Weather of U.S. Cities

Weather of U.S. Cities

A GUIDE TO THE RECENT WEATHER HISTORIES OF 268 KEY CITIES AND WEATHER OBSERVATION STATIONS IN THE UNITED STATES AND ITS ISLAND TERRITORIES

FIFTH EDITION

Richard A. Wood, Ph.D.
Editor

Gale Research
An ITP Information/Reference Group Company

ITP
Changing the Way the World Learns

NEW YORK • LONDON • BONN • BOSTON • DETROIT
MADRID • MELBOURNE • MEXICO CITY • PARIS
SINGAPORE • TOKYO • TORONTO • WASHINGTON
ALBANY NY • BELMONT CA • CINCINNATI OH

Richard A. Wood, Ph.D., *Editor*

Gale Research Staff:

Kimberley A. McGrath, *Project Coordinator*
Mary Beth Trimper, *Production Director*
Evi Seoud, *Assistant Production Manager*
Shanna Heilveil, *Production Assistant*

Cynthia D. Baldwin, *Product Design Manager*
Barbara Yarrow, *Graphic Services Manager*
Erin Martin, *Desktop Publisher*
C.J. Jonik, *Keyliner*

Source for the various charts, as well as the climatological narrative which begins each city's report is the National Oceanic and Atmospheric Administration's series of publications, issued from its National Climatic Data Center entitled *Local Climatological Data, (LCD), Annual Summary for 1993.* The summary for each station in that publication comprises eight pages, some elements of which this book omits because of their ephemeral or esoteric nature.

™ The paper used in this publication meets the minimum requirements of American National Standard for Information Sciences—Permanence Paper for Printed Library Materials, ANSI Z39.48-1984.

This book is printed on recycled paper that meets Environmental Protection Agency Standards.

ISBN: 0-8103-5525-6
Printed in the United States of America

I(T)P Gale Research, an International Thomson Publishing Company.
ITP logo is a trademark under license.

10 9 8 7 6 5 4 3 2 1

Weather of U.S. cities : a guide to the weather histories of 268 key cities and weather observation stations in the United States and its island territories. -- 5th ed. / edited by Richard A. Wood.
p. cm.
ISBN 0-8103-5525-6
1. Urban climatology--United States--Handbooks, manuals, etc. 2. United States--Climate--Handbooks, manuals, etc. I. Wood, Rchard A.
QC983.W393 1995
551.6973'09173'2--dc20 95-5114[illegible]

Contents

Introduction

Even before the initial development of the barometer in the seventeenth century, humankind's interest in weather was more than a casual curiosity. In fact, weather—and its various elements—has wielded crucial influence on our development as a nation, from settlement patterns and trade routes to transportation and communication.

Even today, whether planning a relocation, a business or vacation trip, or just a one-day outing, it is important to be aware of what to expect weather- and climate-wise. For example, while planning a trip to the Pacific coast of the United States in July one might expect it to be rather warm, but I have never been so cold as sitting out at an afternoon baseball game in the San Francisco Bay area, as cold air moved inland from the cold waters of the Pacific.

Weather history of 268 stations

Weather of U.S. Cities is designed to provide a one-stop resource for data on local climatic conditions. Included are detailed records on temperature and other significant weather components, including rain and snow, winds, sunshine, degree days, snowfall, and thunder.

Weather of U.S. Cities relies on the cumulative climatic records of an individual area to present long-term climatic tendencies for a location. All 268 National Weather Service (NWS) "First Order" stations have similar information. Local Climatological Data (LCD) are data observed or compiled using automated equipment that has been tested and accepted by the NWS. Almost all population centers of the country, as well as other strategic locations, are included. Some of the station reports (about 10%) are slightly different in appearance, as the NWS is in the process of switching to new observational equipment—the Automated Surface Observing System (ASOS).

Weather tendencies for an area, while fluctuating annually, are fairly repetitious when studied for the 30-year "normal" period. These longer-range tendencies are what can be described as the *climate* for an area. Annual summaries are used to develop longer range normals for all manners of research. Annual summaries can show changing tendencies within the 30-year normal periods—warmer or colder and wetter or drier periods.

This edition of *Weather of U.S. Cities* includes the recently updated 1961-90 normals for all 268 cities. (Records for the 1951-80 period are still the only records on hand in many libraries.) Such a three-decade period is the time on which climatologists base their standardized reports of "Normal, Means, and Extremes" for an area. Monthly statistics are provided for each station for the years 1964-93.

Sources

The records here provide a statistical picture for each station based on instrument readings recorded by the NWS office in that particular area, as well as the narrative description of each area's climate. The National Climatic Data Center (NCDC) in Asheville, North Carolina, is the agency that has collected the various records and issued them under the title *Local Climatological Data (LCD), Annual Summary for 1993.* Both the NWS and NCDC are subagencies of the National Oceanic and Atmospheric Administration (NOAA).

How to read these reports

Immediately preceding the records is provided a section that describes the specific data items in each report and gives point-by-point guidance in finding and interpreting each data element. The information about the weather history of key cities is easy to find. The reports were designed to be informative, yet easy to read. The terminology used and the standard formats are explained within the section.

Richard A. Wood, Ph.D.
Editor

Local Climatological Data Reports

• INTRODUCTION

Local Climatological Data (LCD) are data observed at principal meteorological stations by trained observers or automated equipment that has been tested and accepted by the controlling agency. The stations are located worldwide and are operated by agencies of the United States government. The controlling agencies are the National Weather Service (NWS), the U.S. Air Force/Air Weather Service (AWS), the U.S. Navy/National Oceanographic Command (NAVOCFANCOM), and the Federal Aviation Administration (FAA). The data are collected on a wide range of time scales from one minute for some current automated equipment to three observations per day for intermittent periods at some remote, part-time locations. The majority of stations collect hourly observations and special (between hour) observations when weather conditions require. The data are used initially for aviation guidance and safety, weather forecasting, and severe weather warnings. The hourly data, special observations, and summaries are generally transmitted over global telecommunication circuits.

In 1992, NWS began automating surface weather observations through the Automated Surface Observing System (ASOS). The measurement of all weather elements included in conventional observations have not been automated. Some of those elements that have not been automated include clouds above 12,000 feet (303 m), snowfall and snow depth, sunshine, and the identification of certain types of weather such as tornado, thunderstorms, hail, drizzle, smoke, and blowing phenomena. Until new automated sensors are designed to measure these elements, the data will be obtained from other sources. ASOS is considered to be only one source of the total surface observation. The total surface observation concept includes the ASOS system, augmentation of ASOS data by station personnel, supplementary data from networks distinct from ASOS (e.g. severe weather spotter networks, hydrological networks, and cooperative networks), and complementary data from other observing technologies such as satellite, radar, and lightning detection systems.

• HOW TO READ THESE REPORTS

Weather statistics on 268 U.S. cities

The information about the history of key cities is easy to find. The reports were planned to be informative, yet easy to read. The terminology used and the standard formats are explained below.

Narrative report

Typically, each report begins with a narrative description of the local area climate prepared by a local climatologist.

The narrative describes the area in terms of terrain, water bodies, and other topographical features as these features exercise key influences on the local weather. They are usually the cause if an area's weather differs sharply from weather in areas only a few miles away. For example, if a lake is near a city, it will always influence the city's weather, and, in fact, may even create climatic differences from one part of the city to another. Mountains, swamps, even plowed fields exercise their influences on air masses as these masses move toward a city.

The report generally discusses temperatures in the city, rainfall tendencies, snowfall history, and other points. It closes with notes about the area's agricultural adaptability. The history of first fall frost and last spring freeze is usually described, along with suggestions about the types of crops for which the area is climatologically suited.

Statistics

The statistics are of two different kinds. The first type distills many years of history to give a profile of the city's weather (e.g., *Normals, Means, and Extremes*). The second group offers data for individual years, allowing the user to see what variances and/or patterns have occurred.

There are several meteorological terms used in the table. The following notes are designed to clarify the information included in the tables.

Location

Precise geographical location is provided at the top of each *Normals, Means, and Extremes* table.

Normal

Term applied as to temperature, degree days, and precipitation refers to the value of that particular element averaged over the period of 1961-90. When the station does not have continuous records from an instrument site with the same "exposure," a difference factor between the old site and the new site is used to adjust the observed values to a common series. The difference factor is determined from a period of simultaneous measurements. The base period is revised every ten years by adding the averages for the most recent decade and dropping them for the first decade of the former normals for 1961-90. *Normal* does not refer to "normalcy" or "expectation," but only to the actual averages for a particular thirty-year period.

Means and extremes (note "a")

Means and extremes are based on the period of years in which observations have been made under comparable conditions of instrument exposure. Data are included for dates through 1985 unless otherwise noted. The Date of an Extreme is the most recent one in cases of repeated occurrences.

Length of observational record

Length of observational record for *Means and Extremes* is based on the length of January data for the present instrument site exposure (15 equals 15 years). The table does not give the all time high or low value if it was recorded at a different site within the area. The *Mean* (or average) values for relative humidity, wind sunshine, sky condition, and the mean number of days with the various other weather conditions listed are also based on length of record noted in each instance. Check the first column for each of these rows to read this length of record for each item.

Average of the highest temperature (°F)

Average of the highest temperature on each day of the month and year for the period 1961-90. This value is obtained by taking the sum of the highest temperature for each day of the period adjusted for site exposure if necessary and dividing by the number of days included.

Average of the lowest temperature (°F)

Average of the lowest temperature on each day of the month and year for the period 1961-90.

Average of all daily temperatures (°F)

Average of all daily temperatures for the month and year for the period 1961-90; computed as being the sum of the minimum and maximum values divided by two. The monthly mean maximum or minimum temperatures are the sum of the daily maximums or minimums divided by the number of days in the month. The monthly average temperature is the sum of the monthly mean maximum and minimum temperatures divided by two.

Extremes—highest temperature (°F)

Highest temperature ever recorded during any month at present site exposure.

Extremes-lowest temperature (°F)

Lowest temperature ever recorded during any month at present site exposure.

Average number of heating degree days

The number of heating degree days (average) for each month and year for the period 1961-90. The statistic is based on the amount that the daily mean temperature falls below 65°F (18.3°C). Each degree of mean temperature below 65 is counted as one heating degree day. If the daily mean temperature is 65°F (18.3°C) or higher, the heating degree day value for that day is zero. Monthly and annual sums are calculated for each period and averaged over the appropriate thirty years of record to establish these "normal" values. Compare this with cooling degree days.

Average number of cooling degree days

The number of cooling degree days (average) for each month and year for the period 1961-90. The concept of this statistic is the mirror image of the concept of heating degree days and is based on the amount that the daily mean temperature exceeds 65°F. Each degree of mean temperature above 65°F is counted as one cooling degree day. If the daily mean temperature is 65°F or below, the cooling degree day is zero. It should be noted that heating and cooling degree days are calculated independently and do not cancel each other out.

Sunshine

The average percent of daytime hours subject to direct radiation from the sun at the present site. The percentage is given without regard for the intensity of sunshine. That is, thin clouds, light haze, or other minor obstructions to direct solar rays may be present but would not mitigate the full counting of an hour.

Mean sky cover

Average amount of daytime sky obscured by any type of cover expressed in tenths (e.g. , 4.8 or, 48%).

Activity limiting weather

Average number of days in month with specified weather conditions (precipitation, storms, etc.) based on present exposure. An (*) indicates less than half a day.

Cloudiness

Average number of days in month at the present site with various amounts of cloud cover. Clear indicates average daytime cloudiness of 0.3 or less; partly cloudy indicates average daytime cloudiness between 0.4 and 0.7; cloudy indicates average daytime cloudiness of 0.8 or more.

Maximum (+) temperature days

Average number of days in month and year when the temperatures at the present site 90°F (32.2°C) or above. (70°F [21.1°C] or above at Alaskan stations.)

Minimum temperature days

Average number of days at the present site when the temperatures remained below 32°F (0°C) at all times.

Freezing days

Average number of days at the present site when the temperature dropped to a minimum of 32°F or below.

Minimum (-) days

Average number of days at present site when the minimum temperature was 0°F or below.

Average station pressure

Given in millibars (mb).

Average relative humidity

Relative humidity levels at various hours of the day. The time is expressed in terms of the 24-hour clock (00 is midnight, 06 is 6 A.M., 12 is noon, and 18 is 6 P.M.). Values are for present site only.

Average precipitation

Precipitation in inches of water equivalent for each month and year during the period 1961-90. As in the other precipitation data, the values are expressed in inches of depth of the liquid water content of all forms of precipitation even if initially frozen. As in all "normal" values, when the station does not have continuous records from the same instrument site, a "ratio factor" between the new and old exposure is used to adjust the observed values to a common series.

Maximum precipitation (month)

Greatest amount (in inches) of water equivalent ever recorded during *any month* at present site.

Minimum precipitation (month)

Least amount (in inches) of water equivalent ever recorded during any month at present site.

Maximum precipitation (day)

Greatest amount (in inches) of water equivalent ever recorded during *any day* at present site.

Summaries for snow, ice pellets, hail

Figures include sleet and are similar to those for total precipitation. The values are expressed in inches of actual snow or ice fall. The water equivalent can be estimated roughly by using the rule-of-thumb that 10 inches of snow equal 1 inch (2.54 cm) of water.

Wind

Average speed of wind is expressed in miles-per-hour without regard of direction.

Prevailing wind direction

The most common single wind direction without regard for wind speed or any minimum amount of persistence. The aggregate total of wind from the other directions may be very much greater than that from the "prevailing" direction. Direction is coded in two different ways, some reports use letters, some use numbers. When letters are used, they have the usual meaning, such as WSW, indicating west-southwest. When numbers are used, they are given in tens of degrees clockwise from true north, so that 09 is 90° clockwise from north (east), 18 is 180° (south), 27 is 270° (west), and 36 is 360° (north).

High wind

The greatest speed in miles per hour of any "mile" of wind passing the station. The accompanying direction and year of occurrence are also given. A mile of wind passing the station one minute has an average speed of 60 mph (97 kph); two minutes—30 mph (48 kph); five minutes—12 mph (19 kph), etc. The wind cups of the particular instrument involved operate much like the wheel of a car in actuating the car's odometer. The instrument does not record the strength of individual wind gusts which usually last less than 20 seconds and which may be very much greater than the value given. The fastest mile however does give some idea of the extremes of wind that can be encountered.

Special symbols

Symbols that appear on many of the individual summaries: (*) less than one half; (**T**) trace, an amount too small to measure; (-) below zero temperatures are preceded by a minus sign.

Weather averages year-by-year

PRECIPITATION refers to the inches of water equivalent in the total of all forms of liquid or frozen precipitation that fell during each month. Snowfall refers to the actual amount of snow in inches that fell during the month. T (trace) is a precipitation amount of less than 0.005 inches. [Note: In estimating the water equivalent of snow, a ratio of 10 inches (25.4 cm) of snow equals 1 inch (2.54 cm) of water is customarily employed.]

AVERAGE TEMPERATURE equals the average of the maximum and minimum temperatures for each day of the month for the given year; afternoon temperatures were typically higher than these values and late night/early morning temperatures were typically below them.

HEATING AND COOLING DEGREE DAYS Heating Degree Days (HDD) provide a well-established index of relative fuel consumption for space heating in a given place—a month of 2,000 HDD requires about twice the amount of space heating energy as one of 1,000 HDD, while 100 HDD will require about the same fuel whether accumulated in two or four days. Regional differences in the Heating Degree Day Index are only partially useful in estimating comparative fuel requirements because the building construction and cultural expectations tend to be different in different parts of the country (e.g., subjective ideas of comfort in relation to temperature will vary). For example, the average standards of efficiency in heating equipment and insulation are generally lower in warmer climates so that fuel requirements tend not to decrease as rapidly as the heating degree days decrease.

The complementary index of Cooling Degree Days provides only a rough guide to relative energy consumption in air conditioning. A proper air–conditioning index will almost certainly require a factor for humidity variation and possibly factors for cloudiness or other weather variables. However, the Cooling Degree Days Index will have some usefulness in indicating relative outdoor comfort and relative indoor air conditioning requirements.

BIRMINGHAM (MUNICIPAL AIRPORT), ALABAMA

Birmingham is located in a hilly area of north-central Alabama in the foothills of the Appalachians about 300 miles inland from the Gulf of Mexico. There is a series of southwest to northeast valleys and ridges in the area.

The city is far enough inland to be protected from destructive tropical hurricanes, yet close enough that the Gulf has a pronounced modifying effect on the climate.

Although summers are long and hot, they are not generally excessively hot. On a typical mid-summer day, the temperature will be nearly 70 degrees at daybreak, approach 90 degrees at mid-day, and level off in the low 90s during the afternoon. It is not unusual for the temperature to remain below 100 degrees for several years in a row. However, every few years an extended heat wave will bring temperatures over 100 degrees. July is normally the hottest month but there is little difference from mid-June to mid-August. Rather persistent high humidity adds to the summer discomfort.

January is normally the coldest month but there is not much difference from mid-December to mid-February. Overall, winters are relatively mild. Even in cold spells, it is unusual for the temperature to remain below freezing all day. Sub-zero cold is extremely rare, occurring only a very few times this century. Extremely low temperatures almost always occur under clear skies after a snowfall.

Snowfall is erratic. Sometimes there is a two- or three-year span with no measurable snow. On rare occasions, there may be a 2 to 4 inch snowstorm. The snow usually melts quickly. Even 1 or 2 inches of snow can effectively shut down this sunbelt city because of the hilly terrain, the wetness of the snow and the unfamiliarity of motorists driving on snow and ice.

Birmingham is blessed with abundant rainfall. It is fairly well distributed throughout the year. However, some of the wetter winter months, plus March and July, have twice the rainfall of October, the driest month. Summer rainfall is almost entirely from scattered afternoon and early evening thunderstorms. Serious droughts are rare and most dry spells are not severe.

The stormiest time of the year with the greatest risk of severe thunderstorms and tornadoes is in spring, especially in March and April.

In a normal year, the last 32 degree minimum temperature in the spring is in mid to late March and the first in autumn is in early November.

NORMALS, MEANS, AND EXTREMES

BIRMINGHAM (MUNICIPAL AIRPORT), ALABAMA

LATITUDE: 33°34'N LONGITUDE: 86°45' W ELEVATION: FT. GRND 620 BARO 622 TIME ZONE: CENTRAL WBAN: 13876

	(a)	JAN	FEB	MAR	APR	MAY	JUNE	JULY	AUG	SEP	OCT	NOV	DEC	YEAR
TEMPERATURE °F:														
Normals														
-Daily Maximum		51.7	56.9	66.1	74.6	81.0	87.4	89.9	89.1	83.9	74.7	64.6	55.7	73.0
-Daily Minimum		31.3	34.5	42.3	49.2	57.7	65.2	69.5	68.8	62.9	50.2	41.6	34.8	50.7
-Monthly		41.5	45.7	54.2	62.0	69.4	76.3	79.8	79.0	73.4	62.5	53.1	45.2	61.8
Extremes														
-Record Highest	50	81	83	89	92	99	102	106	103	100	94	84	80	106
-Year		1949	1962	1982	1987	1962	1954	1980	1990	1990	1954	1961	1951	JUL 1980
-Record Lowest	50	-6	3	2	26	35	42	51	51	37	27	5	1	-6
-Year		1985	1958	1993	1973	1944	1966	1967	1946	1967	1956	1950	1989	JAN 1985
NORMAL DEGREE DAYS:														
Heating (base 65°F)		729	540	350	135	31	0	0	0	7	148	364	614	2918
Cooling (base 65°F)		0	0	15	45	168	339	459	434	259	71	7	0	1797
% OF POSSIBLE SUNSHINE	34	42	50	55	63	66	65	59	63	61	66	55	46	58
MEAN SKY COVER (tenths)														
Sunrise - Sunset	37	6.9	6.5	6.5	5.8	6.0	5.9	6.4	5.8	5.6	4.6	5.7	6.4	6.0
MEAN NUMBER OF DAYS:														
Sunrise to Sunset														
-Clear	37	6.9	7.2	7.4	8.8	8.0	7.1	4.7	6.8	9.5	14.0	10.2	8.4	98.8
-Partly Cloudy	37	6.2	6.4	7.9	8.0	10.9	12.6	14.4	14.5	9.3	7.6	6.8	6.6	111.2
-Cloudy	37	17.9	14.8	15.8	13.2	12.2	10.4	12.0	9.7	11.2	9.4	13.0	15.9	155.3
Precipitation .01 inches or more	50	11.2	10.2	11.0	9.2	9.8	9.6	12.4	9.6	8.0	6.2	9.0	10.7	116.9
Snow,Ice pellets,hail 1.0 inches or more	50	0.3	0.1	0.1	0.*	0.0	0.0	0.0	0.0	0.0	0.0	0.*	0.1	0.6
Thunderstorms	50	1.6	2.3	4.5	5.1	6.9	8.4	11.6	8.8	4.3	1.2	1.8	1.2	57.7
Heavy Fog Visibility 1/4 mile or less	50	1.3	0.6	0.6	0.3	0.3	0.7	0.4	0.5	0.5	0.7	1.0	1.2	8.2
Temperature °F														
-Maximum														
90° and above	30	0.0	0.0	0.0	0.1	1.7	10.8	17.3	15.3	6.3	0.1	0.0	0.0	51.7
32° and below	30	1.4	0.4	0.1	0.0	0.0	0.0	0.0	0.0	0.0	0.0	0.*	0.5	2.5
-Minimum														
32° and below	30	17.0	13.2	6.1	1.2	0.0	0.0	0.0	0.0	0.0	0.5	6.8	14.2	59.0
0° and below	30	0.2	0.0	0.0	0.0	0.0	0.0	0.0	0.0	0.0	0.0	0.0	0.0	0.2
AVG. STATION PRESS.(mb)	21	998.3	997.0	994.6	994.2	993.4	994.0	995.1	995.2	995.4	997.1	997.6	998.5	995.9
RELATIVE HUMIDITY (%)														
Hour 00	30	76	73	72	77	84	84	86	86	85	84	80	77	80
Hour 06 (Local Time)	30	81	79	79	83	86	86	89	90	88	87	84	81	84
Hour 12	30	61	55	52	51	55	56	60	59	59	54	57	60	57
Hour 18	30	65	57	52	52	59	60	65	65	69	70	69	68	63
PRECIPITATION (inches):														
Water Equivalent														
-Normal		5.10	4.72	6.19	4.96	4.85	3.73	5.25	3.59	3.93	2.81	4.33	5.12	54.58
-Maximum Monthly	50	11.00	17.67	15.80	13.75	11.10	8.44	13.70	10.85	10.43	7.52	15.25	13.98	17.67
-Year		1949	1961	1980	1979	1969	1963	1950	1967	1977	1977	1948	1961	FEB 1961
-Minimum Monthly	50	1.09	1.20	1.71	0.42	1.15	0.67	0.30	0.38	T	0.07	0.42	0.81	T
-Year		1981	1968	1985	1986	1951	1968	1983	1989	1955	1991	1949	1980	SEP 1955
-Maximum in 24 hrs	44	5.81	6.57	7.05	5.08	4.63	3.85	5.47	5.13	5.03	3.75	4.87	5.29	7.05
-Year		1949	1961	1970	1966	1969	1957	1985	1952	1977	1977	1948	1961	MAR 1970
Snow,Ice pellets,hail														
-Maximum Monthly	50	6.6	2.3	13.0	5.0	T	T	T	0.0	T	T	1.4	8.0	13.0
-Year		1982	1960	1993	1987	1993	1992	1990		1992	1993	1950	1963	MAR 1993
-Maximum in 24 hrs	44	4.5	2.3	13.0	5.0	T	T	T	0.0	T	T	1.4	8.4	13.0
-Year		1948	1960	1993	1987	1993	1992	1990		1992	1993	1950	1963	MAR 1993
WIND:														
Mean Speed (mph)	50	8.1	8.7	9.1	8.3	6.8	6.1	5.7	5.4	6.3	6.2	7.3	7.8	7.1
Prevailing Direction through 1963		S	N	S	S	S	SSW	SSW	NE	ENE	ENE	N	NNW	S
Fastest Mile														
-Direction (!!)	34	W	SE	SW	SW	NW	SW	SW	NW	SE	W	N	SE	SW
-Speed (MPH)	34	49	59	65	56	65	56	57	50	50	43	52	41	65
-Year		1975	1960	1955	1956	1951	1957	1960	1956	1951	1955	1944	1954	MAR 1955
Peak Gust														
-Direction (!!)														
-Speed (mph)														
-Date														

PRECIPITATION (inches)

BIRMINGHAM (MUNICIPAL AIRPORT), ALABAMA

YEAR	JAN	FEB	MAR	APR	MAY	JUNE	JULY	AUG	SEP	OCT	NOV	DEC	ANNUAL
1964	5.37	4.11	9.44	9.90	3.20	4.08	4.34	2.78	3.26	2.95	3.24	5.09	57.76
1965	3.21	6.22	6.10	2.54	1.37	8.17	4.87	3.55	2.60	0.67	2.80	2.06	44.16
1966	4.74	8.67	3.77	8.37	3.30	2.87	4.99	6.48	5.12	3.13	2.28	2.34	56.06
1967	2.84	4.74	1.79	1.35	9.32	4.37	6.60	10.85	2.84	4.23	6.42	11.49	66.84
1968	5.56	1.20	6.17	6.23	3.51	0.67	9.39	1.81	3.42	1.20	4.66	7.38	51.20
1969	7.78	3.17	5.19	5.32	11.10	3.75	2.91	1.76	6.85	2.51	2.66	6.07	59.07
1970	2.47	2.35	11.36	5.56	2.27	3.55	3.37	7.01	1.05	7.04	2.31	3.32	51.66
1971	3.58	9.28	6.65	4.25	2.64	6.57	8.90	3.68	3.35	1.21	1.76	5.92	57.79
1972	9.30	2.15	4.79	2.56	3.82	2.70	3.55	2.01	8.09	3.35	4.47	5.76	52.55
1973	6.85	2.33	9.71	5.33	8.29	3.74	8.36	5.41	2.64	0.96	4.91	7.58	66.11
1974	6.85	4.94	2.43	5.43	5.43	1.42	4.69	8.28	4.94	1.49	4.13	5.97	56.00
1975	7.23	4.96	7.57	3.19	4.15	2.44	7.33	3.33	3.69	3.74	4.15	3.49	55.27
1976	4.12	1.80	14.15	1.99	9.00	2.75	4.92	3.34	4.91	1.59	2.23	4.35	55.15
1977	5.08	3.89	8.70	6.73	3.51	0.96	6.24	0.87	10.43	7.52	4.10	2.01	60.04
1978	4.54	1.31	3.07	2.64	8.51	5.04	5.09	2.09	1.14	0.22	2.71	5.43	41.79
1979	5.94	4.70	5.69	13.75	6.64	1.19	9.98	2.30	10.40	2.05	5.51	1.55	69.70
1980	6.63	2.36	15.80	9.10	7.30	3.01	2.11	2.84	5.26	3.21	3.04	0.81	61.47
1981	1.09	4.87	7.23	2.45	2.81	2.49	3.88	5.30	0.93	3.34	1.67	5.82	41.88
1982	5.19	6.29	2.71	7.86	3.19	5.39	3.53	2.68	0.66	3.73	7.11	9.51	57.85
1983	3.26	6.42	5.06	8.28	9.57	3.81	0.30	0.99	2.83	3.67	9.14	12.63	65.96
1984	3.96	2.79	4.07	8.61	6.07	1.41	5.06	3.86	0.16	3.91	5.42	2.30	47.62
1985	5.22	5.79	1.71	2.86	4.36	5.34	10.07	4.07	1.97	4.12	2.62	2.54	50.67
1986	1.21	1.79	2.45	0.42	3.66	3.87	1.61	5.56	2.52	5.24	9.66	3.08	41.07
1987	5.89	5.82	4.77	1.03	6.03	4.59	2.30	3.96	3.52	1.16	3.17	3.08	45.32
1988	5.55	2.52	3.18	3.18	1.22	0.79	2.95	3.43	8.57	3.41	6.33	2.84	43.97
1989	4.76	4.31	5.70	3.40	3.82	8.00	6.42	0.38	7.38	1.52	4.63	3.39	53.71
1990	7.38	7.43	5.81	2.38	4.12	2.08	3.16	0.59	2.04	2.98	4.02	5.47	47.46
1991	3.19	4.27	5.42	4.87	8.90	7.52	4.00	4.59	3.02	0.07	4.00	3.64	53.49
1992	3.22	3.96	3.36	2.61	1.18	5.34	7.41	7.43	5.70	2.18	7.94	5.27	55.60
1993	6.11	2.35	4.40	2.99	3.93	1.47	1.16	2.72	4.62	3.22	2.22	4.01	39.20
Record Mean	4.94	4.81	5.97	4.70	4.25	3.95	5.27	4.18	3.53	2.74	3.79	5.02	53.15

AVERAGE TEMPERATURE (deg. F)

BIRMINGHAM (MUNICIPAL AIRPORT), ALABAMA

YEAR	JAN	FEB	MAR	APR	MAY	JUNE	JULY	AUG	SEP	OCT	NOV	DEC	ANNUAL
1964	42.8	42.7	53.9	65.3	71.7	78.2	77.7	78.1	76.0	59.6	55.3	48.0	62.5
1965	45.8	44.6	49.8	64.8	71.8	75.0	77.7	78.0	73.7	60.0	54.8	46.0	61.9
1966	37.7	46.2	52.3	62.2	66.9	73.6	81.3	77.0	71.8	58.8	54.1	44.3	60.5
1967	44.0	41.8	58.3	67.2	67.7	76.2	75.7	74.2	68.4	60.7	48.4	48.5	60.9
1968	42.5	38.7	52.8	62.9	68.3	77.9	78.8	79.4	72.1	63.2	50.3	41.5	60.7
1969	42.6	44.5	46.6	62.7	69.0	76.2	81.7	77.8	70.9	63.9	50.1	43.3	60.8
1970	37.9	44.5	52.6	65.6	71.3	75.7	80.6	78.7	77.0	63.3	48.4	47.9	62.0
1971	43.6	44.2	48.5	60.2	65.3	77.9	77.5	77.8	75.0	67.2	50.7	54.8	61.9
1972	47.2	46.9	53.7	63.1	67.2	74.2	76.9	78.8	75.1	61.7	48.9	47.7	61.8
1973	40.7	42.3	58.6	58.1	67.5	75.9	79.4	77.2	77.2	66.6	56.2	45.0	62.1
1974	52.8	47.3	61.2	61.5	71.0	72.7	79.2	78.4	70.1	60.6	52.7	47.3	62.9
1975	49.1	50.1	53.9	60.5	72.4	76.4	78.1	79.1	70.2	63.5	54.1	44.6	62.7
1976	39.1	54.3	58.7	61.5	64.2	74.4	78.3	77.0	71.9	57.4	45.5	41.2	60.3
1977	31.6	44.8	57.3	65.2	72.4	79.8	83.4	82.1	75.7	59.9	55.0	44.1	62.6
1978	33.5	37.4	50.2	62.9	68.2	77.1	81.5	79.8	76.9	60.6	57.9	46.1	61.0
1979	37.7	44.0	55.2	63.4	69.8	75.1	79.9	78.4	71.4	63.9	52.8	45.9	61.5
1980	45.2	42.3	51.4	60.9	70.0	77.6	84.4	83.0	78.4	60.3	52.4	44.0	62.5
1981	39.2	48.2	51.3	67.1	67.8	80.6	82.2	79.2	71.9	61.8	54.6	42.7	62.2
1982	41.9	47.3	58.9	59.1	72.6	75.7	80.7	79.6	73.1	64.2	54.1	51.2	63.2
1983	40.6	44.8	50.8	56.6	67.3	74.1	80.3	81.7	71.7	63.0	51.9	39.8	60.2
1984	38.5	46.8	52.4	59.0	67.5	77.1	78.2	77.5	72.1	71.1	51.4	54.3	62.2
1985	35.5	43.2	56.7	63.7	69.5	76.2	78.1	78.2	72.0	67.3	61.0	40.2	61.8
1986	41.9	49.9	55.5	61.8	71.2	78.6	82.5	77.8	76.9	63.9	57.4	44.0	63.5
1987	42.1	46.9	54.6	59.5	73.8	76.7	80.5	82.0	73.1	56.8	54.5	49.1	62.5
1988	39.7	43.5	54.1	61.2	67.6	77.5	79.8	81.5	74.4	57.7	55.2	45.8	61.5
1989	48.9	45.7	56.4	60.3	67.7	75.5	79.2	79.5	72.5	61.9	52.9	38.0	61.5
1990	49.0	54.5	57.0	60.7	68.9	77.8	79.8	82.3	77.5	63.9	55.7	49.7	64.7
1991	44.8	49.7	55.8	65.7	73.1	76.9	81.4	80.8	75.4	65.1	50.0	49.3	64.0
1992	43.4	50.6	53.8	61.0	67.4	74.7	80.4	75.6	73.4	61.5	51.7	45.8	61.6
1993	47.0	44.9	50.4	58.8	68.7	78.5	83.8	81.2	73.2	61.9	51.7	44.4	62.0
Record Mean	44.8	47.0	54.8	62.7	70.5	77.6	80.1	79.6	75.0	64.2	53.4	46.2	63.0
Max	54.1	56.9	65.7	74.1	81.4	88.2	90.1	89.7	85.4	75.6	64.2	55.6	73.4
Min	35.5	37.0	44.0	51.3	59.5	67.0	70.1	69.5	64.5	52.7	42.6	36.7	52.5

REFERENCE NOTES

BIRMINGHAM (MUNICIPAL AIRPORT), ALABAMA

GENERAL
T - TRACE AMOUNT.
BLANK ENTRIES DENOTE MISSING/UNREPORTED DATA.
# INDICATES A STATION OR INSTRUMENT RELOCATION.

SPECIFIC

PM - INCLUDES LAST DAY OF PREVIOUS MONTH
ASOS - AUTOMATED SURFACE OBSERVING SYSTEM IN OPERATION DURING THESE MONTHS.

(a) - LENGTH OF RECORD IN YEARS, ALTHOUGH INDIVIDUAL MONTHS MAY BE MISSING.
0.* OR * - THE VALUE IS BETWEEN 0.0 AND 0.05
NORMALS - BASED ON THE 1961-1990 RECORD PERIOD.
EXTREMES - DATES ARE THE MOST RECENT OCCURRENCE.
WIND DIR.- NUMERALS SHOW TENS OF DEGRESS CLOCKWISE FROM TRUE NORTH. "00" INDICATES CALM.
RESULTANT DIRECTIONS ARE GIVEN TO WHOLE DEGREES.
BOLD VALUES INDICATE EXTREME VALUES WHICH OCCURRED AFTER THE ASOS SYSTEM WAS COMMISSIONED.

RECORD = PERIOD OF RECORD
RECORD MEAN PRECIPITATION IS THE MEAN OF ALL DAILY PRECIPITATION AMOUNTS DURING THE PERIOD OF RECORD.
RECORD MAX(MIN) TEMPERATURE IS THE MEAN OF ALL DAILY MAX(MIN) TEMPERATURES DURING THE PERIOD OF RECORD.
RECORD MEAN TEMPERATURE IS THE SUM OF THE RECORD MAX AND RECORD MIN DIVIDED BY 2.
AVERAGE TEMPERATURE IS THE SUM OF THE MEAN DAILY MAX AND MIN TEMPERATURE DIVIDED BY 2.

EXCEPTIONS

1. PERCENT OF POSSIBLE SUNSHINE, MEAN SKY COVER, AND DAYS CLEAR-PARTLY CLOUDY-CLOUDY ARE THROUGH 1977.
2. MAXIMUM 24-HOUR PRECIPITATION AND SNOW, AND FASTEST MILE WINDS ARE THROUGH SEPTEMBER 1978.

RECORD MEANS ARE THROUGH THE CURRENT YEAR, BEGINNING IN 1896 FOR TEMPERATURE
1896 FOR PRECIPITATION
1944 FOR SNOWFALL

HEATING DEGREE DAYS Base 65 deg. F BIRMINGHAM (MUNICIPAL AIRPORT), ALABAMA

SEASON	JULY	AUG	SEP	OCT	NOV	DEC	JAN	FEB	MAR	APR	MAY	JUNE	TOTAL
1964-65	0	0	0	196	305	520	591	571	478	85	2	0	2748
1965-66	0	0	13	171	305	583	840	520	393	149	34	9	3017
1966-67	0	0	3	203	331	635	643	645	248	51	65	2	2826
1967-68	0	0	41	162	492	510	690	755	387	108	30	0	3175
1968-69	0	0	1	148	434	723	687	566	563	95	26	4	3247
1969-70	0	0	4	110	443	663	830	568	377	91	23	0	3109
1970-71	0	0	10	104	493	523	657	578	500	171	74	0	3110
1971-72	0	0	0	28	437	320	548	526	348	138	21	7	2373
1972-73	0	0	9	135	486	534	749	629	217	228	48	0	3035
1973-74	0	0	3	73	285	615	374	491	161	147	9	0	2158
1974-75	0	0	19	159	378	542	493	412	361	190	2	0	2556
1975-76	0	0	49	105	349	626	795	314	213	129	65	0	2645
1976-77	0	0	7	241	580	732	1026	566	252	73	5	0	3482
1977-78	0	0	0	176	292	640	967	768	452	120	42	0	3457
1978-79	0	0	0	160	220	578	839	584	303	83	33	0	2800
1979-80	0	0	0	101	365	588	604	655	417	144	14	0	2888
1980-81	0	0	7	181	372	642	795	464	428	46	51	0	2986
1981-82	0	0	19	138	314	682	711	490	250	199	4	0	2807
1982-83	0	0	16	134	331	449	751	558	437	262	39	0	2977
1983-84	0	0	26	108	388	774	817	519	392	202	68	2	3296
1984-85	0	0	10	27	410	330	907	604	278	123	16	3	2708
1985-86	0	0	14	62	165	761	711	415	305	132	21	0	2586
1986-87	0	1	0	112	239	644	705	500	320	201	0	0	2722
1987-88	0	0	5	248	315	488	777	616	347	134	18	0	2948
1988-89	0	0	0	234	289	589	492	545	289	200	65	0	2703
1989-90	0	0	27	140	363	834	491	299	265	172	32	0	2623
1990-91	0	0	16	138	282	474	620	421	304	46	7	0	2308
1991-92	0	0	12	97	455	492	663	409	344	173	46	0	2691
1992-93	0	0	6	116	398	587	550	559	452	196	20	0	2884
1993-94	0	0	16	153	411	633							

COOLING DEGREE DAYS Base 65 deg. F BIRMINGHAM (MUNICIPAL AIRPORT), ALABAMA

YEAR	JAN	FEB	MAR	APR	MAY	JUNE	JULY	AUG	SEP	OCT	NOV	DEC	TOTAL
1969	0	0	0	33	156	346	523	402	188	84	2	0	1734
1970	0	0	2	115	227	330	492	432	377	58	0	0	2033
1971	1	0	0	35	90	394	393	403	309	105	10	13	1753
1972	4	6	7	88	93	291	379	434	318	40	8	2	1670
1973	0	0	24	28	132	333	452	385	376	129	28	0	1887
1974	4	3	52	48	204	238	450	418	180	28	15	0	1640
1975	8	0	26	61	241	351	416	447	211	66	29	2	1858
1976	0	8	25	27	48	290	418	376	219	15	1	0	1427
1977	0	5	21	85	241	450	578	537	329	25	0	1	2272
1978	0	0	0	64	151	370	515	463	365	34	12	1	1975
1979	0	1	8	43	188	308	466	424	198	74	7	2	1719
1980	0	4	6	25	176	386	607	563	413	43	2	0	2225
1981	0	1	9	113	145	475	539	448	231	46	9	0	2016
1982	1	0	65	28	247	327	495	459	265	118	12	27	2044
1983	0	0	7	17	115	281	481	528	233	53	2	0	1717
1984	0	0	11	30	151	373	421	395	230	221	6	6	1844
1985	0	1	27	90	163	349	413	417	233	139	51	1	1884
1986	0	0	16	43	222	413	547	406	364	84	18	0	2113
1987	0	0	5	46	280	357	487	532	254	1	7	3	1972
1988	0	0	13	26	105	382	467	516	292	14	1	0	1816
1989	0	14	27	65	155	323	447	459	257	51	3	0	1801
1990	1	14	21	49	160	393	466	543	398	110	9	4	2168
1991	0	0	28	74	266	364	515	496	333	106	9	10	2201
1992	0	0	2	59	129	298	484	333	265	14	5	0	1589
1993	0	0	5	19	140	410	591	511	271	63	18	0	2028

SNOWFALL (inches) BIRMINGHAM (MUNICIPAL AIRPORT), ALABAMA

SEASON	JULY	AUG	SEP	OCT	NOV	DEC	JAN	FEB	MAR	APR	MAY	JUNE	TOTAL
1964-65	0.0	0.0	0.0	0.0	0.0	T	T	T	T	0.0	0.0	0.0	T
1965-66	0.0	0.0	0.0	0.0	0.0	0.0	1.2	T	T	0.0	0.0	0.0	1.2
1966-67	0.0	0.0	0.0	0.0	T	T	T	0.8	T	0.0	0.0	0.0	0.8
1967-68	0.0	0.0	0.0	0.0	T	T	1.0	0.7	T	0.0	0.0	0.0	1.7
1968-69	0.0	0.0	0.0	0.0	T	T	T	T	T	0.0	0.0	0.0	T
1969-70	0.0	0.0	0.0	0.0	T	T	1.7	0.1	T	0.0	0.0	0.0	1.8
1970-71	0.0	0.0	0.0	0.0	T	0.0	T	0.8	T	T	0.0	0.0	0.8
1971-72	0.0	0.0	0.0	0.0	T	0.0	T	T	T	0.0	0.0	0.0	T
1972-73	0.0	0.0	0.0	0.0	0.0	0.0	T	0.0	0.0	0.0	0.0	0.0	T
1973-74	0.0	0.0	0.0	0.0	0.0	T	0.0	T	0.0	0.0	0.0	0.0	T
1974-75	0.0	0.0	0.0	0.0	T	0.4	T	T	T	0.0	0.0	0.0	0.4
1975-76	0.0	0.0	0.0	0.0	T	T	T	0.0	0.0	0.0	0.0	0.0	T
1976-77	0.0	0.0	0.0	0.0	T	T	1.4	T	0.0	0.0	0.0	0.0	1.4
1977-78	0.0	0.0	0.0	0.0	0.0	T	1.9	T	T	0.0	0.0	0.0	1.9
1978-79	0.0	0.0	0.0	0.0	0.0	0.0	T	T	0.0	0.0	0.0	0.0	T
1979-80	0.0	0.0	0.0	0.0	0.0	0.0	T	T	0.3	0.0	0.0	0.0	0.3
1980-81	0.0	0.0	0.0	0.0	T	T	0.0	T	0.0	0.0	0.0	0.0	T
1981-82	0.0	0.0	0.0	0.0	0.0	T	6.6	0.0	T	0.0	0.0	0.0	6.6
1982-83	0.0	0.0	0.0	0.0	0.0	T	1.0	T	1.5	0.0	0.0	0.0	2.5
1983-84	0.0	0.0	0.0	0.0	0.0	T	T	T	2.0	0.0	0.0	0.0	2.0
1984-85	0.0	0.0	0.0	0.0	0.0	T	T	0.3	0.0	0.0	0.0	0.0	0.3
1985-86	0.0	0.0	0.0	0.0	0.0	T	T	T	0.0	0.0	0.0	0.0	T
1986-87	0.0	0.0	0.0	0.0	0.0	T	2.6	0.0	T	5.0	0.0	0.0	7.6
1987-88	0.0	0.0	0.0	0.0	0.0	0.0	1.0	T	0.0	0.0	0.0	0.0	1.0
1988-89	0.0	0.0	0.0	0.0	0.0	T	0.0	T	T	T	0.0	0.0	T
1989-90	0.0	0.0	0.0	0.0	T	0.4	0.0	T	0.0	0.0	T	T	0.4
1990-91	T	0.0	0.0	0.0	0.0	T	T	T	T	T	0.0	T	T
1991-92	0.0	0.0	0.0	0.0	T	0.0	4.4	0.0	T	0.0	T	T	4.4
1992-93	0.0	0.0	T	T	0.0	0.0	0.0	T	13.0	0.0	T	0.0	13.0
1993-94	0.0	0.0	0.0	T	0.0	T							
Record Mean	T	0.0	T	T	T	0.3	0.7	0.2	0.4	0.1	T	T	1.6

HUNTSVILLE, ALABAMA

Huntsville has a temperate climate. Summers are characterized by warm and humid weather, with rather frequent thunderstorms. Winters are usually rather cool, but vary considerably from one year to the next.

The city of Huntsville is almost surrounded by the foothills of the Appalachian Mountains. The Tennessee River winds its way westward about 7 miles to the south of the city, and the broad, fertile Tennessee River Valley, with flat to gently rolling terrain, extends to the west. The weather station is located at the Huntsville–Madison County Airport, which is 11 miles southwest of the center of Huntsville. Mountain ridges, with elevations from 1,200 to 1,600 feet above sea level, are located some 14 miles to the northeast, east, and southeast of the airport.

Cold air masses from the continent are predominant over the area during the winter season, but, at times, mild air from the Gulf of Mexico spreads northward to Huntsville or beyond, and may persist for several days in succession. The contrast between air masses frequenting the region in winter provides a potential source of energy for producing extensive periods of low cloudiness and rain, the result being that four months, December through March, account for about 43 percent of the normal annual precipitation. Severely cold weather seldom occurs.

In the transition from winter to spring, appearances of warm, moist air in place of the cold air become more frequent, and the greatest variety of weather usually occurs during this season. Spring season thunderstorms in the vicinity of the boundary between warm and cold air masses are more likely to be accompanied by locally severe weather conditions than thunderstorms in other seasons.

Day to day weather changes in the summer season are rather small, other than the occurrence of thunderstorms that provide relief from the heat on about one–third of the days. Temperatures frequently rise to 90 degrees or higher, but reach 100 degrees only on rare occasions.

During the fall the weather is usually dry and pleasant. The air masses are cooler in the lower levels and the thunderstorm activity of summer decreases sharply. The dry air is very favorable for the harvesting of cotton and hay crops, important items in the economy of the area. A major departure from the relatively dry weather of fall is an occasional rainy spell of one or more days associated with a decaying hurricane drifting northward from the Gulf of Mexico.

Precipitation amounts for the drier months of the fall are appreciably less than for the relatively wet season in winter. However, with the exception of an infrequent long dry spell, precipitation distribution is such as to provide adequate moisture for plant growth throughout the year. Precipitation is mostly in the form of rain, but snow can be expected to some extent each winter and seasonal totals have ranged from less than 1 inch to over 20 inches.

The growing season is 214 days. The climate is suitable for truck farming, as well as for staple crops and livestock. The average date for the last occurrence of freezing temperatures in the spring is late March and the average date of the first freeze is late October.

NORMALS, MEANS, AND EXTREMES

HUNTSVILLE, ALABAMA

LATITUDE: 34°39'N LONGITUDE: 86°46'W ELEVATION: FT. GRND 624 BARO 631 TIME ZONE: CENTRAL WBAN: 03856

	(a)	JAN	FEB	MAR	APR	MAY	JUNE	JULY	AUG	SEP	OCT	NOV	DEC	YEAR
TEMPERATURE °F:														
Normals														
-Daily Maximum		48.2	53.5	62.8	72.5	79.3	86.6	89.0	88.8	82.8	73.0	62.4	52.6	71.0
-Daily Minimum		29.2	32.6	40.9	49.0	57.3	64.9	68.9	67.9	61.6	49.3	40.5	33.1	49.6
-Monthly		38.8	43.1	51.9	60.8	68.4	75.8	79.0	78.3	72.2	61.2	51.5	42.9	60.3
Extremes														
-Record Highest	26	76	79	84	90	93	101	104	103	101	90	83	77	104
-Year		1972	1989	1991	1989	1986	1988	1993	1990	1990	1986	1982	1978	JUL 1993
-Record Lowest	26	-11	6	6	26	36	45	54	52	40	29	15	-3	-11
-Year		1985	1981	1980	1992	1971	1972	1972	1992	1981	1993	1976	1989	JAN 1985
NORMAL DEGREE DAYS:														
Heating (base 65°F)		812	613	417	158	41	0	0	0	13	173	411	685	3323
Cooling (base 65°F)		0	0	11	32	147	324	434	412	229	56	6	0	1651
% OF POSSIBLE SUNSHINE														
MEAN SKY COVER (tenths)														
Sunrise - Sunset	26	6.9	6.7	6.7	6.0	6.3	5.8	5.9	5.5	5.7	5.2	6.1	6.7	6.1
MEAN NUMBER OF DAYS:														
Sunrise to Sunset														
-Clear	26	7.2	6.9	6.7	8.9	7.6	8.1	7.3	8.9	9.3	12.4	9.0	7.6	99.9
-Partly Cloudy	26	6.0	5.8	7.7	7.4	9.9	10.7	13.0	12.6	9.0	7.0	6.5	5.8	101.3
-Cloudy	26	17.8	15.5	16.6	13.7	13.5	11.2	10.6	9.5	11.8	11.7	14.5	17.6	164.0
Precipitation														
.01 inches or more	26	11.4	9.5	11.7	9.5	10.6	9.1	10.3	8.5	8.7	7.3	9.6	11.0	117.2
Snow, Ice pellets, hail														
1.0 inches or more	26	0.5	0.4	0.2	0.0	0.0	0.0	0.0	0.0	0.0	0.0	0.0	0.1	1.1
Thunderstorms	26	1.2	2.1	4.3	4.6	7.2	8.4	10.3	8.3	5.0	2.2	2.3	1.2	57.1
Heavy Fog Visibility														
1/4 mile or less	26	2.6	1.5	1.2	0.8	1.2	1.3	1.6	1.7	2.1	2.3	1.8	1.3	19.4
Temperature °F														
-Maximum														
90° and above	26	0.0	0.0	0.0	0.2	0.8	9.2	17.0	14.8	4.9	0.*	0.0	0.0	46.9
32° and below	26	3.0	1.0	0.1	0.0	0.0	0.0	0.0	0.0	0.0	0.0	0.1	1.2	5.3
-Minimum														
32° and below	26	18.8	14.1	6.8	0.9	0.0	0.0	0.0	0.0	0.0	0.5	7.8	15.7	64.6
0° and below	26	0.2	0.0	0.0	0.0	0.0	0.0	0.0	0.0	0.0	0.0	0.0	0.1	0.3
AVG. STATION PRESS. (mb)	21	996.1	996.4	993.8	993.4	992.7	993.3	994.3	994.6	995.0	996.5	996.8	997.8	995.1
RELATIVE HUMIDITY (%)														
Hour 00	26	78	75	73	75	83	85	87	87	87	83	79	78	81
Hour 06 (Local Time)	26	82	80	81	82	87	88	90	91	90	87	84	81	85
Hour 12	26	64	59	56	52	56	56	59	58	59	55	58	63	58
Hour 18	26	68	61	57	52	60	60	64	65	68	65	65	68	63
PRECIPITATION (inches):														
Water Equivalent														
-Normal		5.17	4.87	6.62	4.93	5.08	4.13	4.85	3.47	4.08	3.25	4.86	5.87	57.18
-Maximum Monthly	26	10.92	9.57	17.00	10.29	11.88	14.99	9.44	9.81	9.78	12.06	11.53	18.68	18.68
-Year		1982	1971	1980	1991	1983	1989	1973	1986	1980	1975	1977	1990	DEC 1990
-Minimum Monthly	26	1.32	0.59	2.70	0.42	1.99	0.17	0.79	0.93	0.55	0.77	1.82	0.91	0.17
-Year		1986	1978	1982	1986	1988	1988	1983	1973	1982	1971	1971	1980	JUN 1988
-Maximum in 24 hrs	26	4.90	4.03	7.70	3.85	5.90	4.81	4.47	4.89	3.99	6.04	3.33	10.22	10.22
-Year		1982	1991	1973	1983	1983	1992	1975	1986	1980	1975	1973	1990	DEC 1990
Snow, Ice pellets, hail														
-Maximum Monthly	26	9.6	4.2	7.3	T	T	0.0	0.0	T	0.0	T	0.8	1.0	9.6
-Year		1988	1985	1993	1993	1993			1990		1993	1974	1974	JAN 1988
-Maximum in 24 hrs	26	9.6	3.6	7.3	T	T	0.0	0.0	T	0.0	T	0.8	1.0	9.6
-Year		1988	1985	1993	1993	1993			1990		1993	1974	1974	JAN 1988
WIND:														
Mean Speed (mph)	26	9.3	9.7	10.1	9.3	8.1	7.0	6.3	6.0	6.9	7.5	8.6	9.4	8.2
Prevailing Direction														
Fastest Obs. 1 Min.														
-Direction (!!!)	26	26	08	12	18	24	25	27	02	23	33	27	30	02
-Speed (MPH)	26	44	43	46	44	46	44	43	63	43	32	40	40	63
-Year		1978	1987	1969	1970	1993	1989	1969	1990	1968	1972	1989	1988	AUG 1990
Peak Gust														
-Direction (!!!)	10	SW	E	W	SW	NW	N	SW	N	S	W	W	NW	W
-Speed (mph)	10	48	58	70	58	69	52	62	55	48	47	51	53	70
-Date		1992	1987	1991	1985	1990	1992	1985	1986	1987	1990	1989	1993	MAR 1991

PRECIPITATION (inches)

HUNTSVILLE, ALABAMA

YEAR	JAN	FEB	MAR	APR	MAY	JUNE	JULY	AUG	SEP	OCT	NOV	DEC	ANNUAL
1964	5.34	4.00	10.75	12.55	2.13	4.28	7.42	2.54	4.49	3.25	4.00	3.45	64.20
1965	3.61	5.67	8.20	3.15	3.10	5.79	4.44	3.38	3.60	0.65	0.64	0.76	42.99
1966	4.29	4.93	2.16	4.04	6.23	2.90	3.28	6.30	4.53	3.36	1.72	3.17	46.91
#1967	1.83	4.02	1.63	3.90	6.48	4.20	14.81	6.24	2.87	3.89	4.12	11.06	65.05
1968	6.01	1.14	5.63	4.05	5.87	0.77	2.90	0.94	4.15	1.94	3.84	5.19	42.43
1969	7.38	6.97	3.00	6.79	6.85	6.15	4.36	3.38	4.39	2.49	2.47	9.48	63.71
1970	1.71	2.75	7.40	9.08	3.06	6.61	2.44	4.04	1.80	4.76	2.58	4.73	50.96
1971	5.25	9.57	6.64	3.29	3.59	1.34	4.70	2.97	5.99	0.77	1.82	9.89	55.82
1972	7.97	2.58	4.94	4.02	3.53	1.90	6.57	2.25	3.83	2.20	5.07	6.62	51.48
1973	6.33	4.84	14.77	3.86	4.38	2.24	9.44	0.93	1.78	3.41	6.79	8.44	67.21
1974	10.50	7.73	3.56	5.57	9.03	3.00	5.14	3.91	3.23	1.36	5.25	6.01	64.29
1975	5.38	4.96	12.17	3.17	6.48	4.61	8.32	3.57	5.11	12.06	3.59	3.55	72.97
1976	3.94	3.36	6.84	1.84	6.60	7.30	4.35	1.27	7.40	4.38	2.61	3.47	53.36
1977	3.71	3.40	9.50	6.68	4.13	2.40	3.41	2.46	9.03	5.31	11.53	2.25	63.81
1978	3.86	0.59	4.81	2.89	6.30	3.87	5.64	3.45	2.83	0.79	3.63	7.68	46.34
1979	5.95	5.49	5.74	7.46	4.03	2.60	6.49	3.04	5.38	2.91	6.65	2.20	57.94
1980	5.50	1.79	17.00	5.04	9.08	2.09	1.87	3.55	9.78	2.46	5.83	0.91	64.90
1981	1.80	3.09	5.93	2.87	3.55	1.99	3.72	4.37	0.78	4.07	5.41	4.23	41.81
1982	10.92	6.12	2.70	10.02	3.89	3.10	4.09	3.92	0.55	1.98	5.49	6.87	59.65
1983	2.26	5.17	5.02	9.95	11.88	5.87	0.79	1.35	2.17	2.59	8.60	11.74	67.39
1984	3.66	3.65	5.86	5.85	7.51	3.40	4.51	4.58	0.76	6.67	6.89	2.69	56.03
1985	4.65	4.18	3.48	3.01	6.56	4.72	6.49	7.19	3.26	7.01	1.86	2.65	55.06
1986	1.32	4.43	3.26	0.42	5.68	4.85	3.07	9.81	5.10	3.90	7.87	4.86	54.57
1987	6.48	7.74	4.16	1.31	2.60	4.42	2.49	1.45	8.97	1.30	5.85	5.08	51.85
1988	6.19	2.40	3.02	3.79	1.99	.0.17	5.24	1.11	7.91	3.24	6.60	5.00	46.66
1989	7.64	6.80	6.45	2.96	4.24	14.99	5.44	6.63	5.19	2.65	6.75	3.64	73.58
1990	7.89	9.34	8.92	4.48	4.38	2.66	3.55	1.89	2.34	3.40	4.73	18.68	72.26
1991	2.77	8.46	9.33	10.29	8.68	1.72	2.59	4.53	3.34	2.49	4.05	12.01	70.26
1992	2.76	3.21	4.92	2.49	2.45	10.59	6.13	3.38	5.86	2.61	5.19	6.01	55.60
1993	4.52	2.79	7.90	3.51	8.27	1.56	2.17	5.72	4.48	1.83	3.88	4.93	51.56
Record Mean	5.02	4.76	6.64	4.88	5.16	4.14	4.66	3.62	4.12	3.23	4.73	5.95	56.91

AVERAGE TEMPERATURE (deg. F)

HUNTSVILLE, ALABAMA

YEAR	JAN	FEB	MAR	APR	MAY	JUNE	JULY	AUG	SEP	OCT	NOV	DEC	ANNUAL
1964	40.9	40.8	52.4	65.1	71.7	78.8	77.9	76.4	73.8	57.2	52.0	44.0	60.9
1965	41.8	42.8	47.0	65.6	74.3	76.5	79.0	79.1	75.0	61.9	56.1	46.0	62.1
1966	35.3	45.3	52.4	63.0	68.7	76.9	83.8	77.8	72.2	59.9	52.4	41.2	60.7
#1967	43.5	40.1	58.9	67.1	67.7	77.8	75.7	73.8	68.4	60.4	46.4	45.8	60.5
1968	38.0	34.6	50.9	61.2	67.1	77.5	78.8	81.0	71.4	62.1	50.6	38.7	59.3
1969	39.3	41.6	43.7	62.3	70.7	79.2	83.3	77.8	70.8	62.5	48.0	41.3	60.0
1970	35.6	40.6	49.7	64.4	69.6	73.9	78.0	79.5	77.3	62.9	48.7	47.8	60.7
1971	41.9	43.4	47.8	61.1	64.4	78.0	77.2	76.6	73.5	66.3	49.6	50.5	60.9
1972	45.3	44.3	52.6	61.7	66.5	73.4	76.3	76.3	74.1	58.9	46.7	46.7	60.2
1973	40.0	42.7	58.0	58.6	66.1	76.3	79.4	77.8	76.8	67.0	55.4	43.4	61.8
1974	49.1	44.6	57.7	59.9	69.1	70.5	78.2	78.2	68.3	59.7	51.5	43.9	60.9
1975	45.8	45.6	48.1	59.9	69.7	75.6	77.0	78.1	68.2	62.6	52.6	42.2	60.5
1976	36.3	50.7	55.3	59.0	63.1	73.3	77.1	76.4	68.8	55.1	42.9	40.0	58.2
1977	28.7	42.4	55.3	63.6	71.5	77.2	82.7	80.2	73.6	58.3	52.7	41.5	60.7
1978	30.5	34.2	48.2	61.5	67.5	77.3	80.2	78.6	74.8	58.9	57.6	45.4	59.5
1979	33.3	40.2	52.6	60.1	66.9	74.1	77.7	77.5	71.1	61.0	49.2	41.7	58.8
1980	40.5	37.5	47.7	59.4	69.2	76.0	82.2	81.1	75.6	58.0	49.9	40.6	59.8
1981	35.1	44.9	49.2	65.6	65.6	77.8	79.6	76.6	68.5	59.1	52.1	40.2	59.5
1982	38.9	44.6	55.5	56.8	71.1	74.4	79.1	77.0	70.6	61.6	51.6	49.0	60.9
1983	38.8	43.2	50.0	54.9	66.3	73.7	80.3	82.3	71.6	62.1	50.0	36.6	59.1
1984	35.3	45.1	49.5	58.1	66.4	76.6	77.1	76.2	70.0	68.3	48.2	51.7	60.2
1985	31.7	40.5	54.7	62.3	68.6	76.6	77.9	77.0	70.7	64.9	57.9	36.0	59.9
1986	38.9	46.9	53.3	61.5	70.2	78.1	82.0	77.2	75.6	62.2	54.8	42.3	61.9
1987	39.2	45.5	53.6	58.4	73.6	77.3	80.3	81.9	72.3	55.7	53.5	47.2	61.5
1988	36.7	41.3	51.6	60.4	67.8	77.4	79.9	81.2	73.1	55.8	53.2	43.4	60.2
1989	46.5	43.3	54.5	59.4	66.6	74.6	78.4	78.1	71.4	60.7	50.9	33.8	59.9
1990	46.9	52.0	54.3	59.7	67.4	77.1	79.6	81.4	76.2	61.8	55.2	46.8	63.2
1991	43.0	47.5	54.0	64.4	72.2	76.6	80.9	79.7	74.0	62.7	47.7	46.7	62.5
1992	41.8	47.9	51.6	60.3	66.9	73.5	79.1	74.5	71.7	60.3	49.9	43.1	60.1
1993	44.5	41.8	48.7	57.1	68.6	77.3	83.6	80.6	71.1	59.9	49.7	42.0	60.4
Record Mean	39.3	43.5	51.7	61.1	68.8	76.2	79.4	78.6	72.7	61.4	51.2	42.9	60.5
Max	48.8	53.8	62.7	72.8	79.9	86.9	89.7	89.0	83.2	73.3	62.1	52.5	71.2
Min	29.7	33.1	40.7	49.3	57.6	65.4	69.2	68.1	62.1	49.5	40.2	33.2	49.8

REFERENCE NOTES

HUNTSVILLE, ALABAMA

GENERAL
T - TRACE AMOUNT.
BLANK ENTRIES DENOTE MISSING/UNREPORTED DATA.
# INDICATES A STATION OR INSTRUMENT RELOCATION.

SPECIFIC

EXCEPTIONS

RECORD MEANS ARE THROUGH THE CURRENT YEAR,
BEGINNING IN 1959 FOR TEMPERATURE
1959 FOR PRECIPITATION
1968 FOR SNOWFALL

(a) - LENGTH OF RECORD IN YEARS, ALTHOUGH INDIVIDUAL MONTHS MAY BE MISSING.
0.* OR * - THE VALUE IS BETWEEN 0.0 AND 0.05
NORMALS - BASED ON THE 1961-1990 RECORD PERIOD.
EXTREMES - DATES ARE THE MOST RECENT OCCURRENCE.
WIND DIR. - NUMERALS SHOW TENS OF DEGRESS CLOCKWISE FROM TRUE NORTH. "00" INDICATES CALM.
RESULTANT DIRECTIONS ARE GIVEN TO WHOLE DEGREES.
BOLD VALUES INDICATE EXTREME VALUES WHICH OCCURRED AFTER THE ASOS SYSTEM WAS COMMISSIONED.

RECORD = PERIOD OF RECORD
RECORD MEAN PRECIPITATION IS THE MEAN OF ALL DAILY PRECIPITATION AMOUNTS DURING THE PERIOD OF RECORD.
RECORD MAX(MIN) TEMPERATURE IS THE MEAN OF ALL DAILY MAX(MIN) TEMPERATURES DURING THE PERIOD OF RECORD.
RECORD MEAN TEMPERATURE IS THE SUM OF THE RECORD MAX AND RECORD MIN DIVIDED BY 2.
AVERAGE TEMPERATURE IS THE SUM OF THE MEAN DAILY MAX AND MIN TEMPERATURE DIVIDED BY 2.

HEATING DEGREE DAYS Base 65 deg. F HUNTSVILLE, ALABAMA

SEASON	JULY	AUG	SEP	OCT	NOV	DEC	JAN	FEB	MAR	APR	MAY	JUNE	TOTAL
1964-65	0	0	2	257	389	644	711	614	549	87	0	0	3253
1965-66	0	0	15	135	268	585	916	548	394	152	35	3	3051
1966-67	0	0	1	178	376	728	660	692	255	69	66	0	3025
#1967-68	0	0	42	179	553	586	829	876	445	148	35	0	3693
1968-69	0	0	7	161	429	806	788	649	655	116	21	0	3632
1969-70	0	0	9	142	502	728	902	679	466	99	25	0	3552
1970-71	0	0	10	116	484	529	708	597	524	161	83	0	3212
1971-72	0	0	0	44	469	440	606	590	382	163	36	4	2734
1972-73	0	0	10	195	544	561	767	618	219	213	53	0	3180
1973-74	0	0	6	67	294	664	486	565	232	179	31	2	2526
1974-75	0	0	33	181	414	646	587	537	520	199	3	0	3120
1975-76	0	0	65	126	376	699	883	415	303	181	93	0	3141
1976-77	0	0	19	306	658	766	1120	628	301	105	12	0	3915
1977-78	0	0	4	216	367	721	1062	855	514	129	60	0	3928
1978-79	0	0	0	191	223	602	974	691	385	155	48	0	3269
1979-80	0	0	0	152	469	714	753	793	532	178	21	1	3613
1980-81	0	0	12	239	448	749	922	559	486	57	66	0	3538
1981-82	0	0	36	206	381	762	804	564	314	250	8	0	3325
1982-83	0	0	35	180	401	511	802	604	460	311	46	0	3350
1983-84	0	0	35	115	444	871	916	569	479	225	70	2	3726
1984-85	0	0	26	32	499	407	1027	682	336	145	26	2	3182
1985-86	0	0	25	74	232	893	804	499	363	147	31	0	3068
1986-87	0	2	0	140	311	697	791	539	346	226	1	0	3053
1987-88	0	0	3	283	340	546	870	679	418	158	22	1	3320
1988-89	0	0	2	289	344	661	568	607	334	233	76	0	3114
1989-90	0	0	29	159	420	958	557	361	340	198	49	0	3071
1990-91	0	0	18	173	292	557	677	485	355	59	11	0	2627
1991-92	0	0	24	126	510	559	712	488	409	192	52	0	3072
1992-93	0	0	8	150	448	671	628	644	499	244	24	0	3316
1993-94	0	0	23	188	461	706							

COOLING DEGREE DAYS Base 65 deg. F HUNTSVILLE, ALABAMA

YEAR	JAN	FEB	MAR	APR	MAY	JUNE	JULY	AUG	SEP	OCT	NOV	DEC	TOTAL
1969	0	0	0	40	207	431	572	401	188	75	0	0	1914
1970	0	0	0	88	176	274	408	459	387	55	0	3	1850
1971	0	0	0	50	75	397	386	364	263	90	13	0	1638
1972	0	0	5	71	91	262	359	354	291	15	2	1	1451
1973	0	0	8	29	93	344	454	405	367	138	11	0	1849
1974	0	0	15	30	166	172	415	417	138	25	17	1	1396
1975	0	0	3	53	156	325	378	413	170	59	11	0	1568
1976	0	3	14	8	43	257	380	360	142	8	0	0	1215
1977	0	1	8	70	222	372	557	477	271	13	4	0	1995
1978	0	0	0	34	145	374	477	427	298	10	8	0	1773
1979	0	0	9	16	113	280	400	393	189	36	0	0	1436
1980	0	0	0	17	157	338	542	508	335	28	1	0	1926
1981	0	0	4	83	91	391	456	366	151	28	2	0	1572
1982	0	0	31	11	205	289	446	380	208	81	7	19	1677
1983	0	0	4	13	96	267	479	544	242	36	2	0	1683
1984	0	0	3	23	120	357	383	352	182	140	1	2	1563
1985	0	0	20	70	145	356	407	382	204	76	28	0	1688
1986	0	0	10	48	199	400	535	387	323	62	11	0	1975
1987	0	0	2	34	273	377	486	532	230	0	3	4	1941
1988	0	0	8	24	116	381	467	510	251	13	0	0	1770
1989	0	4	16	71	132	291	423	415	228	34	2	0	1616
1990	0	1	13	45	130	369	463	517	359	82	8	0	1987
1991	0	0	24	46	241	358	499	461	298	64	0	0	1991
1992	0	0	0	57	120	264	444	302	216	13	3	0	1419
1993	0	0	2	12	142	377	581	493	210	38	10	0	1865

SNOWFALL (inches) HUNTSVILLE, ALABAMA

SEASON	JULY	AUG	SEP	OCT	NOV	DEC	JAN	FEB	MAR	APR	MAY	JUNE	TOTAL
1964-65	0.0	0.0	0.0	0.0	T	T	1.6	1.6	2.3	0.0	0.0	0.0	5.5
1965-66	0.0	0.0	0.0	0.0	0.0	T	8.0	T	T	0.0	0.0	0.0	8.0
1966-67	0.0	0.0	0.0	0.0	4.0	0.7	T	1.0	T	0.0	0.0	0.0	5.7
#1967-68	0.0	0.0	0.0	0.0	0.0	1.4	4.3	2.7	2.1	0.0	0.0	0.0	10.5
1968-69	0.0	0.0	0.0	0.0	T	T	T	0.5	T	0.0	0.0	0.0	0.5
1969-70	0.0	0.0	0.0	0.0	0.1	T	3.3	0.9	0.0	0.0	0.0	0.0	4.3
1970-71	0.0	0.0	0.0	0.0	0.0	1.0	0.5	2.4	T	T	0.0	0.0	3.9
1971-72	0.0	0.0	0.0	0.0	T	0.0	0.3	T	0.0	0.0	0.0	0.0	0.3
1972-73	0.0	0.0	0.0	0.0	T	T	0.2	T	0.0	T	0.0	0.0	0.2
1973-74	0.0	0.0	0.0	0.0	0.0	0.1	0.0	T	T	0.0	0.0	0.0	0.1
1974-75	0.0	0.0	0.0	0.0	0.8	1.0	3.7	T	T	0.0	0.0	0.0	5.5
1975-76	0.0	0.0	0.0	0.0	T	T	0.2	0.0	0.0	0.0	0.0	0.0	0.2
1976-77	0.0	0.0	0.0	0.0	0.1	T	0.9	T	0.0	0.0	0.0	0.0	1.0
1977-78	0.0	0.0	0.0	0.0	0.0	T	3.4	0.5	T	0.0	0.0	0.0	3.9
1978-79	0.0	0.0	0.0	0.0	0.0	T	1.1	2.5	T	0.0	0.0	0.0	3.6
1979-80	0.0	0.0	0.0	0.0	0.0	0.0	T	2.8	2.0	0.0	0.0	0.0	4.8
1980-81	0.0	0.0	0.0	0.0	T	T	T	0.5	0.0	0.0	0.0	0.0	0.5
1981-82	0.0	0.0	0.0	0.0	0.0	T	4.3	T	0.3	0.0	0.0	0.0	4.6
1982-83	0.0	0.0	0.0	0.0	0.0	T	T	T	T	0.0	0.0	0.0	T
1983-84	0.0	0.0	0.0	0.0	0.0	0.6	1.0	1.0	T	0.0	0.0	0.0	2.6
1984-85	0.0	0.0	0.0	0.0	0.0	T	4.0	4.2	0.0	0.0	0.0	0.0	8.2
1985-86	0.0	0.0	0.0	0.0	0.0	0.9	T	T	0.0	0.0	0.0	0.0	0.9
1986-87	0.0	0.0	0.0	0.0	T	T	1.1	T	T	T	0.0	0.0	1.1
1987-88	0.0	0.0	0.0	0.0	0.0	0.0	9.6	1.0	T	0.0	0.0	0.0	10.6
1988-89	0.0	0.0	0.0	0.0	0.0	T	0.2	0.3	T	0.0	0.0	0.0	0.5
1989-90	0.0	0.0	0.0	T	T	0.3	0.0	0.0	T	0.0	0.0	0.0	0.3
1990-91	0.0	T	0.0	0.0	0.0	T	T	T	T	0.0	0.0	0.0	T
1991-92	0.0	0.0	0.0	0.0	T	0.0	0.2	0.0	T	0.0	0.0	0.0	0.2
1992-93	0.0	0.0	0.0	0.0	T	T	0.0	T	7.3	T	T	0.0	7.3
1993-94	0.0	0.0	0.0	T	0.0	T							
Record Mean	0.0	T	0.0	T	T	0.2	1.5	0.7	0.4	T	T	0.0	2.9

MOBILE, ALABAMA

Mobile is located at the head of Mobile Bay and approximately 30 miles from the Gulf of Mexico. Its weather is influenced to a considerable extent by the Gulf.

The summers are consistently warm, but temperatures are seldom as high as they are at inland stations. Normally, in summer, the day begins in the low 70s and the temperature rises rapidly before noon to the high 80s or low 90s, when it is checked by the onset of the sea breeze. On the rare occasions when northerly winds prevail throughout the day, temperatures may reach the high 90s or rise slightly above 100 degrees.

Winter weather is usually mild except for occasional invasions of cold air that last about three days. January is the coldest month in the year. Unusual winters may produce readings that require extensive protective measures as some citrus fruit is grown in the area and outdoor nurseries are numerous.

Based on the 1951–1980 period, the average first occurrence of 32 degrees Fahrenheit in the fall is November 26 and the average last occurrence in the spring is February 27.

The yearly rainfall is among the highest in the United States. It is fairly evenly distributed throughout the year with a slight maximum at the height of the summer thunderstorm season and a slight minimum during the late fall. Rainfall is usually of the shower type and long periods of continuous rain are rare.

Frontal thunderstorms may occur in any month of the year. There may be a thunderstorm every other day in July and August. The summer storms are usually not too violent and seldom produce hail.

The area is subject to hurricanes from the West Indies, the western Caribbean, and the Gulf of Mexico.

NORMALS, MEANS, AND EXTREMES

MOBILE, ALABAMA

LATITUDE: 30°41'N LONGITUDE: 88°15' W ELEVATION: FT. GRND 211 BARO 226 TIME ZONE: CENTRAL WBAN: 13894

	(a)	JAN	FEB	MAR	APR	MAY	JUNE	JULY	AUG	SEP	OCT	NOV	DEC	YEAR
TEMPERATURE °F:														
Normals														
-Daily Maximum		59.7	63.6	70.9	78.5	84.6	90.0	91.3	90.5	86.9	79.5	70.3	62.9	77.4
-Daily Minimum		40.0	42.7	50.1	57.1	64.4	70.7	73.2	72.9	68.7	57.3	49.1	43.1	57.4
-Monthly		49.9	53.2	60.5	67.8	74.5	80.4	82.3	81.8	77.9	68.4	59.8	53.0	67.5
Extremes														
-Record Highest	52	84	82	90	94	100	102	104	102	99	93	87	81	104
-Year		1949	1989	1946	1987	1953	1952	1952	1968	1990	1963	1971	1974	JUL 1952
-Record Lowest	52	3	11	21	32	43	49	60	59	42	30	22	8	3
-Year		1985	1951	1993	1987	1960	1984	1947	1956	1967	1993	1950	1983	JAN 1985
NORMAL DEGREE DAYS:														
Heating (base 65°F)		492	344	177	48	0	0	0	0	0	52	196	393	1702
Cooling (base 65°F)		24	14	38	132	295	462	536	521	387	157	40	21	2627
% OF POSSIBLE SUNSHINE														
MEAN SKY COVER (tenths)														
Sunrise - Sunset	45	6.6	6.2	6.1	5.7	5.7	5.8	6.5	6.0	5.7	4.4	5.3	6.3	5.9
MEAN NUMBER OF DAYS:														
Sunrise to Sunset														
-Clear	45	7.7	7.7	8.8	9.3	8.7	7.1	4.1	6.1	8.8	14.6	11.0	8.9	102.8
-Partly Cloudy	45	6.3	6.7	7.8	8.7	11.2	13.8	15.1	14.8	10.1	7.6	7.1	6.3	115.5
-Cloudy	45	17.0	13.9	14.4	12.0	11.1	9.1	11.8	10.1	11.1	8.9	11.8	15.8	146.9
Precipitation														
.01 inches or more	52	10.8	9.7	10.4	7.4	8.5	11.3	16.2	14.0	10.2	5.7	7.9	10.2	122.3
Snow,Ice pellets,hail														
1.0 inches or more	52	0.*	0.1	0.*	0.0	0.0	0.0	0.0	0.0	0.0	0.0	0.0	0.*	0.2
Thunderstorms	52	2.0	2.3	4.9	4.9	7.2	11.8	18.0	14.4	7.5	2.1	2.3	2.2	79.7
Heavy Fog Visibility														
1/4 mile or less	52	6.1	4.5	5.3	4.6	2.7	0.9	1.0	1.3	1.9	2.7	4.3	4.9	40.4
Temperature °F														
-Maximum														
90° and above	31	0.0	0.0	0.0	0.3	3.7	17.0	22.4	20.6	10.4	1.0	0.0	0.0	75.5
32° and below	31	0.2	0.0	0.0	0.0	0.0	0.0	0.0	0.0	0.0	0.0	0.0	0.1	0.4
-Minimum														
32° and below	31	8.1	5.4	1.2	0.*	0.0	0.0	0.0	0.0	0.0	0.*	1.4	5.6	21.7
0° and below	31	0.0	0.0	0.0	0.0	0.0	0.0	0.0	0.0	0.0	0.0	0.0	0.0	0.0
AVG. STATION PRESS.(mb)	21	1012.8	1011.6	1009.3	1008.7	1007.5	1008.0	1009.3	1008.9	1008.5	1010.5	1011.6	1013.0	1010.0
RELATIVE HUMIDITY (%)														
Hour 00	31	79	78	81	83	85	85	87	88	86	83	82	80	83
Hour 06 (Local Time)	31	82	82	85	87	88	88	90	91	89	86	86	84	87
Hour 12	31	61	56	55	52	54	55	60	61	59	53	57	61	57
Hour 18	31	69	63	63	62	63	67	71	74	72	67	71	72	68
PRECIPITATION (inches):														
Water Equivalent														
-Normal		4.76	5.46	6.41	4.48	5.74	5.04	6.85	6.96	5.91	2.94	4.10	5.31	63.96
-Maximum Monthly	52	16.07	11.89	15.58	17.69	15.08	13.07	19.29	15.19	14.04	13.20	13.65	11.38	19.29
-Year		1991	1983	1946	1955	1980	1961	1949	1984	1988	1985	1948	1953	JUL 1949
-Minimum Monthly	52	0.98	1.31	0.59	0.48	0.45	1.19	1.72	1.46	0.58	T	0.25	1.29	T
-Year		1968	1948	1967	1954	1962	1966	1983	1990	1963	1978	1960	1980	OCT 1978
-Maximum in 24 hrs	52	8.34	5.37	10.57	13.36	8.00	7.38	5.34	6.62	8.55	5.65	7.02	5.50	13.36
-Year		1965	1981	1990	1955	1981	1961	1975	1969	1979	1985	1975	1968	APR 1955
Snow,Ice pellets,hail														
-Maximum Monthly	52	3.5	3.6	2.7	T	T	0.0	T	0.0	0.0	0.0	T	3.0	3.6
-Year		1955	1973	1993	1988	1991		1992				1966	1963	FEB 1973
-Maximum in 24 hrs	52	3.5	3.6	2.7	T	T	0.0	T	0.0	0.0	0.0	T	3.0	3.6
-Year		1955	1973	1993	1988	1991		1992				1966	1963	FEB 1973
WIND:														
Mean Speed (mph)	45	10.3	10.6	10.8	10.2	8.9	7.7	7.0	6.8	7.9	8.2	9.3	10.0	9.0
Prevailing Direction through 1963		N	N	N	S	S	S	S	NE	NE	N	N	N	N
Fastest Obs. 1 Min.														
-Direction (!!)	35	18	23	10	01	32	22	21	14	09	36	36	32	09
-Speed (MPH)	35	44	46	40	44	51	44	60	63	63	46	37	38	63
-Year		1959	1960	1985	1964	1963	1989	1992	1969	1979	1964	1959	1959	SEP 1979
Peak Gust														
-Direction (!!)	10	W	S	SW	SE	SW	SW	SW	E	SE	S	NW	NW	SW
-Speed (mph)	10	45	61	55	46	62	60	64	53	60	52	48	43	64
-Date		1993	1987	1991	1993	1985	1989	1992	1991	1985	1985	1988	1990	JUL 1992

PRECIPITATION (inches)

MOBILE, ALABAMA

YEAR	JAN	FEB	MAR	APR	MAY	JUNE	JULY	AUG	SEP	OCT	NOV	DEC	ANNUAL
1964	7.36	2.73	4.23	13.45	4.32	4.27	8.83	6.20	2.19	2.42	3.73	4.12	63.85
1965	9.35	4.94	4.97	0.60	4.33	3.70	8.13	4.72	10.45	1.59	0.73	3.97	57.48
1966	5.52	9.01	3.21	2.37	4.66	1.19	5.20	6.48	4.32	4.44	1.35	4.60	52.35
1967	5.46	6.13	0.59	2.26	5.05	2.88	7.80	8.86	7.76	6.69	0.36	7.50	61.34
1968	0.98	2.88	1.30	2.24	3.17	3.71	3.08	4.59	6.13	1.13	4.05	10.70	43.96
1969	2.79	3.40	8.15	4.53	7.95	1.33	14.14	12.05	1.95	0.67	0.44	5.58	62.98
1970	3.85	5.32	8.60	1.67	8.84	7.68	9.19	9.20	2.56	5.71	1.54	5.77	69.93
1971	2.15	7.15	6.49	0.99	3.16	2.88	7.75	10.46	7.30	0.03	1.23	5.58	55.17
1972	5.94	4.46	5.87	1.81	7.72	3.65	2.16	2.35	3.28	1.26	5.67	5.59	49.76
1973	1.96	3.40	11.63	9.91	2.97	5.33	6.24	5.46	12.68	2.15	4.52	4.57	70.82
1974	3.89	6.47	6.16	5.91	2.31	3.48	6.60	7.03	10.60	0.88	3.91	4.31	61.55
1975	3.43	3.75	7.45	9.05	7.12	3.76	11.82	8.50	7.37	6.72	12.63	4.98	86.58
1976	1.80	2.36	9.60	1.69	11.11	4.71	4.43	4.77	3.51	5.67	4.64	4.21	58.50
1977	5.54	1.86	6.06	2.73	5.30	1.26	7.77	5.42	9.61	4.57	9.51	4.94	64.57
1978	10.40	3.88	4.08	7.03	9.74	6.13	7.74	7.34	3.04	T	5.05	4.36	68.79
1979	5.14	9.14	8.37	4.05	7.02	3.70	10.80	6.10	11.73	2.80	6.46	3.94	79.25
1980	4.95	1.58	13.46	15.43	15.08	2.57	6.54	6.42	5.94	2.90	1.96	1.29	78.12
1981	1.23	8.75	3.00	0.96	12.51	6.50	6.53	5.06	3.97	0.94	0.85	6.82	57.12
1982	3.56	7.42	6.81	4.48	2.84	11.00	13.14	7.00	4.68	1.61	3.66	8.24	74.44
1983	5.82	11.89	6.89	12.53	1.53	8.09	1.72	11.57	5.97	3.81	5.30	8.34	83.46
1984	6.13	4.79	4.75	3.53	4.28	2.07	2.36	15.19	0.74	6.19	1.67	2.12	53.82
1985	5.06	6.39	5.49	1.22	5.77	3.94	8.31	4.32	10.32	13.20	1.61	4.34	69.97
1986	2.67	4.17	4.53	2.16	4.18	7.53	7.13	5.60	4.41	4.83	8.45	3.68	59.34
1987	5.81	8.64	6.18	0.83	10.69	7.68	4.18	10.33	3.62	0.02	5.54	3.60	67.12
1988	4.64	6.26	7.80	4.19	0.58	2.34	6.04	10.43	14.04	1.83	2.30	1.80	62.25
1989	2.13	1.47	5.57	3.55	6.47	9.82	7.16	3.82	4.55	0.90	11.33	7.23	64.00
1990	7.35	8.13	12.24	4.51	6.47	2.68	2.28	1.46	2.65	1.23	1.77	5.20	55.97
1991	16.07	3.35	4.83	10.43	15.03	7.04	8.15	5.17	1.87	0.86	5.78	3.09	81.67
1992	9.89	9.94	3.61	1.90	1.58	5.24	7.77	8.27	3.26	1.03	12.70	5.27	70.46
1993	8.03	4.48	10.45	4.80	5.24	3.01	5.64	3.52	3.35	5.32	2.87	3.69	60.40
Record Mean	4.87	5.15	6.59	5.04	4.84	5.53	7.42	6.66	5.54	3.20	3.85	5.12	63.82

AVERAGE TEMPERATURE (deg. F)

MOBILE, ALABAMA

YEAR	JAN	FEB	MAR	APR	MAY	JUNE	JULY	AUG	SEP	OCT	NOV	DEC	ANNUAL
1964	49.2	48.1	59.4	69.6	74.9	80.4	80.5	81.5	78.2	64.3	61.1	54.2	66.8
1965	52.5	53.4	58.2	71.4	75.7	79.4	82.4	81.2	79.1	68.7	63.9	53.7	68.3
1966	47.2	52.8	58.8	68.4	75.1	79.1	83.8	81.2	77.9	68.0	60.7	53.2	67.2
1967	53.0	52.2	64.5	74.6	74.6	82.5	81.5	79.8	74.2	66.7	59.5	57.9	68.4
1968	50.3	46.7	58.1	70.3	76.0	83.2	82.7	83.1	78.1	70.1	56.3	49.5	67.0
1969	51.5	53.7	53.8	68.9	74.5	82.4	83.5	80.6	77.5	71.2	58.0	53.7	67.4
1970	45.0	51.4	59.2	71.9	75.3	80.2	83.0	83.0	81.7	70.5	55.9	57.4	67.9
1971	53.7	54.3	59.9	67.9	73.0	82.4	82.3	82.1	79.5	72.9	59.2	63.5	69.2
1972	58.4	55.0	61.8	69.4	74.7	80.7	81.4	83.4	81.6	71.0	56.8	55.5	69.2
1973	50.5	51.7	64.6	65.6	74.6	81.6	84.1	81.4	80.2	73.6	65.3	52.9	68.9
1974	64.0	55.2	64.3	66.1	75.5	77.8	82.2	81.4	75.5	65.5	59.2	55.1	68.5
1975	56.6	59.8	60.7	66.2	76.5	80.7	81.8	81.7	75.3	70.3	60.5	52.1	68.5
1976	49.5	59.4	64.0	69.9	72.7	79.8	83.4	81.0	75.5	61.4	50.6	48.0	66.3
1977	40.9	52.2	62.6	68.2	75.6	82.9	83.5	83.4	80.9	65.2	60.9	50.8	67.3
1978	41.2	45.1	56.7	68.5	75.4	81.2	82.8	83.2	81.6	69.8	65.1	54.4	67.1
1979	45.1	50.7	60.8	69.3	73.3	79.7	81.3	81.1	76.8	68.1	57.2	51.3	66.2
1980	56.2	49.7	61.6	65.9	74.6	81.2	84.6	83.2	81.8	66.3	57.8	51.6	67.9
1981	46.0	53.4	59.5	71.0	71.6	82.6	84.0	82.7	76.8	69.4	63.0	49.9	67.5
1982	49.7	53.3	63.5	67.1	74.4	80.7	81.0	81.0	76.0	68.5	60.4	56.9	67.7
1983	47.1	51.1	55.0	61.8	72.1	76.4	81.8	81.4	73.6	67.4	57.7	47.7	64.4
1984	46.1	52.4	58.6	65.7	72.5	78.1	79.9	78.8	75.6	72.7	56.3	60.4	66.4
1985	43.5	50.7	64.5	67.7	73.7	79.9	79.7	81.0	76.0	71.4	66.1	48.8	66.9
1986	49.7	56.4	59.7	66.1	74.4	81.0	83.3	80.5	79.5	68.5	64.3	51.1	67.9
1987	48.8	53.6	59.2	64.6	75.7	78.9	81.8	82.2	77.2	62.5	59.2	56.5	66.7
1988	46.8	50.6	58.6	67.3	72.8	79.5	81.8	81.3	78.3	65.3	62.7	53.8	66.6
1989	57.1	54.0	61.7	65.9	74.1	79.0	81.0	81.6	76.7	66.1	59.1	44.4	66.7
1990	54.6	59.0	61.7	65.0	73.4	81.1	81.7	83.1	78.6	68.0	60.9	56.7	68.7
1991	52.4	55.9	61.8	69.3	76.0	79.7	81.9	81.3	77.1	69.4	54.1	54.5	67.8
1992	49.4	57.0	60.3	65.4	72.2	79.3	82.4	79.4	77.2	67.4	56.4	55.6	66.8
1993	55.1	52.7	57.0	62.1	71.9	79.9	82.6	82.8	78.3	68.1	56.8	49.8	66.4
Record Mean	51.5	54.2	60.1	67.1	74.2	80.3	81.8	81.5	78.0	68.8	59.0	53.0	67.5
Max	60.4	63.3	69.2	76.2	83.2	88.9	90.1	89.8	86.4	78.5	68.7	61.9	76.4
Min	42.5	45.1	51.0	57.9	65.1	71.6	73.5	73.2	69.5	59.0	49.2	44.0	58.5

REFERENCE NOTES

MOBILE, ALABAMA

GENERAL
T - TRACE AMOUNT.
BLANK ENTRIES DENOTE MISSING/UNREPORTED DATA.
# INDICATES A STATION OR INSTRUMENT RELOCATION.

SPECIFIC

PM - INCLUDES LAST DAY OF PREVIOUS MONTH
ASOS - AUTOMATED SURFACE OBSERVING SYSTEM IN OPERATION DURING THESE MONTHS.

(a) - LENGTH OF RECORD IN YEARS, ALTHOUGH INDIVIDUAL MONTHS MAY BE MISSING.
0.* OR * - THE VALUE IS BETWEEN 0.0 AND 0.05
NORMALS - BASED ON THE 1961-1990 RECORD PERIOD.
EXTREMES - DATES ARE THE MOST RECENT OCCURRENCE
WIND DIR.- NUMERALS SHOW TENS OF DEGRESS CLOCKWISE FROM TRUE NORTH. "00" INDICATES CALM.
RESULTANT DIRECTIONS ARE GIVEN TO WHOLE DEGREES.
BOLD VALUES INDICATE EXTREME VALUES WHICH OCCURRED AFTER THE ASOS SYSTEM WAS COMMISSIONED.

RECORD = PERIOD OF RECORD
RECORD MEAN PRECIPITATION IS THE MEAN OF ALL DAILY PRECIPITATION AMOUNTS DURING THE PERIOD OF RECORD.
RECORD MAX(MIN) TEMPERATURE IS THE MEAN OF ALL DAILY MAX(MIN) TEMPERATURES DURING THE PERIOD OF RECORD.
RECORD MEAN TEMPERATURE IS THE SUM OF THE RECORD MAX AND RECORD MIN DIVIDED BY 2.
AVERAGE TEMPERATURE IS THE SUM OF THE MEAN DAILY MAX AND MIN TEMPERATURE DIVIDED BY 2.

EXCEPTIONS

RECORD MEANS ARE THROUGH THE CURRENT YEAR,
BEGINNING IN 1871 FOR TEMPERATURE
1871 FOR PRECIPITATION
1942 FOR SNOWFALL

HEATING DEGREE DAYS Base 65 deg. F MOBILE, ALABAMA

SEASON	JULY	AUG	SEP	OCT	NOV	DEC	JAN	FEB	MAR	APR	MAY	JUNE	TOTAL
1964-65	0	0	0	78	167	347	387	342	240	4	0	0	1565
1965-66	0	0	0	43	100	345	554	333	208	27	1	0	1611
1966-67	0	0	0	32	173	374	370	363	100	0	3	0	1415
1967-68	0	0	23	46	202	273	448	525	251	12	0	0	1780
1968-69	0	0	0	50	284	474	420	320	342	8	0	0	1898
1969-70	0	0	0	20	220	345	618	375	190	14	10	0	1792
1970-71	0	0	0	7	273	265	361	308	190	57	6	0	1467
1971-72	0	0	0	8	204	136	244	295	131	35	0	0	1053
1972-73	0	0	0	33	284	308	442	368	86	75	0	0	1596
1973-74	0	0	0	18	91	377	108	288	106	49	0	0	1037
1974-75	0	0	0	58	215	336	276	185	183	70	0	0	1323
1975-76	0	0	9	11	211	402	475	174	100	8	0	0	1390
1976-77	0	0	0	149	430	520	738	355	141	28	0	0	2361
1977-78	0	0	0	73	149	439	731	551	268	16	0	0	2227
1978-79	0	0	0	20	53	355	613	401	152	8	4	0	1606
1979-80	0	0	0	43	246	424	267	454	159	52	0	0	1645
1980-81	0	0	0	63	235	420	581	323	180	9	4	0	1815
1981-82	0	0	5	42	113	463	485	323	161	48	0	0	1640
1982-83	0	0	5	63	185	296	545	383	306	119	4	0	1906
1983-84	0	0	7	60	243	529	582	363	218	69	10	0	2081
1984-85	0	0	2	17	270	172	665	397	84	50	1	0	1658
1985-86	0	0	1	22	69	503	469	251	188	40	0	0	1543
1986-87	0	0	0	43	105	433	500	311	193	106	0	0	1691
1987-88	0	0	0	108	194	283	565	421	214	33	0	0	1818
1988-89	0	0	0	53	122	363	254	335	170	81	2	0	1380
1989-90	0	0	6	71	204	630	326	186	128	71	0	0	1622
1990-91	0	0	2	58	145	287	382	258	156	13	0	0	1301
1991-92	0	0	3	32	344	342	477	236	165	77	16	0	1692
1992-93	0	0	0	18	276	297	307	340	260	110	3	0	1611
1993-94	0	0	0	64	278	464							

COOLING DEGREE DAYS Base 65 deg. F MOBILE, ALABAMA

YEAR	JAN	FEB	MAR	APR	MAY	JUNE	JULY	AUG	SEP	OCT	NOV	DEC	TOTAL
1969	5	10	4	134	303	527	582	490	380	220	17	4	2676
1970	6	2	16	229	335	467	564	567	505	184	6	36	2917
1971	18	13	37	151	260	531	541	534	441	261	37	94	2918
1972	47	15	40	173	309	478	513	575	505	224	46	20	2945
1973	0	1	80	98	303	508	598	518	462	293	110	10	2981
1974	85	19	91	88	332	392	539	517	323	81	45	36	2548
1975	21	44	57	115	363	475	529	524	325	183	85	11	2732
1976	0	17	74	160	245	454	578	504	321	48	4	0	2405
1977	0	4	74	132	333	543	579	576	482	85	31	7	2846
1978	0	0	17	127	329	496	560	573	507	177	64	34	2884
1979	0	6	28	144	267	448	512	509	361	146	16	5	2442
1980	2	17	61	89	305	495	618	571	512	110	25	9	2814
1981	0	4	15	195	217	538	595	557	366	185	60	4	2736
1982	17	2	119	118	296	475	502	503	342	177	54	53	2658
1983	0	0	4	27	229	351	532	514	271	140	29	2	2099
1984	0	3	24	96	249	401	469	435	327	261	16	34	2315
1985	4	2	77	133	275	452	461	504	339	227	108	7	2589
1986	0	15	31	79	295	486	573	489	444	158	89	8	2667
1987	4	0	21	99	341	422	527	544	373	41	27	28	2427
1988	7	10	21	108	251	442	529	513	404	67	61	20	2433
1989	16	33	74	116	289	431	498	523	366	114	31	0	2491
1990	7	23	34	80	267	493	525	569	417	155	29	36	2635
1991	1	6	66	149	348	448	528	513	371	176	23	22	2651
1992	0	9	31	97	249	437	548	454	373	102	20	13	2333
1993	5	2	21	33	224	455	554	559	408	165	37	0	2463

SNOWFALL (inches) MOBILE, ALABAMA

SEASON	JULY	AUG	SEP	OCT	NOV	DEC	JAN	FEB	MAR	APR	MAY	JUNE	TOTAL
1970-71	0.0	0.0	0.0	0.0	0.0	0.0	0.0	0.0	T	0.0	0.0	0.0	T
1971-72	0.0	0.0	0.0	0.0	0.0	0.0	0.0	0.0	0.0	0.0	0.0	0.0	0.0
1972-73	0.0	0.0	0.0	0.0	0.0	0.0	T	3.6	0.0	0.0	0.0	0.0	3.6
1973-74	0.0	0.0	0.0	0.0	0.0	T	0.0	0.0	0.0	0.0	0.0	0.0	T
1974-75	0.0	0.0	0.0	0.0	0.0	0.0	0.0	0.0	0.0	0.0	0.0	0.0	0.0
1975-76	0.0	0.0	0.0	0.0	0.0	0.0	0.0	0.0	0.0	0.0	0.0	0.0	0.0
1976-77	0.0	0.0	0.0	0.0	0.0	T	1.9	0.0	0.0	0.0	0.0	0.0	1.9
1977-78	0.0	0.0	0.0	0.0	0.0	0.0	0.4	T	0.0	0.0	0.0	0.0	0.4
1978-79	0.0	0.0	0.0	0.0	0.0	0.0	T	0.0	0.0	0.0	0.0	0.0	T
1979-80	0.0	0.0	0.0	0.0	0.0	0.0	0.0	T	T	0.0	0.0	0.0	T
1980-81	0.0	0.0	0.0	0.0	0.0	0.0	T	T	0.0	0.0	0.0	0.0	T
1981-82	0.0	0.0	0.0	0.0	0.0	0.0	T	0.0	0.0	0.0	0.0	0.0	T
1982-83	0.0	0.0	0.0	0.0	0.0	0.0	0.0	0.0	0.0	0.0	0.0	0.0	0.0
1983-84	0.0	0.0	0.0	0.0	0.0	T	0.0	T	0.0	0.0	0.0	0.0	T
1984-85	0.0	0.0	0.0	0.0	0.0	0.0	T	T	0.0	0.0	0.0	0.0	T
1985-86	0.0	0.0	0.0	0.0	0.0	0.0	0.0	0.0	0.0	0.0	0.0	0.0	0.0
1986-87	0.0	0.0	0.0	0.0	0.0	0.0	T	0.0	0.0	T	0.0	0.0	T
1987-88	0.0	0.0	0.0	0.0	0.0	0.0	0.0	1.7	0.0	T	0.0	0.0	1.7
1988-89	0.0	0.0	0.0	0.0	0.0	0.0	0.0	0.0	T	0.0	0.0	0.0	T
1989-90	T	0.0	0.0	0.0	0.0	T	0.0	0.0	0.0	0.0	0.0	0.0	T
1990-91	0.0	0.0	0.0	0.0	0.0	0.0	0.0	0.0	0.0	0.0	T	0.0	T
1991-92	0.0	0.0	0.0	0.0	0.0	0.0	0.0	0.0	0.0	0.0	0.0	0.0	0.0
1992-93	T	0.0	0.0	0.0	0.0	0.0	0.0	0.0	2.7	0.0	0.0	0.0	2.7
1993-94	0.0	0.0	0.0	0.0	0.0	T							
Record Mean	T	0.0	0.0	0.0	T	0.1	0.1	0.2	0.1	T	T	0.0	0.4

MONTGOMERY, ALABAMA

Montgomery is located in a gently rolling area of Alabama with no local topographic features which appreciably influence weather and climate. The National Weather Service Office is on the north side of Dannelly Field which is located about 6 airline miles south-southwest of the downtown bend in the Alabama River. Surrounding terrain is rather level with long gentle slopes toward the northeast and east.

During the months of June through September, inclusive, temperature and humidity conditions generally show little change from day to day. During the coldest months, December, January, and February, there are frequent shifts between mild and moist air from the Gulf of Mexico and dry, cool continental air.

From late June through the first half of August, nearly all precipitation is from local, mostly afternoon, thunderstorms, and there are apt to be considerable differences in day-to-day amounts of rainfall in different parts of the Montgomery area. In late August and in September, summer conditions of temperature and humidity persist as air continues to drift in from the Gulf, but local thunderstorms become less frequent because of the shortening of the days and the decrease in heat received from the sun. As this late summer season progresses, the local thunderstorms give way to thunderstorms which occur with cold fronts and occasional general rains associated with storms on the Gulf.

All types and intensities of rain, except the local thunderstorms of summer, may occur at any time from December through March or early April. Floods in the rivers are correspondingly most frequent during this period.

Most rain from late April through early June is in the form of showers or thunderstorms occurring in advance of approaching cool fronts, which become weaker and less frequent as summer approaches. It is during this spring season, and during the late summer and early autumn, that droughts sometimes occur.

Snow in Montgomery is important only as a curiosity.

NORMALS, MEANS, AND EXTREMES

MONTGOMERY, ALABAMA

LATITUDE: 32°18'N LONGITUDE: 86°24' W ELEVATION: FT. GRND 192 BARO 210 TIME ZONE: CENTRAL WBAN: 13895

	(a)	JAN	FEB	MAR	APR	MAY	JUNE	JULY	AUG	SEP	OCT	NOV	DEC	YEAR
TEMPERATURE °F:														
Normals														
-Daily Maximum		56.3	60.8	68.6	76.4	82.9	89.4	91.1	90.4	87.0	78.2	68.7	60.2	75.8
-Daily Minimum		35.8	38.8	45.7	52.9	60.8	67.9	71.5	70.8	66.1	53.3	44.6	38.7	53.9
-Monthly		46.1	49.8	57.2	64.7	71.9	78.7	81.3	80.6	76.5	65.8	56.6	49.5	64.9
Extremes														
-Record Highest	49	83	85	89	91	98	105	105	104	101	100	87	85	105
-Year		1949	1962	1954	1988	1953	1954	1952	1983	1980	1954	1986	1982	JUN 1954
-Record Lowest	49	0	10	17	28	40	49	59	56	39	26	13	5	0
-Year		1985	1951	1993	1987	1971	1984	1947	1992	1967	1952	1950	1983	JAN 1985
NORMAL DEGREE DAYS:														
Heating (base 65°F)		594	426	263	78	6	0	0	0	0	91	274	492	2224
Cooling (base 65°F)		8	0	21	69	220	411	505	484	345	116	22	11	2212
% OF POSSIBLE SUNSHINE	43	47	53	59	65	63	63	62	63	62	65	55	49	59
MEAN SKY COVER (tenths)														
Sunrise - Sunset	49	6.6	6.3	6.2	5.6	5.7	5.6	6.2	5.6	5.5	4.7	5.5	6.2	5.8
MEAN NUMBER OF DAYS:														
Sunrise to Sunset														
-Clear	49	7.4	7.9	8.2	9.7	9.0	8.3	5.6	8.0	9.8	13.8	10.9	9.1	107.6
-Partly Cloudy	49	6.4	6.1	7.8	8.4	10.0	11.6	14.1	13.9	9.4	7.5	6.3	5.9	107.3
-Cloudy	49	17.2	14.2	15.1	11.9	12.1	10.1	11.3	9.1	10.8	9.7	12.8	16.0	150.3
Precipitation														
.01 inches or more	49	10.8	9.3	10.0	8.2	8.6	9.0	11.8	9.0	7.7	5.6	7.8	10.1	107.8
Snow,Ice pellets,hail														
1.0 inches or more	49	0.1	0.*	0.*	0.0	0.0	0.0	0.0	0.0	0.0	0.0	0.0	0.*	0.2
Thunderstorms	49	1.6	2.3	4.7	5.2	6.4	8.8	11.9	8.8	4.0	1.4	2.0	1.7	58.7
Heavy Fog Visibility														
1/4 mile or less	49	3.4	1.9	1.5	1.2	1.1	0.7	1.0	1.0	1.6	2.0	2.9	3.3	21.7
Temperature °F														
-Maximum														
90° and above	30	0.0	0.0	0.0	0.2	3.3	17.0	21.7	20.5	12.5	1.3	0.0	0.0	76.4
32° and below	30	0.4	0.0	0.0	0.0	0.0	0.0	0.0	0.0	0.0	0.0	0.0	0.2	0.6
-Minimum														
32° and below	30	13.2	8.7	2.6	0.1	0.0	0.0	0.0	0.0	0.0	0.3	4.3	10.0	39.3
0° and below	30	0.*	0.0	0.0	0.0	0.0	0.0	0.0	0.0	0.0	0.0	0.0	0.0	*
AVG. STATION PRESS.(mb)	21	1013.9	1012.6	1010.2	1009.6	1008.4	1008.8	1010.0	1009.9	1010.0	1011.9	1012.9	1014.2	1011.0
RELATIVE HUMIDITY (%)														
Hour 00	30	77	74	74	78	83	84	88	88	86	84	82	78	81
Hour 06 (Local Time)	30	82	80	83	86	88	88	91	92	90	89	87	83	87
Hour 12	30	60	55	53	51	54	55	60	59	56	52	54	59	56
Hour 18	30	64	58	55	55	60	59	66	68	67	65	66	67	63
PRECIPITATION (inches):														
Water Equivalent														
-Normal		4.68	5.48	6.26	4.49	3.92	3.90	5.19	3.69	4.09	2.45	4.08	5.20	53.43
-Maximum Monthly	49	10.16	13.38	12.40	15.64	12.01	14.44	9.99	10.43	10.62	9.06	21.32	11.35	21.32
-Year		1990	1961	1990	1964	1978	1989	1988	1984	1953	1959	1948	1961	NOV 1948
-Minimum Monthly	49	0.72	1.47	1.93	0.52	1.14	0.33	1.58	0.78	0.44	0.01	0.32	1.36	0.01
-Year		1954	1947	1985	1986	1962	1979	1952	1958	1954	1978	1949	1955	OCT 1978
-Maximum in 24 hrs	49	4.73	5.95	8.62	4.59	6.36	6.99	4.52	5.68	8.81	4.25	8.17	4.89	8.81
-Year		1965	1982	1990	1957	1978	1946	1970	1984	1953	1964	1948	1992	SEP 1953
Snow,Ice pellets,hail														
-Maximum Monthly	49	6.0	3.1	3.8	0.8	0.0	0.0	0.0	0.0	0.0	T	T	1.0	6.0
-Year		1977	1973	1993	1987						1993	1971	1993	JAN 1977
-Maximum in 24 hrs	49	3.0	3.1	3.8	0.8	T	0.0	0.0	0.0	0.0	T	T	1.0	3.8
-Year		1977	1973	1993	1987	1990					1993	1971	1993	MAR 1993
WIND:														
Mean Speed (mph)	49	7.7	8.2	8.3	7.3	6.1	5.8	5.7	5.2	5.9	5.7	6.5	7.1	6.6
Prevailing Direction through 1963		NW	S	NW	S	S	S	SW	ENE	NE	NE	NW	NW	S
Fastest Obs. 1 Min.														
-Direction (!!)	8	30	26	24	32	30	29	36	12	19	23	23	30	29
-Speed (MPH)	8	28	38	35	39	35	44	35	26	24	31	30	33	44
-Year		1990	1990	1992	1990	1991	1985	1991	1992	1989	1990	1989	1990	JUN 1985
Peak Gust														
-Direction (!!)	10	NW	SW	NW	NW	NW	N	N	NE	NW	SW	SW	W	SW
-Speed (mph)	10	36	66	54	60	60	60	55	44	41	37	56	48	66
-Date		1993	1990	1993	1985	1991	1992	1993	1987	1989	1990	1991	1993	FEB 1990

PRECIPITATION (inches)

MONTGOMERY, ALABAMA

YEAR	JAN	FEB	MAR	APR	MAY	JUNE	JULY	AUG	SEP	OCT	NOV	DEC	ANNUAL
1964	6.49	4.49	5.01	15.64	1.92	2.72	5.19	2.55	5.38	6.27	3.46	5.29	64.41
1965	6.10	6.37	8.08	0.88	1.33	2.76	4.99	3.98	6.47	1.14	1.34	3.57	47.01
1966	6.20	8.06	4.26	1.24	2.36	3.20	2.58	1.79	5.85	6.08	4.65	5.11	51.38
1967	2.77	4.46	2.00	1.05	5.48	3.69	7.07	2.67	3.75	4.38	2.58	8.51	48.41
1968	2.78	1.87	2.80	3.31	1.94	1.89	4.96	2.50	1.06	0.80	5.90	5.65	35.46
1969	1.85	3.64	4.73	1.82	4.97	1.75	3.24	5.06	6.06	1.60	0.73	3.88	39.33
1970	2.83	3.77	6.20	2.01	3.49	7.83	5.89	3.70	2.02	4.98	1.50	4.18	48.40
1971	2.53	6.61	10.77	3.81	3.73	3.18	4.94	3.22	5.86	0.29	2.27	6.24	53.45
1972	6.35	3.98	5.25	1.45	1.70	5.49	5.84	3.03	1.61	1.56	3.71	7.14	47.11
1973	3.78	4.62	9.84	7.74	3.97	5.37	7.87	2.73	4.07	0.46	4.31	3.70	58.46
1974	5.63	2.09	2.69	3.69	3.30	1.57	5.41	10.42	5.38	1.45	3.59	4.67	49.89
1975	6.75	7.80	7.26	6.63	4.39	4.76	7.58	4.18	8.56	6.42	4.01	4.64	72.98
1976	3.03	1.75	9.41	1.65	6.78	2.74	3.03	2.21	2.79	1.78	5.32	4.94	45.43
1977	4.87	3.19	7.16	1.53	1.62	1.82	6.62	3.55	5.64	2.60	2.78	3.10	44.48
1978	6.95	2.29	2.61	4.57	12.01	3.87	4.02	3.52	2.18	0.01	3.09	4.24	49.36
1979	5.74	7.73	4.23	10.15	5.35	0.33	5.16	1.54	5.25	0.60	4.10	2.84	53.02
1980	4.06	3.30	10.29	5.47	7.50	2.49	3.25	1.02	3.13	3.10	2.07	1.39	47.07
1981	1.23	6.89	5.95	4.95	3.75	3.06	4.10	4.12	2.04	2.99	0.51	8.50	48.09
1982	4.73	9.87	3.81	8.97	2.87	4.81	7.21	3.17	1.96	1.93	7.71	8.80	65.84
1983	6.12	7.68	8.96	11.50	3.22	6.46	2.83	4.24	5.99	3.58	7.27	7.62	75.47
1984	4.89	4.33	3.68	3.51	6.04	2.50	4.51	10.43	0.81	3.91	4.65	4.60	53.86
1985	3.31	9.00	1.93	3.39	2.67	2.11	6.75	3.61	4.72	3.53	2.60	4.72	48.34
1986	1.12	6.10	9.26	0.52	4.78	1.30	5.01	2.67	5.03	3.54	10.18	3.02	52.53
1987	7.78	9.18	4.82	0.71	4.84	7.84	2.12	4.64	3.96	0.43	4.40	4.30	55.02
1988	5.90	5.54	5.20	5.62	3.35	2.28	9.99	1.96	8.21	3.50	9.43	4.45	65.43
1989	2.03	2.88	8.06	7.26	3.85	14.44	8.79	4.13	4.72	3.79	6.36	9.18	75.49
1990	10.16	7.63	12.40	3.68	4.25	0.86	5.59	0.94	1.07	2.07	1.43	3.30	53.38
1991	8.26	2.80	8.38	6.61	11.57	5.57	5.94	2.48	4.02	0.59	5.85	5.53	67.60
1992	7.75	8.22	3.82	2.21	2.00	7.34	7.81	4.26	1.37	3.08	9.95	7.74	65.55
1993	5.19	4.45	7.69	3.32	1.62	2.92	3.52	6.83	2.47	4.13	6.12	4.68	52.94
Record Mean	4.76	5.19	6.24	4.65	3.80	4.06	4.95	3.88	3.56	2.35	3.59	4.86	51.89

AVERAGE TEMPERATURE (deg. F)

MONTGOMERY, ALABAMA

YEAR	JAN	FEB	MAR	APR	MAY	JUNE	JULY	AUG	SEP	OCT	NOV	DEC	ANNUAL
1964	45.2	45.3	57.2	67.0	72.9	81.2	80.0	80.3	76.3	61.4	57.5	50.4	64.6
1965	48.3	48.9	53.8	67.5	74.8	77.5	80.5	79.5	76.4	63.8	58.3	48.6	64.8
1966	41.9	49.1	55.4	66.6	71.9	76.0	82.7	79.8	76.2	64.3	56.6	48.3	64.1
1967	47.9	48.0	61.8	70.6	71.4	78.5	78.8	78.4	70.7	63.0	53.8	53.2	64.7
1968	45.0	42.5	55.7	66.6	71.7	80.2	80.9	81.8	75.6	65.9	53.3	45.6	63.7
1969	47.5	48.9	50.1	65.4	71.5	79.4	83.1	78.0	73.4	67.0	52.4	45.8	63.6
1970	40.2	46.8	55.3	67.9	72.2	77.0	81.5	80.7	79.2	67.0	51.5	51.8	64.2
1971	48.0	49.0	52.8	62.6	68.2	79.9	79.7	79.5	77.8	70.2	54.1	58.3	65.0
1972	52.6	49.7	57.5	65.7	71.0	78.4	80.1	81.3	78.6	66.3	52.6	52.6	65.6
1973	46.3	46.8	62.1	62.9	71.0	79.3	82.6	80.9	79.3	68.3	59.7	48.4	65.7
1974	58.2	49.9	62.0	63.6	73.6	75.6	80.5	78.4	72.6	62.0	54.8	50.5	65.2
1975	52.6	54.1	57.1	62.8	73.1	77.1	81.3	82.2	74.5	69.1	57.1	46.7	65.6
1976	43.5	54.9	59.2	63.7	66.2	77.0	80.3	78.8	74.3	59.7	48.0	43.5	62.4
1977	36.0	46.8	58.7	66.8	74.7	82.6	83.6	81.6	79.4	64.0	60.3	49.5	65.4
1978	40.5	43.2	54.4	65.7	72.4	80.0	82.4	80.8	79.1	65.3	61.9	51.4	64.8
1979	42.8	48.1	59.0	66.0	71.4	76.3	80.3	80.0	76.0	64.9	55.6	49.7	64.2
1980	51.2	47.9	56.4	63.3	72.1	79.1	83.6	83.3	80.8	61.4	54.4	48.0	65.1
1981	42.9	49.7	54.0	68.4	70.6	83.3	83.5	81.2	75.4	65.7	59.6	46.6	65.1
1982	44.3	49.2	61.7	64.8	74.4	79.3	81.9	81.7	76.2	68.2	59.5	56.6	66.5
1983	44.4	49.0	54.0	60.0	70.5	75.6	81.9	82.9	74.3	67.4	56.3	45.5	63.5
1984	43.0	50.6	56.8	63.8	70.8	79.0	80.1	79.0	74.7	73.4	54.5	58.4	65.3
1985	41.2	49.2	62.4	66.7	73.3	79.3	80.8	80.6	74.9	70.0	64.0	43.6	65.5
1986	45.2	53.0	57.6	64.0	73.4	81.0	84.2	80.3	79.0	66.8	62.0	47.7	66.2
1987	45.6	50.2	57.1	62.4	74.8	78.8	82.4	83.6	76.2	59.7	57.5	52.9	65.1
1988	43.9	47.5	56.3	64.4	70.2	78.9	80.2	81.9	77.1	61.8	59.0	48.4	64.1
1989	52.7	50.7	59.2	63.1	71.4	78.0	80.6	81.4	75.7	64.3	55.1	41.3	64.5
1990	51.0	55.9	59.8	63.0	71.1	80.3	81.9	84.7	78.6	66.8	57.5	53.3	67.0
1991	48.0	53.5	58.6	68.1	75.7	78.3	82.2	81.2	76.6	66.6	52.3	50.9	66.0
1992	45.7	53.8	56.7	62.6	69.5	76.4	81.4	78.4	76.5	64.2	54.2	49.5	64.1
1993	51.2	48.7	54.3	60.6	71.4	80.4	84.6	82.4	77.3	65.2	54.4	46.9	64.8
Record Mean	48.2	51.0	57.7	65.2	72.9	79.7	81.7	81.0	76.7	66.3	55.9	49.4	65.5
Max	57.6	61.0	68.2	76.1	83.6	89.8	91.2	90.5	86.6	77.3	66.8	59.1	75.7
Min	38.7	41.1	47.2	54.2	62.3	69.5	72.1	71.5	66.8	55.2	45.0	39.7	55.3

REFERENCE NOTES

MONTGOMERY, ALABAMA

GENERAL
T - TRACE AMOUNT.
BLANK ENTRIES DENOTE MISSING/UNREPORTED DATA.
# INDICATES A STATION OR INSTRUMENT RELOCATION.

SPECIFIC

(a) - LENGTH OF RECORD IN YEARS, ALTHOUGH INDIVIDUAL MONTHS MAY BE MISSING.
0.* OR * - THE VALUE IS BETWEEN 0.0 AND 0.05
NORMALS - BASED ON THE 1961-1990 RECORD PERIOD.
EXTREMES - DATES ARE THE MOST RECENT OCCURRENCE.
WIND DIR.- NUMERALS SHOW TENS OF DEGRESS CLOCKWISE FROM TRUE NORTH. "00" INDICATES CALM.
RESULTANT DIRECTIONS ARE GIVEN TO WHOLE DEGREES.
BOLD VALUES INDICATE EXTREME VALUES WHICH OCCURRED AFTER THE ASOS SYSTEM WAS COMMISSIONED.

RECORD = PERIOD OF RECORD
RECORD MEAN PRECIPITATION IS THE MEAN OF ALL DAILY PRECIPITATION AMOUNTS DURING THE PERIOD OF RECORD.
RECORD MAX(MIN) TEMPERATURE IS THE MEAN OF ALL DAILY MAX(MIN) TEMPERATURES DURING THE PERIOD OF RECORD.
RECORD MEAN TEMPERATURE IS THE SUM OF THE RECORD MAX AND RECORD MIN DIVIDED BY 2.
AVERAGE TEMPERATURE IS THE SUM OF THE MEAN DAILY MAX AND MIN TEMPERATURE DIVIDED BY 2.

EXCEPTIONS

FASTEST MILE WIND IS THROUGH NOVEMBER 1983.

RECORD MEANS ARE THROUGH THE CURRENT YEAR, BEGINNING IN 1873 FOR TEMPERATURE
1873 FOR PRECIPITATION
1945 FOR SNOWFALL

HEATING DEGREE DAYS Base 65 deg. F MONTGOMERY, ALABAMA

SEASON	JULY	AUG	SEP	OCT	NOV	DEC	JAN	FEB	MAR	APR	MAY	JUNE	TOTAL
1964-65	0	0	0	151	242	454	512	459	364	46	0	0	2228
1965-66	0	0	0	107	209	501	709	438	301	69	7	0	2341
1966-67	0	0	0	77	266	515	521	474	166	21	16	1	2057
1967-68	0	0	30	108	338	379	614	645	308	41	8	0	2471
1968-69	0	0	0	102	349	594	536	456	459	42	8	0	2546
1969-70	0	0	0	55	374	589	760	505	297	51	13	0	2644
1970-71	0	0	1	39	398	410	525	445	376	122	35	0	2351
1971-72	0	0	0	18	343	242	391	443	241	91	0	0	1769
1972-73	0	0	2	67	380	393	575	502	135	119	18	0	2191
1973-74	0	0	0	57	195	506	218	419	144	104	0	0	1643
1974-75	0	0	5	115	315	452	385	303	269	122	2	0	1968
1975-76	0	0	19	31	312	561	660	291	197	67	33	0	2171
1976-77	0	0	0	180	502	657	894	501	216	53	0	0	3003
1977-78	0	0	0	87	164	479	756	601	326	50	2	0	2465
1978-79	0	0	0	72	109	431	681	465	194	39	9	0	2000
1979-80	0	0	0	76	282	466	420	499	274	84	2	0	2103
1980-81	0	0	3	144	313	523	682	421	341	24	7	0	2458
1981-82	0	0	3	63	195	562	637	436	180	76	0	0	2152
1982-83	0	0	5	62	192	301	632	442	339	160	7	0	2140
1983-84	0	0	8	41	273	605	675	412	267	111	20	0	2412
1984-85	0	0	4	16	318	219	735	436	120	65	0	0	1913
1985-86	0	0	5	26	103	656	607	332	238	86	1	0	2054
1986-87	0	1	0	75	152	534	599	410	249	136	0	0	2156
1987-88	0	0	0	173	246	391	647	503	283	70	1	0	2314
1988-89	0	0	0	126	197	510	375	413	217	146	20	0	2004
1989-90	0	0	13	111	301	729	430	259	177	119	12	0	2151
1990-91	0	0	6	92	229	365	520	322	224	29	0	0	1787
1991-92	0	0	1	62	387	447	588	324	260	126	25	0	2220
1992-93	0	0	1	58	323	474	421	453	339	141	8	0	2218
1993-94	0	0	2	106	347	554							

COOLING DEGREE DAYS Base 65 deg. F MONTGOMERY, ALABAMA

YEAR	JAN	FEB	MAR	APR	MAY	JUNE	JULY	AUG	SEP	OCT	NOV	DEC	TOTAL
1969	3	8	3	62	218	441	568	409	258	127	4	0	2101
1970	0	0	4	146	245	366	516	494	434	109	0	7	2321
1971	4	3	2	58	142	453	460	454	388	185	22	43	2214
1972	11	4	13	117	196	408	476	514	417	115	17	15	2303
1973	0	0	54	64	212	438	551	499	438	166	45	0	2467
1974	16	4	56	69	272	325	485	420	239	29	18	8	1941
1975	7	5	32	64	259	369	514	545	311	163	80	0	2349
1976	0	3	23	34	75	367	482	436	287	23	0	0	1730
1977	0	0	27	113	308	533	585	523	442	63	29	7	2630
1978	1	0	6	78	239	458	547	498	430	87	24	20	2388
1979	0	2	13	77	214	346	482	474	337	79	8	1	2033
1980	0	9	15	43	231	429	582	577	486	40	3	4	2419
1981	0	0	5	135	187	556	584	508	321	90	39	2	2427
1982	1	0	85	76	298	437	530	527	349	170	36	50	2559
1983	0	0	8	14	183	326	532	561	293	121	19	8	2065
1984	0	0	21	82	204	425	476	442	305	284	10	20	2269
1985	4	1	45	125	260	435	497	489	309	186	78	0	2429
1986	0	3	16	66	265	489	603	483	428	138	69	1	2561
1987	0	0	8	65	311	419	545	586	344	15	28	21	2342
1988	3	5	22	59	170	420	477	530	373	34	23	0	2116
1989	5	18	47	94	226	397	491	517	342	98	10	0	2245
1990	3	13	22	67	211	464	531	619	419	157	12	11	2529
1991	0	4	33	130	340	407	539	508	355	119	15	18	2468
1992	0	5	10	60	169	351	517	422	354	42	8	3	1941
1993	0	2	16	15	215	468	615	548	377	119	34	0	2409

SNOWFALL (inches) MONTGOMERY, ALABAMA

SEASON	JULY	AUG	SEP	OCT	NOV	DEC	JAN	FEB	MAR	APR	MAY	JUNE	TOTAL
1964-65	0.0	0.0	0.0	0.0	0.0	0.0	T	T	T	0.0	0.0	0.0	T
1965-66	0.0	0.0	0.0	0.0	0.0	0.0	T	T	0.0	0.0	0.0	0.0	T
1966-67	0.0	0.0	0.0	0.0	T	0.0	0.0	T	0.0	0.0	0.0	0.0	T
1967-68	0.0	0.0	0.0	0.0	0.0	0.0	T	T	T	0.0	0.0	0.0	T
1968-69	0.0	0.0	0.0	0.0	T	0.0	0.0	T	0.0	0.0	0.0	0.0	T
1969-70	0.0	0.0	0.0	0.0	0.0	T	T	0.0	0.0	0.0	0.0	0.0	T
1970-71	0.0	0.0	0.0	0.0	0.0	0.0	0.0	T	0.0	0.0	0.0	0.0	T
1971-72	0.0	0.0	0.0	0.0	T	0.0	0.0	0.0	0.0	0.0	0.0	0.0	T
1972-73	0.0	0.0	0.0	0.0	0.0	0.0	T	3.1	0.0	0.0	0.0	0.0	3.1
1973-74	0.0	0.0	0.0	0.0	0.0	T	0.0	0.0	0.0	0.0	0.0	0.0	T
1974-75	0.0	0.0	0.0	0.0	0.0	T	0.0	0.0	0.0	0.0	0.0	0.0	T
1975-76	0.0	0.0	0.0	0.0	0.0	0.0	T	0.0	0.0	0.0	0.0	0.0	T
1976-77	0.0	0.0	0.0	0.0	0.0	0.0	6.0	0.0	0.0	0.0	0.0	0.0	6.0
1977-78	0.0	0.0	0.0	0.0	0.0	T	T	T	T	0.0	0.0	0.0	T
1978-79	0.0	0.0	0.0	0.0	0.0	0.0	T	T	0.0	0.0	0.0	0.0	T
1979-80	0.0	0.0	0.0	0.0	0.0	0.0	0.0	0.0	T	0.0	0.0	0.0	T
1980-81	0.0	0.0	0.0	0.0	0.0	T	0.0	0.0	0.0	0.0	0.0	0.0	T
1981-82	0.0	0.0	0.0	0.0	0.0	0.0	0.3	0.0	0.0	0.0	0.0	0.0	0.3
1982-83	0.0	0.0	0.0	0.0	0.0	0.0	0.0	0.0	T	0.0	0.0	0.0	T
1983-84	0.0	0.0	0.0	0.0	0.0	T	0.0	T	0.0	0.0	0.0	0.0	T
1984-85	0.0	0.0	0.0	0.0	0.0	0.0	T	T	0.0	0.0	0.0	0.0	T
1985-86	0.0	0.0	0.0	0.0	0.0	0.0	0.0	0.0	0.0	0.0	0.0	0.0	0.0
1986-87	0.0	0.0	0.0	0.0	0.0	0.0	T	0.0	0.0	0.8	0.0	0.0	0.8
1987-88	0.0	0.0	0.0	0.0	0.0	0.0	0.0	0.0	0.0	0.0	0.0	0.0	0.0
1988-89	0.0	0.0	0.0	0.0	0.0	T	0.0	T	0.0	0.0	0.0	0.0	T
1989-90	0.0	0.0	0.0	0.0	0.0	T	0.0	0.0	0.0	0.0	0.0	0.0	T
1990-91	0.0	0.0	0.0	0.0	0.0	0.0	0.0	0.0	0.0	T	0.0	0.0	T
1991-92	0.0	0.0	0.0	0.0	0.0	0.0	T	0.0	0.0	0.0	0.0	0.0	T
1992-93	0.0	0.0	0.0	0.0	0.0	0.0	0.0	0.0	3.8	0.0	0.0	0.0	3.8
1993-94	0.0	0.0	0.0	T	0.0	1.0							
Record Mean	0.0	0.0	0.0	T	T	T	0.2	0.1	0.1	T	0.0	0.0	0.4

ANCHORAGE, ALASKA

Anchorage is in a broad valley with adjacent narrow bodies of water. Cook Inlet, including Knik Arm and Turnagain Arm, lies approximately 2 miles to the west, north, and south. The terrain rises gradually to the east for about 10 miles, with marshes interspersed with glacial moraines, shallow depressions, small streams, and knolls. Beyond this area, the Chugach Mountains rise abruptly into a range oriented north-northeast to south-southwest, with average elevation 4,000 to 5,000 feet and some peaks to 8,000 or 10,000 feet. The Chugach Range acts as a barrier to the influx of warm, moist air from the Gulf of Alaska, so the average annual precipitation is only 10 to 15 percent of that at stations located on the Gulf of Alaska side of the Chugach Range. The Alaska Mountain Range lies in a long arc from southwest, through northwest, to northeast, approximately 100 miles distant from Anchorage. During the winter, this range is an effective barrier to the influx of very cold air from the north side of the range.

The four seasons are well marked in Anchorage. In the summer, high temperatures average about 60 degrees and low temperatures nearly 50 degrees. Temperatures in the 70s are considered very warm. On summer days, temperatures on the east side of Anchorage may be about 10 degrees warmer than the official airport readings. Rain increases after mid-June. About two-thirds of the days in July and August are cloudy and one-third have rain.

Autumn is brief, beginning in early September and ending in mid-October. Temperatures begin to fall in September with snow becoming more frequent in October.

Winter can be considered as mid-October to early April when streams and lakes are frozen. Temperatures steadily decrease into January when the highs are near 20 degrees and lows near 5 degrees. The coldest weather is normally in January, when very cold days have high temperatures below zero. Cold days generally have clear skies and calm wind. Mild days do occur with temperatures in the 30s. On cold winter nights, temperatures on the east side of Anchorage may be 10-20 degrees lower than airport readings on the west side. Most winter precipitation is snow, but rain may occur on a few days.

Annual snowfall varies from about 70 inches on the west side to about 90 inches on the east side of Anchorage at low elevations. Along the Chugach Mountains, snow totals increase steadily with increasing elevations and winter arrives a month earlier and stays a month longer at the 1,000 to 2,000 foot level. Most snow is light or dry, i.e., low in water content. Freezing rain is extremely rare. Fog, made of water droplets, occurs on about fifteen days. In general, ice-fog does not occur in Anchorage.

Spring begins in late April and May when days are warm and sunny, nights are cool, and precipitation is exceedingly small. Foliage turns green by late May.

The wind in Anchorage is generally light. However, on several days each winter, strong northerly winds, up to 90 mph, affect the entire Anchorage area. Also during the winter there are about eight occurrences of very strong southeast winds which affect only the east side of Anchorage and the slopes of the Chugach Mountains. These winds occur more often above the 800 feet elevation in the Chugach where winds are funneled thru creek canyons. On the east side of Anchorage, damaging winds of over 100 mph have been recorded.

The average occurrence of the first snow is mid-October, but has occurred as early as mid-September. The average date of the last snow is mid-April, but has occurred as late as early May. The growing season is about 125 days. Average occurrence of the last temperature of 32 degrees in spring is mid-May and the first in fall is mid-September. Daylight varies from about 19 hours in late June to 6 hours in late December with 12 hours of daylight occurring in late September and late March.

NORMALS, MEANS, AND EXTREMES

ANCHORAGE, ALASKA

LATITUDE: 61°13'N　LONGITUDE: 149°53' W　ELEVATION: FT. GRND 114 BARO 157　TIME ZONE: YUKON　WBAN: 26451

	(a)	JAN	FEB	MAR	APR	MAY	JUNE	JULY	AUG	SEP	OCT	NOV	DEC	YEAR
TEMPERATURE °F:														
Normals														
-Daily Maximum		21.4	25.8	33.1	42.8	54.4	61.6	65.2	63.0	55.2	40.5	27.2	22.5	42.7
-Daily Minimum		8.4	11.5	18.1	28.6	38.8	47.2	51.7	49.5	41.6	28.7	15.1	10.0	29.1
-Monthly		14.9	18.7	25.7	35.8	46.6	54.4	58.4	56.3	48.4	34.6	21.2	16.3	35.9
Extremes														
-Record Highest	40	50	48	51	65	77	85	82	82	73	61	53	48	85
-Year		1961	1991	1984	1976	1969	1969	1989	1978	1957	1993	1979	1992	JUN 1969
-Record Lowest	40	-34	-26	-24	-4	17	33	38	31	19	-5	-21	-30	-34
-Year		1975	1956	1971	1985	1964	1961	1964	1984	1992	1956	1956	1964	JAN 1975
NORMAL DEGREE DAYS:														
Heating (base 65°F)		1553	1296	1218	876	570	318	205	270	498	942	1314	1510	10570
Cooling (base 65°F)		0	0	0	0	0	0	0	0	0	0	0	0	0
% OF POSSIBLE SUNSHINE	38	36	43	51	51	52	47	44	39	40	37	33	28	42
MEAN SKY COVER (tenths)														
Sunrise - Sunset	40	7.0	7.1	6.8	7.1	7.6	7.9	7.9	7.9	7.8	7.6	7.4	7.5	7.5
MEAN NUMBER OF DAYS:														
Sunrise to Sunset														
-Clear	40	7.1	6.5	7.6	5.7	4.0	2.7	3.5	3.3	3.8	5.2	5.4	5.9	60.7
-Partly Cloudy	40	4.7	3.6	5.4	6.2	6.7	6.9	5.7	6.0	5.4	4.7	4.8	4.0	64.1
-Cloudy	40	19.3	18.1	18.0	18.1	20.3	20.4	21.8	21.7	20.9	21.1	19.8	21.1	240.5
Precipitation														
.01 inches or more	29	7.9	7.9	7.4	6.1	7.2	8.1	11.6	13.6	14.3	12.2	9.6	11.3	117.4
Snow,Ice pellets,hail														
1.0 inches or more	29	2.8	3.1	2.6	1.5	0.0	0.0	0.0	0.0	0.2	2.3	3.4	4.7	20.6
Thunderstorms	40	0.0	0.0	0.0	0.0	0.1	0.1	0.4	0.3	0.1	0.0	0.0	0.0	1.0
Heavy Fog Visibility														
1/4 mile or less	40	6.0	4.3	1.4	0.6	0.3	0.1	0.2	0.9	1.4	2.0	3.8	5.2	26.1
Temperature °F														
-Maximum														
70° and above	29	0.0	0.0	0.0	0.0	0.6	3.1	6.5	3.2	0.1	0.0	0.0	0.0	13.6
32° and below	29	24.6	19.7	11.3	2.1	0.0	0.0	0.0	0.0	0.0	4.7	20.6	24.3	107.2
-Minimum														
32° and below	29	30.5	27.2	28.2	20.9	2.9	0.0	0.0	0.1	3.3	20.0	28.0	30.1	191.2
0° and below	29	9.7	7.2	2.2	0.*	0.0	0.0	0.0	0.0	0.0	0.1	3.2	7.3	29.8
AVG. STATION PRESS.(mb)	21	1000.0	1001.5	1001.5	1004.0	1005.9	1007.9	1010.0	1008.3	1004.1	999.0	998.8	999.4	1003.4
RELATIVE HUMIDITY (%)														
Hour 03	40	73	73	71	72	72	75	80	82	81	77	78	76	76
Hour 09 (Local Time)	40	74	73	70	66	63	67	73	78	79	78	77	77	73
Hour 15	40	72	67	57	53	49	56	62	64	63	66	73	76	63
Hour 21	40	73	71	68	64	59	62	69	75	77	75	77	77	71
PRECIPITATION (inches):														
Water Equivalent														
-Normal		0.79	0.78	0.69	0.67	0.73	1.14	1.71	2.44	2.70	2.03	1.11	1.12	15.91
-Maximum Monthly	40	2.71	3.07	2.76	1.91	1.93	3.40	4.44	9.77	6.64	4.11	2.84	2.67	9.77
-Year		1987	1955	1979	1977	1989	1962	1958	1989	1990	1986	1976	1955	AUG 1989
-Minimum Monthly	40	0.02	0.07	T	T	0.02	0.17	0.42	0.33	0.76	0.35	0.08	0.11	T
-Year		1982	1958	1983	1969	1957	1993	1972	1969	1973	1960	1985	1982	MAR 1983
-Maximum in 24 hrs	40	1.19	1.16	1.25	0.78	1.18	1.84	2.06	4.12	1.92	1.60	1.66	1.62	4.12
-Year		1961	1956	1986	1989	1980	1962	1956	1989	1961	1986	1964	1955	AUG 1989
Snow,Ice pellets,hail														
-Maximum Monthly	40	27.5	48.5	31.0	27.6	3.9	0.0	0.0	0.0	4.6	27.1	32.4	41.6	48.5
-Year		1990	1955	1979	1963	1963				1965	1982	1956	1955	FEB 1955
-Maximum in 24 hrs	40	10.5	12.4	14.5	9.1	3.9	0.0	0.0	0.0	3.5	11.2	16.4	17.7	17.7
-Year		1955	1956	1959	1955	1963				1965	1991	1964	1955	DEC 1955
WIND:														
Mean Speed (mph)	40	6.4	6.8	6.9	7.2	8.4	8.3	7.3	6.9	6.7	6.7	6.5	6.2	7.0
Prevailing Direction through 1963		NNE	N	N	N	S	S	S	S	NNE	N	NNE	NNE	N
Fastest Obs. 1 Min.														
-Direction (!!)	36	03	04	03	15	35	17	16	02	22	03	04	05	03
-Speed (MPH)	36	61	52	51	35	33	30	29	31	35	40	41	41	61
-Year		1971	1979	1989	1964	1964	1971	1957	1987	1993	1966	1978	1964	JAN 1971
Peak Gust														
-Direction (!!)	14	E	NE	NE	SE	S	SE	SE	N	S	S	NE	SE	NE
-Speed (mph)	14	64	61	75	43	43	46	40	44	48	55	55	55	75
-Date		1986	1993	1989	1987	1988	1985	1980	1987	1985	1987	1990	1992	MAR 1989

PRECIPITATION (inches)

ANCHORAGE, ALASKA

YEAR	JAN	FEB	MAR	APR	MAY	JUNE	JULY	AUG	SEP	OCT	NOV	DEC	ANNUAL
#1964	0.35	1.15	1.07	0.89	0.97	1.73	1.08	2.16	0.83	2.31	2.71	0.64	15.89
1965	0.57	0.67	0.83	0.30	0.51	0.96	1.74	1.58	4.60	1.44	1.86	1.44	16.50
1966	0.63	0.80	0.44	0.70	0.75	0.27	0.71	2.47	2.45	0.86	1.11	1.06	12.25
1967	1.25	1.01	0.98	0.49	1.07	1.44	2.47	2.96	2.86	0.51	1.72	2.40	19.16
1968	0.83	1.67	0.29	0.85	1.60	0.62	1.34	0.69	1.05	1.61	1.08	0.45	12.08
1969	0.28	0.73	0.10	T	0.86	0.18	2.14	0.33	0.78	0.90	0.84	0.94	8.08
1970	0.86	0.57	0.29	0.27	0.43	0.85	2.03	2.23	1.11	1.62	1.21	1.62	13.09
1971	0.24	1.49	0.70	0.63	0.52	0.37	2.86	2.58	1.79	2.16	0.67	0.87	14.88
1972	0.56	0.63	0.68	0.73	0.81	0.61	0.42	1.40	4.42	2.89	0.76	0.72	14.63
1973	0.72	0.11	0.65	0.33	0.14	1.07	0.60	3.40	0.76	1.74	0.78	0.38	10.68
1974	0.02	1.15	0.60	0.61	0.34	0.69	1.22	1.62	1.53	2.63	1.01	2.00	13.42
1975	0.43	0.77	0.54	1.71	0.40	0.47	1.33	1.19	4.52	0.69	0.10	0.89	13.04
1976	0.98	0.33	1.77	0.74	0.16	0.33	0.60	0.97	3.50	1.29	2.84	1.03	14.54
1977	1.35	0.52	0.84	1.91	0.46	0.49	1.37	1.35	4.08	1.92	0.53	0.69	15.51
1978	0.39	1.19	0.45	0.02	0.03	3.09	1.78	0.54	2.16	1.65	0.85	2.60	14.75
1979	0.23	0.69	2.76	0.94	0.15	1.79	3.84	1.56	2.73	2.54	2.77	1.15	21.15
1980	1.28	1.18	0.30	0.19	1.68	2.73	2.27	3.06	2.53	3.05	0.49	0.41	19.17
1981	0.93	0.97	0.41	0.19	0.81	0.83	4.39	4.96	2.15	3.49	1.85	0.36	21.34
1982	0.02	0.69	0.42	0.27	0.54	1.56	2.41	2.33	4.66	2.95	1.72	0.11	17.68
1983	0.21	0.23	T	1.36	0.59	0.66	0.55	2.89	2.29	2.67	0.23	0.48	12.16
1984	1.30	1.08	0.08	0.93	0.96	1.10	1.11	3.21	2.59	1.38	0.15	1.08	14.97
1985	0.70	0.67	0.86	0.50	1.45	1.01	0.99	3.54	3.17	1.07	0.08	1.47	15.51
1986	0.20	0.55	1.70	0.42	0.50	0.33	2.02	3.62	2.85	4.11	1.23	1.42	18.95
1987	1.72	0.20	0.17	0.24	0.67	1.09	1.89	0.43	1.91	2.60	1.90	1.12	14.93
1988	0.38	0.32	0.65	0.37	0.56	0.79	0.64	3.77	1.26	2.96	1.11	1.51	14.32
1989	0.26	0.17	0.22	0.98	1.93	1.14	2.89	9.77	3.92	3.63	1.01	1.63	27.55
1990	1.42	1.46	0.46	0.27	0.71	1.52	0.81	1.90	6.64	0.73	1.31	1.78	19.01
1991	0.62	0.42	0.65	0.23	0.12	0.18	2.82	3.54	3.41	1.93	1.57	1.82	17.31
1992	1.17	1.04	0.31	0.08	0.58	1.21	0.79	2.49	2.83	2.08	1.17	0.69	14.44
1993	0.94	1.17	0.29	0.09	1.17	0.17	0.57	4.02	4.27	1.90	2.00	0.30	16.89
Record Mean	0.83	0.79	0.61	0.54	0.64	1.08	1.87	2.55	2.69	1.80	1.06	1.07	15.52

AVERAGE TEMPERATURE (deg. F)

ANCHORAGE, ALASKA

YEAR	JAN	FEB	MAR	APR	MAY	JUNE	JULY	AUG	SEP	OCT	NOV	DEC	ANNUAL
#1964	14.1	20.1	17.3	33.7	41.2	55.9	57.1	54.6	48.7	35.0	21.3	1.0	33.3
1965	10.0	10.3	36.5	39.7	45.4	52.9	59.0	56.3	53.7	30.7	20.5	13.1	35.7
1966	9.7	14.7	16.7	35.3	44.2	55.4	57.9	54.4	46.3	32.3	17.5	10.8	32.9
1967	7.1	14.4	23.2	35.1	47.1	56.1	59.3	58.3	48.1	35.7	30.0	16.0	35.9
1968	12.5	23.1	28.4	35.0	48.2	54.8	59.9	58.2	47.1	33.1	22.1	6.3	35.7
1969	4.6	17.9	28.6	39.4	47.7	57.7	59.3	54.2	48.9	41.2	23.7	28.4	37.7
1970	9.2	29.6	35.4	36.4	48.1	54.8	57.1	54.6	46.4	32.5	24.9	14.9	37.0
1971	2.7	20.6	14.2	33.4	41.6	51.2	55.4	55.0	46.1	32.3	18.3	16.2	32.3
1972	6.4	13.5	15.7	26.8	43.3	51.9	59.0	56.6	44.5	31.7	21.3	12.4	31.9
1973	2.9	13.1	24.2	35.8	43.6	51.4	57.8	53.8	45.7	32.2	13.6	18.3	32.7
1974	6.8	14.1	23.3	37.9	47.9	55.5	57.3	56.3	49.8	34.5	22.6	18.8	35.4
1975	11.9	12.9	22.5	32.9	46.3	53.0	58.6	56.6	49.3	34.6	14.2	11.6	33.7
1976	17.1	12.8	24.1	34.8	44.9	53.7	58.9	56.3	47.4	33.4	30.6	23.1	36.4
1977	32.0	32.7	24.7	35.7	46.9	57.8	62.6	60.3	50.7	38.3	15.3	11.3	39.0
1978	21.2	26.3	29.3	39.1	49.0	54.5	58.8	59.8	51.5	39.3	26.3	21.4	39.7
1979	22.3	10.6	31.6	38.8	50.2	55.9	60.4	58.8	52.0	41.1	33.5	10.0	38.8
1980	14.3	27.4	27.2	39.3	45.8	53.2	57.0	54.4	46.7	37.2	27.6	0.8	35.9
1981	31.5	24.8	34.4	36.0	50.7	53.8	57.4	54.8	47.9	36.0	21.8	15.9	38.8
1982	6.4	15.5	26.3	33.1	44.5	52.9	56.2	54.7	47.5	26.6	21.0	21.5	33.8
1983	16.2	21.4	28.7	37.4	48.7	55.9	58.5	56.1	45.3	34.4	24.9	16.7	37.0
1984	18.8	19.4	36.4	38.8	49.6	58.8	60.8	56.6	49.3	35.5	19.8	18.9	38.6
1985	30.3	13.5	26.7	28.4	45.1	51.9	58.5	55.2	47.6	30.3	14.0	27.5	35.8
1986	25.6	21.8	24.1	31.0	46.6	54.6	58.0	54.3	48.6	39.0	25.0	28.3	38.1
1987	22.8	25.3	26.8	37.9	47.2	51.9	57.1	57.3	48.0	38.9	26.9	18.2	38.2
1988	18.0	22.7	31.3	37.1	48.5	55.2	58.8	56.0	48.0	33.3	20.5	22.0	37.6
1989	3.5	17.6	23.6	39.3	46.3	55.3	59.4	59.0	50.6	34.0	17.2	24.2	35.8
1990	15.5	3.8	28.6	39.9	49.9	57.1	58.6	57.8	49.6	32.3	9.9	14.8	34.8
1991	15.9	19.6	23.7	37.7	46.6	55.7	57.5	55.5	51.0	33.0	25.0	20.5	36.8
1992	20.3	15.0	24.9	35.2	46.1	55.9	59.5	55.8	40.3	31.2	27.1	15.0	35.5
1993	14.5	21.0	28.8	40.6	50.7	56.3	61.1	58.8	48.8	38.7	25.2	24.0	39.0
Record Mean	13.9	17.8	24.6	35.4	46.4	54.4	58.1	56.0	47.9	34.7	21.5	15.4	35.5
Max	21.0	25.6	32.9	43.2	54.7	62.2	65.4	63.2	55.0	41.0	27.8	21.9	42.8
Min	6.8	9.9	16.2	27.5	38.1	46.5	50.8	48.7	40.7	28.4	15.3	8.9	28.2

REFERENCE NOTES

ANCHORAGE, ALASKA

GENERAL
T - TRACE AMOUNT.
BLANK ENTRIES DENOTE MISSING/UNREPORTED DATA.
# INDICATES A STATION OR INSTRUMENT RELOCATION.

SPECIFIC

PM - INCLUDES LAST DAY OF PREVIOUS MONTH
ASOS - AUTOMATED SURFACE OBSERVING SYSTEM IN OPERATION DURING THESE MONTHS.

(a) - LENGTH OF RECORD IN YEARS, ALTHOUGH INDIVIDUAL MONTHS MAY BE MISSING.
0.* OR * - THE VALUE IS BETWEEN 0.0 AND 0.05
NORMALS - BASED ON THE 1961-1990 RECORD PERIOD.
EXTREMES - DATES ARE THE MOST RECENT OCCURRENCE
WIND DIR.- NUMERALS SHOW TENS OF DEGRESS CLOCKWISE FROM TRUE NORTH. "00" INDICATES CALM.
RESULTANT DIRECTIONS ARE GIVEN TO WHOLE DEGREES.
BOLD VALUES INDICATE EXTREME VALUES WHICH OCCURRED AFTER THE ASOS SYSTEM WAS COMMISSIONED.

RECORD = PERIOD OF RECORD
RECORD MEAN PRECIPITATION IS THE MEAN OF ALL DAILY PRECIPITATION AMOUNTS DURING THE PERIOD OF RECORD.
RECORD MAX(MIN) TEMPERATURE IS THE MEAN OF ALL DAILY MAX(MIN) TEMPERATURES DURING THE PERIOD OF RECORD.
RECORD MEAN TEMPERATURE IS THE SUM OF THE RECORD MAX AND RECORD MIN DIVIDED BY 2.
AVERAGE TEMPERATURE IS THE SUM OF THE MEAN DAILY MAX AND MIN TEMPERATURE DIVIDED BY 2.

EXCEPTIONS

RECORD MEANS ARE THROUGH THE CURRENT YEAR, BEGINNING IN 1943 FOR TEMPERATURE
1943 FOR PRECIPITATION
1943 FOR SNOWFALL

HEATING DEGREE DAYS Base 65 deg. F

ANCHORAGE, ALASKA

SEASON	JULY	AUG	SEP	OCT	NOV	DEC	JAN	FEB	MAR	APR	MAY	JUNE	TOTAL
1964-65	240	315	485	922	1303	1980	1701	1530	879	754	598	357	11064
1965-66	178	261	333	1055	1328	1603	1713	1402	1493	882	641	285	11174
1966-67	214	324	553	1008	1419	1674	1791	1411	1292	890	547	261	11384
1967-68	172	200	501	901	1042	1513	1625	1209	1129	894	510	301	9997
1968-69	153	208	530	982	1280	1816	1869	1312	1124	761	527	217	10779
1969-70	168	330	478	732	1235	1128	1722	983	910	852	515	298	9351
1970-71	239	312	552	1003	1201	1550	1934	1238	1574	940	717	410	11670
1971-72	291	302	561	1008	1396	1508	1814	1488	1521	1138	666	384	12077
1972-73	185	252	608	1025	1308	1627	1925	1448	1258	866	654	399	11555
1973-74	216	342	573	1012	1532	1440	1797	1416	1285	805	526	279	11223
1974-75	235	263	452	937	1263	1425	1643	1454	1313	954	575	354	10868
1975-76	192	252	463	937	1517	1654	1485	1511	1260	897	615	332	11115
1976-77	184	262	521	972	1028	1294	1017	897	1241	872	554	208	9050
1977-78	75	144	421	820	1486	1659	1349	1077	1100	771	491	308	9701
1978-79	186	160	400	792	1153	1344	1321	1515	1029	781	454	268	9403
1979-80	138	184	384	735	937	1704	1568	1083	1164	764	592	347	9600
1980-81	243	320	542	855	1115	1990	1032	1122	943	863	438	329	9792
1981-82	230	307	507	893	1290	1516	1813	1382	1191	949	625	356	11059
1982-83	261	313	520	1184	1315	1342	1507	1216	1117	821	500	267	10363
1983-84	194	269	585	945	1194	1491	1425	1319	880	778	471	179	9730
1984-85	129	254	464	906	1350	1423	1070	1437	1182	1091	610	388	10304
1985-86	193	298	516	1065	1523	1155	1215	1206	1260	1013	564	307	10315
1986-87	215	325	486	800	1194	1133	1303	1104	1176	805	543	386	9470
1987-88	243	232	506	801	1136	1444	1450	1221	1037	830	504	285	9689
1988-89	184	270	503	975	1331	1326	1908	1322	1277	765	573	286	10720
1989-90	173	181	423	956	1428	1255	1533	1715	1121	746	465	237	10233
1990-91	191	222	457	1006	1648	1552	1518	1265	1273	813	563	273	10781
1991-92	226	287	414	988	1193	1373	1380	1444	1240	891	579	268	10283
1992-93	161	280	735	1039	1131	1543	1563	1226	1117	725	436	252	10208
1993-94	125	187	477	808	1191	1267							

COOLING DEGREE DAYS Base 65 deg. F

ANCHORAGE, ALASKA

YEAR	JAN	FEB	MAR	APR	MAY	JUNE	JULY	AUG	SEP	OCT	NOV	DEC	TOTAL
1969	0	0	0	0	0	5	1	2	0	0	0	0	8
1970	0	0	0	0	0	0	0	0	0	0	0	0	0
1971	0	0	0	0	0	0	1	0	0	0	0	0	1
1972	0	0	0	0	0	0	5	0	0	0	0	0	5
1973	0	0	0	0	0	0	0	0	0	0	0	0	0
1974	0	0	0	0	0	0	1	0	0	0	0	0	1
1975	0	0	0	0	0	0	2	0	0	0	0	0	2
1976	0	0	0	0	0	0	3	0	0	0	0	0	3
1977	0	0	0	0	0	0	8	3	0	0	0	0	11
1978	0	0	0	0	0	0	1	7	0	0	0	0	8
1979	0	0	0	0	0	0	4	0	0	0	0	0	4
1980	0	0	0	0	0	0	0	0	0	0	0	0	0
1981	0	0	0	0	0	0	0	0	0	0	0	0	0
1982	0	0	0	0	0	0	0	0	0	0	0	0	0
1983	0	0	0	0	0	0	0	0	0	0	0	0	0
1984	0	0	0	0	0	0	5	1	0	0	0	0	6
1985	0	0	0	0	0	0	0	0	0	0	0	0	0
1986	0	0	0	0	0	0	4	0	0	0	0	0	4
1987	0	0	0	0	0	0	2	0	0	0	0	0	2
1988	0	0	0	0	0	0	0	0	0	0	0	0	0
1989	0	0	0	0	0	0	5	2	0	0	0	0	7
1990	0	0	0	0	0	3	1	2	0	0	0	0	6
1991	0	0	0	0	0	0	0	0	0	0	0	0	0
1992	0	0	0	0	0	0	0	0	0	0	0	0	0
1993	0	0	0	0	0	0	11	0	0	0	0	0	11

SNOWFALL (inches)

ANCHORAGE, ALASKA

SEASON	JULY	AUG	SEP	OCT	NOV	DEC	JAN	FEB	MAR	APR	MAY	JUNE	TOTAL
1964-65	0.0	0.0	0.0	10.6	25.8	16.2	8.5	11.5	3.0	2.8	0.2	0.0	78.6
1965-66	0.0	0.0	4.6	16.7	17.9	23.9	9.7	10.7	5.2	6.1	T	0.0	94.8
1966-67	0.0	0.0	0.0	1.9	13.5	13.4	16.7	16.3	8.8	4.0	0.0	0.0	74.6
1967-68	0.0	0.0	0.0	4.8	2.5	26.5	7.3	26.1	3.2	7.8	T	0.0	78.2
1968-69	0.0	0.0	1.3	15.1	14.7	7.9	6.5	10.6	1.1	T	0.0	0.0	57.2
1969-70	0.0	0.0	0.0	0.2	2.6	8.7	16.1	8.7	1.4	4.1	0.0	0.0	41.8
1970-71	0.0	0.0	T	4.2	4.2	11.2	1.2	18.5	11.1	8.3	T	0.0	58.7
1971-72	0.0	0.0	0.0	11.9	8.2	11.4	9.6	8.9	12.0	12.6	T	0.0	74.6
1972-73	0.0	0.0	1.5	3.3	10.7	6.5	8.1	1.0	16.1	1.3	0.0	0.0	48.5
1973-74	0.0	0.0	0.0	6.6	10.6	6.7	0.5	23.3	8.2	1.9	0.0	0.0	57.8
1974-75	0.0	0.0	0.0	4.4	8.4	29.2	5.7	15.4	8.3	16.1	0.4	0.0	87.9
1975-76	0.0	0.0	0.0	T	2.0	11.5	9.7	1.8	30.7	5.6	T	0.0	61.3
1976-77	0.0	0.0	0.0	11.4	11.1	13.8	6.1	2.1	9.5	14.0	0.0	0.0	68.0
1977-78	0.0	0.0	1.0	13.2	12.6	10.6	7.3	20.8	9.5	T	0.0	0.0	75.0
1978-79	0.0	0.0	0.0	3.9	8.5	35.2	3.6	6.2	31.0	2.8	0.0	0.0	91.2
1979-80	0.0	0.0	0.0	4.3	13.7	16.0	12.0	18.7	3.4	0.8	0.0	0.0	68.9
1980-81	0.0	0.0	0.0	10.2	4.2	1.4	5.0	6.6	4.4	1.1	T	0.0	32.9
1981-82	0.0	0.0	1.5	6.3	20.0	7.6	0.5	0.6	5.6	3.5	0.7	0.0	46.3
1982-83	0.0	0.0	0.0	27.1	23.4	1.9	3.7	4.3	T	11.0	0.0	0.0	71.4
1983-84	0.0	0.0	T	23.7	2.1	10.5	15.0	18.9	0.2	9.8	0.0	0.0	80.2
1984-85	0.0	0.0	0.0	3.3	1.8	18.0	9.7	7.9	12.8	7.3	1.3	0.0	62.1
1985-86	0.0	0.0	0.0	0.8	1.5	6.1	5.1	6.1	21.0	5.4	0.1	0.0	46.1
1986-87	0.0	0.0	0.0	T	3.8	10.1	18.5	2.2	2.5	1.6	0.0	0.0	38.7
1987-88	0.0	0.0	0.0	T	29.2	26.3	4.7	9.2	8.5	2.0	0.0	0.0	79.9
1988-89	0.0	0.0	0.0	12.0	15.3	18.6	10.1	2.3	5.1	T	0.2	0.0	63.6
1989-90	0.0	0.0	0.0	16.3	10.1	20.0	27.5	23.0	4.7	0.8	T	0.0	102.4
1990-91	0.0	0.0	0.0	1.6	16.9	21.4	7.7	5.4	12.7	T	0.0	0.0	65.7
1991-92	0.0	0.0	0.0	11.6	19.3	26.2	21.4	18.3	2.7	T	0.2	0.0	99.7
1992-93	0.0	0.0	3.0	13.0	9.1	12.1	13.7	18.3	5.7	0.0	0.0	0.0	74.9
1993-94	0.0	0.0	T	4.4	11.9	5.1							
Record Mean	0.0	0.0	0.3	7.3	10.7	14.8	10.7	11.5	9.0	4.8	0.4	0.0	69.3

ANNETTE, ALASKA

Annette is located on Annette Island, one of the islands at the southern tip of the Alexander Archipelago of southeastern Alaska and is 20 airline miles south of Ketchikan.

Due to the proximity of Pacific Ocean waters, particularly to the south and west, the climate of Annette is maritime. Temperatures are relatively mild and daily variations between high and low readings are confined within rather narrow limits, with the ranges between average maximum and minimum readings for all months of the year averaging 8 to 13 degrees. Periods of sub–freezing temperatures seldom extend beyond ten days duration. A minimum reading below zero is a rare occurrence. Conversely, extreme high temperatures occur infrequently during the summer months. Temperatures reach the middle 80s nearly every summer. It is not unusual for daily maximum readings to remain in the 50s during the summer months.

Because it is located almost directly in the path of easterly–moving storms crossing the Gulf of Alaska, Annette experiences frequent and relatively heavy precipitation with annual amounts closely approximating that of the immediate coastal regions of Washington and Oregon. October and November are usually the wettest months and the greatest percentage of precipitation, even in the winter months, occurs in the form of rain. Some snow, mixed with rain, may occur as early as October. However, appreciable snowfall seldom occurs from late March to late November. Accumulated snow depths of a foot or more are infrequent and, because of moderating temperatures, snow cover seldom persists beyond a week or two. Topography has a pronounced effect on precipitation in this region and Annette averages 65 percent as much as Ketchikan, only 20 miles to the north. Despite relatively heavy precipitation, flood damage is seldom of consequence because of the small drainage areas and steep topography which provide rapid escape of surface runoff into the sea.

Strong southeasterly winds are frequent from October through March. Thick fogs are infrequent and of short duration. Thunderstorms average one or two a year and generally are the result of frontal activity. The considerable amount of cloudiness prevalent over this area results in a corresponding low percentage of possible sunshine.

NORMALS, MEANS, AND EXTREMES

ANNETTE, ALASKA

LATITUDE: 55 °02'N LONGITUDE: 131 °34' W ELEVATION: FT. GRND 109 BARO 114 TIME ZONE: YUKON WBAN: 25308

	(a)	JAN	FEB	MAR	APR	MAY	JUNE	JULY	AUG	SEP	OCT	NOV	DEC	YEAR
TEMPERATURE °F:														
Normals														
-Daily Maximum		38.9	41.9	44.7	49.5	55.7	60.8	64.5	64.9	60.2	51.8	43.8	40.0	51.4
-Daily Minimum		29.7	32.1	33.9	37.0	42.3	47.8	52.0	52.3	48.1	41.9	34.7	31.4	40.3
-Monthly		34.3	37.0	39.3	43.3	49.0	54.3	58.3	58.6	54.2	46.9	39.3	35.7	45.9
Extremes														
-Record Highest	46	61	65	62	82	88	89	89	90	82	71	67	62	90
-Year		1981	1993	1969	1976	1963	1969	1976	1960	1989	1964	1970	1962	AUG 1960
-Record Lowest	46	1	2	1	21	31	37	40	40	33	18	-3	1	-3
-Year		1982	1956	1955	1954	1971	1965	1973	1973	1983	1984	1985	1964	NOV 1985
NORMAL DEGREE DAYS:														
Heating (base 65°F)		952	784	797	651	496	325	214	204	324	561	771	908	6987
Cooling (base 65°F)		0	0	0	0	0	0	7	6	0	0	0	0	13
% OF POSSIBLE SUNSHINE														
MEAN SKY COVER (tenths)														
Sunrise - Sunset	41	8.1	8.2	8.0	7.9	7.7	8.1	7.9	7.7	7.8	8.6	8.3	8.4	8.0
MEAN NUMBER OF DAYS:														
Sunrise to Sunset														
-Clear	41	4.2	3.0	3.7	3.5	3.8	2.5	3.5	4.0	3.7	2.1	3.1	2.9	40.0
-Partly Cloudy	41	3.7	4.1	4.7	5.8	6.7	6.5	6.1	6.0	5.7	4.0	3.4	3.8	60.3
-Cloudy	41	23.1	21.2	22.5	20.7	20.5	21.1	21.4	21.1	20.6	24.8	23.5	24.3	264.9
Precipitation														
.01 inches or more	46	19.7	18.6	19.9	18.4	16.9	14.7	14.0	14.7	17.6	23.6	22.3	22.8	223.1
Snow, Ice pellets, hail														
1.0 inches or more	46	4.0	3.2	2.3	0.7	0.*	0.0	0.0	0.0	0.0	0.1	1.4	3.0	14.8
Thunderstorms	31	0.1	0.0	0.*	0.0	0.0	0.2	0.2	0.1	0.2	0.3	0.4	0.1	1.5
Heavy Fog Visibility														
1/4 mile or less	31	0.6	0.6	0.7	0.4	1.0	1.3	2.2	2.3	3.4	1.5	0.9	0.4	15.2
Temperature °F														
-Maximum														
70° and above	33	0.0	0.0	0.0	0.3	1.7	3.4	6.8	6.7	1.8	0.*	0.0	0.0	20.8
32° and below	33	5.0	1.7	0.5	0.0	0.0	0.0	0.0	0.0	0.0	0.1	1.4	3.8	12.6
-Minimum														
32° and below	33	17.4	13.3	11.8	4.1	0.3	0.0	0.0	0.0	0.0	1.6	10.0	14.6	73.1
0° and below	33	0.0	0.0	0.0	0.0	0.0	0.0	0.0	0.0	0.0	0.0	0.*	0.0	*
AVG. STATION PRESS. (mb)	6	1008.3	1006.3	1006.6	1010.7	1011.2	1012.7	1013.7	1013.1	1011.8	1007.7	1006.7	1004.6	1009.4
RELATIVE HUMIDITY (%)														
Hour 03	28	81	80	80	82	85	87	89	90	89	87	83	82	85
Hour 09 (Local Time)	28	81	80	78	77	77	79	82	84	86	85	83	82	81
Hour 15	28	76	72	68	66	66	68	70	72	73	78	78	80	72
Hour 21	16	78	78	78	77	75	77	78	82	84	84	81	81	79
PRECIPITATION (inches):														
Water Equivalent														
-Normal		10.07	8.79	7.89	7.79	6.03	4.67	4.27	6.15	9.28	15.47	11.67	11.20	103.28
-Maximum Monthly	46	20.69	18.06	23.57	21.35	14.68	10.40	10.85	20.72	17.40	34.87	28.09	28.90	34.87
-Year		1958	1965	1960	1949	1966	1956	1959	1958	1987	1958	1959	1959	OCT 1958
-Minimum Monthly	46	0.77	0.12	2.13	1.23	1.57	0.61	0.56	0.71	2.52	7.04	3.61	3.86	0.12
-Year		1950	1989	1983	1948	1968	1982	1971	1954	1965	1987	1970	1983	FEB 1989
-Maximum in 24 hrs	46	4.65	5.58	4.02	4.75	4.29	3.79	2.63	4.68	5.04	7.59	5.19	5.20	7.59
-Year		1962	1965	1966	1956	1958	1962	1967	1956	1952	1958	1957	1959	OCT 1958
Snow, Ice pellets, hail														
-Maximum Monthly	46	36.3	33.4	46.9	22.2	5.0	T	0.0	0.0	0.0	4.2	18.3	44.8	46.9
-Year		1974	1974	1971	1971	1964	1988				1956	1973	1964	MAR 1971
-Maximum in 24 hrs	46	12.3	14.2	11.8	14.0	5.0	T	0.0	0.0	0.0	3.2	9.6	12.8	14.2
-Year		1972	1948	1971	1971	1964	1988				1956	1973	1964	FEB 1948
WIND:														
Mean Speed (mph)	31	12.0	12.3	11.0	11.2	9.3	9.0	8.0	8.3	9.3	12.0	12.4	12.6	10.6
Prevailing Direction through 1963		ESE	SE	SE	SSE	SSE	SSE	SSE	SSE	SE	SE	ESE	ESE	SSE
Fastest Obs. 1 Min.														
-Direction (!!)	41	16	16	14	14	14	16	16	16	11	16	13	16	14
-Speed (MPH)	41	58	50	48	60	44	44	35	40	51	55	51	58	60
-Year		1960	1954	1977	1952	1958	1959	1954	1955	1963	1958	1984	1959	APR 1952
Peak Gust														
-Direction (!!)	14	SE	SE	SE	SE	SE	SE	SE	E	SE	SE	SE	SE	SE
-Speed (mph)	14	63	66	60	59	55	49	39	51	55	74	71	69	74
-Date		1988	1991	1986	1985	1985	1993	1988	1991	1980	1990	1980	1986	OCT 1990

PRECIPITATION (inches)

ANNETTE, ALASKA

YEAR	JAN	FEB	MAR	APR	MAY	JUNE	JULY	AUG	SEP	OCT	NOV	DEC	ANNUAL
1964	14.94	13.22	11.06	8.11	4.96	3.82	6.23	10.49	9.76	18.62	9.87	9.81	120.89
1965	15.46	18.06	3.00	9.87	8.24	4.94	3.34	3.91	2.52	27.15	11.47	13.72	121.68
1966	11.19	11.72	17.13	3.22	14.68	2.23	2.77	4.81	9.40	17.90	6.17	9.45	110.67
1967	9.25	15.32	3.01	2.33	7.97	2.98	8.78	10.25	15.51	17.90	6.02	12.11	111.43
1968	11.05	5.80	10.43	13.63	1.57	4.13	4.81	6.42	15.19	18.81	14.09	4.87	110.80
1969	3.39	3.40	3.50	10.68	4.56	2.83	6.09	11.25	4.90	9.56	24.89	16.25	101.30
1970	7.76	8.03	7.83	8.11	11.21	5.50	5.76	7.88	11.82	15.29	3.61	6.82	99.62
1971	7.59	10.70	9.49	9.96	4.24	3.61	0.56	8.57	7.30	14.04	13.64	8.33	98.03
1972	8.20	8.41	10.54	8.78	5.18	4.95	4.36	9.90	4.78	13.46	10.11	9.43	98.10
1973	9.78	9.66	8.83	8.85	9.50	5.96	5.46	4.67	12.81	15.79	5.37	10.63	107.31
1974	6.42	11.91	3.07	8.02	7.70	5.19	3.10	2.54	7.46	22.49	16.05	15.95	109.90
1975	11.30	7.51	4.43	7.95	3.92	5.56	4.94	6.57	9.93	8.82	9.20	14.88	95.01
1976	16.61	9.94	11.92	8.79	9.78	4.44	6.61	7.56	13.15	11.40	10.02	15.49	125.71
#1977	5.88	16.50	7.46	7.31	3.08	6.46	4.15	2.01	5.66	15.31	11.68	5.94	91.44
1978	3.01	6.10	7.67	5.84	3.92	0.95	1.78	6.00	11.66	15.95	14.39	11.35	88.62
1979	6.51	7.18	9.69	2.24	9.81	6.86	2.44	2.77	12.42	10.35	10.20	12.34	92.81
1980	4.89	8.14	9.72	16.88	4.55	0.81	6.72	6.56	8.09	18.35	17.37	14.11	116.19
1981	10.21	6.41	10.57	10.33	4.69	5.31	2.19	7.31	13.17	8.25	16.61	9.26	104.31
1982	8.80	3.53	6.16	7.48	7.56	0.61	4.73	1.85	5.54	15.55	8.57	7.31	77.69
1983	10.73	6.30	2.13	3.97	4.61	3.50	6.65	11.29	8.37	15.53	8.51	3.86	85.45
1984	11.82	11.41	8.35	5.71	5.30	9.37	6.77	7.23	5.45	13.26	8.35	7.85	100.87
1985	11.86	9.65	8.18	7.55	4.17	3.27	1.74	5.18	6.52	12.40	4.77	9.81	85.10
1986	12.82	3.20	9.93	7.18	6.51	3.60	4.11	3.28	5.71	23.51	11.74	10.21	101.80
1987	10.81	8.65	4.69	11.32	8.31	7.46	1.07	3.71	17.40	7.04	17.82	12.16	110.44
1988	9.49	6.06	13.72	7.14	8.91	5.36	7.16	8.01	8.58	11.29	13.97	8.51	108.20
1989	15.28	0.12	4.41	1.93	2.68	3.00	1.77	2.33	5.22	13.90	17.43	17.80	85.87
1990	10.04	5.78	9.91	3.98	3.77	6.21	2.67	5.76	6.66	14.76	7.09	12.53	89.16
1991	8.44	10.20	6.79	4.45	5.78	4.29	8.16	9.21	10.81	17.72	23.32	22.89	132.06
1992	15.53	6.24	8.01	10.17	7.80	4.77	3.69	4.19	17.19	11.30	14.11	8.94	111.94
1993	12.06	16.53	6.36	5.85	4.71	3.53	4.16	3.43	5.47	11.56	14.59	17.05	105.30
Record Mean	10.23	8.95	8.59	8.26	6.12	4.81	4.86	6.59	9.67	16.21	12.69	12.00	108.98

AVERAGE TEMPERATURE (deg. F)

ANNETTE, ALASKA

YEAR	JAN	FEB	MAR	APR	MAY	JUNE	JULY	AUG	SEP	OCT	NOV	DEC	ANNUAL
1964	35.6	38.7	36.7	40.9	47.4	56.4	55.6	55.7	52.5	46.6	36.4	26.0	42.0
1965	32.1	34.5	38.4	42.7	45.9	51.5	57.0	59.3	55.7	46.9	38.9	33.0	44.7
1966	28.9	36.4	39.2	42.8	46.3	53.6	59.9	57.4	54.6	45.7	38.6	38.9	45.2
1967	35.1	39.2	36.0	44.1	49.3	58.5	57.9	61.0	55.1	48.3	40.7	36.3	46.8
1968	32.2	38.6	42.0	41.9	53.1	55.1	60.5	59.7	53.4	46.6	42.6	31.2	46.4
1969	21.2	35.9	40.0	44.2	53.7	62.2	59.0	58.0	57.8	50.8	42.5	40.4	47.2
1970	33.7	41.9	44.3	44.6	50.2	56.9	56.7	56.8	52.6	48.2	39.3	31.4	46.4
1971	28.8	34.9	35.9	41.6	45.7	52.5	61.7	57.7	53.1	43.9	39.2	30.5	43.8
1972	26.7	30.1	36.9	38.0	49.0	52.7	58.9	58.3	52.3	44.0	41.8	33.1	43.5
1973	33.2	36.3	38.1	42.5	47.2	51.2	54.6	55.4	52.8	44.0	30.8	36.2	43.6
1974	28.1	35.4	36.8	43.0	46.8	50.6	55.1	58.7	55.9	46.6	38.9	38.9	44.6
1975	30.5	30.3	35.1	40.6	46.9	50.6	56.2	55.9	53.8	46.0	37.2	37.3	43.4
1976	38.1	34.2	36.7	44.5	48.4	54.2	58.3	58.5	55.4	47.0	42.6	39.4	46.4
#1977	37.2	43.7	39.5	44.6	49.9	53.3	56.3	62.9	53.5	46.7	35.6	30.0	46.1
1978	34.2	39.0	38.9	45.1	47.6	57.6	58.0	58.5	53.7	48.1	38.7	34.2	46.1
1979	30.7	31.6	41.3	44.0	47.7	52.4	59.1	60.5	55.4	49.0	43.1	37.2	46.0
1980	30.8	40.8	38.3	44.4	50.2	56.5	56.9	58.2	53.5	49.4	43.8	34.2	46.4
1981	45.6	39.6	42.7	42.4	52.2	53.0	60.1	60.2	54.1	46.8	43.2	36.2	48.0
1982	29.4	33.0	38.2	40.0	46.7	57.0	57.8	56.4	54.6	45.7	36.8	38.0	44.5
1983	40.3	43.2	42.9	47.1	50.9	55.3	58.1	57.7	52.0	46.7	41.5	32.9	47.4
1984	39.6	40.4	43.6	43.6	48.1	52.4	55.7	57.0	52.9	44.6	37.8	33.4	45.8
1985	42.0	34.8	38.4	40.8	48.3	51.7	59.9	56.2	52.3	43.4	27.6	38.3	44.5
1986	40.9	33.9	43.3	42.1	48.0	54.0	57.5	58.4	54.8	51.3	39.8	40.5	47.0
1987	38.8	41.2	38.9	43.5	48.8	53.4	58.9	60.2	53.9	47.3	43.7	38.2	47.2
1988	34.3	38.9	40.5	43.4	49.6	53.2	55.8	57.7	52.6	48.9	42.2	35.8	46.1
1989	34.0	31.2	38.0	46.7	50.6	56.5	59.8	60.0	56.7	46.4	40.6	43.1	47.0
1990	35.9	33.8	41.4	47.0	52.2	56.2	62.1	61.2	55.2	44.8	36.0	33.1	46.6
1991	32.0	40.8	36.9	43.7	49.8	54.6	57.4	59.2	54.5	45.1	43.9	41.5	46.6
1992	41.2	41.0	41.8	44.9	50.6	57.1	59.6	59.2	50.9	46.2	43.2	33.8	47.5
1993	33.5	41.0	41.0	47.5	55.7	57.2	60.8	61.3	56.7	49.5	41.0	41.5	48.9
Record Mean	34.2	36.8	38.9	43.1	49.4	53.9	58.0	58.4	54.0	46.8	39.8	36.0	45.8
Max	38.7	41.6	44.2	49.2	56.0	59.7	64.1	64.6	59.8	51.6	44.2	40.3	51.2
Min	29.6	31.9	33.5	37.0	42.7	48.0	51.8	52.2	48.1	41.9	35.3	31.8	40.3

REFERENCE NOTES

ANNETTE, ALASKA

GENERAL
T - TRACE AMOUNT.
BLANK ENTRIES DENOTE MISSING/UNREPORTED DATA.
# INDICATES A STATION OR INSTRUMENT RELOCATION.

SPECIFIC

(a) - LENGTH OF RECORD IN YEARS, ALTHOUGH INDIVIDUAL MONTHS MAY BE MISSING.
0.* OR * - THE VALUE IS BETWEEN 0.0 AND 0.05
NORMALS - BASED ON THE 1961-1990 RECORD PERIOD.
EXTREMES - DATES ARE THE MOST RECENT OCCURRENCE
WIND DIR.- NUMERALS SHOW TENS OF DEGRESS CLOCKWISE FROM TRUE NORTH. "00" INDICATES CALM.
RESULTANT DIRECTIONS ARE GIVEN TO WHOLE DEGREES.
BOLD VALUES INDICATE EXTREME VALUES WHICH OCCURRED AFTER THE ASOS SYSTEM WAS COMMISSIONED.

RECORD = PERIOD OF RECORD
RECORD MEAN PRECIPITATION IS THE MEAN OF ALL DAILY PRECIPITATION AMOUNTS DURING THE PERIOD OF RECORD.
RECORD MAX(MIN) TEMPERATURE IS THE MEAN OF ALL DAILY MAX(MIN) TEMPERATURES DURING THE PERIOD OF RECORD.
RECORD MEAN TEMPERATURE IS THE SUM OF THE RECORD MAX AND RECORD MIN DIVIDED BY 2.
AVERAGE TEMPERATURE IS THE SUM OF THE MEAN DAILY MAX AND MIN TEMPERATURE DIVIDED BY 2.

EXCEPTIONS

MEAN WIND SPEED, THUNDERSTORMS, AND HEAVY FOG ARE THROUGH 1976 PLUS 1980.

RECORD MEANS ARE THROUGH THE CURRENT YEAR, BEGINNING IN 1941 FOR TEMPERATURE
1942 FOR PRECIPITATION
1948 FOR SNOWFALL

HEATING DEGREE DAYS Base 65 deg. F ANNETTE, ALASKA

SEASON	JULY	AUG	SEP	OCT	NOV	DEC	JAN	FEB	MAR	APR	MAY	JUNE	TOTAL
1964-65	285	282	369	565	855	1200	1014	846	819	661	588	398	7882
1965-66	240	175	272	552	778	985	1114	795	796	659	575	333	7274
1966-67	161	227	307	591	784	802	921	716	890	620	480	191	6690
1967-68	214	128	290	513	723	884	1011	758	708	688	358	292	6567
1968-69	136	161	342	565	664	1038	1354	809	770	619	348	125	6931
1969-70	181	211	210	432	668	758	963	640	638	604	452	235	5992
1970-71	248	249	367	514	762	1032	1111	836	892	694	590	367	7662
1971-72	145	220	349	648	766	1062	1181	1006	862	803	488	363	7893
1972-73	198	201	376	644	690	982	981	797	827	668	543	408	7315
1973-74	314	289	359	646	1017	885	1138	824	867	656	554	424	7973
1974-75	298	200	266	564	776	803	1062	965	920	727	550	423	7554
1975-76	271	276	328	584	825	852	827	888	867	610	509	315	7152
#1976-77	216	194	283	550	666	787	856	591	786	604	462	344	6339
1977-78	262	119	338	560	876	1077	951	721	800	590	534	216	7044
1978-79	213	196	333	519	780	947	1055	930	728	621	532	373	7227
1979-80	176	137	280	490	652	854	1050	698	819	612	454	252	6474
1980-81	247	217	337	475	629	946	596	705	683	672	389	353	6249
1981-82	156	158	321	560	648	887	1096	893	824	740	561	232	7076
1982-83	224	263	305	592	839	828	755	605	679	529	432	287	6338
1983-84	209	222	387	561	699	987	779	709	654	639	518	371	6735
1984-85	278	242	354	627	807	970	705	837	819	719	509	393	7260
1985-86	159	269	377	661	1113	820	742	866	667	681	520	322	7197
1986-87	225	199	299	420	748	751	805	659	804	639	492	350	6391
1987-88	189	154	328	540	631	823	945	751	757	643	468	349	6578
1988-89	279	219	366	491	678	898	953	940	829	538	439	249	6879
1989-90	158	152	248	570	725	675	895	871	728	533	389	254	6198
1990-91	119	127	288	617	862	983	1018	673	864	635	463	310	6959
1991-92	228	197	309	612	626	721	732	690	713	595	441	240	6104
1992-93	164	176	415	576	648	961	970	666	735	516	291	229	6347
1993-94	125	134	247	471	712	719							

COOLING DEGREE DAYS Base 65 deg. F ANNETTE, ALASKA

YEAR	JAN	FEB	MAR	APR	MAY	JUNE	JULY	AUG	SEP	OCT	NOV	DEC	TOTAL
1969	0	0	0	0	7	49	0	0	0	0	0	0	56
1970	0	0	0	0	0	0	0	3	0	0	0	0	3
1971	0	0	0	0	0	0	51	2	0	0	0	0	53
1972	0	0	0	0	0	0	15	0	1	0	0	0	16
1973	0	0	0	0	0	0	0	0	0	0	0	0	0
1974	0	0	0	0	0	0	0	12	0	0	0	0	12
1975	0	0	0	0	0	0	8	0	0	0	0	0	8
1976	0	0	0	0	0	0	16	1	0	0	0	0	17
#1977	0	0	0	0	0	0	0	60	0	0	0	0	60
1978	0	0	0	0	0	1	5	0	0	0	0	0	6
1979	0	0	0	0	0	0	1	8	0	0	0	0	9
1980	0	0	0	0	0	3	0	12	0	0	0	0	15
1981	0	0	0	0	0	0	11	17	0	0	0	0	28
1982	0	0	0	0	0	0	10	0	0	0	0	0	10
1983	0	0	0	0	0	2	1	0	0	0	0	0	3
1984	0	0	0	0	0	0	0	0	0	0	0	0	0
1985	0	0	0	0	0	0	7	2	0	0	0	0	9
1986	0	0	0	0	0	0	0	3	1	0	0	0	4
1987	0	0	0	0	0	10	5	10	0	0	0	0	25
1988	0	0	0	0	0	0	0	1	2	0	0	0	3
1989	0	0	0	0	0	1	5	2	5	0	0	0	13
1990	0	0	0	0	0	1	36	17	0	0	0	0	54
1991	0	0	0	0	0	6	3	23	0	0	0	0	32
1992	0	0	0	0	0	10	5	5	0	0	0	0	20
1993	0	0	0	0	10	0	4	26	5	0	0	0	45

SNOWFALL (inches) ANNETTE, ALASKA

SEASON	JULY	AUG	SEP	OCT	NOV	DEC	JAN	FEB	MAR	APR	MAY	JUNE	TOTAL
1964-65	0.0	0.0	0.0	T	1.6	44.8	22.2	24.9	2.5	T	0.6	0.0	96.6
1965-66	0.0	0.0	0.0	T	1.7	12.0	9.9	6.5	20.3	2.3	0.0	0.0	52.7
1966-67	0.0	0.0	0.0	1.9	8.7	5.8	23.2	17.7	9.4	0.4	0.0	0.0	67.1
1967-68	0.0	0.0	0.0	0.0	9.3	17.7	25.2	6.4	3.0	8.1	0.0	0.0	69.7
1968-69	0.0	0.0	0.0	0.0	3.3	14.7	23.2	29.5	0.6	T	0.0	0.0	71.3
1969-70	0.0	0.0	0.0	0.0	T	T	7.1	1.0	1.8	0.2	0.0	0.0	10.1
1970-71	0.0	0.0	0.0	T	1.4	23.2	16.6	13.3	46.9	22.2	0.0	0.0	123.6
1971-72	0.0	0.0	0.0	T	T	37.4	34.6	25.4	25.3	14.0	0.0	0.0	136.7
1972-73	0.0	0.0	0.0	0.0	T	11.4	20.6	22.0	4.5	T	0.0	0.0	58.5
1973-74	0.0	0.0	0.0	T	18.3	1.2	36.3	33.4	8.9	T	0.0	0.0	98.1
1974-75	0.0	0.0	0.0	0.0	4.6	7.1	30.6	13.8	6.0	7.0	T	0.0	69.1
1975-76	0.0	0.0	0.0	0.0	4.1	25.2	18.9	22.5	18.4	5.6	T	0.0	94.7
#1976-77	0.0	0.0	0.0	1.5	T	1.8	0.7	1.2	1.1	3.1	0.0	0.0	9.4
1977-78	0.0	0.0	0.0	0.0	2.8	24.2	0.8	0.2	9.7	T	0.0	0.0	37.7
1978-79	0.0	0.0	0.0	0.0	7.4	10.3	17.5	30.4	4.9	1.2	T	0.0	71.7
1979-80	0.0	0.0	0.0	0.0	0.4	13.1	25.6	3.6	5.4	T	T	0.0	48.1
1980-81	0.0	0.0	0.0	0.0	2.4	9.0	0.0	0.1	0.7	T	0.0	0.0	12.2
1981-82	0.0	0.0	0.0	T	T	17.6	17.8	17.6	6.7	T	T	0.0	59.7
1982-83	0.0	0.0	0.0	T	T	1.4	9.8	0.2	T	0.0	0.0	0.0	11.4
1983-84	0.0	0.0	0.0	T	0.1	8.6	0.3	3.8	T	0.3	0.2	0.0	13.3
1984-85	0.0	0.0	0.0	0.4	6.5	7.4	2.3	23.2	7.9	2.3	0.4	T	50.4
1985-86	0.0	0.0	0.0	1.5	5.4	1.2	1.1	1.8	0.2	5.2	T	0.0	16.4
1986-87	0.0	0.0	0.0	0.0	6.4	T	5.6	4.9	13.2	3.4	T	0.0	33.5
1987-88	0.0	0.0	0.0	0.0	1.5	7.3	3.1	3.2	3.6	4.7	T	T	23.4
1988-89	0.0	0.0	0.0	0.0	5.6	8.4	14.2	T	11.5	T	0.0	0.0	39.7
1989-90	0.0	0.0	0.0	0.0	T	0.0	4.4			0.0	0.0	0.0	
1990-91	0.0	0.0	0.0	0.0			T			0.8	T	0.0	
1991-92	0.0	0.0	0.0	0.6	0.0	2.0	1.1	6.5	T	0.0	0.0	0.0	10.2
1992-93	0.0	0.0	0.0	0.0	0.0		20.5	4.7	T	0.0	0.0	0.0	
1993-94	0.0	0.0	0.0	0.0	0.6	T							
Record Mean	0.0	0.0	0.0	0.2	3.1	10.9	12.8	11.0	8.8	2.2	0.1	T	49.2

BARROW, ALASKA

Barrow is the most northerly First-Order station operated by the National Weather Service. Although this station generally records one of the lowest mean temperatures for the winter months, the surrounding topography prevents the establishment of the lowest minima for the state. With the Arctic Ocean to the north, east, and west, and level tundra stretching 200 miles to the south, there are no natural wind barriers to assist in stilling the wind, permitting the lowering of temperatures by radiation, and no downslope drainage area to aid the flow of cold air to lower levels. Consequently, temperature inversions in the lower levels of the atmosphere are not as marked as those observed at stations in the central interior.

Temperatures at this northern station remain below the freezing point through most of the year, with the daily maxima reaching higher than 32 degrees on an average of only 109 days a year. Freezing temperatures have been observed every month of the year. February is generally the coldest month and March temperatures are but little higher than those observed in the winter months. In April, temperatures begin a general upward trend, with May becoming the definite transitional period from winter to the summer season. July is the warmest month of the year and the frequency of minimum temperatures of 32 degrees or less are about one day out of two for July and August. During late July or early August, the Arctic Ocean is usually ice-free for the first time in summer. The end of the short summer is reached in September. By November about half of the daily mean temperatures are zero or below, and Barrow definitely returns to the clutches of winter cold.

At 1250 p.m. on November 18, the sun dips below the horizon and is not seen again until 1151 a.m. on January 24. Then the amount of possible sunshine each day increases by never less than 9 minutes per day. By 106 a.m. on May 10th the possible sunshine has increased to 24 hours per day. The sun remains visible from that time to August 2, when it again sets for 1 hour and 25 minutes. The decrease in hours of sunshine is as rapid as the increase.

The amount of sunshine appears to have a direct relationship to the occurrence of cloudiness, precipitation, and heavy fog. All three build up to a maximum along with the hours of sunshine. Maximum cloudiness does continue into the fall months, although the amount of sunshine, precipitation, and fog are on the decrease. Since an accurate estimate of cloudiness cannot be made under conditions of darkness, the record of cloudiness for that time is not summarized. However, average cloudiness probably approximates that observed during late winter and spring months.

Variation of wind speed during the year is small, with the fall months being windiest. Extreme winds in the upper 40s and low 50s have been recorded for all months.

NORMALS, MEANS, AND EXTREMES

BARROW, ALASKA

LATITUDE: 71 °18'N LONGITUDE: 156 °47' W ELEVATION: FT. GRND 31 BARO 38 TIME ZONE: YUKON WBAN: 27502

	(a)	JAN	FEB	MAR	APR	MAY	JUNE	JULY	AUG	SEP	OCT	NOV	DEC	YEAR
TEMPERATURE °F:														
Normals														
-Daily Maximum		-7.4	-11.8	-9.0	4.7	24.2	38.3	45.0	42.3	33.8	18.1	3.5	-5.2	14.7
-Daily Minimum		-19.3	-23.7	-21.1	-9.1	14.4	29.7	33.6	33.3	27.0	8.8	-6.9	-17.2	4.1
-Monthly		-13.4	-17.8	-15.1	-2.2	19.3	34.0	39.3	37.9	30.5	13.5	-1.7	-11.2	9.4
Extremes														
-Record Highest	73	36	36	33	42	46	71	79	76	62	43	39	34	79
-Year		1974	1982	1967	1936	1992	1990	1993	1968	1957	1954	1937	1932	JUL 1993
-Record Lowest	73	-53	-56	-52	-42	-19	4	22	20	1	-32	-40	-55	-56
-Year		1975	1924	1971	1924	1984	1969	1936	1925	1975	1970	1948	1924	FEB 1924
NORMAL DEGREE DAYS:														
Heating (base 65°F)		2430	2318	2483	2016	1417	930	797	840	1035	1597	2001	2362	20226
Cooling (base 65°F)		0	0	0	0	0	0	0	0	0	0	0	0	0
% OF POSSIBLE SUNSHINE														
MEAN SKY COVER (tenths)														
Sunrise - Sunset	4	1.3	5.2	4.8	5.5	8.3	8.0	7.8	8.9	9.2	8.6	3.9	0.0	6.0
MEAN NUMBER OF DAYS:														
Sunrise to Sunset														
-Clear	50	5.2	11.7	13.8	11.0	3.3	3.6	3.6	1.4	1.2	2.3	5.4	3.0	65.5
-Partly Cloudy	50	1.6	5.7	6.9	7.0	4.9	5.8	7.1	3.9	2.6	3.9	2.7	0.0	52.1
-Cloudy	50	2.5	10.8	10.3	12.0	22.8	20.6	20.3	25.7	26.2	24.8	10.6	0.0	186.6
Precipitation														
.01 inches or more	73	4.5	4.3	3.7	4.2	4.2	4.8	8.7	10.6	10.6	11.2	6.1	5.0	77.9
Snow,Ice pellets,hail														
1.0 inches or more	49	0.5	0.5	0.3	0.7	0.3	0.2	0.1	0.2	1.1	1.7	0.8	0.5	6.9
Thunderstorms	73	0.0	0.0	0.0	0.0	0.0	0.*	0.*	0.*	0.0	0.0	0.0	0.0	0.1
Heavy Fog Visibility														
1/4 mile or less	52	2.1	1.4	1.2	2.6	7.3	10.3	12.0	11.6	5.0	3.3	2.2	1.2	60.3
Temperature °F														
-Maximum														
70° and above	20	0.0	0.0	0.0	0.0	0.0	0.1	0.8	0.3	0.0	0.0	0.0	0.0	1.2
32° and below	73	30.9	28.2	31.0	29.5	26.2	4.7	0.3	1.7	13.2	28.6	29.9	31.0	255.1
-Minimum														
32° and below	73	31.0	28.2	31.0	30.0	30.7	24.0	14.1	15.5	25.7	30.8	30.0	31.0	322.2
0° and below	73	29.1	27.1	30.3	22.8	3.3	0.0	0.0	0.0	0.0	6.7	21.7	28.9	169.9
AVG. STATION PRESS.(mb)	19	1016.4	1019.4	1018.9	1019.1	1017.3	1014.2	1013.8	1012.1	1011.8	1013.4	1015.6	1015.5	1015.6
RELATIVE HUMIDITY (%)														
Hour 03	43	69	66	67	75	88	93	94	94	92	85	78	71	81
Hour 09 (Local Time)	43	70	66	67	75	87	89	88	92	91	85	78	71	80
Hour 15	43	69	66	68	74	83	85	84	86	87	84	78	71	78
Hour 21	43	69	66	68	75	86	89	88	91	91	85	77	71	80
PRECIPITATION (inches):														
Water Equivalent														
-Normal		0.17	0.15	0.17	0.20	0.16	0.28	0.94	0.96	0.60	0.45	0.25	0.16	4.49
-Maximum Monthly	73	1.04	0.81	1.49	1.36	0.81	1.15	3.19	2.81	1.56	1.65	1.15	0.76	3.19
-Year		1962	1959	1963	1963	1933	1955	1989	1963	1958	1925	1965	1967	JUL 1989
-Minimum Monthly	73	0.00	0.00	0.00	0.00	T	T	T	T	0.01	0.12	T	0.00	0.00
-Year		1939	1936	1928	1938	1939	1937	1937	1934	1969	1936	1989	1936	JAN 1939
-Maximum in 24 hrs	72	0.70	0.36	0.71	0.42	0.30	0.82	1.32	0.83	0.56	1.00	0.41	0.26	1.32
-Year		1937	1959	1963	1963	1969	1955	1987	1960	1959	1926	1925	1930	JUL 1987
Snow,Ice pellets,hail														
-Maximum Monthly	73	11.9	9.4	15.8	15.4	12.9	6.6	9.0	4.0	16.2	21.2	19.0	9.7	21.2
-Year		1962	1944	1963	1963	1933	1933	1922	1969	1987	1925	1925	1925	OCT 1925
-Maximum in 24 hrs	73	5.4	3.6	7.1	4.2	4.5	3.2	6.0	2.5	5.1	15.0	6.0	5.0	15.0
-Year		1962	1959	1963	1963	1923	1981	1922	1936	1987	1926	1925	1922	OCT 1926
WIND:														
Mean Speed (mph)	60	11.6	11.1	11.1	11.5	11.8	11.5	11.7	12.4	13.2	13.3	12.4	11.5	11.9
Prevailing Direction through 1963		ESE	E	ENE	NE	ENE	E	E	E	E	E	E	E	E
Fastest Obs. 1 Min.														
-Direction (!!)	30	09	25	27	25	25	23	27	27	23	27	26	24	27
-Speed (MPH)	30	49	55	58	40	39	35	43	37	49	55	54	55	58
-Year		1962	1989	1960	1961	1968	1961	1993	1991	1986	1963	1966	1973	MAR 1960
Peak Gust														
-Direction (!!)	13	E	SW	E	NE	NE	W	W	W	SW	W	E	SW	SW
-Speed (mph)	13	58	74	56	47	41	43	55	47	66	54	53	61	74
-Date		1988	1989	1985	1992	1991	1992	1993	1991	1986	1993	1981	1992	FEB 1989

PRECIPITATION (inches)

BARROW, ALASKA

YEAR	JAN	FEB	MAR	APR	MAY	JUNE	JULY	AUG	SEP	OCT	NOV	DEC	ANNUAL
1964	0.07	0.38	T	0.07	0.38	0.36	0.08	0.36	0.10	0.81	0.29	0.17	3.07
1965	0.04	T	0.29	0.66	0.37	0.05	0.79	0.89	0.43	0.68	1.15	0.51	5.86
1966	0.12	0.06	0.20	0.28	0.15	0.37	2.01	0.57	0.48	0.49	0.51	0.23	5.47
1967	0.19	0.02	0.36	0.13	0.25	0.28	1.43	0.32	0.21	0.32	0.43	0.76	4.70
1968	0.41	0.11	0.15	0.05	0.16	0.45	0.19	0.09	0.61	0.30	0.54	0.25	3.31
1969	0.12	0.13	0.05	0.17	0.51	0.29	0.34	0.88	0.01	0.45	0.24	0.12	3.31
1970	0.07	0.09	0.06	0.15	0.09	0.02	0.15	0.35	0.14	0.16	0.47	0.09	1.84
1971	0.18	0.13	0.16	0.02	0.25	0.12	0.98	0.35	0.17	0.36	0.17	0.18	3.07
1972	0.05	0.19	T	0.01	0.06	0.05	0.11	1.12	1.33	1.40	0.43	0.17	4.92
1973	0.07	0.16	0.04	0.52	0.19	0.78	1.06	2.20	1.14	0.56	0.36	0.09	7.17
1974	0.38	0.06	0.01	T	0.06	0.52	0.55	0.59	0.49	0.20	0.16	0.04	3.06
1975	0.17	0.29	0.17	0.18	0.08	0.79	1.00	1.16	0.51	0.40	0.07	0.01	4.83
1976	0.04	0.07	0.04	0.10	0.03	0.30	0.37	0.17	0.71	0.58	0.45	0.02	2.88
1977	0.17	0.17	0.18	0.05	0.13	0.19	0.09	0.79	0.79	0.48	0.05	0.18	3.27
1978	0.08	0.17	0.08	0.13	0.03	0.37	0.73	0.50	1.11	0.13	0.32	0.22	3.87
1979	0.05	T	0.09	0.11	0.04	0.10	0.81	0.76	0.28	0.38	0.21	0.18	3.01
1980	0.15	0.15	0.02	0.06	0.01	0.56	0.77	1.41	0.73	0.36	0.12	0.08	4.42
1981	0.22	0.06	0.03	0.19	0.06	0.51	1.77	0.56	0.52	0.32	0.06	0.09	4.39
1982	0.17	0.43	0.24	0.34	0.39	0.21	0.78	0.86	0.59	0.56	0.02	0.13	4.72
1983	0.03	0.09	T	0.20	0.07	0.11	0.10	1.04	0.93	0.36	0.25	0.05	3.23
1984	0.19	0.16	0.11	0.27	0.07	0.03	0.83	1.64	0.15	0.33	0.12	0.08	3.98
1985	0.05	0.10	0.15	0.05	0.25	0.64	0.61	0.51	0.58	0.45	0.25	0.16	3.80
1986	0.16	0.14	0.09	0.03	0.07	0.07	0.79	0.69	1.45	0.43	0.14	0.09	4.15
1987	0.13	0.08	0.03	T	0.13	0.06	1.94	1.00	1.37	0.17	0.05	0.18	5.14
1988	0.02	0.04	0.10	0.03	0.02	0.15	0.74	1.57	0.41	0.24	0.01	0.26	3.59
1989	0.01	0.29		0.42	0.02	0.36	3.19	1.69	0.69	0.20	T	0.20	
1990	0.03	0.06	0.13	0.08	0.13	0.38	1.35	1.19	0.55	0.42	0.17	0.12	4.61
1991	0.07	0.08	0.02	0.07	0.17	0.08	0.22	0.20	0.35	0.39	0.07	0.03	1.75
1992	0.04	0.12	0.13	0.11	0.08	0.16	0.26	0.66	0.47	0.21	0.22	0.24	2.70
1993	0.45	0.17	0.11	0.03	0.11	0.44	0.67	0.98	1.50	0.50	0.25	0.12	5.33
Record Mean	0.18	0.16	0.13	0.15	0.13	0.31	0.88	0.91	0.59	0.50	0.25	0.18	4.37

AVERAGE TEMPERATURE (deg. F)

BARROW, ALASKA

YEAR	JAN	FEB	MAR	APR	MAY	JUNE	JULY	AUG	SEP	OCT	NOV	DEC	ANNUAL
1964	-20.2	-28.6	-23.3	-10.6	17.7	31.9	38.1	34.9	27.0	12.9	-1.5	-22.6	4.6
1965	-21.0	-26.3	-8.7	2.2	16.9	30.8	38.3	38.1	29.3	9.3	12.7	-12.5	9.1
1966	-15.7	-22.4	-22.9	-5.1	16.1	35.1	37.4	34.9	31.0	18.5	6.4	-12.2	8.5
1967	-12.0	-16.3	-8.6	6.0	22.1	33.5	37.5	36.1	25.3	12.7	-2.0	-8.6	10.5
1968	-9.6	-25.3	-11.8	-5.4	21.8	33.0	42.3	41.6	32.8	17.1	-3.7	-12.9	10.0
1969	-12.7	-19.3	-11.6	2.0	20.5	33.4	34.7	31.8	29.8	15.9	-9.8	-7.7	9.0
1970	-12.5	-17.0	-18.4	-3.0	19.1	32.9	37.8	35.0	25.7	0.5	-4.2	-10.2	7.2
1971	-19.5	-26.1	-18.2	-4.5	17.4	35.1	40.4	33.5	31.7	14.2	-0.5	-11.4	7.7
1972	-16.2	-19.2	-19.7	-4.4	18.1	32.5	42.9	40.6	31.2	21.1	1.4	-2.9	10.5
1973	-13.5	-13.2	-20.8	-2.7	19.1	33.2	39.7	39.8	34.0	19.4	7.9	-5.4	11.5
1974	-10.7	-28.3	-20.1	-7.9	18.1	30.0	39.1	42.3	32.9	6.2	-7.6	-26.6	5.6
1975	-26.4	-13.5	-7.3	-4.5	19.1	33.6	37.4	34.7	23.9	5.2	-11.3	-21.8	5.8
1976	-18.9	-22.9	-15.4	-2.1	17.2	33.5	38.5	37.5	32.2	13.7	2.5	-16.4	8.3
1977	-8.0	-16.4	-22.2	-6.4	19.0	34.1	38.8	44.4	35.7	20.2	-4.9	-7.6	10.6
1978	-6.0	-14.2	-10.5	1.6	16.4	33.0	39.8	36.6	33.5	7.1	5.4	-13.9	10.7
1979	-2.5	-20.0	-16.6	-0.9	19.8	32.4	43.0	46.1	36.5	19.0	9.5	-12.0	12.9
1980	-13.9	-10.3	-11.5	-3.8	17.0	37.0	35.7	33.9	25.1	14.5	-5.0	-15.6	8.6
1981	-0.9	-15.7	-10.9	1.1	23.6	34.6	39.7	33.6	25.5	14.2	-0.6	-8.1	11.4
1982	-11.1	-6.5	-12.4	-1.0	16.6	33.9	38.0	36.7	29.6	6.9	-10.1	-9.3	9.3
1983	-19.2	-15.4	-13.4	2.7	16.9	34.6	38.1	34.3	23.9	7.0	1.3	0.6	9.3
1984	-15.3	-33.0	-16.5	-10.6	16.6	37.9	41.0	38.4	34.2	17.2	-8.1	-13.6	7.4
1985	-7.3	-17.2	-12.9	-6.6	22.7	35.8	39.0	38.6	28.9	10.4	2.7	-6.9	10.6
1986	-15.0	-8.8	-17.7	-7.7	20.1	34.5	42.0	39.5	36.9	16.4	0.4	-6.4	11.2
1987	-13.0	-20.0	-11.7	-4.7	20.2	34.1	38.9	38.9	28.3	22.9	-5.4	-8.9	10.0
1988	-10.5	-14.4	-12.7	1.4	19.8	33.6	38.9	35.7	28.2	2.0	-13.6	-9.3	8.3
1989	-24.0	9.3		6.0	17.4	36.5	45.5	46.8	35.5	18.0	-12.6	-9.5	
1990	-23.0	-23.2	-11.4	7.5	26.5	37.7	42.3	37.3	31.5	16.9	-6.9	-15.7	10.0
1991	-13.4	-18.2	-17.7	3.3	28.0	36.9	37.9	36.1	29.9	17.5	-8.0	-15.8	9.7
1992	-18.6	-20.3	-9.2	0.8	20.7	35.9	39.7	39.0	25.0	12.5	-1.1	-8.7	9.6
1993	-12.6	-11.1	-12.6	6.5	23.0	36.4	45.2	36.8	32.4	23.1	5.2	-8.1	13.7
Record Mean	-14.7	-18.1	-15.0	-0.8	19.1	34.0	39.5	38.1	30.4	15.4	-0.9	-11.2	9.6
Max	-8.3	-11.9	-8.4	6.3	24.4	38.5	45.5	43.0	34.0	20.2	4.8	-5.2	15.2
Min	-21.0	-24.2	-21.6	-8.0	13.9	29.4	33.5	33.2	26.8	10.5	-6.7	-17.2	4.1

REFERENCE NOTES

BARROW, ALASKA

GENERAL
T - TRACE AMOUNT.
BLANK ENTRIES DENOTE MISSING/UNREPORTED DATA.
# INDICATES A STATION OR INSTRUMENT RELOCATION.

SPECIFIC

(a) - LENGTH OF RECORD IN YEARS, ALTHOUGH INDIVIDUAL MONTHS MAY BE MISSING.
0.* OR * - THE VALUE IS BETWEEN 0.0 AND 0.05
NORMALS - BASED ON THE 1961-1990 RECORD PERIOD.
EXTREMES - DATES ARE THE MOST RECENT OCCURRENCE.
WIND DIR.- NUMERALS SHOW TENS OF DEGRESS CLOCKWISE FROM TRUE NORTH. "00" INDICATES CALM.
RESULTANT DIRECTIONS ARE GIVEN TO WHOLE DEGREES.
BOLD VALUES INDICATE EXTREME VALUES WHICH OCCURRED AFTER THE ASOS SYSTEM WAS COMMISSIONED.

RECORD = PERIOD OF RECORD
RECORD MEAN PRECIPITATION IS THE MEAN OF ALL DAILY PRECIPITATION AMOUNTS DURING THE PERIOD OF RECORD.
RECORD MAX(MIN) TEMPERATURE IS THE MEAN OF ALL DAILY MAX(MIN) TEMPERATURES DURING THE PERIOD OF RECORD.
RECORD MEAN TEMPERATURE IS THE SUM OF THE RECORD MAX AND RECORD MIN DIVIDED BY 2.
AVERAGE TEMPERATURE IS THE SUM OF THE MEAN DAILY MAX AND MIN TEMPERATURE DIVIDED BY 2.

EXCEPTIONS

THE SUN IS BELOW THE HORIZON FOR THE PERIOD NOVEMBER 19 TO JANUARY 23. SKY COVER DATA IS CALCULATED FOR PERIOD SUN IS ABOVE HORIZON.

RECORD MEANS ARE THROUGH THE CURRENT YEAR, BEGINNING IN 1921 FOR TEMPERATURE
1921 FOR PRECIPITATION
1921 FOR SNOWFALL

HEATING DEGREE DAYS Base 65 deg. F BARROW, ALASKA

SEASON	JULY	AUG	SEP	OCT	NOV	DEC	JAN	FEB	MAR	APR	MAY	JUNE	TOTAL
1964-65	831	926	1134	1607	1993	2723	2676	2562	2289	1885	1482	1019	21127
1965-66	821	825	1067	1720	1566	2408	2508	2453	2734	2108	1509	891	20610
1966-67	850	928	1012	1438	1752	2401	2395	2280	2282	1771	1321	939	19369
1967-68	848	887	1185	1617	2011	2287	2311	2628	2388	2119	1333	953	20567
1968-69	697	718	958	1478	2061	2422	2412	2364	2377	1888	1372	942	19689
1969-70	935	1023	1049	1516	2248	2258	2406	2303	2591	2041	1416	959	20745
1970-71	835	921	1173	1999	2074	2338	2624	2555	2585	2090	1470	891	21555
1971-72	755	970	990	1568	1966	2372	2523	2447	2630	2081	1448	968	20718
1972-73	681	750	1009	1350	1909	2106	2436	2193	2667	2035	1416	950	19502
1973-74	778	772	923	1407	1709	2186	2352	2619	2645	2192	1449	1045	20077
1974-75	795	695	956	1818	2180	2848	2841	2202	2246	2085	1417	934	21017
1975-76	850	931	1226	1852	2293	2699	2609	2554	2501	2014	1476	943	21948
1976-77	816	844	976	1587	1873	2532	2269	2285	2711	2141	1419	922	20375
1977-78	802	632	871	1381	2099	2256	2202	2223	2346	1904	1501	954	19171
1978-79	772	874	936	1793	1786	2448	2091	2390	2534	1978	1395	970	19967
1979-80	676	579	851	1419	1660	2391	2450	2186	2380	2065	1480	835	18972
1980-81	902	961	1191	1558	2102	2501	2043	2265	2360	1917	1275	906	19981
1981-82	777	968	1176	1569	1966	2269	2363	2002	2403	1983	1495	925	19896
1982-83	833	871	1057	1796	2261	2307	2617	2255	2435	1871	1485	906	20694
1983-84	830	941	1225	1798	1907	1998	2495	2848	2532	2273	1498	806	21151
1984-85	738	816	917	1479	2197	2442	2248	2304	2423	2153	1303	870	19890
1985-86	800	810	1078	1691	1867	2233	2485	2066	2573	2183	1384	909	20079
1986-87	707	784	840	1500	1941	2216	2424	2388	2382	2093	1380	920	19575
1987-88	801	802	1095	1297	2114	2295	2344	2305	2415	1906	1394	938	19706
1988-89	803	898	1097	1956	2363	2307	2763	1556		1767	1468	851	
1989-90	595	557	879	1449	2335	2311	2730	2477	2372	1720	1184	816	19425
1990-91	696	850	997	1487	2161	2511	2433	2336	2568	1849	1138	834	19860
1991-92	837	887	1046	1469	2194	2513	2598	2481	2307	1930	1365	866	20493
1992-93	778	800	1193	1619	1982	2285	2407	2130	2410	1751	1294	850	19499
1993-94	609	868	970	1291	1795	2267							

COOLING DEGREE DAYS Base 65 deg. F BARROW, ALASKA

YEAR	JAN	FEB	MAR	APR	MAY	JUNE	JULY	AUG	SEP	OCT	NOV	DEC	TOTAL
1969	0	0	0	0	0	0	0	0	0	0	0	0	0
1970	0	0	0	0	0	0	0	0	0	0	0	0	0
1971	0	0	0	0	0	0	0	0	0	0	0	0	0
1972	0	0	0	0	0	0	0	0	0	0	0	0	0
1973	0	0	0	0	0	0	0	0	0	0	0	0	0
1974	0	0	0	0	0	0	0	0	0	0	0	0	0
1975	0	0	0	0	0	0	0	0	0	0	0	0	0
1976	0	0	0	0	0	0	0	0	0	0	0	0	0
1977	0	0	0	0	0	0	0	0	0	0	0	0	0
1978	0	0	0	0	0	0	0	0	0	0	0	0	0
1979	0	0	0	0	0	0	0	0	0	0	0	0	0
1980	0	0	0	0	0	0	0	0	0	0	0	0	0
1981	0	0	0	0	0	0	0	0	0	0	0	0	0
1982	0	0	0	0	0	0	0	0	0	0	0	0	0
1983	0	0	0	0	0	0	0	0	0	0	0	0	0
1984	0	0	0	0	0	0	0	0	0	0	0	0	0
1985	0	0	0	0	0	0	0	0	0	0	0	0	0
1986	0	0	0	0	0	0	0	0	0	0	0	0	0
1987	0	0	0	0	0	0	0	0	0	0	0	0	0
1988	0	0	0	0	0	0	0	0	0	0	0	0	0
1989	0	0		0	0	0	0	0	0	0	0	0	
1990	0	0	0	0	0	0	0	0	0	0	0	0	0
1991	0	0	0	0	0	0	0	0	0	0	0	0	0
1992	0	0	0	0	0	0	0	0	0	0	0	0	0
1993	0	0	0	0	0	0	0	0	0	0	0	0	0

SNOWFALL (inches) BARROW, ALASKA

SEASON	JULY	AUG	SEP	OCT	NOV	DEC	JAN	FEB	MAR	APR	MAY	JUNE	TOTAL
1964-65	0.1	1.3	1.5	13.3	3.3	2.4	0.5	T	4.5	8.8	4.1	0.3	40.1
1965-66	0.0	0.2	3.9	9.0	12.8	5.7	1.6	0.6	2.2	2.9	3.0	T	41.9
1966-67	T	0.9	1.8	7.6	6.4	2.8	2.4	0.4	4.1	1.7	3.2	1.9	33.2
1967-68	0.3	1.0	2.4	5.2	5.8	9.2	6.2	1.6	2.0	0.7	2.8	0.8	38.0
1968-69	0.2	0.3	5.4	3.3	7.0	3.4	1.8	2.0	0.6	2.4	7.4	0.6	34.4
1969-70	0.8	4.0	0.1	3.8	2.6	1.3	1.1	1.1	1.0	1.8	0.9	T	18.5
1970-71	T	T	2.1	2.3	4.6	0.9	1.8	1.6	1.7	0.2	2.4	0.2	17.8
1971-72	T	2.4	2.0	5.0	2.0	2.2	0.6	2.3	0.5	0.7	1.3	0.7	19.7
1972-73	0.0	0.2	12.9	15.0	4.9	1.9	0.7	1.8	1.0	8.8	3.3	2.9	53.4
1973-74	1.1	0.6	6.5	8.4	4.1	1.7	2.6	0.8	0.2	T	0.9	4.1	31.0
1974-75	T	0.0	3.3	2.7	2.4	0.7	2.3	3.6	2.4	2.7	0.7	T	20.8
1975-76	0.3	1.5	3.2	4.3	0.7	0.1	0.4	0.7	0.6	1.6	0.7	0.1	14.2
1976-77	T	T	2.1	8.3	7.2	0.4	2.2	1.9	2.1	0.7	1.3	T	26.2
1977-78	0.1	0.1	0.8	7.0	1.1	2.7	1.2	2.6	1.3	0.7	0.6	T	18.2
1978-79	0.0	0.2	6.1	3.2	4.8	2.6	0.7	T	0.9	1.1	0.4	1.0	21.0
1979-80	0.5	T	2.1	4.4	2.1	2.1	2.8	2.8	0.4	1.1	0.1	T	18.4
1980-81	T	1.0	2.4	7.9	2.4	1.6	4.5	1.2	0.4	2.8	0.7	3.7	28.6
1981-82	T	0.6	5.5	5.7	0.6	0.9	1.7	4.3	2.4	3.4	4.4	0.2	29.7
1982-83	1.7	0.4	5.1	5.8	0.2	1.3	0.3	0.9	T	1.5	0.7	0.7	18.6
1983-84	0.4	1.8	6.9	3.4	3.1	0.5	1.9	1.6	1.1	2.7	0.7	T	24.1
1984-85	T	2.4	0.7	3.7	1.2	0.8	0.5	1.1	1.6	0.6	2.4	0.4	15.4
1985-86	0.0	0.7	6.1	4.7	2.7	1.7	1.6	1.4	0.9	0.3	0.9	T	21.0
1986-87	T	0.9	4.2	9.4	3.4	3.0	2.9	2.8	0.7	0.4	1.4	0.8	29.9
1987-88	T	0.4	16.2	2.7	0.5	2.7	0.5	1.3	1.7	0.4	1.3	0.6	28.3
1988-89	T	2.5	4.1	2.7	0.1	2.6	0.1	3.3		4.2	0.6	0.2	
1989-90	T	0.0	0.4	6.6	0.4	4.0	0.7	0.8	3.0	1.9	1.1	1.2	20.1
1990-91	0.4	T	4.7	11.0	5.6	2.4	1.9	2.0	0.6	2.0	2.7	1.1	34.4
1991-92	T	1.1	4.6	7.9	1.5	1.2	1.1	2.7	4.6	3.1	1.3	0.5	29.6
1992-93	T	2.1	8.9	7.5	6.8	4.7	8.6	3.2	4.2	3.4	1.4	0.1	50.9
1993-94	0.0	1.5	11.8	8.5	5.5	3.1							
Record Mean	0.5	0.7	3.6	6.7	3.4	2.5	2.2	2.1	1.8	2.1	1.8	0.6	28.2

BETHEL, ALASKA

The two main topographical features affecting the climate of Bethel are the Bering Sea, which is about 100 miles to the west and southwest, and the Kilbuck Range of mountains located about 40 miles to the east and southeast of the station. This range, averaging about 4,000 feet in height, extends, roughly, in a north–south direction in that portion nearest to Bethel. Some 160 miles southeast of the Kilbuck Range the Aleutians, extending in a northeast–southwest direction, provide an additional natural barrier to many of the storms originating on the outward end of the Aleutian Chain and moving out through the Gulf of Alaska. Both ranges tend to direct some of the storms northeastward into the Bering Sea, and thus directly affect the Bethel area. During invasions of such storms, it is not uncommon for wind velocities to exceed 50 mph. Maximum speeds usually accompany northeast winds in the winter and southeast winds in the summer. During the winter season, strong southerly winds tend to be considerably affected by the mountains to the south, producing, at times, a pronounced foehn (chinook) effect. Temperatures have risen almost 50 degrees in less than 24 hours under these conditions.

The climate is somewhat more maritime than continental in character, which tends to modify daily temperature extremes during most of the year. However, there are usually two periods during the year when the area becomes affected by continental climatic influences. In June and July, temperatures in the area rise noticeably under the influence of warmer continental air. Around the latter part of December and early January, cold, clear continental air becomes quite dominant, and the climate of Bethel becomes quite similar to other areas located farther inland. Average temperatures through the entire winter season, however, are considerably higher than those experienced in the Alaskan interior, and temperatures for the entire summer season average considerably cooler than in the Alaskan interior. The last date of freezing temperature in spring averages late May, and the average of the first freezing temperature in autumn falls in early September, resulting in a growing season slightly over 100 days. Cabbages, potatoes, cauliflower, beets, turnips, lettuce, and carrots are successfully grown. August is usually the wettest month of the year.

Thunderstorms are rare. The few thunderstorms that do occur are generally short in duration, but rather severe. They usually develop and move out of the northeast during the months of June and July.

NORMALS, MEANS, AND EXTREMES

BETHEL, ALASKA

LATITUDE: 60 °47'N LONGITUDE: 161 °48' W ELEVATION: FT. GRND 125 BARO 131 TIME ZONE: YUKON WBAN: 26615

	(a)	JAN	FEB	MAR	APR	MAY	JUNE	JULY	AUG	SEP	OCT	NOV	DEC	YEAR
TEMPERATURE °F:														
Normals														
-Daily Maximum		12.9	12.8	21.1	31.3	48.2	58.8	62.3	59.5	52.0	35.0	22.6	14.7	35.9
-Daily Minimum		0.4	-0.8	5.4	15.7	31.7	42.3	47.7	46.3	38.3	23.8	10.9	2.2	22.0
-Monthly		6.7	6.0	13.3	23.6	39.9	50.5	55.0	52.9	45.2	29.4	16.8	8.5	29.0
Extremes														
-Record Highest	35	48	46	46	55	80	86	83	84	72	56	48	45	86
-Year		1963	1970	1986	1993	1993	1959	1993	1977	1979	1979	1991	1963	JUN 1959
-Record Lowest	35	-48	-38	-39	-22	4	28	31	28	18	-5	-24	-37	-48
-Year		1989	1990	1964	1964	1965	1962	1959	1984	1970	1963	1990	1975	JAN 1989
NORMAL DEGREE DAYS:														
Heating (base 65°F)		1807	1652	1603	1242	778	435	310	375	594	1104	1446	1752	13098
Cooling (base 65°F)		0	0	0	0	0	0	0	0	0	0	0	0	0
% OF POSSIBLE SUNSHINE														
MEAN SKY COVER (tenths)														
Sunrise - Sunset	33	6.6	5.9	6.3	6.9	7.6	8.2	8.5	8.7	8.3	8.0	7.3	6.8	7.4
MEAN NUMBER OF DAYS:														
Sunrise to Sunset														
-Clear	33	8.1	9.8	9.3	6.5	3.7	1.6	1.9	1.5	2.4	3.7	5.7	7.8	62.0
-Partly Cloudy	33	4.8	4.2	5.2	6.3	7.8	6.8	5.5	4.0	5.0	5.2	5.0	4.9	64.7
-Cloudy	33	18.2	14.3	16.5	17.2	19.5	21.5	23.6	25.4	22.7	22.1	19.3	18.3	238.6
Precipitation														
.01 inches or more	35	8.6	6.8	9.2	8.8	10.3	12.8	15.0	17.6	16.1	12.7	11.5	11.1	140.5
Snow,Ice pellets,hail														
1.0 inches or more	35	1.8	2.0	2.8	1.8	0.7	0.1	0.0	0.0	0.1	1.4	2.9	3.2	16.7
Thunderstorms	26	0.0	0.0	0.0	0.0	0.2	0.5	0.6	0.4	0.1	0.0	0.0	0.0	1.7
Heavy Fog Visibility														
1/4 mile or less	26	3.5	3.0	4.3	4.1	4.9	3.0	3.9	6.0	4.5	4.1	4.3	4.6	50.3
Temperature °F														
-Maximum														
70° and above	35	0.0	0.0	0.0	0.0	0.5	3.2	6.2	2.4	0.1	0.0	0.0	0.0	12.4
32° and below	35	24.9	22.5	22.3	13.4	1.5	0.0	0.0	0.0	0.1	11.2	21.1	24.4	141.4
-Minimum														
32° and below	35	30.2	27.8	30.6	27.7	16.4	0.7	0.1	0.1	5.8	25.6	28.6	30.4	223.9
0° and below	35	15.6	14.5	12.5	5.4	0.0	0.0	0.0	0.0	0.0	0.5	6.9	15.3	70.7
AVG. STATION PRESS.(mb)	21	998.7	1002.3	1000.7	1004.0	1004.1	1006.3	1008.6	1006.1	1002.0	999.6	998.1	997.1	1002.3
RELATIVE HUMIDITY (%)														
Hour 03	43	77	75	80	84	88	89	92	94	92	88	83	77	85
Hour 09 (Local Time)	43	77	75	79	82	80	80	86	91	91	88	83	77	82
Hour 15	44	76	73	74	72	63	61	67	73	70	76	80	77	72
Hour 21	43	77	75	80	81	72	68	74	82	84	85	83	77	78
PRECIPITATION (inches):														
Water Equivalent														
-Normal		0.58	0.43	0.59	0.70	0.78	1.44	1.98	2.91	2.04	1.45	1.07	1.02	14.99
-Maximum Monthly	35	2.53	2.12	3.44	3.89	1.73	3.33	4.00	5.81	4.79	2.51	3.31	3.49	5.81
-Year		1993	1959	1991	1979	1985	1980	1980	1963	1993	1972	1979	1992	AUG 1963
-Minimum Monthly	35	0.06	T	T	0.02	0.10	0.25	0.56	0.99	0.42	0.11	0.04	0.11	T
-Year		1966	1984	1986	1985	1967	1974	1988	1976	1968	1965	1969	1973	MAR 1986
-Maximum in 24 hrs	35	0.85	0.77	0.64	0.94	0.53	1.37	1.22	1.44	2.02	1.55	1.63	1.34	2.02
-Year		1982	1959	1991	1979	1992	1981	1966	1990	1971	1974	1990	1970	SEP 1971
Snow,Ice pellets,hail														
-Maximum Monthly	35	15.3	18.1	30.6	14.3	6.2	2.2	T	0.0	4.0	12.8	21.7	36.0	36.0
-Year		1993	1959	1991	1975	1992	1963	1974		1992	1978	1990	1992	DEC 1992
-Maximum in 24 hrs	35	7.8	6.5	8.2	4.7	5.7	1.2	T	0.0	3.6	5.5	10.0	8.0	10.0
-Year		1987	1959	1974	1975	1992	1963	1974		1992	1982	1990	1992	NOV 1990
WIND:														
Mean Speed (mph)	35	14.6	15.1	13.9	13.0	11.7	11.3	10.9	11.0	11.7	12.6	13.5	14.2	12.8
Prevailing Direction through 1963		NE	NNE	NNE	NW	S	NW	SSW	SSW	SSW	NNE	NE	NNE	NNE
Fastest Obs. 1 Min.														
-Direction (!!)	39	18	16	19	15	18	19	19	22	18	19	16	17	16
-Speed (MPH)	39	54	62	49	44	41	43	40	46	55	52	60	58	62
-Year		1979	1951	1977	1979	1960	1978	1974	1978	1960	1992	1958	1977	FEB 1951
Peak Gust														
-Direction (!!)	14	S	NE	S	SE	S	S	S	S	SE	S	W	S	S
-Speed (mph)	14	61	59	56	49	53	59	46	51	69	77	66	67	77
-Date		1993	1988	1991	1983	1985	1980	1982	1993	1982	1992	1990	1982	OCT 1992

PRECIPITATION (inches)

BETHEL, ALASKA

YEAR	JAN	FEB	MAR	APR	MAY	JUNE	JULY	AUG	SEP	OCT	NOV	DEC	ANNUAL
1964	0.32	0.36	0.83	0.88	0.71	0.79	0.66	1.71	1.47	1.14	0.66	0.72	10.25
1965	0.59	0.97	1.03	0.64	0.47	0.69	1.13	3.42	3.41	0.11	1.19	0.50	14.15
1966	0.06	0.91	0.12	0.17	0.45	0.94	2.40	2.60	0.54	1.90	1.53	0.55	12.17
1967	0.88	0.24	0.87	1.54	0.10	2.63	2.25	2.01	0.63	0.39	2.49	0.83	14.86
1968	0.48	0.24	0.12	0.34	0.99	1.18	1.60	1.83	0.42	1.07	0.34	0.96	9.57
1969	0.41	0.42	0.47	0.06	0.21	0.78	2.49	3.25	0.77	0.88	0.04	0.55	10.33
1970	0.10	0.30	0.47	0.53	1.28	0.68	3.46	3.03	0.85	2.45	1.62	2.27	17.04
1971	0.15	0.16	0.03	0.14	0.78	0.73	2.67	3.12	3.39	2.21	0.40	2.58	16.36
1972	0.78	0.03	0.56	0.62	0.53	1.11	0.93	2.64	2.17	2.51	0.18	0.64	12.70
1973	0.77	0.65	0.49	0.12	1.00	0.45	1.03	3.42	1.69	2.07	1.20	0.11	13.00
1974	0.24	0.10	1.11	0.29	0.23	0.25	3.55	3.10	2.81	1.95	0.93	0.28	14.84
1975	0.29	0.25	0.22	0.88	0.48	1.52	1.35	2.39	3.33	2.08	0.24	0.37	13.40
1976	0.32	0.43	0.36	0.31	0.50	0.57	1.15	0.99	1.37	0.53	0.47	0.29	7.29
1977	0.20	0.32	0.96	0.50	0.75	1.98	1.90	2.01	3.11	1.47	0.60	1.55	15.35
1978	0.17	0.10	0.08	1.32	0.74	2.78	1.65	3.63	1.40	1.43	2.21	1.96	17.47
1979	0.78	0.10	0.58	3.89	0.33	2.63	2.14	2.34	1.25	0.96	3.31	1.22	19.53
1980	0.86	0.74	1.15	0.39	0.84	3.33	4.00	2.72	0.95	0.90	0.20	0.24	16.32
1981	0.61	1.00	0.41	0.59	0.16	1.79	1.51	1.84	0.95	1.62	2.12	0.40	13.00
1982	1.71	0.63	1.59	0.89	0.88	1.24	3.82	3.53	4.66	2.21	1.61	0.50	23.27
1983	0.11	0.07	0.06	2.62	0.41	0.79	2.26	1.52	1.40	2.18	0.62	1.03	13.07
1984	0.17	T	0.10	0.46	0.39	1.57	1.02	2.85	1.10	0.44	0.66	3.33	12.09
1985	0.93	0.24	1.76	0.02	1.73	1.84	1.24	4.17	4.47	2.14	1.33	0.86	20.73
1986	0.65	0.57	T	0.39	0.78	1.25	1.97	4.40	3.10	1.28	1.90	0.45	16.74
1987	0.72	0.30	0.23	0.05	1.20	1.45	1.89	2.43	1.41	1.43	0.53	1.48	13.12
1988	0.42	0.35	0.45	0.45	1.51	1.59	0.56	3.75	0.73	0.61	0.43	1.68	12.53
1989	1.14	1.76	0.38	0.65	1.46	2.57	1.71	3.61	2.85	1.75	0.92	1.04	19.84
1990	0.92	0.96	0.79	0.25	1.08	0.83	2.72	2.84	3.21	2.05	3.01	2.28	20.94
1991	0.23	0.15	3.44	0.30	0.79	1.20	1.51	1.94	1.72	1.55	0.72	0.91	14.46
1992	0.16	0.50	0.53	0.04	0.79	1.89	1.42	3.52	2.19	1.07	2.22	3.49	17.82
1993	2.53	0.50	0.75	0.31	0.80	1.28	0.67	3.44	4.79	2.11	2.60	0.78	20.56
Record Mean	0.83	0.71	0.84	0.62	0.86	1.32	2.11	3.60	2.56	1.54	1.07	1.02	17.06

AVERAGE TEMPERATURE (deg. F)

BETHEL, ALASKA

YEAR	JAN	FEB	MAR	APR	MAY	JUNE	JULY	AUG	SEP	OCT	NOV	DEC	ANNUAL
1964	4.9	-1.2	3.1	19.5	31.0	52.5	56.6	52.1	47.2	31.9	11.9	-3.0	25.5
1965	-0.6	-4.6	26.0	26.5	32.6	48.6	51.6	50.2	48.1	24.2	22.6	2.2	27.3
1966	5.6	12.5	-3.1	23.0	33.2	51.8	53.2	51.6	45.8	26.2	20.9	3.2	27.0
1967	7.4	9.6	20.7	30.9	42.5	54.0	53.6	54.3	45.3	27.4	22.4	10.0	31.5
1968	11.2	4.1	13.2	23.1	41.6	52.3	57.9	53.8	42.1	29.2	18.6	0.9	29.0
1969	9.1	7.7	17.2	27.1	45.5	52.4	53.7	49.0	47.6	35.1	9.5	12.4	30.6
1970	-11.7	16.4	18.7	22.3	43.3	51.9	51.6	50.7	41.3	25.7	27.5	5.4	28.6
1971	-6.3	-1.5	-0.3	17.0	35.2	50.0	52.8	51.9	44.1	30.8	14.9	12.9	25.2
1972	2.3	0.0	-2.2	14.4	37.2	48.6	57.8	55.4	43.7	34.8	18.1	9.6	26.6
1973	-4.3	8.7	6.0	25.3	39.8	49.8	53.3	51.0	45.4	30.2	20.5	9.0	27.9
1974	6.2	-7.1	15.0	28.9	45.3	51.4	54.7	56.4	49.4	26.2	10.9	-3.9	27.8
1975	-1.5	0.8	10.3	20.4	39.7	49.5	56.6	54.3	42.8	30.0	7.0	3.1	26.1
1976	0.6	-7.4	6.3	17.2	37.6	49.5	54.8	54.2	44.9	30.1	17.6	6.7	26.0
1977	21.7	18.7	4.7	13.8	35.5	52.6	56.1	58.6	47.5	28.9	10.9	11.0	30.0
1978	19.0	12.7	13.9	30.5	44.0	45.1	54.0	55.3	44.6	27.6	24.0	21.5	32.7
1979	21.3	-5.0	18.1	31.5	44.3	47.8	53.6	53.1	44.8	34.0	25.3	-1.8	30.6
1980	0.2	12.0	20.3	30.5	42.7	46.3	54.8	50.7	44.9	31.9	18.7	-0.8	29.3
1981	20.9	12.4	29.5	32.3	48.1	52.4	54.2	54.1	44.5	31.8	17.9	7.9	33.8
1982	12.0	12.1	19.2	17.3	38.6	51.9	53.3	55.1	46.7	28.9	20.7	13.2	30.8
1983	4.9	12.0	18.7	27.5	43.0	51.3	53.7	49.5	42.2	26.5	21.1	20.8	31.0
1984	5.7	-13.2	24.8	22.0	38.1	54.0	55.1	49.7	46.7	30.2	16.4	19.4	29.1
1985	25.7	4.9	13.8	8.3	36.5	47.7	58.8	53.6	45.5	27.0	24.4	25.4	31.0
1986	1.7	15.7	13.1	22.8	39.1	52.5	55.1	51.4	45.0	29.0	20.4	21.6	30.6
1987	9.2	11.1	19.8	23.5	39.3	50.4	55.9	56.3	42.2	34.4	12.8	1.8	29.7
1988	15.1	9.2	6.6	25.3	43.8	52.3	60.0	52.3	45.1	27.2	7.0	11.8	29.6
1989	-13.0	26.1	16.6	29.0	38.1	52.2	54.6	54.2	48.3	31.7	8.8	11.8	29.9
1990	2.7	-12.3	13.8	33.7	43.6	52.2	57.5	55.4	43.7	30.3	16.9	9.4	28.9
1991	9.2	5.9	18.1	30.1	43.9	53.5	57.4	53.7	49.6	35.8	13.5	2.0	31.1
1992	5.2	-0.3	10.2	26.0	36.4	52.6	57.4	53.3	37.7	26.7	16.9	9.1	27.6
1993	3.9	19.3	21.6	34.9	45.8	53.6	57.7	54.2	44.6	34.5	24.9	16.3	34.3
Record Mean	6.3	7.3	12.2	25.2	40.4	51.7	54.9	52.8	45.0	30.6	17.1	7.3	29.2
Max	13.4	14.6	20.7	33.4	49.0	60.6	62.5	59.5	52.0	36.5	23.4	14.1	36.6
Min	-0.7	0.0	3.7	16.9	31.8	42.7	47.3	46.0	38.1	24.6	10.8	0.5	21.8

REFERENCE NOTES

BETHEL, ALASKA

GENERAL
T - TRACE AMOUNT.
BLANK ENTRIES DENOTE MISSING/UNREPORTED DATA.
# INDICATES A STATION OR INSTRUMENT RELOCATION.

SPECIFIC

(a) - LENGTH OF RECORD IN YEARS, ALTHOUGH INDIVIDUAL MONTHS MAY BE MISSING.
0.* OR * - THE VALUE IS BETWEEN 0.0 AND 0.05
NORMALS - BASED ON THE 1961-1990 RECORD PERIOD.
EXTREMES - DATES ARE THE MOST RECENT OCCURRENCE.
WIND DIR. - NUMERALS SHOW TENS OF DEGRESS CLOCKWISE FROM TRUE NORTH. "00" INDICATES CALM.
RESULTANT DIRECTIONS ARE GIVEN TO WHOLE DEGREES.
BOLD VALUES INDICATE EXTREME VALUES WHICH OCCURRED AFTER THE ASOS SYSTEM WAS COMMISSIONED.

RECORD = PERIOD OF RECORD
RECORD MEAN PRECIPITATION IS THE MEAN OF ALL DAILY PRECIPITATION AMOUNTS DURING THE PERIOD OF RECORD.
RECORD MAX(MIN) TEMPERATURE IS THE MEAN OF ALL DAILY MAX(MIN) TEMPERATURES DURING THE PERIOD OF RECORD.
RECORD MEAN TEMPERATURE IS THE SUM OF THE RECORD MAX AND RECORD MIN DIVIDED BY 2.
AVERAGE TEMPERATURE IS THE SUM OF THE MEAN DAILY MAX AND MIN TEMPERATURE DIVIDED BY 2.

EXCEPTIONS

RECORD MEANS ARE THROUGH THE CURRENT YEAR, BEGINNING IN 1924 FOR TEMPERATURE
1924 FOR PRECIPITATION
1959 FOR SNOWFALL

HEATING DEGREE DAYS Base 65 deg. F BETHEL, ALASKA

SEASON	JULY	AUG	SEP	OCT	NOV	DEC	JAN	FEB	MAR	APR	MAY	JUNE	TOTAL
1964-65	254	394	525	1020	1589	2108	2033	1952	1203	1148	997	486	13709
1965-66	406	453	500	1254	1264	1944	1842	1465	2109	1253	979	389	13858
1966-67	365	408	569	1195	1318	1914	1783	1546	1366	1015	692	321	12492
1967-68	354	325	584	1160	1273	1707	1662	1767	1603	1250	721	377	12783
1968-69	221	343	681	1102	1385	1986	1729	1605	1475	1131	600	372	12630
1969-70	343	486	515	921	1660	1626	2383	1355	1430	1272	663	386	13040
1970-71	408	437	707	1210	1118	1848	2210	1863	2022	1430	917	443	14613
1971-72	374	398	621	1050	1500	1611	1947	1883	2088	1516	856	486	14330
1972-73	228	293	632	929	1399	1721	2149	1576	1829	1182	775	447	13160
1973-74	355	425	581	1072	1329	1734	1824	2023	1549	1079	603	399	12973
1974-75	318	258	460	1197	1619	2139	2060	1796	1695	1332	778	456	14108
1975-76	255	324	658	1078	1741	1917	1996	2102	1821	1427	843	459	14621
1976-77	309	328	598	1075	1414	1808	1335	1291	1869	1533	906	363	12829
1977-78	269	199	520	1112	1619	1673	1421	1463	1577	1026	645	590	12114
1978-79	334	293	605	1151	1223	1338	1347	1962	1447	998	634	508	11840
1979-80	345	361	600	952	1186	2073	2007	1532	1382	1029	682	552	12701
1980-81	315	438	594	1017	1383	2044	1362	1472	1094	977	517	371	11584
1981-82	330	335	609	1023	1407	1768	1641	1482	1410	1427	808	387	12627
1982-83	355	302	544	1112	1324	1601	1861	1477	1430	1119	676	402	12203
1983-84	343	474	679	1187	1307	1360	1839	2275	1242	1282	830	326	13144
1984-85	300	464	544	1070	1450	1414	1211	1680	1583	1697	876	511	12800
1985-86	189	352	578	1169	1214	1223	1962	1379	1602	1261	794	368	12091
1986-87	303	417	594	1106	1333	1341	1732	1506	1397	1239	791	430	12189
1987-88	279	261	676	944	1565	1961	1544	1614	1804	1185	652	374	12859
1988-89	152	387	590	1164	1739	1645	2419	1084	1495	1075	826	379	12955
1989-90	319	327	495	1024	1682	1648	1934	2164	1586	935	658	378	13150
1990-91	228	290	631	1069	1438	1722	1729	1656	1448	1039	648	340	12238
1991-92	231	342	454	899	1539	1952	1854	1894	1696	1164	882	369	13276
1992-93	230	357	812	1182	1434	1732	1893	1277	1339	896	591	338	12081
1993-94	228	326	606	939	1197	1507							

COOLING DEGREE DAYS Base 65 deg. F BETHEL, ALASKA

YEAR	JAN	FEB	MAR	APR	MAY	JUNE	JULY	AUG	SEP	OCT	NOV	DEC	TOTAL
1969	0	0	0	0	1	1	0	0	0	0	0	0	2
1970	0	0	0	0	0	0	0	0	0	0	0	0	0
1971	0	0	0	0	0	0	0	0	0	0	0	0	0
1972	0	0	0	0	0	0	17	1	0	0	0	0	18
1973	0	0	0	0	0	0	0	0	0	0	0	0	0
1974	0	0	0	0	0	0	3	0	0	0	0	0	3
1975	0	0	0	0	0	0	1	0	0	0	0	0	1
1976	0	0	0	0	0	0	0	0	0	0	0	0	0
1977	0	0	0	0	0	0	0	8	0	0	0	0	8
1978	0	0	0	0	0	0	0	0	0	0	0	0	0
1979	0	0	0	0	0	0	0	0	0	0	0	0	0
1980	0	0	0	0	0	0	6	0	0	0	0	0	6
1981	0	0	0	0	0	0	0	2	0	0	0	0	2
1982	0	0	0	0	0	0	0	0	0	0	0	0	0
1983	0	0	0	0	0	0	0	0	0	0	0	0	0
1984	0	0	0	0	0	0	0	0	0	0	0	0	0
1985	0	0	0	0	0	0	1	5	0	0	0	0	6
1986	0	0	0	0	0	2	3	0	0	0	0	0	5
1987	0	0	0	0	0	0	2	0	0	0	0	0	2
1988	0	0	0	0	0	0	4	0	0	0	0	0	4
1989	0	0	0	0	0	0	3	0	0	0	0	0	3
1990	0	0	0	0	0	0	0	0	0	0	0	0	0
1991	0	0	0	0	0	2	0	0	0	0	0	0	2
1992	0	0	0	0	0	0	0	0	0	0	0	0	0
1993	0	0	0	0	2	0	7	0	0	0	0	0	9

SNOWFALL (inches) BETHEL, ALASKA

SEASON	JULY	AUG	SEP	OCT	NOV	DEC	JAN	FEB	MAR	APR	MAY	JUNE	TOTAL
1964-65	0.0	0.0	0.0	8.2	6.6	8.3	6.4	10.1	7.8	4.3	2.0	0.0	53.7
1965-66	0.0	0.0	T	2.6	8.1	7.3	1.0	14.8	2.3	0.3	4.8	0.0	41.2
1966-67	0.0	0.0	0.0	2.1	13.2	6.9	8.8	2.8	7.9	5.1	0.0	0.0	46.8
1967-68	0.0	0.0	0.0	2.4	10.1	5.0	2.9	5.9	5.6	8.2	1.7	0.0	41.8
1968-69	0.0	0.0	T	4.5	9.5	8.3	13.5	10.7	9.5	0.6	T	0.0	56.6
1969-70	0.0	0.0	0.0	0.6	4.3	11.8	2.8	1.9	5.9	10.5	1.7	0.0	39.5
1970-71	0.0	0.0	T	10.2	7.5	13.1	6.3	4.8	2.4	6.2	6.2	T	56.7
1971-72	0.0	0.0	T	8.0	9.2	17.5	3.8	0.3	5.6	9.4	1.6	1.1	56.5
1972-73	0.0	0.0	0.2	2.7	6.0	2.7	11.5	9.1	7.0	1.8	5.6	T	46.6
1973-74	0.0	0.0	T	4.4	14.5	1.7	0.3	2.9	18.0	4.8	T	0.0	46.6
1974-75	T	0.0	0.0	1.2	7.1	7.2	6.7	5.7	4.4	14.3	3.5	0.0	50.1
1975-76	0.0	0.0	0.0	1.4	3.4	6.5	6.2	5.6	6.7	5.8	0.9	T	36.5
1976-77	0.0	0.0	0.0	1.8	6.1	6.5	2.1	6.6	16.4	10.1	2.5	0.0	52.1
1977-78	0.0	0.0	T	4.8	7.6	13.8	2.1	1.2	0.8	10.7	T	0.0	41.0
1978-79	0.0	0.0	0.0	12.8	2.7	19.1	6.3	1.0	1.8	5.2	T	0.0	48.9
1979-80	0.0	0.0	0.0	1.7	10.3	12.7	6.0	7.1	12.6	2.1	0.4	0.0	52.9
1980-81	0.0	0.0	0.0	4.0	2.0	2.6	3.7	8.8	2.9	2.8	1.0	0.0	27.8
1981-82	0.0	0.0	T	0.9	21.6	4.3	10.2	T	14.7	6.0	T	0.0	57.7
1982-83	0.0	0.0	0.0	9.0	5.6	4.7	1.0	1.5	0.6	9.5	T	0.0	31.9
1983-84	0.0	0.0	T	T	6.4	3.5	2.9	T	1.1	2.6	T	0.0	16.5
1984-85	0.0	0.0	T	3.4	6.8	19.6	2.5	0.7	16.3	0.4	5.1	0.8	55.6
1985-86	0.0	0.0	T	5.3	8.7	2.3	8.7	1.7	T	2.6	4.2	0.0	33.5
1986-87	0.0	0.0	T	0.3	1.8	2.1	13.5	4.0	1.7	0.6	0.2	T	24.2
1987-88	0.0	0.0	0.0	4.7	7.3	14.3	1.7	5.1	5.3	3.1	T	0.0	41.5
1988-89	0.0	0.0	T	5.2	7.9	15.1	13.1	9.2	4.7	2.7	3.4	0.0	61.3
1989-90	0.0	0.0	0.0	4.3	11.0	8.8	10.4	12.8	8.7	2.5	T	0.0	58.5
1990-91	0.0	0.0	3.1	1.7	21.7	6.6	2.6	4.6	30.6	1.5	0.3	T	72.7
1991-92	0.0	0.0	0.0	2.2	9.1	19.2	3.6	10.3	8.3	0.7	6.2	0.0	59.6
1992-93	0.0	0.0	4.0	2.0	19.3	36.0	15.3	6.2	11.6	0.9	T	T	95.3
1993-94	0.0	0.0	T	1.4	13.0	13.2							
Record Mean	T	0.0	0.3	4.1	8.7	9.7	5.8	5.5	7.6	5.3	1.9	0.1	49.0

BETTLES, ALASKA

Bettles Airport is located on the south side of and adjacent to the Koyukuk River The foothills of the Endicott Mountains are found to the west, north, and east of the station. The Koyukuk River valley extends for about 20 miles to the south where it then curves to the southwest. Changes in elevation for a distance of about 15 miles on all sides of the airport are small, with a very gradual rise from south to north. The land is timbered with low spruce and birch. Bettles Airport is one of four Weather Service stations located north of the Arctic Circle.

The climate of the Bettles area is typical of a continental regime. Temperatures during the long summer days are mild, with maximums mostly in the high 60s and low 70s, and occasionally in the 80s. The sun does not set during the period June 2 to July 9. The freeze-free period averages 89 days, extending from May to late August. There is no commercial agriculture in this area. Bettles provides a center for wilderness guided and unguided tours, hunting and fishing and gold mining.

Winters are typical of interior Alaska. Minimum temperatures average below zero from November through March, and readings in the −45 to −55 degree range are experienced each winter. Here, as in most of the interior, the transition from summer to winter and vice versa is rapid, resulting in short spring and fall seasons.

Annual precipitation amounts are slightly heavier than at most interior locations, but still fall well within what is expected for a continental climate. It also follows the pattern of nearly all Alaskan stations, with precipitation amounts building up to a maximum during late summer and fall months. Snow has occurred during all months except July. The total seasonal snowfall has ranged from less than 40 inches to more than 130 inches. Because of the cold temperatures, much of the snow remains on the ground during the winter.

Surface winds are seldom strong during any season of the year, nor do they show much seasonal variation. Wind directions prevail from the north ten months of the year.

NORMALS, MEANS, AND EXTREMES

BETTLES, ALASKA

LATITUDE: 66°55'N LONGITUDE: 151°31' W ELEVATION: FT. GRND 644 BARO 651 TIME ZONE: YUKON WBAN: 26533

	(a)	JAN	FEB	MAR	APR	MAY	JUNE	JULY	AUG	SEP	OCT	NOV	DEC	YEAR
TEMPERATURE °F:														
Normals														
-Daily Maximum		-4.9	0.3	14.4	31.5	52.9	67.5	69.6	62.7	49.0	24.9	5.1	-1.8	30.9
-Daily Minimum		-20.4	-18.3	-8.6	9.7	33.5	47.0	49.2	43.8	32.4	12.1	-8.8	-16.7	12.9
-Monthly		-12.7	-9.0	2.9	20.7	43.2	57.2	59.4	53.3	40.8	18.5	-1.8	-9.3	21.9
Extremes														
-Record Highest	43	48	40	44	56	86	92	93	87	79	53	45	38	93
-Year		1980	1977	1965	1976	1983	1969	1986	1977	1957	1969	1976	1960	JUL 1986
-Record Lowest	43	-70	-61	-56	-37	-10	27	29	22	0	-35	-57	-59	-70
-Year		1975	1993	1964	1986	1952	1960	1955	1974	1992	1992	1974	1974	JAN 1975
NORMAL DEGREE DAYS:														
Heating (base 65°F)		2409	2072	1925	1329	676	238	192	373	726	1442	2004	2303	15689
Cooling (base 65°F)		0	0	0	0	0	0	18	11	0	0	0	0	29
% OF POSSIBLE SUNSHINE														
MEAN SKY COVER (tenths)														
Sunrise - Sunset	5	4.7	5.8	6.1	6.0	6.2	6.6	7.6	8.0	6.7	7.6	6.0	6.4	6.5
MEAN NUMBER OF DAYS:														
Sunrise to Sunset														
-Clear	5	14.8	9.8	9.4	7.2	6.8	3.8	2.0	2.2	6.6	4.6	9.4	9.0	85.6
-Partly Cloudy	5	5.2	4.2	5.4	9.8	11.0	13.4	9.8	8.2	7.0	5.8	6.2	6.6	92.6
-Cloudy	5	11.0	14.0	16.2	13.0	13.2	12.8	19.2	20.6	16.4	20.6	14.4	15.4	186.8
Precipitation														
.01 inches or more	34	7.6	6.8	7.4	5.3	6.4	9.9	10.9	13.3	11.2	11.4	10.1	10.1	110.4
Snow,Ice pellets,hail														
1.0 inches or more	25	3.6	3.1	3.4	1.7	0.4	0.0	0.0	0.*	0.8	4.5	4.5	5.7	27.8
Thunderstorms	21	0.0	0.0	0.0	0.0	0.5	2.7	3.0	0.4	0.1	0.0	0.0	0.0	6.7
Heavy Fog Visibility 1/4 mile or less	21	0.8	0.2	0.2	0.3	0.3	0.1	0.6	1.3	0.7	0.9	0.4	0.3	6.0
Temperature °F														
-Maximum														
70° and above	41	0.0	0.0	0.0	0.0	1.3	12.8	15.8	5.5	0.4	0.0	0.0	0.0	35.7
32° and below	41	30.1	27.9	28.1	14.4	0.5	0.0	0.0	0.0	1.1	21.8	29.3	30.8	184.1
-Minimum														
32° and below	41	31.0	28.3	30.9	29.1	13.4	0.3	0.1	2.1	15.0	29.5	30.0	31.0	240.7
0° and below	41	25.7	23.1	21.6	8.5	0.3	0.0	0.0	0.0	0.*	7.0	20.2	24.8	131.2
AVG. STATION PRESS.(mb)	13	987.6	991.1	988.8	986.5	986.9	986.8	989.2	988.2	986.7	985.0	987.7	987.6	987.7
RELATIVE HUMIDITY (%)														
Hour 03	44	69	67	68	73	75	78	85	89	84	80	74	71	76
Hour 09 (Local Time)	46	68	66	66	67	63	62	70	80	80	79	73	71	70
Hour 15	47	67	62	61	61	51	48	54	62	64	73	72	71	62
Hour 21	44	67	66	67	67	58	55	64	75	77	79	73	71	68
PRECIPITATION (inches):														
Water Equivalent														
-Normal		0.69	0.64	0.68	0.64	0.61	1.44	1.94	2.38	1.72	1.20	0.90	0.90	13.74
-Maximum Monthly	43	3.42	2.94	3.60	1.88	2.23	3.59	5.42	5.91	4.72	3.82	3.85	3.09	5.91
-Year		1973	1964	1963	1965	1991	1965	1963	1963	1993	1972	1967	1992	AUG 1963
-Minimum Monthly	43	T	T	T	0.01	0.04	T	0.00	0.41	0.13	0.12	0.08	0.13	0.00
-Year		1961	1979	1960	1986	1959	1959	1959	1958	1984	1974	1984	1964	JUL 1959
-Maximum in 24 hrs	43	1.40	0.86	0.87	0.98	0.73	1.93	1.46	1.87	1.31	1.32	1.35	0.78	1.93
-Year		1973	1989	1963	1982	1991	1958	1970	1951	1954	1972	1992	1992	JUN 1958
Snow,Ice pellets,hail														
-Maximum Monthly	43	55.8	32.8	35.2	34.7	12.0	T	0.0	2.6	9.4	28.3	41.6	40.2	55.8
-Year		1973	1989	1991	1984	1952	1974		1969	1968	1972	1967	1987	JAN 1973
-Maximum in 24 hrs	43	21.7	9.4	10.6	10.8	6.7	T	0.0	2.6	5.6	10.8	14.0	11.1	21.7
-Year		1973	1989	1991	1975	1952	1987		1969	1972	1976	1992	1984	JAN 1973
WIND:														
Mean Speed (mph)	18	6.0	6.5	7.2	7.5	7.5	7.1	6.7	6.3	6.7	6.6	6.1	5.9	6.7
Prevailing Direction through 1963		NNW	NNW	NNW	NNW	NNW	NNW	SSE	SSE	NNW	NNW	NNW	NNW	NNW
Fastest Obs. 1 Min.														
-Direction (!!)	6	11	08	08	36	11	01	24	25	24	24	26	24	24
-Speed (MPH)	6	30	25	29	26	31	28	28	32	25	25	38	40	40
-Year		1970	1970	1971	1972	1970	1970	1973	1975	1975	1970	1970	1970	DEC 1970
Peak Gust														
-Direction (!!)														
-Speed (mph)														
-Date														

PRECIPITATION (inches)

BETTLES, ALASKA

YEAR	JAN	FEB	MAR	APR	MAY	JUNE	JULY	AUG	SEP	OCT	NOV	DEC	ANNUAL
1964	0.57	2.94	0.11	1.39	0.69	1.22	0.60	1.94	1.87	2.69	0.47	0.13	14.62
1965	1.44	0.60	2.51	1.88	0.66	3.59	1.78	4.86	3.90	0.61	1.55	1.48	24.86
1966	0.05	0.29	0.47	0.87	0.55	1.37	1.06	2.12	0.47	0.63	1.43	0.54	9.85
1967	0.28	0.94	1.87	1.82	0.11	0.66	2.80	2.17	1.23	0.20	3.85	1.68	17.61
1968	1.13	1.29	0.80	0.71	0.30	0.65	1.12	0.55	0.89	0.38	0.08	1.09	8.99
1969	0.13	0.37	0.10	0.01	0.13	0.53	0.83	2.26	0.36	0.99	0.67	0.50	6.88
1970	0.07	0.52	0.77	0.47	0.53	1.53	3.18	2.34	0.43	1.50	1.91	1.97	15.22
1971	0.37	1.21	0.43	0.13	1.18	0.76	2.14	1.80	1.44	1.60	1.23	1.48	13.77
1972	1.25	0.11	0.50	0.03	0.25	1.78	0.41	2.70	0.39	3.82	0.67	1.46	13.37
1973	3.42	0.20	0.30	0.32	0.51	0.64	1.17	2.31	1.02	1.31	0.92	0.39	12.51
1974	0.35	0.09	0.26	0.12	0.44	1.60	1.53	1.32	2.05	0.12	1.51	0.27	9.66
1975	0.68	0.54	0.29	0.38	0.30	1.25	1.05	0.68	2.37	0.63	0.51	0.58	9.26
1976	0.49	1.43	0.60	0.17	1.20	1.34	2.95	1.92	1.87	1.48	1.45	0.21	15.11
1977	0.40	0.49	0.66	0.71	1.45	1.31	1.37	1.71	3.65	0.80	0.10	1.36	14.01
1978	0.40	0.41	0.43	0.46	0.11	2.45	0.44	0.83	1.22	0.41	1.30	1.04	9.50
1979	0.09	T	0.44	1.07	0.65	1.94	2.25	1.11	0.87	1.13	2.66	0.51	12.72
1980	0.56	0.67	0.10	0.24	1.13	2.49	1.69	1.76	0.46	1.99	0.30	0.39	11.78
1981	0.94	1.21	0.31	0.58	0.52	1.56	4.06	2.79	0.65	1.19	0.81	0.43	15.05
1982	0.35	0.32	1.24	1.51	0.36	1.70	4.29	2.51	4.14	0.38	0.72	1.01	18.53
1983	0.05	0.12	0.11	0.33	0.22	0.18	1.77	4.07	1.81	1.23	0.22	0.17	10.28
1984	0.55	0.35	0.10	0.75	0.42	1.65	3.94	3.23	0.13	0.15	0.08	0.82	12.17
1985	0.87	0.27	0.96	0.12	1.01	1.56	1.45	2.35	3.91	0.81	0.87	0.69	14.87
1986	0.34	0.29	0.16	0.01	0.73	2.62	1.35	2.12	2.47	1.46	0.41	1.06	13.02
1987	0.65	0.30	0.04	0.10	1.02	2.28	2.17	1.52	3.20	1.61	0.51	1.46	14.86
1988	1.28	0.59	0.16	0.28	0.91	0.58	1.44	3.61	1.50	1.05	0.45	1.15	13.00
1989	0.66	2.24	0.08	1.26	1.44	1.26	2.00	3.57	1.01	0.99	0.50	1.01	16.02
1990	0.39	0.59	1.89	0.32	0.68	1.39	1.80	1.54	2.38	1.03	0.53	1.01	13.55
1991	1.61	0.42	2.25	0.44	2.23	0.63	1.51	0.83	1.00	1.16	0.22	0.66	12.96
1992	0.50	0.81	0.51	0.25	0.22	3.11	2.86	2.58	0.59	0.97	3.36	3.09	18.85
1993	3.20	0.49	1.15	0.06	0.83	1.36	1.53	2.59	4.72	1.28	1.91	0.88	20.00
Record Mean	0.80	0.66	0.70	0.56	0.59	1.42	1.85	2.41	1.81	1.16	0.91	0.89	13.75

AVERAGE TEMPERATURE (deg. F)

BETTLES, ALASKA

YEAR	JAN	FEB	MAR	APR	MAY	JUNE	JULY	AUG	SEP	OCT	NOV	DEC	ANNUAL
1964	-19.7	-7.1	-10.6	14.9	33.0	57.4	59.7	54.1	41.8	26.1	1.3	-29.1	18.5
1965	-11.0	-17.1	17.7	20.3	35.3	55.6	57.7	49.1	44.1	9.1	5.1	-12.6	21.1
1966	-23.7	-10.5	-6.1	16.8	40.3	58.6	57.4	52.3	44.2	19.8	-2.6	-17.6	19.1
1967	-17.2	-8.7	5.3	23.8	39.0	56.8	56.1	51.6	44.3	19.4	-0.7	-6.4	21.9
1968	-18.1	-9.5	7.4	18.2	38.2	57.2	63.0	57.0	36.9	15.9	-0.1	-17.9	20.7
1969	-23.6	-9.7	2.8	29.4	46.5	60.8	56.3	45.2	44.8	29.1	-6.5	10.1	23.8
1970	-17.7	1.2	8.8	23.7	44.9	54.2	58.5	51.8	32.4	9.9	-0.6	-13.5	21.1
1971	-34.1	-16.8	-8.5	19.1	41.2	61.2	57.9	52.2	38.8	14.1	-4.7	-7.7	17.7
1972	-24.8	-19.8	-13.0	13.2	42.6	57.7	62.3	55.9	37.9	21.1	-0.7	-6.0	18.9
1973	-24.7	-7.4	0.5	23.2	45.4	58.0	56.5	50.2	42.3	17.4	-8.8	-0.2	21.1
1974	-19.5	-27.6	0.4	23.2	47.3	55.4	59.4	56.8	46.5	16.9	-10.2	-19.1	19.2
1975	-21.7	-6.1	5.9	18.0	45.2	57.8	62.4	53.9	38.5	18.0	-10.1	-18.6	20.3
1976	-13.8	-20.2	4.4	26.2	43.8	56.8	58.4	58.4	44.3	19.0	12.4	-9.3	23.4
1977	5.3	4.8	-5.6	14.6	43.3	58.2	63.7	60.9	42.8	22.1	-8.4	-14.7	23.9
1978	2.2	-1.8	6.9	26.2	45.2	53.3	62.5	56.4	45.6	15.7	8.4	-4.2	26.4
1979	-5.1	-26.7	4.4	22.1	48.0	55.7	58.8	57.8	42.9	27.6	17.0	-22.6	23.3
1980	-14.6	4.9	8.3	24.1	48.0	55.7	60.7	51.5	35.9	23.9	4.9	-27.4	23.0
1981	16.1	-1.7	17.5	21.9	48.5	55.6	53.9	51.3	39.7	23.4	2.5	-8.4	26.7
1982	-14.9	-8.9	6.0	17.2	39.6	56.9	60.4	53.3	43.0	9.6	-0.6	-2.1	21.6
1983	-17.2	-2.3	4.7	26.9	43.3	58.4	61.2	48.6	33.9	14.0	8.5	-7.1	22.8
1984	-16.2	-24.2	9.1	14.8	40.5	60.3	58.2	48.9	42.4	20.8	-5.9	-8.1	20.1
1985	10.3	-14.4	6.9	10.7	41.2	56.2	60.3	53.5	38.7	14.6		6.9	23.2
1986	-5.5	0.3		10.2	41.5	59.4	59.7	49.3	40.8	17.1	-4.9	5.3	22.6
1987	-4.7	-5.0	6.1	21.1	45.2	58.3	60.7	53.1	37.2	25.9	3.8	-5.1	24.7
1988	-5.1	0.0	8.5	25.9	47.6	59.3	61.9	52.9	40.2	9.3	-5.3	0.5	24.6
1989	-31.5	6.3	0.7	26.6	41.8	56.5	58.9	55.8	43.2	19.7	-9.6	0.7	22.4
1990	-16.6	-29.4	11.1	29.3	51.0	59.1	62.5	57.2	38.7	18.2	-12.9	-14.5	21.1
1991	-8.0	-6.3	2.6	26.8	48.6	62.0	58.9	51.7	44.8	20.4	-5.6	-7.5	24.0
1992	-4.7	-12.3	9.7	19.5	34.0	59.9	62.9	52.8	29.3	13.7	-0.2	-13.2	21.0
1993	-11.4	-0.6	10.5	31.5	46.0	60.8	63.6	51.2	36.8	24.7	0.2	1.8	26.3
Record Mean	-12.6	-9.1	4.0	20.8	43.0	57.4	59.1	53.0	40.1	18.6	-1.4	-9.8	21.9
Max	-4.7	0.1	14.6	31.9	52.8	67.8	69.3	62.3	48.3	25.1	5.6	-2.3	30.9
Min	-20.5	-18.3	-6.6	9.6	33.3	46.9	48.8	43.6	31.9	12.1	-8.3	-17.3	12.9

REFERENCE NOTES

BETTLES, ALASKA

GENERAL
T - TRACE AMOUNT.
BLANK ENTRIES DENOTE MISSING/UNREPORTED DATA.
# INDICATES A STATION OR INSTRUMENT RELOCATION.

SPECIFIC

(a) - LENGTH OF RECORD IN YEARS, ALTHOUGH INDIVIDUAL MONTHS MAY BE MISSING.
0.* OR * - THE VALUE IS BETWEEN 0.0 AND 0.05
NORMALS - BASED ON THE 1961-1990 RECORD PERIOD.
EXTREMES - DATES ARE THE MOST RECENT OCCURRENCE
WIND DIR.- NUMERALS SHOW TENS OF DEGRESS CLOCKWISE FROM TRUE NORTH. "00" INDICATES CALM.
RESULTANT DIRECTIONS ARE GIVEN TO WHOLE DEGREES.
BOLD VALUES INDICATE EXTREME VALUES WHICH OCCURRED AFTER THE ASOS SYSTEM WAS COMMISSIONED.

RECORD = PERIOD OF RECORD
RECORD MEAN PRECIPITATION IS THE MEAN OF ALL DAILY PRECIPITATION AMOUNTS DURING THE PERIOD OF RECORD.
RECORD MAX(MIN) TEMPERATURE IS THE MEAN OF ALL DAILY MAX(MIN) TEMPERATURES DURING THE PERIOD OF RECORD.
RECORD MEAN TEMPERATURE IS THE SUM OF THE RECORD MAX AND RECORD MIN DIVIDED BY 2.
AVERAGE TEMPERATURE IS THE SUM OF THE MEAN DAILY MAX AND MIN TEMPERATURE DIVIDED BY 2.

EXCEPTIONS

STATION PRESSURE, WIND DIRECTION AND SPEED WERE NOT REPORTED FOR OCT. 23 - DEC 31.

1. DAYS OF PRECIPITATION .01 INCH OR MORE ARE 1943-1953 AND 1969 TO DATE.
2. MEAN WIND SPEED IS APRIL 1950-1954 AND 1969 TO DATE.
3. MEAN SKY COVER, AND DAYS CLEAR-PARTLY CLOUDY-CLOUDY ARE THROUGH 1975.
4. FASTEST OBSERVED WIND IS THROUGH MARCH 1976.
5. THUNDERSTORMS AND HEAVY FOG MAY BE INCOMPLETE, DUE TO PART-TIME OPERATIONS.

RECORD MEANS ARE THROUGH THE CURRENT YEAR, BEGINNING IN 1951 FOR TEMPERATURE
1951 FOR PRECIPITATION
1951 FOR SNOWFALL

HEATING DEGREE DAYS Base 65 deg. F BETTLES, ALASKA

SEASON	JULY	AUG	SEP	OCT	NOV	DEC	JAN	FEB	MAR	APR	MAY	JUNE	TOTAL
1964-65	163	331	689	1198	1913	2923	2359	2304	1461	1335	913	277	15866
1965-66	223	486	620	1732	1797	2408	2755	2116	2207	1441	760	187	16732
1966-67	226	400	618	1398	2028	2562	2559	2066	1850	1231	802	270	16010
1967-68	281	415	614	1406	1972	2217	2579	2161	1783	1397	822	228	15875
1968-69	98	262	836	1517	1952	2575	2755	2093	1929	1065	568	170	15820
1969-70	263	608	599	1108	2146	1699	2572	1789	1738	1233	616	317	14688
1970-71	196	401	972	1706	1968	2438	3074	2295	2279	1374	728	180	17611
1971-72	221	391	782	1573	2090	2254	2785	2467	2420	1549	685	254	17471
1972-73	123	272	804	1354	1968	2199	2783	2029	2004	1246	600	211	15593
1973-74	255	453	674	1468	2217	2023		2598	2004	1249	540	300	
1974-75	192	260	543	1480	2257	2615	2690	1997	1829	1403	605	210	16081
1975-76	134	337	790	1454	2260	2595	2450	2474	1877	1156	651	240	16418
1976-77	219	204	613	1420	1575	2310	1848	1682	2191	1510	665	196	14433
1977-78	88	157	663	1321	2205	2471	1948	1872	1793	1158	607	341	14624
1978-79	94	257	574	1521	1696	2144	2173	2574	1878	1280	521	271	14983
1979-80	194	218	655	1154	1435	2721	2472	1742	1751	1221	521	279	14363
1980-81	146	420	864	1269	1800	2868	1513	1870	1466	1288	502	283	14289
1981-82	335	416	753	1283	1876	2276	2481	2073	1827	1427	780	247	15774
1982-83	170	356	654	1716	1969	2079	2549	1883	1866	1135	677	223	15277
1983-84	144	498	925	1575	1692	2240	2525	2592	1730	1499	750	144	16314
1984-85	215	490	673	1365	2132	2271	1689	2228	1800	1623	729	269	15484
1985-86	143	359	782	1557		1800	2186	1811		1641	722	170	
1986-87	202	482	719	1479	2097	1850	2158	1963	1827	1310	608	228	14923
1987-88	140	364	828	1206	1836	2174	2175	1888	1748	1169	532	176	14236
1988-89	118	370	738	1725	2113	1998	2996	1638	1993	1145	712	258	15804
1989-90	194	279	646	1398	2240	1994	2532	2649	1669	1064	436	198	15299
1990-91	115	241	784	1448	2341	2470	2268	1996	1935	1140	501	171	15410
1991-92	199	406	599	1373	2122	2249	2159	2246	1707	1358	955	167	15540
1992-93	82	372	1064	1586	1953	2431	2364	1834	1680	999	583	133	15081
1993-94	84	418	838	1240	1939	1959							

COOLING DEGREE DAYS Base 65 deg. F BETTLES, ALASKA

YEAR	JAN	FEB	MAR	APR	MAY	JUNE	JULY	AUG	SEP	OCT	NOV	DEC	TOTAL
1969	0	0	0	0	0	47	2	0	0	0	0	0	49
1970	0	0	0	0	0	0	2	0	0	0	0	0	2
1971	0	0	0	0	0	74	9	0	0	0	0	0	83
1972	0	0	0	0	0	42	47	0	0	0	0	0	89
1973	0	0	0	0	0	9	0	0	0	0	0	0	9
1974	0	0	0	0	0	20	30	11	0	0	0	0	61
1975	0	0	0	0	0	0	59	0	0	0	0	0	59
1976	0	0	0	0	0	2	20	5	0	0	0	0	27
1977	0	0	0	0	0	0	54	36	0	0	0	0	90
1978	0	0	0	0	0	0	21	0	0	0	0	0	21
1979	0	0	0	0	0	0	11	1	0	0	0	0	12
1980	0	0	0	0	0	8	17	11	0	0	0	0	36
1981	0	0	0	0	0	9	0	0	0	0	0	0	9
1982	0	0	0	0	0	13	31	0	0	0	0	0	44
1983	0	0	0	0	9	30	33	0	0	0	0	0	72
1984	0	0	0	0	0	9	11	0	0	0	0	0	20
1985	0	0	0	0	0	9	3	10	0	0		0	22
1986	0	0		0	0	10	46	0	0	0	0	0	56
1987	0	0	0	0	0	34	13	0	0	0	0	0	47
1988	0	0	0	0	0	15	29	0	0	0	0	0	44
1989	0	0	0	0	0	8	10	2	0	0	0	0	20
1990	0	0	0	0	7	27	46	7	0	0	0	0	87
1991	0	0	0	0	0	88	18	0	0	0	0	0	106
1992	0	0	0	0	0	22	21	1	0	0	0	0	44
1993	0	0	0	0	0	12	46	0	0	0	0	0	58

SNOWFALL (inches) BETTLES, ALASKA

SEASON	JULY	AUG	SEP	OCT	NOV	DEC	JAN	FEB	MAR	APR	MAY	JUNE	TOTAL
1964-65	0.0	0.0	T	15.4	5.2	1.3	28.2	2.0	28.8	23.6	1.9	0.0	106.4
1965-66	0.0	T	1.6	6.5	14.6	15.9	0.5	3.5	4.9	8.7	0.7	0.0	56.9
1966-67	0.0	0.0	T	3.2	14.6	5.5	2.8	9.9	18.5	20.7	T	T	75.2
1967-68	0.0	0.0	0.0	2.4	41.6	16.8	10.2	12.9	8.0	7.0	0.3	0.0	99.2
1968-69	0.0	0.0	9.4	5.0	1.5	12.0	1.6	4.7	1.1	0.5	0.2	0.0	36.0
1969-70	0.0	2.6	0.0	1.2	18.3	11.3	1.3	8.1	14.5	17.1	1.6	0.0	76.0
1970-71	0.0	0.0	1.0	20.1	35.9	31.2	7.6	25.3	7.2	3.0	1.7	0.0	133.0
1971-72	0.0	0.0	2.9	22.6	15.0	23.3	20.4	1.1	7.0	2.4	0.7	0.0	95.4
1972-73	0.0	0.0	5.6	28.3	6.8	21.6	55.8	4.0	5.5	3.8	T	0.0	131.4
1973-74	0.0	T	2.3	15.2	16.9	4.5	3.8	2.2	3.0	1.6	0.0	T	49.5
1974-75	0.0	0.0	0.3	4.9	18.6	5.1	11.4	8.5	3.7	11.9	0.3	0.0	64.7
1975-76	0.0	0.0	T	7.2	7.7	12.2	13.1	14.8	8.4	3.2	T	0.0	66.6
1976-77	0.0	0.0	T	27.9	23.4	5.1	10.5	8.7	16.2	6.4	0.2	0.0	98.4
1977-78	0.0	0.0	T	12.6	1.2	29.7	9.4	8.0	13.4	5.5	T	0.0	79.8
1978-79	0.0	0.0	3.2	8.0	21.9	33.1	3.5	0.3	8.6	8.1	T	0.0	86.7
1979-80	0.0	0.0	8.0	15.0	36.3	13.0	8.7	11.0	4.2	6.1	0.0	0.0	102.3
1980-81	0.0	0.0	4.4	17.6	9.2	7.2	18.7	21.1	6.3	6.2	0.0	0.0	90.7
1981-82	0.0	0.0	0.6	8.8	25.8	7.9	4.1	1.8	24.2	5.3	1.1	0.0	79.6
1982-83	0.0	0.0	T	10.0	11.0	20.5	1.5	3.5	1.7	5.9	T	0.0	54.1
1983-84	0.0	T	3.7	21.7	7.5	6.6	12.4	11.0	3.0	34.7	2.0	0.0	102.6
1984-85	0.0	0.0	0.5	4.2	1.4	22.5	20.9	9.9	21.8	2.8	3.7	0.0	87.7
1985-86	0.0	0.0	0.6	8.3	10.7	10.8	7.6	5.3	1.2	0.8	1.2	0.0	46.5
1986-87	0.0	0.0	1.3	12.9	7.2	18.3	12.0	4.9	0.6	4.1	0.0	0.0	61.3
1987-88	0.0	0.0	2.0	19.6	9.6	40.2	18.2	13.5	6.3	2.8	2.6	0.0	114.8
1988-89	0.0	0.0	0.6	13.8	7.0	25.0	13.0	32.8	2.2	10.9	2.2	0.0	107.5
1989-90	0.0	0.0	T	8.2	10.8	18.0	10.8	7.1	31.0	3.3	0.0	0.0	89.2
1990-91	0.0	0.0	5.6	17.1	13.0	21.8	29.3	5.5	35.2	3.0	T	0.0	130.5
1991-92	0.0	0.0	0.0	6.2	4.1	11.0	8.2	11.6	7.9	4.0	3.3	0.0	56.3
1992-93	0.0	0.0	4.6	15.2	36.3	36.6	40.8	10.6	25.0	1.5	T	0.0	170.6
1993-94	0.0	T	8.9	12.4	27.9	20.4							
Record Mean	0.0	0.1	1.8	11.9	13.2	15.1	12.5	9.4	10.5	7.2	1.2	T	82.8

BIG DELTA, ALASKA

Big Delta Airport is located approximately 1 1/2 miles from the east bank of the Delta River on the Richardson Highway and 3 miles south of the junction of the Richardson Highway with the Alaska Highway. The nearest community is Delta Junction, at the highway junction. The community of Big Delta is about 10 miles to the north where the Delta River and Tanana River join. Terrain in all directions is comparatively flat with a very gradual decrease in elevation to the north, and an increase to the south, reaching an elevation of 2,000 feet at a distance of 14 miles. Beyond this point the rise in elevation is more rapid. Isabell Pass, through the highest portion of the Alaska Range, is 60 miles to the south. From a large scale viewpoint, even though the airport is located near the Delta River, its climate is influenced more by the Tanana River valley lying between two northwest-southeast oriented mountain ranges.

The climate in this portion of Alaska is continental. Summer temperatures are mild, with maximums generally in the 65 to 80 degree range, but reaching the 90 degree level on rare occasions. There are 18 to 21 hours of sunshine daily. A number of thunderstorms occur every summer. The average freeze-free period is about 114 days, extending from mid-May to early September. Potatoes are grown commercially in the area.

Winters are cold, generally with four months of minimum temperatures below zero. The days are short, the nights long. The transition periods between summer and winter and vice versa are rapid, with the daily change almost perceptible.

The annual precipitation of less then 12 inches is low for crop growth. However, well over half occurs during the summer, at the time most needed. Winter snowfall is light and generally stays on the ground throughout the winter.

Surface winds follow a normal pattern of strongest speeds in winter, lightest in summer. The direction, east-southeast, follows the orientation of the Tanana Valley from early fall to early spring, and follows the orientation of the Delta River, southwest, during the months of May through July. Wind averages and extremes are high when compared to other interior Alaska locations. With strong pressure gradients both the Tanana and Delta River valleys experience a venturi effect accentuating the already high speeds.

NORMALS, MEANS, AND EXTREMES

BIG DELTA, ALASKA

LATITUDE: 64 °00' N LONGITUDE: 145 °44' W ELEVATION: FT. GRND 1268 BARO 1274 TIME ZONE: YUKON WBAN: 26415

	(a)	JAN	FEB	MAR	APR	MAY	JUNE	JULY	AUG	SEP	OCT	NOV	DEC	YEAR
TEMPERATURE °F:														
Normals														
-Daily Maximum		3.1	10.5	24.5	40.3	56.6	66.4	69.6	64.7	53.2	31.2	12.7	5.4	36.5
-Daily Minimum		-11.0	-6.2	2.5	20.5	36.9	47.2	50.7	46.0	35.7	17.8	-1.7	-8.5	19.2
-Monthly		-4.0	2.2	13.5	30.4	46.8	56.8	60.2	55.4	44.4	24.5	5.5	-1.6	27.8
Extremes														
-Record Highest	49	48	51	58	72	90	92	91	88	79	66	52	55	92
-Year		1981	1943	1981	1979	1947	1969	1958	1977	1963	1969	1979	1985	JUN 1969
-Record Lowest	49	-63	-60	-49	-37	-1	30	32	22	-2	-39	-47	-62	-63
-Year		1947	1947	1964	1944	1945	1974	1971	1948	1983	1975	1986	1946	JAN 1947
NORMAL DEGREE DAYS:														
Heating (base 65°F)		2139	1758	1597	1038	564	250	158	307	618	1256	1785	2065	13535
Cooling (base 65°F)		0	0	0	0	0	0	9	9	0	0	0	0	18
% OF POSSIBLE SUNSHINE														
MEAN SKY COVER (tenths)														
Sunrise - Sunset	3	4.4	7.0	6.1	7.9	7.4	7.2	8.1	7.8	7.2	7.9	6.8	6.6	7.0
MEAN NUMBER OF DAYS:														
Sunrise to Sunset														
-Clear	3	14.3	4.7	8.3	1.7	1.7	3.7	0.3	2.3	5.3	3.7	7.0	7.3	60.3
-Partly Cloudy	3	8.0	6.7	8.0	8.7	11.0	9.3	8.3	8.0	6.7	4.7	6.3	7.0	92.7
-Cloudy	3	8.7	16.7	14.7	19.7	18.3	17.0	22.3	20.7	18.0	22.7	16.7	16.7	212.0
Precipitation														
.01 inches or more	33	5.7	4.2	4.1	3.8	6.6	11.8	13.7	12.2	8.7	8.7	7.1	5.8	92.3
Snow,Ice pellets,hail														
1.0 inches or more	22	2.2	1.5	1.5	1.0	0.3	0.0	0.0	0.0	0.7	4.0	3.9	2.2	17.2
Thunderstorms	7	0.0	0.0	0.0	0.0	0.0	1.8	1.5	0.1	0.0	0.0	0.0	0.0	3.5
Heavy Fog Visibility														
1/4 mile or less	7	2.9	0.3	0.0	0.2	0.0	0.0	0.3	0.9	1.9	1.1	0.7	0.9	9.1
Temperature °F														
-Maximum														
70° and above	49	0.0	0.0	0.0	0.*	2.0	10.4	14.9	8.1	0.6	0.0	0.0	0.0	35.9
32° and below	49	29.2	24.7	21.4	7.3	0.3	0.0	0.0	0.0	0.6	16.3	26.5	29.1	155.4
-Minimum														
32° and below	48	31.0	28.0	30.3	26.3	8.0	0.1	0.*	0.9	9.9	27.6	29.7	30.9	222.7
0° and below	48	21.1	17.5	14.3	3.0	0.1	0.0	0.0	0.0	0.*	3.5	14.5	20.5	94.5
AVG. STATION PRESS.(mb)														
RELATIVE HUMIDITY (%)														
Hour 03	16	66	70	68	70	68	73	77	79	79	77	70	70	72
Hour 09 (Local Time)	34	66	68	66	59	52	60	67	71	73	77	72	70	67
Hour 15	34	65	62	54	46	39	46	51	53	55	69	70	69	57
Hour 21	33	66	68	63	59	51	56	64	69	71	77	72	70	66
PRECIPITATION (inches):														
Water Equivalent														
-Normal		0.33	0.28	0.25	0.29	0.92	2.56	2.72	1.92	1.06	0.71	0.52	0.40	11.96
-Maximum Monthly	48	1.35	1.33	1.12	1.98	3.07	5.42	6.18	3.72	3.01	2.07	1.38	2.57	6.18
-Year		1957	1957	1967	1948	1988	1985	1945	1975	1952	1989	1989	1955	JUL 1945
-Minimum Monthly	48	T	T	T	T	0.02	0.51	0.71	0.62	0.14	T	0.03	T	T
-Year		1961	1975	1987	1976	1944	1944	1970	1981	1984	1969	1953	1973	MAR 1987
-Maximum in 24 hrs	45	0.73	0.44	0.38	0.94	1.16	2.14	2.06	1.26	1.17	0.54	0.43	0.90	2.14
-Year		1957	1957	1963	1963	1962	1943	1975	1961	1952	1952	1964	1955	JUN 1943
Snow,Ice pellets,hail														
-Maximum Monthly	48	20.9	15.9	24.9	19.0	10.9	T	0.0	T	14.0	19.8	27.1	29.0	29.0
-Year		1950	1967	1967	1948	1989	1974		1993	1981	1978	1993	1955	DEC 1955
-Maximum in 24 hrs	45	8.5	5.5	6.4	10.0	9.3	T	0.0	T	8.0	10.4	8.6	10.0	10.4
-Year		1957	1957	1950	1982	1989	1963		1993	1981	1981	1993	1965	OCT 1981
WIND:														
Mean Speed (mph)	20	10.8	9.9	8.2	7.6	7.9	6.5	6.0	6.6	7.3	8.5	9.5	9.5	8.2
Prevailing Direction through 1964		ESE	ESE	ESE	W	SW	SW	SW	S	ESE	ESE	ESE	ESE	ESE
Fastest Obs. 1 Min.														
-Direction (!!)	23	29	18	20	18	20	16	18	16	18	18	11	11	29
-Speed (MPH)	23	74	60	63	60	55	51	63	47	66	58	56	63	74
-Year														
Peak Gust														
-Direction (!!)														
-Speed (mph)														
-Date														

PRECIPITATION (inches) BIG DELTA, ALASKA

YEAR	JAN	FEB	MAR	APR	MAY	JUNE	JULY	AUG	SEP	OCT	NOV	DEC	ANNUAL
1964	0.09	0.10	0.25	0.35	0.23	0.90	2.47	0.75	1.45	0.17	1.12	0.84	8.72
1965	0.07	0.14	0.29	0.18	0.06	0.96	2.91	2.97	1.23	0.84	0.50	1.01	11.16
1966	0.01	0.43	0.20	0.35	1.92	2.55	0.91	0.88	0.78	0.83	0.94	0.12	9.92
1967	0.41	0.60	1.12	0.21	1.24	1.93	4.59	1.56	1.49	0.34	0.46	0.32	14.27
1968	0.91	0.14	0.01	0.03	1.44	2.74	1.38	1.95	0.38	0.91	0.44	0.10	10.43
1969	0.36	T	0.27	0.13	1.10	0.84	3.54	0.71	0.45	T	0.29	T	7.69
1970	0.14	0.07	0.25	0.31	0.74	1.62	0.71	1.52	1.58	0.91	0.78	0.59	9.22
1971	0.29	0.49	0.16	0.34	0.29	1.26	1.71	2.50	1.58	0.89	0.21	0.51	10.23
1972	0.24	0.28	0.22	0.29	0.35	2.59	2.36	0.94	1.03	0.88	0.08	0.22	9.48
1973	0.21	0.13	0.04	0.05	0.53	4.40	3.54	1.79	0.26	0.53	0.99	T	12.47
1974	0.11	0.42	0.09	0.05	0.43	2.65	2.57	1.74	0.26	0.36	0.36	0.37	9.41
1975	0.74	T	0.36	0.28	1.59	3.78	4.70	3.72	1.55	0.39	0.45	0.01	17.57
1976	0.07	0.08	0.47	T	1.41	1.28	3.87	2.25	0.89	0.16	0.13	0.14	10.75
1977	0.22	0.05	0.27	0.21	0.87	4.04	1.91	1.64	1.69	0.65	0.07	0.15	11.77
1978	0.04	0.14	0.03	0.38	0.80	1.58	1.43	2.50	0.54	0.34	0.36	0.18	8.32
1979	0.15	0.07	0.06	0.15	0.10	2.97	3.55	2.08	0.51	0.08	0.07	0.19	9.98
1980	0.17	0.09	0.04	0.01	0.80	1.98	1.32	2.44	0.86	0.23	0.22	0.02	8.18
1981	0.01	0.17	T	0.12	1.62	2.43	2.94	0.62	0.75	0.68	0.22	0.18	9.74
1982	0.12	0.39	0.17	0.55	1.09	2.28	3.57	1.30	0.98	0.85	0.44	0.11	11.85
1983	0.22	0.18	0.13	0.09	0.31	1.28	4.62	2.16	1.09	1.31	0.08	0.18	11.65
1984	0.35	0.13	T	0.02	1.72	3.74	5.98	2.74	0.14	0.89	0.19	0.81	16.71
1985	0.80	0.70	0.19	0.12	0.06	5.42	1.79	2.17	1.94	1.45	1.26	0.30	16.20
1986	0.59	0.38	0.45	0.37	0.27	1.63	3.00	1.67	1.16	0.79	1.22	0.49	12.02
1987	0.31	0.58	T	0.56	0.15	1.96	2.96	1.14	1.26	0.86	1.12	0.99	11.89
1988	0.41	0.49	0.46	0.63	3.07	3.25	3.00	1.23	0.31	1.13	0.73	0.53	15.24
1989	1.05	0.27	0.32	0.04	1.65	3.73	2.76	2.33	1.56	2.07	1.38	0.57	17.73
1990	0.44												
1991													
1992													
1993							1.70	2.23	1.55	0.75	1.07	0.15	
Record Mean	0.35	0.27	0.27	0.26	0.91	2.42	2.71	1.93	1.12	0.64	0.45	0.38	11.73

AVERAGE TEMPERATURE (deg. F) BIG DELTA, ALASKA

YEAR	JAN	FEB	MAR	APR	MAY	JUNE	JULY	AUG	SEP	OCT	NOV	DEC	ANNUAL
1964	-5.0	7.0	3.2	26.1	37.3	57.8	58.7	54.2	44.4	29.8	7.8	-22.5	24.9
1965	-6.8	-8.4	31.6	31.4	43.4	54.0	58.9	53.6	48.0	14.9	6.2	-7.5	26.6
1966	-11.4	1.1	2.1	28.8	42.0	58.1	59.1	54.3	46.4	21.3	3.1	-7.9	24.8
1967	-7.3	0.2	12.5	31.3	44.0	58.8	58.3	57.7	45.7	24.6	14.2	1.0	28.4
1968	-8.0	6.7	21.6	31.5	46.6	56.3	63.0	56.9	41.1	21.1	7.7	-10.8	27.8
1969	-22.6	1.0	16.3	36.1	47.6	63.2	57.3	48.4	46.7	35.1	6.3	15.6	29.3
1970	-12.9	14.3	25.8	30.1	48.1	53.9	59.2	54.9	37.7	18.1	12.6	-5.4	28.0
1971	-24.4	2.5	0.7	26.4	44.3	59.1	58.7	53.6	42.5	24.1	1.5	-4.8	23.7
1972	-16.5	-9.0	-1.7	18.9	45.3	56.6	61.7	58.1	39.5	27.6	9.5	-1.1	24.1
1973	-16.5	2.9	12.2	34.5	47.2	56.0	59.8	52.1	44.8	23.5	-2.3	4.3	26.6
1974	-13.7	-14.6	7.3	33.0	47.3	55.0	59.7	56.0	49.2	22.2	4.5	-6.8	25.0
1975	-12.0	-0.7	13.1	28.9	49.4	56.5	60.4	53.8	45.5	21.8	-7.6	-11.7	24.8
1976	-2.9	-9.4	13.3	35.6	46.0	58.0	62.3	59.1	46.8	25.1	23.3	4.1	30.1
1977	17.1	19.3	9.0	30.5	47.7	58.7	62.0	61.6	46.2	27.8	-4.9	-11.5	30.3
1978	9.1	13.0	19.3	35.6	51.8	56.1	63.4	57.9	46.2	25.4	12.2	3.2	32.8
1979	0.8	-19.7	15.8	33.8	51.3	56.6	61.3	60.5	48.3	34.9	24.3	-11.1	29.8
1980	-8.2	18.8	21.3	39.0	52.6	57.6	61.6	55.9	43.2	31.2	14.6	-21.2	30.6
1981	24.1	10.8	30.1	28.9	54.2	55.9	57.9	55.4	43.3	27.5	11.2	-1.5	33.1
1982	-13.2	-4.9	12.5	26.0	43.7	56.5	59.6	53.0	47.2	13.1	5.9	5.7	25.4
1983	-8.4	7.0	9.6	34.5	47.3	57.5	59.6	51.2	37.2	21.7	12.7	-1.3	27.4
1984	-0.8	-8.7	24.3	30.7	45.6	57.3	57.8	51.8	45.8	23.9	0.7	-0.3	27.3
1985	20.8	-7.5	17.4	19.8	44.5	53.3	59.8	53.5	43.1	17.2	-0.4	16.0	28.1
1986	7.4	7.7	5.9	20.2	44.1	58.0	60.0	53.4	44.4	28.6	3.2	14.9	29.0
1987	6.1	10.6	14.5	32.4	47.9	57.3	60.1	55.5	43.4	33.2	7.3	2.4	30.9
1988	3.4	9.1	19.5	32.9	48.8	57.9	61.3	56.1	43.1	18.5	0.8	7.8	29.9
1989	-20.6	8.0	6.4	36.4	46.6	57.4	62.7	59.6	47.2	24.7	-2.8	11.8	28.1
1990	-4.9												
1991													
1992													
1993								55.6	44.9	31.0	10.5	9.3	
Record Mean	-4.1	1.9	12.6	29.7	46.6	56.8	59.8	55.1	44.0	25.0	6.5	-2.3	27.7
Max	3.1	10.7	23.8	39.6	56.4	66.4	69.2	64.4	52.5	31.7	13.5	4.9	36.4
Min	-11.2	-6.8	1.5	19.8	36.7	47.2	50.5	45.8	35.4	18.2	-0.5	-9.5	18.9

REFERENCE NOTES BIG DELTA, ALASKA

GENERAL
T - TRACE AMOUNT.
BLANK ENTRIES DENOTE MISSING/UNREPORTED DATA.
# INDICATES A STATION OR INSTRUMENT RELOCATION.

SPECIFIC

(a) - LENGTH OF RECORD IN YEARS, ALTHOUGH INDIVIDUAL MONTHS MAY BE MISSING.
0.* OR * - THE VALUE IS BETWEEN 0.0 AND 0.05
NORMALS - BASED ON THE 1961-1990 RECORD PERIOD.
EXTREMES - DATES ARE THE MOST RECENT OCCURRENCE
WIND DIR.- NUMERALS SHOW TENS OF DEGRESS CLOCKWISE FROM TRUE NORTH. "00" INDICATES CALM.
RESULTANT DIRECTIONS ARE GIVEN TO WHOLE DEGREES.
BOLD VALUES INDICATE EXTREME VALUES WHICH OCCURRED AFTER THE ASOS SYSTEM WAS COMMISSIONED.

RECORD = PERIOD OF RECORD
RECORD MEAN PRECIPITATION IS THE MEAN OF ALL DAILY PRECIPITATION AMOUNTS DURING THE PERIOD OF RECORD.
RECORD MAX(MIN) TEMPERATURE IS THE MEAN OF ALL DAILY MAX(MIN) TEMPERATURES DURING THE PERIOD OF RECORD.
RECORD MEAN TEMPERATURE IS THE SUM OF THE RECORD MAX AND RECORD MIN DIVIDED BY 2.
AVERAGE TEMPERATURE IS THE SUM OF THE MEAN DAILY MAX AND MIN TEMPERATURE DIVIDED BY 2.

EXCEPTIONS

1. DAYS OF PRECIPITATION .01 INCH OR MORE ARE 1943-1953 AND 1969 TO DATE.
2. MEAN WIND SPEED IS JULY 1944-AUGUST 1964.
3. FASTEST OBSERVED WIND IS JULY 1944-AUGUST 1964 AND JANUARY 1969-MAY 1972, YEARS OF EXTREMES NOT AVAILABLE.
4. THUNDERSTORMS AND HEAVY FOG ARE THROUGH 1973 AND MAY BE INCOMPLETE, DUE TO PART-TIME OPERATIONS.
5. MEAN SKY COVER, AND DAYS CLEAR-PARTLY CLOUDY-CLOUDY ARE 1969-1971.

RECORD MEANS ARE THROUGH THE CURRENT YEAR, BEGINNING IN 1945 FOR TEMPERATURE
1943 FOR PRECIPITATION
1945 FOR SNOWFALL

HEATING DEGREE DAYS Base 65 deg. F — BIG DELTA, ALASKA

SEASON	JULY	AUG	SEP	OCT	NOV	DEC	JAN	FEB	MAR	APR	MAY	JUNE	TOTAL
1964-65	192	327	611	1086	1711	2719	2227	2060	1029	1001	662	321	13946
1965-66	188	346	505	1553	1758	2250	2370	1789	1950	1079	704	203	14695
1966-67	177	326	552	1350	1856	2261	2246	1816	1623	1005	644	200	14056
1967-68	212	227	574	1246	1521	1984	2265	1690	1335	998	564	258	12874
1968-69	88	245	708	1356	1714	2352	2726	1790	1507	860	532	104	13982
1969-70	234	509	541	918	1760	1526	2421	1416	1206	1044	517	325	12417
1970-71	179	305	812	1450	1571	2182	2777	1751	1993	1152	637	186	14995
1971-72	197	348	664	1261	1904	2162	2530	2149	2069	1374	601	269	15528
1972-73	129	209	755	1153	1659	2049	2528	1737	1635	905	542	268	13569
1973-74	163	391	601	1280	2019	1880	2444	2232	1789	952	539	296	14586
1974-75	165	275	464	1319	1815	2231	2387	1840	1605	1075	481	251	13908
1975-76	148	346	577	1334	2184	2379	2105	2161	1597	877	581	204	14493
1976-77	98	189	540	1229	1247	1884	1478	1277	1729	1026	528	184	11409
1977-78	114	138	558	1149	2101	2372	1728	1454	1406	875	403	262	12560
1978-79	83	216	558	1222	1579	1914	1991	2375	1520	929	420	244	13051
1979-80	118	146	492	925	1216	2360	2267	1335	1347	775	380	218	11579
1980-81	109	295	647	1041	1509	2679	1259	1516	1075	1075	331	268	11804
1981-82	216	293	648	1156	1611	2061	2432	1956	1623	1165	651	261	14073
1982-83	172	363	528	1607	1773	1838	2275	1622	1711	909	547	238	13583
1983-84	168	423	825	1334	1563	2057	2044	2140	1254	1022	594	227	13651
1984-85	219	405	573	1267	1929	2023	1361	2030	1469	1348	630	343	13597
1985-86	155	351	650	1480	1962	1516	1783	1603	1831	1343	640	208	13522
1986-87	171	353	615	1119	1855	1548	1826	1516	1563	972	521	244	12303
1987-88	149	286	641	982	1726	1942	1908	1620	1403	957	495	207	12316
1988-89	121	272	651	1434	1927	1769	2655	1593	1813	851	566	222	13874
1989-90	103	166	526	1242	2033	1643	2168						
1990-91													
1991-92													
1992-93													
1993-94		295	598	1047	1631	1722							

COOLING DEGREE DAYS Base 65 deg. F — BIG DELTA, ALASKA

YEAR	JAN	FEB	MAR	APR	MAY	JUNE	JULY	AUG	SEP	OCT	NOV	DEC	TOTAL
1969	0	0	0	0	0	57	1	0	0	0	0	0	58
1970	0	0	0	0	0	0	2	0	0	0	0	0	2
1971	0	0	0	0	0	15	7	0	0	0	0	0	22
1972	0	0	0	0	0	22	34	1	0	0	0	0	57
1973	0	0	0	0	0	3	11	0	0	0	0	0	14
1974	0	0	0	0	0	1	7	4	0	0	0	0	12
1975	0	0	0	0	0	3	14	4	0	0	0	0	21
1976	0	0	0	0	0	4	20	12	0	0	0	0	36
1977	0	0	0	0	0	0	26	39	0	0	0	0	65
1978	0	0	0	0	0	0	39	4	0	0	0	0	43
1979	0	0	0	0	0	0	13	15	0	0	0	0	28
1980	0	0	0	0	1	4	11	16	0	0	0	0	32
1981	0	0	0	0	2	0	0	0	0	0	0	0	2
1982	0	0	0	0	0	13	11	0	0	0	0	0	24
1983	0	0	0	0	5	20	5	0	0	0	0	0	30
1984	0	0	0	0	0	0	0	4	0	0	0	0	4
1985	0	0	0	0	0	0	1	0	0	0	0	0	1
1986	0	0	0	0	0	4	22	2	0	0	0	0	28
1987	0	0	0	0	0	19	5	0	0	0	0	0	24
1988	0	0	0	0	0	2	15	5	0	0	0	0	22
1989	0	0	0	0	0	1	39	5	0	0	0	0	45
1990	0												
1991													
1992													
1993								9	0	0	0	0	

SNOWFALL (inches) — BIG DELTA, ALASKA

SEASON	JULY	AUG	SEP	OCT	NOV	DEC	JAN	FEB	MAR	APR	MAY	JUNE	TOTAL
1964-65	0.0	0.0	T	1.7	19.7	13.2	0.7	4.4	2.8	1.4	T	0.0	43.9
1965-66	0.0	0.0	0.9	16.9	8.4	20.9	T	6.9	3.2	3.1	0.6	0.0	60.9
1966-67	0.0	0.0	0.0	15.9	15.6	5.5	17.2	15.9	24.9	3.2	1.5	0.0	99.7
1967-68	0.0	0.0	T	6.4	9.5	5.6	14.3	2.5	0.1	1.5	T	0.0	39.9
1968-69	0.0	0.0	T	12.2	7.1	1.3	3.0	0.2	5.3	1.2	T	0.0	30.3
1969-70	0.0	T	0.0	T	5.6	T	4.3	1.2	2.9	7.2	0.0	0.0	21.2
1970-71	0.0	0.0	7.1	16.2	13.5	8.1	5.2	8.8	2.9	6.0	1.6	0.0	69.4
1971-72	0.0	0.0	T	14.0	5.6	9.9	4.4	3.4	3.7	5.9	2.0	0.0	48.9
1972-73	0.0	0.0	8.7	15.9	3.5	6.8	6.1	3.9	1.0	0.8	7.0	0.0	53.7
1973-74	0.0	0.0	0.0	6.9	12.6	T	2.2	6.7	1.7	0.8	0.0	T	30.9
1974-75	0.0	T	T	7.6	7.7	9.3	6.2	T	5.9	3.6	T	0.0	40.3
1975-76	0.0	0.0	T	9.4	10.4	0.7	2.5	1.5	8.3	T	T	0.0	32.8
1976-77	0.0	0.0	T	6.0	3.0	2.5	6.9	1.8	6.1	2.9	0.0	0.0	29.2
1977-78	0.0	0.0	6.0	15.7	3.2	3.8	2.0	4.0	1.0	4.8	0.0	0.0	40.5
1978-79	0.0	0.0	T	19.8	13.5	7.5	3.7	2.3	1.3	1.0	0.0	0.0	49.1
1979-80	0.0	0.0	T	1.6	5.9	5.9	6.5	7.0	1.6	T	0.0	0.0	28.5
1980-81	0.0	0.0	0.8	7.1	6.8	1.5	T	3.8	0.3	5.0	0.0	0.0	25.3
1981-82	0.0	T	14.0	18.4	3.4	4.0	1.5	4.5	4.3	12.8	T	0.0	62.9
1982-83	0.0	0.0	T	9.6	6.7	0.9	3.5	2.3	2.4	0.2	0.0	0.0	25.6
1983-84	0.0	0.0	3.5	13.6	2.4	6.2	8.1	2.2	T	2.0	T	0.0	38.0
1984-85	0.0	T	0.0	16.0	5.6	17.2	9.7	10.2	3.0	2.3	0.5	0.0	64.5
1985-86	0.0	0.0	2.0	10.5	24.7	6.7	9.7	8.9	6.4	5.3	1.4	0.0	75.6
1986-87	0.0	0.0	3.2	4.0	16.3	5.6	3.8	2.4	0.0	2.9	0.0	0.0	38.2
1987-88	0.0	0.0	T	8.4	16.4	8.6	4.8	6.4	5.4	0.6	T	0.0	50.6
1988-89	0.0	0.0	0.1	13.0	10.1	6.0	11.1	2.9	3.1	T	10.9	0.0	57.2
1989-90	0.0	0.0	T	19.7	11.5	6.1							
1992-93													
1993-94	0.0	T	5.6	5.4	27.1	2.5							
Record Mean	0.0	T	1.7	9.0	7.8	5.9	5.7	4.4	4.3	3.0	0.7	T	42.6

COLD BAY, ALASKA

The station at Cold Bay is located approximately 30 miles from the end of the Alaskan Peninsula on the northwest side of Cold Bay. Ten miles south–southwest of the station, Frosty Peak rises to an elevation of 6,700 feet. Across the bay to the east several mountains rise to elevations in excess of 5,000 feet. The mountains to the east and southwest provide a sheltering effect from winds and precipitation approaching from these directions. Winds reaching the station from southwesterly or easterly directions rarely exceed 15 mph. The open bay area to the south–southeastward tends to provide a funneling effect upon all winds approaching the Cold Bay area from the southwest to the southeast. From west to the northeast the land is relatively flat with numerous lakes and swamps. Winds from northerly directions are influenced very little by this flat terrain.

The high frequency of cyclonic storms crossing the Northern Pacific and the Bering Sea are the dominant factors in the weather at Cold Bay. These storms account for the high winds and the frequent occurrences of low ceilings and low visibilities encountered at this station. The winds generally result from the strong pressure gradient developing between the Pacific High and the cyclonic storms in the Northern Pacific and Bering Sea.

The climate at Cold Bay is basically maritime, due to the nearness to extensive open ocean areas, and temperature extremes, both seasonal and diurnal, are generally confined to fairly narrow limits. Differences between maximum and minimum temperatures for all individual months average less than 10 degrees. Although it is practically impossible for cold, continental air masses to reach the Cold Bay area by moving overland along the somewhat narrow Alaskan Peninsula, air overlying the frozen ocean surface of the Bering Sea may take on continental characteristics and bring rather cold temperatures to the area. Although below–zero readings have been recorded from December to March, inclusive, below–zero readings are infrequent.

Due to the moderating effects of nearby ocean areas, it is difficult to define the seasonal periods at Cold Bay. The beginning of spring is late. The vegetation does not begin to grow until late May or early June. August is regarded as the midsummer period and autumn arrives in early October. The greatest frequency of fog usually comes in the summer season, with the foggy period extending from mid–July to mid–September. During the winter months visibilities are frequently restricted due to blowing snow. Precipitation is frequent but not abundant. The shortest day of the year at Cold Bay has 7 hours and 7 minutes of possible sunshine, the longest day has 17 hours and 27 minutes of possible sunshine.

NORMALS, MEANS, AND EXTREMES

COLD BAY, ALASKA

LATITUDE: 55°12'N LONGITUDE: 162°43' W ELEVATION: FT. GRND 96 BARO 99 TIME ZONE: YUKON WBAN: 25624

	(a)	JAN	FEB	MAR	APR	MAY	JUNE	JULY	AUG	SEP	OCT	NOV	DEC	YEAR
TEMPERATURE °F:														
Normals														
-Daily Maximum		33.1	32.0	34.8	37.9	44.4	50.4	55.1	55.9	52.2	44.3	38.9	35.3	42.9
-Daily Minimum		24.1	22.8	25.0	28.6	34.8	40.9	46.0	47.1	43.2	34.8	29.9	26.6	33.7
-Monthly		28.6	27.4	29.9	33.3	39.6	45.7	50.5	51.5	47.7	39.6	34.4	31.0	38.3
Extremes														
-Record Highest	49	51	50	56	60	67	69	77	78	76	69	59	54	78
-Year		1993	1991	1974	1948	1979	1953	1960	1948	1985	1964	1986	1990	AUG 1948
-Record Lowest	50	-8	-9	-13	4	18	29	33	33	26	10	1	-1	-13
-Year		1989	1947	1971	1976	1973	1952	1982	1946	1992	1976	1963	1979	MAR 1971
NORMAL DEGREE DAYS:														
Heating (base 65°F)		1128	1053	1088	951	787	579	450	419	519	787	918	1054	9733
Cooling (base 65°F)		0	0	0	0	0	0	0	0	0	0	0	0	0
% OF POSSIBLE SUNSHINE														
MEAN SKY COVER (tenths)														
Sunrise - Sunset	38	8.2	8.4	8.3	8.9	9.1	9.2	9.4	9.5	9.1	8.7	8.5	8.5	8.8
MEAN NUMBER OF DAYS:														
Sunrise to Sunset														
-Clear	38	2.4	1.9	1.9	0.6	0.5	0.4	0.2	0.2	0.2	0.7	1.1	1.7	11.9
-Partly Cloudy	38	5.6	4.7	6.0	3.9	2.9	2.6	2.3	1.9	3.5	5.2	5.3	5.3	49.3
-Cloudy	38	22.9	21.7	23.1	25.4	27.6	26.9	28.4	28.9	26.3	25.2	23.6	24.1	304.1
Precipitation														
.01 inches or more	48	19.0	17.2	17.9	16.1	17.0	15.9	16.4	19.8	20.6	22.7	21.6	20.6	224.7
Snow,Ice pellets,hail														
1.0 inches or more	43	3.6	3.7	3.4	1.9	0.5	0.0	0.0	0.0	0.0	1.0	2.5	3.7	20.2
Thunderstorms	38	0.0	0.0	0.0	0.0	0.0	0.0	0.0	0.0	0.*	0.1	0.1	0.1	0.2
Heavy Fog Visibility														
1/4 mile or less	38	1.8	1.4	1.9	1.3	1.5	2.0	4.1	3.5	1.1	0.3	0.6	1.6	21.0
Temperature °F														
-Maximum														
70° and above	49	0.0	0.0	0.0	0.0	0.0	0.0	0.1	0.2	0.*	0.0	0.0	0.0	0.4
32° and below	49	11.3	11.1	9.9	5.6	0.4	0.0	0.0	0.0	0.0	0.5	3.6	9.7	52.1
-Minimum														
32° and below	50	24.3	23.6	25.1	21.4	8.6	0.4	0.0	0.0	0.4	8.5	18.8	23.8	155.0
0° and below	50	0.6	0.4	0.6	0.0	0.0	0.0	0.0	0.0	0.0	0.0	0.0	0.1	1.6
AVG. STATION PRESS.(mb)	21	995.3	998.8	999.3	1004.4	1004.7	1007.8	1011.3	1009.0	1004.3	1000.9	998.2	995.5	1002.5
RELATIVE HUMIDITY (%)														
Hour 03	24	85	85	84	86	88	91	92	92	89	85	83	85	87
Hour 09 (Local Time)	24	85	84	85	84	84	86	90	90	88	84	84	84	86
Hour 15	24	82	80	78	77	76	78	82	82	80	77	80	83	80
Hour 21	24	85	84	84	84	83	84	88	90	87	83	83	85	85
PRECIPITATION (inches):														
Water Equivalent														
-Normal		2.84	2.27	2.16	1.97	2.29	2.10	2.52	3.24	4.41	4.34	4.19	3.67	36.00
-Maximum Monthly	48	8.46	7.87	4.70	6.55	6.37	6.98	6.13	9.97	9.79	8.02	8.94	7.31	9.97
-Year		1948	1944	1977	1979	1958	1952	1982	1951	1965	1968	1960	1983	AUG 1951
-Minimum Monthly	48	0.60	0.08	0.41	0.02	0.54	0.12	0.28	1.10	0.91	1.88	1.15	0.19	0.02
-Year		1956	1950	1972	1948	1992	1962	1950	1975	1952	1961	1975	1956	APR 1948
-Maximum in 24 hrs	48	2.49	2.49	2.06	2.19	2.22	2.10	1.77	2.17	3.43	4.90	3.43	2.44	4.90
-Year		1948	1956	1976	1993	1958	1971	1986	1951	1965	1968	1972	1978	OCT 1968
Snow,Ice pellets,hail														
-Maximum Monthly	43	34.6	54.3	28.6	19.5	9.3	0.5	T	0.0	0.2	15.6	27.4	24.2	54.3
-Year		1982	1984	1985	1976	1971	1971	1976		1972	1968	1983	1976	FEB 1984
-Maximum in 24 hrs	43	18.0	17.7	7.4	6.0	4.0	2.6	T	0.0	0.2	11.4	21.4	9.4	21.4
-Year		1982	1984	1987	1956	1986	1971	1976		1972	1968	1983	1975	NOV 1983
WIND:														
Mean Speed (mph)	38	17.9	17.8	17.4	17.6	16.2	15.9	15.7	16.5	16.4	16.8	17.5	17.5	16.9
Prevailing Direction through 1963		SSE	SSE	NNW	SSE	SSE	WNW	SSE	SSE	SSE	WSW	SSE	NNW	SSE
Fastest Obs. 1 Min.														
-Direction (!!)	38	17	16	17	15	14	11	17	16	17	21	14	11	17
-Speed (MPH)	38	71	73	67	60	60	63	54	64	75	60	66	64	75
-Year		1985	1967	1977	1987	1985	1959	1972	1985	1988	1978	1967	1960	SEP 1988
Peak Gust														
-Direction (!!)	9	SE	SE	E	SE	SE	SE	S	SE	S	SE	SE	SW	S
-Speed (mph)	9	85	83	76	77	72	69	58	81	95	77	75	78	95
-Date		1987	1989	1987	1987	1991	1993	1992	1985	1988	1991	1993	1988	SEP 1988

PRECIPITATION (inches)

COLD BAY, ALASKA

YEAR	JAN	FEB	MAR	APR	MAY	JUNE	JULY	AUG	SEP	OCT	NOV	DEC	ANNUAL
1964	1.33	1.75	1.32	0.21	1.15	1.85	1.74	3.56	5.25	3.33	3.71	1.31	26.51
1965	1.21	2.78	3.05	0.83	2.66	2.54	1.20	1.57	9.79	2.75	2.55	1.77	32.70
1966	2.21	1.59	1.40	1.31	2.49	0.79	4.63	3.73	4.28	2.96	2.80	2.09	30.28
1967	1.60	2.58	2.54	3.06	0.62	2.48	2.89	4.72	2.91	2.71	7.40	4.49	38.00
1968	2.77	1.29	1.21	1.37	1.30	0.84	0.99	3.53	2.55	8.02	3.32	1.17	28.36
1969	3.75	2.33	1.92	1.09	3.44	2.52	2.01	5.02	5.18	3.90	2.97	2.33	36.46
1970	2.11	4.15	3.32	3.83	2.06	2.99	3.86	3.82	5.62	5.82	2.89	5.94	46.41
1971	1.34	2.17	0.59	0.43	3.75	6.67	2.27	2.76	3.86	3.28	5.11	4.87	37.10
1972	4.08	1.09	0.41	3.09	2.06	2.91	1.82	3.84	1.30	3.91	6.96	6.49	37.96
1973	1.97	1.60	1.87	1.30	1.06	0.78	1.31	2.16	3.07	4.80	2.45	2.67	25.04
1974	2.96	2.72	0.72	1.69	3.12	0.92	1.93	2.63	2.55	2.15	3.78	1.73	26.90
1975	3.12	4.93	2.85	2.53	0.88	3.03	2.55	1.10	4.23	3.18	1.15	5.03	34.58
1976	1.88	2.88	3.76	2.09	0.94	2.69	1.92	2.01	2.38	5.63	2.51	2.89	31.58
1977	4.82	2.36	4.70	2.38	1.71	1.14	2.89	3.32	2.63	5.12	6.85	3.55	41.47
1978	3.70	1.74	2.22	5.42	3.63	2.84	5.67	2.88	3.82	7.67	6.67	6.89	53.15
1979	4.10	0.78	4.65	6.55	4.92	1.98	2.02	5.33	5.31	7.14	7.57	2.21	52.56
1980	3.51	1.69	3.52	1.71	4.22	3.67	2.68	3.95	5.23	4.42	2.88	2.24	39.72
1981	2.34	4.45	2.34	1.30	3.09	1.75	2.64	5.73	2.25	6.51	3.11	3.16	38.67
1982	5.41	1.13	3.45	1.33	4.13	2.93	6.13	2.17	6.44	2.41	5.12	3.10	43.75
1983	1.58	0.66	0.88	3.53	1.59	1.31	2.71	4.06	4.41	4.82	5.69	7.31	38.55
1984	2.30	2.82	1.56	1.79	1.20	1.45	1.77	1.48	2.87	3.64	7.61	3.19	31.68
1985	3.29	2.42	2.85	1.01	2.45	2.19	2.27	5.47	7.14	6.59	7.72	4.95	48.35
1986	2.05	2.23	0.55	1.12	2.02	1.91	2.48	2.63	7.37	3.03	5.08	4.94	35.41
1987	3.17	3.15	3.18	1.94	1.52	4.00	1.80	2.56	4.25	5.60	3.17	3.69	38.03
1988	3.70	2.91	0.89	1.81	2.70	1.69	1.12	3.03	3.90	3.28	3.97	6.37	35.37
1989	1.68	4.02	0.52	2.20	2.21	2.48	1.40	3.20	7.77	4.39	2.60	3.81	36.28
1990	3.99	2.18	1.84	1.16	3.23	1.38	2.13	2.89	6.55	4.21	2.06	5.96	37.58
1991	2.74	2.95	4.33	1.96	2.47	3.78	0.45	4.17	5.82	5.83	1.55	3.32	39.37
1992	1.57	0.68	1.22	0.26	0.54	3.27	2.36	2.86	2.21	2.59	4.42	6.20	28.18
1993	5.32	3.04	1.85	3.98	2.63	3.64	2.03	4.94	5.86	4.48	6.79	4.28	48.84
Record Mean	2.79	2.54	2.06	1.86	2.26	2.25	2.31	3.61	4.14	4.47	4.34	3.44	36.07

AVERAGE TEMPERATURE (deg. F)

COLD BAY, ALASKA

YEAR	JAN	FEB	MAR	APR	MAY	JUNE	JULY	AUG	SEP	OCT	NOV	DEC	ANNUAL
1964	30.0	26.2	29.3	32.7	38.1	46.4	47.8	48.6	47.1	39.7	32.9	30.3	37.4
1965	27.3	24.7	35.6	35.2	36.3	44.4	48.7	50.3	48.9	36.7	36.0	28.5	37.7
1966	32.8	31.7	21.4	34.4	37.4	44.2	48.6	49.1	44.6	33.9	32.3	27.7	36.5
1967	26.5	28.5	34.5	38.2	41.5	46.8	52.9	51.3	46.3	38.2	38.0	30.4	39.5
1968	28.3	24.0	30.4	34.2	42.7	45.4	52.9	52.0	46.8	38.9	36.0	30.9	38.6
1969	31.8	28.0	32.5	34.5	41.7	46.9	53.2	53.4	50.2	42.9	33.5	31.3	40.0
1970	22.4	28.1	31.8	31.8	40.8	47.8	51.3	50.2	46.9	40.0	37.6	30.5	38.3
1971	22.1	26.0	20.2	28.0	34.3	40.5	47.8	48.5	44.5	38.7	32.4	32.4	34.6
1972	27.2	21.4	16.3	29.5	38.6	41.6	48.6	50.0	47.9	39.8	34.3	30.4	35.5
1973	22.9	30.6	29.0	31.5	36.0	42.5	47.5	50.0	46.4	38.5	37.5	30.1	36.9
1974	29.7	19.0	28.9	32.2	41.4	45.8	49.9	53.2	49.2	40.4	34.2	27.5	37.6
1975	24.4	23.7	26.8	32.0	38.4	46.2	50.5	52.5	46.4	39.3	30.1	27.9	36.5
1976	25.5	23.9	22.0	27.2	36.6	44.6	49.7	50.8	45.3	39.1	30.5	28.2	35.3
1977	35.3	33.5	30.7	31.3	39.3	49.6	54.0	53.6	50.1	38.9	31.2	29.4	39.8
1978	33.4	28.4	30.8	37.9	40.5	47.0	49.6	54.3	48.8	40.1	38.1	35.2	40.3
1979	35.1	28.4	35.4	40.8	43.4	50.6	52.3	51.9	49.0	41.9	35.0	26.7	40.9
1980	23.5	25.4	33.7	35.6	41.4	45.9	52.9	51.5	48.1	40.3	36.4	32.1	38.9
1981	30.9	29.4	35.9	38.4	44.8	47.6	52.9	52.2	48.1	40.8	33.8	30.7	40.5
1982	29.8	27.1	33.9	32.1	38.0	45.0	46.8	50.2	45.1	37.5	35.3	30.5	37.6
1983	24.6	31.5	33.5	36.8	41.7	48.4	51.6	52.2	47.3	39.7	34.6	37.5	40.0
1984	31.2	18.7	33.7	31.6	38.0	47.0	49.7	54.7	49.7	40.8	37.0	37.3	39.1
1985	36.1	27.9	30.1	26.8	38.3	42.5	50.6	50.8	49.5	39.6	38.9	35.6	38.9
1986	24.4	28.4	27.0	32.2	38.0	44.7	51.7	51.2	49.8	42.2	37.0	34.4	38.4
1987	30.5	31.2	33.6	34.2	38.8	44.5	50.9	52.8	46.9	41.0	30.0	28.3	38.6
1988	31.2	30.2	26.1	31.1	40.3	46.0	51.0	51.1	46.9	40.5	32.5	30.9	38.2
1989	22.3	35.0	31.5	34.3	40.6	46.0	50.9	53.3	49.8	42.3	32.1	31.3	39.1
1990	30.4	26.3	32.1	36.4	41.4	46.9	50.1	52.2	47.6	40.3	34.6	36.4	39.6
1991	31.1	22.8	34.3	35.6	40.0	45.5	50.8	51.6	49.5	42.0	36.0	29.7	39.1
1992	29.1	25.4	29.9	33.1	40.3	48.0	49.9	50.6	45.8	38.9	33.0	32.4	38.0
1993	27.0	31.1	33.1	37.4	41.9	47.2	52.0	52.7	49.0	42.2	38.2	31.7	40.3
Record Mean	28.5	27.8	29.5	33.2	39.6	45.5	50.2	51.5	47.4	39.9	34.3	30.4	38.2
Max	32.9	32.4	34.3	37.8	44.4	50.1	54.6	55.6	51.9	44.5	38.8	34.7	42.7
Min	24.0	23.1	24.6	28.5	34.7	40.9	45.7	47.3	42.9	35.2	29.9	26.0	33.6

REFERENCE NOTES

COLD BAY, ALASKA

GENERAL
T - TRACE AMOUNT.
BLANK ENTRIES DENOTE MISSING/UNREPORTED DATA.
# INDICATES A STATION OR INSTRUMENT RELOCATION.

SPECIFIC

(a) - LENGTH OF RECORD IN YEARS, ALTHOUGH INDIVIDUAL MONTHS MAY BE MISSING.
0.* OR * - THE VALUE IS BETWEEN 0.0 AND 0.05
NORMALS - BASED ON THE 1961-1990 RECORD PERIOD.
EXTREMES - DATES ARE THE MOST RECENT OCCURRENCE.
WIND DIR. - NUMERALS SHOW TENS OF DEGRESS CLOCKWISE FROM TRUE NORTH. "00" INDICATES CALM.
RESULTANT DIRECTIONS ARE GIVEN TO WHOLE DEGREES.
BOLD VALUES INDICATE EXTREME VALUES WHICH OCCURRED AFTER THE ASOS SYSTEM WAS COMMISSIONED.

RECORD = PERIOD OF RECORD
RECORD MEAN PRECIPITATION IS THE MEAN OF ALL DAILY PRECIPITATION AMOUNTS DURING THE PERIOD OF RECORD.
RECORD MAX(MIN) TEMPERATURE IS THE MEAN OF ALL DAILY MAX(MIN) TEMPERATURES DURING THE PERIOD OF RECORD.
RECORD MEAN TEMPERATURE IS THE SUM OF THE RECORD MAX AND RECORD MIN DIVIDED BY 2.
AVERAGE TEMPERATURE IS THE SUM OF THE MEAN DAILY MAX AND MIN TEMPERATURE DIVIDED BY 2.

EXCEPTIONS

RECORD MEANS ARE THROUGH THE CURRENT YEAR, BEGINNING IN 1943 FOR TEMPERATURE
1943 FOR PRECIPITATION
1951 FOR SNOWFALL

HEATING DEGREE DAYS Base 65 deg. F COLD BAY, ALASKA

SEASON	JULY	AUG	SEP	OCT	NOV	DEC	JAN	FEB	MAR	APR	MAY	JUNE	TOTAL
1964-65	522	501	529	778	957	1068	1162	1121	907	886	884	608	9923
1965-66	497	450	477	871	864	1126	991	924	1346	914	850	616	9926
1966-67	500	486	607	956	977	1153	1185	1017	936	798	717	540	9872
1967-68	369	417	555	823	804	1064	1133	1183	1067	917	682	577	9591
1968-69	368	394	538	804	865	1053	1025	1028	999	907	714	537	9232
1969-70	361	353	436	680	939	1039	1311	1025	1024	987	745	512	9412
1970-71	416	452	538	766	815	1059	1323	1086	1381	1103	945	728	10612
1971-72	527	504	608	809	973	1004	1165	1260	1503	1056	814	697	10920
1972-73	500	457	505	774	915	1065	1295	960	1105	998	895	668	10137
1973-74	537	459	549	815	817	1076	1089	1281	1113	977	725	569	10007
1974-75	460	360	466	756	917	1157	1252	1149	1176	984	816	557	10050
1975-76	443	380	553	790	1038	1145	1217	1184	1324	1125	877	603	10679
1976-77	463	434	585	797	1029	1134	914	874	1057	1006	789	456	9538
1977-78	334	344	441	805	1006	1095	972	1017	1053	808	750	535	9160
1978-79	471	325	481	766	799	917	921	1020	910	720	661	425	8416
1979-80	387	400	474	712	893	1179	1279	1143	964	876	726	569	9602
1980-81	368	414	499	757	851	1014	1051	990	898	791	621	512	8766
1981-82	368	390	500	742	929	1058	1083	1056	956	980	828	592	9482
1982-83	559	451	590	848	884	1062	1246	934	969	838	718	493	9592
1983-84	408	388	523	779	907	847	1045	1338	967	995	834	533	9564
1984-85	467	310	452	744	834	854	889	1033	1075	1140	821	669	9288
1985-86	440	434	455	779	775	905	1251	1018	1173	976	832	605	9643
1986-87	406	423	448	699	834	939	1063	943	965	916	806	607	9049
1987-88	431	373	538	737	1042	1131	1043	1003	1203	1014	754	565	9834
1988-89	427	423	537	755	970	1050	1318	834	1034	917	751	564	9580
1989-90	432	353	447	695	978	1037	1063	1077	1010	853	727	535	9207
1990-91	455	390	516	757	905	879	1046	1178	945	873	771	578	9293
1991-92	432	410	457	705	864	1089	1103	1140	1083	947	759	503	9492
1992-93	463	436	570	798	952	1001	1170	941	983	824	706	528	9372
1993-94	398	373	472	699	797	1022							

COOLING DEGREE DAYS Base 65 deg. F COLD BAY, ALASKA

YEAR	JAN	FEB	MAR	APR	MAY	JUNE	JULY	AUG	SEP	OCT	NOV	DEC	TOTAL
1969	0	0	0	0	0	0	0	0	0	0	0	0	0
1970	0	0	0	0	0	0	0	1	0	0	0	0	1
1971	0	0	0	0	0	0	0	0	0	0	0	0	0
1972	0	0	0	0	0	0	0	0	0	0	0	0	0
1973	0	0	0	0	0	0	0	0	0	0	0	0	0
1974	0	0	0	0	0	0	0	0	0	0	0	0	0
1975	0	0	0	0	0	0	0	0	0	0	0	0	0
1976	0	0	0	0	0	0	0	0	0	0	0	0	0
1977	0	0	0	0	0	0	0	0	0	0	0	0	0
1978	0	0	0	0	0	0	0	0	0	0	0	0	0
1979	0	0	0	0	0	0	0	0	0	0	0	0	0
1980	0	0	0	0	0	0	0	0	0	0	0	0	0
1981	0	0	0	0	0	0	0	0	0	0	0	0	0
1982	0	0	0	0	0	0	0	0	0	0	0	0	0
1983	0	0	0	0	0	0	0	0	0	0	0	0	0
1984	0	0	0	0	0	0	0	0	0	0	0	0	0
1985	0	0	0	0	0	0	0	0	0	0	0	0	0
1986	0	0	0	0	0	0	0	0	0	0	0	0	0
1987	0	0	0	0	0	0	0	0	0	0	0	0	0
1988	0	0	0	0	0	0	0	0	0	0	0	0	0
1989	0	0	0	0	0	0	0	0	0	0	0	0	0
1990	0	0	0	0	0	0	0	0	0	0	0	0	0
1991	0	0	0	0	0	0	0	0	0	0	0	0	0
1992	0	0	0	0	0	0	0	0	0	0	0	0	0
1993	0	0	0	0	0	0	0	0	0	0	0	0	0

SNOWFALL (inches) COLD BAY, ALASKA

SEASON	JULY	AUG	SEP	OCT	NOV	DEC	JAN	FEB	MAR	APR	MAY	JUNE	TOTAL
1964-65	0.0	0.0	0.0	3.4	16.2	9.1	16.4	23.4	1.6	3.2	5.3	0.0	78.6
1965-66	0.0	0.0	0.1	7.7	9.7	9.7	9.3	8.8	14.0	1.4	0.7	0.0	61.4
1966-67	0.0	0.0	0.0	10.7	6.0	12.8	11.4	15.4	4.5	1.0	0.0	0.0	61.8
1967-68	0.0	0.0	T	4.5	4.8	12.2	6.8	10.3	7.8	5.1	2.1	0.0	53.6
1968-69	0.0	0.0	T	15.6	6.7	7.8	10.1	21.0	13.4	6.2	4.8	0.0	85.6
1969-70	0.0	0.0	0.0	2.2	4.5	7.2	9.4	8.6	10.6	6.8	6.8	0.0	56.1
1970-71	0.0	0.0	0.0	1.0	3.3	8.8	19.9	10.5	3.9	6.1	9.3	0.5	63.3
1971-72	0.0	0.0	0.0	T	2.1	9.3	15.6	9.8	4.2	9.9	1.4	T	52.3
1972-73	0.0	0.0	0.2	T	4.0	14.6	10.6	11.5	9.8	10.1	3.5	T	64.3
1973-74	0.0	0.0	T	3.9	2.4	13.0	6.5	17.1	5.9	4.5	1.8	0.0	55.1
1974-75	0.0	0.0	0.0	0.4	10.8	11.7	6.2	13.5	20.5	8.6	2.5	0.0	74.2
1975-76	0.0	0.0	0.0	3.7	12.3	20.4	16.2	16.9	13.9	19.5	0.8	0.0	103.7
1976-77	T	0.0	0.0	6.9	10.4	24.2	4.8	7.5	27.0	7.1	3.6	0.0	91.5
1977-78	0.0	0.0	0.0	9.4	15.0	4.8	14.4	13.9	16.4	3.5	0.1	0.0	77.5
1978-79	0.0	0.0	T	6.4	2.7	9.7	12.4	4.4	11.5	2.1	0.3	0.0	49.5
1979-80	0.0	0.0	T	0.3	15.2	10.5	14.1	11.1	23.7	13.6	1.1	0.0	89.6
1980-81	0.0	0.0	T	2.2	7.6	9.6	10.4	12.0	11.8	2.1	1.4	0.0	57.1
1981-82	0.0	0.0	T	5.5	13.2	14.0	34.6	5.8	15.7	5.6	1.0	0.0	95.4
1982-83	0.0	0.0	T	0.4	3.3	17.6	14.2	10.7	9.1	5.7	T	0.0	61.0
1983-84	0.0	0.0	T	4.0	27.4	2.6	14.3	54.3	4.3	7.6	1.4	0.0	115.9
1984-85	0.0	0.0	0.0	5.3	10.6	6.9	4.2	8.2	28.6	8.8	1.5	0.4	74.5
1985-86	0.0	0.0	T	2.8	1.1	12.1	24.2	11.3	6.1	5.3	6.2	0.0	69.1
1986-87	0.0	0.0	0.0	T	7.6	7.2	12.4	10.5	20.3	7.6	1.0	T	66.6
1987-88	0.0	0.0	0.1	0.7	8.1	14.3	11.0	5.9	7.5	11.6	0.1	0.0	59.3
1988-89	0.0	0.0	0.0	0.3	15.2	22.1	11.2	13.7	3.1	10.4	0.3	0.0	76.3
1989-90	0.0	0.0	0.0	0.1	14.0	6.9	16.4	18.7	8.8	3.9	0.6	0.0	69.4
1990-91	0.0	0.0	T	4.7	10.6	6.3	6.5	13.3	8.4	2.8	0.5	T	53.1
1991-92	0.0	0.0	0.0	0.8	3.3	19.9	7.2	5.0	8.0	1.9	1.8	0.0	47.9
1992-93	0.0	0.0	T	3.4	9.0	6.9	6.8	8.6	2.3	2.6	T	0.0	39.6
1993-94	0.0	0.0	T	T	5.9	16.0							
Record Mean	T	0.0	T	2.9	7.7	10.3	10.5	11.4	10.3	5.9	1.7	T	60.8

FAIRBANKS, ALASKA

Fairbanks is located in the Tanana Valley, in the interior of Alaska. It has a distinctly continental climate, with large variation of temperature from winter to summer.

The climate in Fairbanks is conditioned mainly by the response of the land mass to large changes in solar heat received by the area during the year. The sun is above the horizon from 18 to 21 hours during June and July. During this period, daily average maximum temperatures reach the lower 70s. Temperatures of 80 degrees or higher occur on about 10 days each summer. In contrast, from November to early March, when the period of daylight ranges from 10 to less than 4 hours per day, the lowest temperature readings normally fall below zero quite regularly. Low temperatures of −40 degrees or colder occur each winter. The range of temperatures in summer is comparatively low, from the lower 30s to the mid 90s. In winter, this range is larger, from about 65 below to 45 degrees above. This large winter range of temperature reflects the great difference between frigid weather associated with dry northerly airflow from the Arctic to mild temperatures associated with southerly airflow from the Gulf of Alaska, accompanied by chinook winds off the Alaska Range, 80 miles to the south of Fairbanks.

Snow cover is persistent in Fairbanks, without interruption, from October through April. Snowfalls of 4 inches or more in a day occur only three times during winter. Blizzard conditions are almost never seen, as winds in Fairbanks are above 20 miles an hour less than 1 percent of the time. Precipitation normally reaches a minimum in spring, and a maximum in August, when rainfall is common. During summer, thunderstorms occur in Fairbanks on an average of about eight days. Thunderstorms are about three times more frequent over the hills to the north and east of Fairbanks. Damaging hail or wind rarely accompany thunderstorms around Fairbanks.

There are rolling hills reaching elevations up to 2,000 feet above Fairbanks to the north and east of the city. During winter, the uplands are often warmer than Fairbanks, as cold air settles into the valley. In some months, temperatures in the uplands will average more than 10 degrees warmer than Fairbanks. During summer, the uplands are a few degrees cooler than the city. Precipitation in the uplands around Fairbanks is heavier than it is in the city by roughly 20 to 50 percent. Fairbanks exhibits an urban heat island, especially during winter. Low lying areas nearby, such as the community of North Pole, are often colder than the city, sometimes by as much as 15 degrees.

During winter, with temperatures of −20 degrees or colder, ice fog frequently forms in the city. Cold snaps accompanied by ice fog generally last about a week, but can last three weeks in unusual situations. The fog is almost always less than 300 feet deep, so that the surrounding uplands are usually in the clear, with warmer temperatures. Visibility in the ice fog is sometimes quite low, and this can hinder aircraft operations for as much as a day in severe cases. Aside from the low visibility in winter ice fog, flying weather in Fairbanks is quite favorable, especially from February through May, when crystal clear weather is common and the length of daylight is rapidly increasing.

Hardy vegetables and grains grow luxuriantly. Freezing of local rivers normally begins in the first week of October. The date when ice will normally support a persons weight is October 27. Rivers remain frozen and safe for travel until early April. Breakup of the river ice usually occurs in the first week of May.

NORMALS, MEANS, AND EXTREMES

FAIRBANKS, ALASKA

LATITUDE: 64°49'N LONGITUDE: 147°52'W ELEVATION: FT. GRND 436 BARO 452 TIME ZONE: YUKON WBAN: 26411

	(a)	JAN	FEB	MAR	APR	MAY	JUNE	JULY	AUG	SEP	OCT	NOV	DEC	YEAR
TEMPERATURE °F:														
Normals														
-Daily Maximum		-1.6	7.2	23.8	41.0	59.3	70.1	72.3	66.3	54.8	32.0	10.9	1.8	36.5
-Daily Minimum		-18.5	-14.4	-1.7	20.4	38.0	49.5	52.6	47.2	36.2	18.1	-5.6	-14.8	17.3
-Monthly		-10.1	-3.6	11.0	30.7	48.6	59.8	62.5	56.8	45.5	25.1	2.7	-6.5	26.9
Extremes														
-Record Highest	42	50	47	53	74	89	96	94	90	84	65	46	44	96
-Year		1981	1987	1993	1960	1960	1969	1975	1976	1957	1969	1970	1985	JUN 1969
-Record Lowest	42	-61	-58	-49	-24	-1	31	35	27	3	-27	-46	-62	-62
-Year		1969	1993	1956	1986	1964	1963	1959	1987	1992	1992	1990	1961	DEC 1961
NORMAL DEGREE DAYS:														
Heating (base 65°F)		2328	1921	1674	1029	508	182	123	267	585	1237	1869	2217	13940
Cooling (base 65°F)		0	0	0	0	0	26	45	13	0	0	0	0	84
% OF POSSIBLE SUNSHINE														
MEAN SKY COVER (tenths)														
Sunrise - Sunset	42	6.3	6.2	5.9	6.6	6.9	7.3	7.4	7.8	7.6	7.9	6.9	7.0	7.0
MEAN NUMBER OF DAYS:														
Sunrise to Sunset														
-Clear	42	9.0	8.2	9.9	6.6	4.5	2.9	3.4	3.0	4.2	4.0	6.8	6.8	69.5
-Partly Cloudy	42	6.0	5.9	6.8	7.6	11.0	10.2	8.9	6.7	6.1	5.0	5.2	5.7	85.0
-Cloudy	42	16.1	14.2	14.3	15.8	15.5	16.8	18.7	21.2	19.7	22.0	18.0	18.5	210.7
Precipitation														
.01 inches or more	42	7.7	6.7	5.9	4.7	7.0	10.6	12.3	12.4	9.6	10.7	10.3	9.0	106.9
Snow,Ice pellets,hail														
1.0 inches or more	42	3.4	2.5	2.3	1.1	0.2	0.0	0.0	0.0	0.5	4.2	4.4	4.4	22.9
Thunderstorms	42	0.0	0.0	0.0	0.*	0.4	3.0	2.3	0.9	0.1	0.0	0.0	0.0	6.7
Heavy Fog Visibility														
1/4 mile or less	42	4.6	2.4	0.5	0.3	0.2	0.4	0.8	1.7	1.3	1.7	1.1	3.2	18.1
Temperature °F														
-Maximum														
70° and above	30	0.0	0.0	0.0	0.*	3.9	17.1	21.2	10.8	1.3	0.0	0.0	0.0	54.3
32° and below	30	29.8	26.0	20.8	5.6	0.2	0.0	0.0	0.0	0.4	16.1	27.9	29.6	156.5
-Minimum														
32° and below	30	31.0	28.3	30.9	26.5	6.2	0.0	0.0	0.6	9.1	28.4	30.0	31.0	221.8
0° and below	30	25.9	22.3	16.0	2.2	0.*	0.0	0.0	0.0	0.0	3.7	18.5	24.3	113.0
AVG. STATION PRESS. (mb)	21	993.5	995.7	992.9	993.3	992.9	993.5	995.8	995.3	992.8	990.4	991.5	992.2	993.3
RELATIVE HUMIDITY (%)														
Hour 03	42	69	66	67	68	68	75	81	86	82	79	73	71	74
Hour 09 (Local Time)	42	69	66	66	60	53	60	69	77	78	79	74	71	69
Hour 15	43	69	63	53	46	38	43	50	54	55	67	73	71	57
Hour 21	42	69	66	63	55	47	50	60	70	72	77	74	71	65
PRECIPITATION (inches):														
Water Equivalent														
-Normal		0.47	0.40	0.37	0.32	0.61	1.37	1.87	1.96	0.95	0.90	0.80	0.85	10.87
-Maximum Monthly	42	2.40	1.75	2.24	0.93	1.67	3.52	4.87	6.20	3.05	2.19	3.32	3.23	6.20
-Year		1993	1966	1991	1982	1955	1955	1990	1967	1960	1983	1970	1984	AUG 1967
-Minimum Monthly	42	0.01	0.01	T	T	0.07	0.19	0.35	0.40	0.15	0.08	T	T	T
-Year		1966	1976	1987	1991	1957	1966	1993	1957	1968	1954	1953	1969	APR 1991
-Maximum in 24 hrs	42	0.75	0.97	1.17	0.47	0.88	1.52	1.73	3.42	1.21	2.22	0.84	1.25	3.42
-Year		1993	1966	1991	1979	1955	1955	1990	1967	1954	1976	1970	1968	AUG 1967
Snow,Ice pellets,hail														
-Maximum Monthly	42	40.2	43.1	30.4	11.5	14.1	T	T	T	24.4	25.9	54.0	50.7	54.0
-Year		1993	1966	1991	1992	1992	1993	1990	1969	1992	1982	1970	1984	NOV 1970
-Maximum in 24 hrs	42	10.1	20.1	12.6	6.3	9.4	T	T	T	9.0	10.4	14.6	14.7	20.1
-Year		1993	1966	1963	1992	1992	1993	1990	1969	1992	1974	1970	1968	FEB 1966
WIND:														
Mean Speed (mph)	42	3.1	4.0	5.2	6.6	7.7	7.1	6.6	6.2	6.2	5.4	3.9	3.2	5.4
Prevailing Direction through 1963		N	N	N	N	N	SW	SW	N	N	N	N	N	N
Fastest Obs. 1 Min.														
-Direction (!!!)	42	25	27	22	24	23	25	27	27	22	25	25	24	25
-Speed (MPH)	42	31	33	40	32	32	40	32	34	30	40	35	37	40
-Year		1991	1955	1970	1983	1984	1974	1989	1954	1975	1958	1970	1970	JUN 1974
Peak Gust														
-Direction (!!!)	9	SW	SW	SW	SW	W	E	NE	S	W	W	W	SW	NE
-Speed (mph)	9	47	40	46	40	44	48	63	38	51	40	35	38	63
-Date		1993	1989	1985	1992	1992	1992	1990	1985	1985	1993	1992	1985	JUL 1990

PRECIPITATION (inches) FAIRBANKS, ALASKA

YEAR	JAN	FEB	MAR	APR	MAY	JUNE	JULY	AUG	SEP	OCT	NOV	DEC	ANNUAL
1964	0.21	0.46	0.18	0.68	0.97	1.33	1.28	2.37	0.85	0.53	0.86	0.34	10.06
1965	0.07	0.32	0.27	0.47	0.14	1.16	1.39	1.48	2.11	0.74	1.21	1.92	11.28
1966	0.01	1.75	0.34	0.32	0.38	0.19	0.83	0.59	0.15	0.29	2.06	0.16	7.07
1967	0.40	0.25	1.90	0.84	0.43	1.13	3.34	6.20	0.25	0.32	0.93	1.34	17.33
1968	1.19	0.15	T	0.29	0.67	1.52	0.84	0.96	0.15	0.31	0.27	1.38	7.73
1969	0.55	0.10	0.60	T	0.95	0.39	1.33	2.04	0.28	0.10	0.54	T	6.88
1970	0.10	0.32	0.25	0.45	0.42	2.57	1.81	1.98	0.65	1.84	3.32	2.29	16.00
1971	0.33	0.63	0.20	0.11	0.16	0.31	2.08	2.32	2.45	1.35	0.54	1.83	12.31
1972	0.73	0.16	0.27	0.20	0.35	0.55	0.63	1.09	2.08	0.86	0.46	1.13	8.51
1973	0.44	0.11	0.40	0.05	0.99	0.97	1.92	2.19	0.19	0.91	0.80	0.15	9.12
1974	0.14	0.33	0.27	0.21	0.11	1.22	1.17	1.14	0.47	1.08	1.03	0.55	7.72
1975	0.60	0.04	0.22	0.47	0.49	0.99	1.81	2.10	0.20	0.79	0.44	0.31	8.46
1976	0.22	0.01	0.55	0.08	0.94	1.08	1.60	0.69	1.05	0.89	0.13	0.08	7.32
1977	0.31	0.81	0.26	0.36	1.63	3.01	1.58	0.41	2.51	1.11	0.19	0.80	12.98
1978	0.39	0.19	0.09	0.16	0.44	1.71	1.19	1.24	0.98	0.59	1.02	1.40	9.40
1979	0.58	0.02	0.47	0.83	0.88	1.54	2.54	1.22	0.19	0.94	0.63	0.49	10.33
1980	0.52	0.22	0.13	0.10	0.31	1.38	1.37	1.68	0.76	0.39	0.70	0.32	7.88
1981	0.31	0.78	0.07	0.32	0.73	1.91	2.41	1.35	0.80	0.91	0.91	0.58	11.08
1982	0.34	0.38	0.39	0.93	0.96	1.96	2.33	1.67	0.77	1.48	1.49	0.23	12.93
1983	0.24	0.18	0.09	0.27	0.14	0.57	1.71	3.33	0.92	2.19	0.08	0.65	10.37
1984	0.89	0.64	0.03	0.47	1.17	0.48	2.95	1.15	0.22	0.70	0.42	3.23	12.35
1985	0.52	0.48	0.57	0.36	0.41	1.80	1.13	1.88	2.59	1.00	0.90	0.08	11.72
1986	0.13	0.19	0.32	0.07	0.54	0.87	2.12	2.36	0.65	1.79	0.48	0.34	9.86
1987	0.68	0.10	T	0.05	0.21	1.02	1.70	0.56	0.57	0.39	0.64	0.51	6.43
1988	0.32	0.13	0.13	0.21	1.51	2.26	1.02	1.95	0.73	1.07	0.68	0.46	10.47
1989	0.52	0.98	0.13	0.05	0.99	2.53	0.91	0.78	0.72	1.28	0.97	0.57	10.43
1990	0.52	0.72	0.11	0.07	0.40	1.73	4.87	3.60	1.74	0.31	1.51	2.94	18.52
1991	1.17	0.17	2.24	T	0.10	0.36	0.81	1.18	1.16	0.71	0.48	1.02	9.40
1992	0.85	0.66	0.07	0.47	1.23	2.15	2.32	0.59	1.34	0.91	0.93	1.21	12.73
1993	2.40	0.31	0.26	0.03	0.63	1.24	0.35	1.58	2.63	0.61	0.86	0.43	11.33
Record Mean	0.71	0.46	0.41	0.28	0.66	1.38	1.84	2.11	1.06	0.87	0.70	0.70	11.18

AVERAGE TEMPERATURE (deg. F) FAIRBANKS, ALASKA

YEAR	JAN	FEB	MAR	APR	MAY	JUNE	JULY	AUG	SEP	OCT	NOV	DEC	ANNUAL
1964	-15.7	0.7	-2.2	25.8	38.6	60.1	59.7	56.5	44.9	28.4	2.4	-26.5	22.8
1965	-18.0	-18.1	24.5	30.9	42.9	55.2	60.2	52.7	48.2	13.9	3.9	-14.1	23.5
1966	-27.4	-7.6	-2.5	27.1	45.4	63.3	62.5	57.1	50.1	24.3	0.6	-20.1	22.7
1967	-15.3	-6.9	9.8	31.7	45.7	61.8	59.8	58.3	46.6	24.8	9.5	-1.6	27.1
1968	-11.0	-5.0	12.8	29.2	47.6	59.5	65.8	58.5	42.8	22.1	2.3	-17.7	25.6
1969	-26.7	-7.3	10.1	36.3	49.4	64.9	59.4	49.8	49.1	34.0	1.2	4.0	27.0
1970	-16.2	8.0	20.9	32.0	51.8	58.0	62.4	56.9	40.8	18.9	10.7	-9.8	27.9
1971	-31.7	-4.6	-0.4	26.7	47.3	63.4	61.0	56.1	44.7	27.9	0.5	-5.8	23.8
1972	-16.3	-10.1	-2.8	20.8	47.4	59.4	64.5	58.9	40.1	26.8	7.0	-2.5	24.5
1973	-18.2	-1.5	11.9	35.3	50.6	60.3	62.0	55.0	47.3	25.1	-0.7	-3.4	27.0
1974	-16.7	-17.8	7.6	34.9	51.4	58.7	63.5	59.1	51.4	21.4	0.6	-11.3	25.2
1975	-15.5	-3.4	12.6	30.5	53.5	63.4	68.4	56.1	45.8	23.9	-8.0	-16.1	25.9
1976	-11.5	-13.7	12.1	36.0	47.8	59.6	61.8	59.2	45.4	23.9	15.9	-3.9	27.7
1977	9.8	8.6	4.6	27.8	48.7	59.6	62.8	62.6	45.6	25.6	-7.6	-14.9	27.8
1978	0.1	3.9	14.0	34.8	50.2	54.6	63.5	59.5	46.8	23.3	8.6	3.3	30.2
1979	-7.7	-25.3	12.0	30.9	49.9	57.4	61.3	60.4	46.5	32.4	20.2	-10.2	27.3
1980	-9.5	16.0	17.3	35.9	50.7	56.5	60.9	53.6	43.0	33.0	11.5	-24.0	28.7
1981	18.1	5.1	27.1	31.4	51.3	58.9	56.5	53.6	44.1	29.5	12.3	-4.2	32.0
1982	-18.0	-3.9	13.1	27.7	46.8	58.5	63.1	56.6	49.3	18.5	4.4	2.2	26.6
1983	-11.0	3.4	13.8	37.4	50.4	62.3	64.2	53.5	41.2	23.6	8.6	-3.7	28.6
1984	-5.9	-13.3	21.6	30.3	47.2	61.6	60.9	53.8	46.8	25.9	0.1	-3.1	27.2
1985	11.1	-9.4	14.6	20.8	46.7	57.8	63.1	56.3	42.9	18.8	-4.7	7.7	27.1
1986	-2.1	4.7	6.0	24.0	47.8	62.6	63.6	54.7	46.3	27.1	0.2	7.3	28.5
1987	0.7	1.5	13.5	35.0	50.9	61.9	64.2	57.7	44.1	33.0	6.0	-3.2	30.4
1988	-5.3	3.9	17.8	33.6	52.8	62.9	65.8	58.4	44.4	17.4	-3.1	4.2	29.4
1989	-21.3	3.4	6.7	36.2	47.8	60.1	64.6	60.8	48.6	26.2	-7.1	4.5	27.5
1990	-12.9	-21.7	18.5	38.1	55.1	61.6	65.3	60.0	44.8	24.1	-4.9	-6.3	26.8
1991	-4.7	-1.2	11.9	35.4	51.2	63.8	60.6	54.4	48.0	24.9	0.6	-2.9	28.5
1992	-4.7	-8.7	14.0	26.1	41.8	60.0	64.1	56.5	31.7	17.5	10.4	-7.6	25.1
1993	-3.9	2.4	17.2	41.1	53.7	62.0	65.6	56.1	44.1	29.3	8.4	0.6	31.4
Record Mean	-10.1	-3.4	10.1	30.4	48.0	59.3	61.3	55.6	44.5	25.6	3.1	-7.7	26.4
Max	-1.2	7.7	23.5	41.6	59.1	70.5	71.9	65.7	54.0	33.1	11.3	0.6	36.5
Min	-19.0	-14.5	-3.4	19.2	36.8	48.0	50.7	45.6	35.0	18.1	-5.1	-16.0	16.3

REFERENCE NOTES FAIRBANKS, ALASKA

GENERAL
T - TRACE AMOUNT.
BLANK ENTRIES DENOTE MISSING/UNREPORTED DATA.
# INDICATES A STATION OR INSTRUMENT RELOCATION.

SPECIFIC

PM - INCLUDES LAST DAY OF PREVIOUS MONTH
ASOS - AUTOMATED SURFACE OBSERVING SYSTEM IN OPERATION DURING THESE MONTHS.

(a) - LENGTH OF RECORD IN YEARS, ALTHOUGH INDIVIDUAL MONTHS MAY BE MISSING.
0.* OR * - THE VALUE IS BETWEEN 0.0 AND 0.05
NORMALS - BASED ON THE 1961-1990 RECORD PERIOD.
EXTREMES - DATES ARE THE MOST RECENT OCCURRENCE.
WIND DIR. - NUMERALS SHOW TENS OF DEGREES CLOCKWISE FROM TRUE NORTH. "00" INDICATES CALM.
RESULTANT DIRECTIONS ARE GIVEN TO WHOLE DEGREES.
BOLD VALUES INDICATE EXTREME VALUES WHICH OCCURRED AFTER THE ASOS SYSTEM WAS COMMISSIONED.

RECORD = PERIOD OF RECORD
RECORD MEAN PRECIPITATION IS THE MEAN OF ALL DAILY PRECIPITATION AMOUNTS DURING THE PERIOD OF RECORD.
RECORD MAX(MIN) TEMPERATURE IS THE MEAN OF ALL DAILY MAX(MIN) TEMPERATURES DURING THE PERIOD OF RECORD.
RECORD MEAN TEMPERATURE IS THE SUM OF THE RECORD MAX AND RECORD MIN DIVIDED BY 2.
AVERAGE TEMPERATURE IS THE SUM OF THE MEAN DAILY MAX AND MIN TEMPERATURE DIVIDED BY 2.

EXCEPTIONS

RECORD MEANS ARE THROUGH THE CURRENT YEAR, BEGINNING IN 1930 FOR TEMPERATURE
1930 FOR PRECIPITATION
1952 FOR SNOWFALL

HEATING DEGREE DAYS Base 65 deg. F FAIRBANKS, ALASKA

SEASON	JULY	AUG	SEP	OCT	NOV	DEC	JAN	FEB	MAR	APR	MAY	JUNE	TOTAL
1964-65	165	256	595	1133	1876	2841	2574	2334	1249	1019	680	287	15009
1965-66	152	374	496	1580	1833	2455	2872	2039	2092	1132	600	63	15688
1966-67	92	247	441	1255	1929	2643	2496	2014	1704	991	594	138	14544
1967-68	178	213	545	1239	1661	2064	2358	2029	1613	1069	533	169	13671
1968-69	50	208	657	1320	1881	2567	2849	2024	1698	855	475	80	14664
1969-70	170	467	472	955	1914	1892	2524	1595	1365	981	401	203	12939
1970-71	96	244	722	1425	1630	2318	3002	1948	2030	1141	542	117	15215
1971-72	134	271	600	1143	1932	2195	2524	2178	2102	1318	539	205	15141
1972-73	63	184	738	1177	1739	2091	2582	1864	1637	883	439	150	13547
1973-74	111	302	523	1231	1968	2124	2535	2322	1778	895	414	188	14391
1974-75	85	195	402	1342	1935	2370	2497	1918	1620	1028	347	69	13808
1975-76	33	270	570	1270	2195	2513	2372	2285	1631	861	527	156	14683
1976-77	116	188	583	1269	1466	2136	1709	1574	1871	1107	500	155	12674
1977-78	101	124	573	1216	2184	2480	2013	1712	1576	898	454	304	13635
1978-79	65	176	542	1286	1689	1912	2260	2533	1638	1018	463	220	13802
1979-80	124	143	548	1004	1336	2335	2312	1415	1475	868	436	248	12244
1980-81	127	351	654	985	1599	2766	1447	1676	1168	999	418	188	12378
1981-82	255	347	622	1094	1573	2150	2581	1929	1602	1113	555	216	14037
1982-83	86	252	465	1434	1816	1946	2356	1725	1581	823	451	133	13068
1983-84	62	351	705	1280	1688	2126	2201	2277	1338	1035	549	120	13732
1984-85	140	344	538	1205	1950	2111	1666	2086	1558	1321	558	215	13692
1985-86	72	267	654	1430	2095	1776	2079	1686	1825	1224	527	113	13748
1986-87	110	312	559	1169	1943	1787	1994	1776	1594	893	428	128	12693
1987-88	61	218	620	987	1768	2111	2185	1768	1455	934	371	96	12574
1988-89	39	202	611	1469	2045	1883	2676	1722	1804	859	529	149	13988
1989-90	73	134	484	1195	2164	1875	2420	2433	1431	798	310	127	13444
1990-91	74	178	600	1261	2097	2212	2161	1849	1640	877	421	130	13500
1991-92	143	321	504	1234	1935	2105	2163	2140	1577	1160	711	157	14150
1992-93	45	259	995	1463	1636	2253	2134	1751	1478	711	343	103	13171
1993-94	43	273	620	1099	1694	1995							

COOLING DEGREE DAYS Base 65 deg. F FAIRBANKS, ALASKA

YEAR	JAN	FEB	MAR	APR	MAY	JUNE	JULY	AUG	SEP	OCT	NOV	DEC	TOTAL
1969	0	0	0	0	0	83	2	0	0	0	0	0	85
1970	0	0	0	0	0	0	22	0	0	0	0	0	22
1971	0	0	0	0	0	73	16	0	0	0	0	0	89
1972	0	0	0	0	0	40	55	0	0	0	0	0	95
1973	0	0	0	0	0	13	24	2	0	0	0	0	39
1974	0	0	0	0	1	9	44	20	0	0	0	0	74
1975	0	0	0	0	0	28	146	1	0	0	0	0	175
1976	0	0	0	0	0	2	23	14	0	0	0	0	39
1977	0	0	0	0	0	0	44	54	0	0	0	0	98
1978	0	0	0	0	0	0	27	10	0	0	0	0	37
1979	0	0	0	0	0	0	16	7	0	0	0	0	23
1980	0	0	0	0	0	0	8	6	0	0	0	0	14
1981	0	0	0	0	1	11	0	2	0	0	0	0	14
1982	0	0	0	0	0	27	36	0	0	0	0	0	63
1983	0	0	0	0	5	61	40	0	0	0	0	0	106
1984	0	0	0	0	0	22	21	2	0	0	0	0	45
1985	0	0	0	0	0	8	20	4	0	0	0	0	32
1986	0	0	0	0	0	46	74	0	0	0	0	0	120
1987	0	0	0	0	0	42	42	0	0	0	0	0	84
1988	0	0	0	0	0	41	72	2	0	0	0	0	115
1989	0	0	0	0	0	10	67	11	0	0	0	0	88
1990	0	0	0	0	11	32	91	35	0	0	0	0	169
1991	0	0	0	0	0	100	13	0	0	0	0	0	113
1992	0	0	0	0	0	17	21	4	0	0	0	0	42
1993	0	0	0	0	0	18	70	3	0	0	0	0	91

SNOWFALL (inches) FAIRBANKS, ALASKA

SEASON	JULY	AUG	SEP	OCT	NOV	DEC	JAN	FEB	MAR	APR	MAY	JUNE	TOTAL
1964-65	0.0	0.0	0.0	4.7	18.1	14.0	3.0	7.6	1.4	5.8	0.7	0.0	55.3
1965-66	0.0	0.0	T	15.7	18.8	33.5	0.7	43.1	8.9	1.5	1.7	0.0	123.9
1966-67	0.0	0.0	0.0	5.5	36.6	2.9	13.0	5.2	28.4	11.1	T	0.0	102.7
1967-68	0.0	0.0	T	5.8	14.4	16.8	22.6	4.8	0.1	5.0	2.7	0.0	72.2
1968-69	0.0	0.0	0.9	6.7	8.0	20.2	10.8	1.4	11.3	0.4	T	0.0	59.7
1969-70	0.0	T	0.0	3.7	11.4	T	3.6	6.4	4.2	7.1	T	0.0	36.4
1970-71	0.0	0.0	4.1	21.9	54.0	32.5	8.5	16.3	6.4	0.9	1.1	0.0	145.7
1971-72	0.0	0.0	3.5	20.4	9.8	29.6	12.6	4.5	5.7	4.3	0.0	0.0	90.4
1972-73	0.0	0.0	7.8	6.9	12.2	26.9	14.2	1.8	8.7	1.0	T	0.0	79.5
1973-74	0.0	0.0	1.2	8.2	17.5	2.5	3.0	8.7	7.0	1.3	0.0	T	49.4
1974-75	0.0	0.0	0.3	24.4	22.5	17.9	14.1	1.6	5.0	4.4	T	0.0	90.2
1975-76	0.0	0.0	T	14.4	11.4	6.7	6.4	0.5	10.4	1.2	T	0.0	51.0
1976-77	0.0	0.0	T	14.7	4.3	2.6	6.7	19.0	5.5	6.8	T	0.0	59.6
1977-78	0.0	0.0	1.0	17.3	5.1	14.4	4.9	3.0	1.6	0.7	0.1	0.0	48.1
1978-79	0.0	0.0	0.3	7.6	15.1	21.7	9.7	0.5	7.7	3.1	0.0	0.0	65.7
1979-80	0.0	0.0	T	6.5	4.4	11.0	10.8	4.4	3.2	0.8	0.0	0.0	41.1
1980-81	0.0	0.0	3.4	5.1	10.6	4.8	6.2	10.2	0.9	1.9	0.0	0.0	43.1
1981-82	0.0	0.0	0.3	7.8	16.2	10.6	7.6	7.0	7.2	11.4	0.3	0.0	68.4
1982-83	0.0	0.0	0.6	25.9	27.8	3.8	4.8	3.6	2.1	1.9	0.4	0.0	70.9
1983-84	0.0	0.0	0.4	17.3	2.4	14.4	13.8	11.1	0.8	6.7	T	0.0	66.9
1984-85	0.0	0.0	0.0	11.3	8.5	50.7	8.1	8.0	7.4	5.9	1.0	0.0	100.9
1985-86	0.0	0.0	2.1	5.4	14.7	1.8	2.3	2.6	3.7	1.0	0.0	0.0	33.6
1986-87	0.0	0.0	T	11.8	7.4	6.0	12.0	1.5	T	1.0	T	0.0	39.7
1987-88	0.0	0.0	T	3.5	14.1	10.4	6.3	2.5	2.6	0.2	0.3	0.0	39.9
1988-89	0.0	0.0	0.2	12.9	15.2	10.5	10.8	13.4	3.9	0.4	0.8	0.0	68.1
1989-90	0.0	0.0	0.5	19.7	18.1	11.1	10.4	16.0	2.6	T	0.0	T	78.4
1990-91	T	0.0	1.6	6.9	37.3	47.5	20.6	3.0	30.4	T	0.0	0.0	147.3
1991-92	0.0	0.0	0.0	12.2	9.7	18.9	15.5	15.0	1.8	11.5	14.1	0.0	98.7
1992-93	0.0	0.0	24.4	16.7	18.7	28.5	40.2	5.4	5.2	T	0.0	T	139.1
1993-94	0.0	0.0	7.5	7.3	16.3	8.5							
Record Mean	T	T	1.6	10.9	13.6	13.8	10.8	8.6	6.6	3.4	0.8	T	70.1

GULKANA, ALASKA

Gulkana Airport is located in the Copper River basin 2 miles west of the river and 9 miles south of the village of Gulkana. The airport is approximately 150 airline miles northeast of Anchorage and 200 miles south–southeast of Fairbanks. The facility can be reached by both highway and air, and is near the junction point of the Glenn Highway, leading to Anchorage and Canada, and the Richardson Highway, leading to Fairbanks. Terrain surrounding the station experiences no rapid changes in elevation. To the east the ground drops about 400 feet to the river. East of the river, the land rises gradually to about 3,000 feet at a distance of 18 miles, and then rises abruptly to 12,000 feet at the top of Mt. Drum, about 25 miles away. Elevation changes in all other directions are small. Numerous small river and creek valleys give the appearance of a fairly rough terrain despite the small elevation changes. From a much larger scale viewpoint, mountain ranges exist in all directions. The Alaska Range lies about 75 miles to the north. Mt. Drum to the east is the western end of the Wrangell Mountains. Fifty miles to the south are the Chugach Mountains. To the west at a distance of about 100 miles, the north–south oriented Talkeetna Mountains, with their southern extension joining the Chugach Range, form a barrier between Gulkana and Anchorage. Elevations of the various mountain ranges are from 6,000 to 12,000 feet. The mountains exert a significant influence on the climate of the Copper River basin.

There is no doubt that Gulkana Airport is under a dominant continental influence. Typical of this are the extremes of temperature in summer and winter, with a range as large as 156 degrees. Because the sun makes only a brief appearance during the middle of the day in winter, the average minimum temperature usually remains below zero. In contrast to the cold winter months with short days, summer is pleasantly cool, but warm enough for outdoor activities, with 18 to 20 hours of sunshine.

Mountains surrounding the Gulkana area capture a large portion of the moisture which might otherwise reach the valley, particularly from the Gulf of Alaska which deposits annual amounts in excess of 60 inches on the windward slopes of the Chugach Range. There is no commercial agriculture in the Copper River basin, but with well over half of the annual total precipitation occurring during the four months of June through September, there is adequate moisture for gardening. The average length of the growing season is 78 days.

Cloud data are available for a short period, but the heaviest precipitation occurrence in summer may be indicative of maximum cloudiness occurring during these months.

Surface wind directions are prevailing southeasterly during spring, summer, and early fall, and from the north during late fall and winter. Monthly wind speeds are highest in summer. However, the infrequent occurrences of strong winds in excess of 40 mph have always been associated with the lighter wind months from October through April.

NORMALS, MEANS, AND EXTREMES

GULKANA, ALASKA

LATITUDE: 62°09'N LONGITUDE: 145°27' W ELEVATION: FT. GRND 1572 BARO 1578 TIME ZONE: YUKON WBAN: 26425

	(a)	JAN	FEB	MAR	APR	MAY	JUNE	JULY	AUG	SEP	OCT	NOV	DEC	YEAR
TEMPERATURE °F:														
Normals														
-Daily Maximum		2.5	13.6	28.1	41.5	54.8	64.2	68.3	64.7	54.1	35.3	12.7	4.6	37.0
-Daily Minimum		-13.9	-7.0	2.6	19.9	32.6	42.2	46.4	42.2	33.4	19.5	-2.5	-10.9	17.0
-Monthly		-5.7	3.3	15.4	30.7	43.7	53.2	57.4	53.5	43.8	27.4	5.1	-3.2	27.1
Extremes														
-Record Highest	46	46	46	50	67	85	90	91	86	74	65	48	44	91
-Year		1948	1943	1981	1958	1960	1969	1953	1977	1989	1969	1964	1943	JUL 1953
-Record Lowest	46	-60	-65	-48	-42	5	28	29	20	2	-23	-44	-58	-65
-Year		1947	1947	1972	1944	1964	1985	1970	1984	1992	1982	1989	1964	FEB 1947
NORMAL DEGREE DAYS:														
Heating (base 65°F)		2192	1728	1538	1029	660	354	236	357	636	1166	1797	2114	13807
Cooling (base 65°F)		0	0	0	0	0	0	0	0	0	0	0	0	0
% OF POSSIBLE SUNSHINE														
MEAN SKY COVER (tenths)														
Sunrise - Sunset	5	4.7	7.0	6.0	6.8	7.2	7.4	7.2	7.4	6.8	7.8	7.5	6.7	6.9
MEAN NUMBER OF DAYS:														
Sunrise to Sunset														
-Clear	5	14.8	6.2	9.4	5.0	3.0	2.0	4.4	2.6	5.6	3.0	4.6	6.8	67.4
-Partly Cloudy	5	4.4	4.4	6.4	9.8	9.8	10.0	9.0	9.4	7.8	6.8	4.6	6.8	89.2
-Cloudy	5	11.8	17.8	15.2	15.2	18.2	18.0	17.6	19.0	16.6	21.2	20.8	17.4	208.8
Precipitation														
.01 inches or more	35	7.2	5.7	5.0	2.9	5.1	9.5	12.7	11.6	10.8	8.5	7.3	8.4	94.7
Snow,Ice pellets,hail														
1.0 inches or more	25	2.5	2.4	2.3	1.3	0.2	0.*	0.0	0.0	0.5	3.0	3.5	3.7	19.5
Thunderstorms	6	0.0	0.0	0.0	0.0	0.6	1.3	2.0	0.6	0.0	0.0	0.0	0.0	4.5
Heavy Fog Visibility														
1/4 mile or less	6	0.2	1.2	1.3	0.3	0.2	0.0	0.8	2.0	2.6	3.6	3.4	2.2	17.8
Temperature °F														
-Maximum														
70° and above	45	0.0	0.0	0.0	0.0	1.4	9.1	12.7	6.6	0.2	0.0	0.0	0.0	30.0
32° and below	45	28.6	24.0	18.1	3.7	0.1	0.0	0.0	0.0	0.3	11.4	26.6	28.4	141.3
-Minimum														
32° and below	45	30.8	28.0	30.5	28.6	14.2	0.9	0.1	3.2	13.4	27.1	29.7	30.8	237.3
0° and below	45	23.3	18.5	14.8	2.2	0.0	0.0	0.0	0.0	0.0	3.6	17.2	23.1	102.7
AVG. STATION PRESS.(mb)														
RELATIVE HUMIDITY (%)														
Hour 03	14	73	77	73	77	78	80	84	85	85	83	80	77	79
Hour 09 (Local Time)	34	72	72	70	64	58	61	69	73	76	81	77	75	71
Hour 15	34	71	67	54	46	40	43	49	50	53	67	76	74	58
Hour 21	34	71	72	66	61	54	54	61	68	72	79	78	75	68
PRECIPITATION (inches):														
Water Equivalent														
-Normal		0.45	0.53	0.36	0.24	0.61	1.60	1.78	1.50	1.38	0.89	0.61	0.92	10.87
-Maximum Monthly	50	1.56	1.65	1.32	0.83	1.68	4.07	3.32	4.19	4.34	2.42	4.11	3.07	4.34
-Year		1943	1968	1972	1992	1989	1962	1975	1947	1951	1965	1956	1955	SEP 1951
-Minimum Monthly	50	T	0.03	T	0.00	T	0.08	0.48	0.14	0.07	0.11	0.01	0.07	0.00
-Year		1974	1987	1987	1954	1950	1946	1976	1982	1969	1967	1975	1972	APR 1954
-Maximum in 24 hrs	49	0.76	0.96	0.78	0.78	1.25	1.51	2.04	2.01	2.06	1.15	2.02	0.99	2.06
-Year		1958	1978	1972	1961	1946	1945	1972	1971	1951	1959	1976	1955	SEP 1951
Snow,Ice pellets,hail														
-Maximum Monthly	50	25.5	26.6	23.4	16.4	5.0	1.0	T	4.0	15.7	22.7	36.2	30.0	36.2
-Year		1989	1990	1944	1992	1945	1986	1993	1955	1992	1965	1956	1955	NOV 1956
-Maximum in 24 hrs	46	14.1	14.1	10.6	14.3	4.0	1.0	T	4.0	12.0	13.0	8.8	12.0	14.3
-Year		1989	1990	1972	1992	1977	1986	1993	1955	1992	1978	1993	1992	APR 1992
WIND:														
Mean Speed (mph)	8	5.1	5.6	6.5	8.6	8.8	8.8	8.2	8.0	7.6	6.3	4.8	3.6	6.8
Prevailing Direction through 1956		N	N	WNW	SE	SE	SE	SE	SE	SE	SE	N	N	SE
Fastest Obs. 1 Min.														
-Direction (!!)	5	04	19	15	18	17	15	15	16	04	19	19	17	04
-Speed (MPH)	5	52	35	35	46	29	38	40	30	31	46	33	40	52
-Year		1971	1970	1970	1970	1972	1971	1971	1970	1972	1970	1969	1969	JAN 1971
Peak Gust														
-Direction (!!)														
-Speed (mph)														
-Date														

PRECIPITATION (inches)

GULKANA, ALASKA

YEAR	JAN	FEB	MAR	APR	MAY	JUNE	JULY	AUG	SEP	OCT	NOV	DEC	ANNUAL
1964	0.31	0.59	0.44	0.61	0.64	2.02	1.72	0.73	0.98	1.40	0.07	1.22	10.73
1965	0.69	0.50	0.41	0.43	0.89	1.70	2.29	1.36	1.59	2.42	0.78	0.42	13.48
1966	0.37	0.39	0.19	0.11	0.99	1.92	1.76	1.07	0.97	0.47	0.87	0.37	9.48
1967	0.61	0.83	0.43	0.03	1.51	0.47	1.33	1.91	1.56	0.11	0.88	0.85	10.52
1968	0.59	1.65	0.17	0.26	1.40	1.52	0.95	1.26	0.66	0.87	0.31	0.11	9.75
1969	0.12	0.08	0.33	T	0.66	0.47	2.40	0.59	0.07	0.21	0.26	0.07	5.26
1970	0.22	0.21	0.13	0.26	1.05	1.91	1.34	1.17	1.73	1.39	0.31	0.50	10.22
1971	0.06	0.20	0.05	0.09	0.10	1.29	1.85	3.92	1.82	1.75	0.73	1.68	13.54
1972	0.80	0.11	1.32	0.27	0.57	1.57	3.05	0.74	0.81	1.18	0.41	0.07	10.90
1973	0.41	0.28	0.03	0.26	0.12	2.42	1.64					0.41	
1974	T	1.54	0.23	T	0.41	1.38	1.99	1.39	1.25	0.87	0.41	1.16	10.63
1975	0.22	0.26	0.05	0.37	0.41	2.56	3.32	1.84	1.28	0.23	0.01	0.30	10.85
1976	0.41	0.15	0.32	0.28	0.52	0.57	0.48	0.90	1.17	0.22	2.87	1.26	9.15
1977	0.42	0.41	0.45	0.45	1.43	2.10	1.22	0.55	1.30	0.48	0.48	0.23	9.52
1978	0.44	1.14	0.07	0.01	0.63	1.20	1.29	0.46	0.47	1.68	0.46	2.57	10.42
1979	0.10	0.37	1.15	0.21	0.37	0.91	1.51	1.61	0.32	0.95	0.61	1.11	9.22
1980	0.73	0.27	0.24	T	0.15	1.73	2.60	1.67	0.77	0.69	0.10	0.62	9.57
1981	0.43	0.14	0.35	T	0.14	2.22	2.17	1.79	0.92	0.57	0.98	1.15	10.86
1982	0.09	0.72	0.93	0.19	0.14	0.48	1.40	0.14	1.57	1.04	0.36	0.09	7.15
1983	0.38	0.26	T	0.67	0.06	0.69	3.02	2.22	2.34	0.25	0.17	0.35	10.41
1984	0.73	0.95	T	0.04	0.05	1.22	2.51	1.95	0.43	0.27	0.99	0.81	9.95
1985	0.20	0.44	0.12	0.77	0.08	3.09	1.99	2.64	2.34	0.33	0.32	1.44	13.76
1986	0.66	0.42	0.80	0.69	0.36	0.38	1.10	1.57	0.48	1.31	0.36	0.37	8.50
1987	0.34	0.03	T	0.09	0.48	2.35	2.19	1.73	1.57	2.29	0.75	1.25	13.07
1988	0.15	0.76	0.63	0.02	1.30	1.51	2.54	2.71	1.26	2.09	0.77	0.82	14.56
1989	1.41	0.15	0.16	0.37	1.68	2.22	0.84	0.69	3.15	0.66	1.06	2.01	14.40
1990	0.44	1.62	0.30	0.01	0.24	0.80	0.66	2.23	3.93	0.53	0.95	1.05	12.76
1991	0.54	0.50	0.29	0.31	0.29	1.86	3.20	1.89	0.89	1.19	1.05	0.66	12.67
1992	0.74	0.79	0.80	0.83	0.06	2.46	1.70	1.80	2.73	0.54	0.58	1.35	14.38
1993	0.61	0.46	0.08	0.03	0.54	1.68	1.07	2.08	1.49	1.40	1.55	0.75	11.74
Record Mean	0.53	0.51	0.35	0.22	0.55	1.44	1.88	1.62	1.61	0.90	0.73	0.86	11.21

AVERAGE TEMPERATURE (deg. F)

GULKANA, ALASKA

YEAR	JAN	FEB	MAR	APR	MAY	JUNE	JULY	AUG	SEP	OCT	NOV	DEC	ANNUAL
1964	-9.3	13.7	4.4	28.5	35.7	54.5	55.4	52.0	44.7	30.4	10.3	-15.3	25.4
1965	-11.1	-6.7	30.3	32.8	41.6	49.6	56.9	52.1	47.8	21.7	-3.4	-8.7	25.2
1966	-20.0	1.5	7.4	31.3	40.7	55.8	57.7	52.2	43.0	22.9	2.1	-9.1	23.8
1967	-11.9	-0.6	14.8	21.5	42.5	56.2	58.0	55.1	43.9	25.2	16.1	-1.1	26.6
1968	-7.9	10.2	19.0	28.1	43.8	51.8	57.1	52.2	41.7	24.5	4.6	-18.9	25.5
1969	-20.5	-1.4	19.1	34.9	45.4	59.1	56.0	47.8	43.6	32.6	4.2	8.9	27.5
1970	-13.2	18.6	25.0	31.7	44.8	51.0	53.9	51.6	39.5	21.1	8.3	-10.8	26.8
1971	-19.5	8.0	9.1	29.3	40.8	53.8	56.8	52.5	41.9	24.3	-0.8	-5.7	24.2
1972	-18.6	-6.7	4.4	18.4	42.1	51.1	59.3	55.3	40.4	26.9	4.7	-8.4	22.4
1973	-15.3	-4.1	18.4	34.7	43.4	52.2	57.0					-5.3	
1974	-16.8	-2.6	10.4	32.4			56.4	54.0	45.7	30.1	10.9	1.2	
1975	-8.3	-3.3	10.3	30.2	44.5	51.8	58.1	52.6	46.2	28.8	0.1	-7.0	25.4
1976	-1.9	-5.0	15.0	32.8	42.3	55.0	57.8		43.3	29.5	21.1	4.5	
1977	16.3	20.5	14.2	32.2	42.5	52.9	58.0	57.6	45.6	31.1	-5.0	-12.8	29.4
1978	-4.5	7.2	16.4	31.0	43.1	50.5			46.1	32.6	10.8	0.3	
1979	-4.1	-17.8	20.8			52.2	57.9	57.7	46.7	34.3	24.1	-9.3	
1980	-4.6	16.2	19.7	36.5	46.1	54.8	57.3	52.1	42.3	31.0	15.3	-24.9	28.5
1981	19.7	14.4	23.2	25.9	48.3	52.1	55.4	52.3	40.1	29.2	7.6	-5.2	30.3
1982	-27.0	1.5	17.0	27.9	42.5	54.6	57.6	53.3	45.5	21.2	3.9	1.5	25.0
1983	-7.7	3.4	12.9	34.7		55.6	57.6	52.5	40.0	28.9	8.6	-2.8	
1984	3.2	4.9	25.9	34.9	44.4	54.8	56.4	54.0	44.3	24.5	2.2	-3.2	28.9
1985	14.7	-5.9	16.5	23.4	42.6	49.1	57.7	51.3	44.2	22.6	-3.6	16.7	27.4
1986	5.7	10.6	11.2	21.3	42.4	52.1	56.6	50.4	43.6	29.1	7.4	15.0	28.8
1987	10.2	11.5	14.0	33.5	44.7	51.9	57.6	53.8	43.7	34.4	15.7	3.8	31.2
1988	-4.5	8.4	22.0	33.3	44.4	54.4	57.7	53.2	42.5	26.1	-0.7	6.2	28.6
1989	-12.7	-4.1	9.6	34.3	46.0	53.3	61.1	58.1	47.0	29.5	-0.6	13.0	27.9
1990	-1.3	-8.4	20.2	35.1	49.0	57.1	59.9	55.5	46.9	23.8	-6.6	-2.3	27.4
1991	-3.9	7.4	18.2	34.4	46.0	55.5	55.2	51.8	46.2	22.4	5.4	3.3	28.5
1992	6.6	1.0	17.7	26.9	41.3	54.9	57.7	52.6	33.0	23.2	14.7	-6.9	26.9
1993	-3.7	9.4	18.4	36.3	49.1	55.2	60.9	55.2	44.8	31.9	12.8	3.3	31.1
Record Mean	-5.7	3.1	15.1	30.2	44.0	53.8	57.3	53.4	43.5	27.4	6.7	-3.1	27.1
Max	2.8	13.7	28.2	41.6	55.2	65.2	68.4	64.5	53.6	35.7	14.2	4.7	37.3
Min	-14.2	-7.5	1.9	18.8	32.8	42.3	46.2	42.2	33.4	19.1	-0.9	-10.9	16.9

REFERENCE NOTES

GULKANA, ALASKA

GENERAL
T - TRACE AMOUNT.
BLANK ENTRIES DENOTE MISSING/UNREPORTED DATA.
# INDICATES A STATION OR INSTRUMENT RELOCATION.

SPECIFIC

(a) - LENGTH OF RECORD IN YEARS, ALTHOUGH INDIVIDUAL MONTHS MAY BE MISSING.
0.* OR * - THE VALUE IS BETWEEN 0.0 AND 0.05
NORMALS - BASED ON THE 1961-1990 RECORD PERIOD.
EXTREMES - DATES ARE THE MOST RECENT OCCURRENCE
WIND DIR.- NUMERALS SHOW TENS OF DEGRESS CLOCKWISE FROM TRUE NORTH. "00" INDICATES CALM.
RESULTANT DIRECTIONS ARE GIVEN TO WHOLE DEGREES.
BOLD VALUES INDICATE EXTREME VALUES WHICH OCCURRED AFTER THE ASOS SYSTEM WAS COMMISSIONED.

RECORD = PERIOD OF RECORD
RECORD MEAN PRECIPITATION IS THE MEAN OF ALL DAILY PRECIPITATION AMOUNTS DURING THE PERIOD OF RECORD.
RECORD MAX(MIN) TEMPERATURE IS THE MEAN OF ALL DAILY MAX(MIN) TEMPERATURES DURING THE PERIOD OF RECORD.
RECORD MEAN TEMPERATURE IS THE SUM OF THE RECORD MAX AND RECORD MIN DIVIDED BY 2.
AVERAGE TEMPERATURE IS THE SUM OF THE MEAN DAILY MAX AND MIN TEMPERATURE DIVIDED BY 2.

EXCEPTIONS

1. DAYS OF PRECIPITATION .01 INCH OR MORE IS 1943-1952 AND 1968 TO DATE.
2. MEAN WIND SPEED IS 1950-1954 AND 1968-1970.
3. PRIOR TO 1968, MAXIMUM 24-HOUR PRECIPITATION AND SNOW ARE FOR CALENDAR DAY (MIDNIGHT TO MIDNIGHT) INSTEAD OF FOR ANY CONSECUTIVE 24 HOURS.
4. THUNDERSTORMS AND HEAVY FOG ARE THROUGH 1972 AND MAY BE INCOMPLETE, DUE TO PART-TIME OPERATIONS.
5. MEAN SKY COVER, AND DAYS CLEAR-PARTLY CLOUDY-CLOUDY ARE THROUGH 1972.
6. FASTEST OBSERVED WIND IS THROUGH JULY 1973.

RECORD MEANS ARE THROUGH THE CURRENT YEAR, BEGINNING IN 1943 FOR TEMPERATURE
1943 FOR PRECIPITATION
1943 FOR SNOWFALL

HEATING DEGREE DAYS Base 65 deg. F — GULKANA, ALASKA

SEASON	JULY	AUG	SEP	OCT	NOV	DEC	JAN	FEB	MAR	APR	MAY	JUNE	TOTAL
1964-65	291	397	602	1065	1636	2496	2360	2013	1066	960	720	456	14062
1965-66	247	394	507	1336	2055	2289	2643	1778	1784	1004	745	272	15054
1966-67	221	387	653	1301	1888	2301	2386	1836	1552	1057	688	257	14527
1967-68	211	301	624	1227	1467	2051	2262	1586	1420	1098	651	392	13290
1968-69	238	388	689	1247	1807	2606	2659	1862	1419	897	600	182	14594
1969-70	269	528	635	998	1824	1739	2427	1294	1233	994	620	414	12975
1970-71	340	410	758	1357	1699	2354	2624	1594	1730	1063	743	330	15002
1971-72	257	381	686	1254	1973	2195	2596	2082	1881	1391	704	410	15810
1972-73	172	295	729	1176	1808	2277	2490	1939	1437	902	664	377	14266
1973-74	240					2180	2543	1892	1692	973			
1974-75	261	331	573	1077	1620	1976	2271	1909	1693	1039	629	389	13768
1975-76	230	378	555	1117	1945	2233	2074	2034	1544	960	697	294	14061
1976-77	222		641	1093	1310	1874	1507	1242	1569	977	689	355	
1977-78	217	221	575	1046	2105	2415	2152	1620	1502	1013	671	429	13966
1978-79			559	995	1624	2009	2140	2324	1362			378	
1979-80	215	218	540	944	1220	2307	2164	1412	1397	849	577	302	12145
1980-81	231	393	675	1049	1483	2797	1398	1413	1289	1168	514	380	12790
1981-82	291	387	740	1100	1722	2177	2858	1782	1483	1107	688	309	14644
1982-83	222	354	577	1351	1833	1971	2253	1725	1609	904		273	
1983-84	224	380	743	1110	1688	2104	1919	1741	1206	894	630	299	12938
1984-85	263	339	615	1248	1883	2115	1552	1984	1497	1241	689	472	13898
1985-86	220	419	617	1309	2058	1495	1832	1519	1665	1307	692	378	13511
1986-87	255	449	636	1108	1726	1547	1696	1493	1578	938	625	386	12437
1987-88	220	339	633	942	1471	1897	2155	1639	1327	945	629	309	12506
1988-89	220	359	667	1199	1970	1819	2412	1936	1714	915	584	345	14140
1989-90	121	208	532	1092	1968	1604	2057	2055	1382	893	491	231	12634
1990-91	167	297	538	1272	2147	2089	2138	1611	1449	909	583	285	13485
1991-92	295	402	559	1316	1785	1911	1811	1858	1461	1135	725	299	13557
1992-93	221	376	954	1285	1503	2233	2130	1552	1439	852	488	289	13322
1993-94	137	296	601	1018	1562	1915							

COOLING DEGREE DAYS Base 65 deg. F — GULKANA, ALASKA

YEAR	JAN	FEB	MAR	APR	MAY	JUNE	JULY	AUG	SEP	OCT	NOV	DEC	TOTAL
1969	0	0	0	0	0	11	0	0	0	0	0	0	11
1970	0	0	0	0	0	0	2	0	0	0	0	0	2
1971	0	0	0	0	0	1	11	0	0	0	0	0	12
1972	0	0	0	0	0	0	5	0	0	0	0	0	5
1973	0	0	0	0	0	0	0					0	
1974	0	0	0	0	0		1	0	0	0	0	0	
1975	0	0	0	0	0	0	22	0	0	0	0	0	22
1976	0	0	0	0	0	0	8		0	0	0	0	
1977	0	0	0	0	0	0	4	1	0	0	0	0	5
1978	0	0	0	0	0	0			0	0	0	0	
1979	0	0	0			0	1	0	0	0	0	0	
1980	0	0	0	0	0	2	1	0	0	0	0	0	3
1981	0	0	0	0	0	0	0	0	0	0	0	0	0
1982	0	0	0	0	0	0	0	0	0	0	0	0	0
1983	0	0	0	0		1	2	0	0	0	0	0	
1984	0	0	0	0	0	0	0	3	0	0	0	0	3
1985	0	0	0	0	0	2	2	0	0	0	0	0	4
1986	0	0	0	0	0	0	0	0	0	0	0	0	0
1987	0	0	0	0	0	0	0	0	0	0	0	0	0
1988	0	0	0	0	0	0	0	0	0	0	0	0	0
1989	0	0	0	0	0	0	8	0	0	0	0	0	8
1990	0	0	0	0	0	1	15	7	0	0	0	0	23
1991	0	0	0	0	0	8	0	0	0	0	0	0	8
1992	0	0	0	0	0	2	5	0	0	0	0	0	7
1993	0	0	0	0	0	0	20	0	0	0	0	0	20

SNOWFALL (inches) — GULKANA, ALASKA

SEASON	JULY	AUG	SEP	OCT	NOV	DEC	JAN	FEB	MAR	APR	MAY	JUNE	TOTAL
1964-65	0.0	0.0	0.0	6.9	0.1	12.4	8.1	7.9	6.2	3.1	0.3	0.0	45.0
1965-66	0.0	0.0	1.5	22.7	9.3	3.7	3.8	3.5	2.3	0.1	T	0.0	46.9
1966-67	0.0	0.0	0.0	4.0	10.1	3.9	6.2	10.7	4.4	0.3	1.5	0.0	41.1
1967-68	0.0	0.0	0.0	3.0	3.7	7.3	6.5	12.6	2.5	3.8	T	0.0	39.4
1968-69	0.0	0.0	0.0	12.6	4.6	2.0	2.4	1.6	5.0	T	T	0.0	28.2
1969-70	0.0	T	0.0	1.1	6.4	3.1	4.9	5.3	2.9	6.3	1.0	0.0	31.0
1970-71	0.0	0.0	6.4	14.5	7.4	6.7	0.7	4.3	1.4	1.5	1.5	0.0	44.4
1971-72	0.0	0.0	T	19.3	12.5	21.3	10.9	1.8	18.6	4.0	0.4	0.0	88.8
1972-73	0.0	0.0	6.0	11.5	6.2	1.0	4.7	4.4	0.3	1.9	0.2	0.0	36.2
1973-74	0.0					4.8	T	19.2	4.1	T	T	0.0	
1974-75	0.0	T	0.0	3.8	7.9	17.7	3.2	4.4	0.3	5.2	T	0.0	42.5
1975-76	0.0	0.0	T	0.5	1.0	7.7	5.3	2.0	8.7	4.0	T	0.0	29.2
1976-77	0.0	0.0	0.0	2.0	8.0	29.3	3.4	1.0	7.0	6.9	4.0	0.0	61.6
1977-78	0.0	0.0	T	2.5	10.0	1.0	6.9	15.0	1.0	T	T	0.0	36.4
1978-79	0.0	0.0	0.0	18.2	8.0	29.4	2.1	10.4	12.9	3.1	1.0	0.0	85.1
1979-80	0.0	0.0	0.0	4.7	5.1	16.5	10.9	4.7	10.8	T	T	0.0	52.7
1980-81	0.0	0.0	T	5.4	1.6	13.0	3.5	2.4	5.1	0.1	0.0	0.0	31.1
1981-82	0.0	0.0	3.3	9.5	12.5	17.2	2.2	3.7	8.4	2.0	T	0.0	58.8
1982-83	0.0	0.0	0.0	12.5	6.5	2.0	7.8	4.2	T	5.8	T	0.0	38.8
1983-84	0.0	0.0	1.0	2.8	1.3	5.5	15.1	20.1	T	1.1	0.0	0.0	46.9
1984-85	0.0	0.0	0.0	4.1	11.1	10.1	3.0	15.5	6.6	15.3	T	0.0	65.7
1985-86	0.0	0.0	1.6	3.4	8.1	8.0	10.5	9.0	10.7	11.3	T	1.0	63.6
1986-87	0.0	0.0	T	3.8	5.5	5.6	4.9	0.5	T	1.5	T	0.0	21.8
1987-88	0.0	0.0	T	2.7	9.7	9.4	1.1	9.3	8.0	0.3	T	0.0	40.5
1988-89	0.0	0.0	0.0	22.4	12.1	10.5	25.5	2.5	1.7	0.7	4.0	0.0	79.4
1989-90	0.0	0.0	T	6.9	14.5	22.3	5.6	26.6	4.0	T	T	0.0	79.9
1990-91	0.0	0.0	0.0	3.6	13.3	13.5	9.0	5.8	6.3	3.1	0.5	T	55.1
1991-92	0.0	0.0	0.0	17.6	17.8	8.9	12.8	9.8	10.6	16.4	T	0.0	93.9
1992-93	0.0	0.0	15.7	10.4	7.1	19.7	11.5	8.0	0.8	T	T	T	73.2
1993-94	T	0.0	1.5	9.8	20.6	9.0							
Record Mean	T	0.1	1.1	7.5	8.8	10.5	7.4	6.9	5.0	2.6	0.5	T	50.6

HOMER, ALASKA

Homer Airport is located at the head of Coal Bay, on the north shore of Kachemak Bay just to the east of its confluence with Cook Inlet. The shoreline to either side of the station curves toward the north, such that water lies within a mile to the south, west and east. To the northwest through northeast the ground has a gradual rise to 500 feet at a distance of 1 1/2 miles and then rises abruptly to 1,000 feet at a distance of 2 miles from the station. The nearest 1,500-foot elevation in this direction is Lookout Mountain, 4 miles to the north and northeast. The nearest 2,000-foot elevation lies about 12 miles to the southeast across Kachemak Bay, in the foothills of the Kenai Mountains. The width of the bay in this direction is approximately 9 miles. Continuing southeastward beyond Kachemak Bay, at a distance of 15 to 20 miles, is the ridgeline of the northeast-southwest oriented Kenai Mountains with elevations of 4,000 to 6,000 feet.

The climate of Homer is marine but with precipitation amounts modified by the Kenai Mountains. The annual precipitation is reduced when air being lifted over mountains leaves most of its moisture on the windward side. For this reason the usual Gulf Coast amount of near 60 inches is reduced to less than half that amount. The relatively low annual snowfall is a reflection of the mild winter temperatures. Often precipitation will begin as snow but turn to rain shortly afterwards. The occurrence of the heaviest monthly amounts during the fall and winter months is the result of the increased frequency of storms into the Western Gulf of Alaska during those months.

Temperatures experienced at Homer are more nearly representative of marine climate than is precipitation. Winters are mild, seldom getting colder than zero, and summers are cool with the maximum temperature seldom going above 70 degrees. The growing season is about 100 days.

Surface winds at the station are seldom strong even in winter. However, a short distance to the southeast, over Kachemak Bay, and to the west over Cook Inlet, wind speeds requiring warnings to small craft are fairly common in winter and summer.

The occurrence of a thunderstorm is rare. Heavy fog is infrequent and of short duration, but patchy ground fog is common in spring and fall.

NORMALS, MEANS, AND EXTREMES

HOMER, ALASKA

LATITUDE: 59°38'N LONGITUDE: 151°30' W ELEVATION: FT. GRND 67 BARO 89 TIME ZONE: YUKON WBAN: 25507

	(a)	JAN	FEB	MAR	APR	MAY	JUNE	JULY	AUG	SEP	OCT	NOV	DEC	YEAR
TEMPERATURE °F:														
Normals														
-Daily Maximum		28.5	31.2	35.9	42.2	50.0	56.3	60.5	60.4	54.7	43.8	34.5	30.1	44.0
-Daily Minimum		16.8	18.1	21.9	28.6	35.6	42.2	46.2	46.2	40.5	31.2	22.6	18.4	30.7
-Monthly		22.7	24.7	29.0	35.4	42.8	49.3	53.4	53.3	47.6	37.5	28.6	24.3	37.4
Extremes														
-Record Highest	50	51	51	53	63	71	80	81	78	68	64	52	50	81
-Year		1961	1954	1974	1965	1993	1953	1993	1944	1965	1954	1986	1969	JUL 1993
-Record Lowest	50	-24	-18	-21	-9	6	29	34	31	20	2	-7	-16	-24
-Year		1989	1947	1971	1944	1949	1950	1959	1955	1956	1975	1990	1964	JAN 1989
NORMAL DEGREE DAYS:														
Heating (base 65°F)		1311	1128	1116	888	688	471	360	363	522	853	1092	1262	10054
Cooling (base 65°F)		0	0	0	0	0	0	0	0	0	0	0	0	0
% OF POSSIBLE SUNSHINE														
MEAN SKY COVER (tenths)														
Sunrise - Sunset	15	6.8	6.9	7.0	7.2	7.7	7.3	7.6	7.1	7.3	7.4	7.1	6.8	7.2
MEAN NUMBER OF DAYS:														
Sunrise to Sunset														
-Clear	15	7.7	6.8	6.1	5.4	3.7	3.9	3.3	5.3	4.8	5.2	5.9	7.4	65.4
-Partly Cloudy	15	4.2	4.4	6.3	6.3	6.8	8.1	8.6	7.7	7.0	5.2	5.3	5.3	75.1
-Cloudy	15	19.1	17.1	18.7	18.3	20.5	17.9	19.1	18.1	18.1	20.6	18.9	18.3	224.7
Precipitation														
.01 inches or more	37	13.5	11.4	11.3	9.2	9.5	9.0	10.9	12.8	15.4	15.1	12.3	15.1	145.5
Snow,Ice pellets,hail														
1.0 inches or more	25	3.6	4.1	2.9	1.4	0.2	0.0	0.0	0.0	0.0	0.8	2.5	4.4	20.0
Thunderstorms	24	0.0	0.0	0.0	0.0	0.*	0.0	0.2	0.*	0.0	0.*	0.0	0.0	0.3
Heavy Fog Visibility														
1/4 mile or less	24	0.6	0.9	0.6	0.4	0.4	0.3	1.0	1.8	0.8	0.3	0.5	0.7	8.3
Temperature °F														
-Maximum														
70° and above	53	0.0	0.0	0.0	0.0	0.*	0.3	0.8	0.8	0.0	0.0	0.0	0.0	1.9
32° and below	53	17.1	12.5	8.3	1.5	0.1	0.0	0.0	0.0	0.0	1.4	10.5	16.5	67.8
-Minimum														
32° and below	53	28.7	25.0	27.4	22.2	9.3	0.5	0.0	0.*	3.7	17.6	25.1	27.7	187.2
0° and below	53	4.0	2.4	1.3	0.*	0.0	0.0	0.0	0.0	0.0	0.0	0.2	2.9	10.9
AVG. STATION PRESS.(mb)	15	1000.3	1002.8	1003.1	1004.4	1007.1	1010.4	1013.1	1010.7	1006.4	1000.8	1000.1	998.8	1004.8
RELATIVE HUMIDITY (%)														
Hour 03	43	79	78	77	80	84	87	90	90	87	82	79	79	83
Hour 09 (Local Time)	44	79	79	75	74	72	74	79	82	83	81	79	79	78
Hour 15	44	76	72	67	65	65	67	71	72	70	69	74	77	70
Hour 21	44	78	77	74	74	73	73	77	82	82	79	79	78	77
PRECIPITATION (inches):														
Water Equivalent														
-Normal		2.40	2.13	1.70	1.28	1.15	1.03	1.49	2.24	3.29	3.24	2.62	2.82	25.39
-Maximum Monthly	58	6.68	5.62	6.02	4.09	2.28	3.37	3.79	5.56	6.18	8.55	8.72	8.01	8.72
-Year		1981	1978	1981	1988	1980	1941	1948	1939	1990	1969	1983	1939	NOV 1983
-Minimum Monthly	58	0.39	0.12	0.03	0.01	0.08	0.09	0.16	0.47	0.83	0.83	0.12	0.00	0.00
-Year		1969	1979	1989	1954	1974	1957	1953	1978	1969	1992	1955	1933	DEC 1933
-Maximum in 24 hrs	57	2.32	1.68	1.83	1.48	1.06	1.22	1.38	1.80	1.66	2.24	3.20	2.12	3.20
-Year		1984	1941	1987	1980	1977	1978	1980	1936	1990	1946	1983	1991	NOV 1983
Snow,Ice pellets,hail														
-Maximum Monthly	58	33.8	46.0	38.1	17.4	6.6	T	0.0	0.0	0.5	21.9	37.4	54.7	54.7
-Year		1962	1974	1976	1972	1971	1963			1962	1976	1945	1979	DEC 1979
-Maximum in 24 hrs	58	24.0	19.2	15.2	7.4	6.0	T	0.0	0.0	0.4	6.0	24.5	20.2	24.5
-Year		1939	1945	1954	1972	1971	1963			1968	1940	1945	1979	NOV 1945
WIND:														
Mean Speed (mph)	19	8.4	8.2	8.0	8.1	8.2	7.9	7.4	6.6	7.2	7.4	8.1	8.1	7.8
Prevailing Direction through 1963		NE	NE	NE	NE	SW	WSW	WSW	WSW	NE	NE	NE	NE	NE
Fastest Obs. 1 Min.														
-Direction (!!)	21	28	10	22	12	11	13	29	28	36	09	08	24	36
-Speed (MPH)	21	39	38	35	31	40	31	25	32	49	37	44	37	49
-Year		1991	1986	1975	1988	1975	1972	1989	1984	1992	1987	1983	1992	SEP 1992
Peak Gust														
-Direction (!!)														
-Speed (mph)														
-Date														

PRECIPITATION (inches)

HOMER, ALASKA

YEAR	JAN	FEB	MAR	APR	MAY	JUNE	JULY	AUG	SEP	OCT	NOV	DEC	ANNUAL
1964	1.02	4.07	0.84	1.22	1.39	0.82	0.95	2.37	1.78	3.27	2.88	0.48	21.09
1965	0.98	0.90	2.50	1.18	1.10	1.75	2.31	2.04	4.51	1.66	1.55	2.62	23.10
1966	1.27	1.08	1.56	0.60	1.47	0.73	2.79	5.25	4.40	2.52	1.32	1.30	24.29
1967	0.94	1.54	0.30	0.73	0.29	1.70	0.86	3.65	5.39	2.39	2.83	3.35	23.97
1968	0.69	1.77	1.75	0.62	1.96	0.39	0.18	1.16	1.80	2.16	1.17	0.69	14.34
1969	0.39	2.34	1.28	0.36	1.22	0.43	1.21	0.98	0.83	8.55	3.81	7.85	29.25
1970	0.42	2.65	2.66	2.05	0.36	0.48	1.25	2.37	1.29	3.69	1.77	1.19	20.18
1971	0.69	3.25	1.37	1.02	1.90	1.17	1.95	1.95	1.59	4.23	1.48	1.69	22.29
1972	0.76	1.03	0.79	1.20	1.03	0.57	0.75	3.41	4.76	3.18	1.89	0.22	19.59
1973	0.84	1.26			2.30	2.03	0.32	2.10	2.82	1.59	1.77	2.01	
1974	0.74	4.30	0.55	1.32	0.08	0.43	1.56	1.94	4.08	3.33	4.56	3.58	26.47
1975	2.50	3.38	1.64	2.17	1.94	0.93	0.99	0.51	3.48	1.90	1.02	4.07	24.53
1976	2.67	0.37	4.04	1.98	0.56	0.65	0.35	2.28	4.95	3.78	6.90	3.86	32.39
1977	4.40	2.98	2.13	1.60	2.19	0.95	0.31	2.42	1.69	3.51	0.39	0.12	22.69
1978	1.88	5.62	0.97	0.63	0.68	2.47	1.57	0.47	2.22	4.24	2.79	5.00	28.54
1979	2.21	0.12	1.75	0.95	0.70	1.10	1.83	1.18	3.45	5.04	8.52	5.26	32.11
1980	4.39	4.40	2.13	3.27	2.28	1.34	2.99	2.82	2.59	4.57	3.04	1.19	35.01
1981	6.68	2.22	6.02	0.10	1.72	0.57	3.73	3.73	2.20	2.36	4.52	4.25	38.10
1982	0.94	0.38	1.44	0.96	0.28	0.93	1.05	1.10	5.19	1.58	1.61	4.13	19.59
1983	1.46	1.57	0.67	1.05	0.54	0.49	1.00	2.20	1.96	2.43	8.72	0.22	22.31
1984	4.81	2.24	1.22	1.59	0.42	0.87	1.02	4.17	3.86	1.21	0.88	1.92	24.21
1985	4.23	1.02	3.69	1.48	1.49	0.85	2.93	2.46	3.06	0.92	0.90	2.36	25.39
1986	3.77	2.20	0.07	0.22	1.09	0.29	2.20	2.93	4.16	5.56	2.54	7.78	32.81
1987	6.26	2.95	2.45	0.77	1.06	1.07	1.08	0.59	5.12	6.87	2.33	1.99	32.54
1988	4.71	3.95	4.34	4.09	0.91	0.89	0.77	3.22	1.29	2.43	1.85	5.51	33.96
1989	1.90	0.27	0.03	1.13	1.12	0.66	2.10	2.80	4.09	4.08	1.20	2.65	22.03
1990	3.07	1.54	0.33	0.40	1.10	1.20	0.64	2.39	6.18	1.66	0.16	1.48	20.15
1991	1.40	1.40	1.95	0.81	1.54	1.30	1.50	2.20	3.40	0.92	1.73	5.87	24.02
1992	4.33	1.15	3.76	0.49	0.33	0.45	1.36	2.63	1.06	0.83	4.50	2.56	23.45
1993	2.13	1.60	2.00	1.01	0.37	0.60	1.19	2.99	3.95	1.69	3.11	4.37	25.01
Record Mean	2.31	1.81	1.56	1.17	1.00	0.94	1.60	2.45	3.05	3.21	2.84	2.74	24.68

AVERAGE TEMPERATURE (deg. F)

HOMER, ALASKA

YEAR	JAN	FEB	MAR	APR	MAY	JUNE	JULY	AUG	SEP	OCT	NOV	DEC	ANNUAL
1964	24.5	25.6	24.0	34.1	39.0	50.1	51.8	52.7	48.7	38.1	28.6	12.2	35.8
1965	19.4	18.9	35.7	36.0	40.4	45.9	51.1	51.1	50.4	33.0	27.0	18.6	35.1
1966	22.6	23.1	19.1	35.6	39.6	49.4	52.5	52.2	48.5	34.7	27.4	20.6	35.4
1967	18.2	23.4	30.0	36.7	44.3	50.2	55.2	55.0	48.3	37.5	33.0	22.7	37.9
1968	18.7	27.4	30.4	34.2	43.8	49.6	53.4	54.3	45.3	35.3	31.3	15.2	36.6
1969	13.2	22.3	31.4	38.3	43.8	50.9	53.3	50.7	47.1	42.1	27.3	33.4	37.8
1970	16.3	33.3	35.3	35.0	43.6	48.9	51.3	51.3	45.1	34.3	32.9	19.7	37.3
1971	9.7	24.4	18.2	32.9	38.4	46.6	51.6	53.2	45.9	36.8	25.9	23.0	33.9
1972	15.2	19.1	16.0	27.1	40.0	46.6	52.1	53.4	46.5	37.7	30.0	21.7	33.8
1973	11.8	21.8	29.6		41.5	48.2	52.4	51.3	45.8	35.9	22.9	24.6	
1974	14.1	19.5	26.0	36.5	42.7	49.4	52.9	53.9	49.7	38.4	29.7	24.1	36.4
1975	18.6	18.7	26.0	32.7	42.1	47.9	52.9	52.3	47.9	36.9	22.2	19.1	34.8
1976	22.2	20.9	26.6	35.0	41.7	48.0	53.8	52.7	46.7	36.6	35.5	31.4	37.6
1977	37.1	35.6	27.0	35.1	41.8	51.2	54.3	55.9	48.5	38.6	21.1	17.2	38.6
1978	29.4	30.6	31.9	38.1	44.3	49.0	53.0	54.8	47.4	40.2	32.0	29.7	40.1
1979	30.7	15.6	32.9	38.4	43.8	48.7	55.0	54.6	50.7	42.5	36.0	18.6	38.9
1980	18.8	31.5	32.1	39.2	44.2	50.0	53.1	51.3	48.0	39.8	32.0	12.5	37.7
1981	35.5	28.9	36.2	37.3	46.9	49.3	54.5	53.7	47.5	39.4	30.2	25.2	40.4
1982	21.2	21.0	30.6	33.4	41.9	49.0	52.7	52.6	48.2	32.8	32.4	32.3	37.3
1983	23.3	30.3	36.1	38.4	46.1	51.8	54.6	54.1	46.8	38.8	34.6	29.2	40.4
1984	26.7	22.6	38.0	36.5	43.9	52.3	54.6	54.9	49.1	39.6	30.3	30.7	39.9
1985	36.0	20.4	30.0	27.9	41.4	46.9	53.5	51.8	47.1	32.8	24.9	35.1	37.3
1986	29.8	26.9	26.8	32.1	42.1	50.0	53.1	52.5	47.7	40.4	31.7	33.8	38.9
1987	29.6	32.1	31.6	36.7	44.3	50.1	54.7	54.9	46.4	39.7	31.2	24.7	39.7
1988	27.8	30.5	33.5	36.6	45.0	51.2	55.3	53.9	46.0	37.5	26.4	27.4	39.3
1989	11.0	25.8	27.4	37.9	43.5	50.2	55.2	56.7	50.9	38.1	25.8	30.4	37.7
1990	22.6	14.1	32.4	39.9	46.7	51.8	54.7	55.2	48.9	37.1	20.4	25.5	37.4
1991	24.3	24.9	29.6	38.6	43.9	50.3	54.1	52.7	50.3	38.9	32.5	27.3	39.0
1992	27.9	20.7	28.2	36.6	43.3	52.3	55.5	53.9	42.8	36.2	31.7	23.5	37.7
1993	24.2	27.3	33.5	40.5	48.0	51.1	56.5	55.8	49.1	40.8	32.5	32.4	41.0
Record Mean	22.5	24.5	28.4	35.3	42.7	49.2	53.1	53.0	47.3	37.6	28.9	23.7	37.2
Max	28.7	31.3	35.5	42.4	50.1	56.6	60.5	60.3	54.5	44.2	34.9	29.6	44.0
Min	16.3	17.8	21.2	28.1	35.3	41.7.	45.7	45.7	40.0	31.0	22.9	17.7	30.3

REFERENCE NOTES

HOMER, ALASKA

GENERAL
T - TRACE AMOUNT.
BLANK ENTRIES DENOTE MISSING/UNREPORTED DATA.
# INDICATES A STATION OR INSTRUMENT RELOCATION.

SPECIFIC

(a) - LENGTH OF RECORD IN YEARS, ALTHOUGH INDIVIDUAL MONTHS MAY BE MISSING.
0.* OR * - THE VALUE IS BETWEEN 0.0 AND 0.05
NORMALS - BASED ON THE 1961-1990 RECORD PERIOD.
EXTREMES - DATES ARE THE MOST RECENT OCCURRENCE.
WIND DIR.- NUMERALS SHOW TENS OF DEGRESS CLOCKWISE FROM TRUE NORTH. "00" INDICATES CALM.
RESULTANT DIRECTIONS ARE GIVEN TO WHOLE DEGREES.
BOLD VALUES INDICATE EXTREME VALUES WHICH OCCURRED AFTER THE ASOS SYSTEM WAS COMMISSIONED.

RECORD = PERIOD OF RECORD
RECORD MEAN PRECIPITATION IS THE MEAN OF ALL DAILY PRECIPITATION AMOUNTS DURING THE PERIOD OF RECORD.
RECORD MAX(MIN) TEMPERATURE IS THE MEAN OF ALL DAILY MAX(MIN) TEMPERATURES DURING THE PERIOD OF RECORD.
RECORD MEAN TEMPERATURE IS THE SUM OF THE RECORD MAX AND RECORD MIN DIVIDED BY 2.
AVERAGE TEMPERATURE IS THE SUM OF THE MEAN DAILY MAX AND MIN TEMPERATURE DIVIDED BY 2.

EXCEPTIONS

1. MAXIMUM PRECIPITATION IS FOR CALENDAR DAY (MIDNIGHT TO MIDNIGHT) INSTEAD OF FOR ANY CONSECUTIVE 24 HOURS. PRIOR TO 1968 AND BASED ON 6 HOURLY DATA THEREAFTER.
2. MEAN WIND SPEED IS APRIL 1950-1954 AND 1968 TO DATE.
3. DAYS OF PRECIPITATION .01 INCH OR MORE ARE 1940-1951 AND 1968 TO DATE.
4. THUNDERSTORMS AND HEAVY FOG MAY BE INCOMPLETE, DUE TO PART-TIME OPERATIONS.

RECORD MEANS ARE THROUGH THE CURRENT YEAR, BEGINNING IN 1943 FOR TEMPERATURE
1943 FOR PRECIPITATION
1943 FOR SNOWFALL

HEATING DEGREE DAYS Base 65 deg. F HOMER, ALASKA

SEASON	JULY	AUG	SEP	OCT	NOV	DEC	JAN	FEB	MAR	APR	MAY	JUNE	TOTAL
1964-65	401	374	482	825	1087	1631	1409	1284	901	861	754	568	10577
1965-66	424	427	430	987	1133	1431	1310	1166	1417	876	780	459	10840
1966-67	380	392	491	932	1118	1368	1444	1157	1077	843	638	438	10278
1967-68	296	302	496	844	953	1307	1423	1085	1066	920	649	452	9793
1968-69	356	327	586	915	1004	1535	1601	1189	1033	796	652	416	10410
1969-70	355	439	530	704	1123	972	1500	884	912	894	658	475	9446
1970-71	419	418	593	943	960	1396	1711	1132	1447	956	820	545	11340
1971-72	409	360	568	866	1166	1297	1538	1324	1509	1128	764	542	11471
1972-73	396	351	546	841	1045	1332	1645	1201			720	498	
1973-74	385	416	568	894	1258	1247	1571	1269	1201	848	686	464	10807
1974-75	368	336	455	817	1052	1261	1439	1293	1199	960	703	506	10389
1975-76	367	387	504	863	1274	1413	1321	1271	1184	892	716	502	10694
1976-77	342	376	541	875	880	1033	856	816	1171	890	712	409	8901
1977-78	322	273	490	813	1311	1476	1097	957	1022	800	635	472	9668
1978-79	366	310	522	761	984	1088	1056	1375	987	792	651	481	9373
1979-80	302	316	422	692	861	1432	1426	968	1014	768	638	446	9285
1980-81	362	415	507	773	984	1617	906	1003	885	822	554	485	9293
1981-82	319	339	516	787	1040	1228	1350	1230	1057	941	710	471	9988
1982-83	373	377	499	990	972	1006	1286	966	892	789	577	389	9116
1983-84	317	331	538	801	904	1103	1177	1223	828	851	645	376	9094
1984-85	315	303	469	779	1037	1056	892	1247	1078	1106	727	538	9547
1985-86	349	400	531	994	1199	921	1082	1064	1177	980	704	447	9848
1986-87	361	383	511	758	995	958	1090	914	1028	841	633	441	8913
1987-88	312	306	548	778	1007	1240	1146	997	969	845	613	406	9167
1988-89	292	334	566	846	1150	1160	1672	1089	1159	808	659	437	10172
1989-90	299	251	418	828	1170	1069	1306	1417	1005	744	559	390	9456
1990-91	312	297	476	856	1331	1218	1254	1116	1093	785	647	433	9818
1991-92	329	372	434	801	969	1166	1141	1277	1137	846	666	371	9509
1992-93	287	339	657	887	990	1281	1259	1051	971	729	522	411	9384
1993-94	258	279	472	741	966	1002							

COOLING DEGREE DAYS Base 65 deg. F HOMER, ALASKA

YEAR	JAN	FEB	MAR	APR	MAY	JUNE	JULY	AUG	SEP	OCT	NOV	DEC	TOTAL
1969	0	0	0	0	0	0	0	0	0	0	0	0	0
1970	0	0	0	0	0	0	0	0	0	0	0	0	0
1971	0	0	0	0	0	0	0	1	0	0	0	0	1
1972	0	0	0	0	0	0	0	0	0	0	0	0	0
1973	0	0			0	0	0	0	0	0	0	0	
1974	0	0	0	0	0	0	0	0	0	0	0	0	0
1975	0	0	0	0	0	0	0	0	0	0	0	0	0
1976	0	0	0	0	0	0	0	0	0	0	0	0	0
1977	0	0	0	0	0	0	0	0	0	0	0	0	0
1978	0	0	0	0	0	0	0	0	0	0	0	0	0
1979	0	0	0	0	0	0	0	0	0	0	0	0	0
1980	0	0	0	0	0	0	0	0	0	0	0	0	0
1981	0	0	0	0	0	0	0	0	0	0	0	0	0
1982	0	0	0	0	0	0	0	0	0	0	0	0	0
1983	0	0	0	0	0	0	0	0	0	0	0	0	0
1984	0	0	0	0	0	0	0	0	0	0	0	0	0
1985	0	0	0	0	0	0	0	0	0	0	0	0	0
1986	0	0	0	0	0	0	0	0	0	0	0	0	0
1987	0	0	0	0	0	0	0	0	0	0	0	0	0
1988	0	0	0	0	0	0	0	0	0	0	0	0	0
1989	0	0	0	0	0	0	0	1	0	0	0	0	1
1990	0	0	0	0	0	0	0	0	0	0	0	0	0
1991	0	0	0	0	0	0	0	0	0	0	0	0	0
1992	0	0	0	0	0	0	0	0	0	0	0	0	0
1993	0	0	0	0	0	0	0	0	0	0	0	0	0

SNOWFALL (inches) HOMER, ALASKA

SEASON	JULY	AUG	SEP	OCT	NOV	DEC	JAN	FEB	MAR	APR	MAY	JUNE	TOTAL
1964-65	0.0	0.0	0.0	1.6	10.3	6.4	4.4	17.2	3.1	1.5	0.8	0.0	45.3
1965-66	0.0	0.0	T	3.7	11.4	12.0	6.1	14.8	13.0	2.0	T	0.0	63.0
1966-67	0.0	0.0	0.0	2.5	2.7	15.0	8.1	11.8	T	2.0	0.0	0.0	42.1
1967-68	0.0	0.0	0.0	4.9	0.8	8.3	4.6	4.5	14.8	4.0	1.9	0.0	43.8
1968-69	0.0	0.0	0.4	2.0	5.6	19.3	6.7	28.0	9.8	0.3	0.0	0.0	72.1
1969-70	0.0	0.0	0.0	0.0	8.8	24.1	6.3	5.5	5.0	1.0	0.0	0.0	50.7
1970-71	0.0	0.0	0.0	3.6	0.7	16.4	8.6	21.2	28.6	10.8	6.6	0.0	96.5
1971-72	0.0	0.0	0.0	1.7	16.9	17.1	15.8	12.2	12.5	17.4	1.6	0.0	95.2
1972-73	0.0	0.0	0.0	T	6.5	1.9	9.6	5.3			0.0	0.0	
1973-74	0.0	0.0	0.0	T	20.5	6.0	8.9	46.0	6.1	4.5	0.0	0.0	92.0
1974-75	0.0	0.0	0.0	T	18.0	38.3	16.3	18.9	17.1	9.6	1.0	0.0	119.2
1975-76	0.0	0.0	0.0	0.0	4.4	16.5	11.3	1.2	38.1	13.0	0.0	0.0	84.5
1976-77	0.0	0.0	0.0	21.9	16.5	14.1	T	T	18.8	4.4	0.5	0.0	76.2
1977-78	0.0	0.0	0.2	3.2	4.6	0.4	9.7	27.9	6.7	T	0.0	0.0	52.7
1978-79	0.0	0.0	0.0	T	1.4	20.3	11.4	T	5.3	3.0	0.0	0.0	41.4
1979-80	0.0	0.0	0.0	1.2	18.8	54.7	13.1	21.1	3.1	T	0.0	0.0	112.0
1980-81	0.0	0.0	0.0	4.9	2.8	8.5	11.1	8.3	6.9	0.1	0.0	0.0	42.6
1981-82	0.0	0.0	T	7.9	13.8	10.7	6.8	1.9	2.9	3.2	T	0.0	47.2
1982-83	0.0	0.0	0.0	1.7	5.0	1.3	11.0	1.1	T	1.0	0.0	0.0	21.1
1983-84	0.0	0.0	0.0	7.7	13.7	2.9	8.3	32.2	0.6	7.1	T	0.0	72.5
1984-85	0.0	0.0	0.0	0.0	0.9	5.5	3.6	13.1	29.8	9.9	0.4	0.0	63.2
1985-86	0.0	0.0	0.0	3.7	6.1	1.7	15.5	6.8	0.5	0.7	T	0.0	35.0
1986-87	0.0	0.0	0.0	0.0	13.8	8.2	11.0	6.3	T	0.8	0.0	0.0	40.1
1987-88	0.0	0.0	0.0	0.0	6.5	15.3	8.2	9.8	4.5	T	0.0	0.0	44.3
1988-89	0.0	0.0	0.0	0.2	11.4	14.5	9.0	T	T	T	T	0.0	35.1
1989-90	0.0	0.0	T	0.6	3.6	10.6	17.7	17.8	1.0	T	0.0	0.0	51.3
1990-91	0.0	0.0	0.0	1.2	2.4	5.2	5.3	13.7	17.9	0.7	T	0.0	46.4
1991-92	0.0	0.0	0.0	1.0	7.9	30.7	12.0	19.8	19.5	T	0.0	0.0	90.9
1992-93	0.0	0.0	T	0.3	9.6	14.7	9.3	6.0	5.1	0.0	0.0	0.0	45.0
1993-94	0.0	0.0	0.0	T	0.6	9.7							
Record Mean	0.0	0.0	T	2.1	7.3	13.0	10.3	12.1	9.4	3.2	0.4	T	57.9

JUNEAU, ALASKA

Juneau lies well within the area of maritime influences which prevail over the coastal areas of southeastern Alaska, and is in the path of most storms that cross the Gulf of Alaska. Consequently, the area has little sunshine, generally moderate temperatures, and abundant precipitation. In contrast with the characteristic lack of sunshine there are greatly appreciated intervals, sometimes lasting for several days at a stretch, during which clear skies prevail. The rugged terrain exerts a fundamental influence upon local temperatures and the distribution of precipitation, creating considerable variations in both weather elements within relatively short distances.

Temperature variations, both daily and seasonal, are usually confined to relatively narrow limits by the dominant maritime influences. There are, however, periods of comparatively severe cold, which usually start with strong northerly winds, and are most often caused by the flow of cold air from northwestern Canada through nearby mountain passes and over the Juneau ice field. These are generally of brief duration. During such periods strong, gusty winds, known locally as Taku Winds, often occur especially in downtown Juneau, Douglas, and other local areas, but generally they are not felt in the Mendenhall Valley. At times these are strong enough to cause considerable damage. During periods of calm or light winds, temperature differences within short distances are frequently very pronounced. Variations in local sunlight and air drainage patterns produce wide differences in temperatures particularly between upland or sloping areas and areas of low, flat terrain. Juneau International Airport, located on low, flat terrain formed by the Mendenhall River delta, and in the path of drainage air from the Mendenhall Glacier, averages about 10 days a year with minimum readings below zero. Downtown Juneau, located on a sloping portion of a rugged mountain area, experiences on the average only about one day each year with minimum readings below zero. At the airport the growing season averages 146 days, from May 4 to September 28, while the downtown average is 181 days, from April 22 to October 21.

The months of February to June mark the period of lightest precipitation, with monthly averages of about 3 inches. After June the monthly amounts increase gradually, reaching an average of 7.71 inches in October. Due to the rugged topography, precipitation throughout the year tends to vary greatly within short distances. At the Juneau Airport, yearly precipitation is 53 inches while downtown, only 8 miles away, it is 93 inches. The maximum yearly amount received in the city is almost double the maximum received at the airport.

Although a trace of snow has fallen as early as September 9, first falls usually occur in the latter part of October, and sometimes not until the first part of December. On the average there is very little accumulation on the ground at low levels until the last of November, although at higher elevations, and particularly on mountain tops, a cover is usually established in early October. Snow accumulation usually reaches its greatest depth during the middle of February. Individual storms may produce heavy falls as late as the first half of May. However, snow cover is usually gone before the middle of April. Ice accumulations due to alternating thawing and freezing of snow or due to freezing precipitation are frequent problems in the Juneau area during the winter months.

NORMALS, MEANS, AND EXTREMES

JUNEAU, ALASKA

LATITUDE: 58°22'N LONGITUDE: 134°35' W ELEVATION: FT. GRND 12 BARO 20 TIME ZONE: YUKON WBAN: 25309

	(a)	JAN	FEB	MAR	APR	MAY	JUNE	JULY	AUG	SEP	OCT	NOV	DEC	YEAR
TEMPERATURE °F:														
Normals														
-Daily Maximum		29.4	34.1	38.7	47.2	55.1	60.9	63.9	62.7	55.9	47.1	36.7	31.6	46.9
-Daily Minimum		19.0	22.7	26.7	32.1	38.9	45.0	48.1	47.3	42.9	37.2	27.2	22.6	34.1
-Monthly		24.2	28.4	32.7	39.7	47.0	53.0	56.0	55.0	49.4	42.2	32.0	27.1	40.6
Extremes														
-Record Highest	49	57	57	59	71	82	86	90	83	72	61	56	54	90
-Year		1958	1992	1981	1989	1947	1969	1975	1977	1989	1987	1949	1944	JUL 1975
-Record Lowest	49	-22	-22	-15	6	25	31	36	27	23	11	-5	-21	-22
-Year		1972	1968	1972	1963	1972	1971	1950	1948	1972	1984	1966	1949	JAN 1972
NORMAL DEGREE DAYS:														
Heating (base 65°F)		1265	1025	1001	759	558	360	279	310	468	707	990	1175	8897
Cooling (base 65°F)		0	0	0	0	0	0	0	0	0	0	0	0	0
% OF POSSIBLE SUNSHINE	33	32	32	37	39	39	34	31	32	26	19	23	20	30
MEAN SKY COVER (tenths)														
Sunrise - Sunset	43	7.8	8.0	8.1	8.1	8.0	8.1	8.3	8.0	8.5	8.8	8.5	8.5	8.2
MEAN NUMBER OF DAYS:														
Sunrise to Sunset														
-Clear	43	5.6	4.2	4.3	3.5	3.7	3.5	3.0	3.9	2.7	2.3	3.3	3.4	43.3
-Partly Cloudy	43	2.7	2.9	3.2	4.1	4.2	4.1	4.6	4.7	3.2	2.1	2.2	2.0	39.8
-Cloudy	43	22.7	21.2	23.5	22.4	23.1	22.4	23.4	22.4	24.1	26.6	24.5	25.6	282.1
Precipitation														
.01 inches or more	49	18.4	16.7	17.8	16.9	17.0	15.4	16.6	17.6	20.4	23.5	20.1	21.0	221.5
Snow,Ice pellets,hail														
1.0 inches or more	49	7.0	5.3	4.1	0.9	0.0	0.0	0.0	0.0	0.0	0.3	3.5	6.0	27.0
Thunderstorms	49	0.*	0.0	0.0	0.0	0.0	0.1	0.1	0.*	0.1	0.0	0.*	0.0	0.3
Heavy Fog Visibility 1/4 mile or less	49	2.0	2.4	1.8	1.0	0.8	0.3	0.2	1.1	2.8	3.1	3.2	2.4	20.9
Temperature °F														
-Maximum														
70° and above	49	0.0	0.0	0.0	0.*	1.3	5.1	7.2	5.6	0.2	0.0	0.0	0.0	19.4
32° and below	49	15.9	9.3	3.8	0.2	0.0	0.0	0.0	0.0	0.0	0.2	6.4	12.2	48.0
-Minimum														
32° and below	49	25.4	22.3	23.0	14.7	3.7	0.1	0.0	0.1	1.5	7.8	18.1	23.4	140.1
0° and below	49	4.2	1.7	0.5	0.0	0.0	0.0	0.0	0.0	0.0	0.0	0.2	2.0	8.6
AVG. STATION PRESS.(mb)	12	1011.0	1011.4	1009.2	1012.2	1013.5	1014.5	1015.8	1014.7	1012.7	1008.0	1008.2	1008.9	1011.7
RELATIVE HUMIDITY (%)														
Hour 03	27	81	83	84	87	87	86	87	90	93	89	86	85	87
Hour 09 (Local Time)	27	81	83	82	80	79	79	83	87	91	89	86	85	84
Hour 15	27	78	75	68	63	63	65	70	74	78	79	81	83	73
Hour 21	27	81	82	79	76	72	72	75	81	87	88	86	84	80
PRECIPITATION (inches):														
Water Equivalent														
-Normal		4.54	3.75	3.28	2.77	3.42	3.15	4.16	5.32	6.73	7.84	4.91	4.44	54.31
-Maximum Monthly	49	9.11	8.48	6.37	5.32	9.20	6.02	7.88	12.31	15.14	15.25	11.22	9.89	15.25
-Year		1993	1964	1992	1980	1992	1987	1969	1961	1991	1974	1956	1956	OCT 1974
-Minimum Monthly	49	0.94	0.07	0.59	0.27	1.25	1.08	1.15	0.56	2.34	2.71	1.15	0.49	0.07
-Year		1969	1989	1983	1948	1946	1950	1972	1979	1965	1950	1983	1983	FEB 1989
-Maximum in 24 hrs	49	2.74	2.71	1.81	2.03	2.30	1.92	1.92	2.62	3.17	4.66	3.40	3.56	4.66
-Year		1948	1993	1992	1991	1992	1953	1984	1974	1952	1946	1988	1956	OCT 1946
Snow,Ice pellets,hail														
-Maximum Monthly	49	69.2	86.3	52.6	46.3	1.2	T	0.0	0.0	T	15.6	48.8	54.7	86.3
-Year		1982	1965	1948	1963	1964	1970			1974	1956	1990	1964	FEB 1965
-Maximum in 24 hrs	49	20.1	23.7	31.0	24.2	0.7	T	0.0	0.0	T	8.8	16.5	25.6	31.0
-Year		1975	1949	1948	1963	1945	1970			1974	1956	1963	1962	MAR 1948
WIND:														
Mean Speed (mph)	48	8.3	8.5	8.5	8.6	8.3	7.7	7.5	7.5	8.0	9.5	8.5	9.0	8.3
Prevailing Direction through 1963		ESE	ESE	ESE	ESE	ESE	N	N	N	N	ESE	ESE	ESE	ESE
Fastest Obs. 1 Min.														
-Direction (!!)	23	12	13	11	11	12	13	12	12		12	11	11	
-Speed (MPH)	23	40	44	40	40	40	32	32	38	120	49	58	55	120
-Year		1992	1970	1963	1962	1965	1965	1970	1992	1993	1965	1968	1963	SEP 1993
Peak Gust														
-Direction (!!)														
-Speed (mph)														
-Date														

PRECIPITATION (inches)

JUNEAU, ALASKA

YEAR	JAN	FEB	MAR	APR	MAY	JUNE	JULY	AUG	SEP	OCT	NOV	DEC	ANNUAL
1964	3.19	8.48	4.38	4.04	4.35	3.37	6.94	3.48	2.59	7.35	4.89	5.22	58.28
1965	7.75	5.10	1.66	3.33	4.45	3.11	2.26	4.17	2.34	7.99	1.46	4.26	47.88
1966	4.34	3.13	6.36	2.08	6.33	1.74	3.91	6.37	8.20	6.97	4.39	4.48	58.30
1967	4.04	4.74	1.34	1.12	2.94	2.87	4.26	5.46	8.53	5.71	5.81	3.25	50.07
1968	3.25	5.30	3.85	3.25	1.45	1.95	4.60	2.39	10.14	4.60	5.34	1.90	48.02
1969	0.94	0.68	4.17	1.74	3.38	2.41	7.88	7.54	5.44	3.77	8.69	4.36	51.00
1970	2.37	3.35	4.08	3.69	3.92	2.97	5.01	7.47	9.86	5.87	2.01	2.58	53.18
1971	5.56	3.93	3.33	2.44	4.30	1.74	1.67	6.89	5.36	5.80	4.38	3.23	48.63
1972	3.73	2.71	4.19	3.62	4.03	3.98	1.15	8.62	6.24	8.49	3.35	3.56	53.67
1973	4.37	3.94	3.01	2.41	4.09	2.80	3.65	6.64	4.95	6.07	1.63	2.30	45.86
1974	2.37	6.23	1.15	2.59	1.66	4.92	3.12	5.78	5.96	15.25	7.79	7.03	63.85
1975	4.10	3.76	2.17	3.04	3.59	2.48	4.96	2.78	7.25	3.55	2.83	5.81	46.32
1976	8.19	4.82	3.61	2.14	3.42	3.37	2.48	3.16	8.32	6.19	5.15	5.56	56.41
1977	4.59	4.56	3.31	4.02	1.56	3.47	3.19	3.03	5.57	7.14	4.58	2.16	47.18
1978	1.71	1.50	1.84	2.19	2.86	3.18	3.98	4.39	3.07	13.00	3.90	4.46	46.08
1979	2.19	0.91	3.98	0.98	2.45	2.74	5.44	0.56	4.89	9.06	8.36	7.73	49.29
1980	3.44	2.83	2.75	5.32	2.53	4.37	6.49	5.61	7.91	11.26	7.10	2.27	61.88
1981	4.66	2.57	1.88	2.11	3.27	2.44	4.25	6.19	11.61	6.18	6.93	2.24	54.33
1982	3.74	1.42	2.52	2.44	5.10	1.86	1.73	5.97	5.10	7.97	2.10	1.17	41.12
1983	4.00	1.69	0.59	2.53	5.37	2.69	3.16	9.52	6.13	4.24	1.15	0.49	41.56
1984	6.06	5.40	3.75	2.11	1.84	4.17	6.92	6.26	3.39	6.69			46.59
1985												8.33	8.33
1986	7.00	3.25	6.08	2.98	2.54	2.76	2.38	6.89	2.40	12.33	5.96	6.42	60.99
1987	3.99	3.13	2.12	2.08	2.60	6.02	2.54	4.54	8.92	10.36	7.17	5.32	58.79
1988	2.58	6.55	4.15	2.25	3.91	2.05	5.21	5.53	5.46	9.71	8.62	4.75	60.77
1989	6.77	0.07	1.33	0.87	3.44	1.10	3.81	2.82	7.29	6.37	6.23	6.78	46.88
1990	3.72	4.54	4.86	1.06	1.72	3.32	4.65	5.35	10.63	6.59	4.89	6.03	57.36
1991	4.16	6.55	4.41	4.73	4.72	3.41	4.85	9.60	15.14	8.63	9.63	9.32	85.15
1992	8.69	7.24	6.37	3.63	9.20	2.98	5.18	5.02	11.45	5.90	7.91	5.73	79.30
1993	9.11	8.09	3.50	1.94	2.19	4.92	2.25	3.20	8.44	9.00	11.06	7.89	71.59
Record Mean	4.23	3.66	3.44	2.79	3.46	3.06	4.24	5.23	6.97	7.89	5.62	4.63	55.21

AVERAGE TEMPERATURE (deg. F)

JUNEAU, ALASKA

YEAR	JAN	FEB	MAR	APR	MAY	JUNE	JULY	AUG	SEP	OCT	NOV	DEC	ANNUAL
1964	29.3	35.9	28.8	38.6	45.1	53.6	54.3	53.6	49.7	45.0	32.1	16.6	40.2
1965	23.1	23.6	35.0	37.7	42.4	48.5	55.4	55.1	50.3	43.7	29.9	26.6	39.3
1966	8.6	25.9	30.2	38.1	43.3	52.9	56.2	52.4	49.2	38.6	25.1	26.5	37.2
1967	23.1	30.4	24.1	37.3	45.6	54.8	53.8	55.1	50.0	43.1	32.3	27.3	39.7
1968	18.7	28.8	33.1	37.9	48.4	52.2	56.8	55.7	47.6	39.0	34.1	19.4	39.3
1969	6.8	21.2	30.7	40.6	49.9	57.8	53.7	50.3	47.4	41.4	32.3	35.1	38.9
1970	22.0	35.1	36.5	39.1	45.3	50.7	52.3	51.7	46.3	39.6	27.8	21.6	39.0
1971	13.0	28.1	28.9	38.6	43.5	53.2	57.5	55.4	48.1	38.5	31.4	20.9	38.1
1972	15.8	19.5	26.5	34.6	44.8	50.4	58.0	55.4	47.0	38.7	34.5	23.6	37.4
1973	18.9	24.6	32.9	39.6	45.9	51.3	53.7	51.8	48.0	41.2	23.0	28.0	38.2
1974	14.8	28.8	24.7	39.3	46.8	50.2	53.5	54.7	50.1	42.5	36.4	33.8	39.6
1975	23.1	24.5	30.6	38.3	47.4	51.4	55.9	53.9	51.4	41.8	28.5	24.7	39.3
1976	28.3	25.8	32.4	41.1	45.5	52.0	55.7	55.9	50.6	42.8	40.9	34.4	42.1
1977	35.0	40.1	36.0	42.3	47.7	54.1	57.0	58.5	50.5	42.4	29.4	18.9	42.7
1978	25.1	31.8	33.9	42.0	47.8	54.2	55.2	56.3	50.6	45.1	30.3	28.3	41.7
1979	20.6	11.0	35.6	41.1	47.7	52.2	56.7	58.2	51.0	45.2	37.2	26.5	40.2
1980	19.5	33.8	34.1	42.2	49.4	55.6	55.6	54.7	49.0	44.6	38.7	21.7	41.6
1981	37.6	32.7	39.4	39.1	52.1	54.3	56.1	55.9	49.2	42.8	36.8	26.9	43.6
1982	13.8	21.3	31.8	37.1	45.4	56.3	57.7	54.8	50.3	42.2	30.6	31.6	39.4
1983	30.2	31.8	34.8	42.6	49.7	55.6	57.1	54.6	48.0	42.1	31.8	18.9	41.4
1984	32.0		39.6	43.1	49.1	53.5	55.5	55.3	50.0	40.5			45.5
1985												32.5	32.5
1986	34.3	28.7	35.3	37.3	46.6	54.3	56.8	54.4	50.7	45.8	30.5	36.0	42.6
1987	33.1	34.7	31.8	41.6	47.8	51.6	58.6	57.5	50.1	44.4	40.0	34.3	43.8
1988	27.0	32.6	37.6	41.7	48.5	54.0	53.8	53.9	48.2	44.1	36.2	31.3	42.4
1989	25.5	23.9	29.4	42.7	49.0	55.4	60.1	58.2	52.4	41.7	32.9	36.0	42.3
1990	26.4	25.1	36.4	42.7	49.7	55.0	59.3	58.1	51.2	40.7	26.4	23.7	41.2
1991	24.9	35.3	33.1	41.7	48.2	55.0	55.2	55.0	50.3	40.1	36.3	33.6	42.4
1992	35.7	32.7	36.9	41.1	46.4	55.2	57.0	56.0	46.8	40.5	38.4	25.6	42.7
1993	24.5	30.2	35.8	44.4	52.1	55.8	59.6	57.1	51.2	45.8	36.8	35.9	44.1
Record Mean	24.1	28.0	32.5	39.5	47.1	53.3	56.0	54.8	49.4	42.0	32.8	27.7	40.5
Max	29.5	33.7	38.6	46.9	55.2	61.6	63.9	62.6	55.9	47.0	37.4	32.2	47.0
Min	18.8	22.3	26.4	32.0	39.0	45.0	48.0	47.0	42.8	37.0	28.2	23.1	34.1

REFERENCE NOTES

JUNEAU, ALASKA

GENERAL
T - TRACE AMOUNT.
BLANK ENTRIES DENOTE MISSING/UNREPORTED DATA.
# INDICATES A STATION OR INSTRUMENT RELOCATION.

SPECIFIC

PM - INCLUDES LAST DAY OF PREVIOUS MONTH
ASOS - AUTOMATED SURFACE OBSERVING SYSTEM IN OPERATION DURING THESE MONTHS.

(a) - LENGTH OF RECORD IN YEARS, ALTHOUGH INDIVIDUAL MONTHS MAY BE MISSING.
0.* OR * - THE VALUE IS BETWEEN 0.0 AND 0.05
NORMALS - BASED ON THE 1961-1990 RECORD PERIOD.
EXTREMES - DATES ARE THE MOST RECENT OCCURRENCE
WIND DIR.- NUMERALS SHOW TENS OF DEGRESS CLOCKWISE FROM TRUE NORTH. "00" INDICATES CALM.
RESULTANT DIRECTIONS ARE GIVEN TO WHOLE DEGREES.
BOLD VALUES INDICATE EXTREME VALUES WHICH OCCURRED AFTER THE ASOS SYSTEM WAS COMMISSIONED.

RECORD = PERIOD OF RECORD
RECORD MEAN PRECIPITATION IS THE MEAN OF ALL DAILY PRECIPITATION AMOUNTS DURING THE PERIOD OF RECORD.
RECORD MAX(MIN) TEMPERATURE IS THE MEAN OF ALL DAILY MAX(MIN) TEMPERATURES DURING THE PERIOD OF RECORD.
RECORD MEAN TEMPERATURE IS THE SUM OF THE RECORD MAX AND RECORD MIN DIVIDED BY 2.
AVERAGE TEMPERATURE IS THE SUM OF THE MEAN DAILY MAX AND MIN TEMPERATURE DIVIDED BY 2.

EXCEPTIONS

1. PERCENT OF POSSIBLE SUNSHINE, MEAN SKY COVER, AND DAYS CLEAR-PARTLY CLOUDY-CLOUDY ARE THROUGH 1978.
2. FASTEST OBSERVED WIND IS THROUGH MAY 1979.

ALL PAGES
STATION REDUCED ELEMENTS RECORDED.

RECORD MEANS ARE THROUGH THE CURRENT YEAR, BEGINNING IN 1944 FOR TEMPERATURE
1944 FOR PRECIPITATION
1944 FOR SNOWFALL

HEATING DEGREE DAYS Base 65 deg. F JUNEAU, ALASKA

SEASON	JULY	AUG	SEP	OCT	NOV	DEC	JAN	FEB	MAR	APR	MAY	JUNE	TOTAL
1964-65	324	347	453	614	980	1496	1291	1154	923	813	693	487	9575
1965-66	292	297	435	654	1045	1182	1746	1090	1071	805	667	355	9639
1966-67	265	382	466	812	1191	1188	1291	963	1261	824	592	300	9535
1967-68	340	303	444	671	973	1162	1432	1043	984	808	508	374	9042
1968-69	248	281	516	802	920	1405	1801	1219	1054	727	464	218	9655
1969-70	343	448	521	724	975	921	1329	830	876	770	601	422	8760
1970-71	387	405	553	783	1110	1343	1607	1029	1112	784	658	346	10117
1971-72	227	290	504	811	1001	1360	1519	1315	1186	906	618	432	10169
1972-73	207	293	535	810	907	1275	1423	1126	986	752	584	404	9302
1973-74	343	404	505	732	1253	1143	1550	1006	1242	765	556	437	9936
1974-75	349	315	437	690	851	957	1296	1129	1063	791	541	402	8821
1975-76	281	337	402	712	1088	1244	1132	1131	1006	706	597	384	9020
1976-77	280	275	427	679	717	938	918	690	893	673	531	320	7341
1977-78	243	196	428	695	1062	1423	1233	922	954	683	525	317	8681
1978-79	298	262	427	612	1037	1134	1370	1505	904	712	528	378	9167
1979-80	251	205	415	609	830	1187	1404	895	949	678	477	278	8178
1980-81	283	308	472	628	783	1333	843	899	786	772	392	316	7815
1981-82	269	275	469	682	841	1175	1579	1214	1021	830	601	257	9213
1982-83	220	310	435	699	1027	1029	1073	924	931	663	470	275	8056
1983-84	237	317	502	701	991	1423	1014		780	649	486	338	
1984-85	286	291	444	754									
1985-86						1001	943	1011	913	822	564	316	
1986-87	249	319	423	587	1028	891	982	841	1020	697	526	394	7957
1987-88	199	222	440	635	744	944	1169	935	844	691	508	324	7655
1988-89	338	338	497	641	855	1040	1217	1144	1097	663	491	283	8604
1989-90	159	210	370	713	959	890	1191	1109	879	661	467	295	7903
1990-91	180	210	407	748	1152	1274	1238	823	981	694	516	298	8521
1991-92	294	303	435	764	855	966	902	930	865	712	571	292	7889
1992-93	240	274	540	750	791	1217	1250	966	899	612	394	271	8204
1993-94	166	240	406	588	838	891							

COOLING DEGREE DAYS Base 65 deg. F JUNEAU, ALASKA

YEAR	JAN	FEB	MAR	APR	MAY	JUNE	JULY	AUG	SEP	OCT	NOV	DEC	TOTAL
1969	0	0	0	0	0	7	0	0	0	0	0	0	7
1970	0	0	0	0	0	0	0	0	0	0	0	0	0
1971	0	0	0	0	0	0	0	0	0	0	0	0	0
1972	0	0	0	0	0	0	0	0	0	0	0	0	0
1973	0	0	0	0	0	0	0	0	0	0	0	0	0
1974	0	0	0	0	0	0	0	0	0	0	0	0	0
1975	0	0	0	0	0	0	7	0	0	0	0	0	7
1976	0	0	0	0	0	0	3	3	0	0	0	0	6
1977	0	0	0	0	0	0	0	1	0	0	0	0	1
1978	0	0	0	0	0	0	0	0	0	0	0	0	0
1979	0	0	0	0	0	0	0	1	0	0	0	0	1
1980	0	0	0	0	0	1	0	0	0	0	0	0	1
1981	0	0	0	0	0	0	0	0	0	0	0	0	0
1982	0	0	0	0	0	2	0	0	0	0	0	0	2
1983	0	0	0	0	0	0	0	0	0	0	0	0	0
1984	0		0	0	0	0	0	0	0	0			0
1985												0	0
1986	0	0	0	0	0	2	0	0	0	0	0	0	2
1987	0	0	0	0	0	0	5	0	0	0	0	0	5
1988	0	0	0	0	0	0	0	0	0	0	0	0	0
1989	0	0	0	0	0	0	14	0	0	0	0	0	14
1990	0	0	0	0	0	1	8	3	0	0	0	0	12
1991	0	0	0	0	0	6	0	0	0	0	0	0	6
1992	0	0	0	0	0	6	0	0	0	0	0	0	6
1993	0	0	0	0	0	0	5	0	0	0	0	0	5

SNOWFALL (inches) JUNEAU, ALASKA

SEASON	JULY	AUG	SEP	OCT	NOV	DEC	JAN	FEB	MAR	APR	MAY	JUNE	TOTAL
1964-65	0.0	0.0	0.0	T	3.1	54.7	45.2	86.3	4.0	0.8	0.2	0.0	194.3
1965-66	0.0	0.0	0.0	0.3	9.6	16.9	54.8	17.6	49.0	1.5	T	0.0	149.7
1966-67	0.0	0.0	0.0	4.3	20.0	19.6	38.2	32.0	6.6	3.0	0.0	0.0	123.7
1967-68	0.0	0.0	0.0	0.0	9.5	21.1	31.1	17.0	8.2	6.1	T	0.0	93.0
1968-69	0.0	0.0	0.0	T	8.7	35.6	28.2	17.1	27.9	T	0.0	0.0	117.5
1969-70	0.0	0.0	0.0	0.0	18.2	0.7	15.8	2.0	1.1	3.5	0.0	T	41.3
1970-71	0.0	0.0	T	0.4	24.3	26.2	51.0	21.5	50.6	1.1	T	0.0	175.1
1971-72	0.0	0.0	0.0	6.9	20.5	37.1	45.1	31.1	27.1	10.3	0.0	0.0	178.1
1972-73	0.0	0.0	0.0	2.2	2.8	31.6	63.8	20.9	9.6	0.5	0.0	0.0	131.4
1973-74	0.0	0.0	0.0	T	18.6	15.7	36.0	32.8	15.3	0.5	0.0	0.0	118.9
1974-75	0.0	0.0	T	0.9	3.5	17.3	41.5	16.5	18.9	4.5	0.0	0.0	103.1
1975-76	0.0	0.0	0.0	5.3	32.5	51.0	32.9	34.1	25.5	2.3	0.0	0.0	183.6
1976-77	0.0	0.0	0.0	1.2	1.1	26.3	5.0	T	12.7	T	0.0	0.0	46.3
1977-78	0.0	0.0	0.0	T	27.4	16.6	5.2	1.1	1.7	0.4	0.0	0.0	52.4
1978-79	0.0	0.0	0.0	0.0	14.9	21.6	24.2	21.4	5.4	T	0.0	0.0	87.5
1979-80	0.0	0.0	0.0	0.0	1.6	48.4	41.6	4.1	2.4	T	0.0	0.0	98.1
1980-81	0.0	0.0	0.0	0.0	0.5	40.5	2.4	16.4	0.5	2.2	0.0	0.0	62.5
1981-82	0.0	0.0	0.0	0.0	4.0	6.0	69.2	29.6	8.4	1.1	T	0.0	118.3
1982-83	0.0	0.0	0.0	2.0	0.4	10.8	40.1	15.7	0.2	T	0.0	0.0	69.2
1983-84	0.0	0.0	0.0	0.0	8.1	13.3	43.1	0.7	1.0	T	T	0.0	66.2
1984-85	0.0	0.0	0.0	0.0									
1985-86						2.0	10.3	7.4	30.4	4.4	T	0.0	
1986-87	0.0	0.0	0.0	T	22.1	1.4	3.3	1.4	7.3	T	0.0	0.0	35.5
1987-88	0.0	0.0	0.0	T	4.6	6.8	3.5	8.0	1.0	0.5	0.0	0.0	24.4
1988-89	0.0	0.0	0.0	0.0	4.8	11.3	44.7	0.2	10.0	T	0.0	0.0	71.0
1989-90	0.0	0.0	0.0	0.6	32.5	6.4	36.5	39.4	0.6	0.0	0.0	0.0	116.0
1990-91	0.0	0.0	0.0	0.0	48.8	33.2	31.8	15.5	9.4	0.8	T	0.0	139.5
1991-92	0.0	0.0	0.0	5.4	7.7	49.3	14.3	12.4	4.1	T	T	0.0	93.2
1992-93	0.0	0.0	0.0	1.1	4.4	25.0	32.5	36.8	2.6	T	0.0	0.0	102.4
1993-94	0.0	0.0	0.0	0.0	4.3	10.4							
Record Mean	0.0	0.0	T	1.1	11.9	22.6	26.0	19.0	14.9	3.5	T	T	98.9

KING SALMON, ALASKA

King Salmon is located in that area of southwestern Alaska which joins the Alaskan Peninsula and the Alaskan mainland. It is located about 1/4 mile from the Naknek River and lies 18 miles inland from the shores of Kvichak Bay, an arm of the much larger Bristol Bay. The terrain surrounding the station for a radius of 30 to 60 miles to the north through east to south-southwest is gently rolling, barren tundra. Some 60 miles to the east and southeast, however, the Aleutian Range rises to peaks a little above the 7,000-foot level. This mountain range extends in a northeast-southwest direction. The southern end of the Kuskokwim Range reaches southward to an area roughly 100 miles directly west of King Salmon. Nearness to the ocean tends to provide King Salmon with a climate that is predominantly maritime in character, with diurnal and seasonal temperature ranges normally confined to rather narrow limits. However, the area occasionally experiences definite continental influences that cause temperature extremes which tend to exaggerate the climatic conditions generally prevailing. Extreme temperatures range from upper 80s to readings near -40 degrees, but days in summer with maximum readings reaching 80 degrees are extremely rare. In fact, July, the warmest month, has an average of only five days with temperatures reaching 70 degrees or above.

Cloud coverage in the King Salmon area is generally quite high, averaging about eight-tenths the year around. Mountain ranges to the south, east, and west tend to provide uplift for air moving toward King Salmon from these directions and produce considerable cloudiness which is carried out across the local area. When the wind movement is inland from the southwest, the air arrives carrying a high moisture content to condense in low level cloudiness, and this action contributes to the frequent fog occurrences all months of the year. Fog development is most frequent during the months of July and August. During the winter months the high moisture content of the air causes substantial accumulations of frost on outside objects.

Seasonal snowfall averages about 45 inches, with the maximum depth on the ground during the winter season averaging about 10 inches. This indicates the extent of melting that takes place with the snow accumulation. Although most of the snow is received during periods of general snowfall over most of the southwestern mainland, a considerable amount of snow is brought in as snow showers which move inland from the Bristol Bay area. These showers are generally quite local and usually of short duration, but they often follow in rapid succession to bring sizeable accumulations of snow within relatively short periods of time. December has the greatest monthly average snowfall amount.

From December through March the area experiences rather strong winds, due to the passage of eastward-moving Aleutian lows. The strongest winds are usually from a northerly direction, developing after the low centers have passed on east of the local area. Winds of 50 mph or more have occurred in all months with extremes above 90 mph.

Ice in the bay near King Salmon usually becomes safe for man around November 11, with the Naknek River becoming safe for man around November 25. Break-up on the bay averages about April 6, with the break-up on the river averaging about April 18.

The average date of the last freeze is late May and the average date of the first freeze is early September. The average growing season is 100 days.

NORMALS, MEANS, AND EXTREMES

KING SALMON, ALASKA

LATITUDE: 58°41'N LONGITUDE: 156°39'W ELEVATION: FT. GRND 49 BARO 46 TIME ZONE: YUKON WBAN: 25503

	(a)	JAN	FEB	MAR	APR	MAY	JUNE	JULY	AUG	SEP	OCT	NOV	DEC	YEAR
TEMPERATURE °F:														
Normals														
-Daily Maximum		22.1	22.9	30.7	39.0	50.9	58.6	63.0	61.4	54.7	39.7	29.6	23.4	41.3
-Daily Minimum		7.6	6.7	14.1	23.4	33.9	41.4	46.4	46.4	39.7	25.1	15.0	8.3	25.7
-Monthly		14.9	14.8	22.4	31.2	42.4	50.0	54.7	53.9	47.2	32.4	22.3	15.9	33.5
Extremes														
-Record Highest	51	53	57	56	65	79	88	86	84	74	67	56	48	88
-Year		1963	1991	1943	1948	1993	1953	1951	1968	1974	1954	1986	1970	JUN 1953
-Record Lowest	51	-48	-41	-42	-15	4	29	33	25	15	-12	-28	-38	-48
-Year		1989	1993	1971	1985	1945	1987	1993	1984	1983	1983	1988	1942	JAN 1989
NORMAL DEGREE DAYS:														
Heating (base 65°F)		1553	1406	1321	1014	701	450	319	344	534	1011	1281	1522	11456
Cooling (base 65°F)		0	0	0	0	0	0	0	0	0	0	0	0	0
% OF POSSIBLE SUNSHINE														
MEAN SKY COVER (tenths)														
Sunrise - Sunset	38	6.8	6.6	6.8	7.6	8.0	8.6	8.6	8.6	8.3	7.5	7.3	7.1	7.6
MEAN NUMBER OF DAYS:														
Sunrise to Sunset														
-Clear	38	7.7	7.4	7.1	3.9	2.6	1.1	1.1	1.2	1.7	4.1	5.7	6.4	50.2
-Partly Cloudy	38	5.1	4.9	5.6	6.5	6.0	5.0	4.9	5.1	6.0	7.0	5.2	5.6	67.0
-Cloudy	38	18.2	16.0	18.2	19.6	22.3	23.8	24.9	24.7	22.3	19.8	19.2	19.0	248.0
Precipitation														
.01 inches or more	48	10.9	9.5	11.1	10.2	12.2	13.2	14.5	17.3	16.5	13.4	12.0	12.4	153.2
Snow,Ice pellets,hail														
1.0 inches or more	44	2.8	2.2	2.7	1.7	0.3	0.*	0.0	0.0	0.0	1.0	2.2	3.0	15.8
Thunderstorms	38	0.0	0.0	0.0	0.0	0.2	0.4	0.4	0.2	0.1	0.*	0.1	0.*	1.4
Heavy Fog Visibility														
1/4 mile or less	38	2.4	1.5	1.7	1.6	2.4	3.0	4.7	4.3	2.7	2.4	3.4	2.9	32.9
Temperature °F														
-Maximum														
70° and above	30	0.0	0.0	0.0	0.0	0.8	2.0	5.6	3.7	0.2	0.0	0.0	0.0	12.3
32° and below	30	18.6	16.6	12.3	5.2	0.3	0.0	0.0	0.0	0.0	6.2	14.8	17.6	91.5
-Minimum														
32° and below	30	28.1	25.6	27.2	24.9	11.6	0.5	0.0	0.3	5.4	21.3	25.4	28.1	198.3
0° and below	30	11.3	11.9	5.8	1.1	0.0	0.0	0.0	0.0	0.0	0.8	5.2	11.1	47.2
AVG. STATION PRESS.(mb)	21	999.7	1002.6	1002.3	1005.3	1006.5	1009.3	1012.0	1009.8	1005.0	1001.1	1000.0	999.2	1004.4
RELATIVE HUMIDITY (%)														
Hour 03	45	78	77	79	81	83	87	90	89	87	85	82	78	83
Hour 09 (Local Time)	45	78	77	78	76	73	77	83	86	85	85	82	78	80
Hour 15	45	76	71	68	62	57	59	64	67	66	69	77	77	68
Hour 21	45	78	76	76	73	69	70	75	80	80	81	82	78	77
PRECIPITATION (inches):														
Water Equivalent														
-Normal		1.05	0.81	1.07	1.13	1.34	1.58	2.23	2.95	2.74	2.07	1.48	1.37	19.82
-Maximum Monthly	51	3.02	3.00	2.41	2.99	2.95	3.78	5.08	6.44	7.30	6.35	3.35	3.65	7.30
-Year		1957	1943	1967	1963	1988	1950	1990	1953	1961	1946	1985	1978	SEP 1961
-Minimum Monthly	51	0.16	0.11	0.04	T	0.11	0.00	0.32	1.05	0.89	0.20	T	0.12	0.00
-Year		1959	1973	1960	1948	1948	1948	1951	1975	1984	1959	1963	1958	JUN 1948
-Maximum in 24 hrs	51	1.08	1.29	1.03	1.41	0.98	0.87	1.28	2.00	1.69	1.77	1.56	1.17	2.00
-Year		1987	1969	1953	1963	1977	1951	1986	1963	1960	1977	1985	1978	AUG 1963
Snow,Ice pellets,hail														
-Maximum Monthly	44	30.6	20.3	20.0	16.0	6.1	1.3	0.0	0.0	0.6	15.7	16.1	28.4	30.6
-Year		1993	1990	1977	1968	1985	1972			1956	1990	1969	1993	JAN 1993
-Maximum in 24 hrs	44	18.6	9.3	11.0	4.5	3.8	1.2	0.0	0.0	0.6	8.7	8.6	8.5	18.6
-Year		1993	1962	1981	1971	1985	1972			1956	1990	1948	1990	JAN 1993
WIND:														
Mean Speed (mph)	38	10.7	11.2	11.4	11.1	11.2	10.9	10.0	10.2	10.7	10.5	10.7	10.5	10.8
Prevailing Direction through 1963		N	N	N	NNW	S	SW	S	S	S	N	N	N	N
Fastest Obs. 1 Min.														
-Direction (!!)	32	09	14	09	16	09	14	08	08	09	14	09	09	09
-Speed (MPH)	32	48	53	69	46	48	46	49	46	62	53	71	55	71
-Year		1962	1969	1965	1960	1971	1970	1966	1964	1958	1963	1958	1963	NOV 1958
Peak Gust														
-Direction (!!)	14	E	E	E	SE	S	E	E	SW	SE	E	E	E	E
-Speed (mph)	14	69	69	62	55	63	58	47	56	61	67	67	66	69
-Date		1992	1986	1993	1983	1985	1987	1982	1980	1991	1991	1983	1986	JAN 1992

PRECIPITATION (inches)

KING SALMON, ALASKA

YEAR	JAN	FEB	MAR	APR	MAY	JUNE	JULY	AUG	SEP	OCT	NOV	DEC	ANNUAL
1964	0.73	1.92	1.03	0.75	1.87	0.47	1.74	3.45	2.40	2.07	1.41	1.02	18.86
1965	0.79	0.45	2.25	1.29	1.67	1.55	1.93	2.51	3.92	0.39	1.60	1.62	19.97
1966	0.43	1.54	0.81	0.49	2.40	0.99	2.45	5.69	3.73	2.88	2.17	0.54	24.12
1967	0.93	0.54	2.41	1.33	0.22	2.90	3.15	3.15	1.69	0.52	1.90	2.03	20.77
1968	1.10	0.57	0.80	1.41	1.26	1.35	1.14	2.20	2.70	0.51	0.91	1.25	15.20
1969	0.65	1.94	1.19	0.30	0.79	0.56	2.19	3.42	1.28	2.33	1.84	0.57	17.06
1970	0.50	0.45	1.81	1.80	0.41	1.13	2.87	4.31	1.59	2.24	0.79	1.33	19.23
1971	0.45	1.62	0.27	0.84	1.43	1.48	3.25	4.30	3.40	2.72	1.13	3.42	24.31
1972	1.30	0.21	0.17	1.37	1.29	1.62	1.08	1.95	2.95	2.57	1.35	0.59	16.45
1973	0.62	0.11	1.25	0.43	1.83	1.48	2.43	3.80	1.41	1.52	0.97	1.10	16.95
1974	0.86	0.55	1.27	1.18	0.57	2.40	2.01	3.19	1.56	2.90	1.20	1.23	18.92
1975	2.14	0.76	0.93	2.65	0.86	2.69	0.74	1.05	3.90	2.10	0.46	1.38	19.66
1976	1.24	0.97	0.78	0.58	1.47	1.34	2.60	1.71	2.64	0.81	2.06	1.77	17.97
1977	0.85	1.35	1.99	1.68	1.72	0.99	1.60	3.16	2.58	3.29	0.58	1.04	20.83
1978	0.70	0.28	0.26	0.58	0.98	2.81	1.66	2.03	1.87	2.84	1.77	3.65	19.43
1979	1.00	0.29	0.39	1.20	0.46	1.80	2.24	2.50	0.91	2.71	2.89	1.09	17.48
1980	1.46	0.83	1.51	0.42	1.61	2.19	2.97	2.36	2.00	2.46	1.19	0.49	19.49
1981	1.76	2.26	1.83	0.49	0.73	2.27	2.17	3.93	1.82	1.59	1.31	0.59	20.75
1982	1.48	0.15	1.37	1.20	1.55	3.04	1.98	1.99	5.14	1.41	0.83	1.37	21.51
1983	0.42	0.25	0.22	2.22	1.37	1.20	1.53	2.33	2.36	2.82	0.98	0.48	16.18
1984	1.17	0.55	0.44	0.43	1.08	1.59	1.30	2.41	0.89	0.57	1.00	1.79	13.22
1985	0.95	0.73	1.27	0.34	1.16	1.23	1.31	3.24	2.64	2.29	3.35	1.58	20.09
1986	1.33	0.19	0.24	0.98	1.01	0.93	2.44	3.22	4.03	2.50	1.91	0.65	19.43
1987	2.38	0.54	0.55	0.81	1.74	1.49	1.94	2.73	2.99	2.47	2.75	1.07	21.46
1988	0.56	0.75	0.74	1.02	2.95	1.11	2.73	2.88	2.17	1.68	1.52	1.60	19.71
1989	0.84	0.93	0.19	0.99	2.32	1.10	3.04	3.15	5.90	2.86	1.58	1.31	24.21
1990	1.44	1.61	1.71	0.89	1.52	1.22	5.08	2.02	2.75	2.38	2.10	3.26	25.98
1991	0.55	0.58	1.56	0.86	1.24	1.63	1.02	1.79	2.10	1.99	1.34	1.26	15.92
1992	0.79	0.92	1.40	0.19	0.74	2.53	3.02	4.73	1.35	1.11	1.45	1.77	20.00
1993	1.48	0.35	0.26	0.50	0.70	0.50	1.01	3.21	4.53	1.98	3.00	2.15	19.67
Record Mean	1.05	0.87	1.07	0.93	1.20	1.53	2.16	3.13	2.87	2.08	1.50	1.28	19.67

AVERAGE TEMPERATURE (deg. F)

KING SALMON, ALASKA

YEAR	JAN	FEB	MAR	APR	MAY	JUNE	JULY	AUG	SEP	OCT	NOV	DEC	ANNUAL
#1964	15.0	14.4	16.2	28.2	38.2	52.2	54.1	52.9	48.1	35.0	21.3	2.8	31.6
1965	12.3	7.2	33.8	31.2	37.0	46.0	52.7	51.4	50.3	25.6	22.6	5.3	31.3
1966	20.1	17.4	4.8	30.1	37.4	50.7	52.6	50.2	47.0	27.2	20.5	8.9	30.6
1967	7.6	15.0	26.6	34.6	44.0	50.6	54.6	54.6	46.3	30.0	29.3	12.0	33.8
1968	10.0	12.8	25.5	30.1	43.8	50.7	55.5	54.6	43.3	28.4	26.2	3.4	32.0
1969	6.9	12.8	26.3	34.4	44.1	52.5	54.3	50.9	48.1	38.8	17.0	26.2	34.3
1970	-0.3	26.4	30.5	29.8	44.8	51.1	52.7	51.7	44.8	29.3	29.3	11.6	33.5
1971	-2.6	12.2	-7.8	26.8	37.7	47.3	54.5	54.9	46.8	34.3	21.6	18.6	30.0
1972	6.7	6.2	1.8	22.1	40.9	46.6	55.2	54.4	45.5	36.0	25.4	16.2	29.8
1973	1.8	19.5	19.3	35.9	42.9	51.4	55.6	54.6	47.2	34.1	24.7	17.9	33.8
1974	9.5	0.4	23.2	35.6	45.5	51.2	55.4	57.0	50.6	33.4	20.1	8.0	32.5
1975	4.7	3.9	14.5	25.0	39.4	47.1	54.7	53.6	47.1	32.4	12.7	10.2	28.8
1976	12.3	7.3	15.3	29.5	39.5	46.9	53.2	53.1	45.3	31.5	24.2	19.3	31.5
1977	34.4	30.1	18.8	25.7	39.5	50.5	54.3	56.8	47.0	31.7	14.1	10.6	34.5
1978	28.6	24.8	25.6	37.5	45.2	49.5	54.2	57.1	47.7	36.5	30.0	28.0	38.7
1979	30.1	6.2	30.3	39.6	47.3	52.0	57.8	56.0	50.0	39.4	29.4	4.5	37.3
1980	9.0	20.7	27.6	36.4	41.7	48.9	55.1	51.1	47.0	35.2	26.3	5.3	33.7
1981	29.8	21.9	34.4	35.8	46.8	50.3	55.1	54.8	44.9	33.2	23.4	13.3	37.0
1982	17.0	12.8	23.9	25.5	40.3	48.9	51.5	52.3	46.2	28.1	26.1	24.0	33.1
1983	11.9	18.7	33.2	36.5	46.6	53.8	57.4	54.1	45.5	28.8	30.1	27.2	37.0
1984	17.4	-2.1	36.3	29.2	43.0	52.3	53.7	53.5	48.0	30.1	22.5	24.7	34.1
1985	32.6	10.6	22.6	20.8	39.9	47.4	54.3	52.4	47.4	26.7	25.1	34.2	34.5
1986	16.9	22.1	21.5	28.1	42.1	49.9	53.7	52.2	48.8	36.1	26.3	30.6	35.7
1987	21.1	24.3	29.8	32.3	42.8	49.3	55.9	57.0	45.4	37.5	16.5	9.4	35.1
1988	25.6	26.6	24.8	31.1	44.5	52.8	56.8	53.5	45.8	30.9	13.9	20.8	35.6
1989	-2.9	28.8	23.6	36.1	42.0	51.6	56.3	57.1	51.7	36.7	18.1	19.5	34.9
1990	16.8	-1.8	25.4	39.3	45.8	51.4	56.0	55.9	47.5	31.5	17.3	20.4	33.8
1991	17.5	14.2	25.7	36.4	44.5	50.4	55.2	53.7	50.7	37.2	23.1	15.1	35.3
1992	17.7	3.1	22.0	32.4	42.7	52.6	55.6	53.9	41.0	31.7	23.5	19.2	33.0
1993	15.0	22.7	31.1	41.0	48.3	53.1	57.9	56.0	48.6	38.1	29.6	24.6	38.8
Record Mean	14.2	15.0	21.5	31.6	42.7	50.5	54.8	54.1	47.2	33.5	22.5	14.5	33.5
Max	21.6	23.1	29.8	39.5	51.4	59.3	63.0	61.4	54.7	40.7	29.6	22.2	41.4
Min	6.7	6.9	13.2	23.6	34.0	41.6	46.5	46.7	39.6	26.2	15.3	6.8	25.6

REFERENCE NOTES

KING SALMON, ALASKA

GENERAL
T - TRACE AMOUNT.
BLANK ENTRIES DENOTE MISSING/UNREPORTED DATA.
# INDICATES A STATION OR INSTRUMENT RELOCATION.

SPECIFIC

(a) - LENGTH OF RECORD IN YEARS, ALTHOUGH INDIVIDUAL MONTHS MAY BE MISSING.
0.* OR * - THE VALUE IS BETWEEN 0.0 AND 0.05
NORMALS - BASED ON THE 1961-1990 RECORD PERIOD.
EXTREMES - DATES ARE THE MOST RECENT OCCURRENCE
WIND DIR.- NUMERALS SHOW TENS OF DEGRESS CLOCKWISE FROM TRUE NORTH. "00" INDICATES CALM.
RESULTANT DIRECTIONS ARE GIVEN TO WHOLE DEGREES.
BOLD VALUES INDICATE EXTREME VALUES WHICH OCCURRED AFTER THE ASOS SYSTEM WAS COMMISSIONED.

RECORD = PERIOD OF RECORD
RECORD MEAN PRECIPITATION IS THE MEAN OF ALL DAILY PRECIPITATION AMOUNTS DURING THE PERIOD OF RECORD.
RECORD MAX(MIN) TEMPERATURE IS THE MEAN OF ALL DAILY MAX(MIN) TEMPERATURES DURING THE PERIOD OF RECORD.
RECORD MEAN TEMPERATURE IS THE SUM OF THE RECORD MAX AND RECORD MIN DIVIDED BY 2.
AVERAGE TEMPERATURE IS THE SUM OF THE MEAN DAILY MAX AND MIN TEMPERATURE DIVIDED BY 2.

EXCEPTIONS

RECORD MEANS ARE THROUGH THE CURRENT YEAR, BEGINNING IN 1942 FOR TEMPERATURE
1943 FOR PRECIPITATION
1950 FOR SNOWFALL

HEATING DEGREE DAYS Base 65 deg. F KING SALMON, ALASKA

SEASON	JULY	AUG	SEP	OCT	NOV	DEC	JAN	FEB	MAR	APR	MAY	JUNE	TOTAL
1964-65	329	364	503	924	1303	1929	1631	1615	964	1009	861	566	11998
1965-66	373	410	437	1215	1265	1852	1387	1328	1863	1040	851	425	12446
1966-67	381	450	533	1168	1329	1736	1777	1396	1182	907	643	428	11930
1967-68	315	313	556	1078	1065	1640	1700	1512	1216	1039	648	423	11505
1968-69	286	314	647	1128	1155	1908	1800	1463	1193	914	642	370	11820
1969-70	323	428	499	805	1435	1197	2024	1078	1066	1049	616	410	10930
1970-71	374	404	602	1099	1065	1657	2094	1476	1775	1140	838	523	13047
1971-72	324	306	541	945	1296	1436	1806	1706	1963	1281	739	545	12888
1972-73	298	319	576	888	1177	1508	1962	1268	1411	864	679	402	11352
1973-74	284	318	529	953	1204	1452	1719	1812	1292	875	595	409	11442
1974-75	288	245	427	971	1339	1765	1869	1710	1560	1192	785	529	12680
1975-76	311	347	526	1004	1568	1699	1629	1673	1533	1059	784	536	12669
1976-77	359	362	585	1032	1218	1413	941	969	1427	1173	783	431	10693
1977-78	325	246	531	1025	1524	1687	1123	1121	1214	820	605	458	10679
1978-79	326	237	514	876	1042	1141	1076	1648	1069	754	544	380	9607
1979-80	214	274	441	788	1063	1872	1733	1276	1157	852	716	479	10865
1980-81	301	424	533	917	1155	1849	1086	1206	940	869	559	435	10274
1981-82	298	311	595	981	1242	1603	1483	1461	1263	1180	760	475	11652
1982-83	412	386	557	1134	1162	1266	1644	1293	981	848	566	328	10577
1983-84	229	333	578	1117	1045	1165	1472	1948	880	1067	680	374	10888
1984-85	345	350	504	1075	1270	1246	996	1520	1308	1318	770	523	11225
1985-86	322	384	521	1182	1189	946	1485	1199	1343	1101	701	443	10816
1986-87	344	388	483	892	1156	1058	1353	1133	1087	973	681	467	10015
1987-88	273	241	580	846	1451	1722	1214	1106	1239	1008	627	363	10670
1988-89	246	350	571	1049	1531	1365	2104	1008	1278	860	707	398	11467
1989-90	264	239	392	871	1403	1405	1492	1870	1220	766	590	402	10914
1990-91	270	274	519	1030	1426	1379	1469	1423	1208	852	629	432	10911
1991-92	295	341	420	856	1247	1538	1460	1794	1330	973	687	367	11308
1992-93	282	335	712	1026	1238	1414	1552	1183	1046	715	511	349	10363
1993-94	216	276	487	826	1054	1247							

COOLING DEGREE DAYS Base 65 deg. F KING SALMON, ALASKA

YEAR	JAN	FEB	MAR	APR	MAY	JUNE	JULY	AUG	SEP	OCT	NOV	DEC	TOTAL
1969	0	0	0	0	0	0	0	0	0	0	0	0	0
1970	0	0	0	0	0	0	0	0	0	0	0	0	0
1971	0	0	0	0	0	0	5	0	0	0	0	0	5
1972	0	0	0	0	0	0	0	0	0	0	0	0	0
1973	0	0	0	0	0	0	0	3	0	0	0	0	3
1974	0	0	0	0	0	0	0	0	0	0	0	0	0
1975	0	0	0	0	0	0	0	0	0	0	0	0	0
1976	0	0	0	0	0	0	0	0	0	0	0	0	0
1977	0	0	0	0	0	0	0	0	0	0	0	0	0
1978	0	0	0	0	0	0	0	0	0	0	0	0	0
1979	0	0	0	0	0	0	0	0	0	0	0	0	0
1980	0	0	0	0	0	0	3	0	0	0	0	0	3
1981	0	0	0	0	0	0	0	2	0	0	0	0	2
1982	0	0	0	0	0	0	0	0	0	0	0	0	0
1983	0	0	0	0	0	0	0	0	0	0	0	0	0
1984	0	0	0	0	0	0	0	0	0	0	0	0	0
1985	0	0	0	0	0	0	0	0	0	0	0	0	0
1986	0	0	0	0	0	0	0	0	0	0	0	0	0
1987	0	0	0	0	0	0	0	0	0	0	0	0	0
1988	0	0	0	0	0	0	0	0	0	0	0	0	0
1989	0	0	0	0	0	0	0	1	0	0	0	0	1
1990	0	0	0	0	0	0	0	0	0	0	0	0	0
1991	0	0	0	0	0	0	0	0	0	0	0	0	0
1992	0	0	0	0	0	0	0	0	0	0	0	0	0
1993	0	0	0	0	0	0	1	2	0	0	0	0	3

SNOWFALL (inches) KING SALMON, ALASKA

SEASON	JULY	AUG	SEP	OCT	NOV	DEC	JAN	FEB	MAR	APR	MAY	JUNE	TOTAL
1964-65	0.0	0.0	0.0	4.8	7.2	10.7	6.0	5.2	4.3	7.9	5.6	0.0	51.7
1965-66	0.0	0.0	0.3	2.8	4.0	16.0	1.4	11.3	9.9	0.8	2.6	0.0	49.1
1966-67	0.0	0.0	0.0	5.3	8.1	6.1	7.7	5.1	5.7	2.9	0.0	0.0	40.9
1967-68	0.0	0.0	0.0	1.2	4.1	3.5	9.3	5.3	7.5	16.0	2.2	0.0	49.1
1968-69	0.0	0.0	0.3	0.3	8.9	10.8	6.1	13.9	12.3	1.1	T	0.0	53.7
1969-70	0.0	0.0	0.0	T	16.1	1.5	11.1	4.6	6.4	5.5	T	0.0	45.2
1970-71	0.0	0.0	T	8.3	1.4	8.6	4.3	15.2	1.6	8.9	2.5	T	50.8
1971-72	0.0	0.0	0.0	7.9	3.9	15.1	11.7	2.1	1.9	8.7	0.1	1.3	52.7
1972-73	0.0	0.0	T	0.8	8.0	2.1	3.0	0.8	8.1	2.2	0.6	0.0	25.6
1973-74	0.0	0.0	T	2.0	2.1	12.7	11.9	5.3	4.6	5.1	T	0.0	43.7
1974-75	0.0	0.0	0.0	T	4.3	10.9	19.1	6.3	8.7	14.3	2.9	0.0	66.5
1975-76	0.0	0.0	0.0	0.8	3.9	13.9	12.0	3.2	6.7	6.2	3.2	0.0	49.9
1976-77	0.0	0.0	0.0	2.0	10.9	11.0	2.1	11.9	20.0	4.6	T	0.0	62.5
1977-78	0.0	0.0	T	4.3	5.3	4.5	3.9	3.7	2.2	0.6	T	0.0	24.5
1978-79	0.0	0.0	0.0	1.0	2.2	14.1	4.4	0.2	1.1	T	T	0.0	23.0
1979-80	0.0	0.0	0.0	T	8.5	9.7	11.5	11.1	9.0	T	0.8	0.0	50.6
1980-81	0.0	0.0	0.0	0.3	6.1	6.8	10.5	11.3	15.8	0.6	T	T	51.4
1981-82	0.0	0.0	0.5	0.3	4.8	5.9	5.7	T	8.3	8.3	T	0.0	33.8
1982-83	0.0	0.0	0.0	2.8	2.0	2.9	4.0	2.0	T	6.0	0.1	0.0	19.8
1983-84	0.0	0.0	T	9.9	2.3	2.8	8.4	5.5	T	4.0	0.3	0.0	33.2
1984-85	0.0	0.0	0.0	3.4	7.3	3.8	3.7	6.4	8.9	3.4	6.1	0.0	43.0
1985-86	0.0	0.0	0.0	2.5	9.3	3.6	13.5	1.8	2.5	9.8	1.3	0.0	44.3
1986-87	0.0	0.0	0.0	2.3	2.5	4.8	24.7	2.7	2.7	9.4	T	0.0	49.1
1987-88	0.0	0.0	T	0.1	13.2	8.9	3.3	10.1	9.4	4.4	1.2	0.0	50.6
1988-89	0.0	0.0	T	3.4	12.7	9.2	14.9	3.7	5.1	1.5	2.1	0.0	52.6
1989-90	0.0	0.0	T	0.4	12.3	12.4	14.9	20.3	13.5	3.4	0.2	0.0	77.4
1990-91	0.0	0.0	T	15.7	6.7	18.9	3.1	4.3	14.0	2.8	0.0	0.0	65.5
1991-92	0.0	0.0	0.0	T	9.0	9.4	7.2	8.6	8.7	0.5	T	T	43.4
1992-93	0.0	0.0	T	0.9	7.9	8.0	30.6	5.5	5.2	1.8	T	T	59.9
1993-94	0.0	0.0	T	2.0	5.1	28.4							
Record Mean	0.0	0.0	T	3.0	6.3	9.1	8.3	6.8	7.4	4.5	0.9	T	46.3

KODIAK, ALASKA

Kodiak Island is located on the western side of the Gulf of Alaska, 90 miles southwest of the Kenai Peninsula. Oriented northeast–southwest, the island lies 25 miles southeast of the Alaska Peninsula, separated from it by the Shelikof Strait. Afognak Island lies northeast of Kodiak, across Kupreanof Strait, which averages less than half a mile in width. The two islands are generally considered to be a single landmass, approximately 145 miles long by 50 miles wide. The terrain is rugged, with the mountains averaging from 2,000 to 4,000 feet in height. The highest mountains on Kodiak extend to roughly 5,000 feet. The island has many lakes, ponds, interconnecting waterways, and drainage streams. The irregular shoreline is indented by numerous bays, many of which are deep and narrow.

The National Weather Service Office is located on U. S. Coast Guard Base Kodiak, adjacent to Womens Bay, a small U shaped bay extending westward from the main body of Chiniak Bay.

Kodiak has primarily a marine climate which is exemplified by the limited daily and annual temperature ranges. During the summer, the mean air temperature closely approximates the mean sea surface temperature, rising slightly above it during August but falling below again in September. In winter, the mean maximum air temperature more closely resembles the mean sea surface temperature curve. The absolute temperature range is nearly 100 degrees. Summer maximum temperatures will vary 10 to 20 degrees, depending on whether the northwest gradient is strong enough to maintain a flow of air from over the island, or whether it is weak enough that the sea breeze predominates. The highest daily maximum temperatures occur with northwest winds in summer.

Precipitation is normally abundant throughout the year. All months have a wide variation in the amount of precipitation. A very high percentage of the precipitation falls during northeast to southeast winds. Small amounts of snow may fall as late as May or as early as September with good ground cover anticipated in November. Precipitation measurement is often difficult due to strong, gusty surface winds which frequently accompany precipitation. Drifting and blowing snow occasionally close the field for periods of up to 24 hours.

Although the prevailing wind direction is northwesterly every month except May, June, and July, and the average speed is about 10 knots, these data may be misleading because of the extreme variability in both direction and speed. Maximum gusts of over 90 knots has been recorded. Coast Guard Cutters docked in Womens Bay have reported williwaw winds off Old Womens Mountain in excess of 120 knots. Gusts of over 50 knots have occurred during each month of the year, but are most likely to occur in the winter months.

NORMALS, MEANS, AND EXTREMES

KODIAK, ALASKA

LATITUDE: 57°45'N LONGITUDE: 152°30' W ELEVATION: FT. GRND 15 BARO 112 TIME ZONE: YUKON WBAN: 25501

	(a)	JAN	FEB	MAR	APR	MAY	JUNE	JULY	AUG	SEP	OCT	NOV	DEC	YEAR
TEMPERATURE °F:														
Normals														
-Daily Maximum		35.0	36.1	38.9	43.7	49.5	55.6	60.5	61.9	56.6	47.2	39.9	36.4	46.8
-Daily Minimum		24.7	24.9	26.9	31.4	37.5	43.6	48.2	48.4	43.4	34.2	28.8	25.2	34.8
-Monthly		29.9	30.5	32.9	37.5	43.5	49.6	54.4	55.2	50.0	40.7	34.4	30.8	40.8
Extremes														
-Record Highest	45	54	56	57	64	80	86	82	83	73	62	54	56	86
-Year		1963	1957	1963	1965	1968	1953	1989	1968	1985	1983	1986	1984	JUN 1953
-Record Lowest	45	-16	-12	-6	7	20	30	37	36	26	10	0	-1	-16
-Year		1989	1971	1971	1977	1949	1968	1966	1973	1992	1975	1971	1970	JAN 1989
NORMAL DEGREE DAYS:														
Heating (base 65°F)		1088	966	995	825	667	462	329	304	450	753	918	1060	8817
Cooling (base 65°F)		0	0	0	0	0	0	0	0	0	0	0	0	0
% OF POSSIBLE SUNSHINE														
MEAN SKY COVER (tenths)														
Sunrise - Sunset	26	7.3	7.2	7.2	7.5	8.0	8.0	8.0	7.6	7.5	6.9	7.0	7.2	7.4
MEAN NUMBER OF DAYS:														
Sunrise to Sunset														
-Clear	23	5.3	5.8	6.0	4.6	3.5	3.4	3.2	4.1	4.1	6.0	6.7	6.2	58.9
-Partly Cloudy	23	5.9	4.8	6.7	6.4	5.6	5.7	5.8	6.8	7.5	7.7	6.4	5.8	75.0
-Cloudy	23	19.8	17.6	18.3	19.0	22.0	20.9	22.0	20.0	18.4	17.3	16.9	19.0	231.3
Precipitation														
.01 inches or more	44	17.1	15.9	16.5	15.9	17.6	15.2	14.5	14.4	15.6	16.1	16.5	17.2	192.4
Snow, Ice pellets, hail														
1.0 inches or more	30	4.1	5.1	4.1	2.3	0.1	0.0	0.0	0.0	0.0	0.7	2.0	3.8	22.2
Thunderstorms	23	0.0	0.0	0.0	0.0	0.0	0.1	0.*	0.*	0.0	0.0	0.0	0.*	0.3
Heavy Fog Visibility 1/4 mile or less	23	0.9	1.2	0.9	0.3	1.2	1.6	2.6	2.6	0.7	0.3	0.2	0.7	13.0
Temperature °F														
-Maximum														
70° and above	31	0.0	0.0	0.0	0.0	0.4	1.2	2.6	3.1	0.3	0.0	0.0	0.0	7.6
32° and below	31	8.1	7.2	3.8	0.9	0.*	0.0	0.0	0.0	0.0	0.8	3.6	7.2	31.7
-Minimum														
32° and below	31	21.6	20.6	20.8	13.8	3.4	0.1	0.0	0.0	0.7	12.2	18.5	22.4	134.1
0° and below	31	0.5	0.6	0.3	0.0	0.0	0.0	0.0	0.0	0.0	0.0	0.*	0.2	1.6
AVG. STATION PRESS.(mb)	21	997.0	999.0	1000.0	1003.0	1005.3	1008.5	1011.1	1009.0	1003.9	998.2	997.0	996.9	1002.4
RELATIVE HUMIDITY (%)														
Hour 03	46	79	78	78	78	83	86	88	87	85	80	78	76	81
Hour 09 (Local Time)	46	79	79	77	73	76	79	82	81	81	78	78	77	78
Hour 15	46	75	73	69	68	72	75	77	74	73	69	72	74	73
Hour 21	46	79	78	76	75	78	79	82	83	82	78	77	77	79
PRECIPITATION (inches):														
Water Equivalent														
-Normal		7.38	5.28	4.63	4.20	5.52	4.78	3.70	5.15	6.99	7.18	5.96	6.81	67.58
-Maximum Monthly	45	15.77	12.43	9.94	8.74	12.67	16.88	10.21	11.13	12.60	14.53	15.36	19.82	19.82
-Year		1977	1977	1984	1993	1983	1987	1985	1977	1957	1977	1983	1985	DEC 1985
-Minimum Monthly	45	0.24	1.41	0.67	1.13	1.00	0.70	0.84	0.65	1.20	1.56	0.19	1.21	0.19
-Year		1969	1956	1991	1954	1964	1978	1980	1987	1977	1956	1950	1977	NOV 1950
-Maximum in 24 hrs	45	3.20	4.07	2.08	1.92	3.63	3.85	3.12	3.93	2.76	7.44	7.81	4.41	7.81
-Year		1960	1977	1990	1993	1979	1987	1985	1993	1988	1991	1991	1985	NOV 1991
Snow, Ice pellets, hail														
-Maximum Monthly	45	40.1	38.8	74.5	34.8	6.0	T	0.0	0.0	0.4	14.9	30.0	46.4	74.5
-Year		1971	1984	1956	1985	1971	1965			1977	1961	1989	1990	MAR 1956
-Maximum in 24 hrs	45	12.0	19.0	17.8	11.3	4.0	T	0.0	0.0	0.4	10.0	14.0	15.7	19.0
-Year		1976	1992	1956	1985	1968	1965			1977	1970	1965	1990	FEB 1992
WIND:														
Mean Speed (mph)	40	12.8	12.4	12.3	11.6	10.6	9.3	7.6	8.2	9.7	11.2	12.4	12.4	10.9
Prevailing Direction														
Fastest Obs. 1 Min.														
-Direction (!!)	20	31	30	30	10	30	06	30	02	31	31	09	25	31
-Speed (MPH)	20	55	48	55	46	40	36	32	45	60	49	53	50	60
-Year		1993	1974	1989	1988	1986	1989	1984	1969	1986	1985	1968	1968	SEP 1986
Peak Gust														
-Direction (!!)	14	NW	NW	NW	E	W	NE	NW	NW	NW	NW	NW	NE	NW
-Speed (mph)	14	75	67	82	67	59	52	52	67	78	70	74	69	82
-Date		1993	1984	1989	1988	1986	1987	1984	1984	1986	1985	1990	1989	MAR 1989

PRECIPITATION (inches) KODIAK, ALASKA

YEAR	JAN	FEB	MAR	APR	MAY	JUNE	JULY	AUG	SEP	OCT	NOV	DEC	ANNUAL
1964	6.72	4.76	2.23	1.40	1.00	4.49	1.23	9.30	5.80	3.87	3.19	2.85	46.84
1965	6.16	2.54	8.12	3.88	5.42	11.78	6.33	2.24	7.30	6.72	4.82	3.18	68.49
1966	8.82	1.83	6.23	3.80	1.98	8.59	3.22	7.50	7.90	7.25	1.81	5.35	64.28
1967	3.34	4.16	2.21	3.09	2.13	5.67	1.01	8.25	7.75	4.86	6.16	4.58	53.21
1968	1.97	9.61	3.32	4.42	1.97	2.07	7.60	3.68	5.85	4.04	7.34	2.99	54.86
1969	0.24	4.13	3.89	5.46	3.30	7.56	2.00	3.25	9.35	12.38	5.96	12.19	69.71
1970	3.62	8.39	5.96	1.43	3.10	2.94	4.04	7.44	6.39	4.48	2.52	4.75	55.06
1971	6.74	8.82	4.28	4.31	11.89	8.50	7.96	4.86	6.59	4.70	4.52	2.30	75.47
1972	3.70	4.92		1.30	4.57	2.12	1.11	5.93	2.54	5.94	4.13		
1973	0.31	3.59	3.21	5.27	10.75	1.70	6.06	2.43	6.61	6.37	0.68	7.23	54.21
1974	2.86	3.94	6.69	6.15	3.66	4.57	3.32	4.25	10.36	13.07	7.15	5.38	71.40
1975	10.52	4.71	4.12	3.98	11.37	4.55	3.86	5.97	5.65	7.99	7.82	9.31	79.85
1976	10.54	2.97	2.11	3.53	4.26	4.34	4.09	8.89	12.14	5.53	14.79	11.14	84.33
1977	15.77	12.43	1.36	4.04	5.11	4.10	6.01	11.13	1.20	14.53	3.91	1.21	80.80
1978	11.58	11.56	3.63	6.49	11.05	0.70	4.04	3.53	5.85	12.18	6.69	9.95	87.25
1979	8.66	1.50	6.23	3.39	9.29	4.83	3.02	3.71	10.34	10.84	5.19	1.37	68.37
1980	6.09	9.58	5.11	5.85	6.34	2.15	0.84	1.78	8.62	9.38	7.12	4.64	67.50
1981	13.65	6.43	8.05	2.86	7.26	1.87	2.53	4.35	7.47	3.15	7.29	5.22	70.13
1982	7.22	2.12	1.99	3.60	5.55	7.29	3.81	2.23	10.10	3.54	7.58	13.49	68.52
1983	9.19	7.56	8.34	2.25	12.67	7.88	2.16	0.73	2.93	5.36	15.36	2.43	76.86
1984	10.11	5.75	9.94	6.27	4.90	5.15	3.04	1.39	8.37	3.15	4.98	5.50	68.55
1985	14.41	1.91	5.68	5.81	1.53	6.64	10.21	2.96	8.05	3.26	4.44	19.82	84.72
1986	11.63	6.31	4.59	2.69	2.27	13.16	2.72	6.42	2.22	11.37	5.35	11.64	80.37
1987	10.21	4.75	6.31	4.93	3.67	16.88	1.25	0.65	7.83	8.08	4.98	2.49	72.03
1988	6.91	11.50	8.59	6.65	2.42	1.59	4.11	8.28	6.55	10.20	7.93	11.48	86.21
1989	2.53	1.44	3.61	4.13	2.02	7.22	3.86	5.92	6.10	7.04	5.58	12.59	62.04
1990	6.79	2.90	5.56	6.63	4.75	2.14	7.93	7.86	8.96	5.40	2.32	4.91	66.15
1991	10.68	10.80	0.67	8.03	9.74	8.65	2.29	4.77	10.19	11.49	12.96	5.82	96.09
1992	9.07	2.14	5.49	2.10	7.52	5.59	2.25	6.02	2.69	4.84	4.83	4.97	57.51
1993	2.97	7.11	8.17	8.74	5.18	3.51	2.33	8.41	5.44	9.80	5.56	11.68	78.90
Record Mean	6.55	5.01	4.63	4.00	5.23	4.82	3.71	4.81	6.55	6.77	5.99	6.32	64.39

AVERAGE TEMPERATURE (deg. F) KODIAK, ALASKA

YEAR	JAN	FEB	MAR	APR	MAY	JUNE	JULY	AUG	SEP	OCT	NOV	DEC	ANNUAL
1964	31.9	30.8	30.2	35.2	43.5	50.3	54.1	54.0	50.8	41.9	35.5	27.5	40.5
1965	30.8	28.0	38.6	40.6	40.8	46.2	51.9	53.1	53.3	37.5	33.2	26.6	40.1
1966	31.3	28.6	21.8	35.0	40.3	46.8	53.6	53.1	49.2	34.4	34.3	29.8	38.2
1967	27.6	28.7	34.3	38.0	46.1	49.0	57.9	54.6	49.3	38.6	36.8	29.6	40.9
1968	27.9	31.2	33.3	36.5	47.4	49.6	54.6	56.8	48.5	38.5	35.8	25.3	40.5
1969	25.8	30.3	33.8	38.7	42.3	50.7	57.1	56.2	50.1	44.0	33.3	36.7	41.6
1970	26.2	36.0	37.8	35.8	45.2	51.9	54.5	51.4	52.4	39.3	37.8	25.3	41.1
1971	19.9	27.3	23.6	34.8	40.6	48.4	51.8	54.9	48.7	40.6	31.1	28.6	37.5
1972	25.4	27.3		31.1	41.8	45.7	53.8	53.5	47.3	38.0	33.2		
1973	22.3	31.3	30.0	38.0	41.2	46.4	50.4	51.5	46.2	37.9	32.7	32.2	38.4
1974	27.2	24.8	31.5	36.9	43.0	49.7	52.3	55.9	50.6	40.3	33.1	27.9	39.5
1975	25.5	23.2	27.6	33.4	39.0	45.3	51.1	51.9	47.8	39.6	26.8	25.6	36.4
1976	30.1	27.3	28.4	36.2	43.9	51.0	55.0	55.3	49.0	39.3	38.7	32.1	40.5
1977	37.0	36.1	31.5	34.8	41.9	49.1	54.4	53.1	49.9	39.4	30.6	27.2	40.4
1978	35.9	33.2	35.4	39.7	43.8	50.2	52.3	57.1	50.9	44.1	39.3	35.8	43.1
1979	38.1	27.1	37.9	44.2	45.9	53.3	58.6	57.1	52.9	45.6	38.5	26.8	43.8
1980	29.6	34.5	36.4	40.8	45.1	52.7	55.9	56.1	52.2	43.1	37.5	29.1	42.8
1981	38.2	34.0	39.3	41.9	47.2	52.2	57.5	56.1	50.2	43.0	36.3	31.2	43.9
1982	32.7	28.1	35.7	36.0	43.5	49.2	55.0	55.1	49.9	39.1	36.5	37.2	41.5
1983	30.7	37.0	40.5	41.8	46.6	51.9	57.1	58.0	51.5	42.8	38.1	38.1	44.5
1984	32.7	25.5	38.5	36.7	43.7	51.2	55.7	57.5	51.3	41.6	33.9	37.6	42.2
1985	39.2	28.8	35.4	31.8	41.5	46.2	52.3	54.2	51.1	36.5	32.8	36.4	40.5
1986	33.3	30.3	32.4	34.2	43.5	47.4	54.2	53.0	50.5	44.3	36.3	35.8	41.3
1987	34.0	36.2	35.4	38.1	44.8	47.7	55.9	56.5	48.4	41.8	33.3	28.0	41.7
1988	32.2	34.3	34.7	36.8	45.0	51.6	54.1	55.1	48.7	41.2	32.7	30.4	41.4
1989	21.7	33.0	33.1	39.4	44.7	50.1	57.2	57.1	52.0	41.1	32.1	34.7	41.4
1990	29.4	24.3	35.2	41.0	45.7	52.7	53.7	57.1	51.5	39.8	30.2	31.1	41.0
1991	30.3	27.5	33.6	39.0	44.9	49.7	53.8	54.3	50.3	41.7	37.5	30.2	41.1
1992	33.3	27.7	32.3	39.0	44.5	50.9	55.4	53.7	46.4	40.2	35.4	29.9	40.7
1993	27.3	30.9	35.3	40.3	47.1	50.9	57.1	57.1	50.7	42.4	37.5	35.0	42.6
Record Mean	30.4	30.3	32.9	37.0	43.5	49.7	54.3	54.8	49.9	40.7	34.8	30.9	40.8
Max	34.9	35.2	38.0	42.3	48.7	55.0	59.8	60.5	55.6	46.3	39.6	35.7	46.0
Min	25.9	25.4	27.7	31.8	38.3	44.3	48.8	49.1	44.2	35.0	30.0	26.0	35.6

REFERENCE NOTES KODIAK, ALASKA

GENERAL
T - TRACE AMOUNT.
BLANK ENTRIES DENOTE MISSING/UNREPORTED DATA.
# INDICATES A STATION OR INSTRUMENT RELOCATION.

SPECIFIC

(a) - LENGTH OF RECORD IN YEARS, ALTHOUGH INDIVIDUAL MONTHS MAY BE MISSING.
0.* OR * - THE VALUE IS BETWEEN 0.0 AND 0.05
NORMALS - BASED ON THE 1961-1990 RECORD PERIOD.
EXTREMES - DATES ARE THE MOST RECENT OCCURRENCE.
WIND DIR. - NUMERALS SHOW TENS OF DEGRESS CLOCKWISE FROM TRUE NORTH. "00" INDICATES CALM.
RESULTANT DIRECTIONS ARE GIVEN TO WHOLE DEGREES.
BOLD VALUES INDICATE EXTREME VALUES WHICH OCCURRED AFTER THE ASOS SYSTEM WAS COMMISSIONED.

RECORD = PERIOD OF RECORD
RECORD MEAN PRECIPITATION IS THE MEAN OF ALL DAILY PRECIPITATION AMOUNTS DURING THE PERIOD OF RECORD.
RECORD MAX(MIN) TEMPERATURE IS THE MEAN OF ALL DAILY MAX(MIN) TEMPERATURES DURING THE PERIOD OF RECORD.
RECORD MEAN TEMPERATURE IS THE SUM OF THE RECORD MAX AND RECORD MIN DIVIDED BY 2.
AVERAGE TEMPERATURE IS THE SUM OF THE MEAN DAILY MAX AND MIN TEMPERATURE DIVIDED BY 2.

EXCEPTIONS

1. MEAN WIND SPEED IS 1949-1968 AND 1973 TO DATE.
2. THUNDERSTORMS, HEAVY FOG, AND DAYS CLEAR-PARTLY CLOUDY-CLOUDY ARE 1970-1971 AND 1973 TO DATE.

RECORD MEANS ARE THROUGH THE CURRENT YEAR, BEGINNING IN 1949 FOR TEMPERATURE
1949 FOR PRECIPITATION
1963 FOR SNOWFALL

HEATING DEGREE DAYS Base 65 deg. F KODIAK, ALASKA

SEASON	JULY	AUG	SEP	OCT	NOV	DEC	JAN	FEB	MAR	APR	MAY	JUNE	TOTAL
1964-65	337	342	425	715	884	1163	1060	1037	817	733	751	565	8829
1965-66	406	368	350	851	954	1190	1046	1019	1340	900	766	547	9737
1966-67	355	370	474	948	922	1091	1160	1017	953	811	587	481	9169
1967-68	230	323	472	818	846	1098	1150	981	982	855	547	461	8763
1968-69	322	260	496	823	875	1231	1204	963	960	780	695	423	9032
1969-70	242	267	438	648	946	871	1196	806	834	870	605	385	8108
1970-71	319	414	372	788	807	1224	1391	1049	1277	903	751	491	9786
1971-72	403	311	482	749	1011	1121	1224	1086		1011	713	571	
1972-73	340	348	523	831	948		1309	936	1080	803	730	550	
1973-74	447	412	558	837	960	1009	1164	1117	1033	836	672	456	9501
1974-75	386	276	427	758	952	1143	1220	1165	1148	946	800	586	9807
1975-76	424	396	507	782	1138	1215	1074	1087	1128	856	646	415	9668
1976-77	308	292	475	787	782	1013	861	803	1031	899	710	468	8429
1977-78	321	360	446	785	1026	1163	894	884	911	752	650	436	8628
1978-79	383	239	417	640	765	902	829	1055	837	620	586	344	7617
1979-80	200	236	357	597	785	1177	1090	877	879	720	611	361	7890
1980-81	274	270	378	675	820	1105	825	867	788	686	542	379	7609
1981-82	225	272	438	674	854	1038	996	1028	900	864	658	466	8413
1982-83	302	298	447	798	847	855	1053	780	754	688	561	385	7768
1983-84	237	213	400	681	802	826	993	1137	813	844	655	408	8009
1984-85	283	229	401	717	927	845	795	1009	909	989	719	559	8382
1985-86	386	325	410	873	958	879	976	966	1004	916	659	523	8875
1986-87	330	364	427	638	856	899	957	801	914	801	619	514	8120
1987-88	287	259	490	715	944	1139	1011	884	935	840	616	394	8514
1988-89	331	298	484	729	963	1066	1336	892	980	763	621	442	8905
1989-90	236	237	384	735	983	937	1098	1137	916	712	589	363	8327
1990-91	346	239	398	772	1039	1043	1067	1045	965	773	617	453	8757
1991-92	342	324	432	713	820	1073	978	1075	1009	773	627	415	8581
1992-93	289	345	553	764	880	1082	1163	947	910	733	549	418	8633
1993-94	248	246	422	692	817	921							

COOLING DEGREE DAYS Base 65 deg. F KODIAK, ALASKA

YEAR	JAN	FEB	MAR	APR	MAY	JUNE	JULY	AUG	SEP	OCT	NOV	DEC	TOTAL
1973	0	0	0	0	0	0	0	0	0	0	0	0	0
1974	0	0	0	0	0	3	0	0	0	0	0	0	3
1975	0	0	0	0	0	0	0	0	0	0	0	0	0
1976	0	0	0	0	0	0	5	0	0	0	0	0	5
1977	0	0	0	0	0	0	0	0	0	0	0	0	0
1978	0	0	0	0	0	0	0	0	0	0	0	0	0
1979	0	0	0	0	0	0	8	0	0	0	0	0	8
1980	0	0	0	0	0	0	0	1	0	0	0	0	1
1981	0	0	0	0	0	0	0	0	0	0	0	0	0
1982	0	0	0	0	0	0	0	0	0	0	0	0	0
1983	0	0	0	0	0	0	0	3	0	0	0	0	3
1984	0	0	0	0	0	0	3	1	0	0	0	0	4
1985	0	0	0	0	0	0	0	0	0	0	0	0	0
1986	0	0	0	0	0	0	1	0	0	0	0	0	1
1987	0	0	0	0	0	0	7	0	0	0	0	0	7
1988	0	0	0	0	0	0	0	0	0	0	0	0	0
1989	0	0	0	0	0	1	1	0	0	0	0	0	2
1990	0	0	0	0	0	0	0	0	0	0	0	0	0
1991	0	0	0	0	0	0	0	0	0	0	0	0	0
1992	0	0	0	0	0	0	0	0	0	0	0	0	0
1993	0	0	0	0	0	0	8	9	0	0	0	0	17

SNOWFALL (inches) KODIAK, ALASKA

SEASON	JULY	AUG	SEP	OCT	NOV	DEC	JAN	FEB	MAR	APR	MAY	JUNE	TOTAL
1964-65	0.0	0.0	0.0	10.4	1.9	10.7	17.2	18.9	13.1	21.9	1.3	T	95.4
1965-66	0.0	0.0	T	2.1	24.3	13.8	26.7	11.6	51.7	17.5	T	0.0	147.7
1966-67	0.0	0.0	0.0	3.8	5.2	14.0	18.7	24.7	16.7	T	0.0	0.0	83.1
1967-68	0.0	0.0	0.0	3.3	1.0	10.7	11.2	22.1	3.0	6.0	4.0	0.0	61.3
1968-69	0.0	0.0	T	0.3	T	28.2	0.6	32.1	28.9	0.5	T	0.0	90.6
1969-70	0.0	0.0	0.0	T	3.6	3.0	20.2	22.5	2.7	5.2	T	0.0	57.2
1970-71	0.0	0.0	0.0	12.6	T	23.1	40.1	16.8	34.6	25.7	6.0	0.0	158.9
1971-72	0.0	0.0	0.0	T	7.3	12.2	32.8	35.5		9.9	2.8	0.0	
1972-73	0.0	0.0	T	T	5.2		1.9	24.9	12.6	9.3	1.5	0.0	
1973-74	0.0	0.0	0.0	10.6	1.9	14.3	18.6	16.7	12.9	3.3	0.4	0.0	78.7
1974-75	0.0	0.0	0.0	1.4	13.0	14.3	32.2	27.1	23.6	5.2	0.2	0.0	117.0
1975-76	0.0	0.0	0.0	1.2	20.5	33.1	39.7	13.8	13.2	17.0	T	0.0	138.5
1976-77	0.0	0.0	0.0	0.6	8.5	10.2	T	12.2	8.7	1.4	0.2	0.0	41.8
1977-78	0.0	0.0	0.4	T	12.0	14.1	3.9	16.7	7.0	1.6	0.0	0.0	55.7
1978-79	0.0	0.0	0.0	0.0	3.9	8.6	4.2	10.6	18.0	0.1	T	0.0	45.4
1979-80	0.0	0.0	0.0	T	9.7	4.7	15.3	13.4	7.1	1.0	T	0.0	51.2
1980-81	0.0	0.0	0.0	T	0.8	10.6	3.3	10.9	T	1.7	T	0.0	27.3
1981-82	0.0	0.0	0.0	0.2	3.1	9.6	14.2	4.0	9.5	19.8	1.4	0.0	61.8
1982-83	0.0	0.0	0.0	6.0	0.4	2.4	6.4	1.0	0.6	1.8	T	0.0	18.6
1983-84	0.0	0.0	0.0	6.1	7.2	T	19.2	38.8	2.0	9.0	T	0.0	82.3
1984-85	0.0	0.0	0.0	T	2.9	6.0	0.1	12.9	30.0	34.8	0.8	0.0	87.5
1985-86	0.0	0.0	0.0	4.5	9.0	11.0	23.5	17.0	6.4	11.9	0.8	0.0	84.1
1986-87	0.0	0.0	0.0	0.0	4.2	2.1	8.3	2.0	9.9	4.3	T	0.0	30.8
1987-88	0.0	0.0	0.0	0.0	5.8	6.9	7.6	24.4	15.8	16.6	T	0.0	77.1
1988-89	0.0	0.0	0.0	T	8.6	14.4	18.2	9.3	5.7	0.4	0.0	0.0	56.6
1989-90	0.0	0.0	0.0	0.5	30.0	8.3	35.8	35.1	15.9	3.6	0.0	0.0	129.2
1990-91	0.0	0.0	0.0	3.7	5.0	46.4	20.1	13.2	5.0	11.0	T	0.0	104.4
1991-92	0.0	0.0	0.0	0.0	24.8	24.0	7.0	27.6	15.0	0.8	0.8	0.0	100.0
1992-93	0.0	0.0	T	T	1.2	29.4	6.9	4.2	3.8	1.0	T	0.0	46.5
1993-94	0.0	0.0	0.0	0.2	0.5	6.4							
Record Mean	0.0	0.0	T	2.3	7.7	13.2	15.4	17.8	13.0	8.0	0.7	T	78.0

KOTZEBUE, ALASKA

Kotzebue is located 26 miles inside the Arctic Circle and very near the north end of a long narrow peninsula bounded on the north and west by Kotzebue Sound and on the east by Hotham Inlet, known locally as Kobuk Lake. These water bodies produce a maritime type of climate when the water is ice-free, which is roughly from late May to late October, although the western portion of the sound is not completely frozen until about December and not completely free of ice again until the middle of July. Local topography is nearly uniform with a general low relief, so that there are no significant terrain barriers in the immediate area to impede surface air flow or produce pronounced local variations in temperature and precipitation. The mountainous Seward Peninsula to the south, however, does deflect some low pressure systems which originate in or beyond the Bering Sea area and move toward this region.

During the ice-free period cloudy skies prevail, fog occurs, daily temperatures are relatively uniform, relative humidity is high, and westerly winds predominate. These normal conditions are altered only by cyclonic storms or by pressure systems strong enough to overcome local circulation tendencies.

When the water surrounding the peninsula becomes frozen, the climatic characteristics approach the continental type. The change from maritime to approximately continental conditions becomes progressively more pronounced as the ice cover advances across the sound toward the Arctic Ocean. A similar, but inverse, change occurs as the ice diminishes.

Average winter temperatures are not as severe as might be expected at this latitude. Cyclonic storms and the influence of the Arctic Ocean, which is often relatively free of ice, moderate the winter temperatures.

Precipitation is very light. Snow generally falls in every month of the year except July and August. The total for a normal year is 8 inches.

Cyclonic storms are frequent, especially from October to April, and many of them are accompanied by high winds and blizzard conditions during the winter months. The absence of pronounced sheltering terrain results in unimpeded air movement throughout the year. Windy weather, somewhat characteristic of the area, results largely from the numerous cyclonic storms and uneven heat radiation of adjoining land and water areas.

NORMALS, MEANS, AND EXTREMES

KOTZEBUE, ALASKA

LATITUDE: 66°52' N LONGITUDE: 162°38' W ELEVATION: FT. GRND 10 BARO 20 TIME ZONE: YUKON WBAN: 26616

	(a)	JAN	FEB	MAR	APR	MAY	JUNE	JULY	AUG	SEP	OCT	NOV	DEC	YEAR
TEMPERATURE °F:														
Normals														
-Daily Maximum		5.6	2.3	8.7	20.3	37.9	49.7	59.1	57.0	46.9	27.6	13.1	5.5	27.8
-Daily Minimum		-7.5	-12.0	-8.0	2.3	24.5	37.8	48.5	47.3	37.0	18.2	2.4	-7.4	15.3
-Monthly		-1.0	-4.8	0.4	11.3	31.2	43.8	53.8	52.1	42.0	22.9	7.8	-0.9	21.6
Extremes														
-Record Highest	51	39	40	39	48	74	85	85	80	69	51	38	37	85
-Year		1961	1989	1973	1988	1947	1991	1958	1968	1978	1984	1983	1982	JUN 1991
-Record Lowest	51	-49	-52	-48	-44	-18	20	30	30	13	-19	-36	-47	-52
-Year		1989	1968	1956	1943	1945	1948	1976	1980	1992	1975	1948	1974	FEB 1968
NORMAL DEGREE DAYS:														
Heating (base 65°F)		2046	1954	2003	1611	1048	636	353	407	690	1305	1716	2043	15812
Cooling (base 65°F)		0	0	0	0	0	0	6	7	0	0	0	0	13
% OF POSSIBLE SUNSHINE														
MEAN SKY COVER (tenths)														
Sunrise - Sunset	49	5.8	5.2	5.5	5.8	6.3	6.6	7.3	7.8	7.3	6.9	6.5	5.9	6.4
MEAN NUMBER OF DAYS:														
Sunrise to Sunset														
-Clear	49	11.2	12.1	11.8	9.9	7.7	5.6	4.3	3.5	5.9	7.6	8.7	10.8	99.1
-Partly Cloudy	49	4.1	3.4	5.5	6.7	8.9	10.3	8.0	6.4	5.0	4.6	4.0	4.4	71.3
-Cloudy	49	15.6	12.7	13.7	13.4	14.4	14.0	18.7	21.1	19.2	18.8	17.3	15.9	194.7
Precipitation														
.01 inches or more	50	7.9	7.0	7.7	7.0	5.8	6.4	10.5	14.1	12.4	10.7	9.8	9.3	108.5
Snow,Ice pellets,hail														
1.0 inches or more	50	2.1	1.6	2.1	1.6	0.5	0.*	0.0	0.0	0.4	2.2	3.1	2.6	16.2
Thunderstorms	43	0.0	0.0	0.0	0.0	0.0	0.2	0.2	0.2	0.*	0.0	0.0	0.0	0.6
Heavy Fog Visibility														
1/4 mile or less	50	1.1	1.1	0.6	1.6	3.6	4.7	2.2	0.9	0.8	0.6	0.5	0.9	18.6
Temperature °F														
-Maximum														
70° and above	50	0.0	0.0	0.0	0.0	0.1	0.8	2.7	1.2	0.0	0.0	0.0	0.0	4.8
32° and below	50	30.4	27.6	30.5	24.1	8.0	0.*	0.0	0.0	0.6	20.1	28.5	30.4	200.3
-Minimum														
32° and below	50	31.0	28.3	31.0	29.8	25.4	6.7	0.*	0.1	7.8	28.4	30.0	31.0	249.4
0° and below	50	21.2	21.0	22.3	13.8	1.3	0.0	0.0	0.0	0.0	1.9	13.1	21.5	116.1
AVG. STATION PRESS.(mb)	20	1011.3	1014.3	1012.6	1012.9	1011.5	1011.1	1011.9	1010.4	1008.7	1007.8	1008.4	1009.0	1010.8
RELATIVE HUMIDITY (%)														
Hour 03	30	72	71	72	79	86	89	87	88	85	82	76	74	80
Hour 09 (Local Time)	30	72	72	72	77	81	84	82	85	83	81	77	74	78
Hour 15	30	72	71	71	75	78	79	76	77	73	76	76	74	75
Hour 21	30	72	71	73	79	82	81	80	82	81	81	76	75	78
PRECIPITATION (inches):														
Water Equivalent														
-Normal		0.43	0.32	0.35	0.37	0.33	0.52	1.46	1.78	1.58	0.73	0.59	0.52	8.98
-Maximum Monthly	51	1.77	1.23	1.23	1.41	1.05	1.43	3.51	5.18	4.31	3.20	2.22	1.40	5.18
-Year		1973	1989	1954	1989	1989	1979	1981	1951	1977	1993	1990	1987	AUG 1951
-Minimum Monthly	51	T	0.00	T	T	T	0.01	0.01	0.08	0.03	0.04	0.09	0.02	0.00
-Year		1966	1979	1986	1943	1979	1956	1977	1991	1967	1954	1968	1956	FEB 1979
-Maximum in 24 hrs	51	0.81	0.68	0.47	0.43	0.56	0.80	1.78	1.48	1.65	0.98	0.53	0.40	1.78
-Year		1963	1951	1954	1989	1963	1953	1946	1954	1978	1993	1982	1944	JUL 1946
Snow,Ice pellets,hail														
-Maximum Monthly	50	23.9	14.0	21.9	18.1	12.0	2.4	0.1	0.3	7.4	18.0	24.3	23.6	24.3
-Year		1973	1956	1954	1967	1964	1963	1959	1962	1957	1971	1993	1988	NOV 1993
-Maximum in 24 hrs	50	10.0	6.8	8.6	4.6	4.7	2.1	0.1	0.3	5.0	7.4	6.4	7.5	10.0
-Year		1973	1951	1954	1972	1984	1963	1959	1962	1960	1973	1982	1957	JAN 1973
WIND:														
Mean Speed (mph)	47	14.4	12.9	12.1	12.3	11.1	12.2	12.8	13.2	13.3	13.7	14.6	13.2	13.0
Prevailing Direction through 1963		E	E	E	ESE	W	W	W	W	E	ENE	E	NE	E
Fastest Obs. 1 Min.														
-Direction (!!)	40	09	11	11	11	10	29	16	30	29	09	16	09	11
-Speed (MPH)	40	64	93	55	62	40	42	51	49	52	47	88	66	93
-Year		1950	1951	1954	1950	1985	1950	1951	1969	1962	1958	1950	1950	FEB 1951
Peak Gust														
-Direction (!!)	14	E	E	E	NW	NE	SE	SE	S	NE	SE	SE	E	E
-Speed (mph)	14	72	63	66	47	49	46	45	53	54	60	63	68	72
-Date		1982	1990	1985	1992	1993	1982	1984	1993	1993	1992	1993	1992	JAN 1982

PRECIPITATION (inches)

KOTZEBUE, ALASKA

YEAR	JAN	FEB	MAR	APR	MAY	JUNE	JULY	AUG	SEP	OCT	NOV	DEC	ANNUAL
1964	0.14	0.48	0.02	0.10	0.66	0.14	0.45	1.32	1.76	0.59	0.58	0.39	6.63
1965	0.37	0.26	0.86	0.34	0.49	0.13	1.79	3.75	2.65	0.30	0.37	0.79	12.10
1966	T	0.18	0.07	0.17	0.04	0.18	1.25	0.91	1.40	0.48	0.64	0.15	5.47
1967	0.28	0.11	0.34	0.86	0.24	0.34	2.61	1.32	0.03	0.33	0.45	0.37	7.28
1968	0.31	0.37	0.14	0.49	0.94	0.53	1.47	1.38	0.44	0.38	0.09	0.15	6.69
1969	0.22	0.18	0.26	0.18	0.14	0.82	0.66	1.32	0.05	0.55	0.34	0.35	5.07
1970	0.25	0.18	0.22	0.08	0.04	0.05	1.27	1.44	0.60	0.77	0.81	0.29	6.00
1971	0.16	0.17	0.10	0.16	0.26	0.24	2.98	0.38	0.74	0.76	0.24	0.77	6.96
1972	0.51	0.02	0.12	0.67	0.35	0.59	0.46	1.17	1.34	1.00	0.30	0.27	6.80
1973	1.77	0.41	0.74	0.12	0.09	0.52	1.73	3.27	1.51	1.15	1.31	0.22	12.84
1974	0.18	0.39	0.04	0.09	0.04	0.13	1.90	1.78	1.28	0.05	0.78	0.19	6.85
1975	0.34	0.65	0.16	0.43	0.14	0.66	1.44	1.51	0.88	0.86	0.60	0.35	8.02
1976	0.31	0.21	0.82	0.05	0.12	0.68	1.16	0.48	1.34	0.45	0.53	0.44	6.59
1977	0.45	0.26	0.31	0.42	0.31	0.02	0.01	0.71	4.31	1.13	0.12	0.53	8.58
1978	0.07	0.10	0.10	0.37	T	0.82	1.43	0.59	2.49	0.30	0.69	0.71	7.67
1979	0.01	0.00	0.10	0.38	T	1.43	0.79	2.79	1.61	0.46	1.79	0.51	9.87
1980	0.71	0.60	0.62	0.35	0.03	1.12	1.52	0.92	0.54	0.52	0.15	0.33	7.41
1981	0.89	0.48	0.60	0.44	0.16	0.09	3.51	1.12	0.39	0.58	0.43	0.61	9.30
1982	1.29	0.60	0.52	0.33	0.22	0.70	1.24	1.38	1.76	1.12	1.12	0.36	10.64
1983	0.29	0.15	0.27	0.72	0.13	0.55	1.63	3.36	1.68	0.99	1.04	0.51	11.32
1984	0.54	0.01	0.11	0.24	0.64	0.38	2.11	2.98	1.70	0.35	0.50	0.69	10.25
1985	0.39	0.11	0.77	0.46	0.50	0.60	0.06	2.03	2.18	2.39	0.69	0.56	10.74
1986	0.43	0.44	T	0.22	0.01	0.34	1.42	2.39	3.29	0.26	0.30	0.87	9.97
1987	0.47	0.57	0.21	0.09	0.05	0.71	1.15	0.66	1.99	0.41	0.34	1.40	8.05
1988	0.19	0.31	0.77	0.05	0.83	0.15	0.91	2.35	0.99	0.64	0.28	1.37	8.84
1989	0.27	1.23	0.14	1.41	1.05	0.25	2.41	3.17	1.80	1.04	0.36	0.55	13.68
1990	0.36	0.55	0.75	0.45	0.37	0.85	0.92	2.30	3.01	1.91	2.22	1.07	14.76
1991	1.01	0.56	1.00	0.51	0.22	0.12	1.99	0.08	1.07	1.69	0.12	0.29	8.66
1992	0.26	0.80	0.88	0.10	0.40	0.50	0.37	3.00	0.59	0.84	1.16	0.92	9.82
1993	0.59	0.25	0.64	0.08	0.18	0.93	0.69	1.78	2.45	3.20	2.00	0.62	13.41
Record Mean	0.39	0.34	0.38	0.33	0.34	0.53	1.47	2.09	1.54	0.76	0.56	0.46	9.20

AVERAGE TEMPERATURE (deg. F)

KOTZEBUE, ALASKA

YEAR	JAN	FEB	MAR	APR	MAY	JUNE	JULY	AUG	SEP	OCT	NOV	DEC	ANNUAL
1964	-6.9	-21.2	-10.2	5.9	24.5	39.3	50.2	52.9	41.7	26.6	7.8	-12.3	16.5
1965	-7.0	-14.1	11.4	14.6	24.5	44.0	51.1	47.3	43.5	16.6	16.5	-5.6	20.2
1966	-3.6	-4.3	-12.0	9.1	23.8	39.1	52.9	52.3	43.4	22.0	13.5	-7.9	19.0
1967	1.5	-2.6	10.5	19.5	30.4	46.0	52.8	51.1	42.7	23.7	6.1	2.4	23.7
1968	1.4	-17.5	-0.3	10.6	29.4	39.9	57.6	56.0	39.6	22.8	5.8	-5.0	20.0
1969	1.6	0.7	1.0	16.8	38.8	49.7	51.8	45.4	45.5	30.9	1.4	9.0	24.4
1970	-13.2	3.9	6.5	10.9	29.7	44.8	53.8	50.5	36.0	14.3	12.9	-2.1	20.7
1971	-13.7	-18.9	-7.2	9.6	27.1	47.4	56.0	48.7	41.4	21.7	6.3	0.1	18.2
1972	-4.4	-8.6	-13.3	7.3	34.0	43.2	59.0	55.0	41.1	29.2	9.8	1.7	21.2
1973	-9.8	4.3	-5.2	9.5	30.6	42.1	49.2	48.7	42.6	21.5	5.6	2.0	20.1
1974	-3.7	-19.3	-0.3	12.9	30.1	39.6	53.3	55.2	47.7	19.4	2.8	-17.1	18.4
1975	-9.8	-3.8	5.1	9.0	28.3	37.5	50.3	51.5	36.9	20.4	-1.8	-6.6	18.1
1976	-5.1	-17.6	0.8	6.0	30.0	39.4	50.9	55.2	44.6	24.8	12.0	-4.3	19.7
1977	12.9	4.3	-13.3	3.6	30.5	43.8	58.8	58.9	43.9	24.6	3.5	1.1	22.7
1978	12.9	0.5	1.0	17.8	36.4	46.8	58.5	56.3	46.4	20.3	15.8	2.1	26.2
1979	11.9	-9.6	7.8	18.8	33.5	48.7	54.0	55.6	42.0	27.1	18.1	-11.3	24.7
1980	-6.3	7.3	10.8	15.9	38.5	47.4	55.6	50.3	38.9	28.8	10.8	-10.2	24.0
1981	15.2	1.0	10.9	14.6	34.9	48.5	51.4	51.4	40.7	24.2	8.9	-1.3	25.1
1982	0.0	4.6	3.1	6.6	28.8	47.6	56.3	53.5	42.8	20.1	12.2	7.3	23.6
1983	-5.2	-2.0	7.5	22.2	37.6	42.3	54.1	47.8	38.7	19.4	15.6	12.1	24.2
1984	-3.3	-21.0	5.8	0.4	24.1	42.2	51.3	47.0	45.0	28.2	8.0	5.1	19.4
1985	16.2	-7.0	-4.3	-5.4	26.6	39.3	54.1	53.4	40.3	20.8	10.9	12.9	21.5
1986	-3.0	7.1	-4.9	4.1	29.7	43.6	53.3	50.0	41.2	21.3	9.6	9.8	21.8
1987	-2.0	-0.4	4.2	7.8	32.7	47.9	56.6	53.4	37.7	30.7	5.3	-2.6	22.6
1988	7.5	1.4	-2.3	17.7	34.4	44.7	52.2	52.0	42.9	18.1	1.5	3.9	22.8
1989	-19.9	17.7	5.3	21.0	29.6	44.1	54.2	53.9	45.6	23.6	0.4	5.6	23.4
1990	-8.6	-24.1	1.8	15.9	38.3	45.7	58.8	56.8	40.1	24.5	4.0	-6.9	20.5
1991	-3.0	-4.7	-2.9	14.5	33.1	50.3	54.4	51.8	45.8	29.3	7.6	-4.7	22.6
1992	1.2	-8.7	-4.9	9.3	23.2	43.8	57.4	49.9	32.2	18.8	6.6	-2.4	18.9
1993	-9.0	4.5	0.7	19.5	33.5	47.8	59.0	49.4	38.4	29.0	11.2	8.9	24.4
Record Mean	-2.7	-4.7	-0.2	11.9	31.1	43.9	53.8	51.4	41.2	23.5	7.9	-1.9	21.3
Max	3.9	2.7	8.3	21.0	37.9	49.9	59.2	56.2	46.2	28.3	13.4	4.5	27.6
Min	-9.4	-12.0	-8.7	2.7	24.3	37.9	48.3	46.5	36.2	18.8	2.5	-8.3	14.9

REFERENCE NOTES

KOTZEBUE, ALASKA

GENERAL
T - TRACE AMOUNT.
BLANK ENTRIES DENOTE MISSING/UNREPORTED DATA.
# INDICATES A STATION OR INSTRUMENT RELOCATION.

SPECIFIC

EXCEPTIONS

RECORD MEANS ARE THROUGH THE CURRENT YEAR,
BEGINNING IN 1943 FOR TEMPERATURE
1943 FOR PRECIPITATION
1943 FOR SNOWFALL

(a) - LENGTH OF RECORD IN YEARS, ALTHOUGH INDIVIDUAL MONTHS MAY BE MISSING.
0.* OR * - THE VALUE IS BETWEEN 0.0 AND 0.05
NORMALS - BASED ON THE 1961-1990 RECORD PERIOD.
EXTREMES - DATES ARE THE MOST RECENT OCCURRENCE
WIND DIR.- NUMERALS SHOW TENS OF DEGRESS CLOCKWISE FROM TRUE NORTH. "00" INDICATES CALM.
RESULTANT DIRECTIONS ARE GIVEN TO WHOLE DEGREES.
BOLD VALUES INDICATE EXTREME VALUES WHICH OCCURRED AFTER THE ASOS SYSTEM WAS COMMISSIONED.

RECORD = PERIOD OF RECORD
RECORD MEAN PRECIPITATION IS THE MEAN OF ALL DAILY PRECIPITATION AMOUNTS DURING THE PERIOD OF RECORD.
RECORD MAX(MIN) TEMPERATURE IS THE MEAN OF ALL DAILY MAX(MIN) TEMPERATURES DURING THE PERIOD OF RECORD.
RECORD MEAN TEMPERATURE IS THE SUM OF THE RECORD MAX AND RECORD MIN DIVIDED BY 2.
AVERAGE TEMPERATURE IS THE SUM OF THE MEAN DAILY MAX AND MIN TEMPERATURE DIVIDED BY 2.

HEATING DEGREE DAYS Base 65 deg. F — KOTZEBUE, ALASKA

SEASON	JULY	AUG	SEP	OCT	NOV	DEC	JAN	FEB	MAR	APR	MAY	JUNE	TOTAL
1964-65	451	366	694	1184	1715	2400	2235	2217	1662	1506	1247	625	16302
1965-66	424	540	639	1495	1449	2191	2129	1941	2391	1675	1270	770	16914
1966-67	372	386	640	1327	1539	2259	1967	1894	1685	1358	1067	562	15056
1967-68	373	430	663	1274	1765	1941	1971	2397	2026	1631	1097	746	16314
1968-69	247	282	753	1299	1774	2171	1967	1803	1986	1439	803	450	14974
1969-70	402	603	580	1051	1909	1733	2434	1710	1813	1626	1086	598	15545
1970-71	342	444	860	1565	1559	2081	2444	2354	2236	1658	1168	538	17249
1971-72	276	499	700	1335	1757	2014	2154	2141	2434	1725	953	646	16634
1972-73	192	304	710	1104	1651	1964	2324	1701	2177	1663	1060	680	15530
1973-74	483	500	663	1342	1779	1953	2129	2363	2025	1559	1074	756	16626
1974-75	353	297	512	1406	1868	2550	2323	1928	1855	1678	1129	819	16718
1975-76	451	415	834	1382	2006	2222	2176	2403	1993	1771	1078	762	17493
1976-77	430	296	606	1237	1586	2152	1610	1701	2436	1843	1063	627	15587
1977-78	204	183	622	1247	1845	1984	1612	1807	1985	1408	883	537	14317
1978-79	199	264	550	1379	1470	1953	1639	2093	1772	1380	969	485	14153
1979-80	332	282	685	1168	1400	2372	2207	1674	1679	1464	812	520	14595
1980-81	284	451	778	1119	1618	2336	1536	1793	1668	1505	925	489	14502
1981-82	419	416	721	1258	1676	2055	2013	1692	1920	1753	1120	526	15569
1982-83	269	348	660	1384	1580	1783	2177	1879	1779	1278	841	675	14653
1983-84	333	527	786	1408	1472	1638	2117	2500	1834	1937	1262	675	16489
1984-85	418	550	591	1133	1706	1858	1509	2019	2153	2115	1184	765	16001
1985-86	333	352	737	1363	1617	1610	2112	1619	2171	1825	1088	634	15461
1986-87	354	458	709	1345	1661	1710	2080	1830	1883	1712	993	508	15243
1987-88	263	356	811	1054	1787	2094	1784	1846	2089	1418	944	603	15049
1988-89	390	395	658	1447	1905	1894	2635	1321	1844	1313	1090	619	15511
1989-90	331	336	577	1275	1936	1843	2283	2499	1961	1468	822	580	15911
1990-91	199	249	739	1249	1827	2229	2107	1952	2106	1509	986	451	15603
1991-92	321	403	568	1098	1719	2164	1977	2135	2166	1667	1291	627	16136
1992-93	236	460	977	1423	1752	2090	2297	1693	1995	1361	970	511	15765
1993-94	184	476	788	1110	1609	1733							

COOLING DEGREE DAYS Base 65 deg. F — KOTZEBUE, ALASKA

YEAR	JAN	FEB	MAR	APR	MAY	JUNE	JULY	AUG	SEP	OCT	NOV	DEC	TOTAL
1969	0	0	0	0	0	0	0	0	0	0	0	0	0
1970	0	0	0	0	0	0	3	0	0	0	0	0	3
1971	0	0	0	0	0	14	2	0	0	0	0	0	16
1972	0	0	0	0	0	0	14	1	0	0	0	0	15
1973	0	0	0	0	0	0	0	0	0	0	0	0	0
1974	0	0	0	0	0	0	0	0	0	0	0	0	0
1975	0	0	0	0	0	0	0	0	0	0	0	0	0
1976	0	0	0	0	0	0	0	0	0	0	0	0	0
1977	0	0	0	0	0	0	18	1	0	0	0	0	19
1978	0	0	0	0	0	0	1	0	0	0	0	0	1
1979	0	0	0	0	0	0	1	0	0	0	0	0	1
1980	0	0	0	0	0	0	0	0	0	0	0	0	0
1981	0	0	0	0	0	0	5	0	0	0	0	0	5
1982	0	0	0	0	0	12	3	0	0	0	0	0	15
1983	0	0	0	0	0	0	0	0	0	0	0	0	0
1984	0	0	0	0	0	0	0	0	0	0	0	0	0
1985	0	0	0	0	0	0	3	1	0	0	0	0	4
1986	0	0	0	0	0	0	0	0	0	0	0	0	0
1987	0	0	0	0	0	5	9	0	0	0	0	0	14
1988	0	0	0	0	0	0	0	0	0	0	0	0	0
1989	0	0	0	0	0	0	0	0	0	0	0	0	0
1990	0	0	0	0	0	4	15	1	0	0	0	0	20
1991	0	0	0	0	0	17	1	0	0	0	0	0	18
1992	0	0	0	0	0	0	8	0	0	0	0	0	8
1993	0	0	0	0	0	0	9	0	0	0	0	0	9

SNOWFALL (inches) — KOTZEBUE, ALASKA

SEASON	JULY	AUG	SEP	OCT	NOV	DEC	JAN	FEB	MAR	APR	MAY	JUNE	TOTAL
1964-65	0.0	0.0	T	8.8	17.1	5.4	6.7	10.3	10.5	6.7	3.1	T	68.6
1965-66	0.0	T	5.1	16.0	10.3	19.3	T	6.8	2.9	4.6	1.3	0.0	66.3
1966-67	0.0	T	0.0	3.7	15.4	4.4	8.7	3.5	8.8	18.1	2.1	T	64.7
1967-68	0.0	0.0	0.0	6.2	9.5	9.3	6.8	9.8	4.8	11.6	3.3	0.0	61.3
1968-69	0.0	0.0	1.2	4.9	4.5	4.6	4.9	3.1	6.9	4.2	0.2	0.0	34.5
1969-70	0.0	T	T	1.0	7.7	7.4	4.0	4.0	4.3	1.2	T	0.0	29.6
1970-71	0.0	0.0	0.6	6.2	14.6	6.0	2.9	2.8	1.7	4.3	3.7	0.2	43.0
1971-72	0.0	0.0	T	18.0	5.9	16.4	13.4	0.1	2.7	11.0	0.9	0.2	68.6
1972-73	0.0	0.0	T	11.1	6.7	6.5	23.9	5.2	10.2	1.9	1.3	T	66.8
1973-74	0.0	T	3.0	14.5	14.3	5.5	1.1	6.2	0.8	2.2	0.4	0.3	48.3
1974-75	0.0	0.0	0.8	0.8	12.0	3.6	5.5	10.1	2.3	7.2	0.5	T	42.8
1975-76	0.0	0.0	3.1	10.9	2.6	8.8	4.4	2.7	11.9	2.6	1.8	T	48.8
1976-77	T	0.0	T	6.0	12.8	6.7	8.2	5.8	4.1	4.3	2.5	0.0	50.4
1977-78	0.0	0.0	0.2	0.5	2.0	5.6	1.1	1.3	2.2	2.9	T	0.0	15.8
1978-79	0.0	0.0	T	5.1	7.4	9.6	0.1	0.0	1.0	1.9	T	T	25.1
1979-80	0.0	0.0	T	4.0	15.0	4.4	7.2	3.8	6.0	3.3	0.0	T	43.7
1980-81	0.0	0.0	2.1	4.1	2.5	3.3	9.6	5.2	7.3	6.3	0.9	0.0	41.3
1981-82	0.0	0.0	0.4	5.3	4.5	6.1	12.9	2.3	6.4	3.3	T	T	41.2
1982-83	0.0	0.0	0.4	12.7	12.3	3.0	3.0	1.8	2.7	2.8	T	T	38.7
1983-84	0.0	T	1.6	4.1	12.3	6.1	6.1	0.2	2.8	5.2	6.7	0.0	45.1
1984-85	0.0	0.0	T	5.6	7.8	9.4	5.3	1.6	9.3	5.3	4.4	0.0	48.7
1985-86	0.0	0.0	1.5	4.1	8.0	6.5	10.1	3.3	T	3.8	T	0.0	37.3
1986-87	0.0	0.0	2.0	4.1	2.9	11.6	6.0	9.2	3.7	1.0	T	T	40.5
1987-88	0.0	0.0	T	3.8	4.7	21.7	2.8	4.8	9.4	0.6	T	T	47.8
1988-89	0.0	0.0	0.4	6.6	4.9	23.6	4.3	11.3	2.0	16.0	3.4	T	72.5
1989-90	0.0	0.0	T	8.8	4.7	8.6	5.8	6.7	9.5	5.1	0.4	0.0	49.6
1990-91	0.0	0.0	3.1	9.6	22.8	15.4	12.7	6.2	13.4	6.9	0.4	T	90.5
1991-92	0.0	0.0	0.0	6.6	1.1	4.4	5.7	6.8	14.0	1.7	3.7	0.0	44.0
1992-93	0.0	0.0	0.4	10.0	14.8	10.8	7.2	3.1	10.2	1.3	0.3	T	58.1
1993-94	0.0	0.0	5.0	4.1	24.3	11.0							
Record Mean	T	T	1.1	6.5	8.5	7.8	6.1	5.0	5.9	4.8	1.6	0.1	47.5

MCGRATH, ALASKA

McGrath is located in western interior Alaska near the western end of a relatively flat drainage basin in the upper portion of the Kuskokwim River valley. The Kuskokwim Mountains extend in a northeast-southwest direction through the McGrath area, but the most pronounced ridges of this range lie to the west of the station. Consequently, the area is, for practical purposes, a portion of the sheltered continental interior. The characteristic continental climate is more pronounced during the winter season when temperatures become quite cold and precipitation relatively light, with pronounced north or northwesterly winds prevailing. During the summer months, when prevailing winds become southerly, the climate is at least partly affected by maritime influences, and the transition periods between the seasons are relatively short. Average summertime precipitation is more abundant than that received by most stations farther inland, but this may be due as much to topographic influences as to maritime influences. High temperatures in the summer reflect the continental influences of the interior, and high temperatures in the fall are just a little below the values reached in other areas farther inland like Fairbanks.

Over 40 percent of the normal precipitation falls during July, August, and September. The winter months have relatively little precipitation in comparison. However, due to the general nature of the dry snow, the accumulated snowfall is usually quite large. Spring is the driest season of the year with considerable clear and mild weather, and this period usually lasts well into June. Most thunderstorms occur during the months of June and July. Small hail occurs several times a year.

Break-up of the Kuskokwim River usually occurs in the middle of May and the ground is normally thawed enough for cultivation by the first of June. The normal growing season is approximately 120 days and the average occurrence of the last temperature of 32 degrees in the spring is late May. The first temperature of 32 degrees in the autumn normally occurs in the middle of September.

The summer months are relatively warm for Alaska. On approximately 15 days during the summer the daily maximum rises to the 80s. Relatively high temperatures have been observed during the winter months resulting from influxes of warm maritime air from the ocean. These thaws, usually of one or two days duration, occur several times during the winter season.

During the winter months the minimum temperatures fall to well below zero, at times reaching at least -50 degrees. Long periods of five to ten or more days of extremely low temperatures occur during the winter months. Skies remain clear and atmospheric pressures quite high. In the coldest of these periods high inversion ice-fog and ice crystals are a common occurrence. Months with the least cloudiness are January, February, March, and April.

NORMALS, MEANS, AND EXTREMES

MCGRATH, ALASKA

LATITUDE: 62°58'N LONGITUDE: 155°37' W ELEVATION: FT. GRND 344 BARO 340 TIME ZONE: YUKON WBAN: 26510

	(a)	JAN	FEB	MAR	APR	MAY	JUNE	JULY	AUG	SEP	OCT	NOV	DEC	YEAR
TEMPERATURE °F:														
Normals														
-Daily Maximum		0.4	8.9	23.4	37.0	54.5	65.5	68.4	63.5	52.9	31.3	12.4	2.7	35.1
-Daily Minimum		-17.8	-14.2	-2.9	16.0	34.5	45.1	49.0	45.0	35.4	18.0	-3.7	-14.6	15.8
-Monthly		-8.7	-2.6	10.2	26.5	44.5	55.3	58.7	54.3	44.2	24.7	4.4	-6.0	25.5
Extremes														
-Record Highest	51	54	55	51	67	82	90	89	89	76	59	49	49	90
-Year		1945	1943	1962	1958	1993	1969	1977	1977	1957	1969	1967	1977	JUN 1969
-Record Lowest	51	-75	-64	-51	-40	-2	30	33	25	2	-28	-53	-67	-75
-Year		1989	1947	1966	1985	1945	1974	1981	1984	1992	1982	1990	1961	JAN 1989
NORMAL DEGREE DAYS:														
Heating (base 65°F)		2285	1893	1699	1155	636	291	207	338	624	1249	1818	2201	14396
Cooling (base 65°F)		0	0	0	0	0	0	12	6	0	0	0	0	18
% OF POSSIBLE SUNSHINE														
MEAN SKY COVER (tenths)														
Sunrise - Sunset	51	6.3	6.1	6.2	6.5	7.3	7.8	7.9	8.2	7.9	8.1	7.3	6.9	7.2
MEAN NUMBER OF DAYS:														
Sunrise to Sunset														
-Clear	51	9.4	8.9	9.1	7.4	4.1	1.9	2.2	2.2	3.5	3.9	6.3	7.6	66.5
-Partly Cloudy	51	4.8	4.4	6.2	7.1	8.3	8.6	7.6	5.4	5.0	4.8	4.4	4.5	71.1
-Cloudy	51	16.8	14.9	15.8	15.5	18.6	19.5	21.2	23.3	21.6	22.3	19.2	18.9	227.6
Precipitation														
.01 inches or more	51	9.7	8.4	9.3	7.1	9.5	12.7	14.6	17.3	13.9	12.4	11.6	11.8	138.3
Snow,Ice pellets,hail														
1.0 inches or more	44	4.6	3.7	3.7	2.2	0.4	0.0	0.0	0.0	0.4	3.5	5.4	5.6	29.5
Thunderstorms	51	0.0	0.0	0.0	0.0	0.5	3.0	2.5	0.8	0.1	0.0	0.0	0.0	6.8
Heavy Fog Visibility														
1/4 mile or less	51	1.9	1.1	0.6	0.4	0.4	0.4	0.8	1.4	1.0	1.3	1.5	2.4	13.1
Temperature °F														
-Maximum														
70° and above	51	0.0	0.0	0.0	0.0	1.5	10.3	13.9	6.0	0.4	0.0	0.0	0.0	32.2
32° and below	51	29.3	25.6	22.3	8.0	0.5	0.0	0.0	0.0	0.3	15.4	26.9	29.3	157.6
-Minimum														
32° and below	51	30.9	28.2	30.7	28.3	11.4	0.2	0.0	1.0	10.4	28.0	29.7	30.9	229.6
0° and below	51	24.1	21.0	18.3	5.1	0.*	0.0	0.0	0.0	0.0	3.3	16.9	23.5	112.1
AVG. STATION PRESS.(mb)	20	995.6	997.3	995.5	997.0	997.1	998.8	1001.0	999.8	996.3	993.9	994.3	994.6	996.8
RELATIVE HUMIDITY (%)														
Hour 03	41	72	71	73	74	77	81	85	88	85	83	78	74	78
Hour 09 (Local Time)	41	72	71	72	67	63	66	74	83	84	83	78	74	74
Hour 15	43	71	63	53	49	45	48	55	61	61	69	75	74	60
Hour 21	42	72	70	65	59	53	55	64	74	76	79	78	74	68
PRECIPITATION (inches):														
Water Equivalent														
-Normal		0.75	0.66	0.80	0.82	0.87	1.47	2.05	2.47	1.98	1.40	1.29	1.40	15.96
-Maximum Monthly	51	3.74	3.05	2.82	2.78	3.07	4.36	5.36	6.26	6.18	4.59	4.34	5.02	6.26
-Year		1993	1944	1991	1964	1988	1950	1975	1945	1993	1946	1979	1990	AUG 1945
-Minimum Monthly	51	0.10	0.00	T	0.01	0.18	0.42	0.38	0.80	0.19	0.14	0.02	0.11	0.00
-Year		1961	1979	1986	1943	1979	1957	1993	1968	1969	1965	1963	1966	FEB 1979
-Maximum in 24 hrs	51	1.03	1.15	1.17	1.10	1.11	2.39	2.56	2.10	1.64	1.54	1.51	1.90	2.56
-Year		1957	1956	1982	1960	1960	1950	1975	1955	1945	1945	1985	1990	JUL 1975
Snow,Ice pellets,hail														
-Maximum Monthly	51	42.2	38.0	35.8	32.8	10.7	0.1	T	T	13.7	32.1	35.5	54.2	54.2
-Year		1993	1944	1990	1964	1992	1975	1990	1948	1992	1971	1987	1990	DEC 1990
-Maximum in 24 hrs	50	11.7	13.5	14.3	13.2	7.0	0.1	T	T	7.0	10.5	15.2	15.5	15.5
-Year		1957	1976	1982	1964	1971	1975	1990	1948	1992	1971	1946	1978	DEC 1978
WIND:														
Mean Speed (mph)	43	3.2	4.3	5.3	6.4	6.7	6.3	6.0	5.8	5.9	5.4	3.8	3.1	5.2
Prevailing Direction through 1963		N	NW	N	N	N	WNW	S	S	N	N	ESE	N	N
Fastest Obs. 1 Min.														
-Direction (!!)	38	18	18	18	20	15	20	19	18	18	20	19	18	18
-Speed (MPH)	38	32	75	51	38	30	31	31	43	40	36	39	47	75
-Year		1991	1951	1958	1951	1985	1963	1966	1951	1959	1951	1965	1951	FEB 1951
Peak Gust														
-Direction (!!)	14	S	S	SE	S	S	NW	NW	S	S	E	S	SW	NW
-Speed (mph)	14	59	45	46	39	45	62	45	46	49	40	40	52	62
-Date		1993	1989	1991	1989	1985	1985	1987	1985	1983	1980	1985	1992	JUN 1985

PRECIPITATION (inches)

MCGRATH, ALASKA

YEAR	JAN	FEB	MAR	APR	MAY	JUNE	JULY	AUG	SEP	OCT	NOV	DEC	ANNUAL
1964	0.13	1.11	0.86	2.78	0.87	2.44	1.42	1.78	0.69	1.35	0.77	0.30	14.50
1965	0.40	0.43	1.03	1.11	0.41	0.83	1.60	3.41	3.57	0.14	1.23	0.29	14.45
1966	0.48	0.72	0.12	0.31	0.40	0.86	0.55	2.06	0.48	0.64	0.27	0.11	7.00
1967	0.42	0.49	1.67	2.48	0.29	1.93	2.94	3.41	0.26	0.24	2.99	1.72	18.84
1968	0.98	0.76	0.39	0.29	0.78	1.31	1.38	0.80	0.45	0.83	0.25	0.93	9.15
1969	0.32	0.22	0.35	0.03	0.44	1.20	3.33	2.67	0.19	0.64	0.42	0.63	10.44
1970	0.19	0.80	1.03	0.87	1.02	1.97	1.62	2.35	0.92	2.21	2.05	3.02	18.05
1971	0.46	0.71	0.12	0.32	1.31	1.03	1.31	2.07	1.61	2.75	1.22	4.37	17.28
1972	1.21	0.05	0.32	1.19	0.40	1.80	1.64	2.61	2.53	2.02	0.50	1.27	15.54
1973	1.27	0.41	1.13	0.35	0.97	1.60	0.99	3.47	1.61	1.67	2.01	0.36	15.84
1974	0.37	0.36	0.94	0.37	0.30	0.59	1.34	2.02	1.06	2.27	1.86	0.35	11.83
1975	1.41	0.51	0.48	0.62	0.49	1.31	5.36	2.33	3.14	1.96	0.26	1.16	19.03
1976	1.03	1.33	1.41	0.09	1.63	0.96	1.82	0.91	0.50	1.36	1.76	0.30	13.10
1977	0.31	0.44	1.20	1.58	1.35	0.61	0.91	1.46	4.72	0.71	0.67	1.62	15.58
1978	0.17	0.23	0.15	0.13	0.99	2.96	1.41	2.25	1.19	1.17	1.97	3.43	16.05
1979	1.10	0.00	1.22	2.64	0.18	1.12	1.99	1.87	1.82	0.92	4.34	1.14	18.34
1980	1.65	0.33	0.29	0.08	1.25	3.47	2.96	2.46	1.32	0.66	0.73	0.18	15.38
1981	0.39	1.65	0.13	0.56	0.74	1.99	2.99	2.65	0.71	2.08	1.49	0.70	16.08
1982	0.51	0.50	2.38	0.93	1.17	1.36	3.21	2.73	5.00	2.71	1.23	1.02	22.75
1983	0.18	0.11	0.01	2.55	1.34	1.18	1.34	3.79	2.05	1.92	0.51	0.99	15.97
1984	1.43	0.78	1.08	0.20	1.43	1.24	2.11	2.28	0.57	0.40	0.32	3.54	15.38
1985	1.36	1.21	2.43	0.23	1.07	1.96	1.08	3.06	5.56	2.05	2.83	1.19	24.03
1986	0.23	0.08	T	0.36	0.67	1.80	3.45	2.66	3.29	1.75	1.79	0.95	17.03
1987	1.51	0.59	0.12	0.09	1.43	0.51	4.41	1.66	3.34	1.06	2.08	1.76	18.56
1988	0.62	0.87	0.52	0.38	3.07	2.10	0.88	2.71	0.84	1.31	0.35	1.51	15.16
1989	1.78	2.48	0.24	1.01	1.18	1.18	2.65	3.92	3.32	2.31	1.68	1.02	22.77
1990	1.24	1.99	2.76	0.66	1.01	0.61	2.16	3.37	3.19	1.99	1.68	5.02	25.68
1991	0.74	0.32	2.82	0.59	0.26	0.57	2.73	2.96	1.25	1.58	0.62	1.09	15.53
1992	0.19	0.69	0.70	0.30	1.50	0.96	3.04	1.93	2.45	0.67	2.87	2.19	17.49
1993	3.74	0.79	0.54	0.04	1.14	2.51	0.38	3.96	6.18	2.71	1.84	0.53	24.36
Record Mean	0.95	0.82	0.87	0.66	0.92	1.57	2.20	2.95	2.32	1.41	1.22	1.24	17.13

AVERAGE TEMPERATURE (deg. F)

MCGRATH, ALASKA

YEAR	JAN	FEB	MAR	APR	MAY	JUNE	JULY	AUG	SEP	OCT	NOV	DEC	ANNUAL
1964	-13.7	-0.6	1.1	23.7	35.3	57.1	60.0	54.6	45.2	28.2	5.9	-21.7	23.0
1965	-11.4	-11.9	27.2	29.4	39.5	52.7	55.5	52.0	48.1	14.2	8.7	-13.8	24.2
1966	-20.4	0.2	-3.7	23.1	40.0	57.5	55.4	51.6	45.6	23.8	8.8	-14.2	22.3
1967	-10.8	-0.6	17.5	31.8	45.0	54.5	56.3	54.9	45.1	23.4	14.0	-0.7	27.6
1968	-4.9	-3.3	11.7	26.6	44.1	55.5	60.9	56.4	40.3	21.8	2.8	-15.1	24.7
1969	-12.6	-3.2	15.0	32.1	48.3	57.6	56.1	48.4	45.9	33.2	0.3	4.9	27.2
1970	-22.9	6.4	15.8	25.5	47.1	53.9	56.9	52.6	39.6	21.0	13.3	-10.0	25.0
1971	-27.5	-5.3	-3.2	22.5	41.7	58.7	56.8	55.9	44.8	27.8	0.6	0.4	22.8
1972	-15.6	-7.2	-4.1	17.7	42.8	55.2	61.4	57.6	43.0	31.3	7.9	-2.9	23.9
1973	-18.3	-3.0	8.5	30.6	46.5	56.3	57.7	51.7	45.6	24.6	-2.5	-6.3	24.3
1974	-11.3	-12.5	7.0	29.5	47.4	54.4	59.2	57.6	49.4	20.6	0.5	-13.6	24.0
1975	-14.7	-6.4	6.9	22.5	44.3	54.1	59.8	54.0	42.5	23.6	-8.3	-16.8	21.8
1976	-14.2	-16.8	8.2	28.0	43.4	55.4	58.7	56.1	44.2	22.9	11.1	-6.0	24.3
1977	10.5	6.6	0.5	18.1	41.3	57.0	61.7	61.0	46.3	27.9	-5.4	-8.2	26.5
1978	4.8	6.6	13.2	33.1	48.8	51.5	59.1	57.6	45.3	23.1	14.2	9.3	30.6
1979	-0.7	-19.9	14.1	30.1	48.5	52.5	57.6	56.9	46.3	33.9	21.4	-14.7	27.2
1980	-13.5	11.8	16.5	34.0	47.6	51.8	57.2	52.0	42.3	29.6	8.0	-25.5	26.0
1981	14.4	6.0	26.9	30.2	51.1	54.1	55.6	52.9	41.3	28.1	8.7	-6.9	30.2
1982	-11.6	0.2	13.3	23.9	41.2	55.2	57.8	54.7	45.2	16.9	5.8	-3.9	24.9
1983	-15.2	-0.5	9.1	30.2	47.2	56.7	59.6	50.5	39.7	22.2	14.0	0.4	26.2
1984	-8.0	-16.0	18.7	21.0	40.3	57.2	57.7	51.7	44.4	24.8	1.8	1.4	24.6
1985	12.2	-7.9	13.6	10.9	40.6	53.0	60.3	52.0	42.9	20.4	-1.8	10.2	25.5
1986	-5.9	4.5	3.5	21.0	43.6	58.1	59.6	52.2	45.2	24.9	3.9	8.4	26.6
1987	-0.2	0.6	12.7	27.5	45.7	56.2	59.8	56.6	42.4	32.7	6.4	-7.1	27.8
1988	-3.6	2.0	13.6	27.9	49.5	58.0	63.3	54.5	43.7	17.4	-3.8	6.9	27.5
1989	-27.6			33.5	43.2	56.8	60.4	56.8	46.9	25.6	-4.3	1.2	
1990	-11.2	-21.9	16.8	36.5	48.9	58.5	62.4	57.6	42.6	25.4	2.4	-6.7	25.9
1991	-2.9	-0.8	13.4	31.6	47.0	59.4	57.9	53.6	48.9	28.0	2.8	-7.4	27.6
1992	-3.3	-4.4	12.1	27.0	39.8	57.7	61.1	55.3	33.6	18.7	11.8	-4.2	25.4
1993	-1.1	4.7	18.4	36.9	50.5	58.0	63.2	55.1	42.8	30.9	12.2	5.3	31.4
Record Mean	-8.3	-1.9	9.6	26.6	44.5	55.8	58.8	54.1	43.8	25.3	4.8	-6.8	25.5
Max	1.0	9.5	22.9	37.8	54.8	66.2	68.4	63.0	52.2	32.1	12.7	1.7	35.2
Min	-17.5	-13.4	-3.7	15.5	34.3	45.3	49.1	45.2	35.3	18.5	-3.1	-15.2	15.8

REFERENCE NOTES

MCGRATH, ALASKA

GENERAL
T - TRACE AMOUNT.
BLANK ENTRIES DENOTE MISSING/UNREPORTED DATA.
# INDICATES A STATION OR INSTRUMENT RELOCATION.

SPECIFIC

EXCEPTIONS

RECORD MEANS ARE THROUGH THE CURRENT YEAR, BEGINNING IN 1941 FOR TEMPERATURE
1939 FOR PRECIPITATION
1943 FOR SNOWFALL

(a) - LENGTH OF RECORD IN YEARS, ALTHOUGH INDIVIDUAL MONTHS MAY BE MISSING.
0.* OR * - THE VALUE IS BETWEEN 0.0 AND 0.05
NORMALS - BASED ON THE 1961-1990 RECORD PERIOD.
EXTREMES - DATES ARE THE MOST RECENT OCCURRENCE
WIND DIR.- NUMERALS SHOW TENS OF DEGRESS CLOCKWISE FROM TRUE NORTH. "00" INDICATES CALM.
RESULTANT DIRECTIONS ARE GIVEN TO WHOLE DEGREES.
BOLD VALUES INDICATE EXTREME VALUES WHICH OCCURRED AFTER THE ASOS SYSTEM WAS COMMISSIONED.

RECORD = PERIOD OF RECORD
RECORD MEAN PRECIPITATION IS THE MEAN OF ALL DAILY PRECIPITATION AMOUNTS DURING THE PERIOD OF RECORD.
RECORD MAX(MIN) TEMPERATURE IS THE MEAN OF ALL DAILY MAX(MIN) TEMPERATURES DURING THE PERIOD OF RECORD.
RECORD MEAN TEMPERATURE IS THE SUM OF THE RECORD MAX AND RECORD MIN DIVIDED BY 2.
AVERAGE TEMPERATURE IS THE SUM OF THE MEAN DAILY MAX AND MIN TEMPERATURE DIVIDED BY 2.

HEATING DEGREE DAYS Base 65 deg. F　　MCGRATH, ALASKA

SEASON	JULY	AUG	SEP	OCT	NOV	DEC	JAN	FEB	MAR	APR	MAY	JUNE	TOTAL
1964-65	152	316	584	1132	1772	2692	2368	2155	1168	1064	786	363	14552
1965-66	290	400	500	1569	1686	2445	2654	1813	2132	1250	768	217	15724
1966-67	290	407	577	1273	1684	2457	2350	1838	1466	987	613	306	14248
1967-68	272	304	589	1286	1527	2035	2167	1980	1646	1145	638	278	13867
1968-69	142	275	735	1331	1863	2490	2407	1908	1545	982	510	226	14414
1969-70	270	505	566	980	1939	1862	2731	1639	1521	1179	548	326	14066
1970-71	243	378	757	1360	1546	2326	2873	1968	2116	1269	713	210	15759
1971-72	251	274	597	1146	1929	2002	2500	2097	2144	1415	680	290	15325
1972-73	129	219	657	1040	1705	2107	2585	1902	1749	1024	566	259	13942
1973-74	218	409	574	1244	2022	2214	2365	2174	1799	1062	537	313	14931
1974-75	173	230	461	1368	1936	2442	2470	2002	1798	1270	633	321	15104
1975-76	173	334	668	1280	2203	2538	2458	2377	1758	1102	660	280	15831
1976-77	193	269	617	1301	1612	2204	1683	1633	2002	1403	729	234	13880
1977-78	111	134	551	1142	2115	2273	1862	1632	1597	949	497	398	13261
1978-79	176	224	585	1291	1520	1720	2038	2383	1574	1040	507	367	13425
1979-80	221	242	555	958	1299	2473	2434	1539	1499	923	533	391	13067
1980-81	242	393	672	1092	1704	2807	1562	1653	1173	1036	426	323	13083
1981-82	287	368	704	1140	1688	2232	2376	1816	1599	1228	733	292	14463
1982-83	222	314	588	1486	1778	2133	2491	1831	1728	1037	545	242	14395
1983-84	166	447	752	1318	1523	2003	2262	2357	1432	1316	761	226	14563
1984-85	221	407	609	1239	1895	1973	1636	2046	1591	1621	751	354	14343
1985-86	157	398	656	1381	2004	1697	2203	1692	1908	1317	660	206	14279
1986-87	191	391	586	1236	1837	1754	2020	1805	1616	1122	593	256	13407
1987-88	162	255	674	994	1755	2236	2128	1830	1588	1105	472	207	13406
1988-89	70	320	630	1470	2067	1798	2869			937	666	238	
1989-90	159	248	534	1214	2076	1975	2364	2436	1489	849	492	200	14036
1990-91	103	232	665	1223	1876	2226	2105	1842	1595	995	550	198	13610
1991-92	225	344	478	1140	1866	2246	2117	2015	1637	1132	773	213	14186
1992-93	128	291	932	1429	1592	2147	2045	1686	1438	837	443	203	13171
1993-94	89	299	659	1051	1583	1848							

COOLING DEGREE DAYS Base 65 deg. F　　MCGRATH, ALASKA

YEAR	JAN	FEB	MAR	APR	MAY	JUNE	JULY	AUG	SEP	OCT	NOV	DEC	TOTAL
1969	0	0	0	0	0	9	0	0	0	0	0	0	9
1970	0	0	0	0	0	0	0	0	0	0	0	0	0
1971	0	0	0	0	0	27	2	0	0	0	0	0	29
1972	0	0	0	0	0	0	24	0	0	0	0	0	24
1973	0	0	0	0	0	2	0	2	0	0	0	0	4
1974	0	0	0	0	0	3	4	4	0	0	0	0	11
1975	0	0	0	0	0	0	18	0	0	0	0	0	18
1976	0	0	0	0	0	0	2	0	0	0	0	0	2
1977	0	0	0	0	0	0	17	19	0	0	0	0	36
1978	0	0	0	0	0	0	0	2	0	0	0	0	2
1979	0	0	0	0	0	0	1	1	0	0	0	0	2
1980	0	0	0	0	0	0	9	0	0	0	0	0	9
1981	0	0	0	0	1	2	0	0	0	0	0	0	3
1982	0	0	0	0	0	4	7	0	0	0	0	0	11
1983	0	0	0	0	0	0	4	0	0	0	0	0	4
1984	0	0	0	0	0	0	1	1	0	0	0	0	2
1985	0	0	0	0	0	0	18	0	0	0	0	0	18
1986	0	0	0	0	0	7	32	0	0	0	0	0	39
1987	0	0	0	0	0	1	6	0	0	0	0	0	7
1988	0	0	0	0	0	3	23	0	0	0	0	0	26
1989	0			0	0	1	24	1	0	0	0	0	
1990	0	0	0	0	0	12	28	9	0	0	0	0	49
1991	0	0	0	0	0	33	12	0	0	0	0	0	45
1992	0	0	0	0	0	0	11	0	0	0	0	0	11
1993	0	0	0	0	0	0	41	0	0	0	0	0	41

SNOWFALL (inches)　　MCGRATH, ALASKA

SEASON	JULY	AUG	SEP	OCT	NOV	DEC	JAN	FEB	MAR	APR	MAY	JUNE	TOTAL
1964-65	0.0	0.0	0.0	12.9	15.3	14.2	12.1	24.0	9.7	12.1	0.6	0.0	100.9
1965-66	0.0	0.0	4.1	5.5	20.1	18.4	15.2	21.8	4.2	0.8	T	0.0	90.1
1966-67	0.0	0.0	0.0	15.4	19.6	8.1	9.0	11.7	24.1	24.3	T	0.0	112.2
1967-68	0.0	0.0	0.0	4.9	17.4	19.4	14.2	12.6	8.3	4.9	T	0.0	81.7
1968-69	0.0	0.0	2.5	10.8	10.1	16.4	10.1	8.0	8.9	T	T	0.0	66.8
1969-70	0.0	0.0	0.0	4.3	7.8	12.8	3.8	12.0	9.1	9.1	T	T	58.9
1970-71	0.0	0.0	0.6	12.7	30.6	23.2	11.4	15.1	2.3	5.6	7.6	0.0	109.1
1971-72	0.0	0.0	0.0	32.1	21.6	50.9	14.0	1.1	6.8	30.0	0.3	T	156.8
1972-73	0.0	0.0	4.1	7.5	9.0	12.4	13.4	8.0	22.5	2.2	T	0.0	79.1
1973-74	0.0	0.0	0.3	14.9	33.5	6.8	5.8	6.3	13.7	3.8	0.0	0.0	85.1
1974-75	0.0	0.0	T	16.4	26.6	9.3	18.5	10.4	6.9	8.5	1.3	0.1	98.0
1975-76	0.0	0.0	0.0	16.5	5.3	19.1	16.5	15.9	19.6	1.4	T	0.0	94.3
1976-77	0.0	0.0	0.0	13.3	17.1	6.2	3.5	8.3	17.9	20.3	0.4	0.0	87.0
1977-78	0.0	0.0	0.6	4.4	12.0	22.0	2.4	3.8	2.1	1.3	0.0	0.0	48.6
1978-79	0.0	0.0	T	13.7	20.6	54.0	17.3	0.0	12.6	10.4	T	0.0	128.6
1979-80	0.0	0.0	T	4.1	24.3	19.9	22.2	2.2	4.7	T	T	0.0	77.4
1980-81	0.0	0.0	0.5	7.2	16.8	3.1	9.1	21.5	1.7	5.1	T	0.0	65.0
1981-82	0.0	0.0	4.1	15.7	21.7	16.2	12.8	2.1	30.9	5.6	T	0.0	109.1
1982-83	0.0	0.0	0.0	24.0	21.6	10.4	4.8	2.4	0.3	12.6	T	0.0	76.1
1983-84	0.0	0.0	0.7	7.1	8.9	17.6	18.5	12.9	15.1	2.9	1.1	0.0	84.8
1984-85	0.0	0.0	0.0	6.5	7.5	49.6	20.2	20.6	35.0	3.1	4.2	0.0	146.7
1985-86	0.0	0.0	0.5	9.3	33.9	15.9	6.3	2.0	T	8.1	1.2	0.0	77.2
1986-87	0.0	0.0	T	11.8	22.6	17.4	29.3	9.7	2.2	2.3	2.0	0.0	97.3
1987-88	0.0	0.0	0.1	4.2	35.5	27.4	10.6	15.1	9.6	1.3	0.0	0.0	103.8
1988-89	0.0	0.0	0.8	17.3	7.7			17.3	4.5	2.5	0.6	0.0	
1989-90	0.0	0.0	0.0	21.0	26.8	16.6	21.4	33.4	35.8	5.4	1.8	0.0	162.2
1990-91	T	0.0	3.5	13.1	28.0	54.2	9.6	6.8	31.2	2.3	T	T	148.7
1991-92	0.0	0.0	0.0	6.6	11.5	22.1	7.4	15.4	12.5	2.2	10.7	0.0	88.4
1992-93	0.0	0.0	13.7	8.5	31.4	27.6	42.2	11.4	8.8	0.4	0.1	0.0	144.1
1993-94	0.0	0.0	2.2	9.3	19.3	11.4							
Record Mean	T	T	1.1	10.2	17.1	18.7	14.4	12.0	12.4	6.9	0.9	T	93.8

NOME, ALASKA

The weather station at Nome is located at Nome Field, approximately 1 mile northwest of the city. Low, marshy flats lie between the station and Norton Sound to the south, exposing the station to winds from the southeast through the west. A series of foothills, with heights of 500 to 1,200 feet, extend from northwest through north to east at a distance of from 4 to 8 miles. The terrain increases in ruggedness and height farther north, with the Kigluaik Mountains reaching a height of 5,000 feet at a distance of 30 miles. The ground along the coastal flats is swampy during the summer months, but is permanently frozen below a depth of 2 to 3 feet. Vegetation in the Nome area consists mostly of grass and numerous small flowering plants.

The moderating influence of the open water of Norton Sound is effective only from early June to about the middle of November. Storms moving through this area during these months result in extended periods of cloudiness and rain. There is a nearly continuous cloud cover during July and August. During the summer months the daily temperature range is very slight. The freezing of Norton Sound in November causes a rather abrupt change from a maritime to a continental climate. The majority of low pressure systems during this period take a path south of Nome, resulting in strong easterly winds, accompanied by frequent blizzards, with the winds later becoming northerly and reaching Nome across the colder frozen areas of northern Alaska.

Temperatures generally remain well below freezing from the middle of November to the latter part of April, with January usually the coldest month of the year. Temperatures usually begin to rise near the end of February and continue to rise until they reach a maximum in July.

Precipitation reaches its maximum during the late summer months and drops to a minimum in April and May. Snow begins to fall in September, but usually does not accumulate on the ground until the first part of November. The snow cover decreases rapidly in April and May, and normally disappears by the middle of June. Snow depths in Nome have exceeded 70 inches.

Severe windstorms do occur with winds over 70 mph recorded several times. Strong winds during the winter months when there is snow cover produce blowing snow conditions that severely hinder transportation in the area.

NORMALS, MEANS, AND EXTREMES

NOME, ALASKA

LATITUDE: 64 °30′N LONGITUDE: 165 °26′ W ELEVATION: FT. GRND 13 BARO 19 TIME ZONE: YUKON WBAN: 26617

	(a)	JAN	FEB	MAR	APR	MAY	JUNE	JULY	AUG	SEP	OCT	NOV	DEC	YEAR
TEMPERATURE °F:														
Normals														
-Daily Maximum		14.7	12.2	17.4	25.5	42.0	52.9	57.7	56.1	48.7	33.7	22.5	15.1	33.2
-Daily Minimum		-0.7	-4.5	-0.3	9.7	29.2	38.8	45.2	44.1	36.3	22.2	9.3	-0.5	19.1
-Monthly		7.0	3.9	8.6	17.6	35.6	45.9	51.5	50.2	42.5	28.0	15.9	7.3	26.2
Extremes														
-Record Highest	47	43	48	43	51	78	81	86	81	71	59	45	43	86
-Year		1977	1986	1984	1988	1981	1957	1977	1977	1979	1954	1983	1969	JUL 1977
-Record Lowest	47	-54	-42	-46	-30	-11	23	31	27	9	-10	-39	-41	-54
-Year		1989	1978	1971	1968	1949	1974	1976	1984	1992	1966	1948	1961	JAN 1989
NORMAL DEGREE DAYS:														
Heating (base 65°F)		1798	1711	1748	1422	911	573	423	459	675	1147	1473	1789	14129
Cooling (base 65°F)		0	0	0	0	0	0	0	0	0	0	0	0	0
% OF POSSIBLE SUNSHINE	39	40	55	53	54	50	42	37	32	36	35	31	35	42
MEAN SKY COVER (tenths)														
Sunrise - Sunset	47	6.2	5.4	5.8	6.2	6.6	6.8	7.7	8.1	7.5	6.9	6.9	6.2	6.7
MEAN NUMBER OF DAYS:														
Sunrise to Sunset														
-Clear	47	10.0	11.5	11.1	9.0	7.3	5.7	3.9	3.0	4.6	7.1	7.3	10.0	90.7
-Partly Cloudy	47	4.3	3.6	5.0	6.3	7.5	8.8	6.7	5.7	5.7	5.4	4.2	3.9	67.4
-Cloudy	47	16.6	13.1	14.9	14.7	16.3	15.4	20.3	22.3	19.6	18.4	18.5	17.1	207.2
Precipitation														
.01 inches or more	47	10.9	8.2	9.6	9.0	8.2	9.0	12.5	15.5	13.7	10.7	12.1	10.4	129.8
Snow, Ice pellets, hail														
1.0 inches or more	47	3.0	2.1	2.6	2.1	0.7	0.*	0.0	0.0	0.2	1.4	3.6	3.3	19.1
Thunderstorms	47	0.0	0.0	0.0	0.0	0.*	0.2	0.3	0.1	0.0	0.0	0.0	0.0	0.7
Heavy Fog Visibility														
1/4 mile or less	47	1.9	1.4	1.5	1.3	3.0	3.8	3.0	1.4	0.5	0.4	1.1	1.6	21.0
Temperature °F														
-Maximum														
70° and above	27	0.0	0.0	0.0	0.0	0.3	2.2	3.7	1.1	0.*	0.0	0.0	0.0	7.3
32° and below	27	28.0	25.2	27.6	20.3	4.1	0.0	0.0	0.0	0.2	11.6	24.3	27.1	168.6
-Minimum														
32° and below	27	31.0	28.1	30.9	29.1	19.3	3.1	0.3	1.5	10.2	25.2	29.1	30.9	238.6
0° and below	27	15.6	16.7	14.8	8.3	0.4	0.0	0.0	0.0	0.0	0.6	7.7	15.6	79.7
AVG. STATION PRESS. (mb)	21	1006.8	1011.0	1008.6	1010.4	1009.7	1010.4	1011.5	1009.5	1007.0	1005.5	1004.6	1004.8	1008.3
RELATIVE HUMIDITY (%)														
Hour 03	30	74	73	74	79	82	84	88	89	84	80	77	75	80
Hour 09 (Local Time)	30	74	72	73	77	76	76	82	83	80	79	77	74	77
Hour 15	30	74	71	70	74	73	73	78	79	73	72	75	74	74
Hour 21	30	75	73	74	78	77	75	80	84	82	79	77	74	77
PRECIPITATION (inches):														
Water Equivalent														
-Normal		0.79	0.60	0.54	0.68	0.62	1.12	2.17	2.71	2.43	1.35	1.04	0.83	14.88
-Maximum Monthly	47	2.10	2.11	1.95	2.15	2.02	4.15	4.66	7.82	7.46	3.84	4.39	2.16	7.82
-Year		1957	1989	1954	1961	1989	1978	1990	1951	1986	1972	1979	1951	AUG 1951
-Minimum Monthly	47	T	T	T	0.01	0.10	0.04	0.25	0.40	0.39	T	0.03	0.03	T
-Year		1970	1979	1971	1992	1974	1964	1964	1971	1968	1974	1962	1956	FEB 1979
-Maximum in 24 hrs	47	1.23	0.77	0.65	0.65	0.75	2.03	1.77	2.99	1.49	2.28	0.80	1.09	2.99
-Year		1963	1975	1954	1978	1989	1953	1954	1976	1978	1957	1979	1951	AUG 1976
Snow, Ice pellets, hail														
-Maximum Monthly	47	23.7	23.2	19.7	23.3	10.0	2.5	0.0	0.1	3.5	13.6	25.9	28.7	28.7
-Year		1950	1989	1954	1961	1977	1982		1965	1992	1958	1970	1993	DEC 1993
-Maximum in 24 hrs	47	7.8	9.0	7.0	6.3	5.3	1.7	0.0	0.1	2.6	6.6	8.7	8.4	9.0
-Year		1947	1975	1993	1961	1952	1982		1965	1948	1948	1979	1951	FEB 1975
WIND:														
Mean Speed (mph)	46	11.3	10.9	10.2	10.3	10.0	9.8	9.8	10.4	11.1	10.8	11.8	10.4	10.6
Prevailing Direction through 1963		E	NE	E	N	N	WSW	WSW	SW	N	N	N	E	N
Fastest Obs. 1 Min.														
-Direction (!!)	36	09	04	02	05	09	04	24	26	18	20	24	05	24
-Speed (MPH)	36	54	51	44	45	44	35	35	40	44	52	55	54	55
-Year		1967	1971	1991	1964	1966	1971	1973	1964	1973	1960	1965	1958	NOV 1965
Peak Gust														
-Direction (!!)	10	SE	NE	NE	NE	E	W	S	SE	NE	SE	E	E	NE
-Speed (mph)	10	59	66	59	52	46	35	39	56	52	59	59	53	66
-Date		1987	1988	1985	1984	1985	1991	1984	1993	1991	1992	1989	1992	FEB 1988

PRECIPITATION (inches)

NOME, ALASKA

YEAR	JAN	FEB	MAR	APR	MAY	JUNE	JULY	AUG	SEP	OCT	NOV	DEC	ANNUAL
1964	0.82	0.90	0.74	1.09	0.44	0.04	0.25	2.11	5.72	1.24	0.42	0.64	14.41
1965	0.56	0.13	1.59	1.34	1.15	1.00	2.95	4.69	4.08	0.61	1.46	0.53	20.09
1966	0.31	1.46	0.32	0.61	0.12	0.71	2.68	2.19	2.04	2.30	1.08	0.14	13.96
1967	0.44	0.26	0.39	1.00	0.25	0.45	4.22	2.80	0.45	0.79	0.52	1.02	12.59
1968	1.26	0.46	0.39	0.47	0.76	0.22	1.65	2.59	0.39	1.68	0.61	0.40	10.88
1969	0.55	0.32	0.24	0.02	0.44	0.87	1.84	2.29	0.56	1.00	0.18	0.05	8.36
1970	T	0.25	0.28	0.21	0.27	0.29	2.73	2.00	0.83	0.85	1.84	0.38	9.93
1971	0.05	0.09	T	0.03	0.28	0.85	1.54	0.40	1.97	2.28	0.68	0.77	8.94
1972	0.53	T	0.15	0.43	0.27	1.42	2.46	1.39	1.98	3.84	1.37	1.13	14.97
1973	1.70	0.23	0.13	0.10	0.27	0.92	2.10	3.97	2.49	2.36	1.90	0.96	17.13
1974	0.82	0.09	0.49	0.21	0.10	0.42	3.28	3.53	0.79	T	0.84	0.09	10.66
1975	0.61	1.69	0.66	1.06	0.12	1.15	2.20	2.64	1.58	0.53	0.36	0.64	13.24
1976	0.39	0.30	0.76	0.28	0.22	1.19	0.96	2.99	1.21	1.39	0.58	0.17	10.44
1977	0.91	0.47	0.80	0.30	0.90	0.63	0.32	1.51	5.29	0.51	0.18	1.81	13.63
1978	0.43	0.40	0.22	1.09	0.39	4.15	1.67	2.49	3.54	0.75	1.73	1.20	18.06
1979	1.07	T	0.34	1.63	0.12	2.88	2.53	4.08	3.49	0.74	4.39	0.79	22.06
1980	1.45	1.23	0.71	0.43	1.08	3.78	2.51	1.50	1.21	1.65	0.94	0.47	16.96
1981	1.46	1.44	0.82	1.03	0.66	0.80	3.37	2.52	0.92	1.63	0.68	1.19	16.52
1982	1.51	1.59	0.84	0.69	1.23	1.55	1.62	1.72	3.72	1.39	2.20	1.19	19.25
1983	0.22	0.07	0.12	1.18	0.51	0.31	1.47	3.34	2.25	1.87	1.64	1.11	14.09
1984	0.61	0.01	0.22	0.87	0.45	0.80	3.30	4.45	1.58	0.92	0.40	1.31	14.92
1985	0.61	0.56	1.55	0.07	1.12	1.38	1.12	3.90	4.81	3.71	1.12	1.29	21.24
1986	0.40	0.66	0.03	0.51	0.31	0.24	2.84	2.76	7.46	0.35	1.09	1.17	17.82
1987	0.95	0.56	0.49	0.11	0.37	0.60	0.97	1.91	2.42	1.83	0.65	1.17	12.03
1988	0.45	1.41	0.86	0.29	0.97	1.41	0.32	4.12		0.28	0.44	1.39	
1989	0.84	2.11	0.39	1.73	2.02	0.29	4.66	3.29	2.26	1.23	1.20	0.72	20.74
1990	0.89	0.52	0.70	1.09	2.00	1.17	4.66	4.30	1.86	2.35	1.56	1.28	22.38
1991	1.13	0.43	0.55	0.64	0.35	2.28	2.44	0.75	2.69	2.31	0.14	0.46	14.17
1992	0.37	0.53	1.18	0.01	0.23	0.42	0.61	3.94	1.75	1.51	0.95	1.29	12.79
1993	1.14	1.14	1.68	0.18	0.52	0.39	0.92	3.15	2.57	2.78	2.83	1.88	19.18
Record Mean	0.96	0.79	0.74	0.69	0.63	1.13	2.39	3.34	2.66	1.52	1.05	0.95	16.84

AVERAGE TEMPERATURE (deg. F)

NOME, ALASKA

YEAR	JAN	FEB	MAR	APR	MAY	JUNE	JULY	AUG	SEP	OCT	NOV	DEC	ANNUAL
1964	2.8	-8.1	-1.5	13.4	28.1	47.8	52.4	52.1	43.2	31.0	14.7	-2.5	22.8
1965	0.8	-6.0	19.6	20.4	27.5	44.5	46.5	45.6	43.2	23.1	22.4	1.9	24.1
#1966	6.7	8.2	-3.1	15.2	29.0	43.2	49.2	49.8	42.5	25.6	17.1	1.9	23.8
1967	12.0	6.2	16.8	23.0	31.2	46.5	47.1	49.5	43.7	27.0	15.1	10.0	27.3
1968	8.5	-3.1	7.9	14.4	29.1	43.4	54.1	52.0	39.5	28.1	14.5	-0.2	24.0
1969	11.6	5.2	8.2	21.8	42.8	47.7	50.2	45.2	46.5	33.5	8.7	14.4	28.0
1970	-4.3	12.3	11.8	15.1	36.5	46.0	49.6	47.8	36.8	22.7	20.2	5.3	25.0
1971	-5.7	-11.6	-3.5	12.9	29.7	48.4	51.0	48.5	41.6	28.0	15.6	8.0	21.9
1972	3.0	4.2	-7.4	11.9	35.4	46.3	56.1	53.1	41.7	32.3	20.3	10.5	25.7
1973	-5.3	10.7	1.6	18.3	35.2	42.6	45.5	45.3	42.1	26.2	17.6	7.4	24.0
1974	5.1	-9.4	10.5	20.9	38.2	43.1	50.7	51.3	45.2	24.4	10.0	-6.4	23.7
1975	-0.7	2.1	6.9	13.4	33.9	42.4	51.4	49.5	37.5	25.4	7.5	1.9	22.6
1976	0.0	-14.0	3.0	9.7	33.1	42.4	48.6	51.3	45.1	29.2	16.3	9.3	22.8
1977	24.2	17.0	-5.1	9.4	32.9	45.2	56.3	56.4	44.9	27.0	10.9	8.3	27.3
1978	21.6	11.8	10.6	24.9	42.1	44.7	54.5	54.3	46.6	27.0	23.9	14.3	31.4
1979	20.0	-1.0	15.5	25.5	41.8	47.3	50.6	53.1	44.6	34.7	24.5	-3.8	29.4
1980	0.9	13.1	17.9	23.8	43.2	43.8	53.8	49.7	42.6	31.8	19.9	1.4	28.5
1981	19.5	8.1	22.2	24.3	42.7	50.2	49.7	50.9	41.1	28.5	16.4	7.5	30.1
1982	11.0	11.7	9.9	12.4	33.6	49.2	55.4	53.0	45.1	25.1	19.8	13.8	28.3
1983	2.7	11.1	16.9	25.0	43.8	49.2	56.2	47.6	40.3	26.7	22.0	20.9	30.2
1984	2.0	-14.2	18.3	12.4	31.1	49.7	51.1	44.9	45.0	29.5	16.0	12.0	24.8
1985	24.4	-0.6	5.8	1.3	31.8	41.1	51.5	50.9	40.8	25.8	20.7	24.3	26.5
1986	5.2	14.5	6.3	12.1	33.6	51.9	52.6	48.6	41.9	28.0	21.0	19.4	27.9
1987	6.1	10.5	16.8	18.5	36.9	49.1	53.0	52.1	38.8	33.2	11.8	4.6	27.6
1988	13.6	7.6	6.5	23.0	43.2	47.9	53.0	51.4		25.6	7.9	8.4	
1989	-15.2	22.5	13.2	24.7	31.4	46.8	49.8	50.6	46.3	30.1	9.7	12.5	26.9
1990	2.4	-17.3	10.8	26.0	40.0	47.2	56.0	54.4	40.3	30.4	14.6	3.8	25.7
1991	6.9	5.9	8.6	25.0	38.8	51.1	55.1	51.6	48.0	33.3	17.3	0.1	28.5
1992	8.3	-0.1	5.2	21.7	27.5	47.7	52.0	48.3	33.7	23.7	12.8	4.1	23.7
1993	-4.6	14.8	13.2	27.6	40.2	51.8	56.2	49.9	38.9	32.5	21.6	15.3	29.8
Record Mean	5.3	5.0	8.3	19.1	34.9	46.0	50.5	49.7	42.0	28.8	16.0	6.8	26.0
Max	12.7	13.0	17.0	26.9	41.5	53.0	56.5	55.5	48.1	34.5	22.3	13.9	32.9
Min	-2.1	-3.1	-0.4	11.4	28.3	38.9	44.5	43.8	35.9	23.0	9.8	-0.4	19.1

REFERENCE NOTES

NOME, ALASKA

GENERAL
T - TRACE AMOUNT.
BLANK ENTRIES DENOTE MISSING/UNREPORTED DATA.
# INDICATES A STATION OR INSTRUMENT RELOCATION.

SPECIFIC

(a) - LENGTH OF RECORD IN YEARS, ALTHOUGH INDIVIDUAL MONTHS MAY BE MISSING.
0.* OR * - THE VALUE IS BETWEEN 0.0 AND 0.05
NORMALS - BASED ON THE 1961-1990 RECORD PERIOD.
EXTREMES - DATES ARE THE MOST RECENT OCCURRENCE
WIND DIR.- NUMERALS SHOW TENS OF DEGRESS CLOCKWISE FROM TRUE NORTH. "00" INDICATES CALM.
RESULTANT DIRECTIONS ARE GIVEN TO WHOLE DEGREES.
BOLD VALUES INDICATE EXTREME VALUES WHICH OCCURRED AFTER THE ASOS SYSTEM WAS COMMISSIONED.

RECORD = PERIOD OF RECORD
RECORD MEAN PRECIPITATION IS THE MEAN OF ALL DAILY PRECIPITATION AMOUNTS DURING THE PERIOD OF RECORD.
RECORD MAX(MIN) TEMPERATURE IS THE MEAN OF ALL DAILY MAX(MIN) TEMPERATURES DURING THE PERIOD OF RECORD.
RECORD MEAN TEMPERATURE IS THE SUM OF THE RECORD MAX AND RECORD MIN DIVIDED BY 2.
AVERAGE TEMPERATURE IS THE SUM OF THE MEAN DAILY MAX AND MIN TEMPERATURE DIVIDED BY 2.

EXCEPTIONS

RECORD MEANS ARE THROUGH THE CURRENT YEAR, BEGINNING IN 1917 FOR TEMPERATURE
1917 FOR PRECIPITATION
1947 FOR SNOWFALL
SEPTEMBER 1988 DATA WERE NOT RECEIVED.

HEATING DEGREE DAYS Base 65 deg. F — NOME, ALASKA

SEASON	JULY	AUG	SEP	OCT	NOV	DEC	JAN	FEB	MAR	APR	MAY	JUNE	TOTAL
1964-65	384	396	645	1046	1504	2090	1992	1992	1400	1333	1157	606	14545
1965-66	568	595	649	1294	1275	1958	1806	1586	2111	1488	1107	649	15086
# 1966-67	480	468	670	1214	1429	1957	1639	1643	1487	1252	1039	547	13825
1967-68	548	475	635	1171	1492	1704	1748	1976	1769	1511	1106	641	14776
1968-69	335	396	760	1137	1508	2026	1653	1675	1760	1295	681	515	13741
1969-70	451	607	550	971	1684	1563	2150	1475	1647	1494	878	566	14036
1970-71	471	527	841	1305	1334	1849	2194	2150	2124	1558	1087	495	15935
1971-72	427	503	697	1140	1478	1767	1923	1759	2250	1584	911	555	14994
1972-73	286	365	691	1006	1334	1686	2180	1517	1968	1398	918	665	14014
1973-74	598	604	679	1197	1417	1783	1857	2083	1685	1316	823	652	14694
1974-75	435	419	590	1255	1649	2214	2038	1760	1799	1545	960	673	15337
1975-76	414	476	818	1222	1724	1956	2015	2297	1922	1656	983	674	16157
1976-77	502	417	591	1100	1458	1725	1255	1338	2173	1667	988	589	13803
1977-78	274	272	598	1169	1619	1756	1338	1487	1682	1195	705	605	12700
1978-79	321	323	543	1171	1225	1567	1387	1847	1526	1176	715	522	12323
1979-80	435	360	606	935	1207	2134	1990	1501	1459	1229	668	631	13155
1980-81	341	468	669	1022	1346	1969	1404	1595	1317	1215	682	437	12465
1981-82	468	428	709	1123	1454	1779	1669	1494	1705	1575	969	476	13849
1982-83	288	367	590	1233	1350	1583	1930	1507	1490	1194	648	465	12645
1983-84	267	533	734	1179	1287	1362	1956	2305	1445	1575	1043	455	14141
1984-85	422	617	593	1094	1463	1640	1251	1840	1834	1910	1023	711	14398
1985-86	424	433	717	1207	1325	1254	1855	1412	1816	1583	963	392	13381
1986-87	379	503	687	1142	1310	1410	1823	1521	1490	1389	863	467	12984
1987-88	364	392	780	978	1592	1876	1588	1665	1813	1256	670	506	13480
1988-89	363	415		1212	1711	1751	2488	1183	1601	1206	1030	539	
1989-90	466	443	558	1074	1654	1623	1940	2308	1679	1164	768	527	14204
1990-91	284	320	734	1066	1508	1897	1797	1654	1743	1198	804	410	13415
1991-92	301	407	504	975	1423	2014	1753	1887	1850	1295	1154	513	14076
1992-93	398	510	934	1272	1560	1886	2159	1404	1599	1115	764	390	13991
1993-94	284	461	774	999	1298	1536							

COOLING DEGREE DAYS Base 65 deg. F — NOME, ALASKA

YEAR	JAN	FEB	MAR	APR	MAY	JUNE	JULY	AUG	SEP	OCT	NOV	DEC	TOTAL
1969	0	0	0	0	0	0	0	0	0	0	0	0	0
1970	0	0	0	0	0	0	0	0	0	0	0	0	0
1971	0	0	0	0	0	4	0	0	0	0	0	0	4
1972	0	0	0	0	0	0	19	7	0	0	0	0	26
1973	0	0	0	0	0	0	0	0	0	0	0	0	0
1974	0	0	0	0	0	1	0	0	0	0	0	0	1
1975	0	0	0	0	0	0	1	0	0	0	0	0	1
1976	0	0	0	0	0	0	0	1	0	0	0	0	1
1977	0	0	0	0	0	0	11	10	0	0	0	0	21
1978	0	0	0	0	0	0	0	0	0	0	0	0	0
1979	0	0	0	0	0	0	0	0	0	0	0	0	0
1980	0	0	0	0	0	0	0	0	0	0	0	0	0
1981	0	0	0	0	0	0	0	0	0	0	0	0	0
1982	0	0	0	0	0	7	0	0	0	0	0	0	7
1983	0	0	0	0	0	0	0	0	0	0	0	0	0
1984	0	0	0	0	0	2	0	0	0	0	0	0	2
1985	0	0	0	0	0	0	12	0	0	0	0	0	12
1986	0	0	0	0	0	4	2	0	0	0	0	0	6
1987	0	0	0	0	0	0	0	0	0	0	0	0	0
1988	0	0	0	0	0	0	0	0		0	0	0	
1989	0	0	0	0	0	0	0	0	0	0	0	0	0
1990	0	0	0	0	0	0	7	0	0	0	0	0	7
1991	0	0	0	0	0	0	0	0	0	0	0	0	0
1992	0	0	0	0	0	0	0	0	0	0	0	0	0
1993	0	0	0	0	0	0	19	0	0	0	0	0	19

SNOWFALL (inches) — NOME, ALASKA

SEASON	JULY	AUG	SEP	OCT	NOV	DEC	JAN	FEB	MAR	APR	MAY	JUNE	TOTAL
1964-65	0.0	0.0	0.0	2.6	5.2	7.4	6.2	1.7	9.5	11.4	4.5	T	48.5
1965-66	0.0	0.1	0.2	5.9	13.2	5.3	3.1	14.5	3.2	6.2	0.8	0.0	52.5
1966-67	0.0	0.0	0.0	6.5	10.7	1.6	6.2	4.8	3.8	11.9	0.6	T	46.1
1967-68	0.0	0.0	0.0	1.8	2.9	9.3	6.9	4.4	5.0	4.8	3.8	0.0	38.9
1968-69	0.0	0.0	0.4	8.5	7.9	4.0	5.5	4.2	5.0	0.2	0.4	0.0	36.1
1969-70	0.0	0.0	0.0	T	2.0	1.3	T	6.9	3.4	2.6	0.7	0.0	16.9
1970-71	0.0	0.0	T	5.7	25.9	3.9	2.9	2.2	0.1	0.8	4.4	0.0	45.9
1971-72	0.0	0.0	T	10.9	12.0	8.3	10.6	T	3.1	8.9	2.4	0.8	57.0
1972-73	0.0	0.0	1.0	4.9	16.2	4.5	19.7	4.0	4.7	1.1	1.4	0.8	58.3
1973-74	0.0	0.0	0.3	13.4	15.1	12.9	3.6	1.5	7.1	3.5	1.3	T	58.7
1974-75	0.0	0.0	T	0.4	14.2	2.3	9.6	21.3	8.0	13.8	1.1	0.0	70.7
1975-76	0.0	0.0	T	4.2	5.7	9.4	6.4	4.6	10.4	4.9	1.5	T	47.1
1976-77	0.0	0.0	0.0	2.9	7.4	3.6	7.1	5.5	9.4	4.1	10.0	1.6	51.6
1977-78	0.0	0.0	0.2	2.0	6.5	23.4	5.7	4.7	3.0	9.5	0.6	0.0	55.6
1978-79	0.0	0.0	T	8.1	15.6	14.0	11.8	T	4.1	12.2	0.0	T	65.8
1979-80	0.0	0.0	T	3.2	24.1	9.6	16.3	10.2	8.1	5.0	1.4	0.0	77.9
1980-81	0.0	T	T	5.2	9.6	4.9	14.7	11.6	5.9	10.4	4.6	0.0	66.9
1981-82	0.0	0.0	1.4	4.0	7.8	14.8	17.3	2.6	8.0	6.8	2.7	2.5	67.9
1982-83	0.0	0.0	T	8.2	23.6	13.0	2.4	1.4	1.7	10.2	1.5	0.0	62.0
1983-84	0.0	0.0	0.1	1.4	14.8	8.8	6.6	0.1	1.1	9.8	4.8	0.0	47.5
1984-85	0.0	T	0.0	2.5	5.5	13.6	4.5	5.9	11.2	0.7	8.5	T	52.4
1985-86	0.0	0.0	T	1.2	10.9	9.1	4.4	6.1	0.3	4.2	1.8	0.0	38.0
1986-87	0.0	0.0	1.8	2.3	4.6	10.6	9.7	5.7	4.9	1.5	0.9	T	42.0
1987-88	0.0	0.0	0.5	6.2	6.8	11.0	4.8	14.2	8.6	2.6	1.2	0.0	55.9
1988-89	0.0	0.0		1.4	5.7	15.0	10.6	23.2	5.8	16.4	6.4	T	
1989-90	0.0	0.0	T	4.8	22.2	8.4	11.1	7.6	8.2	8.8	0.8	0.0	71.9
1990-91	0.0	0.0	0.1	4.9	17.1	14.2	11.3	5.5	7.7	8.6	T	T	69.4
1991-92	0.0	0.0	0.0	1.8	1.4	6.8	5.8	7.4	19.2	0.1	3.5	0.0	46.0
1992-93	0.0	0.0	3.5	11.0	10.6	14.9	12.2	11.5	19.6	0.8	1.5	0.0	85.6
1993-94	0.0	0.0	2.9	3.2	21.5	28.7							
Record Mean	0.0	T	0.5	4.6	10.8	9.4	9.2	6.5	7.5	6.6	2.2	0.1	57.5

ST. PAUL ISLAND, ALASKA

St. Paul Island, one of the Pribilof group, is located in the central–southeast Bering Sea area. The climate is typically maritime, resulting in considerable cloudiness, heavy fog, high humidity, and rather well restricted daily temperature ranges. Humidities remain uniformly high from May to late September, and during the summer period there is almost continuous low cloudiness and occasional heavy fog. The differences between the high and low temperatures for the entire year are only slightly above 7 degrees, and the greatest monthly variation in March is slightly less than 12 degrees. Temperatures remain on the cool side even during the summer with extreme highs usually around the middle 50s. Although record low readings fall well below the zero mark, such extremely cold days are rather rare. There are only five days each winter with temperatures falling below the zero mark. The climatic environment makes the Pribilofs ideal for their numerous summer inhabitants, the Alaskan Fur Seals.

In spite of an environment of high humidities, precipitation on St. Paul Island is surprisingly light. The annual average of near 24 inches is slightly below the average for Alaska as a whole. April is generally the driest month, with a gradual increase of precipitation until a monthly total of over 3 inches is reached during August, September, and October. This is followed by a gradual decrease during the succeeding months until the return of April.

Frequent windy periods are characteristic of the island area throughout the year. Frequent storms occur from October to April, and these often are accompanied by gale–force winds to produce general blizzard conditions. Under the influence of prolonged north and northeasterly winds between January and April, the ice pack occasionally moves south to surround the island. During recent years, the southward limit of this movement has been between St. Paul and St. George Islands, some 40 miles to the southeast of St. Paul.

Thunderstorms are extremely rare on St. Paul Island.

NORMALS, MEANS, AND EXTREMES

ST. PAUL ISLAND, ALASKA

LATITUDE: 57°09'N LONGITUDE: 170°13' W ELEVATION: FT. GRND 22 BARO 28 TIME ZONE: YUKON WBAN: 25713

	(a)	JAN	FEB	MAR	APR	MAY	JUNE	JULY	AUG	SEP	OCT	NOV	DEC	YEAR
TEMPERATURE °F:														
Normals														
-Daily Maximum		30.7	26.9	28.9	32.7	39.3	45.8	49.6	51.0	48.8	42.0	37.0	33.2	38.8
-Daily Minimum		22.4	17.9	19.4	23.9	31.2	37.1	42.5	44.4	40.3	33.6	28.9	24.8	30.5
-Monthly		26.6	22.4	24.2	28.3	35.2	41.5	46.1	47.7	44.6	37.8	33.0	29.0	34.7
Extremes														
-Record Highest	76	48	44	50	49	58	62	63	66	61	54	50	52	66
-Year		1916	1917	1980	1993	1979	1926	1992	1987	1979	1916	1915	1936	AUG 1987
-Record Lowest	76	-26	-16	-19	-8	8	16	28	29	22	12	4	-5	-26
-Year		1919	1991	1971	1976	1971	1985	1961	1981	1992	1983	1988	1916	JAN 1919
NORMAL DEGREE DAYS:														
Heating (base 65°F)		1190	1193	1265	1101	924	705	586	536	612	843	960	1116	11031
Cooling (base 65°F)		0	0	0	0	0	0	0	0	0	0	0	0	0
% OF POSSIBLE SUNSHINE														
MEAN SKY COVER (tenths)														
Sunrise - Sunset	60	8.3	8.1	7.9	8.3	8.9	9.1	9.3	9.3	8.8	8.5	8.4	8.5	8.6
MEAN NUMBER OF DAYS:														
Sunrise to Sunset														
-Clear	64	2.3	2.5	3.2	2.1	1.2	1.2	0.6	0.7	0.7	0.7	1.2	1.5	18.0
-Partly Cloudy	64	5.7	6.0	7.2	5.8	3.9	3.1	2.3	2.7	5.4	7.3	6.4	5.5	61.2
-Cloudy	64	23.0	19.8	20.6	22.0	26.0	25.8	28.1	27.6	23.8	23.0	22.4	24.0	286.0
Precipitation														
.01 inches or more	75	18.3	14.9	15.8	14.2	14.1	12.5	15.3	18.6	19.7	22.1	21.5	19.8	206.8
Snow,Ice pellets,hail														
1.0 inches or more	67	3.6	3.5	3.0	1.4	0.8	0.*	0.0	0.0	0.*	0.4	2.2	3.4	18.4
Thunderstorms	49	0.0	0.0	0.0	0.0	0.0	0.*	0.0	0.0	0.0	0.0	0.*	0.0	*
Heavy Fog Visibility														
1/4 mile or less	49	2.0	2.7	3.2	3.2	7.1	9.8	12.8	10.3	3.9	0.7	0.8	1.1	57.5
Temperature °F														
-Maximum														
70° and above	76	0.0	0.0	0.0	0.0	0.0	0.0	0.0	0.0	0.0	0.0	0.0	0.0	0.0
32° and below	74	15.0	16.5	16.9	10.8	2.1	0.0	0.0	0.0	0.0	0.4	5.5	12.6	79.8
-Minimum														
32° and below	76	26.3	26.3	29.3	27.0	18.9	3.2	0.2	0.1	1.7	10.7	19.1	25.2	187.9
0° and below	76	1.1	2.7	1.5	0.2	0.0	0.0	0.0	0.0	0.0	0.0	0.0	0.1	5.5
AVG. STATION PRESS. (mb)	8	998.9	1005.0	1003.0	1006.9	1007.0	1010.4	1013.4	1008.9	1005.7	1004.9	1002.1	996.8	1005.3
RELATIVE HUMIDITY (%)														
Hour 03	16	84	85	87	87	90	95	96	95	91	84	82	85	88
Hour 09 (Local Time)	15	83	85	87	87	89	92	95	95	90	84	83	85	88
Hour 15	16	83	84	83	82	82	83	89	90	83	79	80	84	84
Hour 21	9	84	87	88	87	87	89	93	94	90	84	81	84	87
PRECIPITATION (inches):														
Water Equivalent														
-Normal		1.85	1.33	1.22	1.34	1.28	1.23	1.90	2.75	2.59	2.81	2.80	2.22	23.32
-Maximum Monthly	71	4.99	5.69	3.28	3.21	3.11	3.59	5.85	9.32	6.02	6.21	5.40	4.18	9.32
-Year		1964	1964	1973	1979	1931	1958	1950	1953	1924	1987	1993	1949	AUG 1953
-Minimum Monthly	71	0.25	0.02	0.08	0.08	0.21	0.18	0.32	0.17	0.62	0.96	0.67	0.08	0.02
-Year		1918	1984	1960	1992	1948	1954	1931	1977	1977	1977	1939	1933	FEB 1984
-Maximum in 24 hrs	71	1.38	1.51	1.26	1.00	1.27	1.48	1.92	2.00	1.74	1.93	1.76	1.15	2.00
-Year		1964	1932	1973	1966	1931	1949	1950	1984	1993	1949	1925	1930	AUG 1984
Snow,Ice pellets,hail														
-Maximum Monthly	70	40.6	55.8	31.4	19.1	12.7	2.0	0.0	0.0	1.0	14.0	27.3	22.7	55.8
-Year		1931	1964	1973	1928	1971	1927			1965	1948	1964	1930	FEB 1964
-Maximum in 24 hrs	69	13.8	13.6	12.4	10.0	4.0	2.0	0.0	0.0	1.0	10.9	13.4	8.0	13.8
-Year		1964	1964	1973	1966	1935	1927			1965	1978	1964	1930	JAN 1964
WIND:														
Mean Speed (mph)	19	20.3	20.7	19.0	17.8	15.6	13.8	12.3	14.1	15.7	18.2	20.6	21.0	17.4
Prevailing Direction														
Fastest Obs. 1 Min.														
-Direction (!!)	27	20	03	05	17	23	05	18	25	14	28	35	26	35
-Speed (MPH)	27	63	69	72	51	54	44	39	46	53	60	82	62	82
-Year		1969	1964	1971	1973	1985	1965	1973	1978	1964	1964	1964	1970	NOV 1964
Peak Gust														
-Direction (!!)	14		N	N	SE	SW	S	SE	N	N	W	SW	N	SW
-Speed (mph)	14	63	72	67	67	74	53	47	58	61	70	84	75	84
-Date		1989	1987	1987	1987	1985	1982	1990	1980	1985	1989	1990	1980	NOV 1990

PRECIPITATION (inches)

ST. PAUL ISLAND, ALASKA

YEAR	JAN	FEB	MAR	APR	MAY	JUNE	JULY	AUG	SEP	OCT	NOV	DEC	ANNUAL
1964	4.99	5.69	2.65	1.78	1.84	0.68	1.51	2.39	4.45	4.55	3.84	2.24	36.61
1965	3.65	1.59	1.92	1.52	1.35	0.94	2.67	3.97	2.87	1.45	3.84	2.65	28.42
1966	2.87	2.65	1.64	2.09	2.24	0.35	1.41	2.21	2.49	2.37	3.44	1.05	24.81
1967	2.02	1.64	1.85	3.10	0.71	1.21	1.45	2.59	1.59	1.89	3.20	2.70	23.95
1968	1.63	0.49	0.84	0.94	0.79	0.35	2.81	2.29	1.86	3.77	2.59	2.74	21.10
1969	2.25	1.68	0.71	1.00	1.28	0.68	1.32	4.01	1.44	2.30	1.92	0.98	19.57
1970	0.88	0.96	1.24	1.36	0.76	1.45	1.91	1.52	2.24	3.01	2.34	3.63	21.30
1971	0.83	1.03	1.14	0.83	0.94	0.73	2.77	2.78	2.64	2.11	2.03	3.30	21.13
1972	1.45	0.80	1.22	1.83	0.98	1.80	1.28	1.69	3.21	3.52	2.50	1.44	21.72
1973	1.79	3.34	3.28	0.60	1.08	1.07	1.18	2.40	1.64	2.75	3.23	2.22	24.58
1974	2.11	1.33	0.75	0.88	1.26	1.42	2.40	3.29	1.39	1.42	1.63	1.55	19.43
1975	1.47	1.54	1.02	1.30	0.50	1.07	1.20	1.40	1.45	2.07	0.88	0.73	14.63
1976	0.74	0.80	0.74	1.22	1.48	0.90	1.58	1.50	1.44	1.57	0.94	0.76	13.67
1977	0.57	0.89	1.41	1.09	1.19	0.31	0.86	0.17	0.62	0.96	1.10	0.65	9.82
1978	0.61	0.47	0.53	0.92	0.74	1.33	2.52	2.72	1.87	4.45	2.10	1.95	20.21
1979	2.62	1.01	1.22	3.21	1.52	2.18	2.67	2.57	3.51	3.79	4.78	1.19	30.27
1980	1.13	0.37	0.99	0.76	0.83	3.05	1.20	3.78	2.82	2.85	2.06	0.84	20.68
1981	1.58	0.52	0.79	0.54	1.45	1.15	0.85	2.50	1.07	2.48	2.63	1.01	16.57
1982	0.95	0.38	1.47	0.65	1.40	1.08	1.82	1.64	2.31	1.99	3.01	2.17	18.87
1983	0.28	1.09	0.14	2.43	1.08	0.55	2.39	3.44	2.34	2.22	3.32	3.98	23.26
1984	1.22	0.02	0.31	0.48	0.82	0.85	0.91	4.09	2.93	1.76	3.77	3.54	20.70
1985	2.86	0.84	1.56	1.23	2.04	1.91	1.04	2.27	3.75	4.22	4.56	2.41	28.69
1986	1.11	0.39	0.09	0.83	1.17	1.23	3.25	5.50	4.81	2.45	3.61	3.19	27.63
1987	3.36	1.99	2.02	1.53	1.47	2.94	3.13	3.46	4.69	6.21	1.24	3.28	35.32
1988	1.38	1.40	0.28	1.32	1.74	1.73	0.36	3.93	2.59	2.58	3.53	3.37	24.21
1989	2.12	2.56	0.53	2.33	1.91	0.95	2.00	1.81	3.61	2.94	3.92	1.77	26.45
1990	2.50	2.45	1.25	0.31	1.56	0.53	2.01	3.43	3.38	3.69	4.39	2.87	28.37
1991	1.45	1.16	2.88	0.22	1.00	1.10	2.02	3.99	4.95	2.47	1.60	2.17	25.01
1992	1.18	1.81	1.05	0.08	0.88	2.51	2.80	5.12	3.50	2.89	2.51	2.72	27.05
1993	2.28	1.68	1.27	0.89	1.52	0.86	2.05	3.31	4.92	3.12	5.40	0.97	28.27
Record Mean	1.75	1.24	1.22	1.10	1.23	1.25	2.15	3.17	3.07	2.97	2.65	1.94	23.74

AVERAGE TEMPERATURE (deg. F)

ST. PAUL ISLAND, ALASKA

YEAR	JAN	FEB	MAR	APR	MAY	JUNE	JULY	AUG	SEP	OCT	NOV	DEC	ANNUAL
1964	26.5	15.5	22.9	29.3	34.3	41.9	44.9	46.7	43.1	37.1	31.3	29.6	33.6
1965	22.6	16.7	24.4	29.6	32.7	39.7	44.7	46.4	43.9	35.5	34.3	27.6	33.2
1966	32.5	30.4	21.8	27.8	33.6	40.7	45.5	46.8	44.1	35.0	34.8	27.7	35.1
1967	28.9	27.9	33.2	35.6	38.2	44.0	48.2	49.5	45.4	37.8	35.3	31.2	37.9
1968	30.1	17.6	23.8	25.2	33.9	39.2	45.8	47.4	42.9	37.0	33.3	29.9	33.9
1969	31.7	24.2	25.5	29.6	36.7	42.1	48.3	49.9	46.5	39.1	32.5	29.4	36.3
1970	19.5	26.3	25.3	26.4	34.2	41.2	45.0	45.5	43.9	38.8	36.9	29.2	34.4
1971	23.3	15.3	14.9	21.5	27.8	36.2	41.9	44.5	42.3	37.8	32.4	30.0	30.6
1972	25.9	14.6	10.9	25.0	32.8	37.7	42.5	45.9	44.0	37.8	33.2	30.0	31.7
1973	23.1	25.5	20.1	26.3	30.4	39.4	44.3	45.9	43.9	37.5	37.2	27.4	33.4
1974	26.4	14.9	25.9	25.0	35.3	39.4	45.4	47.3	44.8	36.2	29.6	21.6	32.7
1975	17.6	15.3	19.7	27.4	32.2	38.1	43.7	47.4	42.5	36.2	29.3	29.3	31.6
1976	22.0	9.0	12.8	17.3	30.7	39.9	45.1	47.6	43.2	37.8	31.3	23.6	30.0
1977	30.6	27.4	23.2	26.3	36.1	43.3	48.7	51.1	47.8	37.0	30.0	31.1	36.1
1978	33.6	27.4	25.5	34.9	38.9	44.2	46.5	48.7	47.5	39.2	38.0	32.1	38.1
1979	34.3	30.0	30.4	36.1	40.6	46.7	48.7	50.5	48.1	43.3	36.7	31.5	39.8
1980	27.2	23.7	31.6	30.9	38.1	44.5	47.5	46.5	44.2	37.1	33.9	29.9	36.3
1981	27.9	23.9	30.7	34.3	40.4	43.0	47.8	47.3	44.3	37.7	30.8	28.1	36.4
1982	26.4	27.6	29.5	28.9	35.2	41.4	44.2	48.4	42.8	38.3	34.8	29.6	35.6
1983	22.4	28.3	27.8	31.8	36.9	43.7	47.3	48.1	44.1	36.9	30.8	34.4	36.1
1984	26.8	9.9	24.7	26.0	33.5	43.3	46.2	48.9	46.1	39.0	34.5	32.8	34.3
1985	33.8	23.5	21.9	20.2	34.8	38.7	45.8	47.5	44.4	37.2	37.2	33.9	34.9
1986	22.1	25.7	20.4	27.9	35.9	42.5	47.3	47.8	46.5	39.1	35.2	31.7	35.2
1987	27.4	27.9	29.4	29.5	36.6	42.9	47.7	49.3	43.4	38.5	28.1	25.9	35.6
1988	27.3	24.2	17.8	25.8	35.1	42.2	47.7	48.7	44.9	39.2	29.5	27.5	34.2
1989	22.3	34.1	30.4	33.8	37.8	42.6	47.3	49.8	47.1	40.2	30.2	29.4	37.1
1990	27.0	19.6	26.3	30.6	37.7	43.3	47.8	48.9	44.1	38.2	33.5	31.9	35.7
1991	25.0	9.6	27.6	27.9	36.4	41.9	48.5	49.3	45.8	40.7	33.7	26.7	34.4
1992	24.6	25.1	20.7	27.8	33.5	42.4	45.8	47.7	42.2	37.3	30.9	27.0	33.8
1993	22.5	26.3	27.1	33.4	37.4	43.1	49.4	50.9	46.0	41.0	36.4	27.8	36.8
Record Mean	25.7	22.8	24.3	28.7	35.1	41.4	45.9	47.7	44.8	38.4	33.2	28.6	34.7
Max	29.7	27.2	29.0	32.9	39.2	45.7	49.6	51.0	48.6	42.2	36.9	32.5	38.7
Min	21.7	18.4	19.5	24.4	31.0	37.0	42.2	44.4	40.9	34.5	29.5	24.8	30.7

REFERENCE NOTES

ST. PAUL ISLAND, ALASKA

GENERAL
T - TRACE AMOUNT.
BLANK ENTRIES DENOTE MISSING/UNREPORTED DATA.
# INDICATES A STATION OR INSTRUMENT RELOCATION.

SPECIFIC

(a) - LENGTH OF RECORD IN YEARS, ALTHOUGH INDIVIDUAL MONTHS MAY BE MISSING.
0.* OR * - THE VALUE IS BETWEEN 0.0 AND 0.05
NORMALS - BASED ON THE 1961-1990 RECORD PERIOD.
EXTREMES - DATES ARE THE MOST RECENT OCCURRENCE
WIND DIR.- NUMERALS SHOW TENS OF DEGRESS CLOCKWISE FROM TRUE NORTH. "00" INDICATES CALM.
RESULTANT DIRECTIONS ARE GIVEN TO WHOLE DEGREES.
BOLD VALUES INDICATE EXTREME VALUES WHICH OCCURRED AFTER THE ASOS SYSTEM WAS COMMISSIONED.

RECORD = PERIOD OF RECORD
RECORD MEAN PRECIPITATION IS THE MEAN OF ALL DAILY PRECIPITATION AMOUNTS DURING THE PERIOD OF RECORD.
RECORD MAX(MIN) TEMPERATURE IS THE MEAN OF ALL DAILY MAX(MIN) TEMPERATURES DURING THE PERIOD OF RECORD.
RECORD MEAN TEMPERATURE IS THE SUM OF THE RECORD MAX AND RECORD MIN DIVIDED BY 2.
AVERAGE TEMPERATURE IS THE SUM OF THE MEAN DAILY MAX AND MIN TEMPERATURE DIVIDED BY 2.

EXCEPTIONS

MEAN WIND SPEED IS THROUGH 1974.

RECORD MEANS ARE THROUGH THE CURRENT YEAR, BEGINNING IN 1915 FOR TEMPERATURE
1915 FOR PRECIPITATION
1924 FOR SNOWFALL

HEATING DEGREE DAYS Base 65 deg. F ST. PAUL ISLAND, ALASKA

SEASON	JULY	AUG	SEP	OCT	NOV	DEC	JAN	FEB	MAR	APR	MAY	JUNE	TOTAL
1964-65	615	562	650	857	1004	1091	1307	1347	1252	1054	995	750	11484
1965-66	622	569	628	907	917	1153	1002	963	1334	1111	965	724	10895
1966-67	599	559	619	926	901	1151	1113	1033	980	876	824	622	10203
1967-68	511	474	581	836	883	1040	1079	1367	1271	1184	954	769	10949
1968-69	589	539	652	858	946	1079	1022	1135	1216	1056	872	681	10645
1969-70	512	459	548	798	967	1097	1404	1079	1224	1154	949	707	10898
1970-71	611	598	624	806	838	1104	1288	1388	1552	1300	1148	859	12116
1971-72	713	628	677	835	972	1079	1205	1458	1670	1193	993	813	12236
1972-73	691	585	625	840	947	1077	1292	1102	1385	1156	1064	760	11524
1973-74	634	586	625	848	827	1157	1188	1397	1204	1194	914	758	11332
1974-75	604	539	598	882	1053	1337	1462	1389	1399	1120	1013	799	12195
1975-76	653	539	670	886	1064	1102	1327	1622	1612	1422	1055	745	12697
1976-77	610	532	649	838	1006	1280	1059	1048	1291	1156	889	642	11000
1977-78	499	420	508	862	1043	1045	968	1049	1216	896	803	616	9925
1978-79	564	496	518	792	802	1011	946	976	1065	860	747	548	9325
1979-80	497	444	501	667	843	1030	1165	1190	1030	1016	823	608	9814
1980-81	532	568	617	861	924	1079	1142	1145	1057	916	757	651	10249
1981-82	527	541	613	839	1018	1138	1188	1044	1093	1074	915	702	10692
1982-83	638	507	661	819	900	1092	1313	1019	1147	988	868	632	10584
1983-84	541	519	620	862	1018	942	1178	1592	1245	1166	968	646	11297
1984-85	579	492	561	796	909	993	959	1154	1326	1338	926	783	10816
1985-86	589	537	612	852	824	957	1323	1093	1378	1104	895	670	10834
1986-87	543	527	546	796	885	1029	1160	1034	1098	1057	871	656	10202
1987-88	526	480	642	816	1097	1205	1160	1179	1454	1170	916	677	11322
1988-89	530	499	595	792	1059	1155	1319	859	1067	930	834	666	10305
1989-90	543	463	532	763	1037	1093	1172	1265	1193	1027	840	648	10576
1990-91	525	494	619	825	935	1023	1232	1548	1153	1105	880	685	11024
1991-92	505	483	570	746	935	1181	1243	1150	1367	1109	966	672	10927
1992-93	587	529	680	853	1014	1172	1309	1077	1170	945	849	648	10833
1993-94	477	431	564	737	850	1146							

COOLING DEGREE DAYS Base 65 deg. F ST. PAUL ISLAND, ALASKA

YEAR	JAN	FEB	MAR	APR	MAY	JUNE	JULY	AUG	SEP	OCT	NOV	DEC	TOTAL
1969	0	0	0	0	0	0	0	0	0	0	0	0	0
1970	0	0	0	0	0	0	0	0	0	0	0	0	0
1971	0	0	0	0	0	0	0	0	0	0	0	0	0
1972	0	0	0	0	0	0	0	0	0	0	0	0	0
1973	0	0	0	0	0	0	0	0	0	0	0	0	0
1974	0	0	0	0	0	0	0	0	0	0	0	0	0
1975	0	0	0	0	0	0	0	0	0	0	0	0	0
1976	0	0	0	0	0	0	0	0	0	0	0	0	0
1977	0	0	0	0	0	0	0	0	0	0	0	0	0
1978	0	0	0	0	0	0	0	0	0	0	0	0	0
1979	0	0	0	0	0	0	0	0	0	0	0	0	0
1980	0	0	0	0	0	0	0	0	0	0	0	0	0
1981	0	0	0	0	0	0	0	0	0	0	0	0	0
1982	0	0	0	0	0	0	0	0	0	0	0	0	0
1983	0	0	0	0	0	0	0	0	0	0	0	0	0
1984	0	0	0	0	0	0	0	0	0	0	0	0	0
1985	0	0	0	0	0	0	0	0	0	0	0	0	0
1986	0	0	0	0	0	0	0	0	0	0	0	0	0
1987	0	0	0	0	0	0	0	0	0	0	0	0	0
1988	0	0	0	0	0	0	0	0	0	0	0	0	0
1989	0	0	0	0	0	0	0	0	0	0	0	0	0
1990	0	0	0	0	0	0	0	0	0	0	0	0	0
1991	0	0	0	0	0	0	0	0	0	0	0	0	0
1992	0	0	0	0	0	0	0	0	0	0	0	0	0
1993	0	0	0	0	0	0	0	0	0	0	0	0	0

SNOWFALL (inches) ST. PAUL ISLAND, ALASKA

SEASON	JULY	AUG	SEP	OCT	NOV	DEC	JAN	FEB	MAR	APR	MAY	JUNE	TOTAL
1964-65	0.0	0.0	0.0	2.9	27.3	6.5	28.3	11.0	9.0	4.0	2.7	1.6	93.3
1965-66	0.0	0.0	1.0	3.7	8.1	8.9	5.7	16.0	16.3	14.3	1.2	0.0	75.2
1966-67	0.0	0.0	0.0	8.3	4.0	5.9	10.2	15.2	9.0	2.9	T	0.0	55.5
1967-68	0.0	0.0	0.0	7.2	7.9	14.6	6.0	3.9	12.7	9.7	4.0	0.2	66.2
1968-69	0.0	0.0	T	10.1	10.2	10.0	11.7	16.4	12.1	7.3	3.3	0.0	81.1
1969-70	0.0	0.0	0.0	0.8	4.9	8.7	14.7	8.7	12.0	8.5	1.3	0.1	59.7
1970-71	0.0	0.0	T	0.4	2.0	17.5	6.3	9.6	12.3	13.6	12.7	T	74.4
1971-72	0.0	0.0	0.0	4.3	5.1	13.1	12.9	9.3	10.8	15.6	3.2	T	74.3
1972-73	0.0	0.0	0.2	1.8	6.7	5.5	13.0	27.6	31.4	6.3	7.9	0.2	100.6
1973-74	0.0	0.0	0.0	3.8	3.9	12.4	7.7	11.0	5.6	8.9	2.2	T	55.5
1974-75	0.0	0.0	0.0	2.2	4.6	15.7	14.4	11.4	10.1	5.6	4.5	1.2	69.7
1975-76	0.0	0.0	0.4	2.0	5.9	4.7	9.4	7.9	7.8	6.8	1.0	0.0	45.9
1976-77	0.0	0.0	0.0	2.9	7.1	10.3	2.1	7.1	16.0	5.6	1.4	0.0	52.5
1977-78	0.0	0.0	0.0	1.6	7.9	3.9	2.9	5.2	5.9	0.4	T	0.0	27.8
1978-79	0.0	0.0	0.0	12.5	2.4	9.0	17.9	1.6	3.7	0.9	0.5	0.0	48.5
1979-80	0.0	0.0	T	T	14.6	8.7	8.5	5.0	12.1	9.6	1.7	0.0	60.2
1980-81	0.0	0.0	T	2.6	6.1	6.1	10.5	5.5	7.2	2.0	1.3	0.0	41.3
1981-82	0.0	0.0	0.0	0.9	9.1	4.8	2.1	1.5	5.8	3.0	T	0.0	27.2
1982-83	0.0	0.0	T	0.2	4.0	10.7	2.9	8.4	1.2	5.3	0.2	0.0	32.9
1983-84	0.0	0.0	T	4.8	7.4	1.7	6.5	0.2	3.0	3.0	3.3	0.0	29.9
1984-85	0.0	0.0	0.0	2.9	3.4	3.7	4.5	7.8	10.1	13.3	2.2	0.7	48.6
1985-86	0.0	0.0	0.0	1.7	2.5	3.7	12.6	3.2	1.1	4.8	1.9	0.0	31.5
1986-87	0.0	0.0	0.0	T	3.6	4.9	13.5	16.1	13.2	3.3	2.4	0.0	57.0
1987-88	0.0	0.0	T	0.9	10.3	18.6	6.5	7.0	4.9	11.7	1.9	0.0	61.8
1988-89	0.0	0.0	0.0	2.8	25.7			6.8	3.6	5.5	1.4	0.0	
1989-90	0.0	0.0	0.0	0.7	18.2	10.6	23.4	25.1	16.6	2.9	0.1	0.0	97.6
1990-91	0.0	0.0	0.2	3.3	9.7	7.2	7.3	3.9	12.1	2.7	1.7	0.0	48.1
1991-92	0.0	0.0	0.0	0.8	4.3	19.0	13.1	7.8	15.2	1.5	0.5	0.0	62.2
1992-93	0.0	0.0	T	2.0	10.6	6.0	10.0	16.7	6.3	4.6	0.7	0.0	56.9
1993-94	0.0	0.0	T	T	4.1	13.3							
Record Mean	0.0	0.0	0.1	2.6	6.5	9.4	11.8	9.6	9.0	5.6	2.1	0.1	56.9

TALKEETNA, ALASKA

Talkeetna is located 80 airline miles north of Anchorage, and can be reached by both road and air. It lies in the upper end of a broad valley, on the southeast shore of the junction point of the Talkeetna and Susitna Rivers. The junction point of the Chulitna and Susitna Rivers occurs about 3 miles to the northwest of the station. The valley at this point has a gradual slope from station elevation to the 1,000-foot level at a distance of 15 miles to the west-northwest, and 5 to 7 miles from the northeast through the southwest. Elevation changes in these directions are more rapid beyond this point reaching heights of 4,000 to 6,000 feet at a distance of 25 to 30 miles. Except for isolated peaks (none close) of 2,500 to 4,000 feet there are no significant changes in elevation in the remaining directions. The area immediately surrounding the station is a composite of swampland and slightly higher ground which supports growth of birch and spruce trees.

The climate of Talkeetna varies in character between continental and modified maritime. Because of topography the annual precipitation is about half that measured on the Gulf of Alaska coast, and over twice that of the continental regime found in the interior. Temperature is definitely continental in nature. Like most areas of Alaska, Talkeetna experiences a January thaw with several days of above freezing temperatures. Although ground conditions become icy, the thaw conditions lower the total snow depth enough to prevent the accumulation of excessive amounts during the winter snow season. The warmest period in summer is from June through mid-July, with cooler weather after mid-July because of the cloudiness and precipitation maximums from late July through September.

Surface winds predominate north or south, depending on the season, because of the orientation of the valley. Velocity averages are relatively light.

Talkeetna is not a part of the agricultural area of Alaska. Normal moisture distribution results in a maximum rainfall at a time that crops should be maturing, and the freeze-free period, which on the average is from mid-June to mid-August, results in too short a growing period.

NORMALS, MEANS, AND EXTREMES

TALKEETNA, ALASKA

LATITUDE: 62°18'N LONGITUDE: 150°06' W ELEVATION: FT. GRND 345 BARO 354 TIME ZONE: YUKON WBAN: 26528

	(a)	JAN	FEB	MAR	APR	MAY	JUNE	JULY	AUG	SEP	OCT	NOV	DEC	YEAR
TEMPERATURE °F:														
Normals														
-Daily Maximum		19.2	25.4	33.4	43.4	56.0	64.4	67.7	64.5	55.6	39.3	25.2	19.7	42.8
-Daily Minimum		0.6	4.0	9.8	22.4	33.8	44.4	49.0	45.7	36.6	23.1	7.9	2.9	23.4
-Monthly		9.9	14.7	21.6	32.9	45.0	54.5	58.4	55.1	46.1	31.2	16.5	11.3	33.1
Extremes														
-Record Highest	54	45	51	53	69	82	91	90	85	78	68	50	47	91
-Year		1961	1991	1954	1958	1958	1969	1956	1977	1957	1954	1981	1960	JUN 1969
-Record Lowest	54	-48	-46	-43	-37	-5	28	33	25	11	-21	-41	-45	-48
-Year		1972	1947	1956	1944	1945	1940	1941	1974	1992	1956	1956	1970	JAN 1972
NORMAL DEGREE DAYS:														
Heating (base 65°F)		1708	1408	1345	963	620	315	205	307	567	1048	1455	1665	11606
Cooling (base 65°F)		0	0	0	0	0	0	0	0	0	0	0	0	0
% OF POSSIBLE SUNSHINE														
MEAN SKY COVER (tenths)														
Sunrise - Sunset	9	7.4	6.8	6.5	7.3	7.2	7.4	7.4	7.1	7.4	7.0	7.0	7.9	7.2
MEAN NUMBER OF DAYS:														
Sunrise to Sunset														
-Clear	9	6.0	7.3	9.4	5.0	6.0	1.0	5.0	3.0	5.0	6.7	7.4	5.4	67.2
-Partly Cloudy	9	4.3	3.8	4.0	5.0	6.0	12.0	7.0	11.0	4.1	4.5	3.4	2.4	67.6
-Cloudy	9	20.4	16.9	17.6	18.3	19.0	17.0	19.0	17.0	19.1	19.2	19.1	22.1	224.8
Precipitation														
.01 inches or more	39	9.1	8.6	9.0	7.1	11.4	12.6	14.7	16.4	15.9	13.1	10.5	11.6	140.0
Snow,Ice pellets,hail														
1.0 inches or more	26	5.9	5.0	4.9	3.2	0.3	0.0	0.0	0.0	0.2	3.8	5.6	7.3	36.1
Thunderstorms	26	0.0	0.0	0.0	0.0	0.4	1.3	1.5	0.9	0.1	0.0	0.0	0.0	4.2
Heavy Fog Visibility														
1/4 mile or less	26	0.2	0.2	0.2	0.2	0.2	0.2	0.1	0.3	0.8	1.1	0.5	0.5	4.5
Temperature °F														
-Maximum														
70° and above	53	0.0	0.0	0.0	0.0	1.9	9.1	12.5	7.6	0.4	0.0	0.0	0.0	31.5
32° and below	53	26.1	19.5	12.8	1.8	0.*	0.0	0.0	0.0	0.0	5.1	21.4	26.2	112.8
-Minimum														
32° and below	53	30.7	28.0	30.5	28.2	13.5	0.3	0.0	0.8	8.4	25.1	29.2	30.8	225.7
0° and below	53	14.7	11.4	9.3	1.3	0.*	0.0	0.0	0.0	0.0	1.4	9.5	14.0	61.6
AVG. STATION PRESS.(mb)	5	989.1	992.5	993.6	996.3	998.9	1000.2	1002.9	1000.3	998.4	997.1	993.9	994.4	996.5
RELATIVE HUMIDITY (%)														
Hour 03	31	73	73	75	81	86	88	92	95	94	85	78	75	83
Hour 09 (Local Time)	43	73	74	71	67	65	72	79	85	86	83	77	75	76
Hour 15	45	69	65	59	52	50	54	61	64	64	68	71	73	63
Hour 21	43	73	73	71	68	61	64	71	83	87	83	77	74	74
PRECIPITATION (inches):														
Water Equivalent														
-Normal		1.30	1.44	1.46	1.59	1.67	2.63	3.63	4.52	4.23	3.10	1.73	1.91	29.21
-Maximum Monthly	54	5.56	5.57	4.03	4.51	3.48	6.44	8.74	11.92	10.22	9.98	7.04	6.16	11.92
-Year		1948	1951	1946	1977	1946	1970	1981	1945	1990	1986	1979	1990	AUG 1945
-Minimum Monthly	54	T	T	0.06	0.04	0.27	0.21	1.08	0.96	0.91	0.81	0.07	0.34	T
-Year		1974	1979	1987	1958	1949	1952	1977	1941	1969	1953	1975	1972	FEB 1979
-Maximum in 24 hrs	53	1.95	1.80	1.52	1.37	1.39	1.58	2.93	2.54	3.12	5.14	1.59	1.89	5.14
-Year		1948	1990	1990	1977	1941	1955	1981	1945	1942	1986	1979	1986	OCT 1986
Snow,Ice pellets,hail														
-Maximum Monthly	73	60.3	71.2	56.3	40.1	11.0	T	0.0	T	18.0	32.9	62.7	79.0	79.0
-Year		1990	1951	1946	1964	1952	1952		1952	1923	1976	1967	1990	DEC 1990
-Maximum in 24 hrs	72	24.1	36.0	20.7	18.0	10.0	T	0.0	T	18.0	15.0	19.5	21.9	36.0
-Year		1990	1928	1946	1924	1952	1952		1952	1923	1927	1967	1984	FEB 1928
WIND:														
Mean Speed (mph)	11	6.2	5.5	5.3	4.8	5.0	5.2	4.2	3.7	3.7	3.7	5.2	5.1	4.8
Prevailing Direction through 1963		N	N	N	N	S	S	S	S	N	NNW	N	NNW	N
Fastest Obs. 1 Min.														
-Direction (!!)	26	04	03	03	34	14	05	17	02	19	03	02	36	03
-Speed (MPH)	26	38	35	39	29	23	23	22	26	25	31	31	35	39
-Year		1972	1987	1989	1986	1988	1993	1980	1987	1993	1974	1978	1972	MAR 1989
Peak Gust														
-Direction (!!)														
-Speed (mph)														
-Date														

PRECIPITATION (inches)

TALKEETNA, ALASKA

YEAR	JAN	FEB	MAR	APR	MAY	JUNE	JULY	AUG	SEP	OCT	NOV	DEC	ANNUAL
1964	0.39	3.31	1.66	3.47	0.84	2.12	2.42	2.74	2.17	3.66	3.91	0.54	27.23
1965	0.83	0.68	1.63	1.16	2.10	2.36	2.81	6.68	8.20	1.97	1.02	1.13	30.57
1966	1.71	0.62	0.45	1.94	1.35	1.78	3.47	3.52	5.45	4.10	1.20	1.70	27.29
1967	0.60	0.92	3.07	3.12	1.04	4.20	5.72	5.88	2.66	0.26	7.25	4.07	38.79
1968	1.08	3.33	0.31	1.50	2.94	0.54	2.57	1.99	1.27	1.16	0.71	0.98	18.38
1969	0.44	0.92	0.30	0.29	2.10	0.64	3.75	1.91	0.91	2.18	0.87	1.19	15.50
1970	0.27	1.18	1.48	2.33	1.62	6.44	4.62	4.62	3.56	3.00	0.69	2.66	32.47
1971	1.17	2.72	0.84	0.81	1.18	2.38	4.23	7.77	3.99	2.65	1.39	2.09	31.22
1972	1.72	0.79	1.38	1.40	2.03	2.21	1.76	3.09	8.47	4.00	0.57	0.34	27.76
1973	1.74	0.40	1.85	0.38	1.04	2.25	1.57	7.14	1.51	2.59	0.61	0.74	21.82
1974	T	0.85	0.88	1.63	0.38	1.31	2.11	2.36	2.69	3.28	2.99	1.14	19.62
1975	1.95	0.54	1.18	2.18	1.07	2.92	3.97	2.40	6.44	2.45	0.07	1.23	26.40
1976	0.88	1.09	2.71	0.37	0.99	1.38	2.07	2.16	2.85	3.68	4.82	1.31	24.31
1977	2.65	0.86	1.45	4.51	2.52	2.32	1.08	1.34	8.16	1.54	0.24	1.56	28.23
1978	0.29	0.86	0.61	0.33	1.80	4.73	3.18	1.02	2.09	2.98	2.26	3.00	23.15
1979	0.86	T	3.82	2.96	0.77	3.31	4.65	1.77	3.00	5.07	7.04	1.32	34.57
1980	2.78	1.14	0.89	0.52	3.23	5.73	4.68	3.43	2.33	2.14	1.08	0.56	28.51
1981	1.19	2.79	0.41	0.12	1.13	5.25	8.74	7.39	2.02	4.17	1.34	0.52	35.07
1982	0.03	0.79	1.70	0.39	1.31	4.20	5.74	4.55	7.54	2.07	1.70	1.80	31.82
1983	0.46	0.46	0.09	2.58	1.27	1.77	1.75	5.69	3.29	4.63	0.25	0.57	22.81
1984	1.90	0.89	0.21	0.94	1.40	1.40	3.06	6.63	1.73	1.79	0.44	2.69	23.08
1985	2.54	2.63	2.27	0.90	2.89	1.99	2.68	4.92	7.78	1.91	0.86	2.85	34.22
1986	0.71	0.96	0.52	0.59	1.64	0.56	5.67	5.91	4.77	9.98	2.73	3.05	37.09
1987	1.98	1.01	0.06	0.63	1.23	1.86	4.26	3.37	4.32	2.82	1.75	2.38	25.67
1988	0.49	2.00	0.96	0.90	1.75	2.37	2.28	4.98	3.77	3.53	0.86	2.34	26.23
1989	0.87	1.99	0.94	2.25	3.03	2.12	4.53	8.53	5.78	4.41	2.02	3.73	40.20
1990	3.63	4.31	3.94	0.93	1.60	1.37	1.37	4.31	10.22	1.26	0.89	6.16	39.99
1991	2.34	0.62	3.29	1.02	1.05	0.49	3.89	3.19	3.47	2.60	0.86	3.03	25.85
1992	1.67	1.77	0.63	0.38	1.38	0.74	2.37	3.99	2.35	3.41	4.09	3.28	26.06
1993	4.20	1.23	0.43	0.19	1.68	2.53	1.35	7.51	9.57	4.06	4.04	1.03	37.82
Record Mean	1.62	1.57	1.46	1.09	1.47	2.17	3.33	4.86	4.49	3.02	1.77	1.79	28.64

AVERAGE TEMPERATURE (deg. F)

TALKEETNA, ALASKA

YEAR	JAN	FEB	MAR	APR	MAY	JUNE	JULY	AUG	SEP	OCT	NOV	DEC	ANNUAL
1964	11.6	14.0	11.0	29.7	38.0	55.3	58.2	53.9	47.3	31.7	18.4	-1.8	30.6
1965	7.9	7.0	32.3	34.5	42.4	50.2	56.7	52.7	48.9	25.2	13.2	3.8	31.2
1966	8.4	9.7	14.3	33.9	43.0	56.2	57.2	53.8	45.5	30.3	15.9	11.5	31.6
1967	3.5	12.8	20.7	31.5	44.5	56.4	58.7	57.5	47.4	31.6	21.1	8.1	32.8
1968	6.4	19.9	24.1	30.0	45.2	54.8	59.6	56.8	44.1	30.9	17.5	-1.2	32.4
1969	-0.9	14.5	23.3	38.3	47.6	57.8	57.5	50.6	46.0	36.9	19.5	24.3	34.6
1970	4.2	27.5	31.4	32.6	45.5	51.8	56.3	52.6	43.1	26.6	17.6	5.7	32.9
1971	-7.5	12.6	10.3	31.1	40.2	52.6	56.5	55.9	44.9	27.7	10.5	7.8	28.6
1972	0.0	9.0	9.8	21.1	42.0	52.3	61.1	56.4	42.3	28.6	18.3	9.4	29.2
1973	-2.9	10.0	21.0	34.3	43.7	53.3	57.5	52.1	44.2	26.8	7.6	14.3	30.2
1974	1.2	11.5	19.4	35.8	47.6	56.3	56.7	56.0	48.7	32.3	18.2	10.2	32.8
1975	4.1	8.6	16.8	30.9	44.1	52.5	57.9	55.2	47.0	31.1	8.7	4.9	30.2
1976	8.5	7.0	18.8	33.7	44.2	55.9	59.0	56.3	45.7	29.6	28.0	19.3	33.8
1977	27.4	27.8	16.1	30.8	43.2	56.2	60.3	59.2	46.5	33.9	7.7	6.0	34.6
1978	17.9	23.3	25.6	35.6	47.6	52.4	57.2	57.8	47.5	34.0	20.5	14.5	36.1
1979	15.2	5.9	27.8	34.9	47.0	54.0	59.9	57.9	49.2	37.4	27.1	2.1	34.9
1980	7.4	24.3	23.7	37.6	45.4	53.5	58.5	53.7	45.8	35.8	25.7	-4.2	34.0
1981	28.7	21.1	31.3	31.8	50.0	54.0	56.3	54.4	45.8	34.0	19.8	10.9	36.5
1982	1.2	14.0	23.1	32.0	43.5	53.0	56.6	55.7	46.1	23.0	16.7	19.1	32.0
1983	12.5	18.5	25.7	35.5	48.0	56.4	58.7	53.4	42.6	31.3	22.7	13.8	34.9
1984	11.1	14.8	33.0	35.0	45.5	55.7	57.9	54.5	48.5	33.5	17.1	14.2	35.1
1985	27.9	7.6	21.2	24.2	42.3	52.4	59.6	53.7	45.0	27.0	7.1	25.3	32.8
1986	21.6	19.2	21.5	28.8	46.0	55.8	58.1	52.2	46.8	36.9	20.7	24.1	36.0
1987	19.3	22.4	24.4	35.7	45.9	53.5	57.9	56.7	45.6	36.7	24.1	11.7	36.2
1988	13.3	20.5	28.4	36.1	48.5	56.1	60.9	56.2		31.5	16.1	18.2	36.0
1989	-6.0	11.6	20.1	36.3	45.9	55.4	60.4	57.6	48.7	30.1	13.2	19.9	32.8
1990	8.6	-4.2	26.8	38.2	49.3	59.3	60.8	57.5	46.6	28.4	4.7	11.6	32.3
1991	11.6	17.1	19.2	36.4	45.5	58.0	57.9	54.9	48.8	29.8	21.8	15.9	34.7
1992	16.7	13.4	22.1	32.2	43.4	56.1	59.4	54.4	37.3	29.5	22.1	5.8	32.7
1993	8.2	20.7	25.3	38.5	49.7	57.3	61.7	56.4	44.8	36.0	18.8	21.5	36.6
Record Mean	9.6	15.2	20.9	33.4	45.0	54.9	58.2	55.0	46.0	32.2	17.8	10.6	33.2
Max	19.2	25.8	33.4	44.6	56.5	65.8	68.2	64.7	55.5	40.4	26.2	19.4	43.3
Min	-0.1	4.6	8.3	22.2	33.4	43.9	48.2	45.2	36.4	23.9	9.4	1.8	23.1

REFERENCE NOTES

TALKEETNA, ALASKA

GENERAL
T - TRACE AMOUNT.
BLANK ENTRIES DENOTE MISSING/UNREPORTED DATA.
# INDICATES A STATION OR INSTRUMENT RELOCATION.

SPECIFIC

(a) - LENGTH OF RECORD IN YEARS, ALTHOUGH INDIVIDUAL MONTHS MAY BE MISSING.
0.* OR * - THE VALUE IS BETWEEN 0.0 AND 0.05
NORMALS - BASED ON THE 1961-1990 RECORD PERIOD.
EXTREMES - DATES ARE THE MOST RECENT OCCURRENCE
WIND DIR.- NUMERALS SHOW TENS OF DEGRESS CLOCKWISE FROM TRUE NORTH. "00" INDICATES CALM.
RESULTANT DIRECTIONS ARE GIVEN TO WHOLE DEGREES.
BOLD VALUES INDICATE EXTREME VALUES WHICH OCCURRED AFTER THE ASOS SYSTEM WAS COMMISSIONED.

RECORD = PERIOD OF RECORD
RECORD MEAN PRECIPITATION IS THE MEAN OF ALL DAILY PRECIPITATION AMOUNTS DURING THE PERIOD OF RECORD.
RECORD MAX(MIN) TEMPERATURE IS THE MEAN OF ALL DAILY MAX(MIN) TEMPERATURES DURING THE PERIOD OF RECORD.
RECORD MEAN TEMPERATURE IS THE SUM OF THE RECORD MAX AND RECORD MIN DIVIDED BY 2.
AVERAGE TEMPERATURE IS THE SUM OF THE MEAN DAILY MAX AND MIN TEMPERATURE DIVIDED BY 2.

EXCEPTIONS

1. DAYS OF PRECIPITATION .01 INCH OR MORE ARE 1941-1953 AND 1968 TO DATE.
2. MEAN WIND SPEED IS 1950-1954 PLUS 1968.
3. THUNDERSTORMS AND HEAVY FOG MAY BE INCOMPLETE, DUE TO PART-TIME OPERATIONS.

RECORD MEANS ARE THROUGH THE CURRENT YEAR, BEGINNING IN 1931 FOR TEMPERATURE
1931 FOR PRECIPITATION
1931 FOR SNOWFALL

HEATING DEGREE DAYS Base 65 deg. F TALKEETNA, ALASKA

SEASON	JULY	AUG	SEP	OCT	NOV	DEC	JAN	FEB	MAR	APR	MAY	JUNE	TOTAL
1964-65	201	336	524	1027	1392	2074	1768	1621	1007	908	694	439	11991
1965-66	254	376	474	1228	1549	1894	1752	1545	1572	926	675	254	12499
1966-67	239	342	573	1069	1465	1656	1908	1455	1367	998	625	250	11947
1967-68	189	226	520	1027	1313	1761	1814	1305	1260	1042	608	299	11364
1968-69	165	251	621	1048	1418	2056	2046	1412	1284	796	532	222	11851
1969-70	227	441	564	864	1358	1254	1884	1045	1036	969	601	388	10631
1970-71	268	379	651	1184	1419	1835	2249	1465	1693	1010	762	365	13280
1971-72	260	272	594	1148	1631	1772	2013	1621	1706	1310	703	376	13406
1972-73	131	257	675	1124	1396	1719	2103	1537	1354	914	651	342	12203
1973-74	229	394	617	1176	1720	1566	1978	1493	1407	869	530	256	12235
1974-75	249	273	481	1006	1399	1696	1888	1576	1487	1019	638	369	12081
1975-76	222	296	535	1044	1690	1862	1749	1679	1426	931	640	266	12340
1976-77	179	265	572	1091	1105	1409	1159	1034	1509	1020	669	256	10268
1977-78	145	178	551	958	1714	1830	1455	1159	1214	877	531	372	10984
1978-79	238	226	515	952	1328	1564	1536	1649	1148	894	553	319	10922
1979-80	152	216	470	852	1131	1949	1790	1177	1273	816	597	334	10757
1980-81	202	345	569	899	1172	2144	1116	1224	1038	993	454	326	10482
1981-82	263	320	567	956	1348	1674	1979	1428	1293	982	659	355	11824
1982-83	255	281	561	1296	1441	1419	1621	1300	1213	880	520	251	11038
1983-84	186	352	663	1038	1262	1584	1665	1449	981	892	598	270	10940
1984-85	213	324	491	970	1429	1571	1143	1609	1351	1217	695	374	11387
1985-86	162	345	592	1172	1735	1225	1338	1276	1341	1079	582	274	11121
1986-87	208	388	536	863	1320	1262	1410	1185	1251	873	583	341	10220
1987-88	221	249	578	871	1218	1647	1601	1283	1126	862	505	259	10420
1988-89	123	270		1031	1461	1449	2202	1492	1385	856	585	284	
1989-90	144	224	482	1075	1550	1395	1746	1940	1176	799	485	174	11190
1990-91	140	234	545	1129	1804	1655	1650	1338	1416	850	596	210	11567
1991-92	213	307	479	1087	1288	1518	1490	1488	1324	975	666	261	11096
1992-93	168	320	821	1094	1282	1833	1761	1238	1224	786	469	223	11219
1993-94	108	258	600	892	1380	1341							

COOLING DEGREE DAYS Base 65 deg. F TALKEETNA, ALASKA

YEAR	JAN	FEB	MAR	APR	MAY	JUNE	JULY	AUG	SEP	OCT	NOV	DEC	TOTAL
1969	0	0	0	0	0	11	1	0	0	0	0	0	12
1970	0	0	0	0	0	0	8	0	0	0	0	0	8
1971	0	0	0	0	0	0	3	0	0	0	0	0	3
1972	0	0	0	0	0	0	19	0	0	0	0	0	19
1973	0	0	0	0	0	0	0	0	0	0	0	0	0
1974	0	0	0	0	0	0	0	0	0	0	0	0	0
1975	0	0	0	0	0	0	10	0	0	0	0	0	10
1976	0	0	0	0	0	1	0	4	0	0	0	0	5
1977	0	0	0	0	0	0	6	2	0	0	0	0	8
1978	0	0	0	0	0	0	2	6	0	0	0	0	8
1979	0	0	0	0	0	0	0	1	0	0	0	0	1
1980	0	0	0	0	0	0	5	0	0	0	0	0	5
1981	0	0	0	0	0	0	0	0	0	0	0	0	0
1982	0	0	0	0	0	0	1	0	0	0	0	0	1
1983	0	0	0	0	0	0	0	0	0	0	0	0	0
1984	0	0	0	0	0	0	1	2	0	0	0	0	3
1985	0	0	0	0	0	3	0	0	0	0	0	0	3
1986	0	0	0	0	0	4	2	0	0	0	0	0	6
1987	0	0	0	0	0	0	8	0	0	0	0	0	8
1988	0	0	0	0	0	0	3	1		0	0	0	4
1989	0	0	0	0	0	1	8	3	0	0	0	0	12
1990	0	0	0	0	0	10	16	10	0	0	0	0	36
1991	0	0	0	0	0	6	0	0	0	0	0	0	6
1992	0	0	0	0	0	0	1	0	0	0	0	0	1
1993	0	0	0	0	0	0	12	0	0	0	0	0	12

SNOWFALL (inches) TALKEETNA, ALASKA

SEASON	JULY	AUG	SEP	OCT	NOV	DEC	JAN	FEB	MAR	APR	MAY	JUNE	TOTAL
1964-65	0.0	0.0	0.0	15.3	31.1	5.4	8.3	6.8	5.5	4.6	1.7	0.0	78.7
1965-66	0.0	0.0	0.0	15.6	10.2	13.5	5.5	8.6	5.9	2.6	T	0.0	61.9
1966-67	0.0	0.0	0.0	0.9	17.1	18.9	7.5	13.2	33.4	24.7	0.0	0.0	115.7
1967-68	0.0	0.0	0.0	2.0	62.7	11.1	13.1	31.7	2.9	13.8	0.0	0.0	137.3
1968-69	0.0	0.0	0.3	1.7	10.2	13.7	8.3	26.0	8.1	T	T	0.0	68.3
1969-70	0.0	0.0	0.0	0.3	5.9	12.4	8.3	10.7	8.1	11.5	T	0.0	57.2
1970-71	0.0	0.0	0.0	13.2	7.0	41.4	7.7	52.2	12.7	12.2	1.7	0.0	148.1
1971-72	0.0	0.0	0.0	25.1	24.3	37.5	27.5	11.4	30.1	28.1	T	0.0	184.0
1972-73	0.0	0.0	1.6	28.4	13.8	4.7	41.3	8.1	32.9	2.7	0.0	0.0	133.5
1973-74	0.0	0.0	0.0	23.3	13.7	16.2	T	19.9	21.1	10.0	0.0	0.0	104.2
1974-75	0.0	0.0	0.0	19.0	51.7	19.6	31.7	11.2	24.9	25.5	0.4	0.0	184.0
1975-76	0.0	0.0	0.0	15.5	2.3	24.7	14.0	19.1	49.7	5.1	0.0	0.0	130.4
1976-77	0.0	0.0	0.0	32.9	14.2	29.5	12.3	3.3	24.4	19.0	T	0.0	135.6
1977-78	0.0	0.0	T	12.9	4.8	28.3	5.0	17.1	17.1	4.3	0.0	0.0	89.5
1978-79	0.0	0.0	0.0	15.0	32.0	48.9	18.2	T	24.4	16.1	2.6	0.0	157.2
1979-80	0.0	0.0	0.0	5.9	25.2	23.3	23.2	15.9	10.5	2.9	0.0	0.0	106.9
1980-81	0.0	0.0	0.0	4.3	15.5	8.1	13.2	19.8	2.7	3.1	0.0	0.0	66.7
1981-82	0.0	0.0	0.0	5.1	23.2	8.9	0.4	1.8	24.7	7.6	T	0.0	71.7
1982-83	0.0	0.0	0.0	28.5	27.0	10.3	11.9	11.0	2.1	26.7	0.0	0.0	117.5
1983-84	0.0	0.0	2.2	27.0	3.9	7.6	31.6	19.4	1.1	12.3	0.0	0.0	105.1
1984-85	0.0	0.0	0.0	3.0	7.2	47.5	25.8	34.4	31.4	12.9	9.4	0.0	171.6
1985-86	0.0	0.0	T	6.6	11.8	5.6	16.2	6.7	10.1	12.1	6.5	0.0	75.6
1986-87	0.0	0.0	0.0	0.0	30.9	22.5	37.9	19.5	1.5	9.0	0.0	0.0	121.3
1987-88	0.0	0.0	0.0	T	37.4	45.3	7.5	29.2	18.9	5.0	0.0	0.0	143.3
1988-89	0.0	0.0	0.0	18.3	23.3	45.1	19.9	29.8	23.5	T	T	0.0	159.9
1989-90	0.0	0.0	0.0	20.8	29.7	40.5	60.3	60.3	43.6	2.1	0.0	0.0	257.3
1990-91	0.0	0.0	T	11.3	15.2	79.0	46.8	9.3	44.9	2.3	0.2	0.0	209.0
1991-92	0.0	0.0	0.0	15.6	15.5	55.1	39.4	31.6	7.9	2.3	0.7	0.0	168.1
1992-93	0.0	0.0	6.8	15.1	44.6	37.1	57.0	6.0	7.3	T	0.0	0.0	173.9
1993-94	0.0	0.0	T	10.7	45.3	17.6							
Record Mean	0.0	T	0.2	10.3	17.3	21.9	20.0	18.5	17.8	7.9	0.8	T	114.7

VALDEZ, ALASKA

Valdez is located on the Valdez Arm, a rather well sheltered extension of Prince William Sound. Snow-capped mountains, containing extensive glacier areas, extend almost continuously from southeast of Valdez through north to west-southwest, with rugged but normally unglaciated mountains to the south and southwest. Active glaciers extend to within 5 to 10 miles of Valdez to the north and reach down to the level of the glacial plain on which Valdez is located. This level glacial plain is for the most part a well forested area except for the tidal marshes east and the glacial drainage area further east.

The terrain surrounding Valdez exerts a pronounced influence on practically all aspects of the local weather and climate. The effects of the surrounding mountains are to channel the local winds. From October through April the prevailing direction is northeast, and from May through September the prevailing direction is from the southwest. During the winter, high pressure in the interior and low pressure in the gulf may cause east to north winds of about 100 knots to flow out of passes and river canyons.

Precipitation is abundant the year around, but builds up noticeably during late summer and fall. The heaviest precipitation usually occurs in September and October, and almost one-third of the total annual rainfall occurs in these two months. Snowfall during the winter months is very heavy.

There is considerable cloudiness during the entire year, but slightly less than is realized at Alaskan points farther southeast.

Although the high mountain ridges to the north provide a considerable barrier to the flow of cold, continental air from the interior during the winter months, there is a definite offsetting factor in the downslope drainage from the snowfields and glacier areas on the southern slopes of these mountains. The coldest temperatures realized at Valdez appear to be related to the downslope flow of cold air, although temperatures only rarely dip below zero. The nearby snow and ice fields combine with the ocean areas to provide a moderating effect on the summertime high temperatures which have seldom reached the middle 80s. Considerable variations occur in practically all weather elements within relatively short distances.

The growing season averages slightly over 100 days, extending from May 26 to September 12. In addition, the glacier nature of the plain, the ruggedness of other surrounding terrain, and the cold water runoff from glacier melt tends to keep most available agricultural soil at temperatures too cool for desirable vegetation development during the growing season.

NORMALS, MEANS, AND EXTREMES

VALDEZ, ALASKA

LATITUDE: 61°08'N LONGITUDE: 146°21' W ELEVATION: FT. GRND 37 BARO 52 TIME ZONE: YUKON WBAN: 26442

	(a)	JAN	FEB	MAR	APR	MAY	JUNE	JULY	AUG	SEP	OCT	NOV	DEC	YEAR
TEMPERATURE °F:														
Normals														
-Daily Maximum		25.9	30.0	35.9	44.2	52.5	59.1	62.3	60.8	54.1	43.4	32.4	27.8	44.0
-Daily Minimum		15.0	18.1	22.4	29.9	37.8	44.4	47.5	46.2	40.4	32.7	22.4	18.0	31.2
-Monthly		20.5	24.1	29.2	37.1	45.2	51.8	54.9	53.5	47.2	38.1	27.4	22.9	37.7
Extremes														
-Record Highest	22	46	52	51	61	78	74	85	81	73	57	50	52	85
-Year		1991	1991	1974	1983	1993	1992	1979	1990	1979	1991	1977	1983	JUL 1979
-Record Lowest	22	-20	-5	-6	5	21	31	33	32	25	8	1	-6	-20
-Year		1972	1990	1972	1972	1972	1972	1972	1984	1983	1975	1989	1980	JAN 1972
NORMAL DEGREE DAYS:														
Heating (base 65°F)		1380	1145	1110	837	614	396	313	357	534	834	1128	1305	9953
Cooling (base 65°F)		0	0	0	0	0	0	0	0	0	0	0	0	0
% OF POSSIBLE SUNSHINE														
MEAN SKY COVER (tenths)														
Sunrise - Sunset	14	7.7	7.4	7.1	7.4	7.9	8.2	8.2	7.8	8.1	7.9	7.4	8.1	7.8
MEAN NUMBER OF DAYS:														
Sunrise to Sunset														
-Clear	14	5.4	6.1	7.4	5.9	3.8	2.9	3.6	4.5	3.5	5.1	5.7	4.9	58.8
-Partly Cloudy	14	3.4	2.5	3.6	4.9	5.1	5.5	4.3	5.0	4.8	2.9	3.9	1.9	47.6
-Cloudy	14	22.2	19.7	20.0	19.2	22.1	21.6	23.1	21.5	21.7	23.1	20.4	24.2	258.9
Precipitation														
.01 inches or more	21	17.0	14.5	15.7	13.9	16.2	15.1	16.7	17.4	20.3	19.6	15.6	18.0	200.0
Snow,Ice pellets,hail														
1.0 inches or more	21	10.7	9.1	10.3	4.6	0.2	0.0	0.0	0.0	0.*	2.7	8.7	12.6	58.9
Thunderstorms	16	0.0	0.1	0.0	0.0	0.0	0.2	0.1	0.1	0.3	0.1	0.1	0.0	0.8
Heavy Fog Visibility														
1/4 mile or less	16	0.9	1.3	0.6	0.2	0.1	0.1	2.9	4.6	2.4	1.3	1.5	0.9	16.7
Temperature °F														
-Maximum														
70° and above	21	0.0	0.0	0.0	0.0	0.5	1.5	3.8	2.7	0.1	0.0	0.0	0.0	8.5
32° and below	21	19.0	13.8	5.9	0.9	0.0	0.0	0.0	0.0	0.0	1.2	12.8	18.5	72.0
-Minimum														
32° and below	21	29.3	26.9	28.8	17.6	1.0	0.1	0.0	0.*	1.3	11.7	26.9	29.8	173.3
0° and below	21	1.5	0.4	0.1	0.0	0.0	0.0	0.0	0.0	0.0	0.0	0.0	0.8	2.8
AVG. STATION PRESS.(mb)	14	1004.3	1007.6	1007.5	1008.9	1012.5	1013.8	1015.9	1013.9	1010.7	1005.8	1003.0	1005.9	1009.2
RELATIVE HUMIDITY (%)														
Hour 03	18	78	75	77	80	86	90	92	93	90	81	77	77	83
Hour 09 (Local Time)	21	76	74	74	72	75	78	84	85	86	80	74	78	78
Hour 15	21	75	69	66	62	63	65	71	73	74	72	71	76	70
Hour 21	17	78	75	75	74	74	76	83	87	86	78	75	77	78
PRECIPITATION (inches):														
Water Equivalent														
-Normal		5.60	5.13	4.70	3.16	3.83	3.08	3.84	5.96	8.37	8.07	5.50	6.80	64.04
-Maximum Monthly	22	12.53	9.76	9.99	8.11	5.99	6.05	8.96	18.23	16.69	15.43	20.59	17.34	20.59
-Year		1981	1988	1979	1977	1989	1987	1981	1981	1982	1979	1976	1989	NOV 1976
-Minimum Monthly	22	0.01	0.57	0.81	0.57	0.71	0.93	1.25	2.08	2.78	2.49	0.42	1.34	0.01
-Year		1974	1989	1983	1981	1984	1983	1993	1987	1973	1990	1975	1983	JAN 1974
-Maximum in 24 hrs	16	3.75	2.69	2.20	2.03	1.13	1.71	1.98	3.42	3.26	3.96	3.42	3.05	3.96
-Year		1981	1986	1981	1989	1989	1989	1981	1981	1982	1983	1992	1985	OCT 1983
Snow,Ice pellets,hail														
-Maximum Monthly	22	148.5	100.8	113.9	71.4	5.8	0.0	0.0	0.0	4.6	39.0	76.1	137.1	148.5
-Year		1990	1984	1985	1977	1985				1992	1983	1989	1991	JAN 1990
-Maximum in 24 hrs	16	47.5	29.9	32.3	33.3	5.8	0.0	0.0	0.0	4.0	15.3	21.6	29.2	47.5
-Year		1990	1993	1982	1992	1985				1992	1982	1989	1981	JAN 1990
WIND:														
Mean Speed (mph)	13	7.3	7.7	6.0	5.2	5.7	5.8	4.9	4.2	4.5	6.4	7.6	6.6	6.0
Prevailing Direction														
Fastest Obs. 1 Min.														
-Direction (!!)	16	36	34	02	29	36	26	28	36	01	01	01	35	36
-Speed (MPH)	16	58	56	46	33	28	23	24	32	46	40	53	54	58
-Year		1989	1979	1990	1977	1992	1989	1989	1987	1988	1989	1978	1979	JAN 1989
Peak Gust														
-Direction (!!)	14	N	NE	N	N	N	W	N	N	SW	NE	E	N	N
-Speed (mph)	14	94	81	76	55	44	38	41	56	69	66	69	75	94
-Date		1980	1987	1989	1992	1992	1987	1987	1984	1992	1984	1990	1990	JAN 1980

PRECIPITATION (inches)

VALDEZ, ALASKA

YEAR	JAN	FEB	MAR	APR	MAY	JUNE	JULY	AUG	SEP	OCT	NOV	DEC	ANNUAL
1972	1.76	2.66	3.61		4.35	2.78	1.44	3.19			6.85	1.52	
1973	4.63	3.07	3.43	3.75	2.51	3.22	1.68	9.35	2.78	4.35	2.43	3.35	44.55
1974	0.01	5.21	1.55	5.93	0.79	2.08	2.34	4.22	7.59	12.90	9.78	6.42	58.82
1975	5.24	3.98	2.00	5.42	1.95	2.74	4.30	3.41	12.83	5.51	0.42	6.36	54.16
1976	7.00	2.27	3.80	5.47	2.53	1.00	1.87	3.03	12.58	9.53	20.59	8.29	77.96
1977	11.76	7.95	3.45	8.11	2.89	1.81	2.24	2.50	12.53	5.81	0.70	2.66	62.41
1978	4.18	6.18	2.14	1.25	2.14	3.28	3.81	2.17	4.94	15.38	3.27	6.07	54.81
1979	2.66	1.00	9.99	1.09	1.09	2.91	6.04	6.20	8.05	15.43	13.82	5.14	73.42
1980	8.56	9.64	3.27	2.90	3.32	3.22	7.70	6.72	4.78	9.21	4.73	5.08	69.13
1981	12.53	7.13	6.87	0.57	1.87	4.57	8.96	18.23	6.24	11.80	6.76	7.77	93.30
1982	0.99	1.77	5.42	2.44	1.22	3.27	4.83	2.73	16.69	8.10	5.15	8.96	61.57
1983	3.10	3.24	0.81	7.15	2.06	0.93	6.32	10.53	7.91	14.15	1.94	1.34	59.48
1984	10.55	7.37	3.89	3.06	0.71	3.63	3.99	6.12	5.00	6.01	4.21	6.51	61.05
1985	10.30	5.38	8.20	2.78	4.74	2.09	1.86	7.36	10.59	3.77	2.49	16.87	76.43
1986	7.21	7.34	3.59	1.72	1.86	1.40	5.27	7.64	4.93	8.57	7.15	15.73	72.41
1987	12.42	5.65	1.24	1.63	3.85	6.05	2.84	2.08	10.61	15.33	6.49	8.69	76.88
1988	4.33	9.76	9.52	5.29	2.76	5.04	3.46	12.78	9.31	11.02	2.59	15.11	90.97
1989	6.01	0.57	1.49	5.30	5.99	5.88	3.18	6.66	16.43	9.21	8.06	17.34	86.12
1990	9.51	5.90	7.88	3.08	2.34	3.04	3.14	5.66	16.05	2.49	0.97	7.31	67.37
1991	6.47	6.13	4.50	4.82	4.30	2.02	4.01	5.68	11.07	5.12	3.89	9.58	67.59
1992	9.55	9.01	9.88	2.80	1.88	3.11	3.63	8.96	6.01	10.47	14.25	5.59	85.14
1993	7.06	7.93	2.62	1.57	2.86	1.59	1.25	13.18	16.16	11.90	8.64	10.28	85.04
Record Mean	6.63	5.42	4.51	3.63	2.64	2.98	3.83	6.75	9.67	9.34	6.14	8.00	69.52

AVERAGE TEMPERATURE (deg. F)

VALDEZ, ALASKA

YEAR	JAN	FEB	MAR	APR	MAY	JUNE	JULY	AUG	SEP	OCT	NOV	DEC	ANNUAL
1972	12.0	19.0	20.7	28.8	41.6	49.3	58.8	58.0			28.4	19.2	
1973	13.5	21.2	31.0	38.2	44.0	50.2	53.9	51.3	45.9	36.8	22.8	23.1	36.0
1974	14.6	21.1	26.9	37.5	47.2	52.5	54.7	55.2	49.5	38.1	33.8	25.1	37.7
1975	19.6	21.1	27.6	35.3	44.0	50.2	55.3	54.5	46.5	37.3	23.4	19.7	36.2
1976	22.3	19.4	29.0	36.5	43.6	52.8	55.6	53.6	46.2	36.7	33.7	28.2	38.1
1977	30.5	32.0	30.1	36.2	42.9	51.3	55.7	54.9	48.3	40.9	24.4	19.0	38.9
1978	26.7	28.8	31.9	39.7	46.2	50.5	52.8	55.3	48.4	38.3	29.4	24.7	39.4
1979	25.1	15.5	31.8	38.4	46.2	51.3	55.6	54.4	49.8	41.5	33.1	19.8	38.6
1980	21.1	29.8	31.5	39.5	44.9	51.3	54.6	51.9	47.0	38.8	33.9	14.5	38.2
1981	33.5	26.8	34.2	37.5	49.7	52.3	53.3	51.4	46.4	39.4	27.3	24.5	39.7
1982	15.4	23.6	29.3	35.5	45.0	51.7	54.1	54.1	46.4	34.1	27.3	26.0	36.9
1983	20.7	28.1	31.8	37.3	46.2	53.7	55.2	52.0	45.0	37.9	30.1	24.8	38.6
1984	24.1	26.3	35.1	38.6	47.2	52.6	54.5	54.3	47.8	39.3	27.0	26.1	39.4
1985	32.9	20.1	29.3	32.6	42.4	48.5	53.9	50.9	46.1	34.4	21.7	29.8	36.9
1986	28.0	27.5	27.7	32.1	46.1	52.0	55.3	50.8	48.3	40.6	27.7	31.1	38.9
1987	28.2	30.9	28.8	38.1	45.9	49.4	55.6	55.1	46.6	39.7	33.5	26.8	39.9
1988	25.2	28.1	33.5	38.2	45.6	52.2	54.6	52.5	46.5	38.8	28.3	27.2	39.2
1989	15.7	23.7	30.2	39.6	46.4	52.2	57.8	54.6	48.3	38.1	23.5	30.2	38.4
1990	21.0	15.5	32.6	40.3	47.5	53.9	55.6	54.5	47.7	38.7	22.0	23.6	37.7
1991	23.7	28.8	30.6	39.0	45.6	53.8	53.8	53.8	47.3	39.0	30.4	27.6	39.5
1992	29.4	25.2	29.5	38.4	45.9	53.4	55.3	51.4	43.3	35.6	31.1	23.0	38.5
1993	21.9	28.4	32.2	41.8	50.2	54.4	58.7	54.2	46.1	40.8	30.7	28.7	40.7
Record Mean	23.0	24.6	30.2	37.2	45.6	51.8	55.2	53.6	47.0	38.3	28.2	24.6	38.3
Max	27.6	29.9	36.3	44.0	52.6	58.6	62.0	60.5	53.1	43.0	32.5	29.0	44.1
Min	18.3	19.3	24.1	30.4	38.6	44.9	48.4	46.7	40.9	33.6	23.9	20.3	32.4

REFERENCE NOTES

VALDEZ, ALASKA

GENERAL
T - TRACE AMOUNT.
BLANK ENTRIES DENOTE MISSING/UNREPORTED DATA.
# INDICATES A STATION OR INSTRUMENT RELOCATION.

SPECIFIC

(a) - LENGTH OF RECORD IN YEARS, ALTHOUGH INDIVIDUAL MONTHS MAY BE MISSING.
0.* OR * - THE VALUE IS BETWEEN 0.0 AND 0.05
NORMALS - BASED ON THE 1961-1990 RECORD PERIOD.
EXTREMES - DATES ARE THE MOST RECENT OCCURRENCE.
WIND DIR.- NUMERALS SHOW TENS OF DEGRESS CLOCKWISE FROM TRUE NORTH. "00" INDICATES CALM.
RESULTANT DIRECTIONS ARE GIVEN TO WHOLE DEGREES.
BOLD VALUES INDICATE EXTREME VALUES WHICH OCCURRED AFTER THE ASOS SYSTEM WAS COMMISSIONED.

RECORD = PERIOD OF RECORD
RECORD MEAN PRECIPITATION IS THE MEAN OF ALL DAILY PRECIPITATION AMOUNTS DURING THE PERIOD OF RECORD.
RECORD MAX(MIN) TEMPERATURE IS THE MEAN OF ALL DAILY MAX(MIN) TEMPERATURES DURING THE PERIOD OF RECORD.
RECORD MEAN TEMPERATURE IS THE SUM OF THE RECORD MAX AND RECORD MIN DIVIDED BY 2.
AVERAGE TEMPERATURE IS THE SUM OF THE MEAN DAILY MAX AND MIN TEMPERATURE DIVIDED BY 2.

EXCEPTIONS

RECORD MEANS ARE THROUGH THE CURRENT YEAR, BEGINNING IN 1972 FOR TEMPERATURE
1972 FOR PRECIPITATION
1972 FOR SNOWFALL

HEATING DEGREE DAYS Base 65 deg. F VALDEZ, ALASKA

SEASON	JULY	AUG	SEP	OCT	NOV	DEC	JAN	FEB	MAR	APR	MAY	JUNE	TOTAL
1972-73	222	213			1085	1411	1533	1212	1054	795	647	435	
1973-74	332	413	795	863	1258	1302	1558	1226	1176	825	565	369	10682
1974-75	312	298	457	828	1036	1230	1402	1223	1154	887	647	437	9911
1975-76	292	317	548	854	1241	1401	1319	1315	1107	848	655	358	10255
1976-77	285	346	558	873	934	1137	1060	920	1076	858	680	403	9130
1977-78	280	305	496	741	1210	1422	1181	1008	1021	754	575	429	9422
1978-79	374	294	491	820	1059	1244	1229	1379	1020	792	575	402	9679
1979-80	287	326	449	724	948	1397	1353	1016	1032	762	617	405	9316
1980-81	315	401	535	804	925	1561	970	1065	948	819	470	370	9183
1981-82	359	415	551	784	1122	1250	1531	1151	1101	880	612	392	10148
1982-83	330	332	552	951	1124	1200	1369	1029	1022	823	578	334	9644
1983-84	297	397	595	830	1039	1237	1260	1119	919	787	547	366	9393
1984-85	316	326	510	789	1134	1202	989	1252	1101	967	693	485	9764
1985-86	338	429	561	942	1294	1085	1140	1041	1150	979	577	384	9920
1986-87	295	433	494	752	1112	1046	1134	950	1115	799	589	464	9183
1987-88	295	298	546	778	939	1180	1229	1066	967	798	596	379	9071
1988-89	313	379	548	802	1097	1166	1521	1151	1072	757	569	377	9752
1989-90	219	316	496	829	1236	1072	1357	1378	995	733	534	327	9492
1990-91	288	320	510	805	1282	1277	1271	1009	1058	771	595	329	9515
1991-92	340	343	525	802	1028	1152	1100	1152	1093	792	583	343	9253
1992-93	296	412	644	903	1009	1296	1329	1018	1010	689	453	312	9371
1993-94	199	329	559	744	1022	1116							

COOLING DEGREE DAYS Base 65 deg. F VALDEZ, ALASKA

YEAR	JAN	FEB	MAR	APR	MAY	JUNE	JULY	AUG	SEP	OCT	NOV	DEC	TOTAL
1972	0	0	0	0	0	0	0	0	0	0	0	0	0
1973	0	0	0	0	0	0	0	0	0	0	0	0	0
1974	0	0	0	0	0	0	0	0	0	0	0	0	0
1975	0	0	0	0	0	0	0	0	0	0	0	0	0
1976	0	0	0	0	0	0	0	0	0	0	0	0	0
1977	0	0	0	0	0	0	0	0	0	0	0	0	0
1978	0	0	0	0	0	0	0	0	0	0	0	0	0
1979	0	0	0	0	0	0	4	0	0	0	0	0	4
1980	0	0	0	0	0	0	0	0	0	0	0	0	0
1981	0	0	0	0	0	0	0	0	0	0	0	0	0
1982	0	0	0	0	0	0	0	0	0	0	0	0	0
1983	0	0	0	0	0	0	0	1	0	0	0	0	1
1984	0	0	0	0	0	0	0	0	0	0	0	0	0
1985	0	0	0	0	0	0	0	0	0	0	0	0	0
1986	0	0	0	0	0	0	0	0	0	0	0	0	0
1987	0	0	0	0	0	0	12	0	0	0	0	0	12
1988	0	0	0	0	0	0	0	0	0	0	0	0	0
1989	0	0	0	0	0	0	1	0	0	0	0	0	1
1990	0	0	0	0	0	0	0	0	0	0	0	0	0
1991	0	0	0	0	0	0	0	0	0	0	0	0	0
1992	0	0	0	0	0	0	0	0	0	0	0	0	0
1993	0	0	0	0	0	0	11	3	0	0	0	0	14

SNOWFALL (inches) VALDEZ, ALASKA

SEASON	JULY	AUG	SEP	OCT	NOV	DEC	JAN	FEB	MAR	APR	MAY	JUNE	TOTAL
1972-73	0.0	0.0	0.0	4.4	64.6	25.4	83.2	48.7	46.6	10.1	T	0.0	283.0
1973-74	0.0	0.0	0.0	1.0	26.7	40.3	T	77.8	22.4	15.0	0.0	0.0	183.2
1974-75	0.0	0.0	0.0	10.8	38.2	90.6	90.3	65.8	37.4	29.0	1.8	0.0	363.9
1975-76	0.0	0.0	0.0	14.7	6.8	84.1	96.1	33.2	75.5	40.6	T	0.0	351.0
1976-77	0.0	0.0	0.0	25.8	44.7	101.5	54.0	26.2	52.9	71.4	2.8	0.0	379.3
1977-78	0.0	0.0	0.0	0.6	17.5	41.2	32.4	75.9	27.8	0.7	0.0	0.0	196.1
1978-79	0.0	0.0	0.0	19.8	28.0	82.2	45.6	12.1	112.6	12.0	T	0.0	312.3
1979-80	0.0	0.0	0.0	T	31.8	105.5	37.3	86.5	36.4	3.9	0.0	0.0	301.4
1980-81	0.0	0.0	0.0	4.7	27.4	60.2	24.5	61.3	57.9	4.6	0.0	0.0	240.6
1981-82	0.0	0.0	T	2.5	75.6	70.1	17.0	4.6	74.6	13.2	T	0.0	257.6
1982-83	0.0	0.0	0.0	35.0	59.4	56.9	60.2	31.1	10.9	51.0	0.0	0.0	304.5
1983-84	0.0	0.0	0.0	39.0	17.8	24.7	80.5	100.8	15.6	17.6	0.0	0.0	296.0
1984-85	0.0	0.0	0.0	0.5	29.3	60.8	37.9	59.5	113.9	31.1	5.8	0.0	338.8
1985-86	0.0	0.0	0.0	2.4	32.0	55.1	91.9	47.7	50.1	20.7	0.0	0.0	299.9
1986-87	0.0	0.0	0.0	1.2	55.9	95.2	128.0	84.2	10.1	10.1	0.0	0.0	384.7
1987-88	0.0	0.0	0.0	0.8	48.1	85.6	33.4	79.2	86.5	19.9	0.0	0.0	353.5
1988-89	0.0	0.0	0.0	8.6	37.5	110.0	109.6	8.9	22.6	15.6	0.0	0.0	312.8
1989-90	0.0	0.0	0.0	19.2	76.1	123.3	158.5	84.0	79.0	20.6	0.0	0.0	560.7
1990-91	0.0	0.0	0.0	5.1	16.7	91.9	66.5	68.3	69.1	9.5	T	0.0	327.1
1991-92	0.0	0.0	0.0	6.4	59.9	137.1	86.5	85.7	100.2	38.9	1.9	0.0	516.6
1992-93	0.0	0.0	4.6	9.4	57.0	47.7	90.4	85.4	47.6	1.3	0.0	0.0	343.4
1993-94	0.0	0.0	0.0	4.4	75.6	86.5							
Record Mean	0.0	0.0	0.2	10.5	42.1	76.2	65.4	57.4	54.3	21.7	0.6	0.0	328.5

YAKUTAT, ALASKA

The Yakutat area is surrounded on three sides by the waters of the Gulf of Alaska and Yakutat Bay. Consequently, the climate is maritime in character. Although the area in the immediate vicinity of the station is relatively flat, rather rough, hilly terrain exists within short distances. At distances of 40 to 75 miles to the north and northeast, peaks of the St. Elias Range rise to heights of from 14,000 to almost 20,000 feet. The up-slope terrain, combined with the exposure of the station to moisture-laden air from the Gulf, tends to provide Yakutat with abundant rainfall. The annual precipitation of around 130 inches is one of the greatest in the state, and annual amounts have always been in excess of 85 inches. Thunderstorms seldom occur, with about one per year. June has the lowest precipitation of any month with around 5 inches. October, with almost 20 inches, has the heaviest monthly rainfall. In spite of abundant rainfall, runoff from heavy rain seldom creates a problem of any consequence. This is particularly true in the vicinity of the station where runoff not easily reaching drainage ditches is quite readily absorbed by the porous gravel which is exposed as a surface layer over much of the area. The heavy precipitation produces copious growth of various types of vegetation in the surrounding woods, including several types of edible berries. However, the soil is not suitable for agriculture and a great deal of time is required to prepare the soil to produce even small quantities of garden produce. Agricultural activity is of minor importance. Heavy stands of timber in the area are harvested for lumber and pulp. Fishing is a main source of income in the area.

Daily and seasonal temperatures are held within fairly well-confined limits. Differences between readings range from a little over 12 degrees in October to around 16 degrees in April and May. Normal monthly temperatures range from slightly above 26 degrees in January to around 53 degrees in July and August. Although Yakutat has experienced temperatures below -20 degrees, readings approaching this figure are extremely rare. Yakutat averages about 20 days each year with temperatures below zero. The higher mountain areas to the north and northeast of Yakutat, with extensive glaciation, provide down-slope cold air drainage which results in wide variations of temperature within short distances. Temperatures above the 80 degree mark have occurred in June, July, and August.

Snowfall has occurred in all months of the year except June, July, and August.

Cloudiness is abundant with the annual sunrise to sunset cloud cover exceeding eight-tenths. During the spring, fall, and winter months the Yakutat area is subjected to numerous storms, usually accompanied by high winds. The St. Elias Mountain Range, which borders the area on the northeast and contains numerous glaciers, exerts a pronounced effect upon the local weather, particularly when a steep pressure gradient develops with low pressure in the Gulf to the southwest of Yakutat. Under these conditions cold winds move down from the glacier slopes and skies are generally cloudless.

NORMALS, MEANS, AND EXTREMES

YAKUTAT, ALASKA

LATITUDE: 59°31'N LONGITUDE: 139°40' W ELEVATION: FT. GRND 28 BARO 38 TIME ZONE: YUKON WBAN: 25339

	(a)	JAN	FEB	MAR	APR	MAY	JUNE	JULY	AUG	SEP	OCT	NOV	DEC	YEAR
TEMPERATURE °F:														
Normals														
-Daily Maximum		31.5	35.0	38.2	43.5	50.0	55.6	59.3	59.8	55.3	47.1	37.3	33.1	45.5
-Daily Minimum		18.6	21.1	23.7	29.1	36.5	43.3	47.8	46.5	41.0	34.7	25.2	21.1	32.4
-Monthly		25.1	28.0	31.0	36.3	43.3	49.4	53.6	53.2	48.2	40.9	31.3	27.1	39.0
Extremes														
-Record Highest	47	55	54	59	68	79	82	84	86	77	63	55	52	86
-Year		1981	1963	1981	1976	1963	1991	1955	1957	1957	1967	1976	1960	AUG 1957
-Record Lowest	47	-22	-20	-20	3	21	29	35	29	21	6	-6	-24	-24
-Year		1952	1989	1972	1948	1972	1971	1968	1974	1971	1966	1966	1964	DEC 1964
NORMAL DEGREE DAYS:														
Heating (base 65°F)		1237	1036	1054	861	673	468	353	366	504	747	1011	1175	9485
Cooling (base 65°F)		0	0	0	0	0	0	0	0	0	0	0	0	0
% OF POSSIBLE SUNSHINE														
MEAN SKY COVER (tenths)														
Sunrise - Sunset	45	7.8	7.9	7.8	8.0	8.4	8.6	8.6	8.4	8.5	8.5	8.2	8.3	8.2
MEAN NUMBER OF DAYS:														
Sunrise to Sunset														
-Clear	45	5.1	4.2	4.4	3.8	2.3	1.6	2.0	3.1	2.6	3.2	3.8	4.0	40.2
-Partly Cloudy	45	3.5	3.1	4.4	4.7	4.7	4.6	4.4	4.0	3.5	2.7	2.9	2.7	45.3
-Cloudy	45	22.4	20.9	22.2	21.5	23.9	23.7	24.6	23.9	23.8	25.1	23.3	24.3	279.7
Precipitation														
.01 inches or more	47	19.3	18.2	19.3	18.0	19.1	16.7	17.8	18.5	20.9	24.0	21.4	22.2	235.4
Snow, Ice pellets, hail														
1.0 inches or more	45	8.7	8.9	8.7	4.6	0.5	0.0	0.0	0.0	0.0	1.5	5.5	9.4	47.8
Thunderstorms	45	0.1	0.1	0.*	0.*	0.*	0.0	0.1	0.2	0.4	0.7	0.4	0.3	2.3
Heavy Fog Visibility														
1/4 mile or less	45	2.0	2.5	2.3	1.8	2.2	2.9	3.6	5.3	3.6	0.7	0.9	1.8	29.5
Temperature °F														
-Maximum														
70° and above	29	0.0	0.0	0.0	0.0	0.3	0.9	1.3	1.1	0.1	0.0	0.0	0.0	3.8
32° and below	29	15.4	8.9	3.0	0.2	0.0	0.0	0.0	0.0	0.0	0.3	6.2	11.7	45.7
-Minimum														
32° and below	29	24.8	23.2	24.0	20.8	7.3	0.4	0.0	0.2	4.5	10.9	21.4	24.4	162.1
0° and below	29	4.8	2.7	1.0	0.0	0.0	0.0	0.0	0.0	0.0	0.0	0.5	2.3	11.3
AVG. STATION PRESS. (mb)	21	1005.6	1006.4	1007.0	1010.1	1012.3	1014.2	1015.9	1014.5	1011.2	1005.1	1003.7	1005.2	1009.3
RELATIVE HUMIDITY (%)														
Hour 03	29	83	85	86	90	93	94	96	96	95	91	87	85	90
Hour 09 (Local Time)	29	84	85	84	79	79	81	85	87	89	89	87	85	85
Hour 15	29	81	77	72	69	72	74	78	79	79	80	81	84	77
Hour 21	29	83	84	85	86	86	86	89	93	94	90	86	84	87
PRECIPITATION (inches):														
Water Equivalent														
-Normal		12.18	10.67	10.72	9.92	9.66	7.30	8.18	11.54	18.65	22.97	14.52	14.94	151.25
-Maximum Monthly	52	31.81	32.13	37.28	19.12	18.95	18.34	21.49	27.74	48.33	48.81	43.88	35.21	48.81
-Year		1985	1964	1992	1977	1965	1987	1959	1991	1987	1987	1956	1989	OCT 1987
-Minimum Monthly	52	1.59	0.21	2.06	0.75	2.73	0.52	1.70	2.42	2.44	6.68	3.23	3.79	0.21
-Year		1950	1989	1958	1948	1960	1946	1957	1987	1986	1950	1985	1983	FEB 1989
-Maximum in 24 hrs	46	5.40	6.74	8.03	3.93	5.53	6.09	7.12	5.56	8.13	6.97	7.13	10.43	10.43
-Year		1992	1992	1992	1966	1991	1979	1990	1965	1991	1963	1956	1988	DEC 1988
Snow, Ice pellets, hail														
-Maximum Monthly	45	116.2	87.3	111.0	55.6	15.0	T	0.0	0.0	0.8	36.0	77.1	91.6	116.2
-Year		1989	1965	1959	1985	1965	1984			1992	1966	1975	1957	JAN 1989
-Maximum in 24 hrs	45	23.5	20.7	32.4	18.6	10.0	T	0.0	0.0	0.8	14.7	17.3	23.1	32.4
-Year		1971	1959	1960	1982	1965	1984			1992	1956	1961	1961	MAR 1960
WIND:														
Mean Speed (mph)	45	7.5	7.6	7.2	7.2	7.6	7.2	6.7	6.5	7.0	8.0	7.6	8.0	7.3
Prevailing Direction through 1963		E	E	E	E	ESE	ESE	ESE	ESE	E	E	E	E	E
Fastest Obs. 1 Min.														
-Direction (!!)	41	14	12	13	11	14	11	11	11	14	11	11	14	11
-Speed (MPH)	41	63	53	44	51	52	48	35	43	56	75	63	63	75
-Year		1958	1968	1970	1956	1956	1949	1959	1957	1950	1955	1958	1963	OCT 1955
Peak Gust														
-Direction (!!)	14	SE	SE	SE	SE	SE	SE	SE	SE	E	SE	SE	SE	SE
-Speed (mph)	14	81	62	64	64	48	43	44	60	52	60	70	63	81
-Date		1981	1992	1984	1983	1991	1992	1984	1993	1983	1990	1992	1992	JAN 1981

PRECIPITATION (inches)

YAKUTAT, ALASKA

YEAR	JAN	FEB	MAR	APR	MAY	JUNE	JULY	AUG	SEP	OCT	NOV	DEC	ANNUAL
1964	13.50	32.13	7.85	12.93	11.74	9.05	16.35	9.68	11.31	15.67	16.58	11.55	168.34
1965	17.07	8.54	17.16	11.59	18.95	16.24	14.58	14.43	13.60	36.89	6.58	12.16	187.79
1966	2.40	8.61	11.76	8.04	11.15	1.84	3.01	16.94	18.79	17.15	8.92	10.62	119.23
1967	8.79	10.40	3.90	3.34	5.98	4.97	7.70	14.15	19.48	17.03	18.83	12.26	126.83
1968	8.09	18.11	9.46	10.84	5.35	4.37	5.40	3.87	17.95	12.55	13.91	6.59	116.49
1969	3.56	3.58	14.40	5.68	10.42	2.36	8.16	6.05	9.19	18.79	15.50	16.94	114.63
1970	8.50	10.90	17.19	9.01	6.64	4.27	7.09	14.84	11.37	18.83	8.99	8.90	126.53
1971	11.59	15.48	6.18	10.90	17.00	2.40	10.76	6.36	15.78	20.41	9.40	10.00	136.26
1972	8.68	4.66	10.25	5.48	8.11	4.34	2.96	20.93	10.05	17.17	12.85	9.48	114.96
1973	7.14	8.76	11.71	10.89	9.00	3.91	7.36	14.92	11.98	15.17	4.42	9.09	114.35
1974	3.70	12.31	2.58	11.67	4.65	4.38	2.80	2.96	20.29	35.43	25.23	20.23	146.23
1975	10.67	10.45	5.65	12.97	9.20	8.15	10.51	4.46	29.23	16.99	9.82	11.33	139.43
1976	15.59	5.82	8.56	9.80	18.15	2.45	3.91	10.12	29.16	19.67	42.73	15.74	181.70
1977	17.09	19.17	8.33	19.12	6.57	4.30	5.24	5.14	20.34	24.80	6.71	7.86	144.67
1978	9.18	6.73	11.59	5.64	12.34	8.27	9.57	6.34	13.25	28.65	11.31	17.62	140.49
1979	8.63	1.63	27.35	1.92	10.54	13.84	10.84	7.02	13.78	27.09	15.59	11.13	149.36
1980	9.13	11.01	7.48	17.99	7.23	3.74	11.89	11.64	13.74	29.19	17.77	7.76	148.57
1981	22.43	8.25	12.73	7.04	3.81	5.14	11.04	26.93	24.06	29.77	16.89	8.45	176.54
1982	3.19	2.21	5.07	14.71	8.29	8.74	3.33	8.57	17.63	17.86	17.46	13.66	120.72
1983	13.03	11.27	4.55	17.20	11.55	2.54	7.19	17.24	23.36	18.53	6.10	3.79	136.35
1984	17.65	16.61	12.04	7.25	4.92	4.15	10.73	11.12	14.14	15.90	12.01	11.27	137.79
1985	31.81	8.95	12.03	9.66	10.59	15.95	2.58	7.80	23.44	13.34	3.23	29.85	169.23
1986	25.64	9.56	9.45	9.48	8.19	8.43	5.08	24.54	2.44	25.27	18.19	35.02	181.29
1987	22.95	14.50	11.36	15.90	15.07	18.34	1.86	2.42	48.33	48.81	28.54	22.16	250.24
1988	5.87	16.24	17.84	16.14	8.90	4.62	10.70	22.62	16.99	33.90	17.91	30.18	201.91
1989	11.39	0.21	3.34	2.85	13.20	11.39	7.67		23.57	26.43	19.16	35.21	
1990	11.46	14.38	18.17	6.05	5.46	8.07	17.03	18.68	34.80	21.49	7.10	9.07	171.76
1991	12.92	18.81	11.55	14.66	14.26	5.91	14.32	27.74	42.45	18.70	18.28	19.88	219.48
1992	29.84	19.47	37.28	5.48	12.93	13.64	7.84	20.85	16.87	27.11	28.12	12.90	232.33
1993	10.58	11.31	6.01	6.15	4.82	2.99	2.46	12.08	22.72	23.25	16.07	19.73	138.17
Record Mean	11.90	9.93	10.34	8.84	8.78	6.34	8.14	11.82	17.96	21.52	15.26	14.28	145.12

AVERAGE TEMPERATURE (deg. F)

YAKUTAT, ALASKA

YEAR	JAN	FEB	MAR	APR	MAY	JUNE	JULY	AUG	SEP	OCT	NOV	DEC	ANNUAL
#1964	29.1	32.7	27.8	36.1	41.2	52.4	52.9	53.2	47.4	40.5	29.2	15.8	38.2
1965	22.1	21.3	32.1	35.8	39.2	44.3	50.3	49.3	47.5	39.8	26.9	21.4	35.8
1966	13.0	24.7	24.1	33.2	38.3	48.6	52.0	49.9	46.9	35.7	24.2	25.1	34.7
1967	18.8	24.4	22.3	33.3	42.5	50.5	52.2	52.8	47.8	40.0	32.7	24.2	36.8
1968	18.2	29.1	31.9	33.0	42.2	49.0	53.3	52.6	45.3	35.9	32.3	19.1	36.8
1969	6.8	21.9	29.7	36.5	43.6	49.7	52.0	48.5	48.0	41.4	32.4	34.5	37.1
1970	24.2	33.5	34.6	35.6	42.1	49.2	53.5	52.6	47.4	39.3	29.7	19.4	38.4
1971	15.5	25.0	23.3	32.1	37.6	47.9	52.6	53.4	44.5	39.0	30.9	22.0	35.3
1972	15.6	19.7	24.9	28.8	40.0	48.0	56.0	53.7	46.7	39.0	34.3	24.1	35.9
1973	20.3	26.1	31.6	37.4	43.6	48.5	52.1	50.9	46.1	40.3	25.1	29.9	37.7
1974	16.6	27.4	26.2	36.7	43.7	48.7	52.1	53.5	50.0	42.0	35.2	31.7	38.7
1975	23.6	24.3	26.7	34.6	42.9	47.2	53.4	51.8	49.6	39.7	27.4	24.3	37.1
1976	27.0	23.6	30.7	36.2	41.1	49.2	54.4	53.5	47.5	40.4	40.4	33.9	39.8
1977	36.1	37.9	33.5	38.3	43.3	51.1	55.3	55.4	48.8	42.6	27.4	18.6	40.7
1978	25.8	34.5	34.1	39.9	44.8	50.0	53.2	54.9	48.5	43.6	32.0	29.3	40.9
1979	23.1	16.5	34.6	37.1	44.7	49.5	55.6	54.8	50.8	45.1	38.1	27.3	39.8
1980	24.5	35.7	35.2	41.5	45.5	51.5	55.5	53.5	50.1	45.9	38.9	24.9	41.9
1981	40.8	34.5	39.5	37.7	50.9	52.3	56.3	55.6	49.4	44.1	35.0	27.4	43.6
1982	20.0	24.0	30.1	33.4	42.2	49.0	53.0	52.2	49.0	40.0	32.0	32.0	38.1
1983	30.5	33.1	35.0	38.8	45.7	52.9	55.1	54.2	45.2	40.3	31.6	22.2	40.4
1984	31.0	33.2	37.8	38.6	43.6	48.5	52.5	54.0	48.6	40.7	31.6	28.1	40.7
1985	37.6	27.6	32.5	34.0	41.5	46.6	52.4	51.4	47.2	37.5	20.6	33.7	38.6
1986	34.0	29.0	33.6	33.8	43.6	49.2	54.0	51.9	47.6	43.0	32.3	35.7	40.6
1987	32.0	33.7	30.3	38.8	45.0	48.0	53.7	54.0	48.2	42.7	37.1	30.0	41.1
1988	26.5	32.0	35.3	38.8	43.8	50.4	52.4	52.8	46.5	42.7	33.4	28.9	40.3
1989	21.7	20.3	27.2	38.9	45.2	50.5	54.9		51.0	41.0	29.3	36.7	
1990	26.2	22.5	34.5	40.5	46.9	53.1	55.1	55.9	51.2	39.6	24.3	24.5	39.5
1991	23.2	32.6	29.9	38.2	43.6	51.5	54.6	53.5	50.5	39.8	36.5	34.2	40.7
1992	34.3	30.6	33.9	38.7	44.5	51.5	54.8	52.0	43.8	37.1	36.3	24.7	40.2
1993	20.1	29.6	32.9	41.2	47.8	52.7	56.6	54.8	48.6	44.6	34.6	33.2	41.4
Record Mean	25.4	28.2	30.9	36.5	43.6	49.8	53.7	53.1	48.4	41.0	32.3	27.6	39.2
Max	31.9	35.2	38.4	43.9	50.7	56.3	59.6	59.8	55.4	47.3	38.3	33.5	45.8
Min	18.8	21.2	23.4	29.1	36.4	43.3	47.7	46.4	41.3	34.7	26.3	21.8	32.5

REFERENCE NOTES

YAKUTAT, ALASKA

GENERAL
T - TRACE AMOUNT.
BLANK ENTRIES DENOTE MISSING/UNREPORTED DATA.
# INDICATES A STATION OR INSTRUMENT RELOCATION.

SPECIFIC

EXCEPTIONS

RECORD MEANS ARE THROUGH THE CURRENT YEAR,
BEGINNING IN 1941 FOR TEMPERATURE
1941 FOR PRECIPITATION
1949 FOR SNOWFALL

(a) - LENGTH OF RECORD IN YEARS, ALTHOUGH INDIVIDUAL MONTHS MAY BE MISSING.
0.* OR * - THE VALUE IS BETWEEN 0.0 AND 0.05
NORMALS - BASED ON THE 1961-1990 RECORD PERIOD.
EXTREMES - DATES ARE THE MOST RECENT OCCURRENCE
WIND DIR.- NUMERALS SHOW TENS OF DEGRESS CLOCKWISE FROM TRUE NORTH. "00" INDICATES CALM.
RESULTANT DIRECTIONS ARE GIVEN TO WHOLE DEGREES.
BOLD VALUES INDICATE EXTREME VALUES WHICH OCCURRED AFTER THE ASOS SYSTEM WAS COMMISSIONED.

RECORD = PERIOD OF RECORD
RECORD MEAN PRECIPITATION IS THE MEAN OF ALL DAILY PRECIPITATION AMOUNTS DURING THE PERIOD OF RECORD.
RECORD MAX(MIN) TEMPERATURE IS THE MEAN OF ALL DAILY MAX(MIN) TEMPERATURES DURING THE PERIOD OF RECORD.
RECORD MEAN TEMPERATURE IS THE SUM OF THE RECORD MAX AND RECORD MIN DIVIDED BY 2.
AVERAGE TEMPERATURE IS THE SUM OF THE MEAN DAILY MAX AND MIN TEMPERATURE DIVIDED BY 2.

HEATING DEGREE DAYS Base 65 deg. F YAKUTAT, ALASKA

SEASON	JULY	AUG	SEP	OCT	NOV	DEC	JAN	FEB	MAR	APR	MAY	JUNE	TOTAL
#1964-65	364	360	522	754	1068	1520	1322	1218	1012	870	791	614	10415
1965-66	448	477	520	775	1135	1348	1608	1121	1264	947	822	487	10952
1966-67	396	462	537	899	1216	1231	1423	1132	1316	943	691	424	10670
1967-68	388	370	507	764	965	1259	1445	1033	1020	956	700	475	9882
1968-69	354	381	587	896	974	1418	1802	1201	1087	848	652	450	10650
1969-70	396	506	502	724	973	937	1258	873	931	877	704	472	9153
1970-71	348	378	522	788	1053	1408	1527	1115	1288	982	840	507	10756
1971-72	375	356	608	799	1017	1326	1527	1308	1236	1079	766	505	10902
1972-73	274	342	542	800	913	1258	1381	1082	1029	822	657	485	9585
1973-74	394	430	559	761	1190	1081	1490	1049	1196	840	652	483	10125
1974-75	392	348	442	705	884	1023	1278	1134	1183	905	678	529	9501
1975-76	352	402	455	778	1120	1256	1170	1193	1057	858	733	469	9843
1976-77	321	350	516	755	730	959	888	752	969	795	666	411	8112
1977-78	292	293	478	687	1121	1430	1208	847	950	748	617	445	9116
1978-79	356	306	487	657	986	1100	1294	1350	935	834	624	458	9387
1979-80	284	308	420	609	800	1160	1249	844	916	698	598	398	8284
1980-81	288	351	443	587	779	1235	745	850	786	811	432	376	7683
1981-82	263	283	460	643	893	1161	1388	1145	1075	940	702	475	9428
1982-83	365	389	473	770	983	1017	1065	888	925	778	596	357	8606
1983-84	302	330	588	757	995	1320	1046	916	839	784	655	485	9017
1984-85	382	333	481	747	993	1137	842	1041	1000	928	723	546	9153
1985-86	382	414	527	844	1327	966	955	1002	969	928	657	467	9438
1986-87	336	398	518	675	974	905	1015	871	1068	779	614	504	8657
1987-88	344	335	496	683	830	1077	1184	951	915	782	650	432	8679
1988-89	382	373	546	685	939	1113	1336	1246	1162	778	607	427	9594
1989-90	307		414	736	1063	868	1196	1184	939	727	554	351	
1990-91	299	279	409	781	1211	1249	1290	900	1081	797	656	400	9352
1991-92	316	348	425	771	849	950	944	991	956	782	629	398	8359
1992-93	311	398	628	858	852	1243	1387	984	988	706	525	361	9241
1993-94	248	308	482	628	906	980							

COOLING DEGREE DAYS Base 65 deg. F YAKUTAT, ALASKA

YEAR	JAN	FEB	MAR	APR	MAY	JUNE	JULY	AUG	SEP	OCT	NOV	DEC	TOTAL
1969	0	0	0	0	0	0	0	0	0	0	0	0	0
1970	0	0	0	0	0	0	0	0	0	0	0	0	0
1971	0	0	0	0	0	0	0	0	0	0	0	0	0
1972	0	0	0	0	0	0	0	0	0	0	0	0	0
1973	0	0	0	0	0	0	0	0	0	0	0	0	0
1974	0	0	0	0	0	0	0	0	0	0	0	0	0
1975	0	0	0	0	0	0	0	0	0	0	0	0	0
1976	0	0	0	0	0	0	0	0	0	0	0	0	0
1977	0	0	0	0	0	0	0	0	0	0	0	0	0
1978	0	0	0	0	0	0	0	0	0	0	0	0	0
1979	0	0	0	0	0	0	0	0	0	0	0	0	0
1980	0	0	0	0	0	0	0	0	0	0	0	0	0
1981	0	0	0	0	0	0	0	0	0	0	0	0	0
1982	0	0	0	0	0	0	0	0	0	0	0	0	0
1983	0	0	0	0	0	0	0	0	0	0	0	0	0
1984	0	0	0	0	0	0	0	0	0	0	0	0	0
1985	0	0	0	0	0	0	0	0	0	0	0	0	0
1986	0	0	0	0	0	0	0	0	0	0	0	0	0
1987	0	0	0	0	0	0	0	0	0	0	0	0	0
1988	0	0	0	0	0	0	0	0	0	0	0	0	0
1989	0	0	0	0	0	0	0		0	0	0	0	
1990	0	0	0	0	0	0	0	0	0	0	0	0	0
1991	0	0	0	0	0	1	0	0	0	0	0	0	1
1992	0	0	0	0	0	0	0	0	0	0	0	0	0
1993	0	0	0	0	0	0	0	0	0	0	0	0	0

SNOWFALL (inches) YAKUTAT, ALASKA

SEASON	JULY	AUG	SEP	OCT	NOV	DEC	JAN	FEB	MAR	APR	MAY	JUNE	TOTAL
1964-65	0.0	0.0	0.0	1.0	41.2	36.1	38.0	87.3	19.9	14.2	15.0	0.0	252.7
1965-66	0.0	0.0	0.0	0.8	12.9	64.1	24.3	54.3	44.0	18.5	0.5	0.0	219.4
1966-67	0.0	0.0	0.0	36.0	28.8	34.2	68.4	64.2	32.3	28.8	T	0.0	292.7
1967-68	0.0	0.0	0.0	1.6	23.0	31.1	33.7	23.3	14.3	47.7	2.7	0.0	177.4
1968-69	0.0	0.0	0.0	5.3	10.6	70.5	39.9	33.9	74.3	2.0	0.3	0.0	236.8
1969-70	0.0	0.0	0.0	0.0	8.4	5.3	48.6	9.4	31.8	26.8	T	0.0	130.3
1970-71	0.0	0.0	0.0	12.1	19.6	53.2	67.7	51.6	49.8	47.5	11.4	0.0	312.9
1971-72	0.0	0.0	0.0	26.9	20.9	62.9	75.2	41.7	41.4	45.5	2.1	0.0	316.6
1972-73	0.0	0.0	0.0	5.5	12.4	42.3	70.9	35.6	61.9	10.0	T	0.0	238.6
1973-74	0.0	0.0	0.0	0.2	35.5	20.8	33.3	62.9	16.7	9.0	T	0.0	178.4
1974-75	0.0	0.0	0.0	7.0	27.6	50.5	75.5	68.2	61.3	30.3	6.3	0.0	326.7
1975-76	0.0	0.0	0.0	17.0	77.1	64.9	68.1	75.0	77.5	23.2	0.4	0.0	403.2
1976-77	0.0	0.0	0.0	12.6	18.0	25.0	0.8	17.7	65.4	26.6	1.7	0.0	167.8
1977-78	0.0	0.0	0.0	2.4	41.1	42.0	7.2	3.2	23.4	5.1	T	0.0	124.4
1978-79	0.0	0.0	0.0	T	8.8	42.0	51.7	6.8	23.4	6.0	T	0.0	138.7
1979-80	0.0	0.0	0.0	T	4.6	48.3	33.5	12.9	28.1	1.7	T	0.0	129.1
1980-81	0.0	0.0	T	T	1.0	34.5	0.4	23.6	8.4	3.5	0.0	0.0	71.4
1981-82	0.0	0.0	T	0.4	12.9	21.8	35.2	24.4	32.6	47.2	0.2	0.0	174.7
1982-83	0.0	0.0	T	3.7	7.6	8.9	45.1	12.6	0.6	7.2	T	0.0	85.7
1983-84	0.0	0.0	T	0.6	7.9	12.4	39.9	68.6	2.5	4.5	0.0	T	136.4
1984-85	0.0	0.0	0.0	8.6	30.1	29.1	2.3	59.7	84.1	55.6	5.5	0.0	275.0
1985-86	0.0	0.0	0.0	18.4	13.3	7.8	39.6	21.7	26.7	38.9	T	0.0	166.4
1986-87	0.0	0.0	0.0	T	42.5	5.4	33.6	10.3	12.7	9.7	T	0.0	114.2
1987-88	0.0	0.0	T	T	5.1	44.0	20.5	41.5	13.8	10.5	T	0.0	135.4
1988-89	0.0	0.0	0.0	0.3	15.1	42.7	116.2	1.8	28.8	0.0	T	0.0	204.9
1989-90	0.0		0.0	T	48.1	4.9	27.0	68.7	18.5	0.4	0.0	0.0	
1990-91	0.0	0.0	0.0	2.9	41.2	51.3	27.2	56.8	40.4	2.9	0.0	0.0	222.7
1991-92	0.0	0.0	T	4.7	11.4	42.5	10.1	32.5	61.9	1.0	0.2	0.0	164.3
1992-93	0.0	0.0	0.8	7.2	8.5	52.1	41.6	21.9	10.6	1.5	0.0	0.0	144.2
1993-94	0.0	0.0	0.0	T	12.6	15.9							
Record Mean	0.0	0.0	T	5.3	21.8	38.8	39.0	39.1	38.4	17.5	1.5	T	201.3

FLAGSTAFF, ARIZONA

Flagstaff, elevation 7,000 feet, is situated on a volcanic plateau at the base of the highest mountains in Arizona. The climate may be classified as vigorous with cold winters, mild, pleasantly cool summers, moderate humidity, and considerable diurnal temperature change. Only limited farming exists due to the short growing season. The stormy months are January, February, March, July, and August.

Based on the 1951–1980 period, the average first occurrence of 32 degrees Fahrenheit in the fall is September 21 and the average last occurrence in the spring is June 13.

Temperatures in Flagstaff are characteristic of high altitude climates. The average daily range of temperature is relatively high, especially in the winter months, October to March, as a result of extensive snow cover and clear skies. Winter minimum temperatures frequently reach zero or below and temperatures of –25 degrees or less have occurred. Summer maximum temperatures are often above 80 degrees and occasionally, temperatures have exceeded 95 degrees.

The Flagstaff area is semi-arid. Several months have recorded little or no precipitation. Over 90 consecutive days without measurable precipitation have occurred. Annual precipitation ranges from less than 10 inches to more than 35 inches. Winter snowfalls can be heavy, exceeding 100 inches during one month and over 200 inches during the winter season. However, accumulations are quite variable from year to year. Some winter months may experience little or no snow and the winter season has produced total snow accumulations of less than 12 inches.

NORMALS, MEANS, AND EXTREMES

FLAGSTAFF, ARIZONA

LATITUDE: 35°08'N LONGITUDE: 111°40' W ELEVATION: FT. GRND 7006 BARO 6997 TIME ZONE: MOUNTAIN WBAN: 03103

	(a)	JAN	FEB	MAR	APR	MAY	JUNE	JULY	AUG	SEP	OCT	NOV	DEC	YEAR
TEMPERATURE °F:														
Normals														
-Daily Maximum		42.2	45.3	49.2	57.8	67.4	78.2	81.9	79.3	73.2	63.4	51.1	43.3	61.0
-Daily Minimum		15.2	17.7	21.3	26.7	33.3	41.4	50.5	48.9	41.2	31.0	22.4	15.8	30.5
-Monthly		28.7	31.5	35.3	42.3	50.4	59.8	66.3	64.1	57.3	47.2	36.8	29.6	45.8
Extremes														
-Record Highest	44	66	71	73	80	87	96	97	92	90	85	74	68	97
-Year		1971	1986	1988	1992	1974	1970	1973	1978	1950	1980	1977	1950	JUL 1973
-Record Lowest	44	-22	-23	-16	-2	14	22	32	24	23	-2	-13	-23	-23
-Year		1971	1985	1966	1975	1975	1955	1955	1968	1971	1971	1958	1990	DEC 1990
NORMAL DEGREE DAYS:														
Heating (base 65°F)		1125	938	921	681	453	187	29	65	237	552	846	1097	7131
Cooling (base 65°F)		0	0	0	0	0	31	70	38	6	0	0	0	145
% OF POSSIBLE SUNSHINE	14	76	74	76	83	88	86	75	76	81	79	75	73	79
MEAN SKY COVER (tenths)														
Sunrise - Sunset	42	5.3	5.3	5.3	4.7	4.0	3.0	5.3	5.1	3.7	3.6	4.2	4.8	4.5
MEAN NUMBER OF DAYS:														
Sunrise to Sunset														
-Clear	42	12.3	10.9	11.6	12.5	15.3	18.5	9.1	9.8	15.7	17.1	15.4	13.9	162.0
-Partly Cloudy	42	6.3	6.1	7.9	8.8	9.2	7.7	13.1	13.1	9.6	7.0	6.6	6.5	101.9
-Cloudy	42	12.4	11.3	11.5	8.6	6.5	3.9	8.8	8.1	4.7	6.9	8.0	10.7	101.3
Precipitation														
.01 inches or more	44	7.7	6.9	8.5	5.5	4.5	2.8	11.5	11.5	6.3	4.9	5.3	6.4	81.9
Snow, Ice pellets, hail														
1.0 inches or more	43	4.5	4.0	5.1	2.3	0.6	0.0	0.0	0.0	0.*	0.5	2.1	3.6	22.7
Thunderstorms	30	0.*	0.3	0.6	1.3	2.6	3.7	16.6	15.7	6.7	2.2	0.7	0.2	50.5
Heavy Fog Visibility 1/4 mile or less	30	1.8	1.8	1.6	1.2	0.2	0.*	0.1	0.3	0.5	0.9	1.1	1.8	11.4
Temperature °F														
-Maximum														
90° and above	44	0.0	0.0	0.0	0.0	0.0	1.3	1.5	0.3	0.*	0.0	0.0	0.0	3.2
32° and below	44	4.4	2.5	1.6	0.2	0.0	0.0	0.0	0.0	0.0	0.1	1.1	4.5	14.4
-Minimum														
32° and below	44	30.4	27.7	30.0	24.8	13.7	2.9	0.1	0.1	2.8	18.0	27.9	30.3	208.5
0° and below	44	3.2	1.4	0.6	0.*	0.0	0.0	0.0	0.0	0.0	0.*	0.4	2.2	8.0
AVG. STATION PRESS. (mb)	5	786.7	786.9	783.0	784.8	786.4	789.2	791.7	791.6	790.6	789.7	788.2	787.6	788.0
RELATIVE HUMIDITY (%)														
Hour 05	36	74	74	72	66	64	55	68	76	73	71	70	72	70
Hour 11 (Local Time)	38	54	51	45	34	29	23	34	40	37	38	44	51	40
Hour 17	30	51	45	41	31	27	21	38	43	36	36	43	52	39
Hour 23	24	68	65	63	54	49	40	59	69	64	63	64	72	61
PRECIPITATION (inches):														
Water Equivalent														
-Normal		2.04	2.09	2.55	1.48	0.72	0.40	2.78	2.75	2.03	1.61	1.95	2.40	22.80
-Maximum Monthly	44	9.55	10.05	6.75	5.62	4.14	2.92	6.62	8.06	6.75	9.86	6.64	7.30	10.05
-Year		1993	1993	1970	1965	1992	1955	1986	1986	1983	1972	1985	1967	FEB 1993
-Minimum Monthly	44	0.00	T	T	T	T	0.00	T	0.26	T	T	T	T	0.00
-Year		1972	1967	1972	1991	1974	1971	1993	1962	1992	1952	1989	1958	JAN 1972
-Maximum in 24 hrs	44	2.71	4.48	2.96	1.79	1.11	2.79	2.55	3.04	3.43	2.73	3.69	3.11	4.48
-Year		1993	1993	1970	1985	1965	1956	1964	1986	1965	1972	1978	1951	FEB 1993
Snow, Ice pellets, hail														
-Maximum Monthly	43	63.4	45.5	79.4	58.3	8.2	T	T	0.3	2.0	24.7	40.7	86.0	86.0
-Year		1980	1990	1991	1965	1975	1993	1992	1992	1965	1971	1985	1967	DEC 1967
-Maximum in 24 hrs	43	23.1	23.1	26.3	17.2	6.6	T	T	0.3	2.0	13.5	19.6	27.3	27.3
-Year		1980	1987	1970	1977	1965	1993	1992	1992	1965	1974	1991	1967	DEC 1967
WIND:														
Mean Speed (mph)	26	6.8	6.7	7.2	7.6	7.3	6.9	5.5	5.1	5.7	5.8	6.8	6.7	6.5
Prevailing Direction through 1963		NE	S	SSW	SSW	SSW	SSW	SSW	S	S	N	NNE	NE	SSW
Fastest Mile														
-Direction (!!!)	11	SW	SW	SW	SW	SW	SW	NW	SW	W	NW	SW	NE	SW
-Speed (MPH)	11	38	34	37	40	46	35	39	30	33	34	39	38	46
-Year		1975	1980	1974	1974	1975	1984	1976	1978	1974	1978	1978	1982	MAY 1975
Peak Gust														
-Direction (!!!)														
-Speed (mph)														
-Date														

PRECIPITATION (inches) FLAGSTAFF, ARIZONA

YEAR	JAN	FEB	MAR	APR	MAY	JUNE	JULY	AUG	SEP	OCT	NOV	DEC	ANNUAL
1964	1.07	0.14	3.08	2.17	0.84	0.17	5.23	1.32	0.99	0.02	1.27	2.74	19.04
1965	3.05	2.34	3.33	5.62	1.88	0.30	2.34	1.01	4.85	0.27	4.97	6.63	36.59
1966	1.10	1.06	0.95	0.27	T	0.21	1.62	3.55	2.03	0.99	2.33	6.17	20.28
1967	0.93	T	1.11	1.90	0.41	1.05	3.80	2.68	2.25	0.30	0.54	7.30	22.27
1968	1.55	1.29	1.15	2.09	0.55	0.16	3.61	1.13	0.04	1.38	0.87	2.71	16.53
1969	4.63	3.91	3.00	0.11	1.06	0.01	3.81	1.90	1.34	1.14	2.04	0.46	23.41
1970	0.51	0.41	6.75	1.16	T	0.07	2.58	5.15	3.79	0.11	1.37	2.12	24.02
1971	0.08	1.48	0.25	0.55	1.23	0.00	1.97	4.48	2.02	4.37	0.40	4.18	21.01
1972	0.00	0.02	T	0.72	0.14	1.93	1.90	2.82	0.81	9.86	2.34	4.13	24.67
1973	1.89	3.69	6.18	1.21	1.17	0.40	1.87	1.25	T	0.03	1.90	0.12	19.71
1974	3.63	0.26	1.01	0.57	T	T	3.00	2.16	0.93	3.64	1.03	1.18	17.41
1975	1.76	1.90	2.92	2.20	1.16	0.05	2.24	0.74	1.89	0.33	2.96	1.95	20.10
1976	0.17	5.96	2.06	3.09	1.65	T	3.82	0.58	1.16	0.73	0.10	0.80	20.12
1977	1.85	0.84	0.92	1.47	0.96	0.91	3.60	3.72	1.52	1.04	0.76	1.18	18.77
1978	4.09	4.67	5.58	1.60	0.27	0.09	1.17	0.68	0.46	0.56	6.16	5.39	30.72
1979	5.54	1.73	2.52	0.31	2.16	0.18	0.79	2.38	0.13	1.30	1.14	1.50	19.68
1980	6.52	7.81	4.16	1.21	1.79	0.25	2.49	2.19	0.65	1.08	T	1.15	29.30
1981	1.31	1.16	4.04	1.50	0.72	1.09	2.87	3.73	2.54	1.81	2.43	0.17	23.37
1982	4.62	2.55	5.69	0.25	0.86	T	1.89	2.32	3.17	0.71	5.36	3.67	31.09
1983	1.61	3.04	4.36	2.18	0.06	0.28	2.86	3.53	6.75	0.75	1.53	2.52	29.47
1984	0.36	0.13	0.89	0.73	0.21	0.13	4.20	3.86	1.75	1.43	1.40	5.00	20.09
1985	2.38	1.67	2.54	3.39	0.26	0.09	2.36	1.07	3.68	2.44	6.64	0.15	26.67
1986	0.31	1.76	2.60	1.23	0.72	1.16	6.62	8.06	4.80	2.02	1.68	1.43	32.39
1987	2.51	2.54	1.69	0.21	0.70	0.25	1.93	3.74	1.49	4.64	2.91	1.37	23.98
1988	1.64	2.30	0.14	3.83	0.14	1.86	3.48	4.77	0.15	1.23	1.14	1.00	21.68
1989	1.84	1.35	2.08	0.01	0.62	0.23	2.28	3.40	0.30	1.28	T	1.05	14.44
1990	1.54	3.20	2.17	2.32	0.73	0.24	4.32	1.71	6.18	0.49	1.09	1.68	25.67
1991	1.76	2.08	6.00	T	0.14	0.59	1.04	1.64	0.26	1.12	4.47	2.73	21.83
1992	2.03	3.69	4.40	0.78	4.14	0.32	2.67	5.80	T	3.64	0.46	6.78	34.71
1993	9.55	10.05	1.54	0.26	0.44	0.55	T	4.19	1.95	3.29	3.02	0.76	35.60
Record Mean	2.06	2.07	2.11	1.31	0.71	0.51	2.73	2.89	1.78	1.56	1.53	1.95	21.20

AVERAGE TEMPERATURE (deg. F) FLAGSTAFF, ARIZONA

YEAR	JAN	FEB	MAR	APR	MAY	JUNE	JULY	AUG	SEP	OCT	NOV	DEC	ANNUAL
1964	25.3	26.6	30.5	39.7	49.4	57.4	66.9	63.3	56.7	51.3	31.3	30.3	44.1
1965	31.4	29.7	33.0	39.2	46.7	53.0	65.0	63.7	53.8	49.7	39.7	30.2	44.6
1966	23.5	25.1	37.2	44.7	53.7	59.1	66.4	65.2	58.6	46.9	39.5	29.7	45.8
1967	29.1	35.3	39.0	37.6	48.7	56.2	66.6	64.4	57.9	50.0	40.4	23.1	45.7
1968	25.5	35.8	37.0	40.3	50.2	59.9	64.9	59.5	55.9	46.9	35.7	25.0	44.7
1969	31.8	25.4	27.3	41.7	52.4	57.9	66.2	65.1	58.2	40.5	33.9	33.3	44.5
1970	30.9	37.1	33.7	37.5	52.6	60.4	67.7	66.3	54.4	43.9	39.5	28.5	46.1
1971	29.3	31.1	37.0	41.9	46.8	58.7	67.5	64.0	53.0	38.6	33.5	22.5	43.6
1972	28.4	31.8	40.0	39.8	46.4	57.6	66.5	62.9	56.7	45.9	29.6	21.9	43.9
1973	22.9	28.7	26.8	38.0	52.8	60.5	65.7	64.6	57.7	50.0	37.1	33.0	44.8
1974	27.8	30.5	40.4	43.2	54.8	66.5	66.3	64.7	58.7	47.9	36.8	26.8	47.0
1975	27.7	27.6	33.2	36.2	47.5	57.1	65.9	63.5	57.9	46.5	37.3	29.2	44.2
1976	30.6	35.1	35.4	41.4	52.8	58.9	66.5	62.8	57.1	47.0	39.5	31.6	46.5
1977	26.8	34.2	33.5	44.7	47.4	62.2	67.2	66.2	60.2	50.8	41.0	37.6	47.6
1978	31.5	30.7	40.3	42.2	49.4	61.4	66.6	63.7	57.3	49.8	34.6	24.3	46.0
1979	22.6	25.9	32.3	40.7	47.9	57.8	64.1	60.5	58.8	47.8	31.2	31.2	43.4
1980	30.7	32.8	32.3	41.9	45.6	60.9	69.0	66.3	59.7	49.0	41.4	39.9	47.5
1981	36.2	36.3	36.6	48.5	52.1	66.1	68.2	65.4	58.7	46.7	41.8	36.4	49.4
1982	28.3	30.5	35.3	43.5	49.9	57.1	63.8	65.6	57.6	44.0	35.2	28.1	44.9
1983	31.0	32.3	36.2	37.0	49.9	57.3	65.5	63.8	60.7	48.6	36.9	34.0	46.1
1984	31.7	32.8	38.3	40.7	56.8	58.5	65.7	63.8	59.3	42.4	35.3	29.0	46.2
1985	27.5	28.9	35.5	46.2	51.5	62.6	67.2	65.3	53.1	47.7	33.8	32.4	46.0
1986	37.0	34.2	39.7	44.1	52.0	61.2	64.0	65.7	53.0	43.5	37.6	30.2	46.9
1987	27.6	31.1	32.8	45.9	50.7	61.3	62.8	62.8	56.5	50.4	36.2	27.1	45.4
1988	29.1	34.2	37.4	44.0	50.8	61.5	67.4	64.3	55.9	52.5	36.9	27.9	46.8
1989	26.2	32.3	41.8	50.4	54.2	60.8	68.1	63.6	58.7	47.5	38.4	31.6	47.8
1990	28.6	29.3	38.3	46.1	50.3	64.4	66.9	62.6	59.8	48.9	37.6	25.1	46.5
1991	30.5	37.2	32.2	43.1	49.4	57.3	67.1	66.5	59.6	50.8	36.9	30.2	46.7
1992	27.3	35.2	38.1	49.1	53.3	59.0	65.1	64.2	59.3	50.4	33.8	26.3	46.8
1993	32.2	32.5	40.2	46.4	53.9	60.4	65.5	64.8	57.9	47.4	35.0	30.1	47.2
Record Mean	28.1	31.1	35.8	42.9	50.4	59.4	65.8	63.9	57.4	47.1	36.8	29.7	45.7
Max	41.5	44.2	49.3	57.9	66.8	77.4	81.1	78.6	73.3	63.0	51.4	43.2	60.7
Min	14.7	17.9	22.3	27.8	33.9	41.3	50.4	49.2	41.5	31.2	22.1	16.1	30.7

REFERENCE NOTES FLAGSTAFF, ARIZONA

GENERAL
T - TRACE AMOUNT.
BLANK ENTRIES DENOTE MISSING/UNREPORTED DATA.
# INDICATES A STATION OR INSTRUMENT RELOCATION.

SPECIFIC

PM - INCLUDES LAST DAY OF PREVIOUS MONTH
ASOS - AUTOMATED SURFACE OBSERVING SYSTEM IN OPERATION DURING THESE MONTHS.

(a) - LENGTH OF RECORD IN YEARS, ALTHOUGH INDIVIDUAL MONTHS MAY BE MISSING.
0.* OR * - THE VALUE IS BETWEEN 0.0 AND 0.05
NORMALS - BASED ON THE 1961-1990 RECORD PERIOD.
EXTREMES - DATES ARE THE MOST RECENT OCCURRENCE
WIND DIR.- NUMERALS SHOW TENS OF DEGRESS CLOCKWISE FROM TRUE NORTH. "00" INDICATES CALM.
RESULTANT DIRECTIONS ARE GIVEN TO WHOLE DEGREES.
BOLD VALUES INDICATE EXTREME VALUES WHICH OCCURRED AFTER THE ASOS SYSTEM WAS COMMISSIONED.

RECORD = PERIOD OF RECORD
RECORD MEAN PRECIPITATION IS THE MEAN OF ALL DAILY PRECIPITATION AMOUNTS DURING THE PERIOD OF RECORD.
RECORD MAX(MIN) TEMPERATURE IS THE MEAN OF ALL DAILY MAX(MIN) TEMPERATURES DURING THE PERIOD OF RECORD.
RECORD MEAN TEMPERATURE IS THE SUM OF THE RECORD MAX AND RECORD MIN DIVIDED BY 2.
AVERAGE TEMPERATURE IS THE SUM OF THE MEAN DAILY MAX AND MIN TEMPERATURE DIVIDED BY 2.

EXCEPTIONS

1. PERCENT OF POSSIBLE SUNSHINE IS 1973-1976 AND 1983 TO DATE.
2. MEAN WIND SPEED, THUNDERSTORMS, AND HEAVY FOG ARE THROUGH 1978.

RECORD MEANS ARE THROUGH THE CURRENT YEAR, BEGINNING IN 1900 FOR TEMPERATURE
1900 FOR PRECIPITATION
1950 FOR SNOWFALL

HEATING DEGREE DAYS Base 65 deg. F FLAGSTAFF, ARIZONA

SEASON	JULY	AUG	SEP	OCT	NOV	DEC	JAN	FEB	MAR	APR	MAY	JUNE	TOTAL
1964-65	9	72	245	417	1001	1070	1034	983	985	768	560	353	7497
1965-66	22	69	326	467	753	1071	1281	1109	854	603	343	171	7069
1966-67	14	32	185	556	757	1089	1111	824	799	816	498	265	6946
1967-68	11	33	207	461	733	1292	1218	839	861	733	453	172	7013
1968-69	38	164	267	554	872	1234	1021	1105	1164	693	383	207	7702
1969-70	21	37	200	753	926	978	1051	775	962	821	378	174	7076
1970-71	6	16	310	643	758	1125	1098	943	862	683	557	210	7211
1971-72	24	40	352	809	938	1310	1126	956	768	749	573	220	7865
1972-73	22	78	241	584	1055	1330	1297	1006	1176	801	370	164	8124
1973-74	39	45	212	458	829	987	1147	960	754	650	310	67	6458
1974-75	19	34	195	522	841	1178	1150	1041	978	856	536	231	7581
1975-76	23	66	207	566	821	1101	1059	862	912	702	370	190	6879
1976-77	17	71	230	553	757	1027	1177	858	970	603	540	102	6905
1977-78	9	19	157	432	715	843	1032	954	756	677	478	128	6200
1978-79	33	73	227	465	907	1254	1307	1089	1005	722	524	219	7825
1979-80	68	157	186	526	1008	1042	1056	930	1009	684	596	158	7420
1980-81	6	43	153	491	703	774	886	794	873	486	398	50	5657
1981-82	1	39	182	558	689	880	1130	963	911	640	458	230	6681
1982-83	65	22	218	643	888	1136	1046	911	888	835	461	222	7335
1983-84	26	64	134	502	837	952	1025	929	820	722	247	204	6462
1984-85	21	51	165	695	884	1109	1155	1005	911	557	411	102	7066
1985-86	26	31	351	532	931	1005	862	855	777	619	392	119	6500
1986-87	58	28	353	661	816	1069	1150	942	990	566	435	114	7182
1987-88	82	86	246	447	859	1167	1103	885	848	621	434	124	6902
1988-89	9	41	266	381	836	1141	1196	910	710	432	330	139	6391
1989-90	7	67	184	540	792	1030	1124	996	822	558	449	84	6653
1990-91	10	93	167	494	816	1231	1060	772	1008	652	476	225	7004
1991-92	7	12	159	438	835	1071	1159	854	830	470	356	180	6371
1992-93	38	80	163	444	932	1191	1014	903	761	551	339	158	6574
1993-94	29	45	207	538	893	1073							

COOLING DEGREE DAYS Base 65 deg. F FLAGSTAFF, ARIZONA

YEAR	JAN	FEB	MAR	APR	MAY	JUNE	JULY	AUG	SEP	OCT	NOV	DEC	TOTAL
1969	0	0	0	0	0	1	64	48	0	0	0	0	113
1970	0	0	0	0	0	44	99	64	1	0	0	0	208
1971	0	0	0	0	0	28	108	17	0	0	0	0	153
1972	0	0	0	0	0	4	76	20	0	0	0	0	100
1973	0	0	0	0	0	34	69	40	0	0	0	0	143
1974	0	0	0	0	0	120	66	33	13	0	0	0	232
1975	0	0	0	0	0	0	60	28	0	0	0	0	88
1976	0	0	0	0	0	15	70	13	0	0	0	0	98
1977	0	0	0	0	0	28	82	61	20	0	0	0	191
1978	0	0	0	0	0	25	87	38	2	0	0	0	152
1979	0	0	0	0	0	10	47	24	4	0	0	0	85
1980	0	0	0	0	0	43	133	91	2	0	0	0	269
1981	0	0	0	0	0	91	108	58	0	0	0	0	257
1982	0	0	0	0	0	1	36	46	6	0	0	0	89
1983	0	0	0	0	0	0	45	31	12	0	0	0	88
1984	0	0	0	0	2	14	49	23	2	0	0	0	90
1985	0	0	0	0	0	36	102	47	0	0	0	0	185
1986	0	0	0	0	0	11	33	55	0	0	0	0	99
1987	0	0	0	0	0	13	21	26	0	0	0	0	60
1988	0	0	0	0	0	28	94	28	1	0	0	0	151
1989	0	0	0	0	0	22	111	29	0	0	0	0	162
1990	0	0	0	0	0	76	76	28	18	0	0	0	198
1991	0	0	0	0	0	0	79	65	4	0	0	0	148
1992	0	0	0	0	0	6	48	62	0	0	0	0	116
1993	0	0	0	0	0	26	50	46	1	0	0	0	123

SNOWFALL (inches) FLAGSTAFF, ARIZONA

SEASON	JULY	AUG	SEP	OCT	NOV	DEC	JAN	FEB	MAR	APR	MAY	JUNE	TOTAL
1964-65	0.0	0.0	0.0	0.0	14.3	13.7	15.1	23.9	34.4	58.3	7.0	0.0	166.7
1965-66	0.0	0.0	2.0	2.2	2.9	38.5	14.5	10.8	7.4	5.1	0.0	0.0	83.4
1966-67	0.0	0.0	0.0	T	11.3	13.4	9.8	T	10.9	17.7	0.0	0.0	63.1
1967-68	0.0	0.0	0.0	0.0	5.1	86.0	15.2	9.3	10.1	22.7	2.0	0.0	150.4
1968-69	0.0	0.0	0.0	T	4.3	27.5	12.0	42.1	43.6	0.6	4.6	0.0	134.7
1969-70	0.0	0.0	0.0	1.5	3.0	4.6	5.2	2.8	67.3	11.3	0.0	0.0	95.7
1970-71	0.0	0.0	0.0	0.0	2.0	24.5	0.8	15.1	3.1	5.0	6.1	0.0	56.6
1971-72	0.0	0.0	T	24.7	4.9	18.8	0.0	0.4	T	1.5	0.0	0.0	50.3
1972-73	0.0	0.0	0.0	11.8	23.2	28.9	21.0	33.8	77.4	10.9	3.0	0.0	210.0
1973-74	0.0	0.0	0.0	T	20.7	1.2	35.3	2.4	8.8	1.6	0.0	0.0	70.0
1974-75	0.0	0.0	0.0	16.6	8.2	15.6	20.1	18.2	29.1	25.1	8.2	0.0	141.1
1975-76	0.0	0.0	0.0	T	25.2	18.9	1.5	31.2	20.4	31.2	3.2	0.0	131.6
1976-77	0.0	0.0	0.0	T	0.7	9.0	21.1	11.2	8.8	17.8	1.6	0.0	70.2
1977-78	0.0	0.0	0.0	0.0	7.0	1.6	40.4	32.1	24.5	9.3	1.3	0.0	116.2
1978-79	0.0	0.0	0.0	0.0	16.5	19.8	59.4	18.1	22.8	4.1	4.8	0.0	145.5
1979-80	0.0	0.0	0.0	0.5	5.5	20.7	63.4	32.9	42.5	11.6	T	0.0	177.1
1980-81	0.0	0.0	0.0	6.9	T	6.8	11.7	11.9	45.6	9.5	0.0	0.0	92.4
1981-82	0.0	0.0			0.0	0.0	47.5	20.4	26.7	1.9	0.4	0.0	
1982-83	0.0	0.0	T	T	22.6	27.1	15.3	25.0	38.5	13.6	0.5	0.0	142.6
1983-84	0.0	0.0	0.0	0.0	14.3	5.8	4.3	1.8	0.3	5.5	0.0	0.0	32.0
1984-85	0.0	0.0	0.0	0.6	9.4	28.7	26.2	31.3	21.3	18.5	0.0	0.0	136.0
1985-86	0.0	0.0	0.0	0.0	40.7	2.6	0.4	26.9	32.8	1.6	0.4	0.0	105.4
1986-87	0.0	0.0	0.9	0.6	4.8	9.5	38.6	40.7	25.0	1.5	0.0	0.0	121.6
1987-88	0.0	0.0	0.0	0.0	2.9	16.7	28.9	21.0	1.5	33.1	0.4	0.0	104.5
1988-89	0.0	0.0	0.0	0.0	11.9	15.0	21.7	12.0	16.6	T	0.5	0.0	77.7
1989-90	0.0	T	T	T	T	13.1	24.2	45.5	25.0	4.2	1.4	0.0	113.4
1990-91	T	T	T	T	9.6	22.3	3.1	13.5	79.4	T	T	0.0	127.9
1991-92	0.0	0.0	T	5.9	39.5	24.0	24.7	24.9	35.9	4.0	T	T	158.9
1992-93	T	0.3	0.0	0.5	2.1	41.7	55.7	35.2	11.9	2.3	0.3	T	150.0
1993-94	0.0	T	T	0.0	23.0	13.4							
Record Mean	T	T	0.1	2.0	10.2	15.9	20.7	18.3	22.3	9.5	1.8	T	100.7

PHOENIX, ARIZONA

Phoenix is located in the Salt River Valley at an elevation of about 1,100 feet. The valley is oval shaped and flat except for scattered precipitous mountains rising a few hundred to as much as 1,500 feet above the valley floor. Sky Harbor Airport, where the weather observations are taken, is in the southern part of the city. Six miles to the south of the airport are the South Mountains rising to 2,500 feet. Eighteen miles southwest, the Estrella Mountains rise to 4,500 feet, and 30 miles to the west are the White Tank Mountains rising to 4,100 feet. The Superstition Mountains, over 30 miles to the east, rise to as much as 5,000 feet. The valley, though located in the Sonora Desert, supports large acreages of cotton, citrus, and other agriculture along with one of the largest urban populations in the United States. The water supply for this complex desert community is partly from reservoirs on the impounded Salt and Verde Rivers, and partly from a large underground water table.

Temperatures range from very hot in summer to mild in winter. Many winter days reach over 70 degrees and typical high temperatures in the middle of the winter are in the 60s. The climate becomes less attractive in the summer. The normal high temperature is over 90 degrees from early May through early October, and over 100 degrees from early June through early September. Many days each summer will exceed 110 degrees in the afternoon and remain above 85 degrees all night. When temperatures are extremely high, the low humidity does not provide much comfort.

Indeed, the climate is very dry. Annual precipitation is only about 7 inches, and afternoon humidities range from about 30 percent in winter to only about 10 percent in June. Rain comes mostly in two seasons. From about Thanksgiving to early April there are periodic rains from Pacific storms. Moisture from the south and southeast results in a summer thunderstorm peak in July and August. Usually the break from extreme dryness in June to the onset of thunderstorms in early July is very abrupt. Afternoon humidities suddenly double to about 20 percent, which with the great heat, gives a feeling of mugginess. Fog is rare, occurring about once per winter, and is unknown in the other seasons.

The valley is characterized by light winds. High winds associated with thunderstorms occur periodically in the summer. These occasionally create duststorms which move large distances across the deserts. Strong thunderstorm winds occur any month of the year, but are rare outside the summer months. Persistent strong winds of 30 mph or more are rare except for two or three events in an average spring due to Pacific storms. Winter storms rarely bring high winds due to the relatively stable air in the valley during that season.

Based on the 1951–1980 period, the average first occurrence of 32 degrees Fahrenheit in the fall is December 13 and the average last occurrence in the spring is February 7.

NORMALS, MEANS, AND EXTREMES

PHOENIX, ARIZONA

LATITUDE: 33°26'N LONGITUDE: 112°01'W ELEVATION: FT. GRND 1110 BARO 1109 TIME ZONE: MOUNTAIN WBAN: 23183

	(a)	JAN	FEB	MAR	APR	MAY	JUNE	JULY	AUG	SEP	OCT	NOV	DEC	YEAR
TEMPERATURE °F:														
Normals														
-Daily Maximum		65.9	70.7	75.5	84.5	93.6	103.5	105.9	103.7	98.3	88.1	74.9	66.2	85.9
-Daily Minimum		41.2	44.7	48.8	55.3	63.9	72.9	81.0	79.2	72.8	60.8	48.9	41.8	59.3
-Monthly		53.6	57.7	62.2	69.9	78.8	88.2	93.5	91.5	85.6	74.5	61.9	54.1	72.6
Extremes														
-Record Highest	56	88	92	100	105	113	122	118	116	118	107	93	88	122
-Year		1971	1986	1988	1992	1984	1990	1989	1975	1950	1980	1988	1950	JUN 1990
-Record Lowest	56	17	22	25	32	40	50	61	60	47	34	25	22	17
-Year		1950	1948	1966	1945	1967	1944	1944	1942	1965	1971	1938	1948	JAN 1950
NORMAL DEGREE DAYS:														
Heating (base 65°F)		362	227	182	75	8	0	0	0	0	17	134	345	1350
Cooling (base 65°F)		8	22	95	222	436	696	884	822	618	311	41	7	4162
% OF POSSIBLE SUNSHINE	98	78	80	84	89	93	94	85	85	89	88	84	78	86
MEAN SKY COVER (tenths)														
Sunrise - Sunset	48	4.7	4.6	4.3	3.4	2.7	1.9	3.7	3.3	2.3	2.8	3.4	4.1	3.4
MEAN NUMBER OF DAYS:														
Sunrise to Sunset														
-Clear	56	13.9	12.6	14.4	17.2	20.8	23.1	16.6	17.6	21.6	20.3	17.7	15.3	211.0
-Partly Cloudy	56	7.0	6.7	8.1	7.3	6.8	4.7	10.3	9.6	5.4	6.2	6.2	6.3	84.6
-Cloudy	56	10.1	8.9	8.5	5.6	3.5	2.2	4.2	3.8	3.0	4.4	6.1	9.4	69.7
Precipitation														
.01 inches or more	54	4.1	3.9	3.7	1.7	1.0	0.7	4.2	4.9	2.9	2.7	2.6	3.9	36.5
Snow,Ice pellets,hail														
1.0 inches or more	56	0.0	0.0	0.0	0.0	0.0	0.0	0.0	0.0	0.0	0.0	0.0	0.0	0.0
Thunderstorms	54	0.4	0.6	0.9	0.7	1.0	1.0	6.1	7.1	3.5	1.4	0.5	0.6	23.7
Heavy Fog Visibility														
1/4 mile or less	56	0.6	0.1	0.1	0.0	0.0	0.0	0.0	0.0	0.0	0.*	0.2	0.5	1.6
Temperature °F														
-Maximum														
90° and above	33	0.0	0.1	2.0	9.8	23.1	29.2	30.9	30.6	27.6	15.2	0.5	0.0	169.0
32° and below	33	0.0	0.0	0.0	0.0	0.0	0.0	0.0	0.0	0.0	0.0	0.0	0.0	0.0
-Minimum														
32° and below	33	3.3	1.3	0.4	0.0	0.0	0.0	0.0	0.0	0.0	0.0	0.2	1.8	7.0
0° and below	33	0.0	0.0	0.0	0.0	0.0	0.0	0.0	0.0	0.0	0.0	0.0	0.0	0.0
AVG. STATION PRESS.(mb)	21	978.3	977.5	974.4	972.6	970.6	969.7	971.2	971.6	971.7	974.0	976.5	978.3	973.9
RELATIVE HUMIDITY (%)														
Hour 05	33	67	60	57	43	35	31	44	51	49	50	57	67	51
Hour 11 (Local Time)	33	46	39	34	23	19	16	28	33	31	30	36	46	32
Hour 17	33	33	27	24	16	13	12	20	23	23	22	27	34	23
Hour 23	33	56	48	43	28	23	20	33	38	38	40	48	57	39
PRECIPITATION (inches):														
Water Equivalent														
-Normal		0.67	0.68	0.88	0.22	0.12	0.13	0.83	0.96	0.86	0.65	0.66	1.00	7.66
-Maximum Monthly	56	5.22	2.23	4.16	2.10	1.06	1.70	5.15	5.56	4.23	4.40	3.04	3.98	5.56
-Year		1993	1944	1941	1941	1976	1972	1984	1951	1939	1972	1952	1967	AUG 1951
-Minimum Monthly	56	0.00	0.00	0.00	0.00	0.00	0.00	T	T	0.00	0.00	0.00	0.00	0.00
-Year		1972	1967	1959	1962	1983	1983	1993	1975	1973	1973	1980	1981	MAY 1983
-Maximum in 24 hrs	56	1.84	1.49	2.04	1.38	0.96	1.64	2.75	3.07	2.43	2.32	2.16	1.89	3.07
-Year		1993	1987	1983	1941	1976	1972	1984	1943	1970	1988	1993	1967	AUG 1943
Snow,Ice pellets,hail														
-Maximum Monthly	56	T	0.6	T	T	T	0.0	0.0	0.0	0.0	T	0.0	0.4	0.6
-Year		1993	1939	1991	1949	1992					1992		1990	FEB 1939
-Maximum in 24 hrs	56	T	0.6	T	T	T	0.0	0.0	0.0	0.0	T	0.0	0.4	0.6
-Year		1993	1939	1991	1949	1992					1992		1990	FEB 1939
WIND:														
Mean Speed (mph)	48	5.3	5.9	6.7	6.9	7.0	6.8	7.1	6.6	6.3	5.8	5.3	5.1	6.2
Prevailing Direction through 1963		E	E	E	E	E	E	W	E	E	E	E	E	E
Fastest Obs. 1 Min.														
-Direction (!!)	8	27	14	26	27	11	03	08	10	05	21	29	09	26
-Speed (MPH)	8	32	26	35	28	35	31	35	35	35	28	29	26	35
-Year		1988	1992	1992	1986	1992	1986	1991	1989	1990	1986	1990	1988	MAR 1992
Peak Gust														
-Direction (!!)	56	W	W	W	W	SSE	NE	SE	E	SW	W	W	W	SE
-Speed (mph)	56	60	54	51	49	59	73	86	78	75	61	60	68	86
-Date		1983	1980	1989	1981	1954	1978	1976	1978	1950	1981	1982	1953	JUL 1976

PRECIPITATION (inches)

PHOENIX, ARIZONA

YEAR	JAN	FEB	MAR	APR	MAY	JUNE	JULY	AUG	SEP	OCT	NOV	DEC	ANNUAL
1964	0.22	0.01	0.37	0.10	T	0.00	0.60	1.29	1.80	0.17	0.35	1.09	6.00
1965	1.22	0.91	1.39	1.35	0.16	0.91	0.16	0.18	0.60	0.20	0.92	3.19	11.19
1966	0.35	0.95	0.34	T	T	0.22	0.09	2.17	2.00	0.25	0.38	0.52	7.27
1967	0.25	0.00	0.43	0.08	0.05	0.47	0.99	0.02	0.13	0.67	1.27	3.98	8.34
1968	0.19	1.20	1.04	T	T	0.00	1.70	0.59	0.00	0.35	0.91	0.69	6.67
1969	1.37	0.78	0.56	0.03	0.26	0.00	0.28	0.14	2.11	0.08	0.65	0.68	6.94
1970	T	0.30	2.26	T	T	0.00	0.48	1.02	2.85	0.44	0.02	0.26	7.63
1971	0.22	0.35	T	0.13	T	0.00	0.24	0.99	0.92	0.27	T	0.47	3.59
1972	0.00	T	T	T	T	1.70	0.72	1.20	0.28	4.40	1.01	1.56	10.87
1973	0.13	1.36	1.69	0.07	0.10	T	1.30	T	0.00	0.00	1.36	0.00	6.01
1974	0.57	0.02	1.37	0.01	0.00	0.00	0.84	1.15	1.07	2.12	0.44	0.59	8.18
1975	0.02	0.33	0.63	0.43	T	T	0.38	T	0.82	0.23	0.55	1.12	4.51
1976	T	0.47	0.40	0.67	1.06	0.09	1.48	0.12	1.69	0.70	0.43	0.85	7.96
1977	0.35	0.06	0.27	0.06	0.16	0.10	0.30	0.18	0.53	0.61	T	0.54	3.16
1978	2.33	2.21	2.14	0.20	T	0.01	1.44	1.79	T	0.35	2.30	2.46	15.23
1979	2.16	0.09	1.78	0.02	0.76	0.04	0.34	1.18	0.09	0.09	0.12	0.13	6.80
1980	1.58	2.09	0.86	0.44	0.21	0.03	0.56	0.06	0.13	0.02	0.00	0.08	6.06
1981	0.71	1.08	0.98	0.20	0.03	T	1.14	0.11	0.18	1.34	0.95	0.00	6.72
1982	0.81	0.67	1.30	T	0.50	T	0.43	1.97	0.12	T	2.50	1.64	9.94
1983	0.70	1.17	3.17	0.18	0.00	0.00	0.38	2.48	2.43	0.71	0.43	1.16	12.81
1984	0.31	0.00	0.00	0.91	0.18	0.18	5.15	0.87	3.36	0.31	0.71	2.93	14.91
1985	0.95	0.18	0.46	0.17	T	0.00	0.98	0.21	1.60	0.92	1.59	0.86	7.92
1986	0.07	1.19	1.58	0.01	T	0.01	1.19	1.27	0.47	0.41	0.03	1.38	7.61
1987	0.67	2.06	0.28	0.09	0.06	0.01	1.08	0.45	0.57	0.47	1.04	1.62	8.40
1988	0.90	0.23	0.17	1.09	0.00	0.02	0.87	0.63	0.00	2.38	0.78	0.14	7.21
1989	1.19	T	1.25	0.00	T	0.00	0.13	1.11	0.47	0.46	0.14	0.19	4.94
1990	0.80	0.70	0.35	0.17	0.16	0.04	1.05	2.70	1.11	0.04	0.15	0.46	7.73
1991	0.63	0.56	2.05	0.00	0.00	T	0.14	0.12	0.81	1.16	1.25	1.63	8.35
1992	1.62	0.90	2.49	0.49	1.05	0.04	2.95	1.30	0.03	0.26	0.03	3.08	14.24
1993	5.22	1.72	1.62	0.00	0.08	0.01	T	0.55	0.06	1.27	2.79	0.02	13.34
Record Mean	0.82	0.74	0.77	0.33	0.14	0.09	0.93	1.01	0.78	0.53	0.65	0.91	7.71

AVERAGE TEMPERATURE (deg. F)

PHOENIX, ARIZONA

YEAR	JAN	FEB	MAR	APR	MAY	JUNE	JULY	AUG	SEP	OCT	NOV	DEC	ANNUAL
1964	46.7	49.3	56.5	65.2	73.7	82.6	90.6	86.2	80.9	74.9	55.5	52.0	67.8
1965	52.7	52.4	56.1	63.4	71.8	79.0	91.0	89.0	79.2	73.8	62.1	52.9	68.6
1966	48.2	49.7	61.2	69.8	80.1	86.8	93.0	90.9	82.9	70.9	60.5	52.0	70.5
1967	50.7	55.7	62.8	62.4	75.1	81.1	91.6	91.0	84.8	73.5	63.9	48.2	70.1
1968	52.4	59.7	59.9	66.7	76.6	86.2	90.2	86.5	83.6	72.7	59.2	49.5	70.3
1969	54.9	53.0	56.9	68.5	78.3	84.2	93.1	94.4	86.0	69.5	62.1	54.8	71.3
1970	52.1	60.2	59.5	64.7	79.6	88.1	95.0	92.5	82.2	69.1	61.4	52.6	71.4
1971	52.2	56.3	63.3	66.5	73.3	85.3	94.9	89.6	85.6	69.3	59.7	50.2	70.5
1972	51.4	59.1	70.6	71.4	78.3	87.8	94.4	89.9	84.8	71.9	58.1	52.1	72.5
1973	51.2	57.5	56.6	67.2	80.9	88.1	93.5	93.4	84.7	74.4	60.8	55.4	72.0
1974	54.0	56.7	64.5	70.6	80.2	92.2	92.4	91.2	87.2	75.9	61.5	50.6	73.1
1975	52.3	54.0	59.0	62.6	76.7	86.6	94.3	91.9	86.2	72.9	60.9	54.8	71.0
1976	55.4	60.7	61.5	68.7	80.7	87.9	91.6	90.7	83.0	74.0	64.1	55.6	72.8
1977	53.8	61.7	60.8	73.5	75.7	91.4	95.0	94.1	87.6	78.7	65.8	59.9	74.9
1978	56.6	58.7	65.6	69.2	78.5	90.9	94.6	91.4	86.3	78.6	61.5	51.7	73.6
1979	50.1	55.7	60.4	70.1	78.1	89.5	93.8	89.4	90.2	77.2	58.2	55.9	72.4
1980	56.6	60.6	60.7	69.8	76.0	88.9	95.6	92.2	87.3	75.6	64.1	61.3	74.0
1981	59.2	61.4	63.8	76.0	80.5	93.4	95.2	95.8	89.2	73.6	66.1	58.6	76.0
1982	53.9	60.1	62.4	72.5	80.4	88.1	93.7	93.7	86.7	73.5	61.9	54.1	73.4
1983	56.0	58.4	62.2	66.6	80.6	88.6	95.5	92.6	91.0	77.2	62.4	57.2	74.0
1984	57.4	60.1	67.6	70.7	87.0	88.9	91.7	91.2	87.5	71.4	61.9	53.7	74.1
1985	54.3	57.4	62.8	75.1	84.2	92.4	94.9	94.5	82.3	75.1	61.3	55.9	74.2
1986	61.4	61.0	69.3	74.2	82.3	92.8	92.3	94.5	84.1	74.7	65.0	56.4	75.7
1987	54.7	59.7	63.4	77.9	82.6	93.0	93.1	92.2	86.9	80.9	63.1	52.7	75.0
1988	55.1	62.5	66.3	73.0	81.4	93.1	96.2	93.9	87.4	82.4	64.4	55.7	76.0
1989	54.4	61.9	70.1	80.1	83.1	92.1	97.4	93.7	89.9	77.3	66.4	57.0	77.0
1990	55.6	56.6	67.2	76.2	81.1	93.8	93.6	90.8	87.6	78.7	65.9	53.6	75.1
1991	55.9	66.0	60.3	72.2	79.7	87.8	95.1	94.5	88.5	80.2	63.5	57.3	75.1
1992	56.4	62.1	64.7	77.0	83.1	90.1	92.8	92.3	90.5	79.8	61.5	53.8	75.3
1993	58.2	58.2	65.7	73.8	83.7	89.6	92.9	91.6	87.9	76.7	61.4	56.2	74.7
Record Mean	52.4	56.3	61.1	68.7	76.9	86.2	91.5	89.7	84.2	72.7	60.5	53.0	71.1
Max	65.3	69.6	74.9	83.6	92.3	102.1	104.5	102.3	98.0	87.3	74.7	66.1	85.1
Min	39.4	43.0	47.3	53.8	61.4	70.3	78.4	77.0	70.4	58.0	46.2	40.0	57.1

REFERENCE NOTES

PHOENIX, ARIZONA

GENERAL
T - TRACE AMOUNT.
BLANK ENTRIES DENOTE MISSING/UNREPORTED DATA.
# INDICATES A STATION OR INSTRUMENT RELOCATION.

SPECIFIC

PM - INCLUDES LAST DAY OF PREVIOUS MONTH
ASOS - AUTOMATED SURFACE OBSERVING SYSTEM IN OPERATION DURING THESE MONTHS.

(a) - LENGTH OF RECORD IN YEARS, ALTHOUGH INDIVIDUAL MONTHS MAY BE MISSING.
0.* OR * - THE VALUE IS BETWEEN 0.0 AND 0.05
NORMALS - BASED ON THE 1961-1990 RECORD PERIOD.
EXTREMES - DATES ARE THE MOST RECENT OCCURRENCE
WIND DIR.- NUMERALS SHOW TENS OF DEGRESS CLOCKWISE FROM TRUE NORTH. "00" INDICATES CALM.
RESULTANT DIRECTIONS ARE GIVEN TO WHOLE DEGREES.
BOLD VALUES INDICATE EXTREME VALUES WHICH OCCURRED AFTER THE ASOS SYSTEM WAS COMMISSIONED.

RECORD = PERIOD OF RECORD
RECORD MEAN PRECIPITATION IS THE MEAN OF ALL DAILY PRECIPITATION AMOUNTS DURING THE PERIOD OF RECORD.
RECORD MAX(MIN) TEMPERATURE IS THE MEAN OF ALL DAILY MAX(MIN) TEMPERATURES DURING THE PERIOD OF RECORD.
RECORD MEAN TEMPERATURE IS THE SUM OF THE RECORD MAX AND RECORD MIN DIVIDED BY 2.
AVERAGE TEMPERATURE IS THE SUM OF THE MEAN DAILY MAX AND MIN TEMPERATURE DIVIDED BY 2.

EXCEPTIONS

1. PEAK GUST WINDS ARE AS OBSERVED JANUARY 1938 THROUGH OCTOBER 1953 AND FROM RECORDER THEREAFTER.
2. PERCENT OF POSSIBLE SUNSHINE IS FROM CITY OFFICE AUGUST 1895 THROUGH OCTOBER 1953 AND FROM SKY HARBOR AIRPORT THEREAFTER.
3. MEAN SKY COVER IS 1940, 1941, AND 1948 TO DATE.

RECORD MEANS ARE THROUGH THE CURRENT YEAR, BEGINNING IN 1896 FOR TEMPERATURE
1896 FOR PRECIPITATION
1938 FOR SNOWFALL

HEATING DEGREE DAYS Base 65 deg. F PHOENIX, ARIZONA

SEASON	JULY	AUG	SEP	OCT	NOV	DEC	JAN	FEB	MAR	APR	MAY	JUNE	TOTAL
1964-65	0	0	0	0	281	396	375	346	269	133	14	0	1814
1965-66	0	0	4	7	116	370	516	423	145	12	0	0	1593
1966-67	0	0	0	8	139	397	437	256	102	93	10	0	1442
1967-68	0	0	0	6	72	512	384	151	167	39	0	0	1331
1968-69	0	0	0	0	173	473	306	327	265	12	13	0	1569
1969-70	0	0	0	12	95	307	393	134	166	60	0	0	1167
1970-71	0	0	0	19	119	376	396	241	123	53	0	0	1327
1971-72	0	0	0	79	185	455	414	174	22	12	0	0	1341
1972-73	0	0	0	38	205	395	422	200	254	39	0	0	1553
1973-74	0	0	0	2	156	291	333	229	77	5	0	0	1093
1974-75	0	0	0	21	112	439	388	301	191	107	4	0	1563
1975-76	0	0	0	15	159	310	296	123	134	52	0	0	1089
1976-77	0	0	0	2	112	285	339	122	149	33	0	0	1042
1977-78	0	0	0	0	42	155	254	172	67	25	0	0	715
1978-79	0	0	0	1	148	405	455	254	143	30	0	0	1436
1979-80	0	0	0	11	204	277	254	130	129	35	0	0	1040
1980-81	0	0	0	12	108	122	181	131	74	8	0	0	636
1981-82	0	0	0	1	56	196	335	151	99	4	0	0	842
1982-83	0	0	0	1	103	331	272	181	120	53	0	0	1061
1983-84	0	0	0	0	154	236	228	139	16	23	0	0	796
1984-85	0	0	0	7	126	345	328	222	102	5	0	0	1135
1985-86	0	0	0	0	149	274	110	158	66	2	1	0	760
1986-87	0	0	0	0	43	260	318	172	95	4	0	0	892
1987-88	0	0	0	0	98	375	311	100	60	20	2	0	966
1988-89	0	0	0	0	135	284	321	133	46	0	0	0	919
1989-90	0	0	0	1	36	243	291	253	76	0	0	0	900
1990-91	0	0	0	0	65	348	275	27	161	5	0	0	881
1991-92	0	0	0	34	107	233	260	89	56	7	0	0	786
1992-93	0	0	0	0	127	340	205	184	61	0	0	0	917
1993-94	0	0	0	0	125	264							

COOLING DEGREE DAYS Base 65 deg. F PHOENIX, ARIZONA

YEAR	JAN	FEB	MAR	APR	MAY	JUNE	JULY	AUG	SEP	OCT	NOV	DEC	TOTAL
1969	0	0	22	123	433	582	878	918	638	158	16	0	3768
1970	0	4	4	58	459	700	938	862	527	151	18	0	3721
1971	7	2	76	107	265	614	934	773	623	220	30	0	3651
1972	0	11	200	212	419	691	919	780	599	259	4	0	4094
1973	0	0	0	109	499	701	894	885	598	302	36	0	4024
1974	0	2	69	182	477	825	858	821	673	365	13	0	4285
1975	0	0	12	42	374	654	913	839	640	265	45	1	3785
1976	6	4	34	169	495	692	833	804	548	289	91	0	3965
1977	0	36	25	295	334	797	936	907	683	434	73	1	4521
1978	3	1	92	158	422	787	928	828	644	431	49	0	4343
1979	0	0	11	191	411	741	901	763	764	397	7	0	4186
1980	0	5	2	187	344	724	956	852	675	346	88	13	4192
1981	5	36	40	345	489	857	943	961	731	277	95	5	4784
1982	0	21	24	234	481	697	899	897	658	272	12	0	4195
1983	2	1	38	112	489	715	951	861	787	388	85	0	4429
1984	0	2	107	203	688	724	836	821	681	208	41	0	4311
1985	0	17	40	316	603	826	934	920	525	319	47	0	4547
1986	3	52	209	282	543	844	853	921	582	307	51	1	4648
1987	3	30	51	396	553	846	879	850	665	499	48	0	4820
1988	10	31	108	265	520	851	972	904	678	543	124	3	5009
1989	1	49	210	459	566	820	1013	897	751	392	87	0	5245
1990	5	25	150	339	506	873	895	806	683	431	101	0	4814
1991	0	61	22	227	465	691	937	920	712	514	70	0	4619
1992	0	13	54	373	567	762	868	851	771	464	30	0	4753
1993	1	1	89	271	585	746	871	832	695	369	24	0	4484

SNOWFALL (inches) PHOENIX, ARIZONA

SEASON	JULY	AUG	SEP	OCT	NOV	DEC	JAN	FEB	MAR	APR	MAY	JUNE	TOTAL
1964-65	0.0	0.0	0.0	0.0	0.0	0.0	0.0	0.0	0.0	0.0	0.0	0.0	0.0
1965-66	0.0	0.0	0.0	0.0	0.0	0.0	0.0	0.0	0.0	0.0	0.0	0.0	0.0
1966-67	0.0	0.0	0.0	0.0	0.0	0.0	0.0	0.0	0.0	0.0	0.0	0.0	0.0
1967-68	0.0	0.0	0.0	0.0	0.0	T	0.0	0.0	0.0	0.0	0.0	0.0	T
1968-69	0.0	0.0	0.0	0.0	0.0	T	0.0	0.0	0.0	0.0	0.0	0.0	T
1969-70	0.0	0.0	0.0	0.0	0.0	0.0	0.0	0.0	0.0	0.0	0.0	0.0	0.0
1970-71	0.0	0.0	0.0	0.0	0.0	0.0	0.0	0.0	0.0	0.0	0.0	0.0	0.0
1971-72	0.0	0.0	0.0	0.0	0.0	0.0	0.0	0.0	0.0	0.0	0.0	0.0	0.0
1972-73	0.0	0.0	0.0	0.0	0.0	0.0	0.0	0.0	0.0	0.0	0.0	0.0	0.0
1973-74	0.0	0.0	0.0	0.0	0.0	0.0	0.0	0.0	0.0	0.0	0.0	0.0	0.0
1974-75	0.0	0.0	0.0	0.0	0.0	T	0.0	0.0	0.0	0.0	0.0	0.0	T
1975-76	0.0	0.0	0.0	0.0	0.0	0.0	0.0	0.0	T	0.0	0.0	0.0	T
1976-77	0.0	0.0	0.0	0.0	0.0	0.0	0.0	0.0	0.0	0.0	0.0	0.0	0.0
1977-78	0.0	0.0	0.0	0.0	0.0	0.0	0.0	0.0	0.0	0.0	0.0	0.0	0.0
1978-79	0.0	0.0	0.0	0.0	0.0	0.0	0.0	0.0	0.0	0.0	0.0	0.0	0.0
1979-80	0.0	0.0	0.0	0.0	0.0	0.0	0.0	0.0	0.0	0.0	0.0	0.0	0.0
1980-81	0.0	0.0	0.0	0.0	0.0	0.0	0.0	0.0	0.0	0.0	0.0	0.0	0.0
1981-82	0.0	0.0	0.0	0.0	0.0	0.0	0.0	0.0	0.0	0.0	0.0	0.0	0.0
1982-83	0.0	0.0	0.0	0.0	0.0	0.0	0.0	0.0	0.0	0.0	0.0	0.0	0.0
1983-84	0.0	0.0	0.0	0.0	0.0	0.0	0.0	0.0	0.0	0.0	0.0	0.0	0.0
1984-85	0.0	0.0	0.0	0.0	0.0	0.0	0.0	T	0.0	0.0	0.0	0.0	T
1985-86	0.0	0.0	0.0	0.0	0.0	0.1	0.0	0.0	0.0	0.0	0.0	0.0	0.1
1986-87	0.0	0.0	0.0	0.0	0.0	0.0	T	0.0	0.0	0.0	0.0	0.0	T
1987-88	0.0	0.0	0.0	0.0	0.0	0.0	0.0	0.0	0.0	0.0	0.0	0.0	0.0
1988-89	0.0	0.0	0.0	0.0	0.0	0.0	0.0	0.0	0.0	0.0	0.0	0.0	0.0
1989-90	0.0	0.0	0.0	0.0	0.0	0.0	0.0	0.0	0.0	0.0	0.0	0.0	0.0
1990-91	0.0	0.0	0.0	0.0	0.0	0.4	0.0	0.0	T	0.0	0.0	0.0	0.4
1991-92	0.0	0.0	0.0	0.0	0.0	0.0	0.0	0.0	0.0	0.0	T	0.0	T
1992-93	0.0	0.0	0.0	T	0.0	0.0	T	0.0	0.0	0.0	0.0	0.0	T
1993-94	0.0	0.0	0.0	0.0	0.0	0.0							
Record Mean	0.0	0.0	0.0	T	0.0	T	T	T	T	T	T	0.0	T

TUCSON, ARIZONA

Tucson lies at the foot of the Catalina Mountains, north of the airport. The area within about 15 miles of the airport station is flat or gently rolling, with many dry washes. The soil is sandy, and vegetation is mostly brush, cacti, and small trees. Rugged mountains encircle the valley. The mountains to the north, east, and south rise to over 5,000 feet above the airport. The western hills and mountains range from 500 to 4,000 feet.

The climate of Tucson is characterized by a long hot season, from April to October. Temperatures above 90 degrees prevail from May through September. Temperatures of 100 degrees or higher average 41 days annually, including 14 days each for June and July, but these extreme temperatures are moderated by low relative humidities. The temperature range is large, averaging 30 degrees or more a day.

More than 50 percent of the annual precipitation falls between July 1 and September 15, and over 20 percent falls from December through March. During the summer, scattered convective or orographic showers and thunderstorms often fill dry washes to overflowing. On occasion, brief, torrential downpours cause destructive flash floods in the Tucson area. Hail rarely occurs in thunderstorms. The December through March precipitation occurs as prolonged rainstorms that replenish the ground water. During these storms, snow often falls on the higher mountains, but snow in Tucson is infrequent, particularly in accumulations exceeding an inch in depth.

From the first of the year, the humidity decreases steadily until the summer thunderstorm season, when it shows a marked increase. From mid-September, the end of the thunderstorm season, the humidity decreases again until late November. Occasionally during the summer, humidities are high enough to produce discomfort, but only for short periods. During the hot season, humidity values sometimes fall below 5 percent.

Tucson lies in the zone receiving more sunshine than any other section of the United States. Cloudless days are commonplace, and average cloudiness is low.

Surface winds are generally light, with no major seasonal changes in velocity or direction. Occasional duststorms occur in areas where the ground has been disturbed. During the spring, winds may briefly be strong enough to cause some damage to trees and buildings. Wind velocities and directions are influenced by the surrounding mountains, and the general slope of the terrain. Usually local winds tend to be in the southeast quadrant during the night and early morning hours, veering to northwest during the day. Highest velocities usually occur with winds from the southwest and east to south.

While dust and haze are frequently visible, their effect on the general clarity of the atmosphere is not great. Visibility is normally high.

Based on the 1951–1980 period, the average first occurrence of 32 degrees Fahrenheit in the fall is November 29 and the average last occurrence in the spring is February 28.

NORMALS, MEANS, AND EXTREMES

TUCSON, ARIZONA

LATITUDE: 32°07'N LONGITUDE: 110°56' W ELEVATION: FT. GRND 2584 BARO 2589 TIME ZONE: MOUNTAIN WBAN: 23160

	(a)	JAN	FEB	MAR	APR	MAY	JUNE	JULY	AUG	SEP	OCT	NOV	DEC	YEAR
TEMPERATURE °F:														
Normals														
-Daily Maximum		63.9	67.8	72.8	81.2	89.9	99.6	99.4	96.8	93.3	84.3	72.7	64.3	82.2
-Daily Minimum		38.6	41.0	44.6	50.4	58.0	67.9	73.6	72.1	67.5	56.6	45.6	39.8	54.6
-Monthly		51.3	54.4	58.7	65.8	74.0	83.8	86.6	84.5	80.4	70.4	59.2	52.0	68.4
Extremes														
-Record Highest	53	87	92	99	104	107	117	114	112	107	102	90	84	117
-Year		1953	1957	1988	1989	1958	1990	1989	1993	1990	1993	1988	1954	JUN 1990
-Record Lowest	53	16	20	20	27	38	47	59	61	44	26	24	16	16
-Year		1949	1955	1965	1945	1950	1955	1992	1956	1965	1971	1979	1974	DEC 1974
NORMAL DEGREE DAYS:														
Heating (base 65°F)		425	302	229	97	7	0	0	0	0	27	188	403	1678
Cooling (base 65°F)		0	5	33	121	286	564	670	605	462	194	14	0	2954
% OF POSSIBLE SUNSHINE	46	80	82	86	92	93	93	78	80	87	88	85	79	85
MEAN SKY COVER (tenths)														
Sunrise - Sunset	52	4.7	4.6	4.5	3.4	2.8	2.3	5.2	4.6	3.0	2.9	3.5	4.5	3.8
MEAN NUMBER OF DAYS:														
Sunrise to Sunset														
-Clear	53	13.7	12.8	14.7	17.2	20.1	21.5	10.2	12.3	19.3	19.9	17.6	14.8	194.0
-Partly Cloudy	53	7.1	6.4	6.8	7.3	6.8	6.0	12.2	12.1	6.9	6.4	6.1	6.1	90.3
-Cloudy	53	10.3	9.0	9.4	5.5	4.1	2.5	8.6	6.6	3.8	4.8	6.3	10.1	81.0
Precipitation .01 inches or more	53	4.7	3.8	4.3	2.0	1.5	1.7	10.3	9.5	4.7	3.4	2.9	4.6	53.3
Snow,Ice pellets,hail 1.0 inches or more	53	0.2	0.2	0.1	0.*	0.0	0.0	0.0	0.0	0.0	0.0	0.*	0.1	0.5
Thunderstorms	53	0.4	0.3	0.5	0.7	1.6	2.5	13.7	13.6	5.4	2.0	0.5	0.3	41.4
Heavy Fog Visibility 1/4 mile or less	53	0.3	0.2	0.*	0.0	0.0	0.0	0.0	0.0	0.*	0.0	0.2	0.4	1.0
Temperature °F														
-Maximum														
90° and above	53	0.0	0.*	0.5	4.7	17.7	28.2	29.3	28.6	23.9	9.1	0.*	0.0	142.0
32° and below	53	0.0	0.0	0.0	0.0	0.0	0.0	0.0	0.0	0.0	0.0	0.0	0.0	0.0
-Minimum														
32° and below	53	6.2	4.0	1.1	0.*	0.0	0.0	0.0	0.0	0.0	0.*	1.5	5.0	17.8
0° and below	53	0.0	0.0	0.0	0.0	0.0	0.0	0.0	0.0	0.0	0.0	0.0	0.0	0.0
AVG. STATION PRESS.(mb)	21	927.8	927.1	925.0	924.0	922.8	922.7	924.7	925.0	924.4	925.7	926.9	927.9	925.3
RELATIVE HUMIDITY (%)														
Hour 05	53	63	59	54	42	35	32	57	65	55	52	54	62	53
Hour 11 (Local Time)	53	41	35	29	21	17	16	32	38	32	30	32	40	30
Hour 17	53	33	27	23	16	13	13	28	33	26	25	28	35	25
Hour 23	53	58	50	43	31	24	23	47	53	44	43	48	57	43
PRECIPITATION (inches):														
Water Equivalent														
-Normal		0.87	0.70	0.72	0.30	0.18	0.20	2.37	2.19	1.67	1.06	0.67	1.07	12.00
-Maximum Monthly	53	4.81	2.90	2.26	1.66	1.11	1.46	6.17	7.93	5.11	4.98	1.90	5.02	7.93
-Year		1993	1980	1952	1951	1992	1954	1981	1955	1964	1983	1952	1965	AUG 1955
-Minimum Monthly	53	T	0.00	0.00	0.00	0.00	0.00	0.26	0.23	0.00	0.00	0.00	0.00	0.00
-Year		1970	1972	1956	1972	1974	1983	1993	1976	1953	1982	1980	1981	JUN 1983
-Maximum in 24 hrs	53	1.46	1.49	1.19	0.91	0.89	1.27	3.93	2.48	3.05	3.58	1.86	1.54	3.93
-Year		1993	1942	1952	1988	1943	1954	1958	1961	1964	1983	1968	1967	JUL 1958
Snow,Ice pellets,hail														
-Maximum Monthly	53	4.7	3.9	5.7	2.0	T	0.0	0.0	T	T	T	6.4	6.8	6.8
-Year		1987	1965	1964	1976	1992			1993	1990	1991	1958	1971	DEC 1971
-Maximum in 24 hrs	52	4.3	3.9	5.7	2.0	T	0.0	0.0	T	T	T	6.4	6.8	6.8
-Year		1987	1965	1964	1976	1992			1993	1990	1991	1958	1971	DEC 1971
WIND:														
Mean Speed (mph)	48	7.9	8.1	8.5	8.9	8.8	8.7	8.4	7.9	8.3	8.2	8.1	7.8	8.3
Prevailing Direction through 1963		SE	SE	SE	SE	SE	SSE	SE	SE	SE	SE	SE	SE	SE
Fastest Mile														
-Direction (!!!)	45	E	E	SE	SW	SE	SE	SE	NE	SE	SE	E	W	SE
-Speed (MPH)	45	40	59	41	46	43	50	71	54	54	47	55	44	71
-Year		1962	1952	1955	1986	1984	1961	1971	1969	1960	1948	1951	1949	JUL 1971
Peak Gust														
-Direction (!!!)	10	SW	E	SE	SW	SE	SW	SE	SE	SE	NW	E	SE	SE
-Speed (mph)	10	45	46	53	55	55	47	66	71	71	47	46	47	71
-Date		1988	1987	1986	1984	1984	1991	1985	1988	1990	1988	1990	1988	SEP 1990

PRECIPITATION (inches)

TUCSON, ARIZONA

YEAR	JAN	FEB	MAR	APR	MAY	JUNE	JULY	AUG	SEP	OCT	NOV	DEC	ANNUAL
1964	0.14	0.13	0.81	0.67	0.00	0.01	4.82	3.90	5.11	0.91	0.68	0.81	17.99
1965	0.45	0.64	0.27	0.23	T	0.01	2.13	1.12	0.82	0.07	0.77	5.02	11.53
1966	1.74	2.25	0.19	0.12	0.11	0.02	2.57	3.31	3.53	0.32	0.06	0.19	14.41
1967	0.04	0.13	0.41	0.29	0.62	0.42	2.72	2.00	1.35	1.03	0.48	3.44	12.93
1968	0.18	0.99	1.79	0.62	T	0.00	1.97	1.12	T	0.09	1.86	0.32	8.94
1969	0.74	0.50	0.34	0.60	0.46	0.00	1.51	2.57	1.31	0.03	1.06	0.82	9.94
1970	T	0.34	1.13	0.45	0.03	0.33	2.53	1.43	3.58	1.73	0.00	0.43	11.98
1971	0.04	0.50	T	0.56	0.01	T	2.18	3.29	1.75	1.18	0.69	1.97	12.17
1972	0.00	0.00	0.01	0.00	0.24	0.68	3.49	2.93	1.09	4.51	1.30	0.61	14.86
1973	0.06	1.60	2.20	0.02	0.09	0.50	1.74	0.54	T	0.00	0.47	0.00	7.22
1974	0.93	T	0.55	T	0.00	0.01	4.44	1.04	1.69	2.12	0.81	0.33	11.92
1975	0.36	0.13	0.95	0.27	0.11	0.00	2.38	0.32	1.26	T	0.34	0.52	6.64
1976	0.06	0.53	0.38	0.57	0.23	0.10	1.18	0.23	1.68	0.37	0.48	0.47	6.28
1977	1.83	0.04	0.74	0.43	0.08	0.06	0.76	0.80	1.41	2.36	0.33	1.33	10.17
1978	2.05	1.75	0.89	0.01	0.61	0.22	0.78	1.59	1.66	1.86	1.58	2.73	15.73
1979	2.94	0.42	0.64	0.04	0.67	0.53	2.04	2.60	0.02	0.33	0.01	0.15	10.39
1980	0.73	2.90	1.22	0.08	T	0.23	1.78	1.95	2.93	0.22	0.00	0.19	12.23
1981	1.29	0.71	1.98	0.56	0.26	0.16	6.17	0.80	1.10	0.06	0.61	0.00	13.70
1982	1.56	0.06	1.26	0.05	0.51	0.13	2.13	2.51	2.69	0.00	1.30	1.59	13.79
1983	1.70	0.94	1.28	0.14	T	0.00	1.98	4.24	4.28	4.98	1.71	0.61	21.86
1984	0.62	0.00	0.00	0.36	0.06	1.05	2.92	4.19	1.81	0.77	0.45	3.30	15.53
1985	1.71	1.08	0.20	0.45	T	0.07	3.14	1.97	1.13	2.03	0.95	0.15	12.88
1986	0.98	1.13	1.30	T	0.44	0.06	1.82	3.56	0.31	0.50	0.42	1.28	11.80
1987	0.59	1.64	0.83	0.80	0.74	0.16	0.37	2.79	2.30	0.34	0.44	1.50	12.50
1988	0.41	0.53	0.35	1.15	0.02	0.15	1.69	3.64	0.80	2.09	0.75	0.05	11.63
1989	0.96	0.23	0.62	0.00	0.13	0.06	1.42	0.90	0.02	1.84	0.12	0.18	6.48
1990	0.96	0.71	0.38	0.10	0.03	0.64	5.45	2.70	1.63	0.58	0.23	1.54	14.95
1991	1.15	0.91	1.40	0.00	0.00	0.20	0.44	2.17	1.54	0.73	0.80	1.44	10.78
1992	1.21	1.80	2.12	0.19	1.11	0.07	0.93	4.55	0.94	0.03	T	3.47	16.42
1993	4.81	1.50	0.49	0.00	0.59	0.02	0.26	4.93	0.46	0.81	0.98	0.14	14.99
Record Mean	0.90	0.83	0.74	0.34	0.21	0.25	2.20	2.22	1.31	0.72	0.74	1.03	11.49

AVERAGE TEMPERATURE (deg. F)

TUCSON, ARIZONA

YEAR	JAN	FEB	MAR	APR	MAY	JUNE	JULY	AUG	SEP	OCT	NOV	DEC	ANNUAL
1964	47.5	47.7	54.8	63.2	73.2	82.0	86.2	81.6	76.3	72.1	55.2	52.4	66.0
1965	53.6	51.1	55.1	64.5	70.1	77.6	85.0	84.0	76.8	71.9	62.6	52.1	67.1
1966	47.7	47.8	60.1	66.8	76.1	82.8	85.3	82.9	78.3	68.1	61.1	52.4	67.4
1967	51.4	55.6	62.1	62.1	71.9	80.7	85.4	84.6	80.7	71.6	62.9	48.6	68.1
1968	52.4	59.1	58.7	63.2	73.3	83.5	84.9	81.3	80.7	71.7	58.3	50.6	68.1
1969	55.5	53.1	54.3	66.6	74.9	80.7	86.1	86.3	81.2	66.8	58.6	52.4	68.0
1970	50.0	57.0	55.9	61.1	75.2	83.4	87.2	84.8	76.4	65.1	60.1	51.8	67.3
1971	50.5	52.3	59.8	62.8	69.3	81.2	87.5	81.3	79.1	64.2	56.8	47.1	66.0
1972	50.4	55.8	65.0	65.8	72.3	81.6	86.6	82.9	78.6	66.5	53.0	49.0	67.3
1973	47.6	53.4	51.6	59.7	73.0	81.4	84.3	84.7	79.6	70.7	58.4	52.3	66.4
1974	50.2	51.9	60.1	66.1	74.3	86.9	83.5	83.0	77.8	69.1	57.5	47.0	67.3
1975	49.8	50.7	55.3	57.9	69.8	80.5	84.2	85.8	80.0	69.5	59.3	53.0	66.3
1976	52.6	58.4	58.2	64.8	74.5	83.4	83.9	85.3	77.7	67.8	60.0	52.2	68.3
1977	50.7	56.9	55.7	67.0	70.8	84.7	87.0	86.4	82.0	73.3	61.7	56.9	69.4
1978	53.1	53.6	61.8	65.2	73.1	85.8	88.1	84.7	80.9	73.8	58.5	49.7	69.0
1979	48.4	53.8	56.4	65.6	72.2	83.1	87.5	83.4	84.2	73.0	56.6	55.0	68.3
1980	54.3	57.9	57.5	65.6	71.5	84.9	88.6	84.6	80.5	69.6	59.5	58.1	69.4
1981	54.8	57.1	57.1	69.1	73.4	86.1	85.2	86.4	80.7	68.1	62.2	55.0	69.6
1982	50.7	54.7	57.7	66.1	72.3	80.5	84.8	83.9	79.2	67.0	57.7	50.1	67.0
1983	52.9	53.8	57.3	60.4	73.8	81.6	86.9	84.0	82.2	69.5	57.4	53.5	67.8
1984	51.8	53.7	60.5	64.0	79.9	83.1	84.2	82.9	81.5	66.3	57.8	51.5	68.1
1985	50.3	53.1	58.7	68.7	75.9	85.8	87.5	86.1	77.4	70.0	58.0	52.9	68.7
1986	58.7	56.9	63.8	69.0	76.8	86.6	85.5	86.0	79.0	69.6	59.8	52.3	70.3
1987	50.9	54.2	57.9	70.1	74.3	86.3	87.4	85.1	79.9	75.1	58.9	50.3	69.2
1988	53.0	59.4	61.4	68.0	76.4	86.8	87.9	85.9	80.4	75.3	59.2	51.9	70.5
1989	49.9	58.2	65.0	73.8	77.4	85.4	90.0	86.6	84.5	71.1	61.7	53.0	71.4
1990	51.8	52.8	61.8	69.7	75.2	88.7	85.0	82.6	82.2	73.1	61.6	51.1	69.6
1991	52.3	59.8	55.4	65.2	73.5	81.5	87.5	86.6	80.7	74.0	58.9	54.3	69.1
1992	51.6	57.3	59.4	70.8	76.7	84.5	86.8	85.1	83.6	74.2	56.1	51.4	69.8
1993	55.2	54.0	61.3	68.6	78.1	85.0	88.0	85.5	81.4	72.6	58.8	53.4	70.2
Record Mean	50.6	53.6	58.0	64.9	73.0	82.4	86.1	84.1	80.1	69.7	58.3	51.5	67.7
Max	64.4	67.8	73.0	81.1	89.7	99.1	99.3	96.8	94.1	84.9	73.1	65.3	82.4
Min	36.8	39.3	43.0	48.8	56.3	65.7	72.9	71.4	66.1	54.5	43.5	37.7	53.0

REFERENCE NOTES

TUCSON, ARIZONA

GENERAL
T - TRACE AMOUNT.
BLANK ENTRIES DENOTE MISSING/UNREPORTED DATA.
# INDICATES A STATION OR INSTRUMENT RELOCATION.

SPECIFIC

PM - INCLUDES LAST DAY OF PREVIOUS MONTH
ASOS - AUTOMATED SURFACE OBSERVING SYSTEM IN OPERATION DURING THESE MONTHS.

(*) - LENGTH OF RECORD IN YEARS, ALTHOUGH INDIVIDUAL MONTHS MAY BE MISSING.
0.* OR * - THE VALUE IS BETWEEN 0.0 AND 0.05
NORMALS - BASED ON THE 1961-1990 RECORD PERIOD.
EXTREMES - DATES ARE THE MOST RECENT OCCURRENCE.
WIND DIR. - NUMERALS SHOW TENS OF DEGRESS CLOCKWISE FROM TRUE NORTH. "00" INDICATES CALM.
RESULTANT DIRECTIONS ARE GIVEN TO WHOLE DEGREES.
BOLD VALUES INDICATE EXTREME VALUES WHICH OCCURRED AFTER THE ASOS SYSTEM WAS COMMISSIONED.

RECORD = PERIOD OF RECORD
RECORD MEAN PRECIPITATION IS THE MEAN OF ALL DAILY PRECIPITATION AMOUNTS DURING THE PERIOD OF RECORD.
RECORD MAX(MIN) TEMPERATURE IS THE MEAN OF ALL DAILY MAX(MIN) TEMPERATURES DURING THE PERIOD OF RECORD.
RECORD MEAN TEMPERATURE IS THE SUM OF THE RECORD MAX AND RECORD MIN DIVIDED BY 2.
AVERAGE TEMPERATURE IS THE SUM OF THE MEAN DAILY MAX AND MIN TEMPERATURE DIVIDED BY 2.

EXCEPTIONS

RECORD MEANS ARE THROUGH THE CURRENT YEAR, BEGINNING IN 1900 FOR TEMPERATURE
1900 FOR PRECIPITATION
1941 FOR SNOWFALL

HEATING DEGREE DAYS Base 65 deg. F TUCSON, ARIZONA

SEASON	JULY	AUG	SEP	OCT	NOV	DEC	JAN	FEB	MAR	APR	MAY	JUNE	TOTAL
1964-65	0	0	0	5	293	383	348	383	305	114	21	0	1852
1965-66	0	0	8	33	110	396	532	473	166	26	0	0	1744
1966-67	0	0	0	20	126	386	416	256	115	113	20	0	1452
1967-68	0	0	0	14	89	502	384	170	200	91	0	0	1450
1968-69	0	0	0	4	204	440	288	328	339	34	35	0	1672
1969-70	0	0	0	55	188	384	455	224	274	132	8	0	1720
1970-71	0	0	0	58	143	403	445	350	200	111	12	0	1722
1971-72	0	0	0	120	249	548	444	259	73	50	0	0	1743
1972-73	0	0	0	96	358	489	533	320	410	174	19	0	2399
1973-74	0	0	0	23	216	390	451	362	161	49	5	0	1657
1974-75	0	0	0	53	218	552	465	393	299	217	29	0	2226
1975-76	0	0	0	38	191	365	378	180	221	88	5	0	1466
1976-77	0	0	0	45	178	390	435	221	287	65	9	0	1630
1977-78	0	0	0	1	117	242	365	313	144	64	24	0	1270
1978-79	0	0	0	15	213	470	511	311	260	76	20	0	1876
1979-80	0	0	0	26	252	302	323	202	227	84	3	0	1419
1980-81	0	0	0	66	197	210	310	220	244	31	0	0	1278
1981-82	0	0	0	34	106	304	437	291	223	46	10	0	1451
1982-83	0	0	0	41	211	456	371	309	239	168	6	0	1801
1983-84	0	0	0	0	232	348	402	323	140	110	0	0	1555
1984-85	0	0	0	49	221	413	448	328	200	41	0	0	1700
1985-86	0	0	0	9	217	369	193	244	117	22	6	0	1177
1986-87	0	0	0	11	154	387	429	299	225	24	0	0	1529
1987-88	0	0	0	0	188	452	366	171	161	46	12	0	1396
1988-89	0	0	0	0	220	402	461	199	82	9	4	0	1377
1989-90	0	0	0	25	107	361	402	340	156	16	3	0	1410
1990-91	0	0	0	5	152	427	384	140	296	47	3	0	1454
1991-92	0	0	0	56	195	325	408	215	169	24	0	0	1392
1992-93	0	0	0	0	261	418	298	299	129	28	0	0	1433
1993-94	0	0	0	5	186	355							

COOLING DEGREE DAYS Base 65 deg. F TUCSON, ARIZONA

YEAR	JAN	FEB	MAR	APR	MAY	JUNE	JULY	AUG	SEP	OCT	NOV	DEC	TOTAL
1969	0	0	15	87	348	477	658	669	493	118	1	0	2866
1970	0	5	0	25	333	561	693	620	347	68	4	0	2656
1971	6	0	45	51	152	493	706	514	430	101	12	0	2510
1972	0	1	82	82	236	506	678	563	414	150	1	0	2713
1973	0	0	0	21	272	495	603	615	445	206	26	2	2685
1974	0	0	18	87	301	664	581	564	387	185	1	0	2788
1975	0	0	4	11	184	471	604	651	458	182	27	0	2592
1976	2	0	14	89	306	557	597	636	386	139	34	0	2760
1977	0	0	5	133	198	597	691	669	517	266	23	0	3099
1978	0	0	54	76	283	630	721	616	483	293	28	0	3184
1979	0	0	1	101	249	551	706	576	580	282	6	0	3052
1980	0	4	1	109	211	606	742	615	474	216	37	3	3018
1981	0	8	4	159	267	639	633	670	476	137	27	2	3022
1982	0	4	4	82	244	471	622	594	437	112	0	0	2570
1983	0	0	8	36	288	503	688	600	523	145	10	0	2801
1984	0	0	6	87	469	549	601	562	503	96	12	0	2885
1985	0	1	7	159	345	633	704	660	379	173	14	0	3075
1986	2	23	88	150	378	653	643	657	431	158	3	0	3186
1987	0	2	12	184	297	644	702	630	452	325	12	0	3260
1988	2	13	58	142	374	658	716	657	471	327	51	1	3470
1989	0	16	89	281	397	619	780	676	592	221	16	0	3687
1990	0	6	63	164	327	719	625	553	522	262	56	0	3297
1991	0	1	6	58	274	501	703	675	479	345	21	0	3063
1992	0	0	4	204	372	590	683	627	563	291	1	0	3335
1993	1	0	22	142	413	604	721	641	500	250	11	3	3308

SNOWFALL (inches) TUCSON, ARIZONA

SEASON	JULY	AUG	SEP	OCT	NOV	DEC	JAN	FEB	MAR	APR	MAY	JUNE	TOTAL
1964-65	0.0	0.0	0.0	0.0	0.1	0.0	0.0	3.9	0.0	0.0	0.0	0.0	4.0
1965-66	0.0	0.0	0.0	0.0	0.0	0.3	T	1.2	0.0	0.0	0.0	0.0	1.5
1966-67	0.0	0.0	0.0	0.0	0.0	T	0.0	0.0	0.0	T	0.0	0.0	T
1967-68	0.0	0.0	0.0	0.0	0.0	1.6	0.0	0.0	0.0	0.0	0.0	0.0	1.6
1968-69	0.0	0.0	0.0	0.0	0.0	0.4	0.0	0.0	T	0.0	0.0	0.0	0.4
1969-70	0.0	0.0	0.0	0.0	0.0	T	0.0	0.0	T	T	0.0	0.0	T
1970-71	0.0	0.0	0.0	0.0	0.0	0.0	T	T	0.0	0.0	0.0	0.0	T
1971-72	0.0	0.0	0.0	0.0	0.0	6.8	0.0	0.0	0.0	0.0	0.0	0.0	6.8
1972-73	0.0	0.0	0.0	0.0	0.0	0.0	T	0.0	0.0	0.0	0.0	0.0	T
1973-74	0.0	0.0	0.0	0.0	0.0	0.0	0.4	0.0	T	0.0	0.0	0.0	0.4
1974-75	0.0	0.0	0.0	0.0	0.0	T	0.0	T	0.5	0.0	0.0	0.0	0.5
1975-76	0.0	0.0	0.0	0.0	T	T	0.0	0.0	3.8	2.0	0.0	0.0	5.8
1976-77	0.0	0.0	0.0	0.0	0.0	0.0	0.0	0.0	0.0	0.0	0.0	0.0	0.0
1977-78	0.0	0.0	0.0	0.0	0.0	0.0	0.0	0.0	0.0	0.0	0.0	0.0	0.0
1978-79	0.0	0.0	0.0	0.0	0.0	T	1.2	0.0	0.0	0.0	0.0	0.0	1.2
1979-80	0.0	0.0	0.0	0.0	0.0	0.0	0.0	0.0	T	0.0	0.0	0.0	T
1980-81	0.0	0.0	0.0	0.0	0.0	0.0	0.0	0.0	T	0.0	0.0	0.0	T
1981-82	0.0	0.0	0.0	0.0	0.0	0.0	T	0.0	T	0.0	0.0	0.0	T
1982-83	0.0	0.0	0.0	0.0	0.0	T	0.0	0.0	0.0	0.0	0.0	0.0	T
1983-84	0.0	0.0	0.0	0.0	0.0	0.0	0.0	0.0	0.0	0.0	0.0	0.0	0.0
1984-85	0.0	0.0	0.0	0.0	0.0	T	0.0	2.2	0.0	0.0	0.0	0.0	2.2
1985-86	0.0	0.0	0.0	0.0	0.0	T	0.0	T	0.0	0.0	0.0	0.0	T
1986-87	0.0	0.0	0.0	0.0	0.0	0.0	4.7	0.0	T	0.0	0.0	0.0	4.7
1987-88	0.0	0.0	0.0	0.0	0.0	3.6	0.0	0.0	0.0	0.0	0.0	0.0	3.6
1988-89	0.0	0.0	0.0	0.0	0.0	T	0.0	T	0.0	0.0	0.0	0.0	T
1989-90	0.0	0.0	0.0	0.0	0.0	0.0	2.7	2.3	0.0	T	0.0	0.0	5.0
1990-91	0.0	T	T	0.0	0.0	0.6	0.0	T	0.3	0.0	0.0	0.0	0.9
1991-92	0.0	0.0	0.0	T	0.0	T	0.0	0.0	T	0.0	T	0.0	T
1992-93	0.0	T	0.0	0.0	0.0	0.0	0.0	0.0	0.0	0.0	0.0	0.0	T
1993-94	0.0	T	0.0	0.0	0.0	0.0							
Record Mean	0.0	T	T	-	0.1	0.3	0.3	0.2	0.3	0.1	T	0.0	1.3

WINSLOW, ARIZONA

Winslow is located in the Little Colorado River Valley. The adjacent terrain rises gradually in all directions except to the north-northwest along the river. The White Mountain area, 100 miles to the southeast, rises to over 11,000 feet. To the south and west the Mogollon Rim averages very close to 8,000 feet above sea level, while 60 miles to the northwest the San Francisco Peaks rise to 12,655 feet.

The surrounding high terrain has a considerable effect upon the climate and weather of the Winslow area. It acts as a barrier to the movement of low-level moist air currents, as well as to cold wintertime air masses from the plains states. As a consequence, the climate is very dry and relatively mild for the latitude and elevation.

The elevation of Winslow and the generally clear skies tend to create a large diurnal temperature variation during all seasons. Below-zero readings occur during the winter months about one year in three. Daytime temperatures over 70 degrees have been recorded during all winter months. Summer days are warm with temperatures of 90 degrees or higher occurring frequently from late May to mid-September. Because of the extremely low humidity, however, the high daytime temperatures are quite comfortable. The air cools rapidly after sunset so that nights are generally cool during the summer months.

Monthly and annual precipitation is extremely variable in amount. Moist air carried aloft over the surrounding mountains from the Gulf of Mexico and the Pacific Ocean during the summer and early fall helps produce the major portion of the annual precipitation. The lifting of the moist air over the mountains and the intense surface heating of the sparsely covered lower elevations causes considerable thunderstorm activity during this summer period. Snowfall during the winter is generally light, and because of warm daytime temperatures, it soon melts. The annual snowfall is about 10 inches, but occasionally a winter season will pass with only a trace being recorded. With the annual precipitation averaging about 7 inches, agricultural activity in the vicinity of Winslow is restricted to small irrigated tracts. The non-irrigated land is used mostly for winter range purposes.

More than 270 days during the year are clear or only partly cloudy. The average growing period is 186 days.

During the spring months, occasional high winds pick up considerable dust. During the late fall and winter months the prevailing wind direction is from the southeast, while in the spring and summer months the winds blow primarily from the southwest. Destructive weather such as tornadoes and ice storms rarely occur.

NORMALS, MEANS, AND EXTREMES

WINSLOW, ARIZONA

LATITUDE: 35°01'N LONGITUDE: 110°44' W ELEVATION: FT. GRND 4895 BARO 4892 TIME ZONE: MOUNTAIN WBAN: 23194

	(a)	JAN	FEB	MAR	APR	MAY	JUNE	JULY	AUG	SEP	OCT	NOV	DEC	YEAR
TEMPERATURE °F:														
Normals														
-Daily Maximum		44.7	53.4	60.9	70.0	79.6	90.5	93.6	90.4	83.6	72.1	58.4	46.0	70.3
-Daily Minimum		19.1	24.6	30.1	36.5	44.9	53.9	62.8	61.0	52.8	40.2	29.3	20.2	39.6
-Monthly		31.9	39.0	45.5	53.3	62.3	72.2	78.2	75.7	68.2	56.2	43.9	33.1	55.0
Extremes														
-Record Highest	62	75	78	85	92	101	106	109	103	99	93	80	74	109
-Year		1971	1957	1989	1943	1951	1970	1971	1979	1950	1972	1988	1958	JUL 1971
-Record Lowest	62	-18	-7	6	16	23	35	46	41	31	13	-1	-12	-18
-Year		1937	1939	1948	1975	1964	1979	1935	1968	1978	1970	1952	1967	JAN 1937
NORMAL DEGREE DAYS:														
Heating (base 65°F)		1026	728	605	357	128	6	0	0	21	283	633	989	4776
Cooling (base 65°F)		0	0	0	6	44	222	409	332	117	11	0	0	1141
% OF POSSIBLE SUNSHINE														
MEAN SKY COVER (tenths)														
Sunrise - Sunset	30	5.3	4.9	4.7	3.9	3.5	2.9	5.0	4.5	3.1	3.1	3.9	4.7	4.1
MEAN NUMBER OF DAYS:														
Sunrise to Sunset														
-Clear	37	12.2	11.5	12.8	14.8	17.2	19.8	10.7	12.1	18.0	18.6	15.9	13.7	177.4
-Partly Cloudy	37	7.1	6.9	8.8	8.6	8.5	6.6	12.2	12.6	7.9	6.7	6.4	6.8	99.3
-Cloudy	37	11.7	9.8	9.5	6.5	5.3	3.5	8.1	6.3	4.0	5.7	7.6	10.5	88.6
Precipitation														
.01 inches or more	62	4.1	4.5	4.7	3.1	2.9	2.2	7.2	9.1	5.1	3.8	3.1	4.6	54.4
Snow,Ice pellets,hail														
1.0 inches or more	50	1.0	0.6	0.9	0.1	0.0	0.0	0.0	0.0	0.0	0.1	0.3	1.0	4.0
Thunderstorms	31	0.1	0.4	0.5	1.0	2.0	3.1	11.4	10.7	5.4	1.4	0.2	0.1	36.3
Heavy Fog Visibility														
1/4 mile or less	31	1.1	0.5	0.3	0.1	0.0	0.0	0.0	0.0	0.1	0.4	0.6	1.4	4.3
Temperature °F														
-Maximum														
90° and above	33	0.0	0.0	0.0	0.*	2.7	16.5	23.6	18.4	5.2	0.1	0.0	0.0	66.6
32° and below	33	4.8	0.5	0.0	0.0	0.0	0.0	0.0	0.0	0.0	0.0	0.2	3.9	9.5
-Minimum														
32° and below	33	28.5	23.4	18.7	8.3	1.4	0.0	0.0	0.0	0.1	4.9	20.4	28.5	134.2
0° and below	33	2.0	0.*	0.0	0.0	0.0	0.0	0.0	0.0	0.0	0.0	0.*	1.2	3.2
AVG. STATION PRESS. (mb)	6	853.4	852.6	848.8	849.1	849.6	851.2	853.1	853.3	853.0	853.6	853.2	853.6	852.0
RELATIVE HUMIDITY (%)														
Hour 05	18	79	70	62	52	44	38	58	64	64	61	69	77	62
Hour 11 (Local Time)	18	62	48	36	25	20	17	30	35	37	37	46	59	38
Hour 17	18	50	35	26	19	16	14	27	30	28	28	35	50	30
Hour 23	18	73	60	50	39	32	27	47	50	51	50	60	71	51
PRECIPITATION (inches):														
Water Equivalent														
-Normal		0.45	0.52	0.55	0.26	0.31	0.31	1.20	1.39	0.91	0.91	0.57	0.66	8.04
-Maximum Monthly	62	1.62	2.05	2.07	1.59	2.11	3.22	3.53	4.80	2.54	5.61	1.67	3.73	5.61
-Year		1935	1973	1973	1934	1992	1972	1992	1963	1975	1972	1952	1967	OCT 1972
-Minimum Monthly	62	T	T	T	0.00	T	0.00	0.04	0.15	0.00	0.00	0.00	T	0.00
-Year		1984	1972	1988	1989	1977	1971	1993	1968	1957	1952	1932	1981	APR 1989
-Maximum in 24 hrs	49	0.76	0.70	1.07	0.76	0.90	2.12	1.67	2.08	1.25	2.22	1.33	1.51	2.22
-Year		1980	1956	1984	1988	1969	1955	1992	1964	1983	1974	1987	1967	OCT 1974
Snow,Ice pellets,hail														
-Maximum Monthly	62	11.3	10.7	11.0	4.8	0.6	0.0	0.0	0.0	T	8.2	7.4	39.6	39.6
-Year		1987	1973	1973	1977	1978				1945	1961	1952	1967	DEC 1967
-Maximum in 24 hrs	49	6.4	8.0	8.0	2.9	0.6	0.0	0.0	0.0	T	6.6	4.8	17.0	17.0
-Year		1987	1977	1991	1977	1978				1945	1961	1966	1967	DEC 1967
WIND:														
Mean Speed (mph)	35	7.1	8.5	10.6	11.3	10.9	10.6	9.1	8.4	8.2	7.7	7.3	6.7	8.9
Prevailing Direction through 1963		SE	SE	WSW	SW	SW	SW	WSW	SW	SW	SE	SE	SE	SE
Fastest Obs. 1 Min.														
-Direction (!!)	30	23	22	22	25	25	20	16	24	11	22	22	22	22
-Speed (MPH)	30	56	63	58	56	53	52	59	43	40	49	46	52	63
-Year		1951	1971	1975	1957	1950	1953	1954	1966	1950	1970	1964	1966	FEB 1971
Peak Gust														
-Direction (!!)														
-Speed (mph)														
-Date														

PRECIPITATION (inches)

WINSLOW, ARIZONA

YEAR	JAN	FEB	MAR	APR	MAY	JUNE	JULY	AUG	SEP	OCT	NOV	DEC	ANNUAL
1964	0.03	0.14	0.49	0.27	0.04	0.17	1.63	2.81	1.36	0.04	0.33	0.37	7.68
1965	1.42	0.45	0.97	0.44	0.92	0.33	1.47	0.70	0.91	0.16	0.30	1.45	9.52
1966	0.68	0.66	0.26	0.12	T	0.32	1.34	0.96	0.98	0.97	0.75	0.26	7.30
1967	0.10	0.01	0.24	0.10	0.27	1.06	2.67	1.09	0.49	0.11	0.36	3.73	10.23
1968	T	0.27	0.56	0.50	0.03	0.12	0.83	0.15	0.23	0.70	0.70	0.45	4.54
1969	0.12	0.66	0.45	0.09	1.22	T	1.16	0.83	0.26	0.40	0.36	0.40	5.95
1970	0.27	0.07	0.70	0.23	T	0.02	0.57	1.80	0.96	0.25	0.11	0.13	5.11
1971	0.07	0.25	T	0.52	0.07	0.00	0.20	1.40	1.74	0.97	0.06	0.95	6.23
1972	0.00	T	0.00	0.02	0.02	3.22	1.20	0.72	0.15	5.61	0.58	0.73	12.25
1973	0.23	2.05	2.07	0.11	0.09	0.31	0.38	1.40	0.07	T	0.88	0.05	7.64
1974	0.84	0.03	0.15	0.02	0.00	T	0.58	0.69	1.68	3.41	0.12	0.29	7.81
1975	0.41	0.42	0.73	0.29	0.35	T	1.27	0.59	2.54	T	0.38	0.68	7.66
1976	0.09	0.45	0.19	0.19	1.14	0.15	1.55	0.90	1.11	0.40	0.27	0.38	6.82
1977	0.36	0.69	0.21	0.48	T	0.23	2.05	2.19	1.95	0.83	0.05	0.13	9.17
1978	0.36	0.70	0.93	0.07	0.46	0.27	0.80	0.30	0.41	1.70	1.36	1.02	8.38
1979	1.03	0.18	0.36	0.36	1.39	0.04	0.26	1.47	0.06	0.28	0.18	0.54	6.15
1980	1.18	1.36	0.78	0.45	0.03	0.17	1.71	1.29	0.64	0.22	T	0.23	8.06
1981	0.68	0.15	0.30	0.29	0.70	0.80	1.71	1.80	0.53	1.11	0.75	T	8.82
1982	1.18	0.83	0.33	0.11	1.27	T	2.07	2.69	0.69	T	1.54	1.40	12.11
1983	0.35	0.46	0.79	0.18	0.07	0.01	1.09	1.39	2.37	0.69	0.49	0.77	8.66
1984	T	0.21	1.20	0.08	0.05	0.03	2.32	1.53	1.96	1.16	0.31	1.59	10.44
1985	0.47	0.51	1.13	0.71	0.17	T	1.65	1.25	0.96	1.37	0.91	0.44	9.57
1986	0.02	0.75	0.80	0.16	0.04	0.33	2.32	0.42	0.55	1.14	1.28	0.56	8.37
1987	1.21	0.86	0.38	T	0.54	0.29	0.50	1.88	0.15	1.29	1.39	1.28	9.77
1988	0.24	0.82	T	1.28	0.09	1.30	0.64	1.47	0.04	0.61	0.70	0.09	7.28
1989	0.59	0.28	0.43	0.00	0.11	T	1.69	1.73	0.09	0.26	T	0.41	5.59
1990	0.11	0.68	0.25	0.16	0.15	0.02	1.64	0.48	0.88	0.19	0.97	0.31	5.84
1991	0.79	0.18	1.58	T	0.08	0.91	0.10	0.44	0.71	0.27	1.00	1.26	7.32
1992	0.31	0.20	0.84	0.32	2.11	0.06	3.53	2.58	0.15	0.31	0.28	1.80	12.49
1993	1.54	0.56	0.61	T	0.23	0.00	0.04	2.64	T	1.13	0.49	0.06	7.30
Record Mean	0.46	0.48	0.50	0.34	0.34	0.29	1.10	1.42	0.88	0.79	0.45	0.62	7.67

AVERAGE TEMPERATURE (deg. F)

WINSLOW, ARIZONA

YEAR	JAN	FEB	MAR	APR	MAY	JUNE	JULY	AUG	SEP	OCT	NOV	DEC	ANNUAL
1964	30.6	32.3	41.5	50.8	62.6	70.2	80.6	75.7	68.4	60.0	39.4	34.8	53.9
1965	37.3	37.6	42.8	52.8	58.8	67.6	76.6	76.0	65.7	57.2	47.0	36.1	54.6
1966	24.0	33.7	46.4	55.2	65.6	71.9	79.4	77.0	69.8	56.0	46.3	33.1	54.8
1967	31.2	39.4	49.1	50.2	60.1	69.7	78.3	75.6	69.3	57.0	47.1	21.4	54.1
1968	12.9	40.3	46.6	51.2	61.7	72.0	77.8	72.5	67.9	56.6	42.5	27.8	52.5
1969	39.4	38.6	41.4	54.8	65.0	72.1	79.4	78.9	70.5	51.8	41.5	37.4	55.9
1970	32.7	42.5	42.1	47.5	63.4	72.1	79.0	78.0	65.7	52.2	44.8	33.6	54.4
1971	32.8	37.1	46.0	52.2	60.0	73.6	82.5	78.0	68.0	53.9	43.9	30.3	54.9
1972	35.4	40.3	51.7	54.2	62.2	71.8	79.4	76.0	69.2	58.0	42.0	34.0	56.2
1973	32.0	41.2	44.1	50.4	62.8	69.2	75.8	74.7	66.2	55.0	44.3	34.8	54.2
1974	32.3	34.4	48.6	50.7	64.5	76.4	77.2	74.7	67.8	56.6	42.5	29.3	54.6
1975	26.3	37.5	43.4	47.5	58.8	70.5	77.9	75.6	67.3	54.9	40.4	32.7	52.7
1976	32.8	42.4	42.6	52.6	63.3	71.4	78.1	74.4	66.3	52.6	41.2	31.2	54.1
1977	26.5	36.6	41.9	54.3	59.2	73.8	78.5	76.9	70.6	58.8	45.0	41.6	55.3
1978	37.4	39.0	50.3	54.0	60.3	74.3	78.8	76.9	68.8	59.5	45.7	26.8	56.0
1979	29.2	37.8	43.9	54.3	60.6	71.0	78.6	74.7	70.9	57.8	38.0	33.7	54.2
1980	38.3	41.6	42.9	51.8	58.1	72.0	78.5	75.7	68.4	53.7	43.2	40.5	55.4
1981	39.4	41.8	45.2	57.4	62.8	76.2	78.5	75.9	68.5	55.7	47.2	39.0	57.3
1982	32.1	40.2	46.7	55.0	61.2	70.6	75.9	75.4	69.1	52.3	43.5	35.9	54.8
1983	34.1	40.9	47.3	48.3	61.1	69.4	76.6	75.3	71.0	58.1	44.6	40.6	55.6
1984	33.0	35.9	44.9	51.8	68.9	72.5	77.4	75.7	70.2	53.0	45.4	36.0	55.4
1985	35.1	38.8	46.5	56.6	63.2	74.2	77.9	76.1	64.2	56.1	44.4	34.5	55.6
1986	39.8	43.2	49.6	54.6	62.6	73.4	75.9	77.2	65.2	53.7	44.8	34.8	56.2
1987	30.2	39.1	42.7	56.1	62.4	73.4	74.5	73.4	67.5	60.1	42.8	31.0	54.4
1988	33.5	41.3	45.1	53.8	61.6	72.8	79.4	75.1	66.3	61.4	46.4	34.9	56.0
1989	32.3	40.0	51.6	60.6	66.0	72.5	79.0	74.5	68.9	55.4	44.2	33.7	56.6
1990	34.4	37.8	47.7	55.1	60.6	73.3	75.9	72.0	67.6	54.5	41.9	26.4	53.9
1991	32.4	40.5	41.4	51.0	58.5	66.6	74.9	73.7	66.9	56.2	41.3	30.8	52.9
1992	27.8	39.8	45.2	57.2	60.9	68.6	72.3	73.0	68.2	57.7	37.8	24.8	52.8
1993	39.1	40.0	47.4	54.8	63.3	69.7	76.0	73.6	66.6	54.7	40.2	33.3	54.9
Record Mean	32.5	39.0	45.4	53.7	62.3	72.0	77.9	75.7	68.8	56.7	43.0	33.8	55.0
Max	45.3	53.1	60.7	70.1	79.3	89.9	93.2	90.2	84.3	72.4	57.7	46.6	70.2
Min	19.8	25.0	30.1	37.2	45.3	54.0	62.6	61.2	53.2	40.9	28.3	21.1	39.9

REFERENCE NOTES

WINSLOW, ARIZONA

GENERAL
T - TRACE AMOUNT.
BLANK ENTRIES DENOTE MISSING/UNREPORTED DATA.
# INDICATES A STATION OR INSTRUMENT RELOCATION.

SPECIFIC

(a) - LENGTH OF RECORD IN YEARS, ALTHOUGH INDIVIDUAL MONTHS MAY BE MISSING.
0.* OR * - THE VALUE IS BETWEEN 0.0 AND 0.05
NORMALS - BASED ON THE 1961-1990 RECORD PERIOD.
EXTREMES - DATES ARE THE MOST RECENT OCCURRENCE
WIND DIR. - NUMERALS SHOW TENS OF DEGRESS CLOCKWISE FROM TRUE NORTH. "00" INDICATES CALM.
RESULTANT DIRECTIONS ARE GIVEN TO WHOLE DEGREES.
BOLD VALUES INDICATE EXTREME VALUES WHICH OCCURRED AFTER THE ASOS SYSTEM WAS COMMISSIONED.

RECORD = PERIOD OF RECORD
RECORD MEAN PRECIPITATION IS THE MEAN OF ALL DAILY PRECIPITATION AMOUNTS DURING THE PERIOD OF RECORD.
RECORD MAX(MIN) TEMPERATURE IS THE MEAN OF ALL DAILY MAX(MIN) TEMPERATURES DURING THE PERIOD OF RECORD.
RECORD MEAN TEMPERATURE IS THE SUM OF THE RECORD MAX AND RECORD MIN DIVIDED BY 2.
AVERAGE TEMPERATURE IS THE SUM OF THE MEAN DAILY MAX AND MIN TEMPERATURE DIVIDED BY 2.

EXCEPTIONS

RELATIVE HUMIDITY, MEAN WIND SPEED, MEAN SKY COVER, DAYS CLEAR-PARTLY CLOUDY-CLOUDY, THUNDERSTORMS, AND HEAVY FOG ARE THROUGH 1978.

RECORD MEANS ARE THROUGH THE CURRENT YEAR, BEGINNING IN 1932 FOR TEMPERATURE
1932 FOR PRECIPITATION
1932 FOR SNOWFALL

HEATING DEGREE DAYS Base 65 deg. F — WINSLOW, ARIZONA

SEASON	JULY	AUG	SEP	OCT	NOV	DEC	JAN	FEB	MAR	APR	MAY	JUNE	TOTAL
1964-65	0	0	21	164	761	927	849	762	686	363	217	22	4772
1965-66	0	0	77	236	532	890	1264	866	571	287	43	0	4766
1966-67	0	0	4	278	555	983	1038	711	485	440	182	6	4682
1967-68	0	0	11	252	528	1344	1609	710	564	407	135	29	5589
1968-69	0	3	43	253	669	1145	784	738	722	302	99	1	4759
1969-70	0	0	2	404	697	847	996	623	702	518	94	7	4890
1970-71	0	0	84	391	600	964	992	774	583	374	168	10	4940
1971-72	0	0	59	342	623	1068	911	711	403	316	110	2	4545
1972-73	0	0	10	243	681	957	1014	658	640	433	99	31	4766
1973-74	0	0	51	306	614	925	1009	849	505	423	77	9	4768
1974-75	0	0	44	260	667	1099	1192	763	662	518	204	4	5413
1975-76	0	0	28	310	734	995	988	648	687	366	93	8	4857
1976-77	0	0	39	378	711	1041	1186	789	711	319	189	0	5363
1977-78	0	0	5	207	593	716	849	720	448	326	175	0	4039
1978-79	0	0	37	180	573	1176	1102	757	646	314	156	27	4968
1979-80	0	0	6	229	802	966	822	672	678	391	210	14	4790
1980-81	0	0	13	355	645	753	785	642	608	230	105	6	4142
1981-82	0	1	7	289	527	798	1011	690	561	294	129	4	4311
1982-83	0	0	32	390	638	895	950	668	543	494	183	16	4809
1983-84	0	0	23	207	606	748	985	838	617	388	31	0	4443
1984-85	0	0	21	365	580	891	919	726	568	248	85	12	4415
1985-86	0	0	86	268	613	938	774	604	470	308	122	0	4183
1986-87	0	0	72	345	599	931	1075	717	687	262	87	0	4775
1987-88	0	0	12	153	658	1050	972	681	612	330	154	5	4627
1988-89	0	0	44	118	550	926	1005	696	408	156	64	1	3968
1989-90	0	0	21	292	618	964	941	756	530	287	147	9	4565
1990-91	0	4	37	317	686	1187	1002	679	723	415	197	30	5277
1991-92	0	0	13	264	704	1052	1145	723	607	232	125	11	4876
1992-93	2	10	5	220	808	1238	797	696	538	298	103	27	4742
1993-94	0	1	42	318	736	975							

COOLING DEGREE DAYS Base 65 deg. F — WINSLOW, ARIZONA

YEAR	JAN	FEB	MAR	APR	MAY	JUNE	JULY	AUG	SEP	OCT	NOV	DEC	TOTAL
1969	0	0	0	0	109	222	453	441	174	4	0	0	1403
1970	0	0	0	0	53	225	441	408	112	0	0	0	1239
1971	0	0	1	0	20	276	551	408	156	4	0	0	1416
1972	0	0	0	0	30	210	449	347	144	30	0	0	1210
1973	0	0	0	0	38	165	343	310	93	2	0	0	951
1974	0	0	0	0	71	360	385	308	136	5	0	0	1265
1975	0	0	0	0	18	175	408	336	106	3	0	0	1046
1976	0	0	0	0	45	209	413	298	84	0	0	0	1049
1977	0	0	0	2	18	269	423	377	178	21	0	0	1288
1978	0	0	0	2	39	288	435	374	158	16	0	0	1312
1979	0	0	0	2	30	213	430	310	187	13	0	0	1185
1980	0	0	0	0	5	234	426	339	121	9	0	0	1134
1981	0	0	0	12	44	347	423	345	117	6	0	0	1294
1982	0	0	0	2	21	180	346	328	160	0	0	0	1037
1983	0	0	0	0	71	154	365	324	210	0	0	0	1124
1984	0	0	0	0	162	232	388	341	184	2	0	0	1309
1985	0	0	0	1	35	296	406	349	67	0	0	0	1154
1986	0	0	0	0	52	260	344	383	83	0	0	0	1122
1987	0	0	0	0	14	256	302	270	94	9	0	0	945
1988	0	0	0	0	55	247	452	323	93	15	0	0	1185
1989	0	0	0	30	100	231	441	301	143	2	0	0	1248
1990	0	0	0	0	17	264	342	226	122	0	0	0	971
1991	0	0	0	0	3	85	315	277	78	0	0	0	758
1992	0	0	0	5	3	125	236	263	111	0	0	0	743
1993	0	0	0	0	55	174	348	277	97	7	0	0	958

SNOWFALL (inches) — WINSLOW, ARIZONA

SEASON	JULY	AUG	SEP	OCT	NOV	DEC	JAN	FEB	MAR	APR	MAY	JUNE	TOTAL
1964-65	0.0	0.0	0.0	0.0	2.9	3.5	3.5	1.8	4.4	0.2	0.0	0.0	16.3
1965-66	0.0	0.0	0.0	T	0.2	3.2	9.6	4.9	1.3	0.9	0.0	0.0	20.1
1966-67	0.0	0.0	0.0	0.0	4.8	1.4	T	0.1	0.1	0.4	0.0	0.0	6.8
1967-68	0.0	0.0	0.0	0.0	0.2	39.6	0.3	0.0	1.0	2.0	0.0	0.0	43.1
1968-69	0.0	0.0	0.0	0.0	0.0	6.4	T	0.9	6.1	0.0	0.0	0.0	13.4
1969-70	0.0	0.0	0.0	T	0.0	2.4	0.1	T	5.5	1.5	0.0	0.0	9.5
1970-71	0.0	0.0	0.0	0.0	0.0	1.5	1.1	2.3	T	0.8	0.0	0.0	5.7
1971-72	0.0	0.0	0.0	T	T	8.7	0.0	T	0.0	0.2	0.0	0.0	8.9
1972-73	0.0	0.0	0.0	6.0	T	0.6	2.1	10.7	11.0	0.5	0.0	0.0	30.9
1973-74	0.0	0.0	0.0	0.0	3.0	T	2.7	0.6	1.0	0.0	0.0	0.0	7.3
1974-75	0.0	0.0	0.0	T	0.0	3.3	5.3	4.7	3.7	1.8	0.0	0.0	18.8
1975-76	0.0	0.0	0.0	0.0	1.7	4.3	0.0	T	1.3	1.0	0.0	0.0	8.3
1976-77	0.0	0.0	0.0	0.0	2.9	1.0	3.4	8.0	1.8	4.8	0.0	0.0	21.9
1977-78	0.0	0.0	0.0	0.0	T	T	2.0	3.3	T	0.2	0.6	0.0	6.1
1978-79	0.0	0.0	0.0	0.0	T	6.8	7.0	0.4	T	T	T	0.0	14.2
1979-80	0.0	0.0	0.0	0.0	0.1	0.6	6.4	T	6.4	T	0.0	0.0	13.5
1980-81	0.0	0.0	0.0	T	T	2.8	T	T	T	T	0.0	0.0	2.8
1981-82	0.0	0.0	0.0	0.0	T	0.0	9.3	0.0	0.1	T	0.0	0.0	9.4
1982-83	0.0	0.0	0.0	0.0	T	1.8	0.0	1.6	2.3	T	0.0	0.0	5.7
1983-84	0.0	0.0	0.0	0.0	T	0.7	T	T	0.0	T	0.0	0.0	0.7
1984-85	0.0	0.0	0.0	0.0	0.0	5.5	2.6	0.9	3.7	0.6	0.0	0.0	13.3
1985-86	0.0	0.0	0.0	0.0	0.1	5.6	0.0	2.5	0.2	T	0.0	0.0	8.4
1986-87	0.0	0.0	0.0	0.0	0.0	T	11.3	6.6	2.8	0.0	0.0	0.0	20.7
1987-88	0.0	0.0	0.0	0.0	T	11.2	0.5	1.1	T	T	0.0	0.0	12.8
1988-89	0.0	0.0	0.0	0.0	2.9	1.3	1.0	3.3	0.5	0.0	0.0	0.0	9.0
1989-90	0.0	0.0	0.0	0.0	0.0	1.0	0.7	6.9	1.9	0.0	0.0	0.0	10.5
1990-91	0.0	0.0	0.0	0.0	2.2	1.7	0.1	0.0	10.6	T	0.0	0.0	14.6
1991-92	0.0	0.0	0.0	T	2.3	4.0	3.7	T	1.2	0.0	0.0	0.0	11.2
1992-93	0.0	0.0	0.0	0.0	0.2	5.8	0.4	T	0.0	0.0	0.0	0.0	6.4
1993-94	0.0	0.0	0.0	0.0	T	T							
Record Mean	0.0	0.0	T	0.2	0.7	3.1	2.6	1.9	2.0	0.4	T	0.0	10.9

YUMA, ARIZONA

Yuma has a desert climate. Winter is a period of mostly clear skies and abundant sunshine. Yuma records a higher percentage of sunshine than any other place in the United States. Even in December and January, Yuma averages more than eight hours of sunshine a day. Summers in the lower Colorado River Valley are long and hot. Afternoon temperatures reach at least 100 degrees on the average, from June 4 to September 24, and at least 105 degrees from June 22 to August 26. Extremes over 120 degrees have occurred. From mid July to mid September, moisture–laden air from the Gulf of California frequently invades the area. The water content of the air is higher than might be expected over a desert area.

Precipitation in the Yuma area is sparce. Normal annual precipitation is under 3 inches. The wettest years have produced less than 12 inches and the driest years less than 1 inch. Snow is rare in the Yuma area but amounts under 2 inches in a winter season have been recorded.

NORMALS, MEANS, AND EXTREMES

YUMA, ARIZONA

LATITUDE: 32°39'N LONGITUDE: 114°36' W ELEVATION: FT. GRND 194 BARO 213 TIME ZONE: MOUNTAIN WBAN: 23195

	(a)	JAN	FEB	MAR	APR	MAY	JUNE	JULY	AUG	SEP	OCT	NOV	DEC	YEAR
TEMPERATURE °F:														
Normals														
-Daily Maximum		68.7	74.4	78.9	86.3	94.2	103.3	106.6	105.3	100.5	90.3	77.4	68.5	87.9
-Daily Minimum		44.2	46.9	50.8	56.5	63.8	72.0	80.7	80.1	73.2	62.2	50.9	44.3	60.5
-Monthly		56.5	60.7	64.9	71.4	79.0	87.6	93.7	92.7	86.8	76.2	64.2	56.4	74.2
Extremes														
-Record Highest	43	88	97	100	107	116	122	119	120	116	112	98	86	122
-Year		1971	1986	1986	1989	1983	1990	1958	1981	1990	1980	1962	1958	JUN 1990
-Record Lowest	43	24	28	32	41	46	54	63	63	53	35	30	27	24
-Year		1971	1956	1951	1977	1967	1955	1956	1968	1965	1971	1958	1968	JAN 1971
NORMAL DEGREE DAYS:														
Heating (base 65°F)		275	146	98	37	0	0	0	0	0	8	85	278	927
Cooling (base 65°F)		12	26	95	229	434	678	890	859	654	355	61	12	4305
% OF POSSIBLE SUNSHINE	42	84	87	90	94	95	97	91	91	93	92	87	82	90
MEAN SKY COVER (tenths)														
Sunrise - Sunset	32	4.3	3.8	3.6	2.5	1.9	1.3	2.8	2.5	1.7	2.2	3.0	3.8	2.8
MEAN NUMBER OF DAYS:														
Sunrise to Sunset														
-Clear	32	15.2	15.1	17.2	20.6	23.5	25.4	20.1	21.6	24.4	22.8	19.0	16.7	241.6
-Partly Cloudy	32	7.0	6.7	7.4	5.9	5.3	3.6	7.7	6.4	3.8	5.3	5.8	6.5	71.4
-Cloudy	32	8.8	6.5	6.4	3.6	2.2	1.0	3.3	3.0	1.8	2.8	5.2	7.8	52.3
Precipitation														
.01 inches or more	43	2.6	1.7	2.0	0.8	0.3	0.2	1.1	2.2	1.1	1.2	1.4	2.4	17.2
Snow,Ice pellets,hail														
1.0 inches or more	43	0.0	0.0	0.0	0.0	0.0	0.0	0.0	0.0	0.0	0.0	0.0	0.0	0.0
Thunderstorms	29	0.1	0.2	0.2	0.2	0.1	0.3	1.6	2.2	1.3	0.7	0.1	0.2	7.2
Heavy Fog Visibility														
1/4 mile or less	29	0.5	0.2	0.0	0.0	0.0	0.0	0.0	0.0	0.*	0.0	0.2	0.6	1.5
Temperature °F														
-Maximum														
90° and above	29	0.0	0.4	3.2	11.6	23.0	28.7	30.9	30.8	28.0	17.6	1.2	0.0	175.4
32° and below	29	0.0	0.0	0.0	0.0	0.0	0.0	0.0	0.0	0.0	0.0	0.0	0.0	0.0
-Minimum														
32° and below	29	0.6	0.1	0.0	0.0	0.0	0.0	0.0	0.0	0.0	0.0	0.0	0.7	1.4
0° and below	29	0.0	0.0	0.0	0.0	0.0	0.0	0.0	0.0	0.0	0.0	0.0	0.0	0.0
AVG. STATION PRESS.(mb)	6	1010.7	1009.8	1006.3	1004.9	1002.1	1001.5	1001.4	1001.7	1002.0	1005.1	1008.3	1010.5	1005.4
RELATIVE HUMIDITY (%)														
Hour 05	14	57	56	53	48	45	42	50	56	57	54	56	59	53
Hour 11 (Local Time)	14	40	35	30	25	23	22	32	35	35	32	34	41	32
Hour 17	14	28	24	21	17	15	13	22	24	24	23	27	32	23
Hour 23	14	48	46	41	35	32	29	38	42	44	43	48	52	42
PRECIPITATION (inches):														
Water Equivalent														
-Normal		0.35	0.22	0.21	0.14	0.04	0.02	0.26	0.64	0.31	0.29	0.24	0.45	3.17
-Maximum Monthly	43	2.12	1.82	1.56	1.20	0.37	0.27	2.55	3.44	2.47	2.68	1.66	2.09	3.44
-Year		1979	1958	1992	1965	1965	1972	1989	1989	1963	1957	1969	1982	AUG 1989
-Minimum Monthly	43	T	0.00	0.00	0.00	0.00	0.00	0.00	0.00	0.00	0.00	0.00	0.00	0.00
-Year		1975	1974	1972	1982	1983	1983	1983	1976	1980	1981	1980	1981	MAY 1983
-Maximum in 24 hrs	43	1.29	1.34	0.72	1.08	0.37	0.26	2.55	3.44	2.42	2.20	1.42	1.87	3.44
-Year		1979	1958	1992	1965	1965	1972	1989	1989	1963	1972	1969	1982	AUG 1989
Snow,Ice pellets,hail														
-Maximum Monthly	43	0.0	0.0	0.0	0.0	0.0	0.0	0.0	0.0	0.0	0.0	0.0	T	T
-Year													1967	DEC 1967
-Maximum in 24 hrs	43	0.0	0.0	0.0	0.0	0.0	0.0	0.0	0.0	0.0	0.0	0.0	T	T
-Year													1967	DEC 1967
WIND:														
Mean Speed (mph)	28	7.3	7.4	7.9	8.3	8.3	8.5	9.5	8.9	7.3	6.6	6.9	7.2	7.8
Prevailing Direction through 1963		N	N	WNW	W	WNW	SSE	SSE	SSE	SSE	N	N	N	N
Fastest Mile														
-Direction (!!)	40	NW	W	N	NW	NW	SW	NE	SE	E	S	N	W	NE
-Speed (MPH)	40	41	50	43	47	38	42	61	60	57	47	47	47	61
-Year		1964	1964	1956	1954	1957	1966	1989	1973	1976	1964	1957	1959	JUL 1989
Peak Gust														
-Direction (!!)	1	W	SE	S		S	W	S	SE	W	N	N	N	SE
-Speed (mph)	1	43	37	37		39	39	40	52	41	37	43	46	52
-Date		1993	1993	1993		1993	1993	1993	1992	1993	1993	1992	1993	AUG 1992

PRECIPITATION (inches)

YUMA, ARIZONA

YEAR	JAN	FEB	MAR	APR	MAY	JUNE	JULY	AUG	SEP	OCT	NOV	DEC	ANNUAL
1964	T	0.14	T	0.78	T	0.00	0.02	0.49	T	0.24	0.22	0.09	1.98
1965	0.56	0.38	0.04	1.20	0.37	0.00	0.03	0.02	0.00	0.00	0.58	1.67	4.85
1966	0.41	0.17	0.06	T	0.00	T	0.20	0.08	0.04	0.03	T	0.02	1.01
1967	0.25	0.00	0.07	T	T	T	0.01	0.37	1.97	0.00	1.45	0.65	4.77
1968	T	0.27	0.33	T	0.00	0.00	0.91	0.10	0.00	T	T	0.06	1.67
1969	0.68	0.03	0.02	0.08	0.05	0.00	T	T	0.23	0.01	1.66	0.67	3.43
1970	T	0.56	0.82	T	0.00	0.00	T	1.21	0.02	0.02	0.02	0.01	2.66
1971	0.04	0.03	0.00	0.17	T	0.00	T	0.80	1.27	T	0.00	0.15	2.46
1972	0.00	0.00	0.00	T	T	0.27	0.01	0.26	0.00	2.49	0.19	0.06	3.28
1973	0.03	0.50	0.95	T	0.00	0.00	T	0.36	0.00	0.00	0.05	0.00	1.89
1974	0.64	0.00	0.12	0.00	T	0.00	0.06	T	0.10	0.13	0.00	0.14	1.19
1975	T	0.01	0.21	0.26	0.00	0.00	0.10	T	0.15	0.00	0.03	0.44	1.20
1976	0.03	1.22	0.02	0.39	0.09	0.00	0.25	0.00	0.31	0.35	0.39	0.64	3.69
1977	0.03	0.01	0.04	T	0.02	0.02	0.08	2.96	0.46	0.24	T	0.58	4.44
1978	1.38	0.78	0.18	0.01	T	0.00	T	0.24	T	0.58	0.37	0.70	4.24
1979	2.12	0.02	0.17	0.00	0.34	0.00	0.31	1.42	0.00	0.00	0.00	0.08	4.46
1980	0.54	0.73	0.39	0.16	0.10	0.01	0.00	0.05	0.00	0.00	0.00	0.09	2.07
1981	0.20	0.10	0.69	0.01	0.05	0.00	0.03	0.32	0.04	0.00	0.11	0.00	1.55
1982	0.27	0.14	1.18	0.00	T	0.00	0.23	0.93	0.16	T	0.46	2.09	5.46
1983	0.52	0.54	0.19	0.08	0.00	0.00	0.00	2.64	0.23	0.21	T	0.95	5.36
1984	0.14	0.00	0.00	0.55	T	T	2.11	1.08	0.21	0.00	0.53	1.57	6.19
1985	0.07	0.14	0.04	0.00	0.00	0.00	0.06	0.02	1.23	0.40	0.53	0.19	2.68
1986	0.07	0.28	0.14	0.00	0.00	0.14	0.21	0.19	0.35	0.01	0.11	0.23	1.73
1987	0.08	0.32	T	0.00	0.04	0.00	T	0.68	0.04	1.62	0.27	0.62	3.67
1988	0.34	0.03	0.03	0.61	0.00	0.10	T	1.17	T	0.44	T	T	2.72
1989	0.60	0.00	0.13	0.00	T	0.00	2.55	3.44	T	T	0.03	T	6.75
1990	0.18	T	0.08	T	T	T	0.66	0.07	T	0.95	T	0.01	1.95
1991	0.84	0.22	0.77	0.00	0.00	0.00	T	T	0.60	0.11	0.05	0.58	3.17
1992	0.41	0.81	1.56	0.08	0.33	0.00	T	0.28	0.00	0.01	0.00	1.65	5.13
1993	1.69	1.22	0.38	0.00	0.05	0.00	0.00	0.15	0.12	0.03	0.89	T	4.53
Record Mean	0.43	0.37	0.30	0.11	0.03	0.01	0.22	0.58	0.36	0.29	0.23	0.47	3.41

AVERAGE TEMPERATURE (deg. F)

YUMA, ARIZONA

YEAR	JAN	FEB	MAR	APR	MAY	JUNE	JULY	AUG	SEP	OCT	NOV	DEC	ANNUAL
#1964	53.5	56.2	61.6	68.1	75.8	84.8	93.2	91.4	84.7	79.1	58.9	55.8	71.9
1965	56.6	58.4	61.2	69.7	75.8	81.0	92.4	92.5	82.4	77.6	65.0	54.4	72.2
1966	52.2	55.0	65.4	73.6	81.2	86.4	92.8	93.2	86.6	74.2	64.0	57.2	73.5
1967	55.1	60.6	65.3	63.2	77.4	83.1	93.9	93.3	84.4	76.7	66.9	52.4	72.7
1968	55.2	64.3	66.1	70.1	78.7	86.9	91.4	88.8	84.9	74.9	63.4	51.8	73.0
1969	59.6	56.9	62.8	71.0	80.3	83.4	93.4	95.9	88.8	71.6	63.8	56.5	73.7
1970	54.9	61.0	62.4	65.9	79.3	86.6	94.5	93.8	84.4	72.3	63.1	54.1	72.7
1971	54.8	58.2	64.0	67.4	73.1	84.5	92.9	90.6	85.7	69.0	60.1	51.5	71.0
1972	52.9	61.2	71.1	72.2	78.6	87.5	94.7	90.1	84.9	71.7	59.8	53.7	73.2
1973	53.3	58.8	60.0	68.9	80.7	88.4	92.6	91.9	84.3	75.2	62.4	57.5	72.8
1974	55.3	58.0	65.4	70.8	79.1	91.1	92.0	91.7	88.8	76.3	64.5	53.2	73.9
1975	55.3	57.0	61.4	63.5	75.3	85.4	92.9	92.3	87.9	73.1	62.5	56.2	71.9
1976	56.7	61.0	62.9	67.7	80.1	87.1	91.6	90.5	83.1	75.2	65.4	56.8	73.2
1977	55.9	62.3	60.0	72.0	72.5	89.1	94.5	92.1	86.2	78.6	66.4	60.2	74.2
1978	57.0	60.0	67.2	68.8	79.1	90.9	94.4	91.8	85.8	80.6	62.6	52.7	74.3
1979	51.9	58.4	64.0	72.5	78.8	88.9	92.7	91.0	91.7	78.2	62.4	59.6	74.2
1980	60.0	63.9	63.6	71.8	75.5	88.0	95.2	92.6	87.6	78.0	66.3	62.7	75.5
1981	61.9	63.4	65.5	74.8	79.8	92.7	95.2	95.5	89.6	73.8	68.3	61.2	76.8
1982	56.4	63.2	64.4	72.9	78.9	84.9	93.3	93.1	86.2	75.2	62.9	55.8	73.9
1983	59.5	61.1	65.2	67.9	80.1	87.0	94.5	91.2	90.7	78.8	66.1	59.7	75.1
1984	60.0	61.5	67.8	70.8	85.9	87.7	92.6	92.4	90.6	73.5	63.2	56.0	75.2
1985	55.9	59.0	64.3	76.2	81.4	90.9	95.2	94.5	81.8	76.5	62.5	58.2	74.7
1986	63.8	63.6	70.6	73.9	81.3	90.8	92.2	95.6	82.5	74.6	66.0	58.3	76.1
1987	55.9	61.1	64.4	77.1	80.9	90.1	91.9	93.2	87.8	80.6	64.3	54.1	75.1
1988	57.0	63.5	68.1	72.4	79.9	88.9	94.8	93.0	87.6	82.6	65.9	57.3	75.9
1989	56.0	60.9	70.8	79.0	81.4	89.8	96.3	92.1	88.9	77.4	66.9	58.4	76.5
1990	56.6	58.3	68.0	74.7	79.2	91.2	95.7	93.0	89.6	78.3	65.8	53.9	75.4
1991	57.2	66.5	61.6	71.8	77.5	85.4	91.6	94.0	89.0	80.9	65.7	57.6	74.9
1992	57.9	64.7	65.5	77.1	83.3	88.6	93.2	94.0	90.8	79.6	62.8	54.2	76.0
1993	57.9	60.1	69.3	75.4	82.3	88.2	92.3	92.2	88.3	78.8	63.1	57.9	75.5
Record Mean	55.3	59.3	64.3	70.9	77.7	85.9	92.2	91.4	85.9	74.7	63.1	56.0	73.1
Max	67.5	72.8	78.6	86.3	93.6	102.3	106.3	104.9	100.6	89.5	76.8	68.1	87.3
Min	43.0	45.8	50.1	55.4	61.8	69.4	78.1	77.9	71.2	59.9	49.4	43.9	58.8

REFERENCE NOTES

YUMA, ARIZONA

GENERAL
T - TRACE AMOUNT.
BLANK ENTRIES DENOTE MISSING/UNREPORTED DATA.
# INDICATES A STATION OR INSTRUMENT RELOCATION.

SPECIFIC

(a) - LENGTH OF RECORD IN YEARS, ALTHOUGH INDIVIDUAL MONTHS MAY BE MISSING.
0.* OR * - THE VALUE IS BETWEEN 0.0 AND 0.05
NORMALS - BASED ON THE 1961-1990 RECORD PERIOD.
EXTREMES - DATES ARE THE MOST RECENT OCCURRENCE
WIND DIR.- NUMERALS SHOW TENS OF DEGRESS CLOCKWISE FROM TRUE NORTH. "00" INDICATES CALM.
RESULTANT DIRECTIONS ARE GIVEN TO WHOLE DEGREES.
BOLD VALUES INDICATE EXTREME VALUES WHICH OCCURRED AFTER THE ASOS SYSTEM WAS COMMISSIONED.

RECORD = PERIOD OF RECORD
RECORD MEAN PRECIPITATION IS THE MEAN OF ALL DAILY PRECIPITATION AMOUNTS DURING THE PERIOD OF RECORD.
RECORD MAX(MIN) TEMPERATURE IS THE MEAN OF ALL DAILY MAX(MIN) TEMPERATURES DURING THE PERIOD OF RECORD.
RECORD MEAN TEMPERATURE IS THE SUM OF THE RECORD MAX AND RECORD MIN DIVIDED BY 2.
AVERAGE TEMPERATURE IS THE SUM OF THE MEAN DAILY MAX AND MIN TEMPERATURE DIVIDED BY 2.

EXCEPTIONS

1. RELATIVE HUMIDITY, MEAN WIND SPEED, THUNDERSTORMS, AND HEAVY FOG ARE THROUGH 1978.
2. MEAN SKY COVER, AND DAYS CLEAR-PARTLY CLOUDY-CLOUDY ARE THROUGH 1982.

RECORD MEANS ARE THROUGH THE CURRENT YEAR, BEGINNING IN 1879 FOR TEMPERATURE
1876 FOR PRECIPITATION
1951 FOR SNOWFALL

HEATING DEGREE DAYS Base 65 deg. F YUMA, ARIZONA

SEASON	JULY	AUG	SEP	OCT	NOV	DEC	JAN	FEB	MAR	APR	MAY	JUNE	TOTAL
1964-65	0	0	0	0	199	278	258	190	126	64	1	0	1116
1965-66	0	0	0	0	65	321	390	273	71	3	0	0	1123
1966-67	0	0	0	0	82	249	299	126	56	70	1	0	883
1967-68	0	0	0	0	37	385	296	69	41	14	0	0	842
1968-69	0	0	0	0	67	403	167	219	146	3	9	0	1014
1969-70	0	0	0	3	67	257	306	116	108	53	0	0	910
1970-71	0	0	0	4	75	333	331	195	105	36	4	0	1083
1971-72	0	0	0	76	155	415	372	129	11	9	0	0	1167
1972-73	0	0	0	13	147	344	355	167	147	17	0	0	1190
1973-74	0	0	0	0	125	227	293	190	69	3	2	0	909
1974-75	0	0	0	12	45	359	295	220	135	86	5	0	1157
1975-76	0	0	0	7	112	270	257	116	100	48	0	0	910
1976-77	0	0	0	1	72	248	276	106	159	23	0	0	885
1977-78	0	0	0	0	35	142	241	136	34	16	0	0	604
1978-79	0	0	0	0	136	374	399	174	84	1	0	0	1168
1979-80	0	0	0	2	84	169	155	58	54	23	0	0	545
1980-81	0	0	0	2	57	83	93	90	53	4	0	0	382
1981-82	0	0	0	1	33	123	260	108	64	5	0	0	594
1982-83	0	0	0	0	79	278	172	112	54	25	0	0	720
1983-84	0	0	0	0	88	156	147	102	13	22	0	0	528
1984-85	0	0	0	2	91	273	272	181	66	0	0	0	885
1985-86	0	0	0	2	128	204	45	104	30	0	0	0	513
1986-87	0	0	0	0	26	201	279	130	58	3	0	0	697
1987-88	0	0	0	0	70	330	243	70	32	11	3	0	759
1988-89	0	0	0	0	89	234	274	157	21	0	0	0	775
1989-90	0	0	0	3	28	198	251	210	52	0	0	0	742
1990-91	0	0	0	0	54	337	235	12	119	3	0	0	760
1991-92	0	0	0	14	64	223	213	39	30	0	0	0	583
1992-93	0	0	0	0	101	330	214	131	26	0	0	0	802
1993-94	0	0	0	0	95	213							

COOLING DEGREE DAYS Base 65 deg. F YUMA, ARIZONA

YEAR	JAN	FEB	MAR	APR	MAY	JUNE	JULY	AUG	SEP	OCT	NOV	DEC	TOTAL
1969	7	0	83	188	490	555	884	963	722	213	38	0	4143
1970	0	8	32	85	450	653	921	900	590	238	26	0	3903
1971	22	11	84	116	259	589	868	799	627	205	15	0	3595
1972	0	21	206	230	428	683	929	784	603	226	1	0	4111
1973	0	0	0	140	493	709	865	841	587	325	54	0	4014
1974	0	0	88	186	448	790	841	834	719	371	36	0	4313
1975	0	3	32	49	332	619	869	854	694	265	46	3	3766
1976	4	6	42	138	476	668	830	800	549	324	90	0	3927
1977	0	36	14	238	239	731	921	849	644	429	83	0	4184
1978	0	3	109	136	445	783	918	837	630	490	70	0	4421
1979	0	0	60	232	436	724	867	813	809	416	14	10	4381
1980	4	33	17	236	332	696	945	862	686	412	105	19	4347
1981	4	49	74	304	466	836	944	953	745	280	140	12	4807
1982	0	63	52	249	439	602	884	878	641	322	24	0	4154
1983	8	9	66	118	477	665	920	818	778	438	127	0	4424
1984	3	10	104	203	653	686	862	855	774	274	45	0	4469
1985	0	20	52	343	514	784	943	923	514	364	57	0	4514
1986	14	72	211	274	512	784	848	953	533	304	64	0	4569
1987	4	29	43	373	501	760	839	879	689	492	55	0	4664
1988	1	33	134	237	470	724	931	873	683	555	120	2	4763
1989	1	50	209	429	516	750	980	851	725	389	91	1	4992
1990	1	32	153	300	447	793	958	876	742	420	84	0	4806
1991	0	59	21	217	399	618	832	908	726	514	92	0	4386
1992	4	39	52	369	576	716	882	904	783	459	42	0	4826
1993	0	3	169	318	542	701	854	850	703	435	46	1	4622

SNOWFALL (inches) YUMA, ARIZONA

SEASON	JULY	AUG	SEP	OCT	NOV	DEC	JAN	FEB	MAR	APR	MAY	JUNE	TOTAL
1970-71	0.0	0.0	0.0	0.0	0.0	0.0	0.0	0.0	0.0	0.0	0.0	0.0	0.0
1971-72	0.0	0.0	0.0	0.0	0.0	0.0	0.0	0.0	0.0	0.0	0.0	0.0	0.0
1972-73	0.0	0.0	0.0	0.0	0.0	0.0	0.0	0.0	0.0	0.0	0.0	0.0	0.0
1973-74	0.0	0.0	0.0	0.0	0.0	0.0	0.0	0.0	0.0	0.0	0.0	0.0	0.0
1974-75	0.0	0.0	0.0	0.0	0.0	0.0	0.0	0.0	0.0	0.0	0.0	0.0	0.0
1975-76	0.0	0.0	0.0	0.0	0.0	0.0	0.0	0.0	0.0	0.0	0.0	0.0	0.0
1976-77	0.0	0.0	0.0	0.0	0.0	0.0	0.0	0.0	0.0	0.0	0.0	0.0	0.0
1977-78	0.0	0.0	0.0	0.0	0.0	0.0	0.0	0.0	0.0	0.0	0.0	0.0	0.0
1978-79	0.0	0.0	0.0	0.0	0.0	0.0	0.0	0.0	0.0	0.0	0.0	0.0	0.0
1979-80	0.0	0.0	0.0	0.0	0.0	0.0	0.0	0.0	0.0	0.0	0.0	0.0	0.0
1980-81	0.0	0.0	0.0	0.0	0.0	0.0	0.0	0.0	0.0	0.0	0.0	0.0	0.0
1981-82	0.0	0.0	0.0	0.0	0.0	0.0	0.0	0.0	0.0	0.0	0.0	0.0	0.0
1982-83	0.0	0.0	0.0	0.0	0.0	0.0	0.0	0.0	0.0	0.0	0.0	0.0	0.0
1983-84	0.0	0.0	0.0	0.0	0.0	0.0	0.0	0.0	0.0	0.0	0.0	0.0	0.0
1984-85	0.0	0.0	0.0	0.0	0.0	0.0	0.0	0.0	0.0	0.0	0.0	0.0	0.0
1985-86	0.0	0.0	0.0	0.0	0.0	0.0	0.0	0.0	0.0	0.0	0.0	0.0	0.0
1986-87	0.0	0.0	0.0	0.0	0.0	0.0	0.0	0.0	0.0	0.0	0.0	0.0	0.0
1987-88	0.0	0.0	0.0	0.0	0.0	0.0	0.0	0.0	0.0	0.0	0.0	0.0	0.0
1988-89	0.0	0.0	0.0	0.0	0.0	0.0	0.0	0.0	0.0	0.0	0.0	0.0	0.0
1989-90	0.0	0.0	0.0	0.0	0.0	0.0	0.0	0.0	0.0	0.0	0.0	0.0	0.0
1990-91	0.0	0.0	0.0	0.0	0.0	0.0	0.0	0.0	0.0	0.0	0.0	0.0	0.0
1991-92	0.0	0.0	0.0	0.0	0.0	0.0	0.0	0.0	0.0	0.0	0.0	0.0	0.0
1992-93	0.0	0.0	0.0	0.0	0.0	0.0	0.0	0.0	0.0	0.0	0.0	0.0	0.0
1993-94	0.0	0.0	0.0	0.0	0.0	0.0							
Record Mean	0.0	0.0	0.0	0.0	0.0	T	0.0	0.0	0.0	0.0	0.0	0.0	T

FORT SMITH, ARKANSAS

The weather station at Fort Smith, Arkansas, was established on June 1, 1882 by the U. S. Army Signal Service. For the first 63 years, offices were located at several places within a few blocks of each other in downtown Fort Smith. Since 1945 the station has been at the Fort Smith Municipal Airport, about 5 miles southeast of its original location. Fragmentary weather records made by Army Surgeons exist as far back as 1821.

Fort Smith is located on the Arkansas River at its confluence with the Poteau River and at the point where it enters the state from Oklahoma. The river valley is broad and fairly flat, although elevations in the city of Fort Smith range from 390 feet at the river to about 700 feet. Within 20 miles to the north are the Boston Mountains with elevations to about 2,100 feet and about the same distance south are the Ouachita Mountains with a maximum elevation of about 2,600 feet. The general terrain in the area consists of low broken hills separated by creek and river bottom land.

The surrounding terrain has a definite influence on the weather of Fort Smith. Under conditions of light wind, the direction is prevailing northeast throughout the year. When there is a fairly strong inversion these winds may remain northeasterly even though a strong gradient is present. Although infrequent, dense fog will move in from the river to the east and persist longer than would be expected. In the summer this will result in uncomfortably high humidities and in the winter in cooler temperatures than reported at surrounding stations. Summertime temperatures in the mountains to the north are generally several degrees cooler than in the river valley.

Temperature extremes do occur. In summer there is an average of 10 days when the temperature rises to 100 degrees or higher. On the other hand, in about one year in five, the temperature does not reach 100 degrees. Wintertime temperatures rarely fall to zero or below.

Rainfall is well distributed throughout the growing season. January is the driest month, May the wettest. The difference is almost 3 inches, but rainfall is generally adequate for agricultural pursuits. Summer precipitation comes in the form of convective showers. Dry spells occur, but true droughts are infrequent.

Snowfall varies widely from season to season. Although snowfall averages a little over 6 inches, some years go by with no measurable amount being recorded. Ice storms are much more frequent, causing many problems with traffic movement.

Based on the 1951–1980 period, the average first occurrence of 32 degrees Fahrenheit in the fall is October 30 and the average last occurrence in the spring is April 3.

NORMALS, MEANS, AND EXTREMES

FORT SMITH, ARKANSAS

LATITUDE: 35°20'N LONGITUDE: 94°22'W ELEVATION: FT. GRND 447 BARO 464 TIME ZONE: CENTRAL WBAN: 13964

	(a)	JAN	FEB	MAR	APR	MAY	JUNE	JULY	AUG	SEP	OCT	NOV	DEC	YEAR
TEMPERATURE °F:														
Normals														
-Daily Maximum		48.3	53.5	64.0	74.1	80.5	87.9	93.0	92.4	85.2	75.7	62.8	51.2	72.4
-Daily Minimum		25.5	30.2	39.2	48.6	57.7	65.9	70.0	68.8	61.9	48.8	38.4	29.3	48.7
-Monthly		36.9	41.9	51.6	61.4	69.1	76.9	81.5	80.6	73.6	62.3	50.6	40.3	60.6
Extremes														
-Record Highest	48	81	86	94	95	98	105	111	110	106	96	86	82	111
-Year		1952	1962	1974	1987	1951	1953	1954	1964	1947	1963	1987	1951	JUL 1954
-Record Lowest	48	-10	-9	7	22	35	47	50	51	33	22	8	-5	-10
-Year		1977	1951	1948	1975	1954	1972	1972	1986	1967	1952	1976	1989	JAN 1977
NORMAL DEGREE DAYS:														
Heating (base 65°F)		871	647	426	151	36	0	0	0	13	136	432	766	3478
Cooling (base 65°F)		0	0	10	43	164	357	512	484	271	53	0	0	1894
% OF POSSIBLE SUNSHINE	48	50	55	56	60	62	69	73	72	65	65	55	51	61
MEAN SKY COVER (tenths)														
Sunrise - Sunset	48	6.4	6.1	6.3	6.0	6.1	5.5	4.9	4.8	5.1	4.9	5.6	6.1	5.6
MEAN NUMBER OF DAYS:														
Sunrise to Sunset														
-Clear	48	8.9	8.7	8.7	8.9	8.1	9.5	11.9	12.3	11.9	13.5	10.9	9.6	122.9
-Partly Cloudy	48	6.3	5.9	7.2	7.1	9.6	10.3	10.9	10.6	8.1	6.6	6.0	6.6	95.1
-Cloudy	48	15.8	13.7	15.1	14.0	13.3	10.2	8.2	8.2	10.0	10.9	13.1	14.8	147.3
Precipitation														
.01 inches or more	48	7.6	7.9	9.3	9.9	10.7	8.3	7.4	7.0	7.8	6.7	6.9	7.3	96.5
Snow,Ice pellets,hail														
1.0 inches or more	48	0.9	0.7	0.3	0.0	0.0	0.0	0.0	0.0	0.0	0.0	0.2	0.4	2.5
Thunderstorms	48	1.2	1.8	4.8	6.8	8.4	7.7	7.3	6.6	4.7	3.4	2.6	1.6	56.9
Heavy Fog Visibility														
1/4 mile or less	48	2.5	1.4	1.1	0.6	0.9	0.7	0.7	0.7	1.3	2.0	1.6	1.8	15.3
Temperature °F														
-Maximum														
90° and above	29	0.0	0.0	0.2	0.7	1.7	12.6	23.4	21.6	10.1	1.5	0.0	0.0	71.7
32° and below	29	3.4	1.5	0.1	0.0	0.0	0.0	0.0	0.0	0.0	0.0	0.1	1.5	6.5
-Minimum														
32° and below	29	23.4	16.6	7.3	0.9	0.0	0.0	0.0	0.0	0.0	0.8	8.9	19.3	77.1
0° and below	29	0.2	0.*	0.0	0.0	0.0	0.0	0.0	0.0	0.0	0.0	0.0	0.1	0.4
AVG. STATION PRESS.(mb)	21	1004.7	1002.9	999.1	998.3	997.3	998.0	999.3	999.7	1000.6	1002.1	1002.2	1003.9	1000.7
RELATIVE HUMIDITY (%)														
Hour 00	29	77	76	72	75	83	84	81	81	84	82	79	79	79
Hour 06 (Local Time)	29	82	81	79	83	88	89	89	89	89	87	83	83	85
Hour 12	29	61	57	53	51	57	57	54	53	56	51	56	61	56
Hour 18	29	62	55	49	49	57	56	52	52	57	57	61	64	56
PRECIPITATION (inches):														
Water Equivalent														
-Normal		1.90	2.60	3.95	3.97	5.24	3.39	2.99	2.92	3.24	3.68	3.99	3.03	40.90
-Maximum Monthly	48	11.33	7.94	8.52	10.32	13.45	10.40	10.41	6.57	8.96	12.05	14.01	10.09	14.01
-Year		1949	1951	1953	1957	1990	1958	1960	1971	1974	1951	1946	1971	NOV 1946
-Minimum Monthly	48	0.20	0.14	0.80	0.56	0.79	0.38	0.11	0.44	0.06	T	0.59	0.27	T
-Year		1986	1947	1971	1989	1970	1954	1947	1973	1956	1964	1954	1981	OCT 1964
-Maximum in 24 hrs	48	5.42	4.45	3.62	5.12	5.56	3.48	7.13	5.09	3.95	5.90	6.91	5.78	7.13
-Year		1949	1985	1953	1964	1982	1977	1960	1971	1974	1951	1973	1971	JUL 1960
Snow,Ice pellets,hail														
-Maximum Monthly	48	13.0	11.5	5.3	T	T	0.0	0.0	0.0	0.0	T	4.7	7.2	13.0
-Year		1977	1960	1968	1993	1992					1993	1976	1975	JAN 1977
-Maximum in 24 hrs	48	11.0	5.8	5.3	T	T	0.0	0.0	0.0	0.0	T	4.7	7.1	11.0
-Year		1988	1978	1968	1993	1992					1993	1976	1975	JAN 1988
WIND:														
Mean Speed (mph)	48	8.2	8.6	9.5	8.9	7.7	6.7	6.3	6.4	6.6	6.9	7.9	8.1	7.6
Prevailing Direction through 1963		ENE	ENE	ENE	ENE	NE	NE	NE	NE	NE	NE	NE	ENE	NE
Fastest Obs. 1 Min.														
-Direction (!!)	11	33	24	27	32	28	03	09	09	02	29	27	30	09
-Speed (MPH)	11	33	36	38	35	47	48	51	46	38	36	39	35	51
-Year		1990	1988	1991	1986	1986	1992	1989	1992	1992	1992	1988	1990	JUL 1989
Peak Gust														
-Direction (!!)	10	N	NW	NW	NE	N	NE	E	E	NE	W	W	NW	E
-Speed (mph)	10	47	58	55	76	71	61	85	63	55	52	60	46	85
-Date		1990	1991	1986	1984	1985	1992	1989	1989	1992	1991	1988	1990	JUL 1989

PRECIPITATION (inches)

FORT SMITH, ARKANSAS

YEAR	JAN	FEB	MAR	APR	MAY	JUNE	JULY	AUG	SEP	OCT	NOV	DEC	ANNUAL
1964	0.63	3.21	6.08	9.17	5.16	0.63	0.41	5.62	3.18	T	4.16	1.41	39.66
1965	1.69	3.08	2.26	1.97	7.85	2.69	1.96	3.81	2.14	0.90	1.16	1.37	30.88
1966	2.23	4.79	1.17	8.50	3.14	1.23	1.32	2.87	1.26	0.40	2.73	1.96	31.60
1967	0.99	0.57	1.21	6.81	5.79	1.95	3.02	1.64	2.50	6.96	0.81	5.26	37.51
1968	2.81	1.11	8.00	3.71	11.35	2.65	1.61	1.27	2.87	1.79	9.02	5.55	51.74
1969	2.84	2.87	2.60	3.07	1.99	4.57	1.93	2.37	2.19	9.05	0.62	4.16	38.26
1970	0.75	1.98	3.22	9.42	0.79	3.19	0.97	3.11	7.75	8.41	1.21	1.41	42.21
1971	1.65	2.45	0.80	2.88	4.36	3.20	2.77	6.57	1.35	3.74	2.40	10.09	42.26
1972	0.68	1.01	1.28	2.62	2.33	1.51	4.85	1.13	3.47	6.43	6.63	1.61	33.55
1973	3.56	1.10	8.24	6.80	3.76	7.64	1.53	0.44	5.17	4.39	10.92	3.78	57.33
1974	0.96	1.80	4.09	4.96	1.93	4.68	0.64	6.22	8.96	3.82	4.27	2.33	44.66
1975	2.27	3.71	6.70	2.09	6.91	5.86	2.47	4.91	2.49	0.28	3.15	3.66	44.50
1976	0.33	0.70	4.20	3.77	4.91	4.62	1.29	2.82	2.65	5.39	1.05	1.82	33.55
1977	1.93	1.46	6.29	1.04	2.17	4.17	2.61	2.61	4.63	0.87	3.21	1.30	32.29
1978	2.15	2.66	4.20	1.94	5.16	3.05	0.49	0.47	0.84	0.17	7.12	3.26	31.51
1979	2.67	4.55	5.74	4.80	8.04	4.50	5.75	4.44	0.78	3.90	1.87	2.74	49.78
1980	0.99	0.74	2.98	2.24	4.61	3.20	1.43	0.65	3.16	3.24	1.90	1.32	26.46
1981	1.54	2.68	3.04	2.91	5.46	3.79	7.65	2.17	1.88	6.14	3.00	0.27	40.53
1982	4.15	1.33	1.99	1.27	11.23	4.99	4.81	2.35	0.64	3.50	5.72	6.01	47.99
1983	1.72	1.23	2.08	2.94	5.34	3.22	2.61	2.39	1.82	4.97	4.82	1.92	35.06
1984	0.80	2.60	4.13	2.16	4.70	1.87	3.33	2.00	6.18	11.80	7.33	6.16	53.06
1985	1.65	5.71	7.45	4.44	2.78	2.77	2.03	3.57	3.60	4.88	10.73	0.40	50.01
1986	0.20	3.55	1.80	6.97	5.33	4.36	0.20	5.63	1.84	4.20	2.35	1.21	37.64
1987	2.50	3.45	3.90	1.13	6.46	2.79	2.96	2.83	1.71	2.19	4.70	8.12	42.74
1988	1.62	2.09	5.03	5.57	1.26	1.75	2.32	1.05	5.72	1.46	5.51	1.65	35.03
1989	4.43	5.53	5.99	0.56	10.21	5.51	7.09	2.45	3.47	0.83	0.60	0.74	47.41
1990	5.13	5.95	4.61	8.48	13.45	0.63	2.37	3.60	4.87	3.52	3.77	4.84	61.22
1991	2.21	0.82	3.86	7.43	3.43	1.95	2.65	1.80	4.83	8.84	3.99	4.43	46.24
1992	1.43	1.48	2.26	3.90	5.60	8.33	4.51	2.34	6.91	0.89	5.56	5.46	48.67
1993	4.05	3.10	1.90	7.33	5.31	5.03	0.18	2.57	6.17	4.79	4.26	2.87	47.56
Record Mean	2.42	2.73	3.39	4.13	5.02	3.85	3.09	3.04	3.32	3.37	3.28	2.90	40.54

AVERAGE TEMPERATURE (deg. F)

FORT SMITH, ARKANSAS

YEAR	JAN	FEB	MAR	APR	MAY	JUNE	JULY	AUG	SEP	OCT	NOV	DEC	ANNUAL
#1964	40.6	40.5	50.1	64.4	71.3	78.1	83.7	80.6	73.0	59.5	54.1	41.1	61.4
1965	42.0	41.4	41.7	64.4	71.9	77.5	82.2	80.9	74.8	61.7	55.2	46.4	61.7
1966	34.6	41.3	52.5	60.5	68.3	75.7	84.4	77.7	71.3	59.2	54.3	40.4	60.0
1967	41.0	40.0	58.3	65.9	67.4	78.5	78.2	76.7	70.1	62.1	49.3	41.3	60.7
1968	39.6	39.5	51.7	61.3	68.2	78.9	80.3	82.7	71.8	62.0	48.1	39.1	60.3
1969	40.7	42.8	45.3	63.7	70.7	76.0	84.6	80.0	74.4	61.9	49.1	41.2	60.8
1970	34.2	43.3	47.8	63.5	71.3	77.1	81.4	84.3	76.9	60.5	48.9	44.2	61.1
1971	39.1	41.6	48.9	59.7	66.4	78.7	80.3	77.9	74.8	66.8	50.4	46.4	60.9
1972	38.9	43.6	52.6	61.7	67.1	77.0	77.8	79.9	75.6	60.8	43.2	36.1	59.6
1973	36.6	40.2	54.6	57.6	65.6	74.7	80.1	79.4	74.5	64.3	54.5	40.1	60.2
1974	38.3	44.5	56.4	60.5	71.7	72.4	81.8	77.2	65.7	62.0	49.5	40.2	60.0
1975	41.6	39.8	46.7	59.1	69.8	75.7	79.5	79.4	68.6	62.4	51.2	40.9	59.6
1976	36.9	50.7	53.7	61.7	63.9	72.8	78.8	77.4	70.3	55.1	42.0	37.8	58.4
1977	26.5	43.3	53.5	63.9	72.5	80.1	82.6	81.8	77.1	61.6	52.3	39.8	61.2
1978	28.0	31.9	47.4	63.0	68.2	76.7	85.1	83.0	78.8	61.9	53.0	39.9	59.8
1979	25.6	33.8	52.3	60.1	66.7	75.0	79.8	78.5	71.5	63.5	45.9	42.2	57.9
1980	40.0	38.3	47.6	58.2	69.1	79.2	85.5	85.6	78.2	60.2	50.3	42.4	61.2
1981	38.2	43.3	51.8	66.9	65.5	79.0	82.0	78.1	72.1	60.4	51.7	39.1	60.7
1982	34.3	39.5	54.3	56.6	70.2	73.0	80.6	81.2	72.2	61.7	50.2	44.7	59.9
1983	38.1	43.6	50.4	54.8	65.5	74.0	81.0	84.1	74.4	64.3	51.4	28.9	59.2
1984	33.3	45.0	48.7	58.1	67.3	78.5	80.6	80.9	71.4	64.4	49.7	48.4	60.5
1985	31.6	37.1	56.7	62.8	69.3	76.4	82.2	81.1	73.8	63.7	53.0	35.2	60.2
1986	41.2	46.0	55.2	63.5	70.2	79.3	84.7	78.3	76.8	62.8	48.0	41.5	62.3
1987	39.6	46.8	52.1	61.6	74.1	78.0	81.0	83.5	72.8	59.7	52.2	43.0	62.0
1988	35.4	42.2	51.6	61.4	70.7	79.8	83.5	84.2	76.1	60.0	51.2	42.4	61.5
1989	42.6	36.2	53.0	62.9	69.3	75.9	79.6	80.3	70.3	63.9	52.5	33.1	60.0
1990	45.1	48.2	54.7	60.4	67.6	80.7	81.9	82.0	77.2	60.8	55.0	39.1	62.7
1991	36.3	47.7	54.8	63.5	73.4	79.0	82.1	80.0	72.6	63.7	46.9	45.4	62.1
1992	41.4	50.1	53.7	61.9	67.4	75.6	80.8	75.9	71.4	63.1	48.1	42.3	61.0
1993	39.3	40.4	49.6	57.8	68.0	77.7	86.7	85.2	72.6	59.3	46.6	43.7	60.6
Record Mean	38.9	42.9	51.8	61.7	69.5	77.8	82.1	81.2	74.4	63.1	50.8	41.8	61.3
Max	48.7	53.2	62.8	72.9	80.0	88.2	93.0	92.3	85.6	75.1	61.7	51.5	72.1
Min	29.0	32.5	40.7	50.5	59.0	67.3	71.1	70.1	63.2	51.1	39.8	32.1	50.5

REFERENCE NOTES

FORT SMITH, ARKANSAS

GENERAL
T - TRACE AMOUNT.
BLANK ENTRIES DENOTE MISSING/UNREPORTED DATA.
# INDICATES A STATION OR INSTRUMENT RELOCATION.

SPECIFIC

(a) - LENGTH OF RECORD IN YEARS, ALTHOUGH INDIVIDUAL MONTHS MAY BE MISSING.
0.* OR * - THE VALUE IS BETWEEN 0.0 AND 0.05
NORMALS - BASED ON THE 1961-1990 RECORD PERIOD.
EXTREMES - DATES ARE THE MOST RECENT OCCURRENCE.
WIND DIR.- NUMERALS SHOW TENS OF DEGRESS CLOCKWISE FROM TRUE NORTH. "00" INDICATES CALM.
RESULTANT DIRECTIONS ARE GIVEN TO WHOLE DEGREES.
BOLD VALUES INDICATE EXTREME VALUES WHICH OCCURRED AFTER THE ASOS SYSTEM WAS COMMISSIONED.

RECORD = PERIOD OF RECORD
RECORD MEAN PRECIPITATION IS THE MEAN OF ALL DAILY PRECIPITATION AMOUNTS DURING THE PERIOD OF RECORD.
RECORD MAX(MIN) TEMPERATURE IS THE MEAN OF ALL DAILY MAX(MIN) TEMPERATURES DURING THE PERIOD OF RECORD.
RECORD MEAN TEMPERATURE IS THE SUM OF THE RECORD MAX AND RECORD MIN DIVIDED BY 2.
AVERAGE TEMPERATURE IS THE SUM OF THE MEAN DAILY MAX AND MIN TEMPERATURE DIVIDED BY 2.

EXCEPTIONS

FASTEST MILE WINDS ARE THROUGH JULY 1982.

RECORD MEANS ARE THROUGH THE CURRENT YEAR, BEGINNING IN 1882 FOR TEMPERATURE
1882 FOR PRECIPITATION
1946 FOR SNOWFALL

HEATING DEGREE DAYS Base 65 deg. F FORT SMITH, ARKANSAS

SEASON	JULY	AUG	SEP	OCT	NOV	DEC	JAN	FEB	MAR	APR	MAY	JUNE	TOTAL
1964-65	0	0	14	182	338	733	705	658	712	83	0	0	3425
1965-66	0	0	16	144	286	576	938	659	384	152	30	0	3185
1966-67	0	0	8	207	325	770	735	695	272	63	48	0	3123
1967-68	0	0	32	158	464	726	781	734	420	131	22	0	3468
1968-69	0	0	0	161	508	795	745	613	604	75	13	3	3517
1969-70	0	0	0	182	474	731	951	600	528	112	30	5	3613
1970-71	0	0	10	181	479	639	795	647	495	176	37	0	3459
1971-72	0	0	23	26	441	571	804	616	378	145	36	1	3041
1972-73	1	0	14	176	648	887	873	688	315	258	59	0	3919
1973-74	0	0	9	97	316	764	821	565	304	166	6	1	3049
1974-75	0	0	54	113	465	761	718	699	562	217	10	0	3599
1975-76	0	0	62	141	430	741	863	413	353	123	80	0	3206
1976-77	0	0	20	317	682	837	1186	602	352	88	9	0	4093
1977-78	0	0	0	142	377	773	1143	924	537	113	69	0	4078
1978-79	0	0	0	133	359	771	1215	869	393	175	54	0	3969
1979-80	0	0	8	123	567	699	769	772	531	208	27	0	3704
1980-81	0	0	17	190	441	694	822	600	406	54	61	0	3285
1981-82	0	0	25	184	393	796	941	706	353	270	14	4	3686
1982-83	0	0	28	186	448	625	826	594	443	332	58	3	3543
1983-84	0	0	25	91	419	1113	980	574	496	227	32	0	3957
1984-85	0	0	53	99	461	514	1032	774	281	111	5	1	3331
1985-86	0	0	47	112	364	917	731	529	302	87	13	0	3102
1986-87	0	0	0	121	501	724	779	506	405	168	0	0	3204
1987-88	0	0	3	175	390	675	912	656	414	134	4	0	3363
1988-89	0	0	2	179	411	692	688	797	376	160	52	0	3357
1989-90	0	0	39	126	378	982	611	464	333	180	51	0	3164
1990-91	0	0	12	182	301	794	884	481	324	87	17	0	3082
1991-92	0	0	31	107	541	603	728	429	346	148	51	0	2984
1992-93	0	0	19	96	500	700	790	683	479	230	33	0	3530
1993-94	0	0	12	212	547	653							

COOLING DEGREE DAYS Base 65 deg. F FORT SMITH, ARKANSAS

YEAR	JAN	FEB	MAR	APR	MAY	JUNE	JULY	AUG	SEP	OCT	NOV	DEC	TOTAL
1969	1	0	0	42	196	338	613	472	290	91	4	0	2047
1970	0	0	2	76	230	376	516	606	374	47	4	0	2231
1971	0	0	2	25	86	416	482	408	321	86	9	0	1835
1972	0	0	0	58	107	369	404	471	338	55	0	0	1802
1973	0	0	0	42	86	297	472	457	299	80	8	0	1741
1974	0	0	43	40	224	231	527	385	82	28	7	0	1567
1975	0	0	3	46	163	327	457	454	175	67	23	0	1715
1976	0	4	9	33	54	239	435	388	187	16	0	0	1365
1977	0	0	0	60	250	460	554	528	367	40	3	0	2262
1978	0	0	0	60	173	357	632	567	419	46	6	0	2260
1979	0	0	7	33	113	307	464	426	207	85	4	0	1646
1980	0	0	0	11	164	434	643	649	419	50	5	0	2375
1981	0	0	0	117	81	430	534	412	245	49	0	0	1868
1982	0	0	27	26	183	252	493	510	249	91	8	2	1841
1983	0	0	0	30	80	281	501	597	313	76	15	0	1893
1984	0	0	2	25	111	412	490	500	253	86	8	6	1893
1985	0	0	27	52	146	351	538	503	317	80	13	0	2027
1986	0	0	6	49	179	435	617	418	359	58	0	0	2121
1987	0	0	10	72	290	398	502	578	243	15	12	0	2120
1988	0	0	6	31	185	450	583	604	344	31	0	0	2234
1989	0	0	13	102	189	332	460	481	205	102	12	0	1896
1990	0	2	22	52	140	480	533	533	386	59	9	0	2216
1991	0	0	17	49	284	426	538	474	266	72	6	4	2136
1992	0	0	1	62	131	323	496	346	218	43	0	2	1622
1993	0	0	9	17	132	387	678	633	247	45	2	0	2150

SNOWFALL (inches) FORT SMITH, ARKANSAS

SEASON	JULY	AUG	SEP	OCT	NOV	DEC	JAN	FEB	MAR	APR	MAY	JUNE	TOTAL
1964-65	0.0	0.0	0.0	0.0	0.0	T	0.1	T	T	0.0	0.0	0.0	0.1
1965-66	0.0	0.0	0.0	0.0	0.0	0.0	5.8	1.9	T	0.0	0.0	0.0	7.7
1966-67	0.0	0.0	0.0	0.0	0.0	1.0	0.1	T	4.0	0.0	0.0	0.0	5.1
1967-68	0.0	0.0	0.0	0.0	T	2.0	0.3	3.0	5.3	0.0	0.0	0.0	10.6
1968-69	0.0	0.0	0.0	0.0	T	T	0.0	3.7	0.0	0.0	0.0	0.0	3.7
1969-70	0.0	0.0	0.0	0.0	T	2.4	5.2	T	0.2	0.0	0.0	0.0	7.8
1970-71	0.0	0.0	0.0	0.0	0.0	0.0	T	0.3	1.6	T	0.0	0.0	1.9
1971-72	0.0	0.0	0.0	0.0	2.8	2.0	0.3	0.5	0.0	0.0	0.0	0.0	5.6
1972-73	0.0	0.0	0.0	0.0	1.7	0.7	4.4	T	0.0	T	0.0	0.0	6.8
1973-74	0.0	0.0	0.0	0.0	0.0	1.8	T	0.3	0.0	0.0	0.0	0.0	2.1
1974-75	0.0	0.0	0.0	0.0	2.0	T	0.0	5.8	1.8	0.0	0.0	0.0	9.6
1975-76	0.0	0.0	0.0	0.0	2.7	7.2	T	0.0	0.0	0.0	0.0	0.0	9.9
1976-77	0.0	0.0	0.0	0.0	4.7	0.0	13.0	T	0.0	0.0	0.0	0.0	17.7
1977-78	0.0	0.0	0.0	0.0	T	T	11.7	10.0	0.2	0.0	0.0	0.0	21.9
1978-79	0.0	0.0	0.0	0.0	0.0	T	11.8	4.4	0.0	0.0	0.0	0.0	16.2
1979-80	0.0	0.0	0.0	0.0	T	0.0	1.0	1.6	0.1	0.0	0.0	0.0	2.7
1980-81	0.0	0.0	0.0	0.0	0.7	0.0	3.2	T	T	0.0	0.0	0.0	3.9
1981-82	0.0	0.0	0.0	0.0	0.0	T	3.6	6.7	3.0	0.0	0.0	0.0	13.3
1982-83	0.0	0.0	0.0	0.0	0.0	0.0	T	1.0	0.0	0.0	0.0	0.0	1.0
1983-84	0.0	0.0	0.0	0.0	T	0.9	1.9	1.8	0.4	0.0	0.0	0.0	5.0
1984-85	0.0	0.0	0.0	0.0	0.0	3.1	8.1	5.0	0.0	0.0	0.0	0.0	16.2
1985-86	0.0	0.0	0.0	0.0	0.0	2.0	0.0	6.1	0.0	0.0	0.0	0.0	8.1
1986-87	0.0	0.0	0.0	0.0	0.0	0.0	1.4	1.6	0.4	0.0	0.0	0.0	3.4
1987-88	0.0	0.0	0.0	0.0	0.0	0.5	12.3	T	3.0	0.0	0.0	0.0	15.8
1988-89	0.0	0.0	0.0		0.0	0.5	0.5	0.3	3.3	0.0	T	0.0	
1989-90	0.0	0.0	0.0	0.0	0.0	0.2	T	T	0.0	T	T	0.0	0.2
1990-91	0.0	0.0	0.0	0.0	0.0	5.3	0.0	T	T	T	0.0	0.0	5.3
1991-92	0.0	0.0	0.0	0.0	T	0.0	T	T	T	0.0	T	0.0	T
1992-93	0.0	0.0	0.0	0.0	T	T	T	T	T	T	0.0	0.0	T
1993-94	0.0	0.0	0.0	T	0.2	0.0							
Record Mean	0.0	0.0	0.0	T	0.5	0.9	2.5	1.8	0.6	T	T	0.0	6.3

LITTLE ROCK, ARKANSAS

Little Rock is located on the Arkansas River near the geographical center of the state. It is situated on the dividing line between the Ouachita Mountains to the west and the flat lowlands comprising the Mississippi River Valley to the east. Elevations range from 222 feet at the river level to 257 feet over much of the flat land, including the airport in the southeast, to near 600 feet in the hilly residential area of the western portions of the city. Two minor temperature variations are observed due to the terrain; somewhat lower minimum temperatures are observed in the airport vicinity and a slight downslope adiabatic heating effect accompanies airflow from the ridges and hills in the west and northwest.

The modified continental climate of Little Rock includes exposure to all of the North American air mass types. However, with its proximity to the Gulf of Mexico, the summer season is marked by prolonged periods of warm and humid weather. The growing season averages 233 days in which 62 percent of the normal precipitation occurs. Winters are mild, but polar and Arctic outbreaks are not uncommon.

Precipitation is fairly well distributed throughout the year. Summer rainfall is almost completely of the convective type. The driest period usually occurs in the late summer and early fall. Snow is almost negligible. Glaze and ice storms, although infrequent, are at times severe. Warm front weather in the winter and early spring, characterized by shallow surface cold air flow from the north under warm moist Gulf air, results in excellent conditions for the production of freezing precipitation.

NORMALS, MEANS, AND EXTREMES

LITTLE ROCK, ARKANSAS

LATITUDE: 34°44'N LONGITUDE: 92°14'W ELEVATION: FT. GRND 257 BARO 260 TIME ZONE: CENTRAL WBAN: 13963

	(a)	JAN	FEB	MAR	APR	MAY	JUNE	JULY	AUG	SEP	OCT	NOV	DEC	YEAR
TEMPERATURE °F:														
Normals														
-Daily Maximum		49.0	53.9	64.0	73.4	81.3	89.3	92.4	91.4	84.6	75.1	62.7	52.5	72.5
-Daily Minimum		29.1	33.2	42.2	50.7	59.0	67.4	71.5	69.8	63.5	50.9	41.5	33.1	51.0
-Monthly		39.1	43.6	53.1	62.1	70.2	78.4	81.9	80.6	74.1	63.0	52.1	42.8	61.8
Extremes														
-Record Highest	52	83	85	91	95	98	105	112	108	106	97	86	80	112
-Year		1950	1986	1974	1987	1964	1988	1986	1980	1947	1963	1955	1956	JUL 1986
-Record Lowest	52	-4	-5	11	28	40	46	54	52	37	29	17	-1	-5
-Year		1962	1951	1951	1971	1971	1969	1972	1986	1942	1989	1976	1989	FEB 1951
NORMAL DEGREE DAYS:														
Heating (base 65°F)		803	599	384	133	25	0	0	0	8	128	387	688	3155
Cooling (base 65°F)		0	0	15	46	187	402	524	484	281	66	0	0	2005
% OF POSSIBLE SUNSHINE	32	46	54	57	62	68	73	71	73	68	69	56	48	62
MEAN SKY COVER (tenths)														
Sunrise - Sunset	35	6.5	6.0	6.2	6.1	6.0	5.4	5.5	5.0	5.2	4.5	5.5	6.2	5.7
MEAN NUMBER OF DAYS:														
Sunrise to Sunset														
-Clear	35	8.6	9.1	8.6	8.7	8.0	9.5	8.8	11.6	11.2	14.4	11.0	9.2	118.7
-Partly Cloudy	35	6.1	5.7	7.0	7.5	10.8	11.6	12.9	10.9	8.6	7.1	5.9	5.9	99.9
-Cloudy	35	16.3	13.5	15.4	13.8	12.3	8.9	9.3	8.5	10.1	9.5	13.1	16.0	146.6
Precipitation														
.01 inches or more	51	9.5	9.0	10.2	10.2	10.2	8.4	8.2	7.0	7.4	6.8	8.2	9.3	104.5
Snow,Ice pellets,hail														
1.0 inches or more	51	0.9	0.5	0.2	0.0	0.0	0.0	0.0	0.0	0.0	0.0	0.1	0.2	1.9
Thunderstorms	51	1.8	2.4	4.8	6.4	7.3	7.6	8.7	6.5	3.7	2.6	3.0	1.8	56.7
Heavy Fog Visibility														
1/4 mile or less	51	2.7	1.7	1.2	0.8	0.8	0.3	0.5	0.7	1.0	1.7	1.8	2.6	15.9
Temperature °F														
-Maximum														
90° and above	33	0.0	0.0	0.*	0.3	3.6	16.1	22.5	19.5	9.0	1.2	0.0	0.0	72.1
32° and below	33	3.4	1.0	0.1	0.0	0.0	0.0	0.0	0.0	0.0	0.0	0.*	1.5	5.9
-Minimum														
32° and below	33	20.0	13.2	4.4	0.5	0.0	0.0	0.0	0.0	0.0	0.2	5.6	15.2	59.1
0° and below	33	0.1	0.0	0.0	0.0	0.0	0.0	0.0	0.0	0.0	0.0	0.0	0.1	0.2
AVG. STATION PRESS. (mb)	21	1012.2	1010.5	1006.8	1006.0	1005.0	1005.6	1006.8	1007.2	1008.0	1009.6	1010.0	1011.5	1008.3
RELATIVE HUMIDITY (%)														
Hour 00	33	76	74	71	75	83	83	83	84	85	82	78	77	79
Hour 06 (Local Time)	33	80	80	79	83	87	87	88	89	89	86	83	80	84
Hour 12	33	61	59	56	56	58	55	56	56	58	53	58	62	57
Hour 18	33	64	59	55	56	60	58	59	60	64	63	65	66	61
PRECIPITATION (inches):														
Water Equivalent														
-Normal		3.91	4.36	5.31	6.21	7.02	7.84	8.19	8.06	7.41	6.30	5.21	4.28	74.10
-Maximum Monthly	52	12.53	11.02	10.40	14.20	12.74	7.82	7.95	14.46	10.17	15.35	13.14	16.48	16.48
-Year		1950	1956	1990	1973	1968	1974	1988	1966	1978	1984	1988	1987	DEC 1987
-Minimum Monthly	52	0.50	0.51	0.73	0.50	0.69	T	0.14	0.19	0.28	0.01	0.28	1.26	T
-Year		1986	1947	1966	1987	1970	1952	1986	1980	1956	1944	1949	1958	JUN 1952
-Maximum in 24 hrs	46	5.18	5.15	4.56	7.96	7.71	4.61	3.58	7.32	4.05	5.67	7.81	7.01	7.96
-Year		1969	1950	1990	1974	1955	1960	1988	1966	1967	1990	1988	1987	APR 1974
Snow,Ice pellets,hail														
-Maximum Monthly	52	13.6	9.8	7.0	T	0.0	0.0	0.0	0.0	0.0	T	4.8	9.8	13.6
-Year		1988	1979	1971	1991						1993	1971	1963	JAN 1988
-Maximum in 24 hrs	46	12.1	9.6	6.7	T	T	0.0	0.0	0.0	0.0	T	4.8	9.8	12.1
-Year		1988	1966	1971	1991	1988					1993	1971	1963	JAN 1988
WIND:														
Mean Speed (mph)	51	8.5	8.9	9.6	9.0	7.7	7.2	6.7	6.3	6.7	6.8	8.0	8.2	7.8
Prevailing Direction through 1963		S	SW	WNW	S	S	SSW	SW	SW	NE	SW	SW	SW	SW
Fastest Mile														
-Direction (!!)	36	S	SW	SE	NW	NW	NE	NW	NW	NW	SSW	SW	SW	NW
-Speed (MPH)	36	44	57	56	65	61	60	56	54	50	58	49	48	65
-Year		1950	1971	1959	1961	1952	1953	1960	1956	1952	1956	1952	1971	APR 1961
Peak Gust														
-Direction (!!)														
-Speed (mph)														
-Date														

PRECIPITATION (inches)

LITTLE ROCK, ARKANSAS

YEAR	JAN	FEB	MAR	APR	MAY	JUNE	JULY	AUG	SEP	OCT	NOV	DEC	ANNUAL
1964	0.98	2.87	8.22	11.06	1.40	0.31	3.79	3.71	5.46	0.37	3.70	4.37	46.24
1965	4.45	5.73	3.63	1.19	5.42	2.49	2.51	2.03	7.67	0.21	1.54	2.05	38.92
1966	3.03	5.02	0.73	7.29	2.23	0.69	3.54	14.46	1.42	1.95	3.08	4.21	47.65
1967	2.13	2.31	3.11	7.58	8.69	3.02	4.29	1.73	6.25	4.96	1.73	4.95	50.75
1968	4.76	1.08	5.55	4.85	12.74	6.77	5.98	0.26	5.99	2.81	5.30	4.56	60.65
1969	8.06	2.41	3.65	4.30	3.60	2.98	3.40	2.73	2.33	3.60	3.94	8.10	49.10
1970	1.05	4.57	4.87	7.99	0.69	2.30	3.02	2.15	2.82	7.68	2.09	3.85	43.08
1971	2.07	2.21	3.24	1.70	5.37	7.66	4.01	8.62	0.78	2.55	3.38	6.97	48.56
1972	1.71	1.55	3.32	1.81	2.07	2.62	1.77	3.58	6.43	7.63	7.38	5.14	45.01
1973	5.64	2.95	7.89	14.20	3.96	2.66	6.59	1.26	9.09	5.93	9.03	5.19	74.39
1974	5.77	2.60	2.07	9.76	6.26	7.82	4.09	3.20	4.31	3.36	5.73	2.99	57.96
1975	4.64	4.38	7.67	4.14	5.87	1.56	3.98	2.73	1.86	1.62	3.68	2.92	45.05
1976	3.00	5.12	5.43	1.06	4.88	5.69	1.97	0.70	1.82	6.04	1.79	2.30	39.80
1977	2.70	1.96	6.75	4.47	2.89	4.70	5.07	1.37	6.38	0.63	9.34	1.40	47.66
1978	5.44	1.52	3.56	4.22	6.27	5.39	2.70	6.38	10.17	1.01	6.64	11.56	64.86
1979	4.05	5.67	3.10	9.64	11.54	4.45	4.27	6.51	4.35	3.36	4.02	3.53	64.49
1980	2.73	0.89	6.60	5.85	4.57	0.53	0.99	0.19	5.09	2.64	6.28	1.86	38.22
1981	1.11	3.89	4.00	2.75	9.73	7.80	3.15	2.91	1.37	6.11	1.64	1.34	45.80
1982	8.74	3.37	2.87	9.32	5.63	4.10	1.01	4.52	1.47	2.26	9.72	8.28	61.29
1983	2.25	1.49	4.19	6.72	7.58	3.34	1.07	0.79	0.41	3.73	4.47	9.07	45.11
1984	1.31	3.52	5.58	3.77	8.22	1.06	4.15	5.69	3.28	15.35	8.49	3.54	63.96
1985	3.11	2.78	5.27	8.63	2.99	2.40	3.30	3.52	4.36	3.91	5.78	2.97	49.02
1986	0.50	3.45	3.68	7.33	4.07	6.42	0.14	4.56	1.94	6.05	5.67	3.86	47.67
1987	2.07	7.07	3.52	0.50	4.56	4.63	1.60	2.12	7.56	1.37	10.96	16.48	62.44
1988	3.71	3.41	3.50	3.82	2.05	1.04	7.95	2.19	2.54	1.95	13.14	2.91	48.21
1989	3.01	9.55	7.64	2.57	4.04	3.95	7.87	1.21	3.57	1.70	1.95	2.19	49.25
1990	6.50	4.82	10.40	7.73	7.71	0.80	4.63	1.57	4.08	8.75	3.29	6.79	67.07
1991	6.88	3.03	3.56	12.44	2.87	2.28	2.03	6.78	3.01	7.00	5.18	4.59	59.65
1992	1.75	2.05	6.48	1.86	3.67	5.07	6.76	2.14	2.90	0.67	4.71	3.85	41.91
1993	5.06	2.44	3.05	5.40	5.49	2.04	1.24	2.77	1.44	4.10	6.33	4.41	43.77
Record Mean	4.51	3.89	4.70	5.18	5.01	3.64	3.42	3.30	3.36	3.10	4.42	4.30	48.82

AVERAGE TEMPERATURE (deg. F)

LITTLE ROCK, ARKANSAS

YEAR	JAN	FEB	MAR	APR	MAY	JUNE	JULY	AUG	SEP	OCT	NOV	DEC	ANNUAL
1964	40.8	41.6	52.7	64.7	72.3	80.6	83.2	79.3	73.7	60.1	54.0	43.9	62.3
1965	44.2	43.2	44.5	65.8	72.7	78.2	82.9	81.7	74.3	61.6	56.5	46.7	62.7
1966	35.4	42.9	54.6	62.4	68.4	78.0	84.2	77.9	72.0	59.2	54.8	43.2	61.1
1967	41.7	40.4	58.9	66.7	68.6	79.5	77.9	75.5	69.0	60.9	49.1	42.7	60.9
1968	37.7	38.0	50.8	60.9	67.6	77.8	77.8	81.4	70.7	62.9	51.2	42.2	59.9
1969	43.5	43.0	45.5	61.8	69.9	77.6	84.8	79.0	72.9	62.4	49.4	40.7	60.9
1970	35.7	42.1	48.2	63.1	71.9	78.7	79.9	81.2	78.1	61.5	50.3	47.2	61.5
1971	41.0	44.4	50.0	59.4	65.6	79.3	80.0	78.0	76.4	69.3	50.5	49.9	62.0
1972	43.6	46.7	53.3	62.5	69.8	79.4	80.4	81.1	75.8	62.6	47.3	41.0	62.0
1973	39.7	42.1	58.2	59.9	68.2	78.6	81.1	80.4	75.7	67.6	56.6	42.8	62.6
1974	42.4	45.7	58.1	60.7	71.3	74.3	83.2	79.0	69.0	62.3	51.9	44.4	61.9
1975	44.6	44.6	48.7	60.7	72.5	78.6	80.2	79.5	69.2	63.0	51.4	42.9	61.3
1976	39.7	52.5	56.5	61.1	64.6	74.4	80.2	78.7	72.1	57.8	45.9	41.9	60.4
1977	31.3	46.9	56.4	64.7	73.7	80.0	82.1	80.4	77.3	62.6	52.9	42.0	62.5
1978	31.7	34.0	51.0	65.9	71.4	78.9	84.1	83.0	76.7	62.0	54.6	43.3	61.4
1979	29.9	38.7	55.5	62.7	70.2	77.9	81.0	79.0	72.7	65.3	50.3	45.8	60.7
1980	44.0	40.9	50.3	61.5	70.6	79.4	88.1	87.0	78.6	60.4	50.4	43.1	62.9
1981	39.7	44.6	52.5	67.5	67.4	80.1	83.5	79.9	75.3	61.4	55.5	43.2	62.5
1982	37.5	41.3	57.1	58.0	72.7	76.6	83.1	82.1	74.2	64.7	53.2	48.3	62.4
1983	39.2	43.8	51.1	54.4	67.7	77.4	82.5	86.1	76.0	64.0	50.8	30.9	60.3
1984	36.7	46.6	50.1	59.7	68.0	79.8	79.7	78.1	71.0	65.0	49.0	52.1	61.3
1985	33.7	39.3	57.6	63.0	70.0	78.2	81.2	80.8	72.5	66.2	56.1	38.1	61.4
1986	42.5	48.2	55.4	63.6	71.4	79.7	86.3	78.1	77.6	63.1	49.7	42.4	63.2
1987	40.2	47.1	53.4	62.4	76.3	79.9	82.2	84.4	74.9	59.1	53.0	45.2	63.2
1988	35.6	42.9	52.2	61.5	70.4	78.8	81.7	82.4	75.8	60.4	53.3	44.4	61.6
1989	46.3	38.4	52.5	62.6	69.6	76.2	79.3	80.2	71.2	63.3	55.3	35.7	60.9
1990	48.1	51.3	55.4	62.0	68.1	80.6	83.2	82.2	77.5	61.5	56.3	43.1	64.1
1991	39.0	49.2	56.2	64.4	74.1	79.5	82.8	80.0	73.7	64.3	49.2	46.8	63.3
1992	42.7	50.6	54.4	62.6	69.0	76.5	81.0	76.5	72.7	64.4	50.1	43.9	62.0
1993	40.4	43.2	51.3	58.3	68.8	78.7	86.0	83.7	73.5	61.4	48.3	44.9	61.5
Record Mean	41.3	44.7	53.1	62.4	70.2	78.3	81.6	80.6	74.2	63.6	51.8	43.7	62.1
Max	50.0	53.9	62.9	72.5	80.0	88.0	91.2	90.4	84.4	74.4	61.6	52.3	71.8
Min	32.5	35.4	43.2	52.3	60.3	68.5	71.9	70.8	63.9	52.8	42.1	35.0	52.4

REFERENCE NOTES

LITTLE ROCK, ARKANSAS

GENERAL
T - TRACE AMOUNT.
BLANK ENTRIES DENOTE MISSING/UNREPORTED DATA.
# INDICATES A STATION OR INSTRUMENT RELOCATION.

SPECIFIC

PM - INCLUDES LAST DAY OF PREVIOUS MONTH
ASOS - AUTOMATED SURFACE OBSERVING SYSTEM IN OPERATION DURING THESE MONTHS.

(a) - LENGTH OF RECORD IN YEARS, ALTHOUGH INDIVIDUAL MONTHS MAY BE MISSING.
0.* OR * - THE VALUE IS BETWEEN 0.0 AND 0.05
NORMALS - BASED ON THE 1961-1990 RECORD PERIOD.
EXTREMES - DATES ARE THE MOST RECENT OCCURRENCE
WIND DIR.- NUMERALS SHOW TENS OF DEGRESS CLOCKWISE FROM TRUE NORTH. "00" INDICATES CALM.
RESULTANT DIRECTIONS ARE GIVEN TO WHOLE DEGREES.
BOLD VALUES INDICATE EXTREME VALUES WHICH OCCURRED AFTER THE ASOS SYSTEM WAS COMMISSIONED.

RECORD = PERIOD OF RECORD
RECORD MEAN PRECIPITATION IS THE MEAN OF ALL DAILY PRECIPITATION AMOUNTS DURING THE PERIOD OF RECORD.
RECORD MAX(MIN) TEMPERATURE IS THE MEAN OF ALL DAILY MAX(MIN) TEMPERATURES DURING THE PERIOD OF RECORD.
RECORD MEAN TEMPERATURE IS THE SUM OF THE RECORD MAX AND RECORD MIN DIVIDED BY 2.
AVERAGE TEMPERATURE IS THE SUM OF THE MEAN DAILY MAX AND MIN TEMPERATURE DIVIDED BY 2.

EXCEPTIONS

1. PRECIPITATION DATA DECEMBER 15, 1975 THROUGH OCTOBER 1976 IS FROM NORTH LITTLE ROCK.
2. WIND DATA FROM DECEMBER 15, 1975 THROUGH DECEMBER 1976 IS FROM NORTH LITTLE ROCK.
3. MAXIMUM 24-HOUR PRECIPITATION AND SNOW, FASTEST MILE WINDS, MEAN SKY COVER, AND DAYS CLEAR-PARTLY CLOUDY-CLOUDY ARE THROUGH 1977.
4. PERCENT OF POSSIBLE SUNSHINE IS THROUGH 1975

RECORD MEANS ARE THROUGH THE CURRENT YEAR, BEGINNING IN 1880 FOR TEMPERATURE
1880 FOR PRECIPITATION
1943 FOR SNOWFALL

HEATING DEGREE DAYS Base 65 deg. F LITTLE ROCK, ARKANSAS

SEASON	JULY	AUG	SEP	OCT	NOV	DEC	JAN	FEB	MAR	APR	MAY	JUNE	TOTAL
1964-65	0	0	14	175	336	646	639	604	629	70	0	0	3113
1965-66	0	0	19	137	257	561	916	611	327	115	42	0	2985
1966-67	0	0	3	198	309	679	717	682	240	52	33	0	2913
1967-68	0	0	39	162	472	685	840	778	440	147	30	0	3593
1968-69	0	0	3	124	412	700	665	611	595	108	16	1	3235
1969-70	0	0	0	175	458	745	905	636	516	124	23	0	3582
1970-71	0	0	4	153	442	554	737	571	470	182	55	0	3168
1971-72	0	0	9	6	437	466	659	525	360	149	20	0	2631
1972-73	0	0	8	142	530	736	777	637	216	186	28	0	3260
1973-74	0	0	2	61	261	680	690	533	255	163	4	0	2649
1974-75	0	0	23	111	401	634	630	566	499	196	5	0	3065
1975-76	0	0	48	130	414	681	777	359	284	146	70	0	2909
1976-77	0	0	1	257	567	710	1041	502	265	70	10	0	3423
1977-78	0	0	0	105	370	709	1025	862	436	68	48	0	3623
1978-79	0	0	0	118	321	667	1083	732	314	110	13	0	3358
1979-80	0	0	0	80	436	588	645	693	450	142	15	0	3049
1980-81	0	0	16	184	437	673	774	565	388	37	45	0	3119
1981-82	0	0	4	186	278	668	847	656	298	223	6	0	3166
1982-83	0	0	12	119	369	536	795	587	425	332	33	0	3208
1983-84	0	0	19	89	422	1050	872	530	460	190	24	0	3656
1984-85	0	0	44	81	476	408	962	713	251	101	8	0	3044
1985-86	0	0	31	82	283	825	691	467	298	91	7	0	2775
1986-87	0	1	0	112	454	694	762	496	353	145	0	0	3017
1987-88	0	0	0	182	358	609	904	637	388	123	4	0	3205
1988-89	0	0	1	163	358	633	573	738	395	156	39	0	3056
1989-90	0	0	23	112	313	898	516	380	316	152	31	0	2741
1990-91	0	0	8	173	260	675	798	438	294	69	10	0	2725
1991-92	0	0	23	97	483	560	682	413	324	134	40	0	2756
1992-93	0	0	14	69	441	647	755	606	423	215	15	1	3186
1993-94	0	0	13	174	506	613							

COOLING DEGREE DAYS Base 65 deg. F LITTLE ROCK, ARKANSAS

YEAR	JAN	FEB	MAR	APR	MAY	JUNE	JULY	AUG	SEP	OCT	NOV	DEC	TOTAL
1969	4	0	0	21	176	386	622	439	243	100	0	0	1991
1970	3	0	0	77	243	416	471	511	404	51	9	9	2194
1971	0	0	11	19	79	435	470	411	355	146	10	3	1939
1972	2	0	3	81	174	440	484	507	341	76	3	0	2111
1973	0	0	14	41	135	415	506	485	330	147	17	0	2090
1974	0	0	45	37	206	288	572	441	148	36	14	0	1787
1975	6	0	4	73	245	416	475	455	178	73	14	2	1941
1976	0	5	27	37	65	292	480	434	220	42	0	0	1602
1977	0	0	7	67	287	455	537	482	377	40	14	0	2266
1978	0	0	9	104	253	424	599	565	357	31	16	0	2358
1979	0	0	24	48	178	396	502	442	240	95	1	0	1926
1980	0	1	1	42	196	439	725	688	432	50	5	0	2579
1981	0	0	9	117	131	458	580	470	318	78	2	0	2163
1982	0	0	57	21	253	355	570	540	294	114	20	24	2248
1983	0	0	0	21	121	381	550	660	355	63	6	0	2157
1984	0	1	4	38	126	451	462	416	234	88	3	17	1840
1985	0	0	31	48	167	404	508	501	265	125	21	0	2070
1986	0	5	7	56	211	446	668	415	385	63	0	0	2256
1987	0	0	3	74	359	456	540	610	304	8	3	0	2357
1988	0	0	2	24	177	423	523	546	332	25	15	0	2067
1989	0	0	13	91	189	345	450	479	213	70	29	0	1879
1990	0	4	26	67	135	475	571	540	392	73	8	0	2291
1991	0	1	30	56	299	445	558	472	290	82	14	2	2249
1992	0	0	4	70	170	352	505	364	249	60	0	0	1774
1993	0	0	4	22	138	419	656	591	276	69	14	0	2189

SNOWFALL (inches) LITTLE ROCK, ARKANSAS

SEASON	JULY	AUG	SEP	OCT	NOV	DEC	JAN	FEB	MAR	APR	MAY	JUNE	TOTAL
1964-65	0.0	0.0	0.0	0.0	0.0	0.2	T	2.9	4.3	0.0	0.0	0.0	7.4
1965-66	0.0	0.0	0.0	0.0	0.0	0.0	12.0	9.6	0.0	0.0	0.0	0.0	21.6
1966-67	0.0	0.0	0.0	0.0	T	0.2	1.6	T	T	0.0	0.0	0.0	1.8
1967-68	0.0	0.0	0.0	0.0	T	1.1	1.0	4.3	T	0.0	0.0	0.0	6.4
1968-69	0.0	0.0	0.0	0.0	0.0	0.0	T	2.3	T	0.0	0.0	0.0	2.3
1969-70	0.0	0.0	0.0	0.0	T	T	4.0	T	T	0.0	0.0	0.0	4.0
1970-71	0.0	0.0	0.0	0.0	0.0	T	T	0.7	7.0	T	0.0	0.0	7.7
1971-72	0.0	0.0	0.0	0.0	4.8	0.6	0.1	0.3	0.0	0.0	0.0	0.0	5.8
1972-73	0.0	0.0	0.0	0.0	T	0.7	2.6	T	0.0	T	0.0	0.0	3.3
1973-74	0.0	0.0	0.0	0.0	0.0	T	0.3	T	0.0	0.0	0.0	0.0	0.3
1974-75	0.0	0.0	0.0	0.0	T	T	1.4	0.4	2.4	0.0	0.0	0.0	4.2
1975-76	0.0	0.0	0.0	0.0	0.2	1.0	T	0.0	0.0	0.0	0.0	0.0	1.2
1976-77	0.0	0.0	0.0	0.0	1.0	0.0	3.8	0.0	0.0	0.0	0.0	0.0	4.8
1977-78	0.0	0.0	0.0	0.0	0.0	0.0	10.0	3.4	T	0.0	0.0	0.0	13.4
1978-79	0.0	0.0	0.0	0.0	0.0	T	1.4	9.8	0.0	0.0	0.0	0.0	11.2
1979-80	0.0	0.0	0.0	0.0	0.0	0.0	0.9	0.5	T	T	0.0	0.0	1.4
1980-81	0.0	0.0	0.0	0.0	1.8	T	T	0.0	0.0	0.0	0.0	0.0	1.8
1981-82	0.0	0.0	0.0	0.0	0.0	0.0	5.0	6.3	T	0.0	0.0	0.0	11.3
1982-83	0.0	0.0	0.0	0.0	0.0	T	T	T	T	T	0.0	0.0	T
1983-84	0.0	0.0	0.0	0.0	0.0	0.8	1.5	0.2	4.5	0.0	0.0	0.0	7.0
1984-85	0.0	0.0	0.0	0.0	0.0	T	6.3	5.0	0.0	0.0	0.0	0.0	11.3
1985-86	0.0	0.0	0.0	0.0	0.0	T	0.0	1.5	0.0	0.0	0.0	0.0	1.5
1986-87	0.0	0.0	0.0	0.0	T	0.0	1.0	T	0.8	0.0	0.0	0.0	1.8
1987-88	0.0	0.0	0.0	0.0	0.0	T	13.6	2.5	T	0.0	0.0	0.0	16.1
1988-89	0.0	0.0	0.0	0.0	0.0	T	2.0	T	1.0	0.0	0.0	0.0	3.0
1989-90	0.0	0.0	0.0	0.0	0.0	T	T	0.0	0.0	0.0	0.0	0.0	T
1990-91	0.0	0.0	0.0	0.0	0.0	T	T	T	T	T	0.0	0.0	T
1991-92	0.0	0.0	0.0	0.0	T	T	T	0.0	0.0	0.0	0.0	0.0	T
1992-93	0.0	0.0	0.0	0.0	T	T	0.0	T	T	0.0	0.0	0.0	T
1993-94	0.0	0.0	0.0	T	0.0	0.0							
Record Mean	0.0	0.0	0.0	T	0.2	0.7	2.3	1.4	0.5	T	0.0	0.0	5.1

NORTH LITTLE ROCK, ARKANSAS

North Little Rock is located on the Arkansas River near the geographical center of the state. It is situated on the dividing line between the Ouachita Mountains to the west and the flat lowlands comprising the Mississippi River Valley to the east. General elevations range from 250 to 400 feet, but drop as low as 225 feet near the river and rise to nearly 600 feet in the hillier northern and western sections of the city.

The modified continental climate of the area includes exposure to all of the North American air mass types.

The growing season averages 233 days, in which 62 percent of the normal precipitation occurs.

Precipitation is fairly well distributed throughout the year. Summer rainfall is almost completely of the convective type. The driest period usually occurs in the late summer and early fall.

Winters are relatively mild, but outbreaks of polar and Arctic air occur at regular intervals.

Though each winter generally sees several inches of snow, heavy snowfalls are relatively uncommon. In about one winter out of four, snowfall is an inch or less. Freezing rain and sleet generally occur a few times each winter and, on occasion, produce significant ice storms. Warm front weather, characterized by shallow surface cold air flow from the north under warm moist Gulf air, results in excellent conditions for the production of freezing precipitation.

Although severe weather has occurred during every month of the year, the spring months most frequently bring heavy thunderstorms. Hail is not uncommon, but hailstones of large, damaging size are infrequent. Tornadoes threaten the area in about one year out of four.

Hot, humid weather prevails during the summer months. During most years, the temperature reaches or exceeds 100 degrees on a couple of days.

NORMALS, MEANS, AND EXTREMES

NORTH LITTLE ROCK, ARKANSAS

LATITUDE: 34 °50'N LONGITUDE: 92 °15' W ELEVATION: FT. GRND 563 BARO 565 TIME ZONE: CENTRAL WBAN: 03952

	(a)	JAN	FEB	MAR	APR	MAY	JUNE	JULY	AUG	SEP	OCT	NOV	DEC	YEAR
TEMPERATURE °F:														
Normals														
-Daily Maximum		47.6	52.7	62.7	72.9	79.6	87.4	91.5	89.9	82.9	73.1	61.2	50.7	71.0
-Daily Minimum		29.4	33.6	42.6	52.2	59.7	67.5	71.3	70.0	63.7	52.6	42.4	33.7	51.6
-Monthly		38.5	43.2	52.7	62.6	69.7	77.5	81.4	80.0	73.3	62.9	51.8	42.2	61.3
Extremes														
-Record Highest	16	74	83	87	94	93	102	110	104	102	92	83	78	110
-Year		1989	1986	1981	1987	1988	1988	1986	1980	1980	1979	1978	1982	JUL 1986
-Record Lowest	16	-6	5	14	30	40	52	61	53	41	27	19	-2	-6
-Year		1985	1981	1980	1982	1978	1993	1990	1986	1989	1993	1991	1989	JAN 1985
NORMAL DEGREE DAYS:														
Heating (base 65°F)		822	610	395	130	31	0	0	0	11	126	396	707	3228
Cooling (base 65°F)		0	0	13	58	177	375	508	465	260	60	0	0	1916
% OF POSSIBLE SUNSHINE	16	65	65	72	77	72	81	83	81	79	74	61	60	73
MEAN SKY COVER (tenths)														
Sunrise - Sunset														
MEAN NUMBER OF DAYS:														
Sunrise to Sunset														
-Clear														
-Partly Cloudy														
-Cloudy														
Precipitation														
.01 inches or more	16	9.5	9.4	9.8	9.9	11.9	9.3	8.3	7.3	8.0	7.8	8.6	10.2	109.9
Snow,Ice pellets,hail														
1.0 inches or more	16	0.9	0.8	0.3	0.0	0.0	0.0	0.0	0.0	0.0	0.0	0.1	0.2	2.2
Thunderstorms	16	1.6	2.5	4.5	6.7	9.6	9.1	9.1	8.7	4.3	4.0	3.5	2.5	66.1
Heavy Fog Visibility														
1/4 mile or less	16	4.2	2.9	2.5	1.3	1.1	1.0	0.6	1.5	1.3	2.1	3.4	3.9	25.7
Temperature °F														
-Maximum														
90° and above	16	0.0	0.0	0.0	0.3	1.3	12.5	21.4	19.3	7.8	0.3	0.0	0.0	62.8
32° and below	16	3.6	2.2	0.1	0.0	0.0	0.0	0.0	0.0	0.0	0.0	0.1	2.2	8.1
-Minimum														
32° and below	16	18.2	11.6	3.6	0.3	0.0	0.0	0.0	0.0	0.0	0.3	4.6	12.2	50.9
0° and below	16	0.1	0.0	0.0	0.0	0.0	0.0	0.0	0.0	0.0	0.0	0.0	0.3	0.4
AVG. STATION PRESS.(mb)														
RELATIVE HUMIDITY (%)														
Hour 00														
Hour 06 (Local Time)														
Hour 12														
Hour 18														
PRECIPITATION (inches):														
Water Equivalent														
-Normal		3.21	3.53	4.97	5.08	5.24	3.26	3.31	3.10	4.08	3.62	5.16	4.69	49.25
-Maximum Monthly	16	6.65	9.02	10.09	11.01	10.29	6.16	8.45	8.43	8.43	14.03	14.82	13.74	14.82
-Year		1982	1989	1990	1991	1981	1981	1988	1991	1985	1984	1988	1987	NOV 1988
-Minimum Monthly	16	0.55	1.26	2.58	0.66	1.05	0.61	0.64	0.07	0.51	0.89	1.69	0.74	0.07
-Year		1986	1980	1982	1987	1988	1990	1983	1980	1983	1978	1981	1981	AUG 1980
-Maximum in 24 hrs	16	3.63	2.83	5.05	3.42	3.89	2.20	3.82	4.49	5.30	4.70	8.81	6.36	8.81
-Year		1991	1987	1990	1979	1984	1981	1988	1978	1978	1991	1988	1987	NOV 1988
Snow,Ice pellets,hail														
-Maximum Monthly	16	12.4	15.6	5.0	T	T	T	0.0	0.0	0.0	0.2	4.0	2.9	15.6
-Year		1988	1979	1984	1991	1990	1989				1993	1980	1983	FEB 1979
-Maximum in 24 hrs	16	12.0	9.1	5.0	T	T	T	0.0	0.0	0.0	0.2	4.0	2.4	12.0
-Year		1988	1979	1984	1991	1990	1989				1993	1980	1990	JAN 1988
WIND:														
Mean Speed (mph)														
Prevailing Direction														
Fastest Mile														
-Direction (!!)	3	NW	SW	S	NE	SW	NE	W	W	NE	SE	SW	SW	SW
-Speed (MPH)	3	25	25	28	30	30	21	30	28	24	24	27	25	30
-Year		1979	1980	1979	1980	1981	1979	1981	1978	1979	1979	1979	1978	MAY 1981
Peak Gust														
-Direction (!!)														
-Speed (mph)														
-Date														

PRECIPITATION (inches)

NORTH LITTLE ROCK, ARKANSAS

YEAR	JAN	FEB	MAR	APR	MAY	JUNE	JULY	AUG	SEP	OCT	NOV	DEC	ANNUAL
1978	5.73	1.53	3.23	3.62	7.27	4.22	1.91	6.30	6.84	0.89	6.59	8.11	56.24
1979	2.56	5.47	2.93	8.83	9.57	3.38	3.29	4.54	3.98	2.57	2.53	2.93	52.58
1980	2.84	1.26	4.87	7.50	7.21	0.64	0.64	0.07	3.66	3.41	5.38	2.14	39.62
1981	1.28	4.69	3.14	2.98	10.29	6.16	2.07	1.51	1.87	5.53	1.69	0.74	41.95
1982	6.65	3.01	2.58	7.46	5.13	4.68	5.40	5.67	1.43	2.31	9.19	9.48	62.99
1983	2.34	1.62	3.87	6.56	7.58	2.80	0.64	1.94	0.51	2.87	4.31	7.53	42.57
1984	1.43	3.93	5.29	3.22	8.07	1.04	2.87	1.79	4.19	14.03	7.94	5.61	59.41
1985	3.06	3.35	5.26	5.65	2.16	2.52	1.59	3.04	8.43	4.93	7.80	3.14	50.93
1986	0.55	3.00	3.05	6.97	3.43	5.01	1.18	4.31	0.95	6.35	5.72	3.34	43.86
1987	1.47	6.74	3.16	0.66	5.27	3.34	1.00	2.66	5.03	1.61	10.28	13.74	54.96
1988	2.96	3.22	4.18	3.85	1.05	1.60	8.45	1.58	1.56	1.52	14.82	3.15	47.94
1989	2.76	9.02	8.68	1.45	5.44	5.77	5.35	0.98	4.54	1.45	1.90	1.53	48.87
1990	5.57	4.02	10.09	5.93	9.65	0.61	2.09	3.22	3.56	7.58	3.26	8.07	63.65
1991	5.65	2.57	2.77	11.01	3.34	3.37	3.18	8.43	2.05	7.83	4.41	3.84	58.45
1992	1.85	2.30	7.54	2.25	3.30	5.55	5.43	3.98	3.02	1.00	5.32	4.51	46.05
1993	4.22	2.78	3.32	6.31	4.84	2.01	2.65	3.73	1.99	4.67	6.81	4.36	47.69
Record Mean	3.18	3.66	4.62	5.27	5.85	3.29	2.98	3.36	3.35	4.28	6.12	5.14	51.11

AVERAGE TEMPERATURE (deg. F)

NORTH LITTLE ROCK, ARKANSAS

YEAR	JAN	FEB	MAR	APR	MAY	JUNE	JULY	AUG	SEP	OCT	NOV	DEC	ANNUAL
1978	29.8	31.9	48.9	64.6	68.9	77.7	83.4	81.3	75.5	62.0	55.1	42.0	60.1
1979	27.5	36.4	53.1	61.1	67.9	75.6	80.0	78.7	71.1	64.8	49.3	44.4	59.2
1980	41.3	38.5	48.4	60.7	69.9	79.1	88.6	86.4	77.7	61.8	51.2	43.1	62.2
1981	39.8	45.1	53.1	69.2	66.2	79.6	83.0	79.8	73.2	60.6	54.3	40.7	40.7
1982	35.3	40.0	56.3	57.5	72.4	74.6	82.1	80.2	72.2	63.0	51.4	46.9	61.0
1983	38.6	44.5	50.8	54.9	67.0	76.1	82.3	84.5	75.3	64.5	51.5	29.6	60.0
1984	36.7	47.6	49.7	60.1	68.4	80.1	80.2	78.9	71.4	65.5	49.5	51.4	61.6
1985	32.3	38.7	57.0	64.4	70.6	78.0	81.3	80.1	71.8	65.6	55.1	38.2	61.1
1986	43.1	48.0	56.2	64.0	71.0	79.1	86.4	78.1	77.5	63.1	49.0	42.7	63.2
1987	40.6	47.0	54.4	63.8	75.8	78.9	81.7	83.2	74.4	59.6	53.2	44.9	63.1
1988	36.1	42.8	51.9	61.7	71.3	79.2	81.0	82.3	75.6	59.8	53.6	44.8	61.7
1989	46.3	37.2	52.5	63.2	68.9	75.2	78.0	80.0	69.9	64.0	54.8	35.4	60.5
1990	47.4	50.2	54.7	61.1	66.8	80.4	82.2	81.6	77.1	61.8	57.2	42.8	63.6
1991	38.2	49.6	56.3	64.6	73.9	79.2	82.6	80.4	73.6	65.2	48.6	46.8	63.3
1992	42.6	50.4	54.7	62.7	69.0	75.5	81.0	76.1	72.6	64.9	49.5	43.2	61.9
1993	40.0	42.7	50.9	58.7	68.7	78.4	86.3	83.8	72.7	60.7	48.1	44.6	61.3
Record Mean	38.5	43.2	53.1	62.0	69.8	77.9	82.5	80.9	73.8	62.9	52.0	42.6	61.6
Max	46.8	52.0	62.7	72.0	79.1	87.4	92.4	90.8	83.6	72.7	60.9	50.9	71.0
Min	30.1	34.3	43.4	52.0	60.4	68.4	72.6	71.0	64.1	53.1	43.0	34.2	52.2

REFERENCE NOTES

NORTH LITTLE ROCK, ARKANSAS

GENERAL
T - TRACE AMOUNT.
BLANK ENTRIES DENOTE MISSING/UNREPORTED DATA.
# INDICATES A STATION OR INSTRUMENT RELOCATION.

SPECIFIC

EXCEPTIONS

FASTEST MILE WINDS ARE THROUGH AUGUST 1981.

RECORD MEANS ARE THROUGH THE CURRENT YEAR, BEGINNING IN 1978 FOR TEMPERATURE
1978 FOR PRECIPITATION
1978 FOR SNOWFALL

(a) - LENGTH OF RECORD IN YEARS, ALTHOUGH INDIVIDUAL MONTHS MAY BE MISSING.
0.* OR * - THE VALUE IS BETWEEN 0.0 AND 0.05
NORMALS - BASED ON THE 1961-1990 RECORD PERIOD.
EXTREMES - DATES ARE THE MOST RECENT OCCURRENCE
WIND DIR.- NUMERALS SHOW TENS OF DEGRESS CLOCKWISE FROM TRUE NORTH. "00" INDICATES CALM.
RESULTANT DIRECTIONS ARE GIVEN TO WHOLE DEGREES.
BOLD VALUES INDICATE EXTREME VALUES WHICH OCCURRED AFTER THE ASOS SYSTEM WAS COMMISSIONED.

RECORD = PERIOD OF RECORD
RECORD MEAN PRECIPITATION IS THE MEAN OF ALL DAILY PRECIPITATION AMOUNTS DURING THE PERIOD OF RECORD.
RECORD MAX(MIN) TEMPERATURE IS THE MEAN OF ALL DAILY MAX(MIN) TEMPERATURES DURING THE PERIOD OF RECORD.
RECORD MEAN TEMPERATURE IS THE SUM OF THE RECORD MAX AND RECORD MIN DIVIDED BY 2.
AVERAGE TEMPERATURE IS THE SUM OF THE MEAN DAILY MAX AND MIN TEMPERATURE DIVIDED BY 2.

HEATING DEGREE DAYS Base 65 deg. F NORTH LITTLE ROCK, ARKANSAS

SEASON	JULY	AUG	SEP	OCT	NOV	DEC	JAN	FEB	MAR	APR	MAY	JUNE	TOTAL
1977-78							1086	921	499	95	74	0	
1978-79	0	0	0	126	316	707	1158	794	368	149	35	0	3653
1979-80	0	0	5	98	466	633	728	765	505	173	25	0	3398
1980-81	0	0	26	163	419	671	774	555	372	28	63	0	3071
1981-82	0	0	14	183	315	745	912	696	324	237	6	0	3432
1982-83	0	0	21	143	411	567	812	568	434	316	39	0	3311
1983-84	0	0	22	80	409	1092	874	500	469	182	24	0	3652
1984-85	0	0	52	80	456	420	1007	730	265	89	6	0	3105
1985-86	0	0	36	89	308	823	670	481	283	87	9	0	2786
1986-87	0	0	0	106	471	684	750	498	331	134	0	0	2974
1987-88	0	0	0	173	358	618	888	639	404	121	3	0	3204
1988-89	0	0	0	179	345	620	575	774	401	154	48	0	3096
1989-90	0	0	34	110	326	914	540	407	334	168	42	0	2875
1990-91	0	0	5	163	237	681	825	422	290	70	11	0	2704
1991-92	0	0	28	87	491	557	689	415	321	141	47	0	2776
1992-93	0	0	17	71	456	672	766	619	440	211	19	2	3273
1993-94	0	0	13	188	507	628							

COOLING DEGREE DAYS Base 65 deg. F NORTH LITTLE ROCK, ARKANSAS

YEAR	JAN	FEB	MAR	APR	MAY	JUNE	JULY	AUG	SEP	OCT	NOV	DEC	TOTAL
1978	0	0	5	91	199	390	576	513	323	41	26	0	2164
1979	0	0	7	39	131	329	472	431	194	99	1	0	1703
1980	0	0	0	50	184	433	736	670	417	66	13	0	2569
1981	0	4	10	162	111	445	566	467	268	53	0	0	2086
1982	0	3	61	19	241	296	537	476	242	89	10	15	1989
1983	0	0	0	19	112	339	544	609	338	75	10	0	2046
1984	0	0	4	42	136	458	478	437	251	100	1	7	1914
1985	0	0	26	79	187	398	510	477	249	113	17	0	2056
1986	0	10	16	66	205	428	668	412	379	55	0	0	2239
1987	0	0	8	103	342	423	525	572	290	13	12	0	2288
1988	0	0	3	30	205	433	504	542	323	22	10	0	2072
1989	0	0	20	108	175	313	409	471	188	85	25	0	1794
1990	0	1	21	55	104	468	540	524	374	73	10	0	2170
1991	0	0	31	61	290	433	555	486	289	100	8	0	2253
1992	0	0	11	80	178	323	498	351	247	73	0	0	1761
1993	0	0	8	28	142	410	666	588	255	62	9	0	2168

SNOWFALL (inches) NORTH LITTLE ROCK, ARKANSAS

SEASON	JULY	AUG	SEP	OCT	NOV	DEC	JAN	FEB	MAR	APR	MAY	JUNE	TOTAL
1977-78							10.2	4.8	T	0.0	0.0	0.0	
1978-79	0.0	0.0	0.0	0.0	0.0	0.1	3.5	15.6	T	0.0	0.0	0.0	19.2
1979-80	0.0	0.0	0.0	0.0	T	0.0	0.5	2.1	T	T	0.0	0.0	2.6
1980-81	0.0	0.0	0.0	0.0	4.0	T	1.3	0.0	T	0.0	0.0	0.0	5.3
1981-82	0.0	0.0	0.0	0.0	0.0	0.0	5.0	4.2	0.5	0.0	0.0	0.0	9.7
1982-83	0.0	0.0	0.0	0.0	T	T	1.7	1.1	T	T	0.0	0.0	2.8
1983-84	0.0	0.0	0.0	0.0	0.0	2.9	1.0	T	5.0	0.0	0.0	0.0	8.9
1984-85	0.0	0.0	0.0	0.0	0.0	T	8.8	6.6	0.0	0.0	0.0	0.0	15.4
1985-86	0.0	0.0	0.0	0.0	0.0	0.2	0.0	2.7	0.0	0.0	0.0	0.0	2.9
1986-87	0.0	0.0	0.0	0.0	T	0.0	0.3	0.2	1.2	0.0	0.0	0.0	1.7
1987-88	0.0	0.0	0.0	0.0	0.0	T	12.4	1.0	T	0.0	0.0	0.0	13.4
1988-89	0.0	0.0	0.0	0.0	0.0	1.0	2.8	0.5	1.0	0.0	0.0	T	5.3
1989-90	0.0	0.0	0.0	0.0	0.0	0.3	0.0	T	0.0	T	T	0.0	0.3
1990-91	0.0	0.0	0.0	0.0	0.0	2.4	T	T	T	T	0.0	0.0	2.4
1991-92	0.0	0.0	0.0	0.0	1.3	T	0.5	T	T	0.0	0.0	0.0	1.8
1992-93	0.0	0.0	0.0	T	T	T	T	T	T	0.0	0.0	0.0	T
1993-94	0.0	0.0	0.0	0.2	0.0	T							
Record Mean	0.0	0.0	0.0	T	0.3	0.4	3.0	2.4	0.5	T	T	T	6.7

BAKERSFIELD, CALIFORNIA

Bakersfield, situated in the extreme south end of the great San Joaquin Valley, is partially surrounded by a horseshoe-shaped rim of mountains with an open side to the northwest and the crest at an average distance of 40 miles.

The Sierra Nevada mountains to the northeast shut out most of the cold air that flows southward over the continent during winter. They also catch and store snow, which provides irrigation water for use during the dry months. The Tehachapi Mountains, forming the southern boundary, act as an obstruction to northwest wind, causing heavier precipitation on the windward slopes, high wind velocity over the ridges and, at times, continuing cloudiness in the south end of the valley after skies have cleared elsewhere. To the west are the coast ranges, and the ocean shore lies at a distance of 75 to 100 miles.

Because of the nature of the surrounding topography, there are large climatic variations within relatively short distances. These zones of variation may be classified as valley, mountain, and desert areas. The overall climate, however, is warm and semi-arid. There is only one wet season during the year, as 90 percent of all precipitation falls from October through April, inclusive. Snow in the valley is infrequent, with only a trace occurring in about one year out of seven. Thunderstorms seldom occur in the valley.

Summers are cloudless, hot and dry. Cotton, potatoes, grapes, and cattle are the principal agricultural products. There are considerable amounts of deciduous fruits, citrus, grain, and various vegetables. There are actually more than 110 farm crops grown commercially. Certain crops are planted or harvested every month of the year. Severe freezes seldom occur and there are occasional years with no frost at all in certain warm areas.

Winters are mild and semi-arid, yet fairly humid. December and January are characterized by frequent fog, mostly nocturnal, which prevails when marine air is trapped in the valley by a high pressure system. In extreme cases this fog may last continuously for two or three weeks. Its depth is usually less than 3,000 feet and the same condition that produces it also causes clear skies with mild temperatures in the surrounding mountain and desert areas.

Another local characteristic is the occasionally warm, dry, southeast chinook wind that spills through the Tehachapi Pass during winter. This wind usually attains velocities of 30 to 40 miles an hour, sometimes reaching as high as 60 miles an hour.

During summer months northwest sea breezes frequent the Bakersfield area about twice weekly. When above normal temperatures prevail for several days, the gradient builds up sufficiently to draw in cooler air from the coastal section. During prolonged periods of drought this late afternoon breeze may carry varying amounts of dust, and thermal instability sometimes causes the dust to rise as high as 7,000 feet.

Based on the 1951–1980 period, the average first occurrence of 32 degrees Fahrenheit in the fall is December 11 and the average last occurrence in the spring is January 31.

NORMALS, MEANS, AND EXTREMES

BAKERSFIELD, CALIFORNIA

LATITUDE: 35°25'N LONGITUDE: 119°03'W ELEVATION: FT. GRND 496 BARO 499 TIME ZONE: PACIFIC WBAN: 23155

	(a)	JAN	FEB	MAR	APR	MAY	JUNE	JULY	AUG	SEP	OCT	NOV	DEC	YEAR
TEMPERATURE °F:														
Normals														
-Daily Maximum		56.9	63.9	68.9	75.9	84.6	92.4	98.5	96.6	90.1	80.7	66.8	56.5	77.7
-Daily Minimum		38.6	42.6	45.8	50.1	57.3	64.0	69.6	68.5	63.5	54.8	44.7	38.3	53.2
-Monthly		47.8	53.3	57.4	63.0	71.0	78.2	84.1	82.6	76.8	67.8	55.8	47.5	65.4
Extremes														
-Record Highest	56	82	87	92	101	107	114	115	112	112	103	91	83	115
-Year		1984	1989	1969	1981	1982	1976	1950	1981	1955	1990	1949	1979	JUL 1950
-Record Lowest	56	20	25	31	34	37	45	52	52	45	29	28	19	19
-Year		1963	1990	1966	1984	1988	1988	1987	1942	1948	1971	1941	1990	DEC 1990
NORMAL DEGREE DAYS:														
Heating (base 65°F)		533	331	246	144	28	6	0	0	8	60	283	543	2182
Cooling (base 65°F)		0	0	11	84	214	402	592	546	362	147	7	0	2365
% OF POSSIBLE SUNSHINE														
MEAN SKY COVER (tenths)														
Sunrise - Sunset	48	6.7	6.0	5.6	4.6	3.3	1.8	1.3	1.4	1.8	3.0	4.9	6.5	3.9
MEAN NUMBER OF DAYS:														
Sunrise to Sunset														
-Clear	53	6.8	7.6	9.8	12.4	17.7	23.2	26.5	25.9	23.7	19.4	12.0	7.2	192.2
-Partly Cloudy	53	7.8	8.1	9.4	9.1	8.6	4.8	3.1	3.7	4.3	6.5	8.1	7.6	81.1
-Cloudy	53	16.4	12.5	11.8	8.5	4.6	2.1	1.5	1.4	2.0	5.2	9.9	16.1	91.9
Precipitation														
.01 inches or more	56	6.0	6.3	6.6	3.9	1.5	0.5	0.1	0.3	0.9	1.7	3.5	5.2	36.6
Snow,Ice pellets,hail														
1.0 inches or more	56	0.0	0.0	0.*	0.0	0.0	0.0	0.0	0.0	0.0	0.0	0.0	0.0	*
Thunderstorms	55	0.1	0.2	0.5	0.4	0.3	0.3	0.2	0.2	0.6	0.3	0.1	0.1	3.2
Heavy Fog Visibility														
1/4 mile or less	55	8.4	2.6	0.5	0.1	0.*	0.0	0.*	0.*	0.*	0.1	2.6	7.9	22.4
Temperature °F														
-Maximum														
90° and above	30	0.0	0.0	0.1	2.3	10.4	19.5	28.3	26.0	17.6	5.9	0.*	0.0	110.2
32° and below	30	0.0	0.0	0.0	0.0	0.0	0.0	0.0	0.0	0.0	0.0	0.0	0.0	0.0
-Minimum														
32° and below	30	4.8	1.2	0.1	0.0	0.0	0.0	0.0	0.0	0.0	0.1	0.4	5.0	11.6
0° and below	30	0.0	0.0	0.0	0.0	0.0	0.0	0.0	0.0	0.0	0.0	0.0	0.0	0.0
AVG. STATION PRESS.(mb)	10	1002.0	1001.7	999.0	998.6	995.4	994.8	994.7	994.7	995.0	998.4	1001.4	1003.2	998.2
RELATIVE HUMIDITY (%)														
Hour 04	19	83	78	72	65	55	50	48	53	57	63	75	83	65
Hour 10 (Local Time)	30	78	66	58	46	38	35	33	37	41	46	62	75	51
Hour 16	30	63	50	43	33	26	23	22	24	28	33	49	61	38
Hour 22	19	78	71	63	53	40	34	33	38	44	52	69	79	55
PRECIPITATION (inches):														
Water Equivalent														
-Normal		0.86	1.06	1.04	0.57	0.20	0.10	0.01	0.09	0.17	0.29	0.70	0.63	5.72
-Maximum Monthly	56	2.87	4.68	4.61	2.65	2.39	1.11	0.30	1.18	1.06	1.82	3.04	1.81	4.68
-Year		1943	1978	1938	1967	1971	1972	1965	1983	1976	1974	1960	1992	FEB 1978
-Minimum Monthly	56	T	0.03	T	0.00	0.00	0.00	0.00	0.00	0.00	0.00	0.00	0.00	0.00
-Year		1972	1967	1972	1966	1982	1983	1983	1981	1981	1978	1959	1989	DEC 1989
-Maximum in 24 hrs	56	1.38	3.02	1.68	1.00	1.40	1.10	0.30	1.08	0.63	1.51	1.54	1.15	3.02
-Year		1992	1978	1938	1943	1971	1972	1965	1983	1978	1940	1960	1974	FEB 1978
Snow,Ice pellets,hail														
-Maximum Monthly	56	T	T	1.5	0.0	0.0	0.0	0.0	0.0	0.0	0.0	0.0	T	1.5
-Year		1987	1990	1974									1990	MAR 1974
-Maximum in 24 hrs	56	T	T	1.5	0.0	0.0	0.0	0.0	0.0	0.0	0.0	0.0	T	1.5
-Year		1987	1990	1974									1990	MAR 1974
WIND:														
Mean Speed (mph)	43	5.2	5.8	6.5	7.1	7.9	7.9	7.2	6.8	6.2	5.5	5.1	5.0	6.4
Prevailing Direction through 1963		NW	ENE	NW	NW	NW	NW	NW	NW	WNW	NW	ENE	ENE	NW
Fastest Obs. 1 Min.														
-Direction (!!)	45	02	29	36	29	32	15	29	31	14	08	30	13	13
-Speed (MPH)	45	35	44	38	40	40	41	25	30	35	38	35	46	46
-Year		1960	1960	1973	1958	1990	1972	1950	1987	1976	1986	1985	1977	DEC 1977
Peak Gust														
-Direction (!!)	10	SE	SE	SE	NW	NW	NW	SE	S	N	E	NW	SE	SE
-Speed (mph)	10	48	58	49	43	45	35	33	49	39	48	49	56	58
-Date		1986	1986	1987	1984	1990	1990	1986	1987	1989	1986	1985	1987	FEB 1986

PRECIPITATION (inches) BAKERSFIELD, CALIFORNIA

YEAR	JAN	FEB	MAR	APR	MAY	JUNE	JULY	AUG	SEP	OCT	NOV	DEC	ANNUAL
1964	0.27	0.41	0.57	0.56	0.20	0.01	T	0.17	T	0.67	0.46	0.70	4.02
1965	0.74	0.17	1.17	1.65	0.02	T	0.30	T	0.10	0.00	1.05	1.60	6.80
1966	0.70	1.14	0.29	0.00	T	T	T	T	0.03	T	0.88	1.58	4.62
1967	0.96	0.03	0.52	2.65	0.28	0.20	T	0.00	0.11	0.00	1.76	0.54	7.05
1968	0.49	0.56	1.01	0.66	0.06	0.00	T	T	0.00	1.29	0.40	0.67	5.14
1969	2.12	2.83	0.29	1.10	0.08	T	T	T	T	T	0.42	0.16	7.00
1970	0.57	1.56	0.48	0.16	0.00	T	T	0.00	0.00	T	1.70	0.71	5.18
1971	0.53	0.35	0.42	0.56	2.39	0.00	0.00	0.12	0.02	0.09	0.12	1.17	5.77
1972	T	0.27	T	0.08	0.02	1.11	T	T	0.02	0.54	1.55	0.66	4.25
1973	2.07	0.49	2.49	0.18	T	0.00	0.00	0.00	0.00	0.16	0.64	0.79	6.82
1974	1.16	0.13	1.53	0.70	T	0.00	T	0.00	0.00	1.82	0.51	1.19	7.04
1975	0.06	1.60	0.60	0.93	T	0.00	0.00	0.05	T	0.48	0.25	0.13	4.10
1976	0.05	1.64	0.44	0.76	0.55	0.02	T	T	1.06	0.11	0.31	0.13	5.07
1977	0.58	0.07	1.28	T	0.59	0.06	0.02	1.03	0.00	T	0.09	1.80	5.52
1978	1.21	4.68	2.00	0.88	0.02	0.00	0.00	0.00	0.74	0.00	0.21	0.57	10.31
1979	1.80	1.41	1.97	T	T	0.00	0.00	0.00	0.35	0.28	0.16	0.22	6.19
1980	2.60	1.04	1.32	0.66	0.21	0.00	0.00	0.00	0.00	0.03	T	0.15	6.01
1981	0.93	0.78	2.15	0.56	0.18	0.00	0.00	0.00	0.00	0.83	0.41	0.23	6.07
1982	0.53	0.60	2.13	1.07	0.00	0.42	0.00	T	0.70	0.71	1.30	0.33	7.79
1983	2.21	1.49	2.62	0.57	0.01	0.00	0.00	1.18	0.18	0.14	1.31	1.15	10.86
1984	0.05	0.05	0.69	0.50	0.00	0.01	T	0.01	0.02	0.13	1.01	0.95	3.42
1985	0.38	0.48	0.48	T	0.14	0.44	T	0.00	0.24	0.18	1.65	0.27	4.26
1986	1.12	0.80	1.95	0.24	0.02	0.00	T	T	0.03	T	0.56	0.97	5.69
1987	1.61	0.89	1.07	0.10	0.04	0.31	0.00	0.07	0.01	0.18	1.40	0.83	6.51
1988	0.81	0.37	0.41	1.31	0.12	0.04	T	0.00	0.00	0.00	0.64	0.82	4.52
1989	0.16	0.81	0.86	T	0.45	0.00	0.00	T	0.49	0.04	0.07	0.00	2.88
1990	0.85	0.93	0.45	0.18	0.29	T	0.00	T	0.05	0.03	0.47	0.26	3.51
1991	0.62	0.13	4.33	0.06	T	0.00	0.00	T	0.01	0.30	0.01	1.04	6.50
1992	1.56	2.14	1.86	T	0.08	0.00	0.03	0.00	0.00	0.92	0.00	1.81	8.40
1993	2.33	2.02	1.76	T	0.00	0.48	0.00	T	0.00	0.17	0.79	0.62	8.17
Record Mean	1.04	1.01	1.08	0.59	0.28	0.08	0.01	0.03	0.12	0.33	0.55	0.79	5.91

AVERAGE TEMPERATURE (deg. F) BAKERSFIELD, CALIFORNIA

YEAR	JAN	FEB	MAR	APR	MAY	JUNE	JULY	AUG	SEP	OCT	NOV	DEC	ANNUAL
1964	46.4	51.0	54.8	62.0	67.6	76.9	84.1	82.7	74.0	71.0	51.2	50.7	64.4
1965	47.2	50.4	57.1	61.9	69.3	73.3	82.0	82.3	72.2	69.6	57.3	42.8	63.8
1966	46.8	49.0	57.5	67.3	72.6	78.0	81.4	84.7	74.8	67.1	57.7	46.3	65.2
1967	46.6	49.7	56.1	52.7	70.5	75.2	86.7	87.7	80.4	69.0	61.0	45.3	65.1
1968	47.9	58.9	59.8	65.4	70.2	81.2	86.6	79.8	76.9	66.2	55.7	47.0	66.3
1969	48.9	51.1	56.7	63.5	74.3	76.7	86.3	86.2	80.8	64.6	58.3	51.1	66.6
1970	54.0	56.3	58.8	60.0	74.0	80.3	88.3	84.8	77.5	67.5	59.0	49.1	67.4
1971	47.6	49.8	57.8	61.9	67.9	77.7	87.0	86.0	76.6	64.1	53.8	45.4	64.7
1972	41.7	54.9	63.4	63.4	72.4	80.3	85.0	82.8	75.2	65.8	52.6	43.6	65.1
1973	47.9	57.4	54.3	64.5	76.7	82.4	85.0	83.4	76.6	68.4	56.6	50.0	66.9
1974	51.6	52.9	59.6	63.9	73.2	81.6	85.8	84.6	83.0	70.5	56.7	46.8	67.5
1975	46.8	54.4	57.1	58.8	73.5	81.1	84.3	83.0	82.6	66.6	54.7	48.0	65.9
1976	49.9	55.5	57.4	61.3	75.3	79.7	85.5	79.1	78.3	71.0	59.4	51.1	67.0
1977	46.7	56.9	54.2	67.6	67.2	83.9	85.7	85.3	78.9	71.2	59.5	57.1	67.8
1978	54.8	56.2	62.6	61.4	73.2	79.9	85.9	85.0	76.7	75.2	57.2	46.2	67.8
1979	51.6	52.1	58.3	63.2	74.9	81.4	84.6	81.8	81.8	70.6	58.3	54.2	67.8
1980	52.8	55.6	55.2	62.7	67.4	73.9	85.1	82.6	77.4	71.6	57.7	50.1	66.0
1981	51.8	54.7	57.0	65.2	72.1	84.5	86.6	85.0	80.3	65.0	59.6	51.4	67.8
1982	45.7	55.5	57.9	64.2	76.1	79.0	87.1	84.8	77.0	68.7	52.1	46.4	66.2
1983	44.6	53.8	55.8	58.2	69.8	75.8	79.0	82.9	79.8	69.0	57.2	50.6	64.7
1984	48.1	50.6	57.0	57.8	70.6	74.3	85.3	82.4	80.2	61.6	54.5	47.2	64.1
1985	43.4	51.9	53.9	65.9	67.7	80.7	84.6	79.0	70.8	64.6	53.2	43.3	63.3
1986	52.8	54.7	59.3	61.1	69.7	77.9	80.7	83.7	70.1	65.7	56.5	47.1	64.9
1987	45.0	51.8	56.6	67.3	72.2	78.2	76.8	80.9	76.4	71.6	53.8	47.0	64.8
1988	47.8	54.2	58.8	64.1	68.4	75.4	86.1	82.0	77.0	70.0	54.7	47.2	65.5
1989	45.0	50.6	60.1	68.8	69.6	77.0	82.5	79.8	74.9	66.8	56.0	44.2	64.6
1990	47.5	49.3	59.1	66.8	69.2	77.2	84.9	81.1	76.1	69.6	54.2	43.0	64.8
1991	48.5	57.2	53.0	59.9	66.2	76.0	84.6	79.6	81.3	72.1	56.1	48.2	65.2
1992	43.8	56.6	58.9	66.9	76.7	77.5	81.3	84.3	77.7	69.8	55.7	46.5	66.3
1993	48.7	54.0	61.0	61.9	71.0	77.1	80.9	81.2	77.4	69.2	56.1	46.9	65.5
Record Mean	47.4	52.8	57.0	62.8	70.1	77.6	83.7	81.8	76.0	66.8	55.7	47.9	65.0
Max	57.3	64.2	69.1	76.1	84.3	92.7	99.5	97.5	91.1	81.2	68.6	58.3	78.3
Min	37.4	41.3	45.0	49.4	55.9	62.5	67.9	66.0	60.9	52.5	42.8	37.5	51.6

REFERENCE NOTES BAKERSFIELD, CALIFORNIA

GENERAL
T - TRACE AMOUNT.
BLANK ENTRIES DENOTE MISSING/UNREPORTED DATA.
# INDICATES A STATION OR INSTRUMENT RELOCATION.

SPECIFIC

(a) - LENGTH OF RECORD IN YEARS, ALTHOUGH INDIVIDUAL MONTHS MAY BE MISSING.
0.* OR * - THE VALUE IS BETWEEN 0.0 AND 0.05
NORMALS - BASED ON THE 1961-1990 RECORD PERIOD.
EXTREMES - DATES ARE THE MOST RECENT OCCURRENCE
WIND DIR.- NUMERALS SHOW TENS OF DEGRESS CLOCKWISE FROM TRUE NORTH. "00" INDICATES CALM.
RESULTANT DIRECTIONS ARE GIVEN TO WHOLE DEGREES.
BOLD VALUES INDICATE EXTREME VALUES WHICH OCCURRED AFTER THE ASOS SYSTEM WAS COMMISSIONED.

RECORD = PERIOD OF RECORD
RECORD MEAN PRECIPITATION IS THE MEAN OF ALL DAILY PRECIPITATION AMOUNTS DURING THE PERIOD OF RECORD.
RECORD MAX(MIN) TEMPERATURE IS THE MEAN OF ALL DAILY MAX(MIN) TEMPERATURES DURING THE PERIOD OF RECORD.
RECORD MEAN TEMPERATURE IS THE SUM OF THE RECORD MAX AND RECORD MIN DIVIDED BY 2.
AVERAGE TEMPERATURE IS THE SUM OF THE MEAN DAILY MAX AND MIN TEMPERATURE DIVIDED BY 2.

EXCEPTIONS

RECORD MEANS ARE THROUGH THE CURRENT YEAR, BEGINNING IN 1911 FOR TEMPERATURE
1889 FOR PRECIPITATION
1938 FOR SNOWFALL

HEATING DEGREE DAYS Base 65 deg. F **BAKERSFIELD, CALIFORNIA**

SEASON	JULY	AUG	SEP	OCT	NOV	DEC	JAN	FEB	MAR	APR	MAY	JUNE	TOTAL
1964-65	0	0	1	17	407	438	546	405	241	186	45	7	2293
1965-66	0	0	0	18	232	681	557	443	236	46	6	1	2220
1966-67	0	0	1	32	222	572	566	422	271	361	48	7	2502
1967-68	0	0	0	5	137	604	523	177	181	81	24	0	1732
1968-69	0	0	3	34	277	553	492	384	274	88	13	0	2118
1969-70	0	0	0	45	207	426	337	239	195	158	15	0	1622
1970-71	0	0	0	55	184	486	532	420	225	117	30	3	2052
1971-72	0	0	16	163	329	597	717	286	97	82	25	0	2312
1972-73	0	0	0	59	362	657	521	205	324	87	3	0	2218
1973-74	0	0	0	24	263	461	409	333	174	92	20	0	1776
1974-75	0	0	0	26	244	558	559	294	244	192	30	0	2147
1975-76	0	0	0	73	304	521	463	272	243	140	0	0	2016
1976-77	0	0	0	13	193	424	559	229	333	31	37	0	1819
1977-78	0	0	0	12	162	237	311	241	82	124	13	0	1182
1978-79	0	0	0	9	236	578	410	352	211	87	8	0	1891
1979-80	0	0	0	22	196	334	373	266	293	110	49	0	1643
1980-81	0	0	0	38	231	455	404	285	243	86	9	0	1751
1981-82	0	0	0	65	161	418	591	262	213	109	0	0	1819
1982-83	0	0	2	29	381	569	626	310	277	202	59	1	2456
1983-84	0	0	0	2	230	439	515	411	242	226	31	0	2096
1984-85	0	0	0	140	307	545	663	362	333	68	26	2	2446
1985-86	0	0	4	78	350	668	369	281	183	137	36	0	2106
1986-87	0	0	21	45	249	550	614	362	262	55	24	0	2182
1987-88	0	0	0	8	331	551	526	308	192	82	60	11	2069
1988-89	0	0	0	18	308	544	614	400	162	42	22	0	2110
1989-90	0	0	3	63	264	636	538	432	187	22	9	1	2155
1990-91	0	0	0	13	316	677	503	212	368	154	69	0	2312
1991-92	0	0	0	72	271	514	649	237	181	25	0	1	1950
1992-93	0	0	0	7	273	567	501	302	128	113	9	5	1905
1993-94	0	0	2	5	262	553							

COOLING DEGREE DAYS Base 65 deg. F **BAKERSFIELD, CALIFORNIA**

YEAR	JAN	FEB	MAR	APR	MAY	JUNE	JULY	AUG	SEP	OCT	NOV	DEC	TOTAL
1969	0	0	25	49	308	360	668	663	481	43	13	0	2610
1970	2	0	9	13	300	466	727	621	385	140	11	0	2674
1971	0	0	7	32	126	389	691	660	372	144	0	0	2421
1972	0	0	52	42	261	466	627	559	313	90	0	0	2410
1973	0	0	0	79	371	528	624	580	355	137	18	0	2692
1974	0	0	12	63	281	505	651	616	549	206	0	0	2883
1975	0	0	7	14	299	490	606	564	538	130	1	0	2649
1976	0	6	15	34	326	447	643	445	406	204	32	0	2558
1977	0	10	2	115	113	577	651	635	423	211	4	1	2742
1978	0	3	14	23	273	451	655	628	358	332	11	0	2748
1979	0	0	12	40	321	500	614	524	513	205	1	3	2733
1980	4	2	0	49	129	274	629	551	380	248	16	0	2282
1981	3	0	0	100	237	591	678	630	464	71	6	2	2782
1982	0	2	0	92	351	425	694	620	368	155	0	0	2707
1983	0	0	0	6	214	332	443	563	449	135	2	0	2144
1984	0	0	0	18	211	287	638	541	463	42	0	0	2200
1985	0	0	0	100	118	481	614	440	188	71	3	0	2015
1986	0	0	11	25	190	393	493	589	180	71	0	0	1952
1987	0	0	9	130	253	403	374	499	351	218	0	0	2237
1988	0	2	8	61	173	333	660	533	366	180	5	0	2321
1989	0	2	18	163	170	368	550	465	306	126	0	0	2168
1990	0	0	12	83	145	373	626	505	337	163	2	0	2246
1991	0	0	0	9	113	340	614	458	495	301	9	0	2339
1992	0	0	0	91	368	380	512	603	385	161	0	0	2500
1993	0	0	9	28	204	376	497	508	381	143	1	0	2147

SNOWFALL (inches) **BAKERSFIELD, CALIFORNIA**

SEASON	JULY	AUG	SEP	OCT	NOV	DEC	JAN	FEB	MAR	APR	MAY	JUNE	TOTAL
1970-71	0.0	0.0	0.0	0.0	0.0	0.0	0.0	0.0	0.0	0.0	0.0	0.0	0.0
1971-72	0.0	0.0	0.0	0.0	0.0	0.0	0.0	0.0	0.0	0.0	0.0	0.0	0.0
1972-73	0.0	0.0	0.0	0.0	0.0	0.0	T	0.0	0.0	0.0	0.0	0.0	T
1973-74	0.0	0.0	0.0	0.0	0.0	0.0	0.0	0.0	1.5	0.0	0.0	0.0	1.5
1974-75	0.0	0.0	0.0	0.0	0.0	0.0	0.0	0.0	0.0	0.0	0.0	0.0	0.0
1975-76	0.0	0.0	0.0	0.0	0.0	0.0	0.0	0.0	0.0	0.0	0.0	0.0	0.0
1976-77	0.0	0.0	0.0	0.0	0.0	0.0	0.0	0.0	0.0	0.0	0.0	0.0	0.0
1977-78	0.0	0.0	0.0	0.0	0.0	0.0	0.0	0.0	0.0	0.0	0.0	0.0	0.0
1978-79	0.0	0.0	0.0	0.0	0.0	0.0	T	0.0	0.0	0.0	0.0	0.0	T
1979-80	0.0	0.0	0.0	0.0	0.0	0.0	0.0	0.0	0.0	0.0	0.0	0.0	0.0
1980-81	0.0	0.0	0.0	0.0	0.0	0.0	0.0	0.0	0.0	0.0	0.0	0.0	0.0
1981-82	0.0	0.0	0.0	0.0	0.0	0.0	0.0	0.0	T	0.0	0.0	0.0	T
1982-83	0.0	0.0	0.0	0.0	0.0	0.0	0.0	0.0	0.0	0.0	0.0	0.0	0.0
1983-84	0.0	0.0	0.0	0.0	0.0	0.0	0.0	0.0	0.0	0.0	0.0	0.0	0.0
1984-85	0.0	0.0	0.0	0.0	0.0	0.0	0.0	0.0	T	0.0	0.0	0.0	T
1985-86	0.0	0.0	0.0	0.0	0.0	0.0	0.0	0.0	0.0	0.0	0.0	0.0	0.0
1986-87	0.0	0.0	0.0	0.0	0.0	0.0	T	0.0	0.0	0.0	0.0	0.0	T
1987-88	0.0	0.0	0.0	0.0	0.0	0.0	0.0	0.0	0.0	0.0	0.0	0.0	0.0
1988-89	0.0	0.0	0.0	0.0	0.0	0.0	0.0	T	0.0	0.0	0.0	0.0	T
1989-90	0.0	0.0	0.0	0.0	0.0	0.0	0.0	T	0.0	0.0	0.0	0.0	T
1990-91	0.0	0.0	0.0	0.0	0.0	T	0.0	0.0	T	0.0	0.0	0.0	T
1991-92	0.0	0.0	0.0	0.0	0.0	0.0	0.0	0.0	0.0	0.0	0.0	0.0	0.0
1992-93	0.0	0.0	0.0	0.0	0.0	0.0	0.0	0.0	0.0	0.0	0.0	0.0	0.0
1993-94	0.0	0.0	0.0	0.0	0.0	0.0							
Record Mean	0.0	0.0	0.0	0.0	0.0	T	T	T	T	0.0	0.0	0.0	T

BISHOP, CALIFORNIA

The station at Bishop is located at the Municipal Airport 2 1/2 miles east of the town, 1 mile west of the Owens River, on the floor of the Owens Valley, which is orientated northwest to southeast and at this point is 12 miles wide, level, and semi-arid. Peaks of the 12,000 to 14,000 feet Sierra Nevadas are 25 miles west, and the 12,000 to 14,000 feet White Mountains are 10 miles east. The northern end of the valley is partly cut off by 6,000 to 8,000 feet mountains about 45 miles distant. The southern end of the valley makes a gradual descent to the Mojave Desert about 150 miles distant.

During the summer and autumn, the Mojave Desert causes an early morning and late evening northerly wind. Conversely, in the heat of the afternoon, it causes a southerly wind that is occasionally strong. Summer skies are mostly clear with thunderstorms from May through August. The days are hot and dry, the nights cool.

Winter and spring, although seasons of adverse weather, are quite mild. The skies are partly cloudy with the years greatest amounts of precipitation falling during the months of November through April. At times, strong northerly winds blow, especially in the spring during the months of February, March, and April. East and west winds frequently give pronounced foehn effects and turbulence.

During the winter and spring, strong westerly winds aloft, flowing over the Sierra Nevadas, create the Sierra Wave, known to sailplane pilots the world over.

Based on the 1951–1980 period, the average first occurrence of 32 degrees Fahrenheit in the fall is October 13 and the average last occurrence in the spring is May 7.

NORMALS, MEANS, AND EXTREMES

BISHOP, CALIFORNIA

LATITUDE: 37°22'N LONGITUDE: 118°22' W ELEVATION: FT. GRND 4110 BARO 4113 TIME ZONE: PACIFIC WBAN: 23157

	(a)	JAN	FEB	MAR	APR	MAY	JUNE	JULY	AUG	SEP	OCT	NOV	DEC	YEAR
TEMPERATURE °F:														
Normals														
-Daily Maximum		53.5	58.5	63.4	71.1	80.6	90.5	97.2	94.9	86.8	76.3	62.4	53.8	74.1
-Daily Minimum		22.0	26.3	30.2	35.7	43.8	51.2	56.1	54.2	46.8	37.4	27.8	21.7	37.8
-Monthly		37.8	42.4	46.8	53.4	62.2	70.9	76.7	74.6	66.8	56.9	45.2	37.8	56.0
Extremes														
-Record Highest	46	77	81	87	93	101	109	109	107	106	97	84	78	109
-Year		1948	1986	1966	1989	1951	1954	1972	1993	1950	1980	1988	1958	JUL 1972
-Record Lowest	46	-7	-2	9	15	25	29	34	37	26	16	5	-8	-8
-Year		1982	1969	1971	1953	1964	1988	1987	1959	1986	1970	1958	1990	DEC 1990
NORMAL DEGREE DAYS:														
Heating (base 65°F)		843	633	564	361	140	16	0	0	51	265	594	843	4310
Cooling (base 65°F)		0	0	0	13	53	193	363	298	105	14	0	0	1039
% OF POSSIBLE SUNSHINE														
MEAN SKY COVER (tenths)														
Sunrise - Sunset	29	5.3	5.0	4.7	4.3	4.1	2.6	2.4	2.2	1.9	2.9	4.1	4.5	3.7
MEAN NUMBER OF DAYS:														
Sunrise to Sunset														
-Clear	29	11.5	11.0	13.0	13.7	14.8	20.2	21.9	22.9	23.2	19.7	14.9	14.3	201.0
-Partly Cloudy	29	8.0	7.5	8.9	9.0	9.8	6.8	6.7	6.0	4.7	6.7	7.9	7.4	89.3
-Cloudy	29	11.6	9.8	9.0	7.3	6.4	3.0	2.4	2.2	2.1	4.7	7.2	9.3	75.0
Precipitation														
.01 inches or more	46	3.9	3.3	3.2	2.4	2.5	1.5	2.0	1.7	1.9	1.6	2.5	2.9	29.4
Snow,Ice pellets,hail														
1.0 inches or more	46	1.0	0.3	0.3	0.1	0.*	0.0	0.0	0.0	0.0	0.*	0.2	0.5	2.4
Thunderstorms	19	0.0	0.0	0.0	0.2	1.4	1.5	5.2	2.9	1.0	0.5	0.1	0.0	12.7
Heavy Fog Visibility														
1/4 mile or less	19	0.1	0.0	0.0	0.1	0.0	0.0	0.0	0.0	0.0	0.0	0.1	0.1	0.3
Temperature °F														
-Maximum														
90° and above	46	0.0	0.0	0.0	0.4	5.0	18.3	29.1	26.5	13.9	1.4	0.0	0.0	94.8
32° and below	46	0.6	0.1	0.0	0.0	0.0	0.0	0.0	0.0	0.0	0.0	0.0	0.3	1.0
-Minimum														
32° and below	46	29.4	24.2	20.2	8.2	1.1	0.1	0.0	0.0	0.3	6.2	23.5	29.5	142.6
0° and below	46	0.2	0.*	0.0	0.0	0.0	0.0	0.0	0.0	0.0	0.0	0.0	0.2	0.5
AVG. STATION PRESS. (mb)	11	875.2	874.7	871.0	871.5	870.9	871.6	872.9	873.3	873.5	875.1	875.3	876.5	873.5
RELATIVE HUMIDITY (%)														
Hour 04														
Hour 10 (Local Time)	39	49	41	30	23	21	18	19	20	22	25	35	45	29
Hour 16	32	35	27	21	16	16	13	14	13	14	18	26	34	21
Hour 22														
PRECIPITATION (inches):														
Water Equivalent														
-Normal		1.11	0.95	0.39	0.26	0.29	0.18	0.23	0.18	0.24	0.13	0.57	0.84	5.37
-Maximum Monthly	46	8.93	6.01	2.94	2.26	1.30	1.29	1.47	0.64	1.18	1.58	2.59	5.79	8.93
-Year		1969	1969	1991	1956	1962	1982	1976	1983	1975	1957	1960	1966	JAN 1969
-Minimum Monthly	46	0.00	T	0.00	0.00	0.00	0.00	0.00	0.00	0.00	0.00	0.00	0.00	0.00
-Year		1976	1967	1972	1973	1983	1981	1982	1980	1974	1973	1976	1975	MAY 1983
-Maximum in 24 hrs	46	3.32	3.64	1.77	1.58	0.95	0.72	0.86	0.46	0.73	1.05	1.79	3.35	3.64
-Year		1952	1969	1991	1982	1953	1982	1976	1977	1982	1957	1950	1966	FEB 1969
Snow,Ice pellets,hail														
-Maximum Monthly	45	23.2	31.9	14.5	8.8	2.3	0.0	0.0	0.0	T	1.8	3.9	13.2	31.9
-Year		1969	1969	1952	1956	1964				1955	1978	1964	1967	FEB 1969
-Maximum in 24 hrs	45	18.0	14.2	7.5	8.8	2.3	0.0	0.0	0.0	T	1.8	3.9	6.7	18.0
-Year		1969	1976	1952	1956	1964				1955	1978	1964	1967	JAN 1969
WIND:														
Mean Speed (mph)					6.7									
Prevailing Direction														
Fastest Mile														
-Direction (!!)														
-Speed (MPH)	9	51	52	59	60	58	60	55	75	53	56	56	66	75
-Year		1983	1975	1977	1980	1979	1975	1980	1976	1982	1979	1975	1975	AUG 1976
Peak Gust														
-Direction (!!)	10		W											
-Speed (mph)	10	60	63	58	62	62	54	60	70	47	52	66	68	70
-Date		1992	1986	1985	1986	1985	1992	1987	1990	1989	1991	1987	1988	AUG 1990

PRECIPITATION (inches)

BISHOP, CALIFORNIA

YEAR	JAN	FEB	MAR	APR	MAY	JUNE	JULY	AUG	SEP	OCT	NOV	DEC	ANNUAL
1964	0.49	T	0.03	0.49	0.81	T	T	0.05	0.00	0.37	0.22	0.24	2.70
1965	0.48	0.02	0.05	0.49	0.02	0.22	0.59	0.61	0.02	0.00	1.90	2.16	6.56
1966	0.00	0.01	T	T	0.18	T	T	0.10	0.18	0.00	0.27	5.79	6.53
1967	1.64	T	0.50	0.47	0.02	T	0.62	0.03	0.26	0.00	0.30	0.52	4.36
1968	0.01	0.03	0.10	0.01	0.01	T	0.70	0.39	0.00	0.08	0.01	0.48	1.82
1969	8.93	6.01	0.68	0.11	0.27	0.36	0.31	0.04	T	0.02	0.16	0.20	17.09
1970	0.71	0.54	0.05	0.44	T	0.04	0.03	0.01	0.00	0.00	1.64	0.22	3.68
1971	0.01	0.11	0.27	0.06	1.04	0.00	0.14	0.13	0.01	0.01	0.04	1.85	3.67
1972	T	0.00	0.00	T	0.11	0.25	0.04	0.09	0.36	0.90	0.68	0.01	2.44
1973	3.02	1.59	0.32	0.00	0.09	0.14	0.00	0.01	0.00	0.00	1.94	0.60	7.71
1974	1.48	0.00	1.75	0.21	0.33	0.00	0.16	T	0.00	0.80	T	0.64	5.37
1975	T	0.20	0.69	0.20	T	0.07	T	0.10	1.18	0.09	0.02	0.00	2.55
1976	0.00	1.37	0.05	0.05	0.59	0.17	1.47	T	0.94	0.02	0.00	T	4.66
1977	0.77	0.22	0.04	0.02	0.60	0.61	T	0.51	T	T	0.05	2.53	5.35
1978	2.68	3.33	1.64	0.22	0.00	0.02	0.02	0.01	0.51	0.18	0.51	0.50	9.62
1979	0.45	0.64	0.49	T	T	T	0.03	0.02	0.25	0.07	0.13	0.57	2.65
1980	1.56	2.72	0.28	0.43	0.10	T	0.35	0.00	0.14	T	0.08	1.25	6.91
1981	0.65	0.11	0.85	0.68	0.88	0.00	0.00	0.04	0.03	0.09	1.30	0.11	4.74
1982	1.43	0.02	0.50	1.62	0.08	1.29	0.00	0.51	0.74	0.68	0.87	2.67	10.41
1983	1.82	1.29	1.20	0.22	0.00	T	0.05	0.64	0.40	0.08	1.31	1.14	8.15
1984	T	0.36	0.09	0.02	T	0.04	1.04	0.58	T	0.16	1.97	0.85	5.11
1985	0.25	0.01	0.06	0.00	0.00	0.67	0.31	0.00	0.34	0.05	0.95	0.55	3.19
1986	0.86	3.04	1.00	0.65	T	0.00	0.31	0.06	0.12	0.00	0.03	0.08	6.15
1987	0.42	0.31	0.03	0.04	0.54	0.16	0.18	0.03	0.01	0.13	1.67	0.60	4.12
1988	0.87	0.30	0.07	0.63	0.12	0.23	T	T	0.50	0.00	0.12	0.68	3.52
1989	0.06	0.12	0.04	0.00	1.04	0.04	0.00	0.01	0.24	0.00	0.26	0.00	1.81
1990	0.95	0.50	0.00	0.56	0.21	0.15	0.26	0.45	0.28	0.00	0.00	0.00	3.36
1991	T	0.07	2.94	0.07	T	0.02	0.00	0.00	0.21	0.69	0.00	0.58	4.58
1992	0.38	1.31	0.67	0.00	0.06	0.30	0.12	0.06	0.05	0.53	0.00	1.50	4.98
1993	2.03	2.62	0.91	0.00	0.04	0.00	0.00	T	0.00	0.06	0.12	0.08	5.86
Record Mean	1.25	0.95	0.63	0.28	0.28	0.12	0.14	0.13	0.19	0.21	0.47	0.89	5.54

AVERAGE TEMPERATURE (deg. F)

BISHOP, CALIFORNIA

YEAR	JAN	FEB	MAR	APR	MAY	JUNE	JULY	AUG	SEP	OCT	NOV	DEC	ANNUAL
1964	35.3	42.4	44.5	52.7	60.0	70.0	77.0	75.5	65.9	61.6	41.8	39.8	55.5
1965	41.0	43.8	46.7	53.1	59.4	67.1	74.7	74.2	63.2	60.9	46.3	36.7	55.6
1966	34.9	39.2	49.6	57.4	66.7	71.4	75.6	77.3	67.1	58.4	46.5	39.9	57.0
1967	39.8	43.6	47.5	44.9	61.8	67.9	78.5	77.9	68.1	59.3	49.3	34.0	56.0
1968	37.9	47.3	48.2	53.9	63.3	73.5	77.8	71.7	68.0	58.5	45.9	33.9	56.7
1969	38.5	30.1	40.1	54.1	65.6	68.9	76.5	76.8	69.4	52.4	46.5	39.3	54.8
1970	37.9	45.3	46.9	49.5	64.2	70.6	77.8	77.9	65.0	54.9	45.8	35.4	55.9
1971	41.5	43.0	47.4	52.9	57.5	70.4	78.4	78.1	66.5	52.5	43.2	33.5	55.4
1972	37.7	44.8	55.0	53.7	62.9	72.6	78.8	74.6	65.2	53.8	43.0	34.8	56.4
1973	31.7	40.6	42.8	52.2	65.4	72.3	77.2	73.7	66.7	56.2	43.4	40.5	55.3
1974	29.7	41.8	49.0	52.7	64.6	73.9	76.4	73.7	70.5	55.7	45.8	36.9	55.9
1975	39.4	41.4	44.2	46.3	61.6	70.1	77.7	73.0	69.6	54.3	43.5	42.2	55.3
1976	40.0	40.3	46.3	50.7	64.7	69.3	76.4	70.2	66.8	57.6	50.3	39.1	56.0
1977	38.0	46.1	42.8	56.8	55.6	74.0	76.8	76.9	66.7	59.4	49.6	42.0	57.1
1978	39.0	42.7	49.7	50.9	60.7	70.3	76.1	74.6	63.8	60.2	43.0	35.6	55.6
1979	34.8	38.7	48.2	55.1	64.0	70.4	76.5	72.6	70.3	57.9	43.6	38.9	55.9
1980	40.2	42.5	43.2	53.0	57.4	67.4	76.6	74.1	66.5	56.5	45.9	42.7	55.5
1981	39.4	44.7	46.2	56.8	63.6	75.8	77.6	76.7	70.1	53.1	46.8	42.7	57.8
1982	31.9	43.5	43.8	52.1	62.3	68.3	75.3	74.4	64.7	54.5	40.4	37.3	54.0
1983	39.7	43.1	46.2	47.9	61.0	70.0	72.5	72.6	68.4	57.3	45.0	40.6	55.4
1984	42.4	43.1	50.6	52.0	68.6	70.5	78.0	75.1	69.4	52.3	42.4	32.1	56.4
1985	38.1	41.7	44.3	57.8	62.6	74.4	78.1	73.2	61.9	54.4	41.8	38.9	55.6
1986	42.3	44.0	50.0	53.0	62.8	72.1	73.3	75.6	60.6	53.9	45.6	38.1	55.9
1987	36.7	41.1	45.8	56.9	62.2	71.6	72.7	74.9	67.9	59.5	44.9	32.4	55.6
1988	36.8	44.9	49.0	53.8	61.1	70.4	79.2	75.1	66.4	61.8	45.3	34.6	56.5
1989	35.3	38.9	51.2	60.5	63.0	71.0	78.1	72.2	66.2	54.8	46.9	40.9	56.6
1990	38.2	39.1	50.5	58.2	61.3	71.7	77.7	73.3	67.8	57.7	45.8	31.6	56.1
1991	37.3	45.8	41.9	51.6	57.8	69.1	77.2	74.5	69.3	59.6	46.1	38.6	55.7
1992	37.5	44.0	47.5	59.0	66.6	68.6	74.7	76.3	69.1	60.2	43.5	34.3	56.8
1993	30.8	37.9	50.1	55.2	64.1	68.7	75.1	73.6	66.5	57.5	42.5	36.8	54.9
Record Mean	37.1	40.9	46.7	53.9	61.3	69.4	75.9	73.5	66.7	56.7	45.0	38.8	55.5
Max	52.9	56.1	63.6	71.8	79.7	89.2	96.6	94.0	87.1	76.5	63.2	56.3	73.9
Min	21.2	25.7	29.7	35.9	42.9	49.7	55.3	52.9	46.2	36.9	26.9	21.4	37.1

REFERENCE NOTES

BISHOP, CALIFORNIA

GENERAL
T - TRACE AMOUNT.
BLANK ENTRIES DENOTE MISSING/UNREPORTED DATA.
# INDICATES A STATION OR INSTRUMENT RELOCATION.

SPECIFIC

(a) - LENGTH OF RECORD IN YEARS, ALTHOUGH INDIVIDUAL MONTHS MAY BE MISSING.
0.* OR * - THE VALUE IS BETWEEN 0.0 AND 0.05
NORMALS - BASED ON THE 1961-1990 RECORD PERIOD.
EXTREMES - DATES ARE THE MOST RECENT OCCURRENCE
WIND DIR.- NUMERALS SHOW TENS OF DEGRESS CLOCKWISE FROM TRUE NORTH. "00" INDICATES CALM.
RESULTANT DIRECTIONS ARE GIVEN TO WHOLE DEGREES.
BOLD VALUES INDICATE EXTREME VALUES WHICH OCCURRED AFTER THE ASOS SYSTEM WAS COMMISSIONED.

RECORD = PERIOD OF RECORD
RECORD MEAN PRECIPITATION IS THE MEAN OF ALL DAILY PRECIPITATION AMOUNTS DURING THE PERIOD OF RECORD.
RECORD MAX(MIN) TEMPERATURE IS THE MEAN OF ALL DAILY MAX(MIN) TEMPERATURES DURING THE PERIOD OF RECORD.
RECORD MEAN TEMPERATURE IS THE SUM OF THE RECORD MAX AND RECORD MIN DIVIDED BY 2.
AVERAGE TEMPERATURE IS THE SUM OF THE MEAN DAILY MAX AND MIN TEMPERATURE DIVIDED BY 2.

EXCEPTIONS

1. THUNDERSTORMS AND HEAVY FOG ARE THROUGH 1964 AND MAY BE INCOMPLETE, DUE TO PART-TIME OPERATIONS.
2. MEAN SKY COVER, AND DAYS CLEAR-PARTLY CLOUDY-CLOUDY ARE THROUGH 1981.

RECORD MEANS ARE THROUGH THE CURRENT YEAR, BEGINNING IN 1912 FOR TEMPERATURE
1912 FOR PRECIPITATION
1948 FOR SNOWFALL

HEATING DEGREE DAYS Base 65 deg. F — BISHOP, CALIFORNIA

SEASON	JULY	AUG	SEP	OCT	NOV	DEC	JAN	FEB	MAR	APR	MAY	JUNE	TOTAL
1964-65	0	0	27	133	688	777	732	585	562	355	201	38	4098
1965-66	0	0	85	146	553	872	926	718	468	223	36	12	4039
1966-67	0	0	34	205	549	771	772	592	535	595	160	44	4257
1967-68	0	0	9	174	462	953	833	509	514	324	113	19	3910
1968-69	0	15	47	194	567	958	812	971	765	322	53	15	4719
1969-70	0	0	2	388	549	791	833	548	551	457	78	27	4224
1970-71	0	0	59	312	572	911	719	607	539	356	234	29	4338
1971-72	0	0	100	391	648	972	841	578	305	330	103	1	4269
1972-73	0	1	36	343	654	930	1028	676	681	376	66	2	4793
1973-74	0	6	20	264	642	751	1088	642	488	364	87	0	4352
1974-75	0	0	6	277	568	864	787	654	639	552	147	22	4516
1975-76	0	1	0	330	640	702	767	711	573	423	56	21	4224
1976-77	1	5	24	227	432	798	829	524	682	245	304	2	4073
1977-78	0	0	64	169	455	704	796	622	464	419	151	1	3845
1978-79	0	1	93	165	653	906	931	731	514	291	97	18	4400
1979-80	0	0	6	232	634	805	759	646	667	352	236	44	4381
1980-81	0	0	27	298	567	684	787	563	574	249	90	4	3843
1981-82	0	0	8	361	541	685	1021	593	649	380	111	31	4380
1982-83	3	0	109	317	735	853	775	608	574	508	201	6	4689
1983-84	0	0	45	231	594	749	693	627	440	383	44	8	3814
1984-85	0	0	31	387	668	1014	828	645	632	208	84	20	4517
1985-86	0	0	113	326	687	802	699	583	458	354	134	0	4156
1986-87	1	0	199	336	576	824	871	662	588	239	116	4	4416
1987-88	15	0	1	187	597	1003	869	578	491	330	173	41	4285
1988-89	0	0	69	106	582	933	916	723	421	157	107	7	4021
1989-90	0	0	34	305	536	742	823	719	440	205	122	8	3934
1990-91	0	8	37	227	571	1029	852	530	711	391	225	3	4584
1991-92	0	1	6	200	559	813	845	602	536	186	23	36	3807
1992-93	0	0	0	146	638	947	1056	753	454	291	62	54	4401
1993-94	0	0	55	231	669	868							

COOLING DEGREE DAYS Base 65 deg. F — BISHOP, CALIFORNIA

YEAR	JAN	FEB	MAR	APR	MAY	JUNE	JULY	AUG	SEP	OCT	NOV	DEC	TOTAL
1969	0	0	0	0	79	138	363	372	141	3	0	0	1096
1970	0	0	0	0	60	201	404	407	66	5	0	0	1143
1971	0	0	0	0	8	197	423	411	148	12	0	0	1199
1972	0	0	1	0	45	237	437	306	53	0	0	0	1079
1973	0	0	0	1	86	230	385	281	80	0	0	0	1063
1974	0	0	0	0	83	275	358	280	179	0	0	0	1175
1975	0	0	0	0	52	183	402	258	148	5	0	0	1048
1976	0	0	0	0	57	157	361	167	82	3	0	0	827
1977	0	0	0	5	18	281	373	378	120	1	0	0	1176
1978	0	0	0	0	25	167	353	309	64	25	0	0	943
1979	0	0	0	0	70	187	365	243	172	17	0	0	1054
1980	0	0	0	0	6	122	370	289	78	42	0	0	907
1981	0	0	0	9	53	334	398	371	170	0	0	0	1335
1982	0	0	0	0	35	134	331	299	106	0	0	0	905
1983	0	0	0	0	83	164	238	241	156	0	0	0	882
1984	0	0	0	0	159	177	411	320	168	0	0	0	1235
1985	0	0	0	0	17	312	417	260	27	7	0	0	1040
1986	0	0	0	0	73	220	264	334	78	0	0	0	969
1987	0	0	0	1	35	208	264	313	94	24	0	0	939
1988	0	0	0	0	60	209	447	321	117	11	0	0	1165
1989	0	0	0	29	49	197	415	231	77	0	0	0	998
1990	0	0	0	7	11	216	401	273	129	8	0	0	1045
1991	0	0	0	0	10	136	388	302	140	39	0	0	1015
1992	0	0	0	14	81	153	306	359	128	7	0	0	1048
1993	0	0	0	4	42	170	320	275	106	9	0	0	926

SNOWFALL (inches) — BISHOP, CALIFORNIA

SEASON	JULY	AUG	SEP	OCT	NOV	DEC	JAN	FEB	MAR	APR	MAY	JUNE	TOTAL
1964-65	0.0	0.0	0.0	0.0	3.9	0.5	0.2	0.0	0.0	1.4	T	0.0	6.0
1965-66	0.0	0.0	0.0	0.0	T	0.0	0.0	T	0.0	0.0	0.0	0.0	T
1966-67	0.0	0.0	0.0	0.0	0.0	0.0	7.6	T	0.5	2.2	0.0	0.0	10.3
1967-68	0.0	0.0	0.0	0.0	T	13.2	T	0.0	T	0.0	0.0	0.0	13.2
1968-69	0.0	0.0	0.0	0.0	0.0	2.2	23.2	31.9	2.0	0.0	0.0	0.0	59.3
1969-70	0.0	0.0	0.0	0.0	0.0	0.0	0.0	4.0	0.1	T	0.0	0.0	4.1
1970-71	0.0	0.0	0.0	0.0	T	2.3	0.1	T	0.0	0.1	0.0	0.0	2.5
1971-72	0.0	0.0	0.0	0.0	0.0	1.8	T	0.0	0.0	0.0	0.0	0.0	1.8
1972-73	0.0	0.0	0.0	0.0	T	0.0	6.0	T	1.0	0.0	0.0	0.0	7.0
1973-74	0.0	0.0	0.0	0.0	T	1.5	13.8	0.0	1.1	0.0	T	0.0	16.4
1974-75	0.0	0.0	0.0	0.0	0.0	T	0.0	T	T	0.0	0.0	0.0	T
1975-76	0.0	0.0	0.0	0.0	0.1	0.0	0.0	21.0	0.4	0.0	0.0	0.0	21.5
1976-77	0.0	0.0	0.0	0.0	0.0	T	2.6	T	0.4	0.0	T	0.0	3.0
1977-78	0.0	0.0	0.0	0.0	T	4.5	0.4	1.5	T	T	0.0	0.0	6.4
1978-79	0.0	0.0	0.0	1.8	2.1	0.6	4.8	T	0.8	0.0	T	0.0	10.1
1979-80	0.0	0.0	0.0	0.0	0.0	5.5	0.8	0.0	T	1.9	0.4	0.0	8.6
1980-81	0.0	0.0	0.0	0.0	0.0	T	6.0	T	3.8	0.3	0.0	0.0	10.1
1981-82	0.0	0.0	0.0	0.0	2.1	0.0	20.7	T	4.8	0.0	0.0	0.0	27.6
1982-83	0.0	0.0	0.0	0.0	1.2	T	8.1	T	1.1	0.0	0.0	0.0	10.4
1983-84	0.0	0.0	0.0	0.0	1.0	0.0	T	T	0.0	0.0	0.0	0.0	1.0
1984-85	0.0	0.0	0.0	0.0	1.5	7.9	T	T	0.5	0.0	0.0	0.0	9.9
1985-86	0.0	0.0	0.0	0.0	0.6	1.3	0.0	T	T	0.0	0.0	0.0	1.9
1986-87	0.0	0.0	0.0	0.0	0.0	0.0	0.9	0.5	0.0	0.0	T	0.0	1.4
1987-88	0.0	0.0	0.0	0.0	0.0	5.2	8.9	0.0	0.0	0.0	0.4	0.0	14.5
1988-89						0.0	0.0	0.7	0.0	0.0	0.0	0.0	
1989-90	0.0	0.0	0.0	0.0	0.0	0.0	0.1	T	0.0	0.0	0.0	0.0	0.1
1990-91	0.0	0.0	0.0	0.0	0.0	0.0	0.0	0.0		0.0	0.0	0.0	
1991-92	0.0	0.0	0.0	0.0	0.0			0.2		0.0	0.0	0.0	
1992-93	0.0	0.0	0.0	0.0	0.0	1.0	15.8	8.3	0.0	0.0	0.0	0.0	25.1
1993-94	0.0	0.0	0.0	0.0	T	0.0							
Record Mean	0.0	0.0	T	T	0.3	1.3	4.0	1.6	0.8	0.3	0.1	0.0	8.4

EUREKA, CALIFORNIA

Humboldt Bay is one-quarter mile north and one mile west of the station. There are no hills in Eureka of any consequence. The land slopes upward gently from the Bay toward the Coast Range, which begins about 3 miles east of the station and reaches the top of its first ridge approximately 10 miles to the east. The elevation of the ridge is 2,000 feet and extends in a semicircle from a point 20 miles north of Eureka to a point 25 miles south.

The climate of Eureka is completely maritime with high humidity prevailing the entire year. There are definite rainy and dry seasons. The rainy season begins in October and continues through April, accounting for about 90 percent of the annual precipitation. The dry season from May through September is marked by considerable fog or low cloudiness that usually clears in the late morning and sunny weather is generally the case during the early afternoon hours.

Temperatures are moderate the entire year. Although record highs have reached the mid 80s and record lows near 20 degrees, the usual yearly range is from lows in the mid 30s to highs in the mid 70s.

The principal industries are lumbering, fishing, tourism, and dairy farming. There is very little truck farming due to the low temperatures and lack of sunshine, however, the climate is nearly ideal for berries and flowers.

Based on the 1951-1980 period, the average first occurrence of 32 degrees Fahrenheit in the fall is December 10 and the average last occurrence in the spring is February 6.

NORMALS, MEANS, AND EXTREMES

EUREKA, CALIFORNIA

LATITUDE: 40°48'N LONGITUDE: 124°10' W ELEVATION: FT. GRND 58 BARO 00079 TIME ZONE: PACIFIC WBAN: 24213

	(a)	JAN	FEB	MAR	APR	MAY	JUNE	JULY	AUG	SEP	OCT	NOV	DEC	YEAR
TEMPERATURE °F:														
Normals														
-Daily Maximum		54.4	55.6	55.4	55.9	57.9	60.3	61.8	62.6	63.0	60.8	58.1	54.8	58.4
-Daily Minimum		41.5	42.9	43.4	44.3	47.5	50.6	52.3	53.1	51.6	48.6	45.2	42.0	46.9
-Monthly		48.0	49.3	49.4	50.2	52.8	55.5	57.0	57.9	57.3	54.7	51.7	48.4	52.7
Extremes														
-Record Highest	83	78	85	78	80	84	85	76	82	86	87	78	77	87
-Year		1986	1930	1914	1989	1939	1945	1992	1991	1983	1993	1987	1963	OCT 1993
-Record Lowest	83	25	27	29	32	36	41	45	44	41	32	29	21	21
-Year		1937	1990	1917	1929	1954	1966	1924	1935	1946	1971	1935	1972	DEC 1972
NORMAL DEGREE DAYS:														
Heating (base 65°F)		527	440	484	444	378	285	248	223	234	319	399	515	4496
Cooling (base 65°F)		0	0	0	0	0	0	0	0	0	0	0	0	0
% OF POSSIBLE SUNSHINE	83	43	46	52	57	58	59	55	50	54	50	44	41	51
MEAN SKY COVER (tenths)														
Sunrise - Sunset	51	7.2	7.4	7.3	7.0	6.7	6.5	6.5	6.9	6.0	6.5	7.1	7.1	6.9
MEAN NUMBER OF DAYS:														
Sunrise to Sunset														
-Clear	83	6.0	5.3	5.6	6.2	6.5	7.2	6.6	5.4	8.7	8.1	6.3	6.5	78.4
-Partly Cloudy	83	6.2	5.8	7.6	8.3	9.8	9.6	10.9	10.7	8.5	8.3	6.6	6.4	98.8
-Cloudy	83	18.8	17.2	17.7	15.5	14.6	13.2	13.6	14.9	12.8	14.5	17.1	18.1	188.1
Precipitation														
.01 inches or more	83	16.0	14.1	15.6	11.8	8.4	5.3	2.2	2.6	4.4	8.6	13.1	15.4	117.4
Snow,Ice pellets,hail														
1.0 inches or more	83	0.1	0.*	0.0	0.0	0.0	0.0	0.0	0.0	0.0	0.0	0.0	0.*	0.1
Thunderstorms	71	0.7	0.6	0.4	0.2	0.2	0.2	0.2	0.1	0.3	0.4	0.6	0.6	4.4
Heavy Fog Visibility														
1/4 mile or less	71	4.1	2.8	1.9	1.7	1.2	2.1	3.5	5.4	7.7	9.6	6.2	4.4	50.6
Temperature °F														
-Maximum														
90° and above	83	0.0	0.0	0.0	0.0	0.0	0.0	0.0	0.0	0.0	0.0	0.0	0.0	0.0
32° and below	83	0.0	0.0	0.0	0.0	0.0	0.0	0.0	0.0	0.0	0.0	0.0	0.0	0.0
-Minimum														
32° and below	83	1.9	0.8	0.2	0.*	0.0	0.0	0.0	0.0	0.0	0.*	0.2	1.4	4.6
0° and below	83	0.0	0.0	0.0	0.0	0.0	0.0	0.0	0.0	0.0	0.0	0.0	0.0	0.0
AVG. STATION PRESS.(mb)														
RELATIVE HUMIDITY (%)														
Hour 04														
Hour 10 (Local Time)														
Hour 16														
Hour 22														
PRECIPITATION (inches):														
Water Equivalent														
-Normal		6.00	4.73	5.32	2.88	1.44	0.51	0.13	0.48	0.89	2.67	6.44	6.04	37.53
-Maximum Monthly	83	13.92	13.94	13.97	10.68	6.05	2.57	1.34	3.42	3.56	13.04	16.58	14.13	16.58
-Year		1969	1938	1938	1963	1960	1954	1916	1983	1925	1950	1973	1983	NOV 1973
-Minimum Monthly	83	0.66	0.50	0.07	0.31	0.03	0.00	0.00	0.00	0.00	0.00	T	0.52	0.00
-Year		1985	1923	1926	1956	1955	1917	1967	1940	1929	1917	1929	1976	JUL 1967
-Maximum in 24 hrs	83	4.42	4.88	4.02	2.56	2.23	1.73	1.18	2.21	1.54	5.83	4.55	4.17	5.83
-Year		1912	1959	1975	1983	1943	1943	1916	1983	1977	1950	1926	1939	OCT 1950
Snow,Ice pellets,hail														
-Maximum Monthly	83	3.0	3.5	1.0	T	0.0	0.0	0.0	0.0	0.0	0.0	0.1	1.9	3.5
-Year		1935	1989	1966	1993							1977	1972	FEB 1989
-Maximum in 24 hrs	83	3.0	2.0	1.0	T	0.0	0.0	0.0	0.0	0.0	0.0	0.1	1.9	3.0
-Year		1935	1989	1966	1993							1977	1972	JAN 1935
WIND:														
Mean Speed (mph)	54	6.9	7.2	7.6	8.0	7.9	7.4	6.8	5.8	5.5	5.6	6.0	6.4	6.8
Prevailing Direction through 1964		SE	SE	N	N	N	N	N	NW	N	N	SE	SE	N
Fastest Mile														
-Direction (!!)	83	S	SW	SW	N	NW	NW	N	N	N	SW	S	S	SW
-Speed (MPH)	83	54	48	48	49	40	39	35	34	44	56	55	56	56
-Year		1955	1960	1953	1915	1955	1949	1986	1920	1941	1962	1981	1931	OCT 1962
Peak Gust														
-Direction (!!)	9	S	N	S	NE	N	N	N	S	N	SE	S	N	S
-Speed (mph)	9	70	60	72	53	60	51	45	45	49	52	62	62	72
-Date		1993	1987	1991	1991	1990	1991	1986	1990	1988	1988	1991	1990	MAR 1991

PRECIPITATION (inches)

EUREKA, CALIFORNIA

YEAR	JAN	FEB	MAR	APR	MAY	JUNE	JULY	AUG	SEP	OCT	NOV	DEC	ANNUAL
1964	11.13	1.20	5.91	0.67	1.59	0.72	0.83	0.03	0.07	1.82	12.11	10.96	47.04
1965	5.82	1.36	1.23	5.60	0.44	0.35	T	0.36	T	0.70	5.20	5.22	26.28
1966	9.44	3.12	6.57	1.34	0.06	0.30	0.25	0.50	1.33	1.02	9.86	6.52	40.31
1967	8.87	1.47	7.44	5.29	1.52	0.32	0.00	T	1.32	2.15	4.40	4.34	37.12
1968	7.59	2.93	3.85	0.40	1.04	0.20	0.04	1.98	0.60	2.81	5.88	8.32	35.64
1969	13.92	7.82	1.56	3.22	1.01	0.34	0.05	T	0.36	3.20	3.49	9.60	44.57
1970	12.46	3.15	2.70	1.54	1.38	0.29	T	T	0.32	2.11	13.20	10.24	47.39
1971	5.41	3.28	7.91	2.92	1.28	1.51	0.16	0.55	2.08	0.92	6.36	6.38	38.76
1972	7.96	5.93	5.08	2.27	1.11	0.88	0.01	0.07	1.06	1.97	5.41	7.42	39.17
1973	6.47	3.85	7.10	0.35	0.85	0.23	T	0.08	2.35	4.14	16.58	7.02	49.02
1974	6.02	5.98	6.98	3.15	0.42	0.33	0.11	0.32	T	1.76	2.75	6.40	34.22
1975	5.20	7.68	10.73	3.29	1.05	0.58	0.10	0.58	0.01	6.77	4.72	5.38	46.09
1976	1.88	7.51	3.12	2.80	0.54	0.14	0.20	1.70	0.04	0.28	2.98	0.52	21.71
1977	1.90	2.24	4.33	1.20	2.10	0.07	T	0.20	3.35	2.79	4.51	6.60	29.29
1978	4.52	6.06	2.88	4.10	0.82	0.34	0.03	0.59	2.72	0.04	2.39	1.16	25.65
1979	3.82	6.26	1.70	3.94	2.25	0.05	0.31	0.13	1.15	6.14	6.19	3.75	35.69
1980	3.19	4.67	6.14	4.18	1.70	0.42	T	0.07	0.14	1.38	2.49	6.10	30.48
1981	7.67	3.72	4.64	0.71	2.02	0.57	T	0.01	0.97	3.71	9.39	9.88	43.29
1982	4.75	5.76	7.06	5.97	0.07	0.78	0.08	0.03	0.62	4.89	7.83	10.30	48.14
1983	8.48	9.18	10.73	5.47	1.12	0.65	0.89	3.42	0.87	1.87	10.40	14.13	67.21
1984	0.76	5.18	4.70	2.76	2.51	1.07	0.03	0.05	0.55	3.67	15.15	4.27	40.70
1985	0.66	3.69	4.68	0.45	1.14	0.89	0.15	0.52	1.06	4.07	2.98	2.78	23.07
1986	7.19	10.08	6.12	1.46	2.34	0.21	0.02	T	2.70	1.75	1.85	3.83	37.55
1987	6.48	3.38	6.10	1.15	0.41	0.26	0.20	0.06	0.02	1.05	4.23	10.92	34.26
1988	7.13	0.54	1.18	2.06	2.70	2.22	0.05	T	0.12	0.41	8.93	6.26	31.60
1989	4.71	2.88	7.63	2.01	1.67	0.21	0.08	0.13	0.85	2.90	1.60	0.80	25.47
1990	7.20	4.50	3.30	1.41	3.74	0.32	0.22	0.71	0.19	1.73	3.07	2.95	29.34
1991	1.65	2.75	6.94	2.52	2.16	0.26	1.13	0.37	T	1.06	1.95	2.36	23.15
1992	3.99	3.80	3.51	2.42	0.06	1.27	0.25	0.01	0.33	2.08	2.21	9.33	29.26
1993	7.15	5.93	4.72	5.94	4.44	1.23	0.37	0.54	0.03	0.56	1.35	7.12	39.38
Record Mean	6.59	5.68	5.29	3.01	1.86	0.72	0.14	0.25	0.86	2.63	5.36	6.29	38.69

AVERAGE TEMPERATURE (deg. F)

EUREKA, CALIFORNIA

YEAR	JAN	FEB	MAR	APR	MAY	JUNE	JULY	AUG	SEP	OCT	NOV	DEC	ANNUAL
1964	46.6	47.0	45.7	47.1	50.4	54.8	56.6	57.0	55.2	54.9	50.0	48.5	51.1
1965	47.3	45.7	48.5	50.1	49.9	51.9	54.8	58.9	53.9	55.7	54.6	47.0	51.6
1966	47.7	46.5	48.0	50.4	50.4	55.5	56.6	55.9	58.0	53.2	52.6	49.5	52.0
1967	47.4	47.6	46.7	46.2	51.9	54.6	56.5	56.8	59.1	56.9	53.3	45.1	51.8
1968	46.6	53.2	50.6	48.1	52.4	55.5	56.3	59.3	57.5	53.3	51.8	46.5	52.6
1969	44.0	46.2	48.3	49.8	53.4	56.6	55.6	55.8	56.5	55.2	51.3	51.3	52.0
1970	52.1	51.4	49.9	47.4	52.3	54.3	54.8	55.3	55.7	52.6	53.5	47.4	52.2
1971	45.7	46.3	47.1	48.1	50.6	54.3	55.2	60.0	56.7	50.6	49.4	45.0	50.8
1972	44.6	49.0	51.2	49.1	51.2	54.6	57.8	58.2	55.8	54.2	51.8	45.7	52.0
1973	47.3	50.7	47.4	50.1	52.0	55.1	55.6	54.8	57.1	52.1	51.4	51.2	52.1
1974	46.7	46.1	50.1	49.7	51.1	54.1	57.7	57.7	55.6	53.2	51.2	48.7	51.8
1975	45.8	48.0	47.6	46.5	51.4	53.2	56.9	55.3	55.2	54.2	48.5	47.4	50.8
1976	46.4	46.9	45.7	48.4	51.1	52.8	57.5	57.9	56.2	54.4	52.2	47.4	51.4
1977	47.4	50.5	46.2	49.2	51.5	54.2	55.1	58.3	57.1	53.8	51.2	51.0	52.1
1978	51.8	50.3	53.5	51.0	53.3	56.0	55.9	57.0	57.6	54.9	48.0	43.3	52.7
1979	46.6	48.0	50.0	51.0	52.7	54.0	58.1	59.4	62.3	57.3	52.4	51.8	53.6
1980	48.3	53.5	48.6	51.9	52.3	55.5	57.3	55.0	56.6	54.3	51.3	51.2	53.0
1981	52.5	51.3	50.2	50.6	53.3	56.5	55.2	58.1	57.9	54.3	53.6	52.1	53.8
1982	44.9	49.5	48.3	50.8	52.8	56.3	58.3	59.6	58.7	57.3	52.1	49.6	53.2
1983	51.3	53.4	53.4	51.8	54.4	58.0	60.5	61.9	60.7	57.9	53.6	50.6	55.6
1984	49.4	50.0	52.8	51.3	55.0	55.2	57.6	60.2	58.3	55.3	52.1	46.8	53.7
1985	48.4	47.9	47.4	51.6	53.9	56.6	58.6	58.5	56.7	54.5	46.6	47.7	52.4
1986	54.3	52.7	53.1	50.7	53.9	59.0	57.4	57.2	57.2	55.6	53.5	51.3	54.7
1987	49.1	51.4	52.7	54.2	56.6	57.7	59.5	58.5	57.3	57.8	54.8	49.7	54.9
1988	50.4	50.2	50.3	52.8	56.3	57.7	58.8	57.7	55.8	55.2	53.3	47.7	53.9
1989	46.0	45.6	51.8	54.6	55.9	57.6	59.4	59.3	56.6	54.9	52.6	49.2	53.6
1990	48.3	45.5	50.2	52.8	54.0	58.2	59.8	60.4	61.6	54.3	49.7	42.8	53.1
1991	47.8	52.8	48.1	50.9	52.2	53.8	57.3	59.9	57.2	56.1	51.4	48.3	53.0
1992	50.3	53.9	54.1	56.4	56.7	58.3	60.6	59.1	57.5	58.0	51.7	46.9	55.3
1993	46.4	49.4	53.4	53.6	57.9	57.9	57.4	59.8	55.8	56.9	49.9	49.5	54.0
Record Mean	47.4	48.3	48.8	50.3	52.7	55.1	56.4	56.9	56.4	54.2	51.3	48.3	52.2
Max	53.6	54.4	54.7	55.8	57.6	59.8	60.7	61.3	61.8	60.2	57.7	54.6	57.7
Min	41.2	42.2	42.9	44.7	47.7	50.4	52.0	52.5	50.9	48.2	44.9	42.0	46.6

REFERENCE NOTES

EUREKA, CALIFORNIA

GENERAL
T - TRACE AMOUNT.
BLANK ENTRIES DENOTE MISSING/UNREPORTED DATA.
# INDICATES A STATION OR INSTRUMENT RELOCATION.

SPECIFIC

PM - INCLUDES LAST DAY OF PREVIOUS MONTH
ASOS - AUTOMATED SURFACE OBSERVING SYSTEM IN OPERATION DURING THESE MONTHS.

(a) - LENGTH OF RECORD IN YEARS, ALTHOUGH INDIVIDUAL MONTHS MAY BE MISSING.
0.* OR * - THE VALUE IS BETWEEN 0.0 AND 0.05
NORMALS - BASED ON THE 1961-1990 RECORD PERIOD.
EXTREMES - DATES ARE THE MOST RECENT OCCURRENCE.
WIND DIR. - NUMERALS SHOW TENS OF DEGRESS CLOCKWISE FROM TRUE NORTH. "00" INDICATES CALM.
RESULTANT DIRECTIONS ARE GIVEN TO WHOLE DEGREES.
BOLD VALUES INDICATE EXTREME VALUES WHICH OCCURRED AFTER THE ASOS SYSTEM WAS COMMISSIONED.

RECORD = PERIOD OF RECORD
RECORD MEAN PRECIPITATION IS THE MEAN OF ALL DAILY PRECIPITATION AMOUNTS DURING THE PERIOD OF RECORD.
RECORD MAX(MIN) TEMPERATURE IS THE MEAN OF ALL DAILY MAX(MIN) TEMPERATURES DURING THE PERIOD OF RECORD.
RECORD MEAN TEMPERATURE IS THE SUM OF THE RECORD MAX AND RECORD MIN DIVIDED BY 2.
AVERAGE TEMPERATURE IS THE SUM OF THE MEAN DAILY MAX AND MIN TEMPERATURE DIVIDED BY 2.

EXCEPTIONS

1. PRIOR TO 1965, THUNDERSTORMS AND HEAVY FOG MAY BE INCOMPLETE, DUE TO PART-TIME OPERATIONS.
2. THUNDERSTORM AND HEAVY FOG DATA ARE MISSING FROM 1965 THROUGH 1976.
3. MEAN WIND SPEED IS THROUGH 1964.

RECORD MEANS ARE THROUGH THE CURRENT YEAR, BEGINNING IN 1887 FOR TEMPERATURE
1887 FOR PRECIPITATION
1911 FOR SNOWFALL

HEATING DEGREE DAYS Base 65 deg. F EUREKA, CALIFORNIA

SEASON	JULY	AUG	SEP	OCT	NOV	DEC	JAN	FEB	MAR	APR	MAY	JUNE	TOTAL
1964-65	256	243	288	306	444	505	544	533	505	441	459	385	4909
1965-66	309	183	324	285	305	551	531	510	519	433	443	280	4673
1966-67	254	275	209	358	368	472	541	482	561	560	399	307	4786
1967-68	257	249	170	243	345	611	565	337	441	498	384	279	4379
1968-69	260	179	220	354	388	566	644	519	508	449	352	246	4685
1969-70	284	282	247	295	404	419	392	374	462	521	385	313	4378
1970-71	308	292	274	379	338	539	590	518	548	502	438	318	5044
1971-72	299	154	242	440	463	612	624	456	422	472	423	306	4913
1972-73	217	204	269	328	389	590	542	395	537	439	393	293	4596
1973-74	287	308	230	393	399	419	559	523	455	454	423	321	4771
1974-75	222	220	274	360	407	501	587	469	532	547	417	347	4883
1975-76	244	290	286	328	486	542	569	516	590	490	424	360	5125
1976-77	226	213	258	324	375	535	537	400	577	468	415	317	4645
1977-78	302	200	231	342	408	427	403	404	347	415	357	264	4100
1978-79	274	241	215	307	503	667	562	467	459	412	374	323	4804
1979-80	208	165	92	231	369	404	511	330	500	386	388	280	3864
1980-81	230	303	246	328	402	422	384	377	451	423	357	249	4172
1981-82	299	205	203	324	339	396	616	430	512	419	373	258	4374
1982-83	204	158	181	232	381	468	415	317	355	391	320	203	3625
1983-84	133	90	129	215	336	443	475	429	369	403	301	285	3608
1984-85	222	142	195	295	378	556	507	472	532	396	338	243	4276
1985-86	195	194	244	316	546	533	329	337	365	422	338	169	3988
1986-87	227	236	227	290	341	417	487	372	377	316	260	213	3763
1987-88	163	196	226	221	302	470	446	423	453	358	265	215	3738
1988-89	187	218	274	297	345	529	582	535	403	309	278	214	4171
1989-90	164	171	243	306	365	482	513	541	453	356	332	198	4124
1990-91	154	141	95	325	451	680	528	336	516	418	387	330	4361
1991-92	229	164	225	274	401	511	450	316	329	250	248	193	3590
1992-93	135	176	220	211	393	556	570	433	354	337	214	203	3802
1993-94	229	152	268	253	446	472							

COOLING DEGREE DAYS Base 65 deg. F EUREKA, CALIFORNIA

YEAR	JAN	FEB	MAR	APR	MAY	JUNE	JULY	AUG	SEP	OCT	NOV	DEC	TOTAL
1969	0	0	0	0	0	0	0	0	0	0	0	0	0
1970	0	0	0	0	1	0	0	0	1	0	0	0	2
1971	0	0	0	0	0	0	0	2	0	0	0	0	2
1972	0	0	0	0	0	0	0	1	0	0	0	0	1
1973	0	0	0	0	0	1	0	0	0	0	0	0	1
1974	0	0	0	0	0	0	0	0	0	0	0	0	0
1975	0	0	0	0	0	0	0	0	0	0	0	0	0
1976	0	0	0	0	0	0	0	0	0	0	0	0	0
1977	0	0	0	0	0	0	0	0	0	0	0	0	0
1978	0	0	0	0	0	0	0	0	0	1	0	0	1
1979	0	0	0	0	0	0	0	0	15	0	0	2	17
1980	0	2	0	0	0	0	0	0	0	3	0	0	5
1981	4	0	0	0	0	0	0	0	0	0	0	0	4
1982	0	0	0	0	0	3	0	0	0	2	0	0	5
1983	0	0	0	0	0	0	0	2	7	0	0	0	9
1984	0	0	0	0	0	0	0	0	4	1	0	0	5
1985	0	0	0	0	0	0	1	0	0	2	0	0	3
1986	0	0	0	0	0	0	0	0	0	5	0	0	5
1987	0	0	0	0	3	0	0	0	0	5	0	0	8
1988	0	0	0	0	0	0	0	0	4	0	0	0	4
1989	0	0	0	1	0	0	0	0	0	0	0	0	1
1990	0	0	0	0	0	0	0	4	0	0	0	0	4
1991	0	0	0	0	0	0	0	10	0	4	0	0	14
1992	0	0	0	0	0	0	2	0	3	2	0	0	7
1993	0	3	0	0	3	0	0	0	0	8	0	0	14

SNOWFALL (inches) EUREKA, CALIFORNIA

SEASON	JULY	AUG	SEP	OCT	NOV	DEC	JAN	FEB	MAR	APR	MAY	JUNE	TOTAL
1970-71	0.0	0.0	0.0	0.0	0.0	0.0	T	T	0.0	0.0	0.0	0.0	T
1971-72	0.0	0.0	0.0	0.0	0.0	0.0	1.6	0.0	0.0	0.0	0.0	0.0	1.6
1972-73	0.0	0.0	0.0	0.0	0.0	1.9	0.0	0.0	T	0.0	0.0	0.0	1.9
1973-74	0.0	0.0	0.0	0.0	0.0	0.0	T	0.0	0.0	0.0	0.0	0.0	T
1974-75	0.0	0.0	0.0	0.0	0.0	T	0.0	0.0	0.0	T	0.0	0.0	T
1975-76	0.0	0.0	0.0	0.0	0.0	0.0	0.0	0.0	0.1	0.0	0.0	0.0	0.1
1976-77	0.0	0.0	0.0	0.0	0.0	0.0	0.0	0.0	T	0.0	0.0	0.0	T
1977-78	0.0	0.0	0.0	0.0	0.1	0.0	0.0	0.0	0.0	0.0	0.0	0.0	0.1
1978-79	0.0	0.0	0.0	0.0	0.0	0.0	0.0	0.0	0.0	0.0	0.0	0.0	0.0
1979-80	0.0	0.0	0.0	0.0	0.0	0.0	0.0	0.0	0.0	0.0	0.0	0.0	0.0
1980-81	0.0	0.0	0.0	0.0	0.0	0.0	0.0	0.0	0.0	0.0	0.0	0.0	0.0
1981-82	0.0	0.0	0.0	0.0	0.0	0.0	T	0.0	T	T	0.0	0.0	T
1982-83	0.0	0.0	0.0	0.0	0.0	T	0.0	0.0	T	0.0	0.0	0.0	T
1983-84	0.0	0.0	0.0	0.0	T	1.0	0.0	0.0	0.0	0.0	0.0	0.0	1.0
1984-85	0.0	0.0	0.0	0.0	0.0	0.0	0.0	0.0	0.0	0.0	0.0	0.0	0.0
1985-86	0.0	0.0	0.0	0.0	0.0	0.0	0.0	0.0	0.0	0.0	0.0	0.0	0.0
1986-87	0.0	0.0	0.0	0.0	0.0	0.0	0.0	0.0	0.0	0.0	0.0	0.0	0.0
1987-88	0.0	0.0	0.0	0.0	0.0	T	0.0	0.0	0.0	0.0	0.0	0.0	T
1988-89	0.0	0.0	0.0	0.0	0.0	T	0.0	3.5	0.0	0.0	0.0	0.0	3.5
1989-90	0.0	0.0	0.0	0.0	0.0	0.0	0.0	1.0	0.0	0.0	0.0	0.0	1.0
1990-91	0.0	0.0	0.0	0.0	0.0	T	0.0	0.0	T	0.0	0.0	0.0	T
1991-92	0.0	0.0	0.0	0.0	0.0	T	0.0	0.0	0.0	0.0	0.0	0.0	T
1992-93	0.0	0.0	0.0	0.0	0.0	T	0.0	T	0.0	T	0.0	0.0	T
1993-94	0.0	0.0	0.0	0.0	0.0	0.0							
Record Mean	0.0	0.0	0.0	0.0	T	T	0.1	0.1	T	T	0.0	0.0	0.3

FRESNO, CALIFORNIA

Fresno is located about midway and toward the eastern edge of the San Joaquin Valley, which is oriented northwest to southeast and has a length of about 225 miles and an average width of 50 miles. The San Joaquin Valley is generally flat. About 15 miles east of Fresno the terrain slopes upward with the foothills of the Sierra Nevada. The Sierra Nevada attain an elevation of more than 14,000 feet 50 miles east of Fresno. West of the city 45 miles lie the foothills of the Coastal Range.

The climate of Fresno is dry and mild in winter and hot in summer. Nearly nine-tenths of the annual precipitation falls in the six months from November to April.

Due to clear skies during the summer and the protection of the San Joaquin Valley from marine effects, the normal daily maximum temperature reaches the high 90s during the latter part of July. The daily maximum temperature during the warmest month has ranged from 76 to 115 degrees. Low relative humidities and some wind movement substantially lower the sensible temperature during periods of high readings. Humidity readings of 15 percent are common on summer afternoons, and readings as low as 8 percent have been recorded. In contrast to this, humidity readings average 90 percent during the morning hours of December and January.

Winds flow with the major axis of the San Joaquin Valley, generally from the northwest. This feature is especially beneficial since, during the warmest months, the northwest winds increase during the evenings. These refreshing breezes and the normally large temperature variation of about 35 degrees between the highest and lowest readings of the day, generally result in comfortable evening and night temperatures.

Winter temperatures are usually mild with infrequent cold spells dropping the readings below freezing. Heavy frost occurs almost every year, and the first frost usually occurs during the last week of November. The last frost in spring is usually in early March, however, one year in five will have the last frost after the first of April. The growing season is 291 days.

Although the heaviest rains recorded at Fresno for short periods have occurred in June, usually any rainfall during the summer is very light. Snow is a rare occurrence in Fresno.

Fresno enjoys a very high percentage of sunshine, receiving more than 80 percent of the possible amounts during all but the four months of November, December, January, and February. Reduction of sunshine during these months is caused by fog and short periods of stormy weather.

During foggy periods, at times lasting nearly two weeks, sunshine is reduced to a minimum. This fog frequently lifts to a few hundred feet above the surface of the valley and presents the appearance of a heavy, solid cloud layer.

Spring and autumn are very enjoyable seasons in Fresno, with clear skies, light rainfall and winds and mild temperatures.

NORMALS, MEANS, AND EXTREMES

FRESNO, CALIFORNIA

LATITUDE: 36°46'N LONGITUDE: 119°43' W ELEVATION: FT. GRND 328 BARO 330 TIME ZONE: PACIFIC WBAN: 93193

	(a)	JAN	FEB	MAR	APR	MAY	JUNE	JULY	AUG	SEP	OCT	NOV	DEC	YEAR
TEMPERATURE °F:														
Normals														
-Daily Maximum		54.1	61.7	66.6	75.1	84.2	92.7	98.6	96.7	90.1	79.7	64.7	53.7	76.5
-Daily Minimum		37.4	40.5	43.4	47.3	53.7	60.4	65.1	63.8	58.8	50.7	42.5	37.1	50.1
-Monthly		45.7	51.2	55.1	61.2	69.0	76.6	81.9	80.3	74.5	65.2	53.6	45.4	63.3
Extremes														
-Record Highest	44	78	80	90	100	107	110	112	111	111	102	89	76	112
-Year		1986	1991	1972	1981	1984	1964	1991	1990	1955	1980	1949	1958	JUL 1991
-Record Lowest	44	19	24	26	32	36	44	50	49	37	27	26	18	18
-Year		1963	1990	1966	1982	1975	1955	1955	1966	1950	1972	1975	1990	DEC 1990
NORMAL DEGREE DAYS:														
Heating (base 65°F)		598	386	314	182	34	0	0	0	8	84	342	608	2556
Cooling (base 65°F)		0	0	7	68	158	352	524	474	293	91	0	0	1967
% OF POSSIBLE SUNSHINE	44	47	65	78	85	90	95	97	96	94	88	66	46	79
MEAN SKY COVER (tenths)														
Sunrise - Sunset	44	7.3	6.1	5.4	4.4	3.2	1.9	1.2	1.4	1.7	2.8	5.2	6.9	3.9
MEAN NUMBER OF DAYS:														
Sunrise to Sunset														
-Clear	44	5.4	7.8	11.0	14.0	18.8	23.1	26.7	26.2	23.8	20.2	11.7	6.9	195.6
-Partly Cloudy	44	6.9	7.8	8.0	8.2	7.5	4.5	3.0	3.5	3.9	6.1	7.2	6.2	72.8
-Cloudy	44	18.7	12.7	11.9	7.8	4.7	2.4	1.3	1.4	2.3	4.7	11.1	17.9	96.9
Precipitation														
.01 inches or more	44	7.5	7.1	7.1	4.1	1.9	0.7	0.2	0.3	1.0	2.2	5.2	6.7	44.1
Snow, Ice pellets, hail														
1.0 inches or more	44	0.*	0.0	0.0	0.0	0.0	0.0	0.0	0.0	0.0	0.0	0.0	0.*	*
Thunderstorms	44	0.3	0.4	0.9	0.5	0.5	0.5	0.3	0.3	0.7	0.5	0.2	0.3	5.3
Heavy Fog Visibility														
1/4 mile or less	44	11.8	6.0	1.7	0.3	0.1	0.0	0.0	0.*	0.1	0.9	5.8	12.1	38.8
Temperature °F														
-Maximum														
90° and above	30	0.0	0.0	0.*	2.1	10.1	19.6	28.6	26.3	17.4	4.7	0.0	0.0	108.8
32° and below	30	0.0	0.0	0.0	0.0	0.0	0.0	0.0	0.0	0.0	0.0	0.0	0.1	0.1
-Minimum														
32° and below	30	7.5	3.0	0.7	0.1	0.0	0.0	0.0	0.0	0.0	0.1	1.9	8.7	21.9
0° and below	30	0.0	0.0	0.0	0.0	0.0	0.0	0.0	0.0	0.0	0.0	0.0	0.0	0.0
AVG. STATION PRESS. (mb)	21	1008.9	1007.4	1005.2	1004.3	1001.6	1000.4	1000.3	1000.5	1000.8	1004.1	1007.5	1009.1	1004.2
RELATIVE HUMIDITY (%)														
Hour 04	30	91	89	86	80	71	65	61	67	71	78	87	92	78
Hour 10 (Local Time)	30	85	76	65	51	42	39	38	41	45	52	71	84	57
Hour 16	30	67	55	47	34	26	23	22	25	28	35	53	68	40
Hour 22	30	88	82	75	61	49	44	41	46	51	64	81	89	64
PRECIPITATION (inches):														
Water Equivalent														
-Normal		1.96	1.80	1.89	0.97	0.30	0.08	0.01	0.03	0.24	0.53	1.37	1.42	10.60
-Maximum Monthly	44	8.56	5.97	7.24	4.41	1.65	1.61	0.22	0.25	1.19	2.19	3.50	6.73	8.56
-Year		1969	1962	1991	1967	1990	1993	1992	1964	1976	1992	1972	1955	JAN 1969
-Minimum Monthly	44	0.04	T	0.00	0.02	0.00	0.00	0.00	0.00	0.00	0.00	0.00	0.00	0.00
-Year		1976	1964	1972	1991	1982	1983	1983	1981	1981	1978	1959	1989	DEC 1989
-Maximum in 24 hrs	44	2.59	1.99	1.63	1.39	1.42	1.61	0.22	0.25	0.97	1.76	1.35	1.76	2.59
-Year		1969	1969	1958	1983	1990	1993	1992	1964	1978	1992	1953	1955	JAN 1969
Snow, Ice pellets, hail														
-Maximum Monthly	44	2.2	T	T	0.0	0.0	0.0	0.0	0.0	0.0	T	0.0	1.2	2.2
-Year		1962	1992	1991							1974		1968	JAN 1962
-Maximum in 24 hrs	44	1.5	T	T	0.0	0.0	0.0	0.0	0.0	0.0	T	0.0	1.2	1.5
-Year		1962	1992	1991							1974		1968	JAN 1962
WIND:														
Mean Speed (mph)	44	5.2	5.7	6.7	7.4	8.2	8.3	7.4	6.8	6.1	5.2	4.7	4.9	6.4
Prevailing Direction through 1963		SE	NW	NW	NW	NW	NW	NW	NW	NW	NW	NW	SE	NW
Fastest Obs. 1 Min.														
-Direction (!!)	15	14	31	32	30	32	31	31	31	31	12	30	22	14
-Speed (MPH)	15	39	29	29	32	30	28	23	28	29	23	28	28	39
-Year		1993	1977	1982	1984	1991	1992	1983	1984	1978	1992	1993	1991	JAN 1993
Peak Gust														
-Direction (!!)	10	SE	S	SE	NW	NW	NW	NW	NW	NW	N	N	S	SE
-Speed (mph)	10	55	46	43	41	41	35	28	36	33	46	44	48	55
-Date		1987	1986	1987	1984	1991	1992	1991	1984	1991	1991	1991	1991	JAN 1987

PRECIPITATION (inches)

FRESNO, CALIFORNIA

YEAR	JAN	FEB	MAR	APR	MAY	JUNE	JULY	AUG	SEP	OCT	NOV	DEC	ANNUAL
1964	0.66	T	1.27	0.50	0.35	0.06	T	0.25	0.00	1.23	1.49	2.63	8.44
1965	1.05	0.43	2.38	1.74	T	T	T	0.02	0.00	0.30	2.69	1.73	10.34
1966	0.53	0.54	0.01	0.15	0.10	0.07	0.03	0.00	0.03	0.00	1.57	3.04	6.07
1967	2.21	0.22	3.15	4.41	0.19	0.14	T	T	T	0.07	1.55	1.04	12.98
1968	1.05	1.10	1.49	0.70	0.24	0.00	T	T	0.00	1.54	1.94	2.44	10.50
1969	8.56	5.60	1.16	1.64	0.06	0.04	0.04	0.00	0.04	0.06	0.80	1.14	19.14
1970	3.83	1.27	1.65	0.21	0.00	0.08	T	0.00	0.00	0.01	2.30	2.51	11.86
1971	0.40	0.29	0.58	1.04	1.40	0.00	T	T	0.04	0.03	0.65	2.56	6.99
1972	0.37	0.67	0.00	0.27	0.15	0.60	T	0.00	0.29	0.22	3.50	1.40	7.47
1973	1.91	3.69	2.84	0.09	T	T	0.00	T	0.00	1.02	1.39	1.74	12.68
1974	2.82	0.25	2.56	0.64	0.00	0.00	T	T	0.00	1.44	0.34	1.26	9.31
1975	0.69	0.97	2.44	0.55	T	0.00	T	0.05	0.22	1.07	0.20	0.14	6.33
1976	0.04	4.72	0.44	0.93	T	0.37	0.01	0.21	1.19	1.55	0.87	0.71	11.04
1977	0.68	0.09	1.04	0.04	1.16	0.06	T	T	T	0.01	0.46	3.02	6.56
1978	3.16	4.41	4.25	2.85	0.00	0.00	T	T	1.05	0.00	1.34	0.62	17.68
1979	2.71	2.53	2.27	0.07	0.06	0.00	0.08	0.00	T	0.48	1.01	0.74	9.95
1980	3.83	3.30	2.05	0.25	0.18	T	0.01	0.00	0.00	0.03	0.14	0.49	10.28
1981	2.67	1.29	2.59	1.01	T	0.00	0.00	0.00	0.00	0.58	1.22	0.65	10.01
1982	2.11	0.58	4.76	0.89	0.00	0.31	0.00	T	1.10	1.58	3.16	1.59	16.08
1983	5.14	3.70	4.53	2.76	0.01	0.00	0.00	0.09	1.03	0.09	2.51	1.75	21.61
1984	0.15	1.05	0.48	0.25	0.02	0.20	T	T	0.00	0.70	1.94	1.98	6.77
1985	0.43	0.71	1.73	0.12	0.00	0.33	0.04	0.02	0.43	0.85	3.02	0.72	8.40
1986	2.12	3.66	3.42	0.36	0.16	0.00	T	0.00	0.38	0.00	0.01	2.30	12.41
1987	1.93	1.36	2.39	0.07	0.87	0.01	0.00	0.00	T	0.85	0.52	1.19	9.19
1988	1.52	0.83	0.27	2.41	0.45	0.03	0.00	0.00	0.00	0.00	1.42	2.46	9.39
1989	0.48	1.18	2.25	0.05	0.89	0.00	0.00	0.03	1.11	0.42	0.50	0.00	6.91
1990	2.82	1.33	0.67	0.92	1.65	0.00	T	0.00	0.15	0.05	0.46	0.68	8.73
1991	0.13	1.01	7.24	0.02	0.03	T	0.00	T	T	0.80	0.04	1.22	10.49
1992	1.94	4.73	2.14	0.18	T	T	0.22	T	T	2.19	T	2.68	14.08
1993	5.18	2.44	1.76	0.20	0.25	1.61	0.00	0.00	0.00	0.12	1.16	1.03	13.75
Record Mean	1.82	1.66	1.73	0.94	0.36	0.11	0.01	0.01	0.18	0.55	1.00	1.54	9.90

AVERAGE TEMPERATURE (deg. F)

FRESNO, CALIFORNIA

YEAR	JAN	FEB	MAR	APR	MAY	JUNE	JULY	AUG	SEP	OCT	NOV	DEC	ANNUAL
1964	43.8	47.4	51.3	58.8	64.9	73.3	81.0	78.9	71.0	68.3	51.2	49.0	61.6
1965	46.3	49.6	55.5	60.8	67.5	71.4	78.9	78.8	68.6	65.8	54.7	42.0	61.6
1966	43.4	47.2	56.3	65.5	70.9	76.3	78.2	81.0	72.8	64.8	56.9	45.2	63.2
1967	46.1	48.9	54.4	52.6	68.8	74.3	83.8	83.6	77.4	66.0	56.8	42.6	62.9
1968	44.8	55.8	55.8	61.5	68.1	78.0	82.4	77.2	73.7	63.3	51.9	43.3	63.0
1969	44.8	47.5	53.1	59.7	70.4	72.9	80.9	79.7	75.7	59.7	53.1	46.2	62.0
1970	49.1	52.7	55.3	57.0	70.8	76.5	83.3	79.9	73.0	63.4	55.4	46.3	63.5
1971	45.7	47.6	54.4	59.1	64.2	74.4	81.9	81.1	73.4	60.9	50.7	42.9	61.4
1972	40.6	52.5	60.7	61.1	69.9	77.5	81.5	79.7	71.8	62.6	50.2	40.9	62.4
1973	45.1	51.9	50.4	61.2	72.9	78.6	80.4	78.5	72.0	63.3	52.9	47.2	62.9
1974	47.9	49.1	56.3	60.0	69.5	77.7	81.3	79.3	77.5	66.0	53.1	44.5	63.5
1975	43.4	49.9	51.5	53.9	68.4	74.7	78.1	75.9	75.8	61.4	49.5	43.9	60.6
1976	44.3	49.6	52.4	57.2	69.7	72.9	79.4	72.7	72.2	65.1	53.4	46.5	61.3
1977	44.3	53.5	52.4	65.5	63.6	79.8	81.5	80.6	74.0	66.8	54.6	51.3	64.0
1978	51.4	52.6	60.3	58.9	69.9	76.3	82.4	81.4	73.0	70.0	52.1	42.8	64.2
1979	47.0	51.4	57.4	62.7	71.1	77.9	82.2	79.9	79.5	67.8	54.0	46.9	64.8
1980	49.4	53.8	53.7	61.8	67.2	73.7	84.0	80.7	75.6	68.4	54.2	46.8	64.1
1981	47.9	52.0	54.5	63.2	70.9	82.8	84.9	82.9	76.5	61.4	55.5	47.7	65.0
1982	41.7	50.5	51.4	58.0	69.3	72.9	81.0	80.4	72.3	65.0	51.1	45.4	61.6
1983	45.2	53.1	55.9	57.9	69.7	76.3	79.0	82.1	78.8	68.5	54.6	51.1	64.4
1984	47.8	50.7	58.4	60.8	74.8	77.5	87.0	83.5	81.0	62.4	53.6	46.5	65.3
1985	43.3	51.3	53.1	67.2	69.4	81.8	86.0	80.5	72.3	65.0	52.5	43.8	63.9
1986	53.6	55.7	60.3	62.7	71.2	79.4	81.9	84.2	71.3	66.9	56.7	47.5	66.0
1987	45.3	52.8	55.6	66.7	71.8	78.4	77.0	80.2	75.5	70.1	52.3	44.2	64.2
1988	46.0	52.2	56.8	61.6	67.0	75.6	85.5	81.2	76.4	68.7	54.3	44.5	64.2
1989	42.9	48.8	57.9	67.3	69.6	77.0	82.5	79.3	74.3	65.3	54.3	43.8	63.6
1990	45.5	48.0	57.3	65.7	68.1	76.8	84.0	80.6	75.8	67.7	52.9	41.5	63.7
1991	47.0	55.8	51.5	59.5	66.1	74.7	83.8	78.6	79.9	70.5	55.8	47.0	64.2
1992	42.7	55.5	58.8	66.8	76.0	77.0	81.3	83.2	77.0	68.6	54.3	45.3	65.5
1993	47.1	51.9	60.3	61.7	69.9	75.7	80.2	79.7	75.7	67.8	53.9	45.6	64.1
Record Mean	45.9	51.2	55.2	61.1	68.0	75.5	81.8	80.0	74.1	65.0	54.2	46.3	63.2
Max	54.3	61.4	66.5	74.4	82.6	91.3	98.6	96.7	89.9	79.2	66.0	54.8	76.3
Min	37.4	40.9	43.8	47.8	53.4	59.7	64.9	63.3	58.3	50.8	42.4	37.7	50.0

REFERENCE NOTES

FRESNO, CALIFORNIA

GENERAL
T - TRACE AMOUNT.
BLANK ENTRIES DENOTE MISSING/UNREPORTED DATA.
# INDICATES A STATION OR INSTRUMENT RELOCATION.

SPECIFIC

PM - INCLUDES LAST DAY OF PREVIOUS MONTH
ASOS - AUTOMATED SURFACE OBSERVING SYSTEM IN OPERATION DURING THESE MONTHS.

(a) - LENGTH OF RECORD IN YEARS, ALTHOUGH INDIVIDUAL MONTHS MAY BE MISSING.
0.* OR * - THE VALUE IS BETWEEN 0.0 AND 0.05
NORMALS - BASED ON THE 1961-1990 RECORD PERIOD.
EXTREMES - DATES ARE THE MOST RECENT OCCURRENCE
WIND DIR.- NUMERALS SHOW TENS OF DEGRESS CLOCKWISE FROM TRUE NORTH. "00" INDICATES CALM.
RESULTANT DIRECTIONS ARE GIVEN TO WHOLE DEGREES.
BOLD VALUES INDICATE EXTREME VALUES WHICH OCCURRED AFTER THE ASOS SYSTEM WAS COMMISSIONED.

RECORD = PERIOD OF RECORD
RECORD MEAN PRECIPITATION IS THE MEAN OF ALL DAILY PRECIPITATION AMOUNTS DURING THE PERIOD OF RECORD.
RECORD MAX(MIN) TEMPERATURE IS THE MEAN OF ALL DAILY MAX(MIN) TEMPERATURES DURING THE PERIOD OF RECORD.
RECORD MEAN TEMPERATURE IS THE SUM OF THE RECORD MAX AND RECORD MIN DIVIDED BY 2.
AVERAGE TEMPERATURE IS THE SUM OF THE MEAN DAILY MAX AND MIN TEMPERATURE DIVIDED BY 2.

EXCEPTIONS

RECORD MEANS ARE THROUGH THE CURRENT YEAR, BEGINNING IN 1888 FOR TEMPERATURE
1878 FOR PRECIPITATION
1939 FOR SNOWFALL

HEATING DEGREE DAYS Base 65 deg. F — FRESNO, CALIFORNIA

SEASON	JULY	AUG	SEP	OCT	NOV	DEC	JAN	FEB	MAR	APR	MAY	JUNE	TOTAL
1964-65	0	3	3	41	410	492	572	423	287	191	58	8	2488
1965-66	0	0	12	41	302	707	664	492	271	60	7	1	2557
1966-67	0	0	1	65	238	606	579	444	322	366	59	11	2691
1967-68	0	0	0	29	239	686	619	258	278	139	37	2	2287
1968-69	0	0	12	73	387	665	619	480	366	168	30	0	2800
1969-70	0	0	0	166	349	574	485	340	291	232	25	0	2462
1970-71	0	0	0	108	282	573	593	480	322	181	81	8	2628
1971-72	0	0	20	209	423	678	750	357	142	128	37	0	2744
1972-73	0	0	2	108	437	740	610	358	444	140	12	2	2853
1973-74	0	0	0	94	360	544	522	438	260	160	33	0	2411
1974-75	0	0	0	59	350	628	661	419	409	325	53	3	2907
1975-76	0	0	0	154	455	648	636	440	385	242	10	9	2979
1976-77	0	1	5	63	342	566	636	313	386	42	98	0	2452
1977-78	0	0	0	46	302	417	415	343	143	182	19	0	1867
1978-79	0	0	6	30	382	682	549	372	234	96	34	0	2385
1979-80	0	0	0	56	323	555	473	318	343	129	46	0	2243
1980-81	0	0	0	69	318	553	521	359	316	114	9	0	2259
1981-82	0	0	0	118	278	530	711	398	412	217	21	4	2689
1982-83	0	0	13	62	411	602	607	327	276	206	55	0	2559
1983-84	0	0	1	3	304	421	530	408	198	149	6	0	2020
1984-85	0	0	0	128	335	566	664	378	361	39	8	3	2482
1985-86	0	0	0	63	369	651	345	258	156	98	30	0	1970
1986-87	0	0	13	22	242	537	602	337	282	56	26	0	2117
1987-88	0	0	0	7	374	636	583	366	251	124	69	12	2422
1988-89	0	0	0	20	316	629	679	450	213	52	14	0	2373
1989-90	0	0	7	73	310	649	598	470	236	35	19	1	2398
1990-91	0	0	0	17	356	722	549	253	412	163	65	0	2537
1991-92	0	0	0	81	276	551	683	267	183	25	0	1	2067
1992-93	0	0	0	18	316	602	549	359	145	113	9	12	2123
1993-94	0	0	0	12	326	595							

COOLING DEGREE DAYS Base 65 deg. F — FRESNO, CALIFORNIA

YEAR	JAN	FEB	MAR	APR	MAY	JUNE	JULY	AUG	SEP	OCT	NOV	DEC	TOTAL
1969	0	0	4	15	206	241	500	462	331	10	0	0	1769
1970	0	0	0	0	212	353	573	466	245	67	3	0	1919
1971	0	0	0	8	64	296	529	505	279	89	0	0	1770
1972	0	0	17	18	195	383	518	464	213	42	0	0	1850
1973	0	0	0	32	264	419	484	423	218	47	4	0	1891
1974	0	0	0	20	179	384	512	448	381	96	0	0	2020
1975	0	0	0	0	164	303	413	344	329	49	0	0	1602
1976	0	0	2	16	162	254	456	246	228	73	0	0	1437
1977	0	0	0	62	60	451	518	494	275	108	0	0	1968
1978	0	0	3	6	179	342	546	516	250	187	0	0	2029
1979	0	0	2	37	229	396	541	471	442	149	0	0	2267
1980	0	0	0	39	120	265	594	493	326	181	0	0	2018
1981	0	0	0	67	200	545	622	562	352	14	0	0	2362
1982	0	0	0	12	162	251	501	483	240	70	0	0	1719
1983	0	0	0	0	207	343	440	537	422	119	0	0	2068
1984	0	0	1	30	318	382	688	581	487	55	0	0	2542
1985	0	0	0	111	153	516	657	487	227	69	2	0	2222
1986	0	1	18	34	231	440	530	603	206	87	0	0	2150
1987	0	0	0	114	243	409	379	480	323	172	0	0	2120
1988	0	0	3	28	139	338	642	511	349	143	3	0	2156
1989	0	0	4	129	166	366	546	449	291	90	0	0	2041
1990	0	0	2	61	122	360	595	490	333	108	0	0	2071
1991	0	0	0	6	107	298	588	428	454	259	5	0	2145
1992	0	0	0	88	350	366	511	572	365	135	0	0	2387
1993	0	0	3	20	168	342	476	462	331	105	0	0	1907

SNOWFALL (inches) — FRESNO, CALIFORNIA

SEASON	JULY	AUG	SEP	OCT	NOV	DEC	JAN	FEB	MAR	APR	MAY	JUNE	TOTAL
1970-71	0.0	0.0	0.0	0.0	0.0	0.0	T	0.0	0.0	0.0	0.0	0.0	T
1971-72	0.0	0.0	0.0	0.0	0.0	0.0	T	0.0	0.0	0.0	0.0	0.0	T
1972-73	0.0	0.0	0.0	0.0	0.0	T	0.0	0.0	T	0.0	0.0	0.0	T
1973-74	0.0	0.0	0.0	0.0	0.0	0.0	0.0	0.0	0.0	0.0	0.0	0.0	0.0
1974-75	0.0	0.0	0.0	T	0.0	0.0	0.0	0.0	0.0	0.0	0.0	0.0	T
1975-76	0.0	0.0	0.0	0.0	0.0	0.0	0.0	T	0.0	0.0	0.0	0.0	T
1976-77	0.0	0.0	0.0	0.0	0.0	0.0	0.0	0.0	0.0	0.0	0.0	0.0	0.0
1977-78	0.0	0.0	0.0	0.0	0.0	0.0	0.0	0.0	0.0	0.0	0.0	0.0	0.0
1978-79	0.0	0.0	0.0	0.0	0.0	0.0	0.0	T	T	0.0	0.0	0.0	T
1979-80	0.0	0.0	0.0	0.0	0.0	0.0	0.0	0.0	0.0	0.0	0.0	0.0	0.0
1980-81	0.0	0.0	0.0	0.0	0.0	0.0	0.0	0.0	0.0	0.0	0.0	0.0	0.0
1981-82	0.0	0.0	0.0	0.0	0.0	0.0	0.0	0.0	0.0	0.0	0.0	0.0	0.0
1982-83	0.0	0.0	0.0	0.0	0.0	0.0	0.0	0.0	0.0	0.0	0.0	0.0	0.0
1983-84	0.0	0.0	0.0	0.0	0.0	0.0	0.0	0.0	0.0	0.0	0.0	0.0	0.0
1984-85	0.0	0.0	0.0	0.0	0.0	0.0	0.0	0.0	0.0	0.0	0.0	0.0	0.0
1985-86	0.0	0.0	0.0	0.0	0.0	0.0	0.0	0.0	0.0	0.0	0.0	0.0	0.0
1986-87	0.0	0.0	0.0	0.0	0.0	0.0	0.0	0.0	0.0	0.0	0.0	0.0	0.0
1987-88	0.0	0.0	0.0	0.0	0.0	0.0	0.0	0.0	0.0	0.0	0.0	0.0	0.0
1988-89	0.0	0.0	0.0	0.0	0.0	0.0	0.0	T	0.0	0.0	0.0	0.0	T
1989-90	0.0	0.0	0.0	0.0	0.0	0.0	0.0	T	0.0	0.0	0.0	0.0	T
1990-91	0.0	0.0	0.0	0.0	0.0	T	0.0	0.0	T	0.0	0.0	0.0	T
1991-92	0.0	0.0	0.0	0.0	0.0	0.0	0.0	T	0.0	0.0	0.0	0.0	T
1992-93	0.0	0.0	0.0	0.0	0.0	0.0	0.0	0.0	0.0	0.0	0.0	0.0	0.0
1993-94	0.0	0.0	0.0	0.0	0.0	0.0							
Record Mean	0.0	0.0	0.0	T	0.0	T	0.1	T	T	0.0	0.0	0.0	0.1

LONG BEACH, CALIFORNIA

The climate of the Long Beach Airport is considerably influenced by local topography. In fact, the topography plays a greater role in the climatic conditions at this station than the more general movements of pressure systems which dominate other sections of the country.

The Pacific Ocean, 4 miles south and 12 miles west, has a moderating effect on temperatures. The annual range of temperatures at the airport is much less than is experienced at stations further inland in the Los Angeles basin. Low coastal hills lie immediately between the station and the sea, the highest being Signal Hill, 1 5/8 miles southwest and 498 feet above sea level. The Palos Verdes Hills, 11 miles west-southwest of the station, slope upward to 1,480 feet above sea level. These natural barriers between the ocean and the station cause slightly greater ranges of high and low temperatures locally than at stations on the coast. During the winter months high temperatures are usually in the upper 60s, and lows in the 40s. In the summer highs are in the 70s and low 80s, and lows in the high 50s. Fortunately, high temperatures usually occur with low relative humidities, making infrequent heat waves tolerable for most people.

Precipitation is sparse during the summer months, with an average of only about 0.60 inch for the months of May through October. The greatest rainfall occurs during the winter months. Terrain again plays an important role. Precipitation at the station is considerably less than over the San Gabriel Mountains, about 28 miles to the north and the Santa Ana Mountains, 20 miles to the east. Even the coastal hills influence the local precipitation with greater amounts of rainfall occurring just 1 or 2 miles south and southwest of the station. Snow is an extremely rare phenomenon locally, although the San Gabriel Mountains are blanketed in the higher elevations much of the winter, and occasionally have snow down to the 2,500-foot level. Thunderstorms occur only sporadically at Long Beach.

With the Pacific Ocean only 4 miles south, it might be expected that the sea breeze would be from a southerly component. However, the coastal hills to the southwest combine with the lowest mountain passes leading to the interior desert valleys east of the Los Angeles basin to produce a sea breeze from a westerly component in the afternoon and early evening hours. Occasionally, strong dry northeasterly winds descend the mountain slopes in the fall, winter, and early spring months, developing velocities in excess of 50 mph over localized sections of the Los Angeles basin, usually below canyons. However, these strong winds ordinarily by-pass the station. Actually, the highest winds at Long Beach are recorded in association with the winter and spring storms which invade southern California from the Pacific.

During the summer months low clouds are quite common in the late night and morning hours at this station due to its proximity to the ocean. The tourist from the east and midwest usually expects a wet, rainy day, but by late morning or early afternoon the clouds have disappeared and the balance of the day is sunny and comfortable. Here again is a moderating influence on summertime temperatures locally which is not so prominent at stations further inland where the coastal cloudiness arrives later, burns off earlier, and penetrates less frequently.

NORMALS, MEANS, AND EXTREMES

LONG BEACH, CALIFORNIA

LATITUDE: 33°49'N LONGITUDE: 118°09' W ELEVATION: FT. GRND 25 BARO 68 TIME ZONE: PACIFIC WBAN: 23129

	(a)	JAN	FEB	MAR	APR	MAY	JUNE	JULY	AUG	SEP	OCT	NOV	DEC	YEAR
TEMPERATURE °F:														
Normals														
-Daily Maximum		66.8	67.7	68.0	71.5	73.3	77.0	82.7	84.0	82.1	78.4	72.1	67.0	74.2
-Daily Minimum		44.9	46.9	49.0	51.8	56.3	59.8	63.4	64.8	62.7	57.8	50.4	45.0	54.4
-Monthly		55.9	57.3	58.5	61.7	64.8	68.5	73.1	74.4	72.4	68.1	61.3	56.0	64.3
Extremes														
-Record Highest	41	91	91	98	105	103	109	107	105	110	111	101	92	111
-Year		1976	1986	1988	1989	1967	1981	1985	1967	1963	1961	1966	1958	OCT 1961
-Record Lowest	41	25	33	33	38	40	47	51	52	50	39	34	28	25
-Year		1963	1965	1964	1975	1964	1967	1960	1951	1965	1972	1958	1990	JAN 1963
NORMAL DEGREE DAYS:														
Heating (base 65°F)		285	221	214	134	69	39	0	0	15	24	145	284	1430
Cooling (base 65°F)		0	5	13	35	63	144	255	295	237	120	34	0	1201
% OF POSSIBLE SUNSHINE														
MEAN SKY COVER (tenths)														
Sunrise - Sunset	36	5.2	5.4	5.3	4.6	4.9	4.5	3.3	3.2	3.9	4.4	4.5	4.8	4.5
MEAN NUMBER OF DAYS:														
Sunrise to Sunset														
-Clear	36	11.9	9.8	11.0	12.7	10.8	12.1	18.2	18.8	14.6	13.1	13.4	12.9	159.2
-Partly Cloudy	36	8.2	7.5	9.4	9.9	12.5	12.1	10.6	10.6	10.9	11.1	8.1	8.3	119.1
-Cloudy	36	10.9	11.0	10.6	7.4	7.7	5.8	2.2	1.6	4.6	6.8	8.4	9.9	86.9
Precipitation														
.01 inches or more	49	5.2	5.1	5.4	3.2	1.1	0.4	0.2	0.4	1.1	1.9	3.4	4.6	32.0
Snow,Ice pellets,hail														
1.0 inches or more	49	0.0	0.0	0.0	0.0	0.0	0.0	0.0	0.0	0.0	0.0	0.0	0.0	0.0
Thunderstorms	50	0.4	0.6	0.8	0.3	0.1	0.1	0.2	0.3	0.5	0.4	0.4	0.2	4.2
Heavy Fog Visibility														
1/4 mile or less	50	5.1	3.8	3.0	2.3	1.4	1.5	2.0	3.1	3.8	5.0	5.4	5.7	42.0
Temperature °F														
-Maximum														
90° and above	33	0.1	0.1	0.2	0.8	1.1	1.7	3.7	5.3	5.6	3.2	0.8	0.0	22.6
32° and below	33	0.0	0.0	0.0	0.0	0.0	0.0	0.0	0.0	0.0	0.0	0.0	0.0	0.0
-Minimum														
32° and below	33	0.2	0.0	0.0	0.0	0.0	0.0	0.0	0.0	0.0	0.0	0.0	0.3	0.5
0° and below	33	0.0	0.0	0.0	0.0	0.0	0.0	0.0	0.0	0.0	0.0	0.0	0.0	0.0
AVG. STATION PRESS.(mb)	13	1016.8	1016.5	1014.4	1014.2	1012.6	1011.8	1011.8	1011.8	1011.1	1013.5	1015.4	1017.1	1013.9
RELATIVE HUMIDITY (%)														
Hour 04	25	75	77	78	79	80	82	82	81	82	81	79	77	79
Hour 10 (Local Time)	33	58	61	60	57	61	64	62	61	60	59	56	58	60
Hour 16	33	51	53	53	50	54	55	53	53	54	53	52	52	53
Hour 22	33	72	73	73	73	75	77	77	77	77	77	73	74	75
PRECIPITATION (inches):														
Water Equivalent														
-Normal		2.47	2.47	1.96	0.68	0.17	0.04	0.01	0.11	0.29	0.27	1.65	1.68	11.80
-Maximum Monthly	49	11.24	9.40	8.75	4.42	2.32	0.86	0.21	2.03	1.45	2.08	6.05	5.98	11.24
-Year		1969	1980	1983	1965	1977	1993	1986	1977	1976	1941	1965	1941	JAN 1969
-Minimum Monthly	49	0.00	0.00	0.00	0.00	0.00	0.00	0.00	0.00	0.00	0.00	0.00	T	0.00
-Year		1976	1964	1959	1993	1952	1978	1983	1978	1974	1969	1980	1989	APR 1993
-Maximum in 24 hrs	49	6.86	3.59	3.52	1.49	2.06	0.86	0.20	1.90	1.42	1.27	3.14	3.52	6.86
-Year		1956	1963	1983	1958	1977	1993	1986	1977	1986	1983	1967	1992	JAN 1956
Snow,Ice pellets,hail														
-Maximum Monthly	50	T	0.0	0.0	0.0	0.0	0.0	0.0	0.0	0.0	0.0	0.0	0.0	T
-Year		1993												JAN 1993
-Maximum in 24 hrs	49	T	0.0	0.0	0.0	0.0	0.0	0.0	0.0	0.0	0.0	0.0	0.0	T
-Year		1993												JAN 1993
WIND:														
Mean Speed (mph)	26	5.6	6.1	6.9	7.4	7.3	7.0	6.8	6.6	6.2	5.8	5.5	5.1	6.3
Prevailing Direction through 1963		WNW	W	W	W	S	S	WNW	WNW	WNW	WNW	WNW	WNW	WNW
Fastest Obs. 1 Min.														
-Direction (!!)	25	17	18	32	29	27	29	18	16	33	30	25	32	25
-Speed (MPH)	25	37	40	35	44	30	24	23	23	23	37	44	39	44
-Year		1964	1959	1961	1963	1961	1980	1967	1983	1967	1979	1982	1959	NOV 1982
Peak Gust														
-Direction (!!)	10	N	SE		W	W	N	S		W	E	W	NW	W
-Speed (mph)	10	43	40	49	46	43	40	26	28	29	38	54	45	54
-Date		1984	1993	1991	1988	1991	1990	1989	1990	1992	1992	1985	1991	NOV 1985

PRECIPITATION (inches)

LONG BEACH, CALIFORNIA

YEAR	JAN	FEB	MAR	APR	MAY	JUNE	JULY	AUG	SEP	OCT	NOV	DEC	ANNUAL
1964	0.58	0.00	1.08	0.25	0.04	0.14	T	T	T	0.18	1.26	1.43	4.96
1965	0.57	0.42	2.25	4.42	T	0.02	T	0.17	0.20	0.00	6.05	3.44	17.54
1966	1.19	1.65	0.47	T	0.02	T	T	T	0.01	0.01	1.78	3.91	9.04
1967	3.77	0.03	1.51	2.29	0.02	T	T	T	0.53	T	4.61	1.06	13.82
1968	0.52	0.38	2.20	0.38	T	T	T	0.00	T	0.26	0.37	1.55	5.66
1969	11.24	6.07	0.66	0.48	0.03	T	0.05	T	0.02	0.00	1.41	0.06	20.02
1970	1.79	2.02	1.10	T	T	T	T	T	T	0.05	3.50	3.81	12.27
1971	0.58	0.66	0.23	0.62	0.67	T	T	0.00	T	0.57	0.22	5.29	8.84
1972	0.00	0.03	T	0.18	0.02	0.09	0.00	0.22	0.15	0.03	3.94	1.36	6.02
1973	3.39	4.98	2.45	T	T	T	T	T	T	0.21	2.04	0.36	13.43
1974	6.12	0.17	3.28	0.18	0.05	T	T	T	0.00	0.58	0.03	5.21	15.62
1975	0.09	4.44	3.60	1.49	0.01	T	T	0.00	T	0.25	0.13	0.21	10.22
1976	0.00	2.40	0.66	1.18	0.01	0.14	T	0.03	1.45	0.07	0.98	0.43	7.35
1977	1.80	0.35	1.35	T	2.32	T	0.00	2.03	0.02	T	T	3.03	10.90
1978	7.62	8.60	6.17	0.80	T	0.00	0.00	0.00	1.04	0.02	2.00	1.42	27.67
1979	8.41	2.25	4.07	T	T	T	T	T	T	0.37	0.23	0.28	15.61
1980	7.17	9.40	2.86	0.29	0.10	T	T	T	T	T	0.00	1.54	21.36
1981	1.85	1.55	3.41	0.32	T	T	0.00	T	0.07	0.59	2.39	0.98	11.16
1982	1.92	0.20	3.12	0.76	0.16	T	0.00	T	0.40	0.19	3.07	0.92	10.74
1983	3.04	4.17	8.75	2.30	0.18	0.01	0.00	0.57	1.31	1.44	2.93	1.99	26.69
1984	0.25	0.01	0.13	1.06	0.00	0.01	0.05	0.08	0.15	0.35	1.20	5.20	8.49
1985	0.91	1.58	0.61	T	0.21	0.00	T	0.00	0.24	0.14	4.21	0.33	8.23
1986	1.88	4.97	2.68	0.43	0.00	T	0.21	0.00	1.43	0.40	1.12	0.37	13.49
1987	1.88	1.39	0.63	0.06	T	0.10	0.05	0.05	0.02	1.63	0.64	1.79	8.24
1988	1.67	1.05	0.02	1.33	0.00	T	T	0.02	0.04	T	0.75	3.21	8.09
1989	0.37	0.87	0.80	0.01	0.02	T	0.00	T	0.34	0.45	0.14	T	3.00
1990	1.59	2.08	0.09	0.50	1.20	T	0.00	T	T	0.00	0.22	0.02	5.70
1991	1.42	3.41	4.87	0.05	T	T	0.13	0.01	0.02	0.14	0.05	2.06	12.16
1992	1.48	4.54	5.29	0.02	0.02	T	0.06	0.00	T	0.52	T	4.97	16.90
1993	9.12	5.51	2.00	0.00	T	0.86	T	0.00	T	0.04	0.88	0.78	19.19
Record Mean	2.36	2.39	1.87	0.67	0.13	0.05	0.01	0.07	0.19	0.23	1.35	1.65	10.97

AVERAGE TEMPERATURE (deg. F)

LONG BEACH, CALIFORNIA

YEAR	JAN	FEB	MAR	APR	MAY	JUNE	JULY	AUG	SEP	OCT	NOV	DEC	ANNUAL
1964	56.2	55.7	55.9	59.1	59.5	63.1	70.0	70.9	67.4	67.5	56.1	53.8	61.3
1965	53.5	53.8	57.3	60.0	62.0	63.2	68.2	73.5	68.3	70.0	60.7	54.4	62.1
1966	53.4	53.6	59.1	62.8	64.8	69.3	73.6	76.3	72.4	69.2	61.5	56.3	64.4
1967	54.6	57.5	58.1	54.6	65.3	65.3	75.4	80.1	75.5	71.6	66.0	54.7	64.9
1968	55.6	60.4	61.1	63.5	66.1	68.5	73.8	73.5	71.8	65.7	61.1	54.0	64.6
1969	57.3	53.7	56.2	60.6	64.5	66.6	73.0	75.2	71.3	66.3	64.8	58.2	64.0
1970	55.9	59.3	59.6	59.9	66.0	69.0	73.8	75.2	72.2	66.2	60.9	54.5	64.4
1971	54.4	56.0	59.5	60.3	63.2	68.2	74.0	79.4	76.1	68.2	60.4	52.6	64.4
1972	53.5	56.4	59.5	63.0	66.8	70.7	74.4	73.9	72.3	66.1	59.1	54.0	64.1
1973	53.5	57.5	55.2	62.3	65.7	71.5	73.0	73.4	68.6	67.2	58.0	55.4	63.4
1974	53.4	55.7	57.2	62.0	64.7	70.4	73.9	73.3	72.2	66.7	63.2	54.1	63.9
1975	56.7	58.4	56.8	57.2	63.4	66.8	72.8	72.7	74.2	66.7	60.6	57.2	63.6
1976	60.1	59.3	59.9	59.6	66.3	72.0	73.6	74.2	74.3	72.3	66.2	59.0	66.4
1977	57.7	59.4	55.5	62.8	63.2	68.9	74.2	76.3	71.6	70.4	66.4	61.2	65.6
1978	56.2	56.9	64.0	61.4	68.9	69.9	72.6	73.0	74.2	69.7	58.4	52.6	64.8
1979	53.7	55.2	59.4	64.2	65.7	72.1	72.0	73.5	74.9	66.4	59.8	58.6	64.6
1980	58.5	60.9	58.0	62.6	62.5	68.9	74.1	74.4	70.3	67.9	61.7	59.2	64.9
1981	58.4	59.7	58.6	63.0	66.7	75.1	75.8	76.0	73.2	66.6	61.9	58.3	66.1
1982	53.8	58.7	58.0	61.5	64.9	65.5	73.3	75.9	74.0	69.4	60.5	55.7	64.3
1983	58.4	58.8	59.8	61.1	66.2	68.3	74.1	79.1	76.7	71.0	60.4	56.2	65.8
1984	57.8	58.0	61.5	62.3	68.0	69.3	76.7	76.6	79.2	66.0	57.7	54.0	65.6
1985	54.3	56.1	56.2	62.7	63.5	69.5	75.6	73.2	70.0	67.4	58.1	57.7	63.7
1986	60.8	58.4	60.3	62.4	65.1	69.0	71.4	74.5	67.6	66.7	63.1	57.2	64.7
1987	54.2	57.3	59.0	65.5	66.2	68.1	69.6	72.1	73.4	70.5	62.0	53.1	64.3
1988	56.1	60.6	63.3	63.4	66.2	66.9	73.3	73.1	71.1	68.4	60.4	56.0	64.9
1989	55.3	55.8	61.1	66.8	65.4	68.1	72.6	72.2	72.3	67.2	64.8	59.3	65.1
1990	56.8	55.5	59.3	64.2	66.0	71.7	75.8	73.4	73.6	70.3	63.1	55.5	65.4
1991	57.0	61.3	56.1	62.6	63.0	66.0	70.3	72.3	72.4	69.9	63.8	58.0	64.4
1992	57.3	60.4	59.3	67.2	68.1	69.4	74.8	77.5	74.5	68.9	62.9	54.5	66.2
1993	55.9	57.0	61.8	64.3	67.7	70.5	72.2	72.9	72.1	68.8	62.2	57.0	65.2
Record Mean	55.2	56.5	58.0	61.2	64.3	67.7	72.2	73.2	71.5	67.2	61.0	55.5	63.7
Max	66.0	66.7	67.5	70.8	72.9	76.3	81.8	82.8	81.4	77.4	72.2	67.0	73.6
Min	44.4	46.3	48.4	51.6	55.6	59.1	62.6	63.5	61.6	57.0	49.8	43.9	53.7

REFERENCE NOTES

LONG BEACH, CALIFORNIA

GENERAL
T - TRACE AMOUNT.
BLANK ENTRIES DENOTE MISSING/UNREPORTED DATA.
# INDICATES A STATION OR INSTRUMENT RELOCATION.

SPECIFIC

(a) - LENGTH OF RECORD IN YEARS, ALTHOUGH INDIVIDUAL MONTHS MAY BE MISSING.
0.* OR * - THE VALUE IS BETWEEN 0.0 AND 0.05
NORMALS - BASED ON THE 1961-1990 RECORD PERIOD.
EXTREMES - DATES ARE THE MOST RECENT OCCURRENCE
WIND DIR.- NUMERALS SHOW TENS OF DEGRESS CLOCKWISE FROM TRUE NORTH. "00" INDICATES CALM.
RESULTANT DIRECTIONS ARE GIVEN TO WHOLE DEGREES.
BOLD VALUES INDICATE EXTREME VALUES WHICH OCCURRED AFTER THE ASOS SYSTEM WAS COMMISSIONED.

RECORD = PERIOD OF RECORD
RECORD MEAN PRECIPITATION IS THE MEAN OF ALL DAILY PRECIPITATION AMOUNTS DURING THE PERIOD OF RECORD.
RECORD MAX(MIN) TEMPERATURE IS THE MEAN OF ALL DAILY MAX(MIN) TEMPERATURES DURING THE PERIOD OF RECORD.
RECORD MEAN TEMPERATURE IS THE SUM OF THE RECORD MAX AND RECORD MIN DIVIDED BY 2.
AVERAGE TEMPERATURE IS THE SUM OF THE MEAN DAILY MAX AND MIN TEMPERATURE DIVIDED BY 2.

EXCEPTIONS

STATIONS DO NOT REPORT PEAK GUST DIRECTION OR SNOWFALL WHEN THEY ARE OPERATING PART TIME.

MAXIMUM 24-HOUR PRECIPITATION BASED ON SIX HOUR MEASUREMENTS THROUGH NOVEMBER 1976.

RECORD MEANS ARE THROUGH THE CURRENT YEAR, BEGINNING IN 1941 FOR TEMPERATURE
1942 FOR PRECIPITATION
1944 FOR SNOWFALL

HEATING DEGREE DAYS Base 65 deg. F LONG BEACH, CALIFORNIA

SEASON	JULY	AUG	SEP	OCT	NOV	DEC	JAN	FEB	MAR	APR	MAY	JUNE	TOTAL
1964-65	1	0	8	23	264	342	352	305	231	172	99	53	1850
1965-66	0	0	0	10	126	320	353	313	182	83	22	8	1417
1966-67	0	0	0	5	130	265	315	205	206	308	64	30	1528
1967-68	0	0	0	0	37	312	282	135	130	72	25	3	996
1968-69	0	0	0	17	122	335	246	311	270	136	42	7	1486
1969-70	0	0	0	23	56	213	274	155	163	151	32	1	1068
1970-71	0	0	0	21	124	317	332	251	185	146	72	9	1457
1971-72	0	0	1	62	138	379	354	241	164	62	33	0	1434
1972-73	0	0	0	31	169	334	352	203	295	77	29	0	1490
1973-74	0	0	1	9	207	292	354	257	234	96	37	0	1487
1974-75	0	0	0	27	87	331	254	180	246	227	63	2	1417
1975-76	0	0	0	28	145	239	177	161	165	165	8	1	1089
1976-77	0	0	0	0	42	180	224	157	285	74	72	1	1035
1977-78	0	0	0	4	35	125	265	223	88	104	12	5	861
1978-79	0	0	0	4	201	376	344	268	177	41	30	1	1442
1979-80	0	0	0	21	147	204	195	116	209	99	82	11	1084
1980-81	0	0	0	14	103	185	197	156	189	89	8	0	941
1981-82	0	0	0	26	111	198	339	173	211	119	33	10	1220
1982-83	0	0	0	7	136	283	203	170	161	119	24	1	1104
1983-84	0	0	0	0	145	268	220	198	109	101	15	0	1056
1984-85	0	0	0	32	213	333	324	253	266	87	57	5	1570
1985-86	0	0	0	17	205	219	132	191	154	98	31	0	1047
1986-87	0	0	14	11	64	233	330	214	182	50	23	0	1121
1987-88	0	0	0	4	106	365	270	133	112	85	28	14	1117
1988-89	0	0	0	0	136	277	299	264	134	37	24	3	1174
1989-90	0	0	1	11	49	173	244	261	179	39	22	0	979
1990-91	0	0	0	2	82	292	241	107	268	89	86	9	1176
1991-92	0	0	0	23	72	212	236	141	171	17	0	0	872
1992-93	0	0	0	0	79	319	280	217	103	44	3	6	1051
1993-94	0	0	0	3	90	243							

COOLING DEGREE DAYS Base 65 deg. F LONG BEACH, CALIFORNIA

YEAR	JAN	FEB	MAR	APR	MAY	JUNE	JULY	AUG	SEP	OCT	NOV	DEC	TOTAL
1969	13	0	3	10	33	61	252	323	199	69	56	10	1029
1970	0	0	2	8	70	129	280	325	224	67	7	0	1112
1971	12	7	25	12	24	111	286	453	342	171	11	0	1454
1972	0	0	0	12	97	177	299	285	224	72	0	0	1166
1973	0	0	0	3	57	203	254	267	112	86	6	1	989
1974	0	0	0	15	31	167	284	262	220	87	40	0	1106
1975	6	0	0	0	22	62	248	243	282	90	18	3	974
1976	34	0	17	8	56	215	271	295	286	232	85	0	1499
1977	6	4	0	14	26	124	295	360	207	180	83	13	1312
1978	0	0	65	2	138	162	240	254	284	155	11	0	1311
1979	0	0	10	25	61	220	226	272	302	72	1	14	1203
1980	0	4	0	33	10	132	289	299	167	109	7	8	1058
1981	0	15	0	37	66	309	340	350	251	81	28	1	1478
1982	0	1	2	21	37	33	265	345	276	150	7	0	1137
1983	8	0	7	7	67	108	291	443	359	191	14	0	1495
1984	3	0	8	28	116	138	369	369	430	68	0	0	1529
1985	0	8	0	24	15	149	339	259	156	98	4	2	1054
1986	11	15	12	26	40	126	207	301	102	69	15	0	924
1987	1	4	2	71	67	97	151	225	262	185	22	0	1087
1988	3	12	65	43	70	79	264	259	186	111	3	6	1101
1989	3	12	19	98	43	102	243	229	225	85	49	6	1114
1990	0	1	11	21	60	207	343	266	265	174	32	5	1385
1991	0	9	0	23	31	45	170	232	229	182	46	0	967
1992	0	15	0	89	104	139	312	395	293	128	19	0	1494
1993	3	0	12	33	92	178	229	252	219	131	13	3	1165

SNOWFALL (inches) LONG BEACH, CALIFORNIA

SEASON	JULY	AUG	SEP	OCT	NOV	DEC	JAN	FEB	MAR	APR	MAY	JUNE	TOTAL
1970-71	0.0	0.0	0.0	0.0	0.0	0.0	0.0	0.0	0.0	0.0	0.0	0.0	0.0
1971-72	0.0	0.0	0.0	0.0	0.0	0.0	0.0	0.0	0.0	0.0	0.0	0.0	0.0
1972-73	0.0	0.0	0.0	0.0	0.0	0.0	0.0	0.0	0.0	0.0	0.0	0.0	0.0
1973-74	0.0	0.0	0.0	0.0	0.0	0.0	0.0	0.0	0.0	0.0	0.0	0.0	0.0
1974-75	0.0	0.0	0.0	0.0	0.0	0.0	0.0	0.0	0.0	0.0	0.0	0.0	0.0
1975-76	0.0	0.0	0.0	0.0	0.0	0.0	0.0	0.0	0.0	0.0	0.0	0.0	0.0
1976-77	0.0	0.0	0.0	0.0	0.0	0.0	0.0	0.0	0.0	0.0	0.0	0.0	0.0
1977-78	0.0	0.0	0.0	0.0	0.0	0.0	0.0	0.0	0.0	0.0	0.0	0.0	0.0
1978-79	0.0	0.0	0.0	0.0	0.0	0.0	0.0	0.0	0.0	0.0	0.0	0.0	0.0
1979-80	0.0	0.0	0.0	0.0	0.0	0.0	0.0	0.0	0.0	0.0	0.0	0.0	0.0
1980-81	0.0	0.0	0.0	0.0	0.0	0.0	0.0	0.0	0.0	0.0	0.0	0.0	0.0
1981-82	0.0	0.0	0.0	0.0	0.0	0.0	0.0	0.0	0.0	0.0	0.0	0.0	0.0
1982-83	0.0	0.0	0.0	0.0	0.0	0.0	0.0	0.0	0.0	0.0	0.0	0.0	0.0
1983-84	0.0	0.0	0.0	0.0	0.0	0.0	0.0	0.0	0.0	0.0	0.0	0.0	0.0
1984-85	0.0	0.0	0.0	0.0	0.0	0.0	0.0	0.0	0.0	0.0	0.0	0.0	0.0
1985-86	0.0	0.0	0.0	0.0	0.0	0.0	0.0	0.0	0.0	0.0	0.0	0.0	0.0
1986-87	0.0	0.0	0.0	0.0	0.0	0.0	0.0	0.0	0.0	0.0	0.0	0.0	0.0
1987-88	0.0	0.0	0.0	0.0	0.0	0.0	0.0	0.0	0.0	0.0	0.0	0.0	0.0
1988-89	0.0	0.0	0.0	0.0		0.0		0.0	0.0	0.0	0.0	0.0	
1989-90	0.0	0.0	0.0	0.0	0.0	0.0	0.0	0.0	0.0	0.0	0.0	0.0	0.0
1990-91	0.0	0.0	0.0	0.0	0.0	0.0	0.0	0.0	0.0	0.0	0.0	0.0	0.0
1991-92	0.0	0.0	0.0	0.0	0.0	0.0	0.0	0.0	0.0	0.0	0.0	0.0	0.0
1992-93	0.0	0.0	0.0	0.0	0.0	0.0	T	0.0	0.0	0.0	0.0	0.0	T
1993-94	0.0	0.0	0.0	0.0	0.0	0.0							
Record Mean	0.0	0.0	0.0	0.0	0.0	0.0	T	0.0	0.0	0.0	0.0	0.0	T

LOS ANGELES, CALIFORNIA
CIVIC CENTER

The climate of Los Angeles is normally pleasant and mild through the year. The Pacific Ocean is the primary moderating influence. The coastal mountain ranges lying along the north and east sides of the Los Angeles coastal basin act as a buffer against extremes of summer heat and winter cold occurring in desert and plateau regions in the interior. A variable balance between mild sea breezes, and either hot or cold winds from the interior, results in some variety in weather conditions, but temperature and humidity are usually well within the limits of human comfort. An important, and somewhat unusual, aspect of the climate of the Los Angeles metropolitan area is the pronounced difference in temperature, humidity, cloudiness, fog, rain, and sunshine over fairly short distances.

These differences are closely related to the distance from, and elevation above, the Pacific Ocean. Both high and low temperatures become more extreme and the average relative humidity becomes lower as one goes inland and up foothill slopes. Relative humidity is frequently high near the coast, but may be quite low along the foothills. During periods of high temperatures, the relative humidity is usually below normal so that discomfort is rare, except for infrequent periods when high temperatures and high humidities occur together.

Like other Pacific Coast areas, most rainfall comes during the winter with nearly 85 percent of the annual total occurring from November through March, while summers are practically rainless. As in many semi-arid regions, there is a marked variability in monthly and seasonal totals. Precipitation generally increases with distance from the ocean, from a yearly total of around inches in coastal sections to the south of th city to over 20 inches in foothill areas. Destructive flash floods occasionally develop in and below some mountain canyons. Snow is visible on nearby mountains in the winter, but extremely rare in the coastal basin. Thunderstorms are infrequent.

Prevailing winds are from the west during the spring, summer, and early autumn, with northeasterly wind predominating the remainder of the year. At times, the lack of air movement, combined with a frequent and persistent temperature inversion, is associated with concentrations of air pollution in the Los Angeles coastal basin and some adjacent areas. In fall, winter, and early spring months, occasional foehn-like descending Santa Ana winds come from the northeast over ridges and through passes in the coastal mountains. These Santa Ana winds may pick up considerable amounts of dust and reach speeds of 35 to 50 mph in north and east sections of the city, with higher speeds in outlying areas to the north and east, but rarely reach coastal portions of the city.

Sunshine, fog, and clouds depend a great deal on topography and distance from the ocean. Low clouds are common at night and in the morning along the coast during spring and summer, but form later and clear earlier near the foothills so that annual cloudiness and fog frequencies are greatest near the ocean, and sunshine totals are highest on the inland side of the city. The sun shines about 75 percent of daytime hours at the Civic Center. Light fog may accompany the usual night and morning low clouds, but dense fog is more likely to occur during the night and early morning hours of the winter months.

NORMALS, MEANS, AND EXTREMES

LOS ANGELES, CALIFORNIA CIVIC CENTER

LATITUDE: 34°03'N LONGITUDE: 118°14'W ELEVATION: FT. GRND 270 BARO TIME ZONE: PACIFIC WBAN: 93134

	(a)	JAN	FEB	MAR	APR	MAY	JUNE	JULY	AUG	SEP	OCT	NOV	DEC	YEAR
TEMPERATURE °F:														
Normals														
-Daily Maximum		67.7	69.4	69.5	72.3	73.9	78.3	84.0	84.5	82.7	79.0	72.4	67.8	75.1
-Daily Minimum		48.9	50.6	51.8	54.2	57.7	61.1	64.5	65.7	64.6	60.3	53.5	48.8	56.8
-Monthly		58.3	60.1	60.7	63.3	65.8	69.7	74.3	75.1	73.7	69.7	63.0	58.3	66.0
Extremes														
-Record Highest	53	95	94	98	106	102	112	107	105	110	108	100	91	112
-Year		1971	1986	1988	1989	1967	1990	1985	1983	1988	1987	1966	1979	JUN 1990
-Record Lowest	53	28	34	35	39	46	50	54	53	51	41	38	30	28
-Year		1949	1989	1976	1975	1964	1953	1952	1943	1948	1971	1978	1978	JAN 1949
NORMAL DEGREE DAYS:														
Heating (base 65°F)		222	170	169	128	72	35	0	0	10	17	105	226	1154
Cooling (base 65°F)		14	32	36	77	97	176	293	316	271	162	45	18	1537
% OF POSSIBLE SUNSHINE	32	69	72	73	70	66	65	82	83	79	73	74	71	73
MEAN SKY COVER (tenths)														
Sunrise - Sunset	34	4.4	4.7	4.7	4.7	4.8	4.3	2.7	2.6	3.0	3.8	3.7	4.2	4.0
MEAN NUMBER OF DAYS:														
Sunrise to Sunset														
-Clear	34	14.3	12.4	12.9	12.0	11.4	13.6	20.9	22.4	18.4	16.1	16.5	15.0	186.0
-Partly Cloudy	34	8.1	6.9	9.3	9.8	11.8	10.5	8.9	7.4	8.4	9.3	7.4	8.0	105.8
-Cloudy	34	8.5	9.0	8.7	8.2	7.8	5.9	1.1	1.2	3.3	5.6	6.1	8.0	73.5
Precipitation														
.01 inches or more	53	5.9	5.4	6.3	3.5	1.2	0.6	0.2	0.6	1.2	1.9	3.4	4.9	35.2
Snow,Ice pellets,hail														
1.0 inches or more	45	0.0	0.0	0.0	0.0	0.0	0.0	0.0	0.0	0.0	0.0	0.0	0.0	0.0
Thunderstorms	24	0.5	1.1	0.9	0.8	0.2	0.1	0.2	0.4	0.4	0.3	0.6	0.7	6.1
Heavy Fog Visibility 1/4 mile or less	24	1.5	1.8	1.1	1.3	0.5	0.6	0.5	0.8	1.5	2.6	2.5	2.1	16.8
Temperature °F														
-Maximum														
90° and above	53	0.1	0.1	0.2	0.9	1.1	1.5	3.8	4.3	6.0	3.2	0.7	0.*	21.8
32° and below	53	0.0	0.0	0.0	0.0	0.0	0.0	0.0	0.0	0.0	0.0	0.0	0.0	0.0
-Minimum														
32° and below	53	0.1	0.0	0.0	0.0	0.0	0.0	0.0	0.0	0.0	0.0	0.0	0.1	0.2
0° and below	53	0.0	0.0	0.0	0.0	0.0	0.0	0.0	0.0	0.0	0.0	0.0	0.0	0.0
AVG. STATION PRESS.(mb)														
RELATIVE HUMIDITY (%)														
Hour 04	17	63	71	74	78	81	85	84	84	78	76	61	62	75
Hour 10 (Local Time)	10	51	54	52	53	56	59	54	56	52	55	45	45	53
Hour 16	23	50	52	52	54	55	56	53	55	54	56	49	50	53
Hour 22	11	67	70	72	74	75	78	79	79	76	74	62	62	72
PRECIPITATION (inches):														
Water Equivalent														
-Normal		2.92	3.07	2.61	1.03	0.19	0.03	0.01	0.14	0.45	0.31	1.98	2.03	14.77
-Maximum Monthly	53	14.94	12.75	8.37	6.02	3.03	0.76	0.18	2.26	2.82	2.37	9.68	6.57	14.94
-Year		1969	1980	1983	1965	1977	1993	1986	1977	1976	1987	1965	1971	JAN 1969
-Minimum Monthly	53	0.00	T	0.00	0.00	0.00	0.00	0.00	0.00	0.00	0.00	0.00	0.00	0.00
-Year		1976	1951	1959	1979	1981	1982	1983	1982	1980	1980	1980	1990	DEC 1990
-Maximum in 24 hrs	53	6.11	4.02	3.79	2.05	2.41	0.76	0.18	2.22	1.95	1.77	4.07	3.92	6.11
-Year		1956	1944	1978	1956	1977	1993	1986	1977	1986	1983	1970	1965	JAN 1956
Snow,Ice pellets,hail														
-Maximum Monthly	47	0.3	T	0.0	0.0	0.0	0.0	0.0	0.0	0.0	0.0	0.0	T	0.3
-Year		1949	1951										1947	JAN 1949
-Maximum in 24 hrs	44	0.3	T	0.0	0.0	0.0	0.0	0.0	0.0	0.0	0.0	0.0	T	0.3
-Year		1949	1951										1947	JAN 1949
WIND:														
Mean Speed (mph)	24	6.8	6.9	7.0	6.6	6.3	5.7	5.4	5.3	5.3	5.7	6.4	6.6	6.2
Prevailing Direction through 1963		NE	W	W	W	W	W	W	W	W	W	W	NE	W
Fastest Mile														
-Direction (!!!)	36	N	NW	NW	NW	NW	N	W	E	NW	N	N	SE	N
-Speed (MPH)	36	49	40	47	40	39	32	21	24	27	48	42	44	49
-Year		1946	1961	1964	1955	1945	1949	1947	1945	1941	1959	1946	1943	JAN 1946
Peak Gust														
-Direction (!!!)														
-Speed (mph)														
-Date														

PRECIPITATION (inches)

LOS ANGELES, CALIFORNIA CIVIC CENTER

YEAR	JAN	FEB	MAR	APR	MAY	JUNE	JULY	AUG	SEP	OCT	NOV	DEC	ANNUAL
1964	1.43	T	1.79	0.33	0.01	0.32	T	0.00	0.00	0.33	1.72	2.05	7.98
1965	0.84	0.23	2.49	6.02	0.00	0.01	T	0.01	1.80	0.00	9.68	5.73	26.81
1966	0.96	1.51	0.53	0.00	0.22	0.00	T	T	0.30	0.06	4.07	5.26	12.91
1967	5.93	0.11	2.50	3.76	0.01	0.00	0.00	T	1.02	0.00	8.67	1.66	23.66
1968	0.90	0.49	3.34	0.49	0.00	0.01	0.01	0.11	0.03	0.55	0.37	1.28	7.58
1969	14.94	8.03	1.49	0.63	0.03	T	0.03	0.00	T	0.00	1.11	0.06	26.32
1970	1.59	2.58	2.36	0.00	0.00	0.04	0.00	0.00	0.00	0.00	5.05	4.92	16.54
1971	0.43	0.67	0.53	0.50	0.22	0.00	T	0.00	T	0.04	0.30	6.57	9.26
1972	0.00	0.13	T	0.03	0.03	0.07	0.00	0.35	0.02	0.29	3.26	2.36	6.54
1973	4.39	7.89	2.70	0.00	T	0.00	0.00	0.00	0.00	0.12	1.68	0.67	17.45
1974	8.35	0.14	3.78	0.10	0.08	0.00	0.00	0.00	0.00	0.58	0.07	3.59	16.69
1975	0.12	3.54	4.83	1.53	0.09	0.00	0.00	0.00	0.00	0.27	0.00	0.32	10.70
1976	0.00	3.71	1.81	0.84	0.05	0.22	0.00	0.08	2.82	0.24	0.49	0.75	11.01
1977	2.84	0.17	1.89	0.00	3.03	0.00	0.00	2.26	0.00	0.00	0.08	4.70	14.97
1978	7.70	8.91	8.02	1.77	0.00	0.00	0.00	0.00	0.39	0.05	2.28	1.45	30.57
1979	6.59	3.06	5.85	0.00	0.00	0.00	0.00	0.01	T	0.77	0.21	0.51	17.00
1980	7.50	12.75	4.79	0.31	0.13	0.00	0.00	0.00	0.00	0.00	0.00	0.85	26.33
1981	2.02	1.48	4.10	0.53	0.00	0.00	0.00	0.00	0.02	0.49	1.80	0.48	10.92
1982	2.17	0.70	3.54	1.39	0.12	0.00	0.00	0.00	0.84	0.19	4.41	1.05	14.41
1983	6.49	4.37	8.37	5.16	0.36	0.01	0.00	0.79	1.99	0.75	2.52	3.23	34.04
1984	0.17	0.00	0.28	0.69	0.00	0.01	0.00	0.40	0.23	0.15	1.44	5.53	8.90
1985	0.71	2.84	1.29	0.00	0.23	0.00	0.00	0.00	0.19	0.42	2.91	0.33	8.92
1986	2.19	6.10	5.27	0.45	0.00	0.00	0.18	0.00	1.97	0.53	0.94	0.37	18.00
1987	1.39	1.22	0.95	0.06	0.00	0.05	0.01	0.00	0.09	2.37	1.13	1.84	9.11
1988	1.65	1.72	0.26	3.41	0.00	0.00	0.00	0.05	0.04	0.00	0.70	3.80	9.98
1989	0.73	1.90	0.81	0.00	0.05	0.00	0.00	0.00	0.35	0.43	0.29	0.00	4.56
1990	1.24	3.12	0.17	0.58	1.17	0.00	0.00	0.02	0.00	0.00	0.19	0.00	6.49
1991	1.69	4.13	5.92	0.03	0.00	0.01	0.13	0.00	0.09	0.37	0.00	3.22	15.59
1992	1.74	7.96	7.12	0.33	0.04	0.00	0.08	0.00	0.00	0.70	0.00	4.68	22.65
1993	11.77	6.61	2.74	0.00	0.02	0.76	0.00	0.00	0.00	0.16	0.66	0.78	23.50
Record Mean	3.08	3.17	2.66	1.04	0.30	0.06	0.01	0.06	0.25	0.50	1.34	2.51	14.97

AVERAGE TEMPERATURE (deg. F)

LOS ANGELES, CALIFORNIA CIVIC CENTER

YEAR	JAN	FEB	MAR	APR	MAY	JUNE	JULY	AUG	SEP	OCT	NOV	DEC	ANNUAL
1964	56.7	59.0	58.6	60.5	61.7	64.8	71.8	73.4	70.2	70.5	59.8	56.5	63.6
1965	58.3	58.0	59.0	61.7	63.4	64.1	70.1	75.6	69.6	73.1	62.0	56.9	64.3
1966	55.7	56.2	61.3	64.4	64.5	70.2	74.4	76.6	73.4	71.2	63.6	60.7	66.0
1967	59.2	62.9	61.0	56.1	67.3	66.3	75.7	79.2	75.2	72.5	66.6	55.6	66.5
1968	58.5	63.8	62.9	64.0	65.5	69.2	74.7	74.1	73.4	69.2	63.3	56.2	66.2
1969	58.3	54.9	59.7	63.8	66.6	67.2	73.8	77.0	71.7	67.3	65.2	59.2	65.4
1970	57.6	61.4	61.2	60.9	67.4	70.0	75.3	76.2	74.4	68.3	63.3	57.2	66.1
1971	58.8	59.2	60.3	62.0	64.0	68.8	74.2	78.9	74.6	67.4	60.2	52.8	65.1
1972	55.5	60.3	63.7	63.9	67.6	72.2	78.0	77.4	72.3	67.2	62.2	58.1	66.5
1973	56.4	60.0	57.9	63.1	65.8	72.0	72.4	73.6	70.0	68.8	60.0	59.9	65.0
1974	55.2	59.2	59.6	64.7	65.7	72.2	74.1	72.3	73.2	67.6	64.0	56.2	65.3
1975	57.6	55.8	55.7	56.0	62.7	65.7	72.5	71.9	74.0	66.4	61.2	57.0	63.0
1976	59.4	56.4	58.4	57.8	64.3	71.1	72.6	71.6	72.6	70.7	66.9	60.4	65.2
1977	58.1	63.1	56.9	63.7	61.9	69.2	74.2	75.6	71.8	69.0	66.3	60.8	65.9
1978	58.1	58.9	63.2	60.8	68.6	71.8	73.4	73.7	76.0	70.3	58.4	53.2	65.5
1979	53.3	55.0	57.9	62.7	65.4	71.5	72.1	72.9	77.4	68.7	64.6	63.2	65.4
1980	60.9	64.6	60.9	64.8	63.2	71.8	77.1	76.3	72.6	71.5	65.3	63.7	67.7
1981	61.8	64.3	62.0	66.0	68.9	77.4	77.2	78.3	75.0	68.5	65.0	62.1	68.9
1982	57.1	64.0	59.3	62.2	64.4	65.3	74.0	75.1	73.9	70.9	61.7	58.1	65.5
1983	61.9	63.0	63.9	63.2	70.7	70.7	75.9	80.8	79.1	74.2	63.5	59.8	68.9
1984	61.2	61.9	65.6	65.3	72.4	72.2	78.7	76.4	81.3	68.5	61.0	57.2	68.5
1985	57.5	60.4	59.3	66.8	66.3	73.5	79.2	75.7	71.8	71.3	60.4	61.7	67.0
1986	65.9	62.4	64.5	66.4	68.1	71.2	73.2	76.0	68.8	69.4	66.4	60.1	67.7
1987	57.2	60.3	61.2	67.8	68.1	69.7	70.8	73.0	75.2	71.9	62.9	54.4	66.0
1988	58.3	62.9	64.9	64.1	67.2	67.9	74.3	72.9	72.2	69.7	61.9	57.1	66.8
1989	56.3	56.4	62.4	67.9	66.2	69.8	75.1	72.8	74.5	69.2	66.7	62.7	66.7
1990	59.4	58.0	61.7	65.7	66.9	74.3	77.3	74.0	76.0	73.2	65.6	57.5	67.5
1991	59.2	63.5	56.8	64.2	63.9	67.1	71.0	73.1	73.6	72.1	66.2	59.6	65.9
1992	60.3	62.3	60.8	69.6	69.0	70.4	75.9	78.9	76.6	70.4	65.1	56.5	68.0
1993	57.3	58.3	64.5	67.0	68.9	72.4	73.0	74.4	74.3	71.2	64.6	60.8	67.2
Record Mean	56.3	57.5	58.9	61.3	63.8	67.7	71.9	72.7	71.3	67.2	62.4	57.9	64.1
Max	65.6	66.7	68.0	70.6	72.6	77.0	82.2	82.8	81.5	77.2	72.8	67.3	73.7
Min	47.0	48.3	49.8	52.1	55.1	58.3	61.7	62.5	61.0	57.1	52.0	48.4	54.4

REFERENCE NOTES

LOS ANGELES, CALIFORNIA CIVIC CENTER

GENERAL
T - TRACE AMOUNT.
BLANK ENTRIES DENOTE MISSING/UNREPORTED DATA.
# INDICATES A STATION OR INSTRUMENT RELOCATION.

SPECIFIC

PM - INCLUDES LAST DAY OF PREVIOUS MONTH
ASOS - AUTOMATED SURFACE OBSERVING SYSTEM IN OPERATION DURING THESE MONTHS.

(a) - LENGTH OF RECORD IN YEARS, ALTHOUGH INDIVIDUAL MONTHS MAY BE MISSING.
0.* OR * - THE VALUE IS BETWEEN 0.0 AND 0.05
NORMALS - BASED ON THE 1961-1990 RECORD PERIOD.
EXTREMES - DATES ARE THE MOST RECENT OCCURRENCE
WIND DIR.- NUMERALS SHOW TENS OF DEGRESS CLOCKWISE FROM TRUE NORTH. "00" INDICATES CALM.
RESULTANT DIRECTIONS ARE GIVEN TO WHOLE DEGREES.
BOLD VALUES INDICATE EXTREME VALUES WHICH OCCURRED AFTER THE ASOS SYSTEM WAS COMMISSIONED.

RECORD = PERIOD OF RECORD
RECORD MEAN PRECIPITATION IS THE MEAN OF ALL DAILY PRECIPITATION AMOUNTS DURING THE PERIOD OF RECORD.
RECORD MAX(MIN) TEMPERATURE IS THE MEAN OF ALL DAILY MAX(MIN) TEMPERATURES DURING THE PERIOD OF RECORD.
RECORD MEAN TEMPERATURE IS THE SUM OF THE RECORD MAX AND RECORD MIN DIVIDED BY 2.
AVERAGE TEMPERATURE IS THE SUM OF THE MEAN DAILY MAX AND MIN TEMPERATURE DIVIDED BY 2.

EXCEPTIONS

1. RELATIVE HUMIDITY IS THROUGH 1963.
2. THUNDERSTORMS AND HEAVY FOG ARE THROUGH 1964 AND MAY BE INCOMPLETE, DUE TO PART-TIME OPERATIONS.
3. MEAN WIND SPEED IS THROUGH 1964.
4. PERCENT OF POSSIBLE SUNSHINE, MEAN SKY COVER AND DAYS CLEAR-PARTLY CLOUDY-CLOUDY ARE THROUGH 1976.
5. FASTEST MILE WIND IS THROUGH NOVEMBER 1977.

RECORD MEANS ARE THROUGH THE CURRENT YEAR, BEGINNING IN 1878 FOR TEMPERATURE
1878 FOR PRECIPITATION
1941 FOR SNOWFALL

HEATING DEGREE DAYS Base 65 deg. F LOS ANGELES, CALIFORNIA CIVIC CENTER

SEASON	JULY	AUG	SEP	OCT	NOV	DEC	JAN	FEB	MAR	APR	MAY	JUNE	TOTAL
1964-65	0	0	0	5	186	256	224	196	183	168	70	38	1326
1965-66	0	0	0	1	98	264	281	244	126	58	34	0	1106
1966-67	0	0	0	0	88	145	179	81	133	260	43	25	954
1967-68	0	0	0	0	32	287	207	70	99	70	50	7	822
1968-69	0	0	0	4	76	267	219	277	186	68	25	1	1123
1969-70	0	0	0	22	52	182	222	106	128	134	27	1	874
1970-71	0	0	0	8	72	243	255	184	154	127	77	11	1131
1971-72	0	0	0	94	153	369	288	132	61	61	35	0	1193
1972-73	0	0	0	14	97	230	266	136	214	77	32	2	1068
1973-74	0	0	1	8	156	174	300	160	171	54	32	1	1057
1974-75	0	0	0	25	73	268	243	254	283	262	75	21	1504
1975-76	0	0	0	31	132	247	190	246	215	215	50	12	1338
1976-77	0	0	0	0	62	138	215	86	247	57	108	1	914
1977-78	0	0	0	14	51	132	209	174	102	122	24	0	828
1978-79	0	0	0	7	209	361	354	274	226	80	46	6	1563
1979-80	0	0	0	1	59	114	128	60	123	79	75	4	643
1980-81	0	0	0	2	41	85	103	91	97	43	1	0	463
1981-82	0	0	0	11	58	102	238	58	184	113	41	16	821
1982-83	0	0	0	3	117	205	134	73	68	68	2	0	670
1983-84	0	0	0	0	99	158	140	99	29	59	5	0	589
1984-85	0	0	0	4	129	239	225	162	179	40	21	0	999
1985-86	0	0	0	0	163	131	42	125	92	32	7	0	592
1986-87	0	0	8	2	14	151	241	140	131	31	19	0	737
1987-88	0	0	0	3	91	323	216	82	81	88	30	11	925
1988-89	0	0	1	2	98	258	270	271	104	36	27	5	1072
1989-90	0	0	0	2	27	102	173	206	130	26	16	2	684
1990-91	0	0	0	0	42	244	183	63	248	74	72	2	928
1991-92	0	0	0	23	46	168	159	114	125	1	0	0	636
1992-93	0	0	0	1	49	256	235	181	68	9	0	6	805
1993-94	0	0	0	0	45	136							

COOLING DEGREE DAYS Base 65 deg. F LOS ANGELES, CALIFORNIA CIVIC CENTER

YEAR	JAN	FEB	MAR	APR	MAY	JUNE	JULY	AUG	SEP	OCT	NOV	DEC	TOTAL
1969	19	0	30	39	81	73	276	377	209	100	65	10	1279
1970	0	11	17	17	106	155	326	352	287	118	28	10	1427
1971	71	28	16	46	50	131	291	435	296	176	17	0	1557
1972	0	4	27	34	122	223	409	391	225	89	21	25	1570
1973	9	2	0	25	64	220	236	272	157	133	9	21	1148
1974	3	5	8	51	58	223	288	235	254	115	53	2	1295
1975	21	0	0	0	11	48	241	221	277	82	26	4	931
1976	22	3	20	10	32	203	245	212	233	185	123	2	1290
1977	7	39	3	23	18	135	293	334	210	148	96	8	1314
1978	0	8	52	2	145	212	269	277	338	177	17	0	1497
1979	0	0	14	17	67	209	229	252	379	124	53	62	1406
1980	10	54	3	82	26	215	380	357	233	210	56	53	1679
1981	12	75	13	81	132	380	387	422	306	124	67	17	2016
1982	0	36	15	36	33	32	286	322	275	194	25	0	1254
1983	44	23	41	21	185	174	342	495	432	292	60	4	2113
1984	29	14	56	73	240	222	433	360	496	123	13	6	2065
1985	0	41	10	100	68	264	447	339	210	203	31	35	1748
1986	77	56	83	80	110	194	261	349	132	145	65	6	1558
1987	6	18	21	120	121	147	186	257	312	221	36	1	1446
1988	13	30	84	68	107	107	297	252	223	154	14	20	1356
1989	8	37	31	131	73	154	318	251	290	139	85	41	1558
1990	10	16	36	54	81	291	388	287	336	262	68	20	1849
1991	8	28	0	54	43	72	191	260	265	250	88	6	1265
1992	17	44	1	146	132	168	347	435	353	175	59	0	1877
1993	6	0	59	77	129	232	254	299	287	199	38	12	1592

SNOWFALL (inches) LOS ANGELES, CALIFORNIA CIVIC CENTER

SEASON	JULY	AUG	SEP	OCT	NOV	DEC	JAN	FEB	MAR	APR	MAY	JUNE	TOTAL
1970-71	0.0	0.0	0.0	0.0	0.0	0.0	0.0	0.0	0.0	0.0	0.0	0.0	0.0
1971-72	0.0	0.0	0.0	0.0	0.0	0.0	0.0	0.0	0.0	0.0	0.0	0.0	0.0
1972-73	0.0	0.0	0.0	0.0	0.0	0.0	0.0	0.0	0.0	0.0	0.0	0.0	0.0
1973-74	0.0	0.0	0.0	0.0	0.0	0.0	0.0	0.0	0.0	0.0	0.0	0.0	0.0
1974-75	0.0	0.0	0.0	0.0	0.0	0.0	0.0	0.0	0.0	0.0	0.0	0.0	0.0
1975-76	0.0	0.0	0.0	0.0	0.0	0.0	0.0	0.0	0.0	0.0	0.0	0.0	0.0
1976-77	0.0	0.0	0.0	0.0	0.0	0.0	0.0	0.0	0.0	0.0	0.0	0.0	0.0
1977-78	0.0	0.0	0.0	0.0	0.0	0.0	0.0	0.0	0.0	0.0	0.0	0.0	0.0
1978-79	0.0	0.0	0.0	0.0	0.0	0.0	0.0	0.0	0.0	0.0	0.0	0.0	0.0
1979-80	0.0	0.0	0.0	0.0	0.0	0.0	0.0	0.0	0.0	0.0	0.0	0.0	0.0
1980-81	0.0	0.0	0.0	0.0	0.0	0.0	0.0	0.0	0.0	0.0	0.0	0.0	0.0
1981-82	0.0	0.0	0.0	0.0	0.0	0.0	0.0	0.0	0.0	0.0	0.0	0.0	0.0
1982-83	0.0	0.0	0.0	0.0	0.0	0.0	0.0	0.0	0.0	0.0	0.0	0.0	0.0
1983-84	0.0	0.0	0.0	0.0	0.0	0.0	0.0	0.0	0.0	0.0			
1984-85													
Record Mean	0.0	0.0	0.0	0.0	0.0	T	T	T	0.0	0.0	0.0	0.0	T

LOS ANGELES, CALIFORNIA INTERNATIONAL AIRPORT

Predominating influences on the climate of the Los Angeles International Airport are the Pacific Ocean, 3 miles to the west, the southern California coastal mountain ranges which line the inland side of the coastal plain surrounding the airport, and the large scale weather patterns which allow Pacific storm paths to extend as far south as the Los Angeles area only during late fall, winter, and early spring. Marine air covers the coastal plain most of the year but air from the interior reaches the coast at times, especially during the fall and winter months. The coast ranges act as a buffer to the more extreme conditions of the interior. Pronounced differences in temperature, humidity, cloudiness, fog, sunshine, and rain occur over fairly short distances on the coastal plains and the adjoining foothills due to the local topography and the decreased marine effect further inland. In general, temperature ranges are least and humidity highest close to the coast, while precipitation increases with elevation on the foothills.

The most characteristic feature of the climate of the coastal plain around the station is the night and morning low cloudiness and sunny afternoons which prevail during the spring and summer months and occur often during the remainder of the year. The coastal low cloudiness, combined with the westerly sea breeze, produces mild temperatures throughout the year. Daily temperature range is usually less than 15 degrees in spring and summer and about 20 degrees in fall and winter. Hot weather is not frequent at any season along the coast, although readings have exceeded the mid 80s at the airport every month of the year. When high temperatures do occur, the humidity is almost always low so that discomfort is unusual. Nighttime temperatures are generally cool but minimum temperatures below 40 degrees are rare and periods of over 10 years have passed with no readings below freezing at the airport. Prevailing daytime winds are from the west, but night and early morning breezes are usually light and from the east and northeast. Strongest winds observed at the station have been from the west and north following winter storms. During the fall, winter, and spring, gusty dry northeasterly Santa Ana winds blow over southern California mountains and through passes to the coast. These winds rarely reach L.A. International Airport but extremely dry air and dust clouds associated with them can be expected several times each year.

Precipitation occurs mainly in the winter. Measurable rain may fall on about one day in four from late October into early April, but in three years out of four, traces or less are reported for the entire months of July and August. Thunderstorms do not occur often near the coast, but showers and thunderstorms are observed over the coastal ranges at times during the summer when moist air from the south and southeast invades southern California. Annual rainfall at Los Angeles International Airport is somewhat less than that recorded on the Palos Verdes Hills, rising to an elevation of nearly 1,500 feet on a peninsula 12 miles to the south, and on the Hollywood Hills and Santa Monica Mountains which extend east–west 12 miles north of the station with peaks reaching to nearly 2,000 feet. Traces of snow have fallen at Los Angeles International Airport only a few times, melting as they fell.

Visibility at Los Angeles International Airport is frequently restricted by haze, fog, or smoke. Low visibilities are favored by a layer of moist marine air with warm dry air above and light winds but at times a moderate afternoon sea breeze will bring a fog bank ashore and over the airport. Light fog occurs at some time nearly every month, but heavy fog is observed least during the summer and can be expected on about one night or early morning in four during the winter.

NORMALS, MEANS, AND EXTREMES

LOS ANGELES, CALIFORNIA INTERNATIONAL AIRPORT

LATITUDE: 33°56'N LONGITUDE: 118°24'W ELEVATION: FT. GRND 97 BARO 109 TIME ZONE: PACIFIC WBAN: 23174

	(a)	JAN	FEB	MAR	APR	MAY	JUNE	JULY	AUG	SEP	OCT	NOV	DEC	YEAR
TEMPERATURE °F:														
Normals														
-Daily Maximum		65.7	65.9	65.5	67.4	69.0	71.9	75.3	76.6	76.6	74.4	70.3	65.9	70.4
-Daily Minimum		47.8	49.3	50.5	52.8	56.3	59.5	62.8	64.2	63.2	59.2	52.8	47.9	55.5
-Monthly		56.8	57.6	58.0	60.1	62.7	65.7	69.1	70.5	69.9	66.8	61.6	56.9	63.0
Extremes														
-Record Highest	58	88	92	95	102	97	104	97	98	110	106	101	94	110
-Year		1986	1963	1988	1989	1979	1981	1985	1955	1963	1961	1966	1958	SEP 1963
-Record Lowest	58	23	32	34	39	43	48	49	51	47	41	34	32	23
-Year		1937	1942	1939	1942	1938	1950	1942	1948	1948	1942	1939	1968	JAN 1937
NORMAL DEGREE DAYS:														
Heating (base 65°F)		258	215	224	158	96	54	9	5	22	32	127	258	1458
Cooling (base 65°F)		0	8	7	11	25	75	136	176	169	88	25	7	727
% OF POSSIBLE SUNSHINE														
MEAN SKY COVER (tenths)														
Sunrise - Sunset	45	5.2	5.2	5.1	4.8	5.1	4.9	4.0	3.8	4.2	4.5	4.4	4.7	4.6
MEAN NUMBER OF DAYS:														
Sunrise to Sunset														
-Clear	58	12.2	11.2	11.6	11.5	10.4	9.7	12.6	13.8	13.2	13.0	14.6	13.0	146.9
-Partly Cloudy	58	8.1	6.4	8.6	9.2	10.8	11.3	13.1	12.0	10.6	10.0	7.7	8.4	116.0
-Cloudy	58	10.7	10.7	10.8	9.3	9.8	9.0	5.3	5.2	6.3	7.9	7.6	9.6	102.3
Precipitation														
.01 inches or more	58	6.0	6.1	5.8	3.2	1.1	0.5	0.6	0.4	1.1	1.9	3.4	5.2	35.4
Snow, Ice pellets, hail														
1.0 inches or more	58	0.0	0.0	0.0	0.0	0.0	0.0	0.0	0.0	0.0	0.0	0.0	0.0	0.0
Thunderstorms	51	0.4	0.5	0.8	0.4	0.1	0.1	0.2	0.3	0.3	0.3	0.3	0.4	4.1
Heavy Fog Visibility														
1/4 mile or less	61	4.3	3.3	3.1	2.3	1.4	1.3	1.6	2.1	3.5	4.6	4.8	5.0	37.2
Temperature °F														
-Maximum														
90° and above	34	0.0	0.*	0.1	0.3	0.2	0.4	0.2	0.2	1.6	1.5	0.4	0.*	5.1
32° and below	34	0.0	0.0	0.0	0.0	0.0	0.0	0.0	0.0	0.0	0.0	0.0	0.0	0.0
-Minimum														
32° and below	34	0.*	0.0	0.0	0.0	0.0	0.0	0.0	0.0	0.0	0.0	0.0	0.*	0.1
0° and below	34	0.0	0.0	0.0	0.0	0.0	0.0	0.0	0.0	0.0	0.0	0.0	0.0	0.0
AVG. STATION PRESS. (mb)	21	1014.5	1014.0	1012.4	1011.6	1010.2	1009.4	1009.8	1009.7	1009.0	1011.1	1013.2	1014.5	1011.6
RELATIVE HUMIDITY (%)														
Hour 04	34	70	74	79	80	83	85	86	86	83	79	72	68	79
Hour 10 (Local Time)	34	55	59	61	60	65	68	68	68	65	59	53	52	61
Hour 16	34	59	63	65	64	66	67	68	69	67	65	61	59	64
Hour 22	34	69	72	75	76	80	82	83	83	80	77	70	68	76
PRECIPITATION (inches):														
Water Equivalent														
-Normal		2.40	2.51	1.98	0.72	0.14	0.03	0.01	0.15	0.31	0.34	1.76	1.66	12.01
-Maximum Monthly	58	10.63	11.07	6.37	4.52	2.55	0.74	0.32	2.47	4.39	2.34	7.92	6.57	11.07
-Year		1993	1962	1983	1965	1977	1993	1992	1977	1939	1936	1946	1936	FEB 1962
-Minimum Monthly	58	0.00	T	0.00	0.00	0.00	0.00	0.00	0.00	0.00	0.00	0.00	0.00	0.00
-Year		1976	1964	1959	1979	1943	1978	1983	1981	1968	1969	1980	1990	DEC 1990
-Maximum in 24 hrs	58	6.19	4.16	3.54	1.88	1.72	0.74	0.28	2.40	4.20	1.77	5.60	3.01	6.19
-Year		1956	1962	1968	1960	1977	1993	1992	1977	1939	1972	1967	1951	JAN 1956
Snow, Ice pellets, hail														
-Maximum Monthly	58	T	T	T	0.0	0.0	0.0	0.0	0.0	0.0	0.0	0.0	T	T
-Year		1982	1951	1991									1971	MAR 1991
-Maximum in 24 hrs	58	T	T	T	0.0	0.0	0.0	0.0	0.0	0.0	0.0	0.0	T	T
-Year		1982	1951	1991									1971	MAR 1991
WIND:														
Mean Speed (mph)	45	6.7	7.4	8.1	8.5	8.4	8.0	7.9	7.8	7.3	7.0	6.7	6.6	7.5
Prevailing Direction through 1963		W	W	W	WSW	WSW	WSW	WSW	WSW	WSW	W	W	W	WSW
Fastest Obs. 1 Min.														
-Direction (!!)														
-Speed (MPH)														
-Year														
Peak Gust														
-Direction (!!)	43	NE	N	W	N	W	W	SW	SE	E	W	W	NW	W
-Speed (mph)	43	51	57	62	59	49	40	31	33	39	46	60	49	62
-Date		1984	1953	1952	1957	1988	1990	1977	1955	1983	1974	1982	1984	MAR 1952

PRECIPITATION (inches)

LOS ANGELES, CALIFORNIA INTERNATIONAL AIRPORT

YEAR	JAN	FEB	MAR	APR	MAY	JUNE	JULY	AUG	SEP	OCT	NOV	DEC	ANNUAL
1964	1.49	T	1.20	0.20	0.01	0.29	T	T	T	0.30	1.07	1.95	6.51
1965	0.43	0.34	1.63	4.52	T	0.03	T	0.12	0.11	T	6.38	3.25	16.81
1966	0.84	1.40	0.49	0.01	0.02	T	0.01	T	0.19	0.04	2.69	3.67	9.36
1967	2.71	0.05	1.47	2.68	0.03	T	T	0.00	0.44	T	7.47	1.05	15.90
1968	0.84	0.44	3.77	0.49	T	T	0.04	T	0.00	0.32	0.24	1.42	7.56
1969	9.60	3.76	0.42	0.38	T	T	0.15	0.00	0.01	0.00	1.37	0.01	15.70
1970	1.44	1.39	1.29	T	T	0.01	T	0.00	T	0.02	3.68	4.12	11.95
1971	0.66	0.36	0.23	0.68	0.17	0.00	T	0.00	T	0.28	0.22	5.70	8.30
1972	0.00	0.16	T	T	0.01	0.06	T	0.06	0.03	1.79	3.13	1.88	7.12
1973	3.16	4.87	2.42	T	0.01	T	0.00	0.02	T	0.08	1.92	0.45	12.93
1974	5.68	0.13	2.49	0.14	0.02	T	T	T	T	0.54	T	3.76	12.76
1975	0.01	3.21	2.98	0.74	0.04	T	T	T	T	0.24	T	0.10	7.32
1976	0.00	2.15	0.83	0.77	T	0.28	0.02	0.03	1.85	1.50	0.87	0.95	9.25
1977	3.21	0.26	1.23	T	2.55	T	0.00	2.47	T	T	0.04	3.92	13.68
1978	7.48	7.66	5.75	1.23	T	0.00	0.00	T	0.39	0.04	1.20	0.83	24.58
1979	5.26	2.53	4.74	0.00	T	T	0.00	T	0.04	0.31	0.22	0.42	13.52
1980	6.97	9.13	3.69	0.17	0.07	T	0.00	0.00	T	T	0.00	1.57	21.60
1981	1.51	1.58	3.24	0.46	T	T	0.00	0.00	0.05	0.40	2.63	1.52	11.39
1982	2.78	0.66	3.41	1.61	0.11	0.01	0.00	T	0.78	0.18	3.48	0.66	13.68
1983	5.25	5.64	6.37	3.18	0.04	0.03	0.00	1.25	1.91	0.94	2.74	2.11	29.46
1984	0.39	0.01	0.14	1.16	T	T	0.00	0.29	0.09	0.28	1.24	4.21	7.81
1985	0.70	1.91	0.72	T	0.16	0.00	T	0.00	0.28	0.36	4.75	0.44	9.32
1986	2.31	5.36	4.89	0.30	0.00	T	0.09	T	1.44	0.10	1.14	0.30	15.93
1987	1.27	0.64	0.92	0.02	T	0.09	0.08	T	0.08	1.74	0.60	1.79	7.23
1988	1.61	1.79	0.08	1.14	T	T	0.00	0.02	0.07	T	0.73	2.52	7.96
1989	0.59	1.72	0.86	T	0.04	T	T	T	0.26	0.34	0.38	0.00	4.19
1990	1.18	2.60	0.14	0.34	0.83	T	0.00	0.02	T	0.00	0.10	0.03	5.24
1991	1.38	2.53	3.96	T	T	T	0.17	T	0.09	0.06	T	2.86	11.05
1992	1.61	4.70	5.08	0.18	0.04	T	0.32	0.00	0.00	0.50	0.00	4.16	16.59
1993	10.63	5.48	1.83	0.00	T	0.74	T	0.00	T	0.09	0.93	0.97	20.67
Record Mean	2.56	2.81	2.05	0.82	0.11	0.04	0.02	0.08	0.24	0.35	1.37	1.98	12.43

AVERAGE TEMPERATURE (deg. F)

LOS ANGELES, CALIFORNIA INTERNATIONAL AIRPORT

YEAR	JAN	FEB	MAR	APR	MAY	JUNE	JULY	AUG	SEP	OCT	NOV	DEC	ANNUAL
1964	55.3	56.0	55.9	58.2	58.5	62.2	66.8	69.4	66.7	67.0	58.9	55.2	60.9
1965	55.7	55.3	57.5	58.6	60.6	62.9	65.7	71.5	68.0	69.5	60.5	55.1	61.7
1966	53.6	53.0	57.5	61.7	63.5	66.6	68.2	71.9	69.6	69.5	62.5	57.5	63.0
1967	56.6	59.6	59.3	56.0	64.3	62.9	69.3	72.8	72.0	68.5	64.2	54.1	63.3
1968	57.3	60.8	60.5	61.6	63.9	66.2	68.3	68.5	68.6	65.3	61.7	54.3	63.1
1969	56.9	53.9	56.4	60.3	63.1	65.5	69.1	71.0	67.1	66.1	64.3	58.8	62.7
1970	56.9	60.7	59.4	59.3	63.3	65.7	68.7	69.4	69.2	66.1	61.2	56.1	63.0
1971	55.5	56.7	56.8	58.8	61.4	64.7	69.5	74.3	70.9	65.0	59.2	52.7	62.1
1972	53.9	56.9	59.8	61.5	64.1	67.1	70.3	72.8	69.8	66.4	60.2	56.8	63.3
1973	54.6	58.5	55.5	60.1	62.4	66.7	68.0	69.1	67.6	65.7	59.5	58.3	62.1
1974	54.3	57.5	56.6	59.7	62.9	66.7	70.5	69.6	69.0	65.9	62.7	55.8	62.6
1975	56.9	55.6	55.7	56.7	61.1	64.5	68.9	68.3	70.2	65.5	60.6	57.7	61.8
1976	60.0	57.9	58.4	58.0	63.3	68.0	70.1	69.3	70.4	69.2	65.9	60.1	64.2
1977	57.6	60.6	56.0	61.1	61.2	65.1	68.6	70.9	68.7	66.2	64.7	60.9	63.5
1978	58.6	58.3	62.1	59.2	65.5	67.6	67.7	69.8	73.6	68.3	59.0	54.3	63.7
1979	54.6	54.2	57.5	60.2	63.6	69.2	68.4	71.0	73.8	66.1	61.1	60.6	63.4
1980	59.7	61.3	58.3	60.6	60.1	65.7	68.7	70.9	67.0	66.2	62.4	60.2	63.4
1981	59.5	60.6	58.1	61.2	64.8	71.9	71.7	71.3	69.6	65.3	62.1	59.4	64.6
1982	54.6	59.1	57.6	60.2	62.5	63.6	69.3	71.4	71.6	69.2	61.2	56.3	63.1
1983	58.9	57.4	57.8	59.2	62.5	65.4	69.6	73.6	72.4	69.5	61.0	57.1	63.7
1984	58.2	58.5	61.3	61.4	66.1	67.1	71.3	73.1	76.5	65.6	58.7	55.2	64.4
1985	55.5	56.6	55.2	60.7	61.2	66.0	71.4	69.9	68.7	67.1	58.6	59.1	62.5
1986	62.3	58.7	59.2	61.7	63.4	66.3	68.5	68.9	65.8	66.4	65.0	58.4	63.7
1987	55.0	58.3	58.9	63.1	64.2	64.7	66.3	67.7	69.6	69.0	61.9	53.9	62.7
1988	56.7	60.6	62.5	61.3	63.0	63.6	69.2	68.3	67.7	66.1	60.1	56.5	63.5
1989	55.4	54.8	58.9	64.7	62.5	65.2	69.4	67.7	68.3	65.9	65.1	60.5	63.2
1990	57.1	55.0	57.5	62.5	62.7	67.5	71.4	69.7	71.0	69.6	64.2	56.8	63.8
1991	57.2	59.9	55.0	61.0	60.3	63.9	67.0	69.0	67.8	67.3	63.4	58.7	62.5
1992	58.4	60.8	59.2	65.6	66.1	65.9	71.5	72.9	70.7	67.0	63.7	55.8	64.8
1993	56.2	57.0	60.5	62.8	65.1	68.3	69.5	69.5	69.0	68.4	63.6	58.4	64.0
Record Mean	55.5	56.6	57.4	60.0	62.4	65.3	68.6	69.7	69.0	65.9	61.0	56.9	62.3
Max	64.6	65.2	65.3	67.5	69.2	71.7	75.0	76.1	76.1	73.8	70.5	66.3	70.1
Min	46.4	48.0	49.5	52.4	55.6	58.9	62.2	63.2	61.9	57.9	51.5	47.4	54.6

REFERENCE NOTES

LOS ANGELES, CALIFORNIA INTERNATIONAL AIRPORT

GENERAL
T - TRACE AMOUNT.
BLANK ENTRIES DENOTE MISSING/UNREPORTED DATA.
# INDICATES A STATION OR INSTRUMENT RELOCATION.

EXCEPTIONS

RECORD MEANS ARE THROUGH THE CURRENT YEAR,
BEGINNING IN 1937 FOR TEMPERATURE
1936 FOR PRECIPITATION
1936 FOR SNOWFALL

SPECIFIC

(a) - LENGTH OF RECORD IN YEARS, ALTHOUGH INDIVIDUAL MONTHS MAY BE MISSING.
0.* OR * - THE VALUE IS BETWEEN 0.0 AND 0.05
NORMALS - BASED ON THE 1961-1990 RECORD PERIOD.
EXTREMES - DATES ARE THE MOST RECENT OCCURRENCE
WIND DIR.- NUMERALS SHOW TENS OF DEGRESS CLOCKWISE FROM TRUE NORTH. "00" INDICATES CALM.
RESULTANT DIRECTIONS ARE GIVEN TO WHOLE DEGREES.
BOLD VALUES INDICATE EXTREME VALUES WHICH OCCURRED AFTER THE ASOS SYSTEM WAS COMMISSIONED.

RECORD = PERIOD OF RECORD
RECORD MEAN PRECIPITATION IS THE MEAN OF ALL DAILY PRECIPITATION AMOUNTS DURING THE PERIOD OF RECORD.
RECORD MAX(MIN) TEMPERATURE IS THE MEAN OF ALL DAILY MAX(MIN) TEMPERATURES DURING THE PERIOD OF RECORD.
RECORD MEAN TEMPERATURE IS THE SUM OF THE RECORD MAX AND RECORD MIN DIVIDED BY 2.
AVERAGE TEMPERATURE IS THE SUM OF THE MEAN DAILY MAX AND MIN TEMPERATURE DIVIDED BY 2.

HEATING DEGREE DAYS Base 65 deg. F — LOS ANGELES, CALIFORNIA INTERNATIONAL AIRPORT

SEASON	JULY	AUG	SEP	OCT	NOV	DEC	JAN	FEB	MAR	APR	MAY	JUNE	TOTAL
1964-65	9	0	8	21	197	301	291	271	226	198	130	60	1712
1965-66	12	0	1	9	127	304	350	328	225	113	45	8	1522
1966-67	0	0	0	1	102	224	254	150	171	263	64	61	1290
1967-68	1	0	0	3	49	331	236	118	135	115	61	5	1054
1968-69	3	0	1	26	104	328	250	306	266	143	64	7	1498
1969-70	0	0	3	28	59	199	244	121	168	167	71	10	1070
1970-71	0	1	5	20	118	274	311	234	248	191	130	34	1566
1971-72	0	0	3	92	180	371	339	229	161	103	53	0	1531
1972-73	0	0	0	15	138	248	314	177	287	146	89	12	1426
1973-74	0	0	1	24	159	207	323	208	256	158	62	2	1400
1974-75	0	0	0	26	100	279	258	258	278	243	117	23	1582
1975-76	0	0	0	31	145	226	176	202	206	205	49	15	1255
1976-77	0	0	0	5	62	145	227	129	272	111	115	13	1079
1977-78	1	0	0	19	62	120	195	184	116	168	46	3	914
1978-79	1	0	0	5	179	326	316	295	237	137	74	1	1571
1979-80	0	0	0	18	121	150	161	113	203	143	141	34	1084
1980-81	1	0	3	27	77	158	164	147	207	134	29	0	947
1981-82	0	0	0	36	112	172	314	165	225	147	75	45	1291
1982-83	0	0	0	3	119	261	203	205	218	167	80	23	1279
1983-84	0	0	0	0	129	237	206	182	116	117	22	0	1009
1984-85	0	0	0	29	183	299	287	243	296	148	115	14	1614
1985-86	0	0	1	17	192	181	107	184	182	114	52	7	1037
1986-87	0	0	21	15	34	199	306	189	190	84	44	19	1101
1987-88	1	1	0	6	111	337	257	137	127	118	72	61	1228
1988-89	0	0	5	16	145	265	294	291	189	80	72	25	1382
1989-90	0	2	2	15	44	152	237	275	223	74	82	5	1111
1990-91	0	0	0	1	62	261	236	145	303	130	145	41	1324
1991-92	0	0	0	29	79	190	202	130	172	25	0	3	830
1992-93	0	0	0	2	67	277	269	216	147	69	23	10	1080
1993-94	0	0	0	3	59	199							

COOLING DEGREE DAYS Base 65 deg. F — LOS ANGELES, CALIFORNIA INTERNATIONAL AIRPORT

YEAR	JAN	FEB	MAR	APR	MAY	JUNE	JULY	AUG	SEP	OCT	NOV	DEC	TOTAL
1969	7	0	4	9	10	29	137	196	74	70	45	13	594
1970	0	6	0	4	27	38	121	146	139	59	13	2	555
1971	23	8	0	13	23	33	148	292	184	101	14	0	839
1972	0	0	8	3	34	68	173	251	151	65	1	4	758
1973	1	0	0	4	14	70	99	134	83	52	2	8	467
1974	0	3	0	2	5	58	179	150	125	65	40	0	627
1975	13	0	0	0	0	14	128	108	163	51	21	7	505
1976	29	2	7	1	1	111	163	141	167	143	96	3	864
1977	3	13	0	1	0	23	121	193	118	63	62	5	602
1978	0	3	32	0	70	86	92	159	266	115	4	0	827
1979	0	0	9	0	35	133	112	193	271	60	9	23	845
1980	3	9	0	15	0	62	124	190	70	71	9	15	568
1981	0	30	0	25	29	212	214	204	145	53	31	4	947
1982	0	3	1	10	6	7	144	204	205	140	15	0	735
1983	22	0	0	1	11	40	151	274	231	146	17	0	893
1984	4	0	8	14	61	69	202	257	352	54	0	1	1022
1985	0	14	0	25	2	51	203	160	118	91	6	6	676
1986	29	11	9	24	9	52	115	124	52	64	40	2	531
1987	3	11	9	34	27	17	47	94	148	136	22	0	548
1988	9	16	56	14	17	26	134	109	91	56	1	8	528
1989	5	12	6	80	2	34	145	93	105	48	55	18	603
1990	1	1	1	5	16	86	204	151	187	147	43	13	855
1991	1	4	0	17	9	13	70	129	92	107	37	2	481
1992	5	14	0	50	43	36	207	250	177	73	31	0	886
1993	4	0	13	12	32	116	148	147	128	114	23	5	742

SNOWFALL (inches) — LOS ANGELES, CALIFORNIA INTERNATIONAL AIRPORT

SEASON	JULY	AUG	SEP	OCT	NOV	DEC	JAN	FEB	MAR	APR	MAY	JUNE	TOTAL
1970-71	0.0	0.0	0.0	0.0	0.0	0.0	0.0	0.0	0.0	0.0	0.0	0.0	0.0
1971-72	0.0	0.0	0.0	0.0	0.0	T	0.0	0.0	0.0	0.0	0.0	0.0	T
1972-73	0.0	0.0	0.0	0.0	0.0	0.0	0.0	0.0	0.0	0.0	0.0	0.0	0.0
1973-74	0.0	0.0	0.0	0.0	0.0	0.0	T	0.0	0.0	0.0	0.0	0.0	T
1974-75	0.0	0.0	0.0	0.0	0.0	0.0	0.0	0.0	0.0	0.0	0.0	0.0	0.0
1975-76	0.0	0.0	0.0	0.0	0.0	0.0	0.0	0.0	0.0	0.0	0.0	0.0	0.0
1976-77	0.0	0.0	0.0	0.0	0.0	0.0	0.0	0.0	0.0	0.0	0.0	0.0	0.0
1977-78	0.0	0.0	0.0	0.0	0.0	0.0	0.0	0.0	0.0	0.0	0.0	0.0	0.0
1978-79	0.0	0.0	0.0	0.0	0.0	0.0	0.0	0.0	0.0	0.0	0.0	0.0	0.0
1979-80	0.0	0.0	0.0	0.0	0.0	0.0	0.0	0.0	0.0	0.0	0.0	0.0	0.0
1980-81	0.0	0.0	0.0	0.0	0.0	0.0	0.0	0.0	0.0	0.0	0.0	0.0	0.0
1981-82	0.0	0.0	0.0	0.0	0.0	0.0	T	0.0	0.0	0.0	0.0	0.0	T
1982-83	0.0	0.0	0.0	0.0	0.0	0.0	0.0	0.0	0.0	0.0	0.0	0.0	0.0
1983-84	0.0	0.0	0.0	0.0	0.0	0.0	0.0	0.0	0.0	0.0	0.0	0.0	0.0
1984-85	0.0	0.0	0.0	0.0	0.0	0.0	0.0	0.0	0.0	0.0	0.0	0.0	0.0
1985-86	0.0	0.0	0.0	0.0	0.0	0.0	0.0	0.0	0.0	0.0	0.0	0.0	0.0
1986-87	0.0	0.0	0.0	0.0	0.0	0.0	0.0	0.0	0.0	0.0	0.0	0.0	0.0
1987-88	0.0	0.0	0.0	0.0	0.0	0.0	0.0	0.0	0.0	0.0	0.0	0.0	0.0
1988-89	0.0	0.0	0.0	0.0	0.0	0.0	0.0	0.0	0.0	0.0	0.0	0.0	0.0
1989-90	0.0	0.0	0.0	0.0	0.0	0.0	0.0	0.0	T	0.0	0.0	0.0	T
1990-91	0.0	0.0	0.0	0.0	0.0	0.0	0.0	0.0	T	0.0	0.0	0.0	T
1991-92	0.0	0.0	0.0	0.0	0.0	0.0	0.0	0.0	0.0	0.0	0.0	0.0	0.0
1992-93	0.0	0.0	0.0	0.0	0.0	0.0	0.0	0.0	0.0	0.0	0.0	0.0	0.0
1993-94	0.0	0.0	0.0	0.0	0.0	0.0							
Record Mean	0.0	0.0	0.0	0.0	0.0	T	T	T	T	0.0	0.0	0.0	T

REDDING, CALIFORNIA

No narrative available

NORMALS, MEANS, AND EXTREMES

REDDING, CALIFORNIA

LATITUDE: 40°30'N LONGITUDE: 122°18' W ELEVATION: FT. GRND 502 BARO 536 TIME ZONE: PACIFIC WBAN: 24257

	(a)	JAN	FEB	MAR	APR	MAY	JUNE	JULY	AUG	SEP	OCT	NOV	DEC	YEAR
TEMPERATURE °F:														
Normals														
-Daily Maximum		55.3	61.3	62.5	69.9	80.5	90.4	98.3	95.7	89.3	77.6	62.1	54.7	74.8
-Daily Minimum		35.7	40.0	41.7	46.0	52.3	61.8	64.7	63.1	58.8	49.2	41.4	35.2	49.2
-Monthly		45.5	50.7	52.2	58.0	66.4	76.1	81.5	79.5	74.1	63.5	51.8	45.0	62.0
Extremes														
-Record Highest	7	76	83	85	94	104	111	118	115	116	105	87	74	118
-Year		1989	1992	1988	1989	1987	1987	1988	1990	1988	1991	1986	1988	JUL 1988
-Record Lowest	7	19	21	28	35	36	42	54	53	40	33	23	17	17
-Year		1991	1989	1990	1992	1988	1990	1989	1993	1986	1989	1993	1990	DEC 1990
NORMAL DEGREE DAYS:														
Heating (base 65°F)		605	400	397	252	63	0	0	0	0	122	396	620	2855
Cooling (base 65°F)		0	0	0	42	107	333	512	450	278	75	0	0	1797
% OF POSSIBLE SUNSHINE	7	75	83	84	91	92	94	97	97	93	91	84	76	88
MEAN SKY COVER (tenths)														
Sunrise - Sunset	7	6.5	6.0	6.5	6.2	4.8	3.4	1.5	2.1	2.2	3.7	5.1	5.8	4.5
MEAN NUMBER OF DAYS:														
Sunrise to Sunset														
-Clear	7	8.6	8.7	7.6	7.6	12.3	17.3	25.7	23.4	22.3	17.6	12.1	11.3	174.4
-Partly Cloudy	7	5.6	6.7	6.7	9.0	9.0	7.9	3.7	5.7	4.1	6.9	7.8	4.9	77.9
-Cloudy	7	16.9	12.9	16.7	13.4	9.7	4.9	1.4	1.9	3.6	6.5	10.1	14.9	112.8
Precipitation														
.01 inches or more	7	11.7	8.7	12.9	7.0	6.6	3.9	0.7	1.1	2.4	5.0	6.8	8.8	75.4
Snow, Ice pellets, hail														
1.0 inches or more	7	0.3	0.1	0.1	0.0	0.1	0.0	0.0	0.0	0.0	0.0	0.0	0.4	1.1
Thunderstorms	7	0.1	0.7	1.6	1.4	1.9	1.9	1.1	1.6	0.9	1.1	0.3	0.4	12.9
Heavy Fog Visibility														
1/4 mile or less	7	5.6	1.1	0.7	0.1	0.0	0.0	0.0	0.0	0.0	0.5	2.0	4.9	14.9
Temperature °F														
-Maximum														
90° and above	7	0.0	0.0	0.0	1.4	6.6	16.7	27.1	24.9	18.4	7.9	0.0	0.0	103.0
32° and below	7	0.0	0.0	0.0	0.0	0.0	0.0	0.0	0.0	0.0	0.0	0.0	0.1	0.1
-Minimum														
32° and below	7	13.6	6.3	1.9	0.0	0.0	0.0	0.0	0.0	0.0	0.0	3.4	15.3	40.3
0° and below	7	0.0	0.0	0.0	0.0	0.0	0.0	0.0	0.0	0.0	0.0	0.0	0.0	0.0
AVG. STATION PRESS. (mb)	7	1002.8	1000.8	998.8	998.1	995.8	994.3	994.0	994.2	995.1	997.9	1002.4	1002.7	998.1
RELATIVE HUMIDITY (%)														
Hour 04	7	83	79	81	76	70	64	57	60	60	70	79	81	72
Hour 10 (Local Time)	7	74	63	63	51	42	37	32	33	34	44	61	73	51
Hour 16	7	55	44	49	37	31	26	19	19	22	31	45	55	36
Hour 22	7	78	70	72	64	55	48	40	42	44	59	72	76	60
PRECIPITATION (inches):														
Water Equivalent														
-Normal		6.06	4.45	4.38	2.08	1.27	0.56	0.17	0.46	0.91	2.24	5.21	5.51	33.30
-Maximum Monthly	7	10.38	10.15	10.94	3.76	6.72	1.74	0.49	1.06	4.83	6.26	10.11	10.37	10.94
-Year		1993	1992	1989	1989	1993	1988	1990	1990	1989	1992	1988	1992	MAR 1989
-Minimum Monthly	7	0.89	0.14	0.52	0.14	0.01	T	0.00	0.00	0.00	0.11	0.41	0.00	0.00
-Year		1991	1988	1988	1987	1987	1987	1990	1987	1988	1988	1986	1989	JUL 1990
-Maximum in 24 hrs	7	3.96	2.14	2.56	2.33	3.79	1.24	0.49	0.83	3.15	4.09	3.23	2.72	4.09
-Year		1990	1992	1993	1993	1993	1988	1990	1993	1989	1992	1988	1992	OCT 1992
Snow, Ice pellets, hail														
-Maximum Monthly	7	3.2	1.4	1.8	T	1.5	T	0.0	T	0.0	0.0	T	17.0	17.0
-Year		1989	1990	1987	1989	1990	1992		1993			1988	1988	DEC 1988
-Maximum in 24 hrs	7	2.0	1.4	1.8	T	1.5	T	0.0	T	0.0	0.0	T	10.0	10.0
-Year		1989	1990	1987	1989	1990	1992	1987	1993	1986	1986	1988	1988	DEC 1988
WIND:														
Mean Speed (mph)	7	6.2	7.1	7.8	7.3	8.0	8.2	7.5	6.8	6.5	6.4	6.3	7.0	7.1
Prevailing Direction through														
Fastest Obs. 1 Min.														
-Direction (!!!)	4	18	17	17	17	17	02	17	23	35	19	01	17	17
-Speed (MPH)	4	43	46	46	36	29	38	23	23	29	37	29	44	46
-Year		1993	1993	1991	1993	1990	1992	1991	1990	1993	1992	1993	1992	FEB 1993
Peak Gust														
-Direction (!!!)	7	S	S	S	S	S	N	S	S		S	S	S	S
-Speed (mph)	7	63	64	67	46	54	60	35	46	44	66	58	63	67
-Date		1993	1993	1991	1987	1988	1992	1990	1987	1986	1989	1988	1993	MAR 1991

PRECIPITATION (inches) **REDDING, CALIFORNIA**

YEAR	JAN	FEB	MAR	APR	MAY	JUNE	JULY	AUG	SEP	OCT	NOV	DEC	ANNUAL
1986									2.18	0.80	0.41	1.94	5.33
1987	7.01	4.97	7.00	0.14	0.01	T	0.21	0.00	T	0.48	3.53	9.07	32.42
1988	7.25	0.14	0.52	3.29	3.99	1.74	T	T	0.00	0.11	10.11	3.68	30.83
1989	2.14	1.11	10.94	3.76	0.73	0.95	0.00	0.23	4.83	3.69	1.20	0.00	29.58
1990	8.14	1.37	2.40	0.65	6.60	0.82	0.49	1.06	1.17	0.83	0.67	0.56	24.76
1991	0.89	3.97	9.67	0.52	2.13	0.11	0.03	0.00	0.00	1.93	1.27	5.09	25.61
1992	3.03	10.15	3.41	1.91	0.03	1.66	T	T	0.00	6.26	0.92	10.37	37.74
1993	10.38	7.52	6.34	3.66	6.72	0.65	0.00	0.99	0.00	2.92	1.52	3.16	43.86
Record Mean	5.55	4.18	5.75	1.99	2.89	0.85	0.10	0.33	1.02	2.13	2.45	4.23	31.47

AVERAGE TEMPERATURE (deg. F) **REDDING, CALIFORNIA**

YEAR	JAN	FEB	MAR	APR	MAY	JUNE	JULY	AUG	SEP	OCT	NOV	DEC	ANNUAL
1986									67.8	63.8	54.8	45.5	58.0
1987	43.3	49.3	51.6	63.3	72.1	79.3	78.1	81.2	74.1	68.2	52.2	44.9	63.1
1988	45.5	53.9	56.5	59.7	63.5	75.0	86.7	81.9	76.5	69.1	50.7	45.4	63.7
1989	44.5	45.9	52.0	63.0	66.5	76.1	80.2	77.5	71.3	61.0	53.6	46.4	61.5
1990	45.0	45.9	55.2	64.6	65.4	75.0	83.7	79.6	74.9	65.4	51.6	40.3	62.2
1991	45.8	53.7	48.7	56.4	63.6	73.3	85.0	77.8	79.1	68.9	54.4	46.1	62.7
1992	44.5	53.2	55.3	61.5	75.1	76.1	80.3	81.6	74.5	65.9	52.0	43.1	63.6
1993	43.4	47.2	57.0	57.4	65.1	73.5	81.1	78.5	74.0	65.2	50.7	45.5	61.6
Record Mean	44.5	49.8	53.8	60.8	67.3	75.5	82.1	79.7	74.0	65.9	52.5	44.6	62.5
Max	54.8	61.5	64.7	73.4	81.2	89.8	98.2	96.8	91.2	81.5	64.8	55.5	76.1
Min	34.2	38.1	42.8	48.2	53.4	61.1	66.0	62.6	56.8	50.3	40.1	33.7	49.0

REFERENCE NOTES **REDDING, CALIFORNIA**

GENERAL
T - TRACE AMOUNT.
BLANK ENTRIES DENOTE MISSING/UNREPORTED DATA.
\# INDICATES A STATION OR INSTRUMENT RELOCATION.

SPECIFIC

(a) - LENGTH OF RECORD IN YEARS, ALTHOUGH INDIVIDUAL MONTHS MAY BE MISSING.
0.* OR * - THE VALUE IS BETWEEN 0.0 AND 0.05
NORMALS - BASED ON THE 1961-1990 RECORD PERIOD.
EXTREMES - DATES ARE THE MOST RECENT OCCURRENCE
WIND DIR.- NUMERALS SHOW TENS OF DEGRESS CLOCKWISE FROM TRUE NORTH. "00" INDICATES CALM.
RESULTANT DIRECTIONS ARE GIVEN TO WHOLE DEGREES.
BOLD VALUES INDICATE EXTREME VALUES WHICH OCCURRED AFTER THE ASOS SYSTEM WAS COMMISSIONED.

RECORD = PERIOD OF RECORD
RECORD MEAN PRECIPITATION IS THE MEAN OF ALL DAILY PRECIPITATION AMOUNTS DURING THE PERIOD OF RECORD.
RECORD MAX(MIN) TEMPERATURE IS THE MEAN OF ALL DAILY MAX(MIN) TEMPERATURES DURING THE PERIOD OF RECORD.
RECORD MEAN TEMPERATURE IS THE SUM OF THE RECORD MAX AND RECORD MIN DIVIDED BY 2.
AVERAGE TEMPERATURE IS THE SUM OF THE MEAN DAILY MAX AND MIN TEMPERATURE DIVIDED BY 2.

HEATING DEGREE DAYS Base 65 deg. F — REDDING, CALIFORNIA

SEASON	JULY	AUG	SEP	OCT	NOV	DEC	JAN	FEB	MAR	APR	MAY	JUNE	TOTAL
1985-86													
1986-87			87	94	309	598	665	435	408	90	18	0	
1987-88	0	0	0	27	381	615	601	315	257	168	124	43	2531
1988-89	0	0	0	30	421	602	627	530	397	140	49	4	2800
1989-90	0	0	12	135	336	569	611	531	301	52	76	7	2630
1990-91	0	1	0	50	396	760	591	310	499	253	106	0	2966
1991-92	0	0	0	105	314	579	628	337	295	120	0	15	2393
1992-93	0	0	0	57	386	671	663	493	242	242	54	21	2829
1993-94	0	0	0	56	426	599							

COOLING DEGREE DAYS Base 65 deg. F — REDDING, CALIFORNIA

YEAR	JAN	FEB	MAR	APR	MAY	JUNE	JULY	AUG	SEP	OCT	NOV	DEC	TOTAL
1986									177	63	9	0	249
1987	0	0	0	45	245	439	412	512	279	132	1	0	2065
1988	0	0	4	16	82	349	679	531	350	165	0	1	2177
1989	0	0	0	85	103	344	477	394	210	17	0	0	1630
1990	0	0	1	47	97	314	586	463	306	67	0	0	1881
1991	0	0	0	2	72	257	627	404	430	234	3	0	2029
1992	0	0	0	22	318	355	481	521	291	90	0	0	2078
1993	0	0	0	19	65	282	504	425	279	71	4	0	1649

SNOWFALL (inches) — REDDING, CALIFORNIA

SEASON	JULY	AUG	SEP	OCT	NOV	DEC	JAN	FEB	MAR	APR	MAY	JUNE	TOTAL
1985-86													
1986-87			0.0	0.0	0.0	0.0	0.0	0.0	1.8	0.0	0.0	0.0	
1987-88	0.0	0.0	0.0	0.0	0.0	0.5	0.0	0.0	0.0	0.0	0.0	0.0	0.5
1988-89	0.0	0.0	0.0	0.0	T	17.0	3.2	T	T	T	0.0	0.0	20.2
1989-90	0.0	0.0	0.0	0.0	0.0	0.0	0.0	1.4	T	0.0	1.5	0.0	2.9
1990-91	0.0	0.0	0.0	0.0	0.0	0.0	T	0.0	0.0	0.0	0.0	0.0	T
1991-92	0.0	0.0	0.0	0.0	0.0	0.0	T	0.0	0.0	0.0	0.0	T	T
1992-93	0.0	0.0	0.0	0.0	0.0	T	0.4	T	0.0	0.0	0.0	0.0	0.4
1993-94	0.0	T	0.0	0.0	0.0	0.0							
Record Mean	0.0	T	0.0	0.0	T	2.2	0.5	0.2	0.3	T	0.2	T	3.4

SACRAMENTO, CALIFORNIA

Sacramento, and the lower Sacramento Valley, has a mild climate with abundant sunshine most of the year. A nearly cloud-free sky prevails throughout the summer months, and in much of the spring and fall. The summers are usually dry with warm to hot afternoons and mostly mild nights. The rainy season generally is November through March. About 75 percent of the annual precipitation occurs then, but measurable rain falls only on an average of nine days per month during that period. The shielding effect of mountains to the north, east, and west usually modifies winter storms. The Sierra Nevada snow fields, only 70 miles east of Sacramento, usually provide an adequate water supply during the dry season, and an important recreational area in winter. Heavy snowfall and torrential rains frequently fall on the western Sierra slopes, and may produce flood conditions along the Sacramento River and its tributaries. In the valley, however, excessive rainfall as well as damaging winds are rare.

The prevailing wind at Sacramento is southerly every month but November, when it is northerly. Topographic effects, the north-south alignment of the valley, the coast range, and the Sierra Nevada strongly influence the wind flow in the valley. A sea level gap in the coast range permits cool, oceanic air to flow, occasionally, into the valley during the summer season with a marked lowering of temperature through the Sacramento-San Joaquin River Delta to the capital. In the spring and fall, a large north-to-south pressure gradient develops over the northern part of the state. Air flowing over the Siskiyou mountains to the north warms and dries as it descends to the valley floor. This gusty, blustery north wind is a local variation of the chinook. It apparently carries a form of pollen which may cause allergic responses by susceptible individuals.

As is well known, relative humidity has a marked influence on the reactions of plants and animals to temperature. The extremely low relative humidity that ordinarily accompanies high temperatures in this valley should be considered when comparing temperatures here with those of cities in more humid regions. The extreme hot spells, with temperatures exceeding 100 degrees, are usually caused by air flow from a sub-tropical high pressure area that brings light to nearly calm winds and humidities below 20 percent.

Thunderstorms are few in number, usually mild in character, and occur mainly in the spring. An occasional thunderstorm may drift over the valley from the Sierra Nevada in the summer. Snow falls so rarely, and in such small amounts, that its occurrence may be disregarded as a climatic feature. Heavy fog occurs mostly in midwinter, never in summer, and seldom in spring or autumn. An occasional winter fog, under stagnant atmospheric conditions, may continue for several days. Light and moderate fogs are more frequent, and may come anytime during the wet, cold season. The fog is the radiational cooling type, and is usually confined to the early morning hours.

Sacramento is the geographical center of the great interior valley of California that reaches from Red Bluff in the north to Bakersville in the south. This predominantly agricultural region produces an extremely wide and abundant variety of fruits, grains, and vegetables ranging from the semi-tropical to the hardier varieties.

Based on the 1951-1980 period, the average first occurrence of 32 degrees Fahrenheit in the fall is December 1 and the average last occurrence in the spring is February 14.

NORMALS, MEANS, AND EXTREMES

SACRAMENTO, CALIFORNIA

LATITUDE: 38°31'N LONGITUDE: 121°30' W ELEVATION: FT. GRND 17 BARO 20 TIME ZONE: PACIFIC WBAN: 23232

	(a)	JAN	FEB	MAR	APR	MAY	JUNE	JULY	AUG	SEP	OCT	NOV	DEC	YEAR
TEMPERATURE °F:														
Normals														
-Daily Maximum		52.7	60.0	64.0	71.1	80.3	87.8	93.2	92.1	87.3	77.9	63.1	52.7	73.5
-Daily Minimum		37.7	41.4	43.2	45.5	50.3	55.3	58.1	58.0	55.7	50.4	43.4	37.8	48.1
-Monthly		45.2	50.7	53.6	58.3	65.3	71.6	75.7	75.1	71.5	64.2	53.3	45.3	60.8
Extremes														
-Record Highest	43	70	76	88	93	105	115	114	109	108	101	87	72	115
-Year		1991	1992	1988	1990	1984	1961	1972	1990	1988	1970	1960	1989	JUN 1961
-Record Lowest	43	23	23	26	32	36	41	48	49	43	36	26	18	18
-Year		1979	1989	1971	1953	1974	1990	1983	1978	1978	1989	1993	1990	DEC 1990
NORMAL DEGREE DAYS:														
Heating (base 65°F)		614	400	357	230	80	12	0	0	16	78	351	611	2749
Cooling (base 65°F)		0	0	0	29	89	210	332	313	211	53	0	0	1237
% OF POSSIBLE SUNSHINE	45	48	65	74	82	89	94	97	96	93	86	66	49	78
MEAN SKY COVER (tenths)														
Sunrise - Sunset	45	7.0	6.2	5.6	4.8	3.5	2.2	1.1	1.5	1.8	3.2	5.6	6.7	4.1
MEAN NUMBER OF DAYS:														
Sunrise to Sunset														
-Clear	45	6.6	8.0	10.1	12.0	17.3	21.7	27.0	25.4	23.6	19.2	10.2	7.8	188.8
-Partly Cloudy	45	6.0	6.8	8.3	9.6	8.4	5.9	3.1	4.3	4.2	6.1	7.2	5.9	75.8
-Cloudy	45	18.4	13.4	12.6	8.4	5.3	2.4	0.9	1.3	2.2	5.7	12.6	17.3	100.6
Precipitation														
.01 inches or more	54	9.9	8.7	8.8	5.2	2.7	1.1	0.2	0.4	1.3	3.4	6.8	8.9	57.5
Snow,Ice pellets,hail														
1.0 inches or more	45	0.0	0.*	0.0	0.0	0.0	0.0	0.0	0.0	0.0	0.0	0.0	0.0	*
Thunderstorms	45	0.4	0.6	0.8	0.7	0.4	0.3	0.2	0.2	0.5	0.3	0.2	0.2	4.8
Heavy Fog Visibility														
1/4 mile or less	45	10.2	5.2	1.7	0.3	0.2	0.0	0.0	0.*	0.2	1.4	5.2	9.6	34.0
Temperature °F														
-Maximum														
90° and above	43	0.0	0.0	0.0	0.4	5.3	12.0	22.3	19.4	12.5	2.9	0.0	0.0	74.8
32° and below	43	0.*	0.0	0.0	0.0	0.0	0.0	0.0	0.0	0.0	0.0	0.0	0.*	*
-Minimum														
32° and below	43	6.6	1.7	0.5	0.*	0.0	0.0	0.0	0.0	0.0	0.0	1.3	6.6	16.8
0° and below	43	0.0	0.0	0.0	0.0	0.0	0.0	0.0	0.0	0.0	0.0	0.0	0.0	0.0
AVG. STATION PRESS. (mb)	21	1019.3	1017.6	1015.6	1015.1	1012.5	1011.2	1011.0	1011.2	1011.3	1014.5	1018.0	1019.6	1014.7
RELATIVE HUMIDITY (%)														
Hour 04	33	90	88	85	82	81	78	77	78	78	80	86	90	83
Hour 10 (Local Time)	33	85	78	70	58	50	47	47	50	50	57	73	84	62
Hour 16	33	70	60	54	43	35	31	28	29	31	38	57	70	46
Hour 22	33	86	81	78	73	69	65	61	64	64	70	81	87	73
PRECIPITATION (inches):														
Water Equivalent														
-Normal		3.73	2.87	2.57	1.16	0.27	0.12	0.05	0.07	0.37	1.08	2.72	2.51	17.52
-Maximum Monthly	54	9.14	8.77	7.12	4.76	3.13	1.26	0.79	0.65	2.78	7.51	7.41	12.64	12.64
-Year		1978	1962	1982	1941	1948	1993	1974	1976	1989	1962	1970	1955	DEC 1955
-Minimum Monthly	54	0.16	0.15	0.14	0.00	T	0.00	0.00	0.00	0.00	0.00	0.02	0.00	0.00
-Year		1984	1964	1966	1949	1992	1981	1983	1982	1980	1966	1959	1989	DEC 1989
-Maximum in 24 hrs	45	3.41	3.01	2.30	2.22	0.78	1.21	0.78	0.65	1.79	5.59	2.95	3.64	5.59
-Year		1967	1986	1982	1958	1957	1993	1974	1965	1989	1962	1970	1955	OCT 1962
Snow,Ice pellets,hail														
-Maximum Monthly	45	T	2.0	T	0.0	0.0	0.0	0.0	0.0	0.0	0.0	0.0	T	2.0
-Year		1974	1976	1982									1993	FEB 1976
-Maximum in 24 hrs	45	T	2.0	T	0.0	0.0	0.0	0.0	0.0	0.0	0.0	0.0	T	2.0
-Year		1974	1976	1982									1993	FEB 1976
WIND:														
Mean Speed (mph)	44	7.2	7.5	8.6	8.7	9.2	9.7	9.0	8.6	7.5	6.4	6.0	6.7	7.9
Prevailing Direction through 1963		SE	SSE	SW	SW	SW	SW	SSW	SW	SW	SW	NNW	SSE	SW
Fastest Mile														
-Direction (!!)	45	SE	SE	S	SW	SW	SW	SW	SW	NW	SE	SE	SE	SW
-Speed (MPH)	45	60	51	66	45	74	47	36	38	42	68	70	70	74
-Year		1954	1959	1952	1955	1993	1950	1956	1954	1965	1950	1953	1952	MAY 1993
Peak Gust														
-Direction (!!)														
-Speed (mph)														
-Date														

PRECIPITATION (inches) SACRAMENTO, CALIFORNIA

YEAR	JAN	FEB	MAR	APR	MAY	JUNE	JULY	AUG	SEP	OCT	NOV	DEC	ANNUAL
1964	3.83	0.15	1.36	0.17	0.23	0.39	0.01	0.11	0.00	1.72	2.70	6.03	16.70
1965	3.01	0.41	1.47	2.70	0.09	T	T	0.65	T	0.11	2.93	2.44	13.81
1966	1.91	1.56	0.14	0.47	0.25	0.02	0.10	T	0.07	0.00	5.73	3.53	13.78
1967	8.42	0.41	3.91	3.40	0.13	0.60	T	0.00	0.04	0.24	1.18	1.29	19.62
1968	3.77	2.13	2.39	0.42	0.16	0.15	T	0.02	0.00	0.60	2.49	2.77	14.90
1969	8.50	6.98	0.94	1.63	0.04	0.08	T	0.00	0.02	0.72	0.60	4.41	23.92
1970	7.88	1.58	1.62	0.18	T	0.16	0.00	0.00	0.00	0.84	7.41	3.40	23.07
1971	0.90	0.56	2.05	0.44	0.77	0.01	0.00	0.00	T	0.13	0.87	4.05	9.78
1972	0.81	1.28	0.29	1.39	0.28	0.19	0.00	0.00	0.90	1.75	5.14	1.88	13.91
1973	6.87	5.64	2.76	0.05	0.13	0.00	0.00	0.00	0.33	1.64	6.27	2.79	26.48
1974	3.58	1.37	3.27	0.96	0.01	0.50	0.79	T	0.00	1.16	0.66	2.86	15.16
1975	0.73	4.59	4.28	0.81	T	T	0.04	0.23	T	2.03	0.29	0.18	13.18
1976	0.36	1.49	0.44	1.53	0.00	0.04	0.00	0.65	0.52	0.02	0.55	0.65	6.25
1977	1.17	1.17	1.27	0.30	0.73	0.00	T	0.00	0.76	0.12	1.92	4.27	11.71
1978	9.14	4.46	3.38	2.31	T	T	0.00	T	0.30	T	3.20	0.95	23.74
1979	5.66	4.55	2.47	0.76	0.14	0.00	0.25	0.00	T	1.62	1.48	3.41	20.34
1980	5.64	7.12	2.62	1.06	0.49	0.04	0.40	0.00	0.00	0.06	0.12	1.79	19.34
1981	4.56	0.87	3.55	0.66	0.50	0.00	0.00	0.00	0.25	2.57	6.09	3.28	22.33
1982	5.50	2.35	7.12	3.07	T	0.15	0.00	0.00	1.81	2.61	5.74	3.25	31.60
1983	4.92	5.56	6.75	4.21	0.25	0.40	0.00	0.11	0.66	0.40	4.91	5.26	33.43
1984	0.16	1.22	1.35	0.34	0.01	0.10	T	0.01	0.07	1.39	3.61	1.23	9.49
1985	0.66	1.52	2.01	T	0.01	0.15	T	0.06	0.56	0.53	3.72	2.34	11.56
1986	3.67	8.60	3.20	0.91	0.07	0.00	0.00	0.00	0.60	0.19	0.14	0.76	18.14
1987	2.29	3.23	3.05	0.20	T	T	0.00	0.00	0.00	1.28	2.53	3.25	15.83
1988	2.96	0.99	0.17	1.58	0.89	0.19	0.00	0.00	0.00	0.19	1.68	2.73	11.38
1989	0.71	1.25	6.29	0.31	0.06	0.43	0.00	0.20	2.78	1.76	1.32	0.00	15.11
1990	4.97	2.91	0.93	0.73	2.10	0.00	T	0.00	0.00	0.09	0.43	1.60	13.76
1991	0.36	3.10	6.14	0.29	0.25	0.53	T	0.14	0.04	1.25	0.19	1.60	13.89
1992	1.39	5.47	2.05	0.92	T	0.15	0.00	T	0.00	1.31	0.28	4.94	16.51
1993	8.63	4.94	2.39	0.63	1.14	1.26	0.00	0.00	0.00	0.47	2.28	1.75	23.49
Record Mean	3.59	3.00	2.53	1.30	0.44	0.13	0.03	0.05	0.28	0.99	2.20	2.89	17.43

AVERAGE TEMPERATURE (deg. F) SACRAMENTO, CALIFORNIA

YEAR	JAN	FEB	MAR	APR	MAY	JUNE	JULY	AUG	SEP	OCT	NOV	DEC	ANNUAL
1964	45.1	49.4	52.3	58.9	63.0	70.0	75.9	75.8	70.3	66.7	50.3	49.7	60.6
1965	45.4	49.2	53.0	57.3	64.4	67.0	74.4	75.7	67.6	65.9	54.7	41.0	59.6
1966	45.7	47.5	54.3	62.9	66.4	72.6	72.9	76.9	72.0	64.1	54.7	46.1	61.4
1967	45.9	48.4	50.8	49.6	65.2	69.4	78.0	79.0	74.9	65.5	56.2	44.0	60.6
1968	43.4	54.5	55.1	59.8	64.7	73.9	76.2	73.0	72.2	62.0	52.3	43.5	60.9
1969	44.0	47.4	52.4	57.1	66.9	69.0	76.3	77.9	74.0	61.0	53.9	48.0	60.7
1970	49.3	51.9	54.9	56.1	67.6	71.5	76.8	74.7	72.8	62.1	55.3	46.1	61.6
1971	45.3	48.2	52.6	56.4	61.8	70.5	76.2	76.8	72.2	61.2	52.7	43.0	59.8
1972	41.0	51.4	58.6	58.6	66.9	72.5	76.0	75.9	69.5	62.0	49.7	40.6	60.2
1973	44.3	53.1	51.1	60.8	69.4	74.6	76.8	74.2	70.8	63.4	51.9	46.9	61.5
1974	46.3	48.4	54.1	56.6	63.8	70.5	74.1	74.0	72.2	66.3	53.2	46.4	60.5
1975	43.4	49.2	50.3	51.8	68.2	73.2	77.3	76.9	77.4	65.1	54.6	47.4	61.2
1976	47.2	51.9	54.6	57.9	70.1	73.9	76.5	73.4	71.8	66.1	56.7	46.5	62.2
1977	43.8	52.4	51.0	62.2	59.4	72.2	74.2	73.9	68.8	63.9	54.5	49.6	60.5
1978	50.3	51.9	57.2	55.8	66.4	70.0	75.1	75.0	69.6	65.9	49.8	41.7	60.7
1979	45.3	48.8	54.6	56.9	66.7	71.9	75.6	73.1	74.6	64.0	51.7	46.7	60.9
1980	46.9	51.9	51.6	59.6	62.7	66.8	75.0	71.4	69.4	63.7	53.5	45.4	59.9
1981	46.8	50.4	51.2	57.9	64.7	74.8	75.1	74.5	69.7	63.5	60.3	48.6	61.5
1982	42.0	50.5	50.8	55.5	64.6	66.2	72.1	71.7	68.2	61.0	46.9	43.0	57.7
1983	43.1	52.2	53.4	54.7	64.2	70.8	72.2	76.6	74.9	67.5	53.7	51.0	61.2
1984	48.2	50.2	58.1	58.7	70.0	71.7	78.3	75.3	75.5	62.8	53.6	45.1	62.3
1985	42.4	51.4	50.8	61.5	63.2	75.1	77.0	72.9	68.5	63.3	49.8	42.6	59.9
1986	51.4	54.7	58.8	58.4	65.5	71.6	75.0	75.2	66.2	64.8	55.5	45.7	61.9
1987	44.9	51.3	53.8	62.7	69.1	72.4	71.8	74.9	71.8	67.6	53.4	47.2	61.7
1988	48.0	54.2	58.0	60.9	64.7	72.9	80.4	75.9	72.5	66.5	53.8	46.2	62.8
1989	44.1	47.1	55.6	63.2	65.8	71.7	76.2	73.8	69.6	62.4	54.3	44.3	60.7
1990	47.5	48.6	55.4	63.4	65.5	72.4	77.7	76.6	74.0	66.6	53.0	41.0	61.8
1991	47.3	55.0	51.0	58.0	63.3	70.2	77.1	73.2	74.8	68.8	55.9	46.3	61.7
1992	43.6	54.1	56.2	62.1	70.6	70.9	75.3	77.0	72.4	66.6	53.4	44.1	62.2
1993	45.2	49.5	57.9	58.4	64.6	71.7	74.3	74.1	71.5	65.0	51.5	44.3	60.7
Record Mean	45.3	50.4	53.5	58.6	64.9	71.0	75.4	74.4	71.5	63.9	53.0	45.6	60.6
Max	53.0	59.7	64.2	71.4	79.6	87.0	92.8	91.3	87.5	77.7	63.5	53.3	73.4
Min	37.5	41.0	42.9	45.7	50.2	55.0	57.9	57.4	55.5	50.1	42.6	37.9	47.8

REFERENCE NOTES SACRAMENTO, CALIFORNIA

GENERAL
T - TRACE AMOUNT.
BLANK ENTRIES DENOTE MISSING/UNREPORTED DATA.
# INDICATES A STATION OR INSTRUMENT RELOCATION

SPECIFIC

PM - INCLUDES LAST DAY OF PREVIOUS MONTH
ASOS - AUTOMATED SURFACE OBSERVING SYSTEM IN OPERATION DURING THESE MONTHS.

(a) - LENGTH OF RECORD IN YEARS, ALTHOUGH INDIVIDUAL MONTHS MAY BE MISSING.
0.* OR * - THE VALUE IS BETWEEN 0.0 AND 0.05
NORMALS - BASED ON THE 1961-1990 RECORD PERIOD.
EXTREMES - DATES ARE THE MOST RECENT OCCURRENCE
WIND DIR.- NUMERALS SHOW TENS OF DEGRESS CLOCKWISE FROM TRUE NORTH. "00" INDICATES CALM.
RESULTANT DIRECTIONS ARE GIVEN TO WHOLE DEGREES.
BOLD VALUES INDICATE EXTREME VALUES WHICH OCCURRED AFTER THE ASOS SYSTEM WAS COMMISSIONED.

RECORD = PERIOD OF RECORD
RECORD MEAN PRECIPITATION IS THE MEAN OF ALL DAILY PRECIPITATION AMOUNTS DURING THE PERIOD OF RECORD.
RECORD MAX(MIN) TEMPERATURE IS THE MEAN OF ALL DAILY MAX(MIN) TEMPERATURES DURING THE PERIOD OF RECORD.
RECORD MEAN TEMPERATURE IS THE SUM OF THE RECORD MAX AND RECORD MIN DIVIDED BY 2.
AVERAGE TEMPERATURE IS THE SUM OF THE MEAN DAILY MAX AND MIN TEMPERATURE DIVIDED BY 2.

EXCEPTIONS

RECORD MEANS ARE THROUGH THE CURRENT YEAR,
BEGINNING IN 1941 FOR TEMPERATURE
1940 FOR PRECIPITATION
1949 FOR SNOWFALL

HEATING DEGREE DAYS Base 65 deg. F SACRAMENTO, CALIFORNIA

SEASON	JULY	AUG	SEP	OCT	NOV	DEC	JAN	FEB	MAR	APR	MAY	JUNE	TOTAL
1964-65	0	4	10	49	436	469	602	434	363	246	86	21	2720
1965-66	0	0	19	32	303	738	591	485	326	96	36	14	2640
1966-67	1	0	4	60	303	580	584	461	435	456	98	34	3016
1967-68	0	0	0	33	259	643	663	296	299	171	80	2	2446
1968-69	0	3	2	97	374	662	644	486	384	230	41	6	2929
1969-70	0	0	1	129	323	519	478	359	304	258	56	9	2436
1970-71	0	0	4	129	286	578	603	466	379	249	114	14	2822
1971-72	0	0	33	191	363	673	731	390	197	190	63	11	2842
1972-73	0	0	6	115	451	749	636	325	424	141	15	1	2863
1973-74	0	0	0	77	384	553	571	456	332	251	93	9	2726
1974-75	7	0	0	44	347	569	661	435	449	389	69	1	2971
1975-76	0	0	0	72	306	539	547	374	315	211	1	3	2368
1976-77	0	0	1	44	252	567	650	345	424	92	187	9	2571
1977-78	0	0	17	68	309	472	451	362	235	269	46	0	2229
1978-79	0	0	11	51	449	715	606	446	313	236	57	2	2886
1979-80	0	0	0	100	391	558	551	373	408	164	107	29	2681
1980-81	2	0	4	134	339	596	557	405	420	229	81	2	2769
1981-82	0	0	9	66	145	498	708	398	434	282	70	40	2650
1982-83	3	0	31	125	532	675	670	353	354	303	99	4	3149
1983-84	3	0	0	7	333	425	514	421	206	191	22	11	2133
1984-85	0	0	0	115	335	611	693	377	433	122	89	11	2786
1985-86	0	2	15	95	450	689	411	284	192	200	73	0	2411
1986-87	0	0	53	47	277	593	614	377	340	95	37	0	2433
1987-88	1	0	0	11	339	544	522	307	212	138	94	27	2195
1988-89	0	0	3	38	329	576	640	496	285	106	50	3	2526
1989-90	0	0	11	107	316	634	536	453	289	71	53	6	2476
1990-91	0	0	0	24	356	739	543	274	427	205	104	6	2678
1991-92	0	0	0	82	267	572	657	310	265	104	0	9	2266
1992-93	0	0	0	24	340	643	605	426	214	202	55	21	2530
1993-94	0	0	5	33	399	634							

COOLING DEGREE DAYS Base 65 deg. F SACRAMENTO, CALIFORNIA

YEAR	JAN	FEB	MAR	APR	MAY	JUNE	JULY	AUG	SEP	OCT	NOV	DEC	TOTAL
1969	0	0	0	1	108	136	361	409	278	13	0	0	1306
1970	0	0	0	0	143	208	374	305	245	46	0	0	1321
1971	0	0	0	0	22	186	355	375	254	82	0	0	1274
1972	0	0	6	5	129	245	351	349	147	30	0	0	1262
1973	0	0	0	19	156	295	373	293	181	34	0	0	1351
1974	0	0	0	5	61	180	296	285	222	89	0	0	1138
1975	0	0	0	0	177	258	388	375	375	81	0	0	1654
1976	0	0	1	8	167	278	363	270	213	83	8	0	1391
1977	0	0	0	12	19	230	290	284	139	40	0	0	1014
1978	0	0	0	0	98	157	318	315	157	87	0	0	1132
1979	0	0	0	0	117	214	336	260	295	72	0	0	1294
1980	0	0	0	8	42	91	317	207	145	99	0	0	909
1981	0	0	0	26	78	303	318	301	155	28	7	0	1216
1982	0	0	0	2	67	83	230	213	133	9	0	0	737
1983	0	0	0	0	81	183	235	368	304	92	0	0	1263
1984	0	0	0	6	183	216	419	327	320	57	0	0	1528
1985	0	0	0	22	41	319	380	254	128	48	0	0	1192
1986	0	0	10	9	95	207	315	321	95	47	0	0	1099
1987	0	0	0	34	171	234	220	314	212	100	0	0	1285
1988	0	0	5	22	88	269	484	346	233	92	0	0	1539
1989	0	0	1	60	83	211	354	280	158	32	0	0	1179
1990	0	0	0	33	75	236	399	367	276	82	0	0	1468
1991	0	0	0	3	54	171	379	261	300	208	0	0	1376
1992	0	0	0	23	180	193	330	381	231	81	0	0	1419
1993	0	0	1	9	49	227	294	291	207	38	0	0	1116

SNOWFALL (inches) SACRAMENTO, CALIFORNIA

SEASON	JULY	AUG	SEP	OCT	NOV	DEC	JAN	FEB	MAR	APR	MAY	JUNE	TOTAL
1970-71	0.0	0.0	0.0	0.0	0.0	0.0	0.0	0.0	0.0	0.0	0.0	0.0	0.0
1971-72	0.0	0.0	0.0	0.0	0.0	0.0	0.0	0.0	0.0	0.0	0.0	0.0	0.0
1972-73	0.0	0.0	0.0	0.0	0.0	T	0.0	0.0	0.0	0.0	0.0	0.0	T
1973-74	0.0	0.0	0.0	0.0	0.0	0.0	T	0.0	0.0	0.0	0.0	0.0	T
1974-75	0.0	0.0	0.0	0.0	0.0	0.0	0.0	0.0	0.0	0.0	0.0	0.0	0.0
1975-76	0.0	0.0	0.0	0.0	0.0	0.0	0.0	2.0	0.0	0.0	0.0	0.0	2.0
1976-77	0.0	0.0	0.0	0.0	0.0	0.0	0.0	0.0	0.0	0.0	0.0	0.0	0.0
1977-78	0.0	0.0	0.0	0.0	0.0	0.0	0.0	0.0	0.0	0.0	0.0	0.0	0.0
1978-79	0.0	0.0	0.0	0.0	0.0	0.0	0.0	0.0	0.0	0.0	0.0	0.0	0.0
1979-80	0.0	0.0	0.0	0.0	0.0	0.0	0.0	0.0	0.0	0.0	0.0	0.0	0.0
1980-81	0.0	0.0	0.0	0.0	0.0	0.0	0.0	0.0	0.0	0.0	0.0	0.0	0.0
1981-82	0.0	0.0	0.0	0.0	0.0	0.0	0.0	0.0	T	0.0	0.0	0.0	T
1982-83	0.0	0.0	0.0	0.0	0.0	0.0	0.0	0.0	0.0	0.0	0.0	0.0	0.0
1983-84	0.0	0.0	0.0	0.0	0.0	0.0	0.0	0.0	0.0	0.0	0.0	0.0	0.0
1984-85	0.0	0.0	0.0	0.0	0.0	0.0	0.0	0.0	0.0	0.0	0.0	0.0	0.0
1985-86	0.0	0.0	0.0	0.0	0.0	0.0	0.0	0.0	0.0	0.0	0.0	0.0	0.0
1986-87	0.0	0.0	0.0	0.0	0.0	0.0	0.0	0.0	0.0	0.0	0.0	0.0	0.0
1987-88	0.0	0.0	0.0	0.0	0.0	0.0	0.0	0.0	0.0	0.0	0.0	0.0	0.0
1988-89	0.0	0.0	0.0	0.0	0.0	T	0.0	0.0	0.0	0.0	0.0	0.0	T
1989-90	0.0	0.0	0.0	0.0	0.0	0.0	0.0	0.0	0.0	0.0	0.0	0.0	0.0
1990-91	0.0	0.0	0.0	0.0	0.0	0.0	0.0	0.0	0.0	0.0	0.0	0.0	0.0
1991-92	0.0	0.0	0.0	0.0	0.0	0.0	0.0	0.0	0.0	0.0	0.0	0.0	0.0
1992-93	0.0	0.0	0.0	0.0	0.0	T	0.0	0.0	0.0	0.0	0.0	0.0	T
1993-94	0.0	0.0	0.0	0.0	0.0	T							
Record Mean	0.0	0.0	0.0	0.0	0.0	T	T	T	T	0.0	0.0	0.0	T

SAN DIEGO, CALIFORNIA

The city of San Diego is located on San Diego Bay in the southwest corner of southern California. The prevailing winds and weather are tempered by the Pacific Ocean, with the result that summers are cool and winters warm in comparison with other places along the same general latitude. Temperatures of freezing or below have rarely occurred at the station since the record began in 1871, but hot weather, 90 degrees or above, is more frequent.

Dry easterly winds sometimes blow in the vicinity for several days at a time, bringing temperatures in the 90s and at times even in the 100s in the eastern sections of the city and outlying suburbs. At the National Weather Service station itself, however, there have been relatively few days on which 100 degrees or higher was reached.

As these hot winds are predominant in the fall, highest temperatures occur in the months of September and October. Records show that over 60 percent of the days with 90 degrees or higher have occurred in these two months. High temperatures are almost invariably accompanied by very low relative humidities, which often drop below 20 percent and occasionally below 10 percent.

A marked feature of the climate is the wide variation in temperature within short distances. In nearby valleys daytimes are much warmer in summer and nights noticeably cooler in winter, and freezing occurs much more frequently than in the city. Although records show unusually small daily temperature ranges, only about 15 degrees between the highest and lowest readings, a few miles inland these ranges increase to 30 degrees or more.

Strong winds and gales associated with Pacific, or tropical storms, are infrequent due to the latitude.

The seasonal rainfall is about 10 inches in the city, but increases with elevation and distance from the coast. In the mountains to the north and east the average is between 20 and 40 inches, depending on slope and elevation. Most of the precipitation falls in winter, except in the mountains where there is an occasional thunderstorm. Eighty-five percent of the rainfall occurs from November through March, but wide variations take place in monthly and seasonal totals. Infrequent measurable amounts of hail occur in San Diego, but snow is practically unknown at the Weather Service Office location. In each occurrence of snowfall only a trace was recorded officially, but in some locations amounts up to or slightly exceeding a half-inch fell, and remained on the ground for an hour or more.

As on the rest of the Pacific Coast, a dominant characteristic of spring and summer is the nighttime and early morning cloudiness. Low clouds form regularly and frequently extend inland over the coastal valleys and foothills, but they usually dissipate during the morning and the afternoons are generally clear.

Considerable fog occurs along the coast, but the amount decreases with distance inland. The fall and winter months are usually the foggiest. Thunderstorms are rare, averaging about three a year in the city. Visibilities are good as a rule. The sunshine is plentiful for a marine location, with a marked increase toward the interior.

NORMALS, MEANS, AND EXTREMES

SAN DIEGO, CALIFORNIA

LATITUDE: 32°44'N LONGITUDE: 117°10' W ELEVATION: FT. GRND 13 BARO 33 TIME ZONE: PACIFIC WBAN: 23188

	(a)	JAN	FEB	MAR	APR	MAY	JUNE	JULY	AUG	SEP	OCT	NOV	DEC	YEAR
TEMPERATURE °F:														
Normals														
-Daily Maximum		65.9	66.5	66.3	68.4	69.1	71.6	76.2	77.8	77.1	74.6	69.9	66.1	70.8
-Daily Minimum		48.9	50.7	52.8	55.6	59.1	61.9	65.7	67.3	65.6	60.9	53.9	48.8	57.6
-Monthly		57.4	58.6	59.6	62.0	64.1	66.8	71.0	72.6	71.4	67.7	62.0	57.4	64.2
Extremes														
-Record Highest	53	88	88	93	98	96	101	95	98	111	107	97	88	111
-Year		1953	1954	1988	1989	1953	1979	1985	1955	1963	1961	1976	1963	SEP 1963
-Record Lowest	53	29	36	39	41	48	51	55	57	51	43	38	34	29
-Year		1949	1949	1971	1945	1967	1967	1948	1944	1948	1971	1964	1987	JAN 1949
NORMAL DEGREE DAYS:														
Heating (base 65°F)		245	189	177	113	73	51	13	0	19	24	109	243	1256
Cooling (base 65°F)		9	10	9	23	45	105	199	240	211	107	19	7	984
% OF POSSIBLE SUNSHINE	53	72	72	70	67	59	58	68	70	69	68	75	73	68
MEAN SKY COVER (tenths)														
Sunrise - Sunset	53	5.1	5.2	5.3	5.2	5.7	5.5	4.5	4.2	4.1	4.4	4.2	4.7	4.8
MEAN NUMBER OF DAYS:														
Sunrise to Sunset														
-Clear	53	12.5	10.4	10.9	10.3	8.6	9.2	13.0	14.9	14.9	13.7	14.8	13.6	146.8
-Partly Cloudy	53	7.4	7.5	9.5	10.0	11.4	11.7	13.0	11.6	9.5	9.7	7.8	7.8	116.7
-Cloudy	53	11.2	10.4	10.7	9.8	11.1	9.0	5.0	4.5	5.5	7.6	7.4	9.5	101.7
Precipitation														
.01 inches or more	53	6.8	6.0	7.0	4.4	2.1	1.0	0.4	0.5	1.1	2.5	4.5	5.8	42.1
Snow,Ice pellets,hail														
1.0 inches or more	53	0.0	0.0	0.0	0.0	0.0	0.0	0.0	0.0	0.0	0.0	0.0	0.0	0.0
Thunderstorms	53	0.2	0.3	0.4	0.1	0.1	0.1	0.2	0.2	0.3	0.3	0.3	0.4	3.0
Heavy Fog Visibility 1/4 mile or less	53	3.0	2.5	1.5	1.2	0.6	0.7	0.6	0.6	2.3	3.3	3.5	3.9	23.6
Temperature °F														
-Maximum														
90° and above	33	0.0	0.0	0.1	0.2	0.1	0.5	0.2	0.2	1.3	0.9	0.2	0.0	3.7
32° and below	33	0.0	0.0	0.0	0.0	0.0	0.0	0.0	0.0	0.0	0.0	0.0	0.0	0.0
-Minimum														
32° and below	33	0.*	0.0	0.0	0.0	0.0	0.0	0.0	0.0	0.0	0.0	0.0	0.0	*
0° and below	33	0.0	0.0	0.0	0.0	0.0	0.0	0.0	0.0	0.0	0.0	0.0	0.0	0.0
AVG. STATION PRESS.(mb)	21	1017.2	1016.8	1015.4	1014.5	1013.2	1012.2	1012.5	1012.3	1011.6	1013.7	1015.8	1017.1	1014.4
RELATIVE HUMIDITY (%)														
Hour 04	33	71	74	75	76	77	81	82	82	80	77	72	71	77
Hour 10 (Local Time)	33	55	58	61	60	65	69	69	68	66	61	55	54	62
Hour 16	33	56	59	60	59	64	67	66	67	65	64	61	58	62
Hour 22	33	70	72	73	73	75	78	80	79	78	76	73	71	75
PRECIPITATION (inches):														
Water Equivalent														
-Normal		1.80	1.53	1.77	0.79	0.19	0.07	0.02	0.10	0.24	0.37	1.45	1.57	9.90
-Maximum Monthly	53	9.09	5.40	6.96	3.71	1.79	0.87	0.24	2.13	1.90	2.90	5.82	7.60	9.09
-Year		1993	1976	1991	1988	1977	1990	1991	1977	1963	1941	1965	1943	JAN 1993
-Minimum Monthly	53	T	0.00	T	0.00	0.00	0.00	0.00	0.00	0.00	0.00	0.00	0.02	0.00
-Year		1976	1967	1972	1993	1952	1981	1982	1981	1979	1967	1980	1979	APR 1993
-Maximum in 24 hrs	53	2.65	2.61	2.40	1.98	1.50	0.82	0.23	2.13	1.00	1.39	2.44	3.07	3.07
-Year		1978	1979	1952	1988	1977	1990	1991	1977	1986	1986	1944	1945	DEC 1945
Snow,Ice pellets,hail														
-Maximum Monthly	53	T	0.0	T	0.0	0.0	0.0	0.0	0.0	0.0	0.0	T	T	T
-Year		1949		1985								1985	1967	MAR 1985
-Maximum in 24 hrs	53	T	0.0	T	0.0	0.0	0.0	0.0	0.0	0.0	0.0	T	T	T
-Year		1949		1985								1985	1967	MAR 1985
WIND:														
Mean Speed (mph)	53	5.9	6.6	7.5	7.8	7.9	7.8	7.4	7.3	7.1	6.6	5.9	5.6	7.0
Prevailing Direction through 1963		NE	WNW	WNW	WNW	WNW	SSW	WNW	WNW	NW	WNW	NE	NE	WNW
Fastest Mile														
-Direction (!!)	49	SE	S	SW	S	S	S	NW	NW	S	N	SE	NW	SE
-Speed (MPH)	49	56	45	46	37	30	26	23	23	31	31	51	39	56
-Year		1980	1980	1945	1958	1977	1948	1968	1982	1978	1961	1944	1982	JAN 1980
Peak Gust														
-Direction (!!)	10	W	SW	NW	SW	NW	NW	SW	SW	NW	NW	SW	NW	W
-Speed (mph)	10	64	46	41	40	40	28	30	29	31	32	37	44	64
-Date		1988	1993	1985	1988	1988	1988	1985	1991	1989	1991	1985	1991	JAN 1988

PRECIPITATION (inches)

SAN DIEGO, CALIFORNIA

YEAR	JAN	FEB	MAR	APR	MAY	JUNE	JULY	AUG	SEP	OCT	NOV	DEC	ANNUAL
1964	1.30	0.37	0.97	0.20	0.15	0.08	0.00	T	0.00	0.02	1.01	1.17	5.27
1965	0.40	0.52	1.79	3.58	T	0.01	0.02	T	0.29	T	5.82	6.60	19.03
1966	1.29	0.86	0.17	T	0.02	T	T	0.00	T	0.80	0.82	3.22	7.18
1967	2.20	0.00	1.14	2.24	0.05	0.16	0.01	0.14	0.08	0.00	3.53	1.66	11.21
1968	0.35	0.22	1.55	0.34	0.08	T	0.13	T	T	0.04	0.36	0.61	3.68
1969	4.78	4.34	0.94	0.21	0.17	0.02	T	0.01	T	0.04	0.79	0.46	11.76
1970	0.86	2.58	1.50	0.09	0.01	T	T	0.00	T	0.07	2.05	2.22	9.38
1971	0.30	1.27	0.20	0.93	0.95	0.01	T	0.03	T	1.66	0.06	3.27	8.68
1972	0.07	0.10	T	0.02	0.10	0.38	T	0.02	0.44	0.58	3.16	1.61	6.48
1973	1.68	1.63	2.26	0.05	T	T	T	T	0.02	0.01	1.63	0.19	7.47
1974	2.96	0.04	1.70	0.02	0.01	0.02	0.01	T	T	1.03	0.14	2.20	8.13
1975	0.49	0.96	3.79	2.00	0.01	0.02	T	T	T	0.09	0.64	0.37	8.37
1976	T	5.40	0.99	1.33	0.27	0.02	0.02	0.01	1.00	0.38	0.75	1.06	11.23
1977	2.36	0.06	0.61	0.01	1.79	0.03	T	2.13	T	0.50	0.05	1.67	9.21
1978	5.95	2.64	5.00	0.73	0.04	T	0.00	T	0.72	0.05	2.09	2.19	19.41
1979	5.82	0.85	3.71	0.02	0.09	0.01	0.09	0.01	0.00	0.73	0.27	0.02	11.62
1980	5.58	4.47	2.71	1.18	0.65	0.01	T	0.00	T	0.05	0.00	0.31	14.96
1981	1.48	2.26	3.74	0.22	0.04	0.00	T	0.00	0.03	0.14	1.79	0.54	10.24
1982	2.71	0.88	4.74	0.62	0.01	0.04	0.00	T	0.38	0.05	2.10	1.43	12.96
1983	2.10	3.88	6.57	1.74	0.01	T	0.01	0.39	0.21	0.40	1.94	1.53	18.78
1984	0.46	0.09	0.04	0.62	0.00	0.04	0.19	0.06	T	0.29	2.37	4.55	8.71
1985	0.52	0.77	0.58	0.32	T	T	0.00	T	0.20	0.29	4.92	1.06	8.66
1986	0.75	2.59	3.12	1.17	0.00	T	0.01	0.00	1.04	1.39	1.16	0.95	12.18
1987	1.68	1.53	1.04	0.78	0.03	T	0.03	0.01	0.70	1.74	1.33	2.73	11.60
1988	0.89	1.37	0.59	3.71	0.08	0.00	T	T	T	T	1.39	2.23	10.26
1989	0.42	0.70	0.69	0.12	0.04	0.06	0.00	T	0.23	0.47	0.09	1.01	3.83
1990	2.52	1.13	0.25	0.76	0.51	0.87	T	0.01	T	T	0.65	0.59	7.29
1991	1.06	2.46	6.96	0.05	0.01	T	0.24	0.01	0.28	0.69	0.05	1.70	13.51
1992	1.81	3.34	4.42	0.28	0.07	0.04	0.03	0.05	0.00	0.18	0.03	2.56	12.81
1993	9.09	4.73	1.22	0.00	0.01	0.41	0.03	T	T	0.22	0.77	0.78	17.26
Record Mean	1.91	1.91	1.63	0.71	0.26	0.06	0.04	0.09	0.13	0.42	0.99	1.83	9.97

AVERAGE TEMPERATURE (deg. F)

SAN DIEGO, CALIFORNIA

YEAR	JAN	FEB	MAR	APR	MAY	JUNE	JULY	AUG	SEP	OCT	NOV	DEC	ANNUAL
1964	55.3	56.7	57.8	60.2	60.9	64.0	69.2	70.7	67.7	68.6	59.1	55.6	62.1
1965	56.0	55.9	58.6	60.7	62.5	63.7	67.7	72.0	68.5	69.4	60.9	55.1	62.6
1966	53.9	54.6	58.1	61.3	63.5	66.5	69.2	72.6	69.9	67.6	61.9	57.2	63.0
1967	55.0	57.8	59.0	56.5	63.5	63.6	70.4	73.1	72.0	68.1	64.1	55.5	63.2
1968	57.2	60.7	60.7	62.4	63.9	65.8	71.7	72.2	71.3	66.6	61.7	54.9	64.1
1969	58.1	54.9	56.8	61.7	62.9	65.5	69.4	72.8	69.9	66.0	64.1	59.1	63.4
1970	57.0	59.7	60.5	60.1	63.6	65.6	70.4	72.8	69.7	66.3	61.4	55.4	63.5
1971	54.3	55.4	57.8	60.7	61.5	64.9	69.4	75.4	72.2	65.7	59.5	54.2	62.6
1972	54.9	57.8	60.2	62.3	64.7	67.0	72.7	72.2	68.7	65.6	59.8	57.5	63.6
1973	55.6	59.9	58.1	61.5	63.4	68.0	69.1	70.5	68.8	66.8	60.6	58.2	63.4
1974	56.9	58.2	59.1	62.0	63.3	66.9	71.4	70.2	70.3	66.8	62.2	56.3	63.6
1975	56.1	56.4	57.5	58.7	62.2	65.0	69.4	68.9	71.5	65.9	60.4	56.9	62.4
1976	58.9	59.6	60.3	61.0	65.2	69.7	71.1	72.4	73.8	71.2	66.8	60.7	65.9
1977	60.3	61.7	57.5	61.4	61.9	65.8	71.6	73.1	72.2	68.9	64.9	63.3	65.2
1978	61.0	60.9	64.3	63.4	68.2	71.3	71.6	72.9	74.0	70.1	61.7	55.2	66.2
1979	56.9	56.9	60.1	63.4	65.6	70.2	71.8	73.9	76.3	68.7	62.4	60.6	65.6
1980	61.1	63.5	61.5	63.9	63.8	68.5	72.9	74.2	70.4	67.3	62.7	60.8	65.9
1981	61.3	62.2	61.1	64.4	67.3	72.9	75.6	75.8	73.7	67.1	63.5	60.3	67.1
1982	56.6	60.7	60.5	63.8	65.8	66.7	71.9	73.5	73.1	70.1	62.1	57.4	65.2
1983	60.7	60.9	62.0	62.4	66.2	68.1	72.6	77.4	76.8	72.2	64.4	60.6	67.0
1984	61.2	60.2	63.7	64.3	68.1	69.9	77.2	76.6	78.9	68.5	61.4	56.7	67.2
1985	57.0	57.2	58.9	63.6	64.8	69.0	75.3	72.4	69.8	67.9	60.1	58.0	64.5
1986	61.0	58.9	60.5	62.8	64.6	67.4	69.6	71.8	66.9	65.5	62.8	57.6	64.1
1987	55.4	58.0	59.1	63.4	64.7	65.8	67.1	69.9	69.9	69.5	61.8	53.9	63.2
1988	56.7	59.9	61.6	62.4	63.9	64.9	70.4	71.0	70.0	66.7	60.1	56.0	63.6
1989	54.7	56.7	59.8	65.6	63.7	66.0	70.1	71.0	70.4	66.3	63.1	58.7	63.8
1990	56.6	55.2	58.7	63.2	64.3	69.0	72.3	71.6	71.7	68.6	62.7	55.6	64.1
1991	57.4	59.4	56.5	61.7	62.1	64.1	67.4	68.9	69.4	68.0	62.3	57.3	62.9
1992	57.4	61.1	60.4	67.0	68.0	68.1	71.8	74.9	72.4	68.2	62.6	55.3	65.6
1993	56.9	58.0	61.3	63.8	66.0	68.6	69.8	70.2	69.0	67.3	61.6	57.0	64.1
Record Mean	55.5	56.6	58.0	60.3	62.4	65.2	68.8	70.3	68.8	65.2	60.9	57.0	62.4
Max	63.8	64.4	65.1	66.7	67.8	70.4	74.1	75.6	74.9	72.2	69.4	65.5	69.2
Min	47.3	48.8	50.8	53.8	57.0	59.9	63.5	64.9	62.8	58.2	52.3	48.5	55.6

REFERENCE NOTES

SAN DIEGO, CALIFORNIA

GENERAL
T - TRACE AMOUNT.
BLANK ENTRIES DENOTE MISSING/UNREPORTED DATA.
# INDICATES A STATION OR INSTRUMENT RELOCATION.

SPECIFIC

PM - INCLUDES LAST DAY OF PREVIOUS MONTH
ASOS - AUTOMATED SURFACE OBSERVING SYSTEM IN OPERATION DURING THESE MONTHS.

(a) - LENGTH OF RECORD IN YEARS, ALTHOUGH INDIVIDUAL MONTHS MAY BE MISSING.
0.* OR * - THE VALUE IS BETWEEN 0.0 AND 0.05
NORMALS - BASED ON THE 1961-1990 RECORD PERIOD.
EXTREMES - DATES ARE THE MOST RECENT OCCURRENCE
WIND DIR.- NUMERALS SHOW TENS OF DEGRESS CLOCKWISE FROM TRUE NORTH. "00" INDICATES CALM.
RESULTANT DIRECTIONS ARE GIVEN TO WHOLE DEGREES.
BOLD VALUES INDICATE EXTREME VALUES WHICH OCCURRED AFTER THE ASOS SYSTEM WAS COMMISSIONED.

RECORD = PERIOD OF RECORD
RECORD MEAN PRECIPITATION IS THE MEAN OF ALL DAILY PRECIPITATION AMOUNTS DURING THE PERIOD OF RECORD.
RECORD MAX(MIN) TEMPERATURE IS THE MEAN OF ALL DAILY MAX(MIN) TEMPERATURES DURING THE PERIOD OF RECORD.
RECORD MEAN TEMPERATURE IS THE SUM OF THE RECORD MAX AND RECORD MIN DIVIDED BY 2.
AVERAGE TEMPERATURE IS THE SUM OF THE MEAN DAILY MAX AND MIN TEMPERATURE DIVIDED BY 2.

EXCEPTIONS

RECORD MEANS ARE THROUGH THE CURRENT YEAR, BEGINNING IN 1875 FOR TEMPERATURE
1850 FOR PRECIPITATION
1941 FOR SNOWFALL

HEATING DEGREE DAYS Base 65 deg. F SAN DIEGO, CALIFORNIA

SEASON	JULY	AUG	SEP	OCT	NOV	DEC	JAN	FEB	MAR	APR	MAY	JUNE	TOTAL
1964-65	0	0	0	8	187	280	277	249	195	138	73	35	1442
1965-66	3	0	0	9	118	303	335	284	209	107	40	4	1412
1966-67	0	0	0	4	113	236	302	197	183	245	72	48	1400
1967-68	0	0	0	3	42	288	239	119	135	85	47	8	966
1968-69	0	0	0	9	104	306	214	274	248	101	63	9	1328
1969-70	0	0	0	14	44	178	240	142	133	143	58	12	964
1970-71	0	0	0	12	107	290	331	266	215	143	109	29	1502
1971-72	0	0	0	78	160	326	310	203	139	78	34	0	1328
1972-73	0	0	0	29	149	224	286	131	208	107	61	1	1196
1973-74	0	0	0	6	132	205	243	184	176	85	55	4	1090
1974-75	0	0	0	14	97	265	273	237	225	182	83	10	1386
1975-76	0	0	0	19	141	246	196	150	148	115	16	0	1031
1976-77	0	0	0	0	39	129	143	94	224	103	88	3	823
1977-78	0	0	0	0	37	55	117	117	52	43	8	0	429
1978-79	0	0	0	0	102	297	244	219	153	45	20	6	1086
1979-80	0	0	0	4	75	136	117	50	104	61	43	1	591
1980-81	0	0	0	6	75	133	113	101	116	40	1	0	585
1981-82	0	0	0	9	57	136	258	119	139	64	9	2	793
1982-83	0	0	0	1	93	228	137	110	88	83	9	0	749
1983-84	0	0	0	0	66	130	123	134	51	43	4	0	551
1984-85	0	0	0	4	104	250	238	219	183	60	18	2	1078
1985-86	0	0	0	3	145	211	118	173	132	85	29	0	896
1986-87	0	0	7	10	66	223	291	197	178	72	21	6	1071
1987-88	0	0	0	0	98	338	250	147	125	85	53	22	1118
1988-89	0	0	0	4	141	275	313	237	158	37	40	14	1219
1989-90	0	0	1	13	67	188	252	268	185	52	39	1	1066
1990-91	0	0	0	3	88	284	227	152	254	104	96	28	1236
1991-92	0	0	0	24	96	231	228	115	136	8	1	0	839
1992-93	0	0	0	0	84	294	240	191	111	44	7	6	977
1993-94	0	0	0	4	103	242							

COOLING DEGREE DAYS Base 65 deg. F SAN DIEGO, CALIFORNIA

YEAR	JAN	FEB	MAR	APR	MAY	JUNE	JULY	AUG	SEP	OCT	NOV	DEC	TOTAL
1969	5	0	1	9	3	31	144	247	154	53	26	5	678
1970	0	1	1	1	21	40	172	247	145	58	7	0	693
1971	5	3	0	19	7	31	142	327	224	107	1	0	866
1972	0	0	0	4	33	68	247	230	117	53	0	1	753
1973	0	0	0	10	17	97	133	176	121	70	8	1	633
1974	0	0	0	2	9	69	204	169	164	75	19	0	711
1975	0	0	0	0	1	18	142	124	201	54	8	0	548
1976	14	0	10	3	31	147	196	240	269	200	102	0	1212
1977	5	9	0	2	1	34	212	258	224	128	40	8	921
1978	1	7	38	4	115	194	213	251	276	166	11	0	1276
1979	0	0	10	6	46	169	216	283	348	124	5	8	1215
1980	2	13	3	35	15	110	253	289	170	86	15	7	998
1981	7	29	0	26	81	244	335	343	265	80	21	0	1431
1982	0	7	6	32	42	58	219	271	250	164	12	0	1061
1983	11	0	1	9	51	99	242	392	364	231	55	0	1455
1984	13	0	15	31	107	156	387	366	422	119	4	0	1620
1985	0	7	0	22	19	128	325	235	153	104	6	0	999
1986	2	11	4	29	23	78	152	218	73	31	9	0	630
1987	0	6	5	29	17	35	71	158	154	147	10	0	632
1988	0	5	28	16	25	26	176	193	161	64	0	5	699
1989	0	13	2	63	5	48	165	193	168	58	17	0	732
1990	0	0	2	6	21	127	233	211	209	123	25	0	957
1991	0	0	0	9	11	10	80	130	138	122	21	0	521
1992	0	12	1	72	99	100	216	313	228	106	20	0	1167
1993	0	0	4	12	47	119	155	170	126	81	8	0	722

SNOWFALL (inches) SAN DIEGO, CALIFORNIA

SEASON	JULY	AUG	SEP	OCT	NOV	DEC	JAN	FEB	MAR	APR	MAY	JUNE	TOTAL
1970-71	0.0	0.0	0.0	0.0	0.0	0.0	0.0	0.0	0.0	0.0	0.0	0.0	0.0
1971-72	0.0	0.0	0.0	0.0	0.0	0.0	0.0	0.0	0.0	0.0	0.0	0.0	0.0
1972-73	0.0	0.0	0.0	0.0	0.0	0.0	0.0	0.0	0.0	0.0	0.0	0.0	0.0
1973-74	0.0	0.0	0.0	0.0	0.0	0.0	0.0	0.0	0.0	0.0	0.0	0.0	0.0
1974-75	0.0	0.0	0.0	0.0	0.0	0.0	0.0	0.0	0.0	0.0	0.0	0.0	0.0
1975-76	0.0	0.0	0.0	0.0	0.0	0.0	0.0	0.0	0.0	0.0	0.0	0.0	0.0
1976-77	0.0	0.0	0.0	0.0	0.0	0.0	0.0	0.0	0.0	0.0	0.0	0.0	0.0
1977-78	0.0	0.0	0.0	0.0	0.0	0.0	0.0	0.0	0.0	0.0	0.0	0.0	0.0
1978-79	0.0	0.0	0.0	0.0	0.0	0.0	0.0	0.0	0.0	0.0	0.0	0.0	0.0
1979-80	0.0	0.0	0.0	0.0	0.0	0.0	0.0	0.0	0.0	0.0	0.0	0.0	0.0
1980-81	0.0	0.0	0.0	0.0	0.0	0.0	0.0	0.0	0.0	0.0	0.0	0.0	0.0
1981-82	0.0	0.0	0.0	0.0	0.0	0.0	0.0	0.0	0.0	0.0	0.0	0.0	0.0
1982-83	0.0	0.0	0.0	0.0	0.0	0.0	0.0	0.0	0.0	0.0	0.0	0.0	0.0
1983-84	0.0	0.0	0.0	0.0	0.0	0.0	0.0	0.0	0.0	0.0	0.0	0.0	0.0
1984-85	0.0	0.0	0.0	0.0	0.0	0.0	0.0	0.0	T	0.0	0.0	0.0	T
1985-86	0.0	0.0	0.0	0.0	T	0.0	0.0	0.0	0.0	0.0	0.0	0.0	T
1986-87	0.0	0.0	0.0	0.0	0.0	0.0	0.0	0.0	0.0	0.0	0.0	0.0	0.0
1987-88	0.0	0.0	0.0	0.0	0.0	0.0	0.0	0.0	0.0	0.0	0.0	0.0	0.0
1988-89	0.0	0.0	0.0	0.0	0.0	0.0	0.0	0.0	0.0	0.0	0.0	0.0	0.0
1989-90	0.0	0.0	0.0	0.0	0.0	0.0	0.0	0.0	0.0	0.0	0.0	0.0	0.0
1990-91	0.0	0.0	0.0	0.0	0.0	0.0	0.0	0.0	0.0	0.0	0.0	0.0	0.0
1991-92	0.0	0.0	0.0	0.0	0.0	0.0	0.0	0.0	0.0	0.0	0.0	0.0	0.0
1992-93	0.0	0.0	0.0	0.0	0.0	0.0	0.0	0.0	0.0	0.0	0.0	0.0	0.0
1993-94	0.0	0.0	0.0	0.0	0.0	0.0							
Record Mean	0.0	0.0	0.0	0.0	T	T	T	0.0	T	0.0	0.0	0.0	T

SAN FRANCISCO, CALIFORNIA
INTERNATIONAL AIRPORT

The station is located in the central Terminal Building of the San Francisco International Airport, which is on flat filled tideland on the west shore of San Francisco Bay. The bay borders the airport from the north to the south-southeast. San Bruno Mountain, 5 miles to the north-northwest, rises to 1,300 feet. A north-south trending ridge of coastal mountains, 4 miles to the west, varies in elevation from 700 to 1,900 feet, being highest southward along the peninsula. The Pacific Ocean west of the ridge is 6 miles from the airport. A broad gap to the northwest of the station, between San Bruno Mountain and the coastal mountains, allows a strong flow of marine air over the station and dominate the local climate.

San Francisco Airport enjoys a marine-type climate characterized by mild and moderately wet winters and by dry, cool summers. Winter rains, occurring from November through March, account for over 80 percent of the annual rainfall, and measurable precipitation occurs on an average of 10 days per month during this period. However, there are frequent dry periods lasting well over a week. Severe winter storms with gale winds and heavy rains occur only occasionally. Thunderstorms average two a year and may occur in any month.

The daily and annual range in temperature is small. A few frosty mornings occur during the winter but the temperature seldom drops below freezing. Winter temperatures generally rise to the high 50s in the early afternoon.

The summer weather is dominated by a cool sea breeze resulting in an average summer wind speed of nearly 15 mph. Winds are light in the early morning but normally reach 20 to 25 mph in the afternoon.

A sea fog, arriving over the station during the late evening or night as a low cloud, is another persistent feature of the summer weather. This high fog, occasionally producing drizzle or mist, usually disappears during the late forenoon. Despite the morning overcast, summer days are sunny. On the average a total of only 14 days during the four months from June through September are classified as cloudy.

Daytime temperatures are held down both by the morning low overcast and the afternoon strengthening sea breeze, resulting in daily maximum readings averaging about 70 degrees from May through August. However, during these months occasional hot spells, lasting a few days, are experienced without the usual high fog and sea breeze. September, when the sea breeze becomes less pronounced, is the warmest month with highs in the 70s. Low temperatures during the summer are in the mid-50s.

A strong temperature inversion with its base usually about 1,500 feet persists throughout the summer. Inversions close to the ground are infrequent in summer but rather common in fall and winter. As a consequence of these factors and the continued population and economic growth of the area, atmospheric pollution has become a problem of increasing importance.

NORMALS, MEANS, AND EXTREMES

SAN FRANCISCO, CALIFORNIA INTERNATIONAL AIRPORT

LATITUDE: 37°37'N LONGITUDE: 122°23' W ELEVATION: FT. GRND 8 BARO 90 TIME ZONE: PACIFIC WBAN: 2323

	(a)	JAN	FEB	MAR	APR	MAY	JUNE	JULY	AUG	SEP	OCT	NOV	DEC	YEAR
TEMPERATURE °F:														
Normals														
-Daily Maximum		55.6	59.4	60.8	63.9	66.5	70.3	71.6	72.3	73.6	70.1	62.4	56.1	65.
-Daily Minimum		41.8	45.0	45.8	47.2	49.7	52.6	53.9	55.0	55.2	51.8	47.1	42.7	49.
-Monthly		48.7	52.2	53.3	55.6	58.1	61.5	62.7	63.7	64.5	61.0	54.8	49.4	57.
Extremes														
-Record Highest	66	72	78	85	92	97	106	105	100	103	99	85	75	10
-Year		1948	1930	1952	1989	1984	1961	1988	1993	1971	1987	1967	1958	JUN 196
-Record Lowest	66	24	25	30	31	36	41	43	42	38	34	25	20	2
-Year		1928	1929	1929	1929	1929	1932	1928	1935	1929	1929	1931	1932	DEC 193
NORMAL DEGREE DAYS:														
Heating (base 65°F)		505	358	363	287	218	121	92	68	79	135	306	484	301
Cooling (base 65°F)		0	0	0	5	0	16	21	28	64	11	0	0	14
% OF POSSIBLE SUNSHINE														
MEAN SKY COVER (tenths)														
Sunrise - Sunset	52	6.2	6.2	5.8	5.2	4.5	3.8	2.9	3.4	3.2	4.1	5.4	5.9	4.
MEAN NUMBER OF DAYS:														
Sunrise to Sunset														
-Clear	66	8.6	7.9	9.6	10.8	13.7	16.0	20.7	18.8	18.1	15.5	11.2	9.6	160.
-Partly Cloudy	66	7.7	7.4	8.6	9.3	9.7	8.7	7.5	8.8	8.2	8.9	8.3	7.5	100.
-Cloudy	66	14.7	13.0	12.8	9.9	7.6	5.3	2.8	3.4	3.7	6.6	10.6	13.9	104.
Precipitation														
.01 inches or more	66	10.6	9.7	9.9	5.8	2.8	1.2	0.3	0.5	1.0	3.6	7.0	9.7	62.
Snow,Ice pellets,hail														
1.0 inches or more	66	0.*	0.0	0.0	0.0	0.0	0.0	0.0	0.0	0.0	0.0	0.0	0.*	
Thunderstorms	66	0.3	0.4	0.3	0.2	0.1	0.1	0.1	0.1	0.2	0.2	0.1	0.2	2.
Heavy Fog Visibility														
1/4 mile or less	56	3.5	2.6	0.4	0.1	0.1	0.*	0.*	0.2	0.7	1.4	2.1	3.2	14.
Temperature °F														
-Maximum														
90° and above	34	0.0	0.0	0.0	0.1	0.4	0.9	0.7	0.3	1.3	0.4	0.0	0.0	4.
32° and below	34	0.0	0.0	0.0	0.0	0.0	0.0	0.0	0.0	0.0	0.0	0.0	0.0	0.
-Minimum														
32° and below	34	1.1	0.1	0.*	0.0	0.0	0.0	0.0	0.0	0.0	0.0	0.0	0.9	2.
0° and below	34	0.0	0.0	0.0	0.0	0.0	0.0	0.0	0.0	0.0	0.0	0.0	0.0	0.
AVG. STATION PRESS. (mb)	21	1019.2	1018.1	1016.8	1016.9	1015.1	1014.2	1014.2	1014.0	1013.6	1016.0	1018.5	1019.6	1016.
RELATIVE HUMIDITY (%)														
Hour 04	34	86	84	82	82	83	84	86	87	84	82	83	85	8
Hour 10 (Local Time)	34	79	75	70	65	63	63	65	67	66	68	72	77	6
Hour 16	34	66	65	63	59	59	58	59	61	59	59	63	67	6
Hour 22	34	80	79	77	77	78	79	82	82	79	77	77	80	7
PRECIPITATION (inches):														
Water Equivalent														
-Normal		4.35	3.17	3.06	1.37	0.19	0.11	0.03	0.05	0.20	1.22	2.86	3.09	19.70
-Maximum Monthly	66	11.26	9.52	9.01	6.36	3.81	0.86	0.35	0.66	2.30	7.30	7.94	12.30	12.30
-Year		1993	1958	1958	1958	1957	1967	1977	1976	1959	1962	1973	1955	DEC 1955
-Minimum Monthly	66	0.24	T	T	T	T	0.00	0.00	T	T	T	0.00	0.01	0.00
-Year		1991	1953	1934	1977	1992	1928	1930	1993	1993	1978	1929	1989	JUL 1930
-Maximum in 24 hrs	66	5.71	2.64	2.46	2.66	1.54	0.83	0.35	0.36	2.30	3.74	2.39	3.33	5.71
-Year		1982	1987	1982	1958	1957	1967	1977	1976	1959	1962	1973	1955	JAN 1982
Snow,Ice pellets,hail														
-Maximum Monthly	66	1.5	T	T	0.0	0.0	0.0	0.0	0.0	0.0	0.0	0.0	1.0	1.5
-Year		1962	1989	1992									1932	JAN 1962
-Maximum in 24 hrs	66	1.5	T	T	T	0.0	0.0	0.0	0.0	0.0	0.0	0.0	1.0	1.5
-Year		1962	1989	1992	1987								1932	JAN 1962
WIND:														
Mean Speed (mph)	66	7.2	8.7	10.5	12.2	13.4	13.9	13.6	12.8	11.1	9.4	7.5	7.1	10.6
Prevailing Direction through 1963		WNW	WNW	WNW	WNW	W	W	NW	NW	NW	WNW	WNW	WNW	WNW
Fastest Obs. 1 Min.														
-Direction (!!)	44	16	22	19	18	31	28	28	27	28	25	20	130	16
-Speed (MPH)	44	58	55	41	47	41	44	40	36	38	44	47	48	58
-Year		1963	1990	1984	1982	1984	1969	1993	1974	1968	1950	1953	1983	JAN 1963
Peak Gust														
-Direction (!!)	10	S	S	S	NW	NW	W	W	W	W	S	S	N	S
-Speed (mph)	10	53	69	59	54	58	51	52	45	43	58	60	64	69
-Date		1993	1986	1986	1984	1984	1984	1987	1992	1988	1989	1984	1988	FEB 1986

PRECIPITATION (inches)

SAN FRANCISCO, CALIFORNIA INTERNATIONAL AIRPORT

YEAR	JAN	FEB	MAR	APR	MAY	JUNE	JULY	AUG	SEP	OCT	NOV	DEC	ANNUAL
1964	4.38	0.27	1.95	0.01	0.32	0.60	T	0.01	T	1.26	3.32	5.42	17.54
1965	4.37	0.91	1.76	3.47	T	T	T	0.29	T	T	5.40	5.02	21.22
1966	2.70	3.18	0.59	0.40	0.12	0.04	0.03	0.09	0.08	T	4.79	3.96	15.98
1967	10.43	0.09	5.04	5.31	0.26	0.86	T	T	0.01	0.48	1.29	3.50	27.27
1968	5.25	1.44	3.03	0.55	0.28	T	T	0.06	T	0.45	2.47	4.49	18.02
1969	8.92	8.62	1.34	1.87	T	0.06	T	T	0.02	1.96	0.69	4.59	28.07
1970	8.33	2.18	1.22	0.22	0.01	0.36	T	T	T	0.75	6.41	6.21	25.69
1971	1.27	0.26	2.68	0.77	0.25	T	T	T	0.11	0.03	0.99	3.44	9.80
1972	1.09	1.35	0.18	1.20	T	0.06	T	T	0.30	5.24	5.15	2.40	16.97
1973	8.32	6.82	2.93	0.11	0.07	T	T	T	0.04	1.60	7.94	3.55	31.38
1974	3.21	1.70	4.21	2.32	T	0.14	0.23	T	T	0.93	0.50	2.36	15.60
1975	2.60	3.94	5.91	1.66	0.02	0.04	0.13	0.21	T	2.27	0.26	0.21	17.25
1976	0.37	2.13	1.22	0.92	T	0.01	T	0.66	0.30	0.34	1.37	2.70	10.02
1977	2.22	1.04	2.01	T	0.41	T	0.35	T	0.47	0.15	2.20	3.69	12.54
1978	8.90	4.92	4.90	4.50	0.02	T	T	T	0.26	T	1.67	0.64	25.81
1979	6.61	5.87	2.74	0.69	0.13	T	0.09	T	T	2.20	1.94	4.30	24.57
1980	4.85	7.62	2.65	0.90	0.24	0.03	0.10	T	T	0.10	0.12	1.73	18.34
1981	5.92	2.21	3.60	0.24	0.07	T	T	T	0.28	2.35	4.89	3.91	23.47
1982	8.81	2.82	7.63	3.25	T	0.06	T	T	0.96	1.95	5.34	3.99	34.81
1983	6.83	6.64	8.50	3.11	0.32	T	0.01	T	0.57	0.10	6.03	6.23	38.34
1984	0.46	1.47	1.36	0.68	T	0.03	T	0.11	0.05	1.96	6.12	1.89	14.13
1985	0.74	2.35	3.30	0.12	0.05	0.29	0.03	0.02	0.18	0.69	3.19	1.61	12.57
1986	4.04	8.09	5.84	0.39	0.15	T	0.01	T	0.47	0.02	0.06	1.66	20.73
1987	2.80	3.52	1.98	0.16	0.06	T	T	T	T	0.93	1.64	4.51	15.60
1988	3.92	0.38	0.05	2.02	0.29	0.60	T	T	0.03	0.42	2.31	3.65	13.67
1989	1.25	1.28	4.00	0.78	0.04	0.01	T	T	1.24	1.40	1.34	0.01	11.35
1990	3.06	2.28	0.79	0.20	1.55	T	0.01	T	0.20	0.19	0.28	1.79	10.35
1991	0.24	3.76	6.07	0.61	0.21	0.11	T	0.27	0.04	1.73	0.23	2.70	15.97
1992	2.04	6.44	4.12	0.25	T	0.39	0.00	0.14	T	1.12	0.15	6.04	20.69
1993	11.26	4.68	2.34	0.41	0.55	0.16	T	T	T	0.45	1.47	2.19	23.51
Record Mean	4.14	3.34	2.94	1.30	0.32	0.12	0.02	0.04	0.18	0.95	2.10	3.54	18.99

AVERAGE TEMPERATURE (deg. F)

SAN FRANCISCO, CALIFORNIA INTERNATIONAL AIRPORT

YEAR	JAN	FEB	MAR	APR	MAY	JUNE	JULY	AUG	SEP	OCT	NOV	DEC	ANNUAL
1964	48.2	50.8	51.9	54.2	55.8	61.3	63.1	64.3	63.5	60.9	51.6	51.3	56.4
1965	48.5	50.4	52.3	54.5	55.3	58.3	60.3	63.3	61.1	62.5	55.2	44.9	55.6
1966	47.6	48.8	51.6	57.9	57.3	61.4	60.5	62.3	64.0	60.4	54.7	49.0	56.3
1967	49.2	51.4	52.4	50.9	58.5	59.3	62.7	62.8	66.4	63.5	57.8	48.1	56.9
1968	46.5	54.5	54.5	54.9	56.3	61.3	61.9	64.8	64.2	59.5	54.4	47.8	56.7
1969	47.1	48.9	52.1	54.0	59.5	61.1	62.3	63.0	63.9	61.2	56.1	53.7	56.9
1970	52.4	54.8	56.3	53.7	60.5	60.5	62.9	61.6	65.4	58.5	55.9	48.7	57.6
1971	48.3	50.0	52.1	53.0	56.4	60.1	60.7	65.2	66.2	57.0	53.2	46.1	55.7
1972	45.6	52.0	55.3	55.4	57.6	60.9	64.3	64.0	62.6	61.0	53.2	44.9	56.4
1973	48.0	52.9	51.4	56.7	58.7	63.5	62.1	60.7	63.5	60.7	53.9	50.0	56.8
1974	48.6	49.7	53.4	54.8	56.2	60.4	63.0	63.7	62.5	62.0	53.8	48.6	56.4
1975	47.4	50.9	51.4	50.6	58.4	59.8	61.8	63.0	61.4	58.8	52.2	49.3	55.4
1976	48.5	50.5	51.0	53.3	58.3	63.2	62.5	64.3	63.3	61.3	57.0	48.8	56.9
1977	47.0	53.2	50.9	55.5	55.8	60.4	62.4	64.1	63.5	60.5	55.3	52.3	56.7
1978	52.5	52.8	57.0	54.9	60.4	60.4	61.4	63.2	65.8	61.1	52.5	46.0	57.3
1979	47.5	50.3	54.5	55.5	60.4	61.0	63.8	63.7	67.3	62.6	54.1	50.9	57.6
1980	50.5	54.4	53.0	55.9	56.3	59.9	63.0	61.5	63.2	61.2	55.6	50.7	57.1
1981	51.1	54.0	53.2	56.2	59.0	65.0	61.7	63.0	62.7	58.6	56.3	52.2	57.7
1982	45.1	51.7	51.3	54.6	57.4	59.7	61.7	63.3	64.0	61.1	52.3	48.9	55.9
1983	48.0	53.4	54.1	54.7	57.8	61.5	65.1	66.9	68.3	64.4	55.5	53.4	58.6
1984	51.3	52.9	57.0	56.0	61.8	61.3	65.6	64.4	69.7	60.4	53.8	47.7	58.5
1985	46.4	51.6	51.4	59.0	58.6	65.2	64.8	64.0	63.2	60.7	52.0	47.1	57.0
1986	53.7	56.3	57.1	56.2	58.6	62.5	62.4	61.2	62.9	61.4	57.1	50.3	58.3
1987	49.3	53.3	54.9	59.2	61.6	62.4	63.1	65.1	64.0	63.9	57.0	50.5	58.7
1988	50.6	54.5	56.5	58.1	59.5	62.5	65.3	65.0	63.1	61.4	56.5	50.4	58.6
1989	48.3	48.4	54.9	60.8	59.8	62.7	62.8	64.0	61.4	60.8	56.4	50.1	57.5
1990	49.9	49.2	53.3	58.5	59.0	62.4	64.5	66.3	66.4	63.1	56.0	46.4	57.9
1991	50.1	55.3	52.2	55.7	56.9	59.4	63.7	64.4	62.9	62.8	57.5	50.8	57.6
1992	48.9	56.1	57.4	60.6	63.4	63.6	65.8	63.8	65.3	65.4	56.8	49.8	59.7
1993	49.3	52.5	57.5	58.1	62.2	64.8	64.8	67.3	63.1	63.4	56.3	50.1	59.1
Record Mean	48.4	51.4	53.2	55.2	57.8	60.8	62.1	62.6	63.3	60.5	54.6	49.4	56.6
Max	55.7	59.1	61.1	63.5	66.2	69.6	70.9	71.3	73.0	70.1	63.3	56.6	65.1
Min	41.1	43.7	45.3	46.8	49.4	51.9	53.2	53.8	53.5	50.9	45.9	42.2	48.1

REFERENCE NOTES

SAN FRANCISCO, CALIFORNIA INTERNATIONAL AIRPORT

GENERAL
T - TRACE AMOUNT.
BLANK ENTRIES DENOTE MISSING/UNREPORTED DATA.
# INDICATES A STATION OR INSTRUMENT RELOCATION.

SPECIFIC

(a) - LENGTH OF RECORD IN YEARS, ALTHOUGH INDIVIDUAL MONTHS MAY BE MISSING.
0.* OR * - THE VALUE IS BETWEEN 0.0 AND 0.05
NORMALS - BASED ON THE 1961-1990 RECORD PERIOD.
EXTREMES - DATES ARE THE MOST RECENT OCCURRENCE
WIND DIR.- NUMERALS SHOW TENS OF DEGRESS CLOCKWISE FROM TRUE NORTH. "00" INDICATES CALM.
RESULTANT DIRECTIONS ARE GIVEN TO WHOLE DEGREES.
BOLD VALUES INDICATE EXTREME VALUES WHICH OCCURRED AFTER THE ASOS SYSTEM WAS COMMISSIONED.

RECORD = PERIOD OF RECORD
RECORD MEAN PRECIPITATION IS THE MEAN OF ALL DAILY PRECIPITATION AMOUNTS DURING THE PERIOD OF RECORD.
RECORD MAX(MIN) TEMPERATURE IS THE MEAN OF ALL DAILY MAX(MIN) TEMPERATURES DURING THE PERIOD OF RECORD.
RECORD MEAN TEMPERATURE IS THE SUM OF THE RECORD MAX AND RECORD MIN DIVIDED BY 2.
AVERAGE TEMPERATURE IS THE SUM OF THE MEAN DAILY MAX AND MIN TEMPERATURE DIVIDED BY 2.

EXCEPTIONS

RECORD MEANS ARE THROUGH THE CURRENT YEAR,
BEGINNING IN 1928 FOR TEMPERATURE
1928 FOR PRECIPITATION
1928 FOR SNOWFALL

HEATING DEGREE DAYS Base 65 deg. F — SAN FRANCISCO, CALIFORNIA INTERNATIONAL AIRPORT

SEASON	JULY	AUG	SEP	OCT	NOV	DEC	JAN	FEB	MAR	APR	MAY	JUNE	TOTAL
1964-65	76	46	94	135	393	416	505	404	384	311	295	193	3252
1965-66	139	64	113	101	287	617	535	449	407	214	231	123	3280
1966-67	135	89	57	148	301	489	483	376	383	416	208	163	3248
1967-68	85	63	11	75	217	521	564	297	317	293	264	116	2823
1968-69	108	42	56	167	311	525	547	443	391	322	167	113	3192
1969-70	86	70	54	113	261	341	384	277	260	331	168	133	2478
1970-71	92	117	59	203	265	495	513	414	394	354	258	154	3318
1971-72	133	21	47	252	349	579	594	368	294	282	226	135	3280
1972-73	64	42	75	131	350	613	521	334	416	241	197	103	3087
1973-74	100	129	76	128	327	459	501	424	354	298	270	145	3211
1974-75	83	54	101	117	329	499	540	387	415	424	220	151	3320
1975-76	109	75	120	188	377	480	504	415	427	344	222	136	3397
1976-77	72	38	79	127	231	494	549	326	432	278	278	141	3045
1977-78	103	48	55	139	284	385	381	335	238	295	161	135	2559
1978-79	111	65	32	143	371	581	536	406	319	277	148	132	3121
1979-80	55	56	13	85	320	431	441	298	366	269	261	155	2750
1980-81	76	109	74	145	275	436	424	301	358	279	180	53	2710
1981-82	112	65	71	197	252	389	611	364	416	307	241	154	3179
1982-83	100	63	47	130	376	491	521	322	330	301	225	113	3019
1983-84	33	1	18	43	281	354	415	342	242	269	124	115	2237
1984-85	43	34	5	147	328	527	570	370	416	180	192	27	2839
1985-86	49	51	60	158	382	546	343	236	239	260	191	78	2593
1986-87	77	113	62	122	228	447	477	320	309	168	128	85	2536
1987-88	60	16	40	70	233	445	440	296	259	212	184	78	2333
1988-89	40	29	71	128	246	447	511	455	308	162	160	94	2651
1989-90	70	38	103	138	249	454	459	437	356	189	185	94	2772
1990-91	33	13	8	77	262	570	454	265	387	273	244	166	2752
1991-92	57	44	65	96	223	434	494	252	226	133	52	58	2134
1992-93	20	49	34	40	235	466	480	346	225	201	100	54	2250
1993-94	33	11	71	67	256	455							

COOLING DEGREE DAYS Base 65 deg. F — SAN FRANCISCO, CALIFORNIA INTERNATIONAL AIRPORT

YEAR	JAN	FEB	MAR	APR	MAY	JUNE	JULY	AUG	SEP	OCT	NOV	DEC	TOTAL
1969	0	0	0	0	3	3	8	15	26	2	0	0	57
1970	0	0	0	0	34	3	32	21	77	8	0	0	175
1971	0	0	0	0	0	12	5	34	91	11	0	0	153
1972	0	0	1	0	7	14	49	16	8	14	0	0	109
1973	0	0	0	0	7	64	18	2	37	1	0	0	129
1974	0	0	0	0	6	11	27	22	31	30	0	0	127
1975	0	0	0	0	21	4	15	19	19	2	0	0	80
1976	0	0	0	0	21	88	4	23	33	23	0	0	192
1977	0	0	0	0	0	10	30	26	17	5	0	0	88
1978	0	0	0	0	24	0	7	18	62	33	0	0	144
1979	0	0	0	0	11	19	25	21	88	18	0	0	182
1980	0	0	0	0	0	10	22	7	30	33	1	0	103
1981	0	0	0	17	1	61	17	7	7	3	0	0	113
1982	0	0	0	1	12	5	7	15	23	12	0	0	75
1983	0	0	0	0	7	16	42	66	119	32	0	0	282
1984	0	0	0	4	33	10	70	24	152	9	0	0	302
1985	0	0	0	8	1	38	50	28	11	33	0	0	169
1986	0	0	1	2	0	11	5	0	7	16	0	0	42
1987	0	0	0	4	29	15	9	26	17	43	0	0	143
1988	0	0	0	11	19	8	55	34	23	24	0	0	174
1989	0	0	0	40	6	35	8	15	2	16	0	0	122
1990	0	0	0	1	3	23	23	58	58	24	0	0	190
1991	0	0	0	0	0	6	21	29	13	33	2	0	104
1992	0	0	0	9	10	21	53	18	49	58	0	0	218
1993	0	0	0	0	18	58	33	91	19	26	1	0	246

SNOWFALL (inches) — SAN FRANCISCO, CALIFORNIA INTERNATIONAL AIRPORT

SEASON	JULY	AUG	SEP	OCT	NOV	DEC	JAN	FEB	MAR	APR	MAY	JUNE	TOTAL
1970-71	0.0	0.0	0.0	0.0	0.0	0.0	T	0.0	0.0	0.0	0.0	0.0	T
1971-72	0.0	0.0	0.0	0.0	0.0	0.0	T	0.0	0.0	0.0	0.0	0.0	T
1972-73	0.0	0.0	0.0	0.0	0.0	T	0.0	0.0	T	0.0	0.0	0.0	T
1973-74	0.0	0.0	0.0	0.0	0.0	0.0	0.0	0.0	0.0	0.0	0.0	0.0	0.0
1974-75	0.0	0.0	0.0	0.0	0.0	0.0	T	0.0	0.0	0.0	0.0	0.0	T
1975-76	0.0	0.0	0.0	0.0	0.0	0.0	0.0	T	T	0.0	0.0	0.0	T
1976-77	0.0	0.0	0.0	0.0	0.0	0.0	0.0	0.0	0.0	0.0	0.0	0.0	0.0
1977-78	0.0	0.0	0.0	0.0	0.0	0.0	0.0	0.0	0.0	0.0	0.0	0.0	0.0
1978-79	0.0	0.0	0.0	0.0	0.0	0.0	T	0.0	0.0	0.0	0.0	0.0	T
1979-80	0.0	0.0	0.0	0.0	0.0	0.0	0.0	0.0	T	0.0	0.0	0.0	T
1980-81	0.0	0.0	0.0	0.0	0.0	0.0	0.0	T	0.0	0.0	0.0	0.0	T
1981-82	0.0	0.0	0.0	0.0	0.0	0.0	T	0.0	T	0.0	0.0	0.0	T
1982-83	0.0	0.0	0.0	0.0	0.0	0.0	0.0	T	T	0.0	0.0	0.0	T
1983-84	0.0	0.0	0.0	0.0	0.0	0.0	0.0	0.0	0.0	0.0	0.0	0.0	0.0
1984-85	0.0	0.0	0.0	0.0	0.0	0.0	0.0	0.0	0.0	0.0	0.0	0.0	0.0
1985-86	0.0	0.0	0.0	0.0	0.0	0.0	0.0	0.0	0.0	0.0	0.0	0.0	0.0
1986-87	0.0	0.0	0.0	0.0	0.0	0.0	0.0	T	0.0	0.0	0.0	0.0	T
1987-88	0.0	0.0	0.0	0.0	0.0	0.0	0.0	0.0	0.0	0.0	0.0	0.0	0.0
1988-89	0.0	0.0	0.0	0.0	0.0	T	T	T	0.0	0.0	0.0	0.0	T
1989-90	0.0	0.0	0.0	0.0	0.0	0.0	0.0	0.0	0.0	0.0	0.0	0.0	0.0
1990-91	0.0	0.0	0.0	0.0	0.0	0.0	0.0	0.0	0.0	0.0	0.0	0.0	0.0
1991-92	0.0	0.0	0.0	0.0	0.0	0.0	0.0	0.0	T	0.0	0.0	0.0	T
1992-93	0.0	0.0	0.0	0.0	0.0	T	0.0	0.0	0.0	0.0	0.0	0.0	T
1993-94	0.0	0.0	0.0	0.0	0.0	0.0							
Record Mean	0.0	0.0	0.0	0.0	0.0	T	T	T	T	0.0	0.0	0.0	T

SAN FRANCISCO, CALIFORNIA
MISSION DOLORES

San Francisco is located at the northern end of a narrow peninsula which separates San Francisco Bay from the Pacific Ocean. It is known as the air conditioned city with cool pleasant summers and mild winters. Flowers bloom throughout the year and warm clothing may be needed at times during any month.

Precipitation averages about 20 inches a year with pronounced wet and dry seasons, characteristic of its Mediterranean climate. Little or no rain falls from June through September while about 80 percent of the annual total falls from November through March. Snow is extremely rare. Measurable amounts fall about once every 15 years. Freezing temperatures are also extremely rare. On average, thunderstorms occur on only two days each year. The average annual wind speed is about 9 mph with lighter winds, 6 to 7 mph, occurring in the winter and stronger winds, 10 to 11 mph, in the summer.

San Francisco probably has greater climatic variability by far with respect to temperature, cloudiness, and sunshine within its 49 square mile area than any other similarly sized urban area in the country. Likewise, the San Francisco Bay area has considerably more variability than San Francisco itself.

Sea fogs, and the low stratus clouds associated with them are most common in the summertime, but may occur at any time of the year. In the summer the temperature of the Pacific Ocean is much lower than the temperature inland, particularly in the Central Valley of California. This condition tends to enhance the sea breeze effect common to coastal areas. Brisk westerly winds blow throughout the afternoon and evening hours. The fog is carried inland by these westerly winds in the late afternoon and evening and then evaporates during the subsequent forenoon.

The complex topography of San Francisco causes complex patterns of fog and sun as well as temperature. A range of hills with elevations of nearly 1000 feet above sea level, bisects the city from north to south. This range partially blocks the inland movement of the fog, but gaps in the hills permit small masses of fog to pass through, further complicating the pattern. Occasionally, the fog will reach 50 miles south to San Jose, while the area just to the lee of the highest hills is still mostly clear.

Sunshine varies greatly from one part of the city to another, especially in the summer. Spring and fall are the sunniest seasons. In the summer the sunniest area is a triangular shaped area to the lee of the highest hills and extending to the bay. The least sunny area is along the ocean due to the high frequency of fog there. The percent of possible summer sunshine varies from an estimated 25 to 35 percent at the ocean to 70 to 80 percent in the sunniest area.

The extent and behavior of the summertime fog on a particular day depends on several factors. A typical day would find the fog covering the entire city at sunrise and little wind. During the forenoon the skies become sunny in the eastern part of the city with some partial clearing reaching the ocean for a couple of hours in the early afternoon. By early afternoon the winds pick up and by late afternoon the fog is rolling inland again. The wind usually reaches a maximum velocity in the early evening.

In the winter relatively little difference in the climate is observed from one part of the city to another. This is due to the lack of temperature contrast between the ocean and the land and to the relative frequency of passage of Pacific frontal systems. However, those areas near the ocean have more sunshine than areas further inland. The source region for fog is inland during winter, mainly in the Central Valley, rather than the ocean.

Temperature patterns in the city are the same as those of sunshine. In the winter there is little variation, with average maximums from 55 to 60 degrees and average minimums in the mid to upper 40s. Average temperatures rise until June and remain nearly constant through August with average maximums in the lower 60s near the ocean and upper 60s in the sunny eastern half of the city. Summer minimums range from 50 to 55. The warmest time of the year is September and October when the fog diminishes greatly and some of the warmth from the Central Valley flows westward. At this time of year the average maximums are in the mid 60s near the ocean and in the mid 70s in the warmest areas of the city. The average minimums are about the same as they are during the summer.

NORMALS, MEANS, AND EXTREMES

SAN FRANCISCO, CALIFORNIA MISSION DOLORES

LATITUDE: 37°46'N LONGITUDE: 122°26'W ELEVATION: FT. GRND 75 BARO TIME ZONE: PACIFIC WBAN: 23272

	(a)	JAN	FEB	MAR	APR	MAY	JUNE	JULY	AUG	SEP	OCT	NOV	DEC	YEAR
TEMPERATURE °F:														
Normals														
-Daily Maximum		56.3	60.0	60.8	62.1	62.7	64.1	64.6	65.6	68.7	68.7	62.7	56.4	62.7
-Daily Minimum		45.8	48.7	49.0	49.8	50.5	52.6	53.5	54.6	55.9	55.2	51.6	47.0	51.2
-Monthly		51.1	54.4	54.9	56.0	56.6	58.4	59.1	60.1	62.3	62.0	57.2	51.7	57.0
Extremes														
-Record Highest	57	79	81	83	94	96	101	103	98	101	102	86	76	103
-Year		1962	1986	1952	1989	1976	1961	1988	1993	1971	1987	1966	1958	JUL 1988
-Record Lowest	57	30	31	38	40	44	47	47	48	48	45	41	28	28
-Year		1937	1989	1942	1967	1964	1955	1953	1969	1955	1949	1985	1990	DEC 1990
NORMAL DEGREE DAYS:														
Heating (base 65°F)		431	297	313	275	260	198	187	156	116	123	237	412	3005
Cooling (base 65°F)		0	0	0	0	0	0	0	0	35	30	0	0	65
% OF POSSIBLE SUNSHINE	38	56	62	69	73	72	73	66	65	72	70	62	53	66
MEAN SKY COVER (tenths)														
Sunrise - Sunset														
MEAN NUMBER OF DAYS:														
Sunrise to Sunset														
-Clear														
-Partly Cloudy														
-Cloudy														
Precipitation														
.01 inches or more	57	10.7	10.1	10.7	6.1	3.1	1.3	0.5	0.9	1.5	4.3	7.9	10.1	67.2
Snow,Ice pellets,hail														
1.0 inches or more	37	0.0	0.0	0.0	0.0	0.0	0.0	0.0	0.0	0.0	0.0	0.0	0.0	0.0
Thunderstorms	29	0.3	0.2	0.1	0.3	0.2	0.1	0.2	0.*	0.3	0.2	0.1	0.3	2.2
Heavy Fog Visibility														
1/4 mile or less														
Temperature °F														
-Maximum														
90° and above	57	0.0	0.0	0.0	0.1	0.2	0.4	0.2	0.1	1.0	0.4	0.0	0.0	2.3
32° and below	57	0.0	0.0	0.0	0.0	0.0	0.0	0.0	0.0	0.0	0.0	0.0	0.0	0.0
-Minimum														
32° and below	57	0.1	0.*	0.0	0.0	0.0	0.0	0.0	0.0	0.0	0.0	0.0	0.1	0.2
0° and below	57	0.0	0.0	0.0	0.0	0.0	0.0	0.0	0.0	0.0	0.0	0.0	0.0	0.0
AVG. STATION PRESS.(mb)														
RELATIVE HUMIDITY (%)														
Hour 04	7	81	83	81	82	89	89	92	93	87	81	82	80	85
Hour 10 (Local Time)	7	72	70	61	59	65	70	73	73	64	62	69	71	67
Hour 16	7	63	63	61	61	68	72	74	73	66	60	63	63	66
Hour 22	7	76	78	76	80	86	88	90	90	82	74	76	74	81
PRECIPITATION (inches):														
Water Equivalent														
-Normal		4.06	2.95	3.07	1.29	0.25	0.15	0.04	0.07	0.26	1.26	3.21	3.10	19.71
-Maximum Monthly	57	10.69	8.49	9.04	5.47	3.19	1.42	0.62	0.78	2.06	5.51	8.20	11.47	11.47
-Year		1952	1938	1983	1958	1957	1967	1974	1976	1959	1962	1983	1955	DEC 1955
-Minimum Monthly	57	0.31	0.04	0.07	T	0.00	0.00	0.00	0.00	0.00	0.00	T	0.00	0.00
-Year		1976	1953	1988	1949	1982	1983	1982	1982	1980	1980	1959	1989	DEC 1989
-Maximum in 24 hrs	57	4.22	2.34	3.65	2.36	1.47	1.36	0.61	0.49	2.06	3.11	2.72	3.14	4.22
-Year		1982	1940	1940	1953	1990	1967	1974	1965	1959	1962	1973	1945	JAN 1982
Snow,Ice pellets,hail														
-Maximum Monthly	40	T	T	T	0.0	0.0	0.0	0.0	0.0	0.0	0.0	0.0	T	T
-Year		1962	1951	1951									1972	DEC 1972
-Maximum in 24 hrs	36	T	T	T	0.0	0.0	0.0	0.0	0.0	0.0	0.0	0.0	T	T
-Year		1962	1951	1951									1941	JAN 1962
WIND:														
Mean Speed (mph)	28	6.7	7.5	8.5	9.5	10.4	10.9	11.2	10.5	9.1	7.6	6.3	6.5	8.7
Prevailing Direction through 1963		N	W	W	W	W	W	W	W	W	W	W	N	W
Fastest Mile														
-Direction (!!!)	36	SE	SW	S	W	W	W	W	W	W	SE	S	SE	SE
-Speed (MPH)	36	47	47	44	38	38	40	38	34	32	43	41	45	47
-Year		1965	1938	1948	1965	1965	1965	1939	1966	1956	1950	1953	1965	JAN 1965
Peak Gust														
-Direction (!!!)														
-Speed (mph)														
-Date														

PRECIPITATION (inches)

SAN FRANCISCO, CALIFORNIA MISSION DOLORES

YEAR	JAN	FEB	MAR	APR	MAY	JUNE	JULY	AUG	SEP	OCT	NOV	DEC	ANNUAL
1964	3.37	0.19	2.12	0.01	0.22	0.57	T	0.01	T	1.90	3.99	5.35	17.73
1965	3.97	0.94	2.92	3.21	T	T	0.02	0.49	T	0.01	4.79	3.51	19.86
1966	3.27	2.72	0.80	0.36	0.19	0.17	0.06	0.10	0.10	0.01	4.80	3.87	16.45
1967	9.49	0.22	4.35	4.90	0.09	1.42	0.00	T	0.04	0.53	1.10	2.12	24.26
1968	4.54	2.28	3.15	0.48	0.22	T	T	0.03	0.06	0.62	2.67	3.91	17.96
1969	7.74	7.26	1.01	1.74	T	0.05	T	T	0.01	2.61	0.45	6.15	27.02
1970	7.81	1.56	1.55	0.06	0.03	0.57	T	T	0.00	0.84	6.44	5.39	24.25
1971	2.04	0.26	2.91	0.72	0.19	T	0.01	0.01	0.22	0.11	1.92	3.93	12.32
1972	1.32	2.13	0.23	1.07	T	0.11	0.01	0.04	0.54	5.41	6.40	3.53	20.79
1973	9.38	6.32	2.63	0.02	0.08	0.00	0.00	0.00	0.30	1.62	7.80	3.65	31.80
1974	3.40	1.53	4.49	2.34	0.00	0.10	0.62	0.00	0.00	0.85	0.40	1.53	15.26
1975	2.57	3.72	5.15	1.25	0.02	0.04	0.20	0.02	0.00	2.44	0.43	0.18	16.02
1976	0.31	1.83	1.01	0.70	0.01	0.03	0.00	0.78	0.51	0.38	1.04	2.13	8.73
1977	1.65	0.90	2.01	0.05	0.57	0.00	0.00	0.03	0.86	0.17	1.96	3.30	11.50
1978	6.20	3.54	5.20	3.82	0.00	0.00	0.00	0.00	0.20	0.00	1.67	0.89	21.52
1979	6.74	4.96	1.58	0.87	0.15	0.00	0.07	0.00	0.01	1.66	2.98	3.10	22.12
1980	3.77	4.84	1.25	0.97	0.23	0.02	0.04	0.00	0.00	0.00	0.14	2.95	14.21
1981	4.00	1.78	3.71	0.17	0.12	0.00	0.00	0.00	0.22	1.74	3.73	4.15	19.62
1982	6.84	3.26	7.65	3.03	0.00	0.06	0.00	0.00	0.72	2.79	5.62	2.22	32.19
1983	5.77	8.06	9.04	3.48	0.47	0.00	0.01	0.06	0.68	0.26	8.20	7.72	43.75
1984	0.50	2.34	1.32	0.92	0.16	0.30	0.00	0.24	0.10	2.94	7.45	2.10	18.37
1985	0.59	1.98	3.94	0.27	0.09	0.31	0.00	0.00	0.38	0.80	4.83	2.47	15.66
1986	4.77	8.29	6.25	0.76	0.13	0.00	0.03	0.01	1.32	0.11	0.20	1.64	23.51
1987	4.26	3.77	2.31	0.14	0.06	0.01	0.00	0.00	0.00	1.07	3.09	5.09	19.80
1988	4.93	0.40	0.07	1.73	0.66	0.70	0.00	0.00	0.00	0.64	3.70	4.23	17.06
1989	1.26	1.49	5.28	0.70	0.06	0.07	0.00	0.05	0.98	1.18	1.33	0.00	12.40
1990	4.02	2.45	1.34	0.58	2.38	0.01	0.00	0.04	0.12	0.20	0.52	1.94	13.60
1991	0.60	3.29	5.89	1.07	0.36	0.05	0.00	0.42	0.00	2.35	0.50	2.32	16.85
1992	2.09	6.34	4.41	0.38	0.00	0.39	0.00	0.02	0.00	1.16	0.40	6.03	21.22
1993	9.82	4.48	2.90	0.71	0.87	0.27	0.00	0.00	0.00	0.33	2.16	2.25	23.79
Record Mean	4.54	3.55	3.08	1.47	0.58	0.15	0.02	0.03	0.28	1.01	2.48	4.09	21.27

AVERAGE TEMPERATURE (deg. F)

SAN FRANCISCO, CALIFORNIA MISSION DOLORES

YEAR	JAN	FEB	MAR	APR	MAY	JUNE	JULY	AUG	SEP	OCT	NOV	DEC	ANNUAL
1964	51.0	55.0	53.2	53.8	53.4	57.8	58.9	60.0	62.4	63.1	55.3	53.7	56.5
1965	51.4	54.0	54.4	55.7	54.9	56.2	57.4	61.2	61.2	65.0	58.1	48.3	56.5
1966	52.1	51.8	53.8	57.9	55.1	59.4	58.2	58.8	63.6	62.6	57.2	51.3	56.8
1967	52.6	53.2	52.7	50.8	57.9	57.1	58.9	59.2	63.5	65.5	60.0	51.9	57.0
1968	49.8	56.7	56.7	56.2	55.7	59.0	58.0	62.3	63.1	60.5	56.2	49.8	57.0
1969	48.6	50.1	54.3	54.2	57.0	58.7	57.6	59.4	60.9	61.9	59.3	55.8	56.5
1970	54.0	57.4	57.8	53.3	57.7	56.8	57.8	57.2	64.4	58.6	57.9	50.6	57.0
1971	50.8	51.9	53.3	53.1	54.6	57.3	57.5	61.1	64.7	57.8	55.6	49.0	55.6
1972	48.5	54.0	55.8	55.5	55.5	57.5	60.9	60.2	61.5	61.7	54.9	47.2	56.1
1973	50.1	54.9	52.5	57.2	56.3	60.7	58.6	57.1	61.3	61.0	55.3	52.0	56.4
1974	51.1	52.2	53.3	55.4	54.9	58.2	59.6	59.9	60.3	62.2	56.6	51.1	56.3
1975	51.0	53.3	53.1	51.9	57.2	56.9	58.9	59.5	59.5	59.7	55.6	53.4	55.8
1976	53.4	52.8	52.5	54.1	56.8	61.5	59.2	62.5	62.2	62.8	60.4	54.6	57.8
1977	49.9	56.1	53.2	56.1	55.3	57.1	59.0	61.6	62.0	60.6	58.6	54.9	57.0
1978	55.0	55.2	59.0	56.3	60.7	58.9	58.4	60.6	65.5	61.9	55.9	49.6	58.1
1979	51.0	52.9	55.7	56.5	59.2	58.6	60.2	60.8	66.3	63.2	57.7	55.4	58.1
1980	53.0	57.2	56.0	56.9	55.4	57.9	59.5	58.0	61.3	62.0	58.3	53.4	57.4
1981	52.4	56.1	54.9	55.8	56.8	62.2	57.8	59.2	60.4	59.3	58.3	54.0	57.3
1982	48.5	55.0	52.8	55.6	55.8	56.3	57.9	60.1	62.6	62.8	54.4	52.2	56.2
1983	49.4	54.6	55.3	56.8	59.7	61.8	63.4	65.9	67.1	64.0	56.1	52.8	58.9
1984	51.6	52.6	56.7	54.2	59.9	59.7	63.9	62.8	69.4	61.5	56.0	50.9	58.3
1985	50.0	56.0	53.2	59.8	58.1	63.9	64.1	64.1	64.1	63.2	55.0	51.3	58.6
1986	56.6	58.9	60.4	58.6	60.0	63.2	62.8	61.9	62.8	63.6	60.2	52.5	60.1
1987	51.8	56.4	57.1	60.5	61.1	60.5	61.5	63.5	63.8	65.1	58.8	52.3	59.4
1988	52.8	57.7	59.1	58.8	59.1	61.1	64.2	64.0	63.1	61.5	57.3	53.3	59.3
1989	51.3	50.0	55.4	60.9	59.3	61.6	62.4	63.0	61.8	62.0	58.8	52.6	58.3
1990	52.8	52.0	54.9	59.2	59.0	62.4	62.9	65.3	66.0	64.2	58.0	49.1	58.8
1991	53.4	57.9	53.2	57.1	56.8	58.6	61.3	63.0	63.1	64.4	60.1	53.4	58.5
1992	51.5	58.4	59.2	62.6	62.7	62.5	65.1	63.8	65.8	66.7	59.8	51.7	60.8
1993	51.1	53.8	59.0	59.4	62.5	65.9	63.4	66.6	63.4	64.3	58.2	51.5	59.9
Record Mean	50.5	53.3	54.6	55.9	57.1	59.1	59.2	59.8	62.1	61.3	57.0	51.7	56.8
Max	55.7	59.0	60.7	62.3	63.4	65.6	65.2	65.7	69.0	68.4	63.1	56.7	62.9
Min	45.3	47.5	48.5	49.4	50.8	52.6	53.2	54.0	55.2	54.1	50.9	46.7	50.7

REFERENCE NOTES

SAN FRANCISCO, CALIFORNIA MISSION DOLORES

GENERAL
T - TRACE AMOUNT.
BLANK ENTRIES DENOTE MISSING/UNREPORTED DATA.
# INDICATES A STATION OR INSTRUMENT RELOCATION.

SPECIFIC

PM - INCLUDES LAST DAY OF PREVIOUS MONTH
ASOS - AUTOMATED SURFACE OBSERVING SYSTEM IN OPERATION DURING THESE MONTHS.

(a) - LENGTH OF RECORD IN YEARS, ALTHOUGH INDIVIDUAL MONTHS MAY BE MISSING.
0.* OR * - THE VALUE IS BETWEEN 0.0 AND 0.05
NORMALS - BASED ON THE 1961-1990 RECORD PERIOD.
EXTREMES - DATES ARE THE MOST RECENT OCCURRENCE
WIND DIR.- NUMERALS SHOW TENS OF DEGRESS CLOCKWISE FROM TRUE NORTH. "00" INDICATES CALM.
RESULTANT DIRECTIONS ARE GIVEN TO WHOLE DEGREES.
BOLD VALUES INDICATE EXTREME VALUES WHICH OCCURRED AFTER THE ASOS SYSTEM WAS COMMISSIONED.

RECORD = PERIOD OF RECORD
RECORD MEAN PRECIPITATION IS THE MEAN OF ALL DAILY PRECIPITATION AMOUNTS DURING THE PERIOD OF RECORD.
RECORD MAX(MIN) TEMPERATURE IS THE MEAN OF ALL DAILY MAX(MIN) TEMPERATURES DURING THE PERIOD OF RECORD.
RECORD MEAN TEMPERATURE IS THE SUM OF THE RECORD MAX AND RECORD MIN DIVIDED BY 2.
AVERAGE TEMPERATURE IS THE SUM OF THE MEAN DAILY MAX AND MIN TEMPERATURE DIVIDED BY 2.

EXCEPTIONS

1. THUNDERSTORM DATA ARE THROUGH 1964 AND MAY BE INCOMPLETE, DUE TO PART-TIME OPERATIONS.
2. MEAN WIND SPEED ARE THROUGH 1964.
3. RELATIVE HUMIDITY, SNOW, AND DAYS SNOW 1.0 INCH OR MORE ARE THROUGH 1972.
3. FASTEST MILE WINDS ARE THROUGH MARCH 1973.
4. PERCENT OF POSSIBLE SUNSHINE IS THROUGH 1973.

RECORD MEANS ARE THROUGH THE CURRENT YEAR, BEGINNING IN 1875 FOR TEMPERATURE
1850 FOR PRECIPITATION
1939 FOR SNOWFALL

HEATING DEGREE DAYS Base 65 deg. F SAN FRANCISCO, CALIFORNIA MISSION DOLORES

SEASON	JULY	AUG	SEP	OCT	NOV	DEC	JAN	FEB	MAR	APR	MAY	JUNE	TOTAL
1964-65	186	152	139	100	284	343	414	302	319	280	306	259	3084
1965-66	230	123	120	73	199	509	393	362	336	217	300	182	3044
1966-67	205	188	78	102	241	417	375	325	372	420	230	229	3182
1967-68	184	173	67	47	158	398	465	235	254	263	280	174	2698
1968-69	213	98	92	148	258	462	505	412	333	315	242	185	3263
1969-70	221	171	132	107	166	277	334	210	222	341	237	242	2660
1970-71	231	246	95	201	208	439	433	359	354	351	317	230	3464
1971-72	225	120	85	229	275	490	506	312	279	278	287	226	3312
1972-73	144	139	110	115	298	546	455	277	380	225	267	167	3123
1973-74	199	239	137	134	285	396	423	353	354	284	309	202	3315
1974-75	167	153	153	119	243	422	428	320	360	386	257	238	3246
1975-76	185	170	177	164	276	352	353	344	378	321	262	179	3161
1976-77	173	86	114	102	152	315	463	242	359	261	294	233	2794
1977-78	197	116	93	139	187	304	304	269	188	254	154	177	2382
1978-79	196	137	43	138	268	471	431	332	281	250	185	197	2929
1979-80	151	125	23	64	213	293	366	221	275	240	291	218	2480
1980-81	165	212	132	128	201	350	384	243	304	284	248	121	2772
1981-82	220	175	136	174	196	334	506	273	369	278	281	253	3195
1982-83	210	151	82	90	311	391	478	283	296	238	174	112	2816
1983-84	83	14	30	52	257	369	408	354	251	321	172	159	2470
1984-85	70	84	19	117	263	430	459	254	359	173	209	57	2494
1985-86	71	49	35	99	302	419	254	172	146	196	154	67	1964
1986-87	78	93	63	84	152	383	399	233	243	140	154	147	2169
1987-88	106	52	50	60	179	387	370	207	183	197	196	125	2112
1988-89	60	53	79	137	229	358	419	416	289	171	181	129	2521
1989-90	86	68	94	113	183	377	371	357	305	170	188	97	2409
1990-91	74	23	8	56	204	487	351	192	358	234	248	188	2423
1991-92	129	74	74	74	150	350	414	186	171	87	75	76	1860
1992-93	30	51	28	31	163	406	423	310	178	163	90	49	1922
1993-94	73	33	74	63	204	411							

COOLING DEGREE DAYS Base 65 deg. F SAN FRANCISCO, CALIFORNIA MISSION DOLORES

YEAR	JAN	FEB	MAR	APR	MAY	JUNE	JULY	AUG	SEP	OCT	NOV	DEC	TOTAL
1969	0	0	5	0	1	1	0	0	14	20	3	0	44
1970	0	0	5	0	19	1	17	8	82	10	0	0	142
1971	0	0	0	0	0	7	0	6	83	12	0	0	108
1972	0	0	3	0	1	4	22	0	13	20	0	0	63
1973	0	0	0	2	5	43	8	0	34	15	0	0	107
1974	0	0	0	3	4	3	4	3	19	41	0	0	77
1975	0	0	0	0	23	0	4	8	15	5	0	0	55
1976	0	0	0	0	16	82	0	15	37	41	15	0	206
1977	0	0	0	0	0	1	18	17	8	8	0	0	52
1978	0	0	8	2	30	0	0	6	65	46	3	0	160
1979	0	0	0	0	11	13	10	3	72	16	0	0	125
1980	0	0	0	5	0	12	2	3	29	43	3	0	97
1981	0	0	0	13	1	44	6	3	3	5	2	0	77
1982	0	0	0	7	1	0	0	8	16	27	0	0	59
1983	0	0	0	0	16	21	41	50	101	27	0	0	256
1984	0	0	0	5	20	5	42	20	158	14	0	0	264
1985	0	7	0	24	2	28	50	27	16	49	8	0	211
1986	0	7	10	8	6	22	17	1	4	49	12	0	136
1987	0	0	5	12	38	19	5	14	22	68	0	0	183
1988	0	3	6	18	20	12	42	27	30	34	3	0	195
1989	0	0	0	56	9	35	15	12	5	28	2	0	162
1990	0	0	0	5	8	25	18	39	45	40	1	0	181
1991	0	0	0	3	1	5	20	20	26	62	9	0	146
1992	0	2	0	21	13	9	41	23	61	93	10	0	273
1993	0	0	0	1	21	82	32	88	31	48	7	0	310

SNOWFALL (inches) SAN FRANCISCO, CALIFORNIA MISSION DOLORES

SEASON	JULY	AUG	SEP	OCT	NOV	DEC	JAN	FEB	MAR	APR	MAY	JUNE	TOTAL
1964-65	0.0	0.0	0.0	0.0	0.0	0.0	0.0	0.0	0.0	0.0	0.0	0.0	0.0
1965-66	0.0	0.0	0.0	0.0	0.0	0.0	0.0	0.0	0.0	0.0	0.0	0.0	0.0
1966-67	0.0	0.0	0.0	0.0	0.0	0.0	0.0	0.0	0.0	0.0	0.0	0.0	0.0
1967-68	0.0	0.0	0.0	0.0	0.0	0.0	0.0	0.0	0.0	0.0	0.0	0.0	0.0
1968-69	0.0	0.0	0.0	0.0	0.0	0.0	0.0	0.0	0.0	0.0	0.0	0.0	0.0
1969-70	0.0	0.0	0.0	0.0	0.0	0.0	0.0	0.0	0.0	0.0	0.0	0.0	0.0
1970-71	0.0	0.0	0.0	0.0	0.0	0.0	0.0	0.0	0.0	0.0	0.0	0.0	0.0
1971-72	0.0	0.0	0.0	0.0	0.0	0.0	0.0	0.0	0.0	0.0	0.0	0.0	0.0
1972-73	0.0	0.0	0.0	0.0	0.0	T	0.0	0.0	0.0	0.0			
1973-74													
1975-76													
1976-77													
1977-78													
1978-79													
1979-80													
1980-81													
1981-82													
1982-83													
1983-84													
1984-85													
Record Mean	0.0	0.0	0.0	0.0	0.0	T	T	T	T	0.0	0.0	0.0	T

SANTA BARBARA, CALIFORNIA

No narrative available

NORMALS, MEANS, AND EXTREMES

SANTA BARBARA, CALIFORNIA

LATITUDE: 34°26'N LONGITUDE: 119°50' W ELEVATION: FT. GRND 9 BARO 13 TIME ZONE: PACIFIC WBAN: 23190

	(a)	JAN	FEB	MAR	APR	MAY	JUNE	JULY	AUG	SEP	OCT	NOV	DEC	YEAR
TEMPERATURE °F:														
Normals														
-Daily Maximum		63.7	64.6	65.3	67.4	68.6	71.2	73.9	75.4	75.2	73.1	68.5	64.4	69.3
-Daily Minimum		40.3	43.2	45.2	47.1	50.1	53.6	56.9	58.2	56.5	51.5	44.7	40.1	49.0
-Monthly		52.0	53.9	55.2	57.2	59.4	62.4	65.4	66.8	65.8	62.3	56.6	52.3	59.1
Extremes														
-Record Highest	10	86	82	90	96	92	109	109	101	102	103	97	83	109
-Year		1991	1992	1989	1989	1988	1990	1985	1984	1988	1987	1992	1990	JUN 1990
-Record Lowest	10	26	25	34	38	38	42	49	47	45	36	30	20	20
-Year		1987	1989	1991	1984	1991	1988	1987	1988	1988	1989	1989	1990	DEC 1990
NORMAL DEGREE DAYS:														
Heating (base 65°F)		403	311	304	237	178	102	42	36	71	108	252	394	2438
Cooling (base 65°F)		0	0	0	0	0	24	55	91	95	24	0	0	289
% OF POSSIBLE SUNSHINE														
MEAN SKY COVER (tenths)														
Sunrise - Sunset														
MEAN NUMBER OF DAYS:														
Sunrise to Sunset														
-Clear														
-Partly Cloudy														
-Cloudy														
Precipitation														
.01 inches or more	10	5.0	5.8	6.0	1.6	0.6	0.6	0.8	0.3	0.9	1.8	3.3	4.7	31.4
Snow, Ice pellets, hail														
1.0 inches or more	10	0.0	0.0	0.0	0.0	0.0	0.0	0.0	0.0	0.0	0.0	0.0	0.0	0.0
Thunderstorms	10	0.2	0.2	0.3	0.1	0.2	0.4	0.1	0.1	0.2	0.2	0.1	0.5	2.6
Heavy Fog Visibility														
1/4 mile or less	10	1.0	2.4	1.4	1.3	0.8	1.2	1.1	1.8	3.2	3.3	1.3	1.3	20.1
Temperature °F														
-Maximum														
90° and above	10	0.0	0.0	0.2	0.4	0.1	0.2	0.6	0.2	0.7	0.8	0.4	0.0	3.6
32° and below	10	0.0	0.0	0.0	0.0	0.0	0.0	0.0	0.0	0.0	0.0	0.0	0.0	0.0
-Minimum														
32° and below	10	3.4	1.4	0.0	0.0	0.0	0.0	0.0	0.0	0.0	0.0	0.3	3.1	8.2
0° and below	10	0.0	0.0	0.0	0.0	0.0	0.0	0.0	0.0	0.0	0.0	0.0	0.0	0.0
AVG. STATION PRESS. (mb)	4	1018.7	1017.0	1016.4	1014.4	1013.1	1012.6	1013.0	1012.6	1012.1	1014.1	1016.6	1018.0	1014.9
RELATIVE HUMIDITY (%)														
Hour 04	4	77	76	80	72	78	87	89	89	85	79	77	81	81
Hour 10 (Local Time)	10	60	61	65	62	64	70	71	71	69	64	53	55	64
Hour 16	10	56	57	59	58	59	63	64	64	63	62	53	54	59
Hour 22	10	78	76	77	74	79	84	87	88	85	81	74	75	80
PRECIPITATION (inches):														
Water Equivalent														
-Normal		3.21	3.61	2.84	1.06	0.16	0.04	0.02	0.11	0.45	0.52	2.03	2.20	16.25
-Maximum Monthly	10	9.07	7.17	11.45	3.01	0.72	0.46	0.29	0.81	1.45	2.21	3.76	5.10	11.45
-Year		1993	1992	1991	1988	1990	1988	1992	1984	1986	1987	1985	1991	MAR 1991
-Minimum Monthly	10	0.02	0.01	0.02	0.00	0.00	T	T	0.00	0.00	0.00	0.00	0.00	0.00
-Year		1984	1984	1990	1993	1991	1992	1990	1986	1992	1986	1992	1989	APR 1993
-Maximum in 24 hrs	10	2.15	3.97	4.47	1.54	0.69	0.41	0.27	0.47	1.45	1.48	1.17	2.16	4.47
-Year		1993	1985	1991	1988	1990	1988	1992	1984	1986	1987	1985	1991	MAR 1991
Snow, Ice pellets, hail														
-Maximum Monthly		0.0	0.0	0.0	0.0	0.0	0.0	0.0	0.0	0.0	0.0	0.0	0.0	
-Year														
-Maximum in 24 hrs	10	0.0	0.0	0.0	0.0	0.0	0.0	0.0	0.0	0.0	0.0	0.0	0.0	
-Year														
WIND:														
Mean Speed (mph)	4	4.8	6.3	6.7	7.6	7.1	6.8	6.5	6.1	5.7	5.4	5.0	4.7	6.1
Prevailing Direction														
Fastest Obs. 1 Min.														
-Direction (!!)														
-Speed (MPH)														
-Year														
Peak Gust														
-Direction (!!)														
-Speed (mph)														
-Date														

PRECIPITATION (inches)

SANTA BARBARA, CALIFORNIA

YEAR	JAN	FEB	MAR	APR	MAY	JUNE	JULY	AUG	SEP	OCT	NOV	DEC	ANNUAL
1984	0.02	0.01	0.73	0.14	0.02	T	T	0.81	0.29	0.56	2.42	4.43	9.43
1985	0.72	4.09	1.92	0.02	T	T	0.03	T	0.02	0.58	3.76	0.93	12.07
1986	1.98	6.98	7.15	0.50	T	T	T	0.00	1.45	0.00	0.54	0.85	19.45
1987	1.25	2.36	4.40	0.06	T	T	T	T	T	2.21	0.82	3.56	14.66
1988	2.46	1.52	0.02	3.01	T	0.46	0.02	T	T	0.01	1.00	4.02	12.52
1989	0.48	2.35	0.54	0.19	0.15	T	T	T	0.06	0.46	0.10	0.00	4.33
1990	2.56	1.64	0.02	0.20	0.72	T	T	T	0.20	0.00	0.11	0.01	5.46
1991	1.65	3.10	11.45	0.02	0.00	0.20	0.16	0.25	0.02	0.23	0.07	5.10	22.25
1992	2.19	7.17	3.11	0.01	0.02	T	0.29	T	0.00	0.82	0.00	4.34	17.95
1993	9.07	5.42	4.75	0.00	0.04	0.38	0.01	T	0.00	0.05	1.33	1.52	22.57
Record Mean	2.24	3.46	3.41	0.42	0.09	0.10	0.05	0.11	0.20	0.49	1.01	2.48	14.07

AVERAGE TEMPERATURE (deg. F)

SANTA BARBARA, CALIFORNIA

YEAR	JAN	FEB	MAR	APR	MAY	JUNE	JULY	AUG	SEP	OCT	NOV	DEC	ANNUAL
1984	54.1	54.1	58.1	59.1	63.8	63.2	67.2	70.2	72.7	61.6	54.8	50.5	60.8
1985	50.2	51.5	52.6	57.5	57.3	62.7	68.1	66.1	64.5	61.2	53.6	53.2	58.2
1986	55.3	55.4	56.4	57.7	58.5	62.3	64.6	64.5	61.7	62.5	58.6	53.5	59.3
1987	50.4	53.6	55.1	59.6	62.1	61.8	63.3	64.6	65.5	64.9	56.8	50.3	59.0
1988	51.5	53.9	58.2	57.9	60.3	61.7	66.0	65.2	63.2	61.9		52.2	58.9
1989	49.1	50.5	57.7	61.5	60.6	62.7	66.1	64.0	63.5	60.8	56.9	54.1	59.0
1990	51.9	51.6	55.0	59.5	60.3	65.1	68.0	67.6	66.5	63.1	57.6	49.2	59.6
1991	53.1	55.8	52.9	57.4	58.6	59.9	64.4	65.8	64.8	63.3	57.6	53.2	58.9
1992	52.8	57.9	57.0	62.5	63.6	65.6	70.4	70.2	67.6	64.5	60.2	52.1	62.0
1993	54.4	54.8	59.4	62.5	63.6	66.4	67.4	68.4	66.2	64.2	60.0	52.7	61.7
Record Mean	52.3	53.9	56.2	59.5	60.8	63.2	66.5	66.7	65.6	62.8	57.3	52.1	59.7
Max	65.1	65.2	66.8	70.7	71.1	72.6	75.3	75.7	75.9	73.9	71.1	65.4	70.7
Min	39.4	42.6	45.6	48.3	50.6	53.7	57.7	57.6	55.3	51.7	43.6	38.7	48.7

REFERENCE NOTES

SANTA BARBARA, CALIFORNIA

GENERAL
T - TRACE AMOUNT.
BLANK ENTRIES DENOTE MISSING/UNREPORTED DATA.
# INDICATES A STATION OR INSTRUMENT RELOCATION.

SPECIFIC

(a) - LENGTH OF RECORD IN YEARS, ALTHOUGH INDIVIDUAL MONTHS MAY BE MISSING.
0.* OR * - THE VALUE IS BETWEEN 0.0 AND 0.05
NORMALS - BASED ON THE 1961-1990 RECORD PERIOD.
EXTREMES - DATES ARE THE MOST RECENT OCCURRENCE
WIND DIR.- NUMERALS SHOW TENS OF DEGRESS CLOCKWISE FROM TRUE NORTH. "00" INDICATES CALM.
RESULTANT DIRECTIONS ARE GIVEN TO WHOLE DEGREES.
BOLD VALUES INDICATE EXTREME VALUES WHICH OCCURRED AFTER THE ASOS SYSTEM WAS COMMISSIONED.

RECORD = PERIOD OF RECORD
RECORD MEAN PRECIPITATION IS THE MEAN OF ALL DAILY PRECIPITATION AMOUNTS DURING THE PERIOD OF RECORD.
RECORD MAX(MIN) TEMPERATURE IS THE MEAN OF ALL DAILY MAX(MIN) TEMPERATURES DURING THE PERIOD OF RECORD.
RECORD MEAN TEMPERATURE IS THE SUM OF THE RECORD MAX AND RECORD MIN DIVIDED BY 2.
AVERAGE TEMPERATURE IS THE SUM OF THE MEAN DAILY MAX AND MIN TEMPERATURE DIVIDED BY 2.

EXCEPTIONS

RECORD MEANS ARE THROUGH THE CURRENT YEAR, BEGINNING IN 1984 FOR TEMPERATURE
1984 FOR PRECIPITATION
1984 FOR SNOWFALL

HEATING DEGREE DAYS Base 65 deg. F — SANTA BARBARA, CALIFORNIA

SEASON	JULY	AUG	SEP	OCT	NOV	DEC	JAN	FEB	MAR	APR	MAY	JUNE	TOTAL
1983-84							329	309	211	173	63	57	
1984-85	3	1	0	116	300	443	452	371	378	226	234	69	2593
1985-86	7	14	38	134	335	360	294	266	258	211	191	77	2185
1986-87	20	38	99	88	189	348	447	312	302	161	92	92	2188
1987-88	53	20	20	50	238	449	408	314	226	207	149	112	2246
1988-89	20	27	81	109		390	485	400	229	142	141	74	
1989-90	4	38	57	123	235	329	402	371	304	159	147	42	2211
1990-91	5	4	4	64	217	480	370	252	370	227	199	148	2340
1991-92	22	15	22	81	217	358	372	204	237	84	50	18	1680
1992-93	1	0	1	36	162	394	324	277	171	86	64	9	1525
1993-94	0	0	16	40	148	376							

COOLING DEGREE DAYS Base 65 deg. F — SANTA BARBARA, CALIFORNIA

YEAR	JAN	FEB	MAR	APR	MAY	JUNE	JULY	AUG	SEP	OCT	NOV	DEC	TOTAL
1984	0	0	2	2	32	9	81	168	237	14	0	0	545
1985	0	0	0	7	0	9	109	55	29	23	0	0	232
1986	0	1	0	0	0	2	15	30	8	18	0	0	74
1987	0	0	0	5	8	2	10	13	42	53	0	0	133
1988	0	0	21	1	10	19	62	42	35	15		0	205
1989	0	0	9	44	12	12	44	13	19	1	0	0	154
1990	0	0	2	0	7	51	105	94	57	10	4	0	330
1991	8	0	0	6	9	1	9	45	23	34	0	0	135
1992	0	4	0	15	13	43	178	167	87	28	25	0	560
1993	0	0	3	19	24	57	79	112	59	22	3	0	378

SNOWFALL (inches) — SANTA BARBARA, CALIFORNIA

SEASON	JULY	AUG	SEP	OCT	NOV	DEC	JAN	FEB	MAR	APR	MAY	JUNE	TOTAL
1983-84							0.0	0.0	0.0	0.0	0.0	0.0	
1984-85	0.0	0.0	0.0	0.0	0.0	0.0	0.0	0.0	0.0	0.0	0.0	0.0	0.0
1985-86	0.0	0.0	0.0	0.0	0.0	0.0	0.0	0.0	0.0	0.0	0.0	0.0	0.0
1986-87	0.0	0.0	0.0	0.0	0.0	0.0	0.0	0.0	0.0	0.0	0.0	0.0	0.0
1987-88	0.0	0.0	0.0	0.0	0.0	0.0	0.0	0.0	0.0	0.0	0.0	0.0	0.0
1988-89	0.0	0.0	0.0	0.0	0.0	0.0	0.0	0.0	0.0	0.0	0.0	0.0	0.0
1989-90	0.0	0.0	0.0	0.0	0.0	0.0	0.0	0.0	0.0	0.0	0.0	0.0	0.0
1990-91	0.0	0.0	0.0	0.0	0.0	0.0	0.0	0.0	0.0	0.0	0.0	0.0	0.0
1991-92	0.0	0.0	0.0	0.0	0.0	0.0	0.0	0.0	0.0	0.0	0.0	0.0	0.0
1992-93	0.0	0.0	0.0	0.0	0.0	0.0	0.0	0.0	0.0	0.0	0.0	0.0	0.0
1993-94	0.0	0.0	0.0	0.0	0.0	0.0							
Record Mean	0.0	0.0	0.0	0.0	0.0	0.0	0.0	0.0	0.0	0.0	0.0	0.0	0.0

SANTA MARIA, CALIFORNIA

Santa Maria Valley is a flat, fertile valley opening on the Pacific Ocean where it is widest and tapering inland for a distance approximately 30 miles. The valley is 10 miles wide at the site of the station, which is located 13 miles inland at an elevation of 236 feet. It is bounded by the foothills of the San Rafael Mountains, the Solomon Hills, and the Casmalia Hills ranging from 1,300 to 4,000 feet.

Located 150 miles west-northwest of Los Angeles and 250 miles south of San Francisco, Santa Maria has a maritime climate, displaying characteristics of those of both neighbors. Year-round mild temperatures moving through gradual transitions characterize the climate more than do clearly defined seasons. The annual range of temperatures is about 13 degrees, while the daily temperature range is about 20 degrees for May through September and a few degrees higher from October through April.

The area is primarily agricultural, with vegetable and other produce crops thriving successfully the year-round. Temperatures of 32 degrees or slightly lower occur about twenty-three times during the winter months and necessitate the rotation of crops to the hardier varieties during this season. Precipitation, particularly during the summer months, is insufficient for some crops and is supplemented by irrigation from subterranean water reserves. High humidity and moderate temperatures, however, substantially limit the irrigation requirement.

Based on the 1951-1980 period, the average first occurrence of 32 degrees Fahrenheit in the fall is December 5 and the average last occurrence in the spring is March 15.

The rainfall season, typical of the mid-California coast, is in the winter. About three-fourths of the total annual rainfall occurs from December through March in connection with Pacific cold fronts and storm centers passing inland. During the remainder of the year, and particularly from June to October, the northward displacement and intensification of the semipermanent Pacific anticyclone produces a circulation resulting in little or no precipitation here. Thunderstorms are rare.

During most days, clear, sunny afternoons prevail. But under the influence of the Pacific high, considerable advective and radiative cooling frequently produces nightly low stratus clouds, known as California stratus, and early-morning fog. Both clouds and fog, however, are generally dissipated before noon.

The unequal daytime solar heating over land and ocean, in conjunction with the Pacific high, gives rise to a consistent and prevailing westerly sea breeze during most afternoons. The winds generally decrease to a calm by sundown. Thus the two factors of nighttime stratus and daytime sea breezes effectively combine to maintain relatively cool days and warm nights with little diurnal change.

NORMALS, MEANS, AND EXTREMES

SANTA MARIA, CALIFORNIA

LATITUDE: 34°54'N LONGITUDE: 120°27' W ELEVATION: FT. GRND 236 BARO 270 TIME ZONE: PACIFIC WBAN: 23273

	(a)	JAN	FEB	MAR	APR	MAY	JUNE	JULY	AUG	SEP	OCT	NOV	DEC	YEAR
TEMPERATURE °F:														
Normals														
-Daily Maximum		63.8	64.7	64.2	66.9	67.8	71.0	73.3	74.1	74.8	73.9	68.7	64.2	69.0
-Daily Minimum		38.3	40.4	41.3	42.5	46.4	50.3	52.9	53.9	52.3	48.0	42.3	37.8	45.5
-Monthly		51.1	52.6	52.8	54.8	57.1	60.7	63.1	64.0	63.6	61.0	55.5	51.1	57.3
Extremes														
-Record Highest	51	86	87	95	103	100	102	104	103	103	108	93	90	108
-Year		1976	1943	1988	1989	1970	1976	1985	1962	1978	1987	1956	1958	OCT 1987
-Record Lowest	51	20	22	24	31	31	36	43	43	36	26	25	20	20
-Year		1976	1971	1971	1984	1964	1962	1964	1973	1948	1971	1958	1978	DEC 1978
NORMAL DEGREE DAYS:														
Heating (base 65°F)		431	347	378	311	248	138	96	72	100	143	289	431	2984
Cooling (base 65°F)		0	0	0	5	0	9	37	41	58	19	0	0	169
% OF POSSIBLE SUNSHINE														
MEAN SKY COVER (tenths)														
Sunrise - Sunset	45	5.0	5.1	4.9	4.6	4.3	3.8	3.4	3.5	3.6	3.7	4.1	4.7	4.2
MEAN NUMBER OF DAYS:														
Sunrise to Sunset														
-Clear	45	12.7	11.6	12.5	13.0	14.3	15.5	17.3	16.8	16.1	16.7	15.2	13.8	175.6
-Partly Cloudy	45	7.4	6.4	8.5	8.8	9.9	10.6	12.0	12.5	10.3	8.8	7.2	7.3	109.7
-Cloudy	45	11.0	10.2	10.0	8.3	6.9	3.9	1.6	1.7	3.6	5.4	7.6	9.9	79.9
Precipitation														
.01 inches or more	51	7.2	7.4	8.0	4.5	1.7	0.8	0.4	0.4	1.3	2.4	5.0	6.4	45.5
Snow, Ice pellets, hail														
1.0 inches or more	51	0.0	0.0	0.0	0.0	0.0	0.0	0.0	0.0	0.0	0.0	0.0	0.0	0.0
Thunderstorms	23	0.2	0.1	0.3	0.2	0.2	0.*	0.2	0.3	0.4	0.1	0.1	0.1	2.3
Heavy Fog Visibility														
1/4 mile or less	23	4.6	4.4	4.7	5.8	6.3	6.7	7.7	9.9	12.3	12.0	6.5	5.8	86.7
Temperature °F														
-Maximum														
90° and above	30	0.0	0.0	0.1	0.4	0.4	0.8	0.4	0.3	1.5	1.4	0.1	0.0	5.4
32° and below	30	0.0	0.0	0.0	0.0	0.0	0.0	0.0	0.0	0.0	0.0	0.0	0.0	0.0
-Minimum														
32° and below	30	6.0	3.2	1.7	0.5	0.1	0.0	0.0	0.0	0.0	0.2	1.6	6.0	19.2
0° and below	30	0.0	0.0	0.0	0.0	0.0	0.0	0.0	0.0	0.0	0.0	0.0	0.0	0.0
AVG. STATION PRESS. (mb)														
RELATIVE HUMIDITY (%)														
Hour 04	19	81	84	86	88	91	92	88	93	91	85	78	79	86
Hour 10 (Local Time)	27	62	63	63	58	60	61	63	65	63	57	54	58	61
Hour 16	28	59	61	64	60	61	60	60	62	63	62	60	58	61
Hour 22	19	81	83	84	87	87	88	86	91	89	85	77	79	85
PRECIPITATION (inches):														
Water Equivalent														
-Normal		2.16	2.62	2.27	0.99	0.20	0.03	0.01	0.05	0.30	0.49	1.46	1.78	12.36
-Maximum Monthly	51	7.09	9.69	9.41	4.24	2.44	0.26	0.62	0.86	3.05	2.07	4.74	4.82	9.69
-Year		1969	1962	1991	1958	1977	1957	1950	1976	1976	1960	1965	1955	FEB 1962
-Minimum Monthly	51	T	T	T	T	0.00	T	0.00	0.00	0.00	0.00	0.00	0.02	0.00
-Year		1976	1953	1959	1992	1978	1992	1982	1971	1992	1988	1959	1989	SEP 1992
-Maximum in 24 hrs	51	2.55	2.61	3.22	1.60	1.35	0.26	0.62	0.85	1.78	2.07	1.93	3.15	3.22
-Year		1943	1978	1991	1960	1977	1957	1950	1976	1976	1960	1965	1974	MAR 1991
Snow, Ice pellets, hail														
-Maximum Monthly	51	T	T	T	0.0	0.0	0.0	0.0	0.0	0.0	0.0	T	T	T
-Year		1962	1992	1992								1975	1990	FEB 1992
-Maximum in 24 hrs	51	T	T	T	0.0	0.0	0.0	0.0	0.0	0.0	0.0	T	T	T
-Year		1962	1992	1992								1975	1990	FEB 1992
WIND:														
Mean Speed (mph)	15	6.7	7.2	8.3	8.0	8.3	7.9	6.5	6.2	5.9	6.2	6.6	6.4	7.0
Prevailing Direction through 1963		WNW	WNW	WNW	WNW	WNW	WNW	WNW	W	W	W	WNW	WNW	WNW
Fastest Obs. 1 Min.														
-Direction (!!)	4	152	14	152	30	30	30	29	28	28	31	30	26	26
-Speed (MPH)	4	32	29	30	31	31	29	25	28	28	25	29	35	35
-Year		1992	1993	1991	1991	1991	1993	1992	1993	1993	1992	1992	1993	DEC 1993
Peak Gust														
-Direction (!!)	3	N	SE	SE	NW	NW	NW	NW	NW	NW	NW	N	SW	SE
-Speed (mph)	3	40	58	51	47	45	41	37	35	38	36	38	51	58
-Date		1990	1993	1991	1990	1991	1993	1992	1990	1992	1991	1991	1993	FEB 1993

PRECIPITATION (inches)

SANTA MARIA, CALIFORNIA

YEAR	JAN	FEB	MAR	APR	MAY	JUNE	JULY	AUG	SEP	OCT	NOV	DEC	ANNUAL
1964	1.23	0.05	2.38	0.34	0.27	0.09	0.06	0.10	T	1.64	2.43	1.34	9.93
1965	0.74	0.35	1.69	3.42	T	T	0.01	T	T	0.01	4.74	2.47	13.43
1966	1.05	0.92	0.20	0.03	0.01	0.03	0.16	T	0.16	0.02	2.04	2.86	7.48
1967	3.53	0.51	2.31	3.63	0.22	0.23	T	T	0.18	T	1.91	1.33	13.85
1968	0.60	0.90	1.90	0.65	0.05	T	T	T	T	1.89	0.76	1.41	8.16
1969	7.09	7.57	0.66	1.74	0.01	0.01	T	T	0.13	0.14	0.94	0.29	18.58
1970	2.15	2.67	1.98	0.03	T	0.06	T	T	T	0.01	2.88	3.47	13.25
1971	0.61	0.10	0.35	1.01	0.98	T	T	0.00	0.04	0.24	0.49	2.80	6.62
1972	0.16	0.31	0.01	0.19	T	0.01	0.05	T	T	0.60	4.28	1.14	6.75
1973	4.81	6.20	3.02	T	0.03	0.01	T	0.01	0.09	0.38	2.15	3.04	19.74
1974	4.16	0.14	5.08	0.81	T	T	T	T	T	1.00	0.10	4.64	15.93
1975	0.10	3.22	3.02	0.81	T	T	T	T	T	0.79	0.34	0.17	8.45
1976	T	4.40	0.71	1.19	0.03	0.05	T	0.86	3.05	0.25	0.23	1.09	11.86
1977	1.39	0.06	1.55	0.04	2.44	T	T	0.01	0.02	T	0.15	4.22	9.88
1978	5.62	8.56	4.97	2.46	0.00	0.04	T	T	1.61	T	1.45	1.04	25.75
1979	3.69	3.44	3.49	0.04	0.09	T	T	T	0.40	0.48	0.40	1.25	13.28
1980	3.76	5.53	1.79	0.71	0.32	T	0.08	T	T	0.01	0.01	1.26	13.47
1981	3.54	2.71	5.59	0.44	T	T	T	T	T	0.82	1.23	0.78	15.11
1982	2.70	1.13	4.86	2.01	T	0.11	0.00	0.20	0.38	1.26	3.31	1.05	17.01
1983	6.71	5.76	5.26	2.29	0.09	T	T	0.28	0.77	0.43	2.44	2.73	26.76
1984	0.08	0.49	0.60	0.49	T	T	T	T	0.02	0.53	2.03	3.23	7.47
1985	0.74	0.87	1.82	0.07	T	T	T	0.03	0.02	0.38	3.11	0.76	7.80
1986	0.95	3.68	4.99	1.35	T	T	T	T	0.67	T	0.89	1.41	13.94
1987	1.22	1.01	3.47	0.40	T	0.05	0.01	T	T	2.00	0.57	3.09	11.82
1988	1.42	2.39	0.08	2.54	0.23	0.04	T	T	0.01	0.00	0.78	3.74	11.23
1989	0.41	0.94	0.61	0.08	0.06	T	T	0.00	0.64	0.16	0.38	0.02	3.30
1990	2.28	1.65	0.19	0.22	0.47	T	T	0.01	0.36	0.01	0.17	0.69	6.05
1991	0.96	2.29	9.41	0.28	0.00	0.04	0.00	0.04	T	0.31	0.15	3.12	16.60
1992	2.15	5.73	2.25	T	T	T	0.40	T	0.00	0.45	0.01	2.80	13.79
1993	5.46	3.91	3.54	0.04	0.21	0.12	T	T	0.01	0.40	0.76	1.77	16.22
Record Mean	2.23	2.49	2.28	0.99	0.21	0.03	0.03	0.04	0.21	0.48	1.28	1.89	12.16

AVERAGE TEMPERATURE (deg. F)

SANTA MARIA, CALIFORNIA

YEAR	JAN	FEB	MAR	APR	MAY	JUNE	JULY	AUG	SEP	OCT	NOV	DEC	ANNUAL
1964	48.7	49.6	49.7	52.4	53.5	58.0	60.8	62.2	60.4	60.9	51.6	52.0	55.0
1965	52.1	50.5	52.7	55.3	55.8	58.3	61.6	64.7	62.3	64.2	56.3	48.8	56.9
1966	48.1	48.1	53.8	57.4	58.2	60.9	61.1	63.6	63.4	61.4	56.3	49.9	56.9
1967	51.1	51.9	50.9	48.4	57.2	58.7	63.8	64.3	64.9	62.4	58.2	49.5	56.8
1968	50.9	56.8	54.3	55.7	58.0	61.6	64.2	64.7	63.9	61.4	57.5	49.9	58.3
1969	53.2	51.0	51.3	53.8	58.2	60.7	62.9	62.4	62.7	60.3	58.5	52.2	57.2
1970	53.9	54.8	54.3	53.0	59.4	59.2	62.2	60.3	61.3	57.8	55.5	49.1	56.7
1971	49.9	49.6	51.0	50.6	54.2	58.6	61.5	65.1	63.3	55.7	50.8	45.6	54.6
1972	48.3	53.0	55.7	55.5	56.7	61.1	63.7	63.9	62.4	60.4	52.7	48.0	56.8
1973	47.3	53.0	48.9	54.1	59.5	64.1	61.8	62.0	62.3	61.0	52.3	51.2	56.4
1974	49.7	50.5	52.8	54.0	54.8	60.2	63.5	63.9	62.4	59.8	55.2	49.9	56.4
1975	50.2	50.4	50.3	49.8	55.8	60.2	61.5	62.9	62.3	58.1	53.7	50.8	55.5
1976	52.5	52.5	52.4	51.7	57.2	62.7	63.8	64.7	65.3	62.3	58.8	51.8	58.0
1977	51.0	54.7	49.2	54.9	55.2	60.6	63.0	65.2	63.8	60.3	59.9	57.1	57.9
1978	54.4	53.9	57.7	53.9	59.4	59.4	61.1	64.1	64.9	60.4	52.2	47.3	57.4
1979	48.6	49.0	52.8	53.9	59.0	60.9	63.2	63.9	65.6	62.0	54.8	54.1	57.3
1980	54.9	55.9	52.5	56.5	55.6	59.0	63.2	63.4	62.6	60.3	55.3	55.4	57.9
1981	54.3	54.4	53.1	55.5	57.1	64.6	63.8	63.5	63.0	58.4	57.4	52.9	58.1
1982	48.3	54.6	52.0	55.1	57.6	59.8	62.3	63.3	64.9	62.1	54.1	51.5	57.1
1983	53.5	54.1	55.0	55.0	58.9	62.6	66.1	68.8	70.0	66.2	56.3	54.1	60.1
1984	53.2	51.8	55.5	54.4	61.3	61.3	67.4	68.4	71.9	60.4	54.5	50.6	59.2
1985	50.0	52.7	51.8	58.7	58.4	64.1	68.7	64.1	63.8	60.8	53.4	53.5	58.3
1986	56.6	55.4	56.0	55.8	57.2	60.6	63.6	63.9	60.8	60.8	57.5	51.3	58.3
1987	48.0	52.3	52.4	57.4	59.7	60.5	60.9	62.6	62.4	64.1	55.2	48.7	57.0
1988	50.9	55.5	56.6	56.9	57.0	60.2	65.2	64.7	61.6	61.1	55.3	50.5	58.0
1989	49.2	49.3	55.4	59.9	57.5	61.3	62.4	62.6	61.6	62.1	59.2	54.5	57.9
1990	51.3	49.9	53.6	58.8	57.6	60.7	64.4	65.7	63.9	62.1	56.5	48.9	57.8
1991	52.4	55.6	50.3	54.6	54.0	58.2	63.7	63.5	63.4	60.9	57.3	52.6	57.2
1992	52.5	57.4	56.1	60.2	61.3	61.5	64.8	64.7	64.1	63.0	57.4	50.4	59.5
1993	52.6	53.1	57.5	57.9	60.7	62.9	63.9	65.0	63.0	62.4	57.3	51.8	59.0
Record Mean	50.8	52.4	53.0	55.1	57.2	60.3	62.8	63.3	63.2	60.7	55.8	51.5	57.1
Max	63.2	64.3	64.3	66.8	67.8	70.6	72.6	73.1	74.4	73.5	69.2	64.5	68.7
Min	38.3	40.5	41.7	43.4	46.6	50.0	52.9	53.4	52.0	47.9	42.3	38.6	45.6

REFERENCE NOTES

SANTA MARIA, CALIFORNIA

GENERAL
T - TRACE AMOUNT.
BLANK ENTRIES DENOTE MISSING/UNREPORTED DATA.
# INDICATES A STATION OR INSTRUMENT RELOCATION.

SPECIFIC

(a) - LENGTH OF RECORD IN YEARS, ALTHOUGH INDIVIDUAL MONTHS MAY BE MISSING.
0.* OR * - THE VALUE IS BETWEEN 0.0 AND 0.05
NORMALS - BASED ON THE 1961-1990 RECORD PERIOD.
EXTREMES - DATES ARE THE MOST RECENT OCCURRENCE
WIND DIR.- NUMERALS SHOW TENS OF DEGRESS CLOCKWISE FROM TRUE NORTH. "00" INDICATES CALM.
RESULTANT DIRECTIONS ARE GIVEN TO WHOLE DEGREES.
BOLD VALUES INDICATE EXTREME VALUES WHICH OCCURRED AFTER THE ASOS SYSTEM WAS COMMISSIONED.

RECORD = PERIOD OF RECORD
RECORD MEAN PRECIPITATION IS THE MEAN OF ALL DAILY PRECIPITATION AMOUNTS DURING THE PERIOD OF RECORD.
RECORD MAX(MIN) TEMPERATURE IS THE MEAN OF ALL DAILY MAX(MIN) TEMPERATURES DURING THE PERIOD OF RECORD.
RECORD MEAN TEMPERATURE IS THE SUM OF THE RECORD MAX AND RECORD MIN DIVIDED BY 2.
AVERAGE TEMPERATURE IS THE SUM OF THE MEAN DAILY MAX AND MIN TEMPERATURE DIVIDED BY 2.

EXCEPTIONS

1. MEAN WIND SPEED IS THROUGH 1963.
2. THUNDERSTORMS AND HEAVY FOG ARE THROUGH 1964 AND MAY BE INCOMPLETE, DUE TO PART-TIME OPERATIONS.
3. MEAN SKY COVER, AND DAYS CLEAR-PARTLY CLOUDY-CLOUDY ARE THROUGH 1980.

RECORD MEANS ARE THROUGH THE CURRENT YEAR, BEGINNING IN 1934 FOR TEMPERATURE
1934 FOR PRECIPITATION
1943 FOR SNOWFALL

HEATING DEGREE DAYS Base 65 deg. F SANTA MARIA, CALIFORNIA

SEASON	JULY	AUG	SEP	OCT	NOV	DEC	JAN	FEB	MAR	APR	MAY	JUNE	TOTAL
1964-65	127	90	149	142	398	395	394	400	376	286	279	192	3228
1965-66	111	42	79	92	253	495	515	467	342	235	205	124	2960
1966-67	113	44	67	128	268	463	424	361	430	490	244	183	3215
1967-68	40	24	26	97	197	474	431	233	326	276	212	96	2432
1968-69	34	25	54	116	225	462	359	386	419	328	207	124	2739
1969-70	66	78	71	147	190	385	335	279	322	356	188	167	2584
1970-71	99	141	144	218	279	486	471	425	427	426	330	188	3634
1971-72	110	22	76	301	421	596	510	340	284	280	252	116	3308
1972-73	53	56	91	146	363	521	542	326	496	320	180	82	3176
1973-74	95	86	95	126	379	421	468	398	373	322	312	138	3213
1974-75	60	41	78	164	287	462	450	402	446	451	276	136	3253
1975-76	103	63	96	215	334	432	381	354	385	394	242	132	3131
1976-77	44	30	23	97	184	402	427	282	487	297	298	125	2696
1977-78	81	27	44	145	161	238	323	305	216	325	181	160	2206
1978-79	112	45	54	139	379	544	505	442	372	326	189	130	3237
1979-80	69	35	27	97	299	330	307	254	378	252	286	183	2517
1980-81	68	60	85	155	288	290	327	297	360	289	238	57	2514
1981-82	47	52	58	202	221	367	509	288	398	289	224	153	2808
1982-83	76	53	38	104	319	409	349	301	305	291	184	66	2495
1983-84	21	1	0	34	258	330	356	377	286	318	135	106	2222
1984-85	1	6	1	144	310	436	461	339	402	188	195	45	2528
1985-86	5	45	61	150	343	348	254	262	271	270	235	127	2371
1986-87	48	39	125	132	218	420	516	352	385	223	159	132	2749
1987-88	117	71	80	79	287	499	430	272	267	237	244	139	2722
1988-89	44	33	108	132	294	441	483	433	291	196	227	111	2793
1989-90	82	69	98	112	172	319	418	415	347	181	221	128	2562
1990-91	40	11	37	93	253	491	383	259	446	306	334	203	2856
1991-92	39	62	49	131	234	373	380	217	271	144	110	98	2108
1992-93	47	24	34	77	226	445	377	330	227	204	137	86	2214
1993-94	47	23	79	81	224	403							

COOLING DEGREE DAYS Base 65 deg. F SANTA MARIA, CALIFORNIA

YEAR	JAN	FEB	MAR	APR	MAY	JUNE	JULY	AUG	SEP	OCT	NOV	DEC	TOTAL
1969	0	0	0	0	3	1	8	4	8	9	4	0	37
1970	0	0	0	0	21	1	16	0	43	2	0	0	83
1971	7	0	0	0	0	3	7	32	35	20	0	0	104
1972	0	0	2	0	1	2	19	30	19	10	0	0	83
1973	0	0	0	0	18	61	4	0	24	8	4	0	119
1974	0	0	0	0	0	0	19	12	4	11	0	0	46
1975	0	0	0	0	0	0	1	4	19	9	2	0	35
1976	0	0	0	0	7	72	13	25	40	18	6	0	181
1977	0	0	0	0	0	0	24	40	17	7	15	1	104
1978	0	0	0	0	12	0	0	25	59	2	0	0	98
1979	0	0	0	0	8	14	21	7	51	12	0	0	113
1980	0	0	0	3	0	13	20	16	19	19	2	0	92
1981	0	4	0	9	0	52	20	12	5	6	0	0	108
1982	0	0	0	0	0	2	1	8	42	21	0	0	74
1983	0	0	0	0	0	0	62	127	158	77	3	0	427
1984	0	0	0	6	26	0	84	119	212	9	0	0	456
1985	0	3	0	7	0	25	126	24	34	28	2	0	249
1986	0	2	0	2	0	0	11	14	1	7	0	0	37
1987	0	0	0	0	2	4	0	3	8	56	0	0	73
1988	0	3	15	2	2	2	56	28	13	20	11	0	152
1989	0	0	0	51	2	6	10	3	3	28	5	2	110
1990	0	0	2	0	0	6	31	38	10	13	6	0	106
1991	0	0	0	1	0	7	7	23	7	13	7	0	65
1992	0	0	0	4	3	1	50	23	14	23	7	0	125
1993	0	0	0	0	10	32	21	32	23	10	2	0	130

SNOWFALL (inches) SANTA MARIA, CALIFORNIA

SEASON	JULY	AUG	SEP	OCT	NOV	DEC	JAN	FEB	MAR	APR	MAY	JUNE	TOTAL
1970-71	0.0	0.0	0.0	0.0	0.0	0.0	0.0	0.0	0.0	0.0	0.0	0.0	0.0
1971-72	0.0	0.0	0.0	0.0	0.0	0.0	0.0	0.0	0.0	0.0	0.0	0.0	0.0
1972-73	0.0	0.0	0.0	0.0	0.0	0.0	0.0	0.0	0.0	0.0	0.0	0.0	0.0
1973-74	0.0	0.0	0.0	0.0	0.0	0.0	0.0	0.0	0.0	0.0	0.0	0.0	0.0
1974-75	0.0	0.0	0.0	0.0	0.0	0.0	0.0	0.0	0.0	0.0	0.0	0.0	0.0
1975-76	0.0	0.0	0.0	0.0	T	0.0	0.0	0.0	0.0	0.0	0.0	0.0	T
1976-77	0.0	0.0	0.0	0.0	0.0	0.0	0.0	0.0	0.0	0.0	0.0	0.0	0.0
1977-78	0.0	0.0	0.0	0.0	0.0	0.0	0.0	0.0	0.0	0.0	0.0	0.0	0.0
1978-79	0.0	0.0	0.0	0.0	0.0	0.0	0.0	0.0	0.0	0.0	0.0	0.0	0.0
1979-80	0.0	0.0	0.0	0.0	0.0	0.0	0.0	0.0	0.0	0.0	0.0	0.0	0.0
1980-81	0.0	0.0	0.0	0.0	0.0	0.0	0.0	0.0	0.0	0.0	0.0	0.0	0.0
1981-82	0.0	0.0	0.0	0.0	0.0	0.0	0.0	0.0	0.0	0.0	0.0	0.0	0.0
1982-83	0.0	0.0	0.0	0.0	0.0	0.0	0.0	0.0	0.0	0.0	0.0	0.0	0.0
1983-84	0.0	0.0	0.0	0.0	0.0	0.0	0.0	0.0	0.0	0.0	0.0	0.0	0.0
1984-85	0.0	0.0	0.0	0.0	0.0	0.0	0.0	0.0	0.0	0.0	0.0	0.0	0.0
1985-86	0.0	0.0	0.0	0.0	0.0	0.0	0.0	0.0	T	0.0	0.0	0.0	T
1986-87	0.0	0.0	0.0	0.0	0.0	0.0	0.0	T	0.0	0.0	0.0	0.0	T
1987-88	0.0	0.0	0.0	0.0	0.0	0.0	0.0	0.0	0.0	0.0	0.0	0.0	0.0
1988-89	0.0	0.0	0.0	0.0	0.0	0.0	0.0	0.0	0.0	0.0	0.0	0.0	0.0
1989-90	0.0	0.0	0.0	0.0	0.0	0.0	0.0	T	0.0	0.0	0.0	0.0	T
1990-91	0.0	0.0	0.0	0.0	0.0	T	0.0	0.0	T	0.0	0.0	0.0	T
1991-92	0.0	0.0	0.0	0.0	0.0	0.0	0.0	T	T	0.0	0.0	0.0	T
1992-93	0.0	0.0	0.0	0.0	0.0	0.0	0.0	0.0	0.0	0.0	0.0	0.0	0.0
1993-94	0.0	0.0	0.0	0.0	0.0	0.0							
Record Mean	0.0	0.0	0.0	0.0	T	T	T	T	T	0.0	0.0	0.0	T

STOCKTON, CALIFORNIA

Stockton, the county seat of San Joaquin County, is located near the center of the Great Central Valley of California. It is on the southeast corner of the broad delta formed by the confluence of the San Joaquin and Sacramento Rivers. The surrounding terrain is flat, irrigated farm and orchard land, near sea level, with the rivers and canals of the delta controlled by a system of levees.

Approximately 25 miles east and northeast of Stockton lie the foothills of the Sierra Nevada, rising gradually to an elevation of about 1,000 feet. Beyond the foothills, the mountains rise abruptly to the crest of the Sierra, at a distance of about 75 miles, with some peaks here exceeding 9,000 feet in elevation. On a few days during the year, when atmospheric conditions are favorable, the downslope effect of a north or northeast wind can bring unseasonably dry weather to the delta area, but on the whole the Sierra Nevada has little or no effect on the weather of San Joaquin County. The Sierra Nevada does affect the area, however, to the extent that the entire economy of the Great Valley depends upon the water supplied by the melting snows in the mountains.

To the west and southwest, the Coast Range, with peaks above 2,000 feet, form a barrier separating the Great Valley from the marine air which dominates the climate of the coastal communities. Several gaps in the Coast Range in the San Francisco Bay Area, however, permit the passage inland of a sea breeze which fans out into the delta and has a moderating effect on summer heat, with the result that Stockton enjoys slightly cooler summer days than communities in the upper San Joaquin and Sacramento Valleys.

The summer climate in Stockton is characterized by warm, dry days and relatively cool nights with clear skies and no rainfall. Winter brings mild temperatures and relatively light rains with frequent heavy fogs.

The annual rainfall averages about 14 inches, with 90 percent of the precipitation falling from November through April. Thunderstorms are infrequent, occurring on 3 or 4 days a year. Snow is practically unknown in the Stockton area.

In summer, temperatures exceeding 100 degrees can be expected on about 15 days. During these hot afternoons the air is extremely dry, with relative humidities running generally less than 20 percent. Even on these hot days, however, temperatures will fall into the low 60s at night. In winter the nighttime temperature on clear nights will fall to or slightly below freezing, and will rise in the afternoon into the low 50s.

In late autumn and early winter, clear still nights give rise to the formation of dense fogs, which normally settle in during the night and burn off sometime during the day. In December and January, the so-called fog season, under stagnant atmospheric conditions the fog may last for as long as 4 or 5 weeks, with only brief and temporary periods of clearing.

NORMALS, MEANS, AND EXTREMES

STOCKTON, CALIFORNIA

LATITUDE: 37°54'N LONGITUDE: 121°15' W ELEVATION: FT. GRND 22 BARO 37 TIME ZONE: PACIFIC WBAN: 23237

	(a)	JAN	FEB	MAR	APR	MAY	JUNE	JULY	AUG	SEP	OCT	NOV	DEC	YEAR
TEMPERATURE °F:														
Normals														
-Daily Maximum		53.0	60.7	65.5	72.9	81.6	89.1	94.4	92.9	87.9	78.3	63.4	52.8	74.4
-Daily Minimum		37.0	40.3	42.6	45.7	51.8	57.5	60.9	60.7	57.5	50.8	42.7	37.1	48.7
-Monthly		45.0	50.5	54.1	59.3	66.7	73.3	77.7	76.8	72.7	64.6	53.0	45.0	61.6
Extremes														
-Record Highest	34	71	78	87	100	107	111	114	109	108	101	84	72	114
-Year		1981	1977	1988	1981	1984	1961	1972	1983	1979	1980	1966	1987	JUL 1972
-Record Lowest	34	19	22	27	32	38	45	50	50	43	33	25	17	17
-Year		1963	1989	1971	1976	1964	1971	1981	1968	1982	1972	1985	1990	DEC 1990
NORMAL DEGREE DAYS:														
Heating (base 65°F)		620	406	341	211	53	7	0	0	10	79	360	620	2707
Cooling (base 65°F)		0	0	0	40	106	256	394	366	241	67	0	0	1470
% OF POSSIBLE SUNSHINE														
MEAN SKY COVER (tenths)														
Sunrise - Sunset	46	7.3	6.5	5.7	4.9	3.5	2.3	1.2	1.4	1.8	3.3	5.7	7.1	4.2
MEAN NUMBER OF DAYS:														
Sunrise to Sunset														
-Clear	46	5.3	7.1	9.9	11.9	17.5	21.8	26.8	25.7	23.4	19.2	9.2	6.7	184.5
-Partly Cloudy	46	6.3	6.5	8.5	8.7	8.5	5.7	3.3	3.9	4.4	6.0	8.5	6.5	76.8
-Cloudy	46	19.3	14.7	12.8	9.4	5.1	2.5	0.9	1.4	2.2	5.8	12.3	17.8	104.2
Precipitation														
.01 inches or more	51	8.7	8.2	8.5	5.2	2.3	0.8	0.3	0.4	1.0	3.1	6.4	6.4	51.3
Snow, Ice pellets, hail														
1.0 inches or more	39	0.0	0.0	0.0	0.0	0.0	0.0	0.0	0.0	0.0	0.0	0.0	0.0	0.0
Thunderstorms	32	0.2	0.3	0.3	0.7	0.2	0.3	0.2	0.2	0.4	0.2	0.1	0.1	3.1
Heavy Fog Visibility														
1/4 mile or less	37	12.0	7.5	1.6	0.4	0.1	0.*	0.0	0.*	0.1	2.0	7.9	11.4	43.0
Temperature °F														
-Maximum														
90° and above	34	0.0	0.0	0.0	0.7	6.7	14.3	23.8	21.5	13.5	3.2	0.0	0.0	83.7
32° and below	34	0.1	0.0	0.0	0.0	0.0	0.0	0.0	0.0	0.0	0.0	0.0	0.*	0.1
-Minimum														
32° and below	34	8.1	2.6	0.8	0.1	0.0	0.0	0.0	0.0	0.0	0.0	1.6	8.7	21.9
0° and below	34	0.0	0.0	0.0	0.0	0.0	0.0	0.0	0.0	0.0	0.0	0.0	0.0	0.0
AVG. STATION PRESS. (mb)	10	1018.5	1017.7	1015.9	1015.7	1012.7	1011.9	1011.6	1011.7	1011.9	1015.0	1018.7	1020.3	1015.1
RELATIVE HUMIDITY (%)														
Hour 04	19	90	88	83	78	74	69	65	67	69	75	85	91	78
Hour 10 (Local Time)	27	87	79	67	54	45	42	42	44	48	57	75	87	61
Hour 16	22	70	61	51	40	32	28	26	27	30	38	58	72	44
Hour 22	19	86	82	74	69	62	56	50	52	57	64	79	87	68
PRECIPITATION (inches):														
Water Equivalent														
-Normal		2.84	1.97	2.17	1.08	0.26	0.08	0.06	0.07	0.35	0.79	2.17	2.11	13.95
-Maximum Monthly	52	7.06	6.00	6.48	3.55	2.33	0.66	0.61	0.81	3.00	2.97	6.22	8.05	8.05
-Year		1967	1962	1982	1958	1990	1964	1974	1975	1959	1945	1972	1955	DEC 1955
-Minimum Monthly	52	0.14	0.05	T	0.00	0.00	0.00	0.00	0.00	0.00	0.00	T	0.03	0.00
-Year		1976	1964	1956	1949	1982	1981	1983	1982	1980	1978	1959	1989	JUL 1983
-Maximum in 24 hrs	52	3.01	2.28	1.71	1.54	1.66	0.53	0.57	0.81	2.64	1.88	2.23	3.01	3.01
-Year		1967	1945	1968	1958	1990	1964	1974	1975	1959	1991	1950	1955	JAN 1967
Snow, Ice pellets, hail														
-Maximum Monthly	39	0.0	0.3	T	T	0.0	0.0	0.0	0.0	0.0	0.0	0.0	0.2	0.3
-Year			1976	1991	1970								1988	FEB 1976
-Maximum in 24 hrs	39	0.0	0.3	T	T	0.0	0.0	0.0	0.0	0.0	0.0	0.0	0.2	0.3
-Year			1976	1991	1970								1988	FEB 1976
WIND:														
Mean Speed (mph)	35	6.7	6.9	7.7	8.3	9.2	9.2	8.2	7.7	7.1	6.4	5.8	6.2	7.4
Prevailing Direction through 1963		SE	SE	W	W	W	W	WNW	WNW	W	W	W	SE	W
Fastest Obs. 1 Min.														
-Direction (!!)	30	14	35	33	24	35	26	23	27	34	33	13	15	14
-Speed (MPH)	30	46	39	39	35	35	31	29	28	33	37	40	44	46
-Year		1967	1966	1970	1964	1978	1970	1987	1984	1964	1986	1965	1965	JAN 1967
Peak Gust														
-Direction (!!)														
-Speed (mph)														
-Date														

PRECIPITATION (inches)

STOCKTON, CALIFORNIA

YEAR	JAN	FEB	MAR	APR	MAY	JUNE	JULY	AUG	SEP	OCT	NOV	DEC	ANNUAL
1964	1.99	0.05	0.94	0.14	0.47	0.66	0.00	0.29	0.00	1.68	2.55	5.33	14.10
1965	2.62	0.61	2.17	2.08	0.03	0.00	T	0.35	0.01	0.16	3.52	3.01	14.56
1966	1.60	1.75	0.18	0.27	0.20	0.04	0.14	0.00	0.02	0.00	3.63	3.86	11.69
1967	6.80	0.33	2.84	2.81	0.18	0.07	0.00	0.00	0.03	0.15	1.27	1.33	15.81
1968	4.10	1.74	2.68	1.01	0.19	T	0.00	0.03	T	0.62	2.63	4.38	17.38
1969	6.24	5.43	0.91	1.60	0.03	T	0.00	0.00	0.22	1.07	0.60	2.43	18.53
1970	7.06	1.42	2.56	0.95	T	0.07	0.00	0.00	T	0.95	6.12	3.61	22.74
1971	0.95	0.83	1.88	0.98	0.86	0.01	0.00	0.00	0.19	0.28	0.81	3.82	10.61
1972	0.69	0.70	0.07	0.57	0.11	0.15	0.00	0.00	0.66	0.74	6.22	2.38	12.29
1973	6.31	4.19	3.18	0.23	0.04	T	0.00	0.00	0.08	2.08	3.66	3.87	23.64
1974	1.73	0.74	2.59	2.79	0.00	0.34	0.61	0.00	0.00	0.97	0.78	2.15	12.70
1975	0.93	2.15	3.08	0.74	0.00	T	0.02	0.81	0.01	1.36	0.25	0.12	9.47
1976	0.14	1.23	0.63	1.11	0.00	0.03	0.07	0.52	0.39	0.26	0.59	0.63	5.60
1977	0.83	0.79	0.95	0.50	0.80	0.00	0.01	0.00	0.55	0.02	0.86	2.46	7.77
1978	4.33	2.50	5.56	2.09	0.02	T	0.00	0.00	0.47	0.00	1.70	0.58	17.25
1979	4.29	3.53	1.15	0.78	0.05	0.00	0.22	T	0.00	1.36	0.99	1.76	14.13
1980	3.06	2.73	1.12	0.90	0.27	T	0.50	0.00	0.00	0.06	0.04	1.22	9.90
1981	4.30	0.59	3.18	1.02	0.05	0.00	0.00	0.00	0.02	1.48	3.68	1.72	16.04
1982	3.87	2.28	6.48	1.55	0.00	0.18	0.00	0.00	2.47	2.22	3.93	2.60	25.58
1983	5.80	3.49	5.10	2.10	0.16	0.02	0.00	T	1.79	0.50	4.23	3.46	26.65
1984	0.22	1.20	0.65	0.37	0.01	0.06	T	0.01	0.06	1.47	3.53	1.69	9.27
1985	0.67	0.85	2.21	0.13	0.00	0.22	0.05	0.01	0.07	1.25	2.49	1.72	9.67
1986	1.69	5.82	3.64	0.93	0.15	0.00	0.02	0.00	0.80	0.01	0.03	0.67	13.76
1987	2.03	3.28	2.84	0.12	0.02	T	0.00	0.00	0.00	1.04	1.31	2.41	13.05
1988	1.97	0.31	0.19	1.98	0.54	0.23	0.00	0.00	0.00	0.01	1.36	2.28	8.87
1989	0.37	1.07	2.16	0.07	0.07	0.21	0.00	T	2.06	0.87	1.11	0.03	8.02
1990	2.23	1.30	0.86	0.52	2.33	0.00	T	T	T	0.20	0.45	0.91	8.80
1991	0.17	2.32	5.35	0.45	0.12	0.11	0.02	0.07	T	1.98	0.34	0.76	11.69
1992	1.39	5.37	1.30	0.72	0.00	0.12	0.00	T	0.00	0.52	0.20	3.49	13.11
1993	4.91	3.87	3.22	0.26	0.83	0.42	0.00	0.00	0.00	0.31	2.04	1.01	16.87
Record Mean	2.68	2.14	2.20	1.16	0.36	0.08	0.03	0.04	0.28	0.77	1.75	2.32	13.81

AVERAGE TEMPERATURE (deg. F)

STOCKTON, CALIFORNIA

YEAR	JAN	FEB	MAR	APR	MAY	JUNE	JULY	AUG	SEP	OCT	NOV	DEC	ANNUAL
1964	43.7	47.3	51.7	59.1	63.5	70.9	77.7	77.0	70.0	66.8	49.6	50.1	60.6
1965	45.7	49.5	54.5	59.7	67.0	69.5	77.1	77.7	68.4	65.5	53.8	40.4	60.7
1966	44.1	47.0	54.1	63.5	68.0	73.9	74.6	78.4	71.9	65.3	55.8	46.9	62.0
1967	46.2	48.1	51.9	50.1	66.2	70.7	80.3	81.2	75.7	66.0	55.8	42.7	61.3
1968	42.6	53.9	54.1	59.4	64.1	73.9	76.4	72.9	72.8	63.0	51.9	42.0	60.6
1969	43.8	46.4	51.8	58.0	68.5	70.1	78.4	79.1	74.9	60.5	53.2	48.2	61.1
1970	49.2	51.3	54.4	56.4	68.6	73.0	78.7	76.4	73.9	63.2	55.4	46.6	62.3
1971	44.9	47.8	53.1	57.5	63.4	72.3	79.2	79.6	73.8	62.0	52.1	42.9	60.7
1972	40.6	51.2	59.8	59.7	67.7	74.3	78.1	78.1	71.2	62.6	50.0	42.0	61.3
1973	45.3	52.4	51.3	61.8	71.3	77.1	76.8	74.5	70.2	63.2	53.2	47.5	62.1
1974	47.3	48.5	54.8	58.4	66.5	75.0	78.7	76.4	75.5	65.9	51.4	44.2	61.9
1975	41.7	49.3	51.3	53.6	68.0	72.4	75.8	75.0	74.4	61.5	50.8	44.7	59.9
1976	45.1	50.7	51.5	53.5	66.1	72.8	77.4	74.1	73.9	67.8	56.2	46.0	61.3
1977	43.8	53.1	53.6	65.3	63.0	76.7	79.2	78.8	72.8	66.5	55.3	50.3	63.2
1978	50.4	51.9	58.6	57.7	68.9	73.1	78.5	78.4	71.9	68.3	51.5	41.7	62.6
1979	45.3	51.1	56.4	60.6	70.2	75.1	78.1	77.2	77.8	66.2	53.5	48.0	63.3
1980	49.4	54.5	53.8	60.7	66.0	70.8	79.5	76.7	71.2	65.4	54.5	45.0	62.3
1981	48.2	52.6	54.0	61.7	69.2	80.5	78.0	76.7	73.8	61.3	56.0	49.4	63.5
1982	41.9	50.6	51.7	57.3	67.2	69.6	76.0	76.0	71.2	63.6	49.6	45.2	60.0
1983	44.2	52.6	55.2	56.3	66.5	73.8	75.9	79.2	76.2	67.0	52.6	50.2	62.5
1984	45.5	49.0	57.2	58.2	71.0	74.2	81.8	77.8	76.9	61.3	52.0	44.0	62.4
1985	41.4	50.2	49.7	62.0	64.1	75.9	78.3	73.6	68.1	62.9	50.4	41.1	59.8
1986	51.0	54.1	57.6	59.3	66.6	72.8	75.9	76.0	67.1	64.5	55.7	45.9	62.2
1987	44.9	51.2	54.4	64.1	69.7	73.6	72.8	74.9	72.7	67.8	52.8	47.1	62.2
1988	46.9	52.4	57.3	61.0	64.6	72.2	80.0	76.0	72.9	66.5	53.9	45.2	62.4
1989	43.6	47.6	56.2	63.3	66.5	72.1	77.2	74.8	70.2	63.3	53.7	42.2	60.9
1990	46.5	47.4	56.1	64.1	65.9	73.2	79.0	77.7		67.3	53.6	40.9	
1991	47.0	54.6	51.6	58.5	63.5	69.9	77.9	73.9	76.2	69.6	56.3	45.8	62.1
1992	42.8	54.8	57.1	64.0	72.3	72.4	76.8	78.9	74.2	68.0	54.7	44.3	63.4
1993	45.7	50.4	59.2	60.7	66.8	73.9	77.2	76.7	74.0	66.7	53.3	44.5	62.4
Record Mean	45.1	50.3	53.9	59.3	66.0	72.3	77.1	75.8	72.5	64.3	53.2	45.5	61.3
Max	53.3	60.5	65.4	72.6	80.5	88.0	93.9	92.0	88.0	78.2	64.3	53.6	74.2
Min	36.9	40.1	42.3	45.9	51.4	56.7	60.2	59.5	56.9	50.4	42.0	37.3	48.3

REFERENCE NOTES

STOCKTON, CALIFORNIA

GENERAL
T - TRACE AMOUNT.
BLANK ENTRIES DENOTE MISSING/UNREPORTED DATA.
# INDICATES A STATION OR INSTRUMENT RELOCATION.

SPECIFIC

(a) - LENGTH OF RECORD IN YEARS, ALTHOUGH INDIVIDUAL MONTHS MAY BE MISSING.
0.* OR * - THE VALUE IS BETWEEN 0.0 AND 0.05
NORMALS - BASED ON THE 1961-1990 RECORD PERIOD.
EXTREMES - DATES ARE THE MOST RECENT OCCURRENCE
WIND DIR. - NUMERALS SHOW TENS OF DEGRESS CLOCKWISE FROM TRUE NORTH. "00" INDICATES CALM.
RESULTANT DIRECTIONS ARE GIVEN TO WHOLE DEGREES.
BOLD VALUES INDICATE EXTREME VALUES WHICH OCCURRED AFTER THE ASOS SYSTEM WAS COMMISSIONED.

RECORD = PERIOD OF RECORD
RECORD MEAN PRECIPITATION IS THE MEAN OF ALL DAILY PRECIPITATION AMOUNTS DURING THE PERIOD OF RECORD.
RECORD MAX(MIN) TEMPERATURE IS THE MEAN OF ALL DAILY MAX(MIN) TEMPERATURES DURING THE PERIOD OF RECORD.
RECORD MEAN TEMPERATURE IS THE SUM OF THE RECORD MAX AND RECORD MIN DIVIDED BY 2.
AVERAGE TEMPERATURE IS THE SUM OF THE MEAN DAILY MAX AND MIN TEMPERATURE DIVIDED BY 2.

EXCEPTIONS

1. MAXIMUM 24-HOUR SNOW IS BASED ON SIX HOUR MEASUREMENTS THROUGH 1974.
2. MEAN WIND SPEED IS THROUGH 1977.
3. THUNDERSTORMS, AND HEAVY FOG ARE THROUGH 1978.

RECORD MEANS ARE THROUGH THE CURRENT YEAR, BEGINNING IN 1942 FOR TEMPERATURE
1942 FOR PRECIPITATION
1955 FOR SNOWFALL

HEATING DEGREE DAYS Base 65 deg. F STOCKTON, CALIFORNIA

SEASON	JULY	AUG	SEP	OCT	NOV	DEC	JAN	FEB	MAR	APR	MAY	JUNE	TOTAL
1964-65	0	6	8	50	457	452	591	427	320	201	50	10	2572
1965-66	0	0	15	38	328	754	644	497	332	80	16	10	2714
1966-67	0	0	5	47	273	555	576	466	398	442	77	32	2871
1967-68	0	0	0	25	267	687	691	317	331	182	84	7	2591
1968-69	0	4	2	75	385	707	649	516	407	210	26	1	2982
1969-70	0	0	0	149	349	513	482	379	322	251	47	5	2497
1970-71	0	0	0	106	281	564	616	477	363	217	77	10	2711
1971-72	0	0	18	174	382	676	751	394	163	165	53	4	2780
1972-73	0	0	0	112	445	704	602	344	421	119	12	2	2761
1973-74	0	0	0	90	345	533	541	454	309	194	47	0	2513
1974-75	0	0	0	60	400	639	716	434	419	335	56	6	3065
1975-76	0	1	0	140	420	622	613	408	412	339	42	7	3004
1976-77	0	0	0	29	255	581	650	324	348	41	100	0	2328
1977-78	0	0	2	41	284	451	449	358	191	212	16	0	2004
1978-79	0	0	3	33	398	713	605	384	262	126	27	0	2551
1979-80	0	0	0	60	338	518	473	300	345	140	54	3	2231
1980-81	0	0	5	107	309	613	512	341	330	137	23	0	2377
1981-82	0	0	0	116	263	477	708	396	405	228	35	13	2641
1982-83	0	0	9	73	453	607	638	341	297	252	70	0	2740
1983-84	0	0	0	12	365	451	595	456	234	203	20	4	2340
1984-85	0	0	0	136	383	643	724	408	468	105	64	8	2939
1985-86	0	0	22	110	435	734	425	298	221	174	63	0	2482
1986-87	0	0	39	50	272	585	615	380	323	75	36	0	2375
1987-88	0	0	0	15	360	550	555	361	234	134	96	26	2331
1988-89	0	0	2	40	328	607	656	480	268	94	38	2	2515
1989-90	0	0	6	94	335	699	565	487	269	65	42	2	2564
1990-91	0	0		20	336	741	553	284	406	196	99	7	
1991-92	0	0	0	79	256	585	680	289	237	59	0	2	2187
1992-93	0	0	0	20	305	635	594	405	178	141	26	15	2319
1993-94	0	0	4	14	346	628							

COOLING DEGREE DAYS Base 65 deg. F STOCKTON, CALIFORNIA

YEAR	JAN	FEB	MAR	APR	MAY	JUNE	JULY	AUG	SEP	OCT	NOV	DEC	TOTAL
1969	0	0	0	4	142	159	421	445	304	16	0	0	1491
1970	0	0	0	0	168	253	430	360	274	57	0	0	1542
1971	0	0	0	0	37	233	442	459	287	90	0	0	1548
1972	0	0	6	12	144	286	411	413	194	43	0	0	1509
1973	0	0	0	28	213	371	375	303	160	41	0	0	1491
1974	0	0	0	6	103	304	433	358	323	96	0	0	1623
1975	0	0	0	0	157	236	341	318	288	36	0	0	1376
1976	0	0	0	0	81	248	392	288	273	121	0	0	1403
1977	0	0	0	55	46	356	446	436	243	95	0	0	1677
1978	0	0	0	1	142	250	425	426	218	142	0	0	1604
1979	0	0	0	0	192	309	412	386	390	104	0	0	1793
1980	0	0	0	18	91	184	456	370	199	126	0	0	1444
1981	0	0	0	48	159	474	411	371	270	9	0	0	1742
1982	0	0	0	2	111	155	348	345	202	37	0	0	1200
1983	0	0	0	0	122	273	347	446	347	79	1	0	1615
1984	0	0	0	6	214	288	527	402	364	27	0	0	1828
1985	0	0	0	20	47	342	422	270	123	48	0	0	1272
1986	0	0	0	10	118	240	341	348	111	43	0	0	1211
1987	0	0	0	54	189	265	251	314	236	112	0	0	1421
1988	0	0	2	22	86	250	473	348	246	93	1	0	1521
1989	0	0	0	50	94	222	389	312	168	48	0	0	1283
1990	0	0	0	45	79	253	440	401		99	0	0	
1991	0	0	0	9	58	162	407	283	340	227	1	0	1487
1992	0	0	0	35	234	229	373	439	284	118	3	0	1715
1993	0	0	3	20	88	287	388	371	279	73	0	0	1509

SNOWFALL (inches) STOCKTON, CALIFORNIA

SEASON	JULY	AUG	SEP	OCT	NOV	DEC	JAN	FEB	MAR	APR	MAY	JUNE	TOTAL
1970-71	0.0	0.0	0.0	0.0	0.0	0.0	0.0	T	0.0	0.0	0.0	0.0	T
1971-72	0.0	0.0	0.0	0.0	0.0	0.0	0.0	T	0.0	0.0	0.0	0.0	T
1972-73	0.0	0.0	0.0	0.0	0.0	T	0.0	0.0	0.0	0.0	0.0	0.0	T
1973-74	0.0	0.0	0.0	0.0	0.0	0.0	0.0	0.0	0.0	0.0	0.0	0.0	0.0
1974-75	0.0	0.0	0.0	0.0	0.0	0.0	0.0	0.0	0.0	0.0	0.0	0.0	0.0
1975-76	0.0	0.0	0.0	0.0	0.0	0.0	0.0	0.3	T	0.0	0.0	0.0	0.3
1976-77	0.0	0.0	0.0	0.0	0.0	0.0	0.0	0.0	0.0	0.0	0.0	0.0	0.0
1977-78	0.0	0.0	0.0	0.0	0.0	0.0	0.0	0.0	0.0	0.0	0.0	0.0	0.0
1978-79	0.0	0.0	0.0	0.0	0.0	0.0	0.0	0.0	0.0	0.0	0.0	0.0	0.0
1979-80	0.0	0.0	0.0	0.0	0.0	0.0	0.0	0.0	0.0	0.0	0.0	0.0	0.0
1980-81	0.0	0.0	0.0	0.0	0.0	0.0	0.0	0.0	0.0	0.0	0.0	0.0	0.0
1981-82	0.0	0.0	0.0	0.0	0.0	0.0	0.0	0.0	0.0	0.0	0.0	0.0	0.0
1982-83	0.0	0.0	0.0	0.0	0.0	0.0	0.0	0.0	0.0	0.0	0.0	0.0	0.0
1983-84	0.0	0.0	0.0	0.0	0.0	0.0	0.0	0.0	0.0	0.0	0.0	0.0	0.0
1984-85	0.0	0.0	0.0	0.0	0.0	0.0	0.0	0.0	0.0	0.0	0.0	0.0	0.0
1985-86	0.0	0.0	0.0	0.0	0.0	0.0	0.0	0.0	0.0	0.0	0.0	0.0	0.0
1986-87	0.0	0.0	0.0	0.0	0.0	0.0	0.0	0.0	0.0	0.0	0.0	0.0	0.0
1987-88	0.0	0.0	0.0	0.0	0.0	0.0	0.0	0.0	0.0	0.0	0.0	0.0	0.0
1988-89	0.0	0.0	0.0	0.0	0.0	0.2	0.0	0.0	0.0	0.0	0.0	0.0	0.2
1989-90	0.0	0.0	0.0	0.0	0.0	0.0	0.0	0.0	0.0	0.0	0.0	0.0	0.0
1990-91	0.0	0.0	0.0	0.0	0.0	0.0	0.0	0.0	T	0.0	0.0	0.0	T
1991-92	0.0	0.0	0.0	0.0	0.0	0.0	0.0	0.0	0.0	0.0	0.0	0.0	0.0
1992-93	0.0	0.0	0.0	0.0	0.0	T	0.0	0.0	0.0	0.0	0.0	0.0	T
1993-94	0.0	0.0	0.0	0.0	0.0	0.0							
Record Mean	0.0	0.0	0.0	0.0	0.0	T	0.0	T	T	T	0.0	0.0	T

1993 ALAMOSA, COLORADO (ALS)

Alamosa is located in the south-central part of Colorado, near the center of the San Luis Valley which lies in a broad depression between mountain ranges converging to the north. The valley is the first of a series of basins along the Rio Grande River. The mountain ranges to the east reach altitudes over 14,000 feet and those to the west are between 13,000 and 14,000 feet. The length of the valley from north to south is over 80 miles, and its greatest width is about 50 miles. The valley floor ranges in altitude from 7,500 to near 8,000 feet and has a remarkably flat surface, except for a range of low hills across the southern portion. From the lowest areas which lie along an axis near the eastern border, the valley floor rises to the foothills, steeply to the east and more gently to the west.

The climate of the San Luis Valley is marked by cold winters and moderate summers, light precipitation, and much sunshine. At Alamosa about 80 percent of the annual precipitation occurs from April to October, most of it in the form of scattered light showers and thunderstorms that develop over the mountains and move into the valley during the afternoon. More than half of these thunderstorms occur during July and August. Hail frequently falls in some parts of the valley during their movement. Winter snows occur mainly in frequent light falls, with occasional falls as early as September or as late as May. A good snow cover will remain on the ground for several weeks during the coldest months.

All agriculture in the valley is dependent on irrigation, using water supplied by the more abundant precipitation in the surrounding mountains. Summer grazing of cattle and sheep on nearby mountain ranges and smaller valleys is extensive. A wide variety of vegetables, grains and feed crops are grown locally, with potatoes being the main commercial crop.

Summer is characterized by frequent days with maximum temperatures in the middle 80s and minimum temperatures in the low 40s. Relative humidity ranges from about 76 percent in the early mornings to around 40 percent during the afternoons. Winds are light during the coldest weather, but are strong with occasional blowing dust during the spring and early summer months.

Based on the 1951-1980 period, the average first occurrence of 32 degrees Fahrenheit in the fall is September 8 and the average last occurrence in the spring is June 8.

NORMALS, MEANS, AND EXTREMES FOR 1993

ALAMOSA, CO (ALS)

LATITUDE: 37° 26′ 10″ N | LONGTITUDE: 105° 51′ 56″ W | ELEVATION (FT): GRND: 7536 BARO: 7546 | TIME ZONE: MOUNTAIN (UTC+07) | WBAN: 23061

	ELEMENT	POR	JAN	FEB	MAR	APR	MAY	JUN	JUL	AUG	SEP	OCT	NOV	DEC	YEAR
TEMPERATURE °F	NORMAL DAILY MAXIMUM	30	33.3	39.9	48.7	58.6	68.0	77.8	82.0	79.2	72.7	62.5	47.4	35.4	58.8
	MEAN DAILY MAXIMUM	46	33.6	39.9	48.2	58.6	67.8	78.0	81.9	79.3	73.5	62.8	46.9	35.7	58.9
	HIGHEST DAILY MAXIMUM	48	62	66	73	80	85	93	96	90	87	81	71	61	96
	YEAR OF OCCURRENCE		1971	1986	1989	1989	1984	1990	1989	1977	1990	1979	1980	1958	JUL 1989
	MEAN OF EXTREME MAXS.	46	48.1	53.5	63.1	72.2	79.4	87.2	88.8	86.3	82.5	74.7	61.3	50.5	70.6
	NORMAL DAILY MINIMUM	30	-3.9	4.8	15.8	23.5	32.8	41.0	47.8	45.4	36.7	24.6	12.3	-.5	23.4
	MEAN DAILY MINIMUM	46	-2.7	5.0	15.7	23.8	33.2	41.4	47.6	45.4	36.4	24.6	11.4	-0.4	23.5
	LOWEST DAILY MINIMUM	48	-50	-35	-20	-6	11	24	33	29	15	-10	-30	-42	-50
	YEAR OF OCCURRENCE		1948	1948	1964	1973	1967	1990	1993	1964	1985	1945	1952	1978	JAN 1948
	MEAN OF EXTREME MINS.	46	-23.4	-14.3	-0.9	8.8	20.0	30.3	39.2	36.1	24.3	9.9	-6.5	-19.9	8.6
	NORMAL DRY BULB	30	14.7	22.4	32.3	41.1	50.4	59.4	64.9	62.4	54.8	43.5	29.9	17.5	41.1
	MEAN DRY BULB	48	15.5	22.5	32.6	42.1	51.6	60.9	66.1	63.7	56.1	44.6	29.7	18.2	42.0
	MEAN WET BULB	46	23.0	25.9	32.3	38.5	43.9	50.4	56.1	56.0	48.1	37.2	27.6	17.3	38.0
	MEAN DEW POINT	46	12.3	13.2	19.6	20.4	30.0	33.4	46.4	50.0	40.3	26.5	17.4	8.6	26.5
	NORMAL NO. DAYS WITH:														
	MAXIMUM ≥ 90°	30	0.0	0.0	0.0	0.0	0.0	0.3	0.9	*	0.0	0.0	0.0	0.0	1.2
	MAXIMUM ≤ 32°	30	13.5	6.4	1.5	0.1	0.0	0.0	0.0	0.0	0.0	0.1	2.9	11.3	35.8
	MINIMUM ≤ 32°	30	31.0	28.0	30.5	26.9	14.2	2.1	0.0	0.2	7.3	26.5	29.5	31.0	227.2
	MINIMUM ≤ 0°	30	19.3	9.7	1.5	0.1	0.0	0.0	0.0	0.0	0.0	0.1	3.6	16.0	50.3
H/C	NORMAL HEATING DEG. DAYS	30	1559	1193	1014	717	453	174	42	98	306	667	1053	1473	8749
	NORMAL COOLING DEG. DAYS	30	0	0	0	0	0	6	39	17	0	0	0	0	62
RH	NORMAL (PERCENT)														
	HOUR 05 LST	30	80	80	76	70	72	77	85	86	83	78	80	80	78
	HOUR 11 LST	30	60	55	43	32	29	29	39	42	40	39	49	59	43
	HOUR 17 LST	30	58	49	38	28	27	26	36	38	34	33	48	60	39
	HOUR 23 LST	30	82	79	69	60	58	56	67	70	67	67	78	81	70
S	PERCENT POSSIBLE SUNSHINE														
W/O	MEAN NO. DAYS WITH:														
	HEAVY FOG(VISBY ≤ 1/4 MI)	21	3.7	2.1	1.4	0.8	0.7	0.4	0.5	1.0	1.3	0.7	2.0	3.7	18.3
	THUNDERSTORMS	21	0.0	0.2	0.2	1.2	6.3	5.7	11.7	12.1	4.8	1.0	0.1	0.0	43.3
CLOUDINESS	MEAN:														
	SUNRISE-SUNSET (TENTHS)	1		7.0	5.0	6.0	6.0	3.0	3.0	8.0	3.0	5.0	5.0	4.0	
	MIDNIGHT-MIDNIGHT (TENTHS)	1		8.0	5.0	6.0	7.0		3.0	9.0	3.0	5.0	5.0	4.0	
	MEAN NO. DAYS WITH:														
	CLEAR	0		8.0	12.0	8.0	6.0	14.0	16.0	7.0	19.0	12.0	9.0	17.0	
	PARTLY CLOUDY	1	7.0	7.0	5.0	6.0	8.0	7.0	3.0	8.0	3.0	8.0	6.0	10.0	78.0
	CLOUDY	1	13.0	12.0	7.0	7.0	7.0			13.0	3.0	4.0	3.0	3.0	
PR	MEAN STATION PRESSURE(IN.)	12	22.77	22.73	22.67	22.70	22.76	22.85	22.93	22.94	22.89	22.85	22.76	22.76	22.80
	MEAN SEA-LEVEL PRES. (IN.)	1	30.30	30.10	30.10	29.90	30.00	29.90	29.90	30.10	30.10	30.10	30.20	30.30	30.08
WINDS	MEAN SPEED (MPH)	30	5.9	7.1	10.0	11.2	10.7	9.9	8.0	7.7	7.9	7.6	7.1	6.2	8.3
	PREVAIL.DIR(TENS OF DEGS.)	30	18	18	18	22	18	18	18	18	18	18	18	18	18
	MAXIMUM 2-MINUTE:														
	SPEED (MPH)	1	37	46	37	43	36	36	39	33	32	33	36	36	46
	DIR. (TENS OF DEGS.)	1	24	23	23	25	24	27	30	29	02	24	24	24	23
	YEAR OF OCCURRENCE		1993	1993	1993	1993	1993	1993	1993	1993	1993	1993	1993	1993	FEB 1993
	MAXIMUM 5-SECOND:														
	SPEED (MPH)	1	44		40	49	43	43	53	43	37	38	44	40	
	DIR. (TENS OF DEGS.)	1	28		23	25	25	25	29	28	02	23	18	24	
	YEAR OF OCCURRENCE		1993		1993	1993	1993	1993	1993	1993	1993	1993	1993	1993	
PRECIPITATION	NORMAL TOTAL (IN.)	30	0.26	0.29	0.45	0.49	0.64	0.67	1.19	1.12	0.89	0.70	0.43	0.44	7.57
	MAXIMUM TOTAL (IN.)	48	0.75	1.42	1.62	1.72	1.85	2.58	3.50	5.40	1.94	2.37	1.23	1.52	5.40
	YEAR OF OCCURRENCE		1979	1963	1992	1990	1973	1969	1968	1993	1959	1969	1991	1964	AUG 1993
	MINIMUM TOTAL (IN.)	48	T	T	T	T	0.01	T	0.03	0.21	T	T	T	T	T
	YEAR OF OCCURRENCE		1981	1954	1955	1972	1975	1980	1987	1980	1956	1983	1989	1980	DEC 1980
	MAXIMUM IN 24 HOURS (IN.)	48	0.47	1.15	1.15	1.33	1.04	1.04	1.57	1.34	1.82	1.27	0.78	0.93	1.82
	YEAR OF OCCURRENCE		1956	1963	1992	1952	1990	1969	1971	1993	1959	1969	1985	1964	SEP 1959
	NORMAL NO. DAYS WITH:														
	PRECIPITATION ≥ 0.01	30	3.9	4.4	5.9	4.9	5.3	6.0	9.7	10.4	6.5	4.4	4.1	4.5	70.0
	PRECIPITATION ≥ 1.00	30	0.0	0.0	*	*	0.0	*	0.1	0.0	0.0	*	0.0	0.0	.1
SNOWFALL	NORMAL TOTAL (IN.)	30	4.5	4.6	7.0	4.0	1.8	0.*	0.0	0.0	0.3	3.6	4.5	7.1	37.4
	MAXIMUM TOTAL (IN.)	48	13.8	16.0	29.2	16.4	13.5	0.2	T	0.0	4.2	20.3	19.8	27.7	29.2
	YEAR OF OCCURRENCE		1979	1963	1973	1947	1978	1983	1981		1961	1969	1972	1967	MAR 1973
	MAXIMUM IN 24 HOURS (IN.)	48	9.4	11.5	14.0	10.0	12.0	0.2	T	0.0	4.2	15.5	9.2	15.8	15.8
	YEAR OF OCCURRENCE		1987	1963	1962	1957	1990	1983	1981	1993	1961	1969	1985	1967	DEC 1967
	MAXIMUM SNOW DEPTH (IN.)	46	10	11	11	11	4	T	T	0	1	12	12	18	18
	YEAR OF OCCURRENCE		1992	1963	1992	1957	1978	1990	1990		1959	1991	1972	1964	DEC 1964
	NORMAL NO. DAYS WITH:														
	SNOWFALL ≥ 1.0	30	1.5	1.7	2.3	1.1	0.5	0.0	0.0	0.0	0.1	0.9	1.3	2.1	11.5

PRECIPITATION (inches) 1993 ALAMOSA, COLORADO (ALS)

YEAR	JAN	FEB	MAR	APR	MAY	JUN	JUL	AUG	SEP	OCT	NOV	DEC	ANNUAL
1964	0.26	0.27	0.41	0.22	0.50	0.39	0.91	0.73	1.06	T	0.80	1.52	7.07
1965	0.28	0.37	0.52	0.36	0.59	1.77	1.52	0.95	1.59	1.08	0.05	0.76	9.84
1966	0.28	0.23	0.11	0.15	0.30	0.72	0.78	1.42	0.03	0.49	0.10	0.35	4.96
1967	0.07	0.78	0.15	0.58	1.22	0.84	1.78	3.28	0.53	0.42	0.01	1.20	10.86
1968	0.04	0.42	0.21	0.27	0.20	0.06	3.50	2.22	0.41	0.11	0.28	0.38	8.10
1969	0.16	0.12	0.47	0.32	0.49	2.58	1.92	1.31	1.29	2.37	0.11	0.41	11.55
1970	0.06	0.03	0.85	0.54	0.86	0.38	1.35	1.30	1.53	1.09	0.06	0.03	8.08
1971	0.15	0.26	0.03	0.33	1.07	0.08	2.59	1.21	1.45	0.71	0.44	0.45	8.77
1972	0.24	0.09	0.12	T	0.07	0.60	0.80	1.16	1.00	2.16	1.00	0.46	7.70
1973	0.16	0.12	1.42	0.41	1.85	0.69	1.09	0.65	1.06	0.64	0.11	0.19	8.39
1974	0.70	0.08	0.24	0.18	0.09	0.72	1.78	0.72	0.62	0.74	0.15	0.74	6.76
1975	0.38	0.22	0.50	0.33	0.01	0.65	0.51	0.90	1.47	0.78	0.43	0.04	6.22
1976	0.05	0.33	0.39	0.50	0.77	0.07	1.43	1.22	0.67	0.51	0.20	0.07	6.21
1977	0.25	0.27	0.14	0.82	0.35	1.17	2.20	0.63	1.15	0.08	0.63	0.17	7.86
1978	0.33	0.07	0.13	0.20	1.59	1.23	1.04	0.27	0.19	0.51	0.90	0.81	7.27
1979	0.75	0.09	0.29	0.42	0.94	0.72	0.19	1.61	0.22	0.19	0.50	0.55	6.47
1980	0.32	0.31	0.65	1.48	1.21	T	0.54	0.21	0.46	0.52	0.01	T	5.71
1981	T	0.13	0.62	0.01	0.99	0.95	1.43	1.94	1.40	0.34	0.78	0.33	8.92
1982	0.07	0.49	0.40	0.37	0.57	0.22	0.51	0.58	1.85	0.19	0.25	0.49	5.99
1983	0.21	0.25	0.85	0.32	0.87	1.23	0.50	0.87	0.38	T	0.78	0.99	7.25
1984	0.14	0.28	1.12	0.49	0.18	0.55	0.74	1.07	0.36	1.48	0.10	0.59	7.10
1985	0.28	0.28	0.44	0.97	0.37	0.47	1.68	0.91	1.33	2.02	0.68	0.37	9.80
1986	0.05	0.10	0.37	1.08	0.74	0.67	0.54	0.66	1.20	1.18	1.02	0.12	7.73
1987	0.65	0.48	0.29	0.85	1.00	0.14	0.03	1.06	0.22	0.31	0.95	0.51	6.49
1988	0.26	0.25	0.18	0.35	0.51	0.83	0.66	1.08	0.64	0.20	0.35	0.11	5.42
1989	0.31	0.28	0.10	0.09	0.12	0.14	1.46	0.35	0.00	0.09	T	0.15	3.09
1990	0.62	0.20	0.43	1.72	0.78	0.45	1.86	1.28	1.48	0.72	0.90	0.75	11.19
1991	0.14	0.36	0.32	0.16	0.66	0.30	0.59	0.88	0.70	0.95	1.23	0.98	7.27
1992	0.08	0.08	1.62	0.04	1.13	1.23	1.21	1.97	0.50	0.01	0.48	0.79	9.14
1993	0.25	0.39	0.68	0.42	0.93	0.14	0.33	5.40	0.58	0.32	0.35	0.10	9.89
POR= 48 YRS	0.26	0.26	0.40	0.52	0.68	0.56	1.09	1.19	0.75	0.65	0.38	0.37	7.11

AVERAGE TEMPERATURE (°F) 1993 ALAMOSA, COLORADO (ALS)

YEAR	JAN	FEB	MAR	APR	MAY	JUN	JUL	AUG	SEP	OCT	NOV	DEC	ANNUAL
1964	13.2	16.0	25.0	38.2	51.5	58.8	66.8	61.9	55.2	44.9	25.9	6.7	38.7
1965	16.6	18.4	28.1	41.8	49.1	57.5	65.0	60.7	53.2	45.2	34.8	22.4	41.1
1966	13.0	16.0	33.2	41.4	52.5	59.1	67.5	62.6	55.2	43.8	34.5	20.9	41.6
1967	18.9	23.0	37.2	41.8	48.7	57.6	65.2	60.7	54.3	43.4	32.5	11.6	41.2
1968	7.9	22.6	33.3	37.3	48.7	60.1	63.6	60.6	52.5	44.9	28.5	16.2	39.7
1969	24.0	23.5	27.9	43.2	53.2	56.8	66.4	65.9	54.9	38.6	30.3	19.6	42.0
1970	17.0	28.6	28.6	36.2	51.4	57.5	65.8	64.5	52.5	39.5	31.3	22.9	41.3
1971	19.1	22.3	31.1	40.1	47.3	59.2	63.8	63.0	52.7	42.2	28.1	15.8	40.4
1972	17.4	27.9	37.3	42.7	49.9	61.0	64.1	62.2	56.0	46.6	18.6	10.2	41.2
1973	5.6	16.2	31.6	36.4	50.2	59.0	63.4	62.1	53.4	44.3	33.5	20.7	39.7
1974	11.2	14.9	37.7	38.8	53.2	60.1	63.9	59.3	53.5	45.5	29.2	13.2	40.0
1975	6.8	22.0	31.8	37.3	47.4	57.7	64.7	61.6	54.2	42.4	26.3	16.9	39.1
1976	13.8	29.8	32.3	42.4	50.8	57.9	64.9	60.5	54.3	39.7	28.0	13.3	40.6
1977	13.2	23.5	29.3	43.0	50.6	61.4	65.3	63.9	56.7	44.8	33.0	24.5	42.4
1978	22.8	25.3	35.9	43.3	48.1	60.8	65.4	60.2	55.5	44.4	32.6	8.0	41.9
1979	6.0	10.6	30.4	41.4	50.7	58.0	63.6	61.1	55.9	45.7	21.0	18.5	38.6
1980	20.8	29.4	30.2	38.2	48.5	61.9	67.0	61.9	56.0	40.4	30.4	28.1	42.7
1981	23.7	25.6	33.2	45.5	50.1	62.6	65.9	61.9	56.2	43.6	34.6	20.7	43.6
1982	17.8	22.2	33.2	40.2	48.5	57.2	64.1	64.2	55.6	41.4	30.8	20.9	41.3
1983	20.3	26.2	34.2	36.3	46.8	56.5	65.3	64.7	57.7	43.2	27.7	13.8	41.1
1984	1.6	10.8	27.1	38.1	55.2	58.6	65.9	63.2	56.3	40.7	29.9	20.1	39.0
1985	17.7	21.6	34.5	43.9	51.4	60.3	65.3	63.0	52.2	44.3	30.0	17.4	41.8
1986	25.0	29.7	36.8	43.4	50.5	60.3	63.1	62.8	52.6	41.3	31.3	20.4	43.1
1987	13.4	23.4	31.0	42.7	50.7	61.0	63.9	62.3	52.7	45.4	27.1	14.7	40.7
1988	4.5	17.2	31.5	42.8	50.1	61.8	64.5	63.9	53.5	46.1	29.8	17.8	40.3
1989	15.0	21.5	37.2	44.7	53.3	58.9	65.9	62.3	55.8	42.2	31.4	19.6	42.3
1990	14.8	25.9	36.4	43.3	49.3	62.5	63.6	61.1	58.3	44.4	31.7	13.3	42.0
1991	10.9	26.1	34.0	40.0	51.6	59.4	64.8	63.4	55.4	44.5	22.7	5.2	39.8
1992	1.4	14.5	29.6	47.0	53.5	58.8	62.1	60.8	55.0	45.2	22.1	6.7	38.1
1993	11.9	19.1	33.9	41.7	50.8	58.7	63.3	61.1	53.4	41.1	25.8	18.5	39.9
POR= 48 YRS	15.5	22.5	31.9	41.2	50.5	59.7	64.8	62.4	55.0	43.6	29.0	17.8	41.2

REFERENCE NOTES:

THE TEMPERATURE GRAPH SHOWS NORMAL MAXIMUM AND NORMAL MINIMUM DAILY TEMPERATURES (SOLID CURVES) AND THE ACTUAL DAILY HIGH AND LOW TEMPERATURES (VERTICAL BARS).

H/C INDICATES HEATING AND COOLING DEGREE DAYS.
RH INDICATES RELATIVE HUMIDITY
W/O INDICATES WEATHER AND OBSTRUCTIONS
S INDICATES SUNSHINE.
PR INDICATES PRESSURE.
CLOUDINESS IS THE SUM OF THE CEILOMETER AND SATELLITE DATA NOT TO EXCEED TEN TENTHS.
MAXIMUM SNOW DEPTH IS THE DEPTH MEASURED AT 12 UTC.

GENERAL:
T INDICATES TRACE PRECIPITATION, AN AMOUNT GREATER THAN ZERO BUT LESS THAN THE LOWEST REPORTABLE VALUE.
+ INDICATES THE VALUE ALSO OCCURS ON EARLIER DATES.
BLANK ENTRIES DENOTE MISSING OR UNREPORTED DATA.
NORMALS ARE 30-YEAR AVERAGES (1961 - 1990).
ASOS INDICATES AUTOMATED SURFACE OBSERVING SYSTEM.
PM INDICATES THE LAST DAY OF THE PREVIOUS MONTH.
POR (PERIOD OF RECORD) INDICATES THE NUMBER OF YEARS USED TO COMPUTE THE MEAN. INDIVIDUAL MONTHS WITHIN THE POR MAY BE MISSING.
WHEN THE POR FOR A NORMAL IS LESS THAN 30 YEARS, THE NORMAL IS PROVISIONAL AND IS BASED ON THE NUMBER OF YEARS INDICATED.
0.* OR * INDICATES THE VALUE OR MEAN DAYS WITH IS BETWEEN 0.00 AND 0.05.
CLOUDINESS FOR ASOS STATIONS DIFFERS FROM THE NON-ASOS OBSERVATION TAKEN BY A HUMAN OBSERVER. ASOS STATION CLOUDINESS IS BASED ON TIME-AVERAGED CEILOMETER DATA FOR CLOUDS AT OR BELOW 12,000 FEET AND ON SATELLITE DATA FOR CLOUDS ABOVE 12,000 FEET.
THE NUMBER OF DAYS WITH CLEAR, PARTLY CLOUDY, AND CLOUDY CONDITIONS FOR ASOS STATIONS IS THE SUM OF THE CEILOMETER AND SATELLITE DATA FOR THE SUNRISE TO SUNSET PERIOD.

GENERAL CONTINUED:
CLEAR INDICATES 0 - 3 TENTHS, PARTLY CLOUDY INDICATES 4 - 7 TENTHS, AND CLOUDY INDICATES 8 OR MORE TENTHS. WHEN AT LEAST ONE OF THE ELEMENTS (CEILOMETER OR SATELLITE) IS MISSING, THE DAILY CLOUDINESS IS NOT COMPUTED.
WIND DIRECTION IS RECORDED IN TENS OF DEGREES (2 DIGITS) CLOCKWISE FROM TRUE NORTH. "00" INDICATES CALM. "36" INDICATES TRUE NORTH.
RESULTANT WIND IS THE VECTOR AVERAGE OF THE SPEED AND DIRECTION.
AVERAGE TEMPERATURE IS THE SUM OF THE MEAN DAILY MAXIMUM AND MINIMUM TEMPERATURE DIVIDED BY 2.
SNOWFALL DATA COMPRISE ALL FORMS OF FROZEN PRECIPITATION, INCLUDING HAIL.
A HEATING (COOLING) DEGREE DAY IS THE DIFFERENCE BETWEEN THE AVERAGE DAILY TEMPERATURE AND 65° F.
DRY BULB IS THE TEMPERATURE OF THE AMBIENT AIR.
DEW POINT IS THE TEMPERATURE TO WHICH THE AIR MUST BE COOLED TO ACHIEVE 100 PERCENT RELATIVE HUMIDITY.
WET BULB IS THE TEMPERATURE THE AIR WOULD HAVE IF THE MOISTURE CONTENT WAS INCREASED TO 100 PERCENT RELATIVE HUMIDITY.

HEATING DEGREE DAYS (base 65°F) 1993 ALAMOSA, COLORADO (ALS)

YEAR	JUL	AUG	SEP	OCT	NOV	DEC	JAN	FEB	MAR	APR	MAY	JUN	TOTAL
1964-65	8	107	289	621	1166	1807	1495	1299	1136	689	484	219	9320
1965-66	31	128	348	607	899	1316	1604	1364	983	700	383	171	8534
1966-67	8	86	286	651	910	1362	1423	1172	855	690	498	217	8158
1967-68	23	135	313	663	970	1651	1768	1227	975	823	499	147	9194
1968-69	49	134	370	616	1087	1509	1263	1155	1144	648	361	240	8576
1969-70	9	25	295	812	1036	1402	1485	1013	1123	857	414	236	8707
1970-71	11	40	368	783	1002	1298	1417	1188	1042	738	543	170	8600
1971-72	64	63	361	698	1098	1518	1467	1071	855	666	462	111	8434
1972-73	55	97	267	561	1384	1695	1839	1360	1028	854	449	188	9777
1973-74	74	91	342	633	937	1366	1662	1394	839	778	359	173	8648
1974-75	41	170	339	595	1067	1601	1802	1198	1023	826	537	212	9411
1975-76	26	102	319	695	1157	1485	1579	1014	1010	672	432	208	8699
1976-77	24	132	314	779	1104	1596	1598	1155	1097	655	440	104	8998
1977-78	19	51	246	621	951	1252	1302	1103	900	647	516	126	7734
1978-79	18	141	278	632	966	1762	1827	1518	1069	704	438	203	9556
1979-80	57	127	267	590	1312	1438	1363	1029	1071	798	504	107	8663
1980-81	5	102	263	757	1031	1136	1274	1097	979	576	458	102	7780
1981-82	14	108	256	656	904	1366	1457	1192	977	736	508	230	8404
1982-83	59	47	275	724	1016	1361	1380	1080	946	856	556	249	8549
1983-84	28	35	213	674	1112	1582	1964	1567	1166	799	297	188	9625
1984-85	11	56	252	748	1051	1384	1462	1209	937	625	415	146	8296
1985-86	30	66	378	636	1045	1473	1231	983	866	639	446	138	7931
1986-87	63	75	366	728	1004	1377	1594	1160	1049	661	436	115	8628
1987-88	66	96	364	601	1130	1557	1872	1381	1031	658	454	102	9312
1988-89	28	50	337	577	1049	1453	1544	1211	854	600	357	180	8240
1989-90	17	82	270	698	1001	1400	1550	1089	880	640	480	105	8212
1990-91	59	118	201	633	990	1599	1671	1081	954	742	410	172	8630
1991-92	33	51	280	630	1263	1852	1973	1459	1093	535	350	179	9698
1992-93	97	131	295	607	1281	1803	1638	1280	958	692	435	185	9402
1993-	51	118	342	735	1167	1435							
POR= 49 YRS	35	86	294	654	1071	1458	1530	1194	1017	703	441	160	8643

COOLING DEGREE DAYS (base 65°F) 1993 ALAMOSA, COLORADO (ALS)

YEAR	JAN	FEB	MAR	APR	MAY	JUN	JUL	AUG	SEP	OCT	NOV	DEC	ANNUAL
1969	0	0	0	0	0	1	58	58	0	0	0	0	117
1970	0	0	0	0	0	18	45	32	0	0	0	0	95
1971	0	0	0	0	0	3	35	8	0	0	0	0	46
1972	0	0	0	0	0	1	36	18	0	0	0	0	55
1973	0	0	0	0	0	14	30	8	0	0	0	0	52
1974	0	0	0	0	0	31	13	0	0	0	0	0	44
1975	0	0	0	0	0	0	24	4	0	0	0	0	28
1976	0	0	0	0	0	1	25	0	0	0	0	0	26
1977	0	0	0	0	0	2	36	25	2	0	0	0	65
1978	0	0	0	0	0	8	39	0	0	0	0	0	47
1979	0	0	0	0	0	0	21	13	0	0	0	0	34
1980	0	0	0	0	0	22	76	12	0	0	0	0	110
1981	0	0	0	0	0	35	47	19	0	0	0	0	101
1982	0	0	0	0	0	0	38	27	0	0	0	0	65
1983	0	0	0	0	0	4	43	30	1	0	0	0	78
1984	0	0	0	0	2	1	45	8	0	0	0	0	56
1985	0	0	0	0	0	9	47	9	0	0	0	0	65
1986	0	0	0	0	0	3	11	14	0	0	0	0	28
1987	0	0	0	0	0	0	39	22	0	0	0	0	61
1988	0	0	0	0	0	13	17	22	0	0	0	0	52
1989	0	0	0	0	0	6	52	6	0	0	0	0	64
1990	0	0	0	0	0	35	22	3	7	0	0	0	67
1991	0	0	0	0	0	12	33	8	0	0	0	0	53
1992	0	0	0	0	0	1	12	6	0	0	0	0	19
1993	0	0	0	0	0	6	6	4	0	0	0	0	16
POR= 25 YRS	0	0	0	0	0	8	33	13	0	0	0	0	54

SNOWFALL (inches) 1993 ALAMOSA, COLORADO (ALS)

YEAR	JUL	AUG	SEP	OCT	NOV	DEC	JAN	FEB	MAR	APR	MAY	JUN	TOTAL
1964-65	0.0	0.0	0.0	0.0	13.6	27.0	2.7	8.4	9.1	4.5	0.5	0.0	65.8
1965-66	0.0	0.0	0.0	0.5	0.3	10.8	4.7	5.8	3.3	3.6	0.0	0.0	29.0
1966-67	0.0	0.0	0.0	4.8	0.3	4.9	1.2	12.4	4.0	6.4	1.2	0.0	35.2
1967-68	0.0	0.0	0.0	4.3	0.3	27.7	1.3	6.4	4.8	5.0	T	0.0	49.8
1968-69	0.0	0.0	0.0	0.2	2.8	6.3	3.5	2.4	12.5	0.9	0.9	0.0	29.5
1969-70	0.0	0.0	0.0	20.3	1.7	7.6	1.5	1.0	19.1	8.7	0.8	0.0	60.7
1970-71	0.0	0.0	1.8	14.2	0.8	1.2	3.1	7.7	0.8	4.1	7.4	0.0	41.1
1971-72	0.0	0.0	1.2	T	7.7	11.0	6.9	2.0	4.4	T	T	0.0	33.2
1972-73	0.0	0.0	0.0	14.3	19.8	7.6	3.9	3.6	29.2	6.9	12.2	0.0	97.5
1973-74	0.0	0.0	1.0	8.1	1.1	3.1	12.4	1.9	4.3	5.5	0.0	0.0	37.4
1974-75	0.0	0.0	T	0.2	3.3	10.0	7.0	4.2	6.5	4.3	T	T	35.5
1975-76	0.0	0.0	0.0	0.5	5.9	0.8	0.8	3.4	6.4	2.1	0.0	0.0	19.9
1976-77	0.0	0.0	T	6.5	2.6	2.1	3.4	5.7	2.5	7.1	T	0.0	29.9
1977-78	0.0	0.0	0.0	0.0	0.4	3.9	4.4	0.9	1.0	0.1	13.5	0.0	24.2
1978-79	0.0	0.0	0.0	0.2	4.1	12.1	13.8	0.9	3.0	1.8	2.4	T	38.3
1979-80	0.0	0.0	0.0	1.1	5.3	6.8	5.0	2.1	6.3	8.3	2.3	0.0	37.2
1980-81	0.0	0.0	0.0	2.2	0.1	T	T	1.8	6.0	0.0	T	0.0	10.1
1981-82	T	0.0	0.0	T	5.4	4.9	1.2	6.9	2.9	1.4	2.1	0.0	24.8
1982-83	0.0	0.0	0.0	2.1	2.2	6.0	3.4	5.1	10.2	3.5	0.5	0.2	33.2
1983-84	0.0	0.0	0.0	0.0	8.1	11.2	1.4	2.8	10.6	2.8	T	0.0	36.9
1984-85	0.0	0.0	0.0	6.7	0.9	5.6	2.8	2.8	6.1	0.8	1.2	0.0	26.9
1985-86	0.0	0.0	0.0	6.0	9.7	6.7	0.5	1.0	3.5	4.5	2.1	0.0	34.0
1986-87	0.0	0.0	T	2.7	6.9	1.9	12.8	7.0	3.9	8.4	0.0	0.0	43.6
1987-88	0.0	0.0	0.0	0.0	6.8	7.5	6.0	2.9	3.4	0.7	0.0	0.0	27.3
1988-89	0.0	0.0	0.0	0.0	3.0	1.4	5.2	3.0	1.0	0.9	0.0	0.0	14.5
1989-90	0.0	0.0		T	T	2.2	13.0	2.5	3.3	9.2	4.8	0.0	
1990-91	0.0	0.0	0.0	0.2	6.4	9.9	1.4	4.6	3.2	1.4	0.2	T	27.3
1991-92	0.0	0.0	0.0	15.1	10.2	8.9	0.9	1.7	16.2	0.3	1.0	T	54.3
1992-93	0.0	0.0											
1993-													
POR= 48 YRS	T	0.0	0.2	3.2	4.2	5.6	4.5	4.2	6.1	4.4	1.6	T	34.1

1993 COLORADO SPRINGS, COLORADO (COS)

At an elevation near 6,200 feet above sea level, Colorado Springs is located in relatively flat semi-arid country on the eastern slope of the Rocky Mountains. Immediately to the west the mountains rise abruptly to heights ranging from 10,000 to 14,000 feet but generally averaging near 11,000 feet. To the east lie gently undulating prairie lands. The land slopes upward to the north, reaching an average height of about 8,000 feet in 20 miles at the top of Palmer Lake Divide.

Colorado Springs is in the Arkansas River drainage basin. The principal tributary feeding the Arkansas from this area is Fountain Creek which rises in the high mountains west of the city and is fed by Monument Creek originating to the north in the Palmer Lake Divide area.

Other topographical features of the area, and particularly its wide range of elevations, help to give Colorado Springs the various and altogether delightful plains and mountain mixture of climate that has established the locality as a highly desirable place to live. The higher elevations immediately to the west and north of the city produce significant differences in temperature and precipitation. Precipitation amounts at these higher elevations are approximately twice those at nearby lower elevations and the number of rainy days is almost triple.

In Colorado Springs itself, precipitation is relatively sparse. Over 80 percent of it falls between April 1 and September 30, mostly as heavy downpours accompanying summer thunderstorms. Temperatures, in view of the station latitude and elevation, are mild. Uncomfortable extremes, in either summer or winter, are comparatively rare and of short duration. Relative humidity is normally low and wind movement moderately high. This is notably true of the west-to-east movement of the chinook winds, that cause rapid rises in winter temperatures and remind us that the Indian meaning of CHINOOK is SNOW EATER.

Colorado Springs is best known as a resort city, but is also important to the high-tech industry and military community. Several military installations, including the United States Air Force Academy and the Space Command are located within or near the city. The surrounding prairie is also important for cattle raising and a considerable amount of grazing land is used for sheep in the summer months. The growing season varies considerably in length but averages from the first week in May to the first week of October.

NORMALS, MEANS, AND EXTREMES FOR 1993

COLORADO SPRINGS, CO (COS)

LATITUDE:	LONGTITUDE:	ELEVATION (FT):		TIME ZONE:	WBAN: 93037
38° 48' 43" N	104° 42' 40" W	GRND: 6145	BARO: 6093	MOUNTAIN (UTC+07)	

	ELEMENT	POR	JAN	FEB	MAR	APR	MAY	JUN	JUL	AUG	SEP	OCT	NOV	DEC	YEAR
TEMPERATURE °F	NORMAL DAILY MAXIMUM	30	41.4	44.6	50.0	59.8	68.7	79.0	84.4	81.3	73.6	63.5	50.7	42.2	61.6
	MEAN DAILY MAXIMUM	45	42.0	45.1	49.8	59.7	68.6	79.3	86.5	83.9	76.3	65.5	51.6	44.4	62.7
	HIGHEST DAILY MAXIMUM	45	72	76	81	87	93	100	100	99	94	86	78	77	100
	YEAR OF OCCURRENCE		1974	1963	1971	1992	1984	1954	1954	1954	1990	1979	1981	1955	JUL 1954
	MEAN OF EXTREME MAXS.	16	62.4	65.6	69.4	79.6	83.9	92.9	95.6	92.1	88.1	79.8	70.7	63.8	78.7
	NORMAL DAILY MINIMUM	30	16.1	19.3	24.6	33.0	42.1	51.1	57.1	55.2	47.1	36.3	24.9	17.4	35.4
	MEAN DAILY MINIMUM	45	16.1	19.3	24.2	32.8	42.3	51.3	58.1	56.5	48.3	37.3	25.1	18.3	35.8
	LOWEST DAILY MINIMUM	45	-26	-27	-11	-3	21	32	42	39	22	5	-8	-24	-27
	YEAR OF OCCURRENCE		1951	1951	1956	1959	1954	1951	1952	1992	1985	1969	1976	1990	FEB 1951
	MEAN OF EXTREME MINS.	16	-1.4	-1.5	8.0	18.7	30.1	40.4	50.2	48.1	33.4	20.1	7.8	-2.4	21.0
	NORMAL DRY BULB	30	28.8	32.0	37.3	46.4	55.4	65.0	70.8	68.3	60.4	49.9	37.8	29.8	48.5
	MEAN DRY BULB	45	29.1	32.2	37.0	46.2	55.5	65.3	72.3	70.2	62.3	51.5	38.4	31.4	49.3
	MEAN WET BULB	44	22.4	25.1	29.0	36.0	44.1	51.7	56.7	55.8	48.6	39.1	29.4	23.8	38.5
	MEAN DEW POINT	44	10.1	12.6	16.4	22.9	33.0	40.8	47.7	47.4	38.0	26.4	17.3	11.5	27.0
	NORMAL NO. DAYS WITH:														
	MAXIMUM ≥ 90°	12	0.0	0.0	0.0	0.0	0.2	4.3	10.8	3.3	0.8	0.0	0.0	0.0	19.4
	MAXIMUM ≤ 32°	12	8.3	6.6	2.7	0.7	0.0	0.0	0.0	0.0	0.1	0.3	3.3	7.8	29.8
	MINIMUM ≤ 32°	12	30.3	26.2	25.1	12.8	2.1	0.0	0.0	0.0	1.1	9.1	24.4	29.3	160.4
	MINIMUM ≤ 0°	12	2.6	2.3	0.3	0.0	0.0	0.0	0.0	0.0	0.0	0.0	0.0	2.3	7.5
H/C	NORMAL HEATING DEG. DAYS	30	1122	924	859	558	302	87	6	18	164	468	816	1091	6415
	NORMAL COOLING DEG. DAYS	30	0	0	0	0	0	87	186	120	26	0	0	0	419
RH	NORMAL (PERCENT)	30	51	50	50	46	49	48	50	54	51	46	50	51	50
	HOUR 05 LST	30	56	59	61	61	66	66	68	70	66	58	59	56	62
	HOUR 11 LST	30	41	40	40	35	37	36	36	40	39	35	39	42	38
	HOUR 17 LST	30	46	41	38	34	36	34	39	41	37	36	44	48	40
	HOUR 23 LST	30	57	58	58	55	59	57	60	64	60	56	58	57	58
S	PERCENT POSSIBLE SUNSHINE														
W/O	MEAN NO. DAYS WITH:														
	HEAVY FOG(VISBY ≤ 1/4 MI)	45	2.3	2.5	2.5	2.0	1.8	0.8	0.5	1.0	1.9	1.8	2.3	2.2	21.6
	THUNDERSTORMS	45	0.0	0.0	0.5	2.3	8.2	11.0	15.6	13.1	4.6	0.8	0.0	0.0	56.1
CLOUDINESS	MEAN:														
	SUNRISE-SUNSET (TENTHS)	1		6.0	7.0	7.0		5.0	4.0	8.0	4.0	5.0	4.0	3.0	
	MIDNIGHT-MIDNIGHT (TENTHS)	1		6.0	8.0	7.0			4.0	8.0	4.0	6.0	4.0	3.0	
	MEAN NO. DAYS WITH:														
	CLEAR	1	13.0	10.0	11.0	8.0	1.0	8.0	11.0	7.0	15.0	13.0	12.0	20.0	129.0
	PARTLY CLOUDY	1	3.0	8.0	5.0	6.0		8.0	6.0	8.0	5.0	2.0	4.0	8.0	
	CLOUDY	1	4.0	9.0	9.0	8.0	1.0	5.0	3.0	13.0	6.0	9.0	3.0	1.0	71.0
PR	MEAN STATION PRESSURE(IN.)	21	23.89	23.89	23.82	23.88	23.90	23.98	24.06	24.07	24.04	24.00	23.91	23.90	23.94
	MEAN SEA-LEVEL PRES. (IN.)	44	30.09	30.06	29.96	29.92	29.90	29.88	29.95	29.97	29.99	30.05	30.08	30.09	30.00
WINDS	MEAN SPEED (MPH)	29	8.9	9.7	11.2	11.6	11.0	10.1	9.2	9.1	9.5	9.7	9.1	9.0	9.8
	PREVAIL.DIR(TENS OF DEGS.)	29	36	36	36	36	36	36	36	36	36	36	36	2	36
	MAXIMUM 2-MINUTE:														
	SPEED (MPH)	1	43	45	40	39	39	38	44	36	36	34	41	53	53
	DIR. (TENS OF DEGS.)	1	31	26	22	30	35	36	33	33	36	33	34	29	29
	YEAR OF OCCURRENCE		1993	1993	1993	1993	1993	1993	1993	1993	1993	1993	1993	1993	DEC 1993
	MAXIMUM 5-SECOND:														
	SPEED (MPH)	1	54	53	46	45	45	46	49	43	41	40	46	59	59
	DIR. (TENS OF DEGS.)	1	33	34	22	30	03	29	33	32	21	34	34	29	29
	YEAR OF OCCURRENCE		1993	1993	1993	1993	1993	1993	1993	1993	1993	1993	1993	1993	DEC 1993
PRECIPITATION	NORMAL TOTAL (IN.)	30	0.29	0.40	0.94	1.19	2.15	2.25	2.90	3.02	1.33	0.84	0.47	0.46	16.24
	MAXIMUM TOTAL (IN.)	45	1.17	2.45	2.38	5.90	5.67	8.00	5.27	6.06	4.28	5.01	2.21	1.05	8.00
	YEAR OF OCCURRENCE		1987	1987	1979	1957	1957	1965	1968	1986	1976	1984	1957	1988	JUN 1965
	MINIMUM TOTAL (IN.)	45	T	T	0.01	0.01	0.33	0.13	0.67	0.15	T	0.01	T	T	T
	YEAR OF OCCURRENCE		1964	1991	1966	1964	1974	1990	1987	1962	1953	1980	1965	1970	DEC 1970
	MAXIMUM IN 24 HOURS (IN.)	45	0.79	1.49	1.51	2.45	2.57	3.09	3.00	3.73	1.73	1.60	1.45	0.69	3.73
	YEAR OF OCCURRENCE		1987	1987	1987	1957	1955	1954	1951	1976	1959	1960	1979	1981	AUG 1976
	NORMAL NO. DAYS WITH:														
	PRECIPITATION ≥ 0.01	12	6.2	5.8	8.3	7.7	12.5	9.7	11.0	14.0	6.8	5.2	4.3	6.4	97.9
	PRECIPITATION ≥ 1.00	12	0.0	0.1	0.1	0.2	0.4	0.2	0.4	0.5	0.2	0.1	0.0	0.0	2.2
SNOWFALL	NORMAL TOTAL (IN.)	12	6.7	7.1	12.1	4.5	2.5	0.0	0.0	0.0	0.2	2.8	5.3	10.2	51.4
	MAXIMUM TOTAL (IN.)	45	28.7	23.2	23.2	42.7	19.4	1.1	T	T	27.9	25.9	26.3	18.2	42.7
	YEAR OF OCCURRENCE		1987	1987	1984	1957	1978	1975	1992	1992	1959	1984	1991	1983	APR 1957
	MAXIMUM IN 24 HOURS (IN.)	45	22.0	14.8	13.3	18.0	17.4	1.1	T	T	17.1	14.6	14.5	9.6	22.0
	YEAR OF OCCURRENCE		1987	1987	1964	1957	1978	1975	1992	1992	1959	1984	1972	1979	JAN 1987
	MAXIMUM SNOW DEPTH (IN.)	13	16	12	14	8	11	0	0	0	2	12	11	10	16
	YEAR OF OCCURRENCE		1987	1987	1985	1979	1978				1985	1984	1979	1979	JAN 1987
	NORMAL NO. DAYS WITH:														
	SNOWFALL ≥ 1.0	12	1.9	1.9	3.1	1.3	0.4	0.0	0.0	0.0	0.1	0.6	1.5	3.0	13.8

PRECIPITATION (inches) 1993 COLORADO SPRINGS, COLORADO (COS)

YEAR	JAN	FEB	MAR	APR	MAY	JUN	JUL	AUG	SEP	OCT	NOV	DEC	ANNUAL
1964	T	0.22	1.08	0.01	2.54	0.96	1.14	0.60	1.33	0.03	0.46	0.22	8.59
1965	0.14	0.72	1.12	1.61	1.81	8.00	5.02	3.83	2.24	0.49	T	0.45	25.43
1966	0.39	0.49	0.01	0.79	0.95	2.56	2.91	2.00	2.12	0.36	0.16	0.17	12.91
1967	0.31	0.15	0.18	2.04	2.18	2.74	5.26	3.09	0.73	1.68	0.25	0.67	19.28
1968	0.10	0.22	0.37	0.54	0.62	0.15	5.27	2.12	1.03	0.43	1.32	0.24	12.41
1969	0.11	0.12	0.77	1.83	4.46	2.72	3.90	2.38	1.13	2.86	0.39	0.32	20.99
1970	0.05	0.17	1.06	0.91	0.33	3.63	3.79	4.24	1.09	0.95	0.27	T	16.49
1971	0.34	0.53	0.34	1.36	2.24	0.39	2.82	1.99	1.36	0.23	0.03	0.23	11.86
1972	0.27	0.25	0.55	0.42	1.46	2.07	4.08	3.55	4.13	1.34	1.08	0.83	20.03
1973	0.06	0.06	1.16	1.72	4.27	0.47	3.31	0.89	1.03	0.35	0.15	0.64	14.11
1974	0.26	0.18	0.52	1.92	0.33	1.29	1.42	1.14	0.43	1.36	0.23	0.42	9.50
1975	0.13	0.29	0.24	0.68	1.00	2.97	2.65	2.06	0.16	0.52	1.00	0.07	11.77
1976	0.32	0.23	0.63	1.63	2.09	2.46	1.75	5.94	4.28	0.49	0.40	0.12	20.34
1977	0.29	0.20	1.18	2.57	1.12	3.87	3.02	5.11	0.45	0.19	0.60	0.18	18.78
1978	0.25	0.38	0.40	1.15	3.58	0.54	2.14	2.51	0.05	0.90	0.37	1.01	13.28
1979	0.53	0.04	2.38	1.83	3.13	1.58	2.73	2.50	0.92	0.55	1.82	1.02	19.03
1980	0.25	0.54	1.30	3.64	4.99	1.60	1.69	4.59	0.65	0.01	0.35	0.05	19.66
1981	0.07	0.12	0.93	0.13	3.14	1.98	3.64	5.24	0.52	0.37	0.03	0.82	16.99
1982	0.25	0.27	0.73	0.76	3.07	3.81	3.64	5.37	3.02	0.22	0.10	0.70	21.94
1983	0.43	0.09	1.79	0.97	3.08	2.41	0.99	2.59	0.37	0.28	1.09	0.70	14.79
1984	0.32	0.09	1.93	1.66	0.74	1.54	3.97	4.03	0.93	5.01	0.14	0.64	21.00
1985	0.42	0.24	1.68	2.07	3.36	0.78	4.92	1.56	1.49	0.52	0.42	0.55	18.01
1986	0.01	0.30	0.31	0.65	1.89	2.47	1.63	6.06	0.61	1.41	0.64	0.28	16.26
1987	1.17	2.45	1.79	0.50	3.82	2.89	0.67	2.77	0.55	0.54	0.44	0.64	18.23
1988	0.43	0.68	0.90	0.27	1.01	1.69	2.07	2.88	1.19	0.08	0.36	1.05	12.61
1989	0.23	1.23	0.49	1.06	1.11	3.42	2.26	2.63	2.30	0.28	0.02	0.41	15.44
1990	0.53	0.59	1.77	2.04	3.90	0.13	5.13	1.45	1.50	1.46	0.30	0.27	19.07
1991	0.09	T	0.42	1.76	0.80	3.07	2.87	4.57	0.56	0.88	2.05	0.45	17.52
1992	0.06	0.02	2.36	0.92	2.07	3.91	0.76	3.37	0.13	0.30	0.75	0.11	14.76
1993	0.52	0.21	0.79	1.02	1.60	1.27	2.38	2.17	1.44	0.91	0.97	0.11	13.39
POR= 45 YRS	0.30	0.35	0.89	1.23	2.22	2.13	2.82	2.80	1.24	0.80	0.51	0.36	15.67

WBAN : 93037

AVERAGE TEMPERATURE (°F) 1993 COLORADO SPRINGS, COLORADO (COS)

YEAR	JAN	FEB	MAR	APR	MAY	JUN	JUL	AUG	SEP	OCT	NOV	DEC	ANNUAL
1964	30.3	26.1	30.9	46.2	57.6	65.0	75.3	69.3	61.9	52.3	38.4	32.8	48.8
1965	35.4	27.8	25.6	48.3	54.9	62.1	69.4	65.8	55.0	52.9	43.8	34.7	48.0
1966	25.1	26.8	40.8	44.4	57.1	65.7	73.9	66.8	62.1	48.9	40.6	29.4	48.5
1967	33.1	33.0	42.9	48.2	52.4	62.0	69.3	66.1	60.6	52.3	39.6	25.3	48.7
1968	30.7	32.6	38.6	42.5	51.5	66.9	68.4	65.8	60.3	51.5	34.5	29.4	47.7
1969	33.4	33.8	29.6	48.9	56.8	59.7	71.7	70.2	61.9	41.5	37.0	30.8	47.9
1970	28.2	37.2	32.6	42.6	58.2	64.6	71.2	71.4	57.7	44.2	38.2	32.9	48.2
1971	29.8	29.8	37.7	45.3	53.3	67.8	68.8	69.1	56.7	49.3	37.9	30.4	48.0
1972	30.0	36.1	43.3	49.1	55.9	67.8	68.9	67.7	60.8	49.4	29.8	23.5	48.5
1973	25.4	32.3	36.5	40.9	53.3	65.6	68.4	70.6	58.8	52.6	39.6	31.2	47.9
1974	27.0	33.9	42.1	46.0	59.7	66.2	72.6	67.9	58.0	52.6	38.5	28.0	49.4
1975	29.1	29.8	35.6	44.4	53.4	63.8	71.0	70.2	59.4	52.3	36.4	35.3	48.4
1976	30.1	37.8	36.1	47.7	54.7	64.3	72.1	68.1	59.3	45.7	36.2	32.9	48.8
1977	26.8	35.0	37.1	48.3	58.6	68.0	71.4	68.6	64.0	51.5	38.6	34.6	50.2
1978	25.1	27.7	40.9	48.8	52.5	66.2	72.8	67.5	62.8	51.9	36.5	21.9	47.9
1979	16.9	32.5	38.1	48.3	54.0	64.3	70.6	67.5	64.4	51.7	31.2	33.5	47.8
1980	26.7	34.3	35.7	44.3	53.4	69.2	75.3	70.4	62.3	49.9	39.5	39.8	50.1
1981	34.9	34.4	39.3	53.8	54.5	69.4	71.9	67.3	63.4	50.8	43.3	32.7	51.3
1982	29.4	29.0	38.1	45.7	52.7	60.1	70.2	68.5	59.0	47.6	35.4	29.8	47.1
1983	32.5	34.4	35.7	40.1	50.5	61.1	72.3	71.9	63.8	51.3	37.7	18.4	47.5
1984	26.1	33.3	35.3	41.4	58.2	65.1	71.5	68.4	59.5	42.8	38.4	33.1	47.8
1985	25.1	26.3	37.9	48.7	57.2	65.0	70.3	69.8	58.0	49.0	32.1	27.9	47.3
1986	38.2	34.7	44.3	48.5	54.6	65.8	70.4	67.6	59.1	48.0	37.7	29.8	49.9
1987	29.4	33.0	35.3	48.5	56.0	65.1	70.7	66.1	60.0	50.4	39.2	29.0	48.6
1988	24.3	31.8	36.2	48.2	56.4	68.7	70.6	70.4	60.8	53.0	39.2	29.2	49.1
1989	32.8	21.8	43.7	49.0	57.6	62.1	71.8	68.4	61.0	49.7	41.5	27.3	48.9
1990	33.6	31.7	38.9	47.1	53.8	69.5	68.0	68.0	64.5	49.5	42.7	24.3	49.3
1991	28.0	38.0	39.9	45.8	58.0	67.0	69.4	68.1	60.5	50.3	33.0	31.0	49.1
1992	32.6	37.6	41.6	52.1	57.8	62.5	68.4	66.6	62.9	52.4	31.9	29.3	49.6
1993	26.7	29.4	39.8	46.1	55.6	65.0	70.7	67.3	58.1	48.0	32.3	32.3	47.6
POR= 45 YRS	29.0	32.2	37.1	46.3	55.5	65.3	70.8	68.7	60.9	50.3	37.6	30.7	48.7

WBAN : 93037

REFERENCE NOTES:

THE TEMPERATURE GRAPH SHOWS NORMAL MAXIMUM AND NORMAL MINIMUM DAILY TEMPERATURES (SOLID CURVES) AND THE ACTUAL DAILY HIGH AND LOW TEMPERATURES (VERTICAL BARS).

H/C INDICATES HEATING AND COOLING DEGREE DAYS.
RH INDICATES RELATIVE HUMIDITY
W/O INDICATES WEATHER AND OBSTRUCTIONS
S INDICATES SUNSHINE.
PR INDICATES PRESSURE.
CLOUDINESS ON PAGE 3 IS THE SUM OF THE CEILOMETER AND SATELLITE DATA NOT TO EXCEED TEN TENTHS.
MAXIMUM SNOW DEPTH IS THE DEPTH MEASURED AT 12 UTC.

GENERAL:
T INDICATES TRACE PRECIPITATION, AN AMOUNT GREATER THAN ZERO BUT LESS THAN THE LOWEST REPORTABLE VALUE.
+ INDICATES THE VALUE ALSO OCCURS ON EARLIER DATES.
BLANK ENTRIES DENOTE MISSING OR UNREPORTED DATA.
NORMALS ARE 30-YEAR AVERAGES (1961 - 1990).
ASOS INDICATES AUTOMATED SURFACE OBSERVING SYSTEM.
PM INDICATES THE LAST DAY OF THE PREVIOUS MONTH.
POR (PERIOD OF RECORD) INDICATES THE NUMBER OF YEARS USED TO COMPUTE THE MEAN. INDIVIDUAL MONTHS WITHIN THE POR MAY BE MISSING.
WHEN THE POR FOR A NORMAL IS LESS THAN 30 YEARS, THE NORMAL IS PROVISIONAL AND IS BASED ON THE NUMBER OF YEARS INDICATED.
0.* OR * INDICATES THE VALUE OR MEAN DAYS WITH IS BETWEEN 0.00 AND 0.05.
CLOUDINESS FOR ASOS STATIONS DIFFERS FROM THE NON-ASOS OBSERVATION TAKEN BY A HUMAN OBSERVER. ASOS STATION CLOUDINESS IS BASED ON TIME-AVERAGED CEILOMETER DATA FOR CLOUDS AT OR BELOW 12,000 FEET AND ON SATELLITE DATA FOR CLOUDS ABOVE 12,000 FEET.
THE NUMBER OF DAYS WITH CLEAR, PARTLY CLOUDY, AND CLOUDY CONDITIONS FOR ASOS STATIONS IS THE SUM OF THE CEILOMETER AND SATELLITE DATA FOR THE SUNRISE TO SUNSET PERIOD.

GENERAL CONTINUED:
CLEAR INDICATES 0 - 3 TENTHS, PARTLY CLOUDY INDICATES 4 - 7 TENTHS, AND CLOUDY INDICATES 8 OR MORE TENTHS. WHEN AT LEAST ONE OF THE ELEMENTS (CEILOMETER OR SATELLITE) IS MISSING, THE DAILY CLOUDINESS IS NOT COMPUTED.
WIND DIRECTION IS RECORDED IN TENS OF DEGREES (2 DIGITS) CLOCKWISE FROM TRUE NORTH. "00" INDICATES CALM. "36" INDICATES TRUE NORTH.
RESULTANT WIND IS THE VECTOR AVERAGE OF THE SPEED AND DIRECTION.
AVERAGE TEMPERATURE IS THE SUM OF THE MEAN DAILY MAXIMUM AND MINIMUM TEMPERATURE DIVIDED BY 2.
SNOWFALL DATA COMPRISE ALL FORMS OF FROZEN PRECIPITATION, INCLUDING HAIL.
A HEATING (COOLING) DEGREE DAY IS THE DIFFERENCE BETWEEN THE AVERAGE DAILY TEMPERATURE AND 65° F.
DRY BULB IS THE TEMPERATURE OF THE AMBIENT AIR.
DEW POINT IS THE TEMPERATURE TO WHICH THE AIR MUST BE COOLED TO ACHIEVE 100 PERCENT RELATIVE HUMIDITY.
WET BULB IS THE TEMPERATURE THE AIR WOULD HAVE IF THE MOISTURE CONTENT WAS INCREASED TO 100 PERCENT RELATIVE HUMIDITY.

HEATING DEGREE DAYS (base 65°F) 1993 COLORADO SPRINGS, COLORADO (COS)

YEAR	JUL	AUG	SEP	OCT	NOV	DEC	JAN	FEB	MAR	APR	MAY	JUN	TOTAL
1964-65	0	22	151	385	790	992	910	1034	1214	497	306	107	6408
1965-66	10	31	304	366	632	931	1232	1061	744	613	249	63	6236
1966-67	0	33	111	489	728	1099	981	889	679	496	401	95	6001
1967-68	6	52	138	397	758	1223	1057	932	812	669	411	49	6504
1968-69	22	52	149	409	906	1096	969	868	1090	477	260	180	6478
1969-70	1	8	104	720	833	1053	1135	772	998	664	218	95	6601
1970-71	2	3	235	637	797	987	1086	980	837	584	357	32	6537
1971-72	33	5	285	481	806	1067	1080	832	665	469	273	6	6002
1972-73	41	34	136	476	1049	1281	1221	912	877	715	359	76	7177
1973-74	32	1	194	378	754	1041	1172	866	700	566	176	88	5968
1974-75	1	17	229	376	789	1143	1102	980	904	608	350	88	6587
1975-76	0	10	200	391	852	916	1075	782	891	512	314	80	6023
1976-77	0	11	191	593	859	988	1181	837	858	494	192	5	6209
1977-78	2	22	73	413	784	938	1231	1036	741	479	386	98	6203
1978-79	3	44	119	400	848	1329	1484	906	825	494	336	97	6885
1979-80	6	41	88	407	1005	969	1180	883	901	615	351	32	6478
1980-81	0	7	113	463	759	776	928	850	789	335	321	38	5379
1981-82	5	30	70	433	643	993	1095	1001	827	571	374	163	6205
1982-83	8	11	198	532	880	1084	1001	851	904	742	444	159	6814
1983-84	2	0	101	417	811	1438	1198	911	912	700	220	58	6768
1984-85	0	6	200	684	790	982	1233	1077	830	481	242	77	6602
1985-86	5	8	253	487	978	1142	822	840	635	487	315	49	6021
1986-87	4	14	174	519	813	1081	1096	888	912	491	272	50	6314
1987-88	17	74	150	445	767	1108	1256	958	886	499	273	25	6458
1988-89	7	8	154	366	767	1099	989	1207	655	475	247	134	6108
1989-90	0	4	172	473	699	1164	966	928	805	526	345	24	6106
1990-91	28	21	83	473	663	1258	1142	750	773	568	219	33	6011
1991-92	16	16	145	453	954	1048	998	788	717	383	219	96	5833
1992-93	21	53	91	383	990	1101	1179	991	776	558	286	84	6513
1993-	0	40	212	519	972	1008							
POR= 46 YRS	8	18	151	447	816	1054	1106	918	860	556	295	74	6303

WBAN : 93037

COOLING DEGREE DAYS (base 65°F) 1993 COLORADO SPRINGS, COLORADO (COS)

YEAR	JAN	FEB	MAR	APR	MAY	JUN	JUL	AUG	SEP	OCT	NOV	DEC	ANNUAL
1969	0	0	0	0	11	27	216	177	18	0	0	0	449
1970	0	0	0	0	15	89	200	209	22	0	0	0	535
1971	0	0	0	0	1	126	158	138	41	0	0	0	464
1972	0	0	0	2	0	96	168	124	17	0	0	0	407
1973	0	0	0	0	0	104	145	180	13	0	0	0	442
1974	0	0	0	0	18	130	241	109	26	0	0	0	524
1975	0	0	0	0	0	59	195	180	41	5	0	0	480
1976	0	0	0	0	0	66	227	114	28	0	0	0	435
1977	0	0	0	0	0	103	204	142	49	0	0	0	498
1978	0	0	0	0	4	143	255	127	59	1	0	0	589
1979	0	0	0	0	1	84	185	124	77	2	0	0	473
1980	0	0	0	0	0	169	327	180	41	0	0	0	717
1981	0	0	0	4	2	176	226	105	27	0	0	0	540
1982	0	0	0	0	0	23	176	127	26	0	0	0	352
1983	0	0	0	0	1	48	236	219	71	0	0	0	575
1984	0	0	0	0	17	68	207	119	42	0	0	0	453
1985	0	0	0	0	5	83	179	163	51	0	0	0	481
1986	0	0	0	0	1	82	180	102	3	0	0	0	368
1987	0	0	0	0	0	62	199	113	6	0	0	0	380
1988	0	0	0	0	12	143	190	181	33	0	0	0	559
1989	0	0	0	3	25	54	220	117	57	3	0	0	479
1990	0	0	0	0	6	168	128	121	73	0	0	0	496
1991	0	0	0	0	8	101	161	120	15	4	0	0	409
1992	0	0	0	4	3	28	131	106	32	0	0	0	304
1993	0	0	0	0	2	89	183	117	11	1	0	0	403
POR= 25 YRS	0	0	0	0	4	92	197	140	35	0	0	0	468

WBAN : 93037

SNOWFALL (inches) 1993 COLORADO SPRINGS, COLORADO (COS)

YEAR	JUL	AUG	SEP	OCT	NOV	DEC	JAN	FEB	MAR	APR	MAY	JUN	TOTAL
1964-65	0.0	0.0	0.0	0.0	4.9	4.9	3.2	9.0	14.4	2.2	0.0	0.0	38.6
1965-66	0.0	0.0	0.1	T	T	6.1	6.0	9.7	0.1	4.1	4.2	0.0	30.3
1966-67	0.0	0.0	0.0	3.4	1.7	2.9	4.1	3.6	2.1	13.9	T	0.0	31.7
1967-68	0.0	0.0	0.0	T	3.1	11.7	3.0	4.8	5.7	2.8	0.2	0.0	31.3
1968-69	0.0	0.0	0.0	0.2	12.3	4.9	5.7	2.0	11.3	6.6	0.0	T	43.0
1969-70	0.0	0.0	0.0	21.7	5.6	7.3	0.8	2.4	19.9	7.0	T	T	64.7
1970-71	0.0	0.0	0.8	4.6	1.5	T	8.7	11.5	3.0	2.8	T	0.0	32.9
1971-72	0.0	0.0	9.7	2.9	0.3	3.8	5.2	3.6	7.0	3.3	T	0.0	35.8
1972-73	0.0	0.0	0.0	14.4	16.4	11.4	3.9	2.3	15.6	11.2	0.8	0.0	76.0
1973-74	0.0	0.0	0.0	3.9	1.8	9.1	2.4	2.4	6.8	10.0	0.0	T	36.4
1974-75	0.0	0.0	T	T	1.5	6.3	2.8	4.8	3.5	6.9	0.2	1.1	27.1
1975-76	0.0	0.0	0.0	4.3	9.8	0.9	6.9	5.2	8.6	10.1	0.0	0.0	45.8
1976-77	0.0	0.0	T	2.5	4.9	2.6	4.8	2.5	13.8	4.3	0.0	0.0	35.4
1977-78	0.0	0.0	0.0	0.9	1.9	3.0	4.2	8.6	3.3	1.1	19.4	0.0	42.4
1978-79	0.0	0.0	T	0.5	4.0	15.2	9.9	1.2	20.0	14.6	4.1	0.0	69.5
1979-80	0.0	0.0	0.0	1.3	19.1	17.6	4.7	5.9	12.7	11.3	0.0	0.0	72.6
1980-81	0.0	0.0	0.0	0.2	4.4	1.4	1.0	1.7	9.0	0.3	0.2	0.0	18.2
1981-82	0.0	0.0	0.0	0.4	0.5	9.1	3.6	6.2	8.4	2.3	3.9	0.0	34.4
1982-83	0.0	0.0	0.0	0.2	0.9	8.2	4.0	1.1	16.3	4.8	0.8	0.0	36.3
1983-84	0.0	0.0	0.0	0.0	10.3	18.2	7.8	1.4	23.2	9.0	0.8	0.0	70.7
1984-85	0.0	0.0	0.9	25.9	2.0	10.9	8.0	4.7	22.3	0.8	T	0.0	75.5
1985-86	0.0	0.0	1.9	1.7	8.3	6.3	0.2	4.6	2.9	4.0	T	0.0	29.9
1986-87	0.0	0.0	0.0	1.4	7.3	4.4	28.7	23.2	14.9	3.3	T	0.0	83.2
1987-88	0.0	0.0	0.0	0.0	4.8	9.5	4.9	11.5	12.6	1.0	0.3	0.0	44.6
1988-89	0.0	0.0	0.0	0.0	1.6	13.6	3.0	18.9	T	1.0	0.0	T	38.1
1989-90	T	T	T	2.1	0.2	7.5	8.7	9.3	9.7	11.3	4.2	0.0	53.0
1990-91	0.0	0.0	0.0	8.2	2.7	4.1	0.9	T	5.2	5.9	1.5	0.0	28.5
1991-92	0.0	0.0	0.0	7.5	26.3	4.5	1.2	0.2	3.0	T	0.0	T	42.7
1992-93	T	T	0.0	T	11.4	2.6	7.5	3.3	2.3	1.2	T	T	28.3
1993-	T	0.0	0.1	0.5	12.3	0.5							
POR= 46 YRS	T	T	1.0	3.1	5.6	5.5	5.0	5.1	9.3	6.2	1.5	T	42.3

WBAN : 93037

DENVER, COLORADO

Denver enjoys the invigorating climate that prevails over much of the central Rocky Mountain region, without the extremely cold mornings of the high elevations during winter, or the hot afternoons of summer at lower altitudes. Extremely warm or cold weather in Denver is usually of short duration.

Situated a long distance from any moisture source, and separated from the Pacific Ocean by several high mountain barriers, Denver enjoys low relative humidity, light precipitation, and abundant sunshine.

Air masses from four different sources influence Denver weather. These include arctic air from Canada and Alaska, warm, moist air from the Gulf of Mexico, warm, dry air from Mexico and the southwestern deserts, and Pacific air modified by its passage over mountains to the west.

In winter, the high altitude and mountains to the west combine to moderate temperatures in Denver. Invasions of cold air from the north, intensified by the high altitude, can be abrupt and severe. However, many of the cold air masses that spread southward out of Canada never reach the altitude of Denver, but move off over the lower plains to the east. Surges of air from the west are moderated in their descent down the east face of the Rockies, and reach Denver in the form of chinook winds that often raise temperatures into the 60s, even in midwinter.

In spring, polar air often collides with warm, moist air from the Gulf of Mexico and these collisions result in frequent, rapid and drastic weather changes. Spring is the cloudiest, windiest, and wettest season in the city. Much of the precipitation falls as snow, especially in March and early April. Stormy periods are interspersed with stretches of mild, sunny weather that quickly melt previous snow cover.

Summer precipitation falls mainly from scattered thunderstorms during the afternoon and evening. Mornings are usually clear and sunny, with clouds forming during early afternoon to cut off the sunshine at what would otherwise be the hottest part of the day. Severe thunderstorms, with large hail and heavy rain occasionally occur in the city, but these conditions are more common on the plains to the east.

Autumn is the most pleasant season. Few thunderstorms occur and invasions of cold air are infrequent. As a result, there is more sunshine and less severe weather than at any other time of the year.

Based on the 1951–1980 period, the average first occurrence of 32 degrees Fahrenheit in the fall is October 8 and the average last occurrence in the spring is May 3.

NORMALS, MEANS, AND EXTREMES

DENVER, COLORADO

LATITUDE: 39°45'N LONGITUDE: 104°52' W ELEVATION: FT. GRND 5282 BARO 5287 TIME ZONE: MOUNTAIN WBAN: 23062

	(a)	JAN	FEB	MAR	APR	MAY	JUNE	JULY	AUG	SEP	OCT	NOV	DEC	YEAR
TEMPERATURE °F:														
Normals														
-Daily Maximum		43.2	46.6	52.2	61.8	70.8	81.4	88.2	85.8	76.9	66.3	52.5	44.5	64.2
-Daily Minimum		16.1	20.2	25.8	34.5	43.6	52.4	58.6	56.9	47.6	36.4	25.4	17.4	36.2
-Monthly		29.7	33.4	39.0	48.2	57.2	66.9	73.5	71.4	62.3	51.4	39.0	31.0	50.3
Extremes														
-Record Highest	59	73	76	84	90	96	104	104	101	97	89	79	75	104
-Year		1982	1963	1971	1992	1942	1936	1939	1938	1960	1991	1990	1980	JUL 1939
-Record Lowest	59	-25	-30	-11	-2	22	30	43	41	17	3	-8	-25	-30
-Year		1963	1936	1943	1975	1954	1951	1972	1964	1985	1969	1993	1990	FEB 1936
NORMAL DEGREE DAYS:														
Heating (base 65°F)		1094	885	806	504	253	71	0	0	144	429	780	1054	6020
Cooling (base 65°F)		0	0	0	0	11	128	267	203	63	7	0	0	679
% OF POSSIBLE SUNSHINE	44	71	70	69	67	64	70	71	71	74	72	64	67	69
MEAN SKY COVER (tenths)														
Sunrise - Sunset	45	5.5	5.9	6.2	6.1	6.2	5.1	5.0	5.0	4.4	4.5	5.5	5.3	5.4
MEAN NUMBER OF DAYS:														
Sunrise to Sunset														
-Clear	59	10.3	8.0	7.8	6.7	6.1	9.6	9.2	9.7	13.4	13.3	10.1	10.9	115.2
-Partly Cloudy	59	9.3	8.7	10.1	10.7	12.2	12.2	15.6	13.7	9.0	9.1	9.6	9.6	129.7
-Cloudy	59	11.5	11.5	13.2	12.6	12.7	8.2	6.2	7.6	7.6	8.6	10.3	10.5	120.3
Precipitation .01 inches or more	59	5.8	5.9	8.6	8.5	10.8	8.7	9.3	8.8	6.3	5.3	5.6	5.3	89.1
Snow, Ice pellets, hail 1.0 inches or more	59	2.4	2.4	3.6	2.5	0.4	0.0	0.0	0.0	0.3	1.2	2.6	2.5	17.9
Thunderstorms	59	0.*	0.1	0.3	1.5	6.4	9.9	11.1	8.3	3.5	0.8	0.1	0.0	41.9
Heavy Fog Visibility 1/4 mile or less	53	1.1	1.6	1.1	0.8	0.5	0.4	0.4	0.6	0.6	0.6	1.2	1.1	9.9
Temperature °F														
-Maximum														
90° and above	33	0.0	0.0	0.0	0.*	0.3	6.3	15.0	9.3	2.2	0.0	0.0	0.0	33.1
32° and below	33	6.5	4.3	2.8	0.4	0.0	0.0	0.0	0.0	0.*	0.4	2.5	5.3	22.2
-Minimum														
32° and below	33	29.8	26.0	24.3	11.4	1.5	0.0	0.0	0.0	0.8	8.6	24.4	29.2	156.1
0° and below	33	3.8	1.7	0.5	0.*	0.0	0.0	0.0	0.0	0.0	0.0	0.2	2.8	9.0
AVG. STATION PRESS. (mb)	21	834.7	834.7	832.3	833.5	834.4	836.4	838.8	839.1	838.5	837.8	835.3	834.9	835.9
RELATIVE HUMIDITY (%)														
Hour 05	33	64	67	68	67	71	69	69	70	69	65	68	65	68
Hour 11 (Local Time)	33	45	44	42	38	39	37	35	37	38	36	44	45	40
Hour 17	33	49	44	40	35	38	35	34	35	34	35	49	52	40
Hour 23	33	63	65	62	58	61	59	57	59	59	59	65	64	61
PRECIPITATION (inches):														
Water Equivalent														
-Normal		0.50	0.57	1.28	1.71	2.40	1.79	1.91	1.51	1.24	0.98	0.87	0.64	15.40
-Maximum Monthly	59	1.44	1.66	4.56	4.17	7.31	4.69	6.41	5.85	4.67	4.17	2.97	2.84	7.31
-Year		1948	1960	1983	1942	1957	1967	1965	1979	1961	1969	1946	1973	MAY 1957
-Minimum Monthly	59	0.01	0.01	0.13	0.03	0.06	0.09	0.17	0.06	T	0.05	0.01	0.03	T
-Year		1952	1970	1945	1963	1974	1980	1939	1960	1944	1962	1949	1977	SEP 1944
-Maximum in 24 hrs	59	1.02	1.01	2.79	3.25	3.55	3.16	2.42	3.43	2.44	1.71	1.29	2.00	3.55
-Year		1962	1953	1983	1967	1973	1970	1965	1951	1936	1947	1975	1982	MAY 1973
Snow, Ice pellets, hail														
-Maximum Monthly	59	24.3	18.3	30.5	28.3	13.6	0.3	T	T	21.3	31.2	39.1	30.8	39.1
-Year		1992	1960	1983	1935	1950	1951	1992	1991	1936	1969	1946	1973	NOV 1946
-Maximum in 24 hrs	59	14.5	9.5	18.0	17.3	10.7	0.3	T	T	19.4	12.4	15.9	23.6	23.6
-Year		1992	1953	1983	1957	1950	1951	1992	1991	1936	1969	1983	1982	DEC 1982
WIND:														
Mean Speed (mph)	45	8.6	8.8	9.7	10.0	9.3	8.9	8.3	8.0	7.9	7.8	8.2	8.4	8.6
Prevailing Direction through 1963		S	S	S	S	S	S	S	S	S	S	S	S	S
Fastest Obs. 1 Min.														
-Direction (!!!)	12	32	30	30	33	36	21	29	33	29	01	36	32	33
-Speed (MPH)	12	44	36	41	46	43	38	46	33	36	36	36	38	46
-Year		1982	1989	1991	1993	1983	1987	1993	1989	1988	1992	1987	1981	APR 1993
Peak Gust														
-Direction (!!!)	10	W	NW	W	NW	SE	NW	W	NW	W	NW	W	W	NW
-Speed (mph)	10	55	52	59	62	60	60	55	52	56	48	49	51	62
-Date		1993	1990	1989	1986	1991	1988	1993	1989	1984	1990	1990	1990	APR 1986

PRECIPITATION (inches)

DENVER, COLORADO

YEAR	JAN	FEB	MAR	APR	MAY	JUNE	JULY	AUG	SEP	OCT	NOV	DEC	ANNUAL
1964	0.26	1.04	1.38	1.25	2.53	0.82	0.72	0.27	0.41	0.18	0.88	0.40	10.14
1965	1.00	1.27	1.20	1.05	1.82	4.14	6.41	1.06	2.58	0.45	0.36	0.53	21.87
1966	0.30	1.28	0.32	1.46	0.34	1.41	1.04	2.06	1.15	0.96	0.32	0.17	10.81
1967	0.84	0.39	0.79	3.95	4.77	4.69	3.25	0.83	0.60	1.13	1.01	1.06	23.31
1968	0.51	0.74	0.85	2.39	0.71	0.50	1.34	2.53	0.59	0.75	0.71	0.51	12.13
1969	0.17	0.43	1.10	1.33	6.12	2.99	1.81	0.79	1.67	4.17	0.62	0.32	21.52
1970	0.10	0.01	1.34	0.97	0.64	3.83	1.67	0.54	2.47	0.88	1.19	0.09	13.73
1971	0.35	0.78	0.53	1.98	1.34	0.23	1.20	0.85	2.85	0.44	0.16	0.25	10.96
1972	0.36	0.44	0.50	3.52	0.49	2.94	0.63	2.71	2.07	0.82	1.69	0.70	16.87
1973	1.31	0.16	1.76	3.73	5.06	0.20	2.47	1.28	2.85	0.47	0.83	2.84	22.96
1974	1.03	0.82	1.32	2.28	0.06	2.01	2.34	0.16	0.98	1.68	1.06	0.29	14.03
1975	0.23	0.37	1.19	1.14	2.80	2.11	2.78	2.00	0.24	0.30	1.88	0.47	15.51
1976	0.19	0.54	1.34	1.27	1.34	0.63	2.31	2.50	1.88	0.93	0.32	0.16	13.41
1977	0.16	0.27	1.24	2.13	0.34	1.02	2.98	1.00	0.10	0.48	0.59	0.03	10.34
1978	0.27	0.27	1.07	1.82	3.46	1.17	0.54	0.26	0.07	1.45	0.50	0.82	11.70
1979	0.34	0.42	1.25	1.41	3.53	2.39	0.81	5.85	0.36	1.28	1.66	1.06	20.36
1980	0.64	0.45	1.15	2.54	2.73	0.09	2.93	1.65	0.63	0.10	0.66	0.10	13.67
1981	0.29	0.35	2.27	1.01	3.76	0.63	0.90	1.16	0.35	0.79	0.42	0.66	12.59
1982	0.32	0.09	0.18	0.34	3.48	2.26	0.92	1.16	1.38	1.51	0.47	2.34	14.45
1983	0.15	0.07	4.56	2.10	3.62	2.65	1.75	1.51	0.13	0.39	2.63	0.63	20.19
1984	0.18	0.81	1.19	2.42	0.65	1.26	2.11	3.20	0.47	3.47	0.27	0.46	16.49
1985	0.68	0.59	0.69	2.61	1.33	1.46	3.71	0.28	2.33	0.77	1.20	0.66	16.31
1986	0.22	0.65	0.43	2.59	1.30	1.07	1.69	0.53	0.43	1.80	1.07	0.31	12.09
1987	0.69	1.21	1.34	1.03	4.64	3.50	0.76	2.00	0.70	1.24	1.62	1.30	20.03
1988	0.40	0.60	1.28	0.65	4.26	1.28	2.19	1.83	0.90	0.06	0.47	1.04	14.96
1989	1.14	0.66	0.56	1.00	3.83	2.04	1.64	1.28	1.55	0.81	0.15	0.81	15.47
1990	0.74	0.55	3.10	1.01	1.51	0.21	3.57	1.96	1.46	1.03	1.28	0.27	16.69
1991	0.76	0.08	0.76	1.94	2.43	2.20	4.11	3.69	0.79	0.70	2.67	0.19	20.32
1992	1.19	0.09	3.50	0.53	1.13	2.02	2.23	2.33	0.01	0.51	1.46	0.68	15.68
1993	0.25	1.05	0.89	2.08	0.93	1.67	0.91	0.64	2.29	2.27	1.38	0.42	14.78
Record Mean	0.48	0.57	1.16	1.95	2.41	1.51	1.74	1.45	1.09	1.02	0.72	0.63	14.72

AVERAGE TEMPERATURE (deg. F)

DENVER, COLORADO

YEAR	JAN	FEB	MAR	APR	MAY	JUNE	JULY	AUG	SEP	OCT	NOV	DEC	ANNUAL
1964	30.6	27.4	33.0	46.6	58.8	65.0	75.8	70.4	62.5	52.7	40.0	33.2	49.7
1965	35.0	27.4	29.0	51.2	57.1	63.9	72.7	70.2	55.7	55.1	43.3	35.0	49.6
1966	28.6	28.4	42.5	44.6	58.7	64.6	76.9	70.8	65.0	52.2	41.5	31.9	50.5
1967	34.0	35.1	42.9	48.2	52.6	60.6	69.1	68.2	62.1	52.5	40.5	26.5	49.4
1968	29.7	34.2	40.6	43.0	53.9	67.8	71.7	68.1	60.9	51.9	35.7	28.9	48.9
1969	35.0	35.4	32.2	52.2	59.3	61.5	74.7	73.9	64.5	39.0	39.1	32.5	49.9
1970	30.6	38.6	33.5	43.7	58.8	65.2	72.0	73.9	59.5	45.9	39.1	33.3	49.5
1971	32.1	30.6	38.5	47.8	54.2	69.0	70.6	72.8	57.5	49.4	39.1	31.9	49.5
1972	30.5	36.2	44.8	48.5	57.0	68.3	70.2	71.0	62.1	52.1	32.9	24.9	49.9
1973	27.3	35.5	39.9	43.2	55.6	67.5	71.0	73.5	59.9	54.5	39.5	31.6	49.9
1974	23.7	35.2	43.2	47.9	61.6	68.4	74.7	69.5	59.4	52.4	38.0	31.2	50.5
1975	31.7	30.6	37.3	44.1	54.3	64.3	72.7	70.8	59.5	53.2	36.8	37.5	49.4
1976	32.3	39.3	37.1	49.2	56.7	66.3	75.3	70.2	61.8	48.4	39.5	35.5	51.0
1977	29.2	38.0	39.9	51.1	60.7	71.9	74.3	70.2	66.6	53.3	40.3	35.1	52.5
1978	25.8	31.4	43.3	50.3	54.4	66.9	74.7	69.6	65.0	53.1	37.8	24.6	49.7
1979	18.0	34.2	40.5	49.1	54.8	65.8	73.7	69.5	66.3	53.8	33.3	34.5	49.5
1980	26.0	34.5	38.0	47.7	57.1	71.9	76.4	73.2	65.8	52.4	41.9	41.2	52.2
1981	37.3	36.2	41.2	56.4	57.1	70.4	75.9	72.0	68.2	52.6	45.9	35.8	54.1
1982	30.3	32.0	41.1	47.4	55.1	63.1	72.7	73.1	61.7	49.0	35.7	30.9	49.3
1983	31.9	36.6	36.2	41.0	51.4	62.8	73.3	74.4	64.9	52.7	37.0	17.5	48.3
1984	27.3	34.1	37.2	42.3	60.0	66.5	74.9	71.8	60.7	44.8	39.7	32.8	49.3
1985	25.6	27.7	40.8	51.0	60.0	68.0	73.0	72.4	58.8	50.7	29.8	29.4	48.9
1986	40.3	36.1	47.1	49.6	56.7	70.3	73.5	72.2	60.7	49.3	39.0	31.0	52.2
1987	32.2	36.1	38.8	51.9	59.7	69.2	74.4	70.7	62.4	51.7	40.0	28.4	51.3
1988	25.2	34.1	38.5	50.3	59.0	71.9	74.2	73.6	62.3	54.0	40.6	31.1	51.2
1989	33.5	22.4	43.3	51.1	59.0	65.4	75.9	71.7	62.5	51.3	42.8	27.3	50.5
1990	36.4	33.3	39.5	49.1	56.6	72.6	70.8	71.3	66.9	52.3	44.0	25.7	51.5
1991	27.9	40.3	42.8	47.8	60.3	69.4	73.1	71.5	62.5	50.3	34.7	33.1	51.1
1992	31.8	40.1	43.0	54.8	60.6	66.1	70.5	68.4	65.2	53.7	33.9	25.4	51.1
1993	27.3	29.4	42.6	48.5	58.8	66.4	73.0	70.0	60.6	49.2	34.7	34.2	49.6
Record Mean	30.1	33.0	38.9	47.7	56.8	66.8	72.8	71.3	62.8	51.5	39.4	32.1	50.3
Max	42.8	45.4	51.5	60.5	69.6	80.7	86.7	85.0	77.0	65.5	52.4	44.8	63.5
Min	17.4	20.5	26.3	34.9	44.1	52.9	58.9	57.6	48.5	37.5	26.4	19.4	37.0

REFERENCE NOTES

DENVER, COLORADO

GENERAL
T - TRACE AMOUNT.
BLANK ENTRIES DENOTE MISSING/UNREPORTED DATA.
# INDICATES A STATION OR INSTRUMENT RELOCATION.

SPECIFIC

PM - INCLUDES LAST DAY OF PREVIOUS MONTH
ASOS - AUTOMATED SURFACE OBSERVING SYSTEM IN OPERATION DURING THESE MONTHS.

(a) - LENGTH OF RECORD IN YEARS, ALTHOUGH INDIVIDUAL MONTHS MAY BE MISSING.
0.* OR * - THE VALUE IS BETWEEN 0.0 AND 0.05
NORMALS - BASED ON THE 1961-1990 RECORD PERIOD.
EXTREMES - DATES ARE THE MOST RECENT OCCURRENCE.
WIND DIR.- NUMERALS SHOW TENS OF DEGRESS CLOCKWISE FROM TRUE NORTH. "00" INDICATES CALM.
RESULTANT DIRECTIONS ARE GIVEN TO WHOLE DEGREES.
BOLD VALUES INDICATE EXTREME VALUES WHICH OCCURRED AFTER THE ASOS SYSTEM WAS COMMISSIONED.

RECORD = PERIOD OF RECORD
RECORD MEAN PRECIPITATION IS THE MEAN OF ALL DAILY PRECIPITATION AMOUNTS DURING THE PERIOD OF RECORD.
RECORD MAX(MIN) TEMPERATURE IS THE MEAN OF ALL DAILY MAX(MIN) TEMPERATURES DURING THE PERIOD OF RECORD.
RECORD MEAN TEMPERATURE IS THE SUM OF THE RECORD MAX AND RECORD MIN DIVIDED BY 2.
AVERAGE TEMPERATURE IS THE SUM OF THE MEAN DAILY MAX AND MIN TEMPERATURE DIVIDED BY 2.

EXCEPTIONS

1. FASTEST MILE WINDS ARE THROUGH AUGUST 1981.

RECORD MEANS ARE THROUGH THE CURRENT YEAR, BEGINNING IN 1872 FOR TEMPERATURE
1872 FOR PRECIPITATION
1935 FOR SNOWFALL

HEATING DEGREE DAYS Base 65 deg. F DENVER, COLORADO

SEASON	JULY	AUG	SEP	OCT	NOV	DEC	JAN	FEB	MAR	APR	MAY	JUNE	TOTAL
1964-65	0	16	123	375	743	981	921	1044	1108	411	245	63	6030
1965-66	6	7	296	302	645	924	1122	1017	691	604	204	82	5900
1966-67	0	9	61	391	699	1018	954	832	679	498	388	135	5664
1967-68	4	16	108	389	729	1186	1086	885	751	655	343	38	6190
1968-69	10	35	145	399	871	1114	925	821	1011	378	204	144	6057
1969-70	2	0	56	801	769	998	1061	734	969	632	200	78	6300
1970-71	0	0	198	584	770	977	1018	958	817	508	329	25	6184
1971-72	24	0	273	479	771	1019	1063	832	621	486	246	4	5818
1972-73	42	15	107	397	960	1239	1162	820	771	646	290	56	6505
1973-74	8	0	166	321	758	1029	1277	831	671	507	137	67	5772
1974-75	0	9	199	381	803	1043	1024	957	852	621	332	85	6306
1975-76	0	4	195	363	840	843	1006	740	859	469	254	64	5637
1976-77	0	7	142	509	759	907	1105	749	771	414	137	0	5500
1977-78	2	14	38	358	737	920	1206	936	665	435	335	87	5733
1978-79	0	20	96	366	811	1245	1450	854	751	473	313	81	6460
1979-80	0	20	58	347	941	939	1204	876	828	514	247	9	5983
1980-81	0	4	56	386	683	731	853	801	727	260	243	26	4770
1981-82	0	12	19	375	570	898	1071	918	733	522	306	92	5516
1982-83	3	0	151	487	875	1050	1017	789	885	712	419	129	6517
1983-84	3	0	87	372	833	1469	1163	889	854	673	183	51	6577
1984-85	0	1	183	622	753	990	1215	1041	742	412	167	42	6168
1985-86	0	1	241	435	1051	1094	758	802	548	456	260	22	5668
1986-87	0	0	145	477	775	1045	1012	803	805	392	170	22	5646
1987-88	11	21	110	410	743	1125	1227	889	811	437	215	14	6013
1988-89	7	0	129	333	723	1043	969	1193	665	432	213	76	5783
1989-90	0	0	153	424	658	1162	879	882	781	469	265	7	5680
1990-91	12	3	64	388	623	1211	1143	684	682	510	174	16	5510
1991-92	6	4	118	449	902	982	1022	714	673	309	158	35	5372
1992-93	10	35	58	346	926	1219	1162	992	686	489	195	71	6189
1993-94	1	20	152	488	900	948							

COOLING DEGREE DAYS Base 65 deg. F DENVER, COLORADO

YEAR	JAN	FEB	MAR	APR	MAY	JUNE	JULY	AUG	SEP	OCT	NOV	DEC	TOTAL
1969	0	0	0	0	35	44	312	284	46	0	0	0	721
1970	0	0	0	0	16	93	222	282	40	0	0	0	653
1971	0	0	0	0	0	149	203	248	53	0	0	0	653
1972	0	0	0	0	6	110	210	207	28	1	0	0	562
1973	0	0	0	0	2	138	199	270	21	1	0	0	631
1974	0	0	0	0	36	176	307	157	39	0	0	0	715
1975	0	0	0	0	3	69	246	192	39	5	0	0	554
1976	0	0	0	0	3	112	324	176	52	0	0	0	667
1977	0	0	0	2	11	214	297	182	93	0	0	0	799
1978	0	0	0	0	12	152	308	171	103	2	0	0	748
1979	0	0	0	0	2	112	275	163	102	7	0	0	661
1980	0	0	0	2	10	224	358	263	88	1	0	0	946
1981	0	0	0	7	6	195	346	236	121	1	0	0	912
1982	0	0	0	0	6	42	247	257	59	0	0	0	611
1983	0	0	0	0	7	69	264	301	91	0	0	0	732
1984	0	0	0	0	33	104	315	218	60	0	0	0	730
1985	0	0	0	1	19	137	256	238	63	0	0	0	714
1986	0	0	0	0	11	188	271	227	20	0	0	0	717
1987	0	0	0	3	12	153	309	205	36	2	0	0	720
1988	0	0	0	1	35	225	300	277	55	0	0	0	893
1989	0	0	0	19	34	96	345	214	83	5	0	0	796
1990	0	0	0	0	9	244	196	203	129	1	0	0	782
1991	0	0	0	1	34	156	267	211	50	1	0	0	720
1992	0	0	0	9	29	76	187	148	74	1	0	0	524
1993	0	0	0	0	10	121	256	185	30	5	0	0	607

SNOWFALL (inches) DENVER, COLORADO

SEASON	JULY	AUG	SEP	OCT	NOV	DEC	JAN	FEB	MAR	APR	MAY	JUNE	TOTAL
1964-65	0.0	0.0	0.0	T	6.0	4.4	13.2	17.1	14.9	0.3	T	0.0	55.9
1965-66	0.0	0.0	5.5	0.0	5.5	5.6	3.6	14.6	2.8	6.4	2.9	0.0	46.9
1966-67	0.0	0.0	T	8.3	3.0	1.9	9.9	4.4	6.6	3.6	3.0	0.0	40.7
1967-68	0.0	0.0	0.0	1.7	9.4	13.1	3.0	7.3	9.2	15.1	T	0.0	58.8
1968-69	0.0	0.0	0.0	0.4	5.8	6.9	2.8	4.2	13.2	T	0.0	0.0	33.3
1969-70	0.0	0.0	0.0	31.2	5.1	3.1	0.9	0.3	20.5	4.7	T	0.0	65.8
1970-71	0.0	0.0	4.6	5.9	9.2	0.9	8.6	11.9	9.6	6.0	T	0.0	56.7
1971-72	0.0	0.0	17.2	3.1	1.4	8.4	10.9	9.1	7.1	17.2	0.0	0.0	74.4
1972-73	0.0	0.0	0.0	9.7	19.4	9.8	12.1	3.0	15.1	24.8	1.0	0.0	94.9
1973-74	0.0	0.0	0.0	2.3	9.3	30.8	8.2	10.3	12.8	17.8	0.0	T	91.5
1974-75	0.0	0.0	1.8	1.0	11.9	2.1	3.6	4.0	14.3	10.9	6.1	0.0	55.7
1975-76	0.0	0.0	0.0	2.7	15.2	7.3	3.2	6.4	18.7	1.2	0.0	0.0	54.7
1976-77	0.0	0.0	0.0	7.2	4.5	3.1	2.4	3.1	9.6	4.7	0.0	0.0	34.6
1977-78	0.0	0.0	0.0	3.3	4.1	0.7	5.5	6.2	8.6	4.6	13.5	0.0	46.5
1978-79	0.0	0.0	T	2.7	6.9	14.2	9.1	5.8	18.2	8.1	8.2	0.0	73.2
1979-80	0.0	0.0	0.0	2.7	22.3	16.5	12.3	9.6	12.1	10.0	T	0.0	85.5
1980-81	0.0	0.0	0.0	1.5	7.1	1.2	4.1	4.3	24.0	2.9	T	0.0	45.1
1981-82	0.0	0.0	0.0	2.8	3.3	9.9	4.8	1.8	2.1	2.0	T	0.0	26.7
1982-83	0.0	0.0	0.0	1.2	1.8	27.1	1.3	0.8	30.5	11.3	7.6	0.0	81.6
1983-84	0.0	0.0	T	T	29.3	11.5	3.4	7.9	12.0	16.8	T	0.0	80.9
1984-85	0.0	0.0	5.2	13.1	2.3	5.0	12.5	8.7	7.6	0.8	T	0.0	55.2
1985-86	0.0	0.0	8.7	1.9	17.0	10.3	2.4	6.2	2.6	14.0	0.0	0.0	63.1
1986-87	0.0	0.0	0.0	4.3	11.5	4.9	17.0	12.2	11.5	9.9	0.0	0.0	71.3
1987-88	0.0	0.0	0.0	T	11.2	21.5	5.8	7.0	13.5	2.0	1.3	0.0	62.3
1988-89	0.0	0.0	0.0	0.0	2.8	12.3	13.0	8.2	4.8	9.0	T	T	50.1
1989-90	0.0	0.0	2.3	7.8	1.6	11.8	8.4	7.0	21.9	3.6	0.1	0.0	64.5
1990-91	T	T	0.0	7.6	12.0	4.7	14.1	0.8	4.2	14.3	T	T	57.7
1991-92	T	T	0.0	7.3	29.6	1.9	24.3	0.3	15.6	0.0	0.0	T	79.0
1992-93	T	0.0	0.0	0.1	20.1	11.0	3.7	16.9	4.9	3.7	0.0	T	60.4
1993-94	0.0	0.0	5.4	5.4	17.4	5.6							
Record Mean	T	T	1.6	3.8	8.9	7.4	8.2	7.4	12.6	8.9	1.6	T	60.4

GRAND JUNCTION, COLORADO

Grand Junction is located at the junction of the Colorado and Gunnison Rivers. It is on the west slope of the Rockies, in a large mountain valley. The area has a climate marked by the wide seasonal range usual to interior localities at this latitude. Thanks, however, to the protective topography of the vicinity, sudden and severe weather changes are very infrequent. The valley floor slopes from 4,800 feet near Palisade to 4,400 feet at the west end near Fruita. Mountains are on all sides at distances of from 10 to 60 miles and reach heights of 9,000 to over 12,000 feet.

This mountain valley location, with attendant valley breezes, provides protection from spring and fall frosts. This results in a growing season averaging 191 days in the city. This varies considerably in the outlying districts. It is about the same in the upper valley around Palisade, and 3 to 4 weeks shorter near the river west of Grand Junction. The growing season is sufficiently long to permit commercial growth of almost all fruits except citrus varieties. Summer grazing of cattle and sheep on nearby mountain ranges is extensive.

The interior, continental location, ringed by mountains on all sides, results in quite low precipitation in all seasons. Consequently, agriculture is dependent on irrigation. Adequate supplies of water are available from mountain snows and rains. Summer rains occur chiefly as scattered light showers and thunderstorms which develop over nearby mountains. Winter snows are fairly frequent, but are mostly light and quick to melt. Even the infrequent snows of from 4 to 8 inches seldom remain on the ground for prolonged periods. Blizzard conditions in the valley are extremely rare.

Temperatures above 100 degrees are infrequent, and about one-third of the winters have no readings below zero. Summer days with maximum temperatures in the middle 90s and minimums in the low 60s are common. Relative humidity is very low during the summer, with values similar to other dry locations such as the southern parts of New Mexico and Arizona. Spells of cold winter weather are sometimes prolonged due to cold air becoming trapped in the valley. Winds are usually very light during the coldest weather. Changes in winter are normally gradual, and abrupt changes are much less frequent than in eastern Colorado. Cold waves are rare. Sunny days predominate in all seasons.

The prevailing wind is from the east-southeast due to the valley breeze effect. The strongest winds are associated with thunderstorms or with pre-frontal weather. They usually are from the south or southwest.

NORMALS, MEANS, AND EXTREMES

GRAND JUNCTION, COLORADO

LATITUDE: 39°07'N LONGITUDE: 108°32' W ELEVATION: FT. GRND 4843 BARO 4836 TIME ZONE: MOUNTAIN WBAN: 23066

	(a)	JAN	FEB	MAR	APR	MAY	JUNE	JULY	AUG	SEP	OCT	NOV	DEC	YEAR
TEMPERATURE °F:														
Normals														
-Daily Maximum		35.5	45.4	55.6	65.8	76.0	87.7	93.6	90.5	81.1	67.7	51.4	38.7	65.8
-Daily Minimum		14.5	23.5	31.3	38.5	47.9	57.1	63.9	62.2	52.8	41.6	29.4	18.7	40.1
-Monthly		25.0	34.5	43.4	52.2	62.0	72.4	78.8	76.4	67.0	54.7	40.4	28.7	53.0
Extremes														
-Record Highest	47	60	68	81	89	95	105	105	103	98	88	75	64	105
-Year		1971	1986	1971	1992	1956	1990	1976	1969	1977	1963	1977	1980	JUN 1990
-Record Lowest	47	-23	-18	5	11	26	34	44	43	29	18	-2	-17	-23
-Year		1963	1989	1948	1975	1970	1976	1993	1968	1978	1993	1976	1990	JAN 1963
NORMAL DEGREE DAYS:														
Heating (base 65°F)		1240	854	670	389	132	13	0	0	55	332	738	1125	5548
Cooling (base 65°F)		0	0	0	0	39	235	428	353	115	13	0	0	1183
% OF POSSIBLE SUNSHINE	47	61	65	64	70	73	81	79	77	79	74	63	61	71
MEAN SKY COVER (tenths)														
Sunrise - Sunset	47	6.1	6.2	6.2	5.9	5.5	4.0	4.2	4.4	3.6	4.2	5.4	5.8	5.1
MEAN NUMBER OF DAYS:														
Sunrise to Sunset														
-Clear	47	9.3	7.6	7.9	8.1	9.7	15.0	13.8	13.4	16.4	15.0	10.7	9.7	136.6
-Partly Cloudy	47	7.0	7.4	8.5	9.4	10.9	9.4	11.6	11.4	8.2	7.9	7.4	7.9	106.9
-Cloudy	47	14.6	13.3	14.6	12.6	10.4	5.7	5.6	6.2	5.4	8.2	11.8	13.4	121.8
Precipitation														
.01 inches or more	47	7.0	6.0	7.8	6.5	6.5	4.1	5.2	6.6	5.7	5.6	5.7	6.2	72.8
Snow,Ice pellets,hail														
1.0 inches or more	47	2.6	1.2	1.3	0.3	0.*	0.0	0.0	0.0	0.*	0.2	0.9	2.1	8.7
Thunderstorms	47	0.1	0.3	0.8	2.1	4.6	4.9	7.6	7.9	5.1	1.5	0.4	0.1	35.3
Heavy Fog Visibility														
1/4 mile or less	47	2.9	1.9	0.6	0.1	0.*	0.0	0.0	0.0	0.0	0.1	0.8	2.0	8.3
Temperature °F														
-Maximum														
90° and above	30	0.0	0.0	0.0	0.0	1.2	14.0	24.4	19.4	4.2	0.0	0.0	0.0	63.2
32° and below	30	11.9	2.3	0.2	0.0	0.0	0.0	0.0	0.0	0.0	0.0	0.7	7.0	22.1
-Minimum														
32° and below	30	30.2	25.0	16.4	6.5	0.5	0.0	0.0	0.0	0.1	3.1	20.8	29.6	132.2
0° and below	30	3.8	0.7	0.0	0.0	0.0	0.0	0.0	0.0	0.0	0.0	0.*	1.1	5.7
AVG. STATION PRESS.(mb)	21	855.1	853.6	849.6	849.8	849.6	850.9	853.1	853.6	853.5	854.3	853.9	855.1	852.7
RELATIVE HUMIDITY (%)														
Hour 05	30	78	73	64	57	54	45	48	51	52	58	70	77	61
Hour 11 (Local Time)	30	65	54	43	34	31	25	29	32	33	39	51	62	42
Hour 17	30	64	48	36	28	25	19	22	24	26	33	47	60	36
Hour 23	30	77	68	57	47	42	33	37	40	41	51	65	75	53
PRECIPITATION (inches):														
Water Equivalent														
-Normal		0.56	0.48	0.90	0.75	0.87	0.50	0.65	0.81	0.82	0.98	0.71	0.61	8.64
-Maximum Monthly	47	2.46	1.56	2.02	1.95	1.99	2.07	1.92	3.48	2.81	3.45	2.00	1.89	3.48
-Year		1957	1948	1979	1965	1993	1969	1983	1957	1982	1972	1983	1951	AUG 1957
-Minimum Monthly	47	T	T	0.02	0.06	T	T	0.03	0.04	T	0.00	T	0.01	0.00
-Year		1961	1972	1972	1958	1970	1980	1972	1956	1953	1952	1989	1976	OCT 1952
-Maximum in 24 hrs	47	0.71	0.69	1.15	1.33	1.13	1.57	1.42	1.21	1.35	1.24	0.83	1.16	1.57
-Year		1989	1989	1993	1965	1983	1969	1974	1953	1965	1957	1983	1951	JUN 1969
Snow,Ice pellets,hail														
-Maximum Monthly	47	33.7	18.4	14.9	14.3	5.0	0.0	0.0	T	3.1	6.1	12.1	19.0	33.7
-Year		1957	1948	1948	1975	1979			1993	1965	1975	1964	1983	JAN 1957
-Maximum in 24 hrs	47	9.1	9.0	6.1	8.9	5.0	0.0	0.0	T	3.1	6.1	8.4	6.0	9.1
-Year		1957	1989	1948	1975	1979			1993	1965	1975	1954	1967	JAN 1957
WIND:														
Mean Speed (mph)	47	5.7	6.7	8.3	9.4	9.6	9.7	9.3	9.1	9.0	7.9	6.8	5.9	8.1
Prevailing Direction through 1963		ESE	ESE	ESE	ESE	ESE	ESE	ESE	ESE	ESE	ESE	ESE	ESE	ESE
Fastest Obs. 1 Min.														
-Direction (!!)	14	18	24	28	29	20	27	08	25	28	32	31	27	27
-Speed (MPH)	14	35	32	35	46	46	53	40	37	38	35	39	32	53
-Year		1982	1993	1985	1985	1989	1981	1988	1988	1992	1990	1982	1982	JUN 1981
Peak Gust														
-Direction (!!)	10	SW	SW	W	S	W	S	SW	S	W	W	W	NW	S
-Speed (mph)	10	41	43	52	78	62	66	64	74	64	54	49	41	78
-Date		1993	1993	1988	1985	1989	1991	1993	1984	1992	1985	1984	1990	APR 1985

PRECIPITATION (inches)

GRAND JUNCTION, COLORADO

YEAR	JAN	FEB	MAR	APR	MAY	JUNE	JULY	AUG	SEP	OCT	NOV	DEC	ANNUAL
1964	0.47	0.02	0.70	1.37	0.41	0.16	0.65	1.16	0.36	T	1.05	0.53	6.88
1965	0.66	0.94	0.87	1.95	1.35	1.33	0.89	0.67	2.52	1.40	0.82	0.76	14.16
1966	0.65	0.44	0.04	0.83	0.71	0.13	0.31	0.52	0.26	1.23	0.62	1.78	7.52
1967	0.25	0.22	0.18	0.08	1.52	1.34	1.03	0.59	0.37	0.63	0.33	1.14	7.68
1968	0.26	1.13	0.44	0.67	1.14	0.05	0.49	1.37	0.07	0.94	0.42	0.47	7.45
1969	1.03	0.40	0.67	0.33	0.45	2.07	0.21	0.88	1.45	2.01	0.45	0.36	10.31
1970	0.52	0.05	1.75	0.76	T	0.91	0.60	0.44	0.78	1.56	0.54	0.39	8.30
1971	0.19	0.13	0.02	0.42	1.10	0.03	0.15	1.02	0.58	1.13	0.56	0.67	6.00
1972	0.20	T	0.02	0.11	0.44	0.64	0.03	0.29	0.72	3.45	0.69	0.74	7.33
1973	0.79	0.12	0.65	0.86	1.45	0.87	0.52	0.62	0.33	0.20	0.91	0.62	7.94
1974	1.20	0.40	0.81	1.03	0.01	0.14	1.53	0.48	0.38	0.72	1.18	0.32	8.20
1975	0.53	0.49	1.74	1.38	1.23	0.43	1.39	0.09	0.16	0.85	0.39	0.50	9.18
1976	0.13	0.81	0.75	0.40	1.49	0.14	0.20	0.31	0.67	0.32	0.04	0.01	5.27
1977	0.37	0.06	0.50	0.54	0.59	0.04	0.89	0.59	0.52	0.50	0.70	0.38	5.68
1978	1.08	0.64	1.19	1.19	0.55	0.01	0.25	0.54	0.49	0.03	0.62	1.30	7.89
1979	1.36	0.63	2.02	0.42	1.45	0.78	0.08	0.61	0.01	0.25	1.02	0.27	8.90
1980	0.57	1.10	1.77	0.53	1.17	T	0.96	1.39	0.58	1.31	0.52	0.24	10.14
1981	0.44	0.16	1.35	0.56	1.49	0.17	0.41	0.82	0.25	2.06	0.47	0.60	8.78
1982	0.29	0.41	0.79	0.09	0.75	0.21	0.35	0.94	2.81	0.83	0.48	0.27	8.22
1983	0.50	0.64	1.59	0.90	1.68	1.54	1.92	0.73	1.11	0.36	2.00	1.85	14.82
1984	0.28	0.11	1.57	1.21	0.55	1.68	0.62	1.77	0.34	2.65	0.38	0.43	11.59
1985	0.51	0.26	0.92	1.78	1.09	0.39	1.21	0.24	1.67	2.32	1.10	0.73	12.22
1986	0.13	0.33	0.25	0.71	1.15	0.15	0.94	0.97	1.52	1.22	1.02	0.47	8.86
1987	0.30	1.21	1.95	0.46	1.51	0.23	1.51	0.83	0.13	0.65	1.92	0.83	11.53
1988	1.07	0.21	0.72	0.99	1.10	0.21	0.18	1.37	0.76	0.02	1.02	0.20	7.85
1989	0.98	1.33	0.51	0.23	0.39	0.24	0.27	1.01	0.33	0.14	T	0.08	5.51
1990	0.59	0.55	1.07	0.71	0.05	0.26	0.96	0.49	1.23	0.95	0.57	0.98	8.41
1991	0.92	0.13	0.70	0.87	0.20	0.30	0.40	0.57	2.30	1.20	1.10	0.54	9.23
1992	0.24	0.35	1.71	0.15	1.81	0.17	1.03	0.84	0.33	1.45	0.76	0.35	9.19
1993	1.36	1.09	1.72	1.30	1.99	0.03	0.04	1.42	0.41	1.34	0.41	0.57	11.68
Record Mean	0.60	0.58	0.83	0.75	0.78	0.43	0.62	1.01	0.89	0.92	0.63	0.58	8.63

AVERAGE TEMPERATURE (deg. F)

GRAND JUNCTION, COLORADO

YEAR	JAN	FEB	MAR	APR	MAY	JUNE	JULY	AUG	SEP	OCT	NOV	DEC	ANNUAL
1964	25.5	29.5	37.0	50.3	62.3	70.2	81.3	74.1	66.6	58.0	38.1	30.4	51.9
1965	32.3	32.9	40.4	53.9	62.4	70.3	78.9	76.6	63.4	58.1	48.3	33.8	54.3
1966	24.5	30.6	46.1	53.6	65.4	72.0	80.9	77.0	69.1	56.5	43.9	29.5	54.1
1967	23.9	36.4	48.5	51.6	59.5	68.6	77.7	76.2	67.5	55.3	41.0	18.3	52.0
1968	15.4	37.0	43.8	46.3	59.1	72.8	78.0	70.3	64.2	53.8	38.0	22.7	50.1
1969	28.5	33.6	38.1	53.8	66.4	67.7	80.3	79.5	69.3	47.4	39.2	32.7	53.1
1970	29.3	40.9	39.8	45.9	63.3	71.4	79.1	78.7	64.1	48.8	41.0	31.9	52.9
1971	27.7	34.1	41.1	52.2	60.1	74.1	80.2	78.5	63.2	52.3	38.4	25.9	52.3
1972	30.0	36.6	46.6	53.3	63.1	74.3	80.2	77.1	68.1	54.0	37.1	22.7	53.6
1973	11.5	29.1	42.1	48.1	61.5	70.5	78.1	77.4	65.7	56.4	41.2	30.1	51.0
1974	16.9	19.9	48.2	51.0	65.0	74.9	78.3	75.6	66.4	56.2	39.6	27.1	51.6
1975	20.0	33.0	41.0	46.4	57.1	67.5	78.3	75.4	67.1	53.5	36.2	27.4	50.2
1976	21.7	38.2	38.7	51.9	61.8	70.4	79.6	75.3	66.9	51.2	39.1	27.6	51.9
1977	23.9	37.1	40.8	56.9	63.7	79.1	80.2	78.3	70.2	58.2	40.3	33.3	55.2
1978	29.4	34.4	46.7	52.3	58.8	73.0	78.4	74.5	65.7	54.7	40.2	16.0	52.0
1979	16.6	23.5	41.1	52.5	60.3	71.0	78.7	74.6	72.0	58.9	33.2	26.9	50.8
1980	32.6	39.2	40.9	51.4	59.1	74.0	78.6	75.3	67.9	53.8	42.3	40.1	54.6
1981	36.8	37.9	44.0	56.8	60.9	76.4	79.4	76.8	69.2	50.7	41.6	31.3	55.1
1982	26.0	34.7	46.0	51.3	61.4	72.2	79.0	78.2	67.5	51.9	41.2	33.0	53.5
1983	34.3	40.9	45.9	48.7	58.7	69.8	78.3	80.4	71.4	58.2	42.1	30.5	54.9
1984	20.7	31.7	44.4	49.0	66.2	70.1	78.0	76.6	68.3	50.2	40.8	32.6	52.4
1985	31.1	32.0	43.9	54.5	63.6	73.0	77.6	77.1	62.4	52.8	38.8	31.9	53.2
1986	34.2	40.4	49.0	52.5	60.6	74.3	76.0	75.6	63.2	51.5	40.8	32.4	54.2
1987	27.4	36.8	40.1	54.5	61.2	73.5	75.2	72.9	66.2	56.8	39.6	27.9	52.7
1988	17.4	29.2	41.0	53.1	60.9	76.5	80.6	76.2	64.5	58.9	40.7	30.0	52.4
1989	20.2	27.7	47.5	57.0	63.8	71.0	80.5	74.4	68.3	54.9	40.6	29.2	52.9
1990	28.5	35.4	46.8	55.7	61.3	75.2	78.0	76.6	69.7	53.2	39.5	20.6	53.4
1991	17.6	32.0	42.0	48.8	62.0	72.5	77.6	76.2	66.5	55.0	37.7	26.2	51.2
1992	19.9	37.6	47.2	59.0	64.5	71.2	75.1	75.1	67.7	57.8	35.9	24.5	53.0
1993	31.9	36.2	45.6	50.0	61.9	70.1	76.7	73.2	65.9	51.7	35.6	29.2	52.3
Record Mean	25.8	33.5	43.0	52.3	61.8	71.9	78.3	75.7	66.9	54.2	39.8	28.6	52.7
Max	36.2	44.2	54.6	65.2	75.3	86.7	92.4	89.2	80.6	67.2	51.2	38.7	65.1
Min	15.3	22.9	31.3	39.4	48.3	57.2	64.1	62.1	53.2	41.2	28.4	18.4	40.2

REFERENCE NOTES

GRAND JUNCTION, COLORADO

GENERAL
T - TRACE AMOUNT.
BLANK ENTRIES DENOTE MISSING/UNREPORTED DATA.
# INDICATES A STATION OR INSTRUMENT RELOCATION.

SPECIFIC

PM - INCLUDES LAST DAY OF PREVIOUS MONTH
ASOS - AUTOMATED SURFACE OBSERVING SYSTEM IN OPERATION DURING THESE MONTHS.

(a) - LENGTH OF RECORD IN YEARS, ALTHOUGH INDIVIDUAL MONTHS MAY BE MISSING.
0.* OR * - THE VALUE IS BETWEEN 0.0 AND 0.05
NORMALS - BASED ON THE 1961-1990 RECORD PERIOD.
EXTREMES - DATES ARE THE MOST RECENT OCCURRENCE.
WIND DIR.- NUMERALS SHOW TENS OF DEGRESS CLOCKWISE FROM TRUE NORTH. "00" INDICATES CALM.
RESULTANT DIRECTIONS ARE GIVEN TO WHOLE DEGREES.
BOLD VALUES INDICATE EXTREME VALUES WHICH OCCURRED AFTER THE ASOS SYSTEM WAS COMMISSIONED.

RECORD = PERIOD OF RECORD
RECORD MEAN PRECIPITATION IS THE MEAN OF ALL DAILY PRECIPITATION AMOUNTS DURING THE PERIOD OF RECORD.
RECORD MAX(MIN) TEMPERATURE IS THE MEAN OF ALL DAILY MAX(MIN) TEMPERATURES DURING THE PERIOD OF RECORD.
RECORD MEAN TEMPERATURE IS THE SUM OF THE RECORD MAX AND RECORD MIN DIVIDED BY 2.
AVERAGE TEMPERATURE IS THE SUM OF THE MEAN DAILY MAX AND MIN TEMPERATURE DIVIDED BY 2.

EXCEPTIONS

RECORD MEANS ARE THROUGH THE CURRENT YEAR, BEGINNING IN 1892 FOR TEMPERATURE
1892 FOR PRECIPITATION
1947 FOR SNOWFALL

HEATING DEGREE DAYS Base 65 deg. F GRAND JUNCTION, COLORADO

SEASON	JULY	AUG	SEP	OCT	NOV	DEC	JAN	FEB	MAR	APR	MAY	JUNE	TOTAL
1964-65	0	2	39	219	800	1067	1004	893	758	334	158	8	5282
1965-66	0	0	138	209	496	960	1246	959	582	337	59	4	4990
1966-67	0	2	16	256	628	1092	1268	795	508	396	213	21	5195
1967-68	0	0	28	320	714	1442	1532	805	648	552	218	24	6283
1968-69	0	12	86	346	804	1302	1125	874	826	332	52	34	5793
1969-70	0	0	20	545	764	995	1100	671	772	564	115	39	5585
1970-71	0	0	93	495	714	1019	1152	858	734	378	182	4	5629
1971-72	0	0	134	389	792	1204	1076	813	563	346	139	0	5456
1972-73	0	0	31	335	832	1303	1651	999	705	499	139	49	6543
1973-74	0	0	72	266	708	1075	1487	1260	513	415	66	32	5894
1974-75	0	0	60	266	756	1167	1387	888	736	551	249	51	6111
1975-76	0	0	35	358	858	1161	1335	775	807	386	122	25	5862
1976-77	0	0	41	421	769	1153	1267	775	743	250	94	0	5513
1977-78	0	1	17	214	736	975	1098	852	561	373	210	9	5046
1978-79	0	6	95	313	737	1510	1493	1154	732	377	192	37	6646
1979-80	0	3	0	209	945	1175	999	741	740	405	195	4	5416
1980-81	0	2	21	359	674	765	864	754	645	247	153	15	4499
1981-82	0	0	12	439	696	1039	1203	841	581	405	136	6	5358
1982-83	2	0	61	397	704	983	946	668	586	482	238	22	5089
1983-84	0	0	27	208	678	1064	1366	959	631	474	89	44	5540
1984-85	0	0	54	452	719	996	1044	919	646	310	81	12	5233
1985-86	0	0	139	371	779	1018	949	685	489	366	168	3	4967
1986-87	0	0	130	414	718	1001	1159	785	765	314	143	0	5429
1987-88	0	6	34	248	754	1147	1469	1031	741	350	172	8	5960
1988-89	0	0	106	183	724	1078	1379	1038	534	258	113	8	5421
1989-90	0	0	40	316	729	1103	1124	820	557	271	139	20	5119
1990-91	0	0	28	360	759	1371	1464	919	706	478	136	18	6239
1991-92	0	2	37	304	815	1193	1390	788	540	195	53	8	5325
1992-93	0	6	25	222	868	1245	1018	799	597	446	144	33	5403
1993-94	4	0	59	410	875	1102							

COOLING DEGREE DAYS Base 65 deg. F GRAND JUNCTION, COLORADO

YEAR	JAN	FEB	MAR	APR	MAY	JUNE	JULY	AUG	SEP	OCT	NOV	DEC	TOTAL
1969	0	0	0	3	104	124	481	458	158	1	0	0	1329
1970	0	0	0	0	67	238	442	430	72	0	0	0	1249
1971	0	0	0	0	36	284	479	425	86	3	0	0	1313
1972	0	0	0	0	86	288	479	381	130	3	0	0	1367
1973	0	0	0	0	35	222	410	393	101	6	0	0	1167
1974	0	0	0	1	73	335	420	335	109	0	0	0	1273
1975	0	0	0	0	9	133	419	328	106	9	0	0	1004
1976	0	0	0	0	32	195	460	324	103	0	0	0	1114
1977	0	0	0	16	60	429	477	420	180	10	0	0	1592
1978	0	0	0	0	25	258	420	308	123	1	0	0	1135
1979	0	0	0	6	52	225	428	310	215	27	0	0	1263
1980	0	0	0	1	16	280	427	325	115	19	0	0	1183
1981	0	0	0	9	31	367	456	375	143	0	0	0	1381
1982	0	0	0	0	33	229	443	415	144	0	0	0	1264
1983	0	0	0	0	49	171	421	483	226	3	0	0	1353
1984	0	0	0	0	134	200	408	368	159	0	0	0	1269
1985	0	0	0	4	45	261	396	382	67	0	0	0	1155
1986	0	0	0	0	39	289	348	334	82	0	0	0	1092
1987	0	0	0	5	30	262	324	256	76	2	0	0	955
1988	0	0	0	0	51	360	489	357	98	4	0	0	1359
1989	0	0	0	26	85	195	489	300	145	11	0	0	1251
1990	0	0	0	1	34	331	412	368	174	3	0	0	1323
1991	0	0	0	0	50	247	398	356	88	0	0	0	1139
1992	0	0	0	21	43	203	319	328	114	7	0	0	1035
1993	0	0	0	0	56	193	371	260	92	6	0	0	978

SNOWFALL (inches) GRAND JUNCTION, COLORADO

SEASON	JULY	AUG	SEP	OCT	NOV	DEC	JAN	FEB	MAR	APR	MAY	JUNE	TOTAL
1964-65	0.0	0.0	0.0	0.0	12.1	3.0	5.9	7.5	3.8	0.2	T	0.0	32.5
1965-66	0.0	0.0	3.1	0.0	2.8	3.5	11.0	5.9	0.3	0.8	T	0.0	27.4
1966-67	0.0	0.0	0.0	1.9	4.2	6.3	2.8	T	0.2	0.4	T	0.0	15.8
1967-68	0.0	0.0	0.0	T	0.5	16.7	3.7	0.9	0.3	0.6	T	0.0	22.7
1968-69	0.0	0.0	0.0	0.0	0.2	7.8	3.4	3.8	9.5	T	T	0.0	24.7
1969-70	0.0	0.0	0.0	0.6	0.5	4.8	5.2	T	12.4	1.2	0.0	0.0	24.7
1970-71	0.0	0.0	0.0	3.1	T	4.1	3.2	1.5	0.2	1.1	T	0.0	13.2
1971-72	0.0	0.0	0.0	0.5	3.9	10.4	4.0	T	0.0	T	0.0	0.0	18.8
1972-73	0.0	0.0	0.0	5.7	1.3	9.7	12.8	1.2	1.3	2.0	T	0.0	34.0
1973-74	0.0	0.0	0.0	0.0	7.7	5.7	17.0	5.5	T	1.2	0.0	0.0	37.1
1974-75	0.0	0.0	0.0	T	0.1	4.6	7.9	4.4	8.8	14.3	1.3	0.0	41.4
1975-76	0.0	0.0	0.0	6.1	3.9	7.2	1.7	4.0	6.8	0.2	0.0	0.0	29.9
1976-77	0.0	0.0	0.0	0.0	T	0.1	4.2	T	2.3	1.7	0.0	0.0	8.3
1977-78	0.0	0.0	0.0	T	3.3	2.5	12.0	2.5	0.6	T	T	0.0	20.9
1978-79	0.0	0.0	0.0	0.0	2.9	11.8	18.7	9.6	3.4	1.1	5.0	0.0	52.5
1979-80	0.0	0.0	0.0	0.0	8.2	3.5	2.2	0.5	7.3	0.2	0.0	0.0	21.9
1980-81	0.0	0.0	0.0	0.0	T	0.0	3.9	0.8	1.2	T	0.0	0.0	5.9
1981-82	0.0	0.0	0.0	0.5	3.3	3.4	3.4	4.0	0.8	T	0.0	0.0	15.4
1982-83	0.0	0.0	0.0	T	T	1.9	6.1	3.1	1.5	2.2	T	0.0	14.8
1983-84	0.0	0.0	0.0	0.0	4.2	19.0	3.7	0.6	6.1	2.9	0.0	0.0	36.5
1984-85	0.0	0.0	0.0	0.7	2.0	2.7	5.0	2.7	5.6	0.1	0.0	0.0	18.8
1985-86	0.0	0.0	0.0	0.0	4.6	4.4	1.8	0.7	T	0.2	T	0.0	11.7
1986-87	0.0	0.0	0.0	2.2	1.2	1.0	3.0	5.5	9.4	0.6	T	0.0	22.9
1987-88	0.0	0.0	0.0	0.0	1.1	7.1	12.2	2.2	4.3	0.0	0.0	0.0	26.9
1988-89	0.0	0.0	0.0	0.0	0.9	3.1	10.2	16.0	1.1	T	T	0.0	31.3
1989-90	0.0	0.0	0.0	0.0	0.0	1.1	6.2	8.6	1.8	0.8	0.0	0.0	18.5
1990-91	0.0	0.0	0.0	0.0	1.5	5.1	12.7	0.3	3.7	4.7	0.0	0.0	28.0
1991-92	0.0	0.0	T	2.5	1.9	7.6	2.7	1.9	T	0.0	T	0.0	16.6
1992-93	0.0	0.0	0.0	0.0	2.0	4.4	6.0	8.4	0.0	0.4	T	0.0	21.2
1993-94	0.0	T	0.0	0.0	0.6	5.9							
Record Mean	0.0	T	0.1	0.5	2.7	5.3	7.3	4.2	3.8	1.1	0.1	0.0	25.2

1993 PUEBLO, COLORADO (PUB)

The city of Pueblo is located about 40 miles east-southeast of the Royal Gorge, at the junction of the Arkansas and Fountain Rivers. The mountains west of the city extend from within 25 miles to the southwest to about 35 miles to the northwest. Lake Pueblo, the largest body of water in southern Colorado, is located 7 miles west of the city and provides a variety of water sports, fishing, picnicing, and a wildlife preserve. The countryside surrounding Pueblo consists of rolling plains, broken by normally dry arroyos, and is generally treeless, covered mainly with sparse bunchgrass and occasional cacti. The business section of the city is 4,663 feet above sea level. The National Weather Service Office is located at Pueblo Memorial Airport, 6 miles east of the Pueblo Post Office, and about 1 1/2 miles north of the Arkansas River. Terrain at the airport is relatively flat, and from 50 to 100 feet above the river. The air quality in Pueblo is rated the best of large Colorado cities along the front range.

The climate is semi-arid and marked by large daily temperature variations. The temperature reaches 90 degrees or more about half the time during the summer, but thanks to the low relative humidity, the heat is not oppressive. Summer nights are invariably cool since mountain breezes prevail from shortly after sunset to about noon the following day. The sun shines about 76 percent of the time. Winter is comparatively mild due to the abundant sunshine and the protection afforded by the nearby mountains. Temperatures reach 50 degrees or higher in the winter. The temperature drops to zero or below about eight times during the winter. Cold spells are generally broken after a few days by chinook winds, a very dry, warm, downslope westerly wind.

The probability of measurable precipitation in summer is one day out of four and in winter one out of eight. Summer rains usually occur in the form of afternoon thunderstorms. Blowing dust frequently develops during the spring months of abnormally dry years, especially in areas where dry farming has been attempted.

Agriculture consists chiefly of cattle grazing on the dry plains and irrigated farming near streams. Sugar beets, corn, chili peppers, and melons are the most important crops. In addition, a variety of vegetables, from asparagus to zucchini, are grown. Some dry farming is attempted, but the extent of such operations is limited by the annual precipitation of less than 12 inches.

NORMALS, MEANS, AND EXTREMES FOR 1993

PUEBLO, CO (PUB)

LATITUDE: 38° 17' 21" N LONGTITUDE: 104° 29' 46" W ELEVATION (FT): GRND: 4684 BARO: 4661 TIME ZONE: MOUNTAIN (UTC+07) WBAN: 93058

	ELEMENT	POR	JAN	FEB	MAR	APR	MAY	JUN	JUL	AUG	SEP	OCT	NOV	DEC	YEAR
TEMPERATURE °F	NORMAL DAILY MAXIMUM	30	45.3	50.7	57.2	67.9	76.5	87.7	93.0	89.7	81.3	70.5	56.8	46.7	68.6
	MEAN DAILY MAXIMUM	40	44.9	50.3	56.9	67.2	76.4	87.5	92.4	89.5	81.5	70.4	55.7	47.2	68.3
	HIGHEST DAILY MAXIMUM	52	78	81	86	93	98	108	106	104	100	94	84	82	108
	YEAR OF OCCURRENCE		1971	1981	1989	1992	1989	1990	1981	1980	1990	1991	1980	1980	JUN 1990
	MEAN OF EXTREME MAXS.	40	65.6	69.3	77.4	85.4	90.9	99.9	102.1	99.3	95.1	87.4	75.7	69.4	84.8
	NORMAL DAILY MINIMUM	30	14.2	19.5	26.1	35.7	45.6	54.2	61.2	59.0	50.2	36.6	24.3	15.5	36.8
	MEAN DAILY MINIMUM	40	13.9	19.1	25.9	35.5	45.8	54.5	60.7	58.8	49.7	36.7	23.9	16.2	36.7
	LOWEST DAILY MINIMUM	52	-29	-31	-20	2	25	37	44	40	27	5	-17	-25	-31
	YEAR OF OCCURRENCE		1948	1951	1948	1975	1954	1988	1945	1968	1985	1993	1991	1990	FEB 1951
	MEAN OF EXTREME MINS.	40	-7.5	-0.5	8.0	21.0	33.3	43.8	53.2	50.6	36.2	22.6	6.1	-3.3	22.0
	NORMAL DRY BULB	30	29.8	35.1	41.7	51.8	61.1	71.0	77.1	74.3	65.8	53.6	40.5	31.1	52.7
	MEAN DRY BULB	40	28.8	33.9	41.4	51.4	61.1	71.0	76.6	74.2	65.6	53.6	39.8	31.7	52.4
	MEAN WET BULB	40	23.8	27.9	32.6	40.2	48.8	55.8	61.1	60.2	52.4	42.5	32.3	26.3	42.0
	MEAN DEW POINT	40	13.8	16.7	20.2	26.4	36.5	44.4	52.1	51.5	42.0	30.6	22.6	15.9	31.1
	NORMAL NO. DAYS WITH:														
	MAXIMUM ≥ 90°	30	0.0	0.0	0.0	0.1	2.6	14.2	22.8	17.9	6.6	0.3	0.0	0.0	64.5
	MAXIMUM ≤ 32°	30	6.5	3.1	1.2	*	0.0	0.0	0.0	0.0	0.0	0.1	1.5	5.5	17.9
	MINIMUM ≤ 32°	30	30.2	26.3	24.2	9.0	0.8	0.0	0.0	0.0	0.5	9.3	25.5	29.9	155.7
	MINIMUM ≤ 0°	30	4.5	1.4	0.2	0.0	0.0	0.0	0.0	0.0	0.0	0.0	0.2	3.2	9.5
H/C	NORMAL HEATING DEG. DAYS	30	1091	837	722	396	152	10	0	0	62	357	735	1051	5413
	NORMAL COOLING DEG. DAYS	30	0	0	0	0	31	190	375	291	86	0	0	0	973
RH	NORMAL (PERCENT)	30	57	52	49	44	45	45	49	52	50	47	57	57	50
	HOUR 05 LST	30	68	66	66	63	67	69	72	74	71	65	72	67	68
	HOUR 11 LST	30	52	47	42	34	37	34	36	39	39	36	45	51	41
	HOUR 17 LST	30	50	40	34	29	32	28	32	34	32	32	43	50	36
	HOUR 23 LST	30	66	61	56	51	52	52	58	60	58	55	65	65	57
S	PERCENT POSSIBLE SUNSHINE	52	75	74	74	75	75	80	79	78	81	79	73	72	76
W/O	MEAN NO. DAYS WITH:														
	HEAVY FOG(VISBY ≤ 1/4 MI)	45	1.2	1.2	1.0	0.7	0.4	0.2	0.1	0.3	0.5	0.5	1.0	1.5	8.6
	THUNDERSTORMS	45	0.0	0.0	0.3	1.6	6.0	7.9	11.5	9.3	3.1	0.5	0.1	0.0	40.3
CLOUDINESS	MEAN:														
	SUNRISE-SUNSET (TENTHS)	1		6.0	6.0	5.0	7.0	4.0	3.0	6.0	4.0	4.0	4.0	3.0	
	MIDNIGHT-MIDNIGHT (TENTHS)	1		6.0	6.0	6.0	7.0		3.0	6.0	4.0	4.0	4.0	3.0	
	MEAN NO. DAYS WITH:														
	CLEAR	1	11.0	11.0	9.0	8.0	4.0	15.0	17.0	11.0	18.0	13.0	12.0	21.0	150.0
	PARTLY CLOUDY	1	4.0	7.0	4.0	9.0	13.0	3.0	2.0	9.0	3.0	2.0	3.0	7.0	66.0
	CLOUDY	1	4.0	8.0	9.0	5.0	6.0	4.0	2.0	9.0	5.0	7.0	4.0	2.0	65.0
PR	MEAN STATION PRESSURE(IN.)	7	25.23	25.23	25.11	25.16	25.19	25.22	25.30	25.30	25.32	25.30	25.25	25.22	25.24
	MEAN SEA-LEVEL PRES. (IN.)	40	30.09	30.03	29.93	29.86	29.85	29.83	29.91	29.92	29.95	30.00	30.06	30.08	29.96
WINDS	MEAN SPEED (MPH)	30	8.4	9.4	11.0	11.8	11.2	10.5	9.8	9.2	9.2	8.5	8.1	8.1	9.6
	PREVAIL.DIR(TENS OF DEGS.)	30	30	12	12	12	12	12	14	14	12	12	27	27	12
	MAXIMUM 2-MINUTE:														
	SPEED (MPH)	1	37	38	38	37	40	37	40	40	39	32	37	43	43
	DIR. (TENS OF DEGS.)	1	25	30	20	01	36	10	35	34	02	36	36	27	27
	YEAR OF OCCURRENCE		1993	1993	1993	1993	1993	1993	1993	1993	1993	1993	1993	1993	DEC 1993
	MAXIMUM 5-SECOND:														
	SPEED (MPH)	1	43	44	45	43	46	43	51	44	45	38	43	49	51
	DIR. (TENS OF DEGS.)	1	26	30	25	02	36	10	33	34	03	36	36	27	33
	YEAR OF OCCURRENCE		1993	1993	1993	1993	1993	1993	1993	1993	1993	1993	1993	1993	JUL 1993
PRECIPITATION	NORMAL TOTAL (IN.)	30	0.32	0.31	0.78	0.88	1.25	1.25	2.09	1.99	0.90	0.57	0.43	0.42	11.19
	MAXIMUM TOTAL (IN.)	52	1.45	1.39	2.34	6.17	5.43	4.26	5.14	5.85	2.73	4.91	2.48	0.97	6.17
	YEAR OF OCCURRENCE		1948	1965	1948	1942	1957	1961	1990	1955	1976	1957	1991	1979	APR 1942
	MINIMUM TOTAL (IN.)	52	0.00	T	0.05	T	0.30	T	0.09	0.08	T	T	T	T	0.0
	YEAR OF OCCURRENCE		1993	1970	1966	1963	1965	1990	1987	1960	1956	1988	1989	1993	JAN 1993
	MAXIMUM IN 24 HOURS (IN.)	52	0.61	0.60	1.23	2.49	2.53	2.24	1.96	2.95	1.57	3.77	0.92	0.71	3.77
	YEAR OF OCCURRENCE		1990	1987	1983	1957	1957	1979	1977	1955	1982	1957	1972	1979	OCT 1957
	NORMAL NO. DAYS WITH:														
	PRECIPITATION ≥ 0.01	30	4.3	3.9	6.0	5.1	7.4	7.0	9.4	8.8	5.8	3.3	3.5	4.5	69.0
	PRECIPITATION ≥ 1.00	30	0.0	0.0	0.0	0.1	0.1	0.1	0.3	0.4	*	0.0	0.0	0.0	1.0
SNOWFALL	NORMAL TOTAL (IN.)	30	6.0	4.3	6.8	3.1	0.6	0.0	0.0	0.0	0.4	1.1	3.6	5.9	31.8
	MAXIMUM TOTAL (IN.)	52	20.0	14.4	22.4	21.2	10.6	T	T	T	14.0	16.3	29.3	15.3	29.3
	YEAR OF OCCURRENCE		1948	1965	1948	1957	1990	1992	1992	1991	1959	1991	1946	1967	NOV 1946
	MAXIMUM IN 24 HOURS (IN.)	52	12.4	8.2	10.9	16.8	9.4	T	T	T	9.5	12.6	16.3	8.7	16.8
	YEAR OF OCCURRENCE		1990	1989	1972	1990	1990	1992	1992	1991	1959	1991	1946	1961	APR 1990
	MAXIMUM SNOW DEPTH (IN.)	40	12	7	9	9	3	0	0	0	1	7	12	10	12
	YEAR OF OCCURRENCE		1960	1989	1964	1990	1990				1984	1991	1972	1967	NOV 1972
	NORMAL NO. DAYS WITH:														
	SNOWFALL ≥ 1.0	30	1.7	1.4	2.1	0.7	0.2	0.0	0.0	0.0	0.1	0.3	1.0	1.9	9.4

PRECIPITATION (inches) 1993 PUEBLO, COLORADO (PUB)

YEAR	JAN	FEB	MAR	APR	MAY	JUN	JUL	AUG	SEP	OCT	NOV	DEC	ANNUAL
1964	0.04	0.20	1.11	0.03	1.88	0.13	0.73	2.70	1.11	0.02	0.26	0.08	8.29
1965	0.19	0.71	0.42	1.46	0.30	2.79	2.41	2.69	1.12	0.38	0.01	0.07	12.55
1966	0.31	0.17	0.05	0.24	0.38	0.70	1.68	1.34	1.05	0.21	T	0.14	6.27
1967	0.13	0.12	0.23	2.03	1.25	2.03	3.59	1.07	0.83	0.79	0.11	0.93	13.11
1968	0.10	0.54	0.78	0.68	1.62	0.29	4.41	0.90	0.52	0.29	0.43	0.36	10.92
1969	0.04	0.05	0.82	0.54	0.87	1.74	1.73	1.58	1.28	2.04	0.48	0.50	11.67
1970	0.25	T	0.92	1.32	0.44	0.77	0.70	1.64	1.24	1.55	0.45	T	9.28
1971	0.42	0.23	0.55	0.88	0.73	0.66	1.93	0.85	1.20	0.20	0.45	0.18	8.28
1972	0.35	0.13	0.93	0.68	0.84	0.98	1.99	1.96	1.12	1.05	1.49	0.41	11.93
1973	0.11	0.07	1.34	1.64	1.84	0.38	3.23	1.77	1.19	0.76	0.05	0.82	13.20
1974	0.40	0.17	0.45	0.29	1.17	1.70	1.23	1.75	0.65	0.86	0.69	0.32	9.68
1975	0.07	0.24	0.12	0.85	0.69	2.05	2.19	2.24	0.06	0.61	0.71	0.05	9.88
1976	0.15	0.11	0.86	0.88	0.68	1.34	2.08	0.88	2.73	0.98	0.30	0.14	11.13
1977	0.10	0.32	0.10	1.77	0.60	0.25	2.96	3.53	0.08	T	0.14	0.02	9.87
1978	0.16	0.34	0.40	0.44	1.95	1.19	0.99	1.23	0.08	0.68	0.24	0.71	8.41
1979	0.59	0.06	2.34	0.54	2.60	3.51	1.47	2.04	0.63	0.36	0.94	0.97	16.05
1980	0.39	0.14	1.22	2.99	2.20	0.43	0.75	2.35	0.47	T	0.64	0.01	11.59
1981	0.12	0.15	1.24	0.13	0.76	0.16	1.28	2.55	0.32	0.19	0.04	0.67	7.61
1982	0.54	0.31	0.26	0.13	2.28	1.71	2.71	4.35	2.24	0.37	0.12	0.44	15.46
1983	0.60	0.04	2.08	1.05	1.96	2.07	1.49	1.58	0.18	0.22	0.51	0.95	12.73
1984	0.23	0.07	1.15	1.17	1.11	0.35	4.21	4.37	0.27	2.55	0.20	0.44	16.12
1985	0.50	0.36	0.72	1.70	1.46	0.10	4.82	0.95	1.01	0.60	1.13	0.27	13.62
1986	0.25	0.14	0.55	0.43	0.84	2.21	1.71	2.42	0.35	0.90	0.48	0.49	10.77
1987	0.74	1.39	0.45	0.30	2.09	1.29	0.09	2.89	0.31	0.04	0.49	0.74	10.82
1988	0.94	0.38	0.93	0.70	1.33	1.86	2.00	0.67	1.80	T	0.17	0.60	11.38
1989	0.42	0.83	0.14	0.59	1.45	1.27	0.61	1.02	1.03	0.10	T	0.87	8.33
1990	0.76	0.69	1.14	1.57	2.34	0.00	5.14	3.08	1.83	0.59	0.54	0.22	17.90
1991	0.08	0.01	0.74	0.83	0.72	1.97	2.79	2.14	1.36	0.62	2.48	0.52	14.26
1992	0.04	0.19	1.31	0.43	0.40	3.15	1.39	2.48	0.70	0.71	1.30	0.47	12.57
1993	0.00	0.18	1.72	1.40	1.99	1.49	1.37	3.71	0.62	0.54	1.07	T	14.09
POR= 39 YRS	0.33	0.31	0.78	0.93	1.45	1.27	1.99	2.04	0.89	0.74	0.54	0.38	11.63

AVERAGE TEMPERATURE (°F) 1993 PUEBLO, COLORADO (PUB)

YEAR	JAN	FEB	MAR	APR	MAY	JUN	JUL	AUG	SEP	OCT	NOV	DEC	ANNUAL
1964	33.2	27.7	33.9	49.5	61.3	69.4	78.8	73.4	67.1	54.8	41.6	35.5	52.2
1965	36.5	29.5	31.5	53.8	61.3	69.6	76.8	72.2	61.6	56.0	46.0	36.1	52.6
1966	25.5	31.9	44.1	50.4	62.4	70.8	79.9	72.7	67.0	51.5	42.6	29.6	52.4
1967	34.6	36.0	45.4	54.5	59.0	69.1	75.8	71.8	63.6	52.3	38.4	23.5	52.0
1968	27.9	35.3	42.8	50.2	59.3	71.5	74.5	71.5	68.0	57.9	40.3	30.8	52.5
1969	36.5	37.8	36.3	56.1	64.3	68.0	80.2	79.0	69.8	49.9	44.0	36.8	54.9
1970	31.7	41.8	39.2	50.8	65.2	73.1	80.3	79.8	65.7	50.0	45.6	38.4	55.1
1971	34.9	35.3	41.7	49.9	58.5	73.4	75.4	74.6	62.2	52.8	40.7	33.5	52.7
1972	29.8	38.6	45.9	52.9	60.4	73.2	74.4	74.2	66.5	53.8	31.9	23.1	52.1
1973	29.5	35.1	42.7	47.0	59.4	71.2	75.3	76.5	64.7	56.2	42.1	32.0	52.6
1974	26.3	36.8	47.0	51.3	65.6	72.4	78.7	72.8	63.3	56.8	41.5	28.8	53.4
1975	31.6	33.5	40.5	49.3	59.8	70.0	76.8	75.3	64.0	54.6	38.3	36.5	52.5
1976	30.9	40.2	38.5	51.3	59.5	69.4	76.6	72.5	63.8	48.4	36.0	33.0	51.7
1977	27.6	38.2	42.0	54.2	65.3	74.2	77.5	74.1	67.5	53.8	40.6	36.0	54.2
1978	25.2	29.1	43.7	53.2	58.0	70.2	77.9	72.8	66.7	53.6	38.9	24.0	51.1
1979	16.1	34.4	43.1	51.8	57.6	68.9	76.5	71.7	67.4	55.2	35.8	33.9	51.0
1980	28.2	37.2	40.6	48.3	58.3	72.8	80.2	76.0	67.0	52.4	40.8	41.2	53.6
1981	36.6	39.8	45.9	59.6	62.0	75.6	79.2	74.2	68.7	56.3	46.4	34.5	56.6
1982	31.0	33.3	46.1	53.0	60.8	67.8	77.3	76.6	66.3	51.0	38.3	32.2	52.8
1983	33.3	37.5	41.0	46.1	56.9	67.1	77.5	77.4	68.0	54.1	41.8	20.8	51.8
1984	26.6	36.0	40.1	47.9	63.5	71.7	77.5	75.3	64.0	49.5	41.0	35.5	52.4
1985	25.8	29.1	43.4	54.0	61.9	70.9	75.4	75.0	62.7	51.6	31.1	27.4	50.7
1986	39.5	38.9	47.8	53.2	59.9	71.1	75.9	73.3	63.1	50.9	40.1	30.2	53.7
1987	29.8	37.3	40.4	52.9	61.7	71.2	77.0	72.0	65.0	53.3	39.6	29.0	52.4
1988	19.7	33.6	39.7	51.5	60.8	73.2	75.7	75.8	64.8	55.0	41.8	30.5	51.8
1989	33.1	24.0	46.3	52.8	62.8	67.9	76.8	73.6	65.3	53.2	42.3	26.1	52.0
1990	33.6	33.4	42.4	51.6	57.7	74.3	72.8	72.0	68.3	53.3	44.4	24.6	52.4
1991	28.8	38.7	43.4	51.3	64.0	72.3	74.9	73.7	64.7	52.8	33.9	32.1	52.5
1992	33.9	38.6	45.1	54.6	61.9	66.8	73.2	70.5	65.6	52.1	31.1	28.2	51.8
1993	26.5	30.6	42.1	50.5	59.1	68.8	74.9	71.2	60.1	49.0	32.3	31.8	49.7
POR= 40 YRS	29.4	34.7	41.4	51.4	61.1	70.9	76.6	74.2	65.7	53.6	39.8	31.7	52.5

REFERENCE NOTES:

THE TEMPERATURE GRAPH SHOWS NORMAL MAXIMUM AND NORMAL MINIMUM DAILY TEMPERATURES (SOLID CURVES) AND THE ACTUAL DAILY HIGH AND LOW TEMPERATURES (VERTICAL BARS).

H/C INDICATES HEATING AND COOLING DEGREE DAYS.
RH INDICATES RELATIVE HUMIDITY
W/O INDICATES WEATHER AND OBSTRUCTIONS
S INDICATES SUNSHINE.
PR INDICATES PRESSURE.
CLOUDINESS IS THE SUM OF THE CEILOMETER AND SATELLITE DATA NOT TO EXCEED TEN TENTHS.
MAXIMUM SNOW DEPTH IS THE DEPTH MEASURED AT 12 UTC.

GENERAL:
T INDICATES TRACE PRECIPITATION, AN AMOUNT GREATER THAN ZERO BUT LESS THAN THE LOWEST REPORTABLE VALUE.
+ INDICATES THE VALUE ALSO OCCURS ON EARLIER DATES.
BLANK ENTRIES DENOTE MISSING OR UNREPORTED DATA.
NORMALS ARE 30-YEAR AVERAGES (1961 - 1990).
ASOS INDICATES AUTOMATED SURFACE OBSERVING SYSTEM.
PM INDICATES THE LAST DAY OF THE PREVIOUS MONTH.
POR (PERIOD OF RECORD) INDICATES THE NUMBER OF YEARS USED TO COMPUTE THE MEAN. INDIVIDUAL MONTHS WITHIN THE POR MAY BE MISSING.
WHEN THE POR FOR A NORMAL IS LESS THAN 30 YEARS, THE NORMAL IS PROVISIONAL AND IS BASED ON THE NUMBER OF YEARS INDICATED.
0.* OR * INDICATES THE VALUE OR MEAN DAYS WITH IS BETWEEN 0.00 AND 0.05.
CLOUDINESS FOR ASOS STATIONS DIFFERS FROM THE NON-ASOS OBSERVATION TAKEN BY A HUMAN OBSERVER. ASOS STATION CLOUDINESS IS BASED ON TIME-AVERAGED CEILOMETER DATA FOR CLOUDS AT OR BELOW 12,000 FEET AND ON SATELLITE DATA FOR CLOUDS ABOVE 12,000 FEET.
THE NUMBER OF DAYS WITH CLEAR, PARTLY CLOUDY, AND CLOUDY CONDITIONS FOR ASOS STATIONS IS THE SUM OF THE CEILOMETER AND SATELLITE DATA FOR THE SUNRISE TO SUNSET PERIOD.

GENERAL CONTINUED:
CLEAR INDICATES 0 - 3 TENTHS, PARTLY CLOUDY INDICATES 4 - 7 TENTHS, AND CLOUDY INDICATES 8 OR MORE TENTHS. WHEN AT LEAST ONE OF THE ELEMENTS (CEILOMETER OR SATELLITE) IS MISSING, THE DAILY CLOUDINESS IS NOT COMPUTED.
WIND DIRECTION IS RECORDED IN TENS OF DEGREES (2 DIGITS) CLOCKWISE FROM TRUE NORTH. "00" INDICATES CALM. "36" INDICATES TRUE NORTH.
RESULTANT WIND IS THE VECTOR AVERAGE OF THE SPEED AND DIRECTION.
AVERAGE TEMPERATURE IS THE SUM OF THE MEAN DAILY MAXIMUM AND MINIMUM TEMPERATURE DIVIDED BY 2.
SNOWFALL DATA COMPRISE ALL FORMS OF FROZEN PRECIPITATION, INCLUDING HAIL.
A HEATING (COOLING) DEGREE DAY IS THE DIFFERENCE BETWEEN THE AVERAGE DAILY TEMPERATURE AND 65° F.
DRY BULB IS THE TEMPERATURE OF THE AMBIENT AIR.
DEW POINT IS THE TEMPERATURE TO WHICH THE AIR MUST BE COOLED TO ACHIEVE 100 PERCENT RELATIVE HUMIDITY.
WET BULB IS THE TEMPERATURE THE AIR WOULD HAVE IF THE MOISTURE CONTENT WAS INCREASED TO 100 PERCENT RELATIVE HUMIDITY.

HEATING DEGREE DAYS (base 65°F) 1993 PUEBLO, COLORADO (PUB)

YEAR	JUL	AUG	SEP	OCT	NOV	DEC	JAN	FEB	MAR	APR	MAY	JUN	TOTAL
1964-65	0	2	59	317	693	905	876	987	1030	338	143	15	5365
1965-66	3	0	180	273	566	889	1218	917	641	433	119	5	5244
1966-67	0	4	33	413	666	1091	937	803	600	310	213	16	5086
1967-68	0	15	76	392	790	1282	1140	855	683	437	182	5	5857
1968-69	7	10	20	228	731	1052	879	753	883	260	83	42	4948
1969-70	0	0	4	468	626	865	1026	645	796	418	89	20	4957
1970-71	0	0	105	471	572	820	926	825	715	450	200	0	5084
1971-72	3	0	180	371	723	968	1081	756	585	359	170	0	5196
1972-73	16	2	58	340	987	1294	1093	832	685	532	197	25	6061
1973-74	2	0	84	268	682	1017	1192	782	552	409	59	36	5083
1974-75	0	0	103	250	698	1118	1024	876	756	472	167	39	5503
1975-76	0	0	119	320	795	874	1049	714	814	401	175	25	5286
1976-77	0	2	106	507	861	986	1151	747	708	318	44	1	5431
1977-78	0	4	34	343	723	894	1228	1001	653	347	238	54	5519
1978-79	0	6	59	347	778	1264	1509	849	674	391	247	48	6172
1979-80	0	8	45	299	870	959	1135	797	751	492	214	6	5576
1980-81	0	0	46	383	717	731	871	697	584	175	119	3	4326
1981-82	0	0	22	272	554	938	1047	883	582	358	151	35	4842
1982-83	0	0	62	427	794	1010	974	762	740	561	258	50	5638
1983-84	0	0	52	330	689	1365	1183	834	763	505	120	2	5843
1984-85	0	0	127	474	713	907	1209	999	662	325	125	9	5550
1985-86	0	0	172	410	1010	1161	783	728	523	346	169	21	5323
1986-87	0	0	94	428	741	1069	1082	768	756	358	119	10	5425
1987-88	4	17	43	355	754	1111	1399	903	777	399	167	8	5937
1988-89	1	0	84	308	689	1062	980	1145	573	378	134	35	5389
1989-90	0	0	94	373	676	1199	964	877	695	394	233	2	5507
1990-91	1	0	34	360	610	1245	1116	730	667	406	103	3	5275
1991-92	1	0	76	380	929	1014	958	759	608	309	125	41	5200
1992-93	0	15	58	390	1009	1132	1186	959	703	428	195	30	6105
1993-	0	18	155	491	973	1020							
POR= 40 YRS	0	2	76	352	748	1023	1093	849	721	400	152	19	5435

COOLING DEGREE DAYS (base 65°F) 1993 PUEBLO, COLORADO (PUB)

YEAR	JAN	FEB	MAR	APR	MAY	JUN	JUL	AUG	SEP	OCT	NOV	DEC	ANNUAL
1969	0	0	0	0	68	141	480	440	155	8	0	0	1292
1970	0	0	0	2	100	271	481	470	133	14	0	0	1471
1971	0	0	0	0	4	259	329	305	101	0	0	0	998
1972	0	0	0	2	34	255	316	294	109	3	0	0	1013
1973	0	0	0	0	29	219	328	363	83	2	0	0	1024
1974	0	0	0	6	83	264	432	249	60	4	0	0	1098
1975	0	0	0	6	14	191	371	326	97	3	0	0	1008
1976	0	0	0	0	9	163	365	239	76	0	0	0	852
1977	0	0	0	1	57	283	396	293	115	0	0	0	1145
1978	0	0	3	0	27	217	411	255	116	2	0	0	1031
1979	0	0	0	1	25	169	363	221	124	1	0	0	904
1980	0	0	0	1	14	247	482	347	111	0	0	0	1202
1981	0	0	0	20	34	328	446	291	142	9	0	0	1270
1982	0	0	0	3	30	123	395	359	110	0	0	0	1020
1983	0	0	0	0	16	120	392	391	149	0	0	0	1068
1984	0	0	0	0	76	210	395	328	103	0	0	0	1112
1985	0	0	0	0	36	192	329	318	114	1	0	0	990
1986	0	0	0	0	17	212	347	264	45	0	0	0	885
1987	0	0	0	2	26	203	385	240	50	2	0	0	908
1988	0	0	0	0	45	261	342	342	84	4	0	0	1078
1989	0	0	0	16	72	128	373	273	109	11	0	0	982
1990	0	0	0	0	13	286	250	223	140	3	0	0	915
1991	0	0	0	0	78	232	314	276	75	9	0	0	984
1992	0	0	0	3	34	100	262	191	82	0	0	0	672
1993	0	0	0	1	18	152	316	221	16	1	0	0	725
POR= 25 YRS	0	0	0	1	38	208	371	300	99	2	0	0	1019

SNOWFALL (inches) 1993 PUEBLO, COLORADO (PUB)

YEAR	JUL	AUG	SEP	OCT	NOV	DEC	JAN	FEB	MAR	APR	MAY	JUN	TOTAL
1964-65	0.0	0.0	0.0	0.0	2.9	0.9	5.4	14.4	9.8	0.0	0.0	0.0	33.4
1965-66	0.0	0.0	T	0.0	0.1	1.1	9.9	3.0	0.5	1.3	0.2	0.0	16.1
1966-67	0.0	0.0	0.0	0.1	T	2.6	1.7	1.2	3.0	0.5	0.0	0.0	9.1
1967-68	0.0	0.0	0.0	T	1.3	15.3	2.9	6.6	9.3	4.6	T	0.0	40.0
1968-69	0.0	0.0	0.0	0.0	4.0	7.1	1.1	0.6	12.7	0.0	0.0	0.0	25.5
1969-70	0.0	0.0	0.0	3.9	2.8	5.9	3.3	T	15.3	5.6	0.0	0.0	36.8
1970-71	0.0	0.0	0.0	6.3	T	T	6.0	6.6	8.0	5.9	T	0.0	32.8
1971-72	0.0	0.0	7.5	1.9	3.5	2.2	6.5	3.0	10.9	T	0.0	0.0	35.5
1972-73	0.0	0.0	0.0	12.6	15.7	9.2	3.8	0.8	6.2	4.6	0.6	0.0	53.5
1973-74	0.0	0.0	0.0	T	1.7	8.1	4.4	2.7	4.8	3.3	0.0	0.0	25.0
1974-75	0.0	0.0	0.0	0.0	1.5	4.2	1.7	4.1	1.4	8.2	0.0	0.0	21.1
1975-76	0.0	0.0	0.0	0.0	5.4	0.9	2.8	2.2	11.1	T	0.0	0.0	22.4
1976-77	0.0	0.0	0.0	2.4	4.0	2.0	1.8	3.6	1.1	T	0.0	0.0	14.9
1977-78	0.0	0.0	0.0	T	0.1	0.3	3.6	5.3	1.5	0.2	3.5	0.0	14.5
1978-79	0.0	0.0	0.0	T	1.0	11.5	11.0	T	7.7	0.8	1.4	0.0	33.4
1979-80	0.0	0.0	0.0	T	2.9	9.9	4.1	3.0	13.0	9.7	0.0	0.0	42.6
1980-81	0.0	0.0	0.0	0.0	5.3	0.1	1.6	3.0	6.4	0.4	0.0	0.0	16.8
1981-82	0.0	0.0	0.0	0.0	0.4	4.9	9.1	4.3	2.4	0.3	2.2	0.0	23.6
1982-83	0.0	0.0	0.0	0.0	0.4	3.3	5.7	0.4	8.9	3.6	0.0	0.0	22.3
1983-84	0.0	0.0	0.0	0.0	1.9	12.9	5.3	0.2	11.5	8.9	0.3	0.0	41.0
1984-85	0.0	0.0	3.0	4.9	1.5	6.2	10.1	4.1	7.8	0.0	0.0	0.0	37.6
1985-86	0.0	0.0	1.2	T	23.5	4.9	3.9	3.7	2.4	2.5	T	0.0	42.1
1986-87	0.0	0.0	0.0	T	3.6	6.2	13.9	4.9	1.8	1.1	0.0	0.0	31.5
1987-88	0.0	0.0	0.0	0.0	5.2	12.0	18.3	8.3	11.1	8.9	0.4	0.0	64.2
1988-89	0.0	0.0	0.0	0.0	1.2	9.3	9.0	12.6	T	1.2	T	T	33.3
1989-90	0.0	0.0	0.0	T	T	10.4	15.0	12.7	2.5	18.4	10.6	0.0	69.6
1990-91	0.0	0.0	0.0	1.6	5.1	6.5	2.4	0.1	10.5	2.0	T	T	28.2
1991-92	T	T	T	16.3	25.6	4.0	1.1	0.5	1.2	0.0	0.0	T	48.7
1992-93	T	0.0	0.0				6.4						
1993-													
POR= 39 YRS	T	T	0.7	1.5	4.2	5.3	6.1	4.3	6.5	3.4	0.5	T	32.5

BRIDGEPORT, CONNECTICUT

The airport is located on Stratford Point, a peninsula jutting out into Long Island Sound. Station instrumentation is located approximately 1 mile from the sound. Land around the airport is flat, with marshes to the south. The terrain is of glacial origin, rising in a rolling, mostly wooded manner, to the foothills of the Berkshires, 30 miles to the north and northwest.

Cities in close proximity to the station are Bridgeport, Fairfield, and Milford, while Danbury, New Haven, Norwalk and Stamford are located within a 35-mile radius.

The most pronounced topographical effect is the land-sea breeze, an occurrence generally associated with the spring through early autumn months.

Mean monthly temperatures during the summer months average 3 to 5 degrees lower than nearby inland stations because of the sea-breeze effect. Temperatures during the fall and winter months are moderated because of the proximity of Long Island Sound.

Winter snowfall is generally around 10 inches less than areas a few miles inland, also due to the proximity of the station to Long Island Sound.

One of the hazards along the coastal areas is the flooding of low-lying areas (usually during periods of high tide) with the approach of slow-moving deepening low pressure systems, resulting in 3 to 5 feet higher tides than normal.

NORMALS, MEANS, AND EXTREMES

BRIDGEPORT, CONNECTICUT

LATITUDE: 41°10'N   LONGITUDE: 73°08' W   ELEVATION: FT. GRND 7 BARO 28   TIME ZONE: EASTERN   WBAN: 94702

	(a)	JAN	FEB	MAR	APR	MAY	JUNE	JULY	AUG	SEP	OCT	NOV	DEC	YEAR
TEMPERATURE °F:														
Normals														
-Daily Maximum		36.0	37.6	46.2	56.6	66.7	76.0	81.7	80.9	74.2	64.1	53.0	41.0	59.5
-Daily Minimum		21.9	23.2	30.9	39.8	50.0	59.1	65.7	65.1	57.6	47.1	38.2	27.6	43.9
-Monthly		28.9	30.4	38.6	48.2	58.4	67.6	73.7	73.1	65.9	55.6	45.6	34.3	51.7
Extremes														
-Record Highest	45	65	67	84	91	92	96	103	98	99	85	78	65	103
-Year		1974	1976	1990	1990	1987	1974	1957	1953	1953	1959	1993	1971	JUL 1957
-Record Lowest	45	-7	-5	4	18	31	41	49	44	36	26	16	-4	-7
-Year		1984	1963	1967	1982	1966	1967	1988	1982	1963	1988	1972	1980	JAN 1984
NORMAL DEGREE DAYS:														
Heating (base 65°F)		1119	969	818	504	219	18	0	0	54	302	582	952	5537
Cooling (base 65°F)		0	0	0	0	15	96	270	251	81	11	0	0	724
% OF POSSIBLE SUNSHINE														
MEAN SKY COVER (tenths)														
Sunrise - Sunset	44	6.3	6.1	6.3	6.4	6.5	6.0	6.0	5.8	5.7	5.4	6.3	6.3	6.1
MEAN NUMBER OF DAYS:														
Sunrise to Sunset														
-Clear	44	8.3	8.3	8.0	7.1	6.8	7.8	7.5	8.7	9.5	10.8	7.9	8.3	99.1
-Partly Cloudy	44	7.8	6.8	8.5	8.5	10.1	10.1	11.9	10.7	8.9	8.1	7.7	7.8	106.9
-Cloudy	44	14.8	13.2	14.5	14.4	14.1	12.2	11.7	11.6	11.6	12.0	14.3	14.9	159.2
Precipitation														
.01 inches or more	45	10.6	9.7	11.2	10.6	11.0	9.6	8.5	9.3	8.6	7.3	10.0	11.2	117.7
Snow,Ice pellets,hail														
1.0 inches or more	45	2.1	1.9	1.3	0.2	0.0	0.0	0.0	0.0	0.0	0.0	0.2	1.5	7.2
Thunderstorms	32	0.1	0.3	0.9	1.7	2.8	3.8	5.0	4.1	1.9	0.8	0.3	0.1	21.8
Heavy Fog Visibility														
1/4 mile or less	32	3.2	2.8	3.4	3.1	4.3	3.7	1.9	1.1	1.2	1.5	1.5	1.9	29.5
Temperature °F														
-Maximum														
90° and above	28	0.0	0.0	0.0	0.*	0.2	1.0	3.2	1.8	0.4	0.0	0.0	0.0	6.6
32° and below	28	10.0	6.8	1.2	0.*	0.0	0.0	0.0	0.0	0.0	0.0	0.1	4.8	22.9
-Minimum														
32° and below	28	26.1	23.3	17.0	3.4	0.1	0.0	0.0	0.0	0.0	0.9	7.2	21.2	99.1
0° and below	28	0.5	0.1	0.0	0.0	0.0	0.0	0.0	0.0	0.0	0.0	0.0	0.1	0.6
AVG. STATION PRESS. (mb)	7	1016.9	1016.7	1016.2	1013.8	1014.0	1015.9	1014.7	1016.8	1017.7	1017.6	1017.0	1017.4	1016.2
RELATIVE HUMIDITY (%)														
Hour 01	14	69	67	69	70	79	83	82	83	83	77	75	72	76
Hour 07 (Local Time)	27	72	71	72	71	75	78	78	80	82	79	77	73	76
Hour 13	28	60	58	56	54	60	61	61	62	62	59	61	61	60
Hour 19	27	65	63	61	62	67	69	69	71	72	69	68	66	67
PRECIPITATION (inches):														
Water Equivalent														
-Normal		3.24	3.01	3.75	3.75	3.93	3.46	3.78	3.25	3.07	3.11	3.81	3.50	41.66
-Maximum Monthly	45	11.20	6.65	9.40	10.72	9.53	17.70	12.84	13.29	7.42	10.72	10.22	7.87	17.70
-Year		1979	1972	1953	1983	1989	1972	1971	1952	1960	1955	1972	1972	JUN 1972
-Minimum Monthly	45	0.40	0.43	0.69	0.69	0.41	0.07	0.47	0.72	0.43	0.33	0.36	0.33	0.07
-Year		1970	1987	1981	1985	1986	1949	1979	1981	1959	1963	1976	1955	JUN 1949
-Maximum in 24 hrs	45	4.55	2.31	4.60	3.32	3.23	6.89	5.95	4.72	4.67	4.28	4.07	3.69	6.89
-Year		1979	1969	1977	1980	1968	1972	1971	1991	1960	1972	1954	1968	JUN 1972
Snow,Ice pellets,hail														
-Maximum Monthly	45	26.2	24.0	21.8	6.0	T	0.0	T	T	0.0	0.5	6.6	16.2	26.2
-Year		1965	1967	1967	1982	1977		1991	1992		1987	1989	1963	JAN 1965
-Maximum in 24 hrs	45	16.7	16.7	11.1	6.0	T	0.0	T	T	0.0	0.5	6.6	7.8	16.7
-Year		1978	1969	1967	1982	1977		1991	1992		1987	1989	1966	JAN 1978
WIND:														
Mean Speed (mph)	23	13.2	13.6	13.5	13.0	11.6	10.5	10.0	10.1	11.2	11.9	12.7	13.0	12.0
Prevailing Direction through 1963		NW	NW	NW	N	E	SW	SW	SW	NE	NE	NW	NW	NW
Fastest Obs. 1 Min.														
-Direction (!!)	33	34	34	08	32	34	29	29	04	18	09	14	25	18
-Speed (MPH)	33	67	65	58	55	50	38	40	58	74	58	58	53	74
-Year		1964	1969	1977	1962	1974	1978	1973	1976	1985	1980	1964	1962	SEP 1985
Peak Gust														
-Direction (!!)	10	SW	N	E	NW	SW	NW	NW	NW	SW		SW	NW	E
-Speed (mph)	10	59	61	69	48	51	48	59	68	49	53	61	53	69
-Date		1992	1992	1993	1988	1989	1985	1991	1992	1989	1988	1988	1992	MAR 1993

PRECIPITATION (inches)

BRIDGEPORT, CONNECTICUT

YEAR	JAN	FEB	MAR	APR	MAY	JUNE	JULY	AUG	SEP	OCT	NOV	DEC	ANNUAL
1964	2.85	2.24	2.00	4.79	0.98	1.41	1.85	0.99	0.54	1.24	1.90	2.24	23.03
1965	3.34	2.75	1.16	2.71	4.16	1.42	3.99	1.81	0.82	1.72	1.62	1.35	26.85
1966	2.53	3.80	1.62	2.26	3.59	0.99	1.39	1.83	6.01	3.74	3.12	2.80	33.68
1967	1.29	2.52	6.44	3.46	5.12	2.67	4.07	3.65	1.60	2.39	2.93	5.60	41.74
1968	1.92	1.50	4.51	1.75	5.59	5.68	1.00	4.72	3.17	1.79	6.59	7.45	45.67
1969	1.21	3.68	3.35	5.26	2.71	1.14	5.46	1.66	4.83	2.02	3.82	6.55	41.69
1970	0.40	4.47	3.88	3.90	2.03	2.39	0.94	3.37	2.13	1.63	5.97	2.46	33.57
1971	3.09	6.15	4.34	2.11	4.60	0.39	12.84	4.98	5.72	3.24	6.49	3.33	57.28
1972	2.23	6.65	5.33	5.64	7.38	17.70	1.76	0.77	2.02	6.36	10.22	7.87	73.93
1973	4.77	4.00	5.00	8.14	6.04	4.58	6.55	2.05	2.56	2.83	1.84	6.53	54.89
1974	4.45	1.97	6.05	2.70	2.79	2.24	1.43	3.02	5.39	2.52	1.26	5.93	39.75
1975	4.70	3.11	3.05	2.49	3.27	3.39	6.48	1.91	6.18	3.96	4.41	4.50	47.45
1976	5.56	3.20	2.75	2.38	3.15	2.62	2.09	6.16	1.52	6.62	0.36	2.59	39.00
1977	2.43	1.74	7.74	3.60	2.07	2.75	1.03	4.69	7.26	3.94	4.93	5.17	47.35
1978	7.91	1.34	3.95	1.97	5.12	1.59	2.59	5.90	3.75	2.54	1.74	4.76	43.16
1979	11.20	3.65	3.70	4.53	4.88	3.29	0.47	4.35	4.46	2.71	2.54	2.24	48.02
1980	1.02	1.07	7.05	7.03	2.69	2.52	5.97	2.38	2.39	4.12	3.95	0.95	41.14
1981	0.54	4.66	0.69	3.19	1.92	2.10	4.45	0.72	4.50	4.32	1.93	3.65	32.67
1982	5.50	2.47	2.76	3.83	3.02	11.53	3.31	3.14	1.30	1.52	3.13	1.10	42.61
1983	3.72	2.40	9.21	10.72	4.77	3.72	1.66	2.57	2.20	4.63	6.58	4.74	56.92
1984	1.52	4.72	3.49	4.37	8.14	3.53	6.54	1.23	2.24	2.79	1.83	2.56	42.96
1985	1.25	1.72	1.93	0.69	5.11	5.34	5.19	4.62	1.60	1.48	5.67	1.25	35.85
1986	2.66	3.05	2.32	1.65	0.41	3.16	5.74	2.43	0.85	2.14	4.91	4.41	33.73
1987	4.78	0.43	4.77	4.73	1.20	1.55	1.78	3.89	4.09	2.20	2.87	2.08	34.37
1988	2.65	3.64	2.36	1.59	2.65	0.79	8.53	1.86	2.26	3.26	7.58	1.63	38.80
1989	1.44	2.40	4.06	3.15	9.53	5.60	3.44	6.57	3.21	7.02	3.27	0.83	50.52
1990	4.01	1.94	2.10	4.87	6.89	1.91	2.83	6.47	1.75	5.72	1.89	3.53	43.91
1991	2.82	1.83	4.07	3.19	3.83	2.29	2.17	7.84	3.47	1.88	2.82	4.27	40.48
1992	1.92	2.12	3.64	1.89	2.85	5.13	3.76	8.38	5.32	2.42	4.46	4.30	46.19
1993	2.60	2.60	6.75	3.50	2.25	1.42	1.58	1.58	6.60	4.00	1.71	4.54	39.13
Record Mean	3.52	3.21	3.94	3.82	3.74	3.33	3.69	4.06	3.49	3.36	3.78	3.64	43.58

AVERAGE TEMPERATURE (deg. F)

BRIDGEPORT, CONNECTICUT

YEAR	JAN	FEB	MAR	APR	MAY	JUNE	JULY	AUG	SEP	OCT	NOV	DEC	ANNUAL
1964	31.6	29.9	39.1	46.4	60.4	67.4	72.3	68.9	64.9	53.1	46.0	34.1	51.2
#1965	26.4	29.4	36.7	46.5	61.4	66.7	71.2	73.5	67.5	55.0	45.2	38.1	51.4
1966	30.1	29.3	36.6	43.4	53.0	67.9	77.0	74.9	66.3	54.9	46.5	33.8	51.1
1967	33.8	27.4	33.6	44.8	51.1	67.2	73.7	72.3	65.4	56.4	41.1	35.9	50.2
1968	25.5	27.9	40.2	51.8	58.0	67.1	75.1	73.3	68.1	58.8	45.4	32.4	52.0
1969	30.6	29.9	37.0	50.7	59.8	67.9	71.6	73.3	64.9	54.7	45.1	32.7	51.5
1970	23.9	31.5	37.1	49.3	59.5	66.9	74.3	73.6	67.1	57.5	47.7	32.6	51.8
1971	26.2	32.9	38.1	47.0	57.3	69.5	73.5	72.2	69.0	60.2	43.5	38.6	52.3
1972	33.0	30.0	37.2	46.1	59.0	65.3	74.1	72.8	67.3	51.2	41.8	35.1	51.1
1973	30.4	29.2	42.1	49.1	55.5	70.4	75.2	75.8	68.0	59.9	48.0	37.7	53.5
1974	33.1	29.6	40.1	52.2	58.4	68.6	76.2	75.1	65.7	52.9	46.5	37.4	53.0
1975	35.3	32.2	36.8	44.7	61.9	67.2	74.5	74.0	64.5	59.0	50.5	35.5	53.0
1976	26.9	36.3	40.2	51.1	56.9	70.4	71.5	73.1	65.4	53.5	42.4	29.9	51.5
1977	23.7	31.2	41.9	49.1	60.4	65.1	72.6	74.1	67.8	56.9	50.0	33.5	52.2
1978	26.6	24.1	35.6	46.9	56.8	66.4	73.2	75.0	64.2	55.4	47.2	36.2	50.7
1979	30.6	24.6	43.0	49.5	61.7	66.6	73.8	72.8	64.8	53.2	47.3	37.9	52.2
1980	31.7	28.1	37.0	48.1	60.4	67.3	75.8	75.9	68.4	55.4	43.2	31.3	51.9
1981	22.2	33.9	38.1	49.3	58.6	68.0	74.0	71.4	62.8	50.5	44.1	33.8	50.6
1982	22.9	32.8	37.9	46.3	59.6	63.9	72.9	69.3	64.1	52.9	47.0	38.7	50.7
1983	31.4	32.3	40.5	48.1	55.8	67.7	74.3	73.1	67.3	55.1	46.5	33.1	52.1
1984	26.6	36.5	34.1	47.9	57.8	71.0	73.1	74.8	63.3	57.7	45.0	40.2	52.3
1985	26.2	32.3	41.8	50.8	60.9	65.6	73.6	72.8	66.6	56.0	46.6	31.4	52.1
1986	30.6	29.2	39.9	50.6	61.1	66.8	73.1	70.7	64.7	54.3	42.3	35.5	51.6
1987	30.1	30.0	41.3	50.2	58.8	68.6	74.9	71.0	65.2	51.3	44.9	36.8	51.9
1988	26.4	31.2	39.4	48.1	59.4	68.2	75.5	75.6	64.6	50.5	45.9	32.7	51.5
1989	33.8	31.0	38.8	47.9	59.6	68.5	71.8	71.6	65.2	54.9	43.2	23.3	50.8
1990	36.7	35.0	40.1	49.2	56.9	69.1	73.9	73.9	65.6	59.6	46.6	40.0	53.9
1991	31.7	36.2	42.2	51.8	64.5	70.3	75.5	74.8	65.1	56.7	45.9	36.7	54.3
1992	32.2	33.8	37.1	46.5	58.0	67.1	71.1	70.4	65.0	52.1	44.7	36.0	51.2
1993	33.6	27.9	36.7	49.2	61.4	70.0	76.7	75.3	65.7	53.3	45.0	35.5	52.5
Record Mean	28.7	30.6	38.0	48.1	58.5	67.8	73.4	72.1	65.2	54.6	44.2	33.3	51.2
Max	35.2	38.5	46.5	57.6	68.4	77.5	82.7	81.2	74.4	64.2	52.5	40.8	60.0
Min	22.1	22.6	29.5	38.6	48.6	58.1	64.1	62.9	55.9	45.1	35.9	25.8	42.4

REFERENCE NOTES

BRIDGEPORT, CONNECTICUT

GENERAL
T - TRACE AMOUNT.
BLANK ENTRIES DENOTE MISSING/UNREPORTED DATA.
# INDICATES A STATION OR INSTRUMENT RELOCATION.

SPECIFIC

(a) - LENGTH OF RECORD IN YEARS, ALTHOUGH INDIVIDUAL MONTHS MAY BE MISSING.
0.* OR * - THE VALUE IS BETWEEN 0.0 AND 0.05
NORMALS - BASED ON THE 1961-1990 RECORD PERIOD.
EXTREMES - DATES ARE THE MOST RECENT OCCURRENCE
WIND DIR.- NUMERALS SHOW TENS OF DEGRESS CLOCKWISE FROM TRUE NORTH. "00" INDICATES CALM.
RESULTANT DIRECTIONS ARE GIVEN TO WHOLE DEGREES.
BOLD VALUES INDICATE EXTREME VALUES WHICH OCCURRED AFTER THE ASOS SYSTEM WAS COMMISSIONED.

RECORD = PERIOD OF RECORD
RECORD MEAN PRECIPITATION IS THE MEAN OF ALL DAILY PRECIPITATION AMOUNTS DURING THE PERIOD OF RECORD.
RECORD MAX(MIN) TEMPERATURE IS THE MEAN OF ALL DAILY MAX(MIN) TEMPERATURES DURING THE PERIOD OF RECORD.
RECORD MEAN TEMPERATURE IS THE SUM OF THE RECORD MAX AND RECORD MIN DIVIDED BY 2.
AVERAGE TEMPERATURE IS THE SUM OF THE MEAN DAILY MAX AND MIN TEMPERATURE DIVIDED BY 2.

EXCEPTIONS

MEAN WIND SPEED, THUNDERSTORMS, AND HEAVY FOG ARE THROUGH 1980.

RECORD MEANS ARE THROUGH THE CURRENT YEAR, BEGINNING IN 1903 FOR TEMPERATURE
1894 FOR PRECIPITATION
1949 FOR SNOWFALL

HEATING DEGREE DAYS Base 65 deg. F — BRIDGEPORT, CONNECTICUT

SEASON	JULY	AUG	SEP	OCT	NOV	DEC	JAN	FEB	MAR	APR	MAY	JUNE	TOTAL
1964-65	6	9	83	364	562	952	1190	993	872	546	145	50	5772
#1965-66	0	12	50	307	587	827	1078	996	872	643	366	43	5781
1966-67	0	0	55	308	549	961	960	1047	968	597	424	35	5904
1967-68	0	2	69	274	712	895	1216	1068	762	391	214	30	5633
1968-69	0	1	12	209	579	1005	1061	978	862	421	184	13	5325
1969-70	1	4	92	321	593	996	1269	931	860	465	172	32	5736
1970-71	0	0	49	247	512	997	1196	893	825	531	234	22	5506
1971-72	0	4	27	150	643	813	983	1009	853	564	187	33	5266
1972-73	7	2	43	418	691	920	1066	996	704	472	290	8	5617
1973-74	0	0	35	171	501	838	984	988	765	380	219	21	4902
1974-75	0	0	67	368	551	847	916	907	866	604	141	34	5301
1975-76	0	0	59	190	432	908	1174	826	762	420	251	37	5059
1976-77	0	7	59	356	673	1079	1274	940	710	470	160	62	5790
1977-78	4	3	52	248	442	970	1181	1136	904	536	265	43	5784
1978-79	4	0	92	290	524	889	1062	1126	675	460	114	23	5259
1979-80	8	13	84	360	523	833	1025	1064	862	499	159	39	5469
1980-81	0	0	41	297	646	1038	1320	865	831	465	214	18	5735
1981-82	0	1	102	445	619	959	1297	897	832	556	164	72	5944
1982-83	1	19	66	371	530	809	1034	914	754	501	280	29	5308
1983-84	0	8	73	320	550	981	1188	819	952	508	227	18	5644
1984-85	0	0	104	219	593	761	1197	908	713	423	138	43	5099
1985-86	0	2	54	278	541	1032	1060	997	772	425	174	41	5376
1986-87	3	23	68	345	673	908	1077	975	727	434	232	23	5488
1987-88	0	10	53	417	596	867	1186	974	787	499	195	45	5629
1988-89	6	3	62	449	569	995	958	945	804	505	180	19	5495
1989-90	0	7	88	305	648	1285	869	836	766	476	243	12	5535
1990-91	5	1	77	208	546	771	1023	799	700	396	106	14	4646
1991-92	0	0	91	268	566	871	1012	895	859	548	225	23	5358
1992-93	4	3	80	393	599	892	966	1036	871	465	127	27	5463
1993-94	2	0	89	358	597	909							

COOLING DEGREE DAYS Base 65 deg. F — BRIDGEPORT, CONNECTICUT

YEAR	JAN	FEB	MAR	APR	MAY	JUNE	JULY	AUG	SEP	OCT	NOV	DEC	TOTAL
1969	0	0	0	0	29	108	212	269	96	11	0	0	725
1970	0	0	0	0	10	96	293	275	122	22	0	0	818
1971	0	0	0	0	1	161	269	233	154	9	2	0	829
1972	0	0	0	0	8	53	298	253	116	1	0	0	729
1973	0	0	0	0	6	177	324	342	131	21	0	0	1001
1974	0	0	0	6	20	135	353	320	95	0	0	0	929
1975	0	0	0	0	50	108	300	285	46	11	4	0	804
1976	0	0	0	10	6	205	204	266	78	6	0	0	775
1977	0	0	0	0	24	70	249	290	143	5	0	0	781
1978	0	0	0	0	19	93	264	318	76	1	0	0	771
1979	0	0	0	0	16	79	288	259	85	4	0	0	731
1980	0	0	0	0	23	114	338	346	148	6	0	0	975
1981	0	0	0	0	24	116	286	205	44	0	0	0	675
1982	0	0	0	0	9	46	253	158	45	2	0	0	513
1983	0	0	0	0	0	116	296	267	149	20	0	0	848
1984	0	0	0	0	7	207	258	313	61	0	0	0	846
1985	0	0	0	3	19	71	276	251	110	8	0	0	738
1986	0	0	0	0	60	103	263	204	66	18	0	0	714
1987	0	0	0	0	46	140	312	205	64	0	0	0	767
1988	0	0	0	0	30	148	339	341	57	6	0	0	921
1989	0	0	0	0	18	131	219	218	102	0	0	0	688
1990	0	0	0	10	0	142	288	282	104	47	0	0	873
1991	0	0	0	7	98	178	328	310	101	18	0	0	1040
1992	0	0	0	0	17	94	200	179	85	0	0	0	575
1993	0	0	0	0	22	182	371	324	117	1	2	0	1019

SNOWFALL (inches) — BRIDGEPORT, CONNECTICUT

SEASON	JULY	AUG	SEP	OCT	NOV	DEC	JAN	FEB	MAR	APR	MAY	JUNE	TOTAL
1964-65	0.0	0.0	0.0	0.0	0.0	3.9	26.2	6.1	4.7	1.3	0.0	0.0	42.2
1965-66	0.0	0.0	0.0	0.0	T	0.3	10.6	9.5	0.5	T	0.0	0.0	20.9
1966-67	0.0	0.0	0.0	0.0	0.0	14.0	1.8	24.0	21.8	T	0.0	0.0	61.6
1967-68	0.0	0.0	0.0	0.0	2.1	10.1	6.4	1.0	1.6	0.0	0.0	0.0	21.2
1968-69	0.0	0.0	0.0	0.0	1.3	2.4	0.5	21.7	2.0	0.0	0.0	0.0	27.9
1969-70	0.0	0.0	0.0	T	T	11.6	4.3	6.3	8.6	0.7	0.0	0.0	31.5
1970-71	0.0	0.0	0.0	T	T	9.3	8.5	2.9	3.1	1.2	0.0	0.0	25.0
1971-72	0.0	0.0	0.0	0.0	T	0.8	2.3	14.9	5.1	0.1	0.0	0.0	23.2
1972-73	0.0	0.0	0.0	T	0.1	1.5	4.3	2.3	T	T	0.0	0.0	8.2
1973-74	0.0	0.0	0.0	0.0	T	0.2	16.6	8.8	8.4	T	0.0	0.0	34.0
1974-75	0.0	0.0	0.0	0.0	0.4	2.3	2.5	10.4	1.8	0.1	0.0	0.0	17.5
1975-76	0.0	0.0	0.0	0.0	T	8.6	12.3	3.1	5.4	0.0	0.0	0.0	29.4
1976-77	0.0	0.0	0.0	0.0	T	6.5	10.3	5.8	4.7	T	T	0.0	27.3
1977-78	0.0	0.0	0.0	0.0	1.0	3.5	23.2	15.9	9.1	0.0	0.0	0.0	52.7
1978-79	0.0	0.0	0.0	0.0	2.2	1.2	6.1	8.6	T	0.0	0.0	0.0	18.1
1979-80	0.0	0.0	0.0	T	0.0	0.9	0.7	1.7	6.3	0.0	0.0	0.0	9.6
1980-81	0.0	0.0	0.0	0.0	1.0	3.6	5.4	T	1.5	0.0	0.0	0.0	11.5
1981-82	0.0	0.0	0.0	0.0	0.0	3.1	9.0	T	1.6	6.0	0.0	0.0	19.7
1982-83	0.0	0.0	0.0	0.0	0.0	3.2	5.1	14.2	T	0.5	0.0	0.0	23.0
1983-84	0.0	0.0	0.0	0.0	T	0.6	11.5	T	8.4	0.0	0.0	0.0	20.5
1984-85	0.0	0.0	0.0	0.0	T	1.5	11.2	6.5	0.5	0.0	0.0	0.0	19.7
1985-86	0.0	0.0	0.0	0.0	1.0	3.5	2.0	11.0	T	T	0.0	0.0	17.5
1986-87	0.0	0.0	0.0	0.0	1.8	3.3	11.4	2.9	2.7	0.0	0.0	0.0	22.1
1987-88	0.0	0.0	0.0	0.5	0.9	1.8	15.5	2.1	2.1	T	0.0	0.0	22.9
1988-89					0.0	7.5	1.6	2.1	1.8	0.0	0.0	0.0	
1989-90	0.0	0.0	0.0	0.0	6.6	4.1	5.2	9.3	4.4	1.7	0.0	0.0	31.3
1990-91	0.0	0.0	0.0	0.0	0.0	7.1	6.3	5.6	3.0	0.0	0.0	0.0	22.0
1991-92	T	0.0	0.0	0.0	0.0	2.1	1.8	2.4	10.2	0.0	0.0	0.0	16.5
1992-93	0.0	T	0.0	0.0	0.0	2.4	3.3	9.5	13.7	0.0	0.0	0.0	28.9
1993-94	0.0	0.0	0.0	0.0	0.0	1.9							
Record Mean	T	T	0.0	T	0.6	4.5	7.3	7.3	4.8	0.4	T	0.0	25.0

HARTFORD, CONNECTICUT

Bradley International Airport is located about 3 miles west of the Connecticut River on a slight rise of ground in a broad portion of the Connecticut River Valley between north-south mountain ranges whose heights do not exceed 1,200 feet.

The station is in the northern temperate climate zone. The prevailing west to east movement of air brings the majority of weather systems into Connecticut from the west. The average wintertime position of the Polar Front boundary between cold, dry polar air and warm, moist tropical air is just south of New England, which helps to explain the extensive winter storm activity and day to day variability of local weather. In summer, the Polar Front has an average position along the New England-Canada border with this station in a warm and pleasant atmosphere.

The location of Hartford, relative to continent and ocean, is also significant. Rapid weather changes result when storms move northward along the mid-Atlantic coast, frequently producing strong and persistent northeast winds associated with storms known locally as coastals or northeasters. Seasonally, weather characteristics vary from the cold and dry continental-polar air of winter to the warm and humid maritime air of summer.

Summer thunderstorms develop in the Berkshire Mountains to the west and northwest, move over the Connecticut Valley, and when accompanied by wind and hail, sometimes cause considerable damage to crops, particularly tobacco. During the winter, rain often falls through cold air trapped in the valley, creating extremely hazardous ice conditions. On clear nights in the late summer or early autumn, cool air drainage into the valley, and moisture from the Connecticut River, produce steam and/or ground fog which becomes quite dense throughout the valley, hampering ground and air transportation.

The mean date of the last springtime temperature of 32 degrees or lower is April 22, and the mean date of the first autumn temperature of 32 degrees is October 15.

NORMALS, MEANS, AND EXTREMES

HARTFORD, CONNECTICUT

LATITUDE: 41°56'N LONGITUDE: 72°41'W ELEVATION: FT. GRND 169 BARO 201 TIME ZONE: EASTERN WBAN: 14740

	(a)	JAN	FEB	MAR	APR	MAY	JUNE	JULY	AUG	SEP	OCT	NOV	DEC	YEAR
TEMPERATURE °F:														
Normals														
-Daily Maximum		33.2	36.4	46.8	59.9	71.6	80.0	85.0	82.7	74.8	63.7	51.0	37.5	60.2
-Daily Minimum		15.8	18.6	28.1	37.5	47.6	56.9	62.2	60.4	51.8	40.7	32.8	21.3	39.5
-Monthly		24.6	27.5	37.5	48.7	59.6	68.5	73.7	71.6	63.3	52.2	41.9	29.5	49.9
Extremes														
-Record Highest	39	65	73	87	96	97	100	102	101	99	91	81	74	102
-Year		1967	1985	1977	1976	1979	1964	1966	1975	1983	1963	1974	1984	JUL 1966
-Record Lowest	39	-26	-21	-6	9	28	37	44	36	30	17	1	-14	-26
-Year		1961	1961	1967	1970	1985	1986	1962	1965	1979	1978	1989	1980	JAN 1961
NORMAL DEGREE DAYS:														
Heating (base 65°F)		1252	1050	853	489	194	20	0	6	96	397	693	1101	6151
Cooling (base 65°F)		0	0	0	0	27	125	270	210	45	0	0	0	677
% OF POSSIBLE SUNSHINE	39	55	57	56	55	57	60	62	62	59	57	46	48	56
MEAN SKY COVER (tenths)														
Sunrise - Sunset	39	6.4	6.5	6.7	6.7	6.8	6.6	6.4	6.3	6.1	5.9	6.9	6.7	6.5
MEAN NUMBER OF DAYS:														
Sunrise to Sunset														
-Clear	39	7.7	6.8	6.7	6.4	5.6	5.7	5.8	6.7	8.5	9.3	5.7	6.9	81.6
-Partly Cloudy	39	7.9	7.5	8.5	8.4	9.7	10.4	12.1	10.9	8.8	8.7	8.3	7.6	108.8
-Cloudy	39	15.4	13.9	15.8	15.2	15.7	13.9	13.2	13.4	12.7	13.0	16.1	16.5	174.8
Precipitation														
.01 inches or more	39	10.5	10.3	11.4	11.2	11.7	11.3	9.7	9.8	9.5	8.5	11.2	11.9	127.0
Snow,Ice pellets,hail														
1.0 inches or more	39	2.9	2.6	2.5	0.5	0.*	0.0	0.0	0.0	0.0	0.*	0.5	3.2	12.3
Thunderstorms	39	0.1	0.2	0.7	1.2	2.2	3.9	4.5	3.7	2.1	1.0	0.4	0.1	20.2
Heavy Fog Visibility														
1/4 mile or less	39	2.2	2.4	2.2	1.3	1.7	2.3	1.9	2.3	3.2	3.6	2.2	2.9	28.4
Temperature °F														
-Maximum														
90° and above	34	0.0	0.0	0.0	0.3	1.2	3.4	7.6	4.8	1.3	0.*	0.0	0.0	18.8
32° and below	34	13.6	9.1	1.9	0.1	0.0	0.0	0.0	0.0	0.0	0.0	0.5	9.4	34.5
-Minimum														
32° and below	34	28.7	25.6	21.5	8.4	0.7	0.0	0.0	0.0	0.4	6.4	16.5	26.9	135.1
0° and below	34	2.8	1.6	0.*	0.0	0.0	0.0	0.0	0.0	0.0	0.0	0.0	1.1	5.5
AVG. STATION PRESS.(mb)	21	1010.0	1010.4	1009.3	1007.7	1008.4	1007.9	1008.5	1010.3	1011.7	1012.0	1011.1	1011.1	1009.9
RELATIVE HUMIDITY (%)														
Hour 01	34	69	68	68	69	76	81	82	84	86	81	75	73	76
Hour 07 (Local Time)	34	71	72	72	70	73	77	78	83	86	83	79	76	77
Hour 13	34	56	54	50	45	47	51	51	53	55	51	56	59	52
Hour 19	34	62	60	56	53	56	60	62	66	71	67	67	67	62
PRECIPITATION (inches):														
Water Equivalent														
-Normal		3.41	3.23	3.63	3.85	4.12	3.75	3.19	3.65	3.79	3.57	4.04	3.91	44.14
-Maximum Monthly	39	9.61	7.27	6.86	9.90	12.00	13.60	8.43	21.87	9.02	11.61	8.53	8.36	21.87
-Year		1978	1981	1983	1983	1989	1982	1988	1955	1975	1955	1972	1969	AUG 1955
-Minimum Monthly	39	0.38	0.45	0.27	1.38	0.73	0.67	1.07	0.54	0.84	0.35	0.51	0.78	0.27
-Year		1981	1987	1981	1966	1959	1988	1983	1981	1986	1963	1976	1955	MAR 1981
-Maximum in 24 hrs	39	2.56	2.16	2.62	3.01	4.90	6.14	3.48	12.12	5.28	4.45	2.90	3.12	12.12
-Year		1979	1965	1987	1979	1989	1982	1960	1955	1960	1959	1988	1973	AUG 1955
Snow,Ice pellets,hail														
-Maximum Monthly	39	37.0	32.2	43.3	14.3	1.3	T	0.0	0.0	0.0	1.7	8.7	35.4	43.3
-Year		1978	1969	1956	1982	1977	1993				1979	1986	1969	MAR 1956
-Maximum in 24 hrs	39	14.7	21.0	14.8	14.1	1.3	T	0.0	0.0	0.0	1.7	8.6	13.9	21.0
-Year		1978	1983	1993	1982	1977	1993				1979	1980	1969	FEB 1983
WIND:														
Mean Speed (mph)	39	9.0	9.5	10.0	10.0	8.8	8.1	7.4	7.2	7.3	7.8	8.5	8.7	8.5
Prevailing Direction through 1963		NW	NW	NW	S	S	S	S	S	S	N	S	N	S
Fastest Obs. 1 Min.														
-Direction (!!)	8	30	29	18	29	05	30	35	18	17	26	19	31	17
-Speed (MPH)	8	30	37	37	33	28	26	35	40	43	33	40	39	43
-Year		1993	1989	1987	1988	1987	1986	1991	1988	1985	1986	1989	1992	SEP 1985
Peak Gust														
-Direction (!!)	10	NW	NW	S	NW	NW	W	SW	NW	S	S	S	NW	SW
-Speed (mph)	10	49	49	62	48	48	53	89	55	66	72	60	55	89
-Date		1993	1990	1987	1988	1990	1993	1988	1991	1985	1990	1989	1992	JUL 1988

PRECIPITATION (inches) HARTFORD, CONNECTICUT

YEAR	JAN	FEB	MAR	APR	MAY	JUNE	JULY	AUG	SEP	OCT	NOV	DEC	ANNUAL
1964	5.54	3.43	3.63	3.76	0.87	1.68	3.07	2.46	1.42	1.98	2.20	4.51	34.55
1965	2.73	4.43	1.50	2.14	1.20	1.98	1.68	1.09	3.50	4.97	2.21	2.02	29.45
1966	3.05	4.40	3.15	1.38	3.01	2.72	3.96	1.90	6.05	4.23	3.82	3.60	41.27
1967	2.01	2.00	4.43	4.18	6.34	3.82	2.59	4.34	2.85	2.59	3.13	6.77	45.05
1968	1.92	1.14	4.55	2.74	3.93	6.65	1.58	2.53	2.88	2.03	5.46	5.22	40.63
1969	1.19	3.32	3.11	5.63	3.13	2.63	5.79	3.32	3.53	1.45	6.17	8.36	47.63
1970	0.39	5.15	4.28	4.12	3.59	2.49	1.49	3.18	3.20	2.08	4.31	4.15	38.43
1971	2.80	4.60	3.24	2.85	4.08	0.71	3.23	5.31	4.24	4.46	5.68	3.55	44.75
1972	2.02	5.12	6.71	4.61	7.49	9.66	3.84	3.45	1.84	4.20	8.53	7.08	64.55
1973	3.28	3.05	3.22	6.59	5.95	5.07	1.77	4.50	3.73	3.47	2.14	8.31	51.08
1974	4.10	1.95	4.49	3.64	3.03	2.38	2.39	3.36	8.57	2.34	2.62	4.52	43.39
1975	4.30	3.22	3.82	2.99	3.29	3.83	6.11	4.60	9.02	5.28	4.57	4.31	55.34
1976	5.57	3.11	2.86	3.93	4.45	2.86	3.51	5.76	2.55	4.10	0.51	2.97	42.18
1977	2.41	2.81	6.57	4.89	3.70	3.99	3.37	2.44	8.17	5.45	4.38	5.68	53.86
1978	9.61	1.42	3.63	1.51	4.61	2.94	2.51	3.61	2.67	1.75	2.12	4.23	40.61
1979	9.12	2.83	4.25	5.88	3.48	0.91	1.97	4.44	2.95	4.76	3.46	2.57	46.62
1980	0.72	0.98	5.87	5.39	1.65	3.81	2.65	1.60	1.40	2.58	4.22	0.82	31.69
1981	0.38	7.27	0.27	2.92	2.17	1.37	4.21	0.54	4.49	5.19	2.34	4.00	35.15
1982	4.76	2.83	2.23	4.12	3.30	13.60	2.60	4.41	2.41	3.31	3.12	1.32	48.01
1983	4.68	3.83	6.86	9.90	4.82	2.61	1.07	2.55	2.10	5.52	6.09	5.97	56.00
1984	1.80	4.72	3.93	4.24	11.55	2.16	4.22	1.32	1.20	2.76	2.49	2.46	42.85
1985	0.73	1.72	2.16	1.54	2.77	3.55	4.55	6.44	3.83	2.27	6.04	1.28	36.88
1986	5.34	3.02	2.72	1.55	2.28	6.79	4.44	3.44	0.84	2.18	5.57	6.15	44.32
1987	6.20	0.45	4.44	5.23	2.18	3.66	2.27	4.25	7.19	3.67	3.66	1.57	44.77
1988	3.36	3.99	2.06	2.35	3.46	0.67	8.43	2.12	1.88	2.29	7.84	1.35	39.80
1989	0.88	1.85	3.02	3.33	12.00	6.65	3.40	6.81	4.67	7.62	2.89	1.49	54.61
1990	4.03	3.37	2.46	4.55	6.38	3.59	2.09	8.32	2.13	7.63	3.76	4.86	53.17
1991	2.45	1.78	4.52	3.54	5.18	2.37	2.90	8.69	5.67	3.17	4.03	2.96	47.26
1992	2.73	2.23	3.79	3.13	3.21	5.77	4.62	3.60	2.43	1.95	4.19	4.33	41.98
1993	2.63	2.90	6.67	4.71	1.92	2.63	4.90	1.80	5.35	4.15	3.27	4.16	45.09
Record Mean	3.50	3.17	3.74	3.76	3.71	3.59	3.57	3.91	3.62	3.24	3.83	3.71	43.35

AVERAGE TEMPERATURE (deg. F) HARTFORD, CONNECTICUT

YEAR	JAN	FEB	MAR	APR	MAY	JUNE	JULY	AUG	SEP	OCT	NOV	DEC	ANNUAL
1964	27.1	25.9	37.7	48.8	62.6	68.3	75.0	68.1	63.0	51.3	43.3	29.7	50.1
1965	22.5	26.6	36.1	47.5	64.4	68.9	72.4	71.7	64.1	51.1	39.6	33.8	49.9
1966	26.4	28.8	39.1	46.9	56.4	70.7	75.8	73.4	63.4	51.9	46.2	31.8	50.9
1967	32.8	24.5	33.3	47.1	53.7	70.6	74.0	70.2	62.8	52.8	38.0	32.1	49.4
1968	21.1	25.8	39.9	52.4	58.2	67.4	75.1	70.8	65.0	54.7	38.9	25.4	49.6
1969	24.0	26.8	34.5	51.1	58.8	68.2	70.6	73.3	63.8	51.7	41.2	26.9	49.2
1970	16.8	28.2	35.1	49.0	61.5	67.8	75.2	73.8	65.4	54.9	44.1	26.8	49.9
1971	19.4	29.0	35.4	46.3	58.1	71.0	73.7	72.2	67.7	58.3	39.6	32.8	50.3
1972	27.9	26.0	34.6	44.3	60.1	65.9	73.8	71.1	64.3	48.9	38.7	30.8	48.9
1973	29.4	27.2	43.3	50.4	58.0	71.8	75.0	76.4	64.7	54.7	43.6	32.9	52.3
1974	28.2	27.0	36.8	50.7	56.6	67.3	73.7	72.7	63.3	47.3	40.7	30.7	49.6
1975	31.2	29.7	35.9	45.8	64.6	68.3	76.1	71.8	61.7	55.7	48.2	28.5	51.4
1976	19.5	34.8	40.3	53.3	58.4	72.7	72.4	71.0	62.4	49.9	38.3	24.8	49.8
1977	18.7	27.6	42.9	51.2	63.6	68.0	74.5	72.8	64.0	52.0	44.4	28.0	50.6
1978	23.6	22.1	35.1	48.1	59.9	69.2	71.9	70.0	58.6	49.0	38.6	29.3	47.9
1979	26.6	18.0	41.2	49.0	64.1	69.0	74.6	70.8	61.6	50.7	45.5	33.6	50.4
1980	27.6	24.3	35.2	49.2	61.0	66.4	74.2	73.2	64.9	50.3	37.9	24.6	49.1
1981	17.8	35.3	38.1	52.0	61.6	69.6	74.8	70.6	62.5	49.3	43.7	31.0	50.6
1982	18.8	29.2	36.7	45.8	61.4	65.0	74.4	69.5	63.0	51.5	45.8	36.0	49.8
1983	27.1	29.1	39.2	48.9	56.8	69.9	74.9	72.7	66.5	52.5	42.7	28.1	50.7
1984	21.8	34.3	31.4	48.0	56.0	69.8	71.8	73.2	59.8	55.2	41.5	35.7	49.9
1985	21.5	29.9	39.7	50.7	60.6	63.7	72.4	70.2	63.4	51.9	43.2	27.5	49.6
1986	27.4	26.2	38.7	51.0	61.7	66.0	72.3	69.5	61.8	51.4	38.3	33.1	49.8
1987	25.0	26.7	39.8	49.7	60.8	68.8	74.2	69.0	62.9	49.2	41.4	33.2	50.1
1988	23.1	28.3	38.5	47.4	59.7	66.7	75.2	74.5	62.2	47.6	42.3	29.3	49.6
1989	30.8	28.6	37.4	46.5	60.4	68.3	72.6	71.4	63.9	53.4	40.9	18.1	49.4
1990	34.7	33.0	40.2	49.2	56.7	69.0	74.4	73.3	64.0	57.4	44.5	36.7	52.8
1991	27.0	33.9	40.5	53.3	65.8	70.5	73.7	73.1	62.1	55.1	42.7	32.8	52.5
1992	28.6	30.3	34.6	46.4	58.5	66.4	69.9	69.1	62.6	49.2	40.6	31.2	49.0
1993	28.8	23.8	34.5	49.6	61.2	68.6	74.3	73.4	63.0	49.8	40.8	30.9	49.9
Record Mean	26.6	27.9	37.2	48.2	59.3	67.9	73.2	71.0	63.5	53.0	42.1	30.4	50.0
Max	34.8	36.4	46.1	58.5	70.4	78.7	83.7	81.4	74.1	63.7	50.7	38.2	59.7
Min	18.5	19.3	28.2	37.9	48.1	57.1	62.6	60.6	52.8	42.2	33.4	22.6	40.3

REFERENCE NOTES HARTFORD, CONNECTICUT

GENERAL
T - TRACE AMOUNT.
BLANK ENTRIES DENOTE MISSING/UNREPORTED DATA.
# INDICATES A STATION OR INSTRUMENT RELOCATION.

SPECIFIC

PM - INCLUDES LAST DAY OF PREVIOUS MONTH
ASOS - AUTOMATED SURFACE OBSERVING SYSTEM IN OPERATION DURING THESE MONTHS.

(a) - LENGTH OF RECORD IN YEARS, ALTHOUGH INDIVIDUAL MONTHS MAY BE MISSING.
0.* OR * - THE VALUE IS BETWEEN 0.0 AND 0.05
NORMALS - BASED ON THE 1961-1990 RECORD PERIOD.
EXTREMES - DATES ARE THE MOST RECENT OCCURRENCE
WIND DIR.- NUMERALS SHOW TENS OF DEGRESS CLOCKWISE FROM TRUE NORTH. "00" INDICATES CALM.
RESULTANT DIRECTIONS ARE GIVEN TO WHOLE DEGREES.
BOLD VALUES INDICATE EXTREME VALUES WHICH OCCURRED AFTER THE ASOS SYSTEM WAS COMMISSIONED.

RECORD = PERIOD OF RECORD
RECORD MEAN PRECIPITATION IS THE MEAN OF ALL DAILY PRECIPITATION AMOUNTS DURING THE PERIOD OF RECORD.
RECORD MAX(MIN) TEMPERATURE IS THE MEAN OF ALL DAILY MAX(MIN) TEMPERATURES DURING THE PERIOD OF RECORD.
RECORD MEAN TEMPERATURE IS THE SUM OF THE RECORD MAX AND RECORD MIN DIVIDED BY 2.
AVERAGE TEMPERATURE IS THE SUM OF THE MEAN DAILY MAX AND MIN TEMPERATURE DIVIDED BY 2.

EXCEPTIONS

RECORD MEANS ARE THROUGH THE CURRENT YEAR, BEGINNING IN 1905 FOR TEMPERATURE
1905 FOR PRECIPITATION
1955 FOR SNOWFALL

HEATING DEGREE DAYS Base 65 deg. F HARTFORD, CONNECTICUT

SEASON	JULY	AUG	SEP	OCT	NOV	DEC	JAN	FEB	MAR	APR	MAY	JUNE	TOTAL
1964-65	2	16	128	419	644	1084	1311	1068	888	518	99	45	6222
1965-66	2	34	110	426	753	961	1186	1008	792	534	275	31	6112
1966-67	0	0	98	398	554	1019	993	1126	976	530	349	18	6061
1967-68	0	10	112	380	803	1013	1353	1134	771	374	205	46	6201
1968-69	0	16	56	326	775	1219	1262	1063	938	415	221	27	6318
1969-70	8	7	116	409	707	1175	1485	1026	922	483	161	37	6536
1970-71	0	0	89	322	619	1177	1410	1001	911	556	215	17	6317
1971-72	2	11	55	214	756	992	1140	1125	935	619	172	50	6071
1972-73	6	9	82	494	782	1054	1097	1051	665	444	229	13	5926
1973-74	0	0	102	322	635	988	1134	1056	868	435	287	37	5864
1974-75	2	1	121	542	725	1057	1040	986	894	567	111	43	6089
1975-76	0	11	121	292	503	1125	1403	869	759	391	213	20	5707
1976-77	0	16	118	467	794	1242	1429	1038	684	419	130	45	6382
1977-78	1	8	112	399	610	1141	1276	1192	920	500	220	25	6404
1978-79	9	15	209	489	790	1102	1184	1310	730	473	81	26	6418
1979-80	16	30	152	442	578	965	1151	1174	916	466	146	68	6104
1980-81	0	0	99	449	808	1246	1456	824	828	380	149	10	6249
1981-82	0	9	115	481	635	1048	1427	996	871	569	128	64	6343
1982-83	1	30	96	416	575	894	1170	1002	793	483	261	24	5745
1983-84	0	7	106	404	662	1135	1332	884	1035	503	286	32	6386
1984-85	3	3	186	298	698	896	1341	975	776	428	167	76	5847
1985-86	0	14	119	401	648	1157	1159	1081	809	413	174	63	6038
1986-87	14	32	135	422	793	981	1230	1065	773	452	191	29	6117
1987-88	1	31	100	481	700	981	1292	1057	817	523	186	75	6244
1988-89	9	23	112	539	672	1101	1054	1012	847	553	175	31	6128
1989-90	0	22	103	354	715	1444	935	890	763	478	251	21	5976
1990-91	5	0	112	276	608	873	1170	863	755	373	107	16	5158
1991-92	1	0	156	311	663	990	1122	1002	936	553	218	37	5989
1992-93	9	16	138	486	722	1042	1114	1148	935	454	139	43	6246
1993-94	3	4	142	464	722	1049							

COOLING DEGREE DAYS Base 65 deg. F HARTFORD, CONNECTICUT

YEAR	JAN	FEB	MAR	APR	MAY	JUNE	JULY	AUG	SEP	OCT	NOV	DEC	TOTAL
1969	0	0	0	4	35	131	189	273	88	5	0	0	725
1970	0	0	0	8	58	129	322	280	108	16	0	0	921
1971	0	0	0	0	7	201	274	243	142	11	4	0	882
1972	0	0	0	3	30	83	286	203	70	1	0	0	676
1973	0	0	0	11	19	221	318	362	99	8	0	0	1038
1974	0	0	0	11	34	110	282	247	77	0	3	0	764
1975	0	0	0	0	106	147	348	229	27	7	6	0	870
1976	0	0	0	47	18	257	236	208	47	6	0	0	819
1977	0	0	4	13	93	144	303	259	88	1	0	0	905
1978	0	0	0	0	71	159	228	173	26	0	0	0	657
1979	0	0	0	0	60	151	320	218	56	6	0	0	811
1980	0	0	0	0	31	117	296	263	107	1	0	0	815
1981	0	0	0	0	53	152	311	190	48	0	0	0	754
1982	0	0	0	0	22	70	298	176	45	2	3	0	616
1983	0	0	0	5	16	177	313	253	158	23	0	0	945
1984	0	0	0	0	11	182	218	265	38	4	0	0	718
1985	0	0	0	3	37	44	234	182	78	3	0	0	581
1986	0	0	0	0	79	103	249	179	48	7	0	0	665
1987	0	0	0	3	70	150	292	161	42	0	0	0	718
1988	0	0	0	0	27	134	331	326	37	6	0	0	861
1989	0	0	0	0	37	136	240	224	77	0	0	0	714
1990	0	0	0	13	1	146	305	263	89	48	1	0	866
1991	0	0	0	29	139	191	278	257	76	12	0	0	982
1992	0	0	0	0	21	85	170	151	74	2	0	0	503
1993	0	0	0	0	28	155	297	271	89	0	0	0	840

SNOWFALL (inches) HARTFORD, CONNECTICUT

SEASON	JULY	AUG	SEP	OCT	NOV	DEC	JAN	FEB	MAR	APR	MAY	JUNE	TOTAL
1964-65	0.0	0.0	0.0	T	T	12.3	28.7	6.0	7.8	1.7	0.0	0.0	56.5
1965-66	0.0	0.0	0.0	0.0	T	4.5	18.4	19.1	10.6	T	0.0	0.0	52.6
1966-67	0.0	0.0	0.0	0.0	T	20.0	3.0	25.2	33.2	1.4	T	0.0	82.8
1967-68	0.0	0.0	0.0	0.0	3.7	21.7	7.5	1.6	6.8	0.0	0.0	0.0	41.3
1968-69	0.0	0.0	0.0	0.0	7.9	13.9	3.4	32.2	4.4	0.0	0.0	0.0	61.8
1969-70	0.0	0.0	0.0	T	T	35.4	3.2	6.7	14.7	2.0	0.0	0.0	62.0
1970-71	0.0	0.0	0.0	T	T	27.0	17.7	8.4	12.8	4.0	0.0	0.0	69.9
1971-72	0.0	0.0	0.0	0.0	8.2	6.6	2.9	24.9	13.2	2.7	0.0	0.0	58.5
1972-73	0.0	0.0	0.0	0.4	2.1	12.0	14.1	5.9	0.4	0.3	0.0	0.0	35.2
1973-74	0.0	0.0	0.0	0.0	T	3.1	14.3	5.8	4.8	2.1	0.0	0.0	30.1
1974-75	0.0	0.0	0.0	T	0.8	8.5	10.2	16.0	2.5	0.3	0.0	0.0	38.3
1975-76	0.0	0.0	0.0	0.0	0.3	13.4	15.6	5.0	12.3	0.0	0.0	0.0	46.6
1976-77	0.0	0.0	0.0	0.0	0.4	7.3	20.0	9.1	11.0	0.3	1.3	0.0	49.4
1977-78	0.0	0.0	0.0	0.0	1.3	12.6	37.0	18.1	13.3	T	0.0	0.0	82.3
1978-79	0.0	0.0	0.0	0.0	4.3	10.3	8.6	9.2	T	3.6	0.0	0.0	36.0
1979-80	0.0	0.0	0.0	1.7	0.0	0.9	0.2	7.7	5.9	T	0.0	0.0	16.4
1980-81	0.0	0.0	0.0	0.0	8.6	3.9	4.1	0.9	0.2	0.0	0.0	0.0	17.7
1981-82	0.0	0.0	0.0	T	T	13.1	16.7	5.8	6.5	14.3	0.0	0.0	56.4
1982-83	0.0	0.0	0.0	0.0	T	5.7	10.2	29.4	0.2	0.9	0.0	0.0	46.4
1983-84	0.0	0.0	0.0	0.0	T	7.9	14.7	1.3	19.3	T	0.0	0.0	43.2
1984-85	0.0	0.0	0.0	0.0	0.1	3.8	6.9	9.4	2.1	1.4	0.0	0.0	23.7
1985-86	0.0	0.0	0.0	0.0	2.0	5.4	5.1	11.8	0.2	0.8	0.0	0.0	25.3
1986-87	0.0	0.0	0.0	0.0	8.7	4.9	34.0	1.6	1.7	0.4	0.0	0.0	51.3
1987-88	0.0	0.0	0.0	T	8.6	5.8	22.6	17.6	4.9	T	T	0.0	59.5
1988-89	0.0	0.0	0.0	0.0	0.0	6.3	0.6	4.6	3.4	T	T	0.0	14.9
1989-90	0.0	0.0	0.0	0.0	5.3	12.4	10.5	9.0	4.3	1.5	0.0	0.0	43.0
1990-91	0.0	0.0	0.0	0.0	T	8.1	10.2	5.8	5.7	0.0	0.0	0.0	29.8
1991-92	0.0	0.0	0.0	0.0	0.7	6.0	1.7	5.3	7.3	2.6	0.0	0.0	23.6
1992-93	0.0	0.0	0.0	T	T	6.7	10.5	13.8	31.1	T	0.0	T	62.1
1993-94	0.0	0.0	0.0	0.0	T	6.7							
Record Mean	0.0	0.0	0.0	0.1	2.0	10.3	12.0	11.4	9.9	1.6	T	T	47.1

WILMINGTON, DELAWARE

Delaware is part of the Atlantic Coastal Plain consisting mainly of flat low land with many marshes. Small streams and tidal estuaries comprise the drainage of the State. Wilmington, at the northern end of the State, marks the beginning of low rolling hills extending northward and northwestward into Pennsylvania. The Delaware River, the Delaware Bay, and the Atlantic Ocean are along the eastern boundary of the State. The broad Chesapeake Bay lies 35 miles, or less, to the west of the western boundary of nearly the entire State. These large water areas considerably influence the climate of the Wilmington, Delaware region.

Summers are warm and humid, winters are usually mild. During the summer maximum temperatures are usually in the 80s. The temperature reaches 100 degrees on the average once in six years. During January, the coldest month of the year, the daily average temperature is 32 degrees. Temperatures of zero may be expected once in four years. Most of the winter precipitation falls as rain. Seasonal snowfall has been as little as 1 inch, and as much as 50 inches. Snow is frequently mixed with rain and sleet, and seldom remains on the ground more than a few days.

The proximity of large water areas and the inflow of southerly winds cause the relative humidity to be quite high all year. During the summer months the relative humidity is approximately 75 percent. Fog is relatively frequent and may occur in any month. Light southeast winds blowing up the Delaware Bay favor the formation of fog. Light north-northeast winds bring in smoke from Philadelphia and from the heavy industry area located along the Delaware River north of Wilmington.

Rainfall distribution throughout the year is fairly uniform, however, the greatest amounts normally come during the summer months. Mostly, the summer rainfall comes in the form of thunderstorms. Moisture deficiencies for crops occur occasionally, but severe droughts are rare. During the fall, winter, and spring seasons, much of the rainfall comes from storms forming over the southern states or the South Atlantic and moving northward along the coast. During the late summer and early fall, hurricanes occasionally cause heavy rainfall, but winds seldom reach hurricane force in Wilmington. Heavy rains occasionally cause minor flooding, but the streams and rivers of northern Delaware are not subject to major flooding. Strong easterly and southeasterly winds sometimes cause high tides in the Delaware Bay and the Delaware River, resulting in the flooding of lowlands and damage to bay front and river front properties.

Based on the 1951–1980 period, the average first occurrence of 32 degrees Fahrenheit in the fall is October 29 and the average last occurrence in the spring is April 13.

NORMALS, MEANS, AND EXTREMES

WILMINGTON, DELAWARE

LATITUDE: 39°40'N LONGITUDE: 75°36' W ELEVATION: FT. GRND 74 BARO 96 TIME ZONE: EASTERN WBAN: 13781

	(a)	JAN	FEB	MAR	APR	MAY	JUNE	JULY	AUG	SEP	OCT	NOV	DEC	YEAR
TEMPERATURE °F:														
Normals														
-Daily Maximum		38.7	41.9	52.1	62.6	72.9	81.4	85.6	84.1	77.7	66.6	55.5	43.9	63.6
-Daily Minimum		22.4	24.8	33.1	41.8	52.2	61.6	67.1	65.9	58.2	45.7	37.0	27.6	44.8
-Monthly		30.6	33.4	42.7	52.2	62.5	71.5	76.4	75.0	68.0	56.2	46.3	35.8	54.2
Extremes														
-Record Highest	46	75	78	86	94	95	99	102	101	100	91	85	74	102
-Year		1950	1985	1948	1985	1962	1952	1966	1955	1983	1951	1950	1984	JUL 1966
-Record Lowest	46	-14	-6	2	18	30	41	48	43	36	24	14	-7	-14
-Year		1985	1979	1984	1982	1978	1972	1988	1982	1974	1976	1955	1983	JAN 1985
NORMAL DEGREE DAYS:														
Heating (base 65°F)		1066	885	691	384	122	0	0	0	35	288	561	905	4937
Cooling (base 65°F)		0	0	0	0	44	199	353	310	125	15	0	0	1046
% OF POSSIBLE SUNSHINE														
MEAN SKY COVER (tenths)														
Sunrise - Sunset	46	6.6	6.4	6.4	6.4	6.5	6.0	5.9	5.8	5.7	5.5	6.4	6.6	6.2
MEAN NUMBER OF DAYS:														
Sunrise to Sunset														
-Clear	46	7.2	7.3	7.8	7.2	6.6	7.7	8.0	9.1	10.0	10.8	7.6	7.6	96.9
-Partly Cloudy	46	7.2	6.7	8.1	8.3	10.0	10.8	11.3	10.2	8.0	8.3	8.1	7.4	104.3
-Cloudy	46	16.6	14.2	15.2	14.5	14.4	11.5	11.7	11.7	12.0	11.9	14.3	16.0	164.1
Precipitation														
.01 inches or more	46	10.7	9.4	11.0	10.9	11.3	9.6	9.2	8.9	8.2	7.7	9.4	9.9	116.1
Snow, Ice pellets, hail														
1.0 inches or more	46	2.1	1.5	0.9	0.1	0.0	0.0	0.0	0.0	0.0	0.*	0.2	0.9	5.7
Thunderstorms	46	0.2	0.3	1.2	2.1	4.3	5.8	6.2	5.7	2.4	0.9	0.6	0.2	29.8
Heavy Fog Visibility														
1/4 mile or less	46	4.0	3.4	2.9	2.0	2.2	1.7	1.6	2.3	2.5	3.9	3.5	3.6	33.7
Temperature °F														
-Maximum														
90° and above	46	0.0	0.0	0.0	0.2	0.9	3.9	8.2	5.2	1.7	0.*	0.0	0.0	20.1
32° and below	46	7.3	4.4	0.8	0.0	0.0	0.0	0.0	0.0	0.0	0.0	0.1	4.0	16.5
-Minimum														
32° and below	46	25.4	22.0	15.2	3.0	0.*	0.0	0.0	0.0	0.0	1.6	10.3	22.2	99.8
0° and below	46	0.5	0.2	0.0	0.0	0.0	0.0	0.0	0.0	0.0	0.0	0.0	0.1	0.7
AVG. STATION PRESS. (mb)	21	1015.8	1015.7	1014.0	1012.4	1012.7	1012.4	1013.0	1014.5	1015.8	1016.6	1016.4	1016.5	1014.7
RELATIVE HUMIDITY (%)														
Hour 01	46	73	72	71	73	79	82	83	85	85	83	77	75	78
Hour 07 (Local Time)	46	75	74	74	73	76	78	79	83	85	84	80	76	78
Hour 13	46	60	57	52	50	53	53	54	57	56	54	56	60	55
Hour 19	46	68	64	61	60	64	64	66	70	71	70	69	69	66
PRECIPITATION (inches):														
Water Equivalent														
-Normal		3.03	2.91	3.43	3.39	3.84	3.55	4.23	3.40	3.43	2.88	3.27	3.48	40.84
-Maximum Monthly	46	8.41	7.02	7.50	6.80	7.38	7.49	12.63	12.09	9.53	6.41	7.84	7.90	12.63
-Year		1978	1979	1993	1983	1983	1972	1989	1955	1960	1971	1972	1969	JUL 1989
-Minimum Monthly	46	0.52	0.83	0.81	0.35	0.22	0.21	0.16	0.25	0.82	0.21	0.49	0.19	0.16
-Year		1981	1980	1966	1985	1964	1988	1955	1972	1970	1963	1976	1955	JUL 1955
-Maximum in 24 hrs	46	2.12	2.29	3.11	2.56	2.72	4.35	6.83	4.11	5.62	3.88	3.83	2.33	6.83
-Year		1978	1966	1978	1961	1990	1972	1989	1971	1960	1966	1956	1992	JUL 1989
Snow, Ice pellets, hail														
-Maximum Monthly	46	21.4	27.5	20.3	2.6	T	T	T	0.0	0.0	2.5	11.9	21.5	27.5
-Year		1987	1979	1958	1982	1991	1992	1990			1979	1953	1966	FEB 1979
-Maximum in 24 hrs	46	12.1	16.5	15.6	2.4	T	T	T	0.0	0.0	2.5	11.9	12.4	16.5
-Year		1987	1979	1958	1987	1991	1992	1990			1979	1953	1966	FEB 1979
WIND:														
Mean Speed (mph)	45	9.8	10.4	11.1	10.5	9.0	8.4	7.8	7.4	7.8	8.2	9.1	9.3	9.1
Prevailing Direction through 1963		WNW	NW	WNW	WNW	S	S	NW	S	S	NW	NW	WNW	S
Fastest Obs. 1 Min.														
-Direction (!!)	45	29	29	06	29	30	34	27	35	07	20	16	32	20
-Speed (MPH)	45	46	46	43	45	46	43	48	46	40	58	46	46	58
-Year		1957	1956	1984	1963	1984	1991	1963	1971	1956	1954	1950	1988	OCT 1954
Peak Gust														
-Direction (!!)	10	NW	NW	NE	NW	NW	NW	S	NW	NW	W	NW	NW	NW
-Speed (mph)	10	51	51	59	47	71	52	66	60	64	58	59	69	71
-Date		1986	1987	1993	1993	1984	1989	1993	1990	1985	1987	1989	1988	MAY 1984

PRECIPITATION (inches)

WILMINGTON, DELAWARE

YEAR	JAN	FEB	MAR	APR	MAY	JUNE	JULY	AUG	SEP	OCT	NOV	DEC	ANNUAL
1964	4.13	3.37	2.20	5.97	0.22	1.02	3.70	1.83	2.77	1.29	1.62	4.70	32.82
1965	2.38	2.17	3.20	1.76	1.41	1.62	3.84	2.04	2.41	1.59	0.94	1.54	24.90
1966	2.82	4.90	0.81	3.16	3.35	0.70	3.09	1.42	8.53	5.17	1.75	3.81	39.51
1967	1.67	1.90	5.45	2.69	3.79	3.01	4.45	11.16	1.16	2.05	2.08	5.24	44.65
1968	2.29	1.52	4.75	1.57	4.78	2.81	1.83	1.17	1.50	3.28	3.92	2.33	31.75
1969	1.68	1.76	1.71	1.58	3.21	3.62	6.48	2.34	6.84	1.47	1.79	7.90	40.38
1970	1.00	2.13	3.61	5.56	0.94	6.16	6.03	2.31	0.82	2.64	4.09	3.02	38.31
1971	2.22	6.29	2.29	2.15	4.51	2.50	3.65	8.38	6.99	6.41	5.52	1.33	52.24
1972	2.50	5.43	2.40	4.47	3.85	7.49	2.07	0.25	1.64	4.20	7.84	5.99	48.13
1973	3.81	3.42	4.02	6.57	5.56	5.19	2.82	2.44	3.02	2.22	0.67	7.31	47.05
1974	2.92	1.73	4.56	3.08	3.96	3.97	1.49	5.11	5.65	1.77	1.19	4.18	39.61
1975	4.23	2.95	4.63	3.03	5.65	6.16	5.53	2.55	6.19	3.06	2.63	3.00	49.61
1976	4.21	1.70	2.25	1.40	5.05	2.14	4.33	2.00	2.11	6.12	0.49	1.79	33.59
1977	2.18	1.09	4.55	3.91	0.96	4.41	1.38	4.82	1.29	3.59	6.14	5.81	40.13
1978	8.41	1.77	5.59	2.16	6.94	3.00	5.53	5.97	2.18	1.48	2.69	5.56	51.28
1979	7.61	7.02	2.61	4.03	3.10	4.01	4.76	6.11	5.94	3.45	3.23	1.44	53.31
1980	2.44	0.83	6.22	4.55	2.40	4.23	3.49	1.09	1.44	3.99	2.41	0.83	33.92
1981	0.52	3.23	1.26	3.54	5.05	4.50	2.52	3.38	3.82	2.84	0.67	3.95	35.28
1982	3.75	2.71	2.87	5.41	3.72	4.70	2.70	4.68	2.30	1.97	3.87	2.39	41.07
1983	2.98	3.55	6.84	6.80	7.38	3.94	2.33	1.29	3.44	3.87	5.48	6.80	54.70
1984	1.25	4.27	5.40	4.24	5.03	4.54	6.53	1.56	2.02	3.31	1.63	1.94	41.72
1985	1.56	2.05	2.03	0.35	5.52	1.37	6.91	2.28	4.56	1.84	4.46	0.80	33.73
1986	4.21	2.77	1.19	2.77	1.69	4.05	3.99	2.88	2.75	4.04	6.42	6.11	42.87
1987	4.35	1.52	1.16	2.63	3.15	2.31	4.09	4.21	4.85	2.31	3.50	1.90	35.98
1988	2.46	4.14	1.82	2.59	4.95	0.21	8.29	3.03	2.18	1.94	5.29	0.90	37.80
1989	2.48	2.75	3.69	2.76	6.57	5.43	12.63	1.97	4.31	3.92	1.99	1.27	49.77
1990	3.56	1.35	2.15	3.42	7.03	3.94	4.27	6.15	2.64	2.85	1.61	5.16	44.13
1991	4.30	0.97	4.64	3.28	1.98	3.41	3.71	5.38	5.36	1.27	1.26	4.26	39.82
1992	1.05	1.81	4.36	1.76	4.48	3.14	4.34	2.21	4.30	1.11	4.27	4.21	37.04
1993	2.64	3.11	7.50	5.87	3.95	1.60	4.04	2.65	6.26	2.77	2.85	3.51	46.75
Record Mean	3.29	3.04	3.70	3.51	3.69	3.73	4.54	4.50	3.65	2.99	3.24	3.44	43.30

AVERAGE TEMPERATURE (deg. F)

WILMINGTON, DELAWARE

YEAR	JAN	FEB	MAR	APR	MAY	JUNE	JULY	AUG	SEP	OCT	NOV	DEC	ANNUAL
1964	33.2	31.6	43.1	49.4	64.3	72.1	75.8	71.5	67.2	52.8	47.6	37.3	53.8
1965	29.0	33.6	38.4	49.3	66.0	70.2	74.5	73.3	69.4	53.6	45.0	37.7	53.3
1966	29.3	31.9	42.9	47.9	60.3	72.5	77.2	74.8	65.3	53.6	46.6	34.8	53.1
1967	36.2	29.4	39.4	51.5	55.4	71.2	74.4	72.8	65.1	54.7	40.7	36.8	52.3
1968	28.0	30.5	44.6	54.1	59.6	71.4	76.7	77.1	69.7	59.0	47.0	34.2	54.3
1969	30.8	33.6	39.6	55.1	64.2	73.4	75.7	75.4	68.2	56.0	44.6	32.7	54.1
1970	24.5	33.5	38.9	51.9	64.8	71.7	76.6	76.7	72.1	60.4	48.5	35.9	54.6
1971	27.4	35.5	40.7	51.1	60.4	73.4	75.9	74.0	71.1	62.9	45.7	42.1	55.0
1972	36.1	32.6	41.3	50.0	62.5	68.6	76.9	75.5	69.6	53.5	45.2	41.6	54.4
1973	35.6	35.5	49.4	54.7	61.4	75.8	78.1	78.1	70.4	59.9	49.1	37.7	57.2
1974	36.2	33.1	44.3	55.5	62.4	70.4	76.7	76.1	66.5	53.2	46.6	38.9	55.0
1975	37.5	35.9	40.6	47.5	65.2	71.5	75.7	76.3	65.4	60.0	51.0	36.2	55.2
1976	27.7	41.2	46.5	54.7	59.9	72.4	74.2	73.7	66.9	52.5	40.4	31.0	53.4
1977	20.8	33.3	47.2	54.4	63.9	68.9	76.4	75.4	69.3	53.4	46.6	33.4	53.6
1978	27.2	22.8	37.6	50.3	60.3	70.6	73.4	77.1	66.7	54.1	46.8	37.2	52.0
1979	31.4	22.1	45.4	50.6	63.8	67.9	75.3	75.0	68.2	55.0	49.6	38.1	53.5
1980	32.4	29.9	40.0	54.6	64.9	68.9	77.7	78.2	71.3	54.8	43.3	33.1	54.1
1981	25.4	37.9	40.2	55.0	62.5	72.3	77.1	73.5	66.9	53.0	45.3	34.2	53.6
1982	24.2	34.2	41.8	50.6	65.0	69.9	77.3	72.0	67.4	56.0	47.5	41.3	53.9
1983	35.2	35.3	45.9	53.1	61.0	71.8	77.6	77.0	69.3	56.9	46.7	32.1	55.2
1984	24.8	38.6	35.6	50.7	61.2	73.8	75.2	75.2	63.7	61.2	43.3	42.1	53.8
1985	27.5	37.4	47.1	58.0	65.9	70.6	76.6	74.4	69.1	58.3	51.0	32.9	55.7
1986	32.2	31.6	43.6	52.4	65.7	72.1	77.1	72.5	67.5	57.2	44.1	37.3	54.4
1987	31.4	31.9	44.6	52.3	63.1	73.5	79.1	74.3	68.3	51.7	47.4	38.6	54.7
1988	27.4	34.8	44.2	50.8	62.9	71.6	79.4	77.3	65.8	51.0	46.7	35.1	53.9
1989	36.0	34.3	42.1	51.6	62.1	74.3	75.9	74.4	68.4	57.6	44.6	25.0	53.9
1990	40.5	41.1	46.0	53.7	61.5	72.1	77.4	74.6	66.7	60.0	48.4	41.0	56.9
1991	34.3	39.7	44.9	54.7	69.1	73.6	77.3	76.7	67.2	57.6	46.8	39.5	56.8
1992	35.2	37.2	41.5	52.0	60.7	69.6	76.3	72.2	67.3	53.2	47.2	38.3	54.2
1993	37.6	31.0	39.0	52.8	65.3	72.8	79.4	77.8	68.7	55.7	47.4	36.4	55.3
Record Mean	32.3	33.3	42.1	52.3	62.8	71.3	76.1	74.3	67.8	56.5	45.8	35.3	54.2
Max	40.1	41.6	51.5	62.8	73.4	81.6	85.7	83.7	77.6	66.5	54.6	43.1	63.5
Min	24.4	24.9	32.7	41.8	52.1	61.1	66.5	64.8	58.1	46.6	37.0	27.5	44.8

REFERENCE NOTES

WILMINGTON, DELAWARE

GENERAL
T - TRACE AMOUNT.
BLANK ENTRIES DENOTE MISSING/UNREPORTED DATA.
# INDICATES A STATION OR INSTRUMENT RELOCATION.

SPECIFIC

PM - INCLUDES LAST DAY OF PREVIOUS MONTH
ASOS - AUTOMATED SURFACE OBSERVING SYSTEM IN OPERATION DURING THESE MONTHS.

(a) - LENGTH OF RECORD IN YEARS, ALTHOUGH INDIVIDUAL MONTHS MAY BE MISSING.
0.* OR * - THE VALUE IS BETWEEN 0.0 AND 0.05
NORMALS - BASED ON THE 1961-1990 RECORD PERIOD.
EXTREMES - DATES ARE THE MOST RECENT OCCURRENCE.
WIND DIR. - NUMERALS SHOW TENS OF DEGRESS CLOCKWISE FROM TRUE NORTH. "00" INDICATES CALM.
RESULTANT DIRECTIONS ARE GIVEN TO WHOLE DEGREES.
BOLD VALUES INDICATE EXTREME VALUES WHICH OCCURRED AFTER THE ASOS SYSTEM WAS COMMISSIONED.

RECORD = PERIOD OF RECORD
RECORD MEAN PRECIPITATION IS THE MEAN OF ALL DAILY PRECIPITATION AMOUNTS DURING THE PERIOD OF RECORD.
RECORD MAX(MIN) TEMPERATURE IS THE MEAN OF ALL DAILY MAX(MIN) TEMPERATURES DURING THE PERIOD OF RECORD.
RECORD MEAN TEMPERATURE IS THE SUM OF THE RECORD MAX AND RECORD MIN DIVIDED BY 2.
AVERAGE TEMPERATURE IS THE SUM OF THE MEAN DAILY MAX AND MIN TEMPERATURE DIVIDED BY 2.

EXCEPTIONS

RECORD MEANS ARE THROUGH THE CURRENT YEAR,
BEGINNING IN 1895 FOR TEMPERATURE
1894 FOR PRECIPITATION
1948 FOR SNOWFALL

HEATING DEGREE DAYS Base 65 deg. F WILMINGTON, DELAWARE

SEASON	JULY	AUG	SEP	OCT	NOV	DEC	JAN	FEB	MAR	APR	MAY	JUNE	TOTAL
1964-65	0	9	49	372	514	854	1108	873	816	467	63	27	5152
1965-66	0	12	40	349	593	839	1103	921	676	507	186	24	5250
1966-67	0	0	84	346	545	931	888	991	788	402	295	5	5275
1967-68	0	0	80	327	722	866	1139	994	626	319	171	6	5250
1968-69	0	0	7	205	532	949	1052	874	780	298	92	1	4790
1969-70	0	0	41	288	606	993	1250	876	803	389	95	1	5342
1970-71	0	0	25	175	487	895	1159	821	746	411	153	3	4875
1971-72	0	2	23	93	585	702	889	935	732	443	105	27	4536
1972-73	0	0	20	356	586	716	902	820	477	322	143	0	4342
1973-74	0	0	15	179	469	839	886	887	635	300	137	7	4354
1974-75	0	0	65	362	553	805	847	808	753	520	84	5	4802
1975-76	0	0	56	177	418	888	1149	681	567	341	178	31	4486
1976-77	0	4	44	387	734	1047	1361	884	546	333	107	32	5479
1977-78	0	1	31	353	550	975	1165	1179	842	433	191	17	5737
1978-79	6	0	60	337	542	854	1037	1197	605	424	89	28	5179
1979-80	4	7	31	318	458	827	1004	1009	768	307	83	35	4851
1980-81	0	0	20	322	645	985	1222	752	763	299	135	4	5147
1981-82	0	0	57	370	585	947	1259	855	715	426	69	12	5295
1982-83	0	14	29	305	519	724	919	822	587	368	163	7	4457
1983-84	0	0	74	275	542	1013	1240	758	904	422	162	5	5395
1984-85	0	2	113	149	641	701	1154	766	550	248	73	7	4404
1985-86	0	0	45	213	411	986	1011	930	653	373	99	11	4732
1986-87	0	27	36	276	619	848	1032	923	628	374	143	2	4908
1987-88	0	2	22	406	521	811	1159	869	637	419	121	38	5005
1988-89	3	0	52	434	541	923	893	854	710	395	142	0	4947
1989-90	0	2	54	236	605	1231	749	661	593	368	127	6	4632
1990-91	2	1	69	214	494	734	943	700	617	320	61	5	4160
1991-92	0	0	64	244	541	785	914	799	723	386	169	12	4637
1992-93	0	1	57	363	527	817	843	945	799	360	64	12	4788
1993-94	0	0	55	286	526	879							

COOLING DEGREE DAYS Base 65 deg. F WILMINGTON, DELAWARE

YEAR	JAN	FEB	MAR	APR	MAY	JUNE	JULY	AUG	SEP	OCT	NOV	DEC	TOTAL
1969	0	0	0	8	73	258	336	330	143	14	0	0	1162
1970	0	0	0	2	97	211	365	371	244	38	0	0	1328
1971	0	0	0	0	16	261	345	287	214	34	16	0	1173
1972	0	0	2	0	35	143	376	334	165	8	0	0	1063
1973	0	0	0	19	42	332	416	413	183	29	0	0	1434
1974	0	0	0	24	64	175	370	355	113	2	6	0	1109
1975	0	0	0	0	97	207	337	355	73	27	5	0	1101
1976	0	0	0	37	25	260	291	278	106	6	0	0	1003
1977	0	0	5	20	80	156	360	328	166	0	5	0	1120
1978	0	0	0	0	48	188	273	383	117	7	0	0	1016
1979	0	0	4	1	57	123	327	324	138	16	0	0	990
1980	0	0	0	0	83	159	400	417	214	10	0	0	1283
1981	0	0	0	9	62	228	381	270	120	3	0	0	1073
1982	0	0	0	2	75	163	391	238	107	32	1	0	1009
1983	0	0	0	17	47	218	398	378	209	29	0	0	1296
1984	0	0	0	0	50	276	321	327	80	34	0	0	1088
1985	0	0	4	47	106	181	366	300	174	13	0	0	1191
1986	0	0	0	0	129	227	379	267	116	40	0	0	1158
1987	0	0	0	3	91	264	446	295	129	0	0	0	1228
1988	0	0	0	0	62	242	455	389	80	5	0	0	1233
1989	0	0	6	0	61	287	345	299	162	17	0	0	1177
1990	0	0	10	36	23	227	395	304	127	66	0	0	1188
1991	0	0	1	18	197	271	390	371	135	23	2	0	1408
1992	0	0	0	3	39	156	358	230	134	3	0	0	923
1993	0	0	0	1	79	252	452	405	176	3	4	0	1372

SNOWFALL (inches) WILMINGTON, DELAWARE

SEASON	JULY	AUG	SEP	OCT	NOV	DEC	JAN	FEB	MAR	APR	MAY	JUNE	TOTAL
1964-65	0.0	0.0	0.0	0.0	T	1.6	8.4	1.7	5.8	0.6	0.0	0.0	18.1
1965-66	0.0	0.0	0.0	0.0	0.0	0.0	17.2	10.0	T	T	0.0	0.0	27.2
1966-67	0.0	0.0	0.0	0.0	T	21.5	0.6	18.7	2.7	T	0.0	0.0	43.5
1967-68	0.0	0.0	0.0	0.0	6.8	4.0	1.6	0.9	1.3	0.0	0.0	0.0	14.6
1968-69	0.0	0.0	0.0	0.0	T	0.6	2.9	5.5	7.8	0.0	0.0	0.0	16.8
1969-70	0.0	0.0	0.0	T	0.1	6.8	9.7	1.5	0.4	0.0	0.0	0.0	18.5
1970-71	0.0	0.0	0.0	0.0	0.0	3.2	5.8	0.2	2.4	0.6	0.0	0.0	12.2
1971-72	0.0	0.0	0.0	0.0	T	T	2.1	7.0	0.3	0.1	0.0	0.0	9.5
1972-73	0.0	0.0	0.0	T	T	T	T	1.2	T	T	0.0	0.0	1.2
1973-74	0.0	0.0	0.0	0.0	0.0	6.1	2.4	11.5	T	T	0.0	0.0	20.0
1974-75	0.0	0.0	0.0	0.0	T	T	5.1	4.8	1.1	T	0.0	0.0	11.0
1975-76	0.0	0.0	0.0	0.0	0.0	0.3	4.5	1.7	6.7	T	0.0	0.0	13.2
1976-77	0.0	0.0	0.0	0.0	0.1	3.2	14.5	T	0.0	T	0.0	0.0	17.8
1977-78	0.0	0.0	0.0	0.0	0.4	0.9	16.0	18.4	9.9	T	0.0	0.0	45.6
1978-79	0.0	0.0	0.0	0.0	4.5	T	12.0	27.5	0.2	T	0.0	0.0	44.2
1979-80	0.0	0.0	0.0	2.5	0.0	1.4	6.1	0.8	5.1	0.0	0.0	0.0	15.9
1980-81	0.0	0.0	0.0	0.0	0.5	1.4	6.5	T	3.7	0.0	0.0	0.0	12.1
1981-82	0.0	0.0	0.0	0.0	T	2.8	14.6	4.5	0.4	2.6	0.0	0.0	24.9
1982-83	0.0	0.0	0.0	0.0	T	5.8	T	18.5	0.3	0.5	0.0	0.0	25.1
1983-84	0.0	0.0	0.0	0.0	T	T	9.7	T	5.2	T	0.0	0.0	14.9
1984-85	0.0	0.0	0.0	0.0	T	0.3	14.2	0.7	T	0.4	0.0	0.0	15.6
1985-86	0.0	0.0	0.0	0.0	0.0	1.4	3.1	9.7	T	T	0.0	0.0	14.2
1986-87	0.0	0.0	0.0	0.0	T	0.3	21.4	15.7	0.2	2.4	0.0	0.0	40.0
1987-88	0.0	0.0	0.0	0.0	0.7	2.1	10.8	1.1	T	T	0.0	0.0	14.7
1988-89	0.0	0.0	0.0	0.0	T	0.2	6.7	2.9	1.2	0.0	0.0	0.0	11.0
1989-90	0.0	0.0	0.0	0.0	5.6	8.9	1.5	1.0	1.3	1.6	0.0	0.0	19.9
1990-91	T	0.0	0.0	0.0	0.0	6.4	5.2	0.6	0.9	T	T	0.0	13.1
1991-92	0.0	0.0	0.0	0.0	T	0.2	1.5	1.3	0.5	T	0.0	T	3.5
1992-93	0.0	0.0	0.0	0.0	T	0.1	1.4	10.0	13.9	0.0	0.0	0.0	25.4
1993-94	0.0	0.0	0.0	0.0	T	2.4							
Record Mean	T	0.0	0.0	0.1	0.9	3.3	6.6	6.0	3.3	0.2	T	T	20.4

WASHINGTON, D.C .
DULLES INTERNATIONAL AIRPORT

Dulles International Airport is located in the Virginia Piedmont about 23 miles east of the Blue Ridge Mountains and 12 miles east of the Bull Run and Catoctin Mountains. The Blue Ridge rises to about 2,500 feet above sea level and the Bull Run Mountains to about 1,500 feet at their nearest points. Field elevation is 313 feet. The terrain near the airport is mostly low rolling hills about one-fourth wooded. Ponds located on and near the airport, along with poor air drainage contribute to the formation of local ground fog. Easterly winds cause an upslope effect from the Atlantic Ocean, 140 miles east, and from the Chesapeake Bay, about 55 miles east. Westerly winds create a slight foehn effect.

Its location in the middle latitudes, where the general atmospheric flow is from west to east, favors a continental climate with four well defined seasons. Summers are warm and at times humid, Winters are mild. Generally pleasant weather prevails in spring and autumn. The coldest period, when temperatures average 21 degrees, occurs in late January. The warmest period, averaging 88 degrees, occurs in the last half of July.

Precipitation is rather evenly distributed through the year. Annual precipitation has ranged from about 25 inches to more than 55 inches. Rainfalls of over 10 inches in a 24-hour period have been recorded during the passage of tropical storms. The seasonal snowfall is nearly 24 inches, but varies greatly from season to season. Snowfalls of 4 inches or more occur only twice each winter on average. Accumulations of over 20 inches from a single storm are extremely rare.

Storm damage results mainly from heavy snows and freezing rains in winter and from hurricanes and severe thunderstorms during the other seasons. Damage may result from wind, flooding or rain.

Prevailing winds are from the south except during the winter months when they are from the northwest. The windiest period is late winter and early spring. Winds are generally less during the night and early morning hours and increase to a high in the afternoon. Winds may reach 50 to 60 miles per hour or even higher during severe summer thunderstorms, hurricanes, and winter storms.

The growing season averages 169 days. The average date for the last freeze in spring is late April and the average date for the first freeze in the fall is mid October.

NORMALS, MEANS, AND EXTREMES

WASHINGTON, D.C . DULLES INTERNATIONAL AIRPORT

LATITUDE: 38°57'N LONGITUDE: 77°27' W ELEVATION: FT. GRND 290 BARO 310 TIME ZONE: EASTERN WBAN: 93738

	(a)	JAN	FEB	MAR	APR	MAY	JUNE	JULY	AUG	SEP	OCT	NOV	DEC	YEAR
TEMPERATURE °F:														
Normals														
-Daily Maximum		40.1	43.8	54.6	65.1	74.3	82.8	87.0	85.5	78.9	67.7	56.7	45.0	65.1
-Daily Minimum		21.0	23.3	31.8	40:3	50.0	59.2	64.1	62.8	55.4	42.4	34.2	25.8	42.5
-Monthly		30.6	33.6	43.2	52.7	62.2	71.0	75.6	74.2	67.1	55.1	45.4	35.4	53.8
Extremes														
-Record Highest	31	75	79	89	92	97	100	104	104	99	90	84	78	104
-Year		1975	1985	1990	1990	1969	1964	1988	1983	1983	1986	1982	1984	JUL 1988
-Record Lowest	31	-18	-14	-1	17	28	36	41	38	30	15	9	-4	-18
-Year		1984	1979	1993	1969	1970	1977	1988	1982	1974	1969	1989	1989	JAN 1984
NORMAL DEGREE DAYS:														
Heating (base 65°F)		1066	879	676	369	140	7	0	0	44	319	588	918	5006
Cooling (base 65°F)		0	0	0	0	53	187	329	285	107	12	0	0	973
% OF POSSIBLE SUNSHINE														
MEAN SKY COVER (tenths)														
Sunrise - Sunset	31	6.6	6.5	6.7	6.4	6.4	6.2	6.2	6.0	5.9	5.6	6.4	6.7	6.3
MEAN NUMBER OF DAYS:														
Sunrise to Sunset														
-Clear	31	7.5	7.5	6.6	7.1	7.0	6.5	7.8	8.0	8.6	10.5	7.2	7.2	91.4
-Partly Cloudy	31	7.0	6.2	8.4	8.9	9.6	11.8	10.7	10.9	9.3	8.2	8.0	6.8	105.8
-Cloudy	31	16.5	14.5	16.0	14.0	14.4	11.6	12.5	12.1	12.1	12.3	14.9	17.0	168.1
Precipitation														
.01 inches or more	30	10.0	9.0	10.1	10.1	11.9	9.7	10.5	9.8	8.6	7.9	9.1	9.8	116.6
Snow,Ice pellets,hail														
1.0 inches or more	31	2.1	1.5	1.1	0.1	0.0	0.0	0.0	0.0	0.0	0.*	0.3	0.9	6.0
Thunderstorms	31	0.2	0.2	1.1	2.2	3.8	5.6	6.1	4.9	2.0	1.0	0.6	0.1	27.8
Heavy Fog Visibility														
1/4 mile or less	31	3.3	2.4	2.0	1.4	2.3	1.7	2.1	2.6	3.1	3.8	2.1	2.7	29.4
Temperature °F														
-Maximum														
90° and above	31	0.0	0.0	0.0	0.2	0.9	5.4	11.5	8.3	3.0	0.1	0.0	0.0	29.4
32° and below	31	6.8	3.9	0.5	0.0	0.0	0.0	0.0	0.0	0.0	0.0	0.1	3.0	14.3
-Minimum														
32° and below	31	26.1	22.8	17.2	5.7	0.5	0.0	0.0	0.0	0.1	5.2	14.4	23.3	115.2
0° and below	31	1.3	0.4	0.*	0.0	0.0	0.0	0.0	0.0	0.0	0.0	0.0	0.2	1.8
AVG. STATION PRESS.(mb)	21	1007.2	1006.9	1005.3	1003.7	1003.8	1003.9	1004.5	1005.9	1007.1	1007.9	1007.7	1007.9	1006.0
RELATIVE HUMIDITY (%)														
Hour 01	24	75	74	73	73	84	87	88	90	89	86	79	76	81
Hour 07 (Local Time)	24	77	78	78	77	83	84	86	89	90	88	83	79	83
Hour 13	24	58	54	52	48	55	55	55	56	56	54	55	58	55
Hour 19	24	66	61	58	54	64	65	67	70	74	72	68	68	66
PRECIPITATION (inches):														
Water Equivalent														
-Normal		2.70	2.81	3.17	3.11	4.02	3.92	3.49	3.94	3.36	3.20	3.30	3.22	40.24
-Maximum Monthly	30	6.61	5.75	7.65	7.35	10.26	18.19	7.23	10.71	11.26	9.19	7.83	6.74	18.19
-Year		1979	1979	1993	1973	1988	1972	1992	1984	1975	1971	1963	1969	JUN 1972
-Minimum Monthly	30	0.40	0.25	0.99	0.33	0.80	0.52	0.94	0.76	0.62	T	0.24	0.42	T
-Year		1981	1978	1981	1985	1964	1988	1983	1989	1967	1963	1981	1965	OCT 1963
-Maximum in 24 hrs	30	2.13	2.11	2.56	2.43	3.13	11.88	3.02	7.04	5.54	4.13	4.80	3.15	11.88
-Year		1976	1983	1967	1993	1988	1972	1990	1984	1966	1979	1993	1977	JUN 1972
Snow,Ice pellets,hail														
-Maximum Monthly	31	28.8	27.6	15.5	4.0	T	0.0	0.0	T	0.0	1.3	11.4	24.2	28.8
-Year		1987	1979	1993	1990	1963			1993		1979	1967	1966	JAN 1987
-Maximum in 24 hrs	31	15.4	22.8	13.9	3.8	T	0.0	0.0	T	0.0	1.3	11.4	12.1	22.8
-Year		1971	1983	1993	1990	1963			1993		1979	1967	1969	FEB 1983
WIND:														
Mean Speed (mph)	31	8.2	8.6	9.1	8.8	7.4	6.8	6.2	5.9	6.2	6.6	7.7	7.8	7.4
Prevailing Direction														
Fastest Obs. 1 Min.														
-Direction (!!)	30	20	28	29	32	28	31	30	27	25	29	29	30	31
-Speed (MPH)	30	39	36	40	46	40	55	48	40	35	38	35	40	55
-Year		1978	1972	1977	1963	1980	1989	1980	1980	1967	1967	1969	1968	JUN 1989
Peak Gust														
-Direction (!!)	10	NW	W	E	NW	NW	NW	S	NW	SE	NW	NW	NW	NW
-Speed (mph)	10	52	53	58	46	48	74	56	46	43	41	53	56	74
-Date		1985	1990	1993	1993	1985	1989	1985	1993	1989	1992	1986	1988	JUN 1989

PRECIPITATION (inches)

WASHINGTON, D.C . DULLES INTERNATIONAL AIRPORT

YEAR	JAN	FEB	MAR	APR	MAY	JUNE	JULY	AUG	SEP	OCT	NOV	DEC	ANNUAL
1964	5.56	4.50	2.82	3.68	0.80	1.36	3.09	2.49	3.72	1.96	1.80	4.19	35.97
1965	3.24	2.58	4.27	2.53	1.77	1.75	3.94	3.75	2.28	1.87	0.46	0.42	28.86
1966	4.22	3.90	1.24	4.38	3.84	0.94	2.17	1.79	9.39	3.01	1.50	3.67	40.05
1967	1.12	2.18	4.74	0.93	4.37	1.68	3.38	9.28	0.62	2.79	2.61	5.69	39.39
1968	2.31	0.68	3.36	1.29	5.90	5.20	4.08	4.90	2.10	2.77	3.57	2.32	38.48
1969	2.07	1.69	2.04	1.34	0.99	5.55	4.90	5.07	4.89	0.85	2.11	6.74	38.24
1970	1.32	2.76	2.98	4.19	2.86	2.48	6.00	2.65	1.03	2.79	6.00	3.88	38.94
1971	2.27	4.68	2.27	2.56	8.47	2.81	2.15	4.45	3.63	9.19	3.13	1.00	46.61
1972	2.28	5.44	2.37	4.40	4.76	18.19	1.53	2.09	1.40	3.44	7.09	6.06	59.05
1973	2.25	2.84	2.66	7.35	3.97	1.91	4.99	3.18	3.14	2.67	0.81	5.72	41.49
1974	3.07	1.14	3.09	2.14	3.78	5.41	2.69	4.84	4.02	0.71	1.97	5.26	38.12
1975	2.76	1.97	4.38	2.58	2.99	7.08	6.25	5.38	11.26	2.34	1.64	4.01	52.64
1976	2.81	1.42	4.15	1.31	4.18	2.88	2.33	3.13	3.75	7.88	0.57	1.78	36.19
1977	1.10	0.49	3.59	2.58	2.33	3.09	3.25	4.18	1.74	4.19	4.51	4.87	35.92
1978	6.55	0.25	2.85	1.62	5.05	4.36	4.52	4.29	0.78	0.79	3.02	3.58	37.66
1979	6.61	5.75	3.50	2.05	4.89	4.64	2.18	6.05	7.58	8.65	2.65	0.88	55.43
1980	2.95	1.00	4.82	3.64	3.86	1.89	4.41	1.67	2.70	2.77	3.42	0.68	33.81
1981	0.40	4.10	0.99	3.05	4.36	3.86	4.05	3.55	2.07	3.00	0.24	2.46	32.13
1982	2.10	4.09	3.47	2.82	3.57	5.49	2.11	3.36	4.22	2.21	2.87	2.25	38.56
1983	1.40	3.74	4.21	7.24	3.63	4.01	0.94	1.34	2.95	6.00	5.06	5.66	46.18
1984	1.42	4.13	5.81	5.01	4.23	2.19	2.46	10.71	1.49	1.73	3.64	1.25	44.07
1985	2.32	3.73	1.70	0.33	4.82	1.14	2.34	3.35	2.96	4.06	5.27	0.92	32.94
1986	1.58	3.16	1.12	3.01	1.19	1.40	1.86	5.72	1.04	1.30	4.17	4.83	30.38
1987	4.53	2.47	1.46	4.61	2.33	3.38	3.04	0.96	8.11	2.51	5.02	2.35	40.77
1988	2.47	2.06	2.31	2.35	10.26	0.52	7.12	3.92	1.80	1.60	4.48	0.92	39.81
1989	2.65	2.50	4.01	2.70	7.71	5.75	5.99	0.76	3.14	4.73	2.68	1.72	44.34
1990	3.14	1.65	2.78	5.06	4.37	1.77	5.42	5.56	1.49	6.53	2.56	5.00	45.33
1991	2.67	0.81	5.16	1.80	1.51	6.58	2.64	0.99	3.25	1.37	2.70	4.54	34.02
1992	2.13	2.26	3.23	3.15	3.82	2.88	7.23	1.33	5.55	2.81	5.24	4.73	44.36
1993	2.72	2.74	7.65	5.62	2.85	1.77	2.06	4.68	4.66	2.15	6.28	3.78	46.96
Record Mean	2.73	2.69	3.33	3.11	3.90	3.89	3.56	3.87	3.55	3.18	3.38	3.34	40.55

AVERAGE TEMPERATURE (deg. F)

WASHINGTON, D.C . DULLES INTERNATIONAL AIRPORT

YEAR	JAN	FEB	MAR	APR	MAY	JUNE	JULY	AUG	SEP	OCT	NOV	DEC	ANNUAL
1964	33.4	30.9	43.1	50.6	64.6	73.3	76.3	72.2	66.7	51.9	47.5	38.9	54.1
1965	30.7	35.3	39.2	51.3	66.9	70.1	74.9	75.0	70.4	54.0	45.2	39.4	54.4
1966	28.4	31.7	43.3	48.6	60.8	71.5	76.5	74.0	63.6	51.8	44.9	33.0	52.3
1967	36.6	29.1	42.8	54.8	55.9	71.4	74.5	72.8	63.6	53.7	40.6	36.5	52.7
1968	28.2	30.8	46.2	53.1	58.6	70.0	75.6	75.6	66.8	56.4	46.1	31.9	53.3
1969	29.4	32.2	38.7	55.3	64.3	73.6	77.8	71.9	65.3	52.8	41.9	31.0	52.9
1970	23.9	32.7	37.8	52.5	63.5	68.9	74.0	72.2	69.1	56.3	45.7	36.3	52.8
1971	28.5	36.7	39.7	51.4	60.1	72.6	74.9	72.9	69.7	61.4	44.4	42.2	54.5
1972	35.8	32.6	42.8	51.3	61.8	67.4	75.9	73.3	68.0	51.5	42.0	39.3	53.5
1973	33.6	33.1	48.6	53.1	59.8	74.2	75.9	76.4	69.6	56.9	46.0	35.2	55.2
1974	37.8	32.5	44.6	54.3	61.1	67.1	73.5	74.4	66.3	53.3	46.1	36.6	54.0
1975	34.0	35.0	40.3	49.1	65.2	70.9	74.4	75.5	64.4	57.0	47.2	33.2	53.9
1976	27.5	41.1	44.6	52.9	58.1	70.5	72.2	71.5	64.8	50.5	38.3	31.2	51.9
1977	21.0	34.8	48.9	55.4	64.5	68.3	76.4	76.3	69.9	53.9	47.9	33.0	54.2
1978	27.3	26.7	40.6	54.2	63.4	71.6	74.4	77.2	68.9	53.8	46.8	38.2	53.6
1979	31.2	23.3	46.4	52.2	63.3	68.2	74.2	74.4	66.9	54.6	49.7	40.1	53.7
1980	32.7	31.0	41.9	55.6	65.8	69.2	77.3	77.1	70.8	54.1	42.7	34.9	54.4
1981	27.8	38.5	41.4	57.0	60.9	73.5	75.1	71.9	66.7	52.5	45.9	34.1	53.8
1982	26.2	36.5	43.7	51.3	66.8	70.8	77.0	72.3	66.9	57.3	48.4	42.6	55.0
1983	34.7	35.0	46.0	50.2	59.4	70.0	74.6	75.3	66.2	54.7	44.7	30.4	53.4
1984	25.8	38.9	37.5	50.9	59.8	71.2	72.0	74.2	63.0	61.8	42.6	43.3	53.4
1985	28.1	35.8	45.0	57.1	63.8	69.7	75.9	73.4	67.7	58.4	52.6	33.2	55.1
1986	32.5	32.1	44.3	53.6	64.2	73.4	79.0	72.5	68.2	56.8	43.2	36.8	54.7
1987	31.2	33.1	43.5	51.7	64.3	73.7	79.1	75.4	68.7	49.9	47.8	38.6	54.8
1988	27.6	34.5	43.9	52.2	63.5	70.9	78.7	78.4	66.4	49.2	46.0	35.8	53.9
1989	36.3	35.2	43.7	51.7	61.4	73.8	76.0	74.4	68.6	57.3	43.7	23.1	53.8
1990	40.8	42.2	48.1	54.3	61.6	72.0	76.9	73.8	66.1	58.7	48.6	41.3	57.0
1991	34.0	39.5	46.1	56.0	69.3	72.3	77.9	77.0	67.6	56.6	45.7	38.5	56.7
1992	35.0	38.0	41.9	53.1	59.7	68.6	75.9	70.4	66.0	52.2	45.7	36.8	53.6
1993	36.8	31.7	39.3	52.6	64.5	72.0	79.7	78.0	68.7	54.2	45.3	34.0	54.7
Record Mean	31.1	33.9	43.2	52.9	62.5	71.0	75.9	74.3	67.1	54.9	45.5	35.8	54.0
Max	40.7	44.2	54.5	65.3	74.6	82.9	87.3	85.7	78.8	67.5	56.7	45.4	65.3
Min	21.5	23.5	31.9	40.5	50.3	59.1	64.4	62.9	55.4	42.3	34.3	26.1	42.7

REFERENCE NOTES

WASHINGTON, D.C . DULLES INTERNATIONAL AIRPORT

GENERAL
T - TRACE AMOUNT.
BLANK ENTRIES DENOTE MISSING/UNREPORTED DATA.
# INDICATES A STATION OR INSTRUMENT RELOCATION.

SPECIFIC

(*) - LENGTH OF RECORD IN YEARS, ALTHOUGH INDIVIDUAL MONTHS MAY BE MISSING.
0.* OR * - THE VALUE IS BETWEEN 0.0 AND 0.05
NORMALS - BASED ON THE 1961-1990 RECORD PERIOD.
EXTREMES - DATES ARE THE MOST RECENT OCCURRENCE
WIND DIR.- NUMERALS SHOW TENS OF DEGRESS CLOCKWISE FROM TRUE NORTH. "00" INDICATES CALM.
RESULTANT DIRECTIONS ARE GIVEN TO WHOLE DEGREES.
BOLD VALUES INDICATE EXTREME VALUES WHICH OCCURRED AFTER THE ASOS SYSTEM WAS COMMISSIONED.

RECORD = PERIOD OF RECORD
RECORD MEAN PRECIPITATION IS THE MEAN OF ALL DAILY PRECIPITATION AMOUNTS DURING THE PERIOD OF RECORD.
RECORD MAX(MIN) TEMPERATURE IS THE MEAN OF ALL DAILY MAX(MIN) TEMPERATURES DURING THE PERIOD OF RECORD.
RECORD MEAN TEMPERATURE IS THE SUM OF THE RECORD MAX AND RECORD MIN DIVIDED BY 2.
AVERAGE TEMPERATURE IS THE SUM OF THE MEAN DAILY MAX AND MIN TEMPERATURE DIVIDED BY 2.

EXCEPTIONS

RECORD MEANS ARE THROUGH THE CURRENT YEAR, BEGINNING IN 1963 FOR TEMPERATURE
1963 FOR PRECIPITATION
1963 FOR SNOWFALL

HEATING DEGREE DAYS Base 65 deg. F — WASHINGTON, D.C . DULLES INTERNATIONAL AIRPORT

SEASON	JULY	AUG	SEP	OCT	NOV	DEC	JAN	FEB	MAR	APR	MAY	JUNE	TOTAL
1964-65	0	9	55	397	517	800	1057	824	793	406	41	44	4943
1965-66	0	6	35	337	588	785	1128	926	664	491	179	30	5169
1966-67	1	0	105	404	599	981	874	1002	682	317	283	12	5260
1967-68	0	0	96	352	724	876	1136	985	572	350	205	13	5309
1968-69	0	12	16	278	561	1020	1098	913	809	293	91	1	5092
1969-70	0	8	90	376	685	1047	1267	898	837	378	121	17	5724
1970-71	0	0	46	272	573	879	1121	786	776	401	168	5	5027
1971-72	0	2	34	141	631	699	901	935	687	404	117	46	4597
1972-73	2	3	38	414	686	788	964	886	500	371	186	1	4839
1973-74	0	0	24	257	564	917	837	905	631	329	168	21	4653
1974-75	0	0	77	358	574	873	956	837	759	473	75	15	4997
1975-76	0	0	83	247	526	977	1158	688	625	382	232	35	4953
1976-77	0	8	65	447	793	1038	1361	838	508	303	90	41	5492
1977-78	1	2	23	339	521	985	1163	1066	748	317	123	14	5302
1978-79	0	0	58	349	539	824	1040	1163	578	383	115	43	5092
1979-80	7	15	50	337	461	765	993	977	707	272	69	31	4684
1980-81	0	0	33	342	664	928	1145	735	728	254	170	6	5005
1981-82	0	6	63	388	563	952	1196	790	653	407	56	8	5082
1982-83	0	13	41	266	504	687	934	832	581	449	204	16	4527
1983-84	3	1	103	325	600	1066	1209	750	845	414	194	19	5529
1984-85	4	0	137	130	664	664	1136	814	617	261	94	20	4541
1985-86	0	0	73	217	368	977	995	915	636	336	114	10	4641
1986-87	0	31	52	296	648	868	1039	888	660	398	113	4	4997
1987-88	0	5	21	464	508	810	1152	873	646	381	107	49	5016
1988-89	7	0	50	485	564	899	881	825	670	399	167	0	4947
1989-90	0	7	60	260	633	1295	740	634	547	347	124	8	4655
1990-91	1	0	79	236	482	728	952	706	579	290	53	8	4114
1991-92	0	0	69	280	577	814	921	779	708	359	181	20	4708
1992-93	0	4	79	391	572	869	870	929	787	366	79	19	4965
1993-94	0	0	64	330	587	952							

COOLING DEGREE DAYS Base 65 deg. F — WASHINGTON, D.C . DULLES INTERNATIONAL AIRPORT

YEAR	JAN	FEB	MAR	APR	MAY	JUNE	JULY	AUG	SEP	OCT	NOV	DEC	TOTAL
1970	0	0	0	10	83	143	288	228	178	9	0	0	939
1971	0	0	0	0	21	237	316	255	181	35	19	0	1064
1972	0	0	6	3	25	123	346	268	134	4	0	0	909
1973	0	0	0	20	30	281	344	358	168	15	0	0	1216
1974	0	0	4	15	53	90	270	299	127	3	11	0	872
1975	0	0	0	4	89	197	298	330	73	10	0	0	1001
1976	0	0	0	28	24	208	232	215	65	7	0	0	779
1977	0	0	13	22	81	147	361	357	175	3	12	0	1171
1978	0	0	0	0	84	219	298	385	183	10	0	0	1179
1979	0	0	9	5	69	147	297	310	112	21	6	0	976
1980	0	0	0	0	100	165	391	380	211	12	0	0	1259
1981	0	0	2	20	49	268	320	227	120	5	0	0	1011
1982	0	0	0	6	120	191	381	248	103	34	11	2	1096
1983	0	0	0	12	40	173	307	328	149	10	0	0	1019
1984	0	0	0	0	38	215	228	292	82	39	0	0	894
1985	0	1	4	32	65	169	344	267	160	20	5	0	1067
1986	0	0	3	0	93	266	441	273	151	49	0	0	1276
1987	0	0	0	5	97	270	446	335	138	0	0	0	1291
1988	0	0	1	2	71	231	439	423	96	5	0	0	1268
1989	0	0	16	9	63	270	343	305	176	27	1	0	1210
1990	0	0	30	31	26	223	376	279	118	49	0	0	1132
1991	0	0	0	23	196	234	408	379	153	24	3	0	1420
1992	0	0	0	8	24	131	343	180	116	1	0	0	803
1993	0	0	0	2	71	233	464	407	183	2	6	0	1368

SNOWFALL (inches) — WASHINGTON, D.C . DULLES INTERNATIONAL AIRPORT

SEASON	JULY	AUG	SEP	OCT	NOV	DEC	JAN	FEB	MAR	APR	MAY	JUNE	TOTAL
1964-65	0.0	0.0	0.0	0.0	0.2	1.0	10.1	2.1	2.9	T	0.0	0.0	16.3
1965-66	0.0	0.0	0.0	0.0	T	0.3	19.0	11.3	T	T	0.0	0.0	30.6
1966-67	0.0	0.0	0.0	0.0	0.1	24.2	1.4	18.0	0.7	0.0	0.0	0.0	44.4
1967-68	0.0	0.0	0.0	0.0	11.4	4.9	3.3	4.5	6.4	0.0	0.0	0.0	30.5
1968-69	0.0	0.0	0.0	0.0	5.8	T	0.1	7.2	10.4	0.0	0.0	0.0	23.5
1969-70	0.0	0.0	0.0	0.0	T	15.9	9.8	3.6	0.9	T	0.0	0.0	30.2
1970-71	0.0	0.0	0.0	0.0	T	9.2	7.5	0.2	2.4	0.3	0.0	0.0	19.6
1971-72	0.0	0.0	0.0	0.0	2.9	T	T	16.9	0.2	0.4	0.0	0.0	20.4
1972-73	0.0	0.0	0.0	0.1	0.6	0.0	T	0.3	0.2	1.0	0.0	0.0	2.2
1973-74	0.0	0.0	0.0	0.0	T	10.7	1.0	7.0	0.2	0.0	0.0	0.0	18.9
1974-75	0.0	0.0	0.0	T	T	1.0	7.5	8.1	1.2	T	0.0	0.0	17.8
1975-76	0.0	0.0	0.0	0.0	T	0.9	1.6	0.2	6.4	0.0	0.0	0.0	9.1
1976-77	0.0	0.0	0.0	0.0	0.8	0.6	9.2	T	0.0	T	0.0	0.0	10.6
1977-78	0.0	0.0	0.0	T	0.1	1.1	13.4	2.5	10.3	0.0	0.0	0.0	27.4
1978-79	0.0	0.0	0.0	0.0	4.0	T	8.8	27.6	0.2	0.0	0.0	0.0	40.6
1979-80	0.0	0.0	0.0	1.3	0.0	0.2	9.9	5.6	9.8	0.0	0.0	0.0	26.8
1980-81	0.0	0.0	0.0	0.0	T	0.4	4.0	T	T	0.0	0.0	0.0	4.4
1981-82	0.0	0.0	0.0	0.0	0.9	5.7	14.3	5.7	0.3	2.6	0.0	0.0	29.5
1982-83	0.0	0.0	0.0	0.0	0.0	11.9	0.1	27.2	T	T	0.0	0.0	39.2
1983-84	0.0	0.0	0.0	0.0	0.3	T	10.5	0.8	10.8	0.0	0.0	0.0	22.4
1984-85	0.0	0.0	0.0	0.0	T	T	13.8	0.7	T	0.3	0.0	0.0	14.8
1985-86	0.0	0.0	0.0	0.0	0.0	0.2	1.9	16.0	T	T	0.0	0.0	18.1
1986-87	0.0	0.0	0.0	0.0	0.0	T	28.8	12.0	1.4	0.5	0.0	0.0	42.7
1987-88	0.0	0.0	0.0	0.0	5.5	0.1	10.9	T	0.2	T	0.0	0.0	16.7
1988-89	0.0	0.0	0.0	0.0	0.0	1.0	6.6	1.4	0.9	0.0	0.0	0.0	9.9
1989-90	0.0	0.0	0.0	0.0	2.4	11.4	5.0	T	6.4	4.0	0.0	0.0	29.2
1990-91	0.0	0.0	0.0	0.0	0.0	4.7	6.3	0.1	3.5	T	0.0	0.0	14.6
1991-92	0.0	0.0	0.0	0.0	T	T	2.1	3.4	1.2	T	0.0	0.0	6.7
1992-93	0.0	0.0	0.0	0.0	T	2.8	3.2	8.8	15.5	T	0.0	0.0	30.3
1993-94	0.0	T	0.0	0.0	0.0	3.5							
Record Mean	0.0	T	0.0	T	1.1	3.9	7.2	6.8	3.3	0.3	T	0.0	22.6

WASHINGTON, D.C.
NATIONAL AIRPORT

Washington lies at the western edge of the mid Atlantic Coastal Plain, about 50 miles east of the Blue Ridge Mountains and 35 miles west of Chesapeake Bay, adjacent to the Potomac and Anacostia Rivers. Elevations range from a few feet above sea level to about 400 feet in parts of the northwest section of the city.

Observations have been kept continuously since November 1870. Since June 1941 the official observations have been taken at Washington National Airport.

National Airport is located at the center of the urban heat island. As a result, low temperatures are the highest for the area. Differences between the airport and suburban locations are often 10 to 15 degrees. There is less variation in the high temperatures.

Summers are warm and humid and winters are cold, but not severe. Periods of pleasant weather often occur in the spring and fall. The summertime temperature is in the upper 80s and the winter is in the upper 20s. Precipitation is rather uniformly distributed throughout the year.

Thunderstorms can occur at any time but are most frequent during the late spring and summer. The storms are most often accompanied by downpours and gusty winds, but are not usually severe.

Tornadoes, which infrequently occur, have resulted in significant damage. Severe hailstorms have occurred in the spring.

Tropical storms can bring heavy rain, high winds and flooding, but extensive damage from wind and tidal flooding is rare. Wind gusts of nearly 100 mph and rainfall over 7 inches have occurred during the passage of tropical storms and hurricanes.

Major flooding of the Potomac River can result from heavy rains over the basin, occasionally augmented by snowmelt, and above normal tides associated with hurricanes or severe storms along the coast. Flooding may also occur after a cold winter when the Potomac may be blocked with ice.

Although a snowfall of 10 inches or more in 24 hours is unusual, several notable falls of more than 25 inches have occurred. Normal snowfall during the winter season is 18 inches.

The average date of the last freezing temperature in the spring is April 1 and the average date for the first freezing temperature in the fall is November 10.

NORMALS, MEANS, AND EXTREMES

WASHINGTON, D.C. NATIONAL AIRPORT

LATITUDE: 38°51'N LONGITUDE: 77°02'W ELEVATION: FT. GRND 10 BARO 75 TIME ZONE: EASTERN WBAN: 13743

	(a)	JAN	FEB	MAR	APR	MAY	JUNE	JULY	AUG	SEP	OCT	NOV	DEC	YEAR
TEMPERATURE °F:														
Normals														
-Daily Maximum		42.3	45.9	56.5	66.7	76.2	84.7	88.5	86.9	80.1	69.1	58.3	47.0	66.9
-Daily Minimum		26.8	29.1	37.7	46.4	56.6	66.5	71.4	70.0	62.5	50.3	41.1	31.7	49.2
-Monthly		34.6	37.5	47.2	56.5	66.4	75.6	80.0	78.5	71.3	59.7	49.8	39.4	58.0
Extremes														
-Record Highest	52	79	82	89	95	99	101	104	103	101	94	86	75	104
-Year		1950	1948	1990	1976	1991	1988	1988	1988	1980	1954	1974	1984	JUL 1988
-Record Lowest	52	-5	4	11	24	34	47	54	49	39	29	16	1	-5
-Year		1982	1961	1943	1982	1947	1972	1988	1986	1963	1969	1955	1942	JAN 1982
NORMAL DEGREE DAYS:														
Heating (base 65°F)		942	770	552	264	60	0	0	0	14	195	456	794	4047
Cooling (base 65°F)		0	0	0	9	104	318	465	419	203	31	0	0	1549
% OF POSSIBLE SUNSHINE	45	47	51	55	56	58	64	62	62	61	59	51	46	56
MEAN SKY COVER (tenths)														
Sunrise - Sunset	45	6.6	6.5	6.5	6.4	6.4	6.0	6.0	5.8	5.7	5.5	6.2	6.5	6.2
MEAN NUMBER OF DAYS:														
Sunrise to Sunset														
-Clear	45	7.4	7.2	7.4	7.0	7.0	7.5	7.6	8.8	9.7	10.9	8.0	8.1	96.7
-Partly Cloudy	45	7.2	6.6	8.4	9.1	10.0	11.2	11.7	10.3	8.4	7.8	8.0	6.6	105.5
-Cloudy	45	16.3	14.5	15.1	14.0	14.0	11.2	11.7	11.9	11.9	12.3	14.0	16.3	163.1
Precipitation														
.01 inches or more	52	10.2	8.9	11.0	9.8	11.0	9.4	9.9	9.1	7.9	7.4	8.5	9.3	112.3
Snow,Ice pellets,hail														
1.0 inches or more	50	1.6	1.3	0.7	0.0	0.0	0.0	0.0	0.0	0.0	0.0	0.2	0.8	4.6
Thunderstorms	45	0.2	0.2	1.2	2.7	4.6	5.7	6.8	5.1	2.3	1.1	0.6	0.*	30.4
Heavy Fog Visibility														
1/4 mile or less	45	1.7	1.4	0.9	0.8	0.4	0.2	0.2	0.1	0.4	1.4	1.2	1.8	10.3
Temperature °F														
-Maximum														
90° and above	33	0.0	0.0	0.0	0.4	1.7	7.5	14.2	10.2	3.7	0.1	0.0	0.0	37.7
32° and below	33	4.8	2.2	0.2	0.0	0.0	0.0	0.0	0.0	0.0	0.0	0.*	2.1	9.4
-Minimum														
32° and below	33	22.0	18.4	8.5	0.8	0.0	0.0	0.0	0.0	0.0	0.4	4.3	15.7	70.2
0° and below	33	0.1	0.0	0.0	0.0	0.0	0.0	0.0	0.0	0.0	0.0	0.0	0.0	0.1
AVG. STATION PRESS.(mb)	21	1017.1	1016.8	1015.0	1013.3	1013.4	1013.2	1013.8	1015.2	1016.5	1017.4	1017.4	1017.7	1015.6
RELATIVE HUMIDITY (%)														
Hour 01	33	67	66	65	67	74	77	77	80	80	77	72	69	73
Hour 07 (Local Time)	33	70	70	70	70	74	75	76	80	81	80	75	72	74
Hour 13	33	55	52	50	49	52	52	53	55	56	53	54	57	53
Hour 19	33	60	56	54	52	59	60	61	64	66	64	62	62	60
PRECIPITATION (inches):														
Water Equivalent														
-Normal		2.72	2.71	3.17	2.71	3.66	3.38	3.80	3.91	3.31	3.02	3.12	3.12	38.63
-Maximum Monthly	52	7.11	5.71	7.43	6.88	10.69	11.53	11.06	14.31	12.36	8.18	6.70	6.54	14.31
-Year		1978	1961	1953	1983	1953	1972	1945	1955	1975	1942	1963	1969	AUG 1955
-Minimum Monthly	52	0.31	0.42	0.64	0.03	0.75	0.95	0.93	0.55	0.20	T	0.29	0.22	T
-Year		1955	1978	1945	1985	1986	1988	1966	1962	1967	1963	1981	1955	OCT 1963
-Maximum in 24 hrs	50	2.13	1.94	3.43	3.08	4.32	7.19	4.69	6.39	5.31	4.98	4.03	2.86	7.19
-Year		1976	1983	1958	1970	1953	1972	1970	1955	1975	1955	1993	1977	JUN 1972
Snow,Ice pellets,hail														
-Maximum Monthly	50	21.3	30.6	17.1	0.6	T	0.0	T	T	0.0	0.3	11.5	16.2	30.6
-Year		1966	1979	1960	1972	1993		1990	1992		1979	1987	1962	FEB 1979
-Maximum in 24 hrs	50	13.8	18.7	7.9	0.6	T	0.0	T	T	0.0	0.3	11.5	11.4	18.7
-Year		1966	1979	1960	1972	1993		1990	1992		1979	1987	1957	FEB 1979
WIND:														
Mean Speed (mph)	45	10.0	10.4	10.9	10.5	9.3	8.9	8.3	8.1	8.4	8.7	9.3	9.6	9.4
Prevailing Direction through 1963		NW	S	NW	S	S	S	S	S	S	SSW	S	NW	S
Fastest Obs. 1 Min.														
-Direction (!!)	8	29	33	06	33	32	36	31	34	32	23	33	34	32
-Speed (MPH)	8	35	37	38	35	46	38	38	37	33	39	36	38	46
-Year		1992	1991	1993	1992	1993	1986	1990	1988	1985	1990	1989	1993	MAY 1993
Peak Gust														
-Direction (!!)	10	NW	NW	NE	S	NW	N	W	NW	NW	SW	NW	NW	W
-Speed (mph)	10	51	49	55	48	60	53	63	53	54	58	59	56	63
-Date		1985	1987	1993	1988	1984	1986	1991	1988	1985	1990	1992	1985	JUL 1991

PRECIPITATION (inches)

WASHINGTON, D.C. NATIONAL AIRPORT

YEAR	JAN	FEB	MAR	APR	MAY	JUNE	JULY	AUG	SEP	OCT	NOV	DEC	ANNUAL
1964	3.98	3.38	2.53	4.37	1.46	1.30	1.87	1.89	3.07	1.34	1.42	2.87	29.48
1965	2.73	1.89	4.37	1.65	1.72	1.88	2.98	4.44	2.12	2.32	0.37	0.47	26.94
1966	3.95	3.57	1.44	3.33	2.74	2.02	0.93	1.67	6.87	4.72	1.50	3.28	36.02
1967	1.35	2.32	3.49	0.80	4.27	1.51	5.24	9.17	0.20	1.77	2.10	5.93	38.15
1968	1.97	0.80	3.66	1.53	4.23	7.40	1.31	3.95	2.97	3.17	3.62	2.22	36.83
1969	1.69	2.08	1.60	1.71	1.20	3.46	9.44	6.98	5.07	1.14	2.39	6.54	43.30
1970	1.24	2.69	2.82	5.35	2.79	2.80	8.12	1.09	1.57	2.05	5.77	3.33	39.62
1971	1.86	5.44	1.93	2.10	6.80	1.72	4.97	7.18	2.48	6.12	3.76	1.66	46.02
1972	2.45	5.27	2.27	3.99	4.78	11.53	3.43	2.82	1.27	3.56	6.05	4.55	51.97
1973	2.26	2.68	2.97	4.19	3.39	2.11	2.68	4.41	1.58	1.71	0.97	6.03	34.98
1974	2.66	0.95	4.21	2.26	4.37	3.40	1.15	5.77	4.39	1.13	1.24	4.43	35.96
1975	3.09	1.56	5.33	2.13	4.71	2.15	7.16	3.54	12.36	2.38	2.05	4.04	50.50
1976	3.56	1.55	2.51	1.17	3.57	1.21	4.54	2.13	7.23	7.76	0.85	1.99	38.07
1977	1.50	0.66	2.17	2.66	1.73	3.28	4.06	4.74	0.32	5.35	4.81	4.86	36.14
1978	7.11	0.42	4.48	1.38	5.13	2.43	4.28	5.85	1.01	1.16	2.31	4.00	39.56
1979	6.64	5.62	2.45	1.88	3.55	2.99	3.43	5.41	6.64	5.54	2.33	0.85	47.33
1980	2.85	1.16	5.04	3.28	2.64	1.68	3.86	1.11	1.90	2.59	2.56	0.65	29.32
1981	0.38	2.82	1.49	2.63	3.42	2.55	5.69	3.02	1.94	3.64	0.29	2.80	30.67
1982	2.27	3.33	2.64	3.19	5.11	5.41	2.98	2.68	1.71	1.75	2.96	1.74	35.77
1983	1.69	3.09	4.84	6.88	4.62	7.09	1.78	3.11	2.90	4.87	5.09	5.91	51.87
1984	1.71	3.43	6.14	3.71	3.80	2.01	4.09	2.30	2.51	3.18	3.66	1.19	37.73
1985	2.11	3.07	1.88	0.03	5.79	2.05	2.91	2.35	6.67	3.85	4.47	0.68	35.86
1986	2.38	3.49	0.74	1.98	0.75	1.29	3.79	5.33	0.60	2.01	5.23	4.98	32.57
1987	4.90	2.11	1.54	2.28	2.54	3.90	2.59	2.07	5.11	2.53	4.49	2.57	36.63
1988	3.14	2.52	2.27	2.00	4.50	0.95	3.74	2.39	1.85	1.75	5.33	1.30	31.74
1989	2.49	2.80	4.30	3.50	7.77	6.02	5.66	1.15	6.68	5.48	2.37	2.10	50.32
1990	2.95	1.30	2.57	4.09	5.20	3.14	3.78	6.74	0.87	3.30	2.17	4.73	40.84
1991	2.90	0.83	4.42	1.39	1.57	1.27	3.76	2.03	3.50	2.03	0.85	5.07	29.62
1992	2.78	2.23	3.48	2.55	3.41	2.35	5.34	2.48	3.49	2.03	3.38	2.86	36.38
1993	2.90	2.27	6.82	3.62	3.40	1.73	1.36	3.87	3.68	2.62	4.76	4.38	41.41
Record Mean	2.74	2.52	3.35	2.81	3.88	3.30	3.98	4.24	3.31	2.98	3.05	3.10	39.26

AVERAGE TEMPERATURE (deg. F)

WASHINGTON, D.C. NATIONAL AIRPORT

YEAR	JAN	FEB	MAR	APR	MAY	JUNE	JULY	AUG	SEP	OCT	NOV	DEC	ANNUAL
1964	36.3	36.9	47.5	54.1	68.1	76.0	79.1	75.5	69.9	55.1	51.0	39.9	57.5
1965	33.4	36.8	41.4	51.8	69.1	72.6	78.2	77.2	72.6	57.5	49.5	41.5	56.8
1966	32.4	36.2	47.5	52.7	65.0	76.0	80.9	78.7	68.6	57.0	49.5	37.6	56.8
1967	41.0	34.0	45.0	57.6	60.0	74.7	77.2	76.2	68.0	57.9	45.0	39.9	56.4
1968	31.4	34.3	49.7	58.0	63.7	74.1	79.9	79.2	72.0	61.3	50.0	36.6	57.5
1969	34.2	36.9	43.0	58.7	68.4	77.1	79.5	76.3	70.1	58.8	47.2	36.3	57.2
1970	30.0	37.1	41.9	55.3	68.3	75.2	79.2	79.0	75.0	62.5	49.3	39.7	57.7
1971	31.3	39.1	43.2	55.0	63.7	75.9	78.3	76.7	73.0	64.7	48.2	45.5	57.9
1972	38.5	36.5	45.6	54.1	64.6	70.2	77.5	75.9	71.0	56.0	46.8	43.6	56.7
1973	37.6	37.0	51.1	56.0	62.8	77.1	79.2	79.9	74.3	63.3	51.6	41.9	59.3
1974	42.9	39.2	49.2	58.3	65.1	71.5	79.0	78.4	70.2	57.3	50.9	43.1	58.8
1975	40.9	40.6	45.2	53.6	69.7	76.4	79.3	80.1	68.5	63.2	54.4	40.5	59.4
1976	33.9	46.9	51.3	59.9	65.0	77.6	78.4	76.7	70.4	55.4	43.0	35.5	57.9
1977	25.4	38.8	52.7	60.1	69.4	74.3	80.9	78.8	73.9	59.0	51.8	38.1	58.6
1978	32.5	31.4	44.4	57.7	65.8	76.7	78.8	81.3	73.6	59.4	52.2	43.1	58.1
1979	35.1	28.4	51.5	56.0	67.7	72.4	78.6	78.5	71.6	58.6	54.4	43.7	58.1
1980	37.2	36.1	46.2	60.1	69.5	74.8	82.3	82.8	77.1	59.9	48.6	39.8	59.5
1981	33.0	43.7	47.6	62.1	66.2	78.7	80.2	77.0	71.0	58.3	51.4	38.5	59.0
1982	28.1	38.3	45.7	54.0	69.0	72.8	80.3	75.4	70.6	60.2	51.8	45.5	57.7
1983	38.1	38.7	48.8	53.3	64.9	75.0	81.2	81.0	72.6	60.5	50.3	36.0	58.4
1984	32.2	43.8	41.8	54.9	64.9	76.9	76.5	77.8	68.3	65.2	46.0	45.6	57.8
1985	30.8	37.8	47.7	61.6	68.1	72.3	79.0	76.7	71.9	61.2	54.3	36.4	58.2
1986	35.4	35.3	47.4	56.2	68.1	76.6	81.1	74.6	70.9	61.1	46.5	39.8	57.8
1987	34.7	37.0	47.7	54.8	67.2	76.4	82.6	78.7	72.1	54.4	49.9	41.5	58.1
1988	31.0	37.3	47.2	54.4	65.8	74.4	81.9	80.7	68.9	54.4	49.9	38.7	57.1
1989	39.9	37.8	46.1	55.4	64.1	76.8	78.3	77.1	71.4	60.5	48.0	27.9	56.9
1990	43.6	45.2	50.2	56.8	64.3	75.0	79.4	76.5	69.6	62.8	52.0	44.5	60.0
1991	38.6	43.0	48.8	58.2	73.0	76.8	81.4	80.0	71.0	60.4	48.8	42.3	60.2
1992	38.2	41.2	45.0	55.4	62.3	71.7	79.5	74.0	69.2	56.0	48.8	39.6	56.7
1993	39.7	34.3	42.2	54.8	67.4	75.3	83.1	79.6	71.0	58.1	48.8	38.1	57.7
Record Mean	35.7	38.0	46.3	56.3	66.1	74.7	79.0	77.3	70.7	59.5	48.9	38.8	57.6
Max	43.3	46.4	55.4	66.6	75.9	84.0	87.9	86.0	79.5	68.8	57.6	46.3	66.5
Min	28.1	29.6	37.1	46.1	56.3	65.3	70.1	68.7	61.8	50.1	40.2	31.2	48.7

REFERENCE NOTES

WASHINGTON, D.C. NATIONAL AIRPORT

GENERAL
T - TRACE AMOUNT.
BLANK ENTRIES DENOTE MISSING/UNREPORTED DATA.
# INDICATES A STATION OR INSTRUMENT RELOCATION.

SPECIFIC

PM - INCLUDES LAST DAY OF PREVIOUS MONTH
ASOS - AUTOMATED SURFACE OBSERVING SYSTEM IN OPERATION DURING THESE MONTHS.

(a) - LENGTH OF RECORD IN YEARS, ALTHOUGH INDIVIDUAL MONTHS MAY BE MISSING.
0.* OR * - THE VALUE IS BETWEEN 0.0 AND 0.05
NORMALS - BASED ON THE 1961-1990 RECORD PERIOD.
EXTREMES - DATES ARE THE MOST RECENT OCCURRENCE.
WIND DIR.- NUMERALS SHOW TENS OF DEGRESS CLOCKWISE FROM TRUE NORTH. "00" INDICATES CALM.
RESULTANT DIRECTIONS ARE GIVEN TO WHOLE DEGREES.
BOLD VALUES INDICATE EXTREME VALUES WHICH OCCURRED AFTER THE ASOS SYSTEM WAS COMMISSIONED.

RECORD = PERIOD OF RECORD
RECORD MEAN PRECIPITATION IS THE MEAN OF ALL DAILY PRECIPITATION AMOUNTS DURING THE PERIOD OF RECORD.
RECORD MAX(MIN) TEMPERATURE IS THE MEAN OF ALL DAILY MAX(MIN) TEMPERATURES DURING THE PERIOD OF RECORD.
RECORD MEAN TEMPERATURE IS THE SUM OF THE RECORD MAX AND RECORD MIN DIVIDED BY 2.
AVERAGE TEMPERATURE IS THE SUM OF THE MEAN DAILY MAX AND MIN TEMPERATURE DIVIDED BY 2.

EXCEPTIONS

RECORD MEANS ARE THROUGH THE CURRENT YEAR, BEGINNING IN 1872 FOR TEMPERATURE
1871 FOR PRECIPITATION
1944 FOR SNOWFALL

HEATING DEGREE DAYS Base 65 deg. F — WASHINGTON, D.C. NATIONAL AIRPORT

SEASON	JULY	AUG	SEP	OCT	NOV	DEC	JAN	FEB	MAR	APR	MAY	JUNE	TOTAL
1964-65	0	0	29	300	415	771	974	785	724	395	27	24	4444
1965-66	0	1	18	236	458	723	1001	800	535	374	99	11	4256
1966-67	0	0	41	246	462	843	735	859	611	249	178	3	4227
1967-68	0	0	34	240	592	773	1033	886	471	216	87	0	4332
1968-69	0	0	0	162	445	875	949	780	671	208	40	0	4130
1969-70	0	0	18	226	525	883	1077	773	713	294	56	0	4565
1970-71	0	0	17	131	464	777	1034	722	670	294	85	2	4196
1971-72	0	0	12	61	518	597	815	817	599	326	56	21	3822
1972-73	0	0	8	278	543	654	843	777	423	286	109	0	3921
1973-74	0	0	4	103	399	708	677	716	490	228	85	4	3414
1974-75	0	0	26	250	446	674	740	677	608	345	24	0	3790
1975-76	0	0	20	102	328	752	956	524	415	236	80	0	3413
1976-77	0	0	11	306	652	907	1221	729	389	188	32	3	4438
1977-78	0	0	1	196	406	829	1001	933	633	219	86	0	4304
1978-79	0	0	9	192	378	671	918	1019	425	273	30	0	3915
1979-80	0	0	5	231	313	654	857	830	573	149	28	0	3640
1980-81	0	0	4	189	487	774	984	592	536	133	75	0	3774
1981-82	0	0	19	219	399	818	1135	743	592	328	19	3	4275
1982-83	0	2	9	193	402	597	827	730	497	365	77	0	3699
1983-84	0	0	32	177	433	890	1009	610	710	302	95	4	4262
1984-85	0	0	54	59	561	594	1053	757	533	166	30	6	3813
1985-86	0	0	14	147	320	879	913	824	542	267	61	3	3970
1986-87	0	13	18	180	548	775	931	777	527	304	68	0	4141
1987-88	0	0	4	325	448	719	1047	796	544	317	69	25	4294
1988-89	0	0	18	330	442	807	771	755	596	297	112	0	4128
1989-90	0	0	35	167	507	1144	656	550	481	285	64	4	3893
1990-91	0	0	38	153	381	630	810	608	502	237	22	0	3381
1991-92	0	0	27	175	486	696	824	686	614	295	127	5	3935
1992-93	0	0	43	282	477	781	779	855	700	307	32	3	4259
1993-94	0	0	33	217	487	825							

COOLING DEGREE DAYS Base 65 deg. F — WASHINGTON, D.C. NATIONAL AIRPORT

YEAR	JAN	FEB	MAR	APR	MAY	JUNE	JULY	AUG	SEP	OCT	NOV	DEC	TOTAL
1969	0	0	0	24	151	367	458	360	180	39	1	0	1580
1970	0	0	0	10	166	311	449	442	324	60	0	0	1762
1971	0	0	0	0	52	337	422	372	258	60	22	0	1523
1972	0	0	3	5	50	184	393	346	195	8	2	0	1186
1973	0	0	2	21	47	371	448	469	288	57	3	0	1706
1974	0	0	4	33	96	205	441	422	192	17	27	0	1437
1975	1	0	0	12	177	344	448	475	132	50	15	0	1654
1976	0	4	1	92	86	383	424	370	179	15	0	0	1554
1977	0	0	10	49	177	289	496	434	274	18	15	0	1762
1978	0	0	0	10	117	358	434	514	274	25	0	0	1732
1979	0	0	14	9	120	231	431	425	208	39	2	0	1479
1980	0	0	0	9	174	301	546	563	374	38	1	0	2006
1981	0	0	6	49	118	417	478	380	204	18	0	0	1670
1982	0	0	0	6	155	244	479	330	185	51	13	1	1464
1983	0	0	0	21	81	310	510	504	269	42	0	0	1737
1984	0	0	0	4	99	368	365	404	157	73	0	0	1470
1985	0	0	6	70	135	232	444	373	228	37	6	0	1531
1986	0	0	5	10	162	358	503	318	202	70	1	0	1629
1987	0	0	0	8	146	347	554	431	222	0	0	0	1708
1988	0	0	1	4	101	313	534	490	144	11	0	0	1598
1989	0	0	16	14	91	362	417	381	233	33	1	0	1548
1990	0	0	30	46	50	309	451	364	183	88	0	0	1521
1991	0	0	5	38	278	362	517	472	214	41	5	0	1932
1992	0	0	0	16	53	214	457	285	175	8	0	0	1208
1993	0	0	0	5	114	319	569	460	218	9	8	0	1702

SNOWFALL (inches) — WASHINGTON, D.C. NATIONAL AIRPORT

SEASON	JULY	AUG	SEP	OCT	NOV	DEC	JAN	FEB	MAR	APR	MAY	JUNE	TOTAL
1964-65	0.0	0.0	0.0	0.0	0.2	0.5	9.1	1.9	5.4	0.0	0.0	0.0	17.1
1965-66	0.0	0.0	0.0	0.0	0.0	0.2	21.3	6.9	T	0.0	0.0	0.0	28.4
1966-67	0.0	0.0	0.0	0.0	T	16.1	1.3	19.0	0.7	0.0	0.0	0.0	37.1
1967-68	0.0	0.0	0.0	0.0	6.9	6.3	2.8	2.4	3.0	0.0	0.0	0.0	21.4
1968-69	0.0	0.0	0.0	0.0	T	T	0.2	2.2	6.7	0.0	0.0	0.0	9.1
1969-70	0.0	0.0	0.0	0.0	T	6.8	3.6	3.6	T	0.0	0.0	0.0	14.0
1970-71	0.0	0.0	0.0	0.0	T	5.2	4.8	0.3	1.4	T	0.0	0.0	11.7
1971-72	0.0	0.0	0.0	0.0	1.4	0.1	0.3	14.4	T	0.6	0.0	0.0	16.8
1972-73	0.0	0.0	0.0	T	T	0.0	T	0.1	T	T	0.0	0.0	0.1
1973-74	0.0	0.0	0.0	0.0	0.0	11.0	1.5	4.2	T	T	0.0	0.0	16.7
1974-75	0.0	0.0	0.0	T	T	0.1	6.6	5.8	0.3	T	0.0	0.0	12.8
1975-76	0.0	0.0	0.0	0.0	0.0	0.4	0.1	0.9	0.8	0.0	0.0	0.0	2.2
1976-77	0.0	0.0	0.0	0.0	0.8	0.6	9.7	0.0	0.0	T	0.0	0.0	11.1
1977-78	0.0	0.0	0.0	0.0	0.1	0.2	10.3	3.8	8.3	0.0	0.0	0.0	22.7
1978-79	0.0	0.0	0.0	0.0	3.1	T	4.0	30.6	T	0.0	0.0	0.0	37.7
1979-80	0.0	0.0	0.0	0.3	0.0	T	8.6	5.1	6.1	0.0	0.0	0.0	20.1
1980-81	0.0	0.0	0.0	0.0	T	0.3	4.2	T	T	0.0	0.0	0.0	4.5
1981-82	0.0	0.0	0.0	0.0	T	1.7	15.3	5.3	0.2	T	0.0	0.0	22.5
1982-83	0.0	0.0	0.0	0.0	0.0	6.6	T	21.0	0.0	T	0.0	0.0	27.6
1983-84	0.0	0.0	0.0	0.0	0.3	T	6.5	T	1.8	T	0.0	0.0	8.6
1984-85	0.0	0.0	0.0	0.0	T	0.3	10.0	T	T	T	0.0	0.0	10.3
1985-86	0.0	0.0	0.0	0.0	0.0	0.7	1.8	12.9	T	T	0.0	0.0	15.4
1986-87	0.0	0.0	0.0	0.0	0.0	T	20.8	10.3	T	T	0.0	0.0	31.1
1987-88	0.0	0.0	0.0	0.0	11.5	T	13.1	T	0.4	T	0.0	0.0	25.0
1988-89	0.0	0.0	0.0	0.0	0.0	1.2	2.9	1.2	0.4	0.0	0.0	0.0	5.7
1989-90	0.0	0.0	0.0	0.0	3.5	9.0	0.2	T	2.4	0.2	0.0	0.0	15.3
1990-91	T	0.0	0.0	0.0	0.0	3.0	4.8	0.3	T	0.0	T	0.0	8.1
1991-92	0.0	0.0	0.0	0.0	T	0.0	4.0	2.6	T	0.0	0.0	0.0	6.6
1992-93	0.0	T	0.0	0.0	T	1.0	T	4.1	6.6	0.0	T	0.0	11.7
1993-94	0.0	0.0	0.0	0.0	T	2.6							
Record Mean	T	T	0.0	T	0.8	3.1	5.3	5.3	2.1	T	T	0.0	16.6

DAYTONA BEACH, FLORIDA

Daytona Beach is located on the Atlantic Ocean. The Halifax River, part of the Florida Inland Waterway, runs through the city. The terrain in the area is flat and the soil is mostly sandy. Elevations in the area range from 3 to 15 feet above mean sea level near the ocean to about 31 feet at the airport and on a ridge running along the western city limits.

Nearness to the ocean results in a climate tempered by the effect of land and sea breezes. In the summer, while maximum temperatures reach 90 degrees or above during the late morning or early afternoon, the number of hours of 90 degrees or above is relatively small due to the beginning of the sea breeze near midday and the occurrence of local afternoon convective thunderstorms which lower the temperature to the comfortable 80s. Winters, although subject to invasions of cold air, are relatively mild due to the nearness of the ocean and latitudinal location.

The rainy season from June through mid-October produces 60 percent of the annual rainfall. The major portion of the summer rainfall occurs in the form of local convective thunderstorms which are occasionally heavy and produce as much as 2 or 3 inches of rain. The more severe thunderstorms may be attended by strong gusty winds. Almost all rainfall during the winter months is associated with frontal passages.

Long periods of cloudiness and rain are infrequent, usually not lasting over 2 or 3 days. These periods are usually associated with a stationary front, a so-called northeaster, or a tropical disturbance.

Tropical disturbances or hurricanes are not considered a great threat to this area of the state. Generally hurricanes in this latitude tend to pass well offshore or lose much of their intensity while crossing the state before reaching this area. Only in gusts have hurricane-force winds been recorded at this station.

Heavy fog occurs mostly during the winter and early spring. These fogs usually form by radiational cooling at night and dissipate soon after sunrise. On rare occasions sea fog moves in from the ocean and persists for two or three days. There is no significant source in the area for air pollution.

NORMALS, MEANS, AND EXTREMES

DAYTONA BEACH, FLORIDA

LATITUDE: 29°11'N LONGITUDE: 81°03' W ELEVATION: FT. GRND 29 BARO 34 TIME ZONE: EASTERN WBAN: 12834

	(a)	JAN	FEB	MAR	APR	MAY	JUNE	JULY	AUG	SEP	OCT	NOV	DEC	YEAR
TEMPERATURE °F:														
Normals														
-Daily Maximum		68.0	69.5	74.8	80.0	84.5	88.0	89.8	89.1	86.8	81.5	75.5	70.4	79.8
-Daily Minimum		46.9	48.4	53.9	58.6	64.9	70.8	72.5	72.9	71.9	65.2	56.3	49.7	61.0
-Monthly		57.5	59.0	64.3	69.3	74.7	79.4	81.2	81.0	79.4	73.4	65.9	60.1	70.4
Extremes														
-Record Highest	50	87	89	91	96	100	102	102	100	99	95	89	88	102
-Year		1991	1985	1977	1968	1953	1944	1981	1989	1944	1959	1948	1990	JUL 1981
-Record Lowest	50	15	24	26	35	44	52	60	65	52	41	27	19	15
-Year		1985	1958	1980	1950	1971	1984	1981	1984	1956	1993	1950	1983	JAN 1985
NORMAL DEGREE DAYS:														
Heating (base 65°F)		282	205	112	21	0	0	0	0	0	0	82	207	909
Cooling (base 65°F)		50	37	90	150	301	432	502	496	432	265	109	55	2919
% OF POSSIBLE SUNSHINE														
MEAN SKY COVER (tenths)														
Sunrise - Sunset	45	5.8	5.7	5.6	5.0	5.3	6.2	6.2	6.2	6.3	5.7	5.4	5.9	5.8
MEAN NUMBER OF DAYS:														
Sunrise to Sunset														
-Clear	50	9.3	8.7	9.6	10.9	10.1	5.6	4.7	4.9	5.4	9.5	9.9	9.2	97.9
-Partly Cloudy	50	9.1	8.0	9.5	10.6	11.3	12.7	14.5	15.2	12.7	10.3	9.8	9.0	132.6
-Cloudy	50	12.6	11.5	11.8	8.5	9.6	11.7	11.8	10.9	11.9	11.2	10.3	12.8	134.7
Precipitation														
.01 inches or more	50	7.1	7.8	8.0	5.8	8.2	12.3	13.2	13.8	13.1	10.5	7.4	7.3	114.4
Snow,Ice pellets,hail														
1.0 inches or more	50	0.0	0.0	0.0	0.0	0.0	0.0	0.0	0.0	0.0	0.0	0.0	0.0	0.0
Thunderstorms	49	1.1	1.8	3.4	3.7	7.7	13.4	17.5	15.6	8.7	3.1	1.2	1.1	78.2
Heavy Fog Visibility														
1/4 mile or less	49	5.3	3.2	3.2	1.8	1.5	1.0	1.0	1.2	0.6	1.4	2.8	4.5	27.6
Temperature °F														
-Maximum														
90° and above	50	0.0	0.0	0.2	1.6	5.0	11.0	17.0	14.6	6.3	0.8	0.0	0.0	56.5
32° and below	50	0.0	0.0	0.0	0.0	0.0	0.0	0.0	0.0	0.0	0.0	0.0	0.0	0.0
-Minimum														
32° and below	50	2.3	1.1	0.3	0.0	0.0	0.0	0.0	0.0	0.0	0.0	0.2	1.5	5.5
0° and below	50	0.0	0.0	0.0	0.0	0.0	0.0	0.0	0.0	0.0	0.0	0.0	0.0	0.0
AVG. STATION PRESS. (mb)	21	1018.8	1018.0	1016.5	1015.9	1014.8	1015.1	1016.6	1016.0	1014.9	1015.8	1017.6	1019.1	1016.6
RELATIVE HUMIDITY (%)														
Hour 01	49	85	84	84	83	85	88	89	90	88	85	86	86	86
Hour 07 (Local Time)	49	87	86	86	85	85	87	89	91	90	87	88	87	87
Hour 13	49	59	57	55	53	57	63	65	67	67	63	61	60	61
Hour 19	49	77	73	70	69	71	76	78	80	80	77	79	79	76
PRECIPITATION (inches):														
Water Equivalent														
-Normal		2.75	3.11	2.90	2.23	3.45	5.99	5.40	6.16	6.34	4.13	2.84	2.59	47.89
-Maximum Monthly	50	7.16	9.13	8.11	7.12	12.33	15.19	14.58	19.89	15.20	13.00	10.96	11.98	19.89
-Year		1986	1960	1991	1949	1976	1966	1944	1953	1979	1950	1972	1983	AUG 1953
-Minimum Monthly	50	0.15	0.29	0.25	T	0.08	1.03	0.16	2.01	0.42	0.19	T	0.06	T
-Year		1950	1944	1956	1967	1965	1981	1992	1963	1972	1967	1967	1956	APR 1967
-Maximum in 24 hrs	50	5.73	4.39	5.74	4.03	4.22	6.28	4.21	4.76	6.34	9.29	5.83	5.22	9.29
-Year		1989	1971	1953	1982	1947	1966	1986	1974	1964	1953	1979	1983	OCT 1953
Snow,Ice pellets,hail														
-Maximum Monthly	50	T	T	T	0.0	0.0	T	0.0	0.0	0.0	0.0	0.0	T	T
-Year		1977	1951	1993			1989						1989	MAR 1993
-Maximum in 24 hrs	50	T	T	T	0.0	0.0	T	0.0	0.0	0.0	0.0	0.0	T	T
-Year		1977	1951	1993			1989						1989	MAR 1993
WIND:														
Mean Speed (mph)	48	8.9	9.6	9.9	9.6	8.9	8.1	7.4	7.1	8.3	9.2	8.6	8.5	8.7
Prevailing Direction through 1963		NW	NNW	SSW	E	E	SW	SSW	E	E	NE	NW	NW	E
Fastest Obs. 1 Min.														
-Direction (!!)	45	26	20	24	18	24	33	25	11	11	05	27	34	24
-Speed (MPH)	45	43	44	58	46	41	40	40	50	58	53	37	40	58
-Year		1978	1960	1993	1953	1989	1989	1963	1949	1960	1950	1963	1954	MAR 1993
Peak Gust														
-Direction (!!)	10	W	W	SW	SW	SW	NW	SW	W	S	W	SW	NW	SW
-Speed (mph)	10	52	58	77	49	69	67	59	68	48	56	47	43	77
-Date		1986	1992	1993	1988	1989	1989	1989	1989	1984	1985	1988	1989	MAR 1993

PRECIPITATION (inches)

DAYTONA BEACH, FLORIDA

YEAR	JAN	FEB	MAR	APR	MAY	JUNE	JULY	AUG	SEP	OCT	NOV	DEC	ANNUAL
1964	5.29	2.65	4.84	3.61	2.58	4.73	7.67	10.81	11.39	3.54	3.13	2.52	62.76
1965	2.22	3.00	3.05	1.00	0.08	9.00	3.72	2.97	4.33	3.65	0.97	2.14	36.13
1966	2.89	5.58	0.36	2.56	6.77	15.19	7.09	7.93	4.49	4.60	1.19	1.60	60.25
1967	1.26	3.98	0.31	T	0.73	7.51	9.04	3.02	5.56	0.19	T	2.98	34.58
1968	0.42	1.73	1.79	0.40	4.79	14.38	6.25	11.09	6.07	7.44	2.43	1.38	58.17
1969	1.53	2.03	2.74	0.12	6.47	2.47	2.61	9.40	8.89	6.97	1.96	5.03	50.22
1970	3.94	3.79	3.59	2.08	1.68	2.62	3.65	3.61	3.54	3.87	0.31	0.72	33.40
1971	0.61	5.48	2.00	2.57	3.12	4.73	3.20	3.97	7.20	9.53	1.33	2.49	46.23
1972	2.37	3.97	6.66	1.41	4.02	7.06	3.22	8.29	0.42	3.08	10.96	2.48	53.94
1973	4.66	2.02	2.63	3.09	2.41	4.32	4.69	7.58	5.14	4.40	0.75	2.54	44.23
1974	0.30	1.10	3.19	0.44	2.66	8.65	6.31	9.96	10.50	1.42	0.48	2.20	47.21
1975	1.66	2.27	1.52	2.96	2.99	9.00	6.89	3.16	6.61	5.84	1.46	0.83	45.19
1976	0.60	0.70	2.03	4.27	12.33	11.14	1.07	3.80	5.10	1.90	3.38	6.00	52.32
1977	4.69	2.45	1.43	0.41	4.61	1.15	2.23	7.91	6.55	1.46	3.04	4.74	40.67
1978	2.89	5.98	2.31	3.30	0.56	7.48	5.53	7.99	4.63	8.31	0.07	4.89	53.94
#1979	7.10	1.94	4.08	3.96	6.13	3.03	11.69	5.24	15.20	2.13	7.96	0.56	69.02
1980	3.75	0.76	2.41	2.54	3.62	5.57	5.82	4.13	1.83	2.42	3.12	1.39	37.36
1981	0.32	5.54	3.00	0.29	1.74	1.03	4.69	7.19	7.59	1.08	2.57	4.64	39.68
1982	2.46	2.08	5.81	6.04	4.68	8.29	5.31	3.21	4.96	3.23	1.58	2.53	50.18
1983	2.51	5.96	7.71	6.17	3.86	6.37	1.92	6.82	8.57	10.11	2.01	11.98	73.99
1984	1.46	3.44	1.31	5.29	6.04	2.84	6.77	4.02	10.73	1.09	3.52	0.20	46.71
1985	0.79	0.58	1.49	3.14	3.42	6.81	2.16	9.83	10.62	4.08	0.41	2.05	45.38
1986	7.16	1.28	1.85	0.44	0.99	3.50	14.43	3.47	3.58	3.47	5.08	2.76	48.01
1987	2.21	6.64	7.94	0.28	2.65	3.81	2.78	4.89	5.63	2.77	5.87	0.25	45.72
1988	5.36	1.72	4.57	1.68	1.78	2.39	2.94	4.79	6.81	1.24	6.70	0.93	40.91
1989	6.82	0.64	2.01	2.92	2.02	1.84	2.44	4.47	5.04	11.64	0.88	3.93	44.65
1990	1.42	5.61	1.94	1.48	1.45	2.71	5.85	7.00	1.61	5.88	0.83	0.34	36.12
1991	2.25	1.65	8.11	5.57	6.79	12.67	11.97	7.60	5.52	2.94	0.61	1.51	67.19
1992	2.42	1.71	2.28	2.81	3.13	10.64	0.16	8.86	6.57	5.21	2.15	0.47	46.41
1993	4.29	3.02	5.56	0.33	0.65	2.19	5.05	2.66	2.74	5.53	1.83	1.86	35.71
Record Mean	2.38	2.97	3.30	2.53	3.06	6.01	6.14	6.30	6.75	4.91	2.48	2.27	49.11

AVERAGE TEMPERATURE (deg. F)

DAYTONA BEACH, FLORIDA

YEAR	JAN	FEB	MAR	APR	MAY	JUNE	JULY	AUG	SEP	OCT	NOV	DEC	ANNUAL
1964	56.7	56.4	64.2	71.1	74.8	80.6	81.0	80.8	78.3	70.4	67.9	62.9	70.4
1965	57.5	61.3	63.6	70.8	72.3	77.3	78.7	79.9	79.1	71.6	65.5	59.7	69.8
1966	57.0	59.2	61.9	68.2	75.6	77.6	81.6	81.2	79.1	74.2	64.1	58.4	69.8
1967	60.6	58.5	67.0	72.4	75.5	78.0	79.4	78.7	76.0	71.3	64.5	63.2	70.5
1968	58.2	52.9	60.0	72.2	75.1	78.6	80.7	80.8	79.1	73.2	61.6	56.2	69.0
1969	58.0	55.2	57.9	70.5	74.4	80.5	82.8	80.4	79.7	77.1	63.5	56.5	69.7
1970	54.5	56.1	65.5	73.1	76.1	80.8	82.7	83.5	82.4	76.9	60.6	61.5	71.1
1971	58.6	59.9	60.0	68.0	73.3	78.7	80.5	80.7	79.6	76.1	65.9	67.8	70.7
1972	65.5	59.0	64.7	70.8	74.4	79.6	80.7	80.2	78.7	75.0	67.3	63.5	71.6
1973	58.9	56.3	68.2	68.7	75.2	79.8	81.9	80.3	80.8	75.0	69.6	58.4	71.1
1974	69.5	59.5	68.5	69.3	76.0	78.6	79.2	80.3	80.5	72.0	65.5	59.5	71.5
1975	63.6	65.7	65.6	70.1	77.2	79.5	80.5	80.5	79.6	75.2	66.8	58.6	71.9
1976	54.4	61.1	68.1	67.7	73.3	77.4	80.9	80.0	78.3	70.4	60.5	59.4	69.3
1977	50.6	55.5	68.9	70.0	74.8	82.3	82.6	82.6	80.8	70.8	67.1	58.3	70.3
1978	53.9	52.1	62.3	71.3	77.3	81.5	82.7	82.3	80.7	74.2	71.1	65.6	71.3
#1979	56.7	57.1	64.4	72.3	75.5	78.7	82.1	80.0	80.2	72.9	66.1	59.6	70.4
1980	57.7	55.3	66.3	68.8	74.9	79.2	82.8	82.1	80.3	72.7	65.3	57.0	70.2
1981	48.8	59.2	60.4	70.5	73.5	82.0	82.8	81.5	77.7	73.8	62.7	57.1	69.2
1982	56.6	64.4	66.8	69.4	72.6	79.5	80.0	79.9	77.9	71.5	68.8	64.0	70.9
1983	53.9	57.2	60.2	64.3	72.4	77.0	81.7	81.1	77.8	73.5	62.6	58.1	68.3
1984	55.1	58.0	61.8	66.8	72.4	76.3	79.0	81.4	79.5	75.7	66.5	65.2	69.8
1985	53.7	61.0	66.6	69.4	76.1	81.7	80.6	81.4	78.3	76.8	71.2	56.1	71.1
1986	56.7	62.4	63.1	66.3	73.8	79.9	81.4	81.4	79.6	75.3	72.5	64.7	71.4
1987	55.8	59.7	63.3	65.1	74.3	79.7	81.9	82.3	79.6	70.0	66.5	61.9	70.0
1988	55.1	56.8	62.8	69.1	72.6	79.0	81.2	81.5	80.6	70.7	67.5	59.8	69.7
1989	64.8	61.9	67.8	69.5	75.4	80.3	82.7	81.8	80.5	73.4	65.9	53.3	71.4
1990	62.7	67.5	66.4	69.6	77.3	80.7	81.9	81.9	80.5	76.0	67.4	65.1	73.1
1991	63.4	61.9	65.6	73.7	78.5	80.3	82.5	82.4	80.8	73.8	64.4	63.5	72.6
1992	56.5	61.7	63.3	67.2	72.5	80.5	83.4	80.7	79.8	72.3	69.1	61.4	70.7
1993	64.5	57.5	62.9	65.5	72.7	80.0	82.5	81.9	79.6	73.7	66.9	56.4	70.3
Record Mean	58.3	59.6	64.3	69.2	74.7	79.3	81.0	80.9	79.3	73.3	65.7	59.9	70.4
Max	69.1	70.3	75.0	79.8	84.7	88.3	89.9	89.4	87.0	81.5	75.4	70.3	80.1
Min	47.6	48.9	53.5	58.6	64.6	70.2	72.1	72.4	71.7	65.0	56.0	49.4	60.8

REFERENCE NOTES

DAYTONA BEACH, FLORIDA

GENERAL
T - TRACE AMOUNT.
BLANK ENTRIES DENOTE MISSING/UNREPORTED DATA.
# INDICATES A STATION OR INSTRUMENT RELOCATION.

SPECIFIC

PM - INCLUDES LAST DAY OF PREVIOUS MONTH
ASOS - AUTOMATED SURFACE OBSERVING SYSTEM IN OPERATION DURING THESE MONTHS.

(a) - LENGTH OF RECORD IN YEARS, ALTHOUGH INDIVIDUAL MONTHS MAY BE MISSING.
0.* OR * - THE VALUE IS BETWEEN 0.0 AND 0.05
NORMALS - BASED ON THE 1961-1990 RECORD PERIOD.
EXTREMES - DATES ARE THE MOST RECENT OCCURRENCE
WIND DIR.- NUMERALS SHOW TENS OF DEGRESS CLOCKWISE FROM TRUE NORTH. "00" INDICATES CALM.
RESULTANT DIRECTIONS ARE GIVEN TO WHOLE DEGREES.
BOLD VALUES INDICATE EXTREME VALUES WHICH OCCURRED AFTER THE ASOS SYSTEM WAS COMMISSIONED.

RECORD = PERIOD OF RECORD
RECORD MEAN PRECIPITATION IS THE MEAN OF ALL DAILY PRECIPITATION AMOUNTS DURING THE PERIOD OF RECORD.
RECORD MAX(MIN) TEMPERATURE IS THE MEAN OF ALL DAILY MAX(MIN) TEMPERATURES DURING THE PERIOD OF RECORD.
RECORD MEAN TEMPERATURE IS THE SUM OF THE RECORD MAX AND RECORD MIN DIVIDED BY 2.
AVERAGE TEMPERATURE IS THE SUM OF THE MEAN DAILY MAX AND MIN TEMPERATURE DIVIDED BY 2.

EXCEPTIONS

RECORD MEANS ARE THROUGH THE CURRENT YEAR, BEGINNING IN 1935 FOR TEMPERATURE
1935 FOR PRECIPITATION
1944 FOR SNOWFALL

HEATING DEGREE DAYS Base 65 deg. F DAYTONA BEACH, FLORIDA

SEASON	JULY	AUG	SEP	OCT	NOV	DEC	JAN	FEB	MAR	APR	MAY	JUNE	TOTAL
1964-65	0	0	0	18	22	126	246	138	136	9	1	0	696
1965-66	0	0	0	10	57	173	263	192	120	26	0	0	841
1966-67	0	0	0	3	88	217	168	192	37	5	0	0	710
1967-68	0	0	0	6	72	108	227	344	185	0	0	0	942
1968-69	0	0	0	27	152	301	216	275	228	2	0	0	1201
1969-70	0	0	0	0	105	272	330	252	80	2	0	0	1041
1970-71	0	0	0	0	164	143	230	199	194	55	7	0	992
1971-72	0	0	0	0	71	19	85	182	69	15	0	0	441
1972-73	0	0	0	0	62	139	226	251	38	29	0	0	745
1973-74	0	0	0	10	15	239	0	197	30	24	0	0	515
1974-75	0	0	0	0	69	201	108	69	91	30	0	0	568
1975-76	0	0	0	0	102	222	334	143	34	16	0	0	851
1976-77	0	0	0	11	168	209	444	273	53	20	0	0	1178
1977-78	0	0	0	23	63	241	352	356	132	5	0	0	1172
#1978-79	0	0	0	0	4	71	279	244	79	5	0	0	682
1979-80	0	0	0	0	75	183	234	297	84	16	0	0	889
1980-81	0	0	0	11	93	247	497	184	171	0	1	0	1204
1981-82	0	0	0	0	127	284	273	72	63	26	0	0	845
1982-83	0	0	0	24	21	125	345	220	167	74	2	0	978
1983-84	0	0	0	2	126	255	323	215	148	37	3	0	1109
1984-85	0	0	0	0	63	77	372	173	44	21	0	0	750
1985-86	0	0	0	0	24	303	261	119	141	30	0	0	878
1986-87	0	0	0	0	11	84	301	160	99	81	0	0	736
1987-88	0	0	0	10	74	146	316	259	120	23	0	0	948
1988-89	0	0	0	1	39	187	70	154	68	20	1	0	540
1989-90	0	0	0	31	59	369	120	47	37	14	0	0	677
1990-91	0	0	0	9	35	96	114	126	83	9	0	0	472
1991-92	0	0	0	0	110	114	264	139	106	43	15	0	791
1992-93	0	0	0	0	70	136	104	210	107	49	0	0	676
1993-94	0	0	0	12	62	281							

COOLING DEGREE DAYS Base 65 deg. F DAYTONA BEACH, FLORIDA

YEAR	JAN	FEB	MAR	APR	MAY	JUNE	JULY	AUG	SEP	OCT	NOV	DEC	TOTAL
1969	7	4	17	173	300	473	559	483	447	383	66	13	2925
1970	13	9	101	253	351	478	557	580	532	373	37	40	3324
1971	39	65	45	150	270	420	484	495	445	348	106	113	2980
1972	107	16	67	193	299	447	494	477	415	317	138	98	3068
1973	45	14	148	148	324	450	527	482	483	326	156	40	3143
1974	147	50	145	159	349	414	447	480	469	223	90	40	3013
1975	71	95	118	191	386	441	488	487	445	325	163	29	3239
1976	12	36	136	104	264	378	497	471	406	186	39	40	2569
1977	4	12	181	177	310	527	553	554	478	212	131	41	3180
1978	14	0	56	198	388	499	553	543	477	295	192	94	3309
#1979	26	28	68	231	332	419	538	471	462	252	111	23	2961
1980	12	21	131	135	315	435	559	538	467	258	109	5	2985
1981	0	25	37	172	269	516	559	521	385	282	65	47	2878
1982	19	61	127	166	240	440	472	470	392	234	141	97	2859
1983	6	6	28	57	238	369	521	504	391	270	62	46	2498
1984	22	20	55	96	238	345	442	515	441	338	114	91	2717
1985	29	67	101	160	348	506	490	511	405	373	217	35	3242
1986	13	50	89	79	280	452	516	515	444	324	246	82	3090
1987	20	17	52	92	297	449	530	543	442	171	125	58	2796
1988	17	27	62	155	242	425	509	518	474	185	121	32	2767
1989	71	74	162	162	331	468	553	530	474	299	94	11	3229
1990	55	124	85	161	385	478	531	528	470	355	114	107	3393
1991	71	45	113	278	426	465	548	548	480	278	98	74	3424
1992	8	51	60	118	256	472	577	492	448	231	196	35	2944
1993	95	7	48	71	245	456	551	534	446	289	126	23	2891

SNOWFALL (inches) DAYTONA BEACH, FLORIDA

SEASON	JULY	AUG	SEP	OCT	NOV	DEC	JAN	FEB	MAR	APR	MAY	JUNE	TOTAL
1970-71	0.0	0.0	0.0	0.0	0.0	0.0	0.0	0.0	0.0	0.0	0.0	0.0	0.0
1971-72	0.0	0.0	0.0	0.0	0.0	0.0	0.0	0.0	0.0	0.0	0.0	0.0	0.0
1972-73	0.0	0.0	0.0	0.0	0.0	0.0	0.0	0.0	0.0	0.0	0.0	0.0	0.0
1973-74	0.0	0.0	0.0	0.0	0.0	0.0	0.0	0.0	0.0	0.0	0.0	0.0	0.0
1974-75	0.0	0.0	0.0	0.0	0.0	0.0	0.0	0.0	0.0	0.0	0.0	0.0	0.0
1975-76	0.0	0.0	0.0	0.0	0.0	0.0	0.0	0.0	0.0	0.0	0.0	0.0	0.0
1976-77	0.0	0.0	0.0	0.0	0.0	0.0	T	0.0	0.0	0.0	0.0	0.0	T
1977-78	0.0	0.0	0.0	0.0	0.0	0.0	0.0	0.0	0.0	0.0	0.0	0.0	0.0
#1978-79	0.0	0.0	0.0	0.0	0.0	0.0	0.0	0.0	0.0	0.0	0.0	0.0	0.0
1979-80	0.0	0.0	0.0	0.0	0.0	0.0	0.0	0.0	0.0	0.0	0.0	0.0	0.0
1980-81	0.0	0.0	0.0	0.0	0.0	0.0	0.0	0.0	0.0	0.0	0.0	0.0	0.0
1981-82	0.0	0.0	0.0	0.0	0.0	0.0	0.0	0.0	0.0	0.0	0.0	0.0	0.0
1982-83	0.0	0.0	0.0	0.0	0.0	0.0	0.0	0.0	0.0	0.0	0.0	0.0	0.0
1983-84	0.0	0.0	0.0	0.0	0.0	0.0	0.0	0.0	0.0	0.0	0.0	0.0	0.0
1984-85	0.0	0.0	0.0	0.0	0.0	0.0	0.0	0.0	0.0	0.0	0.0	0.0	0.0
1985-86	0.0	0.0	0.0	0.0	0.0	0.0	0.0	0.0	0.0	0.0	0.0	0.0	0.0
1986-87	0.0	0.0	0.0	0.0	0.0	0.0	0.0	0.0	0.0	0.0	0.0	0.0	0.0
1987-88	0.0	0.0	0.0	0.0	0.0	0.0	0.0	0.0	0.0	0.0	0.0	0.0	0.0
1988-89	0.0	0.0	0.0	0.0	0.0	0.0	0.0	0.0	0.0	0.0	0.0	T	T
1989-90	0.0	0.0	0.0	0.0	0.0	T	0.0	0.0	0.0	0.0	0.0	0.0	T
1990-91	0.0	0.0	0.0	0.0	0.0	0.0	0.0	0.0	0.0	0.0	0.0	0.0	0.0
1991-92	0.0	0.0	0.0	0.0	0.0	0.0	0.0	0.0	0.0	0.0	0.0	0.0	0.0
1992-93	0.0	0.0	0.0	0.0	0.0	0.0	0.0	0.0	T	0.0	0.0	0.0	T
1993-94	0.0	0.0	0.0	0.0	0.0	0.0							
Record Mean	0.0	0.0	0.0	0.0	0.0	T	T	T	T	0.0	0.0	T	T

FORT MYERS, FLORIDA

Located on the south bank of the Caloosahatchee River, about 15 miles from the Gulf of Mexico, Fort Myers has a climate characterized as subtropical, with temperature extremes of both summer and winter tempered by the marine influence of the Gulf.

Temperatures generally range from the low 60s in winter to the low 80s in summer. Winters are mild, with many bright, warm days and moderately cool nights. Occasional cold snaps bring temperatures in the 30s, but only rarely do temperatures drop into the 20s. Frost occurs in the farming areas on only a few occasions each year, and usually is light and scattered. In the summer, temperatures have reached 100 degrees, but these occurrences are very rare.

About two-thirds of annual precipitation occurs during June through September. There are frequent long periods during the winter when only very light, or no rain falls. Most rain during the summer occurs as late afternoon or early evening thunderstorms, which bring welcome cooling on hot summer days. These showers seldom last long, even though they yield large amounts of rain. Exceptions are during the late summer or fall when tropical storms or hurricanes may pass near the Fort Myers area. These may result in heavy downpours that may reach torrential proportions. Twenty-four-hour amounts of from 6 to over 10 inches may occur.

The prevailing wind direction is east and, except during the passage of tropical storms, high velocities are not experienced. During winter and spring there are usually a few days with 20 to 30 mph winds and thunderstorms are sometimes accompanied by strong gusts for brief periods. Winds approximating 100 mph have been experienced with the passage of hurricanes during the fall months.

Thunderstorms have occurred during every month, but are infrequent from November to April. From June through September they occur on 2 out of every 3 days on an average, and as a general rule, in the late afternoons or early evenings. Heavy fog is rather infrequent, occurring mostly in winter during the early mornings. There is seldom a day without sunshine at some time.

Relative humidity is high during the night, dropping off in the middle of the day.

NORMALS, MEANS, AND EXTREMES

FORT MYERS, FLORIDA

LATITUDE: 26°35'N LONGITUDE: 81°52' W ELEVATION: FT. GRND 15 BARO 15 TIME ZONE: EASTERN WBAN: 12835

	(a)	JAN	FEB	MAR	APR	MAY	JUNE	JULY	AUG	SEP	OCT	NOV	DEC	YEAR
TEMPERATURE °F:														
Normals														
-Daily Maximum		74.3	75.4	79.7	84.2	88.7	90.3	91.1	91.3	89.8	85.8	80.7	76.0	83.9
-Daily Minimum		53.2	54.2	58.6	62.0	67.9	73.1	74.5	74.7	74.2	68.6	60.9	55.1	64.8
-Monthly		63.8	64.8	69.1	73.1	78.3	81.7	82.8	83.0	82.0	77.2	70.8	65.6	74.4
Extremes														
-Record Highest	54	88	92	93	96	99	103	101	100	96	95	95	90	103
-Year		1990	1962	1980	1986	1989	1981	1942	1942	1990	1990	1986	1978	JUN 1981
-Record Lowest	54	28	30	33	39	50	60	66	65	63	45	34	26	26
-Year		1981	1958	1980	1950	1945	1984	1950	1957	1956	1957	1970	1962	DEC 1962
NORMAL DEGREE DAYS:														
Heating (base 65°F)		153	108	32	0	0	0	0	0	0	0	25	100	418
Cooling (base 65°F)		116	103	160	247	412	501	552	558	510	378	199	119	3855
% OF POSSIBLE SUNSHINE														
MEAN SKY COVER (tenths)														
Sunrise - Sunset	45	5.1	5.0	5.0	4.6	5.0	6.1	6.4	6.3	6.2	5.1	4.8	5.0	5.4
MEAN NUMBER OF DAYS:														
Sunrise to Sunset														
-Clear	52	10.8	10.1	11.0	11.4	9.5	4.5	2.0	2.3	4.0	10.7	11.4	11.1	98.7
-Partly Cloudy	52	11.6	11.0	12.1	13.0	15.0	16.1	18.4	18.9	15.5	12.7	11.7	11.9	167.9
-Cloudy	52	8.5	7.1	7.9	5.6	6.6	9.4	10.6	9.8	10.5	7.6	6.9	8.0	98.7
Precipitation .01 inches or more	53	5.5	5.6	5.8	4.5	7.9	14.8	18.0	17.7	15.3	7.7	4.4	4.8	111.9
Snow, Ice pellets, hail 1.0 inches or more	53	0.0	0.0	0.0	0.0	0.0	0.0	0.0	0.0	0.0	0.0	0.0	0.0	0.0
Thunderstorms	49	0.8	1.4	2.2	2.8	7.0	15.7	22.8	21.2	14.2	4.0	0.9	0.8	93.8
Heavy Fog Visibility 1/4 mile or less	50	4.7	2.9	2.8	1.7	0.9	0.3	0.*	0.1	0.2	0.7	2.2	3.6	20.0
Temperature °F														
-Maximum														
90° and above	33	0.0	0.1	0.5	3.2	13.9	20.5	25.2	25.0	19.7	5.8	0.5	0.1	114.4
32° and below	33	0.0	0.0	0.0	0.0	0.0	0.0	0.0	0.0	0.0	0.0	0.0	0.0	0.0
-Minimum														
32° and below	33	0.5	0.1	0.0	0.0	0.0	0.0	0.0	0.0	0.0	0.0	0.0	0.2	0.8
0° and below	33	0.0	0.0	0.0	0.0	0.0	0.0	0.0	0.0	0.0	0.0	0.0	0.0	0.0
AVG. STATION PRESS. (mb)	20	1019.4	1018.7	1017.3	1016.5	1015.2	1016.0	1017.4	1016.5	1015.1	1015.7	1017.5	1019.5	1017.1
RELATIVE HUMIDITY (%)														
Hour 01	32	86	84	84	84	85	89	89	89	89	87	87	87	87
Hour 07 (Local Time)	32	89	88	89	88	88	89	89	90	91	89	89	89	89
Hour 13	32	57	55	52	47	50	58	59	60	60	56	56	56	56
Hour 19	32	73	69	67	64	66	74	76	77	77	73	74	74	72
PRECIPITATION (inches):														
Water Equivalent														
-Normal		1.84	2.23	3.07	1.06	3.87	9.52	8.26	9.66	7.82	2.94	1.57	1.53	53.37
-Maximum Monthly	54	7.95	10.82	18.58	7.66	10.32	20.10	15.28	16.73	16.60	12.04	8.06	5.42	20.10
-Year		1991	1983	1970	1941	1968	1974	1941	1981	1969	1959	1987	1940	JUN 1974
-Minimum Monthly	54	0.00	T	0.03	T	0.34	1.99	2.28	3.98	1.93	0.05	T	0.02	0.00
-Year		1950	1944	1974	1970	1962	1980	1964	1963	1988	1963	1944	1984	JAN 1950
-Maximum in 24 hrs	54	3.80	2.60	7.92	3.82	7.75	6.67	4.06	6.73	9.34	10.85	3.67	3.00	10.85
-Year		1991	1969	1970	1943	1989	1959	1965	1967	1962	1951	1987	1969	OCT 1951
Snow, Ice pellets, hail														
-Maximum Monthly		0.0	0.0	0.0	0.0	0.0	0.0	0.0	0.0	0.0	0.0	0.0	0.0	
-Year														
-Maximum in 24 hrs	54	0.0	0.0	0.0	0.0	0.0	0.0	0.0	0.0	0.0	0.0	0.0	0.0	
-Year														
WIND:														
Mean Speed (mph)	47	8.4	9.0	9.3	8.8	8.1	7.2	6.6	6.7	7.5	8.4	8.2	8.0	8.0
Prevailing Direction through 1963		E	E	SW	E	E	E	ESE	E	E	NE	NE	NE	E
Fastest Obs. 1 Min.														
-Direction (!!!)	44	25	25	35	20	22	12	18	25	05	23	19	33	05
-Speed (MPH)	44	40	39	46	39	40	46	45	44	92	45	31	35	92
-Year		1958	1958	1970	1958	1965	1966	1952	1986	1960	1953	1988	1967	SEP 1960
Peak Gust														
-Direction (!!!)														
-Speed (mph)														
-Date														

PRECIPITATION (inches)

FORT MYERS, FLORIDA

YEAR	JAN	FEB	MAR	APR	MAY	JUNE	JULY	AUG	SEP	OCT	NOV	DEC	ANNUAL
1964	2.88	3.30	2.12	0.80	0.50	4.58	2.28	4.26	9.45	1.38	0.22	1.06	32.83
1965	1.24	2.99	2.91	2.39	4.70	7.78	12.05	6.57	4.35	4.42	0.58	0.85	50.83
1966	3.39	1.06	0.37	3.03	1.61	12.42	8.22	8.10	4.18	2.14	0.18	0.29	44.99
1967	1.15	2.15	0.72	T	1.46	7.41	6.69	15.86	7.04	3.08	0.92	2.91	49.39
1968	0.40	2.08	0.65	0.57	10.32	15.03	9.85	11.44	8.92	7.99	2.88	0.16	70.29
1969	1.44	2.87	4.74	0.15	4.71	10.63	7.11	8.49	16.60	11.03	0.22	3.95	71.94
1970	4.36	2.20	18.58	T	6.36	7.47	4.74	4.82	8.29	1.19	0.46	0.37	58.84
1971	0.85	1.55	0.55	0.70	3.77	6.18	9.50	8.06	9.21	6.49	0.16	0.30	47.32
1972	0.77	2.14	4.72	0.27	5.20	7.86	9.72	16.22	2.33	2.20	3.85	1.43	56.71
1973	3.14	2.23	3.89	1.71	0.78	3.99	9.57	8.66	8.38	0.16	0.10	1.72	44.33
1974	0.36	0.81	0.03	0.11	2.40	20.10	14.47	7.70	4.31	0.19	1.46	0.89	52.83
1975	0.26	0.27	1.47	0.80	2.78	10.55	10.81	7.74	12.59	3.05	0.49	0.69	51.50
1976	0.21	1.20	0.91	0.90	5.22	10.59	6.14	8.95	8.81	1.96	2.10	1.68	48.67
1977	3.53	0.15	0.09	0.76	6.51	8.96	9.60	10.58	9.21	0.43	1.50	2.74	54.06
1978	2.48	3.36	3.43	2.35	2.52	6.75	10.29	10.90	5.18	1.45	0.04	4.35	53.10
1979	7.45	1.94	0.43	3.12	5.32	8.31	5.96	14.79	13.65	0.39	0.46	5.16	66.98
1980	2.44	1.04	3.59	1.52	8.73	1.99	7.02	8.79	4.64	1.54	3.15	0.55	45.00
1981	0.80	1.65	1.29	0.08	3.07	11.79	8.24	16.73	6.70	0.40	0.71	0.73	52.19
1982	0.78	3.34	3.32	3.91	2.08	15.01	11.33	10.56	9.29	5.00	1.11	0.27	66.00
1983	4.50	10.82	7.41	1.34	0.62	17.92	4.77	6.46	9.72	4.39	3.66	3.24	74.85
1984	0.15	3.18	6.38	1.09	2.80	8.65	8.99	5.50	7.89	0.65	0.71	0.02	46.01
1985	0.68	0.44	2.06	1.47	1.21	3.76	8.78	7.79	11.71	6.78	2.20	0.66	47.54
1986	0.90	1.01	3.59	0.53	4.01	13.94	6.73	11.16	5.43	3.56	0.78	5.22	56.86
1987	2.29	2.86	5.86	0.14	4.11	9.69	14.38	8.54	7.50	5.10	8.06	0.48	69.01
1988	2.19	1.47	2.44	1.36	0.62	7.16	5.13	9.21	1.93	0.40	2.83	0.26	35.00
1989	1.65	0.36	2.89	0.36	8.05	8.67	9.05	8.82	5.18	2.05	0.65	2.16	49.89
1990	0.47	3.37	0.87	0.39	3.66	9.02	6.47	14.97	7.40	2.28	0.01	0.18	49.09
1991	7.95	0.72	1.13	5.01	8.48	11.22	14.51	5.87	7.98	4.01	0.32	0.30	67.50
1992	1.75	3.89	4.87	2.08	0.97	18.73	8.59	7.59	3.68	1.24	1.19	0.87	55.45
1993	5.36	3.35	3.61	2.40	2.71	6.63	7.23	6.57	8.29	6.85	0.78	0.78	54.56
Record Mean	1.78	2.16	2.67	1.90	3.99	9.30	8.76	8.34	8.30	3.87	1.40	1.38	53.84

AVERAGE TEMPERATURE (deg. F)

FORT MYERS, FLORIDA

YEAR	JAN	FEB	MAR	APR	MAY	JUNE	JULY	AUG	SEP	OCT	NOV	DEC	ANNUAL
1964	63.2	62.0	71.9	75.9	77.7	81.8	82.9	82.9	81.5	74.5	72.2	68.4	74.6
1965	63.7	67.7	70.1	74.8	76.8	80.0	81.2	82.7	81.3	76.1	71.3	66.3	74.3
1966	62.5	65.2	67.2	72.4	79.0	80.0	82.6	82.8	81.8	78.1	67.7	62.8	73.5
1967	66.4	63.1	69.9	73.0	79.0	81.4	82.9	82.4	81.6	75.8	69.6	69.1	74.5
1968	64.6	59.9	64.6	74.4	78.7	80.1	82.9	82.6	81.4	76.1	66.8	62.3	72.9
1969	62.8	60.9	63.4	74.1	77.8	82.2	83.0	82.2	80.7	78.3	66.3	61.0	72.8
1970	58.1	60.2	67.2	74.8	75.0	80.6	82.2	82.7	82.1	78.3	66.7	66.6	72.9
1971	65.3	65.8	65.9	70.4	77.1	80.4	82.0	81.5	80.0	78.9	71.0	72.3	74.2
1972	70.9	66.0	70.8	74.0	78.3	81.6	82.9	83.0	82.6	78.3	71.3	67.4	75.6
1973	66.0	60.7	71.8	72.0	78.2	82.0	82.4	82.1	82.5	77.4	73.5	62.7	74.3
1974	73.0	65.6	72.0	74.5	78.9	80.0	81.4	82.9	83.0	75.0	70.6	65.9	75.3
1975	69.5	70.6	71.3	75.1	80.7	82.4	82.0	83.8	81.1	79.0	70.6	63.7	75.8
1976	60.9	65.0	71.5	71.2	77.2	78.8	81.6	81.9	79.3	72.6	66.7	62.8	72.5
1977	55.9	61.0	70.9	71.4	75.9	80.8	81.6	81.5	81.7	74.0	70.2	63.3	72.4
1978	59.2	57.2	65.7	72.8	78.7	81.7	83.4	83.8	82.4	78.1	74.2	69.7	73.9
1979	62.5	64.0	68.4	76.2	79.0	82.6	85.7	84.3	83.6	79.1	73.8	66.9	75.5
1980	65.1	60.7	71.6	72.6	77.9	83.4	84.4	84.0	84.7	78.3	70.8	62.8	74.7
1981	55.7	65.8	67.7	76.1	79.2	85.9	85.7	84.4	83.1	79.8	69.6	65.8	74.9
1982	64.9	72.2	73.0	76.5	76.4	83.7	84.3	82.6	80.5	75.4	72.1	68.9	75.9
1983	62.4	62.7	65.4	70.1	76.0	80.2	82.0	82.0	79.8	77.2	69.0	66.4	72.8
1984	62.1	64.9	67.4	72.1	78.7	81.5	82.4	83.8	81.1	77.8	69.8	69.4	74.3
1985	60.6	67.0	71.6	73.9	80.0	83.7	82.2	84.5	83.1	81.8	75.6	63.5	75.6
1986	64.0	67.8	67.9	72.3	78.4	82.2	83.1	82.9	83.5	79.5	78.4	69.8	75.8
1987	63.4	67.0	69.4	68.8	78.7	82.8	84.1	84.8	83.1	74.8	71.7	67.0	74.6
1988	63.0	63.8	68.5	74.0	77.5	82.8	83.6	83.6	83.8	77.2	74.0	66.6	74.9
1989	69.9	67.9	71.7	75.3	80.4	82.1	82.4	83.1	82.5	76.6	71.7	60.8	75.4
1990	69.6	71.4	71.8	74.4	80.4	83.1	83.6		83.4	79.4	72.7	70.0	
1991	70.3	67.8	71.4	77.7	80.9	82.8	83.0	84.0	82.8	78.0	69.2	68.8	76.4
1992	63.7	67.7	68.9	72.9	76.2	81.8	84.5	82.9	82.6	76.5	73.9	68.1	75.0
1993	70.1	64.9	68.2	70.8	77.7	82.9	83.7	83.7	82.9	78.3	72.9	64.0	75.0
Record Mean	64.4	65.4	69.0	73.4	77.8	81.3	82.5	82.8	81.7	76.8	70.2	65.6	74.2
Max	74.8	76.1	79.7	84.3	88.3	90.4	91.0	91.3	89.8	85.4	79.9	76.0	83.9
Min	53.9	54.7	58.3	62.4	67.2	72.1	73.9	74.2	73.5	68.1	60.4	55.2	64.5

REFERENCE NOTES

FORT MYERS, FLORIDA

GENERAL
T - TRACE AMOUNT.
BLANK ENTRIES DENOTE MISSING/UNREPORTED DATA.
# INDICATES A STATION OR INSTRUMENT RELOCATION.

SPECIFIC

(a) - LENGTH OF RECORD IN YEARS, ALTHOUGH INDIVIDUAL MONTHS MAY BE MISSING.
0.* OR * - THE VALUE IS BETWEEN 0.0 AND 0.05
NORMALS - BASED ON THE 1961-1990 RECORD PERIOD.
EXTREMES - DATES ARE THE MOST RECENT OCCURRENCE
WIND DIR. - NUMERALS SHOW TENS OF DEGRESS CLOCKWISE FROM TRUE NORTH. "00" INDICATES CALM.
RESULTANT DIRECTIONS ARE GIVEN TO WHOLE DEGREES.
BOLD VALUES INDICATE EXTREME VALUES WHICH OCCURRED AFTER THE ASOS SYSTEM WAS COMMISSIONED.

RECORD = PERIOD OF RECORD
RECORD MEAN PRECIPITATION IS THE MEAN OF ALL DAILY PRECIPITATION AMOUNTS DURING THE PERIOD OF RECORD.
RECORD MAX(MIN) TEMPERATURE IS THE MEAN OF ALL DAILY MAX(MIN) TEMPERATURES DURING THE PERIOD OF RECORD.
RECORD MEAN TEMPERATURE IS THE SUM OF THE RECORD MAX AND RECORD MIN DIVIDED BY 2.
AVERAGE TEMPERATURE IS THE SUM OF THE MEAN DAILY MAX AND MIN TEMPERATURE DIVIDED BY 2.

EXCEPTIONS

RECORD MEANS ARE THROUGH THE CURRENT YEAR,
BEGINNING IN 1920 FOR TEMPERATURE
1920 FOR PRECIPITATION
1920 FOR SNOWFALL

HEATING DEGREE DAYS Base 65 deg. F — FORT MYERS, FLORIDA

SEASON	JULY	AUG	SEP	OCT	NOV	DEC	JAN	FEB	MAR	APR	MAY	JUNE	TOTAL
1964-65	0	0	0	0	0	26	96	36	38	0	0	0	196
1965-66	0	0	0	0	7	52	136	74	24	1	0	0	294
1966-67	0	0	0	0	44	105	70	90	2	0	0	0	311
1967-68	0	0	0	0	2	34	67	153	82	0	0	0	338
1968-69	0	0	0	5	70	157	96	130	106	0	0	0	564
1969-70	0	0	0	0	53	136	224	146	40	0	0	0	599
1970-71	0	0	0	0	61	33	110	85	63	13	0	0	365
1971-72	0	0	0	0	3	3	19	60	0	0	0	0	85
1972-73	0	0	0	0	25	82	94	141	2	4	0	0	348
1973-74	0	0	0	0	12	147	0	73	1	1	0	0	234
1974-75	0	0	0	0	11	74	33	13	15	0	0	0	146
1975-76	0	0	0	0	48	102	164	78	5	0	0	0	397
1976-77	0	0	0	2	62	124	281	132	15	0	0	0	616
1977-78	0	0	0	5	34	133	212	216	69	0	0	0	669
1978-79	0	0	0	0	0	19	127	100	13	0	0	0	259
1979-80	0	0	0	0	14	44	85	163	41	0	0	0	347
1980-81	0	0	0	0	33	101	282	70	19	0	0	0	505
1981-82	0	0	0	0	23	103	102	5	11	0	0	0	244
1982-83	0	0	0	2	7	58	130	81	56	8	0	0	342
1983-84	0	0	0	0	15	100	135	76	46	0	0	0	372
1984-85	0	0	0	0	15	37	171	74	7	3	0	0	307
1985-86	0	0	0	0	0	123	87	35	71	0	0	0	316
1986-87	0	0	0	0	0	19	112	41	14	42	0	0	228
1987-88	0	0	0	0	15	62	126	100	36	0	0	0	339
1988-89	0	0	0	0	2	67	8	70	26	0	0	0	173
1989-90	0	0	0	10	7	176	31	7	1	1	0	0	233
1990-91	0		0	2	2	27	21	42	19	0	0	0	
1991-92	0	0	0	0	35	31	92	28	22	9	0	0	217
1992-93	0	0	0	0	20	33	19	46	33	0	0	0	151
1993-94	0	0	0	0	18	86							

COOLING DEGREE DAYS Base 65 deg. F — FORT MYERS, FLORIDA

YEAR	JAN	FEB	MAR	APR	MAY	JUNE	JULY	AUG	SEP	OCT	NOV	DEC	TOTAL
1969	33	20	64	280	401	522	564	541	479	419	96	20	3439
1970	18	18	112	303	322	472	539	556	517	419	121	91	3488
1971	128	111	96	182	383	467	532	517	458	438	188	236	3736
1972	210	97	190	278	420	504	564	570	536	419	222	165	4175
1973	131	27	221	220	416	517	545	538	533	393	271	83	3895
1974	254	92	224	297	441	460	518	561	548	319	187	109	4010
1975	181	177	215	309	495	528	533	588	490	439	224	70	4249
1976	42	86	213	193	386	423	521	531	436	246	122	66	3265
1977	7	25	205	200	346	479	520	521	509	293	196	88	3389
1978	39	3	95	242	429	508	580	588	530	412	284	170	3880
1979	52	80	124	344	444	535	650	604	563	442	287	112	4237
1980	95	45	252	237	408	562	609	594	598	421	216	41	4078
1981	0	98	107	340	446	634	650	606	551	466	169	136	4203
1982	106	213	266	353	360	568	606	554	473	331	228	185	4243
1983	54	25	75	166	349	463	533	535	452	386	144	153	3335
1984	53	81	126	221	435	503	549	588	489	404	164	182	3795
1985	43	137	219	276	471	567	544	607	550	526	325	84	4349
1986	61	121	167	224	422	522	568	562	562	457	406	175	4247
1987	70	103	160	162	431	542	599	621	552	311	223	131	3905
1988	70	73	152	277	397	539	580	585	570	386	279	122	4030
1989	166	158	241	316	485	521	550	570	530	377	217	51	4182
1990	183	192	219	289	484	549	583		560	455	237	192	
1991	193	126	222	391	499	543	568	595	542	412	166	153	4410
1992	61	111	151	251	353	509	612	563	535	364	293	135	3938
1993	183	49	140	180	400	547	587	587	548	419	261	62	3963

SNOWFALL (inches) — FORT MYERS, FLORIDA

SEASON	JULY	AUG	SEP	OCT	NOV	DEC	JAN	FEB	MAR	APR	MAY	JUNE	TOTAL
1970-71	0.0	0.0	0.0	0.0	0.0	0.0	0.0	0.0	0.0	0.0	0.0	0.0	0.0
1971-72	0.0	0.0	0.0	0.0	0.0	0.0	0.0	0.0	0.0	0.0	0.0	0.0	0.0
1972-73	0.0	0.0	0.0	0.0	0.0	0.0	0.0	0.0	0.0	0.0	0.0	0.0	0.0
1973-74	0.0	0.0	0.0	0.0			0.0	0.0	0.0	0.0	0.0	0.0	
1974-75	0.0	0.0	0.0	0.0	0.0	0.0	0.0	0.0	0.0	0.0	0.0	0.0	0.0
1975-76	0.0	0.0	0.0	0.0	0.0	0.0	0.0	0.0	0.0	0.0	0.0	0.0	0.0
1976-77	0.0	0.0	0.0	0.0	0.0	0.0	0.0	0.0	0.0	0.0	0.0	0.0	0.0
1977-78	0.0	0.0	0.0	0.0	0.0	0.0	0.0	0.0	0.0	0.0	0.0	0.0	0.0
1978-79	0.0	0.0	0.0	0.0	0.0	0.0	0.0	0.0	0.0	0.0	0.0	0.0	0.0
1979-80	0.0	0.0	0.0	0.0	0.0	0.0	0.0	0.0	0.0	0.0	0.0	0.0	0.0
1980-81	0.0	0.0	0.0	0.0	0.0	0.0	0.0	0.0	0.0	0.0	0.0	0.0	0.0
1981-82	0.0	0.0	0.0	0.0	0.0	0.0	0.0	0.0	0.0	0.0	0.0	0.0	0.0
1982-83	0.0	0.0	0.0	0.0	0.0	0.0	0.0	0.0	0.0	0.0	0.0	0.0	0.0
1983-84	0.0	0.0	0.0	0.0	0.0	0.0	0.0	0.0	0.0	0.0	0.0	0.0	0.0
1984-85	0.0	0.0	0.0	0.0	0.0	0.0	0.0	0.0	0.0	0.0	0.0	0.0	0.0
1985-86	0.0	0.0	0.0	0.0	0.0	0.0	0.0	0.0		0.0	0.0	0.0	
1986-87	0.0	0.0	0.0	0.0	0.0	0.0	0.0	0.0	0.0	0.0	0.0	0.0	0.0
1987-88	0.0	0.0	0.0	0.0	0.0	0.0	0.0	0.0	0.0	0.0	0.0	0.0	0.0
1988-89	0.0	0.0	0.0	0.0	0.0	0.0	0.0	0.0	0.0	0.0	0.0	0.0	0.0
1989-90	0.0	0.0	0.0	0.0	0.0	0.0	0.0	0.0	0.0	0.0	0.0	0.0	0.0
1990-91	0.0	0.0	0.0	0.0	0.0	0.0	0.0	0.0	0.0	0.0	0.0	0.0	0.0
1991-92	0.0	0.0	0.0	0.0	0.0	0.0	0.0	0.0	0.0	0.0	0.0	0.0	0.0
1992-93	0.0	0.0	0.0	0.0	0.0	0.0	0.0	0.0	0.0	0.0	0.0	0.0	0.0
1993-94	0.0	0.0	0.0	0.0	0.0	0.0							
Record Mean	0.0	0.0	0.0	0.0	0.0	0.0	0.0	0.0	0.0	0.0	0.0	0.0	0.0

GAINESVILLE, FLORIDA

Gainsville lies in the north central part of the Florida peninsula, almost midway between the coasts of the Atlantic Ocean and the Gulf of Mexico. The terrain is fairly level with several nearby lakes to the east and south. Due to its centralized location, maritime influences are somewhat less than they would be along coastlines at the same latitude.

Maximum temperatures in summer average slightly more than 90 degrees. From June to September, the number of days when temperatures exceed 89 degrees is 84 on average. Record high temperatures are in excess of 100 degrees. Minimum temperatures in winter average a little more than 44 degrees. The average number of days per year when temperatures are freezing or below is 18. Record lows occur in the teens. Low temperatures are a consequence of cold winds from the north or nighttime radiational cooling of the ground in contact with rather calm air.

Rainfall is appreciable in every month but is most abundant from showers and thunderstorms in summer. The average number of thunderstorm hours yearly is approximately 160. In winter, large-scale cyclone and frontal activity is responsible for some of the precipitation. Monthly average values range from about 2 inches in November to about 8 inches in August. Snowfall is practically unknown.

Because of its inland location, Gainesville does not have serious problems with hurricanes. An occasional hurricane will cross the Gulf or Atlantic coast and head toward Gainesville, but before it arrives it is weakened by surface friction and a depletion of water vapor.

NORMALS, MEANS, AND EXTREMES

GAINESVILLE, FLORIDA

LATITUDE: 29°41'N LONGITUDE: 82°16' W ELEVATION: FT. GRND 138 BARO 143 TIME ZONE: EASTERN WBAN: 12816

	(a)	JAN	FEB	MAR	APR	MAY	JUNE	JULY	AUG	SEP	OCT	NOV	DEC	YEAR
TEMPERATURE °F:														
Normals														
-Daily Maximum		65.6	68.1	74.9	80:7	86.1	89.5	90.7	90.1	87.3	81.0	74.3	68.1	79.7
-Daily Minimum		42.5	44.1	50.7	55.8	62.6	68.3	70.9	71.0	68.9	59.7	50.5	44.6	57.5
-Monthly		54.1	56.2	62.8	68.3	74.3	79.0	80.8	80.5	78.1	70.4	62.4	56.4	68.6
Extremes														
-Record Highest	10	83	85	89	95	98	102	99	99	95	91	88	84	102
-Year		1993	1991	1991	1991	1989	1985	1990	1986	1993	1989	1986	1991	JUN 1985
-Record Lowest	10	10	23	28	35	42	50	62	62	52	33	30	16	10
-Year		1985	1989	1993	1987	1992	1984	1988	1984	1993	1989	1991	1989	JAN 1985
NORMAL DEGREE DAYS:														
Heating (base 65°F)		373	271	130	30	0	0	0	0	0	20	146	297	1267
Cooling (base 65°F)		35	25	62	129	288	420	490	481	393	188	68	30	2609
% OF POSSIBLE SUNSHINE														
MEAN SKY COVER (tenths)														
Sunrise - Sunset														
MEAN NUMBER OF DAYS:														
Sunrise to Sunset														
-Clear														
-Partly Cloudy														
-Cloudy														
Precipitation .01 inches or more	10	9.8	7.6	8.3	6.2	7.3	13.1	14.7	16.2	12.6	7.0	7.2	5.9	115.9
Snow,Ice pellets,hail 1.0 inches or more	10	0.0	0.0	0.0	0.0	0.0	0.0	0.0	0.0	0.0	0.0	0.0	0.0	0.0
Thunderstorms	10	1.6	2.5	3.5	3.4	6.0	13.9	19.1	17.3	7.8	3.3	1.8	0.6	80.8
Heavy Fog Visibility 1/4 mile or less	10	5.5	3.9	2.6	2.9	2.8	2.3	1.8	2.6	2.4	3.7	5.6	6.4	42.5
Temperature °F														
-Maximum														
90° and above	10	0.0	0.0	0.0	1.4	8.1	17.9	21.2	19.3	10.4	0.9	0.0	0.0	79.2
32° and below	10	0.1	0.0	0.0	0.0	0.0	0.0	0.0	0.0	0.0	0.0	0.0	0.1	0.2
-Minimum														
32° and below	10	4.4	3.0	0.9	0.0	0.0	0.0	0.0	0.0	0.0	0.0	0.5	4.2	13.0
0° and below	10	0.0	0.0	0.0	0.0	0.0	0.0	0.0	0.0	0.0	0.0	0.0	0.0	0.0
AVG. STATION PRESS. (mb)	10	1014.7	1013.5	1012.3	1010.6	1010.7	1010.6	1012.2	1011.2	1011.1	1011.9	1013.5	1015.5	1012.3
RELATIVE HUMIDITY (%)														
Hour 01	10	89	88	89	89	89	93	94	95	95	92	93	91	91
Hour 07 (Local Time)	10	90	90	92	91	91	93	94	96	96	94	94	92	93
Hour 13	10	63	58	55	52	51	60	63	65	65	63	63	63	60
Hour 19	10	77	69	66	63	64	73	78	81	82	81	83	81	75
PRECIPITATION (inches):														
Water Equivalent														
-Normal		3.35	4.21	3.65	2.64	3.76	6.77	6.80	8.01	5.27	1.82	2.26	3.27	51.81
-Maximum Monthly	10	6.66	5.73	9.79	6.02	6.24	12.86	8.84	15.84	11.97	7.98	4.51	5.00	15.84
-Year		1991	1986	1987	1991	1991	1992	1984	1985	1988	1993	1987	1986	AUG 1985
-Minimum Monthly	10	1.14	0.32	1.55	0.38	0.74	2.22	1.52	2.49	2.00	T	0.31	0.21	T
-Year		1989	1991	1985	1987	1986	1988	1992	1987	1993	1987	1991	1984	OCT 1987
-Maximum in 24 hrs	10	2.22	2.47	3.33	2.62	3.42	4.41	2.31	3.45	6.16	5.13	2.03	1.55	6.16
-Year		1986	1993	1991	1985	1985	1990	1990	1993	1988	1992	1986	1987	SEP 1988
Snow,Ice pellets,hail														
-Maximum Monthly	10	0.0	T	T	0.0	0.0	0.0	0.0	T	0.0	0.0	0.0	T	T
-Year			1988	1991					1989				1989	MAR 1991
-Maximum in 24 hrs	10	0.0	T	T	0.0	0.0	0.0	0.0	T	0.0	0.0	0.0	T	T
-Year			1988	1991					1989				1989	MAR 1991
WIND:														
Mean Speed (mph)	10	7.2	7.6	8.0	7.4	7.0	6.2	5.9	5.6	5.9	6.6	6.7	6.3	6.7
Prevailing Direction through														
Fastest Obs. 1 Min.														
-Direction (!!)														
-Speed (MPH)														
-Year														
Peak Gust														
-Direction (!!)														
-Speed (mph)														
-Date														

PRECIPITATION (inches)

GAINESVILLE, FLORIDA

YEAR	JAN	FEB	MAR	APR	MAY	JUNE	JULY	AUG	SEP	OCT	NOV	DEC	ANNUAL
1984	1.29	4.82	3.14	2.68	5.44	3.85	8.84	3.16	3.55	0.81	2.90	0.21	40.69
1985	1.18	1.66	1.55	4.01	3.72	7.10	7.04	15.84	4.00	4.68	1.76	1.56	54.10
1986	5.51	5.73	3.10	0.84	0.74	5.32	5.20	6.91	3.29	3.44	3.07	5.00	48.15
1987	4.07	5.65	9.79	0.38	4.53	2.48	4.86	2.49	3.34	T	4.51	1.95	44.05
1988	5.19	5.01	7.72	1.43	2.35	2.22	2.07	12.07	11.97	0.81	3.32	1.61	55.77
1989	1.14	1.19	2.17	2.93	1.91	9.66	4.44	6.08	4.66	1.02	2.14	3.13	40.47
1990	1.97	3.54	1.82	2.78	0.86	11.27	6.41	4.08	2.99	3.16	1.84	1.28	42.00
1991	6.66	0.32	8.78	6.02	6.24	6.58	7.25	4.02	2.40	1.41	0.31	0.98	50.97
1992	5.20	3.48	4.00	3.78	1.99	12.86	1.52	8.55	4.37	5.74	2.06	0.73	54.28
1993	3.26	4.77	4.61	0.91	1.41	6.07	3.41	5.65	2.00	7.98	1.35	2.23	43.65
Record Mean	3.55	3.62	4.67	2.58	2.92	6.74	5.10	6.88	4.26	2.90	2.33	1.87	47.41

AVERAGE TEMPERATURE (deg. F)

GAINESVILLE, FLORIDA

YEAR	JAN	FEB	MAR	APR	MAY	JUNE	JULY	AUG	SEP	OCT	NOV	DEC	ANNUAL
1984	53.2	57.5	61.7	67.1	74.2	78.3	79.7	80.8	76.7	73.3	59.1	62.5	68.7
1985	49.4	56.6	65.3	67.8	75.5	80.7	80.1	79.8	77.8	76.6	70.9	52.6	69.4
1986	53.2	59.9	61.1	66.6	74.4	80.1	81.9	80.2	79.4	71.9	69.9	59.1	69.8
1987	53.0	57.1	61.3	63.9	74.1	80.1	81.6	82.3	78.4	65.2	63.1	58.3	68.2
1988	50.0	53.3	60.9	68.1	71.7	78.3	80.0	80.4	78.7	67.5	65.6	56.6	67.6
1989	62.0	58.1	65.8	67.8	74.4	79.7	80.7	80.3	78.5	69.9	61.8	47.8	68.9
1990	59.2	63.8	64.7	66.2	76.3	79.9	80.8	80.9	78.3	72.0	63.3	60.8	70.5
1991	57.6	59.2	64.0	72.0	78.8	79.1	80.7	80.7	77.6	70.0	59.2	58.5	69.8
1992	52.8	58.7	61.0	65.1	71.2	79.1	82.1	79.4	77.6	67.4	64.1	56.5	67.9
1993	60.0	54.1	60.0	63.9	72.5	79.1	82.1	81.9	78.7	70.5	62.2	51.7	68.1
Record Mean	55.0	57.8	62.5	66.8	74.3	79.4	80.9	80.7	78.2	70.4	63.9	56.4	68.8
Max	65.8	69.2	74.2	79.0	86.1	89.9	90.7	90.0	87.2	80.3	74.4	67.8	79.5
Min	44.2	46.4	50.9	54.6	62.4	68.9	71.2	71.3	69.1	60.5	53.4	45.0	58.2

REFERENCE NOTES

GAINESVILLE, FLORIDA

GENERAL
T - TRACE AMOUNT.
BLANK ENTRIES DENOTE MISSING/UNREPORTED DATA.
# INDICATES A STATION OR INSTRUMENT RELOCATION.

SPECIFIC

(a) - LENGTH OF RECORD IN YEARS, ALTHOUGH INDIVIDUAL MONTHS MAY BE MISSING.
0.* OR * - THE VALUE IS BETWEEN 0.0 AND 0.05
NORMALS - BASED ON THE 1961-1990 RECORD PERIOD.
EXTREMES - DATES ARE THE MOST RECENT OCCURRENCE
WIND DIR.- NUMERALS SHOW TENS OF DEGRESS CLOCKWISE FROM TRUE NORTH. "00" INDICATES CALM.
RESULTANT DIRECTIONS ARE GIVEN TO WHOLE DEGREES.
BOLD VALUES INDICATE EXTREME VALUES WHICH OCCURRED AFTER THE ASOS SYSTEM WAS COMMISSIONED.

RECORD = PERIOD OF RECORD
RECORD MEAN PRECIPITATION IS THE MEAN OF ALL DAILY PRECIPITATION AMOUNTS DURING THE PERIOD OF RECORD.
RECORD MAX(MIN) TEMPERATURE IS THE MEAN OF ALL DAILY MAX(MIN) TEMPERATURES DURING THE PERIOD OF RECORD.
RECORD MEAN TEMPERATURE IS THE SUM OF THE RECORD MAX AND RECORD MIN DIVIDED BY 2.
AVERAGE TEMPERATURE IS THE SUM OF THE MEAN DAILY MAX AND MIN TEMPERATURE DIVIDED BY 2.

EXCEPTIONS

RECORD MEANS ARE THROUGH THE CURRENT YEAR,
BEGINNING IN 1984 FOR TEMPERATURE
1984 FOR PRECIPITATION
1984 FOR SNOWFALL

HEATING DEGREE DAYS Base 65 deg. F GAINESVILLE, FLORIDA

SEASON	JULY	AUG	SEP	OCT	NOV	DEC	JAN	FEB	MAR	APR	MAY	JUNE	TOTAL
1983-84							368	220	147	43	1	0	
1984-85	0	0	0	4	209	107	493	250	58	39	0	0	1160
1985-86	0	0	0	0	28	398	363	163	174	37	0	0	1163
1986-87	0	0	0	14	32	199	368	220	150	97	0	0	1080
1987-88	0	0	0	59	130	233	465	337	157	38	2	0	1421
1988-89	0	0	0	27	66	273	127	225	98	56	5	0	877
1989-90	0	0	0	56	141	531	194	97	64	50	0	0	1133
1990-91	0	0	0	30	91	183	242	177	119	9	0	0	851
1991-92	0	0	0	18	211	232	374	203	151	83	21	0	1293
1992-93	0	0	0	21	144	264	189	298	182	67	0	0	1165
1993-94	0	0	0	13	137	405							

COOLING DEGREE DAYS Base 65 deg. F GAINESVILLE, FLORIDA

YEAR	JAN	FEB	MAR	APR	MAY	JUNE	JULY	AUG	SEP	OCT	NOV	DEC	TOTAL
1984	7	9	47	115	293	407	463	496	355	271	40	37	2540
1985	16	23	75	129	331	473	478	468	392	369	212	20	2986
1986	2	27	63	93	299	458	528	480	441	238	186	25	2840
1987	3	5	41	68	288	460	522	543	411	71	79	30	2521
1988	7	9	38	138	218	405	471	487	416	114	93	18	2414
1989	39	37	132	145	302	447	493	483	413	215	52	6	2764
1990	24	71	62	93	359	454	496	499	406	259	46	58	2827
1991	23	21	94	224	435	428	491	492	388	182	45	38	2861
1992	4	28	35	94	219	430	538	454	382	104	126	9	2423
1993	39	0	35	42	238	432	538	532	417	190	61	1	2525

SNOWFALL (inches) GAINESVILLE, FLORIDA

SEASON	JULY	AUG	SEP	OCT	NOV	DEC	JAN	FEB	MAR	APR	MAY	JUNE	TOTAL
1983-84							0.0	0.0	0.0	0.0	0.0	0.0	
1984-85	0.0	0.0	0.0	0.0	0.0	0.0	0.0	0.0	0.0	0.0	0.0	0.0	0.0
1985-86	0.0	0.0	0.0	0.0	0.0	0.0	0.0	0.0	0.0	0.0	0.0	0.0	0.0
1986-87	0.0	0.0	0.0	0.0	0.0	0.0	0.0	0.0	0.0	0.0	0.0	0.0	0.0
1987-88	0.0	0.0	0.0	0.0	0.0	0.0	0.0	T	0.0	0.0	0.0	0.0	T
1988-89	0.0	0.0	0.0	0.0	0.0	0.0	0.0	0.0	0.0	0.0	0.0	0.0	0.0
1989-90	0.0	T	0.0	0.0	0.0	T	0.0	0.0	0.0	0.0	0.0	0.0	T
1990-91	0.0	0.0	0.0	0.0	0.0	0.0	0.0	0.0	T	0.0	0.0	0.0	T
1991-92	0.0	0.0	0.0	0.0	0.0	0.0	0.0	0.0	0.0	0.0	0.0	0.0	0.0
1992-93	0.0	0.0	0.0	0.0	0.0	0.0	0.0	0.0	0.0	0.0	0.0	0.0	0.0
1993-94	0.0	0.0	0.0	0.0	0.0	0.0							
Record Mean	0.0	T	0.0	0.0	0.0	T	0.0	T	T	0.0	0.0	0.0	T

JACKSONVILLE, FLORIDA

Jacksonville, a very large metropolitan area covering 840 square miles, extends from the Atlantic Ocean to about 40 miles inland. Downtown Jacksonville is located some 16 miles inland on the St. Johns River. The surrounding terrain is level. Easterly winds blowing about 40 percent of the time produce a maritime influence that modifies to some extent the heat of summer and the cold of winter. Summers are long, warm and relatively humid. Winters, although punctuated with periodic invasions of cool to occasionally cold air from the north, are mild because of the southern latitude and the proximity to the warm Atlantic Ocean waters. Because of the nearness to the ocean, climatic features across the city vary. For example, during the summer months temperatures at Jacksonville International Airport, located 17 miles inland, usually reach into the low and mid-90s before being tempered by sea breezes. Temperatures along the beaches rarely exceed 90 degrees. Summer thunderstorms usually occur before the noon hour along the beaches, while afternoon thunderstorms are the rule inland.

The annual temperature for Jacksonville is between 68 and 69 degrees. June, July, and August are the hottest months, with temperatures averaging near 80 degrees. December, January, and February are the coolest months, with temperatures near the middle 50s. Temperatures exceed 95 degrees only about ten times a year. Night temperatures in summer are usually comfortable, rarely failing to drop below 80 degrees.

The greatest rainfall, mostly in the form of local thundershowers, occurs during the summer months when a measurable amount can be expected one day in two. Rainfall of 1 inch or more in 24 hours normally occurs about fourteen times a year, and very infrequently heavy rains, associated with tropical storms, reach amounts of several inches with durations of more than 24 hours.

The atmosphere is moist, with an average relative humidity of about 75 percent, ranging from about 90 percent in early morning hours to about 55 percent during the afternoon.

Prevailing winds are northeasterly in the fall and winter months, and southwesterly in spring and summer. Wind movement, which averages slightly less than 9 mph, is 2 to 3 mph higher in the early afternoon than the early morning hours, and slightly higher in spring than in other seasons of the year. Although this area is in the Hurricane Belt, this section of the coast has been very fortunate in escaping hurricane-force winds. Most hurricanes reaching this latitude have tended to move parallel to the coastline, keeping well out to sea. Others have lost much of their force moving over land before reaching this area.

NORMALS, MEANS, AND EXTREMES

JACKSONVILLE, FLORIDA

LATITUDE: 30°30'N LONGITUDE: 81°42' W ELEVATION: FT. GRND 26 BARO 29 TIME ZONE: EASTERN WBAN: 13889

	(a)	JAN	FEB	MAR	APR	MAY	JUNE	JULY	AUG	SEP	OCT	NOV	DEC	YEAR
TEMPERATURE °F:														
Normals														
-Daily Maximum		64.2	67.0	73.0	79.1	84.7	89.3	91.4	90.7	87.2	80.2	73.6	66.8	78.9
-Daily Minimum		40.5	43.3	49.2	54.9	62.1	69.1	71.9	71.8	69.0	59.3	50.2	43.4	57.1
-Monthly		52.4	55.2	61.1	67.0	73.4	79.1	81.6	81.2	78.1	69.8	61.9	55.1	68.0
Extremes														
-Record Highest	52	85	88	91	95	100	103	105	102	100	96	88	84	105
-Year		1947	1962	1974	1968	1967	1954	1942	1954	1944	1951	1986	1991	JUL 1942
-Record Lowest	52	7	19	23	34	45	47	61	63	48	36	21	11	7
-Year		1985	1943	1980	1987	1992	1984	1972	1984	1981	1989	1970	1983	JAN 1985
NORMAL DEGREE DAYS:														
Heating (base 65°F)		421	296	169	37	0	0	0	0	0	31	149	331	1434
Cooling (base 65°F)		31	22	48	97	260	423	515	502	393	179	56	25	2551
% OF POSSIBLE SUNSHINE	42	58	61	68	72	71	65	64	63	58	59	59	55	63
MEAN SKY COVER (tenths)														
Sunrise - Sunset	44	6.1	5.9	5.8	5.2	5.5	6.2	6.3	6.3	6.5	5.6	5.4	6.1	5.9
MEAN NUMBER OF DAYS:														
Sunrise to Sunset														
-Clear	45	9.0	8.7	9.1	10.4	9.1	5.3	4.7	4.8	5.4	9.8	10.1	8.7	95.0
-Partly Cloudy	45	8.0	7.2	9.2	10.1	12.0	13.2	14.3	15.1	11.8	9.2	8.6	8.2	127.0
-Cloudy	45	14.0	12.4	12.7	9.4	9.9	11.5	11.9	11.1	12.9	12.0	11.3	14.1	143.2
Precipitation .01 inches or more	52	8.3	7.9	8.1	6.5	8.1	12.2	14.4	14.5	13.0	8.5	6.4	7.7	115.7
Snow,Ice pellets,hail 1.0 inches or more	52	0.0	0.*	0.0	0.0	0.0	0.0	0.0	0.0	0.0	0.0	0.0	0.0	*
Thunderstorms	52	1.0	1.5	3.1	3.7	6.3	11.0	15.7	13.4	6.9	2.1	0.8	0.9	66.5
Heavy Fog Visibility 1/4 mile or less	49	5.4	3.4	3.3	2.7	2.7	1.4	1.0	1.7	1.8	3.3	5.6	5.9	38.1
Temperature °F														
-Maximum														
90° and above	52	0.0	0.0	0.1	1.5	7.9	16.7	23.7	21.3	10.8	1.2	0.0	0.0	83.1
32° and below	52	0.*	0.0	0.0	0.0	0.0	0.0	0.0	0.0	0.0	0.0	0.0	0.*	*
-Minimum														
32° and below	52	5.6	3.4	0.8	0.0	0.0	0.0	0.0	0.0	0.0	0.0	0.9	4.1	14.8
0° and below	52	0.0	0.0	0.0	0.0	0.0	0.0	0.0	0.0	0.0	0.0	0.0	0.0	0.0
AVG. STATION PRESS.(mb)	21	1019.7	1018.8	1017.1	1016.4	1015.4	1015.6	1016.8	1016.6	1016.0	1017.3	1018.9	1020.2	1017.4
RELATIVE HUMIDITY (%)														
Hour 01	57	85	83	83	84	86	87	88	90	90	89	88	86	87
Hour 07 (Local Time)	57	87	86	86	86	85	87	88	91	92	91	90	88	88
Hour 13	57	58	53	50	48	50	57	58	60	62	59	56	58	56
Hour 19	57	74	69	65	64	66	73	75	78	80	80	79	78	73
PRECIPITATION (inches):														
Water Equivalent														
-Normal		3.31	3.93	3.68	2.77	3.55	5.69	5.60	7.93	7.05	2.90	2.19	2.72	51.32
-Maximum Monthly	52	10.20	8.85	10.18	11.61	10.43	12.90	16.21	16.24	19.36	13.44	7.85	7.09	19.36
-Year		1991	1970	1973	1973	1966	1967	1960	1968	1949	1956	1947	1945	SEP 1949
-Minimum Monthly	52	0.06	0.52	0.18	0.14	0.18	1.59	1.97	1.92	1.02	0.16	T	0.04	T
-Year		1950	1962	1945	1987	1990	1990	1977	1942	1961	1942	1970	1956	NOV 1970
-Maximum in 24 hrs	52	3.02	6.22	7.12	8.25	5.40	5.93	10.09	7.93	10.17	7.90	5.44	3.75	10.17
-Year		1963	1970	1970	1973	1975	1968	1966	1968	1950	1992	1969	1983	SEP 1950
Snow,Ice pellets,hail														
-Maximum Monthly	52	T	1.5	0.5	0.0	0.0	0.0	T	0.0	0.0	0.0	0.0	0.8	1.5
-Year		1985	1958	1986				1990					1989	FEB 1958
-Maximum in 24 hrs	52	T	1.5	0.5	0.0	0.0	0.0	T	0.0	0.0	0.0	0.0	0.8	1.5
-Year		1985	1958	1986				1990					1989	FEB 1958
WIND:														
Mean Speed (mph)	44	8.2	9.0	9.0	8.6	8.0	7.7	7.1	6.7	7.5	8.0	7.8	7.8	7.9
Prevailing Direction through 1963		NW	WSW	NW	SE	WSW	SW	SW	SW	NE	NE	NW	NW	NW
Fastest Obs. 1 Min.														
-Direction (!!)	13	30	30	22	32	01	25	23	11	32	29	33	31	32
-Speed (MPH)	13	38	39	44	46	30	35	35	38	32	29	38	40	46
-Year		1987	1991	1993	1992	1980	1987	1991	1982	1993	1993	1983	1983	APR 1992
Peak Gust														
-Direction (!!)	9	NW	NW	SW	NW	N	W	E	N	NW	W	W	W	E
-Speed (mph)	9	55	62	66	67	56	52	69	61	55	43	46	45	69
-Date		1987	1991	1993	1992	1992	1990	1990	1987	1993	1993	1988	1992	JUL 1990

PRECIPITATION (inches)

JACKSONVILLE, FLORIDA

YEAR	JAN	FEB	MAR	APR	MAY	JUNE	JULY	AUG	SEP	OCT	NOV	DEC	ANNUAL
1964	7.29	6.55	1.76	4.65	4.80	4.67	6.12	5.63	10.31	5.09	3.33	4.83	65.03
1965	0.65	5.50	3.91	0.95	0.94	9.79	2.71	9.58	11.02	1.75	1.92	3.75	52.47
1966	4.56	5.97	0.71	2.25	10.43	7.74	11.09	3.88	5.94	1.38	0.21	1.14	55.30
1967	3.05	4.35	0.81	2.00	1.18	12.90	5.22	12.31	1.80	1.13	0.24	4.69	49.68
1968	0.82	3.05	1.20	0.99	2.17	12.25	6.84	16.24	2.68	5.09	1.30	1.09	53.72
1969	0.84	3.39	4.23	0.34	3.78	5.12	5.89	15.10	10.33	9.81	4.56	3.87	67.26
1970	4.18	8.85	9.98	1.77	1.84	2.65	7.60	10.96	3.20	3.95	T	1.57	56.55
#1971	2.01	2.55	2.41	4.07	1.90	5.52	5.07	12.83	4.17	6.46	0.83	5.87	53.69
1972	5.77	3.48	4.43	2.98	8.26	6.75	3.15	9.76	2.60	4.46	4.22	1.43	57.29
1973	4.64	5.07	10.18	11.61	5.33	4.10	5.45	7.49	7.86	4.08	0.44	4.32	70.57
1974	0.28	1.28	3.47	1.53	4.14	5.53	9.83	11.23	8.13	0.34	1.03	1.73	48.52
1975	3.48	2.58	2.46	5.78	7.00	5.21	6.36	6.23	5.24	3.63	0.39	1.79	50.15
1976	2.29	1.05	3.41	0.63	10.02	4.26	5.41	6.37	8.56	1.63	2.43	4.81	50.87
1977	2.96	3.24	1.03	1.76	3.07	2.65	1.97	7.26	7.45	1.68	3.11	3.38	39.56
1978	4.64	4.17	2.83	2.24	9.18	2.62	6.67	2.39	4.40	1.26	0.80	1.84	43.04
1979	6.28	3.75	1.00	4.18	7.54	5.91	4.67	4.78	17.75	0.25	3.64	2.01	61.76
1980	2.61	1.06	6.83	3.91	3.02	4.59	5.29	3.97	3.03	2.69	2.32	0.21	39.53
1981	0.92	4.53	5.41	0.32	1.48	3.31	2.46	6.47	1.22	1.35	4.92	3.38	35.77
1982	3.00	1.67	4.26	3.60	3.55	8.06	3.81	6.93	9.32	3.37	1.93	2.02	51.52
1983	7.19	4.27	8.46	4.65	1.38	6.86	6.11	4.63	4.61	4.29	3.32	6.42	62.19
1984	2.13	4.67	5.77	3.14	1.46	4.76	6.01	3.78	12.28	1.53	3.30	0.13	48.96
1985	1.05	1.45	1.26	2.76	2.08	3.71	6.33	8.93	16.82	8.34	2.07	3.59	58.39
1986	4.19	4.72	5.44	0.93	2.13	2.53	3.27	9.60	1.99	1.80	2.85	4.65	44.10
1987	4.09	6.47	6.27	0.14	0.75	4.18	4.40	4.48	7.13	0.30	5.02	0.16	43.39
1988	6.36	6.08	2.65	3.44	1.35	3.71	4.50	8.48	16.36	2.35	4.27	1.13	60.68
1989	1.73	1.77	2.14	2.79	1.55	3.66	8.98	9.16	14.37	1.39	0.51	3.40	51.45
1990	1.84	4.07	1.59	1.34	0.18	1.59	6.53	3.81	2.60	4.54	1.17	1.94	31.20
1991	10.20	1.52	7.33	6.31	9.35	11.70	15.90	3.48	6.20	6.36	0.71	0.57	79.63
1992	5.79	2.64	4.09	5.33	5.97	7.04	3.32	10.76	7.33	8.34	1.92	0.65	63.18
1993	3.86	2.89	5.98	0.85	1.60	2.52	7.54	2.96	7.60	8.84	3.58	1.90	50.12
Record Mean	2.95	3.19	3.45	2.81	3.84	5.78	6.68	6.60	7.20	4.43	1.98	2.71	51.62

AVERAGE TEMPERATURE (deg. F)

JACKSONVILLE, FLORIDA

YEAR	JAN	FEB	MAR	APR	MAY	JUNE	JULY	AUG	SEP	OCT	NOV	DEC	ANNUAL
1964	53.0	52.4	62.8	70.1	74.5	82.3	81.4	81.2	77.9	67.7	64.9	59.5	69.0
1965	55.3	57.6	62.4	71.3	76.7	77.9	81.8	82.6	79.8	70.4	63.5	55.3	69.6
1966	51.9	54.8	59.3	68.1	74.1	75.8	82.4	80.8	78.2	71.0	60.4	53.9	67.6
1967	56.0	54.6	65.5	73.0	77.2	79.3	82.1	81.4	75.7	69.5	62.2	60.7	69.8
1968	53.4	50.4	60.5	73.0	76.7	81.1	82.8	83.4	79.2	72.4	60.2	54.7	69.0
1969	54.2	53.3	55.9	69.8	73.8	82.4	84.3	81.2	78.3	73.7	58.7	53.5	68.3
1970	48.1	52.8	63.4	71.7	75.3	79.5	83.1	82.4	81.4	73.1	57.7	58.5	68.9
#1971	54.0	57.1	57.4	66.3	71.8	78.8	81.5	81.0	78.9	73.3	61.7	64.4	68.9
1972	61.4	55.1	61.3	68.7	72.9	76.2	80.4	80.6	78.1	71.7	61.6	60.0	69.0
1973	54.2	53.1	66.0	65.5	73.5	79.8	81.9	80.5	79.3	71.1	64.7	53.9	68.6
1974	66.7	55.5	65.2	66.1	74.3	77.5	79.0	79.9	78.2	66.2	60.3	55.0	68.7
1975	58.6	60.9	60.5	66.4	75.5	80.9	80.0	81.2	78.4	72.2	62.2	52.7	69.1
1976	48.7	58.0	65.0	65.3	71.4	76.4	81.8	79.7	76.8	64.7	54.3	52.8	66.2
1977	44.0	50.0	65.0	67.1	73.0	81.3	82.7	81.9	80.1	66.1	62.5	53.3	67.3
1978	48.6	47.5	58.7	68.3	74.8	79.5	81.8	80.8	77.8	67.8	64.6	55.1	67.1
1979	47.9	52.0	60.8	68.6	73.0	77.1	82.0	80.4	79.3	69.3	62.0	54.3	67.2
1980	53.3	51.2	62.4	68.1	75.2	80.2	83.7	83.1	80.9	68.5	61.3	52.7	68.4
1981	46.5	55.4	59.1	70.4	72.5	83.3	84.4	80.8	76.0	69.7	59.9	53.9	67.7
1982	53.9	61.4	65.4	68.2	73.1	81.6	82.6	82.1	77.6	69.0	64.5	59.2	69.9
1983	49.0	52.8	57.9	62.7	72.3	76.5	82.5	82.2	76.4	71.5	60.1	52.4	66.4
1984	50.4	56.0	60.8	66.6	73.2	77.5	79.6	80.3	75.4	72.5	58.2	61.6	67.7
1985	48.2	56.4	64.0	67.3	75.5	80.7	81.1	81.2	77.7	75.3	69.5	51.9	69.1
1986	51.5	59.3	60.7	66.0	73.3	81.3	83.8	81.5	79.7	72.0	68.7	57.3	69.6
1987	51.9	54.4	60.0	64.0	73.2	80.4	82.6	83.9	79.2	63.9	62.3	57.5	67.8
1988	48.7	52.3	59.8	67.8	71.4	78.4	81.6	82.8	79.4	66.8	64.7	54.4	67.3
1989	60.6	58.6	65.4	67.3	73.7	80.8	82.5	82.2	79.5	70.6	62.4	47.7	69.3
1990	58.3	63.1	65.1	66.8	75.3	80.9	83.5	82.8	79.5	72.7	63.4	60.8	71.0
1991	56.3	58.6	64.8	72.3	79.3	80.0	83.8	83.1	79.3	70.9	58.9	57.6	70.4
1992	52.6	58.5	60.7	65.0	72.1	81.1	84.3	82.3	78.9	68.6	64.0	56.4	68.7
1993	59.1	53.3	60.2	63.5	72.8	80.6	84.1	81.9	79.6	69.8	61.7	51.4	68.2
Record Mean	54.5	56.4	61.8	67.4	73.7	78.7	80.7	80.3	77.4	69.0	60.8	54.9	68.0
Max	63.8	65.9	71.5	77.1	82.9	87.2	89.1	88.5	84.8	77.0	69.9	64.2	76.8
Min	45.1	46.9	52.0	57.8	64.5	70.2	72.3	72.2	69.9	60.9	51.7	45.6	59.1

REFERENCE NOTES

JACKSONVILLE, FLORIDA

GENERAL
T - TRACE AMOUNT.
BLANK ENTRIES DENOTE MISSING/UNREPORTED DATA.
# INDICATES A STATION OR INSTRUMENT RELOCATION.

SPECIFIC

(a) - LENGTH OF RECORD IN YEARS, ALTHOUGH INDIVIDUAL MONTHS MAY BE MISSING.
0.* OR * - THE VALUE IS BETWEEN 0.0 AND 0.05
NORMALS - BASED ON THE 1961-1990 RECORD PERIOD.
EXTREMES - DATES ARE THE MOST RECENT OCCURRENCE
WIND DIR.- NUMERALS SHOW TENS OF DEGRESS CLOCKWISE FROM TRUE NORTH. "00" INDICATES CALM.
RESULTANT DIRECTIONS ARE GIVEN TO WHOLE DEGREES.
BOLD VALUES INDICATE EXTREME VALUES WHICH OCCURRED AFTER THE ASOS SYSTEM WAS COMMISSIONED.

RECORD = PERIOD OF RECORD
RECORD MEAN PRECIPITATION IS THE MEAN OF ALL DAILY PRECIPITATION AMOUNTS DURING THE PERIOD OF RECORD.
RECORD MAX(MIN) TEMPERATURE IS THE MEAN OF ALL DAILY MAX(MIN) TEMPERATURES DURING THE PERIOD OF RECORD.
RECORD MEAN TEMPERATURE IS THE SUM OF THE RECORD MAX AND RECORD MIN DIVIDED BY 2.
AVERAGE TEMPERATURE IS THE SUM OF THE MEAN DAILY MAX AND MIN TEMPERATURE DIVIDED BY 2.

EXCEPTIONS

FASTEST MILE WINDS ARE THROUGH APRIL 1980.

RECORD MEANS ARE THROUGH THE CURRENT YEAR, BEGINNING IN 1874 FOR TEMPERATURE
1872 FOR PRECIPITATION
1942 FOR SNOWFALL

HEATING DEGREE DAYS Base 65 deg. F JACKSONVILLE, FLORIDA

SEASON	JULY	AUG	SEP	OCT	NOV	DEC	JAN	FEB	MAR	APR	MAY	JUNE	TOTAL
1964-65	0	0	0	39	66	213	310	231	170	10	0	0	1039
1965-66	0	0	0	38	81	293	402	294	188	37	0	0	1333
1966-67	0	0	0	20	173	346	281	293	92	7	0	0	1212
1967-68	0	0	3	18	141	179	352	419	188	4	0	0	1304
1968-69	0	0	0	41	189	346	328	327	287	12	0	0	1530
1969-70	0	0	0	0	201	353	522	335	123	22	0	0	1556
#1970-71	0	0	0	1	224	215	348	256	259	79	14	0	1396
1971-72	0	0	0	6	144	92	159	289	131	40	3	0	864
1972-73	0	0	0	10	169	204	351	329	67	52	7	0	1189
1973-74	0	0	0	26	82	357	31	287	82	68	0	0	933
1974-75	0	0	0	38	181	321	223	153	193	59	0	0	1168
1975-76	0	0	0	11	176	373	498	210	79	43	9	0	1399
1976-77	0	0	0	87	327	376	643	414	102	46	0	0	1995
1977-78	0	0	0	70	135	366	508	484	221	22	1	0	1807
1978-79	0	0	0	46	68	324	525	371	160	13	3	0	1510
1979-80	0	0	0	19	144	331	356	406	134	24	0	0	1414
1980-81	0	0	0	26	146	379	570	273	202	9	6	0	1611
1981-82	0	0	1	13	180	362	360	126	99	42	0	0	1183
1982-83	0	0	0	49	95	218	490	336	233	109	1	0	1531
1983-84	0	0	0	10	181	412	447	262	172	57	9	1	1551
1984-85	0	0	0	5	235	122	530	269	95	47	0	0	1303
1985-86	0	0	0	2	36	420	411	185	184	53	7	0	1298
1986-87	0	0	0	24	41	249	401	293	189	101	0	0	1298
1987-88	0	0	0	80	147	256	507	374	176	50	7	0	1597
1988-89	0	0	0	34	79	332	157	231	111	63	5	0	1012
1989-90	0	0	0	47	133	531	221	117	70	49	1	0	1169
1990-91	0	0	0	29	96	191	285	199	100	10	0	0	910
1991-92	0	0	0	9	221	261	384	207	165	86	16	0	1349
1992-93	0	0	0	19	147	277	209	323	185	95	1	0	1256
1993-94	0	0	0	14	150	414							

COOLING DEGREE DAYS Base 65 deg. F JACKSONVILLE, FLORIDA

YEAR	JAN	FEB	MAR	APR	MAY	JUNE	JULY	AUG	SEP	OCT	NOV	DEC	TOTAL
1969	2	8	10	164	282	530	604	513	405	275	16	7	2816
1970	5	2	80	227	325	439	567	544	496	256	13	22	2976
#1971	15	42	32	121	231	421	516	504	424	274	51	79	2710
1972	54	8	24	158	255	346	486	491	400	222	73	55	2572
1973	23	0	105	73	278	451	533	488	434	221	80	18	2704
1974	92	29	97	108	295	383	441	467	399	84	47	18	2460
1975	31	42	59	109	334	482	474	508	406	238	99	2	2784
1976	0	16	85	60	213	350	528	462	360	85	13	7	2179
1977	0	1	107	115	255	499	559	531	462	110	68	10	2717
1978	5	0	31	131	311	441	527	497	390	140	62	24	2559
1979	1	13	36	131	259	369	532	484	436	158	61	3	2483
1980	1	15	63	122	322	466	585	568	483	143	41	3	2812
1981	0	6	23	177	245	557	608	497	336	164	33	21	2667
1982	22	32	117	144	257	505	552	535	385	179	84	46	2858
1983	0	0	18	46	236	352	549	541	349	220	38	26	2375
1984	0	8	51	111	271	383	459	481	318	245	39	22	2388
1985	19	35	70	122	332	479	503	506	388	330	177	21	2982
1986	0	31	57	89	272	497	587	520	448	245	158	20	2924
1987	1	0	38	75	261	469	551	589	430	54	73	33	2574
1988	7	10	26	142	210	411	523	557	442	95	77	13	2513
1989	27	58	127	141	281	482	549	541	441	229	59	1	2936
1990	21	66	81	109	327	484	581	557	443	276	56	67	3068
1991	22	25	99	236	453	456	590	568	433	201	43	40	3166
1992	5	27	40	93	243	491	605	542	425	137	124	18	2750
1993	37	0	43	57	252	477	600	534	445	169	55	1	2670

SNOWFALL (inches) JACKSONVILLE, FLORIDA

SEASON	JULY	AUG	SEP	OCT	NOV	DEC	JAN	FEB	MAR	APR	MAY	JUNE	TOTAL
#1970-71	0.0	0.0	0.0	0.0	0.0	0.0	0.0	0.0	0.0	0.0	0.0	0.0	0.0
1971-72	0.0	0.0	0.0	0.0	0.0	0.0	0.0	0.0	0.0	0.0	0.0	0.0	0.0
1972-73	0.0	0.0	0.0	0.0	0.0	0.0	0.0	T	0.0	0.0	0.0	0.0	T
1973-74	0.0	0.0	0.0	0.0	0.0	0.0	0.0	0.0	0.0	0.0	0.0	0.0	0.0
1974-75	0.0	0.0	0.0	0.0	0.0	0.0	0.0	0.0	T	0.0	0.0	0.0	T
1975-76	0.0	0.0	0.0	0.0	0.0	0.0	0.0	0.0	0.0	0.0	0.0	0.0	0.0
1976-77	0.0	0.0	0.0	0.0	0.0	0.0	T	T	0.0	0.0	0.0	0.0	T
1977-78	0.0	0.0	0.0	0.0	0.0	0.0	T	T	0.0	0.0	0.0	0.0	T
1978-79	0.0	0.0	0.0	0.0	0.0	0.0	0.0	0.0	0.0	0.0	0.0	0.0	0.0
1979-80	0.0	0.0	0.0	0.0	0.0	0.0	0.0	0.0	T	0.0	0.0	0.0	T
1980-81	0.0	0.0	0.0	0.0	0.0	0.0	0.0	0.0	0.0	0.0	0.0	0.0	0.0
1981-82	0.0	0.0	0.0	0.0	0.0	0.0	T	0.0	0.0	0.0	0.0	0.0	T
1982-83	0.0	0.0	0.0	0.0	0.0	0.0	0.0	0.0	0.0	0.0	0.0	0.0	0.0
1983-84	0.0	0.0	0.0	0.0	0.0	0.0	0.0	0.0	0.0	0.0	0.0	0.0	0.0
1984-85	0.0	0.0	0.0	0.0	0.0	0.0	T	0.0	0.0	0.0	0.0	0.0	T
1985-86	0.0	0.0	0.0	0.0	0.0	0.0	0.0	0.0	0.5	0.0	0.0	0.0	0.5
1986-87	0.0	0.0	0.0	0.0	0.0	0.0	0.0	0.0	0.0	0.0	0.0	0.0	0.0
1987-88	0.0	0.0	0.0	0.0	0.0	0.0	0.0	T	0.0	0.0	0.0	0.0	T
1988-89	0.0	0.0	0.0	0.0	0.0	0.0	0.0	T	0.0	0.0	0.0	0.0	T
1989-90	0.0	0.0	0.0	0.0	0.0	0.8	0.0	0.0	0.0	0.0	0.0	0.0	0.8
1990-91	T	0.0	0.0	0.0	0.0	0.0	0.0	0.0	T	0.0	0.0	0.0	T
1991-92	0.0	0.0	0.0	0.0	0.0	0.0	0.0	0.0	T	0.0	0.0	0.0	T
1992-93	0.0	0.0	0.0	0.0	0.0	0.0	0.0	0.0	T	0.0	0.0	0.0	T
1993-94	0.0	0.0	0.0	0.0	0.0	T							
Record Mean	T	0.0	0.0	0.0	0.0	T	T	T	T	0.0	0.0	0.0	0.1

KEY WEST, FLORIDA

Key West is located at the end of the Overseas Highway and near the western end of the Florida Keys, which are a chain of islands swinging in a southwesterly arc from the southeast coast of the Florida peninsula. The nearest point of the mainland is about 60 statute miles to the northeast, while Cuba at its closest point is 98 miles south. The city occupies the island of the same name which is 3 1/2 miles long and 1 mile wide. Its mean elevation is around 8 feet. The maximum elevation of 18 feet covers only about one acre in the western portion. Soil is a thin layer of sand, or marlfill, overlying a stratum of Oolitic limestone. Vegetation on the eastern end of the island is scanty, chiefly of low growth. The western end, where settlement and landscaping are older, has a little heavier growth. The airport and Weather Service Office are located on the southeast shore on partially filled mangrove swamp.

The waters surrounding the key are quite shallow up to the mainland on the northeast and for 6 miles to the reef on the south. There is little wave action because the reef disrupts any established wave pattern.

Because of the nearness of the Gulf Stream in the Straits of Florida, about 12 miles south and southeast, and the tempering effects of the Gulf of Mexico to the west and north, Key West has a notably mild, tropical–maritime climate in which the average temperatures during the winter are about 14 degrees lower than in summer. Cold fronts are strongly modified by the warm water as they move in from northerly quadrants in winter. There is no known record of frost, ice, sleet, or snow in Key West. Prevailing easterly tradewinds and sea breezes suppress the usual summertime heating. Diurnal variations throughout the year average only about 10 degrees.

Precipitation is characterized by dry and wet seasons. The period of December through April receives abundant sunshine and slightly less than 25 percent of the annual rainfall. This rainfall usually occurs in advance of cold fronts in a few heavy showers, or occasionally five to eight light showers per month. June through October is normally the wet season, receiving approximately 53 percent of the yearly total in numerous showers and thunderstorms. Early morning is the favored time for diurnal showers. Easterly waves during this season occasionally bring excessive rainfall, while infrequent hurricanes may be accompanied by unusually heavy amounts. Humidity remains relatively high during the entire year.

NORMALS, MEANS, AND EXTREMES

KEY WEST, FLORIDA

LATITUDE: 24 °33' N LONGITUDE: 81 °45' W ELEVATION: FT. GRND 4 BARO 22 TIME ZONE: EASTERN WBAN: 12836

	(a)	JAN	FEB	MAR	APR	MAY	JUNE	JULY	AUG	SEP	OCT	NOV	DEC	YEAR
TEMPERATURE °F:														
Normals														
-Daily Maximum		74.8	75.4	78.6	81.7	85.1	87.6	89.1	89.2	88.0	84.4	80.0	76.1	82.5
-Daily Minimum		65.0	65.6	69.0	72.2	76.1	78.5	79.6	79.3	78.5	75.5	71.2	66.8	73.1
-Monthly		69.9	70.5	73.8	77.0	80.6	83.1	84.4	84.3	83.3	80.0	75.6	71.5	77.8
Extremes														
-Record Highest	41	86	85	87	90	91	94	95	95	94	93	89	86	95
-Year		1991	1991	1991	1991	1989	1952	1951	1957	1951	1962	1988	1978	AUG 1957
-Record Lowest	41	41	46	47	48	64	68	69	68	69	60	49	44	41
-Year		1981	1989	1986	1987	1992	1961	1952	1952	1985	1957	1959	1989	JAN 1981
NORMAL DEGREE DAYS:														
Heating (base 65°F)		44	30	7	0	0	0	0	0	0	0	0	19	100
Cooling (base 65°F)		196	184	279	360	484	543	601	598	549	465	318	221	4798
% OF POSSIBLE SUNSHINE	34	74	77	82	84	81	76	77	76	72	71	71	71	76
MEAN SKY COVER (tenths)														
Sunrise - Sunset	41	5.1	4.7	4.6	4.5	5.1	6.3	6.4	6.4	6.6	5.5	5.2	5.1	5.5
MEAN NUMBER OF DAYS:														
Sunrise to Sunset														
-Clear	41	11.2	11.6	13.0	13.2	9.9	5.0	3.4	3.1	2.7	8.9	10.3	11.0	103.3
-Partly Cloudy	41	11.2	10.1	11.5	11.3	13.4	14.2	16.4	16.8	15.3	12.8	11.2	11.4	155.5
-Cloudy	41	8.6	6.6	6.5	5.5	7.7	10.8	11.2	11.1	12.0	9.3	8.5	8.7	106.4
Precipitation														
.01 inches or more	45	6.4	5.5	5.2	4.6	7.9	11.5	12.2	14.5	15.7	11.1	6.9	6.7	108.3
Snow,Ice pellets,hail														
1.0 inches or more	45	0.0	0.0	0.0	0.0	0.0	0.0	0.0	0.0	0.0	0.0	0.0	0.0	0.0
Thunderstorms	45	1.0	1.2	1.8	1.8	4.3	9.6	13.1	14.4	11.2	4.2	1.2	1.0	64.6
Heavy Fog Visibility														
1/4 mile or less	45	0.4	0.2	0.0	0.0	0.0	0.0	0.0	0.0	0.*	0.1	0.1	0.3	1.1
Temperature °F														
-Maximum														
90° and above	45	0.0	0.0	0.0	0.*	0.7	7.1	15.3	16.5	7.6	0.6	0.0	0.0	47.9
32° and below	45	0.0	0.0	0.0	0.0	0.0	0.0	0.0	0.0	0.0	0.0	0.0	0.0	0.0
-Minimum														
32° and below	45	0.0	0.0	0.0	0.0	0.0	0.0	0.0	0.0	0.0	0.0	0.0	0.0	0.0
0° and below	45	0.0	0.0	0.0	0.0	0.0	0.0	0.0	0.0	0.0	0.0	0.0	0.0	0.0
AVG. STATION PRESS.(mb)	21	1018.4	1017.9	1016.6	1015.8	1014.6	1015.3	1016.8	1015.8	1014.2	1014.5	1016.3	1018.3	1016.2
RELATIVE HUMIDITY (%)														
Hour 01	42	81	78	78	76	77	78	77	78	79	80	80	81	79
Hour 07 (Local Time)	45	82	80	79	76	76	78	77	78	81	82	83	83	80
Hour 13	45	69	67	66	63	65	68	66	67	69	69	69	69	67
Hour 19	45	77	75	73	71	72	73	71	73	75	76	76	77	74
PRECIPITATION (inches):														
Water Equivalent														
-Normal		2.01	1.80	1.71	1.75	3.46	5.09	3.61	5.03	5.85	4.42	2.84	2.02	39.59
-Maximum Monthly	45	17.64	4.46	9.69	12.83	12.90	14.43	11.69	11.34	18.45	21.57	27.67	11.18	27.67
-Year		1983	1965	1987	1948	1960	1972	1970	1945	1963	1969	1980	1986	NOV 1980
-Minimum Monthly	45	T	0.02	T	0.00	0.12	0.57	0.44	2.23	1.70	0.74	0.13	0.07	0.00
-Year		1990	1948	1971	1959	1945	1985	1993	1991	1951	1972	1991	1981	APR 1959
-Maximum in 24 hrs	45	10.32	2.54	5.31	6.55	8.89	6.17	3.05	3.90	6.65	8.47	23.28	6.99	23.28
-Year		1983	1966	1987	1985	1960	1982	1970	1977	1963	1971	1980	1986	NOV 1980
Snow,Ice pellets,hail														
-Maximum Monthly		0.0	0.0	0.0	0.0	0.0	0.0	0.0	0.0	0.0	0.0	0.0	0.0	
-Year														
-Maximum in 24 hrs	45	0.0	0.0	0.0	0.0	0.0	0.0	0.0	0.0	0.0	0.0	0.0	0.0	
-Year														
WIND:														
Mean Speed (mph)	40	11.9	12.2	12.5	12.4	10.9	9.8	9.6	9.4	9.9	11.2	12.1	12.0	11.2
Prevailing Direction through 1963		NE	SE	SE	ESE	ESE	SE	ESE	ESE	ESE	ENE	ENE	NE	ESE
Fastest Obs. 1 Min.														
-Direction (!!)	17	35	18	22	01	13	19	12	19	12	35	12	26	01
-Speed (MPH)	17	38	40	54	58	46	35	33	41	43	46	47	39	58
-Year		1979	1992	1993	1980	1991	1982	1992	1985	1988	1987	1985	1983	APR 1980
Peak Gust														
-Direction (!!)	9	S	W	SW	N	SE	S	SE	S	SW	N	E	E	SW
-Speed (mph)	9	54	52	75	63	52	51	51	56	58	67	69	40	75
-Date		1993	1992	1993	1986	1991	1989	1989	1986	1987	1987	1985	1987	MAR 1993

PRECIPITATION (inches)

KEY WEST, FLORIDA

YEAR	JAN	FEB	MAR	APR	MAY	JUNE	JULY	AUG	SEP	OCT	NOV	DEC	ANNUAL
1964	0.06	1.38	1.67	0.85	0.96	5.42	1.63	3.42	4.10	5.19	1.09	3.59	29.36
1965	0.55	4.46	0.57	0.64	0.69	2.45	2.94	2.26	10.59	6.47	0.46	1.05	33.13
1966	4.01	3.04	1.33	2.15	2.77	13.67	5.63	3.50	8.34	7.12	0.71	1.35	53.62
1967	0.76	1.79	0.59	0.02	0.67	8.57	2.53	3.98	4.96	5.63	1.38	4.00	34.88
1968	0.07	3.46	4.36	0.25	8.00	9.36	3.81	4.77	6.90	6.80	3.92	0.18	51.88
1969	3.85	1.34	0.85	3.10	3.02	9.96	4.19	2.25	10.21	21.57	1.80	0.78	62.92
1970	8.21	2.10	2.20	0.11	2.63	1.96	11.69	4.03	5.35	8.03	0.15	0.36	46.82
1971	0.41	2.76	T	0.24	4.37	2.77	4.80	9.03	5.47	11.12	1.69	1.59	44.25
1972	2.75	2.08	0.59	0.76	2.64	14.43	4.86	3.53	8.70	0.74	2.04	3.16	46.28
1973	2.48	2.43	2.20	0.16	0.82	2.87	4.08	8.55	3.30	2.43	0.18	3.24	32.74
1974	0.57	0.41	0.30	0.21	3.45	2.99	2.38	4.13	3.05	0.95	1.22	0.33	19.99
1975	0.17	1.61	0.59	0.85	1.36	5.22	4.16	4.20	7.11	3.69	0.91	1.90	31.77
1976	0.97	2.18	0.05	4.79	2.65	10.23	0.81	7.10	3.38	3.22	0.50	1.91	37.79
1977	1.35	1.50	0.07	2.95	10.10	1.97	1.91	10.43	3.71	2.38	4.23	4.43	45.03
1978	2.13	1.81	2.43	1.97	9.45	5.06	1.40	2.69	3.03	4.19	1.83	0.43	36.42
1979	1.65	0.95	1.03	4.56	2.20	3.00	3.50	2.51	2.84	3.11	1.61	1.21	28.17
1980	0.98	0.92	0.83	2.84	2.14	2.58	6.65	6.38	6.01	2.95	27.67	0.47	60.42
1981	0.46	2.27	0.84	0.39	0.55	0.90	1.15	8.88	6.68	1.39	3.50	0.07	27.08
1982	0.38	1.68	3.44	1.74	5.78	8.95	1.94	2.61	6.36	2.27	1.20	0.30	36.65
1983	17.64	3.48	6.57	1.88	2.24	3.82	4.64	2.79	2.25	1.20	0.91	4.97	52.39
1984	0.10	4.07	2.18	4.16	5.79	7.09	3.29	2.59	8.11	1.32	1.15	0.10	39.95
1985	0.32	0.29	2.15	10.60	4.39	0.57	3.64	5.92	6.42	3.04	2.62	2.28	42.24
1986	1.26	2.02	0.99	0.93	0.91	5.37	4.48	5.66	4.58	2.52	0.92	11.18	40.82
1987	0.89	0.57	9.69	0.20	2.39	4.35	3.07	3.87	6.88	8.39	5.84	2.76	48.90
1988	4.78	0.46	1.68	0.64	8.12	3.60	5.78	6.29	3.48	0.76	0.78	0.19	36.56
1989	0.44	0.33	1.05	1.37	2.91	1.80	5.90	3.30	3.15	2.70	6.54	1.67	31.16
1990	T	1.05	0.75	1.37	5.54	0.94	4.09	8.01	5.29	5.14	2.93	1.32	36.43
1991	2.24	0.27	2.90	1.45	8.07	8.42	2.21	2.23	5.39	7.12	0.13	0.72	41.15
1992	0.28	1.22	2.81	1.42	1.60	11.77	1.07	3.82	7.15	2.04	3.83	0.08	37.09
1993	7.95	0.33	1.43	1.78	4.41	1.71	0.44	3.81	5.75	6.01	0.22	2.22	36.06
Record Mean	1.87	1.64	1.54	1.71	2.76	4.45	3.60	4.57	6.39	5.43	2.46	1.78	38.19

AVERAGE TEMPERATURE (deg. F)

KEY WEST, FLORIDA

YEAR	JAN	FEB	MAR	APR	MAY	JUNE	JULY	AUG	SEP	OCT	NOV	DEC	ANNUAL
1964	68.9	67.7	75.9	78.7	80.4	83.0	84.6	84.9	83.6	78.3	77.0	73.9	78.1
1965	71.3	75.2	76.6	80.3	81.5	84.1	85.2	85.1	83.2	80.2	76.6	71.9	79.3
1966	69.1	69.1	70.9	75.0	79.7	79.8	83.0	84.1	83.3	79.4	73.0	69.6	76.3
1967	72.5	71.4	75.9	79.1	82.8	84.3	86.8	84.3	83.9	79.7	75.1	74.4	79.2
1968	71.7	68.5	70.8	78.3	79.4	81.7	83.7	84.1	83.1	79.4	72.9	69.0	76.9
1969	70.2	68.2	68.9	77.4	80.6	82.7	85.3	84.9	83.0	80.3	72.6	69.2	77.0
1970	66.9	67.0	72.9	78.9	78.9	82.7	83.8	83.3	81.6	79.3	71.9	72.4	76.6
1971	70.6	71.9	72.4	75.9	80.7	83.0	84.1	83.2	82.7	81.1	75.8	76.2	78.2
1972	75.0	71.9	75.4	77.9	80.9	81.3	83.6	84.4	82.9	81.3	78.0	73.6	78.9
1973	72.3	66.8	75.8	76.8	80.2	82.9	83.5	83.1	83.0	80.3	78.2	70.4	77.8
1974	76.8	71.0	75.8	77.6	80.4	82.9	83.6	84.2	84.4	79.1	75.3	72.1	78.6
1975	74.6	75.4	76.5	79.6	82.4	84.2	83.7	84.5	82.8	81.2	75.4	71.4	79.3
1976	67.9	70.7	77.1	76.3	80.5	80.7	84.5	84.0	83.5	79.4	74.4	70.9	77.5
1977	65.9	68.0	76.4	77.1	79.1	83.2	83.8	83.0	83.3	78.3	74.6	69.4	76.9
1978	64.2	63.1	70.8	77.4	81.8	84.0	85.3	85.3	84.2	80.5	77.9	76.1	77.6
1979	69.2	69.2	72.8	78.7	81.5	84.3	85.3	85.2	83.7	80.9	76.8	72.4	78.4
1980	70.9	66.5	75.2	77.7	81.3	83.8	84.7	85.0	84.4	81.6	75.7	68.5	78.0
1981	61.3	69.1	71.0	78.5	81.0	85.1	85.8	84.0	83.0	81.5	73.9	70.5	77.1
1982	71.3	75.9	77.1	80.8	80.1	83.1	85.5	83.9	82.3	78.9	75.4	73.4	79.0
1983	67.6	68.3	69.7	74.0	78.8	82.1	83.4	83.9	82.7	80.5	75.7	72.7	76.6
1984	68.6	70.8	72.7	76.4	80.8	81.2	82.9	83.5	81.5	80.1	74.7	74.1	77.3
1985	67.0	71.8	75.3	76.3	81.4	85.1	83.3	84.0	82.2	82.1	78.2	69.0	78.0
1986	67.8	72.0	71.5	74.3	80.1	83.5	85.2	83.7	83.5	81.5	80.6	75.0	78.2
1987	68.9	71.8	73.0	71.2	80.2	84.5	85.1	85.3	84.9	77.1	75.9	71.5	77.5
1988	69.1	68.4	71.9	76.6	77.9	82.5	84.2	84.2	83.6	79.8	78.5	71.4	77.3
1989	74.1	72.1	74.3	78.2	81.7	83.9	83.9	85.4	85.1	80.4	77.0	67.3	78.6
1990	73.2	75.5	75.0	76.6	81.7	84.4	84.7	85.6	83.8	80.9	75.6	74.6	79.3
1991	75.5	72.9	75.7	80.5	82.2	83.8	85.7	85.5	83.8	80.2	74.4	72.9	79.4
1992	69.6	72.5	73.4	75.6	77.8	83.5	85.0	85.1	83.1	79.6	77.9	72.8	78.0
1993	74.5	70.7	71.9	74.3	79.2	83.8	85.9	85.3	83.9	81.2	77.1	70.9	78.2
Record Mean	70.0	70.9	73.4	76.6	79.8	82.6	84.0	84.0	82.8	79.4	74.8	71.0	77.4
Max	74.8	75.9	78.4	81.5	84.7	87.5	89.1	89.3	87.9	83.9	79.2	75.6	82.3
Min	65.1	65.9	68.4	71.6	74.9	77.6	78.8	78.7	77.6	74.9	70.5	66.4	72.5

REFERENCE NOTES

KEY WEST, FLORIDA

GENERAL
T - TRACE AMOUNT.
BLANK ENTRIES DENOTE MISSING/UNREPORTED DATA.
# INDICATES A STATION OR INSTRUMENT RELOCATION.

SPECIFIC

EXCEPTIONS

RECORD MEANS ARE THROUGH THE CURRENT YEAR,
BEGINNING IN 1871 FOR TEMPERATURE
1871 FOR PRECIPITATION
1970 FOR SNOWFALL

(a) - LENGTH OF RECORD IN YEARS, ALTHOUGH INDIVIDUAL MONTHS MAY BE MISSING.
0.* OR * - THE VALUE IS BETWEEN 0.0 AND 0.05
NORMALS - BASED ON THE 1961-1990 RECORD PERIOD.
EXTREMES - DATES ARE THE MOST RECENT OCCURRENCE
WIND DIR.- NUMERALS SHOW TENS OF DEGRESS CLOCKWISE FROM TRUE NORTH. "00" INDICATES CALM.
RESULTANT DIRECTIONS ARE GIVEN TO WHOLE DEGREES.
BOLD VALUES INDICATE EXTREME VALUES WHICH OCCURRED AFTER THE ASOS SYSTEM WAS COMMISSIONED.

RECORD = PERIOD OF RECORD
RECORD MEAN PRECIPITATION IS THE MEAN OF ALL DAILY PRECIPITATION AMOUNTS DURING THE PERIOD OF RECORD.
RECORD MAX(MIN) TEMPERATURE IS THE MEAN OF ALL DAILY MAX(MIN) TEMPERATURES DURING THE PERIOD OF RECORD.
RECORD MEAN TEMPERATURE IS THE SUM OF THE RECORD MAX AND RECORD MIN DIVIDED BY 2.
AVERAGE TEMPERATURE IS THE SUM OF THE MEAN DAILY MAX AND MIN TEMPERATURE DIVIDED BY 2.

HEATING DEGREE DAYS Base 65 deg. F KEY WEST, FLORIDA

SEASON	JULY	AUG	SEP	OCT	NOV	DEC	JAN	FEB	MAR	APR	MAY	JUNE	TOTAL
1964-65	0	0	0	0	0	0	10	4	0	0	0	0	14
1965-66	0	0	0	0	0	1	27	26	4	0	0	0	58
1966-67	0	0	0	0	1	6	5	13	0	0	0	0	25
1967-68	0	0	0	0	0	1	5	10	17	0	0	0	33
1968-69	0	0	0	0	4	28	1	13	17	0	0	0	63
1969-70	0	0	0	0	0	15	47	19	2	0	0	0	83
1970-71	0	0	0	0	11	1	25	20	4	0	0	0	61
1971-72	0	0	0	0	0	0	0	8	0	0	0	0	8
1972-73	0	0	0	0	0	8	16	37	0	0	0	0	61
1973-74	0	0	0	0	0	24	0	17	0	0	0	0	41
1974-75	0	0	0	0	0	4	3	0	0	0	0	0	7
1975-76	0	0	0	0	2	13	38	4	0	0	0	0	57
1976-77	0	0	0	0	0	10	65	20	0	0	0	0	95
1977-78	0	0	0	0	0	36	92	85	10	0	0	0	223
1978-79	0	0	0	0	0	0	17	22	0	0	0	0	39
1979-80	0	0	0	0	0	0	12	45	16	0	0	0	73
1980-81	0	0	0	0	0	26	128	18	1	0	0	0	173
1981-82	0	0	0	0	0	25	24	0	0	0	0	0	49
1982-83	0	0	0	0	0	10	24	8	2	0	0	0	44
1983-84	0	0	0	0	0	36	25	8	12	0	0	0	81
1984-85	0	0	0	0	0	2	39	16	0	0	0	0	57
1985-86	0	0	0	0	0	33	31	9	33	0	0	0	106
1986-87	0	0	0	0	0	0	24	7	2	10	0	0	43
1987-88	0	0	0	0	0	5	22	20	17	0	0	0	64
1988-89	0	0	0	0	0	10	0	27	3	0	0	0	40
1989-90	0	0	0	0	0	61	2	0	0	0	0	0	63
1990-91	0	0	0	0	0	3	0	8	2	0	0	0	13
1991-92	0	0	0	0	0	0	12	2	0	0	0	0	14
1992-93	0	0	0	0	0	3	0	5	12	0	0	0	20
1993-94	0	0	0	0	0	9							

COOLING DEGREE DAYS Base 65 deg. F KEY WEST, FLORIDA

YEAR	JAN	FEB	MAR	APR	MAY	JUNE	JULY	AUG	SEP	OCT	NOV	DEC	TOTAL
1969	169	108	144	378	490	539	635	622	547	481	234	150	4497
1970	113	85	254	424	435	538	589	576	506	452	222	239	4433
1971	204	220	241	334	492	544	599	570	536	505	334	355	4934
1972	314	216	331	396	501	496	583	610	544	511	398	284	5184
1973	249	93	339	364	479	545	582	571	549	483	400	201	4855
1974	371	194	343	387	484	544	585	601	587	445	317	230	5088
1975	309	296	361	444	545	584	588	610	538	508	322	217	5322
1976	133	175	382	343	487	476	611	598	565	453	289	198	4710
1977	101	112	360	371	446	551	589	567	556	419	295	177	4544
1978	75	40	195	375	528	580	638	640	583	485	394	353	4886
1979	154	147	250	417	516	588	637	632	566	502	362	235	5006
1980	200	96	340	387	514	575	620	625	589	524	328	140	4938
1981	21	141	195	411	505	613	651	598	548	521	274	201	4679
1982	224	309	384	484	473	549	641	593	525	437	320	277	5216
1983	110	106	156	278	436	519	576	595	537	487	329	282	4411
1984	147	182	258	350	498	492	558	582	502	474	298	291	4632
1985	110	213	326	349	516	610	576	596	521	534	403	163	4917
1986	123	209	239	286	477	563	634	585	560	521	475	315	4987
1987	151	203	258	202	477	594	629	637	603	380	335	211	4680
1988	155	125	239	356	408	531	604	602	562	465	411	216	4674
1989	289	235	298	401	524	572	593	640	611	482	369	138	5152
1990	263	299	315	356	522	588	620	644	575	499	326	309	5316
1991	334	234	344	472	540	573	650	643	569	482	287	253	5381
1992	164	223	266	323	405	561	625	627	548	462	395	252	4851
1993	298	169	231	282	445	569	656	638	574	508	370	197	4937

SNOWFALL (inches) KEY WEST, FLORIDA

SEASON	JULY	AUG	SEP	OCT	NOV	DEC	JAN	FEB	MAR	APR	MAY	JUNE	TOTAL
1970-71	0.0	0.0	0.0	0.0	0.0	0.0	0.0	0.0	0.0	0.0	0.0	0.0	0.0
1971-72	0.0	0.0	0.0	0.0	0.0	0.0	0.0	0.0	0.0	0.0	0.0	0.0	0.0
1972-73	0.0	0.0	0.0	0.0	0.0	0.0	0.0	0.0	0.0	0.0	0.0	0.0	0.0
1973-74	0.0	0.0	0.0	0.0	0.0	0.0	0.0	0.0	0.0	0.0	0.0	0.0	0.0
1974-75	0.0	0.0	0.0	0.0	0.0	0.0	0.0	0.0	0.0	0.0	0.0	0.0	0.0
1975-76	0.0	0.0	0.0	0.0	0.0	0.0	0.0	0.0	0.0	0.0	0.0	0.0	0.0
1976-77	0.0	0.0	0.0	0.0	0.0	0.0	0.0	0.0	0.0	0.0	0.0	0.0	0.0
1977-78	0.0	0.0	0.0	0.0	0.0	0.0	0.0	0.0	0.0	0.0	0.0	0.0	0.0
1978-79	0.0	0.0	0.0	0.0	0.0	0.0	0.0	0.0	0.0	0.0	0.0	0.0	0.0
1979-80	0.0	0.0	0.0	0.0	0.0	0.0	0.0	0.0	0.0	0.0	0.0	0.0	0.0
1980-81	0.0	0.0	0.0	0.0	0.0	0.0	0.0	0.0	0.0	0.0	0.0	0.0	0.0
1981-82	0.0	0.0	0.0	0.0	0.0	0.0	0.0	0.0	0.0	0.0	0.0	0.0	0.0
1982-83	0.0	0.0	0.0	0.0	0.0	0.0	0.0	0.0	0.0	0.0	0.0	0.0	0.0
1983-84	0.0	0.0	0.0	0.0	0.0	0.0	0.0	0.0	0.0	0.0	0.0	0.0	0.0
1984-85	0.0	0.0	0.0	0.0	0.0	0.0	0.0	0.0	0.0	0.0	0.0	0.0	0.0
1985-86	0.0	0.0	0.0	0.0	0.0	0.0	0.0	0.0	0.0	0.0	0.0	0.0	0.0
1986-87	0.0	0.0	0.0	0.0	0.0	0.0	0.0	0.0	0.0	0.0	0.0	0.0	0.0
1987-88	0.0	0.0	0.0	0.0	0.0	0.0	0.0	0.0	0.0	0.0	0.0	0.0	0.0
1988-89	0.0	0.0	0.0	0.0	0.0	0.0	0.0	0.0	0.0	0.0	0.0	0.0	0.0
1989-90	0.0	0.0	0.0	0.0	0.0	0.0	0.0	0.0	0.0	0.0	0.0	0.0	0.0
1990-91	0.0	0.0	0.0	0.0	0.0	0.0	0.0	0.0	0.0	0.0	0.0	0.0	0.0
1991-92	0.0	0.0	0.0	0.0	0.0	0.0	0.0	0.0	0.0	0.0	0.0	0.0	0.0
1992-93	0.0	0.0	0.0	0.0	0.0	0.0	0.0	0.0	0.0	0.0	0.0	0.0	0.0
1993-94	0.0	0.0	0.0	0.0	0.0	0.0							
Record Mean	0.0	0.0	0.0	0.0	0.0	0.0	0.0	0.0	0.0	0.0	0.0	0.0	0.0

MIAMI, FLORIDA

Miami is located on the lower east coast of Florida. To the east of the city lies Biscayne Bay, an arm of the ocean, about 15 miles long and 3 miles wide. East of the bay is the island of Miami Beach, a mile or less wide and about 10 miles long, and beyond Miami Beach is the Atlantic Ocean. The surrounding countryside is level and sparsely wooded.

The climate of Miami is essentially subtropical marine, featured by a long and warm summer, with abundant rainfall, followed by a mild, dry winter. The marine influence is evidenced by the low daily range of temperature and the rapid warming of cold air masses which pass to the east of the state. The Miami area is subject to winds from the east or southeast about half the time, and in several specific respects has a climate whose features differ from those farther inland.

One of these features is the annual precipitation for the area. During the early morning hours more rainfall occurs at Miami Beach than at the airport, while during the afternoon the reverse is true. The airport office is about 9 miles inland.

An even more striking difference appears in the annual number of days with temperatures reaching 90 degrees or higher, with inland stations having about four times more than the beach. Minimum temperature contrasts also are particularly marked under proper conditions, with the difference between inland locations and the Miami Beach station frequently reaching to 15 degrees or more, especially in winter.

Freezing temperatures occur occasionally in the suburbs and farming districts southwest, west, and northwest of the city, but rarely near the ocean.

Hurricanes occasionally affect the area. The months of greatest frequency are September and October. Destructive tornadoes are very rare. Funnel clouds are occasionally sighted and a few touch the ground briefly but significant damage is seldom reported. Waterspouts are often visible from the beaches during the summer months, however, significant damage is seldom reported. June, July, and August have the highest frequency of dangerous lightning events.

NORMALS, MEANS, AND EXTREMES

MIAMI, FLORIDA

LATITUDE: 25°49'N LONGITUDE: 80°17'W ELEVATION: FT. GRND 7 BARO 12 TIME ZONE: EASTERN WBAN: 12839

	(a)	JAN	FEB	MAR	APR	MAY	JUNE	JULY	AUG	SEP	OCT	NOV	DEC	YEAR
TEMPERATURE °F:														
Normals														
-Daily Maximum		75.2	76.5	79.1	82.4	85.3	87.6	89.0	89.0	87.8	84.5	80.4	76.7	82.8
-Daily Minimum		59.2	60.4	64.2	67.8	72.1	75.1	76.2	76.7	75.9	72.1	66.7	61.5	69.0
-Monthly		67.2	68.5	71.7	75.2	78.7	81.4	82.6	82.8	81.9	78.3	73.6	69.1	75.9
Extremes														
-Record Highest	51	88	89	92	96	95	98	98	98	97	95	89	87	98
-Year		1987	1991	1977	1971	1990	1985	1983	1990	1987	1980	1989	1989	AUG 1990
-Record Lowest	51	30	32	32	46	53	60	69	68	68	51	39	30	30
-Year		1985	1947	1980	1971	1945	1984	1985	1950	1983	1943	1950	1989	DEC 1989
NORMAL DEGREE DAYS:														
Heating (base 65°F)		88	51	14	0	0	0	0	0	0	0	6	41	200
Cooling (base 65°F)		156	149	221	306	425	492	546	552	507	412	264	168	4198
% OF POSSIBLE SUNSHINE	17	67	67	74	76	72	69	73	73	72	71	66	63	70
MEAN SKY COVER (tenths)														
Sunrise - Sunset	45	5.3	5.3	5.4	5.2	5.8	6.7	6.5	6.5	6.7	5.9	5.5	5.4	5.9
MEAN NUMBER OF DAYS:														
Sunrise to Sunset														
-Clear	44	9.4	8.5	8.6	8.6	6.4	3.1	2.6	2.5	2.2	6.7	7.5	9.1	75.1
-Partly Cloudy	44	13.2	12.2	14.0	14.7	15.2	14.4	17.3	17.8	15.6	14.2	13.9	12.8	175.3
-Cloudy	44	8.5	7.7	8.4	6.7	9.4	12.5	11.0	10.7	12.2	10.1	8.6	9.1	114.9
Precipitation														
.01 inches or more	51	6.7	6.0	6.1	6.0	10.2	14.9	16.0	17.2	17.4	14.1	8.6	6.4	129.5
Snow,Ice pellets,hail														
1.0 inches or more	44	0.0	0.0	0.0	0.0	0.0	0.0	0.0	0.0	0.0	0.0	0.0	0.0	0.0
Thunderstorms	44	0.9	1.2	1.8	2.6	6.5	12.3	15.1	15.9	11.8	4.6	1.2	0.7	74.5
Heavy Fog Visibility														
1/4 mile or less	45	1.3	0.8	0.6	0.6	0.3	0.0	0.1	0.1	0.1	0.2	0.9	0.9	5.9
Temperature °F														
-Maximum														
90° and above	29	0.0	0.0	0.2	1.4	3.3	9.4	15.0	14.8	9.9	2.4	0.0	0.0	56.3
32° and below	29	0.0	0.0	0.0	0.0	0.0	0.0	0.0	0.0	0.0	0.0	0.0	0.0	0.0
-Minimum														
32° and below	29	0.1	0.0	0.*	0.0	0.0	0.0	0.0	0.0	0.0	0.0	0.0	0.1	0.2
0° and below	29	0.0	0.0	0.0	0.0	0.0	0.0	0.0	0.0	0.0	0.0	0.0	0.0	0.0
AVG. STATION PRESS.(mb)	21	1019.4	1018.7	1017.6	1016.7	1015.6	1016.3	1017.9	1016.8	1015.2	1015.4	1017.4	1019.4	1017.2
RELATIVE HUMIDITY (%)														
Hour 01	29	81	79	77	76	78	83	82	83	84	82	81	80	81
Hour 07 (Local Time)	29	84	83	82	80	80	84	84	85	87	86	85	83	84
Hour 13	29	59	57	56	54	59	65	63	64	66	63	62	59	61
Hour 19	29	69	67	65	64	69	74	72	74	76	73	72	70	70
PRECIPITATION (inches):														
Water Equivalent														
-Normal		2.01	2.08	2.39	2.85	6.21	9.33	5.70	7.58	7.63	5.64	2.66	1.83	55.91
-Maximum Monthly	51	6.66	8.07	10.57	17.29	18.54	22.36	13.51	16.88	24.40	21.64	13.84	6.39	24.40
-Year		1969	1983	1986	1979	1968	1968	1947	1943	1960	1991	1992	1958	SEP 1960
-Minimum Monthly	51	0.04	0.01	0.02	0.05	0.44	1.81	1.77	1.65	2.63	1.25	0.09	0.12	0.01
-Year		1951	1944	1956	1981	1965	1945	1963	1954	1951	1977	1970	1988	FEB 1944
-Maximum in 24 hrs	51	2.68	5.73	7.07	16.21	11.59	8.20	4.55	6.92	7.58	9.95	8.01	4.38	16.21
-Year		1973	1966	1949	1979	1977	1977	1952	1964	1960	1948	1992	1964	APR 1979
Snow,Ice pellets,hail														
-Maximum Monthly		0.0	0.0	0.0	0.0	0.0	0.0	0.0	0.0	0.0	0.0	0.0	0.0	
-Year														
-Maximum in 24 hrs	51	0.0	0.0	0.0	0.0	0.0	0.0	0.0	0.0	0.0	0.0	0.0	0.0	
-Year														
WIND:														
Mean Speed (mph)	44	9.5	10.2	10.6	10.5	9.7	8.4	8.0	7.9	8.2	9.3	9.7	9.2	9.3
Prevailing Direction through 1963		NNW	ESE	SE	ESE	ESE	SE	SE	SE	ESE	ENE	N	N	ESE
Fastest Obs. 1 Min.														
-Direction (!!)	36	24	25	04	24	32	13	25	12	06	05	07	32	12
-Speed (MPH)	36	46	41	46	35	52	37	43	86	69	41	38	38	86
-Year		1978	1983	1966	1993	1980	1967	1990	1992	1965	1966	1985	1967	AUG 1992
Peak Gust														
-Direction (!!)	10	SW	W	SW	NW	SE	S	NE	E	SW	S	E	NE	E
-Speed (mph)	10	45	47	59	55	46	58	56	115	44	47	48	40	115
-Date		1987	1992	1993	1984	1991	1989	1992	1992	1988	1990	1985	1991	AUG 1992

PRECIPITATION (inches) MIAMI, FLORIDA

YEAR	JAN	FEB	MAR	APR	MAY	JUNE	JULY	AUG	SEP	OCT	NOV	DEC	ANNUAL
1964	0.45	2.21	0.50	3.31	4.67	10.48	5.51	9.84	4.22	9.77	3.00	6.24	60.20
1965	1.98	2.98	3.97	1.20	0.44	6.55	6.56	4.97	11.38	16.79	0.96	0.62	58.40
1966	3.97	6.56	3.25	1.80	5.53	21.37	8.50	7.62	8.00	10.88	3.84	0.74	82.06
1967	2.75	1.14	3.60	0.15	1.68	15.98	5.55	8.13	9.18	12.88	3.81	1.37	66.22
1968	1.92	2.77	0.88	1.27	18.54	22.36	6.15	8.34	11.11	8.71	1.21	0.13	83.39
1969	6.66	2.02	1.98	4.63	8.02	11.42	8.48	4.31	8.24	13.57	1.01	1.15	71.49
1970	2.64	1.77	2.61	0.95	10.98	5.53	4.48	3.60	8.89	3.01	0.09	0.17	44.72
1971	0.51	0.80	0.40	0.07	4.13	11.65	4.72	6.02	9.63	7.48	0.98	4.33	50.72
1972	1.60	2.71	3.01	2.67	13.71	10.90	7.13	6.49	5.08	2.86	2.77	4.18	63.11
1973	3.41	2.21	1.76	2.24	1.08	8.93	6.14	14.60	6.59	3.36	0.46	2.46	53.24
1974	2.54	0.10	2.27	2.11	2.63	8.12	6.09	9.29	6.38	3.68	4.62	1.17	49.00
1975	1.39	0.90	0.61	0.53	4.94	6.37	4.99	5.19	4.69	6.25	2.80	0.44	39.10
1976	0.95	3.54	0.23	4.17	10.45	6.81	3.83	9.45	7.75	4.42	2.69	1.61	55.90
1977	1.44	2.10	0.91	1.97	15.82	12.42	5.23	8.28	7.04	1.25	5.94	2.55	64.95
1978	2.07	3.44	2.92	3.50	5.66	5.29	2.69	3.93	3.42	7.68	3.17	2.06	45.83
1979	1.28	0.57	0.30	17.29	5.29	4.06	5.06	4.81	13.36	3.63	1.62	2.84	60.11
1980	1.89	0.88	3.17	10.20	2.14	3.02	9.40	11.32	5.60	6.05	3.47	0.20	57.34
1981	0.61	4.66	1.32	0.05	4.94	5.49	2.78	12.25	14.79	1.62	2.14	0.14	50.79
1982	0.44	1.22	4.22	9.27	8.80	10.82	3.84	5.79	7.62	7.12	7.09	1.18	67.41
1983	5.36	8.07	2.82	1.79	1.44	8.66	6.20	5.88	7.48	3.52	2.01	4.19	57.42
1984	0.18	0.70	6.12	4.51	10.91	7.24	7.38	5.44	10.45	2.35	4.04	0.70	60.02
1985	0.35	0.06	1.35	3.27	3.19	6.33	11.23	11.88	8.59	5.17	1.37	3.47	56.26
1986	5.04	1.72	10.57	0.71	8.24	9.06	7.81	7.67	4.38	3.96	4.75	2.21	66.12
1987	0.87	2.62	3.82	0.38	4.99	5.48	5.17	3.24	10.17	4.33	4.92	4.28	50.27
1988	1.88	0.61	0.39	1.82	5.28	10.36	10.90	7.89	3.09	1.49	0.76	0.12	44.59
1989	0.67	0.71	0.89	2.14	0.99	10.83	3.53	12.78	5.83	2.65	0.99	0.62	42.63
1990	0.24	1.19	2.28	6.96	7.79	6.84	4.31	11.06	3.52	4.82	1.67	1.03	51.71
1991	1.59	2.04	2.32	5.16	2.50	7.51	7.29	8.84	11.17	21.64	1.18	0.18	71.42
1992	1.80	1.49	2.67	2.43	0.55	13.17	4.21	7.22	6.48	2.02	13.84	1.94	57.82
1993	5.04	2.14	5.98	3.08	4.13	3.64	7.28	5.13	12.59	7.23	6.06	0.49	62.79
Record Mean	2.02	1.92	2.35	3.57	5.96	8.58	6.63	7.38	8.40	6.88	2.99	1.75	58.43

AVERAGE TEMPERATURE (deg. F) MIAMI, FLORIDA

YEAR	JAN	FEB	MAR	APR	MAY	JUNE	JULY	AUG	SEP	OCT	NOV	DEC	ANNUAL
#1964	67.4	65.7	74.6	77.1	77.6	81.2	82.8	83.6	82.3	76.9	74.9	72.2	76.4
1965	67.0	70.9	73.1	76.7	78.5	80.7	81.3	81.7	79.9	78.1	74.2	69.3	76.0
1966	66.0	68.9	69.2	72.8	77.6	78.2	81.1	81.7	81.2	77.9	70.2	66.6	74.3
1967	71.4	68.9	72.8	74.7	79.3	79.9	82.4	81.8	81.3	76.2	71.2	70.3	75.8
1968	66.1	62.7	67.0	75.9	77.6	79.6	81.6	83.9	82.2	77.8	71.0	66.6	74.3
1969	67.6	65.2	67.8	77.3	79.6	82.3	84.1	83.5	82.5	80.3	70.4	65.9	75.5
1970	63.8	64.7	71.8	79.0	79.1	82.0	82.7	84.0	82.2	79.5	69.6	70.9	75.8
1971	68.2	70.9	70.3	75.0	79.1	81.0	82.7	81.9	80.7	78.9	73.9	74.2	76.4
1972	73.0	68.4	72.1	75.0	77.6	79.9	80.9	81.7	80.4	77.9	73.3	70.8	75.9
1973	70.3	65.3	74.5	75.6	79.6	81.3	81.8	81.3	81.8	77.6	76.2	67.0	76.1
1974	74.3	68.9	75.6	76.2	80.0	82.1	82.6	84.0	84.1	78.1	72.9	69.0	77.3
1975	72.7	73.1	73.4	77.5	79.4	81.5	81.1	82.6	82.0	79.2	72.3	69.0	77.0
1976	64.7	68.8	75.8	75.1	78.5	79.1	83.1	81.9	80.4	76.3	71.5	68.2	75.3
1977	61.1	66.1	74.9	74.8	77.0	81.7	83.7	83.2	83.0	76.5	74.0	69.1	75.4
1978	64.0	63.2	68.9	74.0	79.2	81.9	82.5	82.6	82.0	78.8	75.7	73.0	75.5
1979	65.0	64.9	69.2	77.8	80.6	81.9	83.2	82.1	80.7	77.9	75.4	70.2	75.7
1980	67.5	64.0	73.2	75.4	79.0	81.4	82.6	82.8	82.1	80.1	74.3	67.3	75.8
1981	59.7	69.5	70.1	77.8	79.6	83.7	85.0	83.2	81.2	79.7	71.4	67.8	75.7
1982	67.8	74.4	74.7	77.9	77.2	82.0	84.3	84.0	82.7	77.9	75.0	72.6	77.6
1983	67.2	67.5	67.6	71.9	78.2	81.8	85.0	83.3	81.6	78.3	72.5	69.8	75.4
1984	67.0	68.6	70.4	73.2	77.1	79.8	81.9	82.6	80.1	78.2	71.5	71.1	75.1
1985	62.1	68.4	72.5	74.2	79.1	82.4	81.0	82.4	80.6	80.5	75.6	66.0	75.4
1986	65.2	69.4	68.6	71.7	77.5	81.3	83.1	83.5	83.3	80.3	79.3	73.6	76.4
1987	66.1	70.8	71.9	70.6	78.7	84.2	84.2	85.4	83.6	77.6	75.3	69.8	76.5
1988	67.9	67.7	70.7	76.1	77.9	82.0	83.1	83.6	84.0	79.1	76.9	70.5	76.6
1989	72.7	70.8	73.6	77.1	81.0	82.7	83.3	84.3	84.0	79.0	76.2	65.0	77.5
1990	73.6	74.0	73.7	75.2	80.3	83.0	83.5	83.7	83.1	80.4	74.4	72.9	78.2
1991	72.9	69.7	73.9	78.4	81.5	82.9	83.5	84.6	82.4	78.9	73.1	72.2	77.8
1992	67.4	70.5	71.9	74.0	77.8	81.5	84.9	84.4	83.2	79.5	76.8	71.6	77.0
1993	73.2	68.9	71.5	74.0	79.2	83.3	84.6	84.8	83.0	80.8	75.9	68.9	77.3
Record Mean	67.3	68.2	71.5	74.9	78.3	81.3	82.6	82.9	81.8	78.2	73.1	68.9	75.7
Max	75.6	76.7	79.5	82.6	85.5	88.1	89.4	89.8	88.3	84.8	80.3	76.9	83.1
Min	59.0	59.7	63.4	67.2	71.0	74.4	75.7	76.0	75.3	71.5	65.9	60.9	68.3

REFERENCE NOTES MIAMI, FLORIDA

GENERAL
T - TRACE AMOUNT.
BLANK ENTRIES DENOTE MISSING/UNREPORTED DATA.
# INDICATES A STATION OR INSTRUMENT RELOCATION.

SPECIFIC

PM - INCLUDES LAST DAY OF PREVIOUS MONTH
ASOS - AUTOMATED SURFACE OBSERVING SYSTEM IN OPERATION DURING THESE MONTHS.

(a) - LENGTH OF RECORD IN YEARS, ALTHOUGH INDIVIDUAL MONTHS MAY BE MISSING.
0.* OR * - THE VALUE IS BETWEEN 0.0 AND 0.05
NORMALS - BASED ON THE 1961-1990 RECORD PERIOD.
EXTREMES - DATES ARE THE MOST RECENT OCCURRENCE.
WIND DIR.- NUMERALS SHOW TENS OF DEGRESS CLOCKWISE FROM TRUE NORTH. "00" INDICATES CALM.
RESULTANT DIRECTIONS ARE GIVEN TO WHOLE DEGREES.
BOLD VALUES INDICATE EXTREME VALUES WHICH OCCURRED AFTER THE ASOS SYSTEM WAS COMMISSIONED.

RECORD = PERIOD OF RECORD
RECORD MEAN PRECIPITATION IS THE MEAN OF ALL DAILY PRECIPITATION AMOUNTS DURING THE PERIOD OF RECORD.
RECORD MAX(MIN) TEMPERATURE IS THE MEAN OF ALL DAILY MAX(MIN) TEMPERATURES DURING THE PERIOD OF RECORD.
RECORD MEAN TEMPERATURE IS THE SUM OF THE RECORD MAX AND RECORD MIN DIVIDED BY 2.
AVERAGE TEMPERATURE IS THE SUM OF THE MEAN DAILY MAX AND MIN TEMPERATURE DIVIDED BY 2.

EXCEPTIONS

RECORD MEANS ARE THROUGH THE CURRENT YEAR, BEGINNING IN 1940 FOR TEMPERATURE
1940 FOR PRECIPITATION
1940 FOR SNOWFALL

HEATING DEGREE DAYS Base 65 deg. F MIAMI, FLORIDA

SEASON	JULY	AUG	SEP	OCT	NOV	DEC	JAN	FEB	MAR	APR	MAY	JUNE	TOTAL
#1964-65	0	0	0	0	0	4	63	25	18	0	0	0	110
1965-66	0	0	0	0	0	16	70	41	16	0	0	0	143
1966-67	0	0	0	0	27	41	28	41	0	0	0	0	137
1967-68	0	0	0	0	0	25	56	101	57	0	0	0	239
1968-69	0	0	0	0	32	80	18	54	49	0	0	0	233
1969-70	0	0	0	0	21	53	117	58	19	0	0	0	268
1970-71	0	0	0	0	42	23	67	45	31	5	0	0	213
1971-72	0	0	0	0	0	0	2	39	0	0	0	0	41
1972-73	0	0	0	0	3	30	41	64	0	0	0	0	138
1973-74	0	0	0	0	1	93	0	37	0	0	0	0	131
1974-75	0	0	0	0	2	32	14	1	10	0	0	0	59
1975-76	0	0	0	0	33	49	93	27	0	0	0	0	202
1976-77	0	0	0	0	9	32	165	62	3	0	0	0	271
1977-78	0	0	0	0	6	58	123	99	34	0	0	0	320
1978-79	0	0	0	0	0	1	84	82	13	0	0	0	180
1979-80	0	0	0	0	6	10	50	95	39	0	0	0	200
1980-81	0	0	0	0	7	59	168	25	12	0	0	0	271
1981-82	0	0	0	0	1	80	65	1	3	0	0	0	150
1982-83	0	0	0	0	0	22	50	25	38	2	0	0	137
1983-84	0	0	0	0	4	69	54	37	17	0	0	0	181
1984-85	0	0	0	0	9	18	135	61	4	1	0	0	228
1985-86	0	0	0	0	2	78	76	22	54	0	0	0	232
1986-87	0	0	0	0	0	0	83	15	6	27	0	0	131
1987-88	0	0	0	0	3	29	49	38	26	0	0	0	145
1988-89	0	0	0	0	0	36	1	49	18	0	0	0	104
1989-90	0	0	0	1	0	110	7	4	0	0	0	0	122
1990-91	0	0	0	0	0	4	2	31	5	0	0	0	42
1991-92	0	0	0	0	7	0	38	7	6	0	0	0	58
1992-93	0	0	0	0	2	10	5	7	21	0	0	0	45
1993-94	0	0	0	0	4	31							

COOLING DEGREE DAYS Base 65 deg. F MIAMI, FLORIDA

YEAR	JAN	FEB	MAR	APR	MAY	JUNE	JULY	AUG	SEP	OCT	NOV	DEC	TOTAL
1969	104	66	145	375	459	526	597	581	532	478	191	88	4142
1970	85	59	239	425	446	518	558	596	522	457	185	213	4303
1971	176	219	202	315	444	488	558	531	476	443	274	292	4418
1972	262	144	227	307	398	454	498	523	471	408	261	217	4170
1973	212	81	301	324	459	499	531	516	511	394	343	163	4334
1974	294	150	335	342	471	518	551	596	578	414	245	163	4657
1975	261	233	276	382	456	501	508	553	517	448	257	178	4570
1976	92	144	336	309	424	429	569	530	470	361	209	141	4014
1977	50	97	318	299	381	508	587	574	549	364	284	191	4202
1978	97	54	163	273	449	515	547	552	513	437	329	254	4183
1979	90	81	149	391	492	516	572	537	481	407	324	178	4218
1980	138	75	296	321	441	501	555	563	519	476	292	135	4312
1981	10	154	177	389	460	568	625	570	492	460	198	173	4276
1982	161	270	311	394	385	518	606	596	537	406	304	264	4752
1983	125	101	124	213	417	514	628	576	503	419	236	221	4077
1984	124	144	194	252	380	452	532	554	460	416	213	214	3935
1985	55	164	244	285	445	529	505	546	476	488	329	114	4180
1986	86	150	175	207	395	495	569	582	556	483	432	272	4402
1987	122	186	227	202	430	580	603	639	565	401	314	182	4451
1988	145	123	209	339	408	516	571	584	578	445	364	216	4498
1989	247	219	292	367	502	540	576	603	578	442	346	114	4826
1990	279	262	276	314	479	547	578	587	552	486	287	254	4901
1991	254	167	288	408	515	547	583	614	531	437	255	231	4830
1992	121	173	226	277	404	503	624	609	553	454	366	222	4532
1993	269	123	227	277	449	557	613	622	550	497	338	159	4681

SNOWFALL (inches) MIAMI, FLORIDA

SEASON	JULY	AUG	SEP	OCT	NOV	DEC	JAN	FEB	MAR	APR	MAY	JUNE	TOTAL
1970-71	0.0	0.0	0.0	0.0	0.0	0.0	0.0	0.0	0.0	0.0	0.0	0.0	0.0
1971-72	0.0	0.0	0.0	0.0	0.0	0.0	0.0	0.0	0.0	0.0	0.0	0.0	0.0
1972-73	0.0	0.0	0.0	0.0	0.0	0.0	0.0	0.0	0.0	0.0	0.0	0.0	0.0
1973-74	0.0	0.0	0.0	0.0	0.0	0.0	0.0	0.0	0.0	0.0	0.0	0.0	0.0
1974-75	0.0	0.0	0.0	0.0	0.0	0.0	0.0	0.0	0.0	0.0	0.0	0.0	0.0
1975-76	0.0	0.0	0.0	0.0	0.0	0.0	0.0	0.0	0.0	0.0	0.0	0.0	0.0
1976-77	0.0	0.0	0.0	0.0	0.0	0.0	0.0	0.0	0.0	0.0	0.0	0.0	0.0
1977-78	0.0	0.0	0.0	0.0	0.0	0.0	0.0	0.0	0.0	0.0	0.0	0.0	0.0
1978-79	0.0	0.0	0.0	0.0	0.0	0.0	0.0	0.0	0.0	0.0	0.0	0.0	0.0
1979-80	0.0	0.0	0.0	0.0	0.0	0.0	0.0	0.0	0.0	0.0	0.0	0.0	0.0
1980-81	0.0	0.0	0.0	0.0	0.0	0.0	0.0	0.0	0.0	0.0	0.0	0.0	0.0
1981-82	0.0	0.0	0.0	0.0	0.0	0.0	0.0	0.0	0.0	0.0	0.0	0.0	0.0
1982-83	0.0	0.0	0.0	0.0	0.0	0.0	0.0	0.0	0.0	0.0	0.0	0.0	0.0
1983-84	0.0	0.0	0.0	0.0	0.0	0.0	0.0	0.0	0.0	0.0	0.0	0.0	0.0
1984-85	0.0	0.0	0.0	0.0	0.0	0.0	0.0	0.0	0.0	0.0	0.0	0.0	0.0
1985-86	0.0	0.0	0.0	0.0	0.0	0.0	0.0	0.0	0.0	0.0	0.0	0.0	0.0
1986-87	0.0	0.0	0.0	0.0	0.0	0.0	0.0	0.0	0.0	0.0	0.0	0.0	0.0
1987-88	0.0	0.0	0.0	0.0	0.0	0.0	0.0	0.0	0.0	0.0	0.0	0.0	0.0
1988-89	0.0	0.0	0.0	0.0	0.0	0.0	0.0	0.0	0.0	0.0	0.0	0.0	0.0
1989-90	0.0	0.0	0.0	0.0	0.0	0.0	0.0	0.0	0.0	0.0	0.0	0.0	0.0
1990-91	0.0	0.0	0.0	0.0	0.0	0.0	0.0	0.0	0.0	0.0	0.0	0.0	0.0
1991-92	0.0	0.0	0.0	0.0	0.0	0.0	0.0	0.0	0.0	0.0	0.0	0.0	0.0
1992-93	0.0	0.0	0.0	0.0	0.0	0.0	0.0	0.0	0.0	0.0	0.0	0.0	0.0
1993-94	0.0	0.0	0.0	0.0	0.0	0.0							
Record Mean	0.0	0.0	0.0	0.0	0.0	0.0	0.0	0.0	0.0	0.0	0.0	0.0	0.0

ORLANDO, FLORIDA

Orlando is located in the central section of the Florida peninsula, surrounded by many lakes. Relative humidities remain high the year-round, with values near 90 percent at night and 40 to 50 percent in the afternoon. On some winter days, the humidity may drop to 20 percent.

The rainy season extends from June through September, sometimes through October when tropical storms are near. During this period, scattered afternoon thunderstorms are an almost daily occurrence, and these bring a drop in temperature to make the climate bearable. Summer temperatures above 95 degrees are rather rare. There is usually a breeze which contributes to the general comfort.

During the winter months rainfall is light. While temperatures, on infrequent occasion, may drop at night to near freezing, they rise rapidly during the day and, in brilliant sunshine, afternoons are pleasant.

Frozen precipitation in the form of snowflakes, snow pellets, or sleet is rare. However, hail is occasionally reported during thunderstorms.

Hurricanes are usually not considered a great threat to Orlando, since, to reach this area, they must pass over a substantial stretch of land and, in so doing, lose much of their punch. Sustained hurricane winds of 75 mph or higher rarely occur. Orlando, being inland, is relatively safe from high water, although heavy rains sometimes briefly flood sections of the city.

NORMALS, MEANS, AND EXTREMES

ORLANDO, FLORIDA

LATITUDE: 28°26'N LONGITUDE: 81°19'W ELEVATION: FT. GRND 96 BARO 94 TIME ZONE: EASTERN WBAN: 12815

	(a)	JAN	FEB	MAR	APR	MAY	JUNE	JULY	AUG	SEP	OCT	NOV	DEC	YEAR
TEMPERATURE °F:														
Normals														
-Daily Maximum		70.8	72.7	78.0	83.0	87.8	90.5	91.5	91.5	89.7	84.6	78.5	72.9	82.6
-Daily Minimum		48.6	49.7	55.2	59.4	65.9	71.8	73.1	73.4	72.4	65.8	57.5	51.3	62.0
-Monthly		59.7	61.2	66.7	71.2	76.9	81.1	82.3	82.5	81.0	75.2	68.0	62.1	72.3
Extremes														
-Record Highest	51	87	90	92	96	102	100	100	100	98	95	89	90	102
-Year		1991	1962	1970	1968	1945	1985	1961	1980	1988	1986	1992	1978	MAY 1945
-Record Lowest	51	19	28	25	38	48	53	64	64	56	43	29	20	19
-Year		1985	1970	1980	1987	1992	1984	1981	1957	1956	1957	1950	1983	JAN 1985
NORMAL DEGREE DAYS:														
Heating (base 65°F)		234	164	65	5	0	0	0	0	0	0	54	164	686
Cooling (base 65°F)		70	58	117	191	369	483	536	543	480	316	144	74	3381
% OF POSSIBLE SUNSHINE														
MEAN SKY COVER (tenths)														
Sunrise - Sunset	45	5.7	5.7	5.6	5.1	5.4	6.4	6.4	6.4	6.5	5.5	5.2	5.7	5.8
MEAN NUMBER OF DAYS:														
Sunrise to Sunset														
-Clear	45	9.0	8.6	9.2	10.2	8.8	4.2	3.4	3.1	3.8	9.6	10.2	9.8	89.9
-Partly Cloudy	45	10.2	8.7	10.4	11.4	13.5	14.3	16.7	17.0	14.6	11.2	10.2	9.2	147.3
-Cloudy	45	11.8	11.0	11.4	8.4	8.7	11.5	10.9	10.9	11.6	10.2	9.6	12.1	128.1
Precipitation														
.01 inches or more	51	6.3	6.9	7.7	5.5	8.5	14.0	17.1	16.0	13.6	8.5	5.7	5.9	115.8
Snow,Ice pellets,hail														
1.0 inches or more	51	0.0	0.0	0.0	0.0	0.0	0.0	0.0	0.0	0.0	0.0	0.0	0.0	0.0
Thunderstorms	49	1.0	1.5	2.9	3.4	7.6	14.6	19.3	17.5	9.7	2.6	1.1	1.1	82.3
Heavy Fog Visibility														
1/4 mile or less	45	5.5	3.2	2.5	1.4	1.4	0.8	0.5	0.9	1.2	1.7	2.7	4.3	26.1
Temperature °F														
-Maximum														
90° and above	30	0.0	0.0	0.4	3.9	10.7	19.6	25.2	25.5	18.7	3.4	0.0	0.*	107.5
32° and below	30	0.0	0.0	0.0	0.0	0.0	0.0	0.0	0.0	0.0	0.0	0.0	0.0	0.0
-Minimum														
32° and below	30	1.5	0.5	0.1	0.0	0.0	0.0	0.0	0.0	0.0	0.0	0.1	0.6	2.8
0° and below	30	0.0	0.0	0.0	0.0	0.0	0.0	0.0	0.0	0.0	0.0	0.0	0.0	0.0
AVG. STATION PRESS. (mb)	21	1016.6	1015.8	1014.3	1013.6	1012.4	1012.9	1014.4	1013.8	1012.6	1013.4	1015.3	1016.8	1014.3
RELATIVE HUMIDITY (%)														
Hour 01	29	85	83	84	84	86	89	90	91	90	87	87	87	87
Hour 07 (Local Time)	30	88	87	88	87	88	89	90	92	92	89	89	88	89
Hour 13	30	56	52	50	46	49	57	58	60	60	56	56	57	55
Hour 19	30	69	63	61	58	62	72	75	77	77	74	73	73	70
PRECIPITATION (inches):														
Water Equivalent														
-Normal		2.30	3.02	3.21	1.80	3.55	7.32	7.25	6.78	6.01	2.42	2.30	2.15	48.11
-Maximum Monthly	51	7.23	8.32	11.38	9.10	10.36	18.28	19.57	16.11	15.87	14.51	10.29	5.33	19.57
-Year		1986	1983	1987	1992	1976	1968	1960	1972	1945	1950	1987	1983	JUL 1960
-Minimum Monthly	51	0.15	0.10	0.16	0.14	0.43	1.97	2.60	2.92	0.43	0.35	0.03	T	T
-Year		1950	1944	1956	1977	1961	1948	1992	1980	1972	1967	1967	1944	DEC 1944
-Maximum in 24 hrs	51	4.19	4.38	5.03	5.65	3.18	8.40	8.19	5.29	9.67	7.74	5.87	3.61	9.67
-Year		1986	1970	1960	1992	1980	1945	1960	1949	1945	1950	1988	1969	SEP 1945
Snow,Ice pellets,hail														
-Maximum Monthly	21	T	0.0	T	T	0.0	0.0	T	T	0.0	0.0	0.0	0.0	T
-Year		1977		1992	1992			1993	1989					JUL 1993
-Maximum in 24 hrs	21	T	0.0	T	T	0.0	0.0	T	T	0.0	0.0	0.0	0.0	T
-Year		1977		1992	1992			1993	1989					JUL 1993
WIND:														
Mean Speed (mph)	45	9.0	9.6	9.9	9.4	8.8	8.0	7.4	7.2	7.6	8.6	8.7	8.6	8.5
Prevailing Direction through 1963		NNE	S	S	SE	SE	SW	S	S	ENE	N	N	NNE	S
Fastest Obs. 1 Min.														
-Direction (!!)	44	25	25	24	02	17	32	14	32	24	05	26	07	32
-Speed (MPH)	44	42	46	46	50	46	64	46	50	46	48	46	32	64
-Year		1953	1969	1993	1956	1981	1970	1961	1957	1969	1950	1968	1968	JUN 1970
Peak Gust														
-Direction (!!)	10	NW	W	W	E	S	W	W	SW	NW	W	NE	W	W
-Speed (mph)	10	48	51	62	53	68	62	68	58	54	40	41	43	68
-Date		1991	1991	1993	1992	1991	1985	1991	1991	1988	1990	1984	1984	JUL 1991

PRECIPITATION (inches)

ORLANDO, FLORIDA

YEAR	JAN	FEB	MAR	APR	MAY	JUNE	JULY	AUG	SEP	OCT	NOV	DEC	ANNUAL
1964	6.18	3.42	4.65	2.14	2.74	6.11	6.68	9.00	9.47	1.64	0.45	1.91	54.39
1965	1.79	3.67	3.02	0.66	0.52	7.36	11.55	5.49	5.99	4.06	1.06	2.23	47.40
1966	4.45	6.31	2.57	1.92	6.57	9.77	6.73	7.76	6.25	1.98	0.09	0.99	55.39
1967	0.84	5.49	1.31	0.28	1.69	11.16	4.63	6.83	5.88	0.35	0.03	2.42	40.91
1968	0.65	2.76	2.27	0.30	3.72	18.28	5.60	3.44	5.91	5.47	2.82	0.88	52.10
1969	2.22	3.30	5.52	2.38	1.40	5.04	6.73	7.17	6.44	9.45	0.87	4.66	55.18
1970	4.05	6.77	3.66	0.45	4.08	4.92	5.97	5.91	3.25	2.60	0.24	2.06	43.96
1971	0.45	2.98	1.46	1.52	4.31	4.39	8.29	7.51	2.98	3.06	1.21	1.93	40.09
1972	0.99	4.96	5.06	1.39	3.76	6.33	3.98	16.11	0.43	2.34	4.11	1.89	51.35
1973	4.82	2.73	4.13	2.82	4.74	6.63	6.24	7.33	11.53	1.10	0.74	2.56	55.37
#1974	0.18	0.63	3.67	1.17	2.69	15.28	6.01	6.56	5.78	0.48	0.31	1.62	44.38
1975	0.98	1.49	1.10	1.36	7.52	9.70	9.26	4.75	4.97	4.74	0.66	0.51	47.04
1976	0.37	0.83	1.72	2.16	10.36	9.93	7.05	3.25	5.87	0.74	2.03	2.77	47.08
1977	1.81	1.76	1.82	0.14	1.47	4.47	6.61	6.28	7.03	0.43	2.60	3.70	38.12
1978	2.49	5.45	2.14	0.61	3.16	10.00	11.92	5.13	4.31	1.51	0.18	3.69	50.59
1979	6.48	1.45	3.24	1.08	7.66	4.00	7.95	5.88	9.19	0.43	1.93	0.94	50.23
1980	2.45	1.64	1.51	4.07	6.96	5.25	5.14	2.92	3.70	0.55	6.55	0.47	41.21
1981	0.21	4.36	1.85	0.18	2.02	12.49	3.53	5.60	8.26	3.13	2.50	2.97	47.10
1982	1.72	1.34	4.85	6.27	5.29	6.06	11.81	5.03	6.96	0.74	0.53	1.01	51.61
1983	2.08	8.32	5.37	3.21	1.77	7.82	6.49	4.83	5.16	3.78	1.36	5.33	55.52
1984	2.01	2.73	1.85	6.21	3.20	5.32	6.19	7.89	6.19	0.56	2.10	0.19	44.44
1985	0.91	1.27	4.59	1.69	3.00	4.54	7.28	11.63	5.45	2.55	0.82	3.46	47.19
1986	7.23	1.84	2.63	0.49	0.88	9.50	5.85	5.99	4.50	5.63	1.69	3.60	49.83
1987	1.27	1.74	11.38	0.59	1.40	3.54	7.95	6.07	8.64	3.41	10.29	0.51	56.79
1988	3.12	1.38	6.07	2.02	2.82	4.17	9.44	7.94	5.67	1.42	7.44	1.00	52.49
1989	3.80	0.15	1.35	2.28	2.38	6.79	4.74	6.20	10.29	1.75	1.44	4.49	45.66
1990	0.23	4.13	1.92	1.73	0.55	6.22	6.68	3.78	2.46	2.10	1.05	0.83	31.68
1991	2.37	0.98	6.66	7.72	9.48	5.98	10.78	7.13	4.53	4.76	0.27	0.24	60.90
1992	1.35	2.42	3.67	9.10	1.19	8.68	2.60	8.03	7.13	5.17	2.74	0.88	52.96
1993	4.89	1.48	6.26	1.78	2.32	4.47	6.49	5.95	5.35	4.61	0.17	0.76	44.53
Record Mean	2.23	2.70	3.53	2.62	3.40	6.98	7.83	6.68	6.74	3.36	1.88	1.91	49.84

AVERAGE TEMPERATURE (deg. F)

ORLANDO, FLORIDA

YEAR	JAN	FEB	MAR	APR	MAY	JUNE	JULY	AUG	SEP	OCT	NOV	DEC	ANNUAL
1964	58.5	58.3	68.1	74.1	77.1	82.4	81.6	82.8	79.8	72.5	70.5	64.4	72.5
1965	60.0	64.1	67.0	74.8	77.1	79.0	80.5	82.2	80.8	74.2	69.5	62.6	72.7
1966	58.7	62.3	64.4	70.5	77.4	78.0	82.3	82.3	80.1	75.8	65.8	60.1	71.5
1967	63.2	60.0	68.3	74.3	78.3	80.2	82.4	82.0	79.7	74.0	67.5	65.9	73.0
1968	59.6	54.8	61.4	73.5	76.7	78.8	81.3	82.3	80.3	74.3	63.4	58.7	70.4
1969	59.8	57.8	60.4	72.5	76.9	82.9	84.4	82.2	81.2	77.9	64.0	58.7	71.6
1970	55.1	58.7	67.0	75.8	77.7	81.8	83.8	82.3	83.6	77.0	63.4	64.6	72.6
1971	62.0	64.1	64.8	72.1	78.2	81.7	83.1	83.3	81.8	79.0	69.5	71.4	74.2
1972	68.9	62.0	68.7	72.7	77.4	82.2	83.2	82.8	81.8	76.8	68.9	66.1	74.3
1973	62.4	59.7	71.1	71.1	78.3	83.1	84.2	81.8	81.4	75.6	70.9	60.4	73.3
#1974	71.6	60.5	70.2	71.4	78.0	80.3	80.7	82.0	81.8	72.6	67.6	60.9	73.1
1975	65.8	67.6	67.4	72.4	79.1	80.8	80.5	82.3	80.7	76.6	67.4	60.2	73.4
1976	56.5	63.7	70.4	71.3	76.8	79.7	82.4	81.9	80.5	72.6	63.0	60.1	71.6
1977	50.6	57.4	69.7	70.6	75.2	82.6	82.0	81.5	82.6	72.9	69.6	61.0	71.3
1978	56.8	55.8	66.3	73.4	79.3	82.9	82.6	82.6	81.7	75.0	72.3	66.8	73.0
1979	58.2	58.4	64.6	73.4	75.4	80.7	83.3	82.4	81.3	74.4	68.3	62.6	71.9
1980	60.5	57.2	68.2	70.4	76.4	80.1	83.6	83.6	81.7	75.4	67.1	59.0	71.9
1981	51.3	61.7	64.0	73.1	76.7	83.2	84.1	82.9	80.0	76.4	65.3	60.5	71.6
1982	60.0	68.4	70.4	72.6	75.3	82.0	82.6	82.2	80.2	74.1	70.8	66.7	73.8
1983	58.0	59.9	63.5	68.6	76.4	80.5	83.2	83.5	80.6	76.5	65.8	61.2	71.5
1984	57.8	61.2	64.7	69.2	75.6	78.4	80.7	81.5	78.9	75.4	65.8	66.0	71.3
1985	54.7	62.2	68.4	70.7	77.2	82.4	82.1	82.3	79.8	79.4	73.0	58.8	72.6
1986	59.8	64.3	65.4	69.3	76.7	81.7	82.3	83.3	81.7	77.5	75.8	67.3	73.8
1987	58.8	62.7	65.9	66.8	76.8	83.1	83.5	85.0	82.7	72.2	69.0	64.2	72.6
1988	58.5	60.4	65.5	72.0	75.5	80.3	80.7	82.8	83.9	73.7	70.5	62.4	72.2
1989	66.9	64.5	69.7	71.9	77.9	81.9	83.2	83.3	82.2	75.3	69.0	55.5	73.4
1990	65.8	69.1	69.3	71.5	79.4	81.9	82.8	83.5	82.0	77.1	69.3	66.3	74.8
1991	66.3	64.2	67.7	75.3	79.5	81.1	82.6	83.0	81.7	75.3	65.8	65.5	74.0
1992	59.7	64.8	66.4	69.9	74.9	81.2	84.5	82.3	81.6	73.4	70.4	63.5	72.7
1993	66.3	60.6	64.8	68.3	75.5	81.9	83.9	83.2	81.2	75.1	68.3	58.9	72.3
Record Mean	60.5	62.2	66.8	71.8	77.2	81.2	82.4	82.6	81.0	74.8	67.6	62.1	72.5
Max	71.6	73.4	78.0	83.1	88.1	90.9	91.8	91.6	89.6	84.0	77.8	72.8	82.7
Min	49.5	50.9	55.5	60.4	66.3	71.5	73.0	73.5	72.4	65.6	57.4	51.3	62.3

REFERENCE NOTES

ORLANDO, FLORIDA

GENERAL
T - TRACE AMOUNT.
BLANK ENTRIES DENOTE MISSING/UNREPORTED DATA.
# INDICATES A STATION OR INSTRUMENT RELOCATION.

SPECIFIC

PM - INCLUDES LAST DAY OF PREVIOUS MONTH
ASOS - AUTOMATED SURFACE OBSERVING SYSTEM IN OPERATION DURING THESE MONTHS.

(a) - LENGTH OF RECORD IN YEARS, ALTHOUGH INDIVIDUAL MONTHS MAY BE MISSING.
0.* OR * - THE VALUE IS BETWEEN 0.0 AND 0.05
NORMALS - BASED ON THE 1961-1990 RECORD PERIOD.
EXTREMES - DATES ARE THE MOST RECENT OCCURRENCE
WIND DIR.- NUMERALS SHOW TENS OF DEGRESS CLOCKWISE FROM TRUE NORTH. "00" INDICATES CALM.
RESULTANT DIRECTIONS ARE GIVEN TO WHOLE DEGREES.
BOLD VALUES INDICATE EXTREME VALUES WHICH OCCURRED AFTER THE ASOS SYSTEM WAS COMMISSIONED.

RECORD = PERIOD OF RECORD
RECORD MEAN PRECIPITATION IS THE MEAN OF ALL DAILY PRECIPITATION AMOUNTS DURING THE PERIOD OF RECORD.
RECORD MAX(MIN) TEMPERATURE IS THE MEAN OF ALL DAILY MAX(MIN) TEMPERATURES DURING THE PERIOD OF RECORD.
RECORD MEAN TEMPERATURE IS THE SUM OF THE RECORD MAX AND RECORD MIN DIVIDED BY 2.
AVERAGE TEMPERATURE IS THE SUM OF THE MEAN DAILY MAX AND MIN TEMPERATURE DIVIDED BY 2.

EXCEPTIONS

RECORD MEANS ARE THROUGH THE CURRENT YEAR, BEGINNING IN 1943 FOR TEMPERATURE
1943 FOR PRECIPITATION
1943 FOR SNOWFALL

HEATING DEGREE DAYS Base 65 deg. F ORLANDO, FLORIDA

SEASON	JULY	AUG	SEP	OCT	NOV	DEC	JAN	FEB	MAR	APR	MAY	JUNE	TOTAL
1964-65	0	0	0	7	14	84	178	89	82	0	0	0	454
1965-66	0	0	0	1	19	112	215	122	72	5	0	0	546
1966-67	0	0	0	2	70	169	119	157	25	0	0	0	542
1967-68	0	0	0	0	29	80	191	293	149	0	0	0	742
1968-69	0	0	0	19	120	237	168	206	169	0	0	0	919
1969-70	0	0	0	0	93	204	316	187	58	0	0	0	858
1970-71	0	0	0	0	120	79	165	115	92	20	0	0	591
1971-72	0	0	0	0	26	9	51	124	24	6	0	0	240
1972-73	0	0	0	0	54	105	160	169	12	9	0	0	509
#1973-74	0	0	0	6	13	193	0	173	15	8	0	0	408
1974-75	0	0	0	0	40	163	73	44	57	10	0	0	387
1975-76	0	0	0	0	85	174	278	104	18	1	0	0	660
1976-77	0	0	0	4	118	197	440	218	41	8	0	0	1026
1977-78	0	0	0	6	38	179	275	255	71	0	0	0	824
1978-79	0	0	0	0	0	56	230	214	71	0	0	0	571
1979-80	0	0	0	0	47	119	161	245	61	4	0	0	637
1980-81	0	0	0	1	67	190	416	119	76	1	0	0	870
1981-82	0	0	0	0	75	205	204	21	33	7	0	0	545
1982-83	0	0	0	14	16	94	233	148	105	13	0	0	623
1983-84	0	0	0	0	63	188	252	137	86	18	0	0	744
1984-85	0	0	0	0	68	71	340	146	22	12	0	0	659
1985-86	0	0	0	0	14	228	180	82	105	4	0	0	613
1986-87	0	0	0	0	0	42	216	97	48	66	0	0	469
1987-88	0	0	0	0	39	97	221	169	71	7	0	0	604
1988-89	0	0	0	0	11	135	32	119	59	4	0	0	360
1989-90	0	0	0	21	27	308	71	34	11	5	0	0	477
1990-91	0	0	0	6	14	69	75	88	52	0	0	0	304
1991-92	0	0	0	0	85	76	187	79	51	19	8	0	505
1992-93	0	0	0	0	47	102	73	131	80	12	0	0	445
1993-94	0	0	0	10	45	201							

COOLING DEGREE DAYS Base 65 deg. F ORLANDO, FLORIDA

YEAR	JAN	FEB	MAR	APR	MAY	JUNE	JULY	AUG	SEP	OCT	NOV	DEC	TOTAL
1969	12	10	32	232	376	544	608	540	495	406	68	16	3339
1970	19	14	128	330	399	511	586	544	565	380	77	72	3625
1971	77	97	90	238	411	505	569	573	510	440	167	214	3891
1972	181	44	146	243	391	524	570	561	509	374	179	148	3870
1973	88	28	207	198	421	548	602	529	501	341	199	58	3720
#1974	213	51	183	207	410	463	492	536	510	241	125	43	3474
1975	105	121	141	237	442	481	489	541	479	366	167	32	3601
1976	18	75	194	196	374	449	549	529	474	247	65	49	3219
1977	1	13	192	182	324	534	537	521	536	257	185	62	3344
1978	26	3	116	259	449	541	550	553	508	321	225	115	3666
1979	26	31	65	260	330	479	575	546	498	299	153	53	3315
1980	27	25	169	172	362	459	586	582	508	331	138	12	3371
1981	0	34	52	253	372	552	602	559	458	359	89	73	3403
1982	56	123	211	241	325	518	550	542	465	303	196	152	3682
1983	22	11	68	129	361	473	573	582	476	362	95	77	3229
1984	37	35	84	151	332	411	490	520	426	331	99	107	3023
1985	27	74	137	191	386	531	539	548	451	454	262	45	3645
1986	25	69	124	139	372	506	543	573	507	392	333	121	3704
1987	32	38	82	127	376	549	582	627	540	230	163	78	3424
1988	26	43	95	223	336	466	496	559	573	275	182	61	3335
1989	101	111	213	216	408	509	573	579	523	346	153	19	3751
1990	102	156	156	206	453	514	559	581	518	388	149	116	3898
1991	121	71	143	315	455	490	553	565	508	326	118	98	3763
1992	28	79	101	175	325	496	612	540	507	265	217	62	3407
1993	120	14	80	118	334	514	591	573	492	331	146	18	3331

SNOWFALL (inches) ORLANDO, FLORIDA

SEASON	JULY	AUG	SEP	OCT	NOV	DEC	JAN	FEB	MAR	APR	MAY	JUNE	TOTAL
1970-71	0.0	0.0	0.0	0.0	0.0	0.0	0.0	0.0	0.0	0.0	0.0	0.0	0.0
1971-72	0.0	0.0	0.0	0.0	0.0	0.0	0.0	0.0	0.0	0.0	0.0	0.0	0.0
1972-73	0.0	0.0	0.0	0.0	0.0	0.0	0.0	0.0	0.0	0.0	0.0	0.0	0.0
#1973-74	0.0	0.0	0.0	0.0	0.0	0.0	0.0	0.0	0.0	0.0	0.0	0.0	0.0
1974-75	0.0	0.0	0.0	0.0	0.0	0.0	0.0	0.0	0.0	0.0	0.0	0.0	0.0
1975-76	0.0	0.0	0.0	0.0	0.0	0.0	0.0	0.0	0.0	0.0	0.0	0.0	0.0
1976-77	0.0	0.0	0.0	0.0	0.0	0.0	T	0.0	0.0	0.0	0.0	0.0	T
1977-78	0.0	0.0	0.0	0.0	0.0	0.0	0.0	0.0	0.0	0.0	0.0	0.0	0.0
1978-79	0.0	0.0	0.0	0.0	0.0	0.0	0.0	0.0	0.0	0.0	0.0	0.0	0.0
1979-80	0.0	0.0	0.0	0.0	0.0	0.0	0.0	0.0	0.0	0.0	0.0	0.0	0.0
1980-81	0.0	0.0	0.0	0.0	0.0	0.0	0.0	0.0	0.0	0.0	0.0	0.0	0.0
1981-82	0.0	0.0	0.0	0.0	0.0	0.0	0.0	0.0	0.0	0.0	0.0	0.0	0.0
1982-83	0.0	0.0	0.0	0.0	0.0	0.0	0.0	0.0	0.0	0.0	0.0	0.0	0.0
1983-84	0.0	0.0	0.0	0.0	0.0	0.0	0.0	0.0	0.0	0.0	0.0	0.0	0.0
1984-85	0.0	0.0	0.0	0.0	0.0	0.0	0.0	0.0	0.0	0.0	0.0	0.0	0.0
1985-86	0.0	0.0	0.0	0.0	0.0	0.0	0.0	0.0	0.0	0.0	0.0	0.0	0.0
1986-87	0.0	0.0	0.0	0.0	0.0	0.0	0.0	0.0	0.0	0.0	0.0	0.0	0.0
1987-88	0.0	0.0	0.0	0.0	0.0	0.0	0.0	0.0	0.0	0.0	0.0	0.0	0.0
1988-89	0.0	0.0	0.0	0.0	0.0	0.0	0.0	0.0	0.0	0.0	0.0	0.0	0.0
1989-90	0.0	T	0.0	0.0	0.0	0.0	0.0	0.0	0.0	0.0	0.0	0.0	T
1990-91	0.0	0.0	0.0	0.0	0.0	0.0	0.0	0.0	0.0	0.0	0.0	0.0	0.0
1991-92	T	0.0	0.0	0.0	0.0	0.0	0.0	0.0	T	T	0.0	0.0	T
1992-93	0.0	0.0	0.0	0.0	0.0	0.0	0.0	0.0	0.0	0.0	0.0	0.0	0.0
1993-94	T	0.0	0.0	0.0	0.0	0.0							
Record Mean	T	T	0.0	0.0	0.0	0.0	T	0.0	T	T	0.0	0.0	T

PENSACOLA, FLORIDA

Pensacola is situated on a somewhat hilly, sandy slope which borders Pensacola Bay, an expanse of deep water several miles in width. The bay is separated from the Gulf of Mexico by a long, narrow island that forms a natural breakwater for the harbor. Elevations in the city range from a few feet above sea level to more than 100 feet in portions of the residential sections, and most of the city is well above storm tides.

The Gulf of Mexico, about 6 miles distant, moderates the climate of Pensacola by tempering the cold Northers of winter and causing cool and refreshing sea breezes during the daytime in summer.

The average temperature for the summer months is around 80 degrees with an average daily range of 12.5 degrees. Temperatures of 90 degrees or higher occur on the average of 39 times yearly. A temperature of 100 degrees or higher occurs occasionally. The average winter temperature is in the low to mid 50s with an average daily range of 15.7 degrees. On the average, the temperature falls to freezing or below on only nine days of the year. The average occurrence of the last temperature as low as 32 degrees in spring is mid-February, and the average earliest occurrence in autumn is early December, making the average growing season 292 days. Severe cold waves are rather infrequent.

Rainfall is usually well distributed through the year with the greatest frequency normally being in July and August. The greatest monthly rainfall occurs, on average, in July and least in October. Much of the rainfall in summer occurs during the daylight hours and comes in the form of thunderstorms, often producing excessive amounts. Winter rains are frequently lighter, but extend over longer periods. Snow has occurred in about 30 percent of the winters but measurable amounts are less frequent.

A moderate sea breeze usually blows off the Gulf of Mexico during most of the day in summer. Seriously destructive hurricanes are occasionally experienced in this vicinity but loss of life is rare. Hurricanes have occurred from early July to mid-October.

NORMALS, MEANS, AND EXTREMES

PENSACOLA, FLORIDA

LATITUDE: 30°28'N LONGITUDE: 87°12'W ELEVATION: FT. GRND 112 BARO 116 TIME ZONE: CENTRAL WBAN: 13899

	(a)	JAN	FEB	MAR	APR	MAY	JUNE	JULY	AUG	SEP	OCT	NOV	DEC	YEAR
TEMPERATURE °F:														
Normals														
-Daily Maximum		59.8	62.9	69.4	76.6	83.2	88.7	89.9	89.2	86.4	79.1	70.1	62.9	76.5
-Daily Minimum		41.4	44.1	51.3	58.5	65.7	71.8	74.2	73.8	70.3	59.4	51.0	44.4	58.8
-Monthly		50.6	53.6	60.4	67.6	74.5	80.3	82.1	81.5	78.4	69.3	60.6	53.7	67.7
Extremes														
-Record Highest	30	80	82	86	96	96	101	106	104	98	92	85	81	106
-Year		1975	1972	1991	1987	1964	1988	1980	1986	1990	1973	1973	1978	JUL 1980
-Record Lowest	30	5	19	22	33	48	56	61	62	43	32	25	11	5
-Year		1985	1970	1980	1987	1992	1984	1967	1992	1967	1993	1976	1989	JAN 1985
NORMAL DEGREE DAYS:														
Heating (base 65°F)		471	331	184	38	0	0	0	0	0	39	183	371	1617
Cooling (base 65°F)		25	12	42	116	295	459	530	512	402	172	51	20	2636
% OF POSSIBLE SUNSHINE	5	48	53	61	63	67	67	57	58	60	71	64	49	60
MEAN SKY COVER (tenths)														
Sunrise - Sunset	27	6.5	5.9	6.0	5.4	5.5	5.5	6.1	6.0	5.4	4.5	5.2	6.2	5.7
MEAN NUMBER OF DAYS:														
Sunrise to Sunset														
-Clear	27	7.7	8.7	9.0	10.1	9.4	7.1	4.5	5.7	9.8	13.7	11.4	8.6	105.7
-Partly Cloudy	27	7.0	6.8	8.0	9.2	11.1	15.0	17.0	15.1	10.7	8.9	7.6	6.6	123.1
-Cloudy	27	16.3	12.7	14.0	10.7	10.5	7.9	9.5	10.1	9.5	8.5	11.0	15.8	136.4
Precipitation														
.01 inches or more	30	10.0	9.1	9.1	6.4	7.4	10.1	13.7	12.7	8.8	4.9	7.7	9.3	109.2
Snow,Ice pellets,hail														
1.0 inches or more	30	0.1	0.*	0.0	0.0	0.0	0.0	0.0	0.0	0.0	0.0	0.0	0.0	0.1
Thunderstorms	24	1.7	2.8	4.1	3.9	5.4	10.0	15.3	14.3	6.8	1.9	2.0	1.4	69.4
Heavy Fog Visibility														
1/4 mile or less	24	5.8	4.8	5.8	3.7	1.7	0.5	0.4	0.2	0.6	1.8	3.6	4.8	33.5
Temperature °F														
-Maximum														
90° and above	30	0.0	0.0	0.0	0.1	2.0	13.2	18.3	15.7	8.7	0.4	0.0	0.0	58.4
32° and below	30	0.1	0.0	0.0	0.0	0.0	0.0	0.0	0.0	0.0	0.0	0.0	0.1	0.2
-Minimum														
32° and below	30	6.5	3.6	0.8	0.0	0.0	0.0	0.0	0.0	0.0	0.*	0.8	4.0	15.8
0° and below	30	0.0	0.0	0.0	0.0	0.0	0.0	0.0	0.0	0.0	0.0	0.0	0.0	0.0
AVG. STATION PRESS. (mb)	21	1016.7	1015.4	1013.1	1012.6	1011.3	1011.8	1013.0	1012.6	1012.2	1014.2	1015.4	1016.9	1013.8
RELATIVE HUMIDITY (%)														
Hour 00	28	80	79	81	82	84	84	86	87	84	80	82	81	83
Hour 06 (Local Time)	28	82	81	84	86	87	86	88	90	88	84	84	84	85
Hour 12	28	62	58	58	56	59	60	64	65	61	55	59	64	60
Hour 18	28	71	68	68	66	67	68	71	74	72	69	75	75	70
PRECIPITATION (inches):														
Water Equivalent														
-Normal		4.68	5.40	5.63	3.77	4.20	6.40	7.42	7.39	5.32	4.21	3.54	4.29	62.25
-Maximum Monthly	30	18.77	11.66	12.96	15.52	10.31	17.68	20.36	14.14	15.71	14.84	11.97	9.58	20.36
-Year		1991	1966	1979	1964	1987	1978	1979	1987	1988	1985	1992	1982	JUL 1979
-Minimum Monthly	30	0.60	1.06	0.87	0.38	0.08	0.86	1.69	2.53	0.39	0.00	0.30	0.57	0.00
-Year		1981	1991	1967	1987	1988	1979	1970	1990	1984	1978	1981	1980	OCT 1978
-Maximum in 24 hrs	30	5.44	4.70	11.10	7.51	5.01	6.77	5.14	5.92	10.02	5.88	4.20	4.52	11.10
-Year		1978	1982	1979	1964	1987	1970	1975	1987	1967	1986	1989	1964	MAR 1979
Snow,Ice pellets,hail														
-Maximum Monthly	30	2.5	1.9	T	0.0	0.0	0.0	T	0.0	0.0	0.0	0.0	T	2.5
-Year		1977	1973	1993				1993					1993	JAN 1977
-Maximum in 24 hrs	30	1.5	1.9	T	0.0	0.0	0.0	T	0.0	0.0	0.0	0.0	T	1.9
-Year		1977	1973	1993				1993					1993	FEB 1973
WIND:														
Mean Speed (mph)	30	9.0	9.4	9.7	9.5	8.6	7.6	7.0	6.7	7.6	7.9	8.4	8.9	8.3
Prevailing Direction														
Fastest Obs. 1 Min.														
-Direction (!!)	23	22	16	31	32	29	33	09	32	10	13	21	10	10
-Speed (MPH)	23	35	35	35	35	32	32	35	35	53	35	35	34	53
-Year		1987	1984	1993	1990	1973	1972	1993	1993	1979	1985	1972	1988	SEP 1979
Peak Gust														
-Direction (!!)														
-Speed (mph)														
-Date														

PRECIPITATION (inches)

PENSACOLA, FLORIDA

YEAR	JAN	FEB	MAR	APR	MAY	JUNE	JULY	AUG	SEP	OCT	NOV	DEC	ANNUAL
1964	11.83	6.78	4.68	15.52	1.64	4.30	10.53	9.37	2.57	6.28	4.19	5.27	82.96
1965	2.35	6.48	3.77	1.59	0.30	9.09	4.88	6.26	9.30	2.82	0.55	3.46	50.85
1966	6.02	11.66	1.83	2.77	4.40	2.00	3.16	6.25	3.59	3.58	1.52	6.01	52.79
1967	4.89	5.47	0.87	1.57	4.23	5.10	4.72	13.09	10.28	5.79	1.07	6.53	63.61
1968	1.22	2.78	2.66	1.39	1.23	4.86	4.97	8.55	4.01	0.93	3.11	5.49	41.20
1969	1.82	4.98	9.26	3.48	8.35	6.93	13.98	8.24	3.09	2.56	1.28	4.39	68.36
1970	4.08	4.99	7.84	2.20	7.66	10.00	1.69	10.32	2.38	12.01	1.26	3.53	67.96
1971	1.61	5.43	3.05	0.67	3.92	5.10	6.53	7.27	4.91	T	2.37	3.71	44.57
1972	3.65	3.48	5.55	2.04	4.58	8.54	3.58	3.10	1.66	4.45	5.67	5.22	51.52
1973	3.93	4.74	11.81	7.88	3.79	3.45	12.92	4.14	6.13	3.97	1.98	7.68	72.42
1974	3.61	3.20	5.34	2.57	3.25	5.44	5.81	5.63	7.10	0.95	2.81	2.47	48.18
1975	4.51	4.28	6.06	5.50	7.07	4.56	16.03	5.04	11.53	7.21	6.54	3.17	81.50
1976	6.11	3.07	6.29	1.77	8.32	6.97	6.09	9.58	4.82	8.36	5.64	4.22	71.24
1977	4.77	1.44	4.20	3.24	2.02	1.52	5.14	6.29	4.82	2.40	4.15	2.41	42.40
1978	13.41	2.66	4.62	5.73	5.68	17.68	8.65	9.86	0.96	0.00	3.70	3.25	76.20
1979	6.42	6.17	12.96	2.58	4.01	0.86	20.36	7.52	10.94	0.57	5.45	1.78	79.62
1980	4.65	1.07	11.33	7.07	5.27	3.17	4.20	6.04	2.20	2.12	2.59	0.57	50.28
1981	0.60	5.18	3.45	1.31	6.67	1.92	4.84	4.55	2.57	3.12	0.30	4.86	39.37
1982	2.84	8.87	5.58	1.90	2.05	4.00	6.99	9.87	1.32	2.51	3.00	9.58	58.51
1983	5.32	8.82	6.68	10.78	3.92	12.20	5.05	3.82	8.29	3.65	4.42	5.33	78.28
1984	4.25	3.85	5.36	4.77	1.23	4.40	9.32	12.34	0.39	1.75	2.62	1.35	51.63
1985	4.54	5.09	4.12	2.07	5.30	7.20	4.02	7.17	5.59	14.84	3.55	5.85	69.34
1986	3.08	9.98	6.01	0.98	4.06	3.41	2.71	5.09	8.83	14.01	6.13	4.26	68.55
1987	6.51	8.18	2.99	0.38	10.31	8.11	5.94	14.14	4.43	0.33	5.29	2.08	68.69
1988	2.99	8.18	5.72	3.79	0.08	2.15	14.97	13.71	15.71	6.34	1.78	1.89	77.31
1989	1.08	1.64	6.23	3.01	6.74	16.97	11.15	2.77	6.22	2.03	7.67	4.44	69.95
1990	4.70	4.96	9.19	4.90	4.61	5.51	2.10	2.53	1.49	8.49	1.09	1.99	51.56
1991	18.77	1.06	8.20	7.05	9.40	5.87	9.93	5.72	1.38	0.67	3.28	1.67	73.00
1992	8.29	6.67	3.36	2.12	1.09	11.15	5.09	12.05	5.89	3.42	11.97	5.49	76.59
1993	3.60	4.77	7.89	4.44	2.12	4.20	5.78	7.65	7.19	6.62	1.07	6.00	61.33
Record Mean	4.39	4.60	5.45	4.32	3.83	5.27	7.07	7.57	5.96	3.97	3.74	4.39	60.56

AVERAGE TEMPERATURE (deg. F)

PENSACOLA, FLORIDA

YEAR	JAN	FEB	MAR	APR	MAY	JUNE	JULY	AUG	SEP	OCT	NOV	DEC	ANNUAL
1964	49.1	48.0	58.3	67.8	74.4	80.1	80.4	81.1	78.2	64.9	62.3	55.6	66.7
1965	52.6	52.8	57.4	70.3	74.7	78.1	80.8	80.1	78.5	68.4	63.1	53.5	67.6
1966	46.9	52.5	58.4	67.6	74.3	77.4	82.4	80.1	77.4	68.3	59.8	52.2	66.4
1967	51.9	51.6	63.0	72.5	73.2	79.9	79.5	78.3	73.2	66.5	58.9	57.5	67.2
1968	49.6	46.5	56.1	68.3	73.9	81.1	81.0	81.5	78.1	69.9	55.8	49.8	66.0
1969	51.2	53.0	53.7	67.4	73.2	81.3	82.3	80.5	77.2	71.3	58.4	53.4	66.9
1970	46.4	51.3	59.8	71.1	76.1	80.6	83.1	82.2	82.2	72.1	57.3	56.7	68.3
1971	53.2	53.8	58.0	66.8	73.2	81.0	81.6	81.0	79.0	73.5	60.2	63.0	68.7
1972	58.7	55.1	61.3	68.8	74.1	79.8	81.7	83.9	82.2	71.6	58.3	56.7	69.3
1973	52.2	52.2	64.6	65.7	75.0	81.9	84.3	82.4	80.6	73.4	66.1	53.9	69.3
1974	65.7	55.8	66.7	67.8	77.3	79.4	82.4	81.1	78.4	68.2	60.4	56.5	70.0
1975	57.3	60.9	61.6	67.2	76.9	81.1	82.1	82.0	76.1	70.7	61.4	52.7	69.2
1976	51.9	59.0	63.3	69.2	71.2	78.0	80.6	80.3	76.5	62.9	51.5	49.2	66.1
1977	41.6	51.8	62.5	67.7	74.5	82.0	82.0	80.4	80.1	66.8	62.9	52.5	67.1
1978	43.0	47.1	57.5	68.2	75.8	82.1	82.9	82.8	82.0	69.9	66.5	55.6	67.8
1979	45.6	50.9	59.9	69.4	73.6	80.4	82.5	81.8	78.1	69.4	58.8	52.1	66.9
1980	55.3	51.1	60.6	65.6	75.1	81.2	85.4	83.7	81.3	65.7	57.6	50.3	67.7
1981	45.1	52.8	57.4	69.1	70.3	81.7	82.8	80.7	75.9	67.2	60.4	49.5	66.1
1982	49.8	53.9	61.4	65.8	73.0	80.4	80.1	80.1	75.7	68.9	60.7	57.1	67.3
1983	47.6	51.2	54.7	61.3	72.1	76.0	83.0	83.0	75.2	69.6	58.9	49.8	65.2
1984	47.9	53.4	59.3	66.3	74.1	80.0	80.6	80.7	78.3	75.2	58.8	61.9	68.0
1985	46.9	54.1	65.3	68.8	75.5	81.7	81.1	81.9	78.1	74.0	67.8	50.8	68.8
1986	51.0	57.5	60.1	66.7	75.0	82.1	84.6	81.6	80.3	70.4	66.8	53.0	69.1
1987	50.7	55.2	60.5	65.8	76.2	80.7	83.8	84.1	78.9	64.9	60.9	58.0	68.3
1988	48.7	51.8	59.6	68.2	74.1	80.3	81.6	81.6	78.8	66.8	63.4	54.5	67.5
1989	58.8	55.5	62.1	66.0	73.9	79.6	81.3	82.1	77.8	68.1	60.5	46.1	67.7
1990	55.2	59.7	62.9	66.2	74.2	81.7	83.1	84.0	79.9	69.6	62.0	57.7	69.7
1991	54.1	57.6	62.7	70.4	77.1	80.4	83.2	81.9	79.4	70.3	55.6	55.8	69.0
1992	51.0	57.5	60.9	66.4	73.7	80.8	83.8	80.0	78.5	68.7	58.8	57.0	68.1
1993	57.2	54.1	58.3	63.3	73.1	81.0	83.9	84.2	79.4	70.0	59.1	51.6	67.9
Record Mean	52.6	55.0	60.4	67.1	74.0	79.8	81.3	81.3	78.3	69.7	60.1	54.2	67.8
Max	60.4	62.8	67.8	74.2	81.0	86.6	88.0	88.0	85.2	77.9	68.5	62.2	75.2
Min	44.8	47.2	52.9	59.9	67.0	73.1	74.7	74.5	71.3	61.5	51.7	46.2	60.4

REFERENCE NOTES

PENSACOLA, FLORIDA

GENERAL
T - TRACE AMOUNT.
BLANK ENTRIES DENOTE MISSING/UNREPORTED DATA.
# INDICATES A STATION OR INSTRUMENT RELOCATION.

SPECIFIC

(a) - LENGTH OF RECORD IN YEARS, ALTHOUGH INDIVIDUAL MONTHS MAY BE MISSING.
0.* OR * - THE VALUE IS BETWEEN 0.0 AND 0.05
NORMALS - BASED ON THE 1961-1990 RECORD PERIOD.
EXTREMES - DATES ARE THE MOST RECENT OCCURRENCE
WIND DIR.- NUMERALS SHOW TENS OF DEGRESS CLOCKWISE FROM TRUE NORTH. "00" INDICATES CALM.
RESULTANT DIRECTIONS ARE GIVEN TO WHOLE DEGREES.
BOLD VALUES INDICATE EXTREME VALUES WHICH OCCURRED AFTER THE ASOS SYSTEM WAS COMMISSIONED.

RECORD = PERIOD OF RECORD
RECORD MEAN PRECIPITATION IS THE MEAN OF ALL DAILY PRECIPITATION AMOUNTS DURING THE PERIOD OF RECORD.
RECORD MAX(MIN) TEMPERATURE IS THE MEAN OF ALL DAILY MAX(MIN) TEMPERATURES DURING THE PERIOD OF RECORD.
RECORD MEAN TEMPERATURE IS THE SUM OF THE RECORD MAX AND RECORD MIN DIVIDED BY 2.
AVERAGE TEMPERATURE IS THE SUM OF THE MEAN DAILY MAX AND MIN TEMPERATURE DIVIDED BY 2.

EXCEPTIONS

PERCENT OF POSSIBLE SUNSHINE IS THROUGH 1968.

RECORD MEANS ARE THROUGH THE CURRENT YEAR, BEGINNING IN 1879 FOR TEMPERATURE
1879 FOR PRECIPITATION
1964 FOR SNOWFALL

HEATING DEGREE DAYS Base 65 deg. F — PENSACOLA, FLORIDA

SEASON	JULY	AUG	SEP	OCT	NOV	DEC	JAN	FEB	MAR	APR	MAY	JUNE	TOTAL
1964-65	0	0	0	73	142	304	380	341	252	9	0	0	1501
1965-66	0	0	0	48	106	350	561	346	211	33	1	0	1656
1966-67	0	0	0	30	193	398	402	381	116	1	4	0	1525
1967-68	0	0	21	47	214	262	473	531	285	17	0	0	1850
1968-69	0	0	0	57	289	465	421	339	346	11	1	0	1929
1969-70	0	0	0	8	210	352	573	377	176	17	3	0	1716
1970-71	0	0	0	1	238	278	373	318	230	66	3	0	1507
1971-72	0	0	0	9	175	130	234	299	143	29	0	0	1019
1972-73	0	0	0	24	254	277	390	350	70	63	0	0	1428
1973-74	0	0	0	17	79	360	68	277	64	32	0	0	897
1974-75	0	0	0	19	185	289	253	158	162	40	0	0	1106
1975-76	0	0	12	9	199	387	404	179	103	12	1	0	1306
1976-77	0	0	0	122	402	484	717	364	144	29	0	0	2262
1977-78	0	0	0	55	108	392	673	494	248	10	0	0	1980
1978-79	0	0	0	17	23	322	592	390	168	6	4	0	1522
1979-80	0	0	0	19	194	392	293	415	169	47	0	0	1529
1980-81	0	0	0	69	231	455	610	338	235	11	0	0	1949
1981-82	0	0	7	55	172	473	475	306	176	60	1	0	1725
1982-83	0	0	3	51	159	278	532	377	316	124	1	0	1841
1983-84	0	0	4	26	211	470	524	328	192	48	3	0	1806
1984-85	0	0	0	4	209	123	566	310	62	29	0	0	1303
1985-86	0	0	0	6	49	449	429	220	179	29	0	0	1361
1986-87	0	0	0	20	67	369	441	268	161	94	0	0	1420
1987-88	0	0	0	60	156	246	506	383	182	28	0	0	1561
1988-89	0	0	0	29	107	331	208	279	154	74	2	0	1184
1989-90	0	0	2	62	167	580	302	157	96	54	0	0	1420
1990-91	0	0	0	43	112	251	332	213	134	10	0	0	1095
1991-92	0	0	0	25	299	305	423	221	157	64	8	0	1502
1992-93	0	0	0	7	212	262	240	304	225	81	0	0	1331
1993-94	0	0	0	34	213	407							

COOLING DEGREE DAYS Base 65 deg. F — PENSACOLA, FLORIDA

YEAR	JAN	FEB	MAR	APR	MAY	JUNE	JULY	AUG	SEP	OCT	NOV	DEC	TOTAL
1969	2	9	4	88	263	495	544	486	375	211	19	0	2496
1970	0	0	19	206	354	474	566	542	522	230	15	28	2956
1971	14	12	20	126	263	487	523	503	428	277	37	75	2765
1972	43	16	35	149	287	449	521	592	524	236	62	27	2941
1973	0	0	65	91	318	510	605	547	473	288	115	22	3034
1974	93	24	121	125	384	440	546	505	408	125	51	35	2857
1975	22	51	64	116	377	490	537	536	350	194	96	12	2845
1976	3	13	59	147	201	394	490	480	353	66	4	1	2211
1977	0	2	75	118	300	519	531	484	462	119	50	10	2670
1978	0	0	21	113	342	522	560	560	517	175	74	38	2922
1979	0	2	16	146	276	467	546	527	401	162	16	1	2560
1980	2	20	41	72	322	493	638	586	496	99	16	4	2789
1981	0	2	7	141	173	507	559	494	340	131	42	1	2397
1982	10	1	72	91	256	469	478	477	333	180	36	43	2446
1983	0	0	1	21	227	334	568	562	316	176	32	7	2244
1984	0	0	23	93	293	459	492	495	405	329	29	34	2652
1985	11	11	79	149	333	507	505	529	398	290	140	17	2969
1986	0	18	34	88	317	520	613	523	465	195	129	7	2909
1987	3	0	29	128	352	479	590	599	423	66	39	37	2745
1988	7	5	21	133	290	466	525	518	421	90	66	12	2554
1989	20	20	70	111	286	448	516	537	395	165	39	0	2607
1990	5	17	39	94	293	510	568	596	453	195	30	31	2831
1991	1	12	69	179	383	466	568	531	441	199	23	23	2895
1992	0	13	37	112	282	481	591	472	410	128	35	18	2579
1993	6	5	24	34	258	487	591	602	439	196	44	0	2686

SNOWFALL (inches) — PENSACOLA, FLORIDA

SEASON	JULY	AUG	SEP	OCT	NOV	DEC	JAN	FEB	MAR	APR	MAY	JUNE	TOTAL
1970-71	0.0	0.0	0.0	0.0	0.0	0.0	0.0	0.0	0.0	0.0	0.0	0.0	0.0
1971-72	0.0	0.0	0.0	0.0	0.0	0.0	0.0	0.0	0.0	0.0	0.0	0.0	0.0
1972-73	0.0	0.0	0.0	0.0	0.0	0.0	T	1.9	0.0	0.0	0.0	0.0	1.9
1973-74	0.0	0.0	0.0	0.0	0.0	0.0	0.0	0.0	0.0	0.0	0.0	0.0	0.0
1974-75	0.0	0.0	0.0	0.0	0.0	0.0	0.0	0.0	0.0	0.0	0.0	0.0	0.0
1975-76	0.0	0.0	0.0	0.0	0.0	0.0	0.0	0.0	0.0	0.0	0.0	0.0	0.0
1976-77	0.0	0.0	0.0	0.0	0.0	0.0	2.5	0.0	0.0	0.0	0.0	0.0	2.5
1977-78	0.0	0.0	0.0	0.0	0.0	0.0	T	T	0.0	0.0	0.0	0.0	T
1978-79	0.0	0.0	0.0	0.0	0.0	0.0	0.0	0.0	0.0	0.0	0.0	0.0	0.0
1979-80	0.0	0.0	0.0	0.0	0.0	0.0	0.0	0.0	T	0.0	0.0	0.0	T
1980-81	0.0	0.0	0.0	0.0	0.0	0.0	0.0	0.0	0.0	0.0	0.0	0.0	0.0
1981-82	0.0	0.0	0.0	0.0	0.0	0.0	T	0.0	0.0	0.0	0.0	0.0	T
1982-83	0.0	0.0	0.0	0.0	0.0	0.0	0.0	0.0	0.0	0.0	0.0	0.0	0.0
1983-84	0.0	0.0	0.0	0.0	0.0	0.0	0.0	0.0	0.0	0.0	0.0	0.0	0.0
1984-85	0.0	0.0	0.0	0.0	0.0	0.0	0.0	0.0	0.0	0.0	0.0	0.0	0.0
1985-86	0.0	0.0	0.0	0.0	0.0	0.0	0.0	0.0	0.0	0.0	0.0	0.0	0.0
1986-87	0.0	0.0	0.0	0.0	0.0	0.0	0.0	0.0	0.0	0.0	0.0	0.0	0.0
1987-88	0.0	0.0	0.0	0.0	0.0	0.0	0.0	T	0.0	0.0	0.0	0.0	T
1988-89	0.0	0.0	0.0	0.0	0.0	0.0	0.0	0.0	0.0	0.0	0.0	0.0	0.0
1989-90	0.0	0.0	0.0	0.0	0.0	T	0.0	0.0	0.0	0.0	0.0	0.0	T
1990-91	0.0	0.0	0.0	0.0	0.0	0.0	0.0	0.0	0.0	0.0	0.0	0.0	0.0
1991-92	0.0	0.0	0.0	0.0	0.0	0.0	0.0	0.0	0.0	0.0	0.0	0.0	0.0
1992-93	0.0	0.0	0.0	0.0	0.0	0.0	0.0	0.0	T	0.0	0.0	0.0	T
1993-94	T	0.0	0.0	0.0	0.0	T							
Record Mean	T	0.0	0.0	0.0	0.0	T	0.1	0.1	T	0.0	0.0	0.0	0.1

TALLAHASSEE, FLORIDA

Located about 20 miles from the Gulf of Mexico, Tallahassee has a mild, moist climate of the Gulf States. In contrast to the southern part of the Florida Peninsula, there is a definite march of the four seasons with considerable winter rainfall and quite a bit less winter sunshine. The annual average temperature is about 68 degrees.

During the winter, topographic effects and cold air drainage into lower elevations produce a wide variation of low temperatures on cold, clear and calm nights. Freezing temperatures at the airport and surrounding suburban areas average about thirty-six occurrences each winter, but freezing temperatures in the city are about half that number. Temperatures of 25 degrees or lower in the suburban areas average about twelve times per winter, with temperatures dropping into the teens on occasions. Below zero temperatures are rarely recorded. Snow in Tallahassee is infrequent. The date for the last occurrence of 32 degrees is February 28, but has been as late as April 8. The date of the first occurrence of 32 degrees in the fall is November 25, but has been as early as October 18. This gives an average growing season of some 270 days.

Summer is the least pleasant time of the year. Thunderstorms occur every other day. Rather high temperatures and very high humidities cause considerable discomfort. Occurrences of temperatures of 90 degrees or higher average about 90 days per year, but only about 22 of these days have readings as high as 95 degrees. Temperatures reach 100 degrees once or twice in less than half the years. In general, summertime cloudiness holds the high temperatures about 90 degrees.

July is the wettest month followed by August, September, and June. The driest months are October, November, and April.

Extended droughts are infrequent, shorter droughts are rather common, but both are significant. Droughts, or rainfall deficiencies, when extended over months or years, cause the disappearance of large lakes and cypress ponds. Droughts of shorter duration create fire danger in the nearby forests.

High winds are infrequent and of short duration, usually associated with strong cold fronts in the late winter and early spring months. The likelihood of a hurricane occurrence in our coastal area is about once every 17 years with fringe effects felt about once every five years.

NORMALS, MEANS, AND EXTREMES

TALLAHASSEE, FLORIDA

LATITUDE: 30°23'N LONGITUDE: 84°22' W ELEVATION: FT. GRND 55 BARO 58 TIME ZONE: EASTERN WBAN: 93805

	(a)	JAN	FEB	MAR	APR	MAY	JUNE	JULY	AUG	SEP	OCT	NOV	DEC	YEAR
TEMPERATURE °F:														
Normals														
-Daily Maximum		62.9	66.3	73.5	80.4	86.3	90.8	91.3	91.0	88.5	81.5	73.0	65.9	79.3
-Daily Minimum		38.1	40.1	46.8	52.2	60.8	68.5	71.1	71.4	68.0	55.9	46.3	40.3	55.0
-Monthly		50.5	53.2	60.2	66.3	73.6	79.6	81.3	81.3	78.3	68.7	59.7	53.2	67.2
Extremes														
-Record Highest	33	82	85	90	95	98	103	103	102	99	94	88	84	103
-Year		1993	1989	1967	1968	1962	1985	1980	1986	1991	1986	1961	1971	JUN 1985
-Record Lowest	33	6	14	20	29	34	46	57	58	40	30	13	10	6
-Year		1985	1971	1986	1987	1971	1984	1967	1992	1967	1993	1970	1962	JAN 1985
NORMAL DEGREE DAYS:														
Heating (base 65°F)		471	344	191	59	0	0	0	0	0	48	203	389	1705
Cooling (base 65°F)		21	13	42	98	267	438	505	505	399	163	44	23	2518
% OF POSSIBLE SUNSHINE														
MEAN SKY COVER (tenths)														
Sunrise - Sunset	32	6.3	5.9	5.8	5.2	5.5	5.9	6.3	6.1	5.7	4.7	5.2	6.0	5.7
MEAN NUMBER OF DAYS:														
Sunrise to Sunset														
-Clear	32	8.5	8.4	9.0	10.6	8.6	5.8	3.8	5.0	8.1	13.7	11.0	8.9	101.7
-Partly Cloudy	32	7.0	7.2	8.8	10.2	12.8	14.6	16.9	16.0	11.9	7.8	8.3	7.9	129.3
-Cloudy	32	15.4	12.7	13.2	9.2	9.6	9.6	10.3	10.0	10.0	9.4	10.7	14.2	134.2
Precipitation														
.01 inches or more	32	9.8	8.7	8.6	6.5	8.5	12.4	16.4	14.7	9.2	4.9	6.8	8.1	114.7
Snow, Ice pellets, hail														
1.0 inches or more	32	0.0	0.0	0.0	0.0	0.0	0.0	0.0	0.0	0.0	0.0	0.0	0.0	0.0
Thunderstorms	32	1.8	2.3	4.0	3.5	7.8	13.5	19.4	16.6	8.2	2.0	1.6	1.6	82.2
Heavy Fog Visibility 1/4 mile or less	32	6.8	5.3	5.6	4.7	4.6	2.6	2.1	1.7	1.6	2.8	5.0	6.4	49.2
Temperature °F														
-Maximum														
90° and above	32	0.0	0.0	0.*	1.4	7.8	19.6	23.1	22.0	15.2	1.8	0.0	0.0	91.0
32° and below	32	0.1	0.0	0.0	0.0	0.0	0.0	0.0	0.0	0.0	0.0	0.0	0.1	0.1
-Minimum														
32° and below	32	10.8	7.7	3.1	0.3	0.0	0.0	0.0	0.0	0.0	0.3	4.1	9.5	35.7
0° and below	32	0.0	0.0	0.0	0.0	0.0	0.0	0.0	0.0	0.0	0.0	0.0	0.0	0.0
AVG. STATION PRESS. (mb)	21	1018.4	1017.3	1015.3	1014.6	1013.4	1013.7	1015.0	1014.7	1014.2	1015.9	1017.3	1018.8	1015.7
RELATIVE HUMIDITY (%)														
Hour 01	32	85	84	86	88	89	90	93	93	90	88	88	87	88
Hour 07 (Local Time)	32	87	87	89	91	90	91	94	95	93	91	90	88	91
Hour 13	32	59	54	51	47	50	55	61	62	58	52	54	57	55
Hour 19	32	72	64	60	57	60	67	74	76	74	72	77	76	69
PRECIPITATION (inches):														
Water Equivalent														
-Normal		4.77	5.56	6.21	3.74	4.75	6.93	8.82	7.53	5.58	2.92	3.87	5.03	65.71
-Maximum Monthly	33	18.94	11.50	13.57	13.13	11.66	17.41	20.12	15.73	15.92	11.79	10.44	12.65	20.12
-Year		1991	1964	1973	1973	1976	1989	1964	1977	1969	1976	1976	1964	JUL 1964
-Minimum Monthly	33	0.40	0.87	1.29	0.33	T	2.09	2.35	2.45	0.11	T	0.55	0.89	T
-Year		1969	1991	1967	1992	1965	1977	1983	1983	1972	1987	1990	1980	OCT 1987
-Maximum in 24 hrs	33	5.36	6.04	8.91	4.73	5.85	6.75	8.94	3.70	9.47	5.95	4.98	9.26	9.47
-Year		1991	1981	1991	1964	1991	1966	1964	1987	1969	1964	1976	1964	SEP 1969
Snow, Ice pellets, hail														
-Maximum Monthly	33	T	0.4	T	0.0	0.0	T	T	0.0	T	0.0	0.0	1.0	1.0
-Year		1985	1973	1993			1989	1993		1990			1989	DEC 1989
-Maximum in 24 hrs	33	T	0.4	T	0.0	0.0	T	T	0.0	T	0.0	0.0	1.0	1.0
-Year		1985	1973	1993			1989	1993		1990			1989	DEC 1989
WIND:														
Mean Speed (mph)	32	6.8	7.3	7.5	6.9	6.3	5.7	5.1	5.0	5.9	6.2	6.2	6.3	6.3
Prevailing Direction through 1963		N	S	S	S	E	S	SW	E	ENE	N	N	N	N
Fastest Obs. 1 Min.														
-Direction (!!!)	34	23	09	27	19	29	03	22	02	08	34	16	15	02
-Speed (MPH)	34	46	40	48	35	40	44	39	58	46	30	40	32	58
-Year		1963	1969	1964	1993	1961	1966	1993	1962	1990	1961	1985	1969	AUG 1962
Peak Gust														
-Direction (!!!)	10	SW	S	W	SE	S	S	S	N	NE	SW	S	S	NE
-Speed (mph)	10	37	51	53	48	41	76	67	49	83	53	68	36	83
-Date		1991	1984	1993	1993	1984	1989	1993	1990	1990	1993	1985	1991	SEP 1990

PRECIPITATION (inches)

TALLAHASSEE, FLORIDA

YEAR	JAN	FEB	MAR	APR	MAY	JUNE	JULY	AUG	SEP	OCT	NOV	DEC	ANNUAL
1964	9.27	11.50	4.06	5.61	2.63	8.16	20.12	9.32	6.51	10.48	3.87	12.65	104.18
1965	3.86	10.03	7.86	7.14	T	12.62	8.09	8.16	10.25	3.49	2.54	4.75	78.79
1966	9.25	9.95	3.24	2.21	8.23	8.05	6.54	8.89	10.67	2.89	1.15	3.99	75.06
1967	6.38	5.60	1.29	1.44	5.43	5.00	8.31	9.03	1.71	3.39	3.06	6.28	56.92
1968	2.15	3.76	2.10	1.06	4.00	2.96	4.89	5.40	8.91	4.70	4.07	6.72	50.72
1969	0.40	5.17	9.13	2.02	4.38	4.88	18.83	4.88	15.92	1.21	1.93	5.77	74.52
1970	6.50	4.62	11.49	3.57	2.41	4.80	16.13	8.93	6.85	4.63	2.31	3.23	75.47
1971	3.04	5.48	4.81	1.85	4.08	7.44	10.80	10.75	1.57	3.46	0.88	4.11	58.27
1972	6.52	7.05	5.80	0.55	9.08	11.13	4.13	5.23	0.11	1.75	9.86	4.85	66.06
1973	4.96	7.16	13.57	13.13	8.38	7.09	4.41	10.78	5.30	2.35	3.21	7.46	87.80
1974	3.36	2.87	3.00	3.99	8.59	3.84	7.60	9.38	10.43	0.93	1.64	3.80	59.43
1975	11.68	2.85	6.18	7.17	10.34	4.77	17.52	6.80	4.88	4.41	1.50	7.83	85.93
1976	5.53	1.21	5.30	1.65	11.66	11.02	4.19	7.35	2.79	11.79	10.44	4.09	77.02
1977	6.40	3.10	6.07	2.73	3.41	2.09	3.19	15.73	5.27	0.80	3.50	6.02	58.31
1978	7.04	4.95	6.41	2.42	3.80	6.13	11.09	7.92	1.55	1.70	3.65	4.63	61.29
1979	10.49	4.09	1.89	10.44	9.24	2.49	15.19	6.75	10.16	0.35	3.64	5.72	80.45
1980	4.38	2.15	11.14	6.01	5.87	5.73	11.02	3.48	6.23	4.88	3.31	0.89	65.09
1981	1.83	8.17	8.83	1.19	1.38	6.17	5.77	2.97	4.86	1.85	2.47	5.15	50.64
1982	3.65	5.55	4.57	3.81	3.61	7.35	11.75	4.10	6.59	1.22	2.34	5.50	60.04
1983	3.89	6.76	13.04	7.62	3.35	9.39	2.35	2.45	4.36	1.03	6.42	6.20	66.86
1984	3.59	6.55	7.11	7.52	6.34	3.33	9.76	4.31	2.49	1.43	2.36	1.41	56.20
1985	3.09	1.49	4.57	0.98	2.57	7.50	9.56	10.15	2.23	7.53	6.05	7.21	62.93
1986	2.69	9.61	2.52	0.39	1.33	9.23	11.79	11.62	4.27	2.01	8.39	7.93	71.78
1987	6.45	6.15	9.46	0.44	6.17	12.54	6.34	8.95	3.96	T	6.35	1.01	67.82
1988	3.52	7.07	7.38	3.77	0.54	2.14	6.74	5.73	3.98	2.53	3.98	1.08	48.46
1989	0.47	2.93	4.35	3.64	5.00	17.41	6.48	6.26	4.51	3.37	4.58	4.59	63.59
1990	3.10	7.33	3.38	3.38	1.94	3.98	3.43	6.83	4.86	2.49	0.55	4.46	45.73
1991	18.94	0.87	12.55	5.96	9.73	6.39	9.60	3.89	0.56	1.54	0.63	1.59	72.25
1992	6.52	6.41	3.84	0.33	5.31	14.96	3.19	6.82	4.20	2.87	6.96	1.37	62.78
1993	9.63	4.50	6.86	0.43	1.79	2.31	8.05	6.13	1.75	5.37	2.12	2.99	51.93
Record Mean	4.26	4.73	5.27	3.83	4.15	6.67	7.96	6.95	5.32	2.80	2.95	4.25	59.16

AVERAGE TEMPERATURE (deg. F)

TALLAHASSEE, FLORIDA

YEAR	JAN	FEB	MAR	APR	MAY	JUNE	JULY	AUG	SEP	OCT	NOV	DEC	ANNUAL
1964	49.5	49.0	60.2	68.4	73.5	80.6	80.5	81.2	78.2	64.6	60.6	55.7	66.8
1965	50.3	53.3	59.8	69.4	75.1	79.0	81.2	81.6	80.1	69.3	63.3	53.9	68.0
1966	49.1	52.6	58.7	66.2	74.0	77.0	82.5	80.2	78.3	70.0	60.1	51.7	66.7
1967	53.8	52.0	64.1	72.1	74.0	80.7	79.9	80.7	74.8	66.4	58.6	58.8	68.0
1968	51.9	47.0	56.6	70.4	74.3	82.1	81.8	82.9	77.9	69.3	54.8	50.0	66.6
1969	51.8	51.9	53.4	68.0	73.4	82.4	82.1	80.2	77.1	72.1	55.6	49.9	66.5
1970	46.0	49.6	60.7	69.1	73.2	78.0	80.6	80.4	79.9	70.9	52.3	53.5	66.2
1971	51.2	53.3	54.6	62.9	69.5	79.4	79.1	80.1	78.5	71.5	58.2	62.8	66.8
1972	59.0	53.4	59.8	68.2	73.0	78.0	80.2	82.0	79.3	70.2	59.2	57.2	68.3
1973	52.1	50.7	64.7	64.1	72.3	79.9	82.7	80.8	80.7	70.0	63.8	52.1	67.8
1974	66.6	53.7	64.2	64.9	74.5	77.9	80.3	80.6	78.8	64.9	56.7	52.3	68.0
1975	54.8	57.8	60.0	65.5	76.4	80.1	79.4	81.3	76.5	70.1	59.8	51.5	67.8
1976	47.4	56.8	62.9	66.8	72.4	78.0	82.1	81.7	76.9	63.4	53.3	50.2	66.0
1977	43.9	49.2	63.9	67.4	72.8	80.0	82.0	80.7	79.1	64.1	60.4	51.1	66.2
1978	44.8	45.2	56.3	66.3	73.9	80.0	80.9	80.7	79.4	68.1	63.6	53.9	66.1
1979	46.1	49.8	58.1	67.2	72.1	77.4	81.0	79.5	77.1	66.3	58.8	52.0	65.5
1980	52.4	50.1	60.9	64.1	73.7	79.7	83.0	82.6	81.3	66.6	58.5	50.3	66.9
1981	44.3	55.0	57.1	68.8	70.8	82.3	82.0	79.8	75.7	68.1	57.6	50.3	66.0
1982	50.8	57.5	63.3	66.5	74.4	80.6	80.8	81.3	76.5	69.9	64.2	59.7	68.8
1983	49.2	53.0	57.9	62.8	72.6	78.2	82.8	82.4	75.7	70.3	57.5	50.7	66.1
1984	49.8	53.9	58.8	65.2	73.5	78.9	79.7	81.1	77.9	73.5	57.9	62.9	67.8
1985	46.5	56.2	64.5	66.4	74.8	80.5	80.5	80.8	77.5	74.4	66.7	48.3	68.1
1986	48.9	56.5	58.7	64.1	73.8	80.7	82.4	80.0	80.0	69.6	67.0	55.0	68.1
1987	49.1	53.2	59.3	62.4	74.9	78.9	82.2	82.5	77.0	62.0	61.1	55.6	66.5
1988	46.5	49.7	57.8	66.4	70.4	78.2	80.5	81.3	79.1	64.5	61.9	51.1	65.6
1989	58.3	56.3	62.9	65.0	72.7	79.3	81.2	81.5	77.6	68.0	59.6	45.5	67.3
1990	55.8	61.1	63.4	66.1	74.8	81.4	82.4	83.2	79.1	71.0	61.0	58.7	69.8
1991	54.5	56.7	62.2	71.5	77.7	79.8	82.2	81.6	79.7	69.5	56.2	55.7	68.9
1992	50.5	56.2	59.4	64.4	71.8	79.5	82.6	80.0	78.5	66.2	61.7	55.5	67.2
1993	58.1	53.5	59.5	62.2	73.5	80.9	83.3	83.4	79.6	69.1	59.3	50.0	67.7
Record Mean	52.8	54.8	61.0	67.2	74.4	79.7	80.9	80.8	77.9	69.1	59.7	53.7	67.7
Max	63.5	66.0	72.6	79.3	86.0	90.1	90.3	90.0	87.2	79.8	71.1	64.6	78.4
Min	42.0	43.6	49.3	55.0	62.8	69.3	71.5	71.6	68.6	58.3	48.3	42.8	56.9

REFERENCE NOTES

TALLAHASSEE, FLORIDA

GENERAL
T - TRACE AMOUNT.
BLANK ENTRIES DENOTE MISSING/UNREPORTED DATA.
# INDICATES A STATION OR INSTRUMENT RELOCATION.

SPECIFIC

(a) - LENGTH OF RECORD IN YEARS, ALTHOUGH INDIVIDUAL MONTHS MAY BE MISSING.
0.* OR * - THE VALUE IS BETWEEN 0.0 AND 0.05
NORMALS - BASED ON THE 1961-1990 RECORD PERIOD.
EXTREMES - DATES ARE THE MOST RECENT OCCURRENCE
WIND DIR.- NUMERALS SHOW TENS OF DEGRESS CLOCKWISE FROM TRUE NORTH. "00" INDICATES CALM.
RESULTANT DIRECTIONS ARE GIVEN TO WHOLE DEGREES.
BOLD VALUES INDICATE EXTREME VALUES WHICH OCCURRED AFTER THE ASOS SYSTEM WAS COMMISSIONED.

RECORD = PERIOD OF RECORD
RECORD MEAN PRECIPITATION IS THE MEAN OF ALL DAILY PRECIPITATION AMOUNTS DURING THE PERIOD OF RECORD.
RECORD MAX(MIN) TEMPERATURE IS THE MEAN OF ALL DAILY MAX(MIN) TEMPERATURES DURING THE PERIOD OF RECORD.
RECORD MEAN TEMPERATURE IS THE SUM OF THE RECORD MAX AND RECORD MIN DIVIDED BY 2.
AVERAGE TEMPERATURE IS THE SUM OF THE MEAN DAILY MAX AND MIN TEMPERATURE DIVIDED BY 2.

EXCEPTIONS

RECORD MEANS ARE THROUGH THE CURRENT YEAR, BEGINNING IN 1885 FOR TEMPERATURE
1885 FOR PRECIPITATION
1962 FOR SNOWFALL

HEATING DEGREE DAYS Base 65 deg. F TALLAHASSEE, FLORIDA

SEASON	JULY	AUG	SEP	OCT	NOV	DEC	JAN	FEB	MAR	APR	MAY	JUNE	TOTAL
1964-65	0	0	0	88	151	304	450	323	218	25	1	0	1560
1965-66	0	0	0	58	93	338	484	347	206	57	0	0	1583
1966-67	0	0	0	33	190	405	344	362	105	7	2	0	1448
1967-68	0	0	12	50	221	222	399	518	266	16	0	0	1704
1968-69	0	0	0	75	306	462	410	372	355	15	3	0	1998
1969-70	0	0	0	2	286	464	589	424	186	41	4	0	1996
1970-71	0	0	1	1	374	357	423	335	338	129	26	0	1984
1971-72	0	0	0	8	226	134	219	337	167	42	0	0	1133
1972-73	0	0	0	13	228	275	409	393	106	89	15	0	1528
1973-74	0	0	0	47	101	409	31	335	94	89	0	0	1106
1974-75	0	0	0	56	272	398	319	217	201	84	0	0	1547
1975-76	0	0	5	22	237	417	537	234	110	32	4	0	1598
1976-77	0	0	0	100	359	456	649	434	116	46	0	0	2160
1977-78	0	0	0	93	175	433	620	548	271	42	0	0	2182
1978-79	0	0	0	41	76	364	581	424	217	28	13	0	1744
1979-80	0	0	0	54	213	399	385	442	169	74	0	0	1736
1980-81	0	0	0	55	212	448	636	288	243	17	5	0	1904
1981-82	0	0	7	36	234	456	442	204	155	50	0	0	1584
1982-83	0	0	1	53	106	208	482	332	249	102	1	0	1534
1983-84	0	0	4	34	231	446	464	320	212	80	3	1	1795
1984-85	0	0	0	14	254	107	573	263	82	61	0	0	1354
1985-86	0	0	0	10	59	525	492	259	222	80	0	0	1647
1986-87	0	0	0	53	62	316	487	324	203	126	0	0	1571
1987-88	0	0	0	114	171	303	567	439	234	54	8	0	1890
1988-89	0	0	0	80	133	426	218	276	135	89	14	0	1371
1989-90	0	0	0	74	186	599	281	128	97	68	0	0	1433
1990-91	0	0	0	58	142	228	327	232	149	19	0	0	1155
1991-92	0	0	0	34	292	299	447	262	195	99	24	0	1652
1992-93	0	0	0	39	182	291	230	315	193	102	0	0	1352
1993-94	0	0	2	41	204	456							

COOLING DEGREE DAYS Base 65 deg. F TALLAHASSEE, FLORIDA

YEAR	JAN	FEB	MAR	APR	MAY	JUNE	JULY	AUG	SEP	OCT	NOV	DEC	TOTAL
1969	7	8	4	113	272	533	540	479	370	227	10	5	2568
1970	4	0	61	174	267	395	490	482	454	190	0	7	2524
1971	3	13	22	70	174	439	443	476	414	214	30	74	2372
1972	40	6	16	146	254	395	477	533	433	181	62	40	2583
1973	15	0	101	69	247	453	557	496	480	209	72	15	2714
1974	88	22	77	94	302	396	482	491	420	60	27	13	2472
1975	6	23	53	101	361	459	455	513	352	189	89	3	2604
1976	0	4	51	94	241	395	537	526	367	57	14	6	2292
1977	0	0	87	123	249	455	536	494	428	73	46	7	2498
1978	0	0	11	87	280	459	498	497	437	143	41	27	2480
1979	0	2	10	103	241	378	503	456	366	102	35	2	2198
1980	0	15	48	54	278	448	566	554	497	110	26	0	2596
1981	0	14	5	136	194	528	537	467	334	140	21	6	2382
1982	11	2	108	102	295	476	497	511	353	214	88	50	2707
1983	0	0	35	43	242	400	558	548	328	203	14	9	2380
1984	0	4	30	94	275	424	464	508	391	285	48	48	2571
1985	8	23	72	111	314	473	486	498	380	310	120	12	2807
1986	0	25	34	56	274	480	544	473	456	206	127	12	2687
1987	0	2	35	56	314	426	541	550	368	30	62	20	2404
1988	0	2	15	104	181	401	488	513	431	73	48	5	2261
1989	15	37	75	94	261	435	505	517	386	174	30	1	2530
1990	0	26	53	108	308	499	544	572	430	251	30	38	2859
1991	7	7	70	222	401	451	542	525	443	184	33	18	2903
1992	3	15	29	85	243	441	549	472	411	85	92	3	2428
1993	23	0	30	26	271	483	570	580	448	173	40	0	2644

SNOWFALL (inches) TALLAHASSEE, FLORIDA

SEASON	JULY	AUG	SEP	OCT	NOV	DEC	JAN	FEB	MAR	APR	MAY	JUNE	TOTAL
1971-72	0.0	0.0	0.0	0.0	0.0	0.0	0.0	0.0	0.0	0.0	0.0	0.0	0.0
1972-73	0.0	0.0	0.0	0.0	0.0	0.0	0.0	0.4	0.0	0.0	0.0	0.0	0.4
1973-74	0.0	0.0	0.0	0.0	0.0	0.0	0.0	0.0	0.0	0.0	0.0	0.0	0.0
1974-75	0.0	0.0	0.0	0.0	0.0	0.0	0.0	0.0	T	0.0	0.0	0.0	T
1975-76	0.0	0.0	0.0	0.0	0.0	0.0	0.0	0.0	0.0	0.0	0.0	0.0	0.0
1976-77	0.0	0.0	0.0	0.0	0.0	T	T	T	0.0	0.0	0.0	0.0	T
1977-78	0.0	0.0	0.0	0.0	0.0	0.0	0.0	0.0	0.0	0.0	0.0	0.0	0.0
1978-79	0.0	0.0	0.0	0.0	0.0	0.0	0.0	0.0	0.0	0.0	0.0	0.0	0.0
1979-80	0.0	0.0	0.0	0.0	0.0	0.0	0.0	0.0	T	0.0	0.0	0.0	T
1980-81	0.0	0.0	0.0	0.0	0.0	0.0	0.0	0.0	0.0	0.0	0.0	0.0	0.0
1981-82	0.0	0.0	0.0	0.0	0.0	0.0	0.0	0.0	0.0	0.0	0.0	0.0	0.0
1982-83	0.0	0.0	0.0	0.0	0.0	0.0	0.0	0.0	0.0	0.0	0.0	0.0	0.0
1983-84	0.0	0.0	0.0	0.0	0.0	0.0	0.0	0.0	0.0	0.0	0.0	0.0	0.0
1984-85	0.0	0.0	0.0	0.0	0.0	0.0	T	0.0	0.0	0.0	0.0	0.0	T
1985-86	0.0	0.0	0.0	0.0	0.0	0.0	0.0	0.0	0.0	0.0	0.0	0.0	0.0
1986-87	0.0	0.0	0.0	0.0	0.0	0.0	0.0	0.0	0.0	0.0	0.0	0.0	0.0
1987-88	0.0	0.0	0.0	0.0	0.0	0.0	0.0	T	0.0	0.0	0.0	0.0	T
1988-89	0.0	0.0	0.0	0.0	0.0	0.0	0.0	0.0	0.0	0.0	0.0	T	T
1989-90	0.0	0.0	0.0	0.0	0.0	1.0	0.0	0.0	0.0	0.0	0.0	0.0	1.0
1990-91	0.0	0.0	T	0.0	0.0	0.0	0.0	0.0	0.0	0.0	0.0	0.0	T
1991-92	0.0	0.0	0.0	0.0	0.0	0.0	0.0	0.0	0.0	0.0	0.0	0.0	0.0
1992-93	0.0	0.0	0.0	0.0	0.0	0.0	0.0	0.0	T	0.0	0.0	0.0	T
1993-94	T	0.0	0.0	0.0	0.0	0.0							
Record Mean	T	0.0	T	0.0	0.0	T	T	T	T	0.0	0.0	T	T

TAMPA, FLORIDA

Tampa is on west central coast of the Florida Peninsula. Very near the Gulf of Mexico at the upper end of Tampa Bay, land and sea breezes modify the subtropical climate. Major rivers flowing into the area are the Hillsborough, the Alafia, and the Little Manatee.

Winters are mild. Summers are long, rather warm, and humid. Low temperatures are about 50 degrees in the winter and 70 degrees during the summer. Afternoon highs range from the low 70s in the winter to around 90 degrees from June through September. Invasions of cold northern air produce an occasional cool winter morning. Freezing temperatures occur on one or two mornings per year during December, January, and February. In some years no freezing temperatures occur. Temperatures rarely fail to recover to the 60s on the cooler winter days. Temperatures above the low 90s are uncommon because of the afternoon sea breezes and thunderstorms. An outstanding feature of the Tampa climate is the summer thunderstorm season. Most of the thunderstorms occur in the late afternoon hours from June through September. The resulting sudden drop in temperature from about 90 degrees to around 70 degrees makes for a pleasant change. Between a dry spring and a dry fall, some 30 inches of rain, about 60 percent of the annual total, falls during the summer months. Snowfall is very rare. Measurable snows under 1/2 inch have occurred only a few times in the last one hundred years.

A large part of the generally flat sandy land near the coast has an elevation of under 15 feet above sea level. This does make the area vulnerable to tidal surges. Tropical storms threaten the area on a few occasions most years. The greatest risk of hurricanes has been during the months of June and October. Many hurricanes, by replenishing the soil moisture and raising the water table, do far more good than harm. The heaviest rains in a 24-hour period, around 12 inches, have been associated with hurricanes.

Fittingly named the Suncoast, the sun shines more than 65 percent of the possible, with the sunniest months being April and May. Afternoon humidities are usually 60 percent or higher in the summer months, but range from 50 to 60 percent the remainder of the year.

Night ground fogs occur frequently during the cooler winter months. Prevailing winds are easterly, but westerly afternoon and early evening sea breezes occur most months of the year. Winds in excess of 25 mph are not common and usually occur only with thunderstorms or tropical disturbances.

Based on the 1951-1980 period, the average first occurrence of 32 degrees Fahrenheit in the fall is December 26 and the average last occurrence in the spring is February 3.

NORMALS, MEANS, AND EXTREMES

TAMPA, FLORIDA

LATITUDE: 27°58' N LONGITUDE: 82°32' W ELEVATION: FT. GRND 19 BARO 41 TIME ZONE: EASTERN WBAN: 12842

	(a)	JAN	FEB	MAR	APR	MAY	JUNE	JULY	AUG	SEP	OCT	NOV	DEC	YEAR
TEMPERATURE °F:														
Normals														
-Daily Maximum		69.8	71.4	76.6	81.7	87.2	89.5	90.2	90.2	89.0	84.3	77.7	72.1	81.6
-Daily Minimum		50.0	51.6	56.5	60.8	67.5	72.9	74.5	74.5	72.8	65.2	57.2	52.3	63.0
-Monthly		59.9	61.5	66.6	71.3	77.4	81.3	82.4	82.4	80.9	74.8	67.5	62.2	72.4
Extremes														
-Record Highest	47	86	88	91	93	98	99	97	98	96	94	90	86	99
-Year		1991	1971	1949	1975	1975	1985	1993	1975	1991	1990	1971	1991	JUN 1985
-Record Lowest	47	21	24	29	40	49	53	63	67	57	40	23	18	18
-Year		1985	1958	1980	1987	1992	1984	1970	1973	1981	1964	1970	1962	DEC 1962
NORMAL DEGREE DAYS:														
Heating (base 65°F)		234	160	81	7	0	0	0	0	0	0	72	171	725
Cooling (base 65°F)		76	62	130	196	384	489	539	539	477	304	147	84	3427
% OF POSSIBLE SUNSHINE	46	63	66	71	75	75	67	62	61	61	65	64	62	66
MEAN SKY COVER (tenths)														
Sunrise - Sunset	47	5.6	5.5	5.4	4.8	5.1	6.1	6.6	6.5	6.2	5.1	5.0	5.5	5.6
MEAN NUMBER OF DAYS:														
Sunrise to Sunset														
-Clear	47	9.6	9.1	10.6	11.4	10.6	5.5	2.8	3.2	5.2	11.6	11.6	10.2	101.5
-Partly Cloudy	47	9.7	9.0	10.0	11.0	12.6	14.4	16.3	16.7	13.9	10.4	9.5	9.8	143.3
-Cloudy	47	11.7	10.1	10.5	7.6	7.8	10.1	11.8	11.1	11.0	8.9	8.9	11.1	120.5
Precipitation														
.01 inches or more	47	6.5	6.8	6.9	4.7	6.4	11.7	15.5	16.5	12.9	6.9	5.5	6.1	106.3
Snow,Ice pellets,hail														
1.0 inches or more	47	0.0	0.0	0.0	0.0	0.0	0.0	0.0	0.0	0.0	0.0	0.0	0.0	0.0
Thunderstorms	47	0.9	1.5	2.7	2.6	5.5	13.7	20.7	20.4	11.7	2.9	1.2	1.2	85.1
Heavy Fog Visibility														
1/4 mile or less	47	5.5	3.0	2.7	1.1	0.4	0.3	0.1	0.2	0.3	1.1	2.6	4.0	21.2
Temperature °F														
-Maximum														
90° and above	30	0.0	0.0	0.0	0.8	8.2	16.6	20.9	21.5	16.3	2.7	0.*	0.0	87.1
32° and below	30	0.0	0.0	0.0	0.0	0.0	0.0	0.0	0.0	0.0	0.0	0.0	0.0	0.0
-Minimum														
32° and below	30	1.7	0.6	0.1	0.0	0.0	0.0	0.0	0.0	0.0	0.0	0.1	0.8	3.3
0° and below	30	0.0	0.0	0.0	0.0	0.0	0.0	0.0	0.0	0.0	0.0	0.0	0.0	0.0
AVG. STATION PRESS.(mb)	20	1020.0	1019.1	1017.7	1016.8	1015.7	1016.1	1017.7	1016.9	1015.6	1016.6	1018.4	1020.2	1017.6
RELATIVE HUMIDITY (%)														
Hour 01	30	85	83	83	82	82	84	86	87	87	86	86	85	85
Hour 07 (Local Time)	30	87	86	87	87	86	87	87	90	91	89	88	87	88
Hour 13	30	59	56	55	51	52	60	63	64	61	57	57	59	58
Hour 19	30	73	69	67	62	63	69	73	76	75	72	74	74	71
PRECIPITATION (inches):														
Water Equivalent														
-Normal		1.99	3.08	3.01	1.15	3.10	5.48	6.58	7.61	5.98	2.02	1.77	2.15	43.92
-Maximum Monthly	47	8.02	7.95	12.64	6.59	17.64	13.75	20.59	18.59	13.98	7.36	6.12	6.66	20.59
-Year		1948	1963	1959	1957	1979	1974	1960	1949	1979	1952	1963	1950	JUL 1960
-Minimum Monthly	47	T	0.21	0.06	T	0.10	1.86	1.65	2.35	1.28	0.09	T	0.07	T
-Year		1950	1950	1956	1981	1992	1951	1981	1952	1972	1988	1960	1984	APR 1981
-Maximum in 24 hrs	47	3.29	3.68	5.20	3.70	11.84	5.53	12.11	5.37	4.99	2.93	4.48	3.28	12.11
-Year		1953	1981	1960	1951	1979	1974	1960	1949	1985	1985	1988	1969	JUL 1960
Snow,Ice pellets,hail														
-Maximum Monthly	47	0.2	T	T	0.0	0.0	0.0	0.0	0.0	0.0	0.0	0.0	T	0.2
-Year		1977	1951	1980									1989	JAN 1977
-Maximum in 24 hrs	47	0.2	T	T	0.0	0.0	0.0	0.0	0.0	0.0	0.0	0.0	T	0.2
-Year		1977	1951	1980									1989	JAN 1977
WIND:														
Mean Speed (mph)	47	8.6	9.2	9.4	9.2	8.7	7.9	7.1	6.9	7.7	8.4	8.4	8.4	8.3
Prevailing Direction through 1963		N	E	S	ENE	E	E	E	ENE	ENE	NNE	NNE	N	E
Fastest Obs. 1 Min.														
-Direction (!!)	41	29	32	29	29	36	31	32	11	34	02	29	36	31
-Speed (MPH)	41	35	50	43	37	46	67	58	38	56	38	40	45	67
-Year		1959	1954	1956	1961	1958	1964	1963	1961	1960	1953	1963	1953	JUN 1964
Peak Gust														
-Direction (!!)	10	W	NW	W	W	N	E	E	S	W	SW	NE	N	E
-Speed (mph)	10	44	46	58	49	51	61	60	48	45	41	47	36	61
-Date		1991	1984	1993	1991	1986	1988	1988	1990	1984	1985	1988	1989	JUN 1988

PRECIPITATION (inches)

TAMPA, FLORIDA

YEAR	JAN	FEB	MAR	APR	MAY	JUNE	JULY	AUG	SEP	OCT	NOV	DEC	ANNUAL
1964	5.08	5.37	3.92	0.53	3.58	6.64	9.75	10.73	5.89	2.86	0.38	3.19	57.92
1965	1.56	2.57	2.28	1.10	0.67	7.30	9.70	5.59	7.36	1.43	0.88	2.34	42.78
1966	4.05	3.08	1.16	1.57	0.71	6.44	7.62	4.75	4.12	1.22	0.39	0.94	36.05
1967	1.32	4.30	0.66	T	0.63	4.70	10.30	10.34	2.20	2.42	0.45	2.04	39.36
1968	0.41	1.52	1.23	0.74	2.08	6.72	8.39	6.79	3.58	4.16	3.14	0.59	39.35
1969	1.78	2.11	5.33	0.05	6.23	4.39	7.39	11.88	4.09	3.12	2.68	5.17	54.22
1970	3.10	4.02	6.12	0.49	4.12	2.05	2.77	8.43	3.95	0.89	1.08	1.25	38.27
1971	0.86	4.25	0.54	1.80	4.09	2.54	7.74	7.46	10.16	4.70	1.40	0.79	46.33
1972	0.54	4.44	3.01	0.38	1.88	5.24	6.65	9.78	1.28	3.29	3.53	2.16	42.18
1973	3.75	2.54	4.21	2.42	0.17	4.19	4.77	9.43	8.91	0.98	2.82	5.52	49.71
1974	0.17	0.89	2.35	0.38	1.11	13.75	3.43	4.67	4.00	0.23	0.12	2.80	33.90
1975	0.91	1.56	1.09	0.91	2.07	8.73	6.65	4.24	11.25	4.94	0.22	0.87	43.44
1976	0.40	0.49	1.64	1.83	8.13	7.22	4.58	7.02	6.04	1.30	1.59	2.05	42.29
1977	2.75	2.41	0.73	0.86	0.73	2.66	5.36	5.98	4.28	0.42	1.89	3.40	31.47
1978	2.82	5.17	2.44	0.94	5.00	2.03	5.85	5.97	3.08	3.42	0.01	3.12	39.85
1979	5.72	2.87	2.43	0.55	17.64	2.07	5.93	12.76	13.98	0.16	0.83	1.52	66.46
1980	1.72	2.01	3.09	4.38	3.94	3.81	5.66	7.62	4.05	1.27	2.68	0.37	40.60
1981	0.44	5.34	1.70	T	1.68	9.37	1.65	7.71	5.87	0.87	0.43	3.58	38.64
1982	1.86	2.09	2.99	1.87	5.90	8.34	10.49	7.20	10.76	2.17	0.85	1.29	55.81
1983	1.25	7.35	7.59	2.76	4.10	7.17	6.37	8.89	6.61	1.74	2.33	4.71	60.87
1984	1.62	3.32	1.31	1.51	3.19	3.24	7.15	5.68	4.21	0.29	0.72	0.07	32.31
1985	2.06	2.07	1.80	0.96	0.22	6.43	6.48	8.65	9.04	4.77	0.99	1.13	44.60
1986	2.37	1.49	4.27	0.95	2.46	5.00	6.24	5.46	3.87	6.21	1.33	1.95	41.60
1987	3.29	1.50	12.01	0.39	2.86	3.39	6.06	8.50	4.76	1.46	4.36	0.50	49.08
1988	2.76	1.44	4.09	1.83	1.27	5.19	3.40	11.09	13.56	0.09	5.97	1.64	52.33
1989	1.54	0.41	1.79	0.71	0.24	7.41	8.86	7.90	6.11	1.89	2.05	4.72	43.63
1990	0.53	4.58	1.71	1.47	1.76	5.16	10.01	3.27	2.42	2.63	0.66	0.19	34.39
1991	2.41	0.41	4.73	1.54	6.88	3.78	9.92	7.35	3.43	0.78	1.26	0.67	43.16
1992	1.47	3.67	0.95	2.17	0.10	7.03	2.80	8.22	2.95	2.20	2.43	0.99	34.98
1993	3.60	2.32	3.93	2.45	1.74	3.18	2.92	5.06	6.60	4.23	0.22	1.28	37.53
Record Mean	2.22	2.77	3.03	2.00	3.04	6.76	7.62	7.92	6.48	2.61	1.64	2.01	48.10

AVERAGE TEMPERATURE (deg. F)

TAMPA, FLORIDA

YEAR	JAN	FEB	MAR	APR	MAY	JUNE	JULY	AUG	SEP	OCT	NOV	DEC	ANNUAL
1964	58.6	57.1	68.0	74.0	76.7	81.8	82.0	81.5	79.3	70.9	67.7	63.2	71.7
1965	58.6	63.8	66.4	72.7	76.3	79.5	80.3	81.5	80.5	72.5	67.0	60.6	71.7
1966	57.0	59.9	63.1	69.2	78.5	79.2	81.7	82.5	80.6	75.3	65.1	59.2	70.9
1967	62.3	59.5	67.8	73.2	77.3	79.9	80.8	80.3	79.2	73.5	65.3	64.8	72.0
1968	59.4	54.3	60.9	72.5	75.7	79.6	80.2	82.0	79.6	73.2	62.3	57.6	69.8
1969	58.5	55.8	58.2	72.3	76.2	82.1	83.1	80.9	80.5	76.9	62.9	58.1	70.5
1970	54.0	57.6	64.9	73.2	76.2	80.1	82.8	82.9	81.9	76.1	61.9	62.4	71.2
1971	60.0	63.1	62.2	69.4	75.8	81.2	82.3	81.8	80.3	77.5	67.7	69.4	72.6
1972	67.0	60.7	66.8	71.4	76.6	81.0	81.9	82.1	81.3	76.1	68.2	65.1	73.2
1973	61.9	57.4	70.2	69.3	76.8	81.8	83.2	81.8	81.7	75.8	70.8	60.1	72.6
1974	71.1	61.1	70.9	70.9	78.2	80.1	81.1	82.8	82.8	72.9	67.9	61.9	73.5
1975	65.1	66.6	67.6	74.0	81.5	82.6	83.2	83.7	81.9	77.9	68.3	60.2	74.4
1976	56.6	63.1	70.5	70.6	76.1	79.0	81.7	81.6	79.3	71.0	62.8	59.6	71.0
1977	51.2	57.5	70.9	71.5	76.5	83.7	82.9	83.0	82.3	72.5	67.7	58.7	71.6
1978	55.0	53.2	64.2	72.3	78.7	82.4	83.0	82.8	81.4	75.1	71.7	66.2	72.2
1979	57.8	59.3	65.4	74.2	75.9	80.8	83.9	82.2	81.9	75.2	68.7	63.0	72.3
1980	62.0	56.6	68.1	70.1	77.2	81.6	84.0	83.0	81.3	74.0	66.4	57.5	71.8
1981	50.4	61.4	62.8	72.4	75.4	81.5	82.5	81.7	78.6	74.5	64.4	59.2	70.4
1982	59.8	67.9	68.1	71.4	74.4	81.5	82.1	82.1	80.2	74.3	70.8	67.6	73.4
1983	58.9	60.3	63.3	68.6	76.8	80.9	82.2	82.2	79.4	75.8	65.9	59.9	71.2
1984	58.0	62.6	66.0	71.0	78.0	80.4	81.5	82.4	79.9	75.7	64.9	67.3	72.3
1985	55.9	63.6	69.4	72.5	79.8	83.7	82.4	83.1	80.5	79.2	73.6	59.0	73.6
1986	59.3	65.0	65.4	69.1	77.4	81.8	83.0	82.6	82.3	76.8	76.3	66.5	73.8
1987	59.2	63.2	66.4	66.4	77.8	82.7	83.1	83.7	81.4	71.3	68.9	64.3	72.4
1988	58.6	59.1	65.6	70.6	75.3	81.0	82.7	82.9	82.0	73.5	70.8	63.0	72.1
1989	67.1	64.9	69.8	72.0	78.4	82.4	83.3	83.0	82.4	75.4	68.9	56.2	73.7
1990	66.1	69.2	69.7	72.1	80.5	82.7	82.5	83.9	82.8	77.6	70.2	66.9	75.4
1991	66.7	64.2	68.4	76.8	81.2	81.3	82.3	83.2	81.9	75.3	65.8	64.6	74.3
1992	59.8	63.6	64.8	69.4	74.3	82.1	83.8	82.4	82.0	72.7	70.0	64.3	72.4
1993	67.0	60.2	64.3	67.2	76.1	81.8	83.8	83.7	81.9	75.8	69.1	59.5	72.5
Record Mean	60.9	62.2	66.7	71.4	77.0	80.8	82.0	82.1	80.7	74.7	67.3	62.0	72.3
Max	70.3	71.7	76.2	81.1	86.4	89.2	89.8	90.0	88.7	83.3	76.7	71.5	81.2
Min	51.4	52.7	57.2	61.7	67.5	72.4	74.1	74.2	72.6	66.0	57.9	52.6	63.3

REFERENCE NOTES

TAMPA, FLORIDA

GENERAL
T - TRACE AMOUNT.
BLANK ENTRIES DENOTE MISSING/UNREPORTED DATA.
# INDICATES A STATION OR INSTRUMENT RELOCATION.

SPECIFIC

PM - INCLUDES LAST DAY OF PREVIOUS MONTH
ASOS - AUTOMATED SURFACE OBSERVING SYSTEM IN OPERATION DURING THESE MONTHS.

(a) - LENGTH OF RECORD IN YEARS, ALTHOUGH INDIVIDUAL MONTHS MAY BE MISSING.
0.* OR * - THE VALUE IS BETWEEN 0.0 AND 0.05
NORMALS - BASED ON THE 1961-1990 RECORD PERIOD.
EXTREMES - DATES ARE THE MOST RECENT OCCURRENCE.
WIND DIR.- NUMERALS SHOW TENS OF DEGRESS CLOCKWISE FROM TRUE NORTH. "00" INDICATES CALM.
RESULTANT DIRECTIONS ARE GIVEN TO WHOLE DEGREES.
BOLD VALUES INDICATE EXTREME VALUES WHICH OCCURRED AFTER THE ASOS SYSTEM WAS COMMISSIONED.

RECORD = PERIOD OF RECORD
RECORD MEAN PRECIPITATION IS THE MEAN OF ALL DAILY PRECIPITATION AMOUNTS DURING THE PERIOD OF RECORD.
RECORD MAX(MIN) TEMPERATURE IS THE MEAN OF ALL DAILY MAX(MIN) TEMPERATURES DURING THE PERIOD OF RECORD.
RECORD MEAN TEMPERATURE IS THE SUM OF THE RECORD MAX AND RECORD MIN DIVIDED BY 2.
AVERAGE TEMPERATURE IS THE SUM OF THE MEAN DAILY MAX AND MIN TEMPERATURE DIVIDED BY 2.

EXCEPTIONS

RECORD MEANS ARE THROUGH THE CURRENT YEAR, BEGINNING IN 1890 FOR TEMPERATURE
1890 FOR PRECIPITATION
1947 FOR SNOWFALL

HEATING DEGREE DAYS Base 65 deg. F — TAMPA, FLORIDA

SEASON	JULY	AUG	SEP	OCT	NOV	DEC	JAN	FEB	MAR	APR	MAY	JUNE	TOTAL
1964-65	0	0	0	22	24	114	204	96	84	2	0	0	546
1965-66	0	0	0	5	37	160	267	172	93	18	0	0	752
1966-67	0	0	0	7	85	198	129	175	23	0	0	0	617
1967-68	0	0	0	0	69	95	193	306	157	0	0	0	820
1968-69	0	0	0	27	138	264	201	252	220	0	0	0	1102
1969-70	0	0	0	0	111	218	343	209	81	0	0	0	962
1970-71	0	0	0	0	145	115	201	134	139	36	0	0	770
1971-72	0	0	0	0	41	13	62	139	26	5	0	0	286
1972-73	0	0	0	0	65	130	166	223	18	21	0	0	623
1973-74	0	0	0	6	24	200	0	159	17	12	0	0	418
1974-75	0	0	0	0	39	138	84	64	61	5	0	0	391
1975-76	0	0	0	0	88	183	268	109	18	2	0	0	668
1976-77	0	0	0	11	122	208	422	214	28	6	0	0	1011
1977-78	0	0	0	18	53	222	320	323	99	4	0	0	1039
1978-79	0	0	0	0	2	75	245	190	53	0	0	0	565
1979-80	0	0	0	0	47	112	136	262	64	8	0	0	629
1980-81	0	0	0	1	65	233	447	127	103	2	0	0	978
1981-82	0	0	0	0	83	223	209	24	53	8	0	0	600
1982-83	0	0	0	12	18	95	218	148	103	20	0	0	614
1983-84	0	0	0	0	57	214	252	115	68	5	0	0	711
1984-85	0	0	0	0	87	61	306	119	17	5	0	0	595
1985-86	0	0	0	0	9	238	185	78	105	7	0	0	622
1986-87	0	0	0	0	0	53	202	88	42	64	0	0	449
1987-88	0	0	0	4	46	107	221	195	85	14	0	0	672
1988-89	0	0	0	0	9	127	41	116	45	7	0	0	345
1989-90	0	0	0	17	27	285	70	32	13	5	0	0	449
1990-91	0	0	0	7	11	70	72	84	46	0	0	0	290
1991-92	0	0	0	0	93	94	179	90	69	32	5	0	562
1992-93	0	0	0	0	57	83	58	137	84	24	0	0	443
1993-94	0	0	0	6	44	185							

COOLING DEGREE DAYS Base 65 deg. F — TAMPA, FLORIDA

YEAR	JAN	FEB	MAR	APR	MAY	JUNE	JULY	AUG	SEP	OCT	NOV	DEC	TOTAL
1969	8	2	14	224	353	521	568	498	471	377	55	10	3101
1970	12	10	85	255	352	464	557	563	514	353	58	40	3263
1971	54	86	58	174	342	490	541	526	463	396	128	155	3413
1972	131	20	90	204	365	487	528	536	497	352	168	141	3519
1973	75	15	188	158	374	510	574	529	510	348	205	55	3541
1974	196	55	204	197	413	460	506	562	540	250	130	48	3561
1975	93	115	151	282	521	536	572	585	512	407	191	39	4004
1976	17	63	199	175	348	424	525	517	440	202	63	45	3018
1977	2	9	218	210	364	567	559	565	526	258	139	36	3453
1978	18	0	79	232	431	529	565	557	500	319	208	121	3559
1979	28	36	73	283	344	482	592	543	515	322	164	55	3437
1980	45	22	164	165	386	506	598	564	493	284	115	7	3349
1981	0	32	43	230	331	501	552	525	414	303	71	49	3051
1982	56	114	156	208	299	499	537	537	467	311	197	182	3563
1983	36	24	57	137	369	487	540	541	439	342	91	64	3127
1984	42	52	104	190	410	468	517	546	454	337	92	135	3347
1985	30	88	163	237	464	569	547	566	475	445	275	58	3917
1986	15	85	123	139	391	510	565	551	526	374	348	107	3734
1987	27	43	91	114	405	538	567	583	497	207	169	91	3332
1988	30	32	110	188	326	489	554	562	517	271	191	74	3344
1989	112	120	202	224	425	528	575	564	529	344	151	18	3792
1990	107	154	164	225	487	537	549	592	541	406	176	139	4077
1991	131	68	158	361	509	498	543	572	515	326	126	88	3895
1992	25	55	70	170	301	519	589	544	518	248	212	72	3323
1993	126	12	72	95	352	511	593	587	513	347	174	21	3403

SNOWFALL (inches) — TAMPA, FLORIDA

SEASON	JULY	AUG	SEP	OCT	NOV	DEC	JAN	FEB	MAR	APR	MAY	JUNE	TOTAL
1970-71	0.0	0.0	0.0	0.0	0.0	0.0	0.0	0.0	0.0	0.0	0.0	0.0	0.0
1971-72	0.0	0.0	0.0	0.0	0.0	0.0	0.0	0.0	0.0	0.0	0.0	0.0	0.0
1972-73	0.0	0.0	0.0	0.0	0.0	0.0	0.0	0.0	0.0	0.0	0.0	0.0	0.0
1973-74	0.0	0.0	0.0	0.0	0.0	0.0	0.0	0.0	0.0	0.0	0.0	0.0	0.0
1974-75	0.0	0.0	0.0	0.0	0.0	0.0	0.0	0.0	0.0	0.0	0.0	0.0	0.0
1975-76	0.0	0.0	0.0	0.0	0.0	0.0	0.0	0.0	0.0	0.0	0.0	0.0	0.0
1976-77	0.0	0.0	0.0	0.0	0.0	0.0	0.2	0.0	0.0	0.0	0.0	0.0	0.2
1977-78	0.0	0.0	0.0	0.0	0.0	0.0	0.0	0.0	0.0	0.0	0.0	0.0	0.0
1978-79	0.0	0.0	0.0	0.0	0.0	0.0	0.0	0.0	0.0	0.0	0.0	0.0	0.0
1979-80	0.0	0.0	0.0	0.0	0.0	0.0	0.0	0.0	T	0.0	0.0	0.0	T
1980-81	0.0	0.0	0.0	0.0	0.0	0.0	0.0	0.0	0.0	0.0	0.0	0.0	0.0
1981-82	0.0	0.0	0.0	0.0	0.0	0.0	0.0	0.0	0.0	0.0	0.0	0.0	0.0
1982-83	0.0	0.0	0.0	0.0	0.0	0.0	0.0	0.0	0.0	0.0	0.0	0.0	0.0
1983-84	0.0	0.0	0.0	0.0	0.0	0.0	0.0	0.0	0.0	0.0	0.0	0.0	0.0
1984-85	0.0	0.0	0.0	0.0	0.0	0.0	0.0	0.0	0.0	0.0	0.0	0.0	0.0
1985-86	0.0	0.0	0.0	0.0	0.0	0.0	0.0	0.0	0.0	0.0	0.0	0.0	0.0
1986-87	0.0	0.0	0.0	0.0	0.0	0.0	0.0	0.0	0.0	0.0	0.0	0.0	0.0
1987-88	0.0	0.0	0.0	0.0	0.0	0.0	0.0	0.0	0.0	0.0	0.0	0.0	0.0
1988-89	0.0	0.0	0.0	0.0	0.0	0.0	0.0	0.0	0.0	0.0	0.0	0.0	0.0
1989-90	0.0	0.0	0.0	0.0	0.0	T	0.0	0.0	0.0	0.0	0.0	0.0	T
1990-91	0.0	0.0	0.0	0.0	0.0	0.0	0.0	0.0	0.0	0.0	0.0	0.0	0.0
1991-92	0.0	0.0	0.0	0.0	0.0	0.0	0.0	0.0	0.0	0.0	0.0	0.0	0.0
1992-93	0.0	0.0	0.0	0.0	0.0	0.0	0.0	0.0	0.0	0.0	0.0	0.0	0.0
1993-94	0.0	0.0	0.0	0.0	0.0	0.0							
Record Mean	0.0	0.0	0.0	0.0	0.0	T	T	T	T	0.0	0.0	0.0	T

VERO BEACH, FLORIDA

Vero Beach is located on the southeast coast of Florida, separated from the Atlantic Ocean by the Inland Waterway and a narrow island offshore. Its climate is strongly influenced by this maritime location. Temperatures in summer rarely reach 100 degrees. The average maximum temperature in July and August is about 90 degrees. In winter the average minimum temperature is slightly above 50 degrees with record lows near 20 degrees. On average, only one day a year experiences freezing temperatures, usually in January.

Rainfall occurs in all seasons but most abundantly in summer when showers are common. Thunderstorms are present approximately 70 to 80 days a year. Monthly precipitation amounts in winter are about half those in summer, and are due in part to cold frontal systems traversing the region. Throughout the year, relative humidity at 7 A.M. tends to range from 80 to 90 percent. The 1 P.M. humidity ranges from 60 to 70 percent with lower values occurring in midafternoon when temperatures are the highest.

Vero Beach lies at the northern boundary of a tropical rainy region. Within that region during summer and fall there may be hurricane activity. Of those hurricanes that pass close to Vero Beach, many move northward offshore, some cross the peninsula of Florida moving generally eastward but being weakened by their passage over land, and some enter the coastal area from the Atlantic Ocean. The frequency of the latter group has been small, about 5 in 114 years.

NORMALS, MEANS, AND EXTREMES

VERO BEACH, FLORIDA

LATITUDE: 27°39'N LONGITUDE: 80°25' W ELEVATION: FT. GRND 24 BARO 39 TIME ZONE: EASTERN WBAN: 12843

	(a)	JAN	FEB	MAR	APR	MAY	JUNE	JULY	AUG	SEP	OCT	NOV	DEC	YEAR
TEMPERATURE °F:														
Normals														
-Daily Maximum		72.5	73.5	77.5	81.5	85.2	88.2	90.1	90.0	88.3	84.0	78.7	74.1	82.0
-Daily Minimum		50.7	51.9	56.9	61.0	66.2	70.8	72.1	72.5	71.9	66.9	60.0	53.3	62.9
-Monthly		61.6	62.7	67.2	71.3	75.8	79.5	81.1	81.3	80.1	75.5	69.3	63.7	72.4
Extremes														
-Record Highest	10	88	88	90	94	95	98	98	98	95	94	92	87	98
-Year		1991	1986	1991	1986	1990	1986	1993	1989	1990	1989	1992	1991	JUL 1993
-Record Lowest	10	21	28	32	45	48	57	67	64	65	46	38	23	21
-Year		1985	1989	1986	1988	1992	1984	1989	1984	1990	1989	1993	1989	JAN 1985
NORMAL DEGREE DAYS:														
Heating (base 65°F)		186	138	56	7	0	0	0	0	0	0	34	127	548
Cooling (base 65°F)		81	74	124	196	335	435	499	505	453	326	163	87	3278
% OF POSSIBLE SUNSHINE														
MEAN SKY COVER (tenths)														
Sunrise - Sunset														
MEAN NUMBER OF DAYS:														
Sunrise to Sunset														
-Clear														
-Partly Cloudy														
-Cloudy														
Precipitation														
.01 inches or more	10	9.0	7.2	8.0	6.5	9.0	13.2	13.2	12.7	15.1	12.4	10.1	6.9	123.3
Snow,Ice pellets,hail														
1.0 inches or more	10	0.0	0.0	0.0	0.0	0.0	0.0	0.0	0.0	0.0	0.0	0.0	0.0	0.0
Thunderstorms	10	1.5	1.8	3.8	3.9	5.0	12.8	15.5	14.8	10.1	4.3	1.4	0.2	75.1
Heavy Fog Visibility														
1/4 mile or less	10	3.3	2.4	1.7	1.8	0.7	0.5	0.1	0.0	0.4	1.1	1.1	1.9	15.0
Temperature °F														
-Maximum														
90° and above	10	0.0	0.0	0.1	1.1	2.6	10.2	17.5	20.0	10.7	2.1	0.2	0.0	64.5
32° and below	10	0.0	0.0	0.0	0.0	0.0	0.0	0.0	0.0	0.0	0.0	0.0	0.0	0.0
-Minimum														
32° and below	10	0.5	0.5	0.2	0.0	0.0	0.0	0.0	0.0	0.0	0.0	0.0	0.4	1.6
0° and below	10	0.0	0.0	0.0	0.0	0.0	0.0	0.0	0.0	0.0	0.0	0.0	0.0	0.0
AVG. STATION PRESS.(mb)	10	1018.8	1017.9	1017.0	1015.3	1015.2	1015.5	1017.3	1016.0	1015.0	1015.3	1017.1	1019.4	1016.7
RELATIVE HUMIDITY (%)														
Hour 01	10	87	84	82	80	81	86	89	89	87	83	85	84	85
Hour 07 (Local Time)	10	88	87	86	83	83	86	89	92	90	87	87	87	87
Hour 13	10	60	56	56	53	57	64	63	64	64	62	63	59	60
Hour 19	10	77	72	70	66	70	76	77	78	78	78	79	79	75
PRECIPITATION (inches):														
Water Equivalent														
-Normal		2.14	2.91	3.09	1.90	4.36	6.46	6.09	6.10	7.15	5.52	3.27	2.17	51.16
-Maximum Monthly	10	8.91	5.76	12.78	6.82	5.75	20.73	11.11	11.49	9.70	12.37	11.76	4.33	20.73
-Year		1993	1991	1993	1991	1984	1992	1986	1991	1985	1986	1984	1986	JUN 1992
-Minimum Monthly	10	1.19	0.32	0.90	0.03	0.98	2.08	1.78	3.35	1.67	1.63	0.34	0.21	0.03
-Year		1985	1985	1990	1986	1992	1993	1992	1993	1988	1988	1989	1987	APR 1986
-Maximum in 24 hrs	10	3.38	3.44	6.80	3.43	3.26	3.11	3.45	3.78	4.39	4.22	4.33	1.51	6.80
-Year		1991	1991	1993	1991	1993	1992	1988	1991	1990	1986	1984	1985	MAR 1993
Snow,Ice pellets,hail														
-Maximum Monthly	10	0.0	0.0	0.0	0.0	0.0	T	T	0.0	0.0	0.0	0.0	0.0	T
-Year							1990	1992						JUL 1992
-Maximum in 24 hrs	10	0.0	0.0	0.0	0.0	0.0	T	T	0.0	0.0	0.0	0.0	0.0	T
-Year		1984	1984	1984	1984	1984	1990	1992	1984	1984	1984	1984	1984	JUL 1992
WIND:														
Mean Speed (mph)	10	8.6	9.1	9.7	9.4	9.2	7.9	7.0	6.5	7.5	8.7	9.2	8.0	8.4
Prevailing Direction through														
Fastest Obs. 1 Min.														
-Direction (!!)														
-Speed (MPH)														
-Year														
Peak Gust														
-Direction (!!)														
-Speed (mph)														
-Date														

PRECIPITATION (inches) VERO BEACH, FLORIDA

YEAR	JAN	FEB	MAR	APR	MAY	JUNE	JULY	AUG	SEP	OCT	NOV	DEC	ANNUAL
1984	1.59	4.79	2.43	0.83	5.75	4.58	3.27	3.70	9.52	1.66	11.76	1.74	51.62
1985	1.19	0.32	1.31	4.69	3.23	5.42	6.06	6.64	9.70	3.56	2.23	2.19	46.54
1986	2.94	1.77	3.76	0.03	1.77	4.47	11.11	4.43	5.30	12.37	4.42	4.33	56.70
1987	1.65	1.31	4.86	0.05	3.07	3.95	7.34	3.57	5.37	6.71	6.28	0.21	44.37
1988	3.35	2.09	5.74	1.77	4.02	4.93	10.71	3.39	1.67	1.63	2.07	3.28	44.65
1989	1.83	1.69	3.88	2.86	1.96	3.40	2.94	5.02	5.45	5.99	0.34	2.45	37.81
1990	1.77	2.51	0.90	1.92	4.14	4.79	5.90	9.71	8.76	3.66	1.60	0.55	46.21
1991	7.05	5.76	7.18	6.82	4.00	5.38	8.90	11.49	3.09	5.21	1.02	0.76	66.66
1992	1.72	3.22	1.07	2.45	0.98	20.73	1.78	5.36	8.05	1.73	5.98	1.78	54.85
1993	8.91	1.99	12.78	1.42	4.85	2.08	4.53	3.35	5.59	5.77	4.56	1.07	56.90
Record Mean	3.20	2.54	4.39	2.28	3.38	5.97	6.25	5.67	6.25	4.83	4.03	1.84	50.63

AVERAGE TEMPERATURE (deg. F) VERO BEACH, FLORIDA

YEAR	JAN	FEB	MAR	APR	MAY	JUNE	JULY	AUG	SEP	OCT	NOV	DEC	ANNUAL
1984	60.9	63.0	65.9	68.5	75.4	77.7	80.2	80.2	78.6	75.7	68.1	66.9	71.8
1985	57.0	63.9	68.3	70.3	74.8	81.5	80.6	81.8	79.3	80.1	73.3	61.6	72.7
1986	61.3	66.4	65.6	68.0	74.9	79.9	81.3	81.5	80.6	77.2	76.8	68.9	73.5
1987	60.3	64.5	66.7		75.7	81.4	82.3	82.6	81.7	74.0	71.1	66.1	72.8
1988	62.1	60.7	65.5	71.4	74.8	80.2	81.5	81.7	82.1	75.0	72.5	64.6	72.7
1989	67.3	65.6	69.9	72.3	76.6	80.1	81.7	82.3	81.8	76.2	70.4	58.1	73.5
1990	68.1	70.3	70.2	72.8	78.9	80.5	81.8	82.2	81.4	78.4	71.1	68.2	75.3
1991	68.2	65.3	68.5	75.4	79.3	81.1	82.4	82.6	81.2	77.0	68.8	67.6	74.8
1992	61.4	66.1	66.9	70.8	74.3	80.6	82.7	81.8	81.2	75.6	73.0	66.1	73.4
1993	68.5	62.6	65.9	68.5	75.5	80.0	82.7	82.1	81.1	77.1	71.3	61.0	73.0
Record Mean	63.5	64.8	67.3	70.8	76.0	80.3	81.7	81.9	80.9	76.6	71.7	64.9	73.4
Max	73.6	74.9	76.9	80.5	84.6	88.5	90.1	90.4	88.7	84.6	80.0	74.9	82.3
Min	53.4	54.7	57.7	61.2	67.4	72.0	73.3	73.3	73.1	68.6	63.3	54.9	64.4

REFERENCE NOTES VERO BEACH, FLORIDA

GENERAL
T - TRACE AMOUNT.
BLANK ENTRIES DENOTE MISSING/UNREPORTED DATA.
# INDICATES A STATION OR INSTRUMENT RELOCATION.

SPECIFIC

(a) - LENGTH OF RECORD IN YEARS, ALTHOUGH INDIVIDUAL MONTHS MAY BE MISSING.
0.* OR * - THE VALUE IS BETWEEN 0.0 AND 0.05
NORMALS - BASED ON THE 1961-1990 RECORD PERIOD.
EXTREMES - DATES ARE THE MOST RECENT OCCURRENCE.
WIND DIR.- NUMERALS SHOW TENS OF DEGRESS CLOCKWISE FROM TRUE NORTH. "00" INDICATES CALM.
RESULTANT DIRECTIONS ARE GIVEN TO WHOLE DEGREES.
BOLD VALUES INDICATE EXTREME VALUES WHICH OCCURRED AFTER THE ASOS SYSTEM WAS COMMISSIONED.

RECORD = PERIOD OF RECORD
RECORD MEAN PRECIPITATION IS THE MEAN OF ALL DAILY PRECIPITATION AMOUNTS DURING THE PERIOD OF RECORD.
RECORD MAX(MIN) TEMPERATURE IS THE MEAN OF ALL DAILY MAX(MIN) TEMPERATURES DURING THE PERIOD OF RECORD.
RECORD MEAN TEMPERATURE IS THE SUM OF THE RECORD MAX AND RECORD MIN DIVIDED BY 2.
AVERAGE TEMPERATURE IS THE SUM OF THE MEAN DAILY MAX AND MIN TEMPERATURE DIVIDED BY 2.

EXCEPTIONS

RECORD MEANS ARE THROUGH THE CURRENT YEAR, BEGINNING IN 1984 FOR TEMPERATURE
1984 FOR PRECIPITATION
1984 FOR SNOWFALL

HEATING DEGREE DAYS Base 65 deg. F VERO BEACH, FLORIDA

SEASON	JULY	AUG	SEP	OCT	NOV	DEC	JAN	FEB	MAR	APR	MAY	JUNE	TOTAL
1983-84							167	105	83	20	0	0	
1984-85	0	0	0	0	34	56	278	132	20	17	0	0	537
1985-86	0	0	0	0	14	175	156	62	109	16	0	0	532
1986-87	0	0	0	0	0	17	191	77	49		0	0	
1987-88	0	0	0	0	28	77	147	159	87	9	0	0	507
1988-89	0	0	0	0	5	90	22	105	41	3	0	0	266
1989-90	0	0	0	15	13	233	37	19	8	2	0	0	327
1990-91	0	0	0	4	5	44	52	76	43	0	0	0	224
1991-92	0	0	0	0	54	39	153	62	55	22	1	0	386
1992-93	0	0	0	0	26	55	37	95	65	14	0	0	292
1993-94	0	0	0	4	28	146							

COOLING DEGREE DAYS Base 65 deg. F VERO BEACH, FLORIDA

YEAR	JAN	FEB	MAR	APR	MAY	JUNE	JULY	AUG	SEP	OCT	NOV	DEC	TOTAL
1984	49	54	119	134	328	389	475	476	415	339	134	121	3033
1985	36	108	128	184	311	500	489	527	434	477	268	77	3539
1986	47	105	137	111	315	454	514	521	474	384	362	143	3567
1987	54	70	109		340	495	542	553	507	286	217	120	3413
1988	62	43	113	207	311	464	517	524	522	318	238	84	3403
1989	101	127	198	228	366	461	523	542	509	369	183	26	3633
1990	138	172	174	243	439	474	527	540	500	429	193	150	3979
1991	156	89	159	316	449	490	547	553	495	377	174	125	3930
1992	48	101	120	198	299	474	556	523	492	335	274	96	3516
1993	153	33	100	127	332	457	554	538	491	385	224	29	3423

SNOWFALL (inches) VERO BEACH, FLORIDA

SEASON	JULY	AUG	SEP	OCT	NOV	DEC	JAN	FEB	MAR	APR	MAY	JUNE	TOTAL
1983-84							0.0	0.0	0.0	0.0	0.0	0.0	
1984-85	0.0	0.0	0.0	0.0	0.0	0.0	0.0	0.0	0.0	0.0	0.0	0.0	0.0
1985-86	0.0	0.0	0.0	0.0	0.0	0.0	0.0	0.0	0.0	0.0	0.0	0.0	0.0
1986-87	0.0	0.0	0.0	0.0	0.0	0.0	0.0	0.0	0.0	0.0	0.0	0.0	0.0
1987-88	0.0	0.0	0.0	0.0	0.0	0.0	0.0	0.0	0.0	0.0	0.0	0.0	0.0
1988-89	0.0	0.0	0.0	0.0	0.0	0.0	0.0	0.0	0.0	0.0	0.0	0.0	0.0
1989-90	0.0	0.0	0.0	0.0	0.0	0.0	0.0	0.0	0.0	0.0	0.0	T	T
1990-91	0.0	0.0	0.0	0.0	0.0	0.0	0.0	0.0	0.0	0.0	0.0	0.0	0.0
1991-92	0.0	0.0	0.0	0.0	0.0	0.0	0.0	0.0	0.0	0.0	0.0	0.0	0.0
1992-93	T	0.0	0.0	0.0	0.0	0.0	0.0	0.0	0.0	0.0	0.0	0.0	T
1993-94	0.0	0.0	0.0	0.0	0.0	0.0							
Record Mean	T	0.0	0.0	0.0	0.0	0.0	0.0	0.0	0.0	0.0	0.0	T	T

WEST PALM BEACH, FLORIDA

West Palm Beach and Palm Beach, both located on the coastal sand ridge of southeastern Florida, are separated by Lake Worth, a portion of the Inland Waterway. The entire coastal ridge is only about 5 miles wide and in early times the Everglades reached to its western edge. Now most of the swampland has been drained and is devoted to agriculture, the peat-like muck soil being very fertile when fortified with certain lacking minerals. The Atlantic Ocean forms the eastern edge of Palm Beach, and the Gulf Stream flows northward about 2 miles offshore, its nearest approach to the Florida coast.

Because of its southerly location and marine influences, the Palm Beach area has a notably equable climate. Cold continental air must either travel over water or flow down the Florida Peninsula to reach the area, and in either case its cold is appreciably modified. Actually, the coldest weather, with infrequent frosts, is experienced the second or third night after the arrival of the cold air, due to the loss of heat through radiation cooling. The frequency of temperatures as low as the freezing mark is about one per three years at the National Weather Service Office, but in the farmlands farther from the coast the frequency of light freezes is much higher.

Summer temperatures are tempered by the ocean breeze, and by the frequent formation of cumulus clouds, which shade the land somewhat without completely obscuring the sun. Temperatures of 89 degrees or higher have occurred in all months of the year, but the 100 degree mark has rarely occurred. August is the warmest month and has an average maximum temperature of about 90 degrees. The occurrence of 90 degree temperatures in August is so common that such can be expected on more than two-thirds of the days. However, temperatures as high as 100 degrees rarely occur.

The moist, unstable air in this area results in frequent showers, usually of short duration. Thunderstorms are frequent during the summer, occurring every other day. Rainfall is heaviest during the summer and fall, the fall rainfall occurring from occasional heavy rains accompanying tropical disturbances. High winds, associated with hurricanes, have been estimated at about 140 mph in the city.

Flying weather is usually very good in this area, with instrument weather occurring only rarely. Heavy fog occurs on an average of only one morning a month in the winter and spring, and almost never in the summer and fall.

NORMALS, MEANS, AND EXTREMES

WEST PALM BEACH, FLORIDA

LATITUDE: 26°41'N LONGITUDE: 80°07' W ELEVATION: FT. GRND 15 BARO 24 TIME ZONE: EASTERN WBAN: 12844

	(a)	JAN	FEB	MAR	APR	MAY	JUNE	JULY	AUG	SEP	OCT	NOV	DEC	YEAR
TEMPERATURE °F:														
Normals														
-Daily Maximum		74.5	75.9	78.8	82.0	85.6	88.1	89.9	90.0	88.6	84.9	80.0	76.0	82.9
-Daily Minimum		55.7	56.5	61.2	64.7	69.7	73.1	74.5	75.0	74.6	70.7	64.5	58.7	66.6
-Monthly		65.1	66.2	70.0	73.4	77.6	80.6	82.2	82.5	81.6	77.8	72.3	67.4	74.7
Extremes														
-Record Highest	57	89	90	94	99	96	98	101	98	97	95	**91**	90	101
-Year		1942	1949	1977	1971	1971	1980	1942	1963	1937	1989	**1992**	1941	JUL 1942
-Record Lowest	57	27	32	30	43	**51**	61	66	65	66	46	36	28	27
-Year		1977	1989	1980	1987	**1992**	1984	1937	1957	1938	1968	1950	1989	JAN 1977
NORMAL DEGREE DAYS:														
Heating (base 65°F)		122	85	27	0	0	0	0	0	0	0	15	74	323
Cooling (base 65°F)		125	119	182	252	391	468	533	543	498	397	234	149	3891
% OF POSSIBLE SUNSHINE														
MEAN SKY COVER (tenths)														
Sunrise - Sunset	45	5.8	5.6	5.6	5.3	5.8	6.7	6.6	6.5	6.8	6.0	5.8	5.6	6.0
MEAN NUMBER OF DAYS:														
Sunrise to Sunset														
-Clear	48	8.0	7.9	8.1	8.8	7.5	4.0	3.7	2.9	2.7	6.4	7.0	8.5	75.4
-Partly Cloudy	48	11.5	10.8	12.9	13.0	13.6	13.4	14.1	16.3	14.2	14.0	13.0	12.1	158.8
-Cloudy	48	11.5	9.5	10.0	8.2	9.9	12.7	13.2	11.8	13.1	10.7	10.0	10.3	131.0
Precipitation														
.01 inches or more	51	7.8	7.1	8.0	6.7	11.0	14.1	14.8	16.2	17.1	13.0	9.2	7.9	133.1
Snow,Ice pellets,hail														
1.0 inches or more	51	0.0	0.0	0.0	0.0	0.0	0.0	0.0	0.0	0.0	0.0	0.0	0.0	0.0
Thunderstorms	51	0.9	1.2	2.4	3.5	7.5	12.8	16.1	16.0	11.2	4.4	1.5	0.9	78.4
Heavy Fog Visibility														
1/4 mile or less	51	1.7	1.1	0.9	0.9	0.2	0.2	0.*	0.1	0.2	0.4	0.6	1.1	7.4
Temperature °F														
-Maximum														
90° and above	29	0.0	0.0	0.3	1.7	3.6	9.8	18.0	19.3	10.7	2.2	0.1	0.0	65.7
32° and below	29	0.0	0.0	0.0	0.0	0.0	0.0	0.0	0.0	0.0	0.0	0.0	0.0	0.0
-Minimum														
32° and below	29	0.4	0.1	0.1	0.0	0.0	0.0	0.0	0.0	0.0	0.0	0.0	0.1	0.7
0° and below	29	0.0	0.0	0.0	0.0	0.0	0.0	0.0	0.0	0.0	0.0	0.0	0.0	0.0
AVG. STATION PRESS. (mb)	21	1018.9	1018.3	1017.1	1016.4	1015.3	1016.0	1017.5	1016.6	1015.0	1015.4	1017.2	1019.0	1016.9
RELATIVE HUMIDITY (%)														
Hour 01	29	81	80	78	76	79	84	85	84	84	81	81	80	81
Hour 07 (Local Time)	29	83	82	81	79	79	83	85	85	87	84	83	83	83
Hour 13	29	59	56	56	54	59	65	64	64	66	62	61	59	60
Hour 19	29	73	69	68	65	70	76	75	75	78	74	74	73	73
PRECIPITATION (inches):														
Water Equivalent														
-Normal		2.80	2.69	3.66	2.91	6.13	8.09	6.14	6.02	8.53	6.60	4.69	2.49	60.75
-Maximum Monthly	55	11.01	8.71	16.78	18.26	15.22	17.91	17.74	13.52	24.86	18.74	14.63	10.10	24.86
-Year		1983	1983	1982	1942	1976	1966	1941	1950	1960	1965	1982	1986	SEP 1960
-Minimum Monthly	55	0.22	0.29	0.33	0.04	0.39	1.07	1.22	1.73	1.77	1.20	0.23	0.06	0.04
-Year		1960	1948	1956	1967	1967	1952	1961	1987	1988	1972	1970	1968	APR 1967
-Maximum in 24 hrs	55	8.45	4.70	8.80	15.23	7.04	9.21	5.83	6.72	8.71	9.58	7.67	5.26	15.23
-Year		1991	1966	1982	1942	1958	1945	1972	1988	1960	1965	1984	1955	APR 1942
Snow,Ice pellets,hail														
-Maximum Monthly	48	T	0.0	0.0	**0.0**	**0.0**	**0.0**	**0.0**	**T**	**0.0**	**0.0**	**0.0**	**0.0**	**T**
-Year		1977							**1993**					**AUG 1993**
-Maximum in 24 hrs	48	T	0.0	0.0	**0.0**	**0.0**	**0.0**	**0.0**	**T**	**0.0**	**0.0**	**0.0**	**0.0**	**T**
-Year		1977							**1993**					**AUG 1993**
WIND:														
Mean Speed (mph)	51	10.1	10.6	11.0	10.9	10.0	8.4	7.7	7.8	8.7	10.2	10.4	10.0	9.7
Prevailing Direction through 1963		NW	SE	SE	E	ESE	ESE	ESE	ESE	ENE	ENE	ENE	NNW	ESE
Fastest Obs. 1 Min.														
-Direction (!!!)	44	29	29	27	32	27	09	34	13	36	16	34	07	13
-Speed (MPH)	44	48	46	51	55	45	71	46	86	58	74	35	36	86
-Year		1955	1956	1957	1958	1954	1957	1974	1964	1979	1964	1959	1958	AUG 1964
Peak Gust														
-Direction (!!!)	10	NW	W	SW	**22**	S	**SW**	SW	NW	NE	**03**	S	SW	SW
-Speed (mph)	10	62	52	74	**63**	48	**66**	59	66	45	**52**	47	48	74
-Date		1987	1991	1993	**1993**	1988	**1992**	1991	1990	1984	**1993**	1988	1989	MAR 1993

PRECIPITATION (inches) — WEST PALM BEACH, FLORIDA

YEAR	JAN	FEB	MAR	APR	MAY	JUNE	JULY	AUG	SEP	OCT	NOV	DEC	ANNUAL
1964	0.98	4.36	1.82	7.45	5.74	10.23	5.29	7.33	11.72	15.23	5.76	3.39	79.30
1965	1.70	4.14	1.15	0.51	1.38	8.80	11.27	2.41	3.84	18.74	2.57	1.75	58.26
1966	6.23	6.88	2.38	2.73	3.75	17.91	8.19	9.76	10.00	8.90	2.00	1.02	79.75
1967	3.18	3.01	2.49	0.04	0.39	11.39	4.88	6.65	6.91	10.00	1.37	1.23	51.54
1968	0.56	4.44	1.48	0.72	13.82	16.78	4.66	8.14	14.01	10.24	2.51	0.06	77.42
1969	3.59	1.67	5.53	0.73	8.32	12.25	7.43	7.75	14.56	13.47	3.29	1.16	79.75
1970	4.57	2.98	11.95	1.00	5.83	8.23	3.17	5.47	7.61	4.13	0.23	0.11	55.28
1971	0.82	1.08	0.55	1.27	5.04	9.03	5.13	4.38	6.24	5.13	10.77	1.87	51.31
1972	2.47	2.55	3.99	12.62	11.19	12.34	13.25	2.76	3.72	1.20	6.98	2.08	75.15
1973	2.62	2.71	2.28	1.45	4.13	7.53	7.74	7.51	7.59	6.97	2.06	2.15	54.74
1974	8.30	0.42	2.44	1.18	2.93	4.87	11.00	5.85	5.62	9.30	3.99	2.56	58.46
1975	0.47	0.68	0.85	0.57	6.23	7.18	5.67	2.49	11.58	5.60	1.81	1.27	44.40
1976	1.09	4.59	1.84	1.26	15.22	4.52	1.67	7.29	7.74	3.75	3.29	3.06	55.32
1977	3.49	1.11	0.53	1.65	14.80	4.03	2.19	8.03	13.21	2.01	5.84	7.37	64.26
1978	3.40	2.49	2.73	0.88	4.55	10.36	5.91	2.27	6.36	6.98	10.98	5.30	62.21
1979	3.06	0.80	1.06	8.71	5.09	6.34	1.89	3.08	19.63	4.78	4.08	2.66	61.18
1980	4.11	3.85	2.53	3.90	8.06	4.26	7.61	5.02	6.76	4.02	5.28	1.27	56.67
1981	0.43	4.22	2.49	0.43	5.12	4.58	3.72	10.33	9.30	2.10	4.52	2.50	49.74
1982	1.29	2.31	16.78	7.67	6.90	10.37	2.57	5.72	6.03	4.83	14.63	1.52	80.62
1983	11.01	8.71	3.42	3.27	5.30	9.39	4.71	8.21	9.62	7.80	4.24	7.03	82.71
1984	1.36	4.49	5.06	5.99	8.44	5.84	7.34	6.39	8.69	1.67	14.39	0.13	69.79
1985	0.84	0.53	3.49	3.35	4.63	4.84	10.77	5.38	9.38	2.89	0.54	1.35	47.99
1986	6.34	1.58	5.48	0.33	1.59	9.82	10.20	4.88	4.08	8.11	6.80	10.10	69.31
1987	1.17	1.20	7.75	2.61	5.63	6.60	4.42	1.73	11.90	4.78	9.43	1.47	58.69
1988	3.38	3.04	5.51	3.65	6.94	10.95	9.01	11.38	1.77	3.54	3.86	1.88	64.91
1989	0.97	1.00	2.58	6.04	0.78	4.96	3.52	5.83	4.05	5.84	1.15	1.94	38.66
1990	1.19	1.39	1.91	2.82	6.74	6.67	10.17	6.60	11.70	3.45	1.23	1.94	55.81
1991	10.43	4.36	3.36	9.54	8.94	7.45	7.16	7.34	6.44	7.58	2.32	4.44	79.36
1992	1.69	3.26	2.61	2.64	1.59	16.08	2.90	6.12	6.99	2.82	12.56	1.85	61.11
1993	8.05	2.06	11.06	2.53	2.22	2.89	5.41	4.78	5.35	11.99	5.78	0.82	62.94
Record Mean	2.89	2.53	3.54	3.57	5.57	7.85	6.36	6.52	9.01	7.13	3.88	2.57	61.43

AVERAGE TEMPERATURE (deg. F) — WEST PALM BEACH, FLORIDA

YEAR	JAN	FEB	MAR	APR	MAY	JUNE	JULY	AUG	SEP	OCT	NOV	DEC	ANNUAL
#1964	65.5	62.8	72.4	76.2	76.4	80.4	82.4	82.6	81.6	76.5	72.7	70.4	75.0
1965	65.5	69.1	71.8	75.5	77.0	79.6	80.9	82.2	82.1	77.2	73.5	68.1	75.2
1966	64.4	67.0	68.1	73.0	78.5	79.5	81.5	81.7	81.3	77.5	69.6	64.8	73.9
1967	68.4	65.5	71.5	73.7	78.1	78.8	81.6	80.9	80.3	76.5	70.5	68.9	74.6
1968	64.8	60.1	64.9	74.7	76.9	78.4	80.4	81.9	80.4	75.3	67.7	63.7	72.4
1969	65.1	61.7	64.7	74.8	77.0	80.4	82.8	81.4	80.5	78.8	67.7	62.7	73.2
1970	60.4	62.1	69.1	75.7	77.8	80.0	82.5	83.5	81.2	78.4	66.5	68.3	73.8
1971	65.0	67.4	65.9	71.8	78.8	80.9	82.3	83.1	81.1	79.0	73.1	73.1	75.1
1972	71.8	67.0	70.9	74.1	78.8	81.8	82.5	83.2	82.6	80.4	74.3	72.0	76.6
1973	68.7	64.0	73.6	74.5	79.2	81.7	82.1	81.7	82.0	77.5	75.0	66.4	75.5
1974	73.2	66.7	73.6	74.3	78.6	80.0	81.8	82.7	82.3	77.0	72.0	66.8	75.7
1975	71.2	72.1	71.8	73.9	77.8	79.5	79.8	81.4	80.5	77.6	71.3	66.9	75.3
1976	62.2	66.3	72.6	72.7	77.4	78.1	81.1	81.2	80.0	75.6	70.6	66.0	73.7
1977	58.5	63.6	73.7	75.0	76.9	81.2	81.9	82.5	81.0	75.3	72.6	67.3	74.1
1978	62.1	60.4	68.4	73.2	78.9	81.6	82.1	82.2	80.7	77.0	73.7	71.3	74.3
1979	62.7	63.7	68.0	74.4	76.0	80.2	82.7	82.5	81.3	77.5	73.2	67.1	74.1
1980	64.1	60.9	70.2	71.8	77.5	82.5	83.5	82.5	81.3	78.6	72.4	65.5	74.2
1981	58.7	67.6	67.9	75.7	78.5	83.4	84.1	82.8	81.1	78.1	67.9	65.2	74.3
1982	65.7	73.6	73.0	75.9	76.8	80.6	83.2	83.0	81.8	78.5	75.6	70.8	76.5
1983	64.8	66.1	66.5	72.2	77.6	80.6	84.0	84.3	81.6	76.9	70.0	68.3	74.4
1984	64.6	66.5	68.7	71.2	77.4	78.7	80.9	81.3	79.6	77.4	71.4	71.1	74.1
1985	60.7	67.5	71.3	73.1	77.3	82.3	81.0	82.9	80.3	80.4	75.7	65.0	74.8
1986	64.2	69.2	68.1	71.6	77.6	80.8	82.1	82.9	82.5	79.4	78.4	72.0	75.7
1987	64.5	68.7	70.3	69.0	77.8	82.3	83.2	84.9	82.7	76.5	73.8	68.4	75.2
1988	66.4	65.7	69.1	74.6	76.7	81.0	82.2	82.5	83.1	77.9	75.0	68.4	75.2
1989	71.3	68.5	72.2	74.8	79.5	81.9	83.0	83.6	83.1	77.6	74.3	62.5	76.0
1990	72.3	73.8	73.2	73.9	79.7	82.2	83.3	83.7	82.7	80.1	73.5	71.3	77.5
1991	71.0	68.2	72.6	77.0	80.2	81.9	83.2	83.6	82.0	78.8	71.5	70.2	76.7
1992	65.3	68.8	69.7	72.5	75.9	80.6	83.4	82.1	81.7	77.8	75.5	69.5	75.2
1993	71.5	66.3	69.0	71.1	77.7	81.6	82.9	83.0	82.3	79.0	74.4	65.6	75.4
Record Mean	65.9	66.9	70.2	73.8	77.8	80.8	82.5	82.8	81.6	77.8	72.3	67.8	75.0
Max	75.1	76.3	79.2	82.4	85.9	88.7	90.4	90.6	88.8	84.9	80.1	76.5	83.2
Min	56.8	57.4	61.2	65.2	69.6	72.9	74.5	74.9	74.4	70.6	64.5	59.1	66.8

REFERENCE NOTES — WEST PALM BEACH, FLORIDA

GENERAL
T - TRACE AMOUNT.
BLANK ENTRIES DENOTE MISSING/UNREPORTED DATA.
# INDICATES A STATION OR INSTRUMENT RELOCATION.

SPECIFIC

EXCEPTIONS

RECORD MEANS ARE THROUGH THE CURRENT YEAR, BEGINNING IN 1939 FOR TEMPERATURE
1939 FOR PRECIPITATION
1939 FOR SNOWFALL

(a) - LENGTH OF RECORD IN YEARS, ALTHOUGH INDIVIDUAL MONTHS MAY BE MISSING.
0.* OR * - THE VALUE IS BETWEEN 0.0 AND 0.05
NORMALS - BASED ON THE 1961-1990 RECORD PERIOD.
EXTREMES - DATES ARE THE MOST RECENT OCCURRENCE
WIND DIR.- NUMERALS SHOW TENS OF DEGRESS CLOCKWISE FROM TRUE NORTH. "00" INDICATES CALM.
RESULTANT DIRECTIONS ARE GIVEN TO WHOLE DEGREES.
BOLD VALUES INDICATE EXTREME VALUES WHICH OCCURRED AFTER THE ASOS SYSTEM WAS COMMISSIONED.

RECORD = PERIOD OF RECORD
RECORD MEAN PRECIPITATION IS THE MEAN OF ALL DAILY PRECIPITATION AMOUNTS DURING THE PERIOD OF RECORD.
RECORD MAX(MIN) TEMPERATURE IS THE MEAN OF ALL DAILY MAX(MIN) TEMPERATURES DURING THE PERIOD OF RECORD.
RECORD MEAN TEMPERATURE IS THE SUM OF THE RECORD MAX AND RECORD MIN DIVIDED BY 2.
AVERAGE TEMPERATURE IS THE SUM OF THE MEAN DAILY MAX AND MIN TEMPERATURE DIVIDED BY 2.

HEATING DEGREE DAYS Base 65 deg. F — WEST PALM BEACH, FLORIDA

SEASON	JULY	AUG	SEP	OCT	NOV	DEC	JAN	FEB	MAR	APR	MAY	JUNE	TOTAL
1964-65	0	0	0	0	0	9	86	36	30	0	0	0	161
1965-66	0	0	0	0	3	31	103	63	26	0	0	0	226
1966-67	0	0	0	0	44	77	56	70	0	0	0	0	247
1967-68	0	0	0	0	1	37	74	155	77	0	0	0	344
1968-69	0	0	0	6	60	129	55	110	88	0	0	0	448
1969-70	0	0	0	0	41	114	176	105	36	0	0	0	472
1970-71	0	0	0	0	66	32	113	73	79	18	0	0	381
1971-72	0	0	0	0	0	0	4	48	0	0	0	0	52
1972-73	0	0	0	0	5	37	56	81	4	5	0	0	188
1973-74	0	0	0	0	2	104	0	65	1	2	0	0	174
1974-75	0	0	0	0	4	67	18	3	15	5	0	0	112
1975-76	0	0	0	0	40	74	132	56	4	0	0	0	306
1976-77	0	0	0	0	15	68	220	93	8	0	0	0	404
1977-78	0	0	0	0	18	77	161	148	41	0	0	0	445
1978-79	0	0	0	0	0	2	124	118	23	0	0	0	267
1979-80	0	0	0	0	11	46	96	153	50	2	0	0	358
1980-81	0	0	0	0	28	76	193	41	27	0	0	0	365
1981-82	0	0	0	0	33	116	86	2	4	0	0	0	241
1982-83	0	0	0	0	0	48	88	36	58	6	0	0	236
1983-84	0	0	0	0	14	82	85	60	46	4	0	0	291
1984-85	0	0	0	0	10	30	176	85	8	7	0	0	316
1985-86	0	0	0	0	6	102	93	28	67	0	0	0	296
1986-87	0	0	0	0	0	2	104	22	10	39	0	0	177
1987-88	0	0	0	0	8	42	62	68	39	0	0	0	219
1988-89	0	0	0	0	0	49	7	66	26	0	0	0	148
1989-90	0	0	0	8	2	146	14	7	0	1	0	0	178
1990-91	0	0	0	2	0	11	18	44	17	0	0	0	92
1991-92	0	0	0	0	23	10	73	21	22	4	0	0	153
1992-93	0	0	0	0	9	27	11	28	28	1	0	0	104
1993-94	0	0	0	0	11	63							

COOLING DEGREE DAYS Base 65 deg. F — WEST PALM BEACH, FLORIDA

YEAR	JAN	FEB	MAR	APR	MAY	JUNE	JULY	AUG	SEP	OCT	NOV	DEC	TOTAL
1969	64	23	87	300	380	471	561	515	468	438	128	50	3485
1970	42	29	171	329	403	459	551	580	495	423	119	140	3741
1971	121	147	116	227	434	484	542	569	486	440	249	257	4072
1972	223	112	192	277	436	511	547	573	534	486	296	261	4448
1973	177	60	279	296	446	509	539	526	517	397	311	154	4211
1974	262	117	274	288	429	457	526	552	528	378	218	130	4159
1975	216	208	230	280	403	442	467	514	472	397	235	139	4003
1976	53	97	251	237	393	404	507	509	457	334	193	105	3540
1977	25	62	287	308	376	493	531	548	484	329	253	155	3851
1978	78	27	152	252	440	506	540	540	477	380	265	203	3860
1979	59	87	123	287	349	461	555	549	496	396	266	117	3745
1980	75	37	218	213	395	530	582	551	492	427	258	100	3878
1981	1	118	124	325	425	557	599	558	487	414	127	129	3864
1982	116	249	263	336	371	476	568	566	510	425	324	234	4438
1983	87	73	110	229	396	474	597	604	506	378	172	191	3817
1984	81	110	165	198	391	417	500	514	444	391	208	226	3645
1985	49	159	210	256	389	523	502	562	464	485	335	107	4041
1986	76	151	170	206	397	481	536	561	534	455	410	224	4201
1987	95	134	180	166	402	522	574	626	539	363	278	152	4031
1988	114	94	173	293	370	488	539	548	550	408	304	162	4043
1989	208	171	257	302	454	515	563	585	553	406	288	75	4377
1990	247	259	262	276	460	520	574	591	542	474	263	211	4679
1991	211	140	263	366	478	514	571	582	518	436	223	180	4482
1992	88	136	175	238	345	476	575	537	507	404	332	172	3985
1993	218	71	158	192	400	506	564	564	527	439	298	91	4028

SNOWFALL (inches) — WEST PALM BEACH, FLORIDA

SEASON	JULY	AUG	SEP	OCT	NOV	DEC	JAN	FEB	MAR	APR	MAY	JUNE	TOTAL
1970-71	0.0	0.0	0.0	0.0	0.0	0.0	0.0	0.0	0.0	0.0	0.0	0.0	0.0
1971-72	0.0	0.0	0.0	0.0	0.0	0.0	0.0	0.0	0.0	0.0	0.0	0.0	0.0
1972-73	0.0	0.0	0.0	0.0	0.0	0.0	0.0	0.0	0.0	0.0	0.0	0.0	0.0
1973-74	0.0	0.0	0.0	0.0	0.0	0.0	0.0	0.0	0.0	0.0	0.0	0.0	0.0
1974-75	0.0	0.0	0.0	0.0	0.0	0.0	0.0	0.0	0.0	0.0	0.0	0.0	0.0
1975-76	0.0	0.0	0.0	0.0	0.0	0.0	0.0	0.0	0.0	0.0	0.0	0.0	0.0
1976-77	0.0	0.0	0.0	0.0	0.0	0.0	T	0.0	0.0	0.0	0.0	0.0	T
1977-78	0.0	0.0	0.0	0.0	0.0	0.0	0.0	0.0	0.0	0.0	0.0	0.0	0.0
1978-79	0.0	0.0	0.0	0.0	0.0	0.0	0.0	0.0	0.0	0.0	0.0	0.0	0.0
1979-80	0.0	0.0	0.0	0.0	0.0	0.0	0.0	0.0	0.0	0.0	0.0	0.0	0.0
1980-81	0.0	0.0	0.0	0.0	0.0	0.0	0.0	0.0	0.0	0.0	0.0	0.0	0.0
1981-82	0.0	0.0	0.0	0.0	0.0	0.0	0.0	0.0	0.0	0.0	0.0	0.0	0.0
1982-83	0.0	0.0	0.0	0.0	0.0	0.0	0.0	0.0	0.0	0.0	0.0	0.0	0.0
1983-84	0.0	0.0	0.0	0.0	0.0	0.0	0.0	0.0	0.0	0.0	0.0	0.0	0.0
1984-85	0.0	0.0	0.0	0.0	0.0	0.0	0.0	0.0	0.0	0.0	0.0	0.0	0.0
1985-86	0.0	0.0	0.0	0.0	0.0	0.0	0.0	0.0	0.0	0.0	0.0	0.0	0.0
1986-87	0.0	0.0	0.0	0.0	0.0	0.0	0.0	0.0	0.0	0.0	0.0	0.0	0.0
1987-88	0.0	0.0	0.0	0.0	0.0	0.0	0.0	0.0	0.0	0.0	0.0	0.0	0.0
1988-89	0.0	0.0	0.0	0.0	0.0	0.0	0.0	0.0	0.0	0.0	0.0	0.0	0.0
1989-90	0.0	0.0	0.0	0.0	0.0	0.0	0.0	0.0	0.0	0.0	0.0	0.0	0.0
1990-91	0.0	0.0	0.0	0.0	0.0	0.0	0.0	0.0	0.0	0.0	0.0	0.0	0.0
1991-92	0.0	0.0	0.0	0.0	0.0	0.0	0.0	0.0	0.0	0.0	0.0	0.0	0.0
1992-93	0.0	0.0	0.0	0.0	0.0	0.0	0.0	0.0	0.0	0.0			
1993-94	0.0	T	0.0	0.0	0.0	0.0							
Record Mean	0.0	T	0.0	0.0	0.0	0.0	T	0.0	0.0	0.0	0.0	0.0	T

ATHENS, GEORGIA

Athens is located in northeast Georgia, in the Piedmont Plateau section of the state. The terrain is rolling to hilly with the elevation within the city averaging about 700 feet above sea level, and that of the county ranging mostly between 600 and 850 feet. The Atlantic Ocean 200 miles to the southeast, the Gulf of Mexico 275 miles to the south, and the southern Appalachian Mountains to the north and northwest, all exert some influence on the climate of Athens, with the total effect being a moderation of both summer and winter weather.

Summers are warm and somewhat humid in Athens, but there is a noticeable absence of prolonged periods of extreme heat. The maximum temperature reaches 90 degrees or higher on about one-half the days during the three months, June through August, but a temperature of 100 degrees or higher occurs during fewer than one-half the years.

With the mountains to the north serving as a partial barrier to the flow of extremely cold air into the area, winters in Athens are not severe. Cold spells are usually short-lived, interspersed with periods of warm southerly air flow, making normal outside activities possible throughout most years.

Average annual precipitation in Athens is fairly evenly distributed throughout the year, with slight maxima in winter, early spring, and in midsummer. In spite of what appears to be an abundant supply of moisture, dry spells of more or less serious consequence occur during most years. Fortunately, they are more frequent in autumn when long periods of clear, mild, weather not only are pleasant, but offer ideal conditions for harvesting operations.

Snowfall is not of much importance in the area. Measurable amounts occur rather infrequently, but occasionally there is sufficient fall to cause some accumulation on the ground.

The average length of the freeze-free growing season in Athens is 220 days, from early April, the average occurrence of the last spring freeze, to early November, the average occurrence of the first fall freeze.

Thunderstorms occur in all months of the year but are most frequent in June and July. Severe thunderstorms are infrequent, but may occur as isolated summer storms, or in squall lines of winter and spring. A few tornadoes have crossed the city and county since 1930, causing several deaths.

NORMALS, MEANS, AND EXTREMES

ATHENS, GEORGIA

LATITUDE: 33°57'N LONGITUDE: 83°19' W ELEVATION: FT. GRND 802 BARO 803 TIME ZONE: EASTERN WBAN: 13873

	(a)	JAN	FEB	MAR	APR	MAY	JUNE	JULY	AUG	SEP	OCT	NOV	DEC	YEAR
TEMPERATURE °F:														
Normals														
-Daily Maximum		51.6	56.3	65.1	73.4	80.5	87.0	89.6	88.1	82.7	73.6	64.4	55.1	72.3
-Daily Minimum		32.0	34.7	42.1	49.8	58.1	65.8	69.5	68.7	63.0	50.7	42.3	35.1	51.0
-Monthly		41.8	45.5	53.6	61.6	69.3	76.4	79.6	78.5	72.9	62.2	53.4	45.1	61.7
Extremes														
-Record Highest	50	80	81	88	93	97	104	104	107	99	98	86	79	107
-Year		1975	1989	1974	1986	1962	1954	1977	1983	1957	1954	1961	1984	AUG 1983
-Record Lowest	50	-4	5	11	26	37	45	55	54	36	24	7	2	-4
-Year		1985	1958	1980	1992	1989	1972	1967	1968	1967	1952	1950	1962	JAN 1985
NORMAL DEGREE DAYS:														
Heating (base 65°F)		719	546	359	131	27	0	0	0	0	140	354	617	2893
Cooling (base 65°F)		0	0	6	29	161	342	453	419	240	53	6	0	1709
% OF POSSIBLE SUNSHINE														
MEAN SKY COVER (tenths)														
Sunrise - Sunset	45	6.2	6.0	6.0	5.4	5.7	5.6	5.9	5.6	5.5	4.7	5.2	6.0	5.7
MEAN NUMBER OF DAYS:														
Sunrise to Sunset														
-Clear	50	8.7	8.6	9.2	10.3	8.8	8.1	6.7	8.0	9.7	13.7	11.7	9.6	112.9
-Partly Cloudy	50	6.9	6.2	7.3	8.4	10.3	11.7	13.0	12.8	9.3	7.3	6.0	6.2	105.5
-Cloudy	50	15.4	13.4	14.5	11.4	11.9	10.1	11.3	10.2	11.0	10.0	12.4	15.2	146.8
Precipitation														
.01 inches or more	50	11.3	9.3	10.7	8.5	8.6	9.1	11.0	9.1	7.8	6.5	8.1	10.3	110.3
Snow,Ice pellets,hail														
1.0 inches or more	50	0.4	0.3	0.1	0.0	0.0	0.0	0.0	0.0	0.0	0.0	0.1	0.1	0.9
Thunderstorms	50	1.1	1.3	3.2	4.0	6.1	8.8	11.7	8.4	3.4	1.0	1.1	0.6	50.8
Heavy Fog Visibility 1/4 mile or less	38	4.9	4.0	3.4	2.2	2.3	1.7	2.4	3.1	3.1	2.7	3.7	4.9	38.5
Temperature °F														
-Maximum														
90° and above	50	0.0	0.0	0.0	0.1	3.0	11.7	16.7	14.4	5.3	0.3	0.0	0.0	51.4
32° and below	50	0.7	0.2	0.*	0.0	0.0	0.0	0.0	0.0	0.0	0.0	0.*	0.3	1.2
-Minimum														
32° and below	50	14.8	11.0	5.3	0.5	0.0	0.0	0.0	0.0	0.0	0.5	5.6	13.8	51.5
0° and below	50	0.1	0.0	0.0	0.0	0.0	0.0	0.0	0.0	0.0	0.0	0.0	0.0	0.1
AVG. STATION PRESS.(mb)	21	990.5	989.5	987.8	987.2	986.9	987.4	988.4	988.8	989.1	990.2	990.7	991.1	989.0
RELATIVE HUMIDITY (%)														
Hour 01	38	75	71	71	72	81	84	87	88	86	83	78	76	79
Hour 07 (Local Time)	38	80	79	81	82	86	87	90	92	91	89	85	82	85
Hour 13	38	58	54	53	50	54	55	58	59	59	54	54	57	55
Hour 19	38	63	58	56	54	61	62	67	70	72	68	65	66	64
PRECIPITATION (inches):														
Water Equivalent														
-Normal		4.60	4.42	5.46	3.99	4.37	3.93	4.88	3.70	3.36	3.28	3.66	4.09	49.74
-Maximum Monthly	50	9.47	9.24	10.93	9.54	11.34	13.21	10.53	7.43	10.30	7.73	14.98	8.45	14.98
-Year		1960	1961	1964	1964	1959	1967	1964	1961	1989	1964	1948	1945	NOV 1948
-Minimum Monthly	50	0.64	0.75	1.15	0.69	0.41	0.87	0.93	0.09	0.52	T	0.33	0.81	T
-Year		1981	1978	1985	1950	1988	1958	1947	1951	1954	1963	1950	1988	OCT 1963
-Maximum in 24 hrs	50	3.86	3.58	4.89	3.87	5.54	9.93	4.14	3.05	5.34	7.61	4.05	4.33	9.93
-Year		1969	1981	1990	1979	1959	1967	1964	1969	1956	1989	1948	1972	JUN 1967
Snow,Ice pellets,hail														
-Maximum Monthly	50	7.1	4.7	8.7	0.0	0.0	0.0	0.0	0.0	0.0	0.0	2.2	3.0	8.7
-Year		1987	1979	1983								1968	1971	MAR 1983
-Maximum in 24 hrs	50	7.1	4.6	8.7	0.0	0.0	0.0	0.0	0.0	0.0	0.0	2.2	3.0	8.7
-Year		1987	1989	1983								1968	1971	MAR 1983
WIND:														
Mean Speed (mph)	38	8.4	8.9	8.8	8.3	7.1	6.6	6.3	5.8	6.4	6.8	7.4	8.0	7.4
Prevailing Direction through 1963		NW	WNW	NW	WNW	ENE	SW	SW	SW	NE	NE	NW	ENE	SW
Fastest Obs. 1 Min.														
-Direction (!!)	38	25	20	24	23	31	18	13	16	15	05	24	29	20
-Speed (MPH)	38	52	52	50	47	35	43	35	40	37	35	41	43	52
-Year		1959	1961	1974	1957	1987	1991	1987	1959	1980	1990	1989	1957	FEB 1961
Peak Gust														
-Direction (!!)	10	W	SW	W	SW	NW	W	SE	NE	SW	NE	SW	NW	W
-Speed (mph)	10	55	54	83	60	52	58	78	48	55	46	59	45	83
-Date		1992	1990	1991	1985	1987	1991	1986	1987	1993	1990	1989	1987	MAR 1991

PRECIPITATION (inches)

ATHENS, GEORGIA

YEAR	JAN	FEB	MAR	APR	MAY	JUNE	JULY	AUG	SEP	OCT	NOV	DEC	ANNUAL
1964	7.41	5.18	10.93	9.54	4.77	3.51	10.53	3.78	1.84	7.73	2.52	3.68	71.42
1965	1.83	4.92	7.00	4.66	0.93	6.02	4.41	1.44	5.27	2.90	1.50	1.03	41.91
1966	8.56	7.52	4.73	4.55	6.50	3.36	2.50	6.27	2.80	3.54	3.27	4.74	58.34
1967	4.27	4.14	2.27	4.87	5.69	13.21	7.98	5.91	1.28	3.60	6.36	6.58	66.16
1968	5.26	1.45	4.34	4.60	6.39	4.94	8.50	3.25	1.94	3.06	6.57	5.94	56.24
1969	4.95	3.38	4.86	5.69	4.61	1.49	2.28	5.92	5.19	1.98	2.02	3.88	46.25
1970	2.28	2.06	6.35	1.70	3.77	1.36	4.07	3.04	7.09	5.54	1.28	2.87	41.41
1971	4.06	5.62	7.91	4.17	3.05	4.40	3.27	4.56	2.75	2.95	4.02	2.81	49.57
1972	6.46	2.94	3.89	1.55	6.02	5.72	4.69	2.64	0.69	2.43	3.57	8.42	49.02
1973	4.28	2.68	9.86	3.58	8.02	3.18	2.15	3.09	4.09	0.66	1.78	6.97	50.34
1974	3.68	4.98	2.32	3.81	9.73	5.04	5.03	5.99	1.32	0.40	3.11	4.58	49.99
1975	5.55	6.43	10.12	2.80	7.21	3.18	6.55	3.66	6.09	3.83	3.86	2.76	62.04
1976	4.06	2.07	8.55	0.74	8.50	2.47	2.97	5.31	3.01	6.22	5.25	5.28	54.43
1977	4.14	1.79	5.99	1.86	0.88	1.76	5.58	6.33	3.35	7.41	4.86	2.19	46.14
1978	6.92	0.75	3.48	2.78	4.82	2.39	6.41	4.35	1.15	0.97	2.50	3.52	40.04
1979	6.54	6.21	2.72	8.10	3.68	2.51	5.94	3.45	2.44	2.56	3.50	1.38	49.03
1980	6.76	1.84	10.00	2.68	7.52	3.43	1.69	2.55	5.55	1.48	2.66	1.90	48.06
1981	0.64	6.73	2.45	2.10	3.67	1.50	2.32	1.00	1.04	2.06	1.81	7.57	32.89
1982	4.83	7.07	1.88	5.74	4.33	3.65	3.13	3.10	2.19	4.56	5.56	3.56	49.60
1983	3.19	5.42	6.10	5.17	2.30	3.17	1.99	2.06	5.62	3.64	6.90	8.22	53.78
1984	4.50	5.94	5.18	4.84	3.40	3.79	10.20	3.59	0.72	2.75	2.32	2.62	49.85
1985	4.11	4.62	1.15	1.73	3.81	2.08	6.36	2.39	0.62	5.15	4.99	1.39	38.40
1986	0.76	1.65	3.27	1.22	2.20	2.13	3.63	3.76	1.94	7.65	4.77	3.03	36.01
1987	6.29	5.17	4.29	0.89	2.30	3.34	4.83	0.98	2.39	0.36	2.61	2.39	35.84
1988	5.35	2.88	2.53	3.39	0.41	0.91	1.67	2.52	5.37	2.33	4.19	0.81	32.36
1989	2.10	3.21	3.96	4.20	3.99	6.21	6.20	1.97	10.30	5.85	3.51	5.29	56.79
1990	5.74	7.47	8.17	2.15	1.96	1.93	6.13	6.21	3.15	5.17	1.38	3.28	52.74
1991	6.32	3.32	6.07	4.52	5.78	10.94	6.17	5.19	2.22	0.12	1.58	3.19	55.42
1992	3.34	4.75	4.14	1.08	1.48	4.39	4.17	6.19	5.72	2.80	8.25	5.71	52.02
1993	4.40	5.01	7.35	2.73	3.24	1.36	1.54	2.03	3.23	3.85	3.43	2.83	41.00
Record Mean	4.67	4.47	5.33	3.88	4.18	3.83	4.79	3.51	3.56	3.03	3.66	4.03	48.93

AVERAGE TEMPERATURE (deg. F)

ATHENS, GEORGIA

YEAR	JAN	FEB	MAR	APR	MAY	JUNE	JULY	AUG	SEP	OCT	NOV	DEC	ANNUAL
1964	41.8	41.1	52.1	61.2	70.6	79.1	76.8	76.4	73.2	58.3	56.0	47.6	61.2
1965	44.5	45.6	49.8	63.0	73.7	74.1	78.8	78.9	72.9	60.2	54.1	46.1	61.8
1966	38.3	44.9	51.0	61.0	67.9	74.5	79.9	77.0	71.7	60.0	53.0	44.5	60.3
1967	45.3	42.1	57.0	64.7	66.1	73.3	75.3	75.1	67.6	60.6	48.6	48.0	60.3
1968	39.6	39.3	52.8	61.3	66.9	75.4	77.9	79.3	70.5	62.0	50.7	40.1	59.6
1969	41.6	43.5	47.4	62.7	69.0	78.4	82.6	76.2	70.7	63.1	50.0	42.3	60.6
1970	36.9	44.7	52.6	63.7	70.2	76.1	80.0	78.4	76.2	64.7	50.0	47.5	61.7
1971	42.8	44.3	47.9	60.9	66.8	77.9	77.3	77.1	74.3	67.1	51.5	52.6	61.7
1972	48.1	43.3	53.4	61.8	66.9	72.6	77.5	78.1	75.6	62.3	51.0	49.7	61.7
1973	43.2	44.0	57.2	57.9	65.4	75.8	79.3	77.4	75.4	63.8	55.2	44.3	61.5
1974	52.6	46.5	57.6	60.9	70.1	72.5	77.5	77.5	70.7	61.3	52.4	45.3	62.1
1975	47.9	48.9	51.3	61.2	71.7	75.1	78.8	79.8	72.0	64.8	54.9	44.4	62.6
1976	39.7	52.4	57.2	62.6	66.2	74.0	77.9	76.9	70.4	57.9	47.1	41.2	60.3
1977	31.2	43.3	56.9	64.4	71.4	78.6	81.8	78.2	74.4	59.5	55.0	43.4	61.5
1978	35.3	39.0	51.7	63.2	68.5	77.8	80.0	78.4	74.9	61.3	57.7	46.1	61.2
1979	38.8	42.1	55.9	62.6	69.9	74.3	77.6	79.0	72.0	61.7	55.1	46.5	61.3
1980	45.0	42.0	51.0	61.0	69.8	76.5	83.2	81.7	76.5	59.5	50.9	45.4	61.9
1981	39.8	46.1	52.2	67.3	67.9	82.1	83.1	77.6	73.0	60.6	53.0	39.7	61.9
1982	38.1	48.4	56.9	60.5	74.7	78.5	82.3	79.8	74.0	63.5	54.7	51.5	63.6
1983	42.1	46.1	53.7	58.3	70.2	75.1	82.3	83.4	72.7	63.7	54.2	42.9	62.1
1984	42.5	48.8	54.0	59.5	69.2	79.7	76.9	79.2	73.1	69.7	50.8	53.9	63.1
1985	37.8	45.8	56.3	64.2	70.2	77.7	78.0	77.3	72.2	65.9	61.8	41.0	62.4
1986	42.0	49.7	54.4	62.8	70.3	79.4	83.9	77.7	74.2	63.5	57.0	44.6	63.3
1987	41.8	45.5	52.5	59.4	72.7	77.9	81.3	81.9	73.8	57.7	55.2	47.9	62.3
1988	39.2	44.7	53.9	61.8	69.4	77.6	80.5	80.5	72.8	57.9	54.0	45.3	61.5
1989	48.1	47.9	55.4	61.2	67.1	76.5	79.3	78.7	71.9	63.0	53.0	38.4	61.7
1990	48.5	53.0	56.4	60.7	69.2	78.0	80.6	79.6	73.9	63.6	55.7	49.0	64.0
1991	43.9	49.3	55.8	64.1	71.8	75.3	80.0	77.3	73.0	63.8	51.3	48.5	62.8
1992	45.2	50.3	52.7	60.5	66.9	74.6	80.6	76.5	73.3	60.7	51.5	43.8	61.4
1993	46.8	44.0	50.1	58.0	69.8	78.8	85.3	81.1	75.4	61.9	52.3	43.7	62.3
Record Mean	42.8	46.0	52.9	61.7	69.6	76.7	79.6	78.6	73.0	62.4	52.5	44.7	61.7
Max	52.6	56.7	64.2	73.6	81.0	87.6	89.8	88.6	83.0	73.8	63.6	54.6	72.4
Min	33.0	35.2	41.6	49.7	58.1	65.8	69.3	68.5	62.9	51.0	41.5	34.7	50.9

REFERENCE NOTES

ATHENS, GEORGIA

GENERAL
T - TRACE AMOUNT.
BLANK ENTRIES DENOTE MISSING/UNREPORTED DATA.
# INDICATES A STATION OR INSTRUMENT RELOCATION.

SPECIFIC

(a) - LENGTH OF RECORD IN YEARS, ALTHOUGH INDIVIDUAL MONTHS MAY BE MISSING.
0.* OR * - THE VALUE IS BETWEEN 0.0 AND 0.05
NORMALS - BASED ON THE 1961-1990 RECORD PERIOD.
EXTREMES - DATES ARE THE MOST RECENT OCCURRENCE
WIND DIR. - NUMERALS SHOW TENS OF DEGREES CLOCKWISE FROM TRUE NORTH. "00" INDICATES CALM.
RESULTANT DIRECTIONS ARE GIVEN TO WHOLE DEGREES.
BOLD VALUES INDICATE EXTREME VALUES WHICH OCCURRED AFTER THE ASOS SYSTEM WAS COMMISSIONED.

RECORD = PERIOD OF RECORD
RECORD MEAN PRECIPITATION IS THE MEAN OF ALL DAILY PRECIPITATION AMOUNTS DURING THE PERIOD OF RECORD.
RECORD MAX(MIN) TEMPERATURE IS THE MEAN OF ALL DAILY MAX(MIN) TEMPERATURES DURING THE PERIOD OF RECORD.
RECORD MEAN TEMPERATURE IS THE SUM OF THE RECORD MAX AND RECORD MIN DIVIDED BY 2.
AVERAGE TEMPERATURE IS THE SUM OF THE MEAN DAILY MAX AND MIN TEMPERATURE DIVIDED BY 2.

EXCEPTIONS

RECORD MEANS ARE THROUGH THE CURRENT YEAR, BEGINNING IN 1944 FOR TEMPERATURE
1944 FOR PRECIPITATION
1944 FOR SNOWFALL

HEATING DEGREE DAYS Base 65 deg. F ATHENS, GEORGIA

SEASON	JULY	AUG	SEP	OCT	NOV	DEC	JAN	FEB	MAR	APR	MAY	JUNE	TOTAL
1964-65	0	0	1	214	273	539	627	540	467	123	0	1	2785
1965-66	0	0	2	171	322	579	821	556	429	159	26	1	3066
1966-67	0	0	4	170	356	627	605	634	262	77	77	18	2830
1967-68	0	0	36	159	485	524	784	735	378	146	40	0	3287
1968-69	0	2	1	152	424	764	717	597	535	100	26	0	3318
1969-70	0	0	13	112	447	699	865	564	379	117	14	0	3210
1970-71	0	0	5	81	443	538	678	573	521	156	50	0	3045
1971-72	0	0	0	33	415	381	519	622	352	152	17	5	2496
1972-73	0	0	3	113	420	472	669	583	248	224	65	0	2797
1973-74	0	0	0	94	295	634	377	515	249	159	19	0	2342
1974-75	0	0	22	142	381	600	528	445	418	162	2	0	2700
1975-76	0	0	14	85	323	635	778	360	245	116	44	5	2605
1976-77	0	0	6	234	532	735	1040	604	248	86	10	0	3495
1977-78	0	0	2	187	298	663	913	721	410	99	38	0	3331
1978-79	0	0	1	132	218	581	804	635	289	92	22	0	2774
1979-80	0	0	6	134	295	567	609	664	430	143	12	0	2860
1980-81	0	0	21	193	413	601	777	525	395	42	40	0	3007
1981-82	0	0	7	171	358	779	824	457	268	166	3	0	3033
1982-83	0	0	3	127	311	428	706	526	350	212	10	0	2673
1983-84	0	0	17	90	315	676	689	460	338	191	30	0	2806
1984-85	0	0	5	23	428	340	838	533	276	107	13	0	2563
1985-86	0	0	20	80	120	736	708	424	329	130	13	0	2560
1986-87	0	8	2	123	257	628	713	541	383	204	5	0	2864
1987-88	0	0	0	224	289	526	795	583	341	123	12	0	2893
1988-89	0	0	0	230	321	602	518	477	312	190	64	0	2714
1989-90	0	0	25	124	357	817	504	333	276	162	25	0	2623
1990-91	0	0	15	111	275	489	648	433	289	82	10	0	2352
1991-92	0	0	12	93	412	513	609	420	379	174	64	4	2680
1992-93	0	0	13	130	397	650	558	581	455	214	17	0	3015
1993-94	0	0	10	140	394	655							

COOLING DEGREE DAYS Base 65 deg. F ATHENS, GEORGIA

YEAR	JAN	FEB	MAR	APR	MAY	JUNE	JULY	AUG	SEP	OCT	NOV	DEC	TOTAL
1969	0	0	0	33	156	408	553	353	193	60	0	0	1756
1970	0	0	0	83	183	340	474	424	347	80	0	1	1932
1971	0	0	0	39	111	394	389	385	286	105	18	5	1732
1972	0	0	0	60	81	239	395	413	329	37	9	4	1567
1973	0	0	13	15	87	332	450	393	320	62	11	0	1683
1974	0	0	25	42	183	231	392	392	199	32	11	0	1507
1975	6	0	3	55	218	312	433	466	232	88	24	0	1837
1976	0	2	11	50	89	282	410	376	177	18	0	0	1415
1977	0	0	6	74	215	415	524	417	292	24	5	1	1973
1978	0	0	3	52	154	392	473	423	306	24	4	1	1832
1979	0	0	15	25	180	286	398	443	221	40	6	0	1614
1980	0	2	1	30	170	354	573	525	373	32	1	0	2061
1981	0	0	5	116	138	517	569	398	256	40	4	0	2043
1982	0	1	28	36	308	413	545	463	279	85	11	15	2184
1983	0	0	8	17	175	312	543	578	256	58	0	0	1947
1984	0	0	4	31	167	450	375	446	254	174	7	2	1910
1985	3	0	14	89	181	391	411	388	242	117	29	0	1865
1986	0	0	8	70	185	441	592	408	283	85	25	0	2097
1987	0	0	2	44	252	395	514	530	273	3	4	2	2019
1988	0	0	4	34	157	387	487	490	239	17	0	0	1815
1989	0	7	23	81	136	352	451	432	238	70	1	0	1791
1990	0	0	16	39	162	397	491	460	290	76	2	0	1933
1991	0	0	11	62	228	315	472	391	259	63	8	11	1820
1992	0	0	3	45	132	297	489	362	269	7	0	0	1604
1993	0	0	3	12	174	421	634	504	330	52	18	0	2148

SNOWFALL (inches) ATHENS, GEORGIA

SEASON	JULY	AUG	SEP	OCT	NOV	DEC	JAN	FEB	MAR	APR	MAY	JUNE	TOTAL
1964-65	0.0	0.0	0.0	0.0	0.0	0.0	2.5	T	T	0.0	0.0	0.0	2.5
1965-66	0.0	0.0	0.0	0.0	0.0	0.0	4.3	1.3	0.0	0.0	0.0	0.0	5.6
1966-67	0.0	0.0	0.0	0.0	0.0	T	0.0	3.0	0.0	0.0	0.0	0.0	3.0
1967-68	0.0	0.0	0.0	0.0	0.0	0.0	1.9	2.1	T	0.0	0.0	0.0	4.0
1968-69	0.0	0.0	0.0	0.0	2.2	0.0	T	2.1	T	0.0	0.0	0.0	4.3
1969-70	0.0	0.0	0.0	0.0	T	T	1.7	T	0.0	0.0	0.0	0.0	1.7
1970-71	0.0	0.0	0.0	0.0	0.0	0.7	T	T	T	0.0	0.0	0.0	0.7
1971-72	0.0	0.0	0.0	0.0	T	3.0	T	T	T	0.0	0.0	0.0	3.0
1972-73	0.0	0.0	0.0	0.0	0.0	0.0	2.7	1.7	0.0	0.0	0.0	0.0	4.4
1973-74	0.0	0.0	0.0	0.0	0.0	1.0	0.0	T	0.0	0.0	0.0	0.0	1.0
1974-75	0.0	0.0	0.0	0.0	0.0	T	0.0	0.0	T	0.0	0.0	0.0	T
1975-76	0.0	0.0	0.0	0.0	1.0	0.0	2.1	0.0	0.0	0.0	0.0	0.0	3.1
1976-77	0.0	0.0	0.0	0.0	T	0.0	1.0	0.0	0.0	0.0	0.0	0.0	1.0
1977-78	0.0	0.0	0.0	0.0	0.0	T	T	0.2	T	0.0	0.0	0.0	0.2
1978-79	0.0	0.0	0.0	0.0	0.0	0.0	T	4.7	0.0	0.0	0.0	0.0	4.7
1979-80	0.0	0.0	0.0	0.0	0.0	0.0	0.0	3.0	3.3	0.0	0.0	0.0	6.3
1980-81	0.0	0.0	0.0	0.0	T	T	T	T	0.0	0.0	0.0	0.0	T
1981-82	0.0	0.0	0.0	0.0	0.0	T	5.1	2.0	0.0	0.0	0.0	0.0	7.1
1982-83	0.0	0.0	0.0	0.0	0.0	0.0	1.8	0.1	8.7	0.0	0.0	0.0	10.6
1983-84	0.0	0.0	0.0	0.0	0.0	T	T	1.0	0.0	0.0	0.0	0.0	1.0
1984-85	0.0	0.0	0.0	0.0	0.0	0.0	0.2	0.5	0.0	0.0	0.0	0.0	0.7
1985-86	0.0	0.0	0.0	0.0	0.0	T	T	T	T	0.0	0.0	0.0	T
1986-87	0.0	0.0	0.0	0.0	0.0	0.0	7.1	0.0	0.4	0.0	0.0	0.0	7.5
1987-88	0.0	0.0	0.0	0.0	0.0	0.0	3.3	0.0	0.0	0.0	0.0	0.0	3.3
1988-89	0.0	0.0	0.0	0.0	0.0	0.0	0.0	4.6	0.0	0.0	0.0	0.0	4.6
1989-90	0.0	0.0	0.0	0.0	0.0	0.3	0.0	0.0	0.0	0.0	0.0	0.0	0.3
1990-91	0.0	0.0	0.0	0.0	0.0	0.0	1.9	T	0.0	0.0	0.0	0.0	1.9
1991-92	0.0	0.0	0.0	0.0	0.0	0.0	2.0	0.0	T	0.0	0.0	0.0	2.0
1992-93	0.0	0.0	0.0	0.0	0.0	0.0	0.0	T	2.7	0.0	0.0	0.0	2.7
1993-94	0.0	0.0	0.0	0.0	0.0	1.5							
Record Mean	0.0	0.0	0.0	0.0	0.1	0.2	0.9	0.7	0.5	0.0	0.0	0.0	2.4

ATLANTA, GEORGIA

Atlanta is located in the foothills of the southern Appalachians in north-central Georgia. The terrain is rolling to hilly and slopes downward toward the east, west, and south so that drainage of the major river systems is generally into the Gulf of Mexico from the western and southern sections of the city and to the Atlantic from the eastern portions of the city.

The Gulf of Mexico and the Atlantic Ocean are approximately 250 miles south and southeast of the city, respectively. Both the Appalachian chain of mountains and the two nearby maritime bodies exert an important influence on the Atlanta climate. Temperatures are moderated throughout the year while abundant precipitation fosters natural vegetation and growth of crops. Summer temperatures in Atlanta are moderated somewhat by elevation but are still rather warm. However, prolonged periods of hot weather are unusual and 100 degree heat is rarely experienced.

With the mountains to the north tending to retard the southward movement of Polar air masses, Atlanta winters are rather mild. Cold spells are not unusual but they are rather short-lived and seldom disrupt outdoor activities for an extended period of time. Late March is the average date of the last temperature of 32 degrees in the spring and mid-November is the average date of the first temperature of 32 degrees in the fall, which gives an average growing season of about 234 days.

Minimum dry precipitation periods occur mainly during the late summer and early autumn. Maximum thunderstorm activity occurs during July, but severe local thunderstorms occur most frequently in March, April, and May, some spawning highly damaging tornadoes.

The average annual snowfall varies widely from year to year. A fall of 4 inches or more occurs about once every five years. Most snows melt in a short period of time due to the rapid warming which often follows the storm. Ice storms, freezing rain or glaze, occur about two out of every three years, causing hazardous travel and disruption of utilities. Severe ice storms occur about once in ten years, causing major disruption of utilities and significant property damage.

The Bermuda High pressure area has a dominant effect on Atlanta weather, particularly in the summer months. East or northeast winds produce the most unpleasant weather although southerly winds are quite humid during the summer. The generally light wind conditions contribute to the formation of an occasional early morning fog.

NORMALS, MEANS, AND EXTREMES

ATLANTA, GEORGIA

LATITUDE: 33°39'N LONGITUDE: 84°25'W ELEVATION: FT. GRND 1010 BARO 1110 TIME ZONE: EASTERN WBAN: 13874

	(a)	JAN	FEB	MAR	APR	MAY	JUNE	JULY	AUG	SEP	OCT	NOV	DEC	YEAR
TEMPERATURE °F:														
Normals														
-Daily Maximum		50.4	55.0	64.3	72.7	79.6	85.8	88.0	87.1	81.8	72.7	63.4	54.0	71.2
-Daily Minimum		31.5	34.5	42.5	50.2	58.7	66.2	69.5	69.0	63.5	51.9	42.8	35.0	51.3
-Monthly		41.0	44.8	53.5	61.5	69.2	76.0	78.8	78.1	72.7	62.3	53.1	44.5	61.3
Extremes														
-Record Highest	45	79	80	85	93	95	101	105	102	98	95	84	79	105
-Year		1949	1989	1991	1986	1953	1952	1980	1980	1954	1954	1961	1991	JUL 1980
-Record Lowest	45	-8	5	10	26	37	46	53	55	36	28	3	0	-8
-Year		1985	1958	1960	1973	1971	1956	1967	1992	1967	1976	1950	1983	JAN 1985
NORMAL DEGREE DAYS:														
Heating (base 65°F)		744	566	365	138	27	0	0	0	10	138	367	636	2991
Cooling (base 65°F)		0	0	8	33	157	330	428	406	241	54	10	0	1667
% OF POSSIBLE SUNSHINE	58	49	54	58	66	68	67	63	64	63	67	58	50	61
MEAN SKY COVER (tenths)														
Sunrise - Sunset	59	6.4	6.2	6.1	5.5	5.7	5.8	6.1	5.8	5.5	4.6	5.4	6.2	5.8
MEAN NUMBER OF DAYS:														
Sunrise to Sunset														
-Clear	59	8.3	8.0	8.8	10.1	9.1	7.8	6.3	7.4	9.7	14.0	11.6	9.2	110.4
-Partly Cloudy	59	6.4	6.2	7.5	8.2	10.5	11.8	13.2	13.1	9.9	7.2	6.0	6.4	106.4
-Cloudy	59	16.3	14.0	14.6	11.8	11.4	10.3	11.6	10.4	10.4	9.8	12.4	15.4	148.5
Precipitation														
.01 inches or more	59	11.5	10.1	11.4	8.9	9.2	9.9	11.9	9.5	7.6	6.4	8.5	10.3	115.1
Snow,Ice pellets,hail														
1.0 inches or more	59	0.3	0.2	0.1	0.0	0.0	0.0	0.0	0.0	0.0	0.0	0.*	0.1	0.6
Thunderstorms	59	1.1	1.9	3.6	4.2	6.0	8.6	10.3	8.0	3.1	1.0	1.0	0.7	49.6
Heavy Fog Visibility														
1/4 mile or less	59	5.0	3.3	2.7	1.4	1.4	1.0	1.5	1.7	1.9	2.2	3.0	4.5	29.6
Temperature °F														
-Maximum														
90° and above	33	0.0	0.0	0.0	0.1	1.0	8.0	12.5	9.5	3.4	0.0	0.0	0.0	34.4
32° and below	33	1.5	0.3	0.*	0.0	0.0	0.0	0.0	0.0	0.0	0.0	0.*	0.4	2.3
-Minimum														
32° and below	33	15.8	12.3	4.9	0.5	0.0	0.0	0.0	0.0	0.0	0.2	4.8	12.8	51.4
0° and below	33	0.2	0.0	0.0	0.0	0.0	0.0	0.0	0.0	0.0	0.0	0.0	0.*	0.2
AVG. STATION PRESS. (mb)	21	982.6	981.6	979.8	979.6	979.2	979.9	981.0	981.3	981.4	982.5	982.7	983.2	981.2
RELATIVE HUMIDITY (%)														
Hour 01	33	74	69	69	69	77	80	85	85	83	78	76	74	77
Hour 07 (Local Time)	33	78	75	78	78	82	84	88	90	88	84	81	79	82
Hour 13	33	59	54	51	49	53	56	60	60	59	53	55	58	56
Hour 19	33	63	56	53	50	58	61	66	67	67	63	63	64	61
PRECIPITATION (inches):														
Water Equivalent														
-Normal		4.75	4.81	5.77	4.26	4.29	3.56	5.01	3.66	3.42	3.05	3.86	4.33	50.77
-Maximum Monthly	59	10.82	12.77	11.66	11.86	8.37	9.99	11.26	8.69	11.64	7.53	15.72	9.92	15.72
-Year		1936	1961	1980	1979	1980	1991	1948	1967	1989	1966	1948	1961	NOV 1948
-Minimum Monthly	59	0.84	0.77	1.86	0.49	0.32	0.16	0.76	0.50	0.04	T	0.41	0.69	T
-Year		1981	1978	1985	1986	1936	1988	1980	1976	1984	1963	1939	1979	OCT 1963
-Maximum in 24 hrs	59	3.91	5.67	5.74	5.58	5.13	4.22	5.44	5.05	5.87	5.41	4.11	3.85	5.87
-Year		1973	1961	1990	1979	1948	1991	1948	1940	1992	1989	1935	1961	SEP 1992
Snow,Ice pellets,hail														
-Maximum Monthly	59	8.3	4.4	7.9	T	0.0	0.0	0.0	0.0	0.0	T	1.0	2.8	8.3
-Year		1940	1979	1983	1990						1993	1968	1993	JAN 1940
-Maximum in 24 hrs	59	8.3	4.2	7.9	T	0.0	0.0	0.0	0.0	0.0	T	1.0	2.8	8.3
-Year		1940	1979	1983	1990						1993	1968	1993	JAN 1940
WIND:														
Mean Speed (mph)	55	10.4	10.8	10.8	10.1	8.7	8.1	7.6	7.2	8.0	8.5	9.1	9.8	9.1
Prevailing Direction through 1963		NW	NW	NW	NW	NW	NW	SW	NW	ENE	NW	NW	NW	NW
Fastest Obs. 1 Min.														
-Direction (!!)	17	23	29	31	31	27	24	30	32	27	09	24	30	30
-Speed (MPH)	17	46	52	49	41	54	51	60	41	37	30	37	33	60
-Year		1978	1990	1993	1989	1984	1989	1984	1986	1980	1985	1984	1984	JUL 1984
Peak Gust														
-Direction (!!)	10	W	W	NW	W	W	SW	NW	N	E	E	NW	NW	NW
-Speed (mph)	10	48	68	61	61	72	60	77	55	45	45	56	47	77
-Date		1992	1990	1993	1985	1984	1989	1984	1990	1985	1985	1986	1993	JUL 1984

PRECIPITATION (inches)

ATLANTA, GEORGIA

YEAR	JAN	FEB	MAR	APR	MAY	JUNE	JULY	AUG	SEP	OCT	NOV	DEC	ANNUAL
1964	6.01	4.17	9.51	8.68	2.59	2.88	7.14	4.10	1.38	5.63	3.11	4.93	60.13
1965	3.74	4.31	5.94	3.10	2.51	7.15	4.59	1.53	3.02	2.35	2.32	1.70	42.26
1966	5.94	7.04	4.02	4.19	5.84	2.95	3.32	3.94	2.15	7.53	4.82	4.60	56.34
1967	4.85	3.75	3.63	2.81	4.95	3.06	6.46	8.69	1.90	2.51	5.52	6.71	54.84
1968	4.07	1.40	4.54	6.33	6.42	2.65	5.34	4.13	1.29	2.99	4.81	4.89	48.86
1969	2.85	3.20	4.00	5.70	7.68	1.00	2.64	6.12	3.74	1.53	2.67	3.27	44.40
1970	2.95	2.99	6.94	3.24	3.01	2.62	3.59	3.26	1.82	6.29	1.86	3.68	42.25
1971	4.40	5.77	8.65	4.09	2.12	3.33	8.22	2.76	3.32	0.31	3.42	2.79	49.18
1972	9.26	3.16	4.49	2.31	4.28	4.04	3.81	2.78	1.86	3.04	3.96	7.62	50.61
1973	8.89	3.44	9.53	4.03	7.14	3.35	2.10	1.35	4.16	0.75	2.31	8.11	55.16
1974	5.36	6.37	2.44	3.72	3.83	3.20	4.64	6.26	1.06	1.22	3.89	5.31	47.30
1975	6.19	8.98	8.31	4.28	4.62	5.52	8.52	3.30	2.99	5.31	4.62	3.36	66.00
1976	5.15	1.84	10.95	1.49	6.99	2.36	4.29	0.50	0.72	3.55	4.11	4.01	45.96
1977	3.49	2.14	6.28	1.77	2.04	3.03	4.26	4.23	4.90	5.00	7.18	2.36	46.68
1978	7.03	0.77	2.63	3.49	7.28	2.86	2.56	5.66	0.94	1.42	2.96	3.75	41.35
1979	5.03	5.71	3.19	11.86	2.43	1.46	3.62	7.28	6.08	2.17	5.19	0.69	54.71
1980	5.69	2.69	11.66	1.88	8.37	4.49	0.76	1.59	4.77	1.61	2.14	1.29	46.94
1981	0.84	6.62	3.93	2.06	3.89	2.69	2.74	2.76	5.27	3.01	1.85	6.25	41.91
1982	4.75	6.99	3.79	6.02	2.60	6.09	6.31	1.45	3.00	5.83	4.15	5.23	56.21
1983	3.09	4.99	6.68	4.79	1.42	1.52	1.85	1.06	7.52	1.97	7.46	9.27	51.62
1984	4.66	5.97	5.83	6.62	6.57	0.74	11.21	6.46	0.04	1.54	2.10	3.65	55.39
1985	4.11	4.98	1.86	2.75	4.69	2.04	9.92	4.57	2.63	5.74	4.23	2.28	49.80
1986	0.88	2.46	4.13	0.49	2.95	2.18	3.27	6.08	3.68	5.15	6.20	3.03	40.50
1987	5.63	6.13	5.44	1.16	2.74	6.36	7.35	1.22	3.02	0.70	2.36	4.13	46.24
1988	4.64	3.32	2.57	6.06	1.71	0.16	5.04	4.92	6.35	5.00	4.87	1.21	45.85
1989	2.57	4.30	3.85	5.24	6.42	9.34	7.65	2.13	11.64	1.71	3.97	4.49	63.31
1990	8.47	9.75	8.36	2.76	5.26	1.39	3.49	4.64	3.01	6.12	1.27	3.04	57.56
1991	4.66	3.10	6.98	5.28	7.35	9.99	5.82	4.37	2.03	0.39	3.19	2.69	55.85
1992	3.58	3.94	3.81	1.03	1.73	4.14	9.03	5.04	8.55	2.84	10.04	6.38	60.11
1993	3.94	4.43	5.73	2.77	4.87	6.01	3.05	2.96	3.91	3.83	4.01	2.54	48.05
Record Mean	4.72	4.69	5.55	4.01	3.71	3.79	4.82	3.95	3.31	2.66	3.32	4.41	48.94

AVERAGE TEMPERATURE (deg. F)

ATLANTA, GEORGIA

YEAR	JAN	FEB	MAR	APR	MAY	JUNE	JULY	AUG	SEP	OCT	NOV	DEC	ANNUAL
1964	40.8	40.3	51.2	60.6	68.9	77.8	76.5	76.0	72.6	58.8	55.6	46.5	60.4
1965	43.2	43.9	48.8	63.3	71.8	71.7	76.6	77.3	72.9	60.3	53.8	44.8	60.7
1966	36.6	43.2	50.2	60.6	67.6	73.4	78.2	75.4	70.6	59.4	52.1	43.4	59.2
1967	43.9	41.8	56.7	65.0	66.2	72.5	74.1	74.0	66.5	59.4	48.5	47.7	59.7
1968	39.2	38.3	52.3	60.9	67.4	75.7	77.7	79.0	71.5	61.7	50.1	40.2	59.5
1969	40.2	42.7	46.9	62.8	69.2	77.4	80.8	76.1	70.9	62.5	50.1	41.7	60.1
1970	35.9	43.9	52.9	64.4	70.1	74.6	78.7	79.1	77.0	65.2	49.5	47.4	61.6
1971	42.9	44.3	47.5	60.8	66.7	77.2	76.3	76.7	73.7	66.8	50.7	52.3	61.3
1972	46.8	42.6	52.4	61.3	66.5	72.2	76.8	77.7	74.6	61.1	49.6	48.3	60.8
1973	41.4	42.8	57.4	57.6	65.0	75.6	78.9	77.4	75.6	64.7	55.6	44.2	61.3
1974	53.2	45.8	57.8	61.1	71.0	72.5	77.9	76.7	70.2	61.3	52.4	44.3	62.0
1975	47.2	47.1	50.5	59.8	71.0	75.3	76.4	77.9	70.3	63.3	54.0	43.4	61.4
1976	38.5	51.5	56.4	61.7	65.4	73.8	76.4	76.0	69.8	56.2	44.2	39.8	59.1
1977	29.3	42.0	55.3	63.0	69.9	77.1	79.5	77.7	73.5	59.6	54.3	42.1	60.3
1978	33.7	39.3	51.6	61.5	67.6	76.3	78.6	78.3	76.3	62.5	58.5	46.1	60.8
1979	37.3	41.7	56.2	62.7	70.1	75.7	78.8	80.1	72.7	62.4	54.3	46.7	61.6
1980	44.9	41.9	52.1	62.6	72.0	79.1	85.1	83.8	78.9	61.5	51.9	44.9	63.2
1981	39.5	46.8	51.8	67.7	67.6	81.3	82.2	77.7	72.4	60.2	54.5	39.1	61.8
1982	38.5	47.4	56.5	58.4	72.5	76.3	79.1	77.5	70.5	62.7	53.7	49.9	61.9
1983	40.4	44.4	51.3	56.4	67.8	74.0	81.4	81.4	70.8	62.1	51.5	39.9	60.1
1984	39.6	47.5	51.7	58.1	67.5	78.3	76.8	77.5	71.3	69.8	50.6	53.7	61.9
1985	36.3	44.2	56.8	64.0	69.9	77.5	78.4	77.6	72.5	66.4	62.0	41.4	62.3
1986	43.4	49.8	54.4	62.9	71.0	80.0	84.1	77.4	74.6	64.0	57.9	45.1	63.7
1987	41.9	45.7	53.2	60.3	73.2	77.8	81.0	82.0	74.1	59.7	55.7	48.8	62.8
1988	39.2	45.5	54.9	63.0	70.0	78.6	80.5	81.0	73.4	59.3	55.0	46.6	62.3
1989	49.7	47.5	56.8	62.9	68.8	76.9	79.8	79.4	72.9	64.2	54.3	39.1	62.7
1990	49.8	54.4	57.7	61.9	70.4	78.6	80.6	80.6	75.7	64.4	56.5	49.1	65.0
1991	44.3	49.2	56.2	65.9	72.8	76.7	81.0	79.2	74.8	64.6	51.0	49.1	63.7
1992	45.1	51.8	54.0	61.9	68.1	74.5	80.2	76.1	73.2	61.9	51.5	44.5	61.9
1993	47.0	45.2	51.7	59.4	71.1	79.3	85.4	82.1	76.7	63.3	53.9	45.3	63.4
Record Mean	43.0	45.7	52.8	61.3	69.6	76.5	78.8	78.0	73.2	62.8	52.3	44.8	61.6
Max	51.5	54.9	62.7	71.6	79.5	86.0	87.8	86.8	82.1	72.3	61.6	53.2	70.8
Min	34.6	36.5	42.9	51.1	59.6	67.0	69.7	69.1	64.2	53.2	43.0	36.3	52.3

REFERENCE NOTES

ATLANTA, GEORGIA

GENERAL
T - TRACE AMOUNT.
BLANK ENTRIES DENOTE MISSING/UNREPORTED DATA.
# INDICATES A STATION OR INSTRUMENT RELOCATION.

SPECIFIC

PM - INCLUDES LAST DAY OF PREVIOUS MONTH
ASOS - AUTOMATED SURFACE OBSERVING SYSTEM IN OPERATION DURING THESE MONTHS.

(*) - LENGTH OF RECORD IN YEARS, ALTHOUGH INDIVIDUAL MONTHS MAY BE MISSING.
0.* OR * - THE VALUE IS BETWEEN 0.0 AND 0.05
NORMALS - BASED ON THE 1961-1990 RECORD PERIOD.
EXTREMES - DATES ARE THE MOST RECENT OCCURRENCE
WIND DIR.- NUMERALS SHOW TENS OF DEGRESS CLOCKWISE FROM TRUE NORTH. "00" INDICATES CALM.
RESULTANT DIRECTIONS ARE GIVEN TO WHOLE DEGREES.
BOLD VALUES INDICATE EXTREME VALUES WHICH OCCURRED AFTER THE ASOS SYSTEM WAS COMMISSIONED.

RECORD = PERIOD OF RECORD
RECORD MEAN PRECIPITATION IS THE MEAN OF ALL DAILY PRECIPITATION AMOUNTS DURING THE PERIOD OF RECORD.
RECORD MAX(MIN) TEMPERATURE IS THE MEAN OF ALL DAILY MAX(MIN) TEMPERATURES DURING THE PERIOD OF RECORD.
RECORD MEAN TEMPERATURE IS THE SUM OF THE RECORD MAX AND RECORD MIN DIVIDED BY 2.
AVERAGE TEMPERATURE IS THE SUM OF THE MEAN DAILY MAX AND MIN TEMPERATURE DIVIDED BY 2.

EXCEPTIONS

RECORD MEANS ARE THROUGH THE CURRENT YEAR,
BEGINNING IN 1879 FOR TEMPERATURE
1879 FOR PRECIPITATION
1934 FOR SNOWFALL

HEATING DEGREE DAYS Base 65 deg. F ATLANTA, GEORGIA

SEASON	JULY	AUG	SEP	OCT	NOV	DEC	JAN	FEB	MAR	APR	MAY	JUNE	TOTAL
1964-65	0	0	2	201	288	567	669	586	500	110	0	3	2926
1965-66	0	0	3	176	332	619	874	605	451	171	26	5	3262
1966-67	0	0	5	177	379	663	645	642	270	67	78	20	2946
1967-68	0	0	52	193	490	530	792	769	389	154	31	0	3400
1968-69	0	2	0	157	441	760	761	620	555	93	28	0	3417
1969-70	0	0	13	125	445	719	895	586	371	95	15	0	3264
1970-71	0	0	3	64	457	537	681	572	533	156	56	0	3059
1971-72	0	0	0	36	436	390	559	643	387	161	21	5	2638
1972-73	0	0	3	136	465	511	725	617	240	230	72	0	2999
1973-74	0	0	1	86	295	639	357	531	241	155	5	0	2310
1974-75	0	0	26	148	381	633	547	493	451	192	2	0	2873
1975-76	0	0	28	113	342	665	814	384	265	124	48	4	2787
1976-77	0	0	10	277	618	775	1099	640	300	102	11	0	3832
1977-78	0	0	4	178	313	701	966	714	412	137	57	0	3482
1978-79	0	0	0	112	194	580	853	646	279	97	16	0	2777
1979-80	0	0	5	122	320	559	616	668	399	113	3	0	2805
1980-81	0	0	18	154	391	618	786	502	410	36	43	0	2958
1981-82	0	0	17	179	314	795	819	486	282	204	2	0	3098
1982-83	0	0	16	139	341	466	755	571	423	261	24	0	2996
1983-84	0	0	32	123	400	770	780	503	409	221	50	0	3288
1984-85	0	0	13	22	426	346	882	576	265	111	14	1	2656
1985-86	0	0	15	71	131	725	663	422	331	133	14	0	2505
1986-87	0	11	2	107	243	609	709	534	359	191	6	0	2771
1987-88	0	0	0	172	279	494	791	559	310	104	6	0	2715
1988-89	0	0	0	188	291	566	468	490	284	160	44	0	2491
1989-90	0	0	29	103	318	797	462	297	250	150	20	0	2426
1990-91	0	0	12	109	252	488	636	437	281	54	8	0	2277
1991-92	0	0	8	76	419	499	611	377	345	161	50	2	2548
1992-93	0	0	12	110	398	627	548	549	417	184	13	0	2858
1993-94	0	0	6	129	346	604							

COOLING DEGREE DAYS Base 65 deg. F ATLANTA, GEORGIA

YEAR	JAN	FEB	MAR	APR	MAY	JUNE	JULY	AUG	SEP	OCT	NOV	DEC	TOTAL
1969	0	0	1	31	162	379	494	348	197	51	0	0	1663
1970	0	0	0	80	178	292	430	442	371	77	0	0	1870
1971	0	0	0	38	115	374	358	371	265	95	13	5	1634
1972	0	0	3	56	72	227	370	396	297	25	7	0	1453
1973	0	0	11	15	78	322	438	388	323	79	20	0	1674
1974	0	1	24	42	198	229	405	368	187	41	11	0	1506
1975	0	0	6	44	195	313	359	406	193	63	21	0	1600
1976	0	1	8	30	67	273	359	346	159	11	0	0	1254
1977	0	0	3	51	171	367	456	403	266	17	1	0	1735
1978	0	0	2	40	144	346	428	420	345	40	7	1	1773
1979	0	0	13	33	181	327	436	475	243	49	5	0	1762
1980	0	4	4	49	227	428	632	589	440	51	0	0	2424
1981	0	0	9	124	131	494	540	398	246	36	4	0	1982
1982	2	0	25	13	243	346	446	394	192	73	8	6	1748
1983	0	0	3	10	118	278	515	512	212	40	0	0	1688
1984	0	0	2	21	132	405	372	397	210	178	1	2	1720
1985	0	0	18	88	172	381	423	401	248	119	49	0	1899
1986	0	0	11	74	208	455	599	401	300	83	34	0	2165
1987	0	0	2	60	266	391	502	531	281	12	6	2	2053
1988	0	0	5	49	169	416	490	502	258	18	0	0	1907
1989	0	7	36	101	170	364	467	452	273	85	6	0	1961
1990	0	5	26	66	194	415	490	488	341	98	2	0	2125
1991	0	0	20	89	258	358	502	446	305	70	4	13	2065
1992	0	1	13	73	155	292	478	349	265	20	0	0	1646
1993	0	0	11	22	208	435	639	536	364	80	19	0	2314

SNOWFALL (inches) ATLANTA, GEORGIA

SEASON	JULY	AUG	SEP	OCT	NOV	DEC	JAN	FEB	MAR	APR	MAY	JUNE	TOTAL
1964-65	0.0	0.0	0.0	0.0	0.0	T	2.4	0.1	0.5	0.0	0.0	0.0	3.0
1965-66	0.0	0.0	0.0	0.0	0.0	0.0	0.7	T	T	0.0	0.0	0.0	0.7
1966-67	0.0	0.0	0.0	0.0	T	0.0	0.0	2.0	0.0	0.0	0.0	0.0	2.0
1967-68	0.0	0.0	0.0	0.0	0.0	0.0	0.7	3.5	T	0.0	0.0	0.0	4.2
1968-69	0.0	0.0	0.0	0.0	1.0	T	T	1.2	0.0	0.0	0.0	0.0	2.2
1969-70	0.0	0.0	0.0	0.0	T	T	0.6	T	0.0	0.0	0.0	0.0	0.6
1970-71	0.0	0.0	0.0	0.0	0.0	T	T	1.0	T	T	0.0	0.0	1.0
1971-72	0.0	0.0	0.0	0.0	T	1.0	T	T	T	0.0	0.0	0.0	1.0
1972-73	0.0	0.0	0.0	0.0	0.0	0.0	1.0	T	0.0	0.0	0.0	0.0	1.0
1973-74	0.0	0.0	0.0	0.0	0.0	T	0.0	T	T	0.0	0.0	0.0	T
1974-75	0.0	0.0	0.0	0.0	0.0	T	T	T	T	0.0	0.0	0.0	T
1975-76	0.0	0.0	0.0	0.0	0.6	0.0	T	T	T	0.0	0.0	0.0	0.6
1976-77	0.0	0.0	0.0	0.0	0.0	0.0	1.0	0.0	0.0	0.0	0.0	0.0	1.0
1977-78	0.0	0.0	0.0	0.0	0.0	T	T	0.3	T	0.0	0.0	0.0	0.3
1978-79	0.0	0.0	0.0	0.0	0.0	0.0	0.2	4.4	0.0	0.0	0.0	0.0	4.6
1979-80	0.0	0.0	0.0	0.0	0.0	0.0	T	1.7	2.7	0.0	0.0	0.0	4.4
1980-81	0.0	0.0	0.0	0.0	0.0	0.0	T	T	0.0	0.0	0.0	0.0	T
1981-82	0.0	0.0	0.0	0.0	0.0	T	7.0	0.7	0.0	0.0	0.0	0.0	7.7
1982-83	0.0	0.0	0.0	0.0	0.0	0.0	1.9	0.5	7.9	0.0	0.0	0.0	10.3
1983-84	0.0	0.0	0.0	0.0	0.0	T	T	1.3	T	0.0	0.0	0.0	1.3
1984-85	0.0	0.0	0.0	0.0	0.0	T	0.4	1.5	0.0	0.0	0.0	0.0	1.9
1985-86	0.0	0.0	0.0	0.0	0.0	T	0.4	T	0.0	0.0	0.0	0.0	0.4
1986-87	0.0	0.0	0.0	0.0	0.0	0.0	3.6	T	1.2	T	0.0	0.0	4.8
1987-88	0.0	0.0	0.0	0.0	0.0	0.0	4.2	T	0.0	0.0	0.0	0.0	4.2
1988-89	0.0	0.0	0.0	0.0	0.0	T	0.0	0.7	0.0	T	0.0	0.0	0.7
1989-90	0.0	0.0	0.0	0.0	0.0	1.3	0.0	0.0	0.0	T	0.0	0.0	1.3
1990-91	0.0	0.0	0.0	0.0	0.0	0.0	2.1	T	T	0.0	0.0	0.0	2.1
1991-92	0.0	0.0	0.0	0.0	0.0	0.0	5.0	0.0	T	0.0	0.0	0.0	5.0
1992-93	0.0	0.0	0.0	0.0	0.0	T	0.0	T	4.2	0.0	0.0	0.0	4.2
1993-94	0.0	0.0	0.0	T	0.0	2.8							
Record Mean	0.0	0.0	0.0	T	T	0.2	0.9	0.5	0.4	T	0.0	0.0	2.1

AUGUSTA, GEORGIA

The boundary between the Piedmont Plateau and the Coastal Plain, known as the Fall Line, crosses the Savannah River basin in a general northeast-southwest direction near Augusta, Georgia. The Weather Service Office at Bush Field is located in the Savannah River Valley approximately 2 miles west of the river and 203 miles above the mouth of the Savannah. Hills some 200 feet higher than the station are found slightly more than 1 mile to the west and approximately 4 miles to the southwest, and some 5 miles to the south and southeast. Swampland is found immediately to the north, east, and south of the station.

The length of the growing season averages 241 days. The average last occurrence in the spring of temperatures of 32 degrees is mid-March, and the first in the fall is mid-November.

Measurable snow is a rarity and then remains on the ground only a short time. Ice storms, damaging winds, and very low temperatures are also of rare occurrence.

Augusta has been protected, to a great extent, from flooding of the Savannah River by the construction of two multipurpose dams. The Clark Hill Dam is located 21.7 miles above the city and Hartwell Dam has been constructed 89 miles above Augusta.

NORMALS, MEANS, AND EXTREMES

AUGUSTA, GEORGIA

LATITUDE: 33°22'N LONGITUDE: 81°58' W ELEVATION: FT. GRND 136 BARO 148 TIME ZONE: EASTERN WBAN: 03820

	(a)	JAN	FEB	MAR	APR	MAY	JUNE	JULY	AUG	SEP	OCT	NOV	DEC	YEAR
TEMPERATURE °F:														
Normals														
-Daily Maximum		55.7	60.1	68.8	76.6	83.7	89.3	91.7	90.3	85.7	77.2	68.3	59.5	75.6
-Daily Minimum		32.0	34.7	42.2	48.6	57.5	65.6	69.9	69.1	63.1	50.3	41.6	34.8	50.8
-Monthly		43.9	47.4	55.5	62.7	70.7	77.5	80.8	79.7	74.5	63.8	55.0	47.2	63.2
Extremes														
-Record Highest	43	80	86	88	96	99	105	107	108	101	97	90	82	108
-Year		1985	1962	1990	1986	1964	1952	1980	1983	1957	1954	1961	1982	AUG 1983
-Record Lowest	43	-1	9	12	26	35	47	55	54	36	22	15	5	-1
-Year		1985	1973	1980	1982	1971	1984	1951	1968	1967	1952	1970	1981	JAN 1985
NORMAL DEGREE DAYS:														
Heating (base 65°F)		654	493	305	106	12	0	0	0	0	129	314	552	2565
Cooling (base 65°F)		0	0	11	37	188	375	490	456	285	92	14	0	1948
% OF POSSIBLE SUNSHINE														
MEAN SKY COVER (tenths)														
Sunrise - Sunset	43	6.3	6.0	6.0	5.4	5.8	5.8	6.1	5.8	5.7	4.8	5.3	6.0	5.8
MEAN NUMBER OF DAYS:														
Sunrise to Sunset														
-Clear	43	9.2	8.8	9.0	10.7	8.4	7.7	6.1	7.2	9.2	13.7	11.7	10.0	111.7
-Partly Cloudy	43	5.6	6.1	8.0	8.0	10.9	11.5	13.2	13.7	9.4	7.3	6.0	6.1	105.9
-Cloudy	43	16.2	13.3	14.1	11.3	11.7	10.8	11.7	10.1	11.4	10.0	12.3	14.9	147.7
Precipitation .01 inches or more	43	10.3	9.1	10.2	7.8	9.2	9.4	11.3	10.3	7.5	5.9	6.9	9.3	107.2
Snow,Ice pellets,hail 1.0 inches or more	43	0.1	0.2	0.*	0.0	0.0	0.0	0.0	0.0	0.0	0.0	0.0	0.*	0.4
Thunderstorms	43	0.8	1.6	2.7	4.0	6.4	9.5	13.0	10.2	3.6	1.2	0.8	0.6	54.4
Heavy Fog Visibility 1/4 mile or less	43	3.1	2.4	1.7	1.7	1.6	1.2	1.4	3.0	3.0	2.7	3.3	3.7	28.7
Temperature °F														
-Maximum														
90° and above	29	0.0	0.0	0.0	0.9	5.4	15.5	22.7	19.0	9.7	0.8	0.0	0.0	74.0
32° and below	29	0.5	0.*	0.*	0.0	0.0	0.0	0.0	0.0	0.0	0.0	0.0	0.1	0.7
-Minimum														
32° and below	29	15.8	12.3	5.1	0.7	0.0	0.0	0.0	0.0	0.0	0.6	6.9	13.8	55.2
0° and below	29	0.*	0.0	0.0	0.0	0.0	0.0	0.0	0.0	0.0	0.0	0.0	0.0	*
AVG. STATION PRESS.(mb)	21	1014.9	1013.9	1011.8	1010.9	1010.3	1010.5	1011.4	1011.9	1012.2	1013.7	1014.6	1015.5	1012.6
RELATIVE HUMIDITY (%)														
Hour 01	28	79	76	76	79	86	86	87	90	90	88	84	81	84
Hour 07 (Local Time)	29	83	82	84	85	87	86	88	92	92	90	88	85	87
Hour 13	29	54	49	48	45	49	51	54	57	55	49	50	53	51
Hour 19	29	66	59	56	54	61	62	67	72	76	76	72	70	66
PRECIPITATION (inches):														
Water Equivalent														
-Normal		4.05	4.27	4.65	3.31	3.77	4.13	4.24	4.50	3.02	2.84	2.48	3.40	44.66
-Maximum Monthly	43	8.91	7.67	11.92	8.43	9.61	8.84	11.43	11.34	9.51	14.82	7.76	8.65	14.82
-Year		1987	1961	1980	1961	1979	1989	1967	1986	1975	1990	1985	1981	OCT 1990
-Minimum Monthly	43	0.75	0.69	0.88	0.60	0.48	0.68	1.02	0.65	0.31	T	0.09	0.32	T
-Year		1981	1968	1968	1970	1951	1984	1987	1980	1984	1953	1960	1955	OCT 1953
-Maximum in 24 hrs	43	3.61	3.69	5.31	3.96	4.44	5.08	3.71	5.98	4.93	8.57	3.82	3.12	8.57
-Year		1960	1985	1967	1955	1981	1981	1979	1964	1969	1990	1985	1970	OCT 1990
Snow,Ice pellets,hail														
-Maximum Monthly	43	2.6	14.0	1.1	T	0.0	0.0	0.0	0.0	0.0	0.0	T	1.0	14.0
-Year		1992	1973	1980	1992							1968	1993	FEB 1973
-Maximum in 24 hrs	43	2.6	13.7	1.1	T	0.0	0.0	0.0	0.0	0.0	0.0	T	1.0	13.7
-Year		1992	1973	1980	1992							1968	1993	FEB 1973
WIND:														
Mean Speed (mph)	42	7.0	7.6	7.9	7.5	6.4	6.1	5.8	5.4	5.5	5.7	6.0	6.6	6.5
Prevailing Direction through 1963		W	WNW	WNW	SE	SE	SE	SE	SE	NE	NW	NW	NW	SE
Fastest Obs. 1 Min.														
-Direction (!!)	42	23	30	23	32	28	08	33	18	32	18	27	25	08
-Speed (MPH)	42	36	40	52	39	48	62	48	45	35	40	40	34	62
-Year		1978	1982	1972	1962	1967	1965	1970	1953	1959	1977	1954	1954	JUN 1965
Peak Gust														
-Direction (!!)	10	NW	W	W	S	W	NW	NW	NE	W	W	N	NW	NW
-Speed (mph)	10	46	51	56	51	45	60	54	48	47	36	45	44	60
-Date		1986	1985	1986	1984	1984	1988	1987	1987	1989	1985	1985	1984	JUN 1988

PRECIPITATION (inches)

AUGUSTA, GEORGIA

YEAR	JAN	FEB	MAR	APR	MAY	JUNE	JULY	AUG	SEP	OCT	NOV	DEC	ANNUAL
1964	7.08	4.84	5.56	5.33	4.27	5.32	9.66	9.91	2.59	6.34	1.24	3.90	66.04
1965	1.34	5.42	7.29	2.21	1.88	5.67	5.39	2.55	1.05	2.62	1.84	1.16	38.42
1966	7.01	5.36	3.73	2.37	5.77	3.50	3.44	5.74	2.10	1.83	0.85	3.32	45.02
1967	3.37	3.86	6.53	1.92	6.98	4.67	11.43	8.00	0.61	0.55	2.61	2.93	53.46
1968	3.77	0.69	0.88	2.44	4.05	5.08	4.44	1.31	4.85	3.13	3.07	2.89	36.60
1969	1.98	2.33	3.23	4.53	4.33	4.53	6.63	4.80	7.03	1.09	1.62	3.76	45.86
1970	2.71	2.38	6.34	0.60	4.13	1.75	6.29	5.39	0.79	3.92	0.63	5.06	39.99
1971	4.62	5.35	9.57	2.38	3.85	3.51	5.11	6.69	2.50	3.48	2.64	2.69	52.39
1972	6.08	3.08	3.06	0.90	4.05	6.25	3.36	2.45	2.60	0.87	2.79	5.27	40.76
1973	5.18	5.22	6.22	3.71	2.55	7.28	2.47	2.63	2.97	2.02	0.57	2.81	43.63
1974	3.99	5.76	2.32	4.02	4.15	3.63	4.05	3.86	2.83	0.09	2.38	4.05	41.13
1975	3.71	5.22	5.23	4.43	5.01	5.10	5.32	3.53	9.51	1.29	2.12	4.58	55.05
1976	3.51	0.95	4.11	2.00	6.12	4.77	2.00	1.81	6.12	5.06	3.61	5.61	45.67
1977	3.66	1.90	8.18	1.22	2.53	1.80	3.07	7.84	3.26	3.48	3.71	3.01	43.66
1978	7.76	1.50	3.54	3.58	2.16	1.59	1.70	4.91	1.34	1.12	2.50	1.26	32.96
1979	3.40	7.34	2.48	5.27	9.61	1.56	6.12	3.56	4.81	1.50	1.95	1.85	49.45
1980	4.07	3.17	11.92	1.28	1.84	4.31	2.12	0.65	5.06	1.62	2.24	0.96	39.24
1981	0.75	5.26	2.62	2.27	5.29	7.08	1.72	6.20	0.72	2.91	0.91	8.65	44.38
1982	3.00	4.60	1.54	5.23	3.78	3.46	3.56	3.09	1.91	3.65	2.34	4.93	41.09
1983	4.47	6.02	6.86	5.47	1.93	3.90	1.44	4.99	5.40	2.31	4.64	5.24	52.67
1984	3.40	4.93	5.88	6.50	7.55	0.68	7.70	4.42	0.31	1.00	0.65	1.25	44.27
1985	3.22	6.63	1.28	0.97	1.75	2.92	3.71	1.79	0.39	6.21	7.76	1.65	38.28
1986	1.46	2.51	3.23	1.02	2.80	1.41	6.26	11.34	0.74	3.92	4.93	4.12	43.74
1987	8.91	7.23	4.27	0.77	1.57	5.75	1.02	3.99	2.04	0.18	4.06	1.38	41.17
1988	4.30	3.30	3.14	5.04	1.60	4.89	1.77	3.47	5.55	5.40	1.30	1.31	41.07
1989	1.51	3.23	4.37	5.24	2.96	8.84	8.15	3.19	3.73	2.14	1.29	4.68	49.33
1990	2.71	2.70	2.03	1.07	1.84	1.44	1.92	6.46	1.11	14.82	2.57	1.98	40.65
1991	7.50	1.97	8.51	3.15	5.76	5.57	11.12	6.98	1.36	0.57	1.39	2.53	56.41
1992	4.57	4.50	3.69	2.70	1.19	4.82	7.04	5.72	3.28	4.94	6.71	2.24	51.40
1993	5.94	3.23	7.57	1.52	1.28	1.54	2.47	2.77	3.45	4.24	1.58	2.70	38.29
Record Mean	3.79	4.09	4.54	3.35	3.27	4.17	4.88	4.65	3.33	2.53	2.54	3.35	44.49

AVERAGE TEMPERATURE (deg. F)

AUGUSTA, GEORGIA

YEAR	JAN	FEB	MAR	APR	MAY	JUNE	JULY	AUG	SEP	OCT	NOV	DEC	ANNUAL
#1964	44.5	43.6	56.0	64.7	72.9	80.6	79.2	79.1	74.4	60.6	58.3	50.9	63.7
1965	46.1	49.5	53.3	65.0	74.8	75.9	79.1	79.3	75.8	63.1	55.6	47.2	63.8
1966	41.1	47.2	52.5	62.2	70.1	74.0	80.6	78.2	72.6	62.8	53.7	45.7	61.7
1967	47.5	44.3	58.5	66.3	68.5	74.4	77.4	76.9	68.7	61.9	51.4	51.2	62.3
1968	41.9	41.4	54.5	64.5	69.1	76.8	80.0	81.7	74.2	64.8	53.6	43.5	62.2
1969	43.9	45.0	50.0	64.2	70.4	79.2	81.6	76.3	71.6	65.1	50.3	43.6	61.8
1970	37.6	45.3	54.6	64.1	70.8	76.5	80.2	78.9	76.4	65.8	50.8	48.4	62.5
1971	45.3	45.8	49.8	60.2	67.4	78.7	79.3	78.6	76.0	68.9	53.6	55.7	63.3
1972	51.8	46.8	54.9	63.1	68.6	74.2	79.7	80.4	75.6	63.5	53.3	51.6	63.6
1973	45.1	44.7	59.9	60.0	68.2	77.4	81.1	78.8	77.4	64.4	56.4	46.8	63.4
1974	56.8	48.6	58.7	61.8	71.8	74.2	78.4	78.4	73.2	60.5	52.0	45.9	63.4
1975	47.4	50.1	53.1	60.7	72.5	75.8	77.5	80.6	74.5	66.7	56.7	47.4	63.6
1976	42.6	53.7	60.4	63.3	67.9	75.3	80.3	78.3	72.4	59.7	48.4	44.3	62.2
1977	35.5	44.6	59.2	65.1	72.4	80.2	82.7	79.7	76.5	60.9	58.5	46.2	63.5
1978	39.1	40.1	52.5	63.2	69.5	78.2	81.8	80.8	75.5	62.6	58.9	48.1	62.5
1979	41.4	44.0	56.5	63.1	70.6	74.8	80.0	79.6	74.3	62.4	55.6	45.9	62.4
1980	45.9	43.1	52.5	63.2	70.7	77.9	83.7	82.3	77.3	61.4	52.5	45.1	63.0
1981	39.6	48.6	52.3	65.2	68.3	81.1	81.5	77.0	71.4	59.3	51.6	41.2	61.4
1982	40.1	48.9	56.3	58.7	70.2	79.8	82.5	78.7	73.0	64.0	56.9	53.1	63.5
1983	41.9	46.7	54.4	58.8	70.6	75.4	82.6	83.3	72.8	65.1	53.9	44.5	62.5
1984	43.0	50.2	55.3	61.3	69.9	78.5	79.8	80.3	72.8	72.0	52.5	55.2	64.2
1985	42.1	48.8	58.7	65.7	73.2	80.4	81.7	80.3	75.2	70.7	64.6	44.2	65.5
1986	42.7	52.6	55.8	63.4	72.5	81.5	86.0	79.6	77.6	66.2	60.9	48.4	65.6
1987	44.6	47.3	54.8	60.6	73.1	79.2	82.5	83.8	75.8	57.8	56.5	50.4	63.9
1988	41.1	46.0	55.5	62.4	69.1	76.3	81.2	81.4	75.4	59.0	55.6	45.7	62.4
1989	50.5	50.4	57.6	62.9	69.2	78.9	80.9	80.0	74.7	65.4	55.8	41.4	64.0
1990	51.8	56.7	59.4	62.4	71.5	79.4	83.5	82.0	76.2	66.6	56.3	51.5	66.4
1991	46.8	52.3	58.5	66.4	75.3	78.5	82.3	80.5	75.1	65.0	51.9	50.4	65.3
1992	46.4	51.0	54.1	61.2	67.6	75.8	81.8	79.1	75.6	62.3	55.5	46.6	63.1
1993	50.1	46.4	53.6	60.7	72.4	81.5	86.3	82.4	77.7	65.1	55.8	45.5	64.8
Record Mean	46.8	49.3	56.3	63.8	72.0	78.7	81.3	80.3	75.6	65.0	54.9	47.9	64.3
Max	57.1	60.1	67.6	75.8	83.6	89.5	91.3	90.2	85.7	76.5	66.6	58.4	75.2
Min	36.6	38.4	45.1	51.7	60.4	67.8	71.2	70.5	65.4	53.4	43.2	37.3	53.4

REFERENCE NOTES

AUGUSTA, GEORGIA

GENERAL
T - TRACE AMOUNT.
BLANK ENTRIES DENOTE MISSING/UNREPORTED DATA.
# INDICATES A STATION OR INSTRUMENT RELOCATION.

SPECIFIC

(a) - LENGTH OF RECORD IN YEARS, ALTHOUGH INDIVIDUAL MONTHS MAY BE MISSING.
0.* OR * - THE VALUE IS BETWEEN 0.0 AND 0.05
NORMALS - BASED ON THE 1961-1990 RECORD PERIOD.
EXTREMES - DATES ARE THE MOST RECENT OCCURRENCE.
WIND DIR.- NUMERALS SHOW TENS OF DEGRESS CLOCKWISE FROM TRUE NORTH. "00" INDICATES CALM.
RESULTANT DIRECTIONS ARE GIVEN TO WHOLE DEGREES.
BOLD VALUES INDICATE EXTREME VALUES WHICH OCCURRED AFTER THE ASOS SYSTEM WAS COMMISSIONED.

RECORD = PERIOD OF RECORD
RECORD MEAN PRECIPITATION IS THE MEAN OF ALL DAILY PRECIPITATION AMOUNTS DURING THE PERIOD OF RECORD.
RECORD MAX(MIN) TEMPERATURE IS THE MEAN OF ALL DAILY MAX(MIN) TEMPERATURES DURING THE PERIOD OF RECORD.
RECORD MEAN TEMPERATURE IS THE SUM OF THE RECORD MAX AND RECORD MIN DIVIDED BY 2.
AVERAGE TEMPERATURE IS THE SUM OF THE MEAN DAILY MAX AND MIN TEMPERATURE DIVIDED BY 2.

EXCEPTIONS

RECORD MEANS ARE THROUGH THE CURRENT YEAR, BEGINNING IN 1875 FOR TEMPERATURE
1871 FOR PRECIPITATION
1951 FOR SNOWFALL

HEATING DEGREE DAYS Base 65 deg. F — AUGUSTA, GEORGIA

SEASON	JULY	AUG	SEP	OCT	NOV	DEC	JAN	FEB	MAR	APR	MAY	JUNE	TOTAL
1964-65	0	0	1	170	216	442	576	432	367	84	0	0	2288
1965-66	0	0	0	119	276	543	733	492	379	141	18	0	2701
1966-67	0	0	0	122	341	590	537	573	227	47	49	5	2491
1967-68	0	0	21	128	404	428	709	676	327	78	19	0	2790
1968-69	0	0	0	113	339	659	645	555	460	70	8	0	2849
1969-70	0	0	3	79	432	655	844	545	320	113	13	0	3004
1970-71	0	0	12	60	418	510	601	532	466	165	39	0	2803
1971-72	0	0	0	18	355	301	406	523	306	123	5	2	2039
1972-73	0	0	0	82	355	410	613	563	181	166	42	0	2412
1973-74	0	0	0	96	263	561	248	456	209	140	8	0	1981
1974-75	0	0	11	152	393	588	539	415	374	171	3	0	2646
1975-76	0	0	2	62	285	540	687	320	167	97	29	2	2191
1976-77	0	0	0	191	490	635	906	565	206	65	13	0	3071
1977-78	0	0	0	153	215	578	797	689	380	93	33	0	2938
1978-79	0	0	0	103	184	531	723	582	265	79	15	2	2484
1979-80	0	0	4	125	294	584	588	631	385	97	17	0	2725
1980-81	0	0	10	143	370	608	778	452	391	67	33	0	2852
1981-82	0	0	13	187	400	730	765	446	286	206	7	0	3040
1982-83	0	0	6	125	253	383	710	504	332	198	8	0	2519
1983-84	0	0	23	66	326	630	674	424	306	146	31	1	2627
1984-85	0	0	4	16	382	296	710	451	222	75	5	0	2161
1985-86	0	0	6	28	83	639	685	342	287	112	12	0	2194
1986-87	0	5	0	82	170	513	625	491	319	177	5	0	2387
1987-88	0	0	0	220	266	448	732	545	294	102	12	0	2619
1988-89	0	0	1	209	282	594	441	411	258	153	41	0	2390
1989-90	0	0	8	97	282	722	403	244	202	118	11	0	2087
1990-91	0	0	7	93	263	417	560	355	223	52	3	0	1973
1991-92	0	0	3	76	404	461	570	398	339	160	45	0	2456
1992-93	0	0	9	103	289	565	457	518	348	148	6	0	2443
1993-94	0	0	7	86	300	599							

COOLING DEGREE DAYS Base 65 deg. F — AUGUSTA, GEORGIA

YEAR	JAN	FEB	MAR	APR	MAY	JUNE	JULY	AUG	SEP	OCT	NOV	DEC	TOTAL
1969	0	0	3	54	183	433	524	357	208	90	1	0	1853
1970	0	0	3	93	201	352	478	438	360	95	0	4	2024
1971	0	0	0	26	122	415	452	428	336	146	19	17	1961
1972	3	2	2	73	126	283	461	486	325	39	10	3	1813
1973	0	0	28	23	149	379	505	436	382	85	11	4	2002
1974	1	4	16	48	226	284	420	421	265	22	11	3	1721
1975	2	3	11	51	242	332	395	488	293	122	43	0	1982
1976	0	0	30	51	127	320	482	419	228	34	0	0	1691
1977	0	0	33	77	250	461	555	465	355	34	27	0	2257
1978	0	0	1	46	177	402	531	497	324	35	9	13	2035
1979	0	0	9	29	197	301	473	460	288	52	20	0	1829
1980	0	0	2	50	201	396	585	543	385	39	0	0	2201
1981	0	0	6	79	143	491	515	378	208	19	4	0	1843
1982	0	0	23	22	174	450	549	431	254	100	16	21	2040
1983	0	0	10	18	187	318	555	573	263	76	0	0	2000
1984	0	0	13	39	188	411	468	481	244	241	16	1	2102
1985	9	4	33	100	263	468	525	480	318	209	78	0	2487
1986	0	0	12	71	248	501	655	464	383	126	56	4	2520
1987	0	0	8	48	264	432	548	590	331	5	17	1	2244
1988	0	1	5	31	147	346	510	518	323	32	8	1	1922
1989	0	8	38	99	178	423	500	471	305	117	12	0	2151
1990	0	18	36	49	219	438	578	536	350	151	9	6	2390
1991	0	6	29	99	329	414	542	483	310	82	16	14	2324
1992	0	1	6	56	135	330	526	444	331	27	14	0	1870
1993	2	0	2	27	240	503	666	547	393	96	31	2	2509

SNOWFALL (inches) — AUGUSTA, GEORGIA

SEASON	JULY	AUG	SEP	OCT	NOV	DEC	JAN	FEB	MAR	APR	MAY	JUNE	TOTAL
1964-65	0.0	0.0	0.0	0.0	0.0	0.0	T	T	T	0.0	0.0	0.0	T
1965-66	0.0	0.0	0.0	0.0	0.0	0.0	0.4	T	0.0	0.0	0.0	0.0	0.4
1966-67	0.0	0.0	0.0	0.0	0.0	T	0.0	3.3	0.0	0.0	0.0	0.0	3.3
1967-68	0.0	0.0	0.0	0.0	0.0	0.0	1.4	1.4	0.0	0.0	0.0	0.0	2.8
1968-69	0.0	0.0	0.0	0.0	T	0.0	T	T	0.0	0.0	0.0	0.0	T
1969-70	0.0	0.0	0.0	0.0	0.0	T	1.5	T	0.0	0.0	0.0	0.0	1.5
1970-71	0.0	0.0	0.0	0.0	0.0	0.0	T	T	T	0.0	0.0	0.0	T
1971-72	0.0	0.0	0.0	0.0	0.0	0.4	0.0	0.0	0.0	0.0	0.0	0.0	0.4
1972-73	0.0	0.0	0.0	0.0	0.0	0.0	0.4	14.0	0.0	0.0	0.0	0.0	14.4
1973-74	0.0	0.0	0.0	0.0	0.0	T	0.0	0.0	0.0	0.0	0.0	0.0	T
1974-75	0.0	0.0	0.0	0.0	0.0	T	0.0	0.0	T	0.0	0.0	0.0	T
1975-76	0.0	0.0	0.0	0.0	0.0	0.0	T	0.0	0.0	0.0	0.0	0.0	T
1976-77	0.0	0.0	0.0	0.0	0.0	0.0	0.5	0.0	0.0	0.0	0.0	0.0	0.5
1977-78	0.0	0.0	0.0	0.0	0.0	0.0	T	1.1	0.0	0.0	0.0	0.0	1.1
1978-79	0.0	0.0	0.0	0.0	0.0	T	0.0	3.4	0.0	0.0	0.0	0.0	3.4
1979-80	0.0	0.0	0.0	0.0	0.0	0.0	0.0	4.2	1.1	0.0	0.0	0.0	5.3
1980-81	0.0	0.0	0.0	0.0	0.0	T	0.0	T	0.0	0.0	0.0	0.0	T
1981-82	0.0	0.0	0.0	0.0	0.0	0.0	1.5	T	0.0	0.0	0.0	0.0	1.5
1982-83	0.0	0.0	0.0	0.0	0.0	0.0	T	T	T	0.0	0.0	0.0	T
1983-84	0.0	0.0	0.0	0.0	0.0	0.0	T	T	0.0	0.0	0.0	0.0	T
1984-85	0.0	0.0	0.0	0.0	0.0	0.0	T	T	0.0	0.0	0.0	0.0	T
1985-86	0.0	0.0	0.0	0.0	0.0	0.0	T	0.0	0.0	0.0	0.0	0.0	T
1986-87	0.0	0.0	0.0	0.0	0.0	0.0	T	T	0.0	0.0	0.0	0.0	T
1987-88	0.0	0.0	0.0	0.0	0.0	0.0	2.3	0.0	0.0	0.0	0.0	0.0	2.3
1988-89	0.0		0.0	0.0	0.0	T	0.0	3.7	0.0	0.0	0.0	0.0	
1989-90	0.0	0.0	0.0	0.0	0.0	0.4	0.0	0.0	0.0	0.0	0.0	0.0	0.4
1990-91	0.0	0.0	0.0	0.0	0.0	0.0	0.0	0.0	T	0.0	0.0	0.0	T
1991-92	0.0	0.0	0.0	0.0	0.0	0.0	2.6	0.0	0.0	T	0.0	0.0	2.6
1992-93	0.0	0.0	0.0	0.0	0.0	0.0	0.0	T	T	0.0	0.0	0.0	T
1993-94	0.0	0.0	0.0	0.0	0.0	1.0							
Record Mean	0.0	0.0	0.0	0.0	T	0.1	0.3	0.8	T	T	0.0	0.0	1.1

COLUMBUS, GEORGIA

Columbus is located on the Chattahoochee River at the western boundary of Georgia, about 225 miles west of the Atlantic Ocean and 170 miles north of the Gulf of Mexico. Elevation of the ground above sea level ranges from 200 to 500 feet and effects of terrain on the weather are negligible. The climate is that of the humid southeast, with pronounced maritime effects at some periods, and equally pronounced continental effects at others.

Annual rainfall is variable, the months of highest rainfall are generally March and July, and the driest are usually October and November. Heavy midsummer rainfall is commonly the result of frequent local thunderstorms. Heavy rains which occasionally come in autumn are likely to be due to Gulf or Caribbean hurricanes moving inland near the Columbus area.

Snow is rare, but almost every winter sees a few snowflakes falling in the area and occasionally a moderate to heavy snowfall is experienced.

The coldest month is usually January and the warmest is usually July.

Most days in summer will have a high temperature of 90 degrees or more, but few will reach 100 degrees. During many winters the minimum temperature does not drop below 20 degrees, but about one year in ten it will drop to 10 degrees or lower.

Based on the 1951–1980 period, the average first occurrence of 32 degrees Fahrenheit in the fall is November 9 and the average last occurrence in the spring is March 21.

NORMALS, MEANS, AND EXTREMES

COLUMBUS, GEORGIA

LATITUDE: 32°31'N LONGITUDE: 84°57' W ELEVATION: FT. GRND 445 BARO 449 TIME ZONE: EASTERN WBAN: 93842

	(a)	JAN	FEB	MAR	APR	MAY	JUNE	JULY	AUG	SEP	OCT	NOV	DEC	YEAR
TEMPERATURE °F:														
Normals														
-Daily Maximum		56.1	61.0	69.1	77.3	83.6	90.4	91.8	91.2	86.2	77.6	68.2	59.0	76.0
-Daily Minimum		35.2	37.6	45.1	52.2	61.0	68.4	71.8	71.4	66.3	54.1	45.0	38.2	53.9
-Monthly		45.7	49.3	57.2	64.8	72.3	79.5	81.9	81.4	76.4	66.0	56.7	48.6	65.0
Extremes														
-Record Highest	48	83	83	89	93	97	104	104	103	100	96	86	82	104
-Year		1949	1962	1982	1986	1962	1978	1986	1986	1990	1954	1961	1977	JUL 1986
-Record Lowest	48	-2	11	16	28	39	44	59	57	38	24	10	4	-2
-Year		1985	1973	1980	1950	1963	1956	1967	1952	1967	1952	1950	1962	JAN 1985
NORMAL DEGREE DAYS:														
Heating (base 65°F)		607	440	260	76	7	0	0	0	0	86	268	517	2261
Cooling (base 65°F)		9	0	18	70	234	435	524	508	342	117	19	8	2284
% OF POSSIBLE SUNSHINE														
MEAN SKY COVER (tenths)														
Sunrise - Sunset	48	6.6	6.2	6.1	5.4	5.7	5.7	6.3	5.7	5.6	4.6	5.4	6.1	5.8
MEAN NUMBER OF DAYS:														
Sunrise to Sunset														
-Clear	48	8.0	8.0	8.8	10.1	9.1	8.2	5.5	7.7	9.9	14.2	11.5	9.4	110.4
-Partly Cloudy	48	6.0	6.1	7.2	8.5	10.2	11.6	13.2	13.5	8.8	6.7	6.1	6.1	104.2
-Cloudy	48	16.9	14.1	15.0	11.3	11.7	10.2	12.4	9.8	11.2	10.1	12.4	15.6	150.6
Precipitation														
.01 inches or more	48	10.5	9.6	10.4	7.9	8.4	9.3	13.1	10.3	7.8	5.4	7.8	9.5	109.8
Snow,Ice pellets,hail														
1.0 inches or more	48	0.1	0.1	0.*	0.0	0.0	0.0	0.0	0.0	0.0	0.0	0.0	0.*	0.2
Thunderstorms	44	1.2	2.1	3.8	4.4	6.6	8.1	13.2	8.9	3.6	1.1	1.2	1.3	55.5
Heavy Fog Visibility														
1/4 mile or less	44	3.4	2.1	1.3	1.1	0.9	0.5	1.0	0.7	0.6	1.0	1.9	2.6	17.0
Temperature °F														
-Maximum														
90° and above	48	0.0	0.0	0.0	0.4	6.0	17.1	21.1	20.7	10.4	0.8	0.0	0.0	76.4
32° and below	48	0.3	0.1	0.0	0.0	0.0	0.0	0.0	0.0	0.0	0.0	0.0	0.1	0.4
-Minimum														
32° and below	48	12.9	9.0	3.7	0.2	0.0	0.0	0.0	0.0	0.0	0.3	4.7	11.2	42.0
0° and below	48	0.*	0.0	0.0	0.0	0.0	0.0	0.0	0.0	0.0	0.0	0.0	0.0	*
AVG. STATION PRESS. (mb)	21	1006.7	1005.5	1003.3	1002.8	1001.9	1002.3	1003.5	1003.5	1003.5	1005.2	1006.1	1007.1	1004.3
RELATIVE HUMIDITY (%)														
Hour 01	28	79	76	77	78	83	82	86	87	86	83	82	80	82
Hour 07 (Local Time)	48	84	83	85	85	85	85	89	91	90	89	87	85	87
Hour 13	48	59	54	52	48	51	52	57	57	56	52	54	58	54
Hour 19	48	68	61	56	54	59	60	68	68	69	69	71	71	65
PRECIPITATION (inches):														
Water Equivalent														
-Normal		4.59	4.85	5.77	4.30	4.17	4.07	5.54	3.73	3.23	2.22	3.56	4.97	51.00
-Maximum Monthly	48	10.22	9.41	12.53	11.67	8.45	10.83	13.24	10.11	6.94	8.09	12.45	9.39	13.24
-Year		1947	1961	1952	1953	1959	1967	1971	1991	1951	1964	1948	1953	JUL 1971
-Minimum Monthly	48	0.72	1.22	1.38	0.10	0.22	0.74	1.74	0.80	0.22	0.00	0.31	0.43	0.00
-Year		1989	1951	1985	1986	1962	1993	1957	1988	1984	1963	1956	1955	OCT 1963
-Maximum in 24 hrs	48	4.27	5.77	7.22	5.74	4.61	3.88	5.45	6.80	4.25	5.63	4.80	4.41	7.22
-Year		1978	1981	1990	1981	1957	1959	1989	1977	1971	1964	1977	1970	MAR 1990
Snow,Ice pellets,hail														
-Maximum Monthly	48	2.0	14.0	2.5	T	T	0.0	0.0	0.0	0.0	0.0	T	1.6	14.0
-Year		1982	1973	1993	1990	1993						1975	1993	FEB 1973
-Maximum in 24 hrs	48	1.4	14.0	2.5	T	T	0.0	0.0	0.0	0.0	0.0	T	1.6	14.0
-Year		1977	1973	1993	1990	1993						1975	1993	FEB 1973
WIND:														
Mean Speed (mph)	35	7.3	7.9	8.0	7.3	6.6	6.1	5.8	5.5	6.5	6.5	6.6	6.9	6.7
Prevailing Direction														
Fastest Obs. 1 Min.														
-Direction (!!)	35	25	20	27	28	23	29	36	18	36	23	31	33	29
-Speed (MPH)	35	37	52	44	40	39	55	52	47	38	26	37	35	55
-Year		1978	1960	1963	1985	1971	1971	1962	1959	1963	1963	1973	1990	JUN 1971
Peak Gust														
-Direction (!!)	10	SW	NW	NW	S	NE	SW	NW	N	N	W	NW	NW	NE
-Speed (mph)	10	45	61	52	67	82	52	53	68	43	35	49	47	82
-Date		1990	1990	1993	1990	1991	1989	1991	1987	1987	1992	1984	1990	MAY 1991

PRECIPITATION (inches)

COLUMBUS, GEORGIA

YEAR	JAN	FEB	MAR	APR	MAY	JUNE	JULY	AUG	SEP	OCT	NOV	DEC	ANNUAL
1964	7.01	5.37	5.56	11.38	3.63	3.91	9.57	5.18	4.80	8.09	2.08	6.64	73.22
1965	2.79	5.46	6.06	1.07	2.54	7.91	5.48	1.88	5.81	2.99	2.13	2.75	46.87
1966	7.87	8.45	7.30	2.76	7.87	4.54	5.10	7.16	3.46	4.70	3.98	6.74	69.93
1967	3.48	3.34	1.40	0.86	6.62	10.83	3.81	5.06	3.84	3.23	2.86	4.90	50.23
1968	1.78	1.54	4.38	2.58	4.73	1.80	3.02	3.64	2.57	0.85	4.34	4.88	36.11
1969	1.22	4.13	4.49	8.30	6.02	2.81	6.55	2.93	6.32	0.31	0.79	5.09	48.96
1970	4.59	4.22	6.85	2.91	4.24	5.35	6.36	2.91	2.94	4.47	2.48	6.89	54.21
1971	5.33	7.13	9.16	3.27	4.42	4.22	13.24	5.11	4.96	0.26	2.41	4.96	64.47
1972	6.10	3.92	5.01	1.07	3.31	7.74	7.75	1.04	2.47	2.31	4.36	9.38	54.46
1973	6.03	4.55	9.82	5.95	4.53	4.42	3.64	6.97	2.23	1.16	3.37	7.14	59.81
1974	7.50	3.98	2.13	5.85	7.54	2.87	5.26	5.20	4.87	0.55	3.88	5.15	54.78
1975	6.64	7.56	7.32	6.66	4.78	3.94	7.87	4.11	2.85	5.42	2.67	3.54	63.36
1976	3.64	1.59	7.88	2.70	3.75	6.88	2.73	3.71	4.97	5.06	5.67	3.87	52.45
1977	4.15	2.02	7.92	1.73	3.49	3.46	3.79	10.07	3.54	2.15	6.12	3.56	52.00
1978	8.35	1.77	4.43	4.32	6.46	2.44	5.14	3.97	0.72	0.02	2.91	3.01	43.54
1979	6.91	7.84	2.72	10.69	4.59	1.24	4.12	2.50	5.50	1.47	4.52	2.52	54.62
1980	3.98	3.53	11.20	5.13	6.90	2.69	3.22	4.49	2.32	1.92	1.77	1.66	48.81
1981	1.27	7.72	4.20	6.88	1.45	3.05	5.40	2.93	2.62	1.90	1.18	8.94	47.54
1982	4.22	4.88	2.17	8.42	4.19	2.46	5.73	2.40	1.12	2.00	5.29	8.74	51.62
1983	3.95	5.75	7.37	4.48	2.02	5.05	3.57	3.16	6.01	1.82	5.07	7.02	55.27
1984	3.60	3.22	5.96	2.16	4.42	1.96	7.50	3.50	0.22	0.42	2.47	2.69	38.12
1985	3.84	6.21	1.38	1.90	4.86	2.03	4.53	3.28	0.98	2.85	2.94	4.85	39.65
1986	1.45	4.67	8.78	0.10	2.55	0.83	3.15	3.37	3.62	3.94	9.85	2.65	44.96
1987	5.47	7.69	4.38	0.62	6.25	9.19	4.31	1.82	2.31	0.37	3.22	2.90	48.53
1988	6.72	2.73	3.35	4.64	1.61	0.94	4.91	0.80	4.25	1.52	3.63	3.72	38.82
1989	0.72	2.64	4.88	6.51	2.42	7.34	11.58	2.27	3.68	2.46	5.11	7.31	56.92
1990	4.45	6.31	9.40	2.64	4.45	1.05	3.58	1.68	0.54	3.00	1.64	2.76	41.50
1991	7.46	1.94	7.86	4.32	7.00	3.90	3.79	10.11	2.38	0.58	4.70	3.34	57.38
1992	6.65	5.07	4.72	2.59	0.80	2.76	8.28	9.15	2.79	2.46	11.63	3.95	60.85
1993	4.12	3.52	8.06	2.21	2.04	0.74	3.41	4.32	0.97	3.72	4.92	4.01	42.04
Record Mean	4.45	4.55	5.85	4.28	4.14	3.85	5.69	4.06	3.18	2.04	3.69	4.75	50.53

AVERAGE TEMPERATURE (deg. F)

COLUMBUS, GEORGIA

YEAR	JAN	FEB	MAR	APR	MAY	JUNE	JULY	AUG	SEP	OCT	NOV	DEC	ANNUAL
1964	44.1	44.1	55.6	65.2	71.6	79.9	79.0	79.5	76.3	61.6	58.4	51.5	63.9
1965	47.7	49.2	54.8	68.1	75.4	76.4	79.7	80.4	76.1	63.7	57.6	48.2	64.8
1966	42.0	47.7	54.0	65.2	72.1	76.3	82.2	79.3	75.5	65.5	56.0	48.6	63.7
1967	49.0	47.1	60.8	69.8	71.6	77.7	78.6	78.6	70.6	62.9	52.6	52.1	64.3
1968	44.2	42.0	53.9	65.1	70.8	79.8	80.5	80.8	74.5	65.3	52.1	43.7	62.7
1969	45.4	46.4	49.8	65.8	71.2	79.6	82.4	78.0	73.4	67.0	51.9	45.4	63.0
1970	39.3	46.6	55.3	67.6	73.7	77.4	81.5	81.0	79.5	69.0	52.5	52.3	64.6
1971	48.2	48.5	52.0	63.0	69.2	80.0	79.2	79.8	78.0	70.2	55.0	57.8	65.1
1972	52.9	48.8	57.4	65.9	70.7	76.7	79.6	81.8	77.9	66.4	54.3	54.6	65.6
1973	46.4	45.7	61.6	61.6	69.1	78.7	82.1	79.8	79.0	67.7	59.1	48.9	65.0
1974	59.6	51.3	61.3	63.9	73.2	75.5	80.5	80.0	73.6	62.6	54.7	49.7	65.5
1975	51.2	52.5	56.0	63.6	74.2	78.4	78.9	80.8	73.4	66.7	57.2	47.6	65.0
1976	43.1	54.6	59.8	65.1	68.2	76.4	80.6	79.6	74.2	59.9	48.7	45.0	62.9
1977	36.6	46.8	60.2	67.6	74.6	81.5	84.1	82.0	78.1	62.7	59.2	47.8	65.1
1978	39.5	42.2	54.3	65.5	72.5	80.7	82.8	81.4	79.2	66.4	62.2	49.9	64.7
1979	42.4	46.1	58.5	66.5	72.4	78.0	82.1	81.6	75.1	66.0	57.0	49.5	64.6
1980	50.6	46.8	55.4	63.1	72.4	79.8	84.3	83.5	80.1	63.9	55.0	47.9	65.2
1981	41.8	50.6	55.1	68.3	70.2	83.3	83.4	80.2	75.6	63.8	57.1	45.1	64.5
1982	44.6	51.4	61.0	62.8	73.7	79.4	81.3	80.8	75.2	67.3	58.4	54.8	65.9
1983	44.9	49.2	55.1	60.6	72.4	78.9	85.1	84.4	74.4	66.4	55.6	45.5	64.4
1984	43.9	50.7	55.3	62.5	70.5	80.4	80.5	80.4	75.5	73.7	54.3	56.8	65.4
1985	40.7	49.1	60.6	66.3	72.2	79.9	80.5	80.5	75.6	71.1	64.7	44.7	65.5
1986	47.4	55.6	58.4	64.5	74.4	82.4	85.8	81.0	78.5	67.3	61.9	49.1	67.2
1987	45.5	49.2	57.1	62.7	74.9	79.3	82.5	84.5	76.5	60.2	58.9	53.3	65.4
1988	43.5	47.8	56.6	64.6	70.8	79.7	81.4	83.3	76.7	62.4	58.4	49.1	64.5
1989	52.8	51.4	59.5	63.6	71.4	78.9	80.6	81.8	75.5	64.8	55.9	41.5	64.8
1990	51.1	55.9	59.6	63.1	71.1	81.1	83.1	84.1	79.3	67.4	58.0	53.3	67.3
1991	48.6	54.2	59.7	68.8	75.4	78.6	82.2	80.7	77.3	67.3	54.7	52.3	66.7
1992	47.7	54.2	56.9	63.4	71.3	78.3	82.7	78.8	75.9	64.2	55.2	48.9	64.8
1993	51.7	48.2	54.4	60.8	71.9	81.5	85.8	82.5	78.4	65.8	56.0	47.2	65.4
Record Mean	47.0	50.0	56.7	64.8	72.4	79.2	81.5	81.0	76.2	65.8	55.8	49.1	64.9
Max	57.5	61.4	68.6	77.4	84.2	90.1	91.5	91.1	86.3	77.6	67.7	59.9	76.1
Min	36.5	38.6	44.7	52.1	60.7	68.3	71.5	70.9	66.0	54.1	44.0	38.2	53.8

REFERENCE NOTES

COLUMBUS, GEORGIA

GENERAL
T - TRACE AMOUNT.
BLANK ENTRIES DENOTE MISSING/UNREPORTED DATA.
# INDICATES A STATION OR INSTRUMENT RELOCATION.

SPECIFIC

(a) - LENGTH OF RECORD IN YEARS, ALTHOUGH INDIVIDUAL MONTHS MAY BE MISSING.
0.* OR * - THE VALUE IS BETWEEN 0.0 AND 0.05
NORMALS - BASED ON THE 1961-1990 RECORD PERIOD.
EXTREMES - DATES ARE THE MOST RECENT OCCURRENCE
WIND DIR.- NUMERALS SHOW TENS OF DEGRESS CLOCKWISE FROM TRUE NORTH. "00" INDICATES CALM.
RESULTANT DIRECTIONS ARE GIVEN TO WHOLE DEGREES.
BOLD VALUES INDICATE EXTREME VALUES WHICH OCCURRED AFTER THE ASOS SYSTEM WAS COMMISSIONED.

RECORD = PERIOD OF RECORD
RECORD MEAN PRECIPITATION IS THE MEAN OF ALL DAILY PRECIPITATION AMOUNTS DURING THE PERIOD OF RECORD.
RECORD MAX(MIN) TEMPERATURE IS THE MEAN OF ALL DAILY MAX(MIN) TEMPERATURES DURING THE PERIOD OF RECORD.
RECORD MEAN TEMPERATURE IS THE SUM OF THE RECORD MAX AND RECORD MIN DIVIDED BY 2.
AVERAGE TEMPERATURE IS THE SUM OF THE MEAN DAILY MAX AND MIN TEMPERATURE DIVIDED BY 2.

EXCEPTIONS

THUNDERSTORMS AND HEAVY FOG ARE THROUGH 1964 AND 1969 TO DATE. DATA PRIOR TO 1964 MAY BE INCOMPLETE, DUE TO PART-TIME OPERATIONS

RECORD MEANS ARE THROUGH THE CURRENT YEAR, BEGINNING IN 1946 FOR TEMPERATURE
1946 FOR PRECIPITATION
1946 FOR SNOWFALL

HEATING DEGREE DAYS Base 65 deg. F COLUMBUS, GEORGIA

SEASON	JULY	AUG	SEP	OCT	NOV	DEC	JAN	FEB	MAR	APR	MAY	JUNE	TOTAL
1964-65	0	0	0	143	219	418	530	441	349	45	0	0	2145
1965-66	0	0	0	105	219	518	707	478	335	91	8	0	2461
1966-67	0	0	0	63	277	505	489	496	172	18	17	3	2040
1967-68	0	0	29	114	370	403	639	660	345	57	6	0	2623
1968-69	0	0	0	104	382	654	602	517	467	41	8	0	2775
1969-70	0	0	0	49	387	601	790	509	294	60	2	0	2692
1970-71	0	0	0	16	369	393	515	458	397	113	25	0	2286
1971-72	0	0	0	11	313	252	371	468	250	84	0	1	1750
1972-73	0	0	1	56	328	336	572	536	143	146	28	0	2146
1973-74	0	0	0	55	200	496	177	382	162	97	1	0	1570
1974-75	0	0	8	113	314	470	429	343	301	116	0	0	2094
1975-76	0	0	20	61	286	532	672	301	179	55	13	0	2119
1976-77	0	0	0	187	483	615	874	501	183	51	0	0	2894
1977-78	0	0	0	110	186	530	781	634	334	63	11	0	2649
1978-79	0	0	0	57	104	468	693	527	219	34	7	0	2109
1979-80	0	0	1	59	255	475	440	529	301	91	3	0	2154
1980-81	0	0	3	93	299	520	709	401	311	25	7	0	2368
1981-82	0	0	6	94	253	608	626	372	194	119	0	0	2272
1982-83	0	0	6	83	222	337	614	437	318	146	3	0	2166
1983-84	0	0	9	59	291	602	647	409	299	123	21	0	2460
1984-85	0	0	0	12	328	251	747	444	158	75	3	0	2018
1985-86	0	0	1	24	88	625	542	269	215	83	0	0	1847
1986-87	0	6	0	66	157	486	597	436	250	142	2	0	2142
1987-88	0	0	0	157	203	370	661	497	270	65	1	0	2224
1988-89	0	0	0	117	197	489	376	390	226	137	21	0	1953
1989-90	0	0	15	101	275	722	427	257	181	112	10	0	2100
1990-91	0	0	5	81	214	363	501	302	203	25	0	0	1694
1991-92	0	0	0	38	322	406	531	311	266	122	23	0	2019
1992-93	0	0	0	56	300	496	405	464	335	135	6	0	2197
1993-94	0	0	0	86	290	543							

COOLING DEGREE DAYS Base 65 deg. F COLUMBUS, GEORGIA

YEAR	JAN	FEB	MAR	APR	MAY	JUNE	JULY	AUG	SEP	OCT	NOV	DEC	TOTAL
1969	0	1	2	72	208	446	547	409	257	118	1	0	2061
1970	0	0	1	144	280	378	520	504	445	147	0	5	2424
1971	0	2	1	61	160	458	444	468	394	179	23	33	2223
1972	6	4	21	118	181	360	458	526	397	105	14	22	2212
1973	0	0	41	51	160	416	539	469	426	146	27	2	2277
1974	18	6	54	69	265	322	490	470	276	46	10	3	2029
1975	5	1	28	79	293	408	440	494	279	122	56	0	2205
1976	0	5	25	63	119	351	493	460	282	36	0	0	1834
1977	0	0	38	135	302	502	600	532	400	45	20	6	2580
1978	0	0	10	84	250	475	557	517	433	109	28	11	2474
1979	0	2	21	86	243	398	536	518	314	99	24	2	2243
1980	0	5	11	39	237	449	607	581	461	67	6	0	2463
1981	0	4	11	131	174	554	577	479	333	65	23	0	2351
1982	1	0	77	57	274	438	512	496	320	159	30	31	2395
1983	0	0	14	24	240	426	634	609	297	108	16	4	2372
1984	0	0	6	58	198	468	488	485	322	287	15	3	2330
1985	1	4	30	122	230	454	488	489	326	220	85	0	2449
1986	0	9	19	75	295	526	654	509	412	145	71	0	2715
1987	0	0	13	80	314	433	549	613	350	12	30	13	2407
1988	2	4	18	62	187	446	515	573	360	42	4	2	2215
1989	4	14	60	99	227	422	488	528	340	102	9	0	2293
1990	2	7	21	61	206	490	569	598	441	163	11	8	2577
1991	0	6	47	145	328	417	541	495	376	116	19	21	2511
1992	0	4	24	80	226	406	553	437	337	34	12	3	2116
1993	0	0	16	17	228	502	651	550	405	118	32	0	2519

SNOWFALL (inches) COLUMBUS, GEORGIA

SEASON	JULY	AUG	SEP	OCT	NOV	DEC	JAN	FEB	MAR	APR	MAY	JUNE	TOTAL
1964-65	0.0	0.0	0.0	0.0	0.0	0.0	T	T	T	0.0	0.0	0.0	T
1965-66	0.0	0.0	0.0	0.0	0.0	0.0	T	T	0.0	0.0	0.0	0.0	T
1966-67	0.0	0.0	0.0	0.0	0.0	0.0	0.0	T	0.0	0.0	0.0	0.0	T
1967-68	0.0	0.0	0.0	0.0	T	0.0	T	T	0.0	0.0	0.0	0.0	T
1968-69	0.0	0.0	0.0	0.0	T	T	0.0	T	0.0	0.0	0.0	0.0	T
1969-70	0.0	0.0	0.0	0.0	0.0	T	T	T	0.0	0.0	0.0	0.0	T
1970-71	0.0	0.0	0.0	0.0	0.0	T	0.0	T	0.0	T	0.0	0.0	T
1971-72	0.0	0.0	0.0	0.0	0.0	0.0	0.0	0.0	0.0	0.0	0.0	0.0	0.0
1972-73	0.0	0.0	0.0	0.0	0.0	0.0	T	14.0	0.0	0.0	0.0	0.0	14.0
1973-74	0.0	0.0	0.0	0.0	0.0	T	0.0	T	0.0	0.0	0.0	0.0	T
1974-75	0.0	0.0	0.0	0.0	0.0	T	0.0	0.0	0.0	0.0	0.0	0.0	T
1975-76	0.0	0.0	0.0	0.0	T	0.0	T	0.0	0.0	0.0	0.0	0.0	T
1976-77	0.0	0.0	0.0	0.0	0.0	0.0	1.4	0.0	0.0	0.0	0.0	0.0	1.4
1977-78	0.0	0.0	0.0	0.0	0.0	T	T	T	T	0.0	0.0	0.0	T
1978-79	0.0	0.0	0.0	0.0	0.0	0.0	T	2.0	0.0	0.0	0.0	0.0	2.0
1979-80	0.0	0.0	0.0	0.0	0.0	0.0	0.0	T	1.0	0.0	0.0	0.0	1.0
1980-81	0.0	0.0	0.0	0.0	0.0	0.0	0.0	T	0.0	0.0	0.0	0.0	T
1981-82	0.0	0.0	0.0	0.0	0.0	T	2.0	0.0	0.0	0.0	0.0	0.0	2.0
1982-83	0.0	0.0	0.0	0.0	0.0	0.0	1.0	0.0	T	0.0	0.0	0.0	1.0
1983-84	0.0	0.0	0.0	0.0	0.0	0.0	T	T	0.0	0.0	0.0	0.0	T
1984-85	0.0	0.0	0.0	0.0	0.0	0.0	T	T	0.0	0.0	0.0	0.0	T
1985-86	0.0	0.0	0.0	0.0	0.0	0.0	T	0.0	0.0	0.0	0.0	0.0	T
1986-87	0.0	0.0	0.0	0.0	0.0	0.0	T	0.0	0.0	0.0	0.0	0.0	T
1987-88	0.0	0.0	0.0	0.0	0.0	0.0	1.0	0.0	0.0	0.0	0.0	0.0	1.0
1988-89	0.0	0.0	0.0	0.0	0.0	T	0.0	T	0.0	0.0	0.0	0.0	T
1989-90	0.0	0.0	0.0	0.0	0.0	T	0.0	0.0	0.0	T	0.0	0.0	T
1990-91	0.0	0.0	0.0	0.0	0.0	0.0	0.0	T	0.0	0.0	0.0	0.0	T
1991-92	0.0	0.0	0.0	0.0	0.0	0.0	T	0.0	T	0.0	0.0	0.0	T
1992-93	0.0	0.0	0.0	0.0	0.0	0.0	0.0	0.0	2.5	0.0	T	0.0	2.5
1993-94	0.0	0.0	0.0	0.0	0.0	1.6							
Fecord Mean	0.0	0.0	0.0	0.0	T	T	0.1	0.3	0.1	T	T	0.0	0.6

MACON, GEORGIA

Located very near the geographical center of Georgia, Macon is well situated to escape rigorous climatic extremes. The climate is a blend of the maritime and continental types. Rarely does either dominate for long unbroken periods. The prevailing northwesterly winds of winter and early spring are frequently superseded by southerly flows of warm, moist tropical air. The southern extremity of the Appalachians presents an effective barrier to the rapid flow of cold air in winter. In summertime the prevailing southerlies frequently give way to the drier westerly and northerly winds. In short, the climate is truly equable.

Severe storms occur occasionally in this locality. Tornadoes occur, about twice each year within the area covered by Bibb and adjacent counties. Thunderstorms occur on approximately two days out of five from June through August. Occasionally, thunderstorms are accompanied by severe squalls, but property damage from this cause has been heavy in only a few instances. As Macon is some 200 miles from both the Atlantic and the Gulf of Mexico, hurricanes offer no direct threat, and secondary effects are generally milder than those produced by the heavier thunderstorms. Property damage of a minor nature occurs occasionally due to gale force winds and heavy rainfall.

Snow occurs at some time during most winters, but amounts of snow are usually quite small. However, on rare occasions heavy snow does occur in this area.

Based on the 1951–1980 period, the average first occurrence of 32 degrees Fahrenheit in the fall is November 8 and the average last occurrence in the spring is March 17.

The National Weather Service Office is surrounded by predominantly flat terrain. Flanking the station on the west, a range of wooded hills about 300 feet in height runs in a general northwest–southeast direction. The nearest point of these hills is about 2 1/2 miles to the southwest. Most of the countryside is well wooded, except for a few farms. Much of the outlying area is swampy, especially in the river and creek bottoms. Besides the swamps, the only bodies of water in the vicinity are the Ocmulgee River, Echeconnee Creek, and Tobesofkee Creek. These have little influence on the climate, except that when other conditions are favorable, they contribute to the formation of fog.

NORMALS, MEANS, AND EXTREMES

MACON, GEORGIA

LATITUDE: 32°42'N LONGITUDE: 83°39' W ELEVATION: FT. GRND 354 BARO 360 TIME ZONE: EASTERN WBAN: 03813

	(a)	JAN	FEB	MAR	APR	MAY	JUNE	JULY	AUG	SEP	OCT	NOV	DEC	YEAR
TEMPERATURE °F:														
Normals														
-Daily Maximum		56.6	60.8	69.5	77.7	84.5	90.2	91.9	91.1	86.4	77.9	68.7	60.3	76.3
-Daily Minimum		34.2	36.8	44.2	50.9	59.3	66.8	70.5	69.8	64.2	51.6	43.1	37.1	52.4
-Monthly		45.4	48.8	56.9	64.3	71.9	78.5	81.2	80.4	75.3	64.8	56.0	48.7	64.4
Extremes														
-Record Highest	45	84	85	95	96	99	106	108	105	102	100	88	82	108
-Year		1949	1989	1949	1986	1967	1954	1980	1986	1980	1954	1961	1972	JUL 1980
-Record Lowest	45	-6	9	14	29	40	46	54	55	35	26	10	5	-6
-Year		1985	1973	1980	1987	1971	1972	1967	1952	1967	1952	1950	1962	JAN 1985
NORMAL DEGREE DAYS:														
Heating (base 65°F)		615	454	267	82	7	0	0	0	5	106	287	511	2334
Cooling (base 65°F)		7	0	15	61	221	405	502	477	314	100	17	6	2125
% OF POSSIBLE SUNSHINE	45	56	61	65	72	73	71	67	71	66	70	64	58	66
MEAN SKY COVER (tenths)														
Sunrise - Sunset	45	6.3	6.1	6.0	5.4	5.7	5.8	6.3	5.7	5.7	4.6	5.2	6.0	5.7
MEAN NUMBER OF DAYS:														
Sunrise to Sunset														
-Clear	45	8.4	8.3	8.9	11.0	9.0	8.0	5.6	8.4	8.9	14.0	12.0	9.5	112.0
-Partly Cloudy	45	6.8	6.6	8.0	7.9	10.6	11.5	12.9	12.4	10.1	6.9	6.1	6.7	106.5
-Cloudy	45	15.8	13.4	14.1	11.2	11.4	10.5	12.5	10.3	11.0	10.1	11.8	14.8	146.8
Precipitation														
.01 inches or more	45	10.6	9.7	10.4	7.4	9.0	9.6	12.5	10.6	7.8	5.7	7.2	9.2	109.6
Snow,Ice pellets,hail														
1.0 inches or more	45	0.2	0.1	0.*	0.0	0.0	0.0	0.0	0.0	0.0	0.0	0.0	0.*	0.4
Thunderstorms	45	1.1	1.9	3.3	4.2	6.3	8.8	13.3	9.6	3.4	0.8	1.1	0.9	54.8
Heavy Fog Visibility														
1/4 mile or less	45	3.8	2.7	2.0	1.1	1.0	0.9	1.0	1.7	2.0	1.8	3.2	3.9	25.0
Temperature °F														
-Maximum														
90° and above	29	0.0	0.0	0.*	1.2	6.0	17.8	22.9	21.2	11.4	1.2	0.0	0.0	81.8
32° and below	29	0.3	0.*	0.*	0.0	0.0	0.0	0.0	0.0	0.0	0.0	0.0	0.1	0.4
-Minimum														
32° and below	29	13.9	10.0	4.0	0.4	0.0	0.0	0.0	0.0	0.0	0.5	5.2	11.2	45.2
0° and below	29	0.*	0.0	0.0	0.0	0.0	0.0	0.0	0.0	0.0	0.0	0.0	0.0	*
AVG. STATION PRESS.(mb)	21	1007.6	1006.5	1004.5	1003.8	1003.0	1003.3	1004.4	1004.6	1004.7	1006.2	1007.1	1008.1	1005.3
RELATIVE HUMIDITY (%)														
Hour 01 (Local Time)	29	79	76	77	78	83	85	87	89	89	85	83	80	83
Hour 07	29	83	83	86	87	87	88	90	92	93	89	87	84	87
Hour 13	29	58	54	52	48	50	52	56	58	57	51	53	56	54
Hour 19	29	65	58	55	52	57	61	66	69	71	66	66	67	63
PRECIPITATION (inches):														
Water Equivalent														
-Normal		4.56	4.74	4.79	3.46	3.57	3.58	4.30	3.63	2.78	2.18	2.73	4.31	44.63
-Maximum Monthly	45	10.87	9.32	11.90	8.42	11.77	9.06	13.60	8.63	8.82	9.39	10.27	10.39	13.60
-Year		1991	1983	1980	1964	1957	1965	1984	1991	1953	1959	1992	1972	JUL 1984
-Minimum Monthly	45	0.69	0.59	1.20	0.11	0.32	0.89	0.37	1.13	0.35	0.00	0.43	0.58	0.00
-Year		1954	1976	1985	1986	1956	1988	1986	1980	1984	1963	1956	1955	OCT 1963
-Maximum in 24 hrs	45	5.07	5.17	4.34	3.65	5.37	4.97	3.97	2.96	4.60	5.35	3.14	4.50	5.37
-Year		1991	1981	1991	1955	1976	1965	1992	1959	1956	1970	1992	1972	MAY 1976
Snow,Ice pellets,hail														
-Maximum Monthly	45	3.7	16.5	2.6	0.0	T	0.0	T	0.0	0.0	0.0	0.2	1.6	16.5
-Year		1955	1973	1993		1993		1993				1950	1993	FEB 1973
-Maximum in 24 hrs	45	3.7	16.5	2.6	0.0	T	0.0	T	0.0	0.0	0.0	0.2	1.6	16.5
-Year		1955	1973	1993		1993		1993				1950	1993	FEB 1973
WIND:														
Mean Speed (mph)	45	8.0	8.6	9.0	8.5	7.5	7.1	6.8	6.3	6.8	6.7	7.0	7.5	7.5
Prevailing Direction through 1963		WNW	WNW	WNW	WNW	NW	NW	SW	NE	NE	NE	WNW	NW	WNW
Fastest Obs. 1 Min.														
-Direction (!!)	11	31	27	08	27	27	32	30	35	16	31	05	30	27
-Speed (MPH)	11	28	46	32	35	35	33	44	29	25	20	23	35	46
-Year		1986	1990	1991	1985	1991	1986	1983	1983	1985	1990	1985	1983	FEB 1990
Peak Gust														
-Direction (!!)	10	NW	W	NW	W	W	NW	NW	NE	S	E	NW	W	W
-Speed (mph)	10	41	87	49	64	71	58	52	54	35	45	44	39	87
-Date		1986	1990	1993	1985	1991	1988	1989	1988	1985	1993	1989	1993	FEB 1990

PRECIPITATION (inches)

MACON, GEORGIA

YEAR	JAN	FEB	MAR	APR	MAY	JUNE	JULY	AUG	SEP	OCT	NOV	DEC	ANNUAL
1964	8.30	5.71	3.79	8.42	3.54	1.53	4.77	2.81	1.46	4.93	1.85	5.74	52.85
1965	1.55	6.29	6.95	3.49	1.62	9.06	7.92	3.19	3.99	1.69	1.99	1.83	49.57
1966	7.25	7.82	5.04	0.97	5.54	4.17	3.20	2.69	2.67	6.61	2.19	3.98	52.13
1967	3.17	3.16	1.51	1.38	2.73	3.92	5.56	4.41	1.44	1.22	2.34	3.86	34.70
1968	2.88	1.39	1.26	1.98	6.29	1.66	4.44	1.39	2.67	2.60	3.78	3.39	33.73
1969	1.85	3.39	3.85	2.54	2.01	4.21	3.64	6.88	1.55	0.81	1.04	4.91	36.68
1970	3.02	2.81	9.25	1.52	3.83	2.17	3.95	4.49	0.41	7.16	0.84	6.96	46.41
1971	4.91	5.61	7.17	3.39	3.50	6.72	6.62	3.81	1.36	0.65	3.40	4.89	52.03
1972	7.46	4.31	2.72	0.53	2.10	6.21	3.22	3.53	1.77	0.92	3.67	10.39	46.83
1973	6.10	5.71	6.49	6.33	4.25	3.29	2.81	7.09	0.98	0.67	1.98	2.78	48.48
1974	5.36	5.37	1.78	4.08	4.24	2.91	1.42	4.55	3.93	0.43	2.20	3.71	39.98
1975	6.09	6.90	6.66	4.90	4.68	3.83	5.84	2.17	4.71	3.53	2.26	3.91	55.48
1976	3.72	0.59	4.81	1.94	8.25	3.05	1.95	3.69	6.74	3.88	3.52	4.20	46.34
1977	3.72	2.02	7.89	1.28	1.55	1.12	5.04	2.96	3.34	1.67	3.38	3.80	37.77
1978	7.48	1.87	3.79	4.42	3.89	4.16	2.14	4.45	1.41	0.20	2.07	2.21	38.09
1979	5.74	8.46	3.73	6.85	3.76	2.54	5.70	4.18	6.10	0.40	3.37	1.88	52.71
1980	3.27	2.43	11.90	2.54	2.79	2.03	4.14	1.13	5.04	1.27	1.37	0.60	38.51
1981	0.97	8.27	3.44	4.18	0.95	7.54	3.19	3.58	1.69	4.51	1.11	8.66	48.09
1982	4.25	5.88	2.10	6.92	3.92	2.69	5.77	2.49	3.00	0.71	4.16	6.85	48.74
1983	4.57	9.32	5.95	3.85	1.44	5.69	2.06	2.98	1.93	0.83	6.96	5.52	51.10
1984	5.17	3.62	4.72	2.97	6.41	1.53	13.60	1.52	0.35	0.66	1.45	2.01	44.01
1985	2.17	5.74	1.20	1.19	3.78	1.36	5.93	2.50	1.79	3.58	3.70	2.98	35.92
1986	1.69	4.04	2.97	0.11	2.36	2.21	0.37	6.61	5.43	2.32	6.70	3.68	38.49
1987	6.92	7.01	4.25	0.68	2.78	5.10	1.38	3.21	1.50	0.05	3.36	1.88	38.12
1988	5.46	2.43	3.21	5.22	2.39	0.89	3.43	6.75	6.05	3.74	1.83	2.27	43.67
1989	1.85	4.58	4.92	5.61	4.38	5.03	6.21	2.34	2.41	2.28	3.01	8.85	51.47
1990	4.59	3.48	4.55	2.08	2.29	0.95	3.83	1.22	1.84	6.31	1.33	3.52	35.99
1991	10.87	1.74	10.01	4.29	2.52	4.11	4.77	8.63	0.68	0.31	0.63	3.62	52.18
1992	5.99	4.49	3.59	1.97	1.14	5.49	9.88	6.32	2.83	2.86	10.27	3.20	58.03
1993	5.46	4.21	10.32	1.91	2.69	3.30	2.19	5.95	1.29	6.37	2.93	2.84	49.46
Record Mean	4.03	4.48	4.93	3.54	3.22	3.65	4.95	4.12	2.99	2.28	2.66	4.03	44.88

AVERAGE TEMPERATURE (deg. F)

MACON, GEORGIA

YEAR	JAN	FEB	MAR	APR	MAY	JUNE	JULY	AUG	SEP	OCT	NOV	DEC	ANNUAL
#1964	45.0	45.2	56.9	66.2	73.0	81.5	78.5	78.7	74.9	60.1	57.7	50.6	64.0
1965	46.7	49.0	53.7	66.2	75.1	75.7	79.4	79.6	74.9	62.4	55.5	47.4	63.8
1966	41.4	47.2	53.4	64.9	70.8	74.9	81.2	78.9	74.1	63.9	54.9	47.9	62.8
1967	48.2	46.5	60.4	68.2	70.5	75.3	76.6	76.5	68.7	61.4	50.8	50.9	62.8
1968	42.3	40.8	54.3	64.8	69.9	78.1	80.7	81.3	73.3	63.7	52.3	43.9	62.1
1969	44.5	45.2	50.4	65.3	70.6	79.5	82.1	77.1	72.1	67.8	53.3	46.4	62.9
1970	40.9	48.5	57.0	67.8	72.9	77.7	82.9	82.0	80.4	69.7	52.6	52.4	65.4
1971	48.4	49.1	52.4	63.8	70.1	80.8	81.2	81.2	79.3	71.0	54.9	56.2	65.7
1972	51.9	47.9	56.5	65.0	69.6	75.5	79.4	80.5	77.3	65.8	54.5	55.1	64.9
1973	46.6	46.0	62.0	62.1	70.1	79.4	82.9	80.9	79.5	67.3	59.6	49.4	65.5
1974	59.3	50.3	61.5	64.2	73.9	76.2	81.1	80.3	73.1	63.3	55.2	49.7	65.7
1975	51.2	52.9	55.6	63.7	73.9	77.3	78.2	80.1	73.1	66.6	56.8	47.4	64.7
1976	43.5	54.9	60.4	65.0	69.0	76.5	81.4	79.6	74.7	61.5	49.7	45.6	63.5
1977	36.8	46.6	59.8	67.3	74.3	82.1	83.2	81.3	78.1	62.2	58.7	47.1	64.8
1978	40.5	42.7	54.7	65.6	72.3	80.9	82.2	81.0	78.3	65.0	61.1	50.0	64.5
1979	42.9	45.5	58.1	64.6	71.3	77.3	81.0	81.3	75.1	65.1	58.6	50.0	64.2
1980	50.6	46.7	54.7	64.1	73.2	80.6	84.8	84.6	80.6	64.6	55.6	48.8	65.8
1981	43.3	52.3	56.2	68.8	70.5	82.6	83.2	78.1	74.6	63.2	56.3	46.4	64.7
1982	44.2	52.6	60.4	62.1	73.6	79.2	81.9	80.8	75.0	66.8	58.6	56.0	65.9
1983	44.9	49.1	56.2	60.9	73.3	78.7	84.0	83.9	74.2	67.2	55.7	45.8	64.5
1984	44.4	51.5	56.6	64.4	71.5	79.5	80.3	82.1	73.2	72.1	54.2	56.0	65.5
1985	41.1	48.9	58.6	65.0	72.9	81.0	81.4	80.0	74.3	69.8	64.2	44.5	65.1
1986	44.7	53.2	57.1	64.9	73.8	82.1	86.6	80.1	77.6	66.4	61.7	49.0	66.4
1987	45.5	48.7	56.0	62.3	75.0	79.1	82.2	84.2	75.3	58.4	57.4	52.0	64.7
1988	42.2	47.1	56.4	63.9	69.8	78.7	81.3	82.2	75.8	61.2	57.7	48.2	63.7
1989	51.7	51.1	59.2	63.3	70.0	78.7	80.4	80.9	75.0	64.3	55.9	41.4	64.3
1990	52.0	56.3	59.7	62.9	72.0	81.0	82.7	83.1	78.0	66.7	56.9	52.8	67.0
1991	47.9	53.4	59.7	68.1	75.4	78.1	82.3	80.3	75.6	66.0	53.5	51.6	66.0
1992	47.3	52.7	55.7	61.8	69.6	77.5	82.4	78.7	75.6	62.7	55.2	47.5	63.9
1993	50.6	47.3	53.8	60.0	71.5	80.5	85.7	82.0	77.5	64.4	54.5	45.6	64.5
Record Mean	47.3	49.6	56.7	64.3	72.3	79.0	81.1	80.5	75.8	65.3	55.3	48.4	64.7
Max	57.7	60.5	68.2	76.3	83.9	89.8	91.2	90.5	86.0	77.0	67.0	58.8	75.6
Min	36.9	38.7	45.2	52.3	60.7	68.2	71.0	70.4	65.5	53.5	43.5	37.9	53.7

REFERENCE NOTES

MACON, GEORGIA

GENERAL
T - TRACE AMOUNT.
BLANK ENTRIES DENOTE MISSING/UNREPORTED DATA.
# INDICATES A STATION OR INSTRUMENT RELOCATION.

SPECIFIC

(a) - LENGTH OF RECORD IN YEARS, ALTHOUGH INDIVIDUAL MONTHS MAY BE MISSING.
0.* OR * - THE VALUE IS BETWEEN 0.0 AND 0.05
NORMALS - BASED ON THE 1961-1990 RECORD PERIOD.
EXTREMES - DATES ARE THE MOST RECENT OCCURRENCE.
WIND DIR.- NUMERALS SHOW TENS OF DEGRESS CLOCKWISE FROM TRUE NORTH. "00" INDICATES CALM.
RESULTANT DIRECTIONS ARE GIVEN TO WHOLE DEGREES.
BOLD VALUES INDICATE EXTREME VALUES WHICH OCCURRED AFTER THE ASOS SYSTEM WAS COMMISSIONED.

RECORD = PERIOD OF RECORD
RECORD MEAN PRECIPITATION IS THE MEAN OF ALL DAILY PRECIPITATION AMOUNTS DURING THE PERIOD OF RECORD.
RECORD MAX(MIN) TEMPERATURE IS THE MEAN OF ALL DAILY MAX(MIN) TEMPERATURES DURING THE PERIOD OF RECORD.
RECORD MEAN TEMPERATURE IS THE SUM OF THE RECORD MAX AND RECORD MIN DIVIDED BY 2.
AVERAGE TEMPERATURE IS THE SUM OF THE MEAN DAILY MAX AND MIN TEMPERATURE DIVIDED BY 2.

EXCEPTIONS

RECORD MEANS ARE THROUGH THE CURRENT YEAR, BEGINNING IN 1899 FOR TEMPERATURE
1899 FOR PRECIPITATION
1949 FOR SNOWFALL

HEATING DEGREE DAYS Base 65 deg. F MACON, GEORGIA

SEASON	JULY	AUG	SEP	OCT	NOV	DEC	JAN	FEB	MAR	APR	MAY	JUNE	TOTAL
1964-65	0	0	0	177	238	446	559	452	373	63	0	0	2308
1965-66	0	0	2	131	279	538	724	493	357	99	13	0	2636
1966-67	0	0	0	94	308	526	514	509	182	33	37	9	2212
1967-68	0	0	33	141	421	440	696	695	338	72	18	0	2854
1968-69	0	0	0	128	379	648	631	550	449	58	12	0	2855
1969-70	0	0	6	45	348	572	742	460	253	55	2	0	2483
1970-71	0	0	0	15	368	397	507	440	387	109	19	0	2242
1971-72	0	0	0	9	321	288	403	493	270	96	0	0	1880
1972-73	0	0	1	50	327	324	562	529	137	128	27	0	2085
1973-74	0	0	0	56	189	480	186	412	160	104	4	0	1591
1974-75	0	0	14	95	299	469	424	338	306	112	0	0	2057
1975-76	0	0	10	57	280	537	658	290	168	62	17	0	2079
1976-77	0	0	0	163	455	593	867	506	190	52	0	0	2826
1977-78	0	0	0	121	202	554	754	617	321	62	12	0	2643
1978-79	0	0	0	70	130	467	675	541	226	62	7	0	2178
1979-80	0	0	3	79	221	461	441	534	321	70	3	0	2133
1980-81	0	0	2	79	284	493	667	349	279	22	6	0	2181
1981-82	0	0	6	106	268	570	639	344	191	129	2	0	2255
1982-83	0	0	3	85	215	310	618	437	281	149	1	0	2099
1983-84	0	0	14	39	284	591	631	386	270	94	18	0	2327
1984-85	0	0	4	23	327	276	737	448	215	87	6	0	2123
1985-86	0	0	7	41	94	625	620	330	253	87	2	0	2059
1986-87	0	8	0	93	162	491	597	451	285	143	2	0	2232
1987-88	0	0	0	202	240	406	700	515	272	87	5	0	2427
1988-89	0	0	0	152	216	513	407	400	229	148	31	0	2096
1989-90	0	0	15	115	280	724	398	250	192	119	10	0	2103
1990-91	0	0	6	96	243	378	522	325	201	32	2	0	1805
1991-92	0	0	0	58	359	429	541	354	302	151	30	0	2224
1992-93	0	0	6	88	307	534	441	487	351	164	9	0	2387
1993-94	0	0	3	109	335	597							

COOLING DEGREE DAYS Base 65 deg. F MACON, GEORGIA

YEAR	JAN	FEB	MAR	APR	MAY	JUNE	JULY	AUG	SEP	OCT	NOV	DEC	TOTAL
1969	0	2	4	75	193	445	538	382	227	139	2	1	2008
1970	0	0	9	143	257	387	560	536	470	167	2	12	2543
1971	0	1	2	80	185	486	506	505	439	205	27	24	2460
1972	2	3	13	102	151	323	455	488	374	80	18	25	2034
1973	0	0	53	47	189	439	562	500	446	136	30	6	2408
1974	17	6	59	87	288	344	505	482	264	49	13	1	2115
1975	3	5	25	80	282	377	415	476	261	113	40	0	2077
1976	0	4	28	70	148	347	514	457	297	63	0	0	1928
1977	0	0	34	127	296	520	573	512	397	40	18	4	2521
1978	0	0	9	87	244	481	544	504	405	76	22	8	2380
1979	0	1	18	57	210	375	505	514	312	86	33	4	2115
1980	0	7	8	54	264	473	622	615	476	76	6	0	2601
1981	0	2	15	142	184	534	574	414	301	56	16	2	2240
1982	4	4	57	48	275	431	532	495	308	149	31	37	2371
1983	0	0	16	33	265	418	594	594	298	112	7	3	2340
1984	0	0	17	83	226	442	482	536	259	250	11	2	2308
1985	5	2	24	94	261	487	516	470	293	195	76	0	2423
1986	0	5	16	92	284	519	673	485	385	144	73	1	2677
1987	0	0	14	70	317	430	540	601	312	6	19	10	2319
1988	1	2	10	58	162	416	516	542	333	42	5	2	2089
1989	2	16	57	101	192	419	486	499	323	101	12	0	2208
1990	2	11	29	61	235	485	553	568	403	157	9	7	2520
1991	0	7	44	132	333	400	546	483	325	97	21	20	2408
1992	0	4	22	63	176	381	548	431	329	25	22	0	2001
1993	0	0	11	22	216	473	647	534	384	97	26	0	2410

SNOWFALL (inches) MACON, GEORGIA

SEASON	JULY	AUG	SEP	OCT	NOV	DEC	JAN	FEB	MAR	APR	MAY	JUNE	TOTAL
1970-71	0.0	0.0	0.0	0.0	T	0.0	0.0	T	0.0	0.0	0.0	0.0	T
1971-72	0.0	0.0	0.0	0.0	0.0	T	0.0	0.0	0.0	0.0	0.0	0.0	T
1972-73	0.0	0.0	0.0	0.0	0.0	0.0	T	16.5	0.0	0.0	0.0	0.0	16.5
1973-74	0.0	0.0	0.0	0.0	0.0	0.0	0.0	0.0	0.0	0.0	0.0	0.0	0.0
1974-75	0.0	0.0	0.0	0.0	0.0	T	0.0	0.0	0.0	0.0	0.0	0.0	T
1975-76	0.0	0.0	0.0	0.0	0.0	0.0	T	0.0	0.0	0.0	0.0	0.0	T
1976-77	0.0	0.0	0.0	0.0	0.0	0.0	3.0	0.0	0.0	0.0	0.0	0.0	3.0
1977-78	0.0	0.0	0.0	0.0	0.0	T	T	0.2	T	0.0	0.0	0.0	0.2
1978-79	0.0	0.0	0.0	0.0	0.0	0.0	T	3.4	0.0	0.0	0.0	0.0	3.4
1979-80	0.0	0.0	0.0	0.0	0.0	0.0	0.0	0.4	1.1	0.0	0.0	0.0	1.5
1980-81	0.0	0.0	0.0	0.0	0.0	T	0.0	T	0.0	0.0	0.0	0.0	T
1981-82	0.0	0.0	0.0	0.0	0.0	0.0	2.2	0.0	T	0.0	0.0	0.0	2.2
1982-83	0.0	0.0	0.0	0.0	0.0	0.0	1.4	T	T	0.0	0.0	0.0	1.4
1983-84	0.0	0.0	0.0	0.0	0.0	T	T	0.1	0.0	0.0	0.0	0.0	0.1
1984-85	0.0	0.0	0.0	0.0	0.0	0.0	T	T	0.0	0.0	0.0	0.0	T
1985-86	0.0	0.0	0.0	0.0	0.0	0.0	T	0.0	T	0.0	0.0	0.0	T
1986-87	0.0	0.0	0.0	0.0	0.0	0.0	T	0.0	T	0.0	0.0	0.0	T
1987-88	0.0	0.0	0.0	0.0	0.0	0.0	2.1	0.0	0.0	0.0	0.0	0.0	2.1
1988-89	0.0	0.0	0.0	0.0	0.0	0.0	0.0	0.2	0.0	0.0	T	0.0	0.2
1989-90	0.0	0.0	0.0	0.0	0.0	T	0.0	0.0	0.0	0.0	0.0	0.0	T
1990-91	0.0	0.0	0.0	0.0	0.0	0.0	0.0	0.4	0.0	0.0	0.0	0.0	0.4
1991-92	0.0	0.0	0.0	0.0	0.0	0.0	T	0.0	0.0	0.0	0.0	0.0	T
1992-93	0.0	0.0	0.0	0.0	0.0	0.0	0.0	0.0	2.6	0.0	T	0.0	2.6
1993-94	T	0.0	0.0	0.0	0.0	1.6							
Record Mean	T	0.0	0.0	0.0	T	T	0.3	0.6	0.1	0.0	T	0.0	1.0

SAVANNAH, GEORGIA

Savannah is surrounded by flat terrain, low and marshy to the north and east, and rising to several feet above sea level to the west and south. About half the land to the west and south is cleared and the other half is wooded and swampy.

The area has a temperate climate, with a seasonal low temperature of 51 degrees in winter, 66 degrees in spring, 80 degrees in summer, and 66 degrees in autumn. The lowest temperatures are below 10 degrees and the highest temperatures are about 100 degrees.

The normal annual rainfall is about 49 inches. About half falls in the thunderstorm season of June 15 through September 15. The remainder, produced principally by squall-line and frontal showers, is spread over the other nine months with a minor peak in March. Considerable periods of fair, mild weather are experienced in October, November, April, and to a less extent, in May. Snow is a rarity and even a trace does not occur on an average of once a year. The heaviest snowfalls are under 5 inches. Severe tropical storms affect this area about once in ten years. Rainfall from these storms constitute the heaviest sustained precipitation. Accumulations exceeding 22 inches have occurred.

The present exposure of the thermometers gives readings more nearly commensurate with those of suburban street levels of Savannah than was the case of previous locations atop various buildings. During that time, especially on still, clear nights, temperatures near the ground and in lower inland areas were as much as 15 degrees lower than the official low temperature. Present differences on comparable nights range from 3 - 8 degrees.

Sunshine is adequate at all seasons and seldom are there two or more days in succession without it. Sea- and land-breeze effect is usually not felt in Savannah, though it is a daily feature on the nearby islands. Dry, continental air masses reach this area in summer mostly by sliding down the Atlantic coast and giving cooler northeast winds. Such masses reaching this area from the northwest or west in summer bring mostly clear skies and high temperatures.

Based on the 1951-1980 period, the average first occurrence of 32 degrees Fahrenheit in the fall is November 15 and the average last occurrence in the spring is March 10.

NORMALS, MEANS, AND EXTREMES

SAVANNAH, GEORGIA

LATITUDE: 32°08'N LONGITUDE: 81°12' W ELEVATION: FT. GRND 46 BARO 52 TIME ZONE: EASTERN WBAN: 03822

	(a)	JAN	FEB	MAR	APR	MAY	JUNE	JULY	AUG	SEP	OCT	NOV	DEC	YEAR
TEMPERATURE °F:														
Normals														
-Daily Maximum		59.7	62.4	70.1	77.5	84.0	88.8	91.1	89.7	85.2	77.5	70.0	62.3	76.5
-Daily Minimum		38.1	41.1	48.3	54.5	62.9	69.2	72.4	72.2	67.8	56.9	48.1	41.0	56.0
-Monthly		48.9	51.8	59.2	66.0	73.5	79.1	81.8	81.0	76.6	67.3	59.1	51.7	66.3
Extremes														
-Record Highest	43	84	86	91	95	100	104	105	104	98	97	89	83	105
-Year		1957	1989	1974	1986	1953	1985	1986	1954	1986	1986	1961	1971	JUL 1986
-Record Lowest	43	3	14	20	32	39	51	61	57	43	28	15	9	3
-Year		1985	1958	1980	1987	1963	1984	1972	1986	1967	1952	1970	1983	JAN 1985
NORMAL DEGREE DAYS:														
Heating (base 65°F)		516	378	204	47	0	0	0	0	0	63	213	426	1847
Cooling (base 65°F)		16	9	24	77	266	423	521	496	348	135	36	14	2365
% OF POSSIBLE SUNSHINE	43	54	57	62	70	68	65	64	62	58	64	62	55	62
MEAN SKY COVER (tenths)														
Sunrise - Sunset	43	6.2	6.1	6.0	5.4	5.7	6.1	6.3	6.2	6.2	5.1	5.4	6.0	5.9
MEAN NUMBER OF DAYS:														
Sunrise to Sunset														
-Clear	43	9.2	8.4	9.0	10.7	9.3	6.9	5.4	5.7	7.0	12.1	11.1	9.3	104.1
-Partly Cloudy	43	6.0	6.3	8.5	8.5	10.0	11.2	13.6	13.7	10.5	7.8	6.9	7.0	110.0
-Cloudy	43	15.8	13.6	13.5	10.8	11.8	11.9	12.0	11.6	12.5	11.1	12.0	14.7	151.1
Precipitation														
.01 inches or more	43	9.4	8.7	9.1	6.9	8.7	11.1	13.5	12.8	10.0	5.9	6.4	8.0	110.6
Snow,Ice pellets,hail														
1.0 inches or more	43	0.*	0.1	0.*	0.0	0.0	0.0	0.0	0.0	0.0	0.0	0.0	0.*	0.2
Thunderstorms	43	1.0	1.2	3.0	3.8	7.2	10.1	14.9	12.6	5.7	1.5	0.5	0.6	62.0
Heavy Fog Visibility														
1/4 mile or less	43	4.7	3.1	3.3	2.6	3.1	2.3	1.2	2.0	3.7	3.4	4.9	4.6	38.8
Temperature °F														
-Maximum														
90° and above	29	0.0	0.0	0.1	1.4	5.0	14.4	22.0	18.2	8.1	0.7	0.0	0.0	69.8
32° and below	29	0.2	0.0	0.*	0.0	0.0	0.0	0.0	0.0	0.0	0.0	0.0	0.1	0.3
-Minimum														
32° and below	29	10.0	6.9	1.9	0.*	0.0	0.0	0.0	0.0	0.0	0.*	2.2	7.4	28.6
0° and below	29	0.0	0.0	0.0	0.0	0.0	0.0	0.0	0.0	0.0	0.0	0.0	0.0	0.0
AVG. STATION PRESS.(mb)	21	1018.9	1018.0	1016.0	1015.3	1014.5	1014.7	1015.8	1015.8	1015.7	1017.2	1018.4	1019.5	1016.7
RELATIVE HUMIDITY (%)														
Hour 01	29	78	76	77	79	84	86	87	89	88	85	83	79	83
Hour 07 (Local Time)	29	81	80	83	83	85	87	88	91	91	88	86	82	85
Hour 13	29	55	50	48	46	51	55	57	60	60	53	52	54	53
Hour 19	29	66	60	60	58	63	68	72	75	75	72	72	69	68
PRECIPITATION (inches):														
Water Equivalent														
-Normal		3.59	3.22	3.78	3.03	4.09	5.66	6.38	7.46	4.47	2.39	2.19	2.96	49.22
-Maximum Monthly	43	8.98	7.92	9.57	10.57	10.08	14.39	20.10	14.94	13.47	12.50	5.26	5.80	20.10
-Year		1991	1964	1959	1991	1957	1963	1964	1971	1953	1990	1993	1977	JUL 1964
-Minimum Monthly	43	0.45	0.26	0.18	0.38	0.51	0.84	1.35	1.02	0.35	0.02	0.15	0.12	0.02
-Year		1989	1991	1955	1986	1953	1954	1972	1980	1991	1963	1966	1984	OCT 1963
-Maximum in 24 hrs	43	3.58	3.46	4.65	5.62	5.67	4.06	6.36	7.04	6.80	5.79	5.02	3.47	7.04
-Year		1984	1964	1959	1976	1976	1963	1957	1971	1979	1990	1969	1964	AUG 1971
Snow,Ice pellets,hail														
-Maximum Monthly	43	2.0	3.6	1.1	0.0	0.0	T	0.0	0.0	0.0	0.0	0.0	3.6	3.6
-Year		1977	1968	1986			1989						1989	DEC 1989
-Maximum in 24 hrs	43	1.3	3.6	1.1	0.0	0.0	T	0.0	0.0	0.0	0.0	0.0	3.4	3.6
-Year		1977	1968	1986			1989						1989	FEB 1968
WIND:														
Mean Speed (mph)	43	8.5	9.2	9.2	8.7	7.7	7.5	7.2	6.6	7.2	7.5	7.5	7.9	7.9
Prevailing Direction through 1963		WNW	NE	WNW	SSE	SW	SW	SW	SW	NE	NNE	NNE	NE	SW
Fastest Obs. 1 Min.														
-Direction (!!)	13	31	29	32	23	22	27	35	25	29	31	23	30	32
-Speed (MPH)	13	30	30	46	35	44	31	37	29	35	35	40	29	46
-Year		1989	1981	1981	1991	1984	1989	1986	1986	1989	1990	1985	1982	MAR 1981
Peak Gust														
-Direction (!!)	10	NW	NW	W	NE	SW	NW	SW	W	W	NW	S	NW	W
-Speed (mph)	10	51	49	68	53	68	66	63	58	54	61	62	48	68
-Date		1989	1991	1993	1991	1984	1991	1993	1990	1989	1990	1985	1993	MAR 1993

PRECIPITATION (inches)

SAVANNAH, GEORGIA

YEAR	JAN	FEB	MAR	APR	MAY	JUNE	JULY	AUG	SEP	OCT	NOV	DEC	ANNUAL
1964	6.29	7.92	2.71	2.64	4.66	2.55	20.10	8.37	3.93	6.94	2.90	4.16	73.17
1965	0.83	4.34	7.75	1.39	2.62	5.63	7.46	4.56	4.82	1.33	2.09	2.99	45.81
1966	6.05	3.66	3.79	1.88	6.73	6.61	7.89	3.73	1.78	1.19	0.15	1.93	45.39
1967	7.18	2.80	0.50	1.38	2.94	4.38	6.23	8.57	2.12	1.36	0.84	2.97	41.27
1968	1.79	1.16	0.89	2.09	4.80	5.86	6.35	3.83	0.48	3.53	2.59	3.97	37.34
1969	1.77	1.59	5.11	0.71	8.74	9.99	7.57	10.33	3.74	4.04	3.82	3.43	60.84
1970	3.11	2.32	8.51	0.95	5.41	5.21	5.76	10.61	5.88	2.29	0.42	3.37	53.84
1971	3.40	2.60	2.63	3.53	3.56	6.98	9.07	14.94	1.67	8.01	1.26	3.69	61.34
1972	3.99	4.61	3.84	1.20	5.84	6.54	1.35	12.62	0.36	0.54	4.91	2.77	48.57
1973	3.61	4.46	5.36	4.43	1.23	9.19	2.89	6.45	3.65	0.19	0.68	3.26	45.40
1974	1.37	2.79	1.87	2.75	7.25	6.00	6.48	7.90	2.61	0.10	0.96	1.85	41.93
1975	3.17	3.01	3.99	4.71	6.00	2.08	11.55	3.13	8.01	1.25	1.09	3.19	51.18
1976	2.19	1.24	2.51	5.62	6.33	7.49	7.56	7.28	10.07	4.75	4.83	3.87	63.74
1977	3.14	1.83	2.72	1.94	1.03	2.00	5.62	8.01	6.52	1.16	2.07	5.80	41.84
1978	4.02	3.14	1.93	3.68	4.50	2.19	3.61	4.43	2.61	0.60	1.85	2.85	35.41
1979	3.96	4.14	2.42	3.83	8.49	7.37	10.78	2.65	12.20	0.70	2.70	2.68	61.92
1980	2.95	1.29	7.75	3.68	4.50	3.47	2.38	1.02	5.81	1.62	2.04	1.33	37.84
1981	1.03	2.94	3.91	1.75	2.10	3.01	5.42	10.91	2.88	1.29	1.65	3.17	40.06
1982	3.47	2.94	1.64	6.25	4.18	9.15	6.70	9.18	2.98	1.74	0.40	3.63	52.26
1983	5.90	5.23	9.01	5.15	1.07	5.81	5.30	3.67	3.39	1.03	4.18	4.77	54.51
1984	8.87	3.21	5.13	3.41	5.29	1.48	7.88	3.46	7.43	1.23	3.15	0.12	50.66
1985	0.51	1.37	1.65	1.37	2.18	6.72	5.00	9.42	0.76	3.37	4.28	2.01	38.64
1986	2.03	5.28	2.85	0.38	2.06	2.98	5.49	12.31	0.49	1.99	4.40	5.07	45.33
1987	8.62	4.39	5.33	0.50	3.82	8.03	4.37	9.46	8.16	0.33	2.06	1.41	56.48
1988	3.44	4.09	2.11	5.05	3.52	2.63	1.80	10.68	9.62	2.81	1.43	0.99	48.17
1989	0.45	0.67	1.41	3.59	3.10	7.30	4.91	6.29	7.98	4.71	1.26	5.20	46.87
1990	3.91	3.08	3.79	1.75	2.07	0.97	1.92	7.25	1.26	12.50	2.48	2.10	43.08
1991	8.98	0.26	5.48	10.57	7.13	5.12	15.41	10.51	0.35	1.60	1.26	1.75	68.42
1992	6.60	2.24	3.97	2.08	2.04	13.01	2.03	7.69	8.42	3.60	5.15	1.53	58.36
1993	5.52	3.35	7.96	3.22	1.31	2.48	4.34	3.04	6.84	2.58	5.26	2.15	48.05
Record Mean	2.98	3.13	3.60	2.95	3.40	5.53	6.65	7.09	5.58	2.77	2.06	2.80	48.55

AVERAGE TEMPERATURE (deg. F)

SAVANNAH, GEORGIA

YEAR	JAN	FEB	MAR	APR	MAY	JUNE	JULY	AUG	SEP	OCT	NOV	DEC	ANNUAL
#1964	48.2	47.8	58.4	66.8	73.2	81.4	79.1	79.6	74.7	63.1	60.1	53.7	65.5
1965	49.9	52.4	56.8	66.4	74.8	76.4	79.8	80.6	76.7	66.2	58.0	50.1	65.7
1966	44.9	51.2	55.6	64.7	71.6	74.6	80.9	79.6	76.4	67.3	57.1	49.5	64.4
1967	51.4	49.6	61.7	69.0	73.2	77.2	79.9	79.4	71.6	64.4	56.4	55.3	65.7
1968	45.7	44.6	57.0	68.0	72.5	79.1	81.6	82.1	76.5	68.7	54.8	47.6	64.8
1969	47.4	47.1	52.4	66.1	71.0	79.7	82.2	78.9	75.4	69.9	53.9	47.7	64.3
1970	42.4	49.2	59.2	68.3	73.0	78.2	81.6	81.8	78.9	70.2	54.8	54.1	66.0
1971	50.1	52.4	54.5	64.6	71.1	80.5	80.9	80.8	78.6	72.4	58.0	60.6	67.0
1972	57.5	51.3	59.3	66.7	71.6	75.7	81.0	81.0	77.3	68.1	58.1	57.1	67.1
1973	49.8	49.6	63.6	64.0	73.5	79.3	82.2	80.0	79.0	68.9	61.6	51.2	66.9
1974	62.9	53.0	63.8	65.6	74.5	76.8	79.2	79.6	76.3	64.6	57.7	53.1	67.3
1975	55.3	57.7	59.5	65.4	76.7	79.5	78.4	81.7	76.5	70.0	59.3	50.5	67.5
1976	45.9	56.4	62.6	64.3	69.7	75.8	81.7	78.2	74.9	62.5	51.5	49.0	64.4
1977	39.9	49.2	62.1	67.6	74.0	82.0	83.3	81.1	78.9	64.1	61.2	50.0	66.1
1978	43.9	43.6	56.3	68.0	73.7	80.0	82.2	82.3	78.2	67.0	64.7	53.4	66.1
1979	45.4	49.0	59.8	67.7	73.9	76.8	82.1	81.4	77.4	67.5	60.4	50.7	66.0
1980	50.7	48.5	57.1	66.2	73.3	79.7	84.4	83.4	80.4	65.4	56.7	48.1	66.1
1981	43.5	52.6	56.8	69.0	71.6	84.5	84.4	79.2	75.2	65.4	57.7	48.3	65.7
1982	48.7	55.9	62.1	64.8	74.3	80.2	81.3	81.0	75.5	67.6	61.8	57.4	67.5
1983	46.1	50.8	58.0	62.8	73.0	78.2	84.1	82.9	75.7	69.7	57.7	48.5	65.6
1984	47.7	53.7	59.3	65.7	72.9	79.2	80.6	81.5	75.0	73.2	55.7	59.5	67.0
1985	45.3	53.0	61.5	66.9	74.6	81.3	82.7	80.7	76.6	72.2	67.5	48 9	67.6
1986	47.6	56.9	59.1	66.8	75.0	82.5	85.7	81.6	79.7	70.0	65.0	53.9	68.7
1987	49.0	50.4	58.2	63.9	74.1	80.6	83.6	84.5	77.8	61.5	60.4	54.7	66.6
1988	45.0	50.2	58.2	66.1	72.4	77.7	82.7	82.5	77.4	64.2	61.3	51.1	65.7
1989	56.6	56.1	60.6	65.6	72.3	81.0	82.9	80.7	76.8	68.6	59.5	43.7	67.0
1990	55.7	60.0	62.6	65.0	74.3	81.5	84.4	82.4	78.9	70.7	60.8	56.5	69.4
1991	52.2	56.2	62.4	70.0	77.5	79.4	83.2	82.2	77.9	68.4	56.6	55.6	68.5
1992	49.9	55.7	59.2	64.5	71.7	78.6	84.3	81.0	77.1	66.6	60.6	52.0	66.8
1993	54.8	50.3	57.0	62.4	73.6	81.3	86.7	83.1	79.6	68.1	59.9	49.7	67.2
Record Mean	51.3	53.3	59.4	66.1	73.6	79.4	81.7	81.0	77.0	67.8	58.8	52.3	66.8
Max	60.8	63.1	69.3	76.1	83.1	88.4	90.4	89.4	85.1	77.1	68.8	62.0	76.2
Min	41.7	43.6	49.5	56.1	64.0	70.3	72.9	72.6	68.9	58.6	48.8	42.6	57.5

REFERENCE NOTES

SAVANNAH, GEORGIA

GENERAL
T - TRACE AMOUNT.
BLANK ENTRIES DENOTE MISSING/UNREPORTED DATA.
# INDICATES A STATION OR INSTRUMENT RELOCATION.

SPECIFIC

PM - INCLUDES LAST DAY OF PREVIOUS MONTH
ASOS - AUTOMATED SURFACE OBSERVING SYSTEM IN OPERATION DURING THESE MONTHS.

(a) - LENGTH OF RECORD IN YEARS, ALTHOUGH INDIVIDUAL MONTHS MAY BE MISSING.
0.* OR * - THE VALUE IS BETWEEN 0.0 AND 0.05
NORMALS - BASED ON THE 1961-1990 RECORD PERIOD.
EXTREMES - DATES ARE THE MOST RECENT OCCURRENCE
WIND DIR.- NUMERALS SHOW TENS OF DEGRESS CLOCKWISE FROM TRUE NORTH. "00" INDICATES CALM.
RESULTANT DIRECTIONS ARE GIVEN TO WHOLE DEGREES.
BOLD VALUES INDICATE EXTREME VALUES WHICH OCCURRED AFTER THE ASOS SYSTEM WAS COMMISSIONED.

RECORD = PERIOD OF RECORD
RECORD MEAN PRECIPITATION IS THE MEAN OF ALL DAILY PRECIPITATION AMOUNTS DURING THE PERIOD OF RECORD.
RECORD MAX(MIN) TEMPERATURE IS THE MEAN OF ALL DAILY MAX(MIN) TEMPERATURES DURING THE PERIOD OF RECORD.
RECORD MEAN TEMPERATURE IS THE SUM OF THE RECORD MAX AND RECORD MIN DIVIDED BY 2.
AVERAGE TEMPERATURE IS THE SUM OF THE MEAN DAILY MAX AND MIN TEMPERATURE DIVIDED BY 2.

EXCEPTIONS

FASTEST MILE WIND IS THROUGH OCTOBER 1980.

RECORD MEANS ARE THROUGH THE CURRENT YEAR, BEGINNING IN 1874 FOR TEMPERATURE
1871 FOR PRECIPITATION
1951 FOR SNOWFALL

HEATING DEGREE DAYS Base 65 deg. F — SAVANNAH, GEORGIA

SEASON	JULY	AUG	SEP	OCT	NOV	DEC	JAN	FEB	MAR	APR	MAY	JUNE	TOTAL
#1964-65	0	0	0	116	157	355	463	359	282	66	0	0	1798
1965-66	0	0	0	92	204	454	612	387	294	88	11	0	2142
1966-67	0	0	0	50	259	480	413	428	159	26	12	1	1828
1967-68	0	0	12	82	269	306	592	582	270	34	3	0	2150
1968-69	0	0	0	73	309	533	539	498	395	48	3	0	2398
1969-70	0	0	0	38	332	530	698	437	199	57	8	0	2299
1970-71	0	0	1	20	303	340	458	359	338	100	15	0	1934
1971-72	0	0	0	10	236	187	256	400	183	74	1	0	1347
1972-73	0	0	0	25	239	261	462	423	98	85	4	0	1597
1973-74	0	0	0	47	148	435	107	340	124	73	1	0	1275
1974-75	0	0	2	79	243	368	315	232	212	86	0	0	1537
1975-76	0	0	0	25	228	446	586	256	134	69	11	0	1755
1976-77	0	0	0	142	405	490	771	437	152	42	2	0	2441
1977-78	0	0	0	96	165	457	645	594	283	35	2	0	2277
1978-79	0	0	0	45	53	378	602	448	181	17	1	0	1725
1979-80	0	0	0	41	183	438	436	489	257	43	7	0	1894
1980-81	0	0	0	72	252	518	659	342	263	25	8	0	2139
1981-82	0	0	3	59	231	513	501	258	149	76	0	0	1790
1982-83	0	0	0	73	139	266	579	392	228	115	0	0	1792
1983-84	0	0	2	19	232	513	531	320	200	68	7	0	1892
1984-85	0	0	1	8	299	185	615	360	157	60	2	0	1687
1985-86	0	0	1	16	51	504	531	240	215	59	4	0	1621
1986-87	0	5	0	48	101	349	491	401	231	110	5	0	1741
1987-88	0	0	0	122	185	332	612	426	218	52	1	0	1948
1988-89	0	0	0	84	141	423	268	289	193	110	13	0	1521
1989-90	0	0	1	59	191	653	286	175	135	81	0	0	1581
1990-91	0	0	0	55	143	279	395	260	139	14	0	0	1285
1991-92	0	0	0	34	270	315	461	274	204	101	22	0	1681
1992-93	0	0	1	45	194	401	320	406	254	113	0	0	1734
1993-94	0	0	1	45	193	470							

COOLING DEGREE DAYS Base 65 deg. F — SAVANNAH, GEORGIA

YEAR	JAN	FEB	MAR	APR	MAY	JUNE	JULY	AUG	SEP	OCT	NOV	DEC	TOTAL
1969	0	4	8	86	194	450	539	438	320	199	8	1	2247
1970	4	0	26	163	262	403	521	528	423	188	4	9	2531
1971	1	13	18	92	210	474	502	495	414	248	34	56	2557
1972	28	8	14	130	214	328	505	503	379	125	38	23	2295
1973	0	0	63	63	275	434	540	473	429	177	52	13	2519
1974	52	9	96	99	302	360	448	462	347	74	32	8	2289
1975	20	33	48	104	374	439	423	528	351	187	65	2	2574
1976	0	12	68	53	164	331	526	416	306	70	7	1	1954
1977	0	1	71	130	289	517	574	508	421	74	58	0	2643
1978	0	0	22	132	281	456	538	543	403	115	51	29	2570
1979	0	5	28	105	282	360	537	516	378	126	51	2	2390
1980	0	16	19	87	270	449	607	579	468	90	8	2	2595
1981	0	1	14	150	219	589	609	444	319	80	17	4	2446
1982	6	10	67	79	292	463	514	503	323	158	48	42	2505
1983	0	0	16	55	253	400	598	562	332	171	18	6	2411
1984	0	0	31	94	261	431	486	517	309	270	24	19	2442
1985	11	31	58	123	307	496	557	496	355	248	133	7	2822
1986	0	18	40	121	321	532	651	525	449	212	107	13	2989
1987	0	0	26	85	292	474	583	611	392	22	50	21	2556
1988	0	3	11	92	238	386	555	547	378	66	35	1	2312
1989	13	45	64	135	248	488	563	493	362	177	33	0	2621
1990	4	41	69	90	296	503	608	549	422	237	25	24	2868
1991	7	18	64	169	397	441	572	540	393	143	25	29	2798
1992	0	11	29	94	234	414	603	504	370	101	69	7	2436
1993	12	0	10	43	271	497	678	567	447	149	46	1	2721

SNOWFALL (inches) — SAVANNAH, GEORGIA

SEASON	JULY	AUG	SEP	OCT	NOV	DEC	JAN	FEB	MAR	APR	MAY	JUNE	TOTAL
1970-71	0.0	0.0	0.0	0.0	0.0	0.0	T	0.0	0.0	0.0	0.0	0.0	T
1971-72	0.0	0.0	0.0	0.0	0.0	0.0	0.0	0.0	0.0	0.0	0.0	0.0	0.0
1972-73	0.0	0.0	0.0	0.0	0.0	0.0	T	3.2	0.0	0.0	0.0	0.0	3.2
1973-74	0.0	0.0	0.0	0.0	0.0	0.0	0.0	0.0	0.0	0.0	0.0	0.0	0.0
1974-75	0.0	0.0	0.0	0.0	0.0	0.0	0.0	0.0	0.0	0.0	0.0	0.0	0.0
1975-76	0.0	0.0	0.0	0.0	0.0	0.0	T	0.0	0.0	0.0	0.0	0.0	T
1976-77	0.0	0.0	0.0	0.0	0.0	0.0	2.0	T	0.0	0.0	0.0	0.0	2.0
1977-78	0.0	0.0	0.0	0.0	0.0	0.0	0.0	0.0	0.0	0.0	0.0	0.0	0.0
1978-79	0.0	0.0	0.0	0.0	0.0	0.0	0.0	T	0.0	0.0	0.0	0.0	T
1979-80	0.0	0.0	0.0	0.0	0.0	0.0	0.0	0.0	T	0.0	0.0	0.0	T
1980-81	0.0	0.0	0.0	0.0	0.0	T	0.0	0.0	0.0	0.0	0.0	0.0	T
1981-82	0.0	0.0	0.0	0.0	0.0	0.0	0.0	0.0	0.0	0.0	0.0	0.0	0.0
1982-83	0.0	0.0	0.0	0.0	0.0	0.0	T	0.0	T	0.0	0.0	0.0	T
1983-84	0.0	0.0	0.0	0.0	0.0	0.0	T	0.0	0.0	0.0	0.0	0.0	T
1984-85	0.0	0.0	0.0	0.0	0.0	0.0	0.0	0.0	0.0	0.0	0.0	0.0	0.0
1985-86	0.0	0.0	0.0	0.0	0.0	0.0	0.3	0.0	1.1	0.0	0.0	0.0	1.4
1986-87	0.0	0.0	0.0	0.0	0.0	0.0	T	0.0	0.0	0.0	0.0	0.0	T
1987-88	0.0	0.0	0.0	0.0	0.0	0.0	T	T	0.0	0.0	0.0	0.0	T
1988-89	0.0	0.0	0.0	0.0	0.0	T	0.0	1.0	0.0	0.0	0.0	T	1.0
1989-90	0.0	0.0	0.0	0.0	0.0	3.6	0.0	0.0	0.0	0.0	0.0	0.0	3.6
1990-91	0.0	0.0	0.0	0.0	0.0	0.0	0.0	T	0.0	0.0	0.0	0.0	T
1991-92	0.0	0.0	0.0	0.0	0.0	0.0	0.0	0.0	0.0	0.0	0.0	0.0	0.0
1992-93	0.0	0.0	0.0	0.0	0.0	0.0	0.0	0.0	0.2	0.0	0.0	0.0	0.2
1993-94	0.0	0.0	0.0	0.0	0.0	T							
Record Mean	0.0	0.0	0.0	0.0	0.0	0.1	0.1	0.2	T	0.0	0.0	T	0.4

HILO, HAWAII

The city of Hilo is located near the midpoint of the eastern shore of the Island of Hawaii. This island is by far the largest of the Hawaiian group, with an area of 4,038 square miles, more than twice that of all the other islands combined. Its topography is dominated by the great volcanic masses of Mauna Loa (13,653 feet), Mauna Kea (13,796 feet), and of Haulalai, the Kohala Mountains, and Kilauea. In fact, the island consists entirely of the slopes of these mountains and of the broad saddles between them. Mauna Loa and Kilauea, which occupy the southern half of the island, are still active volcanoes.

Hawaii lies well within the belt of northeasterly trade winds generated by the semi-permanent Pacific high pressure cell to the north and east. The climate provides equable temperatures from day to day and season to season. In Hilo, July and August are the warmest months, with average daily highs and lows of 83 and 68 degrees. January and February, the coolest months, have highs of 80 degrees and lows of 63 degrees. Greater variations occur in localities with less rain and cloud, but temperatures in the mid-90s and low 50s are uncommon anywhere on the island near sea level.

Over the windward slopes of Hawaii , rainfall occurs principally as orographic showers within the ascending moist trade winds. Mean annual rainfall, except for the semi-sheltered Hamakua district, increases from 100 inches or more along the coasts to a maximum of over 300 inches at elevations of 2,000 to 3,000 feet, and then declines to about 15 inches at the summits of Mauna Kea and Mauna Loa. Leeward areas are topographically sheltered from the trades and are therefore drier, although sea breezes created by daytime heating of the land move onshore and upslope, causing afternoon and evening cloudiness and showers. The driest locality on the island, and in the State, with an annual rainfall of less than 10 inches, is the coastal strip just leeward of the southern portion of the Kohala Mountains and of the saddle between the Kohalas and Mauna Kea.

Within the city of Hilo, average rainfall varies from about 130 inches a year near the shore to as much as 200 upslope. The wettest part of the island, with a mean annual rainfall exceeding 300 inches, lies about 6 miles upslope from the city limits. Relative humidity at Hilo is in the moderate range, however, due to the natural ventilation provided by the prevailing winds, the weather is seldom oppressive.

The trade winds prevail throughout the year and profoundly influence the climate. The islands entire western coast is sheltered from the trades by high mountains, except that unusually strong trade winds may sweep through the saddle between the Kohala Mountains and Mauna Kea and reach the areas to the lee. But even places exposed to the trades may be affected by local mountain circulations. Except for heavy rain, really severe weather seldom occurs. During the winter, cold fronts or the cyclonic storms of subtropical origin may bring blizzards to the upper slopes of Mauna Loa and Mauna Kea, with snow extending at times to 9,000 feet or below and icing nearer the summit.

Storms crossing the Pacific a thousand miles to the north, low pressure or tropical storms, may generate seas that cause heavy swell and surf.

NORMALS, MEANS, AND EXTREMES

HILO, HAWAII

LATITUDE: 19°43'N LONGITUDE: 155°04'W ELEVATION: FT. GRND 27 BARO 34 TIME ZONE: BERING WBAN: 21504

	(a)	JAN	FEB	MAR	APR	MAY	JUNE	JULY	AUG	SEP	OCT	NOV	DEC	YEAR
TEMPERATURE °F:														
Normals														
-Daily Maximum		79.8	79.8	79.5	79.8	81.2	82.7	83.0	83.6	83.8	83.2	81.4	80.0	81.5
-Daily Minimum		63.6	63.6	64.4	65.5	66.5	67.6	68.6	68.9	68.6	68.1	66.8	64.8	66.4
-Monthly		71.7	71.7	72.0	72.7	73.9	75.2	75.8	76.3	76.2	75.7	74.2	72.4	74.0
Extremes														
-Record Highest	47	91	92	93	89	94	90	89	93	92	91	90	93	94
-Year		1979	1968	1972	1978	1966	1969	1986	1950	1951	1979	1985	1980	MAY 1966
-Record Lowest	47	54	53	54	56	58	60	62	63	61	62	58	55	53
-Year		1980	1962	1983	1949	1947	1946	1970	1955	1970	1985	1985	1977	FEB 1962
NORMAL DEGREE DAYS:														
Heating (base 65°F)		0	0	0	0	0	0	0	0	0	0	0	0	0
Cooling (base 65°F)		208	188	217	231	276	306	335	350	336	332	276	229	3284
% OF POSSIBLE SUNSHINE	43	47	46	41	36	37	44	42	42	43	39	34	37	41
MEAN SKY COVER (tenths)														
Sunrise - Sunset	47	6.4	6.6	7.5	8.0	7.8	7.5	7.6	7.3	7.0	7.2	7.3	6.8	7.3
MEAN NUMBER OF DAYS:														
Sunrise to Sunset														
-Clear	47	6.3	5.2	2.8	1.0	1.2	1.6	1.3	1.8	3.1	2.7	3.3	5.1	35.6
-Partly Cloudy	47	11.4	10.2	10.1	8.9	10.7	11.1	11.5	12.2	12.1	11.7	10.4	10.7	131.0
-Cloudy	47	13.3	12.8	18.1	20.1	19.0	17.3	18.2	17.0	14.8	16.6	16.3	15.2	198.7
Precipitation .01 inches or more	51	17.3	17.0	23.4	25.2	25.2	24.5	27.5	26.5	23.4	23.7	23.2	21.4	278.3
Snow, Ice pellets, hail 1.0 inches or more	51	0.0	0.0	0.0	0.0	0.0	0.0	0.0	0.0	0.0	0.0	0.0	0.0	0.0
Thunderstorms	48	0.9	1.2	1.5	1.1	0.6	0.1	0.3	0.3	0.5	1.2	1.1	0.9	9.8
Heavy Fog Visibility 1/4 mile or less	48	0.0	0.0	0.0	0.0	0.0	0.0	0.0	0.0	0.0	0.0	0.0	0.0	0.0
Temperature °F														
-Maximum														
90° and above	48	0.1	0.1	0.*	0.0	0.*	0.*	0.0	0.1	0.2	0.3	0.*	0.1	0.9
32° and below	48	0.0	0.0	0.0	0.0	0.0	0.0	0.0	0.0	0.0	0.0	0.0	0.0	0.0
-Minimum														
32° and below	48	0.0	0.0	0.0	0.0	0.0	0.0	0.0	0.0	0.0	0.0	0.0	0.0	0.0
0° and below	48	0.0	0.0	0.0	0.0	0.0	0.0	0.0	0.0	0.0	0.0	0.0	0.0	0.0
AVG. STATION PRESS. (mb)	21	1014.4	1014.9	1016.9	1016.7	1016.4	1016.3	1015.6	1015.0	1014.1	1014.3	1014.5	1014.8	1015.3
RELATIVE HUMIDITY (%)														
Hour 02 (Local Time)	44	83	83	86	88	88	87	88	88	87	87	87	85	86
Hour 18 (Local Time)	44	79	78	80	81	80	78	81	81	79	80	81	81	80
Hour 14 (Local Time)	44	66	66	67	69	68	65	68	69	68	69	71	69	68
Hour 20 (Local Time)	44	82	81	82	83	82	81	82	83	84	85	86	84	83
PRECIPITATION (inches):														
Water Equivalent														
-Normal		9.88	10.29	13.92	15.26	9.91	6.20	9.71	9.34	8.53	9.60	14.51	12.04	129.19
-Maximum Monthly	51	32.24	45.55	49.93	43.24	25.01	15.50	28.59	26.92	18.47	26.10	45.75	50.82	50.82
-Year		1979	1979	1980	1986	1964	1943	1982	1991	1990	1951	1990	1954	DEC 1954
-Minimum Monthly	51	0.36	0.58	0.88	2.93	1.18	1.80	3.83	2.66	1.59	2.40	1.01	0.28	0.28
-Year		1953	1986	1972	1962	1945	1985	1975	1971	1974	1962	1989	1980	DEC 1980
-Maximum in 24 hrs	51	10.90	22.30	17.05	11.07	10.26	4.21	7.11	11.57	7.23	8.88	15.59	11.45	22.30
-Year		1990	1979	1980	1971	1965	1978	1982	1991	1986	1951	1959	1987	FEB 1979
Snow, Ice pellets, hail														
-Maximum Monthly		0.0	0.0	0.0	0.0	0.0	0.0	0.0	0.0	0.0	0.0	0.0	0.0	
-Year														
-Maximum in 24 hrs	51	0.0	0.0	0.0	0.0	0.0	0.0	0.0	0.0	0.0	0.0	0.0	0.0	
-Year														
WIND:														
Mean Speed (mph)	44	7.5	7.7	7.7	7.5	7.4	7.1	6.9	6.9	6.8	6.7	6.8	7.2	7.2
Prevailing Direction through 1963		SW	SW	SW	WSW	WSW	WSW	WSW	WSW	WSW	SW	WSW	SW	WSW
Fastest Obs. 1 Min.														
-Direction (!!!)	14	36	35	35	34	35	11	05	36	04	34	02	36	36
-Speed (MPH)	14	35	35	28	26	29	25	25	25	25	29	28	29	35
-Year		1987	1987	1987	1987	1987	1982	1984	1993	1987	1983	1987	1989	JAN 1987
Peak Gust														
-Direction (!!!)	10	NW	W	N	N	N	NE	SE	N	SE	SE	S	NW	W
-Speed (mph)	10	47	55	40	40	41	32	36	36	37	33	36	45	55
-Date		1992	1986	1993	1989	1987	1987	1986	1991	1992	1988	1988	1989	FEB 1986

PRECIPITATION (inches) HILO, HAWAII

YEAR	JAN	FEB	MAR	APR	MAY	JUNE	JULY	AUG	SEP	OCT	NOV	DEC	ANNUAL
1964	14.65	18.22	19.58	11.03	25.01	7.01	6.39	7.33	12.62	11.56	23.39	9.65	166.44
1965	9.28	3.71	8.33	18.49	21.05	8.85	7.15	4.79	5.72	5.80	19.18	14.94	127.29
1966	12.56	7.63	5.59	5.24	5.04	7.49	13.26	7.22	8.37	15.69	20.83	15.09	124.01
1967	8.04	10.35	9.46	21.26	9.84	6.26	14.03	19.55	6.78	10.08	21.25	17.10	154.00
1968	4.77	11.46	10.21	29.68	2.71	8.72	7.43	9.62	8.53	5.97	10.22	24.82	134.14
1969	19.66	43.66	30.64	14.57	7.83	2.76	11.75	17.50	7.24	3.19	6.33	8.10	173.23
1970	2.76	2.56	4.89	28.60	20.26	5.60	12.27	20.53	5.61	8.44	7.21	35.25	153.98
1971	13.47	5.31	12.04	27.82	6.49	2.79	4.13	2.66	8.63	7.28	17.88	32.19	140.69
1972	10.96	10.13	0.88	17.79	4.71	4.58	9.07	8.77	5.20	9.52	13.23	4.01	98.85
1973	3.45	5.51	18.84	7.34	8.34	3.69	4.40	3.54	8.07	9.72	26.88	8.19	107.97
1974	5.88	7.57	13.47	19.11	8.07	4.76	7.81	4.25	1.59	6.65	14.56	19.20	112.92
1975	19.62	9.28	10.40	10.23	3.01	4.20	3.83	8.13	2.73	8.88	11.15	8.47	99.93
1976	15.62	11.63	25.00	11.58	6.01	2.97	5.46	5.13	5.31	11.35	7.24	7.37	114.67
1977	1.22	9.56	15.49	10.90	10.86	2.46	6.36	7.60	4.19	10.30	8.78	2.66	90.38
1978	5.41	4.26	12.95	6.53	9.64	10.99	11.19	13.53	5.44	10.12	20.21	8.82	119.09
1979	32.24	45.55	5.32	9.90	4.10	10.45	6.54	7.04	3.64	5.03	21.56	7.40	158.77
1980	0.91	4.14	49.93	11.01	5.88	9.66	9.17	8.24	13.70	7.69	7.13	0.28	127.74
1981	1.51	4.95	5.66	4.63	4.16	2.43	4.32	8.97	12.79	10.23	11.73	18.53	89.91
1982	13.58	1.35	48.50	12.00	6.89	6.03	28.59	25.45	9.92	6.53	4.74	6.78	170.36
1983	0.90	0.83	1.98	10.31	9.60	3.94	7.21	7.48	12.08	8.06	2.33	3.37	68.09
1984	10.76	10.06	3.37	12.08	6.59	4.28	6.63	9.36	4.05	2.52	18.38	12.00	100.08
1985	2.25	16.14	21.28	10.61	17.04	1.80	9.86	6.71	11.78	8.19	4.71	2.59	112.96
1986	4.95	0.58	15.37	43.24	8.61	9.11	11.17	10.64	14.36	11.53	35.72	5.75	171.03
1987	9.02	5.06	4.79	9.24	15.65	12.91	18.26	3.69	11.56	14.21	15.83	22.19	142.41
1988	10.31	9.95	13.09	12.90	7.77	5.11	5.50	16.56	11.30	8.50	25.74	13.46	140.19
1989	27.46	6.54	7.33	37.19	19.80	7.03	22.93	8.82	9.73	13.16	1.01	5.71	166.71
1990	29.13	15.24	10.80	4.02	8.13	10.04	10.78	7.80	18.47	20.96	45.75	30.10	211.22
1991	3.81	9.32	37.88	11.02	8.08	9.85	9.82	26.92	9.41	5.15	6.74	15.04	153.04
1992	1.33	1.29	3.90	6.62	2.99	9.36	17.63	13.62	17.59	3.38	25.16	17.02	119.89
1993	2.17	2.67	11.96	9.04	7.54	6.63	18.43	11.38	4.99	12.83	10.74	16.11	114.49
Record Mean	9.61	11.33	13.82	13.29	9.14	6.40	10.03	10.33	7.91	9.86	14.79	14.00	130.50

AVERAGE TEMPERATURE (deg. F) HILO, HAWAII

YEAR	JAN	FEB	MAR	APR	MAY	JUNE	JULY	AUG	SEP	OCT	NOV	DEC	ANNUAL
1964	72.1	70.2	70.8	71.7	72.0	75.0	75.2	75.4	75.0	73.7	72.7	73.2	73.1
1965	71.7	68.4	69.7	72.4	74.5	74.1	75.2	75.8	76.8	75.4	73.7	69.9	73.1
1966	69.7	69.9	71.6	71.8	75.0	75.1	76.6	77.1	77.4	77.0	75.6	74.3	74.3
1967	71.9	73.9	74.1	74.3	77.1	77.5	77.8	77.1	77.0	77.2	75.4	73.3	75.6
1968	72.9	74.9	74.5	73.8	75.7	76.6	76.8	76.5	76.8	77.1	76.2	73.7	75.5
1969	71.4	72.2	72.5	72.8	74.5	76.2	75.8	76.2	75.0	74.5	73.3	71.5	73.8
1970	72.0	71.1	71.1	72.4	73.7	73.7	74.2	74.9	74.8	75.3	73.9	72.7	73.3
1971	71.3	71.9	69.4	70.9	72.2	75.0	76.6	76.6	76.6	75.4	73.8	70.9	73.4
1972	70.1	70.7	73.8	72.7	73.0	75.3	75.4	76.5	76.4	76.0	73.3	71.4	73.7
1973	72.2	71.1	72.5	72.2	72.9	74.6	75.7	76.3	76.3	75.8	75.4	73.8	74.1
1974	74.5	72.6	73.1	73.5	73.8	75.3	76.1	76.9	77.3	76.9	73.6	72.3	74.6
1975	71.0	71.9	71.2	72.4	73.1	74.4	74.8	75.7	75.5	74.7	73.5	72.2	73.4
1976	71.3	71.2	71.6	72.1	73.0	73.6	74.5	76.2	76.9	76.2	74.5	73.2	73.7
1977	73.9	74.0	73.3	74.2	74.7	76.2	77.1	78.1	77.5	76.9	75.2	73.7	75.4
1978	71.7	72.5	73.2	74.2	76.2	76.5	77.1	76.8	76.2	75.5	74.1	71.1	74.6
1979	69.8	70.4	71.5	73.8	73.8	74.2	74.6	75.6	76.5	76.1	73.0	72.8	73.5
1980	71.6	72.6	72.3	74.5	77.3	77.6	77.8	75.0	75.7	74.8	73.8	74.2	74.8
1981	73.5	72.7	71.6	72.8	74.2	76.0	76.1	76.1	76.2	74.6	73.9	72.0	74.1
1982	71.9	71.8	70.3	71.2	72.9	76.3	76.7	76.9	76.1	74.9	74.6	71.8	73.8
1983	71.4	71.9	72.5	71.9	72.6	74.3	74.8	75.2	74.9	74.1	73.8	72.9	73.3
1984	72.4	71.5	73.8	73.0	74.0	74.7	75.2	75.3	75.4	76.5	73.6	71.1	73.9
1985	69.8	70.5	69.4	69.8	71.4	74.4	75.4	75.7	75.7	74.3	73.0	71.6	72.6
1986	71.1	73.6	74.7	73.6	75.4	76.6	77.8	78.5	77.9	76.4	75.1	72.8	75.3
1987	71.8	70.7	71.6	72.2	72.5	75.4	76.7	77.9	77.8	76.6	74.7	73.1	74.3
1988	71.9	72.3	72.2	72.6	74.2	74.7	75.7	76.0	76.6	77.9	76.3	74.9	74.6
1989	72.2	71.4	72.4	71.1	72.7	74.7	75.2	75.0	74.6	75.6	73.6	71.3	73.3
1990	72.1	70.4	71.2	73.5	74.1	75.0	76.0	77.0	77.2	76.2	75.4	72.5	74.2
1991	72.0	72.8	70.8	72.6	74.2	74.8	76.0	76.9	76.9	76.2	75.8	72.9	74.3
1992	71.2	71.4	72.3	72.4	74.8	76.2	76.2	77.2	77.8	77.7	75.2	73.6	74.7
1993	71.1	70.1	71.6	73.5	73.3	75.4	75.8	77.0	77.1	76.0	73.4	71.7	73.8
Record Mean	71.3	71.3	71.5	72.3	73.4	74.8	75.5	76.0	75.9	75.4	73.7	72.0	73.6
Max	79.3	79.3	79.0	79.4	80.7	82.3	82.6	83.2	83.5	82.9	80.9	79.4	81.1
Min	63.3	63.2	63.9	65.1	66.2	67.3	68.3	68.8	68.3	67.8	66.5	64.5	66.1

REFERENCE NOTES HILO, HAWAII

GENERAL
T - TRACE AMOUNT.
BLANK ENTRIES DENOTE MISSING/UNREPORTED DATA.
# INDICATES A STATION OR INSTRUMENT RELOCATION.

SPECIFIC

PM - INCLUDES LAST DAY OF PREVIOUS MONTH
ASOS - AUTOMATED SURFACE OBSERVING SYSTEM IN OPERATION DURING THESE MONTHS.

(a) - LENGTH OF RECORD IN YEARS, ALTHOUGH INDIVIDUAL MONTHS MAY BE MISSING.
0.* OR * - THE VALUE IS BETWEEN 0.0 AND 0.05
NORMALS - BASED ON THE 1961-1990 RECORD PERIOD.
EXTREMES - DATES ARE THE MOST RECENT OCCURRENCE
WIND DIR.- NUMERALS SHOW TENS OF DEGRESS CLOCKWISE FROM TRUE NORTH. "00" INDICATES CALM.
RESULTANT DIRECTIONS ARE GIVEN TO WHOLE DEGREES.
BOLD VALUES INDICATE EXTREME VALUES WHICH OCCURRED AFTER THE ASOS SYSTEM WAS COMMISSIONED.

RECORD = PERIOD OF RECORD
RECORD MEAN PRECIPITATION IS THE MEAN OF ALL DAILY PRECIPITATION AMOUNTS DURING THE PERIOD OF RECORD.
RECORD MAX(MIN) TEMPERATURE IS THE MEAN OF ALL DAILY MAX(MIN) TEMPERATURES DURING THE PERIOD OF RECORD.
RECORD MEAN TEMPERATURE IS THE SUM OF THE RECORD MAX AND RECORD MIN DIVIDED BY 2.
AVERAGE TEMPERATURE IS THE SUM OF THE MEAN DAILY MAX AND MIN TEMPERATURE DIVIDED BY 2.

EXCEPTIONS

RECORD MEANS ARE THROUGH THE CURRENT YEAR, BEGINNING IN 1947 FOR TEMPERATURE
1943 FOR PRECIPITATION

HEATING DEGREE DAYS Base 65 deg. F HILO, HAWAII

SEASON	JULY	AUG	SEP	OCT	NOV	DEC	JAN	FEB	MAR	APR	MAY	JUNE	TOTAL
1983-84	0	0	0	0	0	0	0	0	0	0	0	0	0
1984-85	0	0	0	0	0	0	0	0	0	0	0	0	0
1985-86	0	0	0	0	0	0	0	0	0	0	0	0	0
1986-87	0	0	0	0	0	0	0	0	0	0	0	0	0
1987-88	0	0	0	0	0	0	0	0	0	0	0	0	0
1988-89	0	0	0	0	0	0	0	0	0	0	0	0	0
1989-90	0	0	0	0	0	0	0	0	0	0	0	0	0
1990-91	0	0	0	0	0	0	0	0	0	0	0	0	0
1991-92	0	0	0	0	0	0	0	0	0	0	0	0	0
1992-93	0	0	0	0	0	0	0	0	0	0	0	0	0
1993-94	0	0	0	0	0	0							

COOLING DEGREE DAYS Base 65 deg. F HILO, HAWAII

YEAR	JAN	FEB	MAR	APR	MAY	JUNE	JULY	AUG	SEP	OCT	NOV	DEC	TOTAL
1969	207	209	240	239	298	341	344	351	308	300	253	208	3298
1970	223	173	195	229	273	270	290	313	303	325	276	246	3116
1971	203	201	142	185	232	306	366	367	354	328	271	188	3143
1972	163	171	281	236	256	316	330	365	349	348	256	207	3278
1973	233	180	239	222	253	294	341	358	345	340	321	278	3404
1974	299	219	261	263	276	315	351	375	375	376	262	235	3607
1975	192	201	197	232	257	288	311	339	323	309	262	233	3144
1976	201	186	214	222	255	268	302	355	364	357	291	261	3276
1977	280	260	264	281	307	343	379	415	382	374	312	274	3871
1978	216	215	263	283	353	351	383	375	341	332	279	195	3586
1979	155	160	210	271	278	280	302	338	351	350	246	248	3189
1980	213	227	234	293	390	385	405	316	328	313	269	295	3668
1981	271	220	210	242	293	338	350	348	345	302	274	225	3418
1982	220	196	170	194	252	348	369	379	340	317	293	219	3297
1983	207	200	239	214	240	287	313	324	303	288	272	250	3137
1984	236	194	282	247	284	298	324	326	320	363	261	195	3330
1985	154	161	142	152	204	290	329	339	329	294	248	211	2853
1986	196	246	308	264	329	356	404	423	396	363	309	250	3844
1987	218	163	212	226	241	319	369	407	389	365	299	259	3467
1988	221	216	233	238	293	298	338	349	353	405	345	315	3604
1989	227	188	238	189	248	297	327	315	294	335	264	202	3124
1990	227	157	200	260	290	308	349	379	376	353	317	237	3453
1991	223	222	188	234	296	301	348	378	365	351	333	251	3490
1992	197	192	235	229	312	343	355	384	387	402	315	275	3626
1993	193	148	213	263	263	318	343	380	370	350	260	217	3318

SNOWFALL (inches) HILO, HAWAII

SEASON	JULY	AUG	SEP	OCT	NOV	DEC	JAN	FEB	MAR	APR	MAY	JUNE	TOTAL
1971-72	0.0	0.0	0.0	0.0	0.0	0.0	0.0	0.0	0.0	0.0	0.0	0.0	0.0
1972-73	0.0	0.0	0.0	0.0	0.0	0.0	0.0	0.0	0.0	0.0	0.0	0.0	0.0
1973-74	0.0	0.0	0.0	0.0	0.0	0.0	0.0	0.0	0.0	0.0	0.0	0.0	0.0
1974-75	0.0	0.0	0.0	0.0	0.0	0.0	0.0	0.0	0.0	0.0	0.0	0.0	0.0
1975-76	0.0	0.0	0.0	0.0	0.0	0.0	0.0	0.0	0.0	0.0	0.0	0.0	0.0
1976-77	0.0	0.0	0.0	0.0	0.0	0.0	0.0	0.0	0.0	0.0	0.0	0.0	0.0
1977-78	0.0	0.0	0.0	0.0	0.0	0.0	0.0	0.0	0.0	0.0	0.0	0.0	0.0
1978-79	0.0	0.0	0.0	0.0	0.0	0.0	0.0	0.0	0.0	0.0	0.0	0.0	0.0
1979-80	0.0	0.0	0.0	0.0	0.0	0.0	0.0	0.0	0.0	0.0	0.0	0.0	0.0
1980-81	0.0	0.0	0.0	0.0	0.0	0.0	0.0	0.0	0.0	0.0	0.0	0.0	0.0
1981-82	0.0	0.0	0.0	0.0	0.0	0.0	0.0	0.0	0.0	0.0	0.0	0.0	0.0
1982-83	0.0	0.0	0.0	0.0	0.0	0.0	0.0	0.0	0.0	0.0	0.0	0.0	0.0
1983-84	0.0	0.0	0.0	0.0	0.0	0.0	0.0	0.0	0.0	0.0	0.0	0.0	0.0
1984-85	0.0	0.0	0.0	0.0	0.0	0.0	0.0	0.0	0.0	0.0	0.0	0.0	0.0
1985-86	0.0	0.0	0.0	0.0	0.0	0.0	0.0	0.0	0.0	0.0	0.0	0.0	0.0
1986-87	0.0	0.0	0.0	0.0	0.0	0.0	0.0	0.0	0.0	0.0	0.0	0.0	0.0
1987-88	0.0	0.0	0.0	0.0	0.0	0.0	0.0	0.0	0.0	0.0	0.0	0.0	0.0
1988-89	0.0	0.0	0.0	0.0	0.0	0.0	0.0	0.0	0.0	0.0	0.0	0.0	0.0
1989-90	0.0	0.0	0.0	0.0	0.0	0.0	0.0	0.0	0.0	0.0	0.0	0.0	0.0
1990-91	0.0	0.0	0.0	0.0	0.0	0.0	0.0	0.0	0.0	0.0	0.0	0.0	0.0
1991-92	0.0	0.0	0.0	0.0	0.0	0.0	0.0	0.0	0.0	0.0	0.0	0.0	0.0
1992-93	0.0	0.0	0.0	0.0	0.0	0.0	0.0	0.0	0.0	0.0	0.0	0.0	0.0
1993-94	0.0	0.0	0.0	0.0	0.0	0.0							
Record Mean	0.0	0.0	0.0	0.0	0.0	0.0	0.0	0.0	0.0	0.0	0.0	0.0	0.0

HONOLULU, HAWAII

Oahu, on which Honolulu is located is the third largest of the Hawaiian Islands. The Koolau Range, at an average elevation of 2,000 feet parallels the northeastern coast. The Waianae Mountains, somewhat higher in elevation, parallel the west coast. Honolulu Airport, the business and Waikiki districts, and a number of the residential areas of Honolulu lie along the southern coastal plain.

The climate of Hawaii is unusually pleasant for the tropics. Its outstanding features are the persistence of the trade winds, the remarkable variability in rainfall over short distances, the sunniness of the leeward lowlands in contrast to the persistent cloudiness over nearby mountain crests, the equable temperature, and the general infrequency of severe storms.

The prevailing wind throughout the year is the northeasterly trade wind, although its average frequency varies from more than 90 percent during the summer to only 50 percent in January.

Heavy mountain rainfall sustains extensive irrigation of cane fields and the water supply for Honolulu. Oahu is driest along the coast west of the Waianaes where rainfall drops to about 20 inches a year. Daytime showers, usually light, often occur while the sun continues to shine, a phenomenon referred to locally as liquid sunshine.

The moderate temperature range is associated with the small seasonal variation in the energy received from the sun and the tempering effect of the surrounding ocean. Honolulu Airport has recorded as high as the lower 90s and as low as the lower 50s.

Because of the trade winds, even the warmest months are usually comfortable. But when the trades diminish or give way to southerly winds, a situation known locally as kona weather, or kona storms when stormy, the humidity may become oppressively high.

Intense rains of the October to April winter season sometimes cause serious, flash flooding. Thunderstorms are infrequent and usually mild and hail seldom occurs. Infrequently, a small tornado or a waterspout may do some damage. Only a few tropical cyclones have struck Hawaii, although others have come near enough for their outlying winds, waves, clouds, and rain to affect the Islands.

NORMALS, MEANS, AND EXTREMES

HONOLULU, HAWAII

LATITUDE: 21°20'N LONGITUDE: 157°56' W ELEVATION: FT. GRND 7 BARO 18 TIME ZONE: BERING WBAN: 22521

	(a)	JAN	FEB	MAR	APR	MAY	JUNE	JULY	AUG	SEP	OCT	NOV	DEC	YEAR
TEMPERATURE °F:														
Normals														
-Daily Maximum		80.1	80.5	81.6	82.8	84.7	86.5	87.5	88.7	88.5	86.9	84.1	81.2	84.4
-Daily Minimum		65.6	65.4	67.2	68.7	70.3	72.2	73.5	74.2	73.5	72.3	70.3	67.0	70.0
-Monthly		72.9	73.0	74.4	75.8	77.5	79.4	80.5	81.4	81.0	79.6	77.2	74.1	77.2
Extremes														
-Record Highest	24	87	88	88	89	93	92	92	93	94	94	93	89	94
-Year		1987	1984	1987	1990	1988	1992	1992	1987	1988	1984	1986	1983	SEP 1988
-Record Lowest	24	53	53	55	57	60	65	66	67	66	61	57	54	53
-Year		1972	1983	1976	1985	1989	1982	1990	1984	1985	1993	1990	1962	FEB 1983
NORMAL DEGREE DAYS:														
Heating (base 65°F)		0	0	0	0	0	0	0	0	0	0	0	0	0
Cooling (base 65°F)		245	224	291	324	388	432	481	508	480	453	366	282	4474
% OF POSSIBLE SUNSHINE	41	64	66	70	69	70	72	75	76	76	69	62	61	69
MEAN SKY COVER (tenths)														
Sunrise - Sunset	47	5.4	5.5	5.8	6.1	5.9	5.5	5.3	5.2	5.2	5.7	5.7	5.6	5.6
MEAN NUMBER OF DAYS:														
Sunrise to Sunset														
-Clear	44	9.4	8.2	7.5	5.7	6.5	6.4	7.6	8.2	8.2	7.4	7.2	8.2	90.4
-Partly Cloudy	44	13.1	12.4	14.0	14.4	15.4	17.2	18.2	16.8	16.1	15.1	13.8	13.6	180.1
-Cloudy	44	8.5	7.7	9.4	9.9	9.1	6.4	5.2	6.0	5.7	8.5	9.1	9.2	94.7
Precipitation .01 inches or more	44	9.5	9.2	8.8	8.8	7.3	5.8	7.3	6.3	7.0	8.8	9.2	10.3	98.4
Snow,Ice pellets,hail 1.0 inches or more	44	0.0	0.0	0.0	0.0	0.0	0.0	0.0	0.0	0.0	0.0	0.0	0.0	0.0
Thunderstorms	44	0.8	1.1	0.9	0.5	0.3	0.1	0.2	0.1	0.5	0.8	0.8	0.7	6.7
Heavy Fog Visibility 1/4 mile or less	44	0.0	0.0	0.0	0.0	0.0	0.0	0.0	0.0	0.0	0.0	0.0	0.0	0.0
Temperature °F														
-Maximum														
90° and above	24	0.0	0.0	0.0	0.0	0.3	1.6	4.7	10.4	10.1	4.0	0.3	0.0	31.3
32° and below	24	0.0	0.0	0.0	0.0	0.0	0.0	0.0	0.0	0.0	0.0	0.0	0.0	0.0
-Minimum														
32° and below	24	0.0	0.0	0.0	0.0	0.0	0.0	0.0	0.0	0.0	0.0	0.0	0.0	0.0
0° and below	24	0.0	0.0	0.0	0.0	0.0	0.0	0.0	0.0	0.0	0.0	0.0	0.0	0.0
AVG. STATION PRESS.(mb)	21	1014.9	1015.5	1017.3	1017.1	1016.9	1016.7	1016.0	1015.4	1014.5	1014.7	1014.9	1015.3	1015.8
RELATIVE HUMIDITY (%)														
Hour 02	24	81	79	76	75	74	73	73	74	75	76	78	80	76
Hour 18 (Local Time)	24	81	78	73	69	67	66	67	68	68	70	75	79	72
Hour 14	24	61	59	57	55	53	52	51	52	52	55	58	60	55
Hour 20	24	73	71	70	69	68	68	68	69	69	70	72	74	70
PRECIPITATION (inches):														
Water Equivalent														
-Normal		3.55	2.21	2.20	1.54	1.13	0.50	0.59	0.44	0.78	2.28	3.00	3.80	22.02
-Maximum Monthly	47	14.74	13.68	20.79	8.92	7.23	2.46	2.33	3.08	2.74	11.15	14.72	17.29	20.79
-Year		1949	1955	1951	1963	1965	1971	1989	1959	1947	1978	1965	1987	MAR 1951
-Minimum Monthly	47	0.18	0.06	0.01	0.01	0.05	T	0.03	T	0.05	0.11	0.03	0.06	T
-Year		1986	1983	1957	1960	1949	1959	1950	1974	1977	1957	1962	1976	AUG 1974
-Maximum in 24 hrs	44	6.72	6.88	17.07	4.21	3.44	2.28	2.20	2.35	1.40	7.57	9.15	8.25	17.07
-Year		1963	1955	1958	1972	1965	1967	1989	1959	1963	1978	1954	1987	MAR 1958
Snow,Ice pellets,hail														
-Maximum Monthly		0.0	0.0	0.0	0.0	0.0	0.0	0.0	0.0	0.0	0.0	0.0	0.0	
-Year														
-Maximum in 24 hrs	47	0.0	0.0	0.0	0.0	0.0	0.0	0.0	0.0	0.0	0.0	0.0	0.0	
-Year														
WIND:														
Mean Speed (mph)	44	9.5	10.2	11.4	11.8	11.8	12.7	13.2	12.8	11.2	10.5	10.7	10.4	11.3
Prevailing Direction through 1963		ENE	ENE	ENE	ENE	ENE	ENE	ENE	ENE	ENE	ENE	ENE	ENE	ENE
Fastest Obs. 1 Min.														
-Direction (!!)	13	07	07	06	06	13	05	07	04	13	07	20	06	20
-Speed (MPH)	13	32	35	30	31	30	26	28	28	38	25	46	33	46
-Year		1982	1990	1990	1980	1985	1981	1980	1982	1992	1981	1982	1993	NOV 1982
Peak Gust														
-Direction (!!)	10	NE	NE	NE	NE	E	NE	NE	NE	SE	SE	NE	NE	SE
-Speed (mph)	10	41	46	46	41	39	35	40	35	49	35	40	46	49
-Date		1987	1990	1985	1986	1985	1993	1984	1990	1992	1991	1986	1993	SEP 1992

PRECIPITATION (inches)

HONOLULU, HAWAII

YEAR	JAN	FEB	MAR	APR	MAY	JUNE	JULY	AUG	SEP	OCT	NOV	DEC	ANNUAL
1964	2.18	0.52	5.21	0.88	0.21	0.08	1.34	0.46	0.97	0.34	2.36	5.57	20.12
1965	3.02	0.80	0.99	1.48	7.23	0.25	1.37	0.87	0.52	3.56	14.72	7.97	42.78
1966	1.39	3.71	0.39	0.46	0.41	0.04	0.43	0.82	0.16	2.95	9.44	2.98	23.18
1967	0.79	2.53	6.78	1.29	2.12	2.43	1.21	2.53	0.42	1.53	2.78	9.93	34.34
1968	8.17	2.91	2.49	3.14	1.22	0.24	0.29	0.10	1.30	2.13	5.64	9.63	37.26
1969	8.20	0.48	3.00	0.10	0.81	0.23	0.63	0.11	0.87	0.96	5.77	1.34	22.50
1970	1.81	0.77	0.07	0.74	0.21	0.23	2.01	0.21	0.39	1.88	5.94	1.23	15.49
1971	6.19	2.37	5.57	2.19	0.43	2.46	0.04	0.26	1.03	2.27	0.95	2.88	26.64
1972	5.28	5.00	2.45	5.15	0.12	0.79	0.20	0.46	0.92	2.39	0.59	3.59	26.94
1973	0.67	0.60	0.40	0.72	0.89	0.09	0.46	0.32	0.64	1.78	3.73	3.94	14.24
1974	4.21	1.28	3.49	4.13	0.82	1.52	0.44	T	2.08	2.77	2.69	0.59	24.02
1975	6.42	2.36	2.02	0.51	0.19	0.03	0.40	0.03	0.11	0.18	11.54	0.60	24.39
1976	1.29	6.08	2.67	0.71	0.26	0.18	0.24	0.17	0.33	0.45	0.46	0.06	12.90
1977	0.52	0.32	2.36	1.81	4.76	0.11	0.14	0.08	0.05	0.15	0.61	1.45	12.36
1978	0.34	0.75	1.37	2.07	3.39	1.06	0.20	0.83	0.28	11.15	1.55	2.06	25.05
1979	4.57	7.21	0.77	0.55	0.21	0.32	0.13	0.15	0.47	0.53	0.52	1.50	16.93
1980	8.91	2.26	3.04	1.13	0.78	1.76	0.37	0.36	0.41	0.30	0.21	7.37	26.90
1981	0.81	0.97	0.71	1.01	0.94	0.14	0.42	0.70	0.39	1.84	1.01	4.47	13.41
1982	12.82	2.16	3.73	1.28	0.13	0.35	0.20	1.98	0.52	7.24	1.32	3.19	34.92
1983	0.32	0.06	0.53	0.42	0.35	0.26	0.22	0.29	1.16	0.23	0.13	1.06	5.03
1984	0.21	0.60	1.08	2.41	0.16	0.08	0.23	0.04	1.36	1.89	3.58	5.44	17.08
1985	1.46	3.87	1.26	0.20	1.11	0.13	0.53	0.16	1.28	5.08	2.11	0.19	17.38
1986	0.18	1.38	0.17	0.35	0.81	0.36	1.54	0.90	2.00	1.23	4.23	0.78	13.93
1987	0.42	0.86	0.31	0.65	0.73	0.46	0.33	0.22	1.13	0.20	0.93	17.29	23.53
1988	3.05	1.31	0.67	0.50	1.25	0.04	0.12	0.34	0.86	0.23	1.39	6.71	16.47
1989	2.07	6.48	2.58	1.23	0.29	0.11	2.33	0.08	0.15	10.37	0.51	1.32	27.52
1990	4.32	4.15	0.86	0.30	0.30	0.08	0.49	0.01	0.98	0.47	2.96	4.92	19.84
1991	0.80	2.09	6.24	1.00	0.48	0.26	0.16	0.16	0.56	3.43	1.52	1.24	17.94
1992	0.43	1.35	0.72	0.11	1.13	0.10	2.01	0.97	2.08	2.57	1.04	6.49	19.00
1993	0.70	0.41	0.02	0.23	0.19	0.11	0.69	1.03	0.10	1.63	0.46	0.27	5.84
Record Mean	3.76	2.50	2.68	1.32	0.96	0.38	0.56	0.59	0.76	2.02	2.66	3.57	21.77

AVERAGE TEMPERATURE (deg. F)

HONOLULU, HAWAII

YEAR	JAN	FEB	MAR	APR	MAY	JUNE	JULY	AUG	SEP	OCT	NOV	DEC	ANNUAL
1964	75.1	74.7	74.7	75.7	76.4	77.9	79.3	80.4	80.2	78.3	76.4	74.9	77.0
1965	71.9	69.8	72.1	75.0	77.0	79.1	79.9	79.7	80.0	78.2	77.1	73.3	76.1
1966	72.6	72.0	75.2	75.1	77.1	80.6	80.8	82.0	82.4	80.9	77.2	74.6	77.6
1967	72.5	73.9	73.8	74.7	78.2	79.7	81.5	82.1	82.4	80.6	77.6	74.0	77.6
1968	73.0	73.4	74.8	76.7	78.4	80.4	81.5	82.9	82.0	80.6	78.8	72.7	77.9
1969	69.1	73.9	73.6	74.7	76.7	79.1	80.9	83.2	82.0	80.5	78.7	76.5	77.4
#1970	74.2	73.3	76.9	78.3	80.4	81.1	82.2	83.8	79.0	78.5	75.8	74.7	78.2
#1971	71.7	74.4	73.9	75.6	76.2	77.3	78.9	79.5	79.1	78.0	75.7	73.3	76.1
1972	70.4	70.6	72.8	75.0	77.3	78.9	80.4	81.1	80.5	79.3	76.7	71.6	76.2
1973	72.9	72.6	76.1	75.5	77.1	79.2	80.5	81.2	81.0	79.4	77.0	73.8	77.2
1974	74.5	74.4	74.0	77.4	78.2	79.3	79.9	81.2	80.0	79.5	75.7	75.8	77.5
1975	72.4	72.8	73.0	74.5	75.7	78.1	79.0	80.1	79.4	79.1	77.1	73.0	76.2
1976	73.7	72.0	73.6	75.1	77.5	78.2	79.8	80.8	80.7	79.1	75.3	75.3	76.8
1977	73.7	75.6	76.2	76.3	77.6	79.5	80.9	82.2	81.6	81.1	78.6	75.1	78.2
1978	74.2	73.2	75.7	76.8	78.2	78.7	79.0	80.5	80.5	77.8	74.7	72.4	76.8
1979	69.9	72.1	72.8	74.8	78.0	80.0	80.9	80.4	81.1	81.0	77.4	75.3	77.0
1980	71.9	72.4	75.0	76.1	78.3	79.5	80.9	81.0	81.6	80.1	78.0	74.4	77.5
1981	73.2	73.6	74.7	75.9	77.3	80.6	79.7	80.1	80.7	78.3	76.7	74.0	77.1
1982	73.2	71.7	74.0	75.4	78.3	79.6	80.6	81.4	81.4	79.4	75.7	72.0	76.9
1983	71.9	71.3	73.5	74.6	75.7	78.9	79.7	82.4	82.3	81.1	80.1	75.1	77.2
1984	74.6	74.6	75.8	77.0	78.7	79.3	81.0	81.7	81.3	80.2	79.0	74.1	78.1
1985	71.4	73.9	74.5	74.5	76.5	79.2	81.6	81.9	81.1	79.8	75.1	73.3	76.9
1986	72.8	72.6	76.5	77.5	78.3	80.0	81.6	82.9	82.1	80.6	79.2	75.1	78.3
1987	73.4	71.2	74.0	76.0	75.7	80.4	82.1	82.7	82.9	81.4	78.8	75.8	77.9
1988	73.1	74.7	76.0	77.3	78.9	80.8	81.8	82.1	82.1	80.1	79.9	75.6	78.5
1989	74.5	73.6	75.3	74.5	78.4	80.9	81.6	81.4	81.9	78.6	76.7	72.9	77.5
1990	74.7	71.5	73.1	76.6	78.1	80.0	80.8	82.3	82.3	80.9	77.3	74.1	77.6
1991	72.4	73.4	72.9	75.9	77.8	79.4	81.2	82.4	81.5	80.0	79.5	76.2	77.7
1992	72.9	73.2	74.9	75.6	77.8	81.3	81.5	82.2	81.3	79.4	77.0	76.7	77.8
1993	70.9	71.1	74.0	77.4	77.2	80.2	80.6	81.3	81.1	79.7	76.3	75.0	77.1
Record Mean	72.6	72.7	73.8	75.3	76.8	78.8	79.8	80.6	80.4	79.0	76.7	74.1	76.8
Max	79.7	79.9	80.7	82.0	83.7	85.6	86.5	87.3	87.4	85.9	83.3	80.7	83.6
Min	65.5	65.4	66.9	68.5	70.0	72.0	73.1	73.9	73.4	72.1	70.0	67.4	69.9

REFERENCE NOTES

HONOLULU, HAWAII

GENERAL
T - TRACE AMOUNT.
BLANK ENTRIES DENOTE MISSING/UNREPORTED DATA.
# INDICATES A STATION OR INSTRUMENT RELOCATION.

SPECIFIC

PM - INCLUDES LAST DAY OF PREVIOUS MONTH
ASOS - AUTOMATED SURFACE OBSERVING SYSTEM IN OPERATION DURING THESE MONTHS.

(a) - LENGTH OF RECORD IN YEARS, ALTHOUGH INDIVIDUAL MONTHS MAY BE MISSING.
0.* OR * - THE VALUE IS BETWEEN 0.0 AND 0.05
NORMALS - BASED ON THE 1961-1990 RECORD PERIOD.
EXTREMES - DATES ARE THE MOST RECENT OCCURRENCE
WIND DIR. - NUMERALS SHOW TENS OF DEGREES CLOCKWISE FROM TRUE NORTH. "00" INDICATES CALM.
RESULTANT DIRECTIONS ARE GIVEN TO WHOLE DEGREES.
BOLD VALUES INDICATE EXTREME VALUES WHICH OCCURRED AFTER THE ASOS SYSTEM WAS COMMISSIONED.

RECORD = PERIOD OF RECORD
RECORD MEAN PRECIPITATION IS THE MEAN OF ALL DAILY PRECIPITATION AMOUNTS DURING THE PERIOD OF RECORD.
RECORD MAX(MIN) TEMPERATURE IS THE MEAN OF ALL DAILY MAX(MIN) TEMPERATURES DURING THE PERIOD OF RECORD.
RECORD MEAN TEMPERATURE IS THE SUM OF THE RECORD MAX AND RECORD MIN DIVIDED BY 2.
AVERAGE TEMPERATURE IS THE SUM OF THE MEAN DAILY MAX AND MIN TEMPERATURE DIVIDED BY 2.

EXCEPTIONS

RECORD MEANS ARE THROUGH THE CURRENT YEAR, BEGINNING IN 1947 THRU 1964 AND
1970 TO DATE FOR TEMPERATURE
1947 FOR PRECIPITATION

HEATING DEGREE DAYS Base 65 deg. F — HONOLULU, HAWAII

SEASON	JULY	AUG	SEP	OCT	NOV	DEC	JAN	FEB	MAR	APR	MAY	JUNE	TOTAL
1983-84	0	0	0	0	0	0	0	0	0	0	0	0	0
1984-85	0	0	0	0	0	0	0	0	0	0	0	0	0
1985-86	0	0	0	0	0	0	0	0	0	0	0	0	0
1986-87	0	0	0	0	0	0	0	0	0	0	0	0	0
1987-88	0	0	0	0	0	0	0	0	0	0	0	0	0
1988-89	0	0	0	0	0	0	0	0	0	0	0	0	0
1989-90	0	0	0	0	0	0	0	0	0	0	0	0	0
1990-91	0	0	0	0	0	0	0	0	0	0	0	0	0
1991-92	0	0	0	0	0	0	0	0	0	0	0	0	0
1992-93	0	0	0	0	0	0	0	0	0	0	0	0	0
1993-94	0	0	0	0	0	0							

COOLING DEGREE DAYS Base 65 deg. F — HONOLULU, HAWAII

YEAR	JAN	FEB	MAR	APR	MAY	JUNE	JULY	AUG	SEP	OCT	NOV	DEC	TOTAL
1969	133	255	269	295	369	428	500	572	517	488	419	365	4610
1970	293	239	378	409	485	490	540	586	429	426	332	306	4913
1971	216	269	284	326	355	376	435	456	431	408	328	265	4149
1972	178	170	249	307	386	425	484	506	470	450	357	209	4191
1973	252	219	353	322	382	432	485	512	484	455	367	279	4542
1974	300	270	285	378	415	434	468	509	458	457	328	341	4643
1975	235	224	256	292	337	400	438	475	438	442	372	257	4166
1976	278	209	275	311	393	402	464	498	479	446	315	325	4395
1977	276	305	355	344	396	441	498	541	505	507	417	320	4905
1978	292	238	336	361	417	418	439	489	473	401	298	239	4401
1979	159	209	250	299	412	458	500	485	489	504	378	326	4469
1980	222	220	317	340	418	442	501	504	503	476	395	295	4633
1981	263	249	311	335	385	474	463	477	477	419	355	284	4492
1982	261	195	288	318	421	442	493	514	499	452	326	225	4434
1983	223	182	270	295	335	425	461	544	525	508	461	318	4547
1984	304	285	340	366	432	438	501	527	494	475	425	291	4878
1985	205	256	300	293	364	437	521	532	491	464	310	264	4437
1986	251	217	366	384	421	457	519	561	521	491	433	318	4939
1987	267	178	285	337	337	465	537	556	544	516	418	342	4782
1988	260	289	346	373	437	482	527	537	520	478	455	336	5040
1989	301	244	325	291	425	482	521	517	512	431	358	252	4659
1990	306	189	258	354	412	456	498	543	525	501	377	289	4708
1991	236	243	250	333	404	442	508	547	501	474	443	353	4734
1992	251	244	316	327	404	495	518	540	495	452	369	368	4779
1993	190	180	285	381	387	463	491	515	491	465	345	314	4507

SNOWFALL (inches) — HONOLULU, HAWAII

SEASON	JULY	AUG	SEP	OCT	NOV	DEC	JAN	FEB	MAR	APR	MAY	JUNE	TOTAL
1971-72	0.0	0.0	0.0	0.0	0.0	0.0	0.0	0.0	0.0	0.0	0.0	0.0	0.0
1972-73	0.0	0.0	0.0	0.0	0.0	0.0	0.0	0.0	0.0	0.0	0.0	0.0	0.0
1973-74	0.0	0.0	0.0	0.0	0.0	0.0	0.0	0.0	0.0	0.0	0.0	0.0	0.0
1974-75	0.0	0.0	0.0	0.0	0.0	0.0	0.0	0.0	0.0	0.0	0.0	0.0	0.0
1975-76	0.0	0.0	0.0	0.0	0.0	0.0	0.0	0.0	0.0	0.0	0.0	0.0	0.0
1976-77	0.0	0.0	0.0	0.0	0.0	0.0	0.0	0.0	0.0	0.0	0.0	0.0	0.0
1977-78	0.0	0.0	0.0	0.0	0.0	0.0	0.0	0.0	0.0	0.0	0.0	0.0	0.0
1978-79	0.0	0.0	0.0	0.0	0.0	0.0	0.0	0.0	0.0	0.0	0.0	0.0	0.0
1979-80	0.0	0.0	0.0	0.0	0.0	0.0	0.0	0.0	0.0	0.0	0.0	0.0	0.0
1980-81	0.0	0.0	0.0	0.0	0.0	0.0	0.0	0.0	0.0	0.0	0.0	0.0	0.0
1981-82	0.0	0.0	0.0	0.0	0.0	0.0	0.0	0.0	0.0	0.0	0.0	0.0	0.0
1982-83	0.0	0.0	0.0	0.0	0.0	0.0	0.0	0.0	0.0	0.0	0.0	0.0	0.0
1983-84	0.0	0.0	0.0	0.0	0.0	0.0	0.0	0.0	0.0	0.0	0.0	0.0	0.0
1984-85	0.0	0.0	0.0	0.0	0.0	0.0	0.0	0.0	0.0	0.0	0.0	0.0	0.0
1985-86	0.0	0.0	0.0	0.0	0.0	0.0	0.0	0.0	0.0	0.0	0.0	0.0	0.0
1986-87	0.0	0.0	0.0	0.0	0.0	0.0	0.0	0.0	0.0	0.0	0.0	0.0	0.0
1987-88	0.0	0.0	0.0	0.0	0.0	0.0	0.0	0.0	0.0	0.0	0.0	0.0	0.0
1988-89	0.0	0.0	0.0	0.0	0.0	0.0	0.0	0.0	0.0	0.0	0.0	0.0	0.0
1989-90	0.0	0.0	0.0	0.0	0.0	0.0	0.0	0.0	0.0	0.0	0.0	0.0	0.0
1990-91	0.0	0.0	0.0	0.0	0.0	0.0	0.0	0.0	0.0	0.0	0.0	0.0	0.0
1991-92	0.0	0.0	0.0	0.0	0.0	0.0	0.0	0.0	0.0	0.0	0.0	0.0	0.0
1992-93	0.0	0.0	0.0	0.0	0.0	0.0	0.0	0.0	0.0	0.0	0.0	0.0	0.0
1993-94	0.0	0.0	0.0	0.0	0.0	0.0							
Record Mean	0.0	0.0	0.0	0.0	0.0	0.0	0.0	0.0	0.0	0.0	0.0	0.0	0.0

KAHULUI, HAWAII

Kahului Airport is located in the relatively broad central valley of Maui near the northern coast of the island. Five miles to the west, the mountains of west Maui rise abruptly, reaching an elevation of 5,788 feet above sea level at the crest of Puu Kukui 10 miles west of the station. To the southeast the terrain rises gradually to the summit of Haleakala at 10,023 feet, located 17 miles from the airport.

The outstanding features of the climate are the equable temperature regime, the marked seasonal variation in rainfall, the persistent surface winds from the northeast quadrant, and the rarity of severe storms.

The extremely equable temperatures at Kahului are associated with the tempering effect of the Pacific Ocean and the small seasonal variation in the amount of energy received from the sun. The range in normal temperature between the warmest month, August, and the coldest month, February, is 7.2 degrees.

Rainfall is relatively light. The contrast between the dry season, which extends from May through October, and the wet season, November through April, is quite pronounced. Major widespread rainstorms, which account for the bulk of the precipitation in the area, usually occur several times during each wet season, but are infrequent in the dry season. Approximately 50 percent of the normal annual rainfall occurs in the three months of December through February, and over 80 percent in the six months of the wet season. June is the driest month, receiving about 1 percent of the annual total. Occasionally, an entire dry season month will go by with no measurable precipitation whatever. At the other extreme, a single wet season storm sometimes contributes more than one-half the total rainfall in an individual year.

Showers constitute the greatest number of rainfall occurrences and although most of these are light and short-lived, very heavy showers do occur at times. Thunderstorms, which are reported rather infrequently, are usually associated with major storms in the wet season.

Violent, damaging, windstorms are rare, but sometimes occur in connection with major storms moving through the region.

Hurricanes, with winds of 75 mph, or more, rarely affect the Kahului area. However, tropical storms, which are similar to hurricanes, except that the wind speed is less than 75 mph, may pass close enough to produce heavy rain and strong wind at Kahului once every several years.

The large Pacific semipermanent high pressure cell, which is usually centered north of the Hawaiian Islands, is one of the important climatic controls affecting the circulation of air in the region. Over the central North Pacific, this cell produces a rather persistent flow of air from the northeast known as the Northeast Trades. Thus, surface wind at Kahului is predominantly from the northeast quadrant. The trade-wind flow is most prevalent during the dry season. Wind is more variable during the wet season although, on the average, the trades still blow more than 50 percent of the time during this period.

The normal trade winds, accentuated by the funneling effect between Haleakala and the west Maui mountains, as well as by the daytime thermally induced low pressure in the valley, often attain a speed of 40 to 45 mph at the airport, but serve to make living conditions in the nearby Kahului-Wailuku community pleasant and comfortable. Air conditioning is used in only a few business establishments and residences.

Humidity at Kahului is usually moderate to high, with wet season humidities averaging slightly higher than those in the dry season. However, due to the system of natural ventilation provided by the prevailing winds, the weather is seldom oppressive even during the warmer months of the year.

NORMALS, MEANS, AND EXTREMES

KAHULUI, HAWAII

LATITUDE: 20°54'N LONGITUDE: 156°26' W ELEVATION: FT. GRND 48 BARO 46 TIME ZONE: BERING WBAN: 22516

	(a)	JAN	FEB	MAR	APR	MAY	JUNE	JULY	AUG	SEP	OCT	NOV	DEC	YEAR
TEMPERATURE °F:														
Normals														
-Daily Maximum		79.9	80.1	81.2	82.2	84.2	85.9	86.8	87.6	87.7	86.4	83.7	81.1	83.9
-Daily Minimum		63.6	63.4	64.6	66.1	67.0	69.0	70.5	70.8	69.8	69.2	67.6	65.1	67.2
-Monthly		71.7	71.8	73.0	74.2	75.7	77.5	78.7	79.3	78.8	77.9	75.7	73.1	75.6
Extremes														
-Record Highest	29	89	88	90	91	92	93	94	96	95	96	93	90	96
-Year		1981	1981	1984	1981	1992	1981	1984	1983	1992	1973	1990	1992	AUG 1983
-Record Lowest	29	48	50	52	54	57	58	58	61	60	58	55	52	48
-Year		1969	1987	1993	1985	1985	1985	1965	1976	1975	1964	1985	1983	JAN 1969
NORMAL DEGREE DAYS:														
Heating (base 65°F)		0	0	0	0	0	0	0	0	0	0	0	0	0
Cooling (base 65°F)		208	190	248	276	332	375	425	443	414	400	321	251	3883
% OF POSSIBLE SUNSHINE	31	64	65	64	62	67	72	70	71	72	67	63	62	67
MEAN SKY COVER (tenths)														
Sunrise - Sunset	35	4.8	4.9	5.3	5.9	5.4	4.9	4.7	4.7	4.7	5.2	5.2	5.0	5.1
MEAN NUMBER OF DAYS:														
Sunrise to Sunset														
-Clear	35	12.8	11.5	10.7	7.7	9.4	10.6	10.9	12.1	11.7	10.5	10.9	11.9	130.9
-Partly Cloudy	35	9.9	9.5	11.2	11.7	13.5	13.4	14.9	13.3	12.5	12.5	10.7	11.1	144.3
-Cloudy	35	8.3	7.2	9.1	10.6	8.0	5.9	5.3	5.5	5.7	8.0	8.4	8.0	90.1
Precipitation														
.01 inches or more	35	10.6	9.8	10.7	10.3	6.3	5.1	6.6	6.2	5.5	7.5	10.3	11.1	100.3
Snow, Ice pellets, hail														
1.0 inches or more	35	0.0	0.0	0.0	0.0	0.0	0.0	0.0	0.0	0.0	0.0	0.0	0.0	0.0
Thunderstorms	35	0.8	0.5	0.5	0.5	0.2	0.0	0.1	0.1	0.2	0.3	0.4	0.5	4.2
Heavy Fog Visibility														
1/4 mile or less	35	0.0	0.0	0.0	0.0	0.0	0.0	0.0	0.0	0.0	0.0	0.0	0.0	0.0
Temperature °F														
-Maximum														
90° and above	29	0.0	0.0	0.1	0.1	1.1	2.0	3.4	5.8	7.2	4.5	1.2	0.1	25.6
32° and below	29	0.0	0.0	0.0	0.0	0.0	0.0	0.0	0.0	0.0	0.0	0.0	0.0	0.0
-Minimum														
32° and below	29	0.0	0.0	0.0	0.0	0.0	0.0	0.0	0.0	0.0	0.0	0.0	0.0	0.0
0° and below	29	0.0	0.0	0.0	0.0	0.0	0.0	0.0	0.0	0.0	0.0	0.0	0.0	0.0
AVG. STATION PRESS. (mb)	11	1012.4	1013.6	1014.9	1014.9	1014.7	1014.2	1013.6	1013.2	1013.4	1012.8	1012.9	1013.2	1013.6
RELATIVE HUMIDITY (%)														
Hour 02	10	85	83	81	81	82	80	80	79	80	80	81	82	81
Hour 18 (Local Time)	29	82	81	77	74	71	69	71	71	71	73	76	80	75
Hour 14	29	62	60	59	58	56	54	55	56	55	57	60	61	58
Hour 20	29	77	75	74	73	72	71	71	72	71	73	75	76	73
PRECIPITATION (inches):														
Water Equivalent														
-Normal		4.14	2.87	2.72	1.84	0.77	0.27	0.38	0.49	0.35	1.23	2.59	3.27	20.92
-Maximum Monthly	39	14.46	8.31	10.90	14.29	4.36	2.50	1.65	1.54	1.43	5.66	9.27	10.19	14.46
-Year		1980	1972	1967	1989	1987	1967	1989	1982	1987	1985	1965	1988	JAN 1980
-Minimum Monthly	39	0.12	0.07	0.09	0.06	T	0.00	0.02	0.02	0.02	T	0.14	0.01	0.00
-Year		1977	1983	1957	1990	1972	1957	1973	1973	1972	1984	1980	1975	JUN 1957
-Maximum in 24 hrs	39	7.01	4.98	5.42	4.83	2.41	2.36	1.04	1.21	1.16	4.85	5.48	5.82	7.01
-Year		1980	1972	1967	1989	1987	1967	1989	1982	1965	1985	1965	1955	JAN 1980
Snow, Ice pellets, hail														
-Maximum Monthly		0.0	0.0	0.0	0.0	0.0	0.0	0.0	0.0	0.0	0.0	0.0	0.0	
-Year														
-Maximum in 24 hrs	39	0.0	0.0	0.0	0.0	0.0	0.0	0.0	0.0	0.0	0.0	0.0	0.0	
-Year														
WIND:														
Mean Speed (mph)	23	10.8	11.1	12.3	13.3	13.2	14.7	15.6	14.8	12.9	12.0	11.8	11.3	12.8
Prevailing Direction through 1963		SSW	S	NE	NE	NE	ENE	NE	NE	NE	NE	NE	NE	NE
Fastest Mile														
-Direction (!!!)	23	SW	NE	N	E	E	E	NE	NE	E	E	SW	E	SW
-Speed (MPH)	23	44	40	43	36	34	33	37	35	33	36	41	36	44
-Year		1980	1971	1968	1976	1986	1986	1978	1975	1977	1975	1982	1971	JAN 1980
Peak Gust														
-Direction (!!!)	10	S	NE	E	NE	E	NE	E	NE	SW	NE	S	E	S
-Speed (mph)	10	54	46	49	45	44	47	46	45	44	46	51	54	54
-Date		1991	1990	1985	1987	1993	1993	1989	1991	1992	1985	1988	1988	JAN 1991

PRECIPITATION (inches)

KAHULUI, HAWAII

YEAR	JAN	FEB	MAR	APR	MAY	JUNE	JULY	AUG	SEP	OCT	NOV	DEC	ANNUAL
1964	1.11	0.35	0.60	0.34	0.18	0.07	0.27	0.20	0.20	0.16	2.88	3.66	10.02
1965	2.45	7.56	3.75	1.08	0.43	0.07	0.74	0.41	1.17	1.93	9.27	2.05	30.91
1966	0.91	4.31	0.75	0.54	0.17	0.08	0.37	0.31	0.62	1.08	2.19	1.16	12.49
1967	3.85	2.13	10.90	1.19	2.05	2.50	1.12	1.33	0.08	0.77	1.69	5.09	32.70
1968	4.31	5.19	6.93	2.00	2.56	0.12	0.42	0.26	0.21	0.80	4.86	7.07	34.73
1969	7.75	3.51	3.17	1.79	0.27	0.37	0.53	0.71	0.49	0.31	0.50	5.81	25.21
1970	3.49	1.64	0.29	1.37	0.02	0.02	0.15	0.23	0.44	0.48	8.71	1.77	18.61
1971	13.66	0.78	2.92	1.13	0.10	0.21	0.09	0.54	0.25	0.07	0.24	0.14	20.13
1972	0.35	8.31	2.00	0.30	T	0.29	0.04	0.84	0.02	1.00	0.39	2.17	15.71
1973	2.14	0.81	1.35	0.44	0.57	0.08	0.02	0.02	0.04	0.34	0.98	3.48	10.27
1974	9.00	0.12	3.33	0.75	0.54	T	0.51	0.12	0.12	1.57	2.25	0.37	18.68
1975	3.30	4.96	1.89	0.10	0.03	0.05	0.23	0.61	0.11	0.37	2.08	0.01	13.74
1976	2.78	2.16	3.71	0.98	0.01	0.02	0.19	0.10	0.15	0.71	1.92	0.10	12.83
1977	0.12	0.92	1.62	3.83	0.20	0.13	0.43	0.71	0.05	0.45	0.23	2.81	11.50
1978	0.18	0.86	2.19	0.26	0.82	0.61	0.40	0.98	0.25	1.99	5.15	5.46	19.15
1979	7.18	7.07	2.80	2.93	0.09	0.07	0.23	0.23	0.21	0.44	0.80	4.77	26.82
1980	14.46	4.07	2.93	2.48	0.53	0.22	0.72	0.30	0.11	0.54	0.14	1.37	27.87
1981	0.46	1.94	0.73	0.89	0.84	0.06	0.06	0.67	0.85	1.48	2.26	2.61	12.85
1982	8.12	3.77	5.20	3.26	0.14	0.22	0.64	1.54	0.56	2.63	1.91	6.05	34.04
1983	0.58	0.07	1.12	0.24	0.94	0.17	0.53	0.67	0.50	1.38	0.98	5.87	13.05
1984	2.45	0.67	1.42	1.07	0.47	0.02	0.09	0.46	0.11	T	1.16	0.64	8.56
1985	1.16	2.03	1.96	0.25	1.20	0.01	0.53	0.52	0.10	5.66	4.61	1.97	20.00
1986	1.30	1.36	3.93	3.95	1.02	0.77	0.25	0.45	0.05	0.96	1.49	2.86	18.39
1987	2.91	1.41	0.57	3.77	4.36	0.12	0.13	0.62	1.43	0.25	3.02	5.72	24.31
1988	7.72	0.93	0.89	1.37	0.17	0.02	0.21	0.46	0.23	0.84	3.76	10.19	26.79
1989	1.59	5.38	3.96	14.29	0.85	0.42	1.65	0.50	0.31	4.71	2.25	4.72	40.63
1990	6.32	7.94	2.98	0.06	1.50	0.90	0.39	0.50	0.50	0.60	6.44	7.07	35.20
1991	2.94	4.94	2.89	0.50	0.24	0.13	0.37	1.06	1.17	0.15	0.23	1.47	16.09
1992	2.40	1.02	0.34	0.90	1.58	0.30	1.09	0.16	1.32	1.47	3.55	2.85	16.98
1993	2.19	0.21	1.55	1.28	0.52	0.09	1.30	0.80	0.78	1.19	1.99	0.79	12.69
Record Mean	3.53	2.75	2.63	1.77	0.69	0.20	0.38	0.43	0.40	1.00	2.08	3.25	19.11

AVERAGE TEMPERATURE (deg. F)

KAHULUI, HAWAII

YEAR	JAN	FEB	MAR	APR	MAY	JUNE	JULY	AUG	SEP	OCT	NOV	DEC	ANNUAL
#1964	74.2	73.3	73.2	74.6	73.6	76.6	77.2	76.7	77.0	74.5	72.2	72.2	74.6
1965	71.2	67.9	69.1	72.6	75.1	76.6	77.0	78.0	77.9	76.4	74.9	70.7	73.9
1966	69.8	69.7	72.0	72.7	76.5	79.0	78.9	79.5	79.2	78.0	75.8	74.1	75.4
1967	71.1	72.1	72.2	72.6	76.7	77.2	78.8	77.8	77.7	77.2	74.3	71.4	74.9
1968	70.8	70.5	72.0	74.1	75.0	77.9	79.0	80.8	79.9	79.0	77.4	72.5	75.8
1969	68.8	72.5	72.2	73.5	75.5	77.9	80.1	79.5	77.9	76.1	74.8	72.7	75.1
1970	71.5	69.7	72.8	75.2	76.9	78.3	79.4	80.3	78.0	78.3	75.5	73.9	75.8
1971	69.8	72.3	73.5	74.3	75.1	77.3	78.0	77.7	78.5	76.7	75.5	72.7	75.1
1972	69.5	70.4	72.0	73.6	74.7	76.0	77.9	79.1	78.6	78.1	75.2	71.2	74.7
1973	69.9	69.7	74.5	73.2	73.3	75.9	77.7	78.5	77.7	77.5	75.5	72.8	74.7
1974	73.8	73.0	72.9	76.2	75.8	76.7	77.0	78.9	77.7	78.0	75.2	74.6	75.8
1975	71.3	72.2	72.0	74.6	73.0	74.7	76.2	77.2	76.2	75.9	73.8	72.3	74.1
1976	71.1	71.5	72.8	74.2	75.0	75.8	78.4	79.5	78.9	78.0	74.8	74.7	75.4
1977	71.7	73.5	74.6	74.4	75.8	76.5	78.0	80.2	79.7	78.9	77.0	73.9	76.2
1978	74.0	72.9	74.8	76.5	78.3	79.1	80.4	81.2	80.9	79.5	76.1	74.1	77.3
1979	71.1	71.8	71.4	73.3	75.9	78.5	79.1	80.5	80.9	80.5	76.5	75.4	76.2
1980	73.3	73.4	75.6	75.9	78.9	80.0	81.2	80.5	81.0	79.8	78.2	76.0	77.8
1981	74.7	74.6	75.0	76.1	78.0	80.7	80.5	81.3	80.4	78.3	76.9	74.5	77.6
1982	73.1	72.4	72.8	73.2	76.0	78.5	80.9	81.5	80.0	79.0	77.0	73.2	76.5
1983	71.4	71.5	72.5	74.1	74.9	77.6	78.1	79.1	78.0	77.3	75.8	73.5	75.3
1984	73.4	73.4	75.9	77.0	78.5	80.5	80.0	79.5	79.4	80.5	79.0	74.2	77.6
1985	72.0	73.0	70.7	70.8	73.1	75.1	77.6	77.8	77.1	76.7	73.0	71.2	74.0
1986	70.4	71.4	73.7	74.3	75.9	77.1	79.4	80.6	79.4	77.5	77.1	72.9	75.8
1987	72.0	69.3	72.0	72.8	72.4	76.5	78.6	78.9	79.5	78.3	75.6	73.8	75.0
1988	71.5	72.3	73.4	74.2	76.5	77.8	78.9	79.2	78.9	77.5	76.8	73.1	75.8
1989	72.3	71.6	73.2	72.0	75.5	77.3	79.0	78.0	78.6	77.6	74.8	71.7	75.1
1990	72.7	70.8	72.0	74.6	75.5	77.9	78.4	79.6	80.3	78.4	76.5	73.0	75.8
1991	71.7	72.6	72.1	74.4	75.8	76.6	78.3	79.6	79.0	78.3	77.3	74.8	75.9
1992	71.4	71.8	72.9	74.1	76.3	78.6	79.2	80.5	80.4	78.5	76.9	75.4	76.3
1993	71.2	69.6	71.9	75.3	76.0	78.7	79.5	80.3	80.0	78.2	75.9	73.8	75.9
Record Mean	71.7	71.6	72.6	74.0	75.4	77.3	78.4	79.1	78.7	77.7	75.7	73.2	75.5
Max	79.9	79.9	80.8	81.9	83.6	85.4	86.2	87.0	87.2	86.1	83.6	81.0	83.6
Min	63.4	63.3	64.4	66.1	67.2	69.1	70.6	71.2	70.1	69.3	67.7	65.3	67.3

REFERENCE NOTES

KAHULUI, HAWAII

GENERAL
T - TRACE AMOUNT.
BLANK ENTRIES DENOTE MISSING/UNREPORTED DATA.
# INDICATES A STATION OR INSTRUMENT RELOCATION.

SPECIFIC

(a) - LENGTH OF RECORD IN YEARS, ALTHOUGH INDIVIDUAL MONTHS MAY BE MISSING.
0.* OR * - THE VALUE IS BETWEEN 0.0 AND 0.05
NORMALS - BASED ON THE 1961-1990 RECORD PERIOD.
EXTREMES - DATES ARE THE MOST RECENT OCCURRENCE
WIND DIR.- NUMERALS SHOW TENS OF DEGRESS CLOCKWISE FROM TRUE NORTH. "00" INDICATES CALM.
RESULTANT DIRECTIONS ARE GIVEN TO WHOLE DEGREES.
BOLD VALUES INDICATE EXTREME VALUES WHICH OCCURRED AFTER THE ASOS SYSTEM WAS COMMISSIONED.

RECORD = PERIOD OF RECORD
RECORD MEAN PRECIPITATION IS THE MEAN OF ALL DAILY PRECIPITATION AMOUNTS DURING THE PERIOD OF RECORD.
RECORD MAX(MIN) TEMPERATURE IS THE MEAN OF ALL DAILY MAX(MIN) TEMPERATURES DURING THE PERIOD OF RECORD.
RECORD MEAN TEMPERATURE IS THE SUM OF THE RECORD MAX AND RECORD MIN DIVIDED BY 2.
AVERAGE TEMPERATURE IS THE SUM OF THE MEAN DAILY MAX AND MIN TEMPERATURE DIVIDED BY 2.

EXCEPTIONS

THUNDERSTORMS AND HEAVY FOG, 1968 TO DATE, MAY BE INCOMPLETE, DUE TO PART-TIME OPERATIONS.

RECORD MEANS ARE THROUGH THE CURRENT YEAR, BEGINNING IN 1954 FOR TEMPERATURE
1901 FOR PRECIPITATION

HEATING DEGREE DAYS Base 65 deg. F　　KAHULUI, HAWAII

SEASON	JULY	AUG	SEP	OCT	NOV	DEC	JAN	FEB	MAR	APR	MAY	JUNE	TOTAL
1983-84	0	0	0	0	0	0	0	0	0	0	0	0	0
1984-85	0	0	0	0	0	0	0	0	0	0	2	0	2
1985-86	0	0	0	0	0	0	0	0	0	0	0	0	0
1986-87	0	0	0	0	0	0	0	1	0	0	0	0	1
1987-88	0	0	0	0	0	0	0	0	0	0	0	0	0
1988-89	0	0	0	0	0	0	0	0	0	0	0	0	0
1989-90	0	0	0	0	0	0	0	1	0	0	0	0	1
1990-91	0	0	0	0	0	0	0	0	0	0	0	0	0
1991-92	0	0	0	0	0	0	0	0	0	0	0	0	0
1992-93	0	0	0	0	0	0	0	0	0	0	0	0	0
1993-94	0	0	0	0	0	0							

COOLING DEGREE DAYS Base 65 deg. F　　KAHULUI, HAWAII

YEAR	JAN	FEB	MAR	APR	MAY	JUNE	JULY	AUG	SEP	OCT	NOV	DEC	TOTAL
1969	137	214	227	259	332	393	474	457	398	352	302	243	3788
1970	212	136	247	310	374	407	454	479	397	418	322	285	4041
1971	157	212	271	285	323	375	409	401	413	370	322	247	3785
1972	149	162	222	264	308	338	408	446	413	410	312	199	3631
1973	160	139	302	253	266	335	402	425	390	396	321	248	3637
1974	277	229	250	342	339	359	378	439	390	408	311	307	4029
1975	205	211	227	297	254	297	355	385	344	350	271	230	3426
1976	196	196	248	283	315	329	423	456	424	411	300	309	3890
1977	216	244	303	290	343	352	409	476	446	440	368	283	4170
1978	286	229	312	349	420	430	487	508	483	456	339	289	4588
1979	197	197	206	255	346	410	446	486	485	487	354	331	4200
1980	266	253	335	334	442	457	512	488	484	462	401	347	4781
1981	307	276	314	337	407	476	489	510	469	417	363	300	4665
1982	259	215	248	254	347	412	498	519	458	438	366	259	4273
1983	205	188	240	278	315	384	414	445	398	389	333	272	3861
1984	264	250	346	367	425	472	470	456	437	486	425	295	4693
1985	223	229	182	178	260	312	397	403	371	370	250	200	3375
1986	174	187	276	287	345	373	454	490	442	392	370	249	4039
1987	226	127	224	242	237	353	430	440	440	418	326	282	3745
1988	210	218	267	285	364	390	438	444	426	395	358	261	4056
1989	234	193	260	218	331	375	440	412	414	400	301	215	3793
1990	246	169	223	294	332	394	422	458	467	425	352	252	4034
1991	211	219	228	290	341	353	421	458	427	417	375	312	4052
1992	207	205	254	282	358	413	448	489	468	427	362	331	4244
1993	199	135	222	316	347	419	454	482	456	417	334	278	4059

SNOWFALL (inches)　　KAHULUI, HAWAII

SEASON	JULY	AUG	SEP	OCT	NOV	DEC	JAN	FEB	MAR	APR	MAY	JUNE	TOTAL
1971-72	0.0	0.0	0.0	0.0	0.0	0.0	0.0	0.0	0.0	0.0	0.0	0.0	0.0
1972-73	0.0	0.0	0.0	0.0	0.0	0.0	0.0	0.0	0.0	0.0	0.0	0.0	0.0
1973-74	0.0	0.0	0.0	0.0	0.0	0.0	0.0	0.0	0.0	0.0	0.0	0.0	0.0
1974-75	0.0	0.0	0.0	0.0	0.0	0.0	0.0	0.0	0.0	0.0	0.0	0.0	0.0
1975-76	0.0	0.0	0.0	0.0	0.0	0.0	0.0	0.0	0.0	0.0	0.0	0.0	0.0
1976-77	0.0	0.0	0.0	0.0	0.0	0.0	0.0	0.0	0.0	0.0	0.0	0.0	0.0
1977-78	0.0	0.0	0.0	0.0	0.0	0.0	0.0	0.0	0.0	0.0	0.0	0.0	0.0
1978-79	0.0	0.0	0.0	0.0	0.0	0.0	0.0	0.0	0.0	0.0	0.0	0.0	0.0
1979-80	0.0	0.0	0.0	0.0	0.0	0.0	0.0	0.0	0.0	0.0	0.0	0.0	0.0
1980-81	0.0	0.0	0.0	0.0	0.0	0.0	0.0	0.0	0.0	0.0	0.0	0.0	0.0
1981-82	0.0	0.0	0.0	0.0	0.0	0.0	0.0	0.0	0.0	0.0	0.0	0.0	0.0
1982-83	0.0	0.0	0.0	0.0	0.0	0.0	0.0	0.0	0.0	0.0	0.0	0.0	0.0
1983-84	0.0	0.0	0.0	0.0	0.0	0.0	0.0	0.0	0.0	0.0	0.0	0.0	0.0
1984-85	0.0	0.0	0.0	0.0	0.0	0.0	0.0	0.0	0.0	0.0	0.0	0.0	0.0
1985-86	0.0	0.0	0.0	0.0	0.0	0.0	0.0	0.0	0.0	0.0	0.0	0.0	0.0
1986-87	0.0	0.0	0.0	0.0	0.0	0.0	0.0	0.0	0.0	0.0	0.0	0.0	0.0
1987-88	0.0	0.0	0.0	0.0	0.0	0.0	0.0	0.0	0.0	0.0	0.0	0.0	0.0
1988-89	0.0	0.0	0.0	0.0	0.0	0.0	0.0	0.0	0.0	0.0	0.0	0.0	0.0
1989-90	0.0	0.0	0.0	0.0	0.0	0.0	0.0	0.0	0.0	0.0	0.0	0.0	0.0
1990-91	0.0	0.0	0.0	0.0	0.0	0.0	0.0	0.0	0.0	0.0	0.0	0.0	0.0
1991-92	0.0	0.0	0.0	0.0	0.0	0.0	0.0	0.0	0.0	0.0	0.0	0.0	0.0
1992-93	0.0	0.0	0.0	0.0	0.0	0.0	0.0	0.0	0.0	0.0	0.0	0.0	0.0
1993-94	0.0	0.0	0.0	0.0	0.0	0.0							
Record Mean	0.0	0.0	0.0	0.0	0.0	0.0	0.0	0.0	0.0	0.0	0.0	0.0	0.0

LIHUE, HAWAII

Lihue Airport, a little more than 100 feet above sea level, is located near the eastern shore of the island of Kauai. The island is 33 miles long and 25 miles wide and has an area of 555 square miles. The eastern one third of Kauai consists of broadly eroded valley lands, the western two thirds is mostly mountainous. Kawaikini, the highest elevation on the island, 5,170 feet above sea level, lies near the center of Kauai and is 20 miles northwest of the airport.

The outstanding features of the climate are the equable temperatures from day to day and season to season, the persistent northeasterly trade winds and the marked variation in rainfall from the wet to the dry season and place to place.

The equable temperatures are associated with the mid-ocean location of the island and to the small seasonal variation in the amount of energy received from the sun. The range in normal temperature from the coolest month, February, to the warmest month, August, less than 8 degrees. The daily range in temperature is also small, less than 15 degrees.

The trade winds blow across the island during most of each year and the dominance of these winds has a marked influence on the climate of the area. Completely cloudless skies are quite rare. On the average, six tenths to seven tenths of the sky is covered by clouds during the daylight hours.

Trade-wind showers are relatively common. Although heavy at times, most of the showers are light and of short duration. The frequency and intensity of the showers increase toward the mountains to the west. Mt. Waialeale receives 486 inches annually, the highest recorded annual average in the world. Mt. Waialeale has recorded annual rainfalls over 620 inches.

Normal annual rainfall is over 40 inches. Three-fourths of this total, on the average, falls during the seven-month wet season which extends from October through April. Widespread rainstorms, which account for much of the precipitation, occur most frequently during this period. Normal precipitation in January, the wettest month, is over 6 inches.

The dry season includes the months of May through September. June, the driest month, receives only about 1 1/2 inches of rain, on the average.

Hurricanes and other severe windstorms are quite rare. Strong winds do occur at times in connection with storm systems moving through the area, but seldom cause extensive damage.

Relative humidity, moderate to high in all seasons, is slightly higher in the wet season than in the dry. However, even during periods when the temperature and humidity are both high, the weather is seldom oppressive. This is due to the trade winds which provide a system of natural ventilation during most of each year.

NORMALS, MEANS, AND EXTREMES

LIHUE, HAWAII

LATITUDE: 21°59'N LONGITUDE: 159°21' W ELEVATION: FT. GRND 103 BARO 118 TIME ZONE: BERING WBAN: 22536

	(a)	JAN	FEB	MAR	APR	MAY	JUNE	JULY	AUG	SEP	OCT	NOV	DEC	YEAR
TEMPERATURE °F:														
Normals														
-Daily Maximum		77.9	78.1	78.5	79.5	81.2	83.1	84.1	84.8	84.8	83.2	80.7	78.6	81.2
-Daily Minimum		65.3	65.1	66.9	68.5	70.3	72.5	73.7	74.2	73.5	72.0	70.3	67.1	70.0
-Monthly		71.6	71.6	72.7	74.0	75.8	77.8	78.9	79.5	79.2	77.6	75.5	72.9	75.6
Extremes														
-Record Highest	43	85	86	88	88	88	89	89	90	90	90	87	86	90
-Year		1981	1986	1986	1981	1981	1969	1981	1987	1993	1957	1980	1981	SEP 1993
-Record Lowest	43	50	52	51	56	59	61	62	66	65	61	57	52	50
-Year		1969	1962	1955	1958	1989	1987	1979	1979	1981	1993	1958	1953	JAN 1969
NORMAL DEGREE DAYS:														
Heating (base 65°F)		0	0	0	0	0	0	0	0	0	0	0	0	0
Cooling (base 65°F)		209	188	239	270	335	384	431	450	426	391	315	245	3883
% OF POSSIBLE SUNSHINE	43	53	56	54	52	58	62	62	64	66	57	49	48	57
MEAN SKY COVER (tenths)														
Sunrise - Sunset	44	5.8	5.8	6.5	6.9	6.6	6.3	6.4	6.2	5.7	6.1	6.5	6.2	6.3
MEAN NUMBER OF DAYS:														
Sunrise to Sunset														
-Clear	43	8.0	7.1	4.7	2.9	2.9	3.3	2.7	3.6	5.5	5.3	4.0	5.8	55.9
-Partly Cloudy	43	12.7	11.9	13.9	13.7	16.3	17.3	18.8	18.3	17.4	15.2	13.9	13.7	183.0
-Cloudy	43	10.3	9.3	12.4	13.4	11.8	9.4	9.6	9.1	7.1	10.5	12.1	11.5	126.4
Precipitation														
.01 inches or more	43	14.8	13.3	16.5	17.3	16.0	16.3	19.3	18.1	15.8	17.9	17.9	17.3	200.4
Snow,Ice pellets,hail														
1.0 inches or more	44	0.0	0.0	0.0	0.0	0.0	0.0	0.0	0.0	0.0	0.0	0.0	0.0	0.0
Thunderstorms	43	1.2	0.7	1.1	0.4	0.6	0.1	0.2	0.2	0.4	1.0	1.0	1.0	7.8
Heavy Fog Visibility														
1/4 mile or less	44	0.0	0.0	0.0	0.0	0.0	0.0	0.0	0.0	0.0	0.0	0.0	0.0	0.0
Temperature °F														
-Maximum														
90° and above	44	0.0	0.0	0.0	0.0	0.0	0.0	0.0	0.1	0.*	0.*	0.0	0.0	0.1
32° and below	44	0.0	0.0	0.0	0.0	0.0	0.0	0.0	0.0	0.0	0.0	0.0	0.0	0.0
-Minimum														
32° and below	44	0.0	0.0	0.0	0.0	0.0	0.0	0.0	0.0	0.0	0.0	0.0	0.0	0.0
0° and below	44	0.0	0.0	0.0	0.0	0.0	0.0	0.0	0.0	0.0	0.0	0.0	0.0	0.0
AVG. STATION PRESS.(mb)	21	1010.5	1011.1	1013.3	1013.2	1012.9	1012.9	1012.3	1011.6	1010.5	1010.6	1010.8	1010.9	1011.7
RELATIVE HUMIDITY (%)														
Hour 02	43	82	81	81	81	81	80	79	80	80	82	81	81	81
Hour 18 (Local Time)	44	82	80	78	77	75	75	75	76	77	78	80	80	78
Hour 14	44	67	66	66	67	66	65	66	66	65	67	69	68	67
Hour 20	43	79	77	77	77	77	76	76	76	76	78	79	79	77
PRECIPITATION (inches):														
Water Equivalent														
-Normal		5.89	3.33	4.17	3.50	3.15	1.69	2.13	1.76	2.37	4.41	5.45	5.15	43.00
-Maximum Monthly	43	17.56	11.35	14.54	10.65	12.59	4.83	8.85	8.13	10.87	18.04	18.45	22.91	22.91
-Year		1956	1989	1951	1972	1977	1991	1954	1959	1980	1982	1955	1968	DEC 1968
-Minimum Monthly	43	0.30	T	0.30	0.32	0.42	0.41	0.75	0.70	0.45	1.02	0.58	0.51	T
-Year		1986	1983	1957	1992	1968	1975	1956	1984	1975	1963	1963	1985	FEB 1983
-Maximum in 24 hrs	43	11.09	7.28	6.42	6.52	6.70	2.17	5.04	5.43	7.16	8.46	11.20	11.54	11.54
-Year		1956	1954	1989	1972	1977	1971	1954	1959	1980	1978	1955	1968	DEC 1968
Snow,Ice pellets,hail														
-Maximum Monthly		0.0	0.0	0.0	0.0	0.0	0.0	0.0	0.0	0.0	0.0	0.0	0.0	
-Year														
-Maximum in 24 hrs	44	0.0	0.0	0.0	0.0	0.0	0.0	0.0	0.0	0.0	0.0	0.0	0.0	
-Year														
WIND:														
Mean Speed (mph)	43	11.0	11.5	12.6	13.2	12.6	13.1	13.6	13.1	11.6	11.4	12.2	11.8	12.3
Prevailing Direction through 1963		ENE	NE	NE	NE	NE	NE	NE	NE	NE	NE	NE	NE	NE
Fastest Obs. 1 Min.														
-Direction (!!)	14	19	23	06	05	06	07	08	05	15	04	18	15	15
-Speed (MPH)	14	37	41	39	36	33	29	29	31	84	33	65	38	84
-Year		1991	1986	1985	1986	1991	1986	1986	1991	1992	1983	1982	1987	SEP 1992
Peak Gust														
-Direction (!!)	10	SW	SW	NE	NE	NE	NE	NE	NE	SW	NE	NE	SE	SW
-Speed (mph)	10	66	59	54	47	40	39	41	38	115	40	51	53	115
-Date		1985	1986	1985	1986	1987	1986	1985	1991	1992	1987	1985	1987	SEP 1992

PRECIPITATION (inches) LIHUE, HAWAII

YEAR	JAN	FEB	MAR	APR	MAY	JUNE	JULY	AUG	SEP	OCT	NOV	DEC	ANNUAL
1964	7.37	1.73	10.16	2.88	2.29	1.02	3.00	2.07	2.17	3.22	9.76	6.08	51.75
1965	7.57	2.52	0.90	7.12	8.48	1.27	1.80	2.84	3.65	6.68	12.10	3.04	57.97
1966	1.62	6.44	0.49	1.63	0.69	1.08	1.95	2.04	1.51	11.41	8.13	3.69	40.68
1967	5.59	6.59	8.91	1.60	4.05	2.09	1.84	1.98	1.56	1.85	4.43	12.79	53.28
1968	7.43	3.69	7.25	1.84	0.42	1.23	1.70	0.70	3.17	9.93	8.62	22.91	68.89
1969	4.97	1.94	1.75	1.52	2.80	0.89	1.47	0.83	1.49	3.15	7.45	5.91	34.17
1970	7.32	0.89	0.93	2.83	5.85	0.77	1.76	1.33	1.42	2.74	11.05	2.29	39.18
1971	11.94	1.69	8.27	6.36	1.03	2.98	0.84	0.87	2.69	3.18	2.25	7.52	49.62
1972	6.44	7.77	1.99	10.65	0.75	1.74	2.25	1.27	1.95	6.07	1.39	1.27	43.54
1973	1.03	1.50	1.26	1.00	1.87	0.51	0.95	1.32	1.77	5.06	12.15	6.85	35.27
1974	5.59	0.90	6.29	5.57	3.67	1.87	4.69	2.23	3.73	2.94	6.51	1.61	45.60
1975	14.07	4.35	3.36	1.17	0.67	0.41	0.94	0.86	0.45	1.03	5.34	2.87	35.52
1976	3.55	4.86	7.39	3.41	0.86	1.09	2.09	2.04	1.73	2.40	1.22	2.19	32.83
1977	2.92	3.19	2.38	4.11	12.59	3.21	1.30	1.33	1.26	1.28	1.86	4.91	40.34
1978	1.37	0.54	1.49	2.11	4.56	4.56	2.74	1.19	1.85	11.02	3.79	3.89	39.11
1979	4.77	10.59	1.26	1.07	1.61	2.38	0.83	1.88	1.19	3.75	3.04	4.72	37.09
1980	6.12	1.65	2.81	3.25	12.20	3.07	2.76	1.27	10.87	4.49	1.89	4.26	54.64
1981	1.12	1.92	2.90	4.13	0.91	0.92	3.76	2.04	1.98	4.17	5.47	8.82	38.14
1982	14.08	5.04	10.56	4.88	3.63	2.02	2.72	3.37	1.71	18.04	3.67	4.68	74.40
1983	0.87	T	0.64	1.29	2.37	1.10	1.79	1.82	1.83	2.57	1.27	0.85	16.40
1984	2.19	1.21	0.96	3.19	0.48	0.87	1.25	0.70	3.23	1.94	9.76	4.34	30.12
1985	2.75	7.40	2.11	1.65	1.42	0.75	0.88	1.71	3.15	3.20	3.38	0.51	28.91
1986	0.30	1.02	2.02	1.95	2.59	1.85	3.20	2.97	2.85	1.98	5.98	1.28	27.99
1987	1.35	2.79	2.13	3.57	1.85	1.43	2.23	1.58	4.41	4.31	2.81	14.49	42.95
1988	9.81	1.46	2.63	1.60	3.79	0.53	1.95	3.28	1.09	1.88	8.18	6.86	43.06
1989	7.09	11.35	8.62	2.12	1.28	2.14	6.90	2.56	1.42	5.24	5.08	2.97	56.77
1990	9.75	1.75	2.94	2.94	2.30	0.48	1.51	1.27	2.00	1.12	9.97	3.34	39.37
1991	0.68	1.68	9.78	3.36	0.73	4.83	1.81	2.44	1.24	4.67	3.37	7.04	41.63
1992	1.13	3.23	4.26	0.32	2.58	1.05	1.68	2.44	3.13	5.86	2.29	22.20	50.17
1993	0.75	0.46	0.32	2.00	2.62	0.60	1.73	1.75	1.77	4.81	2.69	2.77	22.27
Record Mean	5.56	3.48	4.31	3.06	2.70	1.57	2.15	1.94	2.26	4.55	5.36	5.67	42.62

AVERAGE TEMPERATURE (deg. F) LIHUE, HAWAII

YEAR	JAN	FEB	MAR	APR	MAY	JUNE	JULY	AUG	SEP	OCT	NOV	DEC	ANNUAL
1964	73.7	72.9	73.2	74.6	75.7	77.8	78.2	77.1	77.8	75.1	73.4	73.0	75.2
1965	70.5	67.5	68.6	71.7	74.5	76.5	77.6	78.4	77.3	74.9	74.5	71.1	73.6
1966	69.8	67.6	70.3	69.8	74.2	77.7	78.1	78.6	78.6	77.0	74.8	72.3	74.0
1967	69.5	71.8	71.4	71.9	75.8	77.1	79.0	78.8	79.0	77.5	74.9	69.9	74.7
1968	69.5	68.9	70.7	72.4	74.2	77.8	78.2	79.0	78.7	76.2	75.7	70.8	74.3
1969	65.8	70.5	69.9	72.0	75.4	78.3	79.5	80.0	79.5	77.1	75.6	73.3	74.7
1970	71.6	71.3	74.3	75.6	77.2	78.5	79.5	80.8	79.9	78.4	75.4	75.2	76.5
1971	69.7	73.8	73.9	75.9	76.9	77.9	79.4	79.9	80.0	78.3	77.0	74.5	76.4
1972	70.4	70.9	72.4	74.0	76.7	77.6	78.4	79.9	78.8	78.2	76.1	70.5	75.3
1973	70.9	71.9	75.4	74.6	75.3	77.4	78.8	79.2	78.9	77.9	75.5	72.9	75.7
1974	74.4	74.5	73.5	75.9	76.9	78.6	79.0	80.0	79.9	79.3	75.9	75.9	77.0
1975	73.0	71.4	72.6	74.2	75.6	77.3	77.7	78.1	77.8	77.7	76.9	73.5	75.5
1976	73.9	72.8	72.5	75.1	77.8	78.7	79.9	79.3	78.5	76.9	73.8	74.1	76.1
1977	73.0	72.7	74.1	73.9	75.3	76.8	78.6	79.9	80.0	79.3	77.2	73.5	76.2
1978	72.0	72.2	73.6	74.2	75.2	76.9	78.6	79.4	79.5	77.8	73.8	73.1	75.5
1979	73.2	72.7	72.0	72.5	75.1	77.2	78.4	79.1	80.5	78.5	74.8	72.9	75.6
1980	72.3	71.1	73.1	73.5	75.5	76.6	78.2	79.0	79.1	76.7	75.7	73.9	75.4
1981	73.0	72.6	74.8	76.4	77.9	79.8	79.9	81.2	79.5	77.5	74.8	72.4	76.7
1982	71.3	72.0	72.8	74.6	76.3	78.3	80.7	80.2	79.8	77.4	76.4	72.4	76.0
1983	71.3	71.0	72.9	74.5	74.7	78.3	78.6	79.2	78.8	77.6	76.1	72.5	75.5
1984	73.8	73.7	75.2	76.4	78.2	79.3	80.1	80.5	80.0	79.0	77.5	73.7	77.3
1985	70.7	72.8	73.5	72.9	74.8	77.7	79.9	79.9	78.9	77.4	73.5	72.0	75.3
1986	71.1	71.3	73.3	75.4	76.3	77.8	79.5	80.2	80.1	78.7	76.2	73.0	76.1
1987	71.6	69.4	71.3	72.5	72.0	76.8	78.9	80.6	79.3	78.3	76.4	74.5	75.1
1988	71.7	73.8	73.6	75.4	76.4	78.9	78.7	79.2	79.9	77.1	77.2	74.2	76.3
1989	73.1	71.7	73.2	72.2	75.1	77.8	78.2	78.4	79.0	76.7	74.5	71.0	75.1
1990	72.5	69.8	71.4	73.5	74.4	77.6	78.5	80.0	80.0	79.2	74.9	72.1	75.3
1991	71.1	72.3	71.3	73.7	75.6	76.2	78.3	79.0	78.5	78.0	76.7	73.9	75.4
1992	70.4	70.1	72.5	72.5	74.7	78.6	79.3	80.0	79.7	77.5	74.9	74.5	75.4
1993	69.4	70.5	72.5	75.6	75.5	78.5	78.9	79.9	79.8	77.7	75.2	73.9	75.6
Record Mean	70.3	70.3	71.1	72.5	74.3	76.5	77.5	78.1	77.9	76.4	74.1	71.8	74.3
Max	77.7	77.8	78.0	78.9	80.7	82.5	83.4	84.1	84.3	83.1	80.6	78.5	80.8
Min	62.9	62.8	64.2	66.1	68.0	70.4	71.6	72.1	71.4	69.7	67.6	65.1	67.7

REFERENCE NOTES LIHUE, HAWAII

GENERAL
T - TRACE AMOUNT.
BLANK ENTRIES DENOTE MISSING/UNREPORTED DATA.
# INDICATES A STATION OR INSTRUMENT RFLOCATION.

SPECIFIC

(a) - LENGTH OF RECORD IN YEARS, ALTHOUGH INDIVIDUAL MONTHS MAY BE MISSING.
0.* OR * - THE VALUE IS BETWEEN 0.0 AND 0.05
NORMALS - BASED ON THE 1961-1990 RECORD PERIOD.
EXTREMES - DATES ARE THE MOST RECENT OCCURRENCE.
WIND DIR. - NUMERALS SHOW TENS OF DEGRESS CLOCKWISE FROM TRUE NORTH. "00" INDICATES CALM.
RESULTANT DIRECTIONS ARE GIVEN TO WHOLE DEGREES.
BOLD VALUES INDICATE EXTREME VALUES WHICH OCCURRED AFTER THE ASOS SYSTEM WAS COMMISSIONED.

RECORD = PERIOD OF RECORD
RECORD MEAN PRECIPITATION IS THE MEAN OF ALL DAILY PRECIPITATION AMOUNTS DURING THE PERIOD OF RECORD.
RECORD MAX(MIN) TEMPERATURE IS THE MEAN OF ALL DAILY MAX(MIN) TEMPERATURES DURING THE PERIOD OF RECORD.
RECORD MEAN TEMPERATURE IS THE SUM OF THE RECORD MAX AND RECORD MIN DIVIDED BY 2.
AVERAGE TEMPERATURE IS THE SUM OF THE MEAN DAILY MAX AND MIN TEMPERATURE DIVIDED BY 2.

EXCEPTIONS

RECORD MEANS ARE THROUGH THE CURRENT YEAR, BEGINNING IN 1905 FOR TEMPERATURE
1950 FOR PRECIPITATION

HEATING DEGREE DAYS Base 65 deg. F LIHUE, HAWAII

SEASON	JULY	AUG	SEP	OCT	NOV	DEC	JAN	FEB	MAR	APR	MAY	JUNE	TOTAL
1983-84	0	0	0	0	0	0	0	0	0	0	0	0	0
1984-85	0	0	0	0	0	0	3	1	0	0	0	0	4
1985-86	0	0	0	0	0	0	0	0	0	0	0	0	0
1986-87	0	0	0	0	0	0	0	0	0	0	0	0	0
1987-88	0	0	0	0	0	0	0	0	0	0	0	0	0
1988-89	0	0	0	0	0	0	0	0	0	0	0	0	0
1989-90	0	0	0	0	0	0	0	0	0	0	0	0	0
1990-91	0	0	0	0	0	0	0	0	0	0	0	0	0
1991-92	0	0	0	0	0	0	0	0	0	0	0	0	0
1992-93	0	0	0	0	0	0	0	0	0	0	0	0	0
1993-94	0	0	0	0	0	0							

COOLING DEGREE DAYS Base 65 deg. F LIHUE, HAWAII

YEAR	JAN	FEB	MAR	APR	MAY	JUNE	JULY	AUG	SEP	OCT	NOV	DEC	TOTAL
1969	65	160	159	214	332	406	455	472	441	381	325	265	3675
1970	211	184	293	325	387	412	455	498	455	425	320	324	4289
1971	156	253	283	336	376	393	454	471	456	423	365	301	4267
1972	174	178	237	278	368	383	423	472	420	417	341	180	3871
1973	191	199	328	292	326	377	435	445	424	406	321	252	3996
1974	300	272	272	335	378	418	441	473	452	449	336	344	4470
1975	255	187	240	281	336	376	403	413	394	399	365	268	3917
1976	282	234	244	312	404	420	470	450	412	372	271	288	4159
1977	255	224	288	273	326	357	430	469	456	450	376	269	4173
1978	224	205	274	284	323	362	425	454	443	402	269	260	3925
1979	263	224	220	229	320	374	423	445	474	426	303	252	3953
1980	229	186	258	261	330	353	417	438	429	369	328	283	3881
1981	254	217	311	347	406	453	469	508	444	396	300	238	4343
1982	200	202	246	292	357	406	494	480	451	391	346	237	4102
1983	202	173	252	292	307	404	430	445	420	397	338	238	3898
1984	281	260	325	347	412	434	476	487	459	444	380	277	4582
1985	186	224	272	243	309	387	469	472	425	392	260	225	3864
1986	198	183	264	321	358	390	458	477	459	430	344	254	4136
1987	213	130	203	234	226	360	438	490	435	420	349	301	3799
1988	217	263	275	317	363	422	431	450	452	384	375	294	4243
1989	256	193	261	222	321	389	415	426	425	368	293	189	3758
1990	240	142	203	262	298	386	425	473	453	446	305	225	3858
1991	194	210	205	268	334	343	421	445	412	411	356	281	3880
1992	173	153	240	230	306	417	449	471	447	396	304	302	3888
1993	144	160	243	323	330	409	439	469	452	401	314	281	3965

SNOWFALL (inches) LIHUE, HAWAII

SEASON	JULY	AUG	SEP	OCT	NOV	DEC	JAN	FEB	MAR	APR	MAY	JUNE	TOTAL
1971-72	0.0	0.0	0.0	0.0	0.0	0.0	0.0	0.0	0.0	0.0	0.0	0.0	0.0
1972-73	0.0	0.0	0.0	0.0	0.0	0.0	0.0	0.0	0.0	0.0	0.0	0.0	0.0
1973-74	0.0	0.0	0.0	0.0	0.0	0.0	0.0	0.0	0.0	0.0	0.0	0.0	0.0
1974-75	0.0	0.0	0.0	0.0	0.0	0.0	0.0	0.0	0.0	0.0	0.0	0.0	0.0
1975-76	0.0	0.0	0.0	0.0	0.0	0.0	0.0	0.0	0.0	0.0	0.0	0.0	0.0
1976-77	0.0	0.0	0.0	0.0	0.0	0.0	0.0	0.0	0.0	0.0	0.0	0.0	0.0
1977-78	0.0	0.0	0.0	0.0	0.0	0.0	0.0	0.0	0.0	0.0	0.0	0.0	0.0
1978-79	0.0	0.0	0.0	0.0	0.0	0.0	0.0	0.0	0.0	0.0	0.0	0.0	0.0
1979-80	0.0	0.0	0.0	0.0	0.0	0.0	0.0	0.0	0.0	0.0	0.0	0.0	0.0
1980-81	0.0	0.0	0.0	0.0	0.0	0.0	0.0	0.0	0.0	0.0	0.0	0.0	0.0
1981-82	0.0	0.0	0.0	0.0	0.0	0.0	0.0	0.0	0.0	0.0	0.0	0.0	0.0
1982-83	0.0	0.0	0.0	0.0	0.0	0.0	0.0	0.0	0.0	0.0	0.0	0.0	0.0
1983-84	0.0	0.0	0.0	0.0	0.0	0.0	0.0	0.0	0.0	0.0	0.0	0.0	0.0
1984-85	0.0	0.0	0.0	0.0	0.0	0.0	0.0	0.0	0.0	0.0	0.0	0.0	0.0
1985-86	0.0	0.0	0.0	0.0	0.0	0.0	0.0	0.0	0.0	0.0	0.0	0.0	0.0
1986-87	0.0	0.0	0.0	0.0	0.0	0.0	0.0	0.0	0.0	0.0	0.0	0.0	0.0
1987-88	0.0	0.0	0.0	0.0	0.0	0.0	0.0	0.0	0.0	0.0	0.0	0.0	0.0
1988-89	0.0	0.0	0.0	0.0	0.0	0.0	0.0	0.0	0.0	0.0	0.0	0.0	0.0
1989-90	0.0	0.0	0.0	0.0	0.0	0.0	0.0	0.0	0.0	0.0	0.0	0.0	0.0
1990-91	0.0	0.0	0.0	0.0	0.0	0.0	0.0	0.0	0.0	0.0	0.0	0.0	0.0
1991-92	0.0	0.0	0.0	0.0	0.0	0.0	0.0	0.0	0.0	0.0	0.0	0.0	0.0
1992-93	0.0	0.0	0.0	0.0	0.0	0.0	0.0	0.0	0.0	0.0	0.0	0.0	0.0
1993-94	0.0	0.0	0.0	0.0	0.0	0.0							
Record Mean	0.0	0.0	0.0	0.0	0.0	0.0	0.0	0.0	0.0	0.0	0.0	0.0	0.0

BOISE, IDAHO

Boise is situated in the Boise River Valley about 8 miles below the mouth of a mountain canyon where the valley proper begins. Sheltered by large shade trees and averaging 2,710 feet in elevation, the denser part of the city covers a gentle alluvial slope about 2 miles wide, stretching southwest from the foothills of the Boise Mountains to the river. The Boise Mountains immediately north of the city rise 5,000 to 6,000 feet above sea level in about 8 miles, the slopes partly mantled with sagebrush and then chaparral giving way near the summit to ridges of fir, spruce, and pine. Across the river, the land rises in two irregular steps, or benches, for several miles, finally reaching the low divide between the Boise and Snake Rivers. Downstream the valley widens, merging with the valley of the Snake about 40 miles to the northwest. Once semi-arid, the entire area is now irrigated from the upstream reservoirs.

Although air masses from the Pacific are considerably modified by the time they reach Boise, their influence, particularly in winter, alternates with that of atmospheric developments from other directions. The result is almost a typical upland continental type of climate in summer, while winters are usually tempered by periods of cloudy or stormy and mild weather. Autumns have prolonged periods of near ideal weather, while springtime is noted by changeable weather and varied temperatures. The Boise climate in general may be described as dry and temperate, with sufficient variation to be stimulating.

Summer hot periods rarely last longer than a few days. Temperatures of 100 degrees or higher occur nearly every year.

Winter cold spells with temperatures of 10 degrees or lower generally last longer than the summer hot spells. During cold weather, however, there is ordinarily little wind to add to the discomfort.

The normal precipitation pattern in the Boise area shows a winter high and a very pronounced summer low. Total amounts and intensity are generally greatest near the foothills, dwindling to westward and southward.

Tornadoes are very rare as are destructive force winds. Northwesterly winds, drying and rather raw in character, although of moderate velocity, are common from March through May. Diurnal southeasterly winds, descending from nearby foothills at night, frequently have a moderating effect on winter temperatures. There is an occasional, but moderate, duststorm during the warmer months, usually occurring at times of cold frontal passage.

Relative humidity is low but widespread irrigation maintains humidity several percent above the general dryness of western arid conditions in summer. Thunderstorms occur primarily during spring and summer, with less frequency during fall and occasionally during winter. December and January are the months of heavy fog or low stratus cloud conditions. Only a moderate amount of sunshine is received in the average winter, but protracted periods of clear, sunny weather are the rule in summer. Ice storms are practically unknown.

Based on the 1951-1980 period, the average first occurrence of 32 degrees Fahrenheit in the fall is October 9 and the average last occurrence in the spring is May 8.

NORMALS, MEANS, AND EXTREMES

BOISE, IDAHO

LATITUDE: 43°34'N LONGITUDE: 116°13'W ELEVATION: FT. GRND 2838 BARO 2875 TIME ZONE: MOUNTAIN WBAN: 24131

	(a)	JAN	FEB	MAR	APR	MAY	JUNE	JULY	AUG	SEP	OCT	NOV	DEC	YEAR
TEMPERATURE °F:														
Normals														
-Daily Maximum		36.4	44.2	52.9	61.4	71.0	80.9	90.2	88.1	77.0	64.6	48.7	37.7	62.8
-Daily Minimum		21.6	27.5	31.9	36.7	43.9	52.1	57.7	56.8	48.2	39.0	31.1	22.5	39.1
-Monthly		29.0	35.9	42.4	49.1	57.5	66.5	74.0	72.5	62.6	51.8	39.9	30.1	50.9
Extremes														
-Record Highest	54	63	71	81	92	98	109	111	110	102	94	74	65	111
-Year		1953	1992	1978	1987	1986	1940	1960	1961	1945	1992	1988	1964	JUL 1960
-Record Lowest	54	-17	-15	6	19	22	31	35	34	23	11	-3	-25	-25
-Year		1950	1989	1971	1968	1982	1984	1986	1992	1970	1971	1985	1990	DEC 1990
NORMAL DEGREE DAYS:														
Heating (base 65°F)		1116	815	701	477	242	75	6	20	160	414	753	1082	5861
Cooling (base 65°F)		0	0	0	0	9	120	285	252	88	0	0	0	754
% OF POSSIBLE SUNSHINE	51	40	51	62	68	72	76	87	85	81	69	43	39	64
MEAN SKY COVER (tenths)														
Sunrise - Sunset	54	7.7	7.3	6.9	6.5	5.8	4.8	2.8	3.3	3.6	5.1	7.0	7.5	5.7
MEAN NUMBER OF DAYS:														
Sunrise to Sunset														
-Clear	54	4.5	4.5	6.1	6.5	8.5	11.5	20.4	18.4	17.0	12.1	6.1	4.9	120.4
-Partly Cloudy	54	5.0	6.4	7.3	8.8	9.9	10.2	7.2	8.0	7.0	8.2	6.3	5.7	90.1
-Cloudy	54	21.6	17.4	17.6	14.7	12.4	8.3	3.3	4.6	6.0	10.7	17.6	20.4	154.6
Precipitation														
.01 inches or more	54	11.8	10.2	9.7	8.3	7.8	6.1	2.4	2.6	3.7	5.9	10.1	11.1	89.7
Snow, Ice pellets, hail														
1.0 inches or more	54	2.4	1.3	0.5	0.2	0.*	0.0	0.0	0.0	0.0	0.1	0.9	2.2	7.7
Thunderstorms	54	0.*	0.3	0.6	0.9	2.8	2.8	2.5	2.5	1.5	0.6	0.3	0.1	14.8
Heavy Fog Visibility														
1/4 mile or less	54	5.8	3.3	0.8	0.3	0.2	0.1	0.0	0.*	0.1	0.5	3.0	5.8	19.9
Temperature °F														
-Maximum														
90° and above	54	0.0	0.0	0.0	0.1	1.3	5.4	18.3	15.2	3.3	0.1	0.0	0.0	43.6
32° and below	54	10.4	2.9	0.2	0.0	0.0	0.0	0.0	0.0	0.0	0.0	1.0	6.7	21.3
-Minimum														
32° and below	54	26.3	20.8	17.6	8.1	1.8	0.*	0.0	0.0	0.5	5.6	18.0	25.6	124.3
0° and below	54	1.8	0.4	0.0	0.0	0.0	0.0	0.0	0.0	0.0	0.0	0.1	1.1	3.4
AVG. STATION PRESS. (mb)	21	920.0	918.0	914.3	914.5	913.5	913.8	914.4	914.4	915.8	917.8	918.4	920.1	916.2
RELATIVE HUMIDITY (%)														
Hour 05	54	81	80	74	70	69	67	54	52	59	67	78	81	69
Hour 11 (Local Time)	54	74	68	55	47	44	41	33	33	39	47	65	74	52
Hour 17	54	70	60	45	36	34	30	22	23	29	39	60	71	43
Hour 23	54	80	77	68	61	58	53	39	40	49	60	75	80	62
PRECIPITATION (inches):														
Water Equivalent														
-Normal		1.45	1.07	1.29	1.24	1.08	0.81	0.35	0.43	0.80	0.75	1.48	1.36	12.11
-Maximum Monthly	54	3.87	3.70	3.46	3.04	4.07	3.41	1.62	2.37	2.93	2.25	3.36	4.23	4.23
-Year		1970	1986	1989	1955	1990	1941	1982	1968	1986	1956	1988	1983	DEC 1983
-Minimum Monthly	54	0.12	0.19	0.17	0.09	T	0.01	0.00	T	0.00	0.00	0.14	0.09	0.00
-Year		1949	1964	1992	1949	1992	1966	1947	1992	1987	1988	1976	1976	OCT 1988
-Maximum in 24 hrs	54	1.48	1.00	1.65	1.27	2.05	2.24	0.94	1.61	1.74	0.76	0.88	1.16	2.24
-Year		1953	1951	1981	1969	1990	1958	1960	1979	1976	1947	1971	1955	JUN 1958
Snow, Ice pellets, hail														
-Maximum Monthly	54	21.4	25.2	11.9	8.0	4.0	T	T	T	0.0	2.7	18.6	26.2	26.2
-Year		1964	1949	1951	1967	1964	1954	1970	1989		1971	1985	1983	DEC 1983
-Maximum in 24 hrs	54	8.5	13.0	6.4	7.2	4.0	T	T	T	0.0	1.7	6.5	6.7	13.0
-Year		1950	1949	1952	1969	1964	1954	1970	1989		1971	1964	1983	FEB 1949
WIND:														
Mean Speed (mph)	54	8.0	8.9	9.9	9.9	9.5	9.0	8.4	8.2	8.2	8.3	8.4	8.1	8.7
Prevailing Direction through 1963		SE	SE	SE	SE	NW	NW	NW	NW	SE	SE	SE	SE	SE
Fastest Mile														
-Direction (!!!)	54	SE	W	W	W	W	SW	W	SE	SE	SE	NW	NW	W
-Speed (MPH)	54	50	56	52	50	50	50	61	56	50	56	57	56	61
-Year		1941	1954	1957	1942	1954	1948	1944	1963	1960	1950	1953	1950	JUL 1944
Peak Gust														
-Direction (!!!)	10	N	W	NW	W	NW	NW	S	W	W	S	SW	SE	S
-Speed (mph)	10	59	45	48	58	49	54	71	54	43	43	54	44	71
-Date		1986	1989	1984	1986	1992	1987	1987	1984	1992	1992	1984	1992	JUL 1987

PRECIPITATION (inches)

BOISE, IDAHO

YEAR	JAN	FEB	MAR	APR	MAY	JUNE	JULY	AUG	SEP	OCT	NOV	DEC	ANNUAL
1964	2.46	0.19	0.64	1.35	1.76	2.00	0.41	0.53	0.70	0.21	2.33	3.19	15.77
1965	2.89	0.31	0.43	2.81	0.80	1.20	0.25	0.88	0.55	0.28	1.51	0.61	12.52
1966	0.81	0.73	0.60	0.61	0.32	0.01	0.06	0.01	0.19	0.29	1.60	1.41	6.64
1967	1.49	0.35	0.37	1.47	0.49	1.07	0.05	T	0.58	0.42	0.89	0.50	7.68
1968	0.43	1.86	0.71	0.35	0.40	0.60	T	2.37	0.10	0.70	1.50	1.95	10.97
1969	3.50	1.00	0.26	1.35	0.50	2.00	0.02	T	0.68	0.64	0.59	1.77	12.31
1970	3.87	0.30	1.04	0.93	0.73	1.72	0.28	0.10	1.00	0.81	2.03	1.37	14.18
1971	2.04	0.65	1.50	0.40	0.25	1.58	0.12	0.18	0.64	0.53	2.32	1.63	11.84
1972	2.15	0.91	1.50	0.62	0.32	0.90	0.21	0.05	1.11	0.64	1.11	1.79	11.31
1973	1.14	0.42	0.65	1.49	0.74	0.19	0.07	0.03	0.82	1.15	2.44	2.23	11.37
1974	1.35	0.66	1.50	0.67	0.10	0.60	0.53	0.22	T	1.45	0.67	1.71	9.46
1975	0.59	2.62	1.92	1.53	0.88	0.78	0.82	0.48	0.01	1.99	0.78	1.29	13.69
1976	1.49	1.31	0.72	1.60	0.46	1.66	1.15	0.95	2.11	0.52	0.14	0.09	12.20
1977	0.65	0.57	0.86	0.19	1.80	1.26	0.41	0.73	1.20	0.21	1.86	2.46	12.20
1978	2.37	1.50	1.43	2.34	0.36	0.56	0.48	0.24	0.89	T	1.06	0.60	11.83
1979	1.93	1.20	0.48	1.60	1.28	0.18	0.01	1.81	0.04	1.50	1.30	0.74	12.07
1980	1.56	1.29	2.14	1.20	3.77	0.58	0.03	T	1.59	0.30	1.26	1.49	15.21
1981	1.20	1.02	2.76	1.93	0.95	0.77	0.23	0.13	0.36	0.97	2.24	2.72	15.28
1982	1.42	1.54	1.39	0.79	0.39	0.35	1.62	0.19	1.38	1.74	1.10	1.92	13.83
1983	1.67	1.26	2.70	2.29	1.93	0.17	1.16	0.28	0.65	0.56	1.87	4.23	18.77
1984	0.80	0.86	1.43	1.62	1.06	1.47	0.23	1.24	0.69	0.85	2.36	0.63	13.24
1985	0.20	0.55	0.97	0.90	1.52	0.37	0.85	0.04	1.81	0.84	1.85	1.24	11.14
1986	0.98	3.70	2.01	1.55	1.10	0.35	0.17	0.07	2.93	0.33	1.00	0.12	14.31
1987	0.73	1.24	2.01	0.38	0.69	0.58	0.70	0.11	0.00	T	1.00	1.05	8.49
1988	1.30	0.43	1.45	1.80	1.33	0.47	0.02	0.09	0.24	0.00	3.36	0.81	11.30
1989	1.14	1.15	3.46	0.46	0.21	0.08	0.03	0.78	1.20	1.24	0.59	0.10	10.44
1990	0.84	0.79	0.77	2.14	4.07	0.11	0.42	0.39	0.50	0.45	0.61	0.98	12.07
1991	0.96	0.46	0.55	1.65	1.57	0.64	0.37	0.04	0.21	0.91	1.76	0.35	9.47
1992	0.36	0.92	0.17	0.66	T	2.07	0.03	T	0.30	0.90	1.37	0.89	7.67
1993	1.65	0.96	2.45	2.09	0.92	2.10	0.52	0.24	T	0.47	0.38	0.98	12.76
Record Mean	1.44	1.25	1.32	1.21	1.20	0.89	0.29	0.26	0.58	0.92	1.35	1.32	12.03

AVERAGE TEMPERATURE (deg. F)

BOISE, IDAHO

YEAR	JAN	FEB	MAR	APR	MAY	JUNE	JULY	AUG	SEP	OCT	NOV	DEC	ANNUAL
1964	26.0	26.3	37.6	45.3	55.1	62.5	74.9	69.2	59.6	52.9	36.9	35.2	48.5
1965	34.1	36.9	39.2	50.2	54.9	64.7	73.3	71.1	57.1	56.4	44.3	30.1	51.0
1966	32.4	33.3	43.2	49.2	61.5	65.0	73.2	72.1	66.2	49.4	43.0	31.3	51.6
1967	36.2	38.3	42.0	43.9	56.5	66.1	78.5	78.0	68.6	50.7	40.7	27.8	52.3
1968	29.5	41.1	45.9	46.3	57.8	67.9	77.3	68.7	63.0	52.4	42.0	35.1	52.2
1969	34.3	36.2	42.5	50.5	61.1	66.3	72.9	72.4	64.2	45.9	39.7	34.6	51.7
1970	36.2	41.2	41.4	43.8	57.5	68.9	76.3	75.8	57.8	47.5	43.2	33.5	51.9
1971	33.9	37.5	40.2	49.2	60.0	64.9	74.7	78.6	59.2	49.0	39.7	30.2	51.4
1972	30.9	36.5	45.5	46.7	60.6	67.6	73.5	74.5	58.8	51.7	40.5	23.8	50.9
1973	30.7	39.4	43.0	49.1	60.5	67.6	75.6	73.0	63.1	52.4	41.7	38.4	52.9
1974	29.4	38.7	42.6	49.7	55.4	71.7	73.8	71.7	65.3	52.2	41.7	33.2	52.1
1975	28.2	36.9	41.5	44.5	56.2	64.4	78.3	70.0	65.5	52.3	39.5	31.4	50.7
1976	32.2	34.2	37.1	47.3	59.3	62.6	73.1	67.7	64.1	50.8	40.9	29.3	49.9
1977	19.0	33.8	39.8	54.1	53.7	70.2	72.7	73.6	62.0	53.2	39.6	37.3	50.8
1978	37.0	38.2	48.7	48.5	54.3	64.5	72.8	69.9	61.4	52.8	36.7	27.0	51.0
1979	16.2	34.3	43.2	48.7	57.8	66.7	74.1	71.2	67.3	54.4	34.7	35.8	50.3
1980	30.3	39.8	41.1	52.6	57.2	62.6	72.9	67.2	62.6	51.6	39.8	33.1	50.9
1981	33.9	36.8	44.5	50.3	54.7	63.2	71.0	74.2	63.7	48.7	44.0	35.3	51.7
1982	24.8	30.0	41.3	45.2	54.9	65.8	70.1	72.7	60.4	50.9	36.0	31.5	48.6
1983	35.9	41.5	44.6	47.1	56.5	63.8	69.4	74.9	60.8	53.8	42.0	23.2	51.2
1984	21.2	30.2	41.9	45.9	54.8	61.7	74.2	75.5	60.1	46.9	39.1	22.9	47.9
1985	19.1	25.8	36.0	51.6	58.5	67.2	77.7	69.2	56.2	48.4	27.7	12.6	45.8
1986	29.4	41.0	48.0	48.3	59.1	72.0	69.6	75.9	57.4	52.6	40.4	28.0	51.8
1987	27.8	37.6	44.2	56.0	62.2	70.2	71.2	70.0	65.6	54.9	40.7	32.9	52.8
1988	26.1	37.8	42.6	52.7	57.9	70.7	74.6	71.6	61.7	59.9	40.7	27.0	51.9
1989	24.7	23.2	43.7	53.1	56.3	68.5	77.0	70.0	63.3	51.2	39.4	30.4	50.1
1990	34.2	34.1	44.3	54.7	55.7	66.9	76.2	74.0	69.9	51.0	41.2	18.1	51.7
1991	24.4	41.6	43.0	48.3	54.9	62.6	75.7	76.6	65.8	51.8	37.4	31.7	51.2
1992	33.1	42.1	48.9	54.9	64.3	69.8	71.4	73.9	63.4	54.4	34.5	28.5	53.3
1993	24.8	28.9	41.7	48.1	63.1	61.9	65.0	68.6	63.0	53.2	32.8	33.3	48.7
Record Mean	29.4	35.4	42.4	49.8	57.7	65.6	74.2	72.4	62.7	52.3	40.2	31.5	51.1
Max	36.7	43.5	52.5	61.8	70.8	79.4	89.9	87.8	76.9	64.8	49.2	38.7	62.7
Min	22.1	27.3	32.3	37.8	44.6	51.7	58.4	56.9	48.4	39.8	31.1	24.2	39.5

REFERENCE NOTES

BOISE, IDAHO

GENERAL
T - TRACE AMOUNT.
BLANK ENTRIES DENOTE MISSING/UNREPORTED DATA.
# INDICATES A STATION OR INSTRUMENT RELOCATION.

SPECIFIC

PM - INCLUDES LAST DAY OF PREVIOUS MONTH
ASOS - AUTOMATED SURFACE OBSERVING SYSTEM IN OPERATION DURING THESE MONTHS.

(*) - LENGTH OF RECORD IN YEARS, ALTHOUGH INDIVIDUAL MONTHS MAY BE MISSING.
0.* OR * - THE VALUE IS BETWEEN 0.0 AND 0.05
NORMALS - BASED ON THE 1961-1990 RECORD PERIOD.
EXTREMES - DATES ARE THE MOST RECENT OCCURRENCE
WIND DIR. - NUMERALS SHOW TENS OF DEGRESS CLOCKWISE FROM TRUE NORTH "00" INDICATES CALM.
RESULTANT DIRECTIONS ARE GIVEN TO WHOLE DEGREES.
BOLD VALUES INDICATE EXTREME VALUES WHICH OCCURRED AFTER THE ASOS SYSTEM WAS COMMISSIONED.

RECORD = PERIOD OF RECORD
RECORD MEAN PRECIPITATION IS THE MEAN OF ALL DAILY PRECIPITATION AMOUNTS DURING THE PERIOD OF RECORD.
RECORD MAX(MIN) TEMPERATURE IS THE MEAN OF ALL DAILY MAX(MIN) TEMPERATURES DURING THE PERIOD OF RECORD.
RECORD MEAN TEMPERATURE IS THE SUM OF THE RECORD MAX AND RECORD MIN DIVIDED BY 2.
AVERAGE TEMPERATURE IS THE SUM OF THE MEAN DAILY MAX AND MIN TEMPERATURE DIVIDED BY 2.

EXCEPTIONS

RECORD MEANS ARE THROUGH THE CURRENT YEAR,
BEGINNING IN 1900 FOR TEMPERATURE
1900 FOR PRECIPITATION
1940 FOR SNOWFALL

HEATING DEGREE DAYS Base 65 deg. F BOISE, IDAHO

SEASON	JULY	AUG	SEP	OCT	NOV	DEC	JAN	FEB	MAR	APR	MAY	JUNE	TOTAL
1964-65	0	62	166	370	835	916	951	781	793	435	317	82	5708
1965-66	7	30	234	266	613	1075	1004	884	668	467	172	97	5517
1966-67	7	18	68	476	654	1039	885	742	707	627	279	79	5581
1967-68	0	0	49	434	723	1146	1092	688	586	555	235	58	5566
1968-69	0	59	119	382	684	918	945	797	689	432	147	75	5247
1969-70	3	13	91	585	751	935	886	659	724	631	249	79	5606
1970-71	0	0	240	537	650	969	958	765	763	465	176	78	5601
1971-72	10	6	210	492	753	1069	1047	818	598	543	194	45	5785
1972-73	11	0	222	406	727	1270	1056	708	673	470	182	91	5816
1973-74	4	13	103	382	692	817	1099	728	687	452	304	42	5323
1974-75	10	11	53	391	689	983	1132	782	721	607	275	76	5730
1975-76	6	21	74	399	759	1035	1013	886	858	524	189	132	5896
1976-77	3	36	76	434	720	1097	1418	868	772	342	358	7	6131
1977-78	8	32	145	362	758	853	859	744	500	488	329	74	5152
1978-79	5	38	173	370	841	1171	1503	855	668	481	241	72	6418
1979-80	5	2	26	326	903	899	1070	725	736	367	257	133	5449
1980-81	0	41	104	409	750	983	957	783	631	432	315	97	5502
1981-82	15	5	137	497	624	915	1240	974	729	586	312	86	6120
1982-83	27	2	182	432	863	1030	897	653	622	530	309	82	5629
1983-84	38	0	145	338	682	1290	1353	1004	710	566	328	162	6616
1984-85	0	8	204	557	771	1299	1412	1093	895	398	226	53	6916
1985-86	0	26	259	509	1113	1619	1097	668	522	499	280	15	6607
1986-87	35	2	259	376	733	1141	1149	761	639	287	140	41	5563
1987-88	23	18	86	306	722	990	1198	780	686	359	261	59	5488
1988-89	4	5	157	178	724	1169	1242	1166	656	356	276	30	5963
1989-90	0	29	97	421	759	1064	951	858	633	303	286	82	5483
1990-91	6	10	26	430	710	1449	1252	651	676	493	306	100	6109
1991-92	0	0	55	409	822	1026	982	657	492	308	92	54	4897
1992-93	6	40	118	340	907	1124	1239	1004	715	501	140	155	6289
1993-94	53	45	124	367	960	975							

COOLING DEGREE DAYS Base 65 deg. F BOISE, IDAHO

YEAR	JAN	FEB	MAR	APR	MAY	JUNE	JULY	AUG	SEP	OCT	NOV	DEC	TOTAL
1969	0	0	0	0	32	121	256	251	75	0	0	0	735
1970	0	0	0	0	24	201	357	341	30	0	0	0	953
1971	0	0	0	0	31	81	320	435	42	2	0	0	911
1972	0	0	0	0	62	129	283	303	40	0	0	0	817
1973	0	0	0	0	51	177	341	269	51	1	0	0	890
1974	0	0	0	0	13	252	289	226	71	0	0	0	851
1975	0	0	0	0	10	64	426	182	96	11	0	0	789
1976	0	0	0	0	19	66	263	130	55	2	0	0	535
1977	0	0	0	20	10	170	255	306	61	0	0	0	822
1978	0	0	1	0	6	64	254	200	72	0	0	0	597
1979	0	0	0	0	27	129	293	199	101	3	0	0	752
1980	0	0	0	3	25	68	251	117	38	2	0	0	504
1981	0	0	0	1	3	52	205	296	101	0	0	0	658
1982	0	0	0	0	2	117	194	248	50	0	0	0	611
1983	0	0	0	0	55	50	180	313	26	0	0	0	624
1984	0	0	0	0	19	70	291	340	64	2	0	0	786
1985	0	0	0	2	28	125	402	165	4	0	0	0	726
1986	0	0	0	1	103	235	184	348	37	0	0	0	908
1987	0	0	0	23	61	202	223	180	111	0	0	0	800
1988	0	0	0	0	46	237	308	215	66	24	0	0	896
1989	0	0	0	6	14	140	376	191	56	0	0	0	783
1990	0	0	0	3	5	145	357	293	180	6	0	0	989
1991	0	0	0	0	0	36	337	368	85	6	0	0	832
1992	0	0	0	11	76	202	208	321	77	20	0	0	915
1993	0	0	0	0	86	68	61	162	71	8	0	0	456

SNOWFALL (inches) BOISE, IDAHO

SEASON	JULY	AUG	SEP	OCT	NOV	DEC	JAN	FEB	MAR	APR	MAY	JUNE	TOTAL
1964-65	0.0	0.0	0.0	0.0	7.3	8.7	18.4	0.4	0.4	1.3	T	0.0	36.5
1965-66	0.0	0.0	0.0	T	0.7	4.5	3.6	7.1	0.2	0.1	0.0	0.0	16.2
1966-67	0.0	0.0	0.0	0.1	T	9.0	2.0	1.9	0.3	8.0	T	0.0	21.3
1967-68	0.0	0.0	0.0	0.0	1.3	4.3	3.3	0.1	0.8	0.3	T	0.0	10.1
1968-69	0.0	0.0	0.0	0.0	1.3	11.4	14.4	5.5	1.3	7.2	0.0	0.0	41.1
1969-70	0.0	0.0	0.0	1.2	0.3	4.8	2.9	0.1	2.3	1.1	0.5	0.0	13.2
1970-71	T	0.0	0.0	0.3	0.1	4.4	4.0	2.1	1.6	T	0.0	0.0	12.5
1971-72	0.0	0.0	0.0	2.7	0.4	14.8	5.0	3.5	0.6	1.6	0.0	0.0	28.6
1972-73	0.0	0.0	0.0	T	T	12.6	7.3	0.3	0.6	T	0.0	0.0	20.8
1973-74	0.0	0.0	0.0	0.0	8.8	4.5	6.0	2.4	5.7	T	0.0	0.0	27.4
1974-75	0.0	0.0	0.0	0.0	0.9	4.6	5.4	6.3	2.2	2.6	0.9	0.0	22.9
1975-76	0.0	0.0	0.0	T	3.9	4.2	6.3	6.7	3.9	0.2	0.0	0.0	25.2
1976-77	0.0	0.0	0.0	0.0	T	1.1	7.2	3.5	2.9	T	T	0.0	14.7
1977-78	0.0	0.0	0.0	T	4.7	4.4	1.6	7.3	T	T	T	0.0	18.0
1978-79	0.0	0.0	0.0	0.0	0.2	3.1	11.9	4.3	0.8	0.2	T	0.0	20.5
1979-80	0.0	0.0	0.0	0.0	6.6	1.4	3.8	0.8	2.7	T	0.0	0.0	15.3
1980-81	0.0	0.0	0.0	0.0	3.2	1.7	3.6	0.7	T	0.5	0.0	0.0	9.7
1981-82	0.0	0.0	0.0	0.0	2.8	11.1	12.2	1.4	3.6	1.4	0.0	0.0	32.5
1982-83	0.0	0.0	0.0	0.0	2.1	6.4	1.6	0.9	T	T	0.8	0.0	11.8
1983-84	0.0	0.0	0.0	0.0	2.2	26.2	4.3	4.4	T	0.3	T	0.0	37.4
1984-85	0.0	0.0	0.0	T	0.2	7.7	2.6	5.3	1.5	1.0	0.0	0.0	18.3
1985-86	0.0	0.0	0.0	T	18.6	12.6	3.9	4.4	0.0	T	0.0	0.0	39.5
1986-87	0.0	0.0	0.0	0.0	5.9	0.9	0.5	0.6	T	0.0	0.0	0.0	7.9
1987-88	0.0	0.0	0.0	0.0	0.5	3.0	3.9	0.3	2.9	1.2	0.0	0.0	11.8
1988-89	0.0	0.0	0.0	0.0	2.5	10.8	8.0	0.7	T	T	0.0	0.0	22.0
1989-90	0.0	T	0.0	T	0.4	T	5.2	6.5	0.4	T	T	0.0	12.5
1990-91	0.0	0.0	0.0	0.0	0.1	15.7	1.5	T	1.2	T	0.0	0.0	18.5
1991-92	0.0	0.0	0.0	1.1	2.8	T	0.3	T	0.0	T	0.0	0.0	4.2
1992-93	0.0	0.0	0.0	T	3.6	5.7	14.6	10.4	0.2	T	T	0.0	34.5
1993-94	0.0	0.0	0.0	0.0	1.9	0.2							
Record Mean	T	T	0.0	0.1	2.2	5.8	6.8	3.6	1.7	0.6	0.1	T	20.9

LEWISTON, IDAHO

Lewiston is located at the confluence of the Snake and Clearwater Rivers at an elevation of 738 feet above mean sea level. Lower Granite Lake extends from the confluence of the two rivers, 32 miles downstream in the Snake River channel, to Lower Granite Dam. The valley is rather narrow with a range of hills to the north sloping abruptly to about 2,000 feet above the valley floor. To the south the terrain rises more gradually to a more or less flat bench about 700 feet above the valley. The Weather Office is located on the bench at an elevation of 1,413 feet above sea level and about 2 miles south of Lewiston. Although Lewiston is at about the same latitude as Duluth, Minnesota, the climate, especially in the wintertime, is comparatively very mild. This mildness can be explained by its location with respect to the effects of Pacific air masses from the west and by the sheltering effects of the mountains that surround the valley in almost every direction.

Considerable variations in the climate are to be found within relatively short distances from the valley itself. On the prairies surrounding the valley, winter temperatures are much lower and the precipitation is normally almost double that recorded in the valley and at the airport location.

Precipitation normally amounts to about 13 inches annually, which is rather evenly distributed through the year except for the months of July and August, which are characterized by infrequent thunderstorms that usually drop only small amounts of rain. Records show that several times during these two months not more than a trace of rain has been recorded and at times not even a trace. The thunderstorms on the prairie are, at times, accompanied by heavy hail and windstorms. Snowfall in the valley averages about 18 inches during the year, concentrated mostly in the three months of December, January, and February, but in the higher country surrounding the valley the snowfall is much heavier.

Most of the precipitation reaching this vicinity results from strong invasions of moist air from the North Pacific source region. Greatest amounts of both rain and snow occur when this moist air is overrunning a weak front that has become stationary along an east-west line a short distance south of the area.

Temperatures show a wide range from more than 115 degrees to less than -20 degrees. Many winters have gone by without a temperature of zero being recorded in the valley, but the prairie sections usually experience lower temperatures. The summers experience hot and dry periods with as many as 10 consecutive days with afternoon temperatures reaching 100 degrees or more. Considerable cooling after sunset makes the nights very comfortable. Cold waves occur when arctic air, originating in the Yukon Territory, moves southward. Such cold waves are relatively infrequent when compared to the number of arctic outbreaks east of the continental divide in Montana only a short distance away.

Winds are light, usually prevailing from the east, with occasional stronger winds accompanying the well-developed frontal systems from the west.

Relative humidity averages about 70 percent during the winter months and gradually lowers to about 40 percent during July and August.

The growing season of approximately 200 days in this part of the country, makes conditions favorable for the growing of many types of fruits, vegetables, and berries.

NORMALS, MEANS, AND EXTREMES

LEWISTON, IDAHO

LATITUDE: 46°23'N LONGITUDE: 117°01'W ELEVATION: FT. GRND 1413 BARO 1441 TIME ZONE: PACIFIC WBAN: 24149

	(a)	JAN	FEB	MAR	APR	MAY	JUNE	JULY	AUG	SEP	OCT	NOV	DEC	YEAR
TEMPERATURE °F:														
Normals														
-Daily Maximum		39.6	46.6	53.5	61.5	70.2	79.8	89.0	88.3	77.2	63.2	48.1	40.2	63.1
-Daily Minimum		27.6	31.5	34.7	39.7	46.5	53.9	59.2	58.9	50.9	41.4	34.4	28.7	42.3
-Monthly		33.6	39.1	44.1	50.6	58.4	66.9	74.1	73.7	64.1	52.3	41.3	34.5	52.7
Extremes														
-Record Highest	47	66	72	76	97	100	107	110	115	103	89	74	63	115
-Year		1953	1986	1991	1977	1983	1973	1967	1961	1950	1987	1975	1965	AUG 1961
-Record Lowest	47	-22	-15	2	20	23	34	41	41	28	16	-3	-22	-22
-Year		1950	1950	1955	1966	1954	1951	1955	1992	1965	1971	1955	1968	DEC 1968
NORMAL DEGREE DAYS:														
Heating (base 65°F)		973	725	648	436	210	64	6	16	141	394	711	946	5270
Cooling (base 65°F)		0	0	0	0	6	121	288	285	114	0	0	0	814
% OF POSSIBLE SUNSHINE														
MEAN SKY COVER (tenths)														
Sunrise - Sunset	42	8.4	8.2	7.7	7.3	6.7	6.0	3.4	3.9	4.7	6.2	8.2	8.4	6.6
MEAN NUMBER OF DAYS:														
Sunrise to Sunset														
-Clear	42	2.8	2.6	4.2	4.5	6.3	8.3	18.4	16.4	13.1	8.7	2.6	2.8	90.7
-Partly Cloudy	42	4.2	4.7	6.2	6.5	8.7	8.6	7.1	7.6	7.4	7.3	5.6	3.9	77.8
-Cloudy	42	24.0	21.0	20.5	19.0	16.0	13.1	5.6	7.1	9.4	14.9	21.7	24.3	196.4
Precipitation														
.01 inches or more	47	11.2	9.4	10.5	9.9	9.5	8.6	4.8	4.5	5.4	7.6	10.6	10.9	102.7
Snow, Ice pellets, hail														
1.0 inches or more	47	1.9	0.8	0.4	0.*	0.0	0.0	0.0	0.0	0.0	0.*	0.6	1.6	5.3
Thunderstorms	19	0.0	0.1	0.1	0.7	2.5	3.6	3.7	3.2	1.2	0.4	0.1	0.0	15.6
Heavy Fog Visibility 1/4 mile or less	19	3.9	3.5	1.2	0.2	0.4	0.1	0.0	0.0	0.1	2.1	4.1	5.4	21.1
Temperature °F														
-Maximum														
90° and above	47	0.0	0.0	0.0	0.1	1.3	4.7	15.7	14.3	3.5	0.0	0.0	0.0	39.6
32° and below	47	7.6	2.0	0.2	0.0	0.0	0.0	0.0	0.0	0.0	0.0	1.2	4.9	15.9
-Minimum														
32° and below	47	21.8	16.0	11.7	3.6	0.3	0.0	0.0	0.0	0.1	3.0	12.0	20.6	89.1
0° and below	47	1.3	0.3	0.0	0.0	0.0	0.0	0.0	0.0	0.0	0.0	0.1	0.5	2.2
AVG. STATION PRESS. (mb)														
RELATIVE HUMIDITY (%)														
Hour 04	37	81	80	77	75	76	74	61	59	68	80	83	82	75
Hour 10 (Local Time)	47	75	71	62	55	52	48	38	39	49	64	75	76	59
Hour 16	46	70	61	49	42	39	36	25	25	31	48	68	73	47
Hour 22														
PRECIPITATION (inches):														
Water Equivalent														
-Normal		1.28	0.89	1.09	1.13	1.31	1.25	0.67	0.78	0.78	0.90	1.15	1.20	12.43
-Maximum Monthly	47	3.55	1.99	2.70	3.29	4.80	4.70	2.60	2.96	2.36	2.79	2.79	3.28	4.80
-Year		1970	1986	1972	1978	1948	1950	1987	1989	1947	1950	1973	1964	MAY 1948
-Minimum Monthly	47	0.14	0.17	0.25	0.05	0.27	0.24	T	T	T	T	0.23	0.14	T
-Year		1991	1988	1969	1956	1964	1973	1953	1969	1975	1987	1976	1970	OCT 1987
-Maximum in 24 hrs	47	1.35	0.93	0.92	1.06	1.63	1.72	1.40	1.64	1.73	1.19	1.04	1.12	1.73
-Year		1956	1959	1981	1978	1948	1950	1964	1989	1955	1950	1988	1958	SEP 1955
Snow, Ice pellets, hail														
-Maximum Monthly	47	26.1	14.9	9.7	1.1	T	0.0	T	0.0	0.0	2.5	14.4	18.7	26.1
-Year		1957	1956	1955	1972	1990		1991			1971	1961	1968	JAN 1957
-Maximum in 24 hrs	47	12.8	7.5	6.7	1.0	T	0.0	T	0.0	0.0	1.3	8.3	8.4	12.8
-Year		1966	1956	1989	1947	1990		1991			1971	1961	1968	JAN 1966
WIND:														
Mean Speed (mph)														
Prevailing Direction														
Fastest Obs. 1 Min.														
-Direction (!!!)														
-Speed (MPH)														
-Year														
Peak Gust														
-Direction (!!!)	22													
-Speed (mph)	22	72	64	60	58	54	51	46	51	59	59	59	63	72
-Date		1972	1972	1982	1985	1976	1976	1984	1976	1991	1991	1977	1991	JAN 1972

PRECIPITATION (inches)

LEWISTON, IDAHO

YEAR	JAN	FEB	MAR	APR	MAY	JUNE	JULY	AUG	SEP	OCT	NOV	DEC	ANNUAL
1964	0.49	0.21	0.55	0.93	0.27	3.11	2.15	0.66	0.87	0.94	1.35	3.28	14.81
1965	2.99	0.40	0.54	1.93	0.48	0.95	0.82	1.41	0.28	0.34	1.10	0.14	11.38
1966	1.43	0.71	0.97	0.31	0.42	0.68	0.28	0.48	0.22	0.97	1.81	1.70	9.98
1967	1.28	0.29	1.18	1.78	1.17	1.93	0.03	T	0.60	0.88	0.45	1.46	11.05
1968	0.65	1.42	0.62	0.40	0.97	1.34	0.41	1.64	1.27	1.34	2.01	2.13	14.20
1969	2.98	0.76	0.25	2.33	1.25	2.38	0.21	T	1.38	1.10	0.29	1.50	14.43
1970	3.55	0.65	1.14	1.03	1.26	2.28	1.48	0.02	1.29	0.89	1.36	0.14	15.09
1971	1.67	0.73	1.08	0.74	1.92	2.53	0.70	0.96	1.57	1.01	1.45	1.08	15.44
1972	1.36	1.47	2.70	0.97	1.61	0.93	0.74	0.68	0.83	1.20	0.93	1.65	15.07
1973	0.70	0.66	0.50	0.12	1.58	0.24	0.01	0.02	1.12	1.64	2.79	2.99	12.37
1974	1.34	1.64	0.73	1.66	0.76	0.50	0.40	0.01	0.10	0.06	0.55	0.77	8.52
1975	2.81	1.50	0.99	1.25	1.01	1.29	0.68	1.09	T	1.92	0.56	2.09	15.19
1976	0.54	0.71	0.75	1.29	1.41	1.22	0.43	1.76	0.33	1.13	0.23	0.26	10.06
1977	0.34	0.36	0.92	0.10	1.63	0.35	0.39	1.65	2.22	0.55	1.65	2.10	12.26
1978	1.92	1.47	1.09	3.29	1.06	0.30	0.56	1.90	1.06	T	1.06	0.96	14.67
1979	0.97	1.12	0.69	2.17	1.56	0.70	0.21	0.57	0.18	1.57	1.44	0.97	12.15
1980	1.72	1.57	1.23	0.76	1.87	1.31	0.89	0.47	0.97	0.68	1.00	0.88	13.35
1981	0.89	1.22	1.93	0.92	1.11	1.94	0.92	0.01	1.01	1.41	1.54	1.31	14.21
1982	1.57	0.75	1.29	1.14	0.65	0.46	1.74	0.47	0.97	1.98	0.39	1.03	12.44
1983	0.95	1.46	1.48	1.12	1.15	1.70	0.96	0.93	0.74	0.87	1.00	1.14	13.50
1984	0.71	0.46	1.66	1.15	1.68	1.58	0.27	0.93	0.21	0.91	0.89	0.69	11.14
1985	0.24	0.66	0.67	0.93	1.29	0.92	0.57	0.91	1.82	0.60	0.62	0.36	9.59
1986	1.13	1.99	0.63	0.37	1.39	0.41	0.56	0.84	0.94	0.30	1.44	0.53	10.53
1987	0.56	0.44	0.91	0.83	0.84	1.44	2.60	0.34	0.01	T	0.31	0.81	9.09
1988	0.97	0.17	1.04	1.12	0.91	1.69	0.88	0.08	0.82	0.17	2.04	0.53	10.42
1989	1.61	0.33	1.69	0.65	2.57	1.61	0.07	2.96	0.64	0.63	0.67	0.30	13.73
1990	0.84	0.26	1.05	2.08	2.39	0.71	0.35	0.71	0.04	1.18	1.05	0.92	11.58
1991	0.14	0.32	1.11	0.79	3.74	1.86	0.53	0.03	0.24	0.15	2.00	0.40	11.31
1992	0.71	0.74	0.42	1.76	0.49	0.75	1.34	1.37	0.84	0.67	1.25	0.39	10.73
1993	0.99	0.70	1.17	2.78	1.97	1.63	1.19	0.62	0.07	0.67	0.64	0.80	13.23
Record Mean	1.20	0.86	1.04	1.16	1.50	1.44	0.64	0.74	0.78	0.97	1.17	1.14	12.64

AVERAGE TEMPERATURE (deg. F)

LEWISTON, IDAHO

YEAR	JAN	FEB	MAR	APR	MAY	JUNE	JULY	AUG	SEP	OCT	NOV	DEC	ANNUAL
1964	36.8	36.8	41.8	47.7	57.4	64.4	72.7	68.6	60.7	52.3	39.0	31.5	50.8
1965	34.1	39.6	40.1	52.1	56.6	65.2	74.2	72.2	59.2	56.3	44.3	36.5	52.6
1966	36.1	37.9	43.7	50.7	61.2	64.2	72.8	72.9	68.6	51.9	43.2	39.8	53.6
1967	39.7	41.3	42.1	46.5	58.3	68.5	76.3	79.1	69.5	53.4	40.9	33.5	54.1
1968	34.8	43.1	47.4	48.4	58.6	66.4	76.4	69.8	62.9	50.4	41.8	31.1	52.6
1969	23.6	34.7	43.3	50.7	61.4	69.7	72.8	71.9	64.3	47.9	40.8	35.0	51.3
1970	34.4	42.0	42.8	46.2	59.3	70.1	76.1	75.0	58.2	49.9	41.5	34.0	52.4
1971	36.3	39.0	40.8	49.6	60.4	62.3	74.6	78.1	58.6	48.8	39.8	34.2	51.9
1972	32.3	37.6	47.0	47.1	61.0	67.0	73.2	75.1	60.4	51.2	41.3	29.0	51.9
1973	32.2	39.1	45.7	50.5	61.1	67.7	77.1	74.8	64.7	51.3	39.8	39.4	53.6
1974	29.2	41.3	44.3	51.1	56.3	71.2	74.0	73.7	66.3	53.1	43.0	37.4	53.4
1975	34.3	35.3	41.0	46.5	57.8	63.6	77.8	70.4	66.2	51.4	40.3	36.9	51.8
1976	37.0	38.0	40.9	50.1	59.7	63.0	73.9	70.0	69.0	54.3	43.5	35.6	52.9
1977	28.8	41.5	42.2	55.9	56.5	72.3	73.2	77.4	61.3	50.7	39.4	36.8	53.0
1978	35.4	40.7	47.6	50.0	56.0	67.2	74.1	71.1	61.9	52.2	35.1	29.0	51.7
1979	16.9	36.0	46.0	49.9	59.2	67.1	76.0	75.0	69.0	55.9	37.5	39.8	52.3
1980	28.6	39.6	43.1	56.0	58.2	63.2	73.5	68.8	64.0	52.2	41.6	40.4	52.4
1981	38.5	39.7	46.6	51.6	57.0	62.2	70.9	77.7	65.5	49.5	44.2	36.7	53.3
1982	32.1	36.7	44.3	47.7	57.2	69.2	72.7	74.4	63.9	51.4	38.8	34.8	51.9
1983	40.5	42.9	47.1	49.2	59.5	65.2	70.5	76.7	60.7	53.0	43.8	25.3	52.9
1984	36.0	40.8	46.8	49.3	56.1	63.9	75.2	75.0	61.5	48.6	42.2	30.5	52.2
1985	28.0	32.6	42.3	53.5	60.3	66.9	80.0	69.5	56.8	48.6	29.0	23.2	49.2
1986	39.2	39.0	48.8	50.5	60.1	70.9	68.9	77.7	59.4	53.9	40.5	34.7	53.6
1987	32.0	40.4	46.4	57.0	62.3	70.0	71.5	72.0	68.3	54.6	43.6	33.8	54.3
1988	33.3	41.3	44.4	53.5	58.9	66.7	73.8	73.9	64.2	58.7	42.6	34.2	53.8
1989	36.3	26.6	46.3	56.1	59.4	68.6	74.5	71.4	66.1	54.4	47.1	37.9	53.7
1990	41.9	39.7	48.4	57.0	58.8	67.6	77.4	75.6	73.0	53.1	47.1	30.6	55.9
1991	34.5	47.7	45.4	52.0	55.9	60.3	72.3	75.1	65.6	50.6	40.1	37.6	53.1
1992	37.9	42.2	49.1	53.8	63.0	71.2	72.4	73.4	62.4	54.2	40.2	32.7	54.4
1993	27.0	32.9	46.4	50.1	62.9	63.9	65.4	70.2	64.7	54.4	35.2	36.1	50.8
Record Mean	32.2	38.3	43.8	50.8	58.7	66.1	73.5	72.8	63.8	52.0	40.5	34.6	52.3
Max	38.6	46.1	53.6	62.2	71.0	79.0	88.6	87.7	77.6	63.2	47.8	40.7	63.0
Min	25.7	30.4	34.1	39.4	46.4	53.2	58.4	57.8	49.9	40.8	33.3	28.5	41.5

REFERENCE NOTES

LEWISTON, IDAHO

GENERAL
T - TRACE AMOUNT.
BLANK ENTRIES DENOTE MISSING/UNREPORTED DATA.
# INDICATES A STATION OR INSTRUMENT RELOCATION.

SPECIFIC

(a) - LENGTH OF RECORD IN YEARS, ALTHOUGH INDIVIDUAL MONTHS MAY BE MISSING.
0.* OR * - THE VALUE IS BETWEEN 0.0 AND 0.05
NORMALS - BASED ON THE 1961-1990 RECORD PERIOD.
EXTREMES - DATES ARE THE MOST RECENT OCCURRENCE.
WIND DIR.- NUMERALS SHOW TENS OF DEGRESS CLOCKWISE FROM TRUE NORTH. "00" INDICATES CALM.
RESULTANT DIRECTIONS ARE GIVEN TO WHOLE DEGREES.
BOLD VALUES INDICATE EXTREME VALUES WHICH OCCURRED AFTER THE ASOS SYSTEM WAS COMMISSIONED.

RECORD = PERIOD OF RECORD
RECORD MEAN PRECIPITATION IS THE MEAN OF ALL DAILY PRECIPITATION AMOUNTS DURING THE PERIOD OF RECORD.
RECORD MAX(MIN) TEMPERATURE IS THE MEAN OF ALL DAILY MAX(MIN) TEMPERATURES DURING THE PERIOD OF RECORD.
RECORD MEAN TEMPERATURE IS THE SUM OF THE RECORD MAX AND RECORD MIN DIVIDED BY 2.
AVERAGE TEMPERATURE IS THE SUM OF THE MEAN DAILY MAX AND MIN TEMPERATURE DIVIDED BY 2.

EXCEPTIONS

THUNDERSTORMS AND HEAVY FOG ARE THROUGH 1964 AND MAY BE INCOMPLETE, DUE TO PART-TIME OPERATIONS.

RECORD MEANS ARE THROUGH THE CURRENT YEAR, BEGINNING IN 1947 FOR TEMPERATURE
1947 FOR PRECIPITATION
1947 FOR SNOWFALL

HEATING DEGREE DAYS Base 65 deg. F LEWISTON, IDAHO

SEASON	JULY	AUG	SEP	OCT	NOV	DEC	JAN	FEB	MAR	APR	MAY	JUNE	TOTAL
1964-65	3	30	129	386	775	1034	951	705	765	380	260	46	5464
1965-66	10	19	182	266	613	879	890	754	653	421	160	80	4927
1966-67	18	12	19	400	641	769	776	656	702	550	217	26	4786
1967-68	0	0	31	352	719	969	926	629	539	493	200	50	4908
1968-69	0	30	115	444	690	1044	1274	842	666	423	134	38	5700
1969-70	0	10	86	525	719	924	942	638	681	556	187	56	5324
1970-71	3	0	217	465	698	956	885	720	740	457	171	114	5426
1971-72	16	11	206	498	748	948	1007	789	551	530	171	46	5521
1972-73	13	1	178	419	706	1112	1007	719	588	427	173	59	5402
1973-74	1	6	80	415	749	787	1106	658	637	410	268	37	5154
1974-75	9	0	53	362	651	852	945	826	734	547	230	87	5296
1975-76	1	17	48	417	736	863	862	774	738	439	173	118	5186
1976-77	3	17	14	331	638	907	1116	653	701	300	266	10	4956
1977-78	16	17	153	435	761	868	911	673	531	441	272	40	5118
1978-79	6	22	139	390	890	1109	1485	808	582	448	190	61	6130
1979-80	12	0	14	281	819	774	1122	730	673	274	218	86	5003
1980-81	0	35	74	401	695	757	815	700	564	401	249	111	4802
1981-82	13	0	118	472	620	869	1013	787	636	514	239	57	5338
1982-83	15	0	118	416	779	928	755	612	548	470	243	51	4935
1983-84	17	0	143	366	630	1224	893	697	559	470	282	98	5379
1984-85	0	7	158	510	679	1064	1144	901	698	340	198	47	5746
1985-86	0	12	246	499	1076	1285	791	721	496	430	248	25	5829
1986-87	27	3	192	338	729	932	1016	680	570	255	137	31	4910
1987-88	9	3	35	323	635	961	976	681	634	344	215	87	4903
1988-89	12	0	122	208	666	948	884	1071	572	260	184	18	4945
1989-90	3	14	29	322	531	833	709	703	508	233	198	56	4139
1990-91	0	3	0	371	530	1059	940	478	601	383	276	145	4786
1991-92	3	0	45	441	739	842	831	656	488	328	108	40	4521
1992-93	4	39	122	338	737	993	1172	894	568	439	124	81	5511
1993-94	37	25	105	325	886	891							

COOLING DEGREE DAYS Base 65 deg. F LEWISTON, IDAHO

YEAR	JAN	FEB	MAR	APR	MAY	JUNE	JULY	AUG	SEP	OCT	NOV	DEC	TOTAL
1969	0	0	0	0	31	186	247	228	73	0	0	0	765
1970	0	0	0	0	16	220	354	320	19	3	0	0	932
1971	0	0	0	0	35	42	323	424	20	3	0	0	847
1972	0	0	0	0	52	114	277	321	46	0	0	0	810
1973	0	0	0	0	60	144	382	316	79	0	0	0	981
1974	0	0	0	0	7	228	295	277	98	1	0	0	906
1975	0	0	0	0	13	52	405	194	92	5	0	0	761
1976	0	0	0	0	15	62	288	178	139	5	0	0	687
1977	0	0	0	34	10	234	275	410	49	0	0	0	1012
1978	0	0	0	0	0	112	296	217	51	0	0	0	676
1979	0	0	0	0	15	131	360	315	140	7	0	0	968
1980	0	0	0	10	15	37	271	160	51	10	0	0	554
1981	0	0	0	7	8	32	204	403	139	0	0	0	793
1982	0	0	0	0	5	190	260	298	89	1	0	0	843
1983	0	0	0	0	77	63	197	369	20	0	0	0	726
1984	0	0	0	3	11	69	325	322	59	7	0	0	796
1985	0	0	0	0	59	112	468	157	7	0	0	0	803
1986	0	0	0	1	104	210	155	404	31	0	0	0	905
1987	0	0	0	21	62	188	216	224	141	10	0	0	862
1988	0	0	0	2	29	145	293	281	105	19	0	0	874
1989	0	0	0	0	17	131	305	219	66	0	0	0	738
1990	0	0	0	0	12	140	393	337	245	8	0	0	1135
1991	0	0	0	0	0	11	234	317	69	2	0	0	633
1992	0	0	0	0	55	234	243	305	50	8	0	0	895
1993	0	0	0	0	69	55	56	194	105	2	0	0	481

SNOWFALL (inches) LEWISTON, IDAHO

SEASON	JULY	AUG	SEP	OCT	NOV	DEC	JAN	FEB	MAR	APR	MAY	JUNE	TOTAL
1964-65	0.0	0.0	0.0	0.0	4.6	16.2	10.8	1.0	1.8	T	T	0.0	34.4
1965-66	0.0	0.0	0.0	0.0	T	0.9	16.7	4.0	2.0	0.0	0.0	0.0	23.6
1966-67	0.0	0.0	0.0	0.0	0.3	0.1	1.3	1.0	0.3	T	0.0	0.0	3.0
1967-68	0.0	0.0	0.0	0.0	0.8	8.8	2.7	T	T	T	T	0.0	12.3
1968-69	0.0	0.0	0.0	0.0	T	18.7	22.9	3.7	0.0	T	0.0	0.0	45.3
1969-70	0.0	0.0	0.0	0.0	T	5.5	8.2	0.3	3.5	T	0.0	0.0	17.5
1970-71	0.0	0.0	0.0	0.0	1.2	0.4	7.3	1.2	2.4	T	0.0	0.0	12.5
1971-72	0.0	0.0	0.0	2.5	T	5.9	6.0	2.6	T	1.1	0.0	0.0	18.1
1972-73	0.0	0.0	0.0	0.0	T	3.7	3.6	4.5	T	T	T	0.0	11.8
1973-74	0.0	0.0	0.0	0.9	3.2	6.6	5.3	T	0.2	T	T	0.0	16.2
1974-75	0.0	0.0	0.0	0.0	0.0	0.4	5.0	11.3	T	T	T	0.0	16.7
1975-76	0.0	0.0	0.0	0.0	1.1	1.4	2.3	T	0.4	0.3	0.0	0.0	5.5
1976-77	0.0	0.0	0.0	0.0	0.2	0.1	3.5	0.2	2.0	0.0	T	0.0	6.0
1977-78	0.0	0.0	0.0	T	9.3	9.0	3.7	1.2	0.6	0.0	0.0	0.0	23.8
1978-79	0.0	0.0	0.0	0.0	5.2	3.1	11.6	0.1	T	0.1	0.0	0.0	20.1
1979-80	0.0	0.0	0.0	0.0	1.6	T	15.5	2.0	2.4	T	0.0	0.0	21.5
1980-81	0.0	0.0	0.0	0.0	1.9	2.3	T	3.2	0.0	T	0.0	0.0	7.4
1981-82	0.0	0.0	0.0	0.0	T	5.1	8.7	2.2	T	T	0.0	0.0	16.0
1982-83	0.0	0.0	0.0	0.0	T	2.5	T	T	0.0	0.0	0.0	0.0	2.5
1983-84	0.0	0.0	0.0	0.0	T	12.1	0.5	T	0.0	0.0	T	0.0	12.6
1984-85	0.0	0.0	0.0	T	0.8	10.0	3.9	13.9	1.1	T	0.0	0.0	29.7
1985-86	0.0	0.0	0.0	T	6.2	3.0	0.8	3.0	0.0	T	0.0	0.0	13.0
1986-87	0.0	0.0	0.0	0.0	1.6	T	2.3	T	0.0	0.0	0.0	0.0	3.9
1987-88	0.0	0.0	0.0	0.0	T	1.2	3.5	T	T	T	0.0	0.0	4.7
1988-89	0.0	0.0	0.0	0.0	7.3	4.5	4.8	5.2	6.7	0.0	0.0	0.0	28.5
1989-90	0.0	0.0	0.0	0.0	0.0	2.0	T	1.4	T	0.0	T	0.0	3.4
1990-91	0.0	0.0	0.0	T	T	5.7	T	0.0	2.0	T	0.0	0.0	7.7
1991-92	T	0.0	0.0	T	T	0.0	T	0.0	0.0	0.0	0.0	0.0	T
1992-93	0.0	0.0	0.0	0.0	3.5	2.8	11.0	5.1	0.8	T	0.0	0.0	23.2
1993-94	0.0	0.0	0.0	0.0	3.1	T							
Record Mean	T	0.0	0.0	0.1	1.8	4.2	6.1	2.5	1.4	0.1	T	0.0	16.1

POCATELLO, IDAHO

Pocatello is located in the Snake River Valley at the mouth of Portneuf Canyon at an elevation of about 4,500 feet above sea level. A desert composed of sand, lava rock, and craters, extends to the west, while to the east the ground level rises steadily towards the crests of the Continental Divide. Agriculture, which is practiced extensively in the Snake River Valley, depends upon irrigation for all crops because rainfall during the growing season is insufficient.

Except in autumn, which is the season of the finest weather in Pocatello, the main feature of the climate is its variety. In winter there are frequent periods of persistent southwest wind, with a resulting mildness that matches the winters of the north Pacific Coast. There are also periods of several days when the temperature stays below freezing and approaches or falls below zero.

During cold periods, precipitation falling as snow occasionally accumulates to a depth of a foot or more. Cloudy and unsettled weather prevails throughout the winter, with measurable amounts of precipitation on about one-third of the days.

In the spring there is a gradual warming. Normally, spring months are the wettest and windiest. Winds of 20 to 30 mph for days at a time are common.

The summer season begins with a relatively sudden break in the disagreeable spring weather. Home heating, usually discontinued about the first of June, is sometimes needed intermittently until the first part of July. Suitable weather for outside evening activities is very uncertain during June, even though the afternoons may be mild. As night falls, the temperature often drops rapidly into the 40s, accompanied by a chilling wind from snows remaining on the nearby mountains. During the summer, precipitation usually falls as local showers, often accompanied by light to moderate thunderstorms and occasionally by hail. Damage by excessive precipitation, lightning, high winds, or hail is uncommon and quite localized. Long periods of extremely hot weather in July and August are also uncommon. Although afternoon temperatures may run into the 90s, nights are usually cool.

Exceptionally fine weather predominates during the autumn season. The sudden summer showers are gradually replaced by short periods of cloudy and unsettled weather with more general rains. Continuous home heating is not needed until mid-October. Evenings during September are ideal for outdoor activities, and pleasant afternoons are the rule until toward the end of November. The first cold wave may appear during late November but usually not until late December.

Based on the 1951-1980 period, the average first occurrence of 32 degrees Fahrenheit in the fall is September 20 and the average last occurrence in the spring is May 20.

NORMALS, MEANS, AND EXTREMES

POCATELLO, IDAHO

LATITUDE: 42°55'N LONGITUDE: 112°36' W ELEVATION: FT. GRND 4454 BARO 4464 TIME ZONE: MOUNTAIN WBAN: 24156

	(a)	JAN	FEB	MAR	APR	MAY	JUNE	JULY	AUG	SEP	OCT	NOV	DEC	YEAR
TEMPERATURE °F:														
Normals														
-Daily Maximum		32.2	38.4	46.7	57.5	67.5	78.0	88.1	86.3	75.1	62.5	45.2	33.7	59.3
-Daily Minimum		14.4	19.8	25.9	32.3	39.6	47.3	53.0	50.9	42.8	33.5	26.0	15.8	33.4
-Monthly		23.3	29.1	36.3	44.9	53.6	62.7	70.6	68.7	59.0	48.0	35.6	24.8	46.4
Extremes														
-Record Highest	44	57	65	75	86	93	103	102	104	98	91	71	59	104
-Year		1974	1992	1986	1992	1954	1988	1991	1990	1976	1992	1975	1980	AUG 1990
-Record Lowest	44	-30	-33	-12	15	20	30	34	30	19	10	-14	-29	-33
-Year		1962	1985	1985	1970	1972	1989	1981	1992	1985	1971	1993	1990	FEB 1985
NORMAL DEGREE DAYS:														
Heating (base 65°F)		1293	1005	890	603	353	125	9	29	218	527	882	1246	7180
Cooling (base 65°F)		0	0	0	0	0	56	183	144	38	0	0	0	421
% OF POSSIBLE SUNSHINE	44	40	53	61	66	68	75	83	81	79	71	47	40	64
MEAN SKY COVER (tenths)														
Sunrise - Sunset	44	8.0	7.5	7.1	6.7	6.2	4.9	3.5	3.8	4.1	5.1	7.2	7.9	6.0
MEAN NUMBER OF DAYS:														
Sunrise to Sunset														
-Clear	44	2.7	4.0	5.0	6.3	7.5	11.7	17.4	15.4	15.2	12.0	5.1	3.4	105.5
-Partly Cloudy	44	6.7	6.4	8.2	7.9	9.9	9.6	9.0	10.6	8.1	8.4	7.2	6.8	98.8
-Cloudy	44	21.7	17.9	17.8	15.8	13.6	8.7	4.6	5.0	6.8	10.6	17.7	20.8	160.9
Precipitation .01 inches or more	44	12.0	10.3	10.2	8.3	9.2	6.9	4.2	4.5	4.7	5.3	9.2	10.9	95.8
Snow, Ice pellets, hail 1.0 inches or more	44	3.3	2.1	2.1	1.3	0.2	0.0	0.0	0.0	0.*	0.5	1.9	3.0	14.4
Thunderstorms	43	0.*	0.1	0.4	1.0	3.4	4.7	5.6	5.1	2.7	0.5	0.1	0.1	23.8
Heavy Fog Visibility 1/4 mile or less	43	4.6	3.3	1.6	0.4	0.3	0.*	0.*	0.0	0.1	0.4	1.9	4.3	16.9
Temperature °F														
-Maximum														
90° and above	30	0.0	0.0	0.0	0.0	0.0	4.2	14.4	12.3	1.9	0.*	0.0	0.0	32.8
32° and below	30	15.0	7.5	2.0	0.*	0.0	0.0	0.0	0.0	0.0	0.2	4.0	12.7	41.4
-Minimum														
32° and below	30	28.1	25.1	24.5	15.6	4.2	0.3	0.0	0.1	3.3	14.7	23.4	28.5	167.9
0° and below	30	5.6	2.2	0.3	0.0	0.0	0.0	0.0	0.0	0.0	0.0	0.9	3.9	12.8
AVG. STATION PRESS. (mb)	20	865.6	864.4	860.9	861.4	861.1	862.3	863.7	863.9	864.3	865.3	864.7	865.5	863.6
RELATIVE HUMIDITY (%)														
Hour 05	30	79	80	77	70	70	71	65	62	65	70	77	79	72
Hour 11 (Local Time)	30	75	70	60	48	44	42	35	34	38	47	66	75	53
Hour 17	30	71	62	51	38	34	32	24	23	28	37	60	70	44
Hour 23	27	78	76	69	58	56	54	47	44	50	58	73	78	62
PRECIPITATION (inches):														
Water Equivalent														
-Normal		1.04	0.92	1.26	1.20	1.35	1.02	0.65	0.67	0.85	0.91	1.16	1.11	12.14
-Maximum Monthly	44	3.24	2.63	2.95	3.30	3.29	3.30	2.28	3.98	3.43	2.56	2.84	3.39	3.98
-Year		1980	1986	1983	1963	1980	1967	1984	1968	1982	1956	1983	1983	AUG 1968
-Minimum Monthly	44	0.21	0.12	0.10	0.06	0.19	0.02	T	T	0.00	0.00	0.01	0.07	0.00
-Year		1992	1970	1965	1977	1969	1974	1988	1958	1987	1988	1976	1989	OCT 1988
-Maximum in 24 hrs	44	0.97	0.67	1.27	1.25	1.67	1.18	0.98	1.16	1.13	1.82	0.85	0.94	1.82
-Year		1970	1983	1990	1976	1970	1993	1965	1972	1982	1976	1969	1983	OCT 1976
Snow, Ice pellets, hail														
-Maximum Monthly	44	29.6	21.3	16.6	15.5	5.5	0.2	0.0	T	2.0	12.6	27.5	33.7	33.7
-Year		1993	1993	1985	1976	1983	1981		1993	1965	1971	1985	1983	DEC 1983
-Maximum in 24 hrs	44	10.1	6.4	8.4	10.0	5.2	0.2	0.0	T	2.0	8.0	7.8	10.8	10.8
-Year		1950	1984	1985	1976	1983	1981		1993	1965	1980	1992	1988	DEC 1988
WIND:														
Mean Speed (mph)	41	10.6	10.6	11.2	11.7	10.6	10.1	9.1	8.9	9.1	9.3	10.3	9.9	10.1
Prevailing Direction through 1963		SW	SW	SW	SW	SW	SW	SW	SW	SW	SW	SW	SW	SW
Fastest Mile														
-Direction (!!!)	44	SE	W	W	S	W	W	W	SW	W	SW	W	NW	W
-Speed (MPH)	44	61	57	72	61	61	50	57	54	57	54	67	57	72
-Year		1952	1963	1955	1956	1953	1955	1968	1966	1961	1966	1955	1981	MAR 1955
Peak Gust														
-Direction (!!!)	10	W	S	SW	SW	SW	SW	W	SW	NW	S	S	S	W
-Speed (mph)	10	68	60	64	55	61	55	66	68	51	51	58	54	68
-Date		1990	1985	1987	1988	1991	1986	1986	1985	1989	1991	1988	1987	JAN 1990

PRECIPITATION (inches)

POCATELLO, IDAHO

YEAR	JAN	FEB	MAR	APR	MAY	JUNE	JULY	AUG	SEP	OCT	NOV	DEC	ANNUAL
1964	0.76	0.34	0.78	1.46	1.41	2.08	0.30	0.24	T	0.50	0.85	2.95	11.67
1965	0.71	0.38	0.10	1.65	0.84	1.12	1.19	0.78	0.85	0.02	1.21	1.20	10.05
1966	0.27	0.51	0.42	0.31	0.93	0.21	0.10	0.02	0.82	0.40	0.63	0.72	5.34
1967	1.08	0.18	1.21	1.81	1.14	3.30	0.22	0.13	0.14	0.80	0.51	0.91	11.43
1968	0.75	1.11	1.29	0.39	1.70	1.84	0.10	3.98	0.64	0.58	0.92	0.78	14.08
1969	1.80	0.94	0.40	0.34	0.19	2.02	0.14	0.54	0.14	0.56	0.87	1.10	9.04
1970	1.98	0.12	1.11	1.28	2.03	1.18	1.25	0.11	0.80	0.68	1.60	1.13	13.27
1971	1.47	0.62	1.90	2.33	0.50	1.26	0.32	0.58	2.06	2.29	1.81	1.69	16.83
1972	1.45	0.92	0.61	1.36	0.54	1.29	0.56	1.36	1.14	1.39	0.44	1.89	12.95
1973	1.04	1.00	1.45	0.70	0.44	0.87	1.84	0.13	2.29	1.19	1.87	0.84	13.66
1974	1.55	0.66	1.66	1.40	1.28	0.02	0.14	0.15	0.07	1.99	0.77	0.93	10.62
1975	0.65	1.51	1.71	1.51	1.64	0.73	1.61	0.02	0.03	2.54	0.95	0.59	13.49
1976	0.45	1.42	1.08	2.82	0.51	0.98	0.77	0.74	0.45	1.83	0.01	0.20	11.26
1977	0.70	0.37	0.79	0.06	2.09	0.77	0.58	0.32	0.99	0.18	0.94	1.08	8.87
1978	1.17	1.07	1.16	1.24	1.99	0.22	0.03	0.41	0.87	0.06	1.39	0.74	10.35
1979	1.09	1.17	0.79	0.63	0.33	0.96	0.59	0.70	0.23	0.79	1.12	0.40	8.80
1980	3.24	1.12	1.52	0.55	3.29	0.86	0.72	1.25	1.34	1.88	1.13	0.37	17.27
1981	0.72	0.47	1.69	1.27	3.24	0.68	0.29	0.32	0.08	1.52	1.67	2.22	14.17
1982	1.40	0.91	1.98	1.02	1.06	0.47	1.27	2.02	3.43	0.99	1.17	2.00	17.72
1983	0.48	1.12	2.95	0.72	1.99	1.19	1.13	1.67	1.64	1.21	2.84	3.39	20.33
1984	0.49	2.06	0.63	2.13	0.69	1.33	2.28	0.71	0.50	0.89	0.96	0.53	13.20
1985	0.71	0.96	1.55	0.29	1.27	1.04	0.48	0.05	1.42	0.50	2.37	1.17	11.81
1986	1.07	2.63	0.87	2.56	1.90	0.33	0.12	0.14	1.33	0.39	0.89	0.20	12.43
1987	1.03	0.71	0.84	0.47	2.02	0.67	1.93	0.66	0.00	0.25	0.44	1.21	10.23
1988	1.06	0.21	0.83	0.97	0.78	0.37	T	0.20	0.04	0.00	2.27	1.19	7.92
1989	0.59	1.36	2.91	0.33	0.76	0.62	0.11	0.43	0.48	0.84	1.03	0.07	9.53
1990	0.53	0.24	2.28	1.43	1.37	0.75	0.14	0.70	0.33	0.31	0.92	1.15	10.15
1991	0.71	0.28	2.25	2.56	3.02	0.59	0.12	0.85	0.66	0.74	1.65	0.22	13.65
1992	0.21	0.65	0.47	0.72	0.25	1.34	0.71	0.37	0.08	0.98	1.37	1.82	8.97
1993	2.30	2.04	1.68	1.54	1.93	2.95	1.07	1.90	0.27	1.01	0.54	0.65	17.88
Record Mean	1.18	1.01	1.23	1.27	1.36	1.05	0.65	0.70	0.81	0.98	1.03	1.06	12.32

AVERAGE TEMPERATURE (deg. F)

POCATELLO, IDAHO

YEAR	JAN	FEB	MAR	APR	MAY	JUNE	JULY	AUG	SEP	OCT	NOV	DEC	ANNUAL
1964	19.2	16.9	27.8	42.7	53.6	60.0	72.6	67.5	57.3	49.3	35.0	29.1	44.2
1965	29.8	30.8	32.5	46.8	50.5	60.6	68.6	65.8	51.8	50.6	40.6	25.6	46.1
1966	25.7	26.5	37.1	44.2	56.9	62.2	72.2	67.6	61.7	44.6	37.7	24.0	46.7
1967	29.8	33.1	37.5	40.7	52.9	59.5	71.5	71.0	61.7	47.3	35.4	19.7	46.7
1968	21.2	32.6	39.5	41.7	51.7	61.1	71.6	63.0	56.5	46.0	33.4	24.8	45.3
1969	29.1	25.7	29.2	44.9	58.1	59.9	69.7	70.5	62.1	42.7	35.5	28.5	46.3
1970	29.9	37.2	36.1	39.0	53.6	63.1	71.0	72.1	53.7	43.0	37.6	25.0	46.8
1971	27.2	27.8	32.6	42.6	52.7	60.2	68.9	71.1	52.7	42.8	32.7	22.6	44.5
1972	22.9	29.0	41.0	43.0	53.8	62.3	68.6	69.5	55.6	47.8	36.0	18.3	45.6
1973	19.8	27.0	35.9	43.4	55.7	63.4	70.6	69.8	57.3	48.7	36.7	29.6	46.5
1974	22.5	29.8	38.5	45.5	53.3	66.6	71.4	66.8	59.4	48.6	37.1	24.7	47.0
1975	22.4	28.3	35.5	39.9	50.8	61.5	73.8	66.6	60.0	48.1	33.4	32.5	46.1
1976	26.9	29.4	32.6	44.3	57.7	61.6	72.0	65.9	61.1	46.8	37.7	28.4	47.0
1977	17.1	27.5	34.9	49.9	51.7	68.9	70.8	68.9	59.7	50.4	37.1	32.9	47.5
1978	31.1	33.6	42.9	45.9	51.0	61.8	70.1	66.9	58.1	49.5	31.8	20.0	46.9
1979	10.7	27.9	36.1	44.0	54.4	61.7	71.4	69.2	64.7	51.1	28.4	29.3	45.7
1980	24.7	34.5	36.2	48.0	51.4	59.2	68.6	65.2	58.9	46.2	34.7	31.7	46.6
1981	28.5	29.5	39.0	47.5	51.6	61.8	69.6	72.0	61.4	45.4	38.1	30.8	48.0
1982	20.2	23.5	37.7	41.3	51.8	61.8	67.9	70.8	58.3	45.2	31.9	25.3	44.7
1983	31.5	34.0	40.3	42.3	51.5	60.7	67.0	71.1	60.2	49.3	36.3	19.9	47.0
1984	17.7	20.3	34.2	42.8	54.5	59.0	70.1	70.3	58.3	43.8	34.8	19.6	43.8
1985	12.9	18.5	27.5	49.0	56.9	64.1	73.0	66.2	54.2	45.5	26.7	11.4	42.2
1986	23.4	35.5	44.2	45.0	52.8	67.4	67.1	70.5	54.1	48.0	36.0	24.5	47.4
1987	19.2	33.6	39.3	51.5	57.7	65.2	67.7	67.2	60.7	50.4	36.3	26.0	47.9
1988	21.3	32.8	37.6	48.6	54.2	69.6	73.3	68.7	58.0	55.3	35.3	21.8	48.0
1989	19.4	18.9	38.3	49.2	53.4	62.1	73.1	66.8	59.8	46.8	37.0	26.7	46.0
1990	30.0	28.3	40.0	50.1	51.8	62.5	70.9	69.0	65.7	47.8	37.7	14.8	47.4
1991	19.4	36.7	38.5	44.4	51.1	61.1	70.9	72.4	60.0	47.6	32.8	26.4	46.8
1992	26.0	38.0	44.1	51.3	60.2	63.9	67.4	69.6	59.3	51.4	28.8	22.1	48.5
1993	18.1	20.5	36.1	44.4	57.6	57.6	61.1	64.5	57.5	47.4	27.6	26.1	43.2
Record Mean	24.0	29.3	37.0	46.0	54.4	62.8	71.4	69.6	59.7	48.8	36.1	26.6	47.2
Max	32.5	38.0	46.8	57.8	67.5	77.3	87.7	85.7	75.0	62.3	46.0	35.0	59.3
Min	15.5	20.6	27.1	34.1	41.3	48.3	55.1	53.4	44.3	35.3	26.2	18.3	35.0

REFERENCE NOTES

POCATELLO, IDAHO

GENERAL
T - TRACE AMOUNT.
BLANK ENTRIES DENOTE MISSING/UNREPORTED DATA.
# INDICATES A STATION OR INSTRUMENT RELOCATION.

SPECIFIC

(■) - LENGTH OF RECORD IN YEARS, ALTHOUGH INDIVIDUAL MONTHS MAY BE MISSING.
0.* OR * - THE VALUE IS BETWEEN 0.0 AND 0.05
NORMALS - BASED ON THE 1961-1990 RECORD PERIOD.
EXTREMES - DATES ARE THE MOST RECENT OCCURRENCE
WIND DIR.- NUMERALS SHOW TENS OF DEGRESS CLOCKWISE FROM TRUE NORTH. "00" INDICATES CALM.
RESULTANT DIRECTIONS ARE GIVEN TO WHOLE DEGREES.
BOLD VALUES INDICATE EXTREME VALUES WHICH OCCURRED AFTER THE ASOS SYSTEM WAS COMMISSIONED.

RECORD = PERIOD OF RECORD
RECORD MEAN PRECIPITATION IS THE MEAN OF ALL DAILY PRECIPITATION AMOUNTS DURING THE PERIOD OF RECORD.
RECORD MAX(MIN) TEMPERATURE IS THE MEAN OF ALL DAILY MAX(MIN) TEMPERATURES DURING THE PERIOD OF RECORD.
RECORD MEAN TEMPERATURE IS THE SUM OF THE RECORD MAX AND RECORD MIN DIVIDED BY 2.
AVERAGE TEMPERATURE IS THE SUM OF THE MEAN DAILY MAX AND MIN TEMPERATURE DIVIDED BY 2.

EXCEPTIONS

RECORD MEANS ARE THROUGH THE CURRENT YEAR, BEGINNING IN 1900 FOR TEMPERATURE
1900 FOR PRECIPITATION
1950 FOR SNOWFALL

HEATING DEGREE DAYS Base 65 deg. F — POCATELLO, IDAHO

SEASON	JULY	AUG	SEP	OCT	NOV	DEC	JAN	FEB	MAR	APR	MAY	JUNE	TOTAL
1964-65	0	68	223	478	894	1105	1082	952	999	543	443	143	6930
1965-66	16	60	393	440	725	1216	1211	1072	861	618	253	140	7005
1966-67	2	26	133	623	812	1263	1083	888	846	720	372	178	6946
1967-68	0	3	128	545	883	1398	1352	931	783	689	405	154	7271
1968-69	9	143	260	581	942	1240	1105	1094	1104	594	223	167	7462
1969-70	24	17	105	685	878	1124	1080	772	888	771	348	141	6833
1970-71	4	2	333	675	817	1236	1166	1034	997	666	376	168	7474
1971-72	19	13	374	684	964	1309	1300	1039	735	654	346	101	7538
1972-73	30	9	281	527	860	1445	1394	1057	895	643	290	135	7566
1973-74	12	18	237	500	843	1089	1313	979	815	580	359	78	6823
1974-75	11	42	173	500	829	1246	1312	1021	908	745	435	122	7344
1975-76	11	40	168	521	943	1001	1172	1026	999	615	219	140	6855
1976-77	1	34	143	561	811	1127	1478	1043	926	445	406	6	6981
1977-78	1	38	189	447	831	990	1045	875	678	567	431	119	6211
1978-79	10	59	241	473	990	1387	1678	1034	887	623	320	155	7857
1979-80	1	10	66	422	1090	1100	1245	876	888	507	415	184	6804
1980-81	5	57	182	576	899	1026	1125	988	801	516	407	127	6709
1981-82	16	9	130	603	800	1051	1382	1161	838	703	406	139	7238
1982-83	38	0	216	607	987	1227	1029	863	760	675	415	141	6958
1983-84	61	3	163	478	854	1392	1461	1291	949	658	333	205	7848
1984-85	5	7	221	648	897	1400	1612	1298	1157	472	246	85	8048
1985-86	0	44	321	597	1143	1657	1281	818	636	592	389	35	7513
1986-87	28	5	326	519	864	1247	1410	875	792	398	231	64	6759
1987-88	46	31	144	445	854	1203	1348	926	843	485	340	38	6703
1988-89	1	3	229	295	888	1330	1408	1292	820	469	355	113	7203
1989-90	0	47	159	556	831	1183	1079	1022	766	438	399	139	6619
1990-91	3	20	53	526	813	1555	1406	786	813	610	425	122	7132
1991-92	3	1	176	533	960	1188	1201	776	639	407	154	97	6135
1992-93	23	63	175	417	1078	1323	1445	1240	890	612	226	235	7727
1993-94	136	67	219	539	1115	1200							

COOLING DEGREE DAYS Base 65 deg. F — POCATELLO, IDAHO

YEAR	JAN	FEB	MAR	APR	MAY	JUNE	JULY	AUG	SEP	OCT	NOV	DEC	TOTAL
1969	0	0	0	0	15	23	177	195	25	0	0	0	435
1970	0	0	0	0	0	91	197	230	4	0	0	0	522
1971	0	0	0	0	0	29	146	212	9	0	0	0	396
1972	0	0	0	0	7	26	150	152	6	0	0	0	341
1973	0	0	0	0	7	94	194	174	14	0	0	0	483
1974	0	0	0	0	1	130	215	102	12	0	0	0	460
1975	0	0	0	0	2	23	290	98	23	4	0	0	440
1976	0	0	0	0	1	46	227	70	28	0	0	0	372
1977	0	0	0	0	1	129	190	164	35	0	0	0	519
1978	0	0	0	0	0	28	177	123	41	0	0	0	369
1979	0	0	0	0	0	64	207	145	63	0	0	0	479
1980	0	0	0	3	0	17	125	70	6	0	0	0	221
1981	0	0	0	0	0	38	168	232	31	0	0	0	469
1982	0	0	0	0	2	48	133	187	16	0	0	0	386
1983	0	0	0	0	4	19	131	199	25	0	0	0	378
1984	0	0	0	0	15	33	168	178	25	0	0	0	419
1985	0	0	0	0	5	65	257	90	3	0	0	0	420
1986	0	0	0	0	17	110	100	185	4	0	0	0	416
1987	0	0	0	0	8	74	134	104	23	0	0	0	343
1988	0	0	0	0	8	181	261	123	25	0	0	0	598
1989	0	0	0	1	4	34	255	109	10	0	0	0	413
1990	0	0	0	0	0	68	194	152	79	0	0	0	493
1991	0	0	0	0	0	12	192	237	32	0	0	0	473
1992	0	0	0	0	15	73	103	212	11	2	0	0	416
1993	0	0	0	0	3	16	24	57	2	1	0	0	103

SNOWFALL (inches) — POCATELLO, IDAHO

SEASON	JULY	AUG	SEP	OCT	NOV	DEC	JAN	FEB	MAR	APR	MAY	JUNE	TOTAL
1964-65	0.0	0.0	0.0	0.0	3.7	6.9	4.7	3.7	0.9	1.2	0.4	0.0	21.5
1965-66	0.0	0.0	2.0	0.0	2.0	12.1	4.8	8.7	1.5	1.3	T	0.0	32.4
1966-67	0.0	0.0	0.0	1.1	2.5	5.5	3.3	3.5	5.7	13.9	0.3	0.0	35.8
1967-68	0.0	0.0	0.0	T	4.4	13.9	11.6	4.7	5.9	1.1	T	0.0	41.6
1968-69	0.0	0.0	T	0.0	5.1	17.9	7.7	15.2	5.7	T	0.0	0.0	51.6
1969-70	0.0	0.0	0.0	T	T	7.1	1.9	0.5	9.5	11.5	T	0.0	30.5
1970-71	0.0	0.0	T	0.1	1.4	9.9	13.1	5.9	5.2	0.3	T	T	35.9
1971-72	0.0	0.0	T	12.6	9.6	17.6	10.6	6.9	1.0	0.9	0.0	0.0	59.2
1972-73	0.0	0.0	0.0	0.5	2.7	17.4	12.9	7.0	8.7	4.1	0.0	0.0	53.3
1973-74	0.0	0.0	0.0	2.0	8.7	5.9	8.8	5.9	6.5	12.0	T	T	49.8
1974-75	0.0	0.0	T	2.3	3.7	7.7	7.4	8.8	8.3	11.4	4.9	T	54.5
1975-76	0.0	0.0	0.0	7.6	8.6	3.1	4.5	13.6	7.0	15.5	0.0	T	59.9
1976-77	0.0	0.0	0.0	0.0	0.2	1.5	10.3	0.6	7.6	0.6	1.1	0.0	21.9
1977-78	0.0	0.0	0.0	0.3	2.6	4.2	5.8	7.8	3.6	0.4	0.9	0.0	25.6
1978-79	0.0	0.0	T	0.0	11.5	7.6	14.5	9.7	3.1	3.7	0.3	T	50.4
1979-80	0.0	0.0	0.0	0.8	8.9	2.5	12.7	1.9	7.2	1.5	T	0.0	35.5
1980-81	0.0	0.0	0.0	8.0	2.7	2.4	5.9	2.9	6.3	0.3	1.0	0.2	29.7
1981-82	0.0	0.0	0.0	1.9	2.3	17.9	22.3	5.8	13.6	2.4	0.2	0.0	66.4
1982-83	0.0	0.0	T	0.5	7.3	18.3	6.2	4.9	10.5	5.3	5.5	0.0	58.5
1983-84	0.0	0.0	1.0	0.0	10.7	33.7	6.5	20.1	4.3	9.1	0.2	0.0	85.6
1984-85	0.0	0.0	T	3.7	4.3	5.7	9.3	11.9	16.6	1.3	0.0	0.0	52.8
1985-86	0.0	0.0	T	1.4	27.5	11.5	5.7	6.0	3.7	11.5	1.8	0.0	69.1
1986-87	0.0	0.0	T	0.0	8.6	1.7	13.1	4.5	2.7	2.1	0.0	0.0	32.7
1987-88	0.0	0.0	0.0	0.0	1.6	8.4	8.9	1.1	3.3	1.5	2.6	0.0	27.4
1988-89	0.0	0.0	0.0	0.0	7.7	17.0	6.2	12.0	5.9	0.3	0.5	0.0	49.6
1989-90	0.0	0.0	0.0	8.2	2.5	1.2	3.5	3.1	14.7	T	1.6	0.0	34.8
1990-91	0.0	T	0.0	0.1	3.8	15.9	8.0	1.5	8.3	9.8	1.8	0.0	49.2
1991-92	0.0	0.0	0.0	2.5	9.1	1.9	3.1	0.9	0.6	0.4	0.0	0.0	18.5
1992-93	0.0	0.0	0.0	0.1	15.2	19.9	29.6	21.3	1.5	5.5	0.2	T	93.3
1993-94	0.0	T	0.0	0.0	6.2	6.3							
Record Mean	0.0	T	0.1	1.9	5.3	8.9	10.0	6.2	5.9	4.4	0.6	T	43.2

CHICAGO, OHARE INTERNATIONAL AIRPORT, ILLINOIS

Chicago is located along the southwest shore of Lake Michigan and occupies a plain which, for the most part, is only some tens of feet above the lake. Lake Michigan averages 579 feet above sea level. Natural water drainage over most of the city would be into Lake Michigan, and from areas west of the city is into the Mississippi River System. But actual drainage over most of the city is artificially channeled also into the Mississippi system. Topography does not significantly affect air flow in or near the city except that lesser frictional drag over Lake Michigan causes winds to be frequently stronger along the lakeshore, and often permits air masses moving from the north to reach shore areas an hour or more before affecting western parts of the city.

Chicago is in a region of frequently changeable weather. The climate is predominately continental, ranging from relatively warm in summer to relatively cold in winter. However, the continentality is partially modified by Lake Michigan, and to a lesser extent by other Great Lakes. In late autumn and winter, air masses that are initially very cold often reach the city only after being tempered by passage over one or more of the lakes. Similarly, in late spring and summer, air masses reaching the city from the north, northeast, or east are cooler because of movement over the Great Lakes. Very low winter temperatures most often occur in air that flows southward to the west of Lake Superior before reaching the Chicago area. In summer the higher temperatures are with south or southwest flow and are therefore not influenced by the lakes, the only modifying effect being a local lake breeze. Strong south or southwest flow may overcome the lake breeze and cause high temperatures to extend over the entire city.

During the warm season, when the lake is cold relative to land, there is frequently a lake breeze that reduces daytime temperature near the shore, sometimes by 10 degrees or more below temperatures farther inland. When the breeze off the lake is light this effect usually reaches inland only a mile or two, but with stronger on-shore winds the whole city is cooled. On the other hand, temperatures at night are warmer near the lake so that 24-hour averages on the whole are only slightly different in various parts of the city and suburbs.

At the OHare International Airport temperatures of 96 degrees or higher occur in about half the summers, while about half the winters have a minimum as low as -15 degrees. The average occurrence of the first temperature as low as 32 degrees in the fall is mid-October and the average occurrence of the last temperature as low as 32 degrees in the spring is late April.

Precipitation falls mostly from air that has passed over the Gulf of Mexico. But in winter there is sometimes snowfall, light inland but locally heavy near the lakeshore, with Lake Michigan as the principal moisture source. The heavy lakeshore snow occurs when initially colder air moves from the north with a long trajectory over Lake Michigan and impinges on the Chicago lakeshore. In this situation the air mass is warmed and its moisture content increased up to a height of several thousand feet. Snowfall is produced by upward currents that become stronger, because of frictional effects, when the air moves from the lake onto land. This type of snowfall therefore tends to be heavier and to extend farther inland in south-shore areas of the city and in Indiana suburbs, where the angle between wind-flow and shoreline is greatest. The effect of Lake Michigan, both on winter temperatures and lake-produced snowfall, is enhanced by non-freezing of much of the lake during the winter, even though areas and harbors are often ice-choked.

Summer thunderstorms are often locally heavy and variable, parts of the city may receive substantial rainfall and other parts none. Longer periods of continuous precipitation are mostly in autumn, winter, and spring. About one-half the precipitation in winter, and about 10 percent of the yearly total precipitation, falls as snow. Snowfall from month to month and year to year is greatly variable. There is a 50 percent likelihood that the first and last 1-inch snowfall of a season will occur by December 5 and March 20, respectively.

Channeling of winds between tall buildings often causes locally stronger gusts in the central business area. However, the nickname, windy city, is a misnomer as the average wind speed is not greater than in many other parts of the U.S.

NORMALS, MEANS, AND EXTREMES

CHICAGO, OHARE INTERNATIONAL AIRPORT, ILLINOIS

LATITUDE: 42°00'N LONGITUDE: 87°53'W ELEVATION: FT. GRND 658 BARO 697 TIME ZONE: CENTRAL WBAN: 94846

	(a)	JAN	FEB	MAR	APR	MAY	JUNE	JULY	AUG	SEP	OCT	NOV	DEC	YEAR
TEMPERATURE °F:														
Normals														
-Daily Maximum		29.0	33.5	45.8	58.6	70.1	79.6	83.7	81.8	74.8	63.3	48.4	34.0	58.6
-Daily Minimum		12.9	17.2	28.5	38.6	47.7	57.5	62.6	61.6	53.9	42.2	31.6	19.1	39.5
-Monthly		21.0	25.4	37.2	48.6	58.9	68.6	73.2	71.7	64.4	52.8	40.0	26.6	49.0
Extremes														
-Record Highest	35	65	71	88	91	93	104	102	101	99	91	78	71	104
-Year		1989	1976	1986	1980	1977	1988	1988	1991	1985	1963	1978	1982	JUN 1988
-Record Lowest	35	-27	-17	-8	7	24	36	40	41	28	17	1	-25	-27
-Year		1985	1967	1962	1982	1966	1972	1965	1965	1974	1981	1976	1983	JAN 1985
NORMAL DEGREE DAYS:														
Heating (base 65°F)		1364	1109	862	492	235	35	5	19	84	391	750	1190	6536
Cooling (base 65°F)		0	0	0	0	46	143	259	226	66	12	0	0	752
% OF POSSIBLE SUNSHINE	13	46	47	50	50	57	67	65	63	58	56	39	44	54
MEAN SKY COVER (tenths)														
Sunrise - Sunset	35	6.9	6.9	7.2	6.9	6.2	5.9	5.7	5.7	5.8	6.0	7.3	7.2	6.5
MEAN NUMBER OF DAYS:														
Sunrise to Sunset														
-Clear	35	7.0	5.8	4.9	6.1	7.2	7.2	8.2	8.9	8.6	8.7	5.3	5.9	83.8
-Partly Cloudy	35	6.2	6.5	8.3	7.6	9.9	11.5	12.3	11.5	9.7	8.8	6.4	6.2	105.0
-Cloudy	35	17.8	15.9	17.7	16.3	13.9	11.2	10.5	10.7	11.6	13.5	18.3	18.9	176.4
Precipitation														
.01 inches or more	35	11.0	9.4	12.5	12.6	10.8	9.8	9.9	9.3	9.6	9.4	10.8	11.3	126.3
Snow,Ice pellets,hail														
1.0 inches or more	35	3.3	2.6	2.1	0.4	0.0	0.0	0.0	0.0	0.0	0.1	0.6	2.4	11.6
Thunderstorms	35	0.3	0.4	2.1	4.0	4.9	6.4	6.0	5.8	4.6	1.8	1.1	0.6	37.9
Heavy Fog Visibility														
1/4 mile or less	35	1.5	1.8	2.2	0.9	1.2	0.6	0.5	0.7	0.5	0.9	1.3	1.9	13.9
Temperature °F														
-Maximum														
90° and above	35	0.0	0.0	0.0	0.*	0.7	3.7	6.5	4.4	1.7	0.1	0.0	0.0	17.2
32° and below	35	17.7	12.7	4.2	0.1	0.0	0.0	0.0	0.0	0.0	0.0	2.0	11.6	48.3
-Minimum														
32° and below	35	28.7	25.3	21.3	7.7	0.9	0.0	0.0	0.0	0.2	5.2	16.7	26.5	132.4
0° and below	35	6.5	3.0	0.3	0.0	0.0	0.0	0.0	0.0	0.0	0.0	0.0	2.6	12.4
AVG. STATION PRESS.(mb)	21	993.5	993.9	991.2	990.4	990.4	990.3	991.6	992.8	993.3	993.4	992.5	993.5	992.2
RELATIVE HUMIDITY (%)														
Hour 00	35	76	76	76	72	74	75	79	81	81	77	77	78	77
Hour 06 (Local Time)	35	77	78	79	77	77	78	82	86	85	82	81	80	80
Hour 12	35	67	65	61	55	54	55	57	57	57	56	64	70	60
Hour 18	35	72	70	65	58	54	55	58	62	63	64	71	72	64
PRECIPITATION (inches):														
Water Equivalent														
-Normal		1.53	1.36	2.69	3.64	3.32	3.78	3.66	4.22	3.82	2.41	2.92	2.47	35.82
-Maximum Monthly	35	4.11	3.46	5.91	7.69	7.14	9.96	8.33	17.10	11.44	7.36	8.22	8.56	17.10
-Year		1965	1985	1976	1983	1970	1993	1982	1987	1961	1991	1985	1982	AUG 1987
-Minimum Monthly	35	0.10	0.12	0.63	0.97	0.30	0.95	1.18	0.51	0.02	0.16	0.65	0.23	0.02
-Year		1981	1969	1981	1971	1992	1991	1977	1969	1979	1964	1976	1962	SEP 1979
-Maximum in 24 hrs	35	2.00	1.90	2.39	2.78	3.45	3.09	2.90	9.35	3.00	4.62	2.99	4.53	9.35
-Year		1960	1985	1985	1983	1981	1967	1993	1987	1978	1969	1990	1982	AUG 1987
Snow,Ice pellets,hail														
-Maximum Monthly	35	34.3	21.5	24.7	11.1	1.6	T	T	T	T	6.6	10.4	35.3	35.3
-Year		1979	1967	1965	1975	1966	1992	1992	1989	1967	1967	1959	1978	DEC 1978
-Maximum in 24 hrs	35	18.1	9.7	10.6	10.9	1.6	T	T	T	T	6.6	5.8	11.0	18.1
-Year		1967	1990	1970	1975	1966	1992	1992	1989	1967	1967	1975	1969	JAN 1967
WIND:														
Mean Speed (mph)	35	11.7	11.5	12.0	12.0	10.5	9.3	8.3	8.2	8.9	10.0	11.1	11.0	10.4
Prevailing Direction														
Fastest Obs. 1 Min.														
-Direction (!!)	35	28	25	01	24	34	24	36	32	23	20	23	26	23
-Speed (MPH)	35	47	45	54	54	52	41	55	46	58	48	51	46	58
-Year		1971	1967	1964	1965	1962	1970	1980	1960	1959	1971	1958	1970	SEP 1959
Peak Gust														
-Direction (!!)	9	W	N	SW	S	S	S	W	W	N	SW	SW	SW	SW
-Speed (mph)	9	58	54	84	69	55	63	54	64	58	49	51	53	84
-Date		1989	1990	1991	1984	1988	1990	1992	1987	1989	1990	1991	1992	MAR 1991

PRECIPITATION (inches)

CHICAGO, OHARE INTERNATIONAL AIRPORT, ILLINOIS

YEAR	JAN	FEB	MAR	APR	MAY	JUNE	JULY	AUG	SEP	OCT	NOV	DEC	ANNUAL
1964	0.72	0.52	3.45	5.22	2.26	2.86	4.23	1.95	3.96	0.16	2.90	1.51	29.74
1965	4.11	1.18	3.06	3.48	2.36	3.44	3.66	6.40	5.03	1.57	1.47	3.32	39.08
1966	1.09	1.75	2.64	6.28	4.77	2.95	2.19	1.00	0.55	2.16	4.74	1.88	32.00
1967	2.22	1.82	2.30	3.97	1.61	7.94	1.87	2.60	2.45	3.89	2.19	2.41	35.27
1968	1.77	0.87	0.90	2.31	2.99	4.15	2.03	5.32	3.88	1.04	3.70	2.77	31.73
1969	1.62	0.12	1.93	4.02	3.17	7.76	3.43	0.51	3.01	6.55	1.11	1.18	34.41
1970	0.82	0.59	2.12	4.29	7.14	7.14	4.08	1.50	8.69	2.48	2.78	1.77	43.40
1971	0.93	1.94	1.54	0.97	2.23	2.62	3.57	3.97	2.39	0.72	1.32	5.37	27.57
1972	1.01	0.73	3.45	4.77	3.02	3.55	4.97	6.97	8.14	2.92	3.05	2.89	45.47
1973	1.24	1.38	3.91	4.99	3.69	2.87	5.27	0.67	6.01	2.86	1.50	3.71	38.10
1974	3.29	2.11	2.40	4.27	5.09	4.69	2.96	2.60	1.47	1.88	2.47	2.12	35.35
1975	3.69	2.48	2.02	5.50	3.02	5.07	2.19	7.37	0.80	1.90	2.53	3.05	39.62
1976	0.85	1.87	5.91	4.05	4.03	2.93	1.44	1.29	1.49	1.41	0.65	0.64	26.56
1977	0.55	0.71	3.67	2.62	1.88	5.12	1.18	5.39	6.07	1.36	2.05	1.96	32.56
1978	1.48	0.43	1.16	3.94	2.80	6.36	4.61	1.96	6.88	1.08	2.24	4.41	37.35
1979	2.81	1.02	4.49	4.92	2.58	4.63	2.19	7.57	0.02	1.49	2.80	2.58	37.10
1980	1.04	1.24	1.96	3.41	3.22	3.42	3.56	8.54	5.65	2.09	1.10	3.43	38.66
1981	0.10	2.35	0.63	6.14	5.85	4.46	4.50	6.60	3.25	1.80	2.46	1.05	39.19
1982	2.90	0.41	4.15	2.78	2.08	1.56	8.33	3.93	1.15	1.88	6.95	8.56	44.68
1983	0.66	2.06	3.56	7.69	6.26	4.11	4.25	2.08	5.41	4.41	5.87	2.99	49.35
1984	1.15	1.39	3.00	4.11	4.49	2.02	3.19	2.10	3.84	3.15	2.64	2.92	34.00
1985	1.48	3.46	4.73	1.48	2.79	1.97	3.75	3.90	1.82	4.98	8.22	1.49	40.07
1986	0.39	2.58	1.49	1.85	3.11	3.49	4.30	1.15	7.12	3.75	1.41	1.09	31.73
1987	1.67	0.99	1.59	2.34	2.21	2.19	4.19	17.10	0.94	1.59	2.77	3.77	41.35
1988	1.88	1.29	2.15	2.08	1.19	1.05	2.74	3.29	3.79	5.05	6.45	2.40	33.36
1989	0.82	0.77	1.67	1.37	1.59	2.01	5.89	7.31	3.91	1.49	2.16	0.46	29.45
1990	1.97	2.25	3.09	1.79	6.85	4.50	2.25	7.75	1.03	4.10	5.60	1.94	43.12
1991	1.41	0.62	3.54	4.00	5.20	0.95	1.32	2.81	2.51	7.36	3.59	1.71	35.02
1992	0.87	1.39	2.67	2.21	0.30	1.35	3.77	3.56	4.31	1.79	5.41	2.49	30.12
1993	3.83	0.82	4.52	4.57	1.83	9.96	4.45	5.74	4.47	2.19	1.52	1.00	44.90
Record Mean	1.62	1.37	2.74	3.58	3.22	3.76	3.68	4.12	3.69	2.54	2.87	2.29	35.49

AVERAGE TEMPERATURE (deg. F)

CHICAGO, OHARE INTERNATIONAL AIRPORT, ILLINOIS

YEAR	JAN	FEB	MAR	APR	MAY	JUNE	JULY	AUG	SEP	OCT	NOV	DEC	ANNUAL
1964	27.7	26.6	33.7	49.1	62.7	69.0	72.1	67.7	63.3	48.0	41.4	24.7	48.9
1965	21.4	24.3	26.6	46.6	61.7	64.9	69.4	68.0	63.8	53.2	40.3	35.3	47.9
1966	16.3	26.1	39.6	45.2	53.4	68.5	74.5	69.6	62.5	51.4	42.5	27.1	48.1
1967	27.7	19.8	36.5	48.4	53.8	69.8	68.4	66.2	61.7	52.9	37.3	30.3	47.7
1968	23.8	23.6	42.7	52.3	57.0	70.2	72.0	73.7	65.5	54.7	40.0	27.8	50.3
1969	21.1	29.9	34.4	50.8	60.4	64.3	73.0	73.9	65.3	51.8	38.3	28.0	49.3
1970	16.3	26.1	34.8	51.7	61.9	69.4	74.7	72.9	65.2	55.4	40.7	30.8	50.0
1971	18.9	28.2	35.0	48.6	57.2	73.5	71.5	72.0	69.7	61.7	41.7	34.2	51.0
1972	19.6	23.6	34.0	44.8	61.0	65.7	73.6	73.8	63.5	49.3	37.7	23.9	47.6
1973	28.2	28.7	44.0	48.1	54.8	71.1	74.7	74.6	66.0	57.9	41.9	28.1	51.5
1974	24.8	27.4	38.6	52.3	56.8	65.5	73.6	70.0	60.5	52.8	40.6	30.2	49.4
1975	27.3	26.2	34.1	43.3	62.3	70.5	75.5	76.3	61.4	55.8	47.2	31.5	50.9
1976	19.9	35.2	42.8	52.3	55.9	70.1	74.0	70.8	62.7	48.3	32.4	19.4	48.6
1977	10.7	26.9	44.9	55.0	67.2	69.3	77.5	71.9	66.0	51.5	40.0	24.2	50.4
1978	15.7	16.8	31.9	47.5	58.3	67.6	72.0	72.4	68.8	51.4	40.8	25.8	47.4
1979	12.5	16.2	36.4	45.5	59.3	69.2	72.0	71.0	66.1	53.3	40.6	33.7	48.0
1980	23.4	21.5	32.6	46.5	59.7	65.3	75.7	75.7	66.0	48.4	39.9	28.0	48.6
1981	22.6	28.0	37.6	51.8	55.3	69.8	72.5	71.2	61.7	49.1	40.8	24.9	48.8
1982	12.2	21.5	35.1	44.5	64.3	62.1	74.1	68.8	62.1	53.2	39.1	36.0	47.8
1983	26.3	30.5	37.4	43.4	53.2	69.7	76.7	77.3	64.6	52.8	41.1	14.3	49.0
1984	17.1	33.9	29.5	45.8	55.5	70.3	70.3	72.8	61.1	54.7	37.9	31.0	48.3
1985	14.4	20.4	39.4	52.6	60.2	63.6	71.4	69.2	65.4	52.5	37.8	17.0	47.0
1986	22.8	24.0	40.4	51.5	59.5	66.3	74.9	68.5	66.8	53.7	36.0	30.6	49.6
1987	25.9	33.9	40.8	50.6	63.4	72.4	76.7	71.9	65.1	47.3	43.9	32.2	52.0
1988	19.8	22.7	38.1	48.2	61.0	71.7	76.8	76.8	65.9	46.1	41.7	27.7	49.7
1989	32.4	19.6	36.6	46.8	57.8	67.5	73.9	71.4	62.0	54.0	37.7	17.4	48.1
1990	33.9	31.3	41.3	49.9	56.2	69.6	71.7	71.9	65.9	51.6	44.7	28.6	51.4
1991	20.8	31.0	40.4	52.0	65.6	71.9	75.5	73.6	63.7	53.2	35.2	30.3	51.1
1992	28.1	33.3	37.5	46.1	56.9	64.9	69.3	67.0	62.7	50.4	38.3	28.6	48.6
1993	26.2	24.4	34.2	45.0	59.7	66.4	74.3	73.3	59.2	49.5	38.7	29.8	48.4
Record Mean	21.4	25.8	36.6	48.5	59.2	68.3	73.2	71.8	64.3	52.5	39.8	27.0	49.0
Max	29.3	33.7	45.2	58.5	70.3	79.5	83.6	82.0	74.7	63.0	48.0	34.4	58.5
Min	13.5	17.9	28.0	38.5	48.0	57.2	62.7	61.7	53.9	42.1	31.5	19.5	39.5

REFERENCE NOTES

CHICAGO, OHARE INTERNATIONAL AIRPORT, ILLINOIS

GENERAL
T - TRACE AMOUNT.
BLANK ENTRIES DENOTE MISSING/UNREPORTED DATA.
# INDICATES A STATION OR INSTRUMENT RELOCATION.

SPECIFIC

PM - INCLUDES LAST DAY OF PREVIOUS MONTH
ASOS - AUTOMATED SURFACE OBSERVING SYSTEM IN OPERATION DURING THESE MONTHS.

(a) - LENGTH OF RECORD IN YEARS, ALTHOUGH INDIVIDUAL MONTHS MAY BE MISSING.
0.* OR * - THE VALUE IS BETWEEN 0.0 AND 0.05
NORMALS - BASED ON THE 1961-1990 RECORD PERIOD.
EXTREMES - DATES ARE THE MOST RECENT OCCURRENCE.
WIND DIR. - NUMERALS SHOW TENS OF DEGRESS CLOCKWISE FROM TRUE NORTH. "00" INDICATES CALM.
RESULTANT DIRECTIONS ARE GIVEN TO WHOLE DEGREES.
BOLD VALUES INDICATE EXTREME VALUES WHICH OCCURRED AFTER THE ASOS SYSTEM WAS COMMISSIONED.

RECORD = PERIOD OF RECORD
RECORD MEAN PRECIPITATION IS THE MEAN OF ALL DAILY PRECIPITATION AMOUNTS DURING THE PERIOD OF RECORD.
RECORD MAX(MIN) TEMPERATURE IS THE MEAN OF ALL DAILY MAX(MIN) TEMPERATURES DURING THE PERIOD OF RECORD.
RECORD MEAN TEMPERATURE IS THE SUM OF THE RECORD MAX AND RECORD MIN DIVIDED BY 2.
AVERAGE TEMPERATURE IS THE SUM OF THE MEAN DAILY MAX AND MIN TEMPERATURE DIVIDED BY 2.

EXCEPTIONS

RECORD MEANS ARE THROUGH THE CURRENT YEAR, BEGINNING IN 1958 FOR TEMPERATURE
1958 FOR PRECIPITATION
1958 FOR SNOWFALL

HEATING DEGREE DAYS Base 65 deg. F — CHICAGO, OHARE INTERNATIONAL AIRPORT, ILLINOIS

SEASON	JULY	AUG	SEP	OCT	NOV	DEC	JAN	FEB	MAR	APR	MAY	JUNE	TOTAL
1964-65	10	52	148	521	699	1240	1345	1134	1185	545	157	77	7113
1965-66	12	53	110	370	733	915	1502	1079	782	587	371	53	6567
1966-67	1	12	127	420	669	1170	1148	1257	878	491	362	19	6554
1967-68	39	53	160	395	827	1068	1274	1192	682	376	257	28	6351
1968-69	14	12	59	355	740	1146	1355	976	941	419	204	124	6345
1969-70	4	0	75	423	794	1138	1506	1086	929	418	168	44	6585
1970-71	2	0	85	302	725	1055	1422	1026	923	484	262	14	6300
1971-72	7	3	64	154	693	948	1405	1197	954	602	178	80	6285
1972-73	15	10	109	481	811	1269	1135	1012	645	503	311	0	6301
1973-74	0	0	72	244	687	1139	1240	1046	812	383	266	63	5952
1974-75	0	1	176	384	724	1072	1160	1078	951	643	152	30	6371
1975-76	1	0	147	303	531	1033	1392	859	681	411	285	17	5660
1976-77	0	9	119	522	973	1408	1679	1060	616	332	115	41	6874
1977-78	0	8	42	413	741	1254	1521	1346	1020	518	264	46	7173
1978-79	1	4	59	418	718	1206	1622	1360	879	580	233	30	7110
1979-80	16	19	62	382	722	967	1281	1254	995	558	198	83	6537
1980-81	0	3	71	511	746	1140	1308	1031	846	397	313	6	6372
1981-82	8	6	135	489	719	1236	1632	1213	922	608	93	118	7179
1982-83	7	37	152	372	772	891	1194	961	847	643	364	38	6278
1983-84	16	0	125	383	714	1568	1479	894	1095	575	300	18	7167
1984-85	19	1	189	320	807	1046	1563	1245	787	418	183	103	6681
1985-86	0	6	141	380	813	1480	1302	1142	765	417	202	74	6722
1986-87	3	29	64	343	863	1060	1205	866	742	432	162	14	5783
1987-88	4	19	74	541	629	1011	1396	1221	828	503	176	40	6442
1988-89	0	9	63	583	693	1149	1003	1265	882	540	261	43	6491
1989-90	0	5	131	344	813	1471	956	938	733	491	271	33	6186
1990-91	10	5	103	425	605	1120	1365	945	756	393	142	13	5882
1991-92	0	0	163	367	887	1066	1137	913	847	560	284	77	6301
1992-93	9	37	136	449	795	1122	1196	1133	948	595	184	69	6673
1993-94	0	3	185	479	784	1084							

COOLING DEGREE DAYS Base 65 deg. F — CHICAGO, OHARE INTERNATIONAL AIRPORT, ILLINOIS

YEAR	JAN	FEB	MAR	APR	MAY	JUNE	JULY	AUG	SEP	OCT	NOV	DEC	TOTAL
1971	0	0	0	0	27	275	213	228	213	59	0	0	1015
1972	0	0	0	0	64	106	289	289	72	0	0	0	820
1973	0	0	0	5	3	189	308	301	108	32	0	0	946
1974	0	0	0	10	21	83	274	162	48	12	0	0	610
1975	0	0	0	0	76	203	332	358	46	24	1	0	1040
1976	0	0	0	36	6	178	286	196	56	8	0	0	766
1977	0	0	0	39	191	178	395	229	76	0	0	0	1108
1978	0	0	0	0	60	132	227	243	181	2	0	0	845
1979	0	0	0	2	61	164	241	213	99	26	0	0	806
1980	0	0	0	10	43	101	338	342	107	2	0	0	943
1981	0	0	0	9	20	157	248	204	44	0	0	0	682
1982	0	0	0	0	79	38	295	161	69	14	0	0	656
1983	0	0	1	0	4	189	385	388	122	10	0	0	1099
1984	0	0	0	5	11	184	190	254	77	8	0	0	729
1985	0	0	0	53	42	71	204	142	158	0	0	0	670
1986	0	0	7	17	37	118	318	145	123	3	0	0	768
1987	0	0	0	6	116	241	377	238	83	0	1	0	1062
1988	0	0	0	5	59	247	373	383	96	1	0	0	1164
1989	0	0	2	0	44	121	282	207	48	11	0	0	715
1990	0	0	7	43	8	179	226	224	137	11	1	0	836
1991	0	0	0	11	167	226	334	273	132	10	0	0	1153
1992	0	0	0	1	40	79	152	106	75	4	0	0	457
1993	0	0	0	0	28	118	294	266	19	5	0	0	730

SNOWFALL (inches) — CHICAGO, OHARE INTERNATIONAL AIRPORT, ILLINOIS

SEASON	JULY	AUG	SEP	OCT	NOV	DEC	JAN	FEB	MAR	APR	MAY	JUNE	TOTAL
1964-65	0.0	0.0	0.0	0.0	2.3	11.1	11.7	11.5	24.7	T	0.0	0.0	61.3
1965-66	0.0	0.0	0.0	T	0.2	6.6	15.5	4.3	0.7	T	1.6	0.0	28.9
1966-67	0.0	0.0	0.0	T	0.5	8.4	25.1	21.5	8.8	3.4	T	0.0	67.7
1967-68	0.0	0.0	T	6.6	2.4	2.9	10.4	3.8	1.5	0.1	0.0	0.0	27.7
1968-69	0.0	0.0	0.0	0.0	0.7	10.9	3.7	2.3	4.7	0.0	T	0.0	22.3
1969-70	0.0	0.0	0.0	0.0	2.0	19.3	9.5	6.3	11.8	7.2	0.0	0.0	56.1
1970-71	0.0	0.0	0.0	0.0	0.2	2.7	10.0	1.4	8.0	0.8	T	0.0	23.1
1971-72	0.0	0.0	0.0	0.0	1.3	0.2	7.6	7.7	16.8	3.3	0.0	0.0	36.9
1972-73	0.0	0.0	0.0	0.1	0.9	11.2	0.5	9.3	3.4	0.2	T	0.0	25.6
1973-74	0.0	0.0	0.0	0.0	T	18.8	7.4	9.6	1.4	T	0.0	0.0	37.2
1974-75	0.0	0.0	0.0	0.0	1.0	9.4	3.5	8.2	4.5	11.1	0.0	0.0	37.7
1975-76	0.0	0.0	0.0	0.0	6.4	6.8	10.0	1.6	1.9	0.8	T	0.0	27.5
1976-77	0.0	0.0	0.0	1.6	0.5	6.5	7.2	4.0	4.9	T	0.0	0.0	24.7
1977-78	0.0	0.0	0.0	0.0	5.2	12.7	21.9	7.9	4.5	0.2	0.0	0.0	52.4
1978-79	0.0	0.0	0.0	0.0	5.2	35.3	34.3	6.8	2.0	0.1	0.0	0.0	83.7
1979-80	0.0	0.0	0.0	0.0	4.0	0.9	6.2	14.7	11.6	4.2	0.0	0.0	41.6
1980-81	0.0	0.0	0.0	T	5.1	9.7	2.0	15.9	2.3	0.0	0.0	0.0	35.0
1981-82	0.0	0.0	0.0	T	3.6	4.9	21.1	4.8	14.3	10.6	0.0	0.0	59.3
1982-83	0.0	0.0	0.0	0.0	0.4	2.1	5.0	8.9	9.0	1.2	0.0	0.0	26.6
1983-84	0.0	0.0	0.0	0.0	1.0	16.5	17.2	1.9	9.7	2.7	0.0	0.0	49.0
1984-85	0.0	0.0	0.0	0.0	T	6.6	18.9	13.3	0.3	T	0.0	0.0	39.1
1985-86	0.0	0.0	0.0	0.0	1.1	5.2	6.9	10.9	4.1	0.8	0.0	0.0	29.0
1986-87	0.0	0.0	0.0	T	3.8	0.4	17.3	T	4.7	T	0.0	0.0	26.2
1987-88	0.0	0.0	0.0	0.1	1.0	18.7	5.4	15.5	1.9	T	0.0	0.0	42.6
1988-89	0.0	0.0	0.0	T	0.9	5.0	0.4	15.1	2.0	0.6	0.5	0.0	24.5
1989-90	0.0	T	0.0	6.3	3.9	5.4	3.2	13.6	1.3	0.1	T	0.0	33.8
1990-91	0.0	0.0	0.0	T	T	3.2	11.1	3.3	5.9	T	0.0	0.0	23.5
1991-92	0.0	0.0	0.0	T	1.2	7.6	5.6	1.3	11.6	1.1	0.0	T	28.4
1992-93	T	0.0	0.0	0.3	0.2	5.7	15.2	8.0	13.8	3.7	0.0	0.0	46.9
1993-94	0.0	0.0	0.0	T	0.2	1.2							
Record Mean	T	T	T	0.4	1.9	8.3	10.7	8.1	7.0	1.7	0.1	T	38.2

MOLINE, ILLINOIS

The locality is in the heart of the Corn Belt. Agricultural crops include many important staple products in addition to corn. Cattle, hogs, horses, and poultry produced in Iowa and Illinois rank high in the nation. Close to the Mississippi River there is large scale truck gardening and considerable dairying. Field production of grains and livestock attains greater development farther away from the large streams, where the countryside is rolling prairie. Damaging droughts are not common. This, together with the variety of agricultural products, has led to designating the section as the Bread Basket of America.

The climate is favorable for many industries as evidenced by the large number and variety of manufacturing and other enterprises which have located and developed in the community. Among these are some of the largest producers of agricultural machinery in the world.

This area has a temperate continental climate, with a wide temperature range throughout the year. There are some intensely hot, unusually humid, periods in summer and severely cold periods in winter. Maxima of 90 degrees or more have occurred in summer as frequently as 55 days and zero or lower readings have occurred during every winter.

Freezing temperatures have occurred as late in spring as late May and as early in autumn as late September. Precipitation is usually well distributed throughout the year with the greatest amounts falling during the 177-day average crop growing season. Substantial weather changes frequently occur at three or four day intervals, as a direct result of proximity to some of the most important storm tracks.

NORMALS, MEANS, AND EXTREMES

MOLINE, ILLINOIS

LATITUDE: 41 °27'N LONGITUDE: 90 °31' W ELEVATION: FT. GRND 582 BARO 605 TIME ZONE: CENTRAL WBAN: 14923

	(a)	JAN	FEB	MAR	APR	MAY	JUNE	JULY	AUG	SEP	OCT	NOV	DEC	YEAR
TEMPERATURE °F:														
Normals														
-Daily Maximum		28.4	33.7	46.7	61.4	72.8	82.5	85.9	83.5	75.9	64.2	48.4	33.4	59.7
-Daily Minimum		11.3	15.9	28.1	39.4	49.9	59.6	64.5	61.8	53.3	41.8	30.7	17.5	39.5
-Monthly		19.9	24.8	37.4	50.4	61.4	71.1	75.2	72.7	64.7	53.0	39.6	25.4	49.6
Extremes														
-Record Highest	61	69	71	88	93	104	104	105	106	100	92	80	69	106
-Year		1989	1976	1986	1986	1934	1988	1940	1936	1939	1963	1933	1970	AUG 1936
-Record Lowest	61	-27	-25	-19	7	26	39	46	40	24	16	-9	-24	-27
-Year		1979	1979	1960	1982	1966	1993	1971	1986	1942	1988	1976	1989	JAN 1979
NORMAL DEGREE DAYS:														
Heating (base 65°F)		1398	1126	856	438	188	10	0	11	73	384	762	1228	6474
Cooling (base 65°F)		0	0	0	0	76	193	316	250	64	12	0	0	911
% OF POSSIBLE SUNSHINE	50	48	49	49	53	57	63	67	66	62	58	41	41	55
MEAN SKY COVER (tenths)														
Sunrise - Sunset	51	6.5	6.5	6.8	6.6	6.2	5.9	5.4	5.3	5.2	5.3	6.8	6.8	6.1
MEAN NUMBER OF DAYS:														
Sunrise to Sunset														
-Clear	60	7.9	7.2	6.5	6.6	7.5	7.4	10.0	10.3	11.6	11.6	7.3	7.0	100.8
-Partly Cloudy	60	7.0	6.4	7.8	8.6	9.2	10.6	11.8	11.0	7.9	7.7	6.6	6.2	100.6
-Cloudy	60	16.1	14.7	16.7	14.8	14.2	12.0	9.2	9.7	10.5	11.7	16.2	17.8	163.8
Precipitation														
.01 inches or more	61	9.0	8.0	11.0	10.9	11.9	10.1	9.7	9.0	8.8	7.7	8.6	9.3	114.1
Snow,Ice pellets,hail														
1.0 inches or more	61	2.6	2.1	1.7	0.2	0.0	0.0	0.0	0.0	0.0	0.*	0.7	2.3	9.6
Thunderstorms	60	0.3	0.5	2.2	4.5	6.8	8.3	8.4	7.1	4.5	2.5	1.2	0.7	47.1
Heavy Fog Visibility														
1/4 mile or less	59	2.2	2.0	2.0	0.7	0.8	0.6	0.7	1.9	1.6	1.8	1.2	1.9	17.6
Temperature °F														
-Maximum														
90° and above	33	0.0	0.0	0.0	0.1	1.3	5.4	8.6	6.1	1.8	0.1	0.0	0.0	23.3
32° and below	33	17.8	12.2	3.5	0.2	0.0	0.0	0.0	0.0	0.0	0.0	2.3	12.8	48.7
-Minimum														
32° and below	33	29.3	25.5	20.7	7.3	0.6	0.0	0.0	0.0	0.2	6.0	18.3	27.3	135.2
0° and below	33	7.8	4.2	0.2	0.0	0.0	0.0	0.0	0.0	0.0	0.0	0.2	3.7	16.0
AVG. STATION PRESS.(mb)	21	997.6	997.5	994.0	993.0	992.7	992.8	994.2	995.3	995.9	996.3	995.7	997.3	995.2
RELATIVE HUMIDITY (%)														
Hour 00	33	73	74	74	72	75	77	82	85	83	76	76	77	77
Hour 06 (Local Time)	33	74	76	78	79	80	81	85	89	88	81	80	79	81
Hour 12	33	66	64	60	54	53	53	57	58	57	54	63	69	59
Hour 18	33	69	67	62	54	53	53	59	62	66	63	70	74	63
PRECIPITATION (inches):														
Water Equivalent														
-Normal		1.54	1.23	2.98	3.90	4.30	4.27	4.95	4.22	4.02	2.93	2.51	2.23	39.08
-Maximum Monthly	64	4.39	2.80	7.43	11.30	11.43	13.21	11.76	15.23	14.18	9.41	6.77	4.99	15.23
-Year		1974	1985	1973	1973	1974	1993	1992	1987	1970	1941	1992	1982	AUG 1987
-Minimum Monthly	64	0.31	0.17	0.31	0.68	0.31	1.02	0.41	0.35	0.02	0.01	0.29	0.23	0.01
-Year		1961	1933	1958	1946	1992	1963	1991	1971	1979	1964	1933	1929	OCT 1964
-Maximum in 24 hrs	64	2.07	1.90	2.42	5.81	3.68	5.02	5.41	5.13	6.29	4.89	2.73	3.38	6.29
-Year		1960	1948	1985	1973	1970	1967	1963	1987	1961	1954	1961	1942	SEP 1961
Snow,Ice pellets,hail														
-Maximum Monthly	62	26.7	19.3	19.8	13.0	0.3	T	0.0	0.0	0.1	6.6	15.6	21.7	26.7
-Year		1979	1975	1972	1982	1935	1992			1942	1967	1974	1978	JAN 1979
-Maximum in 24 hrs	60	16.4	10.1	11.0	8.2	0.3	T	0.0	0.0	0.1	6.5	8.5	11.4	16.4
-Year		1971	1975	1972	1970	1935	1992			1942	1967	1975	1987	JAN 1971
WIND:														
Mean Speed (mph)	50	10.9	10.8	12.0	12.0	10.2	9.1	7.6	7.2	8.1	9.2	10.8	10.6	9.9
Prevailing Direction through 1963		WNW	WNW	WNW	NW	E	S	E	E	S	S	WNW	WNW	WNW
Fastest Mile														
-Direction (!!)	60	SW	NW	SW	SW	SW	SW	W	N	NW	SW	W	SW	SW
-Speed (MPH)	60	56	58	66	73	68	77	65	63	66	56	60	61	77
-Year		1946	1948	1954	1947	1955	1947	1953	1987	1945	1949	1952	1947	JUN 1947
Peak Gust														
-Direction (!!)	10	W	NW	SW	SW	NW	W	NW	N	SW	SW	W	NW	N
-Speed (mph)	10	59	54	69	69	60	59	52	81	49	61	52	56	81
-Date		1990	1990	1991	1984	1988	1990	1984	1987	1988	1984	1988	1985	AUG 1987

PRECIPITATION (inches)

MOLINE, ILLINOIS

YEAR	JAN	FEB	MAR	APR	MAY	JUNE	JULY	AUG	SEP	OCT	NOV	DEC	ANNUAL
1964	1.59	0.73	4.02	6.37	2.68	3.89	5.31	2.41	3.99	0.01	3.40	1.27	35.67
1965	4.08	0.97	2.52	7.92	4.04	2.07	4.78	9.31	8.27	0.88	1.93	2.82	49.59
1966	2.01	0.89	0.85	3.96	6.39	4.25	7.74	0.77	2.80	4.38	1.12	2.52	37.68
1967	2.00	1.11	1.68	4.91	1.70	8.15	4.15	3.62	4.99	6.23	2.26	1.56	42.36
1968	0.80	0.32	1.96	3.57	2.59	3.70	3.91	2.16	5.70	1.09	3.29	2.76	31.85
1969	3.28	0.21	2.25	3.11	3.40	6.97	11.39	2.00	2.50	4.63	0.53	1.52	41.79
1970	0.54	0.57	2.62	4.66	11.07	3.25	6.27	3.61	14.18	1.75	1.25	1.95	51.72
1971	1.79	1.57	1.99	1.86	3.14	2.36	11.22	0.35	4.54	2.64	2.21	4.77	38.44
1972	1.19	1.24	3.73	5.19	4.24	8.66	3.79	8.20	3.29	2.39	2.75	1.98	46.65
1973	2.53	1.94	7.43	11.30	6.96	6.58	2.63	2.30	6.74	3.06	1.15	3.74	56.36
1974	4.39	1.57	2.94	5.08	11.43	8.83	1.73	3.37	1.08	1.99	2.59	1.83	46.83
1975	1.50	1.75	3.30	2.88	3.23	3.50	2.71	3.23	1.99	0.90	2.98	0.98	28.95
1976	0.88	1.11	3.67	4.21	4.13	2.02	3.34	0.60	1.89	2.13	0.65	0.34	24.97
1977	0.87	0.62	4.61	1.57	2.66	5.18	6.04	6.98	4.80	4.09	2.78	1.76	41.96
1978	0.59	0.67	0.78	3.02	4.65	3.50	3.24	2.47	3.91	1.95	3.32	3.17	31.27
1979	2.72	1.02	2.77	4.01	2.86	2.85	4.40	6.94	0.02	2.51	2.13	2.27	34.50
1980	1.31	1.55	1.74	2.34	4.01	4.97	3.46	9.09	2.94	1.63	0.70	2.75	36.49
1981	0.32	2.45	0.69	4.48	1.48	5.51	3.07	6.90	4.20	4.42	2.42	0.72	36.66
1982	2.08	0.45	3.85	3.33	5.03	3.61	8.83	2.89	1.94	3.38	3.64	4.99	44.02
1983	0.55	1.50	2.88	4.88	2.06	5.21	2.79	0.96	5.30	2.36	5.78	2.55	36.82
1984	0.84	0.79	3.45	4.31	4.97	3.75	4.63	1.31	2.32	6.60	2.91	3.51	39.39
1985	1.04	2.80	4.71	2.14	3.46	2.26	3.33	2.76	1.71	8.54	6.35	2.21	41.31
1986	0.38	2.53	1.35	3.15	5.54	5.10	6.34	3.24	5.63	4.23	1.03	1.95	40.47
1987	0.91	0.89	2.57	1.49	3.35	1.54	1.91	15.23	1.11	0.79	3.33	4.46	37.58
1988	1.74	1.08	2.36	1.36	2.32	1.16	1.79	4.89	1.97	2.37	3.14	1.57	25.75
1989	1.21	1.21	1.85	3.62	3.42	3.31	3.10	6.17	5.42	1.18	0.71	0.73	31.93
1990	1.66	1.99	6.68	3.50	7.68	9.59	3.56	6.42	1.04	2.35	2.87	3.07	50.41
1991	1.19	0.43	6.31	3.69	3.87	1.32	0.41	2.06	2.52	5.80	5.03	1.86	34.49
1992	1.18	1.91	1.07	2.88	0.31	2.11	11.76	1.70	4.80	1.49	6.77	3.96	39.94
1993	2.01	0.92	4.87	5.48	3.61	13.21	5.82	7.76	4.11	0.37	1.50	1.13	50.79
Record Mean	1.53	1.30	2.73	3.64	3.94	4.42	4.25	3.85	3.63	2.76	2.34	1.88	36.26

AVERAGE TEMPERATURE (deg. F)

MOLINE, ILLINOIS

YEAR	JAN	FEB	MAR	APR	MAY	JUNE	JULY	AUG	SEP	OCT	NOV	DEC	ANNUAL
1964	28.4	27.0	34.4	51.5	66.3	72.3	76.6	70.8	65.6	50.7	44.2	24.8	51.1
1965	21.4	25.7	26.9	49.2	65.9	69.6	74.9	71.1	63.5	53.6	40.7	34.5	49.8
1966	14.3	24.8	39.2	47.0	56.3	70.5	77.9	70.6	62.9	52.5	41.7	28.1	48.8
1967	26.0	18.7	39.0	51.6	56.4	70.9	71.4	67.7	62.1	52.1	36.1	28.5	48.4
1968	21.0	22.7	41.9	51.3	57.2	72.1	73.1	73.5	63.1	53.2	37.3	24.5	49.3
1969	17.9	27.9	32.2	51.6	61.7	67.1	75.0	73.7	64.9	51.7	37.9	24.3	48.8
1970	13.1	24.7	33.6	51.8	64.9	70.7	75.7	72.9	65.2	53.9	39.2	28.9	49.5
1971	14.8	25.9	34.7	50.9	58.0	76.6	70.7	71.9	68.7	59.9	40.0	31.6	50.3
1972	18.6	20.7	36.0	48.0	63.0	69.4	74.6	74.0	66.7	50.2	37.3	22.3	48.4
1973	27.1	28.7	45.4	49.7	57.6	72.3	74.9	75.1	65.7	57.3	40.4	23.7	51.5
1974	21.6	26.8	38.5	53.3	58.8	67.5	77.3	70.3	60.4	53.7	40.1	28.6	49.8
1975	24.9	25.0	30.6	45.1	63.8	72.3	75.0	75.1	60.9	56.2	45.0	30.3	50.4
1976	19.1	35.2	41.2	52.9	57.5	70.5	75.8	71.8	63.0	46.8	30.1	19.6	48.6
1977	8.2	27.1	43.6	56.6	68.9	70.0	77.5	71.0	65.5	51.2	38.7	21.7	50.0
1978	12.4	14.1	33.0	50.8	60.2	71.0	74.0	73.2	69.2	51.5	39.2	22.7	47.6
1979	6.3	13.2	36.3	47.4	61.9	71.2	73.5	72.0	64.3	52.1	37.0	31.7	47.2
1980	23.4	19.2	33.5	49.1	62.7	69.2	77.6	75.8	65.7	48.6	39.7	26.6	49.3
1981	23.4	27.1	39.4	54.3	57.5	71.0	73.3	71.2	63.5	49.9	41.1	25.3	49.7
1982	10.6	21.4	34.6	44.7	66.1	66.2	75.4	71.3	64.2	54.7	39.2	35.3	48.7
1983	26.9	32.0	38.9	45.5	57.8	72.9	79.4	79.6	65.6	53.1	42.9	14.0	50.7
1984	19.6	34.5	31.2	49.1	58.2	73.6	73.1	75.2	62.9	55.1	39.3	31.1	50.2
1985	14.3	21.8	42.5	57.1	64.9	69.4	74.2	69.7	66.1	53.6	35.2	14.9	48.6
1986	25.1	22.9	42.5	54.6	63.0	71.7	77.6	68.3	67.3	53.4	34.8	30.C	50.9
1987	25.2	34.4	42.2	52.6	66.8	74.6	79.4	72.9	64.4	47.4	44.0	31.0	52.9
1988	20.2	21.7	38.7	50.6	64.1	73.1	78.2	77.8	67.0	47.3	40.3	29.0	50.7
1989	32.7	17.4	37.0	49.9	58.8	69.2	76.2	72.2	61.9	54.6	38.6	15.6	48.7
1990	32.7	31.3	42.8	50.0	57.3	71.3	74.2	73.7	66.9	51.7	44.2	24.6	51.7
1991	18.4	31.7	41.8	54.2	67.3	74.6	75.3	73.9	64.8	53.3	33.9	31.0	51.7
1992	29.1	34.3	41.3	49.2	61.0	68.6	70.5	67.4	62.5	52.2	37.5	27.4	50.1
1993	24.8	23.5	34.7	48.5	63.0	70.3	75.6	74.9	61.0	51.0	37.2	29.4	49.5
Record Mean	21.7	26.1	36.7	50.3	61.3	71.2	75.3	73.1	65.0	53.5	39.0	26.6	50.0
Max	30.2	34.8	46.0	61.2	72.6	82.3	86.2	84.1	76.5	65.0	48.0	34.5	60.1
Min	13.1	17.5	27.4	39.4	50.1	60.1	64.3	62.1	53.4	42.1	30.0	18.6	39.8

REFERENCE NOTES

MOLINE, ILLINOIS

GENERAL
T - TRACE AMOUNT.
BLANK ENTRIES DENOTE MISSING/UNREPORTED DATA.
# INDICATES A STATION OR INSTRUMENT RELOCATION.

SPECIFIC

(a) - LENGTH OF RECORD IN YEARS, ALTHOUGH INDIVIDUAL MONTHS MAY BE MISSING.
0.* OR * - THE VALUE IS BETWEEN 0.0 AND 0.05
NORMALS - BASED ON THE 1961-1990 RECORD PERIOD.
EXTREMES - DATES ARE THE MOST RECENT OCCURRENCE
WIND DIR.- NUMERALS SHOW TENS OF DEGRESS CLOCKWISE FROM TRUE NORTH. "00" INDICATES CALM.
RESULTANT DIRECTIONS ARE GIVEN TO WHOLE DEGREES.
BOLD VALUES INDICATE EXTREME VALUES WHICH OCCURRED AFTER THE ASOS SYSTEM WAS COMMISSIONED.

RECORD = PERIOD OF RECORD
RECORD MEAN PRECIPITATION IS THE MEAN OF ALL DAILY PRECIPITATION AMOUNTS DURING THE PERIOD OF RECORD.
RECORD MAX(MIN) TEMPERATURE IS THE MEAN OF ALL DAILY MAX(MIN) TEMPERATURES DURING THE PERIOD OF RECORD.
RECORD MEAN TEMPERATURE IS THE SUM OF THE RECORD MAX AND RECORD MIN DIVIDED BY 2.
AVERAGE TEMPERATURE IS THE SUM OF THE MEAN DAILY MAX AND MIN TEMPERATURE DIVIDED BY 2.

EXCEPTIONS

RECORD MEANS ARE THROUGH THE CURRENT YEAR, BEGINNING IN 1930 FOR TEMPERATURE
1927 FOR PRECIPITATION
1933 FOR SNOWFALL

HEATING DEGREE DAYS Base 65 deg. F — MOLINE, ILLINOIS

SEASON	JULY	AUG	SEP	OCT	NOV	DEC	JAN	FEB	MAR	APR	MAY	JUNE	TOTAL
1964-65	3	24	100	437	624	1241	1344	1096	1174	471	86	3	6603
1965-66	0	21	115	363	722	939	1568	1119	794	533	287	31	6492
1966-67	0	6	116	386	695	1137	1202	1289	802	404	305	10	6352
1967-68	21	33	136	424	860	1127	1354	1221	711	409	248	18	6562
1968-69	8	15	96	404	826	1247	1458	1032	1010	398	172	57	6723
1969-70	0	0	80	445	805	1253	1607	1120	967	419	118	21	6835
1970-71	0	2	92	350	764	1111	1556	1088	931	425	242	0	6561
1971-72	17	3	94	195	740	1026	1434	1280	893	504	155	25	6366
1972-73	8	7	80	459	824	1319	1170	1013	598	452	223	1	6154
1973-74	0	0	78	258	733	1274	1338	1061	812	360	235	32	6181
1974-75	0	8	180	349	739	1122	1235	1114	1058	593	117	14	6529
1975-76	9	0	162	297	594	1072	1413	855	731	385	235	2	5755
1976-77	0	5	123	569	1038	1403	1755	1059	656	286	54	15	6963
1977-78	0	5	50	420	780	1337	1625	1417	986	423	229	13	7285
1978-79	1	4	59	415	767	1305	1820	1445	884	521	172	12	7405
1979-80	5	22	95	415	831	1027	1280	1324	966	481	129	27	6602
1980-81	0	1	77	503	755	1180	1281	1055	783	322	253	3	6213
1981-82	12	9	91	459	710	1226	1682	1214	936	604	62	35	7040
1982-83	0	14	110	333	765	916	1175	917	802	578	228	17	5855
1983-84	1	0	110	377	654	1578	1406	880	1039	482	225	2	6754
1984-85	3	4	171	303	761	1042	1568	1203	692	293	66	25	6131
1985-86	0	4	133	347	888	1546	1231	1174	705	338	104	9	6479
1986-87	0	30	55	358	898	1078	1227	850	701	385	101	4	5687
1987-88	0	23	84	540	627	1046	1384	1250	807	427	90	15	6293
1988-89	1	9	45	544	736	1109	998	1326	863	461	240	28	6360
1989-90	0	4	138	334	785	1527	995	937	691	483	241	18	6153
1990-91	2	0	97	422	622	1248	1437	928	719	339	107	0	5921
1991-92	0	0	147	370	927	1049	1107	884	726	473	181	27	5891
1992-93	3	33	147	405	819	1158	1241	1156	936	488	116	53	6555
1993-94	0	1	153	433	827	1096							

COOLING DEGREE DAYS Base 65 deg. F — MOLINE, ILLINOIS

YEAR	JAN	FEB	MAR	APR	MAY	JUNE	JULY	AUG	SEP	OCT	NOV	DEC	TOTAL
1969	0	0	0	1	77	130	320	276	86	41	0	0	931
1970	0	0	0	32	119	198	341	256	105	15	0	0	1066
1971	0	0	0	9	32	355	203	225	212	44	0	0	1080
1972	0	0	0	3	102	165	312	294	138	6	0	0	1020
1973	0	0	0	2	2	226	312	323	103	27	0	0	995
1974	0	0	0	17	50	113	389	180	45	3	0	0	797
1975	0	0	0	1	85	241	325	319	44	29	0	0	1044
1976	0	0	0	26	12	175	341	221	72	13	0	0	860
1977	0	0	0	44	182	173	397	199	73	0	0	0	1068
1978	0	0	1	0	87	202	288	267	195	5	0	0	1045
1979	0	0	0	0	83	206	272	245	79	22	0	0	907
1980	0	0	0	11	65	158	395	343	107	1	0	0	1080
1981	0	0	0	8	27	191	275	209	50	0	0	0	760
1982	0	0	0	0	102	78	327	215	95	20	0	0	837
1983	0	0	0	1	12	260	453	459	138	14	0	0	1337
1984	0	0	0	9	21	268	261	327	116	3	0	0	1005
1985	0	0	0	64	71	165	292	154	172	0	0	0	918
1986	0	0	12	31	48	216	396	143	133	4	0	0	983
1987	0	0	0	20	162	298	453	277	71	0	3	0	1284
1988	0	0	0	0	69	263	416	410	112	3	0	0	1273
1989	0	0	4	13	53	159	356	236	52	20	0	0	893
1990	0	0	10	40	8	216	294	280	159	18	3	0	1028
1991	0	0	6	21	183	297	324	284	149	16	0	0	1280
1992	0	0	0	6	61	143	181	115	77	13	0	0	596
1993	0	0	0	0	60	217	335	317	39	8	0	0	976

SNOWFALL (inches) — MOLINE, ILLINOIS

SEASON	JULY	AUG	SEP	OCT	NOV	DEC	JAN	FEB	MAR	APR	MAY	JUNE	TOTAL
1964-65	0.0	0.0	0.0	0.0	2.1	11.7	10.2	7.4	16.2	0.0	0.0	0.0	47.6
1965-66	0.0	0.0	0.0	0.0	0.3	5.7	6.0	0.8	0.7	0.1	T	0.0	13.6
1966-67	0.0	0.0	0.0	0.0	T	7.2	14.4	13.2	2.2	1.1	0.0	0.0	38.1
1967-68	0.0	0.0	0.0	6.6	0.3	3.2	8.6	2.8	1.1	T	0.0	0.0	22.6
1968-69	0.0	0.0	0.0	T	5.9	5.4	6.2	0.9	5.6	0.0	T	0.0	24.0
1969-70	0.0	0.0	0.0	0.0	1.5	15.4	7.2	6.1	12.7	8.3	0.0	0.0	51.2
1970-71	0.0	0.0	0.0	0.0	0.2	2.2	17.7	2.6	7.8	T	0.0	0.0	30.5
1971-72	0.0	0.0	0.0	0.0	5.9	0.3	11.8	13.7	19.8	2.6	0.0	0.0	54.1
1972-73	0.0	0.0	0.0	2.0	6.7	6.6	0.4	8.6	0.5	0.9	0.0	0.0	25.7
1973-74	0.0	0.0	0.0	0.0	T	11.4	12.7	7.4	3.0	0.6	0.0	0.0	35.1
1974-75	0.0	0.0	0.0	0.0	15.6	8.8	7.6	19.3	12.2	6.2	0.0	0.0	69.7
1975-76	0.0	0.0	0.0	0.0	9.9	2.5	8.6	0.1	0.6	0.0	T	0.0	21.7
1976-77	0.0	0.0	0.0	0.0	7.7	4.1	12.3	2.5	10.8	1.0	0.0	0.0	38.4
1977-78	0.0	0.0	0.0	0.0	4.8	20.8	6.1	6.8	6.5	T	0.0	0.0	45.0
1978-79	0.0	0.0	0.0	0.0	7.1	21.7	26.7	4.5	3.4	0.6	0.0	0.0	64.0
1979-80	0.0	0.0	0.0	T	1.9	0.8	4.2	13.1	7.5	9.5	0.0	0.0	37.0
1980-81	0.0	0.0	0.0	2.4	1.1	4.5	5.3	5.6	T	0.0	0.0	0.0	18.9
1981-82	0.0	0.0	0.0	T	3.3	4.3	13.6	4.5	6.6	13.0	0.0	0.0	45.3
1982-83	0.0	0.0	0.0	0.0	T	0.5	4.7	6.2	12.7	0.7	0.0	0.0	24.8
1983-84	0.0	0.0	0.0	T	0.9	18.9	9.9	1.9	13.6	0.1	0.0	0.0	45.3
1984-85	0.0	0.0	0.0	0.0	T	9.3	13.4	7.5	0.2	T	0.0	0.0	30.4
1985-86	0.0	0.0	0.0	0.0	3.1	11.5	3.7	7.8	0.2	0.4	0.0	0.0	26.7
1986-87	0.0	0.0	0.0	0.0	2.2	0.2	9.1	T	0.1	0.0	0.0	0.0	11.6
1987-88	0.0	0.0	0.0	T	0.6	15.9	2.7	16.4	0.5	0.0	0.0	0.0	36.1
1988-89	0.0	0.0	0.0	T	1.0	2.5	T	13.8	0.4	T	T	0.0	17.7
1989-90	0.0	0.0	0.0	T	0.1	8.2	6.3	10.3	T	T	0.0	T	24.9
1990-91	0.0	0.0	0.0	0.0	T	7.5	11.1	1.5	5.4	0.0	0.0	0.0	25.5
1991-92	0.0	0.0	T	T	1.0	6.7	5.2	1.4	3.6	T	0.0	T	17.9
1992-93	0.0	0.0	0.0	T	0.6	7.2	7.0	6.4	3.0	1.0	T	0.0	25.2
1993-94	0.0	0.0	0.0	T	3.6	5.2							
Record Mean	0.0	0.0	T	0.2	2.2	7.1	7.8	6.0	5.9	1.0	T	T	30.2

PEORIA, ILLINOIS

The airport station is situated on a rather level tableland surrounded by well-drained and gently rolling terrain. It is set back a mile from the rim of the Illinois River Valley and is almost 200 feet above the river bed. Exposures of all instruments are good. The climate of this area is typically continental as shown by its changeable weather and the wide range of temperature extremes.

June and September are usually the most pleasant months of the year. Then during October or the first of November, Indian Summer is often experienced with an extended period of warm, dry weather.

Precipitation is normally heaviest during the growing season and lowest during midwinter.

The earliest snowfalls have occurred in September and the latest in the spring have occurred as late as May. Heavy snowfalls have rarely exceeded 20 inches.

Based on the 1951-1980 period, the average first occurrence of 32 degrees Fahrenheit in the fall is October 20 and the average last occurrence in the spring is April 24.

NORMALS, MEANS, AND EXTREMES

PEORIA, ILLINOIS

LATITUDE: 40°40'N LONGITUDE: 89°41'W ELEVATION: FT. GRND 652 BARO 683 TIME ZONE: CENTRAL WBAN: 14842

	(a)	JAN	FEB	MAR	APR	MAY	JUNE	JULY	AUG	SEP	OCT	NOV	DEC	YEAR
TEMPERATURE °F:														
Normals														
-Daily Maximum		29.9	34.9	48.1	62.0	72.8	82.2	85.7	83.1	76.9	64.8	49.8	34.6	60.4
-Daily Minimum		13.2	17.7	29.8	40.8	50.9	60.7	65.4	63.1	55.2	43.1	32.5	19.3	41.0
-Monthly		21.6	26.3	39.0	51.4	61.9	71.5	75.5	73.1	66.1	54.0	41.2	27.0	50.7
Extremes														
-Record Highest	54	70	72	86	92	93	105	103	103	100	90	81	71	105
-Year		1989	1976	1986	1986	1987	1988	1940	1988	1953	1963	1950	1982	JUN 1988
-Record Lowest	54	-25	-18	-10	14	25	39	47	41	26	19	-2	-23	-25
-Year		1977	1979	1960	1982	1966	1993	1972	1986	1942	1972	1977	1989	JAN 1977
NORMAL DEGREE DAYS:														
Heating (base 65°F)		1345	1084	806	408	183	11	0	9	54	356	714	1178	6148
Cooling (base 65°F)		0	0	0	0	87	206	326	261	87	15	0	0	982
% OF POSSIBLE SUNSHINE	50	47	50	51	55	60	67	68	67	64	61	43	42	56
MEAN SKY COVER (tenths)														
Sunrise - Sunset	50	6.8	6.8	7.0	6.8	6.4	6.0	5.6	5.5	5.3	5.4	6.8	7.0	6.3
MEAN NUMBER OF DAYS:														
Sunrise to Sunset														
-Clear	50	7.4	6.7	6.0	6.2	7.2	7.1	8.9	10.0	10.9	11.1	6.8	6.7	95.0
-Partly Cloudy	50	5.9	5.7	7.3	8.0	9.5	10.6	11.8	10.5	8.6	7.7	5.9	5.8	97.6
-Cloudy	50	17.7	15.8	17.7	15.8	14.3	12.2	10.2	10.5	10.5	12.1	17.3	18.5	172.6
Precipitation .01 inches or more	54	9.3	8.3	11.0	11.8	11.3	9.5	9.0	8.3	8.7	7.9	9.4	9.8	114.4
Snow,Ice pellets,hail 1.0 inches or more	50	2.2	1.7	1.4	0.3	0.0	0.0	0.0	0.0	0.0	0.*	0.6	1.9	8.1
Thunderstorms	50	0.6	0.6	2.7	5.1	6.7	8.3	7.9	6.6	5.0	2.5	1.3	0.6	47.9
Heavy Fog Visibility 1/4 mile or less	50	3.0	2.8	2.3	0.9	0.9	0.6	1.0	1.4	1.4	1.5	2.1	3.4	21.3
Temperature °F														
-Maximum														
90° and above	34	0.0	0.0	0.0	0.1	0.5	4.4	8.0	5.3	1.7	0.*	0.0	0.0	19.9
32° and below	34	16.9	11.5	3.7	0.1	0.0	0.0	0.0	0.0	0.0	0.0	1.9	12.4	46.6
-Minimum														
32° and below	34	29.3	25.4	19.5	5.8	0.4	0.0	0.0	0.0	0.1	4.6	16.8	26.6	128.6
0° and below	34	6.1	3.1	0.2	0.0	0.0	0.0	0.0	0.0	0.0	0.0	0.1	2.7	12.2
AVG. STATION PRESS. (mb)	21	995.4	995.1	991.8	990.8	990.7	991.0	992.3	993.4	993.9	994.4	993.7	995.1	993.1
RELATIVE HUMIDITY (%)														
Hour 00	34	78	79	77	72	75	77	82	84	82	78	80	81	79
Hour 06 (Local Time)	34	79	81	81	79	81	82	87	89	88	85	84	83	83
Hour 12	34	69	67	63	56	56	56	59	60	59	57	67	72	62
Hour 18	34	72	71	64	56	56	56	61	64	64	63	71	76	65
PRECIPITATION (inches):														
Water Equivalent														
-Normal		1.51	1.42	2.91	3.77	3.70	3.99	4.20	3.10	3.87	2.65	2.69	2.44	36.25
-Maximum Monthly	54	8.11	5.18	6.95	8.66	7.96	11.69	10.15	8.61	13.09	10.80	7.62	6.34	13.09
-Year		1965	1942	1973	1947	1957	1974	1993	1965	1961	1941	1985	1949	SEP 1961
-Minimum Monthly	54	0.22	0.33	0.39	0.71	0.82	0.60	0.33	0.25	0.03	0.03	0.43	0.33	0.03
-Year		1986	1947	1958	1971	1992	1988	1988	1992	1979	1964	1953	1962	SEP 1979
-Maximum in 24 hrs	50	4.45	1.92	3.39	5.06	3.62	4.44	4.22	4.32	4.15	3.70	4.32	3.38	5.06
-Year		1965	1954	1944	1950	1956	1974	1993	1955	1961	1969	1990	1949	APR 1950
Snow,Ice pellets,hail														
-Maximum Monthly	50	24.7	15.2	16.9	13.4	0.1	0.0	T	0.0	T	1.8	9.1	21.7	24.7
-Year		1979	1989	1960	1982	1966		1990		1992	1967	1974	1977	JAN 1979
-Maximum in 24 hrs	50	12.2	7.6	9.0	6.1	0.1	0.0	T	0.0	T	1.8	7.2	10.2	12.2
-Year		1979	1944	1946	1982	1966		1990		1992	1967	1951	1973	JAN 1979
WIND:														
Mean Speed (mph)	50	11.1	11.1	12.1	11.8	10.0	8.9	7.8	7.6	8.4	9.4	11.0	10.8	10.0
Prevailing Direction through 1963		S	WNW	WNW	S	S	S	S	S	S	S	S	S	S
Fastest Obs. 1 Min.														
-Direction (!!)	8	30	31	25	36	31	30	26	01	29	27	20	06	20
-Speed (MPH)	8	37	37	46	37	35	44	40	33	35	31	48	46	48
-Year		1985	1990	1991	1989	1991	1987	1990	1387	1992	1985	1988	1987	NOV 1988
Peak Gust														
-Direction (!!)	10	NW	NW	W	N	NW	NW	W	NW	W	W	S	NE	W
-Speed (mph)	10	53	53	68	69	61	63	67	54	75	48	62	59	75
-Date		1985	1990	1991	1989	1992	1987	1992	1990	1991	1985	1988	1987	SEP 1991

PRECIPITATION (inches)

PEORIA, ILLINOIS

YEAR	JAN	FEB	MAR	APR	MAY	JUNE	JULY	AUG	SEP	OCT	NOV	DEC	ANNUAL
1964	1.02	0.54	3.77	6.92	1.04	4.22	1.59	2.06	3.63	0.03	3.01	1.12	28.95
1965	8.11	0.93	3.40	5.58	3.85	1.04	3.31	8.61	8.17	0.69	1.22	3.35	48.26
1966	1.49	2.60	1.78	3.98	4.50	2.29	3.66	2.56	3.06	1.80	2.42	3.00	33.14
1967	1.08	1.07	2.33	5.47	2.80	2.03	5.74	2.34	2.72	5.56	2.18	2.63	35.95
1968	1.12	1.55	0.93	2.20	4.70	6.16	4.32	1.71	4.83	0.58	2.99	2.80	33.89
1969	2.43	0.56	1.20	2.94	2.37	4.93	5.55	2.82	3.31	5.67	0.79	1.13	33.70
1970	0.56	0.64	1.60	7.18	3.89	3.92	5.46	3.21	11.49	4.36	1.11	1.30	44.72
1971	0.59	1.64	1.09	0.71	2.80	0.98	5.21	2.19	3.07	1.71	1.43	4.96	26.38
1972	0.81	0.74	2.48	4.38	1.30	5.97	3.54	4.26	5.21	2.50	2.56	2.48	36.23
1973	1.76	0.99	6.95	4.26	4.51	6.46	6.04	0.90	7.58	5.18	1.48	4.11	50.22
1974	3.09	1.65	2.69	4.11	6.26	11.69	2.63	0.81	1.45	2.07	4.13	1.93	42.51
1975	2.59	2.85	1.73	3.92	5.19	3.90	4.26	5.62	2.74	3.63	2.75	2.04	41.22
1976	0.78	2.56	4.25	4.86	5.11	2.92	2.98	2.30	1.78	2.48	0.83	0.38	31.23
1977	1.22	0.95	4.41	1.24	3.54	2.06	3.43	7.28	6.26	4.00	1.77	2.25	38.41
1978	0.69	0.59	1.56	4.69	7.72	1.96	3.47	1.28	2.32	1.73	2.54	3.54	32.09
1979	2.48	1.37	4.42	4.48	1.96	1.77	4.81	0.87	0.03	1.70	2.76	2.33	28.98
1980	0.59	1.06	2.79	2.78	2.05	8.94	1.43	6.16	4.09	2.44	0.67	2.25	35.25
1981	0.48	2.41	0.92	5.71	5.77	6.22	7.08	5.61	1.31	1.37	1.64	1.24	39.76
1982	2.88	1.13	4.80	5.40	3.15	3.15	7.53	3.97	1.24	1.47	4.95	5.45	45.12
1983	0.53	1.01	2.84	7.06	6.66	4.48	1.99	1.09	5.08	3.01	5.58	2.65	41.98
1984	0.59	2.28	3.95	5.18	4.84	2.90	5.02	0.78	2.38	5.07	3.95	3.82	40.76
1985	0.99	2.62	5.77	1.14	3.14	5.11	3.43	3.70	3.43	4.61	7.62	2.24	43.80
1986	0.22	1.79	0.87	1.39	2.95	6.53	7.00	1.74	6.39	4.64	1.32	2.60	37.44
1987	1.49	0.84	1.98	1.84	1.69	3.27	2.90	4.02	1.62	0.73	2.88	4.15	27.41
1988	1.99	0.71	2.83	1.59	1.68	0.60	0.33	2.11	2.82	1.08	4.19	2.23	22.16
1989	1.00	1.17	1.14	4.39	2.23	1.28	2.22	2.86	2.87	1.57	0.93	0.87	22.53
1990	1.73	3.59	3.95	2.32	6.19	7.99	9.18	5.31	1.03	3.17	7.19	3.70	55.35
1991	1.19	0.57	3.67	2.97	5.94	1.50	0.35	3.41	3.59	7.31	3.57	2.06	36.13
1992	1.06	1.55	2.58	1.61	0.82	0.80	8.19	0.25	5.81	1.33	5.58	2.99	32.57
1993	3.55	1.68	4.08	4.89	3.25	5.70	10.15	7.38	7.56	2.42	2.22	1.19	54.07
Record Mean	1.75	1.77	2.82	3.52	3.91	3.87	3.88	3.10	3.69	2.50	2.41	2.08	35.29

AVERAGE TEMPERATURE (deg. F)

PEORIA, ILLINOIS

YEAR	JAN	FEB	MAR	APR	MAY	JUNE	JULY	AUG	SEP	OCT	NOV	DEC	ANNUAL
1964	28.7	28.1	36.0	52.2	66.3	72.7	76.2	72.4	65.4	50.8	43.5	25.7	51.5
1965	24.3	27.0	28.0	50.9	66.7	71.3	73.7	71.2	64.7	52.5	41.2	35.4	50.6
1966	19.5	27.8	40.7	48.1	57.1	70.4	78.4	70.3	63.8	51.3	42.6	29.2	49.9
1967	27.6	21.8	40.2	52.7	56.4	72.3	72.0	68.1	63.3	52.6	36.8	29.8	49.5
1968	22.1	23.7	43.0	52.4	57.7	72.4	73.2	73.8	64.3	53.9	40.4	27.6	50.4
1969	20.3	30.4	33.0	52.6	61.4	66.4	75.6	73.1	64.8	51.5	37.5	25.8	49.4
1970	15.1	25.9	35.3	52.2	64.2	69.8	74.9	72.3	65.8	53.6	39.1	29.9	49.8
1971	19.2	27.7	35.5	51.9	57.7	76.8	71.2	71.6	68.9	60.0	40.5	34.3	51.3
1972	19.7	24.5	36.9	48.9	63.2	68.0	73.2	72.4	66.2	50.0	36.2	23.8	48.6
1973	27.4	28.9	46.5	50.9	57.9	71.8	75.0	74.7	67.0	57.7	42.4	24.8	52.1
1974	23.1	29.4	41.1	53.6	59.3	66.7	76.9	72.2	61.1	53.6	39.6	29.9	50.5
1975	27.5	26.1	33.8	47.0	64.6	72.1	73.8	74.7	60.9	55.3	44.9	30.2	50.9
1976	19.5	34.9	43.9	54.1	58.2	70.5	75.3	70.3	63.3	47.3	32.0	21.0	49.2
1977	8.6	27.0	44.6	57.1	68.6	70.2	78.2	71.4	66.4	51.3	40.3	22.8	50.5
1978	13.3	15.4	32.4	51.2	60.3	71.3	75.0	73.2	70.5	52.1	41.3	26.8	48.6
1979	9.4	14.7	37.9	47.7	60.5	71.2	73.0	72.5	65.6	52.6	38.0	31.8	47.9
1980	23.6	19.9	35.6	49.2	63.0	69.3	78.5	76.9	67.9	49.6	40.7	28.9	50.2
1981	23.7	27.8	40.7	55.4	58.7	73.2	75.4	72.8	65.9	53.4	45.0	27.3	51.6
1982	15.8	24.8	37.7	46.5	68.4	67.2	75.7	71.3	64.9	55.0	41.5	37.4	50.5
1983	28.2	33.5	40.3	46.8	58.6	72.6	80.2	80.8	67.9	55.3	44.9	15.2	52.0
1984	20.6	35.5	31.3	49.6	58.8	74.3	73.0	74.9	64.0	57.6	40.3	33.9	51.2
1985	16.8	23.2	43.8	57.0	64.3	68.7	73.7	70.2	66.6	55.5	39.0	18.6	49.8
1986	26.6	23.5	43.2	55.4	64.0	72.3	77.4	68.9	68.9	55.2	36.2	30.9	51.9
1987	25.2	35.7	44.1	54.1	68.4	74.0	79.0	73.7	65.7	48.0	44.6	32.4	53.7
1988	22.7	23.5	39.9	51.1	65.4	73.1	78.6	78.5	68.2	48.2	41.5	29.7	51.7
1989	33.9	18.7	39.2	50.7	59.1	69.8	75.3	72.4	62.2	54.9	39.8	16.2	49.4
1990	34.8	33.7	43.4	50.0	57.8	71.0	72.6	72.1	66.1	51.9	45.2	27.7	52.2
1991	21.5	32.4	42.8	55.8	68.0	74.7	75.4	74.3	65.4	53.9	36.0	32.0	52.7
1992	30.1	35.6	42.3	50.0	61.0	69.7	72.4	68.6	64.2	53.1	39.8	29.5	51.4
1993	27.4	25.1	36.4	49.8	63.6	70.4	76.1	75.0	61.4	51.7	39.4	31.4	50.6
Record Mean	24.0	27.7	38.8	51.0	61.8	71.3	75.6	73.6	66.2	54.3	40.3	28.3	51.1
Max	32.1	36.0	48.1	61.5	72.7	82.2	86.4	84.3	77.2	65.2	49.1	35.9	60.9
Min	15.9	19.4	29.4	40.6	50.8	60.5	64.8	62.9	55.1	43.4	31.6	20.7	41.3

REFERENCE NOTES

PEORIA, ILLINOIS

GENERAL
T - TRACE AMOUNT.
BLANK ENTRIES DENOTE MISSING/UNREPORTED DATA.
# INDICATES A STATION OR INSTRUMENT RELOCATION.

SPECIFIC

PM - INCLUDES LAST DAY OF PREVIOUS MONTH
ASOS - AUTOMATED SURFACE OBSERVING SYSTEM IN OPERATION DURING THESE MONTHS.

(a) - LENGTH OF RECORD IN YEARS, ALTHOUGH INDIVIDUAL MONTHS MAY BE MISSING.
0.* OR * - THE VALUE IS BETWEEN 0.0 AND 0.05
NORMALS - BASED ON THE 1961-1990 RECORD PERIOD.
EXTREMES - DATES ARE THE MOST RECENT OCCURRENCE.
WIND DIR.- NUMERALS SHOW TENS OF DEGRESS CLOCKWISE FROM TRUE NORTH. "00" INDICATES CALM.
RESULTANT DIRECTIONS ARE GIVEN TO WHOLE DEGREES.
BOLD VALUES INDICATE EXTREME VALUES WHICH OCCURRED AFTER THE ASOS SYSTEM WAS COMMISSIONED.

RECORD = PERIOD OF RECORD
RECORD MEAN PRECIPITATION IS THE MEAN OF ALL DAILY PRECIPITATION AMOUNTS DURING THE PERIOD OF RECORD.
RECORD MAX(MIN) TEMPERATURE IS THE MEAN OF ALL DAILY MAX(MIN) TEMPERATURES DURING THE PERIOD OF RECORD.
RECORD MEAN TEMPERATURE IS THE SUM OF THE RECORD MAX AND RECORD MIN DIVIDED BY 2.
AVERAGE TEMPERATURE IS THE SUM OF THE MEAN DAILY MAX AND MIN TEMPERATURE DIVIDED BY 2.

EXCEPTIONS

RECORD MEANS ARE THROUGH THE CURRENT YEAR, BEGINNING IN 1905 FOR TEMPERATURE
1856 FOR PRECIPITATION
1944 FOR SNOWFALL

HEATING DEGREE DAYS Base 65 deg. F PEORIA, ILLINOIS

SEASON	JULY	AUG	SEP	OCT	NOV	DEC	JAN	FEB	MAR	APR	MAY	JUNE	TOTAL
1964-65	2	8	108	435	637	1212	1254	1058	1139	416	72	1	6342
1965-66	0	18	100	391	707	914	1406	1037	746	502	260	22	6103
1966-67	0	10	106	419	666	1103	1151	1205	766	371	291	7	6095
1967-68	10	30	113	394	841	1083	1323	1192	673	371	229	13	6272
1968-69	5	7	68	377	732	1150	1377	962	984	367	172	73	6274
1969-70	0	0	77	443	818	1211	1545	1088	914	402	115	28	6641
1970-71	2	3	84	353	774	1080	1418	1035	906	402	241	0	6298
1971-72	17	5	85	186	728	945	1402	1167	866	475	150	42	6068
1972-73	9	19	87	461	855	1271	1161	1005	566	423	216	0	6073
1973-74	0	0	54	253	671	1237	1292	991	736	348	214	42	5838
1974-75	0	1	157	354	756	1081	1156	1085	959	534	92	17	6192
1975-76	13	1	171	306	596	1069	1404	868	648	351	221	5	5653
1976-77	0	9	106	556	981	1357	1747	1061	623	273	60	17	6790
1977-78	0	6	39	418	734	1301	1595	1383	1006	405	222	14	7123
1978-79	0	4	49	390	704	1174	1722	1403	833	510	194	10	6993
1979-80	3	19	70	401	804	1022	1279	1300	907	474	123	26	6428
1980-81	0	0	65	470	722	1112	1273	1037	748	295	221	1	5944
1981-82	1	0	60	360	594	1163	1520	1119	839	548	29	28	6261
1982-83	0	13	94	325	697	849	1133	875	758	537	206	17	5504
1983-84	2	0	92	311	595	1541	1371	849	1038	467	206	1	6473
1984-85	1	1	153	246	734	956	1489	1164	656	284	71	38	5793
1985-86	0	6	111	287	774	1432	1184	1156	683	314	92	7	6046
1986-87	0	26	37	305	858	1048	1228	814	640	340	53	2	5351
1987-88	0	16	68	520	609	1001	1306	1198	772	409	64	12	5975
1988-89	0	4	38	517	698	1090	958	1290	796	442	231	25	6089
1989-90	0	3	134	317	749	1509	929	871	672	475	226	16	5901
1990-91	8	3	99	409	589	1148	1341	909	683	285	86	0	5560
1991-92	0	0	143	347	863	1017	1078	847	698	444	180	19	5636
1992-93	0	23	109	367	748	1094	1159	1110	878	450	96	48	6082
1993-94	0	0	135	408	761	1035							

COOLING DEGREE DAYS Base 65 deg. F PEORIA, ILLINOIS

YEAR	JAN	FEB	MAR	APR	MAY	JUNE	JULY	AUG	SEP	OCT	NOV	DEC	TOTAL
1969	0	0	0	0	67	122	334	260	80	30	0	0	893
1970	0	0	0	24	99	178	317	239	109	6	0	0	972
1971	0	0	0	14	22	359	217	214	210	37	0	0	1073
1972	0	0	0	0	103	140	273	253	130	3	0	0	902
1973	0	0	0	7	3	211	315	306	121	35	0	0	998
1974	0	0	0	12	43	99	377	233	47	6	0	0	817
1975	0	0	0	0	83	237	292	307	55	14	0	0	988
1976	0	0	0	28	14	176	326	180	62	14	0	0	800
1977	0	0	0	42	176	179	416	209	88	0	0	0	1110
1978	0	0	0	0	81	208	316	267	221	0	0	0	1093
1979	0	0	0	0	62	206	259	258	95	23	0	0	903
1980	0	0	0	6	70	160	425	378	156	1	0	0	1196
1981	0	0	0	13	33	250	331	250	93	6	1	0	977
1982	0	0	0	0	141	101	338	215	96	26	0	0	917
1983	0	0	0	0	14	250	479	494	188	17	0	0	1442
1984	0	0	0	12	24	285	256	315	129	22	0	0	1043
1985	0	0	3	48	54	155	279	173	164	0	0	0	876
1986	0	0	15	30	68	234	392	157	161	6	0	0	1063
1987	0	0	0	19	166	278	440	293	92	0	3	0	1291
1988	0	0	0	2	84	266	431	428	140	5	0	0	1356
1989	0	0	2	20	56	177	324	240	57	11	0	0	887
1990	0	0	9	34	8	204	251	230	138	9	0	0	883
1991	0	0	4	16	189	296	331	293	159	11	0	0	1299
1992	0	0	0	1	62	171	237	140	90	6	0	0	707
1993	0	0	0	2	60	218	350	321	33	6	0	0	990

SNOWFALL (inches) PEORIA, ILLINOIS

SEASON	JULY	AUG	SEP	OCT	NOV	DEC	JAN	FEB	MAR	APR	MAY	JUNE	TOTAL
1964-65	0.0	0.0	0.0	0.0	1.0	7.2	5.5	5.8	13.5	0.0	0.0	0.0	33.0
1965-66	0.0	0.0	0.0	0.0	0.1	4.2	0.6	2.0	0.8	T	0.1	0.0	7.8
1966-67	0.0	0.0	0.0	0.0	T	3.8	10.0	3.6	2.7	1.4	0.0	0.0	21.5
1967-68	0.0	0.0	0.0	1.8	2.5	2.2	8.4	1.6	1.1	T	0.0	0.0	17.6
1968-69	0.0	0.0	0.0	0.0	1.5	4.6	4.6	2.2	4.1	0.0	0.0	0.0	17.0
1969-70	0.0	0.0	0.0	0.0	0.6	11.0	7.3	5.9	8.3	4.6	0.0	0.0	37.7
1970-71	0.0	0.0	0.0	0.0	T	3.4	5.4	0.6	5.8	0.0	0.0	0.0	15.2
1971-72	0.0	0.0	0.0	0.0	4.0	T	10.2	6.2	8.0	0.9	0.0	0.0	29.3
1972-73	0.0	0.0	0.0	0.3	7.3	4.7	0.8	2.7	1.2	0.7	0.0	0.0	17.7
1973-74	0.0	0.0	0.0	0.0	0.1	18.9	6.7	1.9	0.9	1.2	0.0	0.0	29.7
1974-75	0.0	0.0	0.0	0.0	9.1	6.2	8.8	12.8	3.8	1.6	0.0	0.0	42.3
1975-76	0.0	0.0	0.0	0.0	8.1	1.4	8.6	2.5	2.0	0.0	0.0	0.0	22.6
1976-77	0.0	0.0	0.0	T	0.4	4.1	16.3	4.1	2.4	1.4	0.0	0.0	28.7
1977-78	0.0	0.0	0.0	0.0	6.4	21.7	7.0	7.0	5.0	0.0	0.0	0.0	47.1
1978-79	0.0	0.0	0.0	0.0	2.6	14.2	24.7	3.6	6.5	T	0.0	0.0	51.6
1979-80	0.0	0.0	0.0	0.0	0.8	0.1	3.7	11.3	5.3	6.3	0.0	0.0	27.5
1980-81	0.0	0.0	0.0	T	4.1	3.3	5.9	10.5	T	0.0	0.0	0.0	23.8
1981-82	0.0	0.0	0.0	0.0	0.1	9.8	11.0	6.1	6.5	13.4	0.0	0.0	46.9
1982-83	0.0	0.0	0.0	0.0	0.9	2.0	5.6	4.8	5.7	0.1	0.0	0.0	19.1
1983-84	0.0	0.0	0.0	0.0	3.3	15.9	6.9	4.1	6.0	0.0	0.0	0.0	36.2
1984-85	0.0	0.0	0.0	0.0	T	0.9	9.8	6.2	T	T	0.0	0.0	16.9
1985-86	0.0	0.0	0.0	0.0	1.0	6.2	1.3	13.9	0.4	0.1	0.0	0.0	22.9
1986-87	0.0	0.0	0.0	0.0	1.0	T	18.0	0.1	T	0.0	0.0	0.0	19.1
1987-88	0.0	0.0	0.0	T	0.3	9.8	1.9	9.7	1.9	0.0	0.0	0.0	23.6
1988-89	0.0	0.0	0.0	0.0	0.7	4.7	0.3	15.2	T	0.9	T	0.0	21.8
1989-90	0.0	0.0	0.0	0.6	T	10.5	4.8	6.2	T	T	0.0	0.0	22.1
1990-91	T	0.0	0.0	T	T	3.8	9.8	1.0	6.9	0.0	0.0	0.0	21.5
1991-92	0.0	0.0	0.0	T	3.6	5.1	6.4	0.4	4.1	T	0.0	0.0	19.6
1992-93	0.0	0.0	T	T	0.7	6.1	6.8	12.4	3.5	0.7	0.0	0.0	30.2
1993-94	0.0	0.0	0.0	T	3.6	2.4							
Record Mean	T	0.0	T	0.1	2.1	5.9	6.6	5.4	4.1	0.9	T	0.0	25.0

ROCKFORD, ILLINOIS

The climate of Rockford is characterized by hot summers and cold winters.

When winter northeasterly winds blow across Lake Michigan, cloudiness often is increased in the Rockford area, and temperatures are somewhat higher than those westward around the Mississippi River. Conversely, in summer, the cooling effect of Lake Michigan sometimes is felt as far westward as Rockford.

While 34 percent of the precipitation occurs in the three summer months of June to August, and 64 percent in the six months, April to September, no month averages less than 4 percent of the annual total.

Though summers may be described as hot, seldom does oppressive heat prevail for extended periods. In general, the summers are pleasant.

Winters are cold. Snow cover is adequate for diversified winter sports, and usually is continuous from late December through February.

Based on the 1951–1980 period, the average first occurrence of 32 degrees Fahrenheit in the fall is October 11 and the average last occurrence in the spring is April 29.

NORMALS, MEANS, AND EXTREMES

ROCKFORD, ILLINOIS

LATITUDE: 42°12'N LONGITUDE: 89°06'W ELEVATION: FT. GRND 724 BARO 732 TIME ZONE: CENTRAL WBAN: 94822

	(a)	JAN	FEB	MAR	APR	MAY	JUNE	JULY	AUG	SEP	OCT	NOV	DEC	YEAR
TEMPERATURE °F:														
Normals														
-Daily Maximum		26.6	31.6	43.8	58.4	70.8	80.2	83.8	81.4	74.1	62.3	46.3	31.8	57.6
-Daily Minimum		9.8	14.2	26.1	36.9	47.4	57.4	62.5	60.3	52.0	40.6	29.5	16.1	37.7
-Monthly		18.2	22.9	35.0	47.7	59.1	68.8	73.2	70.9	63.1	51.5	38.0	24.0	47.7
Extremes														
-Record Highest	43	63	68	85	91	95	101	103	104	102	90	76	66	104
-Year		1989	1976	1986	1980	1975	1988	1955	1988	1953	1971	1961	1984	AUG 1988
-Record Lowest	43	-27	-22	-11	5	24	37	43	41	27	15	-10	-24	-27
-Year		1982	1985	1962	1982	1966	1993	1967	1986	1984	1952	1977	1983	JAN 1982
NORMAL DEGREE DAYS:														
Heating (base 65°F)		1451	1179	930	519	239	24	5	15	99	427	810	1271	6969
Cooling (base 65°F)		0	0	0	0	56	138	259	198	42	9	0	0	702
% OF POSSIBLE SUNSHINE														
MEAN SKY COVER (tenths)														
Sunrise - Sunset	43	6.7	6.5	6.9	6.7	6.3	6.0	5.6	5.7	5.5	5.8	7.1	7.1	6.3
MEAN NUMBER OF DAYS:														
Sunrise to Sunset														
-Clear	43	7.4	7.3	6.0	6.4	7.7	7.4	9.0	9.0	9.8	10.1	6.3	6.7	93.1
-Partly Cloudy	43	6.7	6.1	7.8	7.8	9.2	10.5	11.8	10.9	8.7	7.5	5.7	5.7	98.5
-Cloudy	43	16.9	14.9	17.3	15.8	14.2	12.0	10.2	11.0	11.5	13.4	18.0	18.6	173.7
Precipitation														
.01 inches or more	43	9.3	7.9	11.0	11.8	11.3	10.2	9.8	9.2	9.1	8.7	9.4	10.2	117.9
Snow,Ice pellets,hail														
1.0 inches or more	43	2.7	2.2	2.0	0.4	0.*	0.0	0.0	0.0	0.0	0.*	0.8	2.8	11.1
Thunderstorms	43	0.1	0.4	1.9	4.2	5.5	7.8	7.9	6.3	4.8	2.4	1.1	0.3	42.6
Heavy Fog Visibility														
1/4 mile or less	43	2.6	2.4	2.6	1.2	1.0	0.6	1.2	1.8	1.7	2.0	2.3	2.9	22.2
Temperature °F														
-Maximum														
90° and above	30	0.0	0.0	0.0	0.1	0.6	2.9	6.1	3.6	1.0	0.*	0.0	0.0	14.4
32° and below	30	19.0	13.9	4.7	0.2	0.0	0.0	0.0	0.0	0.0	0.0	3.0	14.1	54.9
-Minimum														
32° and below	30	29.5	26.2	23.0	9.0	1.2	0.0	0.0	0.0	0.6	7.3	19.5	27.7	144.0
0° and below	30	8.7	5.0	0.3	0.0	0.0	0.0	0.0	0.0	0.0	0.0	0.2	3.7	17.9
AVG. STATION PRESS.(mb)	21	991.4	991.8	988.6	987.8	987.8	987.9	989.2	990.5	990.9	991.0	990.2	991.3	989.9
RELATIVE HUMIDITY (%)														
Hour 00	30	79	79	78	75	76	78	83	87	86	81	81	81	80
Hour 06 (Local Time)	30	80	80	81	80	80	81	86	90	90	85	84	83	83
Hour 12	30	70	66	62	56	53	55	57	59	58	56	67	73	61
Hour 18	30	75	70	65	56	54	55	59	62	65	64	73	77	65
PRECIPITATION (inches):														
Water Equivalent														
-Normal		1.28	1.14	2.46	3.65	3.66	4.52	4.12	4.15	3.80	2.88	2.57	2.05	36.28
-Maximum Monthly	43	4.66	2.67	5.62	9.92	6.98	11.85	11.81	13.55	10.68	8.32	5.51	5.04	13.55
-Year		1960	1971	1961	1973	1974	1993	1952	1987	1961	1969	1985	1971	AUG 1987
-Minimum Monthly	43	0.18	0.04	0.52	0.99	0.48	0.46	0.79	0.67	0.05	0.01	0.38	0.37	0.01
-Year		1961	1969	1958	1989	1992	1988	1991	1970	1979	1952	1976	1976	OCT 1952
-Maximum in 24 hrs	43	2.89	1.73	2.50	5.55	3.61	4.15	5.03	6.42	5.56	5.22	3.20	2.24	6.42
-Year		1960	1966	1976	1973	1977	1969	1952	1987	1961	1954	1961	1971	AUG 1987
Snow,Ice pellets,hail														
-Maximum Monthly	43	26.1	19.5	22.7	7.7	1.0	0.0	T	T	0.0	2.2	14.7	25.1	26.1
-Year		1979	1960	1964	1982	1966		1992	1990		1967	1951	1978	JAN 1979
-Maximum in 24 hrs	43	9.9	10.9	10.4	6.7	0.2	0.0	T	T	0.0	2.2	9.5	11.4	11.4
-Year		1979	1960	1972	1970	1990		1992	1990		1967	1951	1987	DEC 1987
WIND:														
Mean Speed (mph)	43	10.6	10.6	11.7	11.8	10.5	9.3	8.1	7.8	8.5	9.5	10.6	10.5	9.9
Prevailing Direction through 1963		WNW	WNW	ENE	WNW	ENE	SSW	SSW	SSW	SSW	SSW	WNW	WNW	SSW
Fastest Obs. 1 Min.														
-Direction (!!)	43	27	25	25	11	27	30	30	26	20	29	20	06	11
-Speed (MPH)	43	40	45	46	54	52	46	53	48	52	37	46	46	54
-Year		1980	1953	1953	1953	1988	1974	1984	1965	1961	1980	1961	1987	APR 1953
Peak Gust														
-Direction (!!)	4			SW			NW							
-Speed (mph)	10	51	54	54	58	81	54	70	58	58	49	59	62	81
-Date		1988	1987	1990	1984	1988	1990	1984	1987	1988	1988	1988	1987	MAY 1988

PRECIPITATION (inches)

ROCKFORD, ILLINOIS

YEAR	JAN	FEB	MAR	APR	MAY	JUNE	JULY	AUG	SEP	OCT	NOV	DEC	ANNUAL
1964	0.97	0.50	3.64	8.17	3.41	6.51	5.20	1.99	3.32	0.11	4.51	0.91	39.24
1965	4.40	0.83	2.30	4.62	5.66	2.56	3.81	9.27	8.23	2.91	1.83	3.03	49.45
1966	1.39	2.04	2.75	4.96	4.76	4.53	1.59	1.50	1.18	2.51	3.53	2.16	32.90
1967	1.41	0.92	1.63	5.30	2.48	9.98	1.32	3.71	3.11	5.06	2.58	1.55	39.05
1968	0.58	0.37	1.00	4.92	3.78	6.26	2.62	6.33	6.76	1.26	2.80	3.31	39.99
1969	1.86	0.04	1.59	4.45	5.75	8.90	5.72	2.14	1.87	8.32	1.45	1.20	43.29
1970	0.51	0.56	1.89	3.75	5.50	4.20	4.29	0.67	10.63	4.03	1.36	1.37	38.76
1971	1.11	2.67	2.11	1.79	1.83	2.37	2.81	1.48	2.36	1.99	2.24	5.04	27.80
1972	0.66	0.60	2.79	6.21	4.57	8.27	8.39	9.10	6.51	4.29	1.39	2.37	55.15
1973	1.67	1.56	4.53	9.92	6.98	5.16	5.81	1.89	8.52	5.15	1.97	3.32	56.48
1974	3.55	1.65	2.03	3.41	6.98	6.30	1.48	2.21	0.35	2.41	2.31	1.63	34.31
1975	2.38	2.03	2.81	2.39	3.03	4.68	1.30	4.08	0.91	0.98	4.24	1.95	30.78
1976	0.59	1.41	4.94	3.60	2.38	2.16	2.05	2.04	1.39	1.94	0.38	0.37	23.25
1977	0.61	0.64	3.52	2.39	5.47	4.67	4.73	5.31	3.91	2.26	2.63	1.87	38.01
1978	0.78	0.47	0.86	3.29	3.84	5.63	7.41	2.65	4.33	0.87	2.46	2.76	35.35
1979	2.43	1.18	3.74	5.20	1.45	4.75	4.34	6.74	0.05	1.55	2.43	2.24	36.10
1980	1.02	1.07	0.97	2.75	2.37	6.06	3.58	5.41	6.13	1.42	0.68	2.71	34.17
1981	0.22	2.40	0.65	5.21	1.84	5.88	1.14	9.18	3.51	2.47	1.63	0.77	34.90
1982	1.58	0.20	3.63	3.47	4.56	4.31	8.89	2.69	1.70	4.22	4.69	3.65	43.59
1983	0.52	1.48	2.70	3.85	4.99	1.55	3.85	2.47	3.80	2.25	4.68	2.31	34.45
1984	0.73	1.29	2.04	3.09	3.95	3.99	2.92	1.63	1.54	5.93	3.44	3.29	33.84
1985	1.00	1.88	3.66	1.10	3.43	3.34	2.98	4.71	2.81	7.39	5.51	2.30	40.11
1986	0.73	2.44	0.93	2.04	4.79	3.77	3.14	2.38	9.89	2.44	1.12	0.66	34.33
1987	0.93	0.82	1.96	1.98	3.73	3.08	5.02	13.55	2.38	0.82	2.57	4.03	40.87
1988	2.45	0.54	2.15	3.14	1.29	0.46	2.39	1.96	2.29	2.10	4.25	1.75	24.77
1989	0.79	0.63	2.41	0.99	2.93	2.46	7.61	6.23	1.86	0.86	0.94	0.51	28.22
1990	1.83	1.23	2.65	2.83	5.10	9.24	4.93	6.73	0.85	3.31	3.64	2.45	44.79
1991	1.12	0.45	4.77	2.15	2.72	3.29	0.79	1.69	4.59	4.77	4.24	1.82	32.40
1992	1.11	1.19	2.02	3.65	0.48	1.18	5.12	2.44	5.44	0.66	4.75	3.36	31.40
1993	2.45	0.87	3.86	6.11	3.05	11.85	3.72	3.07	3.72	0.74	1.77	1.23	42.44
Record Mean	1.64	1.35	2.54	3.39	3.69	4.39	3.79	3.82	3.83	2.77	2.44	1.77	35.44

AVERAGE TEMPERATURE (deg. F)

ROCKFORD, ILLINOIS

YEAR	JAN	FEB	MAR	APR	MAY	JUNE	JULY	AUG	SEP	OCT	NOV	DEC	ANNUAL
1964	26.7	25.7	32.3	48.5	62.6	68.6	72.8	66.9	62.7	48.1	41.1	22.9	48.3
1965	19.5	22.1	25.9	45.6	61.9	67.9	72.9	69.7	62.2	52.4	39.6	32.3	47.7
1966	13.4	23.9	37.6	44.1	51.4	69.2	75.4	69.7	62.5	51.4	39.0	24.8	46.9
1967	24.6	16.9	35.9	47.8	52.9	68.4	69.6	66.1	62.4	51.3	34.4	27.7	46.5
1968	21.2	21.2	40.4	49.7	56.1	69.6	70.9	71.4	63.2	53.2	37.8	25.3	48.4
1969	17.9	28.3	31.0	48.4	59.4	63.6	72.6	71.7	63.1	49.8	36.7	24.2	47.2
1970	12.5	22.4	33.1	48.5	60.4	68.5	73.6	71.4	63.6	52.8	37.7	26.6	47.6
1971	14.8	25.0	32.3	48.0	57.0	74.2	70.7	70.7	68.1	59.2	39.1	31.5	49.2
1972	17.0	20.5	32.4	44.5	62.3	65.8	71.1	71.3	63.3	48.9	36.6	21.6	46.3
1973	26.3	26.9	43.0	47.6	55.8	70.6	73.5	73.2	64.3	56.5	39.9	24.1	50.2
1974	21.2	24.0	36.2	50.8	56.6	66.1	75.3	70.3	60.1	52.0	38.0	26.7	48.1
1975	23.5	23.1	29.2	42.5	62.4	70.5	73.1	73.1	58.0	53.1	42.2	27.0	48.1
1976	16.5	30.8	38.5	49.7	55.3	68.8	73.4	69.4	60.0	45.1	28.1	15.0	45.9
1977	5.9	23.5	42.2	55.1	68.2	67.4	75.4	68.6	63.4	49.1	36.5	19.9	47.9
1978	11.3	12.3	29.3	47.5	58.9	69.1	72.3	71.7	68.1	49.9	37.9	21.9	45.9
1979	7.7	12.9	31.7	42.9	58.7	68.4	71.5	70.0	63.0	50.8	35.9	29.0	45.2
1980	19.3	17.1	30.1	46.4	60.6	66.7	75.4	73.7	64.2	47.3	38.2	25.4	47.0
1981	21.4	25.9	37.3	51.4	56.3	70.0	72.5	71.2	62.5	48.8	39.4	22.9	48.3
1982	8.9	19.6	32.8	44.0	65.5	64.8	74.6	69.0	61.8	53.6	38.1	33.1	47.1
1983	25.3	29.5	36.8	43.9	54.7	70.5	78.0	77.2	63.3	51.5	40.1	12.7	48.6
1984	16.0	31.9	28.6	46.9	55.4	70.9	70.8	72.8	61.6	54.6	37.1	29.2	48.0
1985	12.9	19.3	39.3	54.1	62.2	66.4	71.9	68.3	64.0	51.8	35.3	14.6	46.7
1986	20.9	22.5	40.0	52.2	61.4	68.5	75.3	67.1	64.9	51.7	33.4	28.1	48.8
1987	23.1	32.4	39.5	51.1	63.9	72.5	76.3	70.8	62.8	44.7	41.6	29.4	50.7
1988	16.0	19.7	36.2	47.7	62.5	71.3	76.6	76.4	65.3	45.2	39.6	26.3	48.6
1989	29.9	16.1	34.0	46.9	58.2	68.2	74.7	70.1	60.5	51.8	35.4	14.2	46.7
1990	30.1	28.2	39.9	48.0	55.7	69.7	72.0	71.4	65.4	50.0	41.8	23.7	49.7
1991	17.0	28.4	39.0	51.7	65.5	73.0	74.2	73.8	63.2	51.3	33.5	28.2	49.9
1992	26.7	32.2	37.6	45.9	59.3	67.5	69.1	65.6	61.3	49.5	36.4	25.8	48.1
1993	23.5	22.5	33.2	45.5	60.8	67.3	73.7	72.8	58.4	48.9	36.8	27.5	47.6
Record Mean	20.7	24.5	35.3	48.2	59.5	69.1	73.8	71.8	63.9	52.2	38.2	25.3	48.5
Max	29.2	33.1	44.6	59.2	71.3	80.6	85.4	83.2	75.4	63.3	46.9	33.1	58.8
Min	12.1	15.9	26.1	37.1	47.7	57.5	62.2	60.3	52.3	41.0	29.5	17.5	38.3

REFERENCE NOTES

ROCKFORD, ILLINOIS

GENERAL
T - TRACE AMOUNT.
BLANK ENTRIES DENOTE MISSING/UNREPORTED DATA.
# INDICATES A STATION OR INSTRUMENT RELOCATION.

SPECIFIC

(a) - LENGTH OF RECORD IN YEARS, ALTHOUGH INDIVIDUAL MONTHS MAY BE MISSING.
0.* OR * - THE VALUE IS BETWEEN 0.0 AND 0.05
NORMALS - BASED ON THE 1961-1990 RECORD PERIOD.
EXTREMES - DATES ARE THE MOST RECENT OCCURRENCE
WIND DIR.- NUMERALS SHOW TENS OF DEGRESS CLOCKWISE FROM TRUE NORTH. "00" INDICATES CALM.
RESULTANT DIRECTIONS ARE GIVEN TO WHOLE DEGREES.
BOLD VALUES INDICATE EXTREME VALUES WHICH OCCURRED AFTER THE ASOS SYSTEM WAS COMMISSIONED.

RECORD = PERIOD OF RECORD
RECORD MEAN PRECIPITATION IS THE MEAN OF ALL DAILY PRECIPITATION AMOUNTS DURING THE PERIOD OF RECORD.
RECORD MAX(MIN) TEMPERATURE IS THE MEAN OF ALL DAILY MAX(MIN) TEMPERATURES DURING THE PERIOD OF RECORD.
RECORD MEAN TEMPERATURE IS THE SUM OF THE RECORD MAX AND RECORD MIN DIVIDED BY 2.
AVERAGE TEMPERATURE IS THE SUM OF THE MEAN DAILY MAX AND MIN TEMPERATURE DIVIDED BY 2.

EXCEPTIONS

RECORD MEANS ARE THROUGH THE CURRENT YEAR, BEGINNING IN 1930 FOR TEMPERATURE
1906 FOR PRECIPITATION
1951 FOR SNOWFALL

HEATING DEGREE DAYS Base 65 deg. F ROCKFORD, ILLINOIS

SEASON	JULY	AUG	SEP	OCT	NOV	DEC	JAN	FEB	MAR	APR	MAY	JUNE	TOTAL
1964-65	9	51	145	519	713	1298	1405	1194	1204	579	145	26	7288
1965-66	1	30	134	392	758	1005	1595	1142	844	620	417	45	6983
1966-67	0	7	122	417	773	1238	1244	1341	892	513	385	22	6954
1967-68	42	51	134	439	910	1149	1349	1266	754	453	275	24	6846
1968-69	15	21	95	393	809	1224	1453	1017	1047	490	210	120	6894
1969-70	4	1	115	486	845	1256	1627	1191	983	502	203	44	7257
1970-71	5	2	117	377	812	1182	1549	1114	1007	504	258	5	6932
1971-72	14	5	95	209	771	1031	1483	1287	1006	606	164	79	6750
1972-73	17	26	119	491	845	1338	1193	1061	675	520	281	2	6568
1973-74	0	3	99	281	744	1260	1354	1142	887	427	277	59	6533
1974-75	0	5	175	400	804	1179	1277	1166	1107	668	150	29	6960
1975-76	12	0	225	376	678	1172	1499	986	813	473	296	16	6546
1976-77	2	17	174	615	1100	1545	1830	1158	697	318	73	42	7571
1977-78	0	20	88	488	849	1393	1658	1471	1101	519	259	29	7875
1978-79	2	7	66	461	806	1333	1771	1458	1026	658	245	27	7860
1979-80	9	26	120	450	866	1110	1411	1380	1073	556	184	63	7248
1980-81	0	3	99	538	796	1222	1346	1089	851	402	275	9	6630
1981-82	9	6	109	493	761	1296	1739	1268	990	623	74	61	7429
1982-83	0	32	156	362	801	984	1224	987	867	627	316	34	6390
1983-84	6	0	148	423	742	1615	1515	952	1123	542	302	7	7375
1984-85	9	9	180	322	833	1101	1609	1277	789	368	127	55	6679
1985-86	2	14	157	402	885	1553	1362	1187	771	397	138	35	6903
1986-87	2	42	91	402	942	1135	1290	905	781	418	143	11	6162
1987-88	2	32	108	623	695	1096	1513	1308	887	514	131	23	6932
1988-89	0	8	66	605	755	1193	1081	1364	954	537	246	36	6845
1989-90	0	10	164	403	881	1570	1074	1026	775	534	285	28	6750
1990-91	2	5	115	475	690	1273	1483	1018	798	402	146	6	6413
1991-92	0	1	172	421	940	1134	1181	945	843	567	220	38	6462
1992-93	12	55	165	476	849	1207	1282	1183	977	581	167	71	7025
1993-94	0	4	212	498	836	1156							

COOLING DEGREE DAYS Base 65 deg. F ROCKFORD, ILLINOIS

YEAR	JAN	FEB	MAR	APR	MAY	JUNE	JULY	AUG	SEP	OCT	NOV	DEC	TOTAL
1969	0	0	0	0	45	82	244	213	66	21	0	0	671
1970	0	0	0	16	64	157	276	209	79	5	0	0	806
1971	0	0	0	1	16	288	197	190	193	35	0	0	920
1972	0	0	0	0	89	104	214	229	75	0	0	0	711
1973	0	0	0	5	1	178	272	263	84	25	0	0	828
1974	0	0	0	9	25	98	329	177	35	6	0	0	679
1975	0	0	0	0	75	201	273	258	22	15	0	0	844
1976	0	0	0	21	3	137	270	159	31	5	0	0	626
1977	0	0	0	31	180	120	330	138	48	0	0	0	847
1978	0	0	0	0	75	160	238	220	166	0	0	0	859
1979	0	0	0	0	56	134	217	187	69	16	0	0	679
1980	0	0	0	6	52	117	329	278	80	0	0	0	862
1981	0	0	0	3	15	164	248	202	42	0	0	0	674
1982	0	0	0	0	96	61	303	168	66	15	0	0	709
1983	0	0	0	0	4	207	416	386	104	10	0	0	1127
1984	0	0	0	3	12	189	195	257	87	5	0	0	748
1985	0	0	0	47	45	106	223	120	134	0	0	0	675
1986	0	0	4	22	35	148	329	113	92	1	0	0	744
1987	0	0	0	8	118	244	358	220	47	0	0	0	995
1988	0	0	0	0	58	221	365	369	81	0	0	0	1094
1989	0	0	0	2	43	138	308	173	37	3	0	0	704
1990	0	0	3	32	3	173	225	210	132	13	0	0	791
1991	0	0	0	8	169	254	292	281	126	4	0	0	1134
1992	0	0	0	0	51	119	148	79	62	3	0	0	462
1993	0	0	0	0	42	148	280	253	18	5	0	0	746

SNOWFALL (inches) ROCKFORD, ILLINOIS

SEASON	JULY	AUG	SEP	OCT	NOV	DEC	JAN	FEB	MAR	APR	MAY	JUNE	TOTAL
1964-65	0.0	0.0	0.0	T	2.8	10.3	10.3	5.0	19.2	T	0.0	0.0	47.6
1965-66	0.0	0.0	0.0	0.0	0.1	5.5	11.0	0.3	1.0	T	1.0	0.0	18.9
1966-67	0.0	0.0	0.0	0.0	T	7.8	8.1	11.6	6.5	3.8	T	0.0	37.8
1967-68	0.0	0.0	0.0	2.2	0.7	2.0	2.9	2.3	0.6	0.3	0.0	0.0	11.0
1968-69	0.0	0.0	0.0	0.0	4.5	8.9	4.6	0.8	4.9	T	0.0	0.0	23.7
1969-70	0.0	0.0	0.0	0.0	0.6	16.0	8.2	5.4	6.4	6.7	0.0	0.0	43.3
1970-71	0.0	0.0	0.0	0.0	T	3.2	10.0	3.4	14.6	0.4	0.0	0.0	31.6
1971-72	0.0	0.0	0.0	0.0	5.0	0.2	5.5	7.9	15.7	2.7	0.0	0.0	37.0
1972-73	0.0	0.0	0.0	0.1	4.4	8.7	0.4	6.9	0.9	5.2	0.0	0.0	26.6
1973-74	0.0	0.0	0.0	0.0	T	20.0	12.2	15.1	2.3	T	0.0	0.0	49.6
1974-75	0.0	0.0	0.0	0.0	4.3	11.4	4.2	18.6	8.4	4.6	0.0	0.0	51.5
1975-76	0.0	0.0	0.0	0.0	6.4	7.9	6.6	1.8	0.3	0.3	T	0.0	23.3
1976-77	0.0	0.0	0.0	T	4.5	5.8	8.8	4.3	7.2	0.8	0.0	0.0	31.4
1977-78	0.0	0.0	0.0	0.0	7.3	23.3	10.0	6.9	4.2	T	0.0	0.0	51.7
1978-79	0.0	0.0	0.0	0.0	4.6	25.1	26.1	12.2	4.5	2.0	0.0	0.0	74.5
1979-80	0.0	0.0	0.0	T	4.9	0.9	5.1	13.0	4.6	5.4	0.0	0.0	33.9
1980-81	0.0	0.0	0.0	0.8	1.4	6.7	2.5	9.1	0.6	0.0	0.0	0.0	21.1
1981-82	0.0	0.0	0.0	T	2.8	5.8	14.8	1.8	8.1	7.7	0.0	0.0	41.0
1982-83	0.0	0.0	0.0	T	T	3.6	5.5	9.5	9.4	T	0.0	0.0	28.0
1983-84	0.0	0.0	0.0	0.0	0.5	18.3	8.8	2.2	10.9	0.7	0.0	0.0	41.4
1984-85	0.0	0.0	0.0	0.0	0.2	14.2	13.4	8.3	2.8	2.4	0.0	0.0	41.3
1985-86	0.0	0.0	0.0	0.0	5.1	12.5	8.7	12.3	0.9	1.2	0.0	0.0	40.7
1986-87	0.0	0.0	0.0	T	5.0	2.0	9.5	T	2.8	T	0.0	0.0	19.3
1987-88	0.0	0.0	0.0	T	0.6	21.7	8.7	9.9	3.4	0.7	0.0	0.0	45.0
1988-89	0.0	0.0	0.0	T	1.6	11.8	1.3	13.3	4.2	1.5	T	0.0	33.7
1989-90	0.0	0.0	0.0	1.0	3.0	8.6	9.4	12.0	0.3	0.2	0.2	0.0	34.7
1990-91	T	T	0.0	T	0.7	11.0	15.3	2.8	6.9	T	0.0	0.0	36.7
1991-92	0.0	0.0	0.0	T	0.7	13.0	6.0	1.0	9.7	0.2	0.0	0.0	30.6
1992-93	T	0.0	0.0	0.4	1.7	11.1	13.1	9.9	13.1	1.8	0.0	0.0	51.1
1993-94	0.0	0.0	0.0	T	0.7	4.9							
Record Mean	T	T	0.0	0.1	2.6	9.3	8.6	6.7	6.6	1.4	T	0.0	35.4

SPRINGFIELD, ILLINOIS

The location of Springfield near the center of North America gives it a typical continental climate with warm summers and fairly cold winters. The surrounding country is nearly level. There are no large hills in the vicinity, but rolling terrain is found near the Sangamon River and Spring Creek.

Monthly temperatures range from the upper 20s for January to the upper 70s for July. Considerable variation may take place within the seasons. Temperatures of 70 degrees or higher may occur in winter and temperatures near 50 degrees are sometimes recorded during the summer months.

There are no wet and dry seasons. Monthly precipitation ranges from a little over 4 inches in May and June to about 2 inches in January. There is some variation in rainfall totals from year to year. Thunderstorms are common during hot weather, and these are sometimes locally severe with brief but heavy showers. The average year has about fifty thunderstorms of which two-thirds occur during the months of May through August. Damaging hail accompanies only a few of the thunderstorms and the areas affected are usually small.

Sunshine is particularly abundant during the summer months when days are long and not very cloudy. January is the cloudiest month, with only about a third as much sunshine as July or August. March is the windiest month, and August the month with the least wind. Velocities of more than 40 mph are not unusual for brief periods in most months of the year. The prevailing wind direction is southerly during most of the year with northwesterly winds during the late fall and early spring months.

An overall description of the climate of Springfield would be one indicating pleasant conditions with sharp seasonal changes, but no extended periods of severely cold weather. Summer weather is often uncomfortably warm and humid.

Based on the 1951–1980 period, the average first occurrence of 32 degrees Fahrenheit in the fall is October 19 and the average last occurrence in the spring is April 17.

NORMALS, MEANS, AND EXTREMES

SPRINGFIELD, ILLINOIS

LATITUDE: 39°50'N LONGITUDE: 89°40'W ELEVATION: FT. GRND 588 BARO 597 TIME ZONE: CENTRAL WBAN: 93822

	(a)	JAN	FEB	MAR	APR	MAY	JUNE	JULY	AUG	SEP	OCT	NOV	DEC	YEAR
TEMPERATURE °F:														
Normals														
-Daily Maximum		32.5	37.2	50.0	63.9	74.7	83.9	86.9	84.2	78.7	66.8	51.7	37.3	62.3
-Daily Minimum		15.9	20.2	31.6	42.5	52.3	61.9	66.1	63.3	55.9	44.4	34.0	21.9	42.5
-Monthly		24.2	28.7	40.8	53.2	63.5	72.9	76.5	73.7	67.3	55.6	42.9	29.7	52.4
Extremes														
-Record Highest	46	71	74	87	90	95	103	112	103	101	93	83	74	112
-Year		1950	1972	1981	1986	1967	1954	1954	1964	1984	1954	1950	1984	JUL 1954
-Record Lowest	46	-21	-22	-12	19	28	40	48	43	32	17	-3	-21	-22
-Year		1985	1963	1960	1982	1966	1966	1975	1986	1984	1952	1964	1989	FEB 1963
NORMAL DEGREE DAYS:														
Heating (base 65°F)		1265	1016	750	361	166	7	0	8	44	314	663	1094	5688
Cooling (base 65°F)		0	0	0	7	120	244	357	278	113	22	0	0	1141
% OF POSSIBLE SUNSHINE	45	49	51	51	56	64	68	71	70	68	63	48	44	59
MEAN SKY COVER (tenths)														
Sunrise - Sunset	46	6.7	6.6	7.0	6.6	6.1	5.8	5.4	5.3	5.1	5.2	6.6	6.9	6.1
MEAN NUMBER OF DAYS:														
Sunrise to Sunset														
-Clear	46	7.3	7.2	6.4	7.0	8.0	8.5	9.9	10.9	11.8	12.0	7.8	7.0	103.9
-Partly Cloudy	46	6.5	5.7	6.9	7.8	9.4	9.3	10.7	10.0	7.9	7.2	6.2	6.3	93.9
-Cloudy	46	17.2	15.3	17.7	15.2	13.6	12.2	10.4	10.1	10.4	11.7	16.0	17.7	167.5
Precipitation														
.01 inches or more	46	9.1	8.7	11.9	11.6	10.4	9.7	9.0	8.3	8.3	7.7	9.5	9.9	114.1
Snow, Ice pellets, hail														
1.0 inches or more	46	1.8	1.8	1.3	0.3	0.0	0.0	0.0	0.0	0.0	0.0	0.5	1.6	7.3
Thunderstorms	46	0.5	0.8	2.6	5.4	6.8	7.9	8.7	6.8	4.5	2.4	1.5	0.6	48.5
Heavy Fog Visibility														
1/4 mile or less	46	2.6	2.7	1.9	0.9	0.8	0.3	0.7	1.1	1.1	1.0	1.6	2.5	17.1
Temperature °F														
-Maximum														
90° and above	34	0.0	0.0	0.0	0.*	1.6	7.1	10.5	6.7	3.1	0.1	0.0	0.0	29.2
32° and below	34	14.5	9.4	2.5	0.*	0.0	0.0	0.0	0.0	0.0	0.0	1.2	9.5	37.1
-Minimum														
32° and below	34	28.0	23.5	16.9	4.2	0.1	0.0	0.0	0.0	0.*	3.7	14.6	25.1	116.1
0° and below	34	4.5	2.4	0.1	0.0	0.0	0.0	0.0	0.0	0.0	0.0	0.*	1.9	9.0
AVG. STATION PRESS. (mb)	20	997.6	996.9	993.4	992.4	992.2	992.3	993.7	994.8	995.5	996.2	995.6	997.0	994.8
RELATIVE HUMIDITY (%)														
Hour 00	34	78	78	77	73	76	77	81	84	82	77	79	81	79
Hour 06 (Local Time)	34	79	81	82	80	80	82	85	89	89	83	83	83	83
Hour 12	34	68	68	64	56	54	54	58	60	56	54	65	72	61
Hour 18	34	72	71	65	56	55	55	60	64	63	62	70	76	64
PRECIPITATION (inches):														
Water Equivalent														
-Normal		1.51	1.77	3.24	3.68	3.62	3.43	3.52	3.29	3.33	2.60	2.53	2.73	35.25
-Maximum Monthly	46	5.67	4.89	7.89	9.91	8.84	9.22	10.76	8.37	8.57	6.41	6.94	8.94	10.76
-Year		1949	1990	1973	1964	1990	1990	1981	1981	1986	1991	1985	1982	JUL 1981
-Minimum Monthly	46	0.04	0.51	0.63	0.73	0.52	0.23	0.91	0.46	T	0.16	0.43	0.15	T
-Year		1986	1958	1956	1971	1992	1959	1974	1992	1979	1964	1949	1955	SEP 1979
-Maximum in 24 hrs	46	2.78	2.54	2.84	4.45	3.95	4.73	4.43	4.79	5.12	3.51	2.46	6.12	6.12
-Year		1975	1990	1972	1979	1990	1958	1981	1956	1959	1973	1964	1982	DEC 1982
Snow, Ice pellets, hail														
-Maximum Monthly	46	21.1	16.0	20.3	7.3	T	0.0	0.0	0.0	0.0	0.3	9.2	22.7	22.7
-Year		1977	1993	1960	1980	1989					1989	1951	1973	DEC 1973
-Maximum in 24 hrs	46	8.8	10.3	8.2	6.1	T	0.0	0.0	0.0	0.0	0.3	8.0	10.9	10.9
-Year		1964	1965	1978	1980	1989					1989	1951	1973	DEC 1973
WIND:														
Mean Speed (mph)	46	12.5	12.3	13.5	13.1	11.2	9.7	8.4	7.9	9.0	10.4	12.4	12.4	11.1
Prevailing Direction through 1963		NW	NW	NW	S	SSW	SSW	SSW	SSW	SSW	S	S	S	SSW
Fastest Obs. 1 Min.														
-Direction (!!)	14	28	17	24	24	23	28	15	29	18	24	24	25	24
-Speed (MPH)	14	35	33	56	46	39	41	35	41	39	38	46	36	56
-Year		1990	1988	1991	1988	1987	1993	1992	1987	1988	1983	1988	1990	MAR 1991
Peak Gust														
-Direction (!!)	10	W	NW	SW	SW	NE	N	SE	W	NE	SW	SW	S	SW
-Speed (mph)	10	51	49	71	63	67	59	60	69	54	53	58	60	71
-Date		1990	1990	1991	1984	1990	1993	1992	1987	1989	1988	1988	1984	MAR 1991

PRECIPITATION (inches)

SPRINGFIELD, ILLINOIS

YEAR	JAN	FEB	MAR	APR	MAY	JUNE	JULY	AUG	SEP	OCT	NOV	DEC	ANNUAL
1964	1.64	1.27	4.00	9.91	1.82	2.26	1.32	2.20	1.24	0.16	4.19	1.01	31.02
1965	3.17	1.88	2.62	4.59	1.67	6.54	2.05	5.90	6.43	0.85	0.89	2.49	39.08
1966	0.36	2.20	1.04	5.75	3.54	1.35	0.96	2.71	4.72	2.37	3.08	2.62	30.70
1967	2.41	0.91	2.61	2.11	4.42	2.54	3.39	2.51	4.03	4.20	1.18	6.00	36.31
1968	1.79	1.15	1.25	2.44	5.69	3.25	4.67	0.99	3.29	1.43	3.08	2.64	31.67
1969	2.50	1.96	2.00	5.35	0.96	2.68	4.60	2.34	3.97	5.80	1.13	1.39	34.68
1970	0.54	0.67	1.99	9.10	2.26	4.68	2.55	4.20	7.73	2.50	0.70	1.33	38.25
1971	1.24	2.36	1.18	0.73	3.59	0.96	5.96	1.06	3.76	0.99	1.41	4.38	27.62
1972	1.03	0.82	4.03	3.35	1.88	2.72	1.70	4.52	3.95	1.40	3.27	3.36	32.03
1973	1.31	0.84	7.89	5.29	2.62	7.29	3.36	1.66	3.28	5.46	1.43	3.86	44.29
1974	2.61	3.15	3.39	3.11	6.37	5.00	0.91	7.70	2.17	1.39	3.58	1.44	40.82
1975	4.28	3.63	1.91	2.89	5.90	4.38	2.71	3.34	2.84	1.37	2.50	1.91	37.66
1976	0.98	3.67	5.60	1.07	1.96	1.41	2.29	2.33	2.20	3.00	0.53	0.66	25.70
1977	1.51	1.21	5.09	2.78	5.78	4.26	1.16	5.95	5.94	5.16	1.63	2.24	42.71
1978	0.72	0.83	4.20	2.84	5.81	1.73	2.99	4.04	1.58	1.57	2.13	3.39	31.83
1979	1.90	1.09	3.75	7.17	1.32	0.94	4.63	2.85	T	1.34	1.98	2.36	29.33
1980	0.72	1.42	4.29	2.22	2.22	3.23	2.08	3.91	4.95	1.47	0.57	1.99	29.07
1981	0.43	2.12	2.27	4.57	6.17	5.80	10.76	8.37	1.13	1.94	2.21	2.35	48.12
1982	4.48	1.81	3.04	3.40	4.12	2.54	2.53	3.68	2.75	2.69	4.50	8.94	44.48
1983	0.46	0.96	3.44	5.02	4.53	2.62	1.60	0.84	1.36	3.63	4.71	3.50	32.67
1984	0.70	1.97	4.00	5.45	6.32	2.26	3.46	0.63	4.80	4.74	4.36	3.91	42.60
1985	0.65	2.96	4.19	1.46	1.75	5.82	2.95	6.03	0.64	3.08	6.94	2.43	38.90
1986	0.04	1.80	1.45	1.57	2.56	6.23	5.39	1.13	8.57	3.63	1.95	1.40	35.72
1987	1.46	0.73	2.08	2.59	0.56	4.08	4.12	3.23	0.99	1.26	3.25	5.00	29.35
1988	2.17	1.39	2.69	1.27	1.76	0.62	1.74	1.56	2.84	1.68	4.37	3.22	25.31
1989	0.88	1.27	1.68	5.50	4.18	0.89	3.13	2.57	5.49	1.02	0.84	0.58	28.03
1990	1.49	4.89	3.41	1.28	8.84	9.22	5.48	2.68	1.91	5.03	3.47	4.97	52.67
1991	1.29	0.71	2.64	3.57	5.74	1.26	3.22	4.03	4.35	6.41	3.31	1.38	37.91
1992	1.03	1.59	2.39	2.82	0.52	2.31	6.86	0.46	1.54	1.47	6.14	2.36	29.49
1993	3.97	1.48	2.48	4.59	2.05	7.26	9.46	3.21	6.05	2.92	3.04	1.07	47.58
Record Mean	1.92	2.01	3.08	3.54	4.02	3.96	3.29	3.01	3.38	2.68	2.51	2.20	35.58

AVERAGE TEMPERATURE (deg. F)

SPRINGFIELD, ILLINOIS

YEAR	JAN	FEB	MAR	APR	MAY	JUNE	JULY	AUG	SEP	OCT	NOV	DEC	ANNUAL
1964	30.7	29.1	38.3	55.1	68.7	75.1	77.9	74.5	69.0	53.3	46.5	29.4	53.9
1965	27.3	30.0	31.1	54.6	70.3	73.2	74.2	72.4	68.6	55.1	45.6	38.2	52.4
1966	21.9	29.5	43.1	49.5	57.7	71.5	80.8	72.0	64.0	52.3	44.1	30.5	51.4
1967	29.5	26.4	43.8	56.5	59.3	73.8	72.9	68.6	63.7	54.9	38.5	31.4	51.6
1968	23.8	25.6	42.8	52.9	59.0	74.5	75.8	75.2	66.7	54.8	41.3	29.1	51.8
1969	23.4	31.5	32.6	54.6	64.2	70.7	77.9	74.0	66.4	54.0	39.7	28.2	51.5
1970	17.9	28.7	37.9	54.7	66.5	71.5	76.4	74.0	69.2	56.2	41.7	34.0	52.4
1971	23.7	30.7	39.1	54.4	60.4	79.0	72.4	72.5	71.2	62.9	43.8	38.5	54.1
1972	24.4	29.0	40.8	52.5	64.7	70.5	75.9	73.8	68.7	52.9	38.3	26.7	51.5
1973	29.7	30.7	48.3	51.9	59.6	73.2	76.0	76.0	69.8	60.7	45.5	28.0	54.1
1974	26.1	32.6	44.2	54.8	61.6	68.3	78.8	72.9	61.9	55.8	42.2	32.7	52.7
1975	31.1	29.4	37.4	50.4	66.5	73.6	74.7	76.6	63.8	57.7	47.1	33.4	53.5
1976	23.5	39.6	45.4	55.0	59.7	72.4	77.4	71.6	65.2	49.4	34.6	24.1	51.5
1977	10.3	29.5	47.6	59.9	71.0	71.6	79.4	72.8	68.6	53.8	42.8	27.0	52.9
1978	15.6	16.7	33.6	53.9	62.3	74.2	76.3	73.1	70.9	53.7	45.2	31.4	50.6
1979	12.4	17.1	40.3	50.4	64.0	74.9	75.4	74.0	67.7	55.7	42.0	35.8	50.8
1980	27.8	21.8	37.1	51.1	64.6	72.0	81.4	79.3	68.7	52.8	42.3	31.7	52.6
1981	26.4	32.4	44.1	60.4	60.3	74.9	77.1	73.8	67.2	54.6	45.3	26.8	53.6
1982	17.0	24.2	40.3	48.1	70.0	68.3	77.1	72.5	66.1	55.4	43.0	38.8	51.7
1983	28.7	34.9	40.6	47.8	59.9	73.8	80.4	80.0	69.1	57.2	45.7	16.1	52.9
1984	22.3	35.9	31.8	50.4	60.3	75.2	74.5	76.9	65.7	59.9	41.7	35.9	52.5
1985	18.5	24.9	45.7	57.8	65.2	69.7	74.3	70.8	67.9	58.1	42.8	21.5	51.4
1986	29.1	26.9	45.3	57.4	65.9	74.7	78.8	70.0	70.4	56.1	37.9	32.6	53.8
1987	26.1	36.6	45.4	54.7	70.5	75.4	78.8	75.3	67.6	50.6	46.2	34.6	55.2
1988	25.6	25.6	41.5	52.8	65.8	73.8	78.6	78.7	68.7	49.7	43.0	31.6	53.0
1989	36.0	20.3	40.6	52.3	59.3	70.5	75.4	73.3	63.8	56.9	42.0	18.7	50.8
1990	37.1	36.1	45.5	50.8	59.9	73.1	75.2	74.0	68.4	53.5	47.8	30.0	54.3
1991	24.0	35.3	45.1	57.3	69.2	75.6	75.3	74.5	66.8	55.8	38.5	34.7	54.3
1992	32.1	38.1	43.7	52.2	62.3	69.6	74.3	70.9	67.0	56.1	42.4	32.8	53.5
1993	29.8	27.6	38.6	50.9	64.3	72.0	77.2	76.0	62.4	52.9	40.5	33.1	52.1
Record Mean	26.9	30.2	40.8	53.3	63.7	73.2	77.2	74.9	67.8	56.3	42.5	31.3	53.2
Max	34.7	38.3	49.8	63.3	74.1	83.4	87.5	85.1	78.5	66.7	50.9	38.6	62.6
Min	19.1	22.1	31.9	43.2	53.3	62.9	66.9	64.7	57.2	46.0	34.0	24.0	43.8

REFERENCE NOTES

SPRINGFIELD, ILLINOIS

GENERAL
T - TRACE AMOUNT.
BLANK ENTRIES DENOTE MISSING/UNREPORTED DATA.
# INDICATES A STATION OR INSTRUMENT RELOCATION.

SPECIFIC

PM - INCLUDES LAST DAY OF PREVIOUS MONTH
ASOS - AUTOMATED SURFACE OBSERVING SYSTEM IN OPERATION DURING THESE MONTHS.

(a) - LENGTH OF RECORD IN YEARS, ALTHOUGH INDIVIDUAL MONTHS MAY BE MISSING.
0.* OR * - THE VALUE IS BETWEEN 0.0 AND 0.05
NORMALS - BASED ON THE 1961-1990 RECORD PERIOD.
EXTREMES - DATES ARE THE MOST RECENT OCCURRENCE
WIND DIR.- NUMERALS SHOW TENS OF DEGRESS CLOCKWISE FROM TRUE NORTH. "00" INDICATES CALM.
RESULTANT DIRECTIONS ARE GIVEN TO WHOLE DEGREES.
BOLD VALUES INDICATE EXTREME VALUES WHICH OCCURRED AFTER THE ASOS SYSTEM WAS COMMISSIONED.

RECORD = PERIOD OF RECORD
RECORD MEAN PRECIPITATION IS THE MEAN OF ALL DAILY PRECIPITATION AMOUNTS DURING THE PERIOD OF RECORD.
RECORD MAX(MIN) TEMPERATURE IS THE MEAN OF ALL DAILY MAX(MIN) TEMPERATURES DURING THE PERIOD OF RECORD.
RECORD MEAN TEMPERATURE IS THE SUM OF THE RECORD MAX AND RECORD MIN DIVIDED BY 2.
AVERAGE TEMPERATURE IS THE SUM OF THE MEAN DAILY MAX AND MIN TEMPERATURE DIVIDED BY 2.

EXCEPTIONS

RECORD MEANS ARE THROUGH THE CURRENT YEAR,
BEGINNING IN 1897 FOR TEMPERATURE
1897 FOR PRECIPITATION
1948 FOR SNOWFALL

HEATING DEGREE DAYS Base 65 deg. F SPRINGFIELD, ILLINOIS

SEASON	JULY	AUG	SEP	OCT	NOV	DEC	JAN	FEB	MAR	APR	MAY	JUNE	TOTAL
1964-65	0	10	66	356	554	1100	1158	973	1041	320	36	0	5614
1965-66	0	13	54	321	575	822	1329	987	670	459	242	15	5487
1966-67	0	5	100	388	618	1065	1091	1074	662	289	230	10	5532
1967-68	9	34	104	337	789	1032	1271	1134	684	358	203	8	5963
1968-69	1	2	31	353	703	1107	1282	934	995	314	130	31	5883
1969-70	0	0	59	370	748	1135	1457	1013	830	332	83	26	6053
1970-71	0	1	41	281	690	952	1275	952	793	342	168	0	5495
1971-72	9	1	65	130	633	820	1253	1040	745	375	114	27	5212
1972-73	3	11	58	375	795	1181	1089	957	512	398	174	0	5553
1973-74	0	0	30	191	578	1140	1200	901	642	316	167	29	5194
1974-75	0	2	135	299	680	996	1044	992	848	435	63	7	5501
1975-76	10	0	125	253	528	973	1280	732	601	321	188	2	5013
1976-77	0	3	84	497	905	1260	1693	989	532	222	44	12	6241
1977-78	0	2	21	340	661	1172	1521	1348	968	334	186	4	6557
1978-79	0	0	36	349	591	1035	1627	1336	758	435	119	0	6286
1979-80	0	9	50	323	684	898	1146	1249	857	421	95	8	5740
1980-81	0	0	49	395	675	1028	1193	906	648	181	184	0	5259
1981-82	0	0	51	332	581	1175	1483	1139	760	502	9	20	6052
1982-83	0	5	86	325	656	806	1117	836	749	510	169	14	5273
1983-84	0	0	75	269	574	1512	1320	835	1023	446	170	1	6225
1984-85	0	0	127	194	691	899	1437	1116	601	262	67	32	5426
1985-86	0	11	99	223	658	1340	1105	1060	618	261	74	1	5450
1986-87	0	19	25	284	807	999	1199	788	599	325	34	0	5079
1987-88	0	3	44	440	565	934	1216	1139	720	360	59	11	5491
1988-89	0	5	33	475	654	1027	890	1242	749	405	224	19	5723
1989-90	0	4	106	269	683	1431	856	803	613	459	168	10	5402
1990-91	7	2	72	360	512	1080	1262	827	612	243	69	0	5046
1991-92	0	0	120	295	787	932	1007	771	651	382	154	20	5119
1992-93	0	8	82	296	671	991	1083	1041	813	419	85	35	5524
1993-94	0	0	117	382	730	979							

COOLING DEGREE DAYS Base 65 deg. F SPRINGFIELD, ILLINOIS

YEAR	JAN	FEB	MAR	APR	MAY	JUNE	JULY	AUG	SEP	OCT	NOV	DEC	TOTAL
1969	0	0	0	6	110	206	407	288	108	37	0	0	1162
1970	0	0	0	31	136	229	361	284	174	17	0	0	1232
1971	0	0	0	32	33	425	248	241	254	69	2	0	1304
1972	0	0	0	6	114	199	347	291	178	10	0	0	1145
1973	0	0	0	12	12	251	350	348	184	64	0	0	1221
1974	0	0	7	14	70	136	437	252	49	17	2	0	984
1975	0	0	0	2	115	271	315	367	94	36	0	0	1200
1976	0	0	3	28	34	228	393	214	98	23	0	0	1021
1977	0	0	0	76	238	217	452	251	134	0	1	0	1369
1978	0	0	1	8	107	289	358	258	222	5	5	0	1253
1979	0	0	0	2	93	305	327	295	138	41	0	0	1201
1980	0	0	0	10	88	228	515	452	169	20	0	0	1482
1981	0	0	8	50	43	303	380	280	121	13	0	0	1198
1982	0	0	0	1	173	131	379	244	126	32	1	1	1088
1983	0	0	1	1	17	289	483	471	205	34	0	0	1501
1984	0	0	0	16	30	312	303	375	155	44	0	1	1236
1985	0	0	7	52	80	178	293	202	191	17	0	0	1020
1986	0	0	14	38	109	299	434	182	193	14	0	0	1283
1987	0	0	0	21	212	318	436	329	130	0	8	0	1454
1988	0	0	0	3	93	281	430	437	150	6	0	0	1400
1989	0	0	2	30	57	190	332	268	77	24	0	0	980
1990	0	0	14	39	18	261	328	289	182	12	1	0	1144
1991	0	0	5	20	207	327	326	302	184	18	0	0	1389
1992	0	0	0	4	75	166	294	199	150	28	0	0	916
1993	0	0	0	1	71	253	386	352	46	13	0	0	1122

SNOWFALL (inches) SPRINGFIELD, ILLINOIS

SEASON	JULY	AUG	SEP	OCT	NOV	DEC	JAN	FEB	MAR	APR	MAY	JUNE	TOTAL
1964-65	0.0	0.0	0.0	0.0	4.4	1.6	9.1	14.2	11.1	T	0.0	0.0	40.4
1965-66	0.0	0.0	0.0	0.0	T	1.7	0.4	8.6	3.5	T	T	0.0	14.2
1966-67	0.0	0.0	0.0	0.0	0.3	2.0	6.4	3.7	2.4	0.1	0.0	0.0	14.9
1967-68	0.0	0.0	0.0	0.1	2.8	6.4	13.0	1.7	4.5	T	0.0	0.0	28.5
1968-69	0.0	0.0	0.0	0.0	0.7	4.7	5.0	6.8	7.0	0.0	0.0	0.0	24.2
1969-70	0.0	0.0	0.0	0.0	0.5	10.2	7.0	3.9	4.3	1.1	0.0	0.0	27.0
1970-71	0.0	0.0	0.0	0.0	T	0.9	4.2	3.4	0.7	3.8	0.0	0.0	13.0
1971-72	0.0	0.0	0.0	0.0	4.4	T	8.1	6.0	3.6	T	0.0	0.0	22.1
1972-73	0.0	0.0	0.0	0.0	5.4	3.1	0.6	3.1	0.3	0.9	0.0	0.0	13.4
1973-74	0.0	0.0	0.0	0.0	1.3	22.7	7.3	5.3	4.4	0.4	0.0	0.0	41.4
1974-75	0.0	0.0	0.0	0.0	6.8	2.5	4.5	14.2	4.5	0.4	0.0	0.0	32.9
1975-76	0.0	0.0	0.0	0.0	8.5	4.3	8.5	2.1	2.2	0.0	0.0	0.0	25.6
1976-77	0.0	0.0	0.0	T	T	6.9	21.1	8.9	0.6	1.1	0.0	0.0	38.6
1977-78	0.0	0.0	0.0	0.0	6.4	8.9	8.9	9.4	18.5	T	0.0	0.0	52.1
1978-79	0.0	0.0	0.0	0.0	T	3.2	15.9	4.6	8.1	T	0.0	0.0	31.8
1979-80	0.0	0.0	0.0	0.0	0.8	0.1	5.1	11.7	5.5	7.3	0.0	0.0	30.5
1980-81	0.0	0.0	0.0	T	3.5	2.1	2.7	8.6	0.6	0.0	0.0	0.0	17.5
1981-82	0.0	0.0	0.0	0.0	0.1	21.6	12.0	11.4	0.7	4.6	0.0	0.0	50.4
1982-83	0.0	0.0	0.0	0.0	0.2	0.9	2.4	1.3	5.4	0.2	0.0	0.0	10.4
1983-84	0.0	0.0	0.0	T	T	16.2	3.9	9.8	5.7	0.0	0.0	0.0	35.6
1984-85	0.0	0.0	0.0	0.0	1.8	0.7	9.3	2.5	T	T	0.0	0.0	14.3
1985-86	0.0	0.0	0.0	0.0	T	4.3	0.2	15.1	0.6	0.1	0.0	0.0	20.3
1986-87	0.0	0.0	0.0	0.0	0.5	T	20.3	T	0.0	0.0	0.0	0.0	20.8
1987-88	0.0	0.0	0.0	0.0	0.1	5.7	0.4	10.2	5.0	0.0	0.0	0.0	21.4
1988-89	0.0	0.0	0.0	0.0	T	5.5	0.3	13.7	5.2	T	T	0.0	24.7
1989-90	0.0	0.0	0.0	0.3	T	6.7	0.6	0.5	1.2	T	0.0	0.0	9.3
1990-91	0.0	0.0	0.0	0.0	T	9.4	6.8	0.8	3.1	0.0	0.0	0.0	20.1
1991-92	0.0	0.0	0.0	T	3.2	0.5	4.6	0.9	2.2	T	0.0	0.0	11.4
1992-93	0.0	0.0	0.0	0.0	1.2	2.5	7.0	16.0	2.9	0.4	0.0	0.0	30.0
1993-94	0.0	0.0	0.0	0.2	0.8	1.7							
Record Mean	0.0	0.0	0.0	T	1.7	4.9	5.8	6.4	4.1	0.7	T	0.0	23.5

EVANSVILLE, INDIANA

Evansville, Indiana, is located on the Ohio River. The country around Evansville ranges from level to areas of rolling terrain near the river. Dress Regional Airport, where the observations have been taken since August 31, 1940, is located in a shallow valley with low hills to the east and west which parallel the valley, but slope down to the south. There are hills 5 miles to the north which are about 100 feet higher than the field. The open end of the valley slopes down and south toward the city of Evansville and the Ohio River.

Records of precipitation, temperature, and wind are available from the city office locations prior to August 1940. Both precipitation and temperature records were from roof-top exposures in the city and from ground exposures at the airport. The airport exposure is not subject to the effect of an early morning smoke blanket that was prevalent over the city during the downtown exposure.

Prevailing wind direction is from the south-southwest. The strongest winds occur during a deep winter storm passage through the Lower Ohio Valley. Strong and cold north to northwest winds occur from late autumn to early spring, most often, in January and February, as large domes of arctic high pressure moves into the midwest.

Geographically, Evansville lies in the path of moisture-bearing low pressure formations that move from the western Gulf region, northeastward over the Mississippi and Ohio Valleys to the Great Lakes and northern Atlantic Coast. Much of the precipitation results from these storm systems, especially in the cooler part of the year.

Both temperature and precipitation are closely related to the movement of the polar front and the storms which move along the front. This is especially true in the winter and spring months. In summer and early autumn changes are less severe and periods of polar air invasions are less prolonged. There is considerable variation in seasonal and monthly temperature and precipitation from year to year as these factors depend greatly on the frequency of storm and frontal passages. A comparatively few miles difference in the distance of the paths of these storms, often spells the difference between whether the precipitation is snow, rain, or freezing rain during winter months.

Convective thunderstorms, developing in the maritime tropical air from the Gulf of Mexico and squall line activity, seem to be the factors which combine to supply the summer rainfall. The greatest precipitation intensities for short periods of time come in the months of greatest thunderstorm frequency. The greatest intensities for 24 hours or more are confined to the winter months when storm centers to the south produce a sustained flow of overrunning Gulf air.

Severe storms are rather infrequent but thunderstorms cause some wind damage each year. Hail often occurs with the stronger thunderstorms. Evansville is in tornado alley with the most frequent occurrence in early spring and late fall. The tornado frequency would probably be less than one every ten years for Evansville.

Snowfall varies greatly from season to season, as do rainfall and temperature. Of note is the fact that snowfalls of 2 or more inches are very infrequent, and these amounts are usually melted within a day or two.

The growing season averages 199 days, but has been as long as 250 days and as short as 169 days.

NORMALS, MEANS, AND EXTREMES

EVANSVILLE, INDIANA

LATITUDE: 38°02'N LONGITUDE: 87°32'W ELEVATION: FT. GRND 381 BARO 400 TIME ZONE: CENTRAL WBAN: 93817

	(a)	JAN	FEB	MAR	APR	MAY	JUNE	JULY	AUG	SEP	OCT	NOV	DEC	YEAR
TEMPERATURE °F:														
Normals														
-Daily Maximum		38.9	43.7	55.9	67.4	76.9	86.2	89.1	87.2	80.7	69.6	55.9	43.6	66.3
-Daily Minimum		21.2	25.0	35.7	45.0	54.2	63.3	67.5	64.9	57.6	44.7	36.5	26.7	45.2
-Monthly		30.1	34.4	45.8	56.2	65.5	74.8	78.4	76.1	69.2	57.2	46.2	35.2	55.8
Extremes														
-Record Highest	53	76	79	84	91	95	104	105	102	103	94	83	77	105
-Year		1943	1962	1986	1989	1975	1954	1954	1983	1954	1953	1961	1982	JUL 1954
-Record Lowest	53	-21	-23	-9	23	28	41	47	43	31	21	-3	-15	-23
-Year		1977	1951	1960	1990	1963	1966	1947	1986	1942	1952	1950	1989	FEB 1951
NORMAL DEGREE DAYS:														
Heating (base 65°F)		1082	857	595	273	114	0	0	0	33	266	564	924	4708
Cooling (base 65°F)		0	0	0	9	130	294	415	344	159	25	0	0	1376
% OF POSSIBLE SUNSHINE	53	43	48	54	59	64	72	74	74	69	65	48	42	59
MEAN SKY COVER (tenths)														
Sunrise - Sunset	53	7.1	6.7	6.8	6.5	6.1	5.7	5.4	5.1	5.1	5.0	6.5	7.0	6.1
MEAN NUMBER OF DAYS:														
Sunrise to Sunset														
-Clear	53	6.7	6.6	6.5	6.8	8.5	8.2	9.4	11.2	11.4	12.5	7.8	6.6	102.2
-Partly Cloudy	53	5.4	6.2	8.1	8.3	8.6	11.2	12.3	11.2	8.4	7.5	6.6	6.1	99.9
-Cloudy	53	18.9	15.4	16.5	14.9	13.9	10.5	9.3	8.7	10.1	11.0	15.5	18.3	163.1
Precipitation														
.01 inches or more	53	10.2	9.1	11.8	11.6	11.2	9.6	9.3	7.4	7.5	7.6	9.6	10.5	115.4
Snow, Ice pellets, hail														
1.0 inches or more	53	1.3	1.0	0.7	0.1	0.0	0.0	0.0	0.0	0.0	0.*	0.2	0.8	4.1
Thunderstorms	53	0.9	1.3	3.6	4.8	6.5	7.5	7.7	5.4	3.4	2.1	1.5	0.6	45.3
Heavy Fog Visibility														
1/4 mile or less	53	2.3	1.5	1.0	0.5	0.6	0.5	0.6	0.9	1.3	1.4	1.2	1.9	13.7
Temperature °F														
-Maximum														
90° and above	32	0.0	0.0	0.0	0.1	2.0	9.6	15.5	11.0	4.1	0.3	0.0	0.0	42.5
32° and below	32	9.7	5.7	0.7	0.0	0.0	0.0	0.0	0.0	0.0	0.0	0.5	5.2	21.8
-Minimum														
32° and below	32	25.1	20.4	13.3	2.5	0.1	0.0	0.0	0.0	0.0	3.3	11.7	21.7	98.1
0° and below	32	2.0	1.1	0.1	0.0	0.0	0.0	0.0	0.0	0.0	0.0	0.0	0.8	4.0
AVG. STATION PRESS. (mb)	21	1006.8	1005.7	1002.5	1001.5	1001.0	1001.3	1002.3	1003.2	1004.0	1005.2	1005.2	1006.5	1003.8
RELATIVE HUMIDITY (%)														
Hour 00	32	76	76	75	74	80	80	83	85	85	80	76	78	79
Hour 06 (Local Time)	32	78	79	79	79	82	82	85	87	88	84	80	80	82
Hour 12	32	66	64	60	54	55	54	57	57	57	53	62	68	59
Hour 18	32	69	66	61	55	57	55	59	61	66	64	68	72	63
PRECIPITATION (inches):														
Water Equivalent														
-Normal		2.66	3.12	4.71	4.02	4.75	3.49	4.04	3.11	2.97	2.87	3.73	3.67	43.14
-Maximum Monthly	53	13.50	7.25	12.84	10.26	12.89	9.30	9.69	8.43	9.89	8.33	8.49	8.23	13.50
-Year		1950	1956	1964	1983	1981	1943	1958	1977	1945	1941	1957	1982	JAN 1950
-Minimum Monthly	53	0.51	0.27	0.89	1.10	0.91	0.65	0.18	0.13	0.56	0.01	0.91	0.56	0.01
-Year		1981	1947	1941	1959	1965	1991	1974	1943	1953	1964	1965	1976	OCT 1964
-Maximum in 24 hrs	53	3.73	3.20	5.63	3.95	6.05	3.26	4.09	3.70	3.45	3.00	3.48	2.35	6.05
-Year		1982	1986	1964	1955	1961	1971	1978	1977	1945	1976	1988	1990	MAY 1961
Snow, Ice pellets, hail														
-Maximum Monthly	53	21.3	18.4	20.2	8.6	T	0.0	0.0	0.0	T	4.6	6.9	10.4	21.3
-Year		1977	1993	1960	1971	1993				1990	1993	1958	1973	JAN 1977
-Maximum in 24 hrs	53	8.7	10.9	10.6	8.6	T	0.0	0.0	0.0	T	4.1	6.9	7.0	10.9
-Year		1978	1993	1960	1971	1993				1990	1993	1958	1963	FEB 1993
WIND:														
Mean Speed (mph)	53	9.2	9.4	10.1	9.7	8.0	7.2	6.2	5.8	6.4	6.9	8.7	8.9	8.0
Prevailing Direction through 1963		SSW	NW	WNW	SSW	SSW	SW	SW	SW	SSW	NW	NW	NW	SSW
Fastest Obs. 1 Min.														
-Direction (!!)	8	22	33	23	32	33	33	32	02	32	31	31	24	33
-Speed (MPH)	8	30	31	35	33	46	32	35	35	28	29	41	41	46
-Year		1987	1992	1986	1988	1990	1987	1987	1989	1984	1990	1992	1987	MAY 1990
Peak Gust														
-Direction (!!)	10	NW	SW	SW	SW	W	N	NW	N	SW	SW	NW	SW	NW
-Speed (mph)	10	49	52	52	47	69	58	56	41	55	47	70	56	70
-Date		1985	1988	1986	1991	1987	1987	1986	1989	1990	1985	1992	1987	NOV 1992

PRECIPITATION (inches)

EVANSVILLE, INDIANA

YEAR	JAN	FEB	MAR	APR	MAY	JUNE	JULY	AUG	SEP	OCT	NOV	DEC	ANNUAL
1964	2.19	1.57	12.84	4.85	1.88	1.58	2.35	1.17	3.50	0.01	2.34	4.09	38.37
1965	2.53	5.01	2.99	2.97	0.91	2.03	6.09	3.83	5.18	1.07	0.91	1.51	35.03
1966	2.89	4.67	1.34	5.33	3.48	1.74	1.96	2.24	3.33	1.65	2.38	5.52	36.53
1967	1.09	2.13	3.54	3.93	3.81	1.60	5.44	3.15	1.19	7.92	3.88	5.51	43.19
1968	2.44	1.76	4.70	3.59	4.13	3.07	5.59	2.58	3.05	2.10	4.60	5.60	43.21
1969	7.98	1.39	2.06	3.51	5.36	4.96	8.72	1.91	2.66	4.20	3.29	3.19	49.23
1970	1.00	2.57	6.30	7.74	3.33	5.94	3.49	2.82	3.25	4.70	1.98	2.81	45.93
1971	2.88	4.90	1.65	2.81	3.96	6.89	3.38	2.17	4.47	2.42	1.57	3.15	40.25
1972	1.68	3.05	5.05	6.66	1.84	1.92	5.31	2.49	1.25	3.06	5.47	4.49	42.27
1973	2.45	1.65	6.97	5.47	5.57	5.24	3.41	2.83	0.73	1.99	6.02	3.85	46.18
1974	3.63	1.51	4.54	3.66	6.38	3.81	0.18	6.89	3.90	1.88	3.85	3.04	43.27
1975	4.05	4.06	7.18	6.62	3.63	3.07	3.13	5.47	2.60	2.73	4.21	4.26	51.01
1976	2.11	2.61	2.25	1.30	7.48	4.24	2.14	0.24	3.29	4.82	1.05	0.56	32.09
1977	1.91	1.29	6.17	3.34	2.68	6.57	4.83	8.43	4.58	2.81	4.30	3.17	50.08
1978	2.64	0.76	4.69	3.49	3.93	0.84	7.66	3.64	2.72	1.61	4.86	6.12	42.96
1979	3.60	4.80	6.30	6.07	3.72	2.78	7.22	2.36	2.83	2.68	6.82	3.03	52.21
1980	1.77	1.25	4.38	2.73	4.10	6.01	4.50	2.15	2.51	3.13	2.34	0.89	35.76
1981	0.51	2.89	1.70	2.50	12.89	1.78	5.08	6.04	2.00	2.36	3.40	2.20	43.35
1982	9.15	1.65	5.07	3.24	4.29	2.95	2.62	3.41	6.07	1.75	4.25	8.23	52.68
1983	1.79	0.74	4.33	10.26	8.87	4.59	1.51	0.94	0.73	5.62	5.55	3.55	48.48
1984	0.85	2.55	7.02	5.75	2.89	3.35	1.50	2.70	6.97	5.13	5.05	5.99	49.75
1985	1.76	4.24	6.10	3.80	2.97	4.68	1.18	3.76	3.59	4.46	7.61	1.74	45.89
1986	1.15	5.77	2.64	2.29	2.93	3.77	5.39	2.07	3.84	3.30	2.35	2.18	37.68
1987	0.77	3.51	2.11	2.31	3.90	5.97	3.19	0.47	1.98	1.23	3.36	5.71	34.51
1988	3.28	3.94	2.89	1.77	1.33	1.11	6.63	2.72	1.19	2.86	7.96	2.75	38.43
1989	3.35	7.00	6.40	4.19	3.72	4.00	7.83	3.46	2.21	2.16	1.64	1.38	47.34
1990	4.26	5.60	2.15	3.75	11.34	3.22	1.01	3.47	2.54	4.81	2.92	7.45	52.52
1991	3.02	2.99	4.27	2.56	3.11	0.65	2.58	0.46	2.60	3.05	3.67	3.72	32.68
1992	0.85	1.51	4.50	1.19	3.44	1.44	8.40	4.39	2.89	1.17	4.34	1.69	35.81
1993	3.57	2.61	3.23	4.38	4.20	4.65	2.37	2.17	5.59	3.76	6.62	2.68	45.83
Record Mean	3.50	3.17	4.29	3.91	4.18	3.80	3.54	3.16	3.12	2.77	3.58	3.41	42.43

AVERAGE TEMPERATURE (deg. F)

EVANSVILLE, INDIANA

YEAR	JAN	FEB	MAR	APR	MAY	JUNE	JULY	AUG	SEP	OCT	NOV	DEC	ANNUAL
1964	34.7	33.8	45.4	59.8	67.7	75.6	77.6	75.6	68.4	52.9	47.5	36.3	56.3
1965	32.3	33.9	36.8	57.5	69.7	73.7	76.2	74.6	69.7	55.9	48.1	41.5	55.8
1966	26.4	32.8	46.6	53.3	61.2	73.6	81.8	74.4	65.5	54.1	46.4	35.0	54.3
1967	36.9	31.7	49.2	59.6	61.6	73.6	73.4	70.1	66.1	57.3	42.0	37.4	54.9
1968	28.5	30.4	45.7	56.6	63.2	74.5	77.7	77.4	68.4	56.9	45.9	34.3	54.9
1969	31.5	36.5	38.3	56.1	65.0	74.4	79.2	74.6	68.3	56.7	41.6	31.5	54.5
1970	24.0	32.6	41.6	58.7	68.4	72.0	75.3	75.4	73.5	57.1	44.6	37.5	55.0
1971	29.8	33.6	42.0	56.1	63.4	79.3	77.1	76.6	72.7	63.8	45.7	42.1	56.9
1972	32.5	34.1	44.2	56.4	66.0	71.4	75.6	73.8	70.5	53.4	41.5	34.4	54.5
1973	32.7	34.6	53.6	55.1	62.4	75.8	79.1	77.5	73.2	62.1	49.6	34.1	57.5
1974	37.4	40.0	50.1	57.0	67.0	70.9	79.4	74.2	63.2	56.2	47.4	36.9	56.6
1975	36.8	38.3	42.6	55.1	69.6	76.2	77.8	77.3	66.5	58.4	48.8	36.9	57.0
1976	29.3	43.1	51.6	57.2	61.6	73.3	77.0	73.6	67.1	52.4	38.7	31.8	54.7
1977	14.8	33.9	51.2	61.5	72.0	75.4	80.9	76.9	72.4	55.6	48.8	33.5	56.4
1978	20.3	21.0	39.8	57.5	64.4	76.8	79.1	76.6	72.0	54.5	49.0	36.8	54.0
1979	20.9	24.6	46.7	54.3	63.5	75.1	76.6	74.8	68.3	56.5	44.1	38.6	53.7
1980	33.1	26.8	40.3	52.8	65.4	73.2	82.0	81.6	72.4	55.4	45.1	37.3	55.4
1981	29.6	37.2	44.7	61.7	61.4	76.6	78.5	75.9	67.2	57.0	48.3	34.5	56.0
1982	27.4	32.1	47.8	52.1	70.7	70.7	79.6	74.1	67.5	59.2	48.9	45.2	56.3
1983	35.2	39.3	46.5	51.5	62.7	74.7	81.4	81.9	70.6	59.3	47.7	26.2	56.4
1984	27.1	39.0	40.0	54.4	62.5	78.6	76.0	76.0	66.3	62.9	43.5	43.7	55.8
1985	23.7	29.6	51.9	59.1	66.5	73.7	79.2	75.1	68.5	60.7	50.1	28.2	55.5
1986	32.9	37.5	47.5	58.1	67.6	76.7	80.5	72.9	72.4	58.0	43.7	35.7	57.0
1987	32.2	38.5	47.9	54.2	71.7	76.5	78.3	77.9	71.0	51.2	49.5	39.7	57.4
1988	29.0	33.1	45.4	55.6	67.0	75.7	79.0	78.8	69.5	51.5	46.4	36.5	55.6
1989	40.1	32.5	46.9	56.6	63.2	73.8	77.7	76.8	68.3	58.3	45.6	23.0	55.2
1990	41.9	43.2	49.8	53.8	63.0	74.9	77.2	75.7	71.0	56.1	50.5	38.1	57.9
1991	31.3	39.7	47.9	58.9	71.1	77.0	79.1	77.0	70.2	59.2	43.2	39.2	57.8
1992	35.4	41.6	47.0	57.7	63.9	71.7	79.0	73.4	68.6	57.7	46.9	36.4	56.6
1993	36.5	32.9	43.1	54.1	67.0	75.8	83.1	79.5	67.0	55.5	44.7	36.4	56.3
Record Mean	33.0	35.7	45.8	56.2	66.0	75.1	78.8	77.0	70.4	58.7	46.2	36.1	56.6
Max	41.1	44.2	55.1	66.4	76.3	85.4	88.9	87.3	81.3	70.0	55.3	43.9	66.3
Min	25.0	27.1	36.4	46.0	55.6	64.7	68.7	66.6	59.4	47.4	37.1	28.3	46.9

REFERENCE NOTES

EVANSVILLE, INDIANA

GENERAL
T - TRACE AMOUNT.
BLANK ENTRIES DENOTE MISSING/UNREPORTED DATA.
# INDICATES A STATION OR INSTRUMENT RELOCATION.

SPECIFIC

(a) - LENGTH OF RECORD IN YEARS, ALTHOUGH INDIVIDUAL MONTHS MAY BE MISSING.
0.* OR * - THE VALUE IS BETWEEN 0.0 AND 0.05
NORMALS - BASED ON THE 1961-1990 RECORD PERIOD.
EXTREMES - DATES ARE THE MOST RECENT OCCURRENCE.
WIND DIR.- NUMERALS SHOW TENS OF DEGRESS CLOCKWISE FROM TRUE NORTH. "00" INDICATES CALM.
RESULTANT DIRECTIONS ARE GIVEN TO WHOLE DEGREES.
BOLD VALUES INDICATE EXTREME VALUES WHICH OCCURRED AFTER THE ASOS SYSTEM WAS COMMISSIONED.

RECORD = PERIOD OF RECORD
RECORD MEAN PRECIPITATION IS THE MEAN OF ALL DAILY PRECIPITATION AMOUNTS DURING THE PERIOD OF RECORD.
RECORD MAX(MIN) TEMPERATURE IS THE MEAN OF ALL DAILY MAX(MIN) TEMPERATURES DURING THE PERIOD OF RECORD.
RECORD MEAN TEMPERATURE IS THE SUM OF THE RECORD MAX AND RECORD MIN DIVIDED BY 2.
AVERAGE TEMPERATURE IS THE SUM OF THE MEAN DAILY MAX AND MIN TEMPERATURE DIVIDED BY 2.

EXCEPTIONS

RECORD MEANS ARE THROUGH THE CURRENT YEAR, BEGINNING IN 1897 FOR TEMPERATURE
1878 FOR PRECIPITATION
1941 FOR SNOWFALL

HEATING DEGREE DAYS Base 65 deg. F EVANSVILLE, INDIANA

SEASON	JULY	AUG	SEP	OCT	NOV	DEC	JAN	FEB	MAR	APR	MAY	JUNE	TOTAL
1964-65	0	8	52	367	518	885	1008	867	867	254	18	0	4844
1965-66	0	5	39	283	500	727	1192	894	566	356	149	10	4721
1966-67	0	0	57	341	554	923	864	928	498	218	161	11	4555
1967-68	1	11	72	264	684	844	1128	997	595	256	111	5	4968
1968-69	0	0	15	292	569	946	1029	793	819	269	80	13	4825
1969-70	0	0	35	299	693	1034	1266	896	716	225	57	1	5222
1970-71	2	0	24	255	605	846	1087	874	707	290	90	0	4780
1971-72	0	0	23	85	571	701	1000	887	639	274	64	16	4260
1972-73	2	3	26	356	702	940	997	844	345	312	106	0	4633
1973-74	0	0	11	160	459	954	849	696	480	264	71	6	3950
1974-75	0	0	124	280	537	865	867	742	689	319	23	0	4446
1975-76	0	0	75	223	484	861	1100	628	417	275	139	0	4202
1976-77	0	0	27	391	786	1021	1549	867	428	162	32	4	5267
1977-78	0	0	3	289	495	970	1377	1228	774	233	137	0	5506
1978-79	0	0	10	323	473	865	1360	1125	559	326	106	0	5147
1979-80	0	1	28	290	619	813	982	1103	756	367	81	10	5050
1980-81	0	0	24	329	591	852	1090	771	624	161	155	0	4597
1981-82	0	0	53	256	498	940	1160	914	534	386	16	0	4757
1982-83	0	0	52	233	486	618	918	711	567	406	106	4	4101
1983-84	0	0	61	186	514	1195	1169	747	769	329	131	0	5101
1984-85	0	0	79	108	638	653	1276	985	411	208	55	9	4422
1985-86	0	0	75	185	446	1135	989	762	538	226	70	0	4426
1986-87	0	15	14	240	632	900	1007	735	528	330	19	0	4420
1987-88	0	0	15	423	456	777	1108	917	602	284	46	4	4632
1988-89	0	0	18	418	548	877	765	902	558	308	142	1	4537
1989-90	0	1	54	225	577	1297	707	603	487	358	97	15	4421
1990-91	2	1	35	291	432	828	1037	702	528	191	42	0	4089
1991-92	0	0	88	227	647	791	913	673	549	259	118	10	4275
1992-93	0	0	46	236	538	875	879	892	671	322	46	12	4517
1993-94	0	0	55	296	600	879							

COOLING DEGREE DAYS Base 65 deg. F EVANSVILLE, INDIANA

YEAR	JAN	FEB	MAR	APR	MAY	JUNE	JULY	AUG	SEP	OCT	NOV	DEC	TOTAL
1969	0	0	0	8	85	300	448	307	140	48	0	0	1336
1970	0	0	0	43	169	220	329	331	286	18	0	0	1396
1971	0	0	0	28	49	437	382	367	260	53	0	0	1576
1972	0	0	2	21	99	214	337	285	196	2	4	0	1160
1973	0	0	0	23	35	331	438	396	265	76	3	0	1567
1974	0	0	26	28	141	188	452	292	73	14	15	0	1229
1975	0	0	0	29	176	342	406	388	129	29	1	0	1500
1976	0	0	6	47	40	258	379	274	98	10	0	0	1112
1977	0	0	9	61	255	323	501	376	232	6	16	0	1779
1978	0	0	0	16	125	361	444	366	229	7	2	0	1550
1979	0	0	0	13	65	310	365	312	138	35	0	0	1238
1980	0	0	0	5	102	264	535	521	257	39	3	0	1726
1981	0	0	1	69	50	355	425	343	128	15	3	0	1389
1982	0	0	10	3	198	179	458	290	134	59	9	11	1351
1983	0	0	3	8	42	303	514	532	236	17	0	0	1655
1984	0	0	0	16	60	416	348	349	127	49	0	0	1365
1985	0	3	13	36	108	276	447	319	190	58	5	0	1455
1986	0	0	2	27	156	360	487	265	246	32	0	0	1575
1987	0	0	0	8	235	350	420	408	201	0	1	0	1623
1988	0	0	0	11	113	329	441	436	162	8	0	0	1500
1989	0	0	3	64	96	272	403	369	161	28	0	0	1396
1990	0	0	21	29	43	318	387	336	220	23	3	0	1380
1991	0	0	4	15	241	369	445	379	249	55	0	0	1757
1992	0	0	0	47	90	219	440	268	162	14	0	0	1240
1993	0	0	0	3	115	342	566	456	122	9	0	0	1613

SNOWFALL (inches) EVANSVILLE, INDIANA

SEASON	JULY	AUG	SEP	OCT	NOV	DEC	JAN	FEB	MAR	APR	MAY	JUNE	TOTAL
1964-65	0.0	0.0	0.0	0.0	1.2	T	4.5	9.7	2.6	0.0	0.0	0.0	18.0
1965-66	0.0	0.0	0.0	0.0	T	T	2.9	10.2	0.8	T	0.0	0.0	13.9
1966-67	0.0	0.0	0.0	0.0	2.8	0.6	0.1	5.1	7.2	0.0	0.0	0.0	15.8
1967-68	0.0	0.0	0.0	0.0	T	0.9	10.0	0.7	6.7	0.0	0.0	0.0	18.3
1968-69	0.0	0.0	0.0	0.0	0.3	T	4.5	7.0	0.8	0.0	0.0	0.0	12.6
1969-70	0.0	0.0	0.0	0.0	2.0	7.7	5.4	7.4	15.2	T	0.0	0.0	37.7
1970-71	0.0	0.0	0.0	0.0	T	0.5	0.6	8.5	4.3	8.6	0.0	0.0	22.5
1971-72	0.0	0.0	0.0	0.0	3.4	T	2.0	1.5	0.7	T	0.0	0.0	7.6
1972-73	0.0	0.0	0.0	0.0	0.3	3.5	0.6	1.6	T	0.3	0.0	0.0	6.3
1973-74	0.0	0.0	0.0	0.0	0.0	10.4	0.4	1.0	3.4	T	0.0	0.0	15.2
1974-75	0.0	0.0	0.0	0.0	T	3.0	1.5	1.3	13.3	T	0.0	0.0	19.1
1975-76	0.0	0.0	0.0	0.0	1.7	1.3	2.9	0.8	0.2	0.0	0.0	0.0	6.9
1976-77	0.0	0.0	0.0	0.0	0.3	4.3	21.3	0.9	T	T	0.0	0.0	26.8
1977-78	0.0	0.0	0.0	0.0	3.9	1.7	20.8	5.3	5.7	0.0	0.0	0.0	37.4
1978-79	0.0	0.0	0.0	0.0	0.0	T	15.0	8.5	0.1	0.0	0.0	0.0	23.6
1979-80	0.0	0.0	0.0	0.0	T	T	7.1	5.0	4.2	T	0.0	0.0	16.3
1980-81	0.0	0.0	0.0	0.0	0.5	0.2	2.2	0.5	T	0.0	0.0	0.0	3.4
1981-82	0.0	0.0	0.0	0.0	0.1	0.7	4.1	6.8	0.3	3.0	0.0	0.0	15.0
1982-83	0.0	0.0	0.0	0.0	T	T	1.3	1.2	1.0	0.6	0.0	0.0	4.1
1983-84	0.0	0.0	0.0	0.0	T	1.6	5.5	9.7	2.8	0.0	0.0	0.0	19.6
1984-85	0.0	0.0	0.0	0.0	T	6.7	10.3	9.4	0.0	T	0.0	0.0	26.4
1985-86	0.0	0.0	0.0	0.0	0.0	2.8	1.1	6.8	T	0.0	0.0	0.0	10.7
1986-87	0.0	0.0	0.0	0.0	T	0.1	2.7	3.2	1.7	0.0	0.0	0.0	7.7
1987-88	0.0	0.0	0.0	0.0	T	1.3	4.0	1.4	0.7	0.0	0.0	0.0	7.4
1988-89	0.0	0.0	0.0	0.0	T	3.0	T	0.5	0.1	0.0	0.0	0.0	3.6
1989-90	0.0	0.0	0.0	0.9	T	6.0	1.6	0.2	4.6	0.0	0.0	0.0	13.3
1990-91	0.0	0.0	T	0.0	0.0	7.2	0.8	1.2	0.2	0.0	0.0	0.0	9.4
1991-92	0.0	0.0	0.0	0.0	1.0	T	1.0	T	1.3	T	0.0	0.0	3.3
1992-93	0.0	0.0	0.0	T	0.1	0.7	T	18.4	1.5	T	T	0.0	20.7
1993-94	0.0	0.0	0.0	4.6	0.3	3.4							
Record Mean	0.0	0.0	T	0.1	0.6	2.4	4.1	3.7	2.5	0.3	T	0.0	13.6

FORT WAYNE, INDIANA

Fort Wayne is located at the junction of the St. Marys, St. Joseph, and Maumee Rivers in northeastern Indiana. The surrounding area is generally level south and east of the city. Southwest and west, the terrain is somewhat rolling, while to the northwest and a few miles north from the city, it becomes quite hilly. The highest point in the general area is about 40 miles due north of Fort Wayne, near Angola, Indiana. At this point, the elevation rises to 1,060 feet above sea level.

The climate is representative of northeastern Indiana and is influenced to some extent by the Great Lakes. It does not differ greatly from the climates of other midwestern cities of the same general latitude. Temperature differences between daily highs and lows are invigorating and average about 20 degrees. The average occurrence of the last freeze in the spring is late April, and the first freeze in the fall is mid–October, making the average freeze–free period 173 days. The length of the growing season is favorable for the maturing of all crops and vegetables normally grown in the midwest.

Annual precipitation is well distributed, with somewhat larger monthly amounts falling in late spring and early summer. Damaging hailstorms occur at an average of about twice a year. One of the most notable storms caused severe damage to property, many thousands of trees, and power and telephone lines in the area. Severe flooding has also occurred in the area. Snow usually covers the ground for about 30 days during the winter months, but heavy snowstorms are not frequent.

Except for the considerable cloudiness that occurs during the winter months, Fort Wayne enjoys a good midwestern average sunshine. Heavy fog occurrence is infrequent.

NORMALS, MEANS, AND EXTREMES

FORT WAYNE, INDIANA

LATITUDE: 41°00'N LONGITUDE: 85°12'W ELEVATION: FT. GRND 791 BARO 818 TIME ZONE: EASTERN WBAN: 14827

	(a)	JAN	FEB	MAR	APR	MAY	JUNE	JULY	AUG	SEP	OCT	NOV	DEC	YEAR
TEMPERATURE °F:														
Normals														
-Daily Maximum		30.4	34.0	46.3	59.8	71.2	80.9	84.6	82.2	75.6	63.1	49.1	35.5	59.4
-Daily Minimum		15.3	17.8	28.8	38.5	49.1	59.3	63.2	60.9	54.2	42.5	33.5	21.6	40.4
-Monthly		22.9	25.9	37.6	49.2	60.2	70.2	74.0	71.6	64.9	52.9	41.3	28.6	49.9
Extremes														
-Record Highest	47	69	69	82	88	94	106	103	101	100	90	79	71	106
-Year		1950	1954	1986	1986	1988	1988	1954	1962	1953	1951	1950	1982	JUN 1988
-Record Lowest	47	-22	-18	-10	7	27	38	44	38	29	19	-1	-18	-22
-Year		1985	1982	1967	1982	1966	1993	1967	1965	1951	1988	1958	1989	JAN 1985
NORMAL DEGREE DAYS:														
Heating (base 65°F)		1305	1095	849	474	217	15	0	10	81	388	711	1128	6273
Cooling (base 65°F)		0	0	0	0	68	171	279	215	78	13	0	0	824
% OF POSSIBLE SUNSHINE	47	46	51	55	60	68	73	75	74	68	62	41	38	59
MEAN SKY COVER (tenths)														
Sunrise - Sunset	47	7.5	7.2	7.3	6.9	6.4	6.1	5.8	5.7	5.7	5.9	7.4	7.8	6.6
MEAN NUMBER OF DAYS:														
Sunrise to Sunset														
-Clear	47	4.9	4.6	4.8	5.6	6.9	6.5	8.0	8.8	8.7	9.7	4.9	4.2	77.6
-Partly Cloudy	47	6.2	6.9	7.4	7.4	9.5	11.5	12.0	11.8	9.4	7.9	6.4	6.0	102.4
-Cloudy	47	19.9	16.8	18.8	17.0	14.6	12.0	11.0	10.4	11.8	13.4	18.7	20.9	185.2
Precipitation														
.01 inches or more	47	12.1	10.6	13.4	13.4	11.5	10.1	9.9	9.1	9.0	9.0	11.2	12.9	132.1
Snow,Ice pellets,hail														
1.0 inches or more	47	2.4	2.3	1.6	0.4	0.0	0.0	0.0	0.0	0.0	0.1	1.1	2.1	9.9
Thunderstorms	47	0.3	0.7	2.4	4.1	5.1	7.0	6.8	5.7	3.7	1.7	1.0	0.4	38.7
Heavy Fog Visibility														
1/4 mile or less	47	2.4	2.4	1.9	1.0	0.8	0.7	0.9	1.8	1.6	1.7	1.4	2.7	19.3
Temperature °F														
-Maximum														
90° and above	32	0.0	0.0	0.0	0.0	0.7	3.5	6.1	3.8	1.1	0.0	0.0	0.0	15.3
32° and below	32	16.3	12.0	3.8	0.1	0.0	0.0	0.0	0.0	0.0	0.0	1.6	11.4	45.2
-Minimum														
32° and below	32	28.4	24.8	20.4	8.3	0.7	0.0	0.0	0.0	0.2	5.1	15.6	25.7	129.1
0° and below	32	4.9	3.3	0.3	0.0	0.0	0.0	0.0	0.0	0.0	0.0	0.0	2.0	10.4
AVG. STATION PRESS.(mb)	21	988.2	988.5	986.2	985.3	985.4	985.6	986.9	988.1	988.6	988.8	988.0	988.5	987.3
RELATIVE HUMIDITY (%)														
Hour 01	32	79	78	77	74	75	77	81	85	84	80	80	81	79
Hour 07 (Local Time)	32	80	80	80	79	79	80	84	88	89	85	83	82	82
Hour 13	32	71	69	64	57	54	54	56	58	58	58	68	74	62
Hour 19	32	75	72	67	58	56	56	58	63	65	66	75	79	66
PRECIPITATION (inches):														
Water Equivalent														
-Normal		1.87	1.91	2.90	3.38	3.44	3.59	3.45	3.37	2.67	2.49	2.79	2.89	34.75
-Maximum Monthly	47	9.72	6.84	5.29	7.11	8.80	8.29	11.00	7.69	6.75	9.26	8.01	7.56	11.00
-Year		1950	1990	1955	1957	1991	1958	1986	1975	1972	1954	1992	1990	JUL 1986
-Minimum Monthly	47	0.39	0.30	0.74	1.28	1.04	0.77	0.41	0.42	0.34	0.14	0.62	0.42	0.14
-Year		1966	1978	1981	1962	1977	1988	1974	1969	1979	1964	1976	1962	OCT 1964
-Maximum in 24 hrs	47	2.64	3.03	2.15	2.65	4.35	4.40	3.47	4.05	4.60	2.96	2.50	2.49	4.60
-Year		1950	1990	1953	1963	1991	1989	1955	1957	1950	1947	1982	1990	SEP 1950
Snow,Ice pellets,hail														
-Maximum Monthly	47	29.5	16.9	19.5	11.7	T	T	0.0	T	0.0	8.0	14.1	20.3	29.5
-Year		1982	1980	1964	1961	1989	1992		1989		1989	1950	1973	JAN 1982
-Maximum in 24 hrs	47	10.8	7.7	13.6	6.4	T	T	0.0	T	0.0	6.5	7.0	11.1	13.6
-Year		1982	1952	1964	1957	1989	1992		1989		1989	1950	1973	MAR 1964
WIND:														
Mean Speed (mph)	47	11.5	11.0	11.8	11.5	10.0	9.0	8.0	7.4	8.2	9.1	10.9	11.1	10.0
Prevailing Direction through 1963		W	W	W	SW	SW	SW	SW	SW	SW	SW	W	W	SW
Fastest Mile														
-Direction (!!)	46	SW	W	S	W	S	SE	NW	N	W	SW	SW	SW	S
-Speed (MPH)	46	59	61	65	63	57	65	61	55	52	46	57	52	65
-Year		1949	1967	1948	1962	1960	1948	1954	1965	1960	1988	1957	1953	MAR 1948
Peak Gust														
-Direction (!!)	10	SW	NW	SW	W	W	W	W	SW	NW	SW	W	SW	SW
-Speed (mph)	10	58	51	59	60	59	53	54	59	46	63	58	58	63
-Date		1990	1990	1991	1984	1986	1990	1987	1985	1987	1988	1989	1987	OCT 1988

PRECIPITATION (inches)

FORT WAYNE, INDIANA

YEAR	JAN	FEB	MAR	APR	MAY	JUNE	JULY	AUG	SEP	OCT	NOV	DEC	ANNUAL
1964	2.48	1.12	4.74	5.02	2.55	3.73	2.20	2.04	1.58	0.14	1.17	1.77	28.54
1965	4.49	2.48	2.61	4.13	2.28	2.16	3.44	2.95	2.82	4.62	2.25	3.02	37.25
1966	0.39	1.77	1.87	2.49	3.25	1.15	4.83	4.15	2.82	1.38	4.85	4.94	33.89
1967	1.93	1.82	2.77	2.60	4.19	2.40	2.06	1.61	1.04	4.22	2.70	5.45	32.79
1968	2.05	1.41	1.99	3.12	6.12	5.15	2.54	3.71	1.97	1.00	3.58	3.44	36.08
1969	4.57	0.41	1.68	3.13	4.10	5.87	3.99	0.42	3.49	5.49	3.34	1.00	37.49
1970	0.77	0.85	2.21	6.30	3.74	2.51	5.18	1.11	5.47	2.92	2.93	1.33	35.32
1971	1.11	3.86	1.49	1.33	3.63	3.55	2.73	2.02	4.14	2.22	1.22	4.35	31.65
1972	1.22	0.59	3.19	6.28	3.29	3.83	2.58	3.30	6.75	3.29	3.62	2.84	40.78
1973	1.57	1.31	4.63	2.16	2.19	4.46	3.37	3.75	0.70	3.06	3.39	3.94	34.53
1974	3.25	2.14	2.91	3.44	5.09	2.28	0.41	3.70	2.22	1.29	2.22	2.57	31.52
1975	2.41	2.05	2.03	3.39	2.77	4.40	0.79	7.69	3.48	1.70	3.17	2.85	36.73
1976	1.82	2.30	3.48	2.03	2.18	3.87	3.19	1.54	2.20	2.49	0.62	0.57	26.29
1977	0.50	2.15	4.50	4.01	1.04	3.79	3.99	7.26	3.68	2.11	2.39	2.83	38.25
1978	2.35	0.30	2.32	5.09	3.24	1.67	1.39	2.63	2.26	1.13	3.03	3.75	29.16
1979	1.64	1.04	3.13	2.17	2.64	3.19	4.76	3.49	0.34	2.16	3.87	2.55	30.98
1980	0.71	2.22	4.16	3.03	2.69	4.54	4.03	4.79	2.47	1.96	0.68	2.50	33.78
1981	0.60	2.89	0.74	5.34	4.42	7.96	2.92	1.71	3.34	3.42	1.09	3.18	37.61
1982	5.42	1.50	4.91	2.84	4.74	4.88	2.45	3.44	1.64	1.11	5.77	4.39	43.09
1983	0.98	0.71	1.84	4.83	4.14	2.15	2.36	1.67	1.34	3.58	4.17	4.34	32.11
1984	1.10	1.53	3.71	2.95	4.69	1.48	2.52	1.10	2.98	3.17	2.94	3.20	31.37
1985	2.03	3.58	3.76	1.69	2.16	2.56	1.86	5.15	2.67	2.44	5.41	2.77	36.08
1986	0.64	2.96	3.08	3.27	2.67	4.95	11.00	2.75	4.64	3.47	1.50	1.54	42.47
1987	2.07	0.53	1.34	2.16	3.41	5.41	2.90	6.08	1.35	1.81	3.73	3.46	34.25
1988	1.53	1.97	3.14	2.69	1.99	0.77	6.51	4.29	1.82	2.70	3.43	2.47	33.31
1989	1.99	1.08	1.40	3.39	4.82	5.78	2.48	4.93	4.17	2.05	2.05	1.18	35.32
1990	2.17	6.84	3.11	2.36	5.60	5.49	4.58	6.75	1.50	5.31	3.31	7.56	54.58
1991	1.82	1.28	2.80	5.06	8.80	2.08	3.39	3.15	2.17	8.94	3.24	2.07	44.80
1992	1.67	1.39	3.21	3.50	3.11	3.31	5.35	3.82	6.01	2.45	8.01	2.65	44.48
1993	5.24	1.82	3.64	4.31	3.93	6.09	3.72	1.71	4.85	1.19	4.11	1.49	42.10
Record Mean	2.31	1.86	3.02	3.34	3.61	3.62	3.47	3.12	2.82	2.76	2.64	2.46	35.03

AVERAGE TEMPERATURE (deg. F)

FORT WAYNE, INDIANA

YEAR	JAN	FEB	MAR	APR	MAY	JUNE	JULY	AUG	SEP	OCT	NOV	DEC	ANNUAL
1964	29.3	25.7	36.0	51.0	64.2	70.6	74.5	70.6	64.7	49.4	43.4	27.9	50.6
1965	24.8	26.1	29.4	48.1	65.1	68.6	70.4	69.9	66.8	52.5	42.3	36.6	50.1
1966	21.1	27.8	39.4	45.8	53.0	69.8	73.8	67.8	60.0	47.8	40.3	28.8	48.0
1967	28.1	21.9	36.2	50.2	53.7	70.6	69.8	67.6	61.2	51.1	34.2	29.8	47.9
1968	20.3	23.4	38.8	49.8	55.4	70.4	73.0	72.5	64.4	53.3	41.0	28.3	49.2
1969	23.3	27.8	33.1	50.9	60.3	65.6	74.9	73.7	65.2	52.6	36.8	27.2	49.3
1970	14.6	27.2	34.1	50.4	63.2	70.2	73.5	72.5	67.0	54.9	40.4	31.4	50.0
1971	20.2	27.9	34.3	47.6	57.7	74.5	70.7	69.6	67.4	59.8	40.0	36.0	50.5
1972	24.0	25.1	35.7	47.7	61.7	66.1	72.7	70.7	64.9	49.0	37.9	29.9	48.8
1973	29.2	27.4	45.1	49.2	56.5	72.7	74.1	72.5	66.9	56.9	43.7	27.2	51.8
1974	25.9	25.9	38.8	50.8	58.0	67.0	75.2	72.0	60.5	51.2	40.9	29.7	49.7
1975	29.8	27.8	33.1	43.9	62.9	69.6	72.9	73.0	60.3	53.9	46.6	31.4	50.4
1976	21.6	35.2	43.6	52.1	57.9	71.2	72.9	69.0	62.0	47.4	33.5	22.1	49.1
1977	9.2	25.2	43.4	55.4	67.9	68.3	76.1	70.7	66.5	50.9	41.7	25.4	50.1
1978	16.1	11.8	30.9	48.6	59.5	69.7	72.8	71.9	69.3	51.0	41.9	31.0	47.9
1979	17.4	14.8	41.1	46.8	60.3	69.1	70.9	69.1	64.2	51.1	39.5	31.3	48.0
1980	24.7	20.1	32.5	46.1	59.9	66.5	75.2	74.1	66.2	50.5	39.0	29.6	48.7
1981	20.1	29.0	37.5	51.5	57.1	70.1	73.0	71.2	64.5	49.6	41.0	26.4	49.3
1982	15.3	21.3	35.4	44.4	66.4	65.6	73.8	69.1	62.6	54.2	43.7	39.2	49.2
1983	29.5	33.6	41.0	47.4	58.0	72.0	78.9	78.4	67.8	55.3	44.9	21.2	52.4
1984	19.5	33.5	28.5	47.4	56.8	73.6	71.8	73.4	63.4	58.5	41.5	35.9	50.3
1985	19.8	23.9	43.1	57.0	65.1	68.8	75.0	71.5	65.0	54.4	43.7	20.5	50.7
1986	26.2	25.6	40.1	51.6	61.7	69.1	74.9	67.9	67.1	53.9	37.2	31.1	50.5
1987	24.7	31.4	40.9	51.4	64.5	72.1	75.1	71.8	65.3	46.6	44.1	33.4	51.8
1988	23.6	23.6	38.5	49.8	64.3	73.3	77.1	74.7	64.5	45.9	41.9	28.7	50.5
1989	33.2	23.4	38.3	47.5	57.3	69.7	75.5	70.8	62.2	52.7	39.4	16.9	48.9
1990	34.8	34.1	42.1	49.8	57.6	69.5	71.9	69.9	64.7	52.7	45.3	32.4	52.1
1991	24.8	31.9	41.2	53.1	67.3	74.2	74.8	72.9	63.6	55.4	37.6	32.9	52.5
1992	28.6	34.0	38.1	47.6	59.2	65.9	71.1	67.6	62.5	50.9	40.9	31.6	49.8
1993	28.6	24.0	35.3	48.6	61.8	68.5	76.3	74.4	61.5	51.3	41.1	31.1	50.2
Record Mean	25.0	27.4	37.0	48.9	60.0	69.6	73.9	71.8	64.7	53.3	40.5	29.1	50.1
Max	32.4	35.1	45.7	59.0	70.7	80.2	84.6	82.3	75.5	63.6	48.3	36.0	59.4
Min	17.6	19.6	28.3	38.7	49.3	59.0	63.2	61.2	53.9	43.0	32.7	22.3	40.7

REFERENCE NOTES

FORT WAYNE, INDIANA

GENERAL
T - TRACE AMOUNT.
BLANK ENTRIES DENOTE MISSING/UNREPORTED DATA.
# INDICATES A STATION OR INSTRUMENT RELOCATION.

SPECIFIC

EXCEPTIONS

RECORD MEANS ARE THROUGH THE CURRENT YEAR,
BEGINNING IN 1912 FOR TEMPERATURE
1912 FOR PRECIPITATION
1947 FOR SNOWFALL

(a) - LENGTH OF RECORD IN YEARS, ALTHOUGH INDIVIDUAL MONTHS MAY BE MISSING.
0.* OR * - THE VALUE IS BETWEEN 0.0 AND 0.05
NORMALS - BASED ON THE 1961-1990 RECORD PERIOD.
EXTREMES - DATES ARE THE MOST RECENT OCCURRENCE.
WIND DIR.- NUMERALS SHOW TENS OF DEGRESS CLOCKWISE FROM TRUE NORTH. "00" INDICATES CALM.
RESULTANT DIRECTIONS ARE GIVEN TO WHOLE DEGREES.
BOLD VALUES INDICATE EXTREME VALUES WHICH OCCURRED AFTER THE ASOS SYSTEM WAS COMMISSIONED.

RECORD = PERIOD OF RECORD
RECORD MEAN PRECIPITATION IS THE MEAN OF ALL DAILY PRECIPITATION AMOUNTS DURING THE PERIOD OF RECORD.
RECORD MAX(MIN) TEMPERATURE IS THE MEAN OF ALL DAILY MAX(MIN) TEMPERATURES DURING THE PERIOD OF RECORD.
RECORD MEAN TEMPERATURE IS THE SUM OF THE RECORD MAX AND RECORD MIN DIVIDED BY 2.
AVERAGE TEMPERATURE IS THE SUM OF THE MEAN DAILY MAX AND MIN TEMPERATURE DIVIDED BY 2.

HEATING DEGREE DAYS Base 65 deg. F — FORT WAYNE, INDIANA

SEASON	JULY	AUG	SEP	OCT	NOV	DEC	JAN	FEB	MAR	APR	MAY	JUNE	TOTAL
1964-65	3	31	109	478	642	1145	1238	1083	1098	499	73	22	6421
1965-66	3	31	85	380	673	872	1352	1034	788	569	370	37	6194
1966-67	1	20	188	525	735	1116	1139	1201	885	444	357	13	6624
1967-68	27	31	154	442	914	1083	1382	1199	806	448	298	19	6803
1968-69	7	23	82	393	713	1131	1285	1036	982	423	191	82	6348
1969-70	0	0	94	383	839	1166	1560	1053	951	448	142	32	6668
1970-71	9	1	71	312	731	1033	1382	1031	945	518	239	3	6275
1971-72	5	2	71	181	743	894	1269	1152	900	513	154	66	5950
1972-73	15	17	86	490	809	1081	1102	1046	607	473	259	0	5985
1973-74	0	7	62	266	635	1166	1206	1091	804	423	238	38	5936
1974-75	0	3	167	424	715	1086	1084	1035	981	628	135	48	6306
1975-76	8	4	173	346	546	1036	1338	856	656	404	225	1	5593
1976-77	0	11	121	540	938	1322	1723	1108	662	322	80	42	6869
1977-78	2	11	61	429	694	1218	1509	1482	1051	485	224	26	7192
1978-79	1	5	55	431	686	1050	1466	1401	731	542	198	26	6592
1979-80	8	31	94	440	756	1037	1240	1294	1001	561	190	65	6717
1980-81	0	0	64	454	772	1090	1385	1004	844	402	259	9	6283
1981-82	0	2	88	472	711	1185	1536	1216	908	608	51	50	6827
1982-83	0	17	124	345	632	795	1096	874	736	526	218	25	5388
1983-84	1	0	78	310	596	1352	1405	908	1125	524	272	0	6571
1984-85	4	1	128	204	699	894	1394	1144	673	281	86	23	5531
1985-86	0	3	124	319	632	1373	1197	1097	769	406	152	28	6100
1986-87	2	46	60	351	826	1043	1243	937	737	408	128	10	5791
1987-88	6	26	64	562	620	972	1277	1192	814	453	87	22	6095
1988-89	0	9	62	588	684	1118	978	1158	823	522	268	19	6229
1989-90	0	9	141	384	760	1488	928	858	712	479	231	29	6019
1990-91	2	7	113	389	587	1007	1240	920	730	355	100	1	5451
1991-92	0	0	162	308	816	985	1120	891	830	520	212	49	5893
1992-93	7	19	131	430	714	1028	1121	1141	914	484	128	51	6168
1993-94	0	4	140	425	710	1044							

COOLING DEGREE DAYS Base 65 deg. F — FORT WAYNE, INDIANA

YEAR	JAN	FEB	MAR	APR	MAY	JUNE	JULY	AUG	SEP	OCT	NOV	DEC	TOTAL
1969	0	0	0	8	52	109	312	279	106	4	0	0	870
1970	0	0	0	16	92	194	281	240	134	5	0	0	962
1971	0	0	0	2	19	294	188	150	149	28	0	0	830
1972	0	0	0	0	60	106	262	202	88	0	0	0	718
1973	0	0	0	8	2	240	291	244	124	19	0	0	928
1974	0	0	0	6	27	106	322	225	39	2	0	0	727
1975	0	0	0	2	76	192	256	260	37	9	1	0	833
1976	0	0	0	26	9	193	254	140	38	4	0	0	664
1977	0	0	0	41	177	152	353	197	112	0	0	0	1032
1978	0	0	0	0	60	169	251	225	192	1	0	0	898
1979	0	0	0	4	58	156	198	167	78	16	0	0	677
1980	0	0	0	1	38	115	326	290	107	10	0	0	887
1981	0	0	0	3	21	168	258	202	79	0	0	0	731
1982	0	0	0	0	101	76	280	150	57	15	0	0	679
1983	0	0	0	4	10	241	436	418	168	16	0	0	1293
1984	0	0	0	5	24	266	223	270	88	10	0	0	886
1985	0	0	3	47	94	142	317	211	130	2	0	0	946
1986	0	0	3	12	55	159	316	143	130	13	0	0	831
1987	0	0	0	6	124	233	320	242	82	0	1	0	1008
1988	0	0	0	2	74	281	381	319	53	4	0	0	1114
1989	0	0	1	1	37	169	334	195	64	10	0	0	811
1990	0	0	7	26	7	173	222	164	112	14	1	0	726
1991	0	0	0	6	177	283	308	254	126	16	0	0	1170
1992	0	0	0	3	40	82	203	110	64	0	0	0	502
1993	0	0	0	0	37	162	356	301	41	4	0	0	901

SNOWFALL (inches) — FORT WAYNE, INDIANA

SEASON	JULY	AUG	SEP	OCT	NOV	DEC	JAN	FEB	MAR	APR	MAY	JUNE	TOTAL
1964-65	0.0	0.0	0.0	T	2.7	7.9	6.2	9.2	11.7	0.4	0.0	0.0	38.1
1965-66	0.0	0.0	0.0	0.0	0.2	2.5	3.1	5.0	1.3	2.0	T	0.0	14.1
1966-67	0.0	0.0	0.0	0.0	12.0	7.3	5.6	12.5	8.8	0.7	0.0	0.0	46.9
1967-68	0.0	0.0	0.0	0.0	4.1	3.7	8.9	2.7	9.5	0.4	0.0	0.0	29.3
1968-69	0.0	0.0	0.0	0.0	T	3.3	3.8	2.8	2.4	0.3	0.0	0.0	12.6
1969-70	0.0	0.0	0.0	0.0	4.5	8.0	9.6	4.8	5.2	3.1	0.0	0.0	35.2
1970-71	0.0	0.0	0.0	0.0	2.6	6.5	5.4	6.1	7.3	0.3	0.0	0.0	28.2
1971-72	0.0	0.0	0.0	0.0	6.7	0.5	7.9	5.8	4.8	1.9	0.0	0.0	27.6
1972-73	0.0	0.0	0.0	0.5	4.0	4.0	2.3	6.0	12.5	4.0	0.0	0.0	33.3
1973-74	0.0	0.0	0.0	0.0	T	20.3	8.5	7.9	3.8	1.8	0.0	0.0	42.3
1974-75	0.0	0.0	0.0	1.4	6.5	12.2	5.8	5.2	9.2	0.9	0.0	0.0	41.2
1975-76	0.0	0.0	0.0	0.0	6.0	10.0	6.3	6.1	3.5	T	0.0	0.0	31.9
1976-77	0.0	0.0	0.0	T	2.1	9.5	13.0	9.2	6.0	0.5	0.0	0.0	40.3
1977-78	0.0	0.0	0.0	T	7.1	18.4	25.3	6.1	2.7	1.0	0.0	0.0	60.6
1978-79	0.0	0.0	0.0	0.0	4.6	1.8	14.1	11.3	1.1	T	0.0	0.0	32.9
1979-80	0.0	0.0	0.0	T	2.0	2.0	4.5	16.9	2.0	1.3	0.0	0.0	28.7
1980-81	0.0	0.0	0.0	0.2	6.2	11.0	6.9	9.4	2.0	0.0	0.0	0.0	35.7
1981-82	0.0	0.0	0.0	T	3.6	17.6	29.5	14.0	6.7	9.8	0.0	0.0	81.2
1982-83	0.0	0.0	0.0	0.0	0.1	1.5	2.8	3.6	6.7	0.2	0.0	0.0	14.9
1983-84	0.0	0.0	0.0	0.0	0.9	16.3	10.8	7.9	13.3	0.6	0.0	0.0	49.8
1984-85	0.0	0.0	0.0	0.0	0.7	4.4	15.3	13.7	0.8	0.9	0.0	0.0	35.8
1985-86	0.0	0.0	0.0	0.0	1.9	10.4	4.0	16.3	1.3	0.2	0.0	0.0	34.1
1986-87	0.0	0.0	0.0	0.0	4.6	1.8	17.6	0.2	1.5	T	0.0	0.0	25.7
1987-88	0.0	0.0	0.0	T	3.9	3.7	4.2	16.7	5.6	T	0.0	0.0	34.1
1988-89	0.0	0.0	0.0	T	3.3	10.1	1.7	6.1	1.3	T	T	0.0	22.5
1989-90	0.0	T	0.0	8.0	2.1	8.3	1.7	3.8	T	T	0.0	T	23.9
1990-91	0.0	0.0	0.0	0.0	T	13.5	8.5	6.0	3.2	T	0.0	0.0	31.2
1991-92	0.0	0.0	0.0	T	3.5	2.0	11.8	0.7	9.8	T	0.0	T	27.8
1992-93	0.0	0.0	0.0	1.4	0.5	9.1	9.6	13.0	7.5	0.8	0.0	0.0	41.9
1993-94	0.0	0.0	0.0	1.0	2.0	5.5							
Record Mean	0.0	T	0.0	0.3	3.2	7.1	8.0	7.6	5.3	1.4	T	T	32.8

INDIANAPOLIS, INDIANA

Indianapolis is located in the central part of the state and is situated on level or slightly rolling terrain. The greater part of the city lies east of the White River which flows in a general north to south direction.

The National Weather Service Forecast Office is located approximately 7 miles southwest of the central part of the city at the Indianapolis International Airport. From a field elevation of 797 feet above sea level at the Indianapolis International Airport the terrain slopes gradually downward to a little below 645 feet at the White River, then upward to just over 910 feet in the northwest corner and eastern sections of the county. The street elevation at the former city office located in the Old Federal Building is 718 feet.

Indianapolis has a temperate climate, with very warm summers and without a dry season. Very cold temperatures may be produced by the invasion of continental polar air in the winter from northern latitudes. The polar air can be quite frigid with very low humidity. The arrival of maritime tropical air from the Gulf in the summer brings warm temperatures and moderate humidity. One of the longest and most severe heat waves brought temperatures of 100 degrees or more for nine consecutive days.

Precipitation is distributed fairly evenly throughout the year, and therefore there is no pronounced wet or dry season. Rainfall in the spring and summer is produced mostly by showers and thunderstorms. A rainfall of about 2 1/2 inches in a 24-hour period can be expected about once a year. Snowfalls of 3 inches or more occur on an average of two or three times in the winter.

Local levees and/or channel improvements now protect some formerly flood-prone areas.

Based on the 1951-1980 period, the average first occurrence of 32 degrees Fahrenheit in the fall is October 20 and the average last occurrence in the spring is April 22.

NORMALS, MEANS, AND EXTREMES

INDIANAPOLIS, INDIANA

LATITUDE: 39°42'N LONGITUDE: 86°16' W ELEVATION: FT. GRND 790 BARO 837 TIME ZONE: EASTERN WBAN: 93819

	(a)	JAN	FEB	MAR	APR	MAY	JUNE	JULY	AUG	SEP	OCT	NOV	DEC	YEAR
TEMPERATURE °F:														
Normals														
-Daily Maximum		33.7	38.3	50.9	63.3	73.8	82.7	85.5	83.6	77.6	65.8	51.9	38.5	62.1
-Daily Minimum		17.2	20.9	31.9	41.5	51.7	61.0	65.2	62.8	55.6	43.5	34.1	23.2	42.4
-Monthly		25.5	29.6	41.4	52.4	62.8	71.9	75.4	73.2	66.6	54.7	43.0	30.9	52.3
Extremes														
-Record Highest	54	71	74	85	89	93	102	104	102	100	90	81	74	104
-Year		1950	1972	1981	1970	1988	1988	1954	1988	1954	1954	1950	1982	JUL 1954
-Record Lowest	54	-22	-21	-7	16	28	37	44	41	28	17	-2	-23	-23
-Year		1985	1982	1980	1940	1966	1992	1942	1965	1942	1942	1958	1989	DEC 1989
NORMAL DEGREE DAYS:														
Heating (base 65°F)		1225	991	732	378	165	5	0	6	58	338	660	1057	5615
Cooling (base 65°F)		0	0	0	0	96	212	322	260	106	18	0	0	1014
% OF POSSIBLE SUNSHINE	49	41	49	50	54	60	66	66	68	65	61	41	38	55
MEAN SKY COVER (tenths)														
Sunrise - Sunset	51	7.3	7.0	7.2	6.9	6.5	6.1	5.8	5.6	5.5	5.5	7.1	7.4	6.5
MEAN NUMBER OF DAYS:														
Sunrise to Sunset														
-Clear	62	5.9	5.6	5.6	6.0	7.1	7.1	8.5	9.1	10.5	11.0	6.4	5.1	87.9
-Partly Cloudy	62	5.9	6.3	7.0	7.3	8.9	10.5	12.4	11.7	8.7	7.5	6.5	6.2	98.9
-Cloudy	62	19.2	16.4	18.4	16.7	15.0	12.4	10.1	10.2	10.8	12.5	17.2	19.8	178.6
Precipitation														
.01 inches or more	54	11.8	10.2	13.1	12.2	12.1	9.8	9.7	8.7	7.9	8.2	10.5	11.9	126.0
Snow,Ice pellets,hail														
1.0 inches or more	52	2.0	2.0	1.2	0.2	0.0	0.0	0.0	0.0	0.0	0.1	0.6	1.8	7.7
Thunderstorms	51	0.7	0.7	2.7	4.5	6.3	7.3	7.8	6.2	3.7	1.8	1.1	0.5	43.3
Heavy Fog Visibility														
1/4 mile or less	51	3.3	2.5	1.6	0.6	0.8	0.7	1.1	1.7	1.5	1.3	1.6	2.9	19.5
Temperature °F														
-Maximum														
90° and above	34	0.0	0.0	0.0	0.0	0.5	3.6	7.0	4.3	1.9	0.0	0.0	0.0	17.3
32° and below	34	13.3	8.9	2.3	0.*	0.0	0.0	0.0	0.0	0.0	0.0	1.0	8.6	34.1
-Minimum														
32° and below	34	27.4	23.5	17.1	5.4	0.3	0.0	0.0	0.0	0.*	4.3	14.2	24.9	117.1
0° and below	34	4.0	2.0	0.2	0.0	0.0	0.0	0.0	0.0	0.0	0.0	0.0	1.8	8.0
AVG. STATION PRESS.(mb)	21	989.9	989.5	986.9	986.2	986.2	986.5	987.7	988.8	989.3	989.9	989.2	990.0	988.4
RELATIVE HUMIDITY (%)														
Hour 01	34	78	77	75	73	77	79	84	86	85	80	80	81	80
Hour 07 (Local Time)	34	81	81	80	79	82	82	87	90	90	87	84	83	84
Hour 13	34	70	67	62	56	56	56	60	61	58	57	67	72	62
Hour 19	34	73	70	65	58	57	58	62	66	66	64	72	76	66
PRECIPITATION (inches):														
Water Equivalent														
-Normal		2.32	2.46	3.79	3.70	4.00	3.49	4.47	3.64	2.87	2.63	3.23	3.34	39.94
-Maximum Monthly	54	12.69	5.35	10.74	8.09	10.10	9.74	11.79	8.34	8.06	8.36	8.50	7.72	12.69
-Year		1950	1971	1963	1964	1943	1942	1992	1980	1989	1941	1985	1990	JAN 1950
-Minimum Monthly	54	0.21	0.36	1.03	0.98	1.06	0.36	0.99	0.68	0.24	0.17	0.82	0.45	0.17
-Year		1944	1978	1941	1976	1988	1988	1941	1964	1963	1963	1976	1976	OCT 1963
-Maximum in 24 hrs	51	3.47	2.50	3.05	2.56	3.53	3.80	5.32	4.72	3.07	3.90	4.15	2.83	5.32
-Year		1950	1977	1963	1961	1961	1963	1987	1976	1961	1959	1993	1990	JUL 1987
Snow,Ice pellets,hail														
-Maximum Monthly	62	30.6	18.0	10.5	4.0	0.2	T	0.0	T	0.0	9.3	8.3	27.5	30.6
-Year		1978	1979	1975	1940	1989	1992		1989		1989	1966	1973	JAN 1978
-Maximum in 24 hrs	51	12.2	12.5	5.6	3.1	0.2	T	0.0	T	0.0	7.5	8.2	11.5	12.5
-Year		1978	1965	1948	1953	1989	1992		1989		1989	1966	1973	FEB 1965
WIND:														
Mean Speed (mph)	45	10.9	10.8	11.7	11.2	9.5	8.5	7.5	7.2	8.0	8.8	10.4	10.5	9.6
Prevailing Direction through 1963		NW	WNW	WNW	SW	SW	SW	SW	SW	SW	SW	SW	SW	SW
Fastest Obs. 1 Min.														
-Direction (!!!)	14	29	33	27	33	32	29	36	32	27	23	31	22	33
-Speed (MPH)	14	39	38	38	46	37	46	40	40	31	35	33	40	46
-Year		1980	1980	1991	1988	1986	1980	1987	1980	1990	1988	1992	1987	APR 1988
Peak Gust														
-Direction (!!!)	10	SW	SW	W	NW	NW	SW	SW	NW	NW	SW	SW	SW	NW
-Speed (mph)	10	48	53	54	74	62	62	55	70	46	48	67	64	74
-Date		1990	1988	1991	1988	1986	1992	1992	1985	1993	1991	1991	1987	APR 1988

PRECIPITATION (inches)

INDIANAPOLIS, INDIANA

YEAR	JAN	FEB	MAR	APR	MAY	JUNE	JULY	AUG	SEP	OCT	NOV	DEC	ANNUAL
1964	2.04	2.01	7.20	8.09	1.42	2.73	4.09	0.68	1.27	0.64	3.13	3.05	36.35
1965	3.86	4.33	2.17	5.80	1.44	3.49	3.18	3.25	5.16	1.05	1.41	2.97	38.11
1966	1.13	2.91	1.31	3.32	1.47	1.28	2.71	1.31	5.73	1.60	4.72	5.23	32.72
1967	1.81	1.84	3.33	3.00	5.00	1.07	2.34	2.38	0.80	5.72	2.54	4.92	34.75
1968	2.96	1.51	3.73	2.86	9.25	2.51	2.19	4.45	1.54	1.13	4.74	4.18	41.05
1969	6.19	1.23	1.33	4.42	1.82	4.16	8.02	2.98	2.89	4.83	2.86	2.04	42.77
1970	1.17	1.86	2.51	6.53	2.43	1.97	4.43	1.82	2.41	3.66	2.12	2.07	32.98
1971	1.98	5.35	1.49	1.16	4.25	3.39	5.68	1.93	3.10	1.84	1.29	6.02	37.48
1972	1.57	1.15	2.48	5.81	1.89	6.04	2.01	2.94	5.65	2.25	5.65	2.83	40.27
1973	2.27	1.11	5.63	2.76	1.79	5.91	6.67	2.74	2.43	3.11	3.62	4.27	42.31
1974	3.39	2.58	3.60	3.45	6.27	5.15	1.20	5.63	3.25	0.99	2.99	2.81	41.31
1975	4.37	4.13	4.16	4.14	2.42	5.73	4.63	4.68	2.32	2.80	3.63	3.71	46.72
1976	2.29	2.90	3.46	0.98	3.10	3.97	3.09	7.95	2.02	2.79	0.82	0.45	33.82
1977	1.50	3.62	3.83	1.91	2.78	3.86	2.57	4.47	3.40	2.79	3.01	4.31	38.05
1978	3.80	0.36	3.54	3.59	4.21	4.43	5.04	6.89	0.85	3.82	2.38	4.03	42.94
1979	3.24	2.86	2.43	3.14	2.23	3.93	11.06	6.09	0.36	2.32	4.37	2.57	44.60
1980	1.67	1.84	4.26	2.10	2.26	4.15	2.87	8.34	3.31	1.87	1.41	0.78	34.86
1981	0.36	2.88	1.22	5.81	9.23	1.64	5.75	1.69	2.04	2.35	1.12	3.40	37.49
1982	5.64	1.62	4.73	2.40	5.94	5.16	3.44	1.00	1.20	0.91	4.16	5.78	41.98
1983	1.05	1.03	2.94	4.47	4.68	4.53	1.58	2.79	1.28	3.87	4.55	3.43	36.20
1984	0.97	3.16	3.14	3.90	4.35	1.51	4.83	3.27	4.69	2.60	5.38	4.33	42.13
1985	1.37	3.73	5.94	2.60	4.60	3.06	4.06	5.29	2.71	1.82	8.50	3.30	46.98
1986	0.73	2.84	3.93	4.34	7.37	3.58	4.88	1.18	5.68	7.84	2.32	1.71	46.40
1987	1.55	1.28	1.84	2.68	1.77	4.11	9.22	0.86	1.41	1.36	2.60	4.77	33.45
1988	2.35	3.04	3.22	4.02	1.06	0.36	4.71	1.46	1.14	3.07	4.39	2.50	31.32
1989	1.75	1.32	3.72	4.32	5.79	3.80	6.15	8.05	8.06	2.92	2.79	1.90	50.57
1990	1.79	5.17	3.93	2.44	7.59	3.11	3.68	4.46	2.68	4.64	3.23	7.72	50.44
1991	1.59	1.94	6.41	4.34	4.64	0.91	2.17	3.54	1.12	5.47	3.95	1.45	37.53
1992	1.40	1.15	2.61	4.17	1.56	4.07	11.79	1.42	3.40	2.84	7.88	1.89	44.18
1993	3.29	2.31	3.72	3.75	2.76	5.15	4.94	6.42	5.69	2.25	8.30	2.18	50.76
Record Mean	2.85	2.52	3.78	3.67	3.93	4.00	3.97	3.33	3.14	2.72	3.30	2.95	40.18

AVERAGE TEMPERATURE (deg. F)

INDIANAPOLIS, INDIANA

YEAR	JAN	FEB	MAR	APR	MAY	JUNE	JULY	AUG	SEP	OCT	NOV	DEC	ANNUAL
1964	30.7	29.0	40.6	54.7	65.6	73.4	74.6	72.5	67.2	51.7	45.3	32.2	53.1
1965	27.5	29.5	33.7	54.4	68.3	71.7	72.9	71.1	67.5	53.7	44.7	38.6	52.8
1966	22.4	30.4	43.3	51.3	58.4	72.4	79.2	73.2	65.2	51.9	43.8	32.7	52.0
1967	32.6	27.7	44.0	53.6	59.3	73.9	73.8	71.1	65.9	55.2	39.4	33.7	52.5
1968	25.1	26.4	42.8	53.7	58.6	72.3	75.3	74.5	65.8	54.4	42.9	30.6	51.9
1969	25.7	31.7	35.7	54.0	63.4	69.8	75.8	72.7	65.2	53.8	37.5	28.2	51.1
1970	17.9	28.6	38.1	55.1	65.6	71.5	75.2	74.2	69.1	55.2	40.9	34.7	52.2
1971	23.4	28.9	37.3	51.0	58.9	75.6	72.3	71.3	69.4	62.0	43.1	38.4	52.7
1972	26.5	28.4	39.8	52.1	65.0	68.8	74.6	73.1	68.3	51.5	39.9	31.9	51.7
1973	30.7	31.5	49.4	51.3	59.5	73.3	76.0	74.4	69.9	59.4	46.4	30.9	54.4
1974	31.6	32.4	45.3	55.0	62.1	68.6	76.0	71.8	60.6	52.5	42.5	32.9	52.6
1975	32.0	32.0	36.9	48.9	65.5	71.5	73.5	76.1	62.6	55.9	46.4	32.8	52.8
1976	23.9	38.8	46.6	53.4	58.8	71.3	73.9	71.3	63.6	48.7	34.7	24.6	50.8
1977	10.3	28.2	46.5	57.1	70.6	71.2	78.0	74.0	69.6	54.2	45.9	29.2	52.9
1978	18.2	17.8	36.7	55.2	63.5	74.0	77.2	75.0	70.4	52.7	45.7	34.4	51.7
1979	18.0	18.8	43.4	49.8	61.1	71.1	73.0	72.4	64.9	53.1	41.3	34.9	50.2
1980	28.5	22.5	36.0	49.1	64.0	69.1	78.5	76.6	67.9	50.9	40.3	31.8	51.3
1981	23.5	33.0	39.9	57.4	59.9	73.3	75.4	72.8	64.6	53.1	44.2	27.8	52.1
1982	20.1	26.4	42.5	48.6	68.6	67.8	76.2	71.7	64.4	55.7	44.2	40.2	52.2
1983	30.6	35.5	42.9	48.2	58.2	71.8	79.7	80.1	69.4	57.5	44.1	20.2	53.2
1984	22.8	37.3	33.0	50.0	58.8	75.3	72.4	74.3	64.2	61.3	43.0	38.9	52.6
1985	20.4	26.1	44.6	57.1	64.9	70.6	74.3	71.2	66.2	57.6	46.6	22.5	51.8
1986	28.5	31.4	43.9	53.8	63.5	72.8	77.5	70.1	69.8	55.6	39.8	32.3	53.3
1987	27.6	35.5	44.5	52.5	68.1	73.7	75.8	73.6	68.1	48.6	46.3	35.8	54.2
1988	25.9	27.2	41.4	51.9	64.4	73.4	78.3	77.5	67.3	48.2	44.2	31.7	52.6
1989	36.3	27.2	42.5	51.4	59.4	71.4	75.7	71.8	64.4	55.4	40.9	18.8	51.3
1990	37.3	37.6	46.2	51.4	60.1	71.3	73.9	72.5	66.9	53.9	47.5	34.6	54.4
1991	26.9	35.3	43.7	55.3	68.3	74.6	76.7	75.2	67.9	57.5	40.2	35.7	54.8
1992	31.5	38.4	43.2	52.1	60.5	68.1	73.5	69.5	65.1	53.4	43.2	33.3	52.7
1993	31.6	27.4	39.1	50.7	63.5	71.0	77.2	75.3	62.2	51.6	42.2	32.4	52.0
Record Mean	28.0	30.7	40.4	52.0	62.5	71.8	75.8	73.6	66.8	55.3	42.2	31.8	52.5
Max	35.8	38.8	49.3	61.8	72.6	81.7	85.7	83.6	77.2	65.4	50.4	39.0	61.8
Min	20.3	22.6	31.5	42.2	52.4	61.8	65.8	63.6	56.5	45.1	33.9	24.5	43.3

REFERENCE NOTES

INDIANAPOLIS, INDIANA

GENERAL
T - TRACE AMOUNT.
BLANK ENTRIES DENOTE MISSING/UNREPORTED DATA.
# INDICATES A STATION OR INSTRUMENT RELOCATION.

SPECIFIC

PM - INCLUDES LAST DAY OF PREVIOUS MONTH
ASOS - AUTOMATED SURFACE OBSERVING SYSTEM IN OPERATION DURING THESE MONTHS.

(a) - LENGTH OF RECORD IN YEARS, ALTHOUGH INDIVIDUAL MONTHS MAY BE MISSING.
0.* OR * - THE VALUE IS BETWEEN 0.0 AND 0.05
NORMALS - BASED ON THE 1961-1990 RECORD PERIOD.
EXTREMES - DATES ARE THE MOST RECENT OCCURRENCE
WIND DIR.- NUMERALS SHOW TENS OF DEGRESS CLOCKWISE FROM TRUE NORTH. "00" INDICATES CALM.
RESULTANT DIRECTIONS ARE GIVEN TO WHOLE DEGREES.
BOLD VALUES INDICATE EXTREME VALUES WHICH OCCURRED AFTER THE ASOS SYSTEM WAS COMMISSIONED.

RECORD = PERIOD OF RECORD
RECORD MEAN PRECIPITATION IS THE MEAN OF ALL DAILY PRECIPITATION AMOUNTS DURING THE PERIOD OF RECORD.
RECORD MAX(MIN) TEMPERATURE IS THE MEAN OF ALL DAILY MAX(MIN) TEMPERATURES DURING THE PERIOD OF RECORD.
RECORD MEAN TEMPERATURE IS THE SUM OF THE RECORD MAX AND RECORD MIN DIVIDED BY 2.
AVERAGE TEMPERATURE IS THE SUM OF THE MEAN DAILY MAX AND MIN TEMPERATURE DIVIDED BY 2.

EXCEPTIONS

RECORD MEANS ARE THROUGH THE CURRENT YEAR, BEGINNING IN 1871 FOR TEMPERATURE
1871 FOR PRECIPITATION
1932 FOR SNOWFALL

HEATING DEGREE DAYS Base 65 deg. F   INDIANAPOLIS, INDIANA

SEASON	JULY	AUG	SEP	OCT	NOV	DEC	JAN	FEB	MAR	APR	MAY	JUNE	TOTAL
1964-65	2	16	73	405	581	1009	1155	989	964	323	39	0	5556
1965-66	0	18	66	355	603	811	1312	962	668	412	223	11	5441
1966-67	0	0	77	402	631	995	997	1040	652	353	219	2	5368
1967-68	3	9	83	327	761	964	1232	1112	680	333	205	11	5720
1968-69	4	13	49	354	656	1057	1211	925	904	329	121	38	5661
1969-70	0	0	85	358	816	1137	1458	1012	829	316	95	11	6117
1970-71	4	0	51	301	715	930	1281	1006	852	417	198	0	5755
1971-72	3	1	49	129	648	813	1186	1054	774	389	89	35	5170
1972-73	8	4	36	413	746	1018	1059	937	477	416	184	0	5298
1973-74	0	0	20	211	552	1052	1028	905	617	314	158	18	4875
1974-75	0	5	163	380	671	988	1016	918	866	481	78	22	5588
1975-76	8	0	137	288	551	992	1265	754	564	363	203	1	5126
1976-77	0	2	79	503	904	1249	1693	1025	567	276	45	19	6362
1977-78	0	0	30	326	575	1104	1443	1313	873	292	150	4	6110
1978-79	0	0	36	377	571	944	1453	1288	665	455	170	5	5964
1979-80	3	13	87	384	705	929	1123	1224	893	474	93	36	5964
1980-81	0	0	45	438	734	1022	1279	889	769	244	180	1	5601
1981-82	1	0	94	368	621	1146	1388	1075	690	486	26	18	5913
1982-83	0	2	110	325	621	764	1062	819	681	498	211	21	5114
1983-84	1	0	64	246	619	1386	1304	796	987	447	211	1	6062
1984-85	1	0	119	138	653	803	1375	1082	631	269	85	16	5172
1985-86	0	1	97	245	544	1311	1126	935	652	344	114	7	5376
1986-87	0	24	30	312	750	1009	1150	820	627	378	51	0	5151
1987-88	2	5	37	504	553	900	1205	1090	726	390	84	15	5511
1988-89	0	2	37	517	618	1024	882	1052	693	419	226	11	5481
1989-90	0	8	106	313	718	1426	851	760	588	432	153	19	5374
1990-91	4	1	86	348	518	932	1175	824	651	292	71	0	4902
1991-92	0	0	103	254	739	902	1035	766	670	390	185	34	5078
1992-93	2	7	88	353	645	973	1028	1046	796	424	103	35	5500
1993-94	0	1	129	409	680	1005							

COOLING DEGREE DAYS Base 65 deg. F   INDIANAPOLIS, INDIANA

YEAR	JAN	FEB	MAR	APR	MAY	JUNE	JULY	AUG	SEP	OCT	NOV	DEC	TOTAL
1969	0	0	0	5	78	190	344	245	95	18	0	0	975
1970	0	0	0	26	122	213	327	292	182	5	0	0	1167
1971	0	0	0	2	17	324	237	202	189	40	0	0	1011
1972	0	0	0	8	95	156	313	266	141	0	0	0	979
1973	0	0	0	11	20	256	349	302	172	44	1	0	1155
1974	0	0	11	19	73	131	346	225	40	3	2	0	850
1975	0	0	0	3	100	222	281	355	71	14	0	0	1046
1976	0	0	0	21	15	198	284	205	43	4	0	0	770
1977	0	0	3	45	226	212	410	286	175	0	6	0	1363
1978	0	0	0	4	110	282	382	318	203	1	0	0	1300
1979	0	0	0	7	57	197	255	250	92	24	0	0	882
1980	0	0	0	3	68	168	425	368	139	6	0	0	1177
1981	0	0	1	23	29	256	332	249	88	5	0	0	983
1982	0	0	0	0	146	109	356	214	98	40	3	1	967
1983	0	0	1	3	9	231	464	474	202	18	0	0	1402
1984	0	0	0	6	25	318	237	291	103	28	0	0	1008
1985	0	0	5	36	90	190	296	199	143	21	0	0	980
1986	0	0	6	12	74	249	395	189	181	24	0	0	1130
1987	0	0	0	6	156	265	343	279	137	0	0	0	1186
1988	0	0	1	3	72	274	422	395	114	2	0	0	1283
1989	0	0	1	20	57	215	338	227	94	21	0	0	973
1990	0	0	13	30	10	215	289	241	147	10	1	0	956
1991	0	0	0	7	178	294	367	323	196	30	0	0	1395
1992	0	0	0	10	53	137	274	152	99	1	0	0	726
1993	0	0	0	0	64	223	389	328	51	0	0	0	1055

SNOWFALL (inches)   INDIANAPOLIS, INDIANA

SEASON	JULY	AUG	SEP	OCT	NOV	DEC	JAN	FEB	MAR	APR	MAY	JUNE	TOTAL
1964-65	0.0	0.0	0.0	0.0	2.6	1.1	12.2	15.3	5.3	T	0.0	0.0	36.5
1965-66	0.0	0.0	0.0	0.0	T	2.9	2.5	6.3	0.8	T	T	0.0	12.5
1966-67	0.0	0.0	0.0	0.0	8.3	3.3	2.4	8.1	3.0	0.0	0.0	0.0	25.1
1967-68	0.0	0.0	0.0	T	6.3	3.5	17.0	1.1	8.8	T	0.0	0.0	36.7
1968-69	0.0	0.0	0.0	0.0	0.6	1.6	9.3	0.8	6.4	T	0.0	0.0	18.7
1969-70	0.0	0.0	0.0	0.0	2.5	12.9	7.8	9.2	5.6	0.2	0.0	0.0	38.2
1970-71	0.0	0.0	0.0	0.0	0.2	0.4	1.5	8.2	2.8	T	0.0	0.0	13.1
1971-72	0.0	0.0	0.0	0.0	4.4	0.4	7.9	6.3	0.3	0.6	0.0	0.0	19.9
1972-73	0.0	0.0	0.0	T	1.9	1.1	0.4	1.4	2.0	1.1	0.0	0.0	7.9
1973-74	0.0	0.0	0.0	0.0	0.4	27.5	3.8	8.0	3.0	2.1	0.0	0.0	44.8
1974-75	0.0	0.0	0.0	0.0	3.8	5.8	6.8	4.8	10.5	0.1	0.0	0.0	31.8
1975-76	0.0	0.0	0.0	0.0	4.5	8.1	5.6	0.6	2.3	0.0	0.0	0.0	21.1
1976-77	0.0	0.0	0.0	0.0	0.4	3.1	20.9	3.6	1.6	0.4	0.0	0.0	30.0
1977-78	0.0	0.0	0.0	0.0	2.8	15.2	30.6	3.9	5.4	0.0	0.0	0.0	57.9
1978-79	0.0	0.0	0.0	0.0	T	0.7	19.1	18.0	0.2	0.4	0.0	0.0	38.4
1979-80	0.0	0.0	0.0	0.0	0.8	0.2	5.0	14.5	3.6	0.7	0.0	0.0	24.8
1980-81	0.0	0.0	0.0	T	3.4	2.1	3.9	7.2	0.7	0.0	0.0	0.0	17.3
1981-82	0.0	0.0	0.0	T	0.4	15.6	21.8	13.6	3.5	3.3	0.0	0.0	58.2
1982-83	0.0	0.0	0.0	0.0	0.1	0.4	2.8	2.5	1.3	T	0.0	0.0	7.1
1983-84	0.0	0.0	0.0	0.0	0.1	8.3	7.2	17.1	9.2	T	0.0	0.0	41.9
1984-85	0.0	0.0	0.0	0.0	2.5	3.6	10.6	11.0	T	0.1	0.0	0.0	27.8
1985-86	0.0	0.0	0.0	0.0	T	8.1	1.7	9.5	1.1	T	0.0	0.0	20.4
1986-87	0.0	0.0	0.0	0.0	T	1.6	11.6	5.3	1.4	T	0.0	0.0	19.9
1987-88	0.0	0.0	0.0	0.0	T	1.1	2.9	4.8	2.5	T	0.0	0.0	11.3
1988-89	0.0	0.0	0.0	T	0.4	6.8	0.1	2.4	1.9	1.7	0.2	0.0	13.5
1989-90	0.0	T	0.0	9.3	0.6	8.2	4.1	2.5	1.3	T	0.0	0.0	26.0
1990-91	0.0	0.0	0.0	0.0	0.0	9.6	5.4	2.4	0.1	0.0	T	0.0	17.5
1991-92	0.0	0.0	0.0	0.0	3.0	0.2	7.1	0.2	2.9	1.3	0.0	T	14.7
1992-93	0.0	0.0	0.0	0.1	1.4	3.7	3.1	17.2	3.0	T	0.0	0.0	28.5
1993-94	0.0	0.0	0.0	2.4	1.1	5.6							
Record Mean	0.0	T	0.0	0.2	1.8	4.9	6.1	5.8	3.4	0.5	T	T	22.7

SOUTH BEND, INDIANA

South Bend is located on the Saint Joseph River in the northern portion of Saint Joseph County, situated on mostly level to gently rolling terrain and some former marshland. Drainage for the area is through the Saint Joseph River and Kankakee River.

South Bend is under the climatic influence of Lake Michigan with its nearest shore 20 miles to the northwest. The lake has a moderating effect on the temperature. Temperatures of 100 degrees or higher are rare and cold waves are less severe than at many locations at the same latitude. This results in favorable conditions for orchard and vegetable growth.

Based on the 1951-1980 period, the average first occurrence of 32 degrees Fahrenheit in the fall is October 18 and the average last occurrence in the spring is May 1.

Precipitation is fairly evenly distributed throughout the year with the greatest amounts during the growing season. The predominant snow season is from November through March, although there are also generally lighter amounts in October and April.

Winter is marked by considerable cloudiness and rather high humidity along with frequent periods of snow. Heavy snowfalls, resulting from a cold northwest wind passing over Lake Michigan are not uncommon.

NORMALS, MEANS, AND EXTREMES

SOUTH BEND, INDIANA

LATITUDE: 41°42'N　LONGITUDE: 86°19' W　ELEVATION: FT. GRND 773 BARO 782　TIME ZONE: EASTERN　WBAN: 14848

	(a)	JAN	FEB	MAR	APR	MAY	JUNE	JULY	AUG	SEP	OCT	NOV	DEC	YEAR
TEMPERATURE °F:														
Normals														
-Daily Maximum		30.4	34.1	45.6	58.7	70.0	79.5	82.9	80.7	74.1	62.3	48.5	35.4	58.5
-Daily Minimum		16.1	18.7	29.1	38.7	48.8	58.6	63.0	61.1	53.7	42.8	33.4	22.3	40.5
-Monthly		23.3	26.4	37.4	48.7	59.4	69.1	72.9	70.9	63.9	52.6	40.9	28.9	49.5
Extremes														
-Record Highest	54	68	69	85	91	95	104	101	103	99	92	82	70	104
-Year		1950	1976	1981	1942	1942	1988	1991	1988	1953	1963	1950	1982	JUN 1988
-Record Lowest	54	-22	-17	-13	11	24	35	44	40	29	20	-7	-16	-22
-Year		1943	1951	1943	1972	1968	1972	1972	1965	1942	1988	1950	1960	JAN 1943
NORMAL DEGREE DAYS:														
Heating (base 65°F)		1293	1081	856	489	233	28	7	22	85	395	723	1119	6331
Cooling (base 65°F)		0	0	0	0	59	151	251	205	52	10	0	0	728
% OF POSSIBLE SUNSHINE														
MEAN SKY COVER (tenths)														
Sunrise - Sunset	48	8.0	7.7	7.5	6.9	6.3	6.1	5.7	5.7	5.9	6.1	7.8	8.2	6.8
MEAN NUMBER OF DAYS:														
Sunrise to Sunset														
-Clear	54	3.4	3.6	4.7	5.9	6.9	7.2	8.3	8.6	8.8	8.6	3.6	2.9	72.5
-Partly Cloudy	54	6.0	5.5	7.2	7.7	9.6	10.4	12.6	12.2	8.9	8.1	6.0	5.5	99.6
-Cloudy	54	21.6	19.1	19.2	16.5	14.5	12.4	10.1	10.1	12.3	14.3	20.4	22.6	193.1
Precipitation														
.01 inches or more	54	15.5	12.4	14.1	13.3	11.3	10.5	9.5	9.4	9.3	10.1	13.0	15.4	143.9
Snow,Ice pellets,hail														
1.0 inches or more	54	5.8	5.0	2.9	0.7	0.0	0.0	0.0	0.0	0.*	0.3	2.5	5.5	22.8
Thunderstorms	54	0.4	0.4	2.4	4.4	5.1	8.0	7.3	6.4	4.4	2.0	1.0	0.4	42.2
Heavy Fog Visibility														
1/4 mile or less	54	2.5	2.3	2.0	1.3	1.4	1.1	1.2	2.2	1.9	2.1	1.9	3.2	23.1
Temperature °F														
-Maximum														
90° and above	30	0.0	0.0	0.0	0.0	0.6	3.0	5.3	3.2	0.7	0.0	0.0	0.0	12.8
32° and below	30	16.4	12.4	3.9	0.1	0.0	0.0	0.0	0.0	0.0	0.0	1.8	10.0	44.7
-Minimum														
32° and below	30	28.0	24.4	20.0	7.7	0.7	0.0	0.0	0.0	0.*	3.3	14.0	25.0	123.1
0° and below	30	3.7	2.4	0.1	0.0	0.0	0.0	0.0	0.0	0.0	0.0	0.0	1.3	7.5
AVG. STATION PRESS.(mb)	21	989.4	989.8	987.4	986.7	986.8	986.9	988.2	989.4	989.8	989.9	988.9	989.5	988.6
RELATIVE HUMIDITY (%)														
Hour 01	30	80	78	77	74	75	77	82	85	84	79	80	81	79
Hour 07 (Local Time)	30	81	81	80	78	79	80	84	88	88	84	83	83	82
Hour 13	30	72	69	62	57	54	55	57	59	60	60	69	76	63
Hour 19	30	75	72	66	60	56	57	60	64	67	68	74	79	67
PRECIPITATION (inches):														
Water Equivalent														
-Normal		2.23	1.90	3.10	3.82	3.22	4.11	3.82	3.67	3.62	3.08	3.27	3.30	39.14
-Maximum Monthly	54	5.28	5.23	7.96	9.20	6.86	10.86	7.47	8.30	9.01	9.75	6.72	5.50	10.86
-Year		1959	1976	1976	1947	1990	1993	1982	1979	1977	1954	1985	1965	JUN 1993
-Minimum Monthly	54	0.44	0.54	0.54	0.50	1.17	0.48	0.02	0.32	0.01	0.42	1.37	0.60	0.01
-Year		1945	1969	1958	1971	1992	1988	1946	1950	1979	1950	1962	1943	SEP 1979
-Maximum in 24 hrs	54	2.81	2.64	2.33	3.14	2.99	4.70	3.64	3.70	3.00	3.49	3.95	3.33	4.70
-Year		1960	1954	1972	1947	1976	1968	1989	1966	1977	1988	1990	1965	JUN 1968
Snow,Ice pellets,hail														
-Maximum Monthly	54	86.1	35.1	33.9	14.0	0.6	T	T	T	1.2	8.8	30.3	41.9	86.1
-Year		1978	1958	1960	1982	1966	1992	1989	1989	1942	1989	1977	1962	JAN 1978
-Maximum in 24 hrs	54	16.7	14.4	14.8	8.7	0.6	T	T	T	1.0	8.8	17.5	13.7	17.5
-Year		1978	1993	1960	1982	1966	1992	1989	1989	1942	1989	1977	1981	NOV 1977
WIND:														
Mean Speed (mph)	45	11.9	11.3	12.0	11.6	10.2	9.1	8.2	7.7	8.6	9.6	11.1	11.3	10.2
Prevailing Direction through 1963		SW	SW	NNW	NNW	SSW	SSW	SSW	SSW	SSW	SSW	SSW	SW	SSW
Fastest Obs. 1 Min.														
-Direction (!!)	44	22	20	20	27	27	27	34	32	31	25	22	23	27
-Speed (MPH)	44	52	47	51	55	68	50	45	63	35	38	58	43	68
-Year		1975	1953	1961	1962	1989	1950	1951	1953	1992	1955	1988	1953	MAY 1989
Peak Gust														
-Direction (!!)	10	NW	NW	SW	SW	W	SW	W	SW	W	SW	SW	W	W
-Speed (mph)	10	59	58	54	66	86	71	66	45	63	59	74	54	86
-Date		1985	1987	1991	1984	1989	1987	1992	1988	1988	1991	1988	1987	MAY 1989

PRECIPITATION (inches)

SOUTH BEND, INDIANA

YEAR	JAN	FEB	MAR	APR	MAY	JUNE	JULY	AUG	SEP	OCT	NOV	DEC	ANNUAL
1964	1.57	1.03	5.01	5.37	1.32	4.45	4.80	3.42	3.39	0.89	2.33	2.54	36.12
1965	4.74	2.70	3.35	4.87	1.67	2.24	2.22	5.43	5.55	2.53	1.65	5.50	42.45
1966	1.54	1.73	2.58	5.72	3.89	1.90	4.40	5.48	1.10	1.32	4.89	4.73	39.28
1967	3.56	2.26	1.73	4.96	1.57	5.49	2.11	2.77	2.59	5.01	3.34	4.08	39.47
1968	2.06	3.21	1.24	2.40	2.44	9.09	2.06	3.17	4.38	1.43	4.42	3.42	39.32
1969	3.61	0.54	1.92	5.51	4.26	5.17	4.88	0.44	1.42	4.70	2.29	1.25	35.99
1970	1.34	0.76	2.81	5.43	3.35	3.92	4.14	3.37	5.37	3.68	3.57	1.78	39.52
1971	1.55	2.12	2.10	0.50	1.64	1.13	4.16	2.80	5.16	2.23	1.83	4.58	29.80
1972	1.82	1.39	3.63	3.23	3.01	2.72	4.52	3.85	7.67	3.76	2.90	4.78	43.28
1973	1.64	1.02	3.85	3.88	3.61	4.85	3.33	1.29	2.11	3.49	1.45	4.30	34.82
1974	3.24	2.20	2.81	4.17	4.82	4.08	1.17	1.70	4.65	2.46	3.21	3.00	37.51
1975	4.58	3.26	2.96	6.02	2.08	5.46	2.58	7.55	1.15	1.31	4.73	3.72	45.40
1976	2.21	5.23	7.96	5.20	6.67	6.60	5.96	2.44	3.34	3.23	3.23	2.21	54.28
1977	1.63	1.27	7.05	2.75	1.92	5.71	2.68	6.03	9.01	3.08	4.27	3.65	49.05
1978	4.03	0.86	2.37	4.35	3.35	3.79	5.21	3.80	3.07	3.99	2.79	4.43	42.04
1979	3.22	1.51	4.03	5.80	3.03	4.66	1.75	8.30	0.01	4.79	4.88	3.66	45.64
1980	1.52	1.51	3.74	3.44	1.65	5.97	3.29	7.84	5.64	3.35	1.47	3.91	43.33
1981	0.68	1.92	0.88	5.28	6.79	6.97	3.71	2.30	3.81	1.23	2.23	1.81	37.61
1982	2.95	1.17	4.54	1.46	5.51	3.12	7.47	2.84	2.51	0.91	4.52	3.40	40.40
1983	0.77	0.79	2.46	5.36	4.83	2.04	2.45	1.28	2.81	1.66	2.60	3.23	30.28
1984	0.86	1.45	2.10	4.22	4.02	3.43	1.76	1.47	4.02	4.38	2.73	4.42	34.86
1985	2.58	4.32	3.86	1.93	1.50	2.88	3.80	3.82	1.88	3.36	6.72	2.51	39.16
1986	1.24	2.46	2.09	1.87	3.42	5.06	6.15	1.90	4.27	3.81	2.90	1.67	36.84
1987	2.31	1.32	1.18	2.67	3.50	3.57	3.61	3.34	3.64	3.20	2.11	4.12	34.57
1988	2.21	1.98	3.03	2.91	1.40	0.48	1.28	5.63	4.42	6.68	5.72	2.91	38.65
1989	1.58	1.05	2.27	2.83	2.72	3.49	5.90	5.65	3.78	1.45	3.55	1.83	36.10
1990	2.36	3.66	2.79	2.91	6.86	4.40	5.45	4.60	3.76	7.09	6.69	5.04	55.61
1991	1.64	1.79	2.79	4.58	4.01	0.62	1.32	3.68	2.71	8.75	2.75	1.67	36.31
1992	1.64	1.73	2.93	2.19	1.17	1.74	5.24	2.07	8.84	1.60	5.54	3.99	38.68
1993	3.35	1.20	2.62	3.64	2.34	10.86	1.51	4.38	7.76	4.09	2.39	1.50	45.64
Record Mean	2.27	1.82	2.87	3.42	3.54	3.67	3.36	3.46	3.43	3.07	2.89	2.65	36.45

AVERAGE TEMPERATURE (deg. F)

SOUTH BEND, INDIANA

YEAR	JAN	FEB	MAR	APR	MAY	JUNE	JULY	AUG	SEP	OCT	NOV	DEC	ANNUAL
1964	29.3	27.3	35.8	50.4	63.5	71.1	74.0	69.0	63.0	48.9	43.2	27.5	50.2
1965	24.7	27.0	28.7	46.9	64.2	67.5	69.9	68.1	63.7	51.6	40.6	36.4	49.1
1966	20.3	28.1	39.1	45.3	52.2	68.8	72.9	67.2	62.3	51.7	42.8	29.9	48.4
1967	29.0	21.1	36.5	48.9	54.3	70.6	69.8	67.9	62.5	51.7	36.0	30.5	48.2
1968	22.3	22.3	39.7	50.0	55.2	68.9	71.0	71.6	64.6	53.7	40.7	27.0	48.9
1969	22.3	27.8	32.9	50.3	59.5	64.9	73.1	73.0	63.6	51.4	37.2	27.5	48.6
1970	16.8	25.7	32.3	48.8	61.0	67.8	72.0	70.4	64.4	54.4	39.4	30.4	48.6
1971	20.0	28.6	33.9	46.2	56.3	73.9	69.5	68.1	67.1	60.4	40.9	36.0	50.1
1972	23.9	26.5	33.4	45.4	60.0	63.6	71.4	69.4	62.9	49.7	38.1	28.8	47.8
1973	29.9	30.0	46.7	50.1	56.1	72.3	74.4	74.1	67.0	58.3	44.4	29.2	52.7
1974	27.9	28.1	39.6	51.7	57.3	66.8	75.0	71.9	62.0	53.0	42.4	32.9	50.7
1975	30.0	27.8	34.4	43.6	63.2	70.3	71.8	73.6	60.1	55.9	48.1	32.7	51.0
1976	21.6	35.3	43.6	52.5	57.0	70.9	72.8	69.0	61.3	47.9	33.2	22.8	49.0
1977	12.3	26.8	44.1	55.4	68.7	67.1	76.2	69.8	65.3	50.5	42.6	25.9	50.4
1978	18.5	14.8	31.2	48.7	59.5	69.2	71.7	71.9	68.9	51.9	42.9	29.5	48.2
1979	17.9	16.3	39.2	46.2	58.5	70.1	72.3	71.3	65.4	54.2	42.3	34.8	49.0
1980	27.1	23.6	35.5	49.0	62.1	67.8	76.6	75.0	66.3	50.4	41.6	31.1	50.5
1981	23.4	32.1	40.7	52.5	56.6	69.2	71.4	70.6	62.8	50.0	41.5	27.6	49.9
1982	15.6	22.7	35.1	44.7	66.0	64.5	73.3	68.9	63.3	54.0	42.5	39.0	49.1
1983	29.3	33.2	40.2	45.5	55.3	72.0	78.7	78.3	66.5	53.8	43.8	18.4	51.3
1984	18.5	35.9	30.0	48.2	56.1	72.3	71.7	74.4	63.8	57.1	41.0	34.4	50.3
1985	19.6	23.7	41.3	55.2	63.1	66.8	73.1	69.8	65.6	54.6	41.6	20.1	49.5
1986	25.8	24.8	40.4	51.3	59.9	67.7	74.8	67.6	65.9	53.0	36.1	31.2	49.9
1987	25.3	30.8	40.0	50.4	64.7	72.8	75.8	71.7	64.6	46.8	43.7	32.8	51.6
1988	21.3	22.7	37.3	48.4	62.2	72.0	76.4	75.9	64.5	45.9	42.5	28.8	49.8
1989	33.4	22.2	37.0	47.3	57.2	68.0	73.9	70.6	61.8	52.9	38.8	17.7	48.4
1990	34.0	31.3	40.9	48.7	56.7	68.8	71.2	69.9	65.0	52.3	45.9	31.8	51.4
1991	24.2	32.3	41.5	52.6	67.5	74.0	75.7	73.2	63.2	54.9	36.6	32.1	52.3
1992	29.1	33.2	37.4	46.1	58.4	65.1	70.0	67.4	62.3	51.4	40.0	30.9	49.3
1993	28.3	24.4	35.0	46.9	60.7	67.7	74.5	73.6	58.9	50.5	39.4	29.7	49.1
Record Mean	24.5	26.1	36.4	48.2	59.3	68.9	73.3	71.5	64.6	53.2	40.1	28.6	49.5
Max	31.9	34.0	45.2	58.5	70.4	79.8	84.1	82.3	75.2	63.2	47.9	35.4	59.0
Min	17.0	18.2	27.6	37.9	48.2	57.9	62.4	60.6	53.9	43.2	32.3	21.7	40.1

REFERENCE NOTES

SOUTH BEND, INDIANA

GENERAL
T - TRACE AMOUNT.
BLANK ENTRIES DENOTE MISSING/UNREPORTED DATA.
# INDICATES A STATION OR INSTRUMENT RELOCATION.

SPECIFIC

PM - INCLUDES LAST DAY OF PREVIOUS MONTH
ASOS - AUTOMATED SURFACE OBSERVING SYSTEM IN OPERATION DURING THESE MONTHS.

(*) - LENGTH OF RECORD IN YEARS, ALTHOUGH INDIVIDUAL MONTHS MAY BE MISSING.
0.* OR * - THE VALUE IS BETWEEN 0.0 AND 0.05
NORMALS - BASED ON THE 1961-1990 RECORD PERIOD.
EXTREMES - DATES ARE THE MOST RECENT OCCURRENCE
WIND DIR.- NUMERALS SHOW TENS OF DEGRESS CLOCKWISE FROM TRUE NORTH. "00" INDICATES CALM.
RESULTANT DIRECTIONS ARE GIVEN TO WHOLE DEGREES.
BOLD VALUES INDICATE EXTREME VALUES WHICH OCCURRED AFTER THE ASOS SYSTEM WAS COMMISSIONED.

RECORD = PERIOD OF RECORD
RECORD MEAN PRECIPITATION IS THE MEAN OF ALL DAILY PRECIPITATION AMOUNTS DURING THE PERIOD OF RECORD.
RECORD MAX(MIN) TEMPERATURE IS THE MEAN OF ALL DAILY MAX(MIN) TEMPERATURES DURING THE PERIOD OF RECORD.
RECORD MEAN TEMPERATURE IS THE SUM OF THE RECORD MAX AND RECORD MIN DIVIDED BY 2.
AVERAGE TEMPERATURE IS THE SUM OF THE MEAN DAILY MAX AND MIN TEMPERATURE DIVIDED BY 2.

EXCEPTIONS

RECORD MEANS ARE THROUGH THE CURRENT YEAR, BEGINNING IN 1894 FOR TEMPERATURE
1894 FOR PRECIPITATION
1940 FOR SNOWFALL

HEATING DEGREE DAYS Base 65 deg. F SOUTH BEND, INDIANA

SEASON	JULY	AUG	SEP	OCT	NOV	DEC	JAN	FEB	MAR	APR	MAY	JUNE	TOTAL
1964-65	5	40	155	493	648	1157	1244	1058	1120	535	99	36	6590
1965-66	3	56	121	411	727	879	1377	1028	796	584	402	47	6431
1966-67	4	31	128	412	658	1079	1109	1220	880	474	337	8	6340
1967-68	31	30	137	429	861	1064	1317	1233	779	444	311	35	6671
1968-69	11	27	69	372	724	1171	1316	1034	988	433	226	105	6476
1969-70	0	0	110	421	826	1158	1485	1094	1007	497	181	68	6847
1970-71	18	11	93	327	766	1066	1387	1012	959	562	280	7	6488
1971-72	16	13	86	171	716	892	1269	1110	972	582	197	109	6133
1972-73	24	35	112	468	801	1115	1080	977	561	446	268	0	5887
1973-74	0	4	60	229	611	1102	1142	1028	780	403	250	51	5660
1974-75	0	3	144	378	672	985	1077	1035	944	634	136	37	6045
1975-76	13	1	175	300	503	992	1338	856	658	407	253	9	5505
1976-77	0	19	139	525	949	1302	1628	1063	640	327	85	68	6745
1977-78	2	18	58	443	669	1206	1436	1401	1041	480	233	41	7028
1978-79	5	4	57	399	656	1095	1453	1356	795	560	241	22	6643
1979-80	3	17	73	353	672	928	1172	1195	908	482	147	58	6008
1980-81	0	1	62	449	694	1047	1282	915	749	374	271	6	5850
1981-82	4	8	132	460	700	1154	1523	1178	922	604	64	72	6821
1982-83	2	30	114	353	668	798	1100	886	760	581	298	30	5620
1983-84	6	0	94	352	628	1440	1431	838	1080	503	290	4	6666
1984-85	7	0	128	244	714	940	1401	1153	727	339	116	47	5816
1985-86	0	4	121	319	694	1381	1208	1117	760	425	192	46	6267
1986-87	3	48	81	369	858	1038	1224	950	766	436	139	16	5928
1987-88	5	25	78	558	638	993	1347	1220	851	498	158	37	6408
1988-89	1	11	72	581	670	1116	972	1190	865	528	271	40	6317
1989-90	0	12	147	381	779	1462	954	936	751	521	257	46	6246
1990-91	6	14	110	408	565	1021	1260	910	722	378	106	2	5502
1991-92	0	1	167	318	844	1012	1104	915	850	561	247	75	6094
1992-93	6	25	139	417	742	1049	1130	1130	925	536	157	55	6311
1993-94	0	7	198	450	765	1087							

COOLING DEGREE DAYS Base 65 deg. F SOUTH BEND, INDIANA

YEAR	JAN	FEB	MAR	APR	MAY	JUNE	JULY	AUG	SEP	OCT	NOV	DEC	TOTAL
1969	0	0	0	0	59	109	257	256	73	3	0	0	757
1970	0	0	0	19	63	158	240	185	82	5	0	0	752
1971	0	0	0	3	17	283	163	114	155	34	0	0	769
1972	0	0	0	0	47	74	226	180	55	0	0	0	582
1973	0	0	0	7	1	226	297	292	126	29	0	0	978
1974	0	0	0	10	17	114	321	222	60	11	0	0	755
1975	0	0	0	0	86	203	232	274	32	25	1	0	853
1976	0	0	1	37	13	192	249	150	36	4	0	0	682
1977	0	0	0	47	206	137	355	173	74	0	2	0	994
1978	0	0	0	0	70	173	218	227	179	0	1	0	868
1979	0	0	0	2	48	181	236	220	93	21	0	0	801
1980	0	0	0	9	65	145	367	319	107	6	0	0	1018
1981	0	0	2	6	15	137	211	191	72	0	0	0	634
1982	0	0	0	3	105	65	266	159	71	17	0	0	686
1983	0	0	0	1	3	247	440	417	146	13	0	0	1267
1984	0	0	0	5	19	228	226	298	98	7	0	0	881
1985	0	0	0	52	64	109	260	159	147	3	0	0	794
1986	0	0	4	22	41	135	312	136	113	2	0	0	765
1987	0	0	0	3	136	256	345	240	71	0	3	0	1054
1988	0	0	0	3	77	254	362	357	65	0	0	0	1118
1989	0	0	1	6	37	137	283	194	55	11	0	0	724
1990	0	0	9	37	8	167	203	171	117	21	0	0	733
1991	0	0	0	13	189	281	338	264	121	11	0	0	1217
1992	0	0	0	2	50	90	168	103	66	2	0	0	481
1993	0	0	0	0	29	144	302	284	18	9	0	0	786

SNOWFALL (inches) SOUTH BEND, INDIANA

SEASON	JULY	AUG	SEP	OCT	NOV	DEC	JAN	FEB	MAR	APR	MAY	JUNE	TOTAL
1964-65	0.0	0.0	0.0	T	13.5	18.5	14.7	20.1	20.5	0.4	0.0	0.0	87.7
1965-66	0.0	0.0	0.0	2.4	7.3	8.6	26.9	10.7	11.4	5.4	0.6	0.0	73.3
1966-67	0.0	0.0	T	T	16.2	18.2	30.4	31.6	11.3	2.9	0.0	0.0	110.6
1967-68	0.0	0.0	0.0	5.0	7.0	12.5	12.6	20.1	8.7	1.5	T	0.0	67.4
1968-69	0.0	0.0	0.0	T	8.8	27.9	24.1	10.0	6.8	T	0.0	0.0	77.6
1969-70	0.0	0.0	0.0	T	10.5	17.1	24.8	11.2	16.7	7.9	0.0	0.0	88.2
1970-71	0.0	0.0	0.0	0.0	11.0	20.0	22.5	7.5	19.1	2.6	0.0	0.0	82.7
1971-72	0.0	0.0	0.0	0.0	16.3	4.4	19.2	21.2	16.1	7.1	0.0	0.0	84.3
1972-73	0.0	0.0	0.0	1.5	12.6	19.7	5.5	10.6	4.7	1.7	0.0	0.0	56.3
1973-74	0.0	0.0	0.0	0.0	1.0	22.6	14.4	11.9	9.4	1.2	0.0	0.0	60.5
1974-75	0.0	0.0	0.0	0.6	7.9	19.9	9.2	13.9	17.8	5.4	0.0	0.0	74.7
1975-76	0.0	0.0	0.0	0.0	10.7	14.0	31.2	13.9	3.8	0.6	0.1	0.0	74.3
1976-77	0.0	0.0	0.0	0.8	21.6	37.6	37.2	13.9	15.8	2.3	0.0	0.0	129.2
1977-78	0.0	0.0	0.0	0.3	30.3	33.6	86.1	16.6	5.1	T	0.0	0.0	172.0
1978-79	0.0	0.0	0.0	0.0	7.5	26.4	45.1	15.9	6.3	0.1	0.0	0.0	101.3
1979-80	0.0	0.0	0.0	T	7.5	13.6	11.5	22.3	9.8	1.7	0.0	0.0	66.4
1980-81	0.0	0.0	0.0	1.1	8.8	24.3	23.8	20.7	6.3	T	0.0	0.0	85.0
1981-82	0.0	0.0	0.0	0.1	9.1	41.3	41.3	19.2	10.2	14.0	0.0	0.0	135.2
1982-83	0.0	0.0	0.0	0.0	2.1	2.5	8.0	9.6	12.0	1.1	0.0	0.0	35.3
1983-84	0.0	0.0	0.0	0.0	1.4	35.6	16.7	15.9	11.1	0.4	0.0	0.0	81.1
1984-85	0.0	0.0	0.0	0.0	0.6	14.1	40.0	28.9	1.6	3.1	0.0	0.0	88.3
1985-86	0.0	0.0	0.0	0.0	2.2	40.4	26.3	11.3	3.6	0.2	0.0	0.0	84.0
1986-87	0.0	0.0	0.0	0.0	9.7	4.8	31.4	5.9	2.0	1.5	0.0	0.0	55.3
1987-88	0.0	0.0	T	T	1.6	13.1	11.4	22.9	12.1	T	0.0	0.0	61.1
1988-89	0.0	0.0	0.0	0.3	7.8	14.8	3.1	16.3	2.5	1.7	T	T	46.5
1989-90	T	T	0.0	8.8	15.2	29.4	1.2	13.9	3.4	1.1	0.0	0.0	73.0
1990-91	0.0	0.0	0.0	T	T	17.5	15.7	23.4	2.4	T	T	0.0	59.0
1991-92	0.0	0.0	0.0	T	7.6	11.7	19.9	10.9	16.7	0.9	0.0	T	67.7
1992-93	0.0	0.0	0.0	1.2	7.7	12.6	8.1	32.2	19.0	1.4	0.0	0.0	82.2
1993-94	0.0	0.0	0.1	T	2.2	11.3							
Record Mean	T	T	T	0.7	8.0	17.7	18.9	14.9	9.1	2.0	T	T	71.3

DES MOINES, IOWA

Located in the heart of North America, Des Moines has a climate which is continental in character. This results in a marked seasonal contrast in both temperature and precipitation. There is a gently rolling terrain in and around the Des Moines metropolitan area. Drainage of the area is generally to the southeast to the Des Moines River and its tributaries.

Since agriculture and services for it are the mainstay of the area, it is convenient to separate the year into arbitrary seasons corresponding to the growing seasons of the principal crops of the section. The winter season, when most plant life is dormant, is from mid-November to late March. The summer season, when corn and soybeans can be grown, lasts from early May to early October. The spring growing season, including part of the growing season of oats and forage crops, and the fall harvest season, each runs about 6 weeks. There is a large variation in annual precipitation from a minimum of about 17 inches to a maximum of about 56 inches. The average annual snowfall is 32 inches. Annual variation of snowfall is also large, ranging from a minimum of about 8 inches to as much as 72 inches.

The winter is a season of cold dry air, interrupted by occasional storms of short duration. At the beginning and the end of the season, the precipitation may occur as rain, but during the major portion of the season it falls as snow. Drifting snow may be extensive and impede transportation. The average precipitation for this season is approximately 20 percent of the annual amount. Although occasional cold waves follow the storms, bitterly cold days on which the temperatures fail to rise above zero occur on an average of only 3 days in 4 years.

The average growing season with temperatures above 32 degrees normally spans 160 to 165 days between late April and mid-October. The growing season is characterized by prevailing southerly winds and precipitation falling primarily as showers and thunderstorms, occasionally with damaging wind, erosive downpours or hail. Some 60 percent of the annual precipitation falls during the crop season with the maximum rate normally in late May and June. The autumn is characteristically sunny with diminishing precipitation, a condition favorable for drying and harvesting crops.

NORMALS, MEANS, AND EXTREMES

DES MOINES, IOWA

LATITUDE: 41°32'N LONGITUDE: 93°39'W ELEVATION: FT. GRND 938 BARO 966 TIME ZONE: CENTRAL WBAN: 14933

	(a)	JAN	FEB	MAR	APR	MAY	JUNE	JULY	AUG	SEP	OCT	NOV	DEC	YEAR
TEMPERATURE °F:														
Normals														
-Daily Maximum		28.1	33.7	46.9	61.8	73.0	82.2	86.7	84.2	75.6	64.3	48.0	32.6	59.8
-Daily Minimum		10.7	15.6	27.6	40.0	51.5	61.2	66.5	63.6	54.5	42.7	29.9	16.1	40.0
-Monthly		19.4	24.7	37.3	50.9	62.3	71.8	76.6	73.9	65.1	53.5	39.0	24.4	49.9
Extremes														
-Record Highest	54	65	73	91	93	98	103	105	108	101	95	76	69	108
-Year		1989	1972	1986	1980	1967	1988	1955	1983	1939	1963	1990	1984	AUG 1983
-Record Lowest	54	-24	-20	-22	9	30	38	47	40	26	14	-4	-22	-24
-Year		1970	1958	1962	1975	1967	1945	1971	1950	1942	1972	1991	1989	JAN 1970
NORMAL DEGREE DAYS:														
Heating (base 65°F)		1414	1128	859	428	165	10	0	11	71	372	780	1259	6497
Cooling (base 65°F)		0	0	0	5	81	214	360	287	74	15	0	0	1036
% OF POSSIBLE SUNSHINE	43	52	54	55	56	61	68	72	70	66	62	49	46	59
MEAN SKY COVER (tenths)														
Sunrise - Sunset	44	6.5	6.5	6.8	6.5	6.4	5.8	5.2	5.2	5.1	5.3	6.5	6.8	6.0
MEAN NUMBER OF DAYS:														
Sunrise to Sunset														
-Clear	44	7.9	7.6	6.8	7.3	7.6	8.2	10.4	10.8	11.8	11.7	7.5	7.1	104.7
-Partly Cloudy	44	7.3	5.8	7.3	7.5	8.5	10.5	11.0	10.4	7.4	7.2	6.8	6.7	96.5
-Cloudy	44	15.8	14.9	16.9	15.1	14.8	11.3	9.6	9.8	10.8	12.1	15.7	17.2	164.0
Precipitation														
.01 inches or more	54	7.6	7.3	10.2	10.6	11.4	10.7	9.5	9.2	8.7	7.6	7.3	8.0	108.1
Snow,Ice pellets,hail														
1.0 inches or more	54	2.4	2.3	1.9	0.5	0.0	0.0	0.0	0.0	0.0	0.1	0.9	2.2	10.2
Thunderstorms	54	0.3	0.4	2.1	4.3	7.2	9.2	8.4	7.4	5.1	2.7	1.1	0.3	48.5
Heavy Fog Visibility														
1/4 mile or less	44	2.1	2.4	2.1	0.9	0.8	0.6	0.6	1.2	1.2	1.2	1.8	2.8	17.6
Temperature °F														
-Maximum														
90° and above	32	0.0	0.0	0.*	0.2	0.5	4.3	9.7	6.8	1.7	0.1	0.0	0.0	23.4
32° and below	32	17.4	12.8	4.5	0.2	0.0	0.0	0.0	0.0	0.0	0.0	3.4	13.7	52.0
-Minimum														
32° and below	32	30.0	25.7	20.4	6.3	0.2	0.0	0.0	0.0	0.2	4.9	18.7	28.7	135.0
0° and below	32	7.9	3.9	0.2	0.0	0.0	0.0	0.0	0.0	0.0	0.0	0.3	3.7	16.0
AVG. STATION PRESS. (mb)	21	984.7	984.0	980.2	979.4	979.1	979.3	980.9	981.8	982.4	982.7	982.0	983.7	981.7
RELATIVE HUMIDITY (%)														
Hour 00	32	74	76	73	69	70	73	76	79	79	73	76	78	75
Hour 06 (Local Time)	32	76	78	78	78	78	80	82	85	85	79	79	80	80
Hour 12	32	67	65	61	55	55	56	58	58	59	55	64	69	60
Hour 18	32	69	66	59	53	53	53	57	58	60	58	67	72	60
PRECIPITATION (inches):														
Water Equivalent														
-Normal		0.96	1.11	2.33	3.36	3.66	4.46	3.78	4.20	3.53	2.62	1.79	1.32	33.12
-Maximum Monthly	54	4.38	2.99	5.82	7.76	7.88	14.19	10.51	13.68	10.19	7.29	6.52	3.43	14.19
-Year		1960	1951	1990	1976	1991	1947	1958	1977	1961	1941	1983	1982	JUN 1947
-Minimum Monthly	54	0.07	0.13	0.37	0.23	1.23	1.02	0.04	0.25	0.41	0.03	0.03	0.12	0.03
-Year		1954	1968	1989	1985	1949	1992	1975	1984	1950	1952	1969	1976	NOV 1969
-Maximum in 24 hrs	54	2.97	1.77	2.42	3.80	2.79	5.50	5.14	6.18	4.47	2.81	3.35	1.69	6.18
-Year		1960	1961	1945	1974	1954	1947	1958	1975	1961	1947	1952	1982	AUG 1975
Snow,Ice pellets,hail														
-Maximum Monthly	54	19.8	21.3	18.8	15.6	0.2	T	T	0.0	T	7.4	14.7	23.9	23.9
-Year		1942	1962	1948	1982	1944	1993	1992		1992	1980	1991	1961	DEC 1961
-Maximum in 24 hrs	54	19.8	12.1	8.5	10.4	0.2	T	T	0.0	T	7.4	11.8	11.0	19.8
-Year		1942	1950	1957	1973	1944	1993	1992		1992	1980	1968	1961	JAN 1942
WIND:														
Mean Speed (mph)	44	11.5	11.4	12.7	12.7	11.1	10.2	8.9	8.6	9.4	10.3	11.3	11.2	10.8
Prevailing Direction through 1963		NW	NW	NW	NW	SE	S	S	S	S	S	NW	NW	NW
Fastest Mile														
-Direction (!!!)	40	NW	W	S	W	W	NW	W	SSE	NW	W	W	SW	W
-Speed (MPH)	40	66	56	66	76	70	76	73	60	55	56	72	61	76
-Year		1953	1952	1953	1965	1955	1953	1968	1960	1953	1952	1952	1951	APR 1965
Peak Gust														
-Direction (!!!)	10	NW	NW	W	SW	E	SW	S	NW	W	NW	SW	NW	S
-Speed (mph)	10	55	62	58	66	54	58	83	63	54	60	62	55	83
-Date		1990	1984	1985	1989	1988	1990	1993	1989	1985	1985	1986	1985	JUL 1993

PRECIPITATION (inches)

DES MOINES, IOWA

YEAR	JAN	FEB	MAR	APR	MAY	JUNE	JULY	AUG	SEP	OCT	NOV	DEC	ANNUAL
1964	0.51	0.28	1.25	3.29	2.90	6.49	3.37	3.94	4.28	0.29	0.95	0.87	28.42
1965	1.62	1.14	3.02	4.18	3.89	4.80	1.93	1.93	7.23	0.79	1.79	1.64	33.96
1966	0.96	0.25	1.56	1.74	5.37	5.22	2.43	2.09	0.75	0.29	0.70	0.49	21.85
1967	0.77	0.28	1.64	2.25	2.22	7.39	0.81	0.79	2.54	1.93	0.47	0.73	21.82
1968	0.78	0.13	0.93	4.19	2.62	3.42	4.48	4.05	2.63	1.47	1.90	1.71	28.31
1969	1.01	0.97	1.41	4.46	4.75	7.32	4.34	1.83	2.43	2.87	0.03	1.01	32.43
1970	0.26	0.24	3.28	2.28	4.21	2.45	1.96	4.95	6.51	5.20	1.46	0.83	33.63
1971	1.75	2.34	0.41	1.54	3.87	4.31	2.16	1.83	2.19	3.51	3.29	1.12	28.32
1972	0.44	0.63	1.05	3.56	3.05	2.58	5.86	6.65	5.45	2.36	2.43	1.96	36.02
1973	2.09	2.21	4.15	4.67	5.01	2.04	9.17	1.37	7.07	3.26	1.49	2.65	45.18
1974	1.51	0.84	1.99	6.31	7.19	4.62	1.33	2.81	2.08	3.96	1.20	1.83	35.67
1975	1.41	1.48	1.90	2.65	3.41	5.98	0.04	9.73	1.70	0.63	2.20	0.48	31.61
1976	0.23	2.43	3.04	7.76	2.84	7.25	1.87	2.26	1.00	1.11	0.10	0.12	30.01
1977	0.50	0.36	3.57	2.45	2.29	1.25	2.63	13.68	2.82	5.10	0.69	1.81	37.15
1978	0.28	1.27	0.90	4.57	3.49	2.74	2.95	3.10	6.39	1.14	3.16	1.37	31.36
1979	1.72	0.52	4.23	3.23	2.50	5.78	2.96	5.07	0.97	3.23	1.43	0.20	31.84
1980	1.80	0.64	1.15	0.86	1.94	5.56	1.52	7.24	1.03	1.90	0.45	1.00	25.09
1981	0.25	0.97	0.39	2.00	2.46	5.02	5.76	6.32	2.30	2.06	2.63	1.14	31.30
1982	2.63	0.78	3.30	5.03	5.79	2.59	7.00	5.25	2.94	3.44	2.62	3.43	44.80
1983	1.17	1.95	3.72	3.80	3.93	3.65	2.44	3.01	3.87	5.54	6.52	1.57	41.17
1984	0.99	0.82	1.65	5.85	5.58	7.81	6.22	0.25	2.76	6.28	1.16	2.41	41.78
1985	0.64	1.98	3.37	0.23	1.56	3.72	2.04	2.83	5.42	3.75	1.65	1.31	28.50
1986	0.12	1.76	2.92	5.66	4.35	7.08	3.90	4.52	6.41	3.89	0.99	0.98	42.58
1987	0.42	1.38	2.99	2.92	3.75	2.10	5.08	10.04	1.40	1.03	3.27	2.59	36.97
1988	0.37	0.59	0.66	0.75	1.46	2.75	4.78	3.05	2.89	0.59	3.38	0.84	22.11
1989	1.30	1.05	0.37	1.95	3.62	2.22	3.65	6.53	5.41	2.28	0.19	0.57	29.14
1990	1.43	0.89	5.82	3.43	4.36	9.52	8.75	1.83	1.40	1.80	2.52	2.18	43.93
1991	0.95	0.17	3.90	7.54	7.88	2.87	1.14	3.65	0.90	4.96	3.61	2.20	39.77
1992	0.97	2.12	2.13	3.99	1.45	1.02	7.76	1.39	4.99	0.51	5.20	1.98	33.51
1993	1.59	1.52	3.22	2.96	7.51	7.68	9.75	12.24	5.79	1.70	1.06	0.86	55.88
Record Mean	1.11	1.14	2.03	2.97	4.19	4.64	3.53	3.82	3.39	2.44	1.68	1.23	32.17

AVERAGE TEMPERATURE (deg. F)

DES MOINES, IOWA

YEAR	JAN	FEB	MAR	APR	MAY	JUNE	JULY	AUG	SEP	OCT	NOV	DEC	ANNUAL
1964	27.8	26.3	30.4	48.0	64.3	69.9	77.2	70.4	63.6	50.8	41.2	23.3	49.4
1965	17.6	20.2	22.7	48.1	66.2	70.3	74.8	72.9	61.0	55.3	41.0	35.3	48.8
1966	14.2	24.7	41.0	46.0	57.7	70.4	78.6	70.8	62.4	52.9	38.4	24.8	48.5
1967	23.7	22.2	40.3	51.1	57.1	69.0	72.1	70.3	62.3	51.5	36.5	28.9	48.8
1968	22.3	23.9	42.6	51.1	57.0	72.4	74.2	73.0	63.3	53.8	34.8	21.3	49.2
1969	16.2	25.9	28.2	52.5	62.9	67.3	76.0	74.7	66.1	50.0	39.0	23.8	48.5
1970	12.7	27.1	32.9	51.7	65.8	72.4	76.6	73.9	65.5	54.1	39.8	28.8	50.1
1971	15.7	23.1	35.2	52.1	58.7	76.8	72.1	73.2	68.1	60.5	39.6	28.2	50.3
1972	16.6	19.9	37.1	48.9	62.5	70.2	73.9	72.4	64.6	48.0	33.9	18.4	47.2
1973	22.0	27.7	45.6	49.2	59.8	73.6	76.4	76.9	65.7	59.0	40.7	22.1	51.6
1974	19.5	28.0	39.8	53.0	60.3	69.1	80.9	71.1	60.9	55.0	39.6	28.2	50.4
1975	22.7	22.4	29.7	46.9	65.9	72.3	77.9	77.1	61.8	57.2	43.5	29.9	50.6
1976	22.9	34.1	39.7	55.4	60.7	71.4	77.2	73.8	65.7	48.5	32.6	21.8	50.3
1977	10.1	29.8	44.7	58.5	69.5	74.7	81.0	72.1	66.5	52.3	39.1	23.2	51.8
1978	11.0	13.3	33.2	50.5	61.6	72.8	75.8	75.1	70.5	52.4	39.0	23.0	48.2
1979	7.5	13.8	35.2	47.0	61.1	70.9	74.5	74.3	66.8	54.3	38.0	31.6	47.9
1980	23.4	21.3	34.7	52.0	63.5	71.2	79.9	76.2	66.8	49.6	41.6	26.6	50.6
1981	25.7	29.8	42.7	57.6	60.4	72.9	76.1	72.4	66.0	51.7	42.8	25.5	51.9
1982	9.6	22.9	35.4	46.9	64.7	67.1	76.9	73.2	64.9	54.6	38.4	31.6	48.8
1983	27.3	32.3	39.4	45.4	58.4	73.2	80.9	83.3	67.7	52.6	40.7	9.8	50.9
1984	19.7	35.5	31.1	48.8	58.7	73.2	75.9	77.5	62.4	52.8	39.5	27.6	50.2
1985	15.8	22.2	42.0	55.2	65.1	68.4	76.5	71.9	64.8	52.6	30.0	13.4	48.2
1986	26.9	21.7	42.5	53.9	62.6	73.3	77.2	69.3	67.4	52.6	33.3	28.8	50.8
1987	26.6	35.7	42.7	54.6	66.8	74.7	78.5	72.1	65.2	48.2	43.1	30.0	53.2
1988	19.6	21.6	40.2	51.4	67.4	75.6	78.5	78.8	66.9	47.9	39.8	28.8	51.4
1989	32.5	15.4	37.1	52.3	61.0	68.9	77.1	73.3	62.2	54.1	36.0	16.9	48.9
1990	31.7	31.1	42.0	50.0	57.9	71.6	73.8	74.4	68.1	52.4	43.7	22.9	51.6
1991	16.8	33.5	42.5	53.6	66.7	74.8	76.1	73.7	65.0	52.5	30.9	30.6	51.4
1992	30.7	34.9	42.4	48.9	62.0	70.3	71.2	68.3	63.1	53.4	34.9	27.1	50.6
1993	21.3	21.5	34.1	47.1	61.0	69.0	73.8	73.8	59.2	50.4	36.4	28.4	48.0
Record Mean	20.8	24.9	36.7	50.6	61.7	71.2	76.2	73.8	65.3	53.8	38.4	25.9	49.9
Max	29.5	33.8	45.9	60.9	71.9	81.2	86.5	84.1	75.9	64.4	47.5	34.1	59.6
Min	12.0	16.0	27.4	40.3	51.4	61.2	65.8	63.5	54.8	43.2	29.2	17.8	40.2

REFERENCE NOTES

DES MOINES, IOWA

GENERAL
T - TRACE AMOUNT.
BLANK ENTRIES DENOTE MISSING/UNREPORTED DATA.
# INDICATES A STATION OR INSTRUMENT RELOCATION.

SPECIFIC

PM - INCLUDES LAST DAY OF PREVIOUS MONTH
ASOS - AUTOMATED SURFACE OBSERVING SYSTEM IN OPERATION DURING THESE MONTHS.

(a) - LENGTH OF RECORD IN YEARS, ALTHOUGH INDIVIDUAL MONTHS MAY BE MISSING.
0.* OR * - THE VALUE IS BETWEEN 0.0 AND 0.05
NORMALS - BASED ON THE 1961-1990 RECORD PERIOD.
EXTREMES - DATES ARE THE MOST RECENT OCCURRENCE
WIND DIR.- NUMERALS SHOW TENS OF DEGRESS CLOCKWISE FROM TRUE NORTH. "00" INDICATES CALM.
RESULTANT DIRECTIONS ARE GIVEN TO WHOLE DEGREES.
BOLD VALUES INDICATE EXTREME VALUES WHICH OCCURRED AFTER THE ASOS SYSTEM WAS COMMISSIONED.

RECORD = PERIOD OF RECORD
RECORD MEAN PRECIPITATION IS THE MEAN OF ALL DAILY PRECIPITATION AMOUNTS DURING THE PERIOD OF RECORD.
RECORD MAX(MIN) TEMPERATURE IS THE MEAN OF ALL DAILY MAX(MIN) TEMPERATURES DURING THE PERIOD OF RECORD.
RECORD MEAN TEMPERATURE IS THE SUM OF THE RECORD MAX AND RECORD MIN DIVIDED BY 2.
AVERAGE TEMPERATURE IS THE SUM OF THE MEAN DAILY MAX AND MIN TEMPERATURE DIVIDED BY 2.

EXCEPTIONS

RECORD MEANS ARE THROUGH THE CURRENT YEAR,
BEGINNING IN 1878 FOR TEMPERATURE
1877 FOR PRECIPITATION
1940 FOR SNOWFALL

HEATING DEGREE DAYS Base 65 deg. F DES MOINES, IOWA

SEASON	JULY	AUG	SEP	OCT	NOV	DEC	JAN	FEB	MAR	APR	MAY	JUNE	TOTAL
1964-65	0	28	124	433	707	1286	1465	1247	1305	509	86	1	7191
1965-66	0	14	154	310	715	915	1568	1122	737	565	256	16	6372
1966-67	0	13	136	375	789	1239	1276	1192	767	421	308	22	6538
1967-68	14	18	109	440	850	1112	1317	1184	685	411	260	24	6424
1968-69	2	11	92	378	898	1351	1508	1086	1135	368	159	50	7038
1969-70	0	0	45	478	773	1270	1617	1055	987	422	99	9	6755
1970-71	0	0	100	352	748	1114	1521	1167	915	393	211	0	6521
1971-72	8	4	96	184	756	1132	1495	1303	857	478	149	23	6485
1972-73	7	10	112	523	925	1442	1326	1039	594	470	173	0	6621
1973-74	0	0	68	209	719	1325	1406	1033	775	363	189	20	6107
1974-75	0	12	168	307	755	1131	1308	1185	1090	539	81	11	6587
1975-76	0	0	148	267	637	1085	1297	890	780	302	160	4	5570
1976-77	0	1	76	527	964	1333	1700	981	624	234	25	2	6467
1977-78	0	3	35	388	769	1289	1667	1442	988	427	181	9	7198
1978-79	0	0	48	385	776	1293	1779	1433	912	532	163	13	7334
1979-80	1	10	57	339	801	1031	1281	1263	932	408	124	9	6256
1980-81	0	0	77	473	695	1182	1214	979	684	241	177	0	5722
1981-82	6	2	57	406	660	1218	1713	1175	911	536	73	30	6787
1982-83	0	6	113	326	791	1026	1162	908	787	587	219	17	5942
1983-84	0	0	96	394	720	1709	1401	851	1043	488	217	1	6920
1984-85	0	0	172	376	759	1154	1520	1192	707	335	59	27	6301
1985-86	0	0	172	378	1046	1596	1181	1208	702	344	116	2	6745
1986-87	0	25	51	378	941	1114	1184	813	687	336	58	6	5593
1987-88	0	24	54	513	648	1083	1399	1254	764	400	33	3	6175
1988-89	0	6	35	524	749	1114	1002	1384	866	417	170	31	6298
1989-90	0	6	140	345	865	1489	1025	943	703	469	221	21	6227
1990-91	3	0	85	394	633	1301	1491	875	699	355	112	0	5948
1991-92	0	1	144	395	1019	1058	1054	868	691	486	152	9	5877
1992-93	0	26	135	363	897	1169	1344	1210	953	528	151	42	6818
1993-94	0	2	185	460	851	1129							

COOLING DEGREE DAYS Base 65 deg. F DES MOINES, IOWA

YEAR	JAN	FEB	MAR	APR	MAY	JUNE	JULY	AUG	SEP	OCT	NOV	DEC	TOTAL
1969	0	0	0	0	99	129	349	308	84	20	0	0	989
1970	0	0	0	31	131	238	368	280	123	20	0	0	1191
1971	0	0	0	13	26	362	234	263	194	52	0	0	1144
1972	0	0	0	3	77	184	289	247	109	0	0	0	909
1973	0	0	0	2	19	267	358	378	98	30	0	0	1152
1974	0	0	0	9	52	149	499	209	52	4	0	0	974
1975	0	0	0	3	116	237	408	383	58	32	0	0	1237
1976	0	0	0	19	36	203	385	283	104	20	0	0	1050
1977	0	0	0	48	172	298	505	232	86	1	0	0	1342
1978	0	0	6	0	83	251	341	321	221	3	0	0	1226
1979	0	0	0	0	48	194	304	305	118	15	0	0	984
1980	0	0	0	22	83	200	469	353	138	2	0	0	1267
1981	0	0	1	27	42	243	358	239	95	2	0	0	1007
1982	0	0	0	0	71	101	374	269	120	11	0	0	946
1983	0	0	0	4	20	272	502	574	183	19	0	0	1574
1984	0	0	0	8	27	254	345	397	101	2	0	0	1134
1985	0	0	0	46	69	134	361	221	174	0	0	0	1005
1986	0	0	11	17	46	258	386	162	130	0	0	0	1010
1987	0	0	0	31	121	304	426	250	65	0	0	0	1197
1988	0	0	0	1	112	329	425	444	101	3	0	0	1415
1989	0	0	6	44	53	156	379	269	61	15	0	0	983
1990	0	0	1	26	5	226	283	297	184	11	2	0	1035
1991	0	0	9	19	172	302	349	279	150	14	0	0	1294
1992	0	0	0	10	64	174	197	134	86	10	0	0	675
1993	0	0	0	0	31	167	281	280	19	15	0	0	793

SNOWFALL (inches) DES MOINES, IOWA

SEASON	JULY	AUG	SEP	OCT	NOV	DEC	JAN	FEB	MAR	APR	MAY	JUNE	TOTAL
1964-65	0.0	0.0	0.0	T	0.5	2.6	7.4	7.6	17.3	0.0	0.0	0.0	35.4
1965-66	0.0	0.0	0.0	0.0	T	2.4	3.6	0.5	1.8	T	T	0.0	8.3
1966-67	0.0	0.0	0.0	T	T	7.1	4.0	2.3	0.5	0.9	T	0.0	14.8
1967-68	0.0	0.0	0.0	2.4	0.7	2.6	8.5	2.1	T	0.1	T	0.0	16.4
1968-69	0.0	0.0	0.0	T	13.5	5.2	6.8	12.3	2.8	0.0	0.0	0.0	40.6
1969-70	0.0	0.0	0.0	T	T	12.0	3.2	1.2	9.8	3.9	0.0	0.0	30.1
1970-71	0.0	0.0	0.0	0.0	0.2	1.4	15.9	13.7	5.3	0.4	0.0	0.0	36.9
1971-72	0.0	0.0	0.0	0.0	10.3	2.9	5.8	9.0	1.6	0.6	0.0	0.0	30.2
1972-73	0.0	0.0	0.0	T	10.2	9.5	15.3	4.6	T	15.1	0.0	0.0	54.7
1973-74	0.0	0.0	0.0	0.0	T	9.6	10.6	6.1	1.9	1.2	0.0	0.0	29.4
1974-75	0.0	0.0	0.0	0.0	9.3	9.1	13.3	17.9	5.6	4.5	0.0	0.0	59.7
1975-76	0.0	0.0	0.0	0.0	7.0	0.5	2.4	11.8	0.9	0.0	T	0.0	22.6
1976-77	0.0	0.0	0.0	T	1.1	2.8	7.6	0.9	4.8	1.9	0.0	0.0	19.1
1977-78	0.0	0.0	0.0	0.0	2.2	22.0	2.7	18.6	9.9	0.3	0.0	0.0	55.7
1978-79	0.0	0.0	0.0	0.0	5.8	11.0	16.6	8.2	4.1	8.0	0.0	0.0	53.7
1979-80	0.0	0.0	0.0	T	0.7	0.4	7.9	5.7	5.8	2.8	0.0	0.0	23.3
1980-81	0.0	0.0	0.0	7.4	T	2.7	3.7	6.6	T	0.0	0.0	0.0	20.4
1981-82	0.0	0.0	0.0	0.4	3.2	10.7	18.5	2.6	11.9	15.6	0.0	0.0	62.9
1982-83	0.0	0.0	0.0	0.8	0.3	3.6	4.2	16.8	13.2	12.6	0.0	0.0	51.5
1983-84	0.0	0.0	0.0	T	9.8	19.6	12.5	1.3	13.7	0.1	0.0	0.0	57.0
1984-85	0.0	0.0	0.0	T	2.7	7.6	7.7	6.9	6.7	T	0.0	0.0	31.6
1985-86	0.0	0.0	T	0.0	8.1	15.9	1.3	6.7	0.2	0.1	0.0	0.0	32.3
1986-87	0.0	0.0	0.0	T	3.1	2.9	5.1	2.8	5.5	0.0	0.0	0.0	19.4
1987-88	0.0	0.0	0.0	T	0.1	13.0	1.2	7.6	0.8	0.0	0.0	0.0	22.7
1988-89	0.0	0.0	0.0	0.0	1.1	1.0	0.1	16.3	1.2	0.6	T	0.0	20.3
1989-90	0.0	0.0	0.0	T	1.7	6.6	11.3	7.4	0.1	T	0.0	0.0	27.1
1990-91	0.0	0.0	0.0	T	1.3	12.1	9.9	0.4	0.2	T	T	0.0	23.9
1991-92	0.0	0.0	0.0	T	14.7	1.2	1.9	4.5	0.2	0.1	0.0	0.0	22.6
1992-93	T	0.0	T	T	10.0	3.7	11.1	10.1	5.1	1.5	T	T	41.5
1993-94	0.0	0.0	0.0	0.2	0.9	2.0							
Record Mean	T	0.0	T	0.2	3.0	6.6	8.1	7.2	6.2	1.9	T	T	33.3

DUBUQUE, IOWA

The terrain around Dubuque varies from gently rolling, 10 to 15 miles to the south and west, to steep hills and bluffs around the city and along the Mississippi River.

The principal feature of the climate in Dubuque is its variety. The Dubuque area is subject to weather ranging from the cold, dry, arctic air masses in the winter, with readings as low as 32 degrees below zero, to the hot, dry weather of the desert southwest in the summer when the temperatures reach about 110. More often the area is covered by mild Pacific air that has lost considerable moisture in crossing the mountains far to the west, or by cool, dry Canadian air, or by warm, moist air from the Gulf regions. Most of the year the latter three types of air masses dominate Dubuque weather, with the invasions of Gulf air rarely occurring in the winter.

The seasons vary widely from year to year at Dubuque. For example, successive invasions of cold air from the north may bring a long, cold winter with snow-covered ground from mid November until March and many days of sub-zero temperatures. Another winter can be mild with bare ground most of the season and only a few sub-zero temperature readings. The summers, too, may vary from hot and humid with considerable thunderstorm activity when the Gulf air prevails, to relatively cool, dry weather when air of northerly origin dominates the season.

All seasons are marked by storms that accompany the changes from one type of air mass to another. In winter, rain changes to sleet and snow, and occasionally a peal of thunder is heard at the height of a snowstorm. In summer, thunderstorms are frequently heavy. They are occasionally accompanied by hail and on rare occasions by tornadoes. Thunderstorms have been sufficiently intense at times to raise the Mississippi River, which is about one-fourth mile wide at Debuque, nearly 5 feet overnight. Flash floods have drowned many people.

Most of the precipitation occurs during the spring and fall seasons. The last occurrence of snow and freezing rain can be in late May, and the first occurrence in late September.

While the climate of Dubuque does not lack for variety, there are times when a particular weather condition may persist for an extended period. Cold weather has lasted as long as 20 days in succession with sub-zero readings. Heat waves have persisted for 10 or more days with readings around 100 degrees each day. Hot, dry spells occasionally plague the crops and livestock in summer, but there are frequent periods of mild, dry weather in the spring and frequently in the autumn.

NORMALS, MEANS, AND EXTREMES

DUBUQUE, IOWA

LATITUDE: 42°24'N LONGITUDE: 90°42'W ELEVATION: FT. GRND 1056 BARO 1070 TIME ZONE: CENTRAL WBAN: 94908

	(a)	JAN	FEB	MAR	APR	MAY	JUNE	JULY	AUG	SEP	OCT	NOV	DEC	YEAR
TEMPERATURE °F:														
Normals														
-Daily Maximum		23.9	29.6	42.5	57.7	69.2	78.4	82.4	79.9	71.7	60.5	44.6	29.1	55.8
-Daily Minimum		7.7	12.7	25.0	37.4	48.0	57.1	62.0	59.6	51.4	40.4	27.9	14.0	36.9
-Monthly		15.9	21.2	33.8	47.6	58.6	67.8	72.3	69.8	61.6	50.5	36.3	21.6	46.4
Extremes														
-Record Highest	41	60	61	85	93	91	100	101	100	97	89	73	64	101
-Year		1989	1984	1986	1980	1991	1988	1988	1988	1955	1953	1978	1970	JUL 1988
-Record Lowest	41	-28	-27	-20	11	24	36	44	40	28	13	-17	-25	-28
-Year		1970	1979	1962	1973	1966	1972	1984	1986	1984	1952	1977	1983	JAN 1970
NORMAL DEGREE DAYS:														
Heating (base 65°F)		1522	1226	967	522	237	34	6	27	124	456	861	1345	7327
Cooling (base 65°F)		0	0	0	0	39	118	232	176	22	6	0	0	593
% OF POSSIBLE SUNSHINE														
MEAN SKY COVER (tenths)														
Sunrise - Sunset	21	6.8	6.5	6.9	6.8	6.5	6.4	5.8	5.6	5.5	5.6	6.9	7.0	6.4
MEAN NUMBER OF DAYS:														
Sunrise to Sunset														
-Clear	21	7.1	7.5	6.1	6.3	6.8	6.4	8.0	9.4	10.5	10.3	6.7	6.7	91.9
-Partly Cloudy	21	6.7	6.1	7.9	7.5	8.6	10.6	12.5	10.5	8.0	8.0	5.9	6.3	98.5
-Cloudy	21	17.2	14.7	17.0	16.2	15.6	13.0	10.5	11.1	11.6	12.7	17.4	18.0	174.9
Precipitation .01 inches or more	41	9.2	7.8	10.9	11.5	11.6	10.3	10.1	9.0	9.2	8.7	9.0	10.1	117.4
Snow, Ice pellets, hail 1.0 inches or more	41	2.9	2.7	2.7	0.8	0.*	0.0	0.0	0.0	0.0	0.1	1.2	3.2	13.7
Thunderstorms	24	0.2	0.1	1.7	3.1	5.2	6.1	6.4	6.0	3.4	2.5	0.9	0.2	35.7
Heavy Fog Visibility 1/4 mile or less	24	3.6	2.8	3.6	1.8	1.7	1.1	2.0	2.0	1.9	2.1	3.0	4.1	29.7
Temperature °F														
-Maximum														
90° and above	26	0.0	0.0	0.0	0.*	0.1	1.4	4.5	2.4	0.6	0.0	0.0	0.0	9.0
32° and below	26	21.8	16.0	5.7	0.3	0.0	0.0	0.0	0.0	0.0	0.0	5.1	17.9	66.8
-Minimum														
32° and below	26	30.6	27.2	22.9	8.6	0.8	0.0	0.0	0.0	0.5	7.2	20.9	29.7	148.5
0° and below	26	9.7	5.5	0.4	0.0	0.0	0.0	0.0	0.0	0.0	0.0	0.2	5.0	20.9
AVG. STATION PRESS. (mb)	6	978.7	977.7	975.5	976.2	974.8	975.4	977.4	978.8	978.6	979.4	978.0	976.9	977.3
RELATIVE HUMIDITY (%)														
Hour 00														
Hour 06 (Local Time)	26	77	78	79	77	79	82	85	88	87	82	81	81	81
Hour 12	26	69	66	63	56	57	59	61	62	62	58	66	72	63
Hour 18	8	73	69	64	58	56	62	64	65	69	67	72	78	66
PRECIPITATION (inches):														
Water Equivalent														
-Normal		1.26	1.32	2.89	3.72	4.26	4.13	4.02	4.69	4.67	2.73	2.71	1.96	38.36
-Maximum Monthly	41	6.04	3.61	6.50	7.69	9.43	10.49	12.23	9.90	15.46	8.58	10.63	4.14	15.46
-Year		1960	1953	1959	1964	1962	1969	1961	1987	1965	1967	1961	1982	SEP 1965
-Minimum Monthly	41	0.31	0.17	0.36	0.81	0.69	0.70	0.87	0.08	0.07	T	0.36	0.30	T
-Year		1964	1958	1958	1985	1992	1988	1991	1969	1979	1964	1955	1975	OCT 1964
-Maximum in 24 hrs	41	3.75	2.24	2.35	2.65	4.60	3.60	6.28	3.90	8.85	2.58	5.09	2.31	8.85
-Year		1960	1953	1959	1964	1962	1966	1961	1970	1967	1959	1961	1971	SEP 1967
Snow, Ice pellets, hail														
-Maximum Monthly	41	29.3	25.1	30.2	19.8	3.1	T	0.0	0.0	T	1.5	13.9	26.4	30.2
-Year		1979	1975	1959	1973	1966	1993			1986	1976	1986	1977	MAR 1959
-Maximum in 24 hrs	41	11.8	11.9	15.5	14.6	3.1	T	0.0	0.0	T	1.5	10.4	15.0	15.5
-Year		1971	1962	1959	1973	1966	1993			1986	1976	1992	1990	MAR 1959
WIND:														
Mean Speed (mph)														
Prevailing Direction														
Fastest Obs. 1 Min.														
-Direction (!!)														
-Speed (MPH)														
-Year														
Peak Gust														
-Direction (!!)	9							NW	NW	NW			E	NW
-Speed (mph)	10	58	52	62	68	74	55	74	62	58	54	55	56	74
-Date		1990	1987	1990	1984	1988	1993	1992	1992	1986	1984	1991	1990	JUL 1992

PRECIPITATION (inches)

DUBUQUE, IOWA

YEAR	JAN	FEB	MAR	APR	MAY	JUNE	JULY	AUG	SEP	OCT	NOV	DEC	ANNUAL
1964	0.31	0.28	2.41	7.69	7.13	6.23	2.43	4.03	2.48	T	1.99	1.16	36.14
1965	2.70	1.06	3.81	6.92	6.30	1.68	4.82	8.33	15.46	3.83	3.18	3.33	61.42
1966	1.81	2.23	5.50	2.88	4.75	9.52	2.95	1.96	1.65	1.47	1.86	2.65	39.23
1967	1.80	1.29	3.06	5.53	3.71	6.75	1.00	3.82	11.88	8.58	3.27	2.28	52.97
1968	1.20	0.23	1.26	7.47	3.76	7.31	3.02	5.26	5.67	1.32	0.76	2.70	39.96
1969	2.56	0.40	1.11	3.46	2.37	10.49	5.39	0.08	1.04	4.31	0.60	1.89	33.70
1970	0.49	0.58	2.55	2.04	5.24	3.88	4.26	4.13	9.16	1.65	1.23	1.56	36.77
1971	1.98	2.96	1.58	1.37	4.35	2.51	3.66	2.60	3.96	3.69	4.39	4.04	37.09
1972	0.47	0.99	2.85	5.50	3.44	2.49	5.02	6.02	5.58	3.69	1.80	1.93	39.78
1973	1.66	1.00	4.72	6.30	5.36	2.13	4.34	1.54	7.27	0.73	2.09	2.42	39.56
1974	2.39	1.38	2.62	4.21	7.39	7.74	1.85	4.06	0.62	3.08	1.78	1.46	38.58
1975	0.90	2.53	3.33	3.78	2.79	3.75	1.15	7.44	1.91	0.41	4.22	0.30	32.51
1976	0.57	2.58	3.71	3.74	1.73	1.48	1.97	3.67	0.88	2.43	0.37	0.45	23.58
1977	0.78	1.10	4.82	2.93	3.31	2.22	6.48	6.02	3.45	2.38	3.05	2.49	39.03
1978	0.81	0.66	1.30	4.14	5.50	2.50	3.51	1.10	4.51	1.68	2.93	2.36	31.00
1979	2.61	1.12	2.54	2.37	2.18	4.33	6.72	7.47	0.07	2.80	1.92	1.34	35.47
1980	1.80	1.23	1.12	1.78	4.31	5.62	1.80	9.11	6.48	2.77	0.99	1.61	38.62
1981	0.34	2.38	0.41	5.72	1.05	6.67	1.69	9.65					
1982											5.30	4.14	
1983	0.70	2.17	2.91	3.39	6.42	1.56	3.38	2.23	3.54	1.78	3.88	2.44	34.40
1984	0.81	1.05	1.68	4.08	4.20	6.19	2.31	1.37	2.22	6.55	1.77	2.77	35.00
1985	1.17	2.51	4.35	0.81	4.51	1.09	2.19	3.44	4.89	4.94	4.22	2.20	36.32
1986	0.82	2.26	1.96	2.74	5.92	5.72	2.75	4.29	13.06	3.60	1.61	0.68	45.41
1987	0.79	0.80	2.11	1.53	5.19	2.05	5.55	9.90	3.31	1.03	3.41	3.16	38.83
1988	1.32	0.74	2.31	2.21	0.97	0.70	1.97	3.73	2.72	1.73	2.69	1.48	22.57
1989	0.90	0.45	1.85	2.32	2.25	1.85	3.01	4.35	2.38	2.18	0.81	0.39	22.74
1990	1.27	1.06	4.63	2.27	4.91	5.68	4.28	8.01	0.57	1.35	2.66	2.86	39.55
1991	1.16	0.45	5.81	3.88	4.24	3.34	0.87	3.90	3.76	3.86	4.28	1.59	37.14
1992	1.49	1.87	1.55	3.82	0.69	1.02	7.52	1.63	4.57	1.09	6.84	2.12	34.21
1993	1.37	0.99	3.10	4.61	5.40	8.04	8.50	8.05	4.50	1.64	1.03	0.98	48.21
Record Mean	1.39	1.31	2.43	3.16	4.11	4.43	3.90	3.79	4.01	2.57	2.15	1.59	34.81

AVERAGE TEMPERATURE (deg. F)

DUBUQUE, IOWA

YEAR	JAN	FEB	MAR	APR	MAY	JUNE	JULY	AUG	SEP	OCT	NOV	DEC	ANNUAL
1964	27.4	25.7	30.6	47.5	62.9	68.5	73.7	67.9	61.1	48.5	39.6	20.0	47.6
#1965	16.1	19.1	24.2	45.4	61.3	66.5	72.1	68.4	60.2	50.9	37.4	31.1	46.0
1966	10.4	21.6	37.4	44.7	53.2	68.0	73.7	68.2	60.7	50.2	37.2	24.6	45.8
1967	21.5	14.7	35.0	48.3	54.0	68.2	69.6	65.7	60.8	49.1	33.2	26.0	45.5
1968	19.1	20.6	40.2	50.0	55.3	67.7	70.7	70.3	60.6	50.7	35.8	20.7	46.8
1969	14.0	24.1	29.4	49.6	58.6	62.2	72.1	71.7	61.9	47.6	34.2	20.4	45.5
1970	8.4	19.0	31.4	48.7	61.6	68.1	72.2	70.6	62.5	52.7	36.7	23.8	46.3
1971	11.2	20.0	29.6	47.7	55.8	73.2	68.0	67.1	64.9	57.1	37.6	27.7	46.6
1972	13.4	17.2	30.8	43.7	62.3	66.4	70.2	71.6	62.1	46.0	32.7	17.3	44.5
1973	22.5	25.7	42.3	45.2	55.5	68.9	72.3	72.3	62.5	55.6	38.0	20.4	48.4
1974	17.6	20.5	34.2	49.2	54.8	63.8	74.0	67.1	57.7	50.8	36.8	25.5	46.0
1975	20.4	19.6	24.6	41.3	61.8	68.5	72.0	71.1	57.3	52.5	40.6	26.2	46.3
1976	16.4	30.0	36.5	49.7	55.4	67.9	73.3	68.9	59.7	45.3	27.8	14.7	45.5
1977	3.4	23.6	41.5	54.2	65.1	67.6	74.4	66.5	61.7	48.7	34.0	17.5	46.5
1978	7.7	11.5	28.6	47.5	57.8	67.3	71.4	71.0	66.2	48.6	35.8	19.2	44.4
1979	4.5	10.8	30.8	43.7	57.9	67.6	71.1	69.7	63.3	50.6	35.7	29.6	44.6
1980	20.1	18.1	29.7	47.9	61.7	68.4	75.9	73.0	63.6	47.8	37.9	23.9	47.3
1981	21.7	25.6	38.4	51.2	57.1	69.3	72.2	69.9					
1982											35.7	30.5	
1983	23.1	28.2	35.3	43.4	54.9	69.8	75.9	76.6	62.9	51.2	39.0	9.6	47.5
1984	15.7	30.7	28.4	47.7	55.1	68.9	70.6	72.4	60.6	53.9	37.4	27.4	47.4
1985	11.9	19.3	39.8	54.0	61.3	65.8	71.3	67.1	61.8	49.8	30.4	8.7	45.1
1986	18.4	19.2	37.3	50.9	59.5	67.2	73.6	65.0	63.3	50.8	30.9	25.7	46.8
1987	23.1	31.9	38.4	51.5	62.4	71.0	75.0	69.1	61.7	44.7	40.8	27.8	49.8
1988	13.9	17.9	36.0	47.5	62.6	71.3	76.1	74.7	64.2	45.1	37.7	25.3	47.7
1989	27.4	14.4	32.3	46.5	58.1	67.0	73.5	67.9	58.5	52.3	34.2	13.3	45.5
1990	28.7	26.8	38.8	47.6	55.1	68.3	71.1	70.8	64.2	49.2	41.2	20.0	48.5
1991	14.5	26.9	37.9	50.9	64.1	72.2	73.2	71.3	61.3	49.6	30.3	26.3	48.2
1992	24.7	30.4	36.8	45.3	59.4	66.7	66.7	64.4	59.4	48.8	32.9	23.5	46.6
1993	20.8	20.3	31.7	44.4	59.3	66.2	71.4	71.1	57.3	48.3	34.7	26.2	46.0
Record Mean	18.6	22.5	34.0	48.3	59.7	68.9	73.7	71.3	63.2	51.7	36.6	24.1	47.8
Max	26.9	30.8	42.5	58.1	69.8	78.9	83.9	81.4	73.2	61.4	44.6	31.5	56.9
Min	10.4	14.2	25.6	38.6	49.5	58.9	63.5	61.2	53.2	42.0	28.6	16.8	38.6

REFERENCE NOTES

DUBUQUE, IOWA

GENERAL
T - TRACE AMOUNT.
BLANK ENTRIES DENOTE MISSING/UNREPORTED DATA.
# INDICATES A STATION OR INSTRUMENT RELOCATION.

SPECIFIC

(a) - LENGTH OF RECORD IN YEARS, ALTHOUGH INDIVIDUAL MONTHS MAY BE MISSING.
0.* OR * - THE VALUE IS BETWEEN 0.0 AND 0.05
NORMALS - BASED ON THE 1961-1990 RECORD PERIOD.
EXTREMES - DATES ARE THE MOST RECENT OCCURRENCE
WIND DIR.- NUMERALS SHOW TENS OF DEGRESS CLOCKWISE FROM TRUE NORTH. "00" INDICATES CALM.
RESULTANT DIRECTIONS ARE GIVEN TO WHOLE DEGREES.
BOLD VALUES INDICATE EXTREME VALUES WHICH OCCURRED AFTER THE ASOS SYSTEM HAS COMMISSIONED.

RECORD = PERIOD OF RECORD
RECORD MEAN PRECIPITATION IS THE MEAN OF ALL DAILY PRECIPITATION AMOUNTS DURING THE PERIOD OF RECORD.
RECORD MAX(MIN) TEMPERATURE IS THE MEAN OF ALL DAILY MAX(MIN) TEMPERATURES DURING THE PERIOD OF RECORD.
RECORD MEAN TEMPERATURE IS THE SUM OF THE RECORD MAX AND RECORD MIN DIVIDED BY 2.
AVERAGE TEMPERATURE IS THE SUM OF THE MEAN DAILY MAX AND MIN TEMPERATURE DIVIDED BY 2.

EXCEPTIONS

1. THUNDERSTORMS AND HEAVY FOG ARE THROUGH 1964 AND MAY BE INCOMPLETE, DUE TO PART-TIME OPERATIONS.
2. MEAN SKY COVER, AND DAYS CLEAR-PARTLY CLOUDY-CLOUDY ARE THROUGH 1970.
3. TOTAL BREAK IN RECORD SEPTEMBER 1981 THROUGH OCTOBER 1982.

RECORD MEANS ARE THROUGH THE CURRENT YEAR, BEGINNING IN 1874 FOR TEMPERATURE
1874 FOR PRECIPITATION
1951 FOR SNOWFALL

HEATING DEGREE DAYS Base 65 deg. F — DUBUQUE, IOWA

SEASON	JULY	AUG	SEP	OCT	NOV	DEC	JAN	FEB	MAR	APR	MAY	JUNE	TOTAL
#1964-65	7	43	169	505	756	1391	1509	1282	1259	582	157	34	7694
1965-66	2	36	181	437	822	1044	1693	1209	848	602	368	50	7292
1966-67	2	22	163	453	826	1246	1341	1405	924	501	360	21	7264
1967-68	39	54	156	502	948	1204	1415	1282	763	446	299	54	7162
1968-69	13	22	149	463	869	1368	1578	1140	1094	452	235	139	7522
1969-70	7	0	140	555	916	1376	1756	1281	1035	502	180	46	7794
1970-71	9	3	140	379	840	1271	1663	1257	1090	515	295	9	7471
1971-72	29	24	127	266	818	1150	1594	1379	1051	632	174	63	7307
1972-73	20	20	141	582	961	1476	1311	1095	696	588	285	7	7182
1973-74	0	2	132	302	805	1375	1460	1242	946	472	327	87	7150
1974-75	0	38	234	433	837	1219	1379	1267	1247	706	149	48	7557
1975-76	24	0	245	393	724	1198	1502	1009	877	461	294	16	6743
1976-77	0	14	180	614	1106	1554	1909	1155	722	343	93	40	7730
1977-78	1	43	121	500	924	1472	1770	1491	1122	521	268	41	8274
1978-79	3	12	90	505	869	1414	1877	1511	1053	632	247	25	8238
1979-80	6	27	110	449	873	1087	1383	1356	1086	516	173	28	7094
1980-81	0	3	113	527	806	1266	1337	1097	819	408	262	9	6647
1981-82	15	13											
1982-83					871	1063	1294	1021	913	641	306	25	
1983-84	5	0	144	429	773	1716	1523	988	1126	513	307	11	7535
1984-85	9	8	198	339	824	1158	1642	1275	776	362	135	64	6790
1985-86	4	23	198	465	1032	1744	1438	1276	856	433	180	45	7694
1986-87	3	56	112	428	1013	1210	1292	922	816	410	149	16	6427
1987-88	3	46	127	619	718	1146	1581	1361	892	519	113	20	7145
1988-89	1	14	87	611	813	1226	1160	1416	1010	549	247	41	7175
1989-90	2	15	208	394	917	1602	1120	1063	805	547	301	38	7012
1990-91	7	1	139	490	707	1388	1561	1063	833	425	159	10	6783
1991-92	3	6	197	477	1033	1191	1241	994	865	585	212	46	6850
1992-93	33	75	201	495	958	1281	1362	1246	1026	610	194	74	7555
1993-94	0	10	241	516	901	1199							

COOLING DEGREE DAYS Base 65 deg. F — DUBUQUE, IOWA

YEAR	JAN	FEB	MAR	APR	MAY	JUNE	JULY	AUG	SEP	OCT	NOV	DEC	TOTAL
1969	0	0	0	0	43	61	236	217	54	22	0	0	633
1970	0	0	0	19	80	149	239	182	70	9	0	0	748
1971	0	0	0	1	15	260	124	97	130	30	0	0	657
1972	0	0	0	0	95	110	189	234	61	0	0	0	689
1973	0	0	0	0	0	133	233	236	65	18	0	0	685
1974	0	0	0	4	18	59	287	109	18	0	0	0	495
1975	0	0	0	0	57	159	247	198	19	12	0	0	692
1976	0	0	0	9	3	112	266	144	29	8	0	0	571
1977	0	0	0	26	103	126	302	98	30	0	0	0	685
1978	0	0	0	0	52	117	210	202	131	1	0	0	713
1979	0	0	0	0	36	106	203	181	66	8	0	0	600
1980	0	0	0	11	76	136	347	257	78	0	0	0	905
1981	0	0	0	1	23	144	247	170					
1982											0	0	
1983	0	0	0	0	0	176	349	367	90	9	0	0	991
1984	0	0	0	1	9	138	190	243	72	1	0	0	654
1985	0	0	0	41	28	93	206	94	112	0	0	0	574
1986	0	0	2	15	15	116	276	64	67	0	0	0	555
1987	0	0	0	11	76	203	322	182	34	0	0	0	828
1988	0	0	0	0	49	216	351	325	68	0	0	0	1009
1989	0	0	1	1	41	110	272	112	21	9	0	0	567
1990	0	0	0	29	1	143	203	191	120	6	0	0	693
1991	0	0	0	8	136	234	264	207	91	4	0	0	944
1992	0	0	0	0	44	105	94	63	39	1	0	0	346
1993	0	0	0	0	27	117	204	209	17	5	0	0	579

SNOWFALL (inches) — DUBUQUE, IOWA

SEASON	JULY	AUG	SEP	OCT	NOV	DEC	JAN	FEB	MAR	APR	MAY	JUNE	TOTAL
1964-65	0.0	0.0	0.0	0.0	3.4	9.5	11.7	5.2	19.5	T	0.0	0.0	49.3
1965-66	0.0	0.0	0.0	0.0	0.1	4.2	11.3	1.5	3.3	T	3.1	0.0	23.5
1966-67	0.0	0.0	0.0	0.0	T	11.1	4.8	12.5	10.0	1.0	T	0.0	39.4
1967-68	0.0	0.0	0.0	0.6	4.1	2.8	5.7	1.6	T	0.2	0.0	0.0	15.0
1968-69	0.0	0.0	0.0	T	0.1	12.9	9.4	4.3	4.8	T	T	0.0	31.5
1969-70	0.0	0.0	0.0	T	0.9	21.4	6.4	6.4	3.5	3.7	0.0	0.0	42.3
1970-71	0.0	0.0	0.0	0.3	T	11.0	14.6	7.5	13.4	1.4	0.0	0.0	48.2
1971-72	0.0	0.0	0.0	0.0	12.3	4.1	4.7	12.4	16.4	4.5	0.0	0.0	54.4
1972-73	0.0	0.0	0.0	0.5	9.5	10.4	2.2	6.0	1.4	19.8	0.0	0.0	49.8
1973-74	0.0	0.0	0.0	0.0	T	6.7	9.1	13.5	3.1	0.5	0.0	0.0	32.9
1974-75	0.0	0.0	0.0	0.0	6.2	11.6	6.2	25.1	17.2	8.7	0.0	0.0	75.0
1975-76	0.0	0.0	0.0	0.0	7.1	1.0	6.2	12.4	0.7	T	0.0	0.0	27.4
1976-77	0.0	0.0	0.0	1.5	3.7	5.7	9.6	1.3	9.5	3.5	0.0	0.0	34.8
1977-78	0.0	0.0	0.0	0.0	11.6	26.4	10.6	9.7	13.0	T	0.0	0.0	71.3
1978-79	0.0	0.0	0.0	0.0	8.0	17.0	29.3	8.7	3.3	4.1	0.0	0.0	70.4
1979-80	0.0	0.0	0.0	0.0	1.9	1.5	7.1	12.3	7.9	5.3	0.0	0.0	36.0
1980-81	0.0	0.0	0.0	1.4	2.1	7.0	3.7	6.5	1.0	0.0	0.0	0.0	21.7
1981-82	0.0	0.0											
1982-83					0.0	0.0	5.9	15.6	18.4	12.3	0.0	0.0	
1983-84	0.0	0.0	0.0	0.0	1.3	18.4	11.5	0.9	10.7	3.1	0.0	0.0	45.9
1984-85	0.0	0.0	0.0	0.0	1.0	14.2	13.7	6.4	8.1	0.1	0.0	0.0	43.5
1985-86	0.0	0.0	0.0	0.0	10.4	18.7	15.8	12.8	0.8	1.3	0.0	0.0	59.8
1986-87	0.0	0.0	T	T	13.9	3.3	8.7	T	8.0	T	0.0	0.0	33.9
1987-88	0.0	0.0	0.0	0.0	1.1	20.9	8.4	8.3	1.5	1.7	0.0	0.0	41.9
1988-89					1.0	4.4	2.4	8.3	7.7	0.2	T	0.0	
1989-90	0.0	0.0	0.0	T	1.8	8.3	7.1	9.7	0.5	0.3	0.0	0.0	27.7
1990-91	0.0	0.0	0.0	0.0	5.3	18.8	12.5	2.6	12.7	0.1	T	0.0	52.0
1991-92	0.0	0.0	0.0	T	3.7	6.5	9.1	10.7	6.0	0.2	0.0	0.0	36.2
1992-93	0.0	0.0	0.0	0.4	12.6	7.4	8.6	8.6	14.9	11.3	0.0	T	63.8
1993-94	0.0	0.0	0.0	T	1.0	5.6							
Record Mean	0.0	0.0	T	0.2	3.7	10.5	9.2	7.8	9.6	2.5	0.1	T	43.6

SIOUX CITY, IOWA

Sioux City is located along the Missouri River at a point where Iowa boarders both Nebraska and South Dakota. Except for the river valleys, the countryside is rolling. The Sioux City business section lies in the river valley and the residential sections, for the most part, are spread over the hills which range from 100 to 200 feet higher than the valley. The local topography causes minor variations in wind and temperature.

Located in the midland of a continent and in the northern half of the Great Plains, the climate of Sioux City is typically continental and is largely determined by the movement and interaction of the large-scale weather systems. Under normal conditions, winters are cold and summers warm, and most of the precipitation comes during the warmer months from April to September. There is considerable fluctuation in temperature and precipitation from season to season and from year to year, as elsewhere in the northern plains. Except for an occasional dry year, the climate is quite favorable for agriculture with corn, the small grains, and grasses producing abundantly.

The grass usually starts to grow about the middle of April. The growing season averages about 160 days. Summers are sunny and most summer rains are associated with showers or thunderstorms. Winds are lightest in the summer months, except for occasional strong gusts with thunderstorms. Winds gradually increase in autumn and winter and usually reach their highest average velocities in April.

NORMALS, MEANS, AND EXTREMES

SIOUX CITY, IOWA

LATITUDE: 42°24'N LONGITUDE: 96°23' W ELEVATION: FT. GRND 1095 BARO 1113 TIME ZONE: CENTRAL WBAN: 14943

	(a)	JAN	FEB	MAR	APR	MAY	JUNE	JULY	AUG	SEP	OCT	NOV	DEC	YEAR
TEMPERATURE °F:														
Normals														
-Daily Maximum		27.6	33.2	45.8	62.2	73.2	82.1	86.5	83.5	74.9	64.0	46.4	30.9	59.2
-Daily Minimum		7.7	13.9	25.7	38.3	50.0	59.6	64.8	62.0	51.9	39.6	26.4	12.7	37.7
-Monthly		17.7	23.6	35.8	50.2	61.6	70.8	75.7	72.8	63.4	51.8	36.4	21.8	48.5
Extremes														
-Record Highest	53	70	71	91	97	102	108	107	104	102	93	81	68	108
-Year		1981	1981	1968	1980	1967	1988	1955	1955	1976	1963	1978	1984	JUN 1988
-Record Lowest	53	-26	-26	-22	-2	25	38	42	37	24	12	-9	-24	-26
-Year		1970	1962	1960	1975	1976	1950	1971	1950	1945	1993	1991	1989	JAN 1970
NORMAL DEGREE DAYS:														
Heating (base 65°F)		1466	1159	905	444	171	19	0	12	103	417	858	1339	6893
Cooling (base 65°F)		0	0	0	0	65	193	332	254	55	8	0	0	907
% OF POSSIBLE SUNSHINE	53	57	56	57	59	62	67	73	70	66	63	51	50	61
MEAN SKY COVER (tenths)														
Sunrise - Sunset	53	6.4	6.5	6.8	6.5	6.4	5.7	4.9	5.0	5.1	5.3	6.6	6.6	6.0
MEAN NUMBER OF DAYS:														
Sunrise to Sunset														
-Clear	53	8.2	7.1	6.6	6.8	6.9	8.3	11.2	11.6	11.6	11.7	7.1	7.4	104.5
-Partly Cloudy	53	7.7	6.8	7.7	8.6	9.5	10.9	11.7	10.9	7.6	7.8	7.2	7.3	103.7
-Cloudy	53	15.1	14.4	16.7	14.5	14.6	10.8	8.1	8.5	10.6	11.6	15.7	16.3	156.8
Precipitation														
.01 inches or more	53	6.7	6.5	8.9	9.6	11.2	10.8	9.0	8.9	8.1	6.4	5.7	6.8	98.6
Snow, Ice pellets, hail														
1.0 inches or more	53	2.0	1.9	2.3	0.5	0.*	0.0	0.0	0.0	0.0	0.2	1.1	2.0	10.0
Thunderstorms	53	0.1	0.3	1.3	3.3	6.9	8.7	8.3	7.4	4.9	1.8	0.4	0.1	43.5
Heavy Fog Visibility														
1/4 mile or less	53	2.3	2.7	2.1	0.6	0.6	0.5	0.7	1.5	1.4	1.5	2.2	2.8	18.9
Temperature °F														
-Maximum														
90° and above	34	0.0	0.0	0.*	0.5	1.3	5.1	9.4	6.1	1.7	0.1	0.0	0.0	24.4
32° and below	34	17.9	13.2	5.3	0.2	0.0	0.0	0.0	0.0	0.0	0.1	4.0	15.4	56.1
-Minimum														
32° and below	34	30.6	27.1	23.0	8.1	0.7	0.0	0.0	0.0	0.6	7.3	22.8	30.1	150.4
0° and below	34	9.6	5.1	0.9	0.*	0.0	0.0	0.0	0.0	0.0	0.0	0.6	5.1	21.3
AVG. STATION PRESS. (mb)	21	979.2	979.1	975.2	974.2	973.7	973.5	975.2	976.0	976.7	977.3	977.1	978.6	976.3
RELATIVE HUMIDITY (%)														
Hour 00	33	76	78	77	71	70	74	79	82	80	74	78	79	77
Hour 06 (Local Time)	34	77	79	81	78	79	82	86	88	87	81	82	81	82
Hour 12	34	68	67	63	52	54	57	59	61	58	54	64	71	61
Hour 18	34	70	68	61	49	49	52	56	59	58	55	67	74	60
PRECIPITATION (inches):														
Water Equivalent														
-Normal		0.55	0.71	1.96	2.34	3.67	3.71	3.27	2.97	2.88	1.94	1.08	0.78	25.86
-Maximum Monthly	53	2.44	2.66	5.90	6.73	8.46	8.78	10.33	7.75	9.69	5.30	4.10	2.23	10.33
-Year		1949	1971	1987	1984	1959	1967	1972	1951	1965	1979	1948	1982	JUL 1972
-Minimum Monthly	53	0.11	0.12	0.24	0.45	0.60	0.53	0.41	0.12	0.07	T	0.01	0.01	T
-Year		1986	1985	1968	1942	1955	1988	1985	1971	1950	1958	1967	1943	OCT 1958
-Maximum in 24 hrs	53	1.09	2.15	2.75	1.95	2.70	3.89	5.50	4.28	2.94	4.55	3.45	1.18	5.50
-Year		1982	1971	1987	1993	1992	1957	1972	1961	1965	1979	1948	1959	JUL 1972
Snow, Ice pellets, hail														
-Maximum Monthly	53	29.1	24.5	26.2	9.7	4.0	T	T	0.0	0.4	9.8	16.5	20.6	29.1
-Year		1982	1993	1962	1983	1945	1991	1992		1961	1991	1983	1968	JAN 1982
-Maximum in 24 hrs	53	17.4	15.2	12.4	6.8	4.0	T	T	0.0	0.4	8.8	12.4	9.0	17.4
-Year		1982	1993	1983	1957	1945	1991	1992		1961	1991	1983	1968	JAN 1982
WIND:														
Mean Speed (mph)	52	11.3	11.2	12.4	13.2	11.9	10.7	9.2	9.1	9.8	10.4	11.3	11.0	11.0
Prevailing Direction through 1963		NW	NW	NW	NNW	SSE	S	SSE	SSE	SSE	SE	NW	NW	NW
Fastest Mile														
-Direction (!!)	49	NW	NW	N	W	W	W	NW	NW	S	W	NW	NW	W
-Speed (MPH)	49	56	54	61	68	80	91	66	56	66	70	59	53	91
-Year		1967	1947	1950	1946	1956	1945	1967	1951	1957	1940	1954	1968	JUN 1945
Peak Gust														
-Direction (!!)	10	NW	NW	NW	S	S	NW	NW	NW	W	NW	NW	NW	S
-Speed (mph)	10	61	58	59	69	60	60	67	60	52	53	58	56	69
-Date		1984	1984	1993	1985	1992	1984	1987	1986	1992	1990	1991	1993	APR 1985

PRECIPITATION (inches)

SIOUX CITY, IOWA

YEAR	JAN	FEB	MAR	APR	MAY	JUNE	JULY	AUG	SEP	OCT	NOV	DEC	ANNUAL
1964	0.34	0.30	1.59	3.53	3.19	2.71	6.00	5.35	3.03	0.34	0.10	0.82	27.30
1965	0.44	1.19	1.67	2.49	5.78	2.01	2.16	1.85	9.69	0.47	0.27	0.33	28.35
1966	0.52	1.18	1.78	0.67	1.15	5.09	2.93	4.88	1.29	1.03	0.11	0.82	21.45
1967	0.41	0.28	0.26	2.02	4.21	8.78	2.02	1.35	1.08	1.26	0.01	0.75	22.43
1968	0.32	0.15	0.24	2.74	2.34	3.85	1.68	1.52	4.12	4.77	0.53	1.96	24.22
1969	1.22	1.32	0.69	0.62	2.49	5.71	4.92	5.91	1.23	2.70	0.21	1.35	28.37
1970	0.21	0.25	1.84	2.96	3.04	2.63	1.89	1.01	7.54	4.57	1.80	1.07	28.81
1971	0.45	2.66	0.46	1.61	2.85	3.65	1.57	0.12	0.60	3.16	1.71	0.79	19.63
1972	0.37	0.89	1.31	3.44	4.29	3.79	10.33	2.09	2.62	1.71	0.96	1.92	33.72
1973	1.23	0.46	4.02	1.33	1.72	3.54	4.90	1.05	4.04	2.22	2.66	0.72	27.89
1974	0.27	0.21	0.68	1.60	4.16	3.28	1.29	3.27	0.87	1.55	0.26	0.52	17.96
1975	1.66	0.29	1.43	3.99	2.66	5.34	1.61	4.56	1.17	0.07	3.24	0.29	26.31
1976	0.14	0.85	2.63	2.23	3.12	0.75	1.50	0.30	1.96	0.42	0.04	0.39	14.33
1977	0.27	0.25	4.05	3.14	3.08	4.80	6.34	2.53	4.90	2.79	1.88	1.00	35.03
1978	0.20	0.97	0.60	3.89	3.39	1.50	3.25	4.25	1.77	0.75	0.71	0.80	22.08
1979	1.09	0.35	3.27	2.34	4.43	3.39	1.91	6.28	2.02	5.30	1.90	0.21	32.49
1980	0.59	0.41	1.03	1.37	3.88	1.21	0.50	5.68	1.40	1.39	0.03	0.22	17.71
1981	0.37	0.32	1.78	0.47	1.98	3.07	4.36	1.72	0.97	2.89	1.90	0.57	20.40
1982	1.58	0.38	1.53	0.64	6.93	1.82	3.63	1.54	3.64	4.83	1.75	2.23	30.50
1983	0.39	0.68	4.91	2.19	4.41	4.58	1.27	1.16	3.07	2.09	3.08	0.45	28.28
1984	0.17	1.14	1.79	6.73	5.54	6.79	2.56	0.79	0.71	4.59	1.85	1.36	34.02
1985	0.29	0.12	2.20	5.04	3.24	1.15	0.41	3.69	4.50	1.20	1.01	0.52	23.37
1986	0.11	0.63	3.03	5.73	3.00	5.35	3.26	1.86	5.06	2.56	1.13	0.21	31.93
1987	0.23	0.41	5.90	0.61	4.20	3.09	3.08	3.94	1.44	0.44	0.90	0.48	24.72
1988	0.73	0.43	0.59	2.50	2.47	0.53	2.93	5.16	5.17	0.08	2.08	0.58	23.25
1989	0.96	0.36	0.86	0.94	1.51	1.83	5.76	2.01	4.33	0.19	0.24	0.40	19.39
1990	0.30	0.31	2.18	1.61	7.48	4.20	3.22	1.34	0.86	1.48	0.87	0.78	24.63
1991	0.46	0.38	2.02	3.48	3.87	5.27	0.77	2.17	2.17	2.04	2.70	0.87	26.20
1992	0.60	0.77	3.52	1.13	3.56	2.74	4.81	4.60	2.82	3.69	0.95	0.80	29.99
1993	0.70	2.08	0.85	3.66	3.06	4.23	6.71	6.43	1.29	1.84	1.03	0.53	32.41
Record Mean	0.64	0.82	1.50	2.42	3.64	3.97	3.24	2.93	2.85	1.72	1.09	0.78	25.59

AVERAGE TEMPERATURE (deg. F)

SIOUX CITY, IOWA

YEAR	JAN	FEB	MAR	APR	MAY	JUNE	JULY	AUG	SEP	OCT	NOV	DEC	ANNUAL
1964	26.0	27.9	30.8	51.2	65.7	69.9	78.5	68.7	63.0	51.9	37.5	18.8	49.2
1965	17.3	18.6	23.8	50.1	65.2	69.4	75.2	72.0	56.9	55.6	38.2	33.0	47.9
1966	11.9	23.9	40.2	45.9	58.7	70.6	78.0	70.1	62.3	52.2	35.5	23.1	47.7
1967	19.4	22.0	39.9	49.6	56.4	68.7	73.2	70.4	62.5	50.4	35.4	24.9	47.8
1968	18.6	23.9	43.8	52.1	57.5	71.9	74.6	74.4	63.0	52.9	35.8	19.4	49.0
1969	12.5	22.6	24.9	52.2	61.8	65.0	76.0	73.6	64.8	46.6	37.7	21.7	46.6
1970	10.4	27.0	30.3	49.7	65.4	72.3	75.1	75.0	63.9	49.5	35.8	22.0	48.1
1971	13.8	24.4	35.2	51.9	59.6	75.4	72.5	73.6	65.0	56.6	37.8	24.4	49.2
1972	15.5	18.3	37.7	49.4	62.7	72.3	73.7	72.9	64.1	48.9	36.3	18.0	47.5
1973	21.5	25.5	42.9	50.3	60.2	72.4	74.7	77.1	63.1	57.1	38.1	21.5	50.4
1974	16.8	30.0	39.6	52.6	62.2	69.6	81.4	70.8	60.2	54.6	37.5	24.8	50.0
1975	18.3	19.2	29.1	45.6	64.6	69.5	76.4	75.1	59.6	55.1	36.4	25.3	47.9
1976	21.1	33.5	36.4	53.1	58.6	70.2	76.3	74.2	64.2	45.1	29.4	19.3	48.5
1977	8.8	28.3	40.7	55.6	67.3	70.9	75.9	69.4	64.8	51.0	35.5	22.6	49.2
1978	7.5	10.1	32.5	48.5	60.0	70.7	73.8	71.7	68.0	50.3	33.2	15.5	45.2
1979	3.9	9.2	31.3	46.6	57.6	70.0	74.4	72.4	65.2	50.1	34.1	30.1	45.4
1980	21.6	20.3	33.6	51.5	61.2	70.7	78.6	73.8	64.3	48.2	39.9	24.4	49.0
1981	24.4	28.3	40.5	56.6	59.2	72.0	74.5	71.0	65.1	49.5	40.1	20.7	50.2
1982	6.4	20.4	34.0	46.5	62.4	65.4	76.4	72.4	63.1	53.2	34.2	26.3	46.7
1983	22.5	28.1	36.5	43.2	56.5	68.1	77.2	80.9	66.3	52.6	38.2	6.0	48.0
1984	20.7	31.7	29.3	46.8	57.8	70.4	73.7	74.9	60.4	51.8	37.9	24.4	48.3
1985	17.8	23.7	41.5	54.7	64.5	67.9	74.5	69.3	61.4	51.4	25.9	13.8	47.2
1986	28.0	23.0	41.0	51.2	61.8	72.9	76.8	68.6	64.0	51.9	31.9	27.0	49.8
1987	26.9	34.8	40.5	54.8	66.4	74.3	76.9	70.1	64.0	46.6	41.6	28.9	52.2
1988	16.0	20.8	39.4	49.0	67.5	76.5	76.4	76.1	64.9	47.4	37.5	27.2	49.9
1989	29.9	15.3	34.0	52.2	61.1	69.4	77.4	72.7	61.9	52.6	32.9	14.3	47.8
1990	30.4	28.4	40.2	49.2	58.3	73.4	73.4	73.6	67.8	51.2	40.0	19.6	50.5
1991	15.7	33.1	40.2	52.9	65.0	73.9	74.5	74.0	64.3	49.0	28.3	28.9	50.0
1992	30.7	34.7	40.4	46.7	60.8	68.1	68.6	66.8	62.1	51.1	33.3	24.0	48.9
1993	17.5	18.5	31.6	46.5	59.1	67.7	73.1	72.5	58.4	49.4	32.2	25.6	46.0
Record Mean	18.8	22.8	34.8	49.4	60.8	70.3	75.5	73.0	64.1	52.3	36.1	23.6	48.5
Max	28.2	32.2	44.5	60.6	71.9	81.1	86.4	83.9	75.6	63.9	45.8	32.5	58.9
Min	9.3	13.4	25.1	38.2	49.7	59.6	64.5	62.1	52.6	40.6	26.4	14.8	38.0

REFERENCE NOTES

SIOUX CITY, IOWA

GENERAL

T - TRACE AMOUNT.
BLANK ENTRIES DENOTE MISSING/UNREPORTED DATA.
# INDICATES A STATION OR INSTRUMENT RELOCATION.

SPECIFIC

(a) - LENGTH OF RECORD IN YEARS, ALTHOUGH INDIVIDUAL MONTHS MAY BE MISSING.
0.* OR * - THE VALUE IS BETWEEN 0.0 AND 0.05
NORMALS - BASED ON THE 1961-1990 RECORD PERIOD.
EXTREMES - DATES ARE THE MOST RECENT OCCURRENCE
WIND DIR.- NUMERALS SHOW TENS OF DEGRESS CLOCKWISE FROM TRUE NORTH. "00" INDICATES CALM.
RESULTANT DIRECTIONS ARE GIVEN TO WHOLE DEGREES.
BOLD VALUES INDICATE EXTREME VALUES WHICH OCCURRED AFTER THE ASOS SYSTEM WAS COMMISSIONED.

RECORD = PERIOD OF RECORD
RECORD MEAN PRECIPITATION IS THE MEAN OF ALL DAILY PRECIPITATION AMOUNTS DURING THE PERIOD OF RECORD.
RECORD MAX(MIN) TEMPERATURE IS THE MEAN OF ALL DAILY MAX(MIN) TEMPERATURES DURING THE PERIOD OF RECORD.
RECORD MEAN TEMPERATURE IS THE SUM OF THE RECORD MAX AND RECORD MIN DIVIDED BY 2.
AVERAGE TEMPERATURE IS THE SUM OF THE MEAN DAILY MAX AND MIN TEMPERATURE DIVIDED BY 2.

EXCEPTIONS

RECORD MEANS ARE THROUGH THE CURRENT YEAR, BEGINNING IN 1890 FOR TEMPERATURE
1890 FOR PRECIPITATION
1941 FOR SNOWFALL

HEATING DEGREE DAYS Base 65 deg. F SIOUX CITY, IOWA

SEASON	JULY	AUG	SEP	OCT	NOV	DEC	JAN	FEB	MAR	APR	MAY	JUNE	TOTAL
1964-65	0	41	134	406	818	1424	1474	1294	1270	446	89	2	7398
1965-66	0	11	260	297	795	983	1643	1144	761	570	236	23	6723
1966-67	0	14	131	400	881	1293	1410	1196	771	459	320	19	6894
1967-68	14	21	107	461	880	1235	1430	1187	654	385	252	25	6651
1968-69	5	2	102	388	868	1408	1625	1185	1236	379	165	75	7438
1969-70	0	0	58	570	813	1334	1690	1057	1069	475	110	13	7189
1970-71	4	0	148	489	868	1326	1581	1133	918	390	182	8	7047
1971-72	11	2	126	283	807	1252	1530	1350	838	463	156	16	6834
1972-73	6	5	116	493	854	1452	1342	1099	679	434	179	4	6663
1973-74	0	0	98	259	799	1343	1491	973	783	374	126	25	6271
1974-75	0	7	180	317	820	1241	1439	1277	1107	575	97	21	7081
1975-76	1	0	204	328	850	1223	1355	907	881	361	218	18	6346
1976-77	1	0	108	620	1062	1411	1738	1019	745	296	24	2	7026
1977-78	0	9	55	428	878	1309	1778	1534	1007	487	194	33	7712
1978-79	1	8	63	448	948	1531	1895	1560	1038	546	255	31	8324
1979-80	1	13	81	457	922	1075	1339	1287	966	419	170	11	6741
1980-81	0	0	108	516	746	1254	1253	1023	752	269	200	3	6124
1981-82	10	4	81	471	741	1367	1816	1245	955	547	116	61	7414
1982-83	0	17	132	366	916	1197	1310	1027	879	648	270	44	6806
1983-84	0	0	113	390	799	1825	1365	956	1098	543	235	7	7331
1984-85	0	5	194	405	805	1254	1459	1149	720	336	86	45	6458
1985-86	1	13	221	414	1168	1581	1142	1172	737	409	124	1	6983
1986-87	0	27	79	396	984	1172	1178	840	752	332	76	6	5842
1987-88	1	44	80	563	693	1111	1514	1276	786	475	47	6	6596
1988-89	0	9	75	542	818	1165	1083	1387	956	428	164	35	6662
1989-90	0	7	161	389	956	1567	1065	1018	759	500	218	21	6661
1990-91	0	2	85	431	744	1401	1521	887	759	379	148	2	6359
1991-92	0	0	150	488	1092	1115	1056	872	757	551	173	21	6275
1992-93	15	45	140	437	944	1262	1466	1297	1032	550	194	49	7431
1993-94	0	12	211	499	978	1217							

COOLING DEGREE DAYS Base 65 deg. F SIOUX CITY, IOWA

YEAR	JAN	FEB	MAR	APR	MAY	JUNE	JULY	AUG	SEP	OCT	NOV	DEC	TOTAL
1969	0	0	0	0	72	85	349	276	59	8	0	0	849
1970	0	0	0	23	131	237	323	318	121	15	0	0	1168
1971	0	0	0	7	19	329	251	273	134	30	0	0	1043
1972	0	0	0	3	93	241	284	259	98	1	0	0	979
1973	0	0	0	0	38	232	308	381	50	19	0	0	1028
1974	0	0	0	9	47	173	513	193	41	4	0	0	980
1975	0	0	0	0	91	162	362	320	50	28	0	0	1013
1976	0	0	0	13	26	181	359	294	89	7	0	0	969
1977	0	0	0	21	102	186	346	152	55	0	0	0	862
1978	0	0	5	0	49	210	280	225	159	0	0	0	928
1979	0	0	0	3	33	186	299	249	95	0	0	0	865
1980	0	0	0	24	57	185	430	279	90	4	0	0	1069
1981	0	0	0	25	26	222	309	199	89	0	0	0	870
1982	0	0	0	3	41	79	358	254	85	7	0	0	827
1983	0	0	0	0	16	144	385	502	158	9	0	0	1214
1984	0	0	0	4	16	180	279	320	61	3	0	0	863
1985	0	0	0	35	77	134	302	153	120	1	0	0	822
1986	0	0	2	3	34	246	367	142	55	0	0	0	849
1987	0	0	0	34	132	292	377	210	56	0	0	0	1101
1988	0	0	0	3	132	356	363	361	80	0	0	0	1295
1989	0	0	0	53	49	175	391	254	71	12	0	0	1005
1990	0	0	0	33	14	278	267	272	176	10	0	0	1050
1991	0	0	0	22	156	276	302	287	135	1	0	0	1179
1992	0	0	0	7	49	119	133	108	57	13	0	0	486
1993	0	0	0	0	18	135	261	247	17	18	0	0	696

SNOWFALL (inches) SIOUX CITY, IOWA

SEASON	JULY	AUG	SEP	OCT	NOV	DEC	JAN	FEB	MAR	APR	MAY	JUNE	TOTAL
1964-65	0.0	0.0	0.0	0.0	0.4	10.1	6.4	11.9	17.2	1.0	0.0	0.0	47.0
1965-66	0.0	0.0	0.0	0.0	1.1	1.2	5.5	0.9	13.1	T	0.0	0.0	21.8
1966-67	0.0	0.0	0.0	4.0	0.2	8.6	3.9	3.0	0.5	T	T	0.0	20.2
1967-68	0.0	0.0	0.0	1.7	T	1.3	3.4	1.2	T	0.3	T	0.0	7.9
1968-69	0.0	0.0	0.0	T	2.8	20.6	12.1	13.0	2.6	0.0	0.0	0.0	51.1
1969-70	0.0	0.0	0.0	T	1.2	15.8	4.1	2.7	15.9	T	T	0.0	39.7
1970-71	0.0	0.0	0.0	5.1	0.4	7.0	4.8	3.5	1.7	0.2	0.0	0.0	22.7
1971-72	0.0	0.0	0.0	0.0	3.4	6.7	5.7	9.1	1.6	0.8	0.0	0.0	27.3
1972-73	0.0	0.0	0.0	0.2	1.0	9.7	11.3	3.8	T	0.4	0.0	0.0	26.4
1973-74	0.0	0.0	0.0	0.0	0.6	7.0	4.2	3.3	4.0	2.1	0.0	0.0	21.2
1974-75	0.0	0.0	0.0	0.0	2.4	6.2	18.2	3.0	6.8	3.4	0.0	0.0	40.0
1975-76	0.0	0.0	0.0	0.0	6.2	0.3	0.8	3.8	7.1	0.0	T	0.0	18.2
1976-77	0.0	0.0	0.0	0.2	0.4	4.3	4.1	0.8	9.4	0.2	0.0	0.0	19.4
1977-78	0.0	0.0	0.0	T	5.7	5.3	1.8	10.8	3.0	T	0.0	0.0	26.6
1978-79	0.0	0.0	0.0	0.0	4.7	7.5	11.3	3.3	8.4	1.2	0.0	0.0	36.4
1979-80	0.0	0.0	0.0	0.1	4.0	1.5	4.7	4.3	5.8	1.3	0.0	0.0	21.7
1980-81	0.0	0.0	0.0	3.0	0.2	2.5	5.5	5.5	0.4	0.0	0.0	0.0	17.1
1981-82	0.0	0.0	0.0	2.0	6.8	6.8	29.1	3.7	4.5	3.9	0.0	0.0	56.8
1982-83	0.0	0.0	0.0	8.0	0.3	12.4	3.2	6.8	19.1	9.7	0.0	0.0	59.5
1983-84	0.0	0.0	0.0	0.0	16.5	9.4	1.8	10.2	17.8	5.1	0.0	0.0	60.8
1984-85	0.0	0.0	0.0	T	2.2	6.2	3.7	0.9	10.8	T	0.0	0.0	23.8
1985-86	0.0	0.0	T	T	9.4	9.0	1.4	1.9	5.4	2.1	0.0	0.0	29.2
1986-87	0.0	0.0	0.0	T	5.1	2.1	3.5	3.4	7.7	T	0.0	0.0	21.8
1987-88	0.0	0.0	0.0	T	5.4	1.9	7.5	5.5	1.0	4.4	0.0	0.0	25.7
1988-89	0.0	0.0	0.0	0.0	9.1	1.9	3.3	6.2	6.3	0.3	T	0.0	27.1
1989-90	T	0.0	0.0	0.0	2.4	7.1	5.5	3.8	3.0	0.1	T	T	21.9
1990-91	0.0	0.0	0.0	0.4	5.7	7.7	6.7	3.9	8.5	T	0.0	T	32.9
1991-92	0.0	0.0	0.0	9.8	8.4	0.5	1.2	1.8	2.1	6.2	T	0.0	30.0
1992-93	T	0.0	0.0	T	1.7	6.9	7.8	24.5	2.4	0.3	0.0	0.0	43.6
1993-94	0.0	0.0	T	T	10.2	9.1							
Record Mean	T	0.0	T	0.7	3.9	6.0	6.3	5.8	7.5	1.5	0.1	T	31.7

WATERLOO, IOWA

Waterloo is situated on the banks of the Cedar River in northeast Iowa, and has a continental humid climate. A wide variation is experienced in both temperature and precipitation during the four distinct seasons.

The distribution of precipitation through the year is very favorable for agriculture with an average 72 percent of the annual total falling in the April to September crop season. The annual temperature range is large. January, the coldest month, averages near 14 degrees and July, the warmest month, averages about 73 degrees. Extreme temperatures range from about −35 to 112 degrees.

It is sometimes convenient to divide the year into periods corresponding to the growing season of the area. Winter extends from November through March, based on a mean daily temperature of 40 degrees. The winter period is a season of cold, dry weather occasionally broken by storms of short duration. Precipitation during the winter is mainly snow with rain dominant at the beginning and end of the season. Annual snowfall varies considerably from year to year. Temperatures of zero degrees or below occur on average about 29 days per year. Bitterly cold days with high temperatures of zero degrees or lower average about 3 days per year. During the winter, prevailing winds are from the northwest.

The spring growing season is marked by an increase in both frequency and intensity of rainfall and by a rapid increase in the mean daily temperature. Spring extends from the first of April to mid May, when daily mean temperatures range between 40 and 59 degrees.

The summer growing season extends from mid May to mid September, based on a mean daily temperature of 60 degrees. Precipitation increases during the spring and reaches a maximum monthly amount in July. In summer, precipitation falls mainly from thunderstorms, three–fourths of which occur during the summer growing season. The prevailing summer wind is southerly, supplying moisture from the Gulf of Mexico. Daily temperatures reach their highest level in July or early August.

The fall growing season extends from mid September to the first part of November, by which time the mean daily temperature has fallen to 40 degrees. Precipitation declines and frequent periods of warm days, cool nights, and cloudless, but hazy, skies persist.

NORMALS, MEANS, AND EXTREMES

WATERLOO, IOWA

LATITUDE: 42°33'N LONGITUDE: 92°24' W ELEVATION: FT. GRND 868 BARO 872 TIME ZONE: CENTRAL WBAN: 94910

	(a)	JAN	FEB	MAR	APR	MAY	JUNE	JULY	AUG	SEP	OCT	NOV	DEC	YEAR
TEMPERATURE °F:														
Normals														
-Daily Maximum		23.7	29.4	43.3	59.3	71.6	80.5	83.9	81.5	73.7	61.8	45.2	28.9	56.9
-Daily Minimum		5.4	10.6	25.0	37.1	48.7	57.6	62.2	59.0	50.0	38.4	26.2	11.7	36.0
-Monthly		14.6	20.1	34.2	48.3	60.2	69.1	73.1	70.3	61.9	50.1	35.7	20.3	46.5
Extremes														
-Record Highest	45	58	66	87	100	94	103	105	105	98	95	79	65	105
-Year		1981	1981	1986	1980	1967	1988	1988	1988	1955	1963	1950	1973	JUL 1988
-Record Lowest	45	-31	-29	-34	-4	25	38	42	38	22	11	-17	-27	-34
-Year		1970	1962	1962	1982	1971	1964	1984	1967	1967	1988	1977	1963	MAR 1962
NORMAL DEGREE DAYS:														
Heating (base 65°F)		1562	1257	955	501	203	27	10	29	127	470	879	1386	7406
Cooling (base 65°F)		0	0	0	0	55	150	261	194	34	8	0	0	702
% OF POSSIBLE SUNSHINE														
MEAN SKY COVER (tenths)														
Sunrise - Sunset	33	6.5	6.7	7.0	6.8	6.5	6.1	5.7	5.6	5.7	5.9	7.2	6.9	6.4
MEAN NUMBER OF DAYS:														
Sunrise to Sunset														
-Clear	33	7.5	6.8	6.3	6.1	7.0	7.0	8.5	9.5	9.7	9.8	5.2	7.0	90.3
-Partly Cloudy	33	7.2	6.3	6.8	7.7	9.6	10.9	11.8	10.7	8.0	7.7	6.9	6.1	99.8
-Cloudy	33	16.2	15.2	17.9	16.2	14.5	12.1	10.7	10.8	12.2	13.6	17.8	18.0	175.1
Precipitation														
.01 inches or more	43	7.2	6.4	9.2	10.0	11.1	10.0	9.5	8.5	8.6	7.1	7.4	7.7	102.8
Snow,Ice pellets,hail														
1.0 inches or more	40	2.1	2.2	2.0	0.5	0.*	0.0	0.0	0.0	0.0	0.1	1.0	2.3	10.1
Thunderstorms	39	0.1	0.3	1.5	3.8	6.3	7.4	7.3	6.7	4.7	2.3	0.5	0.2	41.1
Heavy Fog Visibility														
1/4 mile or less	38	2.2	2.3	2.9	1.4	1.3	0.6	0.9	1.7	1.7	1.0	2.0	2.9	20.9
Temperature °F														
-Maximum														
90° and above	34	0.0	0.0	0.0	0.1	0.8	3.6	6.1	4.0	1.2	0.1	0.0	0.0	16.0
32° and below	34	21.3	15.3	6.6	0.3	0.0	0.0	0.0	0.0	0.0	0.0	4.4	17.6	65.5
-Minimum														
32° and below	34	30.6	27.1	24.2	11.0	1.4	0.0	0.0	0.0	1.0	9.8	22.3	29.7	157.1
0° and below	34	11.9	7.3	1.2	0.*	0.0	0.0	0.0	0.0	0.0	0.0	0.5	6.7	27.6
AVG. STATION PRESS.(mb)	21	986.8	987.0	983.6	982.6	982.4	982.5	984.1	985.1	985.5	985.7	985.0	986.5	984.7
RELATIVE HUMIDITY (%)														
Hour 00	34	77	79	79	75	76	79	84	86	84	78	80	80	80
Hour 06 (Local Time)	34	77	80	82	82	81	83	86	90	89	84	83	82	83
Hour 12	34	70	68	65	55	55	55	59	59	59	56	66	72	62
Hour 18	34	73	72	67	56	54	55	59	62	64	63	72	76	64
PRECIPITATION (inches):														
Water Equivalent														
-Normal		0.80	1.08	2.30	3.30	4.08	4.47	4.83	3.64	3.51	2.57	1.82	1.30	33.70
-Maximum Monthly	45	2.78	3.54	5.43	8.53	7.72	10.11	12.60	9.65	11.38	5.45	5.61	3.77	12.60
-Year		1949	1971	1961	1991	1991	1993	1968	1993	1965	1984	1992	1982	JUL 1968
-Minimum Monthly	45	0.10	0.02	0.25	0.95	0.82	0.86	0.43	0.37	0.42	T	T	0.22	T
-Year		1962	1969	1981	1971	1992	1955	1954	1955	1953	1952	1954	1989	NOV 1954
-Maximum in 24 hrs	45	1.64	1.53	1.93	2.86	3.62	4.35	9.31	5.28	3.52	2.65	2.76	1.75	9.31
-Year		1971	1976	1966	1979	1980	1980	1968	1966	1978	1961	1991	1982	JUL 1968
Snow,Ice pellets,hail														
-Maximum Monthly	43	18.1	24.3	20.1	10.3	1.0	T	0.0	0.0	0.0	1.2	15.9	20.1	24.3
-Year		1982	1962	1959	1973	1991	1993				1991	1991	1961	FEB 1962
-Maximum in 24 hrs	38	14.8	9.1	12.2	6.2	1.0	T	0.0	0.0	0.0	1.2	10.2	12.3	14.8
-Year		1971	1993	1959	1973	1991	1993				1991	1991	1985	JAN 1971
WIND:														
Mean Speed (mph)	37	11.5	11.4	12.4	12.7	11.1	10.0	8.6	8.4	9.2	10.2	11.2	11.2	10.7
Prevailing Direction through 1963		NW	NW	NW	NW	S	S	ESE	S	S	S	S	NW	S
Fastest Obs. 1 Min.														
-Direction (!!)	33	29	28	23	25	18	36	35	21	28	32	33	32	35
-Speed (MPH)	33	46	44	46	52	52	52	58	46	38	36	43	39	58
-Year		1963	1971	1982	1963	1962	1993	1980	1967	1980	1979	1964	1969	JUL 1980
Peak Gust														
-Direction (!!)	10	NW	N	W	NW	W	N	NW	W	W	N	SW	NW	N
-Speed (mph)	10	58	52	53	63	58	83	64	51	49	49	59	53	83
-Date		1992	1987	1985	1984	1988	1993	1984	1989	1985	1987	1991	1985	JUN 1993

PRECIPITATION (inches)

WATERLOO, IOWA

YEAR	JAN	FEB	MAR	APR	MAY	JUNE	JULY	AUG	SEP	OCT	NOV	DEC	ANNUAL
1964	0.34	0.21	1.24	8.11	5.01	4.01	3.48	4.86	5.17	0.46	1.11	1.93	35.93
1965	1.11	1.16	3.44	4.81	5.24	5.09	4.41	7.08	11.38	1.93	3.02	2.12	50.79
1966	1.70	1.47	3.40	3.89	4.79	8.34	6.06	5.97	0.55	3.21	0.47	0.80	40.65
1967	1.53	0.52	2.47	4.49	1.89	5.52	2.25	3.95	1.40	2.80	1.03	0.83	28.68
1968	0.73	0.04	1.30	5.15	1.95	4.66	12.60	8.07	3.61	2.97	0.74	2.80	44.62
1969	1.56	0.02	0.99	4.47	6.76	4.82	8.90	0.54	1.35	3.04	0.12	1.95	34.52
1970	0.18	0.42	2.48	1.21	7.24	2.16	4.13	0.99	5.20	3.50	1.41	1.13	30.05
1971	1.76	3.54	1.36	0.95	5.51	3.14	3.14	0.43	2.56	2.43	2.76	1.78	29.36
1972	0.26	1.13	1.78	3.29	3.35	4.33	5.29	3.71	4.86	3.77	1.98	1.84	35.59
1973	1.16	1.49	3.22	3.45	7.23	3.07	3.29	2.72	6.88	0.94	2.75	2.26	38.46
1974	0.98	1.07	1.63	2.25	5.35	3.72	3.32	2.87	2.14	4.08	0.91	1.28	29.60
1975	0.92	1.14	2.35	3.24	1.93	4.00	1.06	4.53	0.80	0.04	3.68	0.40	24.09
1976	0.18	2.16	3.14	4.66	2.21	2.83	2.83	0.68	1.36	1.75	0.06	0.46	22.32
1977	0.37	0.68	3.04	2.26	1.47	4.11	3.93	5.45	6.09	2.94	1.89	1.73	33.96
1978	0.51	0.45	0.50	4.75	2.69	5.70	3.74	3.60	7.91	1.02	3.70	0.79	35.36
1979	1.41	0.32	3.06	4.37	3.83	6.07	6.68	5.33	0.75	4.18	1.41	0.71	38.12
1980	1.34	0.96	0.86	1.22	5.36	8.63	1.75	5.38	3.74	1.56	0.41	0.52	31.73
1981	0.23	2.44	0.25	4.20	2.05	5.23	3.52	4.13	3.13	1.76	1.86	0.83	29.63
1982	1.51	0.38	2.51	2.58	6.66	3.25	3.90	3.29	1.44	2.69	3.08	3.77	35.06
1983	0.84	1.44	3.80	2.43	5.28	5.31	3.89	1.30	3.19	4.35	3.67	1.03	36.53
1984	0.62	0.86	1.40	5.09	4.77	4.87	6.04	0.38	1.57	5.45	1.45	1.95	34.45
1985	0.61	1.23	2.86	1.45	2.32	2.72	3.38	4.22	5.27	3.09	1.78	1.25	30.18
1986	0.40	2.41	1.77	2.57	6.33	7.40	3.24	2.27	4.48	3.74	1.27	0.82	36.70
1987	0.47	0.91	2.48	1.44	2.93	2.66	5.59	8.51	1.84	0.81	2.69	1.82	32.15
1988	0.75	0.37	1.37	1.72	1.61	3.12	1.51	3.15	1.65	0.68	2.54	0.52	18.99
1989	1.09	0.28	0.93	2.41	1.54	1.40	2.67	2.33	3.07	3.13	0.55	0.22	19.62
1990	0.46	0.37	4.46	2.88	5.06	7.98	9.63	5.22	1.46	1.38	1.34	1.26	41.50
1991	0.76	0.33	4.00	8.53	7.72	2.18	1.60	4.88	2.92	4.61	4.08	1.54	43.15
1992	1.06	1.47	2.27	4.55	0.82	1.93	5.87	2.84	2.23	0.72	5.61	1.60	30.97
1993	0.58	1.30	4.00	3.67	6.11	10.11	11.26	9.65	2.86	1.80	0.97	0.76	53.07
Record Mean	0.98	1.03	2.07	2.96	4.02	4.44	4.21	3.65	3.73	2.37	1.80	1.17	32.43

AVERAGE TEMPERATURE (deg. F)

WATERLOO, IOWA

YEAR	JAN	FEB	MAR	APR	MAY	JUNE	JULY	AUG	SEP	OCT	NOV	DEC	ANNUAL
1964	24.4	25.0	28.4	47.2	63.2	69.1	75.8	68.2	60.9	47.7	37.9	19.1	47.2
1965	13.9	16.5	21.9	44.3	60.2	66.2	70.3	67.8	58.6	51.8	38.3	32.6	45.2
1966	7.8	21.3	38.7	43.7	54.6	70.7	75.7	67.5	61.5	50.9	35.7	22.6	45.9
1967	20.2	16.1	36.6	47.7	54.6	67.3	68.6	65.2	58.1	46.4	32.3	25.1	44.9
1968	18.0	19.3	39.4	50.4	54.1	69.1	70.8	69.5	60.2	50.6	33.7	18.6	46.2
1969	12.9	22.0	26.1	48.5	59.0	61.9	72.6	70.6	61.1	45.9	33.6	17.3	44.3
1970	5.2	17.4	29.1	47.9	61.6	69.6	72.5	70.0	61.8	49.9	34.7	19.4	44.9
1971	7.7	16.2	29.6	46.5	53.5	71.6	66.4	66.7	63.4	54.9	34.7	23.9	44.6
1972	10.2	11.9	30.8	44.3	60.8	67.2	70.2	71.1	61.2	46.4	33.2	16.1	43.6
1973	18.8	24.4	41.8	45.2	56.5	71.3	73.5	73.8	61.8	55.0	37.2	20.3	48.3
1974	17.2	21.2	32.9	49.4	56.0	66.5	75.5	68.3	58.0	51.3	36.7	24.4	46.4
1975	18.5	19.3	24.9	42.8	63.1	71.0	73.9	72.9	57.1	52.2	39.1	25.3	46.7
1976	16.6	30.1	35.9	50.5	56.6	68.8	73.8	69.6	60.7	43.6	26.6	12.7	45.4
1977	-0.1	23.3	40.1	51.7	65.5	68.9	74.3	65.9	60.7	46.0	34.0	16.8	45.6
1978	2.7	7.4	28.2	46.9	60.1	70.1	72.2	71.5	67.1	49.7	34.3	19.5	44.2
1979	3.6	9.7	31.8	44.4	58.2	69.3	72.3	71.8	66.1	51.2	35.6	28.3	45.2
1980	19.0	16.8	30.3	49.1	59.8	69.0	75.6	73.1	63.3	46.2	37.5	22.2	46.8
1981	19.3	25.4	38.5	52.8	57.8	68.5	72.5	69.1	60.7	46.3	39.2	19.6	47.5
1982	3.6	18.1	32.3	42.7	64.4	64.0	74.6	70.3	61.6	51.4	34.9	29.9	45.7
1983	24.1	27.8	36.3	43.4	55.9	69.7	77.0	77.4	63.0	49.4	37.2	6.6	47.3
1984	16.1	30.4	26.4	46.6	57.8	70.1	70.3	72.4	60.3	52.2	37.0	24.0	47.0
1985	12.6	17.7	39.2	53.8	63.3	66.3	73.3	67.6	63.2	50.3	28.1	8.9	45.4
1986	19.4	16.9	38.2	51.9	60.4	70.1	75.1	66.5	63.9	50.3	29.0	24.3	47.2
1987	21.5	32.1	39.2	51.8	64.9	73.3	77.0	69.9	62.3	45.1	40.9	27.0	50.4
1988	13.2	16.7	36.2	47.8	65.1	73.0	75.9	76.4	64.7	44.2	36.7	23.4	47.8
1989	27.2	12.8	32.2	46.8	58.3	68.7	76.0	71.6	59.7	51.8	31.7	13.1	45.8
1990	29.2	26.8	39.5	47.7	56.1	69.8	71.8	71.7	64.8	48.9	40.4	18.2	48.7
1991	11.9	25.8	38.0	50.2	63.7	74.1	73.8	71.6	60.6	49.0	28.8	26.2	47.8
1992	25.2	31.1	37.4	46.3	60.6	68.3	67.9	65.0	60.5	50.8	33.7	23.5	47.5
1993	18.3	19.8	30.5	45.9	61.0	68.5	73.2	72.4	58.2	50.1	34.1	25.3	46.4
Record Mean	17.3	21.4	33.9	48.0	59.8	69.2	73.8	71.3	62.9	51.3	36.0	22.4	47.3
Max	26.5	30.6	43.4	59.5	71.5	80.6	85.4	83.1	74.8	63.1	45.5	31.1	57.9
Min	8.0	12.1	24.3	36.6	48.1	57.8	62.2	59.6	51.0	39.5	26.4	13.7	36.6

REFERENCE NOTES

WATERLOO, IOWA

GENERAL
T - TRACE AMOUNT.
BLANK ENTRIES DENOTE MISSING/UNREPORTED DATA.
# INDICATES A STATION OR INSTRUMENT RELOCATION.

SPECIFIC

(a) - LENGTH OF RECORD IN YEARS, ALTHOUGH INDIVIDUAL MONTHS MAY BE MISSING.
0.* OR * - THE VALUE IS BETWEEN 0.0 AND 0.05
NORMALS - BASED ON THE 1961-1990 RECORD PERIOD.
EXTREMES - DATES ARE THE MOST RECENT OCCURRENCE.
WIND DIR.- NUMERALS SHOW TENS OF DEGRESS CLOCKWISE FROM TRUE NORTH. "00" INDICATES CALM.
RESULTANT DIRECTIONS ARE GIVEN TO WHOLE DEGREES.
BOLD VALUES INDICATE EXTREME VALUES WHICH OCCURRED AFTER THE ASOS SYSTEM WAS COMMISSIONED.

RECORD = PERIOD OF RECORD
RECORD MEAN PRECIPITATION IS THE MEAN OF ALL DAILY PRECIPITATION AMOUNTS DURING THE PERIOD OF RECORD.
RECORD MAX(MIN) TEMPERATURE IS THE MEAN OF ALL DAILY MAX(MIN) TEMPERATURES DURING THE PERIOD OF RECORD.
RECORD MEAN TEMPERATURE IS THE SUM OF THE RECORD MAX AND RECORD MIN DIVIDED BY 2.
AVERAGE TEMPERATURE IS THE SUM OF THE MEAN DAILY MAX AND MIN TEMPERATURE DIVIDED BY 2.

EXCEPTIONS

RECORD MEANS ARE THROUGH THE CURRENT YEAR, BEGINNING IN 1895 FOR TEMPERATURE
1895 FOR PRECIPITATION
1955 FOR SNOWFALL

HEATING DEGREE DAYS Base 65 deg. F — WATERLOO, IOWA

SEASON	JULY	AUG	SEP	OCT	NOV	DEC	JAN	FEB	MAR	APR	MAY	JUNE	TOTAL
1964-65	0	53	181	526	809	1417	1582	1353	1327	612	183	31	8074
1965-66	4	60	207	412	793	999	1776	1219	809	632	336	25	7272
1966-67	0	33	152	438	869	1306	1384	1365	878	520	357	27	7329
1967-68	47	75	208	574	976	1231	1450	1319	786	431	338	41	7476
1968-69	14	35	163	464	935	1430	1610	1196	1201	492	217	132	7889
1969-70	9	4	150	595	935	1473	1851	1330	1104	536	182	32	8201
1970-71	15	6	175	466	900	1410	1776	1364	1095	549	354	15	8125
1971-72	37	40	146	328	903	1266	1694	1536	1053	611	196	56	7866
1972-73	23	28	158	571	945	1509	1428	1130	712	589	264	2	7359
1973-74	0	1	149	313	827	1380	1480	1218	985	469	296	52	7170
1974-75	0	27	239	420	842	1251	1435	1277	1238	662	139	28	7558
1975-76	21	1	262	411	770	1224	1495	1005	897	441	263	14	6804
1976-77	2	26	178	664	1142	1615	2017	1162	764	399	86	22	8077
1977-78	2	56	144	584	923	1490	1932	1610	1133	539	224	23	8660
1978-79	2	9	78	468	914	1404	1903	1547	1024	613	238	20	8220
1979-80	2	13	70	431	875	1131	1418	1394	1067	493	203	24	7121
1980-81	0	3	124	578	819	1323	1408	1104	814	368	234	9	6784
1981-82	10	26	152	573	767	1400	1903	1311	1006	661	97	60	7966
1982-83	0	29	166	421	895	1079	1263	1033	885	646	277	35	6729
1983-84	0	0	159	488	826	1811	1509	1000	1190	548	236	4	7771
1984-85	12	10	215	392	833	1264	1621	1320	794	372	109	53	6995
1985-86	0	19	192	448	1105	1738	1410	1344	826	406	160	11	7659
1986-87	0	43	97	448	1070	1255	1342	915	792	403	105	7	6477
1987-88	1	42	109	609	718	1173	1602	1400	887	511	85	14	7151
1988-89	2	18	81	638	843	1281	1163	1459	1009	542	236	39	7311
1989-90	0	8	189	413	994	1604	1105	1063	786	540	273	37	7012
1990-91	5	4	133	500	730	1447	1642	1094	833	449	158	4	6999
1991-92	2	6	217	488	1078	1192	1228	977	848	554	183	33	6806
1992-93	23	75	178	441	934	1278	1442	1260	1060	567	167	50	7475
1993-94	0	3	212	472	919	1223							

COOLING DEGREE DAYS Base 65 deg. F — WATERLOO, IOWA

YEAR	JAN	FEB	MAR	APR	MAY	JUNE	JULY	AUG	SEP	OCT	NOV	DEC	TOTAL
1969	0	0	0	0	41	47	250	185	40	12	0	0	575
1970	0	0	0	28	84	175	254	170	86	3	0	0	800
1971	0	0	0	2	5	217	87	100	104	21	0	0	536
1972	0	0	0	0	74	128	188	224	50	0	0	0	664
1973	0	0	0	0	8	197	268	283	60	12	0	0	828
1974	0	0	0	9	23	105	331	136	38	0	0	0	642
1975	0	0	0	0	86	215	304	253	30	21	0	0	909
1976	0	0	0	10	10	131	282	175	55	5	0	0	668
1977	0	0	0	9	107	145	297	92	23	0	0	0	673
1978	0	0	0	0	79	179	230	216	149	0	0	0	853
1979	0	0	0	0	35	153	237	230	106	9	0	0	770
1980	0	0	0	24	50	151	337	262	76	0	0	0	900
1981	0	0	0	8	19	120	252	158	27	0	0	0	584
1982	0	0	0	0	87	37	307	199	73	8	0	0	711
1983	0	0	0	4	3	182	380	390	106	11	0	0	1076
1984	0	0	0	3	20	164	183	246	80	0	0	0	696
1985	0	0	0	41	61	100	264	106	146	0	0	0	718
1986	0	0	3	18	26	171	321	97	69	0	0	0	705
1987	0	0	0	14	109	260	382	200	38	0	0	0	1003
1988	0	0	0	0	96	262	347	378	78	0	0	0	1161
1989	0	0	1	3	38	154	349	221	37	10	0	0	813
1990	0	0	0	27	6	186	223	218	132	9	0	0	801
1991	0	0	3	12	125	285	285	219	93	1	0	0	1023
1992	0	0	0	2	53	136	121	81	50	7	0	0	450
1993	0	0	0	0	48	164	262	240	16	18	0	0	748

SNOWFALL (inches) — WATERLOO, IOWA

SEASON	JULY	AUG	SEP	OCT	NOV	DEC	JAN	FEB	MAR	APR	MAY	JUNE	TOTAL
1964-65	0.0	0.0	0.0	0.0	2.1	4.8	6.6	6.6	16.1	0.0	0.0	0.0	36.2
1965-66	0.0	0.0	0.0	0.0	T	1.0	8.1	0.6	1.7	0.4	0.3	0.0	12.1
1966-67	0.0	0.0	0.0	0.0	T	6.4	2.6	7.0	3.9	0.4	0.0	0.0	20.3
1967-68	0.0	0.0	0.0	T	2.6	2.2	6.2	0.6	T	T	0.0	0.0	11.6
1968-69	0.0	0.0	0.0	T	1.4	14.5	5.2	0.3	4.4	T	0.0	0.0	25.8
1969-70	0.0	0.0	0.0	T	0.2	19.5	2.8	4.8	8.4	0.0	0.0	0.0	35.7
1970-71	0.0	0.0	0.0	T	1.0	7.3	16.7	14.1	6.5	T	0.0	0.0	45.6
1971-72	0.0	0.0	0.0	0.0	11.7	7.0	2.9	15.4	5.8	3.4	0.0	0.0	46.2
1972-73	0.0	0.0	0.0	0.3	9.6	10.6	8.0	4.8	0.1	10.3	0.0	0.0	43.7
1973-74	0.0	0.0	0.0	0.0	T	8.4	4.3	8.3	3.1	0.6	0.0	0.0	24.7
1974-75	0.0	0.0	0.0	0.0	4.1	5.3	7.0	12.5	10.5	6.6	0.0	0.0	46.0
1975-76	0.0	0.0	0.0	0.0	4.9	T	1.7	7.2	1.4	T	T	0.0	15.2
1976-77	0.0	0.0	0.0	0.1	0.7	5.3	4.9	1.8	3.1	1.5	0.0	0.0	17.4
1977-78	0.0	0.0	0.0	0.0	11.2	16.9	5.9	6.4	4.6	0.4	0.0	0.0	45.4
1978-79	0.0	0.0	0.0	0.0	9.1	8.1	17.2	2.9	4.3	4.7	0.0	0.0	46.3
1979-80	0.0	0.0	0.0	T	2.5	0.2	6.2	10.5	7.9	0.9	0.0	0.0	28.2
1980-81	0.0	0.0	0.0	T	T	3.7	3.9	14.3	T	T	0.0	0.0	21.9
1981-82	0.0	0.0	0.0	T	3.3	5.2	18.1	1.9	3.9	7.5	0.0	0.0	39.9
1982-83	0.0	0.0	0.0	1.2	0.1	3.4	4.0	15.9	9.2	5.1	0.0	0.0	38.9
1983-84	0.0	0.0	0.0	T	3.2	12.5	8.8	2.1	16.1	1.8	0.0	0.0	44.5
1984-85	0.0	0.0	0.0	0.0	0.3	8.7	10.1	4.4	7.3	0.3	0.0	0.0	31.1
1985-86	0.0	0.0	0.0	0.0	6.6	14.5	5.1	11.7	T	2.2	0.0	0.0	40.1
1986-87	0.0	0.0	0.0	0.0	12.9	4.6	4.5	1.5	6.1	T	0.0	0.0	29.6
1987-88	0.0	0.0	0.0	T	0.5	12.8	6.1	4.8	0.8	T	0.0	0.0	25.0
1988-89	0.0	0.0	0.0	T	3.7	1.3	0.2	5.0	4.8	T	T	T	15.0
1989-90	0.0	0.0	0.0	0.4	1.1	5.4	4.9	5.2	T	T	0.0	T	17.0
1990-91	0.0	0.0	0.0	T	2.0	15.2	9.8	1.8	4.3	0.7	1.0	0.0	34.8
1991-92	0.0	0.0	0.0	1.2	15.9	4.5	7.1	6.6	2.8	0.3	0.0	T	38.4
1992-93	0.0	0.0	0.0	T	9.1	6.5	8.2	13.6	13.1	7.1	T	T	57.6
1993-94	0.0	0.0	0.0	T	1.5	5.5							
Record Mean	0.0	0.0	0.0	0.1	3.6	7.2	6.9	6.6	6.0	1.7	T	T	32.1

1993 CONCORDIA, KANSAS (CNK)

A wide variety of weather occurs in the Concordia area which makes possible a great range in crop production. Wheat is ideally suited to the climate of north-central Kansas where a complete crop failure is unknown. Equally well suited to the climate are alfalfa, sweet clover, and sorghum. Corn is generally a successful crop although dry summers and hot winds occasionally prove disastrous. Adequate moisture throughout the year under normal conditions provides fine grazing conditions for a flourishing livestock industry.

Precipitation is light during the winter months, increasing in the spring until June and dropping off during the autumn months. Summer months with less than 1 inch of precipitation are common, even though monthly summer rainfall has exceeded 13 inches. Thunderstorms are frequent in May, June, July and August. Although heavy winter snowfalls are not uncommon, severe storms that paralyze industry and agriculture are very rare. Some periods have been distinguished by very dry or very wet cycles. Sustained periods of hot, dry, and windy weather frequently occur in July and August with temperatures of 100 degrees or more recorded for a week or more at a time. The average last occurrence of temperatures as low as 32 degrees in the spring is mid April. The average first occurrence of 32 degrees in the autumn is late October.

Winds are southerly most of the year except for a short period of northerly winds during the winter. Velocities are nearly constant throughout the year except for a slight increase in the spring months.

The variety of weather in north-central Kansas is invigorating and healthful. Winters are usually mild, and summers are seldom oppressively hot. Spring and autumn, although very different in most respects, are very pleasant. A period of mild, dry Indian Summer weather usually occurs in October and early November before the winter snow and cold begin.

NORMALS, MEANS, AND EXTREMES FOR 1993

CONCORDIA, KS (CNK)

LATITUDE: 39° 32′ 57″ N LONGTITUDE: 97° 39′ 07″ W ELEVATION (FT): GRND: 1470 BARO: 1472 TIME ZONE: CENTRAL (UTC+06) WBAN: 13984

	ELEMENT	POR	JAN	FEB	MAR	APR	MAY	JUN	JUL	AUG	SEP	OCT	NOV	DEC	YEAR
TEMPERATURE °F	NORMAL DAILY MAXIMUM	30	36.0	41.7	52.5	64.4	73.8	85.0	91.2	88.4	79.0	68.1	52.1	39.3	64.3
	MEAN DAILY MAXIMUM	31	36.0	42.0	53.6	65.2	74.4	87.6	94.0	91.1	81.7	70.1	53.1	41.0	65.8
	HIGHEST DAILY MAXIMUM	32	74	86	88	98	99	109	109	108	106	96	84	82	109
	YEAR OF OCCURRENCE		1990	1972	1986	1989	1967	1980	1980	1983	1990	1963	1980	1964	JUL 1980
	MEAN OF EXTREME MAXS.	32	58.8	65.8	78.5	84.5	88.8	98.5	102.8	101.2	94.9	86.7	72.2	62.0	82.9
	NORMAL DAILY MINIMUM	30	15.7	20.3	30.0	41.2	51.9	61.9	67.8	65.5	55.7	44.0	30.8	19.8	42.0
	MEAN DAILY MINIMUM	31	16.1	20.6	30.4	41.5	51.9	63.5	69.6	67.1	57.6	45.4	31.6	21.2	43.0
	LOWEST DAILY MINIMUM	32	-17	-15	-7	14	26	41	48	45	29	14	-4	-26	-26
	YEAR OF OCCURRENCE		1985	1979	1978	1975	1967	1964	1972	1964	1984	1993	1976	1989	DEC 1989
	MEAN OF EXTREME MINS.	32	-5.4	0.1	10.7	25.3	37.0	48.7	56.2	53.3	39.0	26.8	13.2	0.0	25.4
	NORMAL DRY BULB	30	25.9	31.0	41.3	52.8	62.8	73.5	79.5	77.0	67.4	56.1	41.5	29.6	53.2
	MEAN DRY BULB	31	26.1	31.3	42.1	53.4	63.1	75.6	81.8	79.1	69.6	57.8	42.4	31.1	54.4
	MEAN WET BULB	30	22.9	26.9	36.3	46.5	56.1	64.7	68.6	67.0	59.5	48.7	36.8	26.6	46.7
	MEAN DEW POINT	30	16.6	20.5	29.1	39.8	51.0	60.1	63.5	62.1	54.5	42.1	30.9	20.7	40.9
	NORMAL NO. DAYS WITH:														
	MAXIMUM ≥ 90°	29	0.0	0.0	0.0	0.4	1.3	8.8	18.2	14.2	5.0	0.6	0.0	0.0	48.5
	MAXIMUM ≤ 32°	29	11.9	8.1	2.0	0.1	0.0	0.0	0.0	0.0	0.0	0.0	1.9	8.9	32.9
	MINIMUM ≤ 32°	29	28.8	24.2	18.4	5.1	0.2	0.0	0.0	0.0	0.1	3.0	17.4	28.2	125.4
	MINIMUM ≤ 0°	29	4.2	2.1	0.2	0.0	0.0	0.0	0.0	0.0	0.0	0.0	0.1	2.0	8.6
H/C	NORMAL HEATING DEG. DAYS	30	1212	952	735	373	149	8	0	0	53	290	705	1097	5574
	NORMAL COOLING DEG. DAYS	30	0	0	0	7	81	263	450	377	125	14	0	0	1317
RH	NORMAL (PERCENT)	29	71	69	65	64	68	67	62	65	68	64	70	72	67
	HOUR 00 LST	29	75	75	71	71	76	76	71	72	75	70	75	76	73
	HOUR 06 LST	29	78	79	79	80	84	84	81	83	84	80	81	79	81
	HOUR 12 LST	29	64	62	56	54	57	55	51	54	56	52	60	64	57
	HOUR 18 LST	29	66	61	53	50	54	51	47	49	54	55	64	68	56
S	PERCENT POSSIBLE SUNSHINE	31	65	63	64	66	68	76	79	77	70	69	60	59	68
W/O	MEAN NO. DAYS WITH:														
	HEAVY FOG(VISBY ≤ 1/4 MI)	31	2.4	2.5	1.8	1.1	1.0	0.5	0.3	1.0	1.3	1.3	1.8	2.4	17.4
	THUNDERSTORMS	31	0.2	0.4	1.9	5.5	8.7	10.6	10.4	9.1	6.3	2.5	0.9	0.2	56.7
CLOUDINESS	MEAN:														
	SUNRISE-SUNSET (TENTHS)	1		8.0	8.0	9.0	7.0		6.0	4.0	5.0	5.0	5.0	5.0	
	MIDNIGHT-MIDNIGHT (TENTHS)	1		8.0	7.0	9.0	7.0		6.0	4.0	5.0	5.0	5.0	5.0	
	MEAN NO. DAYS WITH:														
	CLEAR	1	7.0	9.0	9.0	5.0	11.0	10.0	7.0	18.0	15.0	13.0	10.0	12.0	126.0
	PARTLY CLOUDY	1	1.0	2.0	3.0	6.0	2.0	6.0	5.0	4.0	2.0	2.0	5.0	9.0	47.0
	CLOUDY	1	12.0	14.0	13.0	11.0	8.0	3.0	8.0	6.0	7.0	8.0	7.0	9.0	106.0
PR	MEAN STATION PRESSURE(IN.)	21	28.52	28.50	28.38	28.36	28.35	28.36	28.41	28.43	28.45	28.47	28.46	28.51	28.43
	MEAN SEA-LEVEL PRES. (IN.)	30	30.14	30.11	29.99	29.92	29.90	29.89	29.93	29.95	29.99	30.03	30.05	30.11	30.00
WINDS	MEAN SPEED (MPH)	30	10.7	10.7	11.4	11.3	10.4	9.9	9.8	9.5	10.0	9.6	10.4	10.8	10.4
	PREVAIL.DIR(TENS OF DEGS.)	30	30	36	36	18	18	18	18	18	18	18	18	18	18
	MAXIMUM 2-MINUTE:														
	SPEED (MPH)	1	32	28	40	38	40	51	49	38	34	31	36	37	51
	DIR. (TENS OF DEGS.)	1	03	12	35	34	14	22	33	27	19	19	33	18	22
	YEAR OF OCCURRENCE		1993	1993	1993	1993	1993	1993	1993	1993	1993	1993	1993	1993	JUN 1993
	MAXIMUM 5-SECOND:														
	SPEED (MPH)	1	37	32	46	45	45	62	56	47	41	37	40	45	62
	DIR. (TENS OF DEGS.)	1	02	12	35	35	14	22	33	29	31	18	32	17	22
	YEAR OF OCCURRENCE		1993	1993	1993	1993	1993	1993	1993	1993	1993	1993	1993	1993	JUN 1993
PRECIPITATION	NORMAL TOTAL (IN.)	30	0.58	0.75	2.20	2.31	4.29	4.49	3.65	3.54	3.01	1.99	1.13	0.84	28.78
	MAXIMUM TOTAL (IN.)	32	1.84	2.50	8.32	5.98	9.74	14.14	16.75	10.72	8.49	4.89	4.88	3.60	16.75
	YEAR OF OCCURRENCE		1983	1971	1987	1984	1970	1967	1993	1977	1973	1992	1971	1984	JUL 1993
	MINIMUM TOTAL (IN.)	32	0.00	0.01	0.02	0.37	0.26	1.27	0.11	0.38	0.21	0.04	T	T	0.0
	YEAR OF OCCURRENCE		1986	1974	1968	1989	1966	1980	1984	1971	1991	1988	1989	1976	JAN 1986
	MAXIMUM IN 24 HOURS (IN.)	32	0.80	1.24	3.49	2.64	5.15	5.46	3.94	4.56	3.18	4.15	3.20	2.61	5.46
	YEAR OF OCCURRENCE		1985	1966	1987	1987	1970	1967	1971	1968	1973	1979	1971	1984	JUN 1967
	NORMAL NO. DAYS WITH:														
	PRECIPITATION ≥ 0.01	29	5.3	4.9	7.8	9.2	10.8	9.7	8.4	8.6	7.9	6.0	5.0	4.9	88.5
	PRECIPITATION ≥ 1.00	29	0.0	0.1	0.6	0.5	1.1	1.2	1.0	1.0	0.8	0.3	0.2	0.1	6.9
SNOWFALL	NORMAL TOTAL (IN.)	29	5.2	5.5	3.8	0.7	-1.1	0.0	0.0	0.0	-1.1	0.2	2.0	4.8	22.2
	MAXIMUM TOTAL (IN.)	32	15.8	20.6	10.7	6.1	T	T	T	0.0	T	4.6	10.1	16.7	20.6
	YEAR OF OCCURRENCE		1993	1971	1987	1983	1991	1989	1992		1985	1970	1972	1983	FEB 1971
	MAXIMUM IN 24 HOURS (IN.)	32	9.3	13.2	9.1	3.8	T	T	T	0.0	T	4.6	7.3	12.5	13.2
	YEAR OF OCCURRENCE		1985	1965	1987	1970	1991	1989	1992	1993	1985	1970	1983	1966	FEB 1965
	MAXIMUM SNOW DEPTH (IN.)	30	13	15	9	3	0	0	0	0	0	3	7	14	15
	YEAR OF OCCURRENCE		1979	1983	1987	1970						1970	1983	1983	FEB 1983
	NORMAL NO. DAYS WITH:														
	SNOWFALL ≥ 1.0	29	1.9	1.6	1.4	0.3	0.0	0.0	0.0	0.0	0.0	0.*	0.6	1.6	7.4

PRECIPITATION (inches) 1993 CONCORDIA, KANSAS (CNK)

YEAR	JAN	FEB	MAR	APR	MAY	JUN	JUL	AUG	SEP	OCT	NOV	DEC	ANNUAL
1964	0.07	0.44	1.14	3.73	2.52	8.51	3.01	3.03	3.67	0.22	1.50	0.77	28.61
1965	0.87	1.98	0.88	1.34	5.19	8.48	3.21	5.24	4.72	0.22	0.03	1.08	33.24
1966	0.50	1.61	0.30	0.84	0.26	2.38	2.10	3.47	2.36	0.40	0.05	0.89	15.16
1967	0.38	0.20	1.11	3.56	2.70	14.14	4.63	1.91	6.61	1.48	0.37	1.15	38.24
1968	0.17	0.44	0.02	3.15	2.61	3.09	3.81	8.31	3.03	3.12	1.29	0.97	30.01
1969	0.71	1.75	1.67	1.89	5.82	2.89	6.17	3.38	1.30	3.26	0.12	0.57	29.53
1970	0.22	0.44	1.13	2.02	9.74	2.76	3.59	1.53	4.42	2.46	0.20	0.04	28.55
1971	0.76	2.50	0.77	1.18	6.47	3.98	10.17	0.38	0.74	2.74	4.88	0.71	35.28
1972	0.23	0.30	0.79	4.29	4.07	1.70	6.90	6.65	1.03	2.11	3.51	1.13	32.71
1973	1.01	0.65	7.37	2.41	3.14	1.78	8.50	2.44	8.49	4.30	1.27	3.06	44.42
1974	0.33	0.01	0.68	2.26	2.35	3.78	0.21	2.37	0.55	1.98	0.84	0.46	15.82
1975	0.71	1.51	1.46	1.99	2.63	6.96	1.64	2.80	1.72	0.49	2.66	0.55	25.12
1976	0.17	0.31	2.58	4.41	1.61	1.61	1.16	0.77	2.84	1.89	0.08	T	17.43
1977	0.64	0.03	2.06	2.37	6.01	5.03	2.33	10.72	2.05	2.22	1.73	0.04	35.23
1978	0.20	0.71	0.96	2.70	6.33	2.13	5.94	3.54	5.08	0.52	2.04	0.21	30.36
1979	1.12	0.11	5.05	1.54	2.10	3.54	4.16	2.10	0.75	4.82	0.92	0.36	26.57
1980	1.24	1.10	3.20	1.77	2.82	1.27	0.73	3.04	1.25	1.98	0.07	0.97	19.44
1981	0.14	0.07	1.30	1.85	6.98	4.68	8.02	3.61	2.75	0.84	3.22	0.21	33.67
1982	0.82	0.45	2.81	2.06	7.53	8.38	7.04	3.17	1.79	0.85	0.57	1.52	36.99
1983	1.84	1.09	3.42	1.92	2.98	3.59	0.16	2.58	1.72	3.60	1.84	0.85	25.59
1984	0.27	0.55	3.09	5.98	3.98	4.36	0.11	1.85	2.50	2.20	0.24	3.60	28.73
1985	1.14	0.86	1.04	2.85	3.78	3.19	2.06	4.46	2.46	2.96	0.55	0.42	25.77
1986	0.00	1.29	2.21	3.11	3.63	5.67	4.69	7.59	6.81	3.04	0.72	1.42	40.18
1987	0.55	0.93	8.32	2.90	6.06	3.51	3.45	3.42	1.55	1.94	0.72	0.79	34.14
1988	0.72	0.62	0.55	0.99	2.91	3.16	1.65	1.67	1.06	0.04	0.54	0.63	14.54
1989	0.56	0.56	0.55	0.37	4.06	5.80	1.40	5.77	4.22	1.11	T	0.42	24.82
1990	0.26	0.43	4.05	1.15	6.37	3.78	3.96	1.48	1.23	1.06	1.45	0.45	25.67
1991	0.58	0.15	2.14	1.98	3.02	3.77	0.86	2.60	0.21	1.49	1.50	1.65	19.95
1992	0.73	0.88	2.06	2.58	3.49	6.31	11.43	2.63	2.69	4.89	2.17	1.66	41.52
1993	1.52	1.33	2.63	2.06	3.41	6.66	16.75	4.78	3.12	0.86	0.57	1.10	44.79
POR= 31 YRS	0.61	0.75	2.18	2.36	4.07	4.61	4.26	3.53	2.78	1.96	1.13	0.90	29.14

AVERAGE TEMPERATURE (˚F) 1993 CONCORDIA, KANSAS (CNK)

YEAR	JAN	FEB	MAR	APR	MAY	JUN	JUL	AUG	SEP	OCT	NOV	DEC	ANNUAL
1964	32.8	31.8	37.9	54.1	67.3	71.6	82.9	73.5	67.3	55.9	43.4	29.0	54.0
1965	29.7	26.4	30.9	55.4	67.1	71.2	77.5	75.7	63.1	59.0	45.4	37.8	53.3
1966	22.9	30.4	45.3	48.8	63.6	74.2	83.4	73.7	65.5	56.5	41.0	28.8	52.8
1967	28.4	31.8	45.2	54.7	59.4	70.7	73.7	72.9	63.9	54.9	41.0	31.6	52.4
1968	26.2	29.8	46.7	52.8	58.1	74.1	78.0	76.4	66.9	57.6	39.2	25.5	52.6
1969	22.6	29.4	33.8	54.0	62.9	68.3	78.9	75.9	68.9	51.2	42.6	30.1	51.5
1970	23.2	35.3	36.5	52.5	67.9	73.3	78.3	80.8	66.5	51.9	39.6	33.1	53.2
1971	21.8	26.6	39.4	53.9	59.4	76.3	74.4	76.3	68.2	58.2	42.4	33.1	52.5
1972	25.2	31.4	45.3	52.1	61.9	73.4	75.8	74.9	68.2	51.6	38.0	24.9	51.9
1973	27.0	33.2	44.8	51.0	60.4	74.6	77.7	78.2	64.4	59.0	42.8	26.9	53.3
1974	22.0	36.1	45.2	54.6	65.1	71.0	84.4	72.7	63.0	58.2	41.8	32.4	53.9
1975	29.1	26.1	34.8	51.5	64.3	71.7	79.3	80.2	64.1	59.6	41.3	33.6	53.0
1976	28.8	42.2	41.9	54.8	60.0	72.9	79.6	79.2	69.0	50.8	35.8	31.1	53.8
1977	18.8	37.3	46.5	57.2	67.9	74.9	81.5	73.8	68.8	55.6	41.3	29.8	54.4
1978	16.1	18.4	39.1	54.3	61.6	74.6	78.7	77.1	71.3	55.9	40.8	28.2	51.3
1979	12.7	18.7	41.7	51.2	60.8	72.9	76.8	76.3	70.7	58.1	38.5	35.5	51.2
1980	26.6	25.1	37.5	52.8	62.2	75.6	85.2	80.5	70.1	55.6	45.5	32.5	54.1
1981	31.8	35.5	45.3	60.7	59.9	74.5	78.1	73.8	68.3	53.7	45.0	30.7	54.8
1982	19.2	27.3	40.8	50.1	64.0	69.4	80.9	78.6	67.2	54.9	37.6	31.8	51.8
1983	28.6	32.9	40.7	44.9	57.9	69.4	82.1	84.5	70.2	55.6	40.3	12.1	51.6
1984	27.0	40.0	36.7	47.9	61.7	75.0	80.1	79.0	64.8	55.3	42.9	32.4	53.6
1985	20.3	24.7	44.9	55.7	64.8	69.6	79.5	73.3	66.1	55.6	32.8	24.6	51.0
1986	36.5	32.7	49.7	55.9	62.6	74.9	80.0	72.9	70.0	55.4	38.6	33.6	55.2
1987	31.6	40.1	44.4	56.3	67.8	75.5	81.1	76.3	69.4	52.6	45.4	33.7	56.2
1988	25.0	29.8	43.4	53.0	68.2	79.1	78.7	80.6	69.5	53.8	44.0	36.8	55.2
1989	37.2	21.9	43.2	57.8	64.2	70.5	78.8	76.5	65.0	57.8	42.7	23.2	53.2
1990	36.9	35.7	45.3	51.9	60.3	76.4	79.3	78.7	73.0	57.2	47.4	26.4	55.7
1991	24.5	41.1	46.5	56.2	67.2	77.5	82.4	77.5	70.3	56.3	34.5	36.5	55.9
1992	37.1	41.0	46.6	53.5	62.2	69.1	74.5	71.0	67.4	54.4	35.8	29.3	53.5
1993	22.0	23.8	38.8	49.4	61.8	71.5	77.3	76.4	62.6	53.6	38.2	35.2	50.9
POR= 31 YRS	26.1	31.3	42.1	53.4	63.2	73.2	79.2	76.7	67.4	55.9	41.0	30.2	53.3

REFERENCE NOTES:

H/C INDICATES HEATING AND COOLING DEGREE DAYS.
RH INDICATES RELATIVE HUMIDITY
W/O INDICATES WEATHER AND OBSTRUCTIONS
S INDICATES SUNSHINE.
PR INDICATES PRESSURE.
CLOUDINESS IS THE SUM OF THE CEILOMETER AND SATELLITE DATA NOT TO EXCEED TEN TENTHS.
MAXIMUM SNOW DEPTH IS THE DEPTH MEASURED AT 12 UTC.

GENERAL:
T INDICATES TRACE PRECIPITATION, AN AMOUNT GREATER THAN ZERO BUT LESS THAN THE LOWEST REPORTABLE VALUE.
+ INDICATES THE VALUE ALSO OCCURS ON EARLIER DATES.
BLANK ENTRIES DENOTE MISSING OR UNREPORTED DATA.
NORMALS ARE 30-YEAR AVERAGES (1961 - 1990).
ASOS INDICATES AUTOMATED SURFACE OBSERVING SYSTEM.
PM INDICATES THE LAST DAY OF THE PREVIOUS MONTH.
POR (PERIOD OF RECORD) INDICATES THE NUMBER OF YEARS USED TO COMPUTE THE MEAN. INDIVIDUAL MONTHS WITHIN THE POR MAY BE MISSING.
WHEN THE POR FOR A NORMAL IS LESS THAN 30 YEARS, THE NORMAL IS PROVISIONAL AND IS BASED ON THE NUMBER OF YEARS INDICATED.
0.* OR * INDICATES THE VALUE OR MEAN DAYS WITH IS BETWEEN 0.00 AND 0.05.
CLOUDINESS FOR ASOS STATIONS DIFFERS FROM THE NON-ASOS OBSERVATION TAKEN BY A HUMAN OBSERVER. ASOS STATION CLOUDINESS IS BASED ON TIME-AVERAGED CEILOMETER DATA FOR CLOUDS AT OR BELOW 12,000 FEET AND ON SATELLITE DATA FOR CLOUDS ABOVE 12,000 FEET.
THE NUMBER OF DAYS WITH CLEAR, PARTLY CLOUDY, AND CLOUDY CONDITIONS FOR ASOS STATIONS IS THE SUM OF THE CEILOMETER AND SATELLITE DATA FOR THE SUNRISE TO SUNSET PERIOD.

GENERAL CONTINUED:
CLEAR INDICATES 0 - 3 TENTHS, PARTLY CLOUDY INDICATES 4 - 7 TENTHS, AND CLOUDY INDICATES 8 OR MORE TENTHS. WHEN AT LEAST ONE OF THE ELEMENTS (CEILOMETER OR SATELLITE) IS MISSING, THE DAILY CLOUDINESS IS NOT COMPUTED.
WIND DIRECTION IS RECORDED IN TENS OF DEGREES (2 DIGITS) CLOCKWISE FROM TRUE NORTH. "00" INDICATES CALM. "36" INDICATES TRUE NORTH.
RESULTANT WIND IS THE VECTOR AVERAGE OF THE SPEED AND DIRECTION.
AVERAGE TEMPERATURE IS THE SUM OF THE MEAN DAILY MAXIMUM AND MINIMUM TEMPERATURE DIVIDED BY 2.
SNOWFALL DATA COMPRISE ALL FORMS OF FROZEN PRECIPITATION, INCLUDING HAIL.
A HEATING (COOLING) DEGREE DAY IS THE DIFFERENCE BETWEEN THE AVERAGE DAILY TEMPERATURE AND 65˚ F.
DRY BULB IS THE TEMPERATURE OF THE AMBIENT AIR.
DEW POINT IS THE TEMPERATURE TO WHICH THE AIR MUST BE COOLED TO ACHIEVE 100 PERCENT RELATIVE HUMIDITY.
WET BULB IS THE TEMPERATURE THE AIR WOULD HAVE IF THE MOISTURE CONTENT WAS INCREASED TO 100 PERCENT RELATIVE HUMIDITY.

HEATING DEGREE DAYS (base 65°F) 1993 CONCORDIA, KANSAS (CNK)

YEAR	JUL	AUG	SEP	OCT	NOV	DEC	JAN	FEB	MAR	APR	MAY	JUN	TOTAL
1964-65	0	12	69	283	645	1107	1088	1074	1050	305	51	2	5686
1965-66	0	3	140	205	580	838	1297	963	602	481	123	13	5245
1966-67	0	6	76	285	711	1119	1129	923	617	317	253	20	5456
1967-68	7	7	82	344	712	1028	1196	1012	567	369	233	10	5567
1968-69	0	6	47	267	769	1217	1307	992	960	322	128	55	6070
1969-70	0	0	12	450	664	1075	1288	824	876	391	59	23	5662
1970-71	1	0	102	424	752	984	1333	1071	787	330	191	0	5975
1971-72	3	0	99	227	669	982	1229	965	606	388	151	12	5331
1972-73	5	2	68	422	807	1236	1171	883	622	414	154	5	5789
1973-74	0	0	95	208	661	1176	1329	803	606	322	86	19	5305
1974-75	0	4	121	223	692	1005	1106	1082	930	414	90	17	5684
1975-76	0	0	135	208	701	968	1115	657	707	310	175	2	4978
1976-77	0	0	57	461	868	1043	1424	769	567	240	15	0	5444
1977-78	0	0	23	291	703	1085	1512	1299	803	326	162	9	6213
1978-79	0	0	41	288	719	1132	1615	1294	720	411	164	19	6403
1979-80	6	10	22	233	788	909	1184	1154	846	381	143	6	5682
1980-81	0	0	50	311	578	998	1026	822	605	177	196	2	4765
1981-82	1	0	33	343	594	1055	1417	1050	743	453	71	27	5787
1982-83	0	1	78	313	815	1022	1121	889	748	596	236	38	5857
1983-84	0	0	77	317	734	1634	1171	718	871	503	146	2	6173
1984-85	0	0	157	311	656	1001	1379	1123	617	287	66	33	5630
1985-86	0	2	162	288	957	1244	875	898	484	284	94	0	5288
1986-87	0	5	37	294	784	967	1027	692	633	301	37	0	4777
1987-88	0	6	22	389	586	963	1230	1014	663	355	40	0	5268
1988-89	4	1	29	347	624	869	857	1202	676	310	108	19	5046
1989-90	0	1	113	255	660	1293	865	813	606	404	162	4	5176
1990-91	0	0	42	265	527	1191	1247	661	571	275	79	0	4858
1991-92	0	0	62	330	909	877	859	688	565	351	137	14	4792
1992-93	0	11	67	338	867	1102	1324	1145	808	459	133	31	6285
1993-	0	3	123	376	799	919							
POR= 32 YRS	0	2	73	300	710	1073	1199	944	706	357	124	12	5500

COOLING DEGREE DAYS (base 65°F) 1993 CONCORDIA, KANSAS (CNK)

YEAR	JAN	FEB	MAR	APR	MAY	JUN	JUL	AUG	SEP	OCT	NOV	DEC	ANNUAL
1969	0	0	0	0	70	161	438	345	138	31	0	0	1183
1970	0	0	0	23	154	277	421	497	153	26	0	0	1551
1971	0	0	0	5	24	344	303	359	201	25	0	0	1261
1972	0	0	2	10	59	270	347	315	171	12	0	0	1186
1973	0	0	0	1	21	298	399	414	86	29	0	0	1248
1974	0	0	0	18	94	207	607	248	69	19	0	0	1262
1975	0	0	0	14	74	221	447	480	115	49	0	0	1400
1976	0	0	0	10	29	246	458	443	182	32	0	0	1400
1977	0	0	0	14	113	304	518	280	143	4	0	0	1376
1978	0	0	6	10	64	303	431	383	235	12	2	0	1446
1979	0	0	1	5	39	261	381	365	201	27	0	0	1280
1980	0	0	0	18	58	334	634	485	212	26	4	0	1771
1981	0	0	0	51	42	294	414	279	141	0	0	0	1221
1982	0	0	0	11	46	166	501	431	150	9	0	0	1314
1983	0	0	0	0	23	177	534	610	240	32	0	0	1616
1984	0	0	0	0	51	309	475	441	156	16	0	0	1448
1985	0	0	0	16	65	175	457	267	201	3	0	0	1184
1986	0	0	17	19	26	306	474	256	192	1	0	0	1291
1987	0	0	0	51	132	319	507	364	161	12	5	0	1551
1988	0	0	0	4	150	430	437	492	168	5	0	0	1686
1989	0	0	8	99	93	190	435	364	120	39	0	0	1348
1990	0	0	0	15	23	354	448	433	290	28	7	0	1598
1991	0	0	5	19	154	384	546	396	227	68	0	0	1799
1992	0	0	2	15	59	145	303	200	147	19	0	0	890
1993	0	0	0	0	40	235	388	365	56	27	0	0	1111
POR= 25 YRS	0	0	0	16	67	267	451	380	165	21	0	0	1367

SNOWFALL (inches) 1993 CONCORDIA, KANSAS (CNK)

YEAR	JUL	AUG	SEP	OCT	NOV	DEC	JAN	FEB	MAR	APR	MAY	JUN	TOTAL
1964-65	0.0	0.0	0.0	0.0	1.2	1.4	0.3	20.3	4.6	0.0	0.0	0.0	27.8
1965-66	0.0	0.0	0.0	0.0	T	2.1	1.7	1.7	2.1	T	0.0	0.0	7.6
1966-67	0.0	0.0	0.0	0.0	T	15.1	2.9	0.7	0.3	0.1	T	0.0	19.1
1967-68	0.0	0.0	0.0	0.0	0.6	7.6	2.0	5.1	T	0.1	0.0	0.0	15.4
1968-69	0.0	0.0	0.0	0.0	0.1	6.6	5.8	14.5	7.4	0.0	0.0	0.0	34.4
1969-70	0.0	0.0	0.0	0.0	T	4.0	2.5	T	6.6	3.0	0.0	0.0	16.1
1970-71	0.0	0.0	0.0	4.6	0.3	0.2	9.1	20.6	9.2	T	0.0	0.0	44.0
1971-72	0.0	0.0	0.0	0.0	7.5	4.3	2.8	2.6	T	T	0.0	0.0	17.2
1972-73	0.0	0.0	0.0	0.0	10.1	5.8	7.1	2.8	0.0	1.8	0.0	0.0	27.6
1973-74	0.0	0.0	0.0	0.0	0.1	11.3	7.7	0.6	0.5	2.2	0.0	0.0	22.4
1974-75	0.0	0.0	0.0	0.0	T	4.5	10.4	10.6	8.6	1.4	0.0	0.0	35.5
1975-76	0.0	0.0	0.0	0.0	7.9	T	3.0	4.0	6.6	0.0	0.0	0.0	21.5
1976-77	0.0	0.0	0.0	0.5	1.5	T	9.1	0.2	T	1.2	0.0	0.0	12.5
1977-78	0.0	0.0	0.0	0.0	1.4	0.2	4.1	11.7	5.4	0.0	0.0	0.0	22.8
1978-79	0.0	0.0	0.0	0.0	0.8	6.1	13.7	2.2	3.1	2.5	0.0	0.0	28.4
1979-80	0.0	0.0	0.0	0.6	T	4.5	4.5	12.7	6.0	T	0.0	0.0	28.3
1980-81	0.0	0.0	0.0	T	T	0.4	3.8	2.2	T	0.0	0.0	0.0	6.4
1981-82	0.0	0.0	0.0	0.0	0.1	3.6	5.8	6.8	2.8	1.5	0.0	0.0	20.6
1982-83	0.0	0.0	0.0	0.0	1.0	1.3	12.2	8.7	5.3	6.1	0.0	0.0	34.6
1983-84	0.0	0.0	0.0	0.0	7.3	16.7	3.3	T	6.2	T	0.0	0.0	33.5
1984-85	0.0	0.0	0.0	0.0	0.7	2.5	12.4	4.3	6.2	0.0	0.0	0.0	26.1
1985-86	0.0	0.0	T	0.0	6.4	6.0	0.0	2.9	T	0.0	0.0	0.0	15.3
1986-87	0.0	0.0	0.0	0.0	0.4	7.1	5.9	0.2	10.7	0.0	0.0	0.0	24.3
1987-88	0.0	0.0	0.0	0.0	5.3	5.4	3.7	8.7	2.7	0.0	0.0	0.0	25.8
1988-89	0.0	0.0	0.0	0.0	1.9	0.9	T	5.1	0.2	0.0	T	T	8.1
1989-90	0.0	0.0	0.0	T	T	6.9	3.0	2.8	4.2	0.0	0.0	0.0	16.9
1990-91	0.0	0.0	0.0	0.0	0.1	5.5	7.8	T	1.4	T	T	0.0	14.8
1991-92	0.0	0.0	0.0	3.6	7.1	0.4	1.0	5.4	1.3	0.5	0.0	0.0	19.3
1992-93	T	0.0			4.4	2.3	15.8	7.6	0.9	0.4	0.0	0.0	
1993-	0.0	0.0	0.0	T	0.2	1.4							
POR= 32 YRS	T	0.0	T	0.3	2.1	4.5	5.5	5.4	3.5	0.7	T	T	22.0

1993 DODGE CITY, KANSAS (DDC)

The climate of Dodge City and southwestern Kansas is classified as semi-arid. Dodge City is nearly 300 miles east of the Rocky Mountains, but the weather reflects the influence of the mountains. The mountains form a barricade against all except high level moisture from the southwest, west, and northwest. Chinook winds occur occasionally but with less frequency and effect than at stations farther to the west. Relatively dry air predominating with an abundance of sunshine contribute to broad diurnal temperature ranges.

Thunderstorms during the growing season contribute most of the moisture. In general, the thunderstorms are widely scattered, occurring during the late afternoons and evenings. They are occasionally accompanied by hail and strong winds, but due to the local nature of the storms, damage to crops and buildings is spotty and variable. Winter is the dry season. However, the moisture accumulated during the winter months is important for the hard winter wheat. The duration of snow cover is generally brief due to mild temperatures and an abundance of sunshine. The exception results from the occasional blizzard that spreads across the flat treeless prairies of the high plains.

Afternoon temperatures in the 90s prevail during the summer months. Temperatures above 100 degrees are the exception. Due to low humidity and a continual breeze, these high temperatures are moderated. Temperatures normally drop sharply after sunset, allowing cool comfortable nights. During the winter months, large temperature changes are frequent, but the duration of extreme cold spells is brief.

The visibility at Dodge City is generally unrestricted as the terrain is favorable for unrestricted movement of air and air masses. Western Kansas is noted for clear skies and an abundance of sunshine.

Based on the 1951-1980 period, the average first occurrence of 32 degrees Fahrenheit in the fall is October 23 and the average last occurrence in the spring is April 21.

NORMALS, MEANS, AND EXTREMES FOR 1993

DODGE CITY, KS (DDC)

LATITUDE: 37° 45′ 44″ N LONGTITUDE: 99° 57′ 52″ W ELEVATION (FT): GRND: 2582 BARO: 2599 TIME ZONE: CENTRAL (UTC+06) WBAN: 13985

	ELEMENT	POR	JAN	FEB	MAR	APR	MAY	JUN	JUL	AUG	SEP	OCT	NOV	DEC	YEAR
TEMPERATURE °F	NORMAL DAILY MAXIMUM	30	41.5	47.3	56.6	68.0	76.3	87.0	93.1	90.7	81.3	70.7	55.2	43.8	67.6
	MEAN DAILY MAXIMUM	51	41.8	47.4	55.5	67.2	75.8	86.6	94.1	92.4	83.3	72.0	55.7	45.2	68.1
	HIGHEST DAILY MAXIMUM	51	80	85	93	100	102	108	109	107	106	96	91	86	109
	YEAR OF OCCURRENCE		1989	1972	1989	1989	1967	1985	1986	1983	1947	1968	1980	1955	JUL 1986
	MEAN OF EXTREME MAXS.	46	66.6	72.0	81.4	87.9	92.9	100.4	102.8	101.8	96.6	90.1	75.8	67.6	86.3
	NORMAL DAILY MINIMUM	30	18.1	22.7	30.4	41.4	52.0	61.8	67.5	65.7	56.7	44.3	31.1	21.2	42.7
	MEAN DAILY MINIMUM	51	18.7	23.2	29.8	41.2	51.7	61.7	68.4	66.9	57.6	45.3	30.9	22.3	43.1
	LOWEST DAILY MINIMUM	51	-13	-15	-15	15	26	41	46	47	29	14	0	-21	-21
	YEAR OF OCCURRENCE		1984	1951	1948	1975	1967	1954	1990	1950	1985	1993	1958	1989	DEC 1989
	MEAN OF EXTREME MINS.	46	-2.1	3.5	10.6	25.1	37.3	49.5	57.7	55.2	40.5	28.4	13.0	3.4	26.8
	NORMAL DRY BULB	30	29.8	35.0	43.5	54.7	64.2	74.4	80.3	78.2	69.0	57.5	43.2	32.5	55.2
	MEAN DRY BULB	51	30.3	35.3	42.7	54.2	63.8	74.2	81.3	79.7	70.4	58.7	43.3	33.7	55.6
	MEAN WET BULB	44	25.0	29.3	35.3	45.3	55.1	63.3	66.7	65.7	58.3	47.9	36.0	27.9	46.3
	MEAN DEW POINT	44	17.7	21.8	26.4	36.4	48.8	57.1	60.3	59.5	51.5	39.5	28.2	20.5	39.0
	NORMAL NO. DAYS WITH:														
	MAXIMUM ≥ 90°	30	0.0	0.0	0.1	0.6	3.0	12.6	22.1	18.8	7.6	1.5	*	0.0	66.3
	MAXIMUM ≤ 32°	30	8.5	5.4	1.9	0.1	0.0	0.0	0.0	0.0	0.0	0.1	1.7	6.6	24.3
	MINIMUM ≤ 32°	30	29.2	23.1	18.1	4.8	0.2	0.0	0.0	0.0	0.1	2.6	17.2	28.1	123.4
	MINIMUM ≤ 0°	30	2.8	1.1	*	0.0	0.0	0.0	0.0	0.0	0.0	0.0	0.0	1.4	5.3
H/C	NORMAL HEATING DEG. DAYS	30	1091	840	667	323	119	7	0	0	39	253	654	1008	5001
	NORMAL COOLING DEG. DAYS	30	0	0	0	14	94	289	474	414	159	21	0	0	1465
RH	NORMAL (PERCENT)	30	66	65	60	58	62	60	55	58	62	58	64	67	61
	HOUR 00 LST	30	72	71	68	67	72	70	64	67	71	66	72	72	69
	HOUR 06 LST	30	76	76	76	75	80	79	76	78	80	74	77	76	77
	HOUR 12 LST	30	57	55	50	46	51	48	43	46	50	46	52	57	50
	HOUR 18 LST	30	58	52	45	42	47	44	39	42	46	46	56	61	48
S	PERCENT POSSIBLE SUNSHINE	51	67	65	65	68	68	76	80	78	75	74	67	66	70
W/O	MEAN NO. DAYS WITH:														
	HEAVY FOG(VISBY ≤ 1/4 MI)	51	2.9	3.4	3.1	1.8	1.6	0.8	0.6	0.9	1.6	2.2	2.2	2.9	24.0
	THUNDERSTORMS	51	0.2	0.4	1.4	3.7	8.2	10.0	10.3	8.8	4.6	1.9	0.5	0.2	50.2
CLOUDINESS	MEAN:														
	SUNRISE-SUNSET (TENTHS)	1			5.0	7.0	5.0		4.0	4.0	3.0	5.0	5.0	4.0	
	MIDNIGHT-MIDNIGHT (TENTHS)	1			6.0	7.0	5.0		4.0	4.0	4.0	6.0	5.0	4.0	
	MEAN NO. DAYS WITH:														
	CLEAR	1	5.0	1.0	10.0	9.0	12.0	11.0	13.0	19.0	18.0	11.0	12.0	19.0	140.0
	PARTLY CLOUDY	1	2.0		6.0	4.0	2.0	4.0	4.0	4.0	2.0	4.0	3.0	4.0	
	CLOUDY	1	13.0		6.0	7.0	7.0	3.0	4.0	5.0	5.0	8.0	6.0	6.0	
PR	MEAN STATION PRESSURE(IN.)	21	27.36	27.34	27.23	27.24	27.23	27.26	27.32	27.33	27.34	27.34	27.32	27.35	27.31
	MEAN SEA-LEVEL PRES. (IN.)	44	30.12	30.08	29.96	29.91	29.88	29.86	29.92	29.93	29.97	30.02	30.06	30.10	29.98
WINDS	MEAN SPEED (MPH)	30	10.9	11.1	11.7	11.7	10.9	10.5	10.2	9.7	10.5	10.0	10.8	11.2	10.8
	PREVAIL.DIR(TENS OF DEGS.)	30	30	36	36	18	18	18	18	18	18	18	18	18	18
	MAXIMUM 2-MINUTE:														
	SPEED (MPH)	1	37		38	40	36	45	52	55	34	43	34	43	
	DIR. (TENS OF DEGS.)	1	36		35	30	16	33	36	26	17	33	22	01	
	YEAR OF OCCURRENCE		1993		1993	1993	1993	1993	1993	1993	1993	1993	1993	1993	
	MAXIMUM 5-SECOND:														
	SPEED (MPH)	1	42		46	48	41	53	71	75	40	51	41	49	
	DIR. (TENS OF DEGS.)	1	33		35	35	16	28	35	25	17	33	21	01	
	YEAR OF OCCURRENCE		1993		1993	1993	1993	1993	1993	1993	1993	1993	1993	1993	
PRECIPITATION	NORMAL TOTAL (IN.)	30	0.49	0.62	1.56	2.05	3.03	3.10	3.24	2.73	1.91	1.28	0.83	0.65	21.49
	MAXIMUM TOTAL (IN.)	51	1.96	2.87	8.84	6.26	8.69	7.95	9.13	7.44	6.80	4.88	3.75	2.41	9.13
	YEAR OF OCCURRENCE		1949	1993	1973	1976	1951	1951	1962	1977	1973	1968	1971	1984	JUL 1962
	MINIMUM TOTAL (IN.)	51	0.00	T	0.02	0.07	0.43	0.12	0.17	0.68	0.01	T	T	T	0.0
	YEAR OF OCCURRENCE		1986	1991	1966	1963	1985	1952	1946	1970	1980	1952	1989	1957	JAN 1986
	MAXIMUM IN 24 HOURS (IN.)	51	1.35	1.82	2.54	4.64	5.57	3.28	4.17	3.23	3.27	4.55	2.42	2.00	5.57
	YEAR OF OCCURRENCE		1990	1993	1973	1978	1978	1944	1944	1986	1959	1968	1971	1984	MAY 1978
	NORMAL NO. DAYS WITH:														
	PRECIPITATION ≥ 0.01	30	4.8	4.8	7.0	6.5	9.7	8.5	8.1	8.0	6.7	4.9	4.3	5.1	78.4
	PRECIPITATION ≥ 1.00	30	*	*	0.3	0.4	0.6	0.9	0.8	0.6	0.4	0.3	0.2	0.1	4.6
SNOWFALL	NORMAL TOTAL (IN.)	30	4.4	4.0	4.3	0.9	0.*	0.0	0.0	0.0	-1.1	0.2	1.7	3.6	19.1
	MAXIMUM TOTAL (IN.)	51	15.7	19.6	24.0	9.0	0.9	T	T	T	T	4.5	16.7	10.8	24.0
	YEAR OF OCCURRENCE		1990	1993	1970	1983	1978	1990	1992	1992	1985	1991	1992	1942	MAR 1970
	MAXIMUM IN 24 HOURS (IN.)	51	11.8	12.0	12.8	7.4	0.9	T	T	T	T	4.1	6.7	6.6	12.8
	YEAR OF OCCURRENCE		1990	1993	1970	1983	1978	1990	1992	1992	1985	1991	1948	1958	MAR 1970
	MAXIMUM SNOW DEPTH (IN.)	44	12	9	14	6	T	0	0	0	0	3	8	7	14
	YEAR OF OCCURRENCE		1988	1978	1957	1983	1979					1976	1948	1966	MAR 1957
	NORMAL NO. DAYS WITH:														
	SNOWFALL ≥ 1.0	30	1.4	1.2	1.3	0.3	0.0	0.0	0.0	0.0	0.0	0.1	0.6	1.0	5.9

PRECIPITATION (inches) 1993 DODGE CITY, KANSAS (DDC)

YEAR	JAN	FEB	MAR	APR	MAY	JUN	JUL	AUG	SEP	OCT	NOV	DEC	ANNUAL
1964	0.04	0.57	0.77	0.69	3.67	2.17	5.30	2.99	1.75	0.83	1.89	0.58	21.25
1965	0.49	0.60	0.15	0.63	5.65	4.76	3.37	1.62	2.63	3.04	T	2.01	24.95
1966	0.45	0.43	0.02	2.18	0.77	1.61	1.48	4.36	1.89	0.32	T	0.63	14.14
1967	0.32	0.01	0.14	1.54	2.24	5.49	5.06	3.16	2.38	0.31	0.14	0.51	21.30
1968	0.14	0.41	0.26	0.93	3.93	4.78	4.08	6.31	0.23	4.88	1.41	0.43	27.79
1969	0.03	1.77	2.31	1.39	2.83	0.75	2.17	4.21	1.40	1.66	0.08	0.19	18.79
1970	0.07	0.01	2.62	1.91	0.81	1.18	1.72	0.68	2.64	0.52	0.06	0.04	12.26
1971	1.10	2.04	0.09	1.09	2.81	3.93	4.49	0.99	1.46	2.87	3.75	0.66	25.28
1972	0.17	0.07	0.27	2.06	5.44	5.94	7.26	3.97	2.48	0.71	2.28	0.35	31.00
1973	0.62	0.35	8.84	3.24	0.73	1.06	3.88	2.44	6.80	1.17	1.34	1.95	32.42
1974	0.30	0.02	1.42	0.46	2.91	4.09	0.79	6.87	0.19	1.93	0.51	0.34	19.83
1975	0.21	0.55	0.60	3.26	3.94	4.49	2.17	1.33	0.15	0.01	1.83	0.29	18.83
1976	0.03	0.58	0.74	6.26	2.94	0.52	0.51	1.20	3.03	0.79	0.17	0.03	16.80
1977	0.45	0.14	0.53	3.32	3.60	1.09	2.59	7.44	0.87	0.94	0.38	0.76	22.11
1978	0.54	1.35	0.97	4.74	4.55	3.45	0.24	0.95	2.26	0.04	1.25	0.42	20.76
1979	1.54	0.09	3.86	1.96	3.48	0.89	5.20	1.49	0.13	3.30	0.97	0.71	23.62
1980	1.06	1.46	2.87	1.89	3.60	3.85	2.00	2.06	0.01	0.23	0.01	0.76	19.80
1981	0.31	0.04	2.27	0.70	5.73	1.39	5.30	2.26	3.07	1.40	2.26	0.39	25.12
1982	0.18	1.30	0.81	0.66	2.74	3.65	5.54	1.01	1.31	0.84	0.44	1.04	19.52
1983	0.56	1.29	2.80	2.88	3.52	5.05	0.57	1.44	2.64	1.87	1.15	0.62	24.39
1984	0.68	0.20	2.73	4.38	1.38	1.89	0.63	1.35	0.35	2.87	0.10	2.41	18.97
1985	0.92	1.28	1.08	2.18	0.43	2.78	3.42	2.91	3.63	2.69	1.05	0.11	22.48
1986	0.00	0.44	0.18	2.19	1.31	4.70	2.12	5.60	1.22	0.93	0.63	0.96	20.28
1987	0.58	1.38	4.34	0.92	2.54	3.79	4.06	2.79	1.53	0.49	0.52	0.72	23.66
1988	0.89	0.13	0.53	3.08	3.56	0.15	1.95	1.11	2.80	0.56	0.10	0.10	14.96
1989	0.25	0.27	0.85	0.41	3.40	6.59	3.42	2.80	2.36	0.05	T	0.55	20.95
1990	1.60	1.17	1.64	3.64	4.91	1.01	3.07	1.19	1.45	0.18	0.42	0.60	20.88
1991	0.31	T	1.05	1.12	2.02	0.91	0.78	1.43	0.20	1.28	1.52	2.10	12.72
1992	0.39	0.35	1.43	0.21	3.06	6.98	2.67	2.90	0.48	0.71	1.58	0.90	21.66
1993	0.81	2.87	2.01	2.89	2.13	3.22	6.33	3.08	0.82	1.01	0.63	1.02	26.82
POR= 51 YRS	0.54	0.66	1.45	1.91	3.13	3.10	3.01	2.61	1.69	1.39	0.76	0.63	20.87

AVERAGE TEMPERATURE (°F) 1993 DODGE CITY, KANSAS (DDC)

YEAR	JAN	FEB	MAR	APR	MAY	JUN	JUL	AUG	SEP	OCT	NOV	DEC	ANNUAL
1964	35.5	31.8	40.8	56.8	69.0	73.9	83.1	77.4	68.2	57.2	43.8	30.9	55.7
1965	34.4	32.4	33.8	57.9	66.4	72.3	79.2	76.2	65.3	60.0	49.1	36.9	55.3
1966	25.3	30.9	48.0	51.2	64.7	75.9	84.2	74.2	67.0	56.9	43.1	28.9	54.2
1967	34.5	36.6	46.9	57.1	61.5	72.8	74.9	73.6	67.3	58.3	42.4	33.2	54.9
1968	32.1	33.4	48.1	54.9	59.3	74.5	77.8	78.3	69.0	59.6	40.8	28.4	54.7
1969	31.7	35.3	31.8	56.9	65.5	70.9	82.5	78.9	70.3	52.6	45.4	35.9	54.8
1970	30.0	40.8	36.5	52.7	68.2	73.7	80.8	81.6	68.2	51.3	41.7	36.4	55.2
1971	27.2	31.9	42.3	53.6	62.0	77.1	77.7	76.6	68.2	57.9	43.5	34.9	54.4
1972	29.0	36.8	49.2	55.1	63.0	73.8	76.0	75.5	67.6	54.4	37.0	27.3	53.7
1973	28.6	35.4	41.7	46.9	58.8	76.4	79.8	81.6	65.3	61.2	44.1	32.0	54.3
1974	26.2	40.5	48.4	57.4	67.7	74.2	83.4	74.9	63.4	60.1	43.6	33.7	56.1
1975	33.8	29.9	40.2	53.6	64.7	72.4	78.3	81.4	67.1	60.6	41.4	37.6	55.1
1976	32.9	45.0	44.5	57.0	59.9	74.3	80.4	79.9	70.1	50.9	39.6	36.4	55.9
1977	23.7	41.9	47.8	58.9	68.8	79.5	84.3	77.6	72.4	58.6	44.5	35.6	57.8
1978	21.9	21.3	44.7	57.3	61.9	75.2	83.4	78.1	72.8	58.0	42.8	30.8	54.0
1979	14.8	28.6	45.0	54.1	60.5	73.7	78.0	74.1	70.7	58.4	37.5	34.4	52.5
1980	26.9	30.4	39.6	53.2	62.6	78.4	87.2	82.1	73.5	58.3	46.7	38.3	56.4
1981	36.1	38.8	44.0	61.0	61.6	78.5	82.2	78.0	71.6	56.4	47.5	34.1	57.5
1982	28.2	29.9	44.4	52.4	64.7	69.8	80.7	81.1	72.1	58.4	41.4	35.5	54.9
1983	33.2	37.0	43.1	48.0	61.1	71.8	84.1	86.5	74.2	59.5	44.3	18.4	55.1
1984	28.6	40.7	38.7	49.0	64.4	76.7	82.4	82.5	70.0	56.1	45.7	35.1	55.8
1985	25.9	32.4	45.9	58.1	67.2	73.7	80.1	77.2	67.0	54.4	35.5	29.8	53.9
1986	40.1	37.4	51.0	56.8	64.3	75.5	81.3	74.6	70.1	55.6	40.7	35.5	56.9
1987	33.0	40.3	43.2	55.3	65.3	73.2	76.9	76.1	68.8	55.7	45.2	33.6	55.5
1988	27.0	35.1	43.8	52.1	65.4	77.2	79.7	80.3	68.2	55.4	45.2	37.9	55.6
1989	37.4	25.3	45.3	58.9	65.1	68.0	76.7	75.0	65.1	58.7	44.1	27.2	53.9
1990	35.3	36.3	45.5	52.2	60.1	77.5	77.4	78.6	72.9	57.3	48.4	27.9	55.8
1991	28.7	43.6	47.0	56.2	69.2	77.4	81.6	78.7	70.4	57.2	36.3	37.3	57.0
1992	38.0	42.1	48.4	56.1	62.7	68.1	75.5	71.9	68.4	56.6	35.7	26.4	54.2
1993	25.1	28.6	41.1	50.3	62.8	72.0	78.4	76.5	65.5	53.0	38.2	36.1	52.3
POR= 51 YRS	30.3	35.4	42.7	54.2	63.8	74.2	79.7	78.2	69.0	57.5	42.5	33.1	55.0

REFERENCE NOTES:

H/C INDICATES HEATING AND COOLING DEGREE DAYS.
RH INDICATES RELATIVE HUMIDITY
W/O INDICATES WEATHER AND OBSTRUCTIONS
S INDICATES SUNSHINE.
PR INDICATES PRESSURE.
CLOUDINESS IS THE SUM OF THE CEILOMETER AND SATELLITE DATA NOT TO EXCEED TEN TENTHS.
MAXIMUM SNOW DEPTH IS THE DEPTH MEASURED AT 12 UTC.

GENERAL:
T INDICATES TRACE PRECIPITATION, AN AMOUNT GREATER THAN ZERO BUT LESS THAN THE LOWEST REPORTABLE VALUE.
\+ INDICATES THE VALUE ALSO OCCURS ON EARLIER DATES.
BLANK ENTRIES DENOTE MISSING OR UNREPORTED DATA.
NORMALS ARE 30-YEAR AVERAGES (1961 - 1990).
ASOS INDICATES AUTOMATED SURFACE OBSERVING SYSTEM.
PM INDICATES THE LAST DAY OF THE PREVIOUS MONTH.
POR (PERIOD OF RECORD) INDICATES THE NUMBER OF YEARS USED TO COMPUTE THE MEAN. INDIVIDUAL MONTHS WITHIN THE POR MAY BE MISSING.
WHEN THE POR FOR A NORMAL IS LESS THAN 30 YEARS, THE NORMAL IS PROVISIONAL AND IS BASED ON THE NUMBER OF YEARS INDICATED.
0.* OR * INDICATES THE VALUE OR MEAN DAYS WITH IS BETWEEN 0.00 AND 0.05.
CLOUDINESS FOR ASOS STATIONS DIFFERS FROM THE NON-ASOS OBSERVATION TAKEN BY A HUMAN OBSERVER. ASOS STATION CLOUDINESS IS BASED ON TIME-AVERAGED CEILOMETER DATA FOR CLOUDS AT OR BELOW 12,000 FEET AND ON SATELLITE DATA FOR CLOUDS ABOVE 12,000 FEET.
THE NUMBER OF DAYS WITH CLEAR, PARTLY CLOUDY, AND CLOUDY CONDITIONS FOR ASOS STATIONS IS THE SUM OF THE CEILOMETER AND SATELLITE DATA FOR THE SUNRISE TO SUNSET PERIOD.

GENERAL CONTINUED:
CLEAR INDICATES 0 - 3 TENTHS, PARTLY CLOUDY INDICATES 4 - 7 TENTHS, AND CLOUDY INDICATES 8 OR MORE TENTHS. WHEN AT LEAST ONE OF THE ELEMENTS (CEILOMETER OR SATELLITE) IS MISSING, THE DAILY CLOUDINESS IS NOT COMPUTED.
WIND DIRECTION IS RECORDED IN TENS OF DEGREES (2 DIGITS) CLOCKWISE FROM TRUE NORTH. "00" INDICATES CALM. "36" INDICATES TRUE NORTH.
RESULTANT WIND IS THE VECTOR AVERAGE OF THE SPEED AND DIRECTION.
AVERAGE TEMPERATURE IS THE SUM OF THE MEAN DAILY MAXIMUM AND MINIMUM TEMPERATURE DIVIDED BY 2.
SNOWFALL DATA COMPRISE ALL FORMS OF FROZEN PRECIPITATION, INCLUDING HAIL.
A HEATING (COOLING) DEGREE DAY IS THE DIFFERENCE BETWEEN THE AVERAGE DAILY TEMPERATURE AND 65° F.
DRY BULB IS THE TEMPERATURE OF THE AMBIENT AIR.
DEW POINT IS THE TEMPERATURE TO WHICH THE AIR MUST BE COOLED TO ACHIEVE 100 PERCENT RELATIVE HUMIDITY.
WET BULB IS THE TEMPERATURE THE AIR WOULD HAVE IF THE MOISTURE CONTENT WAS INCREASED TO 100 PERCENT RELATIVE HUMIDITY.

HEATING DEGREE DAYS (base 65°F) 1993 DODGE CITY, KANSAS (DDC)

YEAR	JUL	AUG	SEP	OCT	NOV	DEC	JAN	FEB	MAR	APR	MAY	JUN	TOTAL
1964-65	0	3	81	254	632	1051	943	905	960	226	65	4	5124
1965-66	0	3	127	175	473	862	1221	948	521	405	116	5	4856
1966-67	0	5	54	273	651	1110	938	789	561	259	205	23	4868
1967-68	0	6	45	265	669	974	1015	908	526	306	215	4	4933
1968-69	0	1	17	225	718	1127	1028	826	1021	241	70	36	5310
1969-70	0	0	8	415	578	894	1079	672	878	382	58	33	4997
1970-71	4	0	85	441	696	880	1163	919	701	337	142	0	5368
1971-72	2	0	110	238	639	929	1111	812	485	301	129	3	4759
1972-73	6	0	70	345	831	1164	1122	822	717	537	199	0	5813
1973-74	0	0	99	168	622	1018	1199	682	510	254	62	2	4616
1974-75	0	1	118	180	633	963	961	975	762	357	78	25	5053
1975-76	0	0	92	190	701	845	986	574	631	249	181	3	4452
1976-77	0	0	48	464	753	879	1274	645	529	198	8	0	4798
1977-78	0	0	3	219	609	903	1330	1216	630	252	169	12	5343
1978-79	0	2	41	228	667	1054	1549	1013	615	343	175	25	5712
1979-80	0	7	27	237	816	940	1171	996	780	364	130	3	5471
1980-81	0	0	26	257	548	822	889	728	644	180	164	0	4258
1981-82	0	0	24	284	517	954	1131	979	632	381	67	26	4995
1982-83	0	0	30	221	701	907	980	778	670	509	164	20	4980
1983-84	0	0	49	209	617	1442	1122	699	805	473	96	3	5515
1984-85	0	0	117	295	570	916	1206	907	585	217	48	10	4871
1985-86	0	0	154	331	878	1086	766	765	438	269	66	0	4753
1986-87	0	6	36	285	722	911	988	685	669	321	63	3	4689
1987-88	4	8	24	290	585	964	1171	861	653	382	88	0	5030
1988-89	0	1	48	301	589	835	847	1105	616	270	110	50	4772
1989-90	0	0	130	236	623	1165	914	795	596	384	190	1	5034
1990-91	4	0	35	258	499	1147	1116	591	555	275	58	0	4538
1991-92	0	0	55	311	853	851	828	654	505	282	135	30	4504
1992-93	0	8	40	289	875	1188	1228	1013	732	439	129	23	5964
1993-	0	11	87	392	797	887							
POR= 52 YRS	0	2	58	260	669	983	1070	832	687	335	121	14	5031

COOLING DEGREE DAYS (base 65°F) 1993 DODGE CITY, KANSAS (DDC)

YEAR	JAN	FEB	MAR	APR	MAY	JUN	JUL	AUG	SEP	OCT	NOV	DEC	ANNUAL
1969	0	0	0	8	92	219	551	438	174	37	0	0	1519
1970	0	0	0	18	161	301	498	520	187	25	0	0	1710
1971	0	0	2	1	53	369	404	366	213	23	0	0	1431
1972	0	0	4	11	72	275	356	334	156	26	0	0	1234
1973	0	0	0	0	15	352	467	521	114	56	0	0	1525
1974	0	0	1	33	152	287	579	313	77	36	0	0	1478
1975	0	0	0	22	76	252	419	515	162	59	0	0	1505
1976	0	0	4	19	32	286	484	466	205	35	0	0	1531
1977	0	0	2	23	132	441	604	396	231	25	0	0	1854
1978	0	0	8	28	77	327	577	416	282	17	6	0	1738
1979	0	0	2	23	46	292	411	296	206	40	0	0	1316
1980	0	0	0	18	62	410	695	538	288	56	5	0	2072
1981	0	0	0	64	65	411	540	407	231	21	0	0	1739
1982	0	0	0	9	64	177	494	504	253	24	0	0	1525
1983	0	0	0	3	50	233	600	674	332	44	3	0	1939
1984	0	0	0	0	86	365	547	552	272	28	0	0	1850
1985	0	0	2	18	121	277	472	383	224	8	0	0	1505
1986	0	0	8	32	49	324	511	311	196	1	0	0	1432
1987	0	0	0	38	79	256	379	359	144	9	1	0	1265
1988	0	0	0	3	108	373	463	480	149	12	0	0	1588
1989	0	0	12	95	119	145	369	317	138	47	0	0	1242
1990	0	0	0	9	44	383	395	429	278	24	4	0	1566
1991	0	0	4	17	195	376	522	430	226	75	0	0	1845
1992	0	0	0	20	70	132	335	232	149	34	0	0	972
1993	0	0	0	3	67	239	420	372	106	26	0	0	1233
POR= 25 YRS	0	0	1	20	83	299	483	421	199	30	0	0	1536

SNOWFALL (inches) 1993 DODGE CITY, KANSAS (DDC)

YEAR	JUL	AUG	SEP	OCT	NOV	DEC	JAN	FEB	MAR	APR	MAY	JUN	TOTAL
1964-65	0.0	0.0	0.0	0.0	2.5	5.4	4.9	5.5	0.3	0.0	0.0	0.0	18.6
1965-66	0.0	0.0	0.0	0.0	0.0	9.1	4.7	1.1	0.2	T	0.0	0.0	15.1
1966-67	0.0	0.0	0.0	0.0	T	8.2	4.1	0.1	0.6	0.0	0.0	0.0	13.0
1967-68	0.0	0.0	0.0	T	1.0	2.9	0.2	3.7	0.3	T	0.0	0.0	8.1
1968-69	0.0	0.0	0.0	0.0	T	2.6	T	8.9	19.4	0.0	0.0	0.0	30.9
1969-70	0.0	0.0	0.0	0.0	0.1	0.3	0.5	T	24.0	1.8	0.0	0.0	26.7
1970-71	0.0	0.0	0.0	1.1	T	T	9.9	14.1	0.1	T	0.0	0.0	25.2
1971-72	0.0	0.0	0.0	T	1.6	5.2	2.2	0.7	1.0	T	0.0	0.0	10.7
1972-73	0.0	0.0	0.0	T	6.3	3.2	8.8	2.7	1.0	2.4	0.0	0.0	24.4
1973-74	0.0	0.0	0.0	0.0	2.7	10.3	4.2	0.2	T	0.4	0.0	0.0	17.8
1974-75	0.0	0.0	0.0	0.0	T	2.8	2.9	6.1	4.0	T	0.0	0.0	15.8
1975-76	0.0	0.0	0.0	0.0	3.2	0.2	0.6	5.7	2.7	0.0	0.0	0.0	12.4
1976-77	0.0	0.0	0.0	3.8	1.7	T	5.5	0.0	0.3	2.7	0.0	0.0	14.0
1977-78	0.0	0.0	0.0	0.0	0.7	T	5.6	16.5	5.0	0.0	0.9	0.0	28.7
1978-79	0.0	0.0	0.0	0.0	T	5.1	10.3	0.9	5.1	3.8	0.3	0.0	25.5
1979-80	0.0	0.0	0.0	T	T	3.4	5.4	15.1	10.2	1.5	0.0	0.0	35.6
1980-81	0.0	0.0	0.0	0.1	0.1	3.9	2.9	0.4	4.4	0.0	0.0	0.0	11.8
1981-82	0.0	0.0	0.0	0.0	T	3.6	0.6	13.6	1.4	0.0	0.0	0.0	19.2
1982-83	0.0	0.0	0.0	0.0	5.0	2.6	4.7	4.1	7.6	9.0	0.0	0.0	33.0
1983-84	0.0	0.0	0.0	0.0	2.8	6.2	6.8	0.1	11.0	0.3	0.0	0.0	27.2
1984-85	0.0	0.0	T	T	0.8	2.8	8.5	1.6	5.5	0.0	0.0	0.0	19.2
1985-86	0.0	0.0	T	0.0	9.8	1.1	0.0	2.8	T	0.0	0.0	0.0	13.7
1986-87	0.0	0.0	0.0	T	T	5.6	7.2	1.4	16.3	T	0.0	0.0	30.5
1987-88	0.0	0.0	0.0	0.0	2.5	7.2	11.8	T	3.5	0.0	0.0	0.0	25.0
1988-89	0.0	0.0	0.0	0.0	1.0	T	0.4	2.7	T	5.7	0.0	0.0	9.8
1989-90	T	0.0	0.0	0.0	T	5.0	15.7	7.7	0.5	0.4	0.0	0.0	29.3
1990-91	0.0	0.0	0.0	0.0	0.7	2.8	3.3	T	3.4	0.0	T	0.0	10.2
1991-92	0.0	0.0	0.0	4.5	1.0	T	T	0.0	0.6	0.0	T	0.0	6.1
1992-93	T	T			16.7	8.1	10.4	19.6	6.5	T	0.0	0.0	
1993-	0.0	0.0	0.0	0.2	T	2.8							
POR= 52 YRS	T	T	T	0.2	2.2	3.6	4.6	3.9	4.8	0.9	T	0.0	20.1

1993 GOODLAND, KANSAS (GLD)

Goodland is situated on an intermediate plain with few native trees. The terrain rises from east to west with only minor variations from north to south. The rate of rise is about 1,600 feet per 150 miles east of Goodland and about 2,500 feet per 150 miles west. This gradual slope in terrain makes conditions favorable for upslope fog, low clouds, and drizzle with easterly winds.

This is a typical steppe climate with wide variations in precipitation from year to year. Evaporation generally exceeds precipitation during the summer months. The number of subnormal years of precipitation nearly equals the above normal years. The mean monthly rainfall increases in the spring to a maximum in June. General storms provide the main source of precipitation during the spring months, while thunderstorms are the major factor during the summer months. Inadequate moisture received from March through June, often results in drought conditions throughout the summer months with thunderstorms providing only local relief. The frequency of thunderstorms increases to a maximum in July with a marked decrease in September. Hail is most frequent in May and June. Winds during thunderstorms have been recorded with gusts up to 80 mph.

Snow is an important factor in the production of winter wheat, and residual soil moisture often offsets the effects of subnormal spring precipitation. When snow is accompanied by strong winds it can become a dangerous enemy. As little as 1 inch of snow accompanied by strong winds can result in serious blocking of roads and highways. The heaviest snowfall is most likely to occur in March, although heavy snows have been recorded in every month from October through May. Snow may cover the ground about one third of the time from November through March.

Temperatures are typical of continental climates with January normally the coldest month and July the warmest. Winters are often modified by persistent foehn winds but polar outbreaks have been known to drop the temperature as much as 70 degrees in a 24-hour period. Low relative humidity during the summer months makes most nights comfortable even in the hottest weather.

Based on the 1951-1980 period, the average first occurrence of 32 degrees Fahrenheit in the fall is October 7 and the average last occurrence in the spring is May 4. The growing season is 156 days.

NORMALS, MEANS, AND EXTREMES FOR 1993

GOODLAND, KS (GLD)

LATITUDE: 39° 22′ 14″ N LONGTITUDE: 101° 41′ 55″ W ELEVATION (FT): GRND: 3654 BARO: 3649 TIME ZONE: MOUNTAIN (UTC+07) WBAN: 23065

	ELEMENT	POR	JAN	FEB	MAR	APR	MAY	JUN	JUL	AUG	SEP	OCT	NOV	DEC	YEAR
TEMPERATURE °F	NORMAL DAILY MAXIMUM	30	41.5	45.3	52.7	63.6	72.3	83.4	89.8	87.3	78.0	66.5	51.6	42.0	64.5
	MEAN DAILY MAXIMUM	50	40.7	45.0	51.7	63.2	72.3	83.7	89.9	87.9	79.2	67.7	51.3	43.5	64.7
	HIGHEST DAILY MAXIMUM	73	79	81	89	96	99	109	111	110	105	96	87	83	111
	YEAR OF OCCURRENCE		1951	1970	1963	1992	1962	1936	1940	1947	1939	1926	1927	1964	JUL 1940
	MEAN OF EXTREME MAXS.	46	64.4	68.6	76.7	85.6	90.3	99.4	101.9	99.9	95.9	87.4	73.7	66.8	84.2
	NORMAL DAILY MINIMUM	30	15.3	19.0	25.4	35.3	45.6	55.2	61.4	59.1	50.0	37.6	25.7	17.0	37.2
	MEAN DAILY MINIMUM	50	14.5	18.6	24.3	34.6	45.3	55.2	61.2	59.4	49.6	37.4	24.7	17.4	36.8
	LOWEST DAILY MINIMUM	73	-26	-22	-20	0	21	31	42	38	19	7	-12	-27	-27
	YEAR OF OCCURRENCE		1959	1982	1960	1936	1967	1951	1952	1964	1985	1993	1925	1989	DEC 1989
	MEAN OF EXTREME MINS.	46	-6.7	-1.3	4.5	19.6	31.7	44.0	51.6	49.8	34.6	22.5	7.8	-2.4	21.3
	NORMAL DRY BULB	30	28.4	32.2	39.1	49.5	59.0	69.3	75.6	73.2	64.0	52.1	38.7	29.5	50.9
	MEAN DRY BULB	50	27.7	31.8	38.0	48.9	58.8	69.5	75.6	73.6	64.4	52.5	38.0	30.5	50.8
	MEAN WET BULB	46	23.0	26.7	31.8	41.0	50.6	59.0	63.4	62.1	53.6	43.2	32.0	25.1	42.6
	MEAN DEW POINT	46	15.4	19.1	23.2	31.7	43.8	52.0	56.7	55.6	45.7	33.8	24.1	17.4	34.9
	NORMAL NO. DAYS WITH:														
	MAXIMUM ≥ 90°	30	0.0	0.0	0.0	0.2	1.5	8.8	18.1	14.9	5.6	0.4	0.0	0.0	49.5
	MAXIMUM ≤ 32°	30	8.1	5.8	3.1	0.2	0.0	0.0	0.0	0.0	0.0	0.1	2.8	7.4	27.5
	MINIMUM ≤ 32°	30	30.1	26.4	24.8	11.0	1.3	0.0	0.0	0.0	0.7	7.5	24.0	29.8	155.6
	MINIMUM ≤ 0°	30	4.5	1.7	0.4	0.0	0.0	0.0	0.0	0.0	0.0	0.0	0.2	2.8	9.6
H/C	NORMAL HEATING DEG. DAYS	30	1135	918	803	465	211	33	0	11	108	400	789	1101	5974
	NORMAL COOLING DEG. DAYS	30	0	0	0	0	25	162	329	265	78	0	0	0	859
RH	NORMAL (PERCENT)	30	66	66	63	59	64	60	58	60	60	57	65	66	62
	HOUR 05 LST	30	75	78	79	78	83	82	81	82	80	74	77	75	79
	HOUR 11 LST	30	56	56	52	46	51	47	44	46	47	43	52	55	49
	HOUR 17 LST	30	59	53	47	41	46	41	38	40	41	43	57	60	47
	HOUR 23 LST	30	73	74	71	69	75	72	69	70	70	67	72	72	71
S	PERCENT POSSIBLE SUNSHINE														
W/O	MEAN NO. DAYS WITH:														
	HEAVY FOG(VISBY ≤ 1/4 MI)	50	2.3	3.0	3.7	2.3	2.7	1.4	1.3	2.1	2.2	2.0	2.7	2.0	27.7
	THUNDERSTORMS	50	0.0	0.3	0.8	2.5	7.6	10.4	11.6	8.9	4.0	1.3	0.2	0.1	47.7
CLOUDINESS	MEAN:														
	SUNRISE-SUNSET (TENTHS)	1		7.0	7.0	8.0			5.0	6.0	5.0	5.0	5.0	4.0	
	MIDNIGHT-MIDNIGHT (TENTHS)	1		7.0	7.0	8.0			5.0	6.0	5.0	5.0	5.0	4.0	
	MEAN NO. DAYS WITH:														
	CLEAR	1	4.0	10.0	9.0	6.0	7.0	10.0	11.0	14.0	18.0	13.0	10.0	19.0	131.0
	PARTLY CLOUDY	1	4.0	2.0	6.0	2.0	2.0	4.0	3.0	6.0	1.0	4.0	4.0	3.0	41.0
	CLOUDY	1	11.0	12.0	9.0	12.0	4.0	3.0	5.0	8.0	6.0	8.0	6.0	8.0	92.0
PR	MEAN STATION PRESSURE(IN.)	21	26.23	26.22	26.14	26.15	26.16	26.20	26.26	26.27	26.27	26.27	26.22	26.23	26.22
	MEAN SEA-LEVEL PRES. (IN.)	44	30.10	30.06	29.96	29.90	29.88	29.86	29.92	29.93	29.97	30.02	30.05	30.08	29.98
WINDS	MEAN SPEED (MPH)	30	9.2	9.2	10.0	10.9	10.3	10.1	9.6	9.3	9.7	9.0	9.2	9.8	9.7
	PREVAIL.DIR(TENS OF DEGS.)	30	33	33	33	33	15	18	18	18	18	18	33	33	18
	MAXIMUM 2-MINUTE:														
	SPEED (MPH)	1	33	33	40	39	39	38	64	38	34	39	40	49	64
	DIR. (TENS OF DEGS.)	1	25	36	35	34	16	19	30	33	01	31	31	30	30
	YEAR OF OCCURRENCE		1993	1993	1993	1993	1993	1993	1993	1993	1993	1993	1993	1993	JUL 1993
	MAXIMUM 5-SECOND:														
	SPEED (MPH)	1	38	43	46	46	47	44	71	41	41	44	48	57	71
	DIR. (TENS OF DEGS.)	1	32	36	35	33	17	19	30	08	36	31	31	30	30
	YEAR OF OCCURRENCE		1993	1993	1993	1993	1993	1993	1993	1993	1993	1993	1993	1993	JUL 1993
PRECIPITATION	NORMAL TOTAL (IN.)	30	0.41	0.39	1.18	1.30	3.49	3.19	2.87	1.80	1.57	0.90	0.69	0.41	18.20
	MAXIMUM TOTAL (IN.)	73	1.59	2.07	3.60	5.69	8.21	9.46	10.10	9.29	5.39	4.94	2.63	2.90	10.10
	YEAR OF OCCURRENCE		1988	1939	1981	1944	1981	1982	1985	1993	1973	1930	1946	1924	JUL 1985
	MINIMUM TOTAL (IN.)	73	0.00	T	0.03	T	0.31	0.10	0.30	0.13	0.01	T	T	T	0.0
	YEAR OF OCCURRENCE		1933	1970	1929	1963	1927	1976	1924	1964	1992	1988	1959	1981	JAN 1933
	MAXIMUM IN 24 HOURS (IN.)	73	1.27	1.79	1.67	3.13	3.49	4.15	3.74	3.45	2.61	2.85	1.47	1.54	4.15
	YEAR OF OCCURRENCE		1988	1939	1981	1981	1972	1989	1985	1991	1940	1930	1975	1924	JUN 1989
	NORMAL NO. DAYS WITH:														
	PRECIPITATION ≥ 0.01														
	PRECIPITATION ≥ 1.00														
SNOWFALL	NORMAL TOTAL (IN.)														
	MAXIMUM TOTAL (IN.)	53	19.4	25.4	27.4	22.0	6.5	3.5	T	T	5.6	17.6	23.3	17.3	27.4
	YEAR OF OCCURRENCE		1988	1984	1940	1988	1990	1989	1992	1990	1985	1979	1983	1979	MAR 1940
	MAXIMUM IN 24 HOURS (IN.)	53	14.0	11.2	15.2	13.0	6.5	3.5	T	T	5.6	12.0	15.1	11.7	15.2
	YEAR OF OCCURRENCE		1988	1984	1940	1988	1990	1989	1992	1990	1985	1979	1983	1979	MAR 1940
	MAXIMUM SNOW DEPTH (IN.)	1	6	8	4	T	0	T	0	0	0	1	5	T	8
	YEAR OF OCCURRENCE		1993	1993	1993	1993		1993				1993	1993	1993	FEB 1993
	NORMAL NO. DAYS WITH:														
	SNOWFALL ≥ 1.0														

PRECIPITATION (inches) 1993 GOODLAND, KANSAS (GLD)

YEAR	JAN	FEB	MAR	APR	MAY	JUN	JUL	AUG	SEP	OCT	NOV	DEC	ANNUAL
1964	T	0.75	0.56	1.36	2.68	1.50	1.97	0.13	1.56	0.02	0.26	0.02	10.81
1965	0.30	0.51	0.48	0.10	3.43	3.46	1.72	2.94	4.31	3.06	0.01	0.55	20.87
1966	0.50	0.23	0.52	0.56	0.55	2.12	2.26	2.15	1.00	1.23	0.04	0.24	11.40
1967	0.16	0.02	0.46	0.74	2.66	3.23	1.20	1.76	4.13	0.52	0.41	0.43	15.72
1968	0.06	0.12	0.20	0.46	1.89	3.48	2.85	2.09	0.46	0.44	0.34	1.31	13.70
1969	0.11	0.41	0.97	1.86	3.83	2.22	2.46	1.78	0.36	4.10	0.31	0.26	18.67
1970	0.03	T	0.91	1.21	2.27	1.68	1.92	2.37	1.11	1.10	0.57	0.03	13.20
1971	0.53	1.36	0.57	1.93	4.24	2.35	1.62	0.65	1.38	1.32	1.31	0.21	17.47
1972	0.32	0.06	0.15	0.60	6.04	4.76	2.64	2.46	0.73	0.82	1.56	0.94	21.08
1973	0.73	0.02	2.90	1.90	2.96	2.03	1.82	1.16	5.39	0.47	0.66	0.90	20.94
1974	0.17	0.42	0.99	1.52	1.10	3.95	1.15	1.63	0.02	0.96	0.86	0.37	13.14
1975	0.20	0.14	0.64	1.03	4.75	5.25	2.43	0.37	0.19	0.02	1.92	0.06	17.00
1976	0.48	0.21	0.31	0.87	1.17	0.10	2.61	0.60	2.03	0.43	0.40	0.01	9.22
1977	0.38	0.05	1.79	1.94	6.11	1.30	1.28	5.45	1.02	0.15	0.45	0.14	20.06
1978	0.38	0.85	0.29	1.33	3.82	2.25	1.71	1.85	0.12	1.29	0.68	0.43	15.00
1979	0.88	0.08	3.11	1.09	4.48	5.08	4.53	3.17	0.21	2.00	0.78	1.15	26.56
1980	0.61	0.49	2.75	2.67	2.80	1.92	7.25	3.38	2.24	0.19	0.12	T	24.42
1981	1.04	0.03	3.60	3.86	8.21	0.29	1.99	0.77	0.33	0.13	1.91	T	22.16
1982	0.21	0.61	0.39	0.87	5.03	9.46	3.26	0.81	3.42	0.43	0.41	1.58	26.48
1983	0.29	0.87	2.20	1.76	2.43	2.39	3.68	1.39	0.43	0.15	1.78	0.26	17.63
1984	0.74	1.54	1.82	3.35	1.51	3.43	3.24	1.15	0.22	3.25	0.28	0.61	21.14
1985	0.43	0.25	0.33	1.03	4.56	0.71	10.10	0.70	1.74	1.16	0.51	0.44	21.96
1986	0.00	0.44	0.76	1.27	3.37	3.78	1.38	0.84	1.30	1.61	0.50	0.11	15.36
1987	0.34	1.01	2.50	0.48	3.54	5.39	2.96	0.50	1.52	0.03	1.01	0.46	19.74
1988	1.59	0.24	0.41	1.99	4.82	1.85	3.44	3.47	2.19	T	0.39	0.47	20.86
1989	0.23	0.14	0.26	0.89	2.76	8.18	1.09	2.84	1.06	0.32	0.05	0.22	18.04
1990	0.70	0.44	1.63	0.76	4.14	1.74	4.06	1.97	1.49	0.86	1.47	0.14	19.40
1991	0.24	0.11	1.08	0.92	4.03	5.36	3.29	4.07	0.76	0.88	1.32	1.18	23.24
1992	0.20	0.50	1.91	0.01	1.38	6.70	4.03	5.59	0.01	0.90	0.52	0.18	21.93
1993	0.31	0.73	0.81	0.42	2.91	1.94	7.96	9.29	0.30	2.33	0.93	0.01	27.94
POR= 50 YRS	0.42	0.42	1.11	1.31	3.13	3.08	2.91	2.19	1.25	0.88	0.69	0.39	17.76

AVERAGE TEMPERATURE (˚F) 1993 GOODLAND, KANSAS (GLD)

YEAR	JAN	FEB	MAR	APR	MAY	JUN	JUL	AUG	SEP	OCT	NOV	DEC	ANNUAL
1964	32.3	28.0	34.7	47.5	60.6	69.0	79.8	72.1	65.0	53.1	39.1	30.4	51.0
1965	32.6	27.8	28.2	53.2	62.1	67.9	76.0	71.5	55.7	55.1	43.2	32.8	50.5
1966	21.7	26.9	42.8	46.9	62.5	70.8	79.6	71.8	66.0	50.7	40.3	27.4	50.6
1967	33.5	33.1	42.3	50.3	54.4	66.0	72.1	71.2	63.5	53.5	39.1	27.7	50.6
1968	31.2	33.5	43.4	48.7	55.0	72.3	75.2	71.4	65.0	55.0	38.0	24.4	51.1
1969	29.2	33.2	31.2	53.0	60.6	64.7	77.3	75.1	68.1	45.2	41.4	33.0	51.0
1970	29.6	37.4	33.4	47.0	62.3	69.5	76.1	76.9	62.4	48.9	39.9	33.0	51.4
1971	29.8	30.6	39.8	49.4	56.3	73.2	73.1	74.2	61.8	52.3	39.6	32.6	51.1
1972	28.0	35.1	44.9	49.9	59.0	70.5	72.0	71.3	63.8	50.3	32.1	21.2	49.8
1973	26.8	33.3	39.9	44.7	56.6	69.5	73.7	75.7	59.7	53.0	38.7	29.9	50.1
1974	23.5	35.2	41.7	49.8	61.1	68.8	78.2	68.2	59.9	54.3	37.8	29.5	50.7
1975	30.3	29.3	34.8	47.5	57.8	66.8	74.4	75.7	64.0	55.8	37.5	35.4	50.8
1976	32.6	39.9	38.5	51.0	57.1	69.2	76.2	73.6	62.9	45.9	35.4	33.0	51.3
1977	22.9	37.1	38.8	52.1	61.8	71.7	76.3	71.0	67.0	52.6	38.9	31.4	51.8
1978	18.7	19.7	39.6	49.7	55.1	69.4	76.5	71.4	67.3	52.9	35.7	25.1	48.4
1979	17.5	31.4	41.9	50.7	57.0	68.7	76.0	71.6	67.7	54.2	34.5	35.5	50.6
1980	24.4	32.4	35.1	47.1	56.9	71.9	79.2	75.2	66.6	52.0	41.1	36.4	51.5
1981	33.0	32.1	40.5	55.5	55.5	71.6	75.2	70.8	65.8	51.9	43.2	33.8	52.4
1982	25.7	28.4	40.3	47.0	59.0	64.6	74.5	74.7	64.5	52.4	37.5	31.8	50.0
1983	31.0	36.5	38.8	44.1	54.8	66.4	78.9	82.6	68.4	53.0	37.8	14.6	50.6
1984	25.3	34.1	36.0	42.6	58.5	67.8	74.3	75.0	61.8	48.6	41.0	30.0	49.6
1985	25.1	28.1	42.0	53.3	62.2	69.1	76.7	73.3	61.0	50.6	30.8	26.5	49.9
1986	38.9	33.9	46.2	52.3	59.5	71.7	77.2	73.6	66.2	52.0	39.0	32.9	53.6
1987	31.6	37.5	37.5	51.8	62.6	71.3	76.1	72.5	64.5	52.4	39.7	30.2	52.3
1988	23.8	31.5	39.0	48.5	60.9	75.1	76.1	75.9	64.7	52.7	40.9	34.4	52.0
1989	34.0	22.7	40.8	51.4	60.3	66.1	75.6	72.7	63.4	53.5	42.0	25.9	50.7
1990	32.7	31.7	41.5	50.3	56.9	74.4	74.1	74.2	68.6	53.8	42.7	25.4	52.2
1991	26.1	40.0	41.6	49.7	62.3	72.9	75.7	73.2	65.8	52.7	35.6	33.8	52.4
1992	31.3	40.0	43.8	52.6	60.3	66.7	71.5	68.7	64.2	52.1	32.7	22.8	50.6
1993	22.8	22.5	39.6	46.6	58.3	67.5	72.6	70.4	59.6	49.9	33.5	32.9	48.0
POR= 50 YRS	27.7	31.8	38.0	48.9	58.8	69.5	75.5	73.6	64.4	52.5	38.0	30.0	50.7

REFERENCE NOTES:

H/C INDICATES HEATING AND COOLING DEGREE DAYS.
RH INDICATES RELATIVE HUMIDITY
W/O INDICATES WEATHER AND OBSTRUCTIONS
S INDICATES SUNSHINE.
PR INDICATES PRESSURE.
CLOUDINESS IS THE SUM OF THE CEILOMETER AND SATELLITE DATA NOT TO EXCEED TEN TENTHS.
MAXIMUM SNOW DEPTH IS THE DEPTH MEASURED AT 12 UTC.

GENERAL:
T INDICATES TRACE PRECIPITATION, AN AMOUNT GREATER THAN ZERO BUT LESS THAN THE LOWEST REPORTABLE VALUE.
+ INDICATES THE VALUE ALSO OCCURS ON EARLIER DATES.
BLANK ENTRIES DENOTE MISSING OR UNREPORTED DATA.
NORMALS ARE 30-YEAR AVERAGES (1961 - 1990).
ASOS INDICATES AUTOMATED SURFACE OBSERVING SYSTEM.
PM INDICATES THE LAST DAY OF THE PREVIOUS MONTH.
POR (PERIOD OF RECORD) INDICATES THE NUMBER OF YEARS USED TO COMPUTE THE MEAN. INDIVIDUAL MONTHS WITHIN THE POR MAY BE MISSING.
WHEN THE POR FOR A NORMAL IS LESS THAN 30 YEARS, THE NORMAL IS PROVISIONAL AND IS BASED ON THE NUMBER OF YEARS INDICATED.
0.* OR * INDICATES THE VALUE OR MEAN DAYS WITH IS BETWEEN 0.00 AND 0.05.
CLOUDINESS FOR ASOS STATIONS DIFFERS FROM THE NON-ASOS OBSERVATION TAKEN BY A HUMAN OBSERVER. ASOS STATION CLOUDINESS IS BASED ON TIME-AVERAGED CEILOMETER DATA FOR CLOUDS AT OR BELOW 12,000 FEET AND ON SATELLITE DATA FOR CLOUDS ABOVE 12,000 FEET.
THE NUMBER OF DAYS WITH CLEAR, PARTLY CLOUDY, AND CLOUDY CONDITIONS FOR ASOS STATIONS IS THE SUM OF THE CEILOMETER AND SATELLITE DATA FOR THE SUNRISE TO SUNSET PERIOD.

GENERAL CONTINUED:
CLEAR INDICATES 0 - 3 TENTHS, PARTLY CLOUDY INDICATES 4 - 7 TENTHS, AND CLOUDY INDICATES 8 OR MORE TENTHS. WHEN AT LEAST ONE OF THE ELEMENTS (CEILOMETER OR SATELLITE) IS MISSING, THE DAILY CLOUDINESS IS NOT COMPUTED.
WIND DIRECTION IS RECORDED IN TENS OF DEGREES (2 DIGITS) CLOCKWISE FROM TRUE NORTH. "00" INDICATES CALM. "36" INDICATES TRUE NORTH.
RESULTANT WIND IS THE VECTOR AVERAGE OF THE SPEED AND DIRECTION.
AVERAGE TEMPERATURE IS THE SUM OF THE MEAN DAILY MAXIMUM AND MINIMUM TEMPERATURE DIVIDED BY 2.
SNOWFALL DATA COMPRISE ALL FORMS OF FROZEN PRECIPITATION, INCLUDING HAIL.
A HEATING (COOLING) DEGREE DAY IS THE DIFFERENCE BETWEEN THE AVERAGE DAILY TEMPERATURE AND 65˚ F.
DRY BULB IS THE TEMPERATURE OF THE AMBIENT AIR.
DEW POINT IS THE TEMPERATURE TO WHICH THE AIR MUST BE COOLED TO ACHIEVE 100 PERCENT RELATIVE HUMIDITY.
WET BULB IS THE TEMPERATURE THE AIR WOULD HAVE IF THE MOISTURE CONTENT WAS INCREASED TO 100 PERCENT RELATIVE HUMIDITY.

HEATING DEGREE DAYS (base 65˚F) 1993 GOODLAND, KANSAS (GLD)

YEAR	JUL	AUG	SEP	OCT	NOV	DEC	JAN	FEB	MAR	APR	MAY	JUN	TOTAL
1964-65	0	29	121	369	770	1065	1001	1035	1134	349	133	20	6026
1965-66	0	11	300	303	646	992	1337	1060	682	535	149	31	6046
1966-67	0	11	66	436	731	1160	970	889	699	433	359	47	5801
1967-68	10	12	87	367	771	1148	1043	904	662	484	316	20	5824
1968-69	5	22	82	314	803	1249	1106	885	1040	353	171	104	6134
1969-70	0	0	25	608	701	985	1092	768	971	532	137	61	5880
1970-71	5	1	174	497	747	986	1084	957	777	463	269	5	5965
1971-72	11	0	193	388	756	998	1138	862	615	445	222	10	5638
1972-73	28	24	117	454	980	1355	1177	881	770	602	265	28	6681
1973-74	4	0	188	365	780	1082	1281	825	715	454	148	51	5893
1974-75	0	24	198	329	811	1099	1069	997	929	524	226	43	6249
1975-76	0	4	129	290	823	909	998	724	815	414	259	14	5379
1976-77	0	3	121	587	884	987	1296	776	803	387	100	0	5944
1977-78	2	10	40	377	776	1033	1431	1263	782	456	315	50	6535
1978-79	0	18	63	375	876	1229	1470	936	710	424	269	58	6428
1979-80	0	21	49	337	909	907	1251	936	921	534	254	16	6135
1980-81	0	2	61	402	714	878	985	915	751	289	297	18	5312
1981-82	1	9	59	405	649	958	1216	1019	761	534	197	73	5881
1982-83	1	12	106	383	820	1022	1049	791	805	623	326	72	6010
1983-84	0	0	76	366	807	1556	1226	892	893	664	208	39	6727
1984-85	0	0	178	503	712	1079	1232	1026	704	351	129	48	5962
1985-86	0	4	237	439	1020	1188	804	867	577	375	185	9	5705
1986-87	0	2	58	399	773	987	1026	762	845	410	111	10	5383
1987-88	14	24	76	396	749	1072	1270	966	797	487	180	8	6039
1988-89	3	4	102	375	713	942	953	1180	745	435	189	70	5711
1989-90	0	1	152	361	685	1203	994	928	723	444	266	9	5766
1990-91	13	1	66	347	662	1224	1203	694	718	453	138	0	5519
1991-92	5	8	87	396	878	964	1039	716	650	379	188	38	5348
1992-93	8	31	88	397	963	1301	1301	1184	779	541	218	57	6868
1993-	0	29	188	474	938	987							
POR= 50 YRS	2	8	113	385	801	1079	1151	930	828	479	217	40	6033

COOLING DEGREE DAYS (base 65˚F) 1993 GOODLAND, KANSAS (GLD)

YEAR	JAN	FEB	MAR	APR	MAY	JUN	JUL	AUG	SEP	OCT	NOV	DEC	ANNUAL
1969	0	0	0	0	43	102	384	318	127	1	0	0	975
1970	0	0	0	0	59	202	355	376	103	6	0	0	1101
1971	0	0	0	0	5	254	270	291	106	2	0	0	928
1972	0	0	0	1	41	183	251	224	85	4	0	0	789
1973	0	0	0	0	13	171	281	338	34	0	0	0	837
1974	0	0	0	5	37	173	416	130	50	3	0	0	814
1975	0	0	0	4	10	106	298	344	104	15	0	0	881
1976	0	0	0	0	22	149	357	275	64	1	0	0	868
1977	0	0	0	7	9	206	360	199	106	0	0	0	887
1978	0	0	0	3	16	192	365	223	137	8	0	0	944
1979	0	0	0	3	29	176	346	235	139	11	0	0	939
1980	0	0	0	4	11	230	447	324	117	6	0	0	1139
1981	0	0	0	11	8	223	323	200	90	3	0	0	858
1982	0	0	0	1	17	71	301	320	98	0	0	0	808
1983	0	0	0	0	17	122	438	551	184	3	0	0	1315
1984	0	0	0	0	12	133	292	316	87	0	0	0	840
1985	0	0	0	6	47	178	372	267	127	1	0	0	998
1986	0	0	0	3	21	213	387	271	104	1	0	0	1000
1987	0	0	0	18	44	205	364	264	69	9	0	0	973
1988	0	0	0	0	59	315	355	350	100	0	0	0	1179
1989	0	0	0	30	49	109	334	248	107	11	0	0	888
1990	0	0	0	9	25	296	305	292	180	5	0	0	1112
1991	0	0	0	3	63	244	342	268	117	23	0	0	1060
1992	0	0	0	10	48	95	216	151	70	5	0	0	595
1993	0	0	0	0	17	137	243	203	32	13	0	0	645
POR= 25 YRS	0	0	0	3	28	179	335	278	101	4	0	0	928

SNOWFALL (inches) 1993 GOODLAND, KANSAS (GLD)

YEAR	JUL	AUG	SEP	OCT	NOV	DEC	JAN	FEB	MAR	APR	MAY	JUN	TOTAL
1964-65	0.0	0.0	0.0	0.0	2.8	0.8	4.6	10.1	6.4	0.0	0.0	0.0	24.7
1965-66	0.0	0.0	0.0	0.0	T	5.4	10.8	2.4	7.0	0.9	0.8	0.0	27.3
1966-67	0.0	0.0	T	8.4	0.3	7.0	4.3	0.4	4.0	0.4	0.3	0.0	25.1
1967-68	0.0	0.0	0.0	T	3.7	8.4	1.4	2.1	3.0	0.7	0.4	0.0	19.7
1968-69	0.0	0.0	0.0	0.0	1.6	16.4	1.1	4.9	13.3	0.0	0.0	0.0	37.3
1969-70	0.0	0.0	0.0	14.7	2.5	6.1	0.7	T	11.5	6.8	0.0	0.0	42.3
1970-71	0.0	0.0	0.0	5.7	3.4	T	8.3	11.4	8.6	T	0.0	0.0	37.4
1971-72	0.0	0.0	T	1.3	3.6	4.1	5.5	1.7	0.8	T	0.0	0.0	17.0
1972-73	0.0	0.0	0.0	6.3	11.2	16.3	11.7	0.5	15.0	10.1	1.3	0.0	72.4
1973-74	0.0	0.0	0.0	1.7	6.9	12.9	3.2	6.4	8.5	7.2	0.0	0.0	46.8
1974-75	0.0	0.0	0.0	0.0	5.9	5.1	3.6	3.3	3.7	12.0	0.0	0.0	33.6
1975-76	0.0	0.0	0.0	T	15.7	2.2	7.7	3.5	5.8	T	0.0	0.0	34.9
1976-77	0.0	0.0	0.0	5.8	5.2	0.1	5.0	0.3	14.1	5.4	0.0	0.0	35.9
1977-78	0.0	0.0	0.0	T	2.2	0.6	6.2	11.8	4.3	T	2.6	0.0	27.7
1978-79	0.0	0.0	0.0	1.3	3.6	9.1	18.1	1.3	17.1	6.5	T	0.0	57.0
1979-80	0.0	0.0	0.0	17.6	7.9	17.3	13.2	8.7	27.4	9.9	0.0	0.0	102.0
1980-81	0.0	0.0	0.0	0.3	2.6	T	16.6	0.9	20.6	0.8	0.0	0.0	41.8
1981-82	0.0	0.0	0.0	0.0	3.4	0.1	5.6	10.5	2.2	0.8	1.8	0.0	24.4
1982-83	0.0	0.0	0.0	0.2	2.4	14.4	4.3	6.8	13.8	6.3	0.0	0.0	48.2
1983-84	0.0	0.0	T	0.0	23.3	7.2	12.1	17.2	20.0	22.0	T	0.0	101.8
1984-85	0.0	0.0	0.7	6.4	1.1	6.7	6.7	7.5	3.5	T	1.0	0.0	33.6
1985-86	0.0	0.0	5.6	0.0	5.6	6.8	0.0	7.5	5.8	0.7	0.0	0.0	32.0
1986-87	0.0	0.0	0.0	5.5	4.6	1.5	6.1	7.5	21.5	4.5	0.0	0.0	51.2
1987-88	0.0	0.0	0.0	0.0	7.2	9.0	19.4	3.6	7.3	13.4	2.2	0.0	62.1
1988-89	0.0	0.0	0.0	0.0	3.1	2.1	1.0	2.2	2.8	12.5	T	3.5	27.2
1989-90	0.0	0.0	0.0	1.0	0.3	7.0	10.3	6.3	5.8	1.4	6.5	0.0	38.6
1990-91	0.0	0.0	0.0	6.1	13.4	2.1	4.2	0.1	11.4	T	1.4	T	38.7
1991-92	T	0.0	0.0	7.5	3.3	8.3	1.9	0.5	1.3	T	0.0	T	22.8
1992-93	T	0.0			10.5	4.3	11.3	12.5	5.8	0.3		T	
1993-	T	0.0	0.0	1.3	6.2	0.7							
POR= 50 YRS	T	0.0	0.2	2.0	4.9	5.4	6.5	5.2	9.3	4.2	0.5	0.1	38.3

1993 TOPEKA, KANSAS (TOP)

Topeka, is located near the geographical center of the United States, and the middle of the temperate zone. The city straddles the Kansas River about 60 miles above its junction with the Missouri River. The Kansas River flows in an easterly direction through northeastern Kansas. Near Topeka, the river valley ranges from 2 to 4 miles wide, and is bordered on both sides by rolling prairie uplands of some 200 to 300 feet. The city is built on both banks of the Kansas River and along two tributaries, Soldier Creek in north Topeka and Shunganunga Creek in the south and east part of town. Flooding is always a threat following periods of heavy rains but protective construction has reduced the problem.

Seventy percent of the annual precipitation normally falls during the six crop-growing months, April through September. The rains of this period are usually of short duration, predominantly of the thunderstorm type. They occur more frequently during the nighttime and early morning hours than at other times of the day. Excessive precipitation rates may occur with warm-season thunderstorms. Rainfall accumulations over 8 inches in 24 hours have occurred in Topeka. Tornadoes have occurred in the area on several occasions and caused severe damage and numerous injuries.

Individual summers show wide departures from average conditions. Hottest summers may produce temperatures of 100 degrees or higher on more than 50 days. On the other hand, 25 percent of the summers pass with two or fewer 100 degree days. Similarly, precipitation has shown a wide range for June, July, and August, varying from under 3 inches to more than 27 inches during the 3 months. Summers are hot with low relative humidity and persistent southerly winds. Oppressively warm periods with high relative humidity are usually of short duration.

Winter temperatures average about 45 degrees cooler than summer. Cold spells are seldom prolonged. Only on rare occasions do daytime temperatures fail to rise above freezing. Winter precipitation is often in the form of snow, sleet, or glaze, but storms of such severity to prevent normal movement of traffic or to interfere with scheduled activity are not common.

In the transitional spring and fall seasons, the numerous days of fair weather are interspersed with short intervals of stormy weather. Strong, blustery winds are quite common in late winter and spring. Autumn is characteristically a season of warm days, cool nights, and infrequent precipitation, with cold air invasions gradually increasing in intensity as the season progresses.

Nearly all crops of the temperate zone can be produced in the vicinity of Topeka. Wheat and other small grains, clover, soybeans, fruit, and berries do well, and the area supports an extensive dairy industry.

Based on the 1951-1980 period, the average first occurrence of 32 degrees Fahrenheit in the fall is October 14 and the average last occurrence in the spring is April 21.

NORMALS, MEANS, AND EXTREMES FOR 1993

TOPEKA, KS (TOP)

LATITUDE: 39° 04' 08" N LONGTITUDE: 95° 37' 20" W ELEVATION (FT): GRND: 877 BARO: 884 TIME ZONE: CENTRAL (UTC+06) WBAN: 13996

	ELEMENT	POR	JAN	FEB	MAR	APR	MAY	JUN	JUL	AUG	SEP	OCT	NOV	DEC	YEAR
TEMPERATURE °F	NORMAL DAILY MAXIMUM	30	37.0	42.6	55.0	66.9	75.8	84.2	89.3	87.5	79.7	69.0	54.0	40.5	65.1
	MEAN DAILY MAXIMUM	47	37.4	42.1	53.8	66.5	75.9	84.6	89.3	88.2	82.3	71.3	54.8	42.7	65.7
	HIGHEST DAILY MAXIMUM	47	73	84	89	95	97	107	110	110	109	96	85	73	110
	YEAR OF OCCURRENCE		1967	1972	1986	1987	1975	1953	1980	1984	1947	1963	1980	1984	AUG 1984
	MEAN OF EXTREME MAXS.	46	61.1	65.0	78.3	86.2	89.7	95.7	99.2	99.3	94.6	87.1	74.4	63.5	82.9
	NORMAL DAILY MINIMUM	30	16.3	21.7	32.1	42.8	53.3	62.9	67.6	64.9	55.7	43.6	32.0	21.0	42.8
	MEAN DAILY MINIMUM	47	16.9	21.5	31.2	42.7	53.3	63.1	67.4	65.3	57.2	45.2	32.4	22.3	43.2
	LOWEST DAILY MINIMUM	47	-20	-23	-7	10	26	43	43	41	29	19	2	-26	-26
	YEAR OF OCCURRENCE		1974	1979	1978	1975	1963	1993	1972	1988	1984	1993	1976	1989	DEC 1989
	MEAN OF EXTREME MINS.	46	-3.8	0.7	12.4	26.1	36.9	49.1	55.1	51.8	38.0	27.0	14.6	1.3	25.8
	NORMAL DRY BULB	30	26.7	32.2	43.6	54.9	64.6	73.6	78.5	76.2	67.8	56.3	43.0	30.8	54.0
	MEAN DRY BULB	47	27.2	31.8	42.5	54.6	64.6	73.8	78.3	76.8	69.8	58.3	43.6	32.5	54.5
	MEAN WET BULB	44	24.2	28.9	37.2	48.0	57.9	66.8	70.4	68.7	60.9	50.2	38.2	28.5	48.3
	MEAN DEW POINT	44	17.9	22.6	30.3	41.3	53.0	62.9	66.8	64.8	56.4	44.4	32.2	22.6	42.9
	NORMAL NO. DAYS WITH:														
	MAXIMUM ≥ 90°	30	0.0	0.0	0.0	0.4	1.1	6.7	15.5	12.8	4.7	0.6	0.0	0.0	41.8
	MAXIMUM ≤ 32°	30	10.9	6.8	1.4	*	0.0	0.0	0.0	0.0	0.0	0.0	1.1	7.7	27.9
	MINIMUM ≤ 32°	30	29.0	23.1	16.7	4.3	0.3	0.0	0.0	0.0	0.2	4.0	16.1	27.1	120.8
	MINIMUM ≤ 0°	30	4.1	2.0	0.1	0.0	0.0	0.0	0.0	0.0	0.0	0.0	0.0	1.9	8.1
H/C	NORMAL HEATING DEG. DAYS	30	1187	918	663	312	121	5	0	0	56	283	660	1060	5265
	NORMAL COOLING DEG. DAYS	30	0	0	0	9	109	263	419	351	140	13	0	0	1304
RH	NORMAL (PERCENT)	30	70	69	66	64	68	71	70	71	72	68	71	72	69
	HOUR 00 LST	30	75	75	73	72	78	81	78	80	82	77	78	77	77
	HOUR 06 LST	30	77	78	79	80	84	87	86	87	88	83	81	80	82
	HOUR 12 LST	30	63	62	57	54	57	60	59	58	58	54	60	65	59
	HOUR 18 LST	30	64	60	53	51	55	58	56	57	60	57	64	68	58
S	PERCENT POSSIBLE SUNSHINE	44	57	56	57	58	61	66	71	71	67	64	55	52	61
W/O	MEAN NO. DAYS WITH:														
	HEAVY FOG(VISBY ≤ 1/4 MI)	47	2.1	1.8	1.0	0.9	0.9	0.6	0.5	1.1	1.0	1.5	1.1	2.0	14.5
	THUNDERSTORMS	47	0.4	0.7	2.6	5.5	9.1	10.1	8.9	8.0	6.1	3.3	1.2	0.4	56.3
CLOUDINESS	MEAN:														
	SUNRISE-SUNSET (TENTHS)	1		8.0	8.0	9.0	8.0		8.0	4.0	5.0	6.0	6.0	6.0	
	MIDNIGHT-MIDNIGHT (TENTHS)	1		8.0	8.0	9.0	8.0		8.0	4.0	5.0	7.0	5.0	6.0	
	MEAN NO. DAYS WITH:														
	CLEAR	1	5.0	11.0	6.0	6.0	9.0	8.0	5.0	15.0	14.0	10.0	9.0	14.0	112.0
	PARTLY CLOUDY	1	6.0	2.0	9.0	4.0	6.0	5.0	5.0	6.0	1.0	1.0	2.0	6.0	53.0
	CLOUDY	1	9.0	11.0	10.0	11.0	8.0	5.0	9.0	6.0	9.0	10.0	10.0	10.0	108.0
PR	MEAN STATION PRESSURE(IN.)	21	29.19	29.15	29.02	29.00	28.99	29.00	29.04	29.06	29.09	29.11	29.10	29.17	29.08
	MEAN SEA-LEVEL PRES. (IN.)	44	30.16	30.12	30.01	29.95	29.92	29.91	29.96	29.98	30.01	30.06	30.08	30.12	30.02
WINDS	MEAN SPEED (MPH)	30	9.7	9.9	10.5	10.4	9.5	8.9	8.6	8.3	8.9	8.5	9.4	9.9	9.4
	PREVAIL.DIR(TENS OF DEGS.)	30	32	36	36	18	18	18	18	18	18	18	18	30	18
	MAXIMUM 2-MINUTE:														
	SPEED (MPH)	1	25	24	31	37	32	37	44	31	34	26	31	33	44
	DIR. (TENS OF DEGS.)	1	01	31	14	26	16	36	34	34	16	25	31	32	34
	YEAR OF OCCURRENCE		1993	1993	1993	1993	1993	1993	1993	1993	1993	1993	1993	1993	JUL 1993
	MAXIMUM 5-SECOND:														
	SPEED (MPH)	1	32	30	36	46	39	45	52	36	40	31	36	41	52
	DIR. (TENS OF DEGS.)	1	02	09	14	25	16	36	35	34	16	34	31	30	35
	YEAR OF OCCURRENCE		1993	1993	1993	1993	1993	1993	1993	1993	1993	1993	1993	1993	JUL 1993
PRECIPITATION	NORMAL TOTAL (IN.)	30	0.95	1.04	2.46	3.08	4.45	5.54	3.59	3.89	3.81	3.06	1.93	1.43	35.23
	MAXIMUM TOTAL (IN.)	47	5.24	3.49	8.44	8.12	9.39	15.20	12.02	11.18	12.71	7.24	6.27	4.30	15.20
	YEAR OF OCCURRENCE		1949	1971	1973	1967	1982	1967	1950	1977	1973	1980	1964	1973	JUN 1967
	MINIMUM TOTAL (IN.)	47	T	0.02	0.10	0.62	0.41	0.56	0.59	0.26	0.66	0.04	T	0.05	T
	YEAR OF OCCURRENCE		1986	1991	1966	1989	1966	1980	1983	1971	1952	1 952	1989	1979	NOV 1989
	MAXIMUM IN 24 HOURS (IN.)	47	1.55	2.33	3.76	3.59	3.62	5.52	4.19	4.48	4.80	4.10	4.66	2.65	5.52
	YEAR OF OCCURRENCE		1988	1971	1987	1967	1978	1967	1951	1962	1989	1985	1964	1980	JUN 1967
	NORMAL NO. DAYS WITH:														
	PRECIPITATION ≥ 0.01	30	5.9	5.9	8.7	9.8	11.0	10.3	7.9	8.3	8.3	6.8	6.5	6.6	96.0
	PRECIPITATION ≥ 1.00	30	0.1	0.1	0.5	0.7	1.3	1.8	1.2	1.3	1.2	1.1	0.3	0.2	9.8
SNOWFALL	NORMAL TOTAL (IN.)	30	5.8	4.7	2.8	0.6	0.0	0.0	0.0	0.0	0.0	0.*	1.2	5.6	20.7
	MAXIMUM TOTAL (IN.)	47	23.0	22.4	22.1	6.8	T	T	T	0.0	0.0	0.8	9.4	18.8	23.0
	YEAR OF OCCURRENCE		1993	1971	1960	1970	1991	1992	1992			1970	1972	1983	JAN 1993
	MAXIMUM IN 24 HOURS (IN.)	47	15.2	15.2	8.4	7.6	T	T	T	0.0	0.0	0.8	7.4	9.0	15.2
	YEAR OF OCCURRENCE		1993	1971	1960	1970	1991	1992	1992	1993	1993	1970	1975	1973	FEB 1971
	MAXIMUM SNOW DEPTH (IN.)	44	12	12	18	4	0	T	0	T	0	T	8	9	18
	YEAR OF OCCURRENCE		1979	1971	1960	1970		1993		1949		1957	1975	1983	MAR 1960
	NORMAL NO. DAYS WITH:														
	SNOWFALL ≥ 1.0	30	1.9	1.5	1.0	0.2	0.0	0.0	0.0	0.0	0.0	0.0	0.4	1.8	6.8

PRECIPITATION (inches) 1993 TOPEKA, KANSAS (TOP)

YEAR	JAN	FEB	MAR	APR	MAY	JUN	JUL	AUG	SEP	OCT	NOV	DEC	ANNUAL
1964	0.54	0.29	1.73	4.70	2.09	8.10	1.81	8.24	1.13	0.14	6.27	0.94	35.98
1965	1.60	1.27	1.58	2.31	3.41	10.14	3.33	3.53	7.22	0.92	0.20	2.46	37.97
1966	0.16	0.54	0.10	1.98	0.41	8.83	0.75	3.62	1.46	0.42	0.24	0.79	19.30
1967	1.06	0.21	1.91	8.12	5.07	15.20	3.06	1.84	4.64	6.01	0.41	3.11	50.64
1968	0.89	0.56	0.46	4.20	3.37	3.18	10.17	7.40	2.50	4.19	1.56	2.10	40.58
1969	0.84	0.42	1.37	7.14	3.77	8.46	3.26	0.87	2.03	3.98	0.10	1.24	33.48
1970	0.19	0.34	1.03	3.49	5.46	5.82	1.39	0.83	7.70	2.49	1.23	1.65	31.62
1971	1.20	3.49	0.64	1.08	4.83	3.10	4.07	0.26	1.35	3.87	3.03	1.83	28.75
1972	0.47	0.56	1.37	3.93	2.90	1.14	4.81	3.26	4.89	2.11	3.99	1.78	31.21
1973	2.67	1.71	8.44	4.03	4.37	2.96	10.16	2.83	12.71	4.57	2.14	4.30	60.89
1974	0.99	1.20	1.22	2.78	3.59	3.72	2.90	4.89	1.40	5.16	2.19	1.18	31.22
1975	1.50	1.67	1.66	3.26	3.88	4.85	0.68	1.69	4.35	0.05	4.44	1.12	29.15
1976	0.41	0.51	1.38	4.85	4.63	1.69	2.04	0.86	1.12	3.01	0.04	0.21	20.75
1977	0.90	0.22	2.06	2.46	7.83	10.91	1.37	11.18	3.22	4.92	3.38	0.26	48.71
1978	0.19	0.84	1.63	2.35	5.75	4.57	2.26	2.89	6.65	0.36	3.22	0.55	31.26
1979	1.81	0.63	3.95	2.37	2.25	5.63	5.84	4.05	2.17	4.15	1.80	0.05	34.70
1980	1.34	0.91	4.15	1.03	4.85	0.56	0.87	5.86	1.19	7.24	0.25	3.86	32.11
1981	0.32	0.21	1.61	1.98	5.93	9.40	7.63	3.92	2.03	3.72	3.63	0.22	40.60
1982	1.67	0.59	1.14	1.58	9.39	5.99	5.08	4.53	1.17	1.25	2.26	3.61	38.26
1983	0.69	0.63	4.39	6.29	4.93	6.08	0.59	0.62	2.25	5.19	3.61	1.34	36.61
1984	0.11	1.35	4.57	4.26	3.45	10.17	1.66	1.04	4.24	4.10	0.72	2.36	38.03
1985	0.70	2.02	2.38	3.60	3.79	5.15	2.90	7.97	8.16	5.20	2.02	0.71	44.60
1986	T	1.55	1.35	3.15	7.53	2.51	4.21	5.50	6.21	3.30	0.87	1.20	37.38
1987	1.09	2.71	5.92	2.33	3.89	4.86	2.78	5.90	1.81	1.86	1.94	1.87	36.96
1988	2.04	0.48	0.73	2.93	3.08	3.13	1.74	1.34	1.94	0.26	0.86	0.86	19.39
1989	1.24	0.86	3.11	0.62	4.05	4.76	5.21	6.22	8.65	3.44	T	0.61	38.77
1990	1.22	2.31	3.75	1.01	4.45	5.57	3.01	5.69	0.83	2.71	2.91	0.97	34.43
1991	0.76	0.00	2.98	3.63	7.09	1.49	1.47	1.76	2.15	3.20	2.20	2.44	29.17
1992	0.89	1.18	5.29	3.25	1.75	3.35	6.37	1.24	3.92	1.41	5.27	2.01	35.93
1993	1.11	1.61	2.56	5.43	6.95	2.18	10.98	5.32	7.03	1.37	1.12	0.90	46.56
POR= 47 YRS	0.97	1.11	2.47	3.14	4.35	5.13	4.26	3.88	3.41	2.78	1.77	1.40	34.66

WBAN : 13996

AVERAGE TEMPERATURE (˚F) 1993 TOPEKA, KANSAS (TOP)

YEAR	JAN	FEB	MAR	APR	MAY	JUN	JUL	AUG	SEP	OCT	NOV	DEC	ANNUAL
1964	34.2	34.3	40.0	56.4	68.9	72.1	81.2	74.5	68.1	54.6	46.2	29.8	55.0
1965	31.0	30.8	33.0	57.2	68.6	73.1	77.0	74.9	66.9	57.3	45.7	39.9	54.6
1966	26.0	31.9	47.0	50.8	63.8	72.7	83.1	74.8	65.8	55.4	44.1	30.7	53.8
1967	31.5	33.4	47.0	57.4	60.8	72.1	75.0	72.1	63.6	54.7	41.7	33.0	53.5
1968	26.6	31.6	46.4	53.5	59.1	74.1	76.4	75.5	66.6	57.3	40.0	29.0	53.0
1969	25.8	33.9	37.1	55.5	65.3	68.7	79.4	75.5	69.0	52.7	42.5	29.9	52.9
1970	23.3	35.9	39.5	53.5	68.1	71.8	77.8	80.6	69.1	54.2	41.2	34.3	54.1
1971	24.1	26.8	41.1	56.3	61.4	77.5	74.5	75.9	71.6	61.6	45.4	34.8	54.2
1972	25.6	31.8	46.7	54.7	63.6	74.3	74.4	74.7	68.5	54.6	39.4	27.7	53.0
1973	27.5	33.9	47.5	52.6	61.3	74.9	77.4	77.1	66.6	60.4	44.8	29.9	54.5
1974	22.3	35.9	46.9	56.8	67.2	70.0	80.6	74.1	61.9	58.4	43.1	32.7	54.2
1975	30.7	28.8	37.8	54.4	67.3	74.2	77.3	79.3	64.0	59.3	45.5	34.5	54.4
1976	27.8	42.7	45.4	57.0	60.4	72.7	78.0	76.9	69.0	50.3	35.4	28.5	53.7
1977	15.2	37.4	49.6	60.2	70.1	75.2	79.4	76.4	71.6	56.7	42.7	30.1	55.4
1978	17.3	20.4	38.4	55.9	63.0	74.6	77.3	75.7	72.9	54.6	43.0	30.0	51.9
1979	11.8	19.2	42.6	51.6	63.1	72.4	77.8	76.9	68.0	57.0	40.0	35.5	51.3
1980	28.6	26.0	40.8	53.7	62.8	76.5	86.4	80.7	70.0	53.9	45.0	32.6	54.8
1981	31.4	35.5	46.1	60.6	60.9	75.5	79.5	73.1	68.0	56.1	47.2	30.1	55.3
1982	21.9	28.5	43.2	50.2	63.7	69.0	78.7	75.5	66.5	55.9	42.0	35.8	52.6
1983	32.5	36.1	44.9	49.4	62.5	73.5	81.1	83.0	72.2	58.7	45.8	14.4	54.5
1984	26.0	40.2	38.1	51.7	62.4	73.9	77.0	78.0	66.5	56.6	45.5	36.8	54.4
1985	19.9	25.6	48.6	58.7	66.5	72.0	79.7	72.8	66.8	56.6	36.7	25.1	52.4
1986	35.8	32.5	49.8	57.7	65.9	77.0	80.4	72.3	71.6	56.6	38.3	34.6	56.0
1987	29.7	40.3	46.7	57.1	70.4	76.2	78.1	75.5	68.2	52.6	47.4	35.9	56.5
1988	28.1	30.8	43.4	53.9	68.8	75.1	76.7	79.5	70.3	52.8	45.2	35.3	55.0
1989	38.0	22.9	44.4	57.9	64.2	71.4	77.6	74.8	62.3	57.1	42.3	21.0	52.8
1990	37.3	36.2	45.5	51.9	60.3	77.2	77.7	76.5	71.6	57.0	49.1	29.6	55.8
1991	25.2		48.2	57.7	69.4	77.1	80.2	77.3	69.3	58.6	37.9	37.4	
1992	37.2	41.5	47.8	54.7	62.5	69.1	75.9	71.7	67.9	56.5	39.1	32.6	54.7
1993	26.4	29.9	40.8	50.2	63.1	72.9	78.2	77.8	63.4	54.0	39.3	34.8	52.6
POR= 47 YRS	27.2	32.5	42.5	54.6	64.7	73.9	78.3	76.8	68.3	57.1	42.7	31.9	54.2

WBAN : 13996

REFERENCE NOTES:

H/C INDICATES HEATING AND COOLING DEGREE DAYS.
RH INDICATES RELATIVE HUMIDITY
W/O INDICATES WEATHER AND OBSTRUCTIONS
S INDICATES SUNSHINE.
PR INDICATES PRESSURE.
CLOUDINESS IS THE SUM OF THE CEILOMETER AND SATELLITE DATA NOT TO EXCEED TEN TENTHS.
MAXIMUM SNOW DEPTH IS THE DEPTH MEASURED AT 12 UTC.

GENERAL:
T INDICATES TRACE PRECIPITATION, AN AMOUNT GREATER THAN ZERO BUT LESS THAN THE LOWEST REPORTABLE VALUE.
+ INDICATES THE VALUE ALSO OCCURS ON EARLIER DATES.
BLANK ENTRIES DENOTE MISSING OR UNREPORTED DATA.
NORMALS ARE 30-YEAR AVERAGES (1961 - 1990).
ASOS INDICATES AUTOMATED SURFACE OBSERVING SYSTEM.
PM INDICATES THE LAST DAY OF THE PREVIOUS MONTH.
POR (PERIOD OF RECORD) INDICATES THE NUMBER OF YEARS USED TO COMPUTE THE MEAN. INDIVIDUAL MONTHS WITHIN THE POR MAY BE MISSING.
WHEN THE POR FOR A NORMAL IS LESS THAN 30 YEARS, THE NORMAL IS PROVISIONAL AND IS BASED ON THE NUMBER OF YEARS INDICATED.
0.* OR * INDICATES THE VALUE OR MEAN DAYS WITH IS BETWEEN 0.00 AND 0.05.
CLOUDINESS FOR ASOS STATIONS DIFFERS FROM THE NON-ASOS OBSERVATION TAKEN BY A HUMAN OBSERVER. ASOS STATION CLOUDINESS IS BASED ON TIME-AVERAGED CEILOMETER DATA FOR CLOUDS AT OR BELOW 12,000 FEET AND ON SATELLITE DATA FOR CLOUDS ABOVE 12,000 FEET.
THE NUMBER OF DAYS WITH CLEAR, PARTLY CLOUDY, AND CLOUDY CONDITIONS FOR ASOS STATIONS IS THE SUM OF THE CEILOMETER AND SATELLITE DATA FOR THE SUNRISE TO SUNSET PERIOD.

GENERAL CONTINUED:
CLEAR INDICATES 0 - 3 TENTHS, PARTLY CLOUDY INDICATES 4 - 7 TENTHS, AND CLOUDY INDICATES 8 OR MORE TENTHS. WHEN AT LEAST ONE OF THE ELEMENTS (CEILOMETER OR SATELLITE) IS MISSING, THE DAILY CLOUDINESS IS NOT COMPUTED.
WIND DIRECTION IS RECORDED IN TENS OF DEGREES (2 DIGITS) CLOCKWISE FROM TRUE NORTH. "00" INDICATES CALM. "36" INDICATES TRUE NORTH.
RESULTANT WIND IS THE VECTOR AVERAGE OF THE SPEED AND DIRECTION.
AVERAGE TEMPERATURE IS THE SUM OF THE MEAN DAILY MAXIMUM AND MINIMUM TEMPERATURE DIVIDED BY 2.
SNOWFALL DATA COMPRISE ALL FORMS OF FROZEN PRECIPITATION, INCLUDING HAIL.
A HEATING (COOLING) DEGREE DAY IS THE DIFFERENCE BETWEEN THE AVERAGE DAILY TEMPERATURE AND 65˚F.
DRY BULB IS THE TEMPERATURE OF THE AMBIENT AIR.
DEW POINT IS THE TEMPERATURE TO WHICH THE AIR MUST BE COOLED TO ACHIEVE 100 PERCENT RELATIVE HUMIDITY.
WET BULB IS THE TEMPERATURE THE AIR WOULD HAVE IF THE MOISTURE CONTENT WAS INCREASED TO 100 PERCENT RELATIVE HUMIDITY.

HEATING DEGREE DAYS (base 65°F) 1993 TOPEKA, KANSAS (TOP)

YEAR	JUL	AUG	SEP	OCT	NOV	DEC	JAN	FEB	MAR	APR	MAY	JUN	TOTAL
1964-65	0	9	67	316	572	1087	1044	955	984	263	30	0	5327
1965-66	0	0	74	253	572	770	1202	922	552	423	107	6	4881
1966-67	0	2	64	316	620	1053	1032	875	570	247	205	15	4999
1967-68	1	7	87	356	690	985	1184	965	572	338	202	5	5392
1968-69	0	1	35	282	745	1108	1210	866	857	282	94	42	5522
1969-70	0	0	17	408	666	1083	1284	811	785	376	45	29	5504
1970-71	0	0	57	344	706	945	1262	1065	735	274	144	0	5532
1971-72	4	0	69	143	580	927	1216	958	567	324	109	10	4907
1972-73	10	0	59	337	764	1152	1158	864	537	378	129	0	5388
1973-74	0	0	58	191	603	1082	1317	807	558	258	64	7	4945
1974-75	0	3	134	213	649	991	1056	1008	839	352	46	7	5298
1975-76	2	0	137	230	581	941	1148	639	599	269	178	5	4729
1976-77	0	0	45	471	881	1126	1537	767	469	180	11	0	5487
1977-78	0	0	6	263	662	1075	1473	1240	824	280	156	6	5985
1978-79	0	0	34	319	655	1078	1643	1277	693	401	129	9	6238
1979-80	0	4	45	267	741	908	1123	1123	744	344	129	3	5431
1980-81	0	0	65	344	591	1001	1035	822	579	175	176	0	4788
1981-82	0	2	46	283	529	1076	1329	1014	664	449	76	32	5500
1982-83	0	0	93	303	683	896	1002	804	615	466	120	13	4995
1983-84	0	0	56	223	570	1565	1204	713	830	405	137	0	5703
1984-85	0	0	145	276	578	871	1389	1098	501	228	35	8	5129
1985-86	0	0	127	259	844	1228	899	906	491	252	49	0	5055
1986-87	0	9	27	263	792	934	1084	688	560	292	16	0	4665
1987-88	0	3	24	376	531	893	1136	988	662	331	16	5	4965
1988-89	2	4	24	383	587	912	832	1174	641	296	125	5	4985
1989-90	0	2	155	276	672	1360	851	801	600	413	176	4	5310
1990-91	1	1	39	276	477	1093	1227		523	233	48	0	
1991-92	0	0	95	262	808	849	855	673	528	326	132	7	4535
1992-93	0	2	68	278	770	995	1189	979	744	440	101	22	5588
1993-	0	1	108	356	763	930							
POR= 47 YRS	0	1	61	272	662	1022	1164	911	693	328	100	8	5222

WBAN : 13996

COOLING DEGREE DAYS (base 65°F) 1993 TOPEKA, KANSAS (TOP)

YEAR	JAN	FEB	MAR	APR	MAY	JUN	JUL	AUG	SEP	OCT	NOV	DEC	ANNUAL
1969	0	0	0	4	107	158	456	330	146	32	0	0	1233
1970	0	0	0	35	149	239	407	490	188	16	0	0	1524
1971	0	0	3	21	42	381	309	345	273	42	0	0	1416
1972	0	0	5	22	74	297	308	309	169	20	0	0	1204
1973	0	0	0	13	21	304	394	384	115	52	0	0	1283
1974	0	0	6	21	140	165	490	292	47	12	0	0	1173
1975	0	0	0	38	129	289	390	448	116	61	3	0	1474
1976	0	0	1	34	40	242	410	376	171	20	0	0	1294
1977	0	0	0	40	176	311	453	360	209	14	0	0	1563
1978	0	0	6	15	101	298	390	339	277	5	3	0	1434
1979	0	0	4	7	76	237	401	379	144	27	0	0	1275
1980	0	0	0	9	69	356	670	496	220	9	0	0	1829
1981	0	0	0	53	58	321	457	260	143	17	0	0	1309
1982	0	0	0	11	43	157	432	334	147	28	0	0	1152
1983	0	0	0	7	50	274	509	564	278	33	2	0	1717
1984	0	0	0	14	67	274	379	407	196	20	0	3	1360
1985	0	0	0	46	88	225	461	249	188	6	0	0	1263
1986	0	0	26	42	85	363	488	243	233	9	0	0	1489
1987	0	0	0	61	192	344	410	335	126	0	9	0	1477
1988	0	0	0	4	140	314	375	458	191	11	0	0	1493
1989	0	0	11	90	107	206	399	311	81	41	0	0	1246
1990	0	0	1	26	39	377	403	366	241	37	7	0	1497
1991	0	0	11	22	192	371	478	387	229	69	0	0	1759
1992	0	0	0	25	61	134	344	217	162	20	0	0	963
1993	0	0	0	0	48	269	417	405	64	22	0	0	1225
POR= 25 YRS	0	0	2	25	91	275	424	362	173	24	0	0	1376

WBAN : 13996

SNOWFALL (inches) 1993 TOPEKA, KANSAS (TOP)

YEAR	JUL	AUG	SEP	OCT	NOV	DEC	JAN	FEB	MAR	APR	MAY	JUN	TOTAL
1964-65	0.0	0.0	0.0	0.0	1.3	4.3	2.9	7.3	5.0	0.0	0.0	0.0	20.8
1965-66	0.0	0.0	0.0	0.0	T	3.6	0.5	0.6	T	T	0.0	0.0	4.7
1966-67	0.0	0.0	0.0	0.0	T	8.3	3.4	1.9	0.8	0.0	0.0	0.0	14.4
1967-68	0.0	0.0	0.0	0.0	T	11.1	4.7	5.7	T	T	0.0	0.0	21.5
1968-69	0.0	0.0	0.0	0.0	T	0.8	6.2	3.4	2.0	0.0	0.0	0.0	12.4
1969-70	0.0	0.0	0.0	0.0	0.8	9.4	2.1	T	4.9	6.8	0.0	0.0	24.0
1970-71	0.0	0.0	0.0	0.8	T	7.9	2.1	22.4	7.5	T	0.0	0.0	40.7
1971-72	0.0	0.0	0.0	0.0	2.4	1.2	3.0	6.7	3.0	T	0.0	0.0	16.3
1972-73	0.0	0.0	0.0	0.0	9.4	4.0	13.5	1.7	0.0	0.1	0.0	0.0	28.7
1973-74	0.0	0.0	0.0	0.0	T	15.2	7.8	2.1	1.5	1.3	0.0	0.0	27.9
1974-75	0.0	0.0	0.0	0.0	1.3	1.4	5.0	6.7	7.8	3.6	0.0	0.0	25.8
1975-76	0.0	0.0	0.0	0.0	8.3	2.7	6.4	0.7	2.9	0.0	0.0	0.0	21.0
1976-77	0.0	0.0	0.0	T	0.3	T	13.6	0.1	T	0.2	0.0	0.0	14.2
1977-78	0.0	0.0	0.0	0.0	0.1	0.2	3.9	12.4	6.2	0.0	0.0	0.0	22.8
1978-79	0.0	0.0	0.0	0.0	T	11.1	20.1	3.1	7.5	1.1	0.0	0.0	42.9
1979-80	0.0	0.0	0.0	0.0	T	T	3.5	11.4	3.4	0.0	0.0	0.0	18.3
1980-81	0.0	0.0	0.0	T	0.0	3.8	2.6	2.5	0.0	0.0	0.0	0.0	8.9
1981-82	0.0	0.0	0.0	0.0	T	1.4	3.2	8.0	0.3	0.5	0.0	0.0	13.4
1982-83	0.0	0.0	0.0	0.0	1.1	5.0	6.1	10.1	0.6	4.5	0.0	0.0	27.4
1983-84	0.0	0.0	0.0	0.0	4.1	18.8	2.6	T	4.2	0.0	0.0	0.0	29.7
1984-85	0.0	0.0	0.0	0.0	T	9.8	18.2	7.9	0.5	0.0	0.0	0.0	36.4
1985-86	0.0	0.0	0.0	0.0	3.3	5.8	T	1.5	T	0.0	0.0	0.0	10.6
1986-87	0.0	0.0	0.0	T	0.7	1.7	15.1	2.3	0.5	0.0	0.0	0.0	20.3
1987-88	0.0	0.0	0.0	0.0	0.9	9.6	0.6	6.0	4.7	0.0	0.0	0.0	21.8
1988-89	0.0	0.0	0.0	0.0	0.7	0.8	T	9.0	1.6	0.0	T	T	12.1
1989-90	0.0	0.0	0.0	0.0	T	9.5	1.0	0.1	7.6	0.0	0.0	0.0	18.2
1990-91	0.0	0.0	0.0	0.0	0.0	2.9	9.6		T	T	T	0.0	
1991-92	0.0	0.0	0.0	T	6.2	0.1	T	T	0.9	T	0.0	T	7.2
1992-93	T	0.0	0.0	T	4.5	0.9	23.0	14.2	0.6	T		T	
1993-	0.0	0.0	0.0	T	T	3.3							
POR= 47 YRS	T	0.0	0.0	T	1.3	4.8	6.1	4.8	3.7	0.5	T	T	21.2

WBAN : 13996

1993 WICHITA, KANSAS (ICT)

Wichita is in the Central Great Plains where masses of warm, moist air from the Gulf of Mexico collide with cold, dry air from the Arctic region to create a wide range of weather the year around. Summers are usually warm and humid, and can be very hot and dry. The winters are usually mild, with brief periods of very cold weather.

The elevation is just over 1,300 feet above sea level. The terrain is basically flat with natural tree areas mainly along the Arkansas River and its tributaries.

The temperature extremes for the period of weather records at Wichita range from more than 110 degrees to less than -20 degrees. Temperatures above 90 degrees occur an average of 63 days per year, while very cold temperatures below zero occur about 2 days per year.

Precipitation averages about 30 inches per year, with 70 percent of that falling from April through September during the growing season. The wettest years have recorded over 50 inches. The driest years less than 15 inches.

Thunderstorms occur mainly during the spring and early summer. They can be severe and cause damage from heavy rain, large hail, strong winds and tornadoes.

The city of Wichita is protected against floods from the Arkansas River and its local tributaries by the Wichita-Vally Center Flood Control Project, which is designed to protect against floods up to the 75 to 100 year frequency class.

Snowfall normally is 15 inches per year, falling from December through March. Monthly snowfalls in excess of 20 inches and 24-hour snowfalls of more than 13 inches have occurred.

The prevailing wind direction is south with the windiest months March and April. July has the least wind. Strong north winds often occur with the passage of cold fronts from late fall through early spring. Extremely low wind chill factors are experienced with very cold outbreaks during the mid winter. On rare occasions during the summer, strong, hot, dry southwest winds can do considerable damage to crops.

NORMALS, MEANS, AND EXTREMES FOR 1993

WICHITA, KS (ICT)

LATITUDE: 37° 39′ 00″ N LONGTITUDE: 97° 25′ 58″ W ELEVATION (FT): GRND: 1321 BARO: 1342 TIME ZONE: CENTRAL (UTC+06) WBAN: 03928

	ELEMENT	POR	JAN	FEB	MAR	APR	MAY	JUN	JUL	AUG	SEP	OCT	NOV	DEC	YEAR
TEMPERATURE °F	NORMAL DAILY MAXIMUM	30	39.8	45.9	57.2	68.3	76.9	86.8	92.8	90.7	81.4	70.6	55.3	43.0	67.4
	MEAN DAILY MAXIMUM	40	40.0	46.2	56.6	68.2	76.8	87.0	93.1	91.3	81.9	70.6	55.0	45.0	67.7
	HIGHEST DAILY MAXIMUM	41	75	84	89	96	100	110	113	110	107	95	85	83	113
	YEAR OF OCCURRENCE		1967	1976	1989	1972	1967	1980	1954	1984	1990	1979	1980	1955	JUL 1954
	MEAN OF EXTREME MAXS.	41	62.2	69.0	79.5	85.9	90.8	99.7	104.0	103.4	97.2	88.1	73.7	64.4	84.8
	NORMAL DAILY MINIMUM	30	19.2	23.7	33.6	44.5	54.3	64.6	69.9	67.9	59.2	46.6	33.9	23.0	45.0
	MEAN DAILY MINIMUM	40	19.8	24.4	33.3	44.7	54.7	64.5	69.8	67.9	59.3	46.8	33.6	24.6	45.3
	LOWEST DAILY MINIMUM	41	-12	-21	-2	15	31	43	51	48	31	18	1	-16	-21
	YEAR OF OCCURRENCE		1962	1982	1960	1975	1976	1969	1975	1967	1984	1993	1975	1989	FEB 1982
	MEAN OF EXTREME MINS.	41	0.9	5.0	14.8	27.8	39.3	51.5	59.2	56.6	42.6	30.4	17.3	6.3	29.3
	NORMAL DRY BULB	30	29.5	34.8	45.4	56.4	65.6	75.7	81.4	79.3	70.3	58.6	44.7	33.0	56.2
	MEAN DRY BULB	40	29.9	35.3	45.0	56.5	65.8	75.7	81.5	79.7	70.6	58.7	44.3	34.8	56.5
	MEAN WET BULB	38	26.3	30.7	38.9	49.2	58.4	66.3	69.8	68.5	61.8	51.4	39.4	30.0	49.2
	MEAN DEW POINT	38	20.1	24.1	31.4	42.2	53.2	61.4	64.0	62.8	56.5	45.1	33.2	23.9	43.2
	NORMAL NO. DAYS WITH:														
	MAXIMUM ≥ 90°	30	0.0	0.0	0.0	0.3	2.0	12.1	21.9	18.9	7.0	0.8	0.0	0.0	63.0
	MAXIMUM ≤ 32°	30	9.4	5.5	1.0	*	0.0	0.0	0.0	0.0	0.0	0.0	0.7	6.0	22.6
	MINIMUM ≤ 32°	30	28.4	22.5	14.6	2.8	0.1	0.0	0.0	0.0	*	1.3	14.3	26.8	110.8
	MINIMUM ≤ 0°	30	2.2	1.0	0.1	0.0	0.0	0.0	0.0	0.0	0.0	0.0	0.0	0.9	4.2
H/C	NORMAL HEATING DEG. DAYS	30	1101	846	608	278	102	5	0	0	29	221	609	992	4791
	NORMAL COOLING DEG. DAYS	30	0	0	0	20	121	326	508	443	188	22	0	0	1628
RH	NORMAL (PERCENT)	30	70	68	64	63	67	64	59	61	67	65	70	72	66
	HOUR 00 LST	30	75	74	71	71	76	74	67	69	75	73	76	76	73
	HOUR 06 LST	30	78	78	77	78	82	82	78	79	83	80	80	80	80
	HOUR 12 LST	30	62	60	55	52	56	53	48	50	56	53	59	63	56
	HOUR 18 LST	30	64	59	52	50	54	49	44	46	53	54	64	67	55
S	PERCENT POSSIBLE SUNSHINE	40	61	61	62	64	65	70	76	76	69	66	59	58	65
W/O	MEAN NO. DAYS WITH:														
	HEAVY FOG(VISBY ≤ 1/4 MI)	40	2.9	2.8	1.5	0.9	0.7	0.3	0.2	0.2	1.0	1.3	2.0	3.2	17.0
	THUNDERSTORMS	40	0.3	0.8	2.9	5.3	8.9	9.8	7.7	7.3	5.8	3.2	1.2	0.3	53.5
CLOUDINESS	MEAN:														
	SUNRISE-SUNSET (TENTHS)	1		8.0	7.0	8.0	7.0	5.0	3.0	3.0	4.0	6.0	5.0	5.0	
	MIDNIGHT-MIDNIGHT (TENTHS)	1		8.0	8.0	7.0	7.0	5.0	3.0	3.0	4.0	6.0	5.0	5.0	
	MEAN NO. DAYS WITH:														
	CLEAR	1	6.0	7.0	8.0	7.0	8.0	7.0	12.0	18.0	12.0	10.0	11.0	16.0	122.0
	PARTLY CLOUDY	1	2.0	6.0	5.0	2.0	7.0	7.0	5.0	6.0	6.0	6.0	3.0	5.0	60.0
	CLOUDY	1	11.0	12.0	13.0	10.0	7.0	4.0	2.0	3.0	5.0	7.0	7.0	9.0	90.0
PR	MEAN STATION PRESSURE(IN.)	21	28.69	28.66	28.53	28.53	28.51	28.52	28.57	28.59	28.61	28.63	28.62	28.68	28.60
	MEAN SEA-LEVEL PRES. (IN.)	38	30.15	30.10	29.99	29.93	29.91	29.90	29.94	29.95	29.99	30.04	30.07	30.12	30.01
WINDS	MEAN SPEED (MPH)	30	9.4	9.8	10.8	10.4	9.2	8.8	9.0	8.1	8.7	8.6	9.7	9.5	9.3
	PREVAIL.DIR(TENS OF DEGS.)	30	36	36	18	18	18	18	18	18	18	18	18	36	18
	MAXIMUM 2-MINUTE:														
	SPEED (MPH)	1	31	31	38	37	38	38	59	33	39	49	38	39	59
	DIR. (TENS OF DEGS.)	1	32	31	31	28	30	01	03	36	18	31	34	17	03
	YEAR OF OCCURRENCE		1993	1993	1993	1993	1993	1993	1993	1993	1993	1993	1993	1993	JUL 1993
	MAXIMUM 5-SECOND:														
	SPEED (MPH)	1	39		47	41	46	49	101	37	46	53	46	44	
	DIR. (TENS OF DEGS.)	1	33		30	28	30	21	04	36	20	31	31	36	
	YEAR OF OCCURRENCE		1993		1993	1993	1993	1993	1993	1993	1993	1993	1993	1993	
PRECIPITATION	NORMAL TOTAL (IN.)	30	0.79	0.96	2.43	2.38	3.81	4.31	3.13	3.02	3.49	2.22	1.59	1.20	29.33
	MAXIMUM TOTAL (IN.)	40	2.73	3.33	9.17	5.57	9.62	10.46	9.22	7.91	9.46	6.13	5.88	4.71	10.46
	YEAR OF OCCURRENCE		1973	1987	1973	1976	1993	1957	1962	1960	1973	1959	1964	1984	JUN 1957
	MINIMUM TOTAL (IN.)	40	T	T	0.01	0.22	0.52	0.94	0.05	0.31	0.03	T	T	0.03	T
	YEAR OF OCCURRENCE		1986	1991	1971	1963	1973	1954	1975	1976	1956	1958	1989	1955	NOV 1989
	MAXIMUM IN 24 HOURS (IN.)	40	1.72	1.53	2.65	2.51	4.70	4.98	3.86	4.50	3.29	5.03	4.33	2.60	5.03
	YEAR OF OCCURRENCE		1980	1973	1961	1988	1963	1965	1983	1991	1989	1985	1964	1984	OCT 1985
	NORMAL NO. DAYS WITH:														
	PRECIPITATION ≥ 0.01	30	5.1	5.3	7.9	8.0	10.6	9.3	7.1	7.7	7.9	6.2	5.2	5.8	86.1
	PRECIPITATION ≥ 1.00	30	0.1	0.1	0.7	0.5	1.1	1.3	0.9	0.8	1.1	0.6	0.5	0.2	7.9
SNOWFALL	NORMAL TOTAL (IN.)	30	4.7	4.5	2.1	0.3	0.0	0.0	0.0	0.0	0.0	-1.1	1.4	3.8	16.8
	MAXIMUM TOTAL (IN.)	40	19.7	16.7	16.5	4.6	T	T	T	0.0	T	1.5	7.1	13.8	19.7
	YEAR OF OCCURRENCE		1987	1971	1970	1979	1991	1992	1993		1992	1991	1972	1983	JAN 1987
	MAXIMUM IN 24 HOURS (IN.)	40	13.0	11.9	13.5	4.6	T	T	T	0.0	T	1.5	6.8	9.0	13.5
	YEAR OF OCCURRENCE		1962	1971	1970	1979	1991	1992	1993	1993	1992	1991	1984	1983	MAR 1970
	MAXIMUM SNOW DEPTH (IN.)	39	17	13	13	3	0	0	0	0	0	T	5	8	17
	YEAR OF OCCURRENCE		1962	1971	1970	1979						1993	1987	1987	JAN 1962
	NORMAL NO. DAYS WITH:														
	SNOWFALL ≥ 1.0	30	1.4	1.3	0.4	0.1	0.0	0.0	0.0	0.0	0.0	0.0	0.4	1.2	4.8

PRECIPITATION (inches) 1993 WICHITA, KANSAS (ICT)

YEAR	JAN	FEB	MAR	APR	MAY	JUN	JUL	AUG	SEP	OCT	NOV	DEC	ANNUAL
1964	0.71	0.53	0.89	2.97	5.84	3.73	2.23	6.10	2.66	1.64	5.88	1.03	34.21
1965	0.56	1.39	0.48	2.63	6.26	8.00	3.62	4.91	8.44	0.32	0.11	2.25	38.97
1966	0.23	1.44	0.26	2.21	0.76	2.67	1.78	1.09	0.72	0.47	0.09	0.43	12.15
1967	0.28	0.09	0.57	1.30	1.42	5.62	4.28	1.91	3.19	2.98	0.39	1.41	23.44
1968	0.14	0.20	1.36	2.16	4.37	2.38	3.65	6.41	5.91	3.06	2.47	1.31	33.42
1969	0.45	1.35	1.73	4.30	3.28	6.82	6.23	1.07	4.77	2.80	0.01	1.36	34.17
1970	0.28	0.21	2.70	4.49	1.58	6.72	0.47	2.37	4.04	1.88	0.05	0.49	25.28
1971	0.98	1.70	0.01	2.35	3.02	2.70	6.65	1.49	1.73	5.54	2.49	0.95	29.61
1972	0.15	0.28	0.56	3.32	2.47	2.02	3.86	3.31	1.31	2.00	3.06	0.97	23.31
1973	2.73	1.20	9.17	3.78	0.52	1.21	6.07	0.68	9.46	3.43	0.91	2.80	41.96
1974	0.56	0.25	2.36	4.29	4.65	2.79	0.09	4.11	1.08	3.44	2.69	2.22	28.53
1975	1.28	2.12	1.72	1.57	8.60	6.88	0.05	2.77	1.19	0.08	2.89	0.48	29.63
1976	0.04	0.25	1.50	5.57	2.69	3.12	6.13	0.31	2.02	1.82	0.06	0.07	23.58
1977	0.54	0.08	1.42	3.32	8.85	3.15	3.98	6.31	4.35	1.19	2.38	0.19	35.76
1978	0.49	1.71	2.10	2.71	2.24	3.19	1.49	1.90	3.58	0.05	2.21	0.59	22.26
1979	1.57	0.23	4.47	1.46	3.05	6.54	2.18	0.67	1.54	2.96	2.05	1.99	28.71
1980	1.82	0.81	3.99	1.07	2.66	1.34	0.47	3.76	0.67	1.25	0.54	2.11	20.49
1981	0.25	0.22	2.15	0.38	6.33	4.25	1.27	2.65	2.25	4.69	2.93	0.29	27.66
1982	1.68	0.77	2.05	0.73	7.82	8.28	0.56	1.51	1.08	0.41	0.73	1.51	27.13
1983	1.66	1.23	4.26	3.80	4.08	7.38	3.86	1.39	2.53	2.97	2.39	1.13	36.68
1984	0.20	1.23	7.57	3.71	1.15	2.30	0.30	0.75	2.18	2.78	1.44	4.71	28.32
1985	0.26	2.07	1.64	2.28	2.01	4.79	3.97	2.86	5.97	5.58	1.60	0.61	33.64
1986	T	1.26	1.22	1.80	2.98	5.39	3.42	6.00	3.81	3.61	0.58	1.22	31.29
1987	1.40	3.33	4.13	0.61	8.01	4.50	2.14	7.69	2.10	0.90	1.50	2.25	38.56
1988	0.51	0.18	2.91	4.46	2.40	1.86	0.91	1.10	0.53	0.94	0.77	0.50	17.07
1989	0.79	0.39	2.38	0.23	4.96	7.96	4.07	5.72	7.38	0.37	T	0.44	34.69
1990	1.73	2.19	2.68	0.80	1.29	1.91	1.72	2.01	1.95	0.64	2.01	0.78	19.71
1991	0.58	T	0.71	2.27	4.09	1.34	2.65	7.57	2.08	1.00	2.38	2.07	26.74
1992	0.67	0.47	3.60	1.42	3.44	7.26	4.77	1.94	2.94	2.80	4.91	1.16	35.38
1993	1.12	2.25	1.75	2.11	9.62	4.44	6.21	1.31	1.77	1.68	0.55	0.30	33.11
POR= 40 YRS	0.75	0.94	2.27	2.23	4.04	4.23	3.34	2.99	3.30	2.43	1.51	1.11	29.13

WBAN : 03928

AVERAGE TEMPERATURE (°F) 1993 WICHITA, KANSAS (ICT)

YEAR	JAN	FEB	MAR	APR	MAY	JUN	JUL	AUG	SEP	OCT	NOV	DEC	ANNUAL
1964	36.8	36.1	42.1	59.2	70.2	76.9	85.0	76.8	70.8	58.7	47.2	31.2	57.6
1965	35.2	34.4	36.8	60.4	68.8	76.3	81.9	78.2	68.5	60.4	50.5	41.2	57.7
1966	28.3	33.6	49.1	53.5	65.7	76.2	84.9	76.1	68.6	58.0	48.3	32.7	56.2
1967	34.3	36.3	50.5	61.5	64.0	74.5	76.2	74.7	65.9	57.7	43.2	34.6	56.1
1968	32.5	33.8	48.0	55.5	61.0	76.1	80.3	77.9	67.9	59.3	41.7	30.7	55.4
1969	30.1	36.1	36.1	56.4	65.2	71.0	82.9	78.9	71.9	54.6	44.5	34.6	55.2
1970	27.3	38.5	39.7	55.1	69.3	74.5	81.3	83.2	69.7	53.9	41.5	36.9	55.9
1971	28.8	28.8	44.0	57.0	63.9	78.9	78.0	76.8	70.8	60.6	43.9	35.8	55.6
1972	27.0	34.8	47.9	55.3	64.1	76.7	77.1	78.4	71.0	55.6	39.8	28.5	54.7
1973	27.4	35.6	48.4	52.2	61.9	76.8	80.3	79.8	67.8	61.1	46.7	31.9	55.8
1974	24.9	38.7	47.9	56.4	68.3	72.5	84.4	76.7	64.1	60.5	44.9	35.1	56.2
1975	33.4	28.5	40.4	54.7	64.0	73.6	79.2	81.4	65.8	60.3	44.8	36.0	55.2
1976	32.0	45.5	46.3	57.5	60.2	73.7	78.4	79.7	69.8	52.4	38.3	33.7	55.6
1977	24.5	41.7	50.2	59.8	69.2	78.3	83.6	77.9	73.2	59.8	46.1	35.0	58.3
1978	20.5	23.8	43.6	59.0	65.0	76.8	85.1	81.2	76.3	58.6	45.0	32.0	55.6
1979	16.7	24.0	46.9	54.1	64.1	74.5	80.4	79.8	72.7	61.9	41.8	37.7	54.6
1980	31.4	28.2	41.5	54.3	63.5	79.9	90.5	85.3	75.2	58.8	46.9	36.9	57.7
1981	34.1	40.1	47.6	63.7	62.6	77.9	83.6	78.1	72.0	55.9	47.0	32.9	58.0
1982	25.5	28.0	46.0	53.5	65.3	70.4	81.5	82.0	71.9	58.0	43.0	36.1	55.1
1983	31.7	35.5	43.3	48.0	60.5	71.5	81.6	85.0	72.7	58.5	45.5	16.4	54.2
1984	26.7	41.4	40.7	51.8	63.6	77.7	81.6	82.8	70.3	58.3	45.5	37.2	56.5
1985	25.2	31.2	49.1	59.9	67.7	74.0	81.7	77.5	69.8	57.4	39.4	28.8	55.1
1986	38.1	37.8	51.9	58.7	66.7	78.7	83.0	75.9	73.6	57.6	39.9	35.8	58.1
1987	29.3	42.3	47.0	57.4	69.7	76.4	80.1	78.7	70.4	55.7	47.7	34.9	57.5
1988	27.0	34.1	44.2	53.9	68.3	78.6	80.9	83.0	72.0	56.7	47.5	38.5	57.1
1989	38.5	27.5	47.0	59.6	66.0	72.0	79.0	77.0	65.7	60.7	45.5	25.2	55.3
1990	39.5	38.8	46.9	54.3	63.7	81.7	81.7	80.8	74.3	58.4	50.4	30.2	58.4
1991	28.9	44.6	49.7	58.9	70.4	79.3	83.8	80.5	69.6	59.4	39.5	39.1	58.6
1992	39.0	44.9	50.0	56.7	63.3	71.2	78.9	73.8	71.0	59.1	41.3	33.3	56.9
1993	28.6	32.2	43.1	52.5	63.5	74.5	81.7	81.0	67.4	55.3	40.8	37.8	54.9
POR= 40 YRS	29.9	35.3	45.0	56.5	65.7	75.8	81.5	79.6	70.6	58.7	44.3	34.0	56.4

WBAN : 03928

REFERENCE NOTES:

H/C INDICATES HEATING AND COOLING DEGREE DAYS.
RH INDICATES RELATIVE HUMIDITY
W/O INDICATES WEATHER AND OBSTRUCTIONS
S INDICATES SUNSHINE.
PR INDICATES PRESSURE.
CLOUDINESS ON IS THE SUM OF THE CEILOMETER AND SATELLITE DATA NOT TO EXCEED TEN TENTHS.
MAXIMUM SNOW DEPTH IS THE DEPTH MEASURED AT 12 UTC.

GENERAL:
T INDICATES TRACE PRECIPITATION, AN AMOUNT GREATER THAN ZERO BUT LESS THAN THE LOWEST REPORTABLE VALUE.
+ INDICATES THE VALUE ALSO OCCURS ON EARLIER DATES.
BLANK ENTRIES DENOTE MISSING OR UNREPORTED DATA.
NORMALS ARE 30-YEAR AVERAGES (1961 - 1990).
ASOS INDICATES AUTOMATED SURFACE OBSERVING SYSTEM.
PM INDICATES THE LAST DAY OF THE PREVIOUS MONTH.
POR (PERIOD OF RECORD) INDICATES THE NUMBER OF YEARS USED TO COMPUTE THE MEAN. INDIVIDUAL MONTHS WITHIN THE POR MAY BE MISSING.
WHEN THE POR FOR A NORMAL IS LESS THAN 30 YEARS, THE NORMAL IS PROVISIONAL AND IS BASED ON THE NUMBER OF YEARS INDICATED.
0.* OR * INDICATES THE VALUE OR MEAN DAYS WITH IS BETWEEN 0.00 AND 0.05.
CLOUDINESS FOR ASOS STATIONS DIFFERS FROM THE NON-ASOS OBSERVATION TAKEN BY A HUMAN OBSERVER. ASOS STATION CLOUDINESS IS BASED ON TIME-AVERAGED CEILOMETER DATA FOR CLOUDS AT OR BELOW 12,000 FEET AND ON SATELLITE DATA FOR CLOUDS ABOVE 12,000 FEET.
THE NUMBER OF DAYS WITH CLEAR, PARTLY CLOUDY, AND CLOUDY CONDITIONS FOR ASOS STATIONS IS THE SUM OF THE CEILOMETER AND SATELLITE DATA FOR THE SUNRISE TO SUNSET PERIOD.

GENERAL CONTINUED:
CLEAR INDICATES 0 - 3 TENTHS, PARTLY CLOUDY INDICATES 4 - 7 TENTHS, AND CLOUDY INDICATES 8 OR MORE TENTHS. WHEN AT LEAST ONE OF THE ELEMENTS (CEILOMETER OR SATELLITE) IS MISSING, THE DAILY CLOUDINESS IS NOT COMPUTED.
WIND DIRECTION IS RECORDED IN TENS OF DEGREES (2 DIGITS) CLOCKWISE FROM TRUE NORTH. "00" INDICATES CALM. "36" INDICATES TRUE NORTH.
RESULTANT WIND IS THE VECTOR AVERAGE OF THE SPEED AND DIRECTION.
AVERAGE TEMPERATURE IS THE SUM OF THE MEAN DAILY MAXIMUM AND MINIMUM TEMPERATURE DIVIDED BY 2.
SNOWFALL DATA COMPRISE ALL FORMS OF FROZEN PRECIPITATION, INCLUDING HAIL.
A HEATING (COOLING) DEGREE DAY IS THE DIFFERENCE BETWEEN THE AVERAGE DAILY TEMPERATURE AND 65°F.
DRY BULB IS THE TEMPERATURE OF THE AMBIENT AIR.
DEW POINT IS THE TEMPERATURE TO WHICH THE AIR MUST BE COOLED TO ACHIEVE 100 PERCENT RELATIVE HUMIDITY.
WET BULB IS THE TEMPERATURE THE AIR WOULD HAVE IF THE MOISTURE CONTENT WAS INCREASED TO 100 PERCENT RELATIVE HUMIDITY.

HEATING DEGREE DAYS (base 65°F) 1993 WICHITA, KANSAS (ICT)

YEAR	JUL	AUG	SEP	OCT	NOV	DEC	JAN	FEB	MAR	APR	MAY	JUN	TOTAL
1964-65	0	1	34	200	531	1038	918	851	868	176	23	0	4640
1965-66	0	1	73	179	428	733	1131	871	490	342	100	2	4350
1966-67	0	0	33	247	499	996	945	798	463	160	143	5	4289
1967-68	2	1	63	280	645	936	998	899	525	285	153	1	4788
1968-69	0	5	22	224	691	1054	1076	804	887	254	77	20	5114
1969-70	0	0	2	361	610	939	1163	738	778	312	43	26	4972
1970-71	1	0	64	358	696	866	1118	1005	643	253	95	0	5099
1971-72	4	0	77	165	628	897	1171	871	524	310	109	0	4756
1972-73	2	0	46	319	750	1123	1159	816	506	386	125	0	5232
1973-74	0	0	58	159	541	1021	1237	732	529	263	42	1	4583
1974-75	0	0	92	156	596	920	974	1016	757	333	72	9	4925
1975-76	0	0	98	201	596	892	1015	562	575	238	173	2	4352
1976-77	0	0	38	409	794	966	1253	646	456	170	12	0	4744
1977-78	0	0	1	176	558	926	1375	1149	663	194	112	6	5160
1978-79	0	0	18	210	598	1016	1491	1143	560	333	104	3	5476
1979-80	0	0	10	156	690	838	1038	1063	723	318	116	0	4952
1980-81	0	0	28	239	535	864	954	692	533	104	126	0	4075
1981-82	0	0	24	292	537	990	1214	1033	583	356	54	17	5100
1982-83	0	0	37	239	653	889	1022	818	664	507	168	20	5017
1983-84	0	0	47	221	582	1504	1180	680	747	394	95	0	5450
1984-85	0	0	103	237	576	856	1224	938	487	184	33	8	4646
1985-86	0	0	111	230	762	1116	826	755	416	220	41	0	4477
1986-87	0	3	11	233	747	899	1099	631	551	263	14	0	4451
1987-88	0	2	7	282	523	924	1170	891	637	330	33	0	4799
1988-89	0	0	16	265	519	813	817	1044	556	238	90	8	4366
1989-90	0	0	105	193	578	1228	783	728	555	332	112	0	4614
1990-91	0	0	18	238	445	1074	1114	567	475	201	53	0	4185
1991-92	0	0	71	242	759	793	802	574	461	259	113	10	4084
1992-93	0	2	25	209	703	975	1122	913	670	368	100	10	5097
1993-	0	0	59	321	718	837							
POR= 40 YRS	0	0	41	228	613	955	1081	831	616	274	82	5	4726

WBAN : 03928

COOLING DEGREE DAYS (base 65°F) 1993 WICHITA, KANSAS (ICT)

YEAR	JAN	FEB	MAR	APR	MAY	JUN	JUL	AUG	SEP	OCT	NOV	DEC	ANNUAL
1969	0	0	0	1	90	208	563	437	214	45	0	0	1558
1970	0	0	2	24	187	315	513	573	212	22	0	0	1848
1971	0	0	0	19	65	425	414	372	259	34	0	0	1588
1972	0	0	2	26	88	358	385	421	234	34	0	0	1548
1973	0	0	0	7	39	360	482	468	149	47	0	0	1552
1974	0	0	4	12	149	233	608	368	69	23	0	0	1466
1975	0	0	0	32	49	275	450	515	128	63	0	0	1512
1976	0	0	2	19	32	270	420	464	187	23	0	0	1417
1977	0	0	3	19	152	404	581	404	254	24	0	0	1841
1978	0	0	6	20	122	366	631	510	364	23	5	0	2047
1979	0	0	5	14	81	294	488	465	249	67	0	0	1663
1980	0	0	0	3	75	456	796	635	340	52	1	0	2358
1981	0	0	0	72	56	393	582	412	240	17	0	0	1772
1982	0	0	1	16	70	186	516	534	253	29	0	0	1605
1983	0	0	0	2	34	220	521	628	286	28	5	0	1724
1984	0	0	0	6	61	388	520	558	272	35	0	0	1840
1985	0	0	0	36	122	285	523	394	262	1	0	0	1623
1986	0	0	17	40	102	419	563	349	275	10	0	0	1775
1987	0	0	0	42	166	350	473	434	177	3	10	0	1655
1988	0	0	0	3	140	415	497	566	235	15	0	0	1871
1989	0	0	5	81	129	225	442	379	132	70	0	0	1463
1990	0	0	0	17	79	508	525	497	306	41	11	0	1984
1991	0	0	9	25	225	433	592	484	214	77	0	0	2059
1992	0	0	0	17	66	203	441	281	213	37	0	0	1258
1993	0	0	0	0	59	304	526	502	134	27	0	0	1552
POR= 25 YRS	0	0	1	22	97	330	521	465	226	33	0	0	1695

WBAN : 03928

SNOWFALL (inches) 1993 WICHITA, KANSAS (ICT)

YEAR	JUL	AUG	SEP	OCT	NOV	DEC	JAN	FEB	MAR	APR	MAY	JUN	TOTAL
1964-65	0.0	0.0	0.0	0.0	0.1	5.4	3.3	2.7	T	0.0	0.0	0.0	11.5
1965-66	0.0	0.0	0.0	0.0	0.0	0.9	4.7	0.8	T	0.0	0.0	0.0	6.4
1966-67	0.0	0.0	0.0	0.0	T	5.0	2.1	T	0.7	0.0	0.0	0.0	7.8
1967-68	0.0	0.0	0.0	0.0	2.2	9.7	0.6	1.0	T	0.0	0.0	0.0	13.5
1968-69	0.0	0.0	0.0	0.0	0.3	2.8	T	8.3	5.1	0.0	0.0	0.0	16.5
1969-70	0.0	0.0	0.0	0.0	T	3.5	2.8	T	16.5	0.1	0.0	0.0	22.9
1970-71	0.0	0.0	0.0	T	T	T	T	16.7	T	0.0	0.0	0.0	16.7
1971-72	0.0	0.0	0.0	0.0	2.3	2.4	1.7	4.4	0.8	0.0	0.0	0.0	11.6
1972-73	0.0	0.0	0.0	0.0	7.1	2.7	17.7	0.2	T	2.3	0.0	0.0	30.0
1973-74	0.0	0.0	0.0	0.0	T	8.7	4.1	0.7	2.0	0.3	0.0	0.0	15.8
1974-75	0.0	0.0	0.0	0.0	1.8	2.2	7.6	15.2	7.6	T	0.0	0.0	34.4
1975-76	0.0	0.0	0.0	0.0	5.5	T	0.6	0.9	T	0.0	0.0	0.0	7.0
1976-77	0.0	0.0	0.0	T	0.3	T	3.6	T	T	T	0.0	0.0	3.9
1977-78	0.0	0.0	0.0	0.0	T	T	7.4	7.8	0.3	0.0	0.0	0.0	15.5
1978-79	0.0	0.0	0.0	0.0	T	6.7	13.9	1.9	1.6	4.6	0.0	0.0	28.7
1979-80	0.0	0.0	0.0	0.0	T	T	T	12.3	0.4	0.0	0.0	0.0	12.7
1980-81	0.0	0.0	0.0	T	T	0.4	T	2.5	0.2	0.0	0.0	0.0	3.1
1981-82	0.0	0.0	0.0	0.0	T	1.2	0.0	12.7	T	0.0	0.0	0.0	13.9
1982-83	0.0	0.0	0.0	0.0	T	1.4	13.0	8.9	1.5	0.7	0.0	0.0	25.5
1983-84	0.0	0.0	0.0	0.0	4.1	13.8	4.3	T	6.9	0.0	0.0	0.0	29.1
1984-85	0.0	0.0	0.0	0.0	6.8	7.6	3.5	3.6	0.2	0.0	0.0	0.0	21.7
1985-86	0.0	0.0	0.0	0.0	1.0	3.0	0.0	7.5	0.0	0.0	0.0	0.0	11.5
1986-87	0.0	0.0	0.0	0.0	T	0.4	19.7	6.0	T	0.0	0.0	0.0	26.1
1987-88	0.0	0.0	0.0	0.0	6.2	12.6	8.5	1.1	11.0	0.0	0.0	0.0	39.4
1988-89	0.0	0.0	0.0	0.0	3.0	0.7	0.3	0.8	T	T	T	T	4.8
1989-90	0.0	0.0	0.0	0.0	T	4.4	0.3	7.0	0.7	0.1	0.0	0.0	12.5
1990-91	0.0	0.0	0.0	0.0	0.0	4.8	2.1	0.0	T	T	T	T	6.9
1991-92	0.0	0.0	0.0	1.5	0.3	0.0	T	T	T	T	0.0	T	1.8
1992-93	0.0	0.0	T	T	3.8	3.3	7.7	5.1	2.2	T	0.0	0.0	22.1
1993-	T	0.0	0.0	T	T	0.1							
POR= 40 YRS	T	0.0	T	T	1.3	3.2	4.6	4.2	2.4	0.3	T	T	16.0

WBAN : 03928

JACKSON, KENTUCKY

Jackson, County Seat of Breathitt County, is located on the leading edge of the Eastern Kentucky Coal Fields. The topography of Breathitt County is mountainous, with 80 to 90 percent of the county area on a greater than 20 percent slope. The county contains a minimal portion of level land. Almost 80 percent of the land area is composed of slopes greater than 50 percent. The highest elevation is 1,547 feet above sea level. The terrain slopes gently westward into the beautiful Kentucky Bluegrass Region. To the east the mountains rise swiftly to heights of 4,000 to 5,000 feet above sea level.

The major industry in Breathitt County is coal. In addition, the county is rich in such natural resources as timber, petroleum, natural gas, sand, and clay. The roughness of the terrain effectively limits farming to small yields of tobacco and garden vegetables.

The climate of Jackson and Eastern Kentucky is temperate and well suited to a variety of plant and animal life. There are no large bodies of water close enough to have any significant effect on the climate. The North Fork of the Kentucky River flows through Breathitt County westward into the Kentucky River and eventually into the Ohio River. There are numerous small creeks and streams in the county. The steep slopes and narrow valleys make these creeks and streams especially prone to flash flooding during periods of heavy rainfall.

Jackson is subject to sudden and large changes in temperature. Extremes of cold and heat are rare and usually of short duration. Temperatures above 100 degrees or below zero are extremely rare. Average daily temperatures range from about 32 degrees in the winter to the low 70s in the summer, and in the low 50s during the spring and fall months. January is the coldest month with an average temperature of 31 degrees. The warmest month is July, with an average temperature of 73 degrees.

Total annual precipitation for the Jackson area averages nearly 44 inches and is fairly evenly distributed throughout the year. The spring and summer seasons average nearly 12 inches each, while winter averages 11 inches and fall slightly over 8 inches. July is the wettest month with an average of nearly 5 inches of precipitation. Snowfall amounts are variable and snow cover is normally limited to only a few days at a time.

NORMALS, MEANS, AND EXTREMES

JACKSON, KENTUCKY

LATITUDE: 37°35'N LONGITUDE: 83°13'W ELEVATION: FT. GRND 1365 BARO 1356 TIME ZONE: EASTERN WBAN: 03889

	(a)	JAN	FEB	MAR	APR	MAY	JUNE	JULY	AUG	SEP	OCT	NOV	DEC	YEAR
TEMPERATURE °F:														
Normals														
-Daily Maximum		41.4	45.5	57.2	67.1	74.5	81.2	83.9	83.2	77.2	67.3	56.4	46.0	65.1
-Daily Minimum		24.2	27.5	37.4	46.0	54.2	61.9	65.2	64.4	58.6	47.0	39.0	29.8	46.3
-Monthly		32.8	36.5	47.3	56.6	64.3	71.6	74.6	73.8	67.9	57.2	47.7	37.9	55.7
Extremes														
-Record Highest	13	72	74	87	92	90	99	101	101	95	86	81	79	101
-Year		1993	1982	1989	1986	1986	1988	1988	1988	1991	1991	1993	1982	JUL 1988
-Record Lowest	13	-18	-4	8	20	32	44	52	45	34	26	13	-13	-18
-Year		1985	1981	1993	1982	1989	1993	1983	1989	1989	1987	1986	1989	JAN 1985
NORMAL DEGREE DAYS:														
Heating (base 65°F)		998	798	549	262	125	0	0	0	38	264	519	840	4393
Cooling (base 65°F)		0	0	0	10	103	202	298	273	125	22	0	0	1033
% OF POSSIBLE SUNSHINE														
MEAN SKY COVER (tenths)														
Sunrise - Sunset	12	7.7	7.8	7.3	6.8	7.0	6.5	6.2	6.2	6.1	6.2	7.2	7.7	6.9
MEAN NUMBER OF DAYS:														
Sunrise to Sunset														
-Clear	12	4.4	4.3	5.3	6.0	5.5	6.0	6.6	6.7	8.1	8.0	5.5	4.6	70.9
-Partly Cloudy	12	5.6	5.1	6.9	7.8	8.4	10.8	12.1	12.2	9.0	7.9	5.8	5.0	96.5
-Cloudy	12	21.0	18.9	18.8	16.3	17.1	13.3	12.3	12.2	12.9	15.1	18.7	21.4	197.8
Precipitation .01 inches or more	13	13.5	12.9	13.4	11.5	13.5	11.5	12.7	9.5	9.1	8.9	11.7	13.9	142.2
Snow,Ice pellets,hail 1.0 inches or more	13	1.8	2.0	0.6	0.3	0.1	0.0	0.0	0.0	0.0	0.1	0.2	1.5	6.6
Thunderstorms	13	0.5	1.1	3.2	4.4	8.1	9.5	12.4	8.8	3.5	1.4	1.5	0.3	54.5
Heavy Fog Visibility 1/4 mile or less	13	3.8	4.6	4.1	1.8	6.8	7.2	9.6	9.5	8.2	5.2	4.3	4.8	70.0
Temperature °F														
-Maximum														
90° and above	13	0.0	0.0	0.0	0.2	0.2	2.0	7.1	4.8	1.0	0.0	0.0	0.0	15.2
32° and below	13	5.6	3.7	0.5	0.2	0.0	0.0	0.0	0.0	0.0	0.0	0.2	4.7	14.8
-Minimum														
32° and below	13	22.9	16.8	11.8	2.5	0.1	0.0	0.0	0.0	0.0	1.2	8.5	17.9	81.6
0° and below	13	0.7	0.1	0.0	0.0	0.0	0.0	0.0	0.0	0.0	0.0	0.0	0.7	1.5
AVG. STATION PRESS.(mb)	12	970.5	969.7	968.1	966.8	968.3	968.5	969.9	970.2	971.3	971.5	970.8	971.4	969.7
RELATIVE HUMIDITY (%)														
Hour 01	12	71	69	64	61	75	82	85	86	83	75	70	74	75
Hour 07 (Local Time)	12	77	76	72	70	81	85	89	91	89	82	75	78	80
Hour 13	12	63	60	53	47	56	60	62	63	61	55	58	65	59
Hour 19	12	64	62	55	49	63	68	72	72	71	62	63	67	64
PRECIPITATION (inches):														
Water Equivalent														
-Normal		3.76	3.82	4.77	3.95	4.63	4.25	5.14	3.91	3.66	3.20	4.20	4.38	49.67
-Maximum Monthly	13	5.28	7.61	6.74	5.72	7.36	7.01	9.74	7.70	7.82	7.36	9.32	12.97	12.97
-Year		1982	1989	1989	1984	1984	1991	1985	1993	1988	1989	1986	1990	DEC 1990
-Minimum Monthly	13	0.80	2.00	1.56	0.78	2.25	1.37	2.21	1.55	1.40	0.51	1.45	1.74	0.51
-Year		1981	1988	1986	1985	1987	1988	1989	1981	1985	1987	1981	1985	OCT 1987
-Maximum in 24 hrs	13	1.59	2.38	3.44	2.01	3.39	2.62	2.77	4.11	2.88	4.24	4.30	3.43	4.30
-Year		1988	1990	1989	1984	1984	1981	1985	1993	1988	1989	1986	1990	NOV 1986
Snow,Ice pellets,hail														
-Maximum Monthly	13	12.1	21.0	21.9	17.8	1.0	0.0	0.0	T	0.0	2.3	1.8	11.5	21.9
-Year		1985	1985	1993	1987	1989			1991		1993	1989	1993	MAR 1993
-Maximum in 24 hrs	13	5.9	12.7	19.8	8.5	1.0	0.0	0.0	T	0.0	2.3	2.3	3.8	19.8
-Year		1987	1985	1993	1987	1989			1991		1993	1993	1982	MAR 1993
WIND:														
Mean Speed (mph)	12	8.0	8.0	8.3	8.2	6.7	6.1	5.7	5.4	6.1	6.6	8.0	8.1	7.1
Prevailing Direction														
Fastest Obs. 1 Min.														
-Direction (!!)	12	23	12	27	18	23	36	33	28	24	22	19	25	18
-Speed (MPH)	12	37	29	32	40	29	29	32	25	21	21	30	29	40
-Year		1992	1983	1988	1983	1985	1981	1985	1987	1989	1989	1983	1987	APR 1983
Peak Gust														
-Direction (!!)	10	SW	S	NW	SW	SW	NW	NW	W	W	SW	W	S	NW
-Speed (mph)	10	55	60	48	58	49	60	55	46	39	48	51	60	60
-Date		1992	1988	1989	1993	1990	1993	1985	1987	1989	1990	1992	1985	JUN 1993

PRECIPITATION (inches)

JACKSON, KENTUCKY

YEAR	JAN	FEB	MAR	APR	MAY	JUNE	JULY	AUG	SEP	OCT	NOV	DEC	ANNUAL
1981	0.80	5.96	2.78	4.62	4.23	4.88	4.53	1.55	2.53	2.83	1.45	3.53	39.69
1982	5.28	4.34	5.24	2.19	5.43	6.13	4.12	6.14	1.53	1.62	3.55	3.70	49.27
1983	1.68	2.09	1.90	3.25	7.34	2.01	3.81	3.33	3.12	3.83	3.27	2.76	38.39
1984	1.31	2.88	3.43	5.72	7.36	2.09	6.71	2.58	2.43	4.09	5.54	4.08	48.22
1985	3.50	2.02	3.27	0.78	5.50	4.19	9.74	5.24	1.40	4.97	6.89	1.74	49.24
1986	1.84	5.44	1.56	0.95	2.42	2.15	2.73	2.49	3.27	2.38	9.32	3.06	37.61
1987	2.70	3.46	1.90	3.70	2.25	3.22	6.37	2.64	2.92	0.51	3.15	5.98	38.80
1988	2.59	2.00	3.09	2.97	4.50	1.37	4.56	4.15	7.82	1.85	6.12	4.03	45.05
1989	3.48	7.61	6.74	3.23	6.43	6.96	2.21	5.22	7.37	7.36	4.28	2.40	63.29
1990	2.56	6.27	3.16	2.95	5.08	4.02	4.18	4.21	1.86	4.73	2.91	12.97	54.90
1991	3.16	4.46	6.08	2.67	4.99	7.01	4.46	2.96	2.62	2.13	5.22	9.35	55.11
1992	1.87	3.12	5.80	1.66	4.61	4.08	6.67	3.20	3.66	1.60	3.23	4.92	44.42
1993	2.05	3.54	5.28	3.26	3.74	4.82	4.70	7.70	6.58	4.63	5.08	4.28	55.66
Record Mean	2.52	4.09	3.86	2.92	4.91	4.07	4.98	3.95	3.62	3.27	4.62	4.83	47.67

AVERAGE TEMPERATURE (deg. F)

JACKSON, KENTUCKY

YEAR	JAN	FEB	MAR	APR	MAY	JUNE	JULY	AUG	SEP	OCT	NOV	DEC	ANNUAL
1981	29.3	38.4	42.7	61.4	60.4	74.2	75.0	72.8	66.6	56.3	48.3	35.3	55.1
1982	30.2	37.9	49.8	53.5	70.9	69.5	76.7	72.8	66.5	58.8	49.2	45.3	56.8
1983	34.4	38.0	47.2	51.8	60.8	71.1	75.9	77.5	68.2	58.3	47.7	30.6	55.1
1984	30.9	42.8	42.0	54.0	61.2	75.0	72.8	74.0	65.5	63.7	43.9	47.2	56.1
1985	24.5	33.0	49.3	60.8	65.3	69.9	74.0	71.0	67.4	62.2	54.4	32.3	55.3
1986	34.4	40.6	48.6	60.0	66.8	73.8	78.7	73.6	70.5	59.0	46.6	37.4	57.5
1987	33.5	38.9	48.7	54.3	69.9	73.6	75.5	75.6	68.1	52.9	51.6	40.2	56.9
1988	32.4	36.4	47.6	56.3	65.0	72.6	77.8	78.0	68.3	51.3	49.3	39.2	56.2
1989	42.7	35.4	50.0	56.4	60.4	71.2	75.4	73.2	67.0	57.8	47.3	26.7	55.3
1990	42.8	46.3	52.0	56.9	63.0	72.1	75.6	73.1	68.8	58.1	52.0	42.8	58.6
1991	36.2	41.2	48.8	61.3	70.7	73.3	76.7	74.8	69.3	61.4	45.9	41.8	58.5
1992	37.5	43.4	47.4	58.2	62.7	69.6	75.6	70.7	67.5	55.9	47.1	38.1	56.1
1993	39.7	35.4	42.7	54.5	65.5	71.8	79.1	75.6	66.9	55.6	47.2	36.5	55.9
Record Mean	34.5	39.0	47.5	56.8	64.8	72.1	76.1	74.0	67.7	57.8	48.5	38.0	56.4
Max	42.9	48.2	57.7	67.7	75.1	82.0	85.7	83.8	77.5	68.0	57.5	46.0	66.0
Min	26.1	29.9	37.2	46.0	54.5	62.2	66.4	64.2	57.9	47.6	39.4	29.9	46.8

REFERENCE NOTES

JACKSON, KENTUCKY

GENERAL
T - TRACE AMOUNT.
BLANK ENTRIES DENOTE MISSING/UNREPORTED DATA.
# INDICATES A STATION OR INSTRUMENT RELOCATION.

SPECIFIC

(a) - LENGTH OF RECORD IN YEARS, ALTHOUGH INDIVIDUAL MONTHS MAY BE MISSING.
0.* OR * - THE VALUE IS BETWEEN 0.0 AND 0.05
NORMALS - BASED ON THE 1961-1990 RECORD PERIOD.
EXTREMES - DATES ARE THE MOST RECENT OCCURRENCE.
WIND DIR.- NUMERALS SHOW TENS OF DEGRESS CLOCKWISE FROM TRUE NORTH. "00" INDICATES CALM.
RESULTANT DIRECTIONS ARE GIVEN TO WHOLE DEGREES.
BOLD VALUES INDICATE EXTREME VALUES WHICH OCCURRED AFTER THE ASOS SYSTEM WAS COMMISSIONED.

RECORD = PERIOD OF RECORD
RECORD MEAN PRECIPITATION IS THE MEAN OF ALL DAILY PRECIPITATION AMOUNTS DURING THE PERIOD OF RECORD.
RECORD MAX(MIN) TEMPERATURE IS THE MEAN OF ALL DAILY MAX(MIN) TEMPERATURES DURING THE PERIOD OF RECORD.
RECORD MEAN TEMPERATURE IS THE SUM OF THE RECORD MAX AND RECORD MIN DIVIDED BY 2.
AVERAGE TEMPERATURE IS THE SUM OF THE MEAN DAILY MAX AND MIN TEMPERATURE DIVIDED BY 2.

EXCEPTIONS

RECORD MEANS ARE THROUGH THE CURRENT YEAR, BEGINNING IN 1981 FOR TEMPERATURE
1981 FOR PRECIPITATION
1981 FOR SNOWFALL

HEATING DEGREE DAYS Base 65 deg. F JACKSON, KENTUCKY

SEASON	JULY	AUG	SEP	OCT	NOV	DEC	JAN	FEB	MAR	APR	MAY	JUNE	TOTAL
1980-81							1100	739	689	153	174	0	
1981-82	0	0	68	266	495	911	1075	755	467	348	6	2	4393
1982-83	0	0	61	248	479	621	942	748	557	400	145	19	4220
1983-84	1	0	71	216	512	1056	1049	638	706	353	168	5	4775
1984-85	0	0	113	82	630	548	1250	890	483	195	67	27	4285
1985-86	0	1	75	138	321	1007	945	677	511	217	73	2	3967
1986-87	0	15	9	221	545	849	970	724	499	343	35	1	4211
1987-88	0	0	29	373	401	758	1005	824	540	276	78	31	4315
1988-89	0	0	24	421	467	792	683	822	482	302	199	4	4196
1989-90	0	11	68	240	526	1179	683	520	431	290	106	8	4062
1990-91	0	2	52	243	390	682	884	658	504	161	30	1	3607
1991-92	0	0	75	182	570	712	844	621	548	261	150	14	3977
1992-93	0	3	51	282	529	826	778	820	683	314	61	25	4372
1993-94	0	0	57	292	533	876							

COOLING DEGREE DAYS Base 65 deg. F JACKSON, KENTUCKY

YEAR	JAN	FEB	MAR	APR	MAY	JUNE	JULY	AUG	SEP	OCT	NOV	DEC	TOTAL
1981	0	0	5	50	40	281	316	249	121	4	2	0	1068
1982	0	0	2	8	195	143	368	249	115	63	10	18	1171
1983	0	0	11	11	22	208	350	393	175	17	0	0	1187
1984	0	0	0	31	58	313	247	285	133	48	3	2	1120
1985	0	0	4	76	81	180	284	194	155	60	9	0	1043
1986	0	0	12	71	136	271	428	289	178	42	0	0	1427
1987	0	0	2	29	194	268	332	335	127	2	7	0	1296
1988	0	0	6	19	83	263	404	410	128	6	1	0	1320
1989	0	0	22	48	66	195	326	271	135	24	1	0	1088
1990	0	0	37	52	50	225	335	262	172	36	6	0	1175
1991	0	0	11	58	213	258	372	312	213	78	3	0	1518
1992	0	0	11	66	89	160	337	183	133	8	0	0	987
1993	0	0	0	7	83	233	440	337	122	7	6	0	1235

SNOWFALL (inches) JACKSON, KENTUCKY

SEASON	JULY	AUG	SEP	OCT	NOV	DEC	JAN	FEB	MAR	APR	MAY	JUNE	TOTAL
1980-81							5.1	6.8	1.8	0.0	0.0	0.0	
1981-82	0.0	0.0	0.0	0.0	T	3.8	6.0	7.6	5.0	0.4	0.0	0.0	22.8
1982-83	0.0	0.0	0.0	0.0	T	4.1	2.0	6.9	2.8	0.1	0.0	0.0	15.9
1983-84	0.0	0.0	0.0	0.0	T	3.3	8.8	2.6	1.7	T	0.0	0.0	16.4
1984-85	0.0	0.0	0.0	0.0	T	1.9	12.1	21.0	T	T	0.0	0.0	35.0
1985-86	0.0	0.0	0.0	0.0	0.0	4.6	7.7	11.6	1.0	T	0.0	0.0	24.9
1986-87	0.0	0.0	0.0	0.0	0.8	T	11.6	6.0	0.7	17.8	0.0	0.0	36.9
1987-88	0.0	0.0	0.0	0.0	T	3.7	5.4	2.0	1.9	0.0	0.0	0.0	13.0
1988-89	0.0	0.0	0.0	0.0	0.2	2.5	0.4	5.6	0.1	T	1.0	0.0	9.8
1989-90	0.0	0.0	0.0	0.5	1.8	10.1	2.5	2.3	1.4	T	0.0	0.0	18.6
1990-91	0.0	0.0	0.0	0.0	0.0	0.3	0.9	4.8	3.0	0.0	0.0	0.0	9.0
1991-92	0.0	T	0.0	0.0	T	0.4	2.1	0.7	1.5	1.7	0.0	0.0	6.4
1992-93	0.0	0.0	0.0	0.0	1.1	4.5	0.1	8.3	21.9	0.2	T	0.0	36.1
1993-94	0.0	0.0	0.0	2.3	0.1	11.5							
Record Mean	0.0	T	0.0	0.2	0.3	3.9	5.0	6.6	3.3	1.6	0.1	0.0	21.0

LEXINGTON, KENTUCKY

Lexington, County Seat of Fayette County, is located in the heart of the famed Kentucky Blue Grass Region. Fayette County is a gently rolling plateau with the elevation varying between 900 and 1,050 feet above sea level. It is noted for its beauty, the fertility of its soil, excellent grass, stock farms, and burley tobacco. The soil has a high phosphorus content and this is very valuable in growing pasture grasses for the grazing of cattle and horses. Lexington has a decided continental climate with a rather large diurnal temperature range. The climate is temperate and well suited to a varied plant and animal life. There are no bodies of water close enough to have any effect on the climate. The closest river is the Kentucky which makes an arc about 15 to 20 miles to the southeast, south, and southwest on its course to the Ohio River. There are numerous small creeks that rise in the county and flow into the river. The reservoirs of the Lexington Water Company are about 5 miles southeast of the city and are the largest bodies of water in the area.

Lexington is subject to rather sudden and large changes in temperature with the spells generally of rather short duration. Temperatures above 100 degrees and below zero degrees are relatively rare. The average temperature for the winter is 35 degrees, spring 62 degrees, fall 50 degrees, and summer 74 degrees.

Precipitation is evenly distributed throughout the winter, spring, and summer, with about 12 inches recorded on the average for each of these seasons. The fall season averages nearly 8 1/2 inches. Snowfall amounts are variable and the ground does not retain snow cover more than a few days at a time.

The months of September and October are the most pleasant of the year. They have the least amount of precipitation, the greatest number of clear days, and generally comfortable temperatures are the rule during these months.

Based on the 1951-1980 period, the average first occurrence of 32 degrees Fahrenheit in the fall is October 25 and the average last occurrence in the spring is April 17.

NORMALS, MEANS, AND EXTREMES

LEXINGTON, KENTUCKY

LATITUDE: 38°02'N LONGITUDE: 84°36'W ELEVATION: FT. GRND 966 BARO 990 TIME ZONE: EASTERN WBAN: 93820

	(a)	JAN	FEB	MAR	APR	MAY	JUNE	JULY	AUG	SEP	OCT	NOV	DEC	YEAR
TEMPERATURE °F:														
Normals														
-Daily Maximum		39.1	43.6	55.3	65.5	74.3	82.7	85.8	84.9	78.3	67.2	54.9	44.2	64.7
-Daily Minimum		22.4	25.3	35.3	44.2	53.5	61.5	65.7	64.4	58.0	46.0	37.0	27.6	45.1
-Monthly		30.8	34.5	45.3	54.8	64.0	72.2	75.8	74.7	68.2	56.7	46.0	35.9	54.9
Extremes														
-Record Highest	49	76	76	83	88	92	101	103	103	103	91	83	75	103
-Year		1950	1945	1945	1962	1987	1988	1988	1983	1954	1959	1987	1982	JUL 1988
-Record Lowest	49	-21	-15	-2	18	26	39	47	42	34	20	-3	-19	-21
-Year		1963	1951	1960	1982	1966	1966	1972	1965	1993	1976	1950	1989	JAN 1963
NORMAL DEGREE DAYS:														
Heating (base 65°F)		1060	854	611	312	135	5	0	0	47	287	570	902	4783
Cooling (base 65°F)		0	0	0	6	104	221	335	301	143	30	0	0	1140
% OF POSSIBLE SUNSHINE														
MEAN SKY COVER (tenths)														
Sunrise - Sunset	49	7.3	7.1	7.1	6.6	6.3	6.0	5.8	5.5	5.4	5.3	6.7	7.3	6.4
MEAN NUMBER OF DAYS:														
Sunrise to Sunset														
-Clear	49	5.7	5.9	5.7	6.3	7.0	7.0	8.0	9.3	10.5	11.7	6.8	5.8	89.7
-Partly Cloudy	49	5.8	5.6	7.4	8.5	10.0	11.6	12.3	12.0	8.4	7.2	6.6	5.9	101.1
-Cloudy	49	19.5	16.8	18.0	15.2	14.0	11.4	10.7	9.7	11.1	12.1	16.6	19.4	174.4
Precipitation														
.01 inches or more	49	12.3	11.2	12.9	12.2	11.9	10.5	11.2	9.1	8.1	8.0	10.8	11.6	130.0
Snow,Ice pellets,hail														
1.0 inches or more	49	1.7	1.5	0.7	0.1	0.0	0.0	0.0	0.0	0.0	0.0	0.4	0.8	5.1
Thunderstorms	49	0.8	0.9	2.9	3.9	6.3	7.9	9.2	6.7	3.2	1.4	1.0	0.4	44.6
Heavy Fog Visibility 1/4 mile or less	49	2.4	2.0	1.4	0.7	1.0	1.1	1.5	1.9	2.1	1.8	1.3	2.0	19.2
Temperature °F														
-Maximum														
90° and above	30	0.0	0.0	0.0	0.0	0.2	3.3	7.5	6.2	1.7	0.0	0.0	0.0	18.9
32° and below	30	9.1	5.7	0.9	0.0	0.0	0.0	0.0	0.0	0.0	0.0	0.5	4.7	21.0
-Minimum														
32° and below	30	23.9	20.2	13.6	3.3	0.1	0.0	0.0	0.0	0.0	2.3	10.9	20.0	94.3
0° and below	30	1.6	0.7	0.1	0.0	0.0	0.0	0.0	0.0	0.0	0.0	0.0	0.4	2.8
AVG. STATION PRESS.(mb)	21	983.7	982.9	980.7	980.1	980.1	980.7	981.8	982.5	983.2	983.8	983.4	983.8	982.2
RELATIVE HUMIDITY (%)														
Hour 01	29	76	75	71	69	77	81	83	83	83	77	76	77	77
Hour 07 (Local Time)	30	80	79	77	75	81	83	86	88	88	84	81	81	82
Hour 13	30	68	64	58	54	57	57	58	59	59	56	63	68	60
Hour 19	30	70	66	59	55	60	61	63	64	66	64	68	72	64
PRECIPITATION (inches):														
Water Equivalent														
-Normal		2.86	3.21	4.40	3.88	4.47	3.66	5.00	3.93	3.20	2.57	3.39	3.98	44.55
-Maximum Monthly	49	16.65	10.12	10.38	9.30	10.84	11.69	10.64	11.18	9.69	6.13	6.87	10.17	16.65
-Year		1950	1989	1975	1970	1983	1960	1958	1974	1979	1983	1951	1990	JAN 1950
-Minimum Monthly	49	0.37	0.67	0.99	0.79	1.20	0.61	1.83	0.56	0.24	0.33	0.45	0.61	0.24
-Year		1981	1978	1966	1946	1965	1988	1951	1984	1959	1963	1976	1965	SEP 1959
-Maximum in 24 hrs	49	2.98	3.79	3.85	4.39	3.24	5.88	4.73	3.56	4.35	3.21	2.71	3.77	5.88
-Year		1951	1989	1952	1948	1983	1960	1978	1968	1979	1962	1988	1978	JUN 1960
Snow,Ice pellets,hail														
-Maximum Monthly	49	21.9	16.4	17.7	5.9	T	T	T	T	0.0	0.2	9.7	10.7	21.9
-Year		1978	1960	1960	1987	1989	1993	1989	1989		1972	1950	1967	JAN 1978
-Maximum in 24 hrs	49	9.4	7.3	9.5	4.9	T	T	T	T	0.0	0.2	7.5	7.8	9.5
-Year		1966	1971	1947	1987	1989	1993	1989	1989		1972	1966	1967	MAR 1947
WIND:														
Mean Speed (mph)	46	10.9	10.8	11.2	10.6	8.7	7.9	7.3	6.9	7.6	8.2	10.0	10.6	9.2
Prevailing Direction through 1963		S	SSW	SSW	SSW	S	S	SSW	S	S	S	S	S	S
Fastest Obs. 1 Min.														
-Direction (!!)	32	27	32	27	32	22	34	29	22	32	26	15	19	32
-Speed (MPH)	32	41	46	36	46	35	38	37	39	32	32	39	37	46
-Year		1978	1962	1962	1963	1965	1992	1966	1964	1980	1965	1983	1971	APR 1963
Peak Gust														
-Direction (!!)	10	SW	W	E	S	NW	N	W	NW	NW	SW	W	SW	N
-Speed (mph)	10	53	56	47	61	51	64	56	51	40	40	49	54	64
-Date		1985	1993	1986	1993	1993	1992	1987	1988	1989	1984	1992	1985	JUN 1992

PRECIPITATION (inches)

LEXINGTON, KENTUCKY

YEAR	JAN	FEB	MAR	APR	MAY	JUNE	JULY	AUG	SEP	OCT	NOV	DEC	ANNUAL
1964	2.83	2.52	10.06	2.86	1.68	3.55	3.99	1.96	4.91	0.57	2.37	6.18	43.48
1965	2.83	2.92	5.45	3.24	1.20	3.28	3.64	1.92	3.55	2.44	1.02	0.61	32.10
1966	3.99	3.64	0.99	7.16	4.85	1.36	5.74	4.55	4.03	1.44	4.01	4.60	46.36
1967	1.35	2.29	6.03	3.61	6.94	2.65	5.87	3.63	2.97	2.11	4.15	3.99	45.59
1968	1.44	0.71	6.76	3.23	6.87	2.83	4.54	5.27	2.98	2.57	3.23	2.89	43.32
1969	4.26	1.60	1.50	3.83	3.78	4.61	4.37	5.96	0.49	1.53	3.42	4.15	39.50
1970	0.95	3.60	4.72	9.30	3.18	3.81	3.39	2.61	5.71	2.37	2.35	5.51	47.50
1971	3.29	4.72	2.02	2.10	6.14	7.84	7.64	0.88	3.51	0.52	1.71	4.30	44.67
1972	4.10	5.60	4.04	8.75	3.84	3.61	5.58	3.95	4.30	2.71	4.21	6.92	57.61
1973	1.53	1.58	5.08	5.67	8.22	6.06	5.15	3.58	1.40	2.65	6.58	3.42	50.92
1974	6.39	2.24	5.89	3.33	5.52	7.21	4.82	11.18	4.18	1.53	4.08	3.72	60.09
1975	3.66	5.70	10.38	6.17	2.69	2.23	5.60	3.96	6.46	5.09	2.93	4.24	59.11
1976	3.59	4.67	3.72	1.24	3.16	3.34	6.74	1.26	4.23	3.86	0.45	1.22	37.48
1977	2.30	1.03	4.21	3.42	1.51	4.80	4.59	4.83	2.71	3.77	3.95	3.04	40.16
1978	6.38	0.67	2.87	3.15	5.74	1.94	7.60	10.00	3.10	3.20	3.11	9.97	57.73
1979	4.07	2.92	3.22	4.92	4.17	2.80	4.72	6.20	9.69	2.96	4.52	3.81	54.00
1980	1.63	1.17	6.04	2.82	2.27	1.88	5.55	5.10	2.47	2.07	2.02	1.67	34.69
1981	0.37	4.76	1.76	4.88	5.10	2.29	5.27	2.72	1.97	2.44	1.99	3.10	36.65
1982	5.48	2.16	3.89	2.19	2.51	3.95	3.82	4.01	1.21	1.56	3.45	4.53	38.76
1983	1.29	1.61	1.48	5.18	10.84	2.18	2.41	1.26	1.33	6.13	3.59	3.46	40.76
1984	1.64	3.31	4.09	5.02	5.34	2.20	4.80	0.56	1.36	3.87	5.19	4.89	42.27
1985	1.91	1.11	3.69	2.34	4.34	4.98	3.37	3.76	1.93	4.23	4.96	1.13	37.75
1986	0.53	2.48	2.43	1.65	3.24	1.29	5.64	2.67	3.08	2.06	6.49	3.30	34.86
1987	1.30	3.62	3.13	2.23	1.80	6.59	3.48	4.18	0.91	0.55	2.72	6.17	36.68
1988	2.94	3.06	2.34	2.93	3.02	0.61	3.51	4.18	5.96	1.34	5.39	3.62	38.90
1989	3.99	10.12	6.08	2.60	5.39	4.26	4.20	3.98	4.98	3.38	2.38	1.80	53.16
1990	4.17	3.43	1.89	2.37	5.41	4.59	6.45	4.36	2.12	4.49	2.69	10.17	52.14
1991	2.57	3.91	5.80	2.70	3.95	2.91	3.60	3.08	2.09	2.70	1.27	7.22	41.80
1992	3.63	1.84	4.70	2.11	4.68	7.74	10.27	4.73	3.44	0.65	3.50	1.80	49.09
1993	2.42	4.15	3.77	3.53	2.43	5.46	3.38	4.52	3.00	4.19	5.42	3.31	45.58
Record Mean	3.94	3.24	4.44	3.63	3.90	4.09	4.53	3.54	2.89	2.45	3.27	3.72	43.62

AVERAGE TEMPERATURE (deg. F)

LEXINGTON, KENTUCKY

YEAR	JAN	FEB	MAR	APR	MAY	JUNE	JULY	AUG	SEP	OCT	NOV	DEC	ANNUAL
1964	35.0	32.0	45.3	58.3	66.9	74.3	75.2	75.0	68.2	52.7	48.0	37.4	55.7
1965	32.2	34.3	37.9	55.9	68.9	71.3	73.7	74.5	68.8	53.9	46.2	39.6	54.8
1966	25.3	33.3	44.6	52.6	60.5	71.6	77.4	72.6	65.3	54.2	45.7	35.5	53.2
1967	37.3	30.2	50.0	58.8	62.0	72.1	72.5	70.6	64.9	56.1	40.7	39.0	54.5
1968	29.6	28.0	46.0	55.1	62.1	72.0	75.9	75.5	67.8	56.6	46.4	34.6	54.1
1969	32.1	36.2	38.9	56.3	65.5	72.8	77.8	74.7	67.8	57.9	42.8	32.7	54.6
1970	25.8	32.4	40.5	57.2	66.5	71.3	74.1	74.6	72.7	57.7	43.6	38.3	54.5
1971	29.7	34.8	40.3	52.0	59.2	73.7	73.9	73.9	72.5	64.7	47.2	45.7	55.6
1972	35.4	33.2	42.2	53.7	63.2	67.6	73.8	72.9	69.8	53.0	44.5	40.9	54.2
1973	35.1	35.2	53.8	52.9	60.1	73.6	75.8	74.9	72.2	60.9	48.5	36.4	56.6
1974	40.7	37.4	48.5	55.8	63.6	67.2	74.4	73.3	62.6	54.3	45.7	37.8	55.1
1975	37.3	39.5	41.4	52.2	67.4	73.5	76.4	77.5	63.4	57.4	48.5	35.8	55.9
1976	28.4	43.9	50.0	54.5	60.6	71.1	73.0	71.2	64.1	49.7	37.1	30.9	52.9
1977	17.8	34.9	50.8	59.6	69.8	72.3	78.4	75.7	72.2	56.1	49.0	33.4	55.8
1978	21.6	21.3	40.4	57.0	61.1	73.1	76.0	74.1	70.9	53.7	48.4	37.9	53.0
1979	23.6	26.9	48.0	53.8	63.1	70.9	74.0	74.0	66.9	55.6	45.7	37.9	53.4
1980	32.4	28.3	41.5	52.5	64.8	71.2	78.8	78.2	70.6	53.7	43.6	36.0	54.3
1981	27.5	37.0	42.6	59.6	60.4	73.8	75.8	73.4	65.8	55.6	45.9	33.0	54.2
1982	28.2	34.9	47.1	50.6	69.8	68.5	77.1	72.9	65.6	58.3	48.4	44.2	55.5
1983	33.8	37.2	46.3	50.8	60.7	72.8	79.8	80.5	70.2	59.0	46.5	28.4	55.5
1984	27.6	41.2	39.7	53.1	60.6	76.0	72.9	74.9	66.5	63.5	41.8	45.4	55.3
1985	23.8	30.5	48.5	58.5	64.7	70.3	75.1	72.8	67.6	60.5	53.0	29.6	54.6
1986	33.2	38.6	46.8	57.3	65.5	74.2	78.6	72.9	71.0	58.0	44.9	35.7	56.4
1987	31.9	38.0	46.8	53.7	70.3	75.0	77.1	77.5	70.0	52.0	50.0	38.9	56.8
1988	29.8	33.7	44.8	54.2	64.5	74.3	79.1	77.9	67.8	49.6	46.1	36.5	54.9
1989	40.5	33.1	47.3	54.2	60.6	71.6	76.5	74.1	67.6	57.0	45.0	23.0	54.2
1990	41.6	43.1	48.9	53.2	61.6	72.2	75.3	73.8	68.5	56.4	49.9	40.4	57.1
1991	33.9	39.1	47.4	58.1	70.5	74.3	77.6	75.8	68.8	58.7	43.4	40.1	57.3
1992	35.2	41.3	45.4	56.1	62.0	69.4	75.6	70.9	67.0	55.8	45.9	37.0	55.1
1993	37.4	33.1	42.0	52.7	65.0	72.1	80.1	76.5	66.0	55.1	44.6	35.0	55.0
Record Mean	33.1	35.2	44.2	54.3	64.1	72.7	76.3	74.8	68.8	57.2	45.0	36.0	55.2
Max	41.1	43.8	53.7	64.5	74.2	82.6	86.0	84.7	79.1	67.5	53.7	43.8	64.6
Min	25.1	26.5	34.6	44.1	54.0	62.7	66.5	65.0	58.6	46.9	36.2	28.1	45.7

REFERENCE NOTES

LEXINGTON, KENTUCKY

GENERAL
T - TRACE AMOUNT.
BLANK ENTRIES DENOTE MISSING/UNREPORTED DATA.
# INDICATES A STATION OR INSTRUMENT RELOCATION.

SPECIFIC

PM - INCLUDES LAST DAY OF PREVIOUS MONTH
ASOS - AUTOMATED SURFACE OBSERVING SYSTEM IN OPERATION DURING THESE MONTHS.

(*) - LENGTH OF RECORD IN YEARS, ALTHOUGH INDIVIDUAL MONTHS MAY BE MISSING.
0.* OR * - THE VALUE IS BETWEEN 0.0 AND 0.05
NORMALS - BASED ON THE 1961-1990 RECORD PERIOD.
EXTREMES - DATES ARE THE MOST RECENT OCCURRENCE
WIND DIR.- NUMERALS SHOW TENS OF DEGRESS CLOCKWISE FROM TRUE NORTH. "00" INDICATES CALM.
RESULTANT DIRECTIONS ARE GIVEN TO WHOLE DEGREES.
BOLD VALUES INDICATE EXTREME VALUES WHICH OCCURRED AFTER THE ASOS SYSTEM WAS COMMISSIONED.

RECORD = PERIOD OF RECORD
RECORD MEAN PRECIPITATION IS THE MEAN OF ALL DAILY PRECIPITATION AMOUNTS DURING THE PERIOD OF RECORD.
RECORD MAX(MIN) TEMPERATURE IS THE MEAN OF ALL DAILY MAX(MIN) TEMPERATURES DURING THE PERIOD OF RECORD.
RECORD MEAN TEMPERATURE IS THE SUM OF THE RECORD MAX AND RECORD MIN DIVIDED BY 2.
AVERAGE TEMPERATURE IS THE SUM OF THE MEAN DAILY MAX AND MIN TEMPERATURE DIVIDED BY 2.

EXCEPTIONS

RECORD MEANS ARE THROUGH THE CURRENT YEAR, BEGINNING IN 1871 FOR TEMPERATURE
1871 FOR PRECIPITATION
1945 FOR SNOWFALL

HEATING DEGREE DAYS Base 65 deg. F LEXINGTON, KENTUCKY

SEASON	JULY	AUG	SEP	OCT	NOV	DEC	JAN	FEB	MAR	APR	MAY	JUNE	TOTAL
1964-65	0	11	61	375	502	850	1010	851	834	277	26	8	4805
1965-66	0	9	52	346	558	777	1225	883	631	373	167	21	5042
1966-67	0	3	62	340	570	908	851	967	468	223	138	23	4553
1967-68	1	4	83	289	722	800	1090	1065	581	298	124	8	5065
1968-69	0	5	20	296	552	938	1011	799	800	261	77	17	4776
1969-70	0	0	44	274	657	996	1209	908	750	251	84	1	5174
1970-71	6	0	31	229	637	821	1087	838	759	382	198	0	4988
1971-72	0	0	8	69	536	591	909	917	700	343	96	47	4216
1972-73	10	1	20	366	612	739	920	827	353	371	167	0	4386
1973-74	0	1	21	172	490	880	744	767	514	289	128	37	4043
1974-75	0	0	125	338	578	836	852	705	726	387	51	4	4602
1975-76	0	0	128	249	488	895	1130	606	468	339	158	2	4463
1976-77	1	4	64	474	829	1050	1457	836	444	208	52	19	5438
1977-78	0	0	6	277	498	972	1338	1219	755	254	179	6	5504
1978-79	0	0	20	348	492	834	1277	1061	522	337	110	15	5016
1979-80	0	5	40	307	574	833	1005	1057	721	371	88	17	5018
1980-81	0	0	23	358	633	892	1156	777	687	182	180	0	4888
1981-82	0	0	77	286	568	985	1134	840	549	429	14	9	4891
1982-83	0	1	75	259	500	646	961	772	580	422	151	7	4374
1983-84	0	0	59	201	550	1128	1152	685	778	370	178	3	5104
1984-85	2	0	89	84	689	601	1275	959	510	228	66	23	4526
1985-86	0	0	72	179	360	1092	978	735	561	259	94	2	4332
1986-87	0	15	14	250	595	903	1016	749	559	342	39	0	4482
1987-88	0	0	17	399	447	804	1085	901	620	328	90	18	4709
1988-89	0	3	30	474	560	877	750	887	548	351	196	8	4684
1989-90	0	6	61	267	592	1297	720	608	505	378	128	17	4579
1990-91	0	3	57	288	453	757	955	719	544	215	34	0	4025
1991-92	0	0	77	230	642	765	915	682	600	293	159	17	4380
1992-93	0	5	64	288	566	863	847	884	705	363	64	27	4676
1993-94	0	0	67	313	608	922							

COOLING DEGREE DAYS Base 65 deg. F LEXINGTON, KENTUCKY

YEAR	JAN	FEB	MAR	APR	MAY	JUNE	JULY	AUG	SEP	OCT	NOV	DEC	TOTAL
1969	0	0	0	7	97	260	404	309	134	57	0	0	1268
1970	0	0	0	26	137	199	295	306	266	11	0	0	1240
1971	0	0	0	0	25	266	281	285	239	65	10	0	1171
1972	0	0	0	11	47	130	287	250	171	0	4	0	900
1973	0	0	12	18	21	266	342	314	245	51	0	0	1269
1974	0	0	10	21	94	108	296	264	60	11	4	0	868
1975	0	0	0	11	130	267	357	394	86	18	0	0	1263
1976	0	0	9	30	26	193	257	205	46	4	0	0	770
1977	0	0	11	52	206	241	422	337	232	8	23	0	1532
1978	0	0	0	19	69	257	349	290	202	4	0	0	1190
1979	0	0	2	8	57	199	287	292	102	21	0	0	968
1980	0	0	0	4	87	210	438	415	199	17	0	0	1370
1981	0	0	1	29	43	270	341	267	109	2	0	0	1062
1982	0	0	0	4	171	121	383	252	101	62	9	7	1110
1983	0	0	4	3	27	248	465	487	219	21	0	0	1474
1984	0	0	0	17	50	340	254	312	141	44	1	0	1159
1985	0	0	5	40	67	189	317	245	155	49	4	0	1071
1986	0	0	4	34	115	285	427	269	197	42	0	0	1373
1987	0	0	0	10	212	304	383	395	173	2	5	0	1484
1988	0	0	1	8	81	306	442	407	120	5	0	0	1370
1989	0	0	8	34	66	214	362	296	146	27	0	0	1153
1990	0	0	13	29	32	239	326	285	168	26	5	0	1123
1991	0	0	3	15	210	285	398	341	198	41	0	0	1491
1992	0	0	0	35	76	155	340	192	132	8	0	0	938
1993	0	0	0	2	74	247	474	365	103	13	3	0	1281

SNOWFALL (inches) LEXINGTON, KENTUCKY

SEASON	JULY	AUG	SEP	OCT	NOV	DEC	JAN	FEB	MAR	APR	MAY	JUNE	TOTAL
1964-65	0.0	0.0	0.0	0.0	0.7	0.7	9.9	1.8	2.8	0.0	0.0	0.0	15.9
1965-66	0.0	0.0	0.0	0.0	T	1.3	12.4	6.5	1.2	T	0.0	0.0	21.4
1966-67	0.0	0.0	0.0	0.0	8.4	0.3	1.5	11.1	4.3	0.0	0.0	0.0	25.6
1967-68	0.0	0.0	0.0	0.0	T	10.7	12.1	3.1	8.6	0.0	0.0	0.0	34.5
1968-69	0.0	0.0	0.0	0.0	1.0	1.5	1.9	6.1	1.7	0.0	0.0	0.0	12.2
1969-70	0.0	0.0	0.0	0.0	1.4	5.2	8.4	9.9	6.2	T	0.0	0.0	31.1
1970-71	0.0	0.0	0.0	0.0	T	1.2	4.0	9.1	5.3	0.0	0.0	0.0	19.6
1971-72	0.0	0.0	0.0	0.0	0.7	T	1.4	6.3	2.4	T	0.0	0.0	10.8
1972-73	0.0	0.0	0.0	0.2	1.7	0.4	1.0	1.4	0.4	0.2	0.0	0.0	5.3
1973-74	0.0	0.0	0.0	0.0	0.0	4.3	T	3.6	0.7	T	0.0	0.0	8.6
1974-75	0.0	0.0	0.0	T	1.4	2.8	4.6	1.5	5.6	T	0.0	0.0	15.9
1975-76	0.0	0.0	0.0	0.0	T	0.9	6.5	0.4	2.2	0.0	0.0	0.0	10.0
1976-77	0.0	0.0	0.0	0.0	2.9	1.6	18.5	3.5	0.1	0.8	0.0	0.0	27.4
1977-78	0.0	0.0	0.0	0.0	1.7	3.0	21.9	7.1	8.4	0.0	0.0	0.0	42.1
1978-79	0.0	0.0	0.0	0.0	0.0	0.7	11.4	11.6	0.1	T	0.0	0.0	23.8
1979-80	0.0	0.0	0.0	0.0	0.1	T	11.9	4.0	4.2	0.3	0.0	0.0	20.5
1980-81	0.0	0.0	0.0	0.0	0.1	0.5	2.2	0.4	0.5	0.0	0.0	0.0	3.7
1981-82	0.0	0.0	0.0	0.0	0.4	1.7	5.6	3.9	0.3	0.7	0.0	0.0	12.6
1982-83	0.0	0.0	0.0	0.0	0.0	T	0.2	7.5	0.3	T	0.0	0.0	8.0
1983-84	0.0	0.0	0.0	0.0	T	1.7	8.4	4.6	0.3	0.0	0.0	0.0	15.0
1984-85	0.0	0.0	0.0	0.0	T	4.9	10.2	10.7	T	0.5	0.0	0.0	26.3
1985-86	0.0	0.0	0.0	0.0	0.0	3.5	1.2	8.9	0.7	T	0.0	0.0	14.3
1986-87	0.0	0.0	0.0	0.0	0.2	T	3.6	3.5	2.1	5.9	0.0	0.0	15.3
1987-88	0.0	0.0	0.0	0.0	1.0	1.8	3.3	3.4	0.7	0.0	0.0	0.0	10.2
1988-89	0.0	0.0	0.0	0.0	T	0.7	T	1.5	T	T	T	0.0	2.2
1989-90	T	T	0.0	T	1.1	9.3	0.2	T	3.7	T	0.0	0.0	14.3
1990-91	0.0	0.0	0.0	0.0	0.0	0.8	0.1	3.2	1.7	0.0	0.0	0.0	5.8
1991-92	0.0	0.0	0.0	0.0	T	0.2	0.7	0.6	1.3	0.4	0.0	0.0	3.2
1992-93	0.0	0.0	0.0	T	1.8	2.5	0.4	11.5	7.1	T	0.0	T	23.3
1993-94	0.0	0.0	0.0	T	0.1	7.4							
Record Mean	T	T	0.0	T	0.6	1.9	5.4	4.8	2.7	0.3	T	T	15.7

LOUISVILLE, KENTUCKY

Louisville is located on the south bank of the Ohio River, 604 miles below Pittsburgh, Pennsylvania, and 377 miles above the mouth of the river at Cairo, Illinois. The city is divided by Beargrass Creek and its south fork into two portions with entirely different types of topography. The eastern portion is rolling, containing several creeks, and consists of plateaus and rolling hillsides. The highest elevation in this area is 565 feet. The western portion is mostly flat with an average elevation about 100 feet lower than the eastern area. Much of the western section lies in the flood plain of the Ohio River. Nearly all of the industries in the city are located in the western portion, while the eastern portion is almost entirely residential. A range of low hills about five miles northwest of Louisville, on the Indiana side of the Ohio River, present a partial barrier to arctic blasts in the winter months. During colder months, snow is frequently observed on the summits of these hills when there is no snow in the city of Louisville or in riverside communities on the Indiana side of the Ohio River.

The climate of Louisville, while continental in type, is of a variable nature because of its position with respect to the paths of high and low pressure systems and the occasional influx of warm moist air from the Gulf of Mexico. In winter and summer there are occasional cold and hot spells of short duration. As a whole, winters are moderately cold and summers are quite warm. Temperatures of 100 degrees or more in summer and zero degrees or less in winter are rare.

Thunderstorms with high rainfall intensities are common during the spring and summer months. The precipitation in Louisville is nonseasonal and varies from year to year. The fall months are usually the driest. Generally, March has the most rainfall and October the least. Snowfall usually occurs from November through March. As with rainfall, amounts vary from year to year and month to month. Some snow has also been recorded in the months of October and April. Mean total amounts for the months of January, February, and March are about the same with January showing a slight edge in total amount. Relative humidity remains rather high throughout the summer months. Cloud cover is about equally distributed throughout the year with the winter months showing somewhat of an increase in amount. The percentage of possible sunshine at Louisville varies from month to month with the greatest amount during the summer months as a result of the decreasing sky cover during that season. Heavy fog is unusual and there is only an average of 10 days during the year with heavy fog and these occur generally in the months of September through March.

The average date for the last occurrence in the spring of temperatures as low as 32 degrees is mid-April, and the first occurrence in the fall is generally in late October.

The prevailing direction of the wind has a southerly component and the velocity averages under 10 mph. The strongest winds are usually associated with thunderstorms.

NORMALS, MEANS, AND EXTREMES

LOUISVILLE, KENTUCKY

LATITUDE: 38°11'N LONGITUDE: 85°44' W ELEVATION: FT. GRND 477 BARO 485 TIME ZONE: EASTERN WBAN: 93821

	(a)	JAN	FEB	MAR	APR	MAY	JUNE	JULY	AUG	SEP	OCT	NOV	DEC	YEAR
TEMPERATURE °F:														
Normals														
-Daily Maximum		40.3	44.8	56.3	67.3	76.0	83.5	87.0	85.7	80.3	69.2	56.8	45.1	66.0
-Daily Minimum		23.2	26.5	36.2	45.4	54.7	62.9	67.3	65.8	58.7	45.8	37.3	28.6	46.0
-Monthly		31.7	35.7	46.3	56.3	65.3	73.2	77.2	75.8	69.5	57.6	47.1	36.9	56.1
Extremes														
-Record Highest	46	77	77	86	91	95	102	105	101	104	92	84	76	105
-Year		1950	1972	1981	1960	1959	1952	1954	1988	1954	1959	1958	1982	JUL 1954
-Record Lowest	46	-20	-19	-1	22	31	42	50	46	33	23	-1	-15	-20
-Year		1963	1951	1960	1982	1966	1966	1972	1986	1949	1952	1950	1989	JAN 1963
NORMAL DEGREE DAYS:														
Heating (base 65°F)		1032	820	580	273	105	6	0	0	36	254	537	871	4514
Cooling (base 65°F)		0	0	0	12	115	252	378	335	171	25	0	0	1288
% OF POSSIBLE SUNSHINE	46	42	48	50	55	60	66	67	66	64	61	46	40	55
MEAN SKY COVER (tenths)														
Sunrise - Sunset	46	7.3	7.0	7.0	6.6	6.3	5.9	5.7	5.3	5.4	5.3	6.7	7.2	6.3
MEAN NUMBER OF DAYS:														
Sunrise to Sunset														
-Clear	46	5.7	5.9	5.8	6.3	7.5	7.6	8.0	10.2	10.5	11.5	7.4	6.1	92.7
-Partly Cloudy	46	5.7	6.3	7.5	8.8	9.4	11.2	12.7	11.5	8.8	7.6	6.0	6.1	101.5
-Cloudy	46	19.6	16.1	17.6	14.9	14.0	11.1	10.3	9.3	10.7	11.9	16.4	18.8	170.8
Precipitation														
.01 inches or more	46	11.2	10.6	13.0	11.7	11.7	9.8	10.5	8.4	8.1	7.6	10.4	11.5	124.7
Snow,Ice pellets,hail														
1.0 inches or more	46	1.6	1.2	0.7	0.1	0.0	0.0	0.0	0.0	0.0	0.*	0.3	0.7	4.5
Thunderstorms	46	0.8	1.1	3.2	4.3	6.6	7.3	8.4	6.9	3.3	1.7	1.4	0.5	45.6
Heavy Fog Visibility														
1/4 mile or less	46	0.9	0.9	0.5	0.2	0.3	0.3	0.5	0.9	1.0	1.5	0.7	0.7	8.4
Temperature °F														
-Maximum														
90° and above	33	0.0	0.0	0.0	0.*	0.5	6.1	11.6	9.4	3.0	0.0	0.0	0.0	30.7
32° and below	33	8.6	4.9	0.6	0.0	0.0	0.0	0.0	0.0	0.0	0.0	0.3	4.2	18.7
-Minimum														
32° and below	33	24.0	19.8	11.8	2.2	0.1	0.0	0.0	0.0	0.0	1.5	9.1	19.9	88.5
0° and below	33	1.2	0.2	0.0	0.0	0.0	0.0	0.0	0.0	0.0	0.0	0.0	0.3	1.8
AVG. STATION PRESS.(mb)	21	1002.5	1001.6	998.8	997.9	997.4	997.8	998.7	999.6	1000.4	1001.4	1001.4	1002.3	1000.0
RELATIVE HUMIDITY (%)														
Hour 01	33	72	72	69	68	77	80	81	82	83	78	75	74	76
Hour 07 (Local Time)	33	76	77	76	76	82	83	85	88	88	85	79	78	81
Hour 13	33	64	61	57	52	55	56	58	58	59	55	61	65	58
Hour 19	33	65	62	57	52	57	58	60	61	63	61	65	68	61
PRECIPITATION (inches):														
Water Equivalent														
-Normal		2.86	3.30	4.66	4.23	4.62	3.46	4.51	3.54	3.16	2.71	3.70	3.64	44.39
-Maximum Monthly	46	11.38	9.02	14.91	11.10	11.57	10.11	10.05	8.79	10.49	6.47	9.12	8.86	14.91
-Year		1950	1989	1964	1970	1990	1960	1979	1974	1979	1983	1957	1990	MAR 1964
-Minimum Monthly	46	0.45	0.76	1.02	0.76	1.37	0.49	0.99	0.23	0.27	0.39	0.72	0.65	0.23
-Year		1981	1978	1966	1976	1977	1984	1983	1953	1953	1987	1976	1976	AUG 1953
-Maximum in 24 hrs	46	3.00	3.66	6.97	4.85	4.60	5.14	5.46	3.13	4.97	3.25	3.58	2.79	6.97
-Year		1988	1990	1964	1970	1961	1960	1979	1992	1979	1977	1948	1978	MAR 1964
Snow,Ice pellets,hail														
-Maximum Monthly	46	28.4	15.9	22.9	1.6	T	T	0.0	0.0	0.0	2.4	13.2	9.3	28.4
-Year		1978	1993	1960	1973	1989	1993				1993	1966	1961	JAN 1978
-Maximum in 24 hrs	46	14.1	11.0	12.1	1.6	T	T	0.0	0.0	0.0	2.4	13.0	5.0	14.1
-Year		1978	1966	1968	1973	1989	1993				1993	1966	1961	JAN 1978
WIND:														
Mean Speed (mph)	46	9.6	9.6	10.2	9.7	8.0	7.4	6.8	6.4	6.8	7.2	8.9	9.2	8.3
Prevailing Direction through 1963		S	NW	NW	SW	SE	S	S	N	SE	SE	S	S	S
Fastest Obs. 1 Min.														
-Direction (!!)	8	W	30	22	33	27	27	32	32	29	27	27	29	32
-Speed (MPH)	8	34	32	36	37	35	41	46	26	39	29	40	31	46
-Year		1984	1990	1986	1988	1985	1985	1987	1986	1986	1992	1993	1984	JUL 1987
Peak Gust														
-Direction (!!)	10	W	NW	NW	S	S	W	NW	S	NW	SE	W	S	NW
-Speed (mph)	10	48	52	60	49	60	72	78	53	48	44	51	52	78
-Date		1990	1990	1986	1985	1985	1990	1987	1984	1984	1984	1993	1987	JUL 1987

AVERAGE TEMPERATURE (deg. F)

LOUISVILLE, KENTUCKY

YEAR	JAN	FEB	MAR	APR	MAY	JUNE	JULY	AUG	SEP	OCT	NOV	DEC	ANNUAL
1964	35.9	33.4	45.3	58.4	66.8	75.1	76.4	76.2	69.1	53.4	48.0	38.0	56.3
1965	34.3	35.9	38.9	58.3	70.0	73.4	76.3	75.2	70.5	55.7	47.3	42.3	56.5
1966	27.1	34.1	46.1	53.4	62.3	73.1	81.1	75.6	67.4	54.6	47.2	35.6	54.8
1967	36.3	30.9	50.6	59.7	62.4	73.4	74.5	72.2	65.7	57.2	42.7	39.4	55.4
1968	30.3	30.1	45.7	56.8	63.4	74.0	77.7	77.9	69.0	57.4	48.0	35.7	55.5
1969	33.1	36.9	39.9	58.0	66.2	73.5	78.7	74.7	67.5	57.3	43.3	33.5	55.2
1970	27.9	33.6	42.3	59.3	67.2	72.9	75.8	76.0	73.4	58.1	45.3	39.6	56.0
1971	30.9	35.0	42.0	54.7	61.5	76.5	74.8	74.1	72.2	64.4	47.0	45.1	56.5
1972	35.2	34.9	44.8	56.2	65.5	70.6	77.1	76.1	72.3	55.3	44.0	39.1	55.9
1973	35.0	36.4	53.7	54.4	61.5	75.6	78.4	77.0	73.6	62.3	49.8	37.1	57.9
1974	39.8	39.3	49.8	57.2	65.1	68.7	75.9	75.0	63.2	54.9	47.0	39.1	56.3
1975	38.1	40.2	43.3	54.4	69.0	75.4	77.7	79.3	66.2	59.4	50.6	38.9	57.7
1976	31.3	45.4	52.4	57.5	62.9	72.9	76.8	74.2	66.8	52.5	39.5	33.1	55.4
1977	18.6	36.9	51.7	60.3	71.2	73.9	80.2	77.5	72.5	55.5	49.6	34.6	56.9
1978	22.9	23.8	41.7	58.0	63.8	75.7	78.5	77.1	73.7	55.5	50.0	40.0	55.1
1979	24.6	28.0	48.3	55.0	64.2	73.9	75.3	76.1	69.4	58.2	46.9	39.2	54.9
1980	33.5	29.6	41.8	53.6	66.8	73.4	81.5	81.0	73.5	55.8	46.3	38.3	56.3
1981	30.4	38.8	45.7	62.4	62.9	76.2	78.8	76.1	67.7	56.5	47.4	33.8	56.4
1982	28.6	34.9	47.1	51.3	70.3	69.3	78.0	73.5	66.8	59.0	48.7	44.9	56.0
1983	34.7	37.5	46.7	51.7	62.1	73.4	81.1	81.7	71.0	59.1	47.8	28.4	56.3
1984	28.9	41.5	40.4	55.0	62.6	77.7	75.5	76.0	67.2	63.9	44.0	45.9	56.6
1985	25.4	32.8	50.2	60.3	66.5	72.1	77.2	74.8	69.2	61.4	53.7	30.4	56.2
1986	34.5	39.9	48.3	58.5	67.0	75.7	80.3	74.3	73.1	59.5	45.9	36.7	57.8
1987	33.7	39.5	47.9	55.4	71.5	76.2	78.9	78.2	71.2	52.6	50.8	40.2	58.0
1988	31.0	34.7	46.1	57.0	67.1	75.6	80.3	80.0	70.1	52.3	47.8	38.0	56.7
1989	41.6	34.0	48.4	56.7	62.6	73.5	78.1	76.6	69.4	58.4	46.7	25.3	55.9
1990	43.1	44.3	51.2	55.5	64.2	75.1	78.5	77.5	71.8	58.7	52.0	40.8	59.4
1991	34.1	40.5	49.4	60.3	73.1	78.3	81.3	79.2	71.7	61.5	45.0	41.4	59.7
1992	37.1	43.7	47.9	58.2	63.9	72.1	78.5	73.2	69.1	58.1	47.9	38.6	57.4
1993	38.5	34.0	44.0	54.9	66.8	74.5	82.0	79.0	68.0	55.9	45.8	36.5	56.7
Record Mean	34.2	36.8	45.8	56.4	66.0	74.6	78.4	76.8	70.3	58.7	46.6	37.2	56.8
Max	42.2	45.3	55.2	66.4	76.1	84.4	88.2	86.6	80.5	69.1	55.2	45.0	66.2
Min	26.2	28.2	36.4	46.3	55.9	64.7	68.6	66.9	60.1	48.2	37.9	29.3	47.4

PRECIPITATION (inches)

LOUISVILLE, KENTUCKY

YEAR	JAN	FEB	MAR	APR	MAY	JUNE	JULY	AUG	SEP	OCT	NOV	DEC	ANNUAL
1964	2.45	2.45	14.91	3.06	1.85	2.24	3.03	2.63	4.16	0.62	3.32	5.86	46.58
1965	2.76	4.67	4.82	3.28	1.60	2.27	4.86	2.12	8.41	2.54	1.33	1.14	39.80
1966	5.73	5.01	1.02	9.56	3.91	0.75	2.13	5.18	2.59	1.04	3.67	4.33	44.92
1967	1.11	2.01	4.37	4.39	4.62	4.41	7.33	4.30	1.73	3.06	3.08	3.51	43.92
1968	2.13	0.80	6.23	3.94	5.16	1.70	3.07	3.68	2.61	1.00	3.34	3.62	37.28
1969	5.31	1.65	1.94	3.77	3.91	2.97	4.05	3.65	1.08	1.69	3.08	3.69	36.79
1970	1.40	2.87	4.52	11.10	1.85	5.20	3.33	7.65	3.57	4.79	1.75	4.18	52.21
1971	2.64	6.28	2.12	2.16	6.15	2.64	6.74	1.83	4.72	1.96	2.06	2.98	42.28
1972	2.87	3.94	4.07	8.48	4.46	1.08	3.64	2.45	4.24	2.55	6.31	5.29	49.38
1973	1.96	1.60	6.26	5.77	7.04	6.20	9.38	0.91	2.34	2.28	7.59	2.64	53.97
1974	4.38	1.64	5.41	2.74	3.86	2.58	2.04	8.79	3.52	2.09	3.03	2.85	42.93
1975	4.87	4.53	9.65	6.47	4.50	3.15	1.91	3.89	2.64	6.12	3.69	4.89	56.31
1976	3.85	3.13	2.87	0.76	5.09	4.71	2.10	3.18	3.10	3.99	0.72	0.65	34.15
1977	2.33	1.45	4.69	3.40	1.37	7.59	3.29	6.12	3.67	4.76	6.11	4.32	49.10
1978	5.90	0.76	3.76	3.33	4.76	2.67	3.77	5.50	0.96	2.26	5.14	7.64	46.45
1979	3.81	4.49	2.71	7.32	3.59	3.03	10.05	2.37	10.49	2.27	5.85	3.82	59.80
1980	1.71	1.09	4.80	2.63	4.58	3.70	5.41	3.76	3.17	3.37	2.42	1.25	37.89
1981	0.45	3.23	1.54	4.44	4.63	3.23	3.98	3.21	3.22	1.60	2.40	2.02	33.95
1982	5.28	1.55	5.89	3.05	2.96	3.86	3.72	3.74	3.46	1.26	5.50	5.11	45.38
1983	1.63	1.52	2.16	7.10	10.58	4.42	0.99	2.39	1.13	6.47	5.03	3.96	47.38
1984	0.92	1.68	4.41	5.53	6.78	0.49	6.94	5.08	3.70	2.12	5.87	5.86	49.38
1985	2.20	2.08	4.43	1.69	3.93	4.37	3.45	4.49	1.48	4.24	4.43	0.96	37.75
1986	0.91	3.90	2.69	1.04	4.28	2.32	7.04	2.19	2.75	3.08	4.62	2.69	37.51
1987	0.81	4.42	3.05	2.35	1.61	3.58	5.31	2.66	1.15	0.39	2.62	4.70	32.65
1988	4.00	3.58	2.97	3.52	2.68	0.87	4.68	3.00	1.48	1.54	5.76	3.45	37.53
1989	3.68	9.02	5.50	4.93	4.39	5.26	6.90	2.20	2.42	2.65	2.57	1.45	50.97
1990	3.90	6.72	2.78	3.46	11.57	6.13	1.96	3.21	2.57	3.97	2.34	8.86	57.47
1991	3.29	3.72	4.79	2.61	4.02	1.23	2.99	3.35	2.74	2.31	1.87	5.23	38.15
1992	1.97	1.74	5.88	2.66	3.51	3.04	6.51	4.71	3.50	0.96	4.71	1.60	40.79
1993	3.50	4.20	5.20	3.57	2.80	4.05	4.58	5.74	3.90	4.03	3.26	2.56	47.39
Record Mean	3.74	3.38	4.50	3.91	3.99	3.81	3.85	3.32	2.80	2.60	3.52	3.58	43.01

REFERENCE NOTES

LOUISVILLE, KENTUCKY

GENERAL
T - TRACE AMOUNT.
BLANK ENTRIES DENOTE MISSING/UNREPORTED DATA.
# INDICATES A STATION OR INSTRUMENT RELOCATION.

SPECIFIC

PM - INCLUDES LAST DAY OF PREVIOUS MONTH
ASOS - AUTOMATED SURFACE OBSERVING SYSTEM IN OPERATION DURING THESE MONTHS.

(a) - LENGTH OF RECORD IN YEARS, ALTHOUGH INDIVIDUAL MONTHS MAY BE MISSING.
0.* OR * - THE VALUE IS BETWEEN 0.0 AND 0.05
NORMALS - BASED ON THE 1961-1990 RECORD PERIOD.
EXTREMES - DATES ARE THE MOST RECENT OCCURRENCE
WIND DIR.- NUMERALS SHOW TENS OF DEGRESS CLOCKWISE FROM TRUE NORTH. "00" INDICATES CALM.
RESULTANT DIRECTIONS ARE GIVEN TO WHOLE DEGREES.
BOLD VALUES INDICATE EXTREME VALUES WHICH OCCURRED AFTER THE ASOS SYSTEM WAS COMMISSIONED.

RECORD = PERIOD OF RECORD
RECORD MEAN PRECIPITATION IS THE MEAN OF ALL DAILY PRECIPITATION AMOUNTS DURING THE PERIOD OF RECORD.
RECORD MAX(MIN) TEMPERATURE IS THE MEAN OF ALL DAILY MAX(MIN) TEMPERATURES DURING THE PERIOD OF RECORD.
RECORD MEAN TEMPERATURE IS THE SUM OF THE RECORD MAX AND RECORD MIN DIVIDED BY 2.
AVERAGE TEMPERATURE IS THE SUM OF THE MEAN DAILY MAX AND MIN TEMPERATURE DIVIDED BY 2.

EXCEPTIONS

RECORD MEANS ARE THROUGH THE CURRENT YEAR, BEGINNING IN 1873 FOR TEMPERATURE
1873 FOR PRECIPITATION
1948 FOR SNOWFALL

HEATING DEGREE DAYS Base 65 deg. F LOUISVILLE, KENTUCKY

SEASON	JULY	AUG	SEP	OCT	NOV	DEC	JAN	FEB	MAR	APR	MAY	JUNE	TOTAL
1964-65	0	5	44	349	502	829	943	808	802	223	20	0	4525
1965-66	0	2	45	304	526	697	1170	857	580	353	127	8	4669
1966-67	0	0	35	324	531	907	882	949	453	209	139	13	4442
1967-68	0	0	68	259	660	788	1069	1007	590	247	98	4	4790
1968-69	0	1	10	276	511	903	987	778	771	213	61	14	4525
1969-70	0	0	42	282	645	971	1141	875	697	200	70	0	4923
1970-71	0	0	23	220	582	781	1052	833	707	303	137	0	4638
1971-72	0	0	13	65	537	610	914	866	623	282	61	19	3990
1972-73	0	0	16	298	628	793	927	796	349	343	129	0	4279
1973-74	0	0	13	144	450	860	772	714	487	257	99	19	3815
1974-75	0	0	122	314	543	794	830	688	665	333	22	0	4311
1975-76	0	0	73	205	431	801	1040	562	405	266	111	1	3895
1976-77	0	0	29	393	757	982	1435	780	421	183	36	7	5023
1977-78	0	0	6	295	472	935	1294	1145	720	221	142	1	5231
1978-79	0	0	4	293	442	765	1246	1030	514	301	94	5	4694
1979-80	0	0	19	244	534	792	969	1021	713	342	68	8	4710
1980-81	0	0	12	309	555	821	1065	728	595	142	122	0	4349
1981-82	0	0	61	268	523	960	1124	837	549	408	13	3	4746
1982-83	0	1	56	246	495	624	933	763	571	399	121	5	4214
1983-84	0	0	54	196	509	1128	1115	673	757	315	141	0	4888
1984-85	0	0	73	84	623	584	1222	896	458	180	52	16	4188
1985-86	0	0	53	160	347	1067	941	696	516	224	69	0	4073
1986-87	0	12	5	210	570	869	962	706	526	294	21	0	4175
1987-88	0	0	9	377	423	762	1048	872	580	244	38	7	4360
1988-89	0	0	13	398	510	833	720	860	513	291	156	4	4298
1989-90	0	0	49	230	539	1222	672	574	445	320	82	13	4146
1990-91	0	0	34	229	387	745	949	677	482	167	27	0	3697
1991-92	0	0	52	168	590	725	855	610	523	244	124	8	3899
1992-93	0	0	40	219	505	813	819	859	644	299	44	18	4260
1993-94	0	0	48	289	572	875							

COOLING DEGREE DAYS Base 65 deg. F LOUISVILLE, KENTUCKY

YEAR	JAN	FEB	MAR	APR	MAY	JUNE	JULY	AUG	SEP	OCT	NOV	DEC	TOTAL
1969	0	0	0	9	106	277	431	308	127	49	0	0	1307
1970	0	0	0	36	147	244	343	346	283	15	0	0	1414
1971	0	0	0	2	35	351	310	291	237	58	3	0	1287
1972	0	0	3	25	81	193	386	351	242	2	4	0	1287
1973	0	0	7	29	28	325	422	380	280	71	2	0	1544
1974	0	0	22	31	109	136	345	319	75	8	10	0	1055
1975	0	0	0	24	152	320	402	451	116	36	5	0	1506
1976	0	0	21	47	51	243	372	294	92	10	0	0	1130
1977	0	0	14	50	234	281	479	396	238	5	20	0	1717
1978	0	0	0	20	110	323	425	383	270	6	2	0	1539
1979	0	0	5	10	73	279	326	350	154	39	0	0	1236
1980	0	0	0	8	134	266	519	504	276	31	1	0	1739
1981	0	0	5	68	63	343	435	348	150	10	0	0	1422
1982	0	0	1	2	183	139	408	274	118	68	13	8	1214
1983	0	0	7	8	39	264	504	524	240	19	0	0	1605
1984	0	0	0	20	69	386	333	349	145	56	0	1	1359
1985	0	2	8	48	106	233	387	311	185	55	14	0	1349
1986	0	0	5	37	138	330	481	306	255	46	0	0	1598
1987	0	0	0	14	232	342	439	416	203	1	4	0	1651
1988	0	0	4	10	111	333	481	472	173	10	0	0	1594
1989	0	0	6	48	88	264	412	364	188	30	0	0	1400
1990	0	0	22	44	65	323	427	392	244	42	7	0	1566
1991	0	0	8	31	286	406	514	445	262	68	0	0	2020
1992	0	0	2	48	100	229	424	262	169	14	0	0	1248
1993	0	0	0	4	106	310	534	442	146	12	3	0	1557

SNOWFALL (inches) LOUISVILLE, KENTUCKY

SEASON	JULY	AUG	SEP	OCT	NOV	DEC	JAN	FEB	MAR	APR	MAY	JUNE	TOTAL
1964-65	0.0	0.0	0.0	0.0	0.9	0.2	11.8	3.9	4.6	0.0	0.0	0.0	21.4
1965-66	0.0	0.0	0.0	0.0	0.0	T	7.0	11.4	1.3	0.4	0.0	0.0	20.1
1966-67	0.0	0.0	0.0	0.0	13.2	0.5	0.4	7.8	9.3	0.0	0.0	0.0	31.2
1967-68	0.0	0.0	0.0	0.0	T	2.7	13.8	1.8	12.7	0.0	0.0	0.0	31.0
1968-69	0.0	0.0	0.0	0.0	0.3	1.1	3.2	6.3	2.2	0.0	0.0	0.0	13.1
1969-70	0.0	0.0	0.0	0.0	0.7	7.7	7.9	7.4	10.7	T	0.0	0.0	34.4
1970-71	0.0	0.0	0.0	0.0	0.3	0.8	3.2	11.9	5.2	0.1	0.0	0.0	21.5
1971-72	0.0	0.0	0.0	0.0	5.4	T	1.6	3.4	1.2	T	0.0	0.0	11.6
1972-73	0.0	0.0	0.0	0.0	2.0	2.2	1.1	1.1	0.5	1.6	0.0	0.0	8.5
1973-74	0.0	0.0	0.0	0.0	0.0	4.5	1.0	0.9	2.8	T	0.0	0.0	9.2
1974-75	0.0	0.0	0.0	0.0	1.0	1.2	3.0	1.3	10.0	T	0.0	0.0	16.5
1975-76	0.0	0.0	0.0	0.0	0.1	0.7	2.5	0.1	0.7	0.0	0.0	0.0	4.1
1976-77	0.0	0.0	0.0	0.0	1.6	1.1	19.6	0.8	0.1	0.8	0.0	0.0	24.0
1977-78	0.0	0.0	0.0	0.0	4.8	2.2	28.4	5.3	9.4	T	0.0	0.0	50.1
1978-79	0.0	0.0	0.0	0.0	0.0	T	8.5	10.9	0.9	T	0.0	0.0	20.3
1979-80	0.0	0.0	0.0	0.0	0.1	T	10.7	3.6	3.9	T	0.0	0.0	18.3
1980-81	0.0	0.0	0.0	T	T	T	2.5	0.3	0.1	0.0	0.0	0.0	2.9
1981-82	0.0	0.0	0.0	0.0	0.1	3.6	2.7	2.9	0.3	1.4	0.0	0.0	11.0
1982-83	0.0	0.0	0.0	0.0	T	0.0	0.6	4.5	0.1	T	0.0	0.0	5.2
1983-84	0.0	0.0	0.0	0.0	0.0	0.6	3.1	8.8	1.0	0.0	0.0	0.0	13.5
1984-85	0.0	0.0	0.0	0.0	T	4.8	7.4	6.7	T	T	0.0	0.0	18.9
1985-86	0.0	0.0	0.0	0.0	0.0	1.6	1.1	8.8	0.1	0.0	0.0	0.0	11.6
1986-87	0.0	0.0	0.0	0.0	T	T	2.2	6.7	9.3	T	0.0	0.0	18.2
1987-88	0.0	0.0	0.0	0.0	T	T	3.0	5.0	0.5	0.0	0.0	0.0	8.5
1988-89	0.0	0.0	0.0	0.0	T	0.3	T	0.6	T	0.0	T	0.0	0.9
1989-90	0.0	0.0	0.0	1.4	T	6.5	1.9	0.8	4.1	T	0.0	0.0	14.7
1990-91	0.0	0.0	0.0	0.0	0.0	4.1	0.3	1.5	0.2	0.0	0.0	0.0	6.1
1991-92	0.0	0.0	0.0	0.0	0.5	0.1	0.9	0.1	0.9	0.7	0.0	0.0	3.2
1992-93	0.0	0.0	0.0	0.0	0.9	1.0	T	15.9	1.1	T	0.0	T	18.9
1993-94	0.0	0.0	0.0	2.4	T	3.6							
Record Mean	0.0	0.0	0.0	0.1	1.0	2.2	5.1	4.5	3.2	0.1	T	T	16.2

PADUCAH, KENTUCKY

No narrative available

NORMALS, MEANS, AND EXTREMES

PADUCAH, KENTUCKY

LATITUDE: 37°04'N LONGITUDE: 88°46'W ELEVATION: FT. GRND 394 BARO 400 TIME ZONE: CENTRAL WBAN: 03816

	(a)	JAN	FEB	MAR	APR	MAY	JUNE	JULY	AUG	SEP	OCT	NOV	DEC	YEAR
TEMPERATURE °F:														
Normals														
-Daily Maximum		41.5	46.3	57.5	68.6	77.3	85.8	89.0	87.4	81.0	70.6	57.6	46.1	67.4
-Daily Minimum		23.5	27.4	37.5	47.1	56.0	64.6	68.5	66.1	59.2	46.8	38.2	28.4	46.9
-Monthly		32.6	36.9	47.5	57.9	66.7	75.2	78.8	76.8	70.2	58.7	47.9	37.3	57.2
Extremes														
-Record Highest	10	70	74	84	90	94	103	102	104	99	89	83	71	104
-Year		1986	1986	1986	1987	1987	1988	1988	1988	1990	1989	1987	1984	AUG 1988
-Record Lowest	10	-15	-8	16	24	35	44	55	44	37	24	10	-10	-15
-Year		1985	1985	1984	1990	1989	1993	1990	1986	1993	1993	1986	1989	JAN 1985
NORMAL DEGREE DAYS:														
Heating (base 65°F)		1004	787	550	231	83	0	0	0	24	228	513	859	4279
Cooling (base 65°F)		0	0	8	18	136	306	428	366	180	33	0	0	1475
% OF POSSIBLE SUNSHINE	9	46	47	52	64	59	64	68	66	59	61	48	43	56
MEAN SKY COVER (tenths)														
Sunrise - Sunset	9	6.9	6.9	6.7	6.1	6.3	6.0	5.2	5.2	5.4	5.6	6.5	6.7	6.1
MEAN NUMBER OF DAYS:														
Sunrise to Sunset														
-Clear	9	6.8	5.9	6.7	8.4	7.0	7.0	10.4	11.0	11.0	11.1	8.9	7.5	101.7
-Partly Cloudy	9	6.6	7.0	7.8	7.7	8.9	11.7	11.2	11.4	7.6	7.3	4.2	6.5	97.8
-Cloudy	9	17.7	15.3	16.6	13.9	15.1	11.3	9.3	8.6	11.4	12.6	16.9	17.0	165.7
Precipitation														
.01 inches or more	10	8.4	9.7	10.5	11.0	10.5	8.4	8.3	7.3	7.5	8.7	10.2	10.0	110.5
Snow,Ice pellets,hail														
1.0 inches or more	10	0.8	1.3	0.3	0.0	0.0	0.0	0.0	0.0	0.0	0.1	0.0	0.7	3.2
Thunderstorms	10	0.8	1.9	3.8	5.2	8.4	9.8	9.1	8.3	5.3	4.2	2.6	1.0	60.4
Heavy Fog Visibility														
1/4 mile or less	10	2.6	2.2	0.7	1.0	1.3	0.8	0.7	3.4	2.2	1.6	1.4	2.7	20.6
Temperature °F														
-Maximum														
90° and above	10	0.0	0.0	0.0	0.1	1.6	11.6	20.2	14.3	5.0	0.0	0.0	0.0	52.8
32° and below	10	4.9	2.9	0.3	0.0	0.0	0.0	0.0	0.0	0.0	0.0	0.1	3.3	11.5
-Minimum														
32° and below	10	23.2	16.6	10.4	2.5	0.0	0.0	0.0	0.0	0.0	2.6	10.3	19.9	85.5
0° and below	10	0.7	0.4	0.0	0.0	0.0	0.0	0.0	0.0	0.0	0.0	0.0	0.6	1.7
AVG. STATION PRESS.(mb)	9	1006.3	1005.4	1002.8	1000.7	1000.8	1000.4	1001.7	1002.4	1003.5	1004.5	1004.4	1006.6	1003.3
RELATIVE HUMIDITY (%)														
Hour 00	9	78	78	74	77	85	86	88	90	88	84	77	79	82
Hour 06 (Local Time)	9	81	82	79	83	88	88	90	94	93	88	83	82	86
Hour 12	9	65	64	57	52	57	55	57	57	57	54	60	64	58
Hour 18	9	70	66	59	53	60	59	62	64	68	68	68	72	64
PRECIPITATION (inches):														
Water Equivalent														
-Normal		3.27	3.90	4.92	5.01	4.94	4.05	4.19	3.34	3.69	3.00	4.32	4.68	49.31
-Maximum Monthly	10	5.38	13.33	5.83	8.45	8.51	9.20	7.07	5.89	9.23	7.26	9.56	9.99	13.33
-Year		1990	1989	1984	1984	1986	1989	1989	1985	1985	1985	1988	1984	FEB 1989
-Minimum Monthly	10	0.99	2.68	1.93	1.55	1.43	0.41	0.56	1.05	2.38	1.58	2.17	1.34	0.41
-Year		1987	1992	1987	1986	1987	1988	1993	1988	1990	1987	1991	1985	JUN 1988
-Maximum in 24 hrs	10	2.49	6.34	2.70	4.31	2.21	4.11	3.44	2.84	7.53	3.40	3.55	4.00	7.53
-Year		1990	1989	1989	1985	1986	1989	1984	1992	1985	1992	1988	1987	SEP 1985
Snow,Ice pellets,hail														
-Maximum Monthly	10	16.8	15.3	2.8	T	0.0	T	T	0.0	0.0	2.2	T	7.1	16.8
-Year		1985	1993	1988	1993		1993	1992			1993	1993	1984	JAN 1985
-Maximum in 24 hrs	10	5.8	7.9	2.8	T	0.0	T	T	0.0	0.0	2.1	T	7.1	7.9
-Year		1985	1993	1988	1993	1984	1993	1992	1984	1984	1993	1993	1984	FEB 1993
WIND:														
Mean Speed (mph)	9	9.4	9.3	10.4	9.4	7.8	6.8	6.6	5.6	6.5	7.3	9.3	9.2	8.1
Prevailing Direction through														
Fastest Obs. 1 Min.														
-Direction (!!)	9	30	22	19	17	30	27	30	29	18	14	26	22	30
-Speed (MPH)	9	41	33	37	35	33	40	44	30	35	36	38	35	44
-Year		1989	1988	1985	1993	1988	1993	1989	1989	1985	1992	1992	1987	JUL 1989
Peak Gust														
-Direction (!!)	9	NW	SW	S	NW	W	NW	W	W	NW	S	W	SW	S
-Speed (mph)	9	51	54	60	55	49	58	59	51	52	49	56	58	60
-Date		1989	1988	1985	1993	1988	1993	1987	1989	1990	1989	1992	1987	MAR 1985

PRECIPITATION (inches) PADUCAH, KENTUCKY

YEAR	JAN	FEB	MAR	APR	MAY	JUNE	JULY	AUG	SEP	OCT	NOV	DEC	ANNUAL
1984	1.21	4.74	5.83	8.45	6.50	1.58	5.44	3.96	6.80	5.88	4.75	9.99	65.13
1985	1.82	3.70	3.67	6.85	4.13	4.85	0.85	5.89	9.23	7.26	4.29	1.34	53.88
1986	1.44	3.73	3.16	1.55	8.51	1.50	7.07	4.33	3.69	4.45	3.59	3.11	46.13
1987	0.99	3.93	1.93	2.30	1.43	4.03	2.58	1.31	2.80	1.58	4.29	9.19	36.36
1988	3.50	5.15	4.60	2.13	3.14	0.41	3.08	1.05	3.49	3.81	9.56	3.05	42.97
1989	5.31	13.33	5.36	2.55	2.33	9.20	7.07	1.80	2.64	3.48	2.59	1.78	57.44
1990	5.38	9.05	3.69	4.76	7.49	2.14	4.03	1.34	2.38	4.45	2.33	9.59	56.63
1991	3.77	4.07	3.55	3.81	4.29	1.47	3.23	2.42	3.25	3.57	2.17	3.84	39.44
1992	2.13	2.68	3.38	2.07	2.08	3.57	6.90	3.47	5.81	3.51	3.45	1.79	40.84
1993	3.79	3.99	2.99	5.14	2.59	5.51	0.56	2.89	6.00	3.82	6.45	3.57	47.30
Record Mean	2.93	5.44	3.82	3.96	4.25	3.43	4.08	2.85	4.61	4.18	4.35	4.72	48.61

AVERAGE TEMPERATURE (deg. F) PADUCAH, KENTUCKY

YEAR	JAN	FEB	MAR	APR	MAY	JUNE	JULY	AUG	SEP	OCT	NOV	DEC	ANNUAL
1984	29.2	42.1	43.6	56.7	64.6	78.6	76.7	76.9	68.5	63.1	45.0	45.2	57.5
1985	23.9	32.0	51.3	60.9	66.8	73.3	78.4	74.8	68.8	62.4	52.5	31.3	56.4
1986	35.5	40.3	49.7	60.6	68.7	77.4	81.7	73.8	73.8	60.1	45.1	36.9	58.6
1987	33.5	40.9	50.2	57.4	73.0	78.2	79.5	79.8	71.6	53.4	50.9	41.1	59.1
1988	32.2	35.1	47.5	57.4	67.3	75.8	80.4	80.9	70.8	52.8	48.4	38.2	57.2
1989	41.4	32.8	48.1	57.3	64.6	73.6	78.3	77.7	69.3	59.9	48.6	27.1	56.6
1990	43.8	45.7	51.5	55.9	63.9	76.4	78.8	75.9	72.1	56.8	53.5	39.9	59.5
1991	34.2	41.9	51.4	62.3	72.5	78.0	80.9	78.4	71.1	61.1	45.4	41.9	59.9
1992	38.1	45.6	49.7	59.2	66.0	73.6	79.7	74.0	69.0	58.5	48.2	38.7	58.4
1993	38.3	36.9	46.6	55.9	67.3	76.5	84.3	78.7	67.3	56.1	45.6	38.5	57.7
Record Mean	35.0	39.3	49.0	58.3	67.4	76.2	79.8	77.1	70.2	58.4	48.3	37.8	58.1
Max	43.9	49.0	59.6	70.0	78.7	87.4	90.6	88.7	81.9	70.8	58.4	47.3	68.8
Min	26.0	29.6	38.3	46.6	56.2	64.9	69.0	65.4	58.5	46.0	38.2	28.4	47.3

REFERENCE NOTES PADUCAH, KENTUCKY

GENERAL
T - TRACE AMOUNT.
BLANK ENTRIES DENOTE MISSING/UNREPORTED DATA.
# INDICATES A STATION OR INSTRUMENT RELOCATION.

SPECIFIC

(a) - LENGTH OF RECORD IN YEARS, ALTHOUGH INDIVIDUAL MONTHS MAY BE MISSING.
0.* OR * - THE VALUE IS BETWEEN 0.0 AND 0.05
NORMALS - BASED ON THE 1961-1990 RECORD PERIOD.
EXTREMES - DATES ARE THE MOST RECENT OCCURRENCE.
WIND DIR.- NUMERALS SHOW TENS OF DEGRESS CLOCKWISE FROM TRUE NORTH. "00" INDICATES CALM.
RESULTANT DIRECTIONS ARE GIVEN TO WHOLE DEGREES.
BOLD VALUES INDICATE EXTREME VALUES WHICH OCCURRED AFTER THE ASOS SYSTEM WAS COMMISSIONED.

RECORD = PERIOD OF RECORD
RECORD MEAN PRECIPITATION IS THE MEAN OF ALL DAILY PRECIPITATION AMOUNTS DURING THE PERIOD OF RECORD.
RECORD MAX(MIN) TEMPERATURE IS THE MEAN OF ALL DAILY MAX(MIN) TEMPERATURES DURING THE PERIOD OF RECORD.
RECORD MEAN TEMPERATURE IS THE SUM OF THE RECORD MAX AND RECORD MIN DIVIDED BY 2.
AVERAGE TEMPERATURE IS THE SUM OF THE MEAN DAILY MAX AND MIN TEMPERATURE DIVIDED BY 2.

EXCEPTIONS

RECORD MEANS ARE THROUGH THE CURRENT YEAR,
BEGINNING IN 1984 FOR TEMPERATURE
1984 FOR PRECIPITATION
1984 FOR SNOWFALL

HEATING DEGREE DAYS Base 65 deg. F PADUCAH, KENTUCKY

SEASON	JULY	AUG	SEP	OCT	NOV	DEC	JAN	FEB	MAR	APR	MAY	JUNE	TOTAL
1983-84							1103	659	656	269	82	0	
1984-85	0	0	69	108	591	611	1269	918	431	170	48	10	4225
1985-86	0	0	67	151	378	1037	909	688	475	173	47	0	3925
1986-87	0	8	3	191	589	863	968	671	451	249	7	0	4000
1987-88	0	0	12	356	419	735	1011	861	534	239	38	0	4205
1988-89	0	0	16	383	491	825	726	897	525	292	115	0	4270
1989-90	0	0	42	193	488	1165	651	534	438	313	89	10	3923
1990-91	0	1	28	273	352	771	949	642	429	119	18	0	3582
1991-92	0	0	72	190	585	709	827	556	470	228	86	5	3728
1992-93	0	0	39	203	502	808	818	781	565	279	40	8	4043
1993-94	0	0	54	293	581	812							

COOLING DEGREE DAYS Base 65 deg. F PADUCAH, KENTUCKY

YEAR	JAN	FEB	MAR	APR	MAY	JUNE	JULY	AUG	SEP	OCT	NOV	DEC	TOTAL
1984	0	0	0	29	76	415	369	375	182	53	0	2	1501
1985	0	1	11	52	112	263	419	309	186	77	9	0	1439
1986	0	0	6	50	169	379	524	286	273	47	0	0	1734
1987	0	0	0	29	258	402	458	465	213	3	3	0	1831
1988	0	0	0	19	117	333	482	499	197	11	0	0	1658
1989	0	0	11	68	108	266	419	398	177	42	3	0	1492
1990	0	0	26	46	63	356	435	344	247	28	12	0	1557
1991	0	0	13	40	252	396	499	424	262	75	4	0	1965
1992	0	0	0	58	123	270	463	285	166	12	5	0	1382
1993	0	0	2	11	120	360	604	430	131	23	5	0	1686

SNOWFALL (inches) PADUCAH, KENTUCKY

SEASON	JULY	AUG	SEP	OCT	NOV	DEC	JAN	FEB	MAR	APR	MAY	JUNE	TOTAL
1983-84							4.3	11.2	T	0.0	0.0	0.0	
1984-85	0.0	0.0	0.0	0.0	0.0	7.1	16.8	4.8	0.0	0.0	0.0	0.0	28.7
1985-86	0.0	0.0	0.0	0.0	0.0	2.0	0.6	8.3	T	0.0	0.0	0.0	10.9
1986-87	0.0	0.0	0.0	0.0	T	T	2.4	1.7	1.3	0.0	0.0	0.0	5.4
1987-88	0.0	0.0	0.0	0.0	0.0	3.4	3.0	1.4	2.8	T	0.0	0.0	10.6
1988-89	0.0	0.0	0.0	0.0	0.0	2.2	T	3.6	T	0.0	0.0	0.0	5.8
1989-90	0.0	0.0	0.0	T	T	1.7	0.1	1.0	0.7	T	0.0	T	3.5
1990-91	0.0	0.0	0.0	0.0	0.0	3.4	0.2	T	T	0.0	0.0	0.0	3.6
1991-92	0.0	0.0	0.0	0.0	T	0.0	T	T	T	T	0.0	0.0	T
1992-93	T	0.0	0.0	0.0	T	3.0	0.3	15.3	0.7	T	0.0	T	19.3
1993-94	0.0	0.0	0.0	2.2	T	0.4							
Record Mean	T	0.0	0.0	0.2	T	2.3	2.8	4.7	0.6	T	0.0	T	10.6

BATON ROUGE, LOUISANA

Baton Rouge, the capital city, is located on the east side of the Mississippi River in the southeast section of the state, some 65 miles inland from the coast. The area is near the first evident relief north of the deltaic coastal plains. The NOAA National Weather Service Office is located at Ryan Airport, some 8 miles north of the downtown area. Elevations in East Baton Rouge Parish range from near 25 feet to more than 100 feet above sea level.

The general climate of Baton Rouge is humid subtropical, but the city is subject to significant polar influences during winter. Prevailing wind flow is from the southerly direction during much of the year. This maritime air from the Gulf of Mexico helps to temper summer heat, shorten winter cold spells, and provides abundant moisture and rainfall. Winds are usually rather light.

Rainfall is heavy and amounts are substantial in all seasons, with an early autumn low in September and October. Almost all rainfall is from brief convective showers. Occasionally during winter, slow moving cold fronts may produce rains lasting for a few days. Extremes of precipitation may occur in all seasons.

The winter months are normally mild with short cold spells. The typical pattern is, turning cold with rain on the first day, colder with clear skies on the second day, and warming on the third day. Freezing or sub-freezing temperatures occur several times annually, but temperatures nearly always rise above freezing during the day. The average date of the first freeze in the autumn is late November, and the average date of the last freeze in spring is late February, producing a mean freeze-free period of 273 days. Annual total snowfall averages only a fraction of an inch and many years pass with no measurable snow.

The summer months are consistently quite warm, but high temperatures rarely exceed 100 degrees. This is because of the high humidity of the maritime tropical air mass, the effects of cloudiness, and the scattered showers and thunderstorms which are a primary feature of the weather during these months. Scattered showers normally fall in the area on about one-half of the days in June, July, and August.

Except for three or four days per month, point rainfall totals are usually less than 0.5 inch. Summer relative humidity exceeds 80 percent for about 12 hours per day. High humidity may be experienced at any hour, but occurs mainly at night. Readings of 50 percent or less occur about two hours per day, usually in the afternoons. Temperatures in the spring are usually mild and pleasant and in the autumn they are generally delightful for outdoor activities.

Thunderstorms occur each month, most frequently in July and August. Severe local storms, including hailstorms, tornadoes, and local wind storms, are most frequent during the spring months. Large damaging hail very rarely occurs and tornadoes are unusual. Hurricane centers have occasionally passed very near Baton Rouge.

NORMALS, MEANS, AND EXTREMES

BATON ROUGE, LOUISIANA

LATITUDE: 30°32'N LONGITUDE: 91°08' W ELEVATION: FT. GRND 64 BARO 73 TIME ZONE: CENTRAL WBAN: 13970

	(a)	JAN	FEB	MAR	APR	MAY	JUNE	JULY	AUG	SEP	OCT	NOV	DEC	YEAR
TEMPERATURE °F:														
Normals														
-Daily Maximum		59.8	63.6	72.3	79.8	85.7	90.5	91.4	91.1	87.4	80.0	70.9	63.3	78.0
-Daily Minimum		39.6	42.5	50.2	57.9	64.8	70.4	73.2	72.6	68.8	57.1	49.0	42.5	57.4
-Monthly		49.8	53.1	61.3	68.9	75.3	80.5	82.3	81.9	78.1	68.6	60.0	52.9	67.7
Extremes														
-Record Highest	43	82	85	91	92	98	103	101	102	99	94	87	85	103
-Year		1989	1989	1963	1987	1953	1954	1960	1962	1990	1986	1986	1982	JUN 1954
-Record Lowest	43	9	20	20	32	44	53	58	**59**	43	**30**	21	8	8
-Year		1985	1958	1980	1987	1954	1984	1967	**1992**	1967	**1993**	1976	1989	DEC 1989
NORMAL DEGREE DAYS:														
Heating (base 65°F)		490	345	170	27	0	0	0	0	0	48	197	392	1669
Cooling (base 65°F)		19	12	55	144	319	465	536	524	393	159	47	17	2690
% OF POSSIBLE SUNSHINE														
MEAN SKY COVER (tenths)														
Sunrise - Sunset	42	6.8	6.4	6.3	6.0	5.9	5.7	6.2	5.8	5.5	4.4	5.7	6.4	5.9
MEAN NUMBER OF DAYS:														
Sunrise to Sunset														
-Clear	42	7.2	7.5	8.0	7.9	7.9	7.0	5.0	7.2	9.2	14.4	9.6	8.4	99.2
-Partly Cloudy	42	5.9	6.2	8.5	8.9	11.4	14.4	15.2	14.6	11.0	8.2	7.7	6.7	118.6
-Cloudy	42	17.9	14.6	14.6	13.2	11.7	8.6	10.8	9.3	9.8	8.4	12.7	16.0	147.5
Precipitation														
.01 inches or more	42	9.8	9.0	8.7	7.4	8.3	9.9	13.3	11.9	8.9	5.4	7.7	9.7	110.0
Snow,Ice pellets,hail														
1.0 inches or more	42	0.0	0.1	0.0	0.0	0.0	0.0	0.0	0.0	0.0	0.0	0.0	0.0	0.1
Thunderstorms	42	2.0	3.5	4.1	5.2	6.4	9.8	15.2	12.6	7.0	2.4	2.6	2.2	73.1
Heavy Fog Visibility														
1/4 mile or less	42	4.5	3.2	2.9	3.0	2.8	1.0	1.7	1.5	2.8	4.1	4.1	4.0	35.6
Temperature °F														
-Maximum														
90° and above	34	0.0	0.0	0.*	0.3	5.5	19.0	24.0	21.6	11.9	1.9	0.0	0.0	84.2
32° and below	34	0.2	0.*	0.0	0.0	0.0	0.0	0.0	0.0	0.0	0.0	0.0	0.1	0.4
-Minimum														
32° and below	34	8.7	4.9	1.1	0.*	0.0	0.0	0.0	0.0	0.0	0.*	1.9	6.2	22.9
0° and below	34	0.0	0.0	0.0	0.0	0.0	0.0	0.0	0.0	0.0	0.0	0.0	0.0	0.0
AVG. STATION PRESS.(mb)	21	1018.8	1017.4	1014.5	1013.8	1012.4	1013.2	1014.6	1014.2	1013.9	1016.1	1017.1	1018.6	1015.4
RELATIVE HUMIDITY (%)														
Hour 00	34	81	79	80	82	85	86	87	89	88	86	85	83	84
Hour 06 (Local Time)	34	85	84	86	88	90	91	91	92	91	89	88	86	88
Hour 12	34	64	59	57	55	57	59	62	62	60	54	58	63	59
Hour 18	34	67	60	58	59	60	64	69	70	70	66	69	69	65
PRECIPITATION (inches):														
Water Equivalent														
-Normal		4.91	5.52	4.81	5.37	4.89	4.48	6.74	6.00	4.85	3.48	4.31	5.53	60.89
-Maximum Monthly	43	13.35	14.51	12.73	14.84	14.67	23.18	10.98	14.48	13.95	14.48	13.55	15.94	23.18
-Year		1993	1966	1973	1980	1989	1989	1963	1987	1977	1984	1989	1982	JUN 1989
-Minimum Monthly	43	1.15	0.70	0.54	0.38	0.63	0.12	2.05	1.32	0.09	T	0.25	1.94	T
-Year		1971	1962	1955	1976	1963	1979	1962	1980	1953	1978	1967	1978	OCT 1978
-Maximum in 24 hrs	43	9.02	4.72	6.07	12.08	4.96	9.73	4.26	8.31	6.31	8.38	7.29	8.28	12.08
-Year		1993	1979	1973	1967	1954	1989	1969	1987	1973	1964	1989	1982	APR 1967
Snow,Ice pellets,hail														
-Maximum Monthly	43	0.6	3.2	T	0.0	T	**0.0**	**0.0**	**0.0**	**0.0**	**0.0**	T	T	3.2
-Year		1973	1988	1993		1989						1976	1989	FEB 1988
-Maximum in 24 hrs	43	0.5	3.2	T	0.0	T	**0.0**	**0.0**	**0.0**	**0.0**	**0.0**	T	T	3.2
-Year		1973	1988	1993		1989						1976	1989	FEB 1988
WIND:														
Mean Speed (mph)	42	8.8	9.3	9.4	8.8	7.7	6.7	6.0	5.6	6.7	6.7	7.8	8.2	7.6
Prevailing Direction through 1963		N	NE	SE	SE	SE	SE	W	E	NE	NE	N	SE	SE
Fastest Obs. 1 Min.														
-Direction (!!)	31	27	25	18	25	17	03	03	**14**	06	33	**18**	**16**	06
-Speed (MPH)	31	35	35	37	35	48	40	41	**46**	58	40	**31**	**30**	58
-Year		1983	1970	1992	1964	1967	1964	1980	**1992**	1965	1964	**1993**	**1993**	SEP 1965
Peak Gust														
-Direction (!!)	10	W	N	NW	NW	SW	**14**	**11**	**SE**	**16**	W	S	W	**SE**
-Speed (mph)	10	39	46	52	51	51	**54**	**60**	**70**	**46**	48	47	48	**70**
-Date		1992	1993	1992	1985	1989	**1993**	**1993**	**1992**	**1993**	1988	1987	1987	**AUG 1992**

PRECIPITATION (inches)

BATON ROUGE, LOUISANA

YEAR	JAN	FEB	MAR	APR	MAY	JUNE	JULY	AUG	SEP	OCT	NOV	DEC	ANNUAL
1964	6.18	5.21	8.73	4.76	2.80	1.41	9.37	4.45	3.52	9.46	4.70	3.24	63.83
1965	2.61	5.93	4.50	1.47	3.90	1.83	4.50	7.03	5.20	0.92	2.40	4.71	45.00
1966	9.93	14.51	1.95	7.97	4.48	2.16	6.33	5.92	2.62	2.43	2.79	2.53	63.62
1967	2.75	4.82	1.90	12.64	7.73	2.76	9.32	4.11	3.48	2.17	0.25	5.95	57.88
1968	3.59	2.72	3.34	2.67	2.63	1.90	3.52	9.19	1.68	0.06	6.81	7.04	45.15
1969	1.34	5.44	5.42	10.25	2.92	0.76	9.85	3.74	2.99	6.04	0.52	3.79	53.06
1970	2.20	2.20	7.02	3.52	5.21	3.68	5.96	6.21	4.34	6.19	1.34	6.69	54.56
1971	1.15	4.39	5.22	0.75	3.71	2.32	9.59	5.13	10.94	2.65	3.17	10.04	59.06
1972	8.25	3.43	5.97	1.44	9.15	2.43	6.34	1.98	4.12	3.69	4.75	8.22	59.77
1973	4.01	3.64	12.73	10.10	5.60	2.99	4.34	4.92	13.08	1.89	7.44	8.29	79.03
1974	8.33	6.66	5.66	5.59	5.23	1.11	5.89	6.45	2.19	1.61	4.55	3.70	56.97
1975	7.77	1.42	4.49	10.18	5.49	5.11	9.30	11.69	3.54	2.58	2.16	2.37	66.10
1976	3.72	4.67	5.18	0.38	4.92	4.90	7.63	2.31	1.46	3.10	5.00	5.80	49.07
1977	6.50	3.89	4.85	7.10	3.97	1.46	6.35	13.31	13.95	3.05	10.35	2.92	77.70
1978	6.55	2.24	1.86	2.92	7.43	2.94	7.14	7.54	3.83	T	4.94	1.94	49.33
1979	6.25	10.83	4.26	11.48	5.37	0.12	8.76	4.94	3.55	2.47	5.06	2.82	65.91
1980	4.67	3.56	8.25	14.84	7.28	5.35	7.68	1.32	7.74	5.66	5.57	2.38	74.30
1981	1.20	7.07	1.74	3.09	4.47	4.70	4.25	4.46	3.95	1.42	1.51	5.35	43.21
1982	3.35	6.48	2.81	4.60	4.05	2.91	5.18	3.93	2.47	2.42	3.05	15.94	57.19
1983	6.25	4.63	5.39	12.75	6.17	12.25	3.39	8.39	4.47	1.55	4.33	8.06	77.63
1984	2.77	6.63	1.20	1.79	3.82	3.00	4.95	3.92	2.37	14.48	2.74	3.76	51.43
1985	4.56	5.95	4.15	1.61	2.72	4.13	8.85	6.92	6.31	10.08	0.42	4.68	60.38
1986	1.53	3.50	2.71	2.94	8.21	6.10	3.31	6.38	1.91	4.40	8.52	6.19	55.70
1987	7.04	7.97	6.02	1.40	4.23	4.48	6.42	14.48	0.78	1.54	3.78	3.89	62.03
1988	3.98	12.49	9.00	4.66	0.95	4.16	6.45	11.02	9.48	2.80	2.88	8.17	76.04
1989	4.02	1.51	4.64	2.34	14.67	23.18	6.25	5.16	4.51	2.18	13.55	6.31	88.32
1990	11.41	7.91	5.84	2.71	3.61	7.15	7.37	4.35	5.06	3.15	2.12	4.77	65.45
1991	9.69	7.85	3.21	9.18	10.63	5.21	5.29	10.67	6.31	4.64	2.70	2.36	77.74
1992	9.70	7.53	4.46	2.29	2.16	14.45	6.52	7.64	1.50	1.21	8.09	4.70	70.25
1993	13.35	2.82	5.36	11.58	2.18	3.14	4.48	4.29	1.46	5.49	3.76	3.29	61.20
Record Mean	5.11	4.87	4.98	4.89	4.95	4.61	6.48	5.62	4.23	3.15	3.95	5.23	58.08

AVERAGE TEMPERATURE (deg. F)

BATON ROUGE, LOUISANA

YEAR	JAN	FEB	MAR	APR	MAY	JUNE	JULY	AUG	SEP	OCT	NOV	DEC	ANNUAL
1964	49.7	49.5	60.5	70.8	75.9	80.7	81.8	82.0	77.4	64.6	62.1	53.7	67.4
1965	53.2	53.1	56.8	71.9	75.2	79.1	81.6	79.6	78.1	66.7	64.3	53.5	67.8
1966	45.3	50.4	58.8	67.8	75.0	78.5	82.9	80.0	76.9	66.4	62.0	53.0	66.4
1967	51.5	50.7	64.7	73.0	72.9	80.7	79.9	80.1	74.5	65.2	58.9	54.4	67.2
1968	49.6	45.2	57.5	69.8	74.6	80.6	81.8	81.7	76.7	70.4	54.4	50.8	66.1
1969	53.4	53.5	53.8	68.8	74.6	81.8	83.5	80.9	77.3	70.2	57.6	52.4	67.3
1970	46.2	51.2	59.1	70.8	73.4	79.4	81.1	82.0	79.3	66.4	54.4	56.1	66.6
1971	52.8	53.3	57.3	66.4	72.7	80.8	81.9	81.7	78.7	72.0	57.9	62.7	68.2
1972	56.6	55.3	62.2	69.9	74.7	82.0	81.2	82.8	81.7	72.2	57.9	54.9	69.3
1973	49.3	51.2	65.2	65.5	74.1	81.0	84.4	80.9	79.9	73.3	66.2	52.9	68.7
1974	60.8	55.4	67.1	68.5	76.9	77.8	81.3	80.4	75.8	67.2	58.5	54.1	68.7
1975	55.7	57.1	60.5	66.5	75.4	79.6	81.4	81.0	74.7	69.0	59.7	51.0	67.6
1976	48.6	59.1	63.6	68.4	71.1	77.9	81.1	80.9	76.9	61.9	51.4	49.9	65.9
1977	41.8	53.1	62.7	67.9	75.6	82.5	83.3	81.3	79.7	66.8	61.4	52.7	67.4
1978	42.6	45.3	57.1	68.0	76.6	81.8	83.1	81.7	78.8	68.0	63.4	52.2	66.6
1979	42.7	50.1	60.5	68.8	72.1	79.1	81.5	80.9	76.0	68.0	54.9	50.0	65.4
1980	52.6	50.3	58.8	66.2	76.2	82.0	83.7	82.2	80.2	64.1	55.9	50.7	66.9
1981	46.2	52.8	59.1	72.0	71.9	81.7	83.6	82.6	76.3	68.4	62.2	51.8	67.4
1982	52.7	51.8	63.1	67.9	76.3	82.6	82.3	82.4	77.0	68.6	60.9	57.0	68.5
1983	48.7	51.4	57.3	62.6	72.6	77.0	82.0	82.2	75.2	68.0	58.9	46.5	65.2
1984	45.4	53.4	59.9	68.0	73.8	78.9	80.3	79.8	75.9	73.5	57.3	61.3	67.3
1985	43.8	50.4	64.9	68.7	74.0	80.2	80.7	82.0	76.5	72.1	66.3	48.6	67.4
1986	50.7	57.8	60.5	68.5	76.4	81.6	83.8	82.1	82.0	69.1	63.8	51.0	68.9
1987	49.0	55.0	58.7	65.7	76.9	79.7	82.4	83.1	77.5	64.3	60.5	57.4	67.5
1988	47.0	51.9	60.2	68.0	73.7	80.3	82.7	83.0	79.0	66.9	63.9	55.0	67.6
1989	58.7	53.7	63.2	67.4	76.5	79.9	82.0	82.5	77.2	67.9	60.9	44.6	67.9
1990	56.9	60.6	63.2	68.1	76.6	83.9	82.2	82.9	79.6	66.5	61.5	56.3	69.9
1991	50.9	56.7	63.1	70.9	77.2	81.6	83.5	82.1	77.6	70.7	55.7	57.1	68.9
1992	50.8	58.2	61.8	67.0	73.3	80.3	84.0	79.8	78.5	68.9	55.3	56.3	67.9
1993	54.3	54.4	58.9	63.8	71.0	79.9	82.3	82.8	78.0	67.0	55.6	50.3	66.5
Record Mean	52.1	54.0	61.3	66.8	73.7	80.4	82.0	81.8	78.2	69.0	58.3	52.7	67.5
Max	62.2	64.2	72.0	76.2	83.3	90.4	91.4	91.5	88.4	80.9	69.6	62.7	77.7
Min	42.0	43.7	50.5	57.3	64.1	70.3	72.5	72.1	67.9	57.1	47.1	42.6	57.3

REFERENCE NOTES

BATON ROUGE, LOUISANA

GENERAL
T - TRACE AMOUNT.
BLANK ENTRIES DENOTE MISSING/UNREPORTED DATA.
# INDICATES A STATION OR INSTRUMENT RELOCATION.

SPECIFIC

(a) - LENGTH OF RECORD IN YEARS, ALTHOUGH INDIVIDUAL MONTHS MAY BE MISSING.
0.* OR * - THE VALUE IS BETWEEN 0.0 AND 0.05
NORMALS - BASED ON THE 1961-1990 RECORD PERIOD.
EXTREMES - DATES ARE THE MOST RECENT OCCURRENCE
WIND DIR.- NUMERALS SHOW TENS OF DEGRESS CLOCKWISE FROM TRUE NORTH. "00" INDICATES CALM.
RESULTANT DIRECTIONS ARE GIVEN TO WHOLE DEGREES.
BOLD VALUES INDICATE EXTREME VALUES WHICH OCCURRED AFTER THE ASOS SYSTEM WAS COMMISSIONED.

RECORD = PERIOD OF RECORD
RECORD MEAN PRECIPITATION IS THE MEAN OF ALL DAILY PRECIPITATION AMOUNTS DURING THE PERIOD OF RECORD.
RECORD MAX(MIN) TEMPERATURE IS THE MEAN OF ALL DAILY MAX(MIN) TEMPERATURES DURING THE PERIOD OF RECORD.
RECORD MEAN TEMPERATURE IS THE SUM OF THE RECORD MAX AND RECORD MIN DIVIDED BY 2.
AVERAGE TEMPERATURE IS THE SUM OF THE MEAN DAILY MAX AND MIN TEMPERATURE DIVIDED BY 2.

EXCEPTIONS

RECORD MEANS ARE THROUGH THE CURRENT YEAR, BEGINNING IN 1893 FOR TEMPERATURE
1893 FOR PRECIPITATION
1951 FOR SNOWFALL

HEATING DEGREE DAYS Base 65 deg. F — BATON ROUGE, LOUISANA

SEASON	JULY	AUG	SEP	OCT	NOV	DEC	JAN	FEB	MAR	APR	MAY	JUNE	TOTAL
1964-65	0	0	0	75	155	372	367	343	286	7	0	0	1605
1965-66	0	0	1	64	86	355	606	403	212	38	0	0	1765
1966-67	0	0	0	55	143	390	426	393	97	2	9	0	1515
1967-68	0	0	23	65	214	347	468	569	270	19	0	0	1975
1968-69	0	0	0	44	330	437	369	325	345	13	0	0	1863
1969-70	0	0	0	35	240	382	591	379	198	31	14	0	1870
1970-71	0	0	0	64	320	289	392	333	250	85	3	0	1736
1971-72	0	0	0	11	242	142	284	288	123	32	0	0	1122
1972-73	0	0	0	19	252	338	479	384	77	91	0	0	1640
1973-74	0	0	0	18	86	384	179	285	70	28	0	0	1050
1974-75	0	0	0	27	236	361	308	250	202	74	0	0	1458
1975-76	0	0	6	24	246	446	507	189	118	13	0	0	1549
1976-77	0	0	0	140	401	464	710	331	141	18	0	0	2205
1977-78	0	0	0	56	144	387	694	546	258	29	2	0	2116
1978-79	0	0	0	32	104	427	687	418	178	19	5	0	1870
1979-80	0	0	0	44	308	465	379	433	225	44	0	0	1898
1980-81	0	0	0	96	279	448	576	345	192	10	6	0	1952
1981-82	0	0	4	64	132	410	425	366	181	59	0	0	1641
1982-83	0	0	3	51	184	297	499	375	249	114	2	0	1774
1983-84	0	0	8	51	216	571	598	338	188	46	5	0	2021
1984-85	0	0	8	16	248	179	648	413	72	33	2	0	1619
1985-86	0	0	0	23	76	509	433	230	169	25	0	0	1465
1986-87	0	0	0	31	118	431	490	280	201	90	0	0	1641
1987-88	0	0	0	66	181	264	559	378	186	25	1	0	1660
1988-89	0	0	0	20	129	325	230	357	170	68	0	0	1299
1989-90	0	0	0	62	177	626	258	153	126	57	0	0	1459
1990-91	0	0	1	94	147	311	432	238	146	15	0	0	1384
1991-92	0	0	1	25	332	273	433	207	130	56	12	0	1469
1992-93	0	0	0	6	297	271	329	299	214	104	0	0	1520
1993-94	0	0	1	83	314	455							

COOLING DEGREE DAYS Base 65 deg. F — BATON ROUGE, LOUISANA

YEAR	JAN	FEB	MAR	APR	MAY	JUNE	JULY	AUG	SEP	OCT	NOV	DEC	TOTAL
1969	20	10	2	133	305	509	579	501	375	202	26	0	2662
1970	16	1	22	211	281	438	507	531	435	111	8	20	2581
1971	21	12	21	132	249	481	530	523	421	236	35	77	2738
1972	31	15	44	183	307	519	507	558	506	250	46	33	2999
1973	0	2	92	113	292	486	607	499	455	282	128	15	2971
1974	57	21	142	141	380	390	511	485	329	103	49	33	2641
1975	30	36	68	126	329	446	515	500	302	156	94	16	2618
1976	2	25	82	123	195	394	506	501	367	51	2	0	2248
1977	0	4	77	111	337	535	573	513	447	121	42	12	2772
1978	8	0	18	129	371	512	567	526	424	131	63	32	2781
1979	1	6	44	141	233	430	521	501	338	144	14	6	2379
1980	0	15	40	84	352	514	586	539	461	75	15	10	2691
1981	0	7	17	227	226	509	581	553	350	176	56	4	2706
1982	49	1	130	152	359	536	546	547	370	170	66	56	2982
1983	0	0	14	48	243	368	533	538	325	152	39	4	2264
1984	0	7	38	141	286	423	480	468	343	286	22	69	2563
1985	0	10	76	148	291	459	492	533	352	248	121	11	2741
1986	0	35	36	136	359	503	592	536	517	164	89	5	2972
1987	3	4	14	116	376	447	548	568	384	53	51	33	2597
1988	7	4	44	121	275	467	555	564	428	88	102	21	2676
1989	42	47	119	148	365	453	534	550	373	157	62	0	2850
1990	17	36	81	155	366	572	543	563	445	150	52	46	3026
1991	2	12	92	201	386	504	577	537	385	209	59	36	3000
1992	0	17	38	126	274	463	593	469	412	135	11	9	2547
1993	6	5	33	75	192	452	543	560	398	152	41	5	2462

SNOWFALL (inches) — BATON ROUGE, LOUISANA

SEASON	JULY	AUG	SEP	OCT	NOV	DEC	JAN	FEB	MAR	APR	MAY	JUNE	TOTAL
1970-71	0.0	0.0	0.0	0.0	0.0	0.0	0.0	T	0.0	0.0	0.0	0.0	T
1971-72	0.0	0.0	0.0	0.0	0.0	0.0	T	0.0	0.0	0.0	0.0	0.0	T
1972-73	0.0	0.0	0.0	0.0	0.0	0.0	0.6	1.8	0.0	0.0	0.0	0.0	2.4
1973-74	0.0	0.0	0.0	0.0	0.0	T	0.0	0.0	0.0	0.0	0.0	0.0	T
1974-75	0.0	0.0	0.0	0.0	0.0	0.0	T	0.0	0.0	0.0	0.0	0.0	T
1975-76	0.0	0.0	0.0	0.0	0.0	0.0	T	0.0	0.0	0.0	0.0	0.0	T
1976-77	0.0	0.0	0.0	0.0	T	0.0	T	0.0	0.0	0.0	0.0	0.0	T
1977-78	0.0	0.0	0.0	0.0	0.0	0.0	T	0.0	0.0	0.0	0.0	0.0	T
1978-79	0.0	0.0	0.0	0.0	0.0	0.0	T	0.0	0.0	0.0	0.0	0.0	T
1979-80	0.0	0.0	0.0	0.0	0.0	0.0	0.0	0.0	T	0.0	0.0	0.0	T
1980-81	0.0	0.0	0.0	0.0	0.0	0.0	T	T	0.0	0.0	0.0	0.0	T
1981-82	0.0	0.0	0.0	0.0	0.0	0.0	T	0.0	0.0	0.0	0.0	0.0	T
1982-83	0.0	0.0	0.0	0.0	0.0	0.0	0.0	0.0	0.0	0.0	0.0	0.0	0.0
1983-84	0.0	0.0	0.0	0.0	0.0	T	0.0	0.0	0.0	0.0	0.0	0.0	T
1984-85	0.0	0.0	0.0	0.0	0.0	0.0	T	T	0.0	0.0	0.0	0.0	T
1985-86	0.0	0.0	0.0	0.0	0.0	0.0	0.0	0.0	0.0	0.0	0.0	0.0	0.0
1986-87	0.0	0.0	0.0	0.0	0.0	0.0	0.0	0.0	0.0	0.0	0.0	0.0	0.0
1987-88	0.0	0.0	0.0	0.0	0.0	0.0	0.0	3.2	0.0	0.0	0.0	0.0	3.2
1988-89	0.0	0.0	0.0	0.0	0.0	0.0	0.0	T	T	0.0	T	0.0	T
1989-90	0.0	0.0	0.0	0.0	0.0	T	0.0	0.0	0.0	0.0	0.0	0.0	T
1990-91	0.0	0.0	0.0	0.0	0.0	0.0	0.0	0.0	0.0	0.0	0.0	0.0	0.0
1991-92	0.0	0.0	0.0	0.0	0.0	0.0	T	0.0	T	0.0	0.0	0.0	T
1992-93	0.0	0.0	0.0	0.0	0.0	0.0	0.0	0.0	T	0.0			
1993-94	0.0	0.0	0.0	0.0	0.0	0.0							
Record Mean	0.0	0.0	0.0	0.0	T	T	T	0.2	T	0.0	T	0.0	0.2

LAKE CHARLES, LOUISIANA

Lake Charles is located on the east side of the lake of the same name. The Calcasieu River enters and exits Lake Charles and several other lakes in the area on its way to the Gulf of Mexico. The terrain is flat, level coastal plain. Extensive marshes begin some 10 to 15 miles south and extend to the coast. Area elevations range from near sea level to about 25 feet above sea level. The National Weather Service Office is at the Lake Charles Municipal Airport, about 7 miles south of the downtown area. Calcasieu Lake is only 6 miles southwest of the airport.

The general classification of the Lake Charles climate is humid subtropical with a strong maritime character. The climate is influenced to a large degree by the amount of water surface in the immediate area and the proximity of the Gulf of Mexico.

Prevailing wind flow is southerly during much of the year. The flow of air from the Gulf of Mexico helps to temper extremes of summer heat, shorten the duration of winter cold spells and provide a source of abundant rain. Winds are usually rather light.

Rainfall is heavy, with the normal annual total more than 50 inches. Amounts are substantial in all seasons. Almost all rainfall occurs from brief convective showers, except occasionally during winter when nearly continuous frontal rains may persist for a few days. In spite of the large normal rainfall amounts, dry spells of two or three weeks duration are not uncommon.

The winter months are normally mild with cold spells usually of short duration. Temperatures of 20 degrees and below are extremely rare, occurring only about one year in five.

Snow is a negligible. Many years pass without measurable snowfall. However, on rare occasions, as much as 22 inches of snow have fallen at Lake Charles. Freezing rain and sleet are only a little less uncommon than snow.

The summer weather is consistently quite warm and humid but the temperature rarely reaches the 100 degree mark. The humidity is often above 90 percent at night and seldom falls below 50 percent during the afternoons.

The spring and fall seasons are very mild and pleasant with only brief rains interrupting long periods of dry sunny weather.

Severe local storms may occur during any season but are most frequent in the spring. The area weather is occasionally influenced by tropical storms or hurricanes. Some of these storms may be accompanied by tornadoes.

NORMALS, MEANS, AND EXTREMES

LAKE CHARLES, LOUISIANA

LATITUDE: 30°07'N LONGITUDE: 93°13' W ELEVATION: FT. GRND 9 BARO 18 TIME ZONE: CENTRAL WBAN: 03937

	(a)	JAN	FEB	MAR	APR	MAY	JUNE	JULY	AUG	SEP	OCT	NOV	DEC	YEAR
TEMPERATURE °F:														
Normals														
-Daily Maximum		59.7	63.4	70.7	77.9	84.0	89.0	90.8	90.8	86.8	80.2	70.8	63.3	77.3
-Daily Minimum		41.1	44.0	50.7	58.7	65.7	71.6	73.6	73.0	68.6	57.9	50.0	43.6	58.2
-Monthly		50.4	53.7	60.7	68.3	74.9	80.3	82.2	81.9	77.7	69.1	60.4	53.5	67.8
Extremes														
-Record Highest	29	82	83	86	95	96	99	102	101	98	92	87	82	102
-Year		1989	1972	1974	1987	1989	1990	1980	1990	1987	1990	1989	1978	JUL 1980
-Record Lowest	29	15	22	25	34	49	56	61	59	47	30	23	11	11
-Year		1985	1981	1989	1971	1992	1984	1967	1992	1967	1993	1976	1989	DEC 1989
NORMAL DEGREE DAYS:														
Heating (base 65°F)		467	328	178	40	0	0	0	0	0	35	193	375	1616
Cooling (base 65°F)		15	12	45	139	307	459	533	524	381	162	55	18	2650
% OF POSSIBLE SUNSHINE	12	55	60	71	71	70	77	80	78	76	72	62	53	69
MEAN SKY COVER (tenths)														
Sunrise - Sunset	32	7.0	6.3	6.5	6.4	6.0	5.5	6.0	5.7	5.4	4.7	5.6	6.6	6.0
MEAN NUMBER OF DAYS:														
Sunrise to Sunset														
-Clear	32	6.8	7.8	7.3	6.9	7.3	8.2	6.3	7.2	9.3	13.5	9.8	7.6	98.0
-Partly Cloudy	32	5.5	5.6	8.1	8.3	11.4	13.2	14.2	14.7	11.3	8.7	7.7	6.8	115.3
-Cloudy	32	18.8	14.8	15.7	14.8	12.3	8.6	10.5	9.1	9.5	8.8	12.5	16.5	151.9
Precipitation														
.01 inches or more	32	9.7	8.2	7.9	6.9	7.9	8.8	10.7	10.5	9.4	6.0	7.7	9.2	102.9
Snow,Ice pellets,hail														
1.0 inches or more	32	0.1	0.*	0.0	0.0	0.0	0.0	0.0	0.0	0.0	0.0	0.0	0.0	0.1
Thunderstorms	32	3.2	2.7	4.3	4.5	7.5	10.0	14.1	13.4	8.5	3.1	3.0	2.8	77.1
Heavy Fog Visibility														
1/4 mile or less	32	8.1	5.6	6.6	3.8	2.1	0.7	0.5	0.6	2.3	6.0	5.9	7.0	49.3
Temperature °F														
-Maximum														
90° and above	29	0.0	0.0	0.0	0.2	2.1	13.7	22.9	21.8	10.9	0.9	0.0	0.0	72.6
32° and below	29	0.1	0.*	0.0	0.0	0.0	0.0	0.0	0.0	0.0	0.0	0.0	0.1	0.2
-Minimum														
32° and below	29	5.6	2.8	0.7	0.0	0.0	0.0	0.0	0.0	0.0	0.1	1.2	3.8	14.1
0° and below	29	0.0	0.0	0.0	0.0	0.0	0.0	0.0	0.0	0.0	0.0	0.0	0.0	0.0
AVG. STATION PRESS.(mb)	21	1020.0	1018.6	1015.4	1014.6	1013.2	1013.9	1015.6	1015.1	1014.8	1017.1	1018.2	1019.7	1016.4
RELATIVE HUMIDITY (%)														
Hour 00	29	86	85	87	88	90	91	92	92	90	89	87	87	89
Hour 06 (Local Time)	29	88	87	89	90	93	93	94	94	93	91	89	88	91
Hour 12	29	69	64	62	61	63	63	64	63	62	56	60	66	63
Hour 18	29	76	70	69	67	68	69	71	71	73	71	75	77	71
PRECIPITATION (inches):														
Water Equivalent														
-Normal		4.52	3.59	3.29	3.33	5.67	4.96	5.20	5.33	5.69	3.95	4.26	5.05	54.84
-Maximum Monthly	32	14.29	6.78	9.01	10.95	20.71	25.33	13.19	17.36	19.96	17.28	8.26	13.27	25.33
-Year		1991	1985	1980	1973	1980	1989	1979	1962	1973	1970	1986	1967	JUN 1989
-Minimum Monthly	32	0.78	0.62	0.27	0.47	0.34	0.84	0.48	0.77	0.43	T	0.11	2.07	T
-Year		1971	1989	1971	1987	1978	1969	1962	1976	1989	1963	1967	1975	OCT 1963
-Maximum in 24 hrs	32	5.80	3.28	4.91	5.50	16.88	7.09	6.59	14.10	11.20	7.24	4.02	6.88	16.88
-Year		1991	1963	1973	1973	1980	1981	1987	1962	1979	1970	1993	1971	MAY 1980
Snow,Ice pellets,hail														
-Maximum Monthly	32	4.0	1.6	T	T	T	0.0	0.0	0.0	0.0	0.0	T	0.2	4.0
-Year		1973	1988	1968	1993	1992						1976	1989	JAN 1973
-Maximum in 24 hrs	32	4.0	1.6	T	T	T	0.0	0.0	0.0	0.0	0.0	T	0.2	4.0
-Year		1973	1988	1968	1993	1992						1976	1989	JAN 1973
WIND:														
Mean Speed (mph)	32	10.0	10.3	10.5	10.1	8.9	7.7	6.5	6.2	7.3	7.7	9.1	9.5	8.6
Prevailing Direction through 1963		N	S	S	S	S	SSW	SSW	SSW	ENE	ENE	ENE	NE	S
Fastest Obs. 1 Min.														
-Direction (!!)	32	32	25	18	06	31	19	12	11	36	33	21	30	32
-Speed (MPH)	32	58	40	40	44	43	53	35	46	40	38	46	35	58
-Year		1962	1971	1973	1973	1986	1991	1974	1964	1971	1990	1987	1987	JAN 1962
Peak Gust														
-Direction (!!)	10	NE	W	SE	S	W	S	SE	S	NW	N	S	NW	S
-Speed (mph)	10	49	62	52	46	55	61	49	52	58	49	69	46	69
-Date		1991	1984	1990	1993	1989	1991	1990	1988	1987	1985	1987	1987	NOV 1987

PRECIPITATION (inches)

LAKE CHARLES, LOUISIANA

YEAR	JAN	FEB	MAR	APR	MAY	JUNE	JULY	AUG	SEP	OCT	NOV	DEC	ANNUAL
1964	5.49	3.05	4.02	1.84	2.12	5.87	4.86	5.58	6.14	0.30	2.06	5.84	47.17
1965	2.49	3.86	3.34	0.96	4.05	1.49	1.85	6.79	3.43	0.32	1.22	5.00	34.80
1966	6.48	6.17	0.75	8.59	6.30	5.1C	4.90	6.63	3.54	2.96	4.45	3.78	59.65
1967	1.70	2.26	1.05	5.63	8.59	1.64	5.47	6.21	4.21	4.57	0.11	13.27	54.71
1968	3.89	2.68	2.76	2.39	3.16	8.90	7.82	3.59	3.11	2.38	5.83	4.73	51.24
1969	1.13	6.75	4.27	9.53	6.65	0.84	10.06	1.97	3.23	4.42	0.70	5.57	55.12
1970	2.16	2.68	5.25	1.81	7.11	4.28	0.97	3.56	4.13	17.28	1.87	4.10	55.20
1971	0.78	3.45	0.27	0.93	5.78	2.23	4.45	6.31	5.38	2.64	1.64	9.90	43.76
1972	7.73	2.24	2.23	1.58	4.53	0.93	7.75	6.64	5.82	6.44	4.30	5.47	55.66
1973	4.14	2.94	7.40	10.95	7.48	3.98	2.96	3.29	19.96	4.12	3.01	4.80	75.03
1974	12.69	3.89	3.37	2.95	11.01	2.89	1.28	5.54	3.60	4.15	7.30	7.77	66.44
1975	6.06	1.08	2.41	7.22	7.04	5.43	6.96	5.15	5.37	2.50	4.22	2.07	55.51
1976	2.06	1.52	2.53	1.22	2.43	7.50	2.72	0.77	3.26	4.51	5.16	6.72	40.40
1977	4.88	1.52	2.67	6.17	1.51	6.16	3.20	11.52	3.95	5.62	6.98	3.19	57.37
1978	6.63	2.10	1.48	0.52	0.34	6.00	5.04	7.22	6.04	0.18	4.14	2.37	42.06
1979	4.82	6.47	6.00	6.32	7.35	2.60	13.19	3.81	14.10	3.17	4.74	3.22	75.79
1980	5.43	2.21	9.01	1.59	20.71	1.30	8.24	1.55	5.76	3.78	3.27	2.33	65.18
1981	2.06	3.13	1.70	1.26	5.88	14.42	7.65	3.51	2.56	3.72	1.50	2.53	49.92
1982	2.12	2.39	2.73	4.04	5.61	4.90	3.03	7.13	8.51	3.00	6.41	10.76	60.63
1983	5.85	4.36	2.63	1.73	10.19	4.13	1.87	8.74	8.64	0.23	3.56	3.03	54.96
1984	4.54	5.42	1.89	1.86	8.24	3.84	5.36	4.86	6.81	12.22	2.80	4.09	61.93
1985	3.41	6.78	3.56	1.25	3.74	1.48	4.22	6.81	2.09	12.75	3.00	3.75	52.84
1986	2.72	0.93	1.91	2.80	6.02	5.84	4.14	4.96	6.22	5.01	8.26	9.84	58.65
1987	6.76	5.75	4.49	0.47	3.44	10.95	9.15	3.10	3.80	3.13	6.58	4.92	62.54
1988	2.75	5.52	6.30	3.32	2.95	5.66	6.02	7.66	7.76	2.37	4.02	4.80	59.13
1989	4.58	0.62	4.98	2.24	7.56	25.33	5.26	2.72	0.43	0.75	3.94	2.15	60.56
1990	8.91	5.13	5.31	3.55	3.79	3.83	3.51	0.94	9.55	1.48	3.64	4.04	53.68
1991	14.29	2.61	2.35	6.66	14.76	7.65	2.30	7.36	3.65	3.37	3.16	2.16	70.32
1992	11.08	3.87	4.10	7.16	2.68	8.19	6.07	2.46	3.56	2.26	7.42	3.84	62.69
1993	6.83	4.26	4.69	8.19	6.62	9.85	1.69	2.71	1.27	3.83	6.90	2.49	59.33
Record Mean	4.67	4.00	3.69	4.20	5.53	5.58	5.89	4.98	4.86	3.61	4.30	5.20	56.50

AVERAGE TEMPERATURE (deg. F)

LAKE CHARLES, LOUISIANA

YEAR	JAN	FEB	MAR	APR	MAY	JUNE	JULY	AUG	SEP	OCT	NOV	DEC	ANNUAL
#1964	49.0	48.1	58.7	69.1	74.8	79.0	79.6	82.7	78.2	65.8	63.6	54.2	66.9
1965	55.0	54.4	56.8	71.8	75.5	80.1	82.8	80.0	78.1	67.8	66.8	55.9	68.8
1966	47.9	52.5	60.0	69.8	75.7	79.7	84.0	82.1	79.0	68.4	64.2	54.5	68.1
1967	52.5	53.5	65.9	74.7	74.6	82.0	81.4	80.1	76.1	68.0	61.0	55.0	68.7
1968	50.8	46.6	58.2	70.5	74.8	79.8	81.5	81.9	75.6	69.7	56.9	52.2	66.6
1969	54.0	53.8	54.0	67.9	73.7	80.0	82.8	81.9	77.2	70.8	57.1	52.5	67.1
1970	46.1	52.2	56.9	70.5	74.1	80.2	82.2	83.7	80.2	68.1	56.7	59.6	67.6
1971	55.6	55.6	59.1	66.5	74.3	82.1	83.0	81.7	80.1	74.8	60.2	61.8	69.6
1972	55.8	56.3	63.1	69.8	74.1	81.3	80.3	80.9	79.6	69.0	54.7	53.0	68.2
1973	48.8	51.5	64.4	64.5	72.7	79.9	82.7	79.6	78.7	72.6	66.7	52.4	67.9
1974	58.0	55.6	66.3	68.2	76.0	78.7	82.0	80.8	75.2	68.9	59.9	54.5	68.7
1975	56.7	56.9	61.1	66.6	75.8	79.8	81.4	80.9	75.1	69.4	60.3	51.8	68.0
1976	49.9	59.9	63.1	68.8	72.3	78.1	80.6	81.0	77.5	61.9	52.3	51.3	66.4
1977	43.1	54.2	62.1	68.3	76.0	81.3	82.1	81.3	80.0	69.0	61.8	53.3	67.7
1978	42.9	45.6	57.8	68.2	76.8	80.8	82.9	81.8	78.4	68.7	64.4	53.5	66.8
1979	44.4	50.3	61.0	68.8	72.8	79.6	81.6	81.3	76.0	69.8	55.9	51.9	66.1
1980	54.1	52.2	59.5	65.0	75.7	81.6	84.4	82.7	80.9	65.6	56.0	52.2	67.5
1981	47.8	52.5	59.3	71.6	73.2	81.2	82.5	82.0	76.5	69.7	61.9	52.7	67.6
1982	52.3	51.7	63.5	66.4	74.2	82.0	82.0	82.0	76.4	69.0	61.6	57.2	68.1
1983	49.3	52.6	57.8	63.5	73.4	78.5	83.5	83.5	76.9	70.3	62.5	48.2	66.7
1984	48.1	56.0	62.5	69.7	75.6	79.5	80.9	80.9	75.9	73.7	58.1	62.5	68.6
1985	45.1	49.9	64.7	69.9	75.4	81.1	81.9	83.2	77.5	71.9	66.2	49.4	68.0
1986	51.9	57.5	60.0	68.6	75.5	81.5	84.1	82.0	81.7	69.3	63.8	51.4	68.9
1987	49.3	55.3	58.7	66.1	76.6	80.3	82.6	84.4	78.0	66.2	60.4	57.1	67.9
1988	47.5	53.1	60.3	68.1	73.7	79.2	82.2	83.2	78.9	68.2	64.1	55.0	67.8
1989	57.7	53.1	61.1	67.2	77.3	80.0	82.3	82.4	77.1	69.4	62.1	45.0	67.9
1990	56.0	59.3	61.5	66.8	75.2	82.3	81.3	82.7	78.7	67.1	61.7	55.0	69.0
1991	50.3	56.7	62.0	70.5	76.3	80.5	82.5	81.0	77.3	71.4	55.6	57.0	68.4
1992	51.3	58.6	63.2	68.0	73.9	81.6	83.0	79.7	78.4	69.8	56.2	56.8	68.4
1993	53.5	54.7	59.5	64.8	73.4	80.5	83.3	84.2	79.6	68.9	57.2	53.7	67.8
Record													
Mean	51.3	54.5	60.6	68.2	75.0	80.7	82.4	82.2	78.2	69.7	60.0	54.0	68.1
Max	60.5	64.0	70.4	77.7	84.2	89.4	91.0	91.0	87.6	80.9	70.5	63.7	77.6
Min	42.1	45.0	50.8	58.6	65.8	71.9	73.8	73.3	68.8	58.4	49.4	44.3	58.5

REFERENCE NOTES

LAKE CHARLES, LOUISIANA

GENERAL
T - TRACE AMOUNT.
BLANK ENTRIES DENOTE MISSING/UNREPORTED DATA.
# INDICATES A STATION OR INSTRUMENT RELOCATION.

SPECIFIC

EXCEPTIONS

RECORD MEANS ARE THROUGH THE CURRENT YEAR,
BEGINNING IN 1939 FOR TEMPERATURE
1939 FOR PRECIPITATION
1962 FOR SNOWFALL

(a) - LENGTH OF RECORD IN YEARS, ALTHOUGH INDIVIDUAL MONTHS MAY BE MISSING.
0.* OR * - THE VALUE IS BETWEEN 0.0 AND 0.05
NORMALS - BASED ON THE 1961-1990 RECORD PERIOD.
EXTREMES - DATES ARE THE MOST RECENT OCCURRENCE.
WIND DIR. - NUMERALS SHOW TENS OF DEGRESS CLOCKWISE FROM TRUE NORTH. "00" INDICATES CALM.
RESULTANT DIRECTIONS ARE GIVEN TO WHOLE DEGREES.
BOLD VALUES INDICATE EXTREME VALUES WHICH OCCURRED AFTER THE ASOS SYSTEM WAS COMMISSIONED.

RECORD = PERIOD OF RECORD
RECORD MEAN PRECIPITATION IS THE MEAN OF ALL DAILY PRECIPITATION AMOUNTS DURING THE PERIOD OF RECORD.
RECORD MAX(MIN) TEMPERATURE IS THE MEAN OF ALL DAILY MAX(MIN) TEMPERATURES DURING THE PERIOD OF RECORD.
RECORD MEAN TEMPERATURE IS THE SUM OF THE RECORD MAX AND RECORD MIN DIVIDED BY 2.
AVERAGE TEMPERATURE IS THE SUM OF THE MEAN DAILY MAX AND MIN TEMPERATURE DIVIDED BY 2.

HEATING DEGREE DAYS Base 65 deg. F — LAKE CHARLES, LOUISIANA

SEASON	JULY	AUG	SEP	OCT	NOV	DEC	JAN	FEB	MAR	APR	MAY	JUNE	TOTAL
#1964-65	0	0	0	60	139	347	313	298	276	4	0	0	1437
1965-66	0	0	2	50	54	283	528	345	174	21	0	0	1457
1966-67	0	0	0	41	113	348	395	322	86	0	2	0	1307
1967-68	0	0	10	39	151	322	439	528	238	14	0	0	1741
1968-69	0	0	0	28	255	391	352	312	338	8	0	0	1684
1969-70	0	0	0	24	260	382	584	354	249	37	5	0	1895
1970-71	0	0	0	37	255	216	311	271	204	67	0	0	1361
1971-72	0	0	0	3	191	159	297	267	102	22	0	0	1041
1972-73	0	0	1	56	319	377	498	370	74	99	1	0	1795
1973-74	0	0	0	11	78	385	245	272	81	32	0	0	1104
1974-75	0	0	0	16	206	348	282	242	179	66	0	0	1339
1975-76	0	0	0	19	212	414	458	167	122	9	0	0	1401
1976-77	0	0	0	137	376	419	671	300	134	11	0	0	2048
1977-78	0	0	0	36	136	367	682	539	236	22	2	0	2020
1978-79	0	0	0	26	98	377	636	406	151	19	0	0	1713
1979-80	0	0	0	25	284	400	336	376	194	48	0	0	1663
1980-81	0	0	0	79	286	400	523	344	182	5	1	0	1820
1981-82	0	0	5	71	118	378	423	368	156	83	0	0	1602
1982-83	0	0	4	41	162	286	478	339	229	93	0	0	1632
1983-84	0	0	6	27	140	533	516	265	130	24	1	0	1642
1984-85	0	0	7	14	234	155	610	432	67	23	0	0	1542
1985-86	0	0	0	20	79	478	400	225	168	18	0	0	1388
1986-87	0	0	0	21	128	417	480	266	200	83	0	0	1595
1987-88	0	0	0	43	176	264	538	347	179	22	0	0	1569
1988-89	0	0	0	12	123	309	249	363	198	62	0	0	1316
1989-90	0	0	0	47	165	618	284	173	150	57	0	0	1494
1990-91	0	0	1	80	141	333	449	229	143	14	1	0	1391
1991-92	0	0	0	14	317	275	419	195	104	34	5	0	1363
1992-93	0	0	0	2	281	254	350	282	191	74	0	0	1434
1993-94	0	0	0	62	260	365							

COOLING DEGREE DAYS Base 65 deg. F — LAKE CHARLES, LOUISIANA

YEAR	JAN	FEB	MAR	APR	MAY	JUNE	JULY	AUG	SEP	OCT	NOV	DEC	TOTAL
1969	17	4	2	101	275	460	559	531	373	209	30	0	2561
1970	7	0	6	206	292	464	538	582	463	142	12	51	2763
1971	25	15	28	120	295	521	563	524	457	313	53	63	2977
1972	20	22	53	172	289	499	479	503	445	184	18	11	2695
1973	0	1	65	92	248	453	555	461	418	257	136	2	2688
1974	34	17	126	133	349	418	534	496	310	143	61	29	2650
1975	32	22	63	120	341	452	510	501	311	166	79	12	2609
1976	0	23	70	127	234	400	490	503	383	47	4	0	2281
1977	0	4	49	118	343	499	539	514	456	167	48	13	2750
1978	5	2	20	126	374	480	561	528	407	151	87	26	2767
1979	2	2	36	140	248	444	521	514	336	183	17	1	2444
1980	6	12	31	52	337	505	608	555	483	108	24	9	2730
1981	0	4	13	211	264	495	551	535	356	224	35	6	2694
1982	33	5	115	129	292	470	534	535	350	174	68	52	2757
1983	0	0	15	53	268	410	580	582	368	199	69	15	2559
1984	0	12	59	173	337	440	502	501	343	293	36	84	2780
1985	0	12	60	177	328	495	528	571	381	242	122	3	2919
1986	0	22	20	135	332	503	600	535	509	161	99	4	2920
1987	1	2	11	121	368	469	553	606	397	85	46	24	2683
1988	3	8	43	122	278	432	541	573	423	117	104	7	2651
1989	30	36	84	137	388	455	543	547	370	191	82	4	2867
1990	14	23	49	118	321	526	512	557	421	151	52	31	2775
1991	0	4	58	184	357	470	550	506	375	220	43	34	2801
1992	0	16	56	131	290	502	563	463	411	159	24	8	2623
1993	0	2	26	73	267	469	574	605	445	189	36	18	2704

SNOWFALL (inches) — LAKE CHARLES, LOUISIANA

SEASON	JULY	AUG	SEP	OCT	NOV	DEC	JAN	FEB	MAR	APR	MAY	JUNE	TOTAL
1970-71	0.0	0.0	0.0	0.0	0.0	0.0	T	0.0	0.0	0.0	0.0	0.0	T
1971-72	0.0	0.0	0.0	0.0	0.0	0.0	0.0	0.0	0.0	0.0	0.0	0.0	0.0
1972-73	0.0	0.0	0.0	0.0	0.0	0.0	4.0	T	0.0	0.0	0.0	0.0	4.0
1973-74	0.0	0.0	0.0	0.0	0.0	0.0	0.0	0.0	0.0	0.0	0.0	0.0	0.0
1974-75	0.0	0.0	0.0	0.0	0.0	0.0	0.0	0.0	0.0	0.0	0.0	0.0	0.0
1975-76	0.0	0.0	0.0	0.0	0.0	0.0	T	0.0	0.0	0.0	0.0	0.0	T
1976-77	0.0	0.0	0.0	0.0	T	0.0	T	0.0	0.0	0.0	0.0	0.0	T
1977-78	0.0	0.0	0.0	0.0	0.0	0.0	1.0	T	0.0	0.0	0.0	0.0	1.0
1978-79	0.0	0.0	0.0	0.0	0.0	0.0	T	0.0	0.0	0.0	0.0	0.0	T
1979-80	0.0	0.0	0.0	0.0	0.0	0.0	0.0	T	0.0	0.0	0.0	0.0	T
1980-81	0.0	0.0	0.0	0.0	0.0	0.0	0.0	T	0.0	0.0	0.0	0.0	T
1981-82	0.0	0.0	0.0	0.0	0.0	0.0	0.2	0.0	0.0	0.0	0.0	0.0	0.2
1982-83	0.0	0.0	0.0	0.0	0.0	0.0	0.0	0.0	0.0	0.0	0.0	0.0	0.0
1983-84	0.0	0.0	0.0	0.0	0.0	0.0	0.0	0.0	0.0	0.0	0.0	0.0	0.0
1984-85	0.0	0.0	0.0	0.0	0.0	0.0	T	T	0.0	0.0	0.0	0.0	T
1985-86	0.0	0.0	0.0	0.0	0.0	T	0.0	0.0	0.0	0.0	0.0	0.0	T
1986-87	0.0	0.0	0.0	0.0	0.0	0.0	T	0.0	0.0	0.0	0.0	0.0	T
1987-88	0.0	0.0	0.0	0.0	0.0	0.0	0.0	1.6	0.0	0.0	0.0	0.0	1.6
1988-89	0.0	0.0	0.0	0.0	0.0	0.0	0.0	0.0	0.0	0.0	0.0	0.0	0.0
1989-90	0.0	0.0	0.0	0.0	0.0	0.2	0.0	0.0	0.0	0.0	0.0	0.0	0.2
1990-91	0.0	0.0	0.0	0.0	0.0	0.0	0.0	0.0	0.0	T	T	0.0	T
1991-92	0.0	0.0	0.0	0.0	0.0	0.0	T	0.0	0.0	0.0	T	0.0	T
1992-93	0.0	0.0	0.0	0.0	0.0	0.0	0.0	0.0	0.0	T	0.0	0.0	T
1993-94	0.0	0.0	0.0	0.0	0.0	T						-	
Record Mean	0.0	0.0	0.0	0.0	T	T	0.2	0.1	T	T	T	0.0	0.2

NEW ORLEANS, LOUISIANA

The New Orleans metropolitan area is virtually surrounded by water. Lake Pontchartrain, some 610 square miles in area, borders the city on the north and is connected to the Gulf of Mexico through Lake Borgne on the east. In other directions there are bayous, lakes, and marshy delta land. The proximity of the Gulf of Mexico also has a great influence on the climate. Elevations in the city vary from a few feet below to a few feet above mean sea level. A massive levee system surrounding the city and along the Mississippi River offers protection against flooding from the river and tidal surges. The New Orleans International Airport is located 12 miles west of downtown New Orleans, between the Mississippi River and Lake Pontchartrain.

The climate of the city can best be described as humid with the surrounding water modifying the temperature and decreasing the range between the extremes. Almost daily sporadic afternoon thunderstorms from mid-June through September keep the temperature from rising much above 90 degrees. From about mid-November to mid-March, the area is subjected alternately to the southerly flow of warm tropical air and to the northerly flow of cold continental air in periods of varying lengths. The usual track of winter storms is to the north of New Orleans, but occasionally one moves this far south, bringing large and rather sudden drops in temperature. However, the cold spells seldom last over three or four days. The lowest temperatures observed are below 10 degrees. In about two-thirds of the years, the lowest temperature is about 24 degrees or warmer. The lowest temperatures in some years are entirely above freezing.

During the winter and spring, the cold Mississippi River water enhances the formation of river fogs, particularly when light southerly winds bring warm, moist air into the area from the Gulf of Mexico. The nearby lakes and marshes also contribute to fog formation. Even so, the fog usually does not seriously affect automobile traffic except for brief periods. However, air travel will be suspended for several hours and river traffic, at times, will be unable to move between New Orleans and the Gulf for several days.

Rather frequent and sometimes very heavy rains are typical for this area. There are an average of 120 days of measurable rain per year and an annual average accumulation of over 60 inches. A fairly definite rainy period occurs from mid-December to mid-March. Precipitation during this period is most likely to be steady rain for two to three day periods. April, May, October, and November are generally dry, but there have been some extremely heavy showers in those months. The greatest 24-hour amounts have exceeded 14 inches. Snowfall is rather infrequent and light. However, on rare occasions, snowstorms have produced accumulations over 8 inches.

While thunder occurs with most of the showers in the area, thunderstorms with damaging winds are infrequent. Hail of a damaging nature seldom occurs, and tornadoes are extremely rare. However, waterspouts are observed quite often on nearby lakes. Hurricanes have effected the area.

The lower Mississippi River floods result from runoff upstream. If the water level in the river becomes dangerously high, the spillways upriver can be opened to divert the floodwaters. Rainfall in the New Orleans area is pumped into the surrounding lakes and bayous. Local street and minor urban flooding of short duration result from occasional downpours.

Air pollution is not a serious problem. The area is not highly industrialized, and long periods of air stagnation are rare.

Based on the 1951-1980 period, the average first occurrence of 32 degrees Fahrenheit in the fall is December 5 and the average last occurrence in the spring is February 20.

NORMALS, MEANS, AND EXTREMES

NEW ORLEANS, LOUISIANA

LATITUDE: 29 °59′N LONGITUDE: 90 °15′ W ELEVATION: FT. GRND 4 BARO 20 TIME ZONE: CENTRAL WBAN: 12916

	(a)	JAN	FEB	MAR	APR	MAY	JUNE	JULY	AUG	SEP	OCT	NOV	DEC	YEAR
TEMPERATURE °F:														
Normals														
-Daily Maximum		60.8	64.1	71.6	78.5	84.4	89.2	90.6	90.2	86.6	79.4	71.1	64.3	77.6
-Daily Minimum		41.8	44.4	51.6	58.4	65.2	70.8	73.1	72.8	69.5	58.7	51.0	44.8	58.5
-Monthly		51.3	54.3	61.6	68.5	74.8	80.0	81.9	81.5	78.1	69.1	61.1	54.5	68.1
Extremes														
-Record Highest	47	83	85	89	92	96	100	101	102	101	92	87	84	102
-Year		1982	1972	1982	1987	1953	1954	1981	1980	1980	1990	1986	1978	AUG 1980
-Record Lowest	47	14	19	25	32	41	50	60	60	42	35	24	11	11
-Year		1985	1970	1980	1971	1960	1984	1967	1968	1967	1993	1970	1989	DEC 1989
NORMAL DEGREE DAYS:														
Heating (base 65°F)		450	316	162	28	0	0	0	0	0	30	178	349	1513
Cooling (base 65°F)		25	17	56	133	304	450	524	512	393	157	61	23	2655
% OF POSSIBLE SUNSHINE	20	46	51	58	63	64	64	60	61	61	66	55	48	58
MEAN SKY COVER (tenths)														
Sunrise - Sunset	45	6.8	6.3	6.3	5.8	5.6	5.6	6.4	5.8	5.5	4.4	5.4	6.4	5.9
MEAN NUMBER OF DAYS:														
Sunrise to Sunset														
-Clear	45	6.7	7.6	7.9	8.0	9.0	8.2	4.6	7.2	9.5	14.6	10.3	7.8	101.4
-Partly Cloudy	45	7.3	6.5	8.0	10.4	11.2	12.6	14.7	13.8	10.5	7.8	8.2	7.4	118.3
-Cloudy	45	17.0	14.1	15.1	11.6	10.8	9.2	11.7	10.1	10.0	8.6	11.5	15.8	145.6
Precipitation														
.01 inches or more	45	10.2	9.1	8.8	7.2	8.0	10.7	14.5	13.2	9.8	5.7	7.4	9.9	114.5
Snow, Ice pellets, hail														
1.0 inches or more	45	0.0	0.*	0.0	0.0	0.0	0.0	0.0	0.0	0.0	0.0	0.0	0.*	*
Thunderstorms	45	1.9	2.9	3.9	4.4	6.0	9.4	15.0	12.8	6.9	1.9	2.0	2.1	69.2
Heavy Fog Visibility														
1/4 mile or less	45	6.0	4.1	3.9	1.7	0.8	0.2	0.1	0.1	0.2	1.6	3.6	4.9	27.3
Temperature °F														
-Maximum														
90° and above	47	0.0	0.0	0.0	0.2	3.3	16.2	20.8	20.0	9.3	1.0	0.0	0.0	70.8
32° and below	47	0.1	0.0	0.0	0.0	0.0	0.0	0.0	0.0	0.0	0.0	0.0	0.1	0.2
-Minimum														
32° and below	47	4.9	2.8	0.5	0.*	0.0	0.0	0.0	0.0	0.0	0.0	0.9	3.7	12.8
0° and below	47	0.0	0.0	0.0	0.0	0.0	0.0	0.0	0.0	0.0	0.0	0.0	0.0	0.0
AVG. STATION PRESS. (mb)	21	1020.0	1018.7	1016.0	1015.3	1013.9	1014.5	1016.0	1015.4	1015.1	1017.3	1018.5	1020.0	1016.7
RELATIVE HUMIDITY (%)														
Hour 00	45	82	81	82	84	86	87	89	89	86	84	84	83	85
Hour 06 (Local Time)	45	85	84	85	88	89	90	91	92	89	88	86	86	88
Hour 12	45	66	63	60	59	60	63	66	66	65	59	61	66	63
Hour 18	45	72	67	64	65	65	68	72	73	74	72	75	75	70
PRECIPITATION (inches):														
Water Equivalent														
-Normal		5.05	6.01	4.90	4.50	4.56	5.84	6.12	6.17	5.51	3.05	4.42	5.75	61.88
-Maximum Monthly	47	19.25	12.59	19.09	16.12	14.33	15.01	13.15	16.12	16.74	13.20	19.81	10.77	19.81
-Year		1991	1983	1948	1980	1959	1987	1991	1977	1971	1985	1989	1967	NOV 1989
-Minimum Monthly	47	0.54	0.15	0.24	0.28	0.95	0.23	1.92	1.68	0.24	0.00	0.21	1.46	0.00
-Year		1968	1989	1955	1976	1992	1979	1981	1980	1953	1978	1949	1958	OCT 1978
-Maximum in 24 hrs	47	6.08	5.60	7.87	8.08	9.86	7.40	4.30	4.96	6.50	4.51	12.66	6.81	12.66
-Year		1978	1961	1948	1988	1959	1988	1966	1992	1971	1985	1989	1990	NOV 1989
Snow, Ice pellets, hail														
-Maximum Monthly	47	0.4	2.0	T	T	T	0.0	0.0	0.0	0.0	0.0	T	2.7	2.7
-Year		1985	1958	1993	1991	1989						1950	1963	DEC 1963
-Maximum in 24 hrs	47	0.4	2.0	T	T	T	0.0	0.0	0.0	0.0	0.0	T	2.7	2.7
-Year		1985	1958	1993	1991	1989						1950	1963	DEC 1963
WIND:														
Mean Speed (mph)	45	9.3	9.8	9.9	9.4	8.2	6.9	6.1	6.0	7.3	7.6	8.7	9.1	8.2
Prevailing Direction														
Fastest Obs. 1 Min.														
-Direction (!!!)	34	21	26	16	27	36	05	13	33	09	17	27	28	09
-Speed (MPH)	34	46	43	38	35	55	48	44	42	69	40	33	46	69
-Year		1975	1970	1987	1993	1973	1971	1979	1969	1965	1964	1992	1973	SEP 1965
Peak Gust														
-Direction (!!!)	10	W	S	N	W	W	W	NE	SE	NE	NE	NW	SW	SE
-Speed (mph)	10	48	49	53	62	60	54	51	66	58	49	41	48	66
-Date		1991	1984	1993	1993	1989	1992	1992	1992	1990	1985	1989	1990	AUG 1992

PRECIPITATION (inches)

NEW ORLEANS, LOUISIANA

YEAR	JAN	FEB	MAR	APR	MAY	JUNE	JULY	AUG	SEP	OCT	NOV	DEC	ANNUAL
1964	9.60	5.35	5.45	5.66	1.69	5.52	5.90	3.88	4.93	3.50	3.51	3.10	58.09
1965	4.48	5.25	1.95	0.33	3.62	2.21	5.26	6.34	10.03	1.03	1.49	7.35	49.34
1966	12.62	10.11	1.90	4.92	9.31	2.10	9.42	2.84	5.55	3.15	0.72	5.44	68.08
1967	4.22	6.80	1.60	2.18	3.56	2.40	6.42	7.51	3.73	3.79	0.45	10.77	53.43
1968	0.54	3.02	3.49	3.59	4.13	3.69	4.96	4.78	2.44	1.40	4.97	6.14	43.15
1969	3.12	4.80	7.08	6.04	5.51	2.47	6.64	7.80	1.08	0.51	1.73	5.26	52.04
1970	2.53	2.28	7.22	0.43	4.68	4.97	3.70	10.21	4.25	4.94	0.85	4.28	50.34
1971	1.13	4.87	3.61	1.53	1.38	8.02	4.55	5.75	16.74	0.58	2.63	6.64	57.43
1972	6.98	6.03	6.07	1.64	6.31	3.10	3.90	4.92	3.29	4.64	8.45	8.65	63.98
1973	2.68	5.40	12.17	10.47	4.68	6.08	5.94	3.37	11.07	5.07	4.04	8.31	79.28
1974	8.46	5.53	6.64	5.52	9.84	3.83	5.66	6.70	7.58	2.26	5.88	4.89	72.79
1975	2.95	3.64	5.32	6.69	8.03	12.28	8.35	10.11	3.97	4.00	11.35	3.81	80.50
1976	2.61	3.85	3.08	0.28	5.58	3.36	5.67	1.69	1.57	5.08	5.80	8.81	47.38
1977	5.62	2.75	3.96	6.38	2.59	1.74	2.91	16.12	13.48	4.33	8.77	4.15	72.80
1978	13.63	2.53	2.67	3.44	9.72	7.82	10.34	14.68	2.98	0.00	4.67	4.42	76.90
1979	5.55	12.49	3.31	4.90	4.38	0.23	11.43	4.57	4.55	1.49	4.27	3.07	60.24
1980	6.37	3.09	10.08	16.12	9.65	3.69	4.84	1.68	6.31	5.87	3.85	1.54	73.09
1981	0.94	8.34	2.70	2.28	5.35	8.47	1.92	11.10	4.78	2.03	1.10	5.50	54.51
1982	2.76	7.88	2.56	5.86	1.19	5.43	13.07	1.92	5.40	3.84	5.45	10.26	65.62
1983	3.31	12.59	4.88	14.86	3.71	10.64	2.95	6.29	5.72	4.88	6.32	9.15	85.30
1984	4.10	5.27	4.90	1.72	3.54	7.21	3.86	9.51	3.79	2.84	2.80	2.53	52.07
1985	4.83	9.28	7.07	2.11	1.16	4.56	6.92	6.37	5.74	13.20	0.96	4.78	66.98
1986	3.49	2.93	1.88	1.50	1.61	8.87	3.60	6.74	1.42	2.87	7.90	5.05	47.86
1987	8.88	7.38	4.39	2.27	3.46	15.01	6.38	5.05	1.29	0.72	2.92	2.88	60.63
1988	3.74	11.31	8.90	9.25	1.68	11.28	6.78	7.53	5.86	2.87	1.26	3.94	74.40
1989	2.47	0.15	7.14	3.20	3.50	8.22	8.34	3.31	4.53	0.51	19.81	6.28	67.46
1990	7.59	11.45	5.98	4.59	5.87	1.01	2.30	2.45	4.55	2.38	3.21	9.67	61.05
1991	19.25	5.42	6.27	15.29	14.28	10.71	13.15	7.86	3.44	1.88	2.19	2.63	102.37
1992	9.94	8.73	6.69	2.52	0.95	9.52	5.75	9.64	6.63	0.55	15.27	5.68	81.87
1993	6.21	2.34	5.65	6.82	7.23	4.96	5.77	2.26	2.47	3.67	2.43	2.90	52.71
Record Mean	5.03	5.17	5.30	4.88	4.79	5.33	6.71	6.02	5.33	3.07	4.45	5.09	61.16

AVERAGE TEMPERATURE (deg. F)

NEW ORLEANS, LOUISIANA

YEAR	JAN	FEB	MAR	APR	MAY	JUNE	JULY	AUG	SEP	OCT	NOV	DEC	ANNUAL
1964	50.0	49.8	60.5	70.6	74.6	78.9	80.6	81.7	77.0	64.8	61.8	56.4	67.2
1965	54.9	54.4	58.9	70.9	75.5	78.5	81.0	77.7	76.2	66.7	64.5	54.0	67.8
1966	46.9	52.0	59.0	68.7	75.0	77.1	82.2	80.4	77.0	68.9	61.6	53.3	66.9
1967	52.5	52.5	63.2	72.2	73.8	81.6	80.5	80.0	75.1	65.6	60.3	58.2	68.0
1968	51.6	47.2	55.9	68.1	74.1	80.4	82.4	81.0	75.4	68.5	55.5	50.8	65.9
1969	54.2	54.6	53.8	68.8	73.3	80.3	82.1	80.1	76.7	71.1	58.5	54.5	67.3
1970	47.3	51.7	60.1	70.7	74.2	79.6	81.6	81.5	80.1	68.7	55.4	57.9	67.4
1971	55.1	53.9	59.2	66.9	72.9	80.0	81.4	81.1	78.9	71.7	59.2	63.4	68.6
1972	58.6	56.1	61.9	69.8	73.9	80.8	79.4	81.1	79.6	70.4	56.6	55.3	68.6
1973	50.3	52.6	65.5	64.2	72.5	81.7	84.4	81.7	79.7	73.4	66.6	54.1	68.9
1974	63.3	55.9	67.3	69.0	75.8	78.1	80.3	80.4	77.0	66.9	59.8	55.3	69.1
1975	57.2	58.9	61.4	66.8	75.1	79.4	80.2	80.5	75.1	69.9	61.2	51.9	68.2
1976	50.6	58.2	64.8	68.5	72.3	78.3	81.2	81.5	77.8	64.0	52.7	50.6	66.7
1977	43.4	53.8	65.0	69.0	75.9	82.4	83.9	81.9	80.2	68.2	62.9	54.3	68.4
1978	44.1	45.0	59.9	71.4	76.9	81.2	82.6	83.2	81.1	69.7	67.1	55.9	68.2
1979	45.9	52.9	62.6	71.1	74.6	81.1	83.6	82.9	79.3	71.0	58.0	52.5	68.0
1980	56.1	52.0	62.0	66.2	77.9	83.3	85.8	85.5	83.5	68.8	60.0	53.6	69.6
1981	48.5	55.3	61.9	71.4	74.8	83.8	85.0	83.1	77.9	71.1	64.9	54.5	69.4
1982	54.5	55.0	65.9	69.8	76.5	81.5	81.4	82.2	76.8	70.4	62.5	59.4	69.6
1983	50.2	53.8	58.3	64.3	74.0	77.6	81.5	82.4	75.3	69.1	60.0	49.5	66.3
1984	46.6	53.9	59.3	67.5	73.7	77.4	78.8	79.2	76.3	73.5	58.8	62.4	67.3
1985	45.2	52.3	65.3	69.0	74.7	79.3	80.2	81.6	77.0	72.6	67.3	51.0	68.0
1986	51.2	59.2	60.6	67.2	76.7	81.0	83.2	81.6	81.0	70.4	66.3	53.2	69.3
1987	50.0	56.3	60.3	66.2	76.8	79.9	82.9	83.5	78.2	64.4	61.6	59.0	68.3
1988	49.4	53.2	60.9	68.4	73.3	78.5	81.6	81.4	79.8	68.0	65.6	56.0	68.0
1989	60.2	55.9	62.8	67.0	76.4	79.4	81.4	81.7	76.9	67.5	62.4	46.9	68.2
1990	57.2	61.3	63.3	67.6	76.2	82.6	82.3	83.0	79.6	68.1	62.3	59.0	70.2
1991	52.8	58.2	64.2	71.2	77.5	81.3	83.5	81.7	78.2	71.6	55.9	57.6	69.5
1992	51.1	58.3	62.0	66.4	72.8	80.6	83.1	79.6	78.5	69.5	57.4	58.2	68.1
1993	57.2	54.7	58.8	64.0	71.9	80.4	83.3	83.6	79.7	69.9	58.0	52.2	67.8
Record Mean	53.3	56.0	61.8	68.7	75.3	80.7	82.3	82.1	78.7	70.3	60.9	55.4	68.8
Max	62.1	65.2	71.1	78.0	84.3	89.5	90.6	90.3	86.7	79.7	70.3	64.4	77.7
Min	44.5	46.8	52.5	59.3	66.2	71.9	74.0	73.8	70.7	60.9	51.4	46.3	59.9

REFERENCE NOTES

NEW ORLEANS, LOUISIANA

GENERAL
T - TRACE AMOUNT.
BLANK ENTRIES DENOTE MISSING/UNREPORTED DATA.
# INDICATES A STATION OR INSTRUMENT RELOCATION.

SPECIFIC

PM - INCLUDES LAST DAY OF PREVIOUS MONTH
ASOS - AUTOMATED SURFACE OBSERVING SYSTEM IN OPERATION DURING THESE MONTHS.

(*) - LENGTH OF RECORD IN YEARS, ALTHOUGH INDIVIDUAL MONTHS MAY BE MISSING.
0.* OR * - THE VALUE IS BETWEEN 0.0 AND 0.05
NORMALS - BASED ON THE 1961-1990 RECORD PERIOD.
EXTREMES - DATES ARE THE MOST RECENT OCCURRENCE
WIND DIR.- NUMERALS SHOW TENS OF DEGRESS CLOCKWISE FROM TRUE NORTH. "00" INDICATES CALM.
RESULTANT DIRECTIONS ARE GIVEN TO WHOLE DEGREES.
BOLD VALUES INDICATE EXTREME VALUES WHICH OCCURRED AFTER THE ASOS SYSTEM WAS COMMISSIONED.

RECORD = PERIOD OF RECORD
RECORD MEAN PRECIPITATION IS THE MEAN OF ALL DAILY PRECIPITATION AMOUNTS DURING THE PERIOD OF RECORD.
RECORD MAX(MIN) TEMPERATURE IS THE MEAN OF ALL DAILY MAX(MIN) TEMPERATURES DURING THE PERIOD OF RECORD.
RECORD MEAN TEMPERATURE IS THE SUM OF THE RECORD MAX AND RECORD MIN DIVIDED BY 2.
AVERAGE TEMPERATURE IS THE SUM OF THE MEAN DAILY MAX AND MIN TEMPERATURE DIVIDED BY 2.

EXCEPTIONS

RECORD MEANS ARE THROUGH THE CURRENT YEAR, BEGINNING IN 1947 FOR TEMPERATURE
1947 FOR PRECIPITATION
1947 FOR SNOWFALL

HEATING DEGREE DAYS Base 65 deg. F NEW ORLEANS, LOUISIANA

SEASON	JULY	AUG	SEP	OCT	NOV	DEC	JAN	FEB	MAR	APR	MAY	JUNE	TOTAL
1964-65	0	0	0	73	163	297	315	316	231	11	1	0	1407
1965-66	0	0	0	54	84	339	560	361	201	31	0	0	1630
1966-67	0	0	0	33	150	376	390	353	129	2	3	0	1436
1967-68	0	0	20	56	187	265	407	511	303	29	2	0	1780
1968-69	0	0	0	55	303	435	353	304	339	12	0	0	1801
1969-70	0	0	0	19	224	330	561	367	171	24	5	0	1701
1970-71	0	0	0	24	284	248	329	328	216	88	2	0	1519
1971-72	0	0	0	10	208	137	245	278	126	25	0	0	1029
1972-73	0	0	0	28	293	314	447	351	72	114	9	0	1628
1973-74	0	0	0	18	80	355	117	274	71	16	0	0	931
1974-75	0	0	0	24	194	341	270	210	183	73	0	0	1295
1975-76	0	0	6	16	222	417	445	205	98	21	0	0	1430
1976-77	0	0	0	93	375	438	664	318	117	18	0	0	2023
1977-78	0	0	0	43	113	342	646	556	191	2	0	0	1893
1978-79	0	0	0	16	39	324	586	347	128	8	2	0	1450
1979-80	0	0	0	13	230	396	278	385	154	38	0	0	1494
1980-81	0	0	0	35	195	363	504	275	123	12	0	0	1507
1981-82	0	0	0	36	100	333	365	278	127	29	0	0	1268
1982-83	0	0	0	31	146	234	453	309	217	81	1	0	1472
1983-84	0	0	1	37	183	483	564	321	197	48	2	0	1836
1984-85	0	0	2	14	214	146	605	359	62	28	0	0	1430
1985-86	0	0	0	12	49	443	421	195	160	28	0	0	1308
1986-87	0	0	0	28	85	370	464	242	168	75	0	0	1432
1987-88	0	0	0	58	149	222	490	351	166	23	0	0	1459
1988-89	0	0	0	12	92	301	186	292	155	60	0	0	1098
1989-90	0	0	0	53	142	559	253	136	101	41	0	0	1285
1990-91	0	0	0	62	122	244	371	196	105	8	0	0	1108
1991-92	0	0	0	22	312	262	426	203	128	54	5	0	1412
1992-93	0	0	0	2	240	218	248	285	209	82	0	0	1284
1993-94	0	0	0	42	259	399							

COOLING DEGREE DAYS Base 65 deg. F NEW ORLEANS, LOUISIANA

YEAR	JAN	FEB	MAR	APR	MAY	JUNE	JULY	AUG	SEP	OCT	NOV	DEC	TOTAL
1969	28	17	1	133	266	462	537	474	360	213	34	9	2534
1970	20	0	25	200	297	442	520	520	460	148	2	32	2666
1971	29	25	44	151	252	456	514	508	424	221	40	92	2756
1972	50	26	38	175	281	479	453	507	446	200	48	19	2722
1973	0	9	96	99	247	507	607	524	448	289	136	24	2986
1974	71	27	147	144	345	402	484	484	368	93	45	45	2655
1975	34	45	80	132	321	440	479	491	314	171	114	16	2637
1976	4	18	100	132	234	404	509	518	390	68	13	0	2390
1977	0	10	123	145	345	528	593	532	463	151	56	16	2962
1978	5	0	39	203	380	493	553	569	489	169	110	49	3059
1979	0	14	63	198	307	491	581	559	435	206	25	16	2895
1980	10	13	70	85	409	554	653	640	561	160	51	17	3223
1981	0	12	35	210	311	570	627	565	396	231	102	12	3071
1982	49	6	160	182	366	504	517	541	363	208	78	66	3040
1983	0	0	16	67	286	385	518	545	317	171	42	10	2357
1984	0	6	31	130	281	379	436	448	351	286	33	71	2452
1985	0	10	78	154	308	437	480	521	366	251	124	13	2742
1986	0	40	32	99	370	487	573	524	488	203	127	9	2952
1987	3	4	30	120	373	456	562	580	402	48	53	42	2673
1988	14	15	49	131	263	411	523	513	448	113	118	30	2628
1989	46	43	95	124	363	439	515	525	365	137	70	6	2728
1990	17	40	56	127	353	538	545	567	448	166	50	62	2969
1991	2	12	88	202	396	496	580	524	402	233	47	40	3022
1992	0	18	41	103	258	472	568	459	408	149	20	14	2510
1993	13	4	26	61	222	468	572	585	447	200	57	8	2663

SNOWFALL (inches) NEW ORLEANS, LOUISIANA

SEASON	JULY	AUG	SEP	OCT	NOV	DEC	JAN	FEB	MAR	APR	MAY	JUNE	TOTAL
1970-71	0.0	0.0	0.0	0.0	0.0	0.0	0.0	0.0	0.0	0.0	0.0	0.0	0.0
1971-72	0.0	0.0	0.0	0.0	0.0	0.0	0.0	0.0	0.0	0.0	0.0	0.0	0.0
1972-73	0.0	0.0	0.0	0.0	0.0	0.0	0.1	0.6	0.0	0.0	0.0	0.0	0.7
1973-74	0.0	0.0	0.0	0.0	0.0	T	0.0	0.0	0.0	0.0	0.0	0.0	T
1974-75	0.0	0.0	0.0	0.0	0.0	0.0	0.0	0.0	0.0	0.0	0.0	0.0	0.0
1975-76	0.0	0.0	0.0	0.0	0.0	0.0	0.0	0.0	0.0	0.0	0.0	0.0	0.0
1976-77	0.0	0.0	0.0	0.0	0.0	0.0	T	0.0	0.0	0.0	0.0	0.0	T
1977-78	0.0	0.0	0.0	0.0	0.0	0.0	T	T	0.0	0.0	0.0	0.0	T
1978-79	0.0	0.0	0.0	0.0	0.0	0.0	T	0.0	0.0	0.0	0.0	0.0	T
1979-80	0.0	0.0	0.0	0.0	0.0	0.0	0.0	0.0	T	0.0	0.0	0.0	T
1980-81	0.0	0.0	0.0	0.0	0.0	0.0	0.0	0.0	0.0	0.0	0.0	0.0	0.0
1981-82	0.0	0.0	0.0	0.0	0.0	0.0	T	0.0	0.0	0.0	0.0	0.0	T
1982-83	0.0	0.0	0.0	0.0	0.0	0.0	0.0	0.0	0.0	0.0	0.0	0.0	0.0
1983-84	0.0	0.0	0.0	0.0	0.0	0.0	0.0	0.0	0.0	0.0	0.0	0.0	0.0
1984-85	0.0	0.0	0.0	0.0	0.0	0.0	0.4	0.0	0.0	0.0	0.0	0.0	0.4
1985-86	0.0	0.0	0.0	0.0	0.0	0.0	0.0	0.0	0.0	0.0	0.0	0.0	0.0
1986-87	0.0	0.0	0.0	0.0	0.0	0.0	0.0	0.0	0.0	0.0	0.0	0.0	0.0
1987-88	0.0	0.0	0.0	0.0	0.0	0.0	0.0	T	0.0	0.0	0.0	0.0	T
1988-89	0.0	0.0	0.0	0.0	0.0	0.0	0.0	0.0	T	0.0	T	0.0	T
1989-90	0.0	0.0	0.0	0.0	0.0	0.5	0.0	0.0	0.0	T	0.0	0.0	0.5
1990-91	0.0	0.0	0.0	0.0	0.0	0.0	0.0	0.0	T	T	0.0	0.0	T
1991-92	0.0	0.0	0.0	0.0	0.0	0.0	0.0	0.0	T	0.0	0.0	0.0	T
1992-93	0.0	0.0	0.0	0.0	0.0	0.0	0.0	0.0	T	0.0	0.0	0.0	T
1993-94	0.0	0.0	0.0	0.0	0.0	0.0							
Record Mean	0.0	0.0	0.0	0.0	T	0.1	T	0.1	T	T	T	0.0	0.1

SHREVEPORT, LOUISIANA

Shreveport is located in the northwestern section of Louisiana, some 30 miles south of Arkansas and 15 miles east of Texas. A portion of the city is situated in the Red River bottom lands and the remainder in gently rolling hills that begin about 1 mile west of the river. The NOAA National Weather Service Office is at the Shreveport Regional Airport, about 7 miles southwest of the downtown area. Elevations in the Shreveport area range from about 170 to 280 feet above sea level. The climate of Shreveport is transitional between the subtropical humid type prevalent to the south and the continental climates of the Great Plains and Middle West to the north. During winter, masses of moderate to severely cold air move periodically through the area. Rainfall is abundant with the normal annual total near 45 inches. Amounts are substantial from late autumn to spring and there is a summer-early autumn low amount with monthly averages less than 3 inches in August, September, and October.

The winter months are normally mild with cold spells generally of short duration. Freezing temperatures are recorded on an average of 34 days during the year. The average first occurrence of 32 degrees in the autumn is mid-November, and the last occurrence in the spring is early March Although temperatures have fallen below zero degrees, they normally drop below about 15 degrees in about one-half the years. Temperatures recorded at the NWS Office at the airport on clear, calm nights are normally 2 to 5 degrees warmer than those experienced in the river bottom lands. The summer months are consistently quite warm and humid with temperatures exceeding 100 degrees on about 10 days a year and exceeding 95 degrees about 45 days per year. Late afternoon humidity rarely drops below 55 percent.

Measurable snow occurs only once every other year on average. Many consecutive years may pass with no measurable snow. The heaviest snowstorms in the Shreveport area have produced more than 10 inches. More troublesome than the infrequent heavy snowfall are ice and sleet storms which may cause considerable damage to trees and utility lines, as well as make travel very difficult.

Thunderstorms occur each month, but are most frequent in spring and summer months. Severe local storms, including hailstorms, tornadoes, and local windstorms have occurred over small areas in all seasons, but are most frequent during the spring months, with a secondary peak from late November through early January. Large hail of a damaging nature is infrequent, although hail as large as grapefruit has fallen on a few occasions.

Tropical cyclones are in the dissipating stages by the time they reach this portion of the state and winds from them are usually not a destructive factor. Associated heavy rainfall can contribute to local flooding.

NORMALS, MEANS, AND EXTREMES

SHREVEPORT, LOUISIANA

LATITUDE: 32°28'N LONGITUDE: 93°49' W ELEVATION: FT. GRND 254 BARO 268 TIME ZONE: CENTRAL WBAN: 13957

	(a)	JAN	FEB	MAR	APR	MAY	JUNE	JULY	AUG	SEP	OCT	NOV	DEC	YEAR
TEMPERATURE °F:														
Normals														
-Daily Maximum		55.4	60.6	69.2	77.1	83.2	89.7	93.0	93.1	87.3	78.7	68.0	58.5	76.2
-Daily Minimum		34.8	38.0	45.8	54.1	62.0	69.0	72.3	71.3	66.0	54.3	45.3	37.3	54.2
-Monthly		45.1	49.3	57.5	65.6	72.6	79.4	82.7	82.2	76.7	66.5	56.7	48.0	65.2
Extremes														
-Record Highest	41	84	89	92	94	95	101	106	107	103	97	88	84	107
-Year		1972	1986	1974	1987	1977	1988	1980	1962	1980	1954	1984	1955	AUG 1962
-Record Lowest	41	3	12	20	31	42	52	58	53	42	28	16	5	3
-Year		1962	1978	1980	1989	1960	1977	1972	1992	1984	1993	1976	1989	JAN 1962
NORMAL DEGREE DAYS:														
Heating (base 65°F)		623	448	262	69	0	0	0	0	0	63	264	535	2264
Cooling (base 65°F)		6	8	30	87	239	432	549	533	351	110	15	8	2368
% OF POSSIBLE SUNSHINE	41	49	55	58	59	64	71	74	73	69	70	59	52	63
MEAN SKY COVER (tenths)														
Sunrise - Sunset	41	6.7	6.3	6.3	6.2	6.1	5.4	5.3	5.1	5.1	4.7	5.5	6.2	5.8
MEAN NUMBER OF DAYS:														
Sunrise to Sunset														
-Clear	41	7.8	8.0	8.6	7.9	7.6	8.6	10.2	10.8	11.3	13.3	10.9	9.2	114.3
-Partly Cloudy	41	5.2	5.4	6.2	7.7	10.2	12.6	11.9	12.3	9.1	7.6	5.9	6.3	100.3
-Cloudy	41	17.9	14.8	16.2	14.4	13.2	8.8	8.8	7.9	9.6	10.2	13.2	15.6	150.7
Precipitation .01 inches or more	41	9.8	8.2	9.3	8.7	9.2	7.9	7.8	6.7	6.8	6.6	8.2	9.3	98.5
Snow,Ice pellets,hail 1.0 inches or more	41	0.2	0.2	0.1	0.0	0.0	0.0	0.0	0.0	0.0	0.0	0.*	0.*	0.6
Thunderstorms	41	1.8	2.8	4.9	5.7	7.2	7.3	8.0	6.6	4.2	2.9	3.1	2.1	56.6
Heavy Fog Visibility 1/4 mile or less	41	3.4	2.1	1.4	1.2	1.0	0.5	0.3	0.4	1.0	2.3	2.6	2.9	19.1
Temperature °F														
-Maximum														
90° and above	41	0.0	0.0	0.*	0.3	4.1	18.0	25.6	25.0	14.1	2.4	0.0	0.0	89.5
32° and below	41	1.0	0.3	0.*	0.0	0.0	0.0	0.0	0.0	0.0	0.0	0.0	0.4	1.7
-Minimum														
32° and below	41	12.7	7.7	2.5	0.1	0.0	0.0	0.0	0.0	0.0	0.2	3.4	10.0	36.5
0° and below	41	0.0	0.0	0.0	0.0	0.0	0.0	0.0	0.0	0.0	0.0	0.0	0.0	0.0
AVG. STATION PRESS.(mb)	21	1011.7	1010.1	1006.5	1005.8	1004.6	1005.4	1007.0	1006.8	1007.2	1009.1	1009.7	1011.3	1007.9
RELATIVE HUMIDITY (%)														
Hour 00	41	78	76	75	79	84	84	83	82	82	81	80	80	80
Hour 06 (Local Time)	41	84	83	83	87	90	91	91	91	91	89	87	85	88
Hour 12	41	63	59	56	57	59	58	57	55	57	54	58	62	58
Hour 18	41	65	58	54	56	60	60	58	57	61	62	67	68	61
PRECIPITATION (inches):														
Water Equivalent														
-Normal		3.88	3.92	3.59	3.75	5.18	4.29	3.67	2.43	3.12	3.73	4.45	4.10	46.11
-Maximum Monthly	41	10.09	8.57	7.23	21.84	11.78	17.11	9.46	9.23	9.59	12.05	10.81	10.00	21.84
-Year		1974	1983	1969	1991	1967	1989	1972	1991	1968	1984	1987	1982	APR 1991
-Minimum Monthly	41	0.27	0.90	0.56	0.43	0.42	0.13	0.15	0.35	0.17	0.00	0.71	0.59	0.00
-Year		1971	1954	1966	1987	1988	1988	1964	1985	1956	1963	1967	1981	OCT 1963
-Maximum in 24 hrs	41	4.35	3.53	3.63	10.44	5.27	7.28	4.30	4.64	5.39	3.88	6.51	3.35	10.44
-Year		1990	1965	1979	1991	1978	1993	1972	1955	1961	1957	1987	1965	APR 1991
Snow,Ice pellets,hail														
-Maximum Monthly	41	5.9	4.4	4.0	0.3	T	0.0	0.0	0.0	0.0	T	1.3	5.4	5.9
-Year		1978	1985	1965	1987	1991					1992	1980	1983	JAN 1978
-Maximum in 24 hrs	41	5.6	4.4	4.0	0.3	T	0.0	0.0	0.0	0.0	T	1.3	5.4	5.6
-Year		1982	1985	1965	1987	1991					1992	1980	1983	JAN 1982
WIND:														
Mean Speed (mph)	41	9.2	9.6	10.1	9.7	8.3	7.5	7.1	6.7	7.2	7.4	8.5	8.9	8.3
Prevailing Direction through 1963		S	S	S	S	S	S	S	S	ENE	SSE	S	S	S
Fastest Obs. 1 Min.														
-Direction (!!)	31	25	27	29	28	35	36	29	25	19	31	29	14	35
-Speed (MPH)	31	37	40	41	52	58	46	46	37	44	35	38	37	58
-Year		1992	1965	1964	1975	1991	1992	1982	1963	1965	1966	1975	1965	MAY 1991
Peak Gust														
-Direction (!!)	10	NW	N	NW	W	N	N	NW	W	NW	NW	NW	W	N
-Speed (mph)	10	49	54	58	81	83	66	66	56	54	41	63	64	83
-Date		1993	1990	1986	1991	1991	1992	1989	1991	1992	1985	1992	1987	MAY 1991

PRECIPITATION (inches)

SHREVEPORT, LOUISIANA

YEAR	JAN	FEB	MAR	APR	MAY	JUNE	JULY	AUG	SEP	OCT	NOV	DEC	ANNUAL
1964	2.57	2.74	4.24	7.27	1.41	1.87	0.15	4.71	2.51	0.64	1.65	2.55	32.31
1965	3.77	6.51	3.39	1.16	5.40	3.18	1.49	1.82	6.55	0.36	1.20	6.29	41.12
1966	4.22	3.45	0.56	8.02	3.78	2.05	0.58	1.71	3.27	1.62	0.97	3.63	33.86
1967	1.36	2.91	1.02	2.11	11.78	0.89	6.15	4.67	1.27	1.34	0.71	3.92	38.13
1968	8.33	2.22	1.89	9.38	6.05	2.78	4.68	1.89	9.59	1.90	5.85	3.27	57.83
1969	1.14	4.32	7.23	6.63	5.18	1.16	1.06	0.50	0.97	3.16	7.50	3.95	42.80
1970	1.23	4.70	4.30	5.12	4.36	1.14	3.94	2.04	1.64	7.44	2.09	3.80	41.80
1971	0.27	4.13	2.11	1.06	5.26	0.97	6.15	2.99	1.30	3.86	3.75	3.65	35.50
1972	5.97	0.94	2.45	2.06	4.13	2.76	9.46	1.27	2.10	6.32	5.32	4.18	46.96
1973	5.65	1.52	5.01	6.44	2.00	5.84	7.63	0.77	6.39	5.38	5.16	6.37	58.16
1974	10.09	3.67	3.60	3.09	4.58	6.29	7.73	3.84	6.64	3.79	5.80	2.34	61.46
1975	4.55	4.51	5.84	3.91	5.31	3.48	3.45	1.65	0.98	3.87	4.44	1.88	43.87
1976	2.07	2.45	6.67	1.75	5.95	4.42	3.47	2.96	6.28	2.08	1.63	3.77	43.50
1977	3.00	3.68	4.94	2.05	2.40	2.41	3.89	4.28	0.53	0.31	2.11	2.58	32.18
1978	4.89	1.90	2.66	2.79	7.92	1.21	1.74	3.90	2.40	2.74	4.18	5.13	41.46
1979	9.22	4.98	5.74	7.42	7.99	3.04	7.50	1.86	4.33	3.96	4.76	3.12	63.92
1980	4.67	3.10	3.75	5.34	4.42	2.60	1.83	0.42	1.63	2.48	3.59	0.74	34.57
1981	1.43	3.83	3.33	1.97	9.96	6.45	2.36	0.94	3.32	5.63	1.49	0.59	41.30
1982	3.59	3.19	2.59	2.72	2.32	1.84	4.25	2.20	1.11	5.19	5.72	10.00	44.72
1983	2.45	8.57	3.68	1.47	8.22	6.60	1.18	1.67	3.12	0.79	4.90	7.18	49.83
1984	2.10	5.66	3.58	2.52	5.86	3.56	2.20	0.87	2.61	12.05	4.46	2.88	48.35
1985	2.38	4.42	4.28	3.05	1.96	4.57	8.40	0.35	4.40	9.87	4.25	3.37	51.30
1986	0.49	3.48	0.75	3.50	6.60	14.67	2.92	1.68	3.51	6.63	9.19	4.69	58.11
1987	2.26	7.80	1.48	0.43	6.67	5.43	1.21	3.50	0.94	5.49	10.81	8.12	54.14
1988	2.06	3.59	3.89	3.45	0.42	0.13	3.12	3.52	1.61	4.44	5.44	4.71	36.38
1989	7.20	4.06	3.41	2.41	10.07	17.11	4.46	3.94	1.08	1.50	2.32	3.34	60.90
1990	10.02	6.92	4.90	4.29	10.48	2.56	3.53	2.88	2.93	4.33	8.81	3.99	65.64
1991	7.70	5.13	2.89	21.84	10.71	2.53	3.47	9.23	3.45	3.59	3.94	7.51	81.99
1992	4.63	6.41	5.94	3.26	2.81	3.95	3.36	1.24	5.15	4.13	4.69	5.84	51.41
1993	4.63	4.80	5.94	4.19	3.30	15.73	0.27	4.09	3.51	4.43	4.85	1.44	57.18
Record Mean	4.16	3.73	4.15	4.63	4.64	3.62	3.51	2.61	2.91	3.16	3.99	4.47	45.58

AVERAGE TEMPERATURE (deg. F)

SHREVEPORT, LOUISIANA

YEAR	JAN	FEB	MAR	APR	MAY	JUNE	JULY	AUG	SEP	OCT	NOV	DEC	ANNUAL
1964	46.5	45.6	56.8	68.7	75.1	81.0	84.2	84.0	78.5	64.2	59.3	50.1	66.2
1965	50.8	49.4	49.6	70.1	73.7	78.4	83.0	82.1	77.2	66.1	63.2	52.0	66.3
1966	42.3	47.5	58.1	66.2	72.7	79.0	85.0	81.0	75.3	64.0	60.4	46.9	64.9
1967	47.6	46.5	63.5	71.0	71.1	80.9	79.8	80.0	73.2	66.5	57.1	48.9	65.5
1968	45.4	43.6	55.9	66.5	72.7	80.1	81.0	82.0	73.6	66.6	53.7	46.6	64.0
1969	49.3	49.2	50.1	65.1	72.4	80.5	86.6	84.3	78.2	67.8	54.7	48.3	65.5
1970	42.0	49.4	53.8	67.4	72.7	78.8	81.7	84.3	81.1	65.1	54.7	54.3	65.5
1971	50.6	50.6	55.4	64.3	70.8	81.7	83.3	80.6	78.1	70.7	56.4	55.4	66.5
1972	49.4	51.9	59.8	66.9	72.3	80.8	81.0	82.6	80.2	67.2	50.9	45.5	65.7
1973	44.7	49.5	60.9	62.0	71.7	78.5	81.1	78.2	74.9	68.2	61.5	46.8	64.8
1974	47.8	51.2	63.6	64.3	74.0	76.6	82.1	80.1	71.0	66.2	55.6	47.3	65.0
1975	50.2	48.4	55.6	63.7	72.4	78.2	80.9	80.7	73.7	68.0	56.3	49.3	64.8
1976	47.1	59.2	60.2	67.3	67.9	75.7	78.6	78.7	74.3	59.9	49.3	46.3	63.7
1977	37.3	50.8	59.6	65.5	74.1	79.5	83.5	80.3	78.7	66.3	56.5	47.3	64.9
1978	34.9	38.1	52.8	65.5	73.6	80.5	85.4	83.3	77.8	66.0	60.0	48.0	63.8
1979	37.4	46.6	59.1	66.0	70.2	78.0	81.2	80.3	74.2	67.0	52.8	48.9	63.5
1980	48.3	47.9	54.9	63.0	74.3	83.4	86.9	85.5	82.1	63.5	54.2	49.1	66.1
1981	44.7	49.8	56.0	70.0	69.2	80.2	82.8	81.3	73.9	65.0	56.8	46.9	64.7
1982	46.1	45.6	61.5	63.3	74.4	78.6	83.3	83.2	75.6	64.6	55.4	51.2	65.2
1983	44.6	48.8	55.0	59.7	70.0	77.4	82.3	84.0	75.7	66.7	55.9	37.5	63.1
1984	40.6	50.4	58.0	64.9	72.1	79.3	81.1	82.1	75.0	70.6	56.4	60.0	65.9
1985	40.0	46.2	61.4	67.0	72.8	79.3	83.1	84.8	76.3	68.5	61.5	44.2	65.4
1986	49.0	54.4	59.7	66.5	72.3	79.9	83.7	80.8	79.4	65.2	55.6	46.1	66.1
1987	44.8	51.7	55.7	64.2	75.3	79.1	82.3	85.3	76.7	64.1	56.1	50.2	65.5
1988	42.2	49.3	56.3	65.2	71.8	79.8	83.3	83.7	77.8	64.1	58.6	49.2	65.1
1989	51.5	45.8	56.7	65.6	73.8	76.7	81.2	81.0	73.7	66.5	58.6	40.8	64.3
1990	52.5	56.4	59.4	65.6	72.6	82.7	82.2	83.3	79.7	65.0	58.9	48.5	67.2
1991	44.2	52.0	60.0	68.6	75.3	80.6	82.6	81.2	75.7	68.7	52.1	51.2	66.0
1992	47.2	54.7	59.5	65.1	71.4	78.5	82.9	78.6	76.0	67.0	52.0	49.5	65.2
1993	46.4	49.3	55.2	61.3	70.7	80.2	84.6	84.8	77.3	64.4	52.0	49.1	64.6
Record Mean	47.0	50.5	58.0	66.0	73.2	80.3	83.1	82.7	77.1	66.9	56.1	49.1	65.8
Max	56.1	60.3	68.3	76.4	83.1	90.2	93.0	92.9	87.4	78.1	66.5	58.4	75.9
Min	37.9	40.7	47.6	55.6	63.2	70.3	73.1	72.4	66.8	55.8	45.7	39.7	55.7

REFERENCE NOTES

SHREVEPORT, LOUISIANA

GENERAL
T - TRACE AMOUNT.
BLANK ENTRIES DENOTE MISSING/UNREPORTED DATA.
# INDICATES A STATION OR INSTRUMENT RELOCATION.

SPECIFIC

PM - INCLUDES LAST DAY OF PREVIOUS MONTH
ASOS - AUTOMATED SURFACE OBSERVING SYSTEM IN OPERATION DURING THESE MONTHS.

(a) - LENGTH OF RECORD IN YEARS, ALTHOUGH INDIVIDUAL MONTHS MAY BE MISSING.
0.* OR * - THE VALUE IS BETWEEN 0.0 AND 0.05
NORMALS - BASED ON THE 1961-1990 RECORD PERIOD.
EXTREMES - DATES ARE THE MOST RECENT OCCURRENCE
WIND DIR.- NUMERALS SHOW TENS OF DEGRESS CLOCKWISE FROM TRUE NORTH. "00" INDICATES CALM.
RESULTANT DIRECTIONS ARE GIVEN TO WHOLE DEGREES.
BOLD VALUES INDICATE EXTREME VALUES WHICH OCCURRED AFTER THE ASOS SYSTEM WAS COMMISSIONED.

RECORD = PERIOD OF RECORD
RECORD MEAN PRECIPITATION IS THE MEAN OF ALL DAILY PRECIPITATION AMOUNTS DURING THE PERIOD OF RECORD.
RECORD MAX(MIN) TEMPERATURE IS THE MEAN OF ALL DAILY MAX(MIN) TEMPERATURES DURING THE PERIOD OF RECORD.
RECORD MEAN TEMPERATURE IS THE SUM OF THE RECORD MAX AND RECORD MIN DIVIDED BY 2.
AVERAGE TEMPERATURE IS THE SUM OF THE MEAN DAILY MAX AND MIN TEMPERATURE DIVIDED BY 2.

EXCEPTIONS

RECORD MEANS ARE THROUGH THE CURRENT YEAR,
BEGINNING IN 1875 FOR TEMPERATURE
1872 FOR PRECIPITATION
1953 FOR SNOWFALL

HEATING DEGREE DAYS Base 65 deg. F — SHREVEPORT, LOUISIANA

SEASON	JULY	AUG	SEP	OCT	NOV	DEC	JAN	FEB	MAR	APR	MAY	JUNE	TOTAL
1964-65	0	0	6	93	216	479	446	432	473	28	0	0	2173
1965-66	0	0	3	78	103	400	703	485	238	62	8	0	2080
1966-67	0	0	0	99	181	570	553	511	140	13	13	0	2080
1967-68	0	0	15	72	244	500	604	614	308	49	0	0	2406
1968-69	0	0	0	57	346	560	495	438	455	44	2	0	2397
1969-70	0	0	0	86	314	510	713	429	344	58	14	0	2468
1970-71	0	0	0	97	324	351	459	398	316	94	15	0	2054
1971-72	0	0	4	7	282	304	499	382	185	62	0	0	1725
1972-73	0	0	6	92	419	597	621	429	135	145	9	0	2453
1973-74	0	0	0	40	164	557	533	386	152	78	2	0	1912
1974-75	0	0	14	32	312	541	473	457	305	124	0	0	2258
1975-76	0	0	4	39	286	492	551	186	202	38	17	0	1815
1976-77	0	0	0	199	471	574	851	399	188	46	0	0	2728
1977-78	0	0	0	72	260	549	933	746	374	61	30	0	3025
1978-79	0	0	0	57	181	528	849	517	216	50	11	0	2409
1979-80	0	0	0	52	366	498	508	494	312	96	4	0	2330
1980-81	0	0	3	128	340	488	620	425	279	14	20	0	2317
1981-82	0	0	8	129	246	554	588	537	202	125	4	0	2393
1982-83	0	0	9	120	309	457	624	449	308	186	12	0	2474
1983-84	0	0	15	69	305	848	747	421	247	81	11	0	2744
1984-85	0	0	19	36	286	208	770	528	151	42	1	0	2041
1985-86	0	0	11	49	174	638	490	331	176	44	1	0	1914
1986-87	0	0	0	86	299	579	618	366	286	117	0	0	2351
1987-88	0	0	0	79	279	456	701	453	278	54	1	0	2301
1988-89	0	0	0	76	218	482	418	535	295	93	2	0	2119
1989-90	0	0	17	85	244	743	382	243	216	92	3	0	2025
1990-91	0	0	6	126	208	509	634	357	197	26	6	0	2069
1991-92	0	0	9	36	401	422	544	297	185	84	23	0	2001
1992-93	0	0	2	33	381	480	569	435	307	154	3	0	2364
1993-94	0	0	5	132	404	490							

COOLING DEGREE DAYS Base 65 deg. F — SHREVEPORT, LOUISIANA

YEAR	JAN	FEB	MAR	APR	MAY	JUNE	JULY	AUG	SEP	OCT	NOV	DEC	TOTAL
1969	15	0	0	57	241	470	675	604	402	177	11	0	2652
1970	9	0	5	138	259	420	528	606	492	106	23	28	2614
1971	22	1	24	82	203	510	576	488	403	191	28	9	2537
1972	22	10	31	128	235	480	501	553	467	167	1	0	2595
1973	0	2	16	64	223	412	504	417	305	145	66	0	2154
1974	7	6	115	63	288	355	541	477	200	77	35	0	2164
1975	23	0	21	91	238	403	501	493	271	141	34	12	2228
1976	2	26	58	116	112	326	428	432	288	48	7	0	1843
1977	0	7	28	69	289	443	580	479	419	119	12	9	2454
1978	5	0	3	84	303	472	637	570	391	96	39	8	2608
1979	0	8	39	86	178	395	509	483	284	124	8	2	2116
1980	1	6	6	43	298	560	686	643	522	86	22	1	2874
1981	0	5	10	171	157	463	558	511	284	135	6	0	2300
1982	14	0	99	81	300	413	573	573	333	115	24	32	2557
1983	0	0	7	34	176	381	540	595	343	126	39	0	2241
1984	0	5	38	83	235	436	511	540	329	219	35	61	2492
1985	0	8	49	109	252	436	568	620	356	163	78	0	2639
1986	2	41	17	95	236	454	586	494	438	101	24	0	2488
1987	1	0	7	99	327	431	544	634	357	57	19	5	2481
1988	3	3	14	67	220	449	575	587	390	53	37	1	2399
1989	8	7	43	121	283	358	508	503	286	140	57	0	2314
1990	2	9	50	115	244	538	536	572	453	132	30	3	2684
1991	0	1	47	141	334	475	550	509	336	158	20	5	2576
1992	0	4	21	93	227	412	561	428	339	99	1	4	2189
1993	0	0	10	50	189	461	614	620	381	121	20	4	2470

SNOWFALL (inches) — SHREVEPORT, LOUISIANA

SEASON	JULY	AUG	SEP	OCT	NOV	DEC	JAN	FEB	MAR	APR	MAY	JUNE	TOTAL
1970-71	0.0	0.0	0.0	0.0	0.0	0.0	0.8	0.6	0.3	0.0	0.0	0.0	1.7
1971-72	0.0	0.0	0.0	0.0	T	0.0	T	0.0	0.0	0.0	0.0	0.0	T
1972-73	0.0	0.0	0.0	0.0	0.0	0.0	0.6	T	0.0	0.0	0.0	0.0	0.6
1973-74	0.0	0.0	0.0	0.0	0.0	T	0.0	0.0	0.0	0.0	0.0	0.0	T
1974-75	0.0	0.0	0.0	0.0	0.0	0.0	3.4	T	T	0.0	0.0	0.0	3.4
1975-76	0.0	0.0	0.0	0.0	T	0.0	0.0	0.0	0.0	0.0	0.0	0.0	T
1976-77	0.0	0.0	0.0	0.0	T	0.0	5.4	0.0	0.0	0.0	0.0	0.0	5.4
1977-78	0.0	0.0	0.0	0.0	0.0	0.0	5.9	2.0	0.3	0.0	0.0	0.0	8.2
1978-79	0.0	0.0	0.0	0.0	0.0	0.0	T	1.5	0.0	0.0	0.0	0.0	1.5
1979-80	0.0	0.0	0.0	0.0	T	0.0	0.0	1.4	T	0.0	0.0	0.0	1.4
1980-81	0.0	0.0	0.0	0.0	1.3	0.0	0.7	0.1	0.0	0.0	0.0	0.0	2.1
1981-82	0.0	0.0	0.0	0.0	0.0	T	5.6	T	T	0.0	0.0	0.0	5.6
1982-83	0.0	0.0	0.0	0.0	0.0	T	T	T	0.0	0.0	0.0	0.0	T
1983-84	0.0	0.0	0.0	0.0	0.0	5.4	T	T	0.0	0.0	0.0	0.0	5.4
1984-85	0.0	0.0	0.0	0.0	0.0	0.0	0.4	4.4	0.0	0.0	0.0	0.0	4.8
1985-86	0.0	0.0	0.0	0.0	0.0	0.0	T	T	0.0	0.0	0.0	0.0	T
1986-87	0.0	0.0	0.0	0.0	0.0	T	0.0	T	T	0.3	0.0	0.0	0.3
1987-88	0.0	0.0	0.0	0.0	0.0	0.0	1.2	0.8	0.0	0.0	0.0	0.0	2.0
1988-89	0.0	0.0	0.0	0.0	0.0	0.0	T	T	T	0.0	T	0.0	T
1989-90	0.0	0.0	0.0	0.0	0.0	T	0.0	T	0.0	0.0	0.0	0.0	T
1990-91	0.0	0.0	0.0	0.0	0.0	T	0.0	T	0.0	T	T	0.0	T
1991-92	0.0	0.0	0.0	0.0	0.0	0.0	T	T	T	0.0	0.0	0.0	T
1992-93	0.0	0.0	0.0	T	T	0.0	T	T	1.0	T	0.0	0.0	1.0
1993-94	0.0	0.0	0.0	0.0	0.0	T							
Record Mean	0.0	0.0	0.0	T	T	0.2	0.8	0.5	0.2	T	T	0.0	1.7

CARIBOU, MAINE

The Caribou Municipal Airport is located in Aroostook County, the largest and northernmost county in the state. The airport lies on top of high land which is about on the same level as most of the surrounding gently rolling hills. The Aroostook River, which runs about 1 mile to the east and southeast of the station, has little effect on the local weather. Even though Caribou is located only 150 miles from the Atlantic coast, its climate can be justly classed as a severe typical continental type. Winters are particularly long and windy, and seasonal snowfalls averaging over 100 inches are not unusual. While the extreme low temperatures may be less severe than one might expect, temperatures of zero or lower normally occur over 40 times per year. A study of heating degree day data will show the outstanding part that cold weather plays here.

Summers are cool and generally favored with abundant rainfall, which is one of the most important factors in the high yield of the potato and grain crops throughout the county. Our location high up in the St. Lawrence Valley allows Aroostook County to come under the influence of the Summer Polar Front, resulting in practically no dry periods of more than 3 or 4 days in the growing season. The growing season at Caribou averages more than 120 days, with the average last freeze in the spring in mid-May and the average first freeze in autumn in late September.

Autumn climate is nearly ideal, with mostly sunny warm days and crisp cool nights predominating. Aroostook County, even with its relatively short growing season, provides profitable farming. The principal crops are potatoes, peas, a variety of grains, and some hardy vegetables.

Probably unknown to many victims of hay fever and similar afflictions, the immediate Caribou area offers sparkling visibility and relatively pollen-free air in the late summer months. This latter condition is principally due to the extremely high degree of cultivation of all available land.

NORMALS, MEANS, AND EXTREMES

CARIBOU, MAINE

LATITUDE: 46°52'N LONGITUDE: 68°01' W ELEVATION: FT. GRND 624 BARO 630 TIME ZONE: EASTERN WBAN: 14607

	(a)	JAN	FEB	MAR	APR	MAY	JUNE	JULY	AUG	SEP	OCT	NOV	DEC	YEAR
TEMPERATURE °F:														
Normals														
-Daily Maximum		19.4	23.0	34.3	46.7	61.7	71.9	76.5	73.6	64.0	52.0	37.6	24.0	48.7
-Daily Minimum		-1.6	0.7	14.9	29.0	40.1	49.1	54.5	52.1	43.2	34.4	23.7	5.5	28.8
-Monthly		8.9	11.9	24.6	37.9	50.9	60.5	65.5	62.8	53.7	43.2	30.7	14.8	38.8
Extremes														
-Record Highest	54	52	52	73	86	96	96	95	95	91	79	68	58	96
-Year		1986	1990	1962	1990	1977	1944	1991	1975	1945	1968	1956	1950	MAY 1977
-Record Lowest	54	-32	-41	-21	-2	18	30	36	34	23	14	-5	-31	-41
-Year		1976	1955	1992	1964	1974	1958	1969	1982	1980	1972	1989	1989	FEB 1955
NORMAL DEGREE DAYS:														
Heating (base 65°F)		1739	1487	1252	813	437	143	61	115	343	676	1029	1556	9651
Cooling (base 65°F)		0	0	0	0	0	8	76	47	0	0	0	0	131
% OF POSSIBLE SUNSHINE														
MEAN SKY COVER (tenths)														
Sunrise - Sunset	48	6.8	6.8	6.8	7.2	7.3	7.3	7.0	6.8	6.8	7.2	8.0	7.3	7.1
MEAN NUMBER OF DAYS:														
Sunrise to Sunset														
-Clear	52	6.9	6.2	7.0	5.2	4.1	3.3	3.1	4.6	5.5	4.8	2.9	5.6	59.1
-Partly Cloudy	52	7.0	6.2	6.8	6.8	9.0	9.8	12.9	11.4	9.2	7.8	6.2	6.6	99.7
-Cloudy	52	17.1	15.8	17.2	18.0	17.9	16.9	15.0	14.9	15.4	18.4	20.9	18.9	206.4
Precipitation .01 inches or more	54	14.6	12.4	12.8	12.9	13.4	13.6	13.8	13.2	12.3	12.7	14.5	14.6	160.6
Snow,Ice pellets,hail 1.0 inches or more	52	6.0	5.7	4.9	2.6	0.3	0.0	0.0	0.0	0.*	0.5	3.1	6.3	29.3
Thunderstorms	25	0.*	0.0	0.1	0.5	1.9	4.3	6.7	4.1	1.3	1.0	0.*	0.0	20.1
Heavy Fog Visibility 1/4 mile or less	25	2.0	1.6	1.7	2.2	1.0	1.6	2.6	2.3	3.2	2.3	3.7	2.7	26.9
Temperature °F														
-Maximum														
90° and above	54	0.0	0.0	0.0	0.0	0.1	0.4	1.0	0.5	0.*	0.0	0.0	0.0	2.1
32° and below	54	26.2	22.9	13.1	1.8	0.*	0.0	0.0	0.0	0.0	0.3	9.1	23.5	96.9
-Minimum														
32° and below	54	30.7	28.0	29.0	21.7	5.4	0.1	0.0	0.0	2.9	14.3	25.0	30.3	187.4
0° and below	54	16.1	12.9	4.4	0.*	0.0	0.0	0.0	0.0	0.0	0.0	0.3	10.1	43.8
AVG. STATION PRESS.(mb)	10	987.8	989.3	988.7	988.0	990.0	989.6	989.2	991.8	992.1	991.7	990.2	989.9	989.9
RELATIVE HUMIDITY (%)														
Hour 01	19	74	75	76	79	79	84	84	89	89	86	85	79	82
Hour 07 (Local Time)	49	74	74	76	76	73	78	83	86	87	86	85	80	80
Hour 13	49	66	63	60	57	52	56	58	59	60	62	71	70	61
Hour 19	48	72	69	67	65	60	65	70	73	75	75	79	79	71
PRECIPITATION (inches):														
Water Equivalent														
-Normal		2.42	1.92	2.43	2.45	3.07	2.91	4.01	4.07	3.45	3.10	3.55	3.22	36.60
-Maximum Monthly	54	5.10	4.13	5.13	5.26	6.27	7.11	6.83	12.09	8.14	8.73	8.15	7.97	12.09
-Year		1978	1955	1953	1973	1947	1940	1957	1981	1954	1990	1983	1973	AUG 1981
-Minimum Monthly	54	0.12	0.26	0.66	0.54	0.47	0.88	0.96	0.93	0.86	0.63	0.45	0.74	0.12
-Year		1944	1978	1965	1967	1982	1983	1991	1957	1968	1955	1939	1963	JAN 1944
-Maximum in 24 hrs	54	1.52	1.38	1.70	2.11	2.25	2.37	2.92	6.89	6.23	4.07	2.27	2.80	6.89
-Year		1992	1988	1984	1958	1948	1957	1957	1981	1954	1970	1983	1973	AUG 1981
Snow,Ice pellets,hail														
-Maximum Monthly	54	41.4	41.0	47.1	36.4	10.9	T	T	0.0	2.5	12.1	34.9	59.9	59.9
-Year		1978	1960	1955	1982	1967	1991	1992		1991	1963	1974	1972	DEC 1972
-Maximum in 24 hrs	54	15.9	18.2	28.6	21.1	5.8	T	T	0.0	2.5	9.4	21.0	19.7	28.6
-Year		1986	1952	1984	1982	1967	1991	1992		1991	1963	1986	1989	MAR 1984
WIND:														
Mean Speed (mph)	15	12.4	12.0	12.9	11.7	11.4	10.4	9.8	9.3	10.4	10.9	11.1	11.5	11.2
Prevailing Direction through 1962		NW	NW	NW	NW	NW	WSW	WSW	WSW	WSW	NNW	WSW	WSW	WSW
Fastest Obs 1 Min.														
-Direction (!!!)		25	33	32	34	34	34	32	33	23	26	28	32	34
-Speed (MPH)	4	30	28	29	28	31	35	32	24	25	25	31	32	35
-Year		1993	1992	1993	1993	1991	1991	1992	1993	1993	1990	1990	1992	JUN 1991
Peak Gust														
-Direction (!!!)														
-Speed (mph)														
-Date														

PRECIPITATION (inches)

CARIBOU, MAINE

YEAR	JAN	FEB	MAR	APR	MAY	JUNE	JULY	AUG	SEP	OCT	NOV	DEC	ANNUAL
1964	2.64	0.56	3.12	2.07	2.81	1.80	3.58	3.38	1.33	4.19	2.28	2.98	30.74
1965	1.21	2.33	0.66	1.10	2.35	1.25	3.71	3.60	3.07	4.14	4.75	1.56	29.73
1966	2.06	1.49	2.61	0.73	1.73	2.06	3.27	1.54	3.19	3.47	3.60	2.17	27.92
1967	2.23	1.75	1.07	0.54	3.95	1.69	4.18	4.81	6.61	2.81	2.79	4.58	37.01
1968	2.31	1.46	4.01	2.15	1.33	1.65	4.24	2.01	0.86	3.51	3.98	4.16	31.67
1969	3.00	2.25	1.49	2.60	2.88	3.47	3.73	4.02	6.80	1.55	4.36	3.59	39.74
1970	0.31	2.09	2.25	3.56	4.26	3.25	2.54	3.18	5.46	6.35	1.85	2.58	37.68
1971	1.71	2.16	3.20	2.09	2.06	2.62	2.79	3.39	2.62	3.43	3.31	2.46	31.84
1972	1.32	2.38	4.72	1.09	3.69	4.97	4.21	5.07	3.88	4.03	2.76	5.28	43.40
1973	2.60	2.89	2.48	5.26	5.03	2.13	4.62	2.95	2.62	1.47	2.63	7.97	42.65
1974	1.89	1.37	3.56	3.81	3.61	2.82	3.39	3.98	3.40	1.15	4.16	2.05	35.19
1975	2.71	1.66	1.94	1.95	3.04	2.40	4.28	1.39	2.98	1.51	3.40	4.05	31.31
1976	3.51	3.43	2.57	3.31	5.11	2.85	6.74	6.17	2.52	5.46	2.33	4.63	48.63
1977	3.42	2.99	2.36	1.83	0.74	6.44	1.75	7.86	3.54	5.30	1.63	3.59	41.45
1978	5.10	0.26	2.69	2.33	1.98	3.70	5.56	1.95	2.74	1.94	1.84	3.04	33.13
1979	4.49	2.22	3.70	3.08	4.27	3.39	3.36	4.98	4.68	1.68	2.90	3.05	41.80
1980	1.55	0.82	3.15	2.57	2.05	2.19	5.42	2.28	4.06	2.51	3.21	2.74	32.55
1981	1.68	2.39	3.43	2.17	3.18	4.15	2.62	12.09	2.38	6.28	2.51	3.81	46.69
1982	2.46	2.17	2.72	4.01	0.47	3.08	4.25	4.78	3.70	1.61	5.50	2.51	37.26
1983	2.95	1.77	3.84	4.20	5.28	0.88	5.92	3.86	2.70	1.81	8.15	5.01	46.37
1984	2.10	3.06	2.55	1.74	5.72	5.90	4.52	1.65	1.54	1.81	2.01	3.00	35.60
1985	0.99	2.77	1.87	1.80	2.64	2.89	5.05	1.74	2.30	1.42	3.50	2.24	29.21
1986	4.86	1.13	2.32	2.29	2.13	1.96	4.21	4.97	3.58	1.47	3.96	1.66	34.54
1987	2.29	0.33	1.24	1.75	2.46	3.59	3.16	1.82	4.37	2.18	2.33	2.56	28.08
1988	2.79	2.65	1.23	1.99	1.84	2.37	2.28	5.65	1.82	3.09	4.10	1.00	30.81
1989	1.88	1.43	1.40	2.24	4.13	2.29	2.63	5.41	3.52	1.62	3.88	2.35	32.78
1990	3.36	1.84	1.16	2.28	3.53	4.56	4.15	3.23	3.78	8.73	4.22	5.60	46.44
1991	2.06	1.13	4.71	2.62	3.51	1.98	0.96	6.72	3.45	4.46	1.61	2.10	35.31
1992	3.76	2.68	1.87	2.57	1.83	4.58	4.49	6.28	1.51	3.52	1.60	1.58	36.27
1993	2.18	2.10	1.72	3.48	3.58	4.54	2.48	3.57	5.11	4.31	2.94	3.99	40.00
Record Mean	2.32	2.05	2.41	2.55	3.01	3.42	3.93	4.03	3.36	3.21	3.36	2.93	36.57

AVERAGE TEMPERATURE (deg. F)

CARIBOU, MAINE

YEAR	JAN	FEB	MAR	APR	MAY	JUNE	JULY	AUG	SEP	OCT	NOV	DEC	ANNUAL
1964	13.4	15.9	24.1	37.1	51.8	58.8	65.8	58.6	50.8	41.0	29.3	16.9	38.6
1965	8.7	11.5	25.6	37.0	49.1	61.1	61.5	60.9	53.9	41.0	25.1	15.9	37.6
1966	16.7	15.3	28.2	37.6	49.0	60.4	64.1	62.0	51.7	42.7	34.8	21.1	40.3
1967	14.3	6.7	17.4	34.4	42.7	62.1	68.4	64.2	55.8	44.2	31.4	17.3	38.2
1968	6.1	9.9	25.7	41.0	49.7	59.1	66.3	59.1	58.5	48.4	28.2	18.0	39.2
1969	15.3	17.8	25.4	35.6	48.2	60.7	62.9	64.0	52.8	41.1	34.9	21.9	40.0
1970	4.9	13.1	25.3	38.2	50.6	61.7	69.7	66.0	53.9	47.4	34.6	10.1	39.6
1971	5.8	14.3	25.4	36.1	51.6	59.3	64.0	61.3	56.2	46.4	27.9	11.7	38.3
1972	8.2	6.4	18.3	34.7	52.4	61.5	64.7	60.5	54.2	38.3	27.1	7.6	36.2
1973	9.7	12.6	29.6	37.9	48.9	63.0	68.6	66.4	52.9	43.9	27.5	23.9	40.4
1974	6.7	10.8	21.0	36.8	45.2	62.7	64.7	64.3	52.7	39.3	31.4	18.5	37.9
1975	9.6	11.5	23.1	35.1	53.7	61.4	68.6	64.4	54.5	42.3	32.8	10.9	39.0
1976	5.3	13.5	22.7	38.4	51.2	64.0	63.7	62.6	51.5	39.2	25.6	7.9	37.1
1977	6.0	12.9	32.3	36.4	53.3	58.8	65.2	64.2	51.2	43.5	33.3	16.9	39.5
1978	9.9	12.5	20.8	34.9	55.3	60.8	66.2	64.9	50.4	41.6	27.7	16.3	38.4
1979	15.4	10.4	31.8	41.2	54.6	63.0	68.6	61.9	54.4	45.2	35.8	19.2	41.8
1980	13.3	11.9	24.0	42.6	51.2	59.4	64.8	66.0	50.8	40.9	30.1	9.0	38.7
1981	5.6	27.6	28.0	39.5	54.2	61.2	66.3	64.5	53.1	40.4	31.9	23.1	41.3
1982	4.0	11.3	24.5	35.1	53.6	60.3	66.4	58.7	54.9	44.4	32.7	22.1	39.0
1983	14.9	15.3	27.0	41.1	48.9	61.9	64.5	64.3	57.2	43.6	32.2	14.5	40.4
1984	6.7	22.4	19.3	39.8	49.2	59.2	65.8	65.5	51.7	44.5	33.4	18.4	39.7
1985	6.0	16.2	23.5	35.3	49.5	58.2	66.0	62.2	56.1	43.8	28.3	12.1	38.1
1986	11.8	12.2	24.4	42.8	52.2	56.9	62.6	60.7	50.5	41.2	25.9	16.2	38.1
1987	9.9	12.5	28.0	43.9	50.6	60.3	65.8	61.1	54.5	44.2	29.0	18.7	39.9
1988	12.1	13.6	23.1	39.4	54.6	58.5	67.4	64.4	52.4	40.5	33.5	13.8	39.4
1989	12.3	10.4	19.5	36.7	56.7	60.1	64.7	63.5	55.4	45.0	27.6	3.5	38.0
1990	17.3	11.7	24.7	39.6	49.0	62.9	66.2	66.6	53.5	44.7	31.0	20.5	40.6
1991	7.1	15.1	27.0	39.6	52.5	61.4	65.8	65.3	52.1	44.8	32.8	13.5	39.8
1992	10.4	10.9	20.2	36.4	53.2	60.2	60.3	63.7	56.2	41.1	29.3	19.6	38.5
1993	8.6	4.2	23.3	39.9	51.2	59.9	64.8	65.8	54.1	39.2	29.7	19.4	38.3
Record Mean	10.1	12.9	23.8	37.3	50.6	60.0	65.2	62.8	53.8	43.1	30.8	15.8	38.8
Max	19.5	23.0	33.2	46.1	61.4	70.7	75.9	73.6	64.2	51.9	37.6	24.0	48.4
Min	0.7	2.8	14.3	28.5	39.8	49.2	54.4	52.1	43.3	34.3	24.0	7.6	29.2

REFERENCE NOTES

CARIBOU, MAINE

GENERAL
T - TRACE AMOUNT.
BLANK ENTRIES DENOTE MISSING/UNREPORTED DATA.
# INDICATES A STATION OR INSTRUMENT RELOCATION.

SPECIFIC

PM - INCLUDES LAST DAY OF PREVIOUS MONTH
ASOS - AUTOMATED SURFACE OBSERVING SYSTEM IN OPERATION DURING THESE MONTHS.

(a) - LENGTH OF RECORD IN YEARS, ALTHOUGH INDIVIDUAL MONTHS MAY BE MISSING.
0.* OR * - THE VALUE IS BETWEEN 0.0 AND 0.05
NORMALS - BASED ON THE 1961-1990 RECORD PERIOD.
EXTREMES - DATES ARE THE MOST RECENT OCCURRENCE
WIND DIR.- NUMERALS SHOW TENS OF DEGRESS CLOCKWISE FROM TRUE NORTH. "00" INDICATES CALM.
RESULTANT DIRECTIONS ARE GIVEN TO WHOLE DEGREES.
BOLD VALUES INDICATE EXTREME VALUES WHICH OCCURRED AFTER THE ASOS SYSTEM WAS COMMISSIONED.

RECORD = PERIOD OF RECORD
RECORD MEAN PRECIPITATION IS THE MEAN OF ALL DAILY PRECIPITATION AMOUNTS DURING THE PERIOD OF RECORD.
RECORD MAX(MIN) TEMPERATURE IS THE MEAN OF ALL DAILY MAX(MIN) TEMPERATURES DURING THE PERIOD OF RECORD.
RECORD MEAN TEMPERATURE IS THE SUM OF THE RECORD MAX AND RECORD MIN DIVIDED BY 2.
AVERAGE TEMPERATURE IS THE SUM OF THE MEAN DAILY MAX AND MIN TEMPERATURE DIVIDED BY 2.

EXCEPTIONS

1. RELATIVE HUMIDITY HOUR 01, AND MEAN WIND SPEED ARE THROUGH 1962.
2. THUNDERSTORMS AND HEAVY FOG ARE THROUGH 1964 AND MAY BE INCOMPLETE, DUE TO PART-TIME OPERATIONS.

RECORD MEANS ARE THROUGH THE CURRENT YEAR,
BEGINNING IN 1939 FOR TEMPERATURE
1939 FOR PRECIPITATION
1939 FOR SNOWFALL

HEATING DEGREE DAYS Base 65 deg. F CARIBOU, MAINE

SEASON	JULY	AUG	SEP	OCT	NOV	DEC	JAN	FEB	MAR	APR	MAY	JUNE	TOTAL
1964-65	46	199	419	737	1063	1484	1743	1493	1213	831	485	164	9877
1965-66	115	161	351	737	1191	1514	1489	1388	1135	813	499	160	9553
1966-67	64	106	393	683	899	1354	1567	1629	1468	912	682	121	9878
1967-68	14	55	276	636	1000	1471	1823	1594	1211	715	468	171	9434
1968-69	48	191	199	515	1097	1452	1534	1316	1219	876	512	167	9126
1969-70	109	86	364	732	895	1331	1861	1449	1221	800	440	142	9430
1970-71	17	80	328	540	905	1697	1833	1415	1224	860	411	199	9509
1971-72	59	141	274	569	1106	1648	1759	1698	1440	903	405	134	10136
1972-73	55	145	326	818	1130	1775	1711	1464	1092	807	491	122	9936
1973-74	9	44	373	649	1117	1270	1810	1512	1359	837	609	81	9670
1974-75	40	66	369	789	998	1433	1717	1498	1295	890	343	158	9596
1975-76	19	93	310	700	960	1674	1849	1489	1302	793	426	123	9738
1976-77	84	132	405	793	1175	1765	1828	1453	1002	850	406	203	10096
1977-78	56	93	408	657	944	1489	1702	1467	1364	894	333	156	9563
1978-79	53	86	434	717	1114	1503	1534	1527	1018	708	326	104	9124
1979-80	34	146	327	612	870	1413	1593	1534	1265	664	420	200	9078
1980-81	71	41	425	740	1042	1733	1839	1040	1141	757	333	125	9287
1981-82	37	77	355	757	984	1292	1891	1499	1251	890	355	145	9533
1982-83	60	199	310	632	961	1325	1544	1387	1171	712	493	148	8942
1983-84	75	78	257	656	978	1562	1804	1229	1412	748	493	199	9491
1984-85	38	57	396	630	940	1441	1822	1363	1281	884	472	198	9522
1985-86	43	118	272	650	1094	1636	1646	1473	1255	660	394	246	9487
1986-87	105	147	427	733	1167	1505	1704	1465	1140	629	442	142	9606
1987-88	79	152	314	641	1071	1430	1636	1485	1289	760	321	232	9410
1988-89	47	114	373	752	939	1583	1627	1524	1402	841	257	181	9640
1989-90	66	101	303	613	1116	1905	1471	1490	1241	759	490	107	9662
1990-91	57	47	337	623	1014	1370	1791	1392	1172	755	383	158	9099
1991-92	55	66	383	619	959	1595	1691	1564	1382	852	381	163	9710
1992-93	145	73	273	732	1064	1402	1747	1699	1284	748	423	175	9765
1993-94	73	53	327	791	1051	1407							

COOLING DEGREE DAYS Base 65 deg. F CARIBOU, MAINE

YEAR	JAN	FEB	MAR	APR	MAY	JUNE	JULY	AUG	SEP	OCT	NOV	DEC	TOTAL
1969	0	0	0	0	0	45	48	62	4	0	0	0	159
1970	0	0	0	0	2	47	164	117	1	0	0	0	331
1971	0	0	0	0	2	36	38	33	16	0	0	0	125
1972	0	0	0	0	21	35	51	11	9	0	0	0	127
1973	0	0	0	0	0	67	126	97	17	0	0	0	307
1974	0	0	0	0	2	21	38	50	5	0	0	0	116
1975	0	0	0	0	0	54	137	80	0	0	0	0	271
1976	0	0	0	0	6	104	52	65	4	0	0	0	231
1977	0	0	0	0	52	24	69	77	1	0	0	0	223
1978	0	0	0	0	38	35	97	93	1	0	0	0	264
1979	0	0	0	0	8	50	153	57	16	6	0	0	290
1980	0	0	0	0	0	37	71	78	8	0	0	0	194
1981	0	0	0	0	3	17	83	68	4	0	0	0	175
1982	0	0	0	0	11	10	110	9	12	0	0	0	152
1983	0	0	0	0	0	62	65	66	32	1	0	0	226
1984	0	0	0	0	9	33	72	79	0	0	0	0	193
1985	0	0	0	0	0	0	82	40	10	0	0	0	132
1986	0	0	0	0	5	10	39	20	1	0	0	0	75
1987	0	0	0	0	2	10	111	38	8	0	0	0	169
1988	0	0	0	0	6	45	131	103	1	0	0	0	286
1989	0	0	0	0	9	37	64	63	19	0	0	0	192
1990	0	0	0	2	0	51	102	106	0	0	0	0	261
1991	0	0	0	0	2	55	89	85	4	0	0	0	235
1992	0	0	0	0	23	26	5	40	12	0	0	0	106
1993	0	0	0	0	0	31	72	83	6	0	0	0	192

SNOWFALL (inches) CARIBOU, MAINE

SEASON	JULY	AUG	SEP	OCT	NOV	DEC	JAN	FEB	MAR	APR	MAY	JUNE	TOTAL
1964-65	0.0	0.0	0.0	0.9	12.9	37.2	18.7	14.8	6.1	4.0	0.0	0.0	94.6
1965-66	0.0	0.0	0.0	1.2	27.0	17.8	30.3	20.9	10.7	2.8	8.2	0.0	118.9
1966-67	0.0	0.0	0.0	0.4	7.1	24.2	27.8	26.0	10.5	2.2	10.9	0.0	109.1
1967-68	0.0	0.0	0.0	T	9.3	31.3	33.5	5.3	26.3	T	0.0	0.0	105.7
1968-69	0.0	0.0	0.0	T	28.0	31.2	29.5	29.7	19.5	14.1	0.0	0.0	152.0
1969-70	0.0	0.0	0.0	2.3	4.6	28.6	6.0	15.0	19.8	11.7	T	0.0	88.0
1970-71	0.0	0.0	0.0	7.6	11.0	32.4	16.6	16.3	38.6	12.2	0.0	T	134.7
1971-72	0.0	0.0	0.0	T	18.9	26.3	15.7	25.1	41.4	9.4	T	0.0	136.8
1972-73	0.0	0.0	0.0	1.7	13.3	59.9	20.3	27.8	7.8	22.2	0.0	0.0	153.0
1973-74	0.0	0.0	T	T	14.5	23.4	20.7	11.9	21.6	13.1	4.2	0.0	109.4
1974-75	0.0	0.0	T	0.4	34.9	12.6	31.2	9.8	18.2	15.1	0.0	0.0	122.2
1975-76	0.0	0.0	0.0	T	10.4	36.4	30.1	30.2	23.3	1.5	0.5	0.0	132.4
1976-77	0.0	0.0	0.0	3.7	10.2	31.9	39.1	34.4	16.0	10.6	T	0.0	145.9
1977-78	0.0	0.0	0.0	0.0	5.5	37.7	41.4	4.4	13.7	15.3	0.8	0.0	118.8
1978-79	0.0	0.0	0.0	T	9.2	40.3	32.1	18.1	8.6	14.9	T	0.0	123.2
1979-80	0.0	0.0	0.0	0.7	3.0	18.0	7.9	12.6	27.6	0.8	0.0	T	70.6
1980-81	0.0	0.0	T	0.2	12.0	28.9	35.2	8.6	36.4	1.4	0.2	0.0	122.9
1981-82	0.0	0.0	T	2.6	4.8	34.3	30.9	25.1	23.7	36.4	1.0	0.0	158.8
1982-83	0.0	0.0	0.0	T	14.6	4.0	22.8	22.8	13.9	4.8	T	0.0	82.9
1983-84	0.0	0.0	0.0	T	21.0	26.6	27.3	20.1	35.4	3.9	0.2	0.0	134.5
1984-85	0.0	0.0	0.0	0.8	3.8	30.7	10.5	26.8	11.9	5.0	1.3	0.0	90.8
1985-86	0.0	0.0	0.0	T	14.8	18.3	30.0	11.9	18.6	11.3	T	0.0	104.9
1986-87	0.0	0.0	T	0.4	28.1	8.5	26.4	4.1	12.3	5.2	T	0.0	85.0
1987-88	0.0	0.0	T	T	5.2	22.6	32.1	28.2	3.9	6.4	T	0.0	98.4
1988-89	0.0	0.0	0.0	1.2	13.8	11.5	18.4	16.6	11.9	9.4	0.0	0.0	82.8
1989-90	0.0	0.0	T	0.1	14.3	38.0	28.5	18.3	7.6	10.7	0.6	T	118.1
1990-91	0.0	0.0	0.0	T	12.0	22.6	21.5	12.2	24.0	2.5	0.0	T	94.8
1991-92	0.0	0.0	2.5	T	7.1	19.4	14.8	38.6	4.2	7.4	T	0.0	94.0
1992-93	T	0.0	T	0.4	5.3	15.9	5.6	22.8	23.4	7.1	T	0.0	80.5
1993-94	0.0	0.0	0.0	0.2	10.5	13.9							
Record Mean	T	0.0	T	1.7	12.1	23.0	23.3	21.6	19.2	8.4	0.7	T	109.9

PORTLAND, MAINE

The Portland City Airport is located 2 3/4 miles west of the site of the former city office. The surrounding country is mostly open, rolling and sloping generally toward the Fore River, a body of brackish water about 1,000 feet wide at a distance of about 1/2 mile from the station and forming one boundary (north through east) of the field. The airport is about 5 1/2 miles west-northwest of the open ocean. A slight rise reaching an elevation of 100 feet, lying northwest of the field, cuts down the wind slightly from that direction. The older portion of the city is situated on a hill rising abruptly from sea level to 170 feet, 1 1/2 miles east of the airport and on the opposite side of the Fore River. A line of low hills southeast of the airport, near the ocean, which reach a maximum height of 160 feet, shuts off sight of the ocean from the airport. Sebago Lake with an area of 44 square miles is situated about 15 miles to the northwest and 45 miles farther are the White Mountains, averaging 3,000 to 5,000 feet in height.

As a rule, Portland has very pleasant summers and falls, cold winters with frequent thaws, and disagreeable springs. Very few summer nights are too warm and humid for comfortable sleeping. Autumn has the greatest number of sunny days and the least cloudiness. Winters are quite severe, but begin late and then extend deeply into the normal springtime.

Heavy seasonal snowfalls, over 100 inches, normally occur about each 10 years. True blizzards are very rare. The White Mountains, to the northwest, keep considerable snow from reaching the Portland area and also moderate the temperature. Normal monthly precipitation is remarkably uniform throughout the year.

Winds are generally quite light with the highest velocities being confined mostly to March and November. Even in these months the occasional northeasterly gales have usually lost much of their severity before reaching the coast of Maine.

Temperatures well below zero are recorded frequently each winter. Cold waves sometimes come in on strong winds, but extremely low temperatures are generally accompanied by light winds.

The average freeze-free season at the airport station is 139 days. Mid-May is the average occurrence of the last freeze in spring, and the average occurrence of the first freeze in fall is late September. The freeze-free period is longer in the city proper, but may be even shorter at susceptible places further inland.

Daily maximum temperatures at the present airport site agree closely with those near the former intown office, but minimum temperatures on clear, quiet mornings range as much as 15 degrees lower at the airport.

NORMALS, MEANS, AND EXTREMES

PORTLAND, MAINE

LATITUDE: 43°39'N LONGITUDE: 70°18' W ELEVATION: FT. GRND 43 BARO 78 TIME ZONE: EASTERN WBAN: 14764

	(a)	JAN	FEB	MAR	APR	MAY	JUNE	JULY	AUG	SEP	OCT	NOV	DEC	YEAR
TEMPERATURE °F:														
Normals														
-Daily Maximum		30.3	33.1	41.4	52.3	63.2	72.7	78.8	77.4	69.3	58.7	47.0	35.1	54.9
-Daily Minimum		11.4	13.5	24.5	34.1	43.4	52.1	58.3	57.1	48.9	38.3	30.4	17.8	35.8
-Monthly		20.8	23.3	33.0	43.3	53.3	62.4	68.6	67.3	59.1	48.5	38.7	26.5	45.4
Extremes														
-Record Highest	53	64	64	86	85	94	98	99	103	95	88	74	69	103
-Year		1950	1957	1946	1957	1987	1991	1977	1975	1983	1963	1987	1982	AUG 1975
-Record Lowest	53	-26	-39	-21	8	23	33	40	33	23	15	3	-21	-39
-Year		1971	1943	1950	1954	1956	1944	1965	1965	1941	1976	1989	1963	FEB 1943
NORMAL DEGREE DAYS:														
Heating (base 65°F)		1370	1168	992	651	363	100	11	39	189	512	789	1194	7378
Cooling (base 65°F)		0	0	0	0	0	22	123	111	12	0	0	0	268
% OF POSSIBLE SUNSHINE	53	56	59	56	54	55	59	63	63	62	58	48	53	57
MEAN SKY COVER (tenths)														
Sunrise - Sunset	50	6.1	6.1	6.4	6.7	6.7	6.5	6.3	5.9	5.7	5.8	6.6	6.2	6.2
MEAN NUMBER OF DAYS:														
Sunrise to Sunset														
-Clear	53	10.0	8.6	8.7	7.4	6.3	6.8	7.2	9.1	10.3	10.4	7.5	9.2	101.3
-Partly Cloudy	53	6.5	7.1	7.0	7.1	9.5	10.0	11.2	10.7	7.8	7.6	7.1	7.2	99.0
-Cloudy	53	14.5	12.5	15.3	15.5	15.2	13.2	12.6	11.2	11.9	12.9	15.4	14.6	164.9
Precipitation														
.01 inches or more	53	11.0	10.0	11.4	11.8	12.5	11.5	9.7	9.5	8.5	9.5	11.6	11.5	128.5
Snow,Ice pellets,hail														
1.0 inches or more	53	4.5	3.8	3.4	0.8	0.1	0.0	0.0	0.0	0.0	0.1	1.0	3.7	17.3
Thunderstorms	53	0.*	0.1	0.4	0.5	2.0	4.1	4.5	3.5	1.5	0.6	0.3	0.1	17.7
Heavy Fog Visibility														
1/4 mile or less	53	1.9	2.1	3.3	3.1	5.3	5.2	6.4	5.6	5.3	4.8	3.5	1.9	48.3
Temperature °F														
-Maximum														
90° and above	53	0.0	0.0	0.0	0.0	0.2	1.1	2.0	1.6	0.3	0.0	0.0	0.0	5.2
32° and below	53	16.7	12.2	4.4	0.1	0.0	0.0	0.0	0.0	0.0	0.0	0.8	11.7	46.0
-Minimum														
32° and below	53	29.8	26.8	25.8	13.8	2.3	0.0	0.0	0.0	1.0	9.0	19.5	28.7	156.8
0° and below	53	5.9	4.1	0.6	0.0	0.0	0.0	0.0	0.0	0.0	0.0	0.0	2.8	13.5
AVG. STATION PRESS.(mb)	21	1013.0	1013.6	1012.8	1011.6	1012.6	1011.4	1011.9	1014.0	1015.3	1015.7	1014.4	1014.1	1013.4
RELATIVE HUMIDITY (%)														
Hour 01	53	74	73	75	79	84	87	89	89	89	85	81	77	82
Hour 07 (Local Time)	53	76	76	74	73	75	78	80	83	85	84	82	79	79
Hour 13	53	60	58	57	55	58	60	59	59	60	59	62	61	59
Hour 19	53	68	67	68	69	71	73	74	77	79	77	75	71	72
PRECIPITATION (inches):														
Water Equivalent														
-Normal		3.53	3.33	3.67	4.08	3.62	3.44	3.09	2.87	3.09	3.90	5.17	4.55	44.34
-Maximum Monthly	53	11.92	7.10	9.97	9.90	9.64	6.75	7.48	15.22	9.81	12.27	13.50	9.69	15.22
-Year		1979	1981	1953	1973	1984	1982	1976	1991	1954	1962	1983	1969	AUG 1991
-Minimum Monthly	53	0.76	0.04	0.81	0.71	0.49	0.70	0.61	0.27	0.30	0.26	0.90	0.98	0.04
-Year		1970	1987	1965	1941	1965	1941	1965	1947	1948	1947	1976	1955	FEB 1987
-Maximum in 24 hrs	53	3.56	3.41	3.47	5.26	4.66	5.58	2.68	7.83	7.49	7.71	4.70	3.82	7.83
-Year		1977	1981	1951	1973	1989	1967	1979	1991	1954	1962	1990	1969	AUG 1991
Snow,Ice pellets,hail														
-Maximum Monthly	53	62.4	61.2	49.0	15.9	7.0	0.0	0.0	0.0	T	3.8	15.6	54.8	62.4
-Year		1979	1969	1993	1982	1945				1992	1969	1972	1970	JAN 1979
-Maximum in 24 hrs	53	27.1	21.5	18.6	15.9	7.0	0.0	0.0	0.0	T	3.6	11.1	22.8	27.1
-Year		1979	1969	1993	1982	1945				1992	1969	1972	1970	JAN 1979
WIND:														
Mean Speed (mph)	53	9.2	9.4	10.0	10.0	9.2	8.3	7.6	7.5	7.9	8.4	8.8	9.0	8.8
Prevailing Direction through 1963		N	N	W	S	S	S	S	S	S	N	W	N	S
Fastest Obs. 1 Min.														
-Direction (!!)	8	17	32	07	18	16	28	26	01	25	07	15	26	01
-Speed (MPH)	8	36	29	37	30	30	35	29	40	29	31	35	35	40
-Year		1992	1992	1993	1993	1990	1991	1988	1991	1993	1988	1993	1993	AUG 1991
Peak Gust														
-Direction (!!)	10	S	NW	E	NE	SE	W	NW	N	SE	N	SE	SE	SE
-Speed (mph)	10	53	47	53	49	45	49	55	61	70	53	52	56	70
-Date		1992	1992	1993	1991	1989	1991	1990	1991	1985	1991	1993	1986	SEP 1985

PRECIPITATION (inches)

PORTLAND, MAINE

YEAR	JAN	FEB	MAR	APR	MAY	JUNE	JULY	AUG	SEP	OCT	NOV	DEC	ANNUAL
1964	4.74	2.66	3.64	3.27	1.56	2.01	3.47	1.77	1.02	3.45	3.25	3.75	34.59
1965	1.68	6.36	0.81	2.93	0.49	2.27	0.61	1.84	2.31	2.49	3.96	2.40	28.15
1966	4.97	3.43	3.77	1.00	1.96	3.04	0.96	4.99	3.84	3.61	5.01	3.46	40.04
1967	2.52	4.63	2.16	4.74	5.34	6.23	3.46	2.57	2.68	1.31	2.52	6.00	44.16
1968	3.00	1.26	4.16	3.37	3.81	4.04	0.65	2.00	1.40	2.00	7.74	7.70	41.13
1969	3.63	6.28	3.36	2.86	1.67	3.53	5.45	2.20	5.28	2.51	8.54	9.69	55.00
1970	0.76	4.31	4.22	4.14	3.15	2.91	0.95	5.12	4.49	3.87	2.38	5.14	41.44
1971	2.25	6.76	4.74	1.86	4.09	1.09	3.14	3.24	3.55	3.95	4.13	2.85	41.65
1972	2.09	5.14	6.01	2.53	3.17	4.24	2.05	0.80	4.31	3.92	7.87	6.49	48.62
1973	2.58	2.57	3.33	9.90	6.28	4.87	1.70	3.48	2.23	3.38	2.40	9.57	52.29
1974	3.41	2.07	3.82	3.82	4.20	4.69	3.66	1.45	5.43	1.74	4.85	4.41	43.55
1975	4.40	2.51	3.20	3.71	1.09	4.87	2.06	3.89	4.34	4.50	6.01	8.14	48.72
1976	4.44	2.84	2.47	2.42	3.99	1.53	7.48	4.87	1.86	5.37	0.90	3.22	41.39
1977	6.46	3.75	6.92	3.46	2.04	3.38	2.83	2.79	4.63	8.30	6.46	6.61	57.63
1978	6.91	0.87	4.19	4.46	4.36	2.42	1.67	2.36	0.59	3.23	2.26	3.15	36.47
1979	11.92	3.50	4.17	6.48	5.15	1.97	5.90	5.53	3.28	6.71	3.95	2.59	61.15
1980	0.98	1.36	4.54	5.78	1.83	3.34	1.99	2.14	3.00	2.99	4.75	1.18	33.88
1981	0.93	7.10	1.44	3.46	2.27	4.59	5.44	2.31	6.14	4.71	2.80	4.51	45.70
1982	5.17	2.53	3.20	4.54	2.91	6.75	2.61	3.35	1.90	1.93	3.61	1.18	39.68
1983	4.59	3.94	9.75	6.82	5.98	1.35	4.31	2.58	1.35	3.38	13.50	8.78	66.33
1984	2.56	4.99	5.12	4.81	9.64	3.87	3.86	2.09	0.84	3.26	3.69	3.44	48.17
1985	1.03	1.54	3.15	1.25	2.03	2.74	3.30	3.18	2.97	4.07	6.36	2.39	34.01
1986	6.58	2.61	4.21	3.49	2.51	3.91	3.44	1.87	2.64	2.09	5.18	5.91	44.44
1987	5.21	0.04	4.29	6.33	2.62	5.01	1.79	2.48	4.64	2.54	3.82	2.01	40.78
1988	1.97	3.34	1.85	3.68	4.27	2.36	5.89	5.24	1.50	3.47	8.84	1.21	43.62
1989	1.15	2.37	2.14	2.94	8.74	4.49	2.50	1.73	4.48	4.81	3.97	2.23	41.55
1990	3.19	2.49	1.42	5.16	5.23	4.12	3.21	1.89	3.12	7.46	7.50	7.90	52.69
1991	2.91	1.85	6.19	6.71	3.77	1.47	2.35	15.22	5.44	2.83	4.34	4.06	57.14
1992	4.80	3.42	3.93	2.47	1.15	3.99	4.06	2.59	3.11	2.48	4.39	2.12	38.51
1993	2.80	3.84	6.26	5.69	1.14	2.89	2.73	1.21	4.10	3.74	4.05	5.40	43.85
Record Mean	3.86	3.66	3.95	3.65	3.43	3.27	3.12	3.07	3.21	3.39	4.12	3.96	42.70

AVERAGE TEMPERATURE (deg. F)

PORTLAND, MAINE

YEAR	JAN	FEB	MAR	APR	MAY	JUNE	JULY	AUG	SEP	OCT	NOV	DEC	ANNUAL
1964	24.4	23.0	31.9	40.9	54.8	62.6	67.6	61.1	55.6	47.1	36.2	25.3	44.2
1965	19.8	21.6	33.3	41.1	54.8	63.1	66.3	66.5	58.0	47.6	36.6	27.9	44.7
1966	22.3	22.6	32.7	39.9	50.4	63.1	67.7	66.2	56.2	47.1	41.1	26.7	44.7
1967	24.4	17.2	26.5	41.5	47.3	63.3	67.5	66.8	59.9	49.7	34.8	27.8	43.9
1968	16.6	19.2	33.2	45.1	51.0	60.0	70.0	66.4	61.7	52.8	36.7	25.2	44.8
1969	24.2	25.0	31.1	43.0	52.8	62.6	67.6	71.3	60.9	48.9	41.1	28.7	46.5
1970	16.7	25.4	32.4	43.6	55.7	63.5	70.1	68.8	61.0	51.7	41.3	19.6	45.8
1971	12.2	21.9	31.6	41.6	52.7	64.3	69.0	68.9	61.3	52.5	35.4	27.3	44.9
1972	22.2	21.1	28.9	40.6	52.6	59.6	67.5	65.3	58.7	45.2	35.0	24.0	43.4
1973	23.1	23.4	37.7	45.5	51.3	63.8	70.8	70.9	58.4	49.2	37.7	33.5	47.1
1974	23.2	24.5	33.9	45.1	50.6	61.3	68.2	67.9	58.8	44.7	39.3	30.2	45.6
1975	26.7	25.0	29.8	40.0	55.6	61.5	70.1	66.8	57.2	49.3	43.0	24.3	45.8
1976	15.7	28.2	31.5	45.0	52.3	65.9	65.9	65.4	57.3	43.6	36.1	19.4	43.9
1977	14.6	21.9	36.2	42.4	54.4	59.0	68.2	66.6	57.6	48.1	39.4	25.4	44.5
1978	21.1	19.2	30.2	40.6	52.7	61.0	68.0	68.6	57.8	48.2	36.4	26.1	44.2
1979	23.7	15.6	35.6	42.2	55.3	62.7	69.3	64.8	57.4	47.5	42.4	29.8	45.5
1980	22.7	20.5	32.0	44.3	53.7	61.0	69.6	71.2	61.5	45.9	36.2	21.3	45.0
1981	13.8	32.3	35.7	45.2	55.3	63.9	68.8	66.0	58.6	46.6	38.9	29.2	46.2
1982	15.0	23.3	32.1	42.0	54.6	58.3	69.2	64.4	59.2	48.1	41.3	32.0	45.0
1983	25.2	26.2	35.8	44.4	52.0	63.8	69.7	67.6	63.0	47.9	40.2	26.1	46.8
1984	19.8	32.0	28.2	43.3	53.0	63.9	69.6	69.1	57.8	49.6	39.2	31.2	46.4
1985	16.2	26.4	34.8	44.1	53.7	61.7	69.8	66.6	60.7	50.8	39.9	24.6	45.8
1986	25.0	23.4	34.6	46.5	53.6	60.9	66.3	66.2	57.4	47.9	36.3	29.6	45.6
1987	21.6	23.0	34.0	44.9	54.6	63.8	68.0	66.2	59.6	47.1	38.1	30.2	45.9
1988	21.6	25.7	34.2	43.5	54.7	63.8	71.1	71.1	59.4	46.5	41.0	26.0	46.6
1989	26.8	24.0	31.6	41.1	55.6	63.9	69.4	68.0	60.5	49.8	37.3	14.1	45.2
1990	30.2	25.7	34.7	44.6	51.8	62.4	70.3	69.8	59.8	52.4	41.8	33.7	48.1
1991	23.4	29.8	37.0	45.9	58.1	65.5	70.1	69.9	58.2	50.8	40.4	26.4	48.0
1992	23.9	25.9	30.9	41.4	51.9	63.4	65.8	66.8	58.9	47.3	37.6	28.9	45.2
1993	23.4	16.8	29.8	43.3	54.8	64.5	69.8	69.5	59.6	46.2	39.0	29.9	45.6
Record Mean	22.3	23.5	32.4	42.9	53.3	62.4	68.3	66.8	59.5	49.3	38.7	26.9	45.5
Max	30.8	32.3	40.3	51.3	62.3	71.7	77.5	75.7	68.5	58.1	46.3	34.7	54.1
Min	13.7	14.8	24.5	34.5	44.3	53.1	59.1	57.8	50.4	40.5	31.0	19.0	36.9

REFERENCE NOTES

PORTLAND, MAINE

GENERAL
T - TRACE AMOUNT.
BLANK ENTRIES DENOTE MISSING/UNREPORTED DATA.
# INDICATES A STATION OR INSTRUMENT RELOCATION.

SPECIFIC

PM - INCLUDES LAST DAY OF PREVIOUS MONTH
ASOS - AUTOMATED SURFACE OBSERVING SYSTEM IN OPERATION DURING THESE MONTHS.

(a) - LENGTH OF RECORD IN YEARS, ALTHOUGH INDIVIDUAL MONTHS MAY BE MISSING.
0.* OR * - THE VALUE IS BETWEEN 0.0 AND 0.05
NORMALS - BASED ON THE 1961-1990 RECORD PERIOD.
EXTREMES - DATES ARE THE MOST RECENT OCCURRENCE
WIND DIR.- NUMERALS SHOW TENS OF DEGRESS CLOCKWISE FROM TRUE NORTH. "00" INDICATES CALM.
RESULTANT DIRECTIONS ARE GIVEN TO WHOLE DEGREES.
BOLD VALUES INDICATE EXTREME VALUES WHICH OCCURRED AFTER THE ASOS SYSTEM WAS COMMISSIONED.

RECORD = PERIOD OF RECORD
RECORD MEAN PRECIPITATION IS THE MEAN OF ALL DAILY PRECIPITATION AMOUNTS DURING THE PERIOD OF RECORD.
RECORD MAX(MIN) TEMPERATURE IS THE MEAN OF ALL DAILY MAX(MIN) TEMPERATURES DURING THE PERIOD OF RECORD.
RECORD MEAN TEMPERATURE IS THE SUM OF THE RECORD MAX AND RECORD MIN DIVIDED BY 2.
AVERAGE TEMPERATURE IS THE SUM OF THE MEAN DAILY MAX AND MIN TEMPERATURE DIVIDED BY 2.

EXCEPTIONS

RECORD MEANS ARE THROUGH THE CURRENT YEAR, BEGINNING IN 1874 FOR TEMPERATURE
1871 FOR PRECIPITATION
1941 FOR SNOWFALL

HEATING DEGREE DAYS Base 65 deg. F PORTLAND, MAINE

SEASON	JULY	AUG	SEP	OCT	NOV	DEC	JAN	FEB	MAR	APR	MAY	JUNE	TOTAL
1964-65	28	134	293	549	857	1222	1396	1212	975	710	320	125	7821
1965-66	41	61	242	532	845	1145	1315	1183	994	745	449	108	7660
1966-67	19	29	264	549	710	1181	1254	1330	1187	699	544	92	7858
1967-68	15	27	155	467	900	1147	1496	1320	981	590	424	164	7686
1968-69	8	46	110	375	841	1228	1259	1112	1043	652	377	118	7169
1969-70	30	7	179	494	710	1120	1491	1104	1005	637	283	93	7153
1970-71	7	14	159	408	704	1400	1634	1202	1028	695	373	81	7705
1971-72	3	20	147	381	878	1164	1322	1265	1112	725	379	155	7551
1972-73	27	53	190	607	893	1264	1292	1157	842	575	419	99	7418
1973-74	0	9	231	480	813	970	1290	1126	958	595	444	131	7047
1974-75	15	17	206	624	762	1071	1177	1112	1086	746	286	146	7248
1975-76	13	59	230	480	653	1258	1522	1061	1030	594	389	94	7383
1976-77	45	73	229	660	858	1404	1559	1199	888	674	354	184	8127
1977-78	29	54	233	518	761	1219	1353	1276	1071	724	377	134	7749
1978-79	39	32	230	513	852	1201	1272	1380	905	677	311	97	7509
1979-80	21	82	240	539	672	1083	1305	1284	1018	613	346	163	7366
1980-81	16	6	163	584	855	1349	1578	910	901	588	312	54	7316
1981-82	16	45	189	566	778	1102	1543	1161	1014	684	320	198	7616
1982-83	20	78	185	519	704	1015	1225	1080	895	612	393	101	6827
1983-84	8	38	139	527	738	1198	1397	949	1132	642	368	110	7246
1984-85	11	13	223	469	767	1043	1506	1076	930	620	347	115	7120
1985-86	4	32	157	433	747	1245	1236	1161	935	548	354	138	6990
1986-87	47	52	242	523	855	1092	1336	1172	955	597	343	77	7291
1987-88	20	58	171	548	798	1070	1339	1130	950	641	323	112	7160
1988-89	13	32	180	569	713	1201	1174	1141	1028	708	286	91	7136
1989-90	6	25	167	464	824	1573	1071	1093	935	607	402	107	7274
1990-91	12	24	170	388	690	964	1283	979	861	568	236	76	6251
1991-92	11	16	228	433	730	1191	1267	1127	1051	700	414	84	7252
1992-93	47	30	208	543	814	1112	1280	1344	1084	645	311	81	7499
1993-94	15	5	192	575	774	1080							

COOLING DEGREE DAYS Base 65 deg. F PORTLAND, MAINE

YEAR	JAN	FEB	MAR	APR	MAY	JUNE	JULY	AUG	SEP	OCT	NOV	DEC	TOTAL
1969	0	0	0	0	3	56	118	211	59	0	0	0	447
1970	0	0	0	0	4	55	172	136	41	0	0	0	408
1971	0	0	0	0	0	66	135	149	43	1	0	0	394
1972	0	0	0	0	0	3	114	71	8	0	0	0	196
1973	0	0	0	0	1	71	189	201	40	0	0	0	502
1974	0	0	0	2	4	26	121	115	28	0	0	0	296
1975	0	0	0	0	1	49	179	120	2	0	0	0	351
1976	0	0	0	0	2	128	80	93	5	0	0	0	308
1977	0	0	0	1	32	12	135	109	19	0	0	0	308
1978	0	0	0	0	6	22	138	150	20	0	0	0	336
1979	0	0	0	0	15	34	162	83	19	3	0	0	316
1980	0	0	0	0	1	50	163	205	67	0	0	0	486
1981	0	0	0	0	20	30	138	84	5	0	0	0	277
1982	0	0	0	0	6	4	158	66	17	0	0	0	251
1983	0	0	0	0	0	73	161	125	84	3	0	0	446
1984	0	0	0	0	0	84	162	147	12	0	0	0	405
1985	0	0	0	0	5	25	161	92	35	1	0	0	319
1986	0	0	0	0	8	23	93	96	19	0	0	0	239
1987	0	0	0	0	28	47	121	103	17	0	0	0	316
1988	0	0	0	0	11	85	209	227	17	2	0	0	551
1989	0	0	0	0	2	67	151	126	38	0	0	0	384
1990	0	0	0	0	0	34	181	179	19	7	0	0	420
1991	0	0	0	0	29	96	178	176	31	0	0	0	510
1992	0	0	0	0	13	43	76	93	30	0	0	0	255
1993	0	0	0	0	1	72	170	150	37	0	0	0	430

SNOWFALL (inches) PORTLAND, MAINE

SEASON	JULY	AUG	SEP	OCT	NOV	DEC	JAN	FEB	MAR	APR	MAY	JUNE	TOTAL
1964-65	0.0	0.0	0.0	0.0	T	20.5	15.8	13.2	3.4	2.6	0.0	0.0	55.5
1965-66	0.0	0.0	0.0	T	0.6	9.1	38.2	21.3	2.8	0.7	2.0	0.0	74.7
1966-67	0.0	0.0	0.0	0.0	0.0	14.7	12.4	45.8	17.5	15.7	0.1	0.0	106.2
1967-68	0.0	0.0	0.0	0.0	3.6	18.9	20.0	4.5	11.9	0.0	0.0	0.0	58.9
1968-69	0.0	0.0	0.0	T	15.3	18.8	5.4	61.2	9.3	T	0.0	0.0	110.0
1969-70	0.0	0.0	0.0	3.8	T	24.7	6.2	9.8	20.7	2.9	0.0	0.0	68.1
1970-71	0.0	0.0	0.0	T	0.0	54.8	17.2	35.6	24.7	9.2	0.0	0.0	141.5
1971-72	0.0	0.0	0.0	0.0	6.2	12.5	7.0	38.0	21.7	6.3	0.0	0.0	91.7
1972-73	0.0	0.0	0.0	T	15.6	35.1	9.6	6.6	0.5	2.3	0.0	0.0	69.7
1973-74	0.0	0.0	0.0	0.0	0.0	7.2	15.0	4.3	6.2	8.3	0.0	0.0	41.0
1974-75	0.0	0.0	0.0	T	3.2	8.2	15.4	11.6	6.3	1.7	0.0	0.0	46.4
1975-76	0.0	0.0	0.0	T	3.6	25.3	18.1	4.9	22.2	T	0.0	0.0	74.1
1976-77	0.0	0.0	0.0	0.0	1.5	23.3	35.2	7.9	19.3	1.4	T	0.0	88.6
1977-78	0.0	0.0	0.0	0.0	1.9	23.1	30.7	8.2	12.5	0.8	0.0	0.0	77.2
1978-79	0.0	0.0	0.0	0.0	3.6	18.9	62.4	4.5	T	2.9	0.0	0.0	92.3
1979-80	0.0	0.0	0.0	1.7	T	1.8	6.0	11.2	6.8	0.0	0.0	0.0	27.5
1980-81	0.0	0.0	0.0	0.0	8.9	13.0	9.2	4.6	3.1	T	0.0	0.0	38.8
1981-82	0.0	0.0	0.0	0.0	T	24.0	25.9	11.0	8.5	15.9	0.0	0.0	85.3
1982-83	0.0	0.0	0.0	0.0	0.6	5.7	12.4	24.5	2.1	T	0.0	0.0	45.3
1983-84	0.0	0.0	0.0	0.0	T	12.6	28.3	3.3	26.4	T	0.0	0.0	70.6
1984-85	0.0	0.0	0.0	0.0	T	17.0	12.1	7.2	13.1	2.4	0.0	0.0	51.8
1985-86	0.0	0.0	0.0	0.0	3.1	11.2	18.6	12.0	6.4	T	0.0	0.0	51.3
1986-87	0.0	0.0	0.0	0.0	5.2	4.0	50.7	0.8	14.3	3.4	0.0	0.0	78.4
1987-88	0.0	0.0	T	0.0	5.4	9.1	19.8	20.8	3.0	4.2	0.0	0.0	62.3
1988-89	0.0	0.0	0.0	T	T	3.5	4.0	13.8	8.9	0.7	0.0	0.0	30.9
1989-90	0.0	0.0	0.0	0.0	5.0	15.6	20.4	25.6	3.2	T	0.0	0.0	69.8
1990-91	0.0	0.0	T	0.0	0.2	6.8	13.4	6.3	5.7	T	0.0	0.0	32.4
1991-92	0.0	0.0	0.0	0.0	T	22.5	2.4	10.3	13.4	10.0	0.0	0.0	58.6
1992-93	0.0	0.0	T	T	2.8	2.1	17.1	33.5	49.0	11.1	0.0	0.0	115.6
1993-94	0.0	0.0	0.0	0.2	T	12.3							
Record Mean	0.0	0.0	T	0.2	3.0	14.6	19.0	17.4	13.0	3.1	0.2	0.0	70.5

BALTIMORE, MARYLAND

Baltimore-Washington International Airport lies in a region about midway between the rigorous climates of the North and the mild climates of the South, and adjacent to the modifying influences of the Chesapeake Bay and Atlantic Ocean to the east and the Appalachian Mountains to the west. Since this region is near the average path of the low pressure systems which move across the country, changes in wind direction are frequent and contribute to the changeable character of the weather. The net effect of the mountains to the west and the bay and ocean to the east is to produce a more equable climate compared with other continental locations farther inland at the same latitude.

Rainfall distribution throughout the year is rather uniform, however, the greatest intensities are confined to the summer and early fall months, the season for hurricanes and severe thunderstorms. Moisture deficiencies for crops occur occasionally during the growing season, but severe droughts are rare. Rainfall during the growing season occurs principally in the form of thunderstorms, and rainfall totals during these months vary appreciably.

The average date for the last occurrence in spring of temperatures as low as 32 degrees is mid-April. The average date for the first occurrence in fall of temperatures as low as 32 degrees is late October. The freeze-free period is approximately 194 days.

In summer, the area is under the influence of the large semi-permanent high pressure system commonly known as the Bermuda High and centered over the Atlantic Ocean near 30 degrees N Latitude. This pressure system brings warm humid air to the area. The proximity of large water areas and the inflow of southerly winds contribute to high relative humidities during much of the year.

January is the coldest month, and July, the warmest. Snowfall occurs on about eleven days per year on the average, however, an average of only about six days annually produces snowfalls of 1 inch or greater. Snow is frequently mixed with rain and sleet, and snow seldom remains on the ground more than a few days.

Glaze or freezing rain which is hazardous to highway traffic occurs on an average of two to three times per year, generally in January or February. Some years pass without the occurrence of freezing rain, while in others it occurs on as many as eight to ten days. Sleet is observed on about five days annually with the greatest frequency of occurrence in January.

The annual prevailing wind direction is from the west. Winter and spring months have the highest average wind speed. Destructive velocities are rare and occur mostly during summer thunderstorms Only rarely have hurricanes in the vicinity caused widespread damage, then primarily through flooding.

NORMALS, MEANS, AND EXTREMES

BALTIMORE, MARYLAND

LATITUDE: 39°11'N LONGITUDE: 76°40'W ELEVATION: FT. GRND 148 BARO 197 TIME ZONE: EASTERN WBAN: 93721

	(a)	JAN	FEB	MAR	APR	MAY	JUNE	JULY	AUG	SEP	OCT	NOV	DEC	YEAR
TEMPERATURE °F:														
Normals														
-Daily Maximum		40.2	43.7	54.0	64.3	74.2	83.2	87.2	85.4	78.5	67.3	56.5	45.2	65.0
-Daily Minimum		23.4	25.9	34.1	42.5	52.6	61.8	66.8	65.7	58.4	45.9	37.1	28.2	45.2
-Monthly		31.8	34.8	44.1	53.4	63.4	72.5	77.0	75.6	68.5	56.6	46.8	36.7	55.1
Extremes														
-Record Highest	43	75	79	87	94	98	100	104	105	100	92	83	77	105
-Year		1975	1985	1979	1960	1991	1988	1988	1983	1983	1954	1974	1984	AUG 1983
-Record Lowest	43	-7	-3	6	20	32	40	50	45	35	25	13	0	-7
-Year		1984	1979	1960	1965	1966	1972	1988	1986	1963	1969	1955	1983	JAN 1984
NORMAL DEGREE DAYS:														
Heating (base 65°F)		1029	846	648	348	108	0	0	0	29	276	546	877	4707
Cooling (base 65°F)		0	0	0	0	59	227	372	329	134	16	0	0	1137
% OF POSSIBLE SUNSHINE	40	51	55	56	56	56	62	64	62	60	58	51	49	57
MEAN SKY COVER (tenths)														
Sunrise - Sunset	43	6.3	6.3	6.3	6.2	6.2	5.7	5.6	5.6	5.4	5.2	6.1	6.4	5.9
MEAN NUMBER OF DAYS:														
Sunrise to Sunset														
-Clear	43	8.1	7.7	7.9	7.7	7.7	8.4	9.2	9.4	10.6	11.9	8.3	8.3	105.3
-Partly Cloudy	43	7.7	6.8	8.8	9.0	10.3	11.6	11.8	10.8	8.5	8.0	8.3	7.2	108.7
-Cloudy	43	15.2	13.7	14.3	13.3	13.0	10.0	10.0	10.8	10.9	11.1	13.3	15.6	151.2
Precipitation .01 inches or more	43	10.4	9.0	10.8	10.7	10.9	9.2	9.0	9.5	7.7	7.5	9.0	9.3	113.1
Snow,Ice pellets,hail 1.0 inches or more	43	2.0	1.8	1.2	0.*	0.0	0.0	0.0	0.0	0.0	0.0	0.3	1.0	6.3
Thunderstorms	43	0.3	0.2	0.9	2.4	4.0	5.3	6.0	5.1	2.0	1.0	0.4	0.1	27.6
Heavy Fog Visibility 1/4 mile or less	43	3.0	3.2	2.6	1.8	1.7	0.9	0.8	1.0	1.3	2.7	2.4	3.4	25.0
Temperature °F														
-Maximum														
90° and above	43	0.0	0.0	0.0	0.4	1.5	6.2	11.4	8.0	3.1	0.1	0.0	0.0	30.7
32° and below	43	6.2	3.8	0.6	0.0	0.0	0.0	0.0	0.0	0.0	0.0	0.1	3.6	14.3
-Minimum														
32° and below	43	24.9	21.2	14.4	3.0	0.*	0.0	0.0	0.0	0.0	1.7	10.9	21.2	97.4
0° and below	43	0.4	0.1	0.0	0.0	0.0	0.0	0.0	0.0	0.0	0.0	0.0	0.*	0.5
AVG. STATION PRESS. (mb)	21	1013.3	1013.1	1011.4	1009.7	1009.9	1009.9	1010.5	1011.9	1013.1	1013.9	1013.8	1013.9	1012.0
RELATIVE HUMIDITY (%)														
Hour 01	40	69	67	67	68	77	81	81	83	83	80	74	71	75
Hour 07 (Local Time)	40	72	71	72	72	77	79	80	84	85	83	78	74	77
Hour 13	40	57	54	51	49	52	52	53	55	55	54	55	58	54
Hour 19	40	62	59	55	54	60	62	64	67	69	68	65	65	63
PRECIPITATION (inches):														
Water Equivalent														
-Normal		3.05	3.12	3.38	3.09	3.72	3.67	3.69	3.92	3.41	2.98	3.32	3.41	40.76
-Maximum Monthly	43	7.84	7.16	8.12	8.15	8.71	9.95	8.18	18.35	8.62	8.09	7.68	7.44	18.35
-Year		1979	1979	1993	1952	1989	1972	1960	1955	1975	1976	1952	1969	AUG 1955
-Minimum Monthly	43	0.29	0.56	0.93	0.39	0.37	0.15	0.30	0.77	0.21	T	0.31	0.20	T
-Year		1955	1978	1966	1985	1986	1954	1955	1951	1967	1963	1981	1955	OCT 1963
-Maximum in 24 hrs	43	3.11	3.26	3.18	2.80	3.64	5.23	5.86	8.35	6.04	3.49	3.43	3.39	8.35
-Year		1976	1983	1958	1952	1960	1972	1952	1955	1985	1955	1952	1977	AUG 1955
Snow,Ice pellets,hail														
-Maximum Monthly	43	25.1	33.1	21.6	0.7	T	0.0	T	0.0	0.0	0.3	8.4	20.4	33.1
-Year		1987	1979	1960	1985	1963		1992			1979	1967	1966	FEB 1979
-Maximum in 24 hrs	43	12.3	22.8	13.0	0.7	T	0.0	T	0.0	0.0	0.3	8.4	14.1	22.8
-Year		1987	1983	1962	1985	1963		1992			1979	1967	1960	FEB 1983
WIND:														
Mean Speed (mph)	43	9.7	10.3	10.8	10.5	9.1	8.5	8.0	7.8	8.0	8.6	9.2	9.3	9.2
Prevailing Direction through 1963		WNW	NW	WNW	WNW	W	WNW	W	W	S	NW	WNW	WNW	WNW
Fastest Mile														
-Direction (!!!)	43	NE	W	SE	W	SW	SW	NW	NE	W	SE	E	W	SE
-Speed (MPH)	43	63	68	80	70	65	80	57	54	56	73	58	57	80
-Year		1958	1956	1952	1954	1961	1952	1962	1955	1952	1954	1952	1953	MAR 1952
Peak Gust														
-Direction (!!!)	10	SW	NW	W	NW	NW	NW	NW	SW	NW	S	NW	NW	NW
-Speed (mph)	10	53	51	58	49	55	45	68	55	45	47	64	77	77
-Date		1992	1987	1985	1993	1993	1985	1987	1987	1985	1990	1989	1988	DEC 1988

PRECIPITATION (inches)

BALTIMORE, MARYLAND

YEAR	JAN	FEB	MAR	APR	MAY	JUNE	JULY	AUG	SEP	OCT	NOV	DEC	ANNUAL
1964	5.27	4.36	2.98	4.37	0.43	2.40	2.66	1.96	2.61	1.19	2.51	3.94	34.68
1965	3.09	2.89	4.31	1.72	1.79	1.94	2.61	4.72	1.94	1.90	0.68	0.63	28.22
1966	4.15	4.24	0.93	4.39	4.53	1.18	1.48	1.87	8.50	4.80	2.78	3.53	42.38
1967	0.99	2.25	4.39	1.73	3.79	1.89	3.56	8.87	0.21	1.34	2.60	5.31	36.93
1968	3.42	0.72	4.41	1.61	5.41	3.35	2.75	4.16	4.39	3.13	3.85	2.60	39.80
1969	1.38	1.75	1.63	1.80	1.46	3.65	5.22	3.81	2.60	1.10	1.74	7.44	33.58
1970	0.94	3.34	3.07	4.53	1.69	4.10	4.32	1.33	0.46	3.04	5.11	3.50	35.43
1971	2.02	6.21	1.90	1.75	6.12	2.92	4.03	10.91	5.55	6.88	3.75	1.29	53.33
1972	2.82	6.01	2.38	5.30	4.11	9.95	2.81	2.22	1.15	3.51	7.05	5.02	52.33
1973	2.81	2.82	3.96	6.41	3.73	3.16	4.22	3.35	4.87	2.86	1.28	6.36	45.83
1974	2.92	0.94	4.12	2.59	3.58	2.84	0.85	5.85	5.45	1.53	1.39	5.70	37.76
1975	3.47	2.47	5.17	2.73	4.63	3.82	7.15	4.23	8.62	2.89	2.03	4.61	51.82
1976	4.10	2.16	2.23	1.27	5.03	2.49	5.56	2.98	6.93	8.09	0.56	2.04	43.44
1977	1.36	0.63	3.93	3.05	1.49	3.44	2.62	3.31	0.62	5.17	5.01	5.76	36.39
1978	7.34	0.56	4.74	1.26	5.49	2.81	6.83	3.39	1.03	0.71	2.70	4.63	41.49
1979	7.84	7.16	2.05	3.37	4.15	5.74	3.71	9.38	6.73	5.53	2.45	0.87	58.98
1980	2.58	1.06	5.46	4.24	3.58	3.04	3.25	4.00	1.00	3.08	2.72	0.70	34.71
1981	0.49	2.93	1.14	2.04	3.63	5.40	4.59	1.93	2.89	2.57	0.31	3.30	31.22
1982	3.37	4.04	3.03	3.61	1.85	5.70	2.16	0.95	3.63	2.31	3.13	2.39	36.17
1983	2.21	4.81	6.80	6.55	5.47	5.23	1.31	1.57	1.76	3.58	5.02	6.72	51.03
1984	1.96	3.90	5.79	2.95	4.29	1.65	3.27	4.11	2.38	1.94	3.01	1.71	36.96
1985	2.03	3.03	2.37	0.39	6.01	2.44	2.53	3.72	6.22	2.48	4.71	0.84	36.77
1986	2.16	3.78	0.96	2.64	0.37	1.46	4.12	4.26	0.58	1.86	5.96	5.52	33.67
1987	5.85	2.22	0.99	1.86	4.16	2.63	5.05	1.61	7.34	2.25	5.05	2.07	41.08
1988	3.24	3.25	2.35	2.44	4.37	0.84	3.78	2.64	2.05	1.59	4.78	0.97	32.30
1989	3.07	3.36	4.24	3.16	8.71	5.98	7.35	3.38	3.64	4.90	1.97	2.12	51.88
1990	3.71	1.48	2.54	4.23	4.92	2.55	5.68	6.17	1.07	2.57	2.10	4.86	41.88
1991	3.54	0.73	5.65	1.68	1.16	1.08	1.76	2.54	3.05	3.20	1.69	4.08	30.16
1992	1.27	2.49	4.58	1.76	2.92	1.89	5.07	2.19	5.96	2.73	3.44	4.63	38.93
1993	2.73	2.84	8.12	3.68	3.66	2.56	1.71	2.55	4.09	3.02	3.09	4.45	42.50
Record Mean	2.92	2.99	3.73	3.20	3.60	3.54	3.84	4.10	3.45	2.98	3.20	3.39	40.93

AVERAGE TEMPERATURE (deg. F)

BALTIMORE, MARYLAND

YEAR	JAN	FEB	MAR	APR	MAY	JUNE	JULY	AUG	SEP	OCT	NOV	DEC	ANNUAL
1964	33.8	33.2	44.4	50.9	65.3	72.7	77.0	73.5	68.1	53.3	49.8	39.1	55.1
1965	30.8	34.9	39.2	50.3	66.9	70.1	76.8	75.5	70.8	53.9	45.6	38.3	54.4
1966	29.8	30.5	43.3	49.5	61.7	72.9	78.9	76.5	66.7	53.4	46.4	35.6	53.8
1967	37.4	30.7	41.8	54.0	57.0	72.7	75.2	73.4	65.3	55.3	42.0	37.3	53.5
1968	29.2	32.2	46.6	54.0	59.7	72.6	78.2	78.7	70.3	59.7	48.2	34.6	55.3
1969	31.7	34.9	40.7	56.2	65.5	74.9	77.4	76.5	69.5	57.7	46.1	35.2	55.5
1970	27.8	35.4	40.4	53.3	66.6	73.0	77.2	77.4	73.7	61.5	48.6	38.2	56.1
1971	30.0	37.4	41.8	52.7	61.2	74.0	76.5	74.2	70.9	62.9	46.4	43.7	56.0
1972	37.6	34.3	43.6	51.6	62.7	68.1	76.9	75.4	69.8	53.5	43.2	40.4	54.8
1973	34.6	34.3	48.3	53.1	59.6	73.5	75.9	76.9	69.8	58.2	47.3	37.3	55.7
1974	37.9	33.8	45.2	55.3	61.9	68.5	76.5	75.0	67.5	55.3	48.2	40.3	55.4
1975	38.5	39.1	42.1	50.4	66.3	73.0	76.1	77.9	66.0	60.7	51.9	37.2	56.6
1976	30.8	44.1	48.1	56.9	62.1	74.8	75.0	73.9	67.5	52.9	40.9	32.6	55.0
1977	22.9	36.5	50.0	57.9	66.7	71.4	79.0	77.7	72.1	56.0	49.2	35.6	56.3
1978	29.2	27.3	41.7	54.2	62.4	73.1	75.9	78.1	69.7	56.1	48.7	40.2	54.7
1979	33.1	25.6	48.5	53.1	64.7	70.7	75.9	75.7	68.8	55.7	50.6	40.3	55.2
1980	33.8	31.5	41.5	55.7	65.5	71.3	78.2	78.7	72.2	55.3	44.2	35.5	55.3
1981	27.9	38.8	41.9	57.0	62.2	74.3	77.3	74.4	67.7	53.2	46.2	34.5	54.6
1982	25.5	35.8	42.9	50.7	66.1	69.4	77.1	73.0	67.3	56.3	48.4	42.0	54.6
1983	34.6	34.7	45.4	51.8	61.5	72.1	78.7	78.0	69.5	57.3	47.1	33.2	55.3
1984	28.5	41.7	38.2	51.5	61.3	73.4	73.9	75.0	64.8	62.2	43.9	44.1	54.9
1985	29.3	38.7	46.0	57.9	65.1	70.4	76.4	74.5	69.4	58.8	52.4	33.8	56.1
1986	33.2	32.9	45.0	53.5	66.7	74.4	79.4	73.1	68.9	58.9	44.8	38.2	55.8
1987	32.5	34.3	46.2	53.1	65.0	74.5	80.0	76.1	69.3	51.5	47.8	39.8	55.8
1988	28.7	35.9	45.1	52.0	64.0	73.0	80.3	78.5	66.8	51.3	48.1	36.3	55.0
1989	37.9	36.5	43.8	52.5	62.0	73.9	76.0	74.4	69.0	58.3	44.8	25.4	54.5
1990	42.0	42.3	47.6	54.8	62.3	73.3	78.4	74.6	67.3	60.7	49.6	42.2	57.9
1991	35.5	40.7	46.7	55.9	70.6	74.6	79.5	77.8	69.0	57.8	45.8	38.7	57.7
1992	34.6	37.1	41.3	52.0	60.8	70.1	77.4	72.3	67.7	54.3	47.2	38.9	54.5
1993	37.9	31.4	39.4	52.5	65.0	72.2	80.2	76.7	68.8	55.5	46.5	36.2	55.2
Record Mean	32.8	35.3	43.4	53.8	63.5	72.3	77.1	75.5	68.6	56.8	46.4	36.7	55.2
Max	41.2	44.3	53.2	64.7	74.3	83.0	87.3	85.3	78.7	67.5	56.1	45.2	65.1
Min	24.3	26.4	33.5	42.9	52.7	61.6	66.9	65.6	58.4	46.1	36.7	28.2	45.3

REFERENCE NOTES

BALTIMORE, MARYLAND

GENERAL
T - TRACE AMOUNT.
BLANK ENTRIES DENOTE MISSING/UNREPORTED DATA.
# INDICATES A STATION OR INSTRUMENT RELOCATION.

SPECIFIC

PM - INCLUDES LAST DAY OF PREVIOUS MONTH
ASOS - AUTOMATED SURFACE OBSERVING SYSTEM IN OPERATION DURING THESE MONTHS.

(a) - LENGTH OF RECORD IN YEARS, ALTHOUGH INDIVIDUAL MONTHS MAY BE MISSING.
0.* OR * - THE VALUE IS BETWEEN 0.0 AND 0.05
NORMALS - BASED ON THE 1961-1990 RECORD PERIOD.
EXTREMES - DATES ARE THE MOST RECENT OCCURRENCE
WIND DIR.- NUMERALS SHOW TENS OF DEGRESS CLOCKWISE FROM TRUE NORTH. "00" INDICATES CALM.
RESULTANT DIRECTIONS ARE GIVEN TO WHOLE DEGREES.
BOLD VALUES INDICATE EXTREME VALUES WHICH OCCURRED AFTER THE ASOS SYSTEM WAS COMMISSIONED.

RECORD = PERIOD OF RECORD
RECORD MEAN PRECIPITATION IS THE MEAN OF ALL DAILY PRECIPITATION AMOUNTS DURING THE PERIOD OF RECORD.
RECORD MAX(MIN) TEMPERATURE IS THE MEAN OF ALL DAILY MAX(MIN) TEMPERATURES DURING THE PERIOD OF RECORD.
RECORD MEAN TEMPERATURE IS THE SUM OF THE RECORD MAX AND RECORD MIN DIVIDED BY 2.
AVERAGE TEMPERATURE IS THE SUM OF THE MEAN DAILY MAX AND MIN TEMPERATURE DIVIDED BY 2.

EXCEPTIONS

RECORD MEANS ARE THROUGH THE CURRENT YEAR,
BEGINNING IN 1951 FOR TEMPERATURE
1951 FOR PRECIPITATION
1951 FOR SNOWFALL

HEATING DEGREE DAYS Base 65 deg. F

BALTIMORE, MARYLAND

SEASON	JULY	AUG	SEP	OCT	NOV	DEC	JAN	FEB	MAR	APR	MAY	JUNE	TOTAL
1964-65	0	0	41	358	448	797	1052	838	792	433	55	44	4858
1965-66	0	5	34	336	576	819	1085	961	665	460	157	23	5121
1966-67	0	0	68	353	555	905	846	955	715	338	254	6	4995
1967-68	0	0	75	318	684	851	1100	943	566	324	173	1	5035
1968-69	0	1	4	197	500	934	1028	835	748	273	69	0	4589
1969-70	0	0	26	251	561	916	1148	822	752	346	77	0	4899
1970-71	0	0	20	149	484	824	1080	766	712	364	134	2	4535
1971-72	0	0	24	96	571	652	841	884	663	396	94	42	4263
1972-73	2	0	16	357	649	759	935	854	511	365	191	1	4640
1973-74	0	0	24	221	524	852	830	868	613	309	148	14	4403
1974-75	0	0	49	303	509	759	818	720	702	436	66	2	4364
1975-76	0	0	50	156	397	853	1050	603	518	293	133	11	4064
1976-77	0	0	34	377	716	1001	1296	790	469	245	62	18	5008
1977-78	0	0	9	278	476	904	1101	1048	715	318	141	9	4999
1978-79	0	0	33	280	483	763	984	1100	520	354	75	6	4598
1979-80	2	3	22	311	425	757	962	967	723	273	74	6	4525
1980-81	0	0	20	311	620	908	1145	727	706	252	148	1	4838
1981-82	0	0	51	363	557	940	1218	808	677	422	58	20	5114
1982-83	0	5	42	289	495	707	936	842	602	410	152	6	4486
1983-84	0	0	70	257	530	979	1123	671	825	397	169	9	5030
1984-85	0	1	96	123	625	643	1101	731	589	252	79	10	4250
1985-86	0	0	41	201	378	962	980	892	613	342	86	6	4501
1986-87	0	23	34	236	598	822	1002	853	576	357	106	1	4608
1987-88	0	1	15	412	511	774	1120	838	613	389	96	27	4796
1988-89	2	0	39	424	504	882	834	792	663	374	145	0	4659
1989-90	0	0	51	229	600	1221	707	631	552	341	102	5	4439
1990-91	1	0	63	195	454	701	907	674	562	289	55	4	3905
1991-92	0	0	49	246	570	809	936	802	730	387	161	8	4698
1992-93	0	1	51	328	529	801	834	934	787	369	61	11	4706
1993-94	0	0	52	292	553	886							

COOLING DEGREE DAYS Base 65 deg. F

BALTIMORE, MARYLAND

YEAR	JAN	FEB	MAR	APR	MAY	JUNE	JULY	AUG	SEP	OCT	NOV	DEC	TOTAL
1969	0	0	0	16	93	304	392	364	169	33	0	0	1371
1970	0	0	0	4	134	246	389	390	291	45	0	0	1499
1971	0	0	0	0	24	278	363	293	208	36	20	0	1222
1972	0	0	5	1	29	140	379	331	166	7	0	0	1058
1973	0	0	0	15	29	263	344	376	173	19	0	0	1219
1974	0	0	4	24	57	126	361	317	130	8	11	0	1038
1975	0	0	0	4	112	252	351	404	85	27	10	0	1245
1976	0	1	0	58	51	315	317	284	114	9	0	0	1149
1977	0	0	10	37	124	217	439	401	229	7	10	0	1474
1978	0	0	0	0	63	260	344	413	182	12	0	0	1274
1979	0	0	15	4	72	183	348	341	145	28	1	0	1137
1980	0	0	0	0	97	203	415	431	245	17	0	0	1408
1981	0	0	0	19	69	287	389	296	141	5	0	0	1206
1982	0	0	0	4	99	160	381	259	119	26	4	1	1053
1983	0	0	0	18	51	228	430	410	214	24	0	0	1375
1984	0	0	0	0	59	268	281	316	98	41	0	2	1065
1985	0	2	7	43	89	179	363	298	178	17	5	0	1181
1986	0	0	0	1	143	295	452	281	158	54	0	0	1384
1987	0	0	0	7	115	292	473	352	152	0	0	0	1391
1988	0	0	2	4	71	274	485	427	100	8	0	0	1371
1989	0	0	14	5	58	276	351	298	178	25	1	0	1206
1990	0	0	19	38	26	261	422	303	137	68	0	0	1274
1991	0	0	2	24	233	303	462	402	177	29	2	0	1634
1992	0	0	0	6	39	168	392	232	139	4	0	0	980
1993	0	0	0	0	70	235	476	371	175	3	5	0	1335

SNOWFALL (inches)

BALTIMORE, MARYLAND

SEASON	JULY	AUG	SEP	OCT	NOV	DEC	JAN	FEB	MAR	APR	MAY	JUNE	TOTAL
1964-65	0.0	0.0	0.0	0.0	T	0.8	8.3	1.1	8.4	0.0	0.0	0.0	18.6
1965-66	0.0	0.0	0.0	0.0	0.0	T	21.4	11.4	0.0	T	0.0	0.0	32.8
1966-67	0.0	0.0	0.0	0.0	T	20.4	0.4	20.1	2.5	0.0	0.0	0.0	43.4
1967-68	0.0	0.0	0.0	0.0	8.4	4.6	2.5	2.6	5.3	0.0	0.0	0.0	23.4
1968-69	0.0	0.0	0.0	0.0	4.3	T	0.1	6.4	7.8	0.0	0.0	0.0	18.6
1969-70	0.0	0.0	0.0	0.0	T	9.0	6.1	4.0	1.9	T	0.0	0.0	21.0
1970-71	0.0	0.0	0.0	0.0	0.0	6.3	4.1	0.6	2.0	T	0.0	0.0	13.0
1971-72	0.0	0.0	0.0	0.0	1.0	T	1.1	11.4	0.2	0.3	0.0	0.0	14.0
1972-73	0.0	0.0	0.0	T	T	T	T	1.2	T	T	0.0	0.0	1.2
1973-74	0.0	0.0	0.0	0.0	0.0	8.3	1.2	7.6	T	T	0.0	0.0	17.1
1974-75	0.0	0.0	0.0	0.0	T	0.4	5.1	5.5	1.2	T	0.0	0.0	12.2
1975-76	0.0	0.0	0.0	0.0	0.0	0.7	1.7	1.3	7.8	0.0	0.0	0.0	11.5
1976-77	0.0	0.0	0.0	0.0	1.1	1.5	8.5	T	T	T	0.0	0.0	11.1
1977-78	0.0	0.0	0.0	T	0.6	0.5	12.4	12.3	8.5	T	0.0	0.0	34.3
1978-79	0.0	0.0	0.0	0.0	3.7	0.0	5.7	33.1	T	T	0.0	0.0	42.5
1979-80	0.0	0.0	0.0	0.3	T	0.1	4.7	3.8	5.7	0.0	0.0	0.0	14.6
1980-81	0.0	0.0	0.0	0.0	T	0.2	4.1	T	0.3	0.0	0.0	0.0	4.6
1981-82	0.0	0.0	0.0	0.0	T	2.4	14.8	7.6	0.7	T	0.0	0.0	25.5
1982-83	0.0	0.0	0.0	0.0	0.0	7.2	1.2	27.2	T	T	0.0	0.0	35.6
1983-84	0.0	0.0	0.0	0.0	T	T	8.4	T	6.1	T	0.0	0.0	14.5
1984-85	0.0	0.0	0.0	0.0	T	0.1	9.1	0.4	T	0.7	0.0	0.0	10.3
1985-86	0.0	0.0	0.0	0.0	0.0	0.7	1.9	13.0	T	T	0.0	0.0	15.6
1986-87	0.0	0.0	0.0	0.0	0.0	T	25.1	10.1	T	T	0.0	0.0	35.2
1987-88	0.0	0.0	0.0	0.0	6.0	0.5	13.7	0.2	T	T	0.0	0.0	20.4
1988-89	0.0	0.0	0.0	0.0	0.0	0.9	6.0	1.1	0.3	0.0	0.0	0.0	8.3
1989-90	0.0	0.0	0.0	0.0	3.8	10.2	0.5	T	2.7	0.1	0.0	0.0	17.3
1990-91	0.0	0.0	0.0	0.0	0.0	4.8	4.2	0.1	0.3	0.0	0.0	0.0	9.4
1991-92	T	0.0	0.0	0.0	T	T	2.2	1.9	T	T	0.0	0.0	4.1
1992-93	T	0.0	0.0	0.0	T	1.5	1.4	8.8	12.7	T	0.0	0.0	24.4
1993-94	0.0	0.0	0.0	0.0	T	2.9							
Record Mean	T	0.0	0.0	T	1.0	3.5	5.9	6.5	3.8	0.1	T	0.0	20.6

BOSTON, MASSACHUSETTS

Climate is the composite of numerous weather elements. Three important influences are responsible for the main features of the Boston climate. First, the latitude places the city in the zone of prevailing west to east atmospheric flow. Both polar and tropical air masses influence the region. Secondly, Boston is situated on or near several tracks frequently followed by low pressure storm systems. The weather fluctuates regularly from fair to cloudy to stormy conditions and assures an adequate amount of precipitation. The third factor is the east-coast location of Boston. The ocean has a moderating influence on temperature extremes of winter and summer.

Hot summer afternoons are frequently relieved by the locally celebrated sea breeze, as air flows inland from the cool water surface to displace the warm air over the land. This refreshing east wind is more commonly experienced along the shore than in the interior of the city or the western suburbs. In winter, under appropriate conditions, the severity of cold waves is reduced by the nearness of the relatively warm ocean. The average last occurrence of freezing temperature in spring is early April and the first occurrence of freezing temperature in autumn is early November. In suburban areas, especially away from the coast, these dates are later in spring and earlier in autumn by up to one month in the more susceptible localities.

Boston has no dry season. Most growing seasons have several shorter dry spells during which irrigation for high-value crops may be useful. Much of the rainfall from June to September comes from showers and thunderstorms. During the rest of the year, low pressure systems pass more or less regularly and produce precipitation on an average of roughly one day in three. Coastal storms, or northeasters, are prolific producers of rain and snow. The main snow season extends from December through March. Periods when the ground is bare or nearly bare of snow may occur at any time in the winter.

Relative humidity has been known to fall as low as 5 percent but such desert dryness is very rare. Heavy fog occurs on an average of about two days per month with its prevalence increasing eastward from the interior of Boston Bay to the open waters beyond.

Although winds of 30 mph or higher may be expected on at least one day in every month of the year, gales are both more common and more severe in winter.

NORMALS, MEANS, AND EXTREMES

BOSTON, MASSACHUSETTS

LATITUDE: 42°22'N LONGITUDE: 71°02'W ELEVATION: FT. GRND 15 BARO 30 TIME ZONE: EASTERN WBAN: 14739

	(a)	JAN	FEB	MAR	APR	MAY	JUNE	JULY	AUG	SEP	OCT	NOV	DEC	YEAR
TEMPERATURE °F:														
Normals														
-Daily Maximum		35.7	37.5	45.8	55.9	66.6	76.3	81.8	79.8	72.8	62.7	52.2	40.4	59.0
-Daily Minimum		21.6	23.0	31.3	40.2	49.8	59.1	65.1	64.0	56.8	46.9	38.3	26.7	43.6
-Monthly		28.6	30.3	38.6	48.1	58.2	67.7	73.5	71.9	64.8	54.8	45.3	33.6	51.3
Extremes														
-Record Highest	42	63	70	81	94	95	100	102	102	100	90	78	73	102
-Year		1990	1985	1989	1976	1979	1952	1977	1975	1953	1963	1993	1984	JUL 1977
-Record Lowest	42	-12	-4	6	16	34	45	50	47	38	28	15	-7	-12
-Year		1957	1961	1984	1982	1956	1986	1988	1986	1965	1976	1989	1980	JAN 1957
NORMAL DEGREE DAYS:														
Heating (base 65°F)		1128	972	818	507	221	32	0	6	72	321	591	973	5641
Cooling (base 65°F)		0	0	0	0	10	113	264	220	66	5	0	0	678
% OF POSSIBLE SUNSHINE	58	53	56	57	56	59	63	66	65	63	60	50	52	58
MEAN SKY COVER (tenths)														
Sunrise - Sunset	58	6.2	6.2	6.4	6.6	6.6	6.3	6.2	5.7	5.6	5.6	6.4	6.3	6.2
MEAN NUMBER OF DAYS:														
Sunrise to Sunset														
-Clear	58	9.0	8.2	7.8	7.0	6.3	6.6	6.7	9.1	10.3	10.8	7.9	8.6	98.4
-Partly Cloudy	58	6.8	6.7	8.1	8.1	9.9	10.5	12.2	10.8	8.1	7.8	7.3	7.5	103.7
-Cloudy	58	15.1	13.3	15.1	14.9	14.8	12.9	12.1	11.1	11.7	12.4	14.8	14.9	163.2
Precipitation														
.01 inches or more	42	11.4	10.5	11.9	11.3	11.5	10.6	9.2	9.9	8.7	9.1	10.9	11.6	126.5
Snow,Ice pellets,hail														
1.0 inches or more	58	3.1	2.6	2.1	0.3	0.0	0.0	0.0	0.0	0.0	0.0	0.4	2.1	10.7
Thunderstorms	58	0.2	0.2	0.6	1.0	2.2	3.7	4.3	3.6	1.6	0.7	0.4	0.2	18.6
Heavy Fog Visibility														
1/4 mile or less	58	1.8	1.6	2.0	1.7	2.8	1.9	2.2	1.8	1.9	2.2	1.8	1.3	23.2
Temperature °F														
-Maximum														
90° and above	29	0.0	0.0	0.0	0.1	0.5	2.6	5.4	3.4	0.8	0.0	0.0	0.0	12.9
32° and below	29	11.1	7.7	2.1	0.*	0.0	0.0	0.0	0.0	0.0	0.0	0.3	5.7	27.0
-Minimum														
32° and below	29	26.0	23.3	16.6	2.4	0.0	0.0	0.0	0.0	0.0	0.6	7.0	21.6	97.4
0° and below	29	0.5	0.3	0.0	0.0	0.0	0.0	0.0	0.0	0.0	0.0	0.0	0.2	1.0
AVG. STATION PRESS.(mb)	21	1015.0	1015.4	1014.6	1013.2	1014.1	1013.3	1013.8	1015.7	1017.1	1017.3	1016.3	1016.1	1015.2
RELATIVE HUMIDITY (%)														
Hour 01	29	65	65	67	69	74	77	77	79	80	76	71	68	72
Hour 07 (Local Time)	29	67	67	69	69	72	73	73	77	79	77	74	70	72
Hour 13	29	57	56	57	55	59	58	57	59	60	58	59	59	58
Hour 19	29	61	59	62	62	65	66	66	69	71	68	66	63	65
PRECIPITATION (inches):														
Water Equivalent														
-Normal		3.59	3.62	3.69	3.60	3.25	3.09	2.84	3.24	3.06	3.30	4.22	4.01	41.51
-Maximum Monthly	42	10.55	7.81	11.00	9.46	13.38	13.20	8.12	17.09	8.31	8.68	8.89	9.74	17.09
-Year		1979	1984	1953	1987	1954	1982	1959	1955	1954	1962	1983	1969	AUG 1955
-Minimum Monthly	42	0.61	0.72	0.62	1.24	0.53	0.48	0.52	0.83	0.35	0.96	0.64	0.81	0.35
-Year		1989	1987	1981	1966	1964	1953	1952	1972	1957	1967	1976	1989	SEP 1957
-Maximum in 24 hrs	42	2.72	2.68	4.13	3.32	5.74	4.17	2.43	8.40	5.64	4.26	3.76	5.14	8.40
-Year		1979	1969	1968	1991	1954	1984	1988	1955	1954	1962	1992	1992	AUG 1955
Snow,Ice pellets,hail														
-Maximum Monthly	58	35.9	41.3	38.9	13.3	0.5	0.0	0.0	0.0	0.0	0.2	10.0	27.9	41.3
-Year		1978	1969	1993	1982	1977					1979	1938	1970	FEB 1969
-Maximum in 24 hrs	58	21.0	23.6	17.7	13.2	0.5	0.0	0.0	0.0	0.0	0.2	8.0	13.0	23.6
-Year		1978	1978	1960	1982	1977					1979	1987	1960	FEB 1978
WIND:														
Mean Speed (mph)	36	13.8	13.8	13.7	13.2	12.2	11.5	11.0	10.8	11.3	12.0	12.9	13.6	12.5
Prevailing Direction through 1963		NW	WNW	NW	WNW	SW	SW	SW	SW	SW	SW	SW	WNW	SW
Fastest Obs. 1 Min.														
-Direction (!!)	8	18	30	06	05	09	28	35	08	21	09	31	05	06
-Speed (MPH)	8	40	37	54	38	33	45	37	47	35	41	38	51	54
-Year		1992	1991	1993	1991	1992	1988	1989	1991	1989	1988	1989	1992	MAR 1993
Peak Gust														
-Direction (!!)	10	S	NW	NE	NE	S	W	NW	NW	S	NE	NW	NE	NE
-Speed (mph)	10	59	58	81	55	52	68	54	64	76	55	55	78	81
-Date		1992	1991	1993	1991	1989	1988	1989	1991	1985	1991	1989	1992	MAR 1993

PRECIPITATION (inches)

BOSTON, MASSACHUSETTS

YEAR	JAN	FEB	MAR	APR	MAY	JUNE	JULY	AUG	SEP	OCT	NOV	DEC	ANNUAL
1964	4.56	4.67	3.48	3.69	0.53	1.91	3.12	1.78	2.65	2.82	2.18	5.08	36.47
1965	2.64	3.17	2.22	2.32	0.93	2.99	0.55	1.48	2.01	1.59	2.08	1.73	23.71
1966	5.29	3.48	1.98	1.24	2.66	3.40	3.21	1.25	3.42	2.62	4.43	3.03	36.01
1967	2.28	4.05	4.67	4.83	7.32	3.48	2.47	5.74	2.00	0.96	3.38	6.42	47.60
1968	3.85	1.15	7.86	1.72	3.26	5.65	0.55	1.63	1.79	1.85	6.74	6.23	42.28
1969	2.26	7.08	2.63	4.37	1.96	0.63	2.98	1.89	4.42	1.64	8.18	9.74	47.78
1970	0.89	4.65	4.32	2.79	3.01	4.62	1.27	4.12	2.60	2.63	4.09	6.92	41.91
1971	1.88	5.05	3.08	2.92	3.72	1.74	2.84	1.59	1.55	2.16	6.74	2.40	35.67
1972	2.05	5.29	5.37	3.34	5.26	6.76	2.19	0.83	5.94	2.98	7.02	6.08	53.11
1973	3.12	2.13	2.20	5.65	3.76	4.68	4.83	2.78	1.95	2.71	1.74	7.20	42.75
1974	3.22	3.24	4.01	3.86	2.87	2.29	1.54	3.41	7.03	3.12	1.73	3.92	40.24
1975	5.70	3.37	2.74	2.40	1.78	2.10	2.35	5.52	5.49	4.41	5.13	4.80	45.79
1976	5.29	2.45	2.42	2.00	1.98	0.58	4.30	7.99	1.56	4.16	0.64	3.35	36.72
1977	4.41	2.40	4.76	4.07	3.52	2.49	2.21	2.91	4.03	4.63	2.54	6.20	44.17
1978	8.12	2.87	2.46	1.79	4.50	1.53	1.48	4.62	1.30	3.13	2.21	3.63	37.64
1979	10.55	3.46	3.03	3.19	4.24	0.86	2.36	5.02	3.61	3.14	3.29	1.42	44.17
1980	0.74	0.88	5.37	4.36	2.30	3.05	2.20	1.55	0.82	4.14	3.01	0.97	29.39
1981	0.95	6.65	0.62	3.14	1.17	1.65	3.47	1.04	2.54	3.43	4.78	6.27	35.71
1982	4.69	2.66	2.17	3.42	2.58	13.20	4.22	2.22	1.57	3.19	3.42	1.27	44.61
1983	5.03	5.00	9.72	6.86	2.94	1.07	1.07	3.28	1.06	3.74	8.89	4.94	53.60
1984	2.31	7.81	6.82	4.43	8.77	3.06	4.43	1.60	1.22	5.18	1.68	2.93	50.24
1985	1.12	1.83	2.29	1.62	3.36	3.94	3.51	6.67	3.00	1.65	6.39	1.21	36.59
1986	3.42	2.83	3.42	1.59	1.31	7.74	3.96	3.32	1.08	3.27	6.01	6.38	44.33
1987	7.28	0.72	4.27	9.46	1.75	2.62	0.82	2.93	7.29	2.73	3.49	2.12	45.48
1988	2.50	3.93	3.52	1.47	2.86	1.29	7.62	1.11	1.29	1.60	6.57	1.02	34.78
1989	0.61	2.51	3.07	3.58	3.54	2.84	5.09	5.92	4.61	5.71	4.13	0.81	42.42
1990	3.78	3.60	1.71	5.94	6.53	0.69	4.08	6.57	1.67	7.36	1.39	3.18	46.50
1991	3.24	1.58	4.33	4.84	0.92	2.89	1.95	5.27	6.32	4.27	4.06	2.58	42.25
1992	3.11	2.28	3.59	2.34	1.40	4.61	2.66	4.25	3.46	1.62	6.14	8.26	43.72
1993	2.17	4.94	7.67	4.86	1.04	1.75	1.75	1.32	4.64	3.61	2.86	6.60	43.21
Record Mean	3.62	3.38	3.86	3.61	3.22	3.15	3.15	3.60	3.19	3.29	3.91	3.65	41.62

AVERAGE TEMPERATURE (deg. F)

BOSTON, MASSACHUSETTS

YEAR	JAN	FEB	MAR	APR	MAY	JUNE	JULY	AUG	SEP	OCT	NOV	DEC	ANNUAL
#1964	31.7	29.1	38.7	46.1	60.3	67.1	71.5	66.4	62.0	52.5	44.1	32.4	50.2
1965	25.4	28.0	35.8	44.2	59.5	67.4	71.0	70.5	62.5	52.8	42.1	36.1	49.6
1966	28.8	31.3	39.8	45.9	57.3	69.4	74.9	71.3	63.5	54.5	46.9	34.2	51.5
1967	35.1	26.4	33.2	44.9	51.7	67.2	73.0	70.9	62.7	53.8	40.1	35.0	49.5
1968	25.6	26.1	39.1	49.6	56.1	64.9	75.2	70.7	65.0	57.9	43.8	30.9	50.4
1969	29.3	29.5	35.4	50.6	58.5	69.3	71.0	74.3	63.7	54.3	44.9	33.4	51.2
1970	23.0	32.3	37.4	49.0	59.6	67.0	74.3	73.6	65.6	54.9	44.8	28.9	50.9
1971	23.8	30.5	36.7	45.1	55.7	69.1	73.4	73.4	68.0	59.8	43.1	36.3	51.2
1972	33.0	29.6	36.3	44.9	57.6	65.4	73.8	71.5	65.7	51.8	42.3	33.0	50.4
1973	31.4	30.1	43.3	49.9	57.0	70.0	74.3	74.8	64.4	55.6	45.8	39.6	53.0
1974	31.7	29.1	38.7	50.9	54.7	64.8	72.4	72.0	63.7	50.1	45.3	37.8	50.9
1975	34.9	32.1	36.9	45.1	61.5	67.5	75.9	72.9	63.9	57.3	51.8	34.4	52.8
1976	26.1	37.3	41.2	55.1	60.2	73.4	72.9	72.0	64.9	52.3	41.9	29.0	52.2
1977	23.3	30.7	44.7	51.3	62.6	67.4	74.9	73.4	64.4	55.3	48.1	34.2	52.5
1978	28.5	27.1	36.2	48.8	59.3	68.3	72.1	71.6	61.4	52.5	43.6	35.3	50.4
1979	32.5	23.1	42.5	48.7	61.1	68.2	74.5	71.7	64.9	52.7	48.6	36.7	52.1
1980	29.4	27.9	36.9	48.7	59.4	66.3	75.8	74.2	67.0	52.4	41.2	28.6	50.6
1981	21.4	36.4	39.1	51.7	60.4	70.7	74.6	72.1	63.7	51.2	43.9	33.2	51.5
1982	22.9	30.8	38.7	48.2	57.8	63.3	74.9	70.3	64.1	54.2	47.6	39.6	51.0
1983	31.2	32.8	40.6	49.1	58.2	70.7	78.0	73.6	70.6	55.2	46.1	32.1	53.2
1984	26.7	37.6	31.9	46.1	58.0	70.5	74.7	74.6	62.1	53.3	44.6	39.5	51.6
1985	24.4	32.8	40.4	49.3	59.3	64.8	73.5	70.4	65.4	55.4	45.4	31.3	51.0
1986	31.4	28.9	40.7	48.4	58.4	66.1	71.0	70.5	63.2	54.0	42.3	35.5	50.9
1987	28.9	29.1	38.5	45.1	57.2	65.1	71.7	70.3	65.4	54.3	43.9	36.1	50.5
1988	27.8	32.2	39.2	46.8	57.6	68.5	73.7	75.5	64.6	50.8	46.7	32.8	51.4
1989	34.5	30.5	37.3	45.9	59.4	67.8	72.8	71.6	64.7	55.3	42.8	21.7	50.4
1990	36.4	34.1	40.1	47.6	54.9	66.6	73.1	73.3	64.6	58.3	48.5	40.7	53.2
1991	29.4	36.1	41.6	51.3	63.3	70.0	74.6	73.8	63.7	56.4	45.2	36.0	53.5
1992	31.0	32.4	35.4	46.4	55.6	67.8	69.5	70.4	63.9	52.5	42.9	34.8	50.2
1993	32.4	27.1	36.4	48.3	60.3	69.5	74.7	73.6	64.8	52.3	45.6	34.2	51.6
Record Mean	28.8	29.4	37.1	47.2	57.9	67.2	72.7	71.0	64.1	54.0	43.7	32.8	50.5
Max	36.3	37.1	44.6	55.2	66.5	75.9	81.1	78.9	72.1	62.0	50.8	39.8	58.4
Min	21.3	21.8	29.6	39.1	49.3	58.4	64.3	63.0	56.1	46.1	36.5	25.7	42.6

REFERENCE NOTES

BOSTON, MASSACHUSETTS

GENERAL
T - TRACE AMOUNT.
BLANK ENTRIES DENOTE MISSING/UNREPORTED DATA.
# INDICATES A STATION OR INSTRUMENT RELOCATION.

SPECIFIC

PM - INCLUDES LAST DAY OF PREVIOUS MONTH
ASOS - AUTOMATED SURFACE OBSERVING SYSTEM IN OPERATION DURING THESE MONTHS.

(a) - LENGTH OF RECORD IN YEARS, ALTHOUGH INDIVIDUAL MONTHS MAY BE MISSING.
0.* OR * - THE VALUE IS BETWEEN 0.0 AND 0.05
NORMALS - BASED ON THE 1961-1990 RECORD PERIOD.
EXTREMES - DATES ARE THE MOST RECENT OCCURRENCE
WIND DIR.- NUMERALS SHOW TENS OF DEGRESS CLOCKWISE FROM TRUE NORTH. "00" INDICATES CALM.
RESULTANT DIRECTIONS ARE GIVEN TO WHOLE DEGREES.
BOLD VALUES INDICATE EXTREME VALUES WHICH OCCURRED AFTER THE ASOS SYSTEM WAS COMMISSIONED.

RECORD = PERIOD OF RECORD
RECORD MEAN PRECIPITATION IS THE MEAN OF ALL DAILY PRECIPITATION AMOUNTS DURING THE PERIOD OF RECORD.
RECORD MAX(MIN) TEMPERATURE IS THE MEAN OF ALL DAILY MAX(MIN) TEMPERATURES DURING THE PERIOD OF RECORD.
RECORD MEAN TEMPERATURE IS THE SUM OF THE RECORD MAX AND RECORD MIN DIVIDED BY 2.
AVERAGE TEMPERATURE IS THE SUM OF THE MEAN DAILY MAX AND MIN TEMPERATURE DIVIDED BY 2.

EXCEPTIONS

RECORD MEANS ARE THROUGH THE CURRENT YEAR, BEGINNING IN 1872 FOR TEMPERATURE
1871 FOR PRECIPITATION
1936 FOR SNOWFALL

HEATING DEGREE DAYS Base 65 deg. F — BOSTON, MASSACHUSETTS

SEASON	JULY	AUG	SEP	OCT	NOV	DEC	JAN	FEB	MAR	APR	MAY	JUNE	TOTAL
1964-65	14	26	140	380	620	1004	1220	1032	900	617	195	80	6228
1965-66	2	37	136	371	680	888	1115	936	776	566	258	46	5811
1966-67	0	1	88	322	535	950	921	1075	977	596	403	58	5926
1967-68	0	4	110	347	739	923	1214	1122	797	454	270	76	6056
1968-69	1	9	46	247	630	1050	1099	987	911	430	208	21	5639
1969-70	2	3	107	326	595	973	1295	909	846	473	184	52	5765
1970-71	0	0	68	314	598	1113	1269	962	868	586	287	25	6090
1971-72	0	2	37	169	651	882	985	1021	883	598	250	54	5532
1972-73	3	4	51	405	673	985	1033	971	666	450	258	24	5523
1973-74	0	2	94	289	570	782	1023	1000	809	429	335	77	5410
1974-75	0	2	102	458	587	836	925	918	866	590	162	59	5505
1975-76	0	8	70	239	395	941	1198	800	733	331	166	16	4897
1976-77	1	10	55	393	688	1108	1290	956	623	414	158	43	5739
1977-78	0	4	85	304	498	948	1127	1057	885	480	209	18	5615
1978-79	11	11	150	381	635	916	1002	1169	691	481	149	19	5615
1979-80	2	15	80	390	484	873	1096	1071	866	481	185	66	5609
1980-81	2	5	72	387	706	1120	1344	794	796	393	200	7	5826
1981-82	2	6	91	419	628	979	1300	948	811	496	231	113	6024
1982-83	2	19	71	338	515	783	1040	896	749	478	223	22	5136
1983-84	0	8	42	327	561	1012	1182	790	1020	563	239	36	5780
1984-85	3	0	142	359	605	781	1255	897	758	471	204	71	5546
1985-86	3	11	65	298	580	1035	1035	1008	746	490	258	66	5595
1986-87	21	16	98	344	674	904	1112	997	814	588	285	76	5929
1987-88	8	18	57	326	626	888	1145	945	792	541	253	61	5660
1988-89	9	10	64	443	541	992	938	959	853	565	196	51	5621
1989-90	2	4	88	294	660	1336	880	857	762	524	307	60	5774
1990-91	4	5	84	236	496	744	1096	803	721	407	126	35	4757
1991-92	1	8	111	273	586	894	1049	937	913	552	317	37	5678
1992-93	21	14	109	386	656	930	1002	1056	880	493	167	31	5745
1993-94	3	1	89	387	579	946							

COOLING DEGREE DAYS Base 65 deg. F — BOSTON, MASSACHUSETTS

YEAR	JAN	FEB	MAR	APR	MAY	JUNE	JULY	AUG	SEP	OCT	NOV	DEC	TOTAL
1969	0	0	0	9	13	156	196	297	74	1	0	0	746
1970	0	0	0	0	25	118	294	273	91	9	0	0	810
1971	0	0	0	0	6	155	269	271	132	15	1	0	849
1972	0	0	0	0	26	74	279	213	79	0	0	0	671
1973	0	0	0	7	18	180	296	316	84	3	0	0	904
1974	0	0	0	10	22	81	235	226	68	1	3	0	646
1975	0	0	0	0	60	139	345	261	44	9	4	0	862
1976	0	0	0	43	25	276	251	231	61	8	0	0	895
1977	0	0	1	13	92	124	314	272	75	6	0	0	897
1978	0	0	0	0	40	122	237	221	48	0	0	0	668
1979	0	0	0	0	35	122	304	226	85	17	0	0	789
1980	0	0	0	0	18	114	347	299	137	1	0	0	916
1981	0	0	0	0	67	185	306	232	60	0	0	0	850
1982	0	0	0	0	15	67	314	192	49	10	2	0	649
1983	0	0	0	7	18	200	410	283	217	27	0	0	1162
1984	0	0	0	3	31	207	312	306	62	3	0	0	924
1985	0	0	0	5	30	72	271	183	83	8	0	0	652
1986	0	0	0	0	60	105	211	190	55	10	0	0	631
1987	0	0	0	0	48	87	221	189	76	0	2	0	623
1988	0	0	0	0	31	173	287	342	59	11	0	0	903
1989	0	0	1	0	29	142	248	214	89	0	0	0	723
1990	0	0	0	10	2	116	261	268	77	34	8	0	776
1991	0	0	0	3	79	189	304	287	79	15	0	0	956
1992	0	0	0	0	30	126	165	189	83	5	0	0	598
1993	0	0	0	0	28	173	310	273	89	0	2	0	875

SNOWFALL (inches) — BOSTON, MASSACHUSETTS

SEASON	JULY	AUG	SEP	OCT	NOV	DEC	JAN	FEB	MAR	APR	MAY	JUNE	TOTAL
1964-65	0.0	0.0	0.0	T	T	12.2	22.2	4.7	9.7	1.6	0.0	0.0	50.4
1965-66	0.0	0.0	0.0	0.0	T	2.3	26.4	12.1	3.3	T	T	0.0	44.1
1966-67	0.0	0.0	0.0	0.0	0.0	9.9	0.5	23.5	22.9	3.3	T	0.0	60.1
1967-68	0.0	0.0	0.0	0.0	2.2	14.7	17.7	3.4	6.8	0.0	0.0	0.0	44.8
1968-69	0.0	0.0	0.0	0.0	0.4	5.1	0.9	41.3	6.1	T	0.0	0.0	53.8
1969-70	0.0	0.0	0.0	T	T	12.6	7.4	10.5	18.2	0.1	0.0	0.0	48.8
1970-71	0.0	0.0	0.0	T	T	27.9	12.0	8.1	7.4	1.9	0.0	0.0	57.3
1971-72	0.0	0.0	0.0	0.0	2.8	7.9	7.8	16.5	12.1	0.4	0.0	0.0	47.5
1972-73	0.0	0.0	0.0	T	0.6	3.3	3.6	2.5	0.3	T	0.0	0.0	10.3
1973-74	0.0	0.0	0.0	0.0	0.0	T	16.0	17.8	0.1	3.0	0.0	0.0	36.9
1974-75	0.0	0.0	0.0	0.0	2.0	3.6	2.2	17.0	1.8	1.0	0.0	0.0	27.6
1975-76	0.0	0.0	0.0	T	0.1	19.3	15.0	1.4	10.8	T	0.0	0.0	46.6
1976-77	0.0	0.0	0.0	0.0	1.0	17.2	23.2	5.9	10.7	T	0.5	0.0	58.5
1977-78	0.0	0.0	0.0	0.0	0.7	5.2	35.9	27.2	16.1	T	0.0	0.0	85.1
1978-79	0.0	0.0	0.0	0.0	4.2	5.8	10.5	6.6	T	0.4	0.0	0.0	27.5
1979-80	0.0	0.0	0.0	0.2	T	2.0	0.4	6.5	3.6	T	0.0	0.0	12.7
1980-81	0.0	0.0	0.0	0.0	2.4	5.6	11.9	1.9	0.5	0.0	0.0	0.0	22.3
1981-82	0.0	0.0	0.0	0.0	T	17.6	18.0	7.6	5.3	13.3	0.0	0.0	61.8
1982-83	0.0	0.0	0.0	0.0	T	5.5	4.7	22.3	0.2	T	0.0	0.0	32.7
1983-84	0.0	0.0	0.0	0.0	T	2.6	21.1	0.3	19.0	T	0.0	0.0	43.0
1984-85	0.0	0.0	0.0	0.0	T	3.7	7.0	10.2	3.7	2.0	0.0	0.0	26.6
1985-86	0.0	0.0	0.0	0.0	3.0	1.3	0.8	10.4	2.6	T	0.0	0.0	18.1
1986-87	0.0	0.0	0.0	0.0	3.5	3.4	24.3	3.7	3.5	4.1	0.0	0.0	42.5
1987-88	0.0	0.0	0.0	0.0	9.0	7.5	17.0	14.1	5.0	T	0.0	0.0	52.6
1988-89	0.0	0.0	0.0	T	0.0	3.7	1.5	6.7	3.2	0.4	0.0	0.0	15.5
1989-90	0.0	0.0	0.0	0.0	4.5	6.2	7.0	16.9	4.1	0.5	0.0	0.0	39.2
1990-91	0.0	0.0	0.0	0.0	T	1.2	11.7	2.8	3.4	0.0	0.0	0.0	19.1
1991-92	0.0	0.0	0.0	0.0	T	5.8	0.4	4.0	10.8	1.0	0.0	0.0	22.0
1992-93	0.0	0.0	0.0	0.0	0.6	9.7	12.9	19.6	38.9	2.2	0.0	0.0	83.9
1993-94	0.0	0.0	0.0	0.0	T	11.6							
Record Mean	0.0	0.0	0.0	T	1.3	7.5	12.0	11.3	7.9	0.9	T	0.0	41.0

BLUE HILL OBSERVATORY, MILTON, MASSACHUSETTS

The altitude of the Observatory and its proximity to Massachusetts Bay play major roles in determining the climate of Blue Hill. The elevation of 635 feet marks the summit as the highest point of a wooded range that extends east-northeast to west-southwest. The station lies at the southwest end of this range and has a three-quadrant unrestricted exposure, at approximately 400 feet above the surrounding terrain. The orographic effect created by this difference in elevation is responsible for lower temperatures, more precipitation, higher winds, more frequent occurrences of fog, and longer periods of snow cover than at nearby lower elevations. Eight miles to the northeast lies the nearest approach of Boston Harbor, and thus, the station is within range of the sea breeze.

Summer temperatures are generally comfortable. Winters at the summit are more severe than those experienced at surrounding areas. Average occurrence of last freezing temperature in the spring is late April and the first in the fall is late October. The freeze-free period is about 178 days. Records indicate that the freeze-free period at base stations is from 6 to 7 weeks shorter than at the summit. This seeming paradox is due to temperature inversions, in which the colder air is found at lower elevations. This condition develops on clear, calm nights, and is responsible for the shorter freeze-free period in base areas.

Total precipitation is fairly evenly distributed throughout the year. Precipitation occurrences are most frequent January through March and least frequent August through October. Hourly precipitation occurrences indicate a coastal type distribution for the year as a whole, with maxima in the early morning and minima in the early afternoon. In the summer, however, convective action or the continental influence dominates, causing a late afternoon maximum. Coastal storms or northeasters are prolific producers of rain and snow.

The main snow season extends from November through early April. Nearly 14 percent of the annual total precipitation occurs as snow or sleet.

Wind velocities are higher in winter than in summer. Speeds average greatest from January through March and least in August. Surface contour is a factor in the stations wind force, particularly from the southerly and westerly directions. These are the steepest slopes of the hill, and winds velocities increase. Peak winds have been recorded from these directions. Winds from the east-northeast and northeast are somewhat slowed by striking the lower range first.

NORMALS, MEANS, AND EXTREMES

BLUE HILL OBSERVATORY, MILTON, MASSACHUSETTS

LATITUDE: 42°13'N LONGITUDE: 71°07'W ELEVATION: FT. GRND 629 BARO 640 TIME ZONE: EASTERN WBAN: 14753

	(a)	JAN	FEB	MAR	APR	MAY	JUNE	JULY	AUG	SEP	OCT	NOV	DEC	YEAR
TEMPERATURE °F:														
Normals														
-Daily Maximum		33.1	35.3	44.1	55.2	66.6	75.1	80.6	78.5	70.7	60.6	49.2	37.4	57.2
-Daily Minimum		17.6	19.2	27.5	36.6	46.5	55.5	61.7	60.5	53.0	43.2	34.3	22.7	39.9
-Monthly		25.3	27.3	35.8	45.9	56.6	65.3	71.2	69.6	61.9	51.9	41.7	30.1	48.6
Extremes														
-Record Highest	108	68	68	85	94	93	99	100	101	99	88	81	72	101
-Year		1950	1985	1945	1976	1930	1919	1977	1975	1953	1963	1950	1984	AUG 1975
-Record Lowest	108	-16	-21	-5	6	27	36	44	39	28	21	5	-19	-21
-Year		1957	1934	1950	1923	1911	1945	1988	1965	1914	1936	1932	1933	FEB 1934
NORMAL DEGREE DAYS:														
Heating (base 65°F)		1231	1056	905	573	265	54	0	11	116	406	699	1082	6398
Cooling (base 65°F)		0	0	0	0	0	63	194	154	23	0	0	0	434
% OF POSSIBLE SUNSHINE	107	46	49	48	49	52	55	57	58	56	55	47	46	52
MEAN SKY COVER (tenths)														
Sunrise - Sunset														
MEAN NUMBER OF DAYS:														
Sunrise to Sunset														
-Clear														
-Partly Cloudy														
-Cloudy														
Precipitation														
.01 inches or more	108	12.3	10.9	12.5	11.9	12.3	11.5	10.7	10.3	9.4	9.6	10.9	11.8	134.1
Snow,Ice pellets,hail														
1.0 inches or more	104	4.0	3.7	3.1	0.8	0.*	0.0	0.0	0.0	0.0	0.1	0.8	2.9	15.3
Thunderstorms	1	0.0	0.0	1.0	4.0	1.0	4.0	4.5	3.5	1.5	1.0	1.0	0.0	21.5
Heavy Fog Visibility														
1/4 mile or less	1	4.0	5.0	13.0	13.0	11.0	10.0	7.5	8.0	13.5	10.5	8.0	9.0	112.5
Temperature °F														
-Maximum														
90° and above	108	0.0	0.0	0.0	0.*	0.3	1.0	2.6	1.4	0.3	0.0	0.0	0.0	5.7
32° and below	108	14.5	12.1	4.6	0.1	0.0	0.0	0.0	0.0	0.0	0.0	1.3	10.4	43.1
-Minimum														
32° and below	108	28.5	26.2	23.5	9.0	0.3	0.0	0.0	0.0	0.*	3.1	14.0	25.9	130.8
0° and below	108	1.8	1.3	0.1	0.0	0.0	0.0	0.0	0.0	0.0	0.0	0.0	0.6	3.8
AVG. STATION PRESS. (mb)														
RELATIVE HUMIDITY (%)														
Hour 01	33	72	72	74	75	80	84	85	87	86	81	79	75	79
Hour 07 (Local Time)	45	76	75	75	72	75	79	81	83	84	82	81	77	78
Hour 13	44	62	59	57	52	55	59	57	59	60	57	62	62	58
Hour 19	45	68	66	66	64	67	72	73	76	78	73	72	70	70
PRECIPITATION (inches):														
Water Equivalent														
-Normal		4.15	4.31	4.41	4.05	3.80	3.43	3.49	3.92	3.82	3.94	4.92	4.71	48.95
-Maximum Monthly	108	11.61	9.32	10.96	10.37	9.16	13.73	11.67	18.78	11.04	10.84	9.78	12.60	18.78
-Year		1979	1969	1968	1987	1954	1982	1938	1955	1933	1962	1983	1969	AUG 1955
-Minimum Monthly	108	0.89	0.71	0.06	0.92	0.50	0.53	0.13	0.53	0.45	0.22	0.55	0.92	0.06
-Year		1955	1987	1915	1892	1944	1912	1952	1981	1914	1924	1976	1955	MAR 1915
-Maximum in 24 hrs	107	3.10	4.85	6.62	3.19	5.02	5.06	4.67	9.93	5.86	6.02	5.06	5.68	9.93
-Year		1889	1886	1968	1991	1984	1984	1979	1955	1961	1895	1955	1969	AUG 1955
Snow,Ice pellets,hail														
-Maximum Monthly	108	56.3	65.4	52.0	21.5	7.8	T	0.0	0.0	0.0	6.8	23.0	45.2	65.4
-Year		1948	1969	1956	1894	1977	1991				1979	1889	1945	FEB 1969
-Maximum in 24 hrs	104	20.0	28.2	27.2	13.8	7.8	T	0.0	0.0	0.0	6.8	16.0	21.0	28.2
-Year		1898	1969	1960	1982	1977	1991				1979	1898	1960	FEB 1969
WIND:														
Mean Speed (mph)	56	17.4	17.3	17.4	16.6	14.7	13.8	12.9	12.6	13.5	15.2	16.4	16.8	15.4
Prevailing Direction through 1964		W	NW	NW	NW	S	S	SW	SW	SW	NW	W	W	NW
Fastest Mile														
-Direction (!!!)	31	S	S	ENE	NW	S	NW	NW	SSW	SSE	S	S	SSE	SSE
-Speed (MPH)	31	76	77	72	66	65	61	78	67	92	62	67	68	92
-Year		1980	1981	1984	1977	1974	1973	1972	1971	1960	1980	1963	1983	SEP 1960
Peak Gust														
-Direction (!!!)	10	S	E	ENE	SE	S	NW	W	ENE	SSE	ESE	S	SE	SSE
-Speed (mph)	10	85	81	109	84	64	74	86	78	115	67	78	79	115
-Date		1992	1985	1984	1984	1990	1988	1985	1991	1985	1988	1989	1986	SEP 1985

PRECIPITATION (inches)

BLUE HILL OBSERVATORY, MILTON, MASSACHUSETTS

YEAR	JAN	FEB	MAR	APR	MAY	JUNE	JULY	AUG	SEP	OCT	NOV	DEC	ANNUAL
1964	5.31	5.33	3.66	4.73	0.62	1.39	2.38	1.66	3.49	2.78	2.50	6.37	40.22
1965	3.06	2.89	2.44	2.64	1.33	2.70	0.92	1.62	2.35	2.88	2.18	1.95	26.96
1966	5.70	4.54	2.04	1.38	3.08	1.99	4.23	2.49	3.84	3.36	4.93	3.56	41.14
1967	1.96	4.79	6.17	5.36	7.91	4.30	2.95	4.98	3.06	1.31	3.15	8.11	54.05
1968	4.01	1.77	10.96	1.94	4.00	6.04	0.72	2.10	2.45	2.26	7.27	6.37	49.89
1969	2.27	9.32	3.11	5.17	2.78	1.80	3.04	2.30	4.87	2.53	8.59	12.60	58.38
1970	1.07	5.68	5.07	2.95	4.42	3.75	1.13	6.86	2.20	3.25	5.63	6.29	48.30
1971	1.96	5.51	3.23	3.10	4.18	1.21	3.41	1.84	1.64	2.51	6.78	2.72	38.09
1972	2.21	6.59	7.37	3.72	6.29	7.15	3.10	1.67	6.69	3.97	9.17	7.58	65.51
1973	3.85	3.02	2.59	6.48	3.78	5.80	4.30	4.35	4.24	2.93	2.09	8.22	51.65
1974	4.14	3.84	5.17	5.14	3.48	2.99	1.77	3.63	7.59	3.34	2.22	4.80	48.11
1975	6.15	3.99	3.57	3.32	3.28	4.28	3.54	7.72	7.10	6.39	6.09	6.95	62.38
1976	6.52	3.06	3.29	2.45	2.84	0.91	4.51	8.92	2.41	5.73	0.55	3.52	44.71
1977	4.85	2.67	5.62	4.05	3.41	4.82	2.23	3.11	5.05	6.43	3.17	6.29	51.70
1978	9.85	2.97	3.24	1.88	5.00	1.53	2.04	5.91	1.36	3.24	2.28	4.19	43.49
1979	11.61	4.22	3.08	4.34	4.67	1.00	5.55	6.76	4.73	4.09	4.17	1.68	55.90
1980	0.97	1.07	6.07	5.08	1.58	3.91	2.43	2.09	0.69	4.72	3.19	1.19	32.99
1981	1.25	7.93	0.63	3.16	1.60	1.89	4.09	0.53	3.76	4.14	5.60	7.24	41.82
1982	5.61	3.25	3.14	4.03	3.47	13.73	4.66	1.84	2.37	3.69	4.34	2.11	52.24
1983	5.53	6.15	10.02	7.57	3.33	1.27	1.11	6.59	1.67	3.35	9.78	6.08	62.45
1984	2.81	7.80	9.00	4.34	9.14	4.10	5.26	0.91	1.73	5.02	1.59	3.21	54.91
1985	1.31	2.91	3.17	1.80	3.58	4.66	4.53	4.04	2.74	1.83	8.16	1.32	40.05
1986	4.11	3.55	3.09	1.80	1.92	7.13	4.97	4.65	1.04	3.61	6.37	7.71	49.95
1987	7.94	0.71	5.42	10.37	2.45	2.18	1.23	4.09	9.62	2.38	4.91	2.99	54.29
1988	3.38	5.46	4.23	2.36	3.67	0.91	10.66	1.28	1.78	3.11	8.93	1.52	47.29
1989	1.00	3.45	3.83	5.03	4.74	4.44	7.31	6.96	5.39	8.07	4.77	1.14	56.13
1990	4.85	5.21	2.21	5.62	7.81	1.35	5.04	6.03	1.86	6.23	1.96	3.95	52.12
1991	4.00	2.03	5.64	5.05	1.54	3.30	2.95	6.46	6.72	6.51	5.01	2.80	52.01
1992	3.64	2.31	4.48	3.33	1.50	5.09	2.85	5.36	3.40	1.95	5.06	9.80	48.77
1993	3.15	4.42	8.72	6.59	1.23	2.12	1.77	1.67	4.16	4.36	2.97	8.01	49.17
Record Mean	4.17	3.95	4.38	3.97	3.58	3.50	3.64	4.07	3.92	3.85	4.34	4.27	47.65

AVERAGE TEMPERATURE (deg. F)

BLUE HILL OBSERVATORY, MILTON, MASSACHUSETTS

YEAR	JAN	FEB	MAR	APR	MAY	JUNE	JULY	AUG	SEP	OCT	NOV	DEC	ANNUAL
1964	29.5	26.3	36.6	44.4	59.9	65.3	70.1	65.5	60.8	51.6	42.9	30.3	48.6
1965	22.8	26.2	33.5	43.3	59.5	65.4	70.7	70.2	62.0	51.2	38.9	32.5	48.0
1966	24.8	28.2	37.1	43.0	54.5	66.6	72.3	69.7	60.5	51.9	44.5	30.9	48.7
1967	32.3	23.5	30.9	43.3	50.6	66.2	71.3	69.1	61.2	52.7	37.2	32.0	47.6
1968	22.6	22.9	37.0	48.8	54.6	62.9	72.5	68.8	63.9	54.9	40.1	27.0	48.0
1969	26.5	26.7	32.7	49.0	56.7	66.6	68.8	72.2	61.7	51.6	41.7	28.9	48.6
1970	17.8	28.6	33.4	47.1	58.1	64.7	72.1	71.2	62.6	53.8	43.3	25.6	48.2
1971	20.6	28.3	34.3	43.2	55.2	66.8	71.8	70.9	65.7	57.3	38.8	31.7	48.7
1972	28.6	25.3	32.4	42.0	55.7	62.9	70.6	68.0	62.1	47.9	38.0	31.3	47.1
1973	29.3	26.9	41.3	48.4	55.4	67.8	71.4	72.6	61.9	52.8	41.6	35.2	50.4
1974	28.5	26.6	36.5	48.7	53.4	64.1	70.3	71.2	61.5	47.6	41.5	33.8	48.6
1975	30.9	28.5	34.1	42.7	60.0	64.6	73.0	69.3	59.9	54.4	48.0	30.5	49.7
1976	22.0	34.0	37.5	51.4	57.0	69.3	70.0	68.5	61.3	48.0	37.2	24.2	48.4
1977	18.6	27.5	41.9	48.2	59.5	63.8	71.5	70.8	61.3	51.2	43.5	29.0	48.9
1978	23.6	22.3	32.6	44.2	56.6	65.5	70.5	70.0	59.0	50.3	40.9	31.3	47.2
1979	28.3	18.5	39.2	45.8	58.9	64.4	72.9	68.8	62.0	50.8	46.7	34.2	49.2
1980	26.6	24.5	34.8	47.7	57.2	63.3	73.1	71.1	64.0	49.1	38.1	25.7	48.0
1981	17.7	34.6	36.2	48.7	58.5	67.3	71.8	69.4	60.0	48.5	40.8	29.9	48.6
1982	19.0	27.5	35.5	44.5	56.3	61.1	71.7	66.1	61.7	51.7	44.7	35.5	47.9
1983	27.5	29.8	37.7	46.8	54.0	66.7	73.5	70.0	65.6	51.3	43.5	29.3	49.6
1984	23.9	35.2	30.0	45.0	56.2	67.6	70.6	70.9	60.3	53.6	42.0	36.3	49.3
1985	20.6	30.0	37.7	47.4	57.4	62.1	71.3	68.5	62.9	53.1	43.1	28.1	48.5
1986	28.7	25.5	38.7	47.1	57.4	63.0	68.3	67.2	60.4	50.9	39.4	32.4	48.3
1987	25.9	26.1	36.9	45.3	57.7	66.2	71.1	68.2	62.0	50.5	41.5	32.9	48.7
1988	24.8	29.4	36.9	44.7	56.9	66.0	72.8	73.3	61.5	47.8	44.1	29.4	49.0
1989	31.2	27.3	35.1	44.8	58.6	65.7	70.2	69.4	63.2	53.5	39.8	17.9	48.1
1990	34.2	31.2	38.2	46.2	53.6	65.5	71.0	71.0	62.1	56.0	44.8	36.8	50.9
1991	26.9	33.0	39.0	49.5	62.1	66.7	72.1	71.1	60.7	54.2	42.3	32.4	50.8
1992	27.9	29.2	32.6	44.0	55.2	65.3	67.9	68.2	61.1	49.2	39.7	31.1	47.6
1993	27.8	22.5	32.7	45.8	58.5	66.3	72.4	72.2	62.1	49.5	42.6	30.7	48.6
Record Mean	26.1	26.5	35.0	45.8	56.7	65.3	71.1	69.3	62.3	52.1	41.4	30.0	48.5
Max	33.9	34.6	43.2	55.0	66.8	75.1	80.4	78.2	71.1	60.7	49.0	37.3	57.1
Min	18.3	18.5	26.7	36.6	46.6	55.6	61.8	60.5	53.6	43.5	33.7	22.6	39.8

REFERENCE NOTES

BLUE HILL OBSERVATORY, MILTON, MASSACHUSETTS

GENERAL
T - TRACE AMOUNT.
BLANK ENTRIES DENOTE MISSING/UNREPORTED DATA.
# INDICATES A STATION OR INSTRUMENT RELOCATION.

SPECIFIC

(a) - LENGTH OF RECORD IN YEARS, ALTHOUGH INDIVIDUAL MONTHS MAY BE MISSING.
0.* OR * - THE VALUE IS BETWEEN 0.0 AND 0.05
NORMALS - BASED ON THE 1961-1990 RECORD PERIOD.
EXTREMES - DATES ARE THE MOST RECENT OCCURRENCE
WIND DIR.- NUMERALS SHOW TENS OF DEGRESS CLOCKWISE FROM TRUE NORTH. "00" INDICATES CALM.
RESULTANT DIRECTIONS ARE GIVEN TO WHOLE DEGREES.
BOLD VALUES INDICATE EXTREME VALUES WHICH OCCURRED AFTER THE ASOS SYSTEM WAS COMMISSIONED.

RECORD = PERIOD OF RECORD
RECORD MEAN PRECIPITATION IS THE MEAN OF ALL DAILY PRECIPITATION AMOUNTS DURING THE PERIOD OF RECORD.
RECORD MAX(MIN) TEMPERATURE IS THE MEAN OF ALL DAILY MAX(MIN) TEMPERATURES DURING THE PERIOD OF RECORD.
RECORD MEAN TEMPERATURE IS THE SUM OF THE RECORD MAX AND RECORD MIN DIVIDED BY 2.
AVERAGE TEMPERATURE IS THE SUM OF THE MEAN DAILY MAX AND MIN TEMPERATURE DIVIDED BY 2.

EXCEPTIONS

MEAN WIND SPEED IS THROUGH 1964.

RECORD MEANS ARE THROUGH THE CURRENT YEAR, BEGINNING IN 1885 FOR TEMPERATURE
1885 FOR PRECIPITATION
1885 FOR SNOWFALL

HEATING DEGREE DAYS Base 65 deg. F — BLUE HILL OBSERVATORY, MILTON, MASSACHUSETTS

SEASON	JULY	AUG	SEP	OCT	NOV	DEC	JAN	FEB	MAR	APR	MAY	JUNE	TOTAL
1964-65	28	47	165	409	654	1066	1299	1079	970	645	190	99	6651
1965-66	2	43	138	422	776	998	1239	1020	856	651	330	74	6549
1966-67	1	7	157	397	610	1046	1009	1160	1052	644	440	74	6597
1967-68	1	18	145	384	826	1017	1309	1215	860	478	316	116	6685
1968-69	2	21	81	330	739	1172	1189	1066	994	481	263	46	6384
1969-70	23	13	146	406	692	1112	1453	1014	969	529	228	85	6670
1970-71	2	3	123	354	642	1215	1371	1022	945	647	303	49	6676
1971-72	0	15	68	237	779	1025	1122	1141	1004	686	288	105	6470
1972-73	12	30	115	522	802	1038	1101	1064	729	491	304	35	6243
1973-74	4	10	153	372	694	916	1123	1073	879	488	371	85	6168
1974-75	4	6	148	530	695	959	1048	1016	950	663	190	92	6301
1975-76	3	31	156	327	502	1061	1327	895	847	450	254	50	5903
1976-77	7	37	135	522	826	1257	1428	1045	710	503	214	92	6776
1977-78	7	17	148	419	640	1109	1278	1188	997	618	284	50	6755
1978-79	19	19	196	447	717	1036	1134	1296	790	573	207	66	6500
1979-80	15	41	138	445	546	949	1183	1168	927	512	245	116	6285
1980-81	0	18	112	489	799	1208	1461	843	888	482	230	22	6552
1981-82	3	16	168	501	721	1080	1417	1046	907	611	269	143	6882
1982-83	8	50	132	407	601	911	1159	979	840	543	337	56	6023
1983-84	4	23	108	431	638	1101	1267	857	1079	592	277	58	6435
1984-85	14	5	170	349	679	882	1370	975	839	522	247	113	6165
1985-86	0	20	110	363	651	1138	1119	1100	803	527	278	113	6222
1986-87	42	39	156	435	762	1003	1205	1081	866	584	272	55	6500
1987-88	6	41	121	442	699	987	1238	1025	864	605	267	90	6385
1988-89	17	22	117	531	615	1098	1043	1050	923	598	207	72	6293
1989-90	9	20	123	349	751	1454	948	941	825	563	350	63	6396
1990-91	10	17	128	299	603	866	1171	889	798	461	147	69	5458
1991-92	6	12	168	333	675	1004	1143	1033	998	624	333	64	6393
1992-93	30	29	160	482	756	1043	1146	1184	996	567	204	61	6658
1993-94	4	2	148	474	665	1054							

COOLING DEGREE DAYS Base 65 deg. F — BLUE HILL OBSERVATORY, MILTON, MASSACHUSETTS

YEAR	JAN	FEB	MAR	APR	MAY	JUNE	JULY	AUG	SEP	OCT	NOV	DEC	TOTAL
1969	0	0	0	4	14	101	148	242	54	0	0	0	563
1970	0	0	0	0	20	84	227	204	58	13	0	0	606
1971	0	0	0	0	4	111	216	203	97	6	0	0	637
1972	0	0	0	0	9	46	190	131	38	0	0	0	414
1973	0	0	0	1	14	126	208	249	65	1	0	0	664
1974	0	0	0	7	18	66	177	206	50	0	0	0	524
1975	0	0	0	0	42	84	255	169	10	5	0	0	565
1976	0	0	0	48	13	184	171	150	30	3	0	0	599
1977	0	0	4	4	48	60	218	204	45	0	0	0	583
1978	0	0	0	0	31	72	196	180	23	0	0	0	502
1979	0	0	0	0	26	56	264	167	57	14	0	0	584
1980	0	0	0	0	7	72	258	213	87	0	0	0	637
1981	0	0	0	0	36	95	224	156	27	0	0	0	538
1982	0	0	0	0	7	34	221	89	40	2	0	0	393
1983	0	0	0	2	4	115	274	190	134	13	0	0	732
1984	0	0	0	0	11	142	195	191	36	2	0	0	577
1985	0	0	0	3	17	32	203	135	54	1	0	0	445
1986	0	0	0	0	49	61	149	113	22	4	0	0	398
1987	0	0	0	0	52	100	203	148	36	0	1	0	540
1988	0	0	0	0	23	128	266	289	23	4	0	0	733
1989	0	0	0	0	16	100	178	163	76	1	0	0	534
1990	0	0	0	8	0	84	203	208	48	26	3	0	580
1991	0	0	0	0	64	124	234	208	50	5	0	0	685
1992	0	0	0	0	36	77	123	135	48	0	0	0	419
1993	0	0	0	0	12	108	244	233	66	1	0	0	664

SNOWFALL (inches) — BLUE HILL OBSERVATORY, MILTON, MASSACHUSETTS

SEASON	JULY	AUG	SEP	OCT	NOV	DEC	JAN	FEB	MAR	APR	MAY	JUNE	TOTAL
1964-65	0.0	0.0	0.0	0.0	0.0	11.9	33.0	5.0	11.8	3.7	0.0	0.0	65.4
1965-66	0.0	0.0	0.0	0.0	T	2.6	46.0	20.5	4.8	T	0.0	0.0	73.9
1966-67	0.0	0.0	0.0	0.0	0.0	13.8	2.4	45.2	41.0	7.2	T	0.0	109.6
1967-68	0.0	0.0	0.0	0.0	5.8	22.8	23.2	5.7	10.2	0.0	0.0	0.0	67.7
1968-69	0.0	0.0	0.0	0.0	0.4	5.2	0.9	65.4	11.9	0.0	0.0	0.0	83.8
1969-70	0.0	0.0	0.0	T	0.0	21.0	10.6	7.9	20.8	1.0	0.0	0.0	61.3
1970-71	0.0	0.0	0.0	T	0.0	26.5	12.2	5.3	9.6	4.8	0.0	0.0	58.4
1971-72	0.0	0.0	0.0	0.0	12.0	10.7	6.2	24.1	21.8	4.5	0.0	0.0	79.3
1972-73	0.0	0.0	0.0	1.0	2.0	9.1	6.1	4.0	2.0	T	0.0	0.0	24.2
1973-74	0.0	0.0	0.0	0.0	0.0	0.4	17.4	13.6	0.4	3.0	0.0	0.0	34.8
1974-75	0.0	0.0	0.0	0.0	4.8	7.3	5.1	20.1	4.2	6.3	0.0	0.0	47.8
1975-76	0.0	0.0	0.0	T	2.3	23.9	17.3	2.4	13.7	T	0.0	0.0	59.6
1976-77	0.0	0.0	0.0	0.0	1.6	22.5	27.1	9.5	12.7	T	7.8	0.0	81.2
1977-78	0.0	0.0	0.0	0.0	0.8	7.6	35.6	30.4	19.0	0.1	T	0.0	93.5
1978-79	0.0	0.0	0.0	0.0	7.5	8.0	9.0	9.1	0.1	3.2	0.0	0.0	36.9
1979-80	0.0	0.0	0.0	6.8	0.0	6.2	0.7	6.9	4.7	T	0.0	0.0	25.3
1980-81	0.0	0.0	0.0	0.0	7.3	8.1	16.1	8.3	2.5	T	0.0	0.0	42.3
1981-82	0.0	0.0	0.0	0.0	T	27.3	17.4	10.9	6.0	14.0	0.0	0.0	75.6
1982-83	0.0	0.0	0.0	0.0	T	15.8	11.2	28.5	1.0	T	0.0	0.0	56.5
1983-84	0.0	0.0	0.0	0.0	T	9.5	23.6	0.6	26.7	T	0.0	0.0	60.4
1984-85	0.0	0.0	0.0	0.0	0.2	5.2	11.1	16.7	10.4	3.5	T	0.0	47.1
1985-86	0.0	0.0	0.0	0.0	5.1	2.9	1.6	13.8	3.0	T	T	0.0	26.4
1986-87	0.0	0.0	0.0	0.0	13.5	6.4	38.3	5.6	8.4	13.5	0.0	0.0	85.7
1987-88	0.0	0.0	0.0	0.0	11.4	12.1	20.3	16.3	8.5	T	0.0	0.0	68.6
1988-89	0.0	0.0	0.0	T	0.0	8.4	1.4	10.2	9.3	4.7	0.0	0.0	34.0
1989-90	0.0	0.0	0.0	0.0	5.6	7.3	13.3	25.8	5.7	1.5	0.2	0.0	59.4
1990-91	0.0	0.0	0.0	0.0	0.4	5.6	11.7	5.8	5.5	T	0.0	T	29.0
1991-92	0.0	0.0	0.0	T	0.2	7.5	1.1	5.0	11.9	1.5	0.0	0.0	27.2
1992-93	0.0	0.0	0.0	0.0	0.5	19.0	13.5	12.8	35.5	3.0	0.0	0.0	84.3
1993-94	0.0	0.0	0.0	0.0	0.3	15.0							
Record Mean	0.0	0.0	0.0	0.2	2.7	10.8	15.3	15.8	11.4	3.0	0.1	T	59.4

WORCESTER, MASSACHUSETTS

Worcester Municipal Airport is located on the crest of a hill, 1,000 feet above sea level. It is about 500 feet above and 3 1/2 miles northwest of the city proper. The airport is surrounded by ridges and valleys with many of the valleys containing reservoirs. Only two of the ridges extend above the airport elevation. One is 400 feet higher and 2 1/2 miles to the northwest, and the other is 1,000 feet higher and 15 miles to the north.

The proximity to the Atlantic Ocean, Long Island Sound, and the Berkshire Hills plays an important part in determining the weather and, hence, the climate of Worcester. Rapid weather changes occur when storms move up the east coast after developing off the Carolina Coast. In the majority of these cases, they pass to the south and east, resulting in northeast and easterly winds with rain or snow and fog. Storms developing in the Texas-Oklahoma area normally travel up the St. Lawrence River Valley and, depending on the movement and intensity, usually deposit little precipitation over the area. However, they do bring an influx of warm air into the region. Wintertime cold snaps are quite frequent, but temperatures are usually modified by the passage of the air over land and mountains before reaching the county. Summertime thunderstorms develop over the hills to the west, with a majority moving toward the northeast. From the use of radar, we find many break up just before reaching Worcester, or pass either north or south of the city proper.

Airport site temperatures are moderate. The normal mean for the warmest month, July, is around 70 degrees. Though winters are reasonably cold, prolonged periods of severe cold weather are extremely rare. The three coldest months, December through February, have an average temperature of over 25 degrees. A review of Worcester Cooperative records since 1901 shows maximum temperatures above 100 degrees and minimum temperatures below −24 degrees.

Precipitation is usually plentiful and well distributed throughout the year. The annual snowfall for all Worcester sites since 1901, averages slightly less than 60 inches. The airport location averages slightly higher.

Based on the 1951–1980 period, the average first occurrence of 32 degrees Fahrenheit in the fall is October 17 and the average last occurrence in the spring is April 27.

NORMALS, MEANS, AND EXTREMES

WORCESTER, MASSACHUSETTS

LATITUDE: 42°16'N LONGITUDE: 71°52'W ELEVATION: FT. GRND 986 BARO 1001 TIME ZONE: EASTERN WBAN: 94746

	(a)	JAN	FEB	MAR	APR	MAY	JUNE	JULY	AUG	SEP	OCT	NOV	DEC	YEAR
TEMPERATURE °F:														
Normals														
-Daily Maximum		30.7	33.0	42.4	54.0	65.9	74.5	79.3	77.3	69.7	59.5	47.4	34.7	55.7
-Daily Minimum		15.0	16.5	24.9	34.7	45.1	53.9	60.1	58.6	50.5	40.5	31.3	20.1	37.6
-Monthly		22.8	24.8	33.7	44.4	55.5	64.2	69.7	68.0	60.2	50.0	39.4	27.4	46.7
Extremes														
-Record Highest	38	60	67	81	91	92	94	96	96	91	85	78	70	96
-Year		1974	1985	1990	1976	1962	1988	1988	1975	1983	1963	1982	1984	JUL 1988
-Record Lowest	38	-19	-12	-4	11	28	36	43	38	30	20	6	-13	-19
-Year		1957	1967	1986	1982	1970	1986	1988	1965	1992	1969	1989	1962	JAN 1957
NORMAL DEGREE DAYS:														
Heating (base 65°F)		1308	1126	970	618	301	69	6	25	157	465	768	1166	6979
Cooling (base 65°F)		0	0	0	0	6	45	151	118	13	0	0	0	333
% OF POSSIBLE SUNSHINE														
MEAN SKY COVER (tenths)														
Sunrise - Sunset	37	6.2	6.2	6.5	6.6	6.7	6.4	6.3	6.0	5.9	5.7	6.7	6.4	6.3
MEAN NUMBER OF DAYS:														
Sunrise to Sunset														
-Clear	38	8.8	7.7	7.7	6.6	6.1	6.0	6.0	8.0	9.0	10.1	7.0	7.6	90.4
-Partly Cloudy	38	7.8	6.9	7.8	8.6	9.8	10.7	12.1	11.1	8.5	8.2	7.6	7.8	106.9
-Cloudy	38	14.4	13.7	15.5	14.8	15.2	13.3	12.9	11.9	12.4	12.7	15.4	15.7	167.9
Precipitation														
.01 inches or more	38	11.6	10.9	12.3	11.4	12.2	11.3	10.1	10.1	9.4	9.0	11.8	12.7	132.6
Snow,Ice pellets,hail														
1.0 inches or more	38	4.2	3.7	3.2	1.0	0.1	0.0	0.0	0.0	0.0	0.1	0.8	3.6	16.8
Thunderstorms	23	0.*	0.1	0.7	1.2	3.0	4.2	5.3	4.1	1.5	1.0	0.5	0.1	21.6
Heavy Fog Visibility														
1/4 mile or less	23	5.7	5.6	7.6	7.0	7.1	8.0	6.3	6.2	8.2	7.2	7.9	7.0	83.8
Temperature °F														
-Maximum														
90° and above	38	0.0	0.0	0.0	0.1	0.1	0.7	1.5	0.7	0.*	0.0	0.0	0.0	3.2
32° and below	38	17.6	13.7	5.5	0.2	0.0	0.0	0.0	0.0	0.0	0.0	1.7	13.2	51.8
-Minimum														
32° and below	38	29.6	26.8	25.3	11.1	0.6	0.0	0.0	0.0	0.1	4.9	16.8	27.8	143.1
0° and below	38	3.0	1.8	0.1	0.0	0.0	0.0	0.0	0.0	0.0	0.0	0.0	0.9	5.9
AVG. STATION PRESS. (mb)	21	977.1	977.8	977.4	976.8	978.1	978.1	978.9	980.5	981.4	980.9	979.4	978.7	978.8
RELATIVE HUMIDITY (%)														
Hour 01	32	69	69	68	68	73	78	79	81	83	78	76	73	75
Hour 07 (Local Time)	38	72	72	70	68	69	74	76	79	81	78	77	75	74
Hour 13	38	58	57	55	50	51	56	57	59	61	57	61	62	57
Hour 19	38	65	63	60	57	59	66	68	72	75	69	69	69	66
PRECIPITATION (inches):														
Water Equivalent														
-Normal		3.68	3.46	3.95	3.91	4.33	3.88	3.85	3.82	4.01	4.32	4.49	4.05	47.75
-Maximum Monthly	38	11.16	8.37	7.96	8.79	9.94	12.17	8.11	8.01	13.13	10.19	10.40	9.83	13.13
-Year		1979	1981	1972	1987	1984	1982	1959	1991	1974	1990	1972	1973	SEP 1974
-Minimum Monthly	38	0.89	0.25	0.74	1.26	0.86	0.79	0.74	1.03	0.69	1.46	0.67	0.74	0.25
-Year		1970	1987	1981	1985	1959	1979	1987	1981	1986	1963	1976	1989	FEB 1987
-Maximum in 24 hrs	38	2.97	2.46	4.56	3.15	3.03	3.98	3.87	5.00	4.79	3.91	2.98	3.00	5.00
-Year		1978	1973	1987	1987	1967	1966	1985	1991	1960	1990	1972	1986	AUG 1991
Snow,Ice pellets,hail														
-Maximum Monthly	38	46.8	45.2	44.1	21.0	12.7	0.1	0.0	0.0	T	7.5	20.7	37.0	46.8
-Year		1987	1962	1993	1987	1977	1992			1992	1979	1971	1992	JAN 1987
-Maximum in 24 hrs	38	18.7	24.0	19.7	17.0	12.7	0.1	0.0	0.0	T	7.5	14.8	28.1	28.1
-Year		1961	1962	1993	1987	1977	1992			1992	1979	1971	1992	DEC 1992
WIND:														
Mean Speed (mph)	29	11.9	11.6	11.4	11.0	10.0	8.9	8.4	8.3	8.6	9.4	10.2	10.9	10.0
Prevailing Direction through 1963		WSW	WNW	W	W	SW	SW	SW	SW	SW	WSW	WSW	WSW	SW
Fastest Obs. 1 Min.														
-Direction (!!!)	37	25	32	29	05	27	25	32	36	14	25	20	23	32
-Speed (MPH)	37	60	76	76	54	48	39	43	44	36	43	54	51	76
-Year		1959	1956	1956	1956	1956	1958	1957	1991	1985	1958	1956	1957	FEB 1956
Peak Gust														
-Direction (!!!)	10	W	NW	NW	SW	S	W	NW	N	SE	NW	S	W	SE
-Speed (mph)	10	62	63	66	54	61	58	48	69	71	59	62	62	71
-Date		1992	1990	1993	1990	1989	1993	1984	1991	1985	1992	1989	1993	SEP 1985

PRECIPITATION (inches)

WORCESTER, MASSACHUSETTS

YEAR	JAN	FEB	MAR	APR	MAY	JUNE	JULY	AUG	SEP	OCT	NOV	DEC	ANNUAL
1964	5.42	3.54	3.65	4.01	1.18	1.77	3.02	2.93	1.78	2.26	3.17	4.92	37.65
1965	1.71	4.30	2.37	3.56	1.51	3.14	1.03	3.40	2.83	3.02	2.90	2.21	31.98
1966	4.22	4.05	2.52	1.62	3.21	1.91	3.76	1.95	5.59	4.13	4.93	3.05	40.94
1967	2.38	3.15	3.73	5.10	7.01	3.72	5.91	3.47	5.06	1.83	3.65	5.72	50.73
1968	3.00	1.26	7.67	2.24	6.83	7.78	1.06	1.25	1.94	1.88	5.75	6.16	46.82
1969	1.29	3.04	2.53	5.30	3.09	1.26	3.81	3.34	6.16	1.78	6.81	7.69	46.10
1970	0.89	5.47	3.51	3.44	4.18	3.88	1.08	5.86	2.17	3.47	3.75	3.09	40.79
1971	1.97	5.60	1.93	1.71	5.22	1.88	5.27	4.67	4.23	3.66	5.01	2.68	43.83
1972	2.34	4.91	7.96	4.29	7.83	9.25	6.39	2.89	4.98	4.93	10.40	5.49	71.66
1973	4.04	3.30	3.50	6.33	4.73	6.98	3.82	4.29	3.80	4.63	2.00	9.83	57.25
1974	3.54	2.75	5.05	3.24	5.15	4.35	3.22	3.50	13.13	3.45	3.06	6.04	56.48
1975	5.35	3.52	3.37	2.66	2.07	3.32	3.92	4.59	7.21	6.06	5.28	4.61	51.96
1976	6.03	2.57	2.74	2.45	3.46	2.95	3.28	5.72	2.38	4.99	0.67	3.20	40.44
1977	2.76	2.46	5.75	3.69	2.20	4.00	3.84	2.73	6.54	6.33	3.69	4.99	48.98
1978	9.90	2.08	3.22	2.24	3.69	1.57	3.57	5.00	1.02	3.85	2.07	3.56	41.77
1979	11.16	2.64	3.71	4.49	4.14	0.79	5.74	7.39	3.80	4.36	3.58	1.89	53.69
1980	0.95	0.73	6.86	4.77	2.23	4.55	3.59	1.95	1.82	6.16	4.58	1.06	39.25
1981	0.93	8.37	0.74	3.85	4.48	2.45	7.90	1.03	4.66	5.49	3.13	5.94	48.97
1982	5.00	3.22	3.67	4.30	2.96	12.17	3.61	3.36	2.69	2.67	4.32	1.70	49.67
1983	4.85	4.67	7.84	8.59	5.97	2.56	1.32	6.26	1.38	5.77	8.75	6.37	64.33
1984	2.44	5.78	5.47	4.23	9.94	2.85	5.69	1.17	1.68	3.99	2.71	2.84	48.79
1985	1.16	2.72	2.89	1.26	5.46	5.24	6.35	3.74	3.77	3.12	6.41	1.93	44.05
1986	5.56	3.14	2.93	1.59	3.14	7.21	4.83	3.20	0.69	2.72	5.63	7.25	47.89
1987	5.52	0.25	6.57	8.79	1.55	4.55	0.74	4.61	6.37	4.18	2.77	1.85	47.75
1988	2.71	2.78	3.46	3.45	4.47	1.25	6.27	2.19	2.70	3.66	7.91	1.42	42.27
1989	1.18	2.47	2.66	4.25	6.17	5.27	5.67	5.65	4.71	8.21	4.00	0.74	50.98
1990	3.75	3.88	1.52	4.78	7.65	1.74	2.44	6.84	1.73	10.19	2.41	5.46	52.39
1991	2.98	2.08	4.92	5.04	4.16	3.06	2.78	8.01	6.40	3.44	5.47	2.89	51.23
1992	3.01	2.51	4.15	2.59	2.54	4.68	5.25	4.83	3.58	2.36	4.94	4.61	45.05
1993	2.56	2.38	5.46	4.00	1.79	2.36	3.34	1.90	8.85	3.88	4.85	5.11	46.48
Record Mean	3.65	3.31	4.05	4.03	4.00	3.63	3.74	4.29	3.99	4.23	4.57	4.02	47.52

AVERAGE TEMPERATURE (deg. F)

WORCESTER, MASSACHUSETTS

YEAR	JAN	FEB	MAR	APR	MAY	JUNE	JULY	AUG	SEP	OCT	NOV	DEC	ANNUAL
1964	27.2	23.9	35.2	43.7	58.9	64.3	68.6	63.4	59.3	49.6	41.2	28.4	47.0
1965	20.9	24.3	31.3	42.9	58.9	63.7	68.2	67.4	61.1	49.1	36.7	30.1	46.2
1966	22.2	25.8	34.2	42.0	52.9	65.4	70.7	68.1	58.2	49.0	43.1	28.0	46.7
1967	29.4	20.5	28.8	41.7	48.8	66.1	70.7	67.8	60.6	51.5	34.4	29.8	45.9
1968	19.2	20.0	35.6	48.5	53.8	62.2	71.2	68.0	63.5	53.0	37.1	24.0	46.3
1969	23.2	25.1	30.8	48.8	56.3	66.3	68.4	70.4	60.8	49.7	39.1	25.3	47.0
1970	14.9	24.2	30.4	44.6	56.9	63.7	71.1	70.1	60.6	51.7	40.2	22.2	45.9
1971	17.7	25.9	29.9	42.5	54.2	65.6	70.6	68.2	63.4	56.5	36.6	30.5	46.8
1972	26.4	23.3	30.4	40.4	55.7	62.2	71.1	66.9	59.7	44.7	34.9	27.6	45.3
1973	26.1	24.7	40.9	46.3	54.2	67.4	71.1	71.7	60.1	51.3	39.8	31.9	48.8
1974	26.4	23.9	33.5	47.5	52.4	64.0	69.7	69.7	59.7	45.8	39.4	30.4	46.9
1975	27.5	25.2	31.3	40.8	60.1	63.6	71.8	67.5	57.3	51.5	45.2	27.3	47.4
1976	19.2	31.2	35.1	49.1	55.0	68.0	67.8	67.2	58.9	46.4	34.6	21.9	46.2
1977	16.2	24.9	39.1	45.9	58.1	62.4	69.3	68.6	59.6	49.3	41.1	26.0	46.7
1978	20.9	20.0	30.8	42.4	56.6	64.3	68.2	68.1	57.6	48.4	39.2	28.8	45.4
1979	25.3	16.5	38.2	44.7	57.7	63.6	71.2	67.7	60.0	49.0	44.3	31.9	47.5
1980	25.0	22.2	33.1	46.0	57.0	61.5	71.3	69.9	61.1	47.3	36.4	24.2	46.3
1981	15.9	32.8	34.7	47.4	57.9	66.0	70.7	67.4	58.2	46.7	39.1	27.8	47.1
1982	17.1	25.7	33.8	43.6	58.3	61.5	71.5	65.9	61.3	50.8	43.8	34.8	47.3
1983	26.1	29.1	36.6	46.6	54.7	67.9	72.3	70.2	64.9	50.3	42.0	27.3	49.0
1984	22.7	34.0	29.2	45.6	55.4	68.1	70.0	71.4	59.6	53.9	41.1	34.9	48.8
1985	20.1	28.6	37.2	47.4	57.8	61.4	70.4	67.9	62.0	51.7	40.9	25.3	47.6
1986	25.9	23.1	36.4	48.1	57.9	62.6	68.2	66.8	59.4	49.6	37.1	30.3	47.1
1987	24.0	24.5	36.8	45.7	57.6	65.8	70.7	66.6	60.6	48.8	39.4	30.8	47.6
1988	23.0	26.6	35.0	43.8	57.3	64.6	72.6	72.3	60.2	45.5	41.7	27.2	47.5
1989	28.5	24.4	33.0	42.4	57.0	64.0	68.7	67.3	60.6	50.7	36.5	15.1	45.7
1990	31.3	28.8	36.2	43.9	51.5	64.6	69.2	68.8	59.3	53.0	42.1	34.0	48.6
1991	23.8	30.3	36.5	48.2	61.2	64.7	70.0	69.1	58.7	51.9	40.2	29.7	48.7
1992	25.3	27.1	30.7	42.3	55.5	63.5	65.9	65.5	59.2	46.8	37.8	28.7	45.7
1993	26.0	19.6	31.3	45.7	57.8	64.5	70.4	70.0	59.5	47.2	39.7	28.9	46.7
Record Mean	23.8	25.6	33.4	45.1	56.0	64.7	70.0	68.2	60.2	50.3	39.7	28.0	47.0
Max	31.5	33.7	41.7	54.6	66.3	74.6	79.3	77.2	69.2	59.3	47.3	35.3	55.8
Min	16.0	17.4	25.1	35.6	45.7	54.8	60.7	59.2	51.2	41.3	32.0	20.6	38.3

REFERENCE NOTES

WORCESTER, MASSACHUSETTS

GENERAL
T - TRACE AMOUNT.
BLANK ENTRIES DENOTE MISSING/UNREPORTED DATA.
# INDICATES A STATION OR INSTRUMENT RELOCATION.

SPECIFIC

PM - INCLUDES LAST DAY OF PREVIOUS MONTH
ASOS - AUTOMATED SURFACE OBSERVING SYSTEM IN OPERATION DURING THESE MONTHS.

(a) - LENGTH OF RECORD IN YEARS, ALTHOUGH INDIVIDUAL MONTHS MAY BE MISSING.
0.* OR * - THE VALUE IS BETWEEN 0.0 AND 0.05
NORMALS - BASED ON THE 1961-1990 RECORD PERIOD.
EXTREMES - DATES ARE THE MOST RECENT OCCURRENCE
WIND DIR. - NUMERALS SHOW TENS OF DEGRESS CLOCKWISE FROM TRUE NORTH. "00" INDICATES CALM.
RESULTANT DIRECTIONS ARE GIVEN TO WHOLE DEGREES.
BOLD VALUES INDICATE EXTREME VALUES WHICH OCCURRED AFTER THE ASOS SYSTEM WAS COMMISSIONED.

RECORD = PERIOD OF RECORD
RECORD MEAN PRECIPITATION IS THE MEAN OF ALL DAILY PRECIPITATION AMOUNTS DURING THE PERIOD OF RECORD.
RECORD MAX(MIN) TEMPERATURE IS THE MEAN OF ALL DAILY MAX(MIN) TEMPERATURES DURING THE PERIOD OF RECORD.
RECORD MEAN TEMPERATURE IS THE SUM OF THE RECORD MAX AND RECORD MIN DIVIDED BY 2.
AVERAGE TEMPERATURE IS THE SUM OF THE MEAN DAILY MAX AND MIN TEMPERATURE DIVIDED BY 2.

EXCEPTIONS

THUNDERSTORMS, AND HEAVY FOG ARE THROUGH 1977.

RECORD MEANS ARE THROUGH THE CURRENT YEAR, BEGINNING IN 1948 FOR TEMPERATURE
1948 FOR PRECIPITATION
1956 FOR SNOWFALL

HEATING DEGREE DAYS Base 65 deg. F — WORCESTER, MASSACHUSETTS

SEASON	JULY	AUG	SEP	OCT	NOV	DEC	JAN	FEB	MAR	APR	MAY	JUNE	TOTAL
1964-65	36	80	189	472	704	1122	1362	1136	1040	658	206	118	7123
1965-66	14	61	163	486	843	1076	1320	1093	948	685	376	80	7145
1966-67	4	19	218	489	649	1141	1097	1242	1115	691	493	73	7231
1967-68	4	28	151	422	912	1087	1414	1300	904	490	343	126	7181
1968-69	8	31	79	375	830	1264	1286	1111	1056	487	282	49	6858
1969-70	19	15	163	468	773	1224	1547	1137	1068	604	261	99	7378
1970-71	2	12	174	410	738	1319	1461	1087	1083	667	327	61	7341
1971-72	3	35	98	258	844	1061	1189	1203	1065	733	290	116	6895
1972-73	16	41	171	622	897	1152	1200	1120	738	555	338	37	6887
1973-74	6	8	202	420	750	1017	1191	1145	969	520	391	82	6701
1974-75	6	9	190	592	760	1063	1158	1110	1037	721	180	104	6930
1975-76	2	47	223	413	586	1161	1412	972	920	511	310	60	6617
1976-77	17	46	192	571	905	1331	1508	1115	798	571	244	115	7413
1977-78	16	32	189	481	711	1202	1359	1255	1054	674	288	69	7330
1978-79	33	38	229	511	767	1116	1225	1354	824	601	243	80	7021
1979-80	25	53	180	500	614	1019	1235	1233	983	562	246	148	6798
1980-81	2	19	165	540	853	1259	1516	894	934	521	241	42	6986
1981-82	1	29	204	562	772	1145	1478	1094	961	639	212	127	7224
1982-83	7	55	140	436	632	929	1199	996	871	548	318	44	6175
1983-84	5	20	115	459	682	1163	1303	892	1099	577	300	54	6669
1984-85	9	1	184	338	713	928	1382	1011	855	518	230	121	6290
1985-86	2	25	128	406	713	1223	1206	1166	879	501	262	119	6630
1986-87	48	49	182	471	830	1069	1266	1130	867	571	273	58	6814
1987-88	9	56	152	495	761	1052	1298	1107	922	629	253	115	6849
1988-89	19	32	155	597	693	1166	1123	1129	984	674	257	90	6919
1989-90	7	38	172	436	846	1540	1035	1006	884	632	413	58	7067
1990-91	22	30	192	379	683	957	1270	965	877	501	167	80	6123
1991-92	13	16	222	402	739	1090	1222	1092	1056	676	321	82	6931
1992-93	40	55	203	558	810	1117	1204	1259	1039	574	229	74	7162
1993-94	10	5	201	545	754	1115							

COOLING DEGREE DAYS Base 65 deg. F — WORCESTER, MASSACHUSETTS

YEAR	JAN	FEB	MAR	APR	MAY	JUNE	JULY	AUG	SEP	OCT	NOV	DEC	TOTAL
1969	0	0	0	5	17	98	131	192	44	1	0	0	488
1970	0	0	0	0	16	70	197	177	50	6	0	0	516
1971	0	0	0	0	0	87	184	141	57	2	0	0	471
1972	0	0	0	0	9	41	215	109	20	0	0	0	394
1973	0	0	0	1	11	118	201	224	62	0	0	0	617
1974	0	0	0	4	8	59	159	164	36	0	0	0	430
1975	0	0	0	0	33	69	218	132	0	1	0	0	453
1976	0	0	0	37	7	154	110	119	16	0	0	0	443
1977	0	0	2	4	37	44	156	150	32	0	0	0	425
1978	0	0	0	0	32	57	138	137	14	0	0	0	378
1979	0	0	0	0	24	44	225	142	38	10	0	0	483
1980	0	0	0	0	8	48	206	178	57	0	0	0	497
1981	0	0	0	0	29	78	184	110	10	0	0	0	411
1982	0	0	0	0	8	29	216	92	34	0	0	0	379
1983	0	0	0	4	4	138	238	188	118	11	0	0	701
1984	0	0	0	2	10	156	171	209	30	0	0	0	578
1985	0	0	0	0	15	22	177	120	47	0	0	0	381
1986	0	0	0	0	50	54	151	111	21	2	0	0	389
1987	0	0	0	0	49	90	193	110	25	0	0	0	467
1988	0	0	0	0	24	112	260	266	17	1	0	0	680
1989	0	0	0	0	16	70	131	116	47	0	0	0	380
1990	0	0	0	6	0	52	159	153	25	10	0	0	405
1991	0	0	0	4	57	79	173	149	40	2	0	0	504
1992	0	0	0	0	30	44	75	78	37	0	0	0	264
1993	0	0	0	0	12	67	181	163	44	0	0	0	467

SNOWFALL (inches) — WORCESTER, MASSACHUSETTS

SEASON	JULY	AUG	SEP	OCT	NOV	DEC	JAN	FEB	MAR	APR	MAY	JUNE	TOTAL
1964-65	0.0	0.0	0.0	2.1	T	11.7	18.7	5.9	17.7	6.7	0.0	0.0	62.8
1965-66	0.0	0.0	0.0	T	0.6	2.3	44.0	19.7	5.8	0.8	T	0.0	73.2
1966-67	0.0	0.0	0.0	0.0	0.7	13.6	2.5	35.5	34.6	7.3	T	0.0	94.2
1967-68	0.0	0.0	0.0	0.0	9.8	22.2	18.6	6.4	9.2	T	0.0	0.0	66.2
1968-69	0.0	0.0	0.0	0.0	15.3	12.2	1.8	39.5	6.9	T	0.0	0.0	75.7
1969-70	0.0	0.0	0.0	T	0.3	29.5	7.7	11.4	19.9	3.3	T	0.0	72.1
1970-71	0.0	0.0	0.0	T	T	32.1	16.6	11.4	12.1	7.8	0.0	0.0	80.0
1971-72	0.0	0.0	0.0	0.0	20.7	9.6	6.7	35.0	20.1	7.2	0.0	0.0	99.3
1972-73	0.0	0.0	0.0	T	6.1	13.8	17.9	5.8	0.4	0.4	0.0	0.0	44.4
1973-74	0.0	0.0	0.0	0.0	T	0.9	12.5	15.0	1.7	3.7	0.0	0.0	33.8
1974-75	0.0	0.0	0.0	T	1.2	13.1	22.6	21.9	4.9	1.4	0.0	0.0	65.1
1975-76	0.0	0.0	0.0	T	1.5	18.1	21.6	4.7	16.4	T	T	0.0	62.3
1976-77	0.0	0.0	0.0	T	3.0	13.5	21.7	13.8	21.5	1.0	12.7	0.0	87.2
1977-78	0.0	0.0	0.0	0.0	2.2	13.7	34.2	20.8	15.0	T	T	0.0	85.9
1978-79	0.0	0.0	0.0	0.0	5.4	13.1	16.0	6.5	1.3	5.4	0.0	0.0	47.7
1979-80	0.0	0.0	0.0	7.5	0.0	2.1	0.8	6.5	9.7	T	0.0	0.0	26.6
1980-81	0.0	0.0	0.0	0.0	9.0	6.8	12.5	11.4	3.3	T	T	0.0	43.0
1981-82	0.0	0.0	0.0	T	T	24.6	16.7	6.5	11.0	15.1	0.0	0.0	73.9
1982-83	0.0	0.0	0.0	0.0	0.5	6.4	18.6	32.1	3.5	2.3	T	0.0	63.4
1983-84	0.0	0.0	0.0	T	1.1	17.2	24.1	3.3	30.9	T	T	0.0	76.6
1984-85	0.0	0.0	0.0	0.0	T	7.0	9.7	11.0	7.2	4.9	0.0	0.0	39.8
1985-86	0.0	0.0	0.0	0.0	6.9	9.1	5.8	14.5	2.3	0.1	0.3	0.0	39.0
1986-87	0.0	0.0	0.0	0.0	11.5	4.9	46.8	3.0	6.4	21.0	0.0	0.0	93.6
1987-88	0.0	0.0	0.0	0.0	10.2	12.9	25.2	15.8	6.4	0.6	0.0	0.0	71.1
1988-89	0.0	0.0	0.0	0.4	T	5.0	2.8	7.7	8.5	3.7	0.0	0.0	28.1
1989-90	0.0	0.0	0.0	0.0	7.9	10.2	11.3	15.2	6.4	2.1	0.0	T	53.1
1990-91	0.0	0.0	0.0	0.0	0.7	5.0	11.3	9.1	9.3	0.2	0.0	0.0	35.6
1991-92	0.0	0.0	0.0	0.0	5.8	14.5	2.7	8.4	11.8	2.7	0.0	0.1	46.0
1992-93	0.0	0.0	T	0.0	1.9	37.0	14.6	19.7	44.1	2.8	0.0	T	120.1
1993-94	0.0	0.0	0.0	T	0.2	12.9							
Record Mean	0.0	0.0	T	0.5	3.6	13.4	16.1	15.9	14.1	3.8	0.3	T	67.8

ALPENA, MICHIGAN

The city of Alpena lies on the northwest shore of Thunder Bay, 8 miles from the open waters of Lake Huron. Lake Huron and Thunder Bay lie at an elevation of 580 feet above sea level. Generally, the land slopes up westward from the lakeshore to 689 feet at the airport. Farther to the west and southwest the land becomes higher and more rolling. A range of hills with tops 1,000 to 1,350 feet lies northwest to southeast about 25 miles southwest of the station.

Summer showers moving from the southwest weaken and sometimes dissipate as they approach Alpena. Winter storms often bring winds with an easterly component. Precipitation from these is increased by moisture and instability picked up from Lake Huron and by forced upslope flow.

The climate of Alpena is influenced by its location with respect to major storm tracks and the effects of the Great Lakes. The normal wintertime storm track is south of the city, and most passing storms bring snow. Rain, freezing rain, and sleet are uncommon, but not unknown, in winter. In summer, most storms pass to the north, often bringing brief showers to the area, but occasionally, heavy thunderstorms with damaging winds occur. The Great Lakes modify most climatic extremes. Precipitation amounts are distributed evenly throughout the year. The lake effect is most pronounced in early winter, before ice forms. Minimum temperatures during this season are higher than would be expected at this latitude. But as nearby waters, particularly the Straits of Mackinac, freeze over, sub-zero temperatures become fairly common by February.

Summers in Alpena are warm and sunny. Brief showers usually occur every few days, often falling on only part of the area. Hailstorms average less than one a year. During prolonged heat waves the highest temperatures in Michigan often occur in the forest area southwest of Alpena. Winter months are cloudy and marked by frequent snow flurries. Storms bring heavier snowfall. Snow cover is sufficiently deep and persistent to provide good protection for grasses and winter grains.

The climate along the immediate Lake Huron shore is semi-maritime and lacks the temperature extremes experienced just a few miles inland. Maximum temperatures near the lake shore average 1.6 degrees lower than those at the airport, minimum temperatures average 5 degrees higher. Afternoon lake breezes which are strongest in the late spring and early summer cause lake shore maximum temperatures to average 3.6 degrees lower during the month of May.

Freezing temperatures have occurred as late as late June and as early as late August. Principal crops in the area are hay, potatoes, berries, and apples.

Prevailing winds are from the northwest except during May and June when southeast winds predominate. Southeast winds are common in the afternoon during all the summer months.

NORMALS, MEANS, AND EXTREMES

ALPENA, MICHIGAN

LATITUDE: 45°04'N LONGITUDE: 83°34' W ELEVATION: FT. GRND 689 BARO 692 TIME ZONE: EASTERN WBAN: 94849

	(a)	JAN	FEB	MAR	APR	MAY	JUNE	JULY	AUG	SEP	OCT	NOV	DEC	YEAR
TEMPERATURE °F:														
Normals														
-Daily Maximum		26.4	28.2	37.9	51.6	64.8	74.5	80.2	76.9	68.3	56.8	43.0	30.9	53.3
-Daily Minimum		8.8	8.3	18.0	30.1	39.2	48.0	54.0	52.6	46.3	37.0	28.4	16.5	32.3
-Monthly		17.6	18.3	28.0	40.9	52.0	61.3	67.1	64.8	57.3	46.9	35.7	23.7	42.8
Extremes														
-Record Highest	35	52	65	76	90	94	98	102	102	94	88	76	65	102
-Year		1973	1984	1990	1990	1962	1988	1983	1988	1983	1971	1978	1982	AUG 1988
-Record Lowest	35	-28	-37	-27	0	20	28	34	30	25	16	-1	-18	-37
-Year		1963	1979	1962	1965	1966	1966	1965	1982	1993	1966	1976	1985	FEB 1979
NORMAL DEGREE DAYS:														
Heating (base 65°F)		1469	1308	1147	723	411	146	43	80	237	561	879	1280	8284
Cooling (base 65°F)		0	0	0	0	8	35	108	74	6	0	0	0	231
% OF POSSIBLE SUNSHINE	34	37	45	52	53	60	63	66	59	51	42	29	28	49
MEAN SKY COVER (tenths)														
Sunrise - Sunset	34	7.8	7.4	6.8	6.8	6.3	6.0	5.7	5.8	6.5	7.1	8.2	8.4	6.9
MEAN NUMBER OF DAYS:														
Sunrise to Sunset														
-Clear	34	3.1	4.1	6.5	6.3	7.4	7.5	8.0	8.6	6.3	4.8	2.4	1.8	66.9
-Partly Cloudy	34	7.3	7.1	7.8	7.7	9.5	10.8	13.3	10.9	9.5	8.5	6.1	6.4	104.9
-Cloudy	34	20.6	17.0	16.7	16.0	14.1	11.7	9.7	11.5	14.2	17.7	21.5	22.9	193.5
Precipitation														
.01 inches or more	34	13.9	11.2	12.4	11.7	11.4	10.7	9.9	10.9	12.5	12.5	13.5	15.5	146.0
Snow,Ice pellets,hail														
1.0 inches or more	34	6.6	4.9	3.7	1.6	0.1	0.0	0.0	0.0	0.0	0.2	2.2	6.2	25.6
Thunderstorms	33	0.1	0.*	0.9	2.2	3.8	5.5	6.9	6.3	4.0	1.5	0.4	0.2	31.8
Heavy Fog Visibility														
1/4 mile or less	33	1.3	1.5	2.8	2.5	2.4	2.4	2.2	2.7	2.6	2.8	2.3	1.3	26.8
Temperature °F														
-Maximum														
90° and above	34	0.0	0.0	0.0	0.*	0.4	1.6	2.7	1.4	0.3	0.0	0.0	0.0	6.4
32° and below	34	22.0	18.4	9.3	0.8	0.0	0.0	0.0	0.0	0.0	0.*	3.7	16.8	71.0
-Minimum														
32° and below	34	30.4	27.5	28.3	18.7	7.1	0.6	0.0	0.1	2.4	11.5	21.9	28.9	177.3
0° and below	34	7.9	7.8	3.3	0.*	0.0	0.0	0.0	0.0	0.0	0.0	0.1	3.2	22.3
AVG. STATION PRESS.(mb)	21	989.8	991.7	990.4	989.5	989.5	989.1	990.1	991.5	991.8	991.5	990.1	990.3	990.4
RELATIVE HUMIDITY (%)														
Hour 01	28	80	78	79	77	80	86	86	90	89	84	82	82	83
Hour 07 (Local Time)	34	80	79	82	80	78	81	84	90	91	86	84	83	83
Hour 13	34	70	65	61	54	51	53	53	58	61	61	69	73	61
Hour 19	34	75	71	68	61	57	60	61	69	77	76	78	79	69
PRECIPITATION (inches):														
Water Equivalent														
-Normal		1.64	1.29	2.11	2.25	2.74	3.04	2.92	3.40	3.11	2.10	2.20	2.03	28.83
-Maximum Monthly	35	3.31	3.17	4.44	4.15	8.29	8.37	7.17	5.92	7.12	6.53	7.45	4.44	8.37
-Year		1978	1971	1976	1980	1983	1969	1975	1984	1986	1991	1966	1971	JUN 1969
-Minimum Monthly	35	0.16	0.12	0.34	1.18	0.99	0.20	0.22	0.92	0.28	0.61	0.61	0.41	0.12
-Year		1961	1982	1993	1978	1988	1991	1989	1970	1979	1971	1986	1993	FEB 1982
-Maximum in 24 hrs	35	1.83	1.24	1.84	1.20	2.50	2.65	3.15	3.18	3.02	1.97	1.93	1.63	3.18
-Year		1978	1977	1990	1991	1973	1969	1974	1968	1968	1991	1966	1971	AUG 1968
Snow,Ice pellets,hail														
-Maximum Monthly	35	43.5	33.4	35.8	12.7	3.7	0.0	0.0	T	T	4.3	30.5	46.3	46.3
-Year		1978	1967	1971	1992	1974			1991	1993	1969	1966	1970	DEC 1970
-Maximum in 24 hrs	35	16.3	13.7	17.3	12.2	3.7	0.0	0.0	T	T	3.8	16.1	16.4	17.3
-Year		1978	1981	1985	1992	1974			1991	1993	1969	1966	1989	MAR 1985
WIND:														
Mean Speed (mph)	33	8.8	8.4	8.9	9.2	8.2	7.6	7.0	6.7	7.1	7.8	8.5	8.4	8.1
Prevailing Direction through 1963		SW	W	SE	NW	SE	SW	W	SE	SW	SW	W	SW	SW
Fastest Obs. 1 Min.														
-Direction (!!)	13	12	35	32	28	21	21	14	31	20	26	17	07	35
-Speed (MPH)	13	29	37	31	31	35	35	37	35	30	25	31	31	37
-Year		1982	1987	1985	1982	1988	1992	1982	1988	1980	1990	1988	1990	FEB 1987
Peak Gust														
-Direction (!!)	10	W	N	NW	W	SW	SW	NW	NW	NW	W	SW	NE	NW
-Speed (mph)	10	44	54	47	52	53	58	46	60	45	47	53	54	60
-Date		1990	1987	1985	1984	1988	1992	1988	1988	1989	1992	1988	1987	AUG 1988

PRECIPITATION (inches)

ALPENA, MICHIGAN

YEAR	JAN	FEB	MAR	APR	MAY	JUNE	JULY	AUG	SEP	OCT	NOV	DEC	ANNUAL
1964	1.28	0.36	1.91	3.34	3.04	1.12	4.30	3.03	3.63	1.10	2.34	1.52	26.97
1965	2.79	1.86	1.25	3.43	4.24	2.95	1.04	3.65	4.64	1.37	2.70	1.93	31.85
1966	1.50	1.18	2.80	1.61	1.05	2.41	0.91	1.76	2.15	2.12	7.45	2.38	27.32
1967	3.15	2.11	1.14	3.08	1.94	8.14	3.12	3.17	1.98	2.28	2.67	2.39	35.17
1968	1.19	1.75	0.87	1.57	2.55	3.80	2.37	5.30	6.67	1.98	2.98	3.02	34.05
1969	2.23	0.34	1.48	1.71	3.11	8.37	1.82	2.50	2.18	4.78	1.64	1.26	31.42
1970	1.81	0.27	2.03	2.00	1.99	1.96	4.41	0.92	4.76	1.47	2.18	3.32	27.12
1971	2.42	3.17	2.75	1.71	2.87	5.88	4.00	4.08	1.00	0.61	2.30	4.44	35.23
1972	0.47	1.08	3.35	1.69	2.18	1.44	3.51	4.85	2.60	2.03	1.14	3.83	28.17
1973	1.02	0.60	1.82	2.31	4.83	2.31	5.16	5.67	2.97	4.03	1.70	2.17	34.59
1974	3.04	1.38	0.64	2.97	2.61	3.15	5.17	2.03	2.18	1.43	1.42	1.04	27.06
1975	2.14	1.26	2.26	1.70	4.45	3.23	7.17	5.76	1.97	0.91	2.51	0.93	34.29
1976	1.73	2.24	4.44	1.33	2.43	2.82	2.03	2.54	1.30	2.14	0.79	0.61	24.40
1977	0.72	2.19	2.49	2.15	1.85	1.38	3.38	4.75	4.01	1.22	2.33	2.28	28.75
1978	3.31	0.60	1.07	1.18	2.69	1.13	2.96	2.65	6.09	1.26	1.55	2.26	26.75
1979	1.79	1.06	2.13	3.02	2.50	2.76	4.54	2.14	0.28	2.58	1.94	2.09	26.83
1980	0.95	0.47	1.22	4.15	1.87	3.34	2.11	3.29	2.51	0.99	1.03	2.19	24.12
1981	0.46	1.93	1.01	3.41	1.14	2.11	1.50	2.20	2.38	2.98	1.12	1.17	21.41
1982	2.06	0.12	2.33	1.62	1.68	3.37	1.65	3.07	3.61	1.45	2.73	2.81	26.50
1983	1.80	1.48	2.82	2.70	8.29	1.76	2.48	2.62	3.50	3.90	1.74	1.92	35.01
1984	0.89	0.81	2.22	2.89	1.89	2.88	2.35	5.92	3.76	2.79	2.01	2.22	30.63
1985	1.71	2.77	2.77	2.63	2.70	1.00	3.80	4.74	3.07	2.80	3.24	2.60	33.83
1986	1.04	1.06	2.30	1.86	2.41	3.36	5.28	1.83	7.12	2.83	0.61	0.95	30.65
1987	0.94	0.87	0.94	1.20	1.75	3.38	1.17	4.77	3.29	2.00	2.21	2.17	24.69
1988	1.51	1.04	2.72	3.25	0.99	0.84	2.61	4.48	1.50	3.43	3.54	1.24	27.15
1989	1.42	0.61	2.68	2.20	1.89	2.67	0.22	2.29	1.90	2.20	2.27	1.53	21.88
1990	2.44	1.17	2.89	2.32	2.88	4.18	3.69	2.35	1.58	3.07	2.89	1.79	31.25
1991	0.96	0.49	3.04	2.57	6.26	0.20	5.16	3.38	2.24	6.53	1.50	1.44	33.77
1992	1.70	1.70	1.50	2.61	1.10	1.69	3.37	2.65	2.53	2.34	4.01	1.24	26.44
1993	1.33	0.81	0.34	3.16	2.49	4.00	1.71	5.67	1.51	1.98	2.56	0.41	25.97
Record Mean	1.77	1.48	1.96	2.29	2.89	3.00	2.72	3.02	3.09	2.58	2.37	1.91	29.07

AVERAGE TEMPERATURE (deg. F)

ALPENA, MICHIGAN

YEAR	JAN	FEB	MAR	APR	MAY	JUNE	JULY	AUG	SEP	OCT	NOV	DEC	ANNUAL
1964	21.9	21.9	26.9	41.2	56.3	61.0	67.6	61.4	55.6	45.9	38.9	23.2	43.5
1965	16.3	17.9	22.4	35.9	54.9	59.2	61.9	62.0	56.2	44.8	34.3	29.1	41.3
1966	14.9	22.8	32.0	38.8	46.8	63.3	68.5	64.9	55.8	46.4	33.7	22.6	42.6
1967	21.3	12.5	25.1	40.2	46.1	64.6	64.0	61.9	55.3	46.4	30.8	25.7	41.2
1968	18.3	14.4	32.0	44.7	47.9	61.3	66.0	64.4	60.5	50.0	35.6	22.0	43.1
1969	20.3	20.6	24.4	41.6	50.3	56.8	65.7	68.2	56.9	44.3	34.2	22.6	42.2
1970	13.6	16.5	24.3	41.9	51.7	61.3	69.0	65.9	58.0	49.4	36.2	23.1	42.6
1971	15.1	18.5	23.9	38.3	50.2	63.9	62.8	62.3	60.2	54.0	35.7	27.5	42.7
1972	18.5	16.4	21.9	35.4	52.7	57.4	65.6	63.2	55.1	42.4	34.3	23.3	40.5
1973	22.7	16.9	35.3	41.2	49.1	63.5	65.9	68.5	56.6	51.9	36.1	22.9	44.2
1974	20.8	14.1	27.8	41.7	48.2	60.5	67.1	64.8	53.1	44.5	36.5	28.0	42.3
1975	22.4	21.9	25.0	35.8	58.0	63.7	68.7	65.7	53.6	49.0	40.7	23.8	44.0
1976	12.6	23.6	30.6	45.0	50.9	65.8	67.5	66.1	57.0	43.4	30.4	15.0	42.3
1977	10.2	19.4	35.9	43.3	56.5	59.2	67.2	61.3	58.2	45.7	34.7	22.4	42.8
1978	16.2	14.1	24.5	36.0	55.4	60.7	65.9	65.6	58.5	46.3	36.0	23.7	41.9
1979	13.9	8.6	31.3	41.1	50.8	60.8	66.9	63.1	59.0	44.7	33.9	28.4	41.9
1980	19.8	16.8	26.1	40.4	53.7	56.8	67.6	68.6	57.2	43.7	33.6	18.6	41.9
1981	14.7	24.5	32.1	44.3	52.0	62.4	67.5	64.9	54.1	41.4	35.8	26.3	43.3
1982	11.2	17.3	25.4	37.5	56.9	56.4	67.9	61.6	57.4	48.7	36.0	31.4	42.3
1983	22.2	25.1	30.8	39.8	47.0	61.9	72.2	68.5	59.8	45.6	35.0	18.7	43.9
1984	13.1	28.4	24.6	43.3	48.5	63.2	66.4	67.5	54.4	48.6	35.3	27.9	43.4
1985	16.3	17.6	30.5	44.8	53.7	58.5	65.8	63.9	59.1	46.8	35.3	19.6	42.7
1986	18.3	19.3	30.1	45.0	54.6	58.7	68.2	62.4	57.2	45.9	33.3	29.2	43.5
1987	23.2	22.4	32.5	44.9	55.5	65.3	70.4	65.6	59.8	43.4	38.0	29.4	45.9
1988	18.6	15.4	27.6	41.6	56.0	64.8	71.6	68.1	58.0	43.1	37.4	23.9	43.8
1989	25.7	16.8	23.9	39.2	53.7	60.8	68.7	64.8	56.2	47.8	31.1	12.5	41.8
1990	26.9	22.1	31.1	44.3	50.4	61.9	66.8	65.8	57.1	45.4	38.5	26.7	44.8
1991	18.6	26.2	31.2	45.4	58.4	65.6	67.7	68.3	55.5	47.9	33.9	25.0	45.3
1992	22.3	23.5	26.8	38.5	52.6	58.5	61.3	62.0	55.7	44.3	34.1	27.9	42.3
1993	21.5	16.1	26.8	39.1	51.4	59.8	69.5	67.8	52.2	43.7	34.6	25.9	42.4
Record Mean	19.4	18.8	26.9	39.8	51.0	61.0	66.7	64.8	57.1	47.2	35.3	24.9	42.7
Max	26.8	27.3	35.3	48.5	61.1	71.3	77.0	74.7	65.9	55.6	41.7	31.0	51.3
Min	11.9	10.2	18.5	31.1	41.0	50.6	56.3	54.9	48.3	38.7	28.9	18.7	34.1

REFERENCE NOTES

ALPENA, MICHIGAN

GENERAL
T - TRACE AMOUNT.
BLANK ENTRIES DENOTE MISSING/UNREPORTED DATA.
# INDICATES A STATION OR INSTRUMENT RELOCATION.

SPECIFIC

(a) - LENGTH OF RECORD IN YEARS, ALTHOUGH INDIVIDUAL MONTHS MAY BE MISSING.
0.* OR * - THE VALUE IS BETWEEN 0.0 AND 0.05
NORMALS - BASED ON THE 1961-1990 RECORD PERIOD.
EXTREMES - DATES ARE THE MOST RECENT OCCURRENCE.
WIND DIR.- NUMERALS SHOW TENS OF DEGRESS CLOCKWISE FROM TRUE NORTH. "00" INDICATES CALM.
RESULTANT DIRECTIONS ARE GIVEN TO WHOLE DEGREES.
BOLD VALUES INDICATE EXTREME VALUES WHICH OCCURRED AFTER THE ASOS SYSTEM WAS COMMISSIONED.

RECORD = PERIOD OF RECORD
RECORD MEAN PRECIPITATION IS THE MEAN OF ALL DAILY PRECIPITATION AMOUNTS DURING THE PERIOD OF RECORD.
RECORD MAX(MIN) TEMPERATURE IS THE MEAN OF ALL DAILY MAX(MIN) TEMPERATURES DURING THE PERIOD OF RECORD.
RECORD MEAN TEMPERATURE IS THE SUM OF THE RECORD MAX AND RECORD MIN DIVIDED BY 2.
AVERAGE TEMPERATURE IS THE SUM OF THE MEAN DAILY MAX AND MIN TEMPERATURE DIVIDED BY 2.

EXCEPTIONS

1. PRIOR TO 1966, THUNDERSTORMS AND HEAVY FOG DATA MAY BE INCOMPLETE, DUE TO PART-TIME OPERATIONS.
2. FASTEST MILE WINDS ARE THROUGH JULY 1980.

RECORD MEANS ARE THROUGH THE CURRENT YEAR, BEGINNING IN 1873 FOR TEMPERATURE
1873 FOR PRECIPITATION
1960 FOR SNOWFALL

HEATING DEGREE DAYS Base 65 deg. F ALPENA, MICHIGAN

SEASON	JULY	AUG	SEP	OCT	NOV	DEC	JAN	FEB	MAR	APR	MAY	JUNE	TOTAL
1964-65	40	155	293	584	778	1290	1506	1313	1313	864	324	197	8657
1965-66	126	142	285	620	917	1104	1544	1174	1015	780	559	122	8388
1966-67	27	60	281	574	932	1302	1346	1465	1232	739	575	56	8589
1967-68	84	130	289	576	1017	1213	1441	1462	1017	603	526	165	8523
1968-69	82	119	143	484	875	1326	1377	1239	1251	695	466	262	8319
1969-70	63	32	260	633	916	1306	1589	1354	1256	687	417	161	8674
1970-71	27	64	230	478	855	1294	1541	1298	1269	796	450	112	8414
1971-72	93	122	193	353	871	1154	1433	1403	1329	883	376	232	8442
1972-73	72	108	292	694	911	1287	1306	1338	913	709	486	86	8202
1973-74	42	35	304	406	866	1298	1364	1421	1146	697	517	151	8247
1974-75	46	63	358	628	849	1140	1312	1198	1234	868	251	103	8050
1975-76	43	72	334	488	724	1267	1621	1197	1061	612	430	68	7917
1976-77	27	70	270	663	1031	1547	1696	1270	892	643	294	206	8609
1977-78	49	156	206	587	902	1317	1506	1421	1245	860	331	191	8771
1978-79	63	48	212	577	862	1270	1580	1576	1038	712	443	164	8545
1979-80	45	107	217	630	929	1128	1396	1390	1201	729	352	281	8405
1980-81	25	15	248	652	937	1430	1551	1129	1013	616	397	115	8128
1981-82	43	62	325	725	868	1193	1663	1330	1219	819	249	250	8746
1982-83	29	145	258	501	864	1034	1319	1112	1053	748	551	154	7768
1983-84	24	32	216	600	893	1427	1605	1054	1247	644	505	105	8352
1984-85	42	39	314	502	884	1144	1502	1321	1060	613	353	197	7971
1985-86	50	88	217	554	885	1401	1441	1272	1075	594	338	209	8124
1986-87	43	105	240	588	945	1100	1289	1188	1002	594	332	75	7501
1987-88	45	64	164	660	805	1096	1431	1433	1151	697	290	112	7948
1988-89	8	77	213	673	824	1269	1211	1342	1268	769	345	147	8146
1989-90	16	90	272	525	1010	1621	1173	1194	1045	639	443	131	8159
1990-91	39	38	259	600	787	1181	1433	1080	1040	595	262	63	7377
1991-92	34	40	304	525	927	1233	1317	1197	1177	784	392	213	8143
1992-93	123	136	286	639	919	1144	1341	1362	1178	772	418	170	8488
1993-94	9	39	378	655	902	1207							

COOLING DEGREE DAYS Base 65 deg. F ALPENA, MICHIGAN

YEAR	JAN	FEB	MAR	APR	MAY	JUNE	JULY	AUG	SEP	OCT	NOV	DEC	TOTAL
1969	0	0	0	0	18	23	93	137	24	0	0	0	295
1970	0	0	0	2	11	59	158	97	26	2	0	0	355
1971	0	0	0	0	0	86	33	43	57	19	0	0	238
1972	0	0	0	0	1	9	98	57	0	0	0	0	165
1973	0	0	0	0	0	50	72	150	60	5	0	0	337
1974	0	0	0	4	3	24	118	64	8	0	0	0	221
1975	0	0	0	0	39	71	167	102	1	0	0	0	380
1976	0	0	0	18	0	96	113	112	36	0	0	0	375
1977	0	0	0	0	36	39	125	48	9	0	0	0	257
1978	0	0	0	0	39	68	99	73	26	0	0	0	305
1979	0	0	0	0	10	47	112	55	41	7	0	0	272
1980	0	0	0	0	9	43	114	133	19	0	0	0	318
1981	0	0	0	3	1	44	124	65	3	0	0	0	240
1982	0	0	0	0	6	1	125	43	37	0	0	0	212
1983	0	0	0	0	0	68	254	144	67	3	0	0	536
1984	0	0	0	0	0	58	92	123	5	0	0	0	278
1985	0	0	0	14	9	8	81	61	46	0	0	0	219
1986	0	0	0	0	24	26	150	34	10	0	0	0	244
1987	0	0	0	0	46	91	218	88	13	0	0	0	456
1988	0	0	0	0	18	113	217	179	9	0	0	0	536
1989	0	0	0	0	1	26	134	89	16	0	0	0	266
1990	0	0	0	25	0	45	101	70	30	0	0	0	271
1991	0	0	0	11	65	88	125	150	25	2	0	0	466
1992	0	0	0	0	13	25	14	51	16	4	0	0	123
1993	0	0	0	0	3	21	153	130	5	0	0	0	312

SNOWFALL (inches) ALPENA, MICHIGAN

SEASON	JULY	AUG	SEP	OCT	NOV	DEC	JAN	FEB	MAR	APR	MAY	JUNE	TOTAL
1964-65	0.0	0.0	0.0	1.0	7.0	18.2	32.5	20.8	11.5	6.8	0.0	0.0	97.8
1965-66	0.0	0.0	T	1.0	5.2	6.5	21.1	5.3	9.9	4.8	0.3	0.0	54.1
1966-67	0.0	0.0	0.0	T	30.5	15.2	32.6	33.4	10.4	0.6	T	0.0	122.7
1967-68	0.0	0.0	0.0	0.2	10.9	9.1	13.6	15.2	5.6	0.5	T	0.0	55.1
1968-69	0.0	0.0	0.0	1.5	14.3	35.2	25.7	8.1	7.7	T	T	0.0	92.5
1969-70	0.0	0.0	0.0	4.3	5.8	25.4	28.6	5.7	20.0	4.2	0.2	0.0	94.2
1970-71	0.0	0.0	0.0	T	11.8	46.3	43.2	25.0	35.8	4.2	T	0.0	166.3
1971-72	0.0	0.0	0.0	0.0	12.0	18.3	8.6	15.9	27.3	3.3	0.0	0.0	85.4
1972-73	0.0	0.0	0.0	0.2	4.7	42.5	9.6	8.3	9.6	8.8	T	0.0	83.7
1973-74	0.0	0.0	0.0	T	4.2	24.2	13.9	17.4	6.8	2.8	3.7	0.0	73.0
1974-75	0.0	0.0	T	0.3	7.7	12.6	14.0	13.0	17.7	1.1	0.0	0.0	66.4
1975-76	0.0	0.0	0.0	T	10.9	16.0	34.3	18.0	18.9	3.5	0.1	0.0	101.7
1976-77	0.0	0.0	0.0	2.5	8.0	13.8	17.3	11.3	3.4	4.6	0.0	0.0	60.9
1977-78	0.0	0.0	0.0	T	19.5	26.1	43.5	16.8	8.5	0.1	T	0.0	114.5
1978-79	0.0	0.0	0.0	T	7.8	29.1	23.3	13.5	4.8	9.1	2.2	0.0	89.8
1979-80	0.0	0.0	0.0	0.2	6.2	20.0	12.0	11.8	15.7	12.3	0.0	0.0	78.2
1980-81	0.0	0.0	0.0	1.5	4.6	32.2	14.0	20.5	9.3	T	0.0	0.0	82.1
1981-82	0.0	0.0	0.0	0.3	3.9	18.0	39.8	3.6	20.1	3.6	0.0	0.0	89.3
1982-83	0.0	0.0	0.0	0.2	4.7	4.7	22.4	18.3	16.3	7.1	0.0	0.0	73.7
1983-84	0.0	0.0	0.0	T	5.5	22.1	21.2	8.8	9.2	7.4	0.3	0.0	74.5
1984-85	0.0	0.0	T	0.0	4.5	15.3	40.6	30.7	26.8	4.8	0.0	0.0	122.7
1985-86	0.0	0.0	0.0	T	15.4	27.7	17.2	16.0	12.6	3.6	T	0.0	92.5
1986-87	0.0	0.0	0.0	0.1	4.1	11.1	12.3	17.2	3.5	7.9	0.0	0.0	56.2
1987-88	0.0	0.0	0.0	0.7	6.6	16.2	13.6	22.7	21.5	5.2	0.0	0.0	86.5
1988-89	0.0	0.0	0.0	0.9	3.3	15.3	17.0	12.7	29.8	8.0	0.1	0.0	87.1
1989-90	0.0	0.0	T	0.7	19.1	41.3	25.0	16.4	2.8	10.3	0.3	0.0	115.9
1990-91	0.0	T	0.0	0.3	10.2	19.7	17.1	12.8	6.5	1.7	0.0	0.0	68.3
1991-92	0.0	T	T	T	2.7	26.6	16.5	21.3	5.3	12.7	0.0	0.0	85.1
1992-93	0.0	0.0	0.0	2.8	10.2	8.1	16.3	19.3	7.9	12.5	T	0.0	77.1
1993-94	0.0	0.0	T	0.2	2.7	4.7							
Record Mean	0.0	T	T	0.6	8.1	20.5	21.4	15.8	13.4	4.9	0.3	0.0	84.8

DETROIT, METROPOLITAN AIRPORT
MICHIGAN

Detroit and the immediate suburbs, including nearby urban areas in Canada, occupy an area approximately 25 miles in radius. The waterway, consisting of the Detroit and St. Clair Rivers, Lake St. Clair, and the west end of Lake Erie, lies at an elevation of 568 to 580 feet above sea level. Nearly flat land slopes up gently from the waters edge northwestward for about 10 miles and then gives way to increasingly rolling terrain. The Irish Hills, parallel to and about 40 miles northwest of the waterway, have tops 1,000 to 1,250 feet above sea level. On the Canadian side of the waterway the land is relatively level.

Northwest winds in winter bring snow flurry accumulations to all of Michigan except in the Detroit Metropolitan area while summer showers moving from the northwest weaken and sometimes dissipate as they approach Detroit. On the other hand, much of the heaviest precipitation in winter comes from southeast winds, especially to the northwest suburbs of the city.

The climate of Detroit is influenced by its location with respect to major storm tracks and the influence of the Great Lakes. The normal wintertime storm track is south of the city, which brings on the average, about 3 inch snowfalls. Winter storms can bring combinations of rain, snow, freezing rain, and sleet with heavy snowfall accumulations possible at times. In summer, most storms pass to the north allowing for intervals of warm, humid, sunny skies with occasional thunderstorms followed by days of mild, dry, and fair weather. Temperatures of 90 degrees or higher are reached during each summer.

The most pronounced lake effect occurs in the winter when arctic air moving across the lakes is warmed and moistened. This produces an excess of cloudiness but a moderation of cold wave temperatures.

Local climatic variations are due largely to the immediate effect of Lake St. Clair and the urban heat island. On warm days in late spring or early summer, lake breezes often lower temperatures by 10 to 15 degrees in the eastern part of the city and the northeastern suburbs. The urban heat island effect shows up mainly at night where minimum temperatures at the Metropolitan Airport average 4 degrees lower than downtown Detroit. On humid summer nights or on very cold winter nights, this difference can exceed 10 degrees.

The growing season averages 180 days and has ranged from 145 days to 205 days. On average, the last freezing temperature occurs in late April while the average first freezing temperature occurs in late October. A freeze has occurred as late as mid-May and as early as late September.

Air pollution comes primarily from heavy industry spread along both shores of the waterway from Port Huron to Toledo. However, wind dispersion is usually sufficient to keep it from becoming a major hazard.

NORMALS, MEANS, AND EXTREMES

DETROIT, METROPOLITAN AIRPORT MICHIGAN

LATITUDE: 42°14'N LONGITUDE: 83°20' W ELEVATION: FT. GRND 633 BARO 647 TIME ZONE: EASTERN WBAN: 94847

	(a)	JAN	FEB	MAR	APR	MAY	JUNE	JULY	AUG	SEP	OCT	NOV	DEC	YEAR
TEMPERATURE °F:														
Normals														
-Daily Maximum		30.3	33.3	44.4	57.7	69.6	78.9	83.3	81.3	73.9	61.5	48.1	35.2	58.1
-Daily Minimum		15.6	17.6	27.0	36.8	47.1	56.3	61.3	59.6	52.5	40.9	32.2	21.4	39.0
-Monthly		22.9	25.4	35.7	47.3	58.4	67.6	72.3	70.5	63.2	51.2	40.2	28.3	48.6
Extremes														
-Record Highest	35	62	65	81	89	93	104	102	100	98	91	77	68	104
-Year		1965	1976	1986	1977	1988	1988	1988	1988	1976	1963	1968	1982	JUN 1988
-Record Lowest	35	-21	-15	-4	10	25	36	41	38	29	17	9	-10	-21
-Year		1984	1985	1978	1982	1966	1972	1965	1982	1974	1974	1969	1983	JAN 1984
NORMAL DEGREE DAYS:														
Heating (base 65°F)		1305	1109	908	531	243	38	0	16	102	435	744	1138	6569
Cooling (base 65°F)		0	0	0	0	38	116	231	186	48	7	0	0	626
% OF POSSIBLE SUNSHINE	28	40	46	51	54	61	65	68	68	61	51	35	30	53
MEAN SKY COVER (tenths)														
Sunrise - Sunset	35	7.6	7.2	7.1	6.8	6.3	5.9	5.6	5.7	6.1	6.4	7.7	8.0	6.7
MEAN NUMBER OF DAYS:														
Sunrise to Sunset														
-Clear	35	4.2	4.7	5.7	6.0	6.8	7.8	8.8	9.0	8.0	7.5	3.9	3.4	75.8
-Partly Cloudy	35	6.7	6.9	7.3	7.6	10.4	11.1	12.4	11.2	9.7	8.8	6.6	6.1	104.9
-Cloudy	35	20.1	16.7	18.0	16.4	13.8	11.1	9.7	10.8	12.3	14.7	19.5	21.5	184.6
Precipitation														
.01 inches or more	35	13.2	11.1	13.2	12.7	11.1	10.2	9.5	9.5	9.9	9.8	11.9	13.7	136.0
Snow,Ice pellets,hail														
1.0 inches or more	35	3.2	2.9	2.2	0.5	0.0	0.0	0.0	0.0	0.0	0.1	1.1	3.2	13.2
Thunderstorms	35	0.2	0.5	1.6	3.2	3.9	5.9	5.8	5.5	4.1	1.3	0.6	0.4	33.1
Heavy Fog Visibility														
1/4 mile or less	35	2.1	2.3	2.4	1.0	0.7	0.7	0.6	1.3	1.7	2.2	1.8	3.2	19.9
Temperature °F														
-Maximum														
90° and above	35	0.0	0.0	0.0	0.0	0.4	2.7	5.0	3.0	0.9	0.1	0.0	0.0	12.1
32° and below	35	16.8	13.1	4.4	0.2	0.0	0.0	0.0	0.0	0.0	0.0	1.4	11.4	47.1
-Minimum														
32° and below	35	29.0	25.6	22.6	9.5	0.8	0.0	0.0	0.0	0.1	4.8	16.5	26.1	135.0
0° and below	35	3.5	1.9	0.1	0.0	0.0	0.0	0.0	0.0	0.0	0.0	0.0	1.3	6.8
AVG. STATION PRESS.(mb)	21	992.8	993.7	991.7	990.6	990.7	990.6	991.7	993.0	993.7	993.9	992.9	993.2	992.4
RELATIVE HUMIDITY (%)														
Hour 01	35	78	77	76	74	76	79	81	83	84	80	79	79	79
Hour 07 (Local Time)	35	80	79	79	78	78	79	82	86	87	84	82	81	81
Hour 13	35	69	65	61	55	53	54	54	56	57	57	65	71	60
Hour 19	35	73	70	66	59	56	57	58	63	67	68	73	76	66
PRECIPITATION (inches):														
Water Equivalent														
-Normal		1.76	1.74	2.55	2.95	2.92	3.61	3.18	3.43	2.89	2.10	2.67	2.82	32.62
-Maximum Monthly	35	3.92	5.02	4.48	5.40	6.20	7.04	6.02	7.83	7.52	4.87	5.68	6.00	7.83
-Year		1993	1990	1973	1961	1991	1987	1969	1975	1986	1967	1982	1965	AUG 1975
-Minimum Monthly	35	0.27	0.15	0.82	0.92	0.87	0.97	0.59	0.72	0.43	0.35	0.79	0.46	0.15
-Year		1961	1969	1981	1971	1988	1988	1974	1982	1960	1964	1976	1960	FEB 1969
-Maximum in 24 hrs	35	1.72	2.36	1.69	1.97	2.87	2.84	3.19	3.21	3.52	2.57	2.20	3.71	3.71
-Year		1967	1990	1985	1965	1968	1983	1966	1964	1990	1985	1982	1965	DEC 1965
Snow,Ice pellets,hail														
-Maximum Monthly	35	29.6	20.8	16.1	9.0	T	0.0	0.0	0.0	T	2.9	11.8	34.9	34.9
-Year		1978	1986	1965	1982	1991				1990	1980	1966	1974	DEC 1974
-Maximum in 24 hrs	35	11.1	10.3	9.2	7.4	T	0.0	0.0	0.0	T	2.9	5.6	19.2	19.2
-Year		1992	1965	1973	1982	1991				1990	1980	1977	1974	DEC 1974
WIND:														
Mean Speed (mph)	35	11.9	11.5	11.8	11.6	10.2	9.3	8.5	8.3	8.8	9.9	11.3	11.5	10.4
Prevailing Direction through 1963		WSW	WSW	WSW	WSW	WSW	SW	SW	SW	SW	WSW	SW	SW	SW
Fastest Obs. 1 Min.														
-Direction (!!)	14	22	31	25	28	31	23	33	25	28	26	17	22	22
-Speed (MPH)	14	48	38	44	44	39	37	40	32	35	36	37	48	48
-Year		1990	1990	1989	1982	1988	1990	1980	1984	1992	1993	1988	1987	JAN 1990
Peak Gust														
-Direction (!!)	10	SW	NW	W	SW	S	NW	W	W	NW	W	SW	SW	SW
-Speed (mph)	10	66	51	60	64	58	56	54	47	54	52	53	59	66
-Date		1990	1987	1989	1991	1984	1991	1991	1990	1987	1985	1991	1987	JAN 1990

PRECIPITATION (inches)

DETROIT, METROPOLITAN AIRPORT MICHIGAN

YEAR	JAN	FEB	MAR	APR	MAY	JUNE	JULY	AUG	SEP	OCT	NOV	DEC	ANNUAL
1964	2.20	0.68	2.98	4.13	1.72	3.84	1.11	7.70	1.31	0.35	0.80	1.92	28.74
1965	3.63	2.54	3.59	3.30	1.15	2.28	2.38	6.94	1.91	3.89	1.49	6.00	39.10
1966	0.61	1.64	2.62	2.65	2.18	4.16	5.24	5.03	1.51	1.15	3.13	4.53	34.45
1967	2.34	1.28	1.03	3.67	1.63	4.23	2.85	2.01	1.96	4.87	2.77	5.19	33.83
1968	2.30	1.48	2.04	1.71	5.88	4.99	5.14	1.83	1.87	1.09	3.31	3.59	35.23
1969	2.83	0.15	1.62	3.77	3.74	4.26	6.02	1.06	0.68	1.41	2.46	1.33	29.33
1970	1.11	0.86	2.62	3.32	3.01	3.90	3.30	2.75	1.63	1.91	2.73	1.61	28.75
1971	1.03	2.68	1.59	0.92	1.97	2.17	1.95	1.62	2.72	1.01	1.34	3.79	22.79
1972	1.28	1.00	2.55	3.63	2.68	3.30	2.21	3.07	3.40	2.24	3.19	3.11	31.66
1973	1.65	1.08	4.48	1.42	3.72	4.86	4.66	1.67	1.82	2.01	3.21	3.51	34.09
1974	3.26	2.37	4.20	2.75	3.49	2.38	0.59	2.95	2.22	0.81	2.86	4.00	31.88
1975	2.90	2.65	1.66	2.50	2.82	2.39	1.98	7.83	3.18	1.29	2.39	3.00	34.59
1976	1.91	2.87	4.24	3.15	3.26	3.26	1.47	1.68	3.66	2.01	0.79	0.79	29.09
1977	0.98	1.64	3.57	4.17	2.40	3.16	3.28	2.23	4.23	1.37	2.88	2.97	32.88
1978	3.16	0.45	2.05	2.49	3.58	2.69	1.97	1.73	1.82	2.49	2.41	2.81	27.65
1979	1.52	0.57	2.44	4.97	2.82	4.04	4.96	2.99	0.94	1.24	4.19	2.36	33.04
1980	0.69	1.00	3.88	4.23	3.22	6.42	4.33	6.09	2.94	1.26	0.88	2.30	37.24
1981	0.57	3.13	0.82	3.44	2.60	3.33	4.29	2.32	5.47	3.92	1.26	2.38	33.53
1982	3.43	1.10	3.14	1.60	2.83	4.11	4.78	0.72	2.55	1.01	5.68	3.29	34.24
1983	0.84	0.89	1.87	4.20	5.47	4.88	4.53	1.57	2.49	2.85	4.28	3.78	37.65
1984	0.78	1.31	3.12	2.48	3.62	1.04	0.95	3.00	2.30	2.28	2.49	2.90	26.27
1985	2.63	3.83	4.42	2.11	3.11	1.62	3.96	4.88	2.59	3.91	5.51	1.51	40.08
1986	1.30	3.46	2.29	2.73	1.36	5.75	2.47	3.52	7.52	3.05	1.88	2.28	37.61
1987	2.35	0.53	2.19	2.14	2.50	7.04	2.20	6.87	2.69	2.00	3.17	4.60	38.28
1988	1.30	2.02	1.16	1.50	0.87	0.97	2.43	3.13	3.65	3.57	4.29	1.97	26.86
1989	1.28	0.77	2.16	2.22	4.16	3.79	4.21	2.14	3.03	1.73	2.53	1.24	29.26
1990	1.80	5.02	1.91	2.72	3.74	4.92	1.47	3.85	6.06	4.14	2.64	4.37	42.64
1991	1.44	0.94	1.41	2.66	6.20	1.89	1.23	4.31	0.90	4.14	2.61	1.91	29.64
1992	1.78	1.54	3.34	4.34	1.33	2.35	5.91	2.50	5.55	2.01	4.33	2.35	37.33
1993	3.92	1.27	2.12	3.32	1.24	6.05	2.17	1.60	4.26	2.21	1.69	0.78	30.63
Record Mean	1.87	1.72	2.48	3.00	2.92	3.63	3.11	3.38	2.90	2.22	2.65	2.65	32.52

AVERAGE TEMPERATURE (deg. F)

DETROIT, METROPOLITAN AIRPORT MICHIGAN

YEAR	JAN	FEB	MAR	APR	MAY	JUNE	JULY	AUG	SEP	OCT	NOV	DEC	ANNUAL
1964	28.1	25.9	34.3	47.6	60.9	66.2	72.9	65.3	60.7	46.5	41.2	28.1	48.1
1965	24.7	25.9	30.1	45.1	62.4	66.8	69.2	68.0	63.3	48.2	40.5	35.0	48.3
1966	20.4	27.4	37.1	44.4	51.9	68.4	73.5	69.1	62.2	51.0	41.8	28.7	48.0
1967	29.3	23.8	35.9	48.0	51.9	70.4	69.0	67.3	60.4	50.7	35.2	30.9	47.7
1968	20.9	24.3	38.2	50.9	56.1	68.4	71.7	72.5	65.5	53.5	41.0	28.1	49.3
1969	23.1	28.2	33.5	49.4	57.4	64.7	73.2	73.1	64.8	51.4	37.9	26.4	48.6
1970	16.6	24.4	32.9	49.3	60.9	68.4	72.6	72.1	64.5	54.2	40.0	29.0	48.8
1971	20.7	27.4	32.0	45.2	56.4	70.8	69.6	70.3	66.6	58.5	38.5	33.3	49.1
1972	23.8	24.6	32.6	44.6	60.3	64.2	71.2	69.1	63.0	47.3	37.4	29.4	47.3
1973	28.8	25.3	43.3	48.8	55.5	69.9	72.6	72.9	64.9	56.2	41.4	28.7	50.7
1974	26.5	23.6	35.7	49.2	55.2	65.9	72.5	72.3	59.7	48.8	40.6	28.6	48.2
1975	28.3	27.5	32.5	40.9	62.8	69.0	72.2	72.3	59.1	52.9	46.8	29.1	49.5
1976	19.2	33.3	40.4	50.0	56.4	70.6	72.7	70.2	62.1	47.4	33.5	21.5	48.1
1977	12.8	25.2	41.5	52.4	64.4	65.5	75.8	70.6	65.1	47.9	40.5	25.5	48.9
1978	19.6	16.3	30.0	45.5	59.3	66.8	70.6	71.9	67.5	50.2	40.5	28.9	47.3
1979	18.6	16.5	37.7	44.6	56.5	66.6	70.4	67.9	62.6	50.1	39.5	31.9	46.9
1980	24.5	22.2	31.3	45.9	59.8	63.7	72.7	72.7	63.8	46.3	37.4	26.0	47.2
1981	19.0	28.8	36.5	49.8	55.9	68.0	72.4	70.0	60.9	47.6	41.1	27.8	48.2
1982	17.1	20.7	33.0	43.2	64.3	64.2	72.4	67.7	61.8	52.6	41.6	37.3	48.0
1983	28.7	31.6	38.4	44.2	54.4	68.2	74.5	73.6	64.0	51.6	41.2	20.8	49.3
1984	18.0	33.3	28.9	47.8	54.5	70.8	70.8	72.7	61.2	54.9	38.6	34.0	48.8
1985	20.4	23.5	38.4	51.0	60.1	62.8	71.3	69.2	64.3	53.0	42.4	22.2	48.2
1986	23.9	24.6	37.6	50.6	61.3	67.3	75.0	68.9	65.9	52.6	37.3	31.7	49.7
1987	26.1	29.6	39.8	50.8	63.3	71.3	76.1	71.6	64.6	46.6	43.5	33.6	51.4
1988	23.8	23.4	36.9	48.5	62.0	70.4	77.1	75.1	63.3	46.0	42.2	28.7	49.8
1989	32.8	24.1	35.2	45.1	57.5	67.5	73.0	69.9	61.9	52.1	38.2	18.0	47.9
1990	33.6	30.7	39.5	49.0	56.6	68.5	72.2	71.2	64.5	52.8	44.2	32.8	51.3
1991	25.0	31.2	40.3	52.0	66.5	72.4	74.9	73.4	63.1	54.8	38.5	32.1	52.0
1992	28.3	30.8	35.5	46.3	58.3	65.5	68.8	66.7	61.4	49.7	40.5	33.2	48.8
1993	29.4	24.2	34.7	47.8	60.2	67.5	75.5	74.5	61.0	51.9	41.2	30.8	49.9
Record Mean	23.4	25.8	35.5	47.5	58.7	67.7	72.3	70.7	63.3	51.3	40.1	28.6	48.8
Max	30.6	33.5	44.2	57.8	69.9	78.8	83.2	81.4	73.8	61.5	47.9	35.4	58.2
Min	16.1	18.0	26.8	37.1	47.5	56.5	61.4	60.0	52.7	41.1	32.2	21.8	39.3

REFERENCE NOTES

DETROIT, METROPOLITAN AIRPORT MICHIGAN

GENERAL
T - TRACE AMOUNT.
BLANK ENTRIES DENOTE MISSING/UNREPORTED DATA.
# INDICATES A STATION OR INSTRUMENT RELOCATION.

SPECIFIC

PM - INCLUDES LAST DAY OF PREVIOUS MONTH
ASOS - AUTOMATED SURFACE OBSERVING SYSTEM IN OPERATION DURING THESE MONTHS.

(a) - LENGTH OF RECORD IN YEARS, ALTHOUGH INDIVIDUAL MONTHS MAY BE MISSING.
0.* OR * - THE VALUE IS BETWEEN 0.0 AND 0.05
NORMALS - BASED ON THE 1961-1990 RECORD PERIOD.
EXTREMES - DATES ARE THE MOST RECENT OCCURRENCE.
WIND DIR. - NUMERALS SHOW TENS OF DEGRESS CLOCKWISE FROM TRUE NORTH. "00" INDICATES CALM.
RESULTANT DIRECTIONS ARE GIVEN TO WHOLE DEGREES.
BOLD VALUES INDICATE EXTREME VALUES WHICH OCCURRED AFTER THE ASOS SYSTEM WAS COMMISSIONED.

RECORD = PERIOD OF RECORD
RECORD MEAN PRECIPITATION IS THE MEAN OF ALL DAILY PRECIPITATION AMOUNTS DURING THE PERIOD OF RECORD.
RECORD MAX(MIN) TEMPERATURE IS THE MEAN OF ALL DAILY MAX(MIN) TEMPERATURES DURING THE PERIOD OF RECORD.
RECORD MEAN TEMPERATURE IS THE SUM OF THE RECORD MAX AND RECORD MIN DIVIDED BY 2.
AVERAGE TEMPERATURE IS THE SUM OF THE MEAN DAILY MAX AND MIN TEMPERATURE DIVIDED BY 2.

EXCEPTIONS

RECORD MEANS ARE THROUGH THE CURRENT YEAR,
BEGINNING IN 1959 FOR TEMPERATURE
1959 FOR PRECIPITATION
1959 FOR SNOWFALL

HEATING DEGREE DAYS Base 65 deg. F — DETROIT, METROPOLITAN AIRPORT MICHIGAN

SEASON	JULY	AUG	SEP	OCT	NOV	DEC	JAN	FEB	MAR	APR	MAY	JUNE	TOTAL
1964-65	4	82	175	566	710	1137	1239	1088	1074	591	131	45	6842
1965-66	18	48	128	510	729	927	1377	1046	860	612	406	53	6714
1966-67	0	13	136	427	688	1119	1097	1148	894	505	406	9	6442
1967-68	32	29	158	452	886	1050	1360	1173	823	416	275	29	6683
1968-69	3	17	71	384	714	1137	1289	1024	972	464	259	102	6436
1969-70	0	0	93	418	804	1189	1491	1130	993	479	168	49	6814
1970-71	8	8	108	339	742	1105	1368	1047	1015	588	278	22	6628
1971-72	13	6	72	213	788	977	1272	1165	997	608	174	91	6376
1972-73	24	28	113	539	822	1096	1115	1103	667	480	289	3	6279
1973-74	0	10	98	276	702	1119	1189	1152	901	476	308	54	6285
1974-75	0	2	189	495	726	1123	1129	1043	996	714	142	41	6600
1975-76	4	0	178	375	537	1107	1413	914	757	473	269	6	6033
1976-77	1	15	133	540	938	1341	1609	1106	721	395	122	85	7006
1977-78	1	17	85	524	729	1218	1400	1357	1077	580	235	65	7288
1978-79	17	0	73	452	728	1112	1432	1355	843	604	291	55	6962
1979-80	12	29	126	471	758	1019	1249	1233	1036	568	191	104	6796
1980-81	0	0	110	578	822	1201	1418	1008	878	452	293	19	6779
1981-82	3	9	167	534	710	1144	1477	1237	985	647	75	70	7058
1982-83	2	39	145	383	696	852	1119	928	816	618	323	59	5980
1983-84	6	0	125	418	708	1367	1450	912	1112	507	334	9	6948
1984-85	11	4	164	310	785	955	1377	1154	818	435	177	93	6283
1985-86	2	8	129	366	672	1317	1271	1125	842	435	166	48	6381
1986-87	1	33	76	380	824	1028	1198	984	776	423	158	11	5892
1987-88	4	30	69	566	639	969	1273	1201	864	486	138	46	6285
1988-89	2	3	90	590	679	1118	991	1138	916	591	254	33	6405
1989-90	0	11	151	400	797	1451	966	955	785	506	258	27	6307
1990-91	1	1	112	380	618	994	1234	939	761	394	125	5	5564
1991-92	0	0	151	319	788	1013	1129	985	906	555	224	59	6129
1992-93	15	30	153	469	725	976	1097	1138	931	506	161	51	6252
1993-94	0	2	155	404	710	1050							

COOLING DEGREE DAYS Base 65 deg. F — DETROIT, METROPOLITAN AIRPORT MICHIGAN

YEAR	JAN	FEB	MAR	APR	MAY	JUNE	JULY	AUG	SEP	OCT	NOV	DEC	TOTAL
1969	0	0	0	1	28	98	260	259	91	4	0	0	741
1970	0	0	0	17	50	161	249	236	100	12	0	0	825
1971	0	0	0	0	14	205	161	177	128	17	0	0	702
1972	0	0	0	0	36	73	222	160	59	0	0	0	550
1973	0	0	0	3	2	156	241	261	104	11	0	0	778
1974	0	0	0	8	10	91	237	237	36	1	0	0	620
1975	0	0	0	0	82	171	233	232	7	6	0	0	731
1976	0	0	0	30	10	182	246	182	53	3	0	0	706
1977	0	0	0	25	108	107	341	198	94	0	0	0	873
1978	0	0	0	0	63	122	200	221	154	0	0	0	760
1979	0	0	0	0	32	109	184	124	57	16	0	0	522
1980	0	0	0	3	38	69	246	248	79	3	0	0	686
1981	0	0	0	1	17	118	241	168	51	0	0	0	596
1982	0	0	0	0	58	55	237	129	57	5	0	0	541
1983	0	0	0	2	0	160	306	272	104	6	0	0	850
1984	0	0	0	0	15	189	197	252	55	2	0	0	710
1985	0	0	0	25	32	32	201	146	116	0	0	0	552
1986	0	0	0	10	55	120	319	160	110	3	0	0	777
1987	0	0	0	4	111	207	355	245	64	0	1	0	987
1988	0	0	0	0	52	214	385	322	46	8	0	0	1027
1989	0	0	0	0	29	114	256	171	64	5	0	0	639
1990	0	0	1	32	8	139	234	200	101	11	0	0	726
1991	0	0	0	10	179	233	315	268	104	9	0	0	1118
1992	0	0	0	0	27	80	143	91	51	0	0	0	392
1993	0	0	0	0	21	135	334	302	41	5	0	0	838

SNOWFALL (inches) — DETROIT, METROPOLITAN AIRPORT MICHIGAN

SEASON	JULY	AUG	SEP	OCT	NOV	DEC	JAN	FEB	MAR	APR	MAY	JUNE	TOTAL
1964-65	0.0	0.0	0.0	T	4.3	7.2	11.7	17.3	16.1	2.5	0.0	0.0	59.1
1965-66	0.0	0.0	0.0	T	0.4	1.6	8.4	3.9	2.5	1.8	0.0	0.0	18.6
1966-67	0.0	0.0	0.0	0.0	11.8	14.9	5.4	11.0	5.8	1.7	0.0	0.0	50.6
1967-68	0.0	0.0	0.0	T	1.4	4.5	11.8	2.8	10.0	0.1	0.0	0.0	30.6
1968-69	0.0	0.0	0.0	T	T	5.9	6.3	2.3	2.3	0.3	0.0	0.0	17.1
1969-70	0.0	0.0	0.0	T	3.9	10.0	10.9	9.4	6.7	4.2	T	0.0	45.1
1970-71	0.0	0.0	0.0	0.0	1.7	9.8	8.7	5.9	8.7	0.6	0.0	0.0	35.4
1971-72	0.0	0.0	0.0	0.0	4.2	2.6	7.9	9.3	2.5	2.5	0.0	0.0	29.0
1972-73	0.0	0.0	0.0	T	7.1	12.5	2.4	12.8	10.1	0.1	T	0.0	45.0
1973-74	0.0	0.0	0.0	0.0	0.1	16.4	14.1	11.2	5.7	1.7	T	0.0	49.2
1974-75	0.0	0.0	0.0	T	7.7	34.9	4.9	7.5	4.5	3.6	0.0	0.0	63.1
1975-76	0.0	0.0	0.0	0.0	6.5	19.8	15.1	4.9	7.5	2.1	T	0.0	55.9
1976-77	0.0	0.0	0.0	T	1.4	9.8	14.7	5.0	12.3	0.7	0.0	0.0	43.9
1977-78	0.0	0.0	0.0	0.0	7.4	16.6	29.6	5.3	2.5	0.3	0.0	0.0	61.7
1978-79	0.0	0.0	0.0	0.0	6.1	6.6	13.3	3.9	2.7	3.0	0.0	0.0	35.6
1979-80	0.0	0.0	0.0	T	3.2	2.3	2.8	5.5	11.7	1.4	0.0	0.0	26.9
1980-81	0.0	0.0	0.0	2.9	3.4	10.5	7.6	13.4	0.6	0.0	0.0	0.0	38.4
1981-82	0.0	0.0	0.0	0.1	0.7	17.3	20.0	13.3	13.6	9.0	0.0	0.0	74.0
1982-83	0.0	0.0	0.0	T	1.8	1.4	1.5	4.3	7.6	3.4	0.0	0.0	20.0
1983-84	0.0	0.0	0.0	0.0	3.5	19.9	9.9	8.7	9.7	0.1	0.0	0.0	51.8
1984-85	0.0	0.0	0.0	0.0	4.1	6.2	20.9	16.9	6.1	0.9	0.0	0.0	55.1
1985-86	0.0	0.0	0.0	0.0	2.0	14.1	8.6	20.8	7.4	1.3	0.0	0.0	54.2
1986-87	0.0	0.0	0.0	T	3.3	6.0	24.0	2.0	13.3	1.1	0.0	0.0	49.7
1987-88	0.0	0.0	0.0	T	0.7	15.3	7.0	19.2	2.7	0.2	0.0	0.0	45.1
1988-89	0.0	0.0	0.0	T	1.0	6.3	5.3	9.6	2.4	0.5	T	0.0	25.1
1989-90	0.0	0.0	0.0	2.7	2.4	11.8	4.0	11.1	7.8	2.0	0.0	0.0	41.8
1990-91	0.0	0.0	T	0.0	T	13.2	8.8	9.2	0.2	T	T	0.0	31.4
1991-92	0.0	0.0	0.0	T	2.2	8.6	18.4	2.4	11.7	0.2	0.0	0.0	43.5
1992-93	0.0	0.0	0.0	0.4	0.9	5.0	11.0	15.2	15.7	4.0	0.0	0.0	52.2
1993-94	0.0	0.0	0.0	0.4	0.6	1.9							
Record Mean	0.0	0.0	T	0.2	3.0	10.2	10.2	9.1	7.0	1.7	T	0.0	41.2

FLINT, MICHIGAN

Flint, Michigan, is located in the Flint River Valley, in the center of Genesee County. Lake Huron lies approximately 65 miles to the east, while Saginaw Bay is about 40 miles to the north. The surrounding terrain is generally level with a slight rising tendency to a range of hills 15 to 20 miles southeast of the city.

Flint is generally under the climatic influence of the Great Lakes. Temperatures of 100 degrees or higher are rare and cold waves are less severe then expected. During the winter months, snow showers occur with strong northwesterly winds, and Lake Michigan, lying 120 miles to the west, causes a tempering effect upon cold waves coming from the northwest. The lake effect also results in delaying the coming of spring and prolonging warmer weather in late autumn. This results in conditions favorable for orchards and small fruit.

Precipitation is usually ample for growth and development of vegetation. The wettest periods normally occur in the late spring, early summer, and early fall. The driest period is normally during the winter, and although there is an occasional heavy snowfall, most of the snow occurs in the form of frequent light flurries.

Winter months are marked by considerable cloudiness and rather high relative humidity, while during the summer relative humidity is usually not excessive and sunshine is plentiful.

Violent windstorms associated with thunderstorms and squall lines occasionally hit this area. Tornadoes are infrequent but have caused extensive property damage and loss of life.

Weather changes are frequent throughout the year, since a majority of atmospheric disturbances moving eastward across the country pass near enough to affect the weather in Flint.

NORMALS, MEANS, AND EXTREMES

FLINT, MICHIGAN

LATITUDE: 42°58'N LONGITUDE: 83°45' W ELEVATION: FT. GRND 771 BARO 770 TIME ZONE: EASTERN WBAN: 14826

	(a)	JAN	FEB	MAR	APR	MAY	JUNE	JULY	AUG	SEP	OCT	NOV	DEC	YEAR
TEMPERATURE °F:														
Normals														
-Daily Maximum		28.7	31.3	42.3	55.9	68.0	76.9	81.5	79.2	71.6	59.7	46.6	33.8	56.3
-Daily Minimum		14.2	15.6	25.6	36.0	45.7	54.7	59.7	57.9	51.1	40.5	32.1	20.6	37.8
-Monthly		21.5	23.5	34.0	46.0	56.8	65.9	70.6	68.6	61.4	50.1	39.4	27.2	47.1
Extremes														
-Record Highest	37	60	63	78	87	93	101	101	98	93	89	76	67	101
-Year		1965	1984	1990	1990	1988	1988	1988	1988	1983	1963	1978	1982	JUN 1988
-Record Lowest	37	-25	-22	-12	6	22	33	40	37	26	19	6	-12	-25
-Year		1976	1967	1978	1982	1966	1966	1965	1982	1991	1974	1976	1989	JAN 1976
NORMAL DEGREE DAYS:														
Heating (base 65°F)		1349	1162	961	570	292	64	11	25	137	468	768	1172	6979
Cooling (base 65°F)		0	0	0	0	37	91	184	136	29	6	0	0	483
% OF POSSIBLE SUNSHINE														
MEAN SKY COVER (tenths)														
Sunrise - Sunset	38	7.8	7.4	7.3	7.0	6.6	6.2	5.9	5.9	6.3	6.7	8.0	8.2	6.9
MEAN NUMBER OF DAYS:														
Sunrise to Sunset														
-Clear	52	3.5	3.8	5.2	5.8	6.1	6.1	7.3	8.1	6.9	7.2	2.7	2.9	65.7
-Partly Cloudy	52	6.4	7.0	7.2	7.0	9.6	11.7	13.3	11.8	10.2	8.9	6.1	5.8	104.9
-Cloudy	52	21.1	17.4	18.6	17.2	15.3	12.2	10.5	11.2	12.9	14.9	21.2	22.3	194.7
Precipitation														
.01 inches or more	52	12.9	11.2	12.6	12.6	10.7	10.1	9.3	9.4	10.0	9.4	12.4	13.2	133.9
Snow,Ice pellets,hail														
1.0 inches or more	52	3.6	3.3	2.5	0.6	0.0	0.0	0.0	0.0	0.0	0.1	1.1	3.2	14.3
Thunderstorms	52	0.2	0.2	1.3	2.6	4.1	6.0	6.1	5.8	3.7	1.4	0.8	0.2	32.3
Heavy Fog Visibility														
1/4 mile or less	52	1.6	1.6	1.8	0.8	1.0	0.8	1.0	2.0	2.0	2.2	1.5	1.8	18.1
Temperature °F														
-Maximum														
90° and above	30	0.0	0.0	0.0	0.0	0.2	1.4	3.0	1.5	0.5	0.0	0.0	0.0	6.6
32° and below	30	18.2	14.5	5.4	0.4	0.0	0.0	0.0	0.0	0.0	0.0	2.2	12.4	53.0
-Minimum														
32° and below	30	29.1	25.7	22.9	10.6	1.6	0.0	0.0	0.0	0.3	6.1	16.4	26.5	139.3
0° and below	30	4.3	3.3	0.5	0.0	0.0	0.0	0.0	0.0	0.0	0.0	0.0	1.6	9.7
AVG. STATION PRESS.(mb)	21	988.4	989.6	987.8	986.8	987.0	987.0	988.1	989.4	989.8	989.8	988.7	989.0	988.5
RELATIVE HUMIDITY (%)														
Hour 01	30	78	77	76	74	75	79	82	86	86	81	80	80	80
Hour 07 (Local Time)	30	79	79	79	77	77	80	83	88	89	84	82	81	82
Hour 13	30	71	67	61	56	54	56	56	57	60	60	68	73	62
Hour 19	30	74	70	66	59	56	58	58	63	70	70	74	77	66
PRECIPITATION (inches):														
Water Equivalent														
-Normal		1.39	1.28	2.16	2.94	2.65	3.21	2.71	3.49	3.56	2.18	2.60	2.11	30.28
-Maximum Monthly	52	3.56	5.28	4.33	5.90	7.35	5.87	9.35	11.04	10.86	4.21	4.94	4.66	11.04
-Year		1947	1954	1948	1947	1945	1949	1992	1975	1986	1954	1988	1971	AUG 1975
-Minimum Monthly	52	0.07	0.17	0.25	0.62	0.34	0.63	0.73	0.45	0.32	0.33	0.66	0.44	0.07
-Year		1945	1969	1958	1942	1988	1988	1978	1969	1979	1944	1980	1969	JAN 1945
-Maximum in 24 hrs	52	1.81	2.85	2.33	2.89	2.25	3.55	3.72	4.45	6.04	3.19	2.11	1.77	6.04
-Year		1967	1954	1948	1976	1974	1943	1957	1968	1950	1981	1952	1971	SEP 1950
Snow,Ice pellets,hail														
-Maximum Monthly	52	28.5	20.8	19.4	17.3	0.6	T	T	0.0	T	4.4	16.2	24.9	28.5
-Year		1976	1990	1965	1975	1961	1992	1992		1975	1989	1951	1951	JAN 1976
-Maximum in 24 hrs	52	19.8	11.3	12.6	16.7	0.5	T	T	0.0	T	3.5	13.4	9.7	19.8
-Year		1967	1965	1973	1975	1961	1992	1992		1975	1989	1951	1970	JAN 1967
WIND:														
Mean Speed (mph)	52	11.8	11.2	11.9	11.5	10.1	9.1	8.2	7.8	8.8	9.8	11.2	11.3	10.2
Prevailing Direction through 1963		SW	WNW	WNW	WNW	WSW	SW	SW	SW	S	SW	SSW	SW	SW
Fastest Obs. 1 Min.														
-Direction (!!)	38	26	24	27	24	32	29	29	27	27	29	23	27	32
-Speed (MPH)	38	45	39	58	44	81	52	41	37	46	32	46	40	81
-Year		1980	1971	1956	1968	1956	1963	1991	1986	1978	1981	1963	1991	MAY 1956
Peak Gust														
-Direction (!!)	10	NW	W	W	SW	W	SW	SW	SE	N	S	S	NW	SW
-Speed (mph)	10	48	54	69	68	54	76	69	71	53	67	52	52	76
-Date		1993	1985	1991	1984	1985	1991	1991	1984	1990	1992	1991	1985	JUN 1991

PRECIPITATION (inches)

FLINT, MICHIGAN

YEAR	JAN	FEB	MAR	APR	MAY	JUNE	JULY	AUG	SEP	OCT	NOV	DEC	ANNUAL
1964	1.44	0.55	2.05	2.96	2.68	3.98	2.82	2.65	2.23	0.61	1.20	1.42	24.59
1965	2.60	2.16	1.49	1.68	0.93	1.87	1.64	4.55	2.90	1.06	2.35	2.56	25.79
1966	0.51	1.20	2.36	1.83	1.31	2.31	1.51	2.44	1.62	0.88	2.81	1.76	20.54
1967	2.41	1.21	0.58	5.63	1.02	5.36	1.76	2.34	1.68	3.32	2.19	2.79	30.29
1968	1.54	2.12	1.61	1.68	3.51	5.40	2.73	5.35	2.21	1.51	3.86	2.38	33.90
1969	1.86	0.17	1.31	4.25	4.64	3.29	5.92	0.45	0.33	2.96	2.96	0.44	28.58
1970	0.78	0.56	2.30	2.69	2.55	3.23	3.85	2.36	4.03	3.19	2.90	1.24	29.68
1971	0.46	2.32	1.85	1.62	1.98	2.43	3.90	1.62	2.10	2.12	1.46	4.66	26.52
1972	1.21	0.81	2.69	3.82	3.36	3.10	3.86	4.57	4.54	3.73	2.33	4.01	38.03
1973	1.38	1.20	4.20	2.84	3.74	3.24	1.78	1.76	2.80	2.73	4.93	2.78	33.38
1974	2.61	2.16	3.60	3.35	5.50	1.93	1.72	2.75	2.20	1.44	2.78	2.35	32.39
1975	2.89	2.01	2.78	4.31	3.70	4.68	1.92	11.04	3.48	1.37	3.17	4.03	45.38
1976	1.53	2.27	3.66	5.59	4.24	4.11	1.98	0.81	2.66	3.17	1.22	0.92	32.16
1977	0.57	0.70	2.78	3.28	0.95	4.70	1.91	3.15	4.59	2.09	2.67	1.47	28.86
1978	2.02	0.51	1.99	1.03	2.97	1.66	0.73	2.62	4.57	1.51	1.44	2.64	23.69
1979	1.78	0.34	1.58	2.33	1.58	3.96	2.96	2.53	0.32	1.90	3.64	2.30	25.22
1980	1.25	0.88	2.20	3.40	2.35	2.38	5.61	3.28	4.75	2.17	0.66	2.95	31.88
1981	0.59	1.83	0.76	4.32	3.07	2.95	3.93	3.22	7.65	3.09	1.41	1.24	34.06
1982	1.57	0.66	2.49	1.23	2.15	3.24	3.06	3.24	2.76	0.57	4.76	2.75	28.48
1983	1.01	0.91	2.40	3.63	3.76	3.94	2.24	2.26	4.06	2.62	3.46	1.85	32.14
1984	0.56	0.52	2.83	2.85	3.54	0.92	3.48	3.66	2.77	2.61	2.65	3.76	30.15
1985	2.06	3.01	4.00	3.60	2.81	1.54	2.99	4.33	8.29	3.45	3.16	1.38	40.62
1986	0.70	2.10	1.44	2.68	1.61	5.13	3.12	5.29	10.86	1.96	0.94	1.71	37.54
1987	1.16	0.42	1.45	1.80	2.28	1.60	2.20	5.03	5.57	1.90	3.29	2.54	29.24
1988	1.27	1.52	1.45	2.69	0.34	0.63	3.74	4.00	3.03	2.76	4.94	1.27	27.64
1989	1.31	0.52	2.11	1.50	3.93	4.90	2.65	5.56	4.33	1.67	2.86	1.00	32.34
1990	1.84	2.59	1.45	2.10	3.26	3.77	1.80	2.37	3.06	4.06	4.29	2.49	33.08
1991	1.49	0.68	2.87	4.07	3.23	2.91	3.78	3.33	0.63	3.52	3.01	2.15	31.67
1992	1.15	1.59	2.56	3.80	1.64	2.26	9.35	3.50	2.50	2.46	4.05	2.01	36.87
1993	2.80	1.16	1.21	4.55	1.68	5.23	2.32	3.37	4.66	2.21	1.32	0.51	31.02
Record Mean	1.59	1.49	2.24	2.97	2.99	3.31	3.06	3.43	3.19	2.24	2.51	1.96	30.98

AVERAGE TEMPERATURE (deg. F)

FLINT, MICHIGAN

YEAR	JAN	FEB	MAR	APR	MAY	JUNE	JULY	AUG	SEP	OCT	NOV	DEC	ANNUAL
1964	27.7	25.8	34.1	48.2	60.3	66.6	71.7	65.3	60.1	46.1	41.5	27.6	47.9
1965	22.1	24.8	27.1	42.9	61.6	65.4	68.4	67.6	62.5	49.4	40.9	35.4	47.3
1966	18.5	25.5	36.8	43.5	50.1	66.5	70.5	66.3	57.5	47.7	38.5	25.6	45.6
1967	26.0	18.3	30.9	47.3	50.7	68.5	67.3	64.8	58.0	49.3	34.8	30.7	45.5
1968	21.1	20.8	36.4	48.9	52.9	65.3	68.6	68.9	62.3	50.9	38.0	26.3	46.7
1969	22.6	26.1	31.1	47.1	54.2	60.1	69.0	70.1	62.5	49.2	36.9	25.5	46.2
1970	15.1	23.5	30.1	47.2	58.3	65.7	70.3	68.9	61.7	52.1	39.4	26.8	46.6
1971	19.6	26.3	29.7	43.3	54.6	69.5	67.5	67.4	64.8	57.6	38.8	32.4	47.6
1972	21.8	22.5	30.9	42.5	59.1	62.5	70.0	67.7	61.1	46.2	36.9	28.8	45.9
1973	28.8	24.3	41.3	46.8	52.7	67.4	70.4	71.0	62.6	55.3	40.3	28.2	49.1
1974	24.8	20.9	34.9	48.1	53.9	64.9	71.3	69.7	58.9	48.1	40.2	29.2	47.0
1975	28.0	27.2	30.6	40.3	61.8	67.2	70.5	69.9	57.4	53.6	47.3	28.8	48.6
1976	18.1	31.1	38.7	48.2	54.5	69.4	71.1	68.2	60.6	46.8	33.3	17.4	46.4
1977	10.9	22.4	39.2	49.5	62.6	63.1	73.8	68.2	63.6	48.8	41.3	25.8	47.5
1978	18.2	13.4	26.7	43.8	58.7	65.1	69.6	70.0	65.4	50.3	40.6	26.5	45.7
1979	15.9	12.8	38.1	44.6	56.8	66.7	69.9	66.8	62.7	49.9	39.6	32.1	46.3
1980	23.4	20.3	29.9	44.6	57.7	61.6	70.5	70.5	60.5	44.3	36.2	24.4	45.3
1981	18.1	28.3	35.6	48.2	54.2	65.9	70.7	69.0	57.7	44.7	38.7	26.1	46.5
1982	16.4	20.2	32.7	43.3	65.0	62.9	71.3	66.6	61.0	52.5	41.4	37.2	47.6
1983	27.7	31.8	37.8	43.7	53.5	67.3	74.0	73.3	64.8	51.5	41.1	20.1	48.9
1984	14.4	31.2	26.8	45.3	51.5	69.3	70.0	71.8	60.7	54.0	39.4	33.4	47.3
1985	19.8	22.7	37.0	51.8	59.3	62.5	70.4	67.2	63.3	51.4	39.8	21.3	47.2
1986	21.7	21.1	35.8	49.4	58.9	64.9	72.2	66.3	63.0	50.5	35.5	29.9	47.4
1987	24.7	28.7	38.1	49.9	62.8	70.3	75.6	69.3	62.6	45.2	41.6	32.2	50.1
1988	22.0	20.1	34.1	46.5	60.3	67.7	75.2	72.6	61.2	44.8	41.4	27.3	47.8
1989	30.9	20.8	32.7	44.0	56.2	66.2	71.3	68.0	59.6	51.2	35.9	16.1	46.1
1990	31.8	26.9	38.1	48.4	55.3	66.8	70.0	69.4	61.5	49.9	42.4	29.8	49.2
1991	21.2	29.3	37.7	50.2	64.2	69.7	71.7	70.0	59.6	52.8	36.2	29.4	49.3
1992	25.8	29.4	32.9	43.6	57.3	64.2	66.9	65.2	60.2	48.1	38.3	31.5	47.0
1993	26.2	20.5	32.2	45.3	58.0	65.8	72.9	70.4	56.3	47.3	36.8	27.0	46.6
Record Mean	22.5	24.2	32.9	45.7	56.9	66.3	70.5	69.0	61.9	51.2	39.8	28.1	47.4
Max	29.6	32.1	41.6	55.7	68.3	77.6	82.0	79.8	71.8	60.3	46.4	34.0	56.6
Min	15.3	16.3	24.2	35.7	45.4	55.1	59.0	58.1	52.0	42.0	33.2	22.1	38.2

REFERENCE NOTES

FLINT, MICHIGAN

GENERAL
T - TRACE AMOUNT.
BLANK ENTRIES DENOTE MISSING/UNREPORTED DATA.
# INDICATES A STATION OR INSTRUMENT RELOCATION.

SPECIFIC

(a) - LENGTH OF RECORD IN YEARS, ALTHOUGH INDIVIDUAL MONTHS MAY BE MISSING.
0.* OR * - THE VALUE IS BETWEEN 0.0 AND 0.05
NORMALS - BASED ON THE 1961-1990 RECORD PERIOD.
EXTREMES - DATES ARE THE MOST RECENT OCCURRENCE
WIND DIR.- NUMERALS SHOW TENS OF DEGRESS CLOCKWISE FROM TRUE NORTH. "00" INDICATES CALM.
RESULTANT DIRECTIONS ARE GIVEN TO WHOLE DEGREES.
BOLD VALUES INDICATE EXTREME VALUES WHICH OCCURRED AFTER THE ASOS SYSTEM WAS COMMISSIONED.

RECORD = PERIOD OF RECORD
RECORD MEAN PRECIPITATION IS THE MEAN OF ALL DAILY PRECIPITATION AMOUNTS DURING THE PERIOD OF RECORD.
RECORD MAX(MIN) TEMPERATURE IS THE MEAN OF ALL DAILY MAX(MIN) TEMPERATURES DURING THE PERIOD OF RECORD.
RECORD MEAN TEMPERATURE IS THE SUM OF THE RECORD MAX AND RECORD MIN DIVIDED BY 2.
AVERAGE TEMPERATURE IS THE SUM OF THE MEAN DAILY MAX AND MIN TEMPERATURE DIVIDED BY 2.

EXCEPTIONS

RECORD MEANS ARE THROUGH THE CURRENT YEAR, BEGINNING IN 1942 FOR TEMPERATURE
1942 FOR PRECIPITATION
1942 FOR SNOWFALL

HEATING DEGREE DAYS Base 65 deg. F FLINT, MICHIGAN

SEASON	JULY	AUG	SEP	OCT	NOV	DEC	JAN	FEB	MAR	APR	MAY	JUNE	TOTAL
1964-65	11	83	207	581	698	1152	1326	1117	1167	660	158	71	7231
1965-66	31	66	142	481	717	911	1435	1100	866	645	459	73	6926
1966-67	13	41	241	531	787	1215	1204	1302	1052	526	444	27	7383
1967-68	48	70	216	486	901	1059	1354	1276	881	481	375	77	7224
1968-69	36	52	118	447	804	1191	1309	1085	1042	529	339	187	7139
1969-70	16	11	143	485	836	1216	1539	1156	1075	544	238	80	7339
1970-71	19	26	151	395	761	1177	1400	1079	1088	646	324	41	7107
1971-72	29	28	108	248	782	1005	1328	1226	1052	669	192	117	6784
1972-73	34	40	139	574	835	1118	1116	1136	725	542	374	17	6650
1973-74	8	18	153	300	733	1132	1240	1230	924	504	344	71	6657
1974-75	4	10	211	519	737	1101	1142	1055	1059	736	162	56	6792
1975-76	8	2	225	355	525	1117	1448	978	809	518	322	16	6323
1976-77	1	33	169	559	943	1467	1671	1186	794	468	152	118	7561
1977-78	7	35	97	496	701	1210	1444	1438	1178	632	238	87	7563
1978-79	17	8	89	449	725	1187	1516	1458	829	606	291	64	7239
1979-80	17	36	120	477	753	1014	1282	1288	1079	607	250	154	7077
1980-81	2	8	152	633	856	1252	1445	1023	903	499	336	41	7150
1981-82	10	24	231	621	780	1199	1501	1246	996	648	83	91	7430
1982-83	3	58	163	385	700	851	1153	924	838	633	351	65	6124
1983-84	12	5	106	422	711	1385	1565	977	1178	586	411	23	7381
1984-85	14	12	176	335	761	972	1395	1177	860	418	197	103	6420
1985-86	7	28	153	413	748	1345	1338	1224	899	475	218	69	6917
1986-87	13	54	113	444	881	1081	1242	1010	826	447	174	25	6310
1987-88	11	44	119	605	695	1011	1326	1295	954	547	189	82	6878
1988-89	5	24	144	626	700	1161	1053	1233	993	626	289	49	6903
1989-90	5	28	199	422	864	1507	1020	1061	829	529	299	42	6805
1990-91	18	10	164	469	668	1084	1352	994	838	450	165	22	6234
1991-92	3	9	218	381	855	1098	1206	1024	988	638	266	78	6764
1992-93	26	56	193	516	793	1032	1194	1238	1007	583	218	71	6927
1993-94	1	16	267	544	837	1169							

COOLING DEGREE DAYS Base 65 deg. F FLINT, MICHIGAN

YEAR	JAN	FEB	MAR	APR	MAY	JUNE	JULY	AUG	SEP	OCT	NOV	DEC	TOTAL
1969	0	0	0	0	13	48	150	173	73	1	0	0	458
1970	0	0	0	17	36	108	191	152	57	4	0	0	565
1971	0	0	0	0	11	187	116	111	105	25	0	0	555
1972	0	0	0	0	18	50	199	130	30	0	0	0	427
1973	0	0	0	4	0	97	182	214	84	6	0	0	587
1974	0	0	0	4	6	77	206	163	35	2	0	0	493
1975	0	0	0	0	68	128	187	158	1	9	0	0	551
1976	0	0	0	20	5	155	196	140	41	3	0	0	560
1977	0	0	0	12	86	70	286	140	62	0	0	0	656
1978	0	0	0	0	51	95	165	172	108	0	0	0	591
1979	0	0	0	0	43	118	177	97	57	17	0	0	509
1980	0	0	0	3	27	55	181	183	27	0	0	0	476
1981	0	0	0	0	10	77	196	151	19	0	0	0	453
1982	0	0	0	0	91	35	204	118	49	6	0	0	503
1983	0	0	0	0	0	142	301	268	110	11	0	0	832
1984	0	0	0	0	1	157	179	231	53	1	0	0	622
1985	0	0	0	28	29	33	179	107	108	0	0	0	484
1986	0	0	0	12	38	73	246	103	57	0	0	0	529
1987	0	0	0	1	113	189	345	185	55	0	2	0	890
1988	0	0	0	0	50	171	328	266	34	6	0	0	855
1989	0	0	0	0	19	88	207	128	46	1	0	0	489
1990	0	0	5	40	5	100	177	153	65	9	1	0	555
1991	0	0	0	11	151	171	217	173	67	10	0	0	800
1992	0	0	0	2	34	61	91	74	53	0	0	0	315
1993	0	0	0	0	9	101	251	189	12	1	0	0	563

SNOWFALL (inches) FLINT, MICHIGAN

SEASON	JULY	AUG	SEP	OCT	NOV	DEC	JAN	FEB	MAR	APR	MAY	JUNE	TOTAL
1964-65	0.0	0.0	0.0	T	4.7	10.0	13.3	19.7	19.4	5.8	0.0	0.0	72.9
1965-66	0.0	0.0	0.0	T	1.9	7.2	9.5	9.3	3.1	2.6	T	0.0	33.6
1966-67	0.0	0.0	0.0	0.0	13.5	10.4	27.6	16.3	10.8	T	0.0	0.0	78.6
1967-68	0.0	0.0	T	2.6	3.0	2.4	14.1	7.1	11.4	0.2	T	0.0	40.8
1968-69	0.0	0.0	0.0	0.0	0.6	8.1	10.2	2.9	6.8	T	0.0	0.0	28.6
1969-70	0.0	0.0	0.0	T	6.2	6.1	13.4	8.1	12.5	8.7	T	0.0	55.0
1970-71	0.0	0.0	0.0	0.0	1.5	19.9	9.7	6.1	16.8	0.9	0.0	0.0	54.9
1971-72	0.0	0.0	0.0	T	10.4	7.2	14.6	12.9	5.5	1.5	0.0	0.0	52.1
1972-73	0.0	0.0	0.0	T	8.9	18.3	1.3	16.7	14.3	3.1	0.3	0.0	62.9
1973-74	0.0	0.0	0.0	0.0	T	18.3	9.3	14.2	6.0	0.4	T	0.0	48.2
1974-75	0.0	0.0	0.0	T	9.2	21.3	6.8	12.3	16.0	17.3	0.0	0.0	82.9
1975-76	0.0	0.0	T	0.0	10.5	21.1	28.5	7.8	5.6	3.1	T	0.0	76.6
1976-77	0.0	0.0	0.0	T	2.4	17.6	15.6	3.8	5.1	0.3	0.0	0.0	44.8
1977-78	0.0	0.0	0.0	T	3.4	9.9	23.6	8.4	5.0	0.3	0.0	0.0	50.6
1978-79	0.0	0.0	0.0	0.0	6.2	9.8	26.1	3.0	5.4	1.5	0.0	0.0	52.0
1979-80	0.0	0.0	0.0	T	9.4	2.5	6.2	7.2	10.8	3.6	0.0	0.0	39.7
1980-81	0.0	0.0	0.0	T	0.6	13.8	9.5	8.4	4.1	T	0.0	0.0	36.4
1981-82	0.0	0.0	0.0	T	2.8	11.9	19.3	10.7	9.3	8.2	0.0	0.0	62.2
1982-83	0.0	0.0	0.0	0.0	1.3	3.6	3.8	4.8	13.0	7.1	0.0	0.0	33.6
1983-84	0.0	0.0	0.0	0.0	2.4	20.0	9.0	3.6	6.5	0.5	0.0	0.0	42.0
1984-85	0.0	0.0	0.0	0.0	1.1	11.0	15.3	14.4	6.4	1.2	0.0	0.0	49.4
1985-86	0.0	0.0	0.0	0.0	3.4	13.2	11.2	20.6	4.3	0.6	0.0	0.0	53.3
1986-87	0.0	0.0	0.0	0.0	4.8	8.9	16.9	2.2	4.3	1.5	0.0	0.0	38.6
1987-88	0.0	0.0		0.5	T	8.8	5.9	19.6	4.3	0.1	0.0	0.0	
1988-89	0.0	0.0	0.0	T	1.3	6.2	8.7	8.7	3.3	1.3	0.0	T	29.5
1989-90	0.0	0.0	0.0	4.4	7.8	11.3	5.6	20.8	2.9	2.6	0.0	0.0	55.4
1990-91	0.0	0.0	0.0	0.0	T	10.6	16.6	5.8	1.2	T	0.0	0.0	34.2
1991-92	0.0	0.0	0.0	T	4.9	16.7	14.0	1.4	14.4	3.0	T	T	54.4
1992-93	T	0.0	0.0	1.8	0.6	4.6	11.3	16.1	11.8	7.0	0.0	0.0	53.2
1993-94	0.0	0.0	0.0	0.8	0.8	3.5							
Record Mean	T	0.0	T	0.2	3.8	9.7	11.6	9.8	7.6	2.4	T	T	45.1

GRAND RAPIDS, MICHIGAN

Grand Rapids, Michigan, is located in the west-central part of Kent County, in the picturesque Grand River valley about 30 air miles east of Lake Michigan. The Grand River, the longest stream in Michigan, flows through the city and bisects it into east and west sections. High hills rise on either side of the valley. Elevations range from 602 feet on the valley floor to 1,020 feet in the extreme southern part of Kent County, southwest of the airport.

Grand Rapids is under the natural climatic influence of Lake Michigan. In spring the cooling effect of Lake Michigan helps retard the growth of vegetation until the danger of frost has passed. The warming effect in the fall retards frost until most of the crops have matured. Fall is a colorful time of year in western Michigan, compensating for the late spring. During the winter, excessive cloudiness and numerous snow flurries occur with strong westerly winds. The tempering effect of Lake Michigan on cold waves coming in from the west and northwest is quite evident.

The tempering effect of the lake promotes the growth of a great variety of fruit trees and berries, especially apples, peaches, cherries, and blueberries. The intense cold of winter is modified, thus reducing winter kill of fruit trees. Summer days are pleasantly warm and most summer nights are quite comfortable, although there are about three weeks of hot, humid weather during most summers. Prolonged severe cold waves with below-zero temperatures are infrequent. The temperature usually rises to above zero during the daytime hours regardless of early morning readings.

July is the sunniest month and December is the month with the least sunshine. November through January is usually a period of excessive cloudiness and minimal sunshine.

Precipitation is usually ample for the growth and development of all vegetation. About one-half of the annual precipitation falls during the growing season, May through September. Droughts occur occasionally, but are seldom of protracted length. The snowfall season extends from mid-November to mid-March. Some winters have had continuous snow cover throughout this period, although there is usually a mid-winter thaw. The Grand River flows through the city and reaches critical heights a couple of times each year, generally once in January-February and again in March-April. Overflow is generally limited to the lowlands of the flood plain.

November is one of the windiest months and although violent windstorms are infrequent, gusts have on occasion exceeded 65 mph. Summer thunderstorms occasionally produce gusty winds over 60 mph.

NORMALS, MEANS, AND EXTREMES

GRAND RAPIDS, MICHIGAN

LATITUDE: 42°53'N LONGITUDE: 85°31'W ELEVATION: FT. GRND 784 BARO 819 TIME ZONE: EASTERN WBAN: 94860

	(a)	JAN	FEB	MAR	APR	MAY	JUNE	JULY	AUG	SEP	OCT	NOV	DEC	YEAR
TEMPERATURE °F:														
Normals														
-Daily Maximum		29.0	31.6	42.8	56.6	69.3	78.7	82.8	80.5	72.0	59.8	45.8	33.5	56.9
-Daily Minimum		14.7	15.8	25.4	35.4	45.6	55.3	60.4	58.4	49.9	39.1	30.2	20.7	37.6
-Monthly		21.8	23.7	34.1	46.0	57.5	67.0	71.6	69.5	61.0	49.5	38.0	27.1	47.2
Extremes														
-Record Highest	30	62	67	78	88	92	98	100	100	93	87	77	67	100
-Year		1967	1976	1986	1970	1978	1988	1988	1964	1973	1975	1975	1982	JUL 1988
-Record Lowest	30	-21	-19	-8	3	22	33	41	39	27	18	5	-18	-21
-Year		1979	1973	1978	1982	1966	1972	1983	1976	1991	1988	1977	1983	JAN 1979
NORMAL DEGREE DAYS:														
Heating (base 65°F)		1339	1156	958	570	273	50	0	18	139	485	810	1175	6973
Cooling (base 65°F)		0	0	0	0	40	110	208	157	19	0	0	0	534
% OF POSSIBLE SUNSHINE	30	30	40	45	51	56	61	63	61	54	44	27	21	46
MEAN SKY COVER (tenths)														
Sunrise - Sunset	30	8.3	7.7	7.4	6.9	6.4	6.1	5.9	5.9	6.4	6.9	8.2	8.6	7.1
MEAN NUMBER OF DAYS:														
Sunrise to Sunset														
-Clear	30	2.6	3.5	4.8	6.0	7.1	6.7	7.8	8.3	7.0	5.7	2.7	2.2	64.4
-Partly Cloudy	30	5.1	5.9	6.9	7.0	9.1	11.1	12.3	11.3	9.2	8.3	5.3	4.0	95.4
-Cloudy	30	23.3	18.9	19.3	17.0	14.8	12.2	10.9	11.5	13.8	16.9	22.0	24.8	205.5
Precipitation														
.01 inches or more	30	16.2	11.8	12.7	13.3	10.4	10.0	9.4	9.4	10.8	11.4	13.6	16.1	145.1
Snow, Ice pellets, hail														
1.0 inches or more	30	6.7	4.1	2.9	0.8	0.0	0.0	0.0	0.0	0.0	0.2	2.6	6.1	23.4
Thunderstorms	30	0.3	0.3	1.7	3.4	3.7	5.7	6.2	5.4	4.3	1.7	1.3	0.4	34.3
Heavy Fog Visibility														
1/4 mile or less	30	2.0	2.3	2.5	1.8	1.5	1.5	1.4	2.0	2.1	2.1	2.1	3.3	24.6
Temperature °F														
-Maximum														
90° and above	30	0.0	0.0	0.0	0.0	0.5	2.2	4.9	2.6	0.4	0.0	0.0	0.0	10.7
32° and below	30	18.9	14.8	5.7	0.4	0.0	0.0	0.0	0.0	0.0	0.0	2.4	13.6	55.8
-Minimum														
32° and below	30	29.2	26.3	23.8	11.7	2.1	0.0	0.0	0.0	0.4	6.4	17.5	27.7	145.1
0° and below	30	3.7	2.7	0.4	0.0	0.0	0.0	0.0	0.0	0.0	0.0	0.0	1.4	8.3
AVG. STATION PRESS. (mb)	21	987.2	988.3	986.3	985.4	985.7	985.6	986.9	988.2	988.6	988.4	987.3	987.6	987.1
RELATIVE HUMIDITY (%)														
Hour 01	30	80	79	77	75	76	80	83	86	86	82	81	82	81
Hour 07 (Local Time)	30	82	81	80	79	79	81	84	89	89	85	83	83	83
Hour 13	30	72	68	64	57	53	55	56	58	61	62	70	75	63
Hour 19	30	76	72	67	59	55	57	58	63	71	72	76	79	67
PRECIPITATION (inches):														
Water Equivalent														
-Normal		1.83	1.42	2.63	3.37	3.13	3.68	3.19	3.57	4.24	2.81	3.32	2.85	36.04
-Maximum Monthly	30	4.36	3.34	5.12	6.11	8.29	8.21	8.83	8.46	11.85	6.30	7.81	6.63	11.85
-Year		1975	1986	1974	1981	1981	1967	1992	1987	1986	1969	1966	1971	SEP 1986
-Minimum Monthly	30	0.47	0.33	1.08	1.79	0.94	0.25	0.81	0.14	T	0.60	0.95	0.66	T
-Year		1981	1969	1968	1989	1987	1988	1976	1969	1979	1964	1986	1969	SEP 1979
-Maximum in 24 hrs	30	2.16	1.52	1.78	2.63	5.48	3.28	3.38	3.68	4.55	3.19	3.00	2.79	5.48
-Year		1993	1985	1985	1991	1981	1972	1992	1987	1986	1993	1990	1982	MAY 1981
Snow, Ice pellets, hail														
-Maximum Monthly	30	45.5	25.1	36.0	12.4	0.2	0.0	T	T	T	8.4	25.3	34.8	45.5
-Year		1979	1989	1965	1982	1990		1990	1993	1967	1967	1991	1983	JAN 1979
-Maximum in 24 hrs	30	16.1	9.1	13.2	9.8	0.2	0.0	T	T	T	8.4	10.4	15.1	16.1
-Year		1978	1985	1970	1975	1990		1990	1993	1967	1967	1991	1970	JAN 1978
WIND:														
Mean Speed (mph)	30	11.5	10.7	11.1	11.0	9.7	8.9	8.2	7.9	8.3	9.4	10.5	10.7	9.8
Prevailing Direction														
Fastest Obs. 1 Min.														
-Direction (!!)	14	26	33	25	24	27	24	29	20	36	25	25	24	25
-Speed (MPH)	14	39	37	41	41	39	39	35	31	31	36	44	39	44
-Year		1980	1987	1982	1984	1990	1992	1980	1989	1989	1993	1991	1982	NOV 1991
Peak Gust														
-Direction (!!)	10	W	NW	W	SW	SW	SW	NW	N	N	W	SW	SW	SW
-Speed (mph)	10	51	55	56	63	68	63	51	61	52	45	55	51	68
-Date		1990	1987	1991	1984	1987	1992	1991	1984	1989	1993	1991	1992	MAY 1987

PRECIPITATION (inches)

GRAND RAPIDS, MICHIGAN

YEAR	JAN	FEB	MAR	APR	MAY	JUNE	JULY	AUG	SEP	OCT	NOV	DEC	ANNUAL
1964	1.42	0.73	3.54	5.28	3.96	4.12	2.14	3.19	3.16	0.60	2.13	2.01	32.28
1965	3.99	1.58	3.11	2.49	1.53	2.89	1.96	5.36	6.62	2.10	2.35	4.23	38.21
1966	1.22	2.28	2.65	4.62	2.16	2.36	1.95	3.82	1.92	2.33	7.81	4.01	37.13
1967	1.94	1.13	2.49	4.27	1.86	8.21	2.77	3.34	2.37	4.42	4.35	3.60	40.75
1968	1.55	2.12	1.08	2.40	2.67	5.02	2.74	2.86	4.21	4.05	4.28	3.16	36.14
1969	2.39	0.33	1.29	5.16	3.40	4.74	6.42	0.14	0.93	6.30	3.00	0.66	34.76
1970	1.18	0.51	2.43	3.27	4.24	6.53	6.32	2.89	7.18	3.65	3.42	1.89	43.51
1971	1.04	2.48	1.77	2.27	1.05	2.01	2.46	0.92	6.30	1.39	2.84	6.63	31.16
1972	1.26	0.90	2.11	3.85	1.99	4.64	3.72	5.01	3.96	2.92	2.06	4.96	37.38
1973	1.66	1.15	3.34	3.47	4.31	3.58	2.06	1.45	2.47	4.12	3.54	3.28	34.43
1974	3.23	2.09	5.12	2.93	4.01	4.43	0.97	4.61	2.05	2.44	3.11	1.83	36.82
1975	4.36	1.92	2.28	4.07	2.08	5.97	2.31	7.38	2.00	1.04	3.82	4.02	41.25
1976	1.67	2.13	4.99	4.75	6.63	2.79	0.81	1.03	1.21	2.00	1.51	1.05	30.57
1977	1.59	1.35	3.81	4.04	1.33	3.50	5.16	4.77	4.26	2.28	2.34	3.35	37.78
1978	2.22	0.54	2.01	2.55	2.91	4.65	2.83	5.00	5.62	3.05	1.83	3.32	36.53
1979	2.09	0.61	3.72	3.56	1.37	4.16	2.27	4.33	T	2.10	5.46	2.97	32.64
1980	1.76	1.76	1.74	3.64	3.19	4.00	5.90	3.18	4.57	1.99	1.57	3.60	36.90
1981	0.47	2.03	1.29	6.11	8.29	4.22	3.74	2.95	9.52	2.54	2.58	1.10	44.84
1982	2.98	0.36	3.36	2.11	3.63	2.45	3.81	3.07	1.92	1.42	5.36	6.49	36.96
1983	1.33	1.12	3.30	5.06	4.64	2.09	4.76	1.49	4.87	2.66	3.00	2.79	37.11
1984	0.94	1.15	2.77	2.10	4.77	0.62	2.12	1.49	2.16	3.34	2.85	4.37	28.68
1985	1.94	3.26	4.20	2.54	1.36	1.68	3.09	6.48	4.26	4.64	5.45	2.00	40.90
1986	1.07	3.34	2.35	2.58	3.88	7.14	5.27	5.30	11.85	2.76	0.95	1.04	47.53
1987	0.67	0.37	1.15	2.40	0.94	3.56	2.93	8.46	4.47	2.33	2.49	3.29	33.06
1988	2.39	1.14	2.12	3.11	1.07	0.25	3.69	3.04	7.49	5.37	4.82	1.88	36.37
1989	0.95	1.01	2.47	1.79	4.33	5.02	1.29	4.78	4.90	1.53	4.86	0.97	33.90
1990	2.39	2.08	1.96	2.23	4.39	3.00	3.73	3.40	4.22	5.05	7.14	2.97	42.56
1991	1.32	0.64	3.58	5.58	4.44	1.76	6.24	3.79	2.93	5.61	6.41	2.63	44.93
1992	1.52	1.06	3.51	3.98	1.45	1.61	8.83	3.55	5.60	2.34	5.64	3.27	42.36
1993	4.21	1.13	2.31	4.93	2.17	6.05	1.83	7.73	8.20	4.32	2.12	1.47	46.47
Record Mean	2.18	1.82	2.51	3.01	3.29	3.54	2.98	2.88	3.60	2.80	2.84	2.48	33.93

AVERAGE TEMPERATURE (deg. F)

GRAND RAPIDS, MICHIGAN

YEAR	JAN	FEB	MAR	APR	MAY	JUNE	JULY	AUG	SEP	OCT	NOV	DEC	ANNUAL
1964	27.5	25.0	34.5	48.2	62.0	68.9	73.1	67.3	61.4	47.2	41.8	26.0	48.6
1965	22.5	23.8	26.2	42.8	62.3	65.9	70.0	67.9	62.3	50.0	40.0	33.5	47.3
1966	19.1	25.5	37.1	43.8	52.0	69.2	73.6	68.5	61.4	51.2	42.1	29.2	47.7
1967	28.6	20.4	34.1	46.4	51.9	69.3	68.6	66.0	60.3	49.9	34.5	29.9	46.7
1968	21.0	21.4	38.0	48.8	54.0	66.7	69.8	70.6	63.7	51.8	37.9	24.3	47.3
1969	20.8	24.4	29.3	46.4	55.7	61.8	71.3	72.1	62.4	47.9	33.5	25.0	45.9
1970	17.0	22.4	28.8	46.8	59.4	66.8	72.1	70.1	61.7	52.2	38.3	25.7	46.8
1971	18.3	24.1	29.0	43.9	54.0	70.9	69.2	68.6	65.6	58.3	38.0	32.0	47.6
1972	20.8	21.9	29.7	42.3	59.3	63.5	70.3	68.7	62.0	46.1	35.9	26.6	45.6
1973	27.4	22.8	41.9	46.8	53.7	69.9	72.4	72.7	63.4	55.0	39.9	26.2	49.3
1974	25.7	21.1	33.6	47.2	53.2	63.0	71.1	67.3	56.4	46.9	37.6	28.4	46.0
1975	25.2	24.3	28.8	39.8	60.7	67.6	70.6	70.5	57.0	53.0	44.7	27.3	47.5
1976	18.9	31.0	37.9	48.2	53.9	69.5	72.5	68.9	60.2	45.7	31.5	19.1	46.4
1977	12.7	22.8	40.0	52.1	65.8	64.5	74.0	67.9	62.9	48.6	39.5	25.9	48.1
1978	19.3	14.4	27.9	45.3	59.9	66.8	70.4	68.1	63.7	47.2	37.6	26.6	45.6
1979	17.1	15.0	36.9	44.5	57.7	67.6	71.1	68.5	63.4	51.5	40.5	33.0	47.2
1980	25.1	22.7	32.5	47.0	59.9	64.4	72.7	72.6	62.7	45.8	37.5	25.1	47.3
1981	20.8	30.2	35.1	47.3	55.5	68.6	72.1	70.7	60.9	48.2	39.8	28.7	48.2
1982	17.2	22.1	32.5	41.8	65.0	62.8	73.1	68.6	61.9	53.6	40.8	36.2	48.0
1983	27.6	30.9	36.9	42.6	52.8	67.7	74.7	72.2	62.4	50.7	40.7	19.2	48.2
1984	17.1	34.0	28.9	47.3	53.5	70.1	70.1	73.0	60.5	54.2	40.3	32.2	48.4
1985	18.6	21.3	36.2	51.7	60.4	63.8	70.7	67.4	63.6	50.6	38.4	22.3	47.1
1986	22.7	22.5	36.7	49.7	58.8	64.4	72.5	66.0	62.8	50.2	35.0	29.8	47.6
1987	25.4	29.9	37.3	50.1	62.6	71.2	74.1	69.3	62.2	45.0	41.1	31.5	50.0
1988	20.8	20.7	34.1	47.1	60.9	68.5	74.7	73.4	61.3	44.2	40.7	27.3	47.8
1989	30.5	19.7	31.8	43.9	55.8	65.9	71.9	68.4	59.4	51.1	35.7	17.2	45.9
1990	32.1	28.1	37.1	47.7	55.0	67.1	70.3	69.0	62.8	49.9	43.3	30.0	49.4
1991	22.2	29.2	37.9	50.8	64.6	71.2	72.4	70.5	60.1	52.0	35.7	29.4	49.7
1992	27.5	30.3	33.7	44.0	57.4	64.3	67.2	65.0	59.7	48.2	38.2	30.5	47.2
1993	25.9	21.9	32.4	44.0	58.2	65.2	72.7	71.4	56.5	48.0	37.9	27.8	46.8
Record Mean	24.0	24.4	34.0	46.6	58.0	67.8	72.6	70.4	62.9	51.5	39.1	28.2	48.3
Max	30.5	31.6	42.2	56.5	68.9	78.7	83.3	81.0	73.0	60.8	46.0	34.1	57.2
Min	17.5	17.2	25.8	36.7	47.1	56.8	61.8	59.8	52.7	42.2	32.2	22.3	39.3

REFERENCE NOTES

GRAND RAPIDS, MICHIGAN

GENERAL
T - TRACE AMOUNT.
BLANK ENTRIES DENOTE MISSING/UNREPORTED DATA.
# INDICATES A STATION OR INSTRUMENT RELOCATION.

SPECIFIC

PM - INCLUDES LAST DAY OF PREVIOUS MONTH
ASOS - AUTOMATED SURFACE OBSERVING SYSTEM IN OPERATION DURING THESE MONTHS.

(a) - LENGTH OF RECORD IN YEARS, ALTHOUGH INDIVIDUAL MONTHS MAY BE MISSING.
0.* OR * - THE VALUE IS BETWEEN 0.0 AND 0.05
NORMALS - BASED ON THE 1961-1990 RECORD PERIOD.
EXTREMES - DATES ARE THE MOST RECENT OCCURRENCE
WIND DIR.- NUMERALS SHOW TENS OF DEGRESS CLOCKWISE FROM TRUE NORTH. "00" INDICATES CALM.
RESULTANT DIRECTIONS ARE GIVEN TO WHOLE DEGREES.
BOLD VALUES INDICATE EXTREME VALUES WHICH OCCURRED AFTER THE ASOS SYSTEM WAS COMMISSIONED.

RECORD = PERIOD OF RECORD
RECORD MEAN PRECIPITATION IS THE MEAN OF ALL DAILY PRECIPITATION AMOUNTS DURING THE PERIOD OF RECORD.
RECORD MAX(MIN) TEMPERATURE IS THE MEAN OF ALL DAILY MAX(MIN) TEMPERATURES DURING THE PERIOD OF RECORD.
RECORD MEAN TEMPERATURE IS THE SUM OF THE RECORD MAX AND RECORD MIN DIVIDED BY 2.
AVERAGE TEMPERATURE IS THE SUM OF THE MEAN DAILY MAX AND MIN TEMPERATURE DIVIDED BY 2.

EXCEPTIONS

RECORD MEANS ARE THROUGH THE CURRENT YEAR, BEGINNING IN 1894 FOR TEMPERATURE
1870 FOR PRECIPITATION
1964 FOR SNOWFALL

HEATING DEGREE DAYS Base 65 deg. F — GRAND RAPIDS, MICHIGAN

SEASON	JULY	AUG	SEP	OCT	NOV	DEC	JAN	FEB	MAR	APR	MAY	JUNE	TOTAL
1964-65	12	54	179	545	690	1202	1310	1145	1197	657	142	57	7190
1965-66	5	50	145	462	743	968	1414	1102	860	632	400	47	6828
1966-67	0	12	146	426	680	1102	1120	1242	950	554	410	9	6651
1967-68	38	47	179	477	909	1080	1356	1255	827	482	342	70	7062
1968-69	30	36	90	434	805	1254	1360	1132	1102	550	300	151	7244
1969-70	3	2	145	524	938	1232	1481	1186	1117	559	216	61	7464
1970-71	13	17	150	397	798	1210	1441	1140	1110	626	342	35	7279
1971-72	15	11	107	233	800	1016	1365	1240	1089	675	204	102	6857
1972-73	28	41	126	577	867	1185	1162	1178	708	552	341	4	6769
1973-74	1	7	136	314	748	1196	1212	1225	967	532	364	113	6815
1974-75	5	22	275	555	816	1127	1227	1133	1113	750	191	61	7275
1975-76	16	5	242	382	602	1161	1420	980	835	526	342	18	6529
1976-77	0	32	185	591	999	1415	1616	1177	765	401	106	91	7378
1977-78	3	48	99	501	759	1204	1407	1413	1140	584	213	62	7433
1978-79	15	22	124	545	816	1182	1477	1393	863	607	264	42	7350
1979-80	7	31	109	431	731	984	1230	1221	1000	541	189	106	6580
1980-81	0	4	115	588	821	1227	1363	969	919	525	298	8	6837
1981-82	7	5	173	513	749	1116	1475	1196	998	689	91	98	7110
1982-83	6	34	140	361	717	884	1151	949	866	663	373	61	6205
1983-84	16	2	149	440	721	1413	1480	892	1107	532	353	13	7118
1984-85	16	4	189	333	735	1014	1431	1216	886	428	173	81	6506
1985-86	9	17	157	436	790	1316	1302	1186	871	466	214	75	6839
1986-87	11	56	118	452	894	1084	1220	978	849	440	178	20	6300
1987-88	18	36	118	610	712	1032	1364	1277	950	531	169	60	6877
1988-89	3	21	135	639	722	1162	1062	1263	1023	625	297	51	7003
1989-90	2	23	203	424	874	1477	1014	1026	865	549	306	45	6808
1990-91	11	15	139	475	645	1077	1323	997	830	430	157	18	6117
1991-92	5	4	211	404	871	1098	1155	1001	964	626	262	75	6676
1992-93	25	66	197	512	794	1060	1204	1203	1004	624	227	84	7000
1993-94	1	16	259	524	810	1150							

COOLING DEGREE DAYS Base 65 deg. F — GRAND RAPIDS, MICHIGAN

YEAR	JAN	FEB	MAR	APR	MAY	JUNE	JULY	AUG	SEP	OCT	NOV	DEC	TOTAL
1969	0	0	0	0	21	61	202	229	76	1	0	0	590
1970	0	0	0	19	49	122	242	180	55	6	0	0	673
1971	0	0	0	0	8	218	154	129	132	31	0	0	672
1972	0	0	0	0	37	61	200	163	42	0	0	0	503
1973	0	0	0	10	0	157	238	255	96	12	0	0	768
1974	0	0	0	4	6	60	201	103	23	3	0	0	400
1975	0	0	0	0	65	148	199	184	8	15	0	0	619
1976	0	0	0	27	4	158	239	161	49	0	0	0	638
1977	0	0	0	19	137	85	286	143	44	0	0	0	714
1978	0	0	0	0	62	123	188	125	90	0	0	0	588
1979	0	0	0	0	46	129	204	147	69	19	0	0	614
1980	0	0	0	6	34	96	247	245	53	0	0	0	681
1981	0	0	0	0	13	124	236	190	55	0	0	0	618
1982	0	0	0	0	99	40	263	153	55	12	0	0	622
1983	0	0	0	0	2	146	325	234	76	5	0	0	788
1984	0	0	0	7	5	174	179	259	60	6	0	0	690
1985	0	0	0	36	39	52	193	99	119	0	0	0	538
1986	0	0	0	14	26	64	252	94	61	0	0	0	511
1987	0	0	0	2	110	213	306	174	38	0	0	0	843
1988	0	0	0	0	50	170	310	289	29	2	0	0	850
1989	0	0	0	0	19	84	223	135	40	0	0	0	501
1990	0	0	6	37	6	115	185	144	83	13	0	0	589
1991	0	0	0	12	155	211	240	182	71	7	0	0	878
1992	0	0	0	2	36	60	99	74	44	1	0	0	316
1993	0	0	0	0	21	94	244	222	10	5	0	0	596

SNOWFALL (inches) — GRAND RAPIDS, MICHIGAN

SEASON	JULY	AUG	SEP	OCT	NOV	DEC	JAN	FEB	MAR	APR	MAY	JUNE	TOTAL
1964-65	0.0	0.0	0.0	T	5.8	13.4	24.1	16.7	36.0	5.4	0.0	0.0	101.4
1965-66	0.0	0.0	T	0.0	9.1	9.6	25.9	15.0	4.6	2.8	T	0.0	67.0
1966-67	0.0	0.0	0.0	0.0	16.6	17.6	29.8	17.5	10.3	T	T	0.0	91.8
1967-68	0.0	0.0	T	8.4	11.0	9.6	11.8	8.5	4.2	1.6	0.0	0.0	55.1
1968-69	0.0	0.0	0.0	0.0	4.2	26.2	27.7	6.3	7.9	0.0	0.0	0.0	72.3
1969-70	0.0	0.0	0.0	T	14.3	11.8	23.2	8.2	19.3	7.8	0.0	0.0	84.6
1970-71	0.0	0.0	0.0	0.0	6.1	33.3	27.2	4.4	25.9	4.1	0.0	0.0	101.0
1971-72	0.0	0.0	0.0	0.0	14.9	3.7	22.6	16.9	14.2	7.5	0.0	0.0	79.8
1972-73	0.0	0.0	0.0	0.9	11.0	19.8	7.0	13.2	8.5	5.0	0.1	0.0	65.5
1973-74	0.0	0.0	0.0	0.0	0.4	20.0	13.3	18.4	11.3	1.0	T	0.0	64.4
1974-75	0.0	0.0	0.0	0.4	8.9	16.5	10.7	10.6	11.8	10.0	0.0	0.0	68.9
1975-76	0.0	0.0	0.0	0.0	6.6	23.3	25.0	6.5	3.5	4.2	0.1	0.0	69.2
1976-77	0.0	0.0	0.0	2.0	8.5	17.7	26.1	5.0	9.5	2.0	0.0	0.0	70.8
1977-78	0.0	0.0	0.0	0.0	10.6	23.2	35.8	8.8	6.2	T	0.0	0.0	84.6
1978-79	0.0	0.0	0.0	T	6.2	30.0	45.5	5.3	7.2	1.8	0.0	0.0	96.0
1979-80	0.0	0.0	0.0	T	9.4	2.6	13.3	12.6	6.6	4.0	T	0.0	48.5
1980-81	0.0	0.0	0.0	0.4	5.5	17.3	8.1	18.8	1.4	T	0.0	0.0	51.5
1981-82	0.0	0.0	0.0	T	4.4	8.9	30.3	6.7	11.8	12.4	0.0	0.0	74.5
1982-83	0.0	0.0	0.0	0.0	5.2	8.2	5.7	2.9	13.2	0.7	0.0	0.0	35.9
1983-84	0.0	0.0	0.0	T	4.7	34.8	19.6	1.6	10.6	0.1	T	0.0	71.4
1984-85	0.0	0.0	0.0	0.0	T	15.7	22.6	21.3	6.7	3.3	0.0	0.0	69.6
1985-86	0.0	0.0	0.0	0.0	3.5	30.7	18.4	20.2	6.1	0.2	0.0	0.0	79.1
1986-87	0.0	0.0	0.0	0.0	5.3	12.7	19.2	0.9	5.7	3.8	0.0	0.0	47.6
1987-88	0.0	0.0	0.0	1.6	0.7	18.2	21.9	18.1	3.4	0.3	0.0	0.0	64.2
1988-89	0.0	0.0	0.0	0.2	5.5	14.4	8.7	25.1	6.3	2.2	T	0.0	62.4
1989-90	0.0	0.0	0.0	5.8	19.4	25.2	10.6	23.8	2.7	2.1	0.2	0.0	89.8
1990-91	T	0.0	0.0	T	2.0	18.6	27.7	9.5	2.8	T	0.0	0.0	60.6
1991-92	0.0	0.0	0.0	0.3	25.3	27.9	13.4	3.5	15.1	2.3	0.0	0.0	87.8
1992-93	0.0	0.0	0.0	2.3	4.2	14.2	11.1	18.6	11.6	3.3	0.0	0.0	65.3
1993-94	0.0	T	0.0	T	1.9	17.7							
Record Mean	T	T	T	0.7	7.7	18.1	20.2	11.9	10.0	3.0	T	0.0	71.6

HOUGHTON LAKE, MICHIGAN

Houghton Lake is located in north-central lower Michigan. The present station is on the northeast shore of Houghton Lake, the largest inland lake in Michigan, with a circumference of about 32 miles. The Muskegon River source is Higgins Lake, 8 miles to the north. It flows through Houghton Lake, then southwestward to Lake Michigan. The station lies within an elongated bowl shaped 1,000-foot plateau, which extends roughly 50 miles north, 75 miles southwest, and about 20 miles southeast of Houghton Lake. In the immediate area, the land is level to rolling, but there are hills and ridges from 100 to 300 feet higher in elevation surrounding the station. Soils are generally sand, or sandy loam supporting little agricultural production, but the area is rich in natural resources of forests, lakes, and streams.

The interior location diminishes the influence of the larger Great Lakes, which lie 70 to 80 miles east and west of Houghton Lake. Hence, the daily temperature range is larger, especially in summer, and temperature extremes are greater than are found nearer the shores of either Lake Michigan or Lake Huron. Temperatures reach the 100 degree mark about one summer out of ten, and at the other extreme, fall below zero an average of twenty-two times during the winter season.

Precipitation is normally a little heavier during the summer season. About 60 percent of the annual total falls in the six-month period from April through September. The heaviest precipitation occurs with summertime thunderstorms.

Snowfall averages above 80 inches per year at Houghton Lake, with considerable variation from year to year. Much heavier snows, averaging over 100 inches a season, fall within a 30- to 60-mile radius to the north and west of Houghton Lake. Seasonal totals have ranged from 24 inches to over 124 inches. Measurable amounts of snow have occurred in nine of the twelve months, and the average number of months with measurable snowfall is six.

Cloudiness is greatest in the late fall and early winter, while sunshine percentage is highest in the spring and summer. Cloudiness is increased in the late fall due to the moisture and warmth picked up by the westerly and northwesterly winds while crossing Lake Michigan.

The growing season is normally quite short, averaging about 90 days between spring and fall freezes.

NORMALS, MEANS, AND EXTREMES

HOUGHTON LAKE, MICHIGAN

LATITUDE: 44 °22'N LONGITUDE: 84 °41' W ELEVATION: FT. GRND 1149 BARO 1157 TIME ZONE: EASTERN WBAN: 94814

	(a)	JAN	FEB	MAR	APR	MAY	JUNE	JULY	AUG	SEP	OCT	NOV	DEC	YEAR
TEMPERATURE °F:														
Normals														
-Daily Maximum		25.3	28.1	38.2	52.7	66.1	74.8	79.6	76.4	67.8	55.8	42.0	29.5	53.0
-Daily Minimum		8.4	8.2	18.2	31.6	42.0	50.6	55.4	53.8	46.9	37.5	28.4	15.7	33.1
-Monthly		16.9	18.2	28.2	42.2	54.1	62.7	67.5	65.1	57.3	46.7	35.2	22.6	43.1
Extremes														
-Record Highest	29	53	59	76	86	90	95	98	94	92	85	70	63	98
-Year		1973	1984	1990	1980	1988	1987	1987	1988	1985	1971	1990	1982	JUL 1987
-Record Lowest	29	-26	-34	-23	3	21	29	33	29	21	16	-2	-21	-34
-Year		1981	1979	1967	1982	1966	1972	1965	1982	1989	1969	1989	1976	FEB 1979
NORMAL DEGREE DAYS:														
Heating (base 65°F)		1491	1310	1141	684	358	120	35	70	234	567	894	1314	8218
Cooling (base 65°F)		0	0	0	0	20	51	112	73	0	0	0	0	256
% OF POSSIBLE SUNSHINE														
MEAN SKY COVER (tenths)														
Sunrise - Sunset	28	8.0	7.4	7.0	6.8	6.4	6.1	5.8	5.9	6.6	7.2	8.4	8.4	7.0
MEAN NUMBER OF DAYS:														
Sunrise to Sunset														
-Clear	28	2.9	4.2	5.9	6.2	6.9	6.6	7.5	8.0	6.3	5.1	2.4	2.3	64.3
-Partly Cloudy	28	6.4	6.6	7.1	7.1	9.6	11.5	13.2	11.2	9.1	7.8	4.9	5.8	100.3
-Cloudy	28	21.7	17.4	18.0	16.6	14.2	11.9	10.0	11.8	14.6	18.0	22.6	22.9	199.8
Precipitation														
.01 inches or more	29	14.8	11.3	12.0	11.9	10.1	10.5	9.1	9.9	11.7	11.9	13.5	15.3	142.0
Snow,Ice pellets,hail														
1.0 inches or more	29	6.0	4.6	3.4	1.4	0.1	0.0	0.0	0.0	0.0	0.3	2.9	5.4	24.1
Thunderstorms	29	0.1	0.1	0.8	2.0	4.0	5.2	5.7	5.9	4.0	1.4	0.5	0.2	30.1
Heavy Fog Visibility														
1/4 mile or less	29	1.9	1.5	2.7	1.5	1.4	1.2	2.0	3.6	3.6	2.8	2.8	2.8	27.8
Temperature °F														
-Maximum														
90° and above	29	0.0	0.0	0.0	0.0	0.*	0.8	1.9	0.6	0.*	0.0	0.0	0.0	3.3
32° and below	29	23.5	18.4	9.0	1.0	0.0	0.0	0.0	0.0	0.0	0.0	5.2	18.9	76.0
-Minimum														
32° and below	29	30.8	27.8	27.6	17.0	4.9	0.3	0.0	0.*	1.6	9.6	22.2	29.6	171.3
0° and below	29	8.4	7.9	3.0	0.0	0.0	0.0	0.0	0.0	0.0	0.0	0.1	3.1	22.5
AVG. STATION PRESS.(mb)	7	972.5	972.5	972.2	973.1	971.9	972.4	973.9	975.7	975.5	975.6	973.1	971.9	973.4
RELATIVE HUMIDITY (%)														
Hour 01	14	83	81	80	77	78	84	84	88	88	85	86	86	83
Hour 07 (Local Time)	29	83	82	84	81	78	81	85	91	91	88	87	85	85
Hour 13	29	73	68	63	55	50	55	55	60	63	65	73	77	63
Hour 19	29	77	73	67	58	53	57	58	66	73	74	80	81	68
PRECIPITATION (inches):														
Water Equivalent														
-Normal		1.50	1.16	2.02	2.22	2.57	3.02	2.58	3.37	3.41	2.18	2.27	1.95	28.25
-Maximum Monthly	29	3.13	3.36	5.67	4.73	5.99	6.67	4.96	7.18	9.49	8.08	5.10	4.48	9.49
-Year		1974	1971	1976	1991	1983	1969	1975	1975	1986	1991	1988	1971	SEP 1986
-Minimum Monthly	29	0.60	0.29	0.62	0.97	0.40	0.85	0.55	0.85	0.01	0.47	0.45	0.65	0.01
-Year		1977	1982	1993	1987	1966	1988	1989	1969	1979	1971	1986	1976	SEP 1979
-Maximum in 24 hrs	29	1.39	1.43	2.18	1.81	1.94	2.59	3.83	3.12	2.55	2.31	1.82	1.70	3.83
-Year		1974	1971	1976	1991	1973	1969	1984	1981	1985	1991	1988	1971	JUL 1984
Snow,Ice pellets,hail														
-Maximum Monthly	29	38.0	23.6	28.7	11.6	2.3	0.0	T	T	0.1	4.4	18.9	30.4	38.0
-Year		1982	1971	1971	1979	1979		1970	1993	1967	1980	1968	1968	JAN 1982
-Maximum in 24 hrs	29	15.4	8.5	11.7	7.6	2.3	0.0	T	T	0.1	3.5	14.4	13.2	15.4
-Year		1978	1974	1970	1979	1979		1970	1993	1967	1980	1981	1980	JAN 1978
WIND:														
Mean Speed (mph)	14	10.1	9.2	9.3	9.8	9.0	8.0	7.6	7.2	8.0	9.1	9.9	9.6	8.9
Prevailing Direction														
Fastest Obs. 1 Min.														
-Direction (!!)	16	26	27	22	30	34	27	32	26	23	23	27	27	26
-Speed (MPH)	16	40	35	31	36	36	40	32	26	32	32	40	32	40
-Year		1972	1979	1977	1979	1992	1969	1966	1977	1970	1968	1970	1970	JAN 1972
Peak Gust														
-Direction (!!)	10	W	NW	W	SW	S	NW	NW	SW	NW		NW	NE	S
-Speed (mph)	10	46	48	55	53	60	58	58	59	46	54	59	55	60
-Date		1990	1987	1991	1984	1988	1992	1987	1988	1985	1991	1989	1990	MAY 1988

PRECIPITATION (inches)

HOUGHTON LAKE, MICHIGAN

YEAR	JAN	FEB	MAR	APR	MAY	JUNE	JULY	AUG	SEP	OCT	NOV	DEC	ANNUAL
#1964	1.10	0.39	1.55	2.64	2.56	1.25	3.03	2.77	3.61	1.29	2.78	1.11	24.08
1965	2.47	1.40	1.83	2.85	2.81	2.43	1.26	5.05	5.76	1.97	2.41	2.73	32.97
1966	1.02	1.02	2.30	2.08	0.40	0.95	1.74	1.92	2.25	1.79	4.81	3.46	23.74
1967	2.02	0.88	1.32	4.56	2.41	5.65	1.28	1.66	1.73	2.73	2.82	2.41	29.47
1968	1.03	1.57	0.85	1.28	2.62	5.54	1.84	1.32	4.17	3.09	2.47	2.50	28.28
1969	1.87	0.32	0.86	2.59	3.57	6.67	3.57	0.85	1.63	5.45	2.19	0.94	30.51
1970	1.16	0.70	1.86	1.50	2.67	4.15	3.93	1.26	5.85	2.75	3.79	2.00	31.62
1971	1.30	3.36	1.79	2.19	1.52	2.84	3.14	1.76	1.20	0.47	1.96	4.48	26.01
1972	0.82	0.99	2.36	1.40	1.79	2.00	2.52	4.63	2.67	2.50	0.99	3.48	26.15
1973	1.22	1.33	1.95	1.66	4.88	2.84	2.34	2.29	1.95	3.11	1.34	1.81	26.72
1974	3.13	1.14	1.44	3.47	2.92	4.60	4.70	2.78	2.53	1.40	1.41	1.43	30.95
1975	1.97	1.14	1.50	2.63	2.79	3.79	4.96	7.18	1.52	0.85	2.20	1.33	31.86
1976	1.77	2.49	5.67	1.86	2.86	2.77	1.22	1.07	0.99	1.31	0.75	0.65	23.41
1977	0.60	1.46	2.40	2.29	1.39	0.94	1.72	5.70	4.19	1.89	2.14	2.20	26.92
1978	1.93	0.55	1.20	1.35	2.42	2.38	0.91	4.10	6.70	1.43	1.28	2.08	26.33
1979	1.51	0.63	3.05	3.18	1.85	4.46	0.87	3.90	0.01	2.43	2.47	1.71	26.07
1980	1.61	0.69	1.10	3.30	1.55	3.44	2.29	2.01	3.75	1.91	1.55	1.95	25.15
1981	0.79	2.10	0.88	3.88	1.73	4.02	1.89	7.06	1.89	2.61	2.07	1.08	30.00
1982	2.43	0.29	2.41	2.46	2.98	3.21	3.31	3.25	3.58	1.86	2.52	3.47	31.77
1983	1.20	0.79	3.11	1.86	5.99	0.95	1.40	3.89	4.63	3.66	1.60	1.69	30.77
1984	1.06	0.88	2.28	1.96	2.60	3.01	4.30	2.95	2.74	2.17	1.99	2.93	28.87
1985	1.64	1.99	3.55	2.42	1.87	1.71	2.28	4.76	6.14	1.63	3.56	2.14	33.69
1986	1.06	1.73	2.20	1.73	3.20	5.43	4.38	1.76	9.49	1.75	0.45	0.85	34.03
1987	1.06	0.61	0.78	0.97	1.56	1.04	1.62	6.69	4.35	2.21	2.63	2.45	25.97
1988	2.09	0.75	2.39	2.37	0.56	0.85	2.49	4.50	3.63	3.38	5.10	1.86	29.97
1989	0.97	0.70	2.99	0.98	3.19	2.90	0.55	2.62	1.03	1.30	2.08	0.92	20.23
1990	2.43	1.09	1.47	1.77	4.15	2.57	3.82	3.28	3.00	2.94	2.90	1.55	30.97
1991	1.22	0.62	3.02	4.73	5.03	1.48	4.43	2.36	3.05	8.08	1.88	1.76	37.66
1992	1.29	1.35	1.84	3.13	0.49	1.91	2.86	3.45	3.63	2.75	4.65	1.98	29.33
1993	1.73	0.89	0.62	2.96	2.50	4.84	1.28	5.67	2.10	1.78	1.58	0.74	26.69
Record Mean	1.45	1.23	1.92	2.42	2.78	3.03	2.69	2.96	3.17	2.59	2.36	1.77	28.39

AVERAGE TEMPERATURE (deg. F)

HOUGHTON LAKE, MICHIGAN

YEAR	JAN	FEB	MAR	APR	MAY	JUNE	JULY	AUG	SEP	OCT	NOV	DEC	ANNUAL
#1964	23.7	23.9	30.0	45.3	59.6	65.0	68.2	62.3	56.3	43.5	38.5	21.8	44.8
1965	15.5	18.3	22.5	37.9	57.9	61.4	63.0	63.0	56.7	45.7	35.4	28.9	42.1
1966	12.9	21.7	31.9	39.3	47.6	64.7	69.1	64.7	55.8	45.6	34.4	23.0	42.6
1967	21.7	11.9	27.0	42.1	48.4	66.4	65.2	62.1	55.4	46.2	30.7	24.2	41.8
1968	17.6	14.8	32.3	45.3	50.7	62.6	65.7	65.6	60.2	48.9	34.7	21.0	43.3
1969	18.3	19.4	23.2	43.1	52.9	56.9	67.5	68.4	57.2	45.4	33.4	21.6	42.3
1970	12.5	15.8	22.9	42.2	53.9	63.2	69.2	66.1	58.1	49.5	35.5	21.9	42.6
1971	14.7	18.7	23.4	38.8	51.6	67.0	64.7	63.4	60.8	54.7	34.4	26.2	43.2
1972	16.3	16.4	22.3	36.6	56.9	59.1	66.5	64.4	56.8	41.7	33.6	22.0	41.1
1973	22.1	18.1	38.1	43.1	50.6	64.9	67.7	68.9	57.4	51.3	35.8	22.9	45.1
1974	20.3	14.4	27.0	43.2	50.0	61.6	67.8	65.0	53.9	44.3	36.3	26.1	42.5
1975	22.0	20.7	24.3	37.0	59.7	64.1	68.1	65.0	53.7	49.6	40.8	23.8	44.1
1976	14.0	24.1	30.8	45.5	50.8	66.1	67.0	64.6	55.6	42.3	28.3	13.8	41.9
1977	8.7	17.8	34.8	46.6	60.3	61.0	70.1	62.1	58.3	45.1	35.8	21.4	43.5
1978	14.5	10.8	23.6	39.1	56.9	61.6	65.1	66.4	58.6	44.3	35.9	21.8	41.6
1979	11.5	10.4	30.9	40.0	52.6	62.3	66.9	62.7	59.2	45.7	35.0	27.6	42.0
1980	19.1	15.7	25.0	41.9	56.0	59.2	67.4	68.0	56.7	41.6	33.9	18.9	41.9
1981	14.8	24.0	32.5	44.3	53.0	63.6	66.8	65.6	55.9	42.9	35.6	25.3	43.7
1982	11.4	17.9	25.8	37.3	60.9	57.4	68.6	61.9	57.2	48.9	36.2	31.2	42.9
1983	21.8	25.9	32.4	40.3	48.5	63.3	71.5	68.6	59.7	46.4	36.6	16.8	44.3
1984	12.4	28.2	24.1	44.8	50.1	65.2	66.4	68.8	56.1	49.2	35.4	27.0	44.0
1985	14.9	17.9	30.3	46.5	57.0	59.5	66.1	63.9	59.8	47.5	34.5	18.0	43.0
1986	17.3	18.4	30.9	47.3	56.9	60.5	69.7	62.8	58.1	46.8	32.3	26.8	44.0
1987	21.6	23.4	33.6	47.3	58.1	67.1	71.4	65.6	59.5	43.9	37.7	28.7	46.5
1988	17.3	16.3	28.0	43.7	58.6	64.8	71.4	68.8	57.9	42.4	36.6	23.7	44.1
1989	24.9	15.5	25.0	41.1	54.5	62.2	69.5	64.8	56.2	47.9	30.7	12.9	42.1
1990	26.2	21.3	32.1	44.9	51.9	63.4	66.8	65.0	57.9	45.3	38.9	25.8	45.0
1991	16.1	24.5	33.1	46.9	61.1	67.7	68.6	68.0	55.8	47.9	33.4	25.4	45.7
1992	22.8	24.1	28.6	39.4	54.3	60.4	63.0	62.0	56.8	44.8	33.6	26.1	43.0
1993	21.0	16.2	28.7	40.2	54.6	61.9	70.2	68.8	52.9	44.6	34.7	25.9	43.3
Record Mean	18.7	19.5	28.5	42.3	54.5	63.5	67.6	65.6	58.0	47.6	35.2	23.8	43.7
Max	27.2	29.5	39.0	53.9	67.8	76.9	81.2	78.5	69.8	58.2	42.6	30.8	54.6
Min	10.2	9.5	18.0	30.6	41.1	50.1	54.0	52.6	46.2	36.9	27.8	16.7	32.8

REFERENCE NOTES

HOUGHTON LAKE, MICHIGAN

GENERAL
T - TRACE AMOUNT.
BLANK ENTRIES DENOTE MISSING/UNREPORTED DATA.
# INDICATES A STATION OR INSTRUMENT RELOCATION.

SPECIFIC

PM - INCLUDES LAST DAY OF PREVIOUS MONTH
ASOS - AUTOMATED SURFACE OBSERVING SYSTEM IN OPERATION DURING THESE MONTHS.

(a) - LENGTH OF RECORD IN YEARS. ALTHOUGH INDIVIDUAL MONTHS MAY BE MISSING.
0.* OR * - THE VALUE IS BETWEEN 0.0 AND 0.05
NORMALS - BASED ON THE 1961-1990 RECORD PERIOD.
EXTREMES - DATES ARE THE MOST RECENT OCCURRENCE
WIND DIR.- NUMERALS SHOW TENS OF DEGRESS CLOCKWISE FROM TRUE NORTH. "00" INDICATES CALM.
RESULTANT DIRECTIONS ARE GIVEN TO WHOLE DEGREES.
BOLD VALUES INDICATE EXTREME VALUES WHICH OCCURRED AFTER THE ASOS SYSTEM WAS COMMISSIONED.

RECORD = PERIOD OF RECORD
RECORD MEAN PRECIPITATION IS THE MEAN OF ALL DAILY PRECIPITATION AMOUNTS DURING THE PERIOD OF RECORD.
RECORD MAX(MIN) TEMPERATURE IS THE MEAN OF ALL DAILY MAX(MIN) TEMPERATURES DURING THE PERIOD OF RECORD.
RECORD MEAN TEMPERATURE IS THE SUM OF THE RECORD MAX AND RECORD MIN DIVIDED BY 2.
AVERAGE TEMPERATURE IS THE SUM OF THE MEAN DAILY MAX AND MIN TEMPERATURE DIVIDED BY 2.

EXCEPTIONS

1. MEAN WIND SPEED IS THROUGH 1978.
2. FASTEST OBSERVED WIND IS THROUGH JUNE 1979.
3. THUNDERSTORMS AND HEAVY FOG MAY BE INCOMPLETE, DUE TO PART-TIME OPERATIONS.

RECORD MEANS ARE THROUGH THE CURRENT YEAR, BEGINNING IN 1918 FOR TEMPERATURE
1918 FOR PRECIPITATION
1965 FOR SNOWFALL

HEATING DEGREE DAYS Base 65 deg. F — HOUGHTON LAKE, MICHIGAN

SEASON	JULY	AUG	SEP	OCT	NOV	DEC	JAN	FEB	MAR	APR	MAY	JUNE	TOTAL
1964-65	38	136	283	659	789	1331	1533	1303	1314	807	237	144	8574
1965-66	109	123	266	593	883	1114	1607	1205	1020	763	535	91	8309
1966-67	19	65	280	592	911	1295	1335	1484	1168	678	510	33	8370
1967-68	69	116	288	579	1022	1260	1468	1453	1005	582	438	137	8417
1968-69	71	100	159	510	903	1353	1444	1272	1290	651	378	254	8385
1969-70	38	28	250	601	939	1342	1621	1373	1296	682	348	122	8640
1970-71	30	54	225	478	878	1330	1556	1289	1282	781	415	65	8383
1971-72	70	95	183	327	912	1197	1502	1401	1320	847	254	191	8299
1972-73	77	92	242	716	933	1326	1324	1307	829	653	442	54	7995
1973-74	26	25	280	420	871	1298	1380	1411	1171	649	463	135	8129
1974-75	27	56	337	633	856	1197	1327	1234	1254	836	215	101	8073
1975-76	44	68	333	472	720	1272	1575	1180	1054	590	434	53	7795
1976-77	24	92	296	698	1094	1582	1743	1315	932	544	199	159	8678
1977-78	29	134	200	610	869	1347	1556	1513	1276	769	289	144	8736
1978-79	79	46	220	636	869	1333	1655	1530	1053	742	396	135	8694
1979-80	48	109	196	597	893	1153	1416	1424	1233	690	286	209	8254
1980-81	18	27	258	716	928	1423	1553	1143	1001	613	368	80	8128
1981-82	50	44	280	676	872	1222	1658	1315	1209	824	148	224	8522
1982-83	12	139	255	494	855	1040	1331	1087	1003	733	505	116	7570
1983-84	25	29	209	569	846	1487	1628	1062	1262	599	458	53	8227
1984-85	42	28	270	481	881	1168	1547	1313	1071	571	248	172	7792
1985-86	46	88	209	535	908	1451	1474	1297	1048	532	274	148	8010
1986-87	23	109	212	557	975	1179	1335	1159	963	523	271	57	7363
1987-88	41	60	169	649	813	1117	1473	1406	1140	636	225	106	7835
1988-89	7	75	218	692	847	1276	1236	1381	1234	711	333	126	8136
1989-90	19	79	277	520	1025	1607	1196	1217	1009	627	398	97	8071
1990-91	32	51	248	604	773	1208	1509	1127	985	540	219	33	7329
1991-92	37	30	295	525	939	1219	1298	1178	1123	763	341	161	7909
1992-93	83	132	257	619	934	1196	1359	1358	1119	737	331	123	8248
1993-94	4	37	358	627	904	1205							

COOLING DEGREE DAYS Base 65 deg. F — HOUGHTON LAKE, MICHIGAN

YEAR	JAN	FEB	MAR	APR	MAY	JUNE	JULY	AUG	SEP	OCT	NOV	DEC	TOTAL
1969	0	0	0	0	10	18	120	138	24	0	0	0	310
1970	0	0	0	6	15	74	167	95	24	1	0	0	382
1971	0	0	0	0	5	131	68	50	62	17	0	0	333
1972	0	0	0	0	11	19	131	78	4	0	0	0	243
1973	0	0	0	1	0	58	119	155	60	2	0	0	395
1974	0	0	0	1	6	41	120	62	11	0	0	0	241
1975	0	0	0	0	58	83	146	78	2	2	0	0	369
1976	0	0	0	13	0	92	93	86	24	0	0	0	308
1977	0	0	0	0	61	46	192	50	5	0	0	0	354
1978	0	0	0	0	44	49	85	96	34	0	0	0	308
1979	0	0	0	0	18	58	112	45	31	7	0	0	271
1980	0	0	0	2	17	43	98	124	16	0	0	0	300
1981	0	0	0	0	3	47	114	70	13	0	0	0	247
1982	0	0	0	0	27	3	132	52	29	3	0	0	246
1983	0	0	0	0	0	73	235	148	56	0	0	0	512
1984	0	0	0	1	2	65	90	151	11	0	0	0	320
1985	0	0	0	22	8	11	86	61	60	0	0	0	248
1986	0	0	0	8	29	22	175	46	15	0	0	0	295
1987	0	0	0	1	64	125	250	86	12	0	0	0	538
1988	0	0	0	0	35	108	213	199	11	0	0	0	566
1989	0	0	0	0	10	49	164	83	19	0	0	0	325
1990	0	0	0	29	0	55	94	56	41	0	0	0	275
1991	0	0	0	2	104	120	157	130	26	0	0	0	539
1992	0	0	0	0	15	31	28	46	19	0	0	0	139
1993	0	0	0	0	16	34	175	161	4	0	0	0	390

SNOWFALL (inches) — HOUGHTON LAKE, MICHIGAN

SEASON	JULY	AUG	SEP	OCT	NOV	DEC	JAN	FEB	MAR	APR	MAY	JUNE	TOTAL
1964-65	0.0	0.0	0.0	4.1	3.8	12.5	26.5	12.9	17.9	10.8	0.0	0.0	88.5
1965-66	0.0	0.0	T	0.3	6.5	11.8	19.5	7.3	9.2	7.1	0.1	0.0	61.8
1966-67	0.0	0.0	0.0	T	15.1	15.4	22.6	14.3	9.4	4.5	T	0.0	81.3
1967-68	0.0	0.0	0.1	3.0	12.6	10.6	9.0	15.6	4.9	0.7	0.0	0.0	56.5
1968-69	0.0	0.0	0.0	1.0	18.9	30.4	25.8	8.7	9.5	0.4	0.3	0.0	95.0
1969-70	0.0	0.0	0.0	2.0	17.6	20.0	22.7	12.8	24.5	3.8	T	0.0	103.4
1970-71	T	0.0	0.0	T	17.7	23.1	27.5	23.6	28.7	3.5	T	0.0	124.1
1971-72	0.0	0.0	0.0	T	16.7	16.7	15.0	18.7	26.8	2.9	0.0	0.0	96.8
1972-73	0.0	0.0	0.0	0.6	4.1	29.2	8.3	15.2	5.3	7.9	0.8	0.0	71.4
1973-74	0.0	0.0	0.0	T	4.8	13.0	11.2	17.5	12.3	3.1	1.0	0.0	62.9
1974-75	0.0	0.0	T	T	6.0	18.0	11.8	12.4	14.0	4.5	0.0	0.0	66.7
1975-76	0.0	0.0	T	T	5.9	14.6	29.6	20.8	17.6	6.0	0.6	0.0	95.1
1976-77	0.0	0.0	0.0	0.7	13.6	18.6	16.3	11.4	3.9	2.7	0.0	0.0	67.2
1977-78	0.0	0.0	0.0	T	16.0	14.0	33.3	13.1	12.5	0.3	0.0	0.0	89.2
1978-79	0.0	0.0	0.0	T	10.5	29.0	21.8	8.3	5.7	11.6	2.3	0.0	89.2
1979-80	0.0	0.0	0.0	0.7	5.0	7.1	14.4	12.3	12.1	7.7	0.0	0.0	59.3
1980-81	0.0	0.0	0.0	4.4	3.2	25.6	19.3	16.3	5.4	0.2	0.0	0.0	74.4
1981-82	0.0	0.0	0.0	3.0	16.1	15.7	38.0	6.1	14.3	5.5	0.0	0.0	98.7
1982-83	0.0	0.0	0.0	0.2	7.7	4.8	15.4	6.4	13.3	2.7	1.0	0.0	51.5
1983-84	0.0	0.0	0.0	0.0	5.0	19.8	17.0	6.3	6.5	5.1	0.4	0.0	60.1
1984-85	0.0	0.0	0.0	0.0	0.6	11.7	25.3	18.7	14.1	10.0	0.0	0.0	80.4
1985-86	0.0	0.0	0.0	0.0	12.6	23.1	12.7	14.0	8.9	0.3	0.0	0.0	71.6
1986-87	0.0	0.0	0.0	T	3.4	10.3	14.1	6.4	2.3	2.0	0.0	0.0	38.5
1987-88	0.0	0.0	0.0	2.4	8.7	18.6	13.4	12.9	10.3	1.4	0.0	0.0	67.7
1988-89	0.0	0.0	0.0		7.8	12.3	9.3	13.5	18.7	3.1	T	0.0	
1989-90	0.0	0.0	T	0.6	12.2	15.0	21.7	11.0	2.4	2.1	0.9	0.0	65.9
1990-91	0.0	0.0	0.0	T	9.1	15.3	22.6	8.5	4.5	1.8	0.0	0.0	61.8
1991-92	0.0	0.0	T	T	4.9	17.5	15.0	15.8	9.5	3.6	0.0	0.0	66.3
1992-93	0.0	0.0	0.0	3.2	11.8	14.3	14.7	13.1	8.1	2.9	0.0	0.0	68.1
1993-94	0.0	T	0.0	0.1	2.3	9.0							
Record Mean	T	T	T	0.8	9.5	16.7	19.1	12.9	11.5	4.1	0.3	0.0	74.8

LANSING, MICHIGAN

The climate at Lansing alternates between continental and semi–marine, depending on meteorological conditions. The marine type is due to the influence of the Great Lakes and is governed by the force and direction of the wind. When there is little or no wind, the weather becomes continental in character, which means pronounced fluctuation in temperature, hot weather in summer and severe cold in winter. On the other hand, a strong wind from the Lakes may immediately transform the weather into a semi–marine type.

Since large bodies of water are less responsive to temperature changes, the Great Lakes hold the winter cold longer in the spring and the summer heat longer in the fall than do the land areas. This fact is illustrated by looking at some monthly mean temperatures at Lansing as compared to similar latitudes west of the Lakes. Such a comparison shows cooler summers and milder winters in Lansing because of the lake effect.

Based on the 1951–1980 period, the average first occurrence of 32 degrees Fahrenheit in the fall is September 30 and the average last occurrence in the spring is May 13.

Precipitation is fairly well distributed through the year, and no conspicuous annual variation is noted, although there is about 1 inch less per month in winter than in summer. The heavier amounts in summer occur in thunderstorms. The wettest months are May and June. Snowfall for Lansing is moderate, averaging about 52 inches per year.

There are almost twice as many cloudy days as clear days throughout the year. Much cloudiness prevails during the winter season, but sunshine is abundant during the summer months. Similarly, relative humidity remains rather high during the winter, but is only moderate in summer.

Tornadoes sometimes occur in this area, but their frequency is less than in states farther to the south and west. Destructive thunder and wind storms are not uncommon. Flooding of streams and rivers in the upper grand Basin occurs in about one year out of three, with floods causing considerable damage in about one year out of ten.

NORMALS, MEANS, AND EXTREMES

LANSING, MICHIGAN

LATITUDE: 42°47'N LONGITUDE: 84°36' W ELEVATION: FT. GRND 841 BARO 871 TIME ZONE: EASTERN WBAN: 14836

	(a)	JAN	FEB	MAR	APR	MAY	JUNE	JULY	AUG	SEP	OCT	NOV	DEC	YEAR
TEMPERATURE °F:														
Normals														
-Daily Maximum		28.5	31.6	42.7	56.7	69.1	78.2	82.6	80.4	71.9	59.4	45.8	33.5	56.7
-Daily Minimum		13.3	14.2	24.5	35.1	44.9	54.5	59.0	56.9	49.9	39.2	30.8	19.3	36.8
-Monthly		20.9	22.9	33.6	45.9	57.0	66.4	70.8	68.7	60.9	49.3	38.3	26.4	46.8
Extremes														
-Record Highest	35	66	64	78	86	94	99	100	100	97	89	77	66	100
-Year		1967	1984	1989	1980	1977	1988	1988	1988	1973	1963	1975	1982	JUL 1988
-Record Lowest	35	-29	-24	-15	-2	19	30	37	35	22	15	4	-17	-29
-Year		1981	1967	1978	1982	1966	1966	1972	1976	1991	1966	1976	1976	JAN 1981
NORMAL DEGREE DAYS:														
Heating (base 65°F)		1367	1179	973	573	283	60	8	26	147	487	801	1197	7101
Cooling (base 65°F)		0	0	0	0	35	102	188	141	24	0	0	0	490
% OF POSSIBLE SUNSHINE	39	37	44	49	52	61	65	69	65	59	50	31	29	51
MEAN SKY COVER (tenths)														
Sunrise - Sunset	39	7.9	7.4	7.2	6.9	6.3	6.0	5.7	5.7	6.1	6.4	7.9	8.2	6.8
MEAN NUMBER OF DAYS:														
Sunrise to Sunset														
-Clear	39	3.3	4.2	5.2	6.1	6.8	7.2	8.2	8.6	7.9	7.1	3.6	2.8	70.9
-Partly Cloudy	39	6.6	6.8	7.7	7.0	10.1	11.4	13.1	11.9	9.3	8.7	5.8	5.5	103.9
-Cloudy	39	21.1	17.3	18.1	16.8	14.1	11.4	9.7	10.5	12.8	15.2	20.6	22.7	190.4
Precipitation														
.01 inches or more	39	14.8	11.8	13.6	12.7	10.6	10.4	9.7	9.4	10.5	9.4	12.8	14.7	140.3
Snow,Ice pellets,hail														
1.0 inches or more	39	3.8	3.3	2.6	0.8	0.0	0.0	0.0	0.0	0.0	0.2	1.5	3.8	16.0
Thunderstorms	39	0.3	0.2	1.4	3.1	3.8	6.4	6.3	5.8	4.1	1.3	0.9	0.4	34.0
Heavy Fog Visibility														
1/4 mile or less	39	1.7	1.8	2.3	1.0	0.9	1.3	1.1	2.1	2.0	2.2	1.7	2.3	20.5
Temperature °F														
-Maximum														
90° and above	30	0.0	0.0	0.0	0.0	0.5	2.2	4.6	2.6	0.8	0.0	0.0	0.0	10.7
32° and below	30	18.8	14.5	5.6	0.4	0.0	0.0	0.0	0.0	0.0	0.0	2.8	13.3	55.4
-Minimum														
32° and below	30	29.3	26.5	24.3	12.4	2.4	0.1	0.0	0.0	0.8	8.2	18.3	27.1	149.3
0° and below	30	5.4	4.4	0.8	0.*	0.0	0.0	0.0	0.0	0.0	0.0	0.0	2.2	12.9
AVG. STATION PRESS. (mb)	21	984.8	985.9	984.0	983.1	983.4	983.4	984.7	986.0	986.3	986.2	985.0	985.2	984.8
RELATIVE HUMIDITY (%)														
Hour 01	30	82	81	80	77	78	81	85	88	88	84	83	84	83
Hour 07 (Local Time)	30	83	83	83	80	80	81	86	90	91	87	85	85	85
Hour 13	30	75	69	65	58	55	56	57	59	62	63	71	76	64
Hour 19	30	78	74	69	60	57	58	60	65	73	74	79	81	69
PRECIPITATION (inches):														
Water Equivalent														
-Normal		1.49	1.36	2.30	2.81	2.61	3.71	2.52	3.20	3.56	2.10	2.63	2.33	30.62
-Maximum Monthly	42	3.61	4.21	4.36	5.16	6.57	10.21	6.43	9.81	8.34	5.58	5.40	4.70	10.21
-Year		1950	1954	1974	1981	1989	1986	1992	1975	1986	1990	1990	1949	JUN 1986
-Minimum Monthly	42	0.39	0.22	0.92	1.07	0.62	0.20	0.50	0.17	T	0.28	0.51	0.37	T
-Year		1981	1969	1960	1982	1977	1988	1965	1969	1979	1956	1962	1960	SEP 1979
-Maximum in 24 hrs	41	1.59	2.40	1.59	2.73	3.28	5.01	2.16	3.75	3.43	3.46	2.47	1.62	5.01
-Year		1949	1954	1954	1993	1989	1986	1972	1975	1981	1981	1990	1970	JUN 1986
Snow,Ice pellets,hail														
-Maximum Monthly	39	34.0	23.7	19.8	17.0	0.3	T	0.0	0.0	T	7.5	16.8	27.8	34.0
-Year		1978	1986	1971	1970	1954	1992			1989	1967	1966	1951	JAN 1978
-Maximum in 24 hrs	39	20.4	9.0	15.5	17.0	0.3	T	0.0	0.0	T	7.5	11.0	15.1	20.4
-Year		1967	1965	1973	1970	1954	1992			1989	1967	1951	1970	JAN 1967
WIND:														
Mean Speed (mph)	34	11.8	11.0	11.3	11.2	9.9	9.0	8.0	7.5	8.2	9.3	10.8	11.2	9.9
Prevailing Direction through 1958		SW	SW	NW	SW	SW	S	SW	SW	S	SW	SW	SW	SW
Fastest Mile														
-Direction (!!)	34	SW	SW	W	W	W	SE	NE	SW	N	SW	W	SW	SE
-Speed (MPH)	34	54	56	59	61	46	63	56	47	57	48	56	56	63
-Year		1975	1971	1961	1963	1985	1963	1959	1971	1962	1982	1960	1971	JUN 1963
Peak Gust														
-Direction (!!)	10	W	W	SW	SW	W	W	W	W	W	S	SW	W	SW
-Speed (mph)	10	53	51	61	70	59	67	60	54	47	55	51	54	70
-Date		1990	1985	1991	1984	1985	1989	1991	1991	1991	1992	1991	1987	APR 1984

PRECIPITATION (inches)

LANSING, MICHIGAN

YEAR	JAN	FEB	MAR	APR	MAY	JUNE	JULY	AUG	SEP	OCT	NOV	DEC	ANNUAL
1964	1.31	0.47	2.79	3.24	2.69	2.79	3.07	3.95	1.90	0.65	1.92	1.62	26.40
1965	3.17	2.28	3.17	2.04	1.01	2.18	0.50	3.68	7.63	0.99	1.85	3.54	32.04
1966	0.87	1.52	2.40	3.01	1.88	2.56	1.59	2.38	1.22	0.88	4.60	2.46	25.37
1967	2.18	1.41	1.16	3.81	1.05	6.77	1.15	2.90	2.53	3.45	2.27	3.45	32.13
1968	1.46	1.56	1.49	2.96	4.16	7.94	3.03	2.98	2.73	1.56	3.26	2.62	35.75
1969	1.91	0.22	1.06	4.02	4.47	4.03	4.89	0.17	1.67	2.74	2.32	0.53	28.03
1970	0.91	0.66	2.20	3.13	2.89	3.02	3.37	2.36	4.61	3.77	2.85	2.48	32.25
1971	0.68	2.40	2.33	1.50	1.93	5.13	4.82	2.50	5.25	2.29	1.61	4.24	34.68
1972	1.51	1.19	2.82	4.56	2.86	3.28	3.27	5.06	2.65	3.30	2.72	4.16	37.38
1973	1.17	1.44	3.59	2.22	3.78	3.62	2.04	1.77	2.81	2.07	4.22	3.09	31.82
1974	3.00	2.58	4.36	2.07	4.07	2.81	1.21	2.67	2.60	1.44	2.50	2.39	31.70
1975	2.77	2.08	2.43	4.76	2.96	2.93	2.47	9.81	1.65	0.95	3.43	3.32	39.56
1976	1.59	2.54	3.85	3.55	2.52	4.50	2.56	0.58	1.66	2.32	1.02	0.96	27.65
1977	0.95	0.66	2.65	2.53	0.62	3.50	2.00	1.60	4.46	1.20	2.17	2.30	24.64
1978	2.60	0.46	1.91	1.55	2.06	2.73	1.23	3.69	3.80	1.98	2.58	2.85	27.44
#1979	2.13	0.56	1.78	2.69	1.35	5.53	1.86	2.30	T	1.99	3.25	2.30	25.74
1980	0.70	0.99	1.94	2.41	1.84	3.20	3.56	4.76	3.22	2.02	0.90	3.06	28.60
1981	0.39	1.39	1.10	5.16	4.69	3.22	1.71	1.98	8.01	1.28	1.50	1.10	31.53
1982	1.55	0.55	2.98	1.07	2.52	3.00	3.00	1.99	3.51	0.43	4.21	3.54	28.35
1983	1.00	0.81	3.25	4.10	3.97	4.27	2.54	2.70	3.74	2.57	3.22	1.83	34.00
1984	0.49	0.89	2.54	3.02	4.05	0.32	2.64	3.16	2.69	3.09	2.54	3.79	29.22
1985	2.04	3.03	3.53	2.90	2.14	2.17	3.06	4.36	3.34	3.25	3.08	1.29	34.19
1986	0.89	2.62	1.56	1.96	1.97	10.21	1.69	2.88	8.34	2.66	1.21	1.13	37.12
1987	1.00	0.35	0.97	1.58	1.37	3.30	3.35	5.64	4.88	1.77	2.89	3.40	30.50
1988	1.53	1.10	1.52	3.95	0.63	0.20	2.56	5.08	5.97	3.35	4.26	1.26	31.41
1989	1.15	0.67	2.13	1.44	6.57	3.61	0.93	4.90	3.49	1.29	3.65	0.86	30.69
1990	1.55	2.65	1.40	2.36	3.43	2.50	3.61	2.40	4.00	5.58	5.40	2.79	37.67
1991	1.28	0.79	3.76	4.41	1.75	2.60	2.41	3.83	1.05	3.80	3.02	2.13	30.83
1992	1.38	1.36	2.68	4.64	2.09	2.07	6.43	2.79	3.21	2.18	3.88	2.16	34.87
1993	3.22	1.17	1.74	4.86	1.37	6.50	2.94	3.80	4.80	2.99	1.39	0.70	35.48
Record Mean	1.70	1.59	2.40	2.87	3.25	3.46	2.72	3.00	3.11	2.42	2.41	2.04	30.98

AVERAGE TEMPERATURE (deg. F)

LANSING, MICHIGAN

YEAR	JAN	FEB	MAR	APR	MAY	JUNE	JULY	AUG	SEP	OCT	NOV	DEC	ANNUAL
1964	28.2	25.8	34.5	48.8	62.0	69.1	73.2	66.8	61.3	46.4	42.3	25.9	48.7
1965	21.9	23.9	26.4	42.5	61.2	65.2	68.6	68.3	62.1	48.9	39.4	32.8	46.8
1966	16.7	23.2	36.7	43.0	50.7	69.2	72.9	67.8	59.7	49.0	39.2	25.8	46.2
1967	26.9	17.8	32.2	47.8	52.3	70.5	69.0	65.5	58.1	49.0	33.7	29.1	46.0
1968	20.1	21.2	36.7	48.9	53.2	66.5	70.2	69.5	62.7	51.3	37.5	25.5	46.9
1969	21.8	25.4	30.4	47.2	56.5	61.9	71.1	72.8	64.2	50.0	36.2	25.8	47.0
1970	16.4	23.6	29.1	46.5	58.6	65.5	70.2	68.2	61.5	51.8	39.1	25.8	46.3
1971	19.1	26.5	30.2	44.2	55.0	70.4	67.2	66.9	64.3	58.2	38.3	31.9	47.7
1972	21.3	21.8	29.7	41.6	58.2	61.6	68.1	67.1	59.9	44.9	35.3	26.5	44.7
1973	27.5	21.3	39.8	46.2	52.9	67.7	70.1	73.0	62.0	54.3	39.9	26.6	48.4
1974	24.9	19.4	34.1	48.0	53.6	64.2	71.3	68.4	57.5	47.6	39.4	29.3	46.5
1975	27.1	26.1	30.6	40.3	61.3	68.4	70.6	69.6	56.2	52.2	45.0	27.8	48.0
1976	17.4	29.9	38.2	48.1	54.0	69.1	70.9	66.8	59.3	44.8	31.6	18.0	45.7
1977	11.1	23.2	40.2	50.8	64.3	64.0	73.6	68.5	62.7	48.3	39.5	25.3	47.6
1978	17.6	11.3	25.6	43.8	58.6	65.7	69.5	69.5	65.1	48.4	39.8	26.8	45.2
1979	16.1	13.0	37.8	44.5	57.1	67.3	70.3	67.5	63.0	49.5	38.4	30.6	46.3
1980	22.2	19.6	29.9	45.0	58.5	63.5	72.5	71.8	61.8	45.2	36.7	23.7	45.9
1981	17.6	28.2	35.5	48.6	54.3	67.2	71.1	69.7	58.1	46.1	39.1	27.1	46.9
1982	15.1	19.8	31.2	41.5	63.9	62.3	70.8	66.7	60.4	51.8	40.1	34.8	46.6
1983	26.5	30.9	35.9	42.1	52.4	67.1	74.4	72.3	62.4	48.9	39.5	17.8	47.5
1984	14.9	32.6	27.1	46.5	52.6	69.7	69.6	71.9	59.6	52.7	38.2	31.3	47.2
1985	17.8	20.2	36.0	51.1	59.3	63.3	70.5	67.2	63.0	50.7	39.0	21.2	46.6
1986	21.9	21.1	36.0	49.2	58.1	64.6	72.3	65.4	62.6	50.3	35.0	29.9	47.2
1987	24.7	29.0	37.5	49.2	62.4	70.6	74.6	69.4	62.3	44.7	41.5	31.7	49.8
1988	20.6	19.9	34.0	46.5	59.7	68.1	74.6	72.6	61.2	44.1	40.7	27.1	47.4
1989	30.9	19.9	32.5	43.6	55.1	65.8	71.3	68.4	59.8	50.7	33.6	15.8	45.6
1990	31.8	26.8	37.0	47.8	54.1	66.5	69.3	68.4	62.1	49.7	42.5	29.2	48.8
1991	20.9	29.4	37.6	50.2	63.7	69.3	71.4	69.3	58.6	51.9	35.5	28.8	48.9
1992	25.9	29.5	32.2	43.4	56.2	63.3	66.1	63.9	58.8	47.3	37.7	30.8	46.3
1993	25.7	20.9	32.4	44.7	57.6	65.1	72.1	70.5	56.5	47.8	36.9	27.5	46.5
Record Mean	22.0	22.6	32.7	45.1	56.8	66.2	70.8	68.9	61.5	50.3	38.0	26.5	46.8
Max	29.2	30.8	41.5	55.5	68.3	77.6	82.4	80.3	72.6	60.5	45.7	33.3	56.5
Min	14.8	14.4	23.9	34.7	45.3	54.8	59.2	57.4	50.5	40.0	30.4	19.8	37.1

REFERENCE NOTES

LANSING, MICHIGAN

GENERAL
T - TRACE AMOUNT.
BLANK ENTRIES DENOTE MISSING/UNREPORTED DATA.
# INDICATES A STATION OR INSTRUMENT RELOCATION.

SPECIFIC

(a) - LENGTH OF RECORD IN YEARS, ALTHOUGH INDIVIDUAL MONTHS MAY BE MISSING.
0.* OR * - THE VALUE IS BETWEEN 0.0 AND 0.05
NORMALS - BASED ON THE 1961-1990 RECORD PERIOD.
EXTREMES - DATES ARE THE MOST RECENT OCCURRENCE.
WIND DIR. - NUMERALS SHOW TENS OF DEGRESS CLOCKWISE FROM TRUE NORTH. "00" INDICATES CALM.
RESULTANT DIRECTIONS ARE GIVEN TO WHOLE DEGREES.
BOLD VALUES INDICATE EXTREME VALUES WHICH OCCURRED AFTER THE ASOS SYSTEM WAS COMMISSIONED.

RECORD = PERIOD OF RECORD
RECORD MEAN PRECIPITATION IS THE MEAN OF ALL DAILY PRECIPITATION AMOUNTS DURING THE PERIOD OF RECORD.
RECORD MAX(MIN) TEMPERATURE IS THE MEAN OF ALL DAILY MAX(MIN) TEMPERATURES DURING THE PERIOD OF RECORD.
RECORD MEAN TEMPERATURE IS THE SUM OF THE RECORD MAX AND RECORD MIN DIVIDED BY 2.
AVERAGE TEMPERATURE IS THE SUM OF THE MEAN DAILY MAX AND MIN TEMPERATURE DIVIDED BY 2.

EXCEPTIONS

PREVAILING WIND DIRECTION IS FROM THE EAST LANSING OFFICE 1937-1958.

RECORD MEANS ARE THROUGH THE CURRENT YEAR, BEGINNING IN 1900 FOR TEMPERATURE
1900 FOR PRECIPITATION
1904 FOR SNOWFALL

HEATING DEGREE DAYS Base 65 deg. F — LANSING, MICHIGAN

SEASON	JULY	AUG	SEP	OCT	NOV	DEC	JAN	FEB	MAR	APR	MAY	JUNE	TOTAL
1964-65	10	69	188	569	672	1206	1328	1145	1190	667	162	71	7277
1965-66	29	67	148	493	764	988	1490	1164	872	656	439	55	7165
1966-67	4	25	196	491	767	1209	1174	1318	1012	513	395	13	7117
1967-68	38	66	217	503	931	1109	1385	1266	869	480	366	64	7294
1968-69	24	57	111	451	818	1218	1331	1103	1068	531	286	152	7150
1969-70	9	6	119	462	858	1208	1498	1153	1106	564	235	97	7315
1970-71	26	31	159	409	771	1208	1416	1070	1069	621	318	34	7132
1971-72	35	39	129	234	792	1016	1348	1246	1089	696	219	142	6985
1972-73	56	56	173	616	884	1186	1158	1218	774	564	370	20	7075
1973-74	11	5	179	339	745	1182	1233	1270	949	513	359	94	6879
1974-75	6	27	256	538	762	1099	1166	1085	1057	737	180	57	6970
1975-76	11	13	266	402	593	1145	1469	1011	826	528	338	30	6632
1976-77	4	57	210	619	994	1449	1666	1164	761	439	137	99	7599
1977-78	7	50	118	514	757	1226	1461	1496	1215	627	244	84	7799
1978-79	24	13	105	507	751	1181	1509	1452	834	608	289	58	7331
1979-80	19	41	118	491	792	1058	1323	1308	1081	600	229	124	7184
1980-81	0	5	136	607	842	1271	1464	1026	907	483	330	24	7095
1981-82	12	23	227	579	770	1168	1540	1259	1044	696	108	113	7539
1982-83	7	62	184	406	739	930	1187	951	895	681	386	66	6494
1983-84	14	9	152	500	756	1456	1546	936	1166	554	387	17	7493
1984-85	21	13	208	374	798	1039	1458	1248	891	439	202	94	6785
1985-86	12	26	168	438	772	1350	1330	1224	893	480	237	77	7007
1986-87	11	63	125	451	895	1079	1240	1005	846	469	188	23	6395
1987-88	16	38	124	621	700	1026	1371	1301	955	547	199	74	6972
1988-89	7	35	147	644	721	1165	1049	1255	999	637	320	53	7032
1989-90	6	28	199	437	934	1520	1023	1062	869	546	336	55	7015
1990-91	22	20	154	484	667	1102	1359	990	841	449	171	30	6289
1991-92	7	13	244	407	877	1114	1207	1023	1007	642	294	95	6930
1992-93	33	88	224	543	813	1050	1215	1231	1003	604	234	81	7119
1993-94	3	22	264	533	835	1158							

COOLING DEGREE DAYS Base 65 deg. F — LANSING, MICHIGAN

YEAR	JAN	FEB	MAR	APR	MAY	JUNE	JULY	AUG	SEP	OCT	NOV	DEC	TOTAL
1969	0	0	0	2	31	65	203	256	101	5	0	0	663
1970	0	0	0	15	42	117	193	134	60	5	0	0	566
1971	0	0	0	0	14	202	112	103	113	28	0	0	572
1972	0	0	0	0	12	46	161	127	26	1	0	0	373
1973	0	0	0	7	0	106	176	264	92	12	0	0	657
1974	0	0	0	8	11	80	208	138	37	5	0	0	487
1975	0	0	0	0	73	165	192	162	5	11	0	0	608
1976	0	0	0	25	4	162	198	121	45	1	0	0	556
1977	0	0	0	19	124	76	282	162	56	0	0	0	719
1978	0	0	0	0	53	110	170	157	115	0	0	0	605
1979	0	0	0	0	48	133	192	125	63	17	0	0	578
1980	0	0	0	4	33	86	239	223	46	0	0	0	631
1981	0	0	0	1	7	101	210	178	27	0	0	0	524
1982	0	0	0	0	80	40	193	122	53	6	0	0	494
1983	0	0	0	0	1	134	311	242	79	7	0	0	774
1984	0	0	0	5	10	164	173	234	51	1	0	0	638
1985	0	0	0	32	31	50	189	100	116	0	0	0	518
1986	0	0	0	10	31	68	246	83	59	0	0	0	497
1987	0	0	0	0	114	199	322	181	48	0	1	0	865
1988	0	0	0	0	41	173	307	277	41	4	0	0	843
1989	0	0	0	0	22	85	209	138	49	1	0	0	504
1990	0	0	6	37	7	106	163	134	74	14	0	0	541
1991	0	0	0	14	139	165	213	151	59	9	0	0	750
1992	0	0	0	2	28	48	74	62	45	1	0	0	260
1993	0	0	0	0	13	92	228	200	13	5	0	0	551

SNOWFALL (inches) — LANSING, MICHIGAN

SEASON	JULY	AUG	SEP	OCT	NOV	DEC	JAN	FEB	MAR	APR	MAY	JUNE	TOTAL
1964-65	0.0	0.0	0.0	T	4.2	12.3	6.3	15.2	18.6	6.6	0.0	0.0	63.2
1965-66	0.0	0.0	0.0	T	1.5	8.5	10.7	9.5	1.1	2.0	T	0.0	33.3
1966-67	0.0	0.0	0.0	0.0	16.8	10.0	30.1	15.1	9.4	T	T	0.0	81.4
1967-68	0.0	0.0	T	7.5	5.8	2.4	10.8	2.9	13.4	0.2	0.0	0.0	43.0
1968-69	0.0	0.0	0.0	T	1.8	8.6	13.2	3.1	4.9	T	T	0.0	31.6
1969-70	0.0	0.0	0.0	T	8.5	6.1	14.3	8.5	13.3	17.0	0.0	0.0	67.7
1970-71	0.0	0.0	0.0	0.0	0.9	26.1	8.9	5.7	19.8	1.2	0.0	0.0	62.6
1971-72	0.0	0.0	0.0	0.0	13.0	3.7	16.9	14.6	9.3	5.8	0.0	0.0	63.3
1972-73	0.0	0.0	0.0	T	11.8	18.3	1.9	21.3	16.5	5.5	T	0.0	75.3
1973-74	0.0	0.0	0.0	0.0	T	23.6	14.9	16.5	10.4	0.5	T	0.0	65.9
1974-75	0.0	0.0	0.0	0.5	6.9	20.4	4.4	12.7	10.4	13.2	0.0	0.0	68.5
1975-76	0.0	0.0	0.0	0.0	10.0	20.2	23.5	4.6	6.8	5.5	T	0.0	70.6
1976-77	0.0	0.0	0.0	1.4	4.6	20.5	17.2	3.5	8.7	1.1	0.0	0.0	57.0
1977-78	0.0	0.0	0.0	0.0	8.1	13.3	34.0	6.7	4.6	T	0.0	0.0	66.7
1978-79	0.0	0.0	0.0	0.0	7.0	14.7	27.1	4.1	4.2	1.5	T	0.0	58.6
#1979-80	0.0	0.0	0.0	T	7.2	3.2	6.7	5.5	8.8	3.3	0.0	0.0	34.7
1980-81	0.0	0.0	0.0	T	4.2	14.2	5.9	12.2	2.2	T	0.0	0.0	38.7
1981-82	0.0	0.0	0.0	0.3	2.2	15.9	15.6	8.2	12.9	7.0	0.0	0.0	62.1
1982-83	0.0	0.0	0.0	0.0	1.1	4.3	5.1	3.5	15.2	4.3	0.0	0.0	33.5
1983-84	0.0	0.0	0.0	0.1	3.0	21.4	9.1	3.2	12.1	0.5	0.0	0.0	49.4
1984-85	0.0	0.0	0.0	0.0	1.2	11.7	25.5	17.5	3.8	0.9	0.0	0.0	60.6
1985-86	0.0	0.0	0.0	0.0	3.7	16.4	12.5	23.7	4.9	T	0.0	0.0	61.2
1986-87	0.0	0.0	0.0	T	4.1	6.8	16.8	1.4	5.7	2.5	0.0	0.0	37.3
1987-88	0.0	0.0	0.0	0.5	T	12.9	9.7	18.7	4.9	1.2	0.0	0.0	47.9
1988-89	0.0	0.0	0.0	T	3.8	8.5	7.2	13.5	4.4	2.5	0.0	0.0	39.9
1989-90	0.0	0.0	T	3.9	11.8	10.2	3.6	21.3	5.0	0.6	T	0.0	56.4
1990-91	0.0	0.0	0.0	0.0	0.1	12.9	13.8	6.5	2.8	0.1	0.0	0.0	36.2
1991-92	0.0	0.0	0.0	1.3	3.8	19.1	11.0	4.2	14.1	4.8	0.0	T	58.3
1992-93	0.0	0.0	0.0	1.2	4.3	8.7	17.6	19.2	13.1	3.3	0.0	0.0	67.4
1993-94	0.0	0.0	0.0	T	1.5	5.7							
Record Mean	0.0	0.0	T	0.4	4.5	10.7	12.0	10.0	8.4	2.7	0.2	T	48.9

MARQUETTE COUNTY AIRPORT, MICHIGAN

The Marquette County Airport lies about 7.5 miles southwest of the nearest shoreline of Lake Superior and about 8 miles west of the city of Marquette. Lake Superior is the largest body of fresh water in the world and the deepest and coldest of the Great Lakes. An irregular northwest-southeast ridge line lies just to the east of the airport. There are several water storage basins in the vicinity of the station. One basin, about 20 miles long, is 3 miles northwest and another, about 8 miles in diameter, is 3 miles west.

The climate is influenced considerably by the proximity of Lake Superior. As a consequence of the cool expanse of water in the summer, there is rarely a long period of sweltering hot weather. Periods of drought are extremely rare. In the winter, cold outbreaks are tempered considerably by the waters of Lake Superior if the lake is unfrozen. However, winds blowing across these relatively warmer waters pick up moisture and cause cloudy weather throughout the winter, as well as frequent periods of light snow. Lake-formed snow showers and snow squalls are intensified near the station by upslope winds, especially from the northwest through northeast. With a northeast through east wind, especially in autumn, the upslope condition will cause light snow at the airport, while along the lakeshore, only drizzle or no precipitation may occur.

The growing season averages 117 days. Precipitation is rather evenly distributed throughout the year, with an average precipitation of 4 inches or more in June and September and less than 2 inch averages only in January and February. One hundred inches or more of snow occur in nine of ten winter seasons.

NORMALS, MEANS, AND EXTREMES

MARQUETTE COUNTY AIRPORT, MICHIGAN

LATITUDE: 46°32'N LONGITUDE: 87°34' W ELEVATION: FT. GRND 1415 BARO 1417 TIME ZONE: EASTERN WBAN: 94850

	(a)	JAN	FEB	MAR	APR	MAY	JUNE	JULY	AUG	SEP	OCT	NOV	DEC	YEAR
TEMPERATURE °F:														
Normals														
-Daily Maximum		20.5	23.9	33.5	47.5	61.9	70.9	76.6	73.3	63.8	52.3	36.9	24.6	48.8
-Daily Minimum		3.3	4.0	13.6	27.0	38.4	47.4	53.5	51.7	44.1	34.8	23.3	9.9	29.3
-Monthly		11.9	14.0	23.6	37.3	50.2	59.2	65.1	62.5	54.0	43.6	30.1	17.3	39.1
Extremes														
-Record Highest	15	46	61	68	92	93	95	99	95	92	87	69	59	99
-Year		1981	1981	1989	1980	1986	1988	1988	1988	1983	1992	1990	1982	JUL 1988
-Record Lowest	15	-25	-34	-23	-5	17	28	36	34	24	14	-5	-28	-34
-Year		1982	1979	1982	1979	1983	1986	1989	1992	1993	1984	1989	1983	FEB 1979
NORMAL DEGREE DAYS:														
Heating (base 65°F)		1646	1428	1283	831	471	193	74	122	330	663	1047	1479	9567
Cooling (base 65°F)		0	0	0	0	13	19	78	45	0	0	0	0	155
% OF POSSIBLE SUNSHINE	12	33	39	45	49	56	56	61	57	48	40	28	27	45
MEAN SKY COVER (tenths)														
Sunrise - Sunset														
MEAN NUMBER OF DAYS:														
Sunrise to Sunset														
-Clear														
-Partly Cloudy														
-Cloudy														
Precipitation .01 inches or more	15	17.7	12.9	14.4	12.5	10.7	11.5	10.7	12.5	14.4	15.5	15.8	17.9	166.4
Snow, Ice pellets, hail 1.0 inches or more	15	11.2	7.1	7.1	3.5	0.2	0.0	0.0	0.0	0.1	2.5	6.4	9.2	47.3
Thunderstorms	15	0.0	0.1	0.5	1.2	3.1	5.6	6.1	5.5	4.2	1.5	0.1	0.0	27.9
Heavy Fog Visibility 1/4 mile or less	15	0.9	1.4	2.3	2.7	2.9	1.4	2.3	3.9	3.5	3.1	1.6	1.9	27.7
Temperature °F														
-Maximum														
90° and above	15	0.0	0.0	0.0	0.1	0.1	0.5	1.7	0.7	0.1	0.0	0.0	0.0	3.2
32° and below	15	26.3	20.4	13.1	2.5	0.1	0.0	0.0	0.0	0.0	0.6	9.9	23.5	96.4
-Minimum														
32° and below	15	31.0	28.0	29.2	21.5	8.5	0.6	0.0	0.0	2.8	17.8	27.3	30.5	197.2
0° and below	15	11.4	11.0	4.7	0.1	0.0	0.0	0.0	0.0	0.0	0.0	0.7	7.6	35.5
AVG. STATION PRESS. (mb)														
RELATIVE HUMIDITY (%)														
Hour 01														
Hour 07 (Local Time)														
Hour 13														
Hour 19														
PRECIPITATION (inches):														
Water Equivalent														
-Normal		2.17	1.73	2.77	2.64	3.03	3.48	2.88	3.41	4.08	3.61	2.89	2.61	35.30
-Maximum Monthly	15	4.02	3.68	6.08	6.56	6.49	6.61	5.40	8.59	6.94	7.59	8.25	4.33	8.59
-Year		1988	1984	1979	1985	1983	1981	1991	1988	1980	1979	1988	1981	AUG 1988
-Minimum Monthly	15	0.92	0.59	0.56	1.48	0.06	0.61	0.57	0.81	1.21	1.98	1.00	0.52	0.06
-Year		1991	1982	1980	1989	1986	1992	1981	1991	1989	1992	1990	1986	MAY 1986
-Maximum in 24 hrs	14	2.23	2.05	2.40	3.09	3.44	2.80	2.64	2.34	2.34	3.66	2.97	2.48	3.66
-Year		1988	1983	1986	1985	1983	1989	1985	1988	1993	1985	1988	1985	OCT 1985
Snow, Ice pellets, hail														
-Maximum Monthly	15	68.8	51.4	59.1	29.2	22.6	T	0.0	T	1.7	18.6	48.9	82.6	82.6
-Year		1982	1985	1985	1982	1990	1990		1989	1993	1979	1991	1981	DEC 1981
-Maximum in 24 hrs	14	23.3	20.6	25.4	20.0	17.2	T	0.0	T	1.7	12.7	19.1	25.8	25.8
-Year		1988	1983	1986	1985	1990	1990		1989	1993	1989	1991	1985	DEC 1985
WIND:														
Mean Speed (mph)														
Prevailing Direction														
Fastest Mile														
-Direction (!!!)	6	NW	NW	NW	NW	N	NW	NW	NW	W	SE	NW	SW	NW
-Speed (MPH)	6	44	31	40	44	34	38	35	37	35	38	31	35	44
-Year		1980	1985	1982	1982	1981	1984	1982	1984	1983	1984	1979	1982	APR 1982
Peak Gust														
-Direction (!!!)														
-Speed (mph)														
-Date														

PRECIPITATION (inches)

MARQUETTE COUNTY AIRPORT, MICHIGAN

YEAR	JAN	FEB	MAR	APR	MAY	JUNE	JULY	AUG	SEP	OCT	NOV	DEC	ANNUAL
1964	1.31	1.42	1.74	2.91	3.13	2.50	2.95	5.04	3.91	1.79	2.65	2.13	31.48
1965	1.16	1.61	1.72	1.43	3.61	1.51	2.10	3.02	4.25	2.23	3.51	2.02	28.17
1966	1.33	0.72	4.46	1.58	1.10	2.46	1.80	3.90	3.10	3.18	3.62	2.79	30.04
1967	2.71	1.35	1.33	3.06	1.63	3.67	1.44	2.80	1.20	5.50	1.71	0.73	27.13
1968	1.24	3.08	0.76	3.11	2.86	7.07	1.81	2.44	7.22	2.47	1.68	3.96	37.70
1969	2.39	0.79	1.82	2.49	2.22	2.76	1.48	0.53	3.23	5.20	1.86	2.11	26.88
1970	1.90	1.20	1.17	1.55	4.86	1.46	5.36	0.94	5.47	2.29	2.68	1.75	30.63
1971	3.57	2.54	2.06	1.41	2.53	4.12	2.22	1.45	3.80	4.71	2.33	2.89	33.63
1972	1.53	1.43	3.25	2.62	2.50	1.70	2.85	4.12	5.32	1.96	3.39	3.03	33.70
1973	1.47	0.99	2.37	3.01	7.16	3.42	2.16	2.93	2.03	2.53	1.16	2.43	31.66
1974	1.31	1.42	0.61	3.28	2.49	3.57	1.80	3.87	4.01	3.05	2.87	0.81	29.09
1975	2.85	2.10	2.11	2.41	3.53	4.59	1.06	2.97	2.80	1.34	3.16	1.87	30.79
1976	2.65	2.22	3.95	1.83	2.95	1.63	1.52	0.50	1.32	2.24	1.68	1.79	24.28
1977	1.01	1.57	4.09	3.19	2.04	3.10	3.89	3.55	5.17	2.65	2.52	3.24	36.02
1978	3.03	0.89	0.56	1.49	3.34	2.68	4.10	4.47	4.54	1.72	2.72	2.18	31.72
#1979	2.43	1.99	6.08	1.84	2.70	6.13	4.63	1.41	2.09	7.59	2.47	1.56	40.92
1980	2.95	1.47	0.56	4.11	1.87	2.51	3.18	3.49	6.94	2.71	1.92	2.21	33.92
1981	1.96	2.18	2.30	3.10	2.43	6.61	0.57	2.28	2.11	4.63	2.00	4.33	34.50
1982	3.80	0.59	1.89	4.45	2.76	1.24	4.65	3.45	5.00	3.87	2.53	2.83	37.06
1983	2.67	3.14	4.85	3.17	6.49	1.62	1.45	3.21	4.98	5.75	5.68	3.35	46.36
1984	1.27	3.68	3.99	2.46	0.79	2.81	1.92	4.29	4.29	2.03	2.35	2.60	32.48
1985	2.96	2.91	4.63	6.56	3.59	1.49	3.76	4.78	6.32	4.88	5.74	3.97	51.59
1986	2.94	1.21	4.77	2.36	0.06	2.45	3.07	4.62	2.85	4.50	1.04	0.52	30.39
1987	1.42	1.53	2.07	2.30	3.93	2.01	4.88	3.58	2.66	4.31	4.05	3.56	36.30
1988	4.02	0.95	3.66	1.48	1.34	0.71	1.46	8.59	3.97	5.01	8.25	2.36	41.80
1989	1.82	1.60	2.89	1.48	2.15	5.36	0.91	2.61	1.21	3.49	3.43	2.58	29.53
1990	1.74	1.96	1.88	2.32	4.25	3.13	2.01	2.70	4.76	5.27	1.00	1.71	32.73
1991	0.92	1.78	3.75	3.50	2.64	3.06	5.40	0.81	3.10	4.73	5.87	1.51	37.07
1992	1.28	2.66	2.57	2.12	1.76	0.61	4.93	3.28	3.35	1.98	4.27	2.46	31.27
1993	2.39	0.68	0.99	4.11	4.61	2.42	1.95	3.98	4.41	3.32	2.84	1.31	33.01
Record Mean	2.06	1.71	2.17	2.49	2.92	3.33	3.02	2.87	3.53	2.90	2.95	2.29	32.24

AVERAGE TEMPERATURE (deg. F)

MARQUETTE COUNTY AIRPORT, MICHIGAN

YEAR	JAN	FEB	MAR	APR	MAY	JUNE	JULY	AUG	SEP	OCT	NOV	DEC	ANNUAL
1964	25.3	23.5	25.4	40.9	56.5	58.8	67.6	62.2	56.0	47.4	36.5	21.0	43.4
1965	15.8	15.8	23.3	38.2	54.2	60.1	63.1	62.6	53.1	47.4	33.9	28.9	41.4
1966	15.4	21.1	31.6	37.8	47.1	63.0	69.9	64.0	58.5	46.6	32.9	23.9	42.7
1967	20.4	13.1	28.3	39.2	45.8	60.5	64.6	63.6	59.0	45.5	31.0	24.7	41.3
1968	17.8	16.2	33.6	42.3	48.6	56.4	66.1	64.3	60.3	50.5	35.0	24.1	42.9
1969	21.3	22.6	25.6	41.7	51.9	55.4	65.3	70.8	59.3	44.5	34.6	25.6	43.2
1970	14.9	17.5	25.1	40.2	47.0	60.8	68.8	67.8	58.8	50.7	35.2	24.2	42.6
1971	14.4	19.2	26.3	39.7	48.4	61.3	64.2	63.4	61.1	53.6	34.9	25.4	42.7
1972	13.6	14.9	22.8	34.9	52.3	57.0	63.9	63.3	54.6	43.6	33.3	18.9	39.4
1973	22.1	20.5	35.2	39.8	45.4	60.3	67.0	69.1	58.6	53.4	35.1	23.0	44.1
1974	17.1	16.4	26.7	39.9	47.7	59.7	69.1	64.5	52.9	46.7	36.6	29.4	42.2
1975	20.4	23.2	26.5	35.4	54.7	61.1	70.3	67.1	54.2	50.1	39.4	23.1	43.8
1976	15.6	26.1	28.4	43.1	48.5	64.1	66.5	67.6	56.8	43.4	29.1	14.6	42.0
1977	10.1	19.9	34.3	43.8	58.4	59.2	67.3	61.8	56.6	48.4	35.0	22.0	43.1
1978	17.9	18.6	27.4	37.3	53.3	60.5	65.1	66.4	58.7	47.0	34.2	20.1	42.2
#1979	5.6	6.9	24.4	34.4	46.2	58.1	64.7	60.8	55.6	39.5	28.8	23.5	37.4
1980	13.1	12.8	21.3	39.1	54.5	56.9	65.7	64.5	52.9	38.6	29.8	15.4	38.7
1981	13.9	18.4	27.3	39.0	48.1	59.6	65.7	63.8	52.0	40.1	34.5	18.7	40.1
1982	4.8	12.1	23.1	32.8	54.9	54.0	66.0	59.8	53.8	45.4	29.1	23.3	38.2
1983	18.0	21.9	26.0	33.8	44.1	60.5	70.1	67.9	57.5	43.2	31.9	8.7	40.3
1984	9.0	25.3	17.5	40.3	48.2	62.0	64.4	65.4	52.4	46.7	30.5	18.7	40.0
1985	11.3	11.8	26.7	40.9	52.1	56.8	62.6	61.4	54.7	44.0	26.2	9.7	38.2
1986	14.4	14.8	26.4	42.0	54.8	57.1	66.3	60.0	53.2	42.7	25.2	21.3	39.9
1987	18.0	22.9	30.1	44.6	53.7	63.8	67.8	62.7	56.9	39.0	32.6	23.5	43.0
1988	11.6	10.8	23.1	38.5	54.7	62.3	68.6	65.0	54.9	38.0	31.8	16.3	39.6
1989	19.1	8.4	19.4	34.7	50.8	57.5	66.7	63.2	55.7	45.3	24.8	9.2	37.9
1990	22.0	18.1	28.5	42.3	46.9	59.9	63.7	63.2	53.7	41.3	34.0	18.0	41.0
1991	10.0	20.1	26.7	42.2	55.7	62.9	64.9	65.2	52.4	41.0	27.0	19.8	40.7
1992	17.0	19.8	22.6	34.6	52.6	56.3	58.4	59.2	53.5	41.4	28.4	20.2	38.7
1993	16.9	13.9	25.6	34.4	49.8	57.8	66.1	65.4	49.9	40.3	27.0	20.8	39.0
Record Mean	16.8	17.6	25.9	38.7	49.5	59.3	65.8	64.3	57.0	46.7	33.1	22.4	41.5
Max	23.8	25.3	33.5	46.5	58.5	69.0	75.1	72.9	65.2	54.0	39.0	28.3	49.3
Min	9.8	10.0	18.2	30.8	40.5	49.5	56.5	55.8	48.8	39.3	27.2	16.5	33.6

REFERENCE NOTES

MARQUETTE COUNTY AIRPORT, MICHIGAN

GENERAL
T - TRACE AMOUNT.
BLANK ENTRIES DENOTE MISSING/UNREPORTED DATA.
# INDICATES A STATION OR INSTRUMENT RELOCATION.

SPECIFIC

PM - INCLUDES LAST DAY OF PREVIOUS MONTH
ASOS - AUTOMATED SURFACE OBSERVING SYSTEM IN OPERATION DURING THESE MONTHS.

(a) - LENGTH OF RECORD IN YEARS, ALTHOUGH INDIVIDUAL MONTHS MAY BE MISSING.
0.* OR * - THE VALUE IS BETWEEN 0.0 AND 0.05
NORMALS - BASED ON THE 1961-1990 RECORD PERIOD.
EXTREMES - DATES ARE THE MOST RECENT OCCURRENCE
WIND DIR.- NUMERALS SHOW TENS OF DEGRESS CLOCKWISE FROM TRUE NORTH. "00" INDICATES CALM.
RESULTANT DIRECTIONS ARE GIVEN TO WHOLE DEGREES.
BOLD VALUES INDICATE EXTREME VALUES WHICH OCCURRED AFTER THE ASOS SYSTEM WAS COMMISSIONED.

RECORD = PERIOD OF RECORD
RECORD MEAN PRECIPITATION IS THE MEAN OF ALL DAILY PRECIPITATION AMOUNTS DURING THE PERIOD OF RECORD.
RECORD MAX(MIN) TEMPERATURE IS THE MEAN OF ALL DAILY MAX(MIN) TEMPERATURES DURING THE PERIOD OF RECORD.
RECORD MEAN TEMPERATURE IS THE SUM OF THE RECORD MAX AND RECORD MIN DIVIDED BY 2.
AVERAGE TEMPERATURE IS THE SUM OF THE MEAN DAILY MAX AND MIN TEMPERATURE DIVIDED BY 2.

EXCEPTIONS

RECORD MEANS ARE THROUGH THE CURRENT YEAR, BEGINNING IN 1875 FOR TEMPERATURE
1872 FOR PRECIPITATION
1938 FOR SNOWFALL

HEATING DEGREE DAYS Base 65 deg. F

MARQUETTE COUNTY AIRPORT, MICHIGAN

SEASON	JULY	AUG	SEP	OCT	NOV	DEC	JAN	FEB	MAR	APR	MAY	JUNE	TOTAL
1954-65	43	137	268	540	850	1355	1520	1370	1287	795	344	187	8696
1965-66	105	131	355	547	929	1113	1533	1223	1026	809	559	149	8479
1966-67	24	84	225	565	956	1267	1375	1448	1131	767	587	172	8601
1967-68	85	118	203	598	1010	1243	1459	1411	963	676	501	263	8530
1968-69	81	106	166	465	890	1262	1349	1182	1214	692	436	297	8140
1969-70	93	21	209	628	905	1215	1548	1323	1234	739	557	186	8658
1970-71	49	44	224	443	887	1257	1563	1279	1189	754	513	162	8364
1971-72	90	123	182	350	897	1218	1586	1447	1304	897	403	257	8754
1972-73	103	121	305	653	950	1423	1323	1239	916	752	598	159	8542
1973-74	59	37	244	360	889	1296	1482	1354	1180	747	535	192	8375
1974-75	27	77	360	563	843	1097	1376	1165	1188	880	320	176	8072
1975-76	49	45	317	455	762	1290	1525	1120	1128	653	506	94	7944
1976-77	36	79	293	668	1069	1556	1695	1253	946	630	239	193	8657
1977-78	50	126	244	504	894	1329	1455	1295	1156	826	381	183	8443
#1978-79	64	57	220	554	918	1385	1840	1626	1254	914	577	214	9623
1979-80	93	153	295	785	1080	1282	1604	1508	1350	772	340	271	9533
1980-81	53	68	361	813	1050	1532	1581	1303	1160	776	519	168	9384
1981-82	73	78	384	766	907	1425	1864	1480	1292	960	313	326	9868
1982-83	37	190	350	598	1069	1289	1446	1201	1205	929	638	207	9159
1983-84	35	38	264	672	989	1740	1737	1145	1469	733	517	124	9463
1984-85	71	72	369	560	1028	1431	1658	1486	1178	724	400	249	9226
1985-86	111	146	323	645	1156	1709	1564	1400	1191	684	334	250	9513
1986-87	71	169	349	684	1185	1348	1450	1171	1075	606	387	106	8601
1987-88	58	130	247	800	964	1278	1650	1569	1295	787	347	166	9291
1988-89	37	101	305	833	988	1504	1416	1584	1412	902	432	245	9759
1989-90	65	117	283	610	1199	1727	1324	1308	1123	702	556	176	9190
1990-91	104	111	347	726	921	1454	1703	1250	1182	678	334	115	8925
1991-92	74	112	385	739	1133	1394	1483	1308	1306	905	398	276	9513
1992-93	208	202	340	732	1092	1383	1485	1424	1215	910	474	233	9698
1993-94	38	78	446	760	1130	1366							

COOLING DEGREE DAYS Base 65 deg. F

MARQUETTE COUNTY AIRPORT, MICHIGAN

YEAR	JAN	FEB	MAR	APR	MAY	JUNE	JULY	AUG	SEP	OCT	NOV	DEC	TOTAL
1969	0	0	0	0	35	14	109	210	44	0	0	0	412
1970	0	0	0	0	6	66	176	139	46	7	0	0	440
1971	0	0	0	0	6	57	73	81	73	4	0	0	294
1972	0	0	0	0	14	26	76	73	0	0	0	0	189
1973	0	0	0	0	0	25	127	172	59	6	0	0	389
1974	0	0	0	0	5	40	162	70	3	0	0	0	280
1975	0	0	0	0	9	67	218	118	1	0	0	0	413
1976	0	0	0	3	0	75	92	166	53	4	0	0	393
1977	0	0	0	0	44	24	126	35	0	0	0	0	229
1978	0	0	0	0	25	53	76	108	38	2	0	0	302
#1979	0	0	0	0	0	15	92	29	19	0	0	0	155
1980	0	0	0	2	21	36	83	61	5	0	0	0	208
1981	0	0	0	0	1	13	103	50	2	0	0	0	169
1982	0	0	0	0	7	2	75	35	23	0	0	0	142
1983	0	0	0	0	0	78	200	135	47	1	0	0	461
1984	0	0	0	0	3	42	59	89	0	0	0	0	193
1985	0	0	0	7	7	10	44	41	21	0	0	0	130
1986	0	0	0	1	21	17	119	21	0	0	0	0	179
1987	0	0	0	3	41	77	150	65	13	0	0	0	349
1988	0	0	0	0	32	89	157	106	8	0	0	0	392
1989	0	0	0	0	1	27	123	70	13	1	0	0	235
1990	0	0	0	24	0	29	70	63	14	0	0	0	200
1991	0	0	0	0	51	60	77	126	13	0	0	0	327
1992	0	0	0	0	21	22	8	30	3	5	0	0	89
1993	0	0	0	0	10	22	79	97	0	0	0	0	208

SNOWFALL (inches)

MARQUETTE COUNTY AIRPORT, MICHIGAN

SEASON	JULY	AUG	SEP	OCT	NOV	DEC	JAN	FEB	MAR	APR	MAY	JUNE	TOTAL
1954-65	0.0	0.0	0.0	1.7	14.9	27.9	17.0	20.1	22.4	2.9	T	0.0	106.9
1965-66	0.0	0.0	T	0.1	12.6	7.6	20.6	13.0	31.5	12.4	0.3	0.0	98.1
1966-67	0.0	0.0	0.0	1.1	28.0	32.6	29.8	23.5	10.4	2.4	4.5	0.0	132.3
1967-68	0.0	0.0	0.0	5.4	17.4	6.2	11.6	44.4	4.1	5.5	T	0.0	94.6
1968-69	0.0	0.0	0.0	2.0	15.6	41.1	29.5	12.2	17.7	1.0	0.1	0.0	119.2
1969-70	0.0	0.0	0.0	4.3	7.5	30.9	30.8	18.3	11.1	0.9	1.6	0.0	105.4
1970-71	0.0	0.0	T	0.1	11.0	26.6	52.6	22.7	15.9	2.7	3.9	0.0	135.5
1971-72	0.0	0.0	0.0	0.0	11.2	25.3	22.6	23.2	34.3	14.4	0.0	0.0	131.0
1972-73	0.0	0.0	0.0	2.2	6.5	37.7	14.8	12.8	0.4	2.5	3.1	0.0	80.0
1973-74	0.0	0.0	0.0	T	7.4	27.9	15.5	18.2	5.2	13.2	2.3	0.0	89.7
1974-75	0.0	0.0	5.1	5.8	10.3	10.6	34.7	28.7	29.0	0.2	0.0	0.0	124.4
1975-76	0.0	0.0	T	0.1	16.9	15.3	45.6	25.9	44.3	1.4	4.3	0.0	153.8
1976-77	0.0	0.0	T	17.5	17.6	33.3	16.7	17.5	16.7	13.5	T	0.0	132.8
1977-78	0.0	0.0	0.0	T	12.2	38.3	43.3	16.3	6.1	1.0	0.0	0.0	117.2
#1978-79	0.0	0.0	0.0	0.4	13.0	24.2	39.6	23.1	43.9	11.8	T	0.0	156.0
1979-80	0.0	0.0	0.0	18.6	24.8	18.9	33.9	30.2	7.0	11.3	1.4	0.0	146.1
1980-81	0.0	0.0	0.1	11.9	13.1	41.5	41.9	29.5	34.0	4.1	T	0.0	176.1
1981-82	0.0	0.0	0.0	14.2	15.3	82.6	68.8	9.6	24.1	29.2	0.0	0.0	243.8
1982-83	0.0	0.0	0.0	7.2	15.1	17.4	42.4	42.9	54.1	20.2	T	0.0	199.3
1983-84	0.0	0.0	T	2.5	31.6	54.3	30.2	38.1	46.3	1.1	T	0.0	204.1
1984-85	0.0	0.0	T	T	13.6	30.6	56.2	51.4	59.1	18.1	0.0	0.0	229.0
1985-86	0.0	0.0	0.0	T	28.1	56.4	49.8	17.4	49.1	6.8	0.1	0.0	207.7
1986-87	0.0	T	0.0	2.3	15.7	10.5	21.9	27.7	19.3	11.3	0.1	0.0	108.8
1987-88	0.0	0.0	T	8.3	21.4	36.1	62.8	28.0	49.0	5.8	0.0	0.0	211.4
1988-89	0.0	0.0	0.0	10.9	4.2	41.7	43.1	33.7	44.3	7.8	0.2	0.0	185.9
1989-90	0.0	T	0.5	16.6	41.7	58.4	26.4	35.8	16.3	17.1	22.6	T	235.4
1990-91	0.0	0.0	T	9.6	2.5	27.8	25.7	35.2	21.1	9.0	0.6	0.0	131.5
1991-92	0.0	0.0	T	2.9	48.9	30.7	33.3	49.6	23.3	16.7	T	0.0	205.4
1992-93	0.0	0.0	T	11.5	30.0	25.7	40.9	19.9	18.1	29.1	T	0.0	175.2
1993-94	0.0	0.0	1.7	8.7	29.0	22.9							
Record Mean	0.0	T	0.2	4.0	16.8	26.7	27.1	22.9	21.1	9.0	1.4	T	129.3

MUSKEGON, MICHIGAN

Muskegon is located on the eastern shore of Lake Michigan approximately 100 miles north of the southern tip of the lake. The terrain is generally level with several sand dunes along the shoreline. Much of the soil is sandy and vegetation grows well, as evidenced by the trees and grass which grow on the dunes. Many crops grow in the area. Asparagus and celery are the principal truck-garden vegetables. A variety of fruits is raised and blueberries lead as a principal product. The main industry in this area is manufacturing with emphasis on foundry and machined products. The area is also a resort center due to features such as extensive sandy beaches, both on Lake Michigan and inland lakes.

Lake Michigan has a very decided effect upon the weather and climate of this area. The prevailing westerly winds tend to moderate the temperatures, resulting in warmer winters than further inland. In the summer the effect is just the opposite. The air temperature usually remains below the uncomfortable readings of the high 90s. Spring arrives about three to four weeks later than normal for this latitude. Autumn is also delayed, as is the cold of early winter.

Precipitation is fairly moderate, but snowfall is moderate to heavy. The heaviest snows occur during late December, January, and February. Precipitation is also influenced by the lake, especially during the winter. Instability in snow showers along the lakeshore vary enormously in intensity, resulting in traces of snow to more than a foot in 24 hours. The heavier snow squalls tend to concentrate over small sections of the shoreline, depending on their intensity and the direction of the wind. With strong winds most snowshowers will fall further inland, sometimes as much as 30 to 40 miles. Snowfall is likely to occur every day for weeks at a time. The daily accumulation of lake effect snow varies greatly. However, due to low water content of most of the storms, the snow settles rapidly.

Summertime thunderstorms have a tendency, as they move inland, to follow the Muskegon and Grand River Valleys. Thus, these areas are more often frequented by severe electrical storms which will pass without a drop of rain 2 to 3 miles from the immediate river valleys. Thunderstorms near the shoreline are most frequent at night. The afternoon convection-type storms seldom occur within 5 miles of the lake. Lake Michigan-spawned thunderstorms give shoreline areas a surprising number of occurrences compared with surrounding areas of the same latitude during late summer and autumn.

Many crops are planted before the frost danger is over. The young plants are protected by farmers, thus extending the length of the growing season.

Based on the 1951-1980 period, the average first occurrence of 32 degrees Fahrenheit in the fall is October 11 and the average last occurrence in the spring is May 8.

NORMALS, MEANS, AND EXTREMES

MUSKEGON, MICHIGAN

LATITUDE: 43°10'N LONGITUDE: 86°15' W ELEVATION: FT. GRND 625 BARO 629 TIME ZONE: EASTERN WBAN: 14840

	(a)	JAN	FEB	MAR	APR	MAY	JUNE	JULY	AUG	SEP	OCT	NOV	DEC	YEAR
TEMPERATURE °F:														
Normals														
-Daily Maximum		28.8	30.8	41.2	54.5	66.6	75.6	80.3	78.1	70.8	59.0	46.0	33.7	55.5
-Daily Minimum		17.7	18.0	25.4	35.4	45.1	54.4	60.2	58.8	51.5	41.5	32.7	23.3	38.7
-Monthly		23.3	24.4	33.3	45.0	55.9	65.0	70.3	68.5	61.2	50.3	39.4	28.5	47.1
Extremes														
-Record Highest	54	63	62	80	86	93	97	96	99	95	83	76	64	99
-Year		1950	1976	1981	1970	1962	1953	1988	1964	1954	1971	1961	1982	AUG 1964
-Record Lowest	54	-13	-14	-10	1	22	31	40	36	27	21	-14	-15	-15
-Year		1948	1973	1943	1982	1947	1972	1945	1979	1991	1980	1950	1976	DEC 1976
NORMAL DEGREE DAYS:														
Heating (base 65°F)		1293	1137	983	600	309	78	10	25	133	456	768	1132	6924
Cooling (base 65°F)		0	0	0	0	27	78	174	133	19	0	0	0	431
% OF POSSIBLE SUNSHINE														
MEAN SKY COVER (tenths)														
Sunrise - Sunset	51	8.8	8.1	7.3	6.7	6.1	5.8	5.1	5.2	5.8	6.5	8.4	8.8	6.9
MEAN NUMBER OF DAYS:														
Sunrise to Sunset														
-Clear	51	1.6	2.9	5.3	6.7	8.3	8.7	11.0	10.7	9.0	7.4	2.4	1.6	75.7
-Partly Cloudy	51	3.4	4.7	7.2	7.3	8.9	9.9	10.9	10.8	9.0	8.3	4.5	3.7	88.5
-Cloudy	51	26.0	20.6	18.5	16.0	13.9	11.4	9.1	9.5	12.0	15.3	23.1	25.7	201.1
Precipitation														
.01 inches or more	53	16.9	13.5	13.2	12.1	10.8	9.5	8.4	8.5	10.3	10.6	14.1	16.2	144.0
Snow,Ice pellets,hail														
1.0 inches or more	53	9.6	5.9	3.7	0.8	0.0	0.0	0.0	0.0	0.0	0.2	3.0	7.9	30.9
Thunderstorms	53	0.2	0.4	1.6	3.4	4.5	6.2	6.2	6.1	4.9	2.2	1.1	0.4	37.2
Heavy Fog Visibility														
1/4 mile or less	50	1.5	1.7	2.2	2.0	2.3	2.0	1.5	2.0	1.6	2.1	1.8	1.4	22.4
Temperature °F														
-Maximum														
90° and above	33	0.0	0.0	0.0	0.0	0.*	0.6	1.0	0.7	0.0	0.0	0.0	0.0	2.3
32° and below	33	18.3	15.1	5.9	0.4	0.0	0.0	0.0	0.0	0.0	0.0	2.0	13.2	54.9
-Minimum														
32° and below	33	29.0	26.4	23.7	11.1	1.8	0.1	0.0	0.0	0.5	5.4	15.4	26.6	140.0
0° and below	33	1.8	1.8	0.3	0.0	0.0	0.0	0.0	0.0	0.0	0.0	0.0	0.5	4.3
AVG. STATION PRESS.(mb)	20	993.6	994.8	992.6	991.7	991.7	991.5	992.7	994.1	994.5	994.4	993.2	994.0	993.2
RELATIVE HUMIDITY (%)														
Hour 01	32	80	78	76	72	73	78	82	85	84	79	78	80	79
Hour 07 (Local Time)	33	81	80	79	77	76	80	84	89	88	83	80	81	82
Hour 13	33	75	70	63	57	54	57	59	61	63	64	70	76	64
Hour 19	33	77	73	67	59	55	58	60	66	72	73	74	78	68
PRECIPITATION (inches):														
Water Equivalent														
-Normal		2.34	1.49	2.51	2.90	2.60	2.35	2.10	3.41	3.88	2.80	3.15	3.03	32.56
-Maximum Monthly	54	4.55	3.07	6.59	7.12	6.48	5.46	6.63	9.88	13.55	7.33	6.61	5.42	13.55
-Year		1982	1960	1976	1947	1990	1967	1952	1975	1986	1991	1985	1949	SEP 1986
-Minimum Monthly	54	0.45	0.36	0.55	0.72	0.33	0.19	0.47	0.11	0.17	0.33	0.62	0.91	0.11
-Year		1956	1982	1958	1989	1992	1959	1951	1969	1979	1944	1986	1960	AUG 1969
-Maximum in 24 hrs	54	1.69	1.39	2.36	2.31	2.59	3.19	2.54	3.72	6.00	3.21	2.31	3.04	6.00
-Year		1974	1957	1976	1963	1989	1967	1959	1975	1986	1954	1990	1982	SEP 1986
Snow,Ice pellets,hail														
-Maximum Monthly	54	102.4	45.8	35.7	20.4	0.4	0.0	T	0.0	T	4.9	21.5	82.6	102.4
-Year		1982	1981	1965	1982	1954		1990		1991	1967	1951	1963	JAN 1982
-Maximum in 24 hrs	54	22.0	17.5	9.9	12.2	0.4	0.0	T	0.0	T	4.7	9.1	20.1	22.0
-Year		1982	1965	1961	1982	1954		1990		1991	1967	1964	1963	JAN 1982
WIND:														
Mean Speed (mph)	32	12.6	11.6	11.9	11.7	10.0	9.4	8.8	8.6	9.5	10.9	12.0	12.1	10.7
Prevailing Direction through 1963		W	ENE	ENE	SSW	SSW	SSW	W	SSW	SSW	SSW	NW	WNW	SSW
Fastest Obs. 1 Min.														
-Direction (!!!)	34	31	34	25	21	22	24	33	35	20	34	01	23	21
-Speed (MPH)	34	44	41	41	46	43	44	40	35	40	38	44	40	46
-Year		1971	1987	1976	1970	1964	1971	1987	1988	1961	1963	1966	1982	APR 1970
Peak Gust														
-Direction (!!!)	10	W	N	S	SW	S	SW	SE	N	SW	NW	SW	NW	N
-Speed (mph)	10	53	67	59	61	54	55	58	63	51	55	56	55	67
-Date		1990	1987	1988	1988	1989	1990	1989	1988	1988	1985	1991	1985	FEB 1987

PRECIPITATION (inches)

MUSKEGON, MICHIGAN

YEAR	JAN	FEB	MAR	APR	MAY	JUNE	JULY	AUG	SEP	OCT	NOV	DEC	ANNUAL
1964	1.96	0.83	2.56	3.89	1.17	1.41	1.10	2.41	3.59	0.77	3.01	1.61	24.31
1965	3.47	1.90	3.12	2.86	2.27	1.27	2.08	5.11	7.68	2.81	1.78	3.69	38.04
1966	1.76	1.48	3.18	3.47	1.23	1.62	2.24	2.27	1.36	2.17	5.63	3.01	29.42
1967	2.09	1.49	2.04	6.11	1.24	5.46	1.58	2.70	2.87	4.06	2.80	2.97	35.41
1968	1.87	2.44	0.81	1.66	2.44	2.86	3.05	1.83	3.55	2.68	2.91	3.71	29.81
1969	3.00	0.43	1.39	3.49	2.85	3.41	3.00	0.11	1.57	6.56	2.65	1.28	29.74
1970	2.51	1.03	2.94	3.35	2.06	2.42	1.90	1.85	7.15	2.50	3.01	2.49	33.21
1971	1.87	2.19	1.99	1.67	1.67	0.98	1.74	1.33	1.77	0.71	1.86	5.34	23.12
1972	2.21	1.01	2.30	2.09	1.18	3.29	3.51	6.67	3.28	3.45	1.32	4.08	34.39
1973	1.35	1.64	2.57	3.65	3.77	2.10	3.08	3.01	1.67	3.31	2.30	2.91	31.36
1974	4.30	2.24	4.29	3.56	4.13	2.65	1.48	2.78	0.76	2.06	2.67	1.58	32.50
1975	3.57	1.98	2.54	3.06	2.63	4.97	2.07	9.88	1.32	1.15	4.23	2.79	40.19
1976	1.72	2.81	6.59	3.21	5.11	1.41	1.18	0.95	1.08	1.29	1.51	1.75	28.61
1977	2.91	1.16	3.48	1.71	1.51	2.52	2.41	2.48	4.40	2.46	3.11	3.97	32.12
1978	3.24	0.81	1.69	3.41	2.99	3.19	1.54	4.25	4.53	3.07	2.37	2.81	33.90
1979	2.76	0.90	4.04	2.91	1.96	2.07	1.71	4.15	0.17	4.66	3.26	2.49	31.08
1980	2.35	0.69	1.02	4.41	1.77	4.36	3.62	5.52	4.31	2.17	1.88	3.27	35.37
1981	0.77	2.85	0.95	3.54	2.82	3.93	1.86	3.15	5.41	3.16	2.50	2.17	33.11
1982	4.55	0.36	2.40	2.32	3.40	2.79	2.43	4.67	1.80	1.63	5.34	5.27	36.96
1983	1.20	1.22	2.60	3.26	3.30	1.72	1.61	1.46	7.47	2.53	4.47	4.07	34.91
1984	1.32	1.18	2.06	1.90	5.62	1.90	2.49	2.02	1.97	2.23	3.19	4.82	30.70
1985	2.45	2.83	4.02	2.13	1.55	0.98	3.27	7.08	3.35	5.01	6.61	2.99	42.27
1986	1.03	2.68	1.73	1.77	2.21	1.76	3.15	2.98	13.55	2.21	0.62	1.36	35.05
1987	1.52	0.68	1.48	2.82	2.51	1.15	2.13	8.12	4.56	3.31	3.18	4.21	35.67
1988	2.89	1.66	1.95	3.83	0.43	0.47	1.65	3.11	5.92	4.79	6.58	2.82	36.10
1989	1.61	1.04	1.92	0.72	4.64	0.89	1.36	4.86	2.41	1.58	2.92	1.90	25.85
1990	2.34	1.74	2.42	2.00	6.48	3.63	1.26	1.22	3.79	4.36	5.31	2.60	37.15
1991	1.36	0.75	4.40	3.56	3.52	3.84	3.31	3.88	3.32	7.33	4.25	1.82	41.34
1992	1.36	1.33	2.30	3.30	0.33	1.50	2.68	1.97	3.54	2.46	6.26	2.69	29.72
1993	2.61	0.83	1.66	4.83	2.26	4.40	3.87	6.40	4.59	1.66	2.00	1.10	36.21
Record Mean	2.24	1.65	2.49	3.03	2.78	2.56	2.40	3.23	3.48	2.78	3.19	2.66	32.48

AVERAGE TEMPERATURE (deg. F)

MUSKEGON, MICHIGAN

YEAR	JAN	FEB	MAR	APR	MAY	JUNE	JULY	AUG	SEP	OCT	NOV	DEC	ANNUAL
1964	29.0	28.2	34.4	47.2	60.3	67.6	72.4	67.9	61.8	47.7	42.9	28.9	49.0
1965	24.3	25.6	29.0	44.5	61.4	67.1	70.1	67.6	62.9	51.9	39.8	33.5	48.2
1966	18.9	25.2	36.8	43.5	50.6	67.0	73.1	67.6	60.9	50.1	40.2	29.1	46.9
1967	28.6	21.0	33.5	45.4	52.5	69.4	69.8	66.5	61.1	50.9	37.0	31.5	47.3
1968	22.8	22.1	38.9	49.0	54.5	64.9	68.9	69.7	62.6	52.8	38.9	26.6	47.6
1969	22.9	25.7	30.6	46.8	56.8	61.1	70.9	71.7	62.5	49.7	36.5	26.3	46.8
1970	19.2	22.9	28.6	46.5	58.4	67.0	71.5	69.4	61.8	53.2	39.4	27.4	47.1
1971	20.0	24.8	29.3	42.9	53.6	68.6	67.9	66.7	64.5	57.8	38.1	31.6	47.2
1972	21.8	22.4	30.3	42.6	58.7	61.7	68.0	68.2	62.3	47.7	38.2	27.5	45.8
1973	27.9	24.1	41.4	44.9	51.9	66.5	71.0	71.6	62.0	54.5	40.0	26.9	48.6
1974	25.1	20.3	32.9	45.2	52.1	62.2	70.1	68.7	58.5	47.9	39.7	31.0	46.2
1975	28.1	26.0	31.0	41.0	59.5	67.4	71.1	70.2	56.8	53.3	46.0	30.4	48.4
1976	23.6	32.3	37.6	46.5	52.0	67.7	70.0	67.1	58.3	46.0	33.1	20.6	46.3
1977	14.3	22.0	38.8	50.3	63.1	62.3	72.2	66.6	62.3	47.8	39.5	26.2	47.1
1978	19.5	15.7	29.2	44.7	58.0	63.2	67.3	68.5	63.2	48.1	37.6	26.4	45.1
1979	17.0	14.5	33.6	41.2	54.7	64.2	68.6	66.4	60.7	49.4	39.2	31.3	45.1
1980	23.7	22.0	30.6	44.8	56.4	61.5	70.9	70.7	60.9	45.7	37.8	25.2	45.9
1981	20.6	28.1	34.9	47.1	53.8	65.9	69.7	69.6	58.9	47.2	38.7	29.7	47.0
1982	16.1	21.6	30.5	40.1	62.0	59.7	71.6	67.1	60.4	52.4	40.3	35.7	46.5
1983	26.9	30.2	36.4	41.9	50.5	66.2	74.4	72.2	62.6	50.0	40.7	20.8	47.7
1984	19.3	31.9	27.5	46.3	51.4	66.9	68.3	71.7	59.7	52.7	39.4	32.0	47.3
1985	20.4	21.6	35.1	50.3	58.7	62.3	69.3	67.1	63.3	50.8	38.0	24.6	46.8
1986	24.6	24.0	36.5	49.4	58.1	64.5	72.9	65.6	61.9	49.7	36.4	31.1	47.9
1987	26.5	30.7	37.0	48.6	60.6	68.9	73.4	68.7	62.0	46.0	41.8	33.3	49.8
1988	22.9	22.5	34.2	45.6	59.8	67.6	72.7	73.5	61.2	45.2	41.2	28.8	47.9
1989	31.0	21.8	31.5	43.9	54.7	65.0	71.8	67.9	58.7	49.8	36.6	20.5	46.1
1990	32.1	28.7	36.9	46.3	53.7	64.7	69.9	68.2	62.3	49.3	43.0	31.0	48.8
1991	24.5	29.5	37.7	49.1	63.0	69.5	71.1	70.6	60.5	52.1	36.3	30.5	49.5
1992	28.8	29.9	33.6	43.1	55.4	61.7	66.2	64.5	59.9	49.0	38.3	31.0	46.8
1993	26.1	23.6	33.3	43.3	56.9	64.7	72.2	71.8	57.1	47.8	38.2	29.3	47.0
Record Mean	23.8	24.6	32.8	45.3	55.9	65.3	70.2	68.8	61.2	50.7	39.1	28.5	47.2
Max	29.8	31.4	41.1	54.8	66.5	75.9	80.4	78.9	70.9	59.6	45.6	34.2	55.8
Min	17.7	17.9	24.6	35.7	45.2	54.7	60.0	58.8	51.5	41.8	32.6	22.8	38.6

REFERENCE NOTES

MUSKEGON, MICHIGAN

GENERAL
T - TRACE AMOUNT.
BLANK ENTRIES DENOTE MISSING/UNREPORTED DATA.
# INDICATES A STATION OR INSTRUMENT RELOCATION.

SPECIFIC

(a) - LENGTH OF RECORD IN YEARS, ALTHOUGH INDIVIDUAL MONTHS MAY BE MISSING.
0.* OR * - THE VALUE IS BETWEEN 0.0 AND 0.05
NORMALS - BASED ON THE 1961-1990 RECORD PERIOD.
EXTREMES - DATES ARE THE MOST RECENT OCCURRENCE
WIND DIR.- NUMERALS SHOW TENS OF DEGRESS CLOCKWISE FROM TRUE NORTH. "00" INDICATES CALM.
RESULTANT DIRECTIONS ARE GIVEN TO WHOLE DEGREES.
BOLD VALUES INDICATE EXTREME VALUES WHICH OCCURRED AFTER THE ASOS SYSTEM WAS COMMISSIONED.

RECORD = PERIOD OF RECORD
RECORD MEAN PRECIPITATION IS THE MEAN OF ALL DAILY PRECIPITATION AMOUNTS DURING THE PERIOD OF RECORD.
RECORD MAX(MIN) TEMPERATURE IS THE MEAN OF ALL DAILY MAX(MIN) TEMPERATURES DURING THE PERIOD OF RECORD.
RECORD MEAN TEMPERATURE IS THE SUM OF THE RECORD MAX AND RECORD MIN DIVIDED BY 2.
AVERAGE TEMPERATURE IS THE SUM OF THE MEAN DAILY MAX AND MIN TEMPERATURE DIVIDED BY 2.

EXCEPTIONS

RECORD MEANS ARE THROUGH THE CURRENT YEAR, BEGINNING IN 1940 FOR TEMPERATURE
1940 FOR PRECIPITATION
1940 FOR SNOWFALL

HEATING DEGREE DAYS Base 65 deg. F MUSKEGON, MICHIGAN

SEASON	JULY	AUG	SEP	OCT	NOV	DEC	JAN	FEB	MAR	APR	MAY	JUNE	TOTAL
1964-65	11	50	165	528	656	1113	1251	1098	1108	609	147	23	6759
1965-66	1	42	113	402	748	969	1424	1107	868	638	444	68	6824
1966-67	1	14	155	455	741	1106	1120	1230	970	582	395	5	6774
1967-68	16	32	162	438	832	1030	1301	1236	801	471	325	89	6733
1968-69	35	40	103	402	778	1184	1296	1092	1059	538	272	162	6961
1969-70	5	1	143	471	850	1192	1411	1172	1124	564	248	54	7235
1970-71	9	20	146	367	763	1161	1387	1119	1098	657	352	39	7118
1971-72	31	22	120	246	801	1031	1333	1231	1068	665	233	133	6914
1972-73	41	44	118	529	797	1152	1147	1137	726	602	401	15	6709
1973-74	3	8	155	328	743	1171	1229	1246	989	587	398	116	6973
1974-75	7	13	215	525	750	1046	1137	1085	1050	714	202	62	6806
1975-76	16	5	249	364	565	1066	1275	942	841	561	400	24	6308
1976-77	5	43	219	581	948	1371	1564	1200	807	436	145	117	7436
1977-78	7	54	101	526	758	1196	1402	1374	1105	599	239	100	7461
1978-79	29	15	125	519	815	1190	1481	1408	969	708	340	84	7683
1979-80	19	49	160	482	766	1039	1277	1239	1060	601	277	152	7121
1980-81	0	8	152	592	806	1228	1368	1028	923	529	345	34	7013
1981-82	17	9	196	545	782	1089	1509	1209	1064	738	147	165	7470
1982-83	9	52	172	390	734	899	1173	969	877	685	449	77	6486
1983-84	20	1	137	457	723	1365	1411	954	1156	559	415	19	7217
1984-85	23	11	193	374	761	1016	1376	1206	920	465	214	123	6682
1985-86	13	15	153	431	802	1244	1244	1144	878	481	219	75	6699
1986-87	7	60	129	466	855	1042	1184	954	863	487	204	32	6283
1987-88	17	35	112	582	690	976	1299	1227	948	573	189	55	6703
1988-89	11	19	130	606	708	1115	1044	1203	1032	626	323	72	6889
1989-90	0	27	215	468	843	1371	1012	1008	863	580	341	79	6807
1990-91	6	24	149	484	655	1046	1249	990	840	469	157	12	6081
1991-92	9	2	199	397	855	1061	1116	1013	965	653	304	134	6708
1992-93	24	70	185	493	796	1047	1200	1152	977	643	277	79	6943
1993-94	0	11	244	530	796	1101							

COOLING DEGREE DAYS Base 65 deg. F MUSKEGON, MICHIGAN

YEAR	JAN	FEB	MAR	APR	MAY	JUNE	JULY	AUG	SEP	OCT	NOV	DEC	TOTAL
1969	0	0	0	0	25	52	195	217	75	1	0	0	565
1970	0	0	0	13	49	121	220	162	54	7	0	0	626
1971	0	0	0	0	5	151	125	82	114	28	0	0	505
1972	0	0	0	0	43	41	144	149	41	1	0	0	419
1973	0	0	0	4	0	67	194	220	75	13	0	0	573
1974	0	0	0	0	3	38	173	137	27	1	0	0	379
1975	0	0	0	0	38	141	213	174	6	11	0	0	583
1976	0	0	0	15	2	113	166	118	21	1	0	0	436
1977	0	0	0	4	95	42	237	114	28	0	0	0	520
1978	0	0	0	0	29	57	109	131	79	0	0	0	405
1979	0	0	0	0	29	67	137	103	36	7	0	0	379
1980	0	0	0	1	20	54	191	190	36	0	0	0	492
1981	0	0	0	0	6	67	172	160	20	0	0	0	425
1982	0	0	0	0	61	12	220	126	41	9	0	0	469
1983	0	0	0	0	4	122	319	230	72	2	0	0	749
1984	0	0	0	2	0	81	134	227	40	0	0	0	484
1985	0	0	0	30	24	47	153	88	108	0	0	0	450
1986	0	0	0	18	11	67	258	82	42	0	0	0	478
1987	0	0	0	1	75	155	282	157	32	0	0	0	702
1988	0	0	0	0	36	140	258	288	23	0	0	0	745
1989	0	0	0	0	12	81	217	122	32	0	0	0	464
1990	0	0	0	25	0	74	166	131	73	4	0	0	473
1991	0	0	0	2	103	154	206	184	70	6	0	0	725
1992	0	0	0	2	13	39	69	62	37	1	0	0	223
1993	0	0	0	0	29	76	229	225	15	2	0	0	576

SNOWFALL (inches) MUSKEGON, MICHIGAN

SEASON	JULY	AUG	SEP	OCT	NOV	DEC	JAN	FEB	MAR	APR	MAY	JUNE	TOTAL
1964-65	0.0	0.0	0.0	T	15.9	31.0	44.2	35.4	35.7	6.1	0.0	0.0	168.3
1965-66	0.0	0.0	0.0	T	3.2	15.1	58.9	19.5	17.6	3.7	0.0	0.0	118.0
1966-67	0.0	0.0	0.0	T	20.3	31.6	37.5	26.8	12.9	2.0	T	0.0	131.1
1967-68	0.0	0.0	0.0	4.9	7.2	20.0	22.5	21.2	3.1	0.7	0.0	0.0	79.6
1968-69	0.0	0.0	0.0	T	5.7	37.4	51.4	9.6	11.6	T	0.0	0.0	115.7
1969-70	0.0	0.0	0.0	T	20.6	17.7	59.9	16.0	21.5	5.4	0.0	0.0	141.1
1970-71	0.0	0.0	0.0	0.0	10.5	27.9	45.2	10.5	29.0	3.2	0.0	0.0	126.3
1971-72	0.0	0.0	0.0	0.0	20.2	8.4	39.0	19.4	22.5	2.5	0.0	0.0	112.0
1972-73	0.0	0.0	0.0	4.6	7.8	24.7	7.9	19.5	4.0	8.3	T	0.0	76.8
1973-74	0.0	0.0	0.0	0.0	8.1	25.3	33.5	25.0	18.2	T	0.0	0.0	110.1
1974-75	0.0	0.0	0.0	4.0	8.3	16.3	18.8	21.6	14.1	9.1	0.0	0.0	92.2
1975-76	0.0	0.0	0.0	0.0	6.2	30.1	32.0	11.0	6.8	0.8	T	0.0	86.9
1976-77	0.0	0.0	0.0	T	19.7	36.9	60.5	11.3	9.3	6.2	0.0	0.0	143.9
1977-78	0.0	0.0	0.0	T	14.3	55.3	61.1	24.2	9.9	T	0.0	0.0	164.8
1978-79	0.0	0.0	0.0	0.1	7.8	24.6	65.7	11.8	10.5	2.3	0.0	0.0	122.8
1979-80	0.0	0.0	0.0	T	13.1	4.3	29.6	14.5	8.1	5.8	0.0	0.0	75.4
1980-81	0.0	0.0	0.0	0.2	6.9	28.8	23.7	45.8	2.2	T	0.0	0.0	107.6
1981-82	0.0	0.0	0.0	0.2	6.8	22.8	102.4	8.1	13.2	20.4	0.0	0.0	173.9
1982-83	0.0	0.0	0.0	T	3.4	5.5	7.9	6.0	12.7	T	0.0	0.0	35.5
1983-84	0.0	0.0	T	0.0	2.6	50.9	28.8	2.7	8.7	0.5	0.0	0.0	94.2
1984-85	0.0	0.0	0.0	0.0	T	23.0	46.7	33.7	10.1	2.2	0.0	0.0	115.7
1985-86	0.0	0.0	0.0	0.0	4.3	57.0	23.8	17.7	7.0	T	0.0	0.0	109.8
1986-87	0.0	0.0	0.0	0.0	5.3	16.3	24.6	1.5	6.4	8.3	0.0	0.0	62.4
1987-88	0.0	0.0	0.0	T	3.9	20.4	33.2	26.7	7.2	T	0.0	0.0	91.4
1988-89	0.0	0.0	0.0	0.6	2.3	32.5	10.2	34.7	10.2	0.6	T	0.0	91.1
1989-90	0.0	0.0	0.0	4.1	17.9	77.0	6.5	23.9	5.4	2.0	T	0.0	136.8
1990-91	T	0.0	0.0	0.0	5.1	17.4	25.6	16.1	4.0	0.4	0.0	0.0	68.6
1991-92	0.0	0.0	T	T	14.5	29.5	21.4	8.1	10.7	1.3	0.0	0.0	85.5
1992-93	0.0	0.0	0.0	0.5	9.6	21.4	22.5	24.8	9.7	3.6	0.0	0.0	92.1
1993-94	0.0	0.0	0.0	0.1	0.9	37.6							
Record Mean	T	0.0	T	0.5	8.8	25.1	31.2	17.4	11.3	2.8	T	0.0	97.1

SAULT STE. MARIE, MICHIGAN

Sault Ste. Marie is located at the extreme eastern tip of the Upper Peninsula of Michigan at the intersection of Lake Superior, Michigan, and Huron. Consequently, the regional climate is essentially maritime during ice-free periods of the year. Lake ice development usually begins in December and progresses to maximum coverage in February. As ice cover develops, the character of the regional climate gradually changes to continental polar by the time of maximum lake ice development. Lake Superior, to the northwest, is the largest, deepest, and coldest of the Great Lakes and is the dominant climatic control for the area. Water in the northern Great Lakes remains relatively cool during the summer and seldom freezes over during the winter. Therefore, temperatures are moderated throughout most of the year, whereas cloudiness and precipitation are increased.

Terrain on the Michigan side of the international border is nearly flat and lies 700 to 800 feet above sea level. Very little climatological influence is related to Michigan terrain. However, terrain on the Canadian side of the border rises rather abruptly to about 1,500 feet above sea level and this definite topographic influence increases the rain and snow shower activity over the Canadian hills.

Heavy fog occurrences reach a maximum in August, September, and October and form in response to the passage of relatively cold air masses over the warmer waters of the northern Great Lakes. Destructive tornadoes and thunderstorms have occurred on rare occasions. Occurrences of river and flash flooding are very rare. Summer and fall pollen counts are extremely low and sources of industrial pollutants are almost non-existent.

Changing weather patterns are common because of the high frequency of migratory low pressure systems moving toward the east and the northeast through the northern Great Lakes. Summer rains are most frequent during periods of southeasterly circulation whereas winter snows most frequently accompany post-frontal northwest winds.

Most summers pass without a temperature reaching 90 degrees. Winters are cold and snowy with total seasonal snowfall ranging from about 30 inches to more than 175 inches. November 21 is the average date for the appearance of the permanent winter snow cover which normally lasts until April 7.

Annual percent of possible sunshine is low but is especially low during late fall and early winter because of cloud cover produced by lake moisture evaporated into the cold air. Sunshine amounts increase as ice development increases in the winter season. Daylight during most of June and July lasts almost 16 hours, whereas winter daylight reaches a minimum of less than 9 hours a day in late December.

Based on the 1951-1980 period, the average first occurrence of 32 degrees Fahrenheit in the fall is September 27 and the average last occurrence in the spring is May 26.

NORMALS, MEANS, AND EXTREMES

SAULT STE. MARIE, MICHIGAN

LATITUDE: 46°28'N LONGITUDE: 84°22' W ELEVATION: FT. GRND 718 BARO 724 TIME ZONE: EASTERN WBAN: 14847

	(a)	JAN	FEB	MAR	APR	MAY	JUNE	JULY	AUG	SEP	OCT	NOV	DEC	YEAR
TEMPERATURE °F:														
Normals														
-Daily Maximum		21.1	23.2	32.8	48.0	62.6	70.5	76.3	73.8	65.9	54.3	40.0	26.2	49.6
-Daily Minimum		4.6	4.8	15.3	28.4	38.4	45.5	51.3	51.3	44.3	36.2	25.9	11.8	29.8
-Monthly		12.9	14.0	24.0	38.2	50.5	58.0	63.8	62.6	55.1	45.3	33.0	19.0	39.7
Extremes														
-Record Highest	53	45	47	75	85	89	93	97	98	95	80	67	60	98
-Year		1973	1984	1946	1990	1991	1983	1988	1947	1976	1968	1990	1982	AUG 1947
-Record Lowest	53	-36	-35	-24	-2	18	26	36	29	25	16	-10	-31	-36
-Year		1982	1979	1943	1982	1966	1982	1992	1982	1993	1981	1976	1993	JAN 1982
NORMAL DEGREE DAYS:														
Heating (base 65°F)		1615	1428	1271	804	457	224	92	128	300	611	960	1426	9316
Cooling (base 65°F)		0	0	0	0	8	14	55	54	0	0	0	0	131
% OF POSSIBLE SUNSHINE	52	35	47	54	54	57	58	63	57	45	39	24	28	47
MEAN SKY COVER (tenths)														
Sunrise - Sunset	52	8.0	7.3	6.9	6.8	6.5	6.3	5.9	6.1	7.1	7.4	8.5	8.2	7.1
MEAN NUMBER OF DAYS:														
Sunrise to Sunset														
-Clear	52	3.7	4.8	6.6	6.5	7.1	6.8	8.5	7.5	4.7	4.9	1.9	3.3	66.2
-Partly Cloudy	52	5.8	5.9	6.8	7.5	8.6	9.8	10.7	10.8	8.7	6.7	4.5	4.4	90.4
-Cloudy	52	21.5	17.6	17.5	16.0	15.3	13.3	11.8	12.6	16.6	19.4	23.6	23.3	208.6
Precipitation .01 inches or more	52	19.2	14.5	13.0	11.4	11.2	11.3	10.1	11.0	13.3	13.6	17.4	19.5	165.6
Snow,Ice pellets,hail 1.0 inches or more	52	9.3	6.1	4.4	1.8	0.2	0.0	0.0	0.0	0.*	0.8	5.3	9.4	37.2
Thunderstorms	52	0.1	0.1	0.7	1.4	3.1	5.9	5.8	5.3	4.2	1.8	0.5	0.1	28.8
Heavy Fog Visibility 1/4 mile or less	52	1.7	1.8	3.3	2.7	2.9	3.5	4.7	5.7	5.8	4.8	2.9	2.8	42.6
Temperature °F														
-Maximum														
90° and above	52	0.0	0.0	0.0	0.0	0.0	0.1	0.7	0.6	0.1	0.0	0.0	0.0	1.5
32° and below	52	25.8	22.4	14.1	1.8	0.0	0.0	0.0	0.0	0.0	0.1	7.1	21.6	92.9
-Minimum														
32° and below	52	30.7	28.1	29.3	21.0	7.1	0.6	0.0	0.*	1.8	9.7	22.5	29.8	180.6
0° and below	52	11.3	9.5	3.6	0.*	0.0	0.0	0.0	0.0	0.0	0.0	0.2	5.9	30.5
AVG. STATION PRESS.(mb)	21	988.0	990.2	989.1	988.4	988.2	987.5	988.5	989.8	989.9	989.6	988.3	988.4	988.8
RELATIVE HUMIDITY (%)														
Hour 01 (Local Time)	52	81	80	80	79	80	87	90	91	91	87	85	83	85
Hour 07	52	81	81	82	80	80	85	89	92	92	89	87	84	85
Hour 13	52	74	70	67	60	56	61	62	63	67	67	75	77	67
Hour 19	52	80	77	74	66	61	66	67	72	79	79	82	82	74
PRECIPITATION (inches):														
Water Equivalent														
-Normal		2.42	1.74	2.30	2.35	2.71	3.14	2.71	3.61	3.69	3.23	3.45	2.88	34.23
-Maximum Monthly	52	4.52	3.74	4.97	5.16	7.41	7.35	6.04	9.48	7.78	6.36	7.72	5.10	9.48
-Year		1982	1971	1976	1954	1970	1969	1956	1974	1970	1979	1988	1984	AUG 1974
-Minimum Monthly	52	0.51	0.21	0.35	0.60	0.80	0.52	0.57	0.50	0.86	0.16	0.87	0.73	0.16
-Year		1961	1993	1993	1949	1980	1988	1989	1947	1943	1963	1962	1954	OCT 1963
-Maximum in 24 hrs	52	1.37	1.11	1.81	2.67	5.10	5.06	2.79	5.92	2.71	2.06	2.42	1.32	5.92
-Year		1950	1977	1976	1954	1970	1970	1965	1974	1986	1979	1988	1971	AUG 1974
Snow,Ice pellets,hail														
-Maximum Monthly	52	71.0	41.3	34.7	25.8	4.6	T	0.0	T	2.7	12.5	46.8	69.3	71.0
-Year		1982	1972	1964	1982	1947	1951		1986	1956	1943	1989	1980	JAN 1982
-Maximum in 24 hrs	52	15.3	12.6	11.9	11.5	4.6	T	0.0	T	2.7	10.1	14.3	12.5	15.3
-Year		1988	1968	1964	1979	1947	1951		1986	1956	1943	1943	1984	JAN 1988
WIND:														
Mean Speed (mph)	52	9.7	9.4	10.0	10.3	9.7	8.6	7.8	7.7	8.6	9.2	9.8	9.7	9.2
Prevailing Direction through 1963		E	E	WNW	WNW	WNW	WNW	WNW	WNW	WNW	E	E	E	WNW
Fastest Mile														
-Direction (!!)	25	NW	W	SE	SE	E	S	SE	NW	W	NW	NW	NW	NW
-Speed (MPH)	25	47	47	42	42	49	37	44	35	43	42	60	45	60
-Year		1974	1968	1985	1966	1989	1981	1988	1977	1970	1985	1975	1976	NOV 1975
Peak Gust														
-Direction (!!)	10	W	NW	NW	W	NW	NW	SW	NW	NW	W	NW	W	NW
-Speed (mph)	10	56	56	59	58	55	52	54	40	54	61	61	58	61
-Date		1990	1989	1991	1984	1986	1988	1984	1987	1992	1988	1989	1992	NOV 1989

PRECIPITATION (inches)

SAULT STE. MARIE, MICHIGAN

YEAR	JAN	FEB	MAR	APR	MAY	JUNE	JULY	AUG	SEP	OCT	NOV	DEC	ANNUAL
1964	2.77	0.94	2.38	2.48	2.45	2.14	1.92	3.64	3.70	2.79	4.32	3.18	32.71
1965	2.86	2.32	0.93	1.33	4.42	3.71	5.46	3.91	7.12	3.21	5.08	2.61	42.96
1966	2.22	1.79	2.69	1.68	1.78	2.04	1.24	3.45	2.64	4.92	5.70	2.65	32.80
1967	3.03	1.72	1.58	3.29	1.62	2.86	1.87	4.40	1.78	3.53	3.46	3.29	32.43
1968	1.40	3.25	1.14	2.71	2.49	3.73	3.04	4.65	4.05	4.20	1.99	3.01	35.66
1969	2.93	0.66	1.13	3.32	1.70	7.35	0.93	2.18	2.56	6.29	2.68	1.40	33.13
1970	2.38	0.84	1.58	1.81	7.41	1.68	4.06	2.07	7.78	4.29	3.10	2.16	39.16
1971	3.50	3.74	2.84	1.13	2.99	2.86	3.42	4.03	4.03	2.85	3.41	4.43	39.23
1972	4.08	2.98	2.41	2.83	2.15	1.29	3.51	4.52	3.25	1.79	3.01	3.63	35.45
1973	1.74	1.59	1.79	2.80	5.21	2.62	4.21	4.51	2.75	2.81	3.92	2.50	36.45
1974	2.67	1.13	1.69	3.63	2.30	4.02	2.32	9.48	3.71	3.51	3.13	1.85	39.44
1975	3.27	2.34	2.24	2.38	3.96	3.69	3.57	1.27	5.12	1.16	3.25	1.63	33.88
1976	2.24	2.35	4.97	1.89	2.21	1.87	1.24	1.39	1.71	1.43	2.47	1.93	25.70
1977	2.72	2.89	4.76	1.66	1.78	2.37	4.87	3.54	4.31	1.86	4.71	3.42	38.89
1978	2.51	0.54	1.10	1.74	3.74	3.82	5.19	3.65	6.71	3.35	2.61	3.25	38.21
1979	1.86	1.94	4.60	3.79	2.15	5.58	4.42	5.22	1.63	6.36	3.70	2.10	43.35
1980	3.42	0.82	1.14	3.45	0.80	3.94	1.77	2.64	3.32	2.43	2.11	4.14	29.98
1981	0.68	2.55	1.17	3.28	1.15	5.59	0.80	2.45	2.18	4.50	1.89	2.42	28.66
1982	4.52	0.62	3.20	3.75	1.37	1.72	1.72	3.34	6.12	2.88	3.89	4.40	37.53
1983	2.13	1.21	2.57	2.60	4.70	1.65	3.27	2.73	3.35	3.66	3.17	3.04	34.08
1984	1.52	1.90	2.06	1.67	1.16	3.91	2.24	2.99	5.08	3.50	3.78	5.10	34.91
1985	2.61	2.78	2.77	4.36	4.37	1.63	3.29	4.24	3.37	2.81	3.69	3.45	39.37
1986	1.70	0.95	4.14	1.22	1.80	5.12	4.76	4.57	5.23	3.14	2.57	2.21	37.41
1987	1.74	0.80	1.14	1.32	1.96	2.86	3.14	4.50	1.85	5.06	3.42	3.07	30.86
1988	3.12	0.92	2.91	2.17	1.49	0.52	2.14	4.77	3.01	4.54	7.72	3.63	36.94
1989	2.30	1.78	2.37	1.98	2.12	2.61	0.57	2.64	1.05	2.43	4.31	1.73	25.89
1990	2.48	2.04	2.39	1.79	3.29	4.99	1.67	2.33	3.95	3.75	3.90	2.48	35.06
1991	1.51	0.63	3.70	3.39	1.62	1.09	3.86	1.55	5.85	3.97	3.52	2.51	33.20
1992	2.86	1.07	1.52	2.50	1.46	1.55	4.09	3.27	5.72	3.29	5.03	3.45	35.81
1993	2.23	0.21	0.35	3.82	4.51	3.79	3.23	1.86	3.66	5.22	2.49	2.56	33.93
Record Mean	2.10	1.49	1.91	2.24	2.72	2.99	2.71	3.03	3.65	3.07	3.18	2.44	31.53

AVERAGE TEMPERATURE (deg. F)

SAULT STE. MARIE, MICHIGAN

YEAR	JAN	FEB	MAR	APR	MAY	JUNE	JULY	AUG	SEP	OCT	NOV	DEC	ANNUAL
1964	20.5	17.1	22.6	38.0	53.8	57.7	66.5	59.4	53.9	44.0	35.0	17.5	40.5
1965	12.0	12.9	22.3	36.0	51.8	57.3	59.1	60.6	53.1	43.2	32.3	26.8	38.9
1966	10.6	19.0	28.4	37.8	45.0	61.0	67.1	63.0	55.4	44.2	30.5	18.9	40.1
1967	17.4	7.7	21.6	36.0	44.5	61.0	61.3	60.3	55.7	43.4	29.0	22.3	38.3
1968	11.5	10.8	26.7	41.5	49.2	57.8	61.9	60.4	59.5	48.1	31.6	18.3	39.8
1969	15.4	18.3	23.0	38.0	47.0	53.0	63.3	65.7	55.8	42.7	32.9	19.6	39.5
1970	9.7	10.2	20.2	39.0	47.7	58.7	64.9	63.8	55.0	47.7	32.9	17.3	38.9
1971	9.5	13.1	20.3	35.8	47.4	60.7	60.4	59.1	57.2	50.9	31.6	22.0	39.0
1972	13.8	10.7	19.0	33.9	55.1	57.3	62.6	61.6	53.2	41.4	32.5	18.7	38.3
1973	19.2	14.6	35.0	39.8	46.3	58.7	62.9	68.2	53.9	49.5	33.1	19.3	41.7
1974	14.3	9.6	21.9	37.9	46.7	59.8	65.0	64.2	50.4	41.2	32.8	24.1	39.0
1975	16.2	17.9	20.9	32.5	56.2	61.2	66.2	64.3	52.4	47.4	35.8	19.8	40.9
1976	10.4	18.8	22.9	40.0	47.7	62.4	63.3	64.8	53.6	41.3	27.3	9.9	38.6
1977	6.3	14.3	28.6	40.0	55.2	54.8	62.8	57.3	53.6	44.7	34.4	19.8	39.3
1978	10.8	11.1	20.4	35.1	55.0	56.2	62.4	63.1	55.5	43.9	31.8	17.9	38.6
1979	8.9	5.2	27.1	37.0	49.4	57.3	65.4	61.3	55.2	42.4	32.2	24.0	38.8
1980	15.0	12.4	22.5	41.1	54.1	54.8	63.1	64.9	52.2	37.7	29.2	12.8	38.3
1981	8.8	18.6	27.2	39.3	50.7	56.2	63.4	62.8	50.3	38.9	32.5	21.7	39.2
1982	4.7	11.5	21.6	31.7	56.2	51.5	62.7	57.6	54.2	46.8	32.3	24.3	37.9
1983	15.9	21.0	27.5	37.0	45.3	60.9	69.2	68.0	58.5	44.8	33.0	11.2	41.1
1984	7.3	23.7	20.9	44.5	47.5	59.7	63.4	64.9	51.9	46.7	31.7	20.6	40.2
1985	11.9	12.9	26.5	38.8	51.2	54.6	60.7	62.8	57.6	45.8	32.1	13.9	39.1
1986	12.6	15.1	25.8	43.5	54.3	55.3	65.2	60.3	53.9	44.5	30.3	25.8	40.6
1987	20.0	22.1	29.7	45.9	53.0	61.3	66.8	63.0	57.3	42.2	33.5	26.0	43.4
1988	13.5	12.2	23.0	39.2	53.6	59.1	67.5	64.8	55.3	41.1	36.6	19.2	40.4
1989	19.5	11.2	19.4	36.9	52.3	58.1	66.7	63.3	56.1	45.9	28.1	7.3	38.7
1990	23.5	17.9	26.6	41.1	48.8	57.1	62.9	62.6	55.0	43.4	35.9	22.2	41.4
1991	12.9	18.9	27.0	42.9	55.7	63.0	64.5	66.2	54.0	44.9	32.2	21.5	42.0
1992	16.6	17.7	21.8	35.6	51.9	56.4	57.4	60.4	54.0	43.4	30.9	23.7	39.2
1993	16.8	11.8	26.4	36.9	48.3	56.4	64.2	65.2	50.8	40.9	29.6	21.3	39.1
Record Mean	14.1	13.9	23.6	37.9	49.6	58.5	63.8	63.0	55.5	45.0	32.5	20.4	39.8
Max	22.0	22.8	32.2	46.8	60.2	69.9	75.0	73.0	64.5	52.8	38.5	27.0	48.7
Min	6.2	5.0	14.9	29.0	38.9	47.1	52.6	52.9	46.4	37.3	26.5	13.7	30.9

REFERENCE NOTES

SAULT STE. MARIE, MICHIGAN

GENERAL
T - TRACE AMOUNT.
BLANK ENTRIES DENOTE MISSING/UNREPORTED DATA.
# INDICATES A STATION OR INSTRUMENT RELOCATION.

SPECIFIC

EXCEPTIONS

RECORD MEANS ARE THROUGH THE CURRENT YEAR, BEGINNING IN 1888 FOR TEMPERATURE
1888 FOR PRECIPITATION
1942 FOR SNOWFALL

(a) - LENGTH OF RECORD IN YEARS, ALTHOUGH INDIVIDUAL MONTHS MAY BE MISSING.
0.* OR * - THE VALUE IS BETWEEN 0.0 AND 0.05
NORMALS - BASED ON THE 1961-1990 RECORD PERIOD.
EXTREMES - DATES ARE THE MOST RECENT OCCURRENCE
WIND DIR.- NUMERALS SHOW TENS OF DEGRESS CLOCKWISE FROM TRUE NORTH. "00" INDICATES CALM.
RESULTANT DIRECTIONS ARE GIVEN TO WHOLE DEGREES.
BOLD VALUES INDICATE EXTREME VALUES WHICH OCCURRED AFTER THE ASOS SYSTEM WAS COMMISSIONED.

RECORD = PERIOD OF RECORD
RECORD MEAN PRECIPITATION IS THE MEAN OF ALL DAILY PRECIPITATION AMOUNTS DURING THE PERIOD OF RECORD.
RECORD MAX(MIN) TEMPERATURE IS THE MEAN OF ALL DAILY MAX(MIN) TEMPERATURES DURING THE PERIOD OF RECORD.
RECORD MEAN TEMPERATURE IS THE SUM OF THE RECORD MAX AND RECORD MIN DIVIDED BY 2.
AVERAGE TEMPERATURE IS THE SUM OF THE MEAN DAILY MAX AND MIN TEMPERATURE DIVIDED BY 2.

HEATING DEGREE DAYS Base 65 deg. F SAULT STE. MARIE, MICHIGAN

SEASON	JULY	AUG	SEP	OCT	NOV	DEC	JAN	FEB	MAR	APR	MAY	JUNE	TOTAL
1964-65	51	177	332	643	894	1467	1636	1454	1316	864	409	229	9472
1965-66	180	162	350	671	976	1177	1684	1284	1127	806	615	156	9188
1966-67	35	97	288	638	1028	1421	1473	1600	1339	865	631	132	9547
1967-68	146	160	283	664	1075	1320	1655	1570	1179	697	484	237	9470
1968-69	149	165	172	527	996	1442	1532	1300	1294	805	555	360	9297
1969-70	92	55	292	683	957	1399	1707	1531	1382	774	528	203	9603
1970-71	63	99	301	530	959	1469	1712	1452	1380	868	538	157	9528
1971-72	144	190	262	434	998	1324	1582	1566	1420	926	314	232	9392
1972-73	105	122	348	722	972	1432	1414	1406	923	748	574	192	8958
1973-74	92	22	376	473	949	1413	1566	1550	1330	805	558	164	9298
1974-75	57	72	432	730	959	1260	1507	1314	1358	968	298	151	9106
1975-76	64	70	373	537	871	1394	1686	1333	1299	746	534	100	9007
1976-77	76	90	349	728	1121	1706	1818	1420	1119	744	318	305	9794
1977-78	111	246	335	621	910	1396	1675	1503	1377	889	333	270	9666
1978-79	112	96	282	645	989	1453	1733	1673	1171	835	477	237	9703
1979-80	69	134	305	696	979	1263	1541	1520	1313	711	342	317	9190
1980-81	70	44	383	841	1066	1613	1740	1297	1164	763	438	262	9681
1981-82	89	83	435	799	968	1336	1868	1495	1337	991	273	393	10067
1982-83	97	228	339	556	976	1256	1520	1229	1153	832	604	169	8959
1983-84	34	25	236	619	951	1665	1787	1193	1362	608	535	177	9192
1984-85	82	51	391	560	990	1367	1643	1453	1184	776	424	306	9227
1985-86	135	100	242	590	978	1579	1619	1392	1207	639	325	285	9091
1986-87	60	153	325	625	1036	1209	1386	1195	1086	568	393	142	8178
1987-88	53	105	229	699	938	1202	1586	1525	1295	767	358	197	8954
1988-89	39	103	291	733	847	1413	1403	1502	1409	835	388	211	9174
1989-90	34	119	271	582	1099	1785	1281	1317	1186	719	494	240	9127
1990-91	96	99	303	662	866	1319	1609	1282	1169	657	292	92	8446
1991-92	68	56	338	615	976	1341	1495	1367	1331	875	405	256	9123
1992-93	230	165	326	661	1014	1277	1486	1482	1187	834	512	252	9426
1993-94	66	66	421	740	1054	1346							

COOLING DEGREE DAYS Base 65 deg. F SAULT STE. MARIE, MICHIGAN

YEAR	JAN	FEB	MAR	APR	MAY	JUNE	JULY	AUG	SEP	OCT	NOV	DEC	TOTAL
1969	0	0	0	0	3	7	46	84	24	0	0	0	164
1970	0	0	0	0	0	21	69	67	7	0	0	0	164
1971	0	0	0	0	0	34	7	14	34	4	0	0	93
1972	0	0	0	0	12	7	36	24	0	0	0	0	79
1973	0	0	0	0	0	9	33	129	49	0	0	0	220
1974	0	0	0	0	0	14	61	51	0	0	0	0	126
1975	0	0	0	0	32	42	113	54	2	0	0	0	243
1976	0	0	0	3	2	29	33	88	12	0	0	0	167
1977	0	0	0	0	19	5	49	13	0	0	0	0	86
1978	0	0	0	0	27	9	40	44	5	0	0	0	125
1979	0	0	0	0	0	13	87	30	15	0	0	0	145
1980	0	0	0	0	10	16	19	48	6	0	0	0	99
1981	0	0	0	0	1	5	46	22	0	0	0	0	74
1982	0	0	0	0	6	0	30	5	21	0	0	0	62
1983	0	0	0	0	0	52	171	124	49	0	0	0	396
1984	0	0	0	0	0	24	41	56	3	0	0	0	124
1985	0	0	0	0	4	1	8	38	28	0	0	0	79
1986	0	0	0	1	2	1	75	12	0	0	0	0	91
1987	0	0	0	1	27	35	114	52	2	0	0	0	231
1988	0	0	0	0	12	25	126	104	8	0	0	0	275
1989	0	0	0	0	0	9	97	74	11	0	0	0	191
1990	0	0	0	10	0	7	36	30	9	0	0	0	92
1991	0	0	0	0	12	39	57	98	14	0	0	0	220
1992	0	0	0	0	8	2	0	27	2	0	0	0	39
1993	0	0	0	0	2	3	48	79	3	0	0	0	135

SNOWFALL (inches) SAULT STE. MARIE, MICHIGAN

SEASON	JULY	AUG	SEP	OCT	NOV	DEC	JAN	FEB	MAR	APR	MAY	JUNE	TOTAL
1964-65	0.0	0.0	0.1	5.2	18.7	38.6	42.3	32.9	18.9	7.7	0.0	0.0	164.4
1965-66	0.0	0.0	1.8	1.7	27.7	13.4	35.6	11.5	9.3	5.0	4.5	0.0	110.5
1966-67	0.0	0.0	T	1.2	17.8	20.4	29.2	25.8	6.8	1.9	2.2	0.0	105.3
1967-68	0.0	0.0	T	1.3	24.6	25.9	14.8	39.5	2.3	4.7	T	0.0	113.1
1968-69	0.0	0.0	0.0	1.2	21.1	32.4	32.2	12.0	6.6	3.8	T	0.0	109.3
1969-70	0.0	0.0	T	11.6	16.9	16.8	35.5	12.5	9.9	2.5	1.4	0.0	107.1
1970-71	0.0	0.0	0.0	0.3	12.2	24.2	50.4	32.6	17.5	6.8	0.5	0.0	144.5
1971-72	0.0	0.0	0.0	0.0	13.7	26.7	53.2	41.3	22.8	14.7	0.0	0.0	172.4
1972-73	0.0	0.0	0.0	10.1	10.1	38.3	13.9	17.1	0.2	2.0	0.5	0.0	92.2
1973-74	0.0	0.0	0.0	0.2	24.4	19.2	27.1	17.6	18.1	2.9	0.3	0.0	109.8
1974-75	0.0	0.0	T	0.8	7.0	21.8	41.8	32.8	21.3	0.6	0.0	0.0	126.1
1975-76	0.0	0.0	T	T	7.7	11.7	40.8	31.7	32.1	5.0	1.8	0.0	130.8
1976-77	0.0	0.0	T	7.5	34.2	34.9	45.9	26.3	18.5	11.3	0.0	0.0	178.6
1977-78	0.0	0.0	0.0	T	16.2	41.9	35.5	12.1	14.7	5.3	0.0	0.0	125.7
1978-79	0.0	0.0	0.0	T	14.0	51.3	39.6	21.7	13.6	19.2	0.7	0.0	160.1
1979-80	0.0	0.0	0.0	3.5	12.4	28.0	30.7	14.3	17.5	1.7	0.0	0.0	108.1
1980-81	0.0	0.0	0.0	5.2	14.6	69.3	13.6	29.9	8.7	0.4	0.0	0.0	141.7
1981-82	0.0	0.0	0.0	6.3	11.7	25.9	71.0	10.8	17.1	25.8	0.0	0.0	168.6
1982-83	0.0	0.0	0.0	T	12.2	22.7	22.6	15.7	9.4	4.4	T	0.0	87.0
1983-84	0.0	0.0	T	T	5.6	57.5	23.3	10.8	7.8	2.5	0.9	0.0	108.4
1984-85	0.0	0.0	0.1	T	8.2	44.8	43.8	31.5	16.8	9.7	0.0	0.0	154.9
1985-86	0.0	0.0	0.0	0.4	15.7	56.4	28.7	17.0	29.9	0.4	0.1	0.0	148.6
1986-87	0.0	T	0.0	T	25.4	29.0	29.2	10.2	7.2	2.0	0.0	0.0	103.0
1987-88	0.0	0.0	0.0	4.6	16.7	27.7	43.0	14.5	23.4	5.2	0.0	0.0	135.1
1988-89	0.0	0.0	0.0	6.6	7.4	50.2	25.6	23.7	24.7	4.1	T	0.0	142.3
1989-90	0.0	0.0	0.1	0.4	46.8	28.8	31.5	27.3	7.4	16.2	0.9	0.0	159.4
1990-91	0.0	0.0	T	0.4	16.7	29.2	25.9	11.3	15.2	8.5	T	0.0	107.2
1991-92	0.0	0.0	0.4	2.2	19.1	25.1	37.6	12.8	10.5	5.0	0.4	0.0	113.1
1992-93	0.0	0.0	T	10.8	13.6	35.0	28.9	2.8	1.9	8.7	0.0	0.0	101.7
1993-94	0.0	0.0	0.0	5.0	12.6	28.4							
Record Mean	0.0	T	0.1	2.5	15.3	29.9	28.7	18.7	14.8	5.6	0.5	T	116.2

DULUTH, MINNESOTA

Duluth, Minnesota is located at the western tip of Lake Superior. The city, about 20 miles long, lies at the base of a range of hills that rise abruptly to 600 - 800 feet above the level of Lake Superior. The range runs in a northeast and southwest direction. Two or 3 miles from the lake the land becomes a slightly rolling plateau.

Duluth in the summer is known as the Air Conditioned City. Being situated below high terrain and along the lake, any easterly component winds automatically cool the city. However, with westerly flow in the summer, the wind generally abates at night, thus, allowing cool lake air to move back into the city area near the lake.

An important influence on the climate is the passage of a succession of high and low pressure systems west and east. The proximity of Lake Superior, which is the largest and coldest of the Great Lakes, modifies the local weather. Summer temperatures are cooler and winter temperatures are warmer. The lake effect at Duluth is most prevalent when low pressure systems pass to the south creating easterly winds. In the summer, warm, moist air flowing over the cold lake surface has a stabilizing effect that results in cool, cloudy weather over Duluth. However, during the winter cold air flowing over the warm open lake surface absorbs moisture that is later precipitated over Duluth as snow. The lake effect is further reflected from the low frequency of severe storms such as wind, hail, tornadoes, freezing rain (glaze), and blizzards when compared to other areas that are a further distance from the lake.

Easterly component winds at Duluth occur 40 to 50 percent of the time from March through August and 20 to 25 percent of the time from November through February. During the winter 60 to 70 percent of the winds are from a westerly component.

The climate of Duluth is predominantly continental with significant local Lake Superior effects. Duluth averages 143 days between the last occurrence of 32 degrees in mid-May and the first in early October. At the Duluth Airport about six miles away from the lake, the average first and last occurrences of 32 degrees are late May and late September, giving a freeze-free period of 123 days.

Fall colors throughout this area are outstanding. Reds, yellows, browns, and combinations of these are an experience to see. Recreation is superb from December through March for cross-country and down-hill skiing and snowmobiling. The snow is dry.

Ice in the harbor forms about mid-November and generally is gone by mid-April. The shipping season can vary from year to year depending on temperatures and the winds that move the ice around. In most years there is little or no shipping during February and March on Lake Superior.

NORMALS, MEANS, AND EXTREMES

DULUTH, MINNESOTA

LATITUDE: 46°50'N LONGITUDE: 92°11' W ELEVATION: FT. GRND 1428 BARO 1430 TIME ZONE: CENTRAL WBAN: 14913

	(a)	JAN	FEB	MAR	APR	MAY	JUNE	JULY	AUG	SEP	OCT	NOV	DEC	YEAR
TEMPERATURE °F:														
Normals														
-Daily Maximum		16.2	21.7	32.9	48.2	61.9	71.0	77.1	73.9	63.8	52.3	35.2	20.7	47.9
-Daily Minimum		-2.2	2.8	15.7	28.9	39.6	48.5	55.1	53.3	44.5	35.1	21.5	4.9	29.0
-Monthly		7.0	12.3	24.4	38.6	50.8	59.8	66.1	63.7	54.2	43.7	28.4	12.8	38.5
Extremes														
-Record Highest	52	52	55	78	88	90	93	97	97	95	86	70	55	97
-Year		1942	1976	1946	1952	1986	1980	1988	1947	1976	1953	1978	1962	JUL 1988
-Record Lowest	52	-39	-33	-29	-5	17	27	35	32	22	8	-23	-34	-39
-Year		1972	1988	1989	1975	1967	1972	1988	1986	1942	1976	1964	1983	JAN 1972
NORMAL DEGREE DAYS:														
Heating (base 65°F)		1798	1476	1259	792	445	170	60	113	329	660	1098	1618	9818
Cooling (base 65°F)		0	0	0	0	0	14	94	72	0	0	0	0	180
% OF POSSIBLE SUNSHINE	43	48	53	54	55	57	58	65	60	51	46	35	39	52
MEAN SKY COVER (tenths)														
Sunrise - Sunset	45	6.8	6.5	6.8	6.8	6.6	6.6	6.0	6.0	6.6	6.8	7.6	7.2	6.7
MEAN NUMBER OF DAYS:														
Sunrise to Sunset														
-Clear	45	7.2	7.3	7.1	6.2	6.3	5.1	6.9	7.4	6.4	6.6	4.5	6.0	77.1
-Partly Cloudy	45	6.9	6.1	7.0	7.8	9.5	11.1	13.1	12.2	8.6	7.9	5.7	5.9	101.8
-Cloudy	45	16.8	14.8	17.0	15.9	15.2	13.8	11.0	11.4	15.0	16.5	19.8	19.1	186.4
Precipitation														
.01 inches or more	52	11.7	9.5	10.7	10.5	12.3	12.6	11.4	11.4	12.0	9.5	11.1	11.7	134.3
Snow,Ice pellets,hail														
1.0 inches or more	50	4.5	3.5	3.8	1.8	0.3	0.0	0.0	0.0	0.*	0.4	3.2	4.3	21.8
Thunderstorms	51	0.1	0.*	0.6	1.6	3.5	6.9	8.0	7.2	4.0	1.3	0.4	0.1	33.8
Heavy Fog Visibility														
1/4 mile or less	45	2.3	2.3	3.7	3.6	5.5	6.5	5.3	6.5	5.7	4.5	3.4	3.3	52.5
Temperature °F														
-Maximum														
90° and above	32	0.0	0.0	0.0	0.0	0.1	0.1	1.1	0.7	0.1	0.0	0.0	0.0	2.1
32° and below	32	28.0	22.4	15.1	1.9	0.0	0.0	0.0	0.0	0.0	0.7	12.5	26.3	107.0
-Minimum														
32° and below	32	31.0	28.1	29.0	20.1	5.3	0.5	0.0	0.*	2.5	12.6	25.8	30.8	185.7
0° and below	32	17.2	12.8	4.6	0.1	0.0	0.0	0.0	0.0	0.0	0.0	1.6	12.2	48.4
AVG. STATION PRESS. (mb)	21	963.6	965.1	963.5	963.5	963.2	962.5	964.2	965.1	964.7	964.3	963.4	963.8	963.9
RELATIVE HUMIDITY (%)														
Hour 00	32	75	73	73	70	72	80	82	85	84	77	78	79	77
Hour 06 (Local Time)	32	77	76	78	76	76	81	85	88	87	82	81	80	81
Hour 12	32	70	65	63	56	53	60	59	63	64	62	70	74	63
Hour 18	32	70	64	62	54	52	58	59	64	68	66	72	75	64
PRECIPITATION (inches):														
Water Equivalent														
-Normal		1.22	0.80	1.91	2.25	3.03	3.82	3.61	3.99	3.84	2.49	1.80	1.24	30.00
-Maximum Monthly	52	4.70	2.37	5.12	5.84	7.67	8.04	8.48	10.31	9.38	7.53	5.01	3.70	10.31
-Year		1969	1971	1965	1948	1962	1986	1949	1972	1991	1949	1983	1968	AUG 1972
-Minimum Monthly	52	0.14	0.13	0.22	0.24	0.15	0.83	0.97	0.71	0.19	0.13	0.19	0.16	0.13
-Year		1961	1988	1959	1987	1976	1987	1947	1970	1952	1944	1976	1979	FEB 1988
-Maximum in 24 hrs	44	1.74	1.38	2.38	2.27	3.25	4.05	3.68	5.79	3.77	2.90	2.64	2.12	5.79
-Year		1975	1965	1977	1954	1979	1958	1987	1978	1972	1973	1968	1950	AUG 1978
Snow,Ice pellets,hail														
-Maximum Monthly	50	46.8	31.5	45.5	31.5	8.1	0.2	T	T	2.4	8.1	50.1	44.3	50.1
-Year		1969	1955	1965	1950	1954	1945	1992	1989	1991	1966	1991	1950	NOV 1991
-Maximum in 24 hrs	50	14.7	17.0	19.4	11.6	4.3	0.2	T	T	2.4	7.9	24.1	25.4	25.4
-Year		1982	1948	1965	1983	1954	1945	1992	1989	1991	1966	1991	1950	DEC 1950
WIND:														
Mean Speed (mph)	44	11.6	11.2	11.8	12.4	11.6	10.5	9.4	9.4	10.5	11.1	11.6	11.2	11.0
Prevailing Direction through 1963		NW	NW	WNW	NW	E	E	WNW	E	WNW	WNW	WNW	NW	WNW
Fastest Obs. 1 Min.														
-Direction (!!)	8	30	08	E	08	09	27	10	11	24	30	24	30	E
-Speed (MPH)	8	45	32	57	37	39	46	29	30	32	32	39	32	57
-Year		1986	1987	1985	1986	1989	1986	1992	1993	1985	1987	1986	1992	MAR 1985
Peak Gust														
-Direction (!!)	10	W	NW	E	E	E	W	N	E	SW	NW	NW	NW	E
-Speed (mph)	10	56	47	71	60	59	69	47	45	60	70	54	49	71
-Date		1990	1987	1985	1986	1989	1986	1991	1993	1991	1987	1993	1989	MAR 1985

PRECIPITATION (inches) DULUTH, MINNESOTA

YEAR	JAN	FEB	MAR	APR	MAY	JUNE	JULY	AUG	SEP	OCT	NOV	DEC	ANNUAL
1964	1.10	0.57	1.10	4.10	5.74	3.50	1.47	6.57	6.58	0.59	1.84	1.79	34.95
1965	0.89	1.73	5.12	1.64	2.96	4.51	3.57	3.76	5.56	3.30	3.33	1.67	38.04
1966	0.98	1.31	3.84	1.99	1.53	4.14	6.13	6.42	1.52	3.56	1.17	1.56	34.15
1967	3.12	0.24	0.66	1.99	0.80	5.21	2.91	2.31	1.34	1.44	0.50	0.91	21.43
1968	0.77	0.22	1.89	4.83	4.02	5.39	3.60	2.07	3.42	5.28	3.10	3.70	38.29
1969	4.70	0.26	0.39	1.46	2.82	2.18	3.03	2.18	4.13	2.42	1.25	2.67	27.49
1970	0.51	0.43	1.15	3.16	2.81	1.68	3.58	0.71	2.01	6.07	3.39	1.97	27.47
1971	1.56	2.37	2.02	1.29	3.45	3.25	3.91	4.50	2.67	6.09	2.07	1.22	34.40
1972	2.28	1.47	1.45	2.17	2.00	3.70	6.71	10.31	5.30	0.83	1.37	2.02	39.61
1973	0.67	0.29	1.54	1.35	3.81	2.43	2.36	8.46	4.28	4.56	1.61	0.69	32.05
1974	0.97	0.82	0.78	2.07	3.09	4.07	4.85	3.79	0.98	1.57	1.36	1.15	25.50
1975	3.69	0.76	2.59	2.21	1.44	5.59	2.26	2.52	2.32	1.20	4.19	0.64	29.41
1976	1.57	1.05	3.67	0.73	0.15	6.16	2.60	1.84	1.84	0.48	0.19	0.39	20.67
1977	0.36	0.47	4.43	1.27	3.50	3.97	3.91	3.26	5.97	3.20	2.37	1.31	34.02
1978	0.52	0.35	0.47	1.96	3.49	2.96	7.67	7.49	1.52	0.77	1.27	1.19	29.66
1979	0.76	1.89	3.58	1.15	6.01	4.33	5.45	2.10	2.01	3.01	0.47	0.16	30.92
1980	1.55	0.56	1.02	0.41	0.82	2.35	3.94	5.34	6.61	1.64	0.70	0.63	25.57
1981	0.32	1.50	1.05	4.48	1.15	5.83	3.26	2.84	2.42	3.59	0.96	0.97	28.37
1982	2.02	0.48	2.06	2.06	4.30	1.97	6.21	1.60	4.19	5.07	3.08	1.19	34.23
1983	1.34	0.49	2.05	2.28	2.12	2.00	3.51	3.37	5.57	2.32	5.01	1.97	32.03
1984	0.78	0.61	0.54	2.34	1.83	5.70	1.33	1.96	3.82	5.19	0.82	1.91	26.83
1985	0.39	0.66	1.85	2.35	4.44	3.18	4.16	3.91	6.02	1.76	2.33	0.78	31.83
1986	0.66	0.74	0.88	4.11	2.59	8.04	4.58	5.29	6.26	0.66	2.01	0.45	36.27
1987	0.69	0.31	0.60	0.24	4.02	0.83	5.46	1.87	2.93	0.96	1.26	0.67	19.84
1988	0.78	0.13	2.55	0.44	3.96	4.56	1.14	6.82	6.18	1.05	3.44	1.12	32.17
1989	1.87	0.34	1.49	2.11	3.50	3.81	1.09	5.02	4.40	1.02	1.01	0.63	26.29
1990	0.51	0.51	3.35	3.76	1.48	4.83	2.42	5.39	6.49	3.51	0.65	0.49	33.39
1991	0.52	0.55	1.17	3.90	6.11	5.64	5.33	2.49	9.38	2.85	4.89	0.61	43.44
1992	0.60	0.58	0.84	2.87	2.87	5.04	3.64	4.13	3.90	1.13	1.84	1.23	28.67
1993	1.79	0.38	0.44	2.42	4.74	6.95	5.75	3.63	1.81	0.56	2.60	1.28	32.35
Record Mean	1.11	0.90	1.64	2.20	3.13	4.03	3.73	3.54	3.45	2.20	1.73	1.12	28.77

AVERAGE TEMPERATURE (deg. F) DULUTH, MINNESOTA

YEAR	JAN	FEB	MAR	APR	MAY	JUNE	JULY	AUG	SEP	OCT	NOV	DEC	ANNUAL
1964	16.9	16.4	19.8	38.7	54.3	58.4	66.3	59.1	52.4	44.4	30.2	8.4	38.8
1965	5.2	6.0	15.8	38.0	52.0	58.0	61.9	61.1	48.0	44.8	27.8	21.1	36.6
1966	-1.3	12.3	27.1	34.7	47.3	61.1	68.0	61.5	54.8	43.2	25.2	13.5	37.3
1967	11.2	3.7	24.8	37.0	45.6	58.2	64.1	61.7	56.0	42.4	26.6	16.3	37.3
1968	9.1	9.3	32.0	40.5	48.4	57.4	64.2	61.7	56.6	46.2	28.9	11.9	38.8
1969	7.5	14.4	21.4	40.6	50.6	53.8	64.0	68.5	56.7	40.7	28.9	17.5	38.7
1970	2.4	8.4	21.1	38.7	46.6	60.8	68.8	66.0	55.5	44.4	27.8	12.7	37.8
1971	0.6	12.6	21.8	38.3	48.1	61.7	62.0	62.4	57.3	47.7	28.1	14.7	37.9
1972	0.2	6.4	17.9	33.3	53.5	58.4	61.2	61.9	50.0	38.9	25.9	7.2	34.6
1973	11.6	13.7	31.8	38.3	47.8	59.1	64.8	65.5	53.6	48.2	27.9	11.7	39.5
1974	5.6	11.0	20.6	37.9	46.6	58.6	67.1	60.1	47.9	42.8	29.2	19.5	37.2
1975	9.6	12.0	18.2	31.0	53.3	57.9	68.7	62.0	50.9	46.7	31.3	11.8	37.8
1976	4.7	20.7	22.0	42.6	50.6	63.2	66.1	64.9	55.3	37.1	22.1	4.5	37.8
1977	-0.2	17.2	31.4	44.4	57.4	60.0	65.9	58.6	53.2	44.3	28.0	11.3	39.3
1978	5.1	10.5	25.6	38.3	54.8	60.2	64.7	63.6	57.3	44.9	26.2	10.1	38.5
1979	0.6	5.0	23.0	34.5	47.2	59.8	66.1	62.0	56.9	43.0	29.4	22.2	37.5
1980	8.7	10.7	19.7	41.4	54.9	59.5	67.6	63.6	52.9	38.9	30.1	11.6	38.3
1981	11.9	16.3	28.3	39.2	50.2	58.7	65.5	64.3	52.9	40.5	34.7	14.8	39.8
1982	-3.2	10.6	20.8	35.7	52.8	54.8	64.5	60.8	54.3	45.1	24.4	19.6	36.7
1983	13.2	21.1	25.6	35.1	46.7	59.9	69.6	69.7	56.6	44.9	31.6	1.8	39.6
1984	7.1	23.0	18.7	42.4	50.5	61.4	66.6	67.7	51.9	45.7	28.4	13.2	39.7
1985	6.5	11.6	30.3	42.1	54.9	56.5	64.6	59.7	52.1	42.7	19.9	3.0	37.0
1986	11.7	11.5	28.5	42.1	51.8	58.9	64.7	60.8	52.7	43.4	24.0	18.8	39.1
1987	15.3	23.7	31.6	46.1	53.1	63.1	67.6	63.5	57.2	39.4	32.2	20.4	42.8
1988	4.8	6.2	24.4	40.1	56.0	62.8	70.0	64.5	55.4	38.6	28.9	13.5	38.8
1989	14.3	3.4	18.9	36.8	51.7	58.4	68.6	64.8	55.9	44.4	23.2	4.1	37.0
1990	18.7	14.9	27.1	40.2	48.3	61.5	64.6	64.1	56.2	42.1	31.4	12.6	40.1
1991	5.9	19.4	26.1	42.7	54.7	61.7	63.8	66.3	53.0	40.4	22.0	15.9	39.3
1992	16.7	21.2	25.9	36.2	53.8	56.3	59.4	60.5	53.8	41.7	27.3	15.9	39.1
1993	11.9	14.5	27.1	37.2	50.3	57.1	63.6	65.0	49.5	40.0	24.4	17.0	38.1
Record Mean	8.5	12.6	24.2	38.3	49.5	58.6	65.2	63.7	55.0	44.2	28.6	14.5	38.5
Max	17.3	21.7	32.6	47.2	59.7	69.1	75.5	73.4	64.0	52.5	35.4	22.2	47.5
Min	-0.3	3.5	15.8	29.4	39.3	48.1	54.9	54.0	45.9	35.8	21.7	6.8	29.6

REFERENCE NOTES DULUTH, MINNESOTA

GENERAL
T - TRACE AMOUNT.
BLANK ENTRIES DENOTE MISSING/UNREPORTED DATA.
# INDICATES A STATION OR INSTRUMENT RELOCATION.

SPECIFIC

PM - INCLUDES LAST DAY OF PREVIOUS MONTH
ASOS - AUTOMATED SURFACE OBSERVING SYSTEM IN OPERATION DURING THESE MONTHS.

(a) - LENGTH OF RECORD IN YEARS, ALTHOUGH INDIVIDUAL MONTHS MAY BE MISSING.
0.* OR * - THE VALUE IS BETWEEN 0.0 AND 0.05
NORMALS - BASED ON THE 1961-1990 RECORD PERIOD.
EXTREMES - DATES ARE THE MOST RECENT OCCURRENCE
WIND DIR.- NUMERALS SHOW TENS OF DEGRESS CLOCKWISE FROM TRUE NORTH. "00" INDICATES CALM.
RESULTANT DIRECTIONS ARE GIVEN TO WHOLE DEGREES.
BOLD VALUES INDICATE EXTREME VALUES WHICH OCCURRED AFTER THE ASOS SYSTEM WAS COMMISSIONED.

RECORD = PERIOD OF RECORD
RECORD MEAN PRECIPITATION IS THE MEAN OF ALL DAILY PRECIPITATION AMOUNTS DURING THE PERIOD OF RECORD.
RECORD MAX(MIN) TEMPERATURE IS THE MEAN OF ALL DAILY MAX(MIN) TEMPERATURES DURING THE PERIOD OF RECORD.
RECORD MEAN TEMPERATURE IS THE SUM OF THE RECORD MAX AND RECORD MIN DIVIDED BY 2.
AVERAGE TEMPERATURE IS THE SUM OF THE MEAN DAILY MAX AND MIN TEMPERATURE DIVIDED BY 2.

EXCEPTIONS

RECORD MEANS ARE THROUGH THE CURRENT YEAR, BEGINNING IN 1904 FOR TEMPERATURE
1904 FOR PRECIPITATION
1944 FOR SNOWFALL

HEATING DEGREE DAYS Base 65 deg. F DULUTH, MINNESOTA

SEASON	JULY	AUG	SEP	OCT	NOV	DEC	JAN	FEB	MAR	APR	MAY	JUNE	TOTAL
1964-65	50	215	377	631	1037	1751	1854	1650	1519	802	397	209	10492
1965-66	127	158	506	617	1109	1355	2053	1473	1166	903	546	165	10178
1966-67	36	145	304	670	1187	1589	1662	1714	1243	832	592	210	10184
1967-68	80	144	269	694	1147	1504	1730	1611	1017	729	507	227	9659
1968-69	90	140	254	576	1073	1644	1780	1411	1346	725	449	332	9820
1969-70	99	20	258	746	1075	1466	1939	1581	1355	781	573	162	10055
1970-71	38	67	311	630	1107	1616	1996	1464	1332	794	518	124	9997
1971-72	111	133	273	531	1099	1553	2012	1698	1458	946	357	211	10382
1972-73	134	146	440	802	1167	1792	1654	1429	1022	793	526	171	10076
1973-74	72	66	354	513	1106	1651	1843	1508	1373	806	565	206	10063
1974-75	45	160	507	684	1069	1404	1713	1479	1447	1013	358	226	10105
1975-76	46	129	418	561	1003	1642	1867	1279	1325	667	440	101	9478
1976-77	35	112	310	860	1280	1871	2017	1334	1033	612	242	170	9876
1977-78	46	196	347	636	1101	1662	1852	1520	1212	794	324	176	9866
1978-79	71	99	262	615	1159	1699	1999	1679	1293	910	549	172	10507
1979-80	52	115	252	674	1059	1317	1745	1573	1398	702	326	201	9414
1980-81	39	76	357	800	1043	1650	1644	1358	1133	769	455	185	9509
1981-82	74	62	363	752	903	1549	2117	1523	1363	875	370	303	10254
1982-83	66	161	332	609	1212	1398	1598	1226	1214	887	562	185	9450
1983-84	30	9	285	615	996	1959	1792	1210	1429	671	446	116	9558
1984-85	40	40	391	588	1093	1601	1809	1494	1067	679	314	252	9368
1985-86	57	178	394	686	1349	1925	1646	1492	1127	680	416	196	10146
1986-87	76	151	361	663	1224	1426	1537	1153	1027	561	376	112	8667
1987-88	34	100	234	782	977	1377	1862	1704	1253	741	300	146	9510
1988-89	22	97	287	812	1081	1590	1567	1721	1424	839	405	206	10051
1989-90	27	78	272	633	1249	1887	1428	1398	1166	745	506	130	9519
1990-91	76	94	273	700	1003	1623	1833	1273	1201	661	332	129	9198
1991-92	97	67	363	754	1284	1515	1492	1266	1207	859	355	271	9530
1992-93	174	167	332	717	1121	1518	1638	1408	1171	825	451	239	9761
1993-94	73	71	459	769	1214	1485							

COOLING DEGREE DAYS Base 65 deg. F DULUTH, MINNESOTA

YEAR	JAN	FEB	MAR	APR	MAY	JUNE	JULY	AUG	SEP	OCT	NOV	DEC	TOTAL
1969	0	0	0	0	11	4	76	135	16	0	0	0	242
1970	0	0	0	0	8	42	162	102	30	0	0	0	344
1971	0	0	0	0	0	32	24	60	47	1	0	0	164
1972	0	0	0	0	8	22	24	59	0	0	0	0	113
1973	0	0	0	0	0	1	72	84	18	0	0	0	175
1974	0	0	0	0	0	19	115	15	0	0	0	0	149
1975	0	0	0	0	2	18	168	41	0	0	0	0	229
1976	0	0	0	0	0	53	75	117	26	0	0	0	271
1977	0	0	0	0	12	24	80	6	0	0	0	0	122
1978	0	0	0	0	17	34	70	64	39	0	0	0	224
1979	0	0	0	0	4	25	95	30	15	0	0	0	169
1980	0	0	0	0	25	46	126	40	3	0	0	0	240
1981	0	0	0	0	2	2	97	48	6	0	0	0	155
1982	0	0	0	0	0	0	58	36	18	0	0	0	112
1983	0	0	0	0	0	42	179	165	42	0	0	0	428
1984	0	0	0	0	4	13	96	133	4	0	0	0	250
1985	0	0	0	0	9	4	54	20	15	0	0	0	102
1986	0	0	0	0	13	18	74	26	0	0	0	0	131
1987	0	0	0	0	13	62	121	60	7	0	0	0	263
1988	0	0	0	0	27	83	183	89	4	0	0	0	386
1989	0	0	0	0	0	17	147	80	7	0	0	0	251
1990	0	0	0	7	0	32	70	73	16	0	0	0	198
1991	0	0	0	0	20	35	69	117	8	0	0	0	249
1992	0	0	0	0	13	18	9	37	3	3	0	0	83
1993	0	0	0	0	0	8	36	76	0	0	0	0	120

SNOWFALL (inches) DULUTH, MINNESOTA

SEASON	JULY	AUG	SEP	OCT	NOV	DEC	JAN	FEB	MAR	APR	MAY	JUNE	TOTAL
1964-65	0.0	0.0	T	T	8.6	24.3	10.6	19.3	45.5	1.6	1.0	0.0	110.9
1965-66	0.0	0.0	T	T	25.9	8.6	11.8	4.3	24.5	10.2	1.7	0.0	87.0
1966-67	0.0	0.0	T	8.1	7.6	15.4	36.7	2.7	6.9	1.6	1.3	0.0	80.3
1967-68	0.0	0.0	0.0	0.8	4.2	4.3	10.8	2.7	4.1	8.7	3.7	0.0	39.3
1968-69	0.0	0.0	0.0	T	21.8	37.7	46.8	3.0	3.0	8.7	T	0.0	121.0
1969-70	0.0	0.0	0.0	4.3	9.1	38.8	7.5	8.9	9.9	14.8	1.6	0.0	94.9
1970-71	0.0	0.0	0.0	0.2	16.8	20.9	31.8	27.4	12.7	6.5	0.6	0.0	116.9
1971-72	0.0	0.0	0.0	0.2	6.4	17.5	30.9	20.7	11.0	20.4	T	0.0	107.1
1972-73	0.0	0.0	T	0.5	6.7	20.0	9.2	3.0	2.1	2.8	1.5	0.0	45.8
1973-74	0.0	0.0	0.0	T	7.6	15.7	13.4	15.3	15.1	5.2	1.0	0.0	73.3
1974-75	0.0	0.0	0.3	0.8	4.5	19.1	32.7	12.3	30.3	0.4	T	0.0	100.4
1975-76	0.0	0.0	0.0	0.1	25.7	7.8	20.6	8.8	26.4	0.0	T	0.0	89.4
1976-77	0.0	0.0	0.0	0.8	2.7	7.8	8.8	5.1	15.3	0.1	0.0	0.0	40.6
1977-78	0.0	0.0	0.0	1.7	16.0	12.0	13.5	11.4	8.3	6.8	0.0	0.0	69.7
1978-79	0.0	0.0	0.0	T	10.1	20.8	11.9	23.3	17.0	4.2	1.4	0.0	88.7
1979-80	0.0	0.0	0.0	0.8	6.0	2.0	21.9	10.3	12.6	1.5	T	0.0	55.1
1980-81	0.0	0.0	0.0	1.3	1.6	7.1	4.7	13.3	4.2	4.3	T	0.0	36.5
1981-82	0.0	0.0	T	1.4	12.5	15.5	34.2	8.0	17.8	6.3	0.0	0.0	95.7
1982-83	0.0	0.0	0.0	0.8	16.0	21.4	15.7	9.0	9.9	23.7	T	0.0	96.5
1983-84	0.0	0.0	T	T	37.7	32.1	20.0	4.0	9.9	3.3	0.3	0.0	107.3
1984-85	0.0	0.0	T	3.5	2.4	8.2	15.3	12.0	26.7	0.1	0.0	0.0	68.2
1985-86	0.0	0.0	0.7	2.3	34.1	18.8	11.5	10.3	11.4	0.2	T	0.0	89.3
1986-87	0.0	0.0	0.0	1.8	8.2	7.3	11.2	5.1	6.7	0.3	T	0.0	40.6
1987-88	0.0	0.0	0.0	3.9	5.8	11.3	16.7	2.3	13.7	0.1	0.0	0.0	53.8
1988-89	0.0	0.0	0.0	0.3	24.3	20.7	31.2	5.3	17.6	19.0	0.7	0.0	119.1
1989-90	0.0	T	0.0	0.6	8.4	14.2	5.5	15.0	11.6	2.4	0.6	T	58.3
1990-91	0.0	0.0	0.0	3.2	5.1	13.5	11.4	9.7	9.5	8.5	2.9	0.0	63.8
1991-92	0.0	0.0	2.4	4.3	50.1	12.9	9.3	10.6	0.7	9.7	0.0	0.0	100.0
1992-93	T	0.0	T	5.1	29.4	15.5	24.9	10.0	5.8	3.5	T	0.0	94.2
1993-94	0.0	0.0	T	0.6	28.5	15.1							
Record Mean	T	T	0.1	1.4	12.5	15.4	16.8	11.2	13.3	6.5	0.8	T	77.9

INTERNATIONAL FALLS, MINNESOTA

Situated on the Canadian border, International Falls is subjected to frequent outbreaks of continental polar air throughout most of the year. These are tempered to mildness during June, July, and August, when the land and lake areas to the north and northwest have been warmed by long days of sunshine. Periods of fine, mild weather occur, interspersed with showers and an occasional three or four day period of cloudy, rainy weather. The area of small lakes, covering up to 30 percent of the area to the north and northwest, supplies a good deal of the moisture for the late afternoon and evening showers and stores heat that tempers southward flow of cold air during September and October. This prolongs the fall season until early November. In November the water surfaces freeze and snow returns to International Falls. From December through February, temperatures fall below zero on most days and occasionally fail to rise above zero for a week or more.

In winter, frost penetrates into the ground to depths of 36 to 60 inches. If winter begins abruptly so that a heavy blanket of snow covers the ground before protracted freezing occurs, it may freeze to only a few inches deep. This is very important to loggers, who depend upon deep soil freezing for road foundations into otherwise inaccessible places. The wide expanse of deep snow and ice prolongs winter. The transition to summer is rapid after the spring thaw. Spring lasts only about a month.

By June 1st, the ground generally is warm enough for successful planting, but vigilance against freezing temperatures is required through most of June. Crops that do not mature by September 1st have little chance of providing a harvest. Heaviest precipitation coincides with the growing season.

Based on the 1951–1980 period, the average first occurrence of 32 degrees Fahrenheit in the fall is September 15 and the average last occurrence in the spring is May 26.

Heavy deposits of glaze occur only about once a year at International Falls. Occasional storms that intensify over the southern plateau or plains states and move rapidly northeastward, drawing up moist gulf air, bring the most violent weather changes. They often produce severe thunderstorms and windstorms in early fall and blizzards with heavy snowfall and drifting in winter. Quite often such a storm brings an abrupt end to fall weather. During winter, a variation of 100 miles in the paths of such storms as they approach the border is of tremendous importance to local transportation and road maintenance.

Surrounding terrain is generally level. Forests of varying density and swampland surround the station for many miles to the east, south, and west. Rainy Lake, approximately 300 square miles in area, lies to the north. The lake is 5 miles from the station at its closest point.

NORMALS, MEANS, AND EXTREMES

INTERNATIONAL FALLS, MINNESOTA

LATITUDE: 48°34'N LONGITUDE: 93°23' W ELEVATION: FT. GRND 1179 BARO 1185 TIME ZONE: CENTRAL WBAN: 14918

	(a)	JAN	FEB	MAR	APR	MAY	JUNE	JULY	AUG	SEP	OCT	NOV	DEC	YEAR
TEMPERATURE °F:														
Normals														
-Daily Maximum		11.9	19.3	32.8	50.1	64.6	73.4	78.8	75.6	64.3	51.8	32.8	16.6	47.7
-Daily Minimum		-10.0	-4.0	11.4	27.8	39.6	49.5	54.6	51.7	42.5	32.9	17.0	-2.2	25.9
-Monthly		1.0	7.7	22.1	39.0	52.1	61.4	66.7	63.7	53.4	42.4	24.9	7.2	36.8
Extremes														
-Record Highest	54	48	55	76	93	95	98	98	95	95	88	73	57	98
-Year		1973	1991	1946	1952	1964	1956	1988	1991	1976	1963	1975	1939	JUL 1988
-Record Lowest	54	-46	-44	-38	-14	11	23	35	30	20	2	-32	-41	-46
-Year		1968	1966	1962	1954	1967	1964	1972	1982	1965	1988	1985	1955	JAN 1968
NORMAL DEGREE DAYS:														
Heating (base 65°F)		1984	1604	1330	780	418	156	55	116	348	701	1203	1792	10487
Cooling (base 65°F)		0	0	0	0	18	48	107	76	0	0	0	0	249
% OF POSSIBLE SUNSHINE														
MEAN SKY COVER (tenths)														
Sunrise - Sunset	44	6.7	6.3	6.6	6.5	6.5	6.8	6.2	6.1	6.9	7.2	8.0	7.2	6.8
MEAN NUMBER OF DAYS:														
Sunrise to Sunset														
-Clear	54	7.5	8.0	7.8	6.9	6.8	4.7	6.3	7.2	5.6	5.9	3.6	6.3	76.7
-Partly Cloudy	54	7.3	6.2	7.6	8.3	9.0	11.0	13.4	11.5	8.9	7.1	5.2	5.8	101.5
-Cloudy	54	16.2	14.1	15.6	14.7	15.2	14.3	11.2	12.2	15.5	18.0	21.2	18.9	187.1
Precipitation														
.01 inches or more	54	11.7	9.1	10.0	9.2	11.3	12.9	11.5	11.6	11.7	9.6	11.1	11.8	131.3
Snow,Ice pellets,hail														
1.0 inches or more	54	3.8	3.0	2.9	1.6	0.3	0.0	0.0	0.0	0.1	0.7	3.4	3.8	19.5
Thunderstorms	54	0.0	0.0	0.3	1.0	3.3	6.7	8.7	7.0	3.7	0.9	0.1	0.0	31.6
Heavy Fog Visibility														
1/4 mile or less	54	0.9	1.1	1.2	0.9	0.9	0.9	1.4	2.1	1.9	1.4	1.2	1.5	15.3
Temperature °F														
-Maximum														
90° and above	54	0.0	0.0	0.0	0.1	0.3	0.7	1.9	1.1	0.1	0.0	0.0	0.0	4.2
32° and below	54	28.8	23.7	14.7	1.9	0.1	0.0	0.0	0.0	0.0	0.9	15.3	27.2	112.6
-Minimum														
32° and below	54	31.0	28.2	29.6	21.8	7.4	0.4	0.0	0.1	4.3	16.1	28.1	30.8	197.7
0° and below	54	21.6	16.9	8.2	0.3	0.0	0.0	0.0	0.0	0.0	0.0	3.5	16.4	66.9
AVG. STATION PRESS.(mb)	21	972.6	974.0	972.2	972.0	970.9	969.7	971.2	972.2	972.1	972.0	971.8	972.4	971.9
RELATIVE HUMIDITY (%)														
Hour 00	51	74	72	72	70	73	81	85	87	87	81	82	79	79
Hour 06 (Local Time)	51	74	74	77	77	77	83	88	91	91	85	84	79	82
Hour 12	51	69	64	61	52	50	56	57	60	63	62	73	74	62
Hour 18	51	70	64	59	51	49	56	57	61	68	67	77	76	63
PRECIPITATION (inches):														
Water Equivalent														
-Normal		0.88	0.63	1.06	1.58	2.47	3.93	3.59	3.11	3.15	1.97	1.15	0.84	24.36
-Maximum Monthly	54	3.03	1.81	3.07	3.33	6.67	8.19	9.52	11.26	7.36	4.84	3.49	1.70	11.26
-Year		1975	1955	1966	1986	1985	1941	1966	1942	1961	1971	1977	1992	AUG 1942
-Minimum Monthly	54	0.10	0.10	0.19	0.08	0.20	0.70	1.00	0.58	0.28	0.14	0.10	0.16	0.08
-Year		1973	1993	1969	1987	1976	1961	1941	1991	1952	1992	1939	1940	APR 1987
-Maximum in 24 hrs	54	1.50	1.14	1.80	1.59	2.76	3.80	4.87	4.82	3.37	2.62	1.56	1.25	4.87
-Year		1975	1946	1957	1979	1991	1964	1966	1942	1973	1979	1977	1960	JUL 1966
Snow,Ice pellets,hail														
-Maximum Monthly	54	43.0	32.3	31.5	23.0	13.4	0.3	T	T	1.9	8.5	29.7	43.9	43.9
-Year		1975	1992	1951	1950	1954	1969	1992	1978	1942	1981	1965	1992	DEC 1992
-Maximum in 24 hrs	54	17.7	11.4	17.0	13.9	7.7	0.3	T	T	1.5	5.9	14.7	14.4	17.7
-Year		1975	1955	1966	1950	1954	1969	1992	1978	1951	1981	1991	1990	JAN 1975
WIND:														
Mean Speed (mph)	41	8.9	8.8	9.4	10.0	9.4	8.5	7.6	7.5	8.7	9.3	9.4	8.8	8.9
Prevailing Direction through 1963		W	W	W	NW	NW	SE	W	SE	SE	SE	W	W	W
Fastest Obs. 1 Min.														
-Direction (!!)	38	30	26	29	23	20	18	29	32	23	30	27	31	23
-Speed (MPH)	38	35	36	42	52	52	46	46	40	35	47	35	33	52
-Year		1986	1965	1960	1960	1959	1962	1959	1965	1959	1972	1959	1980	APR 1960
Peak Gust														
-Direction (!!)	10	NW	NW	NE	SE	SE	N	NW	W	NW	NW	NW	NW	NW
-Speed (mph)	10	53	45	52	51	58	56	62	60	46	46	47	41	62
-Date		1986	1988	1985	1988	1989	1988	1989	1985	1992	1987	1989	1993	JUL 1989

PRECIPITATION (inches)

INTERNATIONAL FALLS, MINNESOTA

YEAR	JAN	FEB	MAR	APR	MAY	JUNE	JULY	AUG	SEP	OCT	NOV	DEC	ANNUAL
1964	0.74	0.63	0.65	2.92	2.12	7.39	1.97	4.90	3.32	0.42	0.73	0.84	26.63
1965	0.22	0.43	1.34	1.37	3.45	4.03	2.13	2.89	6.14	3.32	2.63	1.60	29.55
1966	0.74	0.74	3.07	1.30	1.20	1.63	9.52	2.85	1.22	1.57	0.80	0.92	25.56
1967	0.95	0.72	0.99	3.12	1.12	4.17	3.40	1.85	1.76	1.60	0.74	1.38	21.80
1968	0.77	0.19	1.66	2.52	1.50	4.65	4.97	3.14	4.33	3.22	0.48	1.60	29.03
1969	2.79	0.27	0.19	1.15	2.11	3.21	4.80	2.17	1.77	2.87	1.02	1.08	23.43
1970	0.93	0.43	1.11	1.83	3.24	1.97	1.67	2.28	2.80	4.32	0.99	1.07	22.64
1971	0.52	1.22	0.52	1.06	2.30	2.67	3.40	1.39	5.22	4.84	1.21	0.48	24.83
1972	0.76	0.71	0.57	0.89	2.57	1.52	4.52	2.65	1.70	1.60	0.93	0.97	19.39
1973	0.10	0.27	0.93	1.30	1.69	4.18	5.10	4.20	6.81	1.92	1.11	0.61	28.22
1974	0.87	0.65	0.99	2.11	3.17	4.64	1.06	6.38	0.86	1.61	1.12	0.79	24.25
1975	3.03	0.51	1.55	1.82	1.37	6.22	2.02	3.34	1.66	2.46	0.87	0.86	25.71
1976	0.99	0.46	1.84	0.94	0.20	7.01	5.70	1.89	1.19	0.81	0.21	0.60	21.84
1977	0.66	1.01	1.91	1.01	5.81	4.20	2.16	3.01	6.81	0.80	3.49	0.96	31.83
1978	0.76	0.27	0.41	1.12	3.86	2.89	6.29	2.96	3.62	0.39	1.60	0.93	25.10
1979	0.54	1.03	1.66	2.70	1.75	4.26	1.08	3.18	0.69	3.88	0.76	0.41	21.94
1980	0.92	0.55	0.87	0.45	0.83	1.70	2.23	4.03	4.08	1.81	1.62	0.56	19.65
1981	0.26	0.22	1.18	1.49	2.47	3.71	2.33	2.03	4.12	2.86	0.67	0.76	22.10
1982	1.24	0.51	1.58	0.84	3.51	2.68	2.37	2.88	3.63	3.67	1.52	0.29	24.72
1983	0.36	0.98	0.72	0.62	1.21	5.02	2.98	3.66	4.23	2.58	1.95	0.66	24.97
1984	0.30	0.76	0.22	0.89	1.77	6.50	2.14	1.30	1.14	4.11	0.91	1.27	21.31
1985	0.38	0.70	0.72	3.17	6.67	6.15	1.22	4.27	2.97	1.97	1.57	0.51	30.30
1986	0.61	0.95	0.26	3.33	0.50	3.67	2.59	1.52	2.42	0.64	1.27	0.35	18.11
1987	0.37	0.48	1.05	0.08	3.13	1.11	7.86	2.50	1.38	0.66	0.72	0.20	19.54
1988	0.44	0.14	1.71	0.30	1.18	4.39	3.04	6.66	3.93	1.03	1.39	0.76	24.97
1989	1.42	0.28	0.81	0.54	1.93	6.59	1.51	4.82	1.92	1.21	0.99	0.45	22.47
1990	0.62	0.50	1.35	1.47	0.98	5.50	3.13	2.15	1.23	1.42	0.75	1.21	20.31
1991	0.79	0.91	1.10	2.52	4.09	4.58	4.10	0.58	4.94	1.47	1.83	0.94	27.85
1992	0.78	1.57	0.56	1.43	2.08	2.17	4.00	5.48	2.34	0.14	1.17	1.70	23.42
1993	0.68	0.10	0.32	1.67	2.82	2.01	6.03	3.84	2.04	0.73	0.81	0.44	21.49
Record Mean	0.81	0.68	1.05	1.56	2.57	3.92	3.68	3.31	3.16	1.75	1.28	0.85	24.61

AVERAGE TEMPERATURE (deg. F)

INTERNATIONAL FALLS, MINNESOTA

YEAR	JAN	FEB	MAR	APR	MAY	JUNE	JULY	AUG	SEP	OCT	NOV	DEC	ANNUAL
1964	11.0	11.5	15.2	39.1	54.7	59.7	68.2	59.8	51.1	43.0	26.7	2.2	36.8
1965	-2.0	2.2	13.0	38.2	52.1	59.5	63.1	61.9	46.5	44.0	24.9	17.1	35.0
1966	-10.4	3.3	25.6	34.5	46.9	63.0	69.5	62.8	54.0	41.4	19.1	6.9	34.7
1967	3.3	-2.6	20.1	35.9	45.2	60.3	64.5	62.1	56.7	41.0	22.8	8.8	34.8
1968	3.8	5.4	27.9	38.7	48.2	59.5	65.3	61.3	56.0	44.4	26.3	7.4	37.0
1969	0.9	10.4	18.7	43.2	49.5	53.3	65.1	67.6	54.9	37.6	24.8	14.4	36.7
1970	-3.0	4.5	17.5	37.2	46.5	65.0	70.1	65.2	55.6	43.7	24.0	3.5	35.8
1971	-7.1	9.7	21.2	39.4	50.2	64.4	62.1	63.3	55.6	45.1	24.7	8.5	36.4
1972	-5.7	1.3	19.3	36.9	59.5	62.8	63.9	65.6	51.1	39.9	25.5	3.8	35.4
1973	9.5	9.2	32.7	38.1	52.1	64.7	68.4	68.2	53.0	47.7	25.2	8.5	39.8
1974	-0.4	6.3	16.6	37.9	47.4	61.4	70.0	61.7	48.6	43.2	27.9	16.8	36.5
1975	6.3	6.8	15.1	33.0	55.1	61.2	70.2	62.2	50.6	44.8	28.7	6.9	36.7
1976	-0.4	14.5	17.3	42.8	51.2	64.6	66.0	64.9	53.6	36.3	19.5	-3.0	35.6
1977	-5.8	12.6	29.6	42.9	61.0	61.8	66.6	57.0	52.5	43.3	22.6	3.6	37.3
1978	-3.5	2.6	18.5	37.0	55.5	59.6	63.5	63.0	56.0	43.4	20.5	0.7	34.8
1979	-8.7	-0.6	19.7	34.0	46.0	59.5	66.4	60.2	52.9	36.9	22.2	15.1	33.7
1980	2.1	8.1	16.8	44.3	58.8	62.3	68.8	64.9	52.8	38.9	26.1	4.6	38.1
1981	6.2	14.5	28.6	40.9	53.3	61.5	68.6	67.8	54.0	40.1	34.9	10.5	40.1
1982	-10.5	6.5	18.7	35.4	55.8	56.2	67.3	61.1	53.7	45.0	21.6	16.4	35.6
1983	11.2	16.6	27.8	38.2	49.0	61.9	69.6	68.6	55.1	41.7	27.7	-4.3	38.6
1984	0.7	20.7	17.5	44.0	48.9	61.5	65.1	67.7	49.0	44.6	25.6	3.4	37.4
1985	-0.1	4.7	26.7	41.7	54.4	54.5	63.6	60.1	51.2	42.1	14.3	0.2	34.5
1986	7.7	9.4	27.6	43.2	56.0	60.4	66.7	61.2	52.4	41.4	19.9	15.9	38.5
1987	11.0	23.5	29.8	47.9	55.8	64.1	67.9	62.5	56.2	38.6	31.0	18.6	42.2
1988	0.2	1.9	21.7	39.5	58.5	67.6	68.4	65.2	53.8	37.8	26.1	8.8	37.5
1989	8.0	-1.7	16.8	36.0	53.4	60.2	69.7	65.2	55.0	43.4	21.0	-1.5	35.5
1990	13.4	9.6	27.0	38.7	49.7	63.2	65.8	66.0	55.6	40.6	28.1	5.9	38.6
1991	-0.1	15.4	25.3	44.8	58.1	66.1	66.4	67.9	51.8	38.1	20.2	13.0	38.9
1992	12.3	16.5	23.9	35.6	55.0	57.7	59.4	60.3	53.0	40.4	25.4	8.6	37.3
1993	4.7	9.1	25.5	38.6	50.7	58.6	64.4	65.0	48.5	37.4	22.9	13.1	36.5
Record Mean	2.3	8.3	21.6	38.8	51.7	61.0	66.3	63.6	53.0	42.7	24.8	8.4	36.9
Max	13.0	19.9	32.6	49.8	64.1	72.9	78.4	75.4	64.1	52.7	32.4	17.8	47.8
Min	-8.3	-3.4	10.6	27.7	39.3	49.0	54.1	51.7	42.0	32.7	17.2	-0.9	26.0

REFERENCE NOTES

INTERNATIONAL FALLS, MINNESOTA

GENERAL
T - TRACE AMOUNT.
BLANK ENTRIES DENOTE MISSING/UNREPORTED DATA.
# INDICATES A STATION OR INSTRUMENT RELOCATION.

SPECIFIC

(a) - LENGTH OF RECORD IN YEARS, ALTHOUGH INDIVIDUAL MONTHS MAY BE MISSING.
0.* OR * - THE VALUE IS BETWEEN 0.0 AND 0.05
NORMALS - BASED ON THE 1961-1990 RECORD PERIOD.
EXTREMES - DATES ARE THE MOST RECENT OCCURRENCE
WIND DIR.- NUMERALS SHOW TENS OF DEGRESS CLOCKWISE FROM TRUE NORTH. "00" INDICATES CALM.
RESULTANT DIRECTIONS ARE GIVEN TO WHOLE DEGREES.
BOLD VALUES INDICATE EXTREME VALUES WHICH OCCURRED AFTER THE ASOS SYSTEM WAS COMMISSIONED.

RECORD = PERIOD OF RECORD
RECORD MEAN PRECIPITATION IS THE MEAN OF ALL DAILY PRECIPITATION AMOUNTS DURING THE PERIOD OF RECORD.
RECORD MAX(MIN) TEMPERATURE IS THE MEAN OF ALL DAILY MAX(MIN) TEMPERATURES DURING THE PERIOD OF RECORD.
RECORD MEAN TEMPERATURE IS THE SUM OF THE RECORD MAX AND RECORD MIN DIVIDED BY 2.
AVERAGE TEMPERATURE IS THE SUM OF THE MEAN DAILY MAX AND MIN TEMPERATURE DIVIDED BY 2.

EXCEPTIONS

RECORD MEANS ARE THROUGH THE CURRENT YEAR, BEGINNING IN 1940 FOR TEMPERATURE
1940 FOR PRECIPITATION
1940 FOR SNOWFALL

HEATING DEGREE DAYS Base 65 deg. F — INTERNATIONAL FALLS, MINNESOTA

SEASON	JULY	AUG	SEP	OCT	NOV	DEC	JAN	FEB	MAR	APR	MAY	JUNE	TOTAL
1964-65	18	207	416	675	1144	1944	2077	1756	1606	799	390	170	11202
1965-66	88	142	552	643	1196	1479	2340	1728	1214	908	560	146	10996
1966-67	9	112	334	724	1371	1801	1912	1893	1385	866	608	151	11166
1967-68	76	127	260	738	1260	1739	1899	1727	1142	785	511	163	10427
1968-69	70	149	272	630	1154	1778	1990	1524	1427	649	489	353	10485
1969-70	76	33	308	843	1200	1561	2106	1694	1467	829	565	83	10765
1970-71	19	68	314	650	1226	1908	2239	1550	1352	761	451	69	10607
1971-72	111	106	318	609	1201	1745	2195	1848	1413	833	236	109	10724
1972-73	72	74	412	769	1180	1893	1720	1561	993	800	393	60	9927
1973-74	26	20	364	530	1186	1751	2027	1641	1493	806	539	145	10528
1974-75	12	133	485	669	1106	1491	1815	1629	1542	950	322	163	10317
1975-76	41	118	426	622	1085	1801	2025	1463	1475	659	420	83	10218
1976-77	40	94	362	883	1359	2112	2195	1461	1089	658	183	114	10550
1977-78	34	246	367	666	1266	1908	2126	1745	1436	832	312	195	11133
1978-79	95	119	299	661	1329	1992	2287	1837	1402	923	580	177	11701
1979-80	36	161	364	862	1277	1539	1951	1646	1488	618	257	124	10323
1980-81	19	56	365	805	1161	1873	1818	1411	1119	716	359	108	9810
1981-82	32	17	324	764	894	1685	2343	1638	1431	882	280	256	10546
1982-83	21	161	354	612	1298	1500	1664	1352	1145	797	490	153	9547
1983-84	27	23	320	715	1116	2149	1995	1276	1470	621	494	112	10318
1984-85	50	50	474	626	1176	1908	2015	1687	1182	688	324	307	10487
1985-86	74	163	411	704	1520	2011	1773	1555	1152	646	313	161	10483
1986-87	34	145	367	723	1346	1517	1671	1158	1084	513	299	103	8960
1987-88	31	127	257	809	1018	1433	2011	1828	1337	756	237	53	9897
1988-89	29	85	331	834	1162	1743	1765	1868	1490	863	363	170	10703
1989-90	11	73	308	670	1312	2062	1596	1548	1171	796	464	88	10099
1990-91	39	59	284	748	1101	1833	2017	1383	1219	601	261	35	9580
1991-92	44	43	393	826	1338	1610	1631	1404	1267	875	335	233	9999
1992-93	173	155	354	756	1178	1746	1868	1560	1221	787	434	211	10443
1993-94	58	73	492	845	1257	1606							

COOLING DEGREE DAYS Base 65 deg. F — INTERNATIONAL FALLS, MINNESOTA

YEAR	JAN	FEB	MAR	APR	MAY	JUNE	JULY	AUG	SEP	OCT	NOV	DEC	TOTAL
1969	0	0	0	0	13	8	82	120	14	0	0	0	237
1970	0	0	0	0	0	89	185	82	37	0	0	0	393
1971	0	0	0	0	0	56	27	59	42	0	0	0	184
1972	0	0	0	0	70	51	41	100	0	0	0	0	262
1973	0	0	0	0	0	59	136	127	12	0	0	0	334
1974	0	0	0	0	3	43	175	38	0	0	0	0	259
1975	0	0	0	0	23	54	211	38	0	2	0	0	328
1976	0	0	0	0	0	80	76	98	27	0	0	0	281
1977	0	0	0	0	65	26	92	4	0	0	0	0	187
1978	0	0	0	0	24	41	57	63	34	0	0	0	219
1979	0	0	0	0	1	18	86	18	8	0	0	0	131
1980	0	0	0	2	69	51	142	62	5	0	0	0	331
1981	0	0	0	0	5	9	151	112	3	0	0	0	280
1982	0	0	0	2	1	0	101	46	22	0	0	0	172
1983	0	0	0	0	0	67	178	142	30	0	0	0	417
1984	0	0	0	0	4	16	61	139	0	1	0	0	221
1985	0	0	0	0	0	1	41	19	2	0	0	0	63
1986	0	0	0	0	41	32	96	34	0	0	0	0	203
1987	0	0	0	7	20	83	130	56	0	0	0	0	296
1988	0	0	0	0	40	136	141	96	3	0	0	0	416
1989	0	0	0	0	11	32	165	85	14	5	0	0	312
1990	0	0	0	11	0	40	72	96	10	0	0	0	229
1991	0	0	0	0	54	76	90	137	6	0	0	0	363
1992	0	0	0	0	33	22	5	16	1	0	0	0	77
1993	0	0	0	0	0	26	47	81	0	0	0	0	154

SNOWFALL (inches) — INTERNATIONAL FALLS, MINNESOTA

SEASON	JULY	AUG	SEP	OCT	NOV	DEC	JAN	FEB	MAR	APR	MAY	JUNE	TOTAL
1964-65	0.0	0.0	0.3	0.4	4.4	15.8	2.9	4.5	15.1	5.0	0.6	0.0	49.0
1965-66	0.0	0.0	T	T	29.7	14.5	9.8	5.4	24.6	10.5	2.1	0.0	96.6
1966-67	0.0	0.0	0.0	T	6.5	12.0	12.4	8.8	6.5	5.4	2.4	0.0	54.0
1967-68	0.0	0.0	0.0	1.4	7.0	12.9	10.1	3.0	9.5	18.3	1.3	0.0	63.5
1968-69	0.0	0.0	0.0	0.4	7.7	18.0	30.7	4.1	1.7	T	T	0.3	62.9
1969-70	0.0	0.0	T	3.0	12.8	14.0	12.7	4.5	4.8	12.5	2.0	0.0	66.3
1970-71	0.0	0.0	0.0	5.6	9.7	8.7	9.5	11.8	4.2	3.3	0.3	0.0	53.1
1971-72	0.0	0.0	0.0	1.5	10.2	7.1	13.9	14.2	8.0	12.0	T	0.0	66.9
1972-73	0.0	0.0	T	2.3	5.0	12.3	2.3	4.9	1.6	2.0	0.0	0.0	30.4
1973-74	0.0	0.0	0.0	T	6.1	9.0	15.6	8.6	13.8	5.7	1.5	0.0	60.3
1974-75	0.0	0.0	1.0	1.8	7.8	16.2	43.0	6.2	18.2	2.1	0.2	0.0	96.5
1975-76	0.0	0.0	0.0	2.1	5.5	14.8	21.4	9.7	28.1	0.2	0.4	0.0	82.2
1976-77	0.0	0.0	0.0	6.9	3.8	12.4	11.2	14.2	14.0	0.2	0.0	0.0	62.7
1977-78	0.0	0.0	0.0	T	24.4	9.7	13.4	4.6	5.4	9.2	0.0	0.0	66.7
1978-79	0.0	T	0.0	0.3	21.0	15.0	10.8	18.3	11.5	2.1	0.8	0.0	79.8
1979-80	0.0	0.0	0.0	2.1	10.3	6.2	16.3	11.0	12.5	5.8	T	0.0	64.2
1980-81	0.0	0.0	0.0	1.7	19.1	9.3	6.8	3.9	1.7	3.3	T	0.0	45.8
1981-82	0.0	0.0	1.4	8.5	5.3	16.2	28.4	4.5	14.8	10.8	0.0	0.0	89.9
1982-83	0.0	0.0	0.0	T	13.9	3.9	5.7	16.4	2.3	3.7	0.1	0.0	46.0
1983-84	0.0	0.0	T	0.9	26.5	15.4	5.7	6.1	3.9	T	0.2	0.0	58.7
1984-85	0.0	0.0	T	2.1	10.9	12.4	9.8	15.3	13.9	2.0	T	0.0	66.4
1985-86	0.0	0.0	T	4.8	27.1	15.5	16.4	15.4	5.2	0.1	0.3	0.0	84.8
1986-87	0.0	0.0	0.0	0.7	11.5	8.4	14.1	12.6	6.9	0.5	0.0	0.0	54.7
1987-88	0.0	0.0	0.0	1.4	0.1	8.1	14.7	4.9	16.1	0.2	0.0	0.0	45.5
1988-89	0.0	0.0	0.0	5.4	20.8	21.4	28.9	6.7	12.5	7.1	1.9	0.0	104.7
1989-90	0.0	0.0	T	T	8.5	10.8	9.8	11.7	3.0	17.1	T	T	60.9
1990-91	T	0.0	T	2.5	7.8	31.1	18.5	18.3	12.1	5.1	1.8	0.0	97.2
1991-92	0.0	0.0	T	3.9	23.8	22.2	16.7	32.3	2.7	9.3	0.1	T	111.0
1992-93	T	0.0	0.1	1.8	24.9	43.9	17.6	1.6	6.8	2.0	T	0.0	98.7
1993-94	0.0	0.0	1.3	4.7	18.4	5.7							
Record Mean	T	T	0.1	1.8	12.0	12.1	12.4	9.4	9.6	5.8	0.8	T	64.0

MINNEAPOLIS – ST. PAUL, MINNESOTA

The Twin Cities of Minneapolis and St. Paul are located at the confluence of the Mississippi and Minnesota Rivers over the heart of an artesian water basin. Its flat or gently rolling terrain varies little in elevation from that of the official observation station at International Airport. Numerous lakes dot the surrounding area. Minneapolis alone boasts of 22 lakes within the city park system. The largest body of water, nearly 15,000 acres, is Lake Minnetonka, located about 15 miles west of the airport. Most bodies of water are relatively small and shallow and are ice covered during winter.

The climate of the Minneapolis–St. Paul area is predominantly continental. Seasonal temperature variations are quite large. Temperatures range from less than −30 degrees to over 100 degrees. The growing season is 166 days. Because of this favorable growing season, all crops generally mature before the autumn freeze occurs.

The Twin Cities lie near the northern edge of the influx of moisture from the Gulf of Mexico. Severe storms such as blizzards, freezing rain (glaze), tornadoes, wind and hail storms do occur. The total annual precipitation is important. Even more significant is its proper distribution during the growing season. During the five month growing season, May through September, the major crops produced are corn, soybeans, small grains, and hay. During this period, the normal rainfall is over 16 inches, approximately 65 percent of the annual precipitation. Winter snowfall is nearly 48 inches. Winter recreational weather is excellent because of the dry snow. These conditions exist from about Christmas into early March. Snow depths average 6 to 8 inches in the city and 8 to 10 inches in the suburbs during this period.

Floods occur along the Mississippi River due to spring snow melt, excessive rainfall, or both. Occasionally an ice jam forms and creates a local flood condition. The flood problem at St. Paul is complicated because the Minnesota River empties into the Mississippi River between the two cities. Consequently, high water or flooding on the Minnesota River creates a greater flood potential at St. Paul. Flood stage at St. Paul can be expected on the average once in every eight years.

NORMALS, MEANS, AND EXTREMES

MINNEAPOLIS – ST. PAUL, MINNESOTA

LATITUDE: 44°53'N LONGITUDE: 93°13'W ELEVATION: FT. GRND 834 BARO 860 TIME ZONE: CENTRAL WBAN: 14922

	(a)	JAN	FEB	MAR	APR	MAY	JUNE	JULY	AUG	SEP	OCT	NOV	DEC	YEAR
TEMPERATURE °F:														
Normals														
-Daily Maximum		20.7	26.6	39.2	56.5	69.4	78.8	84.0	80.7	70.7	58.8	41.0	25.5	54.3
-Daily Minimum		2.8	9.2	22.7	36.2	47.6	57.6	63.1	60.3	50.3	38.8	25.2	10.2	35.3
-Monthly		11.8	17.9	31.0	46.4	58.5	68.2	73.6	70.5	60.5	48.8	33.2	17.9	44.9
Extremes														
-Record Highest	55	58	60	83	95	96	102	105	102	98	89	75	63	105
-Year		1944	1981	1986	1980	1978	1985	1988	1947	1976	1953	1944	1982	JUL 1988
-Record Lowest	55	-34	-28	-32	2	18	34	43	39	26	15	-17	-29	-34
-Year		1970	1965	1962	1962	1967	1945	1972	1967	1974	1972	1964	1983	JAN 1970
NORMAL DEGREE DAYS:														
Heating (base 65°F)		1649	1319	1054	558	244	41	11	22	167	502	954	1460	7981
Cooling (base 65°F)		0	0	0	0	43	137	278	192	32	0	0	0	682
% OF POSSIBLE SUNSHINE	55	53	59	57	58	61	65	72	69	62	55	39	42	58
MEAN SKY COVER (tenths)														
Sunrise - Sunset	55	6.3	6.3	6.7	6.5	6.4	6.1	5.3	5.3	5.6	5.8	7.1	7.0	6.2
MEAN NUMBER OF DAYS:														
Sunrise to Sunset														
-Clear	55	8.3	7.7	7.1	7.0	7.1	7.2	9.9	10.2	9.8	9.9	5.5	6.4	96.1
-Partly Cloudy	55	7.4	6.9	7.3	7.7	9.1	10.4	11.8	11.1	8.6	7.6	6.5	6.4	100.8
-Cloudy	55	15.3	13.7	16.6	15.2	14.8	12.3	9.3	9.7	11.6	13.5	18.0	18.2	168.3
Precipitation														
.01 inches or more	55	8.7	7.3	10.3	10.3	11.3	11.7	9.8	10.0	9.6	7.9	8.5	9.4	114.7
Snow,Ice pellets,hail														
1.0 inches or more	55	3.2	2.7	3.0	0.7	0.1	0.0	0.0	0.0	0.*	0.1	2.2	3.1	15.1
Thunderstorms	55	0.*	0.2	1.0	2.6	5.1	7.5	7.6	6.5	4.3	1.8	0.6	0.1	37.4
Heavy Fog Visibility														
1/4 mile or less	55	1.1	1.4	1.3	0.4	0.5	0.5	0.3	0.6	0.9	0.9	1.1	1.3	10.5
Temperature °F														
-Maximum														
90° and above	34	0.0	0.0	0.0	0.1	0.8	2.7	6.1	3.6	0.8	0.0	0.0	0.0	14.2
32° and below	34	23.4	17.9	8.7	0.3	0.0	0.0	0.0	0.0	0.0	0.1	7.0	21.2	78.5
-Minimum														
32° and below	34	30.8	27.3	25.2	11.0	1.1	0.0	0.0	0.0	0.5	7.8	23.0	29.9	156.4
0° and below	34	13.5	8.1	1.9	0.0	0.0	0.0	0.0	0.0	0.0	0.0	0.7	8.0	32.3
AVG. STATION PRESS (mb)	21	987.5	988.3	985.2	984.2	983.6	983.0	984.7	985.6	986.0	986.3	985.9	987.4	985.6
RELATIVE HUMIDITY (%)														
Hour 00	34	73	74	72	67	68	72	75	77	79	74	77	76	74
Hour 06 (Local Time)	34	74	76	76	75	76	79	81	84	85	81	80	78	79
Hour 12	34	67	65	61	52	52	54	54	57	59	58	66	70	60
Hour 18	34	68	66	61	51	50	52	53	56	61	60	69	73	60
PRECIPITATION (inches):														
Water Equivalent														
-Normal		0.95	0.88	1.94	2.42	3.39	4.05	3.53	3.62	2.72	2.19	1.55	1.08	28.32
-Maximum Monthly	55	3.63	2.14	4.75	5.88	8.03	9.82	17.90	9.31	7.53	5.68	5.29	4.27	17.90
-Year		1967	1981	1965	1986	1962	1990	1987	1977	1942	1971	1991	1982	JUL 1987
-Minimum Monthly	55	0.10	0.06	0.32	0.16	0.61	0.22	0.58	0.43	0.41	0.01	0.02	T	T
-Year		1990	1964	1958	1987	1967	1988	1975	1946	1940	1952	1939	1943	DEC 1943
-Maximum in 24 hrs	55	1.21	1.10	1.66	2.23	3.03	3.00	10.00	7.36	3.55	2.95	2.91	2.47	10.00
-Year		1967	1966	1965	1975	1965	1986	1987	1977	1942	1966	1940	1982	JUL 1987
Snow,Ice pellets,hail														
-Maximum Monthly	55	46.4	26.5	40.0	21.8	3.0	T	T	T	1.7	8.2	46.9	33.2	46.9
-Year		1982	1962	1951	1983	1946	1989	1993	1992	1942	1991	1991	1969	NOV 1991
-Maximum in 24 hrs	55	18.5	9.3	14.7	13.6	3.0	T	T	T	1.7	8.2	21.0	16.5	21.0
-Year		1982	1939	1985	1983	1946	1989	1993	1992	1942	1991	1991	1982	NOV 1991
WIND:														
Mean Speed (mph)	55	10.5	10.4	11.3	12.2	11.1	10.5	9.4	9.3	10.0	10.5	10.9	10.4	10.5
Prevailing Direction through 1963		NW	NW	NW	NW	SE	SE	S	SE	S	SE	NW	NW	NW
Fastest Obs. 1 Min.														
-Direction (!!)	14	32	34	08	19	23	01	35	20	18	33	25	34	32
-Speed (MPH)	14	51	37	33	41	35	46	43	44	36	33	41	35	51
-Year		1986	1987	1985	1984	1986	1980	1980	1983	1988	1981	1986	1989	JAN 1986
Peak Gust														
-Direction (!!)	10	NW	NW	W	SW	N	S	NW	W	N	NW	W	NW	W
-Speed (mph)	10	67	55	60	61	67	51	51	71	52	53	66	48	71
-Date		1986	1987	1988	1984	1985	1992	1984	1988	1989	1987	1986	1989	AUG 1988

PRECIPITATION (inches)

MINNEAPOLIS – ST. PAUL, MINNESOTA

YEAR	JAN	FEB	MAR	APR	MAY	JUNE	JULY	AUG	SEP	OCT	NOV	DEC	ANNUAL
1964	0.47	0.06	1.35	2.98	3.44	2.18	2.02	5.42	5.21	0.57	1.19	1.08	25.97
1965	0.47	1.59	4.75	3.52	7.86	4.01	4.69	4.04	4.90	0.90	1.98	1.23	39.94
1966	0.95	1.55	2.48	0.89	1.46	3.51	2.47	4.40	1.69	3.53	0.39	1.02	24.34
1967	3.63	1.59	0.96	4.07	0.61	7.53	1.36	2.79	0.63	1.73	0.09	0.45	25.44
1968	0.71	0.13	1.89	2.94	3.74	6.78	6.46	0.75	6.16	5.62	0.54	2.21	37.93
1969	2.05	0.31	0.90	1.55	1.98	2.93	2.95	0.99	0.49	2.53	0.65	2.06	19.39
1970	0.47	0.16	2.05	3.55	4.77	1.27	3.66	2.19	3.19	4.97	3.82	0.43	30.53
1971	1.22	1.74	1.21	1.11	3.14	3.52	3.94	1.78	2.73	5.68	2.67	0.70	29.44
1972	0.84	0.49	1.25	1.69	2.18	3.31	5.12	2.48	1.96	1.77	1.11	1.57	23.77
1973	0.92	0.84	1.12	2.32	2.48	1.06	2.90	3.05	2.08	1.29	1.97	1.10	21.13
1974	0.17	1.06	1.00	2.42	2.08	5.21	1.14	2.75	0.58	1.69	0.66	0.35	19.11
1975	2.82	0.79	1.67	5.40	3.81	7.99	0.58	4.92	1.31	0.27	4.80	0.79	35.15
1976	0.87	0.59	2.83	0.80	1.13	3.86	2.45	1.39	1.42	0.49	0.16	0.51	16.50
1977	0.65	0.93	2.66	1.84	2.86	3.57	3.72	9.31	4.43	2.34	1.42	1.15	34.88
1978	0.38	0.24	0.79	3.63	3.79	7.09	3.19	5.77	2.47	0.19	1.84	0.88	30.26
1979	1.09	1.39	2.55	0.66	4.55	4.78	2.34	7.04	2.20	3.16	0.98	0.33	31.07
1980	0.94	0.67	1.12	0.83	2.29	5.52	2.30	3.26	3.68	0.66	0.26	0.24	21.77
1981	0.30	2.14	0.71	2.17	2.18	4.42	4.09	4.73	1.46	2.69	2.16	0.92	27.97
1982	2.45	0.43	2.09	1.62	4.99	1.44	0.92	3.80	1.50	3.45	3.27	4.27	30.23
1983	0.67	1.19	3.22	3.97	6.20	5.22	3.07	3.12	3.34	2.61	4.93	1.53	39.07
1984	0.88	1.64	1.47	3.86	2.29	7.95	3.03	5.15	2.65	5.48	0.31	2.24	36.95
1985	0.87	0.50	4.48	1.81	3.65	2.18	2.20	5.02	4.37	3.66	1.72	1.20	31.66
1986	0.90	0.84	2.03	5.88	3.48	5.34	4.11	4.44	6.90	1.77	0.62	0.31	36.62
1987	0.63	0.13	0.64	0.16	1.88	1.95	17.90	3.67	1.28	0.60	2.07	1.25	32.16
1988	1.37	0.30	1.33	1.58	1.70	0.22	1.17	4.29	2.79	0.80	2.86	0.67	19.08
1989	0.52	1.04	2.19	2.66	3.38	3.50	3.50	2.92	1.28	0.53	1.38	0.42	23.32
1990	0.10	0.77	3.66	3.80	3.36	9.82	5.06	1.71	1.88	1.23	0.65	1.01	33.05
1991	0.49	1.03	2.29	3.58	6.35	2.57	2.95	3.14	5.43	2.52	5.29	1.05	36.69
1992	0.66	0.57	1.56	1.99	1.15	3.68	5.21	4.54	5.20	2.11	1.95	1.05	29.67
1993	1.25	0.39	1.25	1.99	4.02	6.28	5.58	6.50	2.04	0.79	1.57	0.55	32.21
Record Mean	0.83	0.85	1.60	2.17	3.38	4.17	3.55	3.40	2.89	2.01	1.45	0.94	27.26

AVERAGE TEMPERATURE (deg. F)

MINNEAPOLIS – ST. PAUL, MINNESOTA

YEAR	JAN	FEB	MAR	APR	MAY	JUNE	JULY	AUG	SEP	OCT	NOV	DEC	ANNUAL
1964	20.0	23.9	25.8	46.8	61.5	68.7	76.0	68.5	58.9	48.2	35.0	14.8	45.7
1965	10.0	11.8	19.5	41.8	58.7	66.5	70.5	68.6	52.8	50.7	33.1	28.0	42.7
1966	3.3	16.3	35.8	42.2	53.6	68.4	76.8	68.2	60.3	47.5	30.1	18.1	43.4
1967	14.6	8.7	29.8	44.7	52.3	66.9	68.8	66.2	60.3	46.3	30.7	21.8	42.6
1968	14.3	15.2	38.8	48.5	53.4	67.2	71.1	70.7	61.1	50.7	34.0	16.9	45.2
1969	9.4	19.3	24.1	49.3	60.6	61.8	73.6	74.4	63.0	46.5	33.6	20.3	44.7
1970	5.6	15.4	26.0	46.1	58.5	71.2	75.2	71.9	61.2	49.6	32.7	18.2	44.3
1971	6.5	17.0	28.0	47.0	55.4	71.5	68.8	69.6	62.8	51.4	32.7	18.4	44.1
1972	5.5	10.5	26.5	41.9	61.3	66.0	68.5	69.8	57.9	43.7	32.2	11.3	41.3
1973	17.4	21.6	40.2	44.4	55.2	69.5	73.8	73.4	60.1	53.8	34.3	16.7	46.7
1974	11.9	16.9	29.5	47.1	54.4	65.5	76.6	67.3	55.3	49.8	33.7	24.4	44.4
1975	14.5	15.5	22.1	38.9	60.9	68.8	76.3	71.7	57.7	52.8	37.5	21.3	44.8
1976	11.6	27.8	31.4	51.8	58.9	71.7	76.1	73.3	61.8	44.6	28.3	13.6	45.9
1977	0.3	22.7	37.5	53.0	66.9	68.4	74.8	66.1	60.5	47.1	30.8	14.4	45.2
1978	5.5	11.6	30.0	45.2	61.8	67.8	71.1	72.2	67.3	49.8	32.5	15.2	44.2
1979	3.2	10.0	28.9	44.0	55.5	67.3	73.6	69.9	63.4	46.6	31.7	26.0	43.3
1980	15.3	15.3	27.3	49.2	61.5	67.6	75.2	70.7	59.5	45.1	36.6	19.8	45.2
1981	18.0	23.4	37.7	49.1	57.1	67.0	70.9	69.3	60.0	46.7	38.0	17.5	46.2
1982	2.3	15.8	29.0	43.8	62.5	63.7	75.6	71.8	60.9	50.3	31.5	25.7	44.4
1983	19.6	26.9	34.2	42.3	54.6	68.0	77.2	76.8	62.6	48.4	34.0	3.7	45.7
1984	12.0	27.5	24.8	47.1	56.0	69.7	72.2	73.5	57.2	50.7	33.3	17.9	45.2
1985	10.1	16.5	35.6	52.1	62.2	63.9	73.9	67.6	59.9	47.5	24.8	7.7	43.5
1986	17.5	15.7	33.9	49.6	59.4	68.6	73.9	67.1	59.8	49.2	28.2	24.7	45.6
1987	21.2	31.6	38.7	53.5	63.5	72.8	76.0	69.0	62.5	44.6	37.9	25.0	49.7
1988	10.4	13.9	33.8	47.4	65.4	74.4	78.1	73.9	62.4	44.0	32.7	20.5	46.4
1989	21.2	8.6	26.6	45.3	57.5	68.4	76.4	70.8	60.9	49.9	28.0	10.6	43.7
1990	26.3	23.7	35.7	46.8	56.3	69.5	71.3	70.6	64.4	48.1	37.4	16.9	47.3
1991	12.5	24.4	34.3	49.1	61.9	72.9	72.3	71.1	59.0	47.2	24.5	21.2	45.9
1992	21.9	28.0	33.1	43.6	60.5	65.6	65.8	65.9	59.6	47.4	31.4	21.2	45.3
1993	14.6	17.2	29.5	44.2	57.2	64.5	70.3	70.4	55.0	46.5	30.6	22.2	43.5
Record Mean	13.2	17.3	30.1	46.0	58.3	67.9	73.2	70.6	61.4	49.5	32.8	19.2	45.0
Max	21.8	25.9	38.4	55.8	68.5	77.8	83.2	80.5	71.3	59.0	40.4	26.7	54.1
Min	4.6	8.6	21.8	36.3	48.0	58.0	63.1	60.7	51.5	40.1	25.2	11.7	35.8

REFERENCE NOTES

MINNEAPOLIS – ST. PAUL, MINNESOTA

GENERAL
T - TRACE AMOUNT.
BLANK ENTRIES DENOTE MISSING/UNREPORTED DATA.
# INDICATES A STATION OR INSTRUMENT RELOCATION.

SPECIFIC

PM - INCLUDES LAST DAY OF PREVIOUS MONTH
ASOS - AUTOMATED SURFACE OBSERVING SYSTEM IN OPERATION DURING THESE MONTHS.

(a) - LENGTH OF RECORD IN YEARS, ALTHOUGH INDIVIDUAL MONTHS MAY BE MISSING.
0.* OR * - THE VALUE IS BETWEEN 0.0 AND 0.05
NORMALS - BASED ON THE 1961-1990 RECORD PERIOD.
EXTREMES - DATES ARE THE MOST RECENT OCCURRENCE
WIND DIR.- NUMERALS SHOW TENS OF DEGRESS CLOCKWISE FROM TRUE NORTH. "00" INDICATES CALM.
RESULTANT DIRECTIONS ARE GIVEN TO WHOLE DEGREES.
BOLD VALUES INDICATE EXTREME VALUES WHICH OCCURRED AFTER THE ASOS SYSTEM WAS COMMISSIONED.

RECORD = PERIOD OF RECORD
RECORD MEAN PRECIPITATION IS THE MEAN OF ALL DAILY PRECIPITATION AMOUNTS DURING THE PERIOD OF RECORD.
RECORD MAX(MIN) TEMPERATURE IS THE MEAN OF ALL DAILY MAX(MIN) TEMPERATURES DURING THE PERIOD OF RECORD.
RECORD MEAN TEMPERATURE IS THE SUM OF THE RECORD MAX AND RECORD MIN DIVIDED BY 2.
AVERAGE TEMPERATURE IS THE SUM OF THE MEAN DAILY MAX AND MIN TEMPERATURE DIVIDED BY 2.

EXCEPTIONS

RECORD MEANS ARE THROUGH THE CURRENT YEAR,
BEGINNING IN 1891 FOR TEMPERATURE
1891 FOR PRECIPITATION
1939 FOR SNOWFALL

HEATING DEGREE DAYS Base 65 deg. F — MINNEAPOLIS - ST. PAUL, MINNESOTA

SEASON	JULY	AUG	SEP	OCT	NOV	DEC	JAN	FEB	MAR	APR	MAY	JUNE	TOTAL
1964-65	0	63	224	515	894	1551	1702	1486	1405	690	211	19	8760
1965-66	7	40	368	447	950	1140	1909	1358	899	678	357	41	8194
1966-67	0	40	185	536	1042	1446	1556	1572	1086	600	404	30	8497
1967-68	36	65	166	577	1024	1335	1567	1440	808	491	358	62	7929
1968-69	10	28	143	451	922	1486	1723	1274	1261	461	204	136	8099
1969-70	5	0	131	580	933	1379	1842	1382	1204	577	249	20	8302
1970-71	3	5	190	476	959	1443	1811	1341	1139	537	297	18	8219
1971-72	16	22	164	413	962	1438	1844	1576	1188	687	204	73	8587
1972-73	34	52	218	651	974	1664	1474	1208	761	611	299	13	7959
1973-74	1	3	185	350	915	1493	1642	1344	1092	535	338	72	7970
1974-75	0	48	289	467	933	1252	1561	1379	1324	775	188	39	8255
1975-76	15	7	231	387	818	1346	1650	1074	1031	405	195	11	7170
1976-77	0	4	162	632	1092	1590	2005	1180	844	365	75	17	7966
1977-78	0	35	145	548	1016	1565	1842	1488	1080	584	162	46	8511
1978-79	5	7	89	464	968	1538	1914	1537	1112	623	307	38	8602
1979-80	0	24	105	566	992	1203	1536	1436	1165	484	184	34	7729
1980-81	0	12	194	611	845	1396	1453	1160	838	472	249	28	7258
1981-82	11	11	172	564	803	1466	1945	1374	1111	629	117	71	8274
1982-83	0	14	168	448	997	1212	1400	1061	947	673	313	49	7282
1983-84	2	0	161	514	923	1901	1641	1082	1240	531	284	7	8286
1984-85	5	12	251	435	943	1453	1694	1355	904	403	123	104	7682
1985-86	0	28	240	537	1201	1774	1466	1377	957	454	212	30	8276
1986-87	0	43	177	480	1096	1243	1352	929	809	347	134	13	6623
1987-88	2	29	106	623	804	1236	1688	1479	962	523	76	4	7532
1988-89	1	16	116	646	963	1373	1353	1576	1184	583	251	44	8106
1989-90	0	6	159	470	1105	1683	1194	1151	899	569	274	37	7547
1990-91	2	5	136	516	820	1484	1624	1130	945	481	197	3	7343
1991-92	7	8	228	548	1206	1354	1333	1067	981	636	190	72	7630
1992-93	32	52	182	542	1003	1351	1557	1335	1096	617	243	70	8080
1993-94	3	18	302	566	1025	1322							

COOLING DEGREE DAYS Base 65 deg. F — MINNEAPOLIS - ST. PAUL, MINNESOTA

YEAR	JAN	FEB	MAR	APR	MAY	JUNE	JULY	AUG	SEP	OCT	NOV	DEC	TOTAL
1969	0	0	0	0	76	49	276	298	77	12	0	0	788
1970	0	0	0	17	54	213	323	225	83	5	0	0	920
1971	0	0	0	2	5	218	141	168	106	3	0	0	643
1972	0	0	0	0	94	109	148	208	13	0	0	0	572
1973	0	0	0	1	4	158	280	271	47	8	0	0	769
1974	0	0	0	5	18	93	369	127	6	1	0	0	619
1975	0	0	0	0	66	159	371	220	18	16	0	0	850
1976	0	0	0	14	14	223	351	269	72	7	0	0	950
1977	0	0	0	12	145	129	310	76	19	0	0	0	691
1978	0	0	0	0	72	138	201	236	164	0	0	0	811
1979	0	0	0	0	17	113	275	181	65	0	0	0	651
1980	0	0	0	16	82	121	322	194	38	1	0	0	774
1981	0	0	0	0	10	96	200	151	28	0	0	0	485
1982	0	0	0	0	46	40	338	232	53	0	0	0	709
1983	0	0	0	0	0	145	389	368	98	8	0	0	1008
1984	0	0	0	0	13	155	237	280	24	0	0	0	709
1985	0	0	0	22	43	77	284	118	93	0	0	0	637
1986	0	0	0	1	45	148	286	115	32	0	0	0	627
1987	0	0	0	11	95	253	348	159	37	0	0	0	903
1988	0	0	0	1	96	296	412	302	45	0	0	0	1152
1989	0	0	0	0	26	153	359	192	41	8	0	0	779
1990	0	0	0	28	11	178	206	191	125	1	0	0	740
1991	0	0	0	8	109	246	238	205	51	0	0	0	857
1992	0	0	0	3	56	96	64	88	28	2	0	0	337
1993	0	0	0	0	12	60	176	195	8	0	0	0	451

SNOWFALL (inches) — MINNEAPOLIS - ST. PAUL, MINNESOTA

SEASON	JULY	AUG	SEP	OCT	NOV	DEC	JAN	FEB	MAR	APR	MAY	JUNE	TOTAL
1964-65	0.0	0.0	0.0	T	4.3	8.1	10.5	11.7	37.1	2.0	T	0.0	73.7
1965-66	0.0	0.0	0.0	0.0	1.6	1.2	11.9	6.8	14.2	0.4	T	0.0	36.1
1966-67	0.0	0.0	0.0	0.2	3.4	12.7	35.3	23.7	2.6	0.2	0.3	0.0	78.4
1967-68	0.0	0.0	0.0	0.3	0.8	2.4	10.6	2.2	0.8	0.4	0.0	0.0	17.5
1968-69	0.0	0.0	0.0	T	4.9	28.7	21.6	5.3	7.3	0.3	0.0	0.0	68.1
1969-70	0.0	0.0	0.0	2.4	3.8	33.2	9.8	4.3	8.6	1.3	T	0.0	63.4
1970-71	0.0	0.0	0.0	T	6.3	5.5	19.9	13.9	7.0	1.9	0.2	0.0	54.7
1971-72	0.0	0.0	0.0	0.0	13.2	12.8	12.2	7.6	10.4	8.0	0.0	0.0	64.2
1972-73	0.0	0.0	T	T	1.1	15.3	11.6	11.3	0.4	2.0	0.0	0.0	41.7
1973-74	0.0	0.0	0.0	0.0	0.1	17.9	2.5	15.7	7.7	7.3	0.0	0.0	51.2
1974-75	0.0	0.0	0.0	0.0	1.2	6.1	27.4	9.0	18.3	2.2	0.0	0.0	64.2
1975-76	0.0	0.0	0.0	0.0	16.2	5.6	12.8	5.1	13.6	0.0	1.2	0.0	54.5
1976-77	0.0	0.0	0.0	2.3	1.4	8.3	13.4	1.8	14.6	1.8	0.0	0.0	43.6
1977-78	0.0	0.0	0.0	3.0	11.7	14.2	6.8	4.6	8.5	1.9	0.0	0.0	50.7
1978-79	0.0	0.0	0.0	0.0	16.5	15.1	14.2	13.5	8.4	0.7	0.0	0.0	68.4
1979-80	0.0	0.0	0.0	T	7.7	1.7	12.9	8.8	13.7	8.5	0.0	0.0	53.3
1980-81	0.0	0.0	0.0	T	0.9	2.8	4.6	11.0	0.1	1.7	0.0	0.0	21.1
1981-82	0.0	0.0	0.0	0.9	14.0	10.6	46.4	7.4	10.9	4.8	0.0	0.0	95.0
1982-83	0.0	0.0	0.0	1.4	3.6	19.3	3.2	10.8	14.3	21.8	0.0	0.0	74.4
1983-84	0.0	0.0	0.0	T	30.4	21.0	10.6	9.3	17.3	9.8	0.0	0.0	98.4
1984-85	0.0	0.0	0.0	0.3	2.0	16.3	13.1	4.2	36.8	T	0.0	0.0	72.7
1985-86	0.0	0.0	0.4	T	23.9	13.5	10.3	12.3	8.7	0.4	0.0	0.0	69.5
1986-87	0.0	0.0	0.0	T	4.4	4.2	5.5	1.2	2.1	T	0.0	0.0	17.4
1987-88	0.0	0.0	0.0	0.3	4.5	7.5	19.5	4.5	3.7	2.4	0.0	0.0	42.4
1988-89	0.0	0.0	0.0	0.2	15.8	7.2	6.0	17.3	22.7	0.8	0.1	T	70.1
1989-90	0.0	0.0	0.0	0.0	11.3	7.0	1.1	10.7	3.2	2.2	0.0	0.0	35.5
1990-91	0.0	0.0	0.0	T	5.0	11.7	6.5	14.2	4.4	1.5	0.3	0.0	43.6
1991-92	0.0	T	0.0	8.2	46.9	6.7	5.0	5.9	10.8	0.6	0.0	0.0	84.1
1992-93	0.0	T	T	1.3	12.2	9.2	12.0	5.3	6.9	0.5	0.0	0.0	47.4
1993-94	T	0.0	0.0	T	7.7	4.5							
Record Mean	T	T	T	0.5	7.9	9.3	9.8	8.4	10.7	2.8	0.1	T	49.6

ROCHESTER, MINNESOTA

Rochester, Minnesota, is in the Zumbro River Valley. The south branch of the Zumbro flows through Rochester. Within the city of Rochester three creeks flow into the south branch. Terrain around Rochester is rolling, and the elevation ranges from 1,000 to 1,300 feet above sea level.

The National Weather Service station is located 8 miles south of Rochester on a ridge 300 feet above the city elevation. Temperatures from radiation cooling on clear, calm nights can sometimes be much lower in the city.

The succession of high and low pressure systems over Rochester brings a variety of weather that is changeable and stimulating. The weather pattern is continental with four definite seasons. Winters are cold, but summers are pleasant.

The season-to-season temperature variation is quite large. The average temperature for a warm winter is 20 degrees and for a cold winter it is 12 degrees. The average temperature for a warm summer is 70 degrees and a cold summer is 67 degrees, which indicates that summer temperatures are not as variable as those during the winter. The average growing season is about 140 days.

Rochester lies near the northern edge of the influx of moisture from the Gulf of Mexico. Severe storms such as blizzards, freezing rain (glaze), tornadoes, wind, and hail storms do occur. During the five month growing season, May through September, the major crops of corn, soybeans, small grains, and hay are produced. During this period, the normal rainfall is over 18 inches, approximately 65 percent of the annual precipitation.

Snowfall averages above 45 inches per season. The snow season usually begins in November. About one year in ten the first 1 inch or more of snow will occur the latter part of October.

Rolling terrain and the thunderstorm probability make the south branch of the Zumbro River and its tributaries susceptible to flash flooding. Some flooding can occur with the spring snowmelt. In some instances the snowmelt is complicated with moderate spring rainfall.

NORMALS, MEANS, AND EXTREMES

ROCHESTER, MINNESOTA

LATITUDE: 43°55'N LONGITUDE: 92°30' W ELEVATION: FT. GRND 1297 BARO 1289 TIME ZONE: CENTRAL WBAN: 14925

	(a)	JAN	FEB	MAR	APR	MAY	JUNE	JULY	AUG	SEP	OCT	NOV	DEC	YEAR
TEMPERATURE °F:														
Normals														
-Daily Maximum		20.2	26.0	38.2	55.2	68.1	77.7	81.8	78.8	69.8	58.2	40.9	25.0	53.3
-Daily Minimum		2.6	8.1	21.3	34.6	45.5	55.2	60.1	57.5	48.5	37.6	24.4	9.4	33.7
-Monthly		11.5	17.1	29.8	44.9	56.8	66.5	70.9	68.2	59.2	47.9	32.6	17.2	43.6
Extremes														
-Record Highest	34	55	63	79	91	92	101	102	99	95	88	73	62	102
-Year		1981	1981	1986	1980	1980	1985	1988	1988	1978	1976	1978	1982	JUL 1988
-Record Lowest	34	-32	-29	-31	5	21	35	42	37	23	11	-20	-33	-33
-Year		1970	1979	1962	1982	1967	1990	1967	1964	1967	1988	1977	1983	DEC 1983
NORMAL DEGREE DAYS:														
Heating (base 65°F)		1659	1341	1091	603	282	56	16	33	185	530	972	1482	8250
Cooling (base 65°F)		0	0	0	0	28	101	199	133	11	0	0	0	472
% OF POSSIBLE SUNSHINE														
MEAN SKY COVER (tenths)														
Sunrise - Sunset	33	6.6	6.6	7.0	6.8	6.4	6.2	5.7	5.7	6.0	6.2	7.4	7.2	6.5
MEAN NUMBER OF DAYS:														
Sunrise to Sunset														
-Clear	33	7.4	7.3	6.3	6.2	6.8	6.7	8.5	9.0	9.1	8.5	5.1	5.9	86.7
-Partly Cloudy	33	7.1	6.2	7.0	7.5	8.8	10.6	12.0	10.9	7.4	7.9	5.5	6.4	97.4
-Cloudy	33	16.5	14.8	17.7	16.3	15.5	12.7	10.5	11.1	13.5	14.5	19.4	18.6	181.2
Precipitation														
.01 inches or more	33	8.5	7.6	10.3	11.6	11.5	11.0	10.2	9.7	10.7	8.4	9.1	9.5	118.1
Snow, Ice pellets, hail														
1.0 inches or more	33	3.0	2.5	3.2	1.1	0.0	0.0	0.0	0.0	0.0	0.2	1.8	3.1	14.8
Thunderstorms	33	0.1	0.2	1.3	3.4	5.8	7.2	7.7	6.6	4.9	2.0	0.6	0.1	39.8
Heavy Fog Visibility														
1/4 mile or less	33	3.0	2.8	4.3	2.3	2.0	1.1	1.5	2.3	3.0	2.2	3.4	4.4	32.4
Temperature °F														
-Maximum														
90° and above	33	0.0	0.0	0.0	0.1	0.3	1.7	3.2	1.7	0.3	0.0	0.0	0.0	7.3
32° and below	33	24.3	18.2	9.4	0.6	0.0	0.0	0.0	0.0	0.0	0.1	7.5	22.4	82.5
-Minimum														
32° and below	33	30.8	27.5	26.1	12.8	1.8	0.0	0.0	0.0	1.0	10.2	23.6	30.3	164.2
0° and below	33	13.8	9.0	2.2	0.0	0.0	0.0	0.0	0.0	0.0	0.0	1.1	8.8	35.0
AVG. STATION PRESS. (mb)	21	969.1	969.9	967.0	966.5	966.5	966.5	968.2	969.2	969.3	969.1	968.1	969.0	968.2
RELATIVE HUMIDITY (%)														
Hour 00	33	78	80	80	75	76	79	84	86	84	78	82	82	80
Hour 06 (Local Time)	33	79	81	82	80	80	82	86	88	88	83	84	83	83
Hour 12	33	74	72	69	60	57	57	60	62	63	59	71	77	65
Hour 18	33	77	75	71	59	56	57	61	65	68	67	77	80	68
PRECIPITATION (inches):														
Water Equivalent														
-Normal		0.78	0.74	1.78	2.73	3.40	3.72	4.20	3.88	3.47	2.32	1.61	1.03	29.66
-Maximum Monthly	34	2.53	2.21	3.58	6.47	8.41	9.27	12.33	9.52	10.50	6.08	5.90	2.83	12.33
-Year		1967	1971	1990	1990	1982	1990	1978	1979	1986	1970	1991	1982	JUL 1978
-Minimum Monthly	34	0.07	0.04	0.43	1.02	1.17	0.94	1.02	1.17	0.38	0.27	0.06	0.22	0.04
-Year		1961	1964	1978	1987	1963	1985	1975	1970	1975	1965	1967	1967	FEB 1964
-Maximum in 24 hrs	34	1.42	1.05	2.04	3.97	2.61	3.01	7.47	3.89	6.01	2.81	2.64	1.35	7.47
-Year		1967	1984	1966	1990	1980	1986	1981	1991	1978	1966	1991	1982	JUL 1981
Snow, Ice pellets, hail														
-Maximum Monthly	34	27.3	19.1	25.2	16.4	0.3	T	T	T	0.8	5.4	22.5	30.6	30.6
-Year		1982	1962	1985	1983	1967	1993	1993	1989	1961	1979	1985	1969	DEC 1969
-Maximum in 24 hrs	34	15.4	9.3	11.3	13.7	0.3	T	T	T	0.8	5.4	9.2	9.0	15.4
-Year		1982	1983	1966	1988	1967	1993	1993	1989	1961	1979	1991	1985	JAN 1982
WIND:														
Mean Speed (mph)	33	14.5	13.8	14.4	14.5	13.3	12.5	11.0	10.9	12.0	13.1	13.7	13.9	13.1
Prevailing Direction through 1963		NW	NW	NW	NW	S	S	S	S	S	S	S	NW	S
Fastest Obs. 1 Min.														
-Direction (!!)	31	30	28	23	30	28	13	17	34	24	34	28	32	23
-Speed (MPH)	31	48	45	58	53	52	53	51	46	44	41	44	47	58
-Year		1990	1971	1982	1982	1964	1973	1971	1987	1980	1985	1986	1989	MAR 1982
Peak Gust														
-Direction (!!)	10	NW	NW	E	S	S	SW	W	N	S	N	S	NW	S
-Speed (mph)	10	69	53	68	85	74	69	71	64	64	62	67	56	85
-Date		1986	1987	1985	1984	1985	1992	1984	1987	1992	1985	1986	1989	APR 1984

PRECIPITATION (inches)

ROCHESTER, MINNESOTA

YEAR	JAN	FEB	MAR	APR	MAY	JUNE	JULY	AUG	SEP	OCT	NOV	DEC	ANNUAL
1964	0.37	0.04	0.94	2.61	1.91	1.08	1.18	2.07	5.99	0.52	2.36	0.84	19.91
1965	0.45	1.34	2.85	3.92	4.14	1.53	4.67	3.40	6.26	0.27	1.31	1.41	31.55
1966	0.68	1.06	3.32	1.08	1.54	3.26	3.03	3.47	1.36	3.86	0.40	0.96	24.02
1967	2.53	0.76	1.52	3.36	1.36	8.34	1.05	3.52	1.24	2.34	0.06	0.22	26.30
1968	0.77	0.14	0.51	4.16	4.37	6.07	5.07	2.06	3.93	2.94	0.52	1.86	32.40
1969	1.25	0.14	0.99	1.35	3.04	5.66	5.38	3.36	1.59	3.18	0.70	1.66	28.30
1970	0.38	0.47	1.57	2.29	6.70	3.86	4.30	1.17	4.09	6.08	2.04	0.82	33.77
1971	1.12	2.21	0.97	1.58	3.65	4.95	3.15	1.99	3.93	3.37	2.18	0.98	30.08
1972	0.71	0.29	0.72	1.66	1.76	1.11	5.99	2.11	7.06	3.03	1.55	1.45	27.44
1973	1.05	0.88	2.85	4.26	5.26	3.24	4.85	5.71	5.84	2.60	3.37	0.99	40.90
1974	0.36	0.73	2.42	2.59	4.53	7.04	1.26	2.37	1.01	2.54	0.86	0.56	26.27
1975	1.91	0.76	1.78	3.66	2.34	3.86	1.02	5.97	0.38	0.68	4.61	1.21	28.18
1976	0.38	0.49	2.91	2.77	2.09	1.20	1.96	1.68	0.76	0.62	0.11	0.47	15.44
1977	0.37	0.97	2.94	2.91	3.74	4.65	2.34	2.63	3.65	1.97	1.51	1.57	29.25
1978	0.58	0.33	0.43	2.36	3.84	5.62	12.33	1.92	8.08	0.95	1.99	0.83	39.26
1979	1.28	0.34	2.49	2.10	3.83	2.40	2.74	9.52	0.63	4.95	2.28	0.48	33.04
1980	1.52	0.52	0.82	1.17	3.72	1.75	2.56	7.86	2.97	1.88	0.13	0.42	25.32
1981	0.23	2.00	0.54	2.47	2.69	3.46	10.46	6.44	1.01	2.13	0.85	0.72	33.00
1982	1.70	0.11	1.31	3.13	8.41	1.36	3.97	4.94	4.05	2.64	2.38	2.83	36.83
1983	0.82	1.27	2.01	2.52	3.92	4.55	3.12	4.63	4.72	2.88	3.90	1.00	35.34
1984	0.11	1.96	1.08	3.91	2.89	3.74	3.34	1.93	2.40	3.78	1.68	1.79	28.61
1985	0.63	0.57	2.31	1.58	1.74	0.94	2.57	5.40	6.41	1.53	2.43	1.14	27.25
1986	0.59	0.61	2.15	3.80	3.40	5.04	6.00	3.17	10.50	3.57	0.84	0.32	39.99
1987	0.58	0.23	1.29	1.02	2.12	3.69	7.24	3.85	2.05	1.61	1.94	1.75	27.37
1988	1.16	0.22	1.56	2.43	2.35	1.52	1.12	2.88	3.77	0.40	2.87	1.11	21.39
1989	0.41	0.42	1.65	3.49	1.74	2.39	3.31	5.73	0.61	1.67	1.62	0.38	23.42
1990	0.55	0.71	3.58	6.47	4.52	9.27	8.29	5.30	1.30	1.86	0.44	1.65	43.94
1991	0.67	0.45	2.82	5.25	3.84	2.25	5.32	4.66	2.31	1.99	5.90	1.47	36.93
1992	1.03	0.55	2.53	3.24	1.60	1.59	3.51	1.50	4.93	1.30	4.02	1.30	27.10
1993	1.15	0.83	2.92	4.56	4.32	7.44	5.00	6.88	2.75	0.85	1.00	0.74	38.44
Record Mean	0.90	0.78	1.67	2.59	3.63	4.23	3.73	3.69	3.21	2.02	1.63	0.97	29.05

AVERAGE TEMPERATURE (deg. F)

ROCHESTER, MINNESOTA

YEAR	JAN	FEB	MAR	APR	MAY	JUNE	JULY	AUG	SEP	OCT	NOV	DEC	ANNUAL
1964	21.3	24.1	25.6	45.2	60.2	67.2	73.8	65.8	57.6	46.5	33.8	15.4	44.7
1965	9.7	11.6	18.1	41.9	58.4	65.4	69.1	66.3	54.9	50.0	33.7	28.5	42.3
1966	3.2	15.2	34.0	40.9	51.8	67.2	71.9	66.2	59.0	47.9	31.8	19.4	42.4
1967	15.5	10.5	31.2	44.5	51.3	66.1	66.9	64.2	58.4	46.0	29.8	21.8	42.2
1968	14.6	14.9	36.8	47.9	51.6	65.1	67.7	67.1	58.2	49.4	33.8	17.2	43.7
1969	9.8	18.6	21.3	47.2	57.6	59.6	69.8	69.2	60.0	45.4	33.6	18.7	42.5
1970	4.3	15.1	26.3	45.7	58.1	67.9	70.8	68.5	59.9	49.6	32.6	18.0	43.1
1971	5.8	16.4	26.2	45.9	54.4	71.4	67.0	67.7	62.0	53.5	34.2	20.0	43.7
1972	6.5	12.4	27.9	41.3	60.1	65.6	67.7	68.2	57.9	43.3	31.3	11.0	41.1
1973	15.8	20.1	39.3	42.4	53.4	68.3	71.0	70.6	59.4	53.9	34.6	14.9	45.3
1974	12.8	15.8	29.4	46.8	53.8	64.3	74.6	66.2	54.7	49.3	33.4	22.3	43.6
1975	14.6	15.4	21.0	39.6	60.7	67.5	73.2	70.8	56.2	52.9	37.5	22.5	44.3
1976	12.8	28.2	31.8	49.5	55.8	68.2	73.8	69.8	58.5	42.0	24.9	9.6	43.8
1977	-1.8	21.7	38.7	53.0	64.5	66.7	73.0	65.1	60.9	47.3	32.0	14.3	44.6
1978	3.5	8.3	28.3	44.7	57.8	66.4	69.2	68.8	63.7	46.1	29.1	12.6	41.5
1979	-1.5	5.7	26.7	42.1	55.1	67.8	72.7	68.7	62.5	47.4	32.4	26.0	42.1
1980	15.5	15.1	26.6	47.1	60.4	67.1	73.2	69.7	60.1	43.6	35.4	19.2	44.4
1981	18.8	22.4	35.6	48.1	55.0	65.9	69.9	67.6	57.3	44.7	35.9	16.4	44.8
1982	3.0	16.2	28.6	40.9	60.3	61.5	72.9	68.7	60.6	49.8	32.7	26.1	43.4
1983	17.8	23.9	32.1	39.3	51.8	65.9	73.9	73.6	60.5	47.1	33.5	2.9	43.5
1984	12.4	25.4	23.4	44.8	52.5	65.9	67.7	70.6	56.2	49.6	33.1	19.1	43.4
1985	10.3	14.8	35.9	50.4	60.4	62.6	70.5	65.3	58.7	46.0	23.0	6.3	42.0
1986	15.9	15.3	32.7	48.4	57.5	65.9	71.4	64.0	59.5	47.5	26.3	22.5	43.9
1987	19.2	29.3	36.2	50.8	61.0	70.0	73.7	66.8	60.0	42.3	36.8	23.3	47.5
1988	8.3	12.3	31.9	44.0	61.7	70.4	73.7	72.6	61.0	42.0	32.5	19.2	44.1
1989	21.7	9.4	26.3	44.1	55.8	64.4	72.3	68.5	58.2	48.7	27.1	10.0	42.2
1990	25.8	22.1	35.3	45.5	53.8	67.6	69.4	69.4	62.4	46.5	37.7	15.4	45.9
1991	10.1	23.5	34.1	48.3	61.0	72.2	70.1	69.0	57.4	45.5	24.0	21.6	44.7
1992	21.4	26.7	31.3	42.1	58.2	64.6	64.2	63.2	57.7	46.0	30.8	19.8	43.8
1993	12.6	14.6	25.7	41.7	55.7	63.6	69.0	68.3	53.6	45.5	30.3	21.5	41.8
Record Mean	12.3	16.9	29.4	44.8	56.9	66.3	71.3	68.8	59.8	48.4	32.4	18.6	43.8
Max	21.4	26.2	38.4	55.4	68.3	77.1	82.9	80.2	71.1	59.2	41.0	26.7	54.0
Min	3.1	7.5	20.3	34.1	45.4	55.4	59.8	57.4	48.4	37.5	23.8	10.4	33.6

REFERENCE NOTES

ROCHESTER, MINNESOTA

GENERAL
T - TRACE AMOUNT.
BLANK ENTRIES DENOTE MISSING/UNREPORTED DATA.
# INDICATES A STATION OR INSTRUMENT RELOCATION.

SPECIFIC

(a) - LENGTH OF RECORD IN YEARS, ALTHOUGH INDIVIDUAL MONTHS MAY BE MISSING.
0.* OR * - THE VALUE IS BETWEEN 0.0 AND 0.05
NORMALS - BASED ON THE 1961-1990 RECORD PERIOD.
EXTREMES - DATES ARE THE MOST RECENT OCCURRENCE
WIND DIR.- NUMERALS SHOW TENS OF DEGRESS CLOCKWISE FROM TRUE NORTH. "00" INDICATES CALM.
RESULTANT DIRECTIONS ARE GIVEN TO WHOLE DEGREES.
BOLD VALUES INDICATE EXTREME VALUES WHICH OCCURRED AFTER THE ASOS SYSTEM WAS COMMISSIONED.

RECORD = PERIOD OF RECORD
RECORD MEAN PRECIPITATION IS THE MEAN OF ALL DAILY PRECIPITATION AMOUNTS DURING THE PERIOD OF RECORD.
RECORD MAX(MIN) TEMPERATURE IS THE MEAN OF ALL DAILY MAX(MIN) TEMPERATURES DURING THE PERIOD OF RECORD.
RECORD MEAN TEMPERATURE IS THE SUM OF THE RECORD MAX AND RECORD MIN DIVIDED BY 2.
AVERAGE TEMPERATURE IS THE SUM OF THE MEAN DAILY MAX AND MIN TEMPERATURE DIVIDED BY 2.

EXCEPTIONS

RECORD MEANS ARE THROUGH THE CURRENT YEAR, BEGINNING IN 1886 FOR TEMPERATURE
1886 FOR PRECIPITATION
1961 FOR SNOWFALL

HEATING DEGREE DAYS Base 65 deg. F ROCHESTER, MINNESOTA

SEASON	JULY	AUG	SEP	OCT	NOV	DEC	JAN	FEB	MAR	APR	MAY	JUNE	TOTAL
1964-65	5	92	253	565	930	1535	1711	1494	1444	684	223	39	8975
1965-66	10	64	307	463	933	1123	1914	1393	956	717	411	53	8344
1966-67	5	56	210	527	988	1407	1531	1521	1041	610	439	28	8363
1967-68	49	91	197	584	1048	1334	1556	1446	868	507	411	88	8179
1968-69	33	68	214	494	930	1478	1706	1294	1353	525	262	181	8538
1969-70	10	12	177	614	934	1428	1880	1392	1189	588	248	56	8528
1970-71	17	11	205	480	966	1450	1832	1357	1195	566	324	19	8422
1971-72	29	31	170	361	917	1387	1810	1521	1142	704	219	88	8379
1972-73	39	71	219	667	1005	1671	1525	1251	792	674	353	20	8287
1973-74	3	15	197	349	903	1546	1611	1372	1098	547	350	78	8069
1974-75	0	50	309	478	941	1317	1556	1385	1359	755	196	60	8406
1975-76	19	7	273	380	821	1311	1615	1059	1021	468	283	22	7279
1976-77	1	26	225	710	1195	1715	2071	1208	808	371	90	37	8457
1977-78	1	60	143	543	986	1570	1903	1581	1131	601	254	66	8839
1978-79	11	20	142	577	1069	1624	2064	1658	1183	680	319	31	9378
1979-80	0	47	123	540	974	1201	1533	1442	1185	543	206	53	7847
1980-81	0	11	181	659	882	1416	1424	1185	905	496	305	42	7506
1981-82	21	27	235	621	868	1503	1922	1363	1121	718	164	129	8692
1982-83	0	48	178	464	964	1197	1456	1145	1010	765	403	86	7716
1983-84	8	0	209	555	938	1925	1625	1142	1281	600	381	40	8704
1984-85	23	18	281	470	950	1414	1692	1404	895	456	166	129	7898
1985-86	6	55	268	580	1255	1817	1516	1388	994	492	248	60	8679
1986-87	5	81	187	535	1153	1312	1412	993	888	425	173	31	7195
1987-88	8	64	168	697	837	1290	1755	1527	1020	620	144	22	8152
1988-89	3	27	154	706	970	1414	1335	1555	1190	620	290	99	8363
1989-90	0	16	224	500	1132	1700	1205	1196	912	596	341	55	7877
1990-91	9	14	167	567	815	1534	1701	1154	950	507	222	12	7652
1991-92	15	24	274	597	1225	1342	1343	1104	1036	678	239	80	7957
1992-93	73	104	226	587	1019	1394	1618	1407	1211	693	294	87	8713
1993-94	7	32	341	605	1033	1344							

COOLING DEGREE DAYS Base 65 deg. F ROCHESTER, MINNESOTA

YEAR	JAN	FEB	MAR	APR	MAY	JUNE	JULY	AUG	SEP	OCT	NOV	DEC	TOTAL
1969	0	0	0	0	39	24	165	146	36	10	0	0	420
1970	0	0	0	15	42	151	202	131	58	5	0	0	604
1971	0	0	0	0	2	218	100	123	88	11	0	0	542
1972	0	0	0	0	73	112	130	178	14	0	0	0	507
1973	0	0	0	0	1	123	196	192	37	11	0	0	560
1974	0	0	0	4	12	65	302	97	7	0	0	0	487
1975	0	0	0	0	68	141	279	194	18	12	0	0	712
1976	0	0	0	9	5	126	278	183	36	4	0	0	641
1977	0	0	0	18	83	98	255	68	26	0	0	0	548
1978	0	0	0	0	39	112	147	146	111	0	0	0	555
1979	0	0	0	0	21	119	246	167	57	1	0	0	611
1980	0	0	0	14	69	121	262	164	41	0	0	0	671
1981	0	0	0	0	4	75	179	113	13	0	0	0	384
1982	0	0	0	0	26	30	254	167	53	0	0	0	530
1983	0	0	0	0	0	121	289	272	79	7	0	0	768
1984	0	0	0	0	1	74	115	201	23	0	0	0	414
1985	0	0	0	26	28	64	182	74	86	0	0	0	460
1986	0	0	0	2	21	94	211	59	29	0	0	0	416
1987	0	0	0	6	57	189	283	124	26	0	0	0	685
1988	0	0	0	0	48	191	280	272	38	0	0	0	829
1989	0	0	0	0	10	90	235	133	26	2	0	0	496
1990	0	0	0	18	3	140	153	158	97	0	0	0	569
1991	0	0	0	11	101	233	181	153	52	0	0	0	731
1992	0	0	0	0	34	75	54	52	16	2	0	0	233
1993	0	0	0	0	10	52	136	137	6	4	0	0	345

SNOWFALL (inches) ROCHESTER, MINNESOTA

SEASON	JULY	AUG	SEP	OCT	NOV	DEC	JAN	FEB	MAR	APR	MAY	JUNE	TOTAL
1964-65	0.0	0.0	0.0	T	4.8	6.4	11.1	8.8	12.2	5.2	T	0.0	48.5
1965-66	0.0	0.0	T	0.0	1.1	0.9	10.5	7.3	12.2	0.4	0.2	0.0	32.6
1966-67	0.0	0.0	0.0	0.2	2.5	10.3	12.4	13.4	5.6	T	0.3	0.0	44.7
1967-68	0.0	0.0	0.0	T	0.5	0.7	4.7	2.4	0.4	0.4	0.0	0.0	9.1
1968-69	0.0	0.0	0.0	0.2	4.5	20.7	9.5	2.6	5.6	0.5	T	0.0	43.6
1969-70	0.0	0.0	0.0	2.1	0.6	30.6	8.5	9.0	8.4	3.4	T	0.0	62.6
1970-71	0.0	0.0	0.0	0.1	7.6	10.8	16.2	16.0	9.7	1.6	0.0	0.0	62.0
1971-72	0.0	0.0	0.0	0.0	7.8	8.4	11.0	3.5	7.8	2.7	0.0	0.0	41.2
1972-73	0.0	0.0	0.0	1.1	0.8	17.2	12.0	7.9	1.5	10.4	0.2	0.0	51.1
1973-74	0.0	0.0	0.0	T	0.3	16.3	2.1	12.3	11.2	0.3	0.0	0.0	42.5
1974-75	0.0	0.0	0.0	0.0	2.9	8.3	14.1	10.2	15.0	2.5	0.0	0.0	53.0
1975-76	0.0	0.0	0.0	T	7.8	1.9	6.6	1.6	10.3	T	0.2	0.0	28.4
1976-77	0.0	0.0	0.0	2.9	1.9	7.7	9.2	1.8	7.5	3.5	0.0	0.0	34.5
1977-78	0.0	0.0	0.0	T	10.0	18.8	7.4	5.9	5.4	1.2	0.0	0.0	48.7
1978-79	0.0	0.0	0.0	0.0	11.2	14.2	24.4	5.0	11.1	7.4	T	0.0	73.3
1979-80	0.0	0.0	0.0	5.4	5.5	0.7	13.6	6.3	12.8	10.9	0.0	0.0	55.2
1980-81	0.0	0.0	0.0	T	0.7	5.1	3.5	16.1	T	0.2	0.0	0.0	25.6
1981-82	0.0	0.0	0.0	2.2	8.2	11.1	27.3	2.2	4.0	7.7	0.0	0.0	62.7
1982-83	0.0	0.0	0.0	0.5	2.4	8.3	7.9	15.9	11.2	16.4	0.0	0.0	62.6
1983-84	0.0	0.0	0.0	T	14.0	16.2	2.7	12.0	16.1	5.0	0.0	0.0	66.0
1984-85	0.0	0.0	0.0	T	3.7	14.4	12.2	9.3	25.2	3.8	0.0	0.0	68.6
1985-86	0.0	0.0	T	0.0	22.5	16.0	11.5	8.7	1.2	0.8	0.0	0.0	60.7
1986-87	0.0	0.0	0.0	T	8.4	3.5	8.1	2.3	4.7	T	0.0	0.0	27.0
1987-88	0.0	0.0	0.0	0.9	1.7	16.0	18.2	5.8	3.1	15.6	0.0	0.0	61.3
1988-89	0.0	0.0	0.0	T	10.8	4.8	4.7	9.3	21.2	0.1	0.2	0.0	51.1
1989-90	0.0	T	0.0	2.6	10.5	6.6	5.9	9.3	0.5	0.2	0.0	T	35.6
1990-91	0.0	0.0	0.0	0.8	1.6	20.8	9.9	6.1	5.9	0.8	0.0	0.0	45.9
1991-92	0.0	0.0	0.0	4.5	20.3	7.3	9.0	6.5	15.0	T	T	0.0	62.6
1992-93	0.0	0.0	0.0	0.8	4.3	11.7	15.7	10.8	9.8	9.3	0.0	T	62.4
1993-94	T	0.0	0.0	T	7.7	6.0							
Record Mean	T	T	T	0.7	5.8	10.7	10.1	7.9	9.6	4.0	T	T	48.8

SAINT CLOUD, MINNESOTA

St. Cloud is located in central Minnesota on the banks of the Mississippi River. The topography is gently rolling terrain with numerous lakes and wooded areas.

The climate is influenced by atmospheric moisture flowing into the state from the Gulf of Mexico and the Pacific coast. Air masses carrying moisture which is eventually released as precipitation may travel nearly 1,500 miles. Due to this long trek, a minor change in the wind system can result in the area receiving well below or well above the normal precipitation. Rainfall is generally ample for farm and garden crops. Although the total amount is important, its distribution during the average 140 day growing season from mid-May to the end of September is even more significant. Thunderstorms are the principal source of rainfall during this period.

Spring, summer, and fall are very pleasant. Prolonged periods of hot and humid weather are infrequent. Extremely hot days with temperatures of 100 degrees or higher occur only once every five to ten years and rarely are temperatures this high recorded on successive days. Tornadoes and severe local storms are common.

Winter is cold, but not unpleasant, since strong winds and high humidities are generally absent on the coldest days. Cold Canadian air masses are prevalent throughout the winter season. The normal winter will have five to ten days with temperatures in the -20 to -30 degree range. Heavy snowfalls do occur, but the northern location limits the numerous heavy snowfalls that occur just a short distance to the south. Snowfalls of 3 inches or more in a 24 hour period occur only on an average of four times per year. Snow generally remains on the ground from the onset of the winter season until spring. Blizzards occur on the average of once per year with a severe blizzard once every three or four years. Ice storms are infrequent because temperatures are usually too cold and the transition period from season to season is rather abrupt.

NORMALS, MEANS, AND EXTREMES

SAINT CLOUD, MINNESOTA

LATITUDE: 45°33'N LONGITUDE: 94°04' W ELEVATION: FT. GRND 1028 BARO 1032 TIME ZONE: CENTRAL WBAN: 14926

	(a)	JAN	FEB	MAR	APR	MAY	JUNE	JULY	AUG	SEP	OCT	NOV	DEC	YEAR
TEMPERATURE °F:														
Normals														
-Daily Maximum		18.5	24.8	37.6	55.0	68.4	77.4	82.6	79.4	69.1	57.4	39.0	23.1	52.7
-Daily Minimum		-2.4	3.8	17.6	32.0	43.4	52.2	57.6	54.8	45.4	34.4	20.3	5.0	30.3
-Monthly		8.1	14.3	27.6	43.5	55.9	64.8	70.1	67.1	57.3	45.9	29.7	14.1	41.5
Extremes														
-Record Highest	53	55	55	79	96	97	102	103	103	98	90	74	60	103
-Year		1942	1991	1968	1980	1969	1988	1940	1947	1978	1992	1978	1982	AUG 1947
-Record Lowest	53	-43	-35	-32	-3	19	32	40	33	18	5	-20	-41	-43
-Year		1977	1951	1962	1975	1967	1993	1972	1974	1974	1976	1964	1983	JAN 1977
NORMAL DEGREE DAYS:														
Heating (base 65°F)		1764	1420	1159	645	307	87	23	54	240	592	1059	1578	8928
Cooling (base 65°F)		0	0	0	0	25	81	181	119	9	0	0	0	415
% OF POSSIBLE SUNSHINE														
MEAN SKY COVER (tenths)														
Sunrise - Sunset	39	6.4	6.0	6.6	6.6	6.4	6.0	5.2	5.2	5.7	5.9	7.1	7.0	6.2
MEAN NUMBER OF DAYS:														
Sunrise to Sunset														
-Clear	48	8.0	8.3	7.1	7.0	6.9	6.8	10.0	10.6	9.8	10.1	5.8	6.6	96.9
-Partly Cloudy	48	7.8	6.8	7.3	7.5	9.6	11.1	11.7	11.0	8.9	7.8	6.1	6.5	102.0
-Cloudy	48	15.2	12.9	16.5	15.6	14.5	12.1	9.2	9.5	11.4	13.1	18.1	17.9	166.0
Precipitation .01 inches or more	53	8.6	6.8	8.8	9.5	11.4	11.4	9.7	9.6	8.9	7.4	8.1	8.9	109.0
Snow,Ice pellets,hail 1.0 inches or more	53	2.9	2.3	3.0	0.8	0.*	0.0	0.0	0.0	0.*	0.1	2.2	2.8	14.2
Thunderstorms	30	0.0	0.*	0.4	1.9	4.6	7.5	6.9	7.0	3.5	1.4	0.3	0.0	33.4
Heavy Fog Visibility 1/4 mile or less	30	1.6	2.0	1.8	0.7	1.0	0.9	1.3	2.3	1.7	2.1	1.6	2.5	19.6
Temperature °F														
-Maximum														
90° and above	53	0.0	0.0	0.0	0.*	0.4	1.8	4.2	3.3	0.7	0.*	0.0	0.0	10.5
32° and below	53	25.7	19.8	10.8	0.8	0.0	0.0	0.0	0.0	0.0	0.2	9.4	23.3	89.9
-Minimum														
32° and below	53	31.0	28.0	28.2	16.5	3.1	0.*	0.0	0.0	2.2	12.6	26.3	30.6	178.5
0° and below	53	17.4	12.6	4.3	0.*	0.0	0.0	0.0	0.0	0.0	0.0	2.0	9.0	45.3
AVG. STATION PRESS.(mb)	8	980.6	981.3	977.4	978.4	976.4	976.2	978.0	978.3	979.1	979.1	979.4	979.7	978.6
RELATIVE HUMIDITY (%)														
Hour 00 (Local Time)	19	76	76	78	75	74	82	85	85	85	79	80	79	80
Hour 06	41	77	79	81	81	81	85	88	91	90	85	83	80	83
Hour 12	43	69	66	63	52	51	55	55	58	59	57	67	71	60
Hour 18	35	72	68	65	53	49	54	56	58	63	62	73	76	62
PRECIPITATION (inches):														
Water Equivalent														
-Normal		0.74	0.63	1.41	2.35	3.16	4.60	3.11	3.96	3.16	2.21	1.27	0.83	27.43
-Maximum Monthly	53	2.52	2.76	3.43	5.55	8.01	10.52	8.00	7.55	9.48	6.16	3.74	2.04	10.52
-Year		1969	1951	1965	1986	1962	1990	1955	1956	1985	1971	1977	1969	JUN 1990
-Minimum Monthly	53	0.02	0.04	0.10	0.05	0.32	0.05	0.21	0.46	0.07	0.07	0.01	0.01	0.01
-Year		1942	1964	1959	1987	1948	1988	1975	1950	1952	1952	1941	1943	DEC 1943
-Maximum in 24 hrs	53	0.99	1.83	1.81	3.15	3.70	4.06	2.29	4.62	5.37	3.24	2.22	1.38	5.37
-Year		1949	1951	1957	1954	1979	1983	1972	1956	1985	1950	1977	1984	SEP 1985
Snow,Ice pellets,hail														
-Maximum Monthly	53	29.9	21.6	51.7	11.1	3.2	0.0	0.0	0.0	1.8	4.1	26.9	25.4	51.7
-Year		1975	1971	1965	1950	1971				1942	1959	1940	1968	MAR 1965
-Maximum in 24 hrs	53	9.4	12.2	14.5	6.3	3.2	0.0	0.0	0.0	1.8	3.6	11.4	10.2	14.5
-Year		1975	1951	1985	1953	1971				1942	1969	1975	1968	MAR 1985
WIND:														
Mean Speed (mph)	9	8.2	7.8	9.0	9.8	8.9	8.2	7.0	6.4	7.1	7.8	8.3	7.7	8.0
Prevailing Direction														
Fastest Obs. 1 Min.														
-Direction (!!!)	4	32	27	07	05	20	07	21	07	31	27	28	30	05
-Speed (MPH)	4	25	21	22	26	21	21	21	23	21	25	25	21	26
-Year		1990	1990	1991	1993	1992	1993	1993	1993	1993	1993	1991	1991	APR 1993
Peak Gust														
-Direction (!!!)	10	NW				SW			NE	NW				
-Speed (mph)	10	58	46	46	78	51	58	51	44	45	46	53	46	78
-Date		1990	1987	1988	1985	1993	1987	1984	1993	1991	1987	1986	1992	APR 1985

PRECIPITATION (inches)

SAINT CLOUD, MINNESOTA

YEAR	JAN	FEB	MAR	APR	MAY	JUNE	JULY	AUG	SEP	OCT	NOV	DEC	ANNUAL
1964	0.18	0.04	1.22	3.31	3.62	1.30	1.71	6.66	1.38	0.19	0.98	0.58	21.17
1965	0.48	0.91	3.43	3.44	6.78	6.43	4.66	4.65	4.94	0.94	1.55	1.11	39.32
1966	0.70	1.17	1.53	1.66	2.22	3.18	3.51	4.67	0.95	1.41	0.49	0.79	22.28
1967	1.99	0.75	0.39	1.05	0.82	7.00	0.59	4.72	1.43	1.14	0.14	1.12	21.14
1968	0.86	0.21	1.17	4.51	2.80	6.98	1.95	2.13	4.74	5.80	0.58	1.95	33.68
1969	2.52	0.69	0.47	3.48	2.16	2.27	2.81	2.16	1.71	1.29	0.38	2.04	21.98
1970	0.24	0.18	1.05	3.01	2.52	3.43	3.26	1.73	1.66	5.10	2.73	0.24	25.15
1971	0.86	1.53	0.31	1.66	3.86	6.49	2.28	2.79	3.12	6.16	2.56	0.39	32.01
1972	0.55	0.47	1.56	1.59	3.30	1.91	7.26	4.94	1.64	2.54	0.74	1.31	27.81
1973	0.52	0.31	1.40	1.65	2.89	2.92	2.94	4.27	2.80	3.13	1.64	0.73	25.20
1974	0.09	0.83	0.88	1.16	3.26	4.36	2.25	3.20	1.97	1.58	1.29	0.54	21.41
1975	2.39	0.40	1.75	3.69	3.02	5.78	0.21	4.83	2.27	1.08	3.24	0.28	28.94
1976	0.85	0.83	1.78	0.92	0.93	4.84	1.92	0.60	1.37	0.44	0.14	0.31	14.93
1977	0.58	0.98	3.03	3.17	3.57	3.48	4.27	6.10	2.34	2.93	3.74	1.40	35.59
1978	0.19	0.17	0.81	3.49	3.20	6.04	4.43	2.88	4.59	0.14	0.95	1.02	27.91
1979	1.28	1.67	3.02	0.74	5.17	6.34	1.21	4.88	1.58	4.36	0.62	0.31	31.18
1980	1.17	0.84	0.76	0.48	1.62	6.06	1.28	7.01	5.99	0.71	0.20	0.22	26.34
1981	0.44	1.10	1.05	3.29	1.40						0.45	1.11	
1982	0.97	0.13	1.75	0.97	3.91	2.53	3.90	3.37	4.38	4.52	2.31	1.72	30.46
1983	0.61	0.13	2.60	1.57	2.39	9.52	2.21	3.48	6.55	3.09	3.11	0.92	36.18
1984	0.67	0.87	0.65	4.16	2.02	8.11	2.94	2.57	3.39	5.84	0.17	1.81	33.20
1985	0.43	0.23	1.70	3.83	2.81	5.28	2.80	4.57	9.48	1.28	1.43	0.57	34.41
1986	0.71	0.79	0.94	5.55	2.56	4.02	7.53	5.23	6.28	0.42	1.06	0.33	35.42
1987	0.41	0.12	0.71	0.05	2.17	2.14	4.41	4.47	1.94	0.72	1.64	0.76	19.54
1988	0.51	0.17	1.39	0.73	2.22	0.05	2.64	5.45	3.19	0.68	2.75	0.60	20.38
1989	0.85	0.62	0.84	2.33	3.26	2.20	1.74	4.35	2.09	0.71	0.97	0.33	20.29
1990	0.06	0.66	3.09	2.68	3.74	10.52	5.15	3.38	1.70	2.70	0.30	0.68	34.66
1991	0.39	1.08	1.84	4.68	5.04	5.63	3.74	2.88	3.91	2.04	2.31	0.48	34.02
1992	0.69	0.27	1.49	1.78	2.20	3.03	2.95	1.51	1.00	1.12	1.39	0.94	18.37
1993	0.99	0.33	1.26	1.95	5.64	4.99	3.76	5.49	2.09	1.44	2.15	0.78	30.87
Record Mean	0.75	0.68	1.32	2.17	3.44	4.47	3.41	3.59	3.06	1.97	1.27	0.67	26.81

AVERAGE TEMPERATURE (deg. F)

SAINT CLOUD, MINNESOTA

YEAR	JAN	FEB	MAR	APR	MAY	JUNE	JULY	AUG	SEP	OCT	NOV	DEC	ANNUAL
1964	16.9	20.1	21.8	45.0	59.8	66.4	74.2	65.8	57.4	47.2	31.0	10.1	43.0
1965	5.3	8.4	15.7	40.7	56.8	64.8	68.9	66.7	50.3	48.7	29.6	24.1	40.0
1966	-1.5	12.3	32.9	39.1	51.8	66.6	74.4	66.1	58.4	45.7	26.6	15.0	40.6
1967	12.2	5.4	26.2	42.1	50.4	64.2	69.1	65.9	58.8	44.8	30.2	18.8	40.7
1968	10.7	10.6	36.0	46.0	51.9	65.4	69.4	68.0	59.2	48.0	31.5	12.5	42.4
1969	5.2	15.8	21.6	46.2	57.8	58.5	70.0	72.6	60.2	43.2	31.3	17.8	41.7
1970	2.8	12.1	21.6	41.8	54.7	68.6	72.9	69.1	59.1	46.5	29.4	14.7	41.1
1971	2.9	14.4	24.7	44.4	53.5	69.2	66.6	67.4	59.6	49.7	30.7	15.5	41.5
1972	3.5	6.8	24.1	39.7	59.0	63.9	67.0	67.2	55.1	41.6	29.8	7.7	38.8
1973	12.5	17.2	37.3	42.5	53.5	65.7	68.9	70.3	57.4	51.3	31.8	14.1	43.5
1974	7.5	12.5	24.9	44.1	51.4	63.3	73.3	63.5	52.6	46.8	31.3	20.7	41.0
1975	9.9	11.9	18.8	36.3	58.8	65.3	73.3	67.6	54.6	47.9	31.9	15.5	41.0
1976	5.9	22.1	25.9	48.3	55.0	67.8	70.7	69.9	57.8	40.1	23.2	7.3	41.2
1977	-2.2	19.0	34.8	50.9	64.4	65.9	70.9	63.3	58.6	46.0	28.0	10.9	42.6
1978	3.4	9.1	27.2	42.0	58.3	64.3	68.2	67.9	62.4	46.1	27.6	9.0	40.5
1979	-2.5	5.1	24.3	39.2	51.2	63.5	69.0	65.1	58.5	44.7	28.7	22.9	39.1
1980	10.8	11.7	23.2	47.4	58.9	64.1	71.2	67.2	56.4	42.7	33.4	16.7	42.0
1981	15.3	20.9	35.5	46.2	54.9						35.4	14.0	
1982	-3.1	12.1	24.5	40.9	59.2	59.6	70.6	67.2	57.1	47.5	25.6	21.8	40.2
1983	14.2	23.8	31.2	40.6	51.5	64.3	73.6	73.4	59.4	46.2	31.9	-0.4	42.5
1984	9.2	25.7	23.0	45.7	54.6	66.5	70.1	70.5	53.7	49.0	30.8	13.9	42.7
1985	7.3	13.8	33.0	49.1	59.7	60.7	69.0	63.8	55.8	44.1	21.1	4.0	40.1
1986	14.5	12.0	30.6	47.8	57.3	65.1	70.7	63.9	57.0	46.1	25.7	19.9	42.6
1987	17.8	28.9	35.7	50.9	59.6	68.2	73.4	65.7	59.3	41.8	34.7	21.3	46.4
1988	6.8	10.2	30.8	44.5	63.6	71.1	74.6	70.3	58.8	40.8	29.1	16.0	43.1
1989	16.0	4.6	23.6	42.5	55.3	64.8	72.7	68.5	57.8	46.4	24.9	7.7	40.4
1990	22.0	19.0	32.2	43.9	53.9	66.6	68.1	67.4	59.9	44.1	33.5	13.7	43.7
1991	8.8	22.1	31.6	46.3	60.5	68.3	69.5	69.1	56.5	43.8	22.2	17.4	43.0
1992	18.5	24.6	30.9	41.4	57.8	61.8	63.1	63.2	56.7	43.9	28.8	16.6	42.3
1993	10.8	14.0	25.9	41.6	54.3	62.1	67.6	67.4	51.7	43.3	27.0	17.5	40.3
Record Mean	9.4	13.9	27.4	43.9	56.2	65.4	70.8	68.3	58.9	46.9	30.1	15.8	42.3
Max	19.6	24.4	37.2	55.0	68.0	76.9	82.6	80.1	70.3	57.9	38.8	24.7	53.0
Min	-0.7	3.5	17.5	32.8	44.3	53.9	59.1	56.6	47.4	35.9	21.5	6.9	31.6

REFERENCE NOTES

SAINT CLOUD, MINNESOTA

GENERAL
T - TRACE AMOUNT.
BLANK ENTRIES DENOTE MISSING/UNREPORTED DATA.
# INDICATES A STATION OR INSTRUMENT RELOCATION.

SPECIFIC

(a) - LENGTH OF RECORD IN YEARS, ALTHOUGH INDIVIDUAL MONTHS MAY BE MISSING.
0.* OR * - THE VALUE IS BETWEEN 0.0 AND 0.05
NORMALS - BASED ON THE 1961-1990 RECORD PERIOD.
EXTREMES - DATES ARE THE MOST RECENT OCCURRENCE.
WIND DIR.- NUMERALS SHOW TENS OF DEGRESS CLOCKWISE FROM TRUE NORTH. "00" INDICATES CALM.
RESULTANT DIRECTIONS ARE GIVEN TO WHOLE DEGREES.
BOLD VALUES INDICATE EXTREME VALUES WHICH OCCURRED AFTER THE ASOS SYSTEM WAS COMMISSIONED.

RECORD = PERIOD OF RECORD
RECORD MEAN PRECIPITATION IS THE MEAN OF ALL DAILY PRECIPITATION AMOUNTS DURING THE PERIOD OF RECORD.
RECORD MAX(MIN) TEMPERATURE IS THE MEAN OF ALL DAILY MAX(MIN) TEMPERATURES DURING THE PERIOD OF RECORD.
RECORD MEAN TEMPERATURE IS THE SUM OF THE RECORD MAX AND RECORD MIN DIVIDED BY 2.
AVERAGE TEMPERATURE IS THE SUM OF THE MEAN DAILY MAX AND MIN TEMPERATURE DIVIDED BY 2.

EXCEPTIONS

1. MEAN WIND SPEED IS THROUGH 1957.
2. THUNDERSTORMS AND HEAVY FOG ARE THROUGH 1964 AND MAY BE INCOMPLETE, DUE TO PART-TIME OPERATIONS.
3. MEAN SKY COVER, AND DAYS CLEAR-PARTLY CLOUDY-CLOUDY ARE THROUGH 1977.

RECORD MEANS ARE THROUGH THE CURRENT YEAR, BEGINNING IN 1894 FOR TEMPERATURE
1894 FOR PRECIPITATION
1940 FOR SNOWFALL

HEATING DEGREE DAYS Base 65 deg. F SAINT CLOUD, MINNESOTA

SEASON	JULY	AUG	SEP	OCT	NOV	DEC	JAN	FEB	MAR	APR	MAY	JUNE	TOTAL
1964-65	0	92	256	547	1014	1700	1850	1583	1523	722	247	49	9583
1965-66	14	57	435	499	1054	1260	2062	1469	988	769	410	68	9085
1966-67	1	68	221	591	1143	1543	1634	1668	1193	681	446	74	9263
1967-68	44	70	199	625	1039	1427	1680	1577	892	566	402	82	8603
1968-69	31	53	188	529	996	1627	1856	1374	1341	555	263	209	9022
1969-70	22	2	182	673	1004	1457	1927	1476	1336	692	334	45	9150
1970-71	11	17	231	569	1060	1548	1927	1414	1237	612	352	24	9002
1971-72	42	44	221	466	1023	1528	1907	1690	1265	752	237	96	9271
1972-73	48	76	297	717	1049	1775	1624	1333	849	670	348	48	8834
1973-74	14	11	255	419	988	1572	1779	1466	1235	623	418	97	8877
1974-75	4	99	366	558	1005	1369	1704	1483	1425	851	223	82	9169
1975-76	17	29	312	524	985	1533	1828	1238	1207	501	303	32	8509
1976-77	7	37	250	766	1249	1787	2085	1283	928	419	112	39	8962
1977-78	6	90	191	582	1103	1674	1911	1561	1169	683	238	92	9300
1978-79	31	26	175	579	1115	1736	2093	1677	1260	767	438	87	9984
1979-80	8	79	220	623	1081	1297	1679	1540	1286	529	227	83	8652
1980-81	4	35	279	683	939	1491	1535	1230	906	556	314		
1981-82					883	1577	2115	1479	1251	717	192	168	
1982-83	13	63	255	534	1176	1333	1568	1148	1044	723	412	114	8383
1983-84	8	0	229	579	987	2026	1729	1135	1293	571	325	34	8916
1984-85	16	29	348	492	1021	1580	1785	1431	984	475	185	162	8508
1985-86	8	76	300	642	1314	1892	1559	1480	1061	508	269	82	9191
1986-87	0	94	252	580	1170	1390	1454	1002	899	426	221	50	7538
1987-88	8	65	179	713	905	1348	1801	1585	1057	609	112	23	8405
1988-89	1	39	202	744	1071	1514	1513	1689	1276	669	305	88	9111
1989-90	0	23	231	576	1198	1773	1325	1281	1010	653	346	55	8471
1990-91	18	41	204	642	939	1591	1739	1197	1029	554	219	23	8196
1991-92	16	23	285	651	1279	1472	1437	1167	1048	703	259	133	8473
1992-93	83	110	252	648	1079	1495	1679	1425	1208	695	328	117	9119
1993-94	21	38	400	667	1134	1470							

COOLING DEGREE DAYS Base 65 deg. F SAINT CLOUD, MINNESOTA

YEAR	JAN	FEB	MAR	APR	MAY	JUNE	JULY	AUG	SEP	OCT	NOV	DEC	TOTAL
1969	0	0	0	0	47	21	186	247	48	4	0	0	553
1970	0	0	0	3	21	159	261	151	61	0	0	0	656
1971	0	0	0	0	1	160	102	125	66	0	0	0	454
1972	0	0	0	0	56	70	115	152	7	0	0	0	400
1973	0	0	0	0	1	78	141	186	34	2	0	0	442
1974	0	0	0	2	6	54	271	60	1	0	0	0	394
1975	0	0	0	0	37	97	283	117	7	1	0	0	542
1976	0	0	0	5	0	122	193	197	40	0	0	0	557
1977	0	0	0	5	101	73	199	44	6	0	0	0	428
1978	0	0	0	0	39	80	136	125	107	0	0	0	487
1979	0	0	0	0	14	52	140	87	31	0	0	0	324
1980	0	0	0	6	44	61	202	111	27	0	0	0	451
1981	0	0	0	0	9						0	0	
1982	0	0	0	0	16	14	194	138	24	0	0	0	386
1983	0	0	0	0	0	98	284	268	66	1	0	0	717
1984	0	0	0	0	8	84	180	210	13	1	0	0	496
1985	0	0	0	4	28	42	138	47	32	0	0	0	291
1986	0	0	0	0	37	92	182	64	19	0	0	0	394
1987	0	0	0	8	59	152	279	92	14	0	0	0	604
1988	0	0	0	0	76	211	304	208	22	0	0	0	821
1989	0	0	0	0	14	90	248	142	21	6	0	0	521
1990	0	0	0	25	8	110	121	121	60	0	0	0	445
1991	0	0	0	0	86	130	167	155	37	0	0	0	575
1992	0	0	0	0	42	45	34	62	9	0	0	0	192
1993	0	0	0	0	4	40	113	121	3	0	0	0	281

SNOWFALL (inches) SAINT CLOUD, MINNESOTA

SEASON	JULY	AUG	SEP	OCT	NOV	DEC	JAN	FEB	MAR	APR	MAY	JUNE	TOTAL
1964-65	0.0	0.0	0.0	0.4	4.4	7.6	6.6	11.1	51.7	6.1	T	0.0	87.9
1965-66	0.0	0.0	T	0.0	4.5	1.7	11.2	2.1	4.4	1.6	T	0.0	25.5
1966-67	0.0	0.0	0.0	1.4	5.3	8.3	26.1	12.5	3.3	0.2	T	0.0	57.1
1967-68	0.0	0.0	0.0	1.1	1.2	0.8	9.0	3.3	0.1	0.6	T	0.0	16.1
1968-69	0.0	0.0	0.0	T	5.7	25.4	22.9	6.8	4.0	2.1	T	0.0	66.9
1969-70	0.0	0.0	0.0	4.0	4.4	25.0	2.7	2.8	6.1	2.6	T	0.0	47.6
1970-71	0.0	0.0	0.0	0.4	9.6	2.5	13.1	21.6	4.3	4.4	3.2	0.0	59.1
1971-72	0.0	0.0	0.0	0.5	8.9	6.6	8.3	6.7	8.6	7.2	0.0	0.0	46.8
1972-73	0.0	0.0	0.0	T	0.4	14.0	5.8	4.0	0.3	T	0.0	0.0	24.5
1973-74	0.0	0.0	0.0	0.0	1.6	8.9	0.4	14.1	4.6	0.5	0.0	0.0	30.1
1974-75	0.0	0.0	0.0	T	2.8	7.5	29.9	5.1	19.4	0.7	0.0	0.0	65.4
1975-76	0.0	0.0	0.0	0.0	17.1	3.1	17.3	4.6	10.2	1.4	0.1	0.0	53.8
1976-77	0.0	0.0	0.0	0.9	0.9	4.8	12.0	3.5	8.6	1.3	0.0	0.0	32.0
1977-78	0.0	0.0	0.0	T	9.0	12.0	2.7	2.6	8.7	5.0	0.0	0.0	40.0
1978-79	0.0	0.0	0.0	T	4.7	17.8	17.8	13.8	12.1	0.4	0.3	0.0	66.9
1979-80	0.0	0.0	0.0	0.0	5.0	1.9	14.4	9.1	12.0	1.8	T	0.0	44.2
1980-81	0.0	0.0	0.0	T	0.6	1.8	6.7	5.5	T	1.9	0.0		
1981-82					0.0	0.0	18.1	2.6	12.1	1.6	0.0	0.0	
1982-83	0.0	0.0	0.0	0.5	16.9	13.9	4.9	1.7	9.3	6.1	0.0	0.0	53.3
1983-84	0.0	0.0	0.0	0.7	25.0	11.0	10.0	6.5	10.2	0.2	0.0	0.0	63.6
1984-85	0.0	0.0	T	1.2	0.9	8.0	8.6	4.1	22.8	T	0.0	0.0	45.6
1985-86	0.0	0.0	T	0.0	18.3	10.9	11.2	11.5	6.7	0.0	0.0	0.0	58.6
1986-87	0.0	0.0	0.0	0.1	6.9	3.9	3.9	1.5	0.4	T	0.0	0.0	16.7
1987-88	0.0	0.0	0.0	1.5	5.4	8.2	7.6	2.9	6.3	0.2	0.0	0.0	32.1
1988-89					18.2	9.0	12.6	11.2	13.1	1.0	T	0.0	
1989-90	0.0	0.0	T	0.0	12.9	4.0	0.1	10.3	2.8	2.4	0.0	0.0	32.5
1990-91	0.0	0.0	0.0	T	1.6	7.8	5.1	15.9	6.1	9.4	0.3	0.0	46.2
1991-92	0.0	0.0	0.0	3.1	21.3	8.5	7.2	2.7	5.7	6.0	0.0	0.0	54.5
1992-93	0.0	0.0	0.0	1.4	14.1	9.7	15.6	4.0	7.5	7.2	T	0.0	59.5
1993-94	0.0	0.0	0.0	0.1	13.5	9.3							
Record Mean	0.0	0.0	T	0.5	7.5	8.1	8.8	7.3	9.6	2.6	0.1	0.0	44.5

JACKSON, MISSISSIPPI

Jackson is located on the west bank of the Pearl River, about 45 miles east of the Mississippi River and 150 miles north of the Gulf of Mexico. The nearby terrain is gently rolling with no topographic features that appreciably influence the weather. The National Weather Service Office is nearly 7 miles east-northeast of the Jackson Post Office and over 5 miles southwest of the Ross Barnett Reservoir. Alluvial plains up to 3 miles wide extend along the river near Jackson, where some levees have been built on both sides of the river.

The climate is significantly humid during most of the year, with relatively short mild winters and long warm summers. The Gulf of Mexico has a moderating effect on the climate. Cold spells are fairly frequent in winter, but are usually of short duration. Sub-zero temperatures rarely occur. Temperatures occasionally exceed 80 degrees in mid-winter. In summer, temperatures reach 90 degrees or higher on about two-thirds of the days, 100 degree readings are infrequent. Extended periods of very hot weather are rare. On unusual occasions, temperatures at night may drop into the 50s, even in July or August.

Snowfall averages less than two inches per season, with nearly two-thirds of the seasons having only a trace of snow or none at all. Ice storms occasionally cause major damage to trees and power lines during the winter or early spring season. Rainfall is abundant and fairly well-distributed throughout the year. The area does not have a true dry season. However, the six-month period, June through November, is relatively dry in comparison with the December through May period when 60 percent of the annual precipitation can be expected.

Excessive rainfall may occur in any season. In spite of the normally abundant rainfall, fairly serious droughts occasionally occur during the summer or fall season. Tropical disturbances, including hurricanes and their remnants, are infrequent. However, those that pass near or visit the Mississippi Coast in the summer or early fall may bring several days of heavy rain to the Jackson area.

Thunderstorms can be expected on an average of 65 days a year, usually occurring in each month. They are most frequent in summer when they occur on about one-third of the days. At other times of the year, thunderstorms are usually associated with passing weather systems and are likely to be attended by higher winds than in summer. Severe thunderstorms normally affect portions of the metropolitan area a few times each year.

NORMALS, MEANS, AND EXTREMES

JACKSON, MISSISSIPPI

LATITUDE: 32°19'N LONGITUDE: 90°05' W ELEVATION: FT. GRND 291 BARO 297 TIME ZONE: CENTRAL WBAN: 03940

	(a)	JAN	FEB	MAR	APR	MAY	JUNE	JULY	AUG	SEP	OCT	NOV	DEC	YEAR
TEMPERATURE °F:														
Normals														
-Daily Maximum		55.6	60.1	69.3	77.4	84.0	90.6	92.4	92.0	88.0	79.1	69.2	59.5	76.4
-Daily Minimum		32.7	35.7	44.1	51.9	60.0	67.1	70.5	69.7	63.7	50.3	42.3	36.1	52.0
-Monthly		44.1	47.9	56.7	64.6	72.0	78.8	81.5	80.9	75.9	64.7	55.8	47.8	64.2
Extremes														
-Record H ghest	30	82	85	89	94	99	105	106	102	104	95	88	84	106
-Year		1972	1989	1982	1987	1964	1988	1980	1981	1980	1986	1971	1978	JUL 1980
-Record Lowest	30	2	11	15	27	38	47	51	**55**	35	**26**	17	4	2
-Year		1985	1970	1980	1987	1971	1984	1967	**1992**	1967	**1993**	1976	1989	JAN 1985
NORMAL DEGREE DAYS:														
Heating (base 65°F)		656	485	285	87	7	0	0	0	6	104	295	542	2467
Cooling (base 65°F)		8	6	28	75	224	414	512	493	333	94	19	9	2215
% OF POSSIBLE SUNSHINE	29	48	54	60	65	62	70	65	66	64	68	57	48	61
MEAN SKY COVER (tenths)														
Sunrise - Sunset	30	6.7	6.3	6.2	5.8	5.9	5.5	5.8	5.4	5.3	4.5	5.6	6.4	5.8
MEAN NUMBER OF DAYS:														
Sunrise to Sunset														
-Clear	30	7.8	7.9	8.8	9.4	8.3	8.8	7.5	9.0	10.4	14.8	9.8	8.5	111.2
-Partly Cloudy	30	5.9	6.1	6.9	7.4	10.4	12.0	13.2	12.9	9.2	6.6	7.3	6.3	104.2
-Cloudy	30	17.2	14.3	15.3	13.1	12.3	9.2	10.3	9.1	10.4	9.7	12.9	16.1	149.9
Precipitation														
.01 inches or more	30	10.9	9.3	10.1	8.5	9.5	8.4	10.4	9.6	8.3	6.2	8.3	10.0	109.6
Snow,Ice pellets,hail														
1.0 inches or more	30	0.2	0.1	0.1	0.0	0.0	0.0	0.0	0.0	0.0	0.0	0.0	0.0	0.3
Thunderstorms	30	1.9	2.7	5.5	5.7	7.3	8.7	12.6	10.5	5.3	2.1	2.7	2.5	67.5
Heavy Fog Visibility														
1/4 mile or less	30	3.4	2.1	1.7	1.5	1.0	0.6	1.2	1.5	1.5	2.2	2.3	3.1	22.1
Temperature °F														
-Maximum														
90° and above	30	0.0	0.0	0.0	0.3	4.8	18.7	24.2	22.6	12.2	1.0	0.0	0.0	83.8
32° and below	30	0.8	0.3	0.0	0.0	0.0	0.0	0.0	0.0	0.0	0.0	0.*	0.3	1.4
-Minimum														
32° and below	30	14.7	10.9	4.1	0.4	0.0	0.0	0.0	0.0	0.0	0.5	6.2	12.0	48.8
0° and below	30	0.0	0.0	0.0	0.0	0.0	0.0	0.0	0.0	0.0	0.0	0.0	0.0	0.0
AVG. STATION PRESS. (mb)	21	1009.4	1007.8	1004.9	1004.4	1003.3	1003.9	1005.2	1004.9	1005.1	1007.0	1007.8	1009.2	1006.1
RELATIVE HUMIDITY (%)														
Hour 00	30	84	82	82	85	87	88	91	91	90	89	87	84	87
Hour 06 (Local Time)	30	87	87	87	91	92	92	94	95	94	93	90	88	91
Hour 12	30	65	60	56	54	56	56	59	58	58	53	57	63	58
Hour 18	30	71	63	58	58	61	61	66	67	70	72	73	74	66
PRECIPITATION (inches):														
Water Equivalent														
-Normal		5.24	4.70	5.82	5.57	5.05	3.18	4.51	3.77	3.55	3.26	4.81	5.91	55.37
-Maximum Monthly	30	14.10	10.28	15.13	15.95	10.82	8.17	13.25	**8.33**	9.61	9.13	9.98	17.70	17.70
-Year		1979	1987	1976	1991	1967	1989	1979	**1992**	1965	1970	1977	1982	DEC 1982
-Minimum Monthly	30	0.75	1.43	2.05	1.21	0.29	0.10	1.04	0.61	0.56	0.00	0.51	0.91	0.00
-Year		1986	1976	1966	1987	1988	1988	1987	1990	1969	1963	1985	1980	OCT 1963
-Maximum in 24 hrs	30	5.63	3.46	4.69	8.42	3.43	3.38	5.37	**4.79**	5.86	6.99	4.34	6.71	8.42
-Year		1979	1974	1991	1979	1989	1989	1981	**1992**	1965	1975	1983	1982	APR 1979
Snow,Ice pellets,hail														
-Maximum Monthly	30	6.3	3.6	5.3	1.1	0.0	0.0	**0.0**	**0.0**	**0.0**	**0.0**	0.2	3.1	6.3
-Year		1982	1968	1968	1987							1976	1963	JAN 1982
-Maximum in 24 hrs	30	6.0	3.6	5.3	1.1	0.0	0.0	**0.0**	**0.0**	**0.0**	**0.0**	0.2	1.8	6.0
-Year		1982	1968	1968	1987							1976	1963	JAN 1982
WIND:														
Mean Speed (mph)	30	8.4	8.6	9.2	8.5	7.3	6.4	5.9	5.6	6.4	6.5	7.6	8.3	7.4
Prevailing Direction														
Fastest Obs. 1 Min.														
-Direction (!!)	17	35	19	16	12	22	35	33	17	14	34	14	35	35
-Speed (MPH)	17	46	36	44	35	35	40	44	37	30	26	41	35	46
-Year		1990	1986	1986	1991	1987	1984	1982	1981	1991	1990	1987	1976	JAN 1990
Peak Gust														
-Direction (!!)	10	N	N	SE	SW	W	SW	**NE**	**E**	W	NE	SW	W	SW
-Speed (mph)	10	59	51	51	63	49	55	**51**	**56**	52	40	69	48	69
-Date		1990	1984	1987	1985	1990	1992	**1992**	**1992**	1984	1985	1987	1987	NOV 1987

PRECIPITATION (inches)

JACKSON, MISSISSIPPI

YEAR	JAN	FEB	MAR	APR	MAY	JUNE	JULY	AUG	SEP	OCT	NOV	DEC	ANNUAL
1964	5.21	2.32	10.92	11.88	2.18	2.45	5.17	6.43	3.67	7.83	6.44	6.95	71.45
1965	2.86	7.90	6.66	1.25	1.36	2.54	1.92	3.61	9.61	0.84	1.79	3.65	43.99
1966	8.23	7.84	2.05	5.76	7.96	1.45	3.14	6.65	4.87	1.63	3.66	4.83	58.07
1967	1.86	3.56	2.32	1.66	10.82	3.76	4.05	3.49	1.80	2.78	0.93	8.68	45.71
1968	4.56	2.54	2.83	7.20	7.54	1.02	2.24	3.90	1.95	0.28	5.55	5.58	45.19
1969	0.86	3.02	4.90	6.59	1.52	1.29	5.46	3.35	0.56	2.26	1.87	7.22	38.90
1970	2.09	2.63	5.40	2.75	2.43	2.54	2.34	5.64	6.48	9.13	2.70	4.23	48.36
1971	3.02	5.68	7.68	6.86	8.05	3.40	6.28	2.64	6.00	0.09	2.54	9.82	62.06
1972	5.94	3.09	5.57	2.44	4.52	2.01	3.31	2.84	5.04	2.08	3.52	9.67	50.03
1973	4.59	4.23	6.12	9.44	5.96	0.32	1.99	2.38	4.44	2.72	6.15	6.71	55.05
1974	11.00	6.72	3.50	6.74	3.01	3.39	1.54	6.17	5.06	1.74	4.12	7.22	60.21
1975	4.57	6.18	4.86	5.07	6.53	7.44	9.81	6.21	2.68	8.25	4.34	4.29	70.23
1976	3.64	1.43	15.13	2.08	8.01	2.80	4.96	5.26	3.78	3.52	3.34	3.44	57.39
1977	6.18	2.26	6.41	7.98	0.74	2.17	6.22	1.45	3.93	2.79	9.98	3.47	53.58
1978	5.32	2.36	3.37	3.54	10.48	1.03	3.65	1.80	2.90	1.21	3.30	8.37	47.33
1979	14.10	8.35	4.67	14.38	5.52	4.38	13.25	7.44	5.93	1.76	8.79	4.18	92.75
1980	7.53	3.19	13.57	14.33	6.60	1.74	2.91	1.45	3.25	3.47	4.11	0.91	63.06
1981	1.41	2.63	6.19	1.26	6.64	3.66	6.51	2.81	3.51	5.12	1.97	4.90	46.61
1982	4.48	5.22	5.13	6.59	0.77	6.27	9.29	4.97	1.05	6.73	7.43	17.70	75.63
1983	8.17	6.55	6.00	15.53	9.41	2.93	1.70	3.70	2.70	1.52	8.11	6.95	73.27
1984	2.64	4.64	4.84	3.96	5.61	3.18	3.07	4.56	0.93	7.68	6.48	2.17	49.76
1985	4.05	7.55	3.13	3.31	0.86	1.74	4.43	7.06	3.94	7.17	0.51	3.61	47.36
1986	0.75	1.53	3.34	1.75	10.00	3.72	4.78	2.03	2.63	5.10	9.40	4.98	50.01
1987	4.66	10.28	5.47	1.21	4.98	6.17	1.04	4.03	1.50	0.27	4.20	3.50	47.31
1988	2.25	3.89	7.46	5.37	0.29	0.10	2.73	3.02	2.28	6.14	5.66	4.80	43.99
1989	4.38	2.52	4.53	2.13	7.92	8.17	4.47	1.74	5.40	0.23	6.86	4.20	52.55
1990	12.17	8.30	3.55	3.66	6.34	1.46	2.84	0.61	4.83	1.24	3.33	5.71	54.04
1991	4.98	5.30	7.50	15.95	7.11	3.35	3.32	1.49	3.45	2.95	2.39	5.27	63.06
1992	4.27	3.83	2.11	1.48	1.22	4.75	4.48	8.33	3.80	1.55	8.27	4.32	48.41
1993	4.72	3.40	3.84	4.09	4.31	3.79	7.00	2.47	2.04	4.08	5.39	2.89	48.02
Record Mean	4.95	4.77	5.54	5.39	4.64	3.75	4.52	3.60	2.87	2.52	4.06	5.47	52.09

AVERAGE TEMPERATURE (deg. F)

JACKSON, MISSISSIPPI

YEAR	JAN	FEB	MAR	APR	MAY	JUNE	JULY	AUG	SEP	OCT	NOV	DEC	ANNUAL
#1964	45.3	44.4	56.3	68.1	74.7	80.7	81.3	81.7	77.1	62.5	59.4	50.5	65.2
1965	48.5	47.2	50.2	67.8	74.0	77.7	81.5	79.5	76.3	62.6	60.1	49.4	64.6
1966	40.8	46.3	54.7	64.8	70.9	76.1	82.4	78.3	74.4	61.1	56.2	46.5	62.7
1967	46.4	44.2	60.5	69.3	69.6	78.3	77.3	77.0	69.8	61.4	52.7	50.1	63.1
1968	43.4	39.9	53.9	65.9	71.2	79.6	80.7	80.9	73.3	66.2	52.7	45.7	62.8
1969	48.0	48.1	49.3	65.1	72.3	80.3	83.5	79.6	75.2	66.2	52.4	47.1	63.9
1970	41.2	46.3	54.0	68.1	72.9	79.1	80.8	81.6	80.1	64.7	52.8	52.7	64.5
1971	48.7	49.1	52.4	62.2	67.9	79.9	81.0	80.5	78.3	70.1	54.7	57.9	65.2
1972	51.5	51.2	58.8	66.4	71.9	79.6	80.4	82.8	81.5	68.0	52.0	50.1	66.2
1973	44.3	46.9	62.1	62.5	70.9	81.3	83.7	80.1	77.6	68.6	61.7	49.3	65.7
1974	55.1	50.1	62.4	63.1	73.8	74.8	80.4	79.3	72.0	64.1	55.6	50.4	65.1
1975	51.8	52.9	56.8	63.5	74.1	78.3	81.2	80.6	72.3	66.2	56.6	47.4	65.1
1976	44.3	56.7	60.0	65.4	67.9	76.7	80.8	80.2	74.8	59.9	47.2	45.0	63.2
1977	35.3	49.3	59.7	66.3	75.5	81.7	82.7	82.0	78.4	62.0	57.1	47.5	64.8
1978	36.5	39.6	52.5	65.4	71.9	80.0	83.4	82.0	78.4	63.6	60.3	48.8	63.6
1979	38.3	46.0	57.4	64.8	70.3	76.7	80.2	79.7	73.8	64.1	50.9	46.2	62.4
1980	47.4	45.5	54.3	61.6	72.8	80.7	85.8	84.7	82.4	61.4	53.2	46.2	64.7
1981	41.8	48.7	55.1	71.0	69.9	81.8	83.6	82.6	73.2	65.4	58.6	46.6	64.9
1982	47.2	48.8	62.1	63.8	74.7	79.5	82.4	81.8	74.1	65.1	56.4	52.9	65.7
1983	43.7	47.4	53.9	60.1	70.7	76.6	82.9	82.7	74.1	65.8	55.1	41.8	62.9
1984	39.9	49.7	56.5	63.9	71.0	78.9	80.6	79.9	74.6	71.2	53.6	58.3	64.8
1985	38.0	44.8	61.3	65.2	72.2	78.8	80.5	80.3	74.0	68.0	62.5	42.7	64.0
1986	45.3	52.6	57.0	64.1	72.9	80.4	83.2	80.5	79.8	66.1	60.1	46.6	65.7
1987	44.4	51.3	55.5	62.2	75.9	79.0	82.0	82.9	75.2	59.9	56.6	52.1	64.8
1988	41.7	47.5	55.8	64.6	70.8	79.8	82.5	82.6	77.9	61.6	59.3	49.3	64.5
1989	52.5	48.2	58.8	63.2	72.0	78.4	80.6	80.8	74.1	63.9	57.1	40.4	64.2
1990	51.0	56.2	60.0	63.7	71.8	81.3	81.5	82.3	78.4	63.7	58.0	51.8	66.6
1991	46.6	51.7	59.3	68.0	75.6	79.9	82.9	81.4	76.0	67.2	52.4	52.0	66.1
1992	45.6	53.7	57.6	64.1	71.1	77.7	82.0	77.8	75.6	65.0	52.7	50.1	64.4
1993	49.9	49.2	54.4	61.0	71.0	80.9	82.5	82.9	75.9	62.3	52.1	45.8	64.0
Record Mean	47.4	50.5	57.3	65.1	72.5	79.6	81.9	81.4	76.8	66.0	55.7	49.1	65.3
Max	57.9	61.6	68.9	76.9	84.0	90.8	92.6	92.4	88.4	79.3	68.1	59.8	76.7
Min	36.8	39.4	45.7	53.2	61.0	68.3	71.2	70.4	65.1	52.7	43.2	38.4	53.8

REFERENCE NOTES

JACKSON, MISSISSIPPI

GENERAL
T - TRACE AMOUNT.
BLANK ENTRIES DENOTE MISSING/UNREPORTED DATA.
# INDICATES A STATION OR INSTRUMENT RELOCATION.

SPECIFIC

PM - INCLUDES LAST DAY OF PREVIOUS MONTH
ASOS - AUTOMATED SURFACE OBSERVING SYSTEM IN OPERATION DURING THESE MONTHS.

(a) - LENGTH OF RECORD IN YEARS, ALTHOUGH INDIVIDUAL MONTHS MAY BE MISSING.
0.* OR * - THE VALUE IS BETWEEN 0.0 AND 0.05
NORMALS - BASED ON THE 1961-1990 RECORD PERIOD.
EXTREMES - DATES ARE THE MOST RECENT OCCURRENCE.
WIND DIR.- NUMERALS SHOW TENS OF DEGRESS CLOCKWISE FROM TRUE NORTH. "00" INDICATES CALM.
RESULTANT DIRECTIONS ARE GIVEN TO WHOLE DEGREES.
BOLD VALUES INDICATE EXTREME VALUES WHICH OCCURRED AFTER THE ASOS SYSTEM WAS COMMISSIONED.

RECORD = PERIOD OF RECORD
RECORD MEAN PRECIPITATION IS THE MEAN OF ALL DAILY PRECIPITATION AMOUNTS DURING THE PERIOD OF RECORD.
RECORD MAX(MIN) TEMPERATURE IS THE MEAN OF ALL DAILY MAX(MIN) TEMPERATURES DURING THE PERIOD OF RECORD.
RECORD MEAN TEMPERATURE IS THE SUM OF THE RECORD MAX AND RECORD MIN DIVIDED BY 2.
AVERAGE TEMPERATURE IS THE SUM OF THE MEAN DAILY MAX AND MIN TEMPERATURE DIVIDED BY 2.

EXCEPTIONS

RECORD MEANS ARE THROUGH THE CURRENT YEAR, BEGINNING IN 1909 FOR TEMPERATURE
1909 FOR PRECIPITATION
1964 FOR SNOWFALL

HEATING DEGREE DAYS Base 65 deg. F JACKSON, MISSISSIPPI

SEASON	JULY	AUG	SEP	OCT	NOV	DEC	JAN	FEB	MAR	APR	MAY	JUNE	TOTAL
1964-65	0	0	0	124	219	465	512	500	459	48	2	0	2329
1965-66	0	0	9	129	173	481	747	520	326	98	13	0	2496
1966-67	0	0	0	155	270	581	570	574	192	20	26	0	2388
1967-68	0	0	35	141	368	461	662	721	362	64	12	0	2826
1968-69	0	0	0	93	369	592	523	470	476	65	3	0	2591
1969-70	0	0	0	95	372	550	741	516	340	72	23	0	2709
1970-71	0	0	0	85	367	388	507	444	394	147	40	0	2372
1971-72	0	0	0	19	328	235	431	410	216	76	1	0	1716
1972-73	0	0	4	71	400	466	634	503	135	146	19	0	2378
1973-74	0	0	0	61	173	486	327	419	165	115	1	0	1747
1974-75	0	0	6	79	308	465	429	348	293	138	0	0	2066
1975-76	0	0	26	59	304	543	635	252	201	59	18	0	2097
1976-77	0	0	0	197	528	614	913	435	196	43	1	0	2927
1977-78	0	0	0	136	246	538	881	706	377	73	14	0	2971
1978-79	0	0	0	98	164	519	821	531	262	69	20	0	2484
1979-80	0	0	0	98	421	580	540	570	337	124	3	0	2673
1980-81	0	0	0	158	363	575	711	454	312	20	18	0	2611
1981-82	0	0	16	104	217	568	569	451	199	123	4	0	2251
1982-83	0	0	19	120	286	421	653	487	347	176	12	0	2521
1983-84	0	0	21	83	315	716	770	440	289	113	29	0	2776
1984-85	0	0	12	27	350	249	832	562	158	92	2	0	2284
1985-86	0	0	14	54	148	685	606	360	261	90	8	0	2226
1986-87	0	0	0	79	183	566	632	378	297	150	0	0	2285
1987-88	0	0	0	161	271	408	716	502	292	74	4	0	2428
1988-89	0	0	0	134	208	484	392	495	239	135	20	0	2107
1989-90	0	0	11	109	265	754	429	261	209	125	9	0	2172
1990-91	0	0	5	145	227	427	566	368	229	35	2	0	2004
1991-92	0	0	4	67	404	403	593	323	233	118	18	0	2163
1992-93	0	0	1	45	364	452	461	437	327	163	5	0	2255
1993-94	0	0	10	161	401	588							

COOLING DEGREE DAYS Base 65 deg. F JACKSON, MISSISSIPPI

YEAR	JAN	FEB	MAR	APR	MAY	JUNE	JULY	AUG	SEP	OCT	NOV	DEC	TOTAL
1969	3	0	0	75	235	465	577	460	312	137	3	0	2267
1970	7	0	5	173	274	431	496	522	460	81	7	18	2474
1971	9	7	10	69	138	453	503	487	406	181	29	20	2312
1972	20	17	30	126	221	445	487	561	506	169	15	16	2613
1973	1	0	51	77	210	498	583	477	382	181	81	5	2546
1974	25	8	91	64	279	298	487	451	223	57	34	19	2036
1975	26	17	46	96	290	407	509	487	255	101	59	7	2300
1976	0	17	52	76	116	357	497	481	300	47	0	0	1943
1977	0	3	39	89	331	508	555	534	409	47	16	4	2535
1978	2	0	0	92	235	456	576	536	409	60	31	24	2421
1979	0	2	35	69	190	357	482	461	269	76	4	2	1947
1980	0	11	11	28	253	476	652	619	527	52	14	1	2644
1981	0	3	12	207	174	514	582	555	267	127	30	2	2473
1982	25	3	116	93	310	443	544	530	297	130	35	51	2577
1983	0	0	13	35	194	354	564	555	300	115	24	6	2160
1984	0	5	35	87	221	422	489	470	310	227	19	49	2334
1985	0	3	54	105	233	424	486	480	290	153	79	0	2307
1986	0	21	18	70	259	469	572	487	446	122	44	2	2510
1987	0	0	9	72	344	429	533	560	311	14	22	15	2309
1988	0	1	14	67	193	452	552	552	395	34	45	3	2308
1989	8	30	54	87	245	406	489	498	293	84	36	0	2230
1990	3	25	60	92	226	494	515	544	412	113	24	24	2532
1991	0	3	59	129	337	456	562	515	341	142	37	7	2588
1992	0	1	10	97	213	383	532	403	324	51	2	0	2016
1993	0	1	6	48	197	485	547	562	340	82	19	0	2287

SNOWFALL (inches) JACKSON, MISSISSIPPI

SEASON	JULY	AUG	SEP	OCT	NOV	DEC	JAN	FEB	MAR	APR	MAY	JUNE	TOTAL
1964-65	0.0	0.0	0.0	0.0	T	T	T	T	T	0.0	0.0	0.0	T
1965-66	0.0	0.0	0.0	0.0	0.0	0.0	T	0.0	0.0	0.0	0.0	0.0	T
1966-67	0.0	0.0	0.0	0.0	T	T	0.0	T	0.0	0.0	0.0	0.0	T
1967-68	0.0	0.0	0.0	0.0	0.0	T	T	3.6	5.3	0.0	0.0	0.0	8.9
1968-69	0.0	0.0	0.0	0.0	0.0	0.0	0.0	0.0	0.0	0.0	0.0	0.0	0.0
1969-70	0.0	0.0	0.0	0.0	0.0	T	T	0.0	T	0.0	0.0	0.0	T
1970-71	0.0	0.0	0.0	0.0	0.0	0.0	T	T	T	T	0.0	0.0	T
1971-72	0.0	0.0	0.0	0.0	T	0.0	0.0	T	0.0	0.0	0.0	0.0	T
1972-73	0.0	0.0	0.0	0.0	0.0	0.0	T	T	0.0	0.0	0.0	0.0	T
1973-74	0.0	0.0	0.0	0.0	0.0	0.3	0.0	0.0	0.0	0.0	0.0	0.0	0.3
1974-75	0.0	0.0	0.0	0.0	T	0.0	T	0.0	T	0.0	0.0	0.0	T
1975-76	0.0	0.0	0.0	0.0	T	T	T	0.0	0.0	0.0	0.0	0.0	T
1976-77	0.0	0.0	0.0	0.0	0.2	0.0	5.8	0.0	0.0	0.0	0.0	0.0	6.0
1977-78	0.0	0.0	0.0	0.0	0.0	0.0	1.1	T	0.1	0.0	0.0	0.0	1.2
1978-79	0.0	0.0	0.0	0.0	0.0	T	T	T	0.0	0.0	0.0	0.0	T
1979-80	0.0	0.0	0.0	0.0	0.0	0.0	T	T	T	0.0	0.0	0.0	T
1980-81	0.0	0.0	0.0	0.0	0.0	T	T	T	0.0	0.0	0.0	0.0	T
1981-82	0.0	0.0	0.0	0.0	0.0	T	6.3	T	T	0.0	0.0	0.0	6.3
1982-83	0.0	0.0	0.0	0.0	0.0	0.0	T	T	0.0	T	0.0	0.0	T
1983-84	0.0	0.0	0.0	0.0	0.0	T	T	T	0.0	0.0	0.0	0.0	T
1984-85	0.0	0.0	0.0	0.0	0.0	T	0.3	1.4	0.0	0.0	0.0	0.0	1.7
1985-86	0.0	0.0	0.0	0.0	0.0	T	0.0	T	0.0	0.0	0.0	0.0	T
1986-87	0.0	0.0	0.0	0.0	0.0	0.0	T	0.0	T	1.1	0.0	0.0	1.1
1987-88	0.0	0.0	0.0	0.0	0.0	0.0	T	T	0.0	0.0	0.0	0.0	T
1988-89	0.0	0.0	0.0	0.0	0.0	T	0.0	T	T	T	0.0	0.0	T
1989-90	0.0	0.0	0.0	0.0	T	0.1	0.0	0.0	0.0	0.0	0.0	0.0	0.1
1990-91	0.0	0.0	0.0	0.0	0.0	T	0.0	0.0	0.0	T	0.0	0.0	T
1991-92	0.0	0.0	0.0	0.0	T	0.0	0.2	0.0	0.0	0.0	0.0	0.0	0.2
1992-93	0.0	0.0	0.0	0.0	T	0.0	T	0.0	1.6	0.0	0.0	0.0	1.6
1993-94	0.0	0.0	0.0	0.0	0.0	T							
Record Mean	0.0	0.0	0.0	0.0	T	T	0.5	0.2	0.2	T	0.0	0.0	1.0

MERIDIAN, MISSISSIPPI

Mild winters and warm summers describe the general temperature pattern for Meridian. However, the terrain features exert a pronounced influence, particularly during the winter months. The hills to the north, east, and west leave Meridian in a valley. During periods of near calm winds, cold air drainage brings temperatures which may be as much as 10 degrees lower than for other locations in the area. January is usually the coldest month, followed closely by December and February. Sub-zero temperatures are very rare. Summer temperatures are consistently warm. Prolonged periods with above 100 degrees readings are rare.

Precipitation is distributed evenly throughout the year. The widespread rains of the winter months reach a maximum in March. Spring showers reach a minimum in May, followed by localized summer thunderstorms in July and August. The driest period of the year is in late September and October, followed by the onset of winter-type precipitation in late November. This pattern is ideally suited to agricultural operations since the spring rains are conducive to crop growth in the early stages and the dry period in the fall is ideal for harvesting operations. Summer thunderstorms are highly localized and occur on one in three days during July and August.

The long growing season averages 235 days, nearly eight months. The average date of the first occurrence of a temperature as low as 32 degrees in autumn is November 7, and the occurrence of 32 degrees before October 20 is very rare. The average date of the last occurrence of 32 degrees in spring is March 19, although 32 degrees has been recorded in late April. Some portions of the area not affected by cold air drainage may have slightly longer average growing seasons.

The nearby Gulf of Mexico provides an abundant supply of moisture to the Meridian area and results in high humidities for prolonged periods. Humidities of greater than 90 percent occur nightly during every month except for short periods during the autumn and winter when cool continental air is flowing from the north. Lowest humidities are observed during the early afternoons, but seldom reach below 40 percent except for short periods.

March is generally the windiest month of the year due to the frequent occurrence of late winter and spring storms across the Gulf States. October has the lowest average wind speed. Prevailing winds are from the north and northeast during the autumn and winter months, and from the south and southwest during the spring and summer. Local thunderstorms produce short periods of high winds during the spring and summer months and can be quite destructive. Severe thunderstorms and tornadoes have caused considerable loss of life and property in this area. The highest sustained wind speed recorded was 50 mph, but there have been short periods with winds in excess of 50 mph.

Fifty years of record show that December, January, and February receive the smallest amount of possible sunshine. About 40 to 45 percent of the days during these months are cloudy. Sunshine reaches a maximum during the dry period in the fall, September and October. These months are characterized by long periods of cloudless skies.

Thunderstorms normally occur during every month in the year, but most occur during the summer months. These summer thunderstorms provide most of the precipitation during the crop growing season. Cloudiness associated with these thunderstorms brings relief from the oppressive heat. Although thunderstorm occurrence is high, hail damage is infrequent and usually confined to a small area.

NORMALS, MEANS, AND EXTREMES

MERIDIAN, MISSISSIPPI

LATITUDE: 32°20'N LONGITUDE: 88°45' W ELEVATION: FT. GRND 290 BARO 295 TIME ZONE: CENTRAL WBAN: 13865

	(a)	JAN	FEB	MAR	APR	MAY	JUNE	JULY	AUG	SEP	OCT	NOV	DEC	YEAR
TEMPERATURE °F:														
Normals														
-Daily Maximum		56.4	61.2	69.6	77.4	83.6	90.0	92.1	91.8	86.9	77.7	68.6	60.1	76.3
-Daily Minimum		33.4	36.6	43.5	50.9	58.9	66.1	69.9	69.2	63.9	50.5	42.4	36.6	51.8
-Monthly		45.0	48.9	56.6	64.1	71.3	78.1	81.0	80.6	75.4	64.1	55.5	48.4	64.1
Extremes														
-Record Highest	48	83	85	90	95	99	104	107	104	105	97	87	82	107
-Year		1950	1982	1974	1987	1951	1988	1980	1990	1990	1954	1946	1982	JUL 1980
-Record Lowest	48	0	8	15	28	38	42	55	53	34	24	16	2	0
-Year		1962	1951	1980	1987	1971	1984	1967	1952	1967	1952	1976	1989	JAN 1962
NORMAL DEGREE DAYS:														
Heating (base 65°F)		628	456	292	99	14	0	0	0	6	123	301	525	2444
Cooling (base 65°F)		8	5	31	72	210	393	496	484	318	95	16	10	2138
% OF POSSIBLE SUNSHINE														
MEAN SKY COVER (tenths)														
Sunrise - Sunset	48	6.7	6.3	6.3	5.7	5.8	5.6	6.0	5.4	5.5	4.6	5.6	6.3	5.8
MEAN NUMBER OF DAYS:														
Sunrise to Sunset														
-Clear	48	7.4	7.4	8.4	9.2	8.6	8.2	5.8	9.4	10.4	14.3	10.4	8.7	108.2
-Partly Cloudy	48	6.5	6.4	7.4	8.1	10.6	12.6	15.3	13.0	9.1	7.3	7.0	6.8	110.0
-Cloudy	48	17.1	14.5	15.3	12.7	11.8	9.2	9.8	8.6	10.8	9.5	12.7	15.5	147.4
Precipitation														
.01 inches or more	48	10.6	9.1	9.8	8.6	8.6	8.3	11.1	9.0	7.5	5.4	7.8	9.8	105.7
Snow,Ice pellets,hail														
1.0 inches or more	48	0.2	0.1	0.1	0.*	0.0	0.0	0.0	0.0	0.0	0.0	0.0	0.*	0.4
Thunderstorms	48	1.6	2.7	4.6	5.5	6.4	7.4	11.6	8.5	3.9	1.6	2.3	1.9	57.9
Heavy Fog Visibility														
1/4 mile or less	48	2.9	2.1	2.0	2.5	1.9	1.4	1.9	1.9	1.8	2.6	3.0	2.8	26.8
Temperature °F														
-Maximum														
90° and above	29	0.0	0.0	0.*	0.3	4.0	16.4	23.9	22.6	11.9	0.9	0.0	0.0	80.1
32° and below	29	0.3	0.1	0.0	0.0	0.0	0.0	0.0	0.0	0.0	0.0	0.0	0.2	0.7
-Minimum														
32° and below	29	15.4	11.2	4.9	0.4	0.0	0.0	0.0	0.0	0.0	0.6	6.3	12.5	51.4
0° and below	29	0.0	0.0	0.0	0.0	0.0	0.0	0.0	0.0	0.0	0.0	0.0	0.0	0.0
AVG. STATION PRESS.(mb)	21	1010.2	1008.8	1006.1	1005.6	1004.5	1005.0	1006.3	1006.1	1006.2	1008.2	1008.9	1010.3	1007.2
RELATIVE HUMIDITY (%)														
Hour 00	29	83	81	81	86	88	89	90	90	89	88	86	83	86
Hour 06 (Local Time)	29	86	85	87	90	91	90	92	92	91	90	88	86	89
Hour 12	29	60	55	52	51	55	54	58	56	56	51	54	59	55
Hour 18	29	68	59	55	55	60	60	66	66	70	74	72	71	65
PRECIPITATION (inches):														
Water Equivalent														
-Normal		5.15	5.43	6.75	5.46	4.42	3.63	5.15	3.58	3.52	3.06	4.49	6.07	56.71
-Maximum Monthly	48	12.14	15.95	16.47	16.82	9.79	8.91	15.29	10.28	10.24	10.65	13.93	14.79	16.82
-Year		1947	1990	1976	1964	1980	1989	1959	1992	1957	1970	1948	1973	APR 1964
-Minimum Monthly	48	1.21	1.67	1.27	0.91	0.27	0.71	1.07	0.72	0.10	0.00	0.38	1.10	0.00
-Year		1986	1947	1955	1987	1951	1968	1952	1989	1982	1963	1956	1980	OCT 1963
-Maximum in 24 hrs	48	5.74	9.23	7.00	6.36	5.84	3.12	6.95	5.29	5.21	6.04	4.50	8.13	9.23
-Year		1987	1990	1979	1964	1952	1992	1959	1992	1988	1970	1957	1973	FEB 1990
Snow,Ice pellets,hail														
-Maximum Monthly	48	5.8	3.1	5.7	2.7	T	0.0	T	0.0	0.0	0.0	T	17.6	17.6
-Year		1948	1960	1993	1987	1991		1989				1976	1963	DEC 1963
-Maximum in 24 hrs	48	4.7	3.1	5.7	2.7	T	0.0	T	0.0	0.0	0.0	T	15.0	15.0
-Year		1948	1960	1993	1987	1991		1989				1976	1963	DEC 1963
WIND:														
Mean Speed (mph)	34	7.0	7.5	7.9	7.1	5.9	5.1	4.8	4.6	5.3	5.2	6.2	6.9	6.1
Prevailing Direction through 1963		NE	S	NE	S	S	S	S	S	N	N	N	N	S
Fastest Obs. 1 Min.														
-Direction (!!)	34	25	02	29	27	17	22	03	14	02	02	19	30	22
-Speed (MPH)	34	30	35	35	35	35	46	40	32	45	35	41	35	46
-Year		1975	1963	1989	1985	1967	1989	1964	1968	1979	1964	1992	1988	JUN 1989
Peak Gust														
-Direction (!!)	10	SW	NW	N	NW	SW	SW	W	SE	SE	N	S	W	W
-Speed (mph)	10	49	55	64	49	47	64	60	56	38	43	48	68	68
-Date		1988	1986	1988	1990	1992	1989	1989	1992	1985	1988	1992	1990	DEC 1990

PRECIPITATION (inches) — MERIDIAN, MISSISSIPPI

YEAR	JAN	FEB	MAR	APR	MAY	JUNE	JULY	AUG	SEP	OCT	NOV	DEC	ANNUAL
1964	4.78	4.08	9.11	16.82	0.97	4.87	4.69	3.10	2.15	4.18	4.80	9.00	68.55
1965	3.46	5.88	5.04	1.65	0.30	5.61	4.54	1.82	2.24	1.45	1.14	4.28	37.41
1966	7.26	10.54	3.37	4.32	8.49	1.02	2.93	7.56	3.32	1.05	2.67	5.87	58.40
1967	2.19	3.65	2.45	2.22	6.96	2.00	4.84	6.69	2.23	2.40	0.66	11.20	47.49
1968	3.70	2.21	2.87	5.96	3.03	0.71	3.04	5.95	2.91	1.99	3.85	9.16	45.38
1969	2.71	2.30	4.76	7.13	2.22	1.29	4.60	3.94	3.64	0.95	2.24	5.39	41.17
1970	2.74	2.92	4.72	4.37	2.45	2.80	5.46	6.20	2.11	10.65	2.20	5.17	51.79
1971	4.25	5.72	8.05	4.66	5.17	2.58	11.65	1.99	7.17	0.84	3.09	7.63	62.80
1972	11.02	3.91	4.95	2.11	5.51	4.17	5.70	1.70	2.64	2.84	4.94	8.35	57.84
1973	5.03	3.94	14.29	8.96	6.62	2.66	10.16	2.51	4.46	2.00	3.61	14.79	79.03
1974	7.74	5.28	3.68	10.06	3.53	5.72	1.16	4.21	8.11	0.61	5.26	6.47	61.83
1975	8.39	6.30	7.16	7.68	5.12	5.49	6.77	3.82	3.64	4.30	2.84	4.05	65.56
1976	3.00	1.72	16.47	1.01	6.31	3.72	3.09	2.36	1.87	2.54	3.55	4.28	49.92
1977	5.61	4.80	12.63	7.58	3.39	1.56	10.13	4.71	6.60	6.48	5.89	1.68	71.06
1978	5.78	2.55	2.98	3.70	5.76	2.77	3.83	1.61	0.54	0.08	3.10	6.07	38.77
1979	8.72	7.45	8.44	10.73	5.46	3.24	6.48	1.46	8.15	3.19	7.41	2.33	73.06
1980	7.49	3.39	13.87	10.21	9.79	3.30	4.03	1.15	3.62	6.80	3.72	1.10	68.47
1981	1.59	4.65	11.81	1.18	3.66	2.55	3.15	2.21	0.98	3.30	2.60	5.79	43.47
1982	3.16	6.73	4.50	6.52	2.57	4.53	10.18	3.90	0.10	1.69	9.82	9.08	62.78
1983	4.54	9.44	6.62	10.33	7.85	6.71	2.33	3.16	3.45	0.97	8.67	6.62	70.69
1984	2.86	4.83	3.92	5.58	5.06	1.64	6.20	4.83	0.81	9.43	4.87	3.58	53.61
1985	2.45	6.84	3.10	4.57	1.83	2.76	5.29	7.41	5.41	7.46	0.81	4.33	52.26
1986	1.21	2.19	3.67	1.65	7.34	2.21	3.63	5.52	2.54	5.02	10.24	4.16	49.38
1987	8.76	11.33	3.96	0.91	5.73	7.60	1.25	2.69	4.41	0.01	4.05	4.49	55.19
1988	3.14	3.80	4.96	6.33	1.12	0.87	4.62	3.48	9.32	4.29	7.56	7.47	56.96
1989	3.94	3.07	9.82	3.01	7.37	8.91	11.08	0.72	7.10	2.68	5.97	6.68	70.35
1990	11.23	15.95	6.83	4.72	3.77	4.00	3.20	1.61	1.72	0.74	5.14	3.65	62.56
1991	5.78	7.41	6.52	11.78	9.02	3.57	3.80	6.65	6.04	1.13	5.03	6.83	73.56
1992	5.06	5.98	4.30	5.48	1.66	7.48	5.62	10.28	0.89	2.07	10.68	5.94	65.44
1993	11.37	3.11	7.30	4.20	2.91	3.50	2.80	3.52	1.75	5.67	5.01	3.29	54.43
Record Mean	5.05	5.29	5.97	5.12	4.38	4.24	5.41	3.98	3.24	2.50	3.86	5.47	54.49

AVERAGE TEMPERATURE (deg. F) — MERIDIAN, MISSISSIPPI

YEAR	JAN	FEB	MAR	APR	MAY	JUNE	JULY	AUG	SEP	OCT	NOV	DEC	ANNUAL
#1964	43.3	42.9	55.8	67.3	72.5	79.1	79.5	79.7	75.3	59.8	56.5	49.8	63.5
1965	48.1	47.3	52.0	67.6	72.2	76.6	79.6	79.1	75.9	62.3	58.4	47.9	63.9
1966	40.4	47.4	53.9	64.2	70.4	75.8	81.8	77.9	74.0	61.2	56.7	48.5	62.7
1967	46.9	45.8	61.0	70.4	70.6	79.2	78.2	77.0	68.6	60.5	52.3	51.3	63.5
1968	44.6	42.3	54.6	66.6	72.2	80.3	80.6	80.3	72.3	64.6	51.9	44.8	62.9
1969	46.7	49.2	48.9	65.1	70.8	78.7	82.5	78.1	73.5	65.0	52.3	46.6	63.1
1970	41.3	46.4	54.1	67.9	71.7	77.4	80.6	80.9	79.1	64.1	51.3	51.0	63.8
1971	48.1	48.2	51.7	61.9	67.7	79.7	79.6	79.7	77.0	70.0	53.4	57.5	64.5
1972	51.6	51.1	57.9	65.6	70.5	77.9	79.3	82.0	80.5	66.1	52.2	51.1	65.5
1973	45.4	48.1	62.4	62.3	70.4	79.0	81.8	79.2	78.0	69.4	61.6	49.9	65.6
1974	57.2	51.1	64.3	64.9	74.5	75.7	82.3	81.2	73.4	63.0	55.7	50.0	66.1
1975	50.4	51.7	55.2	60.9	75.7	79.1	81.1	81.1	73.2	63.1	53.9	45.7	64.3
1976	42.6	53.8	58.2	63.2	65.0	73.8	77.9	77.3	72.0	57.8	46.4	43.9	61.0
1977	34.7	47.4	58.0	64.6	72.7	82.0	83.7	82.6	79.2	63.2	58.9	48.2	64.6
1978	38.8	39.3	50.1	62.5	70.1	78.7	81.6	80.8	77.8	62.5	59.1	47.6	62.4
1979	38.0	45.7	56.8	64.4	69.7	75.1	81.7	80.7	75.1	63.8	52.4	47.5	62.6
1980	49.0	46.8	54.7	62.4	72.3	79.0	84.5	83.6	81.1	61.7	54.1	47.4	64.7
1981	43.0	50.7	55.8	70.0	70.0	82.1	84.6	83.4	74.2	64.9	58.5	47.4	65.4
1982	48.2	51.2	61.5	63.6	73.6	79.2	82.3	82.0	75.1	66.9	58.8	56.7	66.6
1983	44.9	48.2	53.9	59.7	69.6	74.8	80.4	81.1	72.3	64.2	53.4	43.8	62.2
1984	40.1	47.9	54.2	60.8	67.9	75.0	78.0	79.5	74.5	71.5	54.5	57.3	63.4
1985	40.2	47.5	62.2	65.5	72.2	79.2	80.7	79.9	74.1	68.4	61.2	42.6	64.5
1986	44.0	52.4	56.5	62.8	72.5	79.6	83.2	79.7	78.3	65.1	59.4	46.5	65.0
1987	44.0	51.0	56.7	62.7	74.8	77.6	81.5	82.0	74.5	58.3	56.0	52.8	64.3
1988	41.7	46.9	55.0	64.5	68.7	78.3	80.1	81.6	76.2	59.7	58.5	48.3	63.3
1989	52.2	49.0	58.1	62.3	70.4	78.1	80.6	80.9	74.6	62.9	56.2	40.1	63.8
1990	50.9	56.6	60.5	63.9	71.7	79.9	81.0	82.5	78.6	64.4	58.1	53.0	66.8
1991	47.2	52.3	59.0	68.3	76.0	79.1	82.1	80.9	76.4	66.9	51.7	51.9	66.0
1992	45.8	53.5	57.1	64.1	70.7	77.9	82.3	78.8	76.4	64.7	53.5	49.9	64.6
1993	50.4	49.1	53.5	60.7	70.8	80.2	83.5	83.5	76.5	64.7	54.1	48.0	64.6
Record Mean	46.9	49.6	56.7	64.2	71.6	78.5	80.7	80.4	76.3	65.3	55.3	48.5	64.5
Max	57.2	60.5	68.2	76.1	83.0	89.5	91.1	90.9	87.4	78.0	67.4	58.8	75.7
Min	36.5	38.7	45.2	52.3	60.1	67.5	70.3	69.8	65.2	52.6	43.2	38.2	53.3

REFERENCE NOTES — MERIDIAN, MISSISSIPPI

GENERAL
T - TRACE AMOUNT.
BLANK ENTRIES DENOTE MISSING/UNREPORTED DATA.
# INDICATES A STATION OR INSTRUMENT RELOCATION.

SPECIFIC

(a) - LENGTH OF RECORD IN YEARS, ALTHOUGH INDIVIDUAL MONTHS MAY BE MISSING.
0.* OR * - THE VALUE IS BETWEEN 0.0 AND 0.05
NORMALS - BASED ON THE 1961-1990 RECORD PERIOD.
EXTREMES - DATES ARE THE MOST RECENT OCCURRENCE
WIND DIR.- NUMERALS SHOW TENS OF DEGRESS CLOCKWISE FROM TRUE NORTH. "00" INDICATES CALM.
RESULTANT DIRECTIONS ARE GIVEN TO WHOLE DEGREES.
BOLD VALUES INDICATE EXTREME VALUES WHICH OCCURRED AFTER THE ASOS SYSTEM WAS COMMISSIONED.

RECORD = PERIOD OF RECORD
RECORD MEAN PRECIPITATION IS THE MEAN OF ALL DAILY PRECIPITATION AMOUNTS DURING THE PERIOD OF RECORD.
RECORD MAX(MIN) TEMPERATURE IS THE MEAN OF ALL DAILY MAX(MIN) TEMPERATURES DURING THE PERIOD OF RECORD.
RECORD MEAN TEMPERATURE IS THE SUM OF THE RECORD MAX AND RECORD MIN DIVIDED BY 2.
AVERAGE TEMPERATURE IS THE SUM OF THE MEAN DAILY MAX AND MIN TEMPERATURE DIVIDED BY 2.

EXCEPTIONS

RECORD MEANS ARE THROUGH THE CURRENT YEAR, BEGINNING IN 1890 FOR TEMPERATURE
1890 FOR PRECIPITATION
1946 FOR SNOWFALL

HEATING DEGREE DAYS Base 65 deg. F — MERIDIAN, MISSISSIPPI

SEASON	JULY	AUG	SEP	OCT	NOV	DEC	JAN	FEB	MAR	APR	MAY	JUNE	TOTAL
1964-65	0	0	0	188	277	480	521	504	418	51	2	0	2441
1965-66	0	0	8	131	208	524	758	486	347	109	15	2	2588
1966-67	0	0	0	149	257	513	555	530	180	18	21	0	2223
1967-68	0	0	40	160	379	434	626	649	341	45	7	0	2681
1968-69	0	0	0	116	390	619	560	443	494	49	8	0	2679
1969-70	0	0	1	95	374	563	732	515	330	68	23	0	2701
1970-71	0	0	1	87	404	432	521	468	408	142	40	0	2503
1971-72	0	0	0	17	359	244	418	404	231	81	3	0	1757
1972-73	0	0	5	79	401	434	602	467	125	147	19	0	2279
1973-74	0	0	0	49	174	464	259	390	125	76	0	0	1537
1974-75	0	0	8	103	295	461	449	370	317	164	0	0	2167
1975-76	0	0	22	108	349	589	686	323	236	80	51	0	2444
1976-77	0	0	3	235	553	649	932	487	231	53	4	0	3147
1977-78	0	0	0	108	200	517	808	714	457	111	26	0	2941
1978-79	0	0	0	110	178	547	830	536	261	69	15	0	2546
1979-80	0	0	0	93	374	536	489	540	325	106	3	0	2466
1980-81	0	0	0	143	325	539	677	398	290	21	15	0	2408
1981-82	0	0	9	96	221	545	541	381	198	108	1	0	2100
1982-83	0	0	8	79	226	318	616	461	340	170	15	0	2233
1983-84	0	0	21	92	350	658	766	490	344	173	42	3	2939
1984-85	0	0	8	30	323	251	764	488	137	80	2	0	2083
1985-86	0	0	11	53	154	689	644	359	274	100	9	0	2293
1986-87	0	0	0	99	196	566	644	385	258	142	0	0	2290
1987-88	0	0	0	215	282	387	714	517	308	73	10	0	2506
1988-89	0	0	1	179	223	511	401	469	248	148	40	0	2220
1989-90	0	0	11	125	282	765	433	255	185	110	12	0	2178
1990-91	0	0	7	121	219	391	543	349	223	27	1	0	1881
1991-92	0	0	3	65	414	411	587	327	246	111	21	0	2185
1992-93	0	0	0	50	351	461	444	441	353	153	7	0	2260
1993-94	0	0	3	125	350	521							

COOLING DEGREE DAYS Base 65 deg. F — MERIDIAN, MISSISSIPPI

YEAR	JAN	FEB	MAR	APR	MAY	JUNE	JULY	AUG	SEP	OCT	NOV	DEC	TOTAL
1969	0	6	0	61	197	417	546	415	263	104	2	0	2011
1970	4	0	0	163	238	377	489	503	432	67	0	4	2277
1971	5	2	4	57	129	450	462	460	367	177	18	19	2150
1972	8	6	21	106	182	393	449	536	475	121	21	10	2328
1973	0	0	49	73	193	426	526	445	399	191	76	2	2380
1974	27	6	106	79	301	329	542	507	269	49	22	3	2240
1975	4	1	23	50	342	430	507	506	274	57	22	0	2216
1976	0	8	33	36	57	268	409	389	220	18	0	0	1438
1977	0	0	23	50	250	517	584	554	433	59	24	4	2498
1978	1	0	0	44	193	418	522	498	390	40	9	15	2130
1979	0	0	13	57	168	309	526	493	307	63	6	0	1942
1980	0	16	10	35	233	425	612	586	487	45	4	0	2453
1981	0	1	12	178	175	517	612	576	292	100	33	5	2501
1982	28	0	94	73	274	434	541	532	317	146	46	70	2555
1983	0	0	1	18	163	297	484	505	248	73	9	5	1803
1984	0	1	17	51	142	310	411	456	299	237	14	19	1957
1985	0	5	56	102	233	435	494	472	289	165	47	4	2302
1986	0	13	14	39	247	444	573	460	405	111	38	1	2345
1987	0	0	6	76	310	388	514	533	289	14	16	13	2159
1988	0	2	7	66	131	408	475	519	345	20	35	0	2008
1989	9	26	42	75	215	399	493	500	305	67	26	0	2157
1990	3	29	51	85	226	454	502	549	423	112	19	26	2479
1991	0	2	47	131	349	430	537	502	352	129	21	13	2513
1992	0	0	6	88	208	392	542	434	349	50	13	0	2082
1993	0	4	2	32	193	459	581	581	356	121	27	0	2356

SNOWFALL (inches) — MERIDIAN, MISSISSIPPI

SEASON	JULY	AUG	SEP	OCT	NOV	DEC	JAN	FEB	MAR	APR	MAY	JUNE	TOTAL
1964-65	0.0	0.0	0.0	0.0	0.0	0.0	T	T	T	0.0	0.0	0.0	T
1965-66	0.0	0.0	0.0	0.0	0.0	0.0	T	0.0	0.0	0.0	0.0	0.0	T
1966-67	0.0	0.0	0.0	0.0	T	T	0.0	T	0.0	0.0	0.0	0.0	T
1967-68	0.0	0.0	0.0	0.0	0.0	T	T	3.0	1.5	0.0	0.0	0.0	4.5
1968-69	0.0	0.0	0.0	0.0	0.0	0.0	0.0	0.0	0.0	0.0	0.0	0.0	0.0
1969-70	0.0	0.0	0.0	0.0	0.0	0.0	T	0.0	0.0	0.0	0.0	0.0	T
1970-71	0.0	0.0	0.0	0.0	0.0	0.0	0.0	T	T	0.0	0.0	0.0	T
1971-72	0.0	0.0	0.0	0.0	T	0.0	0.0	0.0	0.0	0.0	0.0	0.0	T
1972-73	0.0	0.0	0.0	0.0	0.0	0.0	T	T	0.0	0.0	0.0	0.0	T
1973-74	0.0	0.0	0.0	0.0	0.0	0.2	0.0	T	0.0	0.0	0.0	0.0	0.2
1974-75	0.0	0.0	0.0	0.0	T	0.0	0.0	0.0	T	0.0	0.0	0.0	T
1975-76	0.0	0.0	0.0	0.0	T	T	0.0	0.0	0.0	0.0	0.0	0.0	T
1976-77	0.0	0.0	0.0	0.0	T	0.0	5.0	0.0	0.0	0.0	0.0	0.0	5.0
1977-78	0.0	0.0	0.0	0.0	0.0	0.0	1.0	T	T	0.0	0.0	0.0	1.0
1978-79	0.0	0.0	0.0	0.0	0.0	0.0	T	T	0.0	0.0	0.0	0.0	T
1979-80	0.0	0.0	0.0	0.0	0.0	0.0	T	0.0	T	0.0	0.0	0.0	T
1980-81	0.0	0.0	0.0	0.0	0.0	T	T	T	0.0	0.0	0.0	0.0	T
1981-82	0.0	0.0	0.0	0.0	0.0	0.0	1.8	0.0	T	0.0	0.0	0.0	1.8
1982-83	0.0	0.0	0.0	0.0	0.0	0.0	0.0	T	T	0.0	0.0	0.0	T
1983-84	0.0	0.0	0.0	0.0	0.0	T	T	T	T	0.0	0.0	0.0	T
1984-85	0.0	0.0	0.0	0.0	0.0	T	T	T	0.0	0.0	0.0	0.0	T
1985-86	0.0	0.0	0.0	0.0	0.0	0.0	0.0	T	0.0	0.0	0.0	0.0	T
1986-87	0.0	0.0	0.0	0.0	0.0	0.0	2.3	0.0	0.0	2.7	0.0	0.0	5.0
1987-88	0.0	0.0	0.0	0.0	0.0	0.0	0.0	0.0	0.0	0.0	0.0	0.0	0.0
1988-89	0.0	0.0	0.0	0.0	0.0	0.0	0.0	T	0.0	T	0.0	0.0	T
1989-90	T	0.0	0.0	0.0	0.0	T	0.0	0.0	0.0	T	0.0	0.0	T
1990-91	0.0	0.0	0.0	0.0	0.0	0.0	0.0	0.0	0.0	T	T	0.0	T
1991-92	0.0	0.0	0.0	0.0	0.0	0.0	0.2	0.0	0.0	0.0	0.0	0.0	0.2
1992-93	0.0	0.0	0.0	0.0	0.0	0.0	0.0	0.0	5.7	0.0	0.0	0.0	5.7
1993-94	0.0	0.0	0.0	0.0	0.0	0.8							
Record Mean	T	0.0	0.0	0.0	T	0.4	0.5	0.2	0.2	0.1	T	0.0	1.2

TUPELO, MISSISSIPPI

Tupelo is located in the Black Prairie physiographic region of Mississippi. The surface is flat to gently undulating, underlain by soft limestone which has weathered into dark, fertile soils. The Black Prairie is largely devoid of trees, although precipitation in the region is more than sufficient for forest growth. The Black Prairie is bordered by the Fall Line Hills on the east and the Pontotoc Ridge on the west, thus being situated as a low, flat region from 20 to 25 miles wide trending northwest–southeast in the northeastern corner of Mississippi.

Agricultural interests are varied, but the region is a major producer of soybeans and livestock. Tupelo is the dominant urban center of the region, and population is projected to continue to increase in the future. Manufacturing and industry are growing rapidly in the area, enhanced by recent completion of the Tenn–Tom Waterway. Water supply for the growing municipal and industrial demand has been identified as a potential problem in the region as ground water is being used faster than it can be replenished by natural processes.

Average annual precipitation of over 50 inches is well distributed throughout the year. On average, about half the annual total falls between April and September, with March being the wettest month and October the driest month. There is a 90 percent probability that the freeze–free period will be longer than six months,. with a 50 percent chance for a freeze before the end of October and after the first of April. Maximum temperatures over 90 degrees are expected on 78 days between May and September, with freezes expected on 66 days between October and March.

Record temperatures of over 105 degrees and below −10 degrees have been observed at Tupelo. Snow is not infrequent during the winter, although amounts average less than two and one–half inches each year. However, accumulations greater than 10 inches have been recorded. Frontal passages are common features of the climate from fall through spring, and there are seasonal occurrences of thunderstorms. Tornadoes, though quite rare, have occurred in the region, particularly during late spring. Precipitation amounts of one–half inch or more are expected on 37 days each year, with one-tenth inch or more expected on 73 days. Daily totals of precipitation have exceeded six inches on several occasions.

NORMALS, MEANS, AND EXTREMES

TUPELO, MISSISSIPPI

LATITUDE: 34 °16'N LONGITUDE: 88 °46' W ELEVATION: FT. GRND 349 BARO 352 TIME ZONE: CENTRAL WBAN: 93862

	(a)	JAN	FEB	MAR	APR	MAY	JUNE	JULY	AUG	SEP	OCT	NOV	DEC	YEAR
TEMPERATURE °F:														
Normals														
-Daily Maximum		48.8	54.5	64.3	73.9	80.9	87.8	90.7	89.9	84.7	74.9	63.5	52.8	72.2
-Daily Minimum		30.9	34.0	43.2	51.0	59.5	66.5	70.6	69.3	63.3	49.8	41.7	34.1	51.2
-Monthly		39.9	44.3	53.7	62.4	70.2	77.2	80.6	79.6	74.0	62.4	52.6	43.5	61.7
Extremes														
-Record Highest	10	72	79	84	93	94	101	105	102	103	92	84	78	105
-Year		1989	1989	1991	1987	1986	1988	1986	1990	1990	1986	1987	1987	JUL 1986
-Record Lowest	10	-6	8	19	29	43	49	58	52	44	**29**	20	-3	-6
-Year		1985	1988	1993	1992	1992	1984	1990	1986	1989	**1993**	1991	1989	JAN 1985
NORMAL DEGREE DAYS:														
Heating (base 65°F)		778	580	365	130	27	0	0	0	9	144	379	667	3079
Cooling (base 65°F)		0	0	15	52	188	366	484	453	279	64	7	0	1908
% OF POSSIBLE SUNSHINE	10	56	55	61	73	71	73	74	71	70	63	52	47	64
MEAN SKY COVER (tenths)														
Sunrise - Sunset	10	6.5	6.6	6.4	5.6	6.1	5.4	5.2	4.8	4.8	4.8	6.0	6.7	5.7
MEAN NUMBER OF DAYS:														
Sunrise to Sunset														
-Clear	10	8.5	7.4	8.2	10.2	8.0	9.4	11.3	11.5	12.6	14.1	9.4	7.9	118.5
-Partly Cloudy	10	5.8	5.8	7.3	7.8	9.8	11.2	11.2	11.7	8.9	6.1	6.0	5.5	97.1
-Cloudy	10	16.7	15.1	15.5	12.0	13.2	9.3	8.5	7.8	8.5	10.8	14.6	17.6	149.6
Precipitation														
.01 inches or more	10	10.0	9.8	9.6	9.2	11.7	9.5	7.7	8.0	7.1	7.5	8.0	11.2	109.3
Snow,Ice pellets,hail														
1.0 inches or more	10	0.4	0.3	0.2	0.0	0.0	0.0	0.0	0.0	0.0	0.0	0.0	0.2	1.1
Thunderstorms	10	1.0	2.4	4.0	5.2	7.9	8.7	8.8	7.6	4.7	2.9	3.1	2.1	58.5
Heavy Fog Visibility														
1/4 mile or less	10	2.6	1.4	1.7	1.1	1.0	0.9	0.8	1.9	1.4	3.1	1.9	1.9	19.7
Temperature °F														
-Maximum														
90° and above	10	0.0	0.0	0.0	0.5	2.2	14.7	22.5	19.7	8.9	0.4	0.0	0.0	68.9
32° and below	10	1.4	1.0	0.0	0.0	0.0	0.0	0.0	0.0	0.0	0.0	0.0	1.8	4.2
-Minimum														
32° and below	10	16.8	10.0	4.6	0.8	0.0	0.0	0.0	0.0	0.0	0.4	6.3	13.9	52.7
0° and below	10	0.2	0.0	0.0	0.0	0.0	0.0	0.0	0.0	0.0	0.0	0.0	0.3	0.5
AVG. STATION PRESS. (mb)	10	1008.6	1006.6	1004.5	1002.7	1002.9	1002.6	1004.0	1003.9	1005.1	1006.4	1006.5	1008.7	1005.2
RELATIVE HUMIDITY (%)														
Hour 00	10	77	76	73	75	83	84	85	87	86	84	79	78	81
Hour 06 (Local Time)	10	82	82	80	84	88	89	90	92	91	89	85	82	86
Hour 12	10	62	60	54	51	56	55	56	56	56	54	58	63	57
Hour 18	10	64	60	54	51	59	59	59	61	63	63	64	68	60
PRECIPITATION (inches):														
Water Equivalent														
-Normal		4.89	4.72	6.07	5.25	5.72	3.84	4.30	3.05	3.60	3.42	4.85	6.16	55.87
-Maximum Monthly	10	6.98	10.90	6.35	12.16	17.56	11.08	6.27	6.22	6.31	7.90	9.64	14.49	17.56
-Year		1989	1991	1990	1991	1991	1989	1989	1985	1988	1984	1986	1990	MAY 1991
-Minimum Monthly	10	0.30	2.33	2.13	0.46	1.34	0.17	0.48	0.59	1.30	0.36	1.50	1.57	0.17
-Year		1986	1988	1985	1986	1992	1988	1986	1983	1984	1989	1985	1984	JUN 1988
-Maximum in 24 hrs	10	2.20	5.91	2.88	3.09	8.59	2.72	**3.60**	3.64	**3.13**	3.27	3.71	7.17	8.59
-Year		1989	1991	1991	1991	1991	1991	**1993**	1985	**1993**	1985	1987	1983	MAY 1991
Snow,Ice pellets,hail														
-Maximum Monthly	10	7.2	5.3	1.7	0.4	T	0.0	T	0.0	0.0	**T**	0.1	2.2	7.2
-Year		1988	1985	1993	1987	1989		1990			**1993**	1991	1983	JAN 1988
-Maximum in 24 hrs	10	6.9	3.1	1.7	0.4	T	0.0	T	0.0	0.0	**T**	0.1	1.9	6.9
-Year		1988	1985	1993	1987	1989		1990			**1993**	1991	1983	JAN 1988
WIND:														
Mean Speed (mph)	10	7.5	8.3	8.5	7.7	7.0	5.9	5.9	5.5	6.3	6.0	7.5	7.6	7.0
Prevailing Direction														
Fastest Obs. 1 Min.														
-Direction (!!)	10	23	22	24	19	34	22	**36**	**01**	30	**01**	23	**11**	**01**
-Speed (MPH)	10	26	30	39	31	38	40	**34**	**41**	32	**30**	37	**38**	**41**
-Year		1992	1988	1991	1993	1989	1989	**1993**	**1993**	1990	**1993**	1987	**1993**	**AUG 1993**
Peak Gust														
-Direction (!!)	8	S	W	SW	W	W	SW	**01**	**01**	NW	N	SE	**11**	N
-Speed (mph)	8	47	51	51	51	47	58	**49**	**67**	45	79	56	**51**	79
-Date		1988	1991	1991	1989	1991	1989	**1993**	**1993**	1990	1988	1987	**1993**	OCT 1988

PRECIPITATION (inches)

TUPELO, MISSISSIPPI

YEAR	JAN	FEB	MAR	APR	MAY	JUNE	JULY	AUG	SEP	OCT	NOV	DEC	ANNUAL
1983							3.05	0.59	2.56	3.61	7.03	13.69	
1984	2.87	3.39	3.53	5.88	8.44	0.59	1.90	1.79	1.30	7.90	5.54	1.57	44.70
1985	3.75	5.18	2.13	4.94	5.35	3.02	2.82	6.22	2.86	7.35	1.50	1.68	46.80
1986	0.30	3.47	3.38	0.46	5.19	6.57	0.48	3.36	2.86	5.81	9.64	5.18	46.70
1987	2.68	6.37	2.40	1.20	5.49	3.96	1.50	2.08	2.13	1.03	6.26	4.78	39.88
1988	2.41	2.33	3.59	3.37	2.92	0.17	2.32	1.23	6.31	3.88	4.05	5.05	37.63
1989	6.98	9.88	3.68	2.31	3.79	11.08	6.27	2.78	5.48	0.36	4.07	2.89	59.57
1990	5.84	9.40	6.35	2.50	4.26	2.57	2.01	1.02	2.41	3.40	3.07	14.49	57.32
1991	3.12	10.90	6.25	12.16	17.56	6.22	0.76	3.16	3.06	1.58	5.36	10.92	81.05
1992	1.53	3.17	4.05	2.38	1.34	7.11	2.99	4.54	2.46	2.65	6.70	4.13	43.05
1993	2.94	3.24	4.05	4.63	5.78	2.67	3.80	6.05	5.45	2.45	2.88	4.80	48.74
Record Mean	3.24	5.73	3.94	3.98	6.01	4.40	2.54	2.98	3.35	3.64	5.10	6.29	51.21

AVERAGE TEMPERATURE (deg. F)

TUPELO, MISSISSIPPI

YEAR	JAN	FEB	MAR	APR	MAY	JUNE	JULY	AUG	SEP	OCT	NOV	DEC	ANNUAL
1983							82.1	83.1	72.8	64.2	55.6	36.5	
1984	36.2	46.8	51.5	59.9	68.7	79.3	79.7	78.9	72.0	69.0	49.7	53.3	62.1
1985	32.1	40.4	57.0	63.8	70.6	78.1	80.3	78.4	71.7	66.1	59.1	37.7	61.3
1986	40.7	48.8	54.4	63.0	71.9	79.3	84.6	79.0	77.8	63.3	54.1	42.9	63.3
1987	40.9	47.4	55.6	61.1	75.0	78.5	81.4	82.8	74.4	57.6	54.3	48.0	63.1
1988	37.3	42.6	52.6	62.7	69.7	78.1	80.9	83.1	75.9	58.3	55.0	45.0	61.8
1989	48.0	43.3	55.2	61.2	68.8	76.4	79.6	79.5	71.7	62.2	53.3	34.8	61.2
1990	48.2	52.6	56.1	61.1	68.5	79.0	80.9	81.1	77.6	61.7	56.1	46.7	64.1
1991	42.3	47.9	55.1	64.8	73.8	79.1	83.3	80.6	75.4	64.5	49.2	47.7	63.6
1992	42.3	48.5	53.0	61.9	69.2	75.2	80.6	76.0	72.5	62.1	51.0	44.7	61.4
1993	44.9	43.8	50.9	58.5	68.9	77.4	83.6	80.6	71.4	60.2	49.8	43.3	61.1
Record Mean	41.3	46.2	54.1	61.8	70.5	78.0	81.5	80.3	73.9	62.6	53.0	43.7	62.3
Max	50.7	56.2	65.1	73.4	81.2	88.5	91.7	90.8	84.8	74.3	63.5	52.8	72.8
Min	31.9	36.2	43.1	50.1	59.8	67.5	71.3	69.8	63.0	50.9	42.4	34.5	51.7

REFERENCE NOTES

TUPELO, MISSISSIPPI

GENERAL
T - TRACE AMOUNT.
BLANK ENTRIES DENOTE MISSING/UNREPORTED DATA.
# INDICATES A STATION OR INSTRUMENT RELOCATION.

SPECIFIC

(a) - LENGTH OF RECORD IN YEARS, ALTHOUGH INDIVIDUAL MONTHS MAY BE MISSING.
0.* OR * - THE VALUE IS BETWEEN 0.0 AND 0.05
NORMALS - BASED ON THE 1961-1990 RECORD PERIOD.
EXTREMES - DATES ARE THE MOST RECENT OCCURRENCE
WIND DIR.- NUMERALS SHOW TENS OF DEGRESS CLOCKWISE FROM TRUE NORTH. "00" INDICATES CALM.
RESULTANT DIRECTIONS ARE GIVEN TO WHOLE DEGREES.
BOLD VALUES INDICATE EXTREME VALUES WHICH OCCURRED AFTER THE ASOS SYSTEM WAS COMMISSIONED.

RECORD = PERIOD OF RECORD
RECORD MEAN PRECIPITATION IS THE MEAN OF ALL DAILY PRECIPITATION AMOUNTS DURING THE PERIOD OF RECORD.
RECORD MAX(MIN) TEMPERATURE IS THE MEAN OF ALL DAILY MAX(MIN) TEMPERATURES DURING THE PERIOD OF RECORD.
RECORD MEAN TEMPERATURE IS THE SUM OF THE RECORD MAX AND RECORD MIN DIVIDED BY 2.
AVERAGE TEMPERATURE IS THE SUM OF THE MEAN DAILY MAX AND MIN TEMPERATURE DIVIDED BY 2.

EXCEPTIONS

RECORD MEANS ARE THROUGH THE CURRENT YEAR, BEGINNING IN 1983 FOR TEMPERATURE
1983 FOR PRECIPITATION
1983 FOR SNOWFALL

HEATING DEGREE DAYS Base 65 deg. F TUPELO, MISSISSIPPI

SEASON	JULY	AUG	SEP	OCT	NOV	DEC	JAN	FEB	MAR	APR	MAY	JUNE	TOTAL
1983-84	0	0	29	91	405	876	886	524	421	181	37	0	3450
1984-85	0	0	22	36	458	363	1016	685	272	119	7	0	2978
1985-86	0	0	19	63	216	838	749	451	328	115	14	0	2793
1986-87	0	0	0	115	324	676	740	487	289	165	0	0	2796
1987-88	0	0	0	229	323	521	852	642	388	106	6	0	3067
1988-89	0	0	0	216	300	612	518	607	320	193	55	0	2821
1989-90	0	0	27	135	359	931	512	350	292	176	43	0	2825
1990-91	0	0	14	177	276	557	699	473	331	59	9	0	2595
1991-92	0	0	18	106	477	531	699	472	366	159	33	0	2861
1992-93	0	0	12	102	418	624	617	588	432	209	22	3	3027
1993-94	0	0	21	190	464	665							

COOLING DEGREE DAYS Base 65 deg. F TUPELO, MISSISSIPPI

YEAR	JAN	FEB	MAR	APR	MAY	JUNE	JULY	AUG	SEP	OCT	NOV	DEC	TOTAL
1983							537	566	276	73	8	0	
1984	0	0	8	38	159	437	465	438	239	165	4	8	1961
1985	0	1	32	89	188	400	480	423	228	105	44	0	1990
1986	0	1	6	61	237	433	616	443	389	71	5	0	2262
1987	0	0	5	55	318	409	513	559	288	6	10	0	2163
1988	0	0	9	43	159	399	500	568	334	14	4	0	2030
1989	0	5	24	88	182	347	461	458	236	54	12	0	1867
1990	0	6	25	65	158	427	499	506	397	80	14	0	2177
1991	0	0	29	61	287	433	575	491	334	98	7	0	2315
1992	0	0	1	72	170	316	494	346	242	18	5	0	1664
1993	0	0	2	23	151	381	583	490	219	52	12	0	1913

SNOWFALL (inches) TUPELO, MISSISSIPPI

SEASON	JULY	AUG	SEP	OCT	NOV	DEC	JAN	FEB	MAR	APR	MAY	JUNE	TOTAL
1983-84	0.0	0.0	0.0	0.0	0.0	2.2	1.8	0.8	1.7	0.0	0.0	0.0	6.5
1984-85	0.0	0.0	0.0	0.0	0.0	0.0	3.7	5.3	0.0	0.0	0.0	0.0	9.0
1985-86	0.0	0.0	0.0	0.0	0.0	1.2	T	0.9	0.0	0.0	0.0	0.0	2.1
1986-87	0.0	0.0	0.0	0.0	0.0	0.1	0.1	T	T	0.4	0.0	0.0	0.6
1987-88	0.0	0.0	0.0	0.0	0.0	0.0	7.2	1.1	0.0	0.0	0.0	0.0	8.3
1988-89	0.0	0.0	0.0	0.0	0.0	0.0	0.6	0.3	T	0.0	T	0.0	0.9
1989-90	0.0	0.0	0.0	0.0	T	0.3	0.0	0.0	T	0.0	0.0	0.0	0.3
1990-91	T	0.0	0.0	0.0	0.0	0.1	0.0	T	T	0.0	0.0	0.0	0.1
1991-92	0.0	0.0	0.0	0.0	0.1	0.0	0.2	0.0	0.0	0.0	0.0	0.0	0.3
1992-93	0.0	0.0	0.0	0.0	0.0	0.0	0.0	T	1.7	0.0	0.0	0.0	1.7
1993-94	0.0	0.0	0.0	T	0.0	T							
Record Mean	T	0.0	0.0	T	T	0.4	1.4	0.8	0.3	T	T	0.0	2.9

COLUMBIA, MISSOURI

Columbia, Missouri, with its interior continental location, experiences moderately cold winters and warm summers that are often humid.

There are usually a few days of temperatures below zero during the winter months, but there have been several winters when the temperature did not get this cold. Periods of cold weather are usually interrupted by periods of at least a few mild days. It is not uncommon to find some days with temperatures in the 60s in the midst of the winter months. Some snow falls each winter, but it is very unlikely that a snow cover will persist for more than three weeks. Most of the time when snow does fall, it stays on the ground for less than a week. March is the month in which substantial amounts of snowfall are most likely.

Temperatures of 100 degrees or more occur in most summers, but there have been several summers when temperatures failed to reach this high. The late spring and early summer months produce more frequent and larger amounts of rain than the other months of the year. Thus, in addition to being warm, these months are often quite humid. By late summer smaller amounts of rain fall and rains occur less frequently, so by mid-August, the moisture in the top 2 feet of soil is often depleted.

The average occurrence of the last temperature as cold as 32 degrees in spring is early April, and the first 32 degree temperature in the fall occurs in late October.

NORMALS, MEANS, AND EXTREMES

COLUMBIA, MISSOURI

LATITUDE: 38°49'N LONGITUDE: 92°13' W ELEVATION: FT. GRND 887 BARO 893 TIME ZONE: CENTRAL WBAN: 03945

	(a)	JAN	FEB	MAR	APR	MAY	JUNE	JULY	AUG	SEP	OCT	NOV	DEC	YEAR
TEMPERATURE °F:														
Normals														
-Daily Maximum		36.6	41.4	53.3	65.7	74.1	82.8	88.6	86.7	78.8	67.6	53.6	40.3	64.1
-Daily Minimum		18.5	22.8	33.0	43.7	53.1	61.2	66.2	63.8	57.0	45.5	34.6	23.2	43.6
-Monthly		27.6	32.1	43.2	54.7	63.6	72.1	77.4	75.2	67.9	56.5	44.1	31.8	53.9
Extremes														
-Record Highest	24	74	82	85	90	90	103	111	110	101	93	83	76	111
-Year		1989	1972	1986	1987	1991	1988	1980	1984	1971	1981	1978	1991	JUL 1980
-Record Lowest	24	-19	-15	-5	19	29	40	48	42	32	22	0	-20	-20
-Year		1982	1979	1978	1975	1976	1993	1975	1986	1984	1993	1991	1989	DEC 1989
NORMAL DEGREE DAYS:														
Heating (base 65°F)		1159	921	676	316	137	10	0	0	52	285	627	1029	5212
Cooling (base 65°F)		0	0	0	7	94	223	384	321	139	21	0	0	1189
% OF POSSIBLE SUNSHINE	24	51	48	49	55	57	65	67	64	60	59	47	45	56
MEAN SKY COVER (tenths)														
Sunrise - Sunset	24	6.4	6.7	6.9	6.4	6.5	6.0	5.3	5.3	5.5	5.5	6.5	6.7	6.1
MEAN NUMBER OF DAYS:														
Sunrise to Sunset														
-Clear	24	8.5	6.8	6.7	7.8	7.3	7.3	10.9	10.6	10.5	11.6	8.2	7.7	103.9
-Partly Cloudy	24	6.5	6.0	6.8	7.1	7.5	10.9	9.5	9.9	7.8	6.6	6.1	6.3	91.0
-Cloudy	24	15.9	15.5	17.5	15.1	16.2	11.8	10.4	10.5	11.8	12.8	15.7	17.0	170.1
Precipitation														
.01 inches or more	24	7.3	8.4	11.0	10.9	12.1	8.2	8.4	8.3	8.8	9.3	9.8	9.2	111.8
Snow, Ice pellets, hail														
1.0 inches or more	24	1.9	2.2	1.2	0.2	0.0	0.0	0.0	0.0	0.0	0.0	0.7	1.4	7.6
Thunderstorms	24	0.5	0.9	3.3	5.1	8.1	7.6	8.0	7.3	5.3	3.4	1.8	0.9	52.3
Heavy Fog Visibility 1/4 mile or less	24	2.5	3.1	2.3	1.4	1.3	1.1	1.2	1.5	1.8	1.7	1.7	3.3	22.9
Temperature °F														
-Maximum														
90° and above	24	0.0	0.0	0.0	0.1	0.2	4.6	15.2	11.8	4.3	0.1	0.0	0.0	36.3
32° and below	24	11.9	7.2	1.2	0.0	0.0	0.0	0.0	0.0	0.0	0.0	1.5	7.3	29.1
-Minimum														
32° and below	24	27.5	21.7	14.9	3.3	0.*	0.0	0.0	0.0	0.*	2.4	13.8	24.8	108.3
0° and below	24	2.7	1.6	0.5	0.0	0.0	0.0	0.0	0.0	0.0	0.0	0.*	1.5	6.3
AVG. STATION PRESS. (mb)	21	987.3	986.4	982.7	982.1	981.9	982.6	983.8	984.7	985.3	985.9	985.4	986.5	984.5
RELATIVE HUMIDITY (%)														
Hour 00	24	75	76	73	70	80	82	81	82	81	77	77	77	78
Hour 06 (Local Time)	24	79	80	79	79	85	86	87	88	88	84	82	81	83
Hour 12	24	66	64	59	55	60	59	56	57	58	57	63	68	60
Hour 18	24	68	65	59	55	61	60	58	59	64	63	68	71	63
PRECIPITATION (inches):														
Water Equivalent														
-Normal		1.45	1.84	3.17	3.83	5.01	4.32	3.67	3.28	3.86	3.22	2.93	2.47	39.05
-Maximum Monthly	24	3.58	6.18	10.09	9.06	10.51	10.28	12.14	8.96	12.06	6.12	10.42	6.96	12.14
-Year		1974	1985	1973	1983	1990	1985	1981	1982	1993	1976	1985	1982	JUL 1981
-Minimum Monthly	24	0.05	0.11	0.78	1.38	1.57	0.35	0.24	0.21	0.45	1.03	0.42	0.59	0.05
-Year		1986	1991	1971	1971	1992	1980	1976	1976	1979	1992	1989	1976	JAN 1986
-Maximum in 24 hrs	24	1.81	2.88	4.03	3.12	4.64	3.21	5.95	5.43	3.81	2.72	2.97	2.88	5.95
-Year		1985	1985	1990	1984	1990	1993	1989	1993	1986	1986	1983	1982	JUL 1989
Snow, Ice pellets, hail														
-Maximum Monthly	24	23.5	20.2	18.0	7.1	T	T	0.0	0.0	0.0	0.1	8.3	17.8	23.5
-Year		1979	1993	1978	1980	1990	1990				1976	1971	1973	JAN 1979
-Maximum in 24 hrs	24	8.6	12.7	8.8	7.1	T	T	0.0	0.0	0.0	0.1	8.1	12.7	12.7
-Year		1987	1993	1990	1980	1990	1990				1976	1975	1987	FEB 1993
WIND:														
Mean Speed (mph)	23	10.8	11.0	12.0	11.5	9.1	8.7	8.3	7.9	8.6	9.5	10.6	10.8	9.9
Prevailing Direction through 1963		S	NW	WNW	S	SSE	SSE	SSE	SSE	SSE	SSE	S	S	SSE
Fastest Obs. 1 Min.														
-Direction (!!)	5	NW	25	26	26	24	02	31	19	35	30	24	31	19
-Speed (MPH)	5	49	35	46	39	35	41	46	52	38	30	40	33	52
-Year		1984	1988	1991	1988	1988	1990	1986	1987	1986	1990	1990	1989	AUG 1987
Peak Gust														
-Direction (!!)	10	NW	NW	SW	SW	SW	SW	N	S	SW	SW	SW	SW	SW
-Speed (mph)	10	53	63	66	69	58	95	64	62	54	46	61	55	95
-Date		1984	1984	1991	1984	1988	1985	1986	1987	1985	1992	1990	1984	JUN 1985

PRECIPITATION (inches)

COLUMBIA, MISSOURI

YEAR	JAN	FEB	MAR	APR	MAY	JUNE	JULY	AUG	SEP	OCT	NOV	DEC	ANNUAL
1964	0.75	1.64	3.66	7.01	5.73	2.98	3.29	0.77	2.96	0.36	3.12	1.30	33.57
1965	2.52	1.15	3.40	4.49	2.12	7.55	3.19	3.62	8.92	2.28	0.59	2.42	42.25
1966	0.21	2.11	0.84	4.12	2.22	3.28	4.08	1.46	3.32	2.05	1.12	2.29	27.10
1967	1.92	0.71	2.49	3.12	5.32	4.97	2.04	0.87	1.96	6.02	1.64	2.44	33.50
1968	1.79	1.35	0.75	1.74	5.46	7.35	3.18	2.13	4.64	2.17	3.36	2.92	36.84
#1969	3.87	1.52	1.91	4.84	4.62	10.15	5.04	1.94	5.53	8.99	0.58	1.35	50.34
1970	0.23	0.31	1.01	8.00	9.81	3.16	2.11	4.36	8.94	2.71	1.37	1.76	43.77
1971	1.62	2.39	0.78	1.38	4.30	2.70	3.30	1.16	2.65	1.94	2.41	4.21	28.84
1972	0.88	0.42	4.01	4.44	3.40	0.66	3.86	1.10	5.06	2.66	5.26	2.71	34.46
1973	3.11	1.64	10.09	3.58	4.38	4.15	4.48	0.94	5.71	5.10	2.77	4.55	50.50
1974	3.58	2.70	3.03	3.55	7.75	5.89	1.43	7.57	1.77	1.20	3.81	1.65	43.93
1975	3.38	2.96	3.23	4.29	4.00	3.83	0.55	8.18	5.60	2.06	3.57	2.48	44.13
1976	0.94	2.36	4.39	2.21	3.68	1.79	0.24	0.21	0.47	6.12	0.88	0.59	23.88
1977	1.44	1.23	3.64	2.70	4.31	5.24	1.62	2.26	6.66	4.39	1.80	1.23	36.52
1978	0.67	1.57	5.44	5.80	6.77	2.50	4.56	2.01	1.11	1.73	3.24	2.51	37.91
1979	2.43	1.40	2.94	5.06	3.30	2.69	2.97	2.80	0.45	4.40	2.26	1.36	32.06
1980	0.63	1.42	2.71	2.42	3.36	0.35	1.39	2.98	1.93	4.35	1.00	1.12	23.66
1981	0.88	1.09	1.33	5.47	7.71	8.02	12.14	2.48	0.68	4.03	3.60	1.10	48.53
1982	3.16	1.23	3.23	2.51	5.33	5.03	3.31	8.96	4.33	1.82	2.20	6.96	48.07
1983	0.28	0.45	2.74	9.06	4.94	3.82	1.97	3.27	3.20	4.96	8.87	2.72	46.28
1984	0.30	3.18	4.48	7.23	3.46	6.87	5.29	0.88	5.58	4.55	3.89	4.99	50.70
1985	1.13	6.18	3.18	2.89	5.04	10.28	1.91	6.14	2.08	2.74	10.42	3.38	55.37
1986	0.05	4.10	1.16	2.67	5.33	1.98	3.43	3.01	9.58	6.02	2.31	2.05	41.69
1987	2.04	0.90	1.82	1.87	2.78	2.24	5.24	3.17	2.92	1.75	3.35	5.48	33.56
1988	1.57	2.23	4.09	1.97	2.45	0.47	2.26	4.14	2.32	1.25	4.35	3.42	30.52
1989	1.32	1.57	2.95	1.94	4.84	3.23	7.36	7.24	1.54	1.44	0.42	0.78	34.63
1990	1.97	3.86	7.22	2.84	10.51	7.20	6.00	3.95	0.93	2.58	3.07	3.49	53.62
1991	1.56	0.11	2.13	4.45	4.17	2.36	3.23	3.86	3.43	3.81	3.09	2.72	34.92
1992	0.52	1.96	3.82	1.83	1.57	1.64	5.40	1.61	4.25	1.03	8.10	2.64	34.37
1993	2.37	1.99	2.14	5.54	4.81	7.28	10.16	8.37	12.06	2.05	4.20	1.52	62.49
Record Mean	1.76	1.83	3.01	3.69	4.68	4.50	3.61	3.65	4.15	2.92	2.53	2.03	38.38

AVERAGE TEMPERATURE (deg. F)

COLUMBIA, MISSOURI

YEAR	JAN	FEB	MAR	APR	MAY	JUNE	JULY	AUG	SEP	OCT	NOV	DEC	ANNUAL
1964	36.0	33.2	40.7	58.0	69.4	73.0	79.7	76.2	70.6	56.2	49.1	32.4	56.2
1965	31.8	32.4	32.9	58.4	70.7	73.9	76.6	75.1	69.4	58.1	48.3	42.5	55.9
1966	25.4	32.2	47.4	51.5	62.6	72.6	81.7	73.8	65.8	55.6	48.4	34.3	54.3
1967	33.1	31.3	47.9	57.9	60.6	73.0	75.1	72.2	66.6	57.0	42.3	34.8	54.3
1968	29.1	29.6	46.6	56.3	61.4	75.7	76.6	76.3	67.6	57.6	42.6	30.9	54.2
#1969	27.7	34.4	37.3	57.5	64.7	70.3	80.5	76.8	69.0	54.5	42.9	31.0	53.9
1970	22.8	32.0	39.2	55.7	66.9	71.2	76.8	75.6	69.9	56.0	42.9	37.4	53.9
1971	27.2	32.2	41.8	56.4	61.3	77.5	74.4	74.7	71.3	63.0	45.6	39.7	55.4
1972	28.4	33.6	44.3	55.8	65.1	72.9	76.0	76.8	70.5	54.3	39.7	27.2	53.7
1973	29.9	33.0	48.7	52.1	60.9	73.4	77.8	76.4	69.0	60.4	46.2	30.8	54.9
1974	26.6	35.3	46.9	55.7	63.1	68.3	79.3	72.3	60.2	57.0	43.0	32.0	53.3
1975	31.4	28.9	37.3	52.5	65.0	72.3	76.2	75.7	62.6	57.6	47.1	35.2	53.5
1976	28.6	42.6	46.5	55.2	59.2	70.1	77.9	75.3	67.8	49.8	34.8	27.2	52.9
1977	13.6	32.9	47.2	58.3	68.0	73.5	79.3	77.0	70.2	55.6	45.1	30.8	54.3
1978	20.4	21.7	37.7	56.7	62.7	73.5	79.1	76.8	74.2	56.1	46.5	34.2	53.3
1979	15.9	23.1	44.3	53.1	64.1	72.2	76.1	75.4	67.7	57.5	41.5	35.9	52.3
1980	29.3	26.6	39.0	52.2	64.3	75.3	87.0	81.7	69.9	53.5	43.3	33.8	54.7
1981	29.0	33.4	44.5	61.0	59.1	74.4	78.1	74.8	68.3	55.0	46.7	29.6	54.5
1982	22.4	30.6	45.3	51.0	66.2	68.4	78.4	73.6	66.2	56.2	44.2	39.6	53.5
1983	31.9	38.1	43.6	48.2	60.4	71.7	79.7	82.1	69.4	56.5	45.5	17.6	53.7
1984	27.6	40.0	36.7	52.5	60.8	74.4	75.5	78.8	66.2	58.4	43.3	38.5	54.4
1985	19.9	27.5	48.2	58.8	64.5	69.0	76.6	72.5	67.5	58.4	42.3	24.5	52.5
1986	33.6	32.7	48.2	58.2	66.0	76.0	80.2	71.7	71.4	57.2	39.3	34.6	55.8
1987	29.6	39.8	47.4	55.5	69.1	75.6	79.2	77.1	67.7	51.6	46.2	35.9	56.2
1988	27.7	28.1	43.8	54.1	66.1	75.2	78.6	79.0	69.5	51.1	45.3	34.0	54.4
1989	38.1	23.2	43.7	54.9	61.6	70.2	76.4	74.2	63.5	57.7	43.7	21.5	52.4
1990	39.5	38.3	46.3	53.1	59.5	73.2	75.5	75.2	70.6	55.3	50.3	31.7	55.7
1991	25.9	39.9	48.7	57.8	70.2	75.9	77.9	76.7	67.4	57.3	39.1	37.0	56.2
1992	35.2	40.1	46.5	55.0	62.1	70.3	75.2	70.3	66.0	56.8	40.7	33.4	54.3
1993	29.5	29.3	39.9	51.5	63.4	72.4	78.3	76.9	62.7	52.9	40.0	34.6	52.6
Record Mean	29.5	32.9	43.3	55.0	64.3	73.4	78.0	76.4	68.8	57.5	44.0	33.3	54.7
Max	38.6	42.3	53.5	65.6	74.9	83.8	88.8	87.4	79.9	68.8	53.8	41.9	64.9
Min	20.5	23.5	33.1	44.3	53.8	63.0	67.1	65.3	57.6	46.2	34.1	24.7	44.4

REFERENCE NOTES

COLUMBIA, MISSOURI

GENERAL
T - TRACE AMOUNT.
BLANK ENTRIES DENOTE MISSING/UNREPORTED DATA.
# INDICATES A STATION OR INSTRUMENT RELOCATION.

SPECIFIC

(a) - LENGTH OF RECORD IN YEARS, ALTHOUGH INDIVIDUAL MONTHS MAY BE MISSING.
0.* OR * - THE VALUE IS BETWEEN 0.0 AND 0.05
NORMALS - BASED ON THE 1961-1990 RECORD PERIOD.
EXTREMES - DATES ARE THE MOST RECENT OCCURRENCE.
WIND DIR. - NUMERALS SHOW TENS OF DEGRESS CLOCKWISE FROM TRUE NORTH. "00" INDICATES CALM.
RESULTANT DIRECTIONS ARE GIVEN TO WHOLE DEGREES.
BOLD VALUES INDICATE EXTREME VALUES WHICH OCCURRED AFTER THE ASOS SYSTEM WAS COMMISSIONED.

RECORD = PERIOD OF RECORD
RECORD MEAN PRECIPITATION IS THE MEAN OF ALL DAILY PRECIPITATION AMOUNTS DURING THE PERIOD OF RECORD.
RECORD MAX(MIN) TEMPERATURE IS THE MEAN OF ALL DAILY MAX(MIN) TEMPERATURES DURING THE PERIOD OF RECORD.
RECORD MEAN TEMPERATURE IS THE SUM OF THE RECORD MAX AND RECORD MIN DIVIDED BY 2.
AVERAGE TEMPERATURE IS THE SUM OF THE MEAN DAILY MAX AND MIN TEMPERATURE DIVIDED BY 2.

EXCEPTIONS

RECORD MEANS ARE THROUGH THE CURRENT YEAR, BEGINNING IN 1889 FOR TEMPERATURE
1889 FOR PRECIPITATION
1951 FOR SNOWFALL

HEATING DEGREE DAYS Base 65 deg. F COLUMBIA, MISSOURI

SEASON	JULY	AUG	SEP	OCT	NOV	DEC	JAN	FEB	MAR	APR	MAY	JUNE	TOTAL
1964-65	0	6	44	270	488	1004	1023	908	990	248	22	0	5003
1965-66	0	3	52	227	491	692	1221	911	548	403	140	6	4694
1966-67	0	1	64	305	501	946	979	937	543	243	209	9	4737
1967-68	1	9	67	288	672	929	1106	1018	572	263	149	6	5080
1968-69	0	0	16	269	664	1051	1150	852	851	231	101	26	5211
#1969-70	0	0	25	369	656	1049	1302	922	792	304	59	22	5500
1970-71	0	0	37	282	655	850	1164	910	711	275	147	0	5031
1971-72	4	2	71	113	582	775	1129	901	634	304	94	13	4622
1972-73	2	0	43	336	751	1162	1080	890	499	387	142	0	5292
1973-74	0	0	42	191	559	1052	1187	825	566	290	129	20	4861
1974-75	0	2	167	257	653	1015	1034	1006	850	377	67	7	5435
1975-76	2	0	149	254	528	916	1121	644	566	307	193	2	4682
1976-77	0	0	43	487	898	1169	1587	891	548	209	40	0	5872
1977-78	0	0	14	287	591	1053	1378	1206	842	260	161	3	5795
1978-79	0	0	21	280	552	949	1515	1169	633	360	94	0	5573
1979-80	2	4	34	261	698	897	1101	1108	803	389	92	3	5392
1980-81	0	0	56	368	640	962	1110	878	630	159	211	0	5014
1981-82	0	0	32	326	544	1089	1314	955	607	416	40	25	5348
1982-83	0	3	88	297	619	783	1018	750	656	501	159	14	4888
1983-84	0	0	84	271	575	1466	1153	720	871	383	155	0	5678
1984-85	0	0	125	213	646	814	1391	1042	518	225	63	16	5053
1985-86	0	2	99	210	674	1249	964	897	521	227	53	0	4896
1986-87	0	15	13	251	765	937	1093	699	539	307	19	0	4638
1987-88	0	2	29	409	560	898	1150	1064	653	323	37	2	5127
1988-89	4	2	16	434	583	955	828	1164	660	356	169	14	5185
1989-90	0	2	122	244	635	1342	781	740	590	390	172	13	5031
1990-91	7	1	39	311	435	1027	1205	698	511	221	49	0	4504
1991-92	0	0	110	254	771	861	918	717	567	320	136	15	4669
1992-93	0	13	88	264	722	973	1094	995	775	396	87	27	5434
1993-94	0	4	117	375	743	935							

COOLING DEGREE DAYS Base 65 deg. F COLUMBIA, MISSOURI

YEAR	JAN	FEB	MAR	APR	MAY	JUNE	JULY	AUG	SEP	OCT	NOV	DEC	TOTAL
1969	0	0	0	12	99	195	489	372	152	49	0	0	1368
1970	0	0	0	32	128	217	372	335	192	12	0	0	1288
1971	0	0	0	28	41	379	304	306	266	58	5	0	1387
1972	0	0	0	35	106	256	351	375	211	10	0	0	1344
1973	0	0	0	8	21	261	401	361	170	53	2	0	1277
1974	0	0	12	19	74	124	449	232	30	13	0	0	953
1975	0	0	0	9	73	235	355	339	83	32	0	0	1126
1976	0	0	0	22	21	160	409	326	130	22	0	0	1090
1977	0	0	0	14	140	261	450	377	177	4	0	0	1423
1978	0	0	1	16	99	266	444	373	305	11	5	0	1520
1979	0	0	0	10	73	224	357	337	125	39	0	0	1165
1980	0	0	0	13	78	319	688	523	211	17	0	0	1849
1981	0	0	1	44	34	285	413	308	138	19	0	0	1242
1982	0	0	0	4	84	134	422	274	127	32	3	0	1080
1983	0	0	0	3	23	223	460	536	222	14	0	0	1481
1984	0	0	0	15	30	287	331	436	166	14	0	1	1280
1985	0	0	7	46	56	143	367	243	180	13	0	0	1055
1986	0	0	10	31	90	337	479	231	212	15	0	0	1405
1987	0	0	0	29	151	324	447	384	113	0	3	0	1451
1988	0	0	2	1	78	317	433	442	159	12	0	0	1444
1989	0	0	6	63	73	176	364	293	83	25	2	0	1085
1990	0	0	17	37	10	264	342	326	214	18	1	0	1229
1991	0	0	12	11	219	338	405	371	190	22	0	0	1568
1992	0	0	0	27	53	181	326	184	126	18	0	0	915
1993	0	0	2	0	46	259	419	379	56	6	0	0	1167

SNOWFALL (inches) COLUMBIA, MISSOURI

SEASON	JULY	AUG	SEP	OCT	NOV	DEC	JAN	FEB	MAR	APR	MAY	JUNE	TOTAL
1964-65	0.0	0.0	0.0	0.0	1.0	1.9	5.3	9.9	13.5	0.0	0.0	0.0	31.6
1965-66	0.0	0.0	0.0	0.0	0.0	3.4	1.5	9.0	0.8	T	0.0	0.0	14.7
1966-67	0.0	0.0	0.0	0.0	0.8	1.6	3.4	1.7	1.4	0.0	0.0	0.0	8.9
1967-68	0.0	0.0	0.0	T	1.8	6.2	4.6	3.5	1.6	T	0.0	0.0	17.7
1968-69	0.0	0.0	0.0	0.0	0.4	2.4	6.6	3.7	5.0	0.0	0.0	0.0	18.1
#1969-70	0.0	0.0	0.0	0.0	T	8.5	1.3	3.0	4.9	2.2	0.0	0.0	19.9
1970-71	0.0	0.0	0.0	0.0	T	4.0	1.0	2.2	1.9	3.3	0.0	0.0	12.4
1971-72	0.0	0.0	0.0	0.0	8.3	T	4.0	3.2	2.3	T	0.0	0.0	17.8
1972-73	0.0	0.0	0.0	0.0	6.4	5.1	6.8	2.4	T	4.5	0.0	0.0	25.2
1973-74	0.0	0.0	0.0	0.0	0.0	17.8	6.1	6.7	7.5	T	0.0	0.0	38.1
1974-75	0.0	0.0	0.0	0.0	6.3	0.3	3.0	16.1	3.7	T	0.0	0.0	29.4
1975-76	0.0	0.0	0.0	0.0	8.1	4.2	6.3	6.1	5.7	0.0	0.0	0.0	30.4
1976-77	0.0	0.0	0.0	0.1	1.3	3.3	19.9	9.4	0.8	0.2	0.0	0.0	35.0
1977-78	0.0	0.0	0.0	0.0	4.9	5.8	8.7	17.5	18.0	0.0	0.0	0.0	54.9
1978-79	0.0	0.0	0.0	0.0	T	5.0	23.5	7.0	4.9	T	0.0	0.0	40.4
1979-80	0.0	0.0	0.0	0.0	0.1	T	3.8	11.8	8.3	7.1	0.0	0.0	31.1
1980-81	0.0	0.0	0.0	T	5.8	1.6	1.4	8.8	T	0.0	0.0	0.0	17.6
1981-82	0.0	0.0	0.0	0.0	T	10.4	6.8	11.7	T	3.0	0.0	0.0	31.9
1982-83	0.0	0.0	0.0	0.0	0.2	T	1.2	1.9	0.7	T	0.0	0.0	4.0
1983-84	0.0	0.0	0.0	0.0	1.1	11.7	2.5	9.8	7.2	0.0	0.0	0.0	32.3
1984-85	0.0	0.0	0.0	0.0	2.7	1.6	14.8	7.1	T	0.0	0.0	0.0	26.2
1985-86	0.0	0.0	0.0	0.0	0.2	3.0	T	8.5	T	T	0.0	0.0	11.7
1986-87	0.0	0.0	0.0	0.0	T	T	21.2	T	T	0.0	0.0	0.0	21.2
1987-88	0.0	0.0	0.0	0.0	1.0	16.1	1.4	10.5	4.0	0.0	0.0	0.0	33.0
1988-89	0.0	0.0	0.0	0.0	2.3	3.6	3.7	13.3	8.1				
1989-90			0.0		0.5	6.9	0.7	6.3	8.8	0.0	T	T	
1990-91	0.0	0.0	0.0	0.0	0.2	6.1	8.0	0.2	T	0.0	0.0	0.0	14.5
1991-92	0.0	0.0	0.0	0.0	5.5	0.7	2.0	3.5	1.9	T	0.0	0.0	13.6
1992-93	0.0	0.0	0.0	0.0	0.6	1.4	7.4	20.2	2.2	T	0.0	0.0	31.8
1993-94	0.0	0.0	0.0	T	1.2	0.8							
Record Mean	0.0	0.0	0.0	T	1.8	4.2	5.6	6.5	4.3	0.7	T	T	23.1

INTERNATIONAL AIRPORT, KANSAS CITY, MISSOURI

The National Weather Service Office at Kansas City is very near the geographical center of the United States. The surrounding terrain is gently rolling. It has a modified continental climate. There are no natural topographic obstructions to prevent the free sweep of air from all directions. The influx of moist air from the Gulf of Mexico, or dry air from the semi-arid regions of the southwest, determine whether wet or dry conditions will prevail. There is often conflict between the warm moist gulf air and the cold polar continental air from the north in this area.

Early spring brings a period of frequent and rapid fluctuations in weather, with the fluctuations generally less frequent as spring progresses. The summer season is characterized by warm days and mild nights, with moderate humidities. July is the warmest month. The fall season is normally mild and usually includes a period near the middle of the season characterized by mild, sunny days, and cool nights. Winters are not severely cold. January is the coldest month. Falls of snow to a depth of 10 inches or more are comparatively rare. The distribution of measurable snow normally extends from November to April.

Nearly 60 percent of the annual precipitation occurs during the six months from April through September. More than 75 percent of the annual moisture normally falls during the growing season. The frequency and distribution of precipitation over a normal day is also important. The maximum frequency of precipitation, from April through October, occurs during the six hours following midnight and the minimum frequency occurs during the six hours following noon.

NORMALS, MEANS, AND EXTREMES

INTERNATIONAL AIRPORT, KANSAS CITY, MISSOURI

LATITUDE: 39°19'N LONGITUDE: 94°43' W ELEVATION: FT. GRND 973 BARO 975 TIME ZONE: CENTRAL WBAN: 03947

	(a)	JAN	FEB	MAR	APR	MAY	JUNE	JULY	AUG	SEP	OCT	NOV	DEC	YEAR
TEMPERATURE °F:														
Normals														
-Daily Maximum		34.7	40.6	52.8	65.1	74.3	83.3	88.7	86.4	78.1	67.5	52.6	38.8	63.6
-Daily Minimum		16.7	21.8	32.6	43.8	53.9	63.1	68.2	65.7	56.9	45.7	33.6	21.9	43.7
-Monthly		25.7	31.2	42.7	54.5	64.1	73.2	78.5	76.1	67.5	56.6	43.1	30.4	53.6
Extremes														
-Record Highest	21	69	76	86	93	92	105	107	109	102	92	82	70	109
-Year		1989	1981	1986	1987	1985	1980	1974	1984	1990	1976	1990	1980	AUG 1984
-Record Lowest	21	-17	-19	-10	12	30	42	52	43	33	17	1	-23	-23
-Year		1982	1982	1978	1975	1976	1990	1973	1986	1991	1993	1991	1989	DEC 1989
NORMAL DEGREE DAYS:														
Heating (base 65°F)		1218	946	691	325	135	7	0	6	56	279	657	1073	5393
Cooling (base 65°F)		0	0	0	10	107	253	419	350	131	18	0	0	1288
% OF POSSIBLE SUNSHINE	21	59	55	58	63	62	67	72	67	66	61	49	50	61
MEAN SKY COVER (tenths)														
Sunrise - Sunset	21	5.9	6.2	6.5	6.1	6.2	5.5	4.6	4.8	4.9	5.0	6.1	6.0	5.7
MEAN NUMBER OF DAYS:														
Sunrise to Sunset														
-Clear	21	10.3	7.9	7.4	8.6	7.7	9.6	13.2	12.6	12.2	12.8	8.6	9.8	120.7
-Partly Cloudy	21	6.3	6.1	7.6	7.1	9.4	9.9	10.3	10.5	8.0	7.1	6.9	6.2	95.6
-Cloudy	21	14.4	14.2	16.0	14.2	13.9	10.5	7.5	7.9	9.8	11.1	14.5	15.0	149.0
Precipitation														
.01 inches or more	21	7.3	6.9	10.4	10.8	11.5	10.1	8.7	8.9	8.5	7.5	8.1	7.8	106.5
Snow,Ice pellets,hail														
1.0 inches or more	21	2.0	1.7	0.8	0.3	0.0	0.0	0.0	0.0	0.0	0.0	0.5	1.1	6.5
Thunderstorms	21	0.3	0.6	2.6	5.0	8.3	9.0	8.3	7.3	5.7	3.0	1.2	0.4	51.8
Heavy Fog Visibility														
1/4 mile or less	21	2.3	3.0	2.3	1.2	1.0	0.7	0.5	1.3	1.3	1.5	2.0	3.2	20.3
Temperature °F														
-Maximum														
90° and above	21	0.0	0.0	0.0	0.4	0.2	5.7	15.5	11.7	3.7	0.1	0.0	0.0	37.3
32° and below	21	12.1	8.2	1.7	0.1	0.0	0.0	0.0	0.0	0.0	0.0	2.2	7.8	32.2
-Minimum														
32° and below	21	28.1	22.0	14.5	3.7	0.*	0.0	0.0	0.0	0.0	2.4	13.9	26.2	110.8
0° and below	21	3.7	2.3	0.1	0.0	0.0	0.0	0.0	0.0	0.0	0.0	0.0	2.0	8.0
AVG. STATION PRESS.(mb)	21	982.9	982.0	977.9	977.2	976.9	977.4	978.8	979.7	980.4	981.1	980.5	982.2	979.8
RELATIVE HUMIDITY (%)														
Hour 00	21	73	74	72	70	76	78	76	78	79	73	74	75	75
Hour 06 (Local Time)	21	77	78	78	78	83	85	84	86	86	80	79	79	81
Hour 12	21	64	64	59	56	59	59	58	59	59	55	62	65	60
Hour 18	21	65	63	57	53	57	57	55	58	60	59	66	68	60
PRECIPITATION (inches):														
Water Equivalent														
-Normal		1.09	1.10	2.51	3.12	5.04	4.72	4.38	4.01	4.86	3.29	1.92	1.58	37.62
-Maximum Monthly	21	2.66	2.69	9.08	6.82	10.07	7.44	15.47	9.58	11.34	7.67	5.12	5.42	15.47
-Year		1982	1985	1973	1984	1974	1981	1992	1982	1977	1977	1992	1980	JUL 1992
-Minimum Monthly	21	0.02	0.20	1.18	1.02	1.05	1.80	0.25	0.75	1.13	0.21	T	0.05	T
-Year		1986	1991	1974	1980	1992	1988	1975	1984	1974	1988	1989	1979	NOV 1989
-Maximum in 24 hrs	21	1.83	1.39	1.98	4.69	4.26	2.67	5.08	6.19	8.82	4.92	2.08	3.67	8.82
-Year		1982	1987	1992	1975	1974	1976	1986	1982	1977	1973	1990	1980	SEP 1977
Snow,Ice pellets,hail														
-Maximum Monthly	21	14.2	15.7	11.4	7.2	T	T	T	0.0	T	T	7.1	13.2	15.7
-Year		1977	1993	1978	1983	1990	1993	1992		1992	1993	1975	1983	FEB 1993
-Maximum in 24 hrs	21	9.5	10.8	9.2	4.0	T	T	T	0.0	T	T	6.1	10.8	10.8
-Year		1993	1993	1990	1983	1990	1993	1992		1992	1993	1975	1987	FEB 1993
WIND:														
Mean Speed (mph)	21	11.4	11.3	12.6	12.4	10.4	9.9	9.5	9.1	9.6	10.5	11.5	11.2	10.8
Prevailing Direction														
Fastest Obs. 1 Min.														
-Direction (!!)	9	23	32	22	16	33	35	18	31	29	19	18	20	35
-Speed (MPH)	9	35	35	39	32	45	48	46	40	40	38	35	39	48
-Year		1990	1991	1991	1991	1987	1990	1990	1987	1988	1992	1990	1988	JUN 1990
Peak Gust														
-Direction (!!)	10	NW	NW	SW	SW	NW	N	NW	NW	N	NW	NW	NW	NW
-Speed (mph)	10	58	56	63	62	59	67	75	54	63	54	52	55	75
-Date		1984	1984	1991	1984	1991	1990	1992	1987	1987	1990	1991	1993	JUL 1992

PRECIPITATION (inches)

INTERNATIONAL AIRPORT, KANSAS CITY, MISSOURI

YEAR	JAN	FEB	MAR	APR	MAY	JUNE	JULY	AUG	SEP	OCT	NOV	DEC	ANNUAL
1964	1.20	1.50	3.20	5.40	5.50	5.40	2.94	3.79	3.07	0.36	4.73	1.73	38.82
1965	2.35	1.81	3.51	2.21	2.19	7.50	9.83	6.27	7.75	1.83	0.24	2.25	47.74
1966	0.21	0.76	1.88	3.04	2.48	8.63	0.89	4.24	2.67	0.45	1.05	0.91	27.21
1967	2.09	0.35	2.85	6.86	6.04	9.71	1.81	0.31	7.87	8.63	0.45	1.72	48.69
1968	0.27	1.20	0.73	3.91	5.19	2.33	5.13	4.32	1.04	3.52	3.03	1.29	31.96
1969	1.31	0.49	1.36	4.77	3.89	10.57	10.29	8.70	3.67	6.50	0.08	0.72	52.35
1970	0.21	0.46	1.02	3.95	4.75	5.32	0.62	4.02	10.79	2.30	0.94	1.74	36.12
1971	1.48	0.89	0.87	1.29	3.02	3.20	3.13	1.04	2.70	3.47	1.68	3.82	26.59
#1972	0.56	0.41	2.13	3.98	2.18	2.71	3.43	2.90	2.69	2.29	3.00	1.47	27.75
1973	2.05	1.35	9.08	2.91	5.65	2.84	8.71	1.60	10.32	5.80	2.36	2.59	55.26
1974	1.05	1.12	1.18	2.94	10.07	2.16	1.13	4.98	1.13	7.22	1.62	1.52	36.12
1975	2.14	1.59	1.49	6.61	3.45	2.46	0.25	4.85	6.10	0.35	2.75	2.03	34.07
1976	0.53	0.66	2.53	3.30	5.49	5.08	0.77	0.76	1.41	2.84	0.21	0.10	23.68
1977	1.15	0.57	2.59	2.35	5.43	6.18	2.74	7.99	11.34	7.67	1.36	0.37	49.74
1978	0.39	1.32	1.77	5.36	4.98	2.64	5.49	2.66	4.04	0.33	3.93	1.05	33.96
#1979	2.35	0.78	2.96	2.35	3.00	5.37	4.68	3.66	1.16	3.56	1.83	0.05	31.75
1980	1.60	1.44	3.64	1.02	3.06	2.52	1.99	4.89	1.63	4.13	0.45	5.42	31.79
1981	0.49	0.31	1.43	1.94	9.46	7.44	8.43	2.43	2.71	4.14	2.84	0.45	42.07
1982	2.66	1.13	2.94	1.55	9.81	6.04	2.73	9.58	1.58	3.04	2.21	3.94	47.21
1983	0.58	0.57	2.93	5.52	6.03	5.03	0.26	0.86	1.89	3.85	3.94	1.42	32.88
1984	0.14	1.96	4.52	6.82	2.26	4.14	3.91	0.75	3.42	6.04	1.24	3.57	38.77
1985	0.94	2.69	2.05	1.75	7.00	3.56	5.82	6.98	9.23	7.51	3.95	1.24	52.72
1986	0.02	1.25	1.34	2.12	4.76	2.48	8.36	3.16	10.40	3.17	1.18	1.20	39.44
1987	0.77	2.26	2.85	2.24	4.74	4.58	3.00	4.64	3.66	1.32	1.88	2.05	33.99
1988	1.40	0.72	1.43	2.15	2.14	1.80	1.21	1.87	8.48	0.21	1.96	0.85	24.22
1989	0.98	0.59	2.13	1.50	4.56	3.44	4.76	7.38	8.87	2.88	T	0.55	37.64
1990	1.20	2.11	3.90	2.47	7.36	6.27	4.40	5.04	1.28	2.46	3.01	1.11	40.61
1991	1.37	0.20	2.36	4.99	3.69	3.06	1.72	1.35	2.12	3.71	2.05	2.08	28.70
1992	1.21	2.01	3.79	4.92	1.05	3.84	15.47	2.37	5.69	1.38	5.12	3.78	50.63
1993	1.96	1.28	2.21	5.59	7.30	5.67	10.90	3.98	7.63	1.75	2.07	1.12	51.46
Record Mean	1.30	1.42	2.60	3.37	4.78	4.75	4.02	3.94	4.39	2.98	1.92	1.50	36.97

AVERAGE TEMPERATURE (deg. F)

INTERNATIONAL AIRPORT, KANSAS CITY, MISSOURI

YEAR	JAN	FEB	MAR	APR	MAY	JUNE	JULY	AUG	SEP	OCT	NOV	DEC	ANNUAL
1964	37.2	36.2	41.8	57.4	70.1	73.7	82.5	76.0	69.4	57.6	49.9	32.3	57.0
1965	31.4	32.3	34.0	58.5	71.6	74.6	78.5	77.1	69.9	60.5	48.2	44.0	56.7
1966	28.0	33.7	49.1	52.8	65.6	73.3	83.4	74.9	68.8	59.7	49.1	34.8	56.1
1967	33.9	33.6	47.8	58.9	61.7	72.9	76.5	75.1	67.9	58.6	44.4	35.4	55.6
1968	28.4	32.3	48.1	57.0	62.2	77.6	79.4	77.8	69.5	59.6	41.2	29.7	55.3
1969	26.5	34.2	37.6	56.3	66.1	70.2	81.4	78.9	71.8	55.7	46.1	33.1	54.8
1970	26.1	36.7	42.0	57.4	72.1	74.5	80.8	82.4	71.6	58.6	44.4	38.9	57.1
1971	27.7	31.5	43.7	59.7	64.4	79.4	76.4	78.3	72.6	64.0	46.9	37.0	56.8
#1972	27.3	33.2	47.1	57.4	67.1	76.9	77.2	77.3	70.9	53.9	39.6	27.5	54.6
1973	27.3	33.8	47.7	51.8	61.5	74.5	76.6	77.1	66.7	60.7	45.5	30.1	54.5
1974	23.8	35.3	46.9	56.2	65.6	70.5	82.1	73.0	61.6	57.7	42.8	32.2	54.0
1975	29.9	27.8	36.5	52.8	67.1	74.2	80.9	79.8	63.5	58.9	45.9	34.1	54.3
1976	27.5	42.7	45.0	56.3	60.0	71.3	79.0	77.7	69.0	50.8	35.9	29.0	53.7
1977	15.6	35.1	48.0	59.7	69.0	74.5	79.7	74.4	69.0	55.7	42.3	29.5	54.4
1978	16.8	19.5	37.6	55.5	62.1	74.6	79.3	76.5	73.6	56.4	44.4	31.0	52.3
#1979	12.5	20.9	42.2	51.6	63.5	72.2	75.9	75.7	68.6	57.7	40.7	35.2	51.4
1980	28.7	25.2	38.7	54.6	63.9	75.3	85.2	80.3	69.6	54.1	44.5	32.1	54.4
1981	30.3	33.4	45.2	61.1	60.5	74.1	78.3	72.9	68.4	55.3	45.6	29.0	54.5
1982	18.6	27.8	42.5	51.1	65.5	68.8	79.4	75.0	67.5	55.6	41.9	35.5	52.4
1983	30.1	35.9	43.1	46.3	59.6	70.8	81.5	83.5	71.2	57.2	44.3	13.2	53.1
1984	25.0	38.9	36.0	50.3	60.4	74.3	76.1	79.0	66.1	56.8	43.9	35.5	53.5
1985	18.7	25.3	47.4	57.9	66.0	68.8	77.0	72.1	66.7	56.5	36.8	22.9	51.3
1986	34.5	30.5	48.5	57.1	65.2	76.5	79.7	72.0	71.8	56.9	37.9	34.5	55.4
1987	29.7	39.4	47.1	56.8	70.6	76.0	79.9	76.4	67.8	52.0	46.7	35.1	56.5
1988	26.7	27.9	43.2	54.5	69.1	78.1	79.6	81.3	70.5	52.2	44.8	35.2	55.3
1989	37.7	22.8	43.8	56.9	63.2	71.1	77.8	75.5	63.2	57.9	42.3	21.1	52.8
1990	37.9	36.2	45.7	52.7	60.4	75.5	77.3	77.1	72.1	57.1	50.1	29.3	56.0
1991	22.9	39.4	47.2	56.7	67.6	76.1	80.6	77.7	68.8	57.6	36.7	36.1	55.6
1992	35.9	39.8	46.9	53.1	62.6	69.7	74.4	70.7	66.1	56.6	39.0	32.9	54.0
1993	25.9	28.6	39.8	50.4	63.5	72.9	77.6	77.8	62.8	53.6	39.4	34.5	52.2
Record Mean	29.1	32.8	43.3	55.3	65.1	74.4	79.4	77.7	69.7	58.5	44.3	33.1	55.2
Max	37.7	41.8	53.0	65.2	74.6	83.8	88.9	87.4	79.6	68.6	53.2	41.2	64.6
Min	20.5	23.8	33.5	45.4	55.5	65.0	69.8	68.0	59.8	48.3	35.3	25.0	45.8

REFERENCE NOTES

INTERNATIONAL AIRPORT, KANSAS CITY, MISSOURI

GENERAL
T - TRACE AMOUNT.
BLANK ENTRIES DENOTE MISSING/UNREPORTED DATA.
# INDICATES A STATION OR INSTRUMENT RELOCATION.

SPECIFIC

PM - INCLUDES LAST DAY OF PREVIOUS MONTH
ASOS - AUTOMATED SURFACE OBSERVING SYSTEM IN OPERATION DURING THESE MONTHS.

(*) - LENGTH OF RECORD IN YEARS, ALTHOUGH INDIVIDUAL MONTHS MAY BE MISSING.
0.* OR * - THE VALUE IS BETWEEN 0.0 AND 0.05
NORMALS - BASED ON THE 1961-1990 RECORD PERIOD.
EXTREMES - DATES ARE THE MOST RECENT OCCURRENCE
WIND DIR.- NUMERALS SHOW TENS OF DEGRESS CLOCKWISE FROM TRUE NORTH. "00" INDICATES CALM.
RESULTANT DIRECTIONS ARE GIVEN TO WHOLE DEGREES.
BOLD VALUES INDICATE EXTREME VALUES WHICH OCCURRED AFTER THE ASOS SYSTEM WAS COMMISSIONED.

RECORD = PERIOD OF RECORD
RECORD MEAN PRECIPITATION IS THE MEAN OF ALL DAILY PRECIPITATION AMOUNTS DURING THE PERIOD OF RECORD.
RECORD MAX(MIN) TEMPERATURE IS THE MEAN OF ALL DAILY MAX(MIN) TEMPERATURES DURING THE PERIOD OF RECORD.
RECORD MEAN TEMPERATURE IS THE SUM OF THE RECORD MAX AND RECORD MIN DIVIDED BY 2.
AVERAGE TEMPERATURE IS THE SUM OF THE MEAN DAILY MAX AND MIN TEMPERATURE DIVIDED BY 2.

EXCEPTIONS

RECORD MEANS ARE THROUGH THE CURRENT YEAR,
BEGINNING IN 1889 FOR TEMPERATURE
1889 FOR PRECIPITATION
1936 FOR SNOWFALL

HEATING DEGREE DAYS Base 65 deg. F — INTERNATIONAL AIRPORT, KANSAS CITY, MISSOURI

SEASON	JULY	AUG	SEP	OCT	NOV	DEC	JAN	FEB	MAR	APR	MAY	JUNE	TOTAL
1964-65	0	2	49	235	473	1004	1033	911	955	237	15	0	4914
1965-66	0	0	44	175	497	646	1141	869	492	365	95	9	4333
1966-67	0	2	29	203	478	931	956	874	548	213	193	11	4438
1967-68	0	3	48	255	609	909	1125	941	528	245	131	3	4797
1968-69	0	1	13	224	706	1090	1186	857	845	258	86	24	5290
1969-70	0	0	3	330	562	984	1198	785	707	280	26	22	4897
1970-71	0	0	25	231	610	801	1151	931	659	208	92	0	4708
1971-72	4	0	57	87	538	863	1163	920	557	263	71	0	4523
#1972-73	1	0	37	355	751	1155	1160	868	529	394	127	0	5377
1973-74	0	0	53	173	578	1077	1272	823	559	270	84	6	4895
1974-75	0	6	141	227	660	1009	1084	1036	878	381	42	5	5469
1975-76	0	0	142	226	567	954	1155	640	616	277	178	6	4761
1976-77	0	0	45	469	865	1108	1527	832	521	192	22	0	5581
1977-78	0	0	21	293	673	1094	1487	1266	848	291	172	6	6151
#1978-79	0	0	28	272	618	1050	1624	1230	698	400	115	5	6040
1979-80	5	5	35	247	720	918	1118	1148	809	327	98	3	5433
1980-81	0	0	63	347	609	1011	1069	880	607	169	179	2	4936
1981-82	0	2	40	309	573	1112	1432	1037	690	416	51	32	5694
1982-83	0	2	78	307	688	911	1074	810	675	557	180	29	5311
1983-84	0	0	57	271	617	1602	1234	750	891	443	175	1	6041
1984-85	0	0	143	269	624	907	1431	1102	538	256	41	19	5330
1985-86	0	3	131	260	841	1297	940	960	528	267	60	0	5287
1986-87	0	12	23	251	805	938	1088	712	549	298	15	0	4691
1987-88	0	3	30	398	552	922	1180	1069	668	311	19	0	5152
1988-89	2	1	18	394	599	915	836	1176	658	319	135	7	5060
1989-90	0	1	138	267	675	1360	836	800	601	398	167	10	5253
1990-91	1	0	44	278	452	1104	1296	709	553	258	62	0	4757
1991-92	0	0	96	277	841	888	898	724	554	367	137	6	4788
1992-93	0	7	80	271	773	987	1203	1011	775	431	88	26	5652
1993-94	0	2	118	365	760	938							

COOLING DEGREE DAYS Base 65 deg. F — INTERNATIONAL AIRPORT, KANSAS CITY, MISSOURI

YEAR	JAN	FEB	MAR	APR	MAY	JUNE	JULY	AUG	SEP	OCT	NOV	DEC	TOTAL
#1969	0	0	0	4	126	188	513	435	216	52	0	0	1534
1970	0	0	0	58	250	317	493	546	232	39	0	0	1935
1971	0	0	5	54	81	437	366	421	295	64	3	0	1726
1972	0	3	7	41	146	368	386	389	219	16	0	0	1575
1973	0	0	0	4	30	291	368	382	110	46	0	0	1231
1974	0	0	4	13	109	176	538	264	47	8	0	0	1159
1975	0	0	0	22	117	284	498	464	105	43	1	0	1534
1976	0	0	0	22	29	199	444	399	169	34	0	0	1296
1977	0	0	0	39	154	291	463	298	147	11	0	0	1403
1978	0	0	5	14	86	300	452	364	295	11	8	0	1535
#1979	0	0	1	3	77	229	348	340	147	29	0	0	1174
1980	0	0	0	21	69	316	632	483	210	15	0	0	1746
1981	0	0	0	58	44	279	418	253	149	14	0	0	1215
1982	0	0	0	8	75	154	452	320	158	26	0	0	1193
1983	0	0	3	5	19	210	517	582	251	31	1	0	1619
1984	0	0	0	9	41	287	353	445	184	23	0	0	1342
1985	0	0	0	49	77	137	379	230	191	5	0	0	1068
1986	0	0	24	35	73	352	466	237	232	7	0	0	1426
1987	0	0	0	58	196	336	474	364	118	1	9	0	1556
1988	0	0	2	5	151	400	459	514	188	5	0	0	1724
1989	0	0	9	85	88	196	402	334	87	56	0	0	1257
1990	0	0	8	33	33	331	394	384	263	40	11	0	1497
1991	0	0	9	16	154	339	490	399	219	54	0	0	1680
1992	0	0	1	20	71	154	298	191	121	19	0	0	875
1993	0	0	0	0	49	271	398	406	55	19	0	0	1198

SNOWFALL (inches) — INTERNATIONAL AIRPORT, KANSAS CITY, MISSOURI

SEASON	JULY	AUG	SEP	OCT	NOV	DEC	JAN	FEB	MAR	APR	MAY	JUNE	TOTAL
1964-65	0.0	0.0	0.0	0.0	2.1	6.2	4.2	7.7	9.6	0.0	0.0	0.0	29.8
1965-66	0.0	0.0	0.0	0.0	T	2.5	T	3.7	3.5	T	T	0.0	9.7
1966-67	0.0	0.0	0.0	0.0	T	7.2	7.6	0.9	1.1	T	0.0	0.0	16.8
1967-68	0.0	0.0	0.0	T	0.2	7.0	2.8	2.7	T	T	0.0	0.0	12.7
1968-69	0.0	0.0	0.0	0.0	3.2	0.8	6.0	1.6	2.8	0.0	0.0	0.0	14.4
1969-70	0.0	0.0	0.0	0.0	0.6	3.8	2.5	2.2	1.0	4.6	0.0	0.0	14.7
1970-71	0.0	0.0	0.0	T	T	3.6	1.3	8.2	7.4	0.0	0.0	0.0	20.5
1971-72	0.0	0.0	0.0	0.0	0.2	1.0	3.0	2.9	3.3	T	0.0	0.0	10.4
#1972-73	0.0	0.0	0.0	0.0	3.6	3.1	10.9	0.3	0.0	1.3	0.0	0.0	19.2
1973-74	0.0	0.0	0.0	0.0	T	8.0	4.8	0.3	0.5	0.3	0.0	0.0	13.9
1974-75	0.0	0.0	0.0	0.0	1.4	1.2	5.4	4.6	5.9	2.3	0.0	0.0	20.8
1975-76	0.0	0.0	0.0	0.0	7.1	2.8	5.2	5.2	1.5	0.0	0.0	0.0	21.8
1976-77	0.0	0.0	0.0	T	0.8	0.2	14.2	0.3	T	0.6	0.0	0.0	16.1
1977-78	0.0	0.0	0.0	0.0	T	1.2	3.9	12.7	11.4	0.0	0.0	0.0	29.2
#1978-79	0.0	0.0	0.0	0.0	0.2	11.7	13.3	1.5	4.7	2.0	0.0	0.0	33.4
1979-80	0.0	0.0	0.0	0.0	T	T	5.4	12.7	5.4	0.0	0.0	0.0	23.5
1980-81	0.0	0.0	0.0	T	T	3.2	4.0	2.9	0.1	0.0	0.0	0.0	10.2
1981-82	0.0	0.0	0.0	0.0	0.1	5.3	6.0	12.7	4.0	1.3	0.0	0.0	29.4
1982-83	0.0	0.0	0.0	0.0	0.5	0.7	6.3	7.4	1.3	7.2	0.0	0.0	23.4
1983-84	0.0	0.0	0.0	0.0	0.7	13.2	1.3	0.5	8.7	0.0	0.0	0.0	24.4
1984-85	0.0	0.0	0.0	0.0	0.4	7.0	11.8	6.9	0.3	0.0	0.0	0.0	26.4
1985-86	0.0	0.0	0.0	0.0	3.5	5.4	T	4.5	T	T	0.0	0.0	13.4
1986-87	0.0	0.0	0.0	T	0.6	1.2	10.5	5.0	T	0.0	0.0	0.0	17.3
1987-88	0.0	0.0	0.0	T	2.0	11.9	0.9	9.3	2.2	0.0	0.0	0.0	26.3
1988-89	0.0	0.0	0.0	0.0	0.1	0.1	0.2	6.5	T	0.0	T	T	6.9
1989-90	0.0	0.0	0.0	0.0	T	6.8	1.0	2.1	9.6	0.0	T	0.0	19.5
1990-91	0.0	0.0	0.0	0.0	1.7	1.6	12.1	T	1.2	T	0.0	0.0	16.6
1991-92	0.0	0.0	0.0	0.0	4.6	0.2	T	2.0	0.5	2.8	0.0	0.0	10.1
1992-93	T	0.0	T	T	4.1	0.8	12.0	15.7	1.1	0.6	0.0	T	34.3
1993-94	0.0	0.0	0.0	T	0.5	2.7							
Record Mean	T	0.0	T	T	1.1	4.4	5.8	4.5	3.6	0.8	T	T	20.1

ST. LOUIS, MISSOURI

Saint Louis is located at the confluence of the Missouri and Mississippi Rivers and near the geographical center of the United States. Thus, with a somewhat modified continental climate, it is in the enviable position of being able to enjoy the changes of a four-season climate without the undue hardship of prolonged periods of extreme heat or high humidity. To the south is the warm, moist air of the Gulf of Mexico, and to the north, in Canada, is a favored region of cold air masses. The alternate invasion of Saint Louis by air masses from these sources, and the conflict along the frontal zones where they come together, produce a variety of weather conditions, none of which are likely to persist to the point of monotony.

Winters are brisk and stimulating, seldom severe. Records since 1870 show that temperatures drop to zero or below an average of two or three days per year. Temperatures remain as cold as 32 degrees or lower less than 25 days in most years. Snowfall has averaged a little over 18 inches per winter season. Snowfall of an inch or more is received on five to ten days in most years.

The long-term record for Saint Louis (since 1870) indicates that temperatures of 90 degrees or higher occur on about 35-40 days a year. Extremely hot days of 100 degrees or more are expected on no more than five days per year.

Normal annual precipitation for the Saint Louis area, is a little less than 34 inches. The three winter months are the driest, with an average total of about 6 inches of precipitation. The spring months of March through May are normally the wettest with normal total precipitation of just under 10 1/2 inches. It is not unusual to have extended dry periods of one to two weeks during the growing season.

Thunderstorms occur normally on between 40 and 50 days per year. During any year, there are usually a few of these that can be classified as severe storms with hail and damaging winds. Tornadoes have produced extensive damage and loss of life in the Saint Louis area.

NORMALS, MEANS, AND EXTREMES

ST. LOUIS, MISSOURI

LATITUDE: 38°45'N LONGITUDE: 90°22'W ELEVATION: FT. GRND 535 BARO 565 TIME ZONE: CENTRAL WBAN: 13994

	(a)	JAN	FEB	MAR	APR	MAY	JUNE	JULY	AUG	SEP	OCT	NOV	DEC	YEAR
TEMPERATURE °F:														
Normals														
-Daily Maximum		37.7	42.6	54.6	66.9	76.1	85.2	89.3	87.3	79.9	68.5	54.7	41.7	65.4
-Daily Minimum		20.8	25.1	35.5	46.4	56.0	65.7	70.4	67.9	60.5	48.3	37.7	26.0	46.7
-Monthly		29.3	33.9	45.1	56.7	66.1	75.4	79.8	77.6	70.2	58.4	46.2	33.9	56.1
Extremes														
-Record Highest	36	76	85	89	93	93	102	107	107	104	94	85	76	107
-Year		1970	1972	1985	1989	1991	1988	1980	1984	1984	1963	1989	1970	AUG 1984
-Record Lowest	36	-18	-10	-5	22	31	43	51	47	36	23	1	-16	-18
-Year		1985	1979	1960	1975	1976	1969	1972	1986	1974	1976	1964	1989	JAN 1985
NORMAL DEGREE DAYS:														
Heating (base 65°F)		1107	871	617	266	111	0	0	0	21	237	564	964	4758
Cooling (base 65°F)		0	0	0	17	145	312	459	391	177	33	0	0	1534
% OF POSSIBLE SUNSHINE	34	51	52	53	56	59	66	68	64	63	59	46	43	57
MEAN SKY COVER (tenths)														
Sunrise - Sunset	45	6.7	6.6	6.8	6.5	6.3	6.0	5.6	5.4	5.2	5.1	6.3	6.8	6.1
MEAN NUMBER OF DAYS:														
Sunrise to Sunset														
-Clear	45	7.5	6.8	6.5	6.9	7.2	7.2	9.2	10.0	11.4	12.3	8.6	7.2	100.9
-Partly Cloudy	45	6.6	6.4	8.1	8.1	9.5	10.9	11.4	11.1	8.2	7.4	6.5	6.5	100.9
-Cloudy	45	16.9	15.1	16.4	14.9	14.2	11.9	10.4	9.8	10.4	11.3	14.9	17.3	163.5
Precipitation														
.01 inches or more	36	8.4	8.2	11.4	11.3	10.7	9.3	8.9	7.8	8.0	8.4	9.6	9.4	111.1
Snow,Ice pellets,hail														
1.0 inches or more	36	1.8	1.5	1.0	0.2	0.0	0.0	0.0	0.0	0.0	0.0	0.5	1.3	6.2
Thunderstorms	36	0.7	0.9	3.1	5.7	6.5	7.1	7.4	6.4	3.9	2.6	1.7	0.7	46.8
Heavy Fog Visibility														
1/4 mile or less	36	2.1	1.7	1.5	0.6	0.6	0.3	0.2	0.4	0.5	0.8	1.1	1.9	11.5
Temperature °F														
-Maximum														
90° and above	33	0.0	0.0	0.0	0.3	1.3	8.1	15.1	11.7	4.3	0.2	0.0	0.0	41.1
32° and below	33	11.2	6.8	1.2	0.0	0.0	0.0	0.0	0.0	0.0	0.0	0.8	6.6	26.6
-Minimum														
32° and below	33	26.4	21.0	13.2	2.6	0.1	0.0	0.0	0.0	0.0	1.5	11.0	22.9	98.8
0° and below	33	2.1	0.5	0.*	0.0	0.0	0.0	0.0	0.0	0.0	0.0	0.0	1.0	3.6
AVG. STATION PRESS. (mb)	21	1000.0	999.0	995.4	994.5	994.2	994.5	995.7	996.8	997.5	998.3	997.9	999.4	996.9
RELATIVE HUMIDITY (%)														
Hour 00	33	77	77	74	70	75	76	77	79	80	75	77	79	76
Hour 06 (Local Time)	33	81	82	81	78	82	82	84	87	88	84	83	83	83
Hour 12	33	66	63	59	54	55	55	56	56	58	55	63	68	59
Hour 18	33	69	65	59	53	55	55	56	57	60	60	67	72	61
PRECIPITATION (inches):														
Water Equivalent														
-Normal		1.81	2.12	3.58	3.50	3.97	3.72	3.85	2.85	3.12	2.68	3.28	3.03	37.51
-Maximum Monthly	36	5.38	4.68	6.67	9.09	9.59	9.43	10.71	6.44	9.16	7.12	9.95	7.82	10.71
-Year		1975	1986	1978	1970	1990	1985	1981	1970	1993	1984	1985	1982	JUL 1981
-Minimum Monthly	36	0.10	0.25	1.09	0.99	1.02	0.44	0.60	0.08	T	0.21	0.44	0.32	T
-Year		1986	1963	1966	1977	1972	1991	1970	1971	1979	1975	1969	1958	SEP 1979
-Maximum in 24 hrs	36	2.43	2.56	2.95	4.91	3.62	3.29	3.47	3.17	3.50	2.70	3.71	4.03	4.91
-Year		1975	1959	1977	1979	1990	1960	1982	1993	1986	1986	1985	1982	APR 1979
Snow,Ice pellets,hail														
-Maximum Monthly	57	23.9	20.8	22.3	6.5	0.6	T	0.0	0.0	0.0	T	11.3	26.3	26.3
-Year		1977	1993	1960	1971	1993	1993				1993	1951	1973	DEC 1973
-Maximum in 24 hrs	57	13.9	11.7	10.7	6.1	0.6	T	0.0	0.0	0.0	T	10.3	12.0	13.9
-Year		1982	1993	1989	1971	1993	1993				1993	1951	1973	JAN 1982
WIND:														
Mean Speed (mph)	44	10.6	10.7	11.7	11.3	9.5	8.8	8.0	7.6	8.2	8.9	10.1	10.4	9.7
Prevailing Direction through 1963		NW	NW	WNW	WNW	S	S	S	S	S	S	S	WNW	S
Fastest Obs. 1 Min.														
-Direction (!!)	14	29	30	27	27	34	27	36	31	25	28	11	29	28
-Speed (MPH)	14	40	45	48	49	46	48	46	40	41	52	41	39	52
-Year		1984	1980	1984	1982	1990	1982	1987	1987	1986	1981	1987	1985	OCT 1981
Peak Gust														
-Direction (!!)	10	NW	W	W	SW	N	NW	N	S	W	NW	W	SW	W
-Speed (mph)	10	53	54	66	58	59	60	62	53	49	58	64	55	66
-Date		1992	1991	1984	1984	1990	1990	1987	1985	1986	1988	1984	1984	MAR 1984

PRECIPITATION (inches) ST. LOUIS, MISSOURI

YEAR	JAN	FEB	MAR	APR	MAY	JUNE	JULY	AUG	SEP	OCT	NOV	DEC	ANNUAL
1964	1.70	2.30	3.84	4.99	2.68	2.73	4.25	2.39	1.47	0.73	3.84	1.24	32.16
1965	2.51	1.16	2.34	3.67	1.38	3.03	3.17	3.59	3.00	0.46	0.78	3.17	28.26
1966	0.65	4.12	1.09	6.03	4.59	1.59	1.26	3.72	2.15	2.18	2.47	2.49	32.34
1967	2.89	1.72	2.77	3.40	4.73	4.46	3.84	1.36	4.33	3.45	2.15	6.20	41.30
1968	1.86	1.09	2.06	1.48	6.78	0.90	3.92	1.60	3.74	0.69	5.74	2.63	32.49
1969	3.61	2.04	2.47	4.01	2.11	8.65	7.08	0.52	5.03	5.77	0.44	1.99	43.72
1970	0.22	0.64	2.17	9.09	2.04	5.08	0.60	6.44	5.54	2.21	0.77	1.40	36.20
1971	0.66	3.08	1.81	1.65	5.66	2.43	4.70	0.08	3.98	1.51	1.67	6.50	33.73
1972	0.77	0.74	2.93	4.49	1 02	1.19	3.10	2.69	6.21	1.47	5.59	3.54	33.74
1973	1.40	1.04	5.81	4.25	3.92	4.23	2.85	2.46	3.52	2.33	3.65	4.36	39.82
1974	3.51	4.17	2.58	2.40	5.90	3.45	0.90	5.05	2.50	1.51	3.15	1.71	36.83
1975	5.38	3.59	4.08	4.56	3.23	3.78	2.56	5.44	2.48	0.21	2.62	2.28	40.21
1976	0.83	1.08	4.28	1.37	3.90	2.32	2.28	1.27	0.90	3.37	0.73	1.13	23.46
1977	2.38	2.47	6.28	0.99	2.13	5.47	4.28	5.34	3.64	3.76	4.33	2.34	43.41
1978	1.70	1.60	6.67	3.21	3.69	2.39	6.03	0.76	3.10	2.28	4.47	1.81	37.71
1979	1.95	1.48	3.63	7.47	1.62	1.67	3.67	2.26	T	1.81	2.07	1.85	29.48
1980	0.63	1.54	3.98	1.54	3.40	2.19	3.56	2.72	3.12	2.89	1.25	0.66	27.48
1981	0.64	2.18	2.97	3.40	6.79	5.82	10.71	3.31	1.17	3.81	2.71	2.01	45.52
1982	4.90	1.37	2.88	2.55	4.85	5.96	7.91	5.27	5.27	2.30	3.89	7.82	54.97
1983	0.72	0.95	3.54	7.30	6.32	4.32	1.23	2.24	1.24	5.40	7.79	3.75	44.80
1984	0.84	3.43	5.37	6.29	5.19	2.74	0.76	0.64	8.88	7.12	5.50	4.89	51.65
1985	0.53	3.77	5.18	3.60	3.30	9.43	5.23	3.66	0.43	1.96	9.95	3.69	50.73
1986	0.10	4.68	1.22	1.23	2.42	4.43	2.61	2.22	7.99	5.34	1.58	1.06	34.88
1987	1.98	1.40	2.16	1.74	2.00	3.59	5.04	5.56	1.62	1.74	4.09	7.46	38.38
1988	3.30	2.27	4.73	1.15	1.44	1.97	3.02	2.31	1.99	1.86	6.65	3.24	33.93
1989	2.58	1.43	4.53	2.10	4.11	2.34	4.59	3.00	1.69	0.95	0.59	0.69	28.60
1990	1.42	3.53	2.66	3.07	9.59	3.02	3.34	2.84	0.78	4.96	3.36	6.52	45.09
1991	1.52	0.98	3.20	3.27	3.87	0.44	5.18	0.98	2.98	5.70	3.26	2.10	33.48
1992	1.12	1.89	3.45	2.46	1.45	1.19	4.31	3.45	2.98	1.21	6.32	3.66	33.49
1993	3.54	2.75	3.31	6.16	3.94	7.12	5.06	4.78	9.16	2.61	4.85	1.48	54.76
Record Mean	2.14	2.28	3.35	3.68	4.08	4.04	3.45	2.96	3.19	2.73	2.85	2.36	37.11

AVERAGE TEMPERATURE (deg. F) ST. LOUIS, MISSOURI

YEAR	JAN	FEB	MAR	APR	MAY	JUNE	JULY	AUG	SEP	OCT	NOV	DEC	ANNUAL
1964	34.7	34.3	42.3	58.7	69.6	74.8	78.5	76.5	69.3	54.0	47.3	31.9	56.0
1965	31.9	33.8	34.5	58.4	70.5	75.1	76.8	75.8	69.1	56.4	48.5	41.7	56.1
1966	25.0	32.1	46.7	51.7	61.7	73.7	82.9	74.2	65.6	54.2	47.1	34.8	54.1
1967	34.7	31.3	48.3	58.5	60.8	74.0	74.9	72.4	66.5	57.5	42.0	34.8	54.7
1968	28.6	29.3	46.1	55.6	61.9	77.0	77.6	77.5	68.0	56.9	43.5	32.3	54.5
1969	29.2	35.3	37.8	56.9	65.7	72.5	80.5	77.0	69.3	56.1	43.4	32.1	54.7
1970	24.8	33.0	40.6	58.0	69.3	72.3	77.9	76.3	71.8	56.1	43.6	37.0	55.1
1971	27.4	33.9	41.7	55.9	62.3	78.9	75.4	75.7	72.5	63.7	46.0	40.6	56.2
1972	29.9	34.3	45.0	56.1	66.4	73.4	77.5	76.3	71.1	54.8	39.8	30.4	54.6
1973	32.6	34.5	50.9	53.7	61.7	74.6	78.7	76.9	70.0	60.7	47.0	30.0	55.9
1974	29.8	36.1	48.1	57.5	65.0	69.5	79.8	74.5	62.3	57.8	44.0	34.0	54.9
1975	33.2	32.0	39.0	53.4	67.6	74.9	77.8	77.9	64.7	59.0	48.3	35.5	55.3
1976	28.1	43.5	48.8	56.5	60.8	72.7	79.5	74.3	68.2	50.9	37.0	28.5	54.1
1977	15.1	34.8	49.6	60.8	71.1	74.7	81.2	76.4	70.7	55.5	44.7	30.6	55.4
1978	19.6	21.1	37.9	56.2	63.6	74.4	78.5	76.4	73.1	55.6	47.5	35.0	53.2
1979	16.6	23.1	44.1	52.9	65.5	76.5	79.2	78.4	70.8	59.2	44.5	38.7	54.1
1980	31.4	27.9	41.0	54.5	66.9	75.5	85.0	83.5	72.5	55.9	46.4	36.6	56.4
1981	31.2	36.8	46.7	63.1	60.7	75.7	78.7	76.1	69.2	55.7	48.7	31.1	56.1
1982	22.5	28.6	45.3	51.5	70.7	70.6	79.3	75.2	68.0	58.3	46.3	41.6	54.9
1983	32.3	38.1	44.5	50.3	62.3	75.3	83.5	84.2	72.0	59.7	48.2	20.5	55.9
1984	28.3	40.4	37.1	54.1	63.2	79.5	78.3	80.7	68.2	61.8	44.3	40.7	56.4
1985	22.6	30.5	49.5	60.4	67.6	71.6	79.3	74.7	70.9	61.4	46.5	27.3	55.2
1986	34.9	34.5	49.2	60.8	68.2	78.3	82.8	74.0	73.3	58.3	41.5	35.4	57.6
1987	30.6	40.1	48.6	56.9	72.6	77.9	81.0	78.9	70.5	53.8	49.1	38.0	58.2
1988	29.2	30.5	45.2	57.1	69.0	77.7	81.6	82.7	72.5	53.9	47.2	37.2	57.0
1989	41.2	28.2	45.0	57.7	64.3	74.8	79.3	77.8	67.4	61.3	47.1	24.1	55.7
1990	42.9	41.3	49.8	55.7	63.6	77.2	80.2	77.9	74.1	58.1	52.7	34.7	59.0
1991	29.3	41.7	50.1	61.5	73.0	79.9	80.9	79.7	72.4	60.5	42.4	39.2	59.2
1992	37.0	42.7	48.5	57.8	65.1	73.8	79.0	73.4	69.2	59.4	44.3	35.7	57.2
1993	32.1	31.7	41.5	54.2	66.4	75.1	81.9	80.4	66.4	56.1	43.9	37.2	55.6
Record Mean	31.3	34.7	44.5	56.1	65.9	75.2	79.5	77.6	70.1	58.8	45.4	35.0	56.2
Max	39.5	43.3	53.8	65.8	75.6	84.5	88.8	87.0	79.8	68.5	53.9	42.8	65.3
Min	23.0	26.1	35.1	46.4	56.3	65.8	70.2	68.1	60.4	49.0	36.8	27.3	47.1

REFERENCE NOTES ST. LOUIS, MISSOURI

GENERAL
T - TRACE AMOUNT.
BLANK ENTRIES DENOTE MISSING/UNREPORTED DATA.
# INDICATES A STATION OR INSTRUMENT RELOCATION.

SPECIFIC

PM - INCLUDES LAST DAY OF PREVIOUS MONTH
ASOS - AUTOMATED SURFACE OBSERVING SYSTEM IN OPERATION DURING THESE MONTHS.

(*) - LENGTH OF RECORD IN YEARS, ALTHOUGH INDIVIDUAL MONTHS MAY BE MISSING.
0.* OR * - THE VALUE IS BETWEEN 0.0 AND 0.05
NORMALS - BASED ON THE 1961-1990 RECORD PERIOD.
EXTREMES - DATES ARE THE MOST RECENT OCCURRENCE
WIND DIR.- NUMERALS SHOW TENS OF DEGRESS CLOCKWISE FROM TRUE NORTH. "00" INDICATES CALM.
RESULTANT DIRECTIONS ARE GIVEN TO WHOLE DEGREES.
BOLD VALUES INDICATE EXTREME VALUES WHICH OCCURRED AFTER THE ASOS SYSTEM WAS COMMISSIONED.

RECORD = PERIOD OF RECORD
RECORD MEAN PRECIPITATION IS THE MEAN OF ALL DAILY PRECIPITATION AMOUNTS DURING THE PERIOD OF RECORD.
RECORD MAX(MIN) TEMPERATURE IS THE MEAN OF ALL DAILY MAX(MIN) TEMPERATURES DURING THE PERIOD OF RECORD.
RECORD MEAN TEMPERATURE IS THE SUM OF THE RECORD MAX AND RECORD MIN DIVIDED BY 2.
AVERAGE TEMPERATURE IS THE SUM OF THE MEAN DAILY MAX AND MIN TEMPERATURE DIVIDED BY 2.

EXCEPTIONS

RECORD MEANS ARE THROUGH THE CURRENT YEAR, BEGINNING IN 1873 FOR TEMPERATURE
1871 FOR PRECIPITATION
1937 FOR SNOWFALL

HEATING DEGREE DAYS Base 65 deg. F

ST. LOUIS, MISSOURI

SEASON	JULY	AUG	SEP	OCT	NOV	DEC	JAN	FEB	MAR	APR	MAY	JUNE	TOTAL
1964-65	0	2	55	333	526	1019	1016	869	938	250	19	0	5027
1965-66	0	7	49	286	489	717	1233	915	570	397	156	4	4823
1966-67	0	0	78	336	534	931	932	940	530	257	188	12	4738
1967-68	3	7	67	268	682	928	1121	1031	587	290	137	2	5123
1968-69	0	2	14	293	640	1008	1106	826	834	247	85	15	5070
1969-70	0	0	27	313	644	1013	1241	893	751	257	55	20	5214
1970-71	0	0	24	287	635	863	1159	866	718	303	122	0	4977
1971-72	0	0	47	97	574	751	1081	884	619	295	80	10	4438
1972-73	2	0	29	317	751	1069	997	849	430	348	121	0	4913
1973-74	0	0	31	182	538	1077	1083	804	539	253	101	21	4629
1974-75	0	0	127	242	625	954	979	919	803	353	48	1	5051
1975-76	2	0	110	228	498	910	1137	619	505	288	158	0	4455
1976-77	0	0	24	456	832	1125	1541	839	471	190	36	3	5517
1977-78	0	0	11	291	601	1059	1401	1223	840	275	158	5	5864
1978-79	0	0	24	292	528	923	1496	1167	644	364	80	0	5518
1979-80	0	0	16	223	610	810	1035	1071	740	331	54	0	4890
1980-81	0	0	30	305	553	877	1039	784	569	127	168	0	4452
1981-82	0	0	35	298	483	1048	1308	1015	603	407	9	11	5217
1982-83	0	0	49	261	569	721	1008	745	632	437	117	7	4546
1983-84	0	0	58	192	498	1376	1133	705	860	342	115	0	5279
1984-85	0	0	103	151	616	746	1308	960	487	200	42	10	4623
1985-86	0	0	64	145	550	1159	929	850	506	194	44	0	4441
1986-87	0	11	12	221	699	910	1062	691	501	267	10	0	4384
1987-88	0	0	12	346	490	830	1102	995	610	241	17	3	4646
1988-89	0	0	5	354	528	854	730	1029	625	293	128	4	4550
1989-90	0	0	73	183	536	1261	679	657	496	327	85	9	4306
1990-91	3	0	24	250	375	934	1101	648	474	154	21	0	3984
1991-92	0	0	67	191	674	796	859	642	509	264	95	5	4102
1992-93	0	3	46	204	615	902	1014	927	726	327	52	13	4829
1993-94	0	0	58	292	628	852							

COOLING DEGREE DAYS Base 65 deg. F

ST. LOUIS, MISSOURI

YEAR	JAN	FEB	MAR	APR	MAY	JUNE	JULY	AUG	SEP	OCT	NOV	DEC	TOTAL
1969	0	0	0	9	113	246	486	381	162	43	0	0	1440
1970	0	0	0	54	195	247	410	360	236	17	0	0	1519
1971	0	0	2	35	45	427	326	340	279	62	10	0	1526
1972	0	0	5	35	129	268	394	358	219	9	0	0	1417
1973	0	0	1	17	25	294	435	375	187	57	4	0	1395
1974	0	0	24	36	109	164	463	300	52	25	2	0	1175
1975	0	0	2	12	133	308	405	406	110	51	4	0	1431
1976	0	0	8	38	34	239	458	298	129	25	0	0	1229
1977	0	0	2	69	231	302	509	360	190	4	0	0	1667
1978	0	0	7	17	120	295	426	360	276	10	8	0	1519
1979	0	0	2	9	102	354	446	420	195	50	0	0	1578
1980	0	0	0	23	120	320	626	580	262	31	2	0	1964
1981	0	0	7	77	42	327	431	353	166	13	1	0	1417
1982	0	0	0	7	191	186	453	322	146	63	15	4	1387
1983	0	0	3	3	41	322	578	603	274	36	2	0	1862
1984	0	0	0	24	67	442	423	493	202	57	2	1	1711
1985	0	0	14	70	128	214	451	310	245	43	2	0	1477
1986	0	0	25	75	150	407	561	298	267	21	0	0	1804
1987	0	0	0	32	251	393	501	439	183	5	20	0	1824
1988	0	0	4	10	144	389	521	556	238	16	0	0	1878
1989	0	0	11	80	111	305	450	403	151	75	6	0	1592
1990	0	0	30	55	47	382	480	408	304	41	12	0	1759
1991	0	0	19	56	277	452	498	463	295	58	3	0	2121
1992	0	0	7	54	104	277	440	268	182	36	0	0	1368
1993	0	0	1	9	102	320	529	483	107	23	0	0	1574

SNOWFALL (inches)

ST. LOUIS, MISSOURI

SEASON	JULY	AUG	SEP	OCT	NOV	DEC	JAN	FEB	MAR	APR	MAY	JUNE	TOTAL
1964-65	0.0	0.0	0.0	0.0	3.1	0.3	7.3	5.7	8.7	0.0	0.0	0.0	25.1
1965-66	0.0	0.0	0.0	0.0	T	0.1	2.5	7.1	0.9	T	0.0	0.0	10.6
1966-67	0.0	0.0	0.0	0.0	T	0.1	1.4	0.9	1.2	0.0	0.0	0.0	3.6
1967-68	0.0	0.0	0.0	T	1.6	4.2	6.9	0.4	7.7	0.0	0.0	0.0	20.8
1968-69	0.0	0.0	0.0	0.0	T	1.3	2.3	5.8	2.7	0.0	0.0	0.0	12.1
1969-70	0.0	0.0	0.0	0.0	T	10.2	2.1	3.7	5.0	1.0	0.0	0.0	22.0
1970-71	0.0	0.0	0.0	0.0	T	0.9	0.4	1.4	0.2	6.5	0.0	0.0	9.4
1971-72	0.0	0.0	0.0	0.0	1.3	0.0	4.1	1.9	0.3	T	0.0	0.0	7.6
1972-73	0.0	0.0	0.0	0.0	5.2	1.0	2.2	3.0	0.2	0.2	0.0	0.0	11.8
1973-74	0.0	0.0	0.0	0.0	0.0	26.3	4.2	4.5	7.4	0.0	0.0	0.0	42.4
1974-75	0.0	0.0	0.0	0.0	1.2	1.5	4.1	12.1	6.3	T	0.0	0.0	25.2
1975-76	0.0	0.0	0.0	0.0	7.6	3.9	4.5	4.5	4.8	0.0	0.0	0.0	25.3
1976-77	0.0	0.0	0.0	0.0	0.3	5.2	23.9	6.7	0.1	0.1	0.0	0.0	36.3
1977-78	0.0	0.0	0.0	0.0	6.7	11.7	22.9	9.3	15.4	0.0	0.0	0.0	66.0
1978-79	0.0	0.0	0.0	0.0	0.0	1.4	18.4	4.8	2.0	0.0	0.0	0.0	26.6
1979-80	0.0	0.0	0.0	0.0	0.3	T	4.2	7.4	8.7	5.0	0.0	0.0	25.6
1980-81	0.0	0.0	0.0	0.0	8.0	1.1	0.6	8.2	0.2	0.0	0.0	0.0	18.1
1981-82	0.0	0.0	0.0	0.0	T	7.9	16.6	9.0	0.4	2.7	0.0	0.0	36.6
1982-83	0.0	0.0	0.0	0.0	T	T	3.3	0.3	1.4	2.4	0.0	0.0	7.4
1983-84	0.0	0.0	0.0	0.0	T	6.5	2.3	9.9	5.2	0.0	0.0	0.0	23.9
1984-85	0.0	0.0	0.0	0.0	1.7	1.8	5.1	1.3	T	0.0	0.0	0.0	9.9
1985-86	0.0	0.0	0.0	0.0	T	5.7	1.0	6.1	0.2	T	0.0	0.0	13.0
1986-87	0.0	0.0	0.0	0.0	T	T	23.6	0.6	T	0.0	0.0	0.0	24.2
1987-88	0.0	0.0	0.0	0.0	T	7.3	1.4	6.7	2.8	0.0	0.0	0.0	18.2
1988-89	0.0	0.0	0.0	0.0	2.9	5.9	0.1	3.9	11.0	0.0	0.0	0.0	23.8
1989-90	0.0	0.0	0.0	T	T	9.1	0.2	6.9	8.4	T	T	T	24.6
1990-91	0.0	0.0	0.0	0.0	T	13.2	1.9	T	1.7	0.0	0.0	0.0	16.8
1991-92	0.0	0.0	0.0	T	3.9	0.4	3.8	2.3	3.1	0.0	0.0	0.0	13.5
1992-93	0.0	0.0	0.0	0.0	T	0.7	5.5	20.8	3.0	0.3	0.6	T	30.9
1993-94	0.0	0.0	0.0	T	1.6	1.1							
Record Mean	0.0	0.0	0.0	T	1.4	3.8	5.3	4.7	4.2	0.4	T	T	19.8

SPRINGFIELD, MISSOURI

The entire metropolitan area, airport, and surrounding territory consists of comparatively flat or very gently rolling tableland, practically atop the crest of the Missouri Ozark Mountain plateau. The average elevation of the city proper is slightly over 1,300 feet above sea level. There are no serious problems of instrumental exposure.

As a result of this advantageous location, the city and surrounding territory enjoy what is described as a plateau climate. The winter season in the Ozarks has temperatures considerably milder than in the upland, plain or prairie, and in the summer the Ozarks are appreciably cooler.

The city of Springfield also occupies a unique location with regard to natural water drainage. The line separating two major water sheds crosses the north-central part of the city. Drainage north of this line flows north into the Gasconade and Missouri Rivers. To the south of the line, drainage is to the south into the White and Mississippi Rivers.

The average annual temperature range is over 140 degrees with lowest temperatures below -25 and and highest temperatures above 115 degrees.

The growing season extends over a period of 199 days. Agriculture is greatly diversified, practically every farm product of the temperate zone is grown in this area. It is a noted livestock and poultry production and distribution center. The climate permits green pasturage the year around in varying quantity, resulting in ever increasing cattle production for both meat and dairy products.

The air is remarkably free from palls of industrial smoke, and the altitude of the city also tends to prevent other than few amounts of either radiation or advection fogs.

NORMALS, MEANS, AND EXTREMES

SPRINGFIELD, MISSOURI

LATITUDE: 37°14'N LONGITUDE: 93°23' W ELEVATION: FT. GRND 1268 BARO 1272 TIME ZONE: CENTRAL WBAN: 13995

	(a)	JAN	FEB	MAR	APR	MAY	JUNE	JULY	AUG	SEP	OCT	NOV	DEC	YEAR
TEMPERATURE °F:														
Normals														
-Daily Maximum		41.8	46.3	57.4	67.9	76.0	84.4	89.6	88.6	80.3	69.8	56.6	45.3	67.0
-Daily Minimum		20.4	25.0	34.4	44.1	53.2	61.9	66.6	65.0	57.7	45.9	35.5	25.3	44.6
-Monthly		31.1	35.7	46.0	56.0	64.6	73.2	78.1	76.8	69.0	57.8	46.0	35.3	55.8
Extremes														
-Record Highest	48	76	81	87	93	93	101	113	106	104	93	80	77	113
-Year		1950	1972	1974	1963	1972	1954	1954	1984	1947	1981	1989	1991	JUL 1954
-Record Lowest	48	-13	-17	-3	18	30	42	44	44	31	18	4	-16	-17
-Year		1985	1979	1948	1957	1970	1966	1972	1967	1984	1993	1959	1989	FEB 1979
NORMAL DEGREE DAYS:														
Heating (base 65°F)		1051	820	589	280	110	5	0	0	43	249	570	921	4638
Cooling (base 65°F)		0	0	0	10	98	251	406	366	163	26	0	0	1320
% OF POSSIBLE SUNSHINE	48	51	53	56	59	61	65	71	71	68	64	53	48	60
MEAN SKY COVER (tenths)														
Sunrise - Sunset	48	6.4	6.5	6.5	6.2	6.2	5.7	5.1	5.0	5.0	4.9	5.8	6.3	5.8
MEAN NUMBER OF DAYS:														
Sunrise to Sunset														
-Clear	48	8.4	7.4	7.5	8.3	7.5	8.4	11.8	11.9	12.1	13.3	10.0	8.6	115.0
-Partly Cloudy	48	6.9	6.0	7.8	7.6	9.7	10.4	10.3	10.5	7.4	6.5	6.4	6.3	95.8
-Cloudy	48	15.7	14.8	15.8	14.1	13.9	11.2	8.9	8.6	10.5	11.3	13.5	16.1	154.4
Precipitation														
.01 inches or more	48	8.2	8.4	10.5	10.5	11.1	10.0	7.9	8.5	8.4	8.0	8.5	8.6	108.5
Snow,Ice pellets,hail														
1.0 inches or more	48	1.6	1.2	0.9	0.1	0.0	0.0	0.0	0.0	0.0	0.*	0.4	1.1	5.3
Thunderstorms	48	0.8	1.3	3.8	6.1	8.3	9.1	7.7	7.8	5.4	3.4	2.1	1.1	56.9
Heavy Fog Visibility														
1/4 mile or less	48	3.5	2.7	1.4	0.9	1.0	1.1	0.9	1.0	1.4	1.7	1.9	2.9	20.5
Temperature °F														
-Maximum														
90° and above	33	0.0	0.0	0.0	0.1	0.3	6.1	16.6	14.8	3.9	0.3	0.0	0.0	42.1
32° and below	33	7.3	4.4	0.8	0.0	0.0	0.0	0.0	0.0	0.0	0.0	0.6	4.6	17.8
-Minimum														
32° and below	33	25.9	20.5	13.5	3.2	0.3	0.0	0.0	0.0	0.1	2.4	12.5	22.9	101.2
0° and below	33	2.2	0.8	0.*	0.0	0.0	0.0	0.0	0.0	0.0	0.0	0.0	0.9	3.9
AVG. STATION PRESS.(mb)	21	974.0	972.9	969.5	969.2	969.1	970.0	971.5	972.1	972.6	973.2	972.4	973.6	971.7
RELATIVE HUMIDITY (%)														
Hour 00	33	73	74	72	72	81	84	82	81	83	77	76	76	78
Hour 06 (Local Time)	33	77	79	78	79	85	86	87	87	88	82	80	79	82
Hour 12	33	61	60	56	55	59	60	56	54	59	54	59	63	58
Hour 18	33	64	62	55	54	60	61	57	56	64	62	66	68	61
PRECIPITATION (inches):														
Water Equivalent														
-Normal		1.79	2.17	3.89	4.18	4.38	5.09	2.92	3.51	4.62	3.58	3.75	3.16	43.04
-Maximum Monthly	48	6.77	5.68	9.01	10.19	13.41	11.31	18.75	8.60	17.46	8.70	12.24	8.84	18.75
-Year		1950	1990	1973	1983	1990	1975	1958	1982	1993	1967	1985	1982	JUL 1958
-Minimum Monthly	48	0.08	0.35	0.50	0.15	1.54	0.23	0.33	0.50	0.20	0.43	0.19	0.13	0.08
-Year		1961	1947	1956	1989	1969	1991	1953	1955	1952	1963	1989	1950	JAN 1961
-Maximum in 24 hrs	48	3.91	3.73	4.09	3.85	5.81	4.83	6.85	3.41	6.27	5.48	6.29	4.50	6.85
-Year		1950	1966	1974	1970	1989	1975	1958	1976	1993	1949	1987	1982	JUL 1958
Snow,Ice pellets,hail														
-Maximum Monthly	48	23.1	19.2	23.9	7.1	T	T	T	0.0	0.0	1.2	19.5	14.5	23.9
-Year		1979	1980	1970	1971	1991	1993	1990			1993	1951	1983	MAR 1970
-Maximum in 24 hrs	48	9.1	16.1	15.7	6.9	T	T	T	0.0	0.0	1.1	12.5	7.5	16.1
-Year		1978	1980	1970	1971	1991	1993	1990			1993	1951	1966	FEB 1980
WIND:														
Mean Speed (mph)	48	11.5	11.8	12.8	12.1	10.3	9.6	8.5	8.5	9.3	10.1	11.2	11.5	10.6
Prevailing Direction through 1963		SSE	SSE	SSE	SSE	SSE	SSE	SSE	SSE	SSE	SSE	SSE	SSE	SSE
Fastest Obs. 1 Min.														
-Direction (!!)	8	NW	31	25	22	20	32	28	36	23	22	26	20	36
-Speed (MPH)	8	39	32	39	32	32	32	46	48	37	35	29	31	48
-Year		1984	1991	1991	1988	1988	1992	1992	1986	1987	1992	1993	1988	AUG 1986
Peak Gust														
-Direction (!!)	10	S	SW	W	S	SW	N	W	N	NW	NW	SW	S	W
-Speed (mph)	10	46	52	52	53	47	51	72	59	49	48	53	48	72
-Date		1988	1984	1991	1984	1988	1990	1992	1986	1987	1988	1990	1988	JUL 1992

PRECIPITATION (inches)

SPRINGFIELD, MISSOURI

YEAR	JAN	FEB	MAR	APR	MAY	JUNE	JULY	AUG	SEP	OCT	NOV	DEC	ANNUAL
1964	0.53	1.67	3.85	4.43	2.30	6.95	0.98	4.92	2.14	1.99	3.20	1.66	34.62
1965	2.47	2.13	1.96	7.40	2.37	6.77	1.91	5.13	6.13	0.90	0.43	3.84	41.44
1966	1.59	5.03	1.38	4.85	3.54	1.90	5.24	4.90	3.35	1.26	2.82	2.15	38.01
1967	1.17	1.61	1.66	5.30	3.99	7.12	2.43	1.02	3.36	8.70	1.53	5.85	43.74
1968	3.07	1.84	4.05	2.59	4.58	2.35	3.60	6.26	3.09	3.28	7.59	3.53	45.83
1969	3.33	1.32	3.07	4.02	1.54	5.38	1.33	2.62	2.44	6.26	1.02	2.66	34.99
1970	0.58	0.67	3.79	5.14	2.58	6.39	0.74	3.62	9.87	4.68	1.59	2.19	41.84
1971	1.23	2.37	1.22	3.14	3.84	3.64	2.54	2.30	5.07	3.63	2.66	6.24	37.88
1972	0.45	1.23	0.95	4.82	1.96	1.62	2.24	3.34	6.41	5.96	8.08	1.70	38.76
1973	4.41	0.85	9.01	6.46	4.31	5.47	7.96	0.98	4.71	4.27	6.39	4.58	59.40
1974	2.12	1.87	7.20	3.04	4.69	5.79	1.87	6.04	4.82	4.32	5.43	2.68	49.87
1975	3.95	3.93	6.32	3.45	1.96	11.31	1.85	2.57	11.36	1.05	4.24	1.95	53.94
1976	0.52	1.14	2.51	5.42	3.83	3.16	4.55	4.61	1.90	2.43	0.23	0.99	31.29
1977	1.54	1.64	3.43	2.35	2.30	7.13	1.48	3.14	9.03	2.24	3.02	1.64	38.94
1978	1.54	1.49	5.83	3.96	4.13	4.07	4.63	3.42	2.91	2.18	6.13	3.66	43.95
1979	2.35	1.66	2.85	7.19	7.42	8.49	5.54	3.61	2.06	3.26	3.16	1.35	48.94
1980	0.85	2.55	3.94	2.92	2.41	3.59	0.69	1.21	1.85	4.04	1.28	2.03	27.36
1981	1.10	1.55	2.25	3.09	6.20	8.54	3.08	5.02	0.83	7.89	2.62	1.55	43.72
1982	4.51	0.40	2.15	3.42	4.83	5.91	1.01	8.60	1.15	2.59	4.26	8.84	47.67
1983	0.77	0.43	2.58	10.19	3.78	2.31	5.95	1.28	3.46	6.67	5.08	2.55	45.05
1984	1.08	3.46	5.01	5.05	2.11	3.19	2.46	1.34	4.84	5.70	4.67	6.87	45.78
1985	1.55	4.25	6.28	3.92	3.75	7.20	1.61	5.97	1.66	5.05	12.24	3.02	56.50
1986	0.09	1.95	2.07	3.83	3.05	3.23	1.29	4.82	11.65	4.11	2.85	1.25	40.19
1987	2.07	5.25	3.95	2.15	5.39	5.80	2.57	4.35	4.08	3.24	9.16	7.48	55.49
1988	1.34	2.18	5.87	3.75	4.40	5.35	4.80	5.14	5.14	1.91	5.48	3.10	48.46
1989	1.60	1.98	4.95	0.15	6.93	2.67	4.79	3.91	2.40	1.12	0.19	0.81	31.50
1990	5.26	5.68	7.87	2.51	13.41	4.66	3.01	1.46	5.84	5.02	2.41	6.06	63.19
1991	3.00	0.93	1.12	5.16	5.58	0.23	4.20	2.72	3.55	3.78	3.92	3.40	37.59
1992	0.87	2.20	2.24	2.45	2.03	6.03	6.84	3.86	7.36	0.80	9.11	4.25	48.04
1993	2.82	2.67	2.62	3.09	4.42	8.75	4.19	2.71	17.46	1.37	4.05	1.63	55.78
Record Mean	2.16	2.12	3.44	4.00	4.87	4.92	3.62	3.71	4.03	3.26	3.02	2.51	41.65

AVERAGE TEMPERATURE (deg. F)

SPRINGFIELD, MISSOURI

YEAR	JAN	FEB	MAR	APR	MAY	JUNE	JULY	AUG	SEP	OCT	NOV	DEC	ANNUAL
1964	37.0	35.9	43.0	59.7	67.9	72.8	78.9	74.9	69.0	56.1	49.9	35.8	56.8
1965	36.2	36.7	37.1	60.1	68.3	73.3	76.2	75.2	69.7	56.7	49.5	43.2	56.8
1966	29.0	34.2	48.3	52.2	60.9	70.8	81.6	74.0	65.0	55.3	48.7	33.9	54.5
1967	35.1	32.6	50.4	58.5	61.3	73.4	73.5	72.3	66.4	58.0	44.5	38.3	55.4
1968	33.3	31.5	45.9	55.6	60.8	74.0	76.2	76.1	66.5	56.4	42.2	32.6	54.3
1969	33.4	36.6	39.2	56.9	65.7	70.9	80.6	76.6	69.0	55.5	44.1	33.6	55.2
1970	25.8	34.7	38.7	55.8	65.2	72.8	75.4	78.0	72.3	54.0	42.2	39.8	54.6
1971	31.4	34.4	44.0	55.9	62.5	77.6	76.0	76.2	71.6	62.4	46.1	42.8	56.8
1972	32.9	38.3	47.1	58.4	65.3	74.2	77.0	76.6	70.8	56.7	40.6	31.6	55.8
1973	33.2	35.9	51.6	53.2	60.9	72.7	77.0	77.3	71.1	62.2	49.8	34.7	56.7
1974	32.9	39.4	50.8	56.6	66.2	68.3	79.4	74.8	61.4	58.6	44.7	35.8	55.8
1975	36.4	34.3	40.4	54.6	66.5	74.5	77.5	76.5	63.3	58.5	47.6	38.6	55.7
1976	32.6	46.3	50.2	57.3	59.7	70.4	76.6	74.6	68.7	52.4	38.2	33.4	55.0
1977	18.9	37.9	49.4	59.7	69.1	75.8	80.2	79.3	72.6	57.7	47.4	35.5	56.9
1978	19.8	22.9	40.2	56.4	63.8	72.9	81.3	77.0	73.9	56.8	46.9	34.7	53.9
1979	17.6	27.1	45.3	53.6	62.3	70.5	75.1	75.6	68.7	60.9	43.7	41.3	53.5
1980	36.5	32.3	42.7	53.8	64.1	75.4	83.8	82.9	73.3	56.3	46.1	37.8	57.1
1981	33.6	38.5	47.2	62.6	61.8	75.8	80.2	75.4	69.8	57.1	49.1	34.2	57.1
1982	26.2	34.4	49.7	52.8	68.2	70.1	80.2	78.7	68.6	57.9	46.7	42.1	56.3
1983	34.9	40.4	46.7	50.6	61.9	72.2	79.6	83.4	71.0	60.5	48.7	21.5	56.0
1984	29.4	42.8	42.9	54.0	62.6	75.5	77.4	79.5	68.1	60.3	45.3	43.0	56.7
1985	24.4	30.8	50.1	59.0	67.1	72.6	79.5	75.6	69.0	59.4	46.1	28.0	55.1
1986	36.6	37.6	49.6	58.0	65.4	75.8	82.0	73.8	71.9	57.4	41.6	36.3	57.2
1987	32.7	41.5	48.5	56.7	70.8	74.7	78.5	79.1	67.6	52.9	46.9	37.2	57.3
1988	29.6	32.2	45.4	54.8	65.6	74.5	76.7	78.6	68.9	52.9	47.1	37.3	55.3
1989	39.6	27.4	45.5	57.0	62.5	69.6	75.3	75.4	63.7	59.5	46.6	25.7	54.0
1990	41.3	42.0	48.2	54.8	61.9	75.1	77.2	77.1	72.5	55.9	52.2	34.9	57.8
1991	30.5	42.6	48.6	57.8	68.9	75.0	78.5	78.3	68.0	59.3	42.4	40.2	57.5
1992	37.7	43.8	48.6	57.5	63.4	70.5	76.5	71.1	67.6	58.7	44.0	36.4	56.3
1993	33.5	33.9	42.9	52.7	63.8	73.1	80.1	79.2	66.0	54.2	41.8	37.9	54.9
Record Mean	32.9	36.0	45.3	56.0	64.5	73.2	77.6	76.7	69.4	58.4	45.6	36.1	56.0
Max	42.2	45.7	55.5	66.5	74.5	83.0	87.7	87.1	79.9	69.1	55.2	44.9	65.9
Min	23.5	26.4	35.0	45.5	54.5	63.3	67.4	66.2	58.9	47.7	35.9	27.3	46.0

REFERENCE NOTES

SPRINGFIELD, MISSOURI

GENERAL
T - TRACE AMOUNT.
BLANK ENTRIES DENOTE MISSING/UNREPORTED DATA.
# INDICATES A STATION OR INSTRUMENT RELOCATION.

SPECIFIC

(a) - LENGTH OF RECORD IN YEARS, ALTHOUGH INDIVIDUAL MONTHS MAY BE MISSING.
0.* OR * - THE VALUE IS BETWEEN 0.0 AND 0.05
NORMALS - BASED ON THE 1961-1990 RECORD PERIOD.
EXTREMES - DATES ARE THE MOST RECENT OCCURRENCE
WIND DIR. - NUMERALS SHOW TENS OF DEGRESS CLOCKWISE FROM TRUE NORTH. "00" INDICATES CALM.
RESULTANT DIRECTIONS ARE GIVEN TO WHOLE DEGREES.
BOLD VALUES INDICATE EXTREME VALUES WHICH OCCURRED AFTER THE ASOS SYSTEM WAS COMMISSIONED.

RECORD = PERIOD OF RECORD
RECORD MEAN PRECIPITATION IS THE MEAN OF ALL DAILY PRECIPITATION AMOUNTS DURING THE PERIOD OF RECORD.
RECORD MAX(MIN) TEMPERATURE IS THE MEAN OF ALL DAILY MAX(MIN) TEMPERATURES DURING THE PERIOD OF RECORD.
RECORD MEAN TEMPERATURE IS THE SUM OF THE RECORD MAX AND RECORD MIN DIVIDED BY 2.
AVERAGE TEMPERATURE IS THE SUM OF THE MEAN DAILY MAX AND MIN TEMPERATURE DIVIDED BY 2.

EXCEPTIONS

RECORD MEANS ARE THROUGH THE CURRENT YEAR, BEGINNING IN 1888 FOR TEMPERATURE
1888 FOR PRECIPITATION
1946 FOR SNOWFALL

HEATING DEGREE DAYS Base 65 deg. F SPRINGFIELD, MISSOURI

SEASON	JULY	AUG	SEP	OCT	NOV	DEC	JAN	FEB	MAR	APR	MAY	JUNE	TOTAL
1964-65	1	10	49	276	457	901	884	788	857	180	21	0	4424
1965-66	0	0	49	269	460	667	1109	854	517	375	153	11	4464
1966-67	0	2	66	309	488	959	922	902	462	213	177	7	4507
1967-68	4	10	63	261	609	819	979	965	590	281	154	7	4742
1968-69	0	0	33	296	678	997	975	792	795	248	71	31	4916
1969-70	0	0	20	342	620	968	1208	845	805	290	89	7	5194
1970-71	8	0	28	344	677	771	1035	852	642	283	116	0	4756
1971-72	3	0	61	117	561	683	988	767	549	243	79	2	4053
1972-73	9	0	40	266	727	1029	981	808	409	359	145	1	4774
1973-74	0	0	31	147	454	933	987	708	454	268	64	18	4064
1974-75	0	0	144	211	600	899	878	851	757	334	55	3	4732
1975-76	0	0	152	245	515	815	997	533	460	245	186	3	4151
1976-77	0	0	41	415	797	975	1421	752	476	173	27	0	5077
1977-78	0	0	4	233	522	906	1396	1172	763	258	134	7	5395
1978-79	0	0	19	268	535	933	1463	1053	606	342	132	14	5365
1979-80	0	1	31	184	633	726	877	941	685	332	109	1	4520
1980-81	0	0	42	289	562	834	967	735	547	126	134	0	4236
1981-82	0	0	37	266	468	949	1200	850	474	367	33	26	4670
1982-83	0	0	68	265	551	704	925	680	563	435	133	16	4340
1983-84	0	0	59	162	489	1343	1097	641	680	338	133	0	4942
1984-85	0	0	106	168	584	678	1251	949	459	204	33	4	4436
1985-86	0	1	88	187	558	1139	874	760	474	226	64	0	4371
1986-87	0	12	16	249	692	882	998	651	505	291	1	1	4298
1987-88	0	1	38	370	540	857	1090	944	601	304	54	2	4801
1988-89	0	8	26	385	531	852	780	1049	601	303	156	22	4713
1989-90	2	1	130	204	546	1214	724	639	521	336	126	4	4447
1990-91	6	0	24	291	381	929	1059	620	508	226	48	0	4092
1991-92	0	0	102	202	670	762	842	607	503	261	112	8	4069
1992-93	0	6	66	203	622	878	970	867	681	361	94	12	4760
1993-94	0	2	70	340	692	836							

COOLING DEGREE DAYS Base 65 deg. F SPRINGFIELD, MISSOURI

YEAR	JAN	FEB	MAR	APR	MAY	JUNE	JULY	AUG	SEP	OCT	NOV	DEC	TOTAL
1969	0	0	0	9	99	216	490	365	149	54	0	0	1382
1970	0	0	0	22	105	248	337	412	255	13	0	0	1392
1971	0	0	0	19	43	384	353	353	267	45	1	0	1465
1972	0	0	0	53	94	288	385	366	224	16	0	0	1426
1973	0	0	0	11	26	240	382	388	216	66	4	0	1333
1974	0	0	23	24	104	124	453	309	43	15	0	0	1095
1975	0	0	1	27	110	291	394	364	109	52	1	0	1349
1976	0	0	7	20	28	173	367	305	156	29	0	0	1085
1977	0	0	0	18	161	332	479	449	238	15	0	0	1692
1978	0	0	0	6	104	249	511	380	293	22	0	0	1565
1979	0	0	0	10	56	183	319	341	148	65	0	0	1122
1980	0	0	0	3	87	318	589	565	299	27	1	0	1889
1981	0	0	0	57	42	329	479	328	188	26	0	0	1449
1982	0	0	7	7	139	182	477	432	184	51	5	1	1485
1983	0	0	3	8	41	239	462	577	245	29	9	0	1613
1984	0	0	0	14	64	319	389	455	204	31	0	1	1477
1985	0	0	5	30	108	237	456	337	216	20	0	0	1409
1986	0	0	2	22	85	333	535	290	231	20	0	0	1518
1987	0	0	0	47	188	302	424	446	125	4	4	0	1540
1988	0	0	1	4	81	293	369	436	150	13	0	0	1347
1989	0	0	4	70	88	165	326	328	98	40	2	0	1121
1990	0	0	8	38	35	312	392	379	257	14	4	0	1439
1991	0	0	5	18	175	307	426	419	196	34	0	0	1580
1992	0	0	3	41	68	183	367	202	151	17	0	0	1032
1993	0	0	4	2	61	259	476	450	106	13	0	0	1371

SNOWFALL (inches) SPRINGFIELD, MISSOURI

SEASON	JULY	AUG	SEP	OCT	NOV	DEC	JAN	FEB	MAR	APR	MAY	JUNE	TOTAL
1964-65	0.0	0.0	0.0	0.0	T	T	0.4	4.0	8.0	0.0	0.0	0.0	12.4
1965-66	0.0	0.0	0.0	0.0	0.0	0.2	2.0	6.8	0.1	T	0.0	0.0	9.1
1966-67	0.0	0.0	0.0	0.0	T	8.1	0.7	1.9	1.7	0.0	0.0	0.0	12.4
1967-68	0.0	0.0	0.0	0.0	0.8	4.6	1.0	3.5	13.7	0.0	0.0	0.0	23.6
1968-69	0.0	0.0	0.0	0.0	1.2	1.4	1.4	8.8	0.5	0.0	0.0	0.0	13.3
1969-70	0.0	0.0	0.0	T	T	13.9	5.7	0.1	23.9	T	0.0	0.0	43.6
1970-71	0.0	0.0	0.0	0.0	T	2.9	0.1	7.1	2.8	7.1	0.0	0.0	20.0
1971-72	0.0	0.0	0.0	0.0	2.8	T	0.2	3.1	T	0.0	0.0	0.0	6.1
1972-73	0.0	0.0	0.0	0.0	11.5	1.5	7.4	0.1	T	0.9	0.0	0.0	21.4
1973-74	0.0	0.0	0.0	0.0	0.0	9.9	1.5	2.1	6.9	T	0.0	0.0	20.4
1974-75	0.0	0.0	0.0	0.0	7.9	0.1	0.4	13.0	11.5	T	0.0	0.0	32.9
1975-76	0.0	0.0	0.0	0.0	6.9	3.4	1.2	T	0.6	0.0	0.0	0.0	12.1
1976-77	0.0	0.0	0.0	T	0.3	1.9	17.4	1.0	0.0	0.2	0.0	0.0	20.8
1977-78	0.0	0.0	0.0	0.0	1.0	0.4	15.4	7.5	5.0	0.0	0.0	0.0	29.3
1978-79	0.0	0.0	0.0	0.0	T	3.5	23.1	1.8	T	T	0.0	0.0	28.4
1979-80	0.0	0.0	0.0	T	T	0.8	1.3	19.2	0.9	2.5	0.0	0.0	24.7
1980-81	0.0	0.0	0.0	T	9.2	T	2.6	6.4	0.0	0.0	0.0	0.0	18.2
1981-82	0.0	0.0	0.0	0.0	T	3.1	15.8	4.5	1.2	T	0.0	0.0	24.6
1982-83	0.0	0.0	0.0	0.0	T	T	3.7	0.4	1.5	T	0.0	0.0	5.6
1983-84	0.0	0.0	0.0	0.0	0.2	14.5	6.9	9.6	0.4	0.0	0.0	0.0	31.6
1984-85	0.0	0.0	0.0	0.0	1.4	4.5	13.7	6.0	T	0.0	0.0	0.0	25.6
1985-86	0.0	0.0	0.0	0.0	T	6.2	T	7.3	0.0	0.0	0.0	0.0	13.5
1986-87	0.0	0.0	0.0	0.0	T	0.1	11.7	2.1	T	0.0	0.0	0.0	13.9
1987-88	0.0	0.0	0.0	0.0	0.3	8.7	5.8	4.3	7.0	T	0.0	0.0	26.1
1988-89	0.0	0.0	0.0	0.0	3.0	6.8	0.8	2.7	15.0	0.0	0.0	T	28.3
1989-90	0.0	0.0	0.0	T	T	6.4	0.9	8.5	0.7	T	T	T	16.5
1990-91	T	0.0	0.0	0.0	T	8.2	2.1	4.6	T	T	T	0.0	14.9
1991-92	0.0	0.0	0.0	0.0	T	T	4.6	T	0.1	0.0	0.0	0.0	4.7
1992-93	0.0	0.0	0.0	0.0	0.4	3.5	3.2	14.1	1.5	1.0	0.0	T	23.7
1993-94	0.0	0.0	0.0	1.2	0.4	0.4							
Record Mean	T	0.0	0.0	T	1.6	3.0	4.4	4.4	3.3	0.4	T	T	17.1

BILLINGS, MONTANA

Billings, Montana, at an elevation of 3,100 to 3,600 feet above sea level, is situated in the borderline area between the Great Plains and the Rocky Mountains, and has a climate which takes on some of the characteristics of both regions. Its climate may be classified as semi-arid, but with irrigation and the favorable distribution of the precipitation, it is possible to raise a variety of crops in the area.

About a third of the annual precipitation falls during May and June, with June being the wettest month. The period of least precipitation is from November through February. These four months normally produce less than 20 percent of the annual precipitation. The heaviest snows occur during the spring and fall months when the temperature and moisture conditions are most favorable. Heavy snows of 6 inches or more also occur during November and December. The occurrence of thawing periods normally prevents the snow from accumulating to great depths on the ground. Thunderstorms are most frequent during the summer months. These storms are frequently accompanied by strong, gusty winds and occasionally by hail. Destructive hailstorms, however, are rather infrequent.

Winter is usually cold, though not extremely so, and generally affords several mild periods of a week to several weeks in length. The winter cold periods are ushered in by moderately strong north to northeast winds and snow. The coldest temperatures occur after the snow ends and the sky clears. True blizzard conditions are not observed very often in town, but in the surrounding rural areas, blizzard conditions may develop several times during the winter. Cold weather improves with the onset of moderate to strong southwest winds. This wind is sometimes a foehn condition (chinook), but is more often a drainage wind moving down the Yellowstone Valley which transports warmer air of Pacific origin to the area. Occasionally an open winter occurs when cold Arctic outbreaks pass far to the east and temperatures stay above zero degrees.

Spring brings a period of frequent and rapid fluctuations in the weather. It is usually cloudy and cool with frequent periods of rain and/or snow. As the season progresses, snows become less frequent until late May and June when rain is the rule. The last freezing temperatures in spring usually occur before mid-May though they have occurred as late as late June.

The summer season is characterized by warm days with abundant sunshine and low humidities. The nights are cool because of the altitude and the cool air drainage into the valley from the higher terrain. Seldom is there a protracted rainy spell during this season. Frequent thunderstorms bring threatening afternoon cloudiness but usually only small amounts of rain.

The first freezing temperatures of the fall season occur in late September, but they have been noted as early as late August. Over the years, the fall months have been about evenly distributed between cold, wet ones, and mild, dry, pleasant ones. The change to severe winter weather usually arrives after the middle of November. There have been years when the more severe type of winter weather have been delayed until late in December.

NORMALS, MEANS, AND EXTREMES

BILLINGS, MONTANA

LATITUDE: 45°48'N LONGITUDE: 108°32' W ELEVATION: FT. GRND 3567 BARO 3583 TIME ZONE: MOUNTAIN WBAN: 24033

	(a)	JAN	FEB	MAR	APR	MAY	JUNE	JULY	AUG	SEP	OCT	NOV	DEC	YEAR
TEMPERATURE °F:														
Normals														
-Daily Maximum		31.8	38.6	45.8	57.1	66.7	77.6	86.7	84.7	71.6	60.6	44.5	34.4	58.3
-Daily Minimum		13.7	19.4	25.2	34.0	43.3	52.0	58.3	56.7	46.5	37.5	25.6	16.5	35.7
-Monthly		22.8	29.0	35.5	45.6	55.0	64.8	72.5	70.7	59.1	49.1	35.1	25.5	47.1
Extremes														
-Record Highest	59	68	72	79	92	96	105	106	105	103	90	77	69	106
-Year		1953	1961	1986	1939	1936	1984	1937	1961	1983	1992	1983	1980	JUL 1937
-Record Lowest	59	-30	-38	-19	-5	14	32	41	35	22	-7	-22	-32	-38
-Year		1937	1936	1989	1936	1954	1969	1972	1992	1984	1991	1959	1983	FEB 1936
NORMAL DEGREE DAYS:														
Heating (base 65°F)		1308	1008	915	582	316	119	12	42	242	498	897	1225	7164
Cooling (base 65°F)		0	0	0	0	6	113	244	219	65	5	0	0	652
% OF POSSIBLE SUNSHINE	54	48	54	61	61	61	64	76	76	68	61	46	45	60
MEAN SKY COVER (tenths)														
Sunrise - Sunset	54	7.1	7.1	7.2	7.1	6.6	5.9	4.3	4.4	5.2	5.8	6.9	6.8	6.2
MEAN NUMBER OF DAYS:														
Sunrise to Sunset														
-Clear	54	5.6	4.3	4.3	4.3	5.5	7.1	13.4	13.6	10.4	9.1	5.7	6.1	89.3
-Partly Cloudy	54	7.7	8.0	8.6	8.6	10.4	11.8	11.8	10.9	9.2	9.1	7.7	8.1	112.1
-Cloudy	54	17.7	16.0	18.1	17.1	15.1	11.1	5.8	6.5	10.4	12.7	16.6	16.8	163.9
Precipitation														
.01 inches or more	59	7.7	7.4	9.1	9.6	11.2	11.1	7.3	6.5	7.0	6.3	6.1	6.9	96.1
Snow, Ice pellets, hail														
1.0 inches or more	51	3.2	2.5	3.3	2.3	0.5	0.*	0.0	0.0	0.4	1.3	2.2	2.8	18.4
Thunderstorms	54	0.*	0.*	0.1	1.1	4.3	7.4	7.5	5.6	1.9	0.2	0.0	0.*	28.2
Heavy Fog Visibility														
1/4 mile or less	46	1.4	2.1	2.1	2.6	1.2	0.7	0.3	0.3	1.1	1.9	2.1	1.6	17.4
Temperature °F														
-Maximum														
90° and above	34	0.0	0.0	0.0	0.*	0.5	4.1	11.9	10.5	1.9	0.1	0.0	0.0	29.0
32° and below	34	13.8	7.9	4.7	0.7	0.0	0.0	0.0	0.0	0.*	0.8	5.2	11.4	44.4
-Minimum														
32° and below	34	27.5	23.6	23.1	12.6	1.6	0.1	0.0	0.0	1.4	8.3	21.9	27.7	147.8
0° and below	34	8.1	3.4	1.2	0.0	0.0	0.0	0.0	0.0	0.0	0.*	1.2	4.5	18.4
AVG. STATION PRESS. (mb)	21	890.9	890.9	888.4	889.3	889.1	889.9	891.8	892.0	892.5	892.3	890.8	890.7	890.7
RELATIVE HUMIDITY (%)														
Hour 05	34	64	66	68	68	70	71	64	61	65	63	65	64	66
Hour 11 (Local Time)	34	60	57	54	48	48	45	39	39	47	49	57	59	50
Hour 17	34	56	51	46	41	42	39	31	30	37	41	53	56	44
Hour 23	34	62	62	62	58	60	58	50	47	54	56	62	62	58
PRECIPITATION (inches):														
Water Equivalent														
-Normal		0.90	0.64	1.16	1.74	2.57	1.99	0.94	1.01	1.36	1.14	0.84	0.79	15.08
-Maximum Monthly	59	2.35	1.77	2.70	4.42	7.71	7.64	5.08	3.50	4.99	3.80	2.34	2.00	7.71
-Year		1972	1978	1954	1955	1981	1944	1993	1965	1941	1971	1978	1973	MAY 1981
-Minimum Monthly	59	0.04	0.05	0.13	0.06	0.40	0.24	0.04	0.05	0.06	0.01	T	0.05	T
-Year		1941	1977	1936	1962	1993	1961	1988	1955	1964	1987	1954	1957	NOV 1954
-Maximum in 24 hrs	59	1.41	0.65	1.01	3.19	2.83	2.78	2.32	2.47	2.19	1.98	1.37	0.96	3.19
-Year		1972	1986	1973	1978	1952	1937	1993	1965	1966	1974	1959	1978	APR 1978
Snow, Ice pellets, hail														
-Maximum Monthly	59	27.7	22.4	27.6	42.3	15.6	2.0	0.4	T	9.3	23.1	25.2	28.8	42.3
-Year		1963	1978	1935	1955	1981	1950	1993	1992	1984	1949	1978	1955	APR 1955
-Maximum in 24 hrs	55	16.6	9.0	10.5	23.7	15.3	2.0	0.4	T	7.5	11.2	15.3	13.7	23.7
-Year		1972	1944	1964	1955	1981	1950	1993	1992	1983	1980	1959	1978	APR 1955
WIND:														
Mean Speed (mph)	54	13.0	12.2	11.4	11.5	10.7	10.1	9.5	9.5	10.2	11.0	12.1	13.1	11.2
Prevailing Direction through 1963		SW	SW	SW	SW	NE	SW	SW	SW	SW	SW	SW	WSW	SW
Fastest Mile														
-Direction (!!)	50	W	W	NW	NW	NN	NW	N	NW	NW	NW	NW	NW	NW
-Speed (MPH)	50	66	72	61	72	68	79	73	69	61	68	63	66	79
-Year		1953	1963	1956	1947	1939	1968	1947	1983	1949	1949	1948	1953	JUN 1968
Peak Gust														
-Direction (!!)	10	NW	W	NW	NW	NW	W	NE	NW	NW	NW	SW	W	NW
-Speed (mph)	10	59	62	52	59	60	54	58	69	61	61	58	64	69
-Date		1986	1988	1990	1987	1988	1987	1985	1986	1989	1985	1990	1988	AUG 1986

PRECIPITATION (inches)

BILLINGS, MONTANA

YEAR	JAN	FEB	MAR	APR	MAY	JUNE	JULY	AUG	SEP	OCT	NOV	DEC	ANNUAL
1964	0.11	0.35	1.38	4.11	3.91	5.13	0.10	2.08	0.06	0.22	0.72	0.59	18.76
1965	0.66	1.18	0.90	1.18	1.89	2.30	1.28	3.50	2.17	0.09	0.66	0.74	16.55
1966	0.43	0.32	1.58	1.10	0.84	1.56	1.45	1.09	2.46	0.50	1.07	0.95	13.35
1967	0.36	0.39	1.55	1.63	1.84	5.18	0.37	0.54	0.66	1.04	0.50	0.79	14.85
1968	1.22	0.58	0.66	1.50	1.79	3.86	0.25	2.35	1.38	0.51	1.71	0.81	16.62
1969	0.99	0.17	0.57	1.48	0.78	5.74	1.69	0.42	0.36	1.56	0.66	0.31	14.73
1970	0.87	1.10	0.95	3.04	3.48	1.61	0.37	0.21	1.92	0.93	0.82	0.79	16.09
1971	1.30	0.56	0.83	1.60	2.07	0.70	0.40	0.43	1.80	3.80	0.17	1.13	14.79
1972	2.35	0.81	0.80	1.63	2.51	0.91	1.91	1.61	1.31	2.24	1.01	1.08	18.17
1973	1.30	0.43	1.59	3.00	0.73	0.81	0.29	1.23	2.21	1.36	1.21	2.00	16.16
1974	0.78	0.22	1.20	1.65	2.80	1.94	0.91	2.18	1.80	2.24	0.28	0.30	16.30
1975	2.05	0.75	1.30	1.91	3.62	1.62	2.64	0.37	0.24	2.66	1.39	1.96	20.51
1976	0.58	1.12	0.95	3.53	1.85	2.70	0.07	0.60	1.21	0.91	0.68	0.31	14.51
1977	1.44	0.05	1.37	0.64	1.35	0.63	0.80	1.05	0.65	1.20	1.42	1.65	12.25
1978	2.03	1.77	0.18	4.12	6.97	1.55	1.54	0.52	3.78	0.27	2.34	1.73	26.80
1979	0.72	0.56	1.11	1.20	0.92	1.06	0.46	0.87	0.16	0.73	0.53	0.09	8.41
1980	1.11	0.78	1.53	0.46	4.47	1.64	0.39	1.17	0.77	2.45	0.42	0.33	15.52
1981	0.21	0.24	1.75	0.35	7.71	1.58	1.65	0.55	0.14	1.33	0.41	0.53	16.45
1982	0.71	0.34	1.81	1.53	2.63	5.03	1.91	0.45	1.22	1.15	0.42	1.07	18.27
1983	0.11	0.31	0.73	0.56	2.23	0.88	1.52	1.12	2.26	1.32	0.90	0.92	12.86
1984	0.65	0.93	0.84	1.38	1.12	1.65	0.29	0.58	1.32	0.37	0.95	0.84	10.92
1985	0.31	0.39	2.05	0.31	1.27	1.07	1.40	1.66	1.89	0.69	1.43	0.20	12.67
1986	0.37	1.72	1.04	2.72	1.92	2.15	1.01	0.43	1.24	0.33	1.21	0.12	14.26
1987	0.07	0.49	1.36	0.42	3.84	1.03	2.23	1.73	0.68	0.01	0.29	0.31	12.46
1988	0.45	0.71	0.66	1.82	1.84	0.43	0.04	0.12	2.12	1.01	0.60	0.56	10.36
1989	1.27	0.56	2.04	2.36	2.06	1.18	0.55	0.76	0.70	2.05	0.52	1.36	15.41
1990	0.29	0.50	1.70	2.06	2.81	0.66	0.37	0.93	0.08	1.05	0.33	0.49	11.27
1991	0.82	0.49	0.62	3.87	2.25	5.62	1.04	0.35	3.11	1.29	0.96	0.31	20.73
1992	0.09	0.12	0.65	2.35	1.70	2.69	1.67	0.34	0.62	0.42	0.30	0.51	11.46
1993	0.47	0.32	0.50	1.86	0.40	2.05	5.08	0.69	1.76	2.11	0.26	0.20	15.70
Record Mean	0.72	0.62	1.07	1.61	2.24	2.34	1.02	0.89	1.32	1.12	0.74	0.68	14.39

AVERAGE TEMPERATURE (deg. F)

BILLINGS, MONTANA

YEAR	JAN	FEB	MAR	APR	MAY	JUNE	JULY	AUG	SEP	OCT	NOV	DEC	ANNUAL
1964	29.1	31.4	31.8	45.2	56.7	62.7	75.6	68.1	56.4	51.5	31.2	17.9	46.5
1965	28.1	25.2	21.4	45.8	52.6	61.4	71.5	69.5	47.1	54.4	38.0	28.5	45.3
1966	15.1	25.7	37.1	40.1	55.4	61.6	76.2	68.3	65.5	50.4	34.5	29.2	46.6
1967	30.2	34.3	31.3	39.7	50.1	60.8	72.7	71.0	61.9	49.9	37.2	24.7	47.0
1968	23.4	34.5	43.5	45.3	54.0	62.6	72.1	67.6	61.6	52.1	38.3	19.9	47.9
1969	6.3	24.3	31.1	50.2	56.6	58.7	69.3	73.2	62.7	40.0	38.8	31.0	45.2
1970	20.2	33.0	29.5	38.6	54.5	66.0	73.2	74.6	55.8	45.1	33.6	24.7	45.7
1971	18.7	26.8	33.5	44.2	54.5	64.3	69.2	76.7	55.6	43.4	35.9	19.7	45.2
1972	13.2	26.2	40.1	43.7	53.8	67.0	66.6	70.1	54.8	43.6	34.2	18.1	44.3
1973	23.1	29.2	38.6	41.4	54.3	65.0	71.7	72.3	57.7	50.4	29.5	31.4	47.1
1974	22.6	35.8	35.8	47.7	50.5	66.9	75.0	64.8	56.0	50.5	37.4	29.9	47.7
1975	23.6	18.1	30.2	37.1	51.3	60.2	74.5	67.8	59.5	47.9	32.9	29.4	44.4
1976	26.5	32.9	32.0	46.1	56.2	60.9	73.7	71.1	62.0	46.0	35.2	31.5	47.9
1977	18.0	37.6	36.0	50.3	56.6	69.2	72.6	66.1	59.0	49.7	31.8	20.3	47.2
1978	11.0	16.0	34.8	46.6	53.6	64.0	68.0	67.4	60.6	49.1	24.6	16.1	42.6
1979	7.7	18.9	36.2	43.6	53.8	66.5	73.4	71.5	66.2	52.6	33.5	36.3	46.7
1980	17.0	29.2	34.4	54.6	61.0	66.5	75.7	67.4	62.0	50.7	40.6	30.3	49.1
1981	36.0	32.5	41.7	50.5	55.6	64.6	73.8	73.2	63.3	46.0	40.5	28.5	50.5
1982	13.1	28.0	32.9	42.6	52.4	61.4	70.8	75.8	60.4	50.2	33.8	28.7	45.9
1983	35.3	38.4	38.6	42.9	53.2	63.9	72.0	77.5	59.3	53.5	37.8	8.7	48.4
1984	29.3	38.3	38.1	45.0	55.5	64.5	74.5	74.7	54.0	42.1	37.6	19.6	47.8
1985	20.3	24.1	34.1	50.5	60.3	64.0	74.8	65.6	53.0	49.6	15.2	25.9	44.8
1986	37.2	24.2	46.2	44.5	54.4	69.7	69.3	70.5	53.7	50.0	32.0	32.3	48.7
1987	30.3	35.0	37.6	54.1	60.2	67.9	70.4	66.2	61.2	49.6	40.3	29.8	50.2
1988	23.6	28.7	39.8	48.3	59.7	75.9	76.2	72.2	58.7	52.0	37.0	29.3	50.1
1989	27.2	13.4	29.9	45.0	55.0	63.6	75.3	69.3	60.5	47.2	39.7	25.0	45.9
1990	31.2	29.0	38.1	46.1	53.6	65.1	72.1	72.6	66.5	48.7	40.6	19.2	48.6
1991	20.8	41.0	38.5	43.7	54.8	64.3	72.6	75.5	60.6	45.6	32.7	34.1	48.7
1992	35.2	39.4	44.0	48.9	58.4	65.9	65.8	67.2	60.7	49.8	36.9	19.7	49.3
1993	17.9	21.7	40.6	47.0	59.8	62.2	62.7	65.6	57.0	47.5	31.4	33.7	45.6
Record Mean	23.0	27.9	34.9	45.7	55.5	64.0	72.4	70.7	59.7	49.4	35.3	27.4	47.2
Max	32.3	37.7	45.2	57.3	67.5	76.5	86.6	84.9	72.6	61.1	44.7	36.4	58.6
Min	13.6	18.0	24.5	34.0	43.6	51.5	58.2	56.5	46.8	37.7	25.8	18.4	35.7

REFERENCE NOTES

BILLINGS, MONTANA

GENERAL
T - TRACE AMOUNT.
BLANK ENTRIES DENOTE MISSING/UNREPORTED DATA.
# INDICATES A STATION OR INSTRUMENT RELOCATION.

SPECIFIC

(a) - LENGTH OF RECORD IN YEARS, ALTHOUGH INDIVIDUAL MONTHS MAY BE MISSING.
0.* OR * - THE VALUE IS BETWEEN 0.0 AND 0.05
NORMALS - BASED ON THE 1961-1990 RECORD PERIOD.
EXTREMES - DATES ARE THE MOST RECENT OCCURRENCE.
WIND DIR. - NUMERALS SHOW TENS OF DEGREES CLOCKWISE FROM TRUE NORTH. "00" INDICATES CALM.
RESULTANT DIRECTIONS ARE GIVEN TO WHOLE DEGREES.
BOLD VALUES INDICATE EXTREME VALUES WHICH OCCURRED AFTER THE ASOS SYSTEM WAS COMMISSIONED.

RECORD = PERIOD OF RECORD
RECORD MEAN PRECIPITATION IS THE MEAN OF ALL DAILY PRECIPITATION AMOUNTS DURING THE PERIOD OF RECORD.
RECORD MAX(MIN) TEMPERATURE IS THE MEAN OF ALL DAILY MAX(MIN) TEMPERATURES DURING THE PERIOD OF RECORD.
RECORD MEAN TEMPERATURE IS THE SUM OF THE RECORD MAX AND RECORD MIN DIVIDED BY 2.
AVERAGE TEMPERATURE IS THE SUM OF THE MEAN DAILY MAX AND MIN TEMPERATURE DIVIDED BY 2.

EXCEPTIONS

RECORD MEANS ARE THROUGH THE CURRENT YEAR, BEGINNING IN 1935 FOR TEMPERATURE
1935 FOR PRECIPITATION
1935 FOR SNOWFALL

HEATING DEGREE DAYS Base 65 deg. F — BILLINGS, MONTANA

SEASON	JULY	AUG	SEP	OCT	NOV	DEC	JAN	FEB	MAR	APR	MAY	JUNE	TOTAL
1964-65	0	67	253	415	1009	1454	1135	1110	1347	570	381	120	7861
1965-66	3	33	533	327	806	1125	1545	1094	860	742	319	155	7542
1966-67	0	54	83	446	910	1104	1069	852	1039	755	461	152	6925
1967-68	1	17	149	463	827	1240	1285	880	658	589	334	125	6568
1968-69	16	42	147	394	793	1394	1818	1134	1044	438	283	218	7721
1969-70	17	4	116	766	780	1050	1382	891	1095	787	329	77	7294
1970-71	1	0	296	614	936	1244	1429	1063	972	617	322	86	7580
1971-72	25	7	305	659	869	1397	1601	1118	764	633	353	36	7767
1972-73	85	26	311	660	914	1451	1296	994	810	699	333	99	7678
1973-74	4	3	226	446	1058	1035	1313	811	896	511	443	94	6840
1974-75	7	76	269	444	821	1078	1280	1308	1071	829	418	152	7753
1975-76	1	31	185	536	955	1095	1186	926	1015	559	265	138	6892
1976-77	0	6	131	582	884	1030	1451	759	896	437	258	28	6462
1977-78	0	59	208	463	988	1380	1668	1366	933	545	357	97	8064
1978-79	41	45	204	487	1208	1509	1771	1286	887	636	350	71	8495
1979-80	2	7	44	383	937	884	1484	1033	939	324	159	46	6242
1980-81	0	25	127	462	724	1073	891	905	717	427	292	79	5722
1981-82	12	6	124	583	729	1124	1603	1028	987	666	386	142	7390
1982-83	12	0	215	453	926	1118	911	741	810	656	381	82	6305
1983-84	29	5	234	359	811	1741	1101	769	828	592	316	97	6882
1984-85	12	3	351	701	812	1404	1381	1140	950	428	184	103	7469
1985-86	13	65	358	471	1492	1207	853	1136	579	610	347	18	7149
1986-87	8	2	331	457	982	1005	1070	829	841	337	183	44	6089
1987-88	39	56	134	473	734	1083	1276	1047	775	492	200	14	6323
1988-89	0	6	221	395	833	1099	1168	1441	1084	595	308	97	7247
1989-90	0	25	172	546	752	1235	1042	1002	829	560	346	108	6617
1990-91	9	0	73	500	725	1413	1365	665	814	630	311	50	6555
1991-92	3	0	171	612	963	951	918	737	641	481	232	73	5782
1992-93	53	95	166	475	837	1398	1456	1210	751	531	177	134	7283
1993-94	98	60	250	534	1004	963							

COOLING DEGREE DAYS Base 65 deg. F — BILLINGS, MONTANA

YEAR	JAN	FEB	MAR	APR	MAY	JUNE	JULY	AUG	SEP	OCT	NOV	DEC	TOTAL
1969	0	0	0	0	29	34	156	266	56	0	0	0	541
1970	0	0	0	0	10	115	261	305	28	3	0	0	722
1971	0	0	0	0	4	69	162	377	29	0	0	0	641
1972	0	0	0	0	12	102	139	192	8	0	0	0	453
1973	0	0	0	0	13	107	219	235	16	0	0	0	590
1974	0	0	0	1	2	158	324	78	7	2	0	0	572
1975	0	0	0	0	2	15	299	126	26	10	0	0	478
1976	0	0	0	0	1	21	276	200	49	0	0	0	547
1977	0	0	0	0	7	160	242	102	34	0	0	0	545
1978	0	0	0	0	12	71	140	128	78	0	0	0	429
1979	0	0	0	0	12	126	270	216	86	6	0	0	716
1980	0	0	0	20	40	99	339	105	46	26	0	0	675
1981	0	0	0	0	6	74	291	268	76	3	0	0	718
1982	0	0	0	0	0	41	198	342	82	0	0	0	663
1983	0	0	0	0	24	54	256	400	69	7	2	0	812
1984	0	0	0	0	30	91	315	310	29	0	0	0	775
1985	0	0	0	0	42	83	325	92	6	0	0	0	548
1986	0	0	1	2	25	163	152	177	1	0	0	0	521
1987	0	0	0	17	41	134	215	100	30	4	0	0	541
1988	0	0	0	0	41	351	355	234	37	0	0	0	1018
1989	0	0	0	2	2	64	327	164	45	0	0	0	604
1990	0	0	0	0	0	117	239	245	123	0	0	0	724
1991	0	0	0	1	3	35	244	332	45	16	0	0	676
1992	0	0	0	4	36	106	84	166	42	10	0	0	448
1993	0	0	0	0	18	57	34	85	19	2	0	0	215

SNOWFALL (inches) — BILLINGS, MONTANA

SEASON	JULY	AUG	SEP	OCT	NOV	DEC	JAN	FEB	MAR	APR	MAY	JUNE	TOTAL
1964-65	0.0	0.0	0.0	T	7.2	7.8	8.0	13.9	11.4	3.4	3.5	0.0	55.2
1965-66	0.0	0.0	5.6	0.0	6.3	10.0	6.3	3.6	17.3	2.5	T	0.0	51.6
1966-67	0.0	0.0	0.0	T	12.0	10.3	6.1	3.7	16.2	11.4	7.7	0.0	67.4
1967-68	0.0	0.0	0.0	T	5.1	9.0	13.2	1.6	6.8	12.3	T	0.0	48.0
1968-69	0.0	0.0	T	T	4.6	8.6	12.2	2.0	5.5	3.0	T	T	35.9
1969-70	0.0	0.0	0.0	10.2	5.0	2.3	9.8	11.3	7.2	22.3	T	T	68.1
1970-71	0.0	0.0	4.9	T	4.9	8.4	11.5	5.7	8.0	2.5	T	0.0	45.9
1971-72	0.0	0.0	T	13.0	1.1	12.2	27.6	7.6	8.2	7.3	T	0.0	77.0
1972-73	0.0	0.0	4.7	3.2	T	12.9	13.3	5.4	7.4	14.3	1.8	0.0	63.0
1973-74	0.0	0.0	T	8.5	11.9	14.9	8.4	2.5	10.6	2.5	0.8	0.0	60.1
1974-75	0.0	0.0	T	0.6	1.1	3.5	20.8	6.8	14.0	12.9	1.1	0.0	60.8
1975-76	0.0	0.0	0.0	3.3	13.9	20.0	5.9	11.1	9.8	6.1	0.0	0.0	70.1
1976-77	0.0	0.0	0.0	2.1	7.0	6.2	16.1	0.3	13.7	5.7	T	0.0	51.1
1977-78	0.0	0.0	0.0	3.4	14.1	18.1	25.9	22.4	1.3	T	0.0	0.0	85.2
1978-79	0.0	0.0	T	0.0	25.2	22.2	10.1	8.7	5.9	3.7	0.3	0.0	76.1
1979-80	0.0	0.0	0.0	1.4	7.7	1.7	16.9	10.0	17.3	4.2	0.0	0.0	59.2
1980-81	0.0	0.0	T	17.8	4.5	5.5	2.1	3.4	16.3	0.7	15.6	0.0	65.9
1981-82	0.0	0.0	0.0	5.7	3.2	5.4	9.8	5.3	18.2	13.5	2.0	0.0	63.1
1982-83	0.0	0.0	5.7	1.5	5.6	11.2	0.1	1.0	6.4	5.8	11.9	0.0	49.2
1983-84	0.0	0.0	7.5	T	5.5	10.9	5.3	6.8	4.5	9.0	T	0.0	49.5
1984-85	0.0	0.0	9.3	6.5	9.9	16.1	4.8	3.8	21.7	1.9	0.0	0.0	74.0
1985-86	0.0	0.0	3.6	6.0	17.1	2.0	3.3	13.8	6.4	12.9	8.3	0.0	73.4
1986-87	0.0	0.0	0.0	0.0	12.3	1.9	0.6	6.0	13.3	0.3	0.4	0.0	34.8
1987-88	0.0	0.0	0.0	0.3	2.6	3.6	7.4	8.8	1.9	10.7	2.0	0.0	37.3
1988-89	0.0	0.0	T	2.0	5.6	6.2	18.5	6.8	25.1	11.8	T	0.0	76.0
1989-90	0.0	0.0	T	7.2	5.8	17.1	3.3	8.9	13.0	11.2	T	0.0	66.5
1990-91	T	T	0.0	3.5	1.5	6.2	11.4	1.0	3.7	30.0	3.6	T	60.9
1991-92	T	0.0	0.0	15.6	7.6	3.5	0.9	1.1	3.1	3.4	0.0	0.0	35.2
1992-93	0.0	T	0.0	4.0	1.7	10.6	11.4	5.8	2.7	6.9	0.0	T	43.1
1993-94	0.4	0.0	T	7.8	5.6	3.0							
Record Mean	T	T	1.1	3.7	6.8	8.5	9.1	7.8	10.1	7.9	1.5	T	56.6

GLASGOW, MONTANA

Founded in the days of national expansion as a railroad shop town, Glasgow is situated in the valley of the Milk River, about 20 miles upstream from where the Milk River joins the Missouri. It lies on the natural route from the plains to Marias Pass in the northern Rockies. The city is located on the valley floor at an average elevation of about 2,100 feet above sea level. Hills rise sharply from the northern edge of the city to flat tableland about 200 feet higher than the valley. The Weather Service Office is located on this flat land about 1 mile north-northeast of the city. A gradual incline commences 3 to 4 miles to the south and southwest of the city and reaches to the rolling hills which separate the Milk River drainage from the Fort Peck Reservoir on the Missouri. The northern shore of Fort Peck Reservoir lies about 15 miles south of Glasgow. This is a body of water impounded by Fort Peck Dam which was completed in 1939. The dam, at full capacity, backs water up the Missouri Valley for over 180 miles. The shape of the reservoir is very irregular, but its average width south of Glasgow is about 10 miles.

The climate in the Glasgow area is continental with a large annual range in temperature and limited precipitation. Fort Peck Reservoir, to the south, seems to have little climatic effect as far north as Glasgow, except for brief periods of morning fog in the late fall which occasionally drift northward from the lake before it freezes. Seventy-eight percent of the annual precipitation falls from April through September, with May and June accounting for about 38 percent of the annual total. This distribution of precipitation helps to make the climate quite favorable for the growing of small grains. Winter precipitation nearly always falls as snow, but as a rule, although snow seldom accumulates to any great depth, it usually is formed into drifts in the open, unprotected areas. Blizzards during the winter months occur occasionally, but usually are of short duration. However, it is wise for travelers and stockmen to be on the alert for this danger during the winter months. Glasgow itself is well protected from most strong winds and blizzard conditions by hills to the north of the city, but occasionally the unprotected surrounding areas feel the full brunt of these winter storms.

Glasgow has a wide range of temperature. Winters are quite cold, but mild winter weather occasionally does occur, sometimes caused when the chinook or foehn wind, which descends the eastern slopes of the Rocky Mountains, reaches as far east as Glasgow. Very cold spells also occur, at least once each winter, but as a rule, these last only a few days. Summers are characterized by warm, sunny weather which can last for several weeks at a time. Sunny weather predominates during the warmer season, but interruptions in the form of clouds and showers do occur, usually in the afternoons and evenings. A few days of hot weather in July and August occur at times, but hot days are seldom oppressive because they are usually accompanied by low humidity.

As is usually the case with a continental climate in northern latitudes, the transitional fall and spring seasons at Glagow are quite rapid.

NORMALS, MEANS, AND EXTREMES

GLASGOW, MONTANA

LATITUDE: 48°13'N LONGITUDE: 106°37' W ELEVATION: FT. GRND 2284 BARO 2279 TIME ZONE: MOUNTAIN WBAN: 94008

	(a)	JAN	FEB	MAR	APR	MAY	JUNE	JULY	AUG	SEP	OCT	NOV	DEC	YEAR
TEMPERATURE °F:														
Normals														
-Daily Maximum		19.9	27.1	39.8	56.5	67.3	77.5	84.7	83.5	70.4	58.7	39.6	24.9	54.2
-Daily Minimum		1.2	7.9	19.0	32.0	42.5	51.3	56.7	55.3	44.1	33.3	18.8	5.5	30.6
-Monthly		10.6	17.5	29.4	44.3	55.0	64.4	70.8	69.4	57.3	46.1	29.2	15.2	42.4
Extremes														
-Record Highest	38	57	71	79	91	102	108	104	108	103	90	75	59	108
-Year		1992	1992	1993	1980	1988	1988	1983	1983	1983	1992	1975	1979	JUN 1988
-Record Lowest	38	-47	-37	-27	-3	20	33	41	37	20	-6	-26	-38	-47
-Year		1969	1982	1960	1975	1976	1979	1977	1956	1985	1991	1985	1989	JAN 1969
NORMAL DEGREE DAYS:														
Heating (base 65°F)		1686	1330	1104	621	322	121	19	71	267	586	1074	1544	8745
Cooling (base 65°F)		0	0	0	0	12	103	199	208	36	0	0	0	558
% OF POSSIBLE SUNSHINE														
MEAN SKY COVER (tenths)														
Sunrise - Sunset	38	7.1	7.0	7.0	6.9	6.6	6.0	4.7	4.8	5.6	6.3	7.0	7.2	6.3
MEAN NUMBER OF DAYS:														
Sunrise to Sunset														
-Clear	38	5.5	5.1	5.1	5.4	5.6	7.1	12.4	12.1	9.2	7.6	5.9	5.0	85.9
-Partly Cloudy	38	7.4	7.3	9.2	8.5	10.5	11.8	12.0	11.1	9.4	8.9	6.8	8.1	111.1
-Cloudy	38	18.1	15.8	16.7	16.1	14.9	11.2	6.6	7.9	11.4	14.6	17.3	17.9	168.3
Precipitation														
.01 inches or more	38	8.3	6.8	7.3	7.1	9.6	10.6	8.2	7.3	6.3	4.7	6.1	8.0	90.4
Snow,Ice pellets,hail														
1.0 inches or more	38	2.1	1.5	1.2	0.8	0.1	0.0	0.0	0.0	0.1	0.4	1.1	1.6	9.0
Thunderstorms	34	0.0	0.0	0.1	0.5	3.6	7.5	8.0	6.0	1.5	0.1	0.*	0.0	27.4
Heavy Fog Visibility 1/4 mile or less	34	2.3	2.4	2.1	0.5	0.3	0.2	0.1	0.1	0.2	0.6	1.8	2.5	13.1
Temperature °F														
-Maximum														
90° and above	29	0.0	0.0	0.0	0.*	0.8	3.4	8.9	8.7	1.6	0.1	0.0	0.0	23.6
32° and below	29	22.1	15.9	7.8	0.8	0.0	0.0	0.0	0.0	0.0	0.9	8.5	19.1	75.2
-Minimum														
32° and below	29	30.9	27.6	28.0	15.3	2.5	0.0	0.0	0.0	2.0	13.1	27.3	30.8	177.6
0° and below	29	14.6	9.2	2.6	0.1	0.0	0.0	0.0	0.0	0.0	0.2	2.5	9.6	38.7
AVG. STATION PRESS. (mb)	21	934.1	934.4	931.7	932.2	931.0	931.2	933.0	933.3	934.0	934.0	933.4	933.6	933.0
RELATIVE HUMIDITY (%)														
Hour 05	29	75	78	80	75	73	75	73	69	72	73	79	78	75
Hour 11 (Local Time)	29	72	72	64	50	45	46	43	41	48	52	68	73	56
Hour 17	29	72	70	57	42	39	39	34	32	38	46	64	73	51
Hour 23	29	75	78	75	64	62	64	59	55	61	65	75	77	68
PRECIPITATION (inches):														
Water Equivalent														
-Normal		0.37	0.27	0.41	0.69	1.77	2.11	1.72	1.35	1.00	0.61	0.28	0.38	10.96
-Maximum Monthly	38	1.24	0.74	1.27	1.99	3.74	5.36	5.93	5.74	4.14	1.77	1.26	1.03	5.93
-Year		1969	1979	1987	1969	1982	1963	1993	1985	1978	1975	1958	1982	JUL 1993
-Minimum Monthly	38	T	0.04	0.05	0.07	0.03	0.09	0.01	0.03	0.04	T	T	0.03	T
-Year		1973	1992	1957	1956	1958	1985	1984	1983	1960	1965	1969	1987	JAN 1973
-Maximum in 24 hrs	38	0.37	0.27	0.94	1.16	2.07	2.47	3.98	4.99	1.98	1.21	0.42	0.36	4.99
-Year		1969	1982	1987	1969	1974	1972	1962	1985	1978	1981	1993	1982	AUG 1985
Snow,Ice pellets,hail														
-Maximum Monthly	38	24.2	15.9	14.9	13.7	10.7	T	T	T	2.2	7.0	17.2	13.9	24.2
-Year		1971	1979	1987	1970	1983	1993	1989	1993	1983	1975	1958	1972	JAN 1971
-Maximum in 24 hrs	38	8.8	4.7	10.9	8.4	10.1	T	T	T	2.1	5.3	10.5	7.0	10.9
-Year		1971	1979	1987	1970	1983	1993	1989	1993	1983	1975	1993	1972	MAR 1987
WIND:														
Mean Speed (mph)	24	10.0	10.1	11.2	12.4	12.0	11.2	10.6	10.9	10.9	10.6	9.6	9.8	10.8
Prevailing Direction														
Fastest Obs. 1 Min.														
-Direction (!!)	25	33	28	30	32	12	03	30	27	30	32	27	30	30
-Speed (MPH)	25	41	44	41	46	41	46	69	44	46	44	48	40	69
-Year		1987	1974	1979	1978	1980	1989	1988	1975	1971	1968	1978	1988	JUL 1988
Peak Gust														
-Direction (!!)	10	W	NW	NW	NW	SE	SW	NW	W	S	NW	NW	NW	NW
-Speed (mph)	10	61	68	54	60	63	58	76	72	52	60	55	61	76
-Date		1990	1988	1986	1987	1988	1986	1988	1988	1985	1990	1989	1988	JUL 1988

PRECIPITATION (inches)

GLASGOW, MONTANA

YEAR	JAN	FEB	MAR	APR	MAY	JUNE	JULY	AUG	SEP	OCT	NOV	DEC	ANNUAL
1964	0.21	0.19	0.61	0.41	2.52	2.78	1.20	0.85	0.52	0.20	0.26	0.78	10.53
1965	0.46	0.20	0.30	0.51	3.25	4.64	0.91	3.01	0.79	T	0.12	0.12	14.31
1966	0.51	0.14	0.10	0.59	1.91	0.89	3.13	3.65	0.41	0.68	0.34	0.23	12.58
1967	0.84	0.25	0.83	0.84	0.68	2.23	0.12	0.18	2.20	1.13	0.36	0.26	9.92
1968	0.13	0.15	0.10	0.67	0.23	2.24	0.82	2.00	0.75	0.10	0.07	0.39	7.65
1969	1.24	0.14	0.17	1.99	0.25	1.23	3.45	0.05	0.27	1.33	T	0.36	10.48
1970	0.49	0.05	0.27	1.51	1.88	2.43	1.07	0.04	1.32	0.44	0.49	0.26	10.25
1971	0.99	0.21	0.27	0.42	0.91	1.26	0.20	0.78	0.48	1.08	0.04	0.26	6.90
1972	0.55	0.59	0.50	0.84	2.73	3.77	2.30	2.40	0.69	0.22	0.02	0.61	15.22
1973	T	0.09	0.27	1.24	0.54	4.35	0.59	1.21	1.43	0.24	0.46	0.78	11.20
1974	0.18	0.25	0.93	0.49	3.27	2.00	2.31	2.69	0.07	0.42	0.26	0.25	13.12
1975	0.11	0.23	0.61	1.30	1.93	1.20	4.18	1.37	0.42	1.77	0.40	0.38	13.90
1976	0.27	0.13	0.30	0.60	0.64	4.27	5.00	1.16	0.41	0.21	0.26	0.28	13.53
1977	0.42	0.12	0.11	0.17	1.83	1.16	0.78	0.58	2.62	0.40	0.24	0.86	9.29
1978	0.19	0.40	0.25	0.51	3.65	2.72	2.63	0.16	4.14	0.28	0.48	0.45	15.86
1979	0.15	0.74	0.43	1.27	2.46	0.48	0.75	0.76	0.27	0.57	0.18	0.05	8.11
1980	0.50	0.14	0.36	0.22	0.46	2.55	0.44	1.36	0.87	1.35	0.20	0.50	8.95
1981	0.06	0.10	0.31	0.19	2.13	1.83	1.72	0.23	0.08	1.40	0.55	0.19	8.79
1982	0.75	0.43	0.72	0.22	3.74	1.03	0.97	1.19	0.98	1.09	0.06	1.03	12.21
1983	0.18	0.24	0.39	0.10	1.59	0.77	2.62	0.03	0.97	0.24	0.16	0.27	7.56
1984	0.43	0.20	0.68	0.10	0.78	1.99	0.01	0.43	0.81	0.57	0.18	0.56	6.74
1985	0.05	0.05	0.27	0.91	2.34	0.09	0.69	5.74	0.83	0.64	0.74	0.50	12.85
1986	0.31	0.55	0.32	0.41	2.49	1.60	1.55	0.83	3.81	0.86	0.57	0.10	13.40
1987	0.08	0.11	1.27	0.45	1.73	1.07	2.90	0.88	0.66	0.03	0.03	0.03	9.24
1988	0.45	0.24	0.22	0.33	1.33	1.54	1.28	0.47	0.95	0.14	0.11	0.42	7.48
1989	0.63	0.60	0.36	1.62	2.09	0.95	0.95	1.48	0.20	0.61	0.47	0.35	10.31
1990	0.32	0.09	0.29	0.71	0.94	1.13	0.64	1.45	0.12	0.14	0.48	0.53	6.84
1991	0.21	0.32	0.62	1.29	1.29	4.13	1.32	0.49	0.61	0.13	0.42	0.15	10.98
1992	0.25	0.04	0.45	0.39	0.67	3.15	1.35	0.90	0.41	0.74	0.65	0.24	9.24
1993	0.20	0.34	0.82	0.43	0.50	1.96	5.93	4.31	0.60	0.47	0.58	0.13	16.27
Record Mean	0.42	0.35	0.47	0.78	1.67	2.42	1.63	1.49	0.94	0.54	0.36	0.41	11.49

AVERAGE TEMPERATURE (deg. F)

GLASGOW, MONTANA

YEAR	JAN	FEB	MAR	APR	MAY	JUNE	JULY	AUG	SEP	OCT	NOV	DEC	ANNUAL
#1964	20.9	27.7	24.1	44.7	57.0	63.9	74.3	68.0	55.0	49.3	27.1	1.4	42.8
1965	6.4	11.1	10.5	41.8	53.2	61.9	71.0	69.4	48.1	52.2	26.8	22.7	39.6
1966	-1.2	7.8	32.0	37.8	55.3	62.3	71.6	65.4	62.1	45.9	23.9	18.9	40.1
1967	12.5	15.1	21.8	37.0	50.9	60.6	71.0	70.5	62.2	46.4	30.6	15.9	41.2
1968	11.2	18.5	38.1	42.0	52.1	60.4	69.0	65.0	57.6	45.8	32.9	9.9	41.9
1969	-7.0	8.2	18.3	47.6	53.8	58.3	66.1	73.4	61.9	38.7	34.5	22.8	39.7
1970	7.0	18.4	24.4	39.5	54.1	67.4	72.6	72.2	56.1	42.8	26.8	12.0	41.1
1971	5.1	14.3	24.6	44.9	55.8	65.0	68.8	77.2	56.5	44.3	32.3	10.8	41.6
1972	4.3	10.5	30.3	44.3	55.6	65.6	64.7	71.1	53.6	42.1	32.8	6.4	40.1
1973	18.1	23.6	39.0	41.6	56.2	65.6	69.8	72.9	56.6	49.3	21.3	15.7	44.2
1974	10.4	25.5	29.6	47.5	50.7	66.2	73.3	64.0	55.4	49.2	33.2	25.1	44.2
1975	18.5	12.6	24.7	35.9	52.8	62.4	73.1	64.9	56.3	45.0	30.1	18.5	41.2
1976	13.8	29.5	29.3	47.8	58.2	62.5	70.3	70.0	59.8	42.2	26.4	18.9	44.1
1977	3.1	26.5	33.6	48.2	58.3	65.7	69.1	61.2	55.2	46.0	24.5	8.3	41.6
1978	-2.1	7.4	24.5	43.8	54.8	62.7	67.1	65.9	58.5	45.4	21.3	9.2	38.2
1979	-4.3	1.5	23.4	36.3	50.4	64.3	72.2	69.7	62.5	49.0	26.9	26.5	39.9
1980	7.3	16.4	27.9	51.9	61.1	66.1	72.3	64.7	57.8	47.7	36.3	18.2	44.2
1981	25.3	25.8	38.9	48.2	56.8	60.6	71.7	73.1	60.8	44.1	37.1	17.7	46.7
1982	-5.1	9.8	24.6	40.6	51.2	62.3	69.7	69.3	56.8	45.2	25.7	19.0	39.1
1983	20.6	27.1	33.6	43.1	51.7	63.4	71.8	76.5	55.3	47.7	32.9	-3.2	43.4
1984	18.6	34.3	31.9	47.6	55.5	64.6	74.6	74.4	51.8	42.0	30.0	6.9	44.4
1985	8.5	14.9	33.6	48.6	58.9	60.6	72.3	64.6	50.5	43.1	9.7	12.8	39.8
1986	22.1	10.5	40.4	43.6	55.9	69.0	69.1	69.4	52.2	48.3	25.1	24.9	44.2
1987	24.0	31.0	33.3	52.1	59.8	68.1	70.3	64.6	60.5	44.9	36.9	25.2	47.6
1988	10.2	18.3	35.6	47.0	62.5	76.8	73.6	70.5	57.6	46.8	29.1	19.0	45.6
1989	14.3	3.7	22.9	43.5	55.4	63.9	75.1	70.2	58.4	45.5	31.9	14.3	41.6
1990	23.2	22.5	35.3	43.9	55.3	66.1	70.6	72.0	64.0	45.3	30.9	10.0	44.9
1991	8.0	31.9	33.8	45.7	56.3	65.5	71.3	75.0	58.9	41.9	27.7	24.9	45.1
1992	26.6	30.5	39.3	46.1	57.8	65.2	64.1	66.2	56.8	45.9	31.7	13.0	45.3
1993	4.9	7.5	34.4	45.8	57.8	61.2	62.7	64.5	54.4	44.8	25.1	23.3	40.5
Record Mean	10.7	16.9	28.5	44.0	55.4	64.0	70.9	69.4	57.4	46.2	29.1	16.5	42.4
Max	20.7	27.3	39.3	56.5	68.2	76.8	85.0	83.6	70.9	59.2	39.5	26.4	54.4
Min	0.7	6.5	17.6	31.6	42.6	51.2	56.8	55.2	43.9	33.2	18.6	6.6	30.4

REFERENCE NOTES

GLASGOW, MONTANA

GENERAL
T - TRACE AMOUNT.
BLANK ENTRIES DENOTE MISSING/UNREPORTED DATA.
# INDICATES A STATION OR INSTRUMENT RELOCATION.

SPECIFIC

(a) - LENGTH OF RECORD IN YEARS, ALTHOUGH INDIVIDUAL MONTHS MAY BE MISSING.
0.* OR * - THE VALUE IS BETWEEN 0.0 AND 0.05
NORMALS - BASED ON THE 1961-1990 RECORD PERIOD.
EXTREMES - DATES ARE THE MOST RECENT OCCURRENCE
WIND DIR.- NUMERALS SHOW TENS OF DEGRESS CLOCKWISE FROM TRUE NORTH. "00" INDICATES CALM.
RESULTANT DIRECTIONS ARE GIVEN TO WHOLE DEGREES.
BOLD VALUES INDICATE EXTREME VALUES WHICH OCCURRED AFTER THE ASOS SYSTEM WAS COMMISSIONED.

RECORD = PERIOD OF RECORD
RECORD MEAN PRECIPITATION IS THE MEAN OF ALL DAILY PRECIPITATION AMOUNTS DURING THE PERIOD OF RECORD.
RECORD MAX(MIN) TEMPERATURE IS THE MEAN OF ALL DAILY MAX(MIN) TEMPERATURES DURING THE PERIOD OF RECORD.
RECORD MEAN TEMPERATURE IS THE SUM OF THE RECORD MAX AND RECORD MIN DIVIDED BY 2.
AVERAGE TEMPERATURE IS THE SUM OF THE MEAN DAILY MAX AND MIN TEMPERATURE DIVIDED BY 2.

EXCEPTIONS

RECORD MEANS ARE THROUGH THE CURRENT YEAR, BEGINNING IN 1944 FOR TEMPERATURE
1944 FOR PRECIPITATION
1956 FOR SNOWFALL

HEATING DEGREE DAYS Base 65 deg. F GLASGOW, MONTANA

SEASON	JULY	AUG	SEP	OCT	NOV	DEC	JAN	FEB	MAR	APR	MAY	JUNE	TOTAL
1964-65	0	52	298	482	1135	1974	1816	1507	1689	690	359	118	10120
1965-66	11	41	502	389	1137	1305	2053	1599	1016	812	316	132	9313
1966-67	5	78	136	588	1225	1422	1624	1396	1335	835	429	149	9222
1967-68	19	12	136	571	1024	1519	1664	1344	826	683	397	158	8353
1968-69	38	78	235	589	956	1706	2236	1584	1440	515	361	219	9957
1969-70	39	7	150	807	908	1301	1798	1297	1251	758	334	46	8696
1970-71	5	7	296	680	1142	1638	1856	1416	1245	595	285	59	9224
1971-72	27	0	267	635	973	1676	1883	1579	1067	614	305	55	9081
1972-73	71	11	338	703	958	1814	1452	1152	802	691	277	68	8337
1973-74	8	7	255	480	1306	1522	1692	1103	1089	526	436	70	8494
1974-75	3	88	283	481	945	1229	1437	1463	1244	867	373	102	8515
1975-76	5	68	260	613	1042	1437	1583	1022	1099	510	215	109	7963
1976-77	9	10	185	698	1151	1421	1917	1074	967	497	233	59	8221
1977-78	32	140	292	583	1208	1752	2081	1610	1252	628	319	111	10008
1978-79	42	70	233	601	1306	1725	2145	1780	1283	853	455	86	10579
1979-80	6	17	114	489	1139	1187	1788	1404	1145	398	175	49	7911
1980-81	2	63	232	536	854	1446	1225	1091	799	499	250	145	7142
1981-82	10	6	169	639	834	1459	2171	1546	1246	728	418	121	9347
1982-83	16	39	271	607	1171	1417	1369	1055	969	649	412	97	8072
1983-84	16	5	322	528	956	2113	1431	885	1018	515	318	85	8192
1984-85	0	20	406	705	1042	1797	1754	1400	967	485	218	156	8950
1985-86	11	84	429	673	1657	1617	1324	1524	754	638	320	18	9049
1986-87	15	15	379	513	1192	1237	1262	948	975	397	185	40	7158
1987-88	26	71	160	619	833	1227	1698	1353	903	535	143	13	7581
1988-89	0	10	244	557	1072	1422	1568	1712	1299	641	297	99	8921
1989-90	0	24	209	596	982	1569	1287	1186	912	626	316	99	7806
1990-91	10	5	119	605	1017	1704	1767	921	962	572	273	29	7984
1991-92	3	11	215	709	1114	1237	1185	995	788	564	254	94	7169
1992-93	65	94	256	594	990	1612	1861	1610	940	570	240	143	8975
1993-94	91	82	322	617	1194	1286							

COOLING DEGREE DAYS Base 65 deg. F GLASGOW, MONTANA

YEAR	JAN	FEB	MAR	APR	MAY	JUNE	JULY	AUG	SEP	OCT	NOV	DEC	TOTAL
1969	0	0	0	0	20	26	79	276	62	0	0	0	463
1970	0	0	0	0	3	128	250	236	38	0	0	0	655
1971	0	0	0	0	9	67	153	383	22	0	0	0	634
1972	0	0	0	0	20	80	69	205	4	0	0	0	378
1973	0	0	0	0	13	91	165	259	10	0	0	0	538
1974	0	0	0	5	0	113	268	62	3	0	0	0	451
1975	0	0	0	0	1	33	263	75	6	0	0	0	378
1976	0	0	0	0	9	41	180	172	35	0	0	0	437
1977	0	0	0	1	34	85	166	29	6	0	0	0	321
1978	0	0	0	0	7	47	114	107	45	0	0	0	320
1979	0	0	0	0	8	70	236	170	48	1	0	0	533
1980	0	0	0	12	60	88	234	59	23	7	0	0	483
1981	0	0	0	0	2	22	227	265	48	0	0	0	564
1982	0	0	0	0	0	44	168	180	32	0	0	0	424
1983	0	0	0	0	7	54	234	369	40	0	0	0	704
1984	0	0	0	0	30	78	303	318	15	0	0	0	744
1985	0	0	0	0	33	29	242	78	3	0	0	0	385
1986	0	0	0	0	44	142	150	161	0	0	0	0	497
1987	0	0	0	15	29	139	198	63	29	4	0	0	477
1988	0	0	0	0	73	374	273	187	27	0	0	0	934
1989	0	0	0	1	4	74	318	193	17	0	0	0	607
1990	0	0	0	2	22	138	192	231	96	0	0	0	681
1991	0	0	0	0	11	54	205	329	36	1	0	0	636
1992	0	0	0	5	38	107	44	138	17	8	0	0	357
1993	0	0	0	0	21	37	27	75	7	0	0	0	167

SNOWFALL (inches) GLASGOW, MONTANA

SEASON	JULY	AUG	SEP	OCT	NOV	DEC	JAN	FEB	MAR	APR	MAY	JUNE	TOTAL
1964-65	0.0	0.0	T	1.6	2.4	10.8	10.7	6.6	4.9	3.1	T	0.0	40.1
1965-66	0.0	0.0	1.1	0.0	1.0	1.2	8.4	2.5	2.8	1.7	1.7	0.0	20.4
1966-67	0.0	0.0	0.0	1.6	4.6	3.9	14.2	4.3	14.8	12.1	3.1	0.0	58.6
1967-68	0.0	0.0	0.0	0.0	0.4	4.6	2.7	2.1	T	6.2	0.9	0.0	16.9
1968-69	0.0	0.0	0.0	T	0.7	5.7	24.1	2.7	3.3	0.6	0.0	0.0	37.1
1969-70	0.0	0.0	0.0	1.0	T	4.1	5.5	0.6	2.6	13.7	T	0.0	27.5
1970-71	0.0	0.0	0.1	1.1	6.7	3.4	24.2	4.3	1.1	T	0.4	0.0	41.3
1971-72	0.0	0.0	0.0	0.1	0.6	6.3	12.7	10.6	3.3	0.7	T	0.0	34.3
1972-73	0.0	0.0	0.2	2.2	0.3	13.9	T	1.2	0.1	4.0	T	0.0	21.9
1973-74	0.0	0.0	0.0	0.0	5.1	10.6	2.1	2.6	6.1	0.6	T	0.0	27.1
1974-75	0.0	0.0	0.0	0.2	1.0	2.3	1.5	3.5	7.4	4.0	T	0.0	19.9
1975-76	0.0	0.0	0.0	7.0	5.5	4.7	3.9	0.5	4.9	0.2	0.0	0.0	26.7
1976-77	0.0	0.0	0.0	1.6	2.3	4.0	7.7	0.3	0.7	0.4	0.0	0.0	17.0
1977-78	0.0	0.0	0.0	0.2	4.3	13.2	3.8	8.4	3.0	0.5	0.0	0.0	33.4
1978-79	0.0	0.0	T	T	7.6	6.5	2.4	15.9	7.1	7.8	0.5	0.0	47.8
1979-80	0.0	0.0	0.0	T	2.7	0.4	9.4	1.3	2.6	0.7	0.0	0.0	17.1
1980-81	0.0	0.0	0.0	3.1	1.0	11.0	1.1	0.8	0.1	T	0.0	0.0	17.1
1981-82	0.0	0.0	0.0	4.4	1.8	2.9	16.4	7.3	7.5	1.7	T	0.0	42.0
1982-83	0.0	0.0	1.5	0.6	0.4	11.2	1.1	1.3	3.5	0.1	10.7	0.0	30.4
1983-84	0.0	0.0	2.2	T	1.8	4.7	4.2	1.6	4.3	1.1	0.2	0.0	20.1
1984-85	0.0	0.0	2.0	4.2	1.7	12.7	1.0	0.7	4.6	1.7	0.0	0.0	28.6
1985-86	0.0	0.0	0.2	2.2	13.0	9.1	3.5	9.8	2.3	4.9	T	0.0	45.0
1986-87	0.0	0.0	0.0	T	6.3	0.1	1.8	1.5	14.9	1.2	T	0.0	25.8
1987-88	0.0	0.0	0.0	0.2	0.2	0.4	8.6	3.6	0.2	0.1	0.0	0.0	13.3
1988-89	0.0	0.0	0.0	T	1.2	10.2	14.8	9.4	4.4	0.7	T	T	40.7
1989-90	T	T	T	T	2.8	9.2	5.7	1.6	3.6	1.4	T	T	24.3
1990-91	0.0	0.0	0.0	0.1	3.7	8.4	3.8	3.4	9.0	4.1	1.7	0.0	34.2
1991-92	0.0	0.0	0.0	1.2	0.9	2.0	1.0	1.3	0.2	2.3	T	T	8.9
1992-93	0.0	0.0	T	0.6	2.5	5.4	5.5	7.8	1.1	T	0.0	T	22.9
1993-94	0.0	T	0.0	0.1	13.9	1.2							
Record Mean	T	T	0.2	1.3	3.3	5.4	6.2	4.2	3.7	2.4	0.6	T	27.4

GREAT FALLS, MONTANA

The city of Great Falls is located along the main stem of the Missouri River at its confluence with the Sun River. The Weather Service Office is located at the Municipal Airport on a plateau between the Sun and Missouri Rivers. This plateau is about 200 feet higher than most of the immediate valley area, and the airport is about two miles southwest of the Sun and Missouri River Junction. Except to the north and northeast, the valley is encircled by mountain ranges, which lie about 30 miles away from east to south, 40 miles to the southwest, and 60 to 100 miles distant from west to northwest. Topography plays an important part in the climate of Great Falls. The Continental Divide to the west, and Big and Little Belt Ranges to the south, are primary factors in producing the frequent wintertime chinook winds observed in this part of Montana. The combination of valleys and plateaus in the immediate area, contributes to marked temperature differences between the airport and the city proper, either on calm, clear mornings, or when chinook winds reach the airport before they are felt at the lower elevations in town.

Summertime in the area generally is quite pleasant, with cool nights, moderately warm and sunny days, and very little hot, humid weather. Most of the summer rainfall occurs in showers or thunderstorms, and steady rains may occur during late spring or early summer. At the airport, freezing temperatures do not occur in July or August and very rarely in June. Frost occurs frequently in April and October, but more often in the valleys than on the surrounding hills or plateaus. However, frost may occur on rare occasions in nearby low lying areas at any time of the year.

Winters are not as cold as is usually expected of a continental location at this latitude, largely as a result of the chinook winds for which this area is noted. While sub-zero weather is experienced normally several times during a winter, the coldest weather seldom lasts more than a few days at a time, and is usually terminated by southwest chinook winds which can produce sharp temperature rises of 40 degrees or more in 24 hours.

As a result of recurring chinooks throughout the winter season, snow seldom lies on the ground for more than a few days. In fact, the ground usually is bare, or nearly bare, of snow most of the winter, except in the surrounding mountains and higher foothills. On the other hand, invasions of cold air from the polar regions occur a few times each winter, and sharp temperature falls from above freezing to below zero within 24 hours are observed occasionally.

Precipitation generally falls as snow during late fall, winter, and early spring, although rain can occur in any month. Late spring, summer, and early fall precipitation is almost always rain, but some hail is observed occasionally during summer thunderstorms.

Although average annual precipitation at Great Falls would normally classify the area as semi-arid, it is important to note that about 70 percent of the annual total falls normally during the April to September growing season. The combination of ideal temperatures during the peak of the growing season, long hours of summer sunshine, and adequate precipitation during the six critical months, makes the climate very favorable for dryland farming. Heavy fog occurs about one day per month, but each case lasts only a small part of the day. Although the average windspeed is relatively high, strong winds over 70 mph are seldom observed. Visibility normally is excellent.

NORMALS, MEANS, AND EXTREMES

GREAT FALLS, MONTANA

LATITUDE: 47°29'N LONGITUDE: 111°22' W ELEVATION: FT. GRND 3663 BARO 3665 TIME ZONE: MOUNTAIN WBAN: 24143

	(a)	JAN	FEB	MAR	APR	MAY	JUNE	JULY	AUG	SEP	OCT	NOV	DEC	YEAR
TEMPERATURE °F:														
Normals														
-Daily Maximum		30.6	37.5	43.7	55.3	65.2	74.6	83.3	81.6	69.6	59.3	43.5	33.1	56.4
-Daily Minimum		11.6	17.2	22.8	31.9	40.9	48.6	53.2	52.2	43.5	35.8	24.3	14.6	33.1
-Monthly		21.2	27.4	33.3	43.6	53.1	61.6	68.2	66.9	56.6	47.5	33.9	23.9	44.8
Extremes														
-Record Highest	56	67	70	78	89	93	101	105	106	98	91	76	69	106
-Year		1992	1992	1978	1980	1980	1990	1973	1969	1980	1992	1975	1939	AUG 1969
-Record Lowest	56	-37	-35	-29	-6	15	31	40	30	21	-11	-25	-43	-43
-Year		1969	1989	1951	1975	1954	1950	1992	1992	1985	1991	1985	1968	DEC 1968
NORMAL DEGREE DAYS:														
Heating (base 65°F)		1358	1053	983	642	372	156	37	91	299	543	933	1274	7741
Cooling (base 65°F)		0	0	0	0	0	54	137	150	47	0	0	0	388
% OF POSSIBLE SUNSHINE	46	49	56	66	62	62	65	79	76	67	61	46	44	61
MEAN SKY COVER (tenths)														
Sunrise - Sunset	52	7.3	7.4	7.4	7.4	7.0	6.5	4.4	4.9	5.6	6.4	7.1	7.2	6.5
MEAN NUMBER OF DAYS:														
Sunrise to Sunset														
-Clear	56	5.3	4.2	4.0	3.9	4.6	5.5	12.9	12.0	9.7	7.1	5.0	5.0	79.1
-Partly Cloudy	56	6.4	6.5	8.7	8.1	9.7	10.9	11.7	11.0	9.1	9.2	7.4	7.7	106.4
-Cloudy	56	19.3	17.5	18.3	18.0	16.7	13.7	6.4	8.0	11.2	14.8	17.7	18.3	179.8
Precipitation .01 inches or more	56	8.9	7.8	9.4	9.0	11.5	11.8	7.6	7.7	7.3	5.8	7.1	7.8	101.4
Snow, Ice pellets, hail 1.0 inches or more	56	3.4	2.9	3.4	2.1	0.5	0.1	0.0	0.*	0.5	1.1	2.6	3.1	19.7
Thunderstorms	56	0.*	0.1	0.1	0.7	3.6	6.9	7.3	6.0	1.4	0.2	0.*	0.*	26.4
Heavy Fog Visibility 1/4 mile or less	54	1.2	1.5	2.2	1.6	0.8	0.6	0.3	0.4	0.6	1.2	1.9	1.0	13.2
Temperature °F														
-Maximum														
90° and above	32	0.0	0.0	0.0	0.0	0.1	2.3	7.6	7.4	1.4	0.*	0.0	0.0	18.9
32° and below	32	13.5	8.3	5.8	1.3	0.0	0.0	0.0	0.0	0.*	1.1	5.0	11.5	46.6
-Minimum														
32° and below	32	26.9	23.8	25.2	15.6	3.3	0.1	0.0	0.1	2.6	11.2	21.7	26.4	156.9
0° and below	32	9.8	5.0	2.4	0.1	0.0	0.0	0.0	0.0	0.0	0.2	2.2	6.6	26.2
AVG. STATION PRESS. (mb)	21	887.6	887.7	885.5	886.7	886.5	887.3	889.2	889.4	889.7	889.2	887.5	887.5	887.8
RELATIVE HUMIDITY (%)														
Hour 05	32	66	67	68	68	70	71	67	65	67	64	65	65	67
Hour 11 (Local Time)	32	62	58	55	47	46	45	39	39	46	47	55	60	50
Hour 17	32	60	54	49	41	41	40	31	31	38	43	55	60	45
Hour 23	32	65	66	65	60	61	60	52	51	58	59	63	65	60
PRECIPITATION (inches):														
Water Equivalent														
-Normal		0.91	0.57	1.10	1.41	2.52	2.39	1.24	1.54	1.24	0.78	0.66	0.85	15.21
-Maximum Monthly	56	2.05	2.16	2.18	4.63	8.13	5.37	4.68	4.90	3.56	3.43	2.27	1.92	8.13
-Year		1969	1958	1967	1975	1953	1965	1993	1985	1941	1975	1955	1977	MAY 1953
-Minimum Monthly	56	T	0.01	0.10	0.05	0.67	0.52	0.04	0.03	0.09	T	0.02	T	T
-Year		1944	1950	1986	1981	1950	1960	1959	1969	1990	1965	1954	1954	OCT 1965
-Maximum in 24 hrs	56	0.74	0.88	1.14	2.43	3.42	2.74	2.40	2.74	1.82	1.15	0.97	0.82	3.42
-Year		1966	1951	1977	1951	1980	1964	1983	1989	1982	1954	1946	1972	MAY 1980
Snow, Ice pellets, hail														
-Maximum Monthly	56	22.6	26.1	24.2	35.4	11.6	11.1	T	8.3	10.4	16.6	22.1	25.0	35.4
-Year		1969	1958	1989	1967	1989	1950	1993	1992	1984	1975	1955	1945	APR 1967
-Maximum in 24 hrs	56	10.2	11.0	11.5	16.8	11.6	11.0	T	8.3	8.4	8.3	10.8	9.8	16.8
-Year		1984	1951	1987	1973	1989	1950	1993	1992	1988	1957	1946	1945	APR 1973
WIND:														
Mean Speed (mph)	52	15.2	14.2	12.9	12.8	11.3	11.1	10.0	10.2	11.2	13.1	14.6	15.6	12.7
Prevailing Direction through 1963		SW	SW	SW	SW	SW	SW	SW	SW	SW	SW	SW	SW	SW
Fastest Mile														
-Direction (!!!)	47	SW	W	W	W	SW	NW	W	SW	NW	W	SW	SW	SW
-Speed (MPH)	47	65	72	73	70	65	70	73	71	73	73	73	82	82
-Year		1946	1954	1951	1946	1951	1980	1951	1954	1945	1949	1955	1956	DEC 1956
Peak Gust														
-Direction (!!!)	10	SW	W	W	W	W	SE	SW	W	SW	SW	SW	SW	SW
-Speed (mph)	10	60	61	55	52	60	58	77	63	61	64	61	58	77
-Date		1990	1988	1991	1992	1988	1992	1990	1984	1984	1991	1990	1991	JUL 1990

PRECIPITATION (inches) GREAT FALLS, MONTANA

YEAR	JAN	FEB	MAR	APR	MAY	JUNE	JULY	AUG	SEP	OCT	NOV	DEC	ANNUAL
1964	0.65	0.52	1.74	1.91	3.36	4.34	1.50	1.66	0.28	T	0.72	1.23	17.91
1965	0.84	1.18	0.79	2.51	1.47	5.37	1.03	1.58	1.90	T	1.13	0.59	18.39
1966	1.63	0.51	0.79	0.74	1.54	2.17	1.81	0.77	0.21	1.32	1.62	0.99	14.10
1967	1.12	0.28	2.18	3.69	2.17	3.65	0.91	0.23	1.59	1.13	0.26	1.47	18.68
1968	1.29	0.24	0.90	1.11	2.64	2.89	0.06	2.16	2.92	0.11	0.68	1.36	16.36
1969	2.05	0.40	0.44	0.38	1.14	5.33	1.11	0.03	0.13	0.89	0.11	0.40	12.41
1970	0.99	1.02	1.14	1.88	3.16	2.32	1.16	0.77	0.67	1.00	0.53	0.70	15.34
1971	1.22	0.65	1.12	0.66	3.03	0.62	0.27	1.16	0.61	0.30	0.36	1.48	11.48
1972	1.47	0.62	1.01	0.77	1.59	0.94	1.51	1.26	0.85	1.17	0.20	1.68	13.07
1973	0.33	0.26	0.30	2.89	0.95	1.43	0.13	0.88	1.29	0.97	1.36	1.37	12.16
1974	1.44	0.26	1.10	1.03	3.16	1.08	0.48	4.76	0.73	0.36	0.26	0.60	15.26
1975	1.14	0.71	1.34	4.63	3.89	4.47	1.20	2.13	0.74	3.43	1.01	0.55	25.24
1976	0.57	0.53	0.75	2.33	0.88	4.10	2.07	1.91	0.61	0.19	0.65	0.51	15.10
1977	1.04	0.19	1.90	0.26	2.11	0.54	1.87	1.94	2.22	0.51	0.43	1.92	14.93
1978	1.68	1.21	0.41	1.76	3.20	2.56	1.99	1.04	2.56	0.27	1.44	1.05	19.17
1979	0.71	0.57	1.00	2.05	0.69	2.61	0.27	0.29	0.33	0.84	0.29	0.26	9.91
1980	0.67	1.03	0.74	0.62	5.12	3.91	0.27	0.67	0.98	1.75	0.19	0.27	16.22
1981	0.34	0.44	2.09	0.05	5.20	1.32	1.04	1.21	0.39	1.06	0.29	0.43	13.86
1982	1.09	0.99	1.97	1.04	3.63	3.09	0.66	0.41	2.43	0.75	0.63	0.99	17.68
1983	0.10	0.33	1.61	0.26	1.34	3.03	3.78	1.10	1.89	0.77	1.28	0.70	16.19
1984	0.72	0.69	1.31	0.94	1.34	2.10	0.05	1.01	0.71	1.20	0.49	1.25	11.81
1985	0.35	0.22	1.02	0.41	3.28	0.58	0.47	4.90	3.23	1.10	1.16	0.47	17.19
1986	0.57	0.75	0.10	2.83	1.74	1.72	1.67	0.81	1.52	0.90	0.45	0.27	13.33
1987	0.05	0.24	1.81	0.64	2.63	1.33	3.05	2.43	1.30	0.02	0.30	0.24	14.04
1988	0.76	0.47	0.44	0.77	1.60	1.42	1.82	0.26	2.33	0.66	0.30	0.97	11.80
1989	0.96	1.19	1.38	2.41	2.41	1.70	3.03	4.88	1.87	0.41	0.81	1.32	22.37
1990	0.29	0.17	1.69	0.84	3.97	1.23	1.03	3.19	0.09	0.13	0.70	0.73	14.06
1991	0.63	0.21	1.21	1.54	1.54	4.15	0.75	1.35	1.00	0.81	0.77	0.08	14.04
1992	0.48	0.23	0.43	1.32	2.14	3.22	1.81	1.37	0.25	2.61	0.29	0.31	14.46
1993	1.17	0.70	0.86	3.16	2.74	2.58	4.68	3.04	1.71	1.10	0.97	0.30	23.01
Record Mean	0.81	0.66	1.00	1.28	2.44	2.70	1.33	1.37	1.23	0.81	0.72	0.70	15.05

AVERAGE TEMPERATURE (deg. F) GREAT FALLS, MONTANA

YEAR	JAN	FEB	MAR	APR	MAY	JUNE	JULY	AUG	SEP	OCT	NOV	DEC	ANNUAL
1964	28.7	33.0	26.9	42.3	54.8	62.5	72.4	65.4	53.9	53.0	31.9	11.2	44.7
1965	23.9	27.0	20.9	44.7	51.5	60.3	69.6	68.0	45.0	53.4	34.9	29.1	44.0
1966	13.9	27.1	36.4	40.2	55.8	59.9	70.0	65.1	64.2	47.4	29.0	27.9	44.8
1967	26.3	32.1	27.2	35.5	52.5	60.2	73.0	71.0	62.9	50.2	36.2	22.1	45.8
1968	22.0	33.2	40.5	40.6	49.5	59.2	67.8	64.4	56.5	48.0	35.8	16.1	44.5
1969	-2.8	16.6	26.2	50.3	55.8	59.3	68.3	72.1	61.3	38.1	40.1	29.9	42.9
1970	14.7	32.3	29.0	38.5	53.8	66.5	71.4	70.9	54.6	44.7	30.2	23.0	44.1
1971	16.2	29.4	31.6	45.0	54.6	62.0	67.5	76.0	54.7	44.7	36.3	18.0	44.6
1972	12.8	22.5	38.3	42.4	53.5	65.2	64.7	69.6	53.9	43.2	36.3	17.9	43.4
1973	24.9	29.6	39.5	40.2	55.4	63.6	71.4	71.2	58.2	49.2	25.2	28.9	46.4
1974	19.8	33.8	33.3	47.1	49.5	66.9	72.9	62.9	54.8	51.4	38.7	32.5	47.0
1975	22.7	13.1	27.4	30.9	50.0	58.8	71.8	64.8	57.3	45.7	32.9	28.6	42.0
1976	26.4	30.5	31.6	46.1	56.9	61.0	70.3	67.9	61.2	46.5	36.6	31.6	47.2
1977	21.6	39.2	34.0	47.1	51.3	65.4	68.0	62.5	56.1	47.8	30.8	16.5	45.1
1978	7.7	14.5	33.6	44.1	51.3	62.5	67.1	66.5	58.8	48.8	23.9	17.4	41.4
1979	6.5	18.8	34.7	40.7	51.5	62.9	69.0	68.5	62.9	49.4	33.4	34.6	44.4
1980	15.2	28.1	32.4	52.9	57.1	60.9	69.4	62.4	58.0	49.0	39.4	23.8	45.7
1981	33.8	30.8	37.2	46.5	52.8	58.1	66.5	69.8	59.8	45.3	40.8	24.6	47.2
1982	6.3	19.5	27.9	37.9	48.4	60.2	66.8	65.0	53.9	46.6	32.1	26.8	41.0
1983	32.2	36.7	35.8	41.7	50.7	60.1	65.8	72.4	53.6	48.9	35.1	4.0	44.8
1984	29.5	36.9	35.3	44.4	51.7	59.8	69.8	71.6	51.8	40.2	35.5	13.0	45.0
1985	19.2	21.6	33.4	48.5	57.3	62.2	73.0	61.8	48.2	44.5	12.3	24.6	42.2
1986	36.6	18.6	43.9	42.4	53.2	66.0	64.9	69.0	51.5	49.6	32.0	33.2	46.7
1987	32.4	36.0	36.2	52.8	57.8	65.4	66.8	61.6	58.9	47.5	40.4	29.6	48.8
1988	23.6	28.9	37.3	46.8	56.4	69.5	69.1	67.1	56.0	49.7	35.7	29.5	47.5
1989	28.0	10.3	29.0	43.4	50.9	60.8	70.2	64.0	56.1	46.2	36.6	27.5	43.6
1990	30.0	28.0	35.7	44.3	49.8	59.8	67.5	68.5	62.4	46.0	37.7	17.8	45.6
1991	19.0	39.2	35.0	43.1	51.2	57.9	68.0	71.7	57.7	42.4	31.7	35.0	46.0
1992	34.7	36.6	40.9	47.1	55.3	62.8	61.6	62.9	56.8	47.6	35.2	18.4	46.7
1993	14.8	19.3	37.6	43.7	55.5	56.5	58.1	60.5	52.5	46.2	28.1	33.6	42.2
Record Mean	21.8	26.5	32.5	43.9	53.3	61.1	68.9	67.3	57.2	47.8	34.3	26.3	45.1
Max	31.1	36.5	42.8	55.3	65.2	73.2	83.2	81.5	69.8	59.1	43.5	35.1	56.4
Min	12.5	16.5	22.3	32.5	41.4	49.0	54.6	53.2	44.5	36.5	25.1	17.4	33.8

REFERENCE NOTES GREAT FALLS, MONTANA

GENERAL
T - TRACE AMOUNT.
BLANK ENTRIES DENOTE MISSING/UNREPORTED DATA.
# INDICATES A STATION OR INSTRUMENT RELOCATION.

SPECIFIC

(a) - LENGTH OF RECORD IN YEARS, ALTHOUGH INDIVIDUAL MONTHS MAY BE MISSING.
0.* OR * - THE VALUE IS BETWEEN 0.0 AND 0.05
NORMALS - BASED ON THE 1961-1990 RECORD PERIOD.
EXTREMES - DATES ARE THE MOST RECENT OCCURRENCE.
WIND DIR.- NUMERALS SHOW TENS OF DEGRESS CLOCKWISE FROM TRUE NORTH. "00" INDICATES CALM.
RESULTANT DIRECTIONS ARE GIVEN TO WHOLE DEGREES.
BOLD VALUES INDICATE EXTREME VALUES WHICH OCCURRED AFTER THE ASOS SYSTEM WAS COMMISSIONED.

RECORD = PERIOD OF RECORD
RECORD MEAN PRECIPITATION IS THE MEAN OF ALL DAILY PRECIPITATION AMOUNTS DURING THE PERIOD OF RECORD.
RECORD MAX(MIN) TEMPERATURE IS THE MEAN OF ALL DAILY MAX(MIN) TEMPERATURES DURING THE PERIOD OF RECORD.
RECORD MEAN TEMPERATURE IS THE SUM OF THE RECORD MAX AND RECORD MIN DIVIDED BY 2.
AVERAGE TEMPERATURE IS THE SUM OF THE MEAN DAILY MAX AND MIN TEMPERATURE DIVIDED BY 2.

EXCEPTIONS

RECORD MEANS ARE THROUGH THE CURRENT YEAR, BEGINNING IN 1938 FOR TEMPERATURE
1938 FOR PRECIPITATION
1938 FOR SNOWFALL

HEATING DEGREE DAYS Base 65 deg. F

GREAT FALLS, MONTANA

SEASON	JULY	AUG	SEP	OCT	NOV	DEC	JAN	FEB	MAR	APR	MAY	JUNE	TOTAL
1964-65	0	93	337	375	990	1666	1271	1055	1361	602	411	164	8325
1965-66	17	54	594	355	894	1106	1585	1054	880	736	304	182	7761
1966-67	7	85	101	537	1075	1143	1190	914	1166	878	396	168	7660
1967-68	2	8	136	453	858	1327	1329	914	753	724	472	198	7174
1968-69	38	93	261	520	867	1511	2104	1350	1198	432	292	199	8865
1969-70	27	8	172	828	743	1081	1559	908	1108	785	345	95	7659
1970-71	9	8	319	625	1036	1296	1509	993	1031	594	320	134	7874
1971-72	34	5	326	628	854	1455	1616	1226	820	667	368	77	8076
1972-73	109	23	331	668	856	1458	1240	983	785	739	303	125	7620
1973-74	6	27	226	483	1191	1111	1397	865	974	530	477	85	7372
1974-75	6	109	311	419	783	1000	1304	1450	1159	1015	460	190	8206
1975-76	12	60	235	592	961	1122	1192	994	1030	565	250	165	7178
1976-77	3	20	144	572	845	1031	1339	715	953	529	419	70	6640
1977-78	37	119	280	527	1021	1502	1776	1410	966	622	421	106	8787
1978-79	54	57	236	496	1228	1473	1808	1292	931	722	417	111	8825
1979-80	19	15	106	482	939	934	1538	1066	1004	370	267	148	6888
1980-81	16	110	225	504	763	1275	960	953	855	548	373	218	6800
1981-82	34	14	201	603	718	1244	1819	1271	1142	806	511	161	8524
1982-83	44	66	342	565	978	1181	1007	786	899	692	437	154	7151
1983-84	59	2	356	490	891	1888	1094	810	915	620	419	183	7727
1984-85	12	18	415	760	879	1611	1415	1212	971	489	249	134	8165
1985-86	4	147	498	629	1581	1246	872	1297	648	672	390	48	8032
1986-87	50	22	400	471	987	979	1004	803	888	372	238	70	6284
1987-88	66	136	189	540	729	1090	1278	1039	852	538	281	65	6803
1988-89	24	39	294	468	876	1090	1140	1529	1109	642	430	150	7791
1989-90	3	96	269	575	845	1155	1079	1031	902	613	462	204	7234
1990-91	34	37	118	583	813	1460	1425	714	922	649	417	209	7381
1991-92	19	5	233	705	992	925	935	819	740	532	307	128	6340
1992-93	130	171	260	538	886	1441	1556	1274	841	634	291	266	8288
1993-94	221	165	372	578	1101	966							

COOLING DEGREE DAYS Base 65 deg. F

GREAT FALLS, MONTANA

YEAR	JAN	FEB	MAR	APR	MAY	JUNE	JULY	AUG	SEP	OCT	NOV	DEC	TOTAL
1969	0	0	0	0	15	34	136	235	67	0	0	0	487
1970	0	0	0	0	4	144	214	197	15	6	0	0	580
1971	0	0	0	0	6	50	120	351	22	8	0	0	557
1972	0	0	0	0	19	87	108	175	5	0	0	0	394
1973	0	0	0	0	14	87	213	226	30	0	0	0	570
1974	0	0	0	0	0	148	253	54	11	7	0	0	473
1975	0	0	0	0	0	10	231	62	12	0	0	0	315
1976	0	0	0	0	6	51	174	116	37	5	0	0	389
1977	0	0	0	0	0	86	139	48	20	0	0	0	293
1978	0	0	0	0	0	36	125	111	60	0	0	0	332
1979	0	0	0	0	2	55	152	132	50	5	0	0	396
1980	0	0	0	12	30	31	156	37	21	18	0	0	305
1981	0	0	0	0	3	17	85	168	49	0	0	0	322
1982	0	0	0	0	0	24	104	73	15	0	0	0	216
1983	0	0	0	0	4	14	90	241	19	0	0	0	368
1984	0	0	0	5	15	33	169	229	26	0	0	0	477
1985	0	0	0	0	20	58	260	51	0	0	0	0	389
1986	0	0	0	0	32	85	56	153	0	0	0	0	326
1987	0	0	0	13	23	90	128	36	13	4	0	0	307
1988	0	0	0	0	20	206	160	112	30	0	0	0	528
1989	0	0	0	0	0	31	170	72	8	0	0	0	281
1990	0	0	0	0	0	54	119	157	47	2	0	0	379
1991	0	0	0	0	0	1	119	217	19	13	0	0	369
1992	0	0	0	2	10	70	31	113	21	5	0	0	252
1993	0	0	0	0	4	19	13	30	3	0	0	0	69

SNOWFALL (inches)

GREAT FALLS, MONTANA

SEASON	JULY	AUG	SEP	OCT	NOV	DEC	JAN	FEB	MAR	APR	MAY	JUNE	TOTAL
1964-65	0.0	0.0	T	T	7.9	13.8	6.9	12.5	7.7	11.5	T	0.7	61.0
1965-66	0.0	0.0	4.2	0.0	9.7	6.5	16.9	5.8	7.7	6.0	T	0.0	56.8
1966-67	T	0.0	0.0	3.4	15.2	9.8	10.9	2.8	21.3	35.4	8.0	0.0	106.8
1967-68	0.0	0.0	0.0	0.7	2.4	14.9	12.8	1.9	7.6	6.1	T	0.0	46.4
1968-69	0.0	0.0	3.9	0.2	2.6	13.6	22.6	5.2	5.5	0.9	0.1	5.3	59.9
1969-70	0.0	0.0	0.0	5.0	1.1	4.3	9.4	11.4	10.9	18.7	1.4	0.0	62.2
1970-71	0.0	0.0	3.0	6.4	4.9	7.3	16.2	8.9	9.7	4.7	T	0.0	61.1
1971-72	0.0	0.0	T	3.8	4.2	19.2	18.6	6.3	11.4	3.5	0.2	0.0	67.2
1972-73	0.0	0.0	0.5	9.5	2.0	16.2	3.3	1.8	1.8	24.8	0.3	0.0	60.2
1973-74	0.0	0.0	6.0	3.6	12.2	12.8	13.4	2.8	9.5	6.1	T	T	66.4
1974-75	0.0	0.0	1.3	1.1	1.3	5.9	13.2	7.2	12.4	29.2	5.6	0.0	77.2
1975-76	0.0	0.0	T	16.6	9.7	5.7	6.1	5.6	8.9	16.7	0.0	0.0	69.3
1976-77	0.0	0.0	0.0	0.4	8.8	8.3	13.9	1.8	21.5	1.0	2.1	0.0	57.8
1977-78	0.0	0.0	0.0	3.2	4.8	18.2	19.3	16.8	3.3	5.0	T	0.0	70.6
1978-79	0.0	0.0	0.0	T	16.5	11.5	12.0	8.1	14.8	8.6	2.6	T	74.1
1979-80	0.0	0.0	0.0	0.7	3.1	3.0	7.0	9.2	6.9	4.4	T	0.0	34.3
1980-81	0.0	0.0	0.0	7.7	3.3	5.4	4.1	7.1	11.5	0.1	T	0.0	39.2
1981-82	0.0	0.0	0.0	7.9	1.0	5.9	19.7	16.3	23.4	18.5	7.6	T	100.3
1982-83	0.0	0.0	0.7	1.5	8.8	13.0	0.9	4.1	6.6	1.4	8.6	0.0	45.6
1983-84	0.0	0.0	7.8	T	14.4	11.9	16.2	7.7	19.5	5.2	1.0	0.0	83.7
1984-85	0.0	0.0	10.4	10.9	5.5	16.6	5.4	3.8	11.9	2.4	0.0	0.0	66.9
1985-86	0.0	T	2.5	8.5	18.1	7.9	4.4	15.4	0.5	14.1	2.4	0.0	73.8
1986-87	0.0	0.0	0.1	1.2	7.9	4.5	1.0	1.8	16.5	0.6	5.3	0.0	38.9
1987-88	0.0	0.0	0.0	0.1	2.9	4.7	12.6	9.2	3.9	7.4	0.0	0.0	40.8
1988-89	0.0	0.0	9.1	5.3	6.0	10.9	16.0	18.7	24.2	15.7	11.6	T	117.5
1989-90	T	0.0	1.7	1.3	7.4	19.9	5.1	3.0	16.2	5.4	T	T	60.0
1990-91	T	0.0	0.0	0.4	6.7	8.5	9.2	3.1	23.9	6.0	4.0	T	61.8
1991-92	0.0	T	0.0	11.2	9.1	0.9	6.1	2.2	3.5	3.2	0.9	0.0	37.1
1992-93	0.0	8.3	0.0	8.3	3.3	6.4	14.7	10.1	7.0	4.4	T	T	62.5
1993-94	T	T	T	3.8	13.0	4.2							
Record Mean	T	0.1	1.5	3.4	7.5	8.7	9.9	8.3	10.4	7.2	1.8	0.3	59.1

HELENA, MONTANA

Helena is located on the south side of an intermountain valley bounded on the west and south by the main chain of the Continental Divide. The valley is approximately 25 miles in width from north to south and 35 miles long from east to west. The average height of the mountains above the valley floor is about 3,000 feet.

The climate of Helena may be described as modified continental. Several factors enter into modifying the continental climate characteristics. Some of these are invasion by Pacific Ocean air masses, drainage of cool air into the valley from the surrounding mountains, and the protecting mountain shield in all directions.

The mountains to the north and east sometimes deflect shallow masses of invading cold Arctic air to the east. Following periods of extreme cold, when the return circulation of maritime air has brought warming to most of the eastern part of the state, cold air may remain trapped in the valley for several days before being replaced by warmer air. During these periods of transition from cold-to-warm temperatures, inversions are often quite pronounced.

As may be expected in a northern latitude, cold waves may occur from November through February, with temperatures occasionally dropping to zero or lower.

Summertime temperatures are moderate, with maximum readings generally under 90 degrees and very seldom reaching 100 degrees. Like all mountain stations, there is usually a marked change in temperature from day to night. During the summer this tends to produce an agreeable combination of fairly warm days and cool nights.

Most of the precipitation falls from April through July from frequent showers or thunderstorms, but usually with some steady rains in June, the wettest month of the year. Like summer, fall and winter months are relatively dry. During the April to September growing season, precipitation varies considerably.

Thunderstorms are rather frequent from May through August. Snow can be expected from September through May, but amounts during the spring and fall are usually light, and snow on the ground ordinarily lasts only a day or two. During the winter months snow may remain on the ground for several weeks at a time. There is little drifting of snow in the valley, and blizzard conditions are very infrequent.

Severe ice, sleet, and hailstorms are very seldom observed. Since 1880, only a few hailstorms have caused extensive damage in the city of Helena.

In winter, hours of sunshine are more than would be expected at a mountain location.

Due to the sheltering influence of the mountains, Foehn (Chinook) winds are not as pronounced as might be expected for a location on the eastern slopes of the Rocky Mountains. Strong winds can occur at any time throughout the year, but generally do not last more than a few hours at a time.

Based on the 1951-1980 period, the average first occurrence of 32 degrees Fahrenheit in the fall is September 18 and the average last occurrence in the spring is May 18.

NORMALS, MEANS, AND EXTREMES

HELENA, MONTANA

LATITUDE: 46°36'N LONGITUDE: 112°00' W ELEVATION: FT. GRND 3828 BARO 3898 TIME ZONE: MOUNTAIN WBAN: 24144

	(a)	JAN	FEB	MAR	APR	MAY	JUNE	JULY	AUG	SEP	OCT	NOV	DEC	YEAR
TEMPERATURE °F:														
Normals														
-Daily Maximum		29.6	36.9	44.8	56.1	65.4	75.8	85.0	83.2	69.8	58.5	42.4	31.3	56.6
-Daily Minimum		9.6	15.9	22.3	30.6	39.6	48.3	53.4	51.7	41.0	31.6	20.7	11.2	31.3
-Monthly		19.6	26.4	33.6	43.4	52.5	62.1	69.2	67.4	55.4	45.1	31.6	21.2	44.0
Extremes														
-Record Highest	53	62	68	77	86	92	100	102	105	99	85	72	64	105
-Year		1992	1950	1978	1992	1986	1988	1981	1969	1967	1992	1990	1980	AUG 1969
-Record Lowest	53	-42	-33	-30	1	17	30	36	28	18	-8	-39	-38	-42
-Year		1957	1989	1955	1954	1954	1969	1971	1992	1970	1991	1959	1964	JAN 1957
NORMAL DEGREE DAYS:														
Heating (base 65°F)		1407	1081	973	648	388	137	34	65	321	617	1002	1358	8031
Cooling (base 65°F)		0	0	0	0	0	50	164	139	33	0	0	0	386
% OF POSSIBLE SUNSHINE	53	46	55	60	59	60	63	78	74	67	60	44	42	59
MEAN SKY COVER (tenths)														
Sunrise - Sunset	53	7.5	7.3	7.3	7.2	6.8	6.3	4.1	4.6	5.3	6.1	7.2	7.5	6.4
MEAN NUMBER OF DAYS:														
Sunrise to Sunset														
-Clear	53	4.7	4.3	3.9	3.8	4.9	5.7	14.5	12.8	10.5	8.3	4.6	4.0	82.1
-Partly Cloudy	53	6.2	6.7	8.2	8.7	9.9	11.2	10.6	11.0	8.6	8.6	7.5	7.3	104.4
-Cloudy	53	20.1	17.2	18.9	17.5	16.2	13.0	5.9	7.2	10.9	14.1	17.9	19.7	178.7
Precipitation .01 inches or more	53	7.9	6.5	8.5	8.2	11.0	11.3	7.6	7.6	6.6	5.6	7.0	7.8	95.7
Snow, Ice pellets, hail 1.0 inches or more	53	2.7	1.9	2.4	1.4	0.4	0.*	0.0	0.*	0.4	0.8	1.7	2.4	14.2
Thunderstorms	53	0.1	0.1	0.1	0.9	4.2	7.6	8.8	7.9	1.8	0.3	0.1	0.*	32.2
Heavy Fog Visibility 1/4 mile or less	53	1.6	1.2	0.8	0.2	0.2	0.1	0.1	0.2	0.3	0.5	1.2	2.0	8.2
Temperature °F														
-Maximum														
90° and above	30	0.0	0.0	0.0	0.0	0.2	2.2	7.6	6.9	1.1	0.0	0.0	0.0	18.0
32° and below	30	14.8	8.3	3.9	0.5	0.0	0.0	0.0	0.0	0.0	0.7	5.4	14.6	48.1
-Minimum														
32° and below	30	29.5	26.5	27.2	17.6	3.6	0.1	0.0	0.*	3.5	17.0	26.9	29.6	181.5
0° and below	30	8.7	4.5	1.5	0.0	0.0	0.0	0.0	0.0	0.0	0.2	1.9	6.2	23.0
AVG. STATION PRESS. (mb)	21	881.5	881.0	878.4	879.4	879.3	880.2	881.9	882.1	882.7	882.7	881.2	881.5	881.0
RELATIVE HUMIDITY (%)														
Hour 05	28	71	72	72	70	71	72	67	67	72	73	74	73	71
Hour 11 (Local Time)	30	67	63	55	47	44	44	39	42	48	52	62	68	53
Hour 17	30	63	54	46	38	38	37	30	31	36	42	58	66	45
Hour 23	30	70	69	66	60	59	59	52	53	59	64	70	72	63
PRECIPITATION (inches):														
Water Equivalent														
-Normal		0.63	0.41	0.73	0.97	1.78	1.87	1.10	1.29	1.15	0.60	0.48	0.59	11.60
-Maximum Monthly	53	2.78	1.20	1.62	3.00	6.09	4.74	4.70	4.23	3.37	2.68	1.50	1.48	6.09
-Year		1969	1986	1982	1975	1981	1944	1993	1974	1965	1975	1950	1977	MAY 1981
-Minimum Monthly	53	T	0.02	0.02	0.10	0.29	0.08	0.08	0.02	0.08	0.02	0.04	0.04	T
-Year		1987	1991	1959	1977	1979	1985	1973	1988	1972	1978	1969	1976	JAN 1987
-Maximum in 24 hrs	53	0.77	0.58	1.01	1.25	2.31	1.78	2.26	1.86	1.61	0.85	0.82	0.51	2.31
-Year		1969	1953	1957	1951	1981	1979	1983	1974	1980	1954	1959	1982	MAY 1981
Snow, Ice pellets, hail														
-Maximum Monthly	53	35.6	19.7	21.6	20.6	12.7	2.7	T	6.2	13.7	11.0	32.9	22.8	35.6
-Year		1969	1959	1955	1967	1967	1969	1993	1992	1965	1969	1959	1967	JAN 1969
-Maximum in 24 hrs	53	11.5	11.7	8.7	12.9	12.5	2.7	T	6.2	13.3	7.4	21.5	10.7	21.5
-Year		1969	1993	1955	1960	1967	1969	1993	1992	1957	1969	1959	1991	NOV 1959
WIND:														
Mean Speed (mph)	53	6.7	7.3	8.3	9.2	8.9	8.5	7.8	7.4	7.4	7.1	7.1	6.8	7.7
Prevailing Direction through 1963		W	W	W	W	W	W	W	W	W	W	W	W	W
Fastest Mile														
-Direction (!!!)	53	SW	W	SW	W	SW	W	SW	S	NW	W	SW	NW	W
-Speed (MPH)	53	73	73	61	52	56	56	65	65	54	62	56	59	73
-Year		1944	1949	1955	1950	1975	1970	1975	1947	1943	1948	1962	1946	FEB 1949
Peak Gust														
-Direction (!!!)	10	W	W	W	W	W	S	W	S	W	W	SW	W	W
-Speed (mph)	10	64	63	52	55	58	58	64	58	51	60	61	58	64
-Date		1990	1988	1987	1989	1989	1988	1990	1992	1984	1991	1990	1992	JUL 1990

PRECIPITATION (inches) HELENA, MONTANA

YEAR	JAN	FEB	MAR	APR	MAY	JUNE	JULY	AUG	SEP	OCT	NOV	DEC	ANNUAL
1964	0.31	0.27	0.51	1.56	3.52	2.98	0.83	1.91	0.16	0.04	0.53	0.99	13.61
1965	0.36	0.49	0.85	0.98	2.20	3.85	0.60	1.92	3.37	0.13	0.62	0.15	15.52
1966	0.46	0.33	0.28	0.51	0.43	0.96	0.32	0.42	0.34	0.75	1.04	0.62	6.46
1967	0.61	0.62	1.43	2.38	2.08	2.36	0.46	0.58	0.68	1.50	0.31	1.39	14.40
1968	0.59	0.16	0.53	1.21	1.62	2.68	0.26	2.00	2.22	0.23	0.92	0.75	13.17
1969	2.78	0.22	0.57	0.60	1.13	3.50	1.77	0.38	0.33	1.06	0.04	0.31	12.69
1970	0.51	0.67	0.96	0.81	1.20	2.11	0.93	0.63	0.36	0.58	0.44	0.54	9.74
1971	1.38	0.63	0.41	0.58	1.77	0.93	0.56	1.22	0.89	0.39	0.34	1.02	10.12
1972	1.12	0.54	0.63	0.41	0.77	1.12	0.56	1.63	0.08	0.57	0.33	0.46	8.22
1973	0.22	0.13	0.05	0.66	1.08	0.73	0.08	0.56	0.43	0.66	1.03	0.63	6.26
1974	0.66	0.23	0.38	0.76	2.07	0.34	0.49	4.23	0.22	0.51	0.30	0.26	10.45
1975	1.26	0.72	0.88	3.00	1.95	2.83	3.89	2.47	0.47	2.68	0.48	0.31	20.94
1976	0.26	0.38	0.41	1.34	0.87	2.74	0.29	1.58	1.82	0.04	0.30	0.04	10.07
1977	0.65	0.13	1.11	0.10	1.82	1.37	1.37	0.72	1.93	0.17	0.48	1.48	11.33
1978	0.96	0.61	0.31	0.94	1.20	0.44	2.83	0.59	1.11	0.02	1.19	0.76	10.96
1979	0.77	0.72	1.34	2.26	0.29	2.75	0.32	0.79	0.12	0.38	0.06	0.59	10.39
1980	0.62	0.74	0.88	0.63	4.32	3.16	1.92	0.28	2.57	1.21	0.32	0.40	17.05
1981	0.15	0.10	1.10	0.75	6.09	1.15	1.78	0.10	0.90	0.82	0.54	0.33	13.81
1982	0.80	0.58	1.62	0.54	1.77	2.99	0.49	0.74	2.74	0.35	0.31	1.05	13.98
1983	0.24	0.07	0.36	0.29	1.79	2.20	3.48	2.67	1.56	0.35	0.26	0.76	14.03
1984	0.17	0.15	0.49	1.45	1.03	2.14	0.11	1.11	0.73	0.74	0.47	0.41	9.00
1985	0.16	0.38	0.32	0.46	0.75	0.08	0.10	2.64	2.11	0.76	0.84	0.35	8.95
1986	0.32	1.20	0.49	1.08	0.83	1.56	1.37	1.84	2.45	0.03	0.54	0.38	12.09
1987	T	0.03	1.19	0.76	1.90	1.50	2.88	0.38	0.80	0.05	0.12	0.42	10.03
1988	0.27	0.50	0.45	1.32	1.82	1.50	0.36	0.02	2.09	0.69	0.69	0.32	10.03
1989	1.42	0.82	1.35	0.72	1.00	1.43	1.55	1.61	1.31	0.54	0.26	0.48	12.49
1990	0.47	0.14	0.91	0.43	1.54	0.92	0.40	2.57	0.11	0.11	0.36	0.47	8.43
1991	0.27	0.02	0.90	0.75	1.71	3.27	0.72	0.70	1.26	0.65	0.88	0.79	11.92
1992	0.29	0.10	0.60	0.55	0.64	2.36	1.06	1.01	0.09	1.87	0.19	0.57	9.33
1993	0.80	1.03	0.56	1.63	1.71	3.14	4.70	2.79	1.25	0.71	0.36	0.13	18.81
Record Mean	0.70	0.52	0.74	0.97	1.87	2.18	1.13	0.96	1.14	0.75	0.61	0.62	12.20

AVERAGE TEMPERATURE (deg. F) HELENA, MONTANA

YEAR	JAN	FEB	MAR	APR	MAY	JUNE	JULY	AUG	SEP	OCT	NOV	DEC	ANNUAL
1964	22.0	27.2	29.9	41.8	51.7	59.0	70.2	63.4	53.8	47.4	30.5	19.1	43.0
1965	29.6	26.6	24.6	44.4	50.1	59.5	67.7	65.8	45.6	49.3	34.7	27.5	43.8
1966	20.6	27.0	36.5	42.5	56.5	59.3	71.5	66.7	62.7	46.4	33.8	27.4	45.9
1967	28.7	32.7	29.0	37.8	52.1	59.9	70.5	71.4	61.9	47.4	33.0	19.0	45.3
1968	15.7	28.9	39.6	39.7	49.6	58.3	68.2	63.2	55.0	44.4	33.5	16.7	42.8
1969	7.2	14.5	21.3	46.7	55.2	57.8	67.9	70.1	58.0	38.1	33.5	25.9	41.4
1970	18.8	32.5	28.2	36.5	51.5	62.3	68.2	67.9	51.1	40.2	29.3	20.6	42.2
1971	19.0	28.5	31.9	41.5	51.4	57.0	63.2	69.0	49.7	42.0	33.6	15.6	41.9
1972	12.4	26.8	39.1	39.8	50.9	61.8	61.7	66.0	50.9	39.0	31.1	13.6	41.1
1973	17.7	22.8	36.0	40.3	52.9	61.2	68.9	66.8	54.4	45.8	24.5	28.2	43.3
1974	18.1	32.8	33.9	46.0	48.6	64.9	70.4	61.5	53.5	45.8	34.7	27.7	44.8
1975	21.1	13.7	28.8	32.9	48.6	56.9	69.2	61.2	54.1	43.3	29.9	26.0	40.5
1976	24.9	29.5	30.4	43.2	54.1	56.6	67.6	64.8	57.9	44.9	33.8	28.7	44.7
1977	18.0	34.1	32.4	46.9	50.4	63.6	66.3	63.6	56.3	45.6	31.2	20.7	44.1
1978	16.9	22.8	38.8	47.7	53.2	63.2	67.2	65.5	57.9	46.5	22.7	15.2	43.2
1979	1.1	20.1	34.7	42.7	53.2	62.0	69.2	67.7	61.3	47.8	29.0	28.2	43.1
1980	14.3	25.3	31.3	49.0	55.4	59.8	67.3	62.9	56.8	45.4	34.9	26.9	44.1
1981	28.4	29.7	38.0	46.4	52.8	58.9	67.7	69.5	58.8	43.8	36.4	25.3	46.3
1982	16.5	23.7	33.9	40.5	50.9	61.2	68.6	69.1	55.5	44.9	27.4	22.7	42.9
1983	30.6	35.3	38.3	43.1	51.5	60.2	66.0	70.8	53.5	46.0	34.7	5.5	44.6
1984	27.3	32.4	37.2	43.5	52.8	60.0	70.0	69.8	52.3	41.7	33.1	11.6	44.3
1985	12.3	18.8	33.4	46.9	56.2	63.3	75.0	63.1	49.6	42.3	12.6	15.0	40.7
1986	25.5	21.8	42.9	43.3	53.6	66.4	64.2	68.0	51.2	45.3	29.0	18.6	44.2
1987	23.3	31.9	37.0	50.2	55.9	64.4	66.2	62.8	59.9	46.8	34.7	24.7	46.5
1988	18.8	29.1	36.1	47.1	55.5	68.4	71.3	68.6	56.4	50.3	33.8	23.1	46.5
1989	24.5	6.1	27.6	44.5	51.7	62.2	72.0	64.3	55.6	45.1	36.8	23.8	42.9
1990	28.8	26.9	34.9	45.8	51.0	61.5	69.4	68.4	63.6	45.3	37.4	14.0	45.6
1991	19.0	37.5	35.3	43.8	52.8	59.5	70.8	72.8	58.4	43.3	29.8	22.9	45.5
1992	23.7	34.2	42.7	47.7	57.5	64.9	64.3	64.7	56.7	46.7	32.8	14.7	45.9
1993	12.3	15.0	36.4	44.3	57.3	58.5	59.5	61.5	54.0	43.8	26.8	30.0	41.6
Record Mean	19.8	24.5	32.7	43.6	52.3	60.0	67.8	66.3	55.8	45.4	32.4	23.8	43.7
Max	28.8	34.0	42.8	55.0	64.0	72.2	82.0	80.6	68.7	57.0	41.9	32.5	55.0
Min	10.7	15.0	22.6	32.2	40.5	47.9	53.6	52.0	42.8	33.8	22.8	15.1	32.4

REFERENCE NOTES HELENA, MONTANA

GENERAL
T - TRACE AMOUNT.
BLANK ENTRIES DENOTE MISSING/UNREPORTED DATA.
# INDICATES A STATION OR INSTRUMENT RELOCATION.

SPECIFIC

PM - INCLUDES LAST DAY OF PREVIOUS MONTH
ASOS - AUTOMATED SURFACE OBSERVING SYSTEM IN OPERATION DURING THESE MONTHS.

(a) - LENGTH OF RECORD IN YEARS, ALTHOUGH INDIVIDUAL MONTHS MAY BE MISSING.
0.* OR * - THE VALUE IS BETWEEN 0.0 AND 0.05
NORMALS - BASED ON THE 1961-1990 RECORD PERIOD.
EXTREMES - DATES ARE THE MOST RECENT OCCURRENCE
WIND DIR.- NUMERALS SHOW TENS OF DEGRESS CLOCKWISE FROM TRUE NORTH. "00" INDICATES CALM.
RESULTANT DIRECTIONS ARE GIVEN TO WHOLE DEGREES.
BOLD VALUES INDICATE EXTREME VALUES WHICH OCCURRED AFTER THE ASOS SYSTEM WAS COMMISSIONED.

RECORD = PERIOD OF RECORD
RECORD MEAN PRECIPITATION IS THE MEAN OF ALL DAILY PRECIPITATION AMOUNTS DURING THE PERIOD OF RECORD.
RECORD MAX(MIN) TEMPERATURE IS THE MEAN OF ALL DAILY MAX(MIN) TEMPERATURES DURING THE PERIOD OF RECORD.
RECORD MEAN TEMPERATURE IS THE SUM OF THE RECORD MAX AND RECORD MIN DIVIDED BY 2.
AVERAGE TEMPERATURE IS THE SUM OF THE MEAN DAILY MAX AND MIN TEMPERATURE DIVIDED BY 2.

EXCEPTIONS

RECORD MEANS ARE THROUGH THE CURRENT YEAR, BEGINNING IN 1881 FOR TEMPERATURE
1881 FOR PRECIPITATION
1941 FOR SNOWFALL

HEATING DEGREE DAYS Base 65 deg. F — HELENA, MONTANA

SEASON	JULY	AUG	SEP	OCT	NOV	DEC	JAN	FEB	MAR	APR	MAY	JUNE	TOTAL
1964-65	1	101	332	542	1028	1420	1090	1070	1250	612	454	169	8069
1965-66	14	69	578	478	903	1153	1370	1058	874	670	267	188	7622
1966-67	8	51	117	570	930	1157	1120	898	1107	807	396	171	7332
1967-68	0	1	130	539	954	1422	1525	1042	780	751	472	206	7822
1968-69	23	102	294	633	937	1493	1788	1407	1348	543	301	226	9095
1969-70	35	14	219	826	938	1207	1427	904	1132	849	409	153	8113
1970-71	22	12	413	763	1063	1372	1422	1013	1020	699	414	251	8464
1971-72	91	22	454	707	936	1524	1628	1102	796	747	430	120	8557
1972-73	136	49	418	798	1010	1588	1465	1174	892	732	374	155	8791
1973-74	23	47	317	588	1208	1136	1452	897	956	564	500	99	7787
1974-75	16	130	338	588	905	1149	1355	1429	1114	954	501	235	8714
1975-76	14	119	322	666	1045	1199	1236	1023	1064	649	331	257	7925
1976-77	15	45	219	615	928	1120	1449	862	1005	535	443	90	7326
1977-78	52	92	270	593	1008	1367	1485	1175	806	512	361	87	7808
1978-79	32	60	244	564	1263	1540	1979	1250	930	665	359	128	9014
1979-80	11	15	127	528	1072	1138	1566	1148	1039	473	304	164	7585
1980-81	25	81	242	602	899	1175	1127	986	832	552	371	191	7083
1981-82	21	16	195	650	853	1227	1497	1153	959	726	428	136	7861
1982-83	30	16	304	618	1120	1306	1059	828	823	649	417	152	7322
1983-84	76	0	351	584	901	1842	1164	941	856	640	380	174	7909
1984-85	2	7	377	716	954	1654	1625	1291	973	538	266	97	8500
1985-86	3	105	455	696	1571	1545	1218	1202	677	645	380	42	8539
1986-87	66	23	409	602	1077	1432	1288	923	862	437	276	77	7472
1987-88	75	104	163	556	901	1241	1426	1034	889	529	297	63	7278
1988-89	10	13	282	449	934	1292	1251	1650	1156	610	407	107	8161
1989-90	0	92	274	611	839	1268	1116	1058	925	573	426	177	7359
1990-91	15	31	78	604	823	1579	1420	767	914	630	373	159	7393
1991-92	2	0	220	666	1053	1297	1273	884	687	513	237	94	6926
1992-93	68	141	246	564	960	1552	1631	1397	879	617	228	214	8497
1993-94	179	122	323	652	1139	1080							

COOLING DEGREE DAYS Base 65 deg. F — HELENA, MONTANA

YEAR	JAN	FEB	MAR	APR	MAY	JUNE	JULY	AUG	SEP	OCT	NOV	DEC	TOTAL
1969	0	0	0	0	1	18	128	178	20	0	0	0	345
1970	0	0	0	0	0	76	130	109	5	0	0	0	320
1971	0	0	0	0	0	14	42	154	0	0	0	0	210
1972	0	0	0	0	2	30	42	89	0	0	0	0	163
1973	0	0	0	0	4	45	151	108	3	0	0	0	311
1974	0	0	0	0	0	102	190	31	0	0	0	0	323
1975	0	0	0	0	0	1	154	12	0	0	0	0	167
1976	0	0	0	0	0	14	102	45	10	0	0	0	171
1977	0	0	0	0	0	57	101	54	13	0	0	0	225
1978	0	0	0	0	3	37	109	85	37	0	0	0	271
1979	0	0	0	0	1	45	152	103	21	0	0	0	322
1980	0	0	0	0	14	14	104	25	4	0	0	0	161
1981	0	0	0	0	0	15	109	165	16	0	0	0	305
1982	0	0	0	0	0	30	147	151	25	0	0	0	353
1983	0	0	0	0	4	16	115	186	12	0	0	0	333
1984	0	0	0	0	10	31	165	163	4	0	0	0	373
1985	0	0	0	0	2	55	318	54	0	0	0	0	429
1986	0	0	0	0	35	91	45	123	1	0	0	0	295
1987	0	0	0	0	3	66	122	41	15	0	0	0	247
1988	0	0	0	0	8	170	211	132	30	0	0	0	551
1989	0	0	0	0	0	30	222	82	0	0	0	0	334
1990	0	0	0	0	0	77	159	142	42	0	0	0	420
1991	0	0	0	0	0	2	188	250	29	0	0	0	469
1992	0	0	0	0	10	96	55	140	3	0	0	0	304
1993	0	0	0	0	2	24	15	23	0	0	0	0	64

SNOWFALL (inches) — HELENA, MONTANA

SEASON	JULY	AUG	SEP	OCT	NOV	DEC	JAN	FEB	MAR	APR	MAY	JUNE	TOTAL
1964-65	0.0	0.0	0.0	0.0	6.3	13.2	2.3	9.4	12.9	6.3	5.0	0.0	55.4
1965-66	0.0	0.0	13.7	0.0	2.1	1.4	6.3	5.1	2.3	1.9	T	0.0	32.8
1966-67	0.0	0.0	0.0	T	9.5	5.7	7.3	10.5	14.9	20.6	12.7	0.0	81.2
1967-68	0.0	0.0	0.0	T	2.7	22.8	8.7	0.9	0.8	10.0	T	0.0	45.9
1968-69	0.0	0.0	3.0	0.4	7.4	11.0	35.6	3.6	7.0	2.6	0.7	2.7	74.0
1969-70	0.0	0.0	0.0	11.0	0.1	6.1	7.0	9.5	9.0	7.9	1.4	0.0	52.0
1970-71	0.0	0.0	T	1.4	5.4	7.5	16.2	3.5	2.5	0.7	T	0.0	37.2
1971-72	0.0	0.0	T	0.8	4.6	14.5	14.9	3.3	4.5	3.8	T	0.0	46.4
1972-73	T	0.0	0.3	4.7	0.7	7.8	3.2	1.8	0.1	6.5	T	0.0	25.1
1973-74	0.0	0.0	1.3	7.2	12.5	7.2	9.9	5.9	2.2	1.5	0.2	0.0	47.9
1974-75	0.0	0.0	T	1.5	0.8	2.7	15.2	10.7	12.3	15.4	0.2	0.0	58.8
1975-76	0.0	0.0	0.0	6.3	4.9	3.9	3.3	4.4	6.7	10.9	0.0	0.0	40.4
1976-77	0.0	0.0	0.0	T	2.9	0.9	13.8	0.8	14.1	0.5	1.0	0.0	34.0
1977-78	0.0	0.0	0.0	0.4	6.8	19.5	15.7	13.3	2.6	0.9	T	0.0	59.2
1978-79	0.0	0.0	T	T	22.1	13.8	11.7	7.2	12.4	9.3	T	T	76.5
1979-80	0.0	0.0	0.0	0.2	0.6	6.5	9.3	10.5	10.1	3.1	0.0	0.0	40.3
1980-81	0.0	0.0	0.0	3.7	1.2	3.8	2.7	2.1	3.3	0.1	T	0.0	16.9
1981-82	0.0	0.0	0.0	5.2	3.4	6.1	18.4	4.8	13.9	4.1	0.8	0.0	56.7
1982-83	0.0	0.0	6.5	0.5	4.1	11.3	3.2	0.2	2.2	1.1	9.9	0.0	39.0
1983-84	0.0	0.0	6.4	0.0	1.3	13.0	1.3	1.5	5.7	2.3	0.8	0.0	32.3
1984-85	0.0	0.0	6.3	9.0	5.7	7.5	3.9	6.2	4.0	0.8	0.0	0.0	43.4
1985-86	0.0	0.0	2.9	8.8	10.4	8.5	4.2	15.6	1.2	11.6	0.2	0.0	63.4
1986-87	0.0	0.0	0.0	T	7.6	5.0	0.2	0.2	9.1	4.3	3.8	0.0	30.2
1987-88	0.0	0.0	0.0	0.3	0.9	1.2	4.4	8.0	5.2	2.9	0.0	0.0	22.9
1988-89	0.0	0.0	5.9	1.5	6.0	5.0	23.0	13.0	20.7	7.9	3.5	T	86.5
1989-90	0.0	T	T	2.6	1.8	9.4	4.3	1.8	14.0	0.8	0.1	0.0	34.8
1990-91	0.0	0.5	0.0	0.5	8.6	11.6	5.9	0.8	12.2	5.6	0.4	0.5	46.6
1991-92	0.0	T	0.0	7.3	8.8	15.7	4.7	1.3	T	2.2	T	0.0	40.0
1992-93	T	6.2	T	8.6	1.5	11.8	10.2	17.9	2.3	2.5	T	0.2	61.2
1993-94	T	0.0	T	4.6	7.3	1.1							
Record Mean	T	0.1	1.5	2.4	6.4	8.5	8.7	6.2	7.4	4.8	1.4	0.1	47.5

KALISPELL, MONTANA

The climate of the Flathead Valley is influenced by the topography. The high mountains to the east form an effective barrier to many severe winter cold waves that move into areas east of the Rockies from Alberta. The mountains to the east rise abruptly 4,500 feet above the valley floor. The mountain snows and spring rains assure an adequate supply of water for the area.

In addition to Flathead Lake, the valley contains many smaller lakes, three rivers, and numerous streams and sloughs. Until late in the winter when a large portion of the lakes and sloughs become frozen, this water surface tends to limit temperature extremes. This effect is most noticeable in the southern end of the valley, because of the influence of Flathead Lake. Due to its size, Flathead Lake seldom freezes over.

The weather at the airport is considerably different in some respects from the weather in Kalispell. Generally there is more cloudiness at the airport since it is closer to the mountains to the east and north. Moist air moving in from the west and southwest, lifting and cooling as it moves over the mountains, is the major cause. On average there is more precipitation on the east side of the valley than on the west side. Average snowfall during the winter at the airport is 68 inches and in Kalispell it is 49 inches.

The annual prevailing wind direction at Kalispell is from the west. At the airport it is from the south. Wind speeds average considerably stronger at the airport than in Kalispell.

In the winter, when a cold wave moving down the east side of the Continental Divide does come over the mountains, the airport is in direct line of the pass the cold air comes through. During these cold waves the wind is from the northeast and will usually have speeds reaching 30 to 40 mph. The strongest gusts reported during these storms exceed 80 mph. As the cold air moves down the valley it spreads out, decreasing the wind velocity, and mixes with the warmer air of the valley. Unless these cold strong winds persist for 3 or 4 days, the wind in the lower part of the valley will be from the northwest, because of the influence of Flathead Lake and the mountains to the west. This wind is always much stronger in the northeast end of the valley where the airport is located than any other place in the valley. In the northwest corner where Whitefish is located, and in the southeast part of the valley, there is rarely much wind from this storm.

NORMALS, MEANS, AND EXTREMES

KALISPELL, MONTANA

LATITUDE: 48°18'N LONGITUDE: 114°16' W ELEVATION: FT. GRND 2965 BARO 2978 TIME ZONE: MOUNTAIN WBAN: 24146

	(a)	JAN	FEB	MAR	APR	MAY	JUNE	JULY	AUG	SEP	OCT	NOV	DEC	YEAR
TEMPERATURE °F:														
Normals														
-Daily Maximum		28.2	35.0	43.4	55.2	64.2	71.4	80.1	79.4	67.7	54.4	38.3	29.9	53.9
-Daily Minimum		12.7	18.1	23.9	31.1	38.4	44.1	47.1	46.3	38.6	29.4	23.7	15.5	30.7
-Monthly		20.5	26.6	33.7	43.2	51.3	57.8	63.6	62.9	53.2	41.9	31.0	22.7	42.4
Extremes														
-Record Highest	44	53	57	72	84	94	96	104	105	99	86	65	57	105
-Year		1953	1988	1986	1977	1986	1955	1960	1961	1967	1992	1975	1979	AUG 1961
-Record Lowest	44	-38	-36	-29	10	19	28	31	31	16	-3	-28	-35	-38
-Year		1950	1950	1960	1951	1954	1973	1971	1993	1970	1984	1959	1990	JAN 1950
NORMAL DEGREE DAYS:														
Heating (base 65°F)		1380	1075	970	654	425	235	97	128	367	716	1020	1311	8378
Cooling (base 65°F)		0	0	0	0	0	19	54	63	13	0	0	0	149
% OF POSSIBLE SUNSHINE														
MEAN SKY COVER (tenths)														
Sunrise - Sunset	44	8.8	8.2	7.8	7.5	6.9	6.5	4.3	4.8	5.5	6.9	8.5	8.9	7.1
MEAN NUMBER OF DAYS:														
Sunrise to Sunset														
-Clear	44	1.9	2.4	3.5	4.0	5.2	6.4	14.2	12.5	10.3	6.3	2.0	1.6	70.3
-Partly Cloudy	44	3.7	4.6	6.2	6.5	9.0	9.8	9.4	9.4	7.7	7.0	4.8	3.3	81.3
-Cloudy	44	25.5	21.3	21.3	19.6	16.8	13.8	7.4	9.1	12.0	17.7	23.2	26.1	213.7
Precipitation														
.01 inches or more	44	15.3	11.5	11.5	9.7	11.4	11.9	7.2	8.0	8.1	8.9	12.7	15.5	131.6
Snow, Ice pellets, hail														
1.0 inches or more	44	5.9	3.8	2.0	0.8	0.3	0.0	0.0	0.0	0.*	0.5	2.6	5.4	21.3
Thunderstorms	43	0.0	0.*	0.3	0.9	2.9	5.3	5.3	5.2	1.9	0.4	0.1	0.0	22.3
Heavy Fog Visibility														
1/4 mile or less	33	4.9	4.7	2.4	0.6	1.0	0.8	0.9	1.0	2.2	4.5	4.7	5.0	32.6
Temperature °F														
-Maximum														
90° and above	34	0.0	0.0	0.0	0.0	0.2	1.1	5.9	6.1	0.4	0.0	0.0	0.0	13.7
32° and below	34	17.5	8.6	3.1	0.*	0.0	0.0	0.0	0.0	0.0	0.4	6.0	17.1	52.7
-Minimum														
32° and below	34	29.3	25.8	27.3	18.2	5.7	0.6	0.1	0.3	5.7	21.3	25.4	28.9	188.6
0° and below	34	6.7	3.0	1.0	0.0	0.0	0.0	0.0	0.0	0.0	0.1	0.9	4.2	15.9
AVG. STATION PRESS. (mb)	21	913.3	912.4	909.5	910.4	910.0	910.6	912.0	911.9	913.0	913.4	912.6	913.3	911.9
RELATIVE HUMIDITY (%)														
Hour 05	29	81	82	80	77	80	84	83	81	83	84	84	84	82
Hour 11 (Local Time)	34	79	76	64	52	51	52	46	46	54	65	78	81	62
Hour 17	29	75	67	54	43	43	45	37	35	42	52	72	79	54
Hour 23	28	80	80	76	69	70	73	71	69	76	80	82	83	76
PRECIPITATION (inches):														
Water Equivalent														
-Normal		1.53	1.10	1.02	1.10	1.87	2.21	1.12	1.40	1.26	0.87	1.30	1.73	16.51
-Maximum Monthly	44	3.11	1.99	2.96	2.37	4.75	4.72	6.02	3.78	3.97	2.96	4.44	4.38	6.02
-Year		1970	1981	1987	1978	1990	1966	1993	1976	1985	1951	1959	1990	JUL 1993
-Minimum Monthly	44	0.20	0.42	0.31	0.26	0.43	0.43	0.02	T	0.01	0.04	0.26	0.32	T
-Year		1985	1967	1965	1968	1950	1977	1953	1955	1990	1953	1969	1954	AUG 1955
-Maximum in 24 hrs	44	1.09	0.65	0.82	1.74	1.49	2.71	2.09	1.76	1.25	0.78	1.72	1.35	2.71
-Year		1982	1961	1987	1951	1980	1982	1987	1976	1959	1957	1989	1964	JUN 1982
Snow, Ice pellets, hail														
-Maximum Monthly	44	34.8	21.2	18.9	8.1	8.9	0.3	T	T	3.1	11.1	39.0	52.1	52.1
-Year		1970	1975	1987	1961	1964	1962	1992	1992	1968	1984	1959	1990	DEC 1990
-Maximum in 24 hrs	44	11.8	8.0	7.7	10.0	7.5	0.3	T	T	3.0	6.2	13.8	15.4	15.4
-Year		1982	1990	1987	1951	1964	1962	1992	1992	1968	1951	1993	1951	DEC 1951
WIND:														
Mean Speed (mph)	31	5.8	6.0	7.1	8.1	7.6	7.3	6.7	6.6	6.3	5.3	5.6	5.5	6.5
Prevailing Direction														
Fastest Obs. 1 Min.														
-Direction (!!)	27	04	01	03	25	23	03	31	15	04	32	03	03	04
-Speed (MPH)	27	52	40	41	37	40	38	38	43	35	38	35	52	52
-Year		1982	1965	1990	1989	1965	1982	1985	1990	1980	1971	1993	1972	JAN 1982
Peak Gust														
-Direction (!!)	10	NE	SW	NE	W	SW	SW	NW	SE	NW	SW	NE	NE	SE
-Speed (mph)	10	62	49	52	49	52	55	45	69	48	54	51	62	69
-Date		1989	1990	1990	1989	1986	1988	1985	1990	1992	1991	1993	1990	AUG 1990

PRECIPITATION (inches)

KALISPELL, MONTANA

YEAR	JAN	FEB	MAR	APR	MAY	JUNE	JULY	AUG	SEP	OCT	NOV	DEC	ANNUAL
1964	1.24	0.50	1.26	0.60	2.56	3.56	1.69	1.77	1.87	0.95	2.13	4.23	22.36
1965	1.84	0.69	0.31	1.44	0.74	2.73	0.85	3.47	1.17	0.11	0.72	0.57	14.64
1966	1.61	1.02	0.79	0.65	1.63	4.72	0.37	1.05	0.29	0.88	2.52	1.58	17.11
1967	1.75	0.42	1.09	0.59	0.95	2.38	0.07	0.01	0.33	1.60	0.58	1.45	11.22
1968	1.01	1.00	0.43	0.26	2.59	2.16	0.46	3.10	3.33	1.50	1.55	2.56	19.95
1969	2.97	0.50	0.64	1.30	0.68	3.88	0.10	0.09	1.44	1.07	0.26	1.21	14.14
1970	3.11	1.27	0.84	0.68	2.18	2.58	1.59	0.34	0.90	1.20	1.29	1.39	17.37
1971	1.81	0.81	0.95	0.44	2.17	3.59	0.93	1.47	0.46	0.95	1.36	1.53	16.47
1972	1.58	1.57	1.14	0.86	1.50	1.69	1.51	1.03	0.76	0.84	0.55	1.60	14.63
1973	0.69	0.62	0.46	0.47	0.91	1.52	0.05	0.56	0.71	1.19	2.80	1.87	11.85
1974	1.94	1.02	1.39	1.92	1.06	1.75	0.64	0.70	1.18	0.12	1.04	1.21	13.97
1975	1.95	1.52	1.33	0.83	0.98	1.96	0.98	2.79	0.58	1.67	1.20	1.19	16.98
1976	1.36	1.40	0.35	0.97	1.79	1.69	1.64	3.78	0.36	0.38	0.47	0.65	14.84
1977	0.81	0.97	1.18	0.43	1.40	0.43	2.57	1.13	2.18	0.13	1.46	3.53	16.22
1978	2.30	0.71	0.67	2.37	2.47	0.91	1.50	2.64	0.80	0.07	1.43	0.87	16.74
1979	1.42	1.57	0.86	1.49	1.64	0.84	0.67	1.10	0.39	1.45	0.42	1.27	13.12
1980	2.15	1.92	0.86	1.52	3.90	2.96	0.81	1.60	0.74	0.78	0.49	2.54	20.27
1981	1.44	1.99	1.43	0.94	3.37	3.62	0.72	1.32	0.48	0.20	1.30	1.81	18.62
1982	2.66	1.60	0.92	1.32	0.78	4.05	1.59	0.82	1.92	0.52	1.43	1.88	19.49
1983	1.09	0.93	1.50	2.18	0.78	3.09	2.06	0.73	1.31	0.83	1.61	1.69	17.80
1984	0.78	0.66	1.34	1.53	1.56	1.77	0.51	0.88	1.88	1.85	1.77	1.22	15.75
1985	0.20	1.39	0.71	0.58	1.60	1.56	0.23	1.12	3.97	0.92	1.63	0.72	14.63
1986	2.22	1.87	0.32	0.83	2.45	2.17	1.42	0.68	2.78	0.51	1.84	0.52	17.61
1987	0.66	0.61	2.96	1.19	0.88	1.20	3.98	1.35	0.60	0.05	0.48	1.58	15.54
1988	0.97	0.84	0.80	0.94	2.83	1.49	0.87	0.29	2.10	0.46	1.03	2.32	14.94
1989	1.36	1.32	1.45	1.25	2.68	1.47	1.23	3.49	1.55	0.90	3.26	2.24	22.20
1990	1.79	0.95	1.12	1.48	4.75	1.16	2.37	2.27	0.01	2.07	1.58	4.38	23.93
1991	1.74	0.45	0.81	0.85	2.26	3.58	0.42	0.72	0.67	0.84	2.16	0.65	15.15
1992	1.17	0.79	0.94	1.02	1.18	3.65	2.37	0.71	1.02	0.71	1.59	1.99	17.14
1993	1.67	0.61	0.62	2.20	1.93	3.46	6.02	1.49	1.61	1.01	1.20	1.58	23.40
Record Mean	1.39	1.00	0.97	0.99	1.70	2.17	1.14	1.16	1.24	1.02	1.41	1.43	15.62

AVERAGE TEMPERATURE (deg. F)

KALISPELL, MONTANA

YEAR	JAN	FEB	MAR	APR	MAY	JUNE	JULY	AUG	SEP	OCT	NOV	DEC	ANNUAL
1964	24.6	23.9	27.9	40.7	49.7	56.8	64.0	58.4	50.1	42.3	31.0	18.0	40.6
1965	25.9	24.2	24.5	43.5	48.1	56.0	64.5	63.3	45.1	45.3	34.3	26.1	41.7
1966	24.3	23.9	32.5	41.7	53.0	54.8	64.4	61.6	59.3	41.2	30.7	27.3	42.9
1967	28.8	30.3	30.7	39.2	50.3	58.7	65.8	67.7	59.5	44.5	32.3	23.8	44.3
1968	20.4	31.9	39.1	40.2	49.8	58.0	65.5	61.8	53.0	41.0	31.4	17.8	42.5
1969	11.3	20.2	27.0	46.2	53.8	57.9	62.1	62.9	55.7	39.8	33.2	26.2	41.3
1970	19.1	27.2	30.6	40.1	52.1	62.4	66.1	64.3	48.9	39.1	30.6	23.6	42.0
1971	21.7	28.8	32.9	43.6	53.7	54.8	63.1	68.8	49.1	40.0	33.1	18.4	42.3
1972	13.9	26.2	38.4	40.0	52.8	59.6	62.5	67.6	51.1	39.3	32.8	18.3	41.9
1973	19.4	27.0	39.3	42.5	52.4	58.8	67.1	65.7	54.8	43.8	29.0	29.0	44.1
1974	20.2	31.7	33.7	44.8	48.3	63.4	65.9	63.2	53.8	42.8	34.3	29.0	44.2
1975	18.2	19.2	28.0	38.9	50.0	56.2	69.9	61.5	54.5	43.6	30.1	27.3	41.5
1976	24.1	29.0	31.6	44.8	53.4	56.3	64.6	62.8	56.8	41.4	31.3	27.2	43.6
1977	19.4	30.9	34.7	46.4	50.4	61.5	63.5	63.6	52.2	42.0	28.8	19.4	42.8
1978	19.7	25.5	34.6	44.0	48.8	59.5	64.5	61.6	54.0	42.7	25.9	14.2	41.3
1979	-0.2	23.1	33.7	42.8	51.7	60.3	67.7	67.3	58.5	45.6	29.0	32.2	42.7
1980	12.1	27.4	32.8	48.0	55.4	57.6	63.8	59.4	55.5	43.9	33.2	28.5	43.1
1981	29.7	30.5	38.8	45.0	52.6	53.8	63.0	67.9	55.3	40.2	32.6	21.5	44.2
1982	17.0	21.1	35.7	38.2	48.8	60.4	61.5	63.2	53.3	40.9	28.5	23.3	41.0
1983	29.6	32.9	38.0	42.3	51.5	57.3	61.2	66.3	50.6	42.8	33.9	7.9	42.9
1984	25.7	31.7	37.9	42.9	48.6	56.6	65.4	65.6	50.5	38.5	32.1	16.3	42.7
1985	17.6	15.4	28.0	44.7	54.3	57.1	68.5	59.4	47.9	40.4	16.8	18.2	39.0
1986	24.8	23.8	40.3	43.6	54.2	64.3	60.2	67.9	51.2	43.4	29.4	24.9	44.0
1987	20.9	28.3	36.1	48.6	54.8	62.1	64.1	60.7	56.5	43.1	35.3	23.0	44.5
1988	19.5	30.5	37.9	47.6	51.7	62.5	64.9	64.2	54.3	48.5	35.5	24.0	45.1
1989	26.6	13.8	31.1	44.7	50.7	60.5	67.5	61.9	53.3	42.8	35.7	26.7	42.9
1990	29.5	25.9	36.5	45.1	50.1	57.7	66.8	65.8	59.8	41.1	35.1	15.6	44.1
1991	16.4	33.3	33.0	42.6	50.3	54.7	63.9	66.1	54.8	39.9	29.7	29.4	42.8
1992	28.4	34.3	41.0	45.6	54.2	62.8	62.0	62.7	51.8	45.6	32.8	18.2	45.0
1993	15.4	19.2	35.4	44.2	57.9	56.9	57.4	60.4	52.4	44.2	25.9	27.6	41.4
Record Mean	21.0	25.5	33.5	43.8	52.1	58.5	65.1	63.5	53.8	43.5	31.9	24.4	43.0
Max	28.3	33.8	43.0	55.2	64.3	71.2	80.5	79.0	67.4	55.0	39.0	30.8	53.9
Min	13.7	17.1	24.0	32.3	39.8	45.9	49.7	48.0	40.2	32.0	24.7	17.9	32.1

REFERENCE NOTES

KALISPELL, MONTANA

GENERAL
T - TRACE AMOUNT.
BLANK ENTRIES DENOTE MISSING/UNREPORTED DATA.
# INDICATES A STATION OR INSTRUMENT RELOCATION.

SPECIFIC

(a) - LENGTH OF RECORD IN YEARS, ALTHOUGH INDIVIDUAL MONTHS MAY BE MISSING.
0.* OR * - THE VALUE IS BETWEEN 0.0 AND 0.05
NORMALS - BASED ON THE 1961-1990 RECORD PERIOD.
EXTREMES - DATES ARE THE MOST RECENT OCCURRENCE
WIND DIR.- NUMERALS SHOW TENS OF DEGRESS CLOCKWISE FROM TRUE NORTH. "00" INDICATES CALM.
RESULTANT DIRECTIONS ARE GIVEN TO WHOLE DEGREES.
BOLD VALUES INDICATE EXTREME VALUES WHICH OCCURRED AFTER THE ASOS SYSTEM WAS COMMISSIONED.

RECORD = PERIOD OF RECORD
RECORD MEAN PRECIPITATION IS THE MEAN OF ALL DAILY PRECIPITATION AMOUNTS DURING THE PERIOD OF RECORD.
RECORD MAX(MIN) TEMPERATURE IS THE MEAN OF ALL DAILY MAX(MIN) TEMPERATURES DURING THE PERIOD OF RECORD.
RECORD MEAN TEMPERATURE IS THE SUM OF THE RECORD MAX AND RECORD MIN DIVIDED BY 2.
AVERAGE TEMPERATURE IS THE SUM OF THE MEAN DAILY MAX AND MIN TEMPERATURE DIVIDED BY 2.

EXCEPTIONS

RECORD MEANS ARE THROUGH THE CURRENT YEAR, BEGINNING IN 1897 FOR TEMPERATURE
1897 FOR PRECIPITATION
1950 FOR SNOWFALL

HEATING DEGREE DAYS Base 65 deg. F KALISPELL, MONTANA

SEASON	JULY	AUG	SEP	OCT	NOV	DEC	JAN	FEB	MAR	APR	MAY	JUNE	TOTAL
1964-65	69	199	439	700	1011	1454	1204	1138	1249	638	515	270	8886
1965-66	62	115	591	602	914	1201	1252	1144	1002	693	373	301	8250
1966-67	68	128	162	733	1023	1161	1118	964	1059	766	449	190	7821
1967-68	32	22	176	627	972	1272	1379	953	795	735	468	211	7642
1968-69	80	133	354	737	999	1458	1658	1246	1172	556	339	211	8943
1969-70	109	102	290	774	947	1197	1418	1054	1062	743	393	127	8216
1970-71	48	62	478	798	1026	1276	1338	1006	986	635	344	305	8302
1971-72	116	28	470	768	950	1440	1581	1118	817	742	381	181	8592
1972-73	113	26	413	792	959	1444	1410	1059	787	668	383	205	8259
1973-74	47	72	314	649	1076	1107	1385	926	965	598	510	125	7774
1974-75	54	91	328	681	914	1109	1445	1277	1140	777	459	258	8533
1975-76	15	133	306	654	1040	1160	1261	1037	1028	598	354	271	7857
1976-77	64	104	244	723	1002	1162	1405	949	935	553	443	114	7698
1977-78	85	97	375	707	1081	1405	1397	1098	934	624	497	169	8469
1978-79	60	139	331	685	1167	1567	2023	1168	964	662	404	152	9322
1979-80	50	17	191	595	1071	1008	1639	1084	990	503	299	218	7665
1980-81	74	178	279	647	948	1121	1087	962	806	593	376	329	7400
1981-82	84	36	287	760	966	1342	1484	1226	901	797	495	154	8532
1982-83	133	88	343	742	1088	1286	1090	893	831	676	417	227	7814
1983-84	130	32	427	680	927	1768	1209	958	835	657	501	254	8378
1984-85	60	50	433	816	982	1503	1463	1387	1140	601	333	240	9008
1985-86	5	180	507	755	1439	1444	1238	1148	761	637	375	65	8554
1986-87	152	13	408	664	1061	1234	1360	1022	890	484	309	124	7721
1987-88	103	143	251	671	885	1296	1402	994	832	516	405	123	7621
1988-89	69	70	337	506	879	1265	1182	1429	1045	605	438	139	7964
1989-90	34	140	345	684	872	1177	1096	1086	877	589	457	239	7596
1990-91	40	61	157	733	891	1528	1503	880	985	665	446	302	8191
1991-92	70	47	300	772	1052	1096	1124	886	739	574	325	115	7100
1992-93	105	150	390	590	958	1447	1533	1275	909	615	234	238	8444
1993-94	235	148	375	634	1170	1154							

COOLING DEGREE DAYS Base 65 deg. F KALISPELL, MONTANA

YEAR	JAN	FEB	MAR	APR	MAY	JUNE	JULY	AUG	SEP	OCT	NOV	DEC	TOTAL
1969	0	0	0	0	0	6	27	44	17	0	0	0	94
1970	0	0	0	0	0	57	88	48	1	0	0	0	194
1971	0	0	0	0	0	6	66	155	0	0	0	0	227
1972	0	0	0	0	7	26	41	113	0	0	0	0	187
1973	0	0	0	0	1	26	118	100	13	0	0	0	258
1974	0	0	0	0	0	83	89	43	0	0	0	0	215
1975	0	0	0	0	0	0	177	30	0	0	0	0	207
1976	0	0	0	0	0	15	59	44	3	0	0	0	121
1977	0	0	0	2	0	17	48	62	0	0	0	0	129
1978	0	0	0	0	0	9	51	39	6	0	0	0	105
1979	0	0	0	0	0	18	142	96	2	0	0	0	258
1980	0	0	0	0	6	0	44	13	0	0	0	0	63
1981	0	0	0	0	0	0	32	135	2	0	0	0	169
1982	0	0	0	0	0	21	33	39	0	0	0	0	93
1983	0	0	0	0	4	4	17	83	4	0	0	0	112
1984	0	0	0	0	0	9	78	76	5	0	0	0	168
1985	0	0	0	0	6	9	120	12	0	0	0	0	147
1986	0	0	0	0	49	53	10	107	0	0	0	0	219
1987	0	0	0	0	0	43	84	18	2	0	0	0	147
1988	0	0	0	0	0	57	70	52	25	0	0	0	204
1989	0	0	0	0	0	13	118	52	0	0	0	0	183
1990	0	0	0	0	0	24	99	92	9	0	0	0	224
1991	0	0	0	0	0	0	41	86	0	0	0	0	127
1992	0	0	0	0	0	57	16	83	0	0	0	0	156
1993	0	0	0	0	23	3	5	11	0	0	0	0	42

SNOWFALL (inches) KALISPELL, MONTANA

SEASON	JULY	AUG	SEP	OCT	NOV	DEC	JAN	FEB	MAR	APR	MAY	JUNE	TOTAL
1964-65	0.0	0.0	0.0	1.6	10.5	32.0	20.7	10.7	4.9	4.0	0.2	0.0	84.6
1965-66	0.0	0.0	0.4	0.0	3.9	8.0	22.6	17.6	1.9	0.9	0.6	0.0	55.9
1966-67	0.0	0.0	0.0	T	11.9	15.7	8.9	5.8	10.0	2.1	2.8	0.0	57.2
1967-68	0.0	0.0	0.0	T	2.3	21.3	14.1	3.0	0.7	1.9	1.4	0.0	44.7
1968-69	0.0	0.0	3.1	T	11.1	30.3	34.2	7.2	4.9	0.1	0.0	0.0	90.9
1969-70	0.0	0.0	0.0	2.2	0.8	14.4	34.8	5.7	7.6	5.0	4.7	T	75.2
1970-71	0.0	0.0	T	1.0	6.6	21.7	25.7	8.4	10.2	T	T	0.0	73.6
1971-72	0.0	0.0	0.0	2.4	12.4	29.6	24.5	13.3	10.3	3.8	T	0.0	96.3
1972-73	0.0	0.0	0.2	10.6	4.0	9.4	6.5	8.3	0.9	1.8	T	T	41.7
1973-74	0.0	0.0	0.0	0.6	18.2	10.6	13.8	8.2	8.7	2.5	T	0.0	62.6
1974-75	0.0	0.0	0.0	0.2	2.8	11.9	20.6	21.2	10.7	0.1	2.1	0.0	69.6
1975-76	0.0	0.0	0.0	3.8	12.6	9.2	15.9	10.2	3.3	2.0	0.0	0.0	57.0
1976-77	0.0	0.0	0.0	T	2.3	7.2	8.8	6.3	14.8	2.3	T	0.0	41.7
1977-78	0.0	0.0	0.0	T	13.1	32.5	26.7	8.0	2.6	4.8	T	0.0	87.7
1978-79	0.0	0.0	T	0.1	11.8	14.6	19.4	18.6	7.1	4.3	T	0.0	75.9
1979-80	0.0	0.0	0.0	0.4	2.2	12.0	27.4	11.6	11.3	0.2	T	0.0	65.1
1980-81	0.0	0.0	0.0	0.3	0.7	18.1	9.3	12.6	4.3	4.9	0.0	0.0	50.2
1981-82	0.0	0.0	0.0	T	1.7	16.3	32.7	6.6	1.9	6.8	0.2	0.0	66.2
1982-83	0.0	0.0	T	T	10.4	17.6	5.9	5.2	1.0	4.6	T	0.0	44.7
1983-84	0.0	0.0	0.0	0.0	6.8	21.3	8.6	7.4	2.8	0.4	T	0.0	47.3
1984-85	0.0	0.0	0.0	11.1	9.6	14.0	2.5	19.1	6.7	1.0	T	0.0	64.0
1985-86	0.0	0.0	T	1.0	12.7	11.9	19.8	17.2	0.4	0.5	T	0.0	63.5
1986-87	0.0	0.0	0.0	T	20.3	7.2	9.3	4.0	18.9	0.8	T	0.0	60.5
1987-88	0.0	0.0	0.0	T	5.5	14.0	8.7	8.2	2.7	2.3	T	0.0	41.4
1988-89	0.0	0.0	0.0	T	6.1	20.2	13.5	11.5	5.4	1.1	T	0.0	57.8
1989-90	T	0.0	0.0	0.4	3.2	13.6	12.6	11.5	6.4	7.6	0.4	0.0	55.7
1990-91	0.0	T	0.0	0.4	3.2	52.1	25.9	1.6	4.4	4.2	0.0	0.0	91.8
1991-92	0.0	0.0	0.0	6.3	14.7	4.6	10.6	4.4	T	1.7	T	0.0	42.3
1992-93	T	T	0.2	T	7.3	23.1	17.8	9.7	2.5	0.3	T	0.0	60.9
1993-94	0.0	0.0	0.0	0.0	16.6	14.4							
Record Mean	T	T	0.1	1.4	8.5	16.9	17.9	10.4	6.0	2.5	0.7	T	64.4

MISSOULA, MONTANA

Missoula is situated in the heart of the Montana Rocky Mountains in the extreme north portion of the Bitterroot Valley, and about 5 miles east of the confluence of the Bitterroot and Clark Fork Rivers. The Clark Fork Valley begins at Missoula and extends about 20 miles west-northwestward. The Bitterroot Valley extends about 70 miles due southward from Missoula. The Continental Divide is 60 to 80 miles east of Missoula, and the Bitterroot Range is only about 20 miles away to the southwest. These two mountain ranges have a marked effect on the climate of Missoula.

The prevailing flow of air aloft over western Montana is from the west and southwest during spring and summer months, and from the west and northwest during the winter months. Since this air must pass over the Bitterroot Range, it loses much of its moisture on the western slopes of these mountains. As a result, Missoula receives only between 12 inches and 15 inches of precipitation annually. This small amount of precipitation makes for a semi-arid climate. There is sufficient irrigation water, however, from the nearby mountains. The heaviest precipitation, of about 2 inches, is received in each month of May and June.

Generally the spring months are cool and a little damp, with almost daily shower activity during May and June. There are about 137 growing days each year. The summer months are dry with moderate temperatures and cool nights. Seldom does the temperature reach 100 degrees. Oppressively warm nighttime temperatures are unknown.

In the winter, the Continental Divide shields the Missoula area from much of the severely cold air which moves down the continent from arctic regions. Because of this shielding effect, many of the cold waves which sweep down over eastern Montana miss the Missoula area entirely. Under certain conditions, however, the cold Arctic air does break over the Continental Divide, and moves with force into the Bitterroot and Clark Fork Valleys. When this happens, Missoula experiences severe blizzard conditions. The cold air is funnelled to the city through Hell Gate which is the mouth of the Clark Fork River canyon at Missoula. Locally these blizzards are referred to as Hell Gate Blizzards. After the valleys of western Montana are filled with the cold air, prolonged cold spells may occur. January is the coldest month, although periods of sub-zero weather occur occasionally in December and February. Rarely, there are brief periods of sub-zero weather in November and March. During the winter months the sunshine is limited to about 30 percent of the possible amount.

NORMALS, MEANS, AND EXTREMES

MISSOULA, MONTANA

LATITUDE: 46°55'N LONGITUDE: 114°05' W ELEVATION: FT. GRND 3197 BARO 3203 TIME ZONE: MOUNTAIN WBAN: 24153

	(a)	JAN	FEB	MAR	APR	MAY	JUNE	JULY	AUG	SEP	OCT	NOV	DEC	YEAR
TEMPERATURE °F:														
Normals														
-Daily Maximum		30.0	37.4	46.6	57.5	65.7	73.9	83.4	82.2	70.9	57.0	40.6	30.2	56.3
-Daily Minimum		15.4	20.9	24.9	30.9	37.9	46.1	50.1	49.2	40.4	31.3	24.2	16.4	32.3
-Monthly		22.7	29.2	35.8	44.2	51.8	60.0	66.8	65.8	55.7	44.2	32.4	23.4	44.3
Extremes														
-Record Highest	49	59	60	75	87	95	98	105	105	99	85	66	60	105
-Year		1953	1967	1978	1987	1986	1987	1973	1961	1967	1980	1983	1956	JUL 1973
-Record Lowest	49	-33	-27	-13	14	21	31	31	30	20	0	-23	-30	-33
-Year		1957	1982	1955	1951	1985	1991	1971	1992	1985	1971	1955	1983	JAN 1957
NORMAL DEGREE DAYS:														
Heating (base 65°F)		1311	1002	905	624	409	179	65	81	303	645	978	1290	7792
Cooling (base 65°F)		0	0	0	0	0	29	121	106	24	0	0	0	280
% OF POSSIBLE SUNSHINE	48	34	43	53	57	59	62	80	76	69	55	33	29	54
MEAN SKY COVER (tenths)														
Sunrise - Sunset	49	8.4	8.1	7.8	7.5	7.0	6.4	3.9	4.5	5.4	6.7	8.2	8.6	6.9
MEAN NUMBER OF DAYS:														
Sunrise to Sunset														
-Clear	49	2.5	2.8	3.3	3.8	5.3	6.4	15.5	13.5	10.2	6.9	2.4	2.0	74.7
-Partly Cloudy	49	4.1	4.5	6.0	6.7	8.6	9.3	9.8	9.5	8.3	7.3	4.6	4.2	82.8
-Cloudy	49	24.4	20.9	21.7	19.4	17.2	14.3	5.7	8.0	11.5	16.8	23.0	24.8	207.7
Precipitation .01 inches or more	49	13.6	10.4	11.5	10.3	11.7	11.4	7.3	7.3	7.5	7.8	11.4	12.8	123.0
Snow,Ice pellets,hail 1.0 inches or more	49	4.2	2.1	2.0	0.7	0.2	0.0	0.0	0.0	0.0	0.3	2.2	3.8	15.4
Thunderstorms	49	0.*	0.*	0.1	0.6	3.5	5.5	6.3	5.9	1.9	0.3	0.*	0.0	24.2
Heavy Fog Visibility 1/4 mile or less	49	4.7	4.0	1.5	0.2	0.2	0.3	0.2	0.4	1.0	3.2	4.6	6.4	26.8
Temperature °F														
-Maximum														
90° and above	33	0.0	0.0	0.0	0.0	0.3	2.7	8.9	8.9	0.8	0.0	0.0	0.0	21.6
32° and below	33	16.4	7.2	2.2	0.0	0.0	0.0	0.0	0.0	0.0	0.2	5.8	17.4	49.2
-Minimum														
32° and below	33	29.5	26.2	25.9	17.2	5.4	0.2	0.*	0.1	4.1	17.8	25.8	29.6	182.0
0° and below	33	5.1	1.8	0.3	0.0	0.0	0.0	0.0	0.0	0.0	0.*	0.6	3.1	10.8
AVG. STATION PRESS.(mb)	21	906.7	905.7	902.8	903.7	903.4	903.9	905.3	905.1	906.3	907.0	906.1	906.9	905.2
RELATIVE HUMIDITY (%)														
Hour 05	33	85	85	84	80	82	84	78	75	82	85	87	86	83
Hour 11 (Local Time)	33	81	78	67	54	53	53	44	46	56	67	80	83	64
Hour 17	33	76	66	51	42	43	42	32	31	39	50	71	80	52
Hour 23	33	83	81	75	67	68	68	58	57	68	75	84	85	72
PRECIPITATION (inches):														
Water Equivalent														
-Normal		1.24	0.79	0.97	0.96	1.78	1.78	0.91	1.20	1.12	0.74	0.81	1.16	13.46
-Maximum Monthly	49	2.94	2.18	2.10	2.46	7.38	4.19	3.05	3.29	3.60	3.51	2.51	3.15	7.38
-Year		1969	1986	1989	1956	1980	1958	1946	1985	1985	1975	1973	1964	MAY 1980
-Minimum Monthly	49	0.16	0.17	0.20	0.08	0.25	0.35	0.09	T	0.05	0.01	0.22	0.25	T
-Year		1981	1973	1953	1977	1963	1961	1985	1967	1979	1978	1976	1976	AUG 1967
-Maximum in 24 hrs	49	0.88	1.03	0.80	1.65	1.92	1.61	1.80	1.43	1.34	1.49	0.81	0.94	1.92
-Year		1948	1975	1992	1951	1980	1964	1987	1947	1954	1946	1973	1964	MAY 1980
Snow,Ice pellets,hail														
-Maximum Monthly	49	42.5	20.1	16.3	8.2	8.1	T	T	T	0.4	5.4	17.7	31.6	42.5
-Year		1963	1975	1989	1970	1978	1973	1993	1991	1983	1973	1947	1964	JAN 1963
-Maximum in 24 hrs	49	11.3	14.4	6.9	6.9	8.1	T	T	T	0.4	5.4	6.2	9.6	14.4
-Year		1980	1975	1977	1950	1978	1973	1993	1991	1983	1973	1961	1955	FEB 1975
WIND:														
Mean Speed (mph)	49	5.2	5.6	6.7	7.5	7.4	7.1	6.9	6.6	6.0	5.1	5.1	4.8	6.2
Prevailing Direction through 1963		ESE	NW	NW	NW	NW	NW	NW	NW	NW	NW	NW	E	NW
Fastest Mile														
-Direction (!!!)	48	S	NW	SW	NW	SW	S	SE	SW	N	SW	SW	W	SE
-Speed (MPH)	48	52	47	50	51	57	51	72	58	43	51	42	56	72
-Year		1953	1963	1972	1950	1954	1956	1957	1956	1974	1950	1978	1957	JUL 1957
Peak Gust														
-Direction (!!!)	10	SW	NW	W	SW	SW	SW	NW	SE	NW	SW	W	W	SW
-Speed (mph)	10	56	45	55	63	56	61	59	61	53	61	47	49	63
-Date		1989	1988	1986	1987	1985	1987	1987	1993	1991	1991	1993	1988	APR 1987

PRECIPITATION (inches) — MISSOULA, MONTANA

YEAR	JAN	FEB	MAR	APR	MAY	JUNE	JULY	AUG	SEP	OCT	NOV	DEC	ANNUAL
1964	0.70	0.51	0.67	1.49	1.22	3.13	0.94	1.84	0.51	0.50	0.56	3.15	15.22
1965	1.46	0.62	0.67	1.68	0.63	1.55	1.59	2.24	2.11	0.17	1.05	0.43	14.20
1966	1.44	0.96	0.86	0.66	0.50	1.81	0.71	1.01	0.96	0.81	0.59	0.79	11.10
1967	1.22	0.27	1.09	0.93	1.33	2.67	0.40	T	0.49	1.64	0.65	1.71	12.40
1968	0.87	0.73	0.60	0.94	1.05	1.89	0.21	2.38	1.92	0.54	0.55	1.17	12.85
1969	2.94	0.52	0.72	0.64	1.21	4.18	0.25	0.04	0.66	0.67	0.42	0.92	13.17
1970	2.87	0.34	1.06	1.07	1.75	2.84	1.68	0.08	0.48	1.00	0.97	0.94	15.08
1971	1.81	0.56	0.78	2.09	1.35	1.74	0.53	0.91	0.30	0.24	1.18	1.63	13.12
1972	2.04	1.82	1.62	0.96	0.69	1.37	0.64	0.24	1.66	0.78	0.41	1.46	13.69
1973	0.44	0.17	0.23	0.33	0.54	1.57	0.09	0.31	0.60	0.60	2.51	1.62	9.01
1974	2.07	0.68	1.26	0.61	0.44	1.36	1.03	1.18	0.70	0.25	0.50	0.68	10.76
1975	2.03	1.77	0.74	1.01	1.35	2.02	1.51	2.03	0.51	3.51	1.15	0.85	18.48
1976	0.90	1.04	0.40	0.94	0.79	1.52	1.20	0.88	0.58	0.33	0.22	0.25	9.05
1977	0.66	0.18	0.98	0.08	2.13	0.66	0.72	1.28	1.67	0.72	1.02	2.88	12.98
1978	1.15	0.66	0.67	1.08	1.98	0.77	0.57	1.11	1.78	0.01	1.00	0.99	11.77
1979	1.25	1.04	1.22	1.04	0.74	0.67	0.77	1.31	0.05	0.97	0.50	0.81	10.37
1980	1.80	0.60	0.88	0.96	7.38	2.04	1.58	0.62	0.77	0.75	0.63	1.34	19.35
1981	0.16	0.77	1.43	0.74	4.19	2.70	1.07	1.61	1.01	0.62	1.07	1.98	17.35
1982	2.07	1.31	1.52	1.34	2.03	1.83	0.94	0.38	2.09	0.43	0.37	1.07	15.38
1983	0.62	0.95	1.10	0.72	2.65	2.26	2.44	1.27	1.37	0.37	1.17	1.79	16.71
1984	0.86	0.44	1.32	2.04	2.02	1.47	0.38	1.47	0.79	0.96	0.89	0.66	13.30
1985	0.19	0.70	0.44	0.55	1.57	0.38	0.09	3.29	3.60	0.80	0.51	0.38	12.50
1986	0.93	2.18	0.54	0.51	1.69	2.66	0.84	1.68	3.54	0.44	1.07	0.50	16.58
1987	0.28	0.37	1.23	0.41	1.31	1.52	2.47	1.05	0.09	0.02	0.26	1.12	10.13
1988	0.74	0.57	1.04	0.69	3.12	1.68	0.50	0.29	0.51	0.51	0.67	0.75	11.07
1989	0.75	0.46	2.10	1.01	1.35	1.44	1.58	2.08	0.88	0.46	0.84	0.61	13.56
1990	0.92	0.29	0.72	1.37	3.56	0.42	0.75	2.64	0.06	0.81	0.87	1.11	13.52
1991	0.62	0.19	1.22	0.30	2.27	2.92	0.27	0.62	0.31	0.29	1.66	0.95	11.62
1992	0.40	0.18	1.19	1.67	0.68	2.13	1.75	0.60	0.73	0.53	0.96	0.78	11.60
1993	0.86	0.67	0.66	1.83	1.99	1.45	2.02	1.32	0.35	2.00	0.46	0.40	14.01
Record Mean	1.05	0.82	0.90	1.01	1.82	1.95	0.95	0.93	1.17	0.98	1.01	1.11	13.70

AVERAGE TEMPERATURE (deg. F) — MISSOULA, MONTANA

YEAR	JAN	FEB	MAR	APR	MAY	JUNE	JULY	AUG	SEP	OCT	NOV	DEC	ANNUAL
1964	21.9	22.6	28.5	42.1	48.9	57.0	66.2	60.8	51.1	42.1	30.4	19.4	40.9
1965	24.6	24.6	25.6	44.7	48.9	57.2	65.0	64.1	47.8	47.0	36.4	27.2	42.7
1966	26.3	26.5	35.3	43.7	55.1	58.1	69.1	66.8	62.9	44.0	34.0	29.5	45.9
1967	31.0	34.5	34.0	40.9	51.4	60.3	70.0	71.2	62.2	44.6	31.9	22.9	46.2
1968	20.2	32.4	41.7	42.1	52.6	59.8	69.8	64.2	54.7	42.0	33.7	21.8	44.6
1969	17.9	20.7	29.0	46.5	54.3	58.8	64.3	67.0	56.8	40.2	31.8	26.0	42.8
1970	22.9	32.2	34.3	39.8	52.6	63.0	68.2	67.7	50.6	41.5	33.9	22.5	44.1
1971	26.3	32.2	35.5	44.8	54.5	58.3	67.1	71.5	51.9	42.2	33.7	21.1	44.9
1972	20.6	28.1	40.6	42.6	54.2	62.0	64.9	66.4	51.4	42.5	33.6	19.3	43.9
1973	21.5	30.9	39.0	43.1	53.2	60.0	69.6	67.4	56.5	45.3	30.2	30.3	45.6
1974	21.2	31.9	35.4	45.9	48.5	65.0	68.1	65.1	57.1	46.3	34.7	27.8	45.6
1975	22.7	22.4	33.0	39.3	49.2	55.4	71.8	62.1	56.1	43.6	30.4	25.7	42.7
1976	28.0	31.7	34.8	46.3	55.0	56.8	66.8	63.2	57.4	43.9	32.3	24.4	45.1
1977	18.6	31.9	34.2	46.9	50.3	63.4	65.8	67.6	54.6	44.2	31.3	24.5	44.4
1978	24.2	27.3	39.8	46.1	48.3	59.6	65.3	63.5	55.1	45.3	26.7	15.9	43.1
1979	5.6	25.5	35.9	43.9	52.2	62.3	69.2	69.1	61.4	48.1	27.2	31.2	44.3
1980	16.3	29.6	34.6	49.8	54.7	58.6	65.5	61.8	57.0	45.3	34.2	30.5	44.8
1981	29.5	31.4	40.0	46.7	53.4	57.5	65.0	69.4	56.7	41.7	32.7	24.0	45.7
1982	21.2	22.7	38.3	41.6	51.0	63.0	64.6	66.0	56.2	45.0	29.0	22.7	43.5
1983	30.0	34.5	39.9	44.6	51.8	58.6	62.1	68.9	51.1	43.8	34.3	11.8	44.3
1984	25.4	32.1	39.7	43.8	49.3	56.8	67.2	68.4	52.8	42.5	33.9	20.1	44.3
1985	19.2	23.7	36.2	47.7	55.4	62.0	74.8	62.2	50.5	40.3	21.7	14.7	42.4
1986	26.0	28.6	41.7	43.5	54.3	65.5	62.5	69.5	53.0	44.7	31.7	21.4	45.2
1987	20.5	30.6	38.0	50.1	55.9	62.9	64.7	62.2	59.3	45.2	34.2	24.2	45.7
1988	20.5	32.5	38.3	47.7	52.1	64.0	67.0	66.3	56.4	50.3	34.3	22.9	46.0
1989	26.7	15.8	32.6	44.7	52.2	61.6	71.2	63.3	55.9	44.9	37.2	24.1	44.2
1990	31.5	29.6	38.0	47.8	50.4	59.8	68.1	66.0	62.6	42.7	36.1	15.6	45.7
1991	21.5	34.9	35.8	43.9	51.1	56.8	68.4	69.6	58.3	44.4	30.4	26.6	45.1
1992	26.6	36.4	43.3	46.9	56.4	64.4	64.2	65.9	54.7	47.2	31.1	19.4	46.4
1993	15.3	18.4	38.5	44.7	58.8	58.3	59.4	63.2	55.5	46.1	27.0	27.6	42.7
Record Mean	22.0	27.7	35.8	44.9	52.8	60.0	67.4	66.0	56.1	45.0	32.7	24.6	44.6
Max	29.8	36.5	46.2	57.6	66.4	74.3	84.6	83.0	71.2	57.7	40.9	31.7	56.7
Min	14.1	18.9	25.3	32.2	39.2	45.6	50.3	48.9	41.0	32.3	24.4	17.4	32.5

REFERENCE NOTES — MISSOULA, MONTANA

GENERAL
T - TRACE AMOUNT.
BLANK ENTRIES DENOTE MISSING/UNREPORTED DATA.
# INDICATES A STATION OR INSTRUMENT RELOCATION.

SPECIFIC

(a) - LENGTH OF RECORD IN YEARS, ALTHOUGH INDIVIDUAL MONTHS MAY BE MISSING.
0.* OR * - THE VALUE IS BETWEEN 0.0 AND 0.05
NORMALS - BASED ON THE 1961-1990 RECORD PERIOD.
EXTREMES - DATES ARE THE MOST RECENT OCCURRENCE
WIND DIR.- NUMERALS SHOW TENS OF DEGRESS CLOCKWISE FROM TRUE NORTH. "00" INDICATES CALM.
RESULTANT DIRECTIONS ARE GIVEN TO WHOLE DEGREES.
BOLD VALUES INDICATE EXTREME VALUES WHICH OCCURRED AFTER THE ASOS SYSTEM WAS COMMISSIONED.

RECORD = PERIOD OF RECORD
RECORD MEAN PRECIPITATION IS THE MEAN OF ALL DAILY PRECIPITATION AMOUNTS DURING THE PERIOD OF RECORD.
RECORD MAX(MIN) TEMPERATURE IS THE MEAN OF ALL DAILY MAX(MIN) TEMPERATURES DURING THE PERIOD OF RECORD.
RECORD MEAN TEMPERATURE IS THE SUM OF THE RECORD MAX AND RECORD MIN DIVIDED BY 2.
AVERAGE TEMPERATURE IS THE SUM OF THE MEAN DAILY MAX AND MIN TEMPERATURE DIVIDED BY 2.

EXCEPTIONS

RECORD MEANS ARE THROUGH THE CURRENT YEAR, BEGINNING IN 1892 FOR TEMPERATURE
1886 FOR PRECIPITATION
1945 FOR SNOWFALL

HEATING DEGREE DAYS Base 65 deg. F MISSOULA, MONTANA

SEASON	JULY	AUG	SEP	OCT	NOV	DEC	JAN	FEB	MAR	APR	MAY	JUNE	TOTAL
1964-65	30	153	409	698	1031	1406	1246	1124	1218	602	491	230	8638
1965-66	44	88	509	555	849	1165	1194	1075	914	631	304	216	7544
1966-67	23	46	94	645	921	1097	1045	848	952	712	416	145	6944
1967-68	2	4	116	630	987	1299	1383	941	718	678	379	168	7305
1968-69	10	97	305	705	935	1333	1456	1234	1109	546	325	201	8256
1969-70	69	46	262	763	991	1200	1297	911	942	751	377	140	7749
1970-71	32	12	425	722	928	1313	1195	911	908	600	319	211	7576
1971-72	75	12	386	700	934	1352	1371	1062	749	666	338	133	7778
1972-73	81	39	400	688	936	1416	1343	946	800	651	353	186	7839
1973-74	18	47	262	606	1036	1070	1357	919	910	569	505	103	7402
1974-75	27	60	230	575	903	1147	1302	1184	985	765	483	281	7942
1975-76	9	116	263	657	1032	1213	1141	959	929	554	302	258	7433
1976-77	25	89	231	647	974	1253	1431	918	950	536	450	76	7580
1977-78	59	76	310	637	1001	1245	1259	1050	771	564	511	171	7654
1978-79	67	110	308	604	1141	1515	1841	1099	893	625	389	126	8718
1979-80	37	13	121	517	1130	1043	1506	1018	935	448	318	198	7284
1980-81	53	119	243	603	917	1063	1092	934	768	542	353	228	6915
1981-82	39	16	256	718	963	1266	1351	1182	824	698	425	104	7842
1982-83	71	48	267	611	1072	1304	1079	845	772	607	402	189	7267
1983-84	126	10	413	652	913	1647	1222	947	779	631	479	259	8078
1984-85	37	18	370	692	926	1383	1413	1154	885	513	304	130	7825
1985-86	0	121	429	755	1294	1554	1205	1014	716	637	370	53	8148
1986-87	102	7	358	622	993	1346	1371	956	828	439	273	126	7421
1987-88	87	110	171	604	918	1258	1374	938	820	514	398	118	7310
1988-89	48	36	285	451	915	1300	1180	1377	1000	602	386	124	7704
1989-90	9	114	268	618	828	1262	1035	988	832	511	443	207	7115
1990-91	31	71	92	682	862	1525	1341	839	898	628	419	241	7629
1991-92	12	8	202	628	1031	1187	1185	823	667	536	261	103	6643
1992-93	75	105	304	543	1011	1405	1537	1298	812	603	209	213	8115
1993-94	181	105	285	577	1133	1153							

COOLING DEGREE DAYS Base 65 deg. F MISSOULA, MONTANA

YEAR	JAN	FEB	MAR	APR	MAY	JUNE	JULY	AUG	SEP	OCT	NOV	DEC	TOTAL
1969	0	0	0	0	0	25	56	117	22	0	0	0	220
1970	0	0	0	0	0	88	138	104	3	0	0	0	333
1971	0	0	0	0	1	14	146	220	0	0	0	0	381
1972	0	0	0	0	11	47	85	85	0	0	0	0	228
1973	0	0	0	0	0	43	165	124	15	0	0	0	347
1974	0	0	0	0	0	106	128	68	1	0	0	0	303
1975	0	0	0	0	0	0	227	31	0	0	0	0	258
1976	0	0	0	0	0	21	89	41	7	0	0	0	158
1977	0	0	0	0	1	36	92	163	3	0	0	0	295
1978	0	0	0	0	0	14	83	69	19	0	0	0	185
1979	0	0	0	0	0	50	177	146	17	0	0	0	390
1980	0	0	0	0	4	13	74	26	8	0	0	0	125
1981	0	0	0	0	0	10	48	158	10	0	0	0	226
1982	0	0	0	0	0	51	66	83	7	0	0	0	207
1983	0	0	0	0	0	4	43	138	5	0	0	0	190
1984	0	0	0	0	0	16	114	129	12	0	0	0	271
1985	0	0	0	0	11	47	311	39	0	0	0	0	408
1986	0	0	0	0	46	78	32	154	4	0	0	0	314
1987	0	0	0	0	0	69	85	31	7	0	0	0	192
1988	0	0	0	0	5	92	115	85	32	0	0	0	329
1989	0	0	0	0	0	27	209	71	1	0	0	0	308
1990	0	0	0	0	0	58	135	109	26	0	0	0	328
1991	0	0	0	0	0	1	125	157	6	0	0	0	289
1992	0	0	0	0	0	92	55	140	1	0	0	0	288
1993	0	0	0	0	24	18	15	57	4	0	0	0	118

SNOWFALL (inches) MISSOULA, MONTANA

SEASON	JULY	AUG	SEP	OCT	NOV	DEC	JAN	FEB	MAR	APR	MAY	JUNE	TOTAL
1964-65	0.0	0.0	0.0	1.7	3.7	31.6	21.5	11.2	8.4	4.0	T	0.0	82.1
1965-66	0.0	0.0	0.2	0.0	3.2	6.4	14.3	18.6	9.0	5.7	T	0.0	57.4
1966-67	0.0	0.0	0.0	0.2	3.4	11.2	7.4	2.2	8.9	5.7	2.6	0.0	41.6
1967-68	0.0	0.0	0.0	0.3	5.1	12.6	14.5	2.8	1.5	5.8	T	0.0	42.6
1968-69	0.0	0.0	T	T	1.5	19.5	27.5	9.2	9.0	T	0.0	0.0	66.7
1969-70	0.0	0.0	0.0	T	1.9	8.6	23.5	1.5	9.5	8.2	3.0	0.0	56.2
1970-71	0.0	0.0	T	1.1	4.4	12.7	9.9	3.4	6.3	3.5	T	T	41.3
1971-72	0.0	0.0	T	0.7	13.1	16.7	22.5	15.2	9.9	3.0	0.0	0.0	81.1
1972-73	0.0	0.0	0.2	1.2	4.5	8.9	3.0	2.1	0.5	0.2	T	T	20.6
1973-74	0.0	0.0	0.0	5.4	13.4	13.9	11.1	8.2	4.6	0.3	T	0.0	56.9
1974-75	0.0	0.0	0.0	0.2	2.7	7.5	16.9	20.1	7.1	3.0	T	0.0	57.5
1975-76	0.0	0.0	0.0	5.0	10.4	6.1	8.9	9.9	2.5	2.0	0.0	0.0	44.8
1976-77	0.0	0.0	0.0	0.4	1.2	3.0	10.3	1.8	12.5	1.0	T	0.0	30.2
1977-78	0.0	0.0	T	0.1	9.6	20.6	16.1	6.6	5.2	T	8.1	0.0	66.3
1978-79	0.0	0.0	0.0	0.0	12.3	14.1	20.0	7.1	7.3	1.9	0.0	0.0	62.7
1979-80	0.0	0.0	0.0	T	7.4	3.1	25.5	3.7	11.8	2.2	1.0	0.0	54.7
1980-81	0.0	0.0	0.0	T	0.9	2.1	1.4	8.0	2.0	T	0.0	0.0	14.4
1981-82	0.0	0.0	0.0	1.4	1.2	14.4	27.0	9.7	9.0	6.6	T	0.0	69.3
1982-83	0.0	0.0	0.0	0.5	2.8	10.4	4.8	5.2	0.5	T	T	0.0	24.2
1983-84	0.0	0.0	0.4	0.0	5.6	27.0	4.1	3.6	1.7	T	T	0.0	42.4
1984-85	0.0	0.0	T	3.9	3.3	10.0	2.5	11.8	3.3	0.1	T	0.0	34.9
1985-86	0.0	0.0	0.0	3.5	6.1	6.2	6.2	18.3	0.1	0.1	0.4	0.0	40.9
1986-87	0.0	0.0	0.0	0.0	7.2	7.0	3.9	2.3	7.8	0.2	T	0.0	28.4
1987-88	0.0	0.0	0.0	0.0	2.4	7.5	13.3	3.1	5.6	0.5	0.4	0.0	32.8
1988-89	0.0	0.0	0.0	0.0	7.4	9.0	9.0	9.4	16.3	5.2	0.0	0.0	56.3
1989-90	T	T	T	0.1	1.9	8.8	3.9	6.7	2.7	0.3	T	0.0	24.4
1990-91	0.0	T	0.0	0.4	8.3	17.3	9.4	0.2	7.6	1.9	T	0.0	45.1
1991-92	0.0	T	0.0	3.4	9.7	6.5	6.3	T	0.2	0.1	T	0.0	26.2
1992-93	T	0.0	T	T	4.7	11.4	13.5	12.2	1.1	1.2	T	0.0	44.1
1993-94	T	0.0	0.0	0.0	7.2	5.5							
Record Mean	T	T	T	0.9	6.0	10.8	12.3	7.6	5.9	2.1	0.7	T	46.4

1993 GRAND ISLAND, NEBRASKA (GRI)

The Grand Island Weather Service Office is located at the Hall County Regional Airport, 3 miles northeast of downtown Grand Island. It is situated just west of the mid-point of the north-south runway. The site is less than 50 miles from the geographical center of the contiguous United States and in the shallow Platte River valley. The complex of the Loup River and its tributaries converge approximately 15 miles northwest of the station, then flows eastward across the state. The terrain immediately surrounding the station is flat, sandy, loam. Just to the north is the southern boundary of the Nebraska sandhills. The terrain slopes gently upward from the Missouri River valley in eastern Nebraska to the Rocky Mountains of Colorado and Wyoming.

The climate is primarily continental in nature with occasional incursions of maritime tropical air from the Gulf of Mexico and modified maritime polar air from the Pacific Ocean. Wintertime outbreaks of cold, dry, Arctic air from Canada are common, usually accompanied by strong biting winds.

The east-west upslope produces periods of fog and low stratus when winds have an easterly component, and the characteristics of a Chinook, with warm dry air, when the component is westerly. Dry-season dust storms occur occasionally with these Chinook winds. These have been reduced in recent years by increased farm irrigation. Growing season humidities have also been increased by the expanding irrigation projects. Summers are usually hot and dry with temperatures often reaching 100 degrees or more. Late spring and early summer is the peak season for severe thunderstorms with frequent hail and tornados occasionally occurring. Winters are punctuated by occasional severe blizzards and have wide variations in temperatures that range from mild to bitterly cold.

Based on the 1951-1980 period, the average first occurrence of 32 degrees Fahrenheit in the fall is October 9 and the average last occurrence in the spring is April 29.

NORMALS, MEANS, AND EXTREMES FOR 1993

GRAND ISLAND, NE (GRI)

LATITUDE: 40° 58′ 03″ N LONGTITUDE: 98° 18′ 30″ W ELEVATION (FT): GRND: 1841 BARO: 1845 TIME ZONE: CENTRAL (UTC+06) WBAN: 14935

	ELEMENT	POR	JAN	FEB	MAR	APR	MAY	JUN	JUL	AUG	SEP	OCT	NOV	DEC	YEAR
TEMPERATURE °F	NORMAL DAILY MAXIMUM	30	32.6	38.1	49.2	63.5	73.3	83.7	88.5	85.9	76.2	65.5	48.8	35.8	61.8
	MEAN DAILY MAXIMUM	55	33.1	38.3	49.1	63.7	74.2	84.8	90.5	88.4	79.1	67.9	49.8	37.9	63.1
	HIGHEST DAILY MAXIMUM	48	76	77	90	96	101	107	109	110	104	96	82	76	110
	YEAR OF OCCURRENCE		1990	1972	1986	1989	1989	1988	1954	1983	1947	1947	1980	1964	AUG 1983
	MEAN OF EXTREME MAXS.	46	56.3	62.7	75.2	85.7	90.5	98.8	100.5	99.1	94.1	84.3	71.7	61.1	81.7
	NORMAL DAILY MINIMUM	30	11.1	16.4	26.3	38.1	49.4	59.3	64.8	62.0	51.6	39.2	26.2	14.9	38.3
	MEAN DAILY MINIMUM	55	11.6	16.8	26.4	38.8	50.0	60.2	65.6	63.5	52.8	40.6	26.3	16.6	39.1
	LOWEST DAILY MINIMUM	48	-28	-19	-21	7	23	38	42	40	23	15	-11	-26	-28
	YEAR OF OCCURRENCE		1963	1979	1960	1975	1967	1954	1971	1950	1984	1993	1976	1989	JAN 1963
	MEAN OF EXTREME MINS.	46	-10.9	-5.2	4.9	21.7	33.7	45.4	52.9	50.0	35.0	22.1	7.6	-5.5	21.0
	NORMAL DRY BULB	30	21.9	27.3	37.7	50.8	61.4	71.5	76.7	73.9	63.9	52.4	37.6	25.4	50.0
	MEAN DRY BULB	55	22.4	27.5	37.8	51.3	62.1	72.5	78.1	76.0	66.0	54.3	38.1	27.3	51.1
	MEAN WET BULB	44	19.0	23.9	31.8	43.1	53.5	62.3	66.7	65.1	56.1	45.3	32.7	23.2	43.6
	MEAN DEW POINT	44	12.7	17.7	25.1	35.4	47.5	56.9	61.8	60.3	50.4	37.8	26.0	17.0	37.4
	NORMAL NO. DAYS WITH:														
	MAXIMUM ≥ 90°	30	0.0	0.0	0.1	0.7	1.7	8.0	14.4	11.3	3.9	0.3	0.0	0.0	40.4
	MAXIMUM ≤ 32°	30	14.3	10.2	4.3	0.1	0.0	0.0	0.0	0.0	0.0	0.0	3.0	12.5	44.4
	MINIMUM ≤ 32°	30	30.7	26.6	22.6	7.8	0.7	0.0	0.0	0.0	0.5	6.5	22.6	30.1	148.1
	MINIMUM ≤ 0°	30	7.6	3.7	0.5	0.0	0.0	0.0	0.0	0.0	0.0	0.0	0.4	4.0	16.2
H/C	NORMAL HEATING DEG. DAYS	30	1336	1056	846	430	171	18	0	12	107	395	822	1228	6421
	NORMAL COOLING DEG. DAYS	30	0	0	0	0	59	213	363	288	74	0	0	0	997
RH	NORMAL (PERCENT)	30	69	70	66	61	64	63	65	67	67	63	69	71	66
	HOUR 00 LST	30	74	75	73	70	73	74	74	77	77	72	76	76	74
	HOUR 06 LST	30	76	78	79	78	81	82	82	84	84	79	80	78	80
	HOUR 12 LST	30	62	61	57	50	53	50	52	54	53	48	57	63	55
	HOUR 18 LST	30	65	62	54	46	50	47	49	51	53	52	63	68	55
S	PERCENT POSSIBLE SUNSHINE														
W/O	MEAN NO. DAYS WITH:														
	HEAVY FOG(VISBY ≤ 1/4 MI)	55	1.8	2.4	2.0	1.2	0.7	0.6	0.7	1.1	1.2	1.6	2.1	2.3	17.7
	THUNDERSTORMS	55	0.0	0.2	1.4	3.7	7.5	9.9	9.2	7.9	5.3	1.9	0.5	0.1	47.6
CLOUDINESS	MEAN:														
	SUNRISE-SUNSET (TENTHS)	1		9.0	9.0	8.0	8.0		7.0	5.0	5.0	6.0	5.0	6.0	
	MIDNIGHT-MIDNIGHT (TENTHS)	1		9.0	9.0	9.0	8.0		7.0	5.0	5.0	6.0	5.0	6.0	
	MEAN NO. DAYS WITH:														
	CLEAR	1	4.0	9.0	6.0	6.0	8.0	8.0	8.0	14.0	15.0	9.0	10.0	13.0	110.0
	PARTLY CLOUDY	1	4.0	3.0	6.0	8.0	2.0	3.0	5.0	5.0	5.0	4.0	5.0	4.0	54.0
	CLOUDY	1	11.0	13.0	13.0	7.0	10.0	6.0	7.0	8.0	6.0	6.0	7.0	13.0	107.0
PR	MEAN STATION PRESSURE(IN.)	21	28.11	28.10	27.99	27.98	27.97	27.99	28.04	28.06	28.07	28.08	28.06	28.10	28.05
	MEAN SEA-LEVEL PRES. (IN.)	44	30.15	30.11	30.00	29.94	29.91	29.89	29.95	29.96	30.00	30.04	30.06	30.11	30.01
WINDS	MEAN SPEED (MPH)	30	10.8	11.0	12.2	12.5	11.6	10.9	10.1	9.9	10.5	10.1	10.8	11.0	10.9
	PREVAIL.DIR(TENS OF DEGS.)	30	18	36	36	18	18	18	18	18	18	18	36	18	18
	MAXIMUM 2-MINUTE:														
	SPEED (MPH)	1	34	24	39	36	39	68	59	31	38	32	39	45	68
	DIR. (TENS OF DEGS.)	1	32	31	36	31	18	30	29	22	33	34	33	31	30
	YEAR OF OCCURRENCE		1993	1993	1993	1993	1993	1993	1993	1993	1993	1993	1993	1993	JUN 1993
	MAXIMUM 5-SECOND:														
	SPEED (MPH)	1	37	30	46	39	54	77	72	38	46	37	44	49	77
	DIR. (TENS OF DEGS.)	1	32	02	35	36	20	30	29	21	34	33	34	32	30
	YEAR OF OCCURRENCE		1993	1993	1993	1993	1993	1993	1993	1993	1993	1993	1993	1993	JUN 1993
PRECIPITATION	NORMAL TOTAL (IN.)	30	0.46	0.72	1.89	2.50	3.82	3.91	2.83	2.82	2.85	1.35	1.04	0.71	24.90
	MAXIMUM TOTAL (IN.)	55	1.65	3.39	6.63	4.34	8.88	13.96	10.38	8.73	9.00	4.42	3.77	2.17	13.96
	YEAR OF OCCURRENCE		1960	1971	1987	1984	1982	1967	1993	1977	1965	1946	1983	1968	JUN 1967
	MINIMUM TOTAL (IN.)	55	T	0.06	0.01	0.09	0.43	0.50	0.63	0.50	0.12	0.00	T	0.02	0.0
	YEAR OF OCCURRENCE		1986	1991	1967	1989	1964	1978	1970	1940	1939	1958	1939	1943	OCT 1958
	MAXIMUM IN 24 HOURS (IN.)	55	1.38	2.21	3.15	3.30	3.07	4.54	5.41	4.12	5.88	2.75	1.90	1.20	5.88
	YEAR OF OCCURRENCE		1947	1971	1979	1964	1985	1967	1950	1977	1977	1968	1973	1968	SEP 1977
	NORMAL NO. DAYS WITH:														
	PRECIPITATION ≥ 0.01	30	5.3	5.9	7.5	8.9	10.9	9.1	9.0	7.9	8.0	5.4	4.9	5.4	88.2
	PRECIPITATION ≥ 1.00	30	0.0	0.1	0.4	0.5	1.0	1.1	0.6	0.6	0.7	0.2	0.3	0.1	5.6
SNOWFALL	NORMAL TOTAL (IN.)	30	4.9	6.4	6.5	1.4	0.1	0.0	0.0	0.0	0.1	0.5	3.8	7.5	31.2
	MAXIMUM TOTAL (IN.)	55	17.5	21.5	20.5	9.0	4.5	T	T	T	3.8	9.8	17.1	26.0	26.0
	YEAR OF OCCURRENCE		1993	1969	1984	1984	1947	1991	1991	1992	1985	1991	1983	1973	DEC 1973
	MAXIMUM IN 24 HOURS (IN.)	55	9.0	15.0	12.2	5.5	4.5	T	T	T	3.8	9.8	11.2	12.0	15.0
	YEAR OF OCCURRENCE		1988	1984	1984	1984	1947	1991	1991	1992	1985	1991	1983	1968	FEB 1984
	MAXIMUM SNOW DEPTH (IN.)	44	18	17	14	4	4	T	0	T	1	2	14	20	20
	YEAR OF OCCURRENCE		1974	1969	1984	1961	1967	1952		1951	1985	1980	1983	1968	DEC 1968
	NORMAL NO. DAYS WITH:														
	SNOWFALL ≥ 1.0	30	1.7	2.1	1.7	0.4	0.*	0.0	0.0	0.0	0.*	0.1	1.1	1.7	8.8

PRECIPITATION (inches) 1993 GRAND ISLAND, NEBRASKA (GRI)

YEAR	JAN	FEB	MAR	APR	MAY	JUN	JUL	AUG	SEP	OCT	NOV	DEC	ANNUAL
1964	0.07	0.92	1.64	4.97	0.43	4.50	3.00	3.44	1.27	0.13	0.14	0.16	20.67
1965	0.67	1.23	1.26	2.18	5.97	5.21	2.36	3.35	9.00	0.37	0.50	0.53	32.63
1966	0.22	0.78	0.58	1.20	1.03	2.98	3.47	1.68	0.43	0.65	0.09	0.66	13.77
1967	0.59	0.07	0.01	0.99	3.40	13.96	0.98	1.30	1.02	1.37	0.22	0.46	24.37
1968	0.13	0.32	0.39	3.47	2.23	7.13	4.82	4.41	2.33	3.61	0.61	2.17	31.62
1969	0.91	2.48	0.19	2.55	4.13	3.46	3.10	3.75	2.01	3.43	0.19	0.82	27.02
1970	0.04	0.24	0.42	2.72	2.49	0.81	0.63	3.36	5.76	1.44	0.33	0.07	18.31
1971	0.82	3.39	1.12	0.95	5.32	5.62	2.27	0.66	1.35	1.75	1.82	0.50	25.57
1972	0.19	0.17	0.23	3.00	5.91	1.86	4.98	1.43	2.50	1.04	2.40	2.03	25.74
1973	0.77	0.45	5.57	1.63	3.85	0.84	2.90	1.38	8.39	1.51	2.37	2.07	31.73
1974	0.62	0.07	0.51	1.73	2.44	2.67	1.35	1.55	0.50	1.41	0.25	1.33	14.43
1975	0.86	0.46	1.02	2.76	1.50	6.86	2.35	1.18	1.01	0.11	3.26	0.16	21.53
1976	0.39	0.73	2.15	2.79	3.21	2.11	1.12	0.78	2.42	0.07	0.12	0.04	15.93
1977	0.42	0.18	3.30	4.89	6.85	1.37	1.73	8.73	7.77	1.42	1.13	0.43	38.22
1978	0.27	1.18	0.83	6.12	1.87	0.50	2.88	3.05	1.62	0.56	1.35	0.64	20.87
1979	0.82	0.43	5.56	3.27	3.99	2.65	2.66	1.39	2.54	3.02	1.78	0.48	28.59
1980	0.80	0.65	2.23	1.92	2.06	3.62	0.85	4.42	0.83	1.46	0.12	0.18	19.14
1981	0.16	0.19	3.14	1.14	4.28	0.60	3.45	4.38	0.93	1.08	3.21	0.63	23.19
1982	0.65	0.56	2.41	3.15	8.88	4.47	2.65	5.78	2.37	2.06	1.49	1.18	35.65
1983	0.66	0.35	3.40	1.30	4.59	6.29	1.71	2.04	2.67	1.08	3.77	0.76	28.62
1984	0.24	1.65	3.18	7.34	5.75	3.95	1.99	1.21	0.19	3.49	1.37	1.34	31.70
1985	0.27	0.27	1.15	4.37	4.62	3.98	3.20	2.25	5.80	1.83	0.61	0.27	28.62
1986	T	0.50	1.98	2.52	3.05	3.07	3.77	2.51	3.65	3.31	0.23	0.31	24.90
1987	0.06	0.72	6.63	1.37	4.86	1.34	1.32	5.41	1.19	0.79	1.19	0.81	25.69
1988	1.13	0.33	0.10	2.41	1.77	4.39	3.93	2.79	3.26	0.01	0.70	0.27	21.09
1989	0.71	0.64	0.41	0.09	1.90	4.86	2.20	3.26	6.49	0.94	0.03	0.40	21.93
1990	0.37	0.45	3.00	0.46	4.15	8.21	3.64	3.16	0.72	0.90	0.82	0.76	26.64
1991	0.52	0.06	1.79	2.73	6.27	5.25	5.74	1.34	0.79	1.61	1.71	2.11	29.92
1992	1.40	1.36	2.95	0.76	3.58	5.31	4.67	5.20	1.02	3.63	1.08	0.94	31.90
1993	1.15	1.48	0.97	1.66	2.54	4.71	10.38	4.64	1.55	1.55	0.72	0.41	31.76
POR= 55 YRS	0.54	0.74	1.64	2.49	3.86	4.02	3.00	2.64	2.47	1.22	0.90	0.65	24.19

AVERAGE TEMPERATURE (˚F) 1993 GRAND ISLAND, NEBRASKA (GRI)

YEAR	JAN	FEB	MAR	APR	MAY	JUN	JUL	AUG	SEP	OCT	NOV	DEC	ANNUAL
1964	29.2	27.3	34.6	52.0	66.6	70.8	79.4	69.5	62.2	52.1	37.7	23.1	50.4
1965	22.8	22.3	25.1	51.2	64.0	69.7	74.8	73.6	56.8	55.8	41.3	33.9	49.3
1966	16.7	25.8	42.1	45.5	61.1	71.0	81.0	70.7	63.9	53.8	38.1	26.3	49.7
1967	24.4	28.8	42.6	52.2	56.5	68.2	74.8	72.2	63.1	52.8	37.3	27.1	50.0
1968	23.9	27.0	44.4	51.1	58.2	73.2	75.2	74.4	64.2	54.2	36.8	20.4	50.2
1969	16.9	24.7	29.2	53.0	61.8	66.1	77.2	76.0	67.3	46.0	40.1	25.7	48.7
1970	17.9	31.9	33.5	49.5	65.8	73.6	78.5	77.7	64.4	48.8	37.2	29.3	50.7
1971	19.9	26.1	36.1	51.7	59.0	76.0	72.8	75.4	64.4	56.0	39.9	29.9	50.6
1972	21.8	27.2	41.7	49.6	61.2	72.5	73.9	72.8	64.7	48.9	35.0	20.0	49.1
1973	23.7	29.7	41.1	47.4	57.7	72.3	75.9	77.7	62.3	55.1	36.6	22.8	50.2
1974	17.2	32.4	42.4	52.6	64.3	69.7	82.3	69.0	61.2	54.8	38.3	25.2	50.8
1975	22.7	21.1	30.8	47.4	62.2	69.0	76.8	77.1	61.0	56.1	35.8	27.3	48.9
1976	24.4	37.5	38.7	53.3	59.3	71.0	77.3	76.0	65.4	47.4	33.2	28.1	51.0
1977	17.7	34.1	41.3	56.2	67.1	73.7	78.9	71.7	65.7	52.6	39.2	26.5	52.1
1978	12.2	15.0	35.5	51.3	59.8	73.7	76.3	73.8	68.3	52.0	35.2	18.4	47.6
1979	7.5	13.8	36.9	48.3	57.5	70.5	74.0	74.2	67.6	53.0	34.4	33.2	47.6
1980	21.6	23.7	34.2	51.5	61.0	73.1	80.7	76.9	66.9	51.5	41.7	29.8	51.0
1981	28.9	31.4	41.7	58.2	57.0	73.2	75.8	71.8	64.6	50.6	41.0	24.6	51.6
1982	12.9	23.8	35.0	47.0	60.8	65.7	76.5	72.4	63.2	52.5	35.0	28.6	47.8
1983	25.5	31.4	37.9	43.6	56.8	69.6	79.5	82.4	68.3	53.8	38.3	8.4	49.6
1984	23.6	34.6	33.4	47.4	60.4	73.5	76.9	77.2	62.0	51.9	39.3	26.6	50.6
1985	20.5	23.8	43.3	55.3	65.0	68.8	75.6	70.6	61.9	52.0	27.4	21.8	48.8
1986	33.8	27.8	46.2	51.8	62.3	75.1	77.9	70.4	66.2	51.8	35.4	30.9	52.5
1987	30.2	36.4	39.3	54.0	65.8	73.6	79.0	71.2	64.6	47.9	41.2	29.2	52.7
1988	18.3	26.5	40.2	50.3	65.6	77.0	75.2	75.8	65.9	49.5	39.1	32.5	51.3
1989	32.7	17.0	37.0	53.9	61.9	69.0	76.7	73.9	62.7	54.0	38.6	20.8	49.9
1990	33.8	30.9	41.7	50.5	58.2	72.9	74.0	74.4	69.0	53.8	42.9	22.7	52.1
1991	20.9	38.3	42.0	53.2	64.1	75.1	76.1	74.6	66.5	50.8	31.6	33.2	52.2
1992	34.2	38.0	42.7	50.8	61.0	67.5	70.0	69.0	64.6	51.7	32.9	24.7	50.6
1993	17.6	18.7	36.0	47.6	59.8	68.1	73.5	73.3	60.0	50.7	34.4	31.7	47.6
POR= 55 YRS	22.4	27.5	37.1	50.4	61.0	71.2	76.6	74.7	64.8	53.3	37.4	26.8	50.3

REFERENCE NOTES:

H/C INDICATES HEATING AND COOLING DEGREE DAYS.
RH INDICATES RELATIVE HUMIDITY
W/O INDICATES WEATHER AND OBSTRUCTIONS
S INDICATES SUNSHINE.
PR INDICATES PRESSURE.
CLOUDINESS IS THE SUM OF THE CEILOMETER AND SATELLITE DATA NOT TO EXCEED TEN TENTHS.
MAXIMUM SNOW DEPTH IS THE DEPTH MEASURED AT 12 UTC.

GENERAL:
T INDICATES TRACE PRECIPITATION, AN AMOUNT GREATER THAN ZERO BUT LESS THAN THE LOWEST REPORTABLE VALUE.
+ INDICATES THE VALUE ALSO OCCURS ON EARLIER DATES.
BLANK ENTRIES DENOTE MISSING OR UNREPORTED DATA.
NORMALS ARE 30-YEAR AVERAGES (1961 - 1990).
ASOS INDICATES AUTOMATED SURFACE OBSERVING SYSTEM.
PM INDICATES THE LAST DAY OF THE PREVIOUS MONTH.
POR (PERIOD OF RECORD) INDICATES THE NUMBER OF YEARS USED TO COMPUTE THE MEAN. INDIVIDUAL MONTHS WITHIN THE POR MAY BE MISSING.
WHEN THE POR FOR A NORMAL IS LESS THAN 30 YEARS, THE NORMAL IS PROVISIONAL AND IS BASED ON THE NUMBER OF YEARS INDICATED.
0.* OR * INDICATES THE VALUE OR MEAN DAYS WITH IS BETWEEN 0.00 AND 0.05.
CLOUDINESS FOR ASOS STATIONS DIFFERS FROM THE NON-ASOS OBSERVATION TAKEN BY A HUMAN OBSERVER. ASOS STATION CLOUDINESS IS BASED ON TIME-AVERAGED CEILOMETER DATA FOR CLOUDS AT OR BELOW 12,000 FEET AND ON SATELLITE DATA FOR CLOUDS ABOVE 12,000 FEET.
THE NUMBER OF DAYS WITH CLEAR, PARTLY CLOUDY, AND CLOUDY CONDITIONS FOR ASOS STATIONS IS THE SUM OF THE CEILOMETER AND SATELLITE DATA FOR THE SUNRISE TO SUNSET PERIOD.

GENERAL CONTINUED:
CLEAR INDICATES 0 - 3 TENTHS, PARTLY CLOUDY INDICATES 4 - 7 TENTHS, AND CLOUDY INDICATES 8 OR MORE TENTHS. WHEN AT LEAST ONE OF THE ELEMENTS (CEILOMETER OR SATELLITE) IS MISSING, THE DAILY CLOUDINESS IS NOT COMPUTED.
WIND DIRECTION IS RECORDED IN TENS OF DEGREES (2 DIGITS) CLOCKWISE FROM TRUE NORTH. "00" INDICATES CALM. "36" INDICATES TRUE NORTH.
RESULTANT WIND IS THE VECTOR AVERAGE OF THE SPEED AND DIRECTION.
AVERAGE TEMPERATURE IS THE SUM OF THE MEAN DAILY MAXIMUM AND MINIMUM TEMPERATURE DIVIDED BY 2.
SNOWFALL DATA COMPRISE ALL FORMS OF FROZEN PRECIPITATION, INCLUDING HAIL.
A HEATING (COOLING) DEGREE DAY IS THE DIFFERENCE BETWEEN THE AVERAGE DAILY TEMPERATURE AND 65˚ F.
DRY BULB IS THE TEMPERATURE OF THE AMBIENT AIR.
DEW POINT IS THE TEMPERATURE TO WHICH THE AIR MUST BE COOLED TO ACHIEVE 100 PERCENT RELATIVE HUMIDITY.
WET BULB IS THE TEMPERATURE THE AIR WOULD HAVE IF THE MOISTURE CONTENT WAS INCREASED TO 100 PERCENT RELATIVE HUMIDITY.

HEATING DEGREE DAYS (base 65°F) 1993 GRAND ISLAND, NEBRASKA (GRI)

YEAR	JUL	AUG	SEP	OCT	NOV	DEC	JAN	FEB	MAR	APR	MAY	JUN	TOTAL
1964-65	0	44	159	399	810	1294	1301	1190	1229	422	97	8	6953
1965-66	0	11	271	284	705	956	1492	1091	703	577	188	27	6305
1966-67	0	20	108	353	799	1192	1252	1009	699	391	331	36	6190
1967-68	4	15	93	389	820	1168	1272	1095	634	414	227	13	6144
1968-69	1	9	84	346	840	1374	1483	1121	1102	357	154	82	6953
1969-70	0	0	25	586	740	1212	1455	919	973	471	93	14	6488
1970-71	0	0	146	513	828	1102	1392	1080	891	396	194	0	6542
1971-72	6	0	131	284	743	1083	1335	1089	714	463	181	19	6048
1972-73	6	9	116	494	894	1389	1273	980	734	520	240	11	6666
1973-74	2	0	123	306	847	1304	1479	904	691	369	91	40	6156
1974-75	0	27	165	315	793	1229	1302	1221	1052	524	127	31	6786
1975-76	0	0	178	296	870	1166	1251	791	808	354	198	11	5923
1976-77	0	0	95	551	944	1139	1461	861	729	280	15	0	6075
1977-78	0	1	47	380	768	1185	1632	1395	914	409	205	25	6961
1978-79	0	3	64	400	886	1438	1777	1431	866	492	258	38	7653
1979-80	9	10	49	366	915	978	1340	1191	946	420	176	11	6411
1980-81	0	2	71	420	694	1084	1113	934	716	221	255	7	5517
1981-82	9	0	80	439	713	1245	1612	1148	925	537	148	61	6917
1982-83	0	12	139	385	896	1120	1216	933	832	635	269	37	6474
1983-84	0	0	89	349	793	1751	1276	875	974	521	177	3	6808
1984-85	0	0	184	405	764	1185	1372	1147	664	319	76	39	6155
1985-86	0	13	217	399	1120	1334	962	1035	587	392	108	0	6167
1986-87	0	17	49	403	882	1052	1071	793	789	356	68	7	5487
1987-88	0	32	82	527	708	1103	1441	1109	761	435	87	2	6287
1988-89	7	8	78	474	770	997	994	1339	864	393	152	22	6098
1989-90	0	6	152	345	786	1368	956	947	717	456	221	17	5971
1990-91	7	0	76	360	655	1303	1358	742	708	370	129	1	5709
1991-92	0	0	110	442	994	977	947	778	686	433	170	23	5560
1992-93	10	25	92	414	959	1243	1464	1288	890	518	181	45	7129
1993-	0	9	170	460	911	1028							
POR= 56 YRS	0	6	111	369	821	1179	1316	1053	862	443	173	28	6361

COOLING DEGREE DAYS (base 65°F) 1993 GRAND ISLAND, NEBRASKA (GRI)

YEAR	JAN	FEB	MAR	APR	MAY	JUN	JUL	AUG	SEP	OCT	NOV	DEC	ANNUAL
1969	0	0	0	1	61	122	381	348	100	6	0	0	1019
1970	0	0	0	12	128	279	426	401	132	19	0	0	1397
1971	0	0	0	6	15	338	256	327	122	15	0	0	1079
1972	0	0	0	8	70	251	287	257	111	0	0	0	984
1973	0	0	0	0	19	238	345	401	53	9	0	0	1065
1974	0	0	0	6	77	189	541	159	56	6	0	0	1034
1975	0	0	0	2	49	158	373	384	66	28	0	0	1060
1976	0	0	0	9	29	196	387	344	111	12	0	0	1088
1977	0	0	0	24	85	268	437	214	77	0	0	0	1105
1978	0	0	5	5	47	293	355	283	169	2	0	0	1159
1979	0	0	0	0	36	211	294	303	132	2	0	0	978
1980	0	0	0	20	56	261	493	377	133	9	0	0	1349
1981	0	0	0	25	15	259	356	219	74	0	0	0	948
1982	0	0	0	1	24	90	364	248	92	2	0	0	821
1983	0	0	0	0	23	183	460	546	194	8	0	0	1414
1984	0	0	0	2	42	264	374	386	105	4	0	0	1177
1985	0	0	0	33	81	158	335	195	134	0	0	0	936
1986	0	0	10	1	30	306	407	193	95	1	0	0	1043
1987	0	0	0	30	99	273	442	233	76	4	1	0	1158
1988	0	0	0	2	111	366	332	351	111	0	0	0	1273
1989	0	0	2	67	63	147	371	291	94	9	0	0	1044
1990	0	0	0	26	16	260	295	302	203	19	0	0	1121
1991	0	0	0	22	108	312	351	306	160	11	0	0	1270
1992	0	0	0	13	55	103	172	156	87	9	0	0	595
1993	0	0	0	0	25	146	269	273	27	23	0	0	763
POR= 25 YRS	0	0	0	12	53	226	364	299	107	7	0	0	1068

SNOWFALL (inches) 1993 GRAND ISLAND, NEBRASKA (GRI)

YEAR	JUL	AUG	SEP	OCT	NOV	DEC	JAN	FEB	MAR	APR	MAY	JUN	TOTAL
1964-65	0.0	0.0	0.0	0.0	T	0.8	7.9	15.3	6.3	0.0	0.0	0.0	30.3
1965-66	0.0	0.0	0.0	0.0	0.4	1.9	3.0	1.0	4.1	0.4	T	0.0	10.8
1966-67	0.0	0.0	0.0	0.5	0.2	8.5	6.4	1.0	T	4.1	4.3	0.0	25.0
1967-68	0.0	0.0	0.0	T	0.7	4.3	1.1	3.5	0.3	0.8	0.0	0.0	10.7
1968-69	0.0	0.0	0.0	0.0	1.7	21.8	7.6	21.5	1.8	T	0.0	0.0	54.4
1969-70	0.0	0.0	0.0	4.4	1.5	10.7	0.6	3.2	3.6	0.3	0.0	0.0	24.3
1970-71	0.0	0.0	0.0	0.1	0.1	0.4	10.3	12.7	11.7	0.2	0.0	0.0	35.5
1971-72	0.0	0.0	0.0	T	4.1	2.7	3.5	2.6	1.0	0.1	0.0	0.0	14.0
1972-73	0.0	0.0	0.0	0.2	11.5	17.5	2.6	4.7	2.1	3.8	T	0.0	42.4
1973-74	0.0	0.0	0.0	0.0	6.6	26.0	11.0	1.3	1.2	2.4	0.0	0.0	48.5
1974-75	0.0	0.0	0.0	0.0	0.3	16.0	8.3	7.4	3.1	2.6	0.0	0.0	37.7
1975-76	0.0	0.0	0.0	0.0	16.0	1.2	3.6	1.4	11.7	0.0	0.0	0.0	33.9
1976-77	0.0	0.0	0.0	T	1.1	0.4	5.4	1.1	12.5	4.8	0.0	0.0	25.3
1977-78	0.0	0.0	0.0	0.0	2.5	7.7	4.6	17.9	8.3	T	0.0	0.0	41.0
1978-79	0.0	0.0	0.0	0.0	10.4	8.1	11.0	3.9	10.4	2.2	0.0	0.0	46.0
1979-80	0.0	0.0	0.0	0.6	1.9	6.2	8.3	8.2	11.6	T	0.0	0.0	36.8
1980-81	0.0	0.0	0.0	6.6	0.8	1.6	2.0	2.8	5.4	T	0.0	0.0	19.2
1981-82	0.0	0.0	0.0	0.1	3.9	8.4	7.5	6.9	8.8	1.1	0.0	0.0	36.7
1982-83	0.0	0.0	0.0	0.2	2.7	10.7	5.0	5.9	11.8	3.9	0.0	0.0	40.2
1983-84	0.0	0.0	T	0.0	17.1	13.8	3.1	15.1	20.5	9.0	0.0	0.0	78.6
1984-85	0.0	0.0	0.0	T	1.6	7.5	5.9	3.4	4.5	0.0	0.0	0.0	22.9
1985-86	0.0	0.0	3.8	0.0	7.4	4.5	T	6.1	0.9	T	0.0	0.0	22.7
1986-87	0.0	0.0	0.0	T	2.5	1.8	1.7	0.6	18.6	0.0	0.0	0.0	25.2
1987-88	0.0	0.0	0.0	0.8	6.4	7.4	11.4	5.2	1.3	T	0.0	0.0	32.5
1988-89	0.0	0.0	0.0	0.0	4.3	0.8	0.8	9.8	4.4	T	0.0	0.0	20.1
1989-90	0.0	0.0	0.0	0.2	2.0	4.7	4.4	6.8	5.1	1.5	0.0	0.0	24.7
1990-91	0.0	0.0	0.0	0.0	2.8	4.8	7.2	T	4.9	T	T	T	19.7
1991-92	T	0.0	0.0	9.8	7.3	T	6.3	5.8	0.7	1.5	T	0.0	31.4
1992-93	0.0	T	0.0		3.6	8.1	17.5	12.7	6.9	T		0.0	
1993-	T	0.0	0.0	0.8	4.0	3.7							
POR= 56 YRS	T	T	0.1	0.5	3.6	6.2	5.8	6.2	6.2	1.6	0.4	T	30.5

1993 LINCOLN, NEBRASKA (LNK)

Lincoln is near the center of Lancaster County in southeastern Nebraska. The surrounding area is gently rolling prairie. The western edge of the city is in the flat valley of Salt Creek, which receives a number of tributaries in or near the city and flows northeastward to the lower Platte. The terrain slopes upward to the west and is sufficient to cause instability in moist easterly winds in the Lincoln area. The upward slope to the west is a part of the general rise in elevation that begins at the Missouri River 45 miles east of Lincoln and culminates in the Continental Divide about 575 miles to the west.

The maximum temperature has exceeded 110 degrees. Hot winds, combining unusual wind force and high temperatures, occasionally cause serious injury to crops.

The majority of winter outbreaks of severely cold air from northwestern Canada move over the Lincoln area. The temperature has remained below zero degrees for more than 8 consecutive days. The center of some of the cold air masses move southward far enough to the east that their full effect is usually not felt here.

Normally the crop season, April through September, receives over three-fourths of the annual precipitation. Nighttime thunderstorms are predominant in the summer months, so that the needed moisture is received during much of the growing season at a time of least interference with outdoor work.

Annual snowfall is about 25 inches, although the annual snowfall has exceeded 59 inches. Much of the snow is light and melts rapidly. However, at times a considerable amount accumulates on the ground and has exceeded a depth of 21 inches.

In the summer the higher winds are associated with thunderstorms. Hail is a common occurrence during the late spring and early summer months in eastern Nebraska. Even though hail damage in the city itself is usually slight, agricultural damage (i.e., crops) occurs frequently in the outlying areas around Lincoln. Lincoln has been relatively free from tornadoes. There is much sunshine, averaging 64 percent of the possible duration. Moderate to low humidities are at comfortable levels except for short periods during the summer when warm, moist, tropical air occasionally reaches this area.

NORMALS, MEANS, AND EXTREMES FOR 1993

LINCOLN, NE (LNK)

LATITUDE: 40° 51′ 03″ N   LONGTITUDE: 96° 45′ 32″ W   ELEVATION (FT): GRND: 1178 BARO: 1197   TIME ZONE: CENTRAL (UTC+06)   WBAN: 14939

	ELEMENT	POR	JAN	FEB	MAR	APR	MAY	JUN	JUL	AUG	SEP	OCT	NOV	DEC	YEAR
TEMPERATURE °F	NORMAL DAILY MAXIMUM	30	32.4	37.9	50.3	64.4	74.2	84.7	90.0	86.7	77.2	66.7	50.2	35.8	62.5
	MEAN DAILY MAXIMUM	36	33.2	38.8	49.4	63.5	73.4	84.4	89.8	87.1	80.8	69.0	50.6	38.0	63.2
	HIGHEST DAILY MAXIMUM	22	73	84	89	97	99	107	108	107	101	93	82	69	108
	YEAR OF OCCURRENCE		1990	1972	1986	1989	1989	1988	1990	1983	1975	1975	1980	1989	JUL 1990
	MEAN OF EXTREME MAXS.	46	55.4	61.3	76.3	85.5	89.2	98.3	101.2	99.5	93.7	84.4	71.3	58.9	81.2
	NORMAL DAILY MINIMUM	30	10.1	15.1	26.8	38.9	50.0	60.2	66.3	63.3	53.2	40.5	27.3	15.4	38.9
	MEAN DAILY MINIMUM	36	11.7	17.4	27.5	39.2	49.8	60.5	65.7	63.7	54.8	41.9	28.1	16.9	39.7
	LOWEST DAILY MINIMUM	22	-33	-24	-19	3	25	39	42	41	26	11	-5	-27	-33
	YEAR OF OCCURRENCE		1974	1979	1978	1975	1976	1978	1972	1988	1984	1993	1976	1983	JAN 1974
	MEAN OF EXTREME MINS.	46	-9.5	-7.1	5.2	22.5	34.7	45.4	54.0	49.5	35.2	23.0	9.5	-6.5	21.3
	NORMAL DRY BULB	30	21.3	26.6	38.6	51.7	62.1	72.5	78.2	75.0	65.3	53.6	38.8	25.6	50.8
	MEAN DRY BULB	54	22.5	27.9	38.1	51.6	62.3	72.5	79.2	77.0	67.0	55.6	39.5	28.5	51.8
	MEAN WET BULB	44	19.7	24.6	33.5	44.8	55.8	64.4	68.7	67.4	58.5	47.5	34.4	24.1	45.3
	MEAN DEW POINT	44	13.5	19.0	27.3	37.6	50.3	59.5	64.1	63.2	53.4	41.1	28.4	18.3	39.7
	NORMAL NO. DAYS WITH:														
	MAXIMUM ≥ 90°	19	0.0	0.0	0.0	0.6	0.9	8.9	17.4	11.9	4.4	0.2	0.0	0.0	44.3
	MAXIMUM ≤ 32°	19	14.1	10.7	2.9	0.1	0.0	0.0	0.0	0.0	0.0	0.0	2.8	11.7	42.3
	MINIMUM ≤ 32°	19	30.4	26.0	20.8	7.5	0.8	0.0	0.0	0.0	0.5	6.1	21.3	29.9	143.3
	MINIMUM ≤ 0°	19	7.2	4.4	0.5	0.0	0.0	0.0	0.0	0.0	0.0	0.0	0.2	4.3	16.6
H/C	NORMAL HEATING DEG. DAYS	30	1355	1075	818	399	161	11	0	9	81	362	786	1221	6278
	NORMAL COOLING DEG. DAYS	30	0	0	0	0	72	236	409	319	90	8	0	0	1134
RH	NORMAL (PERCENT)	30	70	72	69	64	67	65	65	69	70	67	72	73	69
	HOUR 00 LST	30	75	79	76	72	76	75	76	78	79	76	78	78	77
	HOUR 06 LST	30	77	80	81	80	83	83	83	86	86	82	81	79	82
	HOUR 12 LST	30	63	64	60	52	56	53	53	57	57	53	60	65	58
	HOUR 18 LST	30	66	65	57	50	54	50	50	54	58	57	66	70	58
S	PERCENT POSSIBLE SUNSHINE	37	59	57	57	60	61	70	72	71	66	64	54	53	62
W/O	MEAN NO. DAYS WITH:														
	HEAVY FOG(VISBY ≤ 1/4 MI)	22	1.3	1.5	1.8	0.6	0.6	0.3	0.2	0.8	0.4	1.1	1.7	2.0	12.3
	THUNDERSTORMS	22	0.1	0.2	2.0	4.4	7.1	8.4	8.6	7.4	5.2	2.0	0.6	0.3	46.3
CLOUDINESS	MEAN:														
	SUNRISE-SUNSET (TENTHS)	1		9.0						5.0	6.0	5.0	5.0	6.0	
	MIDNIGHT-MIDNIGHT (TENTHS)	1		9.0					7.0	6.0	6.0	5.0	5.0	6.0	
	MEAN NO. DAYS WITH:														
	CLEAR	1	6.0	8.0	3.0		5.0	8.0	5.0	12.0	11.0	10.0	13.0	10.0	
	PARTLY CLOUDY	1	4.0	2.0	5.0		5.0	4.0	3.0	9.0	4.0	5.0	4.0	9.0	
	CLOUDY	1	10.0	15.0	4.0		8.0	7.0	7.0	5.0	7.0	6.0	9.0	11.0	
PR	MEAN STATION PRESSURE(IN.)	21	28.83	28.82	28.70	28.68	28.66	28.67	28.71	28.74	28.76	28.78	28.77	28.82	28.75
	MEAN SEA-LEVEL PRES. (IN.)	44	30.16	30.12	30.00	29.95	29.91	29.90	29.26	29.96	30.00	30.05	30.06	30.12	29.96
WINDS	MEAN SPEED (MPH)	23	9.5	10.1	11.2	11.3	10.5	10.0	9.8	9.6	9.8	9.2	9.9	10.0	10.1
	PREVAIL.DIR(TENS OF DEGS.)	23	32	36	36	36	18	17	18	18	18	18	36	34	18
	MAXIMUM 2-MINUTE:														
	SPEED (MPH)	1	33	24	37	38	39	46	41	32	28	31	29	36	46
	DIR. (TENS OF DEGS.)	1	36	36	35	35	13	19	12	35	18	36	34	19	19
	YEAR OF OCCURRENCE		1993	1993	1993	1993	1993	1993	1993	1993	1993	1993	1993	1993	JUN 1993
	MAXIMUM 5-SECOND:														
	SPEED (MPH)	1	38	28	44	45	48	56	71	37	34	37	40	43	71
	DIR. (TENS OF DEGS.)	1	36	01	32	34	14	19	28	36	18	36	33	18	28
	YEAR OF OCCURRENCE		1993	1993	1993	1993	1993	1993	1993	1993	1993	1993	1993	1993	JUL 1993
PRECIPITATION	NORMAL TOTAL (IN.)	30	0.54	0.72	2.09	2.76	3.90	3.89	3.20	3.41	3.48	2.12	1.27	0.88	28.26
	MAXIMUM TOTAL (IN.)	22	1.59	1.26	6.65	7.21	7.97	7.67	12.50	8.57	8.28	5.40	3.81	3.42	12.50
	YEAR OF OCCURRENCE		1975	1984	1973	1978	1984	1983	1993	1982	1989	1986	1981	1984	JUL 1993
	MINIMUM TOTAL (IN.)	22	T	0.08	0.13	0.26	0.91	0.63	0.37	0.07	0.29	0.01	0.01	0.04	T
	YEAR OF OCCURRENCE		1986	1977	1988	1989	1989	1976	1983	1976	1974	1975	1989	1976	JAN 1986
	MAXIMUM IN 24 HOURS (IN.)	22	0.79	0.84	2.56	2.34	3.31	4.24	5.42	2.98	4.95	4.29	1.93	2.28	5.42
	YEAR OF OCCURRENCE		1989	1984	1987	1974	1984	1985	1990	1982	1989	1979	1977	1984	JUL 1990
	NORMAL NO. DAYS WITH:														
	PRECIPITATION ≥ 0.01	19	5.6	5.4	8.3	9.3	11.0	8.1	8.1	8.8	8.0	6.6	5.7	5.8	90.7
	PRECIPITATION ≥ 1.00	19	0.0	0.0	0.7	0.6	1.2	0.7	0.7	1.3	1.0	0.5	0.3	0.1	7.1
SNOWFALL	NORMAL TOTAL (IN.)	19	5.3	4.9	5.4	1.5	0.0	0.0	0.0	0.0	0.*	0.2	3.0	5.8	26.1
	MAXIMUM TOTAL (IN.)	22	16.8	13.8	17.0	7.9	T	T	T	T	0.8	3.3	8.8	19.8	19.8
	YEAR OF OCCURRENCE		1993	1978	1984	1992	1992	1993	1989	1989	1985	1980	1991	1973	DEC 1973
	MAXIMUM IN 24 HOURS (IN.)	22	8.0	7.7	8.3	7.9	T	T	T	T	0.8	3.3	7.7	10.4	10.4
	YEAR OF OCCURRENCE		1975	1978	1977	1992	1992	1993	1989	1989	1985	1980	1972	1973	DEC 1973
	MAXIMUM SNOW DEPTH (IN.)	44	18	13	20	3	3	0	0	0	0	5	10	16	20
	YEAR OF OCCURRENCE		1974	1978	1960	1987	1967					1970	1957	1983	MAR 1960
	NORMAL NO. DAYS WITH:														
	SNOWFALL ≥ 1.0	19	1.8	1.8	1.6	0.4	0.0	0.0	0.0	0.0	0.0	0.1	0.9	1.8	8.4

PRECIPITATION (inches) 1993 LINCOLN, NEBRASKA (LNK)

YEAR	JAN	FEB	MAR	APR	MAY	JUN	JUL	AUG	SEP	OCT	NOV	DEC	ANNUAL
1964	0.21	0.76	1.29	2.54	2.72	8.56	3.18	3.61	2.95	0.37	0.32	0.51	27.02
1965	0.34	3.06	1.49	3.69	6.27	10.71	4.23	2.17	6.76	0.41	0.84	1.36	41.33
1966	0.65	1.10	0.70	0.66	1.74	4.88	2.63	3.83	2.11	0.45	0.24	0.71	19.70
1967	0.47	0.14	0.88	1.63	4.47	12.93	3.99	1.91	2.91	1.91	0.41	0.68	32.33
1968	0.38	0.09	0.11	3.28	2.10	3.28	3.71	3.63	6.33	2.93	1.30	1.89	29.03
1969	0.68	0.89	1.45	4.62	4.46	2.78	3.93	2.09	0.77	2.88	0.12	1.23	25.90
1970	0.08	0.65	0.71	2.41	3.28	2.84	3.48	3.97	5.84	3.93	1.27	0.28	28.74
1971	1.30	2.79	0.70	0.78	6.30	1.75	3.71	0.95	0.96	4.29	3.41	0.83	27.77
1972	0.22	0.23	0.49	4.23	4.24	2.58	3.18	3.63	2.78	3.22	3.58	1.37	29.75
1973	1.12	0.62	6.65	2.59	5.86	0.77	4.48	0.75	7.52	4.92	1.78	2.15	39.21
1974	0.56	0.08	0.73	3.88	5.22	0.91	0.46	4.52	0.29	2.96	1.08	0.69	21.38
1975	1.59	1.26	1.35	2.75	2.58	3.08	1.63	1.37	1.53	0.01	2.53	0.66	20.34
1976	0.36	1.15	2.59	3.60	3.03	0.63	2.99	0.07	3.09	0.32	0.03	0.04	17.90
1977	0.63	0.08	3.54	1.83	5.20	0.99	3.75	7.48	6.05	1.86	2.03	0.35	33.79
1978	0.34	1.19	1.11	7.21	3.68	2.37	5.05	1.90	4.27	1.46	1.50	0.38	30.46
1979	1.11	0.48	4.93	2.93	2.95	2.99	3.61	2.80	0.40	5.29	1.28	0.38	29.15
1980	1.12	0.55	1.80	1.93	2.07	3.00	1.82	6.21	0.33	1.96	0.08	0.77	21.64
1981	0.15	0.22	1.96	1.88	3.99	0.85	3.24	5.07	2.51	0.92	3.81	0.71	25.31
1982	0.89	0.30	2.68	3.30	5.48	5.57	4.05	8.57	3.22	1.17	1.24	1.97	38.44
1983	0.92	0.66	3.84	1.09	5.00	7.67	0.37	1.17	2.62	1.73	3.64	0.67	29.38
1984	0.27	1.26	3.04	6.52	7.97	5.94	1.35	1.40	1.43	3.89	0.18	3.42	36.67
1985	0.27	0.55	1.37	3.45	3.34	6.17	6.93	2.02	4.37	1.45	0.67	0.32	30.91
1986	T	0.60	2.86	5.71	2.55	5.06	3.92	5.95	5.59	5.40	0.85	1.32	39.81
1987	0.08	0.53	6.53	2.29	4.01	4.19	1.80	6.39	3.20	1.14	0.82	0.70	31.68
1988	0.47	0.14	0.13	2.43	3.25	0.65	1.16	2.27	6.18	0.03	1.09	0.57	18.37
1989	1.14	0.81	0.24	0.26	0.91	4.71	2.03	4.02	8.28	0.82	0.01	0.59	23.82
1990	0.48	0.20	2.86	0.66	4.99	2.47	7.35	3.46	0.62	1.52	1.07	0.77	26.45
1991	0.92	0.12	2.60	2.65	2.98	5.04	2.59	1.26	0.94	1.99	1.72	2.01	24.82
1992	1.38	0.93	2.46	1.92	2.38	2.43	6.58	2.07	2.61	1.70	1.44	0.87	26.77
1993	0.73	0.49	1.90	2.43	4.96	5.83	12.50	4.32	3.56	1.55	0.54	0.41	39.22
POR= 54 YRS	0.72	0.92	1.93	2.71	3.77	4.31	3.53	3.37	3.13	1.92	1.22	0.84	28.37

AVERAGE TEMPERATURE (°F) 1993 LINCOLN, NEBRASKA (LNK)

YEAR	JAN	FEB	MAR	APR	MAY	JUN	JUL	AUG	SEP	OCT	NOV	DEC	ANNUAL
1964	31.6	31.3	35.3	53.2	67.9	71.7	82.0	72.0	65.7	54.8	42.2	26.5	52.9
1965	24.4	22.6	26.7	52.5	66.7	70.8	75.8	74.8	60.1	57.8	42.9	36.2	50.9
1966	18.4	28.3	43.2	46.7	61.4	72.5	81.4	72.6	63.9	55.6	40.5	28.2	51.1
1967	25.2	29.3	43.6	54.1	59.0	70.1	74.9	72.8	63.5	53.7	39.4	29.9	51.3
1968	24.9	28.2	46.0	52.9	58.8	74.6	77.8	76.0	66.2	57.2	38.1	24.1	52.1
1969	19.6	27.1	31.6	54.1	64.3	68.5	79.1	76.8	68.7	50.4	42.7	27.0	50.8
1970	18.6	33.2	36.0	53.4	68.3	74.9	79.1	78.0	66.0	52.0	39.0	31.2	52.5
1971	19.9	26.3	37.5	54.3	60.4	78.0	74.6	76.0	68.1	58.8	41.8	31.2	52.2
1972	20.8	26.1	41.5	49.7	60.7	71.2	74.1	72.2	65.1	49.0	36.2	21.6	49.0
1973	23.4	28.4	42.6	49.6	59.0	73.4	75.0	77.2	63.2	56.1	39.0	21.8	50.7
1974	15.6	31.1	42.1	51.8	62.1	70.9	83.8	70.7	60.9	56.1	39.4	26.9	51.0
1975	21.7	20.1	30.6	48.4	63.4	70.9	78.5	79.5	61.7	57.3	39.5	29.9	50.1
1976	24.3	36.8	39.2	54.3	59.0	71.6	78.1	76.2	66.9	47.3	32.1	24.4	50.9
1977	13.1	31.7	43.5	56.7	68.4	74.4	80.9	72.3	66.4	51.4	38.3	25.1	51.8
1978	9.9	13.2	33.6	51.3	60.0	71.6	76.8	74.9	69.1	52.2	36.5	22.3	47.6
1979	7.2	13.0	38.4	49.6	59.7	72.1	75.0	75.2	68.7	54.5	37.3	32.3	48.6
1980	23.4	21.9	35.3	52.2	62.8	74.3	82.2	78.8	66.9	51.9	42.6	28.6	51.7
1981	28.2	31.5	42.8	58.8	59.8	75.0	78.7	72.4	65.9	52.1	42.1	24.9	52.7
1982	11.9	23.5	36.3	48.3	63.4	66.7	78.9	74.1	65.1	54.7	36.9	29.9	49.1
1983	26.9	31.5	39.4	45.3	58.1	71.7	81.1	83.5	69.3	54.8	39.4	8.2	50.8
1984	21.8	35.7	33.4	48.4	59.2	73.5	77.6	78.0	63.0	53.2	40.2	27.9	51.0
1985	20.1	22.5	42.6	54.9	64.5	69.0	76.0	71.3	63.6	53.8	29.0	21.4	49.1
1986	32.7	26.2	45.8	53.2	63.2	75.4	78.7	71.0	68.5	53.7	35.7	30.5	52.9
1987	29.6	37.2	41.8	55.2	67.4	75.8	79.4	72.9	65.7	48.5	43.9	32.4	54.2
1988	21.6	26.3	41.3	51.0	67.6	77.9	77.8	78.3	67.6	50.7	41.3	31.7	52.8
1989	33.5	16.4	38.8	54.8	62.2	70.6	78.3	75.1	63.4	54.4	37.8	19.0	50.4
1990	33.8	31.8	42.8	51.1	59.5	75.3	76.5	76.7	70.5	55.0	44.7	23.5	53.4
1991	18.1	37.6	44.2	54.6	66.6	76.6	78.1	76.3	67.6	52.9	31.7	33.7	53.2
1992	34.4	37.7	43.8	51.6	62.2	69.5	72.8	70.2	65.5	53.7	33.9	26.4	51.8
1993	17.3	20.7	36.0	47.6	60.4	68.8	75.9	74.9	59.9	50.3	35.1	30.8	48.1
POR= 54 YRS	22.5	28.0	38.2	51.6	62.3	72.5	77.8	75.6	65.8	54.5	38.8	28.1	51.3

REFERENCE NOTES:

H/C INDICATES HEATING AND COOLING DEGREE DAYS.
RH INDICATES RELATIVE HUMIDITY
W/O INDICATES WEATHER AND OBSTRUCTIONS
S INDICATES SUNSHINE.
PR INDICATES PRESSURE.
CLOUDINESS IS THE SUM OF THE CEILOMETER AND SATELLITE DATA NOT TO EXCEED TEN TENTHS.
MAXIMUM SNOW DEPTH IS THE DEPTH MEASURED AT 12 UTC.

GENERAL:
T INDICATES TRACE PRECIPITATION, AN AMOUNT GREATER THAN ZERO BUT LESS THAN THE LOWEST REPORTABLE VALUE.
+ INDICATES THE VALUE ALSO OCCURS ON EARLIER DATES.
BLANK ENTRIES DENOTE MISSING OR UNREPORTED DATA.
NORMALS ARE 30-YEAR AVERAGES (1961 - 1990).
ASOS INDICATES AUTOMATED SURFACE OBSERVING SYSTEM.
PM INDICATES THE LAST DAY OF THE PREVIOUS MONTH.
POR (PERIOD OF RECORD) INDICATES THE NUMBER OF YEARS USED TO COMPUTE THE MEAN. INDIVIDUAL MONTHS WITHIN THE POR MAY BE MISSING.
WHEN THE POR FOR A NORMAL IS LESS THAN 30 YEARS, THE NORMAL IS PROVISIONAL AND IS BASED ON THE NUMBER OF YEARS INDICATED.
0.* OR * INDICATES THE VALUE OR MEAN DAYS WITH IS BETWEEN 0.00 AND 0.05.
CLOUDINESS FOR ASOS STATIONS DIFFERS FROM THE NON-ASOS OBSERVATION TAKEN BY A HUMAN OBSERVER. ASOS STATION CLOUDINESS IS BASED ON TIME-AVERAGED CEILOMETER DATA FOR CLOUDS AT OR BELOW 12,000 FEET AND ON SATELLITE DATA FOR CLOUDS ABOVE 12,000 FEET.
THE NUMBER OF DAYS WITH CLEAR, PARTLY CLOUDY, AND CLOUDY CONDITIONS FOR ASOS STATIONS IS THE SUM OF THE CEILOMETER AND SATELLITE DATA FOR THE SUNRISE TO SUNSET PERIOD.

GENERAL CONTINUED:
CLEAR INDICATES 0 - 3 TENTHS, PARTLY CLOUDY INDICATES 4 - 7 TENTHS, AND CLOUDY INDICATES 8 OR MORE TENTHS. WHEN AT LEAST ONE OF THE ELEMENTS (CEILOMETER OR SATELLITE) IS MISSING, THE DAILY CLOUDINESS IS NOT COMPUTED.
WIND DIRECTION IS RECORDED IN TENS OF DEGREES (2 DIGITS) CLOCKWISE FROM TRUE NORTH. "00" INDICATES CALM. "36" INDICATES TRUE NORTH.
RESULTANT WIND IS THE VECTOR AVERAGE OF THE SPEED AND DIRECTION.
AVERAGE TEMPERATURE IS THE SUM OF THE MEAN DAILY MAXIMUM AND MINIMUM TEMPERATURE DIVIDED BY 2.
SNOWFALL DATA COMPRISE ALL FORMS OF FROZEN PRECIPITATION, INCLUDING HAIL.
A HEATING (COOLING) DEGREE DAY IS THE DIFFERENCE BETWEEN THE AVERAGE DAILY TEMPERATURE AND 65° F.
DRY BULB IS THE TEMPERATURE OF THE AMBIENT AIR.
DEW POINT IS THE TEMPERATURE TO WHICH THE AIR MUST BE COOLED TO ACHIEVE 100 PERCENT RELATIVE HUMIDITY.
WET BULB IS THE TEMPERATURE THE AIR WOULD HAVE IF THE MOISTURE CONTENT WAS INCREASED TO 100 PERCENT RELATIVE HUMIDITY.

HEATING DEGREE DAYS (base 65°F) 1993 LINCOLN, NEBRASKA (LNK)

YEAR	JUL	AUG	SEP	OCT	NOV	DEC	JAN	FEB	MAR	APR	MAY	JUN	TOTAL
1964-65	0	14	94	319	684	1187	1252	1182	1182	387	60	1	6362
1965-66	0	4	194	227	657	888	1439	1026	672	541	179	18	5845
1966-67	0	8	106	314	729	1133	1232	993	671	338	264	25	5813
1967-68	2	9	83	372	761	1082	1239	1060	587	371	211	11	5788
1968-69	0	4	48	275	799	1263	1400	1054	1026	321	129	53	6372
1969-70	0	0	12	469	664	1171	1433	883	891	379	61	7	5970
1970-71	1	0	108	424	774	1040	1393	1077	847	321	173	0	6158
1971-72	4	0	89	223	693	1042	1361	1121	720	453	182	24	5912
1972-73	8	15	108	492	855	1340	1281	1018	688	458	199	1	6463
1973-74	1	0	99	281	772	1333	1533	941	704	403	139	32	6238
1974-75	0	14	161	278	762	1177	1338	1250	1059	495	117	20	6671
1975-76	0	0	168	269	759	1079	1255	810	792	328	206	11	5677
1976-77	0	0	83	553	979	1254	1604	927	659	256	11	0	6326
1977-78	0	0	42	413	793	1230	1703	1447	972	410	193	27	7230
1978-79	0	5	62	392	848	1315	1787	1454	816	463	197	16	7355
1979-80	3	7	42	324	825	1007	1281	1241	912	387	133	5	6167
1980-81	0	0	80	402	666	1120	1133	933	680	209	192	0	5415
1981-82	1	3	71	393	678	1237	1639	1159	880	501	88	52	6702
1982-83	0	3	118	324	840	1081	1173	931	785	584	238	25	6102
1983-84	0	0	75	326	763	1758	1334	841	970	496	196	3	6762
1984-85	0	0	167	366	737	1142	1385	1186	687	326	72	25	6093
1985-86	0	8	198	343	1071	1345	996	1079	602	353	95	0	6090
1986-87	0	15	34	344	874	1064	1090	773	713	331	51	2	5291
1987-88	0	17	51	507	626	1006	1341	1115	728	416	33	2	5842
1988-89	3	6	44	438	706	1025	968	1359	811	374	140	15	5889
1989-90	0	3	139	343	807	1424	960	923	680	443	183	10	5915
1990-91	0	1	55	319	604	1283	1449	760	641	330	104	0	5546
1991-92	0	0	106	391	994	964	941	784	654	405	150	11	5400
1992-93	0	12	90	355	927	1191	1472	1233	891	519	164	38	6892
1993-	0	4	169	471	890	1050							
POR= 43 YRS	0	4	89	342	771	1154	1312	1035	813	399	135	14	6068

COOLING DEGREE DAYS (base 65°F) 1993 LINCOLN, NEBRASKA (LNK)

YEAR	JAN	FEB	MAR	APR	MAY	JUN	JUL	AUG	SEP	OCT	NOV	DEC	ANNUAL
1969	0	0	0	0	116	165	445	372	130	27	0	0	1255
1970	0	0	0	38	172	309	445	410	146	26	0	0	1546
1971	0	0	0	8	40	397	309	356	188	38	0	0	1336
1972	0	0	0	0	55	217	296	247	119	2	0	0	936
1973	0	0	0	2	23	258	318	387	53	14	0	0	1055
1974	0	0	0	14	57	215	589	197	46	10	0	0	1128
1975	0	0	0	4	76	204	428	459	74	37	0	0	1282
1976	0	0	0	15	26	216	413	354	144	10	0	0	1178
1977	0	0	0	14	124	289	501	231	91	0	0	0	1250
1978	0	0	5	7	46	230	371	320	193	3	0	0	1175
1979	0	0	0	7	42	235	320	330	160	4	0	0	1098
1980	0	0	0	13	73	293	542	433	142	1	0	0	1497
1981	0	0	0	29	37	308	430	237	102	0	0	0	1143
1982	0	0	0	8	45	111	438	292	124	10	0	0	1028
1983	0	0	0	0	30	235	505	580	211	20	0	0	1581
1984	0	0	0	5	24	264	395	412	114	9	0	0	1223
1985	0	0	0	30	63	152	351	210	162	2	0	0	970
1986	0	0	12	9	43	320	430	209	148	0	0	0	1171
1987	0	0	0	41	130	333	451	267	78	2	0	0	1302
1988	0	0	0	4	117	396	408	427	131	3	0	0	1486
1989	0	0	5	75	63	188	418	321	95	21	0	0	1186
1990	0	0	0	34	17	326	363	370	227	17	3	0	1357
1991	0	0	5	22	162	355	414	356	191	25	0	0	1530
1992	0	0	0	9	71	152	250	182	111	12	0	0	787
1993	0	0	0	0	26	157	345	317	24	18	0	0	887
POR= 25 YRS	0	0	0	15	66	253	406	330	127	12	0	0	1209

SNOWFALL (inches) 1993 LINCOLN, NEBRASKA (LNK)

YEAR	JUL	AUG	SEP	OCT	NOV	DEC	JAN	FEB	MAR	APR	MAY	JUN	TOTAL
1964-65	0.0	0.0	0.0	0.0	0.5	1.5	3.7	26.1	10.3	0.0	0.0	0.0	42.1
1965-66	0.0	0.0	0.0	0.0	T	2.0	1.9	1.5	3.6	T	0.0	0.0	9.0
1966-67	0.0	0.0	0.0	0.0	0.2	9.9	5.5	0.5	T	3.7	3.0	0.0	22.8
1967-68	0.0	0.0	0.0	T	T	1.9	4.3	1.0	T	T	0.0	0.0	7.2
1968-69	0.0	0.0	0.0	0.0	1.7	11.7	7.8	13.6	5.0	0.0	0.0	0.0	39.8
1969-70	0.0	0.0	0.0	T	0.2	15.8	1.7	3.2	4.7	0.6	0.0	0.0	26.2
1970-71	0.0	0.0	0.0	6.6	0.6	0.4	15.1	17.6	8.3	0.4	0.0	0.0	49.0
1971-72	0.0	0.0	0.0	0.0	10.0	2.3	3.6	3.9	0.9	0.9	0.0	0.0	21.6
1972-73	0.0	0.0	0.0	T	8.6	6.5	6.7	5.1	0.1	2.2	0.0	0.0	29.2
1973-74	0.0	0.0	0.0	0.0	T	19.8	11.0	0.9	1.6	0.3	0.0	0.0	33.6
1974-75	0.0	0.0	0.0	0.0	1.9	8.3	14.6	10.9	4.4	2.0	0.0	0.0	42.1
1975-76	0.0	0.0	0.0	0.0	6.3	1.3	3.9	4.3	5.3	0.0	0.0	0.0	21.1
1976-77	0.0	0.0	0.0	0.4	0.4	0.6	8.7	T	8.3	3.4	0.0	0.0	21.8
1977-78	0.0	0.0	0.0	0.0	3.3	3.7	4.5	13.8	5.7	T	0.0	0.0	31.0
1978-79	0.0	0.0	0.0	0.0	7.3	3.0	11.5	4.4	3.7	4.5	0.0	0.0	34.4
1979-80	0.0	0.0	0.0	0.4	0.1	2.4	8.1	5.6	6.5	0.2	0.0	0.0	23.3
1980-81	0.0	0.0	0.0	3.3	T	2.3	1.7	1.9	3.8	0.0	0.0	0.0	13.0
1981-82	0.0	0.0	0.0	T	2.1	9.3	3.8	4.4	7.8	4.9	0.0	0.0	32.3
1982-83	0.0	0.0	0.0	T	T	5.2	5.1	7.6	12.9	7.2	0.0	0.0	38.0
1983-84	0.0	0.0	T	0.0	8.6	13.8	2.5	5.1	17.0	0.5	0.0	0.0	47.5
1984-85	0.0	0.0	0.0	0.0	0.5	6.2	4.7	3.1	7.0	0.0	0.0	0.0	21.5
1985-86	0.0	0.0	0.8	0.0	6.9	5.2	T	5.4	T	0.6	0.0	0.0	18.9
1986-87	0.0	0.0	0.0	0.0	1.8	3.9	1.2	1.2	7.6	0.0	0.0	0.0	15.7
1987-88	0.0	0.0	0.0	0.3	5.7	1.1	3.0	2.3	0.6	0.0	0.0	0.0	13.0
1988-89	0.0	0.0	0.0	0.0	2.7	3.1	1.1	11.5	0.7	T	0.0	0.0	19.1
1989-90	T	T	0.0	T	0.2	6.5	3.2	2.8	4.7	1.3	0.0	0.0	18.7
1990-91	0.0	0.0	0.0	0.0	0.0	8.1	14.1	T	3.6	T	0.0	0.0	25.8
1991-92	0.0	0.0	0.0	2.3	8.8	T	0.9	3.0	0.9	7.9	T	0.0	23.8
1992-93	0.0	0.0	0.0	T	3.8	2.9	16.8	6.4	2.2	T	0.0	T	32.1
1993-	0.0	0.0	0.0	T	2.8	4.2							
POR= 51 YRS	T	T	T	0.3	2.8	5.3	6.4	5.8	5.8	1.1	0.1	T	27.5

NORFOLK, NEBRASKA

Norfolk is located in northeastern Nebraska, in the valley of the Elkhorn River. The city of Norfolk lies at an average elevation of 1,550 feet above sea level. The surrounding country is moderately rolling in all directions. The terrain becomes more level to the south and southwest. Norfolk is situated near the western limit of the Corn Belt. To the east the climate and soils are favorable for diversified farming and dairying. To the west precipitation becomes lighter, and the farming country gives way to the grazing lands of the Great Plains. There are no local topographic features of sufficient importance to affect the climate of the area.

Northeast Nebraska has a climate typical of the interior of large continents in middle latitudes. The rainfall is moderate. Summers are hot and winters cold, and there are great variations in temperature and precipitation from day to day and from season to season. Most of the moisture which falls over this area is brought in from the Gulf of Mexico. The rapid changes in temperature are caused by the interchange of warm air from the south and southwest with cold air from the north. The rapid day to day changes in weather conditions produce an invigorating and healthful climate in northeast Nebraska.

Daily temperature ranges of 30 to 40 degrees are not uncommon. Summertime precipitation is almost wholly in the form of showers and thunderstorms. Practically all precipitation in the colder months is in the form of snow. As a rule, nearly 85 percent of the snowfall occurs from December to March and the ground is covered by snow during this period.

Norfolk is subject to the strong and persistent winds which prevail over the Great Plains states. Winds of 40 to 50 mph are not uncommon in this area, and gusts up to 100 mph have been recorded at Norfolk. Prevailing winds are from the south and southwest from May through September, with prevailing northwesterly winds during the remainder of the year.

Based on the 1951–1980 period, the average first occurrence of 32 degrees Fahrenheit in the fall is October 5 and the average last occurrence in the spring is May 1.

NORMALS, MEANS, AND EXTREMES

NORFOLK, NEBRASKA

LATITUDE: 41°59'N LONGITUDE: 97°26' W ELEVATION: FT. GRND 1544 BARO 1547 TIME ZONE: CENTRAL WBAN: 14941

	(a)	JAN	FEB	MAR	APR	MAY	JUNE	JULY	AUG	SEP	OCT	NOV	DEC	YEAR
TEMPERATURE °F:														
Normals														
-Daily Maximum		29.6	34.7	46.3	61.6	72.6	82.1	86.8	83.8	75.1	64.1	46.9	32.7	59.7
-Daily Minimum		8.4	14.1	25.0	37.6	49.2	58.6	64.0	61.2	51.1	38.6	25.5	12.8	37.2
-Monthly		19.0	24.4	35.7	49.7	60.9	70.3	75.4	72.6	63.1	51.4	36.2	22.8	48.5
Extremes														
-Record Highest	48	71	73	88	95	103	106	113	107	101	95	82	71	113
-Year		1981	1992	1986	1992	1967	1988	1954	1983	1971	1990	1945	1962	JUL 1954
-Record Lowest	48	-27	-26	-20	2	24	38	42	40	26	11	-15	-30	-30
-Year		1974	1981	1960	1975	1976	1990	1971	1986	1991	1993	1964	1989	DEC 1989
NORMAL DEGREE DAYS:														
Heating (base 65°F)		1426	1137	908	459	183	22	0	16	124	426	864	1308	6873
Cooling (base 65°F)		0	0	0	0	56	181	322	251	67	0	0	0	877
% OF POSSIBLE SUNSHINE														
MEAN SKY COVER (tenths)														
Sunrise - Sunset	48	6.1	6.3	6.6	6.2	6.1	5.3	4.7	4.8	4.8	5.1	6.2	6.2	5.7
MEAN NUMBER OF DAYS:														
Sunrise to Sunset														
-Clear	48	8.9	7.6	7.3	7.9	7.8	10.2	12.5	12.5	13.0	12.2	8.5	8.8	117.1
-Partly Cloudy	48	7.6	6.9	7.7	8.4	9.8	10.0	11.2	10.4	7.5	7.9	7.4	7.5	102.3
-Cloudy	48	14.5	13.7	16.1	13.8	13.4	9.8	7.2	8.1	9.6	10.9	14.2	14.6	145.8
Precipitation														
.01 inches or more	48	5.7	5.8	8.2	9.0	11.0	9.8	9.0	8.4	7.7	5.5	5.1	5.5	90.6
Snow,Ice pellets,hail														
1.0 inches or more	48	1.9	2.0	2.1	0.6	0.*	0.0	0.0	0.0	0.*	0.2	1.3	1.9	10.0
Thunderstorms	36	0.*	0.1	1.1	3.7	7.9	9.6	9.7	9.2	4.4	2.2	0.5	0.1	48.4
Heavy Fog Visibility														
1/4 mile or less	36	1.3	2.1	1.8	0.7	0.6	0.6	0.6	1.4	1.0	1.1	1.6	1.4	14.2
Temperature °F														
-Maximum														
90° and above	48	0.0	0.0	0.0	0.5	1.3	6.9	11.7	9.5	3.4	0.3	0.0	0.0	33.6
32° and below	48	16.5	12.1	6.1	0.2	0.0	0.0	0.0	0.0	0.0	0.1	3.9	13.6	52.6
-Minimum														
32° and below	48	30.8	27.2	24.3	9.6	0.9	0.0	0.0	0.0	0.7	7.5	23.8	30.4	155.1
0° and below	48	9.2	4.8	0.8	0.0	0.0	0.0	0.0	0.0	0.0	0.0	0.7	4.3	19.8
AVG. STATION PRESS.(mb)	17	962.8	962.8	959.0	958.2	957.9	958.0	959.4	960.3	960.5	961.3	960.8	962.1	960.3
RELATIVE HUMIDITY (%)														
Hour 00	17	74	77	76	70	73	73	77	81	76	73	78	78	76
Hour 06 (Local Time)	48	76	79	81	79	80	82	83	86	84	80	80	78	81
Hour 12	48	63	64	60	50	51	52	53	55	52	49	59	65	56
Hour 18	48	68	67	60	47	49	49	50	53	53	53	64	70	57
PRECIPITATION (inches):														
Water Equivalent														
-Normal		0.52	0.77	1.86	2.29	3.68	4.46	3.22	2.55	2.45	1.60	1.01	0.74	25.15
-Maximum Monthly	48	2.33	3.18	7.27	7.47	8.61	12.22	9.11	5.53	8.13	4.57	3.97	2.25	12.22
-Year		1949	1971	1987	1984	1977	1967	1950	1951	1970	1968	1983	1982	JUN 1967
-Minimum Monthly	48	0.06	0.04	0.06	0.23	1.01	0.54	0.33	0.53	0.30	T	0.03	0.08	T
-Year		1986	1949	1967	1969	1948	1987	1954	1971	1956	1958	1980	1958	OCT 1958
-Maximum in 24 hrs	48	1.30	2.41	2.76	1.96	4.12	5.51	3.46	2.89	3.88	2.79	1.53	1.34	5.51
-Year		1982	1971	1987	1985	1973	1974	1952	1978	1970	1968	1975	1981	JUN 1974
Snow,Ice pellets,hail														
-Maximum Monthly	48	16.4	22.8	20.8	13.2	2.7	T	T	T	1.1	6.5	22.6	19.1	22
-Year		1982	1984	1960	1984	1947	1993	1993	1992	1985	1991	1983	1968	FEB 19
-Maximum in 24 hrs	48	12.8	22.5	9.7	9.7	2.7	T	T	T	1.1	5.6	14.6	12.1	22
-Year		1982	1984	1966	1984	1947	1993	1993	1992	1985	1991	1983	1978	FEB 1984
WIND:														
Mean Speed (mph)	17	12.2	11.7	13.1	13.3	11.8	10.8	10.0	9.9	10.8	11.0	11.9	12.0	11.5
Prevailing Direction														
Fastest Obs. 1 Min.														
-Direction (!!)	12	33	34	30	32	09	28	21	32	30	32	35	35	32
-Speed (MPH)	12	46	43	44	51	41	46	58	60	55	39	45	46	60
-Year		1984	1984	1985	1982	1983	1986	1986	1989	1989	1990	1982	1982	AUG 1989
Peak Gust														
-Direction (!!)	10	NW	NW	NW	NW	W	W	SW	NW	NW	NW	N	NW	NW
-Speed (mph)	10	60	54	63	66	54	59	78	82	71	53	55	60	82
-Date		1984	1991	1989	1986	1991	1986	1986	1989	1989	1988	1988	1993	AUG 1989

PRECIPITATION (inches)

NORFOLK, NEBRASKA

YEAR	JAN	FEB	MAR	APR	MAY	JUNE	JULY	AUG	SEP	OCT	NOV	DEC	ANNUAL
1964	0.22	0.73	1.50	3.30	2.95	7.46	2.52	4.19	2.35	0.22	0.19	0.73	26.36
1965	0.35	0.90	0.44	1.04	6.62	3.61	3.86	2.78	6.88	0.52	0.13	0.45	27.58
1966	0.52	1.52	1.17	0.94	1.38	5.80	2.17	3.78	2.08	1.07	0.04	0.56	21.03
1967	0.48	0.06	0.06	1.37	2.27	12.22	0.82	1.10	1.40	1.68	0.07	0.74	22.27
1968	0.19	0.06	0.72	3.54	1.38	2.49	2.42	3.12	2.38	4.57	0.38	1.75	23.00
1969	1.21	1.86	0.43	0.23	2.93	8.09	3.40	3.69	1.62	2.63	0.07	1.14	27.30
1970	0.10	0.35	1.38	2.83	3.47	2.17	1.54	1.44	8.13	2.55	0.97	0.37	25.30
1971	0.28	3.18	0.47	1.21	2.19	5.95	4.12	0.53	0.85	2.28	1.48	0.69	23.23
1972	0.42	0.14	0.74	3.98	4.64	3.70	4.85	2.67	1.74	0.77	1.02	1.20	25.87
1973	1.12	0.46	5.14	1.58	5.56	1.33	4.14	1.31	5.12	0.77	1.94	0.78	29.25
1974	0.36	0.12	0.46	1.70	3.38	7.09	0.84	2.72	0.55	1.71	0.11	0.67	19.71
1975	1.26	0.39	0.86	4.03	2.56	7.62	3.40	2.71	0.69	0.16	3.67	0.19	27.54
1976	0.29	0.71	2.68	1.79	3.53	2.48	1.23	0.66	2.56	0.38	0.08	0.21	16.60
1977	0.26	0.81	3.58	3.45	8.61	5.04	3.25	4.33	1.82	3.00	1.52	0.51	36.18
1978	0.21	0.92	0.82	4.35	2.84	0.86	4.48	3.86	0.82	0.62	0.79	0.92	21.49
1979	1.09	0.44	4.45	1.92	5.14	3.08	4.08	2.84	1.81	4.26	1.65	0.54	31.30
1980	0.25	0.54	0.94	1.11	1.77	3.79	1.06	4.66	0.86	1.91	0.03	0.24	17.16
1981	0.14	0.39	2.57	0.36	2.34	3.39	4.16	3.63	0.57	1.49	2.10	0.73	21.87
1982	1.63	0.35	1.68	1.47	7.70	1.62	2.48	3.30	2.72	3.53	2.66	2.25	31.39
1983	0.75	0.55	3.68	2.27	3.28	6.82	2.73	1.35	1.61	1.16	3.97	0.68	28.85
1984	0.34	2.34	2.33	7.47	4.14	5.48	2.31	1.50	0.72	4.30	1.62	1.13	33.68
1985	0.25	0.15	1.46	5.49	2.85	4.53	1.62	2.48	4.61	0.96	0.89	0.47	25.76
1986	0.06	1.12	2.67	4.39	3.37	5.32	4.00	2.81	3.57	2.61	0.48	0.38	30.78
1987	0.28	0.77	7.27	0.70	5.60	0.54	3.68	3.38	3.21	1.11	1.19	0.78	28.51
1988	0.74	0.46	0.21	2.30	5.21	1.34	4.56	2.46	5.47	0.03	1.03	0.88	24.69
1989	1.18	0.58	0.90	0.90	1.71	1.96	3.68	0.89	3.61	0.33	0.06	0.75	16.55
1990	0.66	0.46	2.31	0.72	3.31	7.64	5.13	2.36	1.01	1.50	1.26	0.51	26.87
1991	0.87	0.46	1.94	2.72	7.55	6.52	1.19	1.30	1.52	1.96	2.28	0.80	29.11
1992	0.61	1.48	2.48	0.66	2.14	4.04	4.36	5.06	2.87	3.14	1.30	0.96	29.10
1993	1.30	1.70	1.14	3.68	2.41	6.39	8.97	2.86	3.17	1.40	0.82	0.43	34.27
Record Mean	0.59	0.77	1.76	2.23	3.76	4.40	3.36	2.65	2.30	1.52	0.98	0.68	25.00

AVERAGE TEMPERATURE (deg. F)

NORFOLK, NEBRASKA

YEAR	JAN	FEB	MAR	APR	MAY	JUNE	JULY	AUG	SEP	OCT	NOV	DEC	ANNUAL
1964	25.6	24.6	30.1	49.6	64.6	70.2	78.9	68.0	62.6	51.4	36.5	19.2	48.4
1965	20.0	18.7	23.7	49.3	63.6	69.3	74.1	72.0	54.6	54.3	37.4	32.0	47.4
1966	11.2	22.7	38.3	44.0	58.8	69.7	78.7	69.6	61.7	51.7	34.9	23.7	47.1
1967	20.9	23.3	40.3	49.6	55.4	67.5	73.4	70.3	62.1	49.6	35.8	24.1	47.7
1968	19.8	23.6	42.2	49.3	55.2	72.0	74.8	73.3	62.1	52.1	35.3	18.7	48.2
1969	13.4	23.1	25.5	51.2	61.5	65.4	75.4	74.2	65.4	45.2	38.8	23.1	46.9
1970	13.4	28.5	29.7	48.0	64.2	72.2	76.3	74.8	63.1	47.9	34.7	23.7	48.0
1971	16.0	24.4	33.8	50.3	57.7	74.2	71.3	73.7	63.2	54.4	37.3	24.2	48.4
1972	17.1	21.6	38.1	48.1	60.5	71.4	73.0	72.0	62.9	47.2	35.3	17.9	47.1
1973	21.2	26.1	41.1	48.2	58.5	71.6	75.2	77.5	61.0	54.9	36.2	22.1	49.5
1974	16.7	29.5	39.5	51.2	61.2	68.7	81.2	69.0	60.1	54.0	37.5	25.9	49.6
1975	20.7	19.3	29.2	44.6	63.5	69.6	77.5	76.3	60.0	55.0	35.5	24.7	48.0
1976	21.5	33.8	37.2	53.4	59.7	70.9	77.2	75.2	64.5	46.0	30.8	23.2	49.4
1977	12.5	30.9	41.1	56.2	67.8	72.8	78.1	70.2	65.7	51.2	35.5	22.3	50.3
1978	8.3	12.1	33.8	49.1	60.4	72.0	75.4	72.7	69.0	50.8	34.8	17.2	46.3
1979	5.7	11.2	32.9	47.1	58.0	70.7	74.2	72.9	67.2	51.0	32.9	31.0	46.2
1980	21.8	21.4	33.6	51.4	61.1	72.4	78.9	75.5	65.3	49.7	40.2	26.4	49.8
1981	26.7	29.3	40.8	57.8	57.8	72.2	76.0	71.5	63.8	49.8	40.5	23.5	50.8
1982	9.1	22.5	33.9	46.9	61.7	65.4	75.6	72.7	62.8	52.3	34.3	28.3	47.1
1983	24.5	29.5	37.1	43.6	57.2	69.8	78.2	81.0	66.6	52.2	37.0	6.9	48.6
1984	22.8	31.4	29.0	46.0	58.5	72.1	75.3	76.5	60.8	52.8	38.9	24.2	49.0
1985	18.1	23.4	41.3	54.5	64.6	67.4	73.9	70.0	60.7	51.1	25.0	17.4	47.3
1986	30.9	24.1	42.5	50.3	61.6	72.9	76.8	69.1	63.9	51.5	33.3	28.8	50.5
1987	27.3	35.7	39.4	54.2	66.0	73.1	77.7	70.0	64.0	46.2	40.6	29.0	51.9
1988	16.2	22.9	39.5	49.2	66.2	76.2	75.4	75.4	64.8	48.0	38.4	29.3	50.1
1989	30.6	15.2	34.0	53.2	59.9	68.3	76.9	73.5	62.6	52.1	34.6	15.6	48.0
1990	31.4	29.0	40.3	49.4	57.4	72.9	73.6	74.2	67.8	51.7	40.0	20.4	50.7
1991	17.2	34.7	40.4	52.8	64.5	75.0	76.0	74.6	65.4	49.2	28.8	31.0	50.8
1992	32.6	35.7	41.6	48.7	61.1	67.8	68.5	67.6	62.9	51.5	33.2	24.2	49.6
1993	17.7	18.1	33.8	46.7	60.1	67.7	73.3	72.9	59.3	50.2	33.1	29.4	46.9
Record Mean	19.1	24.4	34.8	49.3	60.5	70.6	75.5	73.4	63.3	51.8	35.8	23.8	48.5
Max	29.8	34.8	45.2	61.2	72.2	82.2	87.0	84.8	75.7	64.6	46.5	33.9	59.8
Min	8.5	14.0	24.4	37.3	48.8	58.9	63.9	61.9	51.0	38.9	25.0	13.7	37.2

REFERENCE NOTES

NORFOLK, NEBRASKA

GENERAL
T - TRACE AMOUNT.
BLANK ENTRIES DENOTE MISSING/UNREPORTED DATA.
• INDICATES A STATION OR INSTRUMENT RELOCATION.

SPECIFIC

(a) - LENGTH OF RECORD IN YEARS, ALTHOUGH INDIVIDUAL MONTHS MAY BE MISSING.
0.* OR * - THE VALUE IS BETWEEN 0.0 AND 0.05
NORMALS - BASED ON THE 1961-1990 RECORD PERIOD.
EXTREMES - DATES ARE THE MOST RECENT OCCURRENCE
WIND DIR.- NUMERALS SHOW TENS OF DEGRESS CLOCKWISE FROM TRUE NORTH. "00" INDICATES CALM.
RESULTANT DIRECTIONS ARE GIVEN TO WHOLE DEGREES.
BOLD VALUES INDICATE EXTREME VALUES WHICH OCCURRED AFTER THE ASOS SYSTEM WAS COMMISSIONED.

RECORD = PERIOD OF RECORD
RECORD MEAN PRECIPITATION IS THE MEAN OF ALL DAILY PRECIPITATION AMOUNTS DURING THE PERIOD OF RECORD.
RECORD MAX(MIN) TEMPERATURE IS THE MEAN OF ALL DAILY MAX(MIN) TEMPERATURES DURING THE PERIOD OF RECORD.
RECORD MEAN TEMPERATURE IS THE SUM OF THE RECORD MAX AND RECORD MIN DIVIDED BY 2.
AVERAGE TEMPERATURE IS THE SUM OF THE MEAN DAILY MAX AND MIN TEMPERATURE DIVIDED BY 2.

EXCEPTIONS

THUNDERSTORMS AND HEAVY FOG ARE THROUGH 1964 AND 1977 TO DATE AND MAY BE INCOMPLETE, DUE TO PART-TIME OPERATIONS.
FASTEST MILE WINDS ARE THROUGH 1981.

RECORD MEANS ARE THROUGH THE CURRENT YEAR, BEGINNING IN 1946 FOR TEMPERATURE
1946 FOR PRECIPITATION
1946 FOR SNOWFALL

HEATING DEGREE DAYS Base 65 deg. F — NORFOLK, NEBRASKA

SEASON	JULY	AUG	SEP	OCT	NOV	DEC	JAN	FEB	MAR	APR	MAY	JUNE	TOTAL
1964-65	0	48	146	416	848	1412	1389	1290	1271	473	113	7	7413
1965-66	0	14	318	330	825	1016	1667	1178	820	625	234	34	7061
1966-67	0	22	149	415	894	1275	1362	1160	764	463	352	34	6890
1967-68	15	32	118	483	874	1260	1395	1193	700	469	308	30	6877
1968-69	4	6	129	407	886	1432	1596	1165	1219	404	171	87	7506
1969-70	0	3	44	612	776	1293	1596	1015	1089	511	126	23	7088
1970-71	6	1	165	532	903	1272	1512	1126	962	435	222	5	7141
1971-72	14	0	148	328	824	1257	1481	1251	824	506	204	25	6862
1972-73	12	20	145	545	884	1457	1353	1086	735	497	221	12	6967
1973-74	4	0	141	314	858	1324	1493	987	783	414	147	45	6510
1974-75	0	18	190	340	817	1208	1369	1275	1107	604	105	25	7058
1975-76	2	0	207	326	877	1247	1343	900	854	352	191	15	6314
1976-77	0	1	108	587	1018	1287	1624	951	733	287	6	0	6602
1977-78	0	5	45	423	882	1318	1756	1477	965	471	187	28	7557
1978-79	1	9	49	434	900	1476	1836	1506	988	533	248	33	8013
1979-80	3	14	56	430	954	1046	1332	1258	966	423	174	8	6664
1980-81	0	1	100	473	738	1190	1180	995	741	238	235	5	5896
1981-82	9	1	98	464	727	1283	1730	1188	958	537	123	60	7178
1982-83	0	10	140	389	915	1132	1246	991	858	637	258	38	6614
1983-84	0	0	120	391	838	1798	1302	969	1109	563	220	6	7316
1984-85	0	1	196	375	775	1260	1451	1160	730	344	79	52	6423
1985-86	1	15	241	424	1195	1469	1048	1140	692	434	127	3	6789
1986-87	0	26	85	413	943	1114	1161	817	783	350	81	9	5782
1987-88	2	45	89	575	726	1110	1508	1218	783	469	61	6	6592
1988-89	4	14	96	520	790	1101	1063	1392	956	417	190	45	6588
1989-90	3	7	156	402	906	1531	1035	1002	759	500	238	19	6558
1990-91	4	0	87	425	739	1379	1477	843	755	381	149	1	6240
1991-92	0	0	130	489	1080	1048	1001	842	717	492	176	21	5996
1992-93	13	42	119	426	949	1259	1460	1308	962	545	167	52	7302
1993-94	0	11	188	473	952	1097							

COOLING DEGREE DAYS Base 65 deg. F — NORFOLK, NEBRASKA

YEAR	JAN	FEB	MAR	APR	MAY	JUNE	JULY	AUG	SEP	OCT	NOV	DEC	TOTAL
1969	0	0	0	0	69	105	329	296	63	7	0	0	869
1970	0	0	0	9	107	244	364	311	115	10	0	0	1160
1971	0	0	0	3	3	288	214	276	103	7	0	0	894
1972	0	0	0	2	72	225	269	243	90	0	0	0	901
1973	0	0	0	0	24	218	326	395	31	9	0	0	1003
1974	0	0	0	4	34	164	506	149	50	4	0	0	911
1975	0	0	0	1	67	171	398	359	60	24	0	0	1080
1976	0	0	0	11	32	200	387	324	100	6	0	0	1060
1977	0	0	0	30	101	241	415	174	73	0	0	0	1034
1978	0	0	4	0	56	244	330	257	176	2	0	0	1069
1979	0	0	0	2	39	212	297	268	128	0	0	0	946
1980	0	0	0	23	61	236	440	334	115	5	0	0	1214
1981	0	0	0	29	19	229	357	206	70	0	0	0	910
1982	0	0	0	2	26	76	336	255	80	2	0	0	777
1983	0	0	0	0	23	188	416	501	175	3	0	0	1306
1984	0	0	0	2	23	226	326	365	76	6	0	0	1024
1985	0	0	0	34	73	135	283	177	119	0	0	0	821
1986	0	0	4	1	30	249	372	159	56	0	0	0	871
1987	0	0	0	32	119	259	405	207	67	1	0	0	1090
1988	0	0	0	2	111	349	333	345	97	0	0	0	1237
1989	0	0	2	69	39	150	382	278	89	6	0	0	1015
1990	0	0	0	38	10	263	280	292	178	16	0	0	1077
1991	0	0	0	21	140	309	348	302	148	6	0	0	1274
1992	0	0	0	9	61	114	130	131	60	14	0	0	519
1993	0	0	0	0	25	141	263	259	23	20	0	0	731

SNOWFALL (inches) — NORFOLK, NEBRASKA

SEASON	JULY	AUG	SEP	OCT	NOV	DEC	JAN	FEB	MAR	APR	MAY	JUNE	TOTAL
1964-65	0.0	0.0	0.0	0.0	1.4	8.4	5.9	13.0	4.1	0.2	0.0	0.0	33.0
1965-66	0.0	0.0	0.0	0.0	0.3	1.7	5.7	1.5	11.1	T	0.0	0.0	20.3
1966-67	0.0	0.0	0.0	1.6	T	6.7	5.5	0.6	0.5	T	T	0.0	14.9
1967-68	0.0	0.0	0.0	0.5	0.7	4.9	2.4	0.5	T	T	T	0.0	9.0
1968-69	0.0	0.0	0.0	0.0	1.3	19.1	9.8	19.1	1.4	0.0	0.0	0.0	50.7
1969-70	0.0	0.0	0.0	1.4	0.8	12.8	1.8	3.6	12.8	0.1	0.0	0.0	33.3
1970-71	0.0	0.0	0.0	1.3	0.9	2.5	5.1	2.6	5.5	0.3	0.0	0.0	18.2
1971-72	0.0	0.0	0.0	0.0	3.9	4.5	4.6	2.5	4.2	1.3	0.0	0.0	21.0
1972-73	0.0	0.0	0.0	0.0	1.1	14.8	7.2	5.9	0.2	0.5	0.0	0.0	29.7
1973-74	0.0	0.0	0.0	T	3.1	10.9	8.6	1.8	3.3	1.6	0.0	0.0	29.3
1974-75	0.0	0.0	0.0	0.0	1.0	8.6	15.5	6.2	3.9	4.1	0.0	0.0	39.3
1975-76	0.0	0.0	0.0	T	15.8	0.7	3.0	4.7	9.6	0.0	0.0	0.0	33.8
1976-77	0.0	0.0	0.0	1.6	1.0	3.2	4.2	4.0	8.9	2.3	0.0	0.0	25.2
1977-78	0.0	0.0	0.0	T	7.6	8.9	3.4	15.1	8.4	T	0.0	0.0	43.4
1978-79	0.0	0.0	0.0	0.0	7.3	15.1	12.2	6.1	10.0	1.8	0.0	0.0	52.5
1979-80	0.0	0.0	0.0	0.1	4.2	2.5	4.4	4.7	3.7	2.6	0.0	0.0	22.2
1980-81	0.0	0.0	0.0	2.3	0.6	2.0	1.4	3.8	T	0.0	0.0	0.0	10.1
1981-82	0.0	0.0	0.0	0.5	4.3	8.7	16.4	3.8	4.8	8.6	0.0	0.0	47.1
1982-83	0.0	0.0	0.0	3.1	4.0	11.1	7.1	4.6	15.4	6.3	0.0	0.0	51.6
1983-84	0.0	0.0	T	0.0	22.6	7.8	3.5	22.8	16.2	13.2	0.0	0.0	86.1
1984-85	0.0	0.0	T	0.4	1.1	5.6	3.1	0.9	6.3	0.5	0.0	0.0	17.9
1985-86	0.0	0.0	1.1	T	7.7	6.2	0.6	4.2	5.3	6.5	0.0	0.0	31.6
1986-87	0.0	0.0	0.0	1.2	1.7	2.6	3.3	4.2	5.8	0.0	0.0	0.0	18.8
1987-88	0.0	0.0	0.0	T	3.9	3.4	7.2	4.4	0.7	3.0	0.0	0.0	22.6
1988-89	0.0	0.0	0.0	0.0	5.0	2.1	3.8	5.8	6.7	0.1	T	0.0	23.5
1989-90	T	0.0	0.0	0.0	1.2	6.0	6.5	4.1	0.3	1.1	0.0	T	19.2
1990-91	0.0	T	0.0	0.4	8.6	5.0	9.5	1.9	8.8	0.0	T	T	34.2
1991-92	0.0	0.0	0.0	6.5	9.2	T	1.0	1.7	3.0	3.1	T	0.0	24.5
1992-93	0.0	T	0.0	T	2.6	7.7	13.0	17.8	3.2	0.2	T	T	44.5
1993-94	T	0.0	0.0	T	5.0	4.6							
Record Mean	T	T	T	0.5	3.8	6.0	5.8	6.0	6.6	1.8	0.1	T	30.5

NORTH PLATTE, NEBRASKA

The climate of North Platte is characterized throughout the year by frequent rapid changes in the weather. During the winter, most North Pacific lows cross the country north of North Platte. The passage usually brings little or no snowfall, and only a moderate drop in temperature. Only when there is a major outbreak of cold air from Canada does the temperature fall to zero or below. The duration of below-zero temperature is hardly more than two mornings, and by the third or fourth day the temperature is ordinarily rising to the 40s or higher. Snowfall at the onset of a cold outbreak is usually less than 2 inches.

Only when a low moves from the middle Rockies through Nebraska, allowing easterly winds to draw moist air into the low circulation, does snowfall of appreciable amounts occur. Few of these storms move slowly enough, or are intense enough, to deposit much precipitation in the North Platte area. However, during some winters the cold outbreak and intense low from the mid-Rockies combine to produce severe cold and snow several inches in depth, with blizzard conditions following. During and after these snowfalls and blizzards, rail and highway traffic may be stalled until the snow is cleared. Widespread loss of unsheltered livestock and wild life results from such conditions.

The sudden and frequent weather changes of the winter continue through spring with decreasing intensity of temperature changes but increasing precipitation. The summer and fall months bring frequent changes from hot to cool weather. Most summer and fall precipitation is associated with thunderstorms, so the amounts are extremely variable. The surrounding area is occasionally damaged by locally severe winds and hailstorms.

Temperatures may reach into the upper 90s and lower 100s frequently during the summer months, but the elevation and clear skies bring rapid cooling after sunset to lows in the 60s or below by daybreak. Since the humidity is generally low, the extremely hot days of summer are not uncomfortable.

Based on the 1951-1980 period, the average first occurrence of 32 degrees Fahrenheit in the fall is September 24 and the average last occurrence in the spring is May 11.

NORMALS, MEANS, AND EXTREMES

NORTH PLATTE, NEBRASKA

LATITUDE: 41°08'N LONGITUDE: 100°41'W ELEVATION: FT. GRND 2775 BARO 2782 TIME ZONE: CENTRAL WBAN: 24023

	(a)	JAN	FEB	MAR	APR	MAY	JUNE	JULY	AUG	SEP	OCT	NOV	DEC	YEAR
TEMPERATURE °F:														
Normals														
-Daily Maximum		34.6	40.9	49.9	62.5	71.8	81.7	87.8	86.0	76.6	65.7	49.5	37.2	62.0
-Daily Minimum		8.6	14.3	23.0	33.9	44.6	54.1	60.1	57.6	46.2	33.5	21.3	11.2	34.0
-Monthly		21.6	27.6	36.5	48.2	58.2	67.9	74.0	71.8	61.4	49.7	35.4	24.2	48.0
Extremes														
-Record Highest	42	73	79	86	98	97	107	112	105	102	94	82	75	112
-Year		1990	1962	1986	1992	1953	1952	1954	1954	1990	1990	1980	1980	JUL 1954
-Record Lowest	42	-23	-22	-22	7	19	29	40	35	17	10	-13	-34	-34
-Year		1979	1981	1962	1975	1989	1969	1990	1976	1984	1993	1976	1989	DEC 1989
NORMAL DEGREE DAYS:														
Heating (base 65°F)		1345	1047	884	504	234	49	0	11	158	474	888	1265	6859
Cooling (base 65°F)		0	0	0	0	23	136	282	222	50	0	0	0	713
% OF POSSIBLE SUNSHINE	41	61	60	60	64	65	71	76	75	71	69	60	60	66
MEAN SKY COVER (tenths)														
Sunrise - Sunset	41	6.1	6.3	6.6	6.3	6.4	5.3	4.7	4.9	4.7	5.0	6.0	5.8	5.7
MEAN NUMBER OF DAYS:														
Sunrise to Sunset														
-Clear	41	8.5	7.1	7.3	7.1	6.7	10.2	12.1	11.6	13.0	12.6	8.9	9.6	114.8
-Partly Cloudy	41	8.6	7.5	7.7	9.1	10.2	10.5	12.0	11.4	8.3	8.4	8.0	7.9	109.6
-Cloudy	41	13.9	13.6	16.0	13.8	14.1	9.3	7.0	8.0	8.7	10.0	13.0	13.5	140.8
Precipitation														
.01 inches or more	41	4.9	5.5	7.0	8.0	11.0	9.5	9.8	7.8	6.5	5.0	4.7	4.3	84.0
Snow, Ice pellets, hail														
1.0 inches or more	41	1.7	1.7	2.0	0.8	0.1	0.0	0.0	0.0	0.*	0.5	1.1	1.4	9.2
Thunderstorms	41	0.*	0.1	0.8	2.6	6.6	10.0	10.3	8.2	4.1	1.2	0.2	0.0	44.2
Heavy Fog Visibility														
1/4 mile or less	41	1.3	2.3	2.1	0.9	0.9	0.8	1.0	1.7	1.9	2.0	2.3	1.5	18.8
Temperature °F														
-Maximum														
90° and above	29	0.0	0.0	0.0	0.3	0.8	5.7	12.8	10.8	3.8	0.3	0.0	0.0	34.4
32° and below	29	12.3	8.2	3.8	0.2	0.*	0.0	0.0	0.0	0.0	0.2	3.8	10.5	39.1
-Minimum														
32° and below	29	31.0	27.9	26.6	12.6	2.4	0.*	0.0	0.0	2.3	14.4	27.8	30.8	175.8
0° and below	29	7.9	3.7	0.6	0.0	0.0	0.0	0.0	0.0	0.0	0.0	0.8	4.6	17.6
AVG. STATION PRESS. (mb)	21	918.9	918.6	915.3	915.4	915.4	916.2	918.0	918.5	918.8	919.1	918.0	918.6	917.6
RELATIVE HUMIDITY (%)														
Hour 00	29	77	77	74	71	74	75	74	76	74	73	76	77	75
Hour 06 (Local Time)	29	79	79	80	80	83	84	83	85	83	81	81	79	81
Hour 12	29	63	59	53	47	51	52	50	51	48	46	55	61	53
Hour 18	29	64	56	49	43	48	48	47	47	46	47	58	63	51
PRECIPITATION (inches):														
Water Equivalent														
-Normal		0.36	0.43	1.20	1.99	3.43	3.37	3.06	1.74	1.61	0.98	0.66	0.47	19.30
-Maximum Monthly	42	1.12	1.98	2.98	5.01	8.01	6.81	7.05	6.30	6.03	2.91	2.89	1.22	8.01
-Year		1960	1978	1992	1984	1962	1965	1979	1992	1963	1969	1979	1977	MAY 1962
-Minimum Monthly	42	T	0.01	0.09	0.10	0.77	0.33	0.42	0.06	T	0.05	0.02	T	T
-Year		1964	1954	1967	1989	1966	1952	1955	1967	1953	1988	1989	1988	DEC 1988
-Maximum in 24 hrs	42	0.75	1.15	2.26	2.42	2.95	3.80	3.15	2.93	2.53	1.37	1.48	0.79	3.80
-Year		1992	1971	1959	1971	1962	1965	1964	1957	1963	1982	1979	1978	JUN 1965
Snow, Ice pellets, hail														
-Maximum Monthly	42	17.1	20.6	21.9	14.5	3.6	T	T	T	3.1	15.7	17.5	14.1	21.9
-Year		1976	1978	1980	1984	1967	1993	1993	1992	1985	1969	1979	1973	MAR 1980
-Maximum in 24 hrs	42	11.9	9.7	15.1	8.5	2.3	T	T	T	3.1	8.8	8.8	8.6	15.1
-Year		1976	1955	1980	1984	1967	1993	1993	1992	1985	1969	1979	1968	MAR 1980
WIND:														
Mean Speed (mph)	41	9.3	9.8	11.7	12.6	11.7	10.4	9.6	9.4	9.7	9.5	9.6	9.2	10.2
Prevailing Direction through 1963		NW	NW	N	N	SE	SE	SE	SSE	SSE	SSE	NW	NW	NW
Fastest Obs. 1 Min.														
-Direction (!!!)	14	34	35	02	32	07	32	31	18	18	33	03	02	02
-Speed (MPH)	14	43	39	47	45	46	52	46	41	38	37	41	52	52
-Year		1992	1981	1992	1986	1992	1981	1981	1985	1986	1990	1983	1982	DEC 1982
Peak Gust														
-Direction (!!!)	10	NW	NW	NE	S	NE	NW	W	NE	S	W	N	NW	S
-Speed (mph)	10	60	55	64	76	72	62	64	74	58	55	52	56	76
-Date		1987	1988	1992	1985	1985	1986	1993	1990	1986	1985	1989	1988	APR 1985

PRECIPITATION (inches)

NORTH PLATTE, NEBRASKA

YEAR	JAN	FEB	MAR	APR	MAY	JUNE	JULY	AUG	SEP	OCT	NOV	DEC	ANNUAL
1964	T	0.62	1.26	3.76	2.06	4.78	4.52	2.63	1.02	0.17	0.04	0.10	20.96
1965	0.61	0.35	0.25	1.42	4.18	6.81	6.68	1.82	5.69	0.73	0.05	1.02	29.61
1966	0.30	0.15	1.12	1.05	0.77	5.14	3.35	2.74	2.09	0.31	0.04	0.51	17.57
1967	0.47	0.03	0.09	1.03	3.94	6.05	3.74	0.06	1.12	0.58	0.19	0.15	17.45
1968	0.12	0.36	0.11	3.04	1.72	2.25	1.82	3.97	0.82	1.50	0.46	0.87	17.04
1969	0.94	0.26	0.17	0.15	1.35	4.20	2.52	1.61	0.80	2.91	0.18	0.10	15.19
1970	0.23	0.28	0.97	2.46	1.31	4.33	1.68	0.28	2.27	1.27	0.91	0.22	16.21
1971	0.47	1.28	0.97	3.94	2.54	5.84	3.48	0.91	2.01	1.75	0.90	0.16	24.25
1972	0.16	0.08	0.65	1.19	3.18	2.96	3.58	0.95	1.46	0.62	1.12	0.42	16.37
1973	0.40	0.10	2.45	1.45	3.85	0.88	2.87	2.82	3.98	1.26	0.54	1.13	21.73
1974	0.27	0.08	0.42	1.17	1.64	3.86	2.27	1.15	0.24	0.61	0.04	0.42	12.17
1975	0.26	0.17	0.92	1.77	2.12	6.12	2.51	0.25	0.56	0.14	1.15	0.22	16.19
1976	0.99	0.14	2.03	2.80	2.88	2.43	0.92	2.87	2.03	1.18	0.08	0.01	18.36
1977	0.18	0.27	2.89	4.85	5.90	1.31	4.41	1.79	1.48	0.18	0.39	1.22	24.87
1978	0.52	1.98	0.40	1.96	4.84	1.75	3.70	1.92	0.33	0.43	1.03	0.99	19.85
1979	0.86	0.09	2.78	1.46	2.96	3.37	7.05	1.49	0.45	1.32	2.89	0.28	25.00
1980	0.52	0.82	2.56	0.77	2.59	1.89	0.64	3.04	0.34	1.00	0.13	0.02	14.32
1981	0.07	0.05	2.72	2.47	5.37	2.32	5.09	2.48	0.25	0.60	1.94	0.43	23.79
1982	0.20	0.15	0.99	1.42	6.32	2.35	1.78	1.18	1.26	2.44	0.73	1.08	19.90
1983	0.33	0.25	1.54	2.12	3.20	3.32	3.74	1.98	0.14	0.56	1.56	0.46	19.20
1984	0.36	0.87	1.20	5.01	2.82	4.37	0.94	1.38	0.39	2.41	0.69	0.72	21.16
1985	0.55	0.14	0.44	1.84	4.01	0.87	3.98	1.16	3.23	1.24	1.09	0.79	19.34
1986	0.02	1.10	0.70	3.77	2.80	1.70	2.57	1.22	1.02	1.58	0.19	0.27	16.94
1987	0.16	1.55	1.65	1.01	3.19	3.95	2.81	1.19	1.16	1.67	1.26	0.81	20.41
1988	0.72	0.03	0.37	2.02	3.59	3.12	3.03	3.93	1.59	0.05	0.40	T	18.85
1989	0.55	0.73	0.38	0.10	3.02	3.51	1.86	2.37	1.11	0.08	0.02	0.28	14.01
1990	0.27	0.18	1.75	1.52	3.65	1.90	1.99	1.79	0.31	1.48	0.87	0.09	15.80
1991	0.35	0.21	1.00	3.00	5.39	2.78	1.81	0.53	1.75	2.14	0.92	0.67	20.55
1992	0.89	1.42	2.98	0.18	3.18	2.61	3.75	6.30	0.25	0.92	0.20	0.33	23.01
1993	0.74	1.37	0.61	1.80	2.47	6.12	5.47	3.78	0.71	1.59	1.32	0.22	26.20
Record Mean	0.41	0.50	0.98	2.05	3.03	3.24	2.75	2.20	1.52	1.05	0.55	0.49	18.77

AVERAGE TEMPERATURE (deg. F)

NORTH PLATTE, NEBRASKA

YEAR	JAN	FEB	MAR	APR	MAY	JUNE	JULY	AUG	SEP	OCT	NOV	DEC	ANNUAL
#1964	28.1	24.9	32.3	47.0	60.9	67.2	77.5	69.8	63.1	50.3	35.8	25.2	48.5
1965	26.3	23.4	25.5	51.4	60.1	67.1	73.1	70.5	53.8	54.0	40.9	28.9	47.9
1966	15.6	21.4	38.5	43.8	60.0	68.8	77.6	67.8	62.1	50.5	36.0	24.5	47.2
1967	24.9	30.8	39.9	49.3	51.9	64.6	70.3	68.8	61.2	49.8	33.9	22.8	47.3
1968	22.2	27.1	39.9	46.2	53.3	69.5	72.6	71.6	61.1	51.1	34.4	19.0	47.4
1969	16.2	24.2	29.0	51.4	61.0	62.8	73.0	73.6	65.8	41.7	38.3	28.4	47.1
1970	22.3	31.8	31.3	45.6	61.9	68.1	74.3	75.6	60.8	45.4	34.8	26.8	48.2
1971	22.1	24.5	33.4	48.0	54.7	70.2	69.8	72.2	59.7	49.5	37.1	27.6	47.4
1972	20.7	29.9	40.7	46.6	58.1	68.4	70.4	70.8	61.2	46.4	29.8	17.2	46.7
1973	21.5	30.3	40.4	45.8	56.0	67.7	72.8	73.7	58.4	51.7	35.0	23.2	48.0
1974	16.0	33.0	40.1	50.0	59.5	66.5	77.5	67.1	57.5	52.9	34.8	24.2	48.3
1975	25.7	23.4	31.5	46.6	57.1	65.7	74.6	73.2	58.5	50.9	30.2	26.4	47.0
1976	17.8	32.7	34.0	47.4	53.1	64.6	72.8	70.6	59.8	45.2	30.5	28.2	46.4
1977	16.6	32.0	37.0	51.7	63.1	70.5	74.5	68.5	63.8	50.0	35.8	24.9	49.0
1978	11.2	14.7	35.1	48.4	56.6	68.2	74.5	71.0	65.4	49.2	32.1	14.6	45.1
1979	6.0	17.2	37.0	48.4	57.4	68.1	74.1	72.9	67.1	53.8	32.2	34.0	47.3
1980	25.1	26.8	35.5	50.5	60.2	72.5	78.0	74.0	63.7	48.9	38.5	32.0	50.5
1981	29.7	29.1	40.2	55.8	54.8	68.6	73.8	70.1	63.0	49.0	39.9	26.7	50.1
1982	17.1	28.1	36.3	44.8	57.3	63.4	75.0	73.1	61.7	49.2	33.3	28.1	47.3
1983	27.1	35.1	37.2	42.3	54.0	65.2	75.3	78.0	64.8	51.3	36.7	7.5	47.9
1984	20.4	33.2	34.1	43.3	58.1	68.0	73.7	75.5	58.4	48.0	37.2	22.5	47.7
1985	18.3	23.1	40.5	52.2	60.8	65.4	75.3	70.1	59.8	48.6	24.6	19.2	46.5
1986	31.6	27.0	44.3	48.9	58.0	71.4	75.6	71.3	63.0	50.4	35.5	30.2	50.6
1987	29.5	35.8	36.7	51.3	63.1	70.2	75.7	70.3	61.3	47.0	38.2	27.7	50.6
1988	16.4	26.5	37.8	48.5	60.7	74.6	74.5	73.4	62.0	48.7	38.0	30.3	49.3
1989	30.2	17.8	35.7	51.1	59.0	65.8	74.3	71.7	61.0	50.7	37.6	20.6	48.0
1990	30.6	30.9	40.0	48.5	56.7	71.1	73.7	74.0	67.6	50.2	38.1	21.9	50.3
1991	23.1	37.5	40.5	49.3	61.8	71.0	74.9	73.7	63.6	48.4	32.2	33.4	50.8
1992	32.1	37.3	41.8	50.6	58.7	65.7	68.5	66.5	63.0	49.9	32.4	22.5	49.1
1993	18.5	18.0	37.0	45.5	58.7	65.3	71.5	70.3	58.3	48.5	32.7	30.4	46.2
Record Mean	23.3	27.8	36.5	48.7	58.7	68.6	74.7	72.8	63.3	51.0	36.5	27.0	49.1
Max	35.7	40.3	49.4	61.8	71.2	81.1	87.6	86.1	77.6	66.0	50.1	39.3	62.2
Min	10.9	15.2	23.6	35.5	46.2	56.0	61.7	59.6	49.0	36.0	23.0	14.6	35.9

REFERENCE NOTES

NORTH PLATTE, NEBRASKA

GENERAL
T - TRACE AMOUNT.
BLANK ENTRIES DENOTE MISSING/UNREPORTED DATA.
# INDICATES A STATION OR INSTRUMENT RELOCATION.

SPECIFIC

PM - INCLUDES LAST DAY OF PREVIOUS MONTH
ASOS - AUTOMATED SURFACE OBSERVING SYSTEM IN OPERATION DURING THESE MONTHS.

(a) - LENGTH OF RECORD IN YEARS, ALTHOUGH INDIVIDUAL MONTHS MAY BE MISSING.
0.* OR * - THE VALUE IS BETWEEN 0.0 AND 0.05
NORMALS - BASED ON THE 1961-1990 RECORD PERIOD.
EXTREMES - DATES ARE THE MOST RECENT OCCURRENCE
WIND DIR.- NUMERALS SHOW TENS OF DEGRESS CLOCKWISE FROM TRUE NORTH. "00" INDICATES CALM.
RESULTANT DIRECTIONS ARE GIVEN TO WHOLE DEGREES.
BOLD VALUES INDICATE EXTREME VALUES WHICH OCCURRED AFTER THE ASOS SYSTEM WAS COMMISSIONED.

RECORD = PERIOD OF RECORD
RECORD MEAN PRECIPITATION IS THE MEAN OF ALL DAILY PRECIPITATION AMOUNTS DURING THE PERIOD OF RECORD.
RECORD MAX(MIN) TEMPERATURE IS THE MEAN OF ALL DAILY MAX(MIN) TEMPERATURES DURING THE PERIOD OF RECORD.
RECORD MEAN TEMPERATURE IS THE SUM OF THE RECORD MAX AND RECORD MIN DIVIDED BY 2.
AVERAGE TEMPERATURE IS THE SUM OF THE MEAN DAILY MAX AND MIN TEMPERATURE DIVIDED BY 2.

EXCEPTIONS

RECORD MEANS ARE THROUGH THE CURRENT YEAR,
BEGINNING IN 1875 FOR TEMPERATURE
1875 FOR PRECIPITATION
1953 FOR SNOWFALL

HEATING DEGREE DAYS Base 65 deg. F NORTH PLATTE, NEBRASKA

SEASON	JULY	AUG	SEP	OCT	NOV	DEC	JAN	FEB	MAR	APR	MAY	JUNE	TOTAL
#1964-65	0	43	143	450	870	1229	1189	1158	1217	405	173	26	6903
1965-66	0	11	345	335	718	1113	1530	1213	815	628	199	48	6955
1966-67	0	34	136	442	863	1250	1240	950	771	465	427	70	6648
1967-68	19	29	132	479	926	1305	1321	1096	770	558	363	26	7024
1968-69	17	18	150	428	911	1421	1509	1133	1106	399	178	122	7392
1969-70	1	0	45	716	795	1129	1315	923	1041	582	136	55	6738
1970-71	2	1	214	600	899	1177	1320	1127	974	501	311	13	7139
1971-72	12	0	210	472	829	1157	1368	1013	745	549	250	32	6637
1972-73	24	15	169	567	1047	1479	1343	967	756	567	283	34	7251
1973-74	10	0	219	407	892	1290	1518	889	765	445	190	62	6687
1974-75	0	40	256	373	900	1262	1216	1160	1032	558	247	62	7106
1975-76	6	0	228	437	1035	1191	1460	929	956	521	363	70	7196
1976-77	0	13	178	608	1028	1133	1493	920	858	395	81	2	6709
1977-78	2	34	96	458	869	1236	1662	1400	924	491	275	71	7518
1978-79	5	24	99	488	982	1560	1828	1335	862	491	259	56	7989
1979-80	4	11	52	341	975	957	1233	1102	909	439	166	10	6199
1980-81	0	6	107	491	790	1019	1089	1000	762	283	318	26	5891
1981-82	9	4	101	492	749	1179	1479	1030	885	601	239	111	6879
1982-83	0	18	160	484	946	1138	1167	833	854	672	343	90	6705
1983-84	2	0	128	419	840	1780	1379	915	953	647	236	33	7332
1984-85	0	0	247	519	829	1312	1440	1168	752	393	156	83	6899
1985-86	0	23	252	502	1205	1416	1029	1060	634	479	219	2	6821
1986-87	0	14	98	446	878	1074	1093	810	868	420	102	15	5818
1987-88	13	36	139	551	796	1152	1501	1109	839	490	170	3	6799
1988-89	0	13	128	498	803	1067	1072	1316	902	430	211	67	6507
1989-90	2	7	180	437	815	1374	1061	948	771	502	259	15	6371
1990-91	15	1	84	457	797	1331	1290	762	754	466	149	5	6111
1991-92	3	1	148	508	977	971	1010	797	714	436	219	45	5829
1992-93	18	60	113	466	970	1310	1436	1311	862	578	204	73	7401
1993-94	2	24	218	513	965	1066							

COOLING DEGREE DAYS Base 65 deg. F NORTH PLATTE, NEBRASKA

YEAR	JAN	FEB	MAR	APR	MAY	JUNE	JULY	AUG	SEP	OCT	NOV	DEC	TOTAL
1969	0	0	0	0	58	63	257	276	76	0	0	0	730
1970	0	0	0	4	48	155	300	341	96	0	0	0	944
1971	0	0	0	0	1	177	167	229	59	1	0	0	634
1972	0	0	0	0	43	137	199	202	64	0	0	0	645
1973	0	0	0	0	11	121	260	278	29	3	0	0	702
1974	0	0	0	5	24	117	394	115	40	3	0	0	698
1975	0	0	0	10	11	89	311	260	39	7	0	0	727
1976	0	0	0	0	5	62	247	192	27	0	0	0	533
1977	0	0	0	2	26	174	305	149	64	0	0	0	720
1978	0	0	0	1	21	174	307	218	117	5	0	0	843
1979	0	0	0	0	27	156	294	262	123	0	0	0	862
1980	0	0	0	10	27	243	411	289	74	0	0	0	1054
1981	0	0	0	10	9	141	288	168	47	1	0	0	664
1982	0	0	0	0	8	70	314	276	68	0	0	0	736
1983	0	0	0	0	8	103	331	412	128	1	0	0	983
1984	0	0	0	0	27	129	281	331	55	2	0	0	825
1985	0	0	0	14	32	100	326	189	100	0	0	0	761
1986	0	0	0	3	11	201	334	217	43	0	0	0	809
1987	0	0	0	16	48	176	352	208	35	0	0	0	835
1988	0	0	0	1	41	293	301	282	46	0	0	0	964
1989	0	0	0	21	35	99	295	220	67	2	0	0	739
1990	0	0	0	15	10	205	291	289	165	5	0	0	980
1991	0	0	0	3	59	194	317	276	112	2	0	0	963
1992	0	0	0	13	31	68	135	116	60	1	0	0	424
1993	0	0	0	0	15	88	212	194	24	6	0	0	539

SNOWFALL (inches) NORTH PLATTE, NEBRASKA

SEASON	JULY	AUG	SEP	OCT	NOV	DEC	JAN	FEB	MAR	APR	MAY	JUNE	TOTAL
1964-65	0.0	0.0	0.0	0.0	T	1.8	9.2	7.2	5.0	T	0.0	0.0	23.2
1965-66	0.0	0.0	T	0.0	0.5	6.3	10.5	1.7	9.6	5.0	0.2	0.0	33.8
1966-67	0.0	0.0	0.0	1.4	0.5	8.5	9.3	0.8	0.7	4.5	3.6	0.0	29.3
1967-68	0.0	0.0	0.0	T	2.9	3.4	1.6	3.0	0.8	2.2	0.0	0.0	13.9
1968-69	0.0	0.0	0.0	0.0	0.3	10.3	12.4	3.2	1.8	T	T	0.0	28.0
1969-70	0.0	0.0	0.0	15.7	0.8	1.8	3.6	4.0	18.4	0.4	0.0	0.0	44.7
1970-71	0.0	0.0	0.0	9.0	3.6	1.8	5.6	2.7	19.0	0.8	T	0.0	42.5
1971-72	0.0	0.0	0.0	2.0	2.5	2.1	2.6	1.9	0.1	0.1	0.0	0.0	11.3
1972-73	0.0	0.0	0.0	0.8	8.0	8.7	3.8	0.4	1.6	1.7	0.0	0.0	25.0
1973-74	0.0	0.0	0.0	T	4.3	14.1	4.2	1.0	2.3	0.7	0.0	0.0	26.6
1974-75	0.0	0.0	0.0	0.0	T	4.5	2.2	2.6	3.9	3.0	0.0	0.0	16.2
1975-76	0.0	0.0	0.0	0.8	10.9	1.2	17.1	1.6	5.2	T	0.0	0.0	36.8
1976-77	0.0	0.0	0.0	1.0	0.6	T	2.5	2.2	9.9	8.1	0.0	0.0	24.3
1977-78	0.0	0.0	0.0	T	0.3	6.4	6.2	20.6	2.0	T	T	0.0	35.5
1978-79	0.0	0.0	0.0	T	7.6	10.3	6.3	0.3	5.9	2.3	T	0.0	32.7
1979-80	0.0	0.0	0.0	2.9	17.5	2.3	5.2	9.1	21.9	7.4	0.0	0.0	66.3
1980-81	0.0	0.0	0.0	0.6	1.2	T	0.7	0.5	0.5	0.4	0.0	0.0	3.9
1981-82	0.0	0.0	0.0	T	5.3	6.4	4.4	1.7	6.4	0.9	0.0	0.0	25.1
1982-83	0.0	0.0	0.0	1.0	2.0	9.7	1.6	0.1	5.9	5.4	0.0	0.0	25.7
1983-84	0.0	0.0	T	0.0	12.1	7.3	5.3	9.6	8.8	14.5	0.2	0.0	57.8
1984-85	0.0	0.0	T	0.3	0.9	8.7	8.5	0.8	2.1	0.0	0.0	0.0	21.3
1985-86	0.0	0.0	3.1	T	13.0	8.1	0.2	8.7	3.8	T	0.0	0.0	36.9
1986-87	0.0	0.0	0.0	2.0	1.0	2.6	1.6	9.2	7.2	0.1	0.0	0.0	23.7
1987-88	0.0	0.0	0.0	1.3	8.2	7.2	12.6	0.6	3.8	2.1	0.0	0.0	35.8
1988-89	0.0	0.0	0.0	T	2.5	T	6.1	10.6	4.3	0.3	T	0.0	23.8
1989-90	T	0.0	T	T	T	2.2	5.9	1.9	5.3	0.3	T	T	15.6
1990-91	0.0	0.0	0.0	2.0	9.7	0.8	4.1	1.1	5.6	T	T	0.0	23.3
1991-92	0.0	0.0	0.0	7.3	1.6	1.8	5.4	1.2	5.1	0.3	T	0.0	22.7
1992-93	T	T	0.0	T	2.2	3.9	10.1	17.2	2.7	2.0	T	T	38.1
1993-94	T	0.0	T	T	3.9	3.3							
Record Mean	T	T	0.1	1.4	4.0	4.7	5.4	5.2	6.8	2.7	0.1	T	30.3

OMAHA (EPPLEY AIRFIELD), NEBRASKA

Omaha, Nebraska, is situated on the west bank of the Missouri River. The river level at Omaha is normally about 965 feet above sea level and the rolling hills in and around Omaha rise to about 1,300 feet above sea level. The climate is typically continental with relatively warm summers and cold, dry winters. It is situated midway between two distinctive climatic zones, the humid east and the dry west. Fluctuations between these two zones produce weather conditions for periods that are characteristic of either zone, or combinations of both. Omaha is also affected by most low pressure systems that cross the country. This causes periodic and rapid changes in weather, especially during the winter months.

Most of the precipitation in Omaha falls during sharp showers or thunderstorms, and these occur mostly during the growing season from April to September. Of the total precipitation, about 75 percent falls during this six-month period. The rain occurs mostly as evening or nighttime showers and thunderstorms. Although winters are relatively cold, precipitation is light, with only 10 percent of the total annual precipitation falling during the winter months.

Sunshine is fairly abundant, ranging around 50 percent of the possible in the winter to 75 percent of the possible in the summer.

NORMALS, MEANS, AND EXTREMES

OMAHA (EPPLEY AIRFIELD), NEBRASKA

LATITUDE: 41°18'N LONGITUDE: 95°54' W ELEVATION: FT. GRND 997 BARO 985 TIME ZONE: CENTRAL WBAN: 14942

	(a)	JAN	FEB	MAR	APR	MAY	JUNE	JULY	AUG	SEP	OCT	NOV	DEC	YEAR
TEMPERATURE °F:														
Normals														
-Daily Maximum		31.3	37.1	49.4	63.8	74.0	83.7	87.9	85.2	76.5	65.6	49.3	34.6	61.5
-Daily Minimum		10.9	16.7	27.7	39.9	50.9	60.4	65.9	62.9	53.6	41.2	28.7	15.6	39.5
-Monthly		21.1	26.9	38.6	51.9	62.4	72.1	76.9	74.1	65.1	53.4	39.0	25.1	50.6
Extremes														
-Record Highest	57	69	78	89	97	99	105	114	110	104	96	80	72	114
-Year		1944	1972	1986	1989	1939	1953	1936	1936	1939	1938	1980	1939	JUL 1936
-Record Lowest	57	-23	-21	-16	5	27	38	44	43	25	13	-9	-23	-23
-Year		1982	1981	1948	1975	1980	1983	1972	1967	1984	1972	1964	1989	DEC 1989
NORMAL DEGREE DAYS:														
Heating (base 65°F)		1361	1067	818	398	164	14	0	9	80	372	780	1237	6300
Cooling (base 65°F)		0	0	0	5	84	227	369	292	83	12	0	0	1072
% OF POSSIBLE SUNSHINE														
MEAN SKY COVER (tenths)														
Sunrise - Sunset	48	6.1	6.3	6.7	6.4	6.3	5.7	4.9	4.9	4.9	4.9	6.1	6.4	5.8
MEAN NUMBER OF DAYS:														
Sunrise to Sunset														
-Clear	48	8.8	7.4	6.9	7.3	7.3	8.1	11.3	12.3	12.3	12.9	8.6	8.0	111.2
-Partly Cloudy	48	7.9	7.4	8.1	8.5	9.5	11.3	12.0	10.1	7.6	8.1	7.4	7.5	105.4
-Cloudy	48	14.3	13.5	16.0	14.2	14.1	10.6	7.7	8.6	10.2	10.1	13.9	15.4	148.6
Precipitation														
.01 inches or more	57	6.4	6.6	8.8	9.6	11.6	10.6	9.4	9.1	8.5	6.4	5.8	6.5	99.1
Snow, Ice pellets, hail														
1.0 inches or more	58	2.3	1.9	2.0	0.4	0.*	0.0	0.0	0.0	0.0	0.1	0.9	1.8	9.4
Thunderstorms	58	0.1	0.4	1.6	3.8	7.4	9.4	8.5	7.8	5.3	2.3	0.8	0.2	47.7
Heavy Fog Visibility														
1/4 mile or less	58	1.8	2.0	1.4	0.6	0.8	0.4	0.6	1.5	1.4	1.5	1.6	2.1	15.7
Temperature °F														
-Maximum														
90° and above	29	0.0	0.0	0.0	0.5	1.4	7.0	12.8	8.8	2.9	0.2	0.0	0.0	33.7
32° and below	29	14.3	10.1	3.5	0.1	0.0	0.0	0.0	0.0	0.0	0.*	2.7	11.4	42.1
-Minimum														
32° and below	29	30.0	26.1	21.2	6.6	0.4	0.0	0.0	0.0	0.4	6.2	20.7	29.3	141.0
0° and below	29	6.9	3.5	0.3	0.0	0.0	0.0	0.0	0.0	0.0	0.0	0.3	3.3	14.4
AVG. STATION PRESS. (mb)	11	982.7	983.6	979.1	978.4	977.9	977.6	979.3	980.5	981.2	982.2	981.4	983.2	980.6
RELATIVE HUMIDITY (%)														
Hour 00	29	76	77	73	69	73	76	79	81	82	76	77	78	76
Hour 06 (Local Time)	29	78	79	79	77	80	82	85	87	87	82	81	80	81
Hour 12	29	65	63	57	52	54	55	58	59	59	55	62	67	59
Hour 18	29	67	64	54	48	51	52	56	58	59	56	65	71	58
PRECIPITATION (inches):														
Water Equivalent														
-Normal		0.74	0.77	2.04	2.66	4.52	3.87	3.51	3.24	3.72	2.28	1.49	1.02	29.86
-Maximum Monthly	57	3.70	2.97	5.96	6.45	10.33	10.81	10.34	10.16	13.75	4.99	4.70	5.42	13.75
-Year		1949	1965	1973	1951	1959	1947	1993	1987	1965	1961	1983	1984	SEP 1965
-Minimum Monthly	57	T	0.09	0.12	0.23	0.56	1.03	0.39	0.61	0.41	T	0.03	T	T
-Year		1986	1981	1956	1936	1948	1972	1983	1984	1953	1952	1976	1943	JAN 1986
-Maximum in 24 hrs	51	1.52	2.24	1.45	2.56	4.16	3.48	3.37	5.27	6.47	3.13	2.53	3.03	6.47
-Year		1967	1954	1990	1938	1987	1942	1958	1987	1965	1968	1948	1984	SEP 1965
Snow, Ice pellets, hail														
-Maximum Monthly	58	25.7	25.4	27.2	10.0	2.0	T	0.0	0.0	T	7.2	12.0	19.9	27.2
-Year		1936	1965	1948	1992	1945	1993			1985	1941	1957	1969	MAR 1948
-Maximum in 24 hrs	51	13.1	18.3	13.0	9.9	2.0	T	0.0	0.0	T	7.2	8.7	10.2	18.3
-Year		1949	1965	1948	1992	1945	1993			1985	1941	1957	1969	FEB 1965
WIND:														
Mean Speed (mph)	57	10.9	11.0	12.3	12.7	10.9	10.1	8.9	8.9	9.4	9.8	10.9	10.7	10.5
Prevailing Direction through 1963		NNW	NNW	NNW	NNW	SSE	SSE	SSE	SSE	SSE	SSE	SSE	SSE	SSE
Fastest Obs. 1 Min.														
-Direction (!!)		33	35	35	28	27	24	28	31	19	30	33	34	28
-Speed (MPH)	7	44	37	38	40	52	40	58	40	36	36	35	37	58
-Year		1990	1988	1987	1989	1991	1991	1986	1988	1989	1990	1991	1992	JUL 1986
Peak Gust														
-Direction (!!)														
-Speed (mph)														
-Date														

PRECIPITATION (inches)

OMAHA (EPPLEY AIRFIELD), NEBRASKA

YEAR	JAN	FEB	MAR	APR	MAY	JUNE	JULY	AUG	SEP	OCT	NOV	DEC	ANNUAL
1964	0.54	0.36	1.58	4.40	7.46	6.39	4.61	4.13	2.60	0.50	2.34	0.84	35.75
1965	0.60	2.97	2.63	3.71	6.19	5.15	4.39	2.06	13.75	0.76	1.24	1.40	44.85
1966	0.81	0.41	0.88	0.83	3.67	5.93	4.70	1.85	2.16	0.64	0.22	0.62	22.72
1967	2.00	0.16	0.82	2.61	2.26	9.86	4.33	1.53	4.15	1.69	0.28	0.79	30.48
1968	0.49	0.10	0.57	3.93	4.35	4.44	2.61	3.30	5.74	4.62	1.50	2.04	33.69
1969	1.10	1.33	1.25	3.95	4.48	3.27	4.27	5.06	1.33	2.24	0.11	1.80	30.19
1970	0.20	0.14	0.85	2.84	2.64	2.48	1.98	4.66	4.93	4.88	1.23	0.42	27.25
1971	0.95	2.43	0.49	0.90	7.21	3.30	1.77	1.60	0.57	4.20	3.12	1.05	27.59
1972	0.38	0.36	1.14	4.63	5.13	1.03	7.28	2.60	4.93	3.21	3.25	1.62	35.56
1973	1.44	0.87	5.96	3.60	4.94	1.56	4.98	1.27	8.04	2.60	1.43	1.65	38.34
1974	0.63	0.16	0.80	1.72	2.65	1.79	0.79	4.17	2.54	2.74	1.41	0.81	20.21
1975	2.01	1.06	1.78	3.37	3.72	4.30	0.46	1.80	2.16	0.01	2.85	0.46	23.98
1976	0.11	1.39	2.13	2.98	3.55	2.84	1.52	0.62	1.64	1.35	0.03	0.21	18.37
1977	0.93	0.38	3.72	3.28	6.05	2.40	4.69	8.63	5.05	4.38	2.90	0.35	42.76
1978	0.15	0.76	1.04	4.61	5.05	2.19	5.89	3.87	6.01	0.96	1.09	0.50	32.12
1979	1.11	0.30	4.59	2.58	2.84			2.07	3.36	3.12	1.23	0.14	
1980	0.93	0.52	1.40	1.72	2.50	8.99	3.63	6.98	0.82	2.70	0.11	0.04	30.34
1981	0.20	0.09	0.88	1.33	4.13	2.14	1.87	4.80	1.51	1.92	2.60	0.86	22.33
1982	1.83	0.26	1.90	1.22	9.92	4.16	2.46	3.21	2.27	1.10	1.81	1.17	31.31
1983	0.86	0.68	3.65	1.00	2.81	6.52	0.39	1.24	2.45	2.16	4.70	0.63	27.09
1984	0.38	0.62	2.32	4.77	4.92	5.56	1.58	0.61	2.55	3.87	0.52	5.42	33.12
1985	0.56	1.88	1.36	3.16	2.46	1.73	3.27	1.50	2.71	1.36	0.85	0.37	21.21
1986	T	1.00	2.51	4.96	4.88	2.37	2.77	3.86	8.11	4.86	0.99	0.89	37.20
1987	0.08	0.55	4.14	2.24	8.64	3.29	6.72	10.16	1.56	1.33	1.60	1.01	41.32
1988	0.42	0.18	0.14	1.57	4.68	1.60	2.68	1.78	2.63	0.14	2.55	0.95	19.32
1989	1.10	0.86	0.40	1.80	0.83	5.05	3.06	1.80	6.46	1.55	0.15	0.74	23.80
1990	0.59	0.34	4.01	0.36	5.08	3.88	6.36	0.81	0.81	1.71	1.15	1.18	26.28
1991	1.08	0.26	2.85	4.46	4.07	7.79	2.96	3.67	1.37	3.76	3.51	1.75	37.53
1992	1.41	1.18	3.08	3.19	2.27	1.44	7.31	1.57	6.86	2.22	3.01	1.15	34.69
1993	1.42	0.93	2.67	2.26	4.90	8.03	10.34	7.53	2.29	1.18	0.66	0.51	42.72
Record Mean	0.74	0.87	1.54	2.64	3.89	4.48	3.71	3.43	3.18	2.10	1.29	0.91	28.78

AVERAGE TEMPERATURE (deg. F)

OMAHA (EPPLEY AIRFIELD), NEBRASKA

YEAR	JAN	FEB	MAR	APR	MAY	JUNE	JULY	AUG	SEP	OCT	NOV	DEC	ANNUAL
1964	28.6	30.5	33.7	52.8	68.0	71.9	80.1	70.5	65.3	53.7	42.2	25.2	51.9
1965	23.0	22.2	27.1	53.2	68.1	71.8	75.7	74.1	61.0	57.2	42.4	36.9	51.1
1966	18.4	28.8	44.1	48.2	61.0	72.6	79.4	71.0	63.7	53.9	40.1	28.3	50.8
1967	25.5	28.4	45.5	55.6	59.1	70.8	74.4	71.7	62.5	52.1	38.4	28.4	51.1
1968	22.6	27.3	44.6	52.5	58.0	74.0	76.4	75.0	65.4	55.8	37.4	24.0	51.1
1969	18.7	26.7	31.5	53.1	63.2	67.9	78.2	75.6	67.7	50.1	41.0	24.8	49.9
1970	16.3	32.0	35.9	53.0	68.1	74.3	77.3	76.3	66.2	51.7	38.2	29.8	51.6
1971	17.8	25.3	37.2	54.4	60.4	77.8	74.2	74.7	67.8	58.8	41.2	28.9	51.5
1972	20.0	25.4	40.8	51.2	62.4	72.6	74.6	73.6	65.4	49.1	37.1	21.1	49.5
1973	22.6	28.6	44.3	50.4	59.4	73.3	75.5	77.6	64.2	57.3	39.8	22.7	51.3
1974	18.8	30.6	42.1	52.6	61.9	69.8	82.2	70.5	59.7	55.3	40.7	28.9	51.1
1975	22.5	22.4	31.7	49.4	66.2	72.5	78.9	79.7	62.6	58.2	41.6	30.9	51.4
1976	25.5	37.5	40.4	56.8	60.2	72.7	79.0	76.3	67.4	48.5	33.3	24.1	51.8
1977	13.3	33.2	41.6	59.0	70.2	75.1	80.6	72.5	67.4	53.1	39.7	26.2	53.0
1978	12.0	15.7	35.7	52.6	61.7	73.7	77.3	75.5	71.1	53.7	39.6	24.4	49.4
1979	10.7	17.3	39.8	50.0	61.8	72.7	76.0	75.7	66.7		35.8	30.3	
1980	22.4	20.3	32.9	50.7	61.6	73.1	79.6	76.6	64.7	49.0	40.3	26.1	49.8
1981	24.1	28.4	40.8	57.4	58.6	72.6	76.6	71.0	64.8	50.2	40.6	22.9	50.6
1982	9.4	22.6	34.8	47.7	62.9	65.5	77.1	72.6	64.7	55.0	37.1	28.3	48.1
1983	24.9	30.1	37.3	43.5	56.5	69.6	79.4	81.5	67.0	52.2	38.5	7.3	48.9
1984	19.6	33.2	30.3	46.6	57.7	71.8	75.1	76.5	61.9	52.2	39.4	27.1	49.3
1985	19.1	23.6	43.4	54.9	63.4	67.2	74.1	69.4	62.1	52.7	28.5	16.4	47.9
1986	29.4	22.8	42.8	51.9	61.3	73.9	77.9	70.1	67.9	53.7	34.4	29.4	51.3
1987	28.6	36.8	42.8	55.3	67.3	74.2	77.8	70.9	64.8	48.3	43.4	30.9	53.4
1988	21.1	23.9	40.7	50.5	67.3	76.3	76.4	77.3	66.2	49.1	40.0	29.5	51.5
1989	32.4	16.0	37.9	54.6	62.4	69.4	77.4	74.4	63.0	53.9	36.2	17.7	49.6
1990	33.5	31.3	42.7	50.8	58.5	73.5	74.8	75.3	69.0	53.3	42.9	21.2	52.2
1991	16.3	34.2	42.5	54.2	67.3	74.6	75.8	74.3	66.0	52.1	30.6	31.8	51.6
1992	32.6	36.7	43.6	50.0	61.8	69.4	71.1	68.4	64.0	53.5	35.4	27.9	51.2
1993	19.5	22.2	35.6	47.7	61.4	70.0	75.1	75.2	59.7	50.9	35.0	29.4	48.5
Record Mean	22.0	26.6	37.8	51.7	62.6	72.2	77.3	75.0	66.2	54.5	39.0	27.1	51.0
Max	31.2	35.9	47.6	62.4	73.0	82.5	87.6	85.2	76.9	65.5	48.5	35.7	61.0
Min	12.7	17.2	27.9	41.0	52.2	61.9	67.0	64.8	55.5	43.6	29.5	18.5	41.0

REFERENCE NOTES

OMAHA (EPPLEY AIRFIELD), NEBRASKA

GENERAL
T - TRACE AMOUNT.
BLANK ENTRIES DENOTE MISSING/UNREPORTED DATA.
# INDICATES A STATION OR INSTRUMENT RELOCATION.

SPECIFIC

PM - INCLUDES LAST DAY OF PREVIOUS MONTH
ASOS - AUTOMATED SURFACE OBSERVING SYSTEM IN OPERATION DURING THESE MONTHS.

(a) - LENGTH OF RECORD IN YEARS, ALTHOUGH INDIVIDUAL MONTHS MAY BE MISSING.
0.* OR * - THE VALUE IS BETWEEN 0.0 AND 0.05
NORMALS - BASED ON THE 1961-1990 RECORD PERIOD.
EXTREMES - DATES ARE THE MOST RECENT OCCURRENCE
WIND DIR.- NUMERALS SHOW TENS OF DEGRESS CLOCKWISE FROM TRUE NORTH. "00" INDICATES CALM.
RESULTANT DIRECTIONS ARE GIVEN TO WHOLE DEGREES.
BOLD VALUES INDICATE EXTREME VALUES WHICH OCCURRED AFTER THE ASOS SYSTEM WAS COMMISSIONED.

RECORD = PERIOD OF RECORD
RECORD MEAN PRECIPITATION IS THE MEAN OF ALL DAILY PRECIPITATION AMOUNTS DURING THE PERIOD OF RECORD.
RECORD MAX(MIN) TEMPERATURE IS THE MEAN OF ALL DAILY MAX(MIN) TEMPERATURES DURING THE PERIOD OF RECORD.
RECORD MEAN TEMPERATURE IS THE SUM OF THE RECORD MAX AND RECORD MIN DIVIDED BY 2.
AVERAGE TEMPERATURE IS THE SUM OF THE MEAN DAILY MAX AND MIN TEMPERATURE DIVIDED BY 2.

EXCEPTIONS

1. MEAN SKY COVER, AND DAYS CLEAR-PARTLY CLOUDY-CLOUDY ARE THROUGH 1976.
2. MAXIMUM 24-HOUR PRECIPITATION AND SNOW, AND FASTEST MILE WINDS ARE THROUGH MAY 1977.

RECORD MEANS ARE THROUGH THE CURRENT YEAR, BEGINNING IN 1873 FOR TEMPERATURE
1871 FOR PRECIPITATION
1936 FOR SNOWFALL

HEATING DEGREE DAYS Base 65 deg. F — OMAHA (EPPLEY AIRFIELD), NEBRASKA

SEASON	JULY	AUG	SEP	OCT	NOV	DEC	JAN	FEB	MAR	APR	MAY	JUNE	TOTAL
1964-65	0	24	93	353	676	1227	1298	1191	1168	361	55	0	6446
1965-66	0	5	169	257	670	864	1440	1008	642	501	185	13	5754
1966-67	0	13	107	354	741	1132	1216	1020	615	295	249	17	5759
1967-68	8	15	109	417	791	1129	1307	1085	631	380	228	11	6111
1968-69	2	1	60	313	824	1268	1430	1066	1028	353	140	56	6541
1969-70	0	0	25	478	714	1242	1507	917	894	388	68	2	6235
1970-71	0	0	105	426	799	1085	1458	1106	853	319	172	1	6324
1971-72	5	1	95	228	707	1113	1392	1143	741	408	147	12	5992
1972-73	6	7	106	488	831	1357	1307	1014	636	437	191	0	6380
1973-74	0	0	90	254	750	1302	1427	955	702	379	140	31	6030
1974-75	0	15	191	300	726	1115	1311	1189	1024	469	72	11	6423
1975-76	0	0	141	251	695	1051	1219	791	757	261	177	4	5347
1976-77	0	0	61	522	947	1265	1598	883	579	219	10	1	6085
1977-78	0	1	28	361	754	1196	1637	1375	910	372	160	17	6811
1978-79	0	0	39	350	754	1255	1676	1333	775	451	156	12	6801
1979-80	1	6	65		867	1070	1318	1290	987	440	158	4	
1980-81	0	3	108	491	735	1198	1259	1018	743	241	221	0	6017
1981-82	7	3	85	452	723	1299	1721	1183	930	518	102	56	7079
1982-83	0	13	115	315	829	1131	1240	971	854	638	278	37	6421
1983-84	0	0	102	405	789	1786	1401	916	1071	552	243	7	7272
1984-85	0	3	184	391	766	1166	1416	1153	666	325	88	45	6203
1985-86	0	13	217	378	1089	1501	1095	1176	689	389	134	1	6682
1986-87	0	15	40	338	913	1096	1122	784	685	322	67	7	5389
1987-88	1	32	67	512	639	1048	1353	1185	748	433	29	6	6053
1988-89	1	7	56	488	744	1095	1002	1368	844	380	143	23	6151
1989-90	0	7	140	356	855	1460	973	935	684	460	206	15	6091
1990-91	4	1	75	371	662	1350	1506	859	695	338	108	0	5969
1991-92	0	0	123	402	1027	1022	999	816	656	449	154	11	5659
1992-93	2	26	114	359	881	1141	1403	1192	904	514	140	34	6710
1993-94	0	1	171	448	895	1097							

COOLING DEGREE DAYS Base 65 deg. F — OMAHA (EPPLEY AIRFIELD), NEBRASKA

YEAR	JAN	FEB	MAR	APR	MAY	JUNE	JULY	AUG	SEP	OCT	NOV	DEC	TOTAL
1969	0	0	0	2	95	149	416	338	113	25	0	0	1138
1970	0	0	0	32	172	286	386	361	148	20	0	0	1405
1971	0	0	0	12	34	393	295	308	188	42	0	0	1272
1972	0	0	0	3	74	249	314	280	124	1	0	0	1045
1973	0	0	0	3	23	257	332	395	74	22	0	0	1106
1974	0	0	0	11	52	182	540	193	39	4	0	0	1021
1975	0	0	0	7	115	242	441	464	76	44	0	0	1389
1976	0	0	0	21	34	240	440	358	139	17	0	0	1249
1977	0	0	0	45	179	310	489	236	105	0	0	0	1364
1978	0	0	7	5	67	287	386	333	231	5	0	0	1321
1979	0	0	0	8	64	249	344	345	122		0	0	
1980	0	0	0	15	61	254	459	368	107	0	0	0	1264
1981	0	0	0	24	29	235	372	196	85	0	0	0	941
1982	0	0	0	5	43	78	383	252	113	12	0	0	886
1983	0	0	0	0	20	183	453	519	167	17	0	0	1359
1984	0	0	0	6	22	220	320	366	96	4	0	0	1034
1985	0	0	0	30	44	116	290	156	137	1	0	0	774
1986	0	0	10	5	26	276	408	181	133	0	0	0	1039
1987	0	0	0	39	145	292	407	221	69	2	1	0	1176
1988	0	0	0	5	109	351	364	394	99	3	0	0	1325
1989	0	0	10	77	68	159	395	306	89	19	0	0	1123
1990	0	0	0	41	9	277	316	327	199	12	4	0	1185
1991	0	0	5	21	184	295	345	295	161	9	0	0	1315
1992	0	0	0	7	63	150	198	136	90	11	0	0	655
1993	0	0	0	0	35	188	322	324	21	19	0	0	909

SNOWFALL (inches) — OMAHA (EPPLEY AIRFIELD), NEBRASKA

SEASON	JULY	AUG	SEP	OCT	NOV	DEC	JAN	FEB	MAR	APR	MAY	JUNE	TOTAL
1964-65	0.0	0.0	0.0	0.0	2.6	5.4	6.8	25.4	16.1	0.0	0.0	0.0	56.3
1965-66	0.0	0.0	0.0	0.0	0.1	1.7	2.8	1.3	8.3	T	T	0.0	14.2
1966-67	0.0	0.0	0.0	T	0.1	6.8	7.2	0.8	0.3	0.7	1.0	0.0	16.9
1967-68	0.0	0.0	0.0	1.5	T	5.7	4.6	1.1	T	T	T	0.0	12.9
1968-69	0.0	0.0	0.0	0.0	4.7	8.6	8.3	14.0	3.2	0.0	0.0	0.0	38.8
1969-70	0.0	0.0	0.0	T	0.5	19.9	3.2	1.1	5.8	T	0.0	0.0	30.5
1970-71	0.0	0.0	0.0	3.5	T	0.5	13.1	17.4	4.3	0.5	0.0	0.0	39.3
1971-72	0.0	0.0	0.0	0.0	9.0	4.4	1.9	5.0	3.1	1.5	0.0	0.0	24.9
1972-73	0.0	0.0	0.0	T	9.7	5.9	15.8	4.6	T	2.3	0.0	0.0	38.3
1973-74	0.0	0.0	0.0	0.0	4.1	10.7	11.5	2.4	4.2	0.8	0.0	0.0	33.7
1974-75	0.0	0.0	0.0	0.0	5.4	8.1	22.7	11.9	5.1	4.6	0.0	0.0	57.8
1975-76	0.0	0.0	0.0	0.0	6.3	0.6	1.8	7.1	4.2	0.0	0.0	0.0	20.0
1976-77	0.0	0.0	0.0	1.4	0.5	2.8	14.4	0.1	3.7	1.6	0.0	0.0	24.5
1977-78	0.0	0.0	0.0	0.0	T	4.0	1.4	17.0	6.5	T	0.0	0.0	28.9
1978-79	0.0	0.0	0.0	0.0	2.0	5.6	10.0	1.4	6.0	1.1	0.0	0.0	26.1
1979-80	0.0	0.0	0.0	2.0	T	1.3	5.0	3.0	9.2	T	0.0	0.0	20.5
1980-81	0.0	0.0	0.0	2.0	T	0.5	3.6	3.0	T	0.0	0.0	0.0	9.1
1981-82	0.0	0.0	0.0	T	1.0	7.0	2.4	3.0	8.0	2.9	0.0	0.0	24.3
1982-83	0.0	0.0	0.0	T	T	4.0	6.5	9.0	9.0	3.0	0.0	0.0	31.5
1983-84	0.0	0.0	0.0	0.0	6.9	13.5	3.3	2.0	14.2	T	0.0	0.0	39.9
1984-85	0.0	0.0	0.0	T	T	5.0	4.0	4.0	6.0	T	0.0	0.0	19.0
1985-86	0.0	0.0	T	0.0	5.0	4.5	T	7.7	T	0.3	0.0	0.0	17.5
1986-87	0.0	0.0	0.0	0.0	1.8	5.5	1.2	1.6	10.5	T	0.0	0.0	20.6
1987-88	0.0	0.0	0.0	T	9.0	2.0	2.0	2.2	0.8	0.0	0.0	0.0	16.0
1988-89	0.0	0.0	0.0	0.0	4.3	2.7	1.4	11.8	3.3	T	T	0.0	23.5
1989-90	0.0	0.0	0.0	T	1.2	5.3	5.2	4.0	6.7	0.1	0.0	T	22.5
1990-91	0.0	0.0	0.0	T	1.1	10.1	14.9	0.3	4.1	1.1	0.0	T	31.6
1991-92	0.0	0.0	0.0	2.5	8.8	T	0.3	1.1	0.4	10.0	0.0	0.0	23.1
1992-93	0.0	0.0	0.0	0.0	5.8	3.9	13.0	8.5	4.1	1.5	0.0	T	36.8
1993-94	0.0	0.0	0.0	T	2.8	3.1							
Record Mean	0.0	0.0	T	0.3	2.6	5.5	7.4	6.6	6.4	1.0	0.1	T	29.9

OMAHA (NORTH), NEBRASKA

Omaha, Nebraska, is situated on the west bank of the Missouri River. The river level at Omaha is normally about 965 feet above sea level and the rolling hills in and around Omaha rise to about 1,300 feet above sea level. The climate is typically continental with relatively warm summers and cold, dry winters. It is situated midway between two distinctive climatic zones, the humid east and the dry west. Fluctuations between these two zones produce weather conditions for periods that are characteristic of either zone, or combinations of both. Omaha is also affected by most low pressure systems that cross the country. This causes periodic and rapid changes in weather, especially during the winter months.

Most of the precipitation in Omaha falls during sharp showers or thunderstorms, and these occur mostly during the growing season from April to September. Of the total precipitation, about 75 percent falls during this six–month period. The rain occurs mostly as evening or nighttime showers and thunderstorms. Although winters are relatively cold, precipitation is light, with only 10 percent of the total annual precipitation falling during the winter months.

Sunshine is fairly abundant, ranging around 50 percent of the possible in the winter to 75 percent of the possible in the summer.

NORMALS, MEANS, AND EXTREMES

OMAHA (NORTH), NEBRASKA

LATITUDE: 41°22'N LONGITUDE: 96°01' W ELEVATION: FT. GRND 1309 BARO 1315 TIME ZONE: CENTRAL WBAN: 94918

	(a)	JAN	FEB	MAR	APR	MAY	JUNE	JULY	AUG	SEP	OCT	NOV	DEC	YEAR
TEMPERATURE °F:														
Normals														
-Daily Maximum		29.7	35.0	47.6	62.4	72.8	82.4	86.5	84.0	74.9	64.0	47.7	32.9	60.0
-Daily Minimum		11.2	16.6	27.8	40.3	51.8	61.4	66.5	63.8	54.7	43.0	29.7	15.9	40.2
-Monthly		20.5	25.9	37.7	51.4	62.4	71.9	76.5	73.9	64.8	53.6	38.7	24.4	50.1
Extremes														
-Record Highest	39	66	76	88	96	100	104	107	106	103	93	79	66	107
-Year		1981	1972	1986	1989	1967	1988	1974	1983	1955	1975	1980	1976	JUL 1974
-Record Lowest	39	-22	-20	-16	7	25	41	44	44	28	15	-11	-25	-25
-Year		1982	1981	1960	1975	1967	1956	1971	1986	1984	1993	1964	1989	DEC 1989
NORMAL DEGREE DAYS:														
Heating (base 65°F)		1380	1095	846	414	155	11	0	11	90	363	789	1259	6413
Cooling (base 65°F)		0	0	0	6	75	218	357	287	84	10	0	0	1037
% OF POSSIBLE SUNSHINE	57	55	53	54	58	61	67	74	70	68	65	51	48	60
MEAN SKY COVER (tenths)														
Sunrise - Sunset	18	6.0	6.4	6.5	6.3	6.3	5.5	5.1	5.2	4.8	5.5	6.3	6.3	5.8
MEAN NUMBER OF DAYS:														
Sunrise to Sunset														
-Clear	18	9.1	7.9	7.8	7.8	7.1	8.9	11.1	10.4	12.6	11.0	8.1	8.5	110.2
-Partly Cloudy	18	8.1	5.7	7.6	8.1	10.0	11.3	10.1	11.3	8.1	7.6	7.2	7.3	102.3
-Cloudy	18	13.9	14.7	15.6	14.1	13.9	9.8	9.8	9.3	9.3	12.4	14.7	15.2	152.8
Precipitation														
.01 inches or more	18	5.3	6.2	9.7	10.5	11.7	10.1	9.9	8.7	8.5	7.2	6.3	7.2	101.3
Snow,Ice pellets,hail														
1.0 inches or more	18	2.1	1.8	1.6	0.6	0.0	0.0	0.0	0.0	0.0	0.2	1.2	2.1	9.6
Thunderstorms	18	0.1	0.4	2.2	3.4	7.4	9.5	9.5	8.0	6.1	2.1	0.7	0.2	49.6
Heavy Fog Visibility														
1/4 mile or less	18	1.9	2.6	2.7	0.6	0.8	0.6	0.3	1.2	0.9	1.1	2.2	3.2	18.0
Temperature °F														
-Maximum														
90° and above	39	0.0	0.0	0.0	0.4	0.8	4.6	9.3	6.7	2.3	0.1	0.0	0.0	24.2
32° and below	39	16.9	12.6	4.9	0.2	0.0	0.0	0.0	0.0	0.0	0.1	3.8	13.3	51.8
-Minimum														
32° and below	39	30.2	26.0	21.3	6.6	0.3	0.0	0.0	0.0	0.1	4.3	19.2	29.1	137.0
0° and below	39	7.5	3.8	0.3	0.0	0.0	0.0	0.0	0.0	0.0	0.0	0.3	3.5	15.4
AVG. STATION PRESS.(mb)	9	970.3	971.5	967.2	966.0	966.2	966.3	967.5	968.7	968.8	969.8	968.5	970.7	968.5
RELATIVE HUMIDITY (%)														
Hour 00	9	72	73	71	67	70	69	77	79	76	69	74	75	73
Hour 06 (Local Time)	9	75	77	78	76	79	79	85	86	84	78	79	77	79
Hour 12	9	63	63	59	53	55	56	63	62	60	55	65	67	60
Hour 18	9	64	63	56	50	53	52	59	62	60	57	67	71	60
PRECIPITATION (inches):														
Water Equivalent														
-Normal		0.65	0.78	2.13	2.74	4.36	3.90	3.27	3.22	3.65	2.41	1.35	0.93	29.39
-Maximum Monthly	39	1.85	2.86	5.27	7.12	9.09	8.83	9.77	11.77	14.10	5.34	5.11	4.45	14.10
-Year		1975	1965	1983	1984	1959	1993	1958	1960	1965	1986	1983	1984	SEP 1965
-Minimum Monthly	39	T	0.09	0.06	0.15	0.55	0.95	0.29	0.63	0.96	0.06	0.03	0.02	T
-Year		1986	1968	1956	1962	1989	1972	1975	1971	1990	1958	1989	1958	JAN 1986
-Maximum in 24 hrs	17	0.95	0.64	2.04	2.59	3.10	2.77	3.72	3.74	2.77	2.61	2.16	3.10	3.74
-Year		1982	1978	1982	1986	1987	1988	1977	1987	1989	1986	1983	1984	AUG 1987
Snow,Ice pellets,hail														
-Maximum Monthly	39	21.5	23.2	23.3	10.3	0.7	0.0	T	0.0	0.3	5.7	13.9	19.3	23.3
-Year		1975	1965	1960	1983	1967		1992		1985	1991	1957	1969	MAR 1960
-Maximum in 24 hrs	17	6.0	10.0	13.3	9.3	0.0	0.0	T	0.0	0.3	5.7	8.5	7.5	13.3
-Year		1979	1978	1987	1992			1992		1985	1991	1983	1984	MAR 1987
WIND:														
Mean Speed (mph)	9	9.9	9.2	10.5	10.2	8.9	8.3	7.5	7.6	8.6	9.0	9.7	9.8	9.1
Prevailing Direction														
Fastest Mile														
-Direction (!!)	15	NW	NW	NW	NW	N	SW	SW	NW	NW	NW	NW	NW	SW
-Speed (MPH)	15	41	38	38	46	34	39	50	39	35	34	38	37	50
-Year		1978	1978	1982	1982	1983	1992	1993	1980	1980	1979	1982	1981	JUL 1993
Peak Gust														
-Direction (!!)	9	NW	N	NW	S	NW	NW	SW	NW	NW	NW	N	NW	SW
-Speed (mph)	9	59	53	54	55	55	51	72	55	48	48	49	47	72
-Date		1990	1988	1993	1985	1992	1993	1993	1989	1986	1985	1988	1993	JUL 1993

PRECIPITATION (inches)

OMAHA (NORTH), NEBRASKA

YEAR	JAN	FEB	MAR	APR	MAY	JUNE	JULY	AUG	SEP	OCT	NOV	DEC	ANNUAL
1964	0.30	0.34	1.50	5.42	5.43	6.30	3.90	3.71	3.25	0.36	0.84	0.79	32.14
1965	0.51	2.86	2.40	3.27	6.89	3.72	4.42	2.17	14.10	0.88	1.33	0.83	43.38
1966	0.78	0.42	0.87	0.89	5.04	6.12	3.41	3.94	1.82	0.79	0.13	0.53	24.74
1967	0.91	0.18	0.71	2.41	2.74	7.85	2.00	1.99	3.94	1.80	0.29	0.87	25.69
1968	0.42	0.09	0.82	4.43	4.64	3.55	3.19	3.62	4.33	4.40	1.35	1.78	32.62
1969	1.06	1.47	1.04	2.89	3.64	3.01	6.00	4.60	1.32	2.45	0.10	1.80	29.38
1970	0.25	0.15	0.89	2.43	3.31	2.41	4.25	3.43	3.52	4.09	1.03	0.13	25.89
1971	1.04	2.42	0.82	0.67	7.03	2.29	1.71	0.63	1.26	4.87	2.28	0.72	25.74
1972	0.40	0.29	1.28	4.65	5.37	0.95	5.63	2.78	4.54	3.61	3.22	1.63	34.35
1973	1.61	0.82	5.17	2.44	6.68	1.86	5.66	0.74	6.97	2.88	1.53	1.67	38.03
1974	0.64	0.20	0.70		2.57	5.68	0.10	4.98	1.89	3.17			
1975	1.85	1.06	1.88	3.15	3.98	3.36	0.29	1.96	2.30	0.08	2.73	0.94	23.58
1976	0.15	1.67	2.04	2.60	4.07	2.96	0.86	0.73	2.40	0.88	0.04	0.21	18.61
1977	0.75	0.17	3.84	2.26	5.59	2.19	7.28	7.13	5.33	4.66	1.70	0.62	41.52
1978	0.20	1.35	1.21	4.33	3.39	1.84	3.35	2.46	4.64	0.63	1.12	0.71	25.23
1979	1.22	0.38	4.13	2.70	2.99	3.55	2.22	2.54	2.31	4.46	1.58	0.22	28.30
1980	0.61	0.75	1.55	1.32	1.87	7.54	2.48	7.72	1.40	3.43	0.12	0.33	29.12
1981	0.30	0.17	0.93	1.69	3.90	1.99	4.45	7.92	1.49	1.74	3.61	0.60	28.79
1982	1.33	0.24	3.22	1.79	9.00	3.97	2.15	3.03	3.96	1.84	2.05	1.68	34.26
1983	1.14	1.03	5.27	2.02	4.82	4.93	1.15	1.03	2.84	1.91	5.11	0.69	31.94
1984	0.32	0.82	2.74	7.12	3.96	8.16	1.18	0.88	3.42	4.50	0.83	4.45	38.38
1985	0.35	0.73	1.62	2.44	4.16	2.52	2.71	1.55	2.84	1.94	0.65	0.34	21.85
1986	T	0.97	3.00	6.89	3.98	3.18	2.64	3.41	6.46	5.34	0.59	0.68	37.14
1987	0.03	0.60	4.16	2.43	6.08	1.90	4.68	7.44	1.58	1.65	0.95	0.75	32.25
1988	0.43	0.18	0.07	1.87	5.49	2.89	2.59	1.46	3.58	0.11	1.89	0.78	21.34
1989	0.89	0.64	0.31	2.27	0.55	4.38	3.18	2.22	6.14	1.24	0.03	0.50	22.35
1990	0.63	0.41	3.22	0.34	4.04	4.19	5.59	0.84	0.96	1.86	1.10	0.75	23.93
1991	0.92	0.46	4.24	5.74	3.42	6.52	3.49	3.87	2.06	3.02	1.67	1.87	37.28
1992	1.30	1.21	3.66	2.85	2.28	1.46	9.66	1.85	4.58	2.87	2.36	1.01	35.09
1993	1.05	0.79	2.51	2.33	4.86	8.83	8.89	6.38	3.39	1.65	0.68	0.38	41.74
Record Mean	0.71	0.79	2.18	2.68	4.30	4.16	3.78	3.65	3.54	2.25	1.30	0.86	30.21

AVERAGE TEMPERATURE (deg. F)

OMAHA (NORTH), NEBRASKA

YEAR	JAN	FEB	MAR	APR	MAY	JUNE	JULY	AUG	SEP	OCT	NOV	DEC	ANNUAL
1964	28.0	29.2	32.3	51.1	66.1	69.8	77.7	68.9	63.4	52.7	39.7	22.3	50.1
1965	20.5	20.1	24.1	50.7	64.7	69.6	73.9	72.1	58.0	56.5	40.4	34.4	48.8
1966	15.5	25.8	41.4	45.4	59.8	70.3	77.6	70.1	61.8	53.2	37.7	25.6	48.7
1967	22.8	24.0	41.6	52.0	57.6	69.1	72.8	71.0	62.5	51.4	37.4	27.1	49.1
1968	21.3	25.4	43.8	51.2	56.9	72.5	74.8	73.5	63.7	54.4	36.1	21.7	49.6
1969	16.1	25.5	28.8	51.8	62.4	66.9	76.2	74.0	66.2	48.3	39.7	23.2	48.3
1970	13.7	29.8	33.2	51.1	66.1	72.2	74.9	74.6	64.5	50.4	36.2	27.0	49.5
1971	15.7	23.0	34.4	52.7	58.4	75.8	71.7	73.4	66.4	56.6	39.0	27.3	49.5
1972	18.7	23.8	39.1	49.6	61.6	71.4	73.0	72.2	64.4	49.1	35.7	20.2	48.2
1973	22.6	27.0	42.8	49.2	59.5	72.7	73.9	76.3	62.9	56.8	39.2	22.4	50.4
1974	18.4	29.7	41.2			69.6	81.8	69.5	59.9	53.9			
1975	21.7	20.5	29.8	46.9	64.9	70.7	77.4	77.5	61.3	57.7	40.0	28.3	49.7
1976	23.6	35.7	38.6	54.6	59.6	71.5	77.5	75.9	66.0	48.2	32.7	23.5	50.6
1977	11.7	31.4	43.3	57.2	68.0	72.9	78.7	70.7	65.9	52.0	37.9	23.9	51.1
1978	10.0	13.5	33.9	50.5	60.7	72.4	75.0	74.0	69.7	52.5	37.0	21.0	47.5
1979	7.9	13.6	35.0	47.6	59.8	70.8	74.3	74.1	67.9	53.3	36.9	31.9	47.8
1980	23.9	22.0	34.3	52.5	63.5	72.4	79.3	76.4	67.1	50.8	42.4	27.5	51.0
1981	27.3	30.5	42.9	58.4	60.0	73.7	76.9	71.5	65.7	50.9	42.2	24.5	52.1
1982	10.2	24.1	36.0	47.9	63.1	66.6	76.8	72.5	64.0	54.4	36.8	29.7	48.5
1983	25.8	31.0	37.6	43.9	57.5	70.6	79.5	81.8	68.0	53.2	38.7	7.8	49.6
1984	21.5	33.9	30.8	47.0	58.8	72.3	75.9	77.3	62.8	53.2	40.5	26.8	50.1
1985	18.6	24.1	43.6	55.6	65.2	68.4	75.8	70.8	62.9	53.8	28.9	17.5	48.8
1986	31.0	24.5	44.1	53.1	63.0	74.5	77.6	69.8	67.2	53.4	34.3	29.3	51.8
1987	29.2	37.5	42.4	55.8	66.9	74.8	78.6	71.4	65.1	49.1	43.4	30.9	53.8
1988	20.9	24.4	41.2	51.6	68.0	77.4	76.7	77.9	67.8	50.4	40.3	30.6	52.3
1989	33.1	16.7	37.3	54.6	62.9	69.8	77.7	74.7	63.7	54.6	36.9	17.5	50.0
1990	34.3	31.3	42.3	50.7	59.0	73.7	74.7	75.6	70.0	54.4	43.8	21.6	52.6
1991	17.1	35.3	42.4	54.2	66.9	74.8	75.8	74.6	66.3	51.9	30.5	32.3	51.8
1992	32.5	36.2	43.3	49.5	62.6	69.7	70.3	69.1	64.6	53.9	34.4	27.6	51.1
1993	19.7	21.1	34.9	47.7	61.0	69.6	74.2	74.1	59.9	50.8	35.3	29.6	48.2
Record Mean	20.5	25.5	37.0	49.5	60.6	71.3	75.8	73.7	64.6	53.3	37.8	25.4	49.5
Max	29.5	34.6	46.7	60.2	70.8	81.5	85.6	83.7	74.8	63.9	46.9	34.0	59.3
Min	11.4	16.5	27.2	38.8	50.3	61.0	66.0	63.7	54.3	42.7	28.6	16.8	39.8

REFERENCE NOTES

OMAHA (NORTH), NEBRASKA

GENERAL
T - TRACE AMOUNT.
BLANK ENTRIES DENOTE MISSING/UNREPORTED DATA.
# INDICATES A STATION OR INSTRUMENT RELOCATION.

SPECIFIC

(a) - LENGTH OF RECORD IN YEARS, ALTHOUGH INDIVIDUAL MONTHS MAY BE MISSING.
0.* OR * - THE VALUE IS BETWEEN 0.0 AND 0.05
NORMALS - BASED ON THE 1961-1990 RECORD PERIOD.
EXTREMES - DATES ARE THE MOST RECENT OCCURRENCE
WIND DIR.- NUMERALS SHOW TENS OF DEGRESS CLOCKWISE FROM TRUE NORTH. "00" INDICATES CALM.
RESULTANT DIRECTIONS ARE GIVEN TO WHOLE DEGREES.
BOLD VALUES INDICATE EXTREME VALUES WHICH OCCURRED AFTER THE ASOS SYSTEM WAS COMMISSIONED.

RECORD = PERIOD OF RECORD
RECORD MEAN PRECIPITATION IS THE MEAN OF ALL DAILY PRECIPITATION AMOUNTS DURING THE PERIOD OF RECORD.
RECORD MAX(MIN) TEMPERATURE IS THE MEAN OF ALL DAILY MAX(MIN) TEMPERATURES DURING THE PERIOD OF RECORD.
RECORD MEAN TEMPERATURE IS THE SUM OF THE RECORD MAX AND RECORD MIN DIVIDED BY 2.
AVERAGE TEMPERATURE IS THE SUM OF THE MEAN DAILY MAX AND MIN TEMPERATURE DIVIDED BY 2.

EXCEPTIONS

PERCENT OF POSSIBLE SUNSHINE IS FROM OMAHA (EPPLEY) THROUGH MAY 1977.

RECORD MEANS ARE THROUGH THE CURRENT YEAR,
BEGINNING IN 1954 FOR TEMPERATURE
1954 FOR PRECIPITATION
1954 FOR SNOWFALL

HEATING DEGREE DAYS Base 65 deg. F   OMAHA (NORTH), NEBRASKA

SEASON	JULY	AUG	SEP	OCT	NOV	DEC	JAN	FEB	MAR	APR	MAY	JUNE	TOTAL
1964-65	0	31	132	379	753	1319	1373	1254	1262	433	95	5	7036
1965-66	0	9	230	270	731	944	1531	1092	723	582	222	25	6359
1966-67	0	18	141	375	813	1212	1303	1141	727	387	300	23	6440
1967-68	12	18	105	435	823	1170	1351	1144	657	414	264	24	6417
1968-69	2	3	90	345	858	1338	1514	1098	1114	388	162	63	6975
1969-70	0	3	38	532	750	1290	1585	982	979	440	92	13	6704
1970-71	4	0	131	464	860	1169	1519	1168	941	368	215	2	6841
1971-72	9	3	111	276	776	1162	1431	1188	795	461	171	21	6404
1972-73	7	12	116	491	871	1383	1308	1057	681	468	199	3	6596
1973-74	2	0	112	263	770	1315	1439	981	734				
1974-75	0	10	188	340			1338	1242	1084	541	90	19	
1975-76	1	0	167	261	744	1131	1275	841	811	314	190	12	5747
1976-77	0	0	87	538	962	1279	1646	935	664	262	16	1	6390
1977-78	0	3	41	394	803	1268	1702	1437	963	432	185	23	7251
1978-79	0	3	52	385	837	1358	1767	1435	921	514	199	24	7495
1979-80	4	15	48	354	837	1016	1267	1241	945	396	129	9	6261
1980-81	0	3	67	444	671	1157	1161	961	677	226	189	1	5557
1981-82	8	4	73	430	677	1251	1695	1143	894	512	94	50	6831
1982-83	0	12	123	333	844	1091	1206	946	844	625	244	26	6294
1983-84	0	0	94	376	782	1768	1344	896	1054	538	212	3	7067
1984-85	0	1	165	363	726	1177	1436	1139	656	315	65	35	6078
1985-86	0	7	209	342	1077	1467	1048	1130	662	357	99	1	6399
1986-87	0	17	45	352	917	1102	1103	765	692	316	72	2	5383
1987-88	2	33	65	486	640	1053	1362	1172	734	401	33	5	5986
1988-89	2	4	38	448	733	1061	984	1348	861	380	127	25	6011
1989-90	2	5	131	343	834	1470	945	938	696	460	189	19	6032
1990-91	5	0	70	334	635	1341	1479	825	699	338	118	0	5844
1991-92	0	0	116	412	1027	1006	1004	830	664	469	132	14	5674
1992-93	8	22	95	359	910	1149	1395	1220	927	510	146	38	6779
1993-94	0	5	171	454	885	1090							

COOLING DEGREE DAYS Base 65 deg. F   OMAHA (NORTH), NEBRASKA

YEAR	JAN	FEB	MAR	APR	MAY	JUNE	JULY	AUG	SEP	OCT	NOV	DEC	TOTAL
1977	0	0	0	35	115	246	436	186	76	0	0	0	1094
1978	0	0	5	2	58	252	317	290	202	2	0	0	1128
1979	0	0	0	1	47	206	301	303	142	2	0	0	1002
1980	0	0	0	25	87	237	451	362	137	11	0	0	1310
1981	0	0	0	34	38	269	384	209	101	0	0	0	1035
1982	0	0	0	4	43	104	375	250	99	12	0	0	887
1983	0	0	0	0	21	200	454	526	193	17	0	0	1411
1984	0	0	0	5	27	225	346	389	105	4	0	0	1101
1985	0	0	0	39	79	142	344	196	153	1	0	0	954
1986	0	0	16	8	45	294	399	172	119	0	0	0	1053
1987	0	0	0	47	138	302	430	235	73	3	1	0	1229
1988	0	0	0	4	128	384	374	410	129	4	0	0	1433
1989	0	0	10	77	67	172	400	315	98	30	0	0	1169
1990	0	0	0	38	11	289	311	333	228	15	3	0	1228
1991	0	0	3	21	182	303	341	307	160	13	0	0	1330
1992	0	0	0	12	64	163	178	155	89	24	0	0	685
1993	0	0	0	0	29	181	292	297	26	18	0	0	843

SNOWFALL (inches)   OMAHA (NORTH), NEBRASKA

SEASON	JULY	AUG	SEP	OCT	NOV	DEC	JAN	FEB	MAR	APR	MAY	JUNE	TOTAL
1964-65	0.0	0.0	0.0	0.0	2.8	5.6	6.5	23.2	15.9	0.0	0.0	0.0	54.0
1965-66	0.0	0.0	0.0	0.0	0.2	1.2	3.5	0.4	8.4	T	T	0.0	13.7
1966-67	0.0	0.0	0.0	0.0	T	6.2	7.8	1.1	0.4	0.8	0.7	0.0	17.0
1967-68	0.0	0.0	0.0	0.0	T	6.2	4.9	0.7	T	T	T	0.0	11.8
1968-69	0.0	0.0	0.0	T	5.1	8.5	7.6	14.1	2.1	0.0	0.0	0.0	37.4
1969-70	0.0	0.0	0.0	T	0.2	19.3	2.9	1.6	5.1	T	T	0.0	29.1
1970-71	0.0	0.0	0.0	4.1	T	0.3	12.8	17.6	6.3	0.4	0.0	0.0	41.5
1971-72	0.0	0.0	0.0	T	7.8	3.3	3.6	4.3	3.0	1.2	0.0	0.0	23.2
1972-73	0.0	0.0	0.0	T	8.8	7.2	16.5	6.3	0.0	2.9	0.0	0.0	41.7
1973-74	0.0	0.0	0.0	0.0	3.6	9.0	10.6	2.9	4.6		0.0	0.0	
1974-75	0.0	0.0	0.0	0.0			21.5	12.2	5.0	3.7	0.0	0.0	
1975-76	0.0	0.0	0.0	0.0	8.0	0.6	2.6	7.6	5.0	0.0	0.0	0.0	23.8
1976-77	0.0	0.0	0.0	1.4	0.7	2.8	13.7	T	3.3	3.0	0.0	0.0	24.9
1977-78	0.0	0.0	0.0	0.0	3.0	7.4	4.0	22.0	8.6	T	0.0	0.0	45.0
1978-79	0.0	0.0	0.0	0.0	5.3	8.8	13.6	3.9	6.6	4.8	0.0	0.0	43.0
1979-80	0.0	0.0	0.0	1.9	0.2	1.5	7.2	9.8	8.0	0.2	0.0	0.0	28.8
1980-81	0.0	0.0	0.0	5.2	0.2	0.8	4.3	3.6	0.6	0.0	0.0	0.0	14.7
1981-82	0.0	0.0	0.0	0.9	4.7	7.8	5.9	3.5	5.2	3.9	0.0	0.0	31.9
1982-83	0.0	0.0	0.0	0.8	0.4	5.1	8.9	11.4	14.1	10.3	0.0	0.0	51.0
1983-84	0.0	0.0	0.0	0.0	13.6	15.9	3.1	3.4	18.3	1.5	0.0	0.0	55.8
1984-85	0.0	0.0	0.0	T	1.6	9.7	4.6	3.3	8.7	T	0.0	0.0	27.9
1985-86	0.0	0.0	0.3	0.0	6.9	5.5	T	7.0	0.5	1.1	0.0	0.0	21.3
1986-87	0.0	0.0	0.0	0.0	0.8	5.1	0.7	1.4	14.1	0.1	0.0	0.0	22.2
1987-88	0.0	0.0	0.0	0.4	6.8	2.0	3.3	2.8	0.9	0.0	0.0	0.0	16.2
1988-89	0.0	0.0	0.0	0.0	4.4	3.3	1.3	10.2	2.2	T	0.0	0.0	21.4
1989-90	0.0	0.0	0.0	T	0.4	6.2	5.1	5.8	5.4	T	0.0	0.0	22.9
1990-91	0.0	0.0	0.0	T	4.0	8.2	13.7	0.4	5.1	T	0.0	0.0	31.4
1991-92	0.0	0.0	0.0	5.7	7.6	T	0.2	1.3	0.2	9.3	0.0	0.0	24.3
1992-93	T	0.0	0.0	0.0	4.6	4.2	16.1	9.0	5.0	2.3	0.0	0.0	41.2
1993-94	0.0	0.0	T	T	2.3	3.8							
Record Mean	T	0.0	T	0.5	3.4	5.3	7.2	6.5	6.9	1.4	T	0.0	31.2

SCOTTSBLUFF, NEBRASKA

Scottsbluff is located in the North Platte river valley that extends from central Wyoming southeast across western Nebraska. The valley is approximately 20 miles wide in the vicinity of Scottsbluff with a range of hills both to the north and south, parallel to the river. To the south the hills average 600 to 700 feet above the river with some projections upward to 1,000 feet. To the north, rolling hills range from 300 to 400 feet higher than the river.

Due to the protection of the higher hills to the south, southerly winds in the valley are rare. Prevailing winds are west to northwest during the winter months and east to southeast during the summer months. West to northwest winds are intensified by the funneling action of the valley and velocities of 30 to 50 mph are common during the winter and early spring. Quite often these winds are warmed by the downslope (chinook) effect from the higher elevations to the west and bring rapid warming and melting of the snow. Outbreaks of Arctic air bring cold wave conditions about five times each season. Snow with strong winds causing blowing and drifting snow occur several times each winter with a severe blizzard of extended duration occurring about once every thirty years. Easterly winds during the winter and early spring cause upslope conditions with low cloudiness and precipitation.

The average temperature is in the upper 40s. Summertime highs generally range from the 80s to the 90s with lows around 60. Summer temperatures of 100 degrees are reached or exceeded at least once each summer. In winter, highs average about 40 degrees with lows in the teens. Temperatures of zero or below occur about 15 times each winter.

Most of the precipitation occurs as thunderstorms during the spring and summer months. Severe thunderstorms with destructive hail are quite common during the late spring and summer. Tornadoes are infrequent and usually of short duration.

The Platte River in the vicinity of Scottsbluff is a wide shallow stream and has very little effect on the climate. Water stored in numerous upstream reservoirs is used for extensive irrigation in the valley. Lowland flooding occurs when heavy rains fall upstream and a greater than normal amount of water is being released from the upstream reservoirs.

Based on the 1951–1980 period, the average first occurrence of 32 degrees Fahrenheit in the fall is September 29 and the average last occurrence in the spring is May 7.

NORMALS, MEANS, AND EXTREMES

SCOTTSBLUFF, NEBRASKA

LATITUDE: 41°52'N　LONGITUDE: 103°36' W　ELEVATION: FT. GRND 3957 BARO 3950　TIME ZONE: MOUNTAIN　WBAN: 24028

	(a)	JAN	FEB	MAR	APR	MAY	JUNE	JULY	AUG	SEP	OCT	NOV	DEC	YEAR
TEMPERATURE °F:														
Normals														
-Daily Maximum		37.9	43.6	50.3	61.4	70.9	82.0	89.7	87.2	77.2	65.5	50.2	39.5	63.0
-Daily Minimum		11.8	16.8	22.1	31.5	41.8	52.7	58.7	56.0	45.6	33.7	22.4	12.9	33.8
-Monthly		24.9	30.2	36.2	46.5	56.4	67.3	74.2	71.6	61.4	49.6	36.3	26.2	48.4
Extremes														
-Record Highest	51	74	77	87	93	95	106	109	104	101	92	80	77	109
-Year		1982	1962	1943	1992	1969	1990	1989	1988	1948	1967	1989	1980	JUL 1989
-Record Lowest	51	-32	-28	-27	-8	15	30	40	39	19	-6	-13	-42	-42
-Year		1963	1962	1948	1975	1983	1969	1959	1944	1985	1991	1947	1989	DEC 1989
NORMAL DEGREE DAYS:														
Heating (base 65°F)		1243	974	893	555	278	61	0	9	175	477	861	1203	6729
Cooling (base 65°F)		0	0	0	0	12	130	290	214	67	0	0	0	713
% OF POSSIBLE SUNSHINE														
MEAN SKY COVER (tenths)														
Sunrise - Sunset	43	6.2	6.3	6.7	6.4	6.4	5.2	4.4	4.6	4.4	4.9	6.1	5.9	5.6
MEAN NUMBER OF DAYS:														
Sunrise to Sunset														
-Clear	50	8.5	7.4	6.4	6.7	6.2	10.2	13.2	12.9	14.3	13.0	8.4	8.9	116.0
-Partly Cloudy	50	8.5	8.2	9.5	8.8	10.6	10.8	12.1	11.3	7.5	7.8	8.1	8.6	111.7
-Cloudy	50	14.0	12.7	15.1	14.5	14.2	9.0	5.7	6.8	8.3	10.2	13.5	13.6	137.5
Precipitation														
.01 inches or more	50	5.5	5.1	7.5	8.5	11.7	11.0	8.6	6.9	6.8	4.9	4.9	5.3	86.8
Snow,Ice pellets,hail														
1.0 inches or more	43	2.0	1.9	2.9	1.6	0.3	0.0	0.0	0.0	0.1	0.8	1.7	2.1	13.3
Thunderstorms	49	0.0	0.0	0.3	1.8	7.4	11.1	10.4	7.9	4.0	0.8	0.1	0.*	43.9
Heavy Fog Visibility														
1/4 mile or less	49	1.0	0.8	1.6	0.6	0.7	0.6	0.4	0.7	0.9	0.9	1.2	1.1	10.4
Temperature °F														
-Maximum														
90° and above	29	0.0	0.0	0.0	0.1	1.2	7.2	16.0	12.8	4.2	0.1	0.0	0.0	41.6
32° and below	29	9.2	6.1	3.1	0.2	0.0	0.0	0.0	0.0	0.*	0.4	3.6	8.7	31.3
-Minimum														
32° and below	29	29.9	27.1	26.6	13.3	1.9	0.1	0.0	0.0	1.7	13.1	26.5	30.2	170.4
0° and below	29	5.9	3.4	0.8	0.1	0.0	0.0	0.0	0.0	0.0	0.1	1.0	4.6	15.9
AVG. STATION PRESS.(mb)	20	879.0	879.0	876.2	877.1	877.3	878.6	880.4	880.8	880.9	880.9	879.0	879.2	879.0
RELATIVE HUMIDITY (%)														
Hour 05	28	74	75	76	76	79	80	80	82	79	75	76	75	77
Hour 11 (Local Time)	29	58	53	50	44	45	44	42	44	42	43	52	56	48
Hour 17	29	58	49	44	40	41	39	37	38	36	40	53	58	44
Hour 23	28	72	71	68	66	68	68	67	69	67	66	70	72	69
PRECIPITATION (inches):														
Water Equivalent														
-Normal		0.50	0.47	1.09	1.58	2.77	2.64	2.06	1.07	1.10	0.81	0.62	0.56	15.27
-Maximum Monthly	51	1.26	1.93	2.64	3.89	7.25	8.33	4.82	3.42	4.22	3.02	2.15	1.54	8.33
-Year		1978	1986	1990	1984	1987	1947	1978	1987	1973	1969	1993	1978	JUN 1947
-Minimum Monthly	51	T	T	0.17	0.29	0.27	0.59	0.07	0.09	T	0.04	T	0.02	T
-Year		1989	1954	1966	1962	1966	1950	1946	1973	1953	1956	1943	1991	JAN 1989
-Maximum in 24 hrs	51	0.92	0.88	1.68	2.01	2.62	3.74	2.53	2.01	3.28	1.32	1.43	0.90	3.74
-Year		1976	1987	1974	1988	1988	1953	1948	1987	1951	1948	1993	1975	JUN 1953
Snow,Ice pellets,hail														
-Maximum Monthly	51	23.7	23.4	23.5	18.0	7.5	0.1	T	0.0	5.1	21.6	18.5	18.1	23.7
-Year		1949	1987	1980	1957	1967	1951	1993		1985	1969	1983	1985	JAN 1949
-Maximum in 24 hrs	51	11.2	9.8	15.0	11.3	3.7	0.1	T	0.0	4.8	8.5	8.4	10.9	15.0
-Year		1976	1987	1974	1988	1979	1951	1993		1985	1969	1979	1975	MAR 1974
WIND:														
Mean Speed (mph)	42	10.8	11.1	12.2	12.6	11.8	10.5	9.3	9.1	9.4	9.7	10.3	10.5	10.6
Prevailing Direction through 1963		WNW	WNW	WNW	NW	ESE	ESE	ESE	ESE	ESE	NW	NW	WNW	ESE
Fastest Obs. 1 Min.														
-Direction (!!)	43	34	29	29	32	32	29	05	35	32	29	29	32	29
-Speed (MPH)	43	53	60	62	54	80	80	52	52	46	44	56	47	80
-Year		1975	1953	1963	1964	1951	1954	1967	1980	1959	1966	1970	1954	JUN 1954
Peak Gust														
-Direction (!!)	4	W	W	NW	NW	W	N	NW	W	W	NW	W	W	N
-Speed (mph)	4	68	60	59	59	60	70	68	48	54	54	56	53	70
-Date		1990	1990	1990	1992	1990	1991	1993	1990	1991	1990	1992	1993	JUN 1991

PRECIPITATION (inches) SCOTTSBLUFF, NEBRASKA

YEAR	JAN	FEB	MAR	APR	MAY	JUNE	JULY	AUG	SEP	OCT	NOV	DEC	ANNUAL
1964	0.04	0.25	0.38	2.40	1.45	1.47	0.69	0.19	0.09	0.08	0.01	0.65	7.70
1965	0.46	0.44	0.20	0.71	3.30	6.53	2.06	0.52	3.15	0.98	0.09	0.65	19.09
1966	0.46	0.21	0.17	1.11	0.27	2.09	3.34	1.86	1.59	0.81	0.20	0.17	12.28
1967	0.18	0.04	0.29	2.42	4.42	4.22	2.29	0.99	0.53	0.17	0.17	0.82	16.54
1968	0.15	0.23	0.64	1.76	2.69	2.29	1.17	2.08	0.25	0.86	0.19	0.55	12.86
1969	0.72	0.52	0.51	1.08	2.44	2.01	0.72	1.26	0.79	3.02	0.40	0.32	13.79
1970	0.57	0.18	0.93	2.26	1.38	2.49	1.10	0.40	0.34	1.57	0.38	0.39	11.99
1971	0.23	0.44	1.31	1.71	4.73	2.47	1.46	1.47	2.09	0.69	0.33	0.13	17.06
1972	0.35	0.25	0.69	2.91	1.54	4.43	3.77	1.71	1.64	0.79	1.75	0.77	20.60
1973	0.58	0.43	1.97	2.24	0.80	0.88	3.69	0.09	4.22	0.74	1.36	1.24	18.24
1974	0.57	0.10	1.99	0.35	0.79	0.98	0.66	1.58	0.94	0.48	0.42	0.18	9.04
1975	0.32	0.46	1.73	1.78	2.25	1.47	1.60	0.47	0.26	0.74	0.45	1.18	12.71
1976	0.96	0.29	0.36	2.48	2.27	1.31	0.25	0.78	0.22	0.35	0.35	0.10	9.72
1977	0.52	0.02	2.04	2.12	2.07	4.06	1.22	0.63	0.39	0.13	0.91	0.82	14.93
1978	1.26	1.17	0.68	1.24	4.37	2.41	4.82	1.25	0.09	0.74	0.62	1.54	20.19
1979	0.74	0.10	1.22	0.90	1.33	2.59	3.17	2.51	0.74	1.66	1.60	0.49	17.05
1980	1.21	0.99	2.16	0.57	2.82	0.79	1.07	0.47	0.47	0.76	0.57	0.15	12.03
1981	0.69	0.14	0.59	1.47	2.75	2.54	3.54	1.10	0.39	0.34	0.26	0.19	14.00
1982	0.32	0.20	0.46	0.50	2.93	6.63	4.78	1.66	1.78	1.22	0.80	0.57	21.85
1983	0.29	0.04	1.94	2.33	4.20	1.81	0.69	1.23	0.13	0.68	1.75	0.60	15.69
1984	0.44	0.50	1.47	3.89	1.23	1.23	1.80	0.57	0.45	0.88	0.28	0.50	13.24
1985	0.64	0.20	0.37	1.23	0.86	1.76	0.80	0.18	2.71	1.01	1.28	1.17	12.21
1986	0.07	1.93	0.83	2.49	1.51	5.55	4.00	1.01	1.86	1.42	0.81	0.26	21.74
1987	0.34	1.88	1.70	0.44	7.25	4.13	1.14	3.42	0.90	0.08	0.95	1.01	23.24
1988	0.80	0.11	1.11	2.27	5.19	2.29	0.85	0.80	0.97	0.11	0.46	0.40	15.36
1989	T	1.03	0.77	0.65	1.89	1.15	0.32	1.13	1.63	0.70	0.07	0.65	9.99
1990	0.59	0.72	2.64	1.75	2.94	1.14	3.10	1.23	0.97	0.99	1.25	0.36	17.68
1991	0.46	0.39	0.50	1.16	4.35	4.00	0.56	0.11	0.90	1.17	0.72	0.02	14.34
1992	0.81	0.86	1.22	0.34	2.03	3.00	2.96	1.65	0.17	1.15	0.98	0.66	15.83
1993	0.45	1.64	1.36	1.95	0.98	5.55	3.10	2.53	2.17	2.35	2.15	0.59	24.82
Record Mean	0.40	0.49	0.93	1.75	2.74	2.72	1.88	1.25	1.25	0.89	0.54	0.51	15.34

AVERAGE TEMPERATURE (deg. F) SCOTTSBLUFF, NEBRASKA

YEAR	JAN	FEB	MAR	APR	MAY	JUNE	JULY	AUG	SEP	OCT	NOV	DEC	ANNUAL
#1964	28.7	26.2	30.7	44.9	58.8	66.4	77.9	68.9	59.6	50.0	37.0	25.1	47.9
1965	32.2	26.9	26.1	51.1	57.4	64.4	73.5	69.6	53.1	53.4	41.5	30.9	48.3
1966	19.6	22.4	40.6	42.5	59.8	67.4	77.6	68.9	63.0	48.7	38.1	27.3	48.0
1967	31.0	32.0	41.3	48.7	51.6	63.1	72.7	70.7	61.9	51.3	34.7	19.7	48.2
1968	24.1	33.1	41.0	43.2	52.4	67.3	73.4	68.7	59.9	50.8	35.5	22.7	47.7
1969	23.6	32.4	30.8	50.8	59.2	61.4	74.9	75.1	66.7	41.1	38.3	30.8	48.8
1970	24.5	33.0	30.4	41.4	56.9	64.5	72.9	73.2	57.3	42.6	36.0	25.7	46.5
1971	26.8	28.6	34.7	48.2	54.5	69.3	70.2	72.9	58.4	47.5	37.9	29.4	48.2
1972	24.9	33.4	41.7	46.0	56.5	68.2	70.0	71.0	61.1	48.8	30.1	20.4	47.7
1973	24.6	29.8	38.1	41.9	54.9	67.5	71.6	72.9	57.2	50.7	35.3	25.6	47.5
1974	19.6	33.9	38.9	47.4	58.1	69.5	76.5	68.3	56.4	52.1	36.3	28.1	48.8
1975	26.6	25.3	33.1	44.5	56.5	65.6	77.2	72.2	59.7	50.9	35.6	31.9	48.3
1976	25.1	37.2	36.6	49.1	58.2	68.1	76.4	72.2	63.1	47.3	34.7	32.2	50.0
1977	20.8	36.6	38.1	53.0	63.5	73.1	76.3	70.2	65.4	51.6	38.1	28.3	51.2
1978	18.3	22.0	40.5	49.6	56.6	69.0	74.9	70.5	66.2	51.6	33.8	18.9	47.7
1979	10.4	28.1	41.0	50.1	56.1	68.6	75.0	71.7	67.0	53.1	32.0	34.3	49.0
1980	23.1	29.7	36.1	49.4	58.6	72.3	78.2	73.1	66.3	51.6	40.8	37.0	51.3
1981	32.5	32.8	42.7	56.4	57.4	71.3	76.2	72.8	66.9	50.9	42.7	31.2	52.8
1982	22.5	27.7	37.6	43.9	55.6	62.8	72.8	74.2	62.1	48.5	34.2	27.9	47.5
1983	32.6	36.5	36.8	40.7	51.1	63.9	74.2	76.7	63.9	51.5	34.2	12.4	47.9
1984	27.3	35.5	37.4	42.2	58.8	66.8	74.8	76.4	59.9	46.2	37.6	24.8	49.0
1985	20.8	25.6	39.5	50.7	61.7	66.3	75.2	71.6	57.4	47.5	22.0	20.2	46.5
1986	32.9	27.9	44.3	47.3	55.3	69.5	72.6	70.9	59.6	48.5	35.2	28.7	49.4
1987	29.0	33.8	33.3	51.0	60.4	68.3	74.7	68.9	60.5	47.6	38.5	25.2	49.3
1988	18.1	30.4	35.9	46.8	58.2	72.7	74.6	72.6	60.4	50.4	37.2	28.1	48.8
1989	30.4	18.4	35.8	48.2	58.4	66.4	77.1	72.5	61.2	48.8	39.3	20.5	48.1
1990	32.3	29.8	37.4	46.8	54.3	70.4	71.7	71.1	65.9	48.8	39.1	20.6	49.0
1991	22.7	37.3	38.9	45.9	59.2	68.2	74.0	73.9	62.7	47.6	32.7	31.9	49.6
1992	31.2	37.8	41.2	50.4	59.7	65.9	68.6	67.1	62.4	49.8	30.8	18.6	48.6
1993	20.9	17.6	37.7	45.9	58.8	63.1	69.9	68.6	57.2	46.4	29.4	30.0	45.5
Record Mean	25.5	29.0	36.4	46.8	56.9	66.9	73.6	71.6	61.6	49.7	36.4	27.6	48.5
Max	39.1	42.9	50.5	61.6	71.3	81.7	89.2	87.5	78.1	66.1	50.9	40.9	63.3
Min	11.9	15.1	22.3	32.1	42.4	52.1	57.9	55.7	45.1	33.3	21.8	14.2	33.7

REFERENCE NOTES SCOTTSBLUFF, NEBRASKA

GENERAL
T - TRACE AMOUNT.
BLANK ENTRIES DENOTE MISSING/UNREPORTED DATA.
# INDICATES A STATION OR INSTRUMENT RELOCATION.

SPECIFIC

(a) - LENGTH OF RECORD IN YEARS, ALTHOUGH INDIVIDUAL MONTHS MAY BE MISSING.
0.* OR * - THE VALUE IS BETWEEN 0.0 AND 0.05
NORMALS - BASED ON THE 1961-1990 RECORD PERIOD.
EXTREMES - DATES ARE THE MOST RECENT OCCURRENCE.
WIND DIR.- NUMERALS SHOW TENS OF DEGRESS CLOCKWISE FROM TRUE NORTH. "00" INDICATES CALM.
RESULTANT DIRECTIONS ARE GIVEN TO WHOLE DEGREES.
BOLD VALUES INDICATE EXTREME VALUES WHICH OCCURRED AFTER THE ASOS SYSTEM WAS COMMISSIONED.

RECORD = PERIOD OF RECORD
RECORD MEAN PRECIPITATION IS THE MEAN OF ALL DAILY PRECIPITATION AMOUNTS DURING THE PERIOD OF RECORD.
RECORD MAX(MIN) TEMPERATURE IS THE MEAN OF ALL DAILY MAX(MIN) TEMPERATURES DURING THE PERIOD OF RECORD.
RECORD MEAN TEMPERATURE IS THE SUM OF THE RECORD MAX AND RECORD MIN DIVIDED BY 2.
AVERAGE TEMPERATURE IS THE SUM OF THE MEAN DAILY MAX AND MIN TEMPERATURE DIVIDED BY 2.

EXCEPTIONS

RECORD MEANS ARE THROUGH THE CURRENT YEAR, BEGINNING IN 1889 FOR TEMPERATURE
1889 FOR PRECIPITATION
1944 FOR SNOWFALL

HEATING DEGREE DAYS Base 65 deg. F SCOTTSBLUFF, NEBRASKA

SEASON	JULY	AUG	SEP	OCT	NOV	DEC	JAN	FEB	MAR	APR	MAY	JUNE	TOTAL
#1964-65	0	52	185	458	831	1232	1010	1060	1198	414	245	56	6741
1965-66	0	12	358	353	698	1050	1402	1187	748	667	215	74	6764
1966-67	0	31	106	500	801	1165	1049	917	729	484	428	92	6302
1967-68	0	24	125	431	904	1396	1265	917	738	648	382	39	6869
1968-69	11	35	171	434	878	1306	1278	906	1052	419	219	149	6858
1969-70	0	0	30	737	793	1054	1247	891	1065	703	252	109	6881
1970-71	0	0	261	689	861	1212	1177	1013	937	497	321	16	6984
1971-72	13	0	243	537	807	1096	1237	908	714	561	273	19	6408
1972-73	41	14	146	495	1041	1379	1245	980	825	684	312	57	7219
1973-74	18	0	249	438	883	1216	1407	863	801	523	230	53	6681
1974-75	0	28	265	394	854	1135	1183	1105	981	609	268	75	6897
1975-76	0	11	193	434	873	1019	1235	799	874	469	222	46	6175
1976-77	0	8	120	542	903	1011	1362	789	826	360	84	4	6009
1977-78	0	23	72	407	800	1131	1444	1199	752	458	286	54	6626
1978-79	2	20	99	412	930	1425	1690	1029	737	447	288	50	7129
1979-80	0	10	55	363	984	946	1295	1019	889	463	215	8	6247
1980-81	0	1	61	411	721	859	998	897	684	261	253	16	5162
1981-82	6	0	42	432	662	1044	1307	1039	842	629	289	101	6393
1982-83	4	1	155	503	921	1142	997	792	864	722	429	112	6642
1983-84	7	0	126	412	919	1627	1165	851	850	677	219	60	6913
1984-85	0	0	223	574	812	1238	1367	1096	780	422	139	75	6726
1985-86	0	16	286	534	1284	1378	986	1036	636	524	297	27	7004
1986-87	0	0	162	504	891	1118	1110	868	978	420	156	28	6235
1987-88	10	37	153	532	788	1228	1446	999	894	539	241	17	6884
1988-89	1	4	154	446	825	1138	1065	1303	897	511	214	71	6629
1989-90	0	0	169	497	764	1372	1009	977	850	541	327	30	6536
1990-91	18	4	79	497	771	1372	1307	771	802	570	220	19	6430
1991-92	5	0	137	536	965	1018	1042	783	731	438	194	45	5894
1992-93	22	57	107	465	1019	1430	1362	1323	841	567	204	121	7518
1993-94	3	23	237	568	1061	1078							

COOLING DEGREE DAYS Base 65 deg. F SCOTTSBLUFF, NEBRASKA

YEAR	JAN	FEB	MAR	APR	MAY	JUNE	JULY	AUG	SEP	OCT	NOV	DEC	TOTAL
1969	0	0	0	0	46	48	315	319	88	0	0	0	816
1970	0	0	0	0	7	99	254	263	37	0	0	0	660
1971	0	0	0	0	1	150	179	254	50	0	0	0	634
1972	0	0	0	0	16	121	203	207	36	0	0	0	583
1973	0	0	0	0	4	138	228	250	21	0	0	0	641
1974	0	0	0	0	24	192	363	135	15	0	0	0	729
1975	0	0	0	1	12	101	386	245	43	4	0	0	792
1976	0	0	0	0	18	146	359	240	66	1	0	0	830
1977	0	0	0	6	45	254	356	194	90	0	0	0	945
1978	0	0	0	0	34	180	316	198	141	0	0	0	869
1979	0	0	0	8	19	163	317	224	121	1	0	0	853
1980	0	0	0	1	22	234	417	263	104	3	0	0	1044
1981	0	0	0	10	23	213	361	250	107	3	0	0	967
1982	0	0	0	0	3	40	251	291	74	0	0	0	659
1983	0	0	0	0	6	82	297	369	99	0	0	0	853
1984	0	0	0	0	35	124	311	359	78	0	0	0	907
1985	0	0	0	3	42	121	324	225	66	0	0	0	781
1986	0	0	0	0	1	167	243	192	10	0	0	0	613
1987	0	0	0	5	20	134	320	164	26	0	0	0	669
1988	0	0	0	0	37	256	302	248	22	0	0	0	865
1989	0	0	0	10	18	118	382	242	61	0	0	0	831
1990	0	0	0	0	1	199	232	202	113	3	0	0	750
1991	0	0	0	2	46	125	289	284	73	2	0	0	821
1992	0	0	0	7	36	79	140	131	39	0	0	0	432
1993	0	0	0	0	18	71	159	142	10	0	0	0	400

SNOWFALL (inches) SCOTTSBLUFF, NEBRASKA

SEASON	JULY	AUG	SEP	OCT	NOV	DEC	JAN	FEB	MAR	APR	MAY	JUNE	TOTAL
1964-65	0.0	0.0	0.0	0.0	0.1	8.8	4.3	6.6	3.6	2.0	1.0	0.0	26.4
1965-66	0.0	0.0	0.8	0.0	1.0	4.2	5.0	2.7	2.3	5.6	0.2	0.0	21.8
1966-67	0.0	0.0	0.0	T	2.3	3.4	3.0	0.6	2.1	3.3	7.5	0.0	22.2
1967-68	0.0	0.0	0.0	0.0	1.6	12.9	2.1	2.1	6.7	8.1	T	0.0	33.5
1968-69	0.0	0.0	0.0	0.0	1.1	6.1	8.5	5.4	6.1	0.3	0.0	T	27.5
1969-70	0.0	0.0	0.0	21.6	1.0	2.9	6.6	2.0	10.2	17.6	T	0.0	61.9
1970-71	0.0	0.0	T	11.2	0.4	4.4	2.2	4.4	14.2	3.0	T	0.0	39.8
1971-72	0.0	0.0	T	1.0	3.4	1.3	4.1	2.3	1.0	6.2	0.0	0.0	19.3
1972-73	0.0	0.0	0.0	5.3	14.0	10.2	6.9	5.9	4.8	5.2	T	0.0	52.3
1973-74	0.0	0.0	T	4.6	12.2	11.8	9.9	1.1	17.9	1.7	0.0	0.0	59.2
1974-75	0.0	0.0	3.8	T	T	2.7	5.9	5.8	17.7	10.7	T	0.0	46.6
1975-76	0.0	0.0	0.0	6.1	4.7	15.5	12.0	5.5	5.9	0.9	0.0	T	50.6
1976-77	0.0	0.0	0.0	0.1	5.4	1.3	8.1	0.1	10.9	1.0	0.0	0.0	26.9
1977-78	0.0	0.0	0.0	0.4	6.3	11.8	14.9	15.3	1.6	0.8	3.4	0.0	54.5
1978-79	0.0	0.0	0.0	2.9	7.2	17.5	10.5	1.6	7.7	4.3	6.4	0.0	58.1
1979-80	0.0	0.0	0.0	3.8	13.5	5.3	17.1	10.6	23.5	3.9	0.8	0.0	78.5
1980-81	0.0	0.0	0.0	1.2	3.7	1.7	8.3	2.1	3.2	1.3	0.0	0.0	21.5
1981-82	0.0	0.0	0.0	0.6	0.2	1.8	4.4	3.8	2.8	2.1	T	0.0	15.7
1982-83	0.0	0.0	0.0	5.3	4.3	4.9	2.8	T	16.4	8.9	2.6	0.0	45.2
1983-84	0.0	0.0	T	0.0	18.5	7.9	4.8	2.7	13.5	13.3	1.2	0.0	61.9
1984-85	0.0	0.0	1.6	2.2	2.5	7.1	7.2	2.3	3.2	1.7	0.0	0.0	27.8
1985-86	0.0	0.0	5.1	0.3	17.5	18.1	0.5	16.0	6.0	12.5	1.0	0.0	77.0
1986-87	0.0	0.0	0.0	1.5	5.1	2.8	3.7	23.4	16.6	0.4	0.0	0.0	53.5
1987-88	0.0	0.0	0.0	0.3	5.8	17.0	14.6	1.9	10.6	11.6	T	0.0	61.8
1988-89	0.0	0.0	0.0	T	3.5	4.1	T	13.0	9.2	2.7	T	T	32.5
1989-90	0.0	0.0	T	2.0	0.6	11.6	7.0	8.6	18.6	6.0	1.5	0.0	55.9
1990-91	0.0	0.0	0.0	7.6	10.7	4.4	5.7	2.2	4.2	2.6	1.8	T	39.2
1991-92	0.0	0.0	0.0	5.9	4.7	0.1	7.8	1.0	6.6	T	0.0	T	26.1
1992-93	T	0.0	T	4.1	10.3	10.4	7.9	19.0	11.5	2.0	T	T	65.2
1993-94	T	0.0	2.5	3.3	11.1	6.2							
Record Mean	T	0.0	0.4	2.4	5.4	6.5	6.5	5.6	9.0	4.7	0.9	T	41.4

VALENTINE, NEBRASKA

Valentine, located near the northern edge of the Sandhills and cattle country of Nebraska, is near the extreme northern border of the state. The city lies in the valley of the Niobrara River, a branch of the Missouri River, about 160 miles above the junction with the Missouri. It is the county seat of Cherry County and had its beginning in the fall of 1882. The name, Valentine, was selected for the new town in honor of Congressman E. K. Valentine, who represented the Third Congressional District of which Cherry County was a part.

The inland location offers a wide variety of weather. The high afternoon temperatures during the two warmest months, July and August, average nearly 90 degrees and the corresponding humidity averages about 40 percent. Uncomfortably warm nights are few with low morning temperatures averaging about 60 degrees. The temperature seldom reaches 100 degrees or more during the summer. The two coldest months are January and February. The minimum temperature generally reaches -20 degrees or colder once each winter.

Valentines location frequently places it in the path of cold Canadian air mass outbreaks during the cold season, alternating with mild, dry air moving across the Rockies from the Pacific. One or two bitterly cold days usually occur each winter when the temperature will stay below zero throughout the day. Blizzards are not frequent, but at least one is likely each winter season. An occasional severe blizzard occurs about once in every three or four winters. Blowing and drifting snow reduce visibility to zero and bring outdoor activities and travel to a complete stop. Lives may be lost for anyone caught away from shelter, and there is usually loss of livestock which varies with the intensity and duration of the storm. Temperatures below 32 degrees have occurred as late as mid-June and as early as early September, and low temperature records in the 30s have occurred during the summer with light frost in low places. However, these are rare occurrences and temperatures below 50 degrees are not common in July and August.

About 65 percent of the annual precipitation falls during the growing season, May through September, and is predominantly the nighttime thunderstorm type with June being the wettest month.

The spring and fall seasons have mostly pleasant days, with the fall season having the most uniform character with lighter winds and gradually falling temperatures as the season progresses. In the spring the weather is windy and extremely variable, with summer-like days mixed with some of cold of winter. The widest extremes of temperature occur in March.

Some of the damaging weather elements other than blizzards, are high winds, which dig blow outs in the sand hills, and an occasional hailstorm. Damage from hail is not extensive and confined mostly to buildings and gardens since the area is mostly prairie. Floods are unknown. Tornadoes occasionally occur but seldom do much damage because ranches and towns are widely scattered.

NORMALS, MEANS, AND EXTREMES

VALENTINE, NEBRASKA

LATITUDE: 42°52' N LONGITUDE: 100°33' W ELEVATION: FT. GRND 2587 BARO 2590 TIME ZONE: CENTRAL WBAN: 24032

	(a)	JAN	FEB	MAR	APR	MAY	JUNE	JULY	AUG	SEP	OCT	NOV	DEC	YEAR
TEMPERATURE °F:														
Normals														
-Daily Maximum		32.6	37.8	46.5	60.0	71.1	81.9	88.9	86.6	76.1	64.1	47.3	35.2	60.7
-Daily Minimum		6.6	11.9	20.9	32.9	44.1	54.0	60.4	57.6	46.3	33.9	20.8	9.3	33.2
-Monthly		19.6	24.9	33.7	46.5	57.6	68.0	74.7	72.1	61.2	49.0	34.1	22.3	47.0
Extremes														
-Record Highest	39	72	78	85	100	99	110	114	108	103	96	82	73	114
-Year		1987	1982	1993	1992	1989	1988	1990	1965	1983	1993	1965	1990	JUL 1990
-Record Lowest	39	-30	-28	-29	3	19	30	38	34	17	-1	-22	-39	-39
-Year		1988	1975	1980	1975	1967	1969	1971	1988	1991	1991	1959	1989	DEC 1989
NORMAL DEGREE DAYS:														
Heating (base 65°F)		1407	1123	970	555	248	49	0	16	167	496	927	1324	7282
Cooling (base 65°F)		0	0	0	0	19	139	305	236	53	0	0	0	752
% OF POSSIBLE SUNSHINE	29	63	62	59	59	62	69	76	76	71	68	61	60	66
MEAN SKY COVER (tenths)														
Sunrise - Sunset	38	6.1	6.3	6.5	6.3	6.1	5.1	4.4	4.4	4.4	4.9	6.0	5.9	5.5
MEAN NUMBER OF DAYS:														
Sunrise to Sunset														
-Clear	38	8.8	7.4	7.3	7.6	8.2	10.8	13.6	13.7	14.0	12.5	8.6	9.6	122.3
-Partly Cloudy	38	7.9	7.4	8.0	8.3	9.9	10.5	11.7	10.8	7.2	8.2	7.9	7.5	105.2
-Cloudy	38	14.2	13.3	15.5	14.1	12.9	8.7	5.7	6.5	8.6	10.1	13.3	13.6	136.3
Precipitation														
.01 inches or more	38	4.5	5.0	6.7	8.4	10.5	10.3	9.4	8.0	6.5	4.5	4.5	4.6	82.9
Snow,Ice pellets,hail														
1.0 inches or more	38	1.5	1.9	2.1	1.1	0.*	0.0	0.0	0.0	0.*	0.4	1.6	1.4	10.2
Thunderstorms	10	0.0	0.0	0.2	1.5	7.6	10.6	11.8	9.3	3.7	0.8	0.2	0.0	45.7
Heavy Fog Visibility														
1/4 mile or less	10	0.4	0.7	0.8	0.6	0.5	0.1	0.5	0.4	0.4	0.4	0.1	0.3	5.4
Temperature °F														
-Maximum														
90° and above	38	0.0	0.0	0.0	0.4	1.4	6.7	14.1	12.8	4.9	0.6	0.0	0.0	40.9
32° and below	38	14.0	10.5	5.8	0.5	0.0	0.0	0.0	0.0	0.0	0.3	5.1	11.8	48.1
-Minimum														
32° and below	38	30.8	27.9	27.3	15.0	2.6	0.1	0.0	0.0	2.1	13.6	27.3	30.8	177.4
0° and below	38	10.3	6.0	1.7	0.0	0.0	0.0	0.0	0.0	0.0	0.*	1.5	6.4	26.1
AVG. STATION PRESS. (mb)	7	925.6	924.7	921.2	922.4	921.1	922.0	923.1	923.4	923.9	924.9	924.7	924.2	923.4
RELATIVE HUMIDITY (%)														
Hour 00	26	73	75	75	67	71	71	68	70	67	66	71	72	71
Hour 06 (Local Time)	26	75	78	80	77	79	79	79	80	77	74	75	75	77
Hour 12	26	62	61	56	48	48	47	47	46	44	45	54	59	51
Hour 18	26	63	60	53	44	45	44	41	41	39	44	57	62	49
PRECIPITATION (inches):														
Water Equivalent														
-Normal		0.29	0.43	1.04	1.67	3.16	2.89	3.06	2.28	1.53	0.91	0.60	0.37	18.23
-Maximum Monthly	38	0.82	1.43	4.23	4.01	8.96	7.09	8.96	6.71	5.91	2.06	2.62	1.81	8.96
-Year		1979	1962	1977	1968	1962	1983	1983	1966	1973	1986	1985	1987	JUL 1983
-Minimum Monthly	38	T	0.01	0.14	0.25	0.45	0.44	0.28	0.37	0.11	T	0.01	T	T
-Year		1961	1957	1956	1967	1966	1976	1980	1969	1958	1958	1962	1991	DEC 1991
-Maximum in 24 hrs	38	0.66	0.84	1.64	1.91	2.83	2.96	3.40	3.38	2.45	1.34	1.12	1.23	3.40
-Year		1988	1987	1977	1971	1962	1992	1983	1971	1973	1973	1985	1987	JUL 1983
Snow,Ice pellets,hail														
-Maximum Monthly	38	15.2	17.5	51.0	16.1	3.3	T	0.0	T	18.4	4.7	34.5	22.5	51.0
-Year		1982	1962	1977	1970	1979	1991		1990	1985	1971	1985	1987	MAR 1977
-Maximum in 24 hrs	38	7.1	8.3	24.0	9.0	3.3	T	0.0	T	18.4	3.8	13.5	18.3	24.0
-Year		1988	1977	1977	1968	1979	1991		1990	1985	1971	1985	1987	MAR 1977
WIND:														
Mean Speed (mph)	25	9.3	9.0	10.3	11.1	10.9	9.9	9.1	9.1	9.6	9.2	9.5	9.1	9.7
Prevailing Direction through 1964		W	N	N	N	S	S	S	S	N	S	N	W	N
Fastest Obs. 1 Min.														
-Direction (!!)	11	33	32	31	31	31	29	32	18	31	32	32	32	29
-Speed (MPH)	11	40	38	40	43	44	53	39	33	40	36	40	40	53
-Year		1993	1984	1984	1986	1983	1993	1988	1988	1984	1990	1993	1985	JUN 1993
Peak Gust														
-Direction (!!)														
-Speed (mph)														
-Date														

PRECIPITATION (inches)

VALENTINE, NEBRASKA

YEAR	JAN	FEB	MAR	APR	MAY	JUNE	JULY	AUG	SEP	OCT	NOV	DEC	ANNUAL
1964	0.14	0.23	0.30	2.30	1.49	2.79	3.38	0.68	1.51	0.02	0.04	0.45	13.33
1965	0.23	0.08	0.20	1.33	3.73	4.23	2.40	0.92	3.70	0.78	0.32	0.29	18.21
1966	0.28	0.39	0.39	1.06	0.45	3.65	1.71	6.71	2.62	1.01	0.11	0.44	18.82
1967	0.14	0.08	0.71	0.25	2.73	4.74	2.04	1.82	0.78	0.98	0.28	0.38	14.93
1968	0.20	0.33	0.22	4.01	2.98	5.73	1.70	3.10	0.60	0.56	0.41	0.62	20.46
1969	0.33	0.64	0.51	0.89	0.68	2.02	2.42	0.37	0.63	1.87	0.10	0.27	10.73
1970	0.02	0.11	0.64	2.80	2.26	2.40	2.64	0.59	0.71	1.11	0.26	0.28	13.82
1971	0.49	0.67	0.51	3.26	2.99	1.46	1.60	3.45	0.97	1.76	0.33	0.27	17.76
1972	0.41	0.13	0.34	2.32	3.64	2.55	4.63	1.00	0.46	0.63	1.37	0.32	17.80
1973	0.69	0.23	2.28	2.10	3.51	0.95	2.81	1.93	5.91	1.60	1.12	0.13	23.26
1974	0.13	0.21	0.55	1.30	2.57	1.26	1.26	1.51	0.74	0.84	0.19	0.01	10.57
1975	0.52	0.16	1.04	0.84	0.76	2.40	1.84	1.75	0.32	0.33	1.01	0.23	11.20
1976	0.27	0.32	0.57	2.03	1.56	0.44	2.12	0.99	2.22	0.38	0.09	0.22	11.21
1977	0.34	0.79	4.23	3.54	6.14	2.93	6.12	4.40	1.72	0.88	0.59	1.00	32.68
1978	0.25	1.22	0.19	3.25	3.38	2.52	4.17	3.19	0.56	0.12	0.55	0.68	20.08
1979	0.82	0.29	1.80	1.52	3.16	4.54	2.37	0.80	1.41	1.61	0.61	0.02	18.95
1980	0.38	0.68	1.52	0.46	2.67	2.79	0.28	3.03	0.43	1.37	0.09	0.42	14.12
1981	0.09	0.13	1.30	0.49	2.73	1.28	5.51	3.65	0.88	1.95	0.67	0.76	19.44
1982	0.50	0.07	1.05	0.74	6.70	2.11	3.20	3.37	2.70	1.80	0.88	0.53	23.65
1983	0.10	0.06	1.31	1.39	5.19	7.09	8.96	0.83	1.18	0.66	1.17	0.56	28.50
1984	0.11	0.34	0.90	3.40	2.73	2.09	6.53	1.40	0.57	0.55	0.57	0.12	19.31
1985	0.56	0.06	0.57	1.54	0.70	2.23	2.03	2.90	2.58	0.82	2.62	0.29	16.90
1986	0.11	0.61	1.96	2.63	2.95	2.31	2.36	2.47	2.70	2.06	0.59	T	20.75
1987	0.13	1.33	2.51	0.55	3.78	3.46	3.31	3.38	0.86	0.57	1.97	1.81	23.66
1988	0.78	0.66	0.55	1.92	4.54	3.36	2.25	3.36	1.70	0.08	0.62	0.18	20.00
1989	0.02	0.74	0.74	0.48	1.24	1.33	2.85	2.05	1.95	0.57	0.12	0.33	12.42
1990	0.04	0.51	1.31	1.20	4.43	2.03	3.10	2.36	0.67	0.51	0.64	0.27	17.07
1991	0.15	0.59	1.77	2.65	4.80	5.03	2.33	2.59	0.56	1.52	1.26	T	23.25
1992	0.54	0.21	1.41	0.79	0.81	5.40	2.25	4.44	1.38	0.43	0.48	0.14	18.28
1993	0.43	0.74	0.99	3.54	1.11	3.92	3.23	2.40	1.78	1.08	1.08	0.50	20.80
Record Mean	0.44	0.52	1.11	1.99	2.83	3.07	2.79	2.37	1.37	1.03	0.60	0.45	18.57

AVERAGE TEMPERATURE (deg. F)

VALENTINE, NEBRASKA

YEAR	JAN	FEB	MAR	APR	MAY	JUNE	JULY	AUG	SEP	OCT	NOV	DEC	ANNUAL
1964	26.2	27.2	28.9	46.3	61.1	67.9	78.0	68.7	61.0	49.6	33.2	19.5	47.3
1965	24.0	23.7	21.1	48.2	58.1	67.8	73.7	71.1	51.5	53.3	36.1	30.3	46.6
1966	12.8	16.9	36.8	41.5	58.0	68.3	79.1	67.9	60.9	49.4	33.9	24.9	45.9
1967	25.5	26.2	37.7	47.7	51.7	64.5	71.8	70.2	62.1	50.1	33.6	20.0	46.8
1968	20.8	25.0	39.1	44.3	52.1	67.5	72.5	71.1	60.3	50.5	34.9	16.5	46.2
1969	12.7	22.4	23.5	51.3	60.1	62.4	74.5	75.6	66.7	42.3	38.5	26.1	46.3
1970	17.0	29.4	27.6	42.6	60.7	69.2	75.8	74.9	61.6	45.1	34.4	21.1	46.6
1971	16.6	20.3	32.4	47.8	55.5	71.3	70.9	75.4	60.2	49.0	36.1	24.4	46.7
1972	16.5	24.0	38.0	45.4	58.3	67.7	70.3	71.0	61.4	45.6	29.6	17.1	45.4
1973	21.2	28.6	39.6	45.0	55.8	68.4	73.8	76.8	57.7	52.7	34.8	23.8	48.2
1974	19.3	30.2	37.2	47.9	57.3	67.5	79.1	69.0	57.9	52.0	35.5	27.3	48.3
1975	24.3	19.0	28.4	44.0	57.9	66.8	77.6	73.4	58.7	51.3	31.2	25.6	46.5
1976	21.4	33.5	33.5	48.6	56.5	68.0	75.3	74.5	62.4	45.1	28.3	24.3	47.6
1977	11.1	30.7	32.8	50.6	63.0	71.2	74.7	67.4	62.6	48.7	34.4	19.8	47.2
1978	8.1	10.3	32.8	44.2	57.5	67.9	72.4	70.6	66.0	48.1	29.7	12.5	43.3
1979	4.1	13.3	33.0	45.7	54.9	67.1	73.4	70.7	66.5	49.3	30.1	32.2	45.0
1980	19.4	23.8	30.5	48.7	58.4	70.5	77.7	72.6	63.7	47.6	38.1	28.4	48.3
1981	28.5	26.4	38.7	54.1	54.7	68.3	73.9	71.2	63.3	48.2	39.8	23.6	49.3
1982	10.2	25.6	34.6	42.8	57.0	62.5	75.1	73.9	61.0	47.5	31.4	26.6	45.7
1983	29.5	34.4	34.7	40.5	52.7	65.1	75.4	78.3	63.3	50.0	34.7	3.9	46.9
1984	21.8	32.0	32.2	42.5	56.7	67.8	73.9	75.0	57.3	47.5	35.2	21.3	46.9
1985	17.7	21.6	37.2	51.3	62.5	64.1	75.2	69.5	58.3	47.6	18.0	15.9	44.9
1986	26.7	24.3	41.6	45.3	56.7	70.2	74.3	68.7	59.0	48.3	31.0	29.3	48.0
1987	28.5	33.0	33.2	50.9	62.8	69.7	76.5	68.9	60.2	45.4	37.8	26.2	49.4
1988	13.2	21.0	34.5	45.9	60.8	76.4	75.3	73.2	61.0	47.3	35.7	26.2	47.5
1989	27.8	13.8	32.2	48.1	58.2	65.5	76.8	73.0	61.8	49.5	35.4	16.3	46.5
1990	31.7	28.2	36.9	46.0	56.0	70.3	72.2	72.4	66.2	49.0	38.3	17.2	48.7
1991	17.2	34.0	37.5	47.8	59.7	71.5	74.1	73.7	62.6	46.1	28.9	29.6	48.6
1992	28.8	34.8	39.0	48.1	59.2	64.2	65.7	66.4	62.2	49.1	30.2	20.6	47.4
1993	17.7	16.5	36.0	44.0	57.1	63.9	69.9	69.5	57.3	46.8	29.0	28.7	44.7
Record Mean	20.7	24.2	33.2	46.3	57.0	67.3	74.1	72.1	61.9	49.5	34.5	24.6	47.2
Max	33.0	36.4	45.2	58.9	69.6	79.9	87.4	85.6	76.0	63.8	47.1	36.6	60.0
Min	8.4	11.9	21.2	33.7	44.5	54.7	60.7	58.5	47.7	35.1	21.9	12.7	34.3

REFERENCE NOTES

VALENTINE, NEBRASKA

GENERAL
T - TRACE AMOUNT.
BLANK ENTRIES DENOTE MISSING/UNREPORTED DATA.
# INDICATES A STATION OR INSTRUMENT RELOCATION.

SPECIFIC

(a) - LENGTH OF RECORD IN YEARS, ALTHOUGH INDIVIDUAL MONTHS MAY BE MISSING.
0.* OR * - THE VALUE IS BETWEEN 0.0 AND 0.05
NORMALS - BASED ON THE 1961-1990 RECORD PERIOD.
EXTREMES - DATES ARE THE MOST RECENT OCCURRENCE.
WIND DIR.- NUMERALS SHOW TENS OF DEGRESS CLOCKWISE FROM TRUE NORTH. "00" INDICATES CALM.
RESULTANT DIRECTIONS ARE GIVEN TO WHOLE DEGREES.
BOLD VALUES INDICATE EXTREME VALUES WHICH OCCURRED AFTER THE ASOS SYSTEM WAS COMMISSIONED.

RECORD = PERIOD OF RECORD
RECORD MEAN PRECIPITATION IS THE MEAN OF ALL DAILY PRECIPITATION AMOUNTS DURING THE PERIOD OF RECORD.
RECORD MAX(MIN) TEMPERATURE IS THE MEAN OF ALL DAILY MAX(MIN) TEMPERATURES DURING THE PERIOD OF RECORD.
RECORD MEAN TEMPERATURE IS THE SUM OF THE RECORD MAX AND RECORD MIN DIVIDED BY 2.
AVERAGE TEMPERATURE IS THE SUM OF THE MEAN DAILY MAX AND MIN TEMPERATURE DIVIDED BY 2.

EXCEPTIONS

1. THUNDERSTORMS AND HEAVY FOG ARE THROUGH 1964 AND MAY BE INCOMPLETE, DUE TO PART-TIME OPERATIONS.
2. RELATIVE HUMIDITY AND MEAN WIND SPEED ARE THROUGH 1964 AND 1977 TO DATE.
3. FASTEST MILE WIND IS THROUGH AUGUST 1982.

RECORD MEANS ARE THROUGH THE CURRENT YEAR, BEGINNING IN 1889 FOR TEMPERATURE
1889 FOR PRECIPITATION
1956 FOR SNOWFALL

HEATING DEGREE DAYS Base 65 deg. F — VALENTINE, NEBRASKA

SEASON	JULY	AUG	SEP	OCT	NOV	DEC	JAN	FEB	MAR	APR	MAY	JUNE	TOTAL
1964-65	0	74	188	469	947	1403	1266	1150	1354	496	235	22	7604
1965-66	0	16	410	355	864	1068	1615	1341	865	699	247	67	7547
1966-67	0	51	179	476	928	1237	1221	1081	843	512	439	62	7029
1967-68	26	34	133	466	934	1389	1365	1154	798	614	393	56	7362
1968-69	15	24	179	451	896	1497	1618	1188	1278	401	223	132	7902
1969-70	0	8	46	696	787	1198	1482	991	1152	665	172	47	7244
1970-71	1	0	204	611	910	1353	1494	1244	1002	508	291	7	7625
1971-72	25	0	215	492	861	1251	1500	1180	827	584	257	43	7235
1972-73	30	38	165	596	1056	1483	1356	1013	779	592	297	28	7433
1973-74	7	0	236	376	899	1270	1413	970	855	510	243	65	6844
1974-75	0	39	233	400	876	1163	1256	1284	1126	642	236	67	7322
1975-76	5	5	231	435	1008	1214	1348	907	969	485	268	58	6933
1976-77	5	3	151	613	1094	1255	1672	952	990	431	101	8	7275
1977-78	0	35	122	499	914	1397	1760	1526	993	618	254	72	8190
1978-79	4	32	109	515	1053	1622	1888	1446	986	574	326	67	8622
1979-80	14	24	77	478	1040	1010	1408	1189	1063	495	231	26	7055
1980-81	0	14	112	536	804	1128	1124	1075	806	325	328	37	6289
1981-82	21	7	114	514	748	1274	1697	1100	936	659	250	108	7428
1982-83	1	16	189	536	1000	1183	1096	850	931	730	385	91	7008
1983-84	6	0	168	461	903	1892	1334	952	1008	667	288	43	7722
1984-85	0	0	280	542	886	1351	1461	1211	854	415	136	112	7248
1985-86	13	29	279	531	1404	1517	1182	1134	720	591	264	20	7684
1986-87	0	36	189	511	1013	1098	1126	893	976	424	124	35	6425
1987-88	10	50	170	601	811	1198	1600	1270	936	570	204	4	7424
1988-89	6	21	160	543	871	1195	1144	1428	1013	523	234	91	7229
1989-90	5	11	169	471	882	1505	1026	1025	864	581	292	26	6857
1990-91	25	0	117	501	795	1479	1476	861	845	514	207	1	6821
1991-92	1	2	167	585	1076	1089	1115	867	800	515	211	70	6498
1992-93	49	75	141	491	1039	1370	1462	1354	892	623	244	100	7840
1993-94	5	31	253	561	1076	1116							

COOLING DEGREE DAYS Base 65 deg. F — VALENTINE, NEBRASKA

YEAR	JAN	FEB	MAR	APR	MAY	JUNE	JULY	AUG	SEP	OCT	NOV	DEC	TOTAL
1969	0	0	0	0	80	61	301	343	104	0	0	0	889
1970	0	0	0	0	49	181	341	315	108	0	0	0	994
1971	0	0	0	0	4	201	214	330	78	0	0	0	827
1972	0	0	0	0	56	132	202	231	63	0	0	0	684
1973	0	0	0	0	18	136	285	373	24	1	0	0	837
1974	0	0	0	4	13	144	446	171	25	3	0	0	806
1975	0	0	0	16	24	124	400	272	47	18	0	0	901
1976	0	0	0	1	11	157	330	306	81	2	0	0	888
1977	0	0	0	3	46	203	307	116	57	0	0	0	732
1978	0	0	0	0	27	163	239	213	148	0	0	0	790
1979	0	0	0	4	20	136	280	208	129	0	0	0	777
1980	0	0	0	15	34	198	404	255	78	4	0	0	988
1981	0	0	0	6	16	145	304	207	70	0	0	0	748
1982	0	0	0	0	10	39	321	299	73	0	0	0	742
1983	0	0	0	0	12	99	338	422	125	0	0	0	996
1984	0	0	0	0	38	134	280	318	55	5	0	0	830
1985	0	0	0	9	66	92	337	176	87	0	0	0	767
1986	0	0	0	4	15	185	295	160	15	0	0	0	674
1987	0	0	0	11	61	184	372	178	34	0	0	0	840
1988	0	0	0	4	80	351	331	285	48	0	0	0	1099
1989	0	0	0	19	29	110	378	265	81	0	0	0	882
1990	0	0	0	16	17	192	257	237	160	12	0	0	891
1991	0	0	0	4	52	203	290	281	102	5	0	0	937
1992	0	0	0	14	41	54	74	125	63	6	0	0	377
1993	0	0	0	0	7	73	163	179	28	5	0	0	455

SNOWFALL (inches) — VALENTINE, NEBRASKA

SEASON	JULY	AUG	SEP	OCT	NOV	DEC	JAN	FEB	MAR	APR	MAY	JUNE	TOTAL
1964-65	0.0	0.0	0.0	0.0	1.0	4.5	2.3	2.4	4.0	T	0.0	0.0	14.2
1965-66	0.0	0.0	0.0	0.0	2.2	1.3	4.4	4.5	2.9	6.2	T	0.0	21.5
1966-67	0.0	0.0	0.0	T	3.5	5.6	1.5	0.9	3.0	T	0.1	0.0	14.6
1967-68	0.0	0.0	0.0	0.0	1.1	5.9	5.9	3.2	0.6	9.6	T	0.0	26.3
1968-69	0.0	0.0	0.0	0.0	1.7	9.6	5.3	6.7	4.6	T	0.0	0.0	27.9
1969-70	0.0	0.0	0.0	4.7	0.9	3.3	0.9	2.2	10.1	16.1	0.0	0.0	38.2
1970-71	0.0	0.0	0.0	2.9	1.9	4.2	9.3	9.3	8.0	7.7	0.0	0.0	43.3
1971-72	0.0	0.0	0.0	4.7	1.0	3.7	5.7	2.0	0.4	0.4	T	0.0	17.9
1972-73	0.0	0.0	0.0	2.6	12.0	7.6	9.2	2.6	1.1	0.4	T	0.0	35.5
1973-74	0.0	0.0	T	T	13.3	3.0	2.5	3.0	4.1	0.5	0.0	0.0	26.4
1974-75	0.0	0.0	0.0	0.0	1.5	0.2	5.8	5.7	16.1	3.1	0.0	0.0	32.4
1975-76	0.0	0.0	0.0	1.1	11.5	1.4	5.8	5.2	8.3	T	0.0	0.0	33.3
1976-77	0.0	0.0	0.0	0.8	1.5	3.5	7.8	10.3	51.0	2.5	0.0	0.0	77.4
1977-78	0.0	0.0	0.0	1.1	3.0	8.3	3.8	15.9	2.5	9.7	0.0	0.0	44.3
1978-79	0.0	0.0	0.0	0.0	7.0	14.8	10.8	5.1	7.3	8.0	3.3	0.0	56.3
1979-80	0.0	0.0	0.0	1.9	6.2	0.5	7.6	12.6	21.6	2.9	T	0.0	53.3
1980-81	0.0	0.0	0.0	2.6	0.5	5.8	1.5	3.8	2.2	0.0	0.0	0.0	16.4
1981-82	0.0	0.0	0.0	1.4	8.0	11.2	15.2	1.5	6.0	4.6	0.0	0.0	47.9
1982-83	0.0	0.0	0.0	3.0	2.0	5.3	0.8	0.2	5.7	1.9	0.0	0.0	18.9
1983-84	0.0	0.0	T	0.0	12.0	12.4	1.5	4.0	7.2	8.5	0.1	0.0	45.7
1984-85	0.0	0.0	0.1	0.3	1.5	1.5	5.5	0.7	3.0	0.1	0.0	0.0	12.7
1985-86	0.0	0.0	18.4	T	34.5	4.9	1.6	2.6	6.9	3.6	0.0	0.0	72.5
1986-87	0.0	0.0	0.0	1.0	5.7	T	1.9	12.3	15.7	T	0.0	0.0	36.6
1987-88	0.0	0.0	0.0	0.7	3.2	22.5	9.6	5.9	5.9	13.2	0.0	0.0	61.0
1988-89	0.0	0.0	0.0	0.0	3.2	2.2	0.3	12.0	13.8	T	T	T	31.5
1989-90	0.0	0.0	0.0	T	T	5.0	0.4	8.6	10.9	T	0.0	T	24.9
1990-91	0.0	T	0.0	1.8	6.4	3.8	3.2	6.3	6.6	2.2	T	T	30.3
1991-92	0.0	0.0	0.0	4.6	11.9	T	5.4	1.1	3.2	T	T	0.0	26.2
1992-93	0.0	0.0	0.0	0.0	5.0	1.9	7.3	14.4	1.4	12.6	0.0	0.0	42.6
1993-94	0.0	0.0	T	0.6	6.4	4.8							
Record Mean	0.0	T	0.5	1.0	5.1	4.6	4.6	5.6	7.5	3.9	0.1	T	32.8

ELKO, NEVADA

Elko is located in the Humbolt River Valley of northeastern Nevada. Weather observations are taken at the Flight Service Station which is located at the Municipal Airport on the west side of town. The elevation at the airport is just above 5,000 feet.

The Ruby mountain range, with many peaks near or exceeding 10,000 feet in height, dominates the landscape from about 40 miles northeast through 40 miles southeast of Elko. The immediate terrain consists of sagebrush–covered valleys and hills. The highest hills are approximately 2,500 feet above the valley floors. A few areas, mostly in the higher mountains, are covered with sparse stands of juniper, aspen, pinion pine, and spruce. The only heavily forested area in northeastern Nevada is in the Jarbidge Wilderness Area north of Elko near the Idaho border.

Because of the high elevation and proximity of the mountains, there is a wide range between the normal high and low temperatures. High radiative cooling at night makes cool nights the rule, even in mid summer.

Normal precipitation is light, especially during the summer months when the precipitation falls mostly as light showers which do not contribute much toward crop growth. The precipitation that falls between November and June (rain and snow) is critical to agriculture in the area. Not only is the precipitation that falls directly on the fields a benefit to farmers and ranchers, but the runoff from snowfall that accumulates in the mountains is used for irrigation.

The principal crop in northeast Nevada is hay. Cattle ranching is a major industry within the area. The ranges ordinarily furnish excellent summer pasture for cattle. Hay crops are needed for winter feeding.

Mining is another major industry. Many of the mines are located in the mountains at rather high elevations and are affected by daily weather. This is especially true during the winter when snow and rain may cause poor or impassable road conditions, thereby halting mining operations.

Transportation by air, rail, or road is seldom affected by the weather for more than short periods.

Based on the 1951–1980 period, the average first occurrence of 32 degrees Fahrenheit in the fall is September 8 and the average last occurrence in the spring is June 5.

NORMALS, MEANS, AND EXTREMES

ELKO, NEVADA

LATITUDE: 40 °50'N LONGITUDE: 115 °47' W ELEVATION: FT. GRND 5050 BARO 5080 TIME ZONE: PACIFIC WBAN: 24121

	(a)	JAN	FEB	MAR	APR	MAY	JUNE	JULY	AUG	SEP	OCT	NOV	DEC	YEAR
TEMPERATURE °F:														
Normals														
-Daily Maximum		36.7	43.0	50.2	59.1	69.4	80.2	91.0	88.6	78.3	65.9	49.1	37.4	62.4
-Daily Minimum		13.4	19.9	25.0	29.5	36.8	44.6	50.3	48.6	38.9	29.6	22.5	14.0	31.1
-Monthly		25.1	31.5	37.6	44.3	53.1	62.4	70.7	68.7	58.7	47.7	35.8	25.7	46.8
Extremes														
-Record Highest	63	64	70	77	86	92	104	107	107	99	88	78	64	107
-Year		1990	1986	1966	1992	1977	1981	1981	1978	1950	1980	1980	1940	JUL 1981
-Record Lowest	63	-43	-37	-9	-2	10	23	30	20	9	3	-12	-38	-43
-Year		1937	1933	1952	1936	1965	1976	1932	1992	1934	1991	1931	1932	JAN 1937
NORMAL DEGREE DAYS:														
Heating (base 65°F)		1237	938	849	621	369	138	18	42	235	536	876	1218	7077
Cooling (base 65°F)		0	0	0	0	0	60	194	157	46	0	0	0	457
% OF POSSIBLE SUNSHINE														
MEAN SKY COVER (tenths)														
Sunrise - Sunset	43	6.8	6.6	6.8	6.6	6.0	4.4	3.4	3.5	3.3	4.4	6.2	6.4	5.4
MEAN NUMBER OF DAYS:														
Sunrise to Sunset														
-Clear	56	6.7	6.3	6.3	6.3	8.1	12.9	17.3	17.8	18.3	14.1	8.4	7.8	130.2
-Partly Cloudy	56	7.6	7.4	8.1	9.1	10.2	9.6	9.6	8.9	6.7	8.0	6.9	6.9	99.1
-Cloudy	56	16.7	14.6	16.6	14.6	12.6	7.5	4.0	4.3	5.0	8.9	14.6	16.3	135.9
Precipitation														
.01 inches or more	63	9.0	8.6	8.8	7.4	7.8	5.6	3.4	3.6	3.8	4.8	7.1	8.4	78.3
Snow,Ice pellets,hail														
1.0 inches or more	45	3.3	2.0	2.2	1.0	0.3	0.0	0.0	0.0	0.*	0.3	1.8	3.2	14.2
Thunderstorms	44	0.2	0.3	0.3	1.0	3.3	3.6	4.8	4.5	2.1	0.5	0.2	0.1	20.8
Heavy Fog Visibility														
1/4 mile or less	44	1.7	0.7	0.5	0.3	0.4	0.1	0.*	0.1	0.1	0.3	0.5	1.3	5.9
Temperature °F														
-Maximum														
90° and above	29	0.0	0.0	0.0	0.0	0.4	5.9	19.6	15.2	3.6	0.0	0.0	0.0	44.6
32° and below	29	9.7	4.2	0.8	0.1	0.0	0.0	0.0	0.0	0.0	0.1	1.7	8.7	25.3
-Minimum														
32° and below	29	29.1	25.8	25.5	20.1	8.1	1.0	0.*	0.4	6.6	20.7	25.1	29.2	191.6
0° and below	29	5.6	1.9	0.1	0.0	0.0	0.0	0.0	0.0	0.0	0.0	0.5	4.4	12.6
AVG. STATION PRESS.(mb)	19	847.0	846.3	842.6	842.9	842.9	844.0	845.6	845.9	845.8	846.9	846.3	847.1	845.3
RELATIVE HUMIDITY (%)														
Hour 04	28	79	79	78	73	71	67	57	57	62	67	77	78	70
Hour 10 (Local Time)	28	72	66	55	43	37	32	25	27	32	41	61	70	47
Hour 16	28	60	52	42	33	30	25	19	20	23	29	48	59	37
Hour 22	28	78	76	70	59	54	48	37	39	47	57	72	76	59
PRECIPITATION (inches):														
Water Equivalent														
-Normal		0.98	0.80	0.96	0.82	1.00	0.91	0.33	0.65	0.62	0.65	1.11	1.10	9.93
-Maximum Monthly	63	3.35	2.93	2.39	2.17	4.09	2.61	2.35	4.61	3.22	2.76	2.77	4.21	4.61
-Year		1956	1932	1989	1963	1971	1963	1950	1970	1978	1938	1942	1983	AUG 1970
-Minimum Monthly	63	0.04	0.06	0.04	0.02	T	T	0.00	T	T	T	T	T	0.00
-Year		1961	1988	1988	1992	1974	1974	1963	1969	1951	1988	1959	1976	JUL 1963
-Maximum in 24 hrs	63	1.27	0.89	1.02	1.10	1.73	1.85	1.04	4.13	2.32	1.31	1.31	1.62	4.13
-Year		1951	1936	1975	1943	1971	1968	1950	1970	1978	1939	1950	1950	AUG 1970
Snow,Ice pellets,hail														
-Maximum Monthly	62	27.4	26.1	23.2	15.6	11.3	T	T	T	2.0	5.6	16.8	33.2	33.2
-Year		1950	1932	1967	1975	1971	1982	1990	1993	1982	1984	1985	1983	DEC 1983
-Maximum in 24 hrs	45	16.7	9.1	13.8	10.0	8.6	T	T	T	2.0	5.2	9.0	9.3	16.7
-Year		1951	1949	1967	1975	1971	1982	1990	1993	1982	1963	1965	1992	JAN 1951
WIND:														
Mean Speed (mph)	42	5.3	5.8	6.6	7.2	6.9	6.7	6.2	5.9	5.4	5.0	5.1	5.1	5.9
Prevailing Direction through 1963		SW	SW	SW	SW	SW	SW	SW	SW	SW	SW	SW	SW	SW
Fastest Obs. 1 Min.														
-Direction (!!!)	37	23	27	29	25	34	27	21	16	27	29	20	27	27
-Speed (MPH)	37	40	39	41	48	55	61	37	35	58	35	40	50	61
-Year		1952	1963	1952	1956	1955	1984	1991	1966	1959	1953	1984	1952	JUN 1984
Peak Gust														
-Direction (!!!)														
-Speed (mph)														
-Date														

PRECIPITATION (inches)

ELKO, NEVADA

YEAR	JAN	FEB	MAR	APR	MAY	JUNE	JULY	AUG	SEP	OCT	NOV	DEC	ANNUAL
1964	1.27	0.10	0.97	1.14	1.15	2.24	0.18	0.10	0.14	0.48	1.07	3.30	12.14
1965	0.84	0.31	0.44	1.81	1.08	1.21	0.62	1.31	0.20	0.63	1.96	0.76	11.17
1966	0.24	0.37	0.26	0.64	0.55	0.36	0.69	0.52	0.44	0.02	0.58	1.83	6.50
1967	1.00	0.08	1.79	0.87	0.84	1.19	1.03	0.08	0.21	0.27	0.60	0.66	8.62
1968	1.16	1.45	1.12	0.53	1.15	2.60	0.04	1.94	0.36	0.79	1.56	1.93	14.63
1969	1.24	1.76	0.35	0.28	0.27	2.11	0.24	T	0.17	1.11	0.34	1.83	9.70
1970	2.36	0.40	0.50	0.37	0.37	1.29	0.48	4.61	0.52	0.64	1.32	1.70	14.56
1971	0.58	1.03	0.53	0.96	4.09	1.01	0.21	0.98	0.74	0.74	1.23	1.57	13.67
1972	0.39	0.57	0.56	0.36	0.16	1.46	T	0.29	1.01	1.92	0.98	0.77	8.47
1973	1.17	0.96	0.56	0.88	0.70	0.56	0.46	0.29	0.18	0.64	1.40	1.30	9.10
1974	0.61	0.38	0.86	0.58	T	T	0.19	0.12	0.00	1.16	0.34	0.53	4.77
1975	1.79	1.06	2.37	1.70	0.98	0.40	0.15	0.10	0.18	1.42	0.94	0.25	11.34
1976	0.35	0.68	0.25	0.65	0.50	0.64	0.44	0.91	1.84	0.58	0.26	T	7.10
1977	0.30	0.26	0.13	0.18	1.44	1.03	0.22	0.77	0.26	0.01	0.96	0.90	6.46
1978	0.68	0.97	1.88	1.98	0.25	0.18	0.58	0.02	3.22	0.25	0.61	0.52	11.14
1979	1.91	1.20	0.59	0.43	0.42	0.38	0.32	0.36	0.25	0.43	1.10	0.35	7.74
1980	3.11	1.89	0.77	1.22	3.15	0.80	0.33	0.10	0.42	0.19	0.62	0.21	12.81
1981	0.64	0.33	1.20	0.75	0.80	0.24	0.02	0.19	0.13	0.69	0.60	3.19	8.78
1982	0.82	0.65	1.94	0.50	1.04	0.54	0.69	1.24	2.55	1.11	1.78	0.86	13.72
1983	1.73	1.34	1.91	1.28	0.60	0.47	0.01	1.25	1.57	1.21	2.76	4.21	18.34
1984	0.57	0.80	1.25	1.00	0.24	1.29	1.04	0.46	0.11	1.75	1.40	0.45	10.36
1985	0.54	0.15	1.09	0.23	0.60	0.17	0.25	0.02	1.17	0.16	2.14	0.78	7.30
1986	0.18	1.86	0.52	1.17	0.75	0.39	0.12	0.02	0.81	0.04	0.13	0.09	6.08
1987	0.54	0.68	1.13	0.26	1.80	0.69	0.14	0.01	0.09	0.55	1.97	0.76	8.62
1988	1.27	0.06	0.04	0.46	0.91	0.58	0.08	0.26	0.11	T	1.94	1.01	6.72
1989	0.46	0.93	2.39	0.28	0.36	0.50	0.18	0.52	0.69	0.27	0.79	0.51	7.88
1990	0.97	0.78	1.07	1.51	0.96	0.97	0.19	0.56	0.15	0.07	0.98	1.22	9.43
1991	0.49	0.46	0.62	0.86	1.71	0.06	0.20	0.25	0.58	1.29	1.29	0.04	7.85
1992	0.17	0.75	1.64	0.02	0.40	0.67	0.27	0.17	0.01	0.54	1.03	1.89	7.56
1993	1.98	0.93	0.68	0.24	0.44	1.43	0.36	0.09	0.41	0.76	0.07	0.22	7.61
Record Mean	1.17	0.91	0.93	0.71	0.87	0.71	0.34	0.35	0.40	0.66	0.84	1.06	8.96

AVERAGE TEMPERATURE (deg. F)

ELKO, NEVADA

YEAR	JAN	FEB	MAR	APR	MAY	JUNE	JULY	AUG	SEP	OCT	NOV	DEC	ANNUAL
#1964	18.8	20.1	29.5	40.8	50.0	56.9	69.9	65.3	54.2	49.3	30.5	30.3	42.9
1965	30.3	31.4	33.5	44.9	48.5	57.7	67.7	66.1	52.3	50.2	38.3	19.2	45.0
1966	22.5	24.0	38.0	44.3	57.8	62.7	70.1	69.0	61.3	45.9	39.1	18.7	46.1
1967	25.2	31.5	37.3	40.1	51.4	62.7	74.9	75.0	64.9	49.9	41.6	23.1	48.2
1968	27.0	39.8	42.2	42.6	54.0	64.1	75.4	65.6	59.2	50.6	38.2	28.1	48.9
1969	28.8	25.4	30.9	45.4	57.6	61.0	71.1	70.8	61.6	41.1	34.3	27.1	46.2
1970	27.8	36.8	33.4	36.4	50.9	60.1	70.0	70.0	53.0	41.0	38.0	22.9	45.0
1971	30.3	34.0	36.1	41.8	48.2	58.1	68.7	70.8	53.2	43.7	33.9	18.8	44.8
1972	23.4	32.4	42.0	42.3	54.0	64.1	67.5	67.0	54.3	46.3	31.9	19.1	45.3
1973	20.0	31.4	36.8	43.0	56.1	62.8	70.8	69.0	57.0	47.9	36.8	30.3	46.8
1974	27.9	32.2	41.1	43.5	52.6	65.2	69.8	65.2	58.7	47.9	37.6	26.7	47.4
1975	18.9	29.2	35.8	37.6	51.0	59.7	71.3	64.1	58.8	46.4	32.4	29.4	44.6
1976	25.8	32.8	33.5	41.4	55.6	59.2	69.4	63.0	61.1	46.6	37.7	27.6	46.1
1977	24.1	35.6	34.8	49.3	50.6	69.3	71.7	70.9	60.9	51.0	38.9	34.7	49.3
1978	32.2	34.7	44.9	45.3	52.1	60.6	69.8	69.2	56.6	50.6	35.3	23.6	47.9
1979	22.7	33.0	39.8	46.4	56.5	65.8	73.9	71.2	66.1	54.6	33.6	31.5	49.6
1980	32.2	39.0	37.0	47.3	51.5	60.5	71.6	67.2	61.6	49.6	40.0	34.2	49.3
1981	34.6	35.5	40.8	48.2	54.2	67.3	73.9	72.8	61.6	44.8	40.5	34.1	50.7
1982	23.2	29.6	37.8	42.3	52.5	61.6	69.1	70.6	58.6	45.4	34.1	28.0	46.1
1983	30.4	31.8	41.8	42.7	52.6	62.1	68.7	72.1	62.7	50.8	36.5	28.7	48.4
1984	17.1	23.8	36.0	41.9	54.7	59.7	71.9	70.7	60.7	43.3	37.7	20.9	44.9
1985	21.8	26.8	35.1	48.0	54.0	64.7	75.9	65.9	53.7	46.2	30.1	23.1	45.4
1986	32.3	37.6	43.8	45.0	52.3	65.7	67.0	70.5	53.0	46.0	34.0	24.9	47.7
1987	21.5	31.1	37.5	49.5	55.6		66.8	66.6	59.6	50.2	35.7	25.7	47.0
1988	20.3	29.3	37.2	46.5	51.5	65.2	72.0	67.7	56.8	53.0	33.5	21.7	46.2
1989	11.6	22.3	41.5	49.0	52.3	61.5	70.8	65.5	57.8	46.1	33.4	27.8	45.0
1990	27.8	26.7	40.9	50.0	50.1	61.8	70.5	67.0	63.9	45.8	33.6	14.5	46.1
1991	22.5	37.4	36.7	41.2	48.1	59.0	70.4	68.7	59.7	46.5	35.5	27.2	46.1
1992	24.4	35.6	41.1	48.1	56.5	60.7	65.1	66.7	56.9	47.6	28.2	17.2	45.7
1993	15.5	20.5	36.4	42.9	55.7	56.0	59.7	62.0	55.6	45.6	26.4	25.4	41.8
Record Mean	23.6	29.9	37.0	44.5	52.8	61.2	69.8	67.4	57.9	47.1	35.0	25.8	46.0
Max	36.3	42.0	50.4	60.1	69.8	79.9	90.9	88.7	79.0	65.9	49.6	38.4	62.6
Min	11.0	17.8	23.6	28.8	35.7	42.5	48.8	46.1	36.8	28.3	20.4	13.1	29.4

REFERENCE NOTES

ELKO, NEVADA

GENERAL
T - TRACE AMOUNT.
BLANK ENTRIES DENOTE MISSING/UNREPORTED DATA.
# INDICATES A STATION OR INSTRUMENT RELOCATION.

SPECIFIC

(a) - LENGTH OF RECORD IN YEARS, ALTHOUGH INDIVIDUAL MONTHS MAY BE MISSING.
0.* OR * - THE VALUE IS BETWEEN 0.0 AND 0.05
NORMALS - BASED ON THE 1961-1990 RECORD PERIOD.
EXTREMES - DATES ARE THE MOST RECENT OCCURRENCE
WIND DIR.- NUMERALS SHOW TENS OF DEGRESS CLOCKWISE FROM TRUE NORTH. "00" INDICATES CALM.
RESULTANT DIRECTIONS ARE GIVEN TO WHOLE DEGREES.
BOLD VALUES INDICATE EXTREME VALUES WHICH OCCURRED AFTER THE ASOS SYSTEM WAS COMMISSIONED.

RECORD = PERIOD OF RECORD
RECORD MEAN PRECIPITATION IS THE MEAN OF ALL DAILY PRECIPITATION AMOUNTS DURING THE PERIOD OF RECORD.
RECORD MAX(MIN) TEMPERATURE IS THE MEAN OF ALL DAILY MAX(MIN) TEMPERATURES DURING THE PERIOD OF RECORD.
RECORD MEAN TEMPERATURE IS THE SUM OF THE RECORD MAX AND RECORD MIN DIVIDED BY 2.
AVERAGE TEMPERATURE IS THE SUM OF THE MEAN DAILY MAX AND MIN TEMPERATURE DIVIDED BY 2.

EXCEPTIONS

RECORD MEANS ARE THROUGH THE CURRENT YEAR, BEGINNING IN 1910 FOR TEMPERATURE
1871 FOR PRECIPITATION
1932 FOR SNOWFALL

HEATING DEGREE DAYS Base 65 deg. F ELKO, NEVADA

SEASON	JULY	AUG	SEP	OCT	NOV	DEC	JAN	FEB	MAR	APR	MAY	JUNE	TOTAL
# 1964-65	7	85	317	481	1029	1067	1071	937	969	596	505	215	7279
1965-66	11	43	375	452	793	1416	1312	1142	830	615	226	128	7343
1966-67	8	27	126	584	769	1427	1227	932	852	737	413	123	7225
1967-68	0	0	56	459	695	1291	1168	725	700	664	334	100	6192
1968-69	3	94	195	438	799	1137	1119	1100	1051	581	225	138	6880
1969-70	24	8	120	736	916	1166	1148	782	974	849	431	190	7344
1970-71	5	1	357	736	804	1297	1068	863	891	690	515	205	7432
1971-72	14	6	351	651	928	1426	1284	940	704	673	339	59	7375
1972-73	17	28	317	573	985	1419	1386	931	867	652	273	135	7583
1973-74	5	19	236	522	844	1064	1145	910	734	637	382	87	6585
1974-75	28	55	187	523	815	1184	1422	994	899	815	427	168	7517
1975-76	8	77	180	570	970	1094	1209	927	967	701	286	177	7166
1976-77	12	80	145	563	810	1154	1262	816	927	464	441	16	6690
1977-78	0	27	164	430	773	933	1009	845	617	583	398	135	5914
1978-79	23	47	267	439	888	1277	1306	889	775	553	265	86	6815
1979-80	0	4	32	320	933	1032	1013	747	862	523	410	165	6041
1980-81	0	34	119	467	743	947	933	819	744	497	336	67	5706
1981-82	0	4	131	618	726	952	1289	987	836	676	381	130	6730
1982-83	32	0	225	598	925	1143	1066	924	713	661	390	114	6791
1983-84	29	0	105	434	847	1119	1480	1187	894	686	318	201	7300
1984-85	0	10	163	664	811	1360	1331	1060	921	505	335	69	7229
1985-86	0	42	338	573	1042	1294	1002	759	650	596	399	49	6744
1986-87	18	5	370	583	924	1235	1341	943	844	459	286		
1987-88	50	32	175	451	872	1211	1381	1028	856	550	420	102	7128
1988-89	0	16	255	365	938	1338	1647	1194	723	475	387	111	7449
1989-90	4	59	214	578	940	1147	1147	1067	742	445	457	135	6935
1990-91	12	50	86	587	933	1559	1310	767	869	710	518	176	7577
1991-92	2	10	178	566	879	1164	1250	845	734	503	256	152	6539
1992-93	51	89	242	534	1100	1477	1528	1239	879	658	283	272	8352
1993-94	168	113	279	591	1153	1222							

COOLING DEGREE DAYS Base 65 deg. F ELKO, NEVADA

YEAR	JAN	FEB	MAR	APR	MAY	JUNE	JULY	AUG	SEP	OCT	NOV	DEC	TOTAL
1969	0	0	0	0	1	26	220	194	25	0	0	0	466
1970	0	0	0	0	0	48	166	162	2	0	0	0	378
1971	0	0	0	0	0	7	140	191	4	0	0	0	342
1972	0	0	0	0	3	38	102	96	3	0	0	0	242
1973	0	0	0	0	3	74	192	152	2	0	0	0	423
1974	0	0	0	0	4	97	185	68	6	0	0	0	360
1975	0	0	0	0	0	15	212	56	0	0	0	0	283
1976	0	0	0	0	0	11	157	24	36	0	0	0	228
1977	0	0	0	0	3	152	217	216	48	0	0	0	636
1978	0	0	0	0	0	9	179	186	21	0	0	0	395
1979	0	0	0	0	9	120	284	207	70	4	0	0	694
1980	0	0	0	0	0	36	211	109	25	0	0	0	381
1981	0	0	0	2	9	145	282	254	36	0	0	0	728
1982	0	0	0	0	0	36	165	181	39	0	0	0	421
1983	0	0	0	0	11	33	151	228	40	0	0	0	463
1984	0	0	0	0	3	48	223	197	42	0	0	0	513
1985	0	0	0	0	0	66	342	77	7	0	0	0	492
1986	0	0	0	0	12	75	89	186	16	0	0	0	378
1987	0	0	0	0	3		113	89	20	0	0	0	277
1988	0	0	0	0	6	113	224	105	19	0	0	0	467
1989	0	0	0	0	0	13	188	78	4	0	0	0	283
1990	0	0	0	0	0	47	193	121	61	0	0	0	422
1991	0	0	0	0	0	2	176	131	26	0	0	0	335
1992	0	0	0	0	0	28	65	152	7	0	0	0	252
1993	0	0	0	0	1	9	10	26	5	0	0	0	51

SNOWFALL (inches) ELKO, NEVADA

SEASON	JULY	AUG	SEP	OCT	NOV	DEC	JAN	FEB	MAR	APR	MAY	JUNE	TOTAL
1964-65	0.0	0.0	0.0	0.0	12.2	11.0	9.3	4.9	4.5	7.7	0.8	0.0	50.4
1965-66	0.0	0.0	T	0.8	9.4	14.3	4.8	7.2	2.6	2.0	T	0.0	41.1
1966-67	0.0	0.0	0.0	0.1	0.9	17.4	12.5	1.2	23.2	11.1	1.2	0.0	67.6
1967-68	0.0	0.0	0.0	T	6.4	8.5	16.1	0.5	10.4	3.8	0.4	T	46.1
1968-69	0.0	0.0	T	0.0	3.3	24.9	10.8	17.4	4.9	2.4	0.0	0.0	63.7
1969-70	0.0	0.0	0.0	2.3	T	11.1	1.5	1.0	4.6	4.1	0.4	0.0	25.0
1970-71	0.0	0.0	0.0	4.0	3.1	17.8	9.0	11.8	2.0	10.1	11.3	T	69.1
1971-72	0.0	0.0	T	5.4	8.9	13.4	1.1	9.1	3.6	0.6	T	0.0	42.1
1972-73	0.0	0.0	0.0	1.0	7.3	8.6	10.5	5.5	4.2	1.0	0.0	0.0	38.1
1973-74	0.0	0.0	0.0	0.0	6.8	11.8	5.1	1.0	2.2	2.1	T	0.0	29.0
1974-75	0.0	0.0	0.0	0.0	T	6.7	14.5	5.9	6.3	15.6	8.0	0.0	57.0
1975-76	0.0	0.0	0.0	0.6	7.3	2.5	4.4	5.2	3.5	0.8	0.0	0.0	24.3
1976-77	0.0	0.0	0.0	0.0	0.5	T	4.3	1.8	1.5	2.0	2.2	0.0	12.3
1977-78	0.0	0.0	0.0	0.0	3.0	5.3	5.5	5.5	T	T	T	0.0	19.3
1978-79	0.0	0.0	T	T	1.0	2.0	8.3	13.0	2.4	2.1	0.6	T	29.4
1979-80	0.0	0.0	0.0	1.5	9.2	2.3	12.5	3.4	3.9	0.5	2.0	0.0	35.3
1980-81	0.0	0.0	0.0	1.2	0.5	0.5	4.4	1.4	2.6	2.9	0.2	T	13.7
1981-82	0.0	0.0	0.0	T	0.3	9.8	11.2	0.3	13.8	3.5	T	T	38.9
1982-83	0.0	0.0	2.0	T	8.1	6.6	16.5	10.6	9.6	1.9	0.2	0.0	55.5
1983-84	0.0	0.0	0.0	0.0	13.4	33.2	6.6	5.8	5.1	5.9	T	0.0	70.0
1984-85	0.0	0.0	0.0	5.6	5.4	4.9	5.6	2.0	7.7	0.4	0.0	0.0	31.6
1985-86	0.0	0.0	0.0	0.7	16.8	5.1	0.8	2.0	1.9	1.4	0.1	0.0	28.8
1986-87	0.0	0.0	T	0.0	1.1	1.0	6.9	5.7	3.1	T	0.0	0.0	17.8
1987-88	0.0	0.0	0.0	0.0	0.3	6.1	14.5	0.2	1.0	T	3.7	0.0	25.8
1988-89	0.0	0.0	0.0	0.0	11.3	16.1	11.0	9.6	4.6	T	0.0	0.0	52.6
1989-90	0.0	0.0	0.0	0.1	4.3	0.0	6.9	9.4	3.5	T	2.1	0.0	26.3
1990-91	T	T	0.0	0.0	3.2	12.7	2.1	0.7	5.0	1.5	2.0	0.0	27.2
1991-92	0.0	0.0	0.0	2.9	5.9	0.7	2.4	3.2	1.8	T	0.0	0.0	16.9
1992-93	0.0	0.0	0.0	0.0	4.5	19.9	23.5	15.9	1.2	0.6	T	0.0	65.6
1993-94	0.0	T	0.0	0.0	1.4	0.9							
Record Mean	T	T	0.1	0.8	4.5	8.1	9.6	6.2	5.3	2.4	0.8	T	37.7

ELY, NEVADA

Ely, Nevada, is located within but near the southern rim of the Great Basin. The neighboring terrain consists of alternate mountain ranges and sagebrush covered valleys. Principal cover on the mountains is juniper, pinion, and, at higher elevations, white fir, and white pine. Valley floors in this region are near 6,000 feet above sea level. This high elevation is conducive to sharp nighttime radiation, which produces pleasant summer nights but also reduces the season that is free from freezing temperatures.

The Ely weather station is near the center of Steptoe Valley, which is 5 miles wide at this point. The mountains of the Egan Range to the west and the Schell Creek Range to the east range up to 4,000 feet above the station elevation and prevent strong surface winds from these directions. A very pronounced drainage wind sweeps down the valley during the morning hours. More precipitation is noted near the mountains than is measured in the center of the valley.

Because of low annual precipitation, farming is limited to areas that can be irrigated from mountain streams or wells. The livestock industry is predominant in agriculture. Cultivated crops consist almost entirely of grains and forage.

The mountain ranges provide fairly good summer pastures for cattle and the lowlands provide food for a good portion of the winter in dry or snow-softened desert plants. All stock, however, has to be finished for market in the feed yards. Sheep share the mountain pastures with cattle in the summer, and as winter approaches move out on the wide flat valleys. These browsers eat snow for water and consume a wide variety of desert plants, including the lowly sagebrush. It is not uncommon for bands of sheep to spend an entire winter without supplemental feed.

Based on the 1951–1980 period, the average first occurrence of 32 degrees Fahrenheit in the fall is September 6 and the average last occurrence in the spring is June 16.

NORMALS, MEANS, AND EXTREMES

ELY, NEVADA

LATITUDE: 39°17'N LONGITUDE: 114°51' W ELEVATION: FT. GRND 6253 BARO 6257 TIME ZONE: PACIFIC WBAN: 23154

	(a)	JAN	FEB	MAR	APR	MAY	JUNE	JULY	AUG	SEP	OCT	NOV	DEC	YEAR
TEMPERATURE °F:														
Normals														
-Daily Maximum		39.7	43.6	48.4	57.0	67.3	78.3	87.0	84.4	75.2	63.5	49.2	40.6	61.2
-Daily Minimum		9.4	15.4	20.7	26.0	33.7	40.7	48.0	46.5	37.3	28.2	19.0	10.6	28.0
-Monthly		24.6	29.5	34.6	41.5	50.5	59.6	67.5	65.5	56.3	45.8	34.2	25.6	44.6
Extremes														
-Record Highest	55	68	67	73	82	89	99	100	97	93	84	75	67	100
-Year		1951	1986	1966	1992	1984	1954	1985	1981	1990	1967	1975	1958	JUL 1985
-Record Lowest	55	-27	-30	-13	-5	7	18	30	24	15	-3	-15	-29	-30
-Year		1949	1989	1952	1982	1950	1976	1983	1960	1968	1991	1985	1990	FEB 1989
NORMAL DEGREE DAYS:														
Heating (base 65°F)		1252	994	942	705	450	188	14	62	274	595	924	1221	7621
Cooling (base 65°F)		0	0	0	0	0	26	92	77	13	0	0	0	208
% OF POSSIBLE SUNSHINE	54	67	67	70	69	72	80	80	81	82	75	67	67	73
MEAN SKY COVER (tenths)														
Sunrise - Sunset	50	6.2	6.3	6.4	6.2	6.0	4.3	3.9	3.9	3.4	4.3	5.8	5.9	5.2
MEAN NUMBER OF DAYS:														
Sunrise to Sunset														
-Clear	55	8.7	7.0	7.3	7.5	7.5	13.3	14.8	14.6	17.2	14.4	9.7	9.1	131.2
-Partly Cloudy	55	7.6	7.2	8.3	9.1	11.2	10.2	11.2	11.8	7.7	8.1	8.0	8.1	108.5
-Cloudy	55	14.7	14.0	15.4	13.4	12.2	6.5	5.0	4.6	5.1	8.5	12.3	13.8	125.4
Precipitation														
.01 inches or more	55	6.8	7.0	8.5	7.2	7.3	4.9	5.5	5.5	4.4	5.0	5.2	6.1	73.6
Snow,Ice pellets,hail														
1.0 inches or more	55	2.6	2.4	3.1	1.9	0.9	0.1	0.0	0.0	0.1	0.8	1.4	2.5	15.7
Thunderstorms	55	0.2	0.3	0.5	1.4	4.2	4.7	8.1	8.3	3.3	1.5	0.3	0.2	33.1
Heavy Fog Visibility														
1/4 mile or less	55	0.3	0.2	0.4	0.3	0.1	0.*	0.0	0.*	0.2	0.2	0.3	0.4	2.4
Temperature °F														
-Maximum														
90° and above	55	0.0	0.0	0.0	0.0	0.0	2.3	10.1	5.5	0.6	0.0	0.0	0.0	18.5
32° and below	55	7.7	4.5	1.9	0.3	0.*	0.0	0.0	0.0	0.0	0.2	2.5	6.3	23.5
-Minimum														
32° and below	55	30.6	27.5	29.3	24.2	13.4	3.5	0.1	0.5	8.0	22.7	28.2	30.5	218.6
0° and below	55	7.5	4.0	1.1	0.1	0.0	0.0	0.0	0.0	0.0	0.1	1.4	4.9	19.0
AVG. STATION PRESS.(mb)	21	809.4	808.6	805.9	806.8	807.0	808.9	811.1	811.3	810.7	811.0	809.7	809.8	809.2
RELATIVE HUMIDITY (%)														
Hour 04	41	72	74	72	68	67	59	52	56	59	65	70	71	65
Hour 10 (Local Time)	41	61	58	50	39	35	28	24	27	31	39	51	58	42
Hour 16	41	55	50	43	33	31	23	22	23	24	31	46	54	36
Hour 22	39	71	71	66	57	54	42	39	43	46	56	66	69	57
PRECIPITATION (inches):														
Water Equivalent														
-Normal		0.70	0.65	0.96	1.00	1.15	0.88	0.69	0.83	1.01	0.89	0.67	0.70	10.13
-Maximum Monthly	55	2.08	2.19	2.40	3.41	3.26	3.53	2.30	2.51	4.99	3.67	1.82	2.11	4.99
-Year		1993	1969	1952	1978	1977	1963	1987	1983	1982	1981	1960	1966	SEP 1982
-Minimum Monthly	55	T	0.01	0.07	T	T	T	T	T	T	0.00	T	T	0.00
-Year		1948	1972	1972	1989	1948	1978	1948	1985	1953	1952	1959	1976	OCT 1952
-Maximum in 24 hrs	55	0.95	1.54	0.86	1.04	1.42	1.50	1.47	1.01	2.87	1.39	1.29	1.12	2.87
-Year		1952	1969	1954	1947	1955	1963	1987	1992	1982	1976	1960	1966	SEP 1982
Snow,Ice pellets,hail														
-Maximum Monthly	55	24.8	20.0	24.8	24.5	12.1	5.6	T	T	6.3	12.1	17.3	22.3	24.8
-Year		1967	1976	1958	1963	1975	1939	1992	1993	1982	1981	1985	1968	JAN 1967
-Maximum in 24 hrs	55	13.1	10.4	10.6	10.7	8.0	5.6	T	T	4.7	7.3	12.9	12.7	13.1
-Year		1943	1956	1954	1970	1975	1939	1992	1993	1986	1954	1978	1970	JAN 1943
WIND:														
Mean Speed (mph)	53	10.1	10.3	10.7	10.9	10.7	10.6	10.3	10.4	10.3	10.0	9.9	9.9	10.4
Prevailing Direction through 1963		S	S	S	S	S	S	S	S	S	S	S	S	S
Fastest Mile														
-Direction (!!)	50	SE	S	SW	S	S	SW	S	E	S	S	S	SE	S
-Speed (MPH)	50	66	56	65	59	74	63	50	57	57	65	51	61	74
-Year		1952	1954	1989	1951	1948	1952	1957	1954	1953	1950	1954	1952	MAY 1948
Peak Gust														
-Direction (!!)	10	S	SE	SW	S	S	S	W	E	S	SW	SE	S	SW
-Speed (mph)	10	46	52	64	62	63	60	56	60	58	61	55	46	64
-Date		1993	1986	1989	1987	1984	1988	1992	1986	1989	1985	1985	1988	MAR 1989

PRECIPITATION (inches)

ELY, NEVADA

YEAR	JAN	FEB	MAR	APR	MAY	JUNE	JULY	AUG	SEP	OCT	NOV	DEC	ANNUAL
1964	1.41	0.07	1.24	2.77	1.17	2.44	0.02	0.58	0.09	0.19	0.93	1.79	12.70
1965	0.46	0.64	0.46	0.74	0.54	1.25	1.12	1.52	1.56	0.27	0.93	1.28	10.77
1966	0.23	0.31	0.16	0.16	0.46	0.14	0.16	0.61	1.34	0.10	0.30	2.11	6.08
1967	1.86	0.10	0.37	1.38	3.05	2.83	0.84	0.41	2.23	0.13	0.84	0.69	14.73
1968	0.15	0.92	0.67	1.26	1.00	1.12	1.32	1.04	0.10	1.44	0.22	0.79	10.03
1969	1.24	2.19	0.41	0.98	0.28	2.80	0.55	0.34	0.37	0.91	0.79	0.59	11.45
1970	0.11	0.14	0.59	1.55	0.01	1.09	1.81	1.45	0.45	0.23	1.69	1.57	10.69
1971	0.63	0.57	0.20	1.31	2.89	0.09	0.17	0.25	0.39	1.08	0.59	1.25	9.42
1972	0.17	0.01	0.07	0.88	0.32	0.83	0.17	0.47	1.82	1.02	0.14	0.69	6.59
1973	1.34	0.71	2.17	0.20	0.38	1.14	0.43	2.06	0.07	0.88	1.10	0.75	11.23
1974	0.41	0.29	0.67	0.18	0.30	T	0.29	0.02	0.01	1.54	0.23	0.28	4.22
1975	0.74	0.76	1.59	1.20	1.48	0.31	1.04	0.51	0.55	0.91	0.29	0.39	9.77
1976	0.38	1.51	0.77	0.77	0.45	0.34	1.57	0.16	0.66	1.48	0.16	T	8.25
1977	0.39	0.09	0.74	0.17	3.26	0.49	0.49	1.59	0.50	0.33	0.24	0.90	9.19
1978	0.64	1.27	2.00	3.41	0.45	T	0.19	0.23	1.33	0.82	1.42	0.71	12.47
1979	0.89	0.59	1.07	0.22	1.44	0.15	1.27	0.58	0.07	0.76	0.28	0.07	7.39
1980	1.55	1.08	1.57	0.51	2.55	0.72	0.76	0.35	1.65	0.37	0.55	1.12	12.78
1981	0.77	0.16	1.32	1.10	2.02	0.15	0.24	0.07	0.36	3.67	0.17	0.26	10.29
1982	1.06	0.31	2.07	0.72	1.57	0.05	0.58	1.41	4.99	1.28	1.03	0.46	15.53
1983	1.41	1.33	1.18	1.87	0.38	2.28	0.09	2.51	0.88	0.50	0.96	1.45	14.84
1984	0.36	0.39	1.09	0.94	0.35	0.63	2.18	2.01	3.73	1.41	0.99	0.76	14.84
1985	0.49	0.42	1.07	0.17	1.33	0.43	0.58	T	1.82	1.44	1.55	0.59	9.89
1986	0.29	0.75	1.47	1.32	0.51	0.02	0.09	1.24	1.42	1.24	0.18	0.07	8.60
1987	0.76	0.61	0.91	0.33	2.35	0.15	2.30	1.21	0.05	1.43	1.53	0.67	12.30
1988	1.22	0.12	0.29	1.62	0.62	0.62	0.15	1.41	0.15	0.40	1.24	0.82	8.66
1989	0.35	0.50	0.61	T	1.36	1.01	0.59	1.25	0.46	0.30	0.15	0.02	6.60
1990	0.59	1.31	0.79	1.14	1.55	0.82	0.32	0.20	0.64	0.67	0.42	0.31	8.76
1991	0.11	0.17	1.70	0.57	2.81	0.35	0.31	0.91	1.32	0.98	0.48	0.27	9.98
1992	0.52	0.68	1.35	0.14	0.53	0.83	1.37	1.70	0.25	1.26	0.25	0.90	9.78
1993	2.08	1.42	1.15	0.24	0.88	1.17	0.32	0.78	0.15	1.03	0.69	0.15	10.06
Record Mean	0.69	0.63	0.96	0.95	1.06	0.77	0.64	0.70	0.82	0.83	0.63	0.63	9.31

AVERAGE TEMPERATURE (deg. F)

ELY, NEVADA

YEAR	JAN	FEB	MAR	APR	MAY	JUNE	JULY	AUG	SEP	OCT	NOV	DEC	ANNUAL
1964	19.3	22.3	27.9	39.6	48.6	56.6	68.4	64.7	54.4	49.6	29.3	26.0	42.2
1965	29.4	27.8	32.4	41.9	45.9	55.2	65.5	63.6	50.5	49.2	37.6	24.5	43.6
1966	20.1	22.5	36.7	42.6	55.5	60.6	68.6	66.3	59.1	45.7	38.1	27.0	45.2
1967	26.2	28.4	38.4	34.7	49.2	54.6	68.1	68.1	57.9	47.7	37.6	17.6	44.1
1968	23.1	35.8	36.7	38.0	49.3	59.4	68.4	61.1	54.5	46.7	34.8	23.8	44.3
1969	31.2	25.9	26.2	43.2	56.9	57.4	68.5	69.6	60.7	40.4	35.1	29.8	45.4
1970	29.3	35.0	32.9	34.8	50.7	58.1	67.3	68.1	52.2	41.1	36.0	21.7	43.9
1971	24.2	29.9	34.7	41.3	47.0	59.1	68.2	68.4	52.5	40.3	32.0	18.9	43.1
1972	24.1	32.5	41.2	41.6	51.1	61.9	68.1	63.8	54.1	45.4	30.9	20.1	44.6
1973	20.3	26.9	30.3	39.1	52.7	59.6	66.7	66.1	53.9	45.7	32.6	27.0	43.4
1974	23.3	28.2	38.6	39.9	52.7	63.8	67.7	63.9	57.4	44.8	36.4	24.4	45.1
1975	24.1	28.4	31.8	34.1	47.5	57.2	68.0	63.6	57.5	44.8	32.1	28.5	43.1
1976	26.3	30.8	30.7	39.8	53.6	58.1	67.1	60.7	56.8	44.1	36.7	27.3	44.3
1977	22.1	30.5	28.4	45.0	44.9	62.5	67.2	65.8	57.4	47.6	36.4	30.8	44.9
1978	29.3	30.6	40.8	40.9	47.9	59.2	67.0	64.0	53.9	48.9	29.9	20.8	44.4
1979	17.1	26.6	34.7	42.3	51.8	60.0	67.4	63.6	61.1	49.9	31.3	31.4	44.8
1980	28.0	34.9	33.1	44.2	47.3	58.3	68.2	64.9	57.5	45.1	36.1	31.7	45.8
1981	31.5	31.3	35.6	46.5	50.0	63.3	69.4	68.1	59.7	42.9	38.8	33.8	47.6
1982	22.5	30.6	33.6	38.8	49.6	57.2	65.7	67.7	54.7	41.8	33.1	25.0	43.3
1983	28.9	30.2	36.7	37.6	47.7	57.9	65.8	65.5	59.2	48.1	33.2	27.1	44.8
1984	24.8	29.7	35.2	39.3	54.6	57.4	67.5	65.5	58.1	40.2	34.2	22.1	44.1
1985	19.5	24.4	32.5	46.4	52.9	63.4	68.5	65.5	52.0	44.3	27.5	25.0	43.5
1986	34.4	35.4	41.1	42.4	51.1	63.3	65.7	69.0	51.5	43.5	35.8	28.1	46.8
1987	21.5	29.7	35.3	47.0	51.5	61.9	64.1	65.1	57.5	49.2	34.7	24.2	45.1
1988	21.2	31.2	35.9	44.6	49.8	62.7	69.1	64.7	56.1	52.0	32.3	22.1	45.1
1989	17.5	24.2	41.7	48.8	51.0	59.1	70.0	64.5	56.4	45.6	34.9	28.6	45.2
1990	26.7	23.8	38.6	46.6	50.0	61.6	68.9	65.1	60.5	48.0	34.4	17.9	45.2
1991	26.4	35.9	33.0	38.6	46.1	58.7	68.2	66.6	56.3	45.6	34.0	28.4	44.8
1992	24.7	35.0	39.3	48.6	54.6	58.5	65.5	66.2	57.8	48.4	29.8	22.3	45.9
1993	20.1	20.3	35.4	42.2	53.1	54.9	62.1	63.1	55.3	43.9	28.5	24.6	42.0
Record Mean	23.9	28.4	34.0	41.8	50.3	58.7	67.2	65.5	56.6	45.9	33.8	26.2	44.4
Max	38.7	42.3	47.8	57.3	66.9	77.4	86.7	84.5	75.8	63.4	49.0	40.9	60.9
Min	9.1	14.5	20.3	26.4	33.6	40.0	47.7	46.4	37.3	28.3	18.5	11.5	27.8

REFERENCE NOTES

ELY, NEVADA

GENERAL
T - TRACE AMOUNT.
BLANK ENTRIES DENOTE MISSING/UNREPORTED DATA.
# INDICATES A STATION OR INSTRUMENT RELOCATION.

SPECIFIC

(a) - LENGTH OF RECORD IN YEARS, ALTHOUGH INDIVIDUAL MONTHS MAY BE MISSING.
0.* OR * - THE VALUE IS BETWEEN 0.0 AND 0.05
NORMALS - BASED ON THE 1961-1990 RECORD PERIOD.
EXTREMES - DATES ARE THE MOST RECENT OCCURRENCE
WIND DIR.- NUMERALS SHOW TENS OF DEGRESS CLOCKWISE FROM TRUE NORTH. "00" INDICATES CALM.
RESULTANT DIRECTIONS ARE GIVEN TO WHOLE DEGREES.
BOLD VALUES INDICATE EXTREME VALUES WHICH OCCURRED AFTER THE ASOS SYSTEM WAS COMMISSIONED.

RECORD = PERIOD OF RECORD
RECORD MEAN PRECIPITATION IS THE MEAN OF ALL DAILY PRECIPITATION AMOUNTS DURING THE PERIOD OF RECORD.
RECORD MAX(MIN) TEMPERATURE IS THE MEAN OF ALL DAILY MAX(MIN) TEMPERATURES DURING THE PERIOD OF RECORD.
RECORD MEAN TEMPERATURE IS THE SUM OF THE RECORD MAX AND RECORD MIN DIVIDED BY 2.
AVERAGE TEMPERATURE IS THE SUM OF THE MEAN DAILY MAX AND MIN TEMPERATURE DIVIDED BY 2.

EXCEPTIONS

RECORD MEANS ARE THROUGH THE CURRENT YEAR, BEGINNING IN 1939 FOR TEMPERATURE
1939 FOR PRECIPITATION
1939 FOR SNOWFALL

HEATING DEGREE DAYS Base 65 deg. F ELY, NEVADA

SEASON	JULY	AUG	SEP	OCT	NOV	DEC	JAN	FEB	MAR	APR	MAY	JUNE	TOTAL
1964-65	17	74	314	470	1064	1203	1097	1037	1006	684	586	285	7837
1965-66	27	76	429	485	814	1248	1387	1188	869	664	291	154	7632
1966-67	9	35	177	592	801	1169	1193	1019	817	904	485	313	7514
1967-68	3	10	210	530	814	1462	1293	840	870	802	483	182	7499
1968-69	10	151	316	559	900	1268	1039	1087	1198	649	244	229	7650
1969-70	26	7	127	757	892	1084	1100	834	990	900	435	234	7386
1970-71	12	7	376	734	863	1334	1259	979	933	705	549	183	7934
1971-72	18	4	369	760	983	1422	1257	936	732	693	425	102	7701
1972-73	16	86	320	599	1019	1384	1379	1059	1067	769	376	195	8269
1973-74	18	42	326	591	965	1170	1285	1021	809	747	373	93	7440
1974-75	12	54	237	622	852	1250	1262	1021	1023	921	533	228	8015
1975-76	19	81	217	617	982	1123	1194	985	1058	749	346	220	7591
1976-77	31	130	245	641	842	1162	1324	959	1126	596	614	91	7761
1977-78	9	43	238	535	850	1056	1101	959	742	716	521	172	6942
1978-79	43	103	336	490	1047	1361	1479	1072	933	675	400	179	8118
1979-80	22	95	124	464	1005	1035	1138	866	983	617	539	208	7096
1980-81	11	76	221	608	859	1026	1033	937	905	546	456	107	6785
1981-82	1	3	159	680	781	960	1311	960	970	778	472	236	7311
1982-83	60	3	311	715	951	1236	1111	967	870	815	533	208	7780
1983-84	45	50	180	518	948	1172	1238	1018	917	764	318	240	7408
1984-85	19	21	209	759	917	1321	1404	1129	1000	548	371	98	7796
1985-86	8	42	386	635	1118	1232	942	821	732	670	421	74	7081
1986-87	25	5	406	662	868	1136	1341	985	914	535	408	105	7390
1987-88	63	37	221	487	902	1258	1353	973	896	603	464	120	7377
1988-89	1	46	269	395	972	1323	1466	1139	718	478	425	189	7421
1989-90	7	59	251	596	898	1124	1180	1150	812	547	459	138	7221
1990-91	3	47	163	521	912	1455	1190	807	987	782	580	197	7644
1991-92	8	18	254	594	924	1126	1242	861	789	485	315	191	6807
1992-93	50	74	210	507	1048	1317	1383	1247	908	677	363	298	8082
1993-94	103	81	287	649	1086	1246							

COOLING DEGREE DAYS Base 65 deg. F ELY, NEVADA

YEAR	JAN	FEB	MAR	APR	MAY	JUNE	JULY	AUG	SEP	OCT	NOV	DEC	TOTAL
1969	0	0	0	0	1	6	144	154	7	0	0	0	312
1970	0	0	0	0	0	31	89	110	0	0	0	0	230
1971	0	0	0	0	0	8	122	117	2	0	0	0	249
1972	0	0	0	0	0	16	117	58	0	0	0	0	191
1973	0	0	0	0	0	38	77	85	0	0	0	0	200
1974	0	0	0	0	0	63	102	30	14	0	0	0	209
1975	0	0	0	0	0	2	120	46	0	0	0	0	168
1976	0	0	0	0	0	18	103	2	4	0	0	0	127
1977	0	0	0	0	0	24	86	74	14	0	0	0	198
1978	0	0	0	0	0	7	110	79	9	0	0	0	205
1979	0	0	0	0	0	35	103	56	14	0	0	0	208
1980	0	0	0	0	0	12	116	78	2	0	0	0	208
1981	0	0	0	0	0	64	143	108	9	0	0	0	324
1982	0	0	0	0	0	6	90	93	11	0	0	0	200
1983	0	0	0	0	0	0	78	74	11	0	0	0	163
1984	0	0	0	0	1	18	103	44	11	0	0	0	177
1985	0	0	0	0	0	57	124	65	2	0	0	0	248
1986	0	0	0	0	0	30	55	137	7	0	0	0	229
1987	0	0	0	0	0	17	41	47	1	0	0	0	106
1988	0	0	0	0	0	56	134	45	9	0	0	0	244
1989	0	0	0	0	0	18	168	54	0	0	0	0	240
1990	0	0	0	0	0	45	130	60	35	0	0	0	270
1991	0	0	0	0	0	16	115	76	3	0	0	0	210
1992	0	0	0	0	0	5	69	115	1	0	0	0	190
1993	0	0	0	0	0	2	16	33	1	0	0	0	52

SNOWFALL (inches) ELY, NEVADA

SEASON	JULY	AUG	SEP	OCT	NOV	DEC	JAN	FEB	MAR	APR	MAY	JUNE	TOTAL
1964-65	0.0	0.0	T	T	12.9	17.9	6.6	9.0	4.6	6.8	1.0	0.0	58.8
1965-66	0.0	0.0	T	2.0	5.5	12.7	3.5	4.2	1.6	1.4	T	0.0	30.9
1966-67	0.0	0.0	0.0	T	1.0	7.1	24.8	2.1	4.4	12.8	3.6	0.8	56.6
1967-68	0.0	0.0	0.0	0.0	12.7	11.8	2.4	8.6	9.5	16.7	0.7	0.0	62.4
1968-69	0.0	0.0	T	3.5	1.9	22.3	8.3	19.1	6.3	3.9	0.0	0.0	65.3
1969-70	0.0	0.0	0.0	2.2	2.9	2.6	1.0	0.9	4.2	19.9	0.0	T	33.7
1970-71	0.0	0.0	0.0	T	8.0	17.5	8.8	7.7	4.1	13.1	6.3	0.0	65.5
1971-72	0.0	0.0	2.2	9.7	6.0	13.6	3.3	0.1	T	4.3	1.3	0.0	40.5
1972-73	0.0	0.0	0.0	T	0.7	10.5	14.6	10.0	24.0	1.7	0.4	0.0	61.9
1973-74	0.0	0.0	0.0	3.2	11.4	7.9	7.2	5.1	8.9	0.7	2.2	0.0	46.6
1974-75	0.0	0.0	0.0	1.6	0.9	4.4	10.0	8.8	16.4	15.6	12.1	T	69.8
1975-76	0.0	0.0	0.0	5.6	4.7	6.2	5.9	20.0	10.4	6.9	0.0	0.0	59.7
1976-77	0.0	0.0	0.0	T	1.4	T	5.9	0.6	11.2	2.6	10.9	0.0	32.6
1977-78	0.0	0.0	0.2	T	2.0	7.9	8.0	9.6	5.9	18.5	4.4	0.0	56.5
1978-79	0.0	0.0	1.3	6.8	17.0	9.1	13.6	7.1	13.7	2.8	2.5	1.5	75.4
1979-80	0.0	0.0	0.0	0.7	2.2	0.8	17.8	6.9	19.0	1.4	8.6	0.2	57.6
1980-81	0.0	0.0	0.0	1.6	4.0	11.5	8.0	2.2	15.6	6.3	0.7	0.0	49.9
1981-82	0.0	0.0	0.0	12.1	1.5	1.9	13.1	1.4	16.2	7.8	3.9	0.1	58.0
1982-83	0.0	0.0	6.3	1.0	9.3	3.6	15.3	10.3	11.1	10.4	3.8	0.0	71.1
1983-84	0.0	0.0	0.0	0.0	9.9	13.1	5.1	5.7	6.5	6.4	0.0	T	46.7
1984-85	0.0	0.0	T	3.8	10.4	11.3	6.3	5.5	15.4	1.0	2.0	0.0	55.7
1985-86	0.0	0.0	T	8.7	17.3	4.5	0.6	4.6	7.5	7.0	5.8	0.0	56.0
1986-87	0.0	0.0	4.7	6.1	0.9	1.2	11.9	6.2	5.9	T	T	0.0	36.9
1987-88	0.0	0.0	0.0	0.0	3.3	8.7	17.3	2.8	1.2	3.5	5.1	0.0	41.9
1988-89	0.0	T	T	0.0	17.1	18.0	7.2	10.9	1.3	T	1.8	0.0	56.3
1989-90	T	0.0	0.0	2.6	1.1	T	6.0	14.2	10.3	1.0	0.7	1.0	36.9
1990-91	0.0	0.0	0.0	1.8	4.8	4.1	1.0	0.5	21.4	3.3	4.7	0.0	41.6
1991-92	0.0	0.0	0.0	9.7	3.2	4.2	6.0	3.8	2.8	0.0	T	0.2	29.9
1992-93	T	T	T	0.2	1.9	11.6	24.3	17.0	2.0	2.2	0.8	T	60.0
1993-94	0.0	T	0.0	T	6.0	1.5							
Record Mean	T	T	0.3	2.3	5.2	7.3	9.0	6.9	9.5	5.7	2.4	0.2	48.8

LAS VEGAS, NEVADA

Las Vegas is situated near the center of a broad desert valley, which is almost surrounded by mountains ranging from 2,000 to 10,000 feet higher than the floor of the valley. This Vegas Valley, comprising about 600 square miles, runs from northwest to southeast, and slopes gradually upward on each side toward the surrounding mountains. Weather observations are taken at McCarran Airport, 7 miles south of downtown Las Vegas, and about 5 miles southwest and 300 feet higher than the lower portions of the valley. Since mountains encircle the valley, drainage winds are usually downslope toward the center, or lowest portion of the valley. This condition also affects minimum temperatures, which in lower portions of the valley can be from 15 to 25 degrees colder than recorded at the airport on clear, calm nights.

The four seasons are well defined. Summers display desert conditions, with maximum temperatures usually in the 100 degree range. The proximity of the mountains contributes to the relatively cool summer nights, with the majority of minimum temperatures in the mid 70s. During about 2 weeks almost every summer warm, moist air predominates in this area, and causes scattered thunderstorms, occasionally quite severe, together with higher than average humidity. Soil erosion, especially near the mountains and foothills surrounding the valley, is evidence of the intensity of some of the thunderstorm activity. Winters, on the whole, are mild and pleasant. Daytime temperatures average near 60 degrees with mostly clear skies. The spring and fall seasons are generally considered most ideal, although rather sharp temperature changes can occur during these months. There are very few days during the spring and fall months when outdoor activities are affected in any degree by the weather.

The Sierra Nevada Mountains of California and the Spring Mountains immediately west of the Vegas Valley, the latter rising to elevations over 10,000 feet above the valley floor, act as effective barriers to moisture moving eastward from the Pacific Ocean. It is mainly these barriers that result in a minimum of dark overcast and rainy days. Rainy days average less than one in June to three per month in the winter months. Snow rarely falls in this valley and it usually melts as it falls, or shortly thereafter. Notable exceptions have occurred.

Strong winds, associated with major storms, usually reach this valley from the southwest or through the pass from the northwest. Winds over 50 mph are infrequent but, when they do occur, are probably the most provoking of the elements experienced in the Vegas Valley, because of the blowing dust and sand associated with them.

Based on the 1951–1980 period, the average first occurrence of 32 degrees Fahrenheit in the fall is November 21 and the average last occurrence in the spring is March 7.

NORMALS, MEANS, AND EXTREMES

LAS VEGAS, NEVADA

LATITUDE: 36°05'N LONGITUDE: 115°10' W ELEVATION: FT. GRND 2162 BARO 2179 TIME ZONE: PACIFIC WBAN: 23169

	(a)	JAN	FEB	MAR	APR	MAY	JUNE	JULY	AUG	SEP	OCT	NOV	DEC	YEAR
TEMPERATURE °F:														
Normals														
-Daily Maximum		57.3	63.3	68.8	77.5	87.8	100.3	105.9	103.2	94.7	82.1	67.4	57.5	80.5
-Daily Minimum		33.6	38.8	43.8	50.7	60.2	69.4	76.2	74.2	66.2	54.3	42.6	33.9	53.7
-Monthly		45.5	51.1	56.3	64.1	74.0	84.9	91.1	88.7	80.5	68.3	55.0	45.7	67.1
Extremes														
-Record Highest	45	77	87	91	99	109	115	116	116	113	103	87	77	116
-Year		1975	1986	1966	1981	1951	1970	1985	1979	1950	1978	1988	1980	JUL 1985
-Record Lowest	45	8	16	23	31	40	48	60	56	46	26	21	11	8
-Year		1963	1989	1971	1975	1964	1993	1987	1968	1965	1971	1952	1990	JAN 1963
NORMAL DEGREE DAYS:														
Heating (base 65°F)		605	389	292	143	14	0	0	0	0	62	304	598	2407
Cooling (base 65°F)		0	0	22	116	293	597	809	735	465	164	0	0	3201
% OF POSSIBLE SUNSHINE	44	77	80	83	87	88	93	88	88	91	87	81	79	85
MEAN SKY COVER (tenths)														
Sunrise - Sunset	45	4.8	4.7	4.6	3.8	3.4	2.1	2.8	2.5	2.1	2.8	3.9	4.4	3.5
MEAN NUMBER OF DAYS:														
Sunrise to Sunset														
-Clear	45	13.8	12.4	13.6	15.8	17.9	22.2	20.1	21.6	22.6	20.4	15.9	14.9	211.1
-Partly Cloudy	45	6.4	6.9	8.5	7.6	8.0	5.2	7.5	6.6	5.0	6.3	7.3	6.7	82.0
-Cloudy	45	10.9	9.0	8.9	6.6	5.1	2.6	3.4	2.8	2.4	4.2	6.8	9.4	72.2
Precipitation														
.01 inches or more	45	3.2	2.8	3.2	1.8	1.4	0.7	2.6	2.9	1.6	1.8	2.0	2.6	26.5
Snow,Ice pellets,hail														
1.0 inches or more	45	0.3	0.*	0.0	0.0	0.0	0.0	0.0	0.0	0.0	0.0	0.1	0.*	0.4
Thunderstorms	45	0.*	0.2	0.4	0.5	1.0	1.0	4.0	4.0	1.6	0.5	0.2	0.*	13.4
Heavy Fog Visibility														
1/4 mile or less	45	0.3	0.1	0.1	0.0	0.0	0.0	0.0	0.0	0.*	0.*	0.1	0.2	0.8
Temperature °F														
-Maximum														
90° and above	33	0.0	0.0	0.*	3.3	15.4	25.7	30.5	29.8	22.2	6.0	0.0	0.0	132.9
32° and below	33	0.1	0.0	0.0	0.0	0.0	0.0	0.0	0.0	0.0	0.0	0.0	0.*	0.2
-Minimum														
32° and below	33	12.5	4.3	1.2	0.1	0.0	0.0	0.0	0.0	0.0	0.1	2.2	11.2	31.5
0° and below	33	0.0	0.0	0.0	0.0	0.0	0.0	0.0	0.0	0.0	0.0	0.0	0.0	0.0
AVG. STATION PRESS.(mb)	21	942.3	940.7	937.3	935.9	933.9	933.6	934.9	935.6	936.0	938.7	940.8	942.5	937.7
RELATIVE HUMIDITY (%)														
Hour 04	33	56	51	46	35	32	25	29	35	33	37	46	55	40
Hour 10 (Local Time)	33	42	36	31	22	19	15	19	24	22	25	32	40	27
Hour 16	33	31	27	22	16	14	11	15	17	17	19	26	32	21
Hour 22	33	50	43	37	26	23	17	22	26	26	30	40	49	32
PRECIPITATION (inches):														
Water Equivalent														
-Normal		0.48	0.48	0.42	0.21	0.28	0.12	0.35	0.49	0.28	0.21	0.43	0.38	4.13
-Maximum Monthly	45	2.41	2.52	4.80	2.44	0.96	0.97	2.48	2.59	1.58	1.22	2.22	1.71	4.80
-Year		1949	1993	1992	1965	1969	1990	1984	1957	1963	1992	1965	1992	MAR 1992
-Minimum Monthly	45	T	0.00	0.00	0.00	0.00	0.00	0.00	0.00	0.00	0.00	0.00	0.00	0.00
-Year		1984	1977	1972	1962	1970	1982	1981	1980	1971	1979	1980	1981	JUN 1982
-Maximum in 24 hrs	45	1.09	1.30	1.27	0.97	0.83	0.97	1.36	2.59	1.07	1.09	1.78	0.95	2.59
-Year		1990	1993	1992	1965	1987	1990	1984	1957	1963	1992	1960	1977	AUG 1957
Snow,Ice pellets,hail														
-Maximum Monthly	45	16.7	1.4	0.1	T	0.0	0.0	0.0	T	0.0	T	4.0	2.0	16.7
-Year		1949	1990	1976	1970				1989		1956	1964	1967	JAN 1949
-Maximum in 24 hrs	45	9.0	6.9	0.1	T	0.0	0.0	0.0	T	0.0	T	4.0	2.0	9.0
-Year		1974	1979	1976	1970				1989		1956	1964	1967	JAN 1974
WIND:														
Mean Speed (mph)	45	7.4	8.6	10.2	11.0	11.1	11.1	10.3	9.6	9.0	8.1	7.8	7.3	9.3
Prevailing Direction through 1963		W	SW	SW	SW	SW	SW	SW	SW	SW	WSW	W	W	SW
Fastest Obs. 1 Min.														
-Direction (!!)	8	23	23	23	22	32	34	10	14	22	31	21	30	32
-Speed (MPH)	8	39	50	51	49	53	48	38	40	35	47	43	40	53
-Year		1987	1989	1989	1988	1991	1989	1984	1989	1989	1989	1985	1984	MAY 1991
Peak Gust														
-Direction (!!)	10	SW	NW	NW	W	NW	NE	SW	SE	SW	SW	SW	SW	SE
-Speed (mph)	10	54	67	82	69	72	59	53	90	49	52	68	54	90
-Date		1987	1984	1984	1988	1991	1984	1984	1989	1989	1984	1987	1990	AUG 1989

PRECIPITATION (inches)

LAS VEGAS, NEVADA

YEAR	JAN	FEB	MAR	APR	MAY	JUNE	JULY	AUG	SEP	OCT	NOV	DEC	ANNUAL
1964	0.05	0.02	0.02	0.03	0.05	0.03	0.24	0.05	T	T	0.63	T	1.12
1965	0.05	0.45	0.74	2.44	0.40	T	0.28	0.38	T	T	2.22	1.00	7.96
1966	T	0.07	0.04	0.01	T	0.15	0.30	0.09	0.35	0.09	0.33	0.48	1.91
1967	0.47	0.00	T	0.09	0.21	0.82	0.20	0.38	1.03	0.00	1.52	0.82	5.54
1968	0.01	0.22	0.22	0.10	T	0.31	0.11	0.04	0.01	T	0.02	0.07	1.11
1969	1.57	0.96	0.57	T	0.96	0.23	0.06	0.33	0.08	0.27	0.06	T	5.09
1970	0.01	0.86	0.28	0.04	0.00	0.18	0.58	1.79	0.00	0.02	0.38	0.15	4.29
1971	T	0.03	T	T	0.84	T	0.08	0.90	0.00	0.06	0.12	0.51	2.54
1972	0.00	T	0.00	0.07	0.46	0.32	0.13	0.84	0.63	1.12	1.09	0.19	4.85
1973	0.49	1.64	1.83	0.35	0.09	0.03	T	0.08	T	0.02	0.14	0.01	4.68
1974	2.00	0.11	0.16	T	T	0.00	0.58	0.08	0.16	0.61	0.23	0.59	4.52
1975	0.01	0.05	1.07	0.42	0.35	T	0.26	0.06	1.17	0.03	T	0.05	3.47
1976	0.00	2.49	0.02	0.13	0.34	0.00	1.95	0.00	1.09	0.70	0.02	0.03	6.77
1977	0.21	0.00	0.28	0.01	0.72	0.05	T	1.38	0.19	0.06	0.01	1.06	3.97
1978	1.00	1.51	1.13	0.36	0.54	0.00	0.19	0.53	0.03	0.62	0.59	1.15	7.65
1979	2.18	0.07	0.96	0.06	0.35	0.00	0.78	2.12	T	0.00	0.03	0.24	6.79
1980	1.45	2.25	0.94	0.18	0.15	T	0.43	0.00	0.18	0.04	0.00	0.01	5.63
1981	0.09	0.20	1.44	0.02	0.50	T	0.00	0.20	0.25	0.15	0.29	0.00	3.14
1982	0.09	1.10	0.29	0.01	0.31	0.00	0.05	0.71	0.07	0.04	0.60	0.72	3.99
1983	0.43	0.32	0.90	0.45	0.16	T	0.06	1.25	0.50	0.26	0.10	0.43	4.86
1984	T	0.03	T	0.04	0.00	0.22	2.48	0.99	0.47	T	0.94	1.68	6.85
1985	0.19	0.02	0.06	0.31	T	0.02	0.13	0.00	0.08	0.07	0.37	0.02	1.27
1986	0.23	0.15	0.32	0.10	0.28	T	0.13	0.04	0.05	0.07	0.81	0.47	2.65
1987	1.13	0.45	0.49	0.17	0.90	0.13	0.13	0.01	T	0.49	1.80	0.89	6.59
1988	0.65	0.26	0.00	0.76	T	0.04	0.04	0.46	T	0.00	T	0.08	2.29
1989	0.51	0.06	0.05	T	0.64	T	0.05	0.80	T	T	0.00	T	2.11
1990	1.18	0.37	T	0.18	T	0.97	0.59	T	0.19	0.17	0.10	T	3.75
1991	0.21	0.54	1.01	T	0.05	0.19	0.54	0.78	0.06	0.06	0.38	0.24	4.06
1992	0.45	1.30	4.80	0.02	0.05	0.09	0.03	0.21	0.00	1.22	0.00	1.71	9.88
1993	1.63	2.52	0.14	0.01	0.01	0.08	0.00	0.26	0.00	0.02	0.17	0.21	5.05
Record Mean	0.53	0.48	0.51	0.22	0.18	0.08	0.41	0.48	0.30	0.24	0.37	0.40	4.21

AVERAGE TEMPERATURE (deg. F)

LAS VEGAS, NEVADA

YEAR	JAN	FEB	MAR	APR	MAY	JUNE	JULY	AUG	SEP	OCT	NOV	DEC	ANNUAL
1964	42.0	45.6	52.3	61.8	70.9	80.9	90.7	87.6	78.4	72.0	50.0	45.3	64.8
1965	47.1	49.5	53.2	61.2	69.6	78.0	88.7	87.9	74.8	69.8	55.9	45.0	65.1
1966	42.7	45.8	57.9	66.4	77.5	83.9	89.3	89.6	80.1	66.5	55.4	46.1	66.8
1967	45.3	50.6	59.3	56.2	72.5	79.6	91.7	90.3	80.0	69.1	56.7	41.6	66.1
1968	44.2	55.7	57.5	62.0	73.5	84.0	89.0	83.5	79.7	67.0	54.5	40.8	66.0
1969	47.5	46.3	53.0	64.4	76.8	81.4	89.7	92.2	82.5	62.8	53.5	45.8	66.3
1970	44.0	52.5	54.9	58.6	75.3	83.4	91.1	88.8	77.2	63.8	55.0	44.5	65.8
1971	44.4	49.7	55.8	63.0	68.0	83.3	92.8	89.0	77.6	61.7	50.9	41.4	64.8
1972	42.3	52.0	63.7	65.1	74.5	84.7	93.1	86.5	78.0	63.5	49.7	41.3	66.2
1973	40.9	49.6	50.7	62.2	76.7	85.2	91.7	87.6	78.9	67.7	53.4	46.2	65.9
1974	41.0	48.9	59.5	63.4	77.0	89.1	88.8	87.7	83.4	69.3	54.8	44.4	67.3
1975	45.3	48.8	53.9	56.6	72.5	83.8	90.3	87.5	81.7	66.1	53.0	48.2	65.6
1976	46.9	53.2	53.4	62.6	77.8	81.5	86.9	85.5	78.7	66.5	58.0	46.4	66.5
1977	45.7	54.2	52.6	68.6	67.7	88.0	92.4	90.1	80.6	71.4	57.2	51.9	68.3
1978	47.9	52.1	59.9	63.1	73.1	87.1	91.9	89.0	79.0	73.5	54.2	42.9	67.8
1979	41.1	48.4	56.0	66.1	75.4	85.5	91.1	85.9	85.3	70.7	51.6	47.2	67.1
1980	49.5	53.2	54.2	63.5	69.0	83.9	92.0	90.2	81.4	68.9	56.8	52.7	67.9
1981	51.1	52.5	56.4	70.6	74.3	88.8	92.7	90.0	82.5	64.4	58.0	48.8	69.2
1982	45.6	50.5	55.1	63.8	73.6	81.5	88.1	87.3	77.9	63.0	50.5	44.5	65.1
1983	46.6	51.7	56.4	58.5	72.8	82.8	88.5	83.8	82.5	67.8	55.3	47.9	66.2
1984	47.1	50.1	57.9	63.1	80.7	83.5	88.2	85.4	81.7	63.0	52.7	44.0	66.5
1985	44.4	47.4	54.9	68.2	76.9	87.4	92.0	89.9	75.4	67.3	51.7	48.3	67.0
1986	51.7	55.8	63.0	66.2	76.6	87.8	87.6	91.2	75.4	65.0	55.8	46.0	68.5
1987	44.7	51.4	54.6	68.4	74.5	86.3	86.9	88.2	81.2	71.0	53.4	42.5	66.9
1988	45.1	52.4	58.1	64.2	73.4	85.3	92.6	86.9	79.1	74.9	56.0	46.0	67.8
1989	43.9	50.0	63.4	72.7	75.7	85.3	93.4	86.9	80.0	67.2	57.3	48.0	68.7
1990	45.2	48.8	60.5	68.8	74.5	85.9	90.8	87.8	82.0	69.2	55.1	40.2	67.4
1991	45.5	55.9	52.7	64.2	69.9	82.1	90.2	87.8	81.9	72.2	55.2	47.0	67.1
1992	45.9	54.1	56.8	70.5	77.7	83.2	88.7	90.5	83.7	70.9	52.7	43.6	68.2
1993	45.7	50.1	60.9	67.5	77.0	82.5	89.4	88.5	81.3	69.1	51.5	46.3	67.5
Record Mean	44.6	49.9	55.8	64.6	73.9	83.5	89.8	87.7	80.0	67.3	53.5	45.7	66.3
Max	56.5	62.4	68.9	78.5	88.3	98.6	104.5	102.2	94.8	81.7	66.6	57.7	80.0
Min	32.7	37.4	42.7	50.6	59.4	68.3	75.0	73.2	65.1	53.0	40.4	33.6	52.6

REFERENCE NOTES

LAS VEGAS, NEVADA

GENERAL
T - TRACE AMOUNT.
BLANK ENTRIES DENOTE MISSING/UNREPORTED DATA.
# INDICATES A STATION OR INSTRUMENT RELOCATION.

SPECIFIC

PM INCLUDES LAST DAY OF PREVIOUS MONTH
ASOS - AUTOMATED SURFACE OBSERVING SYSTEM IN OPERATION DURING THESE MONTHS.

(*) - LENGTH OF RECORD IN YEARS, ALTHOUGH INDIVIDUAL MONTHS MAY BE MISSING.
0.* OR * - THE VALUE IS BETWEEN 0.0 AND 0.05
NORMALS - BASED ON THE 1961-1990 RECORD PERIOD.
EXTREMES - DATES ARE THE MOST RECENT OCCURRENCE
WIND DIR.- NUMERALS SHOW TENS OF DEGRESS CLOCKWISE FROM TRUE NORTH. "00" INDICATES CALM.
RESULTANT DIRECTIONS ARE GIVEN TO WHOLE DEGREES.
BOLD VALUES INDICATE EXTREME VALUES WHICH OCCURRED AFTER THE ASOS SYSTEM WAS COMMISSIONED.

RECORD = PERIOD OF RECORD
RECORD MEAN PRECIPITATION IS THE MEAN OF ALL DAILY PRECIPITATION AMOUNTS DURING THE PERIOD OF RECORD.
RECORD MAX(MIN) TEMPERATURE IS THE MEAN OF ALL DAILY MAX(MIN) TEMPERATURES DURING THE PERIOD OF RECORD.
RECORD MEAN TEMPERATURE IS THE SUM OF THE RECORD MAX AND RECORD MIN DIVIDED BY 2.
AVERAGE TEMPERATURE IS THE SUM OF THE MEAN DAILY MAX AND MIN TEMPERATURE DIVIDED BY 2.

EXCEPTIONS

RECORD MEANS ARE THROUGH THE CURRENT YEAR,
BEGINNING IN 1937 FOR TEMPERATURE
1937 FOR PRECIPITATION
1949 FOR SNOWFALL

HEATING DEGREE DAYS Base 65 deg. F — LAS VEGAS, NEVADA

SEASON	JULY	AUG	SEP	OCT	NOV	DEC	JAN	FEB	MAR	APR	MAY	JUNE	TOTAL
1964-65	0	0	0	12	444	606	551	427	358	220	49	0	2667
1965-66	0	0	15	17	266	615	685	529	235	54	0	0	2416
1966-67	0	0	0	47	286	578	606	397	189	261	25	0	2389
1967-68	0	0	0	18	244	716	638	265	231	110	8	0	2230
1968-69	0	0	1	28	304	743	536	518	381	74	16	0	2601
1969-70	0	0	0	112	341	589	643	344	304	208	8	0	2549
1970-71	0	0	0	111	295	631	630	421	306	105	47	0	2546
1971-72	0	0	4	207	417	724	697	373	99	69	6	0	2596
1972-73	0	0	0	108	453	727	744	428	437	132	12	0	3041
1973-74	0	0	0	42	349	576	738	443	188	82	13	0	2431
1974-75	0	0	0	55	300	634	607	446	340	249	37	0	2668
1975-76	0	0	0	73	354	516	553	339	357	124	1	0	2317
1976-77	0	0	0	39	212	569	593	297	374	45	56	0	2185
1977-78	0	0	0	3	226	399	522	356	168	91	16	0	1781
1978-79	0	0	1	2	324	676	737	458	270	66	18	0	2552
1979-80	0	0	0	44	395	546	474	335	328	108	32	0	2262
1980-81	0	0	0	82	255	374	426	344	263	29	2	0	1775
1981-82	0	0	0	74	214	497	594	398	301	98	9	0	2185
1982-83	0	0	10	84	429	631	564	364	263	198	21	0	2564
1983-84	0	0	0	3	297	524	548	424	216	111	0	0	2123
1984-85	0	0	0	127	363	641	629	487	308	41	0	0	2596
1985-86	0	0	1	31	393	512	404	270	125	57	11	0	1804
1986-87	0	0	14	53	268	586	622	375	316	40	1	0	2275
1987-88	0	0	0	18	342	689	612	357	225	83	33	0	2359
1988-89	0	0	0	0	291	581	647	425	118	23	16	0	2101
1989-90	0	0	0	70	224	519	606	449	172	12	0	0	2052
1990-91	0	0	0	23	290	761	597	247	376	57	25	2	2378
1991-92	0	0	0	77	297	552	584	308	248	7	0	0	2073
1992-93	0	0	0	16	364	655	591	410	143	32	3	8	2222
1993-94	0	0	0	33	398	573							

COOLING DEGREE DAYS Base 65 deg. F — LAS VEGAS, NEVADA

YEAR	JAN	FEB	MAR	APR	MAY	JUNE	JULY	AUG	SEP	OCT	NOV	DEC	TOTAL
1969	0	0	13	62	390	500	772	852	532	54	0	0	3175
1970	0	0	0	21	334	560	818	748	371	81	1	0	2934
1971	0	0	24	53	148	556	871	752	390	112	0	0	2906
1972	0	2	66	80	308	597	876	675	398	69	0	0	3071
1973	0	0	0	54	382	612	833	708	424	134	8	0	3155
1974	0	0	24	43	394	731	744	713	559	195	0	0	3403
1975	0	0	2	2	276	570	792	704	508	117	2	0	2973
1976	0	0	2	57	404	500	687	641	419	93	6	0	2809
1977	0	0	0	161	149	694	858	781	476	210	3	0	3332
1978	0	0	17	40	277	672	841	752	425	268	8	0	3300
1979	0	0	0	104	346	625	813	656	614	229	0	0	3387
1980	0	0	0	68	160	575	842	788	498	211	15	0	3157
1981	0	0	5	205	296	721	866	781	531	64	12	0	3481
1982	0	0	2	70	281	501	721	699	404	30	0	0	2708
1983	0	0	2	9	269	541	735	589	534	94	10	0	2783
1984	0	0	3	61	496	563	724	641	508	74	1	0	3071
1985	0	0	0	143	377	678	844	778	319	110	2	0	3251
1986	0	20	69	98	379	693	707	821	332	59	0	0	3178
1987	0	0	0	148	302	645	685	729	495	211	0	0	3215
1988	0	0	16	64	300	615	864	685	434	312	31	0	3321
1989	0	11	74	259	351	614	887	687	456	143	0	0	3482
1990	0	0	42	134	302	634	810	713	516	163	0	0	3314
1991	0	0	0	42	187	524	788	714	515	307	12	0	3089
1992	0	0	0	180	402	552	742	798	571	206	3	0	3454
1993	0	0	21	114	381	537	765	737	494	166	0	0	3215

SNOWFALL (inches) — LAS VEGAS, NEVADA

SEASON	JULY	AUG	SEP	OCT	NOV	DEC	JAN	FEB	MAR	APR	MAY	JUNE	TOTAL
1970-71	0.0	0.0	0.0	0.0	0.0	T	T	0.0	0.0	0.0	0.0	0.0	T
1971-72	0.0	0.0	0.0	0.0	0.0	T	0.0	0.0	0.0	0.0	0.0	0.0	T
1972-73	0.0	0.0	0.0	0.0	0.0	0.0	0.3	0.0	T	0.0	0.0	0.0	0.3
1973-74	0.0	0.0	0.0	0.0	0.0	0.0	13.4	0.0	0.0	0.0	0.0	0.0	13.4
1974-75	0.0	0.0	0.0	0.0	0.0	T	0.0	T	T	0.0	0.0	0.0	T
1975-76	0.0	0.0	0.0	0.0	T	0.0	0.0	0.0	0.1	0.0	0.0	0.0	0.1
1976-77	0.0	0.0	0.0	0.0	0.0	0.0	0.0	0.0	0.0	0.0	0.0	0.0	0.0
1977-78	0.0	0.0	0.0	0.0	0.0	0.0	0.0	0.0	0.0	0.0	0.0	0.0	0.0
1978-79	0.0	0.0	0.0	0.0	0.0	T	9.9	0.3	0.0	0.0	0.0	0.0	10.2
1979-80	0.0	0.0	0.0	0.0	0.0	0.0	0.0	0.0	0.0	0.0	0.0	0.0	0.0
1980-81	0.0	0.0	0.0	0.0	0.0	0.0	0.0	0.0	0.0	0.0	0.0	0.0	0.0
1981-82	0.0	0.0	0.0	0.0	0.0	0.0	0.0	0.0	0.0	0.0	0.0	0.0	0.0
1982-83	0.0	0.0	0.0	0.0	0.0	0.0	0.0	0.0	0.0	0.0	0.0	0.0	0.0
1983-84	0.0	0.0	0.0	0.0	0.0	0.0	0.0	0.0	0.0	0.0	0.0	0.0	0.0
1984-85	0.0	0.0	0.0	0.0	0.0	T	0.0	T	0.0	0.0	0.0	0.0	T
1985-86	0.0	0.0	0.0	0.0	0.0	T	0.0	0.0	0.0	0.0	0.0	0.0	T
1986-87	0.0	0.0	0.0	0.0	0.0	0.0	T	0.6	0.0	0.0	0.0	0.0	0.6
1987-88	0.0	0.0	0.0	0.0	0.0	0.0	0.0	T	0.0	0.0	0.0	0.0	T
1988-89	0.0	0.0	0.0	0.0	0.0	T	0.0	0.3	0.0	0.0	0.0	0.0	0.3
1989-90	0.0	T	0.0	0.0	0.0	0.0	T	1.4	0.0	0.0	0.0	0.0	1.4
1990-91	0.0	0.0	0.0	0.0	0.0	0.0	0.0	0.0	0.0	0.0	0.0	0.0	0.0
1991-92	0.0	0.0	0.0	0.0	0.0	0.0	0.0	0.0	T	0.0	0.0	0.0	T
1992-93	0.0	0.0	0.0	0.0	0.0	T	0.0	0.0	0.0	0.0	0.0	0.0	T
1993-94	0.0	0.0	0.0	0.0	0.0	0.0							
Record Mean	0.0	T	0.0	T	0.1	0.1	1.0	0.1	T	T	0.0	0.0	1.2

RENO, NEVADA

At an elevation of 4,400 feet above mean sea level, Reno is located at the west edge of Truckee Meadows in a semi-arid plateau lying in the lee of the Sierra Nevada Mountain Range. To the west, the Sierras rise to elevations of 9,000 to 11,000 feet. Hills to the east reach 6,000 to 7,000 feet. The Truckee River, flowing from the Sierras eastward through Reno, drains into Pyramid Lake to the northeast of the city.

The daily temperatures on the whole are mild, but the difference between the high and low often exceeds 45 degrees. While the afternoon high may exceed 90 degrees, a light wrap is often needed shortly after sunset. Nights with low temperatures over 60 degrees are rare. Afternoon temperatures in winter are moderate.

Based on the 1951–1980 period, the average first occurrence of 32 degrees Fahrenheit in the fall is September 16 and the average last occurrence in the spring is June 1.

More than half of the precipitation in Reno occurs mainly as mixed rain and snow, and falls from December to March. Although there is an average of about 25 inches of snow a year, it seldom remains on the ground for more than three or four days at a time. Summer rain comes mainly as brief thunderstorms in the middle and late afternoons. While precipitation is scarce, considerable water is available from the high altitude reservoirs in the Sierra Nevada, where precipitation is heavy.

Humidity is very low during the summer months, and moderately low during the winter. Fogs are rare, and are usually confined to the early morning hours of midwinter. Sunshine is abundant throughout the year.

NORMALS, MEANS, AND EXTREMES

RENO, NEVADA

LATITUDE: 39°30'N LONGITUDE: 119°47' W ELEVATION: FT. GRND 4404 BARO 4402 TIME ZONE: PACIFIC WBAN: 23185

	(a)	JAN	FEB	MAR	APR	MAY	JUNE	JULY	AUG	SEP	OCT	NOV	DEC	YEAR
TEMPERATURE °F:														
Normals														
-Daily Maximum		45.1	51.7	56.3	63.7	72.9	83.1	91.9	89.6	79.5	68.6	53.8	45.5	66.8
-Daily Minimum		20.7	24.2	29.2	33.3	40.1	46.9	51.3	49.6	41.3	32.9	26.7	19.9	34.7
-Monthly		32.9	38.0	42.8	48.6	56.5	65.1	71.6	69.6	60.4	50.8	40.3	32.7	50.8
Extremes														
-Record Highest	52	70	75	83	89	96	103	104	105	101	91	77	70	105
-Year		1967	1986	1966	1981	1986	1988	1980	1983	1950	1980	1980	1969	AUG 1983
-Record Lowest	52	-16	-16	-2	13	18	25	33	24	20	8	1	-16	-16
-Year		1949	1989	1945	1956	1964	1954	1976	1962	1965	1971	1958	1972	FEB 1989
NORMAL DEGREE DAYS:														
Heating (base 65°F)		995	756	688	492	274	76	13	20	178	440	741	1001	5674
Cooling (base 65°F)		0	0	0	0	10	79	217	162	40	0	0	0	508
% OF POSSIBLE SUNSHINE	44	65	68	75	81	81	85	92	92	91	83	70	64	79
MEAN SKY COVER (tenths)														
Sunrise - Sunset	51	6.4	6.3	6.1	5.8	4.9	3.6	2.3	2.4	2.6	4.0	5.7	6.2	4.7
MEAN NUMBER OF DAYS:														
Sunrise to Sunset														
-Clear	51	8.3	7.0	8.1	8.4	11.9	16.5	22.4	21.9	20.7	15.7	9.5	8.5	158.8
-Partly Cloudy	51	7.2	7.5	9.2	10.0	9.8	7.9	6.0	6.4	5.8	7.7	8.0	7.8	93.3
-Cloudy	51	15.5	13.8	13.7	11.6	9.3	5.6	2.5	2.6	3.5	7.7	12.5	14.8	113.2
Precipitation														
.01 inches or more	51	6.1	6.0	6.4	3.9	4.3	3.2	2.3	2.3	2.4	3.0	4.8	5.9	50.5
Snow, Ice pellets, hail														
1.0 inches or more	51	2.1	1.6	1.5	0.5	0.2	0.0	0.0	0.0	0.*	0.1	0.7	1.4	8.0
Thunderstorms	51	0.0	0.*	0.1	0.5	1.9	2.8	3.6	3.1	1.3	0.6	0.0	0.0	14.0
Heavy Fog Visibility														
1/4 mile or less	51	2.2	0.9	0.3	0.1	0.1	0.0	0.*	0.1	0.1	0.2	0.6	2.5	7.0
Temperature °F														
-Maximum														
90° and above	30	0.0	0.0	0.0	0.0	1.2	7.1	20.4	17.0	5.2	0.1	0.0	0.0	51.1
32° and below	30	3.2	0.7	0.*	0.0	0.0	0.0	0.0	0.0	0.0	0.0	0.4	3.5	7.9
-Minimum														
32° and below	30	28.3	24.5	23.0	16.8	5.5	0.6	0.0	0.1	3.6	15.7	23.9	28.2	170.3
0° and below	30	0.9	0.2	0.*	0.0	0.0	0.0	0.0	0.0	0.0	0.0	0.0	1.0	2.1
AVG. STATION PRESS. (mb)	21	868.2	866.7	864.3	865.0	864.5	865.3	866.6	866.6	866.6	868.0	867.9	868.6	866.5
RELATIVE HUMIDITY (%)														
Hour 04	30	79	75	69	65	66	64	63	65	68	72	75	77	70
Hour 10 (Local Time)	30	70	58	48	37	33	30	27	29	34	42	57	67	44
Hour 16	30	51	40	33	27	25	22	18	19	21	27	41	51	31
Hour 22	30	73	63	55	48	44	40	35	38	45	55	65	72	53
PRECIPITATION (inches):														
Water Equivalent														
-Normal		1.07	0.99	0.71	0.38	0.69	0.46	0.28	0.32	0.39	0.38	0.87	0.99	7.53
-Maximum Monthly	52	4.13	4.84	2.02	2.04	2.89	1.53	1.06	1.65	2.31	2.14	3.08	5.25	5.25
-Year		1969	1986	1952	1958	1963	1989	1971	1965	1982	1945	1983	1955	DEC 1955
-Minimum Monthly	52	T	T	T	T	T	0.00	0.00	0.00	0.00	T	0.00	T	0.00
-Year		1966	1967	1988	1985	1985	1959	1951	1957	1974	1966	1959	1989	SEP 1974
-Maximum in 24 hrs	52	2.37	1.80	1.21	1.64	1.76	0.79	0.80	0.97	0.91	1.55	1.65	2.16	2.37
-Year		1943	1990	1943	1958	1987	1969	1949	1965	1982	1962	1988	1955	JAN 1943
Snow, Ice pellets, hail														
-Maximum Monthly	52	22.9	23.5	29.0	7.5	14.1	0.2	0.0	0.0	1.5	5.1	16.5	25.6	29.0
-Year		1993	1969	1952	1958	1964	1970			1982	1971	1985	1971	MAR 1952
-Maximum in 24 hrs	52	12.0	18.0	16.9	7.3	9.0	0.2	0.0	0.0	1.5	3.7	15.4	14.9	18.0
-Year		1956	1990	1952	1958	1962	1970			1982	1971	1985	1971	FEB 1990
WIND:														
Mean Speed (mph)	51	5.6	6.2	7.8	8.3	8.0	7.7	7.1	6.6	5.8	5.4	5.5	5.2	6.6
Prevailing Direction through 1963		S	S	WNW	WNW	WNW	WNW	WNW	WNW	WNW	WNW	S	SW	WNW
Fastest Mile														
-Direction (!!)	29	SW	SW	SW	SW	NW	SE	NW	W	W	SW	S	SW	SW
-Speed (MPH)	29	80	66	80	48	48	64	50	47	44	74	61	68	80
-Year		1968	1975	1968	1982	1986	1987	1988	1984	1987	1985	1983	1968	JAN 1968
Peak Gust														
-Direction (!!)	10	SW	W	SW	SW	NW	SW	NW	W	W	SW	SW	S	SW
-Speed (mph)	10	90	66	71	64	49	67	67	55	54	81	62	75	90
-Date		1988	1986	1991	1988	1986	1987	1988	1984	1986	1985	1984	1987	JAN 1988

PRECIPITATION (inches) RENO, NEVADA

YEAR	JAN	FEB	MAR	APR	MAY	JUNE	JULY	AUG	SEP	OCT	NOV	DEC	ANNUAL
1964	0.68	0.01	0.72	0.53	1.79	0.29	0.13	0.02	0.00	0.04	0.63	2.89	7.73
1965	1.64	0.01	0.98	0.27	0.18	1.31	0.35	1.65	0.50	0.02	1.62	1.19	9.72
1966	T	0.20	0.04	0.03	0.33	0.03	T	0.02	0.10	T	1.07	1.45	3.27
1967	2.11	T	1.93	0.95	0.47	0.59	0.57	1.22	0.82	0.04	0.22	0.55	9.47
1968	1.11	0.92	0.84	0.02	0.30	0.16	0.05	0.13	0.15	0.01	0.73	1.03	5.45
1969	4.13	1.74	0.07	0.10	0.20	1.29	0.17	T	0.01	0.40	0.04	2.07	10.22
1970	1.73	0.32	0.19	0.60	T	0.88	0.05	0.02	0.01	0.03	1.47	1.65	6.95
1971	0.75	0.33	1.54	0.59	2.38	0.09	1.06	0.09	0.10	0.44	0.24	2.97	10.58
1972	0.37	0.14	0.03	0.14	1.02	0.18	0.01	0.14	0.30	1.30	1.00	0.89	5.52
1973	1.54	1.66	0.73	0.13	0.75	0.07	0.27	0.48	0.01	0.56	1.74	1.27	9.21
1974	1.60	0.34	1.16	0.23	0.01	T	0.33	0.19	0.00	0.69	0.27	0.56	5.38
1975	0.32	1.74	1.59	0.62	0.21	0.21	0.03	1.03	0.92	0.15	0.12	0.01	6.95
1976	0.16	1.20	0.36	0.20	0.10	T	0.96	0.62	1.10	0.28	0.07	0.01	5.06
1977	0.67	0.71	0.19	T	1.24	1.03	0.07	0.01	0.01	0.14	0.23	2.54	6.84
1978	1.66	0.98	1.49	0.20	0.31	0.07	0.19	0.15	0.68	0.08	1.30	0.82	7.93
1979	0.66	0.82	0.52	0.41	0.16	T	0.58	0.38	T	0.31	0.17	2.02	6.03
1980	2.77	1.90	0.76	0.51	0.78	0.12	0.54	0.32	0.48	0.14	0.28	0.60	9.20
1981	0.85	0.21	0.58	0.21	0.57	T	0.01	0.36	0.07	0.64	2.13	1.05	6.68
1982	1.20	0.41	1.14	0.34	0.10	1.07	0.04	0.09	2.31	1.65	1.71	1.04	11.10
1983	1.72	1.58	1.31	1.35	0.21	0.53	T	0.78	0.84	0.36	3.08	1.47	13.23
1984	0.36	0.22	0.20	0.24	0.06	0.34	0.45	0.02	0.04	0.60	1.68	0.07	4.28
1985	0.24	0.68	1.07	T	T	0.12	T	0.01	0.63	0.46	1.23	0.55	4.99
1986	0.40	4.84	0.88	0.77	0.26	0.31	0.86	0.07	0.28	0.06	0.02	0.19	8.94
1987	0.49	0.78	0.80	0.49	2.29	1.12	0.01	0.01	0.01	0.54	0.37	0.59	7.50
1988	0.50	0.02	T	0.95	0.12	0.59	0.22	0.01	0.04	0.02	1.99	0.84	5.30
1989	0.20	0.80	0.46	0.03	1.33	1.53	0.00	0.82	1.19	0.43	0.55	T	7.34
1990	0.62	1.98	0.07	0.33	0.19	0.03	0.86	0.21	0.31	0.06	0.15	0.45	5.26
1991	0.01	0.21	1.42	0.47	0.50	0.39	0.04	0.24	0.60	0.23	0.89	0.15	5.15
1992	0.13	0.45	0.69	0.06	0.10	1.12	0.15	0.28	T	0.45	0.06	1.87	5.36
1993	2.42	1.27	0.55	0.01	0.27	0.35	T	T	T	1.42	0.13	0.16	6.58
Record Mean	1.29	1.06	0.76	0.45	0.63	0.38	0.27	0.26	0.31	0.42	0.69	1.02	7.55

AVERAGE TEMPERATURE (deg. F) RENO, NEVADA

YEAR	JAN	FEB	MAR	APR	MAY	JUNE	JULY	AUG	SEP	OCT	NOV	DEC	ANNUAL
1964	32.5	34.2	37.7	45.3	51.9	61.3	70.7	66.9	58.1	55.3	38.0	36.0	49.0
1965	33.6	40.1	43.2	49.6	54.0	59.2	66.1	65.8	54.9	52.4	41.6	29.9	49.2
1966	33.4	34.8	43.4	50.2	60.7	63.7	65.7	67.5	58.7	49.6	41.0	34.0	50.2
1967	37.0	40.3	40.6	40.3	54.9	59.9	69.6	71.0	62.6	51.0	42.7	28.3	49.8
1968	31.7	42.4	42.4	44.6	52.8	62.9	73.0	64.5	59.6	50.6	40.8	31.4	49.7
1969	37.0	34.2	40.7	48.4	58.1	62.0	70.8	68.8	63.4	45.3	39.0	36.7	50.4
1970	37.3	42.0	41.4	44.1	58.0	65.1	73.0	71.5	57.2	48.0	45.4	29.2	51.0
1971	33.0	36.9	40.9	45.9	50.9	59.7	71.5	70.7	56.5	45.8	37.5	26.6	48.0
1972	25.5	40.2	47.1	46.2	56.9	65.1	70.7	67.6	56.6	48.5	38.0	25.1	49.0
1973	28.4	37.7	37.9	43.8	58.8	64.3	70.1	67.4	58.5	49.2	42.8	38.3	49.8
1974	31.6	36.2	42.8	46.5	54.9	63.9	67.0	65.7	61.4	49.3	40.4	31.1	49.2
1975	32.3	34.1	38.4	39.9	54.2	62.4	70.4	64.8	64.2	49.5	35.9	33.8	48.3
1976	32.1	35.0	38.3	44.5	57.5	61.1	69.0	62.1	62.0	51.3	42.4	30.5	48.8
1977	32.3	40.0	38.1	51.3	48.5	68.6	69.5	69.8	60.7	52.2	41.9	38.3	51.0
1978	37.1	38.4	47.2	45.3	52.3	61.6	69.5	67.2	57.6	53.5	37.2	24.9	49.3
1979	28.9	36.9	41.9	46.9	57.1	63.9	69.9	67.7	64.2	54.0	38.0	35.5	50.4
1980	36.9	40.6	38.5	49.8	54.4	60.8	71.3	67.5	63.0	50.9	41.8	36.2	51.0
1981	36.1	38.8	41.7	50.7	57.5	68.3	67.9	69.4	64.7	46.9	42.5	39.0	52.0
1982	28.5	40.3	40.2	44.0	55.0	61.8	70.4	68.8	57.0	46.9	36.2	32.3	48.5
1983	34.2	38.6	40.5	43.8	53.3	62.5	67.2	69.9	62.9	54.1	41.2	38.8	50.6
1984	31.9	37.2	44.3	45.8	59.4	61.7	73.4	69.8	63.1	46.2	39.6	30.7	50.3
1985	30.6	37.0	38.7	52.7	56.3	68.6	73.4	68.5	57.4	50.4	34.8	31.2	50.0
1986	40.3	42.8	47.7	49.2	57.4	67.5	69.4	73.0	56.3	50.9	43.0	35.3	52.7
1987	31.6	38.4	43.4	54.8	59.7	67.9	68.2	71.3	65.0	56.3	41.8	31.9	52.5
1988	33.0	40.2	44.1	51.2	56.6	67.0	75.2	73.2	63.4	58.4	42.7	31.2	53.0
1989	30.9	31.1	46.4	54.0	57.0	66.3	72.6	67.7	61.9	51.4	41.8	35.9	51.4
1990	34.3	30.8	45.7	54.5	56.4	65.7	73.7	71.1	65.4	54.7	40.9	25.8	51.6
1991	31.8	43.7	39.9	46.9	51.8	62.5	74.3	71.5	65.8	56.6	43.4	33.7	51.8
1992	34.4	41.8	46.9	55.8	64.9	66.5	70.8	72.7	64.8	55.3	39.9	30.6	53.7
1993	25.7	34.3	48.5	48.7	58.4	61.9	69.0	70.0	65.3	54.9	40.4	36.5	51.1
Record Mean	32.0	36.8	41.5	47.8	55.1	62.9	70.3	68.7	60.9	51.1	40.9	33.3	50.1
Max	43.7	49.3	55.0	62.8	70.6	80.0	89.5	87.8	79.4	68.2	55.3	45.3	65.6
Min	20.2	24.4	28.0	32.7	39.6	45.7	51.1	49.5	42.4	34.0	26.5	21.3	34.6

REFERENCE NOTES RENO, NEVADA

GENERAL
T - TRACE AMOUNT.
BLANK ENTRIES DENOTE MISSING/UNREPORTED DATA.
# INDICATES A STATION OR INSTRUMENT RELOCATION.

SPECIFIC

(a) - LENGTH OF RECORD IN YEARS, ALTHOUGH INDIVIDUAL MONTHS MAY BE MISSING.
0.* OR * - THE VALUE IS BETWEEN 0.0 AND 0.05
NORMALS - BASED ON THE 1961-1990 RECORD PERIOD.
EXTREMES - DATES ARE THE MOST RECENT OCCURRENCE
WIND DIR.- NUMERALS SHOW TENS OF DEGRESS CLOCKWISE FROM TRUE NORTH. "00" INDICATES CALM.
RESULTANT DIRECTIONS ARE GIVEN TO WHOLE DEGREES.
BOLD VALUES INDICATE EXTREME VALUES WHICH OCCURRED AFTER THE ASOS SYSTEM WAS COMMISSIONED.

RECORD = PERIOD OF RECORD
RECORD MEAN PRECIPITATION IS THE MEAN OF ALL DAILY PRECIPITATION AMOUNTS DURING THE PERIOD OF RECORD.
RECORD MAX(MIN) TEMPERATURE IS THE MEAN OF ALL DAILY MAX(MIN) TEMPERATURES DURING THE PERIOD OF RECORD.
RECORD MEAN TEMPERATURE IS THE SUM OF THE RECORD MAX AND RECORD MIN DIVIDED BY 2.
AVERAGE TEMPERATURE IS THE SUM OF THE MEAN DAILY MAX AND MIN TEMPERATURE DIVIDED BY 2.

EXCEPTIONS

PERCENT OF POSSIBLE SUNSHINE IS THROUGH 1979 AND 1984 TO DATE.

RECORD MEANS ARE THROUGH THE CURRENT YEAR, BEGINNING IN 1888 FOR TEMPERATURE
1870 FOR PRECIPITATION
1943 FOR SNOWFALL

HEATING DEGREE DAYS Base 65 deg. F RENO, NEVADA

SEASON	JULY	AUG	SEP	OCT	NOV	DEC	JAN	FEB	MAR	APR	MAY	JUNE	TOTAL
1964-65	6	28	203	296	802	892	968	695	667	454	336	174	5521
1965-66	34	31	298	383	694	1079	974	838	664	439	135	105	5674
1966-67	45	45	184	473	715	951	859	685	750	737	317	177	5938
1967-68	1	3	87	429	663	1132	1025	650	694	606	369	113	5772
1968-69	0	91	180	441	721	1033	861	856	745	494	208	104	5734
1969-70	5	12	79	605	772	871	851	638	724	622	216	90	5485
1970-71	0	0	234	519	585	1102	985	780	739	567	429	164	6104
1971-72	9	13	264	588	819	1184	1216	709	548	558	259	54	6221
1972-73	8	26	249	503	803	1229	1127	760	834	630	199	87	6455
1973-74	7	36	193	483	659	819	1027	800	680	545	310	73	5632
1974-75	52	42	120	478	733	1046	1006	858	818	746	332	113	6344
1975-76	26	65	51	477	868	961	1012	863	821	610	224	138	6116
1976-77	15	116	116	419	670	1062	1002	695	828	405	509	30	5867
1977-78	10	12	154	392	687	821	858	739	544	582	387	113	5299
1978-79	21	62	234	347	826	1236	1113	781	709	536	247	91	6203
1979-80	9	28	56	339	805	908	865	701	813	451	319	152	5446
1980-81	13	35	79	430	688	885	890	727	715	424	228	48	5162
1981-82	12	7	83	554	669	800	1123	687	760	623	307	133	5758
1982-83	15	11	278	556	855	1006	947	732	752	630	371	77	6230
1983-84	40	8	104	332	708	805	1019	801	637	570	183	133	5340
1984-85	0	8	111	575	753	1056	1060	781	810	359	266	45	5824
1985-86	5	12	230	446	896	1039	757	618	528	469	285	32	5317
1986-87	5	0	291	430	654	913	1028	737	661	299	182	34	5234
1987-88	38	5	45	265	690	1017	982	714	643	408	267	88	5162
1988-89	0	0	132	202	663	1042	1049	944	568	321	256	21	5198
1989-90	0	21	99	417	688	895	943	954	590	312	260	64	5243
1990-91	0	20	55	313	715	1209	1021	588	772	540	404	106	5743
1991-92	0	6	41	265	642	965	945	666	555	273	61	83	4502
1992-93	10	9	41	293	748	1056	1212	853	503	480	200	139	5544
1993-94	16	8	77	316	730	874							

COOLING DEGREE DAYS Base 65 deg. F RENO, NEVADA

YEAR	JAN	FEB	MAR	APR	MAY	JUNE	JULY	AUG	SEP	OCT	NOV	DEC	TOTAL
1969	0	0	0	0	2	19	193	136	35	0	0	0	385
1970	0	0	0	0	6	100	256	210	7	0	0	0	579
1971	0	0	0	0	0	11	217	197	15	0	0	0	440
1972	0	0	0	0	12	65	191	113	4	0	0	0	385
1973	0	0	0	0	13	75	173	117	3	0	0	0	381
1974	0	0	0	0	2	47	118	71	20	0	0	0	258
1975	0	0	0	0	4	43	199	66	33	3	0	0	348
1976	0	0	0	0	0	25	144	34	33	0	0	0	236
1977	0	0	0	0	6	141	154	167	32	0	0	0	500
1978	0	0	0	0	0	16	166	139	19	0	0	0	340
1979	0	0	0	0	9	63	169	122	38	3	0	0	404
1980	0	0	0	0	2	32	218	119	25	1	0	0	397
1981	0	0	0	2	4	153	112	151	80	0	0	0	502
1982	0	0	0	0	2	45	188	135	47	0	0	0	417
1983	0	0	0	0	16	9	115	170	49	0	0	0	359
1984	0	0	0	0	16	42	264	162	61	0	0	0	545
1985	0	0	0	0	3	157	273	126	6	0	0	0	565
1986	0	0	0	0	53	112	148	253	39	0	0	0	605
1987	0	0	0	1	27	126	142	210	53	2	0	0	561
1988	0	0	0	0	11	152	323	264	92	2	0	0	844
1989	0	0	0	0	19	66	240	112	13	0	0	0	450
1990	0	0	0	1	0	95	278	216	76	0	0	0	666
1991	0	0	0	0	0	39	296	214	71	13	0	0	633
1992	0	0	0	5	61	135	197	257	45	1	0	0	701
1993	0	0	0	0	4	52	145	171	92	10	0	0	474

SNOWFALL (inches) RENO, NEVADA

SEASON	JULY	AUG	SEP	OCT	NOV	DEC	JAN	FEB	MAR	APR	MAY	JUNE	TOTAL
1964-65	0.0	0.0	0.0	0.0	5.2	8.0	5.2	T	4.6	T	T	0.0	23.0
1965-66	0.0	0.0	0.0	T	2.2	9.8	T	1.4	0.4	T	0.0	0.0	13.8
1966-67	0.0	0.0	0.0	0.0	T	5.8	12.4	T	15.6	6.8	1.5	0.0	42.1
1967-68	0.0	0.0	0.0	0.0	T	1.9	12.4	2.2	3.5	T	T	0.0	20.0
1968-69	0.0	0.0	0.0	0.0	1.3	9.1	10.7	23.5	1.9	0.3	T	0.0	46.8
1969-70	0.0	0.0	0.0	0.0	T	2.8	T	0.6	1.8	2.3	T	0.2	7.7
1970-71	0.0	0.0	0.0	0.2	T	15.4	11.4	4.9	7.0	3.3	6.3	0.0	48.5
1971-72	0.0	0.0	T	5.1	0.2	25.6	5.7	0.7	0.2	1.0	0.0	0.0	38.5
1972-73	0.0	0.0	0.0	T	T	8.8	11.9	8.2	6.5	0.7	T	0.0	36.1
1973-74	0.0	0.0	0.0	0.0	2.5	6.8	8.7	1.0	6.5	2.6	T	0.0	28.1
1974-75	0.0	0.0	0.0	0.0	0.0	4.0	2.7	19.0	12.1	4.8	2.2	T	44.8
1975-76	0.0	0.0	0.0	T	1.7	T	0.3	15.8	7.8	2.7	T	0.0	28.3
1976-77	0.0	0.0	0.0	0.0	0.0	0.5	0.7	1.4	1.2	T	2.9	0.0	6.7
1977-78	0.0	0.0	0.0	0.0	0.5	5.6	0.3	5.1	0.2	0.1	0.4	0.0	12.2
1978-79	0.0	0.0	T	0.8	9.7	9.1	6.4	3.9	0.2	1.1	0.3	0.0	31.5
1979-80	0.0	0.0	0.0	0.3	T	2.1	4.7	6.1	7.6	1.2	T	0.0	22.0
1980-81	0.0	0.0	0.0	T	T	T	3.9	T	2.2	T	0.0	0.0	6.1
1981-82	0.0	0.0	0.0	1.1	2.0	0.1	12.5	2.3	6.7	0.8	0.5	0.0	26.0
1982-83	0.0	0.0	1.5	0.0	8.6	1.8	1.5	1.0	3.0	2.9	3.5	0.0	23.8
1983-84	0.0	0.0	0.0	0.0	5.7	0.5	6.7	1.5	0.1	T	0.0	0.0	14.5
1984-85	0.0	0.0	0.0	3.4	3.0	1.3	4.3	0.8	7.0	T	T	0.0	19.8
1985-86	0.0	0.0	T	1.2	16.5	1.4	0.0	T	1.4	0.4	T	0.0	20.9
1986-87	0.0	0.0	T	0.0	0.2	0.6	1.8	8.0	2.5	T	0.0	0.0	13.1
1987-88	0.0	0.0	0.0	0.0	0.8	6.3	8.2	0.0	T	T	T	0.0	15.3
1988-89	0.0	0.0	0.0	0.0	4.1	11.7	3.3	13.3	2.2	0.8	T	0.0	35.4
1989-90	0.0	0.0	0.0	T	T	T	5.6	21.6	2.0	0.0	0.0	0.0	29.2
1990-91	0.0	0.0	0.0	0.0	0.4	2.7	0.1	0.0	4.5	T	1.6	0.0	9.3
1991-92	0.0	0.0	0.0	T	5.0	1.4	0.4	1.3	T	0.0	0.0	0.0	8.1
1992-93	0.0	0.0	0.0	0.0	T	14.3	22.9	13.0	0.0	0.0	T	0.0	50.2
1993-94	0.0	0.0	T	T	T	0.3							
Record Mean	0.0	0.0	T	0.3	2.2	4.4	6.0	5.2	4.4	1.2	0.9	T	24.6

WINNEMUCCA, NEVADA

Winnemucca lies at an elevation about 4300 feet above sea level and is effectively cut off by the Sierra Nevada Mountains from the moisture source of the Pacific Ocean. Winnemucca has a climate marked by warm days, cool nights, and light precipitation. Sixty-six percent of the annual rainfall occurs as rain and snow between December and May. The winter snow pack in the surrounding mountains is generally sufficient for essential summertime irrigation. Reservoirs along the streams hold surplus water for less favorable years. As a result of the characteristic dryness of the climate, the neighboring valleys and hills are covered with sagebrush, and trees are found only along streams and in other places where water is sufficient the year round. Though it is heavier in the mountains, snowfall at Winnemucca itself has had measurable amounts fall in every month except July, August and September. During the winter months, snow on the ground permits grazing in many desert regions where there is no other source of stock water. Grazing in the summer months is restricted to mountain range tracts where there is sufficient water. Streams in many areas have been stocked with fish and provide good fishing each year.

Temperatures in this plateau area tend to rise sharply right after sunrise and remain comparatively high during the daylight hours, then drop rapidly about sundown. Daily temperature variations of 50 degrees are not uncommon.

Based on the 1951–1980 period, the average first occurrence of 32 degrees Fahrenheit in the fall is September 10 and the average last occurrence in the spring is June 8.

NORMALS, MEANS, AND EXTREMES

WINNEMUCCA, NEVADA

LATITUDE: 40°54'N LONGITUDE: 117°48' W ELEVATION: FT. GRND 4298 BARO 4301 TIME ZONE: PACIFIC WBAN: 24128

	(a)	JAN	FEB	MAR	APR	MAY	JUNE	JULY	AUG	SEP	OCT	NOV	DEC	YEAR
TEMPERATURE °F:														
Normals														
-Daily Maximum		42.1	49.0	55.0	63.0	73.0	83.1	93.0	90.6	80.1	68.1	52.3	42.7	66.0
-Daily Minimum		16.7	22.8	25.2	29.4	37.7	45.8	51.2	48.7	38.8	29.3	23.3	17.0	32.2
-Monthly		29.4	35.9	40.1	46.2	55.4	64.5	72.2	69.7	59.5	48.7	37.8	29.8	49.1
Extremes														
-Record Highest	44	68	74	81	90	96	106	106	108	103	91	77	67	108
-Year		1971	1986	1972	1981	1986	1988	1979	1983	1950	1980	1980	1980	AUG 1983
-Record Lowest	44	-24	-28	-3	6	12	23	29	28	12	7	-8	-37	-37
-Year		1963	1985	1971	1972	1953	1954	1983	1992	1958	1970	1985	1990	DEC 1990
NORMAL DEGREE DAYS:														
Heating (base 65°F)		1104	815	772	564	307	104	6	28	203	505	816	1091	6315
Cooling (base 65°F)		0	0	0	0	9	89	229	173	38	0	0	0	538
% OF POSSIBLE SUNSHINE	42	51	56	60	66	72	77	86	85	82	74	54	52	68
MEAN SKY COVER (tenths)														
Sunrise - Sunset	44	6.9	6.8	6.7	6.4	5.8	4.4	2.8	3.0	3.2	4.4	6.2	6.4	5.3
MEAN NUMBER OF DAYS:														
Sunrise to Sunset														
-Clear	44	6.4	6.2	6.9	6.4	9.0	13.7	20.1	19.4	18.4	14.5	8.4	8.3	137.6
-Partly Cloudy	44	6.7	6.5	7.3	9.1	9.6	8.7	7.2	7.6	6.4	7.2	6.9	6.3	89.4
-Cloudy	44	18.0	15.6	16.8	14.4	12.4	7.5	3.7	4.0	5.2	9.3	14.8	16.5	138.2
Precipitation														
.01 inches or more	44	8.1	7.0	7.7	6.2	6.3	5.1	2.2	2.6	3.1	4.4	7.3	8.2	68.2
Snow,Ice pellets,hail														
1.0 inches or more	43	1.8	1.4	1.6	0.7	0.3	0.0	0.0	0.0	0.*	0.1	0.9	1.7	8.6
Thunderstorms	43	0.*	0.2	0.4	0.8	2.4	2.9	3.1	3.4	1.7	0.5	0.1	0.1	15.7
Heavy Fog Visibility														
1/4 mile or less	43	1.8	0.5	0.2	0.1	0.*	0.1	0.*	0.*	0.*	0.2	0.3	1.2	4.4
Temperature °F														
-Maximum														
90° and above	44	0.0	0.0	0.0	0.*	1.3	8.5	22.2	18.6	5.8	0.1	0.0	0.0	56.5
32° and below	44	5.8	1.3	0.3	0.0	0.0	0.0	0.0	0.0	0.0	0.0	0.6	4.0	12.0
-Minimum														
32° and below	44	28.3	24.5	25.4	19.6	7.9	1.4	0.1	0.2	6.2	20.4	25.1	28.4	187.3
0° and below	44	3.7	0.8	0.1	0.0	0.0	0.0	0.0	0.0	0.0	0.0	0.5	2.4	7.5
AVG. STATION PRESS.(mb)	19	871.0	869.3	866.6	867.5	866.8	867.7	868.5	868.8	869.3	870.7	870.8	871.5	869.0
RELATIVE HUMIDITY (%)														
Hour 04	44	78	76	72	66	63	57	45	46	52	63	74	79	64
Hour 10 (Local Time)	44	68	60	49	38	34	30	22	23	29	38	56	66	43
Hour 16	44	57	47	38	29	26	22	15	16	20	28	45	57	33
Hour 22	38	75	70	62	52	48	42	30	31	39	52	68	76	54
PRECIPITATION (inches):														
Water Equivalent														
-Normal		0.74	0.62	0.78	0.84	0.83	0.86	0.27	0.45	0.40	0.62	0.94	0.88	8.23
-Maximum Monthly	44	2.70	2.17	1.66	2.92	3.38	2.86	1.74	1.74	1.51	2.19	2.66	3.66	3.66
-Year		1956	1962	1952	1978	1987	1958	1984	1979	1976	1951	1950	1983	DEC 1983
-Minimum Monthly	44	0.04	0.08	0.06	0.06	T	T	0.00	0.00	0.00	T	T	0.03	0.00
-Year		1966	1967	1959	1959	1985	1981	1963	1969	1974	1978	1959	1976	SEP 1974
-Maximum in 24 hrs	44	0.74	0.72	0.67	1.01	1.70	1.79	0.86	0.79	0.83	1.64	1.58	0.95	1.79
-Year		1980	1960	1979	1958	1987	1958	1984	1990	1976	1951	1950	1969	JUN 1958
Snow,Ice pellets,hail														
-Maximum Monthly	44	19.2	13.8	23.4	12.0	5.4	T	0.0	T	1.0	7.4	19.6	17.5	23.4
-Year		1993	1969	1952	1964	1965	1991		1990	1986	1984	1985	1971	MAR 1952
-Maximum in 24 hrs	43	8.8	9.9	9.0	7.2	4.3	T	0.0	T	1.0	4.9	5.7	8.4	9.9
-Year		1993	1959	1982	1971	1971	1991		1990	1986	1984	1985	1955	FEB 1959
WIND:														
Mean Speed (mph)	38	7.6	8.0	8.6	8.7	8.6	8.5	8.4	7.8	7.6	7.3	7.3	7.3	8.0
Prevailing Direction through 1963		NE	S	S	W	W	W	W	W	W	S	S	S	W
Fastest Mile														
-Direction (!!!)	44	W	W	W	W	N	W	W	W	W	NE	S	SW	W
-Speed (MPH)	44	56	59	66	52	61	57	56	51	57	54	40	61	66
-Year		1956	1961	1963	1963	1951	1963	1957	1960	1952	1960	1960	1955	MAR 1963
Peak Gust														
-Direction (!!!)	10	SW	SW	SW	NW	SW	W	SW	W	SW	SW	NW	SW	SW
-Speed (mph)	10	59	54	61	58	49	58	56	51	52	43	54	49	61
-Date		1993	1986	1986	1988	1993	1984	1985	1988	1991	1989	1993	1987	MAR 1986

PRECIPITATION (inches) WINNEMUCCA, NEVADA

YEAR	JAN	FEB	MAR	APR	MAY	JUNE	JULY	AUG	SEP	OCT	NOV	DEC	ANNUAL
1964	1.22	0.11	0.89	1.50	0.87	2.03	0.40	0.50	0.03	0.38	0.81	1.73	10.47
1965	0.98	0.38	0.28	0.94	1.08	0.95	0.36	0.72	0.10	0.17	1.20	0.59	7.75
1966	0.04	0.58	0.38	0.42	0.33	0.38	0.02	0.01	0.34	0.01	1.00	1.00	4.51
1967	0.93	0.08	1.32	1.86	0.25	1.55	0.37	T	0.20	0.09	0.73	0.32	7.70
1968	0.81	0.92	0.31	0.47	0.18	0.85	T	0.69	0.00	0.49	2.54	1.22	8.48
1969	1.30	1.48	0.28	1.29	T	1.65	0.01	0.00	0.01	1.43	0.15	2.07	9.67
1970	1.85	0.30	0.47	0.92	0.42	1.97	0.29	0.40	0.05	0.28	1.10	1.04	9.09
1971	0.19	0.37	0.75	1.47	1.33	1.60	0.03	0.23	0.37	0.26	1.10	1.47	9.17
1972	0.16	0.61	0.38	0.48	0.18	0.55	0.02	T	0.66	1.53	1.08	0.97	6.62
1973	1.35	0.33	0.89	0.73	1.33	0.06	0.10	0.05	0.48	0.64	0.78	1.18	7.92
1974	0.67	0.28	1.13	0.69	T	0.00	0.29	0.39	0.00	1.67	0.06	1.15	6.33
1975	1.02	0.85	1.29	0.53	0.68	0.70	0.06	0.79	0.19	1.74	0.47	0.27	8.59
1976	0.41	0.50	0.29	0.82	0.37	1.14	0.36	1.29	1.51	0.43	0.16	0.03	7.31
1977	0.34	0.14	0.37	0.25	1.56	2.36	0.26	0.64	0.22	T	0.75	1.19	8.08
1978	0.88	0.58	1.08	2.92	0.26	0.08	0.17	T	1.12	T	0.62	0.39	8.10
1979	0.86	1.22	0.96	1.04	0.25	0.39	0.74	1.74	T	0.61	0.61	0.31	8.73
1980	1.93	0.44	0.53	0.26	2.05	0.57	0.13	0.23	0.18	0.27	0.52	0.25	7.36
1981	0.44	0.51	0.97	0.19	1.94	T	0.01	0.03	0.23	1.08	1.40	1.64	8.44
1982	0.30	0.14	1.39	0.29	0.18	1.50	0.88	0.15	1.08	1.19	1.41	0.52	9.03
1983	1.18	0.97	1.46	0.82	0.78	0.85	0.73	1.53	0.59	0.65	1.25	3.66	14.47
1984	0.11	0.91	1.56	1.24	1.41	1.38	1.74	0.19	0.78	1.62	1.57	0.36	12.87
1985	0.83	0.68	0.73	0.20	T	0.04	0.45	0.02	0.52	0.19	2.51	0.84	7.01
1986	0.32	0.86	0.39	0.75	1.07	0.12	0.02	0.27	0.81	0.73	0.10	0.07	5.51
1987	0.53	0.51	1.32	0.54	3.38	0.46	0.19	T	T	0.57	0.86	0.68	9.04
1988	1.19	0.22	0.10	1.55	0.46	0.52	0.07	0.17	0.28	0.04	1.62	0.51	6.73
1989	0.19	0.71	0.89	0.21	0.40	0.33	T	0.44	0.93	0.59	0.77	0.10	5.56
1990	0.47	0.45	0.59	1.33	1.22	0.21	0.09	0.97	0.21	0.06	0.17	0.60	5.90
1991	0.35	0.31	1.15	0.47	1.71	0.54	0.10	0.29	0.81	0.63	0.81	0.63	7.80
1992	0.09	0.21	1.02	0.08	0.02	0.43	0.07	0.05	0.07	0.49	0.60	1.01	4.14
1993	2.02	0.50	1.30	0.45	0.68	1.03	0.18	0.04	0.03	0.54	0.16	0.34	7.27
Record Mean	0.95	0.81	0.88	0.81	0.88	0.75	0.24	0.25	0.39	0.66	0.78	0.97	8.39

AVERAGE TEMPERATURE (deg. F) WINNEMUCCA, NEVADA

YEAR	JAN	FEB	MAR	APR	MAY	JUNE	JULY	AUG	SEP	OCT	NOV	DEC	ANNUAL
1964	28.6	31.9	36.1	43.2	51.8	59.2	71.3	66.2	56.0	52.2	34.5	35.1	47.2
1965	32.7	35.2	37.1	47.9	50.3	61.2	70.2	66.5	54.0	51.4	40.9	28.8	48.0
1966	30.7	31.5	40.2	46.6	60.0	63.2	70.0	70.0	61.6	46.9	40.5	30.6	49.3
1967	32.6	37.4	39.0	39.4	53.6	60.6	72.5	73.7	63.5	48.6	40.5	26.9	49.0
1968	29.8	42.4	41.1	42.4	53.7	65.3	74.3	63.8	57.9	49.2	40.2	29.7	49.2
1969	33.2	31.3	37.7	47.3	58.6	62.3	71.5	69.7	63.0	42.6	36.0	31.8	48.8
1970	34.2	39.9	35.9	37.9	55.2	64.8	73.8	71.0	55.0	44.3	41.9	30.0	48.7
1971	33.7	34.9	37.9	44.5	52.7	62.4	74.3	74.8	55.0	44.6	36.1	24.2	47.9
1972	27.6	36.9	44.6	44.2	58.9	68.1	74.1	72.7	56.9	47.9	36.3	20.8	49.1
1973	26.6	37.0	37.2	44.2	58.4	65.7	72.8	69.7	58.6	47.7	40.8	36.3	49.6
1974	30.3	36.0	43.9	45.4	56.9	69.1	71.3	67.1	60.6	48.0	37.9	28.5	49.6
1975	27.4	34.6	40.7	40.4	53.8	61.8	74.0	67.1	63.3	49.6	36.3	34.8	48.6
1976	33.1	39.9	37.7	45.9	60.5	64.7	73.8	65.6	64.0	49.9	41.8	29.1	50.5
1977	26.9	38.5	37.2	50.4	50.4	70.6	73.0	72.4	58.9	50.4	40.5	38.9	50.7
1978	36.0	37.6	46.6	47.4	54.1	62.6	71.7	69.4	57.3	50.5	34.6	26.0	49.5
1979	25.4	35.9	41.8	47.1	59.1	66.5	72.9	69.5	63.2	53.1	35.5	32.9	50.2
1980	32.7	41.8	38.1	49.1	55.7	60.9	73.2	68.0	60.5	49.6	39.2	33.9	50.2
1981	34.6	36.9	38.6	49.2	53.2	65.9	71.4	71.5	63.0	45.5	41.5	37.4	50.8
1982	28.3	35.8	38.5	44.4	54.3	63.0	70.1	70.3	57.0	44.9	33.3	30.7	47.5
1983	34.3	39.0	41.7	42.7	53.3	62.0	67.2	70.6	59.8	50.2	38.6	31.5	49.3
1984	24.2	34.3	40.3	42.6	55.1	59.0	71.7	71.5	60.0	43.9	38.3	26.4	47.3
1985	19.1	27.2	36.9	48.9	54.7	68.0	74.7	66.9	55.1	44.9	28.0	26.7	45.9
1986	38.7	41.1	45.8	46.7	53.7	68.4	69.8	72.9	54.4	49.0	38.3	29.0	50.7
1987	26.7	34.5	40.2	51.3	58.4	66.7	67.9	70.3	62.4	53.1	39.1	28.8	50.0
1988	27.5	37.1	40.0	48.9	53.8	66.7	75.7	70.7	58.9	55.0	38.4	27.8	50.0
1989	22.0	27.3	43.8	50.7	54.4	65.7	73.1	67.1	60.2	48.0	37.4	30.2	48.3
1990	31.7	29.6	42.6	52.7	53.0	64.2	73.9	70.2	65.4	49.7	35.3	16.6	50.3
1991	26.2	40.3	39.1	43.2	50.1	61.2	74.8	71.1	63.3	50.8	38.9	29.2	49.0
1992	31.6	40.0	44.4	52.2	62.6	65.7	70.5	71.9	61.3	50.6	34.8	26.1	51.0
1993	20.8	29.4	46.5	46.8	59.3	60.0	65.3	67.2	61.0	50.2	30.7	30.6	47.3
Record Mean	28.3	34.3	39.8	46.9	54.9	63.4	71.8	69.4	59.7	48.8	37.8	30.2	48.8
Max	39.8	46.0	53.3	62.0	70.8	80.3	90.6	89.1	79.0	66.6	52.3	42.2	64.3
Min	16.8	22.5	26.4	31.7	39.0	46.5	52.9	49.8	40.3	31.0	23.2	18.1	33.2

REFERENCE NOTES WINNEMUCCA, NEVADA

GENERAL
T - TRACE AMOUNT.
BLANK ENTRIES DENOTE MISSING/UNREPORTED DATA.
# INDICATES A STATION OR INSTRUMENT RELOCATION.

SPECIFIC

(a) - LENGTH OF RECORD IN YEARS, ALTHOUGH INDIVIDUAL MONTHS MAY BE MISSING.
0.* OR * - THE VALUE IS BETWEEN 0.0 AND 0.05
NORMALS - BASED ON THE 1961-1990 RECORD PERIOD.
EXTREMES - DATES ARE THE MOST RECENT OCCURRENCE
WIND DIR.- NUMERALS SHOW TENS OF DEGRESS CLOCKWISE FROM TRUE NORTH. "00" INDICATES CALM.
RESULTANT DIRECTIONS ARE GIVEN TO WHOLE DEGREES.
BOLD VALUES INDICATE EXTREME VALUES WHICH OCCURRED AFTER THE ASOS SYSTEM WAS COMMISSIONED.

RECORD = PERIOD OF RECORD
RECORD MEAN PRECIPITATION IS THE MEAN OF ALL DAILY PRECIPITATION AMOUNTS DURING THE PERIOD OF RECORD.
RECORD MAX(MIN) TEMPERATURE IS THE MEAN OF ALL DAILY MAX(MIN) TEMPERATURES DURING THE PERIOD OF RECORD.
RECORD MEAN TEMPERATURE IS THE SUM OF THE RECORD MAX AND RECORD MIN DIVIDED BY 2.
AVERAGE TEMPERATURE IS THE SUM OF THE MEAN DAILY MAX AND MIN TEMPERATURE DIVIDED BY 2.

EXCEPTIONS

RECORD MEANS ARE THROUGH THE CURRENT YEAR, BEGINNING IN 1878 FOR TEMPERATURE
1878 FOR PRECIPITATION
1950 FOR SNOWFALL

HEATING DEGREE DAYS Base 65 deg. F — WINNEMUCCA, NEVADA

SEASON	JULY	AUG	SEP	OCT	NOV	DEC	JAN	FEB	MAR	APR	MAY	JUNE	TOTAL
1964-65	7	69	262	389	907	919	996	829	859	504	448	137	6326
1965-66	10	39	321	417	716	1116	1055	933	759	548	172	121	6207
1966-67	12	31	119	552	726	1059	996	767	800	760	358	170	6350
1967-68	0	0	80	499	730	1174	1083	648	735	672	346	96	6063
1968-69	3	118	217	483	736	1087	979	938	837	521	194	110	6223
1969-70	16	9	85	685	862	1022	949	699	897	805	298	120	6447
1970-71	0	0	295	635	686	1080	962	837	834	604	374	117	6424
1971-72	3	3	306	625	860	1260	1151	808	626	617	218	36	6513
1972-73	1	6	255	525	855	1370	1184	780	859	616	218	97	6766
1973-74	1	26	200	531	719	881	1070	807	646	581	263	52	5777
1974-75	32	41	153	521	806	1127	1156	844	749	730	339	140	6638
1975-76	13	41	69	474	855	925	982	718	841	568	139	71	5696
1976-77	8	45	81	461	691	1104	1176	734	855	430	450	15	6050
1977-78	2	28	203	444	729	798	894	762	566	523	335	96	5380
1978-79	21	53	241	442	907	1202	1221	809	710	532	205	63	6406
1979-80	1	18	75	363	875	988	995	667	827	471	290	162	5732
1980-81	0	23	146	472	766	958	935	779	809	474	365	89	5816
1981-82	2	0	115	600	697	848	1131	811	812	610	327	124	6077
1982-83	29	8	258	618	943	1054	947	721	714	658	377	103	6430
1983-84	35	6	156	451	785	1032	1257	881	757	662	307	210	6539
1984-85	0	11	185	645	794	1188	1415	1049	863	479	314	49	6992
1985-86	3	39	293	617	1101	1179	809	665	590	542	381	40	6259
1986-87	12	4	345	490	792	1108	1181	849	763	407	221	64	6236
1987-88	47	14	111	364	772	1116	1155	804	769	474	343	94	6063
1988-89	0	4	223	305	792	1146	1328	1051	648	419	336	37	6289
1989-90	3	47	139	523	822	1071	1026	986	690	360	363	92	6122
1990-91	4	38	45	468	887	1498	1195	685	796	647	452	135	6850
1991-92	0	4	93	431	777	1104	1029	717	633	379	106	95	5368
1992-93	13	40	124	438	899	1198	1365	992	567	536	192	183	6547
1993-94	45	40	166	448	1024	1061							

COOLING DEGREE DAYS Base 65 deg. F — WINNEMUCCA, NEVADA

YEAR	JAN	FEB	MAR	APR	MAY	JUNE	JULY	AUG	SEP	OCT	NOV	DEC	TOTAL
1969	0	0	0	0	2	37	222	162	33	0	0	0	456
1970	0	0	0	0	2	124	279	191	0	0	0	0	596
1971	0	0	0	0	0	45	295	313	13	0	0	0	666
1972	0	0	0	0	40	136	287	250	17	1	0	0	731
1973	0	0	0	0	18	124	248	180	15	0	0	0	585
1974	0	0	0	0	19	184	236	112	29	0	0	0	580
1975	0	0	0	0	0	50	297	113	25	3	0	0	488
1976	0	0	0	0	7	68	287	71	56	0	0	0	489
1977	0	0	0	0	4	188	257	265	27	0	0	0	741
1978	0	0	0	0	5	31	234	197	18	0	0	0	485
1979	0	0	0	0	29	114	256	165	26	3	0	0	593
1980	0	0	0	0	8	48	262	126	19	2	0	0	465
1981	0	0	0	6	5	124	208	208	61	0	0	0	612
1982	0	0	0	0	2	71	195	177	22	0	0	0	467
1983	0	0	0	0	21	21	108	189	10	0	0	0	349
1984	0	0	0	0	8	37	217	218	40	0	0	0	520
1985	0	0	0	0	3	148	311	105	3	0	0	0	570
1986	0	0	0	0	37	147	166	254	32	0	0	0	636
1987	0	0	0	3	26	123	144	185	41	0	0	0	522
1988	0	0	0	0	5	152	337	187	48	0	0	0	729
1989	0	0	0	0	13	66	264	118	1	0	0	0	462
1990	0	0	0	0	0	74	290	205	62	0	0	0	631
1991	0	0	0	0	0	26	310	200	49	0	0	0	585
1992	0	0	0	1	39	121	194	258	20	2	0	0	635
1993	0	0	0	0	24	39	60	116	50	0	0	0	289

SNOWFALL (inches) — WINNEMUCCA, NEVADA

SEASON	JULY	AUG	SEP	OCT	NOV	DEC	JAN	FEB	MAR	APR	MAY	JUNE	TOTAL
1964-65	0.0	0.0	0.0	0.0	6.3	4.5	3.2	1.8	1.5	3.3	5.4	0.0	26.0
1965-66	0.0	0.0	0.0	0.0	0.5	5.7	0.4	7.4	2.0	0.0	0.0	0.0	16.0
1966-67	0.0	0.0	0.0	T	T	10.8	4.1	0.4	11.1	6.3	T	0.0	32.7
1967-68	0.0	0.0	0.0	0.0	5.3	4.5	8.2	T	0.6	T	0.0	0.0	18.6
1968-69	0.0	0.0	0.0	0.0	6.1	11.3	6.9	13.8	3.5	2.9	0.0	0.0	44.5
1969-70	0.0	0.0	0.0	3.5	0.3	6.2	0.1	0.8	1.6	9.2	0.2	0.0	21.9
1970-71	0.0	0.0	0.0	0.4	0.1	4.9	1.8	2.4	2.0	8.8	4.3	0.0	24.7
1971-72	0.0	0.0	0.2	2.3	5.5	17.5	0.8	6.2	2.7	2.6	T	0.0	37.8
1972-73	0.0	0.0	0.0	0.1	1.2	13.3	9.9	1.1	7.4	3.4	0.0	0.0	36.4
1973-74	0.0	0.0	0.0	0.0	2.8	4.6	5.5	2.5	10.0	3.1	T	0.0	28.5
1974-75	0.0	0.0	0.0	T	T	5.7	7.0	7.2	4.9	2.9	4.3	0.0	32.0
1975-76	0.0	0.0	0.0	T	3.3	4.1	3.0	1.5	3.7	4.8	0.0	0.0	20.4
1976-77	0.0	0.0	0.0	T	T	0.4	4.5	0.9	4.8	0.7	1.6	0.0	12.9
1977-78	0.0	0.0	T	0.0	3.0	3.6	4.4	3.8	0.0	5.8	1.9	0.0	22.5
1978-79	0.0	0.0	T	0.0	1.8	4.2	5.2	12.1	4.6	1.9	0.1	0.0	29.9
1979-80	0.0	0.0	0.0	0.0	1.6	0.8	2.7	T	0.5	0.4	T	0.0	6.0
1980-81	0.0	0.0	0.0	T	T	0.2	3.7	1.2	1.1	T	T	0.0	6.2
1981-82	0.0	0.0	0.0	T	0.5	3.8	4.6	0.8	15.9	0.9	T	0.0	26.5
1982-83	0.0	0.0	0.0	T	5.8	2.3	9.2	2.1	0.9	1.3	0.3	0.0	21.9
1983-84	0.0	0.0	0.0	0.0	2.2	14.6	1.4	2.4	5.5	2.1	T	0.0	28.2
1984-85	0.0	0.0	0.0	7.4	2.6	3.9	8.8	7.2	5.8	0.4	T	0.0	36.1
1985-86	0.0	0.0	T	T	19.6	0.8	0.3	2.0	1.8	0.4	1.3	0.0	26.2
1986-87	0.0	0.0	1.0	0.0	0.8	T	2.7	5.1	6.0	0.1	0.0	0.0	15.7
1987-88	0.0	0.0	0.0	0.0	0.8	3.4		T					
1988-89						7.1	2.8	8.4	0.5	1.9	0.0	T	
1989-90	0.0	0.0	0.0	0.1	1.6	T	3.3	11.7	0.3	T	T	0.0	17.0
1990-91	0.0	T	0.0	0.0	2.0	8.0	3.7	T	8.6	0.7	T	T	23.0
1991-92	0.0	0.0	0.0	2.3	1.7	1.5	0.9	0.3	0.3	0.0	0.0	0.0	7.0
1992-93	0.0	0.0	0.0	0.0	0.4	6.8	19.2	7.8	T	0.7	T	0.0	34.9
1993-94	0.0	0.0	0.0	0.0	1.0	4.1							
Record Mean	0.0	T	T	0.4	2.3	4.7	5.4	3.8	4.4	2.3	0.6	T	23.9

CONCORD, NEW HAMPSHIRE

Concord, the Capital of New Hampshire, is situated near the geographical center of New England at an altitude of approximately 300 feet above sea level on the Merrimack River. Its surroundings are hilly with many lakes and ponds. The countryside is generously wooded, mostly on land reclaimed from fields which were formerly cleared for farming. From the coast about 50 miles to the southeast, the terrain slopes gently upward to the city. West of the city, the land rises some 2,000 feet higher in only half that distance. Mount Washington, at an elevation of 6,288 feet is in the White Mountains 75 miles north of town.

Northwesterly winds are prevalent. They bring cold, dry air during the winter and pleasantly cool, dry air in the summer. Stronger southerly winds occur during July and August, and easterly winds usually accompany summer and winter storms. Winter breezes are somewhat lighter, and winds are frequently calm during the night and early morning hours. Low temperatures, as a rule, do not interrupt normal out-of-doors activity because winds are calm or light, producing a low wind chill factor.

Very hot summer weather is infrequent. During any month, temperatures considerably above the average maxima and much below the normal minima are observed.

The average amount of precipitation for the warmer half of the year differs little from that for the colder half. Precipitation occurrences average approximately one day of three for the year, with a somewhat higher frequency for the April-May period, offsetting the lower frequency of August-October. The more significant rains and heavier snowfalls are associated with easterly winds, especially northeasterly winds. The first snowfall of an inch or more is likely to come between the middle of November and the middle of December. The snow cover normally lasts from mid-December until the last week of March, but bare ground is not rare in the winter, nor is a snowscape rare earlier or later in the season. Rain, sleet, or freezing rain may also occur.

Agriculture is neither intensive nor large-scale in the vicinity of the station. Potatoes and other frost-resistant vegetables, hardy fruits such as apples, forage for the dairy industry, and maple sugar are the principal crops.

Based on the 1951-1980 period, the average first occurrence of 32 degrees Fahrenheit in the fall is September 22 and the average last occurrence in the spring is May 23. Freezing temperatures have occurred as late as June and as early as August.

NORMALS, MEANS, AND EXTREMES

CONCORD, NEW HAMPSHIRE

LATITUDE: 43°12'N LONGITUDE: 71°30' W ELEVATION: FT. GRND 342 BARO 343 TIME ZONE: EASTERN WBAN: 14745

	(a)	JAN	FEB	MAR	APR	MAY	JUNE	JULY	AUG	SEP	OCT	NOV	DEC	YEAR
TEMPERATURE °F:														
Normals														
-Daily Maximum		29.8	33.0	42.8	56.3	68.9	77.3	82.4	79.8	71.6	60.7	47.1	34.2	57.0
-Daily Minimum		7.4	10.4	22.1	31.5	41.4	51.2	56.5	54.7	46.0	34.9	27.0	14.4	33.1
-Monthly		18.6	21.8	32.4	43.9	55.2	64.2	69.5	67.3	58.8	47.8	37.1	24.3	45.1
Extremes														
-Record Highest	52	68	66	85	95	97	98	102	101	98	90	80	68	102
-Year		1950	1957	1977	1976	1962	1980	1966	1975	1953	1963	1950	1982	JUL 1966
-Record Lowest	52	-33	-37	-16	8	21	30	35	29	21	10	-5	-22	-37
-Year		1984	1943	1967	1969	1966	1972	1965	1965	1947	1972	1989	1951	FEB 1943
NORMAL DEGREE DAYS:														
Heating (base 65°F)		1438	1210	1011	633	312	70	13	39	196	533	837	1262	7554
Cooling (base 65°F)		0	0	0	0	8	46	153	111	10	0	0	0	328
% OF POSSIBLE SUNSHINE	52	52	55	53	53	55	59	63	60	56	53	43	47	54
MEAN SKY COVER (tenths)														
Sunrise - Sunset	52	6.1	6.2	6.4	6.6	6.7	6.3	6.2	5.9	6.0	5.9	6.8	6.4	6.3
MEAN NUMBER OF DAYS:														
Sunrise to Sunset														
-Clear	52	9.2	7.6	7.8	7.0	6.1	6.1	6.7	8.0	8.6	9.2	6.3	7.8	90.5
-Partly Cloudy	52	7.1	7.6	8.0	8.0	9.9	11.8	12.5	11.5	9.1	9.0	7.6	7.9	109.8
-Cloudy	52	14.8	13.1	15.2	14.9	14.9	12.2	11.8	11.4	12.3	12.8	16.1	15.3	164.8
Precipitation														
.01 inches or more	52	10.6	9.6	10.9	11.6	11.9	10.9	10.1	9.8	8.9	9.0	11.4	10.9	125.5
Snow,Ice pellets,hail														
1.0 inches or more	52	4.5	3.8	3.1	0.6	0.*	0.0	0.0	0.0	0.0	0.*	1.3	4.0	17.3
Thunderstorms	52	0.*	0.1	0.3	0.8	2.3	4.3	5.3	3.8	1.8	0.6	0.1	0.*	19.4
Heavy Fog Visibility														
1/4 mile or less	52	2.2	1.8	2.8	1.9	3.2	3.6	5.2	6.8	9.0	6.3	3.6	2.7	49.1
Temperature °F														
-Maximum														
90° and above	28	0.0	0.0	0.0	0.2	1.0	2.2	4.8	3.0	0.5	0.0	0.0	0.0	11.5
32° and below	28	17.7	13.2	4.4	0.2	0.0	0.0	0.0	0.0	0.0	0.*	1.7	13.1	50.3
-Minimum														
32° and below	28	30.4	27.3	25.8	16.9	5.2	0.2	0.0	0.1	2.3	14.1	21.5	29.1	172.9
0° and below	28	9.4	6.6	1.1	0.0	0.0	0.0	0.0	0.0	0.0	0.0	0.2	5.1	22.4
AVG. STATION PRESS.(mb)	21	1002.9	1003.4	1002.6	1001.5	1002.3	1001.6	1002.3	1004.2	1005.5	1005.6	1004.3	1003.9	1003.3
RELATIVE HUMIDITY (%)														
Hour 01	27	74	73	73	76	83	88	90	91	91	86	81	78	82
Hour 07 (Local Time)	28	75	76	77	76	77	82	84	88	91	88	84	80	82
Hour 13	28	58	55	52	47	47	52	51	53	55	53	59	61	54
Hour 19	28	66	62	59	55	58	63	65	70	76	73	73	71	66
PRECIPITATION (inches):														
Water Equivalent														
-Normal		2.51	2.53	2.72	2.91	3.14	3.15	3.23	3.32	2.81	3.23	3.66	3.16	36.37
-Maximum Monthly	52	8.09	7.77	7.81	5.88	9.52	10.10	6.53	7.26	7.78	8.78	7.36	7.52	10.10
-Year		1979	1981	1953	1983	1984	1944	1988	1991	1960	1962	1983	1973	JUN 1944
-Minimum Monthly	52	0.40	0.03	0.86	1.02	0.60	0.64	0.96	0.95	0.41	0.59	0.75	0.58	0.03
-Year		1970	1987	1981	1985	1965	1979	1955	1944	1948	1947	1976	1943	FEB 1987
-Maximum in 24 hrs	52	2.12	2.26	2.27	2.27	2.59	4.47	2.54	3.97	4.12	4.24	2.89	3.31	4.47
-Year		1979	1981	1974	1987	1984	1944	1971	1991	1960	1962	1947	1969	JUN 1944
Snow,Ice pellets,hail														
-Maximum Monthly	52	45.4	49.8	38.3	15.3	5.0	T	0.0	0.0	T	2.1	18.4	38.1	49.8
-Year		1987	1969	1956	1982	1945	1993			1992	1969	1971	1956	FEB 1969
-Maximum in 24 hrs	52	19.0	14.2	16.8	13.9	5.0	T	0.0	0.0	T	2.1	9.5	14.6	19.0
-Year		1944	1972	1993	1982	1945	1993			1992	1969	1961	1946	JAN 1944
WIND:														
Mean Speed (mph)	51	7.2	7.8	8.2	7.8	7.0	6.4	5.7	5.4	5.5	6.0	6.7	7.1	6.7
Prevailing Direction through 1963		NW	NW	NW	NW	NW	NW	NW	NW	NW	NW	NW	NW	NW
Fastest Mile														
-Direction (!!)	50	NW	N	NE	NW	NW	SW	SW	E	E	NW	NE	NW	NE
-Speed (MPH)	50	44	42	71	52	48	44	45	56	42	39	72	52	72
-Year		1972	1950	1950	1945	1945	1986	1971	1934	1960	1944	1950	1962	NOV 1950
Peak Gust														
-Direction (!!)	10	W	E	NW	NW	NW	SW	N	NW	SE	SW	NW	W	E
-Speed (mph)	10	52	60	49	48	45	59	53	45	44	44	53	58	60
-Date		1992	1988	1993	1985	1993	1986	1987	1987	1985	1990	1989	1993	FEB 1988

PRECIPITATION (inches)

CONCORD, NEW HAMPSHIRE

YEAR	JAN	FEB	MAR	APR	MAY	JUNE	JULY	AUG	SEP	OCT	NOV	DEC	ANNUAL
1964	3.64	1.93	3.01	2.05	1.15	0.81	3.37	2.74	0.42	2.24	3.47	3.07	27.90
1965	0.97	3.06	0.93	2.49	0.60	2.72	1.73	3.12	1.87	2.71	2.45	1.52	24.17
1966	2.69	2.13	2.19	1.16	2.16	1.78	2.30	4.05	5.40	3.31	3.01	2.42	32.60
1967	1.23	2.36	1.75	3.52	3.92	3.82	5.91	1.97	2.04	0.99	2.86	3.82	34.19
1968	1.79	0.93	3.80	2.85	5.20	5.90	1.62	3.60	2.34	2.23	5.24	5.82	41.32
1969	1.34	3.69	2.36	2.75	1.26	4.70	4.40	2.84	4.43	1.56	5.87	7.10	42.30
1970	0.40	4.27	2.78	3.38	3.04	2.26	2.33	3.06	3.40	3.64	3.03	3.08	34.67
1971	1.63	3.87	2.28	2.19	3.36	1.67	5.14	2.79	1.91	2.69	2.91	2.36	32.80
1972	1.44	2.60	4.16	2.71	4.20	3.54	5.40	2.12	2.55	2.23	6.57	4.55	42.07
1973	2.44	1.91	2.58	4.55	4.20	4.86	1.05	6.88	1.77	2.46	1.82	7.52	42.04
1974	2.80	2.32	3.98	2.58	3.74	1.82	1.41	2.20	4.74	1.64	3.20	4.02	34.45
1975	4.12	2.36	3.12	2.47	1.22	3.87	3.71	3.94	5.15	4.29	4.91	3.12	42.28
1976	3.40	2.36	2.02	2.43	3.90	2.74	3.20	2.66	2.73	4.05	0.75	2.27	32.51
1977	2.16	2.02	4.51	4.04	2.44	3.47	1.26	3.51	5.64	5.52	3.07	4.00	41.64
1978	6.32	0.67	2.16	2.06	2.67	3.18	1.08	2.87	0.46	2.72	1.77	2.91	28.87
1979	8.09	2.29	2.85	3.10	4.86	0.64	3.45	4.20	3.15	3.79	2.92	1.93	41.27
1980	0.43	0.78	3.37	3.72	0.86	2.83	2.35	3.99	2.19	2.63	3.12	0.79	27.06
1981	0.48	7.77	0.86	3.12	3.21	2.81	5.54	3.25	4.61	6.51	3.51	4.17	45.84
1982	3.98	2.88	2.47	3.08	1.91	7.84	2.83	2.54	1.85	1.52	2.93	0.91	34.74
1983	3.92	2.17	7.07	5.88	5.19	2.52	2.07	2.07	1.21	3.28	7.36	5.35	48.09
1984	1.89	5.06	2.92	3.74	9.52	2.83	4.44	0.97	1.08	4.42	2.67	2.70	42.24
1985	0.95	1.99	2.86	1.02	2.05	3.05	2.83	2.51	3.78	3.62	4.58	1.65	30.89
1986	4.78	2.23	3.58	1.85	1.44	4.95	4.77	3.72	2.27	1.71	4.48	4.50	40.28
1987	3.00	0.03	3.47	4.71	1.08	5.77	3.77	2.84	3.94	4.14	2.50	1.55	36.80
1988	1.97	2.24	1.32	2.75	3.35	0.80	6.53	5.44	1.56	1.23	5.06	1.05	33.30
1989	0.74	2.05	2.18	3.40	5.11	4.25	3.62	3.55	4.22	4.86	3.34	0.91	38.23
1990	2.82	2.63	1.64	3.00	5.09	2.51	1.79	7.19	2.31	4.93	3.25	4.12	41.28
1991	1.85	1.42	3.01	2.61	2.51	1.72	1.80	7.26	5.52	3.32	4.84	3.62	39.48
1992	1.97	1.30	2.77	1.78	2.39	3.28	3.50	2.13	2.48	2.41	3.54	2.23	29.78
1993	1.51	1.63	3.04	3.77	0.97	1.95	2.10	1.97	4.01	3.58	3.72	3.36	31.61
Record Mean	2.88	2.58	3.06	3.00	3.14	3.29	3.53	3.43	3.37	3.19	3.47	3.04	37.97

AVERAGE TEMPERATURE (deg. F)

CONCORD, NEW HAMPSHIRE

YEAR	JAN	FEB	MAR	APR	MAY	JUNE	JULY	AUG	SEP	OCT	NOV	DEC	ANNUAL
1964	22.9	22.4	33.6	43.6	58.8	65.6	70.8	62.7	57.4	47.4	37.1	25.1	45.6
#1965	18.9	22.2	31.9	41.3	57.1	62.5	65.9	66.6	58.8	46.8	34.8	27.3	44.5
1966	20.4	22.5	34.6	42.2	53.4	66.8	72.2	69.6	57.0	48.0	41.5	26.2	46.2
1967	24.8	17.5	27.1	42.0	48.8	66.1	70.2	67.2	58.4	49.3	32.7	27.8	44.3
1968	15.8	18.0	34.8	44.8	51.1	62.0	69.3	64.6	61.0	51.2	34.8	22.5	44.2
1969	21.9	23.2	28.3	44.3	52.3	63.0	65.4	68.9	59.7	47.5	38.9	23.7	44.8
1970	11.0	22.6	30.7	45.8	57.3	63.1	70.9	68.7	60.6	51.0	39.3	20.3	45.1
1971	12.5	23.3	30.4	42.1	54.1	65.5	68.3	67.0	61.1	51.9	32.4	26.6	44.6
1972	22.0	21.2	27.9	40.2	56.8	62.8	69.3	64.1	57.5	42.3	31.2	23.4	43.2
1973	21.0	20.2	35.5	45.0	52.9	66.9	70.3	72.3	58.8	48.2	36.1	28.9	46.3
1974	21.4	21.1	31.8	45.8	51.1	62.8	67.5	67.0	58.9	42.5	35.9	26.7	44.4
1975	21.6	21.5	30.1	40.4	61.3	65.0	72.9	66.5	56.1	47.6	39.9	21.9	45.4
1976	10.9	24.7	31.9	46.7	53.6	68.9	66.8	65.3	57.2	45.0	31.7	16.1	43.3
1977	10.6	20.5	36.7	45.1	58.3	62.9	69.0	68.0	58.5	47.0	39.4	21.0	44.8
1978	17.5	13.5	28.0	40.6	57.5	65.2	69.7	69.6	55.9	46.7	35.3	22.7	43.5
1979	23.3	15.1	37.6	44.2	56.3	63.8	71.2	67.5	59.8	47.4	42.2	29.4	46.5
1980	22.4	19.1	31.8	44.4	55.6	62.8	70.6	68.4	58.2	45.1	34.8	19.2	44.4
1981	12.5	30.8	34.2	47.1	57.5	66.1	69.9	66.6	58.7	45.2	37.8	25.4	46.0
1982	10.9	20.8	30.2	41.6	57.3	60.9	69.5	65.4	60.2	47.6	41.7	32.4	44.9
1983	23.1	26.1	35.9	45.2	53.1	65.1	70.1	69.3	62.3	48.2	39.4	23.4	46.8
1984	16.0	30.5	28.2	44.6	52.9	65.9	68.7	69.3	57.8	50.4	38.3	30.3	46.1
1985	15.9	25.8	35.7	45.2	56.1	62.2	70.3	66.7	60.5	49.2	38.5	22.0	45.7
1986	23.0	21.3	35.5	48.2	57.5	61.3	67.3	66.1	57.3	47.4	34.1	29.0	45.7
1987	20.0	22.0	34.5	46.8	56.7	64.4	70.6	65.1	58.7	45.8	36.9	28.1	45.8
1988	18.5	23.5	33.4	43.9	57.3	62.9	72.6	70.5	57.8	44.9	39.1	23.2	45.6
1989	25.7	22.6	31.8	41.4	58.3	65.0	69.5	67.7	60.8	49.1	35.8	11.9	45.0
1990	28.6	24.6	34.7	45.6	52.8	65.3	70.8	69.8	59.7	51.8	40.1	30.9	47.9
1991	20.2	28.3	36.0	47.6	60.2	65.6	69.2	69.9	57.7	50.8	39.7	25.5	47.6
1992	23.3	26.0	30.2	42.5	55.0	64.0	66.0	67.1	59.7	44.9	35.6	26.5	45.1
1993	23.5	15.7	30.5	45.7	56.9	64.9	71.4	70.0	58.8	45.3	36.4	26.8	45.5
Record Mean	21.0	22.8	32.1	44.4	56.2	64.8	69.9	67.3	59.6	48.9	37.5	25.4	45.8
Max	31.3	33.5	41.8	56.0	68.8	77.1	82.1	79.1	71.4	60.5	47.0	34.7	56.9
Min	10.7	12.0	22.4	32.8	43.5	52.5	57.8	55.6	47.8	37.2	28.1	16.1	34.7

REFERENCE NOTES

CONCORD, NEW HAMPSHIRE

GENERAL
T - TRACE AMOUNT.
BLANK ENTRIES DENOTE MISSING/UNREPORTED DATA.
# INDICATES A STATION OR INSTRUMENT RELOCATION.

SPECIFIC

PM - INCLUDES LAST DAY OF PREVIOUS MONTH
ASOS - AUTOMATED SURFACE OBSERVING SYSTEM IN OPERATION DURING THESE MONTHS.

(a) - LENGTH OF RECORD IN YEARS, ALTHOUGH INDIVIDUAL MONTHS MAY BE MISSING.
0.* OR * - THE VALUE IS BETWEEN 0.0 AND 0.05
NORMALS - BASED ON THE 1961-1990 RECORD PERIOD.
EXTREMES - DATES ARE THE MOST RECENT OCCURRENCE
WIND DIR.- NUMERALS SHOW TENS OF DEGRESS CLOCKWISE FROM TRUE NORTH. "00" INDICATES CALM.
RESULTANT DIRECTIONS ARE GIVEN TO WHOLE DEGREES.
BOLD VALUES INDICATE EXTREME VALUES WHICH OCCURRED AFTER THE ASOS SYSTEM WAS COMMISSIONED.

RECORD = PERIOD OF RECORD
RECORD MEAN PRECIPITATION IS THE MEAN OF ALL DAILY PRECIPITATION AMOUNTS DURING THE PERIOD OF RECORD.
RECORD MAX(MIN) TEMPERATURE IS THE MEAN OF ALL DAILY MAX(MIN) TEMPERATURES DURING THE PERIOD OF RECORD.
RECORD MEAN TEMPERATURE IS THE SUM OF THE RECORD MAX AND RECORD MIN DIVIDED BY 2.
AVERAGE TEMPERATURE IS THE SUM OF THE MEAN DAILY MAX AND MIN TEMPERATURE DIVIDED BY 2.

EXCEPTIONS

RECORD MEANS ARE THROUGH THE CURRENT YEAR, BEGINNING IN 1871 FOR TEMPERATURE
1855 FOR PRECIPITATION
1942 FOR SNOWFALL

HEATING DEGREE DAYS Base 65 deg. F CONCORD, NEW HAMPSHIRE

SEASON	JULY	AUG	SEP	OCT	NOV	DEC	JAN	FEB	MAR	APR	MAY	JUNE	TOTAL
#1964-65	8	106	246	539	831	1230	1418	1191	1015	704	257	133	7678
1965-66	50	78	227	556	899	1163	1376	1184	935	681	362	73	7584
1966-67	6	5	244	520	697	1196	1241	1323	1164	683	496	57	7632
1967-68	8	34	208	482	965	1144	1520	1358	928	598	424	118	7787
1968-69	18	92	133	424	899	1311	1330	1165	1128	613	389	119	7621
1969-70	64	40	193	534	777	1275	1668	1179	1055	572	256	108	7721
1970-71	4	25	181	431	760	1379	1622	1165	1064	682	332	73	7718
1971-72	26	49	165	396	970	1185	1327	1267	1142	736	262	112	7637
1972-73	27	82	223	695	1007	1284	1357	1250	905	596	370	78	7874
1973-74	15	9	244	518	860	1112	1345	1223	1025	573	432	99	7455
1974-75	34	26	213	694	865	1182	1339	1218	1075	730	152	98	7626
1975-76	10	78	260	532	747	1330	1672	1162	1019	565	356	60	7791
1976-77	37	84	234	615	992	1506	1683	1242	870	594	259	119	8235
1977-78	37	58	222	551	760	1360	1466	1435	1138	725	270	72	8094
1978-79	45	34	275	563	882	1304	1284	1392	841	617	280	99	7616
1979-80	33	64	199	546	675	1098	1317	1324	1022	610	290	123	7301
1980-81	13	33	245	611	899	1417	1626	953	951	530	267	40	7585
1981-82	12	43	192	608	810	1222	1674	1233	1072	695	246	136	7943
1982-83	25	66	169	535	692	1007	1291	1086	895	588	364	82	6800
1983-84	14	33	167	521	760	1283	1516	993	1135	607	382	85	7496
1984-85	19	27	238	446	792	1072	1514	1092	901	588	286	106	7081
1985-86	9	38	166	485	785	1326	1295	1216	907	499	267	140	7133
1986-87	41	67	251	538	919	1109	1390	1199	939	542	296	77	7368
1987-88	18	89	201	589	837	1138	1436	1194	971	626	254	137	7490
1988-89	19	60	219	622	769	1289	1211	1182	1022	703	219	80	7395
1989-90	6	53	169	484	865	1639	1121	1121	933	585	369	67	7412
1990-91	22	15	183	409	737	1049	1381	1024	893	517	186	58	6474
1991-92	19	7	238	438	754	1216	1286	1123	1070	669	319	90	7229
1992-93	52	33	203	617	875	1190	1277	1372	1062	574	248	85	7588
1993-94	6	15	229	600	851	1177							

COOLING DEGREE DAYS Base 65 deg. F CONCORD, NEW HAMPSHIRE

YEAR	JAN	FEB	MAR	APR	MAY	JUNE	JULY	AUG	SEP	OCT	NOV	DEC	TOTAL
1969	0	0	0	0	4	65	82	168	44	0	0	0	363
1970	0	0	0	1	27	58	196	145	56	6	0	0	489
1971	0	0	0	0	0	95	135	118	55	0	0	0	403
1972	0	0	0	0	17	52	167	60	5	0	0	0	301
1973	0	0	0	3	2	145	184	242	67	2	0	0	645
1974	0	0	0	3	9	40	118	92	40	0	0	0	302
1975	0	0	0	0	48	108	263	134	0	0	0	0	553
1976	0	0	0	18	9	184	100	99	8	1	0	0	419
1977	0	0	0	5	57	64	168	156	38	0	0	0	488
1978	0	0	0	0	46	83	198	186	8	0	0	0	521
1979	0	0	0	1	15	69	232	150	46	6	0	0	519
1980	0	0	0	0	5	64	193	145	51	0	0	0	458
1981	0	0	0	0	38	80	172	101	11	0	0	0	402
1982	0	0	0	0	11	17	171	87	31	0	0	0	317
1983	0	0	0	0	2	93	179	172	93	5	0	0	544
1984	0	0	0	0	15	117	139	165	28	0	0	0	464
1985	0	0	0	0	16	27	184	93	36	1	0	0	357
1986	0	0	0	0	44	37	118	109	23	0	0	0	331
1987	0	0	0	1	45	64	198	96	21	0	0	0	425
1988	0	0	0	0	19	81	262	238	7	4	0	0	611
1989	0	0	0	0	17	87	153	141	50	0	0	0	448
1990	0	0	0	10	0	80	210	170	30	8	0	0	508
1991	0	0	0	2	44	84	157	166	29	4	0	0	486
1992	0	0	0	2	15	66	90	105	52	0	0	0	330
1993	0	0	0	0	7	90	212	176	49	0	0	0	534

SNOWFALL (inches) CONCORD, NEW HAMPSHIRE

SEASON	JULY	AUG	SEP	OCT	NOV	DEC	JAN	FEB	MAR	APR	MAY	JUNE	TOTAL
1964-65	0.0	0.0	0.0	T	T	12.4	13.2	6.5	5.2	1.3	0.0	0.0	38.6
1965-66	0.0	0.0	0.0	T	1.0	4.6	32.9	17.7	3.7	T	1.1	0.0	61.0
1966-67	0.0	0.0	0.0	0.0	T	14.3	8.3	32.6	19.6	6.0	T	0.0	80.8
1967-68	0.0	0.0	0.0	0.0	15.0	15.0	15.8	6.7	7.4	0.0	0.0	0.0	59.9
1968-69	0.0	0.0	0.0	0.0	11.6	13.4	4.7	49.8	5.7	T	0.0	0.0	85.2
1969-70	0.0	0.0	0.0	2.1	T	20.7	5.0	13.8	14.2	2.8	0.0	0.0	58.6
1970-71	0.0	0.0	0.0	0.0	T	30.1	15.6	19.8	24.0	6.3	0.0	0.0	95.8
1971-72	0.0	0.0	0.0	0.0	18.4	17.4	10.8	29.8	13.6	10.0	0.0	0.0	100.0
1972-73	0.0	0.0	0.0	0.0	12.9	23.2	13.1	5.6	0.1	3.4	0.0	0.0	58.3
1973-74	0.0	0.0	0.0	0.0	T	6.6	15.4	7.0	6.5	3.9	0.0	0.0	39.4
1974-75	0.0	0.0	0.0	0.0	2.9	18.3	19.5	17.1	4.7	5.5	0.0	0.0	68.0
1975-76	0.0	0.0	0.0	0.0	2.2	25.1	14.7	8.4	24.3	T	0.0	0.0	74.7
1976-77	0.0	0.0	0.0	T	3.1	11.6	37.1	11.7	22.3	0.5	T	0.0	86.3
1977-78	0.0	0.0	0.0	0.0	1.5	20.2	37.1	13.5	11.5	0.4	T	0.0	84.2
1978-79	0.0	0.0	0.0	0.0	10.5	16.2	42.3	4.5	0.2	5.1	0.0	0.0	78.8
1979-80	0.0	0.0	0.0	1.3	T	2.1	3.1	11.9	8.6	T	0.0	0.0	27.0
1980-81	0.0	0.0	0.0	T	9.4	9.8	9.2	20.9	5.4	T	0.0	0.0	54.7
1981-82	0.0	0.0	0.0	0.0	T	33.0	26.2	9.0	6.5	15.3	0.0	0.0	90.0
1982-83	0.0	0.0	0.0	0.0	1.3	3.6	9.0	20.8	4.0	T	T	0.0	38.7
1983-84	0.0	0.0	0.0	0.0	T	17.5	20.4	12.7	25.0	T	T	0.0	75.6
1984-85	0.0	0.0	0.0	0.0	T	16.5	11.6	11.0	12.4	1.0	0.0	0.0	52.5
1985-86	0.0	0.0	0.0	0.0	8.3	11.2	15.1	11.5	4.4	T	T	0.0	50.5
1986-87	0.0	0.0	0.0	0.0	14.4	7.7	45.4	0.6	7.0	9.4	0.0	0.0	84.5
1987-88	0.0	0.0	0.0	0.0	5.8	12.0	19.3	23.7	4.6	0.1	0.0	0.0	65.5
1988-89	0.0	0.0	0.0	T	0.5	5.0	5.6	7.0	10.2	0.8	T	0.0	29.1
1989-90	0.0	0.0	0.0	0.0	4.7	12.0	23.1	22.0	1.3	T	0.0	0.0	63.1
1990-91	0.0	0.0	0.0	0.0	0.6	8.8	11.2	6.8	6.0	0.2	0.0	0.0	33.6
1991-92	0.0	0.0	0.0	0.0	1.3	13.5	3.8	5.3	6.8	4.5	0.0	0.0	35.2
1992-93	0.0	0.0	T	T	1.1	10.9	19.7	24.0	35.8	6.6	0.0	T	98.1
1993-94	0.0	0.0	0.0	T	0.2	10.0							
Record Mean	0.0	0.0	T	0.1	3.8	13.6	17.7	14.6	10.9	2.4	0.1	T	63.2

MOUNT WASHINGTON OBS.
GORHAM, NEW HAMPSHIRE

The Mount Washington Observatory is located at the summit of Mount Washington, New Hampshire, highest mountain of the Presidential range. The weather is very severe most of the year, conditions approximating those that would be encountered at a much higher latitude. The upper limits of timberline extend to 4,500 to 5,000 feet.

Prevailing winds are from the west and west-northwest, although the most severe storms are usually from the southeast. Winds are stronger at the summit than at the same elevation at a distance from the mountain, due to the Bernouilli effect. Mount Washington is near the mid-point of a 60-mile-long mountain front trending northeast to southwest. Wind speeds in excess of 100 mph are not uncommon, and the stations highest measured wind, 231 mph, still stands as a world record.

The station is in the clouds approximately 55 percent of the time. This is due partly to the effect of orographic uplift and partly due to the fact that the summit is often above the cloud base when there are low clouds in the area.

Minimum temperatures are not extreme compared to some U. S. valley stations. Annual temperature variations are not as great as they are in the surrounding lowlands, which may actually be colder than the summit when there is a strong inversion. Rime or glaze icing occurs often in winter, when the mountain is frequently in supercooled clouds.

Because of its severe climate, Mount Washington has for many years been used as a natural laboratory for cloud physics research and for the development and testing of instruments, aircraft components, and structures which are required to withstand high winds and icing conditions.

NORMALS, MEANS, AND EXTREMES

MOUNT WASHINGTON OBS. GORHAM, NEW HAMPSHIRE

LATITUDE: 44°16'N LONGITUDE: 71°18' W ELEVATION: FT. GRND 6262 BARO 6274 TIME ZONE: EASTERN WBAN: 14755

	(a)	JAN	FEB	MAR	APR	MAY	JUNE	JULY	AUG	SEP	OCT	NOV	DEC	YEAR
TEMPERATURE °F:														
Normals														
-Daily Maximum		12.3	13.1	20.1	28.7	41.0	49.4	53.6	51.8	45.9	36.4	27.3	17.1	33.1
-Daily Minimum		-4.6	-3.2	5.4	16.1	28.5	37.9	43.0	41.6	34.6	24.2	13.9	0.6	19.8
-Monthly		3.9	5.0	12.8	22.4	34.8	43.7	48.3	46.8	40.3	30.3	20.6	8.9	26.5
Extremes														
-Record Highest	61	44	43	52	60	66	71	71	72	67	59	52	45	72
-Year		1950	1981	1990	1976	1977	1933	1953	1975	1960	1938	1982	1992	AUG 1975
-Record Lowest	61	-47	-46	-38	-20	-2	8	25	20	9	-5	-20	-46	-47
-Year		1934	1943	1950	1954	1966	1945	1982	1986	1992	1939	1958	1933	JAN 1934
NORMAL DEGREE DAYS:														
Heating (base 65°F)		1894	1680	1618	1278	936	639	518	564	741	1076	1332	1739	14015
Cooling (base 65°F)		0	0	0	0	0	0	0	0	0	0	0	0	0
% OF POSSIBLE SUNSHINE	55	32	34	34	35	36	32	30	31	35	38	29	30	33
MEAN SKY COVER (tenths)														
Sunrise - Sunset	55	7.7	7.6	7.7	7.7	7.8	8.1	8.2	8.0	7.5	7.2	8.0	7.9	7.8
MEAN NUMBER OF DAYS:														
Sunrise to Sunset														
-Clear	61	4.7	4.4	4.5	4.0	3.4	2.2	1.5	2.6	4.4	6.1	3.6	4.3	45.9
-Partly Cloudy	61	5.4	5.1	5.5	6.0	7.3	6.7	7.5	7.9	6.5	6.0	5.0	4.9	73.8
-Cloudy	61	20.9	18.7	20.9	20.0	20.3	21.1	22.0	20.5	19.1	18.9	21.4	21.8	245.6
Precipitation														
.01 inches or more	61	19.2	17.8	19.0	17.8	17.2	16.0	16.5	15.7	15.1	15.3	19.2	20.1	209.2
Snow, Ice pellets, hail														
1.0 inches or more	50	11.7	10.4	10.8	8.0	3.0	0.4	0.*	0.1	0.7	4.0	8.6	11.6	69.3
Thunderstorms	61	0.1	0.1	0.3	0.9	1.8	3.5	4.3	3.0	1.0	0.6	0.3	0.1	15.9
Heavy Fog Visibility														
1/4 mile or less	61	26.6	24.9	27.1	24.8	24.0	25.7	27.4	27.8	26.1	25.7	26.9	27.6	314.6
Temperature °F														
-Maximum														
90° and above	61	0.0	0.0	0.0	0.0	0.0	0.0	0.0	0.0	0.0	0.0	0.0	0.0	0.0
32° and below	61	29.0	26.5	26.0	18.4	6.2	0.8	0.*	0.2	2.2	10.5	20.3	27.1	167.2
-Minimum														
32° and below	61	31.0	28.1	30.6	28.1	19.6	6.7	1.5	3.1	12.1	23.2	28.2	30.8	243.0
0° and below	61	17.9	16.5	10.9	2.0	0.1	0.0	0.0	0.0	0.0	0.2	3.8	14.0	65.4
AVG. STATION PRESS. (mb)														
RELATIVE HUMIDITY (%)														
Hour 01	33	82	82	86	86	86	91	93	92	86	84	85	85	87
Hour 07 (Local Time)	33	83	81	84	85	84	89	91	90	85	82	84	84	85
Hour 13	33	83	83	84	84	81	84	87	87	85	81	84	84	84
Hour 19	33	83	82	85	86	84	87	91	91	88	83	85	84	86
PRECIPITATION (inches):														
Water Equivalent														
-Normal		7.94	8.56	8.97	8.17	7.51	7.82	7.08	8.24	7.38	7.19	10.38	9.72	98.96
-Maximum Monthly	61	18.23	25.56	15.98	15.21	18.82	16.00	15.53	20.69	14.07	13.55	19.56	17.95	25.56
-Year		1958	1969	1977	1988	1984	1973	1969	1991	1938	1990	1983	1973	FEB 1969
-Minimum Monthly	61	1.29	0.98	2.15	2.19	1.78	2.43	2.69	2.77	2.74	0.75	2.31	1.49	0.75
-Year		1981	1980	1946	1959	1951	1979	1955	1947	1948	1947	1939	1955	OCT 1947
-Maximum in 24 hrs	61	4.85	10.38	4.64	8.30	4.60	6.50	7.37	6.63	5.38	7.03	6.07	8.64	10.38
-Year		1986	1970	1992	1984	1967	1973	1969	1991	1985	1959	1968	1969	FEB 1970
Snow, Ice pellets, hail														
-Maximum Monthly	61	94.6	172.8	98.0	110.9	52.2	8.1	1.1	2.5	7.8	34.4	86.6	103.7	172.8
-Year		1978	1969	1970	1988	1967	1959	1957	1965	1949	1969	1968	1968	FEB 1969
-Maximum in 24 hrs	61	24.0	49.3	27.4	27.2	22.2	5.1	1.1	2.5	7.7	17.0	25.0	37.5	49.3
-Year		1978	1969	1969	1988	1967	1988	1957	1965	1986	1969	1968	1968	FEB 1969
WIND:														
Mean Speed (mph)	58	46.2	44.3	41.7	36.0	29.6	27.8	25.3	25.1	29.0	33.8	39.5	44.8	35.3
Prevailing Direction through 1963		W	W	W	W	W	W	W	W	W	W	W	W	W
Fastest Obs. 1 Min.														
-Direction (!!)														
-Speed (MPH)														
-Year														
Peak Gust														
-Direction (!!)	61	NW	E	W	SE	W	NW	NW	ENE	SE	W	NW	NW	SE
-Speed (mph)	61	173	166	180	231	164	136	110	142	174	161	163	178	231
-Date		1985	1972	1942	1934	1945	1949	1933	1954	1979	1943	1983	1980	APR 1934

PRECIPITATION (inches)

MOUNT WASHINGTON OBS. GORHAM, NEW HAMPSHIRE

YEAR	JAN	FEB	MAR	APR	MAY	JUNE	JULY	AUG	SEP	OCT	NOV	DEC	ANNUAL
1964	9.21	5.64	8.55	5.14	6.83	5.62	5.57	8.27	4.86	5.30	5.88	5.79	76.66
1965	4.55	7.52	2.71	3.97	2.40	8.15	4.42	8.19	10.01	7.88	6.90	4.65	71.35
1966	9.35	6.40	5.19	3.61	4.55	6.88	5.63	10.84	5.57	5.47	5.79	6.66	75.94
1967	4.92	7.28	2.82	6.23	12.68	6.92	6.80	5.36	7.61	6.67	7.45	7.26	82.00
1968	4.12	3.95	7.53	4.74	6.27	11.02	6.18	5.53	3.89	5.75	16.07	16.10	91.15
1969	8.60	25.56	12.22	6.59	6.01	5.06	15.53	7.50	4.17	7.26	14.41	17.23	130.14
1970	5.53	22.29	13.92	8.43	8.36	5.03	5.59	6.76	10.98	6.89	6.15	13.06	112.99
1971	8.18	8.32	10.19	7.57	10.96	4.89	9.25	9.29	6.24	5.71	10.18	9.51	100.29
1972	7.01	13.37	11.75	9.73	6.36	10.40	7.62	8.31	7.34	8.38	15.28	16.06	121.61
1973	11.74	5.72	9.01	13.33	11.59	16.00	4.38	5.16	6.93	7.71	11.87	17.95	121.39
1974	8.26	9.43	15.96	14.62	7.57	7.00	9.84	7.30	9.03	4.34	11.45	16.24	121.04
1975	13.61	8.92	15.95	13.45	4.32	6.75	8.31	6.50	12.32	6.59	6.37	10.48	113.57
1976	11.87	12.81	9.42	6.04	11.08	9.07	6.29	10.63	6.65	10.36	13.61	13.57	121.40
1977	11.11	12.09	15.98	8.90	3.73	11.12	3.99	7.21	10.64	8.89	11.37	12.08	117.11
1978	18.19	2.98	14.16	9.96	8.13	10.84	3.75	4.23	5.67	4.41	3.24	6.83	92.39
1979	12.62	4.34	5.30	4.98	7.62	2.43	3.80	7.53	6.62	6.75	5.97	3.38	71.34
1980	2.59	0.98	7.99	6.65	2.55	5.37	5.92	7.96	9.96	8.41	11.55	4.31	74.24
1981	1.29	19.81	3.72	5.41	8.59	8.90	11.02	9.38	10.82	9.59	5.72	7.09	101.34
1982	8.03	4.43	5.56	8.25	4.49	10.16	5.90	9.89	6.20	5.54	14.64	7.97	91.06
1983	7.93	5.67	10.26	10.73	11.43	3.60	6.96	10.32	5.57	7.29	19.56	17.38	116.70
1984	4.99	11.56	9.31	14.19	18.82	10.71	8.51	4.79	4.48	5.66	9.84	12.92	115.78
1985	5.48	10.88	12.72	6.77	6.97	9.69	7.93	6.51	9.60	6.15	8.07	7.02	97.79
1986	16.89	4.63	8.06	5.34	7.30	7.68	10.12	10.95	7.79	5.83	10.04	9.48	104.11
1987	8.87	5.48	12.07	8.38	7.49	9.15	6.47	5.95	8.09	8.97	7.97	6.85	95.74
1988	6.52	9.59	8.24	15.21	5.22	6.46	6.56	11.68	7.31	7.43	15.68	3.52	103.42
1989	6.20	4.20	6.41	9.45	14.34	12.22	5.74	10.44	8.19	9.28	14.49	6.19	107.15
1990	8.94	5.24	5.98	5.30	8.27	8.20	6.87	12.73	7.29	13.55	9.72	13.10	105.19
1991	6.53	5.66	6.70	7.50	7.78	4.18	5.35	20.69	8.66	9.86	7.33	8.07	98.31
1992	4.56	6.13	13.43	5.24	2.61	4.66	9.63	8.50	8.30	5.99	7.87	3.07	79.99
1993	4.99	6.05	5.21	5.32	4.42	9.45	9.63	4.25	10.24	5.15	9.02	5.13	78.86
Record Mean	6.61	6.95	7.48	7.03	6.63	7.11	6.77	7.63	7.19	6.67	8.55	8.02	86.64

AVERAGE TEMPERATURE (deg. F)

MOUNT WASHINGTON OBS. GORHAM, NEW HAMPSHIRE

YEAR	JAN	FEB	MAR	APR	MAY	JUNE	JULY	AUG	SEP	OCT	NOV	DEC	ANNUAL
1964	7.9	3.4	13.0	22.4	37.4	42.9	50.9	42.2	37.7	28.5	20.4	13.6	26.7
1965	1.3	3.0	7.7	18.2	36.4	42.5	45.2	46.2	43.1	27.2	15.1	11.6	24.8
1966	5.4	6.1	14.1	20.1	30.1	45.3	48.1	47.4	36.8	28.5	27.0	13.4	26.9
1967	8.9	-1.6	9.1	20.4	25.3	48.2	50.1	47.4	42.0	31.8	15.0	11.6	25.7
1968	4.5	-3.4	15.8	26.6	32.6	43.1	49.1	43.8	43.4	32.1	17.0	7.4	26.0
1969	8.0	8.2	9.1	23.7	31.8	45.8	47.3	47.7	43.0	30.8	22.5	10.7	27.4
1970	-3.6	4.8	11.0	22.5	36.9	44.9	51.1	48.3	41.0	34.9	24.2	6.4	26.9
1971	-2.1	8.2	8.6	20.3	34.1	45.0	47.9	46.9	45.2	39.7	18.0	9.6	26.8
1972	5.7	1.3	12.0	16.6	37.9	43.2	49.4	46.0	42.2	26.3	18.6	10.8	25.8
1973	6.0	5.2	22.6	23.5	33.1	46.0	50.4	51.2	38.4	32.6	17.8	16.7	28.6
1974	6.6	2.9	8.5	24.0	29.4	45.0	47.1	48.5	39.0	23.2	23.0	13.5	25.9
1975	7.5	4.3	9.6	16.4	40.3	44.2	51.4	48.3	38.8	32.2	24.8	9.0	27.2
1976	2.6	8.4	13.5	23.7	33.6	48.3	47.1	47.6	38.7	26.5	11.0	0.0	25.1
1977	-3.3	5.3	18.1	22.3	36.2	42.1	47.4	46.9	39.2	29.6	23.7	6.9	26.2
1978	2.9	-1.3	6.6	18.4	37.0	43.5	48.7	48.9	37.7	28.4	19.6	7.1	24.8
1979	7.5	-1.0	18.4	23.9	37.7	45.1	50.9	45.2	40.9	29.7	25.2	11.4	27.9
1980	4.7	1.1	12.2	25.3	35.1	40.9	48.7	48.6	39.0	25.2	16.5	3.5	25.3
1981	-2.5	15.5	11.5	23.2	36.2	44.8	48.9	47.1	38.7	29.8	21.3	11.2	27.2
1982	-2.2	5.1	12.4	18.5	38.3	40.9	47.9	42.1	43.1	32.3	24.1	16.9	26.6
1983	7.9	10.6	17.9	25.6	32.9	46.1	48.4	48.2	43.0	31.2	22.9	7.5	28.5
1984	4.0	16.7	8.0	25.4	32.1	44.5	47.8	50.6	36.5	34.9	21.8	14.1	28.0
1985	-1.8	6.9	10.9	22.1	35.9	39.3	47.9	46.1	43.2	31.0	23.2	3.7	25.7
1986	5.7	5.4	15.6	28.9	38.0	40.9	47.3	45.0	38.0	29.2	16.9	11.4	26.9
1987	6.0	2.4	16.5	28.5	36.1	44.7	49.4	45.2	39.3	28.3	20.2	11.3	27.3
1988	4.8	5.3	11.7	23.8	37.4	40.1	51.6	49.0	37.7	23.0	22.4	5.3	26.0
1989	8.1	3.2	13.8	19.1	39.0	45.3	49.1	46.4	41.7	32.2	15.3	-5.4	25.7
1990	11.9	9.4	15.9	26.2	31.7	45.6	49.3	50.4	39.3	34.9	23.0	14.7	29.4
1991	4.7	8.3	16.9	28.1	39.8	44.3	47.8	48.2	37.8	32.6	23.6	10.7	28.6
1992	4.9	6.7	8.3	20.9	37.1	43.9	44.7	46.3	42.8	27.2	20.2	11.8	26.2
1993	8.3	0.3	14.2	28.7	35.6	42.4	48.4	49.3	39.6	26.8	19.9	11.9	27.1
Record Mean	5.0	5.4	12.2	22.8	35.0	44.3	48.2	47.3	40.6	30.8	20.5	9.2	26.8
Max	13.4	13.7	19.6	29.3	41.2	50.3	54.4	52.7	46.3	36.8	27.1	17.1	33.5
Min	-3.4	-2.8	4.8	16.2	28.7	38.2	42.0	41.8	34.9	24.8	13.8	1.2	20.0

REFERENCE NOTES

MOUNT WASHINGTON OBS. GORHAM, NEW HAMPSHIRE

GENERAL
T - TRACE AMOUNT.
BLANK ENTRIES DENOTE MISSING/UNREPORTED DATA.
# INDICATES A STATION OR INSTRUMENT RELOCATION.

SPECIFIC

(a) - LENGTH OF RECORD IN YEARS, ALTHOUGH INDIVIDUAL MONTHS MAY BE MISSING.
0.* OR * - THE VALUE IS BETWEEN 0.0 AND 0.05
NORMALS - BASED ON THE 1961-1990 RECORD PERIOD.
EXTREMES - DATES ARE THE MOST RECENT OCCURRENCE.
WIND DIR. - NUMERALS SHOW TENS OF DEGRESS CLOCKWISE FROM TRUE NORTH. "00" INDICATES CALM.
RESULTANT DIRECTIONS ARE GIVEN TO WHOLE DEGREES.
BOLD VALUES INDICATE EXTREME VALUES WHICH OCCURRED AFTER THE ASOS SYSTEM WAS COMMISSIONED.

RECORD = PERIOD OF RECORD
RECORD MEAN PRECIPITATION IS THE MEAN OF ALL DAILY PRECIPITATION AMOUNTS DURING THE PERIOD OF RECORD.
RECORD MAX(MIN) TEMPERATURE IS THE MEAN OF ALL DAILY MAX(MIN) TEMPERATURES DURING THE PERIOD OF RECORD.
RECORD MEAN TEMPERATURE IS THE SUM OF THE RECORD MAX AND RECORD MIN DIVIDED BY 2.
AVERAGE TEMPERATURE IS THE SUM OF THE MEAN DAILY MAX AND MIN TEMPERATURE DIVIDED BY 2.

EXCEPTIONS

1. SUNSHINE DATA IS BY VISUAL OBSERVATION.
2. PRECIPITATION IS MEASURED BY GAGE WITH TILTED ORIFICE DURING SUMMER AND ESTIMATED FROM SNOW FALL/DEPTH MEASUREMENTS DURING WINTER.
3. TEMPERATURE AND PRECIPITATION MAY BE DOUBTFUL DURING PERIODS OF EXTREME WIND VELOCITY AND/OR ICING CONDITIONS.

RECORD MEANS ARE THROUGH THE CURRENT YEAR, BEGINNING IN 1933 FOR TEMPERATURE
1933 FOR PRECIPITATION
1933 FOR SNOWFALL

HEATING DEGREE DAYS Base 65 deg. F — MOUNT WASHINGTON OBS. GORHAM, NEW HAMPSHIRE

SEASON	JULY	AUG	SEP	OCT	NOV	DEC	JAN	FEB	MAR	APR	MAY	JUNE	TOTAL
1964-65	431	699	812	1125	1330	1590	1973	1733	1771	1397	878	668	14407
1965-66	606	576	651	1164	1489	1649	1848	1645	1574	1339	1077	587	14205
1966-67	515	539	837	1124	1132	1595	1738	1862	1729	1328	1224	495	14118
1967-68	456	540	684	1026	1497	1650	1876	1986	1521	1147	994	650	14027
1968-69	484	651	640	1011	1434	1782	1766	1587	1733	1230	1022	569	13909
1969-70	540	527	650	1058	1269	1676	2127	1683	1667	1266	867	592	13922
1970-71	424	511	713	924	1217	1812	2079	1583	1746	1334	949	593	13885
1971-72	527	554	587	780	1404	1714	1838	1846	1639	1450	832	646	13817
1972-73	474	579	675	1196	1388	1677	1830	1672	1306	1239	983	567	13586
1973-74	445	422	793	997	1409	1495	1809	1738	1747	1225	1096	591	13767
1974-75	544	505	771	1291	1252	1589	1777	1699	1713	1453	758	616	13968
1975-76	415	513	780	1009	1197	1735	1935	1638	1591	1233	967	496	13509
1976-77	547	534	781	1184	1617	2017	2114	1670	1445	1275	887	686	14757
1977-78	538	551	768	1091	1233	1796	1924	1855	1805	1393	861	634	14449
1978-79	497	493	811	1128	1355	1792	1782	1849	1440	1229	841	593	13810
1979-80	430	607	715	1090	1185	1657	1867	1853	1633	1184	919	717	13857
1980-81	499	500	774	1227	1448	1907	2093	1381	1651	1246	890	598	14214
1981-82	492	545	781	1085	1305	1660	2085	1676	1623	1391	821	718	14182
1982-83	524	703	649	1009	1221	1488	1766	1520	1455	1176	988	559	13058
1983-84	508	513	654	1042	1256	1780	1885	1394	1766	1180	1012	609	13599
1984-85	527	437	849	926	1290	1574	2068	1622	1675	1280	899	766	13913
1985-86	525	579	646	1044	1249	1898	1838	1669	1526	1074	829	717	13594
1986-87	540	611	802	1101	1439	1657	1826	1750	1495	1088	890	603	13802
1987-88	478	607	765	1132	1340	1659	1863	1729	1648	1231	849	742	14043
1988-89	409	490	814	1295	1271	1849	1759	1729	1582	1370	799	584	13951
1989-90	487	569	691	1007	1487	2184	1641	1551	1519	1157	1030	574	13897
1990-91	479	443	762	928	1255	1558	1866	1584	1486	1100	774	614	12849
1991-92	529	515	808	998	1234	1683	1863	1688	1757	1318	857	627	13877
1992-93	624	570	659	1165	1338	1647	1755	1815	1572	1082	903	670	13800
1993-94	507	484	750	1176	1346	1642							

COOLING DEGREE DAYS Base 65 deg. F — MOUNT WASHINGTON OBS. GORHAM, NEW HAMPSHIRE

YEAR	JAN	FEB	MAR	APR	MAY	JUNE	JULY	AUG	SEP	OCT	NOV	DEC	TOTAL
1969	0	0	0	0	0	0	0	0	0	0	0	0	0
1970	0	0	0	0	0	0	0	0	0	0	0	0	0
1971	0	0	0	0	0	0	0	0	0	0	0	0	0
1972	0	0	0	0	0	0	0	0	0	0	0	0	0
1973	0	0	0	0	0	0	0	0	0	0	0	0	0
1974	0	0	0	0	0	0	0	0	0	0	0	0	0
1975	0	0	0	0	0	0	0	1	0	0	0	0	1
1976	0	0	0	0	0	0	0	0	0	0	0	0	0
1977	0	0	0	0	0	0	0	0	0	0	0	0	0
1978	0	0	0	0	0	0	0	0	0	0	0	0	0
1979	0	0	0	0	0	0	0	0	0	0	0	0	0
1980	0	0	0	0	0	0	0	0	0	0	0	0	0
1981	0	0	0	0	0	0	0	0	0	0	0	0	0
1982	0	0	0	0	0	0	0	0	0	0	0	0	0
1983	0	0	0	0	0	0	0	0	0	0	0	0	0
1984	0	0	0	0	0	0	0	0	0	0	0	0	0
1985	0	0	0	0	0	0	0	0	0	0	0	0	0
1986	0	0	0	0	0	0	0	0	0	0	0	0	0
1987	0	0	0	0	0	0	0	0	0	0	0	0	0
1988	0	0	0	0	0	0	0	0	0	0	0	0	0
1989	0	0	0	0	0	0	0	0	0	0	0	0	0
1990	0	0	0	0	0	0	0	0	0	0	0	0	0
1991	0	0	0	0	0	0	0	0	0	0	0	0	0
1992	0	0	0	0	0	0	0	0	0	0	0	0	0
1993	0	0	0	0	0	0	0	0	0	0	0	0	0

SNOWFALL (inches) — MOUNT WASHINGTON OBS. GORHAM, NEW HAMPSHIRE

SEASON	JULY	AUG	SEP	OCT	NOV	DEC	JAN	FEB	MAR	APR	MAY	JUNE	TOTAL
1964-65	0.0	1.2	0.7	23.5	21.1	31.2	31.7	32.7	19.8	23.3	1.1	7.9	194.2
1965-66	0.0	2.5	0.1	28.2	37.8	29.1	63.9	51.5	29.4	22.5	23.6	0.6	289.2
1966-67	0.0	T	3.9	13.3	18.3	37.0	30.9	52.1	22.8	30.1	52.2	0.0	260.6
1967-68	0.0	0.0	6.9	5.9	41.6	40.0	36.8	26.7	34.8	10.3	6.6	1.5	211.1
1968-69	0.0	0.0	1.5	19.0	86.6	103.7	56.8	172.8	95.0	21.1	9.9	0.0	566.4
1969-70	0.0	0.0	T	34.4	20.3	84.9	30.2	95.7	98.0	34.9	9.2	T	407.6
1970-71	T	0.0	4.9	3.9	23.8	84.5	57.2	45.9	86.6	55.9	33.8	0.0	396.5
1971-72	0.0	0.2	0.0	0.4	52.8	60.3	38.9	79.1	63.1	74.4	7.0	2.0	378.2
1972-73	0.0	T	0.0	12.5	80.7	91.6	70.5	35.2	46.1	71.9	25.3	0.1	433.9
1973-74	0.0	0.0	0.6	10.5	46.3	55.4	49.5	53.7	77.3	64.0	10.6	T	367.9
1974-75	T	0.0	3.7	16.3	44.6	77.1	80.7	50.9	86.3	89.3	1.7	1.1	451.7
1975-76	0.0	0.0	4.3	4.7	11.5	61.5	66.2	65.2	52.0	23.0	19.5	0.2	308.1
1976-77	0.0	1.6	1.5	22.9	76.1	82.0	90.8	65.5	97.4	44.8	16.3	0.3	499.2
1977-78	T	T	0.4	14.9	43.0	73.4	94.6	21.4	71.4	44.3	7.4	3.5	374.3
1978-79	T	0.0	T	8.7	19.9	43.5	67.7	22.3	18.0	24.1	1.8	0.6	206.6
1979-80	0.8	T	0.4	11.5	8.7	23.5	15.6	14.2	43.3	18.1	4.8	1.4	142.3
1980-81	T	0.0	0.7	18.7	55.1	35.8	12.2	36.4	32.3	18.3	5.7	0.0	215.2
1981-82	0.0	0.0	5.6	10.3	22.4	54.3	54.0	34.7	40.6	35.2	0.4	T	257.5
1982-83	0.0	0.6	T	9.4	20.2	13.4	40.8	20.6	36.4	39.2	8.2	T	188.8
1983-84	T	0.0	0.3	11.3	69.6	78.1	40.3	62.9	49.5	41.4	26.3	T	379.7
1984-85	0.0	0.0	1.7	18.3	28.3	62.2	45.6	54.7	71.5	44.3	13.3	2.1	342.0
1985-86	0.0	0.0	0.2	5.1	18.5	51.9	87.5	41.0	47.2	17.4	5.7	0.5	275.0
1986-87	T	1.8	7.7	6.2	42.7	41.0	73.8	44.0	78.3	45.9	3.6	0.0	345.0
1987-88	0.0	T	4.7	23.3	29.4	53.5	50.8	60.7	44.2	110.9	10.0	6.7	394.2
1988-89	0.2	T	T	21.0	60.9	34.3	44.0	28.6	29.8	71.9	2.9	T	293.6
1989-90	T	0.0	7.3	13.4	67.7	55.7	58.5	40.0	25.5	26.0	17.4	0.7	312.2
1990-91	0.0	0.0	5.2	8.6	56.3	58.4	51.6	38.3	43.2	25.3	8.9	0.7	296.5
1991-92	0.0	0.0	2.3	8.3	18.8	51.4	18.2	45.8	72.4	32.5	8.3	T	258.0
1992-93	T	T	3.4	15.1	20.0	19.5	36.5	45.5	41.6	20.3	2.5	4.2	208.6
1993-94	T	0.0	1.9	20.0	23.3	36.8							
Record Mean	T	0.2	1.9	12.0	31.7	42.7	40.0	40.8	42.5	31.3	10.3	1.2	254.6

ATLANTIC CITY, NEW JERSEY
N.A.F.E.C.

The Atlantic City National Weather Service Office is located at the National Aviation Facilities Experimental Center, Pomona, which is about 10 miles west–northwest of Atlantic City and the Atlantic Ocean. The surrounding terrain is fairly flat at an elevation of 50 to 60 feet above sea level. Vegetation in the area consists of scrub pine and low underbrush, but clearing for the air facility has been quite extensive. Bays and salt marshes are as near as 6 miles east of the airport. Atlantic City is located on Abescon Island on the southeast coast of New Jersey. Surrounding terrain, composed of tidal marshes and beach sand, is flat and lies slightly above sea level. The climate is principally continental in character. However, the moderating influence of the Atlantic Ocean is apparent throughout the year, being more marked in the city than at the airport. As a result, summers are relatively cooler and winters milder than elsewhere at the same latitude.

Land and sea breezes, local circulations resulting from the differential heating and cooling of the land and sea, often prevail. These winds occur when moderate or intense storms are not present in the area, thus enabling the local circulation to overcome the general wind pattern. During the warm season sea breezes in the late morning and afternoon hours prevent excessive heating. Frequently, the temperature at Atlantic City during the afternoon hours in the summer averages several degrees lower than at the airport and the airport averages several degrees lower than localities farther inland. On occasions, sea breezes have lowered the temperature as much as 15 to 20 degrees within a half hour. However, the major effect of the sea breeze at the airport is preventing the temperature from rising above the 80s. Because the change in ocean temperature lags behind the air temperature from season to season, the weather tends to remain comparatively mild late into the fall, but on the other hand, warming is retarded in the spring. Normal ocean temperatures range from an average near 37 degrees in January to near 72 degrees in August.

Precipitation is moderate and well distributed throughout the year, with June the driest month and August the wettest. Tropical storms or hurricanes occasionally bring excessive rainfall to the area. The bulk of winter precipitation results from storms which move northeastward along or near the east coast of the United States. Snowfall is considerably less than elsewhere at the same latitude and does not remain long on the ground. Precipitation, often beginning as snow, will frequently become mixed with or change to rain while continuing as snow over more interior sections. In addition, ice storms and resultant glaze are relatively infrequent.

NORMALS, MEANS, AND EXTREMES

ATLANTIC CITY, NEW JERSEY N.A.F.E.C.

LATITUDE: 39°27'N LONGITUDE: 74°34' W ELEVATION: FT. GRND 64 BARO 118 TIME ZONE: EASTERN WBAN: 93730

	(a)	JAN	FEB	MAR	APR	MAY	JUNE	JULY	AUG	SEP	OCT	NOV	DEC	YEAR
TEMPERATURE °F:														
Normals														
-Daily Maximum		40.4	42.5	51.6	60.7	71.2	80.0	84.5	83.3	76.6	66.0	55.7	45.3	63.2
-Daily Minimum		21.4	23.5	31.3	39.3	49.6	58.7	64.8	63.5	55.5	43.7	35.8	26.3	42.8
-Monthly		30.9	33.0	41.5	50.0	60.4	69.4	74.7	73.4	66.1	54.9	45.8	35.8	53.0
Extremes														
-Record Highest	50	78	75	87	94	99	106	104	102	99	90	84	75	106
-Year		1967	1985	1945	1969	1969	1969	1966	1948	1983	1959	1950	1984	JUN 1969
-Record Lowest	50	-10	-11	5	12	25	37	42	40	32	20	10	-7	-11
-Year		1977	1979	1984	1969	1966	1980	1988	1976	1969	1988	1989	1950	FEB 1979
NORMAL DEGREE DAYS:														
Heating (base 65°F)		1057	896	729	450	167	12	0	0	53	324	576	905	5169
Cooling (base 65°F)		0	0	0	0	25	144	301	260	86	10	0	0	826
% OF POSSIBLE SUNSHINE	33	50	52	55	55	56	60	61	64	60	59	50	46	56
MEAN SKY COVER (tenths)														
Sunrise - Sunset	35	6.4	6.3	6.3	6.3	6.4	6.1	6.2	6.0	5.7	5.4	6.2	6.4	6.1
MEAN NUMBER OF DAYS:														
Sunrise to Sunset														
-Clear	35	8.0	7.5	7.9	7.1	6.3	6.9	6.9	7.5	9.5	10.8	7.5	7.8	93.7
-Partly Cloudy	35	8.2	6.9	8.0	9.1	10.6	11.5	11.4	11.5	8.4	8.6	9.0	7.9	111.2
-Cloudy	35	14.8	13.8	15.1	13.7	14.1	11.7	12.7	12.0	12.0	11.6	13.5	15.3	160.4
Precipitation														
.01 inches or more	50	10.6	9.8	10.7	10.8	10.2	8.9	8.8	8.7	7.7	7.3	9.3	9.7	112.4
Snow, Ice pellets, hail														
1.0 inches or more	49	1.6	1.4	0.6	0.1	0.0	0.0	0.0	0.0	0.0	0.0	0.2	0.6	4.5
Thunderstorms	35	0.2	0.4	1.1	2.3	3.5	4.7	6.5	5.2	2.1	0.8	0.5	0.3	27.2
Heavy Fog Visibility														
1/4 mile or less	35	3.1	3.2	3.5	3.8	4.5	4.2	3.9	3.6	3.4	4.6	3.3	2.8	43.8
Temperature °F														
-Maximum														
90° and above	29	0.0	0.0	0.0	0.2	0.9	3.6	7.1	4.8	1.4	0.0	0.0	0.0	18.0
32° and below	29	7.2	4.9	0.7	0.0	0.0	0.0	0.0	0.0	0.0	0.0	0.1	3.2	16.1
-Minimum														
32° and below	29	25.6	22.0	17.2	6.4	0.4	0.0	0.0	0.0	0.1	3.7	12.6	22.2	110.1
0° and below	29	0.8	0.4	0.0	0.0	0.0	0.0	0.0	0.0	0.0	0.0	0.0	0.2	1.5
AVG. STATION PRESS. (mb)	21	1015.5	1015.3	1013.9	1012.4	1012.8	1012.6	1013.1	1014.6	1015.9	1016.4	1016.3	1016.2	1014.6
RELATIVE HUMIDITY (%)														
Hour 01	29	76	76	76	78	84	87	87	89	89	86	81	77	82
Hour 07 (Local Time)	29	78	78	78	77	79	81	83	87	88	88	84	78	82
Hour 13	29	58	56	54	52	56	56	57	58	58	56	58	58	56
Hour 19	29	70	68	66	66	69	70	71	75	79	78	75	71	72
PRECIPITATION (inches):														
Water Equivalent														
-Normal		3.46	3.06	3.62	3.56	3.33	2.64	3.83	4.14	2.93	2.82	3.58	3.32	40.29
-Maximum Monthly	50	7.71	5.98	8.80	7.95	11.51	6.36	13.09	11.98	6.27	7.50	9.65	7.33	13.09
-Year		1948	1958	1993	1952	1948	1970	1959	1967	1966	1943	1972	1969	JUL 1959
-Minimum Monthly	50	0.26	0.82	0.62	0.84	0.40	0.10	0.51	0.34	0.41	0.15	0.68	0.62	0.10
-Year		1955	1980	1945	1976	1957	1954	1983	1943	1970	1963	1976	1955	JUN 1954
-Maximum in 24 hrs	50	2.86	2.59	2.66	3.37	4.15	2.91	6.46	6.40	3.98	2.95	3.93	2.75	6.46
-Year		1944	1966	1979	1952	1959	1952	1959	1966	1954	1958	1953	1951	JUL 1959
Snow, Ice pellets, hail														
-Maximum Monthly	49	20.3	35.2	17.6	3.9	T	0.0	T	0.0	0.0	T	7.8	9.3	35.2
-Year		1987	1967	1969	1990	1989		1991			1990	1967	1989	FEB 1967
-Maximum in 24 hrs	49	16.3	17.1	11.5	3.9	T	0.0	T	0.0	0.0	T	7.8	7.5	17.1
-Year		1987	1979	1969	1990	1989		1991			1990	1967	1960	FEB 1979
WIND:														
Mean Speed (mph)	35	10.9	11.3	11.9	11.7	10.1	9.1	8.4	8.0	8.3	8.9	10.4	10.6	9.9
Prevailing Direction through 1963		WNW	W	WNW	S	S	S	S	S	ENE	W	W	WNW	S
Fastest Obs. 1 Min.														
-Direction (!!)	34	29	27	24	07	17	30	26	12	32	29	27	36	32
-Speed (MPH)	34	47	43	46	46	37	41	37	35	60	41	40	55	60
-Year		1971	1960	1973	1961	1990	1991	1970	1971	1960	1961	1960	1960	SEP 1960
Peak Gust														
-Direction (!!)	10	NW	S	NE	NW	S	NW	N	S	NW	W	NW	E	N
-Speed (mph)	10	52	53	66	53	55	64	81	46	69	56	61	58	81
-Date		1992	1989	1993	1993	1989	1991	1990	1988	1985	1990	1989	1992	JUL 1990

PRECIPITATION (inches)

ATLANTIC CITY, NEW JERSEY N.A.F.E.C.

YEAR	JAN	FEB	MAR	APR	MAY	JUNE	JULY	AUG	SEP	OCT	NOV	DEC	ANNUAL
1964	6.35	4.32	2.80	7.59	1.46	0.84	2.79	1.63	5.91	2.67	1.18	3.47	41.01
1965	3.58	2.44	3.75	2.00	2.59	1.24	2.61	2.40	1.60	1.18	0.79	1.09	25.27
1966	3.45	5.17	0.70	2.58	3.17	1.87	2.59	9.04	6.27	3.73	1.91	4.81	45.29
1967	1.16	3.24	3.78	2.76	3.68	1.37	4.19	11.98	1.50	2.86	1.72	5.57	43.81
1968	2.77	1.69	4.99	1.50	5.55	2.86	1.75	2.20	0.43	2.73	3.10	3.89	33.46
1969	1.68	2.38	3.19	3.55	1.68	1.42	12.64	2.56	1.65	2.09	4.28	7.33	44.45
1970	1.50	3.08	3.11	4.66	1.81	6.36	2.83	2.70	0.41	3.73	5.73	3.04	38.96
1971	2.67	5.26	1.64	1.29	1.88	0.69	3.65	10.40	4.39	4.20	5.02	2.08	43.17
1972	2.93	4.31	3.59	4.62	3.51	4.82	2.81	0.44	3.66	5.11	9.65	3.63	49.08
1973	3.26	3.63	3.08	4.39	3.08	4.32	3.28	2.05	4.73	2.74	1.43	5.48	41.47
1974	3.47	2.40	4.62	2.66	2.61	2.52	1.99	5.50	2.95	1.90	1.08	4.76	36.46
1975	5.94	3.08	3.84	3.90	5.44	3.86	6.02	5.01	5.16	1.76	3.76	2.53	50.30
1976	4.52	2.70	1.39	0.84	3.61	0.97	2.23	4.70	3.06	6.60	0.68	2.52	33.82
1977	3.45	1.41	3.42	2.13	0.64	1.53	2.79	6.49	2.97	4.19	4.75	4.69	38.46
1978	5.70	1.11	5.17	1.53	6.71	3.00	5.77	6.82	1.51	1.21	2.96	3.52	45.01
1979	7.13	5.76	3.62	2.98	2.80	3.15	7.63	4.11	3.32	2.19	3.23	2.19	48.11
1980	2.63	0.82	6.38	5.40	1.61	3.58	2.47	2.63	1.74	3.20	3.63	0.75	34.84
1981	0.56	3.72	1.41	6.20	3.18	4.91	1.28	3.25	1.96	2.96	1.12	3.94	34.49
1982	4.11	2.06	2.70	3.85	2.42	3.03	3.62	1.63	1.34	1.14	4.17	2.85	32.92
1983	2.46	3.32	5.85	7.45	5.21	3.01	0.51	2.90	2.22	3.48	6.70	5.06	48.17
1984	2.41	3.70	5.92	4.84	6.58	1.62	4.35	2.44	1.31	1.46	3.02	1.79	39.44
1985	2.07	1.71	2.38	1.02	5.04	1.55	4.36	3.94	2.26	0.90	3.81	0.93	29.97
1986	3.73	3.42	1.88	4.57	0.54	2.41	4.50	3.35	2.39	3.90	5.04	4.85	40.58
1987	6.23	1.54	3.36	5.83	2.96	2.10	6.12	2.64	3.51	2.56	2.27	2.19	41.31
1988	2.99	3.80	2.21	1.77	3.20	1.03	4.58	2.78	3.07	2.76	4.65	0.64	33.48
1989	2.41	3.47	4.67	4.55	4.64	3.74	6.40	5.68	5.92	4.45	2.87	1.61	50.41
1990	2.70	1.00	2.60	3.46	5.71	1.52	3.64	5.96	1.91	2.63	2.01	3.57	36.71
1991	5.12	1.21	5.52	3.07	0.85	1.43	5.28	3.44	2.62	1.36	0.76	3.75	34.41
1992	0.92	2.12	2.94	1.58	2.34	3.59	3.39	6.64	5.04	1.26	3.10	3.96	36.88
1993	2.62	2.38	8.80	2.23	2.24	1.36	1.76	5.40	2.67	3.82	1.20	4.22	38.70
Record Mean	3.31	2.98	3.82	3.39	3.20	2.52	4.28	4.34	2.99	2.92	3.32	3.32	40.38

AVERAGE TEMPERATURE (deg. F)

ATLANTIC CITY, NEW JERSEY N.A.F.E.C.

YEAR	JAN	FEB	MAR	APR	MAY	JUNE	JULY	AUG	SEP	OCT	NOV	DEC	ANNUAL
#1964	35.3	33.2	43.1	49.7	64.0	70.9	76.3	72.2	67.6	53.5	47.0	36.8	54.1
1965	27.6	33.0	37.8	48.7	64.1	67.6	72.8	72.1	68.0	53.1	44.9	36.4	52.2
1966	28.5	30.1	41.4	47.2	56.6	70.2	75.2	73.4	65.2	52.7	47.7	36.2	52.0
1967	38.9	29.2	38.6	51.4	54.1	69.0	73.4	71.5	61.2	52.4	39.5	35.1	51.2
1968	26.1	27.1	43.0	51.0	58.9	70.1	74.2	73.9	66.5	55.0	43.8	31.5	51.8
1969	30.3	31.8	37.4	53.6	62.0	69.5	73.2	73.3	65.5	53.9	43.9	34.5	52.4
1970	26.9	33.8	38.2	49.6	61.5	68.9	73.7	73.9	69.4	58.2	47.6	36.5	53.2
1971	28.8	35.7	39.0	47.6	57.6	69.7	72.0	70.4	67.5	60.7	44.5	41.1	52.9
1972	33.9	32.5	38.4	46.5	59.1	66.1	75.6	73.5	68.2	51.6	44.3	42.5	52.7
1973	36.2	34.7	48.0	51.3	58.8	72.5	75.1	75.9	66.4	57.4	47.4	39.9	55.3
1974	39.3	33.2	44.4	54.5	60.3	68.0	75.0	74.2	65.9	52.2	46.3	39.0	54.3
1975	37.2	36.5	40.1	45.9	63.3	69.6	74.5	73.8	62.8	57.0	47.9	35.5	53.7
1976	28.4	39.5	43.2	51.5	57.4	69.6	72.5	71.4	64.9	51.2	39.6	30.1	51.6
1977	19.7	32.8	45.8	52.5	61.7	67.5	75.3	75.3	69.5	54.9	49.6	35.4	53.4
1978	30.4	23.8	38.1	49.3	57.2	68.7	72.1	76.5	63.4	53.1	47.8	37.9	51.5
1979	31.5	21.7	44.8	51.2	62.9	66.9	74.6	72.7	65.6	54.4	48.7	38.0	52.7
1980	30.8	28.6	38.7	51.6	61.6	64.3	72.4	73.1	67.9	52.3	41.3	31.2	51.2
1981	22.8	34.2	36.7	50.4	57.8	70.2	76.5	73.8	67.6	53.4	44.8	34.4	51.9
1982	26.4	36.8	43.1	49.8	64.1	69.9	76.9	71.8	64.8	55.2	48.5	41.0	54.0
1983	33.4	35.5	44.9	51.7	60.3	70.7	78.7	75.9	67.9	56.5	46.5	33.9	54.7
1984	28.0	40.1	36.6	49.7	60.7	73.6	75.3	77.1	65.9	62.3	44.6	44.0	54.8
1985	26.8	35.4	44.8	55.1	64.5	70.0	76.9	73.7	68.4	58.1	53.4	33.1	55.0
1986	32.6	32.1	43.0	50.0	63.1	70.5	75.9	71.9	66.3	56.6	44.7	37.9	53.7
1987	31.7	31.0	42.0	49.4	61.3	71.9	76.6	72.0	66.8	50.7	47.4	38.4	53.3
1988	27.9	34.3	42.2	48.7	59.5	69.1	77.1	76.1	64.3	49.7	45.9	34.2	52.4
1989	36.3	33.8	40.6	49.4	59.9	72.2	74.0	73.3	67.3	57.2	44.2	24.7	52.7
1990	40.6	40.5	44.5	51.3	59.2	70.2	75.1	73.4	64.5	59.2	48.0	41.4	55.7
1991	33.7	38.4	43.9	53.3	66.0	71.1	76.9	75.6	65.4	56.4	46.5	38.9	55.5
1992	34.2	35.7	39.5	49.3	57.0	66.5	73.9	70.5	66.1	51.8	46.0	37.4	52.3
1993	37.3	30.5	38.7	50.5	61.2	69.1	76.5	73.5	67.0	53.3	45.3	35.2	53.2
Record Mean	31.4	33.3	41.2	50.7	60.7	69.6	75.0	73.6	66.4	55.3	46.1	35.8	53.3
Max	40.6	42.7	51.1	61.3	71.4	80.2	84.6	83.2	76.7	66.2	56.0	45.2	63.3
Min	22.2	23.8	31.3	40.0	49.9	59.0	65.3	63.9	56.2	44.4	36.2	26.3	43.2

REFERENCE NOTES

ATLANTIC CITY, NEW JERSEY N.A.F.E.C.

GENERAL
T - TRACE AMOUNT.
BLANK ENTRIES DENOTE MISSING/UNREPORTED DATA.
# INDICATES A STATION OR INSTRUMENT RELOCATION.

SPECIFIC

(a) - LENGTH OF RECORD IN YEARS, ALTHOUGH INDIVIDUAL MONTHS MAY BE MISSING.
0.* OR * - THE VALUE IS BETWEEN 0.0 AND 0.05
NORMALS - BASED ON THE 1961-1990 RECORD PERIOD.
EXTREMES - DATES ARE THE MOST RECENT OCCURRENCE
WIND DIR.- NUMERALS SHOW TENS OF DEGRESS CLOCKWISE FROM TRUE NORTH. "00" INDICATES CALM.
RESULTANT DIRECTIONS ARE GIVEN TO WHOLE DEGREES.
BOLD VALUES INDICATE EXTREME VALUES WHICH OCCURRED AFTER THE ASOS SYSTEM WAS COMMISSIONED.

RECORD = PERIOD OF RECORD
RECORD MEAN PRECIPITATION IS THE MEAN OF ALL DAILY PRECIPITATION AMOUNTS DURING THE PERIOD OF RECORD.
RECORD MAX(MIN) TEMPERATURE IS THE MEAN OF ALL DAILY MAX(MIN) TEMPERATURES DURING THE PERIOD OF RECORD.
RECORD MEAN TEMPERATURE IS THE SUM OF THE RECORD MAX AND RECORD MIN DIVIDED BY 2.
AVERAGE TEMPERATURE IS THE SUM OF THE MEAN DAILY MAX AND MIN TEMPERATURE DIVIDED BY 2.

EXCEPTIONS

TEMPERATURE AND PRECIPITATION INCLUDE DATA FROM U. S. NAVAL AIR STATION RECORDS.

RECORD MEANS ARE THROUGH THE CURRENT YEAR, BEGINNING IN 1958 FOR TEMPERATURE
1958 FOR PRECIPITATION
1945 FOR SNOWFALL

HEATING DEGREE DAYS Base 65 deg. F — ATLANTIC CITY, NEW JERSEY N.A.F.E.C.

SEASON	JULY	AUG	SEP	OCT	NOV	DEC	JAN	FEB	MAR	APR	MAY	JUNE	TOTAL
#1964-65	0	1	42	351	532	869	1152	890	838	485	109	77	5346
1965-66	3	24	70	368	596	879	1125	971	726	529	275	43	5609
1966-67	0	0	85	373	513	888	801	996	813	411	341	23	5244
1967-68	0	1	149	399	757	922	1195	1094	674	417	196	13	5817
1968-69	0	18	45	314	629	1031	1068	926	847	354	150	29	5411
1969-70	4	16	108	356	626	939	1174	867	820	453	166	13	5542
1970-71	0	0	49	224	515	876	1117	814	798	514	228	29	5164
1971-72	4	15	47	143	620	736	958	936	816	550	187	44	5056
1972-73	5	8	28	410	612	691	885	842	523	414	207	3	4628
1973-74	0	0	64	245	519	773	788	885	632	329	184	27	4446
1974-75	0	0	68	393	557	800	857	790	763	566	121	17	4932
1975-76	0	2	108	251	505	904	1128	733	668	409	242	55	5005
1976-77	0	17	71	424	757	1075	1398	895	587	380	145	42	5791
1977-78	2	0	28	311	457	910	1069	1146	826	463	256	24	5492
1978-79	6	0	116	365	512	831	1032	1208	623	406	110	39	5248
1979-80	6	19	80	341	483	830	1053	1050	808	394	145	102	5311
1980-81	3	5	38	392	705	1042	1299	855	871	436	251	14	5911
1981-82	0	0	47	356	600	940	1187	787	670	448	79	13	5127
1982-83	0	14	64	318	498	740	973	820	617	405	166	16	4631
1983-84	0	6	94	288	547	958	1137	714	874	451	171	8	5248
1984-85	0	0	80	129	604	646	1179	821	621	319	106	14	4519
1985-86	0	0	53	225	346	981	997	915	674	444	157	25	4817
1986-87	0	30	56	289	600	834	1023	947	705	459	194	6	5143
1987-88	0	10	46	436	520	814	1145	885	698	481	200	68	5303
1988-89	5	9	79	476	565	949	883	867	751	464	184	3	5235
1989-90	0	1	67	260	617	1239	747	680	630	421	188	19	4869
1990-91	1	1	90	240	504	722	963	737	649	362	119	25	4413
1991-92	0	0	102	281	550	803	949	839	785	468	269	44	5090
1992-93	3	0	73	407	565	849	852	961	806	429	150	39	5134
1993-94	0	0	79	353	591	918							

COOLING DEGREE DAYS Base 65 deg. F — ATLANTIC CITY, NEW JERSEY N.A.F.E.C.

YEAR	JAN	FEB	MAR	APR	MAY	JUNE	JULY	AUG	SEP	OCT	NOV	DEC	TOTAL
1969	0	0	0	17	65	170	264	282	129	18	0	0	945
1970	0	0	0	0	63	136	279	285	187	20	0	0	970
1971	0	0	0	0	6	178	228	191	129	17	12	0	761
1972	0	0	0	0	11	83	341	282	132	4	0	0	853
1973	0	0	0	10	21	233	321	345	113	20	0	0	1063
1974	0	0	0	19	44	121	319	292	103	2	6	0	906
1975	0	0	0	0	74	164	305	282	48	10	0	0	883
1976	0	0	0	13	15	201	238	221	74	3	0	0	765
1977	0	0	2	15	51	120	328	325	168	7	2	0	1018
1978	0	0	0	0	19	143	233	364	74	4	0	0	837
1979	0	0	5	0	50	103	308	266	103	15	0	0	850
1980	0	0	0	0	47	88	238	262	132	6	0	0	773
1981	0	0	0	6	31	177	364	279	132	1	0	0	990
1982	0	0	0	0	56	165	375	231	64	23	9	0	923
1983	0	0	0	15	30	196	431	350	190	33	0	0	1245
1984	0	0	0	0	43	273	326	382	115	53	0	1	1193
1985	0	0	3	30	99	170	376	278	163	18	3	0	1140
1986	0	0	0	0	104	198	343	250	99	38	0	0	1032
1987	0	0	0	0	87	217	366	231	106	0	0	0	1007
1988	0	0	0	0	38	197	388	360	63	5	0	0	1051
1989	0	0	3	1	35	225	289	267	145	23	2	0	990
1990	0	0	4	15	16	180	319	270	84	67	0	0	955
1991	0	0	2	15	154	216	376	334	119	19	2	0	1237
1992	0	0	0	3	27	98	287	178	110	6	0	0	709
1993	0	0	0	0	44	171	366	271	146	0	5	0	1003

SNOWFALL (inches) — ATLANTIC CITY, NEW JERSEY N.A.F.E.C.

SEASON	JULY	AUG	SEP	OCT	NOV	DEC	JAN	FEB	MAR	APR	MAY	JUNE	TOTAL
1964-65	0.0	0.0	0.0	0.0	T	1.0	8.2	3.3	2.8	3.2	0.0	0.0	18.5
1965-66	0.0	0.0	0.0	0.0	0.0	0.0	15.1	8.0	T	0.0	0.0	0.0	23.1
1966-67	0.0	0.0	0.0	0.0	T	8.5	1.1	35.2	2.1	T	0.0	0.0	46.9
1967-68	0.0	0.0	0.0	0.0	7.8	3.9	0.8	4.2	1.8	0.0	0.0	0.0	18.5
1968-69	0.0	0.0	0.0	0.0	T	4.3	0.5	7.0	17.6	0.0	0.0	0.0	29.4
1969-70	0.0	0.0	0.0	0.0	T	0.6	10.3	5.9	T	0.0	0.0	0.0	16.8
1970-71	0.0	0.0	0.0	0.0	0.0	1.4	7.2	1.9	0.9	T	0.0	0.0	11.4
1971-72	0.0	0.0	0.0	0.0	0.0	0.1	2.9	5.9	T	T	0.0	0.0	8.9
1972-73	0.0	0.0	0.0	T	T	T	0.0	0.4	T	T	0.0	0.0	0.4
1973-74	0.0	0.0	0.0	0.0	0.1	T	0.4	9.9	T	T	0.0	0.0	10.4
1974-75	0.0	0.0	0.0	0.0	0.2	T	3.3	1.6	2.0	T	0.0	0.0	7.1
1975-76	0.0	0.0	0.0	0.0	0.0	0.8	4.9	1.0	3.3	0.0	0.0	0.0	10.0
1976-77	0.0	0.0	0.0	0.0	T	4.0	7.8	0.5	0.0	0.0	T	0.0	12.3
1977-78	0.0	0.0	0.0	0.0	T	T	2.4	14.6	8.1	T	0.0	0.0	25.1
1978-79	0.0	0.0	0.0	0.0	1.2	T	14.2	27.7	T	0.0	0.0	0.0	43.1
1979-80	0.0	0.0	0.0	T	0.0	4.8	6.3	0.8	2.6	T	0.0	0.0	14.5
1980-81	0.0	0.0	0.0	T	T	0.1	3.2	0.0	T	0.0	0.0	0.0	3.3
1981-82	0.0	0.0	0.0	0.0	0.0	0.6	7.8	4.4	T	2.0	0.0	0.0	14.8
1982-83	0.0	0.0	0.0	0.0	0.0	6.7	T	14.9	T	0.7	0.0	0.0	22.3
1983-84	0.0	0.0	0.0	0.0	0.1	0.4	3.8	T	4.0	T	0.0	0.0	8.3
1984-85	0.0	0.0	0.0	0.0	T	T	15.1	0.0	T	1.3	0.0	0.0	16.4
1985-86	0.0	0.0	0.0	0.0	0.0	4.2	3.9	9.6	T	T	0.0	0.0	17.7
1986-87	0.0	0.0	0.0	0.0	0.0	T	20.3	10.7	1.6	0.7	0.0	0.0	33.3
1987-88	0.0	0.0	0.0	0.0	T	T	7.1	0.2	T	T	0.0	0.0	7.3
1988-89	0.0	0.0	0.0	0.0	0.0	0.4	0.9	12.8	3.4	T	T	0.0	17.5
1989-90	0.0	0.0	0.0	0.0	6.0	9.3	T	T	3.8	3.9	0.0	0.0	23.0
1990-91	0.0	0.0	0.0	T	0.0	3.1	4.7	1.7	T	0.0	0.0	0.0	9.5
1991-92	T	0.0	0.0	0.0	T	T	2.2	1.0	T	T	0.0	0.0	3.2
1992-93	0.0	0.0	0.0	0.0	0.0	0.2	0.8	6.0	3.7	T	0.0	0.0	10.7
1993-94	0.0	0.0	0.0	0.0	0.0	2.7							
Record Mean	T	0.0	0.0	T	0.4	2.2	5.2	5.4	2.6	0.3	T	0.0	16.2

ATLANTIC CITY, NEW JERSEY
STATE MARINA

The Atlantic City State Marina is located on Abescon Island on the southeast coast of New Jersey. Surrounding terrain, composed of tidal marshes and beach sand, is flat and lies slightly above sea level. The climate is principally continental in character. However, the moderating influence of the Atlantic Ocean is apparent throughout the year, being more marked in the city than at the airport. As a result, summers are relatively cooler and winters milder than elsewhere at the same latitude.

Land and sea breezes, local circulations resulting from the differential heating and cooling of the land and sea, often prevail. These winds occur when moderate or intense storms are not present in the area, thus enabling the local circulation to overcome the general wind pattern. During the warm season sea breezes in the late morning and afternoon hours prevent excessive heating. Frequently, the temperature at Atlantic City during the afternoon hours in the summer averages several degrees lower than at the airport and the airport averages several degrees lower than localities farther inland. On occasions, sea breezes have lowered the temperature as much as 15 to 20 degrees within a half hour. However, the major effect of the sea breeze at the airport is preventing the temperature from rising above the 80s. Because the change in ocean temperature lags behind the air temperature from season to season, the weather tends to remain comparatively mild late into the fall, but on the other hand, warming is retarded in the spring. Normal ocean temperatures range from an average near 37 degrees in January to near 72 degrees in August.

Precipitation is moderate and well distributed throughout the year, with June the driest month and August the wettest. Tropical storms or hurricanes occasionally bring excessive rainfall to the area. The bulk of winter precipitation results from storms which move northeastward along or near the east coast of the United States. Snowfall is considerably less than elsewhere at the same latitude and does not remain long on the ground. Precipitation, often beginning as snow, will frequently become mixed with or change to rain while continuing as snow over more interior sections. In addition, ice storms and resultant glaze are relatively infrequent.

NORMALS, MEANS, AND EXTREMES

ATLANTIC CITY, NEW JERSEY STATE MARINA

LATITUDE: 39°23'N LONGITUDE: 74°26' W ELEVATION: FT. GRND 11 BARO TIME ZONE: EASTERN WBAN: 13724

	(a)	JAN	FEB	MAR	APR	MAY	JUNE	JULY	AUG	SEP	OCT	NOV	DEC	YEAR
TEMPERATURE °F:														
Normals														
-Daily Maximum														
-Daily Minimum														
-Monthly														
Extremes														
-Record Highest	33	70	72	76	91	94	97	100	102	92	88	78	72	102
-Year		1967	1976	1991	1960	1969	1991	1983	1983	1985	1986	1993	1984	AUG 1983
-Record Lowest	33	-2	1	2	22	36	45	53	50	42	27	19	4	-2
-Year		1985	1961	1991	1982	1966	1972	1988	1986	1974	1974	1987	1983	JAN 1985
NORMAL DEGREE DAYS:														
Heating (base 65°F)														
Cooling (base 65°F)														
% OF POSSIBLE SUNSHINE														
MEAN SKY COVER (tenths)														
Sunrise - Sunset														
MEAN NUMBER OF DAYS:														
Sunrise to Sunset														
-Clear														
-Partly Cloudy														
-Cloudy														
Precipitation														
.01 inches or more	32	10.2	9.5	10.5	10.0	9.9	8.7	8.6	8.5	7.8	7.5	9.1	10.2	110.5
Snow, Ice pellets, hail														
1.0 inches or more														
Thunderstorms														
Heavy Fog Visibility														
1/4 mile or less														
Temperature °F														
-Maximum														
90° and above	32	0.0	0.0	0.0	0.1	0.3	0.8	2.2	1.3	0.3	0.0	0.0	0.0	4.9
32° and below	32	6.8	3.9	0.6	0.0	0.0	0.0	0.0	0.0	0.0	0.0	0.1	2.3	13.6
-Minimum														
32° and below	32	21.9	18.3	10.2	1.3	0.0	0.0	0.0	0.0	0.0	0.3	4.2	15.6	71.6
0° and below	32	0.*	0.0	0.0	0.0	0.0	0.0	0.0	0.0	0.0	0.0	0.0	0.0	*
AVG. STATION PRESS. (mb)														
RELATIVE HUMIDITY (%)														
Hour 01														
Hour 07 (Local Time)														
Hour 13														
Hour 19														
PRECIPITATION (inches):														
Water Equivalent														
-Normal														
-Maximum Monthly	33	8.40	6.29	8.20	6.68	6.01	7.28	15.69	14.77	5.55	5.47	7.57	6.56	15.69
-Year		1987	1979	1993	1987	1984	1973	1959	1967	1964	1959	1972	1969	JUL 1959
-Minimum Monthly	33	0.35	0.79	0.68	0.82	0.28	0.61	0.57	0.93	0.39	0.02	0.82	0.71	0.02
-Year		1981	1980	1966	1976	1986	1971	1981	1964	1959	1963	1965	1985	OCT 1963
-Maximum in 24 hrs	33	3.40	2.40	3.30	2.73	3.13	4.10	6.62	8.60	2.54	2.67	3.90	3.32	8.60
-Year		1962	1966	1993	1991	1978	1968	1969	1967	1971	1972	1977	1993	AUG 1967
Snow, Ice pellets, hail														
-Maximum Monthly		0.0	0.0	0.0	0.0	0.0	0.0	0.0	0.0	0.0	0.0	0.0	0.0	
-Year														
-Maximum in 24 hrs														
-Year														
WIND:														
Mean Speed (mph)														
Prevailing Direction														
Fastest Mile														
-Direction (!!!)	28	W		W	SSW	SW	NW	W	SE	NNW	SE	E	ENE	SE
-Speed (MPH)	28	60	59	54	53	46	49	48	63	63	48	62	58	63
-Year		1966	1972	1964	1970	1974	1972	1965	1971	1961	1965	1968	1974	AUG 1971
Peak Gust														
-Direction (!!!)	6	SSW	WNW	ENE	NE	NE	SW	SE	ENE	WSW	SE	NW	E	ENE
-Speed (mph)	6	58	60	87	63	52	64	52	67	78	77	55	67	87
-Date		1992	1985	1984	1987	1985	1986	1984	1986	1985	1991	1987	1992	MAR 1984

PRECIPITATION (inches)

ATLANTIC CITY, NEW JERSEY STATE MARINA

YEAR	JAN	FEB	MAR	APR	MAY	JUNE	JULY	AUG	SEP	OCT	NOV	DEC	ANNUAL
1964	5.84	4.04	2.65	6.62	1.30	1.21	1.37	0.93	5.55	2.35	1.21	2.62	35.69
1965	3.48	1.99	3.28	2.30	2.39	1.87	4.92	5.50	1.57	0.61	0.82	1.32	30.05
1966	3.12	4.84	0.68	2.58	4.62	2.22	2.70	4.11	4.35	2.81	1.44	4.58	38.05
1967	0.87	2.98	2.82	2.78	3.33	1.33	2.62	14.77	2.01	2.64	1.09	4.93	42.17
1968	2.09	1.41	4.05	1.13	5.04	6.24	1.26	1.07	2.17	2.36	3.08	3.85	33.75
1969	2.24	2.92	2.63	3.76	1.43	1.64	11.11	4.15	2.46	0.51	4.56	6.55	43.96
1970	1.54	3.39	3.26	4.77	2.46	4.60	3.72	1.24	0.47	3.35	4.37	2.30	35.47
1971	2.89	4.31	1.84	1.35	2.14	0.61	4.20	10.37	5.47	3.60	4.29	2.85	43.92
1972	3.25	3.76	3.87	3.97	5.10	5.21	3.73	1.46	4.35	4.43	7.57	4.98	51.68
1973	2.87	3.51	2.64	4.03	2.83	7.28	3.21	1.06	2.96	1.80	1.33	5.23	38.75
1974	2.78	2.35	4.33	1.75	2.44	2.01	2.57	6.31	2.56	1.59	1.00	3.31	33.00
1975	4.29	3.08	3.39	3.94	2.73	3.72	5.63	2.05	3.82	1.74	2.93	2.57	39.89
1976	3.66	2.98	0.91	0.82	2.19	1.59	1.74	5.01	3.08	4.94	0.95	2.42	30.29
1977	2.48	2.10	2.30	1.89	0.90	1.78	2.17	4.10	1.81	3.05	6.77	5.12	34.47
1978	5.59	1.11	3.82	1.33	6.00	2.07	4.20	4.01	0.87	1.35	2.48	3.95	36.78
1979	6.96	6.29	3.28	4.00	3.55	3.12	2.21	5.93	3.68	2.07	3.80	1.56	46.45
1980.	2.97	0.79	6.39	4.38	1.47	2.73	2.25	1.41	2.08	3.63	3.11	1.13	32.34
1981	0.35	4.08	1.98	4.96	2.16	2.49	0.57	3.75	2.09	2.98	1.28	3.83	30.52
1982	2.81	2.03	2.48	3.98	2.24	2.88	4.14	2.08	1.16	1.11	4.05	3.04	32.00
1983	2.22	2.82	5.63	5.17	3.75	2.50	0.71	2.43	1.88	3.41	6.64	3.47	40.63
1984	2.38	4.32	6.50	4.46	6.01	1.44	4.24	1.03	2.08	1.53	2.28	1.53	37.80
1985	2.04	1.73	2.58	0.97	5.01	2.39	3.37	3.43	2.07	1.08	2.16	0.71	27.54
1986	2.46	2.83	1.46	2.79	0.28			6.11	1.88	3.55	4.19	5.01	
1987	8.40	1.55	4.10	6.68	3.84	0.71	3.11	2.91	3.27	2.81	2.41	1.99	41.78
1988	3.07	3.74	2.00	2.37	3.28	1.55	2.95	3.73	2.22				24.91
1990						1.65	4.06	3.70	1.44	1.98	1.95	3.67	
1991	5.81	1.36	4.90	5.41	0.34	0.81	2.59	4.86	2.86	2.70	1.38	4.29	37.31
1992	1.20	2.40	2.97	1.43	2.81	1.75	5.89	5.11	5.01	1.42	3.25	4.25	37.49
1993	2.75	2.39	8.20	1.96	2.24	1.17	1.89	4.53	1.24	3.81	1.29	5.21	36.68
Record Mean	3.23	3.05	3.64	3.24	2.90	2.45	3.94	4.12	2.68	2.64	3.05	3.36	38.30

AVERAGE TEMPERATURE (deg. F)

ATLANTIC CITY, NEW JERSEY STATE MARINA

YEAR	JAN	FEB	MAR	APR	MAY	JUNE	JULY	AUG	SEP	OCT	NOV	DEC	ANNUAL
1964	36.6	34.6	42.6	48.9	61.9	69.7	73.6	72.4	69.2	56.3	50.6	39.4	54.7
1965	31.8	34.4	39.6	48.3	62.8	68.0	73.0	73.3	69.0	55.8	47.2	39.5	53.6
1966	31.8	34.3	42.6	48.2	58.3	68.4	76.6	73.8	67.1	56.4	49.3	37.7	53.7
1967	38.9	32.5	39.1	51.2	55.4	68.6	75.3	73.5	66.0	58.3	45.2	40.6	53.7
1968	32.0	32.5	44.8	52.6	60.2	70.6	76.5	76.2	71.0	60.2	48.4	35.6	55.1
1969	32.4	33.6	38.6	52.8	61.4	70.8	75.2	75.4	69.3	58.6	47.8	37.9	54.5
1970	29.5	36.7	41.0	51.1	61.4	69.3	72.5	74.9	68.3	58.5	48.3	38.2	54.1
1971	31.5	38.1	40.9	48.7	56.9	67.9	71.9	71.0	67.9	60.0	46.2	44.9	53.8
1972	38.6	36.6	40.4	48.3	58.1	64.5	73.1	72.0	67.5	54.3	46.4	43.6	53.6
1973	39.0	38.4	48.4	54.0	57.2	68.8	73.7	75.1	69.8	60.4	48.9	40.8	56.2
1974	39.1	34.1	43.6	52.5	59.8	67.5	75.6	73.4	65.8	53.3	47.4	39.5	54.3
1975	37.4	36.0	39.4	45.9	61.3	68.8	73.5	74.9	64.7	59.3	50.8	38.5	54.2
1976	32.7	42.0	45.0	53.4	58.0	69.1	71.8	71.4	65.9	53.1	41.6	34.5	53.2
1977	24.4	35.6	45.7	52.2	60.9	67.1	75.3	76.1	71.1	56.6	48.9	36.9	54.2
1978	31.2	27.9	39.8	49.6	55.3	65.8	70.6	74.4	65.0	55.2	48.2	39.3	51.9
1979	33.1	27.4	43.3	50.5	59.3	66.6	72.8	73.1	67.5	56.2	51.4	41.9	53.6
1980	35.5	32.7	40.7	52.2	62.3	67.0	72.9	75.1	69.4	55.6		36.7	
1981	29.9	39.1	41.4	53.2	60.7	71.4	77.5	74.7	67.3	54.6	46.9	37.9	54.5
1982	29.1	35.8	42.2	48.7	60.2	66.1	72.8	69.6	66.0	56.3	48.8	40.7	53.0
1983	34.7	39.0	46.0	51.3	59.0	67.5	76.0	74.6	68.7	57.6	48.3	35.4	54.9
1984	30.4	41.3	39.6	51.1	60.7	72.8	73.4	77.4	67.4	62.8	47.6	47.2	56.0
1985	30.6	36.9	46.4	54.1	62.0	68.4	75.0	74.4	70.4	61.7	54.4	36.4	55.9
1986	35.5	34.0	43.6	51.5	62.4	69.4	77.0	73.3	68.3	59.8	48.4	41.1	55.4
1987	35.1	34.8	44.1	49.8		70.4	76.3	74.5	69.0	54.4	49.4	40.6	55.1
1988	30.5	35.2	43.0	49.4	59.4	68.4	73.6	73.8	66.2				55.5
1990						69.9	75.0	75.4	67.9	62.2	50.1	44.0	
1991	37.5	40.9	44.9	53.0	66.0	71.8	76.4	76.7	68.3	59.1	48.6	41.5	57.1
1992	36.9	37.9	41.0	48.7	56.8	66.7	74.9	72.6	67.8	55.7	48.7	40.3	54.0
1993	39.8	32.9	39.2	49.2	61.4	69.6	78.1	75.1	68.8	57.9	49.3	38.4	55.0
Record Mean	33.7	35.3	42.0	50.7	59.7	68.5	74.3	74.1	67.9	57.7	48.4	38.7	54.3
Max	40.0	41.7	48.3	57.2	65.8	74.6	80.2	79.8	73.8	64.3	54.9	45.3	60.5
Min	27.3	28.8	35.6	44.1	53.6	62.4	68.5	68.3	61.9	51.1	41.8	32.1	48.0

REFERENCE NOTES

ATLANTIC CITY, NEW JERSEY STATE MARINA

GENERAL
T - TRACE AMOUNT.
BLANK ENTRIES DENOTE MISSING/UNREPORTED DATA.
# INDICATES A STATION OR INSTRUMENT RELOCATION.

SPECIFIC

PM - INCLUDES LAST DAY OF PREVIOUS MONTH
ASOS - AUTOMATED SURFACE OBSERVING SYSTEM IN OPERATION DURING THESE MONTHS.

(●) - LENGTH OF RECORD IN YEARS, ALTHOUGH INDIVIDUAL MONTHS MAY BE MISSING.
0.* OR * - THE VALUE IS BETWEEN 0.0 AND 0.05
NORMALS - BASED ON THE 1961-1990 RECORD PERIOD.
EXTREMES - DATES ARE THE MOST RECENT OCCURRENCE
WIND DIR.- NUMERALS SHOW TENS OF DEGRESS CLOCKWISE FROM TRUE NORTH. "00" INDICATES CALM.
RESULTANT DIRECTIONS ARE GIVEN TO WHOLE DEGREES.
BOLD VALUES INDICATE EXTREME VALUES WHICH OCCURRED AFTER THE ASOS SYSTEM WAS COMMISSIONED.

RECORD = PERIOD OF RECORD
RECORD MEAN PRECIPITATION IS THE MEAN OF ALL DAILY PRECIPITATION AMOUNTS DURING THE PERIOD OF RECORD.
RECORD MAX(MIN) TEMPERATURE IS THE MEAN OF ALL DAILY MAX(MIN) TEMPERATURES DURING THE PERIOD OF RECORD.
RECORD MEAN TEMPERATURE IS THE SUM OF THE RECORD MAX AND RECORD MIN DIVIDED BY 2.
AVERAGE TEMPERATURE IS THE SUM OF THE MEAN DAILY MAX AND MIN TEMPERATURE DIVIDED BY 2.

EXCEPTIONS

WINDS ARE FASTEST OBSERVED 1-MINUTE WINDS WITH DIRECTIONS IN COMPASS POINTS.

RECORD MEANS ARE THROUGH THE CURRENT YEAR, BEGINNING IN 1959 FOR TEMPERATURE
1959 FOR PRECIPITATION
1959 FOR SNOWFALL

FOR REPRESENTITIVE SNOWFALL DATA FOR THE ATLANTIC CITY AREA, CONSULT THE ANNUAL ISSUE OF THE ATLANTIC CITY AIRPORT OFFICE, ATLANTIC CITY NEW JERSEY, N.A.F.E.C.

HEATING DEGREE DAYS Base 65 deg. F

ATLANTIC CITY, NEW JERSEY STATE MARINA

SEASON	JULY	AUG	SEP	OCT	NOV	DEC	JAN	FEB	MAR	APR	MAY	JUNE	TOTAL
1964-65	0	0	17	272	426	785	1022	851	778	492	126	38	4807
1965-66	0	8	23	284	528	783	1023	854	688	498	217	26	4932
1966-67	0	0	47	263	463	835	800	903	796	405	293	8	4813
1967-68	0	0	49	226	591	751	1016	937	618	371	156	1	4716
1968-69	0	0	1	180	490	903	1005	872	810	358	141	2	4762
1969-70	0	0	25	222	509	833	1093	789	736	411	142	3	4763
1970-71	0	0	33	204	493	823	1029	749	738	480	246	30	4825
1971-72	0	2	27	158	560	615	812	818	755	494	209	47	4497
1972-73	2	0	23	328	553	658	796	738	508	330	245	9	4190
1973-74	0	0	16	163	476	742	798	860	655	379	173	24	4286
1974-75	0	0	57	358	524	784	848	805	786	567	151	11	4891
1975-76	0	0	49	188	422	814	994	664	612	356	212	29	4340
1976-77	0	3	38	366	696	936	1253	816	590	384	154	36	5272
1977-78	0	0	11	267	477	862	1043	1031	776	455	302	41	5265
1978-79	6	0	62	297	497	791	979	1045	666	427	173	22	4965
1979-80	5	3	27	275	404	708	906	931	746	376	115	43	4539
1980-81	0	0	19	288		871	1080	720	723	350	160	3	
1981-82	0	0	51	315	538	834	1106	813	698	482	155	39	5031
1982-83	1	10	36	280	486	749	931	722	584	405	179	17	4400
1983-84	1	9	63	236	496	911	1065	681	781	406	139	5	4793
1984-85	0	0	56	96	514	543	1059	783	571	336	119	16	4093
1985-86	0	0	19	132	313	876	910	860	654	399	144	15	4322
1986-87	0	9	20	201	491	735	922	842	641	452		3	
1987-88	0	0	14	324	460	747	1062	859	675	465	200	51	4857
1988-89	1	2	40										
1989-90												13	
1990-91	0	0	42	154	437	646	847	668	616	357	79	8	3854
1991-92	0	0	36	205	484	722	865	780	738	484	267	23	4604
1992-93	0	0	37	292	482	759	776	890	792	467	136	19	4650
1993-94	0	0	33	214	469	819							

COOLING DEGREE DAYS Base 65 deg. F

ATLANTIC CITY, NEW JERSEY STATE MARINA

YEAR	JAN	FEB	MAR	APR	MAY	JUNE	JULY	AUG	SEP	OCT	NOV	DEC	TOTAL
1969	0	0	0	0	38	181	322	327	160	32	0	0	1060
1970	0	0	0	0	37	138	240	315	141	11	0	0	882
1971	0	0	0	0	2	122	219	194	336	7	2	0	882
1972	0	0	0	0	0	39	258	224	103	2	0	0	626
1973	0	0	0	3	9	233	278	319	168	29	0	0	1039
1974	0	0	0	10	18	125	334	266	91	0	1	0	845
1975	0	0	0	0	42	131	269	313	50	16	0	0	821
1976	0	0	0	14	2	160	219	210	69	2	0	0	676
1977	0	0	0	6	37	105	324	355	200	10	0	0	1037
1978	0	0	0	0	9	72	186	297	69	0	0	0	633
1979	0	0	0	0	5	74	252	263	111	8	0	0	713
1980	0	0	0	0	39	109	250	319	159	2		0	
1981	0	0	0	3	33	200	392	305	125	1	0	0	1059
1982	0	0	0	0	13	77	247	160	70	20	3	0	590
1983	0	0	0	2	0	96	349	313	181	14	0	0	955
1984	0	0	0	0	12	246	269	393	135	36	0	0	1091
1985	0	0	2	12	34	124	315	298	188	35	0	0	1008
1986	0	0	0	0	70	154	378	274	126	47	0	0	1049
1987	0	0	0	0		173	357	299	142	0	0	0	1042
1988	0	0	0	0	34	160	275	280	83				832
1990						166	318	330	136	74	1	0	
1991	0	0	0	3	116	219	359	370	143	30	0	0	1240
1992	0	0	0	0	21	83	315	243	126	8	0	0	796
1993	0	0	0	0	30	164	415	319	155	2	4	0	1089

SNOWFALL (inches)

ATLANTIC CITY, NEW JERSEY STATE MARINA

SEASON	JULY	AUG	SEP	OCT	NOV	DEC	JAN	FEB	MAR	APR	MAY	JUNE	TOTAL
Record Mean													

NEWARK, NEW JERSEY

Terrain in vicinity of the station is flat and rather marshy. To the northwest are ridges oriented roughly in a south-southwest to north-northeast direction. They rise to an elevation of about 200 feet at 4.5 to 5 miles and to 500 to 600 feet at 7 to 8 miles. All winds between west-northwest and north-northwest are downslope and therefore are subject to some adiabatic temperature increase. This effect is evident in the rapid improvement which normally occurs with shift of wind to westerly, following a coastal storm or frontal passage. The drying effect of the downslope winds accounts for the relatively few local thunderstorms occurring at the station, compared to areas to the west. Easterly winds, particularly southeasterly, moderate the temperature because of the influence of the Atlantic Ocean.

Temperature falls of 5 to 15 degrees, depending on the season, are not uncommon when the wind backs from southwesterly to southeasterly. Periods of very hot weather, lasting as long as a week, are associated with a west-southwest air flow which has a long trajectory over land. Extremes of cold are related to rapidly moving outbreaks of cold air traveling southeastward from the Hudson Bay region. Temperatures of zero or below occur in one winter out of four, but are much more common several miles to the west of the station. Average dates of the last occurrence in spring and the first occurrence in autumn of temperatures as low as 32 degrees are in mid-April and the end of October or early November. Areas to the west of the station experience a growing season at least a month shorter than that at the airport.

A considerable amount of precipitation is realized from the Northeasters of the Atlantic coast. These storms, more typical of the fall and winter, generally last for a period of two days and commonly produce between 1 and 2 inches of precipitation. Storms producing 4 inches or more of snow occur from two to five times a winter. Snowstorms producing 8 inches or more have occurred in about one-half the winters. As many as three such storms have been experienced in one winter. The frequency and intensity of snow storms and the duration of snow cover increase dramatically within a few miles to the west of the station.

NORMALS, MEANS, AND EXTREMES

NEWARK, NEW JERSEY

LATITUDE: 40°42'N LONGITUDE: 74°10' W ELEVATION: FT. GRND 7 BARO 29 TIME ZONE: EASTERN WBAN: 14734

	(a)	JAN	FEB	MAR	APR	MAY	JUNE	JULY	AUG	SEP	OCT	NOV	DEC	YEAR
TEMPERATURE °F:														
Normals														
-Daily Maximum		37.7	40.5	50.8	61.9	72.4	82.3	87.0	85.4	77.6	66.7	55.4	42.9	63.4
-Daily Minimum		23.4	25.4	33.4	42.7	53.2	62.8	68.6	67.4	59.9	48.2	39.2	29.1	46.1
-Monthly		30.6	33.0	42.1	52.3	62.8	72.6	77.8	76.4	68.8	57.5	47.3	36.0	54.8
Extremes														
-Record Highest	52	74	76	89	94	98	102	105	103	105	92	85	72	105
-Year		1950	1949	1945	1990	1992	1993	1993	1948	1953	1949	1950	1982	JUL 1993
-Record Lowest	52	-8	-7	6	16	33	43	52	45	35	28	15	-1	-8
-Year		1985	1943	1943	1982	1947	1945	1945	1982	1947	1969	1955	1980	JAN 1985
NORMAL DEGREE DAYS:														
Heating (base 65°F)		1066	896	710	381	127	0	0	0	26	252	531	899	4888
Cooling (base 65°F)		0	0	0	0	59	232	397	353	140	20	0	0	1201
% OF POSSIBLE SUNSHINE														
MEAN SKY COVER (tenths)														
Sunrise - Sunset	47	6.5	6.3	6.3	6.4	6.5	6.1	6.2	6.0	5.7	5.5	6.4	6.5	6.2
MEAN NUMBER OF DAYS:														
Sunrise to Sunset														
-Clear	51	7.8	7.4	8.0	7.2	6.3	6.8	6.6	7.7	9.5	10.8	7.5	7.9	93.5
-Partly Cloudy	51	7.8	7.5	8.5	8.8	10.8	11.0	12.2	11.7	8.9	8.4	8.3	8.0	111.7
-Cloudy	51	15.5	13.4	14.6	14.0	13.9	12.2	12.2	11.7	11.5	11.8	14.2	15.1	160.0
Precipitation														
.01 inches or more	52	10.8	9.6	11.3	11.1	11.7	10.3	10.1	9.5	8.5	7.9	10.0	10.9	121.6
Snow,Ice pellets,hail														
1.0 inches or more	52	2.1	1.9	1.3	0.2	0.0	0.0	0.0	0.0	0.0	0.0	0.2	1.4	7.0
Thunderstorms	52	0.3	0.2	1.0	1.5	3.6	4.8	6.0	4.7	2.3	1.1	0.5	0.2	26.1
Heavy Fog Visibility														
1/4 mile or less	52	2.1	1.7	1.5	1.1	1.7	1.2	0.4	0.5	0.8	1.9	1.8	1.8	16.4
Temperature °F														
-Maximum														
90° and above	28	0.0	0.0	0.0	0.2	1.5	4.8	9.3	7.0	1.6	0.0	0.0	0.0	24.3
32° and below	28	9.3	5.4	0.9	0.*	0.0	0.0	0.0	0.0	0.0	0.0	0.1	3.9	19.5
-Minimum														
32° and below	28	23.7	20.9	12.7	1.5	0.0	0.0	0.0	0.0	0.0	0.5	5.7	18.9	84.0
0° and below	28	0.5	0.2	0.0	0.0	0.0	0.0	0.0	0.0	0.0	0.0	0.0	0.1	0.7
AVG. STATION PRESS. (mb)	20	1017.0	1016.9	1015.9	1013.8	1014.4	1014.0	1014.4	1016.0	1017.6	1018.1	1017.9	1018.0	1016.2
RELATIVE HUMIDITY (%)														
Hour 01	28	70	68	66	66	72	73	73	76	78	76	72	72	72
Hour 07 (Local Time)	28	74	72	70	66	70	70	72	75	78	78	77	74	73
Hour 13	28	58	54	51	48	51	51	51	53	55	53	56	59	53
Hour 19	28	63	60	57	54	59	58	59	62	65	64	64	64	61
PRECIPITATION (inches):														
Water Equivalent														
-Normal		3.39	3.04	3.87	3.84	4.13	3.22	4.50	3.91	3.66	3.05	3.91	3.45	43.97
-Maximum Monthly	52	10.10	4.94	11.14	11.14	10.22	6.40	9.98	11.84	10.28	8.20	11.53	9.47	11.84
-Year		1979	1979	1983	1983	1984	1975	1988	1955	1944	1943	1977	1983	AUG 1955
-Minimum Monthly	52	0.45	1.22	1.10	0.90	0.52	0.07	0.89	0.50	0.95	0.21	0.51	0.27	0.07
-Year		1981	1968	1981	1963	1964	1949	1966	1964	1951	1963	1976	1955	JUN 1949
-Maximum in 24 hrs	40	3.59	2.45	2.83	3.73	4.22	2.97	3.63	7.84	5.27	3.04	7.22	2.77	7.84
-Year		1979	1961	1991	1984	1979	1992	1988	1971	1971	1973	1977	1983	AUG 1971
Snow,Ice pellets,hail														
-Maximum Monthly	52	27.4	26.1	26.0	13.8	T	0.0	0.0	0.0	0.0	0.3	5.7	29.1	29.1
-Year		1978	1979	1956	1982	1977					1952	1989	1947	DEC 1947
-Maximum in 24 hrs	52	17.8	20.0	17.6	12.8	T	0.0	0.0	0.0	0.0	0.3	5.7	26.0	26.0
-Year		1978	1961	1956	1982	1977					1952	1989	1947	DEC 1947
WIND:														
Mean Speed (mph)	49	11.2	11.5	11.9	11.2	10.0	9.5	8.9	8.7	9.0	9.4	10.2	10.8	10.2
Prevailing Direction through 1963		NE	NW	NW	WNW	SW	SW	SW	SW	SW	SW	SW	SW	SW
Fastest Obs. 1 Min.														
-Direction (!!)	45	30	23	27	27	32	26	35	09	05	11	09	32	09
-Speed (MPH)	45	52	46	43	50	50	58	52	46	51	48	82	55	82
-Year		1964	1965	1950	1951	1963	1984	1988	1955	1960	1954	1950	1962	NOV 1950
Peak Gust														
-Direction (!!)	10	SW	NW	NE	NW	NW	W	NW	N	W	SW	NW	NW	W
-Speed (mph)	10	54	58	62	55	58	83	69	68	67	53	63	60	83
-Date		1992	1984	1993	1993	1988	1984	1988	1985	1985	1990	1989	1988	JUN 1984

PRECIPITATION (inches)

NEWARK, NEW JERSEY

YEAR	JAN	FEB	MAR	APR	MAY	JUNE	JULY	AUG	SEP	OCT	NOV	DEC	ANNUAL
1964	5.12	2.59	2.27	5.56	0.52	3.09	4.74	0.50	1.30	1.55	2.08	4.10	33.42
1965	2.86	2.91	2.81	2.60	1.23	1.23	1.73	2.87	2.20	2.31	1.48	1.86	26.09
1966	2.29	4.41	1.12	3.01	4.86	0.49	0.89	3.08	7.86	3.78	3.06	3.01	37.86
1967	1.15	3.00	5.86	2.84	3.57	3.31	7.53	5.53	1.35	2.87	2.35	4.65	44.01
1968	1.71	1.22	3.59	2.24	6.28	4.37	1.87	2.41	2.48	2.02	4.38	4.32	36.89
1969	1.47	2.68	3.53	3.51	2.73	2.53	7.11	2.24	6.63	1.75	2.80	4.97	41.95
1970	0.87	3.29	3.42	3.52	2.64	2.41	3.68	3.91	1.83	2.36	4.41	2.05	34.39
1971	2.74	4.44	3.29	1.35	3.65	1.48	6.98	10.63	7.88	2.96	3.86	1.51	50.77
1972	2.26	4.01	3.09	3.08	6.02	6.02	4.70	2.30	1.03	4.83	8.42	4.10	49.86
1973	3.65	3.39	3.63	5.77	3.56	4.03	3.63	3.36	3.39	3.35	1.29	7.24	46.29
1974	2.84	1.44	4.11	2.37	3.49	3.60	1.31	7.17	5.76	1.85	0.80	4.02	38.76
1975	3.99	2.56	2.94	2.29	3.27	6.40	8.02	4.36	9.00	3.24	3.67	2.91	52.65
1976	5.04	2.52	2.33	2.50	4.12	1.54	3.91	2.98	2.50	5.07	0.51	2.17	35.19
1977	1.55	2.77	5.67	3.16	1.31	3.89	1.51	4.29	3.99	3.53	11.53	4.77	47.97
1978	7.76	2.26	4.58	2.60	7.97	2.05	4.99	7.30	4.23	1.64	2.66	5.37	53.41
1979	10.10	4.94	3.65	3.66	7.78	2.73	3.39	4.38	5.72	4.58	3.09	2.08	56.10
1980	1.66	1.28	9.13	7.28	2.61	3.27	2.78	0.92	1.87	3.37	3.71	0.63	38.51
1981	0.45	4.81	1.10	3.15	3.88	2.61	4.51	0.57	3.42	3.47	1.75	5.32	35.04
1982	6.77	2.36	2.82	6.20	2.96	5.28	2.86	2.78	2.39	1.68	3.16	1.32	40.58
1983	4.37	3.03	11.14	11.14	4.22	2.81	1.59	3.46	2.93	5.80	5.54	9.47	65.50
1984	2.78	4.57	6.96	6.36	10.22	4.77	8.65	1.74	2.46	3.93	2.88	3.69	59.01
1985	1.22	2.58	1.59	1.17	4.23	4.29	4.52	2.58	4.19	1.29	8.32	1.31	37.29
1986	4.44	3.88	1.95	5.88	1.41	1.71	6.62	4.16	1.96	1.93	6.78	5.23	45.95
1987	6.21	1.30	3.81	5.06	2.55	4.13	4.66	5.26	3.87	3.37	2.94	2.37	45.53
1988	3.74	4.15	2.13	1.97	5.86	1.06	9.98	1.82	1.66	2.45	7.71	0.98	43.51
1989	1.98	2.70	4.42	3.25	8.80	5.41	5.23	7.03	6.45	5.40	2.57	0.75	53.99
1990	4.72	1.71	2.81	3.98	6.87	3.68	4.98	7.71	2.72	5.11	2.82	5.19	52.30
1991	3.72	1.81	5.49	3.91	4.80	2.95	5.21	5.63	3.24	1.29	2.04	3.67	43.76
1992	1.27	1.37	3.48	1.35	3.46	4.67	4.79	3.37	2.60	0.73	5.02	4.63	36.74
1993	2.75	2.87	7.22	4.59	1.77	1.21	2.15	2.84	6.29	3.98	1.95	4.89	42.51
Record Mean	3.34	2.88	4.04	3.66	3.86	3.31	4.08	4.16	3.73	3.03	3.61	3.38	43.08

AVERAGE TEMPERATURE (deg. F)

NEWARK, NEW JERSEY

YEAR	JAN	FEB	MAR	APR	MAY	JUNE	JULY	AUG	SEP	OCT	NOV	DEC	ANNUAL
1964	34.3	31.9	42.6	49.1	65.4	71.2	76.0	73.9	68.9	55.9	49.4	35.9	54.6
#1965	28.3	32.4	39.0	50.0	67.3	71.6	75.7	74.5	68.4	54.0	44.4	38.8	53.7
1966	30.4	33.2	41.7	48.2	59.3	73.8	79.6	76.5	66.6	55.5	48.9	36.5	54.2
1967	36.9	29.4	37.6	50.9	54.3	72.0	74.2	73.5	66.6	56.4	42.2	38.3	52.7
1968	27.8	29.9	43.1	54.0	59.6	69.7	78.2	76.9	70.7	59.7	45.7	32.5	54.0
1969	31.3	31.3	38.8	54.6	64.1	72.8	74.2	77.3	67.5	56.2	45.5	33.1	53.9
1970	24.2	33.0	39.0	51.9	64.6	70.9	77.2	77.3	70.6	59.5	49.1	35.3	54.4
1971	27.3	35.2	41.2	51.4	60.6	74.8	77.8	76.0	71.8	63.2	46.2	41.4	55.6
1972	35.4	31.3	40.5	50.0	63.0	68.8	77.9	75.9	69.8	53.3	44.8	39.7	54.2
1973	35.5	33.3	48.6	54.2	60.4	74.6	78.7	79.6	71.0	60.3	48.8	39.4	57.0
1974	35.4	31.9	43.4	56.5	62.7	70.1	77.1	76.5	66.6	53.9	47.5	38.9	55.0
1975	36.9	35.1	39.7	47.3	65.8	71.6	76.9	75.1	64.3	59.1	51.7	35.4	54.9
1976	26.8	39.3	44.0	55.2	61.1	73.6	74.9	74.4	66.5	52.6	39.9	29.1	53.1
1977	20.9	32.8	46.8	53.7	65.4	70.3	78.2	75.1	68.0	54.5	47.1	33.3	53.9
1978	27.2	25.5	38.5	51.0	60.5	71.6	75.1	76.7	66.1	57.5	48.8	38.1	53.1
1979	32.5	23.5	46.2	52.0	64.5	69.4	77.0	76.6	69.1	56.5	51.8	40.2	54.9
1980	34.0	30.8	38.9	52.6	65.9	70.2	78.9	78.6	70.8	55.0	42.9	30.4	54.1
1981	24.1	37.6	40.2	55.3	64.0	74.6	79.3	75.1	67.2	53.1	46.0	34.6	54.3
1982	24.2	36.2	41.8	50.6	63.2	67.9	78.4	72.5	66.7	56.9	48.8	42.8	54.2
1983	35.0	35.9	44.7	52.2	60.8	73.5	79.6	77.6	70.6	57.8	47.8	34.2	55.8
1984	27.8	40.8	36.5	52.7	62.2	75.0	76.6	77.3	65.4	62.3	45.3	40.8	55.2
1985	24.9	33.5	44.5	57.0	67.1	69.4	76.3	75.6	70.2	58.5	49.5	33.3	55.0
1986	33.0	31.1	44.2	53.4	66.7	72.7	76.9	74.2	68.6	58.0	45.0	38.1	55.2
1987	31.5	33.0	45.0	53.9	63.9	74.5	79.4	75.3	68.7	53.7	47.6	38.4	55.4
1988	28.7	34.4	43.9	51.1	63.4	73.0	80.5	79.8	68.0	52.6	48.9	35.5	55.0
1989	37.0	34.2	42.4	52.5	63.2	74.3	77.2	76.3	69.9	59.1	45.0	25.6	54.7
1990	40.4	39.8	44.9	53.3	61.1	73.4	77.8	76.6	68.6	62.4	50.0	42.3	57.6
1991	33.6	38.6	44.4	54.8	68.9	74.2	77.9	77.7	68.0	58.3	47.6	38.8	56.9
1992	35.2	36.0	39.3	50.2	61.7	72.6	76.9	75.1	69.6	55.7	47.8	38.8	54.9
1993	37.6	31.0	40.2	54.3	67.0	75.8	82.6	79.2	69.2	56.4	47.8	37.2	56.5
Record Mean	31.6	32.9	41.1	51.7	62.5	71.6	76.8	75.1	67.8	56.9	46.3	35.4	54.1
Max	38.7	40.6	49.4	60.9	72.1	81.0	85.8	83.9	76.7	66.1	54.2	42.4	62.6
Min	24.5	25.2	32.8	42.4	52.8	62.1	67.7	66.3	58.8	47.7	38.4	28.3	45.6

REFERENCE NOTES

NEWARK, NEW JERSEY

GENERAL
T - TRACE AMOUNT.
BLANK ENTRIES DENOTE MISSING/UNREPORTED DATA.
# INDICATES A STATION OR INSTRUMENT RELOCATION.

SPECIFIC

PM - INCLUDES LAST DAY OF PREVIOUS MONTH
ASOS - AUTOMATED SURFACE OBSERVING SYSTEM IN OPERATION DURING THESE MONTHS.

(*) - LENGTH OF RECORD IN YEARS, ALTHOUGH INDIVIDUAL MONTHS MAY BE MISSING.
0.* OR * - THE VALUE IS BETWEEN 0.0 AND 0.05
NORMALS - BASED ON THE 1961-1990 RECORD PERIOD.
EXTREMES - DATES ARE THE MOST RECENT OCCURRENCE
WIND DIR.- NUMERALS SHOW TENS OF DEGRESS CLOCKWISE FROM TRUE NORTH. "00" INDICATES CALM.
RESULTANT DIRECTIONS ARE GIVEN TO WHOLE DEGREES.
BOLD VALUES INDICATE EXTREME VALUES WHICH OCCURRED AFTER THE ASOS SYSTEM WAS COMMISSIONED.

RECORD = PERIOD OF RECORD
RECORD MEAN PRECIPITATION IS THE MEAN OF ALL DAILY PRECIPITATION AMOUNTS DURING THE PERIOD OF RECORD.
RECORD MAX(MIN) TEMPERATURE IS THE MEAN OF ALL DAILY MAX(MIN) TEMPERATURES DURING THE PERIOD OF RECORD.
RECORD MEAN TEMPERATURE IS THE SUM OF THE RECORD MAX AND RECORD MIN DIVIDED BY 2.
AVERAGE TEMPERATURE IS THE SUM OF THE MEAN DAILY MAX AND MIN TEMPERATURE DIVIDED BY 2.

EXCEPTIONS

RECORD MEANS ARE THROUGH THE CURRENT YEAR, BEGINNING IN 1931 FOR TEMPERATURE
1931 FOR PRECIPITATION
1942 FOR SNOWFALL

HEATING DEGREE DAYS Base 65 deg. F NEWARK, NEW JERSEY

SEASON	JULY	AUG	SEP	OCT	NOV	DEC	JAN	FEB	MAR	APR	MAY	JUNE	TOTAL
#1964-65	1	0	40	278	461	895	1133	905	799	442	55	17	5026
1965-66	0	11	50	339	610	807	1066	882	717	500	212	14	5208
1966-67	0	0	63	286	480	876	864	991	842	425	331	5	5163
1967-68	0	1	58	285	677	823	1148	1012	676	325	167	12	5184
1968-69	0	0	6	193	573	1003	1039	938	804	317	101	2	4976
1969-70	0	0	49	284	575	984	1255	892	796	390	97	5	5327
1970-71	0	0	24	199	472	914	1160	827	732	402	155	7	4892
1971-72	0	1	12	95	569	724	909	969	757	444	93	19	4592
1972-73	0	0	22	356	599	776	906	882	504	339	163	1	4548
1973-74	0	0	18	166	479	787	909	921	661	273	127	12	4353
1974-75	0	0	62	341	521	802	864	832	775	524	84	6	4811
1975-76	0	1	59	195	400	913	1177	738	645	338	141	17	4624
1976-77	0	4	56	381	745	1107	1361	895	563	352	89	24	5577
1977-78	0	0	50	319	527	975	1168	1099	814	411	190	13	5566
1978-79	6	0	66	239	481	830	1001	1155	577	386	68	11	4820
1979-80	2	4	28	289	393	763	953	987	802	366	62	24	4673
1980-81	0	0	28	314	654	1066	1261	762	764	290	96	0	5235
1981-82	0	0	52	360	563	934	1258	802	712	433	85	42	5241
1982-83	0	13	36	267	493	679	923	810	622	395	162	5	4405
1983-84	0	0	52	249	510	949	1144	696	874	366	128	9	4977
1984-85	0	0	83	114	584	745	1235	877	641	268	62	15	4624
1985-86	0	0	21	212	462	971	985	942	642	341	89	7	4672
1986-87	0	11	22	240	594	826	1030	893	616	331	140	3	4706
1987-88	0	1	25	342	518	818	1117	880	647	410	120	28	4906
1988-89	1	0	18	386	476	906	859	853	698	366	132	6	4701
1989-90	0	0	37	190	594	1215	756	699	622	369	122	2	4606
1990-91	1	1	50	163	446	697	967	734	630	330	63	4	4086
1991-92	0	0	55	227	513	804	917	834	790	441	148	4	4733
1992-93	0	0	38	295	510	807	842	946	765	318	42	4	4567
1993-94	0	0	48	263	513	853							

COOLING DEGREE DAYS Base 65 deg. F NEWARK, NEW JERSEY

YEAR	JAN	FEB	MAR	APR	MAY	JUNE	JULY	AUG	SEP	OCT	NOV	DEC	TOTAL
1969	0	0	0	15	80	243	293	390	131	17	0	0	1169
1970	0	0	0	4	94	187	384	387	201	33	0	0	1290
1971	0	0	0	0	25	307	403	350	222	46	12	0	1365
1972	0	0	3	4	41	142	406	347	175	3	0	0	1121
1973	0	0	0	20	26	296	432	459	205	28	0	0	1466
1974	0	0	0	28	64	172	381	361	115	1	3	0	1125
1975	0	0	0	0	117	211	375	321	46	20	10	0	1100
1976	0	0	0	50	30	281	317	305	110	6	0	0	1099
1977	0	0	6	18	111	191	414	321	146	1	0	0	1208
1978	0	0	0	0	59	217	325	367	105	15	0	0	1088
1979	0	0	0	2	59	147	381	372	158	34	3	0	1156
1980	0	0	0	0	97	187	435	427	209	10	0	0	1365
1981	0	0	0	6	75	293	446	319	124	0	0	0	1263
1982	0	0	0	6	39	136	421	249	95	24	12	0	982
1983	0	0	0	19	39	268	458	396	226	36	0	0	1442
1984	0	0	0	2	47	316	365	388	102	36	0	0	1256
1985	0	0	11	36	134	152	357	335	183	19	3	0	1230
1986	0	0	2	2	149	243	380	303	136	30	0	0	1245
1987	0	0	0	6	116	293	453	327	143	0	1	0	1339
1988	0	0	0	0	75	274	488	465	115	10	0	0	1427
1989	0	0	3	1	81	294	385	360	194	16	0	0	1334
1990	0	0	7	23	11	262	403	365	165	89	2	0	1327
1991	0	0	0	28	190	288	406	399	151	28	0	0	1490
1992	0	0	0	4	52	242	373	323	185	15	0	0	1194
1993	0	0	0	5	113	340	553	450	182	9	4	0	1656

SNOWFALL (inches) NEWARK, NEW JERSEY

SEASON	JULY	AUG	SEP	OCT	NOV	DEC	JAN	FEB	MAR	APR	MAY	JUNE	TOTAL
1964-65	0.0	0.0	0.0	0.0	0.0	3.9	16.1	1.8	4.6	0.7	0.0	0.0	27.1
1965-66	0.0	0.0	0.0	T	0.0	T	10.2	8.6	T	0.0	0.0	0.0	18.8
1966-67	0.0	0.0	0.0	0.0	0.0	12.6	1.3	25.4	18.0	T	0.0	0.0	57.3
1967-68	0.0	0.0	0.0	0.0	3.1	3.9	4.6	0.6	1.7	0.0	0.0	0.0	13.9
1968-69	0.0	0.0	0.0	0.0	0.4	4.7	1.1	16.5	5.9	0.0	0.0	0.0	28.6
1969-70	0.0	0.0	0.0	0.0	T	8.5	9.1	5.5	4.3	T	0.0	0.0	27.4
1970-71	0.0	0.0	0.0	0.0	0.0	2.9	13.2	1.1	4.2	2.2	0.0	0.0	23.6
1971-72	0.0	0.0	0.0	0.0	T	0.4	3.1	12.3	1.0	T	0.0	0.0	16.8
1972-73	0.0	0.0	0.0	T	T	T	0.7	0.6	0.6	T	0.0	0.0	1.9
1973-74	0.0	0.0	0.0	0.0	0.0	2.1	6.8	8.1	3.1	0.3	0.0	0.0	20.4
1974-75	0.0	0.0	0.0	0.0	T	1.2	1.4	12.7	1.1	T	0.0	0.0	16.4
1975-76	0.0	0.0	0.0	0.0	T	2.4	7.2	6.1	4.2	T	0.0	0.0	19.9
1976-77	0.0	0.0	0.0	0.0	T	6.7	10.8	5.8	1.7	T	T	0.0	25.0
1977-78	0.0	0.0	0.0	0.0	1.5	0.2	27.4	25.3	10.5	T	0.0	0.0	64.9
1978-79	0.0	0.0	0.0	0.0	2.6	T	9.4	26.1	T	T	0.0	0.0	38.1
1979-80	0.0	0.0	0.0	T	0.0	3.7	2.5	1.8	6.3	T	0.0	0.0	14.3
1980-81	0.0	0.0	0.0	0.0	0.4	3.1	6.9	T	9.1	0.0	0.0	0.0	19.5
1981-82	0.0	0.0	0.0	0.0	T	3.4	12.3	0.5	0.8	13.8	0.0	0.0	30.8
1982-83	0.0	0.0	0.0	0.0	T	2.9	2.3	21.5	0.2	4.1	0.0	0.0	31.0
1983-84	0.0	0.0	0.0	0.0	1.2	2.4	13.7	0.3	11.3	T	0.0	0.0	28.9
1984-85	0.0	0.0	0.0	0.0	T	6.8	8.9	7.4	0.1	T	0.0	0.0	23.2
1985-86	0.0	0.0	0.0	0.0	0.6	4.6	2.8	13.9	T	0.1	0.0	0.0	22.0
1986-87	0.0	0.0	0.0	0.0	T	2.3	21.4	6.5	2.4	0.0	0.0	0.0	32.6
1987-88	0.0	0.0	0.0	0.0	1.5	2.3	15.4	2.7	0.9	T	0.0		
1988-89	0.0	0.0	0.0	0.0	0.0	0.1	4.1	0.6	2.7	0.0	0.0	0.0	7.5
1989-90	0.0	0.0	0.0	0.0	5.7	0.5	2.4	2.8	2.5	0.6	0.0	0.0	14.5
1990-91	0.0	0.0	0.0	0.0	T	7.6	8.5	5.2	0.2	0.0	0.0	0.0	21.5
1991-92	0.0	0.0	0.0	0.0	T	0.5	1.0	1.0	11.4	T	0.0	0.0	13.9
1992-93	0.0	0.0	0.0	0.0	T	0.5	0.8	10.7	16.8	0.0	0.0	0.0	28.8
1993-94	0.0	0.0	0.0	0.0	T	3.9							
Record Mean	0.0	0.0	0.0	T	0.5	5.5	7.5	7.9	4.9	0.7	T	0.0	26.9

ALBUQUERQUE, NEW MEXICO

The Albuquerque metropolitan area is largely situated in the Rio Grande Valley and on the mesas and piedmont slopes which rise either side of the valley floor. The Rio Grande flows from north to south through the area. The Sandia and Manzano Mountains rise abruptly at the eastern edge of the city with Tijeras Canyon separating the two ranges. West of the city the land gradually rises to the Continental Divide, some 90 miles away.

The climate of Albuquerque is best described as arid continental with abundant sunshine, low humidity, scant precipitation, and a wide yet tolerable seasonal range of temperatures. Sunny days and low humidity are renowned features of the climate. More than three-fourths of the daylight hours have sunshine, even in the winter months. The air is normally dry and muggy days are rare. The combination of dry air and plentiful solar radiation allows widespread use of energy-efficient devices such as evaporative coolers and solar collectors.

Precipitation within the valley area is adequate only for native desert vegetation and deep-rooted imports. However, irrigation supports successful farming and fruit growing in the Rio Grande Valley. On the east slopes of the Sandias and Manzanos, precipitation is sufficient for thick stands of timber and good grass cover.

Meager amounts of precipitation fall in the winter, much of it as snow. Snowfalls of an inch or more occur about four times a year in the Rio Grande Valley, while the mountains receive substantial snowfall on occasion. Snow seldom remains on the ground more than 24 hours in the city proper. However, snow cover on the east slopes of the Sandias is sufficient for skiing during most winters.

Nearly half of the annual precipitation in Albuquerque results from afternoon and evening thunderstorms during the summer. Thunderstorm frequency increases rapidly around July 1st, peaks during August, then tapers off by the end of September. Thunderstorms are usually brief, sometimes produce heavy rainfall, and often lower afternoon temperatures noticeably. Hailstorms are infrequent and tornadoes rare.

Temperatures in Albuquerque are those characteristic of a dry, high altitude, continental climate. The average daily range of temperature is relatively high, but extreme temperatures are rare. High temperatures during the winter are near 50 degrees with only a few days on which the temperature fails to rise above the freezing mark. In the summer, daytime maxima are about 90 degrees, but with the large daily range, the nights usually are comfortably cool.

The average number of days between the last freezing temperature in spring and the first freeze in fall varies widely across the Albuquerque metropolitan area. The growing season in Albuquerque and adjacent suburbs ranges from around 170 days in the Rio Grande Valley to about 200 days in parts of the northeast section of the city.

Sustained winds of 12 mph or less occur approximately 80 percent of the time at the Albuquerque International Airport, while sustained winds greater than 25 mph have a frequency less than 3 percent. Late winter and spring storms along with occasional east winds out of Tijeras Canyon are the main sources of strong wind conditions. Blowing dust, the least attractive feature of the climate, often accompanies the occasional strong winds of winter and spring.

NORMALS, MEANS, AND EXTREMES

ALBUQUERQUE, NEW MEXICO

LATITUDE: 35°03'N LONGITUDE: 106°37' W ELEVATION: FT. GRND 5311 BARO 5313 TIME ZONE: MOUNTAIN WBAN: 23050

	(a)	JAN	FEB	MAR	APR	MAY	JUNE	JULY	AUG	SEP	OCT	NOV	DEC	YEAR
TEMPERATURE °F:														
Normals														
-Daily Maximum		46.8	53.5	61.4	70.8	79.7	90.0	92.5	89.0	81.9	71.0	57.3	47.5	70.1
-Daily Minimum		21.7	26.4	32.2	39.6	48.6	58.3	64.4	62.6	55.2	43.0	31.2	23.1	42.2
-Monthly		34.2	40.0	46.9	55.2	64.2	74.2	78.5	75.9	68.6	57.0	44.3	35.3	56.2
Extremes														
-Record Highest	54	69	76	85	89	98	105	105	101	100	91	77	72	105
-Year		1971	1986	1971	1989	1951	1980	1980	1979	1979	1979	1975	1958	JUN 1980
-Record Lowest	54	-17	-5	8	19	28	40	52	50	37	21	-7	-7	-17
-Year		1971	1951	1948	1980	1975	1980	1985	1992	1971	1991	1976	1990	JAN 1971
NORMAL DEGREE DAYS:														
Heating (base 65°F)		955	700	561	301	89	0	0	0	18	259	621	921	4425
Cooling (base 65°F)		0	0	0	7	64	279	419	338	126	11	0	0	1244
% OF POSSIBLE SUNSHINE	54	72	72	73	77	80	83	76	75	79	79	76	71	76
MEAN SKY COVER (tenths)														
Sunrise - Sunset	54	4.9	5.0	5.1	4.6	4.2	3.4	4.5	4.4	3.6	3.5	4.0	4.6	4.3
MEAN NUMBER OF DAYS:														
Sunrise to Sunset														
-Clear	54	13.0	10.9	11.4	12.6	14.2	17.4	12.0	13.3	16.8	17.2	15.1	13.9	167.9
-Partly Cloudy	54	7.7	7.6	9.6	9.5	10.4	8.6	14.1	12.4	7.6	7.8	7.6	7.3	110.5
-Cloudy	54	10.4	9.7	9.9	7.9	6.4	3.9	4.8	5.2	5.6	6.1	7.2	9.8	86.9
Precipitation														
.01 inches or more	54	4.1	4.1	4.6	3.3	4.5	4.0	8.8	9.6	5.7	4.8	3.5	4.3	61.3
Snow,Ice pellets,hail														
1.0 inches or more	54	1.0	0.9	0.6	0.2	0.*	0.0	0.0	0.0	0.0	0.*	0.5	0.9	4.3
Thunderstorms	54	0.1	0.3	0.9	1.5	4.0	5.1	10.8	10.8	4.6	2.3	0.5	0.2	41.2
Heavy Fog Visibility														
1/4 mile or less	54	1.1	1.0	0.6	0.2	0.*	0.*	0.1	0.*	0.1	0.3	0.6	1.6	5.8
Temperature °F														
-Maximum														
90° and above	33	0.0	0.0	0.0	0.0	2.5	17.0	23.0	15.9	3.8	0.1	0.0	0.0	62.3
32° and below	33	2.2	0.7	0.1	0.0	0.0	0.0	0.0	0.0	0.0	0.0	0.2	1.8	5.0
-Minimum														
32° and below	33	28.8	22.5	15.4	4.3	0.2	0.0	0.0	0.0	0.0	2.1	16.5	28.5	118.2
0° and below	33	0.4	0.0	0.0	0.0	0.0	0.0	0.0	0.0	0.0	0.0	0.1	0.1	0.5
AVG. STATION PRESS.(mb)	21	838.9	837.9	835.2	835.7	836.1	837.9	840.2	840.8	840.2	840.0	838.8	839.2	838.4
RELATIVE HUMIDITY (%)														
Hour 05	33	71	65	56	49	48	46	60	67	62	61	65	70	60
Hour 11 (Local Time)	33	51	44	34	26	25	24	34	40	40	38	42	50	37
Hour 17	33	41	33	25	19	19	18	27	31	30	29	36	44	29
Hour 23	33	61	53	44	35	35	33	47	53	52	49	54	61	48
PRECIPITATION (inches):														
Water Equivalent														
-Normal		0.44	0.46	0.54	0.52	0.50	0.59	1.37	1.64	1.00	0.89	0.43	0.50	8.88
-Maximum Monthly	54	1.32	1.82	2.18	1.97	3.07	2.57	3.33	3.30	2.63	3.08	1.93	1.85	3.33
-Year		1978	1993	1973	1942	1941	1986	1968	1967	1988	1972	1991	1959	JUL 1968
-Minimum Monthly	54	T	T	T	T	T	T	0.08	T	T	0.00	0.00	0.00	0.00
-Year		1970	1984	1966	1993	1945	1975	1980	1962	1957	1952	1949	1981	DEC 1981
-Maximum in 24 hrs	54	0.87	0.80	1.11	1.66	1.14	1.64	1.77	1.75	1.92	1.80	1.67	1.35	1.92
-Year		1962	1993	1973	1969	1969	1952	1961	1980	1955	1969	1991	1958	SEP 1955
Snow,Ice pellets,hail														
-Maximum Monthly	54	9.5	10.3	13.9	8.1	1.0	T	T	T	T	3.2	9.3	14.7	14.7
-Year		1973	1986	1973	1973	1979	1992	1990	1993	1971	1986	1940	1959	DEC 1959
-Maximum in 24 hrs	54	5.1	6.0	10.7	10.9	1.0	T	T	T	T	3.2	5.5	14.2	14.2
-Year		1973	1986	1973	1988	1979	1992	1990	1993	1971	1986	1946	1958	DEC 1958
WIND:														
Mean Speed (mph)	54	8.0	8.8	10.0	10.8	10.5	9.9	9.0	8.2	8.5	8.2	7.9	7.7	9.0
Prevailing Direction through 1963		N	N	SE	S	S	S	SE	SE	SE	SE	N	N	SE
Fastest Obs. 1 Min.														
-Direction (!!)	9	09	09	19	17	28	16	36	27	25	10	27	09	09
-Speed (MPH)	9	52	40	49	46	46	40	52	41	40	38	48	47	52
-Year		1990	1989	1991	1985	1986	1991	1990	1990	1985	1991	1988	1987	JAN 1990
Peak Gust														
-Direction (!!)	10	E	W	SW	E	SW	N	N	E	W	NW	W	E	SW
-Speed (mph)	10	70	63	77	64	70	68	72	63	61	51	63	71	77
-Date		1990	1984	1991	1990	1991	1992	1990	1989	1985	1986	1988	1987	MAR 1991

PRECIPITATION (inches)

ALBUQUERQUE, NEW MEXICO

YEAR	JAN	FEB	MAR	APR	MAY	JUNE	JULY	AUG	SEP	OCT	NOV	DEC	ANNUAL
1964	0.07	1.12	0.13	0.61	0.35	T	1.87	0.98	1.57	0.04	0.21	0.49	7.44
1965	0.47	0.60	0.49	0.49	0.19	0.99	1.65	0.61	1.18	0.89	0.33	1.42	9.31
1966	0.42	0.30	T	0.04	0.02	1.66	1.63	1.06	1.04	0.54	0.09	0.01	6.81
1967	0.01	0.44	0.25	T	0.04	1.71	0.61	3.30	0.79	0.18	0.15	0.56	8.04
1968	0.01	0.98	1.48	0.51	0.99	0.05	3.33	1.49	0.30	0.12	0.59	0.82	10.67
1969	0.08	0.34	0.41	1.76	1.31	0.59	0.94	0.95	1.08	2.37	0.01	0.72	10.56
1970	T	0.27	0.42	0.05	0.33	0.40	1.22	2.24	0.79	0.25	0.08	0.23	6.28
1971	0.27	0.21	0.03	0.78	0.16	0.02	1.05	0.87	1.44	1.15	0.67	1.40	8.05
1972	0.12	0.12	0.08	T	0.18	0.55	1.00	2.93	1.00	3.08	0.69	0.36	10.11
1973	0.85	0.33	2.18	0.91	0.66	1.37	1.80	1.19	1.13	0.35	0.08	0.03	10.88
1974	0.88	0.11	0.85	0.14	0.01	0.22	2.40	0.79	1.58	1.96	0.38	0.51	9.83
1975	0.26	0.99	0.95	0.10	0.66	T	1.43	1.40	1.66	T	0.28	0.28	8.01
1976	0.00	0.40	0.09	0.31	0.82	0.60	1.32	0.73	0.45	0.03	0.24	0.20	5.19
1977	0.88	0.13	0.63	1.07	0.10	0.04	0.69	2.28	0.78	0.76	0.42	0.13	7.91
1978	1.32	1.02	0.54	0.05	0.69	1.05	0.24	2.49	0.59	1.22	1.00	0.76	10.97
1979	1.07	0.62	0.14	0.24	2.48	1.02	0.80	1.53	0.40	0.27	0.91	0.87	10.35
1980	0.87	0.58	0.60	0.60	0.56	0.01	0.08	2.61	1.83	0.09	0.30	0.74	8.87
1981	0.05	0.67	0.80	0.30	0.53	0.35	1.07	1.68	0.41	1.43	0.37	0.00	7.66
1982	0.32	0.20	0.84	0.05	0.52	0.09	1.32	1.09	1.34	0.26	0.60	0.78	7.41
1983	1.10	0.71	0.61	0.02	0.32	1.21	0.55	0.27	0.91	1.20	0.44	0.42	7.76
1984	0.33	T	0.62	0.50	0.16	0.48	1.13	2.70	1.13	3.04	0.63	1.36	12.08
1985	0.49	0.54	0.70	1.69	1.12	0.53	1.16	0.49	1.53	2.15	0.19	0.16	10.75
1986	0.22	1.01	0.17	0.33	1.11	2.57	1.51	2.26	0.53	1.54	1.29	0.44	12.98
1987	0.66	0.61	0.07	1.00	0.58	0.13	0.91	2.98	0.20	0.44	0.42	0.34	8.34
1988	0.15	0.07	0.85	1.42	0.62	1.25	2.26	3.29	2.63	0.32	0.22	0.03	13.11
1989	0.57	0.35	0.48	T	0.02	0.02	1.51	0.48	0.31	0.97	T	0.28	4.99
1990	0.21	0.49	0.41	1.71	0.45	0.27	2.36	1.79	0.96	0.15	0.86	0.59	10.25
1991	0.60	0.06	0.14	T	1.14	0.65	2.63	1.26	1.43	0.26	1.93	1.49	11.59
1992	0.60	0.20	0.63	0.22	1.81	0.67	2.01	2.17	0.79	0.70	1.12	1.16	12.08
1993	0.94	1.82	0.22	T	0.20	0.44	0.23	3.05	0.49	0.64	0.97	0.03	9.03
Record Mean	0.40	0.39	0.45	0.54	0.64	0.60	1.41	1.42	0.93	0.83	0.46	0.47	8.53

AVERAGE TEMPERATURE (deg. F)

ALBUQUERQUE, NEW MEXICO

YEAR	JAN	FEB	MAR	APR	MAY	JUNE	JULY	AUG	SEP	OCT	NOV	DEC	ANNUAL
1964	30.0	29.1	41.5	51.7	65.8	73.6	78.2	76.8	69.3	59.4	43.7	35.5	54.5
1965	38.8	39.4	44.6	54.8	61.7	69.4	77.9	75.4	66.6	58.0	48.4	35.8	55.9
1966	30.1	33.2	45.6	54.6	67.2	72.8	79.8	75.7	68.4	56.8	46.7	34.3	55.4
1967	33.2	40.5	52.0	57.8	63.8	71.5	79.2	74.5	68.4	58.2	46.1	32.4	56.5
1968	36.8	43.3	46.7	53.4	62.7	75.2	76.1	72.4	68.0	58.3	42.8	30.0	55.5
1969	38.0	38.5	41.1	57.4	66.2	73.6	80.2	79.0	70.0	53.8	41.4	39.1	56.6
1970	34.5	42.8	44.1	52.5	66.2	72.7	79.6	77.8	67.5	52.6	44.5	36.4	56.0
1971	33.6	38.9	47.7	53.3	61.7	73.8	78.1	73.9	66.4	53.8	45.2	31.9	54.8
1972	36.1	42.5	53.6	56.9	64.0	73.7	78.6	74.1	68.1	57.6	40.1	35.0	56.7
1973	31.8	35.9	45.1	50.2	62.7	73.5	78.4	78.0	67.5	56.4	44.6	34.0	54.8
1974	33.6	37.9	52.8	56.4	68.5	80.1	77.0	72.7	66.1	58.1	45.0	32.0	56.7
1975	30.8	38.0	45.0	49.9	61.0	73.0	76.8	76.1	66.3	56.5	42.6	35.6	54.3
1976	33.2	43.3	44.3	54.6	62.8	73.4	77.0	75.0	68.0	53.1	40.6	33.0	54.9
1977	29.8	40.7	43.2	56.5	64.2	75.5	78.6	77.4	69.4	58.9	46.4	40.4	56.8
1978	36.8	39.3	50.2	57.7	60.5	75.5	81.6	75.5	69.1	60.3	47.5	34.3	57.4
1979	32.9	41.1	48.4	56.9	63.7	73.3	80.6	77.1	72.3	61.5	41.0	37.7	57.2
1980	40.2	44.2	46.1	52.1	61.1	77.2	82.7	77.4	69.9	54.5	43.5	40.5	57.4
1981	38.0	42.9	46.2	59.0	64.5	77.0	79.8	76.4	69.7	55.7	47.0	40.5	58.0
1982	35.9	39.4	47.4	56.1	63.0	74.8	79.1	77.4	69.5	54.8	42.9	34.4	56.2
1983	35.0	39.7	46.9	50.2	63.0	73.4	80.4	79.4	73.4	58.3	45.1	36.7	56.8
1984	34.1	40.1	46.8	52.8	69.9	73.6	78.9	75.7	68.8	51.6	43.7	35.6	56.0
1985	33.8	38.3	47.5	57.4	64.0	74.1	77.1	76.6	65.9	57.5	45.4	37.6	56.3
1986	41.3	43.0	50.9	56.5	63.7	72.7	74.7	76.0	66.5	54.6	42.0	36.3	56.5
1987	32.3	39.2	43.7	54.8	62.8	73.0	77.8	74.7	68.8	61.3	45.2	35.3	55.7
1988	34.6	43.9	47.0	55.1	64.3	74.4	78.1	75.0	66.3	61.1	45.4	33.9	56.6
1989	35.5	41.9	52.8	61.4	68.8	75.6	78.6	74.3	69.4	56.7	46.4	35.1	58.0
1990	34.6	38.5	48.6	57.3	63.6	79.0	76.8	73.8	70.9	58.3	45.0	32.1	56.5
1991	35.7	44.6	46.1	56.0	65.5	73.4	76.9	75.5	68.1	59.6	43.4	37.3	56.8
1992	32.7	42.3	48.9	60.0	64.6	72.4	76.2	75.0	70.3	60.8	39.7	32.8	56.3
1993	39.7	42.5	48.8	57.1	65.7	75.1	79.9	75.6	69.1	56.2	43.3	37.3	57.5
Record Mean	34.6	39.8	46.5	55.0	63.9	73.6	77.4	75.3	68.4	56.8	43.9	35.3	55.9
Max	47.0	53.1	60.9	70.1	79.1	89.1	91.3	88.7	82.2	71.1	57.4	47.5	69.8
Min	22.2	26.5	32.1	39.9	48.7	58.0	63.5	61.9	54.7	42.6	30.4	23.1	42.0

REFERENCE NOTES

ALBUQUERQUE, NEW MEXICO

GENERAL
T - TRACE AMOUNT.
BLANK ENTRIES DENOTE MISSING/UNREPORTED DATA.
# INDICATES A STATION OR INSTRUMENT RELOCATION.

SPECIFIC

PM - INCLUDES LAST DAY OF PREVIOUS MONTH
ASOS - AUTOMATED SURFACE OBSERVING SYSTEM IN OPERATION DURING THESE MONTHS.

(a) - LENGTH OF RECORD IN YEARS, ALTHOUGH INDIVIDUAL MONTHS MAY BE MISSING.
0.* OR * - THE VALUE IS BETWEEN 0.0 AND 0.05
NORMALS - BASED ON THE 1961-1990 RECORD PERIOD.
EXTREMES - DATES ARE THE MOST RECENT OCCURRENCE
WIND DIR.- NUMERALS SHOW TENS OF DEGRESS CLOCKWISE FROM TRUE NORTH. "00" INDICATES CALM.
RESULTANT DIRECTIONS ARE GIVEN TO WHOLE DEGREES.
BOLD VALUES INDICATE EXTREME VALUES WHICH OCCURRED AFTER THE ASOS SYSTEM WAS COMMISSIONED.

RECORD = PERIOD OF RECORD
RECORD MEAN PRECIPITATION IS THE MEAN OF ALL DAILY PRECIPITATION AMOUNTS DURING THE PERIOD OF RECORD.
RECORD MAX(MIN) TEMPERATURE IS THE MEAN OF ALL DAILY MAX(MIN) TEMPERATURES DURING THE PERIOD OF RECORD.
RECORD MEAN TEMPERATURE IS THE SUM OF THE RECORD MAX AND RECORD MIN DIVIDED BY 2.
AVERAGE TEMPERATURE IS THE SUM OF THE MEAN DAILY MAX AND MIN TEMPERATURE DIVIDED BY 2.

EXCEPTIONS

RECORD MEANS ARE THROUGH THE CURRENT YEAR,
BEGINNING IN 1893 FOR TEMPERATURE
1893 FOR PRECIPITATION
1940 FOR SNOWFALL

HEATING DEGREE DAYS Base 65 deg. F ALBUQUERQUE, NEW MEXICO

SEASON	JULY	AUG	SEP	OCT	NOV	DEC	JAN	FEB	MAR	APR	MAY	JUNE	TOTAL
1964-65	2	0	20	173	632	909	805	709	624	300	128	24	4326
1965-66	0	0	56	217	492	895	1074	882	595	305	53	0	4569
1966-67	0	0	15	247	541	942	980	682	396	211	109	0	4123
1967-68	0	0	13	220	557	1003	870	623	559	343	107	8	4303
1968-69	2	0	12	208	660	1080	831	735	735	228	84	0	4575
1969-70	0	0	1	348	701	795	938	612	644	367	63	11	4480
1970-71	0	0	58	380	605	878	968	725	533	343	122	5	4617
1971-72	0	0	101	341	587	1022	889	648	346	244	76	0	4254
1972-73	0	3	14	244	740	925	1020	811	607	440	113	3	4920
1973-74	0	0	43	257	606	955	963	754	373	255	29	4	4239
1974-75	0	2	68	212	593	1020	1051	748	614	449	143	6	4906
1975-76	0	0	47	256	664	905	979	622	634	304	99	1	4511
1976-77	0	0	35	367	726	985	1084	675	669	250	61	0	4852
1977-78	0	0	1	192	551	757	870	713	454	215	175	2	3930
1978-79	0	0	20	167	521	945	988	665	509	241	100	12	4168
1979-80	0	0	23	148	715	840	763	595	577	379	139	2	4181
1980-81	0	0	6	335	640	752	827	611	575	197	62	2	4007
1981-82	0	0	3	280	534	754	895	709	538	268	94	0	4075
1982-83	0	0	23	314	658	941	922	703	556	439	127	0	4683
1983-84	0	0	11	198	592	875	948	714	559	362	22	3	4284
1984-85	0	0	51	411	631	903	960	744	536	220	74	7	4537
1985-86	0	0	61	228	581	842	727	610	431	249	80	8	3817
1986-87	0	0	51	313	680	882	1004	717	653	300	81	2	4683
1987-88	0	0	2	133	589	914	937	605	551	290	103	2	4126
1988-89	0	5	39	118	579	959	909	640	373	133	31	0	3786
1989-90	0	0	10	260	551	918	934	735	501	233	103	0	4245
1990-91	0	0	14	202	595	1013	903	563	581	263	60	12	4206
1991-92	0	0	21	188	645	851	994	651	493	170	53	5	4071
1992-93	0	0	8	128	752	991	778	624	496	238	69	3	4087
1993-94	0	0	15	284	642	853							

COOLING DEGREE DAYS Base 65 deg. F ALBUQUERQUE, NEW MEXICO

YEAR	JAN	FEB	MAR	APR	MAY	JUNE	JULY	AUG	SEP	OCT	NOV	DEC	TOTAL
1969	0	0	0	6	127	263	478	442	158	7	0	0	1481
1970	0	0	0	0	105	246	461	405	141	4	0	0	1362
1971	0	0	5	0	26	277	414	282	149	0	0	0	1153
1972	0	0	0	5	52	267	428	294	113	23	0	0	1182
1973	0	0	0	0	48	267	422	409	124	0	0	0	1270
1974	0	0	0	5	144	464	380	247	107	6	0	0	1353
1975	0	0	0	0	25	256	372	351	96	0	0	0	1100
1976	0	0	0	0	38	260	382	319	137	5	0	0	1141
1977	0	0	0	0	44	324	427	392	141	7	0	0	1335
1978	0	0	0	4	41	324	521	330	151	27	0	0	1398
1979	0	0	0	5	67	269	491	382	249	45	0	0	1508
1980	0	0	0	0	27	375	557	392	160	15	0	0	1526
1981	0	0	0	28	51	368	470	360	152	1	0	0	1430
1982	0	0	0	6	38	301	441	394	163	4	0	0	1347
1983	0	0	0	1	72	260	484	450	267	1	0	0	1535
1984	0	0	0	4	179	266	441	340	169	1	0	0	1400
1985	0	0	0	0	51	289	383	368	97	0	0	0	1188
1986	0	0	0	1	50	245	310	349	103	0	0	0	1058
1987	0	0	0	0	17	251	404	308	120	25	0	0	1125
1988	0	0	0	1	85	288	411	322	86	3	0	0	1196
1989	0	0	0	31	154	323	426	295	150	10	0	0	1389
1990	0	0	0	10	66	426	374	281	200	2	0	0	1359
1991	0	0	0	0	87	269	375	331	120	25	0	0	1207
1992	0	0	0	27	49	235	354	318	171	5	0	0	1159
1993	0	0	0	8	101	312	470	337	145	16	0	0	1389

SNOWFALL (inches) ALBUQUERQUE, NEW MEXICO

SEASON	JULY	AUG	SEP	OCT	NOV	DEC	JAN	FEB	MAR	APR	MAY	JUNE	TOTAL
1964-65	0.0	0.0	0.0	0.0	T	0.3	1.4	3.6	T	T	0.0	0.0	5.3
1965-66	0.0	0.0	0.0	0.0	T	3.0	5.4	1.0	0.0	T	0.0	0.0	9.4
1966-67	0.0	0.0	0.0	0.0	T	T	T	1.0	1.1	T	0.0	0.0	2.1
1967-68	0.0	0.0	0.0	0.2	1.0	2.8	T	2.0	1.4	T	0.0	0.0	7.4
1968-69	0.0	0.0	0.0	0.0	T	7.4	T	1.8	5.5	T	0.0	0.0	14.7
1969-70	0.0	0.0	0.0	0.0	T	1.1	T	2.7	3.3	0.0	0.0	0.0	7.1
1970-71	0.0	0.0	0.0	0.5	T	0.5	3.0	2.3	0.5	T	0.0	0.0	6.8
1971-72	0.0	0.0	T	T	T	6.8	1.2	1.1	0.0	T	0.0	0.0	9.1
1972-73	0.0	0.0	0.0	T	2.9	1.2	9.5	1.8	13.9	8.1	0.0	0.0	37.4
1973-74	0.0	0.0	0.0	0.3	0.6	0.1	9.3	0.6	2.0	0.0	0.0	0.0	12.9
1974-75	0.0	0.0	0.0	0.0	T	4.9	0.9	6.7	3.8	0.2	0.0	0.0	16.5
1975-76	0.0	0.0	0.0	0.0	0.2	2.9	0.0	T	0.5	0.2	0.0	0.0	3.8
1976-77	0.0	0.0	0.0	T	2.4	1.2	8.4	1.4	2.3	2.6	0.0	0.0	18.3
1977-78	0.0	0.0	0.0	0.0	0.0	T	6.0	3.4	2.0	0.0	0.1	0.0	11.5
1978-79	0.0	0.0	0.0	0.0	T	1.0	2.6	6.0	T	0.5	1.0	0.0	11.1
1979-80	0.0	0.0	0.0	0.9	0.8	2.7	T	0.9	3.1	T	T	0.0	8.4
1980-81	0.0	0.0	0.0	T	2.8	7.4	0.5	2.6	0.9	T	0.0	0.0	14.2
1981-82	0.0	0.0	0.0	0.0	0.0	0.0	3.6	1.2	0.7	T	0.0	0.0	5.5
1982-83	0.0	0.0	0.0	0.0	0.9	3.3	7.3	4.2	1.0	T	T	0.0	16.7
1983-84	0.0	0.0	0.0	0.0	0.8	0.8	4.1	T	0.1	3.0	0.0	0.0	8.8
1984-85	0.0	0.0	0.0	T	T	3.4	2.0	2.9	0.6	0.0	0.0	0.0	8.9
1985-86	0.0	0.0	0.0	0.0	0.7	0.9	2.9	10.3	0.3	0.0	T	0.0	15.1
1986-87	0.0	0.0	0.0	3.2	0.6	0.2	4.9	4.9	0.2	2.2	0.0	0.0	16.2
1987-88	0.0	0.0	0.0	0.0	1.1	1.7	1.2	T	7.9	4.2	0.0	0.0	16.1
1988-89	0.0	0.0	0.0	0.0	1.7	0.3	3.4	3.2	3.1	0.0	0.0	0.0	11.7
1989-90	0.0	0.0	0.0	0.0	T	2.5	1.8	4.8	T	0.3	T	T	9.4
1990-91	T	0.0	0.0	0.0	2.2	6.3	0.9	T	0.8	T	0.0	0.0	10.2
1991-92	0.0	0.0	0.0	2.5	1.5	2.1	5.6	T	1.0	0.0	T	T	12.7
1992-93	0.0	T	0.0	0.0	5.9	7.6	0.8	2.0	0.2	0.0	T	0.0	16.5
1993-94	0.0	T	0.0	T	4.1	0.2							
Record Mean	T	T	T	0.1	1.3	2.6	2.5	2.1	1.8	0.6	T	T	11.1

ROSWELL, NEW MEXICO

The climate at Roswell conforms to the basic trend of the four seasons, but shows certain deviations related to geography. Higher landmasses almost surround the valley location, with a long, gradual descent from points southwest through west and north. The topography acts to modify air masses, especially the cold outbreaks in wintertime. Downslope warming of air, as well as air interchange within a tempering environment, often prevents sharp cooling. Moreover, the elevation of 3,600 feet is high enough to moderate the heat and humidity compared to locations to the south and east.

Summer moves into a wet phase that delivers the most important rain of the year. Rather frequent showers and thunderstorms from June through September account for over half of the annual precipitation. Storm clouds that build up from the heat of the day, overspread the sky on many afternoons, retarding a further rise in temperature. At the same time, relative humidity shows moderation, ranging from about 70 percent in early morning to 30 percent in the mid-afternoon. Temperatures are quite warm on most summer days with readings of 100 degrees or higher occurring on 10 days in an average year.

Rainfall tapers off markedly in the fall with decline in storm activity. This leaves usually agreeable conditions because of low wind movement and mostly clear skies. Frosty nights alternate with warm days. Relative humidity reaches rather low levels in autumn, but dryness is not as rigorous as in the spring.

In winter, sub-freezing at night is tempered by considerable warming during the day. Zero or lower temperatures occur on only one day in an average winter. Sub-zero cold spells are of short duration. Winter is the season of least precipitation.

Spring ushers in the driest season of the year with respect to relative humidity. Wind movement shows a large increase, especially from the plateau areas of the west. Most of the 60 days a year with winds of 25 mph or more occur from February to May. Destructive storms seldom strike the city, but minor damage results from thundersqualls or hailstorms about once a year. Rain is most erratic in spring, ranging from none of consequence in some years, to excessive amounts in others.

NORMALS, MEANS, AND EXTREMES

ROSWELL, NEW MEXICO

LATITUDE: 33°18'N LONGITUDE: 104°32' W ELEVATION: FT. GRND 3649 BARO 3653 TIME ZONE: MOUNTAIN WBAN: 23009

	(a)	JAN	FEB	MAR	APR	MAY	JUNE	JULY	AUG	SEP	OCT	NOV	DEC	YEAR
TEMPERATURE °F:														
Normals														
-Daily Maximum		54.3	59.9	67.8	76.7	84.7	93.6	94.6	91.9	86.0	77.3	66.0	56.0	75.7
-Daily Minimum		24.7	29.1	36.3	45.3	54.7	62.1	66.7	64.9	59.2	47.1	35.1	25.6	45.9
-Monthly		39.5	44.5	52.1	61.0	69.7	77.9	80.7	78.4	72.6	62.2	50.6	40.8	60.8
Extremes														
-Record Highest	21	82	85	93	99	104	110	109	104	101	96	87	81	110
-Year		1975	1989	1989	1989	1989	1990	1989	1980	1992	1993	1988	1981	JUN 1990
-Record Lowest	21	-9	3	14	23	34	47	59	54	42	14	4	-8	-9
-Year		1979	1985	1987	1973	1975	1992	1988	1990	1973	1991	1976	1978	JAN 1979
NORMAL DEGREE DAYS:														
Heating (base 65°F)		791	574	409	162	19	0	0	0	8	122	432	750	3267
Cooling (base 65°F)		0	0	9	42	165	387	487	415	236	35	0	0	1776
% OF POSSIBLE SUNSHINE	7	60	68	75	77	80	83	77	73	72	77	73	71	74
MEAN SKY COVER (tenths)														
Sunrise - Sunset	9	5.3	4.7	4.4	4.2	4.3	3.5	4.6	4.5	4.8	3.6	3.9	4.2	4.3
MEAN NUMBER OF DAYS:														
Sunrise to Sunset														
-Clear	9	11.3	12.1	14.2	14.3	13.4	17.2	11.1	12.8	12.8	18.1	15.8	14.9	168.1
-Partly Cloudy	9	8.8	7.8	9.1	8.9	11.8	9.9	14.7	12.2	7.9	5.9	7.1	9.0	113.0
-Cloudy	9	10.9	8.3	7.7	6.8	5.8	2.9	5.2	6.0	9.3	7.0	7.1	7.1	84.1
Precipitation														
.01 inches or more	21	4.7	3.6	2.6	3.1	4.5	5.0	6.5	8.7	7.0	4.1	2.8	3.9	56.4
Snow,Ice pellets,hail														
1.0 inches or more	21	1.0	1.0	0.3	0.2	0.0	0.*	0.0	0.0	0.*	0.1	0.4	1.0	4.1
Thunderstorms	21	0.2	0.2	0.6	2.3	4.9	7.0	7.6	8.4	4.8	1.8	0.6	0.*	38.4
Heavy Fog Visibility														
1/4 mile or less	21	3.6	2.6	0.6	0.2	0.4	0.2	0.2	0.2	0.9	1.9	2.3	2.1	15.2
Temperature °F														
-Maximum														
90° and above	21	0.0	0.0	0.1	1.2	8.3	21.2	24.5	21.4	9.4	1.5	0.0	0.0	87.7
32° and below	21	2.1	0.9	0.1	0.0	0.0	0.0	0.0	0.0	0.0	0.0	0.4	1.9	5.3
-Minimum														
32° and below	21	25.2	16.8	6.9	1.7	0.0	0.0	0.0	0.0	0.0	1.1	12.6	24.5	88.8
0° and below	21	0.2	0.0	0.0	0.0	0.0	0.0	0.0	0.0	0.0	0.0	0.0	0.3	0.5
AVG. STATION PRESS. (mb)	21	891.3	890.3	887.2	887.3	887.0	888.2	890.4	891.1	891.1	891.3	890.7	891.3	889.8
RELATIVE HUMIDITY (%)														
Hour 05	21	72	66	56	53	59	64	70	74	75	69	68	68	66
Hour 11 (Local Time)	21	52	45	33	29	32	35	41	45	48	43	44	47	41
Hour 17	21	42	34	24	22	25	27	32	37	40	36	38	41	33
Hour 23	21	63	55	43	41	44	47	55	60	63	60	58	61	54
PRECIPITATION (inches):														
Water Equivalent														
-Normal		0.35	0.46	0.33	0.46	1.04	1.61	1.71	2.58	2.02	1.05	0.52	0.45	12.58
-Maximum Monthly	21	1.03	2.02	1.48	2.48	4.57	5.02	6.88	6.48	6.58	5.48	1.89	2.67	6.88
-Year		1992	1987	1973	1985	1992	1986	1991	1974	1980	1986	1986	1991	JUL 1991
-Minimum Monthly	21	0.04	T	0.00	0.02	T	0.02	0.01	0.34	0.10	T	0.00	0.00	0.00
-Year		1984	1984	1980	1991	1974	1990	1980	1985	1992	1980	1991	1976	NOV 1991
-Maximum in 24 hrs	21	0.67	1.41	1.43	2.24	1.77	3.05	4.91	3.94	2.71	3.89	1.33	0.83	4.91
-Year		1986	1988	1973	1985	1981	1981	1981	1977	1980	1986	1984	1987	JUL 1981
Snow,Ice pellets,hail														
-Maximum Monthly	21	7.3	16.9	4.8	5.3	0.8	1.0	0.0	0.0	1.0	4.2	12.3	15.3	16.9
-Year		1980	1988	1984	1983	1989	1991			1989	1976	1980	1987	FEB 1988
-Maximum in 24 hrs	21	7.3	16.5	4.8	4.0	2.0	1.0	0.0	0.0	1.0	3.1	6.3	9.7	16.5
-Year		1980	1988	1984	1983	1988	1991	1973	1973	1989	1976	1980	1987	FEB 1988
WIND:														
Mean Speed (mph)	21	7.7	8.5	10.1	10.1	9.8	9.5	8.7	7.8	8.0	7.8	7.8	7.5	8.6
Prevailing Direction														
Fastest Mile														
-Direction (!!)	9	NW	NW	NW	SW	NW	NW	NE	NW	NE	22	NE	SW	NW
-Speed (MPH)	9	47	56	52	48	60	73	42	44	40	44	65	58	73
-Year		1973	1977	1973	1977	1977	1975	1977	1974	1976	1975	1975	1977	JUN 1975
Peak Gust														
-Direction (!!)														
-Speed (mph)														
-Date														

PRECIPITATION (inches)

ROSWELL, NEW MEXICO

YEAR	JAN	FEB	MAR	APR	MAY	JUNE	JULY	AUG	SEP	OCT	NOV	DEC	ANNUAL
1964	0.80	1.25	0.15	0.02	0.30	1.10	0.17	0.57	2.05	T	0.33	0.24	6.98
1965	0.12	0.84	0.21	0.38	0.35	1.09	1.50	0.83	0.76	0.05	0.08	0.47	6.68
1966	0.53	0.03	0.25	1.97	0.54	2.35	0.15	2.89	0.97	T	T	T	9.68
1967	0.00	0.20	0.07	T	0.11	3.55	0.97	4.00	0.85	0.02	0.22	1.07	11.06
1968	1.50	1.17	1.93	0.06	0.57	0.60	5.50	2.67	0.10	0.41	1.11	0.22	15.84
1969	0.01	0.47	1.14	0.44	0.10	0.35	1.32	0.71	2.67	4.34	T	1.78	13.33
1970	0.01	0.28	0.51	0.02	0.48	2.72	2.07	0.52	0.97	0.78	0.09	0.18	8.63
1971	0.18	0.23	0.11	0.26	T	0.18	1.88	3.62	1.57	0.76	0.45	0.80	10.04
#1972	0.20	0.00	0.03	0.00	0.16	2.06	5.43	3.35	3.25	1.27	0.49	0.26	16.50
1973	0.73	0.92	1.48	0.15	0.73	0.97	2.26	1.27	2.55	0.51	0.01	0.02	11.60
1974	0.24	0.01	0.11	0.50	T	0.03	0.31	6.48	6.47	3.81	0.09	0.60	18.65
1975	0.20	1.06	0.27	0.29	0.13	0.57	2.75	1.28	2.83	0.16	T	0.05	9.59
1976	0.12	0.22	0.24	0.79	0.82	1.55	2.44	1.98	2.29	0.69	0.41	0.00	11.55
1977	0.07	0.36	0.27	1.25	2.43	0.25	0.46	4.45	0.29	0.62	0.48	0.02	10.95
1978	0.50	0.48	0.39	0.02	1.81	4.31	0.52	3.49	3.58	1.47	1.25	0.43	18.25
1979	0.41	0.44	0.13	0.32	1.25	1.56	1.44	2.28	0.15	0.18	T	0.37	8.53
1980	0.85	0.19	0.00	1.06	0.85	0.29	0.01	2.45	6.58	T	0.77	0.15	13.20
1981	0.27	0.17	0.10	0.79	3.35	4.55	6.27	4.73	2.70	1.02	0.25	0.13	24.33
1982	0.66	0.20	0.12	0.41	0.20	0.76	1.03	0.93	2.00	0.20	0.92	1.62	9.05
1983	0.50	0.22	0.11	0.64	0.93	0.67	0.37	0.80	0.42	3.43	1.52	0.42	10.03
1984	0.04	T	0.46	0.03	1.62	4.52	0.85	5.03	1.04	2.74	1.57	0.85	18.75
1985	0.37	0.04	0.70	2.48	2.22	2.58	2.71	0.34	1.93	0.98	0.12	0.07	14.54
1986	0.67	0.50	0.12	0.31	1.19	5.02	1.11	3.11	3.93	5.48	1.89	1.47	24.80
1987	0.45	2.02	0.20	0.26	1.54	3.69	0.40	4.71	0.78	0.28	0.46	1.41	16.20
1988	0.22	1.48	0.03	0.27	3.42	1.27	4.45	0.51	1.56	0.01	0.03	0.51	13.76
1989	0.32	0.49	0.23	0.07	0.44	0.07	1.14	1.93	1.22	0.10	0.00	0.07	6.08
1990	0.41	0.22	0.74	0.84	0.11	0.02	1.24	1.20	1.44	0.38	0.57	0.32	7.49
1991	0.29	0.33	0.02	0.02	1.04	1.24	6.88	3.40	4.10	0.12	0.95	2.67	21.06
1992	1.03	0.70	0.20	0.65	4.57	2.12	1.08	1.56	0.10	0.50	0.08	0.68	13.27
1993	0.75	0.20	0.01	1.15	0.38	0.56	3.39	2.19	0.88	0.55	0.13	0.16	10.35
Record Mean	0.43	0.47	0.48	0.69	1.11	1.45	2.00	1.90	1.88	1.13	0.57	0.53	12.64

AVERAGE TEMPERATURE (deg. F)

ROSWELL, NEW MEXICO

YEAR	JAN	FEB	MAR	APR	MAY	JUNE	JULY	AUG	SEP	OCT	NOV	DEC	ANNUAL
1964	34.7	35.3	48.6	58.2	69.8	76.7	81.7	80.4	72.1	60.9	48.1	39.7	58.8
1965	44.2	40.1	45.3	62.0	69.0	77.0	80.9	76.7	70.7	59.0	52.5	42.0	60.0
1966	30.8	38.5	51.6	59.8	70.0	76.6	84.3	76.5	70.9	56.8	50.6	39.5	58.8
1967	41.2	44.3	57.2	65.5	68.5	75.3	80.0	75.0	68.8	59.3	47.5	35.3	59.8
1968	39.1	43.7	47.7	55.5	66.7	77.0	75.6	75.6	67.2	60.3	44.7	38.2	57.6
#1969	45.0	44.6	43.1	61.9	69.1	77.1	82.9	81.4	71.9	58.1	47.1	41.0	60.3
1970	37.0	46.1	46.6	58.0	67.7	74.3	80.3	78.3	71.0	55.7	48.5	43.4	58.9
1971	41.3	42.9	52.9	58.5	68.6	78.1	79.7	74.9	69.3	59.4	48.4	40.8	59.6
#1972	40.6	45.7	56.2	63.4	68.3	77.5	79.0	76.0	70.8	59.1	42.9	39.5	59.9
1973	39.0	43.6	52.3	54.5	66.2	75.3	78.8	77.6	69.7	61.4	51.1	41.9	59.3
1974	40.4	43.3	57.6	61.2	73.5	79.2	80.0	74.5	64.4	59.2	47.5	39.5	60.1
1975	41.3	42.3	49.9	56.6	65.9	76.7	77.0	78.3	68.5	62.5	49.5	43.4	59.3
1976	40.4	53.3	54.1	63.7	68.7	79.3	78.6	80.3	71.2	56.2	42.7	39.3	60.7
1977	38.6	48.2	52.1	62.3	73.3	81.6	84.2	83.0	78.4	64.1	53.1	47.0	63.9
1978	36.0	43.6	55.6	66.2	71.5	79.3	83.4	78.0	69.2	60.3	49.0	37.2	60.8
1979	34.9	43.6	50.5	60.6	67.5	75.1	81.0	76.5	72.2	63.8	44.6	40.6	59.3
1980	39.9	44.6	51.1	57.7	68.0	83.5	85.4	78.7	71.6	57.9	43.8	44.0	60.5
1981	41.6	46.2	51.7	63.5	68.5	79.0	79.9	75.7	69.9	59.8	53.7	45.3	61.3
1982	38.9	42.7	53.4	60.5	68.0	76.7	80.9	80.5	73.6	59.4	47.0	37.8	60.0
1983	38.9	45.5	52.8	54.7	66.3	76.5	81.6	81.4	76.1	62.9	51.8	35.2	60.3
1984	38.5	46.5	51.7	60.2	72.5	75.8	79.1	76.2	69.3	57.7	48.5	42.1	59.8
1985	37.0	43.2	54.2	63.4	70.1	75.8	79.4	80.4	70.7	61.1	53.2	40.3	60.7
1986	44.1	46.8	56.2	64.6	69.1	75.9	79.2	78.6	72.1	59.3	46.0	40.0	61.0
1987	39.7	44.9	49.1	57.3	67.5	75.8	80.2	78.7	70.6	63.9	49.0	39.3	59.7
1988	37.6	44.5	51.4	59.7	68.2	77.5	78.6	78.7	72.2	64.0	53.3	40.6	60.5
1989	43.1	43.8	56.6	65.6	74.3	78.4	81.6	79.6	71.3	63.3	51.3	37.2	62.2
1990	42.5	47.0	53.1	62.7	70.5	85.0	79.9	77.7	74.6	62.8	52.5	41.7	62.5
1991	39.8	49.6	55.3	64.1	74.2	79.8	78.6	78.3	68.1	61.7	46.4	41.5	61.5
1992	38.4	47.2	54.2	62.8	67.0	75.4	80.6	78.9	74.2	64.4	45.9	41.1	60.8
1993	42.3	45.6	53.8	62.3	70.1	79.9	83.0	79.8	72.4	59.3	45.4	42.2	61.3
Record Mean	39.6	44.1	51.3	59.8	68.3	76.8	79.3	77.8	70.9	60.1	48.0	40.1	59.6
Max	54.6	59.7	67.7	76.1	84.0	92.1	92.8	91.3	84.9	75.4	63.5	54.9	74.7
Min	24.6	28.5	35.0	43.5	52.6	61.5	65.7	64.3	56.9	44.7	32.4	25.3	44.6

REFERENCE NOTES

ROSWELL, NEW MEXICO

GENERAL
T - TRACE AMOUNT.
BLANK ENTRIES DENOTE MISSING/UNREPORTED DATA.
# INDICATES A STATION OR INSTRUMENT RELOCATION.

SPECIFIC

(a) - LENGTH OF RECORD IN YEARS, ALTHOUGH INDIVIDUAL MONTHS MAY BE MISSING.
0.* OR * - THE VALUE IS BETWEEN 0.0 AND 0.05
NORMALS - BASED ON THE 1961-1990 RECORD PERIOD.
EXTREMES - DATES ARE THE MOST RECENT OCCURRENCE
WIND DIR.- NUMERALS SHOW TENS OF DEGRESS CLOCKWISE FROM TRUE NORTH. "00" INDICATES CALM.
RESULTANT DIRECTIONS ARE GIVEN TO WHOLE DEGREES.
BOLD VALUES INDICATE EXTREME VALUES WHICH OCCURRED AFTER THE ASOS SYSTEM WAS COMMISSIONED.

RECORD = PERIOD OF RECORD
RECORD MEAN PRECIPITATION IS THE MEAN OF ALL DAILY PRECIPITATION AMOUNTS DURING THE PERIOD OF RECORD.
RECORD MAX(MIN) TEMPERATURE IS THE MEAN OF ALL DAILY MAX(MIN) TEMPERATURES DURING THE PERIOD OF RECORD.
RECORD MEAN TEMPERATURE IS THE SUM OF THE RECORD MAX AND RECORD MIN DIVIDED BY 2.
AVERAGE TEMPERATURE IS THE SUM OF THE MEAN DAILY MAX AND MIN TEMPERATURE DIVIDED BY 2.

EXCEPTIONS

1. PERCENT OF POSSIBLE SUNSHINE, SKY COVER, AND DAYS CLEAR-PARTLY CLOUDY-CLOUDY ARE THROUGH 1981.
2. FASTEST MILE WINDS ARE THROUGH MAY 1982.

RECORD MEANS ARE THROUGH THE CURRENT YEAR, BEGINNING IN 1895 FOR TEMPERATURE
1895 FOR PRECIPITATION
1949 FOR SNOWFALL

HEATING DEGREE DAYS Base 65 deg. F ROSWELL, NEW MEXICO

SEASON	JULY	AUG	SEP	OCT	NOV	DEC	JAN	FEB	MAR	APR	MAY	JUNE	TOTAL
1964-65	0	0	21	145	501	782	637	693	606	126	22	0	3533
1965-66	0	0	34	200	372	708	1052	734	407	168	33	0	3708
1966-67	0	5	5	257	425	788	731	577	244	59	52	0	3143
1967-68	0	2	9	205	517	915	796	609	529	283	52	0	3917
#1968-69	0	0	26	171	603	823	615	564	672	109	38	2	3623
1969-70	0	0	0	257	529	738	861	523	565	214	55	17	3759
1970-71	0	0	56	298	490	662	729	612	384	206	30	0	3467
1971-72	0	0	82	183	495	743	750	551	273	109	25	0	3211
#1972-73	0	0	13	216	660	785	800	594	386	310	73	0	3837
1973-74	0	0	27	130	410	710	755	599	233	151	11	0	3026
1974-75	0	0	104	190	518	784	729	630	462	269	36	1	3723
1975-76	0	0	48	104	462	662	755	334	334	90	36	0	2825
1976-77	0	0	18	275	663	789	811	463	393	109	0	0	3521
1977-78	0	0	0	68	349	551	895	593	294	45	39	3	2837
1978-79	0	0	55	174	468	856	929	591	444	161	51	2	3731
1979-80	0	0	27	115	602	748	773	587	425	222	53	0	3552
1980-81	0	0	19	234	630	645	719	521	404	100	21	0	3293
1981-82	0	0	11	183	333	603	803	616	357	160	27	0	3093
1982-83	0	0	2	195	530	834	802	541	372	322	44	2	3644
1983-84	0	0	7	115	389	920	817	529	404	164	19	0	3364
1984-85	0	0	56	227	488	704	863	604	332	85	10	1	3370
1985-86	0	0	46	131	344	759	641	504	271	77	31	0	2804
1986-87	0	0	0	181	564	770	779	556	484	243	22	0	3599
1987-88	0	0	4	71	476	788	844	588	417	163	36	0	3387
1988-89	0	7	7	56	359	749	673	590	275	90	7	0	2813
1989-90	0	0	23	121	405	854	691	497	366	119	62	0	3138
1990-91	0	0	0	98	369	715	775	426	299	68	8	0	2758
1991-92	0	0	59	152	551	719	815	509	325	117	32	3	3282
1992-93	0	0	0	64	563	734	697	538	341	124	24	0	3085
1993-94	0	0	8	221	583	699							

COOLING DEGREE DAYS Base 65 deg. F ROSWELL, NEW MEXICO

YEAR	JAN	FEB	MAR	APR	MAY	JUNE	JULY	AUG	SEP	OCT	NOV	DEC	TOTAL
1969	0	0	0	21	174	372	560	516	215	52	0	0	1910
1970	0	0	0	9	146	303	482	420	244	16	0	0	1620
1971	0	0	14	17	146	402	463	315	220	17	0	0	1594
#1972	0	0	7	68	133	382	438	350	197	41	0	0	1616
1973	0	0	0	2	118	316	434	400	174	25	0	0	1469
1974	0	0	11	45	283	433	473	299	95	16	0	0	1655
1975	0	0	0	19	67	361	379	416	159	33	3	0	1437
1976	0	0	7	57	159	436	429	478	211	9	0	0	1786
1977	0	0	0	34	269	505	602	564	408	46	0	0	2428
1978	0	0	8	91	244	438	576	410	187	37	0	0	1991
1979	0	0	0	35	133	311	506	364	251	84	0	0	1684
1980	0	0	0	8	152	564	636	430	223	21	1	0	2035
1981	0	0	0	59	133	430	467	336	166	31	0	0	1622
1982	0	0	6	35	124	357	499	488	266	27	0	0	1802
1983	0	0	0	21	89	353	523	512	346	58	0	0	1902
1984	0	0	2	26	256	331	443	356	189	6	0	0	1609
1985	0	0	8	44	175	332	457	485	224	18	0	0	1743
1986	0	0	6	74	167	331	447	429	221	10	0	0	1685
1987	0	0	0	17	108	330	478	432	176	43	0	0	1584
1988	0	0	0	11	140	382	430	441	230	32	15	0	1681
1989	0	4	19	115	303	412	523	460	217	75	0	0	2128
1990	0	0	4	57	241	607	472	403	295	37	2	0	2118
1991	0	0	3	48	300	451	431	419	158	59	0	0	1869
1992	0	0	0	56	101	319	492	436	283	52	0	0	1739
1993	0	0	0	51	189	454	564	463	236	53	0	0	2010

SNOWFALL (inches) ROSWELL, NEW MEXICO

SEASON	JULY	AUG	SEP	OCT	NOV	DEC	JAN	FEB	MAR	APR	MAY	JUNE	TOTAL
1964-65	0.0	0.0	0.0	0.0	0.0	1.9	1.2	7.8	1.7	0.0	0.0	0.0	12.6
1965-66	0.0	0.0	0.0	0.0	0.0	2.7	5.2	T	T	0.0	0.0	0.0	7.9
1966-67	0.0	0.0	0.0	0.0	0.0	T	0.0	2.0	0.6	0.0	0.0	0.0	2.6
1967-68	0.0	0.0	0.0	0.0	0.8	6.4	2.9	4.0	9.9	0.0	0.0	0.0	24.0
1968-69	0.0	0.0	0.0	0.0	8.5	0.0	T	0.0	10.9	0.0	0.0	0.0	19.4
1969-70	0.0	0.0	0.0	0.0	T	13.9	T	2.0	4.9	0.0	0.0	0.0	20.8
1970-71	0.0	0.0	0.0	0.2	0.0	0.0	2.0	2.4	1.0	0.0	0.0	0.0	5.6
1971-72	0.0	0.0	0.0	0.0	0.0	5.5	1.9	0.0	0.0	0.0	0.0	0.0	7.4
#1972-73	0.0	0.0	0.0	0.6	2.6	4.8	6.8	7.5	0.5	1.1	0.0	0.0	23.9
1973-74	0.0	0.0	0.0	0.0	0.0	T	1.0	T	0.2	0.0	0.0	0.0	1.2
1974-75	0.0	0.0	0.0	0.0	0.0	1.4	1.5	3.7	1.4	0.0	0.0	0.0	8.0
1975-76	0.0	0.0	0.0	0.0	T	0.3	1.2	0.0	2.4	0.0	0.0	0.0	3.9
1976-77	0.0	0.0	0.0	4.2	4.1	0.0	0.4	0.4	0.9	0.0	0.0	0.0	10.0
1977-78	0.0	0.0	0.0	0.0	0.0	0.0	3.2	2.3	0.7	0.0	T	0.0	6.2
1978-79	0.0	0.0	0.0	0.0	0.5	7.7	0.7	6.4	0.0	0.0	0.0	0.0	15.3
1979-80	0.0	0.0	0.0	0.4	0.0	3.8	7.3	4.0	0.0	3.0	0.0	0.0	18.5
1980-81	0.0	0.0	0.0	T	12.3	0.6	5.3	0.0	T	0.0	0.0	0.0	18.2
1981-82	0.0	0.0	0.0	0.0	0.0	0.0	6.9	2.6	1.2	T	0.0	0.0	10.7
1982-83	0.0	0.0	0.0	0.0	7.1	8.0	2.4	1.0	0.0	5.3	0.0	0.0	23.8
1983-84	0.0	0.0	0.0	0.0	0.0	3.5	0.5	T	4.8		0.0	0.0	
1984-85	0.0	0.0	0.0	0.0	0.0	5.4	1.9	0.3	0.0	0.0	0.0	0.0	7.6
1985-86	0.0	0.0	0.0	0.0	0.0	1.1	5.4	4.9	0.0	0.0	0.0	0.0	11.4
1986-87	0.0	0.0	0.0	T	0.5	3.4	5.5	8.7	0.5	0.0	0.0	0.0	18.6
1987-88	0.0	0.0	0.0	0.0	0.0	15.3	0.4	16.9	T	T	0.0	0.0	32.6
1988-89	0.0	0.0	0.0	0.0	T	3.9	T	1.2	4.3	T	0.8	0.0	10.2
1989-90	0.0	0.0	1.0	0.0	0.0	T	2.7	T	1.2	0.0	0.0	0.0	4.9
1990-91	0.0	0.0	0.0	0.0	0.5	0.8	1.9	0.0	0.0	T	T	1.0	4.2
1991-92	0.0	0.0	0.0	1.2	T	2.9	4.4	0.0	0.0	T	0.0	T	8.5
1992-93	0.0	0.0	0.0	0.0	T	5.0	T	0.0	T	0.0	0.0	0.0	5.0
1993-94	0.0	0.0	0.0	1.8	T	T							
Record Mean	0.0	0.0	T	0.2	1.2	3.0	2.8	2.7	1.2	0.3	T	T	11.4

ALBANY, NEW YORK

Albany is located on the west bank of the Hudson River some 150 miles north of New York City, and 8 miles south of the confluence of the Mohawk and Hudson Rivers. The river-front portion of the city is only a few feet above sea level, and there is a tidal effect upstream to Troy. Eleven miles west of Albany the Helderberg escarpment rises to 1,800 feet. Between it and the Hudson River the valley floor is gently rolling, ranging some 200 to 500 feet above sea level. East of the city there is more rugged terrain 5 or 6 miles wide with elevations of 300 to 600 feet. Farther to the east the terrain rises more sharply. It reaches a north-south range of hills 12 miles east of Albany with elevations ranging to 2,000 feet.

The climate at Albany is primarily continental in character, but is subjected to some modification by the Atlantic Ocean. The moderating effect on temperatures is more pronounced during the warmer months than in winter when outbursts of cold air sweep down from Canada. In the warmer seasons, temperatures rise rapidly in the daytime. However, temperatures also fall rapidly after sunset so that the nights are relatively cool. Occasionally there are extended periods of oppressive heat up to a week or more in duration.

Winters are usually cold and sometimes fairly severe. Maximum temperatures during the colder winters are often below freezing and nighttime lows are frequently below 10 degrees. Sub-zero readings occur about twelve times a year. Snowfall throughout the area is quite variable and snow flurries are quite frequent during the winter. Precipitation is sufficient to serve the economy of the region in most years, and only occasionally do periods of drought exist. Most of the rainfall in the summer is from thunderstorms. Tornadoes are quite rare and hail is not usually of any consequence.

Wind velocities are moderate. The north-south Hudson River Valley has a marked effect on the lighter winds and in the warm months, average wind direction is usually southerly. Destructive winds rarely occur.

The area enjoys one of the highest percentages of sunshine in the entire state. Seldom does the area experience long periods of cloudy days and long periods of smog are rare.

Based on the 1951-1980 period, the average first occurrence of 32 degrees Fahrenheit in the fall is September 29 and the average last occurrence in the spring is May 7.

NORMALS, MEANS, AND EXTREMES

ALBANY, NEW YORK

LATITUDE: 42°45'N LONGITUDE: 73°48' W ELEVATION: FT. GRND 275 BARO 296 TIME ZONE: EASTERN WBAN: 14735

	(a)	JAN	FEB	MAR	APR	MAY	JUNE	JULY	AUG	SEP	OCT	NOV	DEC	YEAR
TEMPERATURE °F:														
Normals														
-Daily Maximum		30.2	33.2	44.0	57.5	69.7	79.0	84.0	81.4	73.2	61.8	48.7	34.9	58.1
-Daily Minimum		11.0	13.8	24.5	35.1	45.4	54.6	59.6	57.8	49.4	38.6	30.7	18.2	36.6
-Monthly		20.6	23.5	34.3	46.4	57.6	66.9	71.8	69.6	61.3	50.2	39.7	26.5	47.4
Extremes														
-Record Highest	47	62	67	86	92	94	99	100	99	100	89	82	71	100
-Year		1974	1976	1986	1990	1981	1952	1953	1955	1953	1963	1950	1984	JUL 1953
-Record Lowest	47	-28	-21	-21	10	26	36	40	34	24	16	5	-22	-28
-Year		1971	1973	1948	1965	1968	1986	1978	1982	1947	1969	1972	1969	JAN 1971
NORMAL DEGREE DAYS:														
Heating (base 65°F)		1376	1162	952	558	247	34	0	12	141	459	759	1194	6894
Cooling (base 65°F)		0	0	0	0	18	91	213	155	30	0	0	0	507
% OF POSSIBLE SUNSHINE	55	46	52	53	54	56	60	63	60	57	52	36	38	52
MEAN SKY COVER (tenths)														
Sunrise - Sunset	55	7.0	6.9	6.9	6.9	6.9	6.6	6.3	6.2	6.1	6.3	7.5	7.4	6.8
MEAN NUMBER OF DAYS:														
Sunrise to Sunset														
-Clear	55	5.4	5.6	6.1	5.6	5.1	5.2	5.7	6.7	7.6	7.5	3.6	4.8	69.0
-Partly Cloudy	55	8.1	7.2	7.8	8.0	9.3	11.4	13.2	11.8	9.9	9.4	8.1	6.9	111.1
-Cloudy	55	17.5	15.5	17.1	16.4	16.5	13.5	12.1	12.6	12.5	14.1	18.4	19.3	185.2
Precipitation														
.01 inches or more	47	12.2	10.6	11.9	12.2	13.0	11.1	10.2	10.4	9.6	9.0	11.9	12.3	134.5
Snow,Ice pellets,hail														
1.0 inches or more	47	3.9	3.3	2.5	0.7	0.*	0.0	0.0	0.0	0.0	0.*	1.0	4.0	15.4
Thunderstorms	55	0.1	0.1	0.5	1.3	3.3	5.3	6.3	4.7	2.3	0.9	0.3	0.1	25.4
Heavy Fog Visibility														
1/4 mile or less	55	1.2	1.0	1.3	0.8	1.3	1.2	1.4	2.5	3.7	4.1	1.6	1.7	21.7
Temperature °F														
-Maximum														
90° and above	28	0.0	0.0	0.0	0.2	0.4	1.6	4.3	2.1	0.4	0.0	0.0	0.0	8.9
32° and below	28	16.9	13.0	4.0	0.2	0.0	0.0	0.0	0.0	0.0	0.0	1.1	11.3	46.4
-Minimum														
32° and below	28	29.4	26.1	24.2	12.2	1.6	0.0	0.0	0.0	0.6	8.7	18.3	27.4	148.5
0° and below	28	6.5	4.2	0.5	0.0	0.0	0.0	0.0	0.0	0.0	0.0	0.0	2.3	13.4
AVG. STATION PRESS.(mb)	21	1006.7	1007.3	1005.9	1004.2	1004.6	1004.2	1004.8	1006.6	1007.9	1008.3	1007.6	1007.7	1006.3
RELATIVE HUMIDITY (%)														
Hour 01	28	76	74	72	71	78	83	84	88	88	83	79	78	80
Hour 07 (Local Time)	28	77	76	75	72	76	79	81	86	89	86	81	79	80
Hour 13	28	63	58	53	49	52	56	55	58	59	57	62	65	57
Hour 19	28	70	66	61	55	59	63	64	70	75	72	72	73	67
PRECIPITATION (inches):														
Water Equivalent														
-Normal		2.36	2.27	2.93	2.99	3.41	3.62	3.18	3.47	2.95	2.83	3.23	2.93	36.17
-Maximum Monthly	47	6.44	5.02	5.90	7.95	8.96	7.36	6.96	7.33	7.89	8.83	8.07	6.73	8.96
-Year		1978	1981	1977	1983	1953	1973	1975	1950	1960	1955	1972	1973	MAY 1953
-Minimum Monthly	47	0.42	0.24	0.26	1.14	1.05	0.65	0.49	0.73	0.40	0.20	0.91	0.64	0.20
-Year		1980	1987	1981	1963	1980	1964	1968	1947	1964	1963	1978	1958	OCT 1963
-Maximum in 24 hrs	47	1.91	1.74	2.38	2.20	2.17	3.48	3.42	4.52	3.66	3.31	2.26	4.02	4.52
-Year		1978	1990	1986	1968	1968	1952	1993	1971	1960	1987	1991	1948	AUG 1971
Snow,Ice pellets,hail														
-Maximum Monthly	47	47.8	34.5	34.7	17.7	1.6	T	T	0.0	T	6.5	24.6	57.5	57.5
-Year		1987	1962	1956	1982	1977	1991	1989		1989	1987	1972	1969	DEC 1969
-Maximum in 24 hrs	47	21.2	17.9	26.6	17.5	1.6	T	T	0.0	T	6.5	21.9	18.3	26.6
-Year		1983	1958	1993	1982	1977	1991	1989		1989	1987	1971	1966	MAR 1993
WIND:														
Mean Speed (mph)	55	9.8	10.2	10.6	10.4	9.0	8.3	7.5	7.0	7.4	8.1	9.1	9.3	8.9
Prevailing Direction through 1963		WNW	WNW	WNW	WNW	S	S	S	S	S	S	S	S	S
Fastest Obs. 1 Min.														
-Direction (!!)	10	28	27	28	30	30	31	30	30	18	32	36	31	31
-Speed (MPH)	10	36	33	38	33	30	35	32	36	30	29	35	41	41
-Year		1985	1985	1984	1989	1993	1991	1993	1988	1989	1992	1989	1991	DEC 1991
Peak Gust														
-Direction (!!)	10	W	W	W	W	NW	NW	W	NW	S	W	W	NW	NW
-Speed (mph)	10	58	56	53	47	49	59	52	55	47	56	58	56	59
-Date		1992	1985	1993	1992	1992	1991	1993	1990	1989	1992	1988	1985	JUN 1991

PRECIPITATION (inches) — ALBANY, NEW YORK

YEAR	JAN	FEB	MAR	APR	MAY	JUNE	JULY	AUG	SEP	OCT	NOV	DEC	ANNUAL
1964	3.35	1.63	2.93	2.17	1.31	0.65	1.29	2.55	0.40	0.54	1.45	3.28	21.55
1965	1.95	1.92	1.73	2.38	1.22	1.91	3.52	4.32	3.76	2.37	1.89	0.97	27.94
1966	2.29	2.71	3.63	1.46	2.35	2.95	3.88	1.44	5.61	2.22	1.79	3.04	33.37
1967	1.22	1.76	2.56	3.69	3.36	2.85	3.38	2.17	2.23	3.48	2.68	3.90	33.28
1968	1.48	0.36	2.62	2.64	4.79	4.38	0.49	1.77	1.49	2.18	5.48	4.60	32.28
1969	2.13	1.66	1.32	3.51	2.64	5.30	5.08	2.18	2.06	1.55	5.56	6.51	39.50
1970	0.81	1.98	2.87	3.01	1.78	3.14	1.93	3.35	3.79	2.49	1.48	3.89	30.52
1971	1.78	4.10	3.11	2.00	3.48	2.81	3.89	7.04	2.40	2.09	3.78	3.09	39.57
1972	1.21	3.04	4.05	3.63	5.98	6.84	3.10	1.48	1.99	3.60	8.07	4.19	47.18
1973	2.16	1.34	1.99	4.47	5.45	7.36	1.68	2.89	1.33	2.07	1.27	6.73	38.74
1974	2.04	2.12	3.10	2.80	3.47	3.31	4.84	3.53	5.37	1.49	3.83	2.57	38.47
1975	2.75	3.58	2.72	2.18	2.96	3.80	6.96	5.98	4.57	5.88	2.89	2.78	47.05
1976	3.78	2.60	3.57	3.63	4.89	5.37	2.60	5.04	2.61	5.65	1.41	1.39	42.54
1977	1.51	2.63	5.90	3.41	2.29	2.87	2.31	3.66	6.66	4.00	4.85	4.21	44.30
1978	6.44	0.88	1.99	1.68	1.96	4.60	4.04	3.06	1.87	2.95	0.91	3.08	33.46
1979	6.37	1.71	1.83	3.89	4.13	1.94	2.78	2.67	4.05	3.42	3.41	0.94	37.14
1980	0.42	0.89	4.44	3.02	1.05	4.90	2.69	6.45	2.24	2.27	2.99	1.23	32.59
1981	0.59	5.02	0.26	1.99	2.44	2.78	3.50	1.76	3.45	3.55	1.56	3.54	30.44
1982	3.18	2.14	3.23	2.46	2.60	6.48	2.43	2.01	1.42	0.99	3.80	1.33	32.07
1983	3.73	2.03	5.33	7.95	6.26	1.95	1.34	3.41	2.28	2.18	4.73	5.10	46.29
1984	1.28	2.98	3.04	4.29	7.92	1.74	3.97	3.25	1.53	2.50	2.15	2.48	37.13
1985	0.81	1.18	3.67	1.44	2.71	4.12	1.86	2.23	3.07	1.81	5.00	2.05	29.95
1986	3.17	3.00	3.72	1.49	3.11	5.43	6.68	4.09	2.61	2.12	4.62	3.92	43.96
1987	4.23	0.24	1.99	4.25	1.57	3.54	2.50	3.67	6.98	6;90	1.78	1.64	39.29
1988	1.95	3.00	1.62	2.22	2.95	1.42	3.12	4.77	1.50	1.40	4.58	1.02	29.55
1989	0.46	1.60	2.69	2.68	5.92	6.52	5.91	2.90	2.81	5.53	1.90	0.75	39.67
1990	3.84	3.94	3.66	3.87	6.12	2.66	1.68	6.66	1.81	4.60	3.67	3.50	46.01
1991	2.15	1.67	2.53	4.14	2.74	1.69	1.65	4.32	3.33	3.82	4.76	2.92	35.72
1992	1.86	1.30	1.66	2.77	3.61	1.96	4.26	2.05	2.43	2.80	3.66	3.02	31.38
1993	2.14	2.86	5.12	5.39	1.37	2.87	6.55	1.54	3.22	3.31	3.80	3.08	41.25
Record Mean	2.47	2.35	2.77	2.81	3.34	3.67	3.69	3.52	3.27	3.07	2.96	2.64	36.57

AVERAGE TEMPERATURE (deg. F) — ALBANY, NEW YORK

YEAR	JAN	FEB	MAR	APR	MAY	JUNE	JULY	AUG	SEP	OCT	NOV	DEC	ANNUAL
#1964	23.9	22.3	34.9	45.7	61.7	66.5	74.4	66.2	61.0	49.2	41.4	27.9	47.9
1965	18.1	22.3	31.2	42.2	59.6	66.9	68.9	69.4	63.6	51.2	37.6	30.8	46.8
1966	21.5	23.3	34.3	44.0	53.9	67.4	72.2	69.2	58.0	48.5	42.3	27.3	46.8
1967	27.0	18.0	29.0	43.5	50.4	69.9	71.6	69.3	61.3	51.0	34.8	28.9	46.2
1968	14.7	21.1	37.1	51.1	54.9	66.7	72.7	68.6	63.7	53.3	38.5	23.5	47.2
1969	20.9	24.7	31.1	47.6	56.3	66.0	69.7	70.6	62.4	49.0	39.7	21.6	46.6
1970	9.7	23.1	32.0	48.7	60.5	65.9	72.0	69.6	63.3	52.9	41.9	21.6	46.8
1971	13.9	25.4	30.6	42.3	54.9	66.3	68.4	66.8	64.8	54.7	36.9	30.0	46.3
1972	22.9	21.1	30.5	41.2	59.5	63.6	70.9	67.2	60.7	45.7	35.1	28.9	45.6
1973	27.0	22.0	41.9	48.8	55.3	68.7	72.8	72.9	60.5	51.0	39.9	28.2	49.1
1974	23.3	21.3	32.4	48.1	54.1	65.0	69.3	67.9	58.3	44.4	38.6	28.9	46.0
1975	25.7	24.9	30.8	40.7	61.9	65.1	72.8	70.0	59.4	53.3	45.5	26.1	48.0
1976	16.0	31.5	36.7	49.7	55.0	69.4	68.5	67.4	59.0	46.5	34.9	21.4	46.3
1977	15.5	24.5	40.0	46.8	60.2	64.6	71.7	67.8	61.4	49.7	42.6	26.7	47.6
1978	21.5	18.2	30.8	43.4	58.4	64.4	68.9	69.2	56.8	48.6	38.6	28.7	45.6
1979	22.1	14.4	38.9	45.4	60.0	66.0	72.5	69.0	61.2	50.2	44.1	31.4	47.9
1980	24.1	19.8	33.3	48.0	59.5	63.3	72.2	70.7	62.6	47.4	34.8	19.9	46.3
1981	14.0	33.1	34.7	48.1	58.9	66.7	69.3	68.5	58.8	44.8	37.7	25.7	46.7
1982	14.3	23.4	32.8	44.3	59.5	62.9	70.1	65.5	60.5	50.6	43.0	33.7	46.7
1983	24.3	26.8	37.6	46.7	54.9	67.2	72.2	69.8	62.6	49.6	39.2	24.0	47.9
1984	18.1	32.4	29.0	47.6	53.2	66.4	68.9	71.8	60.2	53.8	40.3	33.8	48.0
1985	19.9	26.8	37.3	49.7	60.0	62.2	70.7	68.7	63.3	50.2	40.1	24.5	47.8
1986	23.0	22.8	37.2	50.5	61.3	64.6	71.3	67.8	60.1	48.9	35.7	30.8	47.8
1987	21.7	21.7	37.7	50.4	60.0	68.3	73.5	67.2	60.6	46.6	40.1	30.7	48.2
1988	20.6	24.1	34.2	46.6	59.5	65.1	75.0	72.3	60.0	46.0	41.0	26.6	47.6
1989	27.8	24.2	33.5	44.6	59.5	68.0	71.6	69.8	62.5	51.5	39.3	13.7	47.2
1990	32.8	28.2	37.8	48.9	55.3	67.3	73.0	70.9	61.7	53.1	41.8	33.6	50.4
1991	23.2	30.0	37.4	51.2	63.2	69.0	71.6	71.2	59.9	53.2	40.1	28.9	49.9
1992	24.5	26.9	31.5	44.7	58.5	65.2	67.6	67.4	61.4	46.5	38.9	29.8	46.9
1993	26.6	18.3	31.4	48.4	59.5	66.3	73.1	71.7	60.5	48.6	38.4	27.4	47.5
Record Mean	22.8	23.9	33.7	46.6	58.5	67.4	72.3	70.1	62.4	51.1	39.6	27.6	48.0
Max	31.1	32.6	42.3	56.5	69.3	78.0	82.8	80.4	72.7	60.9	47.3	35.1	57.4
Min	14.4	15.2	25.1	36.7	47.7	56.8	61.7	59.7	52.1	41.3	31.9	20.1	38.5

REFERENCE NOTES — ALBANY, NEW YORK

GENERAL
T - TRACE AMOUNT.
BLANK ENTRIES DENOTE MISSING/UNREPORTED DATA.
# INDICATES A STATION OR INSTRUMENT RELOCATION.

SPECIFIC

PM - INCLUDES LAST DAY OF PREVIOUS MONTH
ASOS - AUTOMATED SURFACE OBSERVING SYSTEM IN OPERATION DURING THESE MONTHS.

(a) - LENGTH OF RECORD IN YEARS, ALTHOUGH INDIVIDUAL MONTHS MAY BE MISSING.
0.* OR * - THE VALUE IS BETWEEN 0.0 AND 0.05
NORMALS - BASED ON THE 1961-1990 RECORD PERIOD.
EXTREMES - DATES ARE THE MOST RECENT OCCURRENCE
WIND DIR.- NUMERALS SHOW TENS OF DEGRESS CLOCKWISE FROM TRUE NORTH. "00" INDICATES CALM.
RESULTANT DIRECTIONS ARE GIVEN TO WHOLE DEGREES.
BOLD VALUES INDICATE EXTREME VALUES WHICH OCCURRED AFTER THE ASOS SYSTEM WAS COMMISSIONED.

RECORD = PERIOD OF RECORD
RECORD MEAN PRECIPITATION IS THE MEAN OF ALL DAILY PRECIPITATION AMOUNTS DURING THE PERIOD OF RECORD.
RECORD MAX(MIN) TEMPERATURE IS THE MEAN OF ALL DAILY MAX(MIN) TEMPERATURES DURING THE PERIOD OF RECORD.
RECORD MEAN TEMPERATURE IS THE SUM OF THE RECORD MAX AND RECORD MIN DIVIDED BY 2.
AVERAGE TEMPERATURE IS THE SUM OF THE MEAN DAILY MAX AND MIN TEMPERATURE DIVIDED BY 2.

EXCEPTIONS

RECORD MEANS ARE THROUGH THE CURRENT YEAR,
BEGINNING IN 1874 FOR TEMPERATURE
1826 FOR PRECIPITATION
1947 FOR SNOWFALL

HEATING DEGREE DAYS Base 65 deg. F ALBANY, NEW YORK

SEASON	JULY	AUG	SEP	OCT	NOV	DEC	JAN	FEB	MAR	APR	MAY	JUNE	TOTAL
#1964-65	1	48	169	484	702	1141	1449	1193	1041	679	197	68	7172
1965-66	11	49	120	421	817	1051	1342	1162	948	623	347	57	6948
1966-67	3	5	216	502	673	1163	1169	1312	1111	639	447	11	7251
1967-68	0	19	153	429	899	1112	1557	1269	857	412	304	46	7057
1968-69	7	45	76	359	787	1281	1360	1122	1043	518	284	55	6937
1969-70	13	22	137	491	749	1339	1708	1168	1016	495	165	75	7378
1970-71	3	7	127	377	686	1336	1580	1104	1059	672	315	50	7316
1971-72	20	45	109	311	838	1080	1298	1269	1060	707	175	97	7009
1972-73	16	38	154	590	890	1113	1168	1198	709	486	299	47	6708
1973-74	2	3	200	431	750	1136	1285	1216	1005	511	343	54	6936
1974-75	17	14	227	631	786	1113	1212	1115	1053	722	145	88	7123
1975-76	0	19	173	357	580	1199	1511	964	871	472	315	43	6504
1976-77	7	40	196	564	895	1345	1526	1127	764	545	205	85	7299
1977-78	7	51	156	471	666	1179	1340	1306	1051	642	245	84	7198
1978-79	43	19	256	503	784	1119	1324	1414	803	579	188	63	7095
1979-80	19	37	163	468	619	1036	1259	1303	974	503	190	106	6677
1980-81	0	7	140	539	900	1393	1575	885	930	502	235	30	7136
1981-82	8	22	204	622	816	1209	1564	1160	992	617	182	87	7483
1982-83	20	65	156	436	657	969	1255	1062	843	539	312	58	6372
1983-84	5	24	150	479	766	1265	1448	939	1109	517	363	60	7125
1984-85	12	8	170	344	737	959	1389	1062	852	458	184	106	6281
1985-86	7	16	123	452	740	1246	1295	1177	859	432	154	75	6576
1986-87	17	46	173	495	872	1053	1332	1207	842	433	210	29	6709
1987-88	2	56	154	567	741	1056	1370	1181	946	546	198	99	6916
1988-89	8	30	160	584	714	1185	1146	1133	968	607	194	35	6764
1989-90	0	22	134	413	766	1584	990	1026	839	500	298	44	6616
1990-91	5	6	148	388	689	964	1290	973	850	417	141	22	5893
1991-92	6	0	197	372	740	1111	1248	1098	1034	605	210	56	6677
1992-93	17	27	167	565	773	1082	1183	1300	1034	492	185	67	6892
1993-94	0	11	185	500	791	1161							

COOLING DEGREE DAYS Base 65 deg. F ALBANY, NEW YORK

YEAR	JAN	FEB	MAR	APR	MAY	JUNE	JULY	AUG	SEP	OCT	NOV	DEC	TOTAL
1969	0	0	0	2	23	95	165	203	68	0	0	0	556
1970	0	0	0	12	36	107	225	160	83	7	0	0	630
1971	0	0	0	0	9	98	132	107	109	1	0	0	456
1972	0	0	0	0	12	58	208	112	31	0	0	0	421
1973	0	0	0	7	6	164	248	255	71	2	0	0	753
1974	0	0	0	11	12	59	157	111	35	0	1	0	386
1975	0	0	0	0	58	97	248	180	12	0	2	0	597
1976	0	0	0	19	11	184	120	120	22	0	0	0	476
1977	0	0	0	8	66	79	222	146	53	0	0	0	574
1978	0	0	0	0	47	70	169	154	16	0	0	0	456
1979	0	0	0	0	39	99	258	168	55	17	0	0	636
1980	0	0	0	0	28	63	230	189	73	0	0	0	583
1981	0	0	0	2	53	87	149	137	25	0	0	0	453
1982	0	0	0	0	19	31	184	88	29	0	4	0	355
1983	0	0	0	0	8	134	236	179	86	6	0	0	649
1984	0	0	0	0	3	107	140	226	35	3	0	0	514
1985	0	0	0	5	37	27	191	140	80	2	0	0	482
1986	0	0	6	4	46	69	220	140	33	1	0	0	519
1987	0	0	0	4	62	136	271	133	29	0	0	0	635
1988	0	0	0	0	36	110	326	263	16	4	0	0	755
1989	0	0	1	0	31	132	213	178	63	0	0	0	618
1990	0	0	2	22	1	119	261	197	55	24	0	0	681
1991	0	0	0	9	92	147	221	198	50	14	0	0	731
1992	0	0	0	2	15	70	106	112	65	0	0	0	370
1993	0	0	0	0	17	116	259	224	55	2	0	0	673

SNOWFALL (inches) ALBANY, NEW YORK

SEASON	JULY	AUG	SEP	OCT	NOV	DEC	JAN	FEB	MAR	APR	MAY	JUNE	TOTAL
1964-65	0.0	0.0	0.0	T	T	11.2	20.4	3.7	8.4	2.1	0.0	0.0	45.8
1965-66	0.0	0.0	0.0	T	0.5	2.7	28.8	24.5	9.2	T	1.4	0.0	67.1
1966-67	0.0	0.0	0.0	0.0	T	29.4	5.7	16.3	26.2	3.1	0.2	0.0	80.9
1967-68	0.0	0.0	0.0	T	9.0	17.8	8.0	1.8	5.6	0.0	0.0	0.0	42.2
1968-69	0.0	0.0	0.0	0.0	13.5	18.1	6.3	20.7	4.5	0.2	0.0	0.0	63.3
1969-70	0.0	0.0	0.0	T	3.2	57.5	7.2	7.4	11.2	1.2	T	0.0	87.7
1970-71	0.0	0.0	0.0	T	T	43.8	15.2	17.6	32.0	3.9	0.0	0.0	112.5
1971-72	0.0	0.0	0.0	0.0	24.0	10.1	8.5	24.8	15.9	6.0	0.0	0.0	89.3
1972-73	0.0	0.0	0.0	T	24.6	22.5	11.2	12.5	T	0.1	0.0	0.0	70.9
1973-74	0.0	0.0	0.0	0.0	0.1	18.9	10.0	12.4	5.6	11.3	0.0	0.0	58.3
1974-75	0.0	0.0	0.0	T	2.2	12.5	14.0	21.2	2.9	1.8	0.0	0.0	54.6
1975-76	0.0	0.0	0.0	0.0	3.6	16.4	15.0	4.4	14.8	T	T	0.0	54.2
1976-77	0.0	0.0	0.0	T	5.7	7.8	22.1	17.9	15.2	0.3	1.6	0.0	70.6
1977-78	0.0	0.0	0.0	0.0	8.4	19.8	40.8	15.8	7.4	0.2	T	0.0	92.4
1978-79	0.0	0.0	0.0	0.0	3.4	19.9	26.5	4.6	0.9	8.2	0.0	0.0	63.5
1979-80	0.0	0.0	0.0	T	0.0	5.8	0.6	10.2	10.8	0.0	0.0	0.0	27.4
1980-81	0.0	0.0	0.0	0.0	11.8	12.8	11.9	6.9	1.5	T	0.0	0.0	44.9
1981-82	0.0	0.0	0.0	0.0	1.1	31.4	18.2	9.6	19.1	17.7	0.0	0.0	97.1
1982-83	0.0	0.0	0.0	0.0	0.6	5.5	27.5	17.4	9.2	14.7	0.1	0.0	75.0
1983-84	0.0	0.0	0.0	0.0	1.7	11.6	16.5	7.2	28.2	T	0.0	0.0	65.2
1984-85	0.0	0.0	0.0	0.0	2.2	11.7	8.4	10.1	8.7	0.2	0.0	0.0	41.3
1985-86	0.0	0.0	0.0	0.0	11.8	11.5	18.0	16.1	3.4	1.7	T	0.0	62.5
1986-87	0.0	0.0	0.0	0.0	8.3	20.3	47.8	2.8	0.8	0.6	0.0	0.0	80.6
1987-88	0.0	0.0	0.0	6.5	6.2	11.4	21.7	26.0	4.8	0.1	0.0	0.0	76.7
1988-89	0.0	0.0	0.0	T	T	7.8	1.3	5.1	4.7	0.1	0.0	0.0	19.0
1989-90	T	0.0	T	0.0	1.9	8.0	20.3	22.8	4.9	T	0.0	0.0	57.9
1990-91	0.0	0.0	0.0	T	0.4	8.5	11.2	5.3	3.3	0.0	0.0	T	28.7
1991-92	0.0	0.0	0.0	0.0	1.5	12.7	3.4	6.3	4.9	1.9	0.0	0.0	30.7
1992-93	0.0	0.0	0.0	T	2.8	12.6	14.3	28.6	34.3	1.6	0.0	0.0	94.2
1993-94	0.0	0.0	0.0	T	0.7	6.1							
Record Mean	T	0.0	T	0.2	4.1	14.9	16.0	14.2	11.1	2.7	0.1	T	63.2

BINGHAMTON, NEW YORK

Binghamton, in south central New York lies in a comparatively narrow valley at the confluence of the Susquehanna and Chenango Rivers. Within a radius of 5 miles, hills rise to elevations of 1,400–1,600 feet above mean sea level. In the spring, melting snow, sometimes supplemented by rainfall, occasionally causes flooding in the city and along the streams. Less frequently, heavy rains in the warmer months produce some flooding.

The climate of Binghamton is representative of the humid area of the north-eastern United States and is primarily continental in type. The area, being adjacent to the so-called St. Lawrence Valley storm track, and also subject to cold air masses approaching from the west and north, has a variable climate, characterized by frequent and rapid changes. Furthermore, diurnal and seasonal changes assist in the production of an invigorating climate. In the warmer months, it is seldom that either high temperatures or humidity become depressing to humans. As a rule, the temperature rises rapidly to moderate daytime levels with readings of 90 degrees or above only a few days in any month. Summer nights are sufficiently cool to provide favorable sleeping conditions and relief from the heat of the day.

Winters are usually cold, but not commonly severe. Highest daytime temperatures average in the high 20s to low 30s, while the lowest nighttime readings average from the mid-teens to low 20s. Ordinarily a few sub-zero readings may be expected in January and February, with a lesser number in November, December, and March. The transitional seasons, spring and autumn, are the most variable of the year.

Most of the precipitation in the Binghamton area derives from moisture laden air transported from the Gulf of Mexico and cyclonic systems moving northward along the Atlantic coast. The annual rainfall is rather evenly distributed over the year. However, the greatest average monthly amounts occur during the growing season, April through September. As a rule, rainfall is ample for good crop growth and comes mostly in the form of thunderstorms. Ordinarily, the requirements for water supplies are adequately met by the precipitation that is received. Annual snowfall is around 50 inches in Binghamton and above 85 inches at Edwin A. Link Field, some 10 miles to the NNW, and about 700 feet higher in elevation. Most of the snow falls during the normal winter months. However, heavy snows can occur as early as November and as late as April. Being adjacent to the track of storms that move through the St. Lawrence Valley, and being under the influence of winds that sweep across Lakes Erie and Ontario to the interior of the state, the area is subject to much cloudiness and winter snow flurries.

Furthermore, the combination of a valley location and surrounding hills produces numerous advection fogs which also reduce the amount of sunshine received.

For the most part, the winds at Binghamton have northerly and westerly components. Tornadoes, although rare, have struck in the Binghamton area.

The growing season averages 150 to 160 days. Usually the last spring frost occurs during early May, and the first frost in autumn during early October.

NORMALS, MEANS, AND EXTREMES

BINGHAMTON, NEW YORK

LATITUDE: 42°13'N LONGITUDE: 75°59' W ELEVATION: FT. GRND 1572 BARO 1618 TIME ZONE: EASTERN WBAN: 04725

	(a)	JAN	FEB	MAR	APR	MAY	JUNE	JULY	AUG	SEP	OCT	NOV	DEC	YEAR
TEMPERATURE °F:														
Normals														
-Daily Maximum		27.9	30.1	40.4	53.5	65.4	73.8	78.6	76.4	68.6	57.3	44.9	32.7	54.1
-Daily Minimum		14.3	15.3	24.7	35.2	46.2	54.6	59.7	57.9	50.6	40.3	31.6	20.3	37.6
-Monthly		21.1	22.7	32.6	44.4	55.8	64.2	69.2	67.2	59.6	48.8	38.3	26.5	45.9
Extremes														
-Record Highest	42	63	66	82	88	89	94	98	94	96	82	77	65	98
-Year		1967	1954	1977	1990	1982	1952	1988	1985	1953	1963	1982	1984	JUL 1988
-Record Lowest	42	-20	-15	-7	9	25	33	39	37	25	17	3	-18	-20
-Year		1957	1979	1993	1982	1978	1980	1963	1965	1974	1976	1976	1980	JAN 1957
NORMAL DEGREE DAYS:														
Heating (base 65°F)		1361	1184	1004	618	305	73	16	35	180	502	801	1194	7273
Cooling (base 65°F)		0	0	0	0	20	49	147	103	18	0	0	0	337
% OF POSSIBLE SUNSHINE	42	37	42	46	50	56	62	64	61	55	49	32	29	49
MEAN SKY COVER (tenths)														
Sunrise - Sunset	42	8.0	7.8	7.6	7.2	7.0	6.6	6.4	6.5	6.6	6.8	8.1	8.3	7.3
MEAN NUMBER OF DAYS:														
Sunrise to Sunset														
-Clear	42	2.5	3.0	4.0	5.0	5.0	4.9	5.7	5.3	5.5	6.0	2.6	2.0	51.6
-Partly Cloudy	42	7.0	6.2	7.3	7.0	8.8	11.2	12.5	11.5	10.0	8.3	5.9	6.3	102.0
-Cloudy	42	21.5	19.1	19.7	18.0	17.2	13.9	12.8	14.2	14.5	16.7	21.5	22.7	211.7
Precipitation														
.01 inches or more	42	16.6	14.5	15.3	13.7	13.3	12.2	10.8	10.9	10.4	11.8	15.0	17.4	161.8
Snow,Ice pellets,hail														
1.0 inches or more	42	5.7	4.9	3.6	1.2	0.1	0.0	0.0	0.0	0.0	0.1	2.2	5.6	23.5
Thunderstorms	42	0.1	0.1	1.1	2.0	3.6	6.4	6.4	5.1	2.8	1.1	0.3	0.1	29.1
Heavy Fog Visibility														
1/4 mile or less	42	3.3	3.2	5.1	4.0	4.0	4.3	4.6	5.5	5.7	4.4	4.4	4.5	53.2
Temperature °F														
-Maximum														
90° and above	42	0.0	0.0	0.0	0.0	0.0	0.3	1.2	0.6	0.2	0.0	0.0	0.0	2.2
32° and below	42	19.9	16.0	8.5	0.7	0.0	0.0	0.0	0.0	0.0	0.1	3.4	15.4	64.0
-Minimum														
32° and below	42	29.7	26.5	25.0	12.0	1.3	0.0	0.0	0.0	0.3	6.1	17.5	27.4	145.9
0° and below	42	3.5	2.4	0.3	0.0	0.0	0.0	0.0	0.0	0.0	0.0	0.0	1.4	7.6
AVG. STATION PRESS.(mb)	21	955.9	956.6	956.0	955.5	956.9	957.4	958.6	960.0	960.3	959.8	958.3	957.1	957.7
RELATIVE HUMIDITY (%)														
Hour 01	42	78	77	75	72	75	80	81	84	86	80	79	80	79
Hour 07 (Local Time)	42	80	79	79	76	78	82	84	89	90	85	82	82	82
Hour 13	42	70	67	62	56	56	58	57	60	63	62	69	73	63
Hour 19	42	74	71	67	61	61	65	65	70	73	70	74	77	69
PRECIPITATION (inches):														
Water Equivalent														
-Normal		2.40	2.33	2.82	3.13	3.36	3.60	3.50	3.36	3.32	2.89	3.28	3.00	36.99
-Maximum Monthly	42	6.39	4.36	6.00	8.57	6.46	9.46	7.40	7.48	9.66	9.43	7.52	6.11	9.66
-Year		1979	1971	1980	1983	1968	1960	1956	1959	1977	1955	1972	1983	SEP 1977
-Minimum Monthly	42	0.76	0.51	0.69	0.98	0.78	0.98	0.83	0.61	0.61	0.26	1.01	0.94	0.26
-Year		1970	1968	1981	1985	1962	1979	1955	1953	1961	1963	1960	1960	OCT 1963
-Maximum in 24 hrs	42	1.80	2.16	1.95	2.86	3.58	3.19	3.24	3.29	3.57	3.88	2.66	2.81	3.88
-Year		1958	1966	1964	1980	1992	1972	1976	1988	1985	1955	1972	1983	OCT 1955
Snow,Ice pellets,hail														
-Maximum Monthly	42	43.6	44.3	37.9	22.9	3.4	T	0.0	0.0	T	11.6	24.4	59.6	59.6
-Year		1987	1972	1993	1983	1966	1993			1993	1993	1954	1969	DEC 1969
-Maximum in 24 hrs	42	18.4	23.0	19.1	11.5	3.4	T	0.0	0.0	T	6.8	11.2	15.6	23.0
-Year		1964	1961	1993	1960	1966	1993			1993	1993	1993	1969	FEB 1961
WIND:														
Mean Speed (mph)	42	11.5	11.6	11.7	11.4	10.0	9.3	8.4	8.2	8.8	9.8	10.9	11.2	10.2
Prevailing Direction through 1963		WSW	SSE	NW	WNW	NNW	NNW	WSW	SSW	SSW	WSW	NNW	WSW	WSW
Fastest Obs. 1 Min.														
-Direction (!!)	8	28	E	34	24	33	26	30	27	27	31	30	29	26
-Speed (MPH)	8	32	33	32	32	32	40	35	31	32	31	32	33	40
-Year		1992	1985	1993	1991	1988	1991	1988	1988	1992	1990	1989	1991	JUN 1991
Peak Gust														
-Direction (!!)	10	W	E	N	W	NW	NW	NW	S	SW	S	NW	W	NW
-Speed (mph)	10	51	52	52	52	54	59	74	51	48	46	58	48	74
-Date		1992	1985	1993	1985	1988	1988	1992	1989	1989	1990	1989	1991	JUL 1992

PRECIPITATION (inches)

BINGHAMTON, NEW YORK

YEAR	JAN	FEB	MAR	APR	MAY	JUNE	JULY	AUG	SEP	OCT	NOV	DEC	ANNUAL
1964	3.00	2.00	4.56	5.09	2.01	1.22	4.80	1.85	0.66	1.06	1.90	3.18	31.33
1965	4.15	1.55	2.23	2.88	1.73	1.90	2.23	4.25	3.72	3.02	2.30	1.69	31.65
1966	3.19	3.62	2.86	2.51	2.56	2.90	1.45	1.41	3.26	1.40	2.87	2.75	30.78
1967	1.60	1.60	2.78	2.09	5.04	2.90	3.45	4.92	3.11	3.29	4.45	2.49	37.72
1968	2.26	0.51	3.25	1.61	6.46	6.96	1.66	2.26	5.49	3.14	5.62	3.17	42.39
1969	2.00	0.97	0.69	2.78	1.60	4.00	4.32	1.96	1.84	2.25	3.71	4.85	30.97
1970	0.76	2.22	2.41	3.58	3.03	1.15	4.50	3.97	3.85	2.07	2.45	3.58	33.57
1971	1.68	4.36	2.77	2.02	3.31	1.73	4.60	2.10	1.66	1.89	3.17	4.16	33.45
1972	1.29	3.74	3.79	2.83	5.17	9.18	1.68	3.79	2.03	2.17	7.52	4.85	48.04
1973	1.60	1.95	2.01	3.74	3.29	2.93	1.93	2.40	3.12	2.72	1.85	5.81	33.35
1974	2.20	1.70	4.05	2.02	3.19	3.48	4.23	1.69	2.98	0.98	3.51	3.05	33.08
1975	2.55	3.97	2.16	1.77	4.42	2.90	6.13	5.33	8.41	3.48	2.49	3.22	46.83
1976	3.69	2.88	2.78	2.69	2.53	4.42	6.40	6.79	3.85	6.30	1.12	1.71	45.16
1977	1.68	1.54	5.11	2.73	1.72	3.17	3.27	2.95	9.66	4.76	5.10	4.84	46.53
1978	6.06	1.26	2.36	1.92	2.55	3.85	2.54	4.61	1.16	3.57	1.29	3.16	34.33
1979	6.39	1.67	2.73	3.13	4.26	0.98	1.45	2.44	5.70	2.46	3.70	1.83	36.74
1980	1.08	1.08	6.00	5.48	1.54	5.68	2.09	1.58	2.81	2.86	2.96	1.60	34.76
1981	0.89	3.88	0.69	3.18	1.94	3.42	1.99	1.99	3.40	4.72	1.67	2.49	30.26
1982	3.40	2.26	2.61	2.29	3.89	7.09	1.87	2.94	1.86	0.93	4.04	1.90	35.08
1983	2.56	1.50	2.57	8.57	4.05	4.08	2.20	3.21	1.53	2.61	3.58	6.11	42.57
1984	1.59	3.34	2.19	5.07	6.09	2.65	5.44	3.07	1.92	1.58	3.55	3.15	39.64
1985	1.30	1.30	3.63	0.98	2.69	2.61	4.14	2.72	4.76	2.47	4.63	2.19	33.42
1986	2.13	4.00	3.01	2.99	3.22	4.80	7.36	3.01	3.27	2.45	5.75	2.48	44.47
1987	3.04	0.67	1.91	4.20	1.29	3.82	4.35	4.17	4.54	2.66	1.79	2.18	34.62
1988	1.57	3.77	1.61	2.76	3.89	1.05	4.94	5.22	2.05	2.58	2.93	1.19	33.56
1989	1.50	1.95	4.15	1.37	5.82	5.89	3.48	3.40	4.29	3.35	1.95	1.61	38.76
1990	3.33	3.23	1.72	3.11	5.00	2.50	3.07	5.57	2.83	7.19	3.20	5.22	45.97
1991	2.11	2.13	3.42	4.22	1.90	2.36	1.96	3.78	2.61	1.65	4.50	2.74	33.38
1992	1.68	1.95	3.29	3.51	6.37	1.30	6.32	2.87	3.52	3.47	3.78	2.97	41.03
1993	2.11	2.71	3.68	7.10	1.99	3.99	2.30	4.28	3.92	3.41	3.59	3.30	42.38
Record Mean	2.40	2.36	2.85	3.31	3.38	3.55	3.59	3.46	3.26	3.04	3.18	2.90	37.28

AVERAGE TEMPERATURE (deg. F)

BINGHAMTON, NEW YORK

YEAR	JAN	FEB	MAR	APR	MAY	JUNE	JULY	AUG	SEP	OCT	NOV	DEC	ANNUAL
1964	24.0	18.9	31.9	43.0	58.4	62.0	70.3	65.2	61.2	48.3	42.9	28.0	46.2
1965	19.8	22.7	28.7	40.3	59.2	63.1	66.7	66.6	60.8	46.8	36.9	30.4	45.2
1966	19.1	22.8	33.2	41.5	50.6	65.9	71.2	69.4	56.8	46.9	41.1	26.5	45.4
1967	29.0	18.5	29.4	43.6	47.6	68.2	67.8	66.2	59.1	48.6	32.9	28.9	45.0
1968	17.1	17.3	34.5	48.4	51.3	62.9	69.4	67.1	61.8	49.9	36.7	23.6	45.0
1969	22.1	23.0	29.9	47.0	56.3	63.8	67.0	67.6	60.1	48.0	37.9	23.5	45.5
1970	16.5	25.0	28.9	45.0	57.0	62.8	67.9	67.2	60.6	50.9	39.2	24.2	45.5
1971	15.6	23.6	28.4	40.7	54.3	66.3	65.9	64.1	62.8	55.4	35.8	30.9	45.3
1972	24.7	20.6	28.2	40.4	58.8	62.2	71.5	65.8	59.9	42.9	33.6	29.4	44.8
1973	25.8	20.1	40.0	45.4	51.5	66.3	70.2	69.7	59.7	52.6	39.8	29.0	47.5
1974	27.3	22.2	32.1	48.7	52.5	63.2	69.7	68.7	56.9	44.9	37.3	27.9	46.0
1975	24.9	24.5	27.2	37.0	62.0	66.5	72.8	67.4	56.5	51.4	44.6	26.3	46.8
1976	16.9	30.3	35.5	47.0	52.8	67.0	65.7	65.5	57.3	43.7	31.1	20.3	44.4
1977	12.0	24.1	38.0	47.4	59.0	61.8	70.0	66.5	60.3	46.7	40.2	24.6	45.9
1978	19.1	15.0	27.6	41.5	58.6	63.7	68.4	69.5	58.3	47.6	39.2	27.9	44.7
1979	21.2	13.6	37.9	42.5	55.0	62.1	69.0	65.7	58.7	48.4	42.6	31.2	45.6
1980	23.5	20.2	31.8	46.5	58.8	62.0	70.4	70.6	62.2	44.9	33.7	21.4	45.5
1981	15.3	30.7	32.3	47.5	57.3	65.2	69.4	67.8	58.4	46.3	38.2	26.5	46.2
1982	14.9	24.4	32.3	42.2	59.7	62.1	70.0	64.4	61.0	50.6	42.1	34.4	46.5
1983	24.7	28.0	35.9	43.2	53.4	66.8	72.3	70.9	63.7	50.6	39.7	23.8	47.8
1984	19.7	33.1	25.9	45.2	53.0	66.4	68.3	70.0	58.1	54.0	38.3	34.2	47.2
1985	18.9	26.1	35.3	49.1	58.9	61.2	68.8	67.4	61.3	50.0	39.9	22.4	46.6
1986	22.7	21.6	35.9	47.3	59.9	62.4	68.5	66.1	60.8	49.3	35.4	31.1	46.8
1987	23.3	23.0	38.5	48.7	57.8	66.4	71.1	66.1	59.1	45.2	39.0	29.4	47.3
1988	20.2	23.0	33.2	43.5	57.7	63.0	73.5	69.8	57.2	43.0	40.1	25.1	45.8
1989	27.2	21.9	31.9	41.5	55.4	65.1	68.6	65.7	59.2	50.5	35.7	14.0	44.7
1990	31.5	28.6	36.7	47.3	52.8	64.4	68.1	66.5	57.9	51.6	40.7	32.0	48.2
1991	23.4	28.4	36.0	49.2	61.8	66.7	70.1	69.2	59.0	51.4	39.2	30.2	48.7
1992	25.7	26.2	30.1	43.4	55.4	62.7	65.6	64.0	58.4	44.3	36.9	28.1	45.1
1993	25.6	18.0	28.5	44.8	56.8	62.9	69.7	68.3	57.3	46.3	37.0	25.4	45.1
Record Mean	21.3	23.4	32.0	44.7	55.8	64.3	69.1	67.3	59.6	48.9	38.3	26.5	46.0
Max	28.4	30.8	39.7	53.8	65.5	74.0	78.6	76.5	68.7	57.5	45.0	32.8	54.3
Min	14.2	16.0	24.2	35.6	46.0	54.6	59.6	58.0	50.5	40.2	31.5	20.2	37.6

REFERENCE NOTES

BINGHAMTON, NEW YORK

GENERAL
T - TRACE AMOUNT.
BLANK ENTRIES DENOTE MISSING/UNREPORTED DATA.
# INDICATES A STATION OR INSTRUMENT RELOCATION.

SPECIFIC

(a) - LENGTH OF RECORD IN YEARS, ALTHOUGH INDIVIDUAL MONTHS MAY BE MISSING.
0.* OR * - THE VALUE IS BETWEEN 0.0 AND 0.05
NORMALS - BASED ON THE 1961-1990 RECORD PERIOD.
EXTREMES - DATES ARE THE MOST RECENT OCCURRENCE.
WIND DIR. - NUMERALS SHOW TENS OF DEGRESS CLOCKWISE FROM TRUE NORTH. "00" INDICATES CALM.
RESULTANT DIRECTIONS ARE GIVEN TO WHOLE DEGREES.
BOLD VALUES INDICATE EXTREME VALUES WHICH OCCURRED AFTER THE ASOS SYSTEM WAS COMMISSIONED.

RECORD = PERIOD OF RECORD
RECORD MEAN PRECIPITATION IS THE MEAN OF ALL DAILY PRECIPITATION AMOUNTS DURING THE PERIOD OF RECORD.
RECORD MAX(MIN) TEMPERATURE IS THE MEAN OF ALL DAILY MAX(MIN) TEMPERATURES DURING THE PERIOD OF RECORD.
RECORD MEAN TEMPERATURE IS THE SUM OF THE RECORD MAX AND RECORD MIN DIVIDED BY 2.
AVERAGE TEMPERATURE IS THE SUM OF THE MEAN DAILY MAX AND MIN TEMPERATURE DIVIDED BY 2.

EXCEPTIONS

RECORD MEANS ARE THROUGH THE CURRENT YEAR,
BEGINNING IN 1952 FOR TEMPERATURE
1952 FOR PRECIPITATION
1952 FOR SNOWFALL

HEATING DEGREE DAYS Base 65 deg. F BINGHAMTON, NEW YORK

SEASON	JULY	AUG	SEP	OCT	NOV	DEC	JAN	FEB	MAR	APR	MAY	JUNE	TOTAL
1964-65	9	61	165	510	657	1142	1392	1176	1119	734	211	140	7316
1965-66	41	76	184	558	836	1063	1417	1173	981	696	447	88	7560
1966-67	8	16	252	552	709	1186	1107	1297	1098	634	531	27	7417
1967-68	26	40	178	505	953	1113	1477	1375	940	493	416	119	7635
1968-69	29	62	106	465	843	1279	1323	1171	1081	536	292	100	7287
1969-70	30	40	195	525	808	1280	1499	1116	1113	599	261	124	7590
1970-71	20	31	179	431	769	1258	1525	1154	1130	723	339	56	7615
1971-72	45	85	144	296	872	1050	1243	1280	1133	732	205	116	7201
1972-73	19	56	173	680	937	1097	1210	1250	770	584	411	47	7234
1973-74	5	27	200	374	749	1109	1160	1192	1016	495	384	90	6801
1974-75	14	5	251	615	825	1144	1237	1127	1164	833	147	65	7427
1975-76	0	39	250	415	604	1192	1487	999	910	554	374	50	6874
1976-77	31	62	241	653	1013	1382	1636	1137	831	537	237	130	7890
1977-78	15	61	166	560	736	1246	1415	1392	1150	699	248	100	7788
1978-79	53	7	208	533	765	1143	1351	1437	832	667	325	127	7448
1979-80	37	73	204	522	669	1040	1278	1294	1022	548	214	144	7045
1980-81	7	11	143	617	929	1349	1537	951	1006	522	266	56	7394
1981-82	19	21	215	573	799	1185	1549	1135	1006	679	178	102	7461
1982-83	19	72	147	439	685	944	1241	1031	897	649	356	68	6548
1983-84	10	13	130	447	750	1271	1399	917	1205	588	373	65	7168
1984-85	19	19	222	341	795	947	1424	1083	914	488	213	129	6594
1985-86	13	31	169	458	747	1315	1304	1209	897	522	191	116	6972
1986-87	29	58	154	480	880	1041	1284	1168	816	486	260	52	6708
1987-88	14	59	185	606	773	1095	1384	1212	981	640	246	142	7337
1988-89	18	45	232	674	739	1231	1164	1199	1019	698	322	69	7410
1989-90	9	62	201	446	875	1576	1031	1013	873	549	373	74	7082
1990-91	35	34	225	415	721	1018	1282	1018	893	483	180	54	6358
1991-92	3	10	220	417	767	1071	1213	1116	1077	638	302	104	6938
1992-93	34	77	215	634	837	1136	1214	1311	1126	599	257	108	7548
1993-94	11	27	249	576	834	1223							

COOLING DEGREE DAYS Base 65 deg. F BINGHAMTON, NEW YORK

YEAR	JAN	FEB	MAR	APR	MAY	JUNE	JULY	AUG	SEP	OCT	NOV	DEC	TOTAL
1969	0	0	0	0	30	71	101	127	55	0	0	0	384
1970	0	0	0	7	22	66	117	109	52	0	0	0	373
1971	0	0	0	0	13	101	82	61	84	5	0	0	346
1972	0	0	0	0	19	36	228	90	27	0	0	0	400
1973	0	0	0	4	0	92	175	180	50	0	0	0	501
1974	0	0	0	12	5	42	168	128	17	0	0	0	372
1975	0	0	0	0	59	119	251	121	1	1	0	0	552
1976	0	0	0	21	3	119	61	82	15	0	0	0	301
1977	0	0	3	14	56	39	174	115	33	0	0	0	434
1978	0	0	0	0	58	68	165	154	14	0	0	0	459
1979	0	0	0	0	23	47	167	98	22	14	0	0	371
1980	0	0	0	0	30	61	183	191	66	1	0	0	532
1981	0	0	0	4	30	68	161	115	22	0	0	0	400
1982	0	0	0	0	20	23	179	60	35	0	8	0	325
1983	0	0	0	1	0	129	240	205	98	7	0	0	680
1984	0	0	0	0	9	111	128	182	21	4	0	0	455
1985	0	0	0	17	31	21	138	116	67	0	0	0	390
1986	0	0	2	0	43	44	144	98	35	0	0	0	366
1987	0	0	0	2	44	101	213	101	15	0	0	0	476
1988	0	0	0	0	26	89	288	201	7	1	0	0	612
1989	0	0	0	0	30	76	128	89	36	3	0	0	362
1990	0	0	2	25	2	60	138	85	21	5	0	0	338
1991	0	0	0	12	86	110	169	149	47	5	0	0	578
1992	0	0	0	0	12	42	59	54	26	0	0	0	193
1993	0	0	0	0	9	54	165	136	25	1	0	0	390

SNOWFALL (inches) BINGHAMTON, NEW YORK

SEASON	JULY	AUG	SEP	OCT	NOV	DEC	JAN	FEB	MAR	APR	MAY	JUNE	TOTAL
1964-65	0.0	0.0	0.0	T	T	15.0	24.9	6.8	14.3	12.7	0.0	0.0	73.7
1965-66	0.0	0.0	0.0	1.3	4.5	5.4	36.6	16.8	11.7	2.0	3.4	0.0	81.7
1966-67	0.0	0.0	0.0	T	0.1	25.4	13.2	19.4	26.3	3.9	0.2	0.0	88.5
1967-68	0.0	0.0	0.0	T	18.6	15.3	11.0	11.7	7.0	T	0.0	0.0	63.6
1968-69	0.0	0.0	0.0	1.1	15.2	11.9	10.0	12.4	1.3	0.1	T	0.0	52.0
1969-70	0.0	0.0	0.0	0.3	9.7	59.6	14.0	12.9	14.1	3.4	T	0.0	114.0
1970-71	0.0	0.0	T	0.4	1.1	36.9	17.2	18.7	33.5	0.8	T	0.0	108.6
1971-72	0.0	0.0	0.0	0.0	17.9	8.1	11.0	44.3	17.5	7.4	0.0	0.0	106.2
1972-73	0.0	0.0	0.0	2.6	16.1	21.5	6.8	11.8	2.5	4.6	1.8	0.0	67.7
1973-74	0.0	0.0	0.0	T	1.7	26.9	10.7	14.2	21.2	10.1	T	0.0	84.8
1974-75	0.0	0.0	0.0	0.5	7.5	13.7	15.7	16.2	9.1	4.4	0.0	0.0	67.1
1975-76	0.0	0.0	0.0	T	4.2	18.3	29.7	10.7	11.6	1.8	T	0.0	76.3
1976-77	0.0	0.0	0.0	0.1	10.4	12.3	22.8	10.7	15.5	0.5	2.1	0.0	74.4
1977-78	0.0	0.0	0.0	T	13.5	31.8	41.0	17.5	10.1	1.4	T	0.0	115.3
1978-79	0.0	0.0	0.0	T	6.2	24.4	26.3	11.0	2.1	7.3	0.0	0.0	77.3
1979-80	0.0	0.0	0.0	0.1	1.5	16.7	7.0	13.2	17.9	0.4	0.0	0.0	56.8
1980-81	0.0	0.0	0.0	0.5	13.9	12.9	16.1	8.8	7.1	T	0.0	0.0	59.3
1981-82	0.0	0.0	0.0	T	4.0	22.6	15.8	12.0	15.5	11.7	0.0	0.0	81.6
1982-83	0.0	0.0	0.0	0.5	3.2	13.3	23.2	8.6	9.1	22.9	0.2	0.0	81.0
1983-84	0.0	0.0	0.0	T	10.6	7.1	18.6	9.7	24.9	T	T	0.0	70.9
1984-85	0.0	0.0	0.0	0.0	6.8	10.5	23.5	10.5	9.7	1.5	0.0	0.0	62.5
1985-86	0.0	0.0	0.0	0.0	1.3	19.9	20.1	24.7	4.1	4.9	T	0.0	75.0
1986-87	0.0	0.0	0.0	0.0	13.4	9.3	43.6	9.2	2.9	0.4	0.0	0.0	78.8
1987-88	0.0	0.0	0.0	T	6.0	18.9	17.4	27.3	9.0	3.0	0.0	0.0	81.6
1988-89	0.0	0.0	0.0	4.9	0.6	8.8	9.5	12.6	5.8	4.2	1.4	0.0	47.8
1989-90	0.0	0.0	T	0.1	7.3	21.4	22.5	9.0	10.9	3.6	T	0.0	74.8
1990-91	0.0	0.0	0.0	0.3	7.2	14.5	15.0	16.1	9.9	4.0	T	0.0	67.0
1991-92	0.0	0.0	0.0	T	3.9	12.2	12.6	12.8	14.1	0.4	T	0.0	56.0
1992-93	0.0	0.0	T	1.9	4.4	18.8	11.9	31.6	37.9	16.2	0.0	T	122.7
1993-94	0.0	0.0	T	11.6	5.7	22.9							
Record Mean	0.0	0.0	T	0.7	7.4	18.0	19.5	17.5	14.1	5.0	0.3	T	82.5

BUFFALO, NEW YORK

The country surrounding Buffalo is comparatively low and level to the west. To the east and south the land is gently rolling, rising to pronounced hills within 12 to 18 miles, and to 1,000 feet above the level of Lake Erie about 35 miles south-southeast of the city. An escarpment of 50 to 100 feet lies east-west 1-1/2 miles to the north. The eastern end of Lake Erie is 9 miles to the west-southwest, while Lake Ontario lies 25 miles to the north, the two being connected by the Niagara River, which flows north-northwestward from the end of Lake Erie.

Buffalo is located near the mean position of the polar front. Its weather is varied and changeable, characteristic of the latitude. Wide seasonal swings of temperature from hot to cold are tempered appreciably by the proximity of Lakes Erie and Ontario. Lake Erie lies to the southwest, the direction of the prevailing wind. Wind flow throughout the year is somewhat higher due to this exposure. The vigorous interplay of warm and cold air masses during the winter and early spring months causes one or more windstorms. Precipitation is moderate and fairly evenly divided throughout the twelve months.

The spring season is more cloudy and cooler than points not affected by the cold lake. Spring growth of vegetation is retarded, protecting it from late spring frosts. With heavy winter ice accumulations in the lake, typical spring conditions are delayed until late May or early June.

Summer comes suddenly in mid-June. Lake breezes temper the extreme heat of the summer season. Temperatures of 90 degrees and above are infrequent. There is more summer sunshine here than in any other section of the state. Due to the stabilizing effects of Lake Erie, thunderstorms are relatively infrequent. Most of them are caused by frontal action. To the north and south of the city thunderstorms occur more often.

Autumn has long, dry periods and is frost free usually until mid-October. Cloudiness increases in November, continuing mostly cloudy throughout the winter and early spring. Snow flurries off the lake begin in mid-November or early December. Outbreaks of Arctic air in December and throughout the winter months produce locally heavy snowfalls from the lake. At the same time, temperatures of well below zero over Canada and the midwest are raised 10 to 30 degrees in crossing the lakes. Only on rare occasions do polar air masses drop southward from eastern Hudson Bay across Lake Ontario without appreciable warming.

NORMALS, MEANS, AND EXTREMES

BUFFALO, NEW YORK

LATITUDE: 42°56'N LONGITUDE: 78°44' W ELEVATION: FT. GRND 705 BARO 715 TIME ZONE: EASTERN WBAN: 14733

	(a)	JAN	FEB	MAR	APR	MAY	JUNE	JULY	AUG	SEP	OCT	NOV	DEC	YEAR
TEMPERATURE °F:														
Normals														
-Daily Maximum		30.2	31.6	41.7	54.2	66.1	75.3	80.2	77.9	70.8	59.4	47.1	35.3	55.8
-Daily Minimum		17.0	17.4	25.9	36.2	47.0	56.5	61.9	60.1	53.0	42.7	33.9	22.9	39.5
-Monthly		23.6	24.5	33.8	45.2	56.6	65.9	71.1	69.0	61.9	51.1	40.5	29.1	47.7
Extremes														
-Record Highest	50	72	65	81	94	90	96	97	99	98	87	80	74	99
-Year		1950	1981	1945	1990	1991	1988	1988	1948	1953	1951	1961	1982	AUG 1948
-Record Lowest	50	-16	-20	-7	12	26	35	43	38	32	20	9	-10	-20
-Year		1982	1961	1984	1982	1947	1945	1945	1982	1991	1965	1971	1980	FEB 1961
NORMAL DEGREE DAYS:														
Heating (base 65°F)		1283	1134	967	594	279	59	5	17	130	431	735	1113	6747
Cooling (base 65°F)		0	0	0	0	19	86	194	141	37	0	0	0	477
% OF POSSIBLE SUNSHINE	50	31	38	45	51	58	65	67	64	57	49	29	27	48
MEAN SKY COVER (tenths)														
Sunrise - Sunset	50	8.4	8.2	7.6	7.1	6.8	6.2	6.0	6.1	6.4	6.7	8.4	8.5	7.2
MEAN NUMBER OF DAYS:														
Sunrise to Sunset														
-Clear	50	1.4	2.2	3.8	4.9	5.6	6.3	6.9	7.0	6.4	6.3	2.1	1.3	54.1
-Partly Cloudy	50	6.1	5.6	7.6	8.0	9.7	11.5	13.1	11.8	9.9	8.4	5.4	6.1	103.1
-Cloudy	50	23.5	20.5	19.6	17.1	15.7	12.2	11.0	12.3	13.7	16.3	22.6	23.5	208.1
Precipitation														
.01 inches or more	50	19.8	17.2	16.1	14.2	12.4	10.4	9.9	10.6	11.0	11.8	16.1	19.5	168.9
Snow, Ice pellets, hail														
1.0 inches or more	50	7.1	5.5	3.5	1.0	0.1	0.0	0.0	0.0	0.0	0.1	3.0	6.1	26.4
Thunderstorms	50	0.2	0.2	1.3	2.3	3.1	5.2	5.7	6.0	3.8	1.6	1.1	0.4	30.9
Heavy Fog Visibility														
1/4 mile or less	50	1.4	1.7	2.6	2.2	2.3	1.2	0.8	0.9	1.0	1.3	1.2	1.2	17.9
Temperature °F														
-Maximum														
90° and above	33	0.0	0.0	0.0	0.*	0.1	0.5	1.5	0.6	0.*	0.0	0.0	0.0	2.9
32° and below	33	17.1	15.2	7.2	0.5	0.0	0.0	0.0	0.0	0.0	0.0	1.9	11.9	53.8
-Minimum														
32° and below	33	28.5	25.8	23.8	10.3	0.6	0.0	0.0	0.0	0.1	3.3	14.0	25.6	132.1
0° and below	33	2.0	1.5	0.2	0.0	0.0	0.0	0.0	0.0	0.0	0.0	0.0	0.6	4.4
AVG. STATION PRESS. (mb)	21	990.5	991.6	990.2	989.1	989.5	989.5	990.4	991.8	992.4	992.4	991.3	991.2	990.8
RELATIVE HUMIDITY (%)														
Hour 01	33	77	79	78	75	76	79	79	83	83	79	79	79	79
Hour 07 (Local Time)	33	79	80	80	76	75	77	78	83	84	82	80	81	80
Hour 13	33	72	70	65	58	55	56	55	58	60	60	69	73	63
Hour 19	33	76	75	72	64	62	61	60	66	72	72	76	77	69
PRECIPITATION (inches):														
Water Equivalent														
-Normal		2.70	2.31	2.68	2.87	3.14	3.55	3.08	4.17	3.49	3.09	3.83	3.67	38.58
-Maximum Monthly	50	6.88	5.90	5.97	5.90	7.22	8.36	8.93	10.67	8.99	9.13	9.75	8.71	10.67
-Year		1982	1990	1991	1961	1989	1987	1992	1977	1977	1954	1985	1990	AUG 1977
-Minimum Monthly	50	1.03	0.81	1.20	1.27	1.21	0.11	0.93	1.10	0.77	0.30	1.44	0.69	0.11
-Year		1946	1968	1967	1946	1965	1955	1989	1948	1964	1963	1944	1943	JUN 1955
-Maximum in 24 hrs	50	2.57	2.31	2.14	2.09	3.52	5.01	3.38	3.88	4.94	3.49	2.51	2.33	5.01
-Year		1982	1954	1954	1991	1986	1987	1963	1963	1979	1945	1949	1990	JUN 1987
Snow, Ice pellets, hail														
-Maximum Monthly	50	68.3	54.2	29.3	15.0	7.9	T	T	T	T	3.1	31.3	68.4	68.4
-Year		1977	1958	1993	1975	1989	1980	1993	1991	1993	1972	1976	1985	DEC 1985
-Maximum in 24 hrs	50	25.3	19.4	17.2	6.8	7.9	T	T	T	T	2.8	19.9	24.3	25.3
-Year		1982	1984	1993	1975	1989	1980	1993	1991	1993	1993	1955	1945	JAN 1982
WIND:														
Mean Speed (mph)	54	14.2	13.5	13.2	12.6	11.4	10.9	10.3	9.8	10.3	11.1	12.7	13.3	11.9
Prevailing Direction through 1963		WSW	SW	SW	SW	SW	SW	SW	SW	S	S	S	WSW	SW
Fastest Mile														
-Direction (!!!)	47	SW	SW	W	W	SW	NW	NW	SW	SW	SW	SW	S	SW
-Speed (MPH)	47	91	70	68	67	63	56	59	56	59	63	66	60	91
-Year		1950	1946	1959	1957	1950	1954	1953	1944	1954	1954	1948	1945	JAN 1950
Peak Gust														
-Direction (!!!)	10	SW	S	W	W	SW	SW	SW	W	S	SW	SW	SW	W
-Speed (mph)	10	71	55	72	74	61	59	53	71	62	61	72	66	74
-Date		1985	1988	1986	1985	1990	1990	1990	1988	1987	1993	1992	1985	APR 1985

PRECIPITATION (inches)

BUFFALO, NEW YORK

YEAR	JAN	FEB	MAR	APR	MAY	JUNE	JULY	AUG	SEP	OCT	NOV	DEC	ANNUAL
1964	2.12	1.09	3.72	3.36	2.91	1.55	2.57	5.02	0.77	1.89	2.09	2.58	29.67
1965	3.27	2.99	1.97	1.99	1.21	1.50	3.69	4.12	2.37	5.07	4.69	2.60	35.47
1966	3.74	2.11	2.78	2.06	1.36	1.97	4.92	3.60	2.65	0.93	4.50	2.25	32.87
1967	1.18	1.39	1.20	2.60	3.69	2.50	1.57	4.04	6.36	4.78	3.13	2.16	34.60
1968	2.18	0.81	2.67	1.78	3.30	4.45	1.19	5.33	5.63	3.03	4.47	3.42	38.26
1969	3.85	0.97	1.62	4.16	3.75	3.51	3.83	2.48	2.04	2.77	4.09	3.09	36.16
1970	2.06	1.74	1.72	2.54	2.87	2.55	4.02	2.01	4.55	4.20	3.20	3.25	34.71
1971	1.46	3.03	2.07	1.48	1.56	4.25	4.50	4.43	1.88	1.57	3.07	3.61	32.91
1972	2.17	3.44	3.99	2.99	3.64	6.06	0.99	4.19	3.06	2.96	4.28	3.86	41.63
1973	2.03	1.98	3.27	3.56	2.99	1.68	3.68	2.98	1.44	4.27	4.07	4.89	36.84
1974	2.44	2.19	3.19	3.15	3.36	3.86	1.80	3.64	2.42	1.75	5.38	3.13	36.31
1975	2.11	2.93	2.92	1.86	3.31	3.65	2.34	8.49	2.44	1.13	2.77	4.58	38.53
1976	3.19	3.43	5.59	4.01	4.70	3.36	5.65	1.65	5.39	3.61	2.11	3.83	46.52
1977	3.38	1.59	2.42	3.60	1.39	2.79	3.64	10.67	8.99	2.61	4.45	8.02	53.55
1978	6.29	1.36	1.72	1.84	3.95	2.42	1.48	3.51	4.40	3.72	1.55	3.50	35.74
1979	5.43	2.03	2.48	3.16	1.63	2.18	3.51	6.26	5.61	3.88	4.14	3.43	43.74
1980	1.97	1.08	4.05	2.43	1.60	5.82	3.55	3.58	4.53	4.69	2.36	2.65	38.31
1981	1.11	3.50	1.70	3.09	2.56	3.68	5.05	3.13	4.24	3.31	2.22	2.87	36.46
1982	6.88	1.28	2.64	2.33	3.66	3.14	1.50	4.62	3.37	2.06	6.31	3.32	41.11
1983	1.44	1.30	3.20	2.55	3.28	2.99	2.01	3.51	2.11	4.62	5.19	7.30	39.50
1984	1.54	3.59	1.77	2.53	4.67	6.86	1.37	4.16	3.73	0.87	2.66	3.67	37.42
1985	4.27	3.34	4.42	1.33	3.46	3.21	1.81	4.63	1.20	3.73	9.75	4.85	46.00
1986	2.31	2.60	1.95	3.33	4.42	4.15	2.82	2.73	3.88	4.34	3.11	4.02	39.66
1987	2.90	0.85	3.66	3.40	1.35	8.36	3.09	3.38	5.32	2.62	4.44	2.78	42.15
1988	1.58	4.07	2.99	2.96	2.74	1.56	6.35	2.69	2.07	6.08	3.37	2.15	38.61
1989	1.77	2.54	3.15	1.88	7.22	7.83	0.93	1.84	3.85	2.98	4.83	2.34	41.16
1990	2.69	5.90	1.50	5.22	6.08	3.55	3.14	3.25	3.65	4.59	2.61	8.71	50.89
1991	2.07	2.06	5.97	5.83	3.10	0.86	3.34	2.84	3.19	3.11	4.02	3.81	40.20
1992	2.01	2.45	2.93	4.68	3.48	2.21	8.93	3.79	5.56	2.80	4.92	3.80	47.56
1993	4.35	1.92	3.02	2.55	1.79	4.99	1.78	3.86	5.53	3.69	3.58	3.60	40.66
Record Mean	3.06	2.69	2.79	2.75	2.95	2.92	2.96	3.23	3.13	3.09	3.34	3.32	36.23

AVERAGE TEMPERATURE (deg. F)

BUFFALO, NEW YORK

YEAR	JAN	FEB	MAR	APR	MAY	JUNE	JULY	AUG	SEP	OCT	NOV	DEC	ANNUAL
1964	29.3	23.5	34.0	46.9	59.2	65.7	73.1	64.9	60.9	48.1	42.1	29.5	48.1
1965	23.6	25.8	30.0	41.2	59.6	64.3	67.6	67.8	63.5	47.8	40.0	34.3	47.1
1966	20.4	24.9	34.7	43.3	52.2	67.4	71.4	68.5	58.7	48.8	41.5	28.6	46.7
1967	29.8	20.6	30.9	46.1	50.1	72.5	71.2	68.1	60.7	51.9	36.3	33.0	47.6
1968	19.9	20.7	35.7	49.2	53.4	64.8	71.2	69.4	66.1	53.5	40.7	26.8	47.6
1969	25.0	24.6	30.9	46.8	54.4	64.4	70.5	71.2	62.2	51.0	39.1	24.8	47.1
1970	17.6	24.8	30.1	46.9	57.3	66.0	71.0	70.2	64.0	54.5	41.6	27.4	47.6
1971	20.9	27.0	29.8	41.8	54.5	67.6	68.7	67.8	65.4	58.7	39.1	33.5	47.9
1972	25.5	22.0	30.1	41.1	59.1	62.6	71.0	67.7	62.8	46.2	36.0	30.8	46.3
1973	27.6	22.9	42.4	46.9	54.5	68.2	72.3	71.8	61.7	54.3	40.8	29.0	49.4
1974	27.1	22.3	33.0	46.2	53.1	65.6	69.9	69.9	59.6	49.2	40.2	31.7	47.3
1975	30.1	29.1	30.8	39.3	62.1	68.0	72.3	69.7	58.3	53.1	46.9	28.3	49.0
1976	19.7	31.8	37.2	46.5	53.4	68.4	67.8	67.5	60.1	46.3	34.1	22.0	46.3
1977	13.8	24.6	39.8	47.0	60.3	64.4	72.0	68.1	62.6	49.6	43.3	27.9	47.8
1978	20.4	15.5	28.2	42.5	57.4	65.1	70.4	70.3	60.8	49.5	40.4	30.4	45.9
1979	20.5	14.9	38.2	44.3	56.9	66.5	71.3	67.5	61.9	50.7	43.5	33.4	47.5
1980	25.8	21.2	31.8	46.1	58.1	61.9	71.7	72.6	62.4	48.7	39.4	25.3	47.1
1981	19.3	32.9	33.9	47.2	56.4	66.2	71.8	70.0	60.9	48.2	40.4	29.0	48.0
1982	17.2	23.2	32.5	41.6	61.0	62.2	71.8	65.0	61.6	52.6	43.0	37.5	47.5
1983	27.0	29.6	36.7	43.6	53.9	67.6	74.2	71.2	63.7	51.7	40.8	22.7	48.6
1984	20.4	33.8	27.1	47.7	52.9	67.8	70.3	70.3	58.5	53.2	39.0	35.6	48.1
1985	21.1	24.8	35.6	49.5	59.5	62.7	69.7	69.2	64.2	52.5	42.0	25.6	48.0
1986	25.5	24.5	36.2	47.8	59.7	64.1	71.1	67.9	61.8	50.9	37.7	32.4	48.3
1987	26.1	25.0	37.7	50.0	60.5	68.9	74.2	68.9	63.4	47.9	42.5	34.3	50.0
1988	26.6	24.3	35.2	46.1	59.7	64.0	74.8	72.4	62.1	46.9	43.0	30.0	48.8
1989	31.3	22.7	33.0	41.9	55.1	65.9	71.5	68.5	60.8	51.5	37.9	17.4	46.5
1990	33.4	29.3	36.9	48.5	54.9	66.7	71.4	70.4	61.7	52.5	43.4	34.4	50.3
1991	26.0	30.6	37.8	50.5	64.3	69.1	71.9	71.0	62.0	53.1	39.3	31.3	50.6
1992	27.1	27.7	31.6	43.8	57.3	63.4	66.8	66.3	61.6	47.9	40.2	31.9	47.1
1993	29.5	20.7	30.7	47.3	57.0	66.0	73.4	72.0	59.4	49.2	39.6	29.6	47.9
Record Mean	25.0	24.7	32.6	43.8	55.1	64.8	70.5	69.0	62.4	51.5	40.0	29.5	47.4
Max	31.4	31.6	39.9	52.0	63.7	72.7	78.3	77.0	70.5	59.1	46.3	35.4	54.8
Min	18.5	17.7	25.3	35.5	46.4	56.9	62.6	60.9	54.3	44.0	33.8	23.6	40.0

REFERENCE NOTES

BUFFALO, NEW YORK

GENERAL
T - TRACE AMOUNT.
BLANK ENTRIES DENOTE MISSING/UNREPORTED DATA.
# INDICATES A STATION OR INSTRUMENT RELOCATION.

SPECIFIC

PM - INCLUDES LAST DAY OF PREVIOUS MONTH
ASOS - AUTOMATED SURFACE OBSERVING SYSTEM IN OPERATION DURING THESE MONTHS.

(a) - LENGTH OF RECORD IN YEARS, ALTHOUGH INDIVIDUAL MONTHS MAY BE MISSING.
0.* OR * - THE VALUE IS BETWEEN 0.0 AND 0.05
NORMALS - BASED ON THE 1961-1990 RECORD PERIOD.
EXTREMES - DATES ARE THE MOST RECENT OCCURRENCE
WIND DIR.- NUMERALS SHOW TENS OF DEGRESS CLOCKWISE FROM TRUE NORTH. "00" INDICATES CALM.
RESULTANT DIRECTIONS ARE GIVEN TO WHOLE DEGREES.
BOLD VALUES INDICATE EXTREME VALUES WHICH OCCURRED AFTER THE ASOS SYSTEM WAS COMMISSIONED.

RECORD = PERIOD OF RECORD
RECORD MEAN PRECIPITATION IS THE MEAN OF ALL DAILY PRECIPITATION AMOUNTS DURING THE PERIOD OF RECORD.
RECORD MAX(MIN) TEMPERATURE IS THE MEAN OF ALL DAILY MAX(MIN) TEMPERATURES DURING THE PERIOD OF RECORD.
RECORD MEAN TEMPERATURE IS THE SUM OF THE RECORD MAX AND RECORD MIN DIVIDED BY 2.
AVERAGE TEMPERATURE IS THE SUM OF THE MEAN DAILY MAX AND MIN TEMPERATURE DIVIDED BY 2.

EXCEPTIONS

RECORD MEANS ARE THROUGH THE CURRENT YEAR, BEGINNING IN 1874 FOR TEMPERATURE
1871 FOR PRECIPITATION
1944 FOR SNOWFALL

HEATING DEGREE DAYS Base 65 deg. F BUFFALO, NEW YORK

SEASON	JULY	AUG	SEP	OCT	NOV	DEC	JAN	FEB	MAR	APR	MAY	JUNE	TOTAL
1964-65	5	68	176	518	680	1097	1277	1092	1080	706	186	100	6985
1965-66	23	46	122	525	742	942	1374	1114	931	648	401	68	6936
1966-67	7	19	199	495	700	1124	1086	1239	1047	560	457	4	6937
1967-68	12	26	162	403	853	985	1393	1281	901	469	352	84	6921
1968-69	11	29	58	374	722	1180	1233	1125	1052	540	325	102	6751
1969-70	13	16	147	433	769	1240	1459	1121	1076	552	255	66	7147
1970-71	6	6	93	328	695	1161	1361	1057	1085	691	327	36	6846
1971-72	11	29	87	202	771	971	1218	1237	1070	707	187	112	6602
1972-73	16	33	113	574	860	1054	1152	1173	696	542	318	24	6555
1973-74	2	14	171	326	720	1107	1167	1187	989	553	365	51	6652
1974-75	2	0	187	483	738	1024	1077	1001	1053	764	175	32	6536
1975-76	3	15	197	368	535	1134	1400	958	853	557	358	40	6418
1976-77	15	35	180	573	921	1328	1580	1123	775	544	207	90	7371
1977-78	5	40	110	473	646	1146	1376	1378	1130	670	282	81	7337
1978-79	14	3	154	472	732	1067	1371	1400	823	619	285	65	7005
1979-80	16	35	134	455	636	973	1208	1265	1022	559	240	142	6685
1980-81	2	0	128	498	759	1224	1411	895	956	527	269	33	6702
1981-82	6	11	170	514	732	1108	1476	1163	1002	698	147	95	7122
1982-83	4	65	140	382	656	848	1172	987	868	636	342	71	6171
1983-84	5	10	125	418	722	1304	1378	899	1167	519	385	35	6967
1984-85	11	22	210	360	774	905	1354	1120	902	476	196	95	6425
1985-86	8	12	114	378	685	1215	1215	1128	885	519	197	80	6436
1986-87	4	42	137	430	811	1003	1199	1115	837	447	213	28	6266
1987-88	3	25	91	527	665	948	1184	1174	916	560	186	113	6392
1988-89	5	17	122	560	654	1078	1038	1177	985	687	321	60	6704
1989-90	1	28	170	411	806	1466	970	995	866	518	311	46	6588
1990-91	5	2	141	395	640	941	1203	956	836	431	141	22	5713
1991-92	1	1	166	376	762	1037	1169	1076	1027	633	254	93	6595
1992-93	28	41	148	525	738	1021	1095	1235	1053	526	257	60	6727
1993-94	0	8	212	486	752	1089							

COOLING DEGREE DAYS Base 65 deg. F BUFFALO, NEW YORK

YEAR	JAN	FEB	MAR	APR	MAY	JUNE	JULY	AUG	SEP	OCT	NOV	DEC	TOTAL
1969	0	0	0	0	1	88	192	212	69	6	0	0	568
1970	0	0	0	16	21	108	197	173	72	12	0	0	599
1971	0	0	0	0	9	119	136	122	107	15	0	0	508
1972	0	0	0	0	12	48	210	123	57	0	0	0	450
1973	0	0	0	6	2	126	233	230	78	3	0	0	678
1974	0	0	0	0	7	71	163	158	29	0	0	0	428
1975	0	0	0	0	90	129	238	171	3	3	0	0	634
1976	0	0	0	8	7	149	109	119	40	0	0	0	432
1977	0	0	0	12	68	78	228	142	45	0	1	0	574
1978	0	0	0	0	52	91	189	173	35	0	0	0	540
1979	0	0	0	6	40	118	217	120	49	20	0	0	570
1980	0	0	0	0	32	56	217	242	58	2	0	0	607
1981	0	0	0	2	13	78	225	173	55	0	0	0	546
1982	0	0	0	3	31	18	221	74	45	2	0	2	396
1983	0	0	0	0	5	157	300	214	90	15	0	0	781
1984	0	0	0	5	16	123	183	193	23	1	0	0	544
1985	0	0	0	18	32	32	161	151	96	0	1	0	491
1986	0	0	0	7	38	60	200	137	46	0	0	0	488
1987	0	0	0	4	79	151	298	152	49	0	0	0	733
1988	0	0	0	0	29	88	315	255	41	8	0	0	736
1989	0	0	0	0	21	97	207	143	50	0	0	0	518
1990	0	0	3	29	4	104	208	176	47	14	0	0	585
1991	0	0	0	3	125	153	221	193	83	13	0	0	791
1992	0	0	0	1	24	53	90	90	55	0	0	0	313
1993	0	0	0	0	14	97	267	231	51	3	0	0	663

SNOWFALL (inches) BUFFALO, NEW YORK

SEASON	JULY	AUG	SEP	OCT	NOV	DEC	JAN	FEB	MAR	APR	MAY	JUNE	TOTAL
1964-65	0.0	0.0	0.0	T	5.4	15.2	19.2	9.4	17.5	4.2	0.0	0.0	70.9
1965-66	0.0	0.0	0.0	1.2	12.2	7.0	48.0	15.2	11.4	3.2	0.1	0.0	98.3
1966-67	0.0	0.0	0.0	0.0	10.0	12.1	11.6	19.8	10.8	0.6	1.2	0.0	66.1
1967-68	0.0	0.0	0.0	T	19.7	10.4	19.1	11.7	10.6	0.1	0.0	0.0	71.6
1968-69	0.0	0.0	0.0	T	11.6	11.7	31.2	12.8	8.0	3.1	0.0	0.0	78.4
1969-70	0.0	0.0	0.0	1.0	22.1	23.4	38.0	21.9	12.6	1.5	T	0.0	120.5
1970-71	0.0	0.0	0.0	0.0	2.6	32.3	17.2	19.4	22.6	2.9	0.0	0.0	97.0
1971-72	0.0	0.0	0.0	0.0	18.7	12.9	27.6	31.4	14.1	5.2	0.0	0.0	109.9
1972-73	0.0	0.0	0.0	3.1	18.9	19.8	9.9	16.1	8.5	2.4	0.1	0.0	78.8
1973-74	0.0	0.0	0.0	0.0	3.0	23.1	19.7	22.8	12.9	7.1	0.1	0.0	88.7
1974-75	0.0	0.0	0.0	T	22.1	23.6	11.0	16.3	7.6	15.0	0.0	0.0	95.6
1975-76	0.0	0.0	0.0	T	5.5	27.3	21.6	8.3	17.3	2.5	T	0.0	82.5
1976-77	0.0	0.0	0.0	0.2	31.3	60.7	68.3	22.7	13.5	2.2	0.5	0.0	199.4
1977-78	0.0	0.0	0.0	T	15.0	53.4	56.5	21.7	5.8	1.8	0.1	0.0	154.3
1978-79	0.0	0.0	0.0	T	3.0	10.1	42.6	28.3	4.6	8.7	0.0	0.0	97.3
1979-80	0.0	0.0	0.0	T	12.6	19.7	10.2	11.7	13.9	0.3	T	T	68.4
1980-81	0.0	0.0	0.0	T	6.7	21.6	14.4	5.0	13.2	T	0.0	0.0	60.9
1981-82	0.0	0.0	0.0	T	1.8	24.8	53.2	12.7	9.0	10.9	0.0	0.0	112.4
1982-83	0.0	0.0	0.0	0.0	15.8	12.9	9.0	5.5	6.9	2.3	T	0.0	52.4
1983-84	0.0	0.0	0.0	T	17.7	52.0	13.4	32.5	16.0	0.9	T	0.0	132.5
1984-85	0.0	0.0	0.0	0.0	1.4	11.2	65.9	20.9	6.3	1.5	0.0	0.0	107.2
1985-86	0.0	0.0	0.0	0.0	5.2	68.4	17.3	17.3	4.8	1.7	T	0.0	114.7
1986-87	0.0	0.0	0.0	0.0	13.7	4.8	28.5	7.7	10.8	2.0	0.0	0.0	67.5
1987-88	0.0	0.0	0.0	T	0.9	9.8	6.9	31.9	6.1	0.8	0.0	0.0	56.4
1988-89	0.0	0.0	0.0	0.5	0.6	10.8	5.4	29.6	10.1	2.5	7.9	0.0	67.4
1989-90	0.0	0.0	0.0	T	7.8	34.8	11.8	28.0	1.4	9.9	T	0.0	93.7
1990-91	0.0	0.0	0.0	T	0.7	15.4	16.6	16.1	8.5	0.2	T	0.0	57.5
1991-92	0.0	T	0.0	0.2	18.0	21.4	18.4	7.0	22.8	5.0	0.0	0.0	92.8
1992-93	0.0	0.0	0.0	0.6	13.7	16.5	13.1	19.5	29.3	0.5	T	0.0	93.2
1993-94	T	0.0	T	2.9	4.8	27.9							
Record Mean	T	T	T	0.3	11.5	22.7	23.3	18.2	11.7	3.1	0.3	T	91.0

ISLIP, NEW YORK

Long Island is the terminal moraine marking the southernmost advance of the ice sheet along the Atlantic Coast during the last ice age. The terrain is generally flat, with only a gradual rise in elevation from Long Island Sound on the northern shore and from the Atlantic Ocean on the southern shore toward the middle of the island. Islip is located about half-way out Long Island on the southern coast. The airport is located about seven miles to the northeast of the city. Islip is protected from flooding during periods of high tides by Fire Island, a natural barrier located about three miles offshore. Most of the air masses affecting Islip are continental in origin, however the ocean has a pronounced influence on the climate of the area.

A cool sea breeze blowing off the ocean during the summer months helps to alleviate the afternoon heat. There are an average of 7 days between June and September when the afternoon temperature exceeds 90 degrees, while farther inland there are 10 to 15 such days.

It is uncommon for the eye of a tropical storm to pass directly over Long Island. Tropical weather systems moving along the Atlantic Coast, however, are capable of producing episodes of heavy rain and strong winds in the late summer or fall.

The winter season is relatively mild. Below zero temperatures are reported on only one or two days in about half the winters. Temperatures of 10 degrees below zero or colder are extremely rare. The seasonal snowfall averages about 29 inches. Almost all of this snow falls between December and March. Coastal low pressure systems, Northeasters, are the principle source of this snow. These weather systems will occasionally produce a heavy snowfall. There are usually extended periods during the winter when the ground is bare of snow.

The average date of the last spring temperature of 32 degrees is April 27 and the average first fall occurrence is October 21. Inland locations would expect a shorter freeze-free season.

NORMALS, MEANS, AND EXTREMES

ISLIP, NEW YORK

LATITUDE: 40°47'N LONGITUDE: 73°06' W ELEVATION: FT. GRND 84 BARO 84 TIME ZONE: EASTERN WBAN: 04781

	(a)	JAN	FEB	MAR	APR	MAY	JUNE	JULY	AUG	SEP	OCT	NOV	DEC	YEAR
TEMPERATURE °F:														
Normals														
-Daily Maximum		35.9	37.8	45.9	57.1	67.2	75.4	80.8	80.8	72.9	63.4	52.1	41.9	59.3
-Daily Minimum		21.3	21.7	30.8	39.5	49.4	58.7	65.7	65.1	60.0	45.7	35.8	26.5	43.4
-Monthly		28.6	29.8	38.4	48.3	58.3	67.1	73.2	73.0	66.5	54.6	44.0	34.3	51.3
Extremes														
-Record Highest	10	63	67	82	86	95	95	101	93	91	83	78	66	101
-Year		1993	1991	1990	1990	1987	1991	1991	1989	1985	1986	1990	1991	JUL 1991
-Record Lowest	10	-7	2	8	26	34	46	50	45	38	28	11	7	-7
-Year		1988	1993	1990	1992	1987	1990	1988	1986	1989	1985	1989	1988	JAN 1988
NORMAL DEGREE DAYS:														
Heating (base 65°F)		1128	986	825	501	221	24	0	0	51	329	630	952	5647
Cooling (base 65°F)		0	0	0	0	14	87	254	248	96	7	0	0	706
% OF POSSIBLE SUNSHINE														
MEAN SKY COVER (tenths)														
Sunrise - Sunset														
MEAN NUMBER OF DAYS:														
Sunrise to Sunset														
-Clear														
-Partly Cloudy														
-Cloudy														
Precipitation														
.01 inches or more	10	10.5	9.1	10.0	11.9	9.9	9.9	9.8	8.2	8.5	8.2	10.3	10.0	116.3
Snow,Ice pellets,hail														
1.0 inches or more	10	2.1	1.6	1.8	0.1	0.0	0.0	0.0	0.0	0.0	0.0	0.1	1.2	6.9
Thunderstorms	10	0.2	0.5	1.1	1.7	3.2	5.2	5.7	3.8	2.0	1.0	0.8	0.3	25.5
Heavy Fog Visibility														
1/4 mile or less	10	2.8	3.0	3.1	3.9	4.7	3.7	2.6	2.4	1.8	4.9	3.3	2.7	38.9
Temperature °F														
-Maximum														
90° and above	10	0.0	0.0	0.0	0.0	0.9	1.7	2.7	1.2	0.3	0.0	0.0	0.0	6.8
32° and below	10	7.0	4.8	1.5	0.0	0.0	0.0	0.0	0.0	0.0	0.0	0.1	4.3	17.7
-Minimum														
32° and below	10	25.0	21.9	17.4	2.9	0.0	0.0	0.0	0.0	0.0	1.3	9.6	21.4	99.5
0° and below	10	0.5	0.0	0.0	0.0	0.0	0.0	0.0	0.0	0.0	0.0	0.0	0.0	0.5
AVG. STATION PRESS.(mb)	10	1014.5	1014.1	1013.6	1011.5	1012.4	1011.0	1012.1	1013.5	1015.7	1015.8	1015.8	1015.2	1013.7
RELATIVE HUMIDITY (%)														
Hour 01	9	75	74	74	78	83	83	85	88	86	83	78	73	80
Hour 07 (Local Time)	9	76	77	76	77	77	77	80	84	85	85	81	76	79
Hour 13	9	61	59	56	56	57	57	61	61	62	59	60	59	59
Hour 19	9	69	68	67	68	70	69	74	76	78	76	72	67	71
PRECIPITATION (inches):														
Water Equivalent														
-Normal		3.69	3.48	4.10	4.23	3.94	3.82	3.46	4.04	3.48	3.55	4.23	4.05	46.07
-Maximum Monthly	10	6.28	5.55	5.53	5.06	10.14	7.86	8.36	13.78	5.06	8.71	8.02	6.06	13.78
-Year		1987	1984	1984	1990	1989	1989	1984	1990	1984	1989	1988	1992	AUG 1990
-Minimum Monthly	10	1.34	1.11	2.38	1.29	0.73	0.58	1.90	0.47	0.81	1.29	1.27	0.90	0.47
-Year		1985	1987	1985	1992	1986	1988	1987	1984	1985	1992	1991	1985	AUG 1984
-Maximum in 24 hrs	10	1.91	2.33	2.52	1.81	4.76	3.85	2.69	6.92	2.23	3.95	2.63	2.80	6.92
-Year		1990	1984	1987	1990	1989	1992	1984	1990	1984	1989	1988	1993	AUG 1990
Snow,Ice pellets,hail														
-Maximum Monthly	10	13.5	10.9	13.3	3.0	T	0.0	0.0	0.0	0.0	0.0	7.6	10.4	13.5
-Year		1985	1993	1993	1990	1991						1989	1988	JAN 1985
-Maximum in 24 hrs	10	6.0	6.7	8.0	3.0	T	0.0	0.0	0.0	0.0	0.0	7.6	9.2	9.2
-Year		1984	1987	1993	1990	1991	1984	1984	1984	1984	1984	1989	1988	DEC 1988
WIND:														
Mean Speed (mph)	10	9.4	10.1	10.6	9.9	8.7	8.5	7.5	7.2	7.7	8.4	9.4	9.5	8.9
Prevailing Direction through														
Fastest Obs. 1 Min.														
-Direction (!!)														
-Speed (MPH)														
-Year														
Peak Gust														
-Direction (!!)														
-Speed (mph)														
-Date														

PRECIPITATION (inches) ISLIP, NEW YORK

YEAR	JAN	FEB	MAR	APR	MAY	JUNE	JULY	AUG	SEP	OCT	NOV	DEC	ANNUAL
1984	2.63	5.55	5.53	4.81	9.43	5.14	8.36	0.47	5.06	2.43	1.69	2.33	53.43
1985	1.34	2.00	2.38	1.79	4.13	6.32	3.41	3.84	0.81	1.31	6.18	0.90	34.41
1986	3.37	3.20	3.10	2.66	0.73	1.69	4.18	3.95	0.82	2.06	6.56	5.46	37.78
1987	6.28	1.11	4.93	3.65	1.53	2.53	1.90	4.46	3.28	1.96	2.55	2.94	37.12
1988	3.17	5.36	3.94	1.97	2.92	0.58	2.45	1.49	3.59	3.35	8.02	2.96	39.80
1989	2.21	4.01	4.68	4.78	10.14	7.86	4.90	7.68	4.56	8.71	4.82	0.97	65.32
1990	5.68	2.13	2.55	5.06	8.94	5.20	3.33	13.78	2.48	8.12	1.57	4.65	63.49
1991	4.17	1.83	3.84	3.29	2.33	1.81	2.22	8.68	4.11	2.50	1.27	4.96	41.01
1992	2.07	2.26	3.56	1.29	3.60	4.89	6.20	6.62	3.22	1.29	6.05	6.06	47.11
1993	2.57	3.61	5.39	2.95	1.38	1.33	1.98	1.60	4.13	4.75	3.30	4.65	37.64
Record Mean	3.35	3.11	3.99	3.22	4.51	3.73	3.89	5.26	3.21	3.65	4.20	3.59	45.71

AVERAGE TEMPERATURE (deg. F) ISLIP, NEW YORK

YEAR	JAN	FEB	MAR	APR	MAY	JUNE	JULY	AUG	SEP	OCT	NOV	DEC	ANNUAL
1984	27.4	38.1	34.7	48.0	57.2	70.2	71.8	73.9	62.6	57.5	44.7	41.3	52.3
1985	26.3	32.9	41.9	50.8	60.5	65.4	73.4	71.8	66.9	55.9	48.4	32.1	52.2
1986	31.7	30.1	40.4	49.8	61.4	67.6	73.4	70.3	63.7	53.9	42.6	36.4	51.8
1987	30.5	30.3	41.1	49.9	58.0	69.2	74.9	70.9	65.1	51.0	45.7	36.9	52.0
1988	26.6	32.4	39.9	47.8	58.8	67.8	75.0	75.0	64.2	49.7	45.3	33.4	51.3
1989	34.6	31.4	38.2	47.4	58.6	69.9	72.7	73.1	65.0	54.4	42.6	24.8	51.1
1990	37.9	36.0	40.5	48.7	57.0	68.0	74.0	74.1	66.0	61.0	47.3	39.9	54.2
1991	32.2	36.7	42.7	52.0	64.1	70.5	74.8	74.5	64.3	56.0	45.8	37.3	54.2
1992	32.5	34.2	37.0	46.0	57.2	66.3	71.5	70.3	64.7	51.9	44.5	36.1	51.0
1993	34.6	28.4	36.8	49.4	60.8	69.0	75.7	73.1	65.4	52.3	44.0	35.2	52.1
Record Mean	31.4	33.0	39.3	49.0	59.4	68.3	73.7	72.7	64.8	54.4	45.1	35.3	52.2
Max	39.0	40.5	47.4	57.0	68.4	76.9	81.3	80.2	72.8	63.2	53.4	43.0	60.3
Min	23.8	25.5	31.2	41.0	50.3	59.8	66.1	65.2	56.8	45.5	36.7	27.6	44.1

REFERENCE NOTES ISLIP, NEW YORK

GENERAL
T - TRACE AMOUNT.
BLANK ENTRIES DENOTE MISSING/UNREPORTED DATA.
# INDICATES A STATION OR INSTRUMENT RELOCATION.

SPECIFIC

EXCEPTIONS

RECORD MEANS ARE THROUGH THE CURRENT YEAR, BEGINNING IN 1984 FOR TEMPERATURE
1984 FOR PRECIPITATION
1984 FOR SNOWFALL

(a) - LENGTH OF RECORD IN YEARS, ALTHOUGH INDIVIDUAL MONTHS MAY BE MISSING.
0.* OR * - THE VALUE IS BETWEEN 0.0 AND 0.05
NORMALS - BASED ON THE 1961-1990 RECORD PERIOD.
EXTREMES - DATES ARE THE MOST RECENT OCCURRENCE.
WIND DIR. - NUMERALS SHOW TENS OF DEGRESS CLOCKWISE FROM TRUE NORTH. "00" INDICATES CALM.
RESULTANT DIRECTIONS ARE GIVEN TO WHOLE DEGREES.
BOLD VALUES INDICATE EXTREME VALUES WHICH OCCURRED AFTER THE ASOS SYSTEM WAS COMMISSIONED.

RECORD = PERIOD OF RECORD
RECORD MEAN PRECIPITATION IS THE MEAN OF ALL DAILY PRECIPITATION AMOUNTS DURING THE PERIOD OF RECORD.
RECORD MAX(MIN) TEMPERATURE IS THE MEAN OF ALL DAILY MAX(MIN) TEMPERATURES DURING THE PERIOD OF RECORD.
RECORD MEAN TEMPERATURE IS THE SUM OF THE RECORD MAX AND RECORD MIN DIVIDED BY 2.
AVERAGE TEMPERATURE IS THE SUM OF THE MEAN DAILY MAX AND MIN TEMPERATURE DIVIDED BY 2.

HEATING DEGREE DAYS Base 65 deg. F — ISLIP, NEW YORK

SEASON	JULY	AUG	SEP	OCT	NOV	DEC	JAN	FEB	MAR	APR	MAY	JUNE	TOTAL
1983-84							1162	773	933	501	237	29	
1984-85	0	0	119	228	601	729	1193	892	707	421	155	49	5094
1985-86	1	2	57	279	493	1012	1026	970	755	449	166	32	5242
1986-87	2	24	92	356	664	880	1064	964	736	447	257	25	5511
1987-88	0	7	59	427	572	867	1182	940	772	509	212	56	5603
1988-89	5	1	69	467	583	975	934	934	826	520	201	9	5524
1989-90	0	3	94	319	666	1239	833	807	751	488	244	26	5470
1990-91	3	2	71	193	524	771	1010	789	681	390	130	13	4577
1991-92	0	0	105	285	569	851	1001	891	865	563	249	26	5405
1992-93	4	4	82	400	607	888	935	1018	868	464	145	33	5448
1993-94	1	0	90	386	626	917							

COOLING DEGREE DAYS Base 65 deg. F — ISLIP, NEW YORK

YEAR	JAN	FEB	MAR	APR	MAY	JUNE	JULY	AUG	SEP	OCT	NOV	DEC	TOTAL
1984	0	0	0	0	2	189	217	282	53	3	0	0	746
1985	0	0	0	5	22	69	269	218	122	7	0	0	712
1986	0	0	0	0	64	115	270	194	60	19	0	0	722
1987	0	0	0	0	47	156	316	196	67	0	0	0	782
1988	0	0	0	0	25	145	320	319	55	2	0	0	866
1989	0	0	0	0	10	161	245	262	100	0	0	0	778
1990	0	0	0	7	1	123	290	287	108	73	0	0	889
1991	0	0	0	6	108	183	312	301	91	15	0	0	1016
1992	0	0	0	0	15	74	212	176	79	0	0	0	556
1993	0	0	0	0	23	161	339	259	110	0	2	0	894

SNOWFALL (inches) — ISLIP, NEW YORK

SEASON	JULY	AUG	SEP	OCT	NOV	DEC	JAN	FEB	MAR	APR	MAY	JUNE	TOTAL
1983-84							11.9	T	13.0	0.0	0.0	0.0	
1984-85	0.0	0.0	0.0	0.0	T	4.7	13.5	8.7	T	T	0.0	0.0	26.9
1985-86	0.0	0.0	0.0	0.0	T	2.1	2.6	10.4	0.1	T	0.0	0.0	15.2
1986-87	0.0	0.0	0.0	0.0	T	3.4	8.8	8.6	1.7	0.0	0.0	0.0	22.5
1987-88	0.0	0.0	0.0	0.0	1.1	4.2	10.7	0.1	3.4	0.0	0.0	0.0	19.5
1988-89	0.0	0.0	0.0	0.0	0.0	10.4	4.4	1.2	3.0	T	0.0	0.0	19.0
1989-90	0.0	0.0	0.0	0.0	7.6	0.2	2.0	2.0	4.2	3.0	0.0	0.0	19.0
1990-91	0.0	0.0	0.0	0.0	0.0	4.0	3.6	4.3	1.9	0.0	T	0.0	13.8
1991-92	0.0	0.0	0.0	0.0	0.0	2.0	2.3	1.5	7.6	T	0.0	0.0	13.4
1992-93	0.0	0.0	0.0	0.0	0.0	2.1	2.3	10.9	13.3	0.0	0.0	0.0	28.6
1993-94	0.0	0.0	0.0	0.0	T	3.3							
Record Mean	0.0	0.0	0.0	0.0	0.9	3.6	6.2	4.8	4.8	0.3	T	0.0	20.6

NEW YORK, CENTRAL PARK, NEW YORK

New York City, in area exceeding 300 square miles, is located on the Atlantic coastal plain at the mouth of the Hudson River. The terrain is laced with numerous waterways, all but one of the five boroughs in the city are situated on islands. Elevations range from less than 50 feet over most of Manhattan, Brooklyn, and Queens to almost 300 feet in northern Manhattan and the Bronx, and over 400 feet in Staten Island. Extensive suburban areas on Long Island, and in Connecticut, New York State and New Jersey border the city on the east, north, and west. About 30 miles to the west and northwest, hills rise to about 1,500 feet and to the north in upper Westchester County to 800 feet. To the southwest and to the east are the low-lying land areas of the New Jersey coastal plain and of Long Island, bordering on the Atlantic.

The New York Metropolitan area is close to the path of most storm and frontal systems which move across the North American continent. Therefore, weather conditions affecting the city most often approach from a westerly direction. New York City can thus experience higher temperatures in summer and lower ones in winter than would otherwise be expected in a coastal area. However, the frequent passage of weather systems often helps reduce the length of both warm and cold spells, and is also a major factor in keeping periods of prolonged air stagnation to a minimum.

Although continental influence predominates, oceanic influence is by no means absent. During the summer local sea breezes, winds blowing onshore from the cool water surface, often moderate the afternoon heat. The effect of the sea breeze diminishes inland. On winter mornings, ocean temperatures which are warm relative to the land reinforce the effect of the city heat island and low temperatures are often 10-20 degrees lower in the inland suburbs than in the central city. The relatively warm water temperatures also delay the advent of winter snows. Conversely, the lag in warming of water temperatures keeps spring temperatures relatively cool. One year-round measure of the ocean influence is the small average daily variation in temperature.

Precipitation is moderate and distributed fairly evenly throughout the year. Most of the rainfall from May through October comes from thunderstorms, usually of brief duration and sometimes intense. Heavy rains of long duration associated with tropical storms occur infrequently in late summer or fall. For the other months of the year precipitation is more likely to be associated with widespread storm areas, so that day-long rain, snow or a mixture of both is more common. Coastal storms, occurring most often in the fall and winter months, produce on occasion considerable amounts of precipitation and have been responsible for record rains, snows, and high winds.

The average annual precipitation is reasonably uniform within the city but is higher in the northern and western suburbs and less on eastern Long Island. Annual snowfall totals also show a consistent increase to the north and west of the city with lesser amounts along the south shores and the eastern end of Long Island, reflecting the influence of the ocean waters.

Local Climatological Data is published for three locations in New York City, Central Park, La Guardia Airport, and John F. Kennedy International Airport. Other nearby locations for which it is published are Newark, New Jersey, and Bridgeport, Connecticut.

Based on the 1951-1980 period, the average first occurrence of 32 degrees Fahrenheit in the fall is November 11 and the average last occurrence in the spring is April 1.

NORMALS, MEANS, AND EXTREMES

NEW YORK, CENTRAL PARK, NEW YORK

LATITUDE: 40°47'N LONGITUDE: 73°58' W ELEVATION: FT. GRND 132 BARO 87 TIME ZONE: EASTERN WBAN: 94728

	(a)	JAN	FEB	MAR	APR	MAY	JUNE	JULY	AUG	SEP	OCT	NOV	DEC	YEAR
TEMPERATURE °F:														
Normals														
-Daily Maximum		37.6	40.3	50.0	61.2	71.7	80.1	85.2	83.7	76.2	65.3	54.0	42.5	62.3
-Daily Minimum		25.3	26.9	34.8	43.8	53.7	63.0	68.4	67.3	60.1	49.7	41.1	30.7	47.1
-Monthly		31.5	33.6	42.4	52.5	62.7	71.6	76.8	75.5	68.2	57.5	47.6	36.6	54.7
Extremes														
-Record Highest	125	72	75	86	96	99	101	106	104	102	94	84	72	106
-Year		1950	1985	1945	1976	1962	1966	1936	1918	1953	1941	1950	1982	JUL 1936
-Record Lowest	125	-6	-15	3	12	32	44	52	50	39	28	5	-13	-15
-Year		1882	1934	1872	1923	1891	1945	1943	1986	1912	1936	1875	1917	FEB 1934
NORMAL DEGREE DAYS:														
Heating (base 65°F)		1039	879	701	375	125	0	0	0	34	250	522	880	4805
Cooling (base 65°F)		0	0	0	0	54	203	366	326	130	17	0	0	1096
% OF POSSIBLE SUNSHINE	107	51	55	57	58	61	64	65	64	62	61	52	49	58
MEAN SKY COVER (tenths)														
Sunrise - Sunset	42	6.0	5.8	5.7	6.0	5.7	5.6	5.5	5.5	5.2	4.9	5.8	5.9	5.6
MEAN NUMBER OF DAYS:														
Sunrise to Sunset														
-Clear	42	8.1	8.3	8.8	7.6	8.0	8.0	8.5	9.2	10.6	11.8	9.0	8.9	106.7
-Partly Cloudy	42	9.2	8.7	10.1	10.5	12.4	12.4	13.0	12.1	10.0	9.7	9.5	9.1	126.7
-Cloudy	42	13.7	11.2	12.1	11.9	10.7	9.6	9.5	9.7	9.4	9.5	11.5	13.0	131.8
Precipitation														
.01 inches or more	124	11.1	9.8	11.4	10.7	11.0	10.1	10.5	9.8	8.4	8.3	9.2	10.4	120.6
Snow,Ice pellets,hail														
1.0 inches or more	123	2.2	2.2	1.5	0.2	0.0	0.0	0.0	0.0	0.0	0.0	0.3	1.5	7.9
Thunderstorms	30	0.1	0.2	0.9	1.0	2.4	3.8	4.1	3.7	1.2	0.7	0.4	0.1	18.7
Heavy Fog Visibility 1/4 mile or less	6	0.0	0.0	0.0	0.0	0.0	0.0	0.0	0.0	0.0	0.0	0.0	0.0	0.0
Temperature °F														
-Maximum														
90° and above	80	0.0	0.0	0.0	0.1	1.0	3.2	6.8	4.6	1.4	0.1	0.0	0.0	17.3
32° and below	80	8.4	5.6	1.2	0.*	0.0	0.0	0.0	0.0	0.0	0.0	0.2	4.8	20.3
-Minimum														
32° and below	80	22.5	20.1	12.4	1.4	0.*	0.0	0.0	0.0	0.0	0.3	4.6	17.9	79.2
0° and below	80	0.2	0.2	0.0	0.0	0.0	0.0	0.0	0.0	0.0	0.0	0.0	0.1	0.4
AVG. STATION PRESS. (mb)	10	1013.9	1013.7	1012.3	1011.5	1011.3	1012.3	1012.5	1014.2	1014.7	1015.1	1014.5	1014.5	1013.4
RELATIVE HUMIDITY (%)														
Hour 01	49	65	64	64	64	70	73	74	76	76	72	69	67	70
Hour 07 (Local Time)	61	68	68	67	67	71	74	75	78	79	76	73	69	72
Hour 13	61	60	58	55	51	53	55	55	57	57	55	59	60	56
Hour 19	61	60	59	57	56	60	61	63	66	66	63	63	62	61
PRECIPITATION (inches):														
Water Equivalent														
-Normal		3.42	3.27	4.08	4.20	4.42	3.67	4.35	4.01	3.89	3.56	4.47	3.91	47.25
-Maximum Monthly	124	10.52	6.87	10.41	8.77	10.24	9.78	11.89	12.36	16.85	13.31	12.41	9.98	16.85
-Year		1979	1869	1980	1874	1989	1903	1889	1990	1882	1903	1972	1973	SEP 1882
-Minimum Monthly	124	0.58	0.46	0.90	0.95	0.30	0.02	0.49	0.24	0.21	0.14	0.34	0.25	0.02
-Year		1981	1895	1885	1881	1903	1949	1910	1964	1884	1963	1976	1955	JUN 1949
-Maximum in 24 hrs	81	3.91	3.04	4.25	4.22	4.88	4.74	3.60	5.78	8.30	11.17	8.09	3.21	11.17
-Year		1979	1973	1876	1984	1968	1884	1971	1971	1882	1903	1977	1909	OCT 1903
Snow,Ice pellets,hail														
-Maximum Monthly	125	27.4	27.9	30.5	13.5	T	0.0	T	0.0	0.0	0.8	19.0	29.6	30.5
-Year		1925	1934	1896	1875	1977		1990			1925	1898	1947	MAR 1896
-Maximum in 24 hrs	125	13.6	17.6	18.1	10.2	T	0.0	T	0.0	0.0	0.8	10.0	26.4	26.4
-Year		1978	1983	1941	1915	1977		1990			1925	1898	1947	DEC 1947
WIND:														
Mean Speed (mph)	58	10.7	10.8	11.0	10.5	8.8	8.1	7.6	7.6	8.1	8.9	9.9	10.4	9.4
Prevailing Direction through 1963		NW	NW	NW	NW	SW	SW	SW	SW	SW	SW	NW	NW	SW
Fastest Obs. 1 Min.														
-Direction (!!)	10	06	05	04	05	18	29	35	05	27	20	31	05	05
-Speed (MPH)	10	25	28	35	26	24	25	29	21	23	25	29	39	39
-Year		1992	1988	1984	1986	1989	1985	1986	1992	1985	1991	1989	1992	DEC 1992
Peak Gust														
-Direction (!!)	10	NE	NE	NE	SE	ENE	N	WSW	S	W	NE	NW	NE	NE
-Speed (mph)	10	43	51	63	46	44	41	37	43	52	46	58	64	64
-Date		1992	1984	1984	1984	1989	1990	1988	1986	1985	1988	1989	1992	DEC 1992

PRECIPITATION (inches)

NEW YORK, CENTRAL PARK, NEW YORK

YEAR	JAN	FEB	MAR	APR	MAY	JUNE	JULY	AUG	SEP	OCT	NOV	DEC	ANNUAL
1964	4.62	2.93	2.57	5.09	0.57	2.67	4.17	0.24	1.69	1.73	2.55	4.16	32.99
1965	3.09	3.66	2.49	2.90	1.58	1.27	1.33	2.73	1.70	2.16	1.46	1.72	26.09
1966	2.63	4.96	0.94	2.69	4.26	1.17	1.25	1.89	8.82	4.64	3.47	3.18	39.90
1967	1.39	2.68	5.97	3.45	4.08	4.64	6.99	5.94	1.84	3.47	2.59	6.08	49.12
1968	2.04	1.13	4.79	2.82	7.06	6.15	2.63	2.88	1.97	2.20	5.75	4.15	43.57
1969	1.10	3.05	3.73	3.99	2.67	3.16	7.37	2.53	8.32	1.97	3.58	7.07	48.54
1970	0.66	4.52	4.18	3.48	3.34	2.27	2.19	2.47	1.74	2.48	5.14	2.82	35.29
1971	2.67	5.33	3.80	2.95	4.24	2.31	7.20	9.37	7.36	4.14	5.64	1.76	56.77
1972	2.41	5.90	4.55	3.92	8.39	9.30	4.54	1.92	1.33	6.27	12.41	6.09	67.03
1973	4.53	4.55	3.60	8.05	4.51	4.55	5.89	3.08	2.75	3.92	1.82	9.98	57.23
1974	3.80	1.49	5.76	3.83	4.29	3.29	1.33	5.99	8.05	2.59	0.94	6.33	47.69
1975	4.76	3.33	3.32	3.04	3.38	7.58	11.77	3.05	9.32	3.70	4.33	3.63	61.21
1976	5.78	3.13	2.99	2.80	4.77	2.78	1.42	6.52	3.15	5.31	0.34	2.29	41.28
1977	2.25	2.51	7.41	3.75	1.71	3.83	1.60	4.57	4.75	5.03	12.26	5.06	54.73
1978	8.27	1.59	2.73	2.38	9.15	1.69	4.48	5.50	4.06	1.50	2.85	5.61	49.81
1979	10.52	4.58	4.40	4.04	6.23	1.56	1.76	4.27	4.83	3.87	3.38	2.69	52.13
1980	1.72	1.04	10.41	8.26	2.33	3.84	5.26	1.16	1.98	3.86	4.11	0.58	44.55
1981	0.58	6.04	1.19	3.42	3.56	2.71	6.21	0.59	3.45	3.49	1.69	5.18	38.11
1982	6.46	2.37	2.56	5.67	2.43	5.12	3.14	4.66	1.77	2.31	3.44	1.47	41.40
1983													
1984	1.87	4.86	6.30	6.62	9.74	5.76	7.03	1.38	2.51	3.63	4.07	3.26	57.03
1985	1.00	2.41	1.91	1.41	5.72	4.41	4.41	2.58	4.75	1.30	8.09	0.83	38.82
1986	4.23	2.86	1.46	3.93	1.68	1.86	5.56	4.24	2.20	1.92	6.85	6.16	42.95
1987	5.81	1.01	4.93	5.90	1.45	3.94	4.12	4.89	5.25	3.89	3.08	2.17	46.44
1988	3.64	3.91	2.10	2.20	5.27	1.29	8.14	2.19	2.34	3.56	8.90	1.13	44.67
1989	2.29	3.03	4.93	4.26	10.24	8.79	5.13	8.44	6.90	7.48	2.79	0.83	65.11
1990	5.34	2.33	3.64	5.12	9.10	2.50	3.51	12.36	2.24	6.38	2.82	5.58	60.92
1991	3.38	1.93	5.16	3.68	3.11	4.16	4.57	7.13	3.71	2.13	1.96	4.26	45.18
1992	1.68	1.87	4.08	1.76	4.02	4.77	4.49	3.49	4.89	1.16	5.64	5.50	43.35
1993	3.44	2.81	6.64	4.28	1.56	1.49	1.70	5.41	5.25	4.55	2.20	4.95	44.28
Record Mean	3.43	3.34	3.88	3.54	3.64	3.46	4.27	4.35	3.73	3.50	3.52	3.48	44.14

AVERAGE TEMPERATURE (deg. F)

NEW YORK, CENTRAL PARK, NEW YORK

YEAR	JAN	FEB	MAR	APR	MAY	JUNE	JULY	AUG	SEP	OCT	NOV	DEC	ANNUAL
1964	35.7	32.9	43.1	49.7	65.4	71.6	75.4	72.9	67.2	55.0	49.4	36.4	54.6
1965	29.7	33.9	40.0	50.6	66.4	70.1	74.3	73.2	67.5	57.3	46.8	40.5	54.2
1966	32.2	35.1	42.7	49.7	61.6	75.4	79.7	76.9	66.5	56.2	48.9	35.7	55.1
1967	37.4	29.2	37.6	49.6	55.2	72.8	75.3	73.9	66.7	57.2	42.5	38.2	53.0
1968	26.7	28.9	43.3	55.0	59.6	69.7	77.3	76.0	70.6	60.5	46.9	34.3	54.1
1969	31.8	32.6	40.1	55.9	65.3	73.1	74.8	77.4	69.0	57.7	46.4	33.4	54.8
1970	25.1	33.0	38.7	52.1	64.0	70.9	77.1	77.6	70.8	58.9	48.5	34.4	54.3
1971	27.0	35.1	40.1	50.8	61.4	74.2	77.8	75.9	71.6	62.7	45.1	40.8	55.2
1972	35.1	31.4	39.8	50.1	63.3	67.9	77.2	75.6	69.5	53.5	44.4	38.5	53.8
1973	35.5	32.5	46.4	53.4	59.5	73.4	77.4	77.6	69.5	60.2	48.3	39.0	56.1
1974	35.3	31.7	42.1	55.2	61.0	69.0	77.2	76.4	66.7	54.1	48.2	39.4	54.7
1975	37.3	35.8	40.2	47.9	65.8	70.5	75.8	74.4	64.2	59.2	52.3	35.9	54.9
1976	27.4	39.9	44.4	55.0	60.2	73.2	74.8	74.3	66.6	52.9	41.7	29.9	53.4
1977	22.1	33.5	46.8	53.7	65.0	70.2	79.0	75.7	68.2	54.9	47.3	35.7	54.3
1978	28.0	27.2	39.0	51.6	61.5	71.3	74.4	76.0	65.0	54.9	47.8	38.9	53.0
1979	33.6	25.5	46.9	52.6	65.3	69.2	76.9	76.8	70.5	57.3	52.5	41.1	55.7
1980	33.7	31.4	41.2	54.5	65.6	70.3	79.3	80.3	70.8	55.2	44.6	32.5	55.0
1981	26.3	39.3	42.3	56.2	64.8	73.0	78.5	76.0	67.6	54.4	47.7	36.5	55.2
1982	26.1	35.3	42.0	51.2	64.1	68.6	77.9	73.2	68.3	58.5	50.4	42.8	54.9
1983	34.5	36.4	44.0	52.3	60.2	73.4	79.5	77.7	71.8	57.9	48.9	35.2	56.0
1984	29.9	40.6	36.7	51.9	61.6	74.5	74.7	76.7	65.9	61.8	47.3	43.8	55.5
1985	28.8	36.6	45.8	55.5	65.3	68.6	76.2	75.4	70.5	59.5	50.0	34.2	55.5
1986	34.1	32.0	45.1	54.5	66.0	71.6	76.0	73.1	67.9	58.0	45.7	39.0	55.3
1987	32.3	33.2	45.2	53.4	63.6	72.8	78.0	74.2	67.7	53.8	47.7	39.5	55.1
1988	29.5	35.0	43.6	51.2	62.7	71.8	79.3	78.8	67.4	52.8	49.4	35.9	54.8
1989	37.4	34.5	42.4	52.2	62.1	72.0	75.0	74.0	68.1	58.2	45.7	25.9	54.0
1990	41.4	39.8	45.1	53.5	60.2	72.1	76.8	75.3	67.5	61.9	50.4	42.6	57.2
1991	34.9	40.0	44.6	55.7	68.7	74.1	77.7	77.1	67.5	58.4	48.3	39.6	57.2
1992	35.7	36.4	40.0	50.5	61.0	70.3	74.2	73.0	67.2	54.5	46.5	37.9	53.9
1993	36.3	30.8	39.7	53.3	65.7	73.3	80.2	77.2	67.3	56.0	48.8	37.3	55.5
Record Mean	32.2	33.1	41.2	51.5	62.2	71.1	76.3	74.8	68.1	57.7	47.0	36.0	54.3
Max	38.5	40.0	48.7	60.0	71.2	79.8	84.8	83.1	76.3	65.6	53.5	42.0	62.0
Min	25.8	26.2	33.6	43.1	53.2	62.3	67.8	66.6	59.8	49.8	40.4	29.9	46.6

REFERENCE NOTES

NEW YORK, CENTRAL PARK, NEW YORK

GENERAL
T - TRACE AMOUNT.
BLANK ENTRIES DENOTE MISSING/UNREPORTED DATA.
# INDICATES A STATION OR INSTRUMENT RELOCATION.

SPECIFIC

PM - INCLUDES LAST DAY OF PREVIOUS MONTH
ASOS - AUTOMATED SURFACE OBSERVING SYSTEM IN OPERATION DURING THESE MONTHS.

(*) - LENGTH OF RECORD IN YEARS, ALTHOUGH INDIVIDUAL MONTHS MAY BE MISSING.
0.* OR * - THE VALUE IS BETWEEN 0.0 AND 0.05
NORMALS - BASED ON THE 1961-1990 RECORD PERIOD.
EXTREMES - DATES ARE THE MOST RECENT OCCURRENCE
WIND DIR.- NUMERALS SHOW TENS OF DEGRESS CLOCKWISE FROM TRUE NORTH. "00" INDICATES CALM.
RESULTANT DIRECTIONS ARE GIVEN TO WHOLE DEGREES.
BOLD VALUES INDICATE EXTREME VALUES WHICH OCCURRED AFTER THE ASOS SYSTEM WAS COMMISSIONED.

RECORD = PERIOD OF RECORD
RECORD MEAN PRECIPITATION IS THE MEAN OF ALL DAILY PRECIPITATION AMOUNTS DURING THE PERIOD OF RECORD.
RECORD MAX(MIN) TEMPERATURE IS THE MEAN OF ALL DAILY MAX(MIN) TEMPERATURES DURING THE PERIOD OF RECORD.
RECORD MEAN TEMPERATURE IS THE SUM OF THE RECORD MAX AND RECORD MIN DIVIDED BY 2.
AVERAGE TEMPERATURE IS THE SUM OF THE MEAN DAILY MAX AND MIN TEMPERATURE DIVIDED BY 2.

EXCEPTIONS

1. MEAN SKY COVER, AND DAYS CLEAR-PARTLY CLOUDY-CLOUDY ARE THROUGH 1966.
2. PERCENT OF POSSIBLE SUNSHINE, AND MEAN WIND SPEED ARE THROUGH 1976.
3. FASTEST MILE WINDS ARE THROUGH MARCH 1977 AND FEBRUARY 1980 THROUGH OCTOBER 1981.
4. RELATIVE HUMIDITY IS THROUGH 1980.
5. LIQUID PRECIPITATION FOR 1983 IS NOT CONSIDERED IN DETERMINING EXTREMES.

RECORD MEANS ARE THROUGH THE CURRENT YEAR, BEGINNING IN 1912 FOR TEMPERATURE
1869 FOR PRECIPITATION
1869 FOR SNOWFALL

HEATING DEGREE DAYS Base 65 deg. F — NEW YORK, CENTRAL PARK, NEW YORK

SEASON	JULY	AUG	SEP	OCT	NOV	DEC	JAN	FEB	MAR	APR	MAY	JUNE	TOTAL
1964-65	3	0	63	308	461	879	1088	867	765	426	64	30	4954
1965-66	0	13	54	239	538	755	1007	830	685	451	166	9	4747
1966-67	0	0	63	270	475	901	849	999	843	462	305	5	5172
1967-68	0	4	55	264	671	825	1179	1042	668	292	170	15	5185
1968-69	0	0	3	183	538	944	1023	902	768	285	74	0	4720
1969-70	0	0	28	240	551	974	1227	890	809	387	109	6	5221
1970-71	0	0	27	210	490	940	1173	830	764	419	135	9	4997
1971-72	0	0	14	106	596	743	920	965	775	445	94	26	4684
1972-73	2	0	25	355	611	812	907	903	572	362	188	2	4739
1973-74	0	0	29	162	493	800	913	925	704	309	165	27	4527
1974-75	1	0	59	333	502	789	852	812	764	507	86	11	4716
1975-76	0	3	62	193	387	898	1163	723	630	360	167	18	4604
1976-77	0	4	44	373	692	1082	1322	877	560	354	100	27	5435
1977-78	0	0	56	307	524	903	1140	1051	797	394	179	13	5364
1978-79	5	0	75	311	510	802	969	1100	554	369	55	14	4764
1979-80	4	4	20	271	373	734	963	969	731	310	67	22	4468
1980-81	0	0	31	305	602	1000	1194	715	698	264	78	3	4890
1981-82	0	0	48	320	513	876	1198	825	707	413	74	36	5010
1982-83	0	5	24	229	446	679	936	793	644	393	161	3	4313
1983-84	0	0	34	249	480	914	1082	698	870	389	137	9	4862
1984-85	0	0	69	114	525	654	1113	789	596	305	79	24	4268
1985-86	0	0	17	188	448	947	950	917	615	312	89	11	4494
1986-87	0	10	27	236	572	797	1008	883	608	348	146	8	4643
1987-88	0	2	29	343	512	780	1093	867	656	409	133	31	4855
1988-89	3	0	23	385	459	896	844	849	696	376	143	14	4688
1989-90	0	1	54	217	572	1205	724	702	612	366	150	4	4607
1990-91	3	2	57	166	436	686	927	696	625	311	61	3	3973
1991-92	0	0	60	222	496	782	902	827	767	434	160	12	4662
1992-93	0	3	54	324	547	834	882	953	779	347	57	14	4794
1993-94	0	0	65	275	483	852							

COOLING DEGREE DAYS Base 65 deg. F — NEW YORK, CENTRAL PARK, NEW YORK

YEAR	JAN	FEB	MAR	APR	MAY	JUNE	JULY	AUG	SEP	OCT	NOV	DEC	TOTAL
1969	0	0	0	20	88	250	310	392	154	20	0	0	1234
1970	0	0	0	8	86	190	385	398	207	30	0	0	1304
1971	0	0	0	0	29	290	404	347	218	40	7	0	1335
1972	0	0	0	5	47	118	384	338	169	3	0	0	1064
1973	0	0	0	20	23	260	390	401	171	22	2	0	1289
1974	0	0	0	19	47	155	385	360	115	1	6	0	1088
1975	0	0	0	0	120	185	341	299	43	22	15	0	1025
1976	0	0	0	65	24	270	310	299	103	5	0	0	1076
1977	0	0	3	22	110	189	442	338	159	0	0	0	1263
1978	0	0	0	0	77	209	301	348	81	4	0	0	1020
1979	0	0	0	4	71	149	378	376	192	43	5	0	1218
1980	0	0	0	1	94	188	448	480	213	11	0	0	1435
1981	0	0	0	4	78	252	425	347	129	0	0	0	1235
1982	0	0	0	7	55	152	405	266	129	36	16	0	1066
1983	0	0	0	19	16	259	460	404	244	35	0	0	1437
1984	0	0	0	3	39	301	306	367	106	26	0	0	1148
1985	0	0	8	28	95	139	353	329	189	21	5	0	1167
1986	0	0	5	4	127	214	348	269	120	27	0	0	1114
1987	0	0	0	5	110	251	406	295	118	0	2	0	1187
1988	0	0	0	0	66	243	455	435	104	12	0	0	1315
1989	0	0	4	0	61	231	313	287	151	10	0	0	1057
1990	0	0	4	25	8	225	375	328	140	77	4	0	1186
1991	0	0	0	38	182	280	403	382	142	24	1	0	1452
1992	0	0	0	5	46	174	292	256	127	8	0	0	908
1993	0	0	0	0	82	269	474	386	140	3	4	0	1358

SNOWFALL (inches) — NEW YORK, CENTRAL PARK, NEW YORK

SEASON	JULY	AUG	SEP	OCT	NOV	DEC	JAN	FEB	MAR	APR	MAY	JUNE	TOTAL
1964-65	0.0	0.0	0.0	0.0	0.0	3.1	14.8	2.5	2.8	1.2	0.0	0.0	24.4
1965-66	0.0	0.0	0.0	T	0.0	T	11.6	9.8	T	0.0	0.0	0.0	21.4
1966-67	0.0	0.0	0.0	0.0	0.0	9.1	1.4	23.6	17.4	T	0.0	0.0	51.5
1967-68	0.0	0.0	0.0	0.0	3.2	5.5	3.6	1.1	6.1	0.0	0.0	0.0	19.5
1968-69	0.0	0.0	0.0	0.0	T	7.0	1.0	16.6	5.6	0.0	0.0	0.0	30.2
1969-70	0.0	0.0	0.0	0.0	T	6.8	8.4	6.4	4.0	T	0.0	0.0	25.6
1970-71	0.0	0.0	0.0	0.0	0.0	2.4	11.4	T	1.3	0.4	0.0	0.0	15.5
1971-72	0.0	0.0	0.0	0.0	T	T	2.8	17.8	2.3	T	0.0	0.0	22.9
1972-73	0.0	0.0	0.0	T	T	T	1.8	0.8	0.2	T	0.0	0.0	2.8
1973-74	0.0	0.0	0.0	0.0	0.0	2.8	7.8	9.4	3.2	0.3	0.0	0.0	23.5
1974-75	0.0	0.0	0.0	0.0	0.1	0.1	2.0	10.6	0.3	T	0.0	0.0	13.1
1975-76	0.0	0.0	0.0	0.0	T	2.3	5.6	5.0	4.4	T	0.0	0.0	17.3
1976-77	0.0	0.0	0.0	0.0	T	5.1	13.0	5.8	0.6	T	T	0.0	24.5
1977-78	0.0	0.0	0.0	0.0	0.2	0.4	20.3	23.0	6.8	T	0.0	0.0	50.7
1978-79	0.0	0.0	0.0	0.0	2.2	0.5	6.6	20.1	T	T	0.0	0.0	29.4
1979-80	0.0	0.0	0.0	T	0.0	3.5	2.0	2.7	4.6	T	0.0	0.0	12.8
1980-81	0.0	0.0	0.0	0.0	T	2.8	8.0	T	8.6	0.0	0.0	0.0	19.4
1981-82	0.0	0.0	0.0	0.0	0.0	2.1	11.8	0.4	0.7	9.6	0.0	0.0	24.6
1982-83	0.0	0.0	0.0	0.0	0.0	3.0	1.9	23.5	T	0.8	0.0	0.0	29.2
1983-84	0.0	0.0	0.0	0.0	T	1.6	11.7	0.2	11.9	0.0	0.0	0.0	25.4
1984-85	0.0	0.0	0.0	0.0	T	5.5	8.4	10.0	0.2	T	0.0	0.0	24.1
1985-86	0.0	0.0	0.0	0.0	T	0.9	2.2	9.9	T	T	0.0	0.0	13.0
1986-87	0.0	0.0	0.0	0.0	T	0.6	13.6	7.0	1.9	0.0	0.0	0.0	23.1
1987-88	0.0	0.0	0.0	0.0	1.1	2.6	13.9	1.5	T	0.0	0.0	0.0	19.1
1988-89	0.0	0.0	0.0	0.0	0.0	0.3	5.0	0.3	2.5	0.0	0.0	0.0	8.1
1989-90	0.0	0.0	0.0	0.0	4.7	1.4	1.8	1.8	3.1	0.6	0.0	0.0	13.4
1990-91	T	0.0	0.0	0.0	0.0	7.2	8.4	9.1	0.2	0.0	0.0	0.0	24.9
1991-92	0.0	0.0	0.0	0.0	T	0.7	1.5	1.0	9.4	T	0.0	0.0	12.6
1992-93	0.0	0.0	0.0	0.0	0.0	0.4	1.5	10.7	11.9	0.0	0.0	0.0	24.5
1993-94	0.0	0.0	0.0	0.0	T	6.9							
Record Mean	T	0.0	0.0	T	0.9	5.4	7.5	8.4	5.0	0.9	T	0.0	28.2

J.F.K INTERNATIONAL AIRPORT
N.Y.C, NEW YORK

New York City, in area exceeding 300 square miles, is located on the Atlantic coastal plain at the mouth of the Hudson River. The terrain is laced with numerous waterways, all but one of the five boroughs in the city are situated on islands. Elevations range from less than 50 feet over most of Manhattan, Brooklyn, and Queens to almost 300 feet in northern Manhattan and the Bronx, and over 400 feet in Staten Island. Extensive suburban areas on Long Island, and in Connecticut, New York State and New Jersey border the city on the east, north, and west. About 30 miles to the west and northwest, hills rise to about 1,500 feet and to the north in upper Westchester County to 800 feet. To the southwest and to the east are the low-lying land areas of the New Jersey coastal plain and of Long Island, bordering on the Atlantic.

The New York Metropolitan area is close to the path of most storm and frontal systems which move across the North American continent. Therefore, weather conditions affecting the city most often approach from a westerly direction. New York City can thus experience higher temperatures in summer and lower ones in winter than would otherwise be expected in a coastal area. However, the frequent passage of weather systems often helps reduce the length of both warm and cold spells, and is also a major factor in keeping periods of prolonged air stagnation to a minimum.

Although continental influence predominates, oceanic influence is by no means absent. During the summer local sea breezes, winds blowing onshore from the cool water surface, often moderate the afternoon heat. The effect of the sea breeze diminishes inland. On winter mornings, ocean temperatures which are warm relative to the land reinforce the effect of the city heat island and low temperatures are often 10–20 degrees lower in the inland suburbs than in the central city. The relatively warm water temperatures also delay the advent of winter snows. Conversely, the lag in warming of water temperatures keeps spring temperatures relatively cool. One year-round measure of the ocean influence is the small average daily variation in temperature.

Precipitation is moderate and distributed fairly evenly throughout the year. Most of the rainfall from May through October comes from thunderstorms It is therefore usually of brief duration and sometimes intense. Heavy rains of long duration associated with tropical storms occur infrequently in late summer or fall. For the other months of the year precipitation is more likely to be associated with widespread storm areas, so that day-long rain, snow or a mixture of both is more common. Precipitation accompanying winter storms sometimes starts as snow, later changes to rain, and perhaps briefly back to snow before ending. Coastal storms, occurring most often in the fall and winter months, produce on occasion considerable amounts of precipitation and have been responsible for record rains, snows, and high winds.

Relative humidity averages about the same over the metropolitan area except again that the immediate coastal areas are more humid than inland locations.

Local Climatological Data is published for three locations in New York City, Central Park, La Guardia Airport, and John F. Kennedy International Airport. Other nearby locations for which it is published are Newark, New Jersey, and Bridgeport, Connecticut.

NORMALS, MEANS, AND EXTREMES

J.F.K INTERNATIONAL AIRPORT N.Y.C. NEW YORK

LATITUDE: 40°39'N LONGITUDE: 73°47' W ELEVATION: FT. GRND 13 BARO 32 TIME ZONE: EASTERN WBAN: 94789

	(a)	JAN	FEB	MAR	APR	MAY	JUNE	JULY	AUG	SEP	OCT	NOV	DEC	YEAR
TEMPERATURE °F:														
Normals														
-Daily Maximum		37.5	39.5	48.0	58.2	67.6	77.0	82.8	81.9	75.0	64.7	53.8	42.7	60.7
-Daily Minimum		24.9	26.3	34.0	42.8	52.4	61.8	68.1	67.3	60.1	49.3	40.5	30.5	46.5
-Monthly		31.2	32.9	41.0	50.5	60.0	69.4	75.5	74.7	67.6	57.0	47.2	36.6	53.6
Extremes														
-Record Highest	33	65	68	85	90	99	99	104	100	98	85	77	70	104
-Year		1974	1991	1990	1977	1969	1964	1966	1983	1983	1986	1993	1982	JUL 1966
-Record Lowest	33	-2	-2	7	20	34	45	55	46	40	25	19	2	-2
-Year		1985	1963	1967	1982	1966	1967	1979	1965	1963	1961	1987	1983	JAN 1985
NORMAL DEGREE DAYS:														
Heating (base 65°F)		1048	899	744	435	178	9	0	0	39	261	534	880	5027
Cooling (base 65°F)		0	0	0	0	23	141	326	301	117	13	0	0	921
% OF POSSIBLE SUNSHINE														
MEAN SKY COVER (tenths)														
Sunrise - Sunset	35	6.3	6.2	6.2	6.2	6.2	6.0	5.9	5.8	5.6	5.3	6.3	6.4	6.0
MEAN NUMBER OF DAYS:														
Sunrise to Sunset														
-Clear	35	8.0	7.8	8.3	7.9	6.9	7.2	7.4	7.8	9.8	11.2	7.7	7.5	97.7
-Partly Cloudy	35	8.4	7.4	9.0	9.0	11.5	11.3	12.2	12.8	9.2	8.7	8.3	8.3	116.2
-Cloudy	35	14.5	13.0	13.7	13.1	12.6	11.4	11.3	10.4	11.0	11.2	13.9	15.1	151.3
Precipitation														
.01 inches or more	35	10.2	9.6	11.3	10.8	10.9	9.9	9.5	9.1	8.5	7.6	9.7	11.1	118.2
Snow, Ice pellets, hail														
1.0 inches or more	35	2.2	1.8	1.1	0.1	0.0	0.0	0.0	0.0	0.0	0.0	0.1	1.0	6.3
Thunderstorms	35	0.1	0.2	1.1	1.6	3.1	4.0	4.9	4.8	1.9	1.0	0.5	0.1	23.3
Heavy Fog Visibility														
1/4 mile or less	35	2.5	2.6	3.3	2.8	4.0	3.7	2.3	1.1	1.1	2.3	2.1	2.5	30.2
Temperature °F														
-Maximum														
90° and above	32	0.0	0.0	0.0	0.*	0.5	1.9	4.3	2.7	0.8	0.0	0.0	0.0	10.3
32° and below	32	8.8	5.5	0.9	0.*	0.0	0.0	0.0	0.0	0.0	0.0	0.1	4.1	19.4
-Minimum														
32° and below	32	23.4	20.6	11.8	1.3	0.0	0.0	0.0	0.0	0.0	0.3	4.3	17.3	78.9
0° and below	32	0.2	0.1	0.0	0.0	0.0	0.0	0.0	0.0	0.0	0.0	0.0	0.0	0.2
AVG. STATION PRESS. (mb)	21	1017.0	1017.1	1015.8	1014.2	1014.7	1014.4	1014.8	1016.5	1017.8	1018.3	1017.9	1017.9	1016.4
RELATIVE HUMIDITY (%)														
Hour 01	32	68	68	68	71	78	79	79	80	80	76	73	70	74
Hour 07 (Local Time)	32	71	70	70	70	73	75	75	78	80	78	75	72	74
Hour 13	32	59	57	56	55	59	60	59	60	60	56	59	60	58
Hour 19	32	63	62	63	65	70	71	72	72	72	69	67	64	68
PRECIPITATION (inches):														
Water Equivalent														
-Normal		3.17	3.02	3.59	3.90	3.80	3.65	3.80	3.41	3.30	2.88	3.65	3.42	41.59
-Maximum Monthly	46	8.33	5.48	8.17	9.51	10.71	9.20	8.48	17.41	9.65	6.58	9.51	6.73	17.41
-Year		1979	1960	1980	1983	1989	1984	1969	1955	1975	1989	1972	1986	AUG 1955
-Minimum Monthly	46	0.21	1.01	0.95	1.12	0.38	T	0.46	0.42	0.70	0.09	0.32	0.61	T
-Year		1956	1987	1981	1963	1955	1949	1954	1972	1951	1963	1976	1989	JUN 1949
-Maximum in 24 hrs	46	3.25	2.87	2.40	3.31	2.88	6.27	5.92	6.59	5.83	3.42	4.09	2.94	6.59
-Year		1979	1958	1977	1980	1968	1984	1984	1955	1960	1972	1972	1992	AUG 1955
Snow, Ice pellets, hail														
-Maximum Monthly	35	20.1	25.3	21.1	8.2	T	0.0	0.0	0.0	0.0	0.5	3.7	16.4	25.3
-Year		1978	1961	1960	1982	1967					1962	1989	1960	FEB 1961
-Maximum in 24 hrs	35	14.2	21.7	9.0	8.0	T	0.0	0.0	0.0	0.0	0.5	3.7	8.2	21.7
-Year		1978	1983	1993	1982	1967					1962	1989	1960	FEB 1983
WIND:														
Mean Speed (mph)	35	13.3	13.5	13.6	13.0	11.5	10.7	10.3	10.1	10.5	11.1	12.3	12.8	11.9
Prevailing Direction														
Fastest Obs. 1 Min.														
-Direction (!!)	30	26	25	06	29	16	29	22	30	28	26	30	06	26
-Speed (MPH)	30	52	46	46	44	44	41	40	46	46	44	44	46	52
-Year		1966	1967	1993	1982	1973	1981	1984	1965	1985	1980	1989	1974	JAN 1966
Peak Gust														
-Direction (!!)	10	W	NW	NE	E	N	NW	SW	W	W	NW	W	E	N
-Speed (mph)	10	52	60	61	51	71	49	54	47	58	49	58	60	71
-Date		1992	1985	1984	1987	1987	1984	1984	1992	1985	1986	1988	1992	MAY 1987

PRECIPITATION (inches)

J.F.K INTERNATIONAL AIRPORT N.Y.C, NEW YORK

YEAR	JAN	FEB	MAR	APR	MAY	JUNE	JULY	AUG	SEP	OCT	NOV	DEC	ANNUAL
1964	4.59	2.84	2.34	6.60	0.62	2.22	4.65	0.87	2.58	2.03	2.02	5.34	36.70
1965	2.99	2.79	2.74	3.18	1.61	1.12	2.45	2.11	0.96	1.85	1.36	2.22	25.38
1966	3.18	4.08	1.35	3.24	3.51	0.21	0.69	2.99	7.47	4.92	2.29	2.78	36.71
1967	1.47	2.48	5.48	3.67	5.12	3.39	4.98	7.36	1.44	1.89	2.13	5.70	45.11
1968	1.99	1.73	4.90	1.97	5.34	4.16	2.58	2.78	2.54	1.85	6.84	4.55	41.23
1969	1.10	3.72	2.03	4.26	1.72	2.85	8.48	2.39	3.80	1.70	2.48	6.16	40.69
1970	0.56	3.33	4.03	3.15	2.01	3.04	0.54	2.57	1.61	1.93	3.84	2.55	29.16
1971	2.31	4.76	3.17	2.90	2.61	1.71	3.19	4.35	3.72	3.28	4.07	1.46	37.53
1972	1.74	4.89	4.09	3.32	5.02	6.70	2.60	0.42	1.49	5.00	9.51	4.26	49.04
1973	3.37	2.61	3.72	6.98	3.97	5.25	2.64	0.84	3.03	2.26	1.41	5.96	42.04
1974	2.83	1.41	4.60	2.33	2.69	2.38	1.29	4.24	5.97	2.19	1.10	6.07	37.10
1975	5.05	3.57	3.50	3.13	3.52	7.06	7.86	3.54	9.65	3.25	4.02	3.03	57.18
1976	4.55	3.16	2.30	2.26	3.56	3.97	3.50	8.30	2.24	4.06	0.32	2.02	40.24
1977	2.67	2.41	4.70	3.62	2.29	3.25	2.30	4.85	6.69	4.41	6.59	4.92	48.70
1978	7.74	1.74	2.38	2.10	6.44	1.24	3.50	3.58	3.41	1.68	1.86	5.29	40.96
1979	8.33	4.58	3.77	3.03	5.02	3.42	1.66	5.65	4.13	3.53	2.00	2.09	47.21
1980	1.55	1.06	8.17	7.53	2.77	3.57	3.93	0.85	1.32	2.79	3.98	0.90	38.42
1981	0.49	4.33	0.95	2.77	2.80	4.28	5.32	1.05	2.67	3.74	1.77	4.40	34.57
1982	4.83	2.08	2.61	4.40	3.43	5.08	2.67	2.15	1.15	1.29	2.85	1.52	34.06
1983	4.14	2.79	6.66	9.51	3.56	2.10	3.69	5.84	3.95	5.12	5.62	6.14	59.12
1984	1.60	4.55	5.99	5.82	8.59	9.20	6.66	1.47	2.02	1.91	2.47	2.94	53.22
1985	1.06	2.13	2.17	2.12	4.77	4.56	2.63	1.80	3.29	0.93	6.68	0.87	33.01
1986	3.86	3.01	2.04	3.99	1.09	1.61	4.75	3.69	1.75	1.51	6.43	6.73	40.46
1987	5.62	1.01	3.26	4.74	1.70	4.22	3.71	3.84	2.98	3.02	2.73	2.24	39.07
1988	3.13	3.85	2.38	2.01	5.07	1.53	6.70	2.01	2.86	3.39	6.65	1.08	40.66
1989	2.08	2.64	4.17	3.71	10.71	8.07	5.99	4.35	4.31	6.58	2.51	0.61	55.73
1990	4.41	1.17	2.32	4.64	6.97	2.37	4.37	6.68	1.80	5.03	1.59	3.89	45.24
1991	3.73	1.58	3.63	3.90	4.20	1.90	2.48	7.59	2.79	1.52	1.77	3.64	38.73
1992	1.34	1.52	3.23	1.37	3.49	2.57	5.75	5.31	1.78	1.56	5.53	4.93	38.38
1993	2.00	2.20	6.83	3.06	1.88	1.15	2.39	2.17	5.55	3.72	0.88	3.78	35.61
Record Mean	2.93	3.00	3.86	3.83	3.73	3.21	3.81	3.98	3.31	2.99	3.71	3.52	41.89

AVERAGE TEMPERATURE (deg. F)

J.F.K INTERNATIONAL AIRPORT N.Y.C, NEW YORK

YEAR	JAN	FEB	MAR	APR	MAY	JUNE	JULY	AUG	SEP	OCT	NOV	DEC	ANNUAL
1964	33.5	31.5	39.9	47.1	63.1	70.4	74.5	72.2	68.3	55.8	48.8	36.5	53.5
1965	28.5	32.3	38.6	48.3	61.8	67.9	72.2	72.3	66.1	55.6	45.4	38.6	52.3
1966	31.2	33.8	41.9	47.5	56.8	70.0	77.0	74.5	66.2	55.2	47.7	35.6	53.1
1967	36.4	27.8	34.9	46.4	51.7	66.7	73.3	71.4	64.4	55.5	41.1	36.7	50.5
1968	25.6	27.9	40.2	50.5	56.8	68.1	77.0	75.8	70.1	58.2	46.6	33.0	52.5
1969	31.3	30.8	38.2	52.3	61.7	71.2	75.0	77.4	68.5	58.1	47.2	35.2	53.9
1970	26.7	33.3	37.5	50.0	61.7	70.3	77.0	77.4	70.0	58.3	48.6	35.0	53.8
1971	27.9	34.2	38.9	47.8	59.5	72.1	77.2	76.5	71.8	63.5	47.2	42.8	55.0
1972	36.4	33.5	40.4	49.8	61.2	68.0	77.0	74.8	69.3	54.5	46.5	41.2	54.4
1973	36.6	35.1	47.7	54.2	57.2	69.2	74.0	75.3	67.2	57.7	45.9	37.2	54.8
1974	35.1	31.6	42.7	52.5	59.6	67.8	76.6	75.7	66.9	54.4	47.9	39.8	54.2
1975	37.9	35.9	39.9	47.1	62.3	69.0	75.0	74.4	64.1	60.0	52.8	36.6	54.6
1976	28.1	38.7	43.1	53.5	58.6	70.2	72.9	72.7	65.6	52.8	40.7	30.1	52.3
1977	22.0	32.1	43.7	51.0	61.7	67.4	75.1	73.3	66.8	53.7	47.2	34.3	52.3
1978	29.2	26.8	38.7	50.6	59.5	70.7	74.5	77.5	66.6	57.3	48.8	39.3	53.3
1979	33.2	25.0	45.4	51.5	62.8	66.9	74.7	73.6	67.1	54.9	48.3	38.1	53.5
1980	32.2	29.2	38.4	51.9	62.9	68.1	76.5	78.0	69.9	56.3	45.4	33.7	53.6
1981	26.5	38.7	42.2	53.7	60.6	70.5	77.3	75.4	68.1	54.6	47.7	37.5	54.4
1982	26.7	35.8	40.9	49.2	61.8	66.9	75.7	72.7	67.1	57.6	50.4	43.6	54.1
1983	35.8	36.2	44.6	51.5	58.0	71.6	78.7	76.7	71.0	58.8	49.6	35.9	55.7
1984	29.3	39.8	36.2	51.1	61.0	73.9	74.8	77.6	66.1	61.7	47.4	43.9	55.2
1985	28.8	34.8	45.1	54.2	63.1	68.4	74.8	73.9	69.2	58.3	50.1	34.0	54.6
1986	33.3	31.9	42.6	52.0	63.3	70.0	75.3	72.9	66.6	56.8	45.2	39.6	54.1
1987	32.4	32.5	43.4	51.3	60.1	71.1	76.2	72.9	67.3	53.6	47.1	39.3	53.9
1988	28.8	34.1	42.2	49.7	60.3	69.9	75.8	76.5	67.1	52.4	48.1	35.9	53.4
1989	36.5	34.0	40.5	50.0	59.9	71.5	74.2	74.1	68.0	57.7	45.5	26.6	53.2
1990	39.4	38.8	42.4	50.0	58.3	69.1	75.5	75.4	67.1	61.3	48.9	41.7	55.7
1991	34.7	39.0	43.4	52.4	66.2	72.3	76.7	76.8	67.2	58.6	47.6	40.0	56.2
1992	35.7	36.2	39.3	48.5	58.8	68.0	73.6	72.2	67.0	54.2	46.6	37.9	53.2
1993	36.8	30.5	38.9	50.5	63.0	70.5	78.7	75.0	67.4	55.6	46.9	36.9	54.2
Record Mean	31.8	33.7	40.8	50.6	60.3	69.7	75.6	74.6	67.6	57.0	47.1	36.7	53.8
Max	38.1	40.3	47.7	58.2	68.2	77.5	83.0	81.9	75.0	64.6	53.7	42.8	60.9
Min	25.5	27.0	33.8	42.9	52.5	61.9	68.1	67.3	60.1	49.5	40.4	30.6	46.6

REFERENCE NOTES

J.F.K INTERNATIONAL AIRPORT N.Y.C, NEW YORK

GENERAL

T - TRACE AMOUNT.
BLANK ENTRIES DENOTE MISSING/UNREPORTED DATA.
# INDICATES A STATION OR INSTRUMENT RELOCATION.

SPECIFIC

(a) - LENGTH OF RECORD IN YEARS, ALTHOUGH INDIVIDUAL MONTHS MAY BE MISSING.
0.* OR * - THE VALUE IS BETWEEN 0.0 AND 0.05
NORMALS - BASED ON THE 1961-1990 RECORD PERIOD.
EXTREMES - DATES ARE THE MOST RECENT OCCURRENCE
WIND DIR.- NUMERALS SHOW TENS OF DEGRESS CLOCKWISE FROM TRUE NORTH. "00" INDICATES CALM.
RESULTANT DIRECTIONS ARE GIVEN TO WHOLE DEGREES.
BOLD VALUES INDICATE EXTREME VALUES WHICH OCCURRED AFTER THE ASOS SYSTEM WAS COMMISSIONED.

RECORD = PERIOD OF RECORD
RECORD MEAN PRECIPITATION IS THE MEAN OF ALL DAILY PRECIPITATION AMOUNTS DURING THE PERIOD OF RECORD.
RECORD MAX(MIN) TEMPERATURE IS THE MEAN OF ALL DAILY MAX(MIN) TEMPERATURES DURING THE PERIOD OF RECORD.
RECORD MEAN TEMPERATURE IS THE SUM OF THE RECORD MAX AND RECORD MIN DIVIDED BY 2.
AVERAGE TEMPERATURE IS THE SUM OF THE MEAN DAILY MAX AND MIN TEMPERATURE DIVIDED BY 2.

EXCEPTIONS

1. FROM AUGUST 1966-MARCH 1969, MAXIMUM 24-HOUR PRECIPITATION IS FOR CALENDAR DAY (MIDNIGHT TO MIDNIGHT) INSTEAD OF FOR ANY CONSECUTIVE 24 HOURS.
2. MAXIMUM 24-HOUR SNOW IS FOR CALENDAR DAY THROUGH MARCH 1969.

RECORD MEANS ARE THROUGH THE CURRENT YEAR, BEGINNING IN 1951 FOR TEMPERATURE
1951 FOR PRECIPITATION
1959 FOR SNOWFALL

HEATING DEGREE DAYS Base 65 deg. F — J.F.K INTERNATIONAL AIRPORT N.Y.C, NEW YORK

SEASON	JULY	AUG	SEP	OCT	NOV	DEC	JAN	FEB	MAR	APR	MAY	JUNE	TOTAL
1964-65	0	0	42	282	478	878	1126	909	809	496	146	26	5192
1965-66	0	14	58	289	584	810	1041	868	707	521	256	14	5162
1966-67	0	0	60	295	513	904	883	1033	930	554	408	22	5602
1967-68	0	4	72	305	710	871	1214	1066	761	427	245	19	5694
1968-69	0	0	3	221	545	983	1038	950	822	374	144	0	5080
1969-70	0	0	37	232	526	917	1180	881	844	444	134	3	5198
1970-71	0	0	25	219	487	921	1144	858	803	508	180	9	5154
1971-72	0	0	11	83	532	685	881	904	755	449	123	9	4432
1972-73	0	0	21	321	549	732	876	831	530	325	242	13	4440
1973-74	0	0	64	234	568	856	920	929	686	373	183	26	4839
1974-75	0	0	47	322	508	775	833	809	770	531	135	13	4743
1975-76	0	3	60	171	367	874	1140	756	672	348	195	28	4614
1976-77	0	6	53	373	726	1076	1326	916	651	418	147	34	5726
1977-78	0	7	68	344	530	947	1104	1062	805	422	197	7	5493
1978-79	2	0	62	237	481	791	978	1115	603	397	88	19	4773
1979-80	2	6	45	313	493	827	1010	1030	823	387	106	39	5081
1980-81	0	0	25	266	585	962	1184	731	700	334	159	12	4958
1981-82	0	0	46	313	511	844	1182	812	741	468	110	44	5071
1982-83	0	6	24	245	434	655	901	801	625	400	212	8	4311
1983-84	0	2	38	224	455	891	1101	726	886	409	134	7	4873
1984-85	0	0	67	119	522	647	1118	841	606	328	95	18	4361
1985-86	0	0	26	213	441	954	976	920	688	383	124	15	4740
1986-87	0	8	38	274	588	781	1001	906	667	405	211	11	4890
1987-88	0	2	23	345	532	789	1118	892	702	454	171	33	5061
1988-89	0	0	28	388	502	893	879	862	753	443	174	6	4928
1989-90	0	0	51	221	581	1181	790	726	696	444	201	11	4902
1990-91	3	2	50	166	474	717	931	723	661	378	83	11	4199
1991-92	0	0	59	213	514	768	896	829	788	490	217	12	4786
1992-93	1	1	47	331	543	835	869	960	801	428	98	11	4925
1993-94	0	0	50	285	541	866							

COOLING DEGREE DAYS Base 65 deg. F — J.F.K INTERNATIONAL AIRPORT N.Y.C, NEW YORK

YEAR	JAN	FEB	MAR	APR	MAY	JUNE	JULY	AUG	SEP	OCT	NOV	DEC	TOTAL
1969	0	0	0	1	48	192	313	390	147	26	0	0	1117
1970	0	0	0	0	42	168	378	389	182	17	0	0	1176
1971	0	0	0	0	16	231	384	365	221	45	8	0	1270
1972	0	0	0	0	11	107	378	309	159	4	0	0	968
1973	0	0	0	9	7	145	287	324	137	14	0	0	923
1974	0	0	0	7	24	116	366	342	111	1	1	0	968
1975	0	0	0	0	58	137	319	301	41	24	7	0	887
1976	0	0	0	9	5	187	250	253	80	2	0	0	786
1977	0	0	0	5	51	114	321	268	127	0	0	0	886
1978	0	0	0	0	37	185	305	391	119	5	0	0	1042
1979	0	0	0	0	26	83	310	283	115	7	0	0	824
1980	0	0	0	0	48	139	365	412	178	6	0	0	1148
1981	0	0	0	3	30	184	390	327	143	0	0	0	1077
1982	0	0	0	0	16	107	336	252	92	21	4	0	828
1983	0	0	0	0	1	213	431	371	223	36	0	0	1275
1984	0	0	0	0	16	285	310	396	106	21	1	0	1135
1985	0	0	0	10	44	126	312	284	157	14	0	0	947
1986	0	0	0	0	80	172	326	262	94	27	0	0	961
1987	0	0	0	0	65	200	355	256	99	0	0	0	975
1988	0	0	0	0	33	186	344	365	99	2	0	0	1029
1989	0	0	0	0	20	208	293	286	147	3	0	0	957
1990	0	0	0	2	0	142	337	332	118	57	0	0	988
1991	0	0	0	5	124	238	371	374	130	23	0	0	1265
1992	0	0	0	0	31	107	276	233	112	3	0	0	762
1993	0	0	0	0	45	182	431	318	130	0	2	0	1108

SNOWFALL (inches) — J.F.K INTERNATIONAL AIRPORT N.Y.C, NEW YORK

SEASON	JULY	AUG	SEP	OCT	NOV	DEC	JAN	FEB	MAR	APR	MAY	JUNE	TOTAL
1964-65	0.0	0.0	0.0	0.0	T	3.0	17.4	4.1	3.6	1.4	0.0	0.0	29.5
1965-66	0.0	0.0	0.0	T	T	T	10.1	5.5	T	T	0.0	0.0	15.6
1966-67	0.0	0.0	0.0	0.0	T	8.8	2.8	19.9	15.5	T	T	0.0	47.0
1967-68	0.0	0.0	0.0	0.0	2.1	4.3	5.1	1.7	4.2	0.0	0.0	0.0	17.4
1968-69	0.0	0.0	0.0	0.0	T	4.4	0.6	22.4	3.4	0.0	0.0	0.0	30.8
1969-70	0.0	0.0	0.0	0.0	T	6.3	5.5	5.3	2.4	0.0	0.0	0.0	19.5
1970-71	0.0	0.0	0.0	0.0	T	2.3	11.6	0.2	2.8	3.2	0.0	0.0	20.1
1971-72	0.0	0.0	0.0	0.0	T	0.1	1.7	10.3	2.8	T	0.0	0.0	14.9
1972-73	0.0	0.0	0.0	0.0	T	T	0.8	0.8	T	T	0.0	0.0	1.6
1973-74	0.0	0.0	0.0	0.0	0.0	0.5	6.7	10.5	4.1	T	0.0	0.0	21.8
1974-75	0.0	0.0	0.0	0.0	0.2	T	0.6	8.8	0.9	T	0.0	0.0	10.5
1975-76	0.0	0.0	0.0	0.0	T	1.2	6.9	7.3	5.3	0.0	0.0	0.0	20.7
1976-77	0.0	0.0	0.0	0.0	T	5.8	13.4	3.5	T	0.0	0.0	0.0	22.7
1977-78	0.0	0.0	0.0	0.0	0.6	1.0	20.1	18.1	8.7	0.0	0.0	0.0	48.5
1978-79	0.0	0.0	0.0	0.0	1.7	0.3	7.4	17.6	0.2	T	0.0	0.0	27.2
1979-80	0.0	0.0	0.0	T	0.0	2.5	3.0	2.3	3.2	T	0.0	0.0	11.0
1980-81	0.0	0.0	0.0	0.0	0.2	1.7	7.7	T	6.9	0.0	0.0	0.0	16.5
1981-82	0.0	0.0	0.0	0.0	T	3.1	12.5	0.8	0.3	8.2	0.0	0.0	24.9
1982-83	0.0	0.0	0.0	0.0	0.0	4.8	1.0	24.7	0.1	1.5	0.0	0.0	32.1
1983-84	0.0	0.0	0.0	0.0	T	1.2	9.9	T	10.9	T	0.0	0.0	22.0
1984-85	0.0	0.0	0.0	0.0	T	5.5	12.4	9.0	0.4	T	0.0	0.0	27.3
1985-86	0.0	0.0	0.0	0.0	T	2.8	3.0	13.5	T	T	0.0	0.0	19.3
1986-87	0.0	0.0	0.0	0.0	0.4	1.0	11.8	7.9	2.0	0.0	0.0	0.0	23.1
1987-88	0.0	0.0	0.0	0.0	0.4	3.0	15.7	0.5	0.1	0.0	0.0	0.0	19.7
1988-89	0.0	0.0	0.0	0.0	0.0	0.7	4.7	0.1	2.7	0.0	0.0	0.0	8.2
1989-90	0.0	0.0	0.0	0.0	3.7	0.8	1.4	1.3	1.9	0.5	0.0	0.0	9.6
1990-91	0.0	0.0	0.0	0.0	0.0	6.1	5.7	7.9	0.7	0.0	0.0	0.0	20.4
1991-92	0.0	0.0	0.0	0.0	T	0.9	1.9	0.8	6.9	T	0.0	0.0	10.5
1992-93	0.0	0.0	0.0	0.0	T	0.8	0.8	6.7	11.7	0.0	0.0	0.0	20.0
1993-94	0.0	0.0	0.0	0.0	T	8.0							
Record Mean	0.0	0.0	0.0	T	0.3	3.5	7.0	7.6	3.9	0.4	T	0.0	22.7

NEW YORK, LA GUARDIA FIELD, NEW YORK

New York City, in area exceeding 300 square miles, is located on the Atlantic coastal plain at the mouth of the Hudson River. The terrain is laced with numerous waterways, all but one of the five boroughs in the city are situated on islands. Elevations range from less than 50 feet over most of Manhattan, Brooklyn, and Queens to almost 300 feet in northern Manhattan and the Bronx, and over 400 feet in Staten Island. Extensive suburban areas on Long Island, and in Connecticut, New York State and New Jersey border the city on the east, north, and west. About 30 miles to the west and northwest, hills rise to about 1,500 feet and to the north in upper Westchester County to 800 feet. To the southwest and to the east are the low-lying land areas of the New Jersey coastal plain and of Long Island, bordering on the Atlantic.

The New York Metropolitan area is close to the path of most storm and frontal systems which move across the North American continent. Therefore, weather conditions affecting the city most often approach from a westerly direction. New York City can thus experience higher temperatures in summer and lower ones in winter than would otherwise be expected in a coastal area. However, the frequent passage of weather systems often helps reduce the length of both warm and cold spells, and is also a major factor in keeping periods of prolonged air stagnation to a minimum.

Although continental influence predominates, oceanic influence is by no means absent. During the summer local sea breezes, winds blowing onshore from the cool water surface, often moderate the afternoon heat. The effect of the sea breeze diminishes inland. On winter mornings, ocean temperatures which are warm relative to the land reinforce the effect of the city heat island and low temperatures are often 10-20 degrees lower in the inland suburbs than in the central city. The relatively warm water temperatures also delay the advent of winter snows. Conversely, the lag in warming of water temperatures keeps spring temperatures relatively cool. One year-round measure of the ocean influence is the small average daily variation in temperature.

Precipitation is moderate and distributed fairly evenly throughout the year. Most of the rainfall from May through October comes from thunderstorms. It is therefore usually of brief duration and sometimes intense. Heavy rains of long duration associated with tropical storms occur infrequently in late summer or fall. For the other months of the year precipitation is more likely to be associated with widespread storm areas, so that day-long rain, snow or a mixture of both is more common. Precipitation accompanying winter storms sometimes starts as snow, later changes to rain, and perhaps briefly back to snow before ending. Coastal storms, occurring most often in the fall and winter months, produce on occasion considerable amounts of precipitation and have been responsible for record rains, snows, and high winds.

The average annual precipitation and snowfall totals are reasonably uniform within the city but show a consistent increase to the north and west with lesser amounts along the south shores and the eastern end of Long Island, reflecting the influence of the ocean waters. Relative humidity averages about the same over the metropolitan area except again that the immediate coastal areas are more humid than inland locations.

Local Climatological Data is published for three locations in New York City, Central Park, La Guardia Airport, and John F. Kennedy International Airport. Other nearby locations for which it is published are Newark, New Jersey, and Bridgeport, Connecticut.

NORMALS, MEANS, AND EXTREMES

NEW YORK, LA GUARDIA FIELD, NEW YORK

LATITUDE: 40°46'N LONGITUDE: 73°54'W ELEVATION: FT. GRND 11 BARO 39 TIME ZONE: EASTERN WBAN: 14732

	(a)	JAN	FEB	MAR	APR	MAY	JUNE	JULY	AUG	SEP	OCT	NOV	DEC	YEAR
TEMPERATURE °F:														
Normals														
-Daily Maximum		37.0	39.3	48.4	59.1	69.8	78.9	83.9	82.3	75.1	64.1	53.3	42.1	61.1
-Daily Minimum		25.5	27.1	34.6	43.9	53.8	63.3	68.9	68.3	61.3	50.6	41.5	31.1	47.5
-Monthly		31.3	33.2	41.5	51.5	61.8	71.1	76.4	75.4	68.2	57.4	47.4	36.6	54.3
Extremes														
-Record Highest	32	68	73	83	91	96	99	107	97	96	87	80	69	107
-Year		1967	1985	1990	1976	1987	1988	1966	1988	1983	1990	1993	1982	JUL 1966
-Record Lowest	32	-3	-2	8	22	38	46	56	51	44	30	18	-1	-3
-Year		1985	1963	1980	1982	1983	1972	1988	1982	1974	1969	1976	1980	JAN 1985
NORMAL DEGREE DAYS:														
Heating (base 65°F)		1045	890	729	405	144	8	0	0	29	252	528	880	4910
Cooling (base 65°F)		0	0	0	0	44	191	353	322	125	17	0	0	1052
% OF POSSIBLE SUNSHINE														
MEAN SKY COVER (tenths)														
Sunrise - Sunset	45	6.4	6.3	6.3	6.3	6.3	5.9	5.9	5.8	5.7	5.4	6.4	6.4	6.1
MEAN NUMBER OF DAYS:														
Sunrise to Sunset														
-Clear	45	7.6	7.6	7.7	7.4	6.6	7.6	7.4	8.1	9.5	10.8	7.5	7.6	95.4
-Partly Cloudy	45	8.4	7.3	9.1	9.3	11.2	11.4	12.6	12.2	9.3	8.8	8.5	8.9	117.0
-Cloudy	45	15.0	13.3	14.2	13.3	13.2	11.1	11.0	10.8	11.2	11.4	14.0	14.5	152.9
Precipitation .01 inches or more	53	11.0	9.5	10.9	10.9	11.2	9.8	9.5	9.3	8.2	7.8	9.8	10.6	118.4
Snow,Ice pellets,hail 1.0 inches or more	49	2.1	2.0	1.1	0.1	0.0	0.0	0.0	0.0	0.0	0.*	0.1	1.2	6.7
Thunderstorms	45	0.2	0.2	0.9	1.7	3.2	4.1	5.2	4.6	2.3	0.8	0.5	0.2	24.0
Heavy Fog Visibility 1/4 mile or less	45	1.6	1.3	1.4	1.2	1.5	1.2	0.6	0.4	0.2	0.9	0.6	1.1	12.2
Temperature °F														
-Maximum														
90° and above	31	0.0	0.0	0.0	0.*	0.8	3.1	5.9	3.7	1.2	0.0	0.0	0.0	14.7
32° and below	31	9.7	6.3	1.1	0.*	0.0	0.0	0.0	0.0	0.0	0.0	0.1	4.5	21.7
-Minimum														
32° and below	31	22.2	19.8	10.7	1.0	0.0	0.0	0.0	0.0	0.0	0.1	3.2	15.9	72.9
0° and below	31	0.1	0.1	0.0	0.0	0.0	0.0	0.0	0.0	0.0	0.0	0.0	0.*	0.2
AVG. STATION PRESS.(mb)	21	1016.9	1017.0	1015.6	1013.9	1014.4	1014.0	1014.5	1016.1	1017.6	1018.2	1017.7	1017.8	1016.1
RELATIVE HUMIDITY (%)														
Hour 01	31	64	62	64	65	71	71	72	74	74	71	68	66	69
Hour 07 (Local Time)	31	66	65	67	67	71	72	72	76	76	74	71	68	70
Hour 13	31	57	55	53	51	53	54	53	55	57	55	58	59	55
Hour 19	31	59	57	57	56	60	60	61	63	64	63	62	61	60
PRECIPITATION (inches):														
Water Equivalent														
-Normal		3.04	2.86	3.60	3.79	3.82	3.57	4.07	3.75	3.39	3.02	3.81	3.40	42.12
-Maximum Monthly	53	8.68	5.76	8.73	11.51	9.27	8.15	12.33	16.05	9.63	9.09	9.92	7.70	16.05
-Year		1979	1960	1953	1983	1984	1972	1975	1955	1975	1943	1972	1973	AUG 1955
-Minimum Monthly	53	0.51	0.78	0.87	0.99	0.43	0.03	0.69	0.24	0.62	0.06	0.31	0.31	0.03
-Year		1981	1987	1966	1985	1964	1949	1954	1964	1941	1963	1976	1955	JUN 1949
-Maximum in 24 hrs	53	3.55	2.90	3.25	3.06	3.02	4.01	3.82	7.11	4.52	3.58	4.46	3.44	7.11
-Year		1979	1941	1953	1984	1968	1987	1971	1955	1969	1972	1977	1941	AUG 1955
Snow,Ice pellets,hail														
-Maximum Monthly	49	18.3	26.4	18.9	8.2	T	0.0	0.0	0.0	0.0	1.2	6.1	26.8	26.8
-Year		1948	1983	1958	1982	1977					1962	1989	1947	DEC 1947
-Maximum in 24 hrs	49	11.3	22.0	15.3	8.2	T	0.0	0.0	0.0	0.0	1.2	6.1	22.8	22.8
-Year		1987	1983	1960	1982	1977					1962	1989	1947	DEC 1947
WIND:														
Mean Speed (mph)	45	13.8	13.9	14.0	13.0	11.6	11.0	10.4	10.3	11.0	11.7	12.7	13.5	12.2
Prevailing Direction through 1963		WNW	WNW	NW	NW	NE	S	S	S	S	SW	WNW	WNW	S
Fastest Obs. 1 Min.														
-Direction (!!!)	14	06	01	06	16	17	29	24	28	30	10	30	05	30
-Speed (MPH)	14	39	43	52	41	41	38	35	46	46	52	55	52	55
-Year		1992	1987	1993	1993	1989	1993	1989	1993	1985	1980	1989	1992	NOV 1989
Peak Gust														
-Direction (!!!)	10	W	N	NE	SE	S	W	W	W	NW	S	NW	NE	NE
-Speed (mph)	10	58	59	71	63	56	56	49	72	64	64	76	77	77
-Date		1992	1987	1993	1987	1989	1993	1993	1993	1985	1987	1989	1992	DEC 1992

PRECIPITATION (inches)

NEW YORK, LA GUARDIA FIELD, NEW YORK

YEAR	JAN	FEB	MAR	APR	MAY	JUNE	JULY	AUG	SEP	OCT	NOV	DEC	ANNUAL
1964	4.22	2.39	2.14	5.12	0.43	2.85	4.56	0.24	1.78	1.68	1.65	4.25	31.31
1965	2.93	3.24	1.94	2.75	1.35	1.14	1.34	2.02	1.26	1.44	1.29	1.47	22.17
1966	2.36	4.29	0.87	2.31	3.78	1.44	1.12	1.74	5.07	3.73	2.44	2.54	31.69
1967	1.06	2.36	5.77	2.91	3.52	3.13	5.02	7.03	1.39	2.66	2.39	5.82	43.06
1968	1.67	1.37	4.77	2.59	4.67	6.16	2.39	3.19	1.84	1.95	5.13	3.80	39.53
1969	0.93	2.59	3.06	4.57	2.52	4.20	8.35	2.64	7.35	1.69	3.06	5.30	46.26
1970	0.76	3.24	3.16	2.36	2.02	2.36	1.18	2.68	1.70	1.45	4.59	1.86	27.36
1971	2.27	4.10	3.08	2.08	3.37	2.21	6.32	8.38	4.93	3.32	4.58	1.55	46.19
1972	2.02	4.80	4.12	3.19	6.85	8.15	3.47	0.94	1.65	4.91	9.92	4.66	54.68
1973	3.47	3.02	3.47	7.17	4.21	6.79	4.90	2.32	2.63	2.73	1.54	7.70	49.95
1974	2.78	1.28	4.45	2.94	3.82	2.51	1.43	6.07	6.90	2.03	1.01	6.38	41.60
1975	4.50	3.20	2.81	2.77	3.79	7.19	12.33	3.60	9.63	3.36	3.92	3.69	60.79
1976	5.42	2.94	2.30	2.45	3.87	2.74	1.00	5.95	2.82	4.62	0.31	2.23	36.65
1977	2.02	2.13	6.57	2.93	1.82	3.56	1.53	4.48	4.80	5.84	8.30	4.86	48.84
1978	6.11	0.92	2.14	1.95	8.15	1.30	3.79	4.01	3.93	1.41	2.24	4.90	40.85
1979	8.68	4.28	3.76	3.55	4.32	1.51	1.37	4.80	4.07	3.83	3.04	2.57	45.78
1980	1.94	0.95	8.65	6.55	2.14	3.43	4.74	1.32	1.16	3.15	4.17	0.61	38.81
1981	0.51	5.42	1.11	3.01	3.32	2.32	5.73	0.31	2.99	3.21	1.63	4.65	34.21
1982	4.81	2.25	2.39	4.14	2.03	4.70	2.97	3.11	1.41	1.65	3.19	1.42	34.07
1983	4.14	2.90	8.22	11.51	3.77	1.95	3.41	2.67	3.47	7.32	4.85	6.63	60.84
1984	1.51	4.31	5.19	5.26	9.27	6.85	5.75	1.19	2.65	3.01	3.13	2.58	50.70
1985	0.76	1.81	1.81	0.99	5.18	4.48	5.77	2.80	4.23	1.18	7.00	0.63	36.64
1986	4.50	2.74	1.91	3.65	1.45	1.43	3.90	4.60	1.84	1.71	5.94	5.19	38.86
1987	5.43	0.78	4.45	4.79	1.12	6.36	4.42	4.32	3.72	4.01	2.60	2.28	44.28
1988	2.58	3.44	1.98	2.09	4.45	0.94	8.47	1.83	2.59	3.08	7.76	1.18	40.39
1989	2.54	2.83	4.23	3.03	8.83	6.90	5.49	7.21	5.40	5.45	2.53	0.78	55.22
1990	4.10	1.56	2.74	5.30	7.63	2.13	2.77	10.31	1.90	5.72	2.18	4.88	51.22
1991	3.03	1.92	3.69	3.06	2.99	3.31	3.39	6.78	3.56	1.22	1.72	3.49	38.16
#1992	1.39	1.43	4.07	1.52	2.87	3.25	4.38	4.12	2.58	1.05	5.27	5.47	37.40
#1993	3.05	3.25	6.45	3.49	2.31	1.71	1.70	6.11	5.22	4.07	1.37	4.43	43.16
Record Mean	3.09	2.93	3.94	3.67	3.74	3.25	4.04	4.37	3.37	3.15	3.80	3.62	42.97

AVERAGE TEMPERATURE (deg. F)

NEW YORK, LA GUARDIA FIELD, NEW YORK

YEAR	JAN	FEB	MAR	APR	MAY	JUNE	JULY	AUG	SEP	OCT	NOV	DEC	ANNUAL
1964	34.4	32.6	42.8	49.7	65.6	71.3	76.1	73.8	68.1	55.4	49.9	36.7	54.7
1965	30.2	34.2	40.2	50.7	67.1	72.5	76.6	75.7	70.3	55.9	45.9	39.6	54.9
1966	31.8	34.5	42.0	48.6	59.0	74.0	80.8	77.7	68.3	56.6	49.2	36.8	55.0
1967	38.2	30.6	37.8	49.7	55.0	71.8	75.3	73.7	66.5	57.2	42.7	38.6	53.1
1968	26.8	29.1	42.3	54.0	59.4	69.2	77.2	75.9	70.6	59.8	46.4	33.5	53.7
1969	30.9	31.4	39.1	53.9	62.9	70.9	73.2	76.0	67.7	56.5	46.1	33.8	53.5
1970	26.0	32.7	38.4	51.1	62.6	70.5	77.1	77.5	70.4	59.2	49.6	36.2	54.3
1971	28.8	36.0	39.8	49.6	59.4	72.1	76.1	74.8	70.9	62.4	44.5	40.0	54.5
1972	34.2	30.6	37.8	47.6	59.7	65.8	74.6	73.6	67.8	52.2	43.1	38.5	52.1
1973	35.1	32.5	45.0	53.0	58.9	72.7	76.8	77.5	69.3	60.1	48.7	39.4	55.7
1974	35.6	31.9	41.7	54.2	59.9	68.9	76.8	75.7	66.2	53.8	47.8	39.5	54.3
1975	37.1	35.5	39.6	47.1	64.3	69.7	75.7	74.1	64.5	59.4	52.6	37.1	54.7
1976	28.8	40.7	44.8	55.6	61.5	74.1	75.8	75.0	67.5	54.5	43.0	30.7	54.3
1977	22.3	33.2	45.7	52.6	64.1	69.4	77.2	74.8	67.2	54.8	47.3	35.7	53.7
1978	28.8	26.9	38.6	50.3	59.6	70.0	73.8	75.4	65.3	55.7	48.2	38.0	52.5
1979	31.6	23.0	44.2	49.9	63.0	68.9	77.1	75.5	68.2	56.0	50.2	39.0	53.9
1980	32.7	30.7	40.2	53.2	64.7	70.1	78.7	78.2	70.4	55.7	43.1	30.8	54.0
1981	24.7	38.4	41.3	54.3	63.8	72.7	78.2	75.7	66.4	53.9	47.0	36.6	54.4
1982	25.3	35.1	41.1	50.2	63.3	66.8	76.3	72.4	66.9	57.4	49.0	41.8	53.8
1983	34.4	35.0	43.2	51.7	58.9	72.5	78.3	76.7	70.7	57.4	48.0	35.1	55.1
1984	29.3	39.2	35.3	50.3	61.3	73.6	73.5	76.3	65.7	62.5	46.4	43.3	54.7
1985	28.4	35.6	44.9	53.9	64.6	68.8	76.6	75.8	70.7	59.5	50.6	34.7	55.3
1986	34.0	31.6	43.8	53.1	65.4	71.5	76.1	73.4	67.9	58.0	45.9	39.3	55.0
1987	32.7	33.1	44.6	52.7	63.3	73.0	77.7	74.0	68.1	54.5	47.9	39.9	55.1
1988	29.8	35.1	43.0	50.8	62.2	72.1	78.6	78.8	67.9	53.0	49.5	36.6	54.8
1989	37.6	34.3	41.7	51.4	62.5	72.9	76.2	75.3	69.7	59.3	46.4	26.1	54.5
1990	41.1	39.9	44.0	53.0	59.7	72.4	77.1	76.3	68.8	63.1	51.0	43.2	57.5
1991	35.4	40.3	44.8	55.6	69.1	74.8	78.8	78.3	69.0	59.6	48.7	40.3	57.9
#1992	35.9	36.5	40.3	50.4	61.2	71.2	75.4	73.9	68.4	55.6	46.5	38.0	54.4
#1993	36.5	30.4	38.1	52.5	65.2	73.2	80.0	77.4	68.4	56.7	48.8	37.8	55.4
Record Mean	32.3	33.8	41.5	51.8	62.0	71.4	76.7	75.3	68.4	58.0	47.6	36.5	54.7
Max	38.2	40.2	48.7	59.7	70.3	79.5	84.4	82.6	75.6	64.8	53.7	42.3	61.7
Min	26.4	27.4	34.2	43.8	53.7	63.3	69.0	68.1	61.3	51.1	41.5	30.8	47.6

REFERENCE NOTES

NEW YORK, LA GUARDIA FIELD, NEW YORK

GENERAL
T - TRACE AMOUNT.
BLANK ENTRIES DENOTE MISSING/UNREPORTED DATA.
# INDICATES A STATION OR INSTRUMENT RELOCATION.

SPECIFIC

EXCEPTIONS

RECORD MEANS ARE THROUGH THE CURRENT YEAR, BEGINNING IN 1941 FOR TEMPERATURE
1941 FOR PRECIPITATION
1945 FOR SNOWFALL

(a) - LENGTH OF RECORD IN YEARS, ALTHOUGH INDIVIDUAL MONTHS MAY BE MISSING.
0.* OR * - THE VALUE IS BETWEEN 0.0 AND 0.05
NORMALS - BASED ON THE 1961-1990 RECORD PERIOD.
EXTREMES - DATES ARE THE MOST RECENT OCCURRENCE.
WIND DIR.- NUMERALS SHOW TENS OF DEGRESS CLOCKWISE FROM TRUE NORTH. "00" INDICATES CALM.
RESULTANT DIRECTIONS ARE GIVEN TO WHOLE DEGREES.
BOLD VALUES INDICATE EXTREME VALUES WHICH OCCURRED AFTER THE ASOS SYSTEM WAS COMMISSIONED.

RECORD = PERIOD OF RECORD
RECORD MEAN PRECIPITATION IS THE MEAN OF ALL DAILY PRECIPITATION AMOUNTS DURING THE PERIOD OF RECORD.
RECORD MAX(MIN) TEMPERATURE IS THE MEAN OF ALL DAILY MAX(MIN) TEMPERATURES DURING THE PERIOD OF RECORD.
RECORD MEAN TEMPERATURE IS THE SUM OF THE RECORD MAX AND RECORD MIN DIVIDED BY 2.
AVERAGE TEMPERATURE IS THE SUM OF THE MEAN DAILY MAX AND MIN TEMPERATURE DIVIDED BY 2.

HEATING DEGREE DAYS Base 65 deg. F — NEW YORK, LA GUARDIA FIELD, NEW YORK

SEASON	JULY	AUG	SEP	OCT	NOV	DEC	JAN	FEB	MAR	APR	MAY	JUNE	TOTAL
1964-65	2	0	52	290	444	872	1072	858	763	426	47	14	4840
1965-66	0	5	26	280	565	780	1021	846	705	484	214	17	4943
1966-67	0	0	39	256	469	866	823	960	839	454	308	5	5019
1967-68	0	1	56	262	662	812	1181	1034	698	323	176	14	5219
1968-69	0	0	3	196	554	969	1051	933	795	336	111	2	4950
1969-70	0	0	49	272	562	959	1204	896	821	414	118	6	5301
1970-71	0	0	26	200	456	886	1117	807	774	455	179	11	4911
1971-72	0	1	17	108	612	768	945	987	837	520	164	43	5002
1972-73	4	0	37	389	653	813	920	904	613	371	195	4	4903
1973-74	0	0	30	167	485	785	903	919	713	331	186	27	4546
1974-75	0	0	63	341	512	783	858	819	778	529	105	14	4802
1975-76	0	4	56	188	372	857	1118	695	618	320	134	14	4376
1976-77	0	2	36	324	654	1061	1316	883	591	377	115	28	5387
1977-78	0	0	59	309	524	901	1119	1061	812	436	203	16	5440
1978-79	5	0	72	287	498	831	1031	1173	637	446	94	15	5089
1979-80	2	5	32	295	440	802	997	988	764	350	74	21	4770
1980-81	0	0	24	292	651	1052	1241	739	728	316	97	4	5144
1981-82	0	0	63	338	532	875	1222	832	737	443	79	49	5170
1982-83	0	6	36	253	482	712	942	832	669	402	187	5	4526
1983-84	0	2	48	259	505	919	1102	740	913	436	147	9	5080
1984-85	0	0	74	101	552	666	1127	820	620	338	87	20	4405
1985-86	0	0	17	183	428	934	955	929	649	350	98	12	4555
1986-87	0	9	25	235	561	787	995	886	623	361	149	7	4638
1987-88	0	2	24	321	508	772	1083	862	674	421	134	31	4832
1988-89	3	0	22	376	455	872	844	854	719	402	137	9	4693
1989-90	0	0	38	182	552	1198	735	698	645	367	159	3	4577
1990-91	2	0	38	145	421	670	911	687	619	312	51	4	3860
1991-92	0	0	45	190	482	758	894	819	756	431	156	10	4541
1992-93	0	1	47	293	546	831	875	964	828	369	61	9	4824
#1993-94	0	0	51	256	484	837							

COOLING DEGREE DAYS Base 65 deg. F — NEW YORK, LA GUARDIA FIELD, NEW YORK

YEAR	JAN	FEB	MAR	APR	MAY	JUNE	JULY	AUG	SEP	OCT	NOV	DEC	TOTAL
1969	0	0	0	8	55	186	258	346	135	14	0	0	1002
1970	0	0	0	2	54	180	382	395	196	27	0	0	1236
1971	0	0	0	0	12	229	352	308	199	32	6	0	1138
1972	0	0	0	2	8	77	311	274	129	0	0	0	801
1973	0	0	0	14	16	242	372	397	164	24	0	0	1229
1974	0	0	0	16	35	149	371	338	104	0	4	0	1017
1975	0	0	0	0	87	160	338	293	46	20	9	0	953
1976	0	0	0	44	31	293	344	322	119	5	0	0	1158
1977	0	0	0	10	93	166	387	311	130	0	0	0	1097
1978	0	0	0	0	44	172	286	326	86	5	0	0	919
1979	0	0	0	0	41	138	382	335	131	22	0	0	1049
1980	0	0	0	0	71	180	430	413	192	8	0	0	1294
1981	0	0	0	0	66	243	413	336	111	0	0	0	1169
1982	0	0	0	5	32	112	358	239	100	25	6	0	877
1983	0	0	0	9	4	233	417	368	222	31	0	0	1284
1984	0	0	0	0	39	274	271	355	100	31	0	0	1070
1985	0	0	1	11	81	139	368	340	193	20	3	0	1156
1986	0	0	0	0	118	213	352	276	118	26	0	0	1103
1987	0	0	0	1	102	251	398	289	124	0	2	0	1167
1988	0	0	0	0	55	248	431	438	116	11	0	0	1299
1989	0	0	2	0	65	253	351	328	183	15	0	0	1197
1990	0	0	0	17	4	234	381	356	158	95	7	0	1252
1991	0	0	0	34	184	303	435	421	174	30	0	0	1581
#1992	0	0	0	1	45	202	327	284	153	10	0	0	1022
#1993	0	0	0	0	74	261	474	390	157	5	4	0	1365

SNOWFALL (inches) — NEW YORK, LA GUARDIA FIELD, NEW YORK

SEASON	JULY	AUG	SEP	OCT	NOV	DEC	JAN	FEB	MAR	APR	MAY	JUNE	TOTAL
1964-65	0.0	0.0	0.0	0.0	0.0	3.1	11.9	2.3	3.1	0.7	0.0	0.0	21.1
1965-66	0.0	0.0	0.0	T	T	T	9.3	9.8	T	T	0.0	0.0	19.1
1966-67	0.0	0.0	0.0	0.0	0.0	7.6	0.9	20.0	14.9	T	0.0	0.0	43.4
1967-68	0.0	0.0	0.0	0.0	2.2	3.4	5.2	1.5	2.2	0.0	0.0	0.0	14.5
1968-69	0.0	0.0	0.0	0.0	T	4.9	1.4	18.5	4.0	0.0	0.0	0.0	28.8
1969-70	0.0	0.0	0.0	0.0	T	8.1	7.4	4.2	4.6	T	0.0	0.0	24.3
1970-71	0.0	0.0	0.0	0.0	0.0	1.9	10.4	0.3	2.3	1.0	0.0	0.0	15.9
1971-72	0.0	0.0	0.0	0.0	T	T	2.2	17.2	2.7	0.1	0.0	0.0	22.2
1972-73	0.0	0.0	0.0	T	T	T	0.9	1.0	T	T	0.0	0.0	1.9
1973-74	0.0	0.0	0.0	0.0	0.0	2.2	6.4	7.8	2.2	0.3	0.0	0.0	18.9
1974-75	0.0	0.0	0.0	0.0	T	0.4	1.8	9.1	0.4	T	0.0	0.0	11.7
1975-76	0.0	0.0	0.0	0.0	T	3.0	6.1	4.8	2.8	0.0	0.0	0.0	16.7
1976-77	0.0	0.0	0.0	0.0	T	4.9	10.9	5.8	0.4	T	T	0.0	22.0
1977-78	0.0	0.0	0.0	0.0	T	0.3	16.6	18.7	7.9	0.0	0.0	0.0	43.5
1978-79	0.0	0.0	0.0	0.0	2.3	0.2	6.0	17.4	T	T	0.0	0.0	25.9
1979-80	0.0	0.0	0.0	T	0.0	3.2	2.3	1.6	3.2	0.0	0.0	0.0	10.3
1980-81	0.0	0.0	0.0	0.0	0.2	1.8	7.7	T	6.4	0.0	0.0	0.0	16.1
1981-82	0.0	0.0	0.0	0.0	T	3.6	13.1	0.4	0.3	8.2	0.0	0.0	25.6
1982-83	0.0	0.0	0.0	0.0	0.0	2.1	1.7	26.4	T	T	0.0	0.0	30.2
1983-84	0.0	0.0	0.0	0.0	T	1.6	9.8	T	12.7	0.0	0.0	0.0	24.1
1984-85	0.0	0.0	0.0	0.0	T	5.5	8.3	8.8	0.3	T	0.0	0.0	22.9
1985-86	0.0	0.0	0.0	0.0	0.4	0.9	2.8	14.3	T	T	0.0	0.0	18.4
1986-87	0.0	0.0	0.0	0.0	T	T	16.3	6.0	0.9	0.0	0.0	0.0	23.2
1987-88	0.0	0.0	0.0	0.0	T	4.2	15.5	1.3	0.1	T	0.0	0.0	21.1
1988-89	0.0	0.0	0.0	0.0	0.0	0.4	6.4	1.6	2.4	0.0	0.0	0.0	10.8
1989-90	0.0	0.0	0.0	0.0	6.1	2.7	3.0	3.8	4.4	0.9	0.0	0.0	20.9
1990-91	0.0	0.0	0.0	0.0	0.0	7.3	6.2	8.3	0.1	0.0	0.0	0.0	21.9
1991-92	0.0	0.0	0.0	0.0	0.0	1.5	1.3	1.3	10.2	T	0.0	0.0	14.3
#1992-93	0.0	0.0	0.0	T	T	0.5	2.2	13.6	15.4	0.0	0.0	0.0	31.7
#1993-94	0.0	0.0	0.0	0.0	0.0	10.4							
Record Mean	0.0	0.0	0.0	T	0.5	4.9	6.7	8.1	4.5	0.6	T	0.0	25.2

ROCHESTER, NEW YORK

Rochester is located at the mouth of the Genesee River at about the mid point of the south shore of Lake Ontario. The river flows northward from northwest Pennsylvania and empties into Lake Ontario. The land slopes from a lakeshore elevation of 246 feet to over 1,000 feet some 20 miles south. The airport is located just south of the city.

Lake Ontario plays a major role in the Rochester weather. In the summer its cooling effect inhibits the temperature from rising much above the low to mid 90s. In the winter the modifying temperature effect prevents temperatures from falling below -15 degrees most of the time, although temperatures at locations more than 15 miles inland do drop below -30 degrees.

The lake plays a major role in winter snowfall distribution. Well inland from the lake and toward the airport, the seasonal snowfall is usually less than in the area north of the airport and toward the lakeshore where wide variations occur. This is due to what is called the lake effect. Snowfalls of one to two feet or more in 24 hours are common near the lake in winter due the lake effect alone. The lake rarely freezes over because of its depth. The area is also prone to other heavy snowstorms and blizzards because of its proximity to the paths of low pressure systems coming up the east coast, out of the Ohio Valley, or, to a lesser extent, from the Alberta area. The climate is favorable for winter sports activities with a continuous snow cover likely from December through March.

Moisture in the air from the lake enhances the climatic conditions for fruit growing. Apples, peaches, pears, cantaloupes, plums, cherries, and grapes are grown abundantly in Greater Rochester and the Western Finger Lakes Region.

Precipitation is rather evenly distributed throughout the year. Excessive rains occur infrequently but may be caused by slowly moving thunderstorms, slowly moving or stalled major low pressure systems, or by hurricanes and tropical storms that move inland. Hail occurs occasionally and heavy fog is rare.

The growing season averages 150 to 180 days. The years first frost usually occurs in late September and the last frost typically occurs in mid-May.

NORMALS, MEANS, AND EXTREMES

ROCHESTER, NEW YORK

LATITUDE: 43°07'N LONGITUDE: 77°40' W ELEVATION: FT. GRND 547 BARO 547 TIME ZONE: EASTERN WBAN: 14768

	(a)	JAN	FEB	MAR	APR	MAY	JUNE	JULY	AUG	SEP	OCT	NOV	DEC	YEAR
TEMPERATURE °F:														
Normals														
-Daily Maximum		30.9	32.5	42.7	55.9	67.8	75.8	80.7	78.1	71.8	60.5	47.8	35.8	56.7
-Daily Minimum		16.3	16.6	25.7	35.9	46.3	54.3	59.6	57.8	51.7	41.6	33.3	22.4	38.5
-Monthly		23.6	24.6	34.3	45.9	57.1	65.1	70.2	68.0	61.7	51.1	40.5	29.1	47.6
Extremes														
-Record Highest	53	74	67	84	93	94	100	98	99	99	91	81	72	100
-Year		1950	1947	1945	1990	1987	1953	1993	1948	1953	1951	1950	1982	JUN 1953
-Record Lowest	53	-16	-19	-6	13	26	35	42	36	28	20	5	-16	-19
-Year		1957	1979	1980	1982	1979	1949	1963	1965	1947	1972	1971	1942	FEB 1979
NORMAL DEGREE DAYS:														
Heating (base 65°F)		1283	1131	952	573	270	62	10	33	137	435	735	1113	6734
Cooling (base 65°F)		0	0	0	0	25	65	171	126	38	0	0	0	425
% OF POSSIBLE SUNSHINE	53	36	40	49	54	59	67	69	66	59	49	31	31	51
MEAN SKY COVER (tenths)														
Sunrise - Sunset	53	8.2	7.9	7.3	6.8	6.6	6.0	5.8	6.0	6.2	6.7	8.2	8.4	7.0
MEAN NUMBER OF DAYS:														
Sunrise to Sunset														
-Clear	53	2.0	2.4	4.5	6.2	6.1	7.2	7.9	7.7	7.0	6.3	2.2	2.0	61.4
-Partly Cloudy	53	6.7	6.6	8.2	7.8	9.6	10.8	12.4	11.7	10.5	8.4	6.2	5.6	104.5
-Cloudy	53	22.2	19.2	18.3	16.1	15.4	11.9	10.6	11.6	12.5	16.2	21.7	23.4	199.3
Precipitation														
.01 inches or more	53	17.6	15.9	14.6	13.4	11.9	10.6	9.6	10.0	10.8	11.7	15.2	17.5	158.8
Snow,Ice pellets,hail														
1.0 inches or more	53	7.3	6.9	4.0	1.0	0.1	0.0	0.0	0.0	0.0	0.1	2.0	6.2	27.6
Thunderstorms	53	0.1	0.1	1.0	2.0	3.5	5.2	6.2	5.7	3.1	1.0	0.4	0.2	28.3
Heavy Fog Visibility														
1/4 mile or less	53	0.9	0.6	1.6	1.1	1.2	1.1	0.6	0.8	1.3	1.7	0.7	1.0	12.7
Temperature °F														
-Maximum														
90° and above	30	0.0	0.0	0.0	0.1	0.3	1.6	4.3	2.0	0.6	0.0	0.0	0.0	8.9
32° and below	30	16.5	14.3	6.5	0.4	0.0	0.0	0.0	0.0	0.0	0.0	1.5	10.5	49.7
-Minimum														
32° and below	30	28.6	25.6	23.1	10.7	0.9	0.0	0.0	0.0	0.1	4.6	15.2	25.5	134.2
0° and below	30	2.7	2.1	0.2	0.0	0.0	0.0	0.0	0.0	0.0	0.0	0.0	0.7	5.8
AVG. STATION PRESS.(mb)	21	996.4	997.5	996.1	994.8	995.0	994.8	995.6	997.1	998.0	998.2	997.2	997.2	996.5
RELATIVE HUMIDITY (%)														
Hour 01	30	77	77	76	76	78	82	83	87	87	82	80	80	80
Hour 07 (Local Time)	30	77	79	79	77	77	80	82	87	88	85	82	81	81
Hour 13	30	69	67	62	56	54	55	54	58	61	61	68	72	61
Hour 19	30	74	73	69	61	60	60	60	67	74	74	77	78	69
PRECIPITATION (inches):														
Water Equivalent														
-Normal		2.08	2.10	2.28	2.61	2.72	3.00	2.71	3.40	2.97	2.44	2.92	2.73	31.96
-Maximum Monthly	53	5.79	5.07	5.42	4.90	6.62	6.77	9.70	6.00	6.30	7.85	6.99	5.05	9.70
-Year		1978	1950	1942	1944	1974	1980	1947	1984	1977	1955	1985	1944	JUL 1947
-Minimum Monthly	53	0.72	0.66	0.47	1.28	0.36	0.22	0.98	0.76	0.28	0.23	0.44	0.62	0.22
-Year		1988	1987	1958	1971	1977	1963	1989	1951	1960	1963	1976	1958	JUN 1963
-Maximum in 24 hrs	53	1.64	2.43	2.21	2.22	3.85	2.86	3.25	2.39	3.54	2.98	3.13	1.60	3.85
-Year		1966	1950	1942	1991	1974	1950	1987	1968	1979	1980	1945	1978	MAY 1974
Snow,Ice pellets,hail														
-Maximum Monthly	53	60.4	64.8	40.3	20.2	10.9	0.0	T	T	T	2.6	17.6	46.1	64.8
-Year		1978	1958	1959	1979	1989		1990	1965	1993	1993	1983	1981	FEB 1958
-Maximum in 24 hrs	53	18.2	22.8	22.2	10.4	10.8	0.0	T	T	T	2.6	11.2	19.1	22.8
-Year		1966	1978	1993	1990	1989		1990	1965	1993	1993	1953	1978	FEB 1978
WIND:														
Mean Speed (mph)	53	11.6	11.3	11.0	10.7	9.3	8.6	8.0	7.7	8.0	8.8	10.2	10.7	9.7
Prevailing Direction through 1963		WSW	WSW	WSW	WSW	WSW	SW	SW	SW	SW	SW	WSW	WSW	WSW
Fastest Obs. 1 Min.														
-Direction (!!!)	8	28	25	24	25	30	31	24	28	25	24	25	26	24
-Speed (MPH)	8	35	35	52	38	35	35	31	40	37	38	46	37	52
-Year		1992	1988	1991	1991	1991	1993	1990	1986	1993	1993	1992	1991	MAR 1991
Peak Gust														
-Direction (!!!)	10	W	SW	SW	SW	NW	N	NW	W	W	SW	SW	W	SW
-Speed (mph)	10	63	52	68	67	64	52	45	62	48	56	67	53	68
-Date		1988	1988	1991	1984	1991	1988	1988	1986	1993	1993	1992	1991	MAR 1991

PRECIPITATION (inches) ROCHESTER, NEW YORK

YEAR	JAN	FEB	MAR	APR	MAY	JUNE	JULY	AUG	SEP	OCT	NOV	DEC	ANNUAL
1964	1.99	0.89	2.89	3.54	2.77	1.13	1.52	2.74	0.58	0.76	1.60	2.04	22.45
1965	3.05	2.24	2.07	1.85	0.50	0.64	1.46	2.94	1.98	2.82	3.40	2.21	25.16
1966	4.10	2.51	1.42	2.04	1.26	1.85	2.77	2.14	2.47	0.68	3.12	1.75	26.11
1967	0.94	1.67	1.31	1.69	2.74	1.57	2.68	4.64	3.84	4.35	2.89	1.52	29.84
1968	1.91	0.74	2.38	1.33	2.84	2.84	1.42	5.95	1.86	2.89	4.28	3.31	31.75
1969	2.46	0.91	1.16	3.48	2.25	4.69	1.83	1.82	1.77	1.69	3.42	3.66	29.14
1970	1.80	2.28	1.49	2.58	3.03	3.74	4.91	3.88	2.49	3.96	3.50	4.12	37.78
1971	2.66	4.21	3.43	1.28	1.71	3.52	5.59	3.18	1.79	1.34	1.96	3.49	34.16
1972	1.50	3.96	2.19	2.68	3.32	6.56	1.43	3.14	3.84	2.25	4.83	2.58	38.28
1973	1.28	1.70	2.92	3.21	2.68	2.84	1.14	1.94	1.41	2.67	3.82	3.62	29.23
1974	1.75	2.06	3.61	2.60	6.62	2.59	2.82	3.64	3.48	1.34	3.23	2.86	36.60
1975	1.83	2.82	2.74	1.43	2.85	5.35	1.18	2.31	3.15	1.83	1.35	3.76	30.60
1976	2.33	1.67	3.54	3.81	2.63	3.37	5.15	3.04	2.13	4.73	0.44	1.48	34.32
1977	1.49	0.97	2.18	2.49	0.36	1.33	3.26	5.65	6.30	2.64	3.78	4.65	35.10
1978	5.79	2.40	1.48	2.25	2.03	1.30	2.17	2.66	3.63	2.56	1.14	4.35	31.76
1979	4.18	2.40	1.76	3.78	3.14	1.85	3.16	2.05	5.32	2.60	1.80	2.86	34.90
1980	1.11	1.16	3.83	2.35	1.49	6.77	1.90	3.44	3.57	3.73	2.52	2.45	34.32
1981	1.24	3.13	1.04	1.95	2.27	2.70	4.60	4.44	5.37	3.29	2.18	2.78	34.99
1982	4.16	1.01	1.73	1.63	1.77	3.92	3.13	3.00	3.57	1.79	3.95	2.17	31.83
1983	1.43	1.23	2.45	3.50	3.44	2.40	1.13	5.43	1.56	3.26	4.91	4.47	35.21
1984	1.62	2.97	2.08	3.05	5.47	1.67	1.90	6.00	3.34	0.76	1.47	3.31	33.64
1985	2.49	1.78	3.47	1.30	2.08	2.63	1.86	1.11	2.49	2.34	6.99	1.46	30.00
1986	1.63	2.46	1.90	3.80	1.64	4.27	3.13	3.29	5.11	3.56	1.93	3.56	36.28
1987	1.89	0.66	1.98	3.68	1.19	3.94	5.85	3.92	4.60	1.65	2.74	1.98	34.08
1988	0.72	2.18	1.62	2.32	1.73	1.10	4.30	3.81	1.69	2.34	1.68	1.11	24.60
1989	1.18	1.55	3.69	1.62	5.99	5.65	0.98	2.46	2.82	3.13	2.01	1.58	32.66
1990	1.61	3.93	1.56	3.58	5.76	2.88	3.05	3.59	3.36	4.37	2.27	4.18	40.14
1991	1.69	1.16	4.70	4.07	2.43	1.19	2.37	1.80	2.86	1.65	2.39	2.92	29.23
1992	1.46	1.87	3.53	3.43	2.83	1.98	6.03	4.45	3.02	1.78	2.90	2.98	36.26
1993	2.32	1.52	2.44	3.07	1.24	2.76	1.67	1.67	4.37	3.21	3.27	1.60	29.14
Record Mean	2.44	2.35	2.62	2.60	2.90	2.98	3.08	2.89	2.78	2.75	2.69	2.59	32.67

AVERAGE TEMPERATURE (deg. F) ROCHESTER, NEW YORK

YEAR	JAN	FEB	MAR	APR	MAY	JUNE	JULY	AUG	SEP	OCT	NOV	DEC	ANNUAL
1964	28.3	24.0	34.4	47.2	60.3	65.8	73.8	66.2	61.2	48.7	42.3	29.6	48.5
1965	21.6	25.7	29.5	41.3	60.1	64.4	67.5	68.4	63.8	48.8	39.8	34.6	47.1
1966	22.6	25.5	36.6	44.4	53.0	67.8	73.0	68.9	59.2	50.1	43.5	30.8	48.0
1967	31.1	21.4	32.5	46.4	49.1	70.4	69.4	67.6	60.4	52.1	36.3	32.3	47.4
1968	19.8	20.8	36.0	49.5	53.2	64.7	70.7	69.4	64.8	53.8	41.2	27.7	47.6
1969	25.2	26.0	32.0	47.6	55.6	65.0	70.7	72.0	63.6	50.9	40.1	25.1	47.8
1970	18.0	23.7	30.8	48.4	59.5	68.1	72.2	70.1	63.0	53.9	41.4	25.5	47.9
1971	19.5	26.9	28.5	41.1	54.0	67.5	68.8	68.8	67.4	59.0	38.4	30.9	47.6
1972	26.0	23.4	30.3	42.2	60.4	65.2	73.0	69.8	64.3	47.6	37.0	33.1	47.7
1973	28.7	22.2	42.5	48.0	56.1	70.7	73.4	73.0	62.5	55.0	43.0	31.2	50.5
1974	27.1	22.5	33.0	49.4	53.9	65.7	71.3	70.6	59.2	47.6	39.5	31.4	47.6
1975	29.5	28.4	31.6	39.3	63.2	67.1	73.0	70.0	58.4	53.1	47.2	27.8	49.0
1976	19.8	33.3	37.2	48.6	55.4	69.8	69.2	68.4	60.9	47.5	35.4	23.6	47.4
1977	15.5	25.4	39.8	47.9	60.7	64.8	72.9	68.7	62.7	49.4	43.6	28.4	48.3
1978	22.9	16.2	29.7	43.6	60.2	67.4	72.6	71.6	62.4	51.0	41.0	30.0	47.4
1979	21.5	13.7	38.6	44.0	56.4	66.2	72.3	67.1	61.3	50.3	43.0	32.3	47.2
1980	24.0	19.7	32.4	47.8	60.0	63.1	72.9	74.3	63.7	48.8	38.8	24.7	47.5
1981	15.7	32.3	34.5	48.0	57.2	67.3	71.9	69.4	59.8	47.3	39.9	28.7	47.7
1982	16.1	23.0	33.5	43.2	60.9	63.6	72.0	66.1	62.8	52.7	43.4	37.4	47.9
1983	27.4	29.1	37.2	43.9	53.8	66.7	73.8	70.7	63.9	52.9	40.7	25.1	48.8
1984	20.4	33.2	26.5	47.5	52.6	66.8	69.2	72.0	60.6	54.9	40.7	35.9	48.4
1985	21.9	25.6	36.7	49.6	58.6	61.7	68.8	68.7	63.8	51.0	41.4	25.0	47.7
1986	25.0	24.5	37.0	47.9	59.8	63.3	69.8	65.7	59.8	49.9	36.9	31.7	47.6
1987	25.3	23.6	37.1	49.7	59.9	67.9	72.7	67.3	61.6	47.1	40.6	32.6	48.8
1988	25.0	23.7	34.7	45.0	58.7	64.2	73.7	71.1	60.1	45.8	42.6	29.4	47.8
1989	30.3	22.5	32.3	42.1	56.3	67.4	72.8	68.5	61.7	52.6	38.1	17.1	46.8
1990	33.6	29.3	37.3	48.8	54.4	67.2	70.7	69.9	60.7	52.1	42.4	33.8	50.0
1991	25.1	30.5	37.0	50.0	62.8	68.3	72.3	70.3	60.5	52.1	39.0	30.7	49.9
1992	26.2	27.2	30.2	44.1	57.1	63.5	66.6	66.3	60.9	46.4	38.9	30.2	46.5
1993	27.5	18.7	30.0	46.9	56.6	65.5	72.4	71.4	59.0	48.0	39.0	28.4	47.0
Record Mean	24.7	24.6	32.9	45.2	56.8	66.4	71.3	69.3	62.5	51.3	39.8	29.0	47.8
Max	31.6	32.0	40.6	54.2	66.8	76.6	81.3	78.9	72.1	60.1	46.6	35.3	56.3
Min	17.7	17.2	25.2	36.2	46.8	56.2	61.3	59.7	52.9	42.4	33.0	22.7	39.3

REFERENCE NOTES ROCHESTER, NEW YORK

GENERAL
T - TRACE AMOUNT.
BLANK ENTRIES DENOTE MISSING/UNREPORTED DATA.
# INDICATES A STATION OR INSTRUMENT RELOCATION.

SPECIFIC

PM - INCLUDES LAST DAY OF PREVIOUS MONTH
ASOS - AUTOMATED SURFACE OBSERVING SYSTEM IN OPERATION DURING THESE MONTHS.

(a) - LENGTH OF RECORD IN YEARS, ALTHOUGH INDIVIDUAL MONTHS MAY BE MISSING.
0.* OR * - THE VALUE IS BETWEEN 0.0 AND 0.05
NORMALS - BASED ON THE 1961-1990 RECORD PERIOD.
EXTREMES - DATES ARE THE MOST RECENT OCCURRENCE.
WIND DIR.- NUMERALS SHOW TENS OF DEGRESS CLOCKWISE FROM TRUE NORTH. "00" INDICATES CALM.
RESULTANT DIRECTIONS ARE GIVEN TO WHOLE DEGREES.
BOLD VALUES INDICATE EXTREME VALUES WHICH OCCURRED AFTER THE ASOS SYSTEM WAS COMMISSIONED.

RECORD = PERIOD OF RECORD
RECORD MEAN PRECIPITATION IS THE MEAN OF ALL DAILY PRECIPITATION AMOUNTS DURING THE PERIOD OF RECORD.
RECORD MAX(MIN) TEMPERATURE IS THE MEAN OF ALL DAILY MAX(MIN) TEMPERATURES DURING THE PERIOD OF RECORD.
RECORD MEAN TEMPERATURE IS THE SUM OF THE RECORD MAX AND RECORD MIN DIVIDED BY 2.
AVERAGE TEMPERATURE IS THE SUM OF THE MEAN DAILY MAX AND MIN TEMPERATURE DIVIDED BY 2.

EXCEPTIONS

RECORD MEANS ARE THROUGH THE CURRENT YEAR,
BEGINNING IN 1872 FOR TEMPERATURE
1829 FOR PRECIPITATION
1941 FOR SNOWFALL

HEATING DEGREE DAYS Base 65 deg. F ROCHESTER, NEW YORK

SEASON	JULY	AUG	SEP	OCT	NOV	DEC	JAN	FEB	MAR	APR	MAY	JUNE	TOTAL
1964-65	3	48	169	500	674	1090	1337	1096	1095	706	192	114	7024
1965-66	24	53	122	496	748	937	1307	1098	871	612	382	57	6707
1966-67	0	16	190	458	639	1053	1046	1216	998	551	486	8	6661
1967-68	25	28	164	401	856	1005	1399	1275	894	463	357	78	6945
1968-69	15	36	67	369	710	1150	1229	1087	1013	517	302	86	6581
1969-70	18	10	126	437	737	1231	1448	1149	1053	506	209	49	6973
1970-71	3	11	126	349	699	1218	1405	1059	1122	707	342	49	7090
1971-72	17	33	74	194	792	1048	1200	1199	1071	677	161	78	6544
1972-73	7	24	92	534	833	982	1118	1189	690	519	279	17	6284
1973-74	2	14	162	305	653	1040	1167	1187	983	475	352	59	6399
1974-75	1	1	209	535	755	1034	1096	1017	1031	764	139	52	6634
1975-76	4	14	194	365	525	1146	1395	914	858	507	300	35	6257
1976-77	11	27	173	538	879	1279	1524	1103	777	523	204	89	7127
1977-78	9	44	113	477	634	1127	1298	1360	1087	634	220	63	7066
1978-79	5	1	136	428	711	1077	1342	1432	813	626	310	79	6960
1979-80	13	37	155	468	655	1006	1264	1306	1003	510	195	125	6737
1980-81	1	0	108	498	782	1243	1522	908	938	507	260	26	6793
1981-82	6	12	201	546	748	1119	1510	1171	972	648	162	67	7162
1982-83	10	54	113	377	643	847	1161	998	854	627	347	78	6109
1983-84	9	8	121	387	723	1228	1376	917	1187	520	395	50	6921
1984-85	14	7	162	307	724	897	1330	1097	869	471	217	119	6214
1985-86	15	23	121	429	700	1231	1235	1129	864	506	206	100	6559
1986-87	16	62	175	462	840	1026	1223	1153	858	454	234	39	6542
1987-88	7	50	139	547	722	997	1232	1192	933	594	220	126	6759
1988-89	6	40	164	596	664	1095	1070	1184	1009	682	288	33	6831
1989-90	0	33	149	383	801	1478	967	993	853	520	327	46	6550
1990-91	7	6	171	406	669	959	1230	957	862	458	170	29	5924
1991-92	2	1	196	408	776	1057	1195	1088	1069	621	259	89	6761
1992-93	26	46	172	571	774	1071	1158	1289	1077	538	263	66	7051
1993-94	0	10	214	525	775	1127							

COOLING DEGREE DAYS Base 65 deg. F ROCHESTER, NEW YORK

YEAR	JAN	FEB	MAR	APR	MAY	JUNE	JULY	AUG	SEP	OCT	NOV	DEC	TOTAL
1969	0	0	0	0	18	94	202	233	92	7	0	0	646
1970	0	0	0	16	47	148	235	176	69	13	0	0	704
1971	0	0	0	0	10	133	143	159	155	15	0	0	615
1972	0	0	0	0	24	94	261	179	79	0	0	0	637
1973	0	0	0	15	6	194	267	269	96	4	0	0	851
1974	0	0	0	13	14	88	204	181	40	0	0	0	540
1975	0	0	0	0	89	121	257	178	5	6	0	0	656
1976	0	0	0	24	9	189	150	138	55	1	0	0	566
1977	0	0	3	16	80	88	260	164	50	0	1	0	662
1978	0	0	0	0	77	141	245	212	66	3	0	0	744
1979	0	0	0	1	52	121	244	112	49	18	0	0	597
1980	0	0	0	0	46	73	253	294	76	2	0	0	744
1981	0	0	0	5	23	102	228	156	50	0	0	0	564
1982	0	0	0	3	40	30	232	95	52	3	1	0	456
1983	0	0	0	0	7	136	289	192	96	20	0	0	740
1984	0	0	0	1	14	113	152	233	35	1	0	0	549
1985	0	0	0	15	23	27	139	145	90	0	0	0	439
1986	0	0	1	0	50	53	168	94	28	0	0	0	394
1987	0	0	0	1	82	131	254	127	42	0	0	0	637
1988	0	0	0	0	34	107	284	232	29	7	0	0	693
1989	0	0	0	0	26	111	248	153	60	3	0	0	601
1990	0	0	3	41	5	122	192	164	45	14	0	0	586
1991	0	0	0	14	108	135	234	175	68	14	0	0	748
1992	0	0	0	1	19	51	84	96	57	0	0	0	308
1993	0	0	0	0	9	86	239	214	42	5	0	0	595

SNOWFALL (inches) ROCHESTER, NEW YORK

SEASON	JULY	AUG	SEP	OCT	NOV	DEC	JAN	FEB	MAR	APR	MAY	JUNE	TOTAL
1964-65	0.0	0.0	0.0	T	5.1	11.6	26.6	10.3	15.6	1.9	0.0	0.0	71.1
1965-66	0.0	T	0.0	0.9	8.0	6.0	60.2	21.0	6.2	0.9	T	0.0	103.2
1966-67	0.0	0.0	0.0	0.0	3.0	14.4	12.7	27.6	16.0	T	0.3	0.0	74.0
1967-68	0.0	0.0	0.0	T	10.0	6.9	24.2	20.4	15.2	T	0.0	0.0	76.7
1968-69	0.0	0.0	0.0	T	8.6	22.2	25.6	17.8	4.6	1.0	0.0	0.0	79.8
1969-70	0.0	0.0	0.0	T	5.8	42.0	37.9	27.7	4.9	1.3	T	0.0	119.6
1970-71	0.0	0.0	0.0	0.2	3.6	44.2	34.1	29.7	29.7	1.2	0.0	0.0	142.7
1971-72	0.0	0.0	0.0	0.0	11.2	13.8	18.1	35.7	19.0	7.3	0.0	0.0	105.1
1972-73	0.0	0.0	0.0	0.2	16.9	22.7	8.9	18.4	4.4	1.5	T	0.0	73.0
1973-74	0.0	0.0	0.0	0.0	4.2	23.4	14.4	26.6	22.3	8.2	T	0.0	99.1
1974-75	0.0	0.0	0.0	0.3	4.6	26.5	10.8	23.2	10.9	14.9	0.0	0.0	91.2
1975-76	0.0	0.0	0.0	T	1.8	28.3	29.9	8.8	15.2	1.8	0.4	0.0	86.2
1976-77	0.0	0.0	0.0	0.5	6.5	24.5	30.2	15.0	13.0	1.8	0.6	0.0	92.1
1977-78	0.0	0.0	0.0	T	12.7	35.2	60.4	40.7	7.5	4.2	0.2	0.0	160.9
1978-79	0.0	0.0	0.0	T	3.3	30.9	36.8	39.1	8.2	20.2	0.0	0.0	138.5
1979-80	0.0	0.0	0.0	0.2	1.2	12.2	13.1	24.0	21.2	0.3	0.0	0.0	72.2
1980-81	0.0	0.0	0.0	T	8.4	31.8	31.5	9.3	12.0	1.4	0.0	0.0	94.4
1981-82	0.0	0.0	0.0	0.1	2.4	46.1	43.6	14.9	8.9	12.4	0.0	0.0	128.4
1982-83	0.0	0.0	0.0	T	3.0	11.6	10.2	13.6	9.3	12.2	T	0.0	59.9
1983-84	0.0	0.0	0.0	0.0	17.6	19.6	23.4	27.8	29.1	0.5	T	0.0	118.0
1984-85	0.0	0.0	0.0	0.0	1.6	11.6	36.8	26.1	8.4	2.6	0.0	0.0	87.1
1985-86	0.0	0.0	0.0	0.0	7.6	18.3	15.5	17.9	9.3	2.1	T	0.0	70.7
1986-87	0.0	0.0	0.0	0.0	7.4	9.3	29.6	13.0	5.3	2.5	0.0	0.0	67.1
1987-88	0.0	0.0	0.0	T	4.6	19.3	9.8	29.4	5.6	1.1	0.0	0.0	69.8
1988-89	0.0	0.0	0.0	0.1	0.2	10.3	15.0	30.6	15.6	3.9	10.9	0.0	86.6
1989-90	0.0	0.0	0.0	T	6.5	32.8	14.0	31.3	5.4	15.8	T	0.0	105.8
1990-91	T	0.0	0.0	T	4.4	18.2	26.5	16.1	2.0	1.1	0.0	0.0	68.3
1991-92	0.0	0.0	0.0	0.0	13.7	23.9	18.3	12.8	38.1	3.8	0.0	0.0	110.6
1992-93	0.0	0.0	0.0	T	9.5	29.3	22.4	31.2	37.1	2.0	T	0.0	131.5
1993-94	0.0	0.0	T	2.6	9.8	14.0							
Record Mean	T	T	T	0.2	6.7	19.6	22.7	22.4	14.4	3.6	0.3	0.0	89.8

SYRACUSE, NEW YORK

Syracuse is located approximately at the geographical center of the state. Gently rolling terrain stretches northward for about 30 miles to the eastern end of Lake Ontario. Oneida Lake is about 8 miles northeast of Syracuse. Approximately 5 miles south of the city, hills rise to 1,500 feet. Immediately to the west, the terrain is gently rolling with elevations 500 to 800 feet above sea level.

The climate of Syracuse is primarily continental in character and comparatively humid. Nearly all cyclonic systems moving from the interior of the country through the St. Lawrence Valley will affect the Syracuse area. Seasonal and diurnal changes are marked and produce an invigorating climate.

In the summer and in portions of the transitional seasons, temperatures usually rise rapidly during the daytime to moderate levels and as a rule fall rapidly after sunset. The nights are relatively cool and comfortable. There are only a few days in a year when atmospheric humidity causes great personal discomfort.

Winters are usually cold and are sometimes severe in part. Daytime temperatures average in the low 30s with nighttime lows in the teens. Low winter temperatures below -25 degrees have been recorded. The autumn, winter, and spring seasons display marked variability.

Based on the 1951-1980 period, the average first occurrence of 32 degrees Fahrenheit in the fall is October 16 and the average last occurrence in the spring is April 28.

Precipitation in the Syracuse area is derived principally from cyclonic storms which pass from the interior of the country through the St. Lawrence Valley. Lake Ontario provides the source of significant winter precipitation. The lake is quite deep and never freezes so cold air flowing over the lake is quickly saturated and produces the cloudiness and snow squalls which are a well-known feature of winter weather in the Syracuse area.

The area enjoys sufficient precipitation in most years to meet the needs of agriculture and water supplies. The precipitation is uncommonly well distributed, averaging about 3 inches per month throughout the year. Snowfall is moderately heavy with an average just over 100 inches. There are about 30 days per year with thunderstorms, mostly during the warmer months.

Wind velocities are moderate, but during the winter months there are numerous days with sufficient winds to cause blowing and drifting snow.

During December, January, and February there is much cloudiness. Syracuse receives only about one-third of possible sunshine during winter months. Approximately two-thirds of possible sunshine is received during the warm months.

NORMALS, MEANS, AND EXTREMES

SYRACUSE, NEW YORK

LATITUDE: 43°07'N LONGITUDE: 76°07' W ELEVATION: FT. GRND 410 BARO 420 TIME ZONE: EASTERN WBAN: 14771

	(a)	JAN	FEB	MAR	APR	MAY	JUNE	JULY	AUG	SEP	OCT	NOV	DEC	YEAR
TEMPERATURE °F:														
Normals														
-Daily Maximum		30.6	32.5	42.7	56.0	68.3	76.7	81.7	79.0	71.6	60.3	48.0	35.4	56.9
-Daily Minimum		14.2	15.4	25.1	35.5	46.0	53.8	59.0	57.7	51.4	41.1	33.0	21.1	37.8
-Monthly		22.4	24.0	33.9	45.7	57.1	65.3	70.4	68.4	61.5	50.7	40.5	28.3	47.4
Extremes														
-Record Highest	44	70	69	87	92	96	98	97	97	97	87	81	70	98
-Year		1967	1981	1986	1990	1977	1953	1990	1987	1953	1963	1950	1966	JUN 1953
-Record Lowest	44	-26	-26	-16	9	25	35	45	40	28	19	5	-22	-26
-Year		1966	1979	1950	1972	1966	1966	1976	1965	1991	1976	1976	1980	FEB 1979
NORMAL DEGREE DAYS:														
Heating (base 65°F)		1321	1148	964	579	268	61	10	28	139	443	735	1138	6834
Cooling (base 65°F)		0	0	0	0	23	70	178	133	34	0	0	0	438
% OF POSSIBLE SUNSHINE	44	34	39	46	50	55	59	64	59	53	44	26	26	46
MEAN SKY COVER (tenths)														
Sunrise - Sunset	44	8.1	7.9	7.4	6.9	6.6	6.2	5.9	6.2	6.3	6.8	8.3	8.4	7.1
MEAN NUMBER OF DAYS:														
Sunrise to Sunset														
-Clear	44	2.6	3.2	4.8	6.2	6.1	7.2	7.8	6.9	6.8	6.4	2.2	2.3	62.6
-Partly Cloudy	44	6.5	5.8	6.9	6.7	9.8	10.5	12.2	11.4	10.2	7.7	5.5	4.9	98.1
-Cloudy	44	21.8	19.3	19.3	17.1	15.0	12.3	11.0	12.7	13.0	16.9	22.3	23.8	204.6
Precipitation														
.01 inches or more	44	19.0	16.0	16.7	14.3	12.9	11.3	10.9	11.0	11.3	12.4	16.4	19.1	171.2
Snow,Ice pellets,hail														
1.0 inches or more	44	8.7	7.6	5.0	1.4	0.*	0.0	0.0	0.0	0.0	0.3	2.8	7.8	33.6
Thunderstorms	44	0.2	0.2	0.8	1.8	3.3	5.3	6.1	5.5	2.6	1.0	0.6	0.1	27.4
Heavy Fog Visibility														
1/4 mile or less	44	0.7	0.7	0.8	0.6	0.7	0.5	0.5	0.7	0.9	1.2	0.6	0.8	8.8
Temperature °F														
-Maximum														
90° and above	30	0.0	0.0	0.0	0.*	0.3	1.4	3.7	1.8	0.3	0.0	0.0	0.0	7.5
32° and below	30	16.5	14.1	5.9	0.3	0.0	0.0	0.0	0.0	0.0	0.0	1.4	11.2	49.4
-Minimum														
32° and below	30	28.6	25.5	23.5	11.7	0.9	0.0	0.0	0.0	0.2	5.3	14.8	26.2	136.7
0° and below	30	4.4	3.0	0.7	0.0	0.0	0.0	0.0	0.0	0.0	0.0	0.0	1.6	9.6
AVG. STATION PRESS.(mb)	21	1001.7	1002.7	1001.3	999.7	1000.0	999.7	1000.4	1002.0	1003.0	1003.4	1002.5	1002.6	1001.6
RELATIVE HUMIDITY (%)														
Hour 01	30	76	76	76	75	79	83	84	87	86	82	80	79	80
Hour 07 (Local Time)	30	77	78	78	76	77	79	81	86	88	85	81	80	81
Hour 13	30	68	65	60	53	55	56	56	59	62	61	67	71	61
Hour 19	30	74	72	67	59	60	63	63	70	76	75	76	77	69
PRECIPITATION (inches):														
Water Equivalent														
-Normal		2.34	2.15	2.77	3.33	3.28	3.79	3.81	3.51	3.79	3.24	3.72	3.20	38.93
-Maximum Monthly	44	5.77	5.38	6.84	8.12	7.41	12.30	9.52	8.41	8.81	8.29	6.79	5.50	12.30
-Year		1978	1951	1955	1976	1976	1972	1974	1956	1975	1955	1972	1983	JUN 1972
-Minimum Monthly	44	1.02	0.63	1.01	1.22	0.75	1.10	0.90	1.33	0.75	0.21	1.25	1.73	0.21
-Year		1970	1987	1981	1985	1977	1962	1969	1980	1964	1963	1978	1958	OCT 1963
-Maximum in 24 hrs	44	1.47	1.99	1.77	2.85	3.13	3.88	4.07	4.27	4.14	3.60	2.09	2.18	4.27
-Year		1958	1961	1993	1976	1969	1972	1974	1954	1975	1955	1967	1952	AUG 1954
Snow,Ice pellets,hail														
-Maximum Monthly	44	72.2	72.6	54.4	16.4	1.2	T	T	0.0	T	5.7	25.9	64.6	72.6
-Year		1978	1958	1993	1983	1973	1992	1992		1992	1988	1976	1989	FEB 1958
-Maximum in 24 hrs	44	24.5	21.4	35.6	7.1	1.2	T	T	0.0	T	2.9	12.1	17.3	35.6
-Year		1966	1961	1993	1975	1973	1992	1992		1992	1988	1973	1991	MAR 1993
WIND:														
Mean Speed (mph)	44	10.8	10.7	10.8	10.5	9.0	8.3	8.0	7.7	8.2	8.8	10.2	10.4	9.4
Prevailing Direction through 1963		WSW	WNW	WNW	WNW	WNW	WNW	WNW	WSW	S	WSW	WSW	WSW	WNW
Fastest Mile														
-Direction (!!)	44	W	W	SE	NW	30	NW	NW	NW	W	SE	E	W	SE
-Speed (MPH)	44	60	62	56	52	52	49	47	43	52	63	59	52	63
-Year		1974	1967	1956	1957	1991	1961	1982	1958	1962	1954	1950	1962	OCT 1954
Peak Gust														
-Direction (!!)	10	W	SW	SW	W	NW	SE	NW	W	W	NW	W	NW	NW
-Speed (mph)	10	58	56	54	61	76	67	64	46	48	45	58	63	76
-Date		1992	1985	1986	1985	1991	1989	1993	1987	1989	1992	1992	1991	MAY 1991

PRECIPITATION (inches)

SYRACUSE, NEW YORK

YEAR	JAN	FEB	MAR	APR	MAY	JUNE	JULY	AUG	SEP	OCT	NOV	DEC	ANNUAL
1964	2.18	1.13	3.83	3.66	2.31	1.41	2.15	3.09	0.75	1.52	2.20	2.87	27.10
1965	2.28	2.82	1.63	3.53	1.61	2.04	1.34	1.95	3.60	2.70	2.97	1.92	28.39
1966	3.98	2.96	2.27	3.05	1.79	2.73	2.09	2.64	4.75	0.90	2.05	3.93	33.14
1967	1.47	1.49	1.34	2.11	3.33	1.56	6.33	5.00	2.73	3.52	4.48	2.66	36.02
1968	2.08	1.10	3.13	2.40	3.46	6.14	3.77	4.17	3.43	5.81	4.07	4.67	44.23
1969	3.37	1.49	1.08	3.95	4.34	3.74	0.90	1.77	1.13	2.30	4.56	3.42	32.05
1970	1.02	1.84	2.45	3.68	2.79	2.93	4.42	4.07	4.33	3.84	3.53	3.33	38.23
1971	1.90	4.07	2.90	2.19	3.40	3.26	6.49	4.01	2.56	1.62	3.52	3.26	39.18
1972	1.10	2.87	2.49	4.03	6.19	12.30	3.45	3.76	4.12	4.36	6.79	3.95	55.41
1973	1.85	1.71	3.45	6.91	5.58	7.07	3.62	2.97	4.57	3.81	6.73	4.38	52.65
1974	2.08	1.70	4.34	3.09	5.78	4.67	9.52	4.60	4.45	1.58	4.95	3.47	50.23
1975	2.54	3.05	2.67	2.01	2.74	4.08	9.32	5.35	8.81	3.69	3.54	4.10	51.90
1976	2.79	2.71	4.62	8.12	7.41	7.42	5.24	6.73	3.27	6.53	1.53	1.80	58.17
1977	1.84	1.62	3.47	3.04	0.75	3.30	4.76	4.93	6.54	4.75	5.31	4.33	44.64
1978	5.77	0.80	3.08	1.87	1.90	3.58	2.78	3.31	3.93	2.68	1.25	4.12	35.07
1979	4.70	2.54	2.73	3.89	3.07	2.33	2.33	3.69	5.25	2.91	3.25	1.84	38.53
1980	1.47	1.38	4.34	3.33	1.34	4.45	2.57	1.33	3.40	2.56	2.64	3.27	32.08
1981	1.34	2.72	1.01	2.04	2.61	1.89	2.68	2.63	5.58	6.66	3.09	2.96	35.21
1982	3.59	1.26	2.63	1.71	2.87	4.64	3.83	2.60	4.22	0.72	4.52	2.55	35.14
1983	1.92	1.07	2.30	6.34	3.33	1.50	2.31	2.80	2.98	1.98	4.30	5.50	36.33
1984	1.30	2.88	2.39	3.16	4.97	2.02	3.66	5.17	2.61	1.95	3.48	4.38	37.97
1985	2.49	1.55	2.61	1.22	3.39	2.80	2.75	1.44	3.88	3.39	5.18	1.80	32.50
1986	2.41	2.27	2.82	3.42	2.67	4.89	5.23	3.36	5.47	3.32	3.74	3.33	42.93
1987	3.03	0.63	1.86	3.31	1.41	5.04	2.16	2.12	5.99	3.13	3.02	1.99	33.69
1988	1.50	2.13	1.79	2.70	3.05	2.46	5.72	3.77	1.88	3.57	3.95	1.92	34.44
1989	1.06	1.71	3.13	1.52	4.27	5.41	2.20	2.68	5.96	4.08	2.78	2.13	36.93
1990	2.13	3.95	3.70	4.09	5.62	2.92	3.72	5.33	3.45	6.09	3.23	5.24	49.47
1991	2.44	1.54	4.07	3.90	3.90	1.67	2.86	4.03	4.20	2.62	2.72	3.10	37.05
1992	2.62	2.46	3.80	3.54	5.21	1.78	8.00	2.64	4.55	2.69	3.75	2.57	43.61
1993	3.08	2.45	3.75	6.55	2.25	2.93	4.76	4.71	3.83	2.91	3.19	3.20	43.61
Record Mean	2.66	2.48	3.08	3.14	3.07	3.54	3.53	3.34	3.19	3.08	3.03	2.96	37.09

AVERAGE TEMPERATURE (deg. F)

SYRACUSE, NEW YORK

YEAR	JAN	FEB	MAR	APR	MAY	JUNE	JULY	AUG	SEP	OCT	NOV	DEC	ANNUAL
1964	25.9	23.7	35.1	46.5	61.6	66.0	73.2	67.5	61.6	49.5	43.6	29.6	48.6
1965	20.5	24.6	30.3	42.3	59.4	63.7	67.5	69.1	62.8	47.9	38.7	32.1	46.6
1966	19.0	23.1	34.4	42.6	51.2	65.7	71.1	70.4	59.5	49.5	42.9	29.2	46.6
1967	30.6	19.2	31.6	44.9	50.2	69.5	67.7	66.6	60.7	51.8	37.6	32.7	46.9
1968	18.5	21.1	33.2	48.1	53.6	64.9	69.9	68.8	64.8	52.5	40.0	27.2	46.9
1969	24.3	23.6	30.4	46.9	55.7	64.5	69.9	71.3	63.8	51.1	40.4	23.8	47.1
1970	16.1	24.1	31.7	47.2	57.5	63.5	69.7	68.0	61.4	52.3	41.7	25.4	46.5
1971	18.5	26.5	31.2	42.8	55.8	67.9	69.0	67.1	65.6	56.6	36.9	33.2	47.6
1972	26.4	22.9	29.4	40.5	58.5	64.5	72.9	69.2	63.5	46.5	37.0	30.8	46.9
1973	28.4	21.4	42.6	46.8	54.3	69.6	72.7	73.5	62.0	53.7	40.7	29.5	49.6
1974	26.0	21.6	32.3	48.8	54.1	65.6	69.1	68.9	59.1	46.5	40.6	30.4	46.9
1975	29.4	28.1	31.7	39.9	62.9	67.1	71.7	68.2	57.1	53.2	46.6	27.6	48.6
1976	18.1	32.5	36.6	48.4	54.2	67.9	66.7	66.0	59.8	46.9	35.8	22.6	46.3
1977	15.7	26.0	40.1	48.2	60.3	62.7	70.8	67.3	62.5	50.6	44.0	27.3	48.0
1978	21.3	17.6	29.4	42.2	58.3	64.8	71.9	71.7	59.9	49.6	40.3	30.6	46.4
1979	22.4	12.9	39.1	45.1	58.6	66.0	71.7	67.9	61.4	50.9	44.5	33.4	47.8
1980	25.6	19.8	32.4	47.8	59.8	63.0	72.5	73.8	63.4	48.8	37.6	22.6	47.3
1981	15.0	33.7	36.4	50.0	59.2	68.0	73.3	70.4	61.6	47.9	39.0	29.0	48.6
1982	14.8	25.1	33.2	43.9	59.4	63.1	70.4	65.3	60.6	50.4	43.9	34.1	47.0
1983	23.4	26.4	35.7	44.3	53.7	66.7	72.0	69.0	62.5	50.3	39.0	22.5	47.1
1984	18.7	32.0	24.5	46.0	52.4	65.4	68.0	68.8	57.7	52.2	38.3	33.5	46.5
1985	22.0	27.3	36.3	47.8	59.5	62.0	69.8	68.9	63.5	51.4	41.2	26.0	48.0
1986	23.9	23.4	37.4	49.2	61.0	64.3	71.0	66.8	60.5	49.7	36.8	31.6	48.0
1987	23.8	21.7	38.0	51.9	60.3	68.3	73.6	68.5	61.1	47.7	40.9	32.3	49.0
1988	23.1	24.6	34.4	45.7	59.7	64.1	74.0	71.8	60.8	46.6	43.0	27.8	48.0
1989	28.6	22.7	32.9	43.5	58.2	67.3	71.1	68.2	61.8	51.7	38.8	14.7	46.6
1990	33.2	29.0	37.5	49.3	54.5	67.3	71.8	70.3	61.2	52.8	42.2	33.5	50.2
1991	24.3	29.8	37.7	51.0	62.8	68.4	72.4	71.8	60.5	53.1	40.0	30.7	50.2
1992	24.7	26.5	29.3	44.4	57.5	64.0	67.3	67.5	61.3	46.6	39.7	31.0	46.7
1993	27.5	17.0	30.1	46.9	58.0	65.2	72.5	70.7	60.0	48.2	38.6	26.9	46.8
Record Mean	23.9	24.2	33.5	45.6	57.1	66.1	71.2	69.2	62.0	51.2	40.2	28.2	47.7
Max	31.7	32.2	41.7	55.0	67.4	76.3	81.2	79.0	71.7	60.3	47.3	35.1	56.6
Min	16.1	16.1	25.3	36.2	46.8	55.9	61.1	59.4	52.4	42.1	33.1	21.3	38.8

REFERENCE NOTES

SYRACUSE, NEW YORK

GENERAL
T - TRACE AMOUNT.
BLANK ENTRIES DENOTE MISSING/UNREPORTED DATA.
# INDICATES A STATION OR INSTRUMENT RELOCATION.

SPECIFIC

PM - INCLUDES LAST DAY OF PREVIOUS MONTH
ASOS - AUTOMATED SURFACE OBSERVING SYSTEM IN OPERATION DURING THESE MONTHS.

(a) - LENGTH OF RECORD IN YEARS, ALTHOUGH INDIVIDUAL MONTHS MAY BE MISSING.
0.* OR * - THE VALUE IS BETWEEN 0.0 AND 0.05
NORMALS - BASED ON THE 1961-1990 RECORD PERIOD.
EXTREMES - DATES ARE THE MOST RECENT OCCURRENCE
WIND DIR.- NUMERALS SHOW TENS OF DEGRESS CLOCKWISE FROM TRUE NORTH. "00" INDICATES CALM.
RESULTANT DIRECTIONS ARE GIVEN TO WHOLE DEGREES.
BOLD VALUES INDICATE EXTREME VALUES WHICH OCCURRED AFTER THE ASOS SYSTEM WAS COMMISSIONED.

RECORD = PERIOD OF RECORD
RECORD MEAN PRECIPITATION IS THE MEAN OF ALL DAILY PRECIPITATION AMOUNTS DURING THE PERIOD OF RECORD.
RECORD MAX(MIN) TEMPERATURE IS THE MEAN OF ALL DAILY MAX(MIN) TEMPERATURES DURING THE PERIOD OF RECORD.
RECORD MEAN TEMPERATURE IS THE SUM OF THE RECORD MAX AND RECORD MIN DIVIDED BY 2.
AVERAGE TEMPERATURE IS THE SUM OF THE MEAN DAILY MAX AND MIN TEMPERATURE DIVIDED BY 2.

EXCEPTIONS

RECORD MEANS ARE THROUGH THE CURRENT YEAR, BEGINNING IN 1902 FOR TEMPERATURE
1902 FOR PRECIPITATION
1950 FOR SNOWFALL

HEATING DEGREE DAYS Base 65 deg. F SYRACUSE, NEW YORK

SEASON	JULY	AUG	SEP	OCT	NOV	DEC	JAN	FEB	MAR	APR	MAY	JUNE	TOTAL
1964-65	2	30	154	475	636	1087	1370	1125	1069	676	215	132	6971
1965-66	27	50	144	521	782	1011	1422	1168	945	665	429	80	7244
1966-67	6	7	186	473	652	1104	1059	1275	1030	595	453	13	6853
1967-68	24	35	154	407	812	995	1438	1266	979	501	345	83	7039
1968-69	27	41	54	391	745	1163	1256	1152	1063	536	306	103	6837
1969-70	22	20	134	425	730	1269	1508	1139	1027	536	244	115	7169
1970-71	7	27	150	388	692	1222	1437	1069	1040	658	295	50	7035
1971-72	13	51	96	256	840	980	1189	1216	1098	731	204	84	6758
1972-73	9	23	98	567	833	1053	1128	1217	687	547	325	31	6518
1973-74	2	12	164	344	723	1094	1200	1206	1004	493	339	52	6633
1974-75	16	3	202	565	726	1069	1100	1026	1026	749	138	46	6666
1975-76	3	32	230	357	545	1154	1449	936	872	509	329	47	6463
1976-77	24	45	179	556	869	1303	1520	1086	767	511	209	111	7180
1977-78	14	60	121	444	624	1162	1348	1322	1097	677	252	92	7213
1978-79	10	1	184	470	735	1062	1315	1457	796	591	242	74	6937
1979-80	19	39	146	454	607	971	1215	1302	1007	511	194	115	6580
1980-81	3	0	120	496	814	1307	1544	869	882	446	221	27	6729
1981-82	2	4	145	523	775	1110	1552	1114	978	626	183	79	7091
1982-83	13	57	152	449	628	951	1280	1073	902	615	351	67	6538
1983-84	11	25	140	457	769	1312	1432	949	1246	563	386	68	7358
1984-85	16	33	227	390	797	971	1329	1048	882	514	193	109	6509
1985-86	10	18	121	415	702	1200	1266	1156	856	471	172	76	6463
1986-87	12	50	155	468	838	1027	1270	1208	831	395	211	35	6500
1987-88	7	27	138	529	717	1007	1290	1167	942	571	187	131	6713
1988-89	9	33	150	574	653	1148	1120	1175	989	639	242	38	6770
1989-90	3	36	151	406	779	1554	976	1001	849	496	319	43	6613
1990-91	4	4	160	386	675	967	1253	980	839	428	153	24	5873
1991-92	1	0	189	378	743	1056	1240	1112	1099	617	245	79	6759
1992-93	15	33	164	562	753	1047	1156	1337	1074	537	230	68	6976
1993-94	2	10	190	515	785	1172							

COOLING DEGREE DAYS Base 65 deg. F SYRACUSE, NEW YORK

YEAR	JAN	FEB	MAR	APR	MAY	JUNE	JULY	AUG	SEP	OCT	NOV	DEC	TOTAL
1969	0	0	0	0	22	94	183	222	102	1	0	0	624
1970	0	0	0	8	22	74	160	127	51	3	0	0	445
1971	0	0	0	0	17	145	145	124	117	4	0	0	552
1972	0	0	0	0	9	78	262	160	61	0	0	0	570
1973	0	0	0	7	0	177	249	281	79	2	0	0	795
1974	0	0	0	14	6	77	148	128	31	1	0	0	405
1975	0	0	0	0	80	114	221	138	1	1	0	0	555
1976	0	0	0	16	2	141	84	83	31	0	0	0	357
1977	0	0	1	12	71	47	202	138	49	0	0	0	520
1978	0	0	0	0	49	92	231	215	36	0	0	0	623
1979	0	0	0	2	50	109	232	134	46	22	0	0	595
1980	0	0	0	0	41	62	243	279	80	1	0	0	706
1981	0	0	3	4	47	125	264	180	49	0	0	0	672
1982	0	0	0	0	18	25	186	72	25	0	3	0	329
1983	0	0	0	0	2	125	236	155	70	7	0	0	595
1984	0	0	0	0	4	88	119	154	14	1	0	0	380
1985	0	0	0	7	30	26	165	144	87	0	0	0	459
1986	0	0	5	1	52	62	201	112	28	0	0	0	461
1987	0	0	0	7	73	142	280	143	29	0	0	0	674
1988	0	0	0	0	33	112	296	251	32	9	0	0	733
1989	0	0	0	0	37	112	198	144	59	0	0	0	550
1990	0	0	5	33	2	118	222	177	51	16	0	0	624
1991	0	0	0	16	89	136	237	218	61	16	0	0	773
1992	0	0	0	5	21	54	94	118	60	0	0	0	352
1993	0	0	0	0	17	81	241	195	48	0	0	0	582

SNOWFALL (inches) SYRACUSE, NEW YORK

SEASON	JULY	AUG	SEP	OCT	NOV	DEC	JAN	FEB	MAR	APR	MAY	JUNE	TOTAL
1964-65	0.0	0.0	0.0	0.3	4.0	18.3	31.8	24.9	13.3	4.7	0.0	0.0	97.3
1965-66	0.0	0.0	0.0	1.8	2.7	7.1	71.0	27.0	7.8	0.5	0.9	0.0	118.8
1966-67	0.0	0.0	0.0	0.0	T	33.0	18.3	21.0	10.4	0.3	T	0.0	83.0
1967-68	0.0	0.0	0.0	T	14.4	14.4	18.5	23.2	10.7	T	0.0	0.0	81.2
1968-69	0.0	0.0	0.0	0.8	16.5	25.4	24.5	21.3	9.4	T	0.0	0.0	97.9
1969-70	0.0	0.0	0.0	1.7	9.7	52.5	21.7	25.8	12.7	1.2	0.2	0.0	125.5
1970-71	0.0	0.0	0.0	0.8	7.0	51.9	30.3	25.2	37.2	4.8	0.0	0.0	157.2
1971-72	0.0	0.0	0.0	0.0	16.7	18.3	18.2	50.0	22.7	7.8	0.0	0.0	133.7
1972-73	0.0	0.0	0.0	0.3	15.8	29.8	11.9	13.3	3.6	5.3	1.2	0.0	81.2
1973-74	0.0	0.0	0.0	T	20.6	24.4	15.5	23.7	31.2	7.8	T	0.0	123.2
1974-75	0.0	0.0	0.0	2.8	4.8	26.2	11.8	27.3	20.6	12.0	0.0	0.0	105.5
1975-76	0.0	0.0	0.0	T	2.8	27.0	35.8	12.7	16.6	0.9	T	0.0	95.8
1976-77	0.0	0.0	0.0	0.3	25.9	25.7	52.3	24.4	13.5	1.9	1.0	0.0	145.0
1977-78	0.0	0.0	0.0	0.0	11.3	40.1	72.2	26.1	11.1	0.4	T	0.0	161.2
1978-79	0.0	0.0	0.0	T	3.9	40.9	27.9	20.7	14.9	10.2	0.0	0.0	118.5
1979-80	0.0	0.0	0.0	0.1	1.5	13.8	24.5	32.8	20.5	0.2	0.0	0.0	93.4
1980-81	0.0	0.0	0.0	T	7.3	28.8	23.4	8.5	10.6	0.4	0.0	0.0	79.0
1981-82	0.0	0.0	0.0	0.5	12.1	37.3	48.2	11.6	14.4	13.0	0.0	0.0	137.1
1982-83	0.0	0.0	0.0	T	1.9	10.9	20.3	8.2	8.3	16.4	T	0.0	66.0
1983-84	0.0	0.0	0.0	0.0	7.6	24.2	21.8	19.7	40.3	T	0.0	0.0	113.6
1984-85	0.0	0.0	0.0	0.0	5.0	23.4	57.3	21.6	7.1	2.0	0.0	0.0	116.4
1985-86	0.0	0.0	0.0	0.0	8.0	28.2	29.9	26.1	11.0	1.7	T	0.0	104.9
1986-87	0.0	0.0	0.0	0.0	16.1	8.8	49.2	15.1	3.0	1.3	0.0	0.0	93.5
1987-88	0.0	0.0	0.0	T	10.8	20.7	18.0	46.1	10.2	5.6	0.0	0.0	111.4
1988-89	0.0	0.0	0.0	5.7	0.2	34.4	19.4	21.7	9.9	6.5	0.0	0.0	97.8
1989-90	0.0	0.0	T	T	12.9	64.6	27.4	33.3	15.2	8.6	0.0	0.0	162.0
1990-91	0.0	0.0	0.0	0.2	7.8	24.5	30.9	27.7	2.8	3.0	0.0	0.0	96.9
1991-92	0.0	0.0	0.0	0.0	5.5	37.9	50.5	27.6	41.3	4.1	0.0	T	166.9
1992-93	T	0.0	T	1.4	10.1	19.8	42.9	51.3	54.4	12.2	0.0	0.0	192.1
1993-94	0.0	0.0	0.0	1.0	17.1	34.0							
Record Mean	T	0.0	T	0.6	9.3	26.8	29.6	26.2	17.4	4.0	0.1	T	113.9

ASHEVILLE, NORTH CAROLINA

The city of Asheville is located on both banks of the French Broad River, near the center of the French Broad Basin. Upstream from Asheville, the valley runs south for 18 miles and then curves toward the south-southwest. Downstream from the city, the valley is oriented toward the north-northwest. Two miles upstream from the principal section of Asheville, the Swannanoa River joins the French Broad from the east. The entire valley is known as the Asheville Plateau, having an average elevation near 2,200 feet above sea level, and is flanked by mountain ridges to the east and west, whose peaks range from 2,000 to 4,400 feet above the valley floor. At the Carolina-Tennessee border, about 25 miles north-northwest of Asheville, a relatively high ridge of mountains blocks the northern end of the valley. Thirty miles south, the Blue Ridge Mountains form an escarpment, having a general elevation of about 2,700 feet above sea level. The tallest peaks near Asheville are Mt. Mitchell, 6,684 feet above sea level, 20 miles northeast of the city, and Big Pisgah Mountain, 5,721 feet above sea level, 16 miles to the southwest.

Asheville has a temperate, but invigorating, climate. Considerable variation in temperature often occurs from day to day in summer, as well as during the other seasons.

While the office was located in the city, the combination of roof exposure conditions and a smoke blanket, caused by inversions in temperature in the valley on quiet nights, resulted in higher early morning temperatures at City Office sites than were experienced nearer ground level in nearby rural areas. The growing season in this area is of sufficient length for commercial crops, the average length of freeze-free period being about 195 days. The average last occurrence in spring of a temperature 32 degrees or lower is mid-April and the average first occurrence in fall of 32 degrees is late October.

The orientation of the French Broad Valley appears to have a pronounced influence on the wind direction. Prevailing winds are from the northwest during all months of the year. Also, the shielding effect of the nearby mountain barriers apparently has a direct bearing on the annual amount of precipitation received in this vicinity. In an area northwest of Asheville, the average annual precipitation is the lowest in North Carolina. Precipitation increases sharply in all other directions, especially to the south and southwest.

Destructive events caused directly by meteorological conditions are infrequent. The most frequent, occurring at approximately 12-year intervals, are floods on the French Broad River. These floods are usually associated with heavy rains caused by storms moving out of the Gulf of Mexico. Snowstorms which have seriously disrupted normal life in this community are infrequent. Hailstorms that cause property damage are extremely rare.

NORMALS, MEANS, AND EXTREMES

ASHEVILLE, NORTH CAROLINA

LATITUDE: 35°26'N LONGITUDE: 82°33' W ELEVATION: FT. GRND 2140 BARO 2161 TIME ZONE: EASTERN WBAN: 03812

	(a)	JAN	FEB	MAR	APR	MAY	JUNE	JULY	AUG	SEP	OCT	NOV	DEC	YEAR
TEMPERATURE °F:														
Normals														
-Daily Maximum		46.5	50.0	59.2	67.8	75.0	80.4	83.0	82.1	76.9	68.3	59.3	50.3	66.6
-Daily Minimum		24.8	27.4	35.4	42.6	50.9	58.3	62.7	61.9	55.5	43.5	35.7	28.6	43.9
-Monthly		35.7	38.7	47.4	55.2	63.0	69.4	72.8	72.0	66.2	55.9	47.5	39.5	55.3
Extremes														
-Record Highest	29	78	77	83	89	91	96	96	100	92	86	81	78	100
-Year		1975	1989	1985	1972	1969	1969	1988	1983	1991	1986	1974	1971	AUG 1983
-Record Lowest	29	-16	-2	2	22	28	35	44	42	30	21	8	-7	-16
-Year		1985	1967	1993	1987	1989	1966	1988	1986	1967	1976	1970	1983	JAN 1985
NORMAL DEGREE DAYS:														
Heating (base 65°F)		908	736	546	298	117	18	5	5	64	295	525	791	4308
Cooling (base 65°F)		0	0	0	0	55	150	247	222	100	13	0	0	787
% OF POSSIBLE SUNSHINE	29	55	59	60	65	61	63	60	54	56	61	58	55	59
MEAN SKY COVER (tenths)														
Sunrise - Sunset	29	6.1	5.9	6.0	5.6	6.2	6.1	6.3	6.4	6.2	5.1	5.5	6.0	6.0
MEAN NUMBER OF DAYS:														
Sunrise to Sunset														
-Clear	29	9.3	9.0	9.0	9.8	7.3	6.4	5.4	4.9	6.9	12.1	10.7	9.8	100.7
-Partly Cloudy	29	7.4	6.3	8.2	8.5	10.3	12.1	13.8	13.7	10.6	8.0	7.1	7.2	113.1
-Cloudy	29	14.2	12.9	13.8	11.7	13.4	11.5	11.8	12.4	12.4	10.9	12.2	14.0	151.4
Precipitation														
.01 inches or more	29	10.2	9.3	11.4	9.4	11.9	11.1	11.8	12.7	9.6	8.0	9.4	9.8	124.6
Snow,Ice pellets,hail														
1.0 inches or more	29	1.3	1.2	0.7	0.2	0.0	0.0	0.0	0.0	0.0	0.0	0.2	0.5	4.1
Thunderstorms	29	0.3	0.8	2.2	3.3	6.9	7.7	9.3	8.8	3.3	0.8	0.8	0.3	44.4
Heavy Fog Visibility														
1/4 mile or less	29	4.0	3.0	2.3	2.4	5.7	7.8	9.1	13.9	12.0	8.1	4.3	4.5	77.2
Temperature °F														
-Maximum														
90° and above	29	0.0	0.0	0.0	0.0	0.1	1.8	5.1	2.4	0.3	0.0	0.0	0.0	9.8
32° and below	29	3.1	1.5	0.3	0.0	0.0	0.0	0.0	0.0	0.0	0.0	0.1	1.0	5.9
-Minimum														
32° and below	29	23.2	20.1	13.3	4.2	0.3	0.0	0.0	0.0	0.*	4.2	13.4	20.8	99.5
0° and below	29	0.5	0.*	0.0	0.0	0.0	0.0	0.0	0.0	0.0	0.0	0.0	0.1	0.6
AVG. STATION PRESS.(mb)	21	941.7	941.2	940.3	940.2	941.0	942.2	943.4	944.0	944.0	944.0	943.3	942.7	942.3
RELATIVE HUMIDITY (%)														
Hour 01	29	81	78	79	78	89	93	94	96	96	91	85	82	87
Hour 07 (Local Time)	29	85	83	85	85	92	94	95	97	97	94	88	85	90
Hour 13	29	59	55	53	50	57	59	62	63	63	56	57	59	58
Hour 19	29	68	62	60	55	67	70	74	78	81	73	70	70	69
PRECIPITATION (inches):														
Water Equivalent														
-Normal		3.25	3.91	4.63	3.36	4.43	4.23	4.52	4.69	3.87	3.59	3.59	3.52	47.59
-Maximum Monthly	29	7.47	8.07	9.86	7.26	8.83	10.73	9.92	11.28	9.12	8.82	7.76	8.48	11.28
-Year		1978	1990	1975	1979	1973	1989	1982	1967	1977	1990	1979	1973	AUG 1967
-Minimum Monthly	29	0.45	0.44	0.77	0.25	1.06	0.90	0.46	0.52	0.16	0.19	1.19	0.16	0.16
-Year		1981	1978	1985	1976	1988	1990	1986	1981	1984	1991	1981	1965	SEP 1984
-Maximum in 24 hrs	29	2.95	3.47	5.13	3.06	4.95	3.93	4.02	5.10	3.41	3.06	4.03	2.66	5.13
-Year		1978	1982	1968	1973	1973	1987	1969	1990	1975	1990	1977	1973	MAR 1968
Snow,Ice pellets,hail														
-Maximum Monthly	29	17.6	25.5	18.2	11.5	T	T	0.0	T	0.0	T	9.6	16.3	25.5
-Year		1966	1969	1993	1987	1993	1991		1990		1993	1968	1971	FEB 1969
-Maximum in 24 hrs	29	14.0	11.7	16.5	11.5	T	T	0.0	T	0.0	T	5.7	16.3	16.5
-Year		1988	1969	1993	1987	1993	1991		1990		1993	1968	1971	MAR 1993
WIND:														
Mean Speed (mph)	29	9.5	9.5	9.4	8.8	7.1	6.1	5.9	5.4	5.7	6.7	8.2	8.9	7.6
Prevailing Direction														
Fastest Obs. 1 Min.														
-Direction (!!)	29	34	34	33	22	34	36	35	34	32	33	32	34	34
-Speed (MPH)	29	40	60	48	44	40	40	43	40	35	35	40	44	60
-Year		1975	1972	1993	1970	1971	1977	1966	1973	1980	1972	1974	1965	FEB 1972
Peak Gust														
-Direction (!!)	10	NW	NW	NW	N	N	N	S	NW	N	NW	N	NW	NW
-Speed (mph)	10	49	54	64	51	44	52	60	43	37	40	49	49	64
-Date		1984	1987	1993	1988	1992	1987	1990	1993	1992	1990	1989	1992	MAR 1993

PRECIPITATION (inches)

ASHEVILLE, NORTH CAROLINA

YEAR	JAN	FEB	MAR	APR	MAY	JUNE	JULY	AUG	SEP	OCT	NOV	DEC	ANNUAL
#1964	2.83	3.58	5.13	5.21	0.94	0.80	3.29	8.88	5.37	8.46	2.51	2.88	49.88
1965	2.16	4.60	5.10	2.62	3.33	4.12	4.47	4.03	4.69	2.92	1.30	0.16	39.50
1966	3.37	6.56	2.59	5.47	4.73	2.46	3.24	7.73	4.55	5.37	3.32	2.36	51.75
1967	2.02	2.20	2.86	1.11	6.79	4.45	6.90	11.28	2.53	3.30	2.54	6.13	52.11
1968	2.93	0.62	6.65	2.37	2.92	5.06	7.18	3.31	2.64	5.02	2.98	3.10	44.78
1969	2.64	5.08	4.01	3.53	3.32	3.82	7.53	6.47	3.04	2.63	1.91	4.63	48.61
1970	1.75	2.42	2.62	2.96	1.72	2.72	5.02	2.46	1.17	5.55	1.83	2.72	32.94
1971	2.53	4.93	3.48	2.06	3.54	5.00	5.47	3.03	3.80	7.05	2.84	4.32	48.05
1972	3.57	2.02	3.19	1.49	6.63	6.54	4.66	1.88	5.29	4.44	4.42	3.89	48.02
1973	4.26	4.23	8.91	5.71	8.83	3.87	6.95	4.57	3.12	2.41	3.57	8.48	64.91
1974	3.44	4.24	3.18	4.99	5.58	3.73	3.93	7.34	4.13	1.28	4.22	2.38	48.44
1975	3.86	4.56	9.86	0.61	8.17	2.12	3.31	3.63	7.53	3.94	4.89	4.44	56.92
1976	3.51	2.20	4.96	0.25	8.67	5.51	3.18	4.23	3.50	5.59	1.58	4.05	47.23
1977	2.09	1.02	7.29	4.05	3.96	5.11	1.03	3.68	9.12	3.79	6.88	2.43	50.45
1978	7.47	0.44	5.22	2.97	4.65	2.29	0.63	6.91	2.57	0.30	2.49	4.32	40.26
1979	6.81	5.14	5.72	7.26	5.35	2.20	5.52	3.63	5.60	1.40	7.76	1.05	57.44
1980	2.85	0.53	8.26	4.77	4.54	4.68	2.21	2.38	4.36	2.62	3.04	0.59	40.83
1981	0.45	4.80	3.24	2.07	7.50	4.41	2.06	0.52	1.36	2.19	1.19	4.79	34.58
1982	5.41	7.02	1.92	3.62	3.78	3.98	9.92	1.73	1.33	3.48	4.59	4.04	50.82
1983	3.39	5.63	6.27	5.27	3.48	3.71	1.06	0.95	5.66	4.43	4.77	8.30	52.92
1984	2.35	6.43	4.82	4.05	6.62	3.69	5.88	5.02	0.16	2.73	2.61	1.34	45.71
1985	2.95	4.74	0.77	2.74	1.59	1.47	4.37	7.04	1.25	3.41	4.91	0.70	35.94
1986	1.11	1.85	2.75	0.57	3.55	1.28	0.46	6.10	3.15	4.19	5.28	4.28	34.57
1987	3.49	6.17	2.85	3.67	1.87	8.94	1.86	1.79	6.79	0.36	3.09	2.33	43.21
1988	3.71	0.88	1.31	3.46	1.06	0.94	2.65	1.78	2.79	3.12	3.47	1.41	26.58
1989	1.65	4.61	2.91	3.17	5.54	10.73	8.33	4.98	8.17	2.98	4.27	3.29	60.63
1990	3.27	8.07	5.95	1.96	5.09	0.90	6.55	7.78	1.43	8.82	1.55	4.50	55.87
1991	3.25	1.66	6.13	5.38	2.41	5.27	6.07	3.83	1.27	0.19	3.34	4.86	43.66
1992	3.08	3.66	3.52	3.99	6.18	6.62	1.10	7.64	3.15	4.15	7.24	3.71	54.04
1993	3.82	2.03	6.16	3.21	4.59	1.12	2.07	5.29	1.56	1.21	3.32	3.59	37.97
Record Mean	3.21	3.74	4.57	3.29	4.69	4.03	4.26	4.52	3.65	3.41	3.63	3.52	46.51

AVERAGE TEMPERATURE (deg. F)

ASHEVILLE, NORTH CAROLINA

YEAR	JAN	FEB	MAR	APR	MAY	JUNE	JULY	AUG	SEP	OCT	NOV	DEC	ANNUAL
#1964	38.1	34.7	46.3	57.3	65.3	72.8	72.9	71.7	66.0	53.0	51.4	42.9	56.1
1965	37.0	37.9	42.4	57.4	65.8	66.8	72.2	71.1	66.8	53.8	46.4	40.3	54.8
1966	30.1	36.2	43.5	52.0	60.1	66.1	71.1	69.9	62.9	51.7	45.0	37.6	52.2
1967	38.7	35.0	49.8	57.6	59.7	66.8	68.7	68.5	60.2	53.5	42.7	41.8	53.6
1968	34.3	32.4	46.6	54.8	61.0	69.5	73.1	74.1	64.1	56.7	46.0	36.2	54.0
1969	36.7	37.8	41.3	56.7	65.2	73.1	75.9	70.7	65.8	56.2	44.0	36.5	55.0
1970	30.9	39.1	46.8	57.6	63.7	70.1	74.4	73.1	70.7	59.0	46.0	42.8	56.2
1971	36.5	39.5	43.1	55.2	61.1	72.4	72.5	72.2	69.5	61.8	45.8	47.7	56.5
1972	42.1	37.6	46.5	55.8	61.2	66.5	72.3	72.5	69.0	55.0	45.5	45.2	55.7
1973	37.5	38.5	52.7	52.8	60.3	71.0	74.1	74.2	70.0	58.8	49.1	39.8	56.5
1974	48.2	40.5	51.1	54.9	64.2	66.7	72.9	72.3	65.7	54.6	47.4	40.3	56.6
1975	41.7	42.5	44.8	54.1	66.0	68.8	72.4	72.9	65.2	57.3	48.2	38.6	56.1
1976	33.6	45.3	50.6	54.9	59.5	68.1	71.2	70.2	63.1	51.5	41.2	36.2	53.8
1977	24.8	37.4	50.7	58.2	64.6	69.7	75.7	73.8	69.1	54.3	49.3	36.8	55.3
1978	29.3	33.4	45.9	56.8	62.0	71.1	73.4	74.1	70.0	55.7	51.8	41.0	55.4
1979	34.2	35.8	50.0	55.9	64.3	68.8	72.2	73.2	66.5	55.3	49.2	42.0	55.6
1980	40.5	35.1	46.2	56.5	64.8	71.7	77.5	74.8	70.2	54.7	47.1	39.6	56.5
1981	33.3	39.9	44.9	60.1	60.7	74.3	75.0	71.7	66.1	54.5	48.2	35.8	55.4
1982	32.3	41.2	50.0	53.6	67.3	71.5	74.6	71.7	64.5	56.3	47.1	44.9	56.3
1983	36.7	38.8	46.7	51.1	61.6	69.0	75.7	76.5	66.6	57.5	47.3	36.4	55.3
1984	34.0	40.5	44.8	51.7	59.9	70.0	70.6	71.6	62.8	62.7	43.1	46.3	54.8
1985	30.5	38.3	48.1	56.6	62.6	69.8	72.2	70.9	64.2	60.5	56.0	34.7	55.4
1986	35.0	42.2	46.0	56.0	63.3	71.7	76.1	70.9	68.0	57.4	50.7	39.8	56.4
1987	35.3	38.9	46.5	52.6	66.7	71.2	74.7	74.7	66.5	50.1	47.0	42.2	55.5
1988	32.1	37.1	47.1	54.6	61.1	69.3	73.8	74.6	66.3	50.2	46.7	38.7	54.3
1989	42.1	39.8	50.3	54.5	59.4	70.0	73.3	71.9	65.8	56.1	46.2	31.6	55.1
1990	42.8	45.6	50.4	54.2	63.3	70.9	73.8	73.9	67.6	57.8	49.9	45.5	58.0
1991	39.2	42.4	49.6	58.4	67.3	70.3	75.2	72.1	67.3	57.0	45.5	43.4	57.3
1992	40.7	44.2	46.3	55.9	59.9	68.1	74.9	70.4	67.8	54.4	46.2	39.2	55.7
1993	42.0	38.0	43.0	52.6	63.2	71.4	78.0	73.9	67.8	55.0	46.0	36.8	55.6
Record Mean	36.3	39.0	47.0	55.3	62.7	69.8	73.7	72.5	66.5	55.8	47.0	39.9	55.5
Max	46.9	50.4	59.1	68.1	74.7	81.2	84.3	82.8	77.2	68.2	58.7	50.7	66.9
Min	25.6	27.6	35.0	42.4	50.7	58.4	63.1	62.1	55.8	43.4	35.3	29.1	44.1

REFERENCE NOTES

ASHEVILLE, NORTH CAROLINA

GENERAL
T - TRACE AMOUNT.
BLANK ENTRIES DENOTE MISSING/UNREPORTED DATA.
# INDICATES A STATION OR INSTRUMENT RELOCATION.

SPECIFIC

PM - INCLUDES LAST DAY OF PREVIOUS MONTH
ASOS - AUTOMATED SURFACE OBSERVING SYSTEM IN OPERATION DURING THESE MONTHS.

(a) - LENGTH OF RECORD IN YEARS, ALTHOUGH INDIVIDUAL MONTHS MAY BE MISSING.
0.* OR * - THE VALUE IS BETWEEN 0.0 AND 0.05
NORMALS - BASED ON THE 1961-1990 RECORD PERIOD.
EXTREMES - DATES ARE THE MOST RECENT OCCURRENCE.
WIND DIR.- NUMERALS SHOW TENS OF DEGRESS CLOCKWISE. FROM TRUE NORTH. "00" INDICATES CALM.
RESULTANT DIRECTIONS ARE GIVEN TO WHOLE DEGREES.
BOLD VALUES INDICATE EXTREME VALUES WHICH OCCURRED AFTER THE ASOS SYSTEM WAS COMMISSIONED.

RECORD = PERIOD OF RECORD
RECORD MEAN PRECIPITATION IS THE MEAN OF ALL DAILY PRECIPITATION AMOUNTS DURING THE PERIOD OF RECORD.
RECORD MAX(MIN) TEMPERATURE IS THE MEAN OF ALL DAILY MAX(MIN) TEMPERATURES DURING THE PERIOD OF RECORD.
RECORD MEAN TEMPERATURE IS THE SUM OF THE RECORD MAX AND RECORD MIN DIVIDED BY 2.
AVERAGE TEMPERATURE IS THE SUM OF THE MEAN DAILY MAX AND MIN TEMPERATURE DIVIDED BY 2.

EXCEPTIONS

RECORD MEANS ARE THROUGH THE CURRENT YEAR, BEGINNING IN 1965 FOR TEMPERATURE
1965 FOR PRECIPITATION
1965 FOR SNOWFALL

HEATING DEGREE DAYS Base 65 deg. F ASHEVILLE, NORTH CAROLINA

SEASON	JULY	AUG	SEP	OCT	NOV	DEC	JAN	FEB	MAR	APR	MAY	JUNE	TOTAL
#1964-65	0	20	46	372	399	679	863	751	691	232	23	27	4103
1965-66	0	7	39	344	550	759	1075	800	660	383	149	42	4808
1966-67	1	1	87	405	593	838	810	834	465	226	185	51	4496
1967-68	7	2	158	351	660	713	947	939	566	306	150	7	4806
1968-69	0	20	42	258	563	884	873	755	729	246	70	9	4449
1969-70	0	8	59	280	623	875	1050	720	557	236	86	3	4497
1970-71	0	0	29	194	565	682	875	707	672	290	137	0	4151
1971-72	0	0	6	129	576	530	704	790	569	294	116	35	3749
1972-73	3	0	8	304	578	605	846	737	374	362	158	0	3975
1973-74	0	0	7	205	473	772	516	680	423	299	83	24	3482
1974-75	0	0	65	316	519	760	715	624	619	331	46	7	4002
1975-76	0	0	77	232	498	812	966	566	439	296	168	33	4087
1976-77	2	3	83	411	706	884	1239	768	437	198	66	25	4822
1977-78	0	0	14	331	466	868	1101	878	586	241	139	0	4624
1978-79	0	0	12	283	390	741	951	810	457	268	71	18	4001
1979-80	5	0	44	299	468	707	753	861	573	258	65	2	4035
1980-81	0	0	37	315	533	778	978	696	615	152	152	0	4256
1981-82	0	1	57	326	499	897	1006	659	458	333	38	0	4274
1982-83	0	0	74	274	531	616	872	725	562	410	127	13	4204
1983-84	0	0	84	229	527	882	955	706	618	391	176	9	4577
1984-85	1	0	107	91	648	576	1064	737	520	249	109	19	4121
1985-86	0	6	111	156	266	932	923	633	581	273	91	2	3974
1986-87	0	32	16	268	419	774	913	725	567	369	40	1	4124
1987-88	0	0	47	452	532	702	1013	802	545	308	132	31	4564
1988-89	5	0	33	453	544	808	702	698	454	331	200	4	4232
1989-90	1	8	74	279	558	1028	679	535	446	321	91	3	4023
1990-91	0	0	55	229	445	601	793	627	472	204	53	16	3495
1991-92	0	1	64	242	578	663	748	596	573	274	169	25	3933
1992-93	0	0	34	324	558	794	708	751	677	365	82	9	4302
1993-94	0	0	44	310	563	866							

COOLING DEGREE DAYS Base 65 deg. F ASHEVILLE, NORTH CAROLINA

YEAR	JAN	FEB	MAR	APR	MAY	JUNE	JULY	AUG	SEP	OCT	NOV	DEC	TOTAL
1969	0	0	0	4	85	262	343	196	92	15	0	0	997
1970	0	0	0	22	52	159	296	259	206	17	0	0	1011
1971	0	0	0	3	25	232	238	231	149	37	8	0	923
1972	0	0	0	24	6	84	236	237	134	0	1	0	722
1973	0	0	0	1	16	190	288	292	163	19	0	0	969
1974	0	0	0	3	65	82	254	234	92	1	0	0	731
1975	0	0	0	11	82	124	237	252	89	0	0	0	795
1976	0	0	0	0	5	135	198	170	35	2	0	0	545
1977	0	0	0	2	59	173	340	279	146	7	1	0	1007
1978	0	0	0	2	53	188	266	292	168	4	0	0	973
1979	0	0	0	1	55	141	234	261	96	4	0	0	792
1980	0	0	0	8	64	210	396	311	198	4	0	0	1191
1981	0	0	0	10	25	286	316	213	98	7	0	0	955
1982	0	0	0	0	117	206	305	215	64	16	0	0	923
1983	0	0	0	0	25	139	335	362	141	5	0	0	1007
1984	0	0	0	0	25	165	180	211	49	27	0	0	657
1985	0	0	5	2	43	170	229	194	90	25	4	0	762
1986	0	0	0	8	43	209	353	222	112	38	0	0	985
1987	0	0	0	7	97	192	310	309	97	0	0	0	1012
1988	0	0	0	0	18	168	282	304	79	3	0	0	854
1989	0	0	5	23	34	159	264	229	107	11	0	0	832
1990	0	0	0	3	48	187	279	283	141	11	0	0	952
1991	0	0	3	13	132	181	324	227	139	3	0	0	1022
1992	0	0	0	7	20	125	313	174	125	0	0	0	764
1993	0	0	0	0	37	210	411	285	133	5	0	0	1081

SNOWFALL (inches) ASHEVILLE, NORTH CAROLINA

SEASON	JULY	AUG	SEP	OCT	NOV	DEC	JAN	FEB	MAR	APR	MAY	JUNE	TOTAL
#1964-65	0.0	0.0	0.0	0.0	T	T	5.5	4.3	5.0	0.0	0.0	0.0	14.8
1965-66	0.0	0.0	0.0	0.0	T	T	17.6	6.2	0.2	T	0.0	0.0	24.0
1966-67	0.0	0.0	0.0	0.0	1.3	0.8	1.5	4.2	T	0.0	0.0	0.0	7.8
1967-68	0.0	0.0	0.0	0.0	0.0	1.9	7.2	6.0	0.1	0.0	0.0	0.0	15.2
1968-69	0.0	0.0	0.0	0.0	9.6	T	0.1	25.5	13.0	0.0	0.0	0.0	48.2
1969-70	0.0	0.0	0.0	0.0	T	10.9	4.8	1.1	0.5	0.0	0.0	0.0	17.3
1970-71	0.0	0.0	0.0	0.0	T	6.1	0.1	0.1	8.9	0.2	0.0	0.0	15.4
1971-72	0.0	0.0	0.0	0.0	0.6	16.3	T	7.4	7.4	0.0	0.0	0.0	31.7
1972-73	0.0	0.0	0.0	0.0	0.6	T	7.1	0.5	1.0	T	0.0	0.0	9.2
1973-74	0.0	0.0	0.0	0.0	0.0	3.0	T	0.3	1.1	T	0.0	0.0	4.4
1974-75	0.0	0.0	0.0	0.0	3.1	3.0	0.4	4.3	3.7	T	0.0	0.0	14.5
1975-76	0.0	0.0	0.0	0.0	5.0	0.4	1.6	3.5	T	0.0	0.0	0.0	10.5
1976-77	0.0	0.0	0.0	0.0	0.1	0.3	11.9	0.7	0.0	0.0	0.0	0.0	13.0
1977-78	0.0	0.0	0.0	T	T	1.5	9.7	5.3	5.3	T	0.0	0.0	21.8
1978-79	0.0	0.0	0.0	0.0	0.0	T	5.2	17.8	T	0.0	T	0.0	23.0
1979-80	0.0	0.0	0.0	0.0	T	T	2.1	6.3	5.4	T	0.0	0.0	13.8
1980-81	0.0	0.0	0.0	0.0	T	T	4.7	T	9.9	0.0	0.0	0.0	14.6
1981-82	0.0	0.0	0.0	0.0	T	2.0	8.6	8.1	0.1	3.0	0.0	0.0	21.8
1982-83	0.0	0.0	0.0	0.0	0.0	0.4	10.5	9.3	4.5	2.0	0.0	0.0	26.7
1983-84	0.0	0.0	0.0	0.0	T	T	0.2	2.9	T	0.0	0.0	0.0	3.1
1984-85	0.0	0.0	0.0	0.0	0.0	T	4.5	3.1	0.1	0.4	0.0	0.0	8.1
1985-86	0.0	0.0	0.0	0.0	0.0	0.4	0.8	3.7	0.1	T	0.0	0.0	5.0
1986-87	0.0	0.0	0.0	0.0	T	T	15.0	2.7	0.3	11.5	0.0	0.0	29.5
1987-88	0.0	0.0	0.0	0.0	0.3	0.5	14.2	T	T	1.2	0.0	0.0	16.2
1988-89	0.0	0.0	0.0	0.0	0.0	T	1.2	6.0	T	1.0	0.0	0.0	8.2
1989-90	0.0	0.0	0.0	T	T	3.0	T	T	T	0.0	0.0	0.0	3.0
1990-91	0.0	T	0.0	0.0	0.0	T	T	0.4	3.1	0.0	0.0	T	3.5
1991-92	0.0	0.0	0.0	0.0	1.0	0.0	T	T	0.5	0.0	T	0.0	1.5
1992-93	0.0	0.0	0.0	0.0	T	T	T	3.0	18.2	T	T	0.0	21.2
1993-94	0.0	0.0	0.0	T	T	8.0							
Record Mean	0.0	T	0.0	T	0.7	2.0	4.6	4.6	3.0	0.7	T	T	15.7

CAPE HATTERAS, NORTH CAROLINA

Hatteras Island is the largest and easternmost Island in North Carolina. The average elevation of the Island is less than 10 feet above mean sea level. It is separated from the mainland by the Pamlico Sound and is part of a chain of islands known as the Outer Banks. The Island is narrow, ranging from a few hundred yards wide to a few miles wide and is about 54 miles long. Much of the island is a National Park and waterfowl reserve.

The Weather Office is located in the village of Buxton about 1 mile west–northwest of the famous Cape Hatteras Lighthouse. Weather observations have been taken continuously since 1874 from locations all within 10 miles of the present stations location.

With its maritime climate, Cape Hatteras is very humid, with cooler summers and warmer winters than mainland North Carolina. Ninety degree temperatures are rare in summer, as are the teens in winter. The average first occurrence of freezing temperatures is early December, and the average last occurrence is late February.

Average rainfall is greater than any other coastal station in the state. Rainfall is rather evenly distributed throughout the year, with the maximum during July, August, and September. Snowfall is rare and generally light, usually melting as it falls.

Winter storms frequently breed offshore where the warm waters of the Gulf Stream and the southermost penetration of the Labrador Current meet some 20 to 50 miles off the coast. Late summer and fall tracks of tropical cyclones occasionally threaten the island. These storms produce strong winds, heavy rains and tidal flooding from both the ocean and Pamlico Sound. Many ships have been lost near, or wrecked on, the beaches of the island and attest to the fury of wind and wave.

More than a million tourists visit the island each year. The proximity of the gulfstream, natural beaches, excellent surf, and offshore fishing make Cape Hatteras a preferred place for vacationers, sportsmen, and campers. The surfing conditions are said to be the best on the east coast.

NORMALS, MEANS, AND EXTREMES

CAPE HATTERAS, NORTH CAROLINA

LATITUDE: 35°16'N LONGITUDE: 75°33' W ELEVATION: FT. GRND 7 BARO 25 TIME ZONE: EASTERN WBAN: 93729

	(a)	JAN	FEB	MAR	APR	MAY	JUNE	JULY	AUG	SEP	OCT	NOV	DEC	YEAR
TEMPERATURE °F:														
Normals														
-Daily Maximum		52.4	53.4	59.6	67.0	74.4	80.8	84.6	84.7	80.7	72.4	64.7	56.8	69.3
-Daily Minimum		36.7	37.6	43.6	50.8	59.6	67.5	71.8	72.0	67.6	58.3	49.3	41.1	54.7
-Monthly		44.6	45.5	51.6	58.9	67.0	74.2	78.3	78.4	74.2	65.4	57.0	49.0	62.0
Extremes														
-Record Highest	36	75	76	81	89	91	95	96	94	92	89	81	78	96
-Year		1985	1971	1990	1990	1991	1978	1992	1968	1978	1986	1986	1991	JUL 1992
-Record Lowest	36	6	14	19	26	39	44	54	56	45	32	22	12	6
-Year		1985	1958	1967	1972	1971	1966	1972	1979	1970	1979	1967	1983	JAN 1985
NORMAL DEGREE DAYS:														
Heating (base 65°F)		639	546	421	196	38	0	0	0	0	90	265	503	2698
Cooling (base 65°F)		6	0	6	13	100	276	412	415	276	102	25	7	1638
% OF POSSIBLE SUNSHINE	31	48	52	61	66	65	64	65	65	64	60	56	48	60
MEAN SKY COVER (tenths)														
Sunrise - Sunset	36	6.2	6.2	6.0	5.4	5.9	6.1	6.4	6.2	5.6	5.4	5.5	6.0	5.9
MEAN NUMBER OF DAYS:														
Sunrise to Sunset														
-Clear	36	8.9	8.5	9.6	10.3	8.6	7.5	6.6	7.8	9.5	11.2	10.2	9.6	108.2
-Partly Cloudy	36	6.8	5.4	7.4	8.8	9.7	10.3	10.2	10.0	9.5	7.6	8.0	7.3	100.9
-Cloudy	36	15.4	14.4	14.1	11.0	12.7	12.2	14.3	13.3	11.0	12.2	11.8	14.1	156.2
Precipitation														
.01 inches or more	36	11.2	10.4	10.6	8.5	9.8	9.3	11.8	11.1	9.0	9.2	9.1	10.0	120.0
Snow,Ice pellets,hail														
1.0 inches or more	36	0.1	0.3	0.1	0.0	0.0	0.0	0.0	0.0	0.0	0.0	0.0	0.1	0.6
Thunderstorms	36	0.8	1.5	2.3	3.3	5.0	5.3	8.8	7.9	3.3	1.8	1.5	0.9	42.3
Heavy Fog Visibility														
1/4 mile or less	36	2.7	2.6	2.4	1.3	1.0	0.3	0.1	0.3	0.2	0.6	1.1	2.2	14.9
Temperature °F														
-Maximum														
90° and above	36	0.0	0.0	0.0	0.0	0.1	0.6	2.3	1.9	0.5	0.0	0.0	0.0	5.4
32° and below	36	0.7	0.3	0.0	0.0	0.0	0.0	0.0	0.0	0.0	0.0	0.0	0.2	1.2
-Minimum														
32° and below	36	10.1	8.7	3.5	0.3	0.0	0.0	0.0	0.0	0.0	0.*	1.1	6.2	30.0
0° and below	36	0.0	0.0	0.0	0.0	0.0	0.0	0.0	0.0	0.0	0.0	0.0	0.0	0.0
AVG. STATION PRESS.(mb)	21	1019.0	1018.6	1017.2	1016.1	1016.1	1016.2	1016.9	1017.5	1017.8	1018.6	1019.4	1019.8	1017.8
RELATIVE HUMIDITY (%)														
Hour 01	36	78	79	80	81	86	87	89	89	85	82	80	79	83
Hour 07 (Local Time)	36	80	80	80	78	82	82	85	86	84	82	82	80	82
Hour 13	36	67	65	63	59	65	67	70	69	67	65	64	66	66
Hour 19	36	78	76	77	75	79	80	82	83	80	79	78	78	79
PRECIPITATION (inches):														
Water Equivalent														
-Normal		5.30	4.12	4.29	3.53	4.00	4.11	4.98	6.00	5.27	4.98	4.97	4.54	56.09
-Maximum Monthly	36	12.45	8.45	11.20	9.57	11.44	10.80	9.99	16.10	20.00	15.05	16.20	8.63	20.00
-Year		1991	1983	1989	1989	1972	1962	1965	1986	1989	1985	1985	1962	SEP 1989
-Minimum Monthly	36	1.75	1.06	0.98	0.59	0.35	0.38	0.45	0.99	0.08	0.53	1.15	0.64	0.08
-Year		1981	1991	1967	1976	1987	1978	1958	1983	1986	1984	1991	1985	SEP 1986
-Maximum in 24 hrs	36	5.95	2.92	5.94	5.60	3.55	6.63	5.53	8.11	5.59	7.67	7.69	3.59	8.11
-Year		1991	1970	1989	1963	1984	1962	1967	1962	1989	1983	1985	1979	AUG 1962
Snow,Ice pellets,hail														
-Maximum Monthly	36	3.5	4.4	8.5	T	0.0	0.0	0.0	0.0	0.0	0.0	T	13.5	13.5
-Year		1962	1978	1960	1989							1987	1989	DEC 1989
-Maximum in 24 hrs	36	3.5	4.4	7.3	T	0.0	0.0	0.0	0.0	0.0	0.0	T	8.2	8.2
-Year		1962	1978	1980	1989							1987	1989	DEC 1989
WIND:														
Mean Speed (mph)	36	12.0	12.2	11.9	11.7	10.8	10.6	10.0	9.6	10.5	11.2	11.0	11.5	11.1
Prevailing Direction through 1963		NNE	NNE	SW	SW	SW	SSW	SW	SW	NE	NNE	NNE	NNE	NNE
Fastest Obs. 1 Min.														
-Direction (!!)	23	14	16	02	21	18	29	22	33	11	02	15	16	33
-Speed (MPH)	23	37	44	48	40	35	37	26	60	60	44	58	46	60
-Year		1987	1985	1980	1989	1972	1972	1982	1993	1985	1982	1990	1986	AUG 1993
Peak Gust														
-Direction (!!)	9	W	SW	SW	S	NE	W	SW	N	SW	NE	S	N	N
-Speed (mph)	9	59	58	63	60	46	55	44	98	87	66	78	60	98
-Date		1987	1984	1993	1989	1990	1993	1990	1993	1985	1990	1990	1989	AUG 1993

PRECIPITATION (inches)

CAPE HATTERAS, NORTH CAROLINA

YEAR	JAN	FEB	MAR	APR	MAY	JUNE	JULY	AUG	SEP	OCT	NOV	DEC	ANNUAL
1964	6.27	5.86	1.11	4.32	2.63	1.87	6.29	10.62	5.51	7.41	2.08	3.19	57.16
1965	1.95	5.21	2.94	1.20	0.92	6.72	9.99	2.80	2.90	3.19	1.46	2.24	41.52
1966	9.07	4.24	2.32	1.21	7.51	6.80	2.16	3.61	6.41	5.04	1.23	6.58	56.18
1967	4.08	4.51	0.98	1.20	5.48	1.04	8.36	8.65	8.73	2.83	2.38	4.71	52.95
1968	5.62	4.06	1.93	4.07	3.63	3.47	7.65	1.80	6.47	9.58	4.21	2.64	55.13
1969	3.22	3.50	4.87	3.37	0.87	2.31	6.69	6.77	4.41	1.41	3.60	5.27	46.29
1970	5.61	5.92	5.31	4.25	2.86	4.66	6.16	4.21	4.07	2.94	8.28	2.44	56.71
1971	4.53	3.76	4.70	3.73	5.52	1.58	6.48	9.74	8.59	11.24	2.40	2.07	64.34
1972	6.42	5.66	2.96	0.72	11.44	1.83	8.62	1.78	2.74	7.26	7.29	4.84	61.56
1973	3.99	4.75	6.15	3.35	4.79	4.29	2.56	3.02	3.04	1.34	1.23	8.52	47.03
1974	2.02	4.07	3.48	2.38	4.31	8.31	2.49	11.04	2.94	8.99	1.72	5.14	56.89
1975	4.38	2.95	1.89	4.37	4.05	4.40	6.69	1.91	5.21	4.04	3.71	4.87	48.47
1976	3.22	1.42	2.73	0.59	5.20	4.71	4.31	11.73	8.58	4.62	3.08	6.90	57.09
1977	4.53	1.64	4.01	1.89	1.81	3.07	2.48	2.74	2.70	9.13	12.00	5.99	51.99
1978	7.61	2.85	5.70	5.73	6.04	0.38	4.47	1.04	0.73	3.15	6.77	3.32	47.79
1979	9.72	4.67	2.96	2.68	6.66	4.04	6.30	3.33	12.78	2.04	8.46	5.19	68.83
1980	7.76	2.38	8.94	4.03	0.87	5.49	3.47	1.48	5.09	2.58	5.29	4.48	51.86
1981	1.75	1.99	2.36	1.84	4.90	3.91	9.32	11.34	1.31	2.69	2.93	7.20	51.54
1982	5.85	7.03	6.04	5.85	0.90	5.50	6.76	2.14	4.53	3.90	2.50	5.84	56.84
1983	9.27	8.45	9.29	3.40	0.81	4.87	1.75	0.99	6.15	9.46	4.58	6.10	65.12
1984	2.34	4.28	2.99	4.26	7.27	3.12	5.23	4.73	3.81	0.53	4.95	1.69	45.20
1985	3.67	6.28	2.28	0.67	3.77	3.92	3.46	7.17	4.06	15.05	16.20	0.64	67.17
1986	5.75	1.38	1.22	1.93	3.19	3.17	3.10	16.10	0.08	2.09	4.21	6.64	48.86
1987	10.56	4.32	9.56	5.33	0.35	2.94	4.80	6.91	7.61	1.19	5.16	2.23	60.96
1988	5.81	3.98	3.08	3.77	3.62	2.33	1.88	7.42	2.15	3.53	3.95	0.75	42.27
1989	4.30	2.13	11.20	9.57	5.36	4.12	5.16	9.07	20.00	8.29	5.17	6.47	90.84
1990	4.66	4.13	6.92	3.46	5.75	1.80	4.23	3.23	0.59	3.97	3.82	5.13	47.69
1991	12.45	1.06	6.57	1.96	0.49	5.33	8.58	13.19	4.41	4.86	1.15	4.46	64.51
1992	4.40	4.25	7.76	0.71	4.17	0.94	5.28	13.89	6.14	4.03	4.85	5.66	62.08
1993	5.96	5.15	6.67	3.96	0.63	3.15	2.03	9.00	2.99	11.54	4.04	4.14	59.26
Record Mean	4.57	4.08	4.37	3.48	3.76	4.32	5.60	6.11	5.49	4.86	4.13	4.66	55.44

AVERAGE TEMPERATURE (deg. F)

CAPE HATTERAS, NORTH CAROLINA

YEAR	JAN	FEB	MAR	APR	MAY	JUNE	JULY	AUG	SEP	OCT	NOV	DEC	ANNUAL
1964	46.8	44.2	52.1	59.1	66.6	74.5	77.3	76.1	73.7	62.3	58.7	52.0	61.9
1965	45.8	45.5	47.5	55.0	68.9	73.2	77.5	78.5	74.4	63.1	56.0	48.2	61.1
1966	43.4	44.6	49.4	56.8	63.9	70.9	78.1	77.6	73.7	64.8	55.2	46.9	60.4
1967	48.5	45.0	50.4	56.8	62.8	70.5	77.6	76.9	68.8	64.2	51.4	51.6	60.4
1968	41.8	38.7	52.2	58.6	64.8	74.9	78.6	80.9	73.0	66.6	57.4	45.6	61.1
1969	42.5	43.5	45.4	60.0	66.3	75.1	78.9	76.8	73.0	67.1	52.1	46.1	60.6
1970	37.7	44.2	49.5	59.5	67.5	75.2	77.5	78.5	74.9	67.5	54.8	48.2	61.3
1971	42.5	47.3	49.4	54.7	64.9	74.6	78.3	77.7	75.6	70.9	56.0	56.2	62.3
1972	52.9	47.0	52.2	56.3	66.0	71.5	77.9	78.3	73.8	64.8	56.7	54.2	62.6
1973	46.1	43.4	55.2	60.5	67.3	76.5	77.4	78.8	75.9	67.8	57.4	52.6	63.3
1974	55.5	47.8	56.3	61.8	68.5	74.2	77.3	77.6	75.2	61.5	55.2	49.9	63.4
1975	51.1	51.5	52.5	55.9	69.1	76.5	77.6	80.2	76.2	68.3	59.0	48.7	63.9
1976	43.7	51.7	57.3	60.0	67.2	73.6	77.7	75.9	71.8	62.2	49.9	46.5	61.5
1977	35.8	44.3	55.2	62.9	69.6	74.6	80.8	79.5	77.3	63.4	59.4	48.8	62.6
1978	42.0	35.7	50.1	60.2	65.1	73.8	78.0	80.6	76.3	63.5	61.3	51.2	61.5
1979	45.4	41.0	50.7	58.9	67.6	71.6	77.4	77.9	75.2	63.4	58.6	49.1	61.4
1980	45.8	39.0	49.3	61.1	68.6	73.1	79.3	79.1	76.0	62.2	52.0	47.1	61.1
1981	36.2	46.1	45.4	62.6	65.4	78.3	78.6	76.5	71.5	62.9	52.9	44.4	60.1
1982	41.3	47.6	52.1	57.4	68.8	73.3	78.3	77.4	72.8	63.7	59.7	54.1	62.2
1983	44.9	45.4	52.7	57.4	66.8	72.6	79.5	78.7	74.3	66.7	55.9	47.5	61.9
1984	41.7	48.8	48.5	57.7	68.2	75.5	77.8	79.4	72.9	68.8	55.4	55.7	62.5
1985	41.0	45.3	55.1	63.7	69.8	75.2	79.6	78.2	75.8	71.4	66.6	47.7	64.1
1986	44.7	49.3	51.8	60.2	68.1	75.9	80.9	77.0	75.1	68.5	62.8	51.6	63.8
1987	46.0	44.3	48.4	56.4	67.8	75.6	80.1	80.1	75.6	61.7	57.8	48.8	61.9
1988	43.4	46.0	52.3	58.0	66.7	72.2	78.3	80.8	74.5	61.5	60.9	47.9	61.9
1989	50.8	50.1	54.8	59.6	67.2	77.3	79.1	79.3	75.0	66.7	59.6	41.3	63.4
1990	52.2	54.4	56.7	61.3	68.0	74.4	79.7	79.7	74.5	69.5	57.8	56.1	65.4
1991	50.0	50.3	56.4	64.1	73.4	76.6	81.7	80.0	75.6	66.6	57.4	52.8	65.4
1992	48.3	49.3	52.6	59.4	64.3	73.5	80.3	78.0	75.9	63.4	60.2	51.1	63.0
1993	51.4	44.4	50.5	57.2	68.3	75.6	81.9	78.1	76.4	68.9	60.7	47.7	63.4
Record Mean	46.3	46.5	51.8	59.0	67.6	74.6	78.4	78.3	74.7	65.8	56.7	48.9	62.4
Max	53.0	53.4	58.6	65.8	73.8	80.2	83.8	83.7	80.0	71.5	63.0	55.5	68.5
Min	39.5	39.6	44.9	52.2	61.4	69.0	72.9	72.9	69.3	60.0	50.3	42.2	56.2

REFERENCE NOTES

CAPE HATTERAS, NORTH CAROLINA

GENERAL
T - TRACE AMOUNT.
BLANK ENTRIES DENOTE MISSING/UNREPORTED DATA.
# INDICATES A STATION OR INSTRUMENT RELOCATION.

SPECIFIC

EXCEPTIONS

RECORD MEANS ARE THROUGH THE CURRENT YEAR,
BEGINNING IN 1882 FOR TEMPERATURE
1875 FOR PRECIPITATION
1958 FOR SNOWFALL

(a) - LENGTH OF RECORD IN YEARS, ALTHOUGH INDIVIDUAL MONTHS MAY BE MISSING.
0.* OR * - THE VALUE IS BETWEEN 0.0 AND 0.05
NORMALS - BASED ON THE 1961-1990 RECORD PERIOD.
EXTREMES - DATES ARE THE MOST RECENT OCCURRENCE
WIND DIR.- NUMERALS SHOW TENS OF DEGRESS CLOCKWISE FROM TRUE NORTH. "00" INDICATES CALM.
RESULTANT DIRECTIONS ARE GIVEN TO WHOLE DEGREES.
BOLD VALUES INDICATE EXTREME VALUES WHICH OCCURRED AFTER THE ASOS SYSTEM WAS COMMISSIONED.

RECORD = PERIOD OF RECORD
RECORD MEAN PRECIPITATION IS THE MEAN OF ALL DAILY PRECIPITATION AMOUNTS DURING THE PERIOD OF RECORD.
RECORD MAX(MIN) TEMPERATURE IS THE MEAN OF ALL DAILY MAX(MIN) TEMPERATURES DURING THE PERIOD OF RECORD.
RECORD MEAN TEMPERATURE IS THE SUM OF THE RECORD MAX AND RECORD MIN DIVIDED BY 2.
AVERAGE TEMPERATURE IS THE SUM OF THE MEAN DAILY MAX AND MIN TEMPERATURE DIVIDED BY 2.

HEATING DEGREE DAYS Base 65 deg. F — CAPE HATTERAS, NORTH CAROLINA

SEASON	JULY	AUG	SEP	OCT	NOV	DEC	JAN	FEB	MAR	APR	MAY	JUNE	TOTAL
1964-65	0	0	3	136	194	409	588	538	537	296	15	0	2716
1965-66	0	0	0	135	267	514	665	565	478	253	104	23	3004
1966-67	0	0	0	77	296	562	505	551	449	245	107	15	2807
1967-68	0	0	11	77	399	418	708	758	391	196	70	0	3028
1968-69	0	0	0	73	228	594	686	594	601	166	41	0	2983
1969-70	0	0	1	51	390	581	842	573	472	167	44	0	3121
1970-71	0	0	10	40	299	515	690	486	476	300	65	0	2881
1971-72	0	0	0	6	289	278	379	517	393	267	39	16	2184
1972-73	0	0	0	61	257	335	578	597	302	151	50	0	2331
1973-74	0	0	0	31	246	389	293	476	276	143	28	0	1882
1974-75	0	0	3	141	306	459	425	378	383	289	26	0	2410
1975-76	0	0	0	32	206	498	653	379	244	190	52	4	2258
1976-77	0	0	1	135	448	566	900	575	309	104	27	0	3065
1977-78	0	0	0	103	196	504	704	813	455	152	60	0	2987
1978-79	0	0	0	90	123	426	601	663	439	198	32	0	2572
1979-80	0	0	0	112	208	487	587	750	477	145	33	0	2799
1980-81	0	0	0	132	385	549	884	524	600	109	82	0	3265
1981-82	0	0	15	104	360	630	727	482	392	237	44	0	2991
1982-83	0	0	0	86	176	341	613	544	373	238	50	2	2423
1983-84	0	0	10	55	272	535	715	466	507	215	47	0	2822
1984-85	0	0	2	22	296	282	736	547	305	113	19	0	2322
1985-86	0	0	0	1	48	532	618	435	408	156	48	0	2246
1986-87	0	0	0	49	116	411	581	575	508	255	42	0	2537
1987-88	0	0	0	118	226	495	664	544	386	214	52	22	2721
1988-89	0	0	0	144	150	525	432	421	325	192	38	0	2227
1989-90	0	0	0	39	200	728	398	300	289	159	30	0	2143
1990-91	0	0	13	70	221	288	460	408	281	81	0	0	1822
1991-92	0	0	0	49	248	394	511	449	381	200	73	3	2308
1992-93	0	0	2	91	182	426	413	570	442	237	16	0	2379
1993-94	0	0	2	22	150	529							

COOLING DEGREE DAYS Base 65 deg. F — CAPE HATTERAS, NORTH CAROLINA

YEAR	JAN	FEB	MAR	APR	MAY	JUNE	JULY	AUG	SEP	OCT	NOV	DEC	TOTAL
1969	0	0	0	22	89	312	439	371	248	123	8	0	1612
1970	0	0	0	7	127	314	393	422	313	122	0	0	1698
1971	0	0	0	1	67	294	419	402	325	194	26	10	1738
1972	7	0	2	12	74	218	406	422	271	63	17	8	1500
1973	0	1	5	25	128	352	394	434	334	126	25	9	1833
1974	7	0	13	54	146	283	390	399	318	42	20	0	1672
1975	1	7	3	20	160	353	398	479	345	142	32	0	1940
1976	0	0	11	48	127	268	401	344	214	53	0	0	1466
1977	0	1	10	49	177	295	499	453	374	57	37	6	1958
1978	0	0	0	13	68	271	411	493	345	48	21	7	1677
1979	0	0	2	23	121	203	394	407	313	69	24	0	1556
1980	0	0	0	35	153	252	449	445	335	54	3	0	1726
1981	0	0	1	45	104	406	429	361	218	46	1	0	1611
1982	0	0	1	15	168	259	419	393	239	53	24	8	1579
1983	0	1	0	18	110	236	458	433	295	112	8	0	1671
1984	0	0	0	6	153	323	403	452	244	146	13	2	1742
1985	0	2	4	82	176	317	461	418	330	207	104	3	2104
1986	0	0	7	19	151	335	497	379	308	165	58	2	1921
1987	0	0	0	4	137	323	475	475	325	20	17	0	1776
1988	0	2	3	10	114	245	419	497	292	41	33	0	1656
1989	0	11	16	36	115	376	443	450	307	98	43	0	1895
1990	4	9	40	56	127	288	463	462	308	218	10	20	2005
1991	0	2	22	59	267	354	525	469	324	106	25	25	2178
1992	1	0	2	42	56	267	483	414	334	46	45	0	1690
1993	0	0	0	12	123	323	532	413	351	151	30	0	1935

SNOWFALL (inches) — CAPE HATTERAS, NORTH CAROLINA

SEASON	JULY	AUG	SEP	OCT	NOV	DEC	JAN	FEB	MAR	APR	MAY	JUNE	TOTAL
1964-65	0.0	0.0	0.0	0.0	0.0	0.0	0.3	0.0	0.0	0.0	0.0	0.0	0.3
1965-66	0.0	0.0	0.0	0.0	0.0	0.0	T	0.0	0.0	0.0	0.0	0.0	T
1966-67	0.0	0.0	0.0	0.0	0.0	T	0.0	0.0	0.0	0.0	0.0	0.0	T
1967-68	0.0	0.0	0.0	0.0	0.0	T	T	3.5	T	0.0	0.0	0.0	3.5
1968-69	0.0	0.0	0.0	0.0	0.0	T	0.2	T	0.0	0.0	0.0	0.0	0.2
1969-70	0.0	0.0	0.0	0.0	0.0	0.0	1.4	1.2	T	0.0	0.0	0.0	2.6
1970-71	0.0	0.0	0.0	0.0	T	2.5	T	0.0	0.0	0.0	0.0	0.0	2.5
1971-72	0.0	0.0	0.0	0.0	0.0	0.0	T	T	T	0.0	0.0	0.0	T
1972-73	0.0	0.0	0.0	0.0	0.0	0.0	1.9	3.3	0.0	0.0	0.0	0.0	5.2
1973-74	0.0	0.0	0.0	0.0	0.0	T	0.0	0.0	0.0	0.0	0.0	0.0	T
1974-75	0.0	0.0	0.0	0.0	0.0	0.0	0.0	0.0	T	0.0	0.0	0.0	T
1975-76	0.0	0.0	0.0	0.0	0.0	0.0	T	0.0	0.0	0.0	0.0	0.0	T
1976-77	0.0	0.0	0.0	0.0	0.0	T	0.0	1.0	0.0	0.0	0.0	0.0	1.0
1977-78	0.0	0.0	0.0	0.0	0.0	0.0	T	4.4	T	0.0	0.0	0.0	4.4
1978-79	0.0	0.0	0.0	0.0	0.0	0.0	T	1.1	0.0	0.0	0.0	0.0	1.1
1979-80	0.0	0.0	0.0	0.0	0.0	0.0	3.0	T	7.3	0.0	0.0	0.0	10.3
1980-81	0.0	0.0	0.0	0.0	0.0	T	2.9	0.0	0.0	0.0	0.0	0.0	2.9
1981-82	0.0	0.0	0.0	0.0	0.0	0.0	T	0.0	0.0	0.0	0.0	0.0	T
1982-83	0.0	0.0	0.0	0.0	0.0	0.0	0.0	T	0.0	0.0	0.0	0.0	T
1983-84	0.0	0.0	0.0	0.0	0.0	0.2	0.0	T	0.0	0.0	0.0	0.0	0.2
1984-85	0.0	0.0	0.0	0.0	0.0	0.0	0.1	0.0	0.0	0.0	0.0	0.0	0.1
1985-86	0.0	0.0	0.0	0.0	0.0	T	0.3	0.2	T	0.0	0.0	0.0	0.5
1986-87	0.0	0.0	0.0	0.0	0.0	0.0	T	0.0	0.0	0.0	0.0	0.0	T
1987-88	0.0	0.0	0.0	0.0	T	0.0	0.7	T	0.0	0.0	0.0	0.0	0.7
1988-89	0.0	0.0	0.0	0.0	0.0	4.4	0.0	0.3	0.0	T	0.0	0.0	4.7
1989-90	0.0	0.0	0.0	0.0	0.0	13.5	0.0	0.0	0.0	0.0	0.0	0.0	13.5
1990-91	0.0	0.0	0.0	0.0	0.0	0.0	0.0	0.0	T	0.0	0.0	0.0	T
1991-92	0.0	0.0	0.0	0.0	0.0	T	0.5	0.0	T	0.0	0.0	0.0	0.5
1992-93	0.0	0.0	0.0	0.0	0.0	0.0	0.0	0.1	T	0.0	0.0	0.0	0.1
1993-94	0.0	0.0	0.0	0.0	0.0	0.0							
Record Mean	0.0	0.0	0.0	0.0	T	0.6	0.4	0.6	0.4	T	0.0	0.0	2.0

CHARLOTTE, NORTH CAROLINA

Charlotte is located in the Piedmont of the Carolinas, a transitional area of rolling country between the mountains to the west and the Coastal Plain to the east. The mountains are to the northwest about 80 miles from Charlotte. The general elevation of the area around Charlotte is about 730 feet. The Atlantic ocean is about 160 miles southeast.

The mountains have a moderating effect on winter temperatures, causing appreciable warming of cold air from the northwest winds. The ocean is too far away to have any immediate effect on summer temperatures but in winter an occasional general and sustained flow of air from the warm ocean waters results in considerable warming.

Charlotte enjoys a moderate climate, characterized by cool winters and quite warm summers. Temperatures fall as low as the freezing point on a little over one-half of the days in the winter months. Winter weather is changeable, with occasional cold periods, but extreme cold is rare. Snow is infrequent, and the first snowfall of the season usually comes in late November or December. Heavy snowfalls have occurred, but any appreciable accumulation of snow on the ground for more than a day or two is rare.

Summers are long and quite warm, with afternoon temperatures frequently in the low 90s. The growing season is also long, the average length of the freeze-free period being 216 days. On the average, the last occurrence in spring with a temperature of 32 degrees is early April. In the fall the average first occurrence of 32 degrees is early November.

Rainfall is generally rather evenly distributed throughout the year, the driest weather usually coming in the fall. Summer rainfall comes principally from thunderstorms with occasional dry spells of one to three weeks duration.

Hurricanes which strike the Carolina coast may produce heavy rain but seldom cause dangerous winds.

NORMALS, MEANS, AND EXTREMES

CHARLOTTE, NORTH CAROLINA

LATITUDE: 35°13'N LONGITUDE: 80°56' W ELEVATION: FT. GRND 737 BARO 738 TIME ZONE: EASTERN WBAN: 13881

	(a)	JAN	FEB	MAR	APR	MAY	JUNE	JULY	AUG	SEP	OCT	NOV	DEC	YEAR
TEMPERATURE °F:														
Normals														
-Daily Maximum		49.0	53.0	62.3	71.2	78.3	85.8	88.9	87.7	81.9	72.0	62.6	52.3	70.4
-Daily Minimum		29.6	31.9	39.4	47.5	56.4	65.6	69.6	68.9	62.9	50.6	41.5	32.8	49.7
-Monthly		39.3	42.5	50.9	59.4	67.4	75.7	79.3	78.3	72.4	61.3	52.1	42.6	60.1
Extremes														
-Record Highest	54	78	81	90	93	100	103	103	103	104	98	85	77	104
-Year		1952	1989	1945	1960	1941	1954	1986	1983	1954	1954	1961	1971	SEP 1954
-Record Lowest	54	-5	5	4	24	32	45	53	53	39	24	11	2	-5
-Year		1985	1958	1980	1960	1963	1972	1961	1965	1967	1962	1950	1962	JAN 1985
NORMAL DEGREE DAYS:														
Heating (base 65°F)		797	630	437	183	42	0	0	0	6	161	391	694	3341
Cooling (base 65°F)		0	0	0	15	116	321	443	412	228	47	0	0	1582
% OF POSSIBLE SUNSHINE	43	55	58	62	68	66	67	67	65	65	66	58	56	63
MEAN SKY COVER (tenths)														
Sunrise - Sunset	44	6.3	6.1	6.1	5.6	6.1	6.0	6.1	5.9	5.7	4.8	5.4	6.0	5.8
MEAN NUMBER OF DAYS:														
Sunrise to Sunset														
-Clear	45	9.2	8.6	9.2	9.9	7.8	7.5	6.8	7.4	9.4	13.2	11.4	10.1	110.6
-Partly Cloudy	45	6.0	6.0	8.0	8.5	10.2	11.0	11.8	12.5	9.4	7.6	6.4	5.8	103.2
-Cloudy	45	15.8	13.6	13.8	11.6	13.0	11.5	12.4	11.1	11.2	10.2	12.2	15.1	151.4
Precipitation														
.01 inches or more	54	10.1	9.6	11.1	8.8	9.6	9.5	11.2	9.7	7.2	6.6	7.8	9.6	111.0
Snow, Ice pellets, hail														
1.0 inches or more	54	0.7	0.4	0.3	0.0	0.0	0.0	0.0	0.0	0.0	0.0	0.*	0.2	1.6
Thunderstorms	54	0.5	0.9	1.9	3.2	5.6	7.2	9.6	6.9	2.9	1.2	0.6	0.4	41.0
Heavy Fog Visibility														
1/4 mile or less	54	3.7	2.9	2.5	1.3	1.0	1.0	1.1	1.4	2.0	1.8	3.1	4.0	25.9
Temperature °F														
-Maximum														
90° and above	33	0.0	0.0	0.0	0.3	1.7	7.7	14.2	11.0	4.0	0.2	0.0	0.0	38.9
32° and below	33	1.6	0.3	0.*	0.0	0.0	0.0	0.0	0.0	0.0	0.0	0.0	0.5	2.4
-Minimum														
32° and below	33	18.4	15.0	7.3	1.0	0.*	0.0	0.0	0.0	0.0	0.7	6.4	15.5	64.4
0° and below	33	0.1	0.0	0.0	0.0	0.0	0.0	0.0	0.0	0.0	0.0	0.0	0.0	0.1
AVG. STATION PRESS. (mb)	21	992.0	991.3	989.6	988.7	988.8	989.2	990.2	990.9	991.4	992.4	992.6	992.9	990.8
RELATIVE HUMIDITY (%)														
Hour 01	33	72	67	68	68	78	80	82	84	83	79	76	73	76
Hour 07 (Local Time)	33	78	75	78	77	83	84	86	89	89	86	83	79	82
Hour 13	33	56	51	50	46	52	55	56	58	57	52	53	56	54
Hour 19	33	60	54	52	49	58	61	64	66	67	65	62	62	60
PRECIPITATION (inches):														
Water Equivalent														
-Normal		3.71	3.84	4.43	2.68	3.82	3.39	3.92	3.73	3.50	3.36	3.23	3.48	43.09
-Maximum Monthly	54	7.44	7.59	8.76	7.64	12.48	8.26	9.12	9.98	10.89	14.72	8.68	7.49	14.72
-Year		1962	1979	1980	1958	1975	1961	1941	1948	1945	1990	1985	1983	OCT 1990
-Minimum Monthly	54	0.45	0.74	0.58	0.30	0.11	0.15	0.53	0.61	0.02	T	0.46	0.43	T
-Year		1981	1978	1985	1976	1941	1993	1983	1972	1954	1953	1973	1965	OCT 1953
-Maximum in 24 hrs	54	3.57	2.92	3.83	3.20	3.67	3.77	3.00	4.52	4.74	5.46	3.27	2.87	5.46
-Year		1962	1973	1977	1962	1975	1949	1949	1978	1959	1990	1985	1972	OCT 1990
Snow, Ice pellets, hail														
-Maximum Monthly	54	12.1	14.9	19.3	0.1	T	0.0	0.0	0.0	0.0	0.0	2.5	7.5	19.3
-Year		1988	1979	1960	1982	1993						1968	1971	MAR 1960
-Maximum in 24 hrs	54	12.1	12.0	10.3	0.1	T	0.0	0.0	0.0	0.0	0.0	2.5	7.5	12.1
-Year		1988	1969	1983	1982	1993						1968	1971	JAN 1988
WIND:														
Mean Speed (mph)	44	7.8	8.3	8.8	8.7	7.5	6.9	6.6	6.4	6.7	6.8	7.1	7.4	7.4
Prevailing Direction through 1963		SW	NE	SW	S	SW	SW	SW	S	NE	NNE	SSW	SW	SW
Fastest Obs. 1 Min.														
-Direction (!!!)	14	31	07	24	24	23	19	20	32	12	21	24	19	12
-Speed (MPH)	14	30	32	30	29	25	30	35	37	46	37	30	35	46
-Year		1989	1984	1992	1991	1993	1987	1993	1986	1989	1979	1992	1989	SEP 1989
Peak Gust														
-Direction (!!!)	10	NW	NE	NW	SW	N	S	SW	NW	E	N	S	NW	E
-Speed (mph)	10	49	53	60	56	46	49	52	77	87	36	51	47	87
-Date		1989	1984	1991	1991	1985	1987	1993	1990	1989	1987	1988	1987	SEP 1989

PRECIPITATION (inches)

CHARLOTTE, NORTH CAROLINA

YEAR	JAN	FEB	MAR	APR	MAY	JUNE	JULY	AUG	SEP	OCT	NOV	DEC	ANNUAL
1964	5.68	4.29	4.18	4.52	0.89	2.11	5.39	4.83	2.32	7.20	2.05	3.18	46.64
1965	2.17	2.69	6.59	3.32	1.96	5.30	6.19	4.87	0.88	1.76	3.19	0.43	39.35
1966	3.79	5.59	2.13	1.73	3.61	2.58	1.86	2.36	5.42	2.29	1.19	2.58	35.13
1967	2.76	3.79	2.48	1.57	4.70	3.08	5.13	9.23	1.60	0.83	2.51	5.53	43.21
1968	4.95	0.87	4.96	2.73	2.45	6.63	5.13	0.88	1.46	3.02	4.68	2.29	40.05
1969	1.93	5.19	4.04	3.49	2.03	2.32	3.48	5.14	4.83	1.33	1.14	4.87	39.79
1970	1.70	3.66	2.93	2.04	2.69	2.62	5.73	4.03	0.55	5.12	1.58	3.14	35.79
1971	3.40	5.19	4.87	2.79	4.47	5.09	4.99	2.89	2.68	6.88	2.96	2.24	48.45
1972	4.47	3.21	2.59	1.75	5.61	3.01	6.59	0.61	3.77	1.37	5.42	5.83	44.23
1973	4.14	4.44	6.96	2.13	4.31	4.91	3.62	2.31	3.15	2.38	0.46	5.32	44.13
1974	5.22	4.90	3.30	3.26	4.49	2.32	4.16	6.35	6.50	0.46	4.50	3.82	49.28
1975	6.14	3.50	7.62	1.69	12.48	1.86	7.58	4.48	6.51	3.58	2.83	3.82	62.09
1976	1.89	1.13	4.36	0.30	4.26	3.84	2.26	0.90	5.55	8.33	3.37	5.60	41.79
1977	2.73	1.48	8.45	2.05	3.16	3.12	0.82	2.44	6.35	4.74	4.20	1.97	41.51
1978	6.80	0.74	4.97	2.69	4.91	4.19	4.03	8.11	1.16	1.18	2.81	3.13	44.72
#1979	5.31	7.59	3.79	6.47	4.54	4.72	4.74	1.27	9.69	2.95	4.61	1.36	57.04
1980	4.67	1.31	8.76	2.31	3.59	2.27	2.63	1.94	5.37	1.67	3.77	0.83	39.12
1981	0.45	3.63	2.12	0.67	4.27	1.81	6.61	2.67	3.42	3.94	0.87	6.23	36.69
1982	4.30	4.87	1.58	3.84	4.97	4.16	4.19	2.03	0.64	3.83	3.05	4.23	41.69
1983	2.53	5.50	6.07	2.66	2.14	3.77	0.53	3.61	0.74	2.43	4.05	7.49	41.52
1984	4.09	5.90	5.89	4.50	4.78	2.95	5.96	3.95	1.74	0.75	2.08	2.40	44.99
1985	5.20	4.05	0.58	1.90	5.14	5.46	4.14	7.35	0.74	5.16	8.68	0.92	49.32
1986	1.02	1.03	3.01	1.20	1.63	0.41	2.26	5.43	0.83	3.49	3.44	3.16	26.91
1987	4.78	5.19	3.65	2.44	0.99	2.98	1.38	2.76	6.87	0.84	4.05	3.39	39.32
1988	3.43	1.11	3.29	2.27	2.20	1.55	3.56	4.56	4.45	4.12	2.11	1.62	34.27
1989	1.61	4.67	4.92	2.58	5.37	3.20	6.30	2.99	7.27	4.08	3.14	3.66	49.79
1990	3.81	5.65	3.57	2.03	4.99	0.90	2.71	3.47	1.75	14.72	2.75	3.23	49.58
1991	6.02	1.32	7.18	5.43	2.94	2.39	3.70	8.18	1.70	0.50	2.83	2.99	45.18
1992	3.28	4.31	4.93	2.51	3.59	6.83	1.78	3.46	8.18	6.21	5.84	2.81	53.73
1993	6.17	2.72	7.61	2.00	2.60	0.15	1.56	3.77	0.92	2.28	2.69	3.34	35.81
Record Mean	3.81	3.92	4.37	3.21	3.54	3.83	4.61	4.58	3.35	3.07	2.77	3.64	44.69

AVERAGE TEMPERATURE (deg. F)

CHARLOTTE, NORTH CAROLINA

YEAR	JAN	FEB	MAR	APR	MAY	JUNE	JULY	AUG	SEP	OCT	NOV	DEC	ANNUAL
1964	41.5	40.1	51.0	60.6	69.8	77.7	76.9	75.0	70.9	55.7	54.8	45.3	59.9
1965	41.5	44.6	46.8	62.4	73.4	73.3	77.8	78.0	73.1	59.6	51.4	44.0	60.5
1966	36.0	42.2	50.2	58.7	67.6	74.0	79.6	77.2	70.7	58.0	50.3	41.5	58.9
1967	43.3	40.0	53.9	62.7	63.7	73.1	76.3	75.2	67.0	60.0	47.8	46.3	59.1
1968	37.9	38.7	51.6	59.0	64.8	73.8	77.0	79.3	71.4	60.6	49.4	38.6	58.5
1969	38.1	39.7	45.1	60.8	67.9	77.3	80.8	75.6	69.0	60.4	47.7	39.5	58.5
1970	34.7	41.1	48.6	61.2	68.2	74.4	79.5	78.9	75.9	63.6	49.7	44.7	60.0
1971	39.9	41.4	46.3	58.3	65.0	76.5	76.2	75.8	73.1	64.7	49.0	50.6	59.7
1972	44.6	40.4	49.6	58.5	64.7	71.3	77.0	77.1	72.7	59.0	48.8	46.5	59.2
1973	39.6	40.0	54.1	57.6	64.7	74.7	78.5	76.8	74.6	62.1	52.7	41.7	59.8
1974	49.8	44.0	55.2	60.0	68.5	72.2	76.9	75.9	68.4	58.3	49.9	42.1	60.1
1975	45.2	45.0	50.4	59.0	70.5	75.0	76.2	78.9	70.8	63.6	53.5	42.4	60.9
1976	39.3	50.2	54.7	59.9	65.0	72.4	76.9	76.1	69.5	55.5	44.1	39.1	58.6
1977	30.1	42.1	54.9	62.5	69.8	75.2	82.4	79.7	74.4	58.7	53.6	42.5	60.5
1978	36.9	36.7	49.4	61.5	66.7	76.2	78.9	79.0	74.2	60.1	56.2	44.8	60.0
#1979	36.7	37.9	53.7	60.2	67.7	71.6	76.6	78.4	70.4	59.3	52.9	43.7	59.1
1980	41.4	38.9	47.4	60.0	68.2	74.0	80.0	80.9	74.8	58.3	48.8	42.6	59.6
1981	36.1	44.2	48.5	64.2	65.1	78.2	78.7	75.1	70.2	58.0	51.3	39.3	59.1
1982	36.0	45.4	52.7	57.6	72.0	75.0	78.4	76.3	70.0	60.5	52.3	47.9	60.3
1983	38.8	41.8	50.6	54.8	66.7	73.9	80.7	80.2	71.1	61.0	50.0	39.6	59.1
1984	38.3	45.4	50.0	56.5	65.9	76.4	76.1	77.1	69.0	67.8	47.8	50.0	60.0
1985	35.6	43.1	54.0	61.4	68.0	75.8	77.4	75.7	70.6	64.8	58.6	39.7	60.4
1986	38.8	46.8	52.8	64.3	69.5	80.6	84.8	76.9	74.1	64.0	53.2	43.4	62.4
1987	40.0	42.9	51.1	58.9	71.6	77.8	82.3	81.4	73.5	56.2	53.5	45.5	61.2
1988	35.8	43.4	52.7	60.6	68.4	75.3	79.6	80.9	71.8	56.3	53.0	43.8	60.1
1989	46.8	45.9	53.4	60.7	66.7	78.2	79.6	77.8	71.8	63.5	52.2	36.5	61.1
1990	48.5	52.3	55.9	61.1	69.0	77.7	81.1	80.1	74.1	63.8	55.4	49.0	64.0
1991	43.1	48.4	55.2	64.4	74.0	77.6	82.2	78.4	73.8	63.5	51.4	48.2	63.4
1992	45.1	48.7	52.7	61.3	66.0	73.9	82.4	76.1	72.1	60.0	52.4	42.9	61.1
1993	45.0	43.1	49.6	58.7	70.5	78.8	85.5	80.2	74.8	62.3	52.7	42.9	62.0
Record Mean	41.5	44.0	51.2	60.0	68.8	76.0	78.8	77.6	72.2	61.5	51.1	43.2	60.5
Max	50.6	53.7	61.7	71.1	79.5	86.2	88.5	87.0	81.8	71.9	61.2	52.2	70.4
Min	32.5	34.2	40.7	49.0	58.1	65.8	69.2	68.2	62.6	51.0	41.0	34.1	50.5

REFERENCE NOTES

CHARLOTTE, NORTH CAROLINA

GENERAL
T - TRACE AMOUNT.
BLANK ENTRIES DENOTE MISSING/UNREPORTED DATA.
# INDICATES A STATION OR INSTRUMENT RELOCATION.

SPECIFIC

(a) - LENGTH OF RECORD IN YEARS, ALTHOUGH INDIVIDUAL MONTHS MAY BE MISSING.
0.* OR * - THE VALUE IS BETWEEN 0.0 AND 0.05
NORMALS - BASED ON THE 1961-1990 RECORD PERIOD.
EXTREMES - DATES ARE THE MOST RECENT OCCURRENCE
WIND DIR.- NUMERALS SHOW TENS OF DEGRESS CLOCKWISE FROM TRUE NORTH. "00" INDICATES CALM.
RESULTANT DIRECTIONS ARE GIVEN TO WHOLE DEGREES.
BOLD VALUES INDICATE EXTREME VALUES WHICH OCCURRED AFTER THE ASOS SYSTEM WAS COMMISSIONED.

RECORD = PERIOD OF RECORD
RECORD MEAN PRECIPITATION IS THE MEAN OF ALL DAILY PRECIPITATION AMOUNTS DURING THE PERIOD OF RECORD.
RECORD MAX(MIN) TEMPERATURE IS THE MEAN OF ALL DAILY MAX(MIN) TEMPERATURES DURING THE PERIOD OF RECORD.
RECORD MEAN TEMPERATURE IS THE SUM OF THE RECORD MAX AND RECORD MIN DIVIDED BY 2.
AVERAGE TEMPERATURE IS THE SUM OF THE MEAN DAILY MAX AND MIN TEMPERATURE DIVIDED BY 2.

EXCEPTIONS

RECORD MEANS ARE THROUGH THE CURRENT YEAR, BEGINNING IN 1879 FOR TEMPERATURE
1878 FOR PRECIPITATION
1940 FOR SNOWFALL

HEATING DEGREE DAYS Base 65 deg. F CHARLOTTE, NORTH CAROLINA

SEASON	JULY	AUG	SEP	OCT	NOV	DEC	JAN	FEB	MAR	APR	MAY	JUNE	TOTAL
1964-65	0	5	14	287	307	604	721	565	554	128	0	7	3192
1965-66	0	0	7	194	403	643	892	631	455	213	46	8	3492
1966-67	0	0	2	231	437	724	664	695	343	128	122	14	3360
1967-68	0	0	40	177	509	572	835	756	413	197	64	0	3563
1968-69	0	0	0	175	462	811	829	702	610	141	45	0	3775
1969-70	0	0	22	180	511	784	933	665	502	154	42	0	3793
1970-71	0	0	12	111	456	622	770	655	574	208	73	0	3481
1971-72	0	0	4	72	490	444	625	710	472	226	47	13	3103
1972-73	0	0	3	195	490	564	780	693	341	241	74	0	3381
1973-74	0	0	2	124	361	715	463	583	320	182	36	0	2786
1974-75	0	0	50	216	454	699	609	554	444	212	12	0	3250
1975-76	0	0	13	103	347	694	789	425	321	192	68	12	2964
1976-77	0	0	5	299	621	799	1075	636	306	120	36	7	3904
1977-78	0	0	3	215	356	690	862	785	473	140	72	0	3596
1978-79	0	0	7	155	255	620	873	750	350	163	30	4	3207
#1979-80	1	0	14	197	357	655	726	750	538	171	27	0	3436
1980-81	0	0	26	230	482	689	890	574	508	87	74	0	3560
1981-82	0	2	18	228	405	790	891	544	376	235	6	0	3495
1982-83	0	0	20	191	380	529	806	645	441	308	46	1	3367
1983-84	0	0	44	149	445	780	820	562	459	266	74	2	3601
1984-85	0	0	47	31	517	458	905	607	358	150	32	0	3105
1985-86	0	0	28	85	198	777	799	503	381	106	24	0	2901
1986-87	0	6	2	119	357	665	767	614	426	217	9	0	3182
1987-88	0	0	1	264	339	597	899	621	378	154	23	6	3282
1988-89	0	0	1	279	354	653	558	541	378	189	77	0	3030
1989-90	0	0	27	111	382	877	503	350	307	158	23	0	2738
1990-91	0	0	15	109	282	491	671	457	313	88	11	0	2437
1991-92	0	0	9	102	410	525	608	465	375	174	83	1	2752
1992-93	0	0	20	162	372	679	612	610	469	199	16	0	3139
1993-94	0	0	12	139	384	677							

COOLING DEGREE DAYS Base 65 deg. F CHARLOTTE, NORTH CAROLINA

YEAR	JAN	FEB	MAR	APR	MAY	JUNE	JULY	AUG	SEP	OCT	NOV	DEC	TOTAL
1969	0	0	0	22	143	373	498	336	151	42	0	0	1565
1970	0	0	0	47	148	289	453	438	346	71	0	0	1792
1971	0	0	0	12	83	354	354	341	251	69	16	5	1485
1972	0	0	0	40	43	208	377	385	241	17	10	0	1321
1973	0	0	11	25	72	292	426	373	296	39	0	0	1534
1974	0	0	20	38	152	223	375	346	158	15	10	0	1337
1975	0	0	0	40	188	307	356	436	196	67	11	0	1601
1976	0	2	11	43	76	238	374	351	146	14	0	0	1255
1977	0	2	0	50	191	319	545	465	294	27	20	0	1913
1978	0	0	0	41	132	343	438	440	287	13	1	1	1696
#1979	0	0	4	28	122	208	369	419	183	30	3	0	1366
1980	0	0	0	28	133	275	473	497	328	26	0	0	1760
1981	0	0	2	69	87	403	431	323	182	21	0	0	1518
1982	0	0	3	22	230	308	424	357	179	59	6	5	1593
1983	0	0	1	6	107	275	494	476	234	32	0	0	1625
1984	0	0	0	18	110	350	348	381	176	124	6	0	1513
1985	0	0	24	48	132	330	395	338	204	87	11	0	1569
1986	0	0	9	94	174	474	618	381	284	93	10	0	2137
1987	0	0	0	38	222	389	542	515	264	2	1	0	1973
1988	0	0	3	28	135	323	461	500	213	18	1	0	1682
1989	0	14	25	72	138	403	458	405	237	75	1	0	1828
1990	0	1	32	50	157	389	507	475	292	81	4	3	1991
1991	0	0	17	79	300	382	542	423	280	66	8	13	2110
1992	0	0	1	71	124	274	546	351	240	12	2	0	1621
1993	0	0	0	15	195	419	643	476	313	62	21	0	2144

SNOWFALL (inches) CHARLOTTE, NORTH CAROLINA

SEASON	JULY	AUG	SEP	OCT	NOV	DEC	JAN	FEB	MAR	APR	MAY	JUNE	TOTAL
1964-65	0.0	0.0	0.0	0.0	T	0.0	11.2	1.0	T	0.0	0.0	0.0	12.2
1965-66	0.0	0.0	0.0	0.0	0.0	0.0	9.6	1.1	0.0	0.0	0.0	0.0	10.7
1966-67	0.0	0.0	0.0	0.0	0.0	T	T	5.4	0.0	0.0	0.0	0.0	5.4
1967-68	0.0	0.0	0.0	0.0	0.0	T	4.3	2.4	T	0.0	0.0	0.0	6.7
1968-69	0.0	0.0	0.0	0.0	2.5	T	T	13.2	3.0	0.0	0.0	0.0	18.7
1969-70	0.0	0.0	0.0	0.0	0.0	T	4.5	T	0.0	0.0	0.0	0.0	4.5
1970-71	0.0	0.0	0.0	0.0	0.0	2.7	0.1	0.8	5.7	0.0	0.0	0.0	9.3
1971-72	0.0	0.0	0.0	0.0	T	7.5	0.0	0.5	3.7	0.0	0.0	0.0	11.7
1972-73	0.0	0.0	0.0	0.0	T	0.0	5.5	1.0	0.0	0.0	0.0	0.0	6.5
1973-74	0.0	0.0	0.0	0.0	0.0	5.4	0.0	T	1.4	0.0	0.0	0.0	6.8
1974-75	0.0	0.0	0.0	0.0	0.0	T	0.0	1.0	0.2	0.0	0.0	0.0	1.2
1975-76	0.0	0.0	0.0	0.0	0.5	T	T	T	0.0	0.0	0.0	0.0	0.5
1976-77	0.0	0.0	0.0	0.0	T	T	3.4	T	0.0	0.0	0.0	0.0	3.4
1977-78	0.0	0.0	0.0	0.0	0.0	T	T	0.7	4.8	0.0	0.0	0.0	5.5
1978-79	0.0	0.0	0.0	0.0	0.0	0.0	0.4	14.9	0.0	0.0	0.0	0.0	15.3
#1979-80	0.0	0.0	0.0	0.0	0.0	0.0	0.4	7.3	6.8	0.0	0.0	0.0	14.5
1980-81	0.0	0.0	0.0	0.0	0.0	0.3	1.8	T	T	0.0	0.0	0.0	2.1
1981-82	0.0	0.0	0.0	0.0	0.0	T	4.8	5.9	0.0	0.1	0.0	0.0	10.8
1982-83	0.0	0.0	0.0	0.0	0.0	T	0.8	1.5	10.3	0.0	0.0	0.0	12.6
1983-84	0.0	0.0	0.0	0.0	T	T	T	5.9	0.0	0.0	0.0	0.0	5.9
1984-85	0.0	0.0	0.0	0.0	0.0	T	1.7	T	0.0	0.0	0.0	0.0	1.7
1985-86	0.0	0.0	0.0	0.0	0.0	T	T	0.3	T	0.0	0.0	0.0	0.3
1986-87	0.0	0.0	0.0	0.0	0.0	0.0	5.6	2.1	0.3	T	0.0	0.0	8.0
1987-88	0.0	0.0	0.0	0.0	T	0.0	12.1	0.0	0.0	0.0	0.0	0.0	12.1
1988-89	0.0	0.0	0.0	0.0	0.0	T	0.0	3.5	T	0.0	0.0	0.0	3.5
1989-90	0.0	0.0	0.0	0.0	T	0.6	0.0	0.0	T	T	0.0	0.0	0.6
1990-91	0.0	0.0	0.0	0.0	0.0	T	1.0	T	T	0.0	0.0	0.0	1.0
1991-92	0.0	0.0	0.0	0.0	T	0.0	T	0.0	0.0	0.0	0.0	0.0	T
1992-93	0.0	0.0	0.0	0.0	0.0	T	0.0	T	1.6	0.0	T	0.0	1.6
1993-94	0.0	0.0	0.0	0.0	0.0	2.6							
Record Mean	0.0	0.0	0.0	0.0	0.1	0.5	2.0	1.7	1.2	T	T	0.0	5.6

GREENSBORO, NORTH CAROLINA

The Greensboro-High Point-Winston-Salem Regional Airport is located in the west-central part of Guilford County, in the northern Piedmont section of North Carolina. The location is near the headwaters of the Haw and Deep Rivers, both branches of the Cape Fear River system. A few miles west is a ridge beyond which lies the Yadkin River Basin. To the north, across a similar ridge, the waters of the Dan River flow northeastward into the Roanoke. West, beyond the Yadkin River Basin, the land gradually rises into the Brushy Mountains. To the northwest, other outcroppings southeast of the Blue Ridge rise into peaks occasionally exceeding 2,500 feet. The Blue Ridge proper forms a northeast-southwest barrier with heights occasionally exceeding 3,000 feet.

Winter temperatures and rainfall are both modified by the mountain barrier, but to a lesser extent than in areas closer to the Appalachian Range. Shallow cold air masses from the west tend to be stopped or turned aside by the mountains, while deeper masses are lifted over the range, losing moisture and warming during the passage. For this reason the lowest temperatures recorded in Forsyth and Guilford Counties usually occur when clear, cold air drifts southward, east of the Appalachian Range. The summer temperatures vary with the cloudiness and shower activity, but are generally mild.

Northwesterly winds seldom bring heavy or prolonged winter rain or snow. Flurries of light snow may fall when cold air blows across the mountains, but the heavier winter precipitation comes with winds blowing from northeast through east and south to southwest. When moist winds blowing from an easterly or southerly direction meet cold air moving out of the north or northwest in the vicinity of North Carolina, snow, sleet, or glaze may occur. Glazing is more common here than in most of North Carolina, but only occasionally becomes severe enough to do much damage in the northern Piedmont area.

Seasonal snowfall has a wide range and there have been a few winters with only a trace of snow. Snow seldom stays on the ground more than a few days.

Summer precipitation is largely from thunderstorms, mostly local in character. The frequency of these showers and the amount of rain received varies greatly from year to year and from place to place. Sizeable areas are sometimes without significant rain in late spring or early summer for two or more weeks, while other areas in the vicinity may be well watered.

Damaging storms are infrequent in the Northern Piedmont area. The highest winds to occur have been associated with thunderstorms, and were of brief duration. Hail is reported within Guilford and Forsyth Counties each year. The occurrence of tornadoes is rare. Hurricanes have produced heavy rainfall here, but no winds of destructive force.

Based on the 1951-1980 period, the average first occurrence of 32 degrees Fahrenheit in the fall is October 27 and the average last occurrence in the spring is April 11.

NORMALS, MEANS, AND EXTREMES

GREENSBORO, NORTH CAROLINA

LATITUDE: 36°05'N LONGITUDE: 79°57' W ELEVATION: FT. GRND 897 BARO 908 TIME ZONE: EASTERN WBAN: 13723

	(a)	JAN	FEB	MAR	APR	MAY	JUNE	JULY	AUG	SEP	OCT	NOV	DEC	YEAR
TEMPERATURE °F:														
Normals														
-Daily Maximum		46.7	50.7	60.2	69.6	77.0	83.6	86.9	85.5	79.9	69.9	60.5	50.5	68.4
-Daily Minimum		26.6	29.3	37.4	45.6	54.7	62.6	66.9	65.9	59.5	47.2	38.5	30.5	47.1
-Monthly		36.7	40.0	48.8	57.6	65.9	73.1	76.9	75.7	69.7	58.6	49.5	40.5	57.8
Extremes														
-Record Highest	65	78	81	90	94	98	102	102	103	100	95	85	78	103
-Year		1975	1977	1945	1930	1941	1954	1977	1988	1954	1954	1974	1971	AUG 1988
-Record Lowest	65	-8	-4	5	21	32	42	48	45	35	20	10	0	-8
-Year		1985	1936	1960	1943	1989	1977	1933	1986	1942	1962	1970	1962	JAN 1985
NORMAL DEGREE DAYS:														
Heating (base 65°F)		877	700	502	235	78	6	0	0	16	226	465	760	3865
Cooling (base 65°F)		0	0	0	13	106	249	369	332	157	27	0	0	1253
% OF POSSIBLE SUNSHINE	65	52	56	60	63	63	64	62	61	62	64	57	53	60
MEAN SKY COVER (tenths)														
Sunrise - Sunset	65	6.3	6.1	6.0	5.7	5.9	5.9	6.1	6.0	5.6	4.8	5.4	6.0	5.8
MEAN NUMBER OF DAYS:														
Sunrise to Sunset														
-Clear	65	8.8	8.6	9.1	9.5	8.1	6.8	6.5	7.2	9.8	13.6	11.1	10.0	109.2
-Partly Cloudy	65	6.8	6.3	7.9	8.5	10.8	12.5	12.6	12.4	8.7	7.3	6.8	6.3	106.9
-Cloudy	65	15.4	13.3	13.9	11.9	12.0	10.7	11.9	11.3	11.5	10.2	12.2	14.7	149.1
Precipitation														
.01 inches or more	65	10.2	9.6	11.1	9.4	10.3	10.1	12.0	10.5	7.7	7.0	8.2	9.4	115.6
Snow,Ice pellets,hail														
1.0 inches or more	65	0.9	0.8	0.4	0.0	0.0	0.0	0.0	0.0	0.0	0.0	0.*	0.4	2.5
Thunderstorms	65	0.4	0.8	2.0	3.4	6.6	8.2	10.8	8.1	3.2	1.1	0.5	0.3	45.3
Heavy Fog Visibility														
1/4 mile or less	65	4.8	3.3	2.8	1.6	1.9	1.3	1.8	2.3	3.0	2.5	3.2	3.9	32.5
Temperature °F														
-Maximum														
90° and above	30	0.0	0.0	0.0	0.4	1.1	6.2	12.0	8.6	2.2	0.2	0.0	0.0	30.6
32° and below	30	2.6	1.0	0.2	0.0	0.0	0.0	0.0	0.0	0.0	0.0	0.*	1.1	4.8
-Minimum														
32° and below	30	21.8	19.1	10.1	2.0	0.*	0.0	0.0	0.0	0.0	1.5	9.5	18.6	82.7
0° and below	30	0.1	0.0	0.0	0.0	0.0	0.0	0.0	0.0	0.0	0.0	0.0	0.0	0.1
AVG. STATION PRESS.(mb)	21	987.2	986.7	985.2	984.3	984.6	985.2	986.1	986.9	987.6	988.2	988.1	988.1	986.5
RELATIVE HUMIDITY (%)														
Hour 01	30	74	70	70	71	82	84	87	88	87	84	77	76	79
Hour 07 (Local Time)	30	79	76	78	78	83	85	88	90	90	87	83	80	83
Hour 13	30	56	52	50	48	55	56	59	60	59	54	53	56	55
Hour 19	30	63	57	54	52	62	65	68	70	72	71	65	66	64
PRECIPITATION (inches):														
Water Equivalent														
-Normal		3.17	3.32	3.72	2.84	4.02	3.81	4.51	3.88	3.52	3.50	2.97	3.36	42.62
-Maximum Monthly	65	8.24	7.04	8.76	8.03	8.35	7.99	12.72	12.53	13.26	12.59	8.26	6.44	13.26
-Year		1937	1929	1975	1987	1982	1965	1984	1939	1947	1990	1985	1973	SEP 1947
-Minimum Monthly	65	0.66	0.73	0.67	0.47	0.37	T	0.98	0.71	T	0.26	0.35	0.33	T
-Year		1981	1978	1985	1986	1936	1990	1953	1972	1985	1963	1981	1955	JUN 1990
-Maximum in 24 hrs	65	3.06	3.00	3.62	4.42	3.25	4.91	4.43	4.47	7.49	6.24	3.32	3.60	7.49
-Year		1936	1934	1991	1987	1989	1972	1944	1949	1947	1954	1962	1958	SEP 1947
Snow,Ice pellets,hail														
-Maximum Monthly	65	22.9	16.3	21.3	T	T	0.0	0.0	0.0	0.0	0.0	5.9	14.3	22.9
-Year		1966	1979	1960	1993	1993						1968	1930	JAN 1966
-Maximum in 24 hrs	65	14.0	9.3	11.1	T	T	0.0	0.0	0.0	0.0	0.0	5.0	14.3	14.3
-Year		1940	1979	1960	1993	1993						1968	1930	DEC 1930
WIND:														
Mean Speed (mph)	65	8.1	8.5	9.1	8.8	7.6	6.9	6.5	6.2	6.6	7.0	7.5	7.6	7.5
Prevailing Direction through 1963		SW	SW	SW	SW	SW	SW	SW	SW	NE	NE	SW	SW	SW
Fastest Obs. 1 Min.														
-Direction (!!)	13	31	09	29	32	24	25	35	20	12	30	20	29	35
-Speed (MPH)	13	32	29	30	35	42	29	50	28	35	46	38	30	50
-Year		1989	1984	1991	1989	1988	1989	1984	1993	1989	1985	1992	1983	JUL 1984
Peak Gust														
-Direction (!!)	9	SW	E	W	W	SW	NW	N	SE	SE	NW	SW	SW	SW
-Speed (mph)	9	63	48	43	46	59	51	59	39	54	60	48	47	63
-Date		1992	1984	1993	1990	1988	1988	1991	1988	1989	1985	1991	1990	JAN 1992

PRECIPITATION (inches)

GREENSBORO, NORTH CAROLINA

YEAR	JAN	FEB	MAR	APR	MAY	JUNE	JULY	AUG	SEP	OCT	NOV	DEC	ANNUAL
1964	4.48	4.09	2.75	2.93	0.56	3.07	6.61	6.75	1.03	8.15	1.57	3.20	45.19
1965	2.27	2.80	5.04	3.07	1.00	7.99	9.08	3.91	4.31	3.72	1.29	0.64	45.12
1966	4.48	5.44	1.53	0.97	4.15	4.04	2.01	3.01	3.59	3.25	1.55	2.26	36.28
1967	1.84	2.76	1.21	2.03	4.03	2.11	3.11	5.43	2.39	1.23	1.40	6.34	33.88
1968	3.74	0.88	3.97	2.64	3.92	1.62	2.60	2.51	0.58	4.60	3.55	2.25	32.86
1969	2.01	3.01	3.72	2.30	3.17	7.86	4.11	3.95	3.44	1.89	1.41	4.93	41.80
1970	1.66	3.25	3.81	4.52	2.76	3.59	2.56	9.13	0.70	5.57	2.87	2.37	42.79
1971	1.80	4.46	3.32	2.74	5.37	2.99	3.88	4.22	2.51	6.71	2.58	1.61	42.19
1972	2.60	3.71	2.65	2.72	6.24	6.37	2.54	0.71	3.72	1.82	4.87	4.49	42.44
1973	3.50	3.53	5.69	4.31	5.71	4.88	5.30	2.72	1.15	2.31	1.24	6.44	46.78
1974	4.58	3.13	3.32	2.15	5.69	2.72	1.71	7.14	7.57	0.81	2.42	4.23	45.47
1975	5.31	3.15	8.76	2.00	6.23	1.71	12.35	1.48	7.29	2.62	1.81	3.80	56.51
1976	2.25	1.47	2.42	0.62	2.94	3.91	1.87	2.63	2.35	8.61	1.69	3.92	34.68
1977	2.27	1.69	2.79	2.25	0.70	1.54	1.52	4.03	6.35	4.67	3.11	3.30	34.22
1978	7.70	0.73	4.38	4.13	5.09	3.72	9.23	9.60	2.70	1.25	3.17	4.45	56.15
1979	5.87	5.03	2.95	3.21	4.46	2.72	4.30	1.13	13.08	2.64	4.78	1.12	51.29
1980	4.00	1.77	5.04	3.24	3.23	3.44	2.68	2.14	2.95	2.13	2.58	0.78	33.98
1981	0.66	3.61	2.59	1.13	2.78	2.92	8.99	2.60	6.35	3.63	0.35	5.60	41.21
1982	2.91	4.64	2.39	2.72	8.35	7.28	3.46	1.46	1.22	4.65	2.73	3.56	45.37
1983	1.31	3.82	5.75	5.16	4.12	5.15	2.78	1.44	2.68	4.09	5.37	4.38	46.05
1984	3.51	5.25	5.31	2.77	5.83	3.10	12.72	5.55	1.12	1.16	2.02	1.90	50.24
1985	4.34	3.77	0.67	0.94	4.32	2.25	4.31	4.86	T	1.92	8.26	1.46	37.10
1986	0.74	1.61	2.00	0.47	1.12	1.10	3.21	8.03	1.05	3.24	3.50	3.60	29.67
1987	3.77	4.72	3.24	8.03	2.48	1.20	5.80	3.55	7.22	0.71	3.66	3.38	47.76
1988	2.35	1.27	1.86	2.55	3.97	3.19	4.47	3.50	3.24	4.09	3.55	1.19	35.23
1989	0.93	4.48	5.71	3.23	6.36	7.09	2.96	2.31	6.08	5.04	2.92	3.53	50.64
1990	3.76	4.32	2.64	2.90	6.54	T	1.50	2.01	1.16	12.59	1.93	4.17	43.52
1991	4.53	1.15	7.75	6.66	3.89	1.96	3.95	1.91	3.66	1.90	1.67	2.80	41.83
1992	3.54	3.02	2.74	5.83	2.34	2.92	5.43	2.97	3.06	3.66	6.05	2.48	44.04
1993	3.65	3.40	7.61	6.08	2.95	3.14	3.28	2.31	2.82	1.25	2.58	3.28	42.35
Record Mean	3.29	3.30	3.80	3.29	3.67	3.62	4.60	4.13	3.54	3.09	2.88	3.23	42.42

AVERAGE TEMPERATURE (deg. F)

GREENSBORO, NORTH CAROLINA

YEAR	JAN	FEB	MAR	APR	MAY	JUNE	JULY	AUG	SEP	OCT	NOV	DEC	ANNUAL
1964	38.1	37.7	48.7	58.4	67.6	75.7	76.4	74.0	67.9	54.4	52.5	42.6	57.8
1965	38.9	40.6	44.3	58.6	72.6	72.8	75.7	77.0	71.6	57.9	50.1	43.2	58.6
1966	33.1	39.2	48.7	55.9	66.7	73.5	79.0	76.4	69.8	57.7	50.2	40.5	57.6
1967	43.0	38.2	53.0	63.3	64.0	73.7	77.1	75.9	66.6	58.5	46.8	45.1	58.8
1968	35.6	36.3	53.4	60.2	65.0	72.9	77.0	78.1	69.4	60.2	48.9	37.5	57.9
1969	36.4	40.1	44.3	59.5	67.3	76.2	80.4	75.7	68.6	59.3	47.0	38.5	57.8
1970	33.4	40.0	47.8	60.9	67.7	76.0	79.5	75.4	73.0	60.0	47.8	41.1	58.6
1971	35.2	38.7	43.7	54.4	62.5	74.7	75.1	73.3	71.0	63.0	47.7	48.7	57.3
1972	42.3	37.8	47.1	56.2	63.0	68.7	76.1	75.5	70.8	54.7	46.2	44.1	56.9
1973	36.1	37.1	52.2	55.4	62.7	75.4	76.5	76.0	72.3	59.8	49.7	38.2	57.6
1974	45.8	41.1	52.0	58.0	66.1	70.7	76.6	75.5	67.9	57.5	49.5	41.8	58.6
1975	42.9	44.1	46.2	56.7	69.1	75.2	77.1	80.0	70.1	62.0	52.0	40.1	59.6
1976	35.9	49.3	53.8	60.5	64.2	71.2	75.9	73.1	66.9	53.2	41.8	37.0	56.9
1977	26.7	39.1	53.3	61.2	69.1	73.6	80.4	78.2	71.7	55.1	49.8	38.2	58.1
1978	31.3	31.6	45.2	57.0	63.1	73.4	76.4	77.1	71.4	57.2	52.1	42.2	56.5
1979	35.1	32.8	50.9	59.1	66.2	69.8	75.7	76.5	69.1	58.0	52.3	41.5	57.3
1980	38.4	36.0	46.5	58.2	66.0	70.8	78.4	78.7	73.1	56.1	46.6	40.2	57.4
1981	33.5	42.3	45.7	61.9	63.2	77.9	78.5	73.6	68.1	55.6	48.9	36.4	57.1
1982	32.7	42.8	49.2	54.9	70.1	72.1	77.1	75.0	68.0	57.6	49.1	45.5	57.9
1983	36.0	38.6	48.4	53.0	63.8	71.6	78.0	78.2	69.1	58.3	48.7	36.5	56.7
1984	35.2	44.1	46.0	54.3	63.5	75.0	74.2	75.4	65.7	65.5	45.5	48.3	57.7
1985	32.8	40.6	51.7	61.0	66.8	73.1	75.7	73.4	67.9	61.1	56.2	37.6	58.2
1986	37.1	42.3	49.7	60.5	66.5	77.3	80.8	73.6	71.1	61.3	50.7	40.2	59.3
1987	36.4	38.8	47.9	55.6	68.9	75.9	80.1	78.9	71.6	53.1	50.6	42.2	58.3
1988	33.0	40.4	49.4	57.0	65.2	72.4	78.1	79.7	68.5	53.2	49.6	39.5	57.2
1989	42.2	40.4	49.4	56.8	63.2	75.4	77.5	74.8	69.5	59.4	48.8	32.7	57.5
1990	45.3	48.5	52.7	58.1	66.0	75.2	79.3	77.3	70.5	61.2	52.4	46.3	61.1
1991	40.8	45.2	52.4	60.7	71.5	75.0	79.5	76.2	70.6	60.3	48.5	44.9	60.5
1992	42.1	45.2	48.5	57.5	62.4	71.1	79.5	73.0	69.3	56.6	48.9	39.7	57.8
1993	40.4	38.2	45.4	56.9	69.1	75.5	82.0	78.2	72.2	58.3	49.0	39.0	58.7
Record Mean	38.3	40.6	48.2	57.8	66.5	74.1	77.4	76.0	69.9	58.7	48.6	40.1	58.0
Max	48.2	51.4	59.7	70.0	78.1	85.0	87.6	86.0	80.4	70.5	59.9	50.2	68.9
Min	28.4	29.8	36.7	45.5	54.9	63.2	67.2	66.0	59.4	46.8	37.2	30.0	47.1

REFERENCE NOTES

GREENSBORO, NORTH CAROLINA

GENERAL
T - TRACE AMOUNT.
BLANK ENTRIES DENOTE MISSING/UNREPORTED DATA.
# INDICATES A STATION OR INSTRUMENT RELOCATION.

SPECIFIC

(a) - LENGTH OF RECORD IN YEARS, ALTHOUGH INDIVIDUAL MONTHS MAY BE MISSING.
0.* OR * - THE VALUE IS BETWEEN 0.0 AND 0.05
NORMALS - BASED ON THE 1961-1990 RECORD PERIOD.
EXTREMES - DATES ARE THE MOST RECENT OCCURRENCE
WIND DIR.- NUMERALS SHOW TENS OF DEGRESS CLOCKWISE FROM TRUE NORTH. "00" INDICATES CALM.
RESULTANT DIRECTIONS ARE GIVEN TO WHOLE DEGREES.
BOLD VALUES INDICATE EXTREME VALUES WHICH OCCURRED AFTER THE ASOS SYSTEM WAS COMMISSIONED.

RECORD = PERIOD OF RECORD
RECORD MEAN PRECIPITATION IS THE MEAN OF ALL DAILY PRECIPITATION AMOUNTS DURING THE PERIOD OF RECORD.
RECORD MAX(MIN) TEMPERATURE IS THE MEAN OF ALL DAILY MAX(MIN) TEMPERATURES DURING THE PERIOD OF RECORD.
RECORD MEAN TEMPERATURE IS THE SUM OF THE RECORD MAX AND RECORD MIN DIVIDED BY 2.
AVERAGE TEMPERATURE IS THE SUM OF THE MEAN DAILY MAX AND MIN TEMPERATURE DIVIDED BY 2.

EXCEPTIONS

FASTEST MILE WINDS ARE THROUGH FEBRUARY 1980.

RECORD MEANS ARE THROUGH THE CURRENT YEAR, BEGINNING IN 1929 FOR TEMPERATURE
1929 FOR PRECIPITATION
1929 FOR SNOWFALL

HEATING DEGREE DAYS Base 65 deg. F GREENSBORO, NORTH CAROLINA

SEASON	JULY	AUG	SEP	OCT	NOV	DEC	JAN	FEB	MAR	APR	MAY	JUNE	TOTAL
1964-65	0	2	28	329	369	688	803	676	637	215	0	11	3758
1965-66	0	3	15	230	441	670	979	714	501	285	71	12	3921
1966-67	0	0	14	237	438	751	676	742	376	130	111	14	3489
1967-68	0	0	49	218	539	610	904	828	374	170	69	2	3763
1968-69	0	1	0	180	476	848	878	689	637	175	48	0	3932
1969-70	0	0	30	211	536	816	973	695	525	160	59	0	4005
1970-71	0	1	19	179	511	736	916	728	653	312	120	0	4175
1971-72	0	0	9	101	523	501	696	783	550	282	76	30	3551
1972-73	1	0	9	314	562	642	889	774	396	292	111	0	3990
1973-74	0	0	4	179	453	823	586	664	403	232	59	0	3403
1974-75	0	0	53	244	477	711	676	579	578	256	26	0	3600
1975-76	0	0	31	121	391	765	895	454	357	199	74	20	3307
1976-77	0	1	19	370	687	860	1184	723	358	159	42	11	4414
1977-78	0	0	7	307	459	821	1039	927	605	245	131	4	4545
1978-79	0	0	16	238	381	700	921	896	440	198	52	8	3850
1979-80	1	2	25	239	377	719	817	834	565	212	65	7	3863
1980-81	0	0	31	278	544	762	970	630	590	135	116	0	4056
1981-82	0	1	39	299	477	879	997	616	484	304	14	0	4110
1982-83	0	0	38	253	467	602	894	734	505	358	97	9	3957
1983-84	0	0	74	222	482	876	913	600	580	331	107	4	4189
1984-85	0	0	96	53	580	512	992	679	425	160	50	6	3553
1985-86	1	0	65	151	263	843	857	629	472	176	68	0	3525
1986-87	0	17	13	174	426	761	880	724	524	291	34	0	3844
1987-88	0	0	4	361	422	699	984	705	478	241	56	23	3973
1988-89	0	0	24	368	455	781	702	685	492	279	133	0	3919
1989-90	0	1	48	206	481	994	601	458	397	237	51	4	3478
1990-91	0	0	31	161	371	573	739	546	391	167	24	2	3005
1991-92	0	0	29	171	495	619	703	567	505	255	132	8	3484
1992-93	0	0	38	258	474	777	756	746	600	244	18	0	3911
1993-94	0	0	23	232	482	803							

COOLING DEGREE DAYS Base 65 deg. F GREENSBORO, NORTH CAROLINA

YEAR	JAN	FEB	MAR	APR	MAY	JUNE	JULY	AUG	SEP	OCT	NOV	DEC	TOTAL
1969	0	0	0	18	125	345	486	339	145	41	0	0	1499
1970	0	0	0	45	151	338	456	335	264	32	0	0	1621
1971	0	0	0	0	50	298	321	261	195	47	13	5	1190
1972	0	0	3	23	21	147	352	328	188	2	7	0	1071
1973	0	0	6	11	46	319	363	343	229	25	0	0	1342
1974	0	0	8	28	101	176	366	333	144	15	18	0	1189
1975	0	0	1	15	159	313	379	473	192	37	7	0	1576
1976	0	4	14	71	55	212	344	257	82	9	0	0	1048
1977	0	2	1	50	175	275	483	417	214	9	10	0	1636
1978	0	0	0	11	80	263	364	384	216	4	0	0	1322
1979	0	0	6	26	95	159	340	365	157	28	4	0	1180
1980	0	0	0	16	102	189	424	434	281	12	0	0	1458
1981	0	0	0	47	69	394	427	276	141	15	0	0	1369
1982	0	0	0	7	179	216	382	317	136	33	0	2	1272
1983	0	0	0	5	68	210	409	415	202	20	0	0	1329
1984	0	0	0	16	66	309	292	328	122	79	0	0	1212
1985	0	1	18	47	109	259	339	268	159	36	7	0	1243
1986	0	0	3	49	123	376	496	291	202	67	5	0	1612
1987	0	0	0	13	162	333	473	438	212	0	0	0	1631
1988	0	0	2	10	71	249	411	461	137	8	0	0	1349
1989	0	3	13	39	79	320	392	313	189	40	0	0	1388
1990	0	0	23	37	86	317	451	390	202	47	0	0	1553
1991	0	0	9	46	232	310	454	352	202	33	5	5	1648
1992	0	0	0	37	58	198	457	252	175	6	0	0	1183
1993	0	0	0	9	154	321	530	414	247	34	10	0	1719

SNOWFALL (inches) GREENSBORO, NORTH CAROLINA

SEASON	JULY	AUG	SEP	OCT	NOV	DEC	JAN	FEB	MAR	APR	MAY	JUNE	TOTAL
1964-65	0.0	0.0	0.0	0.0	0.3	T	9.1	3.3	0.2	0.0	0.0	0.0	12.9
1965-66	0.0	0.0	0.0	0.0	0.0	T	22.9	3.1	0.0	0.0	0.0	0.0	26.0
1966-67	0.0	0.0	0.0	0.0	T	2.6	0.8	2.7	0.0	0.0	0.0	0.0	6.1
1967-68	0.0	0.0	0.0	0.0	T	T	6.0	3.2	T	0.0	0.0	0.0	9.2
1968-69	0.0	0.0	0.0	0.0	5.9	T	0.3	0.7	10.7	0.0	0.0	0.0	17.6
1969-70	0.0	0.0	0.0	0.0	0.0	1.6	6.4	T	T	0.0	0.0	0.0	8.0
1970-71	0.0	0.0	0.0	0.0	0.0	3.1	0.4	0.7	3.5	T	0.0	0.0	7.7
1971-72	0.0	0.0	0.0	0.0	0.9	4.2	T	4.8	3.1	0.0	0.0	0.0	13.0
1972-73	0.0	0.0	0.0	0.0	T	0.0	6.0	T	1.5	0.0	0.0	0.0	7.5
1973-74	0.0	0.0	0.0	0.0	0.0	8.3	0.0	T	4.5	0.0	0.0	0.0	12.8
1974-75	0.0	0.0	0.0	0.0	0.0	T	T	2.0	2.0	0.0	0.0	0.0	4.0
1975-76	0.0	0.0	0.0	0.0	0.0	T	T	0.3	0.0	0.0	0.0	0.0	0.3
1976-77	0.0	0.0	0.0	0.0	T	1.2	8.1	1.2	0.0	0.0	0.0	0.0	10.5
1977-78	0.0	0.0	0.0	0.0	T	0.7	0.4	5.9	3.6	0.0	0.0	0.0	10.6
1978-79	0.0	0.0	0.0	0.0	T	T	T	16.3	0.0	0.0	0.0	0.0	16.3
1979-80	0.0	0.0	0.0	0.0	0.0	0.0	3.5	7.0	7.9	0.0	0.0	0.0	18.4
1980-81	0.0	0.0	0.0	0.0	T	0.5	2.4	0.0	7.6	0.0	0.0	0.0	10.5
1981-82	0.0	0.0	0.0	0.0	0.0	0.1	8.8	4.8	T	T	0.0	0.0	13.7
1982-83	0.0	0.0	0.0	0.0	0.0	0.2	0.4	5.7	T	T	0.0	0.0	6.3
1983-84	0.0	0.0	0.0	0.0	0.0	T	T	3.8	T	0.0	0.0	0.0	3.8
1984-85	0.0	0.0	0.0	0.0	0.0	0.0	5.1	T	0.0	0.0	0.0	0.0	5.1
1985-86	0.0	0.0	0.0	0.0	0.0	0.3	T	1.1	0.0	0.0	0.0	0.0	1.4
1986-87	0.0	0.0	0.0	0.0	0.0	0.0	14.2	9.0	0.2	T	0.0	0.0	23.4
1987-88	0.0	0.0	0.0	0.0	0.3	0.0	8.6	0.0	0.0	0.0	0.0	0.0	8.9
1988-89	0.0	0.0	0.0	0.0	0.0	0.3	T	13.3	T	T	0.0	0.0	13.6
1989-90	0.0	0.0	0.0	0.0	0.0	2.5	0.0	T	T	0.0	0.0	0.0	2.5
1990-91	0.0	0.0	0.0	0.0	0.0	T	T	T	T	0.0	0.0	0.0	T
1991-92	0.0	0.0	0.0	0.0	T	0.0	0.0	T	0.0	T	0.0	0.0	T
1992-93	0.0	0.0	0.0	0.0	0.0	T	T	1.2	5.7	T	T	0.0	6.9
1993-94	0.0	0.0	0.0	0.0	0.0	1.0							
Record Mean	0.0	0.0	0.0	0.0	0.1	1.2	3.1	2.5	1.7	T	T	0.0	8.7

RALEIGH, NORTH CAROLINA

The Raleigh-Durham Airport is located in the zone of transition between the Coastal Plain and the Piedmont Plateau. The surrounding terrain is rolling, with an average elevation of around 400 feet, the range over a 10-mile radius is roughly between 200 and 550 feet. Being centrally located between the mountains on the west and the coast on the south and east, the Raleigh-Durham area enjoys a favorable climate. The mountains form a partial barrier to cold air masses moving eastward from the interior of the nation. As a result, there are few days in the heart of the winter season when the temperature falls below 20 degrees. Tropical air is present over the eastern and central sections of North Carolina during much of the summer season, bringing warm temperatures and rather high humidities to the Raleigh-Durham area. Afternoon temperatures reach 90 degrees or higher on about one-fourth of the days in the middle of summer, but reach 100 degrees less than once per year. Even in the hottest weather, early morning temperatures almost always drop into the lower 70s.

Rainfall is well distributed throughout the year as a whole. July and August have the greatest amount of rainfall, and October and November the least. There are times in spring and summer when soil moisture is scanty. This usually results from too many days between rains rather than from a shortage of total rainfall, but occasionally the accumulated total during the growing season falls short of plant needs. Most summer rain is produced by thunderstorms, which may occasionally be accompanied by strong winds, intense rains, and hail. The Raleigh-Durham area is far enough from the coast so that the bad weather effects of coastal storms are reduced. While snow and sleet usually occur each year, excessive accumulations of snow are rare.

From September 1887 to December 1950, the office was located in the downtown areas of Raleigh. The various buildings occupied were within an area of three blocks. All thermometers were exposed on the roof, and this, plus the smoke over the city, had an effect on the temperature record of that period. Lowest temperatures at the city office were frequently from 2 to 5 degrees higher than those recorded in surrounding rural areas. Maximum temperatures in the city were generally a degree or two lower. These observations are supported by a period of simultaneous record from the Municipal Airport and the city office location between 1937 and 1940.

From September 1946 to May 1954, simultaneous records were kept at a surface location on the North Carolina State College campus in Raleigh, and at the Raleigh-Durham Airport 10 1/2 air miles to the northwest.

Based on the 1951-1980 period, the average first occurrence of 32 degrees Fahrenheit in the fall is October 27 and the average last occurrence in the spring is April 11.

NORMALS, MEANS, AND EXTREMES

RALEIGH, NORTH CAROLINA

LATITUDE: 35°52'N LONGITUDE: 78°47' W ELEVATION: FT. GRND 416 BARO 415 TIME ZONE: EASTERN WBAN: 13722

	(a)	JAN	FEB	MAR	APR	MAY	JUNE	JULY	AUG	SEP	OCT	NOV	DEC	YEAR
TEMPERATURE °F:														
Normals														
-Daily Maximum		48.9	52.6	62.1	71.7	78.6	85.0	88.0	86.8	81.1	71.6	62.6	52.7	70.1
-Daily Minimum		28.8	31.3	38.7	46.2	55.3	63.6	68.1	67.5	61.1	48.4	39.7	32.4	48.4
-Monthly		38.9	42.0	50.4	59.0	67.0	74.3	78.1	77.1	71.1	60.1	51.2	42.6	59.3
Extremes														
-Record Highest	49	79	84	92	95	97	104	105	105	104	98	88	79	105
-Year		1952	1977	1945	1980	1953	1954	1952	1988	1954	1954	1950	1978	AUG 1988
-Record Lowest	49	-9	5	11	23	31	38	48	46	37	19	11	4	-9
-Year		1985	1971	1980	1985	1977	1977	1975	1965	1983	1962	1970	1983	JAN 1985
NORMAL DEGREE DAYS:														
Heating (base 65°F)		809	644	458	193	47	0	0	0	9	189	414	694	3457
Cooling (base 65°F)		0	0	6	13	109	279	406	375	192	37	0	0	1417
% OF POSSIBLE SUNSHINE	39	53	56	60	63	58	61	60	58	58	61	57	53	58
MEAN SKY COVER (tenths)														
Sunrise - Sunset	44	6.2	6.0	6.0	5.6	6.0	5.8	6.0	6.0	5.7	5.0	5.4	5.9	5.8
MEAN NUMBER OF DAYS:														
Sunrise to Sunset														
-Clear	45	9.0	8.7	9.4	9.6	8.2	7.7	7.3	7.2	9.5	12.9	11.4	9.9	110.7
-Partly Cloudy	45	6.9	6.1	7.4	9.0	10.0	11.8	12.0	12.4	9.2	7.0	7.2	7.0	106.0
-Cloudy	45	15.1	13.5	14.2	11.4	12.8	10.5	11.8	11.4	11.3	11.1	11.4	14.1	148.6
Precipitation .01 inches or more	49	10.0	9.7	10.3	9.0	10.3	9.3	11.3	10.0	7.6	7.0	8.2	9.1	111.8
Snow,Ice pellets,hail 1.0 inches or more	49	0.8	0.7	0.3	0.*	0.0	0.0	0.0	0.0	0.0	0.0	0.1	0.3	2.2
Thunderstorms	49	0.4	0.9	2.0	3.4	6.2	7.1	10.6	7.8	3.4	1.3	0.8	0.3	44.0
Heavy Fog Visibility 1/4 mile or less	44	3.5	2.8	2.2	1.5	2.4	1.9	2.6	3.3	3.4	3.5	3.1	3.5	33.6
Temperature °F														
-Maximum														
90° and above	29	0.0	0.0	0.*	0.6	1.3	7.1	12.9	10.2	3.0	0.2	0.0	0.0	35.3
32° and below	29	2.1	0.5	0.1	0.0	0.0	0.0	0.0	0.0	0.0	0.0	0.*	0.8	3.6
-Minimum														
32° and below	29	20.1	17.2	9.8	2.2	0.*	0.0	0.0	0.0	0.0	1.4	9.1	17.2	77.1
0° and below	29	0.2	0.0	0.0	0.0	0.0	0.0	0.0	0.0	0.0	0.0	0.0	0.0	0.2
AVG. STATION PRESS.(mb)	21	1003.7	1003.1	1001.4	1000.2	1000.3	1000.6	1001.4	1002.3	1003.0	1003.9	1004.3	1004.5	1002.4
RELATIVE HUMIDITY (%)														
Hour 01	29	73	70	71	73	84	86	88	89	88	85	79	75	80
Hour 07 (Local Time)	29	79	77	80	81	86	87	89	92	92	89	84	80	85
Hour 13	29	55	52	49	45	54	56	58	60	58	53	52	55	54
Hour 19	29	63	58	56	53	66	67	70	74	77	76	67	66	66
PRECIPITATION (inches):														
Water Equivalent														
-Normal		3.48	3.69	3.77	2.59	3.92	3.68	4.01	4.02	3.19	2.86	2.98	3.24	41.43
-Maximum Monthly	49	7.52	6.42	7.78	6.10	7.67	9.38	10.27	12.18	12.94	7.53	8.22	6.65	12.94
-Year		1954	1989	1983	1978	1974	1973	1991	1986	1945	1971	1948	1983	SEP 1945
-Minimum Monthly	49	0.87	0.69	1.03	0.23	0.92	0.33	0.80	0.81	0.23	0.44	0.61	0.25	0.23
-Year		1981	1991	1985	1976	1964	1993	1953	1950	1985	1963	1973	1965	SEP 1985
-Maximum in 24 hrs	49	3.11	3.22	3.70	4.04	4.40	3.44	3.89	5.20	5.16	4.10	4.70	3.18	5.20
-Year		1984	1973	1983	1978	1957	1967	1952	1955	1944	1954	1963	1958	AUG 1955
Snow,Ice pellets,hail														
-Maximum Monthly	49	14.4	17.2	14.0	1.8	0.0	0.0	T	0.0	0.0	0.0	2.6	10.6	17.2
-Year		1955	1979	1960	1983			1993				1975	1958	FEB 1979
-Maximum in 24 hrs	49	9.0	10.4	9.3	1.8	0.0	0.0	T	0.0	0.0	0.0	2.6	9.1	10.4
-Year		1966	1979	1969	1983			1993				1975	1958	FEB 1979
WIND:														
Mean Speed (mph)	44	8.5	8.9	9.3	9.0	7.7	7.0	6.7	6.4	6.8	7.1	7.6	8.0	7.7
Prevailing Direction through 1963		SW	SW	SW	SW	SW	SW	SW	NE	NE	NNE	SW	SW	SW
Fastest Obs. 1 Min.														
-Direction (!!)	40	27	12	32	14	20	33	17	33	23	29	32	21	17
-Speed (MPH)	40	41	44	44	40	54	39	153	46	35	73	35	35	153
-Year		1971	1984	1967	1961	1972	1977	1993	1969	1972	1954	1969	1968	JUL 1993
Peak Gust														
-Direction (!!)	10	NW	SE	W	NW	W	N	N	SW	S	SE	S	NW	SE
-Speed (mph)	10	48	62	52	56	55	48	45	61	46	36	41	45	62
-Date		1989	1984	1993	1987	1984	1993	1991	1986	1989	1990	1989	1987	FEB 1984

PRECIPITATION (inches)

RALEIGH, NORTH CAROLINA

YEAR	JAN	FEB	MAR	APR	MAY	JUNE	JULY	AUG	SEP	OCT	NOV	DEC	ANNUAL
1964	3.66	4.11	2.93	3.39	0.92	3.41	4.06	5.68	5.29	3.95	1.38	4.13	42.91
1965	1.47	2.40	4.08	1.51	2.20	8.32	5.54	3.00	2.65	1.77	1.23	0.25	34.42
1966	5.42	4.76	1.81	2.02	4.95	3.68	0.91	5.79	3.58	2.01	2.06	2.61	39.60
1967	1.64	3.80	1.62	3.02	4.15	4.57	3.49	6.22	1.74	2.26	2.14	4.93	39.58
1968	2.88	1.00	2.22	3.03	3.82	1.74	5.15	2.50	1.77	5.15	3.59	2.75	35.60
1969	1.55	3.60	3.95	1.43	2.85	4.81	4.40	6.31	6.21	2.09	1.01	3.31	41.52
1970	2.26	3.47	4.04	2.07	3.36	0.87	5.64	4.47	1.20	4.47	1.59	2.57	36.01
1971	3.28	3.85	3.69	2.59	4.68	2.79	4.56	6.26	2.91	7.53	1.81	1.69	45.64
1972	1.97	4.13	2.50	1.92	5.34	4.16	6.80	4.17	5.80	3.96	5.98	5.01	51.74
1973	2.67	5.50	4.06	4.40	3.99	9.38	3.12	4.60	1.13	0.60	0.61	6.38	46.44
1974	4.39	2.87	3.34	1.32	7.67	4.02	1.56	4.82	3.71	1.23	1.79	4.02	40.74
1975	6.09	2.85	6.26	1.64	3.84	1.66	6.74	2.11	5.77	1.23	4.60	4.04	46.83
1976	3.07	1.54	3.17	0.23	4.74	2.55	1.00	1.52	5.99	3.97	1.89	4.04	33.71
1977	2.82	2.13	5.63	1.89	3.94	0.84	0.89	4.12	3.86	5.06	2.22	3.70	37.10
1978	7.03	1.43	4.40	6.10	4.20	4.06	3.63	1.86	1.37	1.46	4.17	3.26	42.97
#1979	5.71	5.55	2.69	2.63	4.71	3.27	4.84	1.66	6.76	1.88	4.73	0.94	45.37
1980	4.39	1.91	5.87	1.97	2.33	4.89	2.11	1.87	3.76	2.25	2.87	1.42	35.64
1981	0.87	3.02	2.35	1.03	4.28	0.55	5.69	5.34	2.70	4.64	0.95	4.96	36.38
1982	3.43	4.97	3.02	3.33	4.20	8.39	3.34	1.83	1.55	3.93	2.34	4.02	44.35
1983	1.79	6.00	7.78	3.54	5.89	3.09	1.10	1.81	2.13	3.59	3.86	6.65	47.23
1984	4.93	5.65	5.40	4.45	5.43	3.08	9.20	1.13	2.31	0.73	1.64	2.32	46.27
1985	4.83	4.44	1.03	0.64	3.95	2.87	6.28	3.73	0.23	1.75	7.61	0.81	38.17
1986	1.88	1.65	3.06	1.01	2.98	1.92	4.32	12.18	0.95	1.28	2.77	2.95	36.95
1987	6.53	5.52	2.88	4.68	1.19	2.11	1.78	5.80	5.48	1.71	1.39	3.02	42.09
1988	3.15	2.42	1.76	3.56	2.85	2.88	2.69	3.40	4.90	5.67	3.34	1.04	37.66
1989	1.35	6.42	5.40	4.91	3.88	7.30	5.46	5.08	3.96	3.44	3.94	3.01	54.15
1990	3.07	3.82	5.02	2.19	6.97	1.03	2.22	2.65	0.30	5.69	1.51	3.08	37.55
1991	4.12	0.69	4.59	1.04	2.89	2.05	10.27	1.87	3.16	1.40	0.73	2.65	35.46
1992	3.80	2.23	2.95	1.93	2.60	5.12	3.45	7.63	2.22	3.79	5.02	2.44	43.18
1993	4.50	2.22	6.13	4.84	3.32	0.33	2.11	1.77	3.50	2.95	2.66	3.72	38.05
Record Mean	3.43	3.67	3.75	3.21	3.88	4.06	5.15	4.99	3.61	2.91	2.63	3.24	44.52

AVERAGE TEMPERATURE (deg. F)

RALEIGH, NORTH CAROLINA

YEAR	JAN	FEB	MAR	APR	MAY	JUNE	JULY	AUG	SEP	OCT	NOV	DEC	ANNUAL
#1964	41.2	39.4	50.6	59.6	68.1	75.6	76.4	74.9	69.3	55.2	52.7	43.3	58.9
1965	40.5	42.9	44.6	58.3	71.9	71.9	76.2	77.2	72.7	58.8	50.9	43.8	59.1
1966	35.8	42.3	49.7	57.0	66.5	73.4	79.1	77.7	71.0	58.6	50.8	41.9	58.7
1967	44.8	38.9	53.7	62.5	63.9	72.0	76.9	76.3	67.5	59.4	46.8	46.3	59.1
1968	37.1	36.4	52.6	58.1	63.9	73.6	76.9	79.2	70.9	62.0	51.7	38.8	58.4
1969	37.4	40.9	43.4	58.7	66.2	74.7	78.0	73.8	67.5	58.6	46.1	37.7	56.9
1970	32.9	38.9	46.2	59.4	66.0	72.7	76.1	75.3	73.3	61.0	49.4	42.7	57.8
1971	37.0	42.1	45.1	56.4	64.4	75.4	76.6	75.3	72.0	64.8	48.7	50.3	59.0
1972	44.7	40.0	49.5	58.0	64.3	69.9	77.1	75.6	70.4	57.4	48.1	46.2	58.5
1973	39.3	40.0	54.8	57.9	64.5	75.1	76.5	76.6	73.2	62.2	54.6	42.5	59.8
1974	49.3	42.9	54.4	60.2	67.4	71.8	76.5	76.0	69.0	56.5	48.7	43.2	59.7
1975	43.8	43.7	47.1	55.8	68.2	73.7	75.6	78.3	71.1	62.2	53.5	42.0	59.6
1976	37.6	50.1	56.2	60.7	66.8	74.7	78.6	75.4	69.7	55.7	42.5	37.1	58.7
1977	26.6	39.3	53.3	62.6	68.4	73.3	80.6	78.1	72.8	56.4	51.8	39.9	58.6
1978	35.3	33.1	48.2	59.2	66.0	75.8	78.0	79.9	73.7	59.5	55.1	44.9	59.1
1979	39.1	36.5	52.2	59.7	66.8	69.8	75.4	77.2	71.2	59.9	52.0	43.4	58.6
1980	40.6	36.5	46.5	62.0	69.8	75.0	78.9	79.6	74.9	58.7	49.1	40.7	59.3
1981	33.4	43.9	46.2	61.7	64.0	78.9	80.8	74.6	68.3	57.2	50.7	39.7	58.3
1982	35.5	45.5	51.7	57.4	71.0	74.7	79.1	76.5	70.5	60.6	51.9	47.5	60.2
1983	38.1	40.7	50.7	55.1	65.4	72.5	79.1	79.1	70.7	60.4	50.9	39.5	58.5
1984	36.3	45.7	47.2	55.9	65.5	75.5	74.9	76.6	67.5	66.3	47.1	49.7	59.0
1985	34.0	41.9	52.7	62.0	67.3	73.9	76.8	75.2	69.7	63.7	58.4	39.4	59.6
1986	38.5	44.5	51.7	61.2	67.4	78.4	81.7	75.6	72.3	63.0	52.9	42.6	60.8
1987	38.3	40.4	49.2	56.9	69.3	76.3	81.2	79.2	73.0	54.6	52.8	44.5	59.6
1988	34.7	41.9	50.6	57.8	66.0	72.4	79.0	80.3	70.3	54.4	52.0	42.4	58.5
1989	44.8	42.9	50.7	58.0	65.0	77.0	78.1	76.2	71.7	61.3	51.4	34.6	59.3
1990	48.0	51.0	54.9	60.3	67.4	75.2	80.0	78.0	72.2	64.1	54.2	48.1	62.8
1991	41.9	46.8	54.3	62.3	72.3	75.8	80.6	77.8	71.6	61.1	50.0	47.0	61.8
1992	43.4	46.2	50.5	59.2	63.2	72.0	80.4	74.1	71.5	58.1	51.9	42.5	59.4
1993	43.3	40.9	47.9	56.9	69.0	76.1	82.5	78.2	73.7	59.6	51.8	40.3	60.0
Record Mean	41.2	42.9	50.5	59.3	67.9	75.2	78.3	77.2	71.7	60.7	50.8	42.8	59.8
Max	50.6	53.0	61.4	70.9	79.0	85.7	88.2	86.7	81.5	71.5	61.3	52.3	70.2
Min	31.7	32.8	39.5	47.6	56.8	64.7	68.5	67.6	61.8	49.9	40.3	33.2	49.5

REFERENCE NOTES

RALEIGH, NORTH CAROLINA

GENERAL
T - TRACE AMOUNT.
BLANK ENTRIES DENOTE MISSING/UNREPORTED DATA.
# INDICATES A STATION OR INSTRUMENT RELOCATION.

SPECIFIC

PM - INCLUDES LAST DAY OF PREVIOUS MONTH
ASOS - AUTOMATED SURFACE OBSERVING SYSTEM IN OPERATION DURING THESE MONTHS.

(a) - LENGTH OF RECORD IN YEARS, ALTHOUGH INDIVIDUAL MONTHS MAY BE MISSING.
0.* OR * - THE VALUE IS BETWEEN 0.0 AND 0.05
NORMALS - BASED ON THE 1961-1990 RECORD PERIOD.
EXTREMES - DATES ARE THE MOST RECENT OCCURRENCE
WIND DIR. - NUMERALS SHOW TENS OF DEGRESS CLOCKWISE FROM TRUE NORTH. "00" INDICATES CALM.
RESULTANT DIRECTIONS ARE GIVEN TO WHOLE DEGREES.
BOLD VALUES INDICATE EXTREME VALUES WHICH OCCURRED AFTER THE ASOS SYSTEM WAS COMMISSIONED.

RECORD = PERIOD OF RECORD
RECORD MEAN PRECIPITATION IS THE MEAN OF ALL DAILY PRECIPITATION AMOUNTS DURING THE PERIOD OF RECORD.
RECORD MAX(MIN) TEMPERATURE IS THE MEAN OF ALL DAILY MAX(MIN) TEMPERATURES DURING THE PERIOD OF RECORD.
RECORD MEAN TEMPERATURE IS THE SUM OF THE RECORD MAX AND RECORD MIN DIVIDED BY 2.
AVERAGE TEMPERATURE IS THE SUM OF THE MEAN DAILY MAX AND MIN TEMPERATURE DIVIDED BY 2.

EXCEPTIONS

RECORD MEANS ARE THROUGH THE CURRENT YEAR,
BEGINNING IN 1887 FOR TEMPERATURE
1887 FOR PRECIPITATION
1945 FOR SNOWFALL

HEATING DEGREE DAYS Base 65 deg. F RALEIGH, NORTH CAROLINA

SEASON	JULY	AUG	SEP	OCT	NOV	DEC	JAN	FEB	MAR	APR	MAY	JUNE	TOTAL
#1964-65	0	1	25	305	365	665	754	616	628	232	0	15	3606
1965-66	0	5	7	208	416	650	898	629	473	263	72	12	3633
1966-67	0	0	11	216	425	709	623	723	361	151	105	13	3337
1967-68	0	0	35	199	539	574	855	824	391	213	87	0	3717
1968-69	0	0	0	151	396	805	848	667	667	195	61	0	3790
1969-70	0	0	45	222	561	841	989	725	576	200	74	0	4233
1970-71	0	0	22	154	460	684	863	636	611	258	87	0	3775
1971-72	0	0	3	61	496	456	623	718	478	237	51	16	3139
1972-73	0	0	9	238	504	576	790	692	334	231	88	0	3462
1973-74	0	0	2	126	312	690	481	614	346	187	48	0	2806
1974-75	0	0	44	268	501	668	651	589	553	293	34	0	3601
1975-76	0	0	17	117	351	705	843	426	300	194	52	6	3011
1976-77	0	0	7	302	668	857	1183	715	358	132	49	14	4285
1977-78	0	0	4	283	411	768	914	883	514	196	83	0	4056
1978-79	0	0	7	184	292	627	793	792	398	183	43	8	3327
1979-80	0	0	13	196	394	661	753	820	564	130	33	0	3564
1980-81	0	0	16	225	477	747	973	583	579	149	99	0	3848
1981-82	0	4	31	253	425	776	907	538	411	244	15	0	3604
1982-83	0	0	14	182	392	542	828	675	438	305	79	7	3462
1983-84	0	0	59	180	417	784	882	553	545	283	83	5	3791
1984-85	0	0	63	42	530	468	954	644	395	146	42	4	3288
1985-86	0	0	36	96	207	789	812	569	415	157	59	0	3140
1986-87	0	11	12	149	370	687	820	681	484	248	29	0	3491
1987-88	0	0	1	319	362	631	932	665	444	228	62	22	3666
1988-89	0	0	8	336	386	695	619	623	459	257	102	0	3485
1989-90	0	3	30	167	404	934	518	390	357	186	37	0	3026
1990-91	0	0	18	124	323	520	709	501	354	153	18	0	2720
1991-92	0	0	24	156	451	562	659	537	446	226	114	3	3178
1992-93	0	0	29	224	390	691	666	670	524	244	15	0	3453
1993-94	0	0	19	198	405	758							

COOLING DEGREE DAYS Base 65 deg. F RALEIGH, NORTH CAROLINA

YEAR	JAN	FEB	MAR	APR	MAY	JUNE	JULY	AUG	SEP	OCT	NOV	DEC	TOTAL
1969	0	0	0	12	105	295	413	283	128	28	0	0	1264
1970	0	0	0	41	113	236	350	327	278	38	0	0	1383
1971	0	0	0	6	78	320	364	329	222	61	17	5	1402
1972	0	0	5	37	36	170	382	336	177	6	6	2	1157
1973	0	0	24	26	81	310	363	365	254	48	7	0	1478
1974	0	0	25	51	130	210	363	347	169	9	21	0	1325
1975	0	0	3	22	141	269	337	421	209	38	12	0	1452
1976	0	3	31	71	116	304	428	330	157	19	0	0	1459
1977	0	2	4	68	162	272	490	414	245	25	19	0	1701
1978	0	0	2	30	120	330	412	468	275	21	3	10	1671
1979	0	0	6	28	105	159	332	384	205	46	10	0	1275
1980	0	0	0	45	190	306	441	460	321	38	6	0	1807
1981	0	0	2	56	75	425	497	309	139	19	0	0	1522
1982	0	0	3	24	208	299	443	363	183	53	5	7	1588
1983	0	0	0	16	97	238	441	447	239	42	0	0	1520
1984	0	0	0	16	108	324	311	366	143	90	0	0	1358
1985	0	3	20	65	121	277	373	323	181	64	14	0	1441
1986	0	0	7	51	142	408	526	349	237	96	15	0	1831
1987	0	0	0	11	170	347	508	447	250	0	2	0	1735
1988	0	3	5	17	98	249	438	482	172	14	3	0	1481
1989	0	11	23	54	110	367	412	359	237	59	4	0	1636
1990	0	3	49	51	117	312	472	410	239	102	4	5	1764
1991	0	0	28	78	253	333	493	403	230	40	9	13	1880
1992	0	0	3	58	65	218	487	285	228	15	3	0	1362
1993	0	0	0	12	147	338	550	417	286	34	17	0	1801

SNOWFALL (inches) RALEIGH, NORTH CAROLINA

SEASON	JULY	AUG	SEP	OCT	NOV	DEC	JAN	FEB	MAR	APR	MAY	JUNE	TOTAL
1964-65	0.0	0.0	0.0	0.0	0.4	0.0	9.7	3.4	T	0.0	0.0	0.0	13.5
1965-66	0.0	0.0	0.0	0.0	0.0	0.0	12.3	T	0.0	0.0	0.0	0.0	12.3
1966-67	0.0	0.0	0.0	0.0	T	1.0	0.5	9.1	0.0	0.0	0.0	0.0	10.6
1967-68	0.0	0.0	0.0	0.0	T	1.4	3.0	1.3	T	0.0	0.0	0.0	5.7
1968-69	0.0	0.0	0.0	0.0	1.2	0.7	T	0.8	9.3	0.0	0.0	0.0	12.0
1969-70	0.0	0.0	0.0	0.0	0.0	0.0	2.0	T	T	0.0	0.0	0.0	2.0
1970-71	0.0	0.0	0.0	0.0	0.0	0.6	T	T	5.3	0.0	0.0	0.0	5.9
1971-72	0.0	0.0	0.0	0.0	T	3.7	0.0	1.4	2.6	0.0	0.0	0.0	7.7
1972-73	0.0	0.0	0.0	0.0	T	0.0	6.4	4.5	0.4	0.0	0.0	0.0	11.3
1973-74	0.0	0.0	0.0	0.0	0.0	2.8	0.0	T	2.9	0.0	0.0	0.0	5.7
1974-75	0.0	0.0	0.0	0.0	0.0	T	T	T	0.6	0.0	0.0	0.0	0.6
1975-76	0.0	0.0	0.0	0.0	2.6	T	0.4	T	0.0	0.0	0.0	0.0	3.0
1976-77	0.0	0.0	0.0	0.0	T	T	2.1	1.5	0.0	0.0	0.0	0.0	3.6
1977-78	0.0	0.0	0.0	0.0	T	T	T	9.0	1.6	0.0	0.0	0.0	10.6
1978-79	0.0	0.0	0.0	0.0	0.0	0.0	0.4	17.2	T	0.0	0.0	0.0	17.6
#1979-80	0.0	0.0	0.0	0.0	0.0	0.0	2.2	5.0	11.1	0.0	0.0	0.0	18.3
1980-81	0.0	0.0	0.0	0.0	0.0	3.1	2.6	0.0	T	0.0	0.0	0.0	5.7
1981-82	0.0	0.0	0.0	0.0	0.0	T	6.0	0.6	0.0	0.0	0.0	0.0	6.6
1982-83	0.0	0.0	0.0	0.0	0.0	T	T	2.7	7.3	1.8	0.0	0.0	11.8
1983-84	0.0	0.0	0.0	0.0	0.0	0.0	T	6.9	T	0.0	0.0	0.0	6.9
1984-85	0.0	0.0	0.0	0.0	0.0	0.0	4.1	T	0.0	0.0	0.0	0.0	4.1
1985-86	0.0	0.0	0.0	0.0	0.0	T	T	0.9	T	0.0	0.0	0.0	0.9
1986-87	0.0	0.0	0.0	0.0	T	0.0	0.6	10.2	T	T	0.0	0.0	10.8
1987-88	0.0	0.0	0.0	0.0	0.6	0.0	7.3	T	0.0	0.0	0.0	0.0	7.9
1988-89	0.0	0.0	0.0	0.0	0.0	0.1	0.0	11.1	0.5	0.3	0.0	0.0	12.0
1989-90	0.0	0.0	0.0	0.0	0.0	2.7	0.0	T	T	0.0	0.0	0.0	2.7
1990-91	0.0	0.0	0.0	0.0	0.0	T	T	T	T	0.0	0.0	0.0	T
1991-92	0.0	0.0	0.0	0.0	T	0.0	T	T	0.0	0.0	0.0	0.0	T
1992-93	0.0	0.0	0.0	0.0	0.0	T	T	1.6	0.9	0.0	0.0	0.0	2.5
1993-94	T	0.0	0.0	0.0	0.0	3.1							
Record Mean	T	0.0	0.0	0.0	0.1	0.8	2.2	2.6	1.3	T	0.0	0.0	7.1

WILMINGTON, NORTH CAROLINA

Wilmington is located in the tidewater section of southeastern North Carolina, near the Atlantic Ocean. The city proper is built adjacent to the east bank of the Cape Fear River. Because of the curvature of the coastline in this area, the ocean lies about 5 miles east and about 20 miles south. The surrounding terrain is typical of coastal Carolina. It is low-lying with an average elevation of less than 40 feet, and is characterized by level to gently rolling land with rivers, creeks, and lakes that frequently have considerable swamp or marshland adjoining them. Large wooded areas alternate with cultivated fields.

The maritime location makes the climate of Wilmington unusually mild for its latitude. All wind directions from the east-northeast through southwest have some moderating effects on temperatures throughout the year, because the ocean is relatively warm in winter and cool in summer. The daily range in temperatures is moderate compared to a continental type of climate. As a rule, summers are quite warm and humid, but excessive heat is rare. Sea breezes, arriving early in the afternoon, tend to alleviate the heat further inland. Long-term averages show afternoon temperatures reach 90 degrees or higher on one-third of the days in midsummer, but several years may pass without 100 degree weather. During the colder part of the year, numerous outbreaks of polar air masses reach the Atlantic Coast, causing sharp drops in temperatures. However, these cold outbreaks are significantly moderated by the long trajectories from the source regions, the effects of passing over the Appalachian Range, and the warming effects of the ocean air. As a result, most winters are short and quite mild. Even in the most severe cold spells, the temperature usually remains above zero. Normally, the temperature fails to rise above the freezing point during a 24-hour period only once each winter.

Rainfall in this area is usually ample and well-distributed throughout the year, the greatest amount occurring in the summer. Summer rainfall comes principally from thunderstorms, and is therefore usually of short duration, but often heavy and unevenly distributed. Thunderstorms occur about one out of three days from June through August. Winter rain is more likely to be of the slow, steady type, lasting one or two days. Generally, the winter rain is evenly distributed and associated with slow-moving, low-pressure systems. Seldom is there a winter without a few flakes of snow, but several years may pass without a measurable amount, and appreciable accumulation on the ground is rare. Hail occurs less than once a year. Sunshine is abundant, with the area receiving about two-thirds of the sunshine hours possible at its latitude.

Because of these many factors, the growing season is long, averaging 244 days, but records show the range is from 180 days to as long as 302 days. This area is exceptionally good for floriculture. Agricultural pursuits, principally field-grown flowers, nursery plantings, and vegetables, are an important part of the economy. Some types of plants continue to grow throughout the year.

In common with most Atlantic Coastal localities, the area is subject to the effects of coastal storms and occasional hurricanes which produce high winds, above normal tides, and heavy rains.

NORMALS, MEANS, AND EXTREMES

WILMINGTON, NORTH CAROLINA

LATITUDE: 34°16'N LONGITUDE: 77°54' W ELEVATION: FT. GRND 30 BARO 34 TIME ZONE: EASTERN WBAN: 13748

	(a)	JAN	FEB	MAR	APR	MAY	JUNE	JULY	AUG	SEP	OCT	NOV	DEC	YEAR
TEMPERATURE °F:														
Normals														
-Daily Maximum		55.3	58.1	65.7	74.0	80.8	85.4	88.5	87.6	85.2	76.9	69.1	59.4	73.8
-Daily Minimum		34.4	36.4	43.1	50.5	59.3	67.5	71.7	71.0	65.3	53.7	44.8	37.5	52.9
-Monthly		44.9	47.3	54.4	62.3	70.1	76.5	80.1	79.4	75.3	65.3	57.0	48.5	63.4
Extremes														
-Record Highest	42	82	85	89	95	98	104	102	102	98	95	87	81	104
-Year		1975	1962	1974	1967	1953	1952	1977	1954	1975	1986	1974	1984	JUN 1952
-Record Lowest	42	5	11	9	30	38	48	55	55	44	27	20	0	0
-Year		1985	1958	1980	1983	1989	1983	1988	1982	1981	1962	1970	1989	DEC 1989
NORMAL DEGREE DAYS:														
Heating (base 65°F)		631	496	338	115	10	0	0	0	0	102	259	519	2470
Cooling (base 65°F)		8	0	10	34	168	345	468	446	309	111	19	8	1926
% OF POSSIBLE SUNSHINE	42	57	59	64	71	67	66	64	61	62	65	63	59	63
MEAN SKY COVER (tenths)														
Sunrise - Sunset	42	6.1	6.0	5.8	5.3	5.9	6.1	6.3	6.3	6.0	5.1	5.1	5.8	5.8
MEAN NUMBER OF DAYS:														
Sunrise to Sunset														
-Clear	42	9.6	9.0	9.9	11.0	8.5	7.4	6.1	6.7	8.4	12.7	11.9	10.2	111.4
-Partly Cloudy	42	6.0	5.8	7.5	8.1	10.6	10.8	12.2	11.8	9.6	7.3	7.1	7.0	103.9
-Cloudy	42	15.4	13.5	13.6	10.9	11.9	11.8	12.7	12.5	12.0	11.0	11.0	13.8	150.0
Precipitation														
.01 inches or more	42	10.7	9.8	10.2	8.0	9.5	9.9	13.1	12.5	9.3	7.0	7.7	9.1	116.8
Snow,Ice pellets,hail														
1.0 inches or more	42	0.2	0.1	0.1	0.0	0.0	0.0	0.0	0.0	0.0	0.0	0.0	0.2	0.6
Thunderstorms	42	0.5	1.0	2.3	3.2	5.5	7.6	11.6	9.4	4.0	1.3	0.8	0.5	47.7
Heavy Fog Visibility														
1/4 mile or less	42	2.5	1.7	2.3	1.5	2.1	1.5	0.9	1.4	2.4	2.5	2.8	2.7	24.4
Temperature °F														
-Maximum														
90° and above	30	0.0	0.0	0.0	0.9	2.2	8.3	16.5	12.9	4.7	0.2	0.0	0.0	45.8
32° and below	30	0.5	0.1	0.*	0.0	0.0	0.0	0.0	0.0	0.0	0.0	0.0	0.1	0.7
-Minimum														
32° and below	30	13.3	10.6	4.0	0.3	0.0	0.0	0.0	0.0	0.0	0.1	3.0	10.8	42.0
0° and below	30	0.0	0.0	0.0	0.0	0.0	0.0	0.0	0.0	0.0	0.0	0.0	0.*	*
AVG. STATION PRESS.(mb)	21	1018.6	1018.0	1016.3	1015.2	1015.0	1015.2	1015.9	1016.5	1016.8	1017.9	1018.7	1019.4	1017.0
RELATIVE HUMIDITY (%)														
Hour 01	30	78	77	79	80	87	89	89	91	90	87	83	80	84
Hour 07 (Local Time)	30	80	78	82	80	85	85	87	90	90	88	85	82	84
Hour 13	30	56	52	52	48	56	59	63	65	62	56	54	55	57
Hour 19	30	71	66	67	64	71	73	76	80	81	80	77	74	73
PRECIPITATION (inches):														
Water Equivalent														
-Normal		3.87	3.70	3.88	2.87	4.43	5.98	8.13	6.94	5.04	2.69	3.11	3.63	54.27
-Maximum Monthly	42	10.22	8.74	8.09	8.21	9.12	12.87	15.12	14.06	18.94	9.81	7.87	7.06	18.94
-Year		1991	1983	1983	1961	1956	1962	1966	1981	1984	1964	1972	1989	SEP 1984
-Minimum Monthly	42	1.09	1.01	0.93	0.33	0.95	0.89	1.65	1.66	0.70	0.17	0.49	0.48	0.17
-Year		1981	1976	1967	1957	1987	1984	1961	1968	1986	1953	1973	1955	OCT 1953
-Maximum in 24 hrs	42	3.08	3.20	3.31	3.52	4.95	7.73	6.58	5.15	8.24	4.34	4.82	3.88	8.24
-Year		1982	1983	1960	1961	1963	1966	1988	1986	1958	1964	1969	1980	SEP 1958
Snow,Ice pellets,hail														
-Maximum Monthly	42	5.4	12.5	6.6	T	0.0	T	0.0	0.0	0.0	0.0	T	15.3	15.3
-Year		1988	1973	1980	1989		1993					1976	1989	DEC 1989
-Maximum in 24 hrs	42	5.0	11.7	5.7	T	0.0	T	0.0	0.0	0.0	0.0	T	9.7	11.7
-Year		1988	1973	1980	1989		1993					1976	1989	FEB 1973
WIND:														
Mean Speed (mph)	42	9.0	9.7	10.1	10.2	9.1	8.5	7.9	7.4	7.8	8.0	8.1	8.4	8.7
Prevailing Direction through 1963		N	NW	SSW	SSW	SSW	SSW	SSW	SW	N	N	N	N	N
Fastest Obs. 1 Min.														
-Direction (!!)	13	32	21	29	21	04	23	23	24	08	01	10	06	29
-Speed (MPH)	13	36	31	58	38	28	31	45	35	46	35	29	31	58
-Year		1989	1984	1981	1982	1992	1993	1986	1982	1984	1982	1985	1980	MAR 1981
Peak Gust														
-Direction (!!)	10	SW	SW	S	NW	SW	SW	SW	S	E	SW	SW	SW	SW
-Speed (mph)	10	49	49	70	53	45	52	78	41	74	41	45	49	78
-Date		1987	1984	1993	1989	1985	1993	1986	1986	1984	1993	1988	1993	JUL 1986

PRECIPITATION (inches)

WILMINGTON, NORTH CAROLINA

YEAR	JAN	FEB	MAR	APR	MAY	JUNE	JULY	AUG	SEP	OCT	NOV	DEC	ANNUAL
1964	7.08	6.17	2.89	1.32	5.11	5.02	7.45	3.69	3.58	9.81	1.17	4.54	57.83
1965	1.29	4.83	5.72	2.40	2.76	9.38	12.09	10.58	2.90	3.44	1.12	0.81	57.32
1966	6.32	5.54	2.89	1.48	7.50	9.78	15.12	5.67	6.09	1.10	0.70	3.44	65.63
1967	3.89	4.14	0.93	1.15	2.58	6.03	5.78	8.13	1.55	1.07	2.24	4.67	42.16
1968	3.71	1.44	0.97	3.50	2.30	2.52	9.31	1.66	1.24	4.46	4.40	2.26	37.77
1969	2.80	2.53	4.60	3.41	7.32	9.31	13.46	4.69	2.59	0.95	5.23	3.86	60.75
1970	1.98	2.45	7.19	1.37	3.92	4.48	5.71	13.98	2.23	4.12	2.22	3.22	52.87
1971	4.97	3.51	4.57	3.46	2.29	5.60	8.08	10.42	5.04	6.05	2.22	1.49	57.70
1972	4.27	4.57	3.36	0.78	3.50	3.50	6.36	4.91	6.73	0.69	7.87	5.29	51.83
1973	4.37	4.85	3.48	5.32	3.48	10.70	9.33	6.42	5.60	1.52	0.49	5.60	61.16
1974	2.81	4.23	2.30	2.18	5.15	5.37	8.36	13.33	5.09	1.12	3.14	3.72	56.80
1975	5.06	5.09	3.15	3.90	3.11	7.11	9.76	4.52	6.06	3.14	3.00	5.20	59.10
1976	2.93	1.01	2.57	0.91	4.33	12.74	8.28	9.53	4.79	3.31	2.14	5.37	57.91
1977	2.94	1.83	5.66	1.43	6.75	4.26	3.25	5.85	4.88	5.73	6.33	2.83	51.74
1978	6.68	1.33	2.94	3.82	3.46	2.34	7.38	4.77	1.58	1.01	3.71	4.68	43.70
#1979	6.23	4.21	4.82	5.60	5.27	4.74	2.82	2.48	15.17	0.38	2.01	2.35	56.08
1980	4.16	1.52	6.03	1.33	4.65	2.46	5.95	3.21	5.97	1.62	1.87	5.81	44.58
1981	1.09	3.08	3.02	1.43	5.02	2.45	5.23	14.06	1.07	1.39	0.78	5.76	44.38
1982	5.50	6.67	1.84	4.03	2.04	7.59	8.59	3.67	7.08	2.56	1.32	6.57	57.46
1983	4.90	8.74	8.09	2.09	1.13	6.71	5.53	5.63	5.59	1.02	4.49	5.20	59.12
1984	2.58	4.82	4.43	3.23	6.45	0.89	9.01	4.79	18.94	0.49	1.16	1.32	58.11
1985	2.01	5.08	1.66	0.71	2.76	4.56	10.34	3.63	2.75	2.43	6.74	1.35	44.02
1986	2.12	2.52	4.13	0.48	7.09	5.58	11.28	11.44	0.70	3.35	4.44	6.28	59.41
1987	6.49	4.42	2.70	2.96	0.95	5.24	5.19	9.35	6.42	0.51	5.67	1.35	51.25
1988	5.41	2.00	4.05	3.56	7.54	2.93	14.49	9.61	2.80	1.81	3.14	0.59	57.93
1989	1.60	2.64	6.70	7.60	3.58	7.55	9.93	3.93	9.54	4.61	1.91	7.06	66.65
1990	2.34	2.31	5.11	2.21	8.04	2.17	6.59	11.42	1.40	7.10	2.09	2.65	53.43
1991	10.22	1.65	6.52	3.35	1.75	2.57	13.35	8.65	3.31	4.19	2.98	2.17	60.71
1992	5.75	3.45	5.65	3.12	7.55	7.00	3.18	13.56	2.23	3.09	7.14	4.48	66.20
1993	5.83	2.96	6.34	5.78	2.82	3.70	7.04	5.66	8.09	8.29	2.58	2.67	61.76
Record Mean	3.46	3.47	3.71	2.87	3.82	5.17	7.50	6.83	5.46	3.18	2.65	3.30	51.42

AVERAGE TEMPERATURE (deg. F)

WILMINGTON, NORTH CAROLINA

YEAR	JAN	FEB	MAR	APR	MAY	JUNE	JULY	AUG	SEP	OCT	NOV	DEC	ANNUAL
1964	46.7	45.8	54.2	63.0	71.3	77.9	78.3	78.1	74.8	62.0	59.2	52.2	63.6
1965	46.2	48.3	51.7	62.4	74.4	76.2	79.1	80.2	75.4	64.8	57.5	48.8	63.8
1966	43.2	48.0	53.3	62.2	69.6	74.8	80.9	80.1	75.8	66.8	56.1	47.8	63.2
1967	50.4	47.5	58.0	66.0	71.3	76.3	81.1	80.1	72.0	64.9	53.0	52.1	64.4
1968	42.2	40.3	54.5	63.2	69.6	78.3	80.5	81.9	73.6	65.2	54.7	45.0	62.4
1969	43.2	45.1	50.0	63.1	68.8	77.4	80.8	76.9	72.2	65.5	52.5	44.9	61.7
1970	39.2	45.5	53.6	64.0	70.8	76.2	80.2	79.7	76.7	67.4	55.1	49.9	63.2
1971	44.8	48.0	50.2	60.4	68.8	78.0	79.8	78.8	75.4	70.8	55.1	56.6	63.9
1972	52.0	47.0	53.8	61.3	67.2	72.7	80.1	79.3	74.7	64.3	55.4	51.9	63.3
1973	46.5	45.3	58.8	60.8	69.5	76.6	79.8	80.1	77.2	67.3	58.2	50.2	64.2
1974	58.8	49.5	59.7	64.4	70.7	75.5	79.1	79.3	74.4	61.7	54.8	49.5	64.8
1975	51.0	52.1	54.4	62.3	73.8	78.7	79.9	82.4	77.0	68.2	59.7	49.4	65.7
1976	45.1	55.3	60.8	63.9	69.6	76.0	81.7	77.7	74.0	60.5	49.7	46.4	63.4
1977	35.8	45.6	57.9	65.9	71.4	77.7	82.6	80.5	77.8	62.0	58.8	47.6	63.6
1978	41.9	38.5	51.3	64.2	69.8	77.1	80.3	82.1	76.8	63.8	60.2	49.4	62.9
1979	43.7	43.2	53.9	65.6	71.4	74.7	80.4	80.6	74.5	63.3	57.8	47.8	63.1
1980	47.2	41.7	50.4	63.2	70.5	76.9	82.9	82.2	78.7	62.9	51.9	45.0	62.8
1981	37.5	46.6	49.9	63.9	66.9	79.8	80.7	76.9	71.4	60.9	52.3	44.4	60.9
1982	41.5	50.7	55.8	58.6	71.1	75.8	79.1	77.7	72.0	62.0	57.5	52.6	62.8
1983	42.6	45.1	53.7	57.4	67.3	75.0	82.8	81.5	74.8	66.1	55.1	46.9	62.4
1984	42.1	50.4	52.9	61.0	70.0	77.3	77.9	78.3	71.0	68.7	55.0	57.5	63.5
1985	43.2	50.4	59.1	66.3	71.8	78.7	81.4	78.9	75.0	71.1	66.6	47.0	65.8
1986	44.2	51.6	55.6	64.3	71.9	79.4	84.0	78.8	76.5	67.3	59.7	49.9	65.3
1987	45.0	45.5	52.4	59.7	70.6	78.6	81.9	81.4	76.6	60.3	57.6	50.0	63.3
1988	41.2	46.4	54.2	61.5	68.8	74.3	79.9	81.1	74.4	59.2	58.0	46.6	62.1
1989	51.7	50.7	55.5	61.3	68.7	79.6	80.5	78.8	75.3	65.8	57.1	38.6	63.6
1990	52.6	55.5	58.5	63.5	71.2	77.9	81.7	80.2	75.2	68.8	57.7	54.8	66.5
1991	48.3	51.4	57.5	66.5	74.9	77.8	82.9	80.2	75.0	64.9	54.0	51.0	65.4
1992	47.0	51.6	53.9	61.5	66.9	74.3	83.0	78.5	75.0	62.2	58.7	48.3	63.4
1993	50.5	45.8	52.7	59.8	71.2	78.1	84.1	79.8	77.7	66.5	57.6	46.1	64.2
Record Mean	47.0	48.5	54.7	62.1	70.1	76.7	79.9	79.0	74.6	64.8	55.7	48.5	63.5
Max	56.5	58.2	64.5	72.0	79.3	85.2	87.9	86.9	82.9	74.4	65.8	58.3	72.7
Min	37.5	38.8	44.8	52.2	60.9	68.2	71.9	71.0	66.2	55.2	45.5	38.7	54.2

REFERENCE NOTES

WILMINGTON, NORTH CAROLINA

GENERAL
 T - TRACE AMOUNT.
BLANK ENTRIES DENOTE MISSING/UNREPORTED DATA.
 # INDICATES A STATION OR INSTRUMENT RELOCATION.

SPECIFIC

(a) - LENGTH OF RECORD IN YEARS, ALTHOUGH
 INDIVIDUAL MONTHS MAY BE MISSING.
0.* OR * - THE VALUE IS BETWEEN 0.0 AND 0.05
NORMALS - BASED ON THE 1961-1990 RECORD PERIOD.
EXTREMES - DATES ARE THE MOST RECENT OCCURRENCE.
WIND DIR. - NUMERALS SHOW TENS OF DEGRESS CLOCKWISE
 FROM TRUE NORTH. "00" INDICATES CALM.
RESULTANT DIRECTIONS ARE GIVEN TO WHOLE DEGREES.
BOLD VALUES INDICATE EXTREME VALUES WHICH OCCURRED
 AFTER THE ASOS SYSTEM WAS COMMISSIONED.

RECORD = PERIOD OF RECORD
RECORD MEAN PRECIPITATION IS THE MEAN OF ALL DAILY
PRECIPITATION AMOUNTS DURING THE PERIOD OF RECORD.
RECORD MAX(MIN) TEMPERATURE IS THE MEAN OF ALL DAILY
 MAX(MIN) TEMPERATURES DURING THE PERIOD OF RECORD.
RECORD MEAN TEMPERATURE IS THE SUM OF THE RECORD
 MAX AND RECORD MIN DIVIDED BY 2.
AVERAGE TEMPERATURE IS THE SUM OF THE MEAN DAILY
 MAX AND MIN TEMPERATURE DIVIDED BY 2.

EXCEPTIONS

FASTEST MILE WIND IS THROUGH JANUARY 1980.

RECORD MEANS ARE THROUGH THE CURRENT YEAR,
BEGINNING IN 1874 FOR TEMPERATURE
 1871 FOR PRECIPITATION
 1952 FOR SNOWFALL

HEATING DEGREE DAYS Base 65 deg. F   WILMINGTON, NORTH CAROLINA

SEASON	JULY	AUG	SEP	OCT	NOV	DEC	JAN	FEB	MAR	APR	MAY	JUNE	TOTAL
1964-65	0	0	0	145	184	415	575	465	408	131	1	0	2324
1965-66	0	0	0	102	217	494	671	470	360	147	35	2	2498
1966-67	0	0	0	57	276	530	442	485	252	79	15	7	2143
1967-68	0	0	4	88	362	402	701	709	342	106	20	0	2734
1968-69	0	0	0	103	312	609	670	554	458	102	27	0	2835
1969-70	0	0	1	83	376	614	793	539	348	113	16	0	2883
1970-71	0	0	10	43	295	461	620	469	450	160	27	0	2535
1971-72	0	0	0	11	318	276	404	516	340	167	29	4	2065
1972-73	0	0	0	81	299	399	567	546	217	154	30	0	2293
1973-74	0	0	0	39	224	459	211	428	209	113	24	0	1707
1974-75	0	0	5	149	333	473	429	362	338	160	0	0	2249
1975-76	0	0	0	40	199	478	613	294	183	122	32	0	1961
1976-77	0	0	0	172	455	569	899	541	241	78	27	1	2983
1977-78	0	0	0	144	231	537	709	736	419	89	30	0	2895
1978-79	0	0	0	91	154	489	653	606	346	60	5	0	2404
1979-80	0	0	2	110	239	525	543	670	445	120	21	0	2675
1980-81	0	0	0	128	388	613	846	510	465	101	64	0	3115
1981-82	0	0	13	149	374	630	726	396	290	204	17	0	2799
1982-83	0	0	3	145	241	386	688	552	345	231	46	2	2639
1983-84	0	0	12	66	299	551	703	417	371	147	37	1	2604
1984-85	0	0	16	18	306	243	682	420	228	80	11	0	2004
1985-86	0	0	4	17	60	559	637	371	307	92	24	0	2071
1986-87	0	0	0	79	199	469	615	542	395	181	25	0	2505
1987-88	0	0	0	152	230	461	730	537	330	139	38	7	2624
1988-89	0	0	0	202	225	563	406	415	312	187	44	0	2354
1989-90	0	0	4	74	259	810	378	282	233	114	7	0	2161
1990-91	0	0	6	77	223	327	513	382	265	63	0	0	1856
1991-92	0	0	0	78	336	447	548	383	343	167	45	0	2347
1992-93	0	0	4	126	238	509	449	530	378	169	5	0	2408
1993-94	0	0	4	57	244	580							

COOLING DEGREE DAYS Base 65 deg. F   WILMINGTON, NORTH CAROLINA

YEAR	JAN	FEB	MAR	APR	MAY	JUNE	JULY	AUG	SEP	OCT	NOV	DEC	TOTAL
1969	0	2	0	51	152	381	496	377	225	107	10	0	1801
1970	0	0	0	90	205	343	482	463	370	125	3	3	2084
1971	0	0	0	31	152	397	465	433	319	199	28	23	2047
1972	5	0	2	64	101	241	477	453	298	66	19	2	1728
1973	1	0	31	34	174	354	465	472	372	116	24	8	2051
1974	25	0	52	101	208	321	443	449	294	53	32	0	1978
1975	1	5	17	84	276	417	469	545	369	145	45	0	2373
1976	2	17	58	94	183	336	525	403	277	39	2	0	1936
1977	0	4	28	114	234	387	553	487	391	56	53	3	2310
1978	0	0	0	68	187	372	482	536	359	63	17	13	2097
1979	0	2	9	84	210	298	487	490	293	65	28	0	1966
1980	0	1	0	71	198	364	560	540	419	70	2	0	2225
1981	0	0	3	74	131	449	491	375	213	32	2	0	1770
1982	0	3	11	18	211	331	445	400	218	58	24	9	1728
1983	0	0	2	10	122	307	560	517	313	107	10	0	1948
1984	0	0	4	35	201	378	405	418	204	138	12	18	1813
1985	11	17	57	126	229	420	513	437	309	211	112	10	2452
1986	0	4	22	79	247	436	599	435	353	158	47	8	2388
1987	0	0	11	32	205	419	529	516	354	15	16	3	2100
1988	0	4	2	43	164	293	466	505	289	26	20	0	1812
1989	1	20	25	80	167	443	489	432	321	104	29	0	2111
1990	4	20	42	74	206	394	525	480	317	201	8	13	2284
1991	0	7	36	118	311	392	565	478	306	82	11	20	2326
1992	0	2	5	67	108	286	563	427	314	47	57	0	1876
1993	6	0	2	20	206	399	600	467	392	109	29	0	2230

SNOWFALL (inches)   WILMINGTON, NORTH CAROLINA

SEASON	JULY	AUG	SEP	OCT	NOV	DEC	JAN	FEB	MAR	APR	MAY	JUNE	TOTAL
1970-71	0.0	0.0	0.0	0.0	T	4.0	T	0.0	T	0.0	0.0	0.0	4.0
1971-72	0.0	0.0	0.0	0.0	0.0	T	0.0	T	0.0	0.0	0.0	0.0	T
1972-73	0.0	0.0	0.0	0.0	0.0	0.0	1.9	12.5	0.0	0.0	0.0	0.0	14.4
1973-74	0.0	0.0	0.0	0.0	0.0	0.4	0.0	0.0	0.0	0.0	0.0	0.0	0.4
1974-75	0.0	0.0	0.0	0.0	0.0	0.0	0.0	T	0.0	0.0	0.0	0.0	T
1975-76	0.0	0.0	0.0	0.0	0.0	0.0	T	0.0	0.0	0.0	0.0	0.0	T
1976-77	0.0	0.0	0.0	0.0	T	T	0.6	0.3	0.0	0.0	0.0	0.0	0.9
1977-78	0.0	0.0	0.0	0.0	0.0	T	T	T	T	0.0	0.0	0.0	T
1978-79	0.0	0.0	0.0	0.0	0.0	0.0	T	0.2	0.0	0.0	0.0	0.0	0.2
#1979-80	0.0	0.0	0.0	0.0	0.0	0.0	0.0	0.3	6.6	0.0	0.0	0.0	6.9
1980-81	0.0	0.0	0.0	0.0	0.0	0.0	1.0	0.0	0.0	0.0	0.0	0.0	1.0
1981-82	0.0	0.0	0.0	0.0	0.0	0.0	T	0.1	0.0	0.0	0.0	0.0	0.1
1982-83	0.0	0.0	0.0	0.0	0.0	0.0	0.0	0.0	4.2	0.0	0.0	0.0	4.2
1983-84	0.0	0.0	0.0	0.0	0.0	0.0	T	0.0	0.0	0.0	0.0	0.0	T
1984-85	0.0	0.0	0.0	0.0	0.0	T	T	0.0	0.0	0.0	0.0	0.0	T
1985-86	0.0	0.0	0.0	0.0	0.0	0.0	T	T	0.0	0.0	0.0	0.0	T
1986-87	0.0	0.0	0.0	0.0	0.0	0.0	T	0.0	0.0	0.0	0.0	0.0	T
1987-88	0.0	0.0	0.0	0.0	0.0	0.0	5.4	0.0	0.0	0.0	0.0	0.0	5.4
1988-89	0.0	0.0	0.0	0.0	0.0	1.7	0.0	0.8	T	T	0.0	0.0	2.5
1989-90	0.0	0.0	0.0	0.0	0.0	15.3	0.0	0.0	0.0	0.0	0.0	T	15.3
1990-91	0.0	0.0	0.0	0.0	0.0	0.0	0.0	T	0.0	0.0	0.0	0.0	T
1991-92	0.0	0.0	0.0	0.0	0.0	0.0	1.4	T	T	0.0	0.0	0.0	1.4
1992-93	0.0	0.0	0.0	0.0	0.0	0.0	0.0	T	T	0.0	0.0	T	T
1993-94	0.0	0.0	0.0	0.0	0.0	0.7							
Record Mean	0.0	0.0	0.0	0.0	T	0.6	0.4	0.5	0.4	T	0.0	T	2.0

BISMARCK, NORTH DAKOTA

Bismarck, the State Capital and County Seat of Burleigh County, is located in south-central North Dakota, near the center of North America. It is on the east bank of the Missouri River in a shallow basin 7 miles wide and 11 miles long.

The Weather Service Forecast Office is located at the Municipal Airport approximately 2 miles southeast of city center. It is almost entirely surrounded by low-lying hills. The closest hills, 3 miles to the north, and other hills 5 miles to the southeast, are about 200 to 300 feet high. West across the Missouri River the land is more hilly and 300 to 600 feet higher.

The climate is semi-arid, typically continental in character, and invigorating. Summers are warm, but there are not many hot days, and very few hot and humid days. Winters tend to be long and quite cold, but there are plenty of mild days to make winter weather pleasant much of the time. Sunshine is abundant, averaging 2,700 hours out of a possible 4,470 hours.

More than 75 percent of annual precipitation falls during the six month period from April through September, and nearly 50 percent during May, June, and July. Snow has been reported in all months except July and August. Three inches or more can be expected on about three days each year.

Most summer precipitation occurs during thunderstorms in the late afternoon and evening. Thunderstorms occur on about 34 days each year, accompanied by hail on two or three of the days. A damaging hailstorm is experienced about once every ten years. Tornadoes are rare, but damaging winds occasionally occur with the heavier thunderstorms.

The winter season usually begins in late November and continues until late March. Winter precipitation is nearly all in the form of snow and is often associated with strong winds and low temperatures. This combination produces winter storms and occasional blizzards that must never be taken lightly. A severe blizzard lasting two or three days may be expected every few years. But several times each winter storms lasting a few hours occur in which drifting snow can make travel difficult and even block roads. A stalled motorist can be in serious trouble if he is not prepared with adequate winter clothing and some kind of emergency provisions. A motorist must never leave his vehicle in a blinding snowstorm as he can easily become lost.

The temperature range from summer to winter is very large and typical of the Northern Great Plains. The average freeze-free period is 134 days, from mid-May to late September.

NORMALS, MEANS, AND EXTREMES

BISMARCK, NORTH DAKOTA

LATITUDE: 46°46'N LONGITUDE: 100°45' W ELEVATION: FT. GRND 1647 BARO 1655 TIME ZONE: CENTRAL WBAN: 24011

	(a)	JAN	FEB	MAR	APR	MAY	JUNE	JULY	AUG	SEP	OCT	NOV	DEC	YEAR
TEMPERATURE °F:														
Normals														
-Daily Maximum		20.2	26.4	38.5	54.9	67.8	77.1	84.4	82.7	70.8	58.7	39.3	24.5	53.8
-Daily Minimum		-1.7	5.1	17.8	31.0	42.2	51.6	56.4	53.9	43.1	32.5	17.8	3.3	29.4
-Monthly		9.2	15.7	28.2	43.0	55.0	64.4	70.4	68.3	57.0	45.7	28.6	14.0	41.6
Extremes														
-Record Highest	54	62	69	81	93	98	107	109	109	105	95	75	65	109
-Year		1981	1992	1946	1992	1941	1988	1973	1941	1959	1963	1978	1979	JUL 1973
-Record Lowest	54	-44	-39	-31	-12	15	30	35	33	11	-10	-30	-43	-44
-Year		1950	1982	1948	1975	1967	1969	1971	1988	1974	1991	1985	1967	JAN 1950
NORMAL DEGREE DAYS:														
Heating (base 65°F)		1730	1380	1141	660	324	116	15	69	262	598	1092	1581	8968
Cooling (base 65°F)		0	0	0	0	14	98	183	171	22	0	0	0	488
% OF POSSIBLE SUNSHINE	54	54	54	59	59	62	64	74	72	65	58	44	47	59
MEAN SKY COVER (tenths)														
Sunrise - Sunset	54	6.7	6.8	7.0	6.7	6.5	6.0	4.9	4.9	5.4	5.9	7.0	6.8	6.2
MEAN NUMBER OF DAYS:														
Sunrise to Sunset														
-Clear	54	6.9	5.6	5.4	5.9	6.4	7.5	11.3	11.6	10.2	9.4	6.1	6.8	93.1
-Partly Cloudy	54	7.6	7.5	8.5	8.8	10.5	10.5	12.5	11.2	8.8	8.1	6.7	6.9	107.4
-Cloudy	54	16.5	15.2	17.1	15.3	14.1	12.1	7.2	8.3	10.9	13.5	17.3	17.3	164.7
Precipitation														
.01 inches or more	54	7.6	6.8	8.0	7.9	9.7	11.4	9.1	8.4	7.1	5.5	6.4	7.7	95.7
Snow,Ice pellets,hail														
1.0 inches or more	54	2.5	2.2	2.2	1.2	0.3	0.0	0.0	0.0	0.1	0.5	1.9	2.1	12.9
Thunderstorms	54	0.0	0.*	0.1	1.0	3.8	8.6	9.5	7.7	2.7	0.5	0.*	0.0	33.9
Heavy Fog Visibility														
1/4 mile or less	54	1.0	1.4	1.4	0.8	0.4	0.5	0.7	0.6	0.7	1.0	1.4	1.4	11.4
Temperature °F														
-Maximum														
90° and above	34	0.0	0.0	0.0	0.1	0.7	2.9	7.9	8.2	1.8	0.1	0.0	0.0	21.6
32° and below	34	22.4	17.1	9.7	0.7	0.1	0.0	0.0	0.0	0.0	0.6	8.9	20.6	80.1
-Minimum														
32° and below	34	30.9	28.1	28.4	17.5	4.1	0.1	0.0	0.0	2.8	15.6	28.2	31.0	186.6
0° and below	34	16.9	10.6	3.6	0.1	0.0	0.0	0.0	0.0	0.0	0.1	2.4	12.3	45.9
AVG. STATION PRESS. (mb)	21	957.2	957.9	955.1	954.8	953.5	953.2	954.8	955.3	955.9	956.2	956.2	956.7	955.6
RELATIVE HUMIDITY (%)														
Hour 00	34	75	77	77	71	70	78	75	72	74	71	78	78	75
Hour 06 (Local Time)	34	75	78	81	79	78	84	84	83	83	79	81	78	80
Hour 12	34	68	68	62	51	48	52	47	46	50	51	65	71	57
Hour 18	34	71	69	59	46	44	48	43	40	45	50	67	73	55
PRECIPITATION (inches):														
Water Equivalent														
-Normal		0.45	0.43	0.77	1.67	2.18	2.72	2.14	1.72	1.49	0.90	0.49	0.51	15.47
-Maximum Monthly	54	1.29	1.65	3.19	5.46	5.18	8.29	13.75	5.05	6.93	4.30	2.56	0.95	13.75
-Year		1969	1987	1975	1975	1965	1947	1993	1944	1977	1982	1944	1967	JUL 1993
-Minimum Monthly	54	0.02	0.03	0.09	T	0.28	0.50	0.18	0.03	0.02	0.02	T	T	T
-Year		1940	1985	1981	1952	1984	1974	1968	1971	1948	1993	1990	1944	NOV 1990
-Maximum in 24 hrs	54	0.67	0.73	1.30	1.97	2.54	3.25	5.27	2.68	3.02	1.81	0.99	0.59	5.27
-Year		1952	1958	1950	1964	1985	1947	1993	1965	1977	1980	1944	1960	JUL 1993
Snow,Ice pellets,hail														
-Maximum Monthly	54	25.0	25.6	31.1	18.7	10.3	T	T	T	5.0	23.7	29.9	17.2	31.1
-Year		1982	1979	1975	1984	1950	1993	1993	1991	1984	1991	1993	1977	MAR 1975
-Maximum in 24 hrs	54	8.2	8.9	15.5	11.9	11.0	T	T	T	4.8	14.3	11.4	10.0	15.5
-Year		1988	1979	1966	1984	1967	1993	1993	1991	1984	1991	1986	1988	MAR 1966
WIND:														
Mean Speed (mph)	54	10.0	9.8	10.9	11.9	11.7	10.4	9.2	9.4	9.9	10.0	9.9	9.5	10.2
Prevailing Direction through 1963		WNW	WNW	WNW	WNW	SSE	WNW	SSE	E	WNW	WNW	WNW	WNW	WNW
Fastest Obs. 1 Min.														
-Direction (!!)	14	29	32	36	33	27	28	32	30	15	31	31	33	30
-Speed (MPH)	14	44	35	43	41	53	44	46	54	46	39	36	45	54
-Year		1990	1990	1993	1984	1980	1981	1982	1980	1986	1980	1986	1992	AUG 1980
Peak Gust														
-Direction (!!)	10	W	NW	W	SW	NW	NW	NW	SW	SE	W	NW	NW	SE
-Speed (mph)	10	67	53	64	56	56	58	53	61	84	52	56	63	84
-Date		1990	1987	1986	1985	1985	1985	1984	1984	1986	1989	1989	1985	SEP 1986

PRECIPITATION (inches)

BISMARCK, NORTH DAKOTA

YEAR	JAN	FEB	MAR	APR	MAY	JUNE	JULY	AUG	SEP	OCT	NOV	DEC	ANNUAL
1964	0.40	0.22	0.52	2.90	0.90	5.71	2.18	0.76	0.66	0.10	0.43	0.65	15.43
1965	0.54	0.21	0.46	1.88	5.18	2.57	3.35	3.39	3.02	0.46	0.18	0.29	21.53
1966	0.29	0.41	1.85	1.23	0.91	3.60	4.27	2.25	0.86	0.59	0.29	0.18	16.73
1967	0.85	0.68	0.28	2.79	1.67	0.85	0.29	1.21	2.09	1.72	0.16	0.95	13.54
1968	0.30	0.12	0.95	1.30	2.51	6.52	0.18	4.56	1.20	0.05	0.56	0.78	19.03
1969	1.29	1.17	0.15	0.77	1.57	2.01	5.24	0.92	0.49	0.30	0.06	0.79	14.76
1970	0.46	0.34	0.55	4.05	2.32	3.78	1.63	0.13	1.47	0.69	1.33	0.16	16.91
1971	0.78	0.41	0.30	1.06	1.34	3.86	1.04	0.03	1.63	3.74	0.59	0.47	15.25
1972	0.68	0.41	1.32	1.81	3.16	1.81	1.60	1.70	0.29	1.60	0.11	0.67	15.16
1973	0.07	0.08	0.55	0.87	1.77	1.15	1.24	0.30	2.32	1.40	0.46	0.83	11.04
1974	0.11	0.29	0.40	2.23	2.68	0.50	1.10	1.52	0.23	0.74	0.36	0.50	10.66
1975	0.53	0.47	3.19	5.46	1.81	4.60	2.63	0.64	0.57	0.71	0.19	0.70	21.50
1976	0.52	0.27	0.52	2.84	1.08	2.66	0.46	0.32	1.59	0.21	0.15	0.55	11.17
1977	0.55	0.33	0.71	0.13	1.09	2.35	1.39	1.92	6.93	1.00	1.36	0.78	18.54
1978	0.14	0.41	0.30	2.00	4.65	1.79	2.58	1.21	1.80	0.45	1.13	0.49	16.95
1979	0.54	1.21	1.15	0.94	1.07	0.76	3.22	1.73	0.82	0.10	0.05	0.22	11.81
1980	0.70	0.28	0.32	0.43	1.08	1.67	3.16	5.03	1.11	2.31	0.09	0.21	16.39
1981	0.12	0.42	0.09	0.58	0.90	1.67	3.69	3.32	1.88	0.52	0.79	0.48	14.46
1982	0.75	0.40	1.08	0.76	3.71	2.04	2.17	1.52	0.45	4.30	0.41	0.48	18.07
1983	0.23	0.44	1.65	0.51	1.46	2.95	2.04	0.87	1.03	0.75	0.73	0.48	13.14
1984	0.38	0.31	1.65	3.65	0.28	3.58	0.81	0.87	0.94	0.99	0.73	0.58	14.77
1985	0.27	0.03	0.80	1.77	4.13	1.80	0.55	4.61	1.29	1.33	0.91	0.35	17.84
1986	0.36	0.25	0.25	3.60	3.11	3.95	4.24	1.61	4.41	0.35	2.09	0.02	24.24
1987	0.14	1.65	1.34	0.13	4.19	1.52	4.59	3.03	0.29	0.10	0.02	0.13	17.13
1988	0.68	0.40	0.92	0.12	1.16	2.18	0.55	2.19	0.63	0.15	0.48	0.71	10.17
1989	0.59	0.21	0.29	1.86	1.92	0.70	1.76	1.62	1.23	0.21	0.64	0.30	11.33
1990	0.26	0.23	0.55	0.31	1.65	4.73	1.53	1.37	1.25	0.29	T	0.50	12.67
1991	0.16	0.24	0.62	1.62	3.34	2.64	0.65	1.77	2.49	2.33	0.74	0.16	16.76
1992	0.31	0.40	0.61	0.22	1.12	3.64	2.46	0.98	1.28	0.39	0.81	0.48	12.70
1993	0.29	0.33	0.38	1.26	2.36	4.57	13.75	1.89	0.26	0.02	1.04	0.84	26.99
Record Mean	0.46	0.45	0.85	1.49	2.23	3.22	2.29	1.79	1.34	0.92	0.57	0.51	16.11

AVERAGE TEMPERATURE (deg. F)

BISMARCK, NORTH DAKOTA

YEAR	JAN	FEB	MAR	APR	MAY	JUNE	JULY	AUG	SEP	OCT	NOV	DEC	ANNUAL
1964	15.9	21.9	22.8	45.1	56.8	63.0	71.7	64.9	53.9	47.2	25.3	3.5	41.0
1965	2.5	10.2	16.4	42.0	54.5	63.0	69.4	66.8	46.3	49.3	30.1	23.6	39.5
1966	-3.0	10.4	33.0	38.0	53.4	64.2	72.6	64.3	58.7	46.1	21.3	15.9	39.6
1967	11.2	9.2	28.1	38.3	49.3	61.1	69.0	67.4	60.4	45.0	28.3	12.8	40.0
1968	7.1	11.9	32.8	42.8	49.8	61.2	67.6	64.6	57.5	45.1	32.0	11.4	40.3
1969	-0.8	14.7	17.5	46.1	55.9	59.5	70.4	73.8	61.9	40.6	32.5	17.6	40.8
1970	4.0	15.6	18.8	39.6	54.3	67.3	71.0	69.9	58.1	44.4	27.3	11.1	40.1
1971	1.5	12.9	28.2	43.8	52.4	65.7	66.2	71.6	56.8	45.1	30.1	10.9	40.4
1972	4.8	9.1	27.5	42.4	58.1	65.2	66.6	70.7	57.0	41.4	27.6	8.1	39.9
1973	15.5	21.9	39.4	40.9	54.1	63.3	68.4	72.0	54.6	46.8	22.6	9.8	42.5
1974	6.5	15.9	27.1	42.3	50.5	62.8	73.8	63.0	52.9	45.9	28.9	17.5	40.6
1975	12.9	12.5	22.5	35.5	53.9	63.2	72.8	66.8	55.0	47.0	31.5	19.1	41.1
1976	11.9	27.2	27.3	46.5	55.6	67.1	71.2	72.9	60.1	41.4	26.4	14.8	43.5
1977	-1.6	23.9	33.8	49.1	61.7	65.6	70.2	62.0	57.0	46.2	26.4	10.5	42.1
1978	-1.7	9.0	27.9	42.7	56.9	63.0	68.8	68.2	61.8	46.0	22.7	10.8	39.6
1979	-2.2	-0.7	22.2	35.1	50.0	64.3	70.2	65.7	59.4	44.6	25.9	23.7	38.2
1980	8.7	15.4	24.8	48.7	60.1	66.2	71.9	65.7	58.1	45.6	35.9	18.5	43.3
1981	19.5	23.1	34.9	46.5	54.1	60.7	71.2	70.1	59.3	46.9	36.9	14.7	44.8
1982	-4.3	8.7	23.5	40.0	54.9	59.2	70.3	67.5	56.7	44.5	24.4	22.1	39.0
1983	23.0	27.4	30.4	39.9	50.9	62.3	72.3	74.8	56.3	44.8	30.9	-1.2	42.7
1984	15.5	28.0	25.4	41.7	51.9	63.1	70.1	71.6	52.6	45.5	30.3	8.6	42.0
1985	6.2	13.2	31.6	46.0	59.1	59.3	69.9	64.3	53.6	44.3	14.0	9.1	39.2
1986	20.1	13.8	38.8	42.3	55.9	66.5	70.1	65.5	54.3	46.1	23.5	22.3	43.3
1987	21.8	28.1	27.9	51.2	59.8	68.2	72.4	64.9	58.7	42.0	34.5	24.1	46.1
1988	8.9	14.9	30.2	45.0	61.0	75.7	73.2	69.5	56.6	43.8	29.0	18.1	43.8
1989	16.0	6.3	24.8	44.3	56.2	63.7	76.3	69.6	58.6	45.3	29.2	8.2	41.5
1990	23.6	20.0	35.2	43.5	54.8	65.7	70.2	71.8	61.7	43.8	31.2	9.4	44.2
1991	8.8	28.1	32.7	46.1	57.7	68.5	71.4	73.7	58.9	40.7	23.2	25.1	44.6
1992	23.3	27.1	35.0	42.1	58.1	62.7	64.0	64.2	56.5	43.9	28.8	11.8	43.1
1993	8.2	10.9	32.4	42.2	55.4	60.2	64.4	66.6	53.2	43.9	26.7	17.5	40.1
Record Mean	8.6	13.1	26.0	43.0	54.8	64.2	70.5	68.5	57.8	45.5	28.4	15.4	41.3
Max	19.2	23.5	36.2	54.8	67.2	76.1	83.6	82.1	71.1	58.0	38.7	25.4	53.0
Min	-2.0	2.6	15.8	31.3	42.5	52.2	57.4	54.9	44.4	32.9	18.1	5.3	29.6

REFERENCE NOTES

BISMARCK, NORTH DAKOTA

GENERAL
T - TRACE AMOUNT.
BLANK ENTRIES DENOTE MISSING/UNREPORTED DATA.
# INDICATES A STATION OR INSTRUMENT RELOCATION.

SPECIFIC

(a) - LENGTH OF RECORD IN YEARS, ALTHOUGH INDIVIDUAL MONTHS MAY BE MISSING.
0.* OR * - THE VALUE IS BETWEEN 0.0 AND 0.05
NORMALS - BASED ON THE 1961-1990 RECORD PERIOD.
EXTREMES - DATES ARE THE MOST RECENT OCCURRENCE
WIND DIR.- NUMERALS SHOW TENS OF DEGRESS CLOCKWISE FROM TRUE NORTH. "00" INDICATES CALM.
RESULTANT DIRECTIONS ARE GIVEN TO WHOLE DEGREES.
BOLD VALUES INDICATE EXTREME VALUES WHICH OCCURRED AFTER THE ASOS SYSTEM WAS COMMISSIONED.

RECORD = PERIOD OF RECORD
RECORD MEAN PRECIPITATION IS THE MEAN OF ALL DAILY PRECIPITATION AMOUNTS DURING THE PERIOD OF RECORD.
RECORD MAX(MIN) TEMPERATURE IS THE MEAN OF ALL DAILY MAX(MIN) TEMPERATURES DURING THE PERIOD OF RECORD.
RECORD MEAN TEMPERATURE IS THE SUM OF THE RECORD MAX AND RECORD MIN DIVIDED BY 2.
AVERAGE TEMPERATURE IS THE SUM OF THE MEAN DAILY MAX AND MIN TEMPERATURE DIVIDED BY 2.

EXCEPTIONS

RECORD MEANS ARE THROUGH THE CURRENT YEAR, BEGINNING IN 1875 FOR TEMPERATURE
1875 FOR PRECIPITATION
1940 FOR SNOWFALL

HEATING DEGREE DAYS Base 65 deg. F BISMARCK, NORTH DAKOTA

SEASON	JULY	AUG	SEP	OCT	NOV	DEC	JAN	FEB	MAR	APR	MAY	JUNE	TOTAL
1964-65	5	108	342	542	1180	1909	1937	1533	1501	684	324	83	10148
1965-66	14	57	556	479	1040	1279	2106	1525	987	804	368	118	9333
1966-67	4	93	228	577	1307	1517	1664	1562	1138	795	491	129	9505
1967-68	55	43	164	614	1095	1613	1795	1535	990	660	462	139	9165
1968-69	34	82	238	612	982	1658	2040	1405	1461	563	314	169	9558
1969-70	6	3	149	748	967	1463	1892	1376	1423	758	331	34	9150
1970-71	8	24	263	632	1123	1667	1967	1458	1134	631	383	41	9331
1971-72	52	20	271	608	1043	1672	1868	1618	1157	671	243	65	9288
1972-73	41	39	256	724	1116	1762	1528	1201	787	714	332	93	8593
1973-74	12	4	316	557	1263	1709	1816	1369	1167	674	440	120	9447
1974-75	6	119	357	582	1078	1466	1612	1467	1309	880	345	112	9333
1975-76	9	39	295	550	1002	1419	1642	1087	1163	550	288	65	8109
1976-77	11	6	193	726	1152	1551	2063	1147	956	473	164	45	8487
1977-78	17	111	246	574	1153	1687	2066	1564	1143	663	257	105	9586
1978-79	18	35	217	585	1263	1679	2084	1837	1320	891	463	81	10473
1979-80	11	55	191	626	1166	1273	1744	1435	1238	492	211	58	8500
1980-81	5	53	217	594	867	1438	1402	1173	925	547	335	144	7700
1981-82	11	11	191	554	836	1556	2151	1577	1279	742	310	176	9394
1982-83	5	73	291	626	1210	1321	1294	1048	1066	746	434	128	8242
1983-84	6	0	300	620	1015	2049	1532	1067	1225	690	421	94	9019
1984-85	17	27	373	602	1033	1749	1821	1450	1027	562	194	192	9047
1985-86	21	85	345	636	1526	1731	1386	1427	807	673	297	44	8978
1986-87	5	65	314	580	1237	1319	1332	1023	1144	416	189	36	7660
1987-88	10	88	199	708	909	1260	1737	1454	1071	597	184	16	8233
1988-89	1	40	251	651	1075	1448	1513	1640	1239	621	274	110	8863
1989-90	0	38	218	606	1066	1761	1278	1256	916	650	318	78	8185
1990-91	10	5	173	649	1006	1722	1740	1028	995	557	253	10	8148
1991-92	7	0	236	748	1247	1231	1286	1093	921	689	248	117	7823
1992-93	74	111	256	645	1077	1645	1757	1513	1004	678	296	177	9233
1993-94	68	52	347	645	1141	1466							

COOLING DEGREE DAYS Base 65 deg. F BISMARCK, NORTH DAKOTA

YEAR	JAN	FEB	MAR	APR	MAY	JUNE	JULY	AUG	SEP	OCT	NOV	DEC	TOTAL
1969	0	0	0	0	39	11	179	282	60	0	0	0	571
1970	0	0	0	0	5	109	199	182	60	0	0	0	555
1971	0	0	0	0	0	67	94	230	32	0	0	0	423
1972	0	0	0	0	37	76	98	226	24	0	0	0	461
1973	0	0	0	0	1	47	125	227	11	0	0	0	411
1974	0	0	0	0	0	61	287	61	0	0	0	0	409
1975	0	0	0	0	6	64	257	104	2	0	0	0	433
1976	0	0	0	0	3	137	212	257	54	0	0	0	663
1977	0	0	0	3	70	72	186	25	11	0	0	0	367
1978	0	0	0	0	13	50	141	143	128	0	0	0	475
1979	0	0	0	0	6	67	178	86	28	0	0	0	365
1980	0	0	0	8	64	103	227	84	16	0	0	0	502
1981	0	0	0	0	3	24	210	177	27	0	0	0	441
1982	0	0	0	0	3	8	175	159	49	0	0	0	394
1983	0	0	0	0	3	53	243	312	47	0	0	0	658
1984	0	0	0	0	21	44	180	241	11	4	0	0	501
1985	0	0	0	0	16	26	180	67	8	0	0	0	297
1986	0	0	0	0	21	96	169	88	0	0	0	0	374
1987	0	0	0	11	33	136	245	92	15	0	0	0	532
1988	0	0	0	1	65	340	264	184	6	0	0	0	860
1989	0	0	0	4	8	78	359	190	33	0	0	0	672
1990	0	0	0	14	9	105	177	224	82	0	0	0	611
1991	0	0	0	0	36	122	212	278	61	0	0	0	709
1992	0	0	0	9	40	52	50	95	9	0	0	0	255
1993	0	0	0	0	7	42	55	110	3	0	0	0	217

SNOWFALL (inches) BISMARCK, NORTH DAKOTA

SEASON	JULY	AUG	SEP	OCT	NOV	DEC	JAN	FEB	MAR	APR	MAY	JUNE	TOTAL
1964-65	0.0	0.0	0.0	1.0	6.7	10.7	10.2	3.2	7.5	7.2	8.0	0.0	54.5
1965-66	0.0	0.0	3.6	0.0	0.4	2.3	4.4	5.4	22.7	4.9	0.6	0.0	44.3
1966-67	0.0	0.0	0.0	0.4	5.1	2.5	12.7	13.9	4.1	14.8	8.2	0.0	61.7
1967-68	0.0	0.0	0.0	T	2.6	13.1	4.3	1.8	7.4	3.6	2.3	0.0	35.1
1968-69	0.0	0.0	0.0	T	2.4	13.0	16.0	17.4	3.2	0.3	T	T	52.3
1969-70	0.0	0.0	0.0	0.9	0.4	13.7	7.0	4.5	7.6	18.2	0.0	0.0	52.3
1970-71	0.0	0.0	0.0	2.1	13.4	2.8	12.7	6.1	3.3	1.8	T	0.0	42.2
1971-72	0.0	0.0	0.0	3.1	6.4	8.7	13.3	5.8	8.5	0.9	T	0.0	46.7
1972-73	0.0	0.0	T	5.8	1.1	10.2	0.9	1.2	1.4	0.5	0.0	0.0	21.1
1973-74	0.0	0.0	0.0	0.6	5.5	11.5	1.4	4.5	3.4	0.6	0.0	0.0	27.5
1974-75	0.0	0.0	T	T	3.6	6.9	6.3	6.0	31.1	4.8	0.0	0.0	58.7
1975-76	0.0	0.0	0.0	T	2.2	8.6	5.3	2.7	6.5	0.8	T	0.0	26.1
1976-77	0.0	0.0	0.0	1.9	3.3	7.7	8.3	5.1	1.6	0.7	0.0	0.0	28.6
1977-78	0.0	0.0	0.0	3.5	16.2	17.2	4.1	11.1	7.6	4.0	0.0	0.0	63.7
1978-79	0.0	0.0	0.0	T	16.6	10.0	10.3	25.6	11.6	7.3	1.6	0.0	83.0
1979-80	0.0	0.0	0.0	T	1.0	1.6	12.6	5.6	5.7	0.1	0.0	0.0	26.6
1980-81	0.0	0.0	0.0	1.1	0.2	4.7	2.1	3.6	T	T	0.0	0.0	11.7
1981-82	0.0	0.0	0.0	0.7	6.3	10.4	25.0	10.3	22.5	5.1	0.0	0.0	80.3
1982-83	0.0	0.0	0.0	4.8	7.2	2.9	2.2	5.3	9.3	0.5	T	0.0	32.2
1983-84	0.0	0.0	T	0.0	6.6	10.6	4.1	5.9	20.6	18.7	T	0.0	66.5
1984-85	0.0	0.0	5.0	1.3	2.0	14.9	3.9	0.5	12.4	0.2	0.0	0.0	40.2
1985-86	0.0	0.0	T	3.2	24.2	7.1	6.1	6.5	2.4	11.7	T	0.0	61.2
1986-87	0.0	0.0	0.0	T	28.5	1.3	2.6	23.2	11.8	0.5	T	0.0	67.9
1987-88	0.0	0.0	0.0	0.3	T	1.2	12.1	8.2	9.2	0.5	0.0	0.0	31.5
1988-89	0.0	0.0	0.0	0.1	3.3	15.9	12.9	7.4	6.4	1.8	T	T	47.8
1989-90	0.0	T	T	1.7	7.6	5.1	3.7	3.6	3.5	3.1	0.8	T	29.1
1990-91	T	T	0.0	2.5	T	10.9	4.1	5.2	4.8	8.4	6.5	T	42.4
1991-92	0.0	T	T	23.7	7.5	1.5	7.0	9.5	5.3	3.0	T	T	57.5
1992-93	0.0	0.0	0.0	4.8	10.5	8.5	9.0	7.9	3.3	T	T	T	44.0
1993-94	T	0.0	0.0	0.5	29.9	14.0							
Record Mean	T	T	0.3	1.7	6.3	6.9	7.2	6.7	8.1	3.8	0.9	T	41.9

FARGO, NORTH DAKOTA

Moorhead, Minnesota, and Fargo are twin cities in the Red River Valley of the north. The Red River of the north flows northward between the two cities and is a part of the Hudson Bay drainage area. The Red River is approximately 2 miles east of the airport at its nearest point and has no significant effect on the weather. In recent years, spring floods due to melting snow have been common. Summer floods caused by heavy rains are infrequent.

The surrounding terrain is flat and open. Northerly winds blowing up the valley occasionally causing low cloudiness and fog. However, this upslope cloudiness is very infrequent. Aside from this, there are no pronounced climatic differences due to geographical features in the immediate area.

The summers are generally comfortable with very few days of hot and humid weather. Nights, with few exceptions, are comfortably cool. The winter months are cold and dry with temperatures rising above freezing only on an average of six days each month, and nighttime lows dropping below zero approximately half of the time.

Precipitation is the most important climatic factor in the area. The Red River Valley lies in an area where lighter amounts fall to the west and heavier amounts to the east. Seventy-five percent of the precipitation occurs during the growing season (April to September) and is often accompanied by electrical storms and heavy falls in a short time. Winter precipitation is light, indicating that heavy snowfall is the exception rather than the rule. The first light snow in the fall occasionally falls in September, but usually very little, if any, occurs until October or November. The latest fall is generally in April.

With the flat terrain, surface friction has little effect on the wind in the area and this fact has led to the legendary Dakota blizzards. Strong winds with even light snowfall cause much drifting and blowing snow, reducing visibility to near zero. Fortunately, these conditions occur only several times during the winter months.

NORMALS, MEANS, AND EXTREMES

FARGO, NORTH DAKOTA

LATITUDE: 46°54'N LONGITUDE: 96°48' W ELEVATION: FT. GRND 896 BARO 911 TIME ZONE: CENTRAL WBAN: 14914

	(a)	JAN	FEB	MAR	APR	MAY	JUNE	JULY	AUG	SEP	OCT	NOV	DEC	YEAR
TEMPERATURE °F:														
Normals														
-Daily Maximum		15.4	21.1	34.6	53.8	68.5	77.4	83.4	81.3	69.4	56.7	36.8	20.1	51.5
-Daily Minimum		-3.6	2.7	17.3	32.1	43.8	53.6	58.8	56.4	45.9	34.6	19.4	3.1	30.3
-Monthly		5.9	12.0	25.9	43.0	56.2	65.5	71.1	68.8	57.7	45.7	28.1	11.6	41.0
Extremes														
-Record Highest	41	52	66	78	100	98	99	106	106	102	93	74	57	106
-Year		1981	1958	1967	1980	1964	1959	1988	1976	1959	1963	1990	1962	JUL 1988
-Record Lowest	41	-35	-34	-23	-7	20	30	36	33	19	7	-24	-32	-35
-Year		1977	1962	1980	1975	1966	1969	1967	1982	1965	1976	1985	1967	JAN 1977
NORMAL DEGREE DAYS:														
Heating (base 65°F)		1832	1484	1212	660	307	93	19	48	239	598	1107	1655	9254
Cooling (base 65°F)		0	0	0	0	35	108	209	165	20	0	0	0	537
% OF POSSIBLE SUNSHINE	51	50	56	57	59	61	61	71	69	59	54	40	42	57
MEAN SKY COVER (tenths)														
Sunrise - Sunset	48	6.7	6.7	7.0	6.6	6.4	6.2	5.2	5.2	5.8	6.2	7.2	7.0	6.3
MEAN NUMBER OF DAYS:														
Sunrise to Sunset														
-Clear	51	6.8	6.2	5.4	6.5	7.0	6.5	10.1	10.4	8.7	8.7	5.5	6.1	88.0
-Partly Cloudy	51	7.5	7.3	8.9	8.8	9.7	10.9	13.4	11.9	9.1	8.3	6.3	7.2	109.4
-Cloudy	51	16.8	14.8	16.6	14.7	14.3	12.6	7.5	8.7	12.1	14.0	18.2	17.7	167.9
Precipitation .01 inches or more	51	8.5	7.0	7.8	8.2	10.0	10.7	9.6	9.0	7.9	6.3	6.3	8.2	99.4
Snow,Ice pellets,hail 1.0 inches or more	51	2.4	1.7	2.1	1.1	0.*	0.0	0.0	0.0	0.0	0.3	1.9	2.3	11.8
Thunderstorms	51	0.0	0.*	0.3	1.3	3.8	7.3	8.4	7.0	2.9	0.9	0.1	0.*	32.0
Heavy Fog Visibility 1/4 mile or less	51	0.8	1.6	1.9	0.5	0.4	0.6	0.8	1.0	0.9	0.9	1.4	1.8	12.5
Temperature °F														
-Maximum														
90° and above	34	0.0	0.0	0.0	0.1	0.7	2.2	5.1	5.3	1.0	0.1	0.0	0.0	14.6
32° and below	34	26.9	21.3	12.3	1.1	0.*	0.0	0.0	0.0	0.0	0.5	10.9	24.9	98.0
-Minimum														
32° and below	34	31.0	28.2	27.6	16.3	4.0	0.*	0.0	0.0	1.8	13.1	27.0	30.9	179.9
0° and below	34	18.4	12.4	4.4	0.1	0.0	0.0	0.0	0.0	0.0	0.0	2.1	13.3	50.7
AVG. STATION PRESS. (mb)	21	985.2	986.1	983.0	982.0	980.5	979.5	981.2	982.0	982.4	983.0	983.5	984.7	982.8
RELATIVE HUMIDITY (%)														
Hour 00	34	75	77	80	73	67	76	78	77	78	74	79	78	76
Hour 06 (Local Time)	34	75	77	82	79	77	82	85	86	85	80	82	78	81
Hour 12	34	72	72	70	57	50	56	55	54	58	58	70	74	62
Hour 18	34	73	74	71	53	46	52	51	50	55	59	73	76	61
PRECIPITATION (inches):														
Water Equivalent														
-Normal		0.67	0.45	1.06	1.82	2.45	2.82	2.70	2.43	1.99	1.68	0.73	0.65	19.45
-Maximum Monthly	52	1.85	1.74	2.27	5.28	7.30	9.40	8.42	8.52	6.13	7.03	4.58	2.19	9.40
-Year		1989	1979	1983	1986	1977	1975	1952	1944	1957	1982	1977	1951	JUN 1975
-Minimum Monthly	52	0.09	0.03	0.03	0.01	0.46	0.58	0.42	0.18	0.13	0.05	0.02	0.04	0.01
-Year		1961	1954	1958	1988	1976	1972	1950	1984	1974	1986	1990	1958	APR 1988
-Maximum in 24 hrs	52	1.00	1.22	1.16	1.91	4.10	4.02	5.10	4.72	3.97	3.22	1.99	0.87	5.10
-Year		1989	1946	1950	1963	1977	1975	1993	1943	1957	1982	1977	1960	JUL 1993
Snow,Ice pellets,hail														
-Maximum Monthly	52	31.5	19.5	18.7	12.8	1.0	T	0.0	T	0.6	8.1	24.3	20.3	31.5
-Year		1989	1979	1975	1970	1950	1991		1989	1942	1951	1985	1951	JAN 1989
-Maximum in 24 hrs	52	19.4	11.2	11.5	8.6	1.0	T	0.0	T	0.6	7.8	12.6	9.3	19.4
-Year		1989	1951	1990	1970	1950	1991		1989	1942	1951	1977	1988	JAN 1989
WIND:														
Mean Speed (mph)	51	12.7	12.4	13.1	13.9	12.9	11.6	10.5	11.0	11.9	12.6	12.8	12.3	12.3
Prevailing Direction through 1963		SSE	N	N	N	N	SSE	S	SSE	SSE	SSE	S	S	SSE
Fastest Obs. 1 Min.														
-Direction (!!)	8	30	33	34	14	32	08	30	29	31	31	31	30	31
-Speed (MPH)	8	40	38	36	37	37	41	37	40	45	46	37	41	46
-Year		1990	1988	1989	1992	1989	1993	1987	1987	1988	1993	1986	1992	OCT 1993
Peak Gust														
-Direction (!!)	10	NW	N	SE	NW	NW	SW	NW	NW	NW	NW	NW	NW	NW
-Speed (mph)	10	60	59	53	51	62	66	69	60	62	54	52	59	69
-Date		1990	1984	1985	1991	1988	1991	1987	1988	1988	1993	1991	1992	JUL 1987

PRECIPITATION (inches)

FARGO, NORTH DAKOTA

YEAR	JAN	FEB	MAR	APR	MAY	JUNE	JULY	AUG	SEP	OCT	NOV	DEC	ANNUAL
1964	0.54	0.27	0.92	3.76	0.87	4.85	0.77	2.85	1.70	0.10	0.72	0.91	18.26
1965	0.10	0.14	1.36	3.04	3.06	3.10	4.81	2.55	3.50	0.55	0.79	1.01	24.01
1966	0.40	0.26	1.92	1.78	1.27	2.91	4.01	3.80	0.54	1.40	0.18	0.50	18.97
1967	1.03	0.21	0.34	4.14	1.00	2.54	0.60	0.41	0.31	1.06	0.04	1.36	13.04
1968	0.37	0.27	1.29	4.09	2.08	3.94	1.49	1.61	2.23	1.75	0.37	1.11	20.60
1969	1.27	0.46	0.54	1.55	2.36	2.03	5.92	0.38	1.55	1.51	0.14	0.81	18.52
1970	0.10	0.20	1.52	2.30	2.83	2.63	0.43	1.24	3.61	1.61	0.96	0.47	17.90
1971	0.81	0.34	0.56	1.10	2.68	3.51	2.80	0.92	4.30	4.42	0.83	0.59	22.86
1972	0.94	0.61	0.74	0.96	3.52	0.58	2.78	3.45	1.22	1.25	0.22	1.51	17.78
1973	0.12	0.13	1.25	0.70	1.65	1.78	3.60	3.85	4.98	1.54	0.90	1.02	21.52
1974	0.35	0.36	0.71	3.40	4.03	0.90	4.75	6.46	0.13	3.10	0.48	0.32	24.99
1975	1.32	0.27	1.48	3.24	1.45	9.40	2.42	2.90	1.24	1.76	0.64	0.18	26.30
1976	1.25	0.35	1.00	1.19	0.46	2.34	0.63	0.41	0.55	0.16	0.26	0.24	8.84
1977	0.65	1.24	1.72	0.84	7.30	1.64	5.36	2.53	3.21	2.46	4.58	0.75	32.28
1978	0.16	0.18	0.43	1.15	1.78	4.40	2.92	3.79	0.92	0.13	1.11	0.47	17.44
1979	0.44	1.74	2.00	3.04	2.02	2.92	3.38	0.90	0.31	2.60	0.48	0.14	19.97
1980	1.23	0.57	0.62	0.02	0.64	2.68	0.76	4.24	2.52	1.06	0.47	0.30	15.11
1981	0.11	0.49	0.67	0.61	3.46	2.56	3.21	1.76	1.11	2.36	0.40	0.85	17.59
1982	1.32	0.54	1.25	0.45	1.82	1.61	2.64	1.12	1.12	7.03	1.13	0.17	20.20
1983	0.46	0.21	2.27	0.42	2.00	2.34	4.16	2.56	1.63	1.62	1.04	0.96	19.67
1984	0.79	0.90	1.12	1.68	0.61	5.38	0.64	0.18	1.23	6.76	0.18	0.90	20.37
1985	0.20	0.18	1.35	0.60	5.03	1.44	3.91	2.30	1.39	1.12	1.06	0.59	19.17
1986	0.85	0.27	0.19	5.28	1.00	3.98	4.78	1.72	3.67	0.05	1.43	0.29	23.51
1987	0.27	0.86	0.49	0.12	3.46	0.66	2.86	3.23	1.70	0.18	0.48	0.69	15.00
1988	1.62	0.22	1.02	0.01	1.82	1.24	0.46	2.14	3.22	0.49	1.18	1.11	14.53
1989	1.85	0.21	1.49	1.03	2.60	1.51	0.62	6.07	2.10	0.31	1.18	0.24	19.21
1990	0.13	0.58	1.54	1.78	1.52	6.05	0.78	0.99	1.75	1.22	0.02	0.77	17.13
1991	0.29	1.27	0.97	3.15	2.38	6.26	1.86	1.87	1.28	0.71	0.46	0.37	20.87
1992	0.89	0.51	1.05	0.89	2.32	6.47	0.83	2.35	2.55	0.26	1.73	0.56	20.41
1993	0.79	0.19	0.83	0.74	2.67	4.28	7.71	1.13	0.49	0.19	1.88	1.00	21.90
Record Mean	0.64	0.59	0.92	1.92	2.45	3.42	3.09	2.74	1.94	1.50	0.87	0.66	20.74

AVERAGE TEMPERATURE (deg. F)

FARGO, NORTH DAKOTA

YEAR	JAN	FEB	MAR	APR	MAY	JUNE	JULY	AUG	SEP	OCT	NOV	DEC	ANNUAL
1964	15.6	18.9	20.8	46.4	61.1	67.2	74.0	64.9	55.4	45.2	29.2	3.8	41.9
1965	-1.2	7.1	13.7	41.6	54.4	63.9	68.5	66.5	48.9	47.3	26.0	19.2	38.0
1966	-6.4	6.2	29.7	37.2	51.4	66.1	73.8	65.5	58.2	44.0	23.1	11.9	38.4
1967	9.3	3.8	26.8	38.3	49.7	62.6	67.9	66.7	60.8	44.0	29.0	15.1	39.5
1968	7.7	9.9	34.1	43.9	52.9	64.1	69.6	68.1	59.3	46.7	31.0	8.9	41.4
1969	-1.6	12.8	15.3	45.4	54.8	57.3	68.4	72.4	59.0	40.3	30.5	15.6	39.2
1970	0.6	10.9	18.6	39.1	51.7	67.9	71.8	69.7	59.7	46.8	27.8	9.3	39.5
1971	-0.7	12.7	27.6	44.5	54.2	67.5	65.1	68.3	58.5	47.4	29.6	12.2	40.6
1972	2.7	4.1	23.9	41.0	59.8	66.9	68.4	70.5	56.9	42.4	28.6	3.8	39.1
1973	10.3	16.3	36.0	41.4	54.2	64.7	68.3	71.7	54.8	50.2	25.1	10.1	41.9
1974	1.7	9.6	22.9	42.2	51.1	64.4	73.7	64.3	53.4	47.5	29.2	20.9	40.1
1975	12.3	10.1	18.4	35.9	56.6	65.2	74.3	68.1	55.4	49.4	31.1	14.8	41.0
1976	7.7	21.4	23.0	47.0	55.8	68.5	71.8	73.6	60.0	39.5	23.2	6.9	41.5
1977	-3.3	17.5	32.0	49.5	66.5	66.6	72.2	62.5	57.9	47.1	25.6	6.5	41.7
1978	-1.4	3.4	23.5	42.5	59.1	64.7	69.5	69.1	63.6	46.4	22.8	7.3	39.2
1979	-4.2	-1.5	20.4	36.0	50.4	65.4	71.9	67.3	62.0	42.6	24.5	20.7	38.0
1980	6.6	8.3	20.7	49.0	61.4	65.7	71.9	67.5	56.9	42.4	33.1	12.7	41.6
1981	11.8	19.6	33.5	45.6	55.5	62.8	71.1	69.6	57.4	44.5	35.4	8.7	43.0
1982	-7.0	8.9	22.9	40.7	58.1	59.1	70.9	68.5	57.5	45.7	24.1	20.9	39.2
1983	16.1	21.8	29.9	40.2	52.1	66.1	73.5	72.9	56.7	44.4	31.3	-0.3	42.1
1984	9.7	24.9	23.4	45.6	54.2	65.8	70.6	73.3	54.4	47.4	29.7	9.6	42.4
1985	5.1	10.9	32.9	46.6	60.2	60.0	69.0	64.5	53.9	44.6	15.4	3.9	38.9
1986	13.8	10.5	31.6	43.9	57.5	67.5	71.5	65.6	56.1	45.3	23.1	20.8	42.3
1987	18.2	27.5	31.4	51.5	61.7	69.1	74.0	66.8	59.6	42.6	33.4	20.6	46.4
1988	5.9	9.3	29.5	44.5	63.9	73.8	75.8	72.2	58.5	42.9	27.5	15.2	43.3
1989	11.4	1.7	20.1	42.2	58.2	64.1	75.9	70.8	58.5	45.8	24.0	4.3	39.8
1990	21.8	17.6	31.4	43.6	55.0	67.0	70.0	71.1	62.3	45.6	32.1	12.2	44.1
1991	6.4	20.5	30.4	48.0	61.5	70.1	70.2	72.7	58.8	42.0	22.0	18.8	43.5
1992	17.3	23.5	32.6	41.4	58.7	62.0	64.3	64.8	56.8	45.1	27.3	10.5	42.0
1993	7.5	9.8	25.6	43.3	56.7	63.1	67.0	69.2	54.7	43.9	26.6	16.0	40.3
Record Mean	5.5	10.2	24.6	42.5	55.2	64.8	70.1	68.1	58.0	45.5	27.4	12.7	40.4
Max	15.0	19.7	33.7	53.2	67.3	76.1	81.9	80.2	69.7	56.4	36.0	21.4	50.9
Min	-4.0	0.7	15.5	31.8	43.1	53.5	58.3	56.0	46.2	34.6	18.7	4.0	29.9

REFERENCE NOTES

FARGO, NORTH DAKOTA

GENERAL
T - TRACE AMOUNT.
BLANK ENTRIES DENOTE MISSING/UNREPORTED DATA.
# INDICATES A STATION OR INSTRUMENT RELOCATION.

SPECIFIC

PM - INCLUDES LAST DAY OF PREVIOUS MONTH
ASOS - AUTOMATED SURFACE OBSERVING SYSTEM IN OPERATION DURING THESE MONTHS.

(a) - LENGTH OF RECORD IN YEARS, ALTHOUGH INDIVIDUAL MONTHS MAY BE MISSING.
0.* OR * - THE VALUE IS BETWEEN 0.0 AND 0.05
NORMALS - BASED ON THE 1961-1990 RECORD PERIOD.
EXTREMES - DATES ARE THE MOST RECENT OCCURRENCE.
WIND DIR.- NUMERALS SHOW TENS OF DEGREES CLOCKWISE FROM TRUE NORTH. "00" INDICATES CALM.
RESULTANT DIRECTIONS ARE GIVEN TO WHOLE DEGREES.
BOLD VALUES INDICATE EXTREME VALUES WHICH OCCURRED AFTER THE ASOS SYSTEM WAS COMMISSIONED.

RECORD = PERIOD OF RECORD
RECORD MEAN PRECIPITATION IS THE MEAN OF ALL DAILY PRECIPITATION AMOUNTS DURING THE PERIOD OF RECORD.
RECORD MAX(MIN) TEMPERATURE IS THE MEAN OF ALL DAILY MAX(MIN) TEMPERATURES DURING THE PERIOD OF RECORD.
RECORD MEAN TEMPERATURE IS THE SUM OF THE RECORD MAX AND RECORD MIN DIVIDED BY 2.
AVERAGE TEMPERATURE IS THE SUM OF THE MEAN DAILY MAX AND MIN TEMPERATURE DIVIDED BY 2.

EXCEPTIONS

RECORD MEANS ARE THROUGH THE CURRENT YEAR, BEGINNING IN 1881 FOR TEMPERATURE
1881 FOR PRECIPITATION
1943 FOR SNOWFALL

HEATING DEGREE DAYS Base 65 deg. F — FARGO, NORTH DAKOTA

SEASON	JULY	AUG	SEP	OCT	NOV	DEC	JAN	FEB	MAR	APR	MAY	JUNE	TOTAL
1964-65	5	94	304	604	1065	1902	2051	1618	1582	695	324	66	10310
1965-66	19	59	477	544	1161	1415	2216	1649	1089	829	430	86	9974
1966-67	0	84	230	644	1253	1642	1723	1716	1178	795	480	96	9841
1967-68	65	49	158	645	1073	1544	1773	1592	952	628	375	79	8933
1968-69	31	66	201	565	1015	1736	2066	1460	1536	582	348	230	9836
1969-70	20	10	229	757	1028	1522	1996	1511	1431	773	407	55	9739
1970-71	16	26	251	560	1109	1723	2035	1461	1153	609	333	30	9306
1971-72	57	36	241	535	1052	1630	1931	1763	1269	718	231	60	9523
1972-73	25	41	261	695	1089	1897	1688	1361	893	701	328	79	9058
1973-74	32	3	309	451	1187	1698	1963	1550	1298	676	431	86	9684
1974-75	3	91	345	537	1066	1362	1630	1535	1438	867	265	79	9218
1975-76	14	22	284	492	1012	1550	1774	1257	1296	533	285	55	8574
1976-77	13	9	227	788	1247	1797	2119	1327	1015	466	74	30	9112
1977-78	7	95	211	549	1178	1817	2061	1721	1284	668	209	90	9890
1978-79	15	39	179	571	1262	1788	2147	1863	1377	861	457	64	10623
1979-80	3	45	139	689	1209	1367	1808	1644	1363	493	206	61	9027
1980-81	3	35	267	696	951	1616	1645	1266	971	574	298	84	8406
1981-82	14	10	250	627	881	1742	2236	1570	1298	725	222	187	9762
1982-83	0	66	257	589	1219	1359	1513	1206	1082	738	390	74	8493
1983-84	16	2	301	631	1004	2023	1714	1154	1280	576	344	52	9097
1984-85	15	13	339	541	1053	1715	1853	1514	988	550	172	179	8932
1985-86	13	72	329	625	1487	1895	1585	1527	1027	627	266	45	9498
1986-87	0	69	268	602	1251	1360	1447	1047	1036	415	163	39	7697
1987-88	15	59	177	688	940	1369	1832	1614	1092	609	131	8	8534
1988-89	3	25	207	677	1118	1537	1658	1771	1386	677	224	96	9379
1989-90	0	17	224	599	1224	1881	1332	1324	1034	666	314	58	8673
1990-91	8	18	173	594	982	1637	1813	1242	1066	505	211	6	8255
1991-92	3	3	234	708	1284	1425	1473	1198	998	709	247	137	8419
1992-93	66	84	252	613	1125	1686	1780	1542	1214	643	269	111	9385
1993-94	35	19	310	652	1144	1514							

COOLING DEGREE DAYS Base 65 deg. F — FARGO, NORTH DAKOTA

YEAR	JAN	FEB	MAR	APR	MAY	JUNE	JULY	AUG	SEP	OCT	NOV	DEC	TOTAL
1969	0	0	0	0	39	5	131	249	59	0	0	0	483
1970	0	0	0	0	2	146	233	180	95	3	0	0	659
1971	0	0	0	0	3	109	65	142	56	0	0	0	375
1972	0	0	0	0	74	125	135	217	23	0	0	0	574
1973	0	0	0	0	0	76	140	219	13	0	0	0	448
1974	0	0	0	0	9	75	281	75	3	1	0	0	444
1975	0	0	0	0	11	92	308	126	1	15	0	0	553
1976	0	0	0	0	4	164	228	283	83	4	0	0	766
1977	0	0	0	6	129	86	235	23	8	0	0	0	487
1978	0	0	0	0	31	86	165	176	146	0	0	0	604
1979	0	0	0	0	12	85	225	124	58	0	0	0	504
1980	0	0	0	18	102	89	222	119	31	1	0	0	582
1981	0	0	0	0	9	25	212	159	26	0	0	0	431
1982	0	0	0	2	11	20	189	179	39	0	0	0	440
1983	0	0	0	0	2	113	288	252	55	0	0	0	710
1984	0	0	0	0	18	81	196	279	24	4	0	0	602
1985	0	0	0	6	31	35	143	63	4	0	0	0	282
1986	0	0	0	0	41	126	208	92	10	0	0	0	477
1987	0	0	0	17	66	169	303	121	25	0	0	0	701
1988	0	0	0	0	102	280	346	252	22	0	0	0	1002
1989	0	0	0	0	19	76	345	201	34	11	0	0	686
1990	0	0	0	29	10	123	172	214	98	1	0	0	647
1991	0	0	0	2	107	166	171	250	54	0	0	0	750
1992	0	0	0	9	58	52	49	85	11	3	0	0	267
1993	0	0	0	0	19	61	107	155	9	4	0	0	355

SNOWFALL (inches) — FARGO, NORTH DAKOTA

SEASON	JULY	AUG	SEP	OCT	NOV	DEC	JAN	FEB	MAR	APR	MAY	JUNE	TOTAL
1964-65	0.0	0.0	0.0	T	2.7	11.9	1.2	1.7	13.1	2.5	T	0.0	33.1
1965-66	0.0	0.0	T	0.0	6.1	6.1	5.0	1.1	15.4	5.0	0.0	0.0	38.7
1966-67	0.0	0.0	0.0	1.2	1.3	3.4	15.1	2.6	4.9	5.0	T	0.0	33.5
1967-68	0.0	0.0	0.0	0.7	0.7	10.3	4.0	2.4	2.0	11.6	0.4	0.0	32.1
1968-69	0.0	0.0	0.0	0.8	3.9	11.4	14.5	7.8	3.0	T	T	0.0	41.4
1969-70	0.0	0.0	0.0	2.0	1.9	9.5	2.3	3.6	9.1	12.8	T	0.0	41.2
1970-71	0.0	0.0	0.0	0.9	6.4	8.3	15.1	4.8	1.8	1.0	T	0.0	38.3
1971-72	0.0	0.0	0.0	3.8	2.3	10.0	16.5	10.9	7.1	3.1	0.0	0.0	53.7
1972-73	0.0	0.0	T	3.8	1.7	18.5	1.7	1.4	1.4	2.4	0.0	0.0	30.9
1973-74	0.0	0.0	0.0	T	3.9	12.3	6.1	7.1	10.5	2.7	0.0	0.0	42.6
1974-75	0.0	0.0	0.0	0.4	1.0	5.1	18.3	5.9	18.7	3.7	T	0.0	53.1
1975-76	0.0	0.0	0.0	0.4	3.8	1.9	14.0	6.2	14.0	T	0.1	0.0	40.4
1976-77	0.0	0.0	0.0	0.1	2.7	5.5	12.6	10.7	4.6	2.1	0.0	0.0	38.3
1977-78	0.0	0.0	0.0	T	24.2	7.2	4.6	3.9	7.1	2.8	0.0	0.0	49.8
1978-79	0.0	0.0	0.0	T	8.5	11.7	7.8	19.5	4.3	2.7	0.8	0.0	55.3
1979-80	0.0	0.0	0.0	1.4	6.0	1.5	17.3	7.2	6.5	T	0.0	0.0	39.9
1980-81	0.0	0.0	0.0	0.5	1.1	4.6	2.1	4.5	0.3	T	0.0	0.0	13.1
1981-82	0.0	0.0	T	2.3	2.2	9.9	30.0	10.9	14.0	0.2	0.0	0.0	69.5
1982-83	0.0	0.0	0.0	0.0	6.8	0.3	3.8	2.0	7.4	2.9	T	0.0	23.2
1983-84	0.0	0.0	T	T	5.3	11.8	11.5	3.1	7.7	0.5	0.0	0.0	39.9
1984-85	0.0	0.0	T	T	1.4	7.4	3.7	3.1	12.6	T	0.0	0.0	28.2
1985-86	0.0	0.0	0.0	T	24.3	10.4	11.2	6.7	0.7	3.7	T	0.0	57.0
1986-87	0.0	0.0	0.0	T	5.3	3.8	2.8	10.4	1.2	T	0.0	0.0	23.5
1987-88	0.0	0.0	0.0	T	3.0	6.6	24.3	4.4	6.2	T	0.0	0.0	44.5
1988-89	0.0	0.0	0.0	T	11.6	14.9	31.5	2.3	12.4	0.9	T	0.0	73.6
1989-90	0.0	T	T	T	16.3	2.6	0.8	7.9	11.5	7.2	T	0.0	46.3
1990-91	0.0	0.0	0.0	1.3	0.2	12.4	4.0	15.3	10.9	4.2	T	T	48.3
1991-92	0.0	0.0	T	0.3	5.2	5.9	10.5	2.1	0.2	3.3	T	0.0	27.5
1992-93	0.0	0.0	T	1.8	16.4	9.2	16.7	3.3	6.4	T	0.0	0.0	53.8
1993-94	0.0	0.0	0.0	T	21.5	13.8							
Record Mean	0.0	T	T	0.7	5.8	7.1	8.6	5.9	6.7	3.1	0.1	T	37.9

WILLISTON, NORTH DAKOTA

Williston lies in a flat valley at the junction of the Missouri River and Little Muddy Creek. The surrounding country is rolling. Hills to the east are highest, ranging from 250 to 300 feet in height at a distance of 5 to 7 miles. Across the Missouri River to the south, the bluffs are about 225 feet high at 4 miles distance.

Great extremes of temperatures are encountered, winters being cold, while summer days are usually warm. In winter, temperatures below zero are common and lows of -50 degrees have been recorded. When temperatures are lowest, however, the air is generally dry, with little or no wind and the weather is fine and invigorating. At the other extreme, temperatures above 100 degrees have been reached in all months from May to September. The low humidity that generally prevails on the hottest summer days keeps them from becoming oppressive.

The climate of Williston and vicinity is continental, semi-arid, characterized by marked season changes. Winter is the relatively dry season, with only about 1/2 inch of monthly precipitation occurring from November to February. There is considerably less than the average amount of snowfall for similar locations in the United States. Ice crystals, which rarely yield more than a trace of precipitation, are common in the cold months. Although snow has been observed every month except July and August, there is usually very little from April to November. Accumulated winter snow remains unmelted on the ground until about March. Summer precipitation is variable from year to year. The amount of rain occurring during the growing period is the most important element of climate for agricultural interests in the vicinity of Williston. Generally, considerably more precipitation occurs in the spring and summer months than in winter, but even so, the rainfall is just adequate for successful farming operations in normal years. A series of dry years, in addition to causing failure of crops, may result in erosion of the fertile topsoil by winds.

The growing season averages 131 days. It has ranged from 94 to 172 days during the period of record.

Clear and partly cloudy skies, nearly equally distributed, occur about 70 percent of the time. Heavy fog occurs on the average about ten times a year. Because of the northern latitude of Williston, it enjoys long hours of daylight in the spring and summer. Relatively little cloudiness occurs then, so that the duration of sunshine averages about two-thirds of the possible amount. These conditions are conducive to rapid growth of vegetation, making successful agricultural pursuits possible in spite of the relatively short growing season.

Summer storms are generally in the form of thunderstorms or rain showers, occasionally accompanied by hail and squally winds. Tornadoes are rare in this area. In the winter, cold waves and occasionally blizzard conditions occur. Cold waves result when extremely cold air advances southward from northwestern Canada. In blizzard conditions the advancing cold wave is accompanied by winds of gale force and the air is filled with fine, wind-driven snow. In extreme instances in the country, it becomes impossible for persons to ascertain their bearings or to remain alive many hours without shelter in such storms.

NORMALS, MEANS, AND EXTREMES

WILLISTON, NORTH DAKOTA

LATITUDE: 48°11'N LONGITUDE: 103°38' W ELEVATION: FT. GRND 1900 BARO 1903 TIME ZONE: CENTRAL WBAN: 94014

	(a)	JAN	FEB	MAR	APR	MAY	JUNE	JULY	AUG	SEP	OCT	NOV	DEC	YEAR
TEMPERATURE °F:														
Normals														
-Daily Maximum		19.6	26.8	39.4	55.8	68.3	78.0	84.8	83.3	70.1	58.1	38.0	23.6	53.8
-Daily Minimum		-1.8	5.3	17.4	30.5	42.2	51.4	56.5	54.1	42.4	31.4	16.3	2.8	29.0
-Monthly		8.9	16.1	28.4	43.2	55.3	64.7	70.7	68.7	56.3	44.8	27.2	13.2	41.5
Extremes														
-Record Highest	32	53	66	78	92	106	106	109	107	104	93	73	58	109
-Year		1981	1992	1978	1980	1980	1988	1980	1983	1983	1963	1981	1979	JUL 1980
-Record Lowest	32	-40	-41	-28	-15	17	30	34	34	17	-3	-24	-50	-50
-Year		1966	1962	1962	1975	1980	1969	1967	1988	1974	1991	1985	1983	DEC 1983
NORMAL DEGREE DAYS:														
Heating (base 65°F)		1739	1369	1135	654	316	115	26	76	294	626	1134	1606	9090
Cooling (base 65°F)		0	0	0	0	15	106	203	191	33	0	0	0	548
% OF POSSIBLE SUNSHINE	30	52	58	62	61	62	66	74	74	65	59	42	48	60
MEAN SKY COVER (tenths)														
Sunrise - Sunset	32	6.7	6.8	6.6	6.6	6.4	5.9	4.8	4.9	5.5	6.0	6.8	6.7	6.1
MEAN NUMBER OF DAYS:														
Sunrise to Sunset														
-Clear	32	6.3	5.7	6.9	6.2	6.4	7.7	11.7	11.4	9.7	8.9	6.4	6.5	93.6
-Partly Cloudy	32	8.3	7.4	8.3	9.2	11.2	11.0	12.4	11.8	8.7	8.2	7.4	7.4	111.3
-Cloudy	32	16.5	15.2	15.8	14.6	13.4	11.3	7.0	7.8	11.6	13.9	16.2	17.1	160.3
Precipitation														
.01 inches or more	32	7.9	6.3	7.7	7.8	9.8	10.4	8.9	7.3	7.0	5.2	6.5	8.4	93.2
Snow,Ice pellets,hail														
1.0 inches or more	32	2.3	1.7	2.0	1.5	0.1	0.0	0.0	0.0	0.2	0.5	1.6	2.5	12.3
Thunderstorms	29	0.0	0.0	0.1	0.9	3.2	7.8	8.6	6.2	2.1	0.1	0.0	0.0	29.0
Heavy Fog Visibility														
1/4 mile or less	29	0.9	1.3	1.0	0.9	0.3	0.2	0.4	0.2	0.6	0.9	1.7	1.1	9.4
Temperature °F														
-Maximum														
90° and above	32	0.0	0.0	0.0	0.1	1.0	3.1	8.0	8.8	1.9	0.1	0.0	0.0	23.0
32° and below	32	22.4	16.3	9.0	0.8	0.0	0.0	0.0	0.0	0.0	0.8	9.8	20.5	79.7
-Minimum														
32° and below	32	30.8	27.8	28.7	17.7	4.1	0.2	0.0	0.0	2.9	16.2	28.5	31.0	187.8
0° and below	32	16.6	10.8	3.6	0.1	0.0	0.0	0.0	0.0	0.0	0.1	3.4	12.8	47.4
AVG. STATION PRESS.(mb)	21	948.0	948.5	945.8	945.9	944.6	944.4	946.2	946.7	947.3	947.3	947.1	947.4	946.6
RELATIVE HUMIDITY (%)														
Hour 00	26	79	80	79	69	67	71	68	64	69	72	79	80	73
Hour 06 (Local Time)	32	78	80	83	80	78	81	80	78	80	79	82	80	80
Hour 12	32	72	71	64	51	47	49	46	45	50	53	68	73	57
Hour 18	32	74	71	60	46	42	44	39	36	43	51	71	76	54
PRECIPITATION (inches):														
Water Equivalent														
-Normal		0.53	0.42	0.69	1.28	1.99	2.28	2.10	1.25	1.33	0.77	0.45	0.58	13.67
-Maximum Monthly	32	1.42	1.48	2.26	3.31	7.38	6.16	6.28	4.66	3.11	3.56	1.15	1.43	7.38
-Year		1967	1967	1975	1967	1965	1991	1993	1993	1986	1971	1986	1982	MAY 1965
-Minimum Monthly	32	0.03	0.04	0.01	0.03	0.15	0.71	0.49	0.07	0.11	T	0.04	0.07	T
-Year		1973	1990	1966	1983	1980	1987	1976	1971	1963	1965	1969	1987	OCT 1965
-Maximum in 24 hrs	32	0.51	0.43	0.92	2.04	2.05	2.20	5.03	3.11	2.24	2.21	0.80	0.88	5.03
-Year		1988	1962	1985	1967	1965	1964	1963	1993	1971	1971	1974	1982	JUL 1963
Snow,Ice pellets,hail														
-Maximum Monthly	32	24.3	16.5	30.9	22.2	15.5	T	T	T	4.0	14.2	16.4	15.2	30.9
-Year		1982	1972	1975	1970	1983	1991	1993	1993	1984	1985	1993	1978	MAR 1975
-Maximum in 24 hrs	32	8.4	4.3	9.7	15.0	14.6	T	T	T	4.0	10.5	7.9	10.1	15.0
-Year		1989	1962	1985	1986	1983	1991	1993	1993	1984	1985	1975	1978	APR 1986
WIND:														
Mean Speed (mph)	29	9.8	9.7	10.3	11.2	11.2	10.4	9.4	9.6	10.0	10.1	9.2	9.7	10.1
Prevailing Direction through 1963		W	NE	NW	SE	SE	SE	SE	SW	SW	SW	SW	SW	SW
Fastest Obs. 1 Min.														
-Direction (!!)	14	32	33	29	25	27	27	26	28	07	31	32	31	26
-Speed (MPH)	14	44	33	46	40	46	52	63	44	38	41	37	45	63
-Year		1987	1989	1986	1981	1984	1991	1991	1985	1986	1988	1989	1985	JUL 1991
Peak Gust														
-Direction (!!)	10	W	W	W	N	W	N	W	W	E	NW	NW	NW	W
-Speed (mph)	10	62	51	58	62	62	69	69	55	60	60	53	63	69
-Date		1990	1987	1986	1984	1992	1991	1991	1985	1986	1991	1989	1985	JUL 1991

PRECIPITATION (inches) WILLISTON, NORTH DAKOTA

YEAR	JAN	FEB	MAR	APR	MAY	JUNE	JULY	AUG	SEP	OCT	NOV	DEC	ANNUAL
1964	0.45	0.24	0.94	1.08	1.10	5.92	2.24	1.14	0.40	0.27	0.65	1.28	15.71
1965	0.71	0.10	0.48	0.99	7.38	2.03	1.52	1.23	2.20	T	0.21	0.61	17.46
1966	0.42	0.31	0.01	0.94	1.75	0.92	1.92	1.80	0.40	0.36	0.30	0.33	9.46
1967	1.42	1.48	0.70	3.31	0.18	0.91	1.05	0.92	0.67	1.15	0.33	0.56	12.68
1968	0.53	0.16	0.10	0.78	1.05	3.68	0.63	3.38	0.85	0.31	0.37	1.09	12.93
1969	1.03	0.60	0.45	0.93	0.42	2.73	3.47	0.54	1.14	0.79	0.04	0.48	12.62
1970	0.56	0.10	0.59	3.20	3.00	2.07	3.11	0.53	2.16	0.56	0.81	0.49	17.18
1971	0.79	0.15	0.45	0.91	2.28	3.88	1.92	0.07	2.48	3.56	0.40	0.42	17.31
1972	0.43	1.37	0.91	1.91	4.10	1.18	2.14	3.21	1.32	0.43	0.15	0.80	17.95
1973	0.03	0.12	0.52	1.79	1.06	2.30	1.01	0.64	3.06	0.28	0.75	0.56	12.12
1974	0.47	0.32	0.91	0.77	3.73	1.88	1.29	2.46	0.17	1.08	0.13	0.30	13.51
1975	0.14	0.13	2.26	2.81	0.86	3.41	2.35	1.50	1.77	1.29	1.14	1.20	18.86
1976	0.58	0.21	0.74	1.08	0.51	2.45	0.49	1.06	0.96	0.28	0.31	0.49	9.16
1977	0.97	0.30	0.02	0.53	2.94	1.30	1.28	1.55	2.30	0.41	0.97	0.86	13.43
1978	0.30	0.52	0.44	0.65	4.17	2.53	2.62	0.51	2.09	0.18	0.89	0.98	15.88
1979	0.12	0.81	1.01	1.69	1.82	2.93	1.78	0.33	0.38	0.23	0.06	0.09	11.25
1980	0.58	0.28	0.16	0.40	0.15	1.80	0.54	1.83	2.24	1.62	0.48	0.72	10.80
1981	0.08	0.21	0.18	0.39	1.18	3.58	2.47	0.87	0.57	0.51	0.51	0.36	10.91
1982	1.28	0.45	1.34	1.02	1.79	2.77	1.17	1.96	1.86	1.91	0.07	1.43	17.05
1983	0.35	0.07	1.07	0.03	1.98	0.87	2.12	0.87	0.51	0.30	0.78	0.57	9.52
1984	0.81	0.12	0.90	1.48	0.47	1.87	0.93	0.65	2.24	0.49	0.33	0.27	10.56
1985	0.13	0.24	1.07	1.53	1.08	1.12	1.40	1.96	1.08	1.86	0.58	0.64	12.69
1986	0.35	0.62	0.83	2.57	3.39	2.19	5.70	1.06	3.11	0.78	1.15	0.09	21.84
1987	0.32	0.19	1.70	0.29	2.02	0.71	4.97	0.47	0.65	0.11	0.18	0.07	11.68
1988	0.75	0.28	0.63	0.15	1.39	3.46	0.52	0.39	1.49	0.28	0.51	0.79	10.64
1989	1.00	0.39	0.53	1.44	1.68	1.48	0.87	1.01	0.54	1.75	0.44	0.41	11.54
1990	0.36	0.04	0.44	0.61	2.08	1.77	1.32	1.73	0.31	0.10	0.14	0.40	9.30
1991	0.18	0.37	0.44	1.60	2.60	6.16	1.04	1.46	2.74	0.38	0.46	0.18	17.61
1992	0.48	0.10	0.56	2.28	1.44	1.92	1.04	2.18	0.76	0.52	0.77	0.44	12.49
1993	0.23	0.23	0.88	0.41	0.81	3.76	6.28	4.66	0.28	0.09	0.95	0.38	18.96
Record Mean	0.53	0.43	0.69	1.16	1.86	3.10	1.98	1.45	1.15	0.77	0.55	0.54	14.22

AVERAGE TEMPERATURE (deg. F) WILLISTON, NORTH DAKOTA

YEAR	JAN	FEB	MAR	APR	MAY	JUNE	JULY	AUG	SEP	OCT	NOV	DEC	ANNUAL
1964	16.2	24.4	21.3	45.4	57.4	62.9	72.7	66.3	53.5	46.4	23.6	0.0	40.8
1965	3.0	9.3	12.4	40.8	52.9	63.4	70.4	67.7	45.4	48.9	26.7	20.9	38.5
1966	-2.7	9.6	32.8	36.8	54.8	63.5	73.0	65.3	60.1	45.7	21.6	17.9	39.9
1967	11.7	10.5	23.3	38.9	51.1	62.5	70.3	69.7	62.0	45.2	29.7	13.1	40.7
1968	8.8	15.6	35.7	42.0	52.0	61.4	69.4	64.8	58.0	45.3	30.6	7.9	40.9
1969	-4.2	12.0	20.8	49.0	55.0	57.8	67.9	72.8	60.8	39.4	32.5	18.6	40.2
1970	3.6	16.9	21.6	37.9	52.5	67.9	72.5	70.0	56.4	41.6	25.3	10.4	39.7
1971	1.6	14.9	28.2	43.1	54.9	64.8	66.3	73.1	56.0	43.0	29.3	9.6	40.4
1972	3.3	8.2	25.9	41.3	56.3	65.8	64.0	69.3	52.6	39.4	28.5	6.7	38.4
1973	16.9	21.5	37.2	41.2	54.8	64.6	68.4	72.8	55.5	47.2	20.2	13.0	42.8
1974	6.8	19.9	26.0	43.0	50.3	65.1	74.3	63.2	53.0	47.0	29.7	23.0	41.8
1975	12.0	12.3	21.9	34.7	52.9	62.1	73.0	65.3	55.0	44.4	28.4	15.7	39.8
1976	10.8	22.6	24.6	46.2	56.3	65.0	70.5	70.5	59.2	40.9	24.3	14.5	42.1
1977	-0.3	25.1	33.9	49.1	62.3	66.4	71.1	61.6	56.1	46.2	24.4	8.3	42.0
1978	-2.4	7.7	27.8	43.0	57.2	63.8	68.6	67.5	60.2	46.4	20.8	8.3	39.1
1979	-2.2	1.7	23.0	34.6	49.2	65.0	70.3	66.9	61.1	46.7	25.1	23.8	38.8
1980	7.7	14.9	25.6	49.7	60.6	67.1	73.2	65.7	58.1	46.6	33.2	16.2	43.6
1981	21.8	23.3	38.1	49.1	59.2	63.8	73.8	74.3	61.8	45.4	34.6	16.1	46.8
1982	-5.8	11.8	26.3	41.8	54.9	64.3	71.7	70.3	56.4	44.4	24.0	16.0	39.7
1983	20.0	25.0	30.0	41.2	51.1	63.7	74.9	78.0	56.3	47.2	30.3	-4.5	42.8
1984	15.2	30.9	28.3	47.4	54.5	63.4	73.0	73.7	50.2	41.0	27.2	5.5	42.5
1985	7.0	13.2	32.4	47.0	58.9	58.8	71.2	64.6	51.2	42.8	13.2	11.2	39.3
1986	22.2	13.3	39.8	41.9	55.8	66.8	67.3	66.8	52.0	45.0	22.6	22.1	43.0
1987	20.7	29.2	32.1	51.6	59.8	68.7	70.0	63.7	59.5	42.3	34.2	22.9	46.2
1988	7.6	15.1	33.2	45.3	61.6	77.3	72.9	69.3	55.7	44.2	27.7	17.1	43.9
1989	12.9	2.2	23.2	43.7	56.3	63.2	75.7	70.2	58.0	45.1	29.3	10.5	40.9
1990	22.9	20.5	34.9	43.2	54.5	65.5	69.9	71.4	61.9	44.6	30.3	9.1	44.1
1991	7.7	29.1	32.6	47.0	57.4	66.4	70.5	73.6	57.3	40.8	25.2	23.3	44.2
1992	22.7	28.2	36.2	43.2	55.6	63.4	64.0	63.9	55.4	44.0	27.6	9.9	42.8
1993	5.7	9.6	33.4	43.9	56.1	58.8	62.3	63.9	52.3	42.1	24.9	20.5	39.5
Record Mean	8.3	12.6	25.6	42.9	54.4	63.5	69.8	67.8	56.5	44.6	27.6	14.8	40.7
Max	18.6	22.9	35.9	54.7	66.7	75.4	83.0	81.6	69.7	56.9	37.5	24.5	52.3
Min	-2.0	2.3	15.2	31.1	42.1	51.6	56.7	54.1	43.3	32.3	17.6	5.0	29.1

REFERENCE NOTES WILLISTON, NORTH DAKOTA

GENERAL
T - TRACE AMOUNT.
BLANK ENTRIES DENOTE MISSING/UNREPORTED DATA.
# INDICATES A STATION OR INSTRUMENT RELOCATION.

SPECIFIC

(a) - LENGTH OF RECORD IN YEARS, ALTHOUGH INDIVIDUAL MONTHS MAY BE MISSING.
0.* OR * - THE VALUE IS BETWEEN 0.0 AND 0.05
NORMALS - BASED ON THE 1961-1990 RECORD PERIOD.
EXTREMES - DATES ARE THE MOST RECENT OCCURRENCE.
WIND DIR. - NUMERALS SHOW TENS OF DEGRESS CLOCKWISE FROM TRUE NORTH. "00" INDICATES CALM.
RESULTANT DIRECTIONS ARE GIVEN TO WHOLE DEGREES.
BOLD VALUES INDICATE EXTREME VALUES WHICH OCCURRED AFTER THE ASOS SYSTEM WAS COMMISSIONED.

RECORD = PERIOD OF RECORD
RECORD MEAN PRECIPITATION IS THE MEAN OF ALL DAILY PRECIPITATION AMOUNTS DURING THE PERIOD OF RECORD.
RECORD MAX(MIN) TEMPERATURE IS THE MEAN OF ALL DAILY MAX(MIN) TEMPERATURES DURING THE PERIOD OF RECORD.
RECORD MEAN TEMPERATURE IS THE SUM OF THE RECORD MAX AND RECORD MIN DIVIDED BY 2.
AVERAGE TEMPERATURE IS THE SUM OF THE MEAN DAILY MAX AND MIN TEMPERATURE DIVIDED BY 2.

EXCEPTIONS

PRIOR TO MARCH 1967, THUNDERSTORMS AND HEAVY FOG MAY BE INCOMPLETE, DUE TO PART-TIME OPERATIONS.

RECORD MEANS ARE THROUGH THE CURRENT YEAR, BEGINNING IN 1879 FOR TEMPERATURE
1879 FOR PRECIPITATION
1962 FOR SNOWFALL

HEATING DEGREE DAYS Base 65 deg. F WILLISTON, NORTH DAKOTA

SEASON	JULY	AUG	SEP	OCT	NOV	DEC	JAN	FEB	MAR	APR	MAY	JUNE	TOTAL
1964-65	0	86	349	567	1236	2017	1920	1557	1624	717	370	87	10530
1965-66	3	62	580	494	1142	1358	2098	1547	992	839	323	128	9566
1966-67	0	94	190	593	1297	1455	1649	1525	1285	775	435	99	9397
1967-68	44	27	151	605	1056	1607	1741	1425	900	683	397	146	8782
1968-69	26	90	225	603	1028	1769	2149	1480	1361	475	326	223	9755
1969-70	25	10	166	786	966	1431	1903	1340	1337	807	382	30	9183
1970-71	2	17	286	719	1182	1689	1964	1400	1135	649	310	62	9415
1971-72	42	5	283	674	1063	1713	1913	1644	1209	702	291	50	9589
1972-73	75	25	380	789	1089	1805	1488	1213	855	708	316	78	8821
1973-74	14	1	291	546	1338	1608	1801	1258	1204	653	448	80	9242
1974-75	1	100	354	553	1052	1294	1635	1470	1329	904	371	117	9180
1975-76	1	60	299	632	1092	1522	1675	1226	1247	556	267	87	8664
1976-77	10	11	203	740	1215	1563	2024	1112	959	469	149	46	8501
1977-78	8	131	267	577	1212	1759	2091	1600	1148	655	252	91	9791
1978-79	21	39	213	569	1322	1756	2085	1774	1294	906	495	76	10550
1979-80	7	42	145	560	1190	1272	1775	1447	1216	460	237	43	8394
1980-81	5	52	221	567	947	1511	1330	1162	825	473	205	70	7368
1981-82	4	9	149	600	903	1509	2198	1485	1191	692	322	79	9141
1982-83	2	46	278	633	1223	1516	1389	1115	1076	708	436	101	8523
1983-84	1	1	300	544	1031	2158	1538	983	1131	520	343	92	8642
1984-85	0	25	444	741	1128	1845	1793	1446	1001	535	210	201	9369
1985-86	21	73	415	683	1553	1664	1321	1446	775	685	303	35	8974
1986-87	14	48	386	612	1266	1324	1369	997	1012	402	190	33	7653
1987-88	31	88	183	698	920	1301	1781	1444	980	585	175	8	8194
1988-89	1	35	279	637	1114	1479	1613	1757	1292	640	275	115	9237
1989-90	0	35	225	609	1065	1690	1294	1241	927	649	333	88	8156
1990-91	19	7	166	626	1034	1728	1775	999	997	533	253	20	8157
1991-92	5	8	262	744	1188	1289	1306	1058	887	649	313	112	7821
1992-93	75	111	296	644	1116	1704	1838	1551	973	630	277	203	9418
1993-94	100	87	377	704	1197	1376							

COOLING DEGREE DAYS Base 65 deg. F WILLISTON, NORTH DAKOTA

YEAR	JAN	FEB	MAR	APR	MAY	JUNE	JULY	AUG	SEP	OCT	NOV	DEC	TOTAL
1969	0	0	0	0	23	14	119	258	50	0	0	0	464
1970	0	0	0	0	3	123	244	176	34	0	0	0	580
1971	0	0	0	0	3	62	90	262	23	0	0	0	440
1972	0	0	0	0	28	80	52	167	15	0	0	0	342
1973	0	0	0	0	7	72	125	250	13	0	0	0	467
1974	0	0	0	2	0	92	295	51	0	0	0	0	440
1975	0	0	0	0	2	39	253	77	5	0	0	0	376
1976	0	0	0	0	4	95	185	185	37	1	0	0	507
1977	0	0	0	0	74	96	204	30	6	0	0	0	410
1978	0	0	0	0	19	60	142	125	75	0	0	0	421
1979	0	0	0	0	12	83	180	107	33	0	0	0	415
1980	0	0	0	7	107	114	265	84	22	4	0	0	603
1981	0	0	0	2	31	41	285	301	60	0	0	0	720
1982	0	0	0	0	13	64	217	215	29	0	0	0	538
1983	0	0	0	0	8	68	314	413	45	0	0	0	848
1984	0	0	0	0	21	54	251	304	8	0	0	0	638
1985	0	0	0	0	26	21	218	68	6	0	0	0	339
1986	0	0	0	0	27	96	94	110	0	0	0	0	327
1987	0	0	0	6	37	150	192	53	22	0	0	0	460
1988	0	0	0	0	75	383	253	175	4	0	0	0	890
1989	0	0	0	5	9	68	338	202	22	0	0	0	644
1990	0	0	0	4	17	110	175	211	75	0	0	0	592
1991	0	0	0	0	25	68	182	282	40	0	0	0	597
1992	0	0	0	3	29	71	50	87	13	2	0	0	255
1993	0	0	0	0	8	28	22	58	2	0	0	0	118

SNOWFALL (inches) WILLISTON, NORTH DAKOTA

SEASON	JULY	AUG	SEP	OCT	NOV	DEC	JAN	FEB	MAR	APR	MAY	JUNE	TOTAL
1964-65	0.0	0.0	0.0	1.3	6.2	12.8	7.1	0.9	5.0	1.6	1.4	0.0	36.3
1965-66	0.0	0.0	3.0	0.0	2.2	6.7	4.3	3.3	0.1	6.2	0.5	0.0	26.3
1966-67	0.0	0.0	T	2.0	3.4	3.7	16.6	15.0	6.2	12.4	1.1	0.0	60.4
1967-68	0.0	0.0	0.0	T	1.6	5.6	8.9	1.2	0.3	2.9	T	0.0	20.5
1968-69	0.0	0.0	0.0	T	2.6	11.6	10.7	6.0	4.5	T	T	0.0	35.4
1969-70	0.0	0.0	0.0	1.5	T	5.3	5.9	1.0	7.6	22.2	0.0	0.0	43.5
1970-71	0.0	0.0	T	T	6.4	5.6	10.5	2.5	1.6	1.9	0.7	0.0	29.2
1971-72	0.0	0.0	T	1.8	4.3	5.4	6.5	16.5	6.4	2.7	0.0	0.0	43.6
1972-73	0.0	0.0	3.0	3.9	1.5	11.5	0.6	1.8	2.3	1.6	0.0	0.0	26.2
1973-74	0.0	0.0	0.0	T	6.4	5.5	3.4	5.7	9.3	1.4	T	0.0	31.7
1974-75	0.0	0.0	0.0	0.1	2.2	4.4	3.6	1.3	30.9	10.3	0.0	0.0	52.8
1975-76	0.0	0.0	0.0	4.5	14.1	14.1	10.3	2.0	14.6	0.4	0.0	0.0	60.0
1976-77	0.0	0.0	0.0	1.9	2.9	6.9	11.1	3.1	T	2.0	0.0	0.0	27.9
1977-78	0.0	0.0	0.0	T	8.4	10.0	3.9	5.7	4.6	1.4	0.0	0.0	34.0
1978-79	0.0	0.0	0.0	T	9.4	15.2	2.9	14.1	10.3	10.1	T	0.0	62.0
1979-80	0.0	0.0	0.0	T	1.2	0.3	10.4	7.9	2.1	3.5	0.0	0.0	25.4
1980-81	0.0	0.0	0.0	1.1	3.3	10.4	1.2	3.1	T	0.0	T	0.0	19.1
1981-82	0.0	0.0	0.0	0.3	4.6	5.7	24.3	8.2	17.0	9.8	0.5	0.0	70.4
1982-83	0.0	0.0	T	T	0.8	13.4	1.7	1.6	9.2	0.1	15.5	0.0	42.3
1983-84	0.0	0.0	1.0	T	6.0	9.2	7.2	0.5	7.9	13.5	0.3	0.0	45.6
1984-85	0.0	0.0	4.0	4.5	3.5	6.5	1.8	3.1	11.4	1.5	0.0	0.0	36.3
1985-86	0.0	0.0	0.1	14.2	11.3	9.2	3.6	8.7	4.4	17.6	T	0.0	69.1
1986-87	0.0	0.0	0.0	T	13.4	1.2	4.2	1.8	14.2	0.4	0.0	0.0	35.2
1987-88	0.0	0.0	0.0	0.2	T	1.2	8.9	3.4	3.7	T	0.0	0.0	17.4
1988-89	0.0	0.0	0.0	T	4.2	13.3	22.8	6.2	1.8	1.3	0.0	T	49.6
1989-90	T	0.0	T	2.2	4.6	5.5	6.0	0.8	4.3	2.4	2.1	T	27.9
1990-91	T	T	0.0	T	1.6	8.7	4.4	5.2	7.2	1.8	2.7	T	31.6
1991-92	T	T	T	5.3	3.3	2.7	6.3	2.2	0.9	3.5	0.0	0.0	24.2
1992-93	0.0	T	0.0	3.8	8.5	10.7	7.3	4.1	2.6	T	0.0	0.0	37.0
1993-94	T	T	T	0.4	16.4	5.2							
Record Mean	T	T	0.3	1.7	5.0	7.3	7.2	4.8	6.5	4.7	0.8	T	38.3

AKRON, OHIO

The station at the Akron-Canton Airport is located about midway between Akron and Canton, a few miles south of the crest separating the Lake Erie and Muskingum River drainage areas. Precipitation at the station and southward drains through the Muskingum River into the Ohio, while northward of the crest the Cuyahoga and other streams flow into Lake Erie. The terrain is rolling with highest elevations near 1,300 feet above sea level and many small lakes provide water for local industry as well as recreational facilities for the densely populated region. The area is mainly industrial, agricultural operations having diminished rapidly in recent years.

Lake Erie has considerable influence on the area weather, tempering cold air masses during the late fall and winter, as well as contributing to the formation of brief, but heavy snow squalls until the lake freezes over.

The arrival of spring is late in this area, but has the good effect of retarding plant growth and allowing growing of normally frost-susceptible fruits. Summers are moderately warm, but quite humid, while the months of September, October, and sometimes November are usually pleasant although with considerable morning fog. The average last occurrence of freezing temperatures in spring is the end of April, and the first occurrence in fall is late October. In past years, growing seasons for most vegetation has varied from 120 to 211 days. Temperatures and occurences of frost vary widely over the area because of the hilly terrain. Due to the influence of Lake Erie, snowfall is usually much heavier north of the station.

NORMALS, MEANS, AND EXTREMES

AKRON, OHIO

LATITUDE: 40°55'N LONGITUDE: 81°26' W ELEVATION: FT. GRND 1208 BARO 1242 TIME ZONE: EASTERN WBAN: 14895

	(a)	JAN	FEB	MAR	APR	MAY	JUNE	JULY	AUG	SEP	OCT	NOV	DEC	YEAR
TEMPERATURE °F:														
Normals														
-Daily Maximum		32.6	35.9	47.3	59.1	69.7	78.5	82.3	80.4	73.7	62.1	49.7	37.7	59.1
-Daily Minimum		16.9	18.9	28.6	37.9	48.2	57.0	61.5	60.1	53.7	42.7	34.2	23.6	40.3
-Monthly		24.8	27.4	38.0	48.5	59.0	67.8	71.9	70.3	63.7	52.4	42.0	30.7	49.7
Extremes														
-Record Highest	45	70	68	81	88	93	100	101	98	99	86	80	76	101
-Year		1950	1961	1986	1986	1991	1988	1988	1953	1953	1952	1961	1982	JUL 1988
-Record Lowest	45	-24	-13	-3	10	24	32	43	41	32	20	-1	-16	-24
-Year		1985	1979	1980	1964	1966	1972	1988	1982	1956	1952	1958	1989	JAN 1985
NORMAL DEGREE DAYS:														
Heating (base 65°F)		1246	1053	837	495	224	37	0	13	103	399	690	1063	6160
Cooling (base 65°F)		0	0	0	0	38	121	217	177	64	8	0	0	625
% OF POSSIBLE SUNSHINE														
MEAN SKY COVER (tenths)														
Sunrise - Sunset	45	8.0	7.8	7.5	7.2	6.6	6.2	5.9	5.9	6.0	6.1	7.7	8.1	6.9
MEAN NUMBER OF DAYS:														
Sunrise to Sunset														
-Clear	45	3.3	3.4	4.4	5.0	6.4	6.6	7.0	7.7	8.4	8.9	3.9	2.8	67.7
-Partly Cloudy	45	5.6	6.0	6.6	7.5	9.1	11.0	12.8	12.2	9.1	7.7	6.2	5.5	99.4
-Cloudy	45	22.1	18.9	20.0	17.6	15.5	12.4	11.1	11.1	12.5	14.4	19.9	22.7	198.2
Precipitation														
.01 inches or more	45	16.0	14.5	15.8	14.3	12.9	10.9	11.0	9.6	9.7	10.4	14.1	15.6	154.7
Snow, Ice pellets, hail														
1.0 inches or more	45	3.9	2.9	2.8	0.8	0.*	0.0	0.0	0.0	0.0	0.2	1.3	3.0	14.9
Thunderstorms	45	0.3	0.4	2.1	3.8	5.7	7.1	7.8	5.8	3.4	1.4	0.8	0.3	38.8
Heavy Fog Visibility														
1/4 mile or less	45	2.6	2.8	2.3	1.6	1.5	1.4	1.7	2.5	2.4	2.2	1.9	3.0	25.9
Temperature °F														
-Maximum														
90° and above	30	0.0	0.0	0.0	0.0	0.3	1.6	3.8	1.7	0.4	0.0	0.0	0.0	7.8
32° and below	30	14.4	10.8	3.7	0.2	0.0	0.0	0.0	0.0	0.0	0.0	1.5	9.1	39.6
-Minimum														
32° and below	30	28.1	23.8	20.8	9.1	0.8	0.*	0.0	0.0	0.0	3.8	14.1	24.7	125.2
0° and below	30	3.0	1.9	0.1	0.0	0.0	0.0	0.0	0.0	0.0	0.0	0.0	0.8	5.7
AVG. STATION PRESS. (mb)	21	972.7	972.9	971.4	970.8	971.3	971.8	972.9	974.1	974.5	974.5	973.5	973.3	972.8
RELATIVE HUMIDITY (%)														
Hour 01	30	76	75	73	70	74	79	82	84	83	78	76	77	77
Hour 07 (Local Time)	30	79	78	77	75	77	80	84	88	87	82	79	79	80
Hour 13	30	69	65	60	54	55	56	56	58	60	58	65	70	61
Hour 19	30	72	69	64	58	59	61	62	66	71	67	71	74	66
PRECIPITATION (inches):														
Water Equivalent														
-Normal		2.16	2.23	3.33	3.16	3.73	3.18	4.08	3.32	3.32	2.35	3.01	2.95	36.82
-Maximum Monthly	45	8.70	5.24	8.83	6.46	9.60	8.42	11.43	8.19	9.02	8.42	9.39	6.72	11.43
-Year		1950	1956	1964	1981	1956	1989	1958	1974	1990	1954	1985	1990	JUL 1958
-Minimum Monthly	45	0.71	0.31	1.04	0.91	1.05	0.37	0.67	0.49	0.20	0.45	0.62	0.31	0.20
-Year		1961	1987	1958	1971	1977	1988	1991	1970	1960	1953	1976	1955	SEP 1960
-Maximum in 24 hrs	45	2.99	2.57	3.29	2.01	3.18	2.87	4.18	3.00	6.30	2.77	2.66	1.86	6.30
-Year		1959	1959	1964	1987	1985	1970	1958	1975	1979	1954	1985	1991	SEP 1979
Snow, Ice pellets, hail														
-Maximum Monthly	45	37.5	21.1	20.9	20.9	3.2	T	0.0	0.0	T	6.8	22.3	29.4	37.5
-Year		1978	1984	1960	1987	1966	1991			1965	1952	1950	1974	JAN 1978
-Maximum in 24 hrs	45	10.9	12.3	10.7	19.7	3.2	T	0.0	0.0	T	4.4	7.4	17.9	19.7
-Year		1966	1984	1973	1987	1966	1991			1965	1993	1950	1974	APR 1987
WIND:														
Mean Speed (mph)	45	11.6	11.1	11.4	10.8	9.2	8.4	7.6	7.4	8.1	9.1	10.9	11.3	9.8
Prevailing Direction through 1963		SW	NW	NW	SW	SW	S	SW	SW	S	S	S	S	S
Fastest Obs. 1 Min.														
-Direction (!!)	32	22	25	22	30	32	29	32	32	36	27	30	24	25
-Speed (MPH)	32	44	51	40	40	46	39	40	35	40	30	35	40	51
-Year		1978	1962	1964	1968	1967	1986	1991	1987	1962	1988	1989	1968	FEB 1962
Peak Gust														
-Direction (!!)	10	W	W	NW	SW	W	NW	NW	W	W	SW	SW	SW	NW
-Speed (mph)	10	54	53	60	60	56	56	68	60	52	48	58	58	68
-Date		1990	1990	1989	1984	1990	1986	1991	1992	1986	1986	1988	1987	JUL 1991

PRECIPITATION (inches) AKRON, OHIO

YEAR	JAN	FEB	MAR	APR	MAY	JUNE	JULY	AUG	SEP	OCT	NOV	DEC	ANNUAL
1964	2.19	1.81	8.83	6.06	3.29	1.17	3.54	5.65	0.58	1.26	1.90	4.54	40.82
1965	5.25	2.66	2.68	1.57	2.55	2.04	1.56	4.16	3.12	3.80	2.41	1.75	33.55
1966	2.03	2.20	1.95	3.32	2.30	1.92	2.82	2.95	2.01	0.98	5.05	2.11	29.64
1967	1.14	2.47	3.07	3.76	4.43	1.01	2.11	3.01	3.69	1.98	2.80	2.41	31.88
1968	2.88	0.48	2.71	2.35	6.54	2.40	4.13	1.61	3.25	1.38	3.86	3.88	35.47
1969	2.57	0.88	1.91	3.88	3.78	2.94	6.08	1.68	1.51	2.15	2.85	2.73	32.96
1970	1.48	1.84	2.46	3.94	3.30	7.06	4.22	0.49	3.46	4.24	3.11	2.63	38.23
1971	1.75	3.89	2.41	0.91	2.65	4.61	2.65	2.86	3.06	1.07	2.02	4.32	32.20
1972	1.33	2.35	4.09	3.96	2.88	3.20	7.26	3.20	6.47	1.54	4.27	3.33	43.88
1973	1.67	2.33	4.36	3.45	3.97	3.62	3.88	2.01	2.09	4.60	2.43	2.38	36.79
1974	2.93	1.85	5.68	2.50	4.85	2.74	2.42	8.19	1.81	1.51	3.41	4.32	42.21
1975	3.79	3.58	3.45	1.76	3.57	2.76	2.22	7.47	5.27	2.57	1.67	2.88	40.99
1976	3.16	2.99	3.49	1.64	1.15	3.94	9.00	2.01	3.90	2.31	0.62	1.42	35.63
1977	1.24	1.13	3.62	4.12	1.05	5.43	4.75	4.58	3.72	2.04	4.19	4.09	39.96
1978	3.69	0.48	2.54	2.21	4.41	3.31	2.76	4.14	1.98	2.58	1.06	4.20	33.36
1979	3.17	2.11	1.74	3.58	5.06	2.48	1.96	4.28	7.85	1.51	3.13	2.44	39.31
1980	1.58	1.64	4.71	2.97	5.42	3.76	5.43	4.94	1.76	1.86	1.94	2.15	38.16
1981	0.80	4.63	1.95	6.46	4.84	5.76	4.13	1.47	3.20	1.66	1.92	3.35	40.17
1982	4.71	1.80	3.81	1.18	3.13	4.23	2.13	1.47	2.98	0.85	4.94	3.45	34.68
1983	1.58	1.38	3.76	5.13	4.53	2.03	3.38	3.06	2.73	3.22	4.16	3.43	38.39
1984	1.15	3.02	3.00	2.82	5.44	1.64	3.02	3.77	2.62	2.66	3.41	2.58	35.13
1985	1.33	1.80	4.64	1.14	6.49	2.72	2.56	3.47	0.72	1.55	9.39	2.80	38.61
1986	1.38	3.17	2.07	1.83	2.11	3.05	3.70	1.24	3.51	3.23	3.41	2.81	31.51
1987	2.05	0.31	2.87	4.02	2.61	3.83	2.85	3.96	2.11	1.99	1.38	2.33	30.31
1988	1.15	2.69	2.11	2.82	1.88	0.37	5.05	3.89	4.92	2.63	3.88	1.83	33.22
1989	2.23	2.11	3.86	2.46	5.98	8.42	2.83	1.13	3.38	2.45	2.49	1.98	39.32
1990	2.18	5.01	1.27	5.12	7.28	3.60	10.03	6.26	9.02	7.10	2.11	6.72	65.70
1991	2.44	1.96	2.66	3.04	1.49	0.81	0.67	1.70	2.50	1.49	1.99	3.33	24.08
1992	1.94	1.97	3.87	3.72	3.02	1.59	10.94	4.96	3.71	2.07	5.10	2.25	45.14
1993	4.03	1.92	3.99	3.96	1.36	5.46	1.86	1.66	3.95	4.43	5.85	2.76	41.23
Record Mean	2.53	2.22	3.12	3.15	3.78	3.59	3.97	3.35	3.25	2.50	2.62	2.58	36.66

AVERAGE TEMPERATURE (deg. F) AKRON, OHIO

YEAR	JAN	FEB	MAR	APR	MAY	JUNE	JULY	AUG	SEP	OCT	NOV	DEC	ANNUAL
1964	28.8	25.0	37.4	49.9	61.3	68.0	72.9	67.1	63.4	50.0	44.6	31.9	50.0
1965	25.2	26.2	32.3	47.3	64.6	66.9	69.8	69.2	66.1	50.5	42.3	36.4	49.7
1966	22.0	28.4	38.7	46.0	53.9	69.4	73.5	69.7	61.5	50.5	41.7	31.0	48.9
1967	32.3	24.9	38.2	50.6	53.2	72.4	70.7	68.2	60.9	53.2	36.7	34.2	49.6
1968	23.0	23.4	40.1	52.2	57.1	69.8	72.9	72.9	66.4	54.6	42.9	30.4	50.5
1969	27.5	30.4	33.9	50.1	59.0	65.9	72.1	70.2	62.5	52.2	39.1	26.9	49.2
1970	19.6	26.5	33.4	50.1	62.1	68.8	71.9	71.5	67.6	56.3	42.7	32.0	50.2
1971	22.6	30.4	34.3	46.3	57.3	71.6	71.0	68.3	66.7	58.7	38.5	35.7	50.1
1972	25.7	24.0	32.6	45.4	59.2	63.2	71.2	69.8	64.5	49.6	39.6	35.6	48.4
1973	29.8	27.5	47.5	49.9	57.2	71.0	73.1	72.5	65.7	57.1	45.3	32.9	52.5
1974	30.7	27.1	39.3	51.8	58.1	66.3	71.9	71.1	60.1	50.7	43.5	31.5	50.2
1975	32.0	31.2	36.0	43.9	63.2	69.5	72.7	73.2	58.9	53.9	47.5	33.4	51.3
1976	23.0	36.8	43.3	49.2	55.2	68.6	68.4	66.3	59.8	46.6	33.2	23.6	47.9
1977	11.4	27.0	44.7	54.0	65.7	66.0	74.8	71.3	64.8	49.9	43.7	28.3	50.1
1978	19.1	15.9	32.2	47.0	58.6	67.8	70.5	72.1	68.2	50.1	42.5	32.6	48.0
1979	19.8	16.8	41.9	46.6	56.7	66.2	69.3	69.2	63.1	50.8	42.6	33.1	48.0
1980	26.8	23.3	34.9	47.4	60.5	63.9	72.6	74.5	65.8	48.6	39.4	29.5	48.9
1981	21.3	31.9	36.0	51.9	58.5	69.8	71.7	70.6	62.8	50.4	43.5	30.7	49.9
1982	20.1	27.7	37.6	45.4	66.1	64.7	73.4	68.2	62.4	55.0	45.0	40.0	50.5
1983	28.8	32.8	41.7	47.3	56.7	68.4	75.0	73.8	65.4	54.0	44.5	24.7	51.1
1984	22.2	36.2	30.0	48.5	54.7	70.5	69.8	70.1	60.2	56.7	40.1	37.1	49.7
1985	19.7	24.8	40.7	53.6	60.6	63.1	70.7	69.3	65.3	54.2	45.5	24.3	49.3
1986	26.3	29.1	39.6	51.7	61.2	67.3	72.8	68.9	66.3	53.4	39.1	31.7	50.6
1987	26.8	30.9	40.2	49.4	62.2	69.8	74.4	70.6	63.4	46.8	44.7	33.9	51.1
1988	24.6	26.8	37.3	47.6	60.4	68.0	76.2	73.5	62.8	45.7	42.7	29.8	49.6
1989	34.2	25.3	39.6	45.4	57.0	67.5	73.0	70.1	63.0	52.3	39.2	18.2	48.7
1990	34.9	34.2	42.1	49.3	55.9	66.7	70.3	69.1	62.8	53.5	44.4	35.5	51.6
1991	27.8	33.4	42.7	54.0	68.7	72.8	75.5	73.0	63.6	54.3	38.7	33.2	53.1
1992	29.0	32.4	36.0	48.1	57.5	64.5	70.8	66.5	61.9	48.7	40.8	33.0	49.1
1993	31.4	25.0	35.2	47.9	59.3	67.3	74.2	73.9	61.2	50.0	41.0	29.5	49.7
Record Mean	26.6	27.4	36.8	48.1	59.2	68.2	72.3	70.6	64.2	52.6	40.8	30.3	49.8
Max	34.2	35.5	45.7	58.3	69.9	78.8	83.1	81.2	74.6	62.4	48.5	37.1	59.1
Min	19.0	19.3	27.8	37.9	48.4	57.5	61.6	60.0	53.7	42.8	33.1	23.5	40.4

REFERENCE NOTES AKRON, OHIO

GENERAL
T - TRACE AMOUNT.
BLANK ENTRIES DENOTE MISSING/UNREPORTED DATA.
# INDICATES A STATION OR INSTRUMENT RELOCATION.

SPECIFIC

(a) - LENGTH OF RECORD IN YEARS, ALTHOUGH INDIVIDUAL MONTHS MAY BE MISSING.
0.* OR * - THE VALUE IS BETWEEN 0.0 AND 0.05
NORMALS - BASED ON THE 1961-1990 RECORD PERIOD.
EXTREMES - DATES ARE THE MOST RECENT OCCURRENCE
WIND DIR. - NUMERALS SHOW TENS OF DEGRESS CLOCKWISE FROM TRUE NORTH. "00" INDICATES CALM.
RESULTANT DIRECTIONS ARE GIVEN TO WHOLE DEGREES.
BOLD VALUES INDICATE EXTREME VALUES WHICH OCCURRED AFTER THE ASOS SYSTEM WAS COMMISSIONED.

RECORD = PERIOD OF RECORD
RECORD MEAN PRECIPITATION IS THE MEAN OF ALL DAILY PRECIPITATION AMOUNTS DURING THE PERIOD OF RECORD.
RECORD MAX(MIN) TEMPERATURE IS THE MEAN OF ALL DAILY MAX(MIN) TEMPERATURES DURING THE PERIOD OF RECORD.
RECORD MEAN TEMPERATURE IS THE SUM OF THE RECORD MAX AND RECORD MIN DIVIDED BY 2.
AVERAGE TEMPERATURE IS THE SUM OF THE MEAN DAILY MAX AND MIN TEMPERATURE DIVIDED BY 2.

EXCEPTIONS

RECORD MEANS ARE THROUGH THE CURRENT YEAR, BEGINNING IN 1887 FOR TEMPERATURE
1887 FOR PRECIPITATION
1949 FOR SNOWFALL

HEATING DEGREE DAYS Base 65 deg. F AKRON, OHIO

SEASON	JULY	AUG	SEP	OCT	NOV	DEC	JAN	FEB	MAR	APR	MAY	JUNE	TOTAL
1964-65	2	46	121	458	605	1020	1226	1081	1005	524	94	56	6238
1965-66	8	40	82	442	675	878	1329	1019	807	568	344	46	6238
1966-67	3	10	148	445	690	1044	1006	1116	825	433	368	8	6096
1967-68	17	22	146	368	841	949	1296	1198	763	379	246	19	6244
1968-69	9	19	48	338	655	1067	1156	963	959	441	215	81	5951
1969-70	1	6	133	401	768	1172	1402	1073	972	458	161	33	6580
1970-71	10	2	61	281	664	1015	1309	962	946	554	248	9	6061
1971-72	0	18	68	202	786	899	1214	1182	999	581	184	110	6243
1972-73	25	16	84	470	755	905	1082	1043	538	455	243	1	5617
1973-74	2	12	78	240	587	988	1059	1059	790	402	241	45	5503
1974-75	0	0	169	438	637	1029	1015	940	895	626	122	38	5909
1975-76	2	1	183	344	519	971	1293	811	666	484	308	24	5606
1976-77	15	56	171	564	948	1272	1656	1057	627	354	105	82	6907
1977-78	4	21	68	458	632	1130	1413	1366	1013	533	247	51	6936
1978-79	9	0	58	456	670	997	1397	1348	712	550	288	62	6547
1979-80	35	28	118	444	664	981	1179	1202	925	520	168	95	6359
1980-81	1	3	79	500	763	1093	1351	920	889	384	222	15	6220
1981-82	6	6	138	444	638	1058	1386	1037	844	583	54	53	6247
1982-83	5	32	129	318	596	770	1113	896	715	527	261	45	5407
1983-84	10	0	103	338	608	1244	1321	831	1078	493	323	7	6356
1984-85	7	16	180	254	740	860	1397	1119	746	371	174	82	5946
1985-86	3	5	116	329	578	1254	1192	999	783	403	158	53	5873
1986-87	2	43	70	361	770	1023	1178	945	765	466	160	32	5815
1987-88	2	23	88	558	606	957	1243	1103	853	515	175	67	6190
1988-89	8	11	95	599	659	1083	949	1106	784	583	280	38	6195
1989-90	3	13	124	391	764	1444	923	856	707	494	280	69	6068
1990-91	14	3	130	352	611	909	1143	880	684	349	76	4	5155
1991-92	0	0	139	345	782	978	1111	939	893	509	251	76	6023
1992-93	7	38	145	497	717	986	1036	1116	914	506	192	56	6210
1993-94	2	1	161	458	715	1093							

COOLING DEGREE DAYS Base 65 deg. F AKRON, OHIO

YEAR	JAN	FEB	MAR	APR	MAY	JUNE	JULY	AUG	SEP	OCT	NOV	DEC	TOTAL
1969	0	0	0	3	36	115	225	179	63	11	0	0	632
1970	0	0	0	17	79	152	231	210	146	18	0	0	853
1971	0	0	0	0	15	217	196	130	123	13	0	0	694
1972	0	0	0	0	13	63	223	169	77	0	0	0	545
1973	0	0	0	9	5	188	259	252	104	5	0	0	822
1974	0	0	0	13	33	90	222	196	29	1	1	0	585
1975	0	0	0	0	74	179	250	260	5	7	0	0	775
1976	0	0	0	16	14	139	132	101	22	0	0	0	424
1977	0	0	5	30	131	118	314	223	71	0	0	0	892
1978	0	0	0	0	54	143	185	226	159	0	0	0	767
1979	0	0	0	5	36	106	173	166	68	10	0	0	564
1980	0	0	0	0	37	69	244	306	109	0	0	0	765
1981	0	0	0	1	26	166	223	187	78	1	0	0	682
1982	0	0	0	3	97	52	272	140	59	14	3	3	643
1983	0	0	0	2	11	153	326	279	122	5	0	0	898
1984	0	0	0	3	12	179	163	180	43	4	0	0	584
1985	0	0	0	36	44	32	186	149	134	1	1	0	583
1986	0	0	2	13	47	131	249	169	119	10	0	0	740
1987	0	0	0	5	84	184	298	203	48	0	1	0	823
1988	0	0	0	0	38	163	362	283	35	6	0	0	887
1989	0	0	5	0	40	121	257	176	70	4	0	0	673
1990	0	0	7	30	5	128	187	137	70	3	0	0	567
1991	0	0	0	25	198	245	335	255	106	21	0	0	1185
1992	0	0	0	7	25	68	195	93	58	0	0	0	446
1993	0	0	0	0	22	129	295	285	57	0	0	0	788

SNOWFALL (inches) AKRON, OHIO

SEASON	JULY	AUG	SEP	OCT	NOV	DEC	JAN	FEB	MAR	APR	MAY	JUNE	TOTAL
1964-65	0.0	0.0	0.0	T	5.1	9.2	15.9	8.9	10.9	T	0.0	0.0	50.0
1965-66	0.0	0.0	T	0.1	0.9	1.8	16.3	6.8	4.0	4.8	3.2	0.0	37.9
1966-67	0.0	0.0	0.0	T	9.7	6.8	4.9	19.7	13.8	0.8	T	0.0	55.7
1967-68	0.0	0.0	0.0	T	8.8	4.4	16.6	6.6	4.1	0.1	0.0	0.0	40.6
1968-69	0.0	0.0	0.0	0.2	2.4	11.1	4.8	5.8	5.9	T	0.0	0.0	30.2
1969-70	0.0	0.0	0.0	0.5	4.7	17.1	12.0	8.4	10.5	0.2	T	0.0	53.4
1970-71	0.0	0.0	0.0	0.0	0.7	4.4	9.0	15.7	15.1	T	T	0.0	44.9
1971-72	0.0	0.0	0.0	0.0	10.8	2.1	7.5	14.7	10.5	1.1	0.0	0.0	46.7
1972-73	0.0	0.0	0.0	0.5	2.9	6.5	3.9	9.9	10.8	2.6	T	0.0	37.1
1973-74	0.0	0.0	0.0	0.0	0.7	8.8	8.3	8.7	8.0	3.9	T	0.0	38.4
1974-75	0.0	0.0	0.0	3.5	6.0	29.4	8.8	10.5	11.9	2.6	0.0	0.0	72.7
1975-76	0.0	0.0	0.0	0.0	3.9	4.8	25.1	9.8	4.8	2.7	T	0.0	51.1
1976-77	0.0	0.0	0.0	0.4	7.0	7.4	19.5	5.6	1.0	1.5	0.0	0.0	42.4
1977-78	0.0	0.0	0.0	T	10.0	19.3	37.5	9.1	6.0	0.1	0.0	0.0	82.0
1978-79	0.0	0.0	0.0	0.0	1.1	4.5	21.9	14.5	3.1	1.6	0.0	0.0	46.7
1979-80	0.0	0.0	0.0	0.4	1.6	5.1	8.5	12.2	6.0	0.4	T	0.0	34.2
1980-81	0.0	0.0	0.0	T	7.0	11.0	13.0	9.3	12.0	T	0.0	0.0	52.3
1981-82	0.0	0.0	0.0	T	1.4	19.0	17.8	5.3	10.4	7.8	0.0	0.0	61.7
1982-83	0.0	0.0	0.0	T	2.4	14.2	4.7	8.9	7.6	1.0	0.0	0.0	38.8
1983-84	0.0	0.0	0.0	0.0	4.4	8.8	10.5	21.1	12.5	0.2	0.0	0.0	57.5
1984-85	0.0	0.0	0.0	0.0	2.6	7.8	20.9	12.4	1.2	5.3	0.0	0.0	50.2
1985-86	0.0	0.0	0.0	0.0	T	10.7	9.5	7.6	5.3	2.6	0.0	0.0	35.7
1986-87	0.0	0.0	0.0	0.0	5.4	0.8	11.0	1.8	9.8	20.9	0.0	0.0	49.7
1987-88	0.0	0.0	0.0	T	2.7	13.3	7.7	8.1	16.7	0.4	0.0	0.0	48.9
1988-89	0.0	0.0	0.0	0.2	1.6	8.0	4.1	5.4	11.4	3.2	1.5	0.0	35.4
1989-90	0.0	0.0	0.0	3.7	2.5	13.2	11.5	6.3	0.5	3.9	0.0	0.0	41.6
1990-91	0.0	0.0	0.0	T	0.3	7.4	7.4	10.3	2.1	0.1	0.0	T	27.6
1991-92	0.0	0.0	0.0	T	6.6	5.3	12.4	5.7	14.0	4.2	T	0.0	48.2
1992-93	0.0	0.0	0.0	0.5	5.3	8.1	5.8	16.3	14.6	1.2	0.0	0.0	51.8
1993-94	0.0	0.0	0.0	6.6	2.3	11.7							
Record Mean	0.0	0.0	T	0.6	4.6	10.0	11.2	9.5	8.9	2.6	0.1	T	47.5

CINCINNATI, (GREATER CINCINNATI AIRPORT) OHIO

Greater Cincinnati Airport is located on a gently rolling plateau about 12 miles southwest of downtown Cincinnati and 2 miles south of the Ohio River at its nearest point. The river valley is rather narrow and steep-sided varying from 1 to 3 miles in width and the river bed is 500 feet below the level of the airport.

The climate is continental with a rather wide range of temperatures from winter to summer. A precipitation maximum occurs during winter and spring with a late summer and fall minimum. On the average, the maximum snowfall occurs during January, although the heaviest 24-hour amounts have been recorded during late November and February.

The heaviest precipitation, as well as the precipitation of the longest duration, is normally associated with low pressure disturbances moving in a general southwest to northeast direction through the Ohio valley and south of the Cincinnati area.

Summers are warm and rather humid. The temperature will reach 100 degrees or more in 1 year out of 3. However, the temperature will reach 90 degrees or higher on about 19 days each year. Winters are moderately cold with frequent periods of extensive cloudiness.

The freeze free period lasts on the average 187 days from mid-April to the latter part of October.

NORMALS, MEANS, AND EXTREMES

CINCINNATI, (GREATER CINCINNATI AIRPORT) OHIO

LATITUDE: 39°03'N LONGITUDE: 84°40'W ELEVATION: FT. GRND 869 BARO 888 TIME ZONE: EASTERN WBAN: 93814

	(a)	JAN	FEB	MAR	APR	MAY	JUNE	JULY	AUG	SEP	OCT	NOV	DEC	YEAR
TEMPERATURE °F:														
Normals														
-Daily Maximum		36.6	40.8	53.0	64.2	74.0	82.0	85.5	84.1	77.9	66.0	53.3	41.5	63.2
-Daily Minimum		19.5	22.7	33.1	42.2	51.8	60.0	64.8	62.9	56.6	44.2	35.3	25.3	43.2
-Monthly		28.1	31.8	43.0	53.2	62.9	71.0	75.1	73.5	67.3	55.1	44.3	33.5	53.2
Extremes														
-Record Highest	32	69	73	84	89	93	102	103	102	98	88	81	75	103
-Year		1967	1972	1986	1976	1962	1988	1988	1962	1964	1963	1987	1982	JUL 1988
-Record Lowest	32	-25	-11	-11	17	27	39	47	43	31	16	1	-20	-25
-Year		1977	1982	1980	1964	1963	1972	1963	1986	1993	1962	1976	1989	JAN 1977
NORMAL DEGREE DAYS:														
Heating (base 65°F)		1144	930	682	354	151	11	0	0	51	327	621	977	5248
Cooling (base 65°F)		0	0	0	0	86	191	313	266	120	20	0	0	996
% OF POSSIBLE SUNSHINE	10	37	43	47	56	56	62	62	62	62	53	37	32	51
MEAN SKY COVER (tenths)														
Sunrise - Sunset	42	7.5	7.3	7.3	6.9	6.7	6.2	6.0	5.7	5.6	5.6	7.1	7.6	6.6
MEAN NUMBER OF DAYS:														
Sunrise to Sunset														
-Clear	42	5.2	5.1	5.3	5.8	6.0	7.0	7.7	8.3	9.7	10.5	5.9	5.1	81.6
-Partly Cloudy	42	6.0	5.9	6.7	7.5	9.7	10.5	11.6	11.9	8.8	7.4	5.9	5.8	97.8
-Cloudy	42	19.7	17.2	19.0	16.6	15.3	12.5	11.6	10.8	11.5	13.1	18.2	20.1	185.9
Precipitation														
.01 inches or more	46	12.0	11.3	13.3	12.5	11.7	10.5	10.2	9.1	8.0	8.3	11.2	12.2	130.2
Snow,Ice pellets,hail														
1.0 inches or more	46	2.1	1.8	1.2	0.2	0.0	0.0	0.0	0.0	0.0	0.1	0.5	1.2	7.2
Thunderstorms	46	0.7	0.8	2.4	4.2	5.7	6.9	8.2	7.5	3.2	1.4	1.2	0.5	42.6
Heavy Fog Visibility														
1/4 mile or less	30	2.3	2.0	1.5	0.8	1.2	1.2	1.7	2.7	3.3	2.5	1.6	2.5	23.4
Temperature °F														
-Maximum														
90° and above	31	0.0	0.0	0.0	0.0	0.5	4.1	8.0	5.3	1.8	0.0	0.0	0.0	19.7
32° and below	31	11.4	7.6	1.3	0.0	0.0	0.0	0.0	0.0	0.0	0.0	0.7	6.4	27.4
-Minimum														
32° and below	31	26.2	22.4	15.9	4.5	0.2	0.0	0.0	0.0	0.*	3.5	12.6	21.9	107.3
0° and below	31	2.9	1.5	0.1	0.0	0.0	0.0	0.0	0.0	0.0	0.0	0.0	1.1	5.7
AVG. STATION PRESS.(mb)	21	987.5	987.0	984.7	984.0	983.9	984.5	985.6	986.5	987.1	987.7	987.2	987.7	986.1
RELATIVE HUMIDITY (%)														
Hour 01	31	76	74	73	71	77	81	83	84	83	78	75	77	78
Hour 07 (Local Time)	31	79	78	78	76	80	82	85	88	88	84	80	80	82
Hour 13	31	68	64	59	54	55	56	57	58	58	55	63	69	60
Hour 19	31	70	65	61	55	58	59	61	64	67	65	69	72	64
PRECIPITATION (inches):														
Water Equivalent														
-Normal		2.59	2.69	4.24	3.75	4.28	3.84	4.24	3.35	2.88	2.86	3.46	3.15	41.33
-Maximum Monthly	46	9.43	6.72	12.18	7.19	9.48	7.36	8.36	7.71	8.61	8.60	7.51	7.90	12.18
-Year		1950	1955	1964	1970	1968	1977	1962	1982	1979	1983	1985	1990	MAR 1964
-Minimum Monthly	46	0.57	0.25	1.14	1.04	1.13	0.95	1.18	0.31	0.18	0.25	0.43	0.51	0.18
-Year		1981	1978	1960	1971	1964	1965	1951	1953	1963	1963	1949	1976	SEP 1963
-Maximum in 24 hrs	46	4.33	2.84	5.21	2.72	3.71	3.45	4.28	3.12	4.54	4.47	3.36	2.96	5.21
-Year		1959	1990	1964	1950	1956	1974	1988	1957	1979	1985	1948	1948	MAR 1964
Snow,Ice pellets,hail														
-Maximum Monthly	46	31.5	19.9	13.0	3.7	0.2	T	T	0.0	0.0	6.2	12.1	12.5	31.5
-Year		1978	1993	1968	1977	1989	1993	1991			1993	1966	1989	JAN 1978
-Maximum in 24 hrs	46	8.1	9.3	9.8	3.6	0.2	T	T	0.0	0.0	5.9	9.0	7.5	9.8
-Year		1978	1966	1968	1977	1989	1993	1991			1993	1966	1990	MAR 1968
WIND:														
Mean Speed (mph)	46	10.7	10.4	11.1	10.7	8.7	8.0	7.2	6.8	7.4	8.2	9.7	10.2	9.1
Prevailing Direction through 1963		SSW	SSW	SSW	SSW	SSW	SSW	SSW	SSW	SSW	SSW	SSW	SSW	SSW
Fastest Obs. 1 Min.														
-Direction (!!)	30	28	29	25	25	31	24	34	31	30	27	21	21	25
-Speed (MPH)	30	46	40	44	46	37	40	35	37	32	35	35	40	46
-Year		1976	1967	1977	1982	1986	1971	1980	1983	1990	1967	1988	1973	APR 1982
Peak Gust														
-Direction (!!)	10	SW	SW	SW	S	W	SW	W	NE	NW	SW	W	SW	SW
-Speed (mph)	10	51	55	64	61	53	60	49	47	45	43	56	59	64
-Date		1990	1988	1986	1985	1987	1992	1990	1990	1990	1986	1988	1987	MAR 1986

PRECIPITATION (inches)

CINCINNATI, (GREATER CINCINNATI AIRPORT) OHIO

YEAR	JAN	FEB	MAR	APR	MAY	JUNE	JULY	AUG	SEP	OCT	NOV	DEC	ANNUAL
1964	2.88	1.98	12.18	6.73	1.13	4.32	2.56	2.25	1.65	0.59	2.69	4.92	43.88
1965	3.11	5.07	2.86	4.90	1.46	0.95	4.42	3.24	6.06	3.81	1.26	1.19	38.33
1966	3.84	3.73	1.22	5.38	2.42	2.52	4.06	4.31	3.13	0.57	4.18	3.31	38.67
1967	0.75	1.86	3.63	3.66	5.64	1.72	4.99	0.77	1.79	2.65	3.84	3.94	35.24
1968	1.79	0.64	4.24	3.47	9.48	2.43	7.50	2.26	2.16	1.35	3.21	4.01	42.54
1969	4.64	1.27	1.42	3.59	2.05	4.91	3.28	2.15	2.87	1.53	3.67	2.59	33.97
1970	1.27	1.68	4.71	7.19	1.88	5.73	3.47	2.96	3.87	2.44	2.29	3.32	40.81
1971	2.47	5.89	2.55	1.04	3.31	5.18	3.70	3.45	6.56	1.44	1.68	3.39	40.66
1972	1.96	2.20	3.68	5.89	6.02	2.41	1.50	2.64	5.96	2.55	6.26	4.23	45.30
1973	1.79	1.58	6.11	5.81	3.46	6.27	7.16	2.62	2.63	4.39	4.95	2.66	49.43
1974	3.65	1.63	4.39	5.08	5.53	4.38	3.82	5.75	4.44	1.07	4.19	2.83	46.76
1975	4.05	3.38	6.76	4.16	3.11	5.09	1.62	1.97	3.64	4.59	2.50	3.36	44.23
1976	3.00	2.37	2.14	1.21	1.80	5.94	2.33	4.36	1.95	3.85	0.83	0.51	30.29
1977	1.90	1.29	4.52	4.16	1.53	7.36	1.90	5.45	1.80	3.74	3.90	4.00	41.55
1978	4.52	0.25	1.99	2.28	5.30	6.63	6.86	4.41	0.43	5.03	2.67	6.46	46.83
1979	3.68	3.77	2.05	4.90	4.00	5.92	5.49	4.80	8.61	1.77	4.86	2.91	52.76
1980	2.26	1.04	4.50	1.96	4.59	4.13	5.51	4.19	1.83	3.28	2.58	1.26	37.13
1981	0.57	3.86	1.72	5.05	5.07	3.34	3.66	2.15	1.47	2.33	2.94	2.39	34.55
1982	7.17	1.17	4.67	2.18	4.60	3.61	2.44	7.71	1.27	0.99	5.08	4.25	45.14
1983	1.56	1.14	2.02	4.84	8.89	2.22	1.96	3.23	1.22	8.60	4.20	2.84	42.72
1984	0.75	2.40	3.61	4.88	4.82	2.11	2.57	3.30	3.50	3.85	6.00	4.21	42.00
1985	1.68	2.25	6.90	1.34	6.18	4.55	3.59	2.02	0.76	5.83	7.51	1.52	44.13
1986	1.01	2.85	3.07	1.57	3.59	1.46	3.33	3.78	3.53	3.08	3.79	2.58	33.64
1987	0.92	1.62	4.65	2.88	2.73	4.62	5.07	2.27	1.17	1.42	1.82	3.43	32.60
1988	2.75	4.94	3.42	3.92	1.99	1.19	6.85	2.44	3.05	1.86	4.78	2.78	39.97
1989	3.21	4.67	6.40	5.19	4.64	3.04	5.97	5.33	2.97	3.18	3.05	1.96	49.61
1990	2.59	5.82	2.75	3.22	9.41	5.01	3.68	5.67	4.13	5.09	2.31	7.90	57.58
1991	2.84	3.99	6.20	3.62	3.41	1.39	2.66	5.04	2.60	1.37	1.89	5.08	40.09
1992	2.99	0.93	4.19	2.71	2.84	3.65	7.00	3.17	3.23	1.11	4.31	1.36	37.49
1993	3.83	3.43	3.60	3.13	2.33	4.80	1.26	4.20	2.68	2.61	4.31	2.53	38.71
Record Mean	3.19	2.90	3.97	3.52	3.97	3.86	4.16	3.21	2.82	2.69	3.42	3.10	40.82

AVERAGE TEMPERATURE (deg. F)

CINCINNATI, (GREATER CINCINNATI AIRPORT) OHIO

YEAR	JAN	FEB	MAR	APR	MAY	JUNE	JULY	AUG	SEP	OCT	NOV	DEC	ANNUAL
1964	32.1	30.2	43.1	56.1	65.8	71.9	74.2	73.7	67.1	52.7	47.1	35.5	54.1
1965	30.6	32.9	36.6	53.3	67.1	71.9	73.2	72.5	67.1	53.4	44.2	38.8	53.5
1966	23.8	31.5	42.9	50.1	60.4	73.1	78.8	74.5	66.5	51.0	44.7	33.6	52.6
1967	35.4	27.5	44.8	55.4	60.5	73.0	72.9	71.1	65.3	55.5	40.3	36.8	53.2
1968	26.8	28.1	45.2	55.5	60.7	72.9	75.7	75.3	67.4	55.9	45.6	33.7	53.6
1969	30.2	33.7	37.3	55.2	65.0	71.9	77.3	74.0	66.5	55.7	40.3	30.6	53.1
1970	24.0	30.6	39.0	56.8	66.8	72.2	75.5	75.1	73.7	58.6	45.0	38.0	54.6
1971	28.3	33.3	40.2	53.1	60.8	75.8	74.8	72.4	70.3	61.4	43.2	41.2	54.6
1972	30.2	29.5	40.6	52.5	63.1	65.3	75.5	73.4	68.7	51.7	40.3	36.1	52.3
1973	31.7	33.0	50.6	51.1	58.4	72.3	74.8	73.7	69.6	59.0	45.4	33.6	54.4
1974	35.7	33.9	45.9	54.3	61.9	67.7	75.2	74.3	61.6	53.2	44.4	34.4	53.5
1975	34.0	37.1	39.8	50.6	67.6	72.3	74.6	76.9	62.4	57.5	49.1	37.4	54.9
1976	27.1	41.7	47.0	56.1	60.0	71.5	73.6	71.2	63.6	48.8	34.9	27.6	52.0
1977	12.0	29.8	46.5	56.3	68.2	68.7	77.9	73.9	69.8	52.3	45.4	28.7	52.5
1978	18.4	18.2	36.4	53.4	60.1	72.2	75.0	73.0	70.8	52.6	46.4	36.0	51.0
1979	21.3	21.4	46.6	50.9	60.2	69.4	73.4	72.3	66.1	54.3	44.2	36.2	51.4
1980	29.9	24.0	38.5	50.3	64.9	70.1	76.6	76.5	68.6	50.6	41.5	32.9	52.0
1981	24.1	34.2	40.1	58.1	59.8	72.3	75.9	73.6	65.2	53.9	43.7	29.0	52.5
1982	23.9	30.6	44.3	49.6	68.1	67.4	77.0	71.3	66.9	59.3	48.4	42.9	54.1
1983	31.6	35.3	44.7	49.4	59.0	71.6	79.2	78.3	67.2	55.7	44.7	24.6	53.4
1984	23.7	38.2	34.4	51.1	58.9	74.1	72.2	74.3	65.7	61.5	42.0	42.4	53.2
1985	22.7	29.4	47.5	57.9	65.4	69.9	75.1	72.5	67.5	59.3	49.9	26.2	53.6
1986	30.9	35.2	45.2	55.3	64.5	72.9	77.6	72.0	70.1	56.4	42.7	34.0	54.7
1987	30.7	37.3	45.0	53.0	69.3	73.6	76.1	75.2	68.3	49.3	48.0	36.8	55.2
1988	27.5	30.5	42.2	52.4	64.4	72.4	78.5	77.5	67.2	48.5	45.0	34.1	53.4
1989	38.6	30.8	45.4	52.9	60.1	71.3	76.7	73.5	66.4	55.7	43.9	21.6	53.1
1990	40.0	40.8	48.2	52.8	61.6	71.8	74.7	73.7	67.7	55.9	48.9	38.3	56.2
1991	31.0	37.3	45.4	56.9	70.4	75.4	77.9	74.8	68.6	58.6	41.4	37.3	56.3
1992	32.9	39.7	43.6	54.1	61.2	68.0	73.5	69.6	64.5	53.3	44.2	35.6	53.4
1993	34.5	29.3	40.2	52.0	63.3	70.9	79.1	76.2	63.9	53.0	43.5	32.8	53.2
Record Mean	29.7	33.1	42.4	53.5	63.3	71.7	75.7	74.1	67.3	55.6	43.9	33.9	53.7
Max	37.9	42.0	52.0	64.2	74.2	82.4	86.1	84.8	78.1	66.4	52.8	41.8	63.6
Min	21.4	24.2	32.7	42.8	52.3	60.9	65.3	63.4	56.4	44.7	35.0	26.0	43.8

REFERENCE NOTES

CINCINNATI, (GREATER CINCINNATI AIRPORT) OHIO

GENERAL
T - TRACE AMOUNT.
BLANK ENTRIES DENOTE MISSING/UNREPORTED DATA.
# INDICATES A STATION OR INSTRUMENT RELOCATION.

SPECIFIC

Pm - INCLUDES LAST DAY OF PREVIOUS MONTH
ASOS - AUTOMATED SURFACE OBSERVING SYSTEM IN OPERATION DURING THESE MONTHS.

(a) - LENGTH OF RECORD IN YEARS, ALTHOUGH INDIVIDUAL MONTHS MAY BE MISSING.
0.* OR * - THE VALUE IS BETWEEN 0.0 AND 0.05
NORMALS - BASED ON THE 1961-1990 RECORD PERIOD.
EXTREMES - DATES ARE THE MOST RECENT OCCURRENCE
WIND DIR.- NUMERALS SHOW TENS OF DEGRESS CLOCKWISE FROM TRUE NORTH. "00" INDICATES CALM.
RESULTANT DIRECTIONS ARE GIVEN TO WHOLE DEGREES.
BOLD VALUES INDICATE EXTREME VALUES WHICH OCCURRED AFTER THE ASOS SYSTEM WAS COMMISSIONED.

RECORD = PERIOD OF RECORD
RECORD MEAN PRECIPITATION IS THE MEAN OF ALL DAILY PRECIPITATION AMOUNTS DURING THE PERIOD OF RECORD.
RECORD MAX(MIN) TEMPERATURE IS THE MEAN OF ALL DAILY MAX(MIN) TEMPERATURES DURING THE PERIOD OF RECORD.
RECORD MEAN TEMPERATURE IS THE SUM OF THE RECORD MAX AND RECORD MIN DIVIDED BY 2.
AVERAGE TEMPERATURE IS THE SUM OF THE MEAN DAILY MAX AND MIN TEMPERATURE DIVIDED BY 2.

EXCEPTIONS

RECORD MEANS ARE THROUGH THE CURRENT YEAR, BEGINNING IN 1948 FOR TEMPERATURE
1948 FOR PRECIPITATION
1948 FOR SNOWFALL

HEATING DEGREE DAYS Base 65 deg. F CINCINNATI, (GREATER CINCINNATI AIRPORT) OHIO

SEASON	JULY	AUG	SEP	OCT	NOV	DEC	JAN	FEB	MAR	APR	MAY	JUNE	TOTAL
1964-65	0	14	68	376	529	907	1058	894	872	350	43	0	5111
1965-66	2	15	64	362	614	803	1270	931	681	442	181	15	5380
1966-67	0	0	70	428	602	969	910	1043	622	305	175	9	5133
1967-68	0	2	82	313	736	867	1178	1063	609	284	151	6	5291
1968-69	0	4	32	316	576	965	1072	872	852	298	84	29	5100
1969-70	0	0	66	309	737	1058	1265	954	797	264	91	2	5543
1970-71	4	0	28	212	591	829	1133	881	758	351	157	0	4944
1971-72	0	0	31	128	649	731	1073	1024	751	374	116	79	4956
1972-73	2	3	24	404	733	891	1025	888	445	425	206	3	5049
1973-74	0	1	27	217	583	964	901	863	591	332	161	31	4671
1974-75	0	0	147	371	614	942	953	774	778	435	50	10	5074
1975-76	5	0	142	244	474	848	1168	670	558	321	176	5	4611
1976-77	0	4	72	498	893	1157	1640	980	571	276	67	36	6194
1977-78	0	2	32	391	586	1118	1440	1303	880	346	207	10	6315
1978-79	0	0	21	381	552	891	1348	1216	563	425	179	15	5591
1979-80	1	14	60	346	616	887	1080	1182	814	434	92	24	5550
1980-81	0	0	48	446	697	988	1261	858	768	230	191	6	5493
1981-82	0	0	87	344	634	1107	1268	956	635	460	28	19	5538
1982-83	0	1	56	244	505	682	1029	825	627	466	199	21	4655
1983-84	1	0	89	288	600	1247	1274	773	939	425	219	4	5859
1984-85	0	0	101	128	684	692	1306	992	543	256	72	22	4796
1985-86	0	0	78	212	450	1195	1056	828	613	305	105	3	4845
1986-87	0	21	25	292	664	955	1058	766	612	365	52	2	4812
1987-88	0	1	39	477	505	868	1156	991	699	374	84	22	5216
1988-89	1	0	38	509	595	949	811	949	608	380	211	14	5065
1989-90	0	4	77	297	630	1335	770	671	531	390	127	21	4853
1990-91	0	1	66	296	477	821	1046	773	602	250	44	0	4376
1991-92	0	0	81	232	700	853	988	727	658	339	172	27	4777
1992-93	0	8	97	358	617	907	937	997	762	384	103	40	5210
1993-94	0	0	101	370	640	992							

COOLING DEGREE DAYS Base 65 deg. F CINCINNATI, (GREATER CINCINNATI AIRPORT) OHIO

YEAR	JAN	FEB	MAR	APR	MAY	JUNE	JULY	AUG	SEP	OCT	NOV	DEC	TOTAL
1969	0	0	0	9	92	240	389	285	116	31	0	0	1162
1970	0	0	0	28	151	225	336	320	292	21	0	0	1373
1971	0	0	0	0	33	335	313	236	195	25	0	0	1137
1972	0	0	0	7	65	96	334	268	142	0	0	0	912
1973	0	0	4	13	7	228	310	278	170	36	0	0	1046
1974	0	0	8	17	70	121	323	297	50	12	2	0	900
1975	0	0	0	8	138	236	309	376	72	19	3	0	1161
1976	0	0	6	59	28	207	276	202	39	5	0	0	822
1977	0	0	7	22	171	152	407	285	181	4	5	0	1234
1978	0	0	0	4	63	231	315	255	200	2	0	0	1070
1979	0	0	2	8	38	154	271	248	102	22	0	0	845
1980	0	0	0	0	98	187	364	363	166	5	0	0	1183
1981	0	0	1	31	34	234	343	275	99	9	0	0	1026
1982	0	0	0	5	129	99	381	203	120	73	13	8	1031
1983	0	0	4	4	18	225	448	417	161	8	0	0	1285
1984	0	0	0	13	38	289	233	295	130	29	0	0	1027
1985	0	0	6	47	93	174	318	241	162	41	5	0	1087
1986	0	0	4	22	97	247	399	243	183	30	0	0	1225
1987	0	0	0	12	193	266	353	325	147	0	4	0	1300
1988	0	0	2	3	70	251	425	392	111	6	0	0	1260
1989	0	0	7	26	67	210	369	275	125	17	0	0	1096
1990	0	0	17	32	27	230	309	276	155	21	3	0	1070
1991	0	0	0	14	218	317	408	310	195	42	0	0	1504
1992	0	0	0	19	59	126	273	158	88	2	0	0	725
1993	0	0	0	0	56	224	443	353	75	5	0	0	1156

SNOWFALL (inches) CINCINNATI, (GREATER CINCINNATI AIRPORT) OHIO

SEASON	JULY	AUG	SEP	OCT	NOV	DEC	JAN	FEB	MAR	APR	MAY	JUNE	TOTAL
1964-65	0.0	0.0	0.0	0.0	1.7	0.9	9.3	5.9	6.2	T	0.0	0.0	24.0
1965-66	0.0	0.0	0.0	0.0	T	2.5	6.8	9.5	1.2	0.1	T	0.0	20.1
1966-67	0.0	0.0	0.0	0.0	12.1	1.6	2.5	7.3	8.1	0.0	0.0	0.0	31.6
1967-68	0.0	0.0	0.0	T	3.1	3.1	8.7	3.6	13.0	0.0	0.0	0.0	31.5
1968-69	0.0	0.0	0.0	0.0	0.4	2.0	1.0	0.3	2.8	0.0	0.0	0.0	6.5
1969-70	0.0	0.0	0.0	0.0	1.1	6.0	6.7	4.3	12.0	T	0.0	0.0	30.1
1970-71	0.0	0.0	0.0	0.0	0.4	0.4	3.2	13.3	9.7	T	0.0	0.0	27.0
1971-72	0.0	0.0	0.0	0.0	3.0	T	1.6	10.8	0.1	0.5	0.0	0.0	16.0
1972-73	0.0	0.0	0.0	T	6.5	1.9	0.5	3.1	3.9	1.8	0.0	0.0	17.7
1973-74	0.0	0.0	0.0	0.0	0.4	2.0	1.7	3.2	3.4	0.5	0.0	0.0	11.2
1974-75	0.0	0.0	0.0	T	5.0	5.6	2.0	2.2	6.8	0.2	0.0	0.0	21.8
1975-76	0.0	0.0	0.0	0.0	3.7	1.7	8.0	0.1	0.6	0.0	0.0	0.0	14.1
1976-77	0.0	0.0	0.0	0.0	2.7	1.0	30.3	4.2	5.4	3.7	0.0	0.0	47.3
1977-78	0.0	0.0	0.0	0.0	4.0	6.3	31.5	4.6	7.5	0.0	0.0	0.0	53.9
1978-79	0.0	0.0	0.0	0.0	T	0.7	17.5	11.7	0.6	0.1	0.0	0.0	30.6
1979-80	0.0	0.0	0.0	0.0	1.0	1.0	8.3	11.9	7.9	T	0.0	0.0	30.1
1980-81	0.0	0.0	0.0	T	1.2	3.7	4.0	2.6	2.5	0.0	0.0	0.0	14.0
1981-82	0.0	0.0	0.0	T	0.3	10.9	7.1	3.9	0.5	1.5	0.0	0.0	24.2
1982-83	0.0	0.0	0.0	0.0	T	T	0.8	5.5	0.3	T	0.0	0.0	6.6
1983-84	0.0	0.0	0.0	0.0	T	1.7	4.1	6.7	4.1	0.0	0.0	0.0	16.6
1984-85	0.0	0.0	0.0	0.0	1.4	7.3	12.2	9.5	0.4	1.7	0.0	0.0	32.5
1985-86	0.0	0.0	0.0	0.0	0.0	5.0	2.8	11.3	0.8	T	0.0	0.0	19.9
1986-87	0.0	0.0	0.0	0.0	T	0.8	1.6	2.4	8.8	2.3	0.0	0.0	15.9
1987-88	0.0	0.0	0.0	0.0	0.1	0.2	4.3	4.7	2.3	T	0.0	0.0	11.6
1988-89	0.0	0.0	0.0	0.0	0.7	2.9	T	3.0	1.2	0.3	0.2	0.0	8.3
1989-90	0.0	0.0	0.0	5.9	0.2	12.5	1.3	3.6	5.6	T	0.0	0.0	29.1
1990-91	0.0	0.0	0.0	0.0	0.0	8.6	4.3	2.6	T	0.0	0.0	0.0	15.5
1991-92	T	0.0	0.0	0.0	1.9	0.5	3.6	1.2	3.6	2.9	0.1	0.0	13.8
1992-93	0.0	0.0	0.0	T	1.6	3.8	0.3	19.9	3.9	T	0.0	T	29.5
1993-94	0.0	0.0	0.0	6.2	0.8	5.4							
Record Mean	T	0.0	0.0	0.3	2.1	3.9	6.7	5.5	4.2	0.5	T	T	23.2

CLEVELAND, OHIO

Cleveland is on the south shore of Lake Erie in northeast Ohio. The metropolitan area has a lake frontage of 31 miles. The surrounding terrain is generally level except for an abrupt ridge on the eastern edge of the city which rises some 500 feet above the shore terrain. The Cuyahoga River, which flows through a rather deep but narrow north-south valley, bisects the city.

Local climate is continental in character but with strong modifying influences by Lake Erie. West to northerly winds blowing off Lake Erie tend to lower daily high temperatures in summer and raise temperatures in winter. Temperatures at Hopkins Airport which is 5 miles south of the lakeshore average from 2-4 degrees higher than the lakeshore in summer, while overnight low temperatures average from 2-4 degrees lower than the lakefront during all seasons.

In this area, summers are moderately warm and humid with occasional days when temperatures exceed 90 degrees. Winters are relatively cold and cloudy with an average of 5 days with sub-zero temperatures. Weather changes occur every few days from the passing of cold fronts.

The daily range in temperature is usually greatest in late summer and least in winter. Annual extremes in temperature normally occur soon after late June and December. Maximum temperatures below freezing occur most often in December, January, and February. Temperatures of 100 degrees or higher are rare. On the average, freezing temperatures in fall are first recorded in October while the last freezing temperature in spring normally occurs in April.

As is characteristic of continental climates, precipitation varies widely from year to year. However, it is normally abundant and well distributed throughout the year with spring being the wettest season. Showers and thunderstorms account for most of the rainfall during the growing season. Thunderstorms are most frequent from April through August. Snowfall may fluctuate widely. Mean annual snowfall increases from west to east in Cuyahoga County ranging from about 45 inches in the west to more than 90 inches in the extreme east.

Damaging winds of 50 mph or greater are usually associated with thunderstorms. Tornadoes, one of the most destructive of all atmospheric storms, occasionally occur in Cuyahoga County.

NORMALS, MEANS, AND EXTREMES

CLEVELAND, OHIO

LATITUDE: 41 °25'N LONGITUDE: 81 °52' W ELEVATION: FT. GRND 777 BARO 779 TIME ZONE: EASTERN WBAN: 14820

	(a)	JAN	FEB	MAR	APR	MAY	JUNE	JULY	AUG	SEP	OCT	NOV	DEC	YEAR
TEMPERATURE °F:														
Normals														
-Daily Maximum		31.9	35.0	46.3	57.9	68.6	78.3	82.4	80.5	73.6	62.1	50.0	37.4	58.7
-Daily Minimum		17.6	19.3	28.2	37.3	47.3	56.8	61.4	60.3	54.2	43.5	35.0	24.5	40.5
-Monthly		24.8	27.2	37.3	47.6	58.0	67.6	71.9	70.4	63.9	52.8	42.6	30.9	49.6
Extremes														
-Record Highest	52	73	69	83	88	92	104	103	102	101	90	82	77	104
-Year		1950	1961	1945	1986	1959	1988	1941	1948	1953	1946	1950	1982	JUN 1988
-Record Lowest	52	-19	-15	-5	10	25	31	41	38	32	19	3	-15	-19
-Year		1963	1963	1984	1964	1966	1972	1968	1982	1942	1988	1976	1989	JAN 1963
NORMAL DEGREE DAYS:														
Heating (base 65°F)		1246	1058	859	522	250	40	0	11	99	387	672	1057	6201
Cooling (base 65°F)		0	0	0	0	33	118	218	178	66	8	0	0	621
% OF POSSIBLE SUNSHINE	50	31	37	44	52	58	65	67	63	59	52	31	26	49
MEAN SKY COVER (tenths)														
Sunrise - Sunset	52	8.3	7.9	7.5	7.1	6.6	6.1	5.7	5.7	6.0	6.3	8.0	8.4	7.0
MEAN NUMBER OF DAYS:														
Sunrise to Sunset														
-Clear	52	2.8	2.9	4.4	5.0	6.2	6.8	8.3	8.7	7.9	8.0	2.9	2.5	66.4
-Partly Cloudy	52	4.8	5.5	6.4	7.9	9.8	11.1	11.9	11.2	9.6	8.0	5.9	4.7	96.7
-Cloudy	52	23.5	19.8	20.2	17.1	15.1	12.2	10.8	11.2	12.5	15.0	21.2	23.8	202.1
Precipitation														
.01 inches or more	52	16.2	14.3	15.4	14.4	12.9	10.8	10.2	9.8	10.0	11.2	14.5	16.3	156.0
Snow,Ice pellets,hail														
1.0 inches or more	52	4.2	4.0	3.3	0.8	0.*	0.0	0.0	0.0	0.0	0.2	1.8	4.2	18.4
Thunderstorms	52	0.1	0.5	1.7	3.4	4.9	6.5	6.4	5.2	3.3	1.6	1.0	0.3	34.8
Heavy Fog Visibility														
1/4 mile or less	52	1.3	1.7	1.9	1.2	1.3	0.7	0.4	0.9	0.6	0.8	0.6	1.1	12.5
Temperature °F														
-Maximum														
90° and above	33	0.0	0.0	0.0	0.0	0.3	1.8	4.0	2.1	0.6	0.0	0.0	0.0	8.7
32° and below	33	15.0	12.4	4.8	0.2	0.0	0.0	0.0	0.0	0.0	0.0	1.1	10.0	43.5
-Minimum														
32° and below	33	27.8	24.3	20.9	9.1	0.8	0.*	0.0	0.0	0.0	2.9	12.5	24.5	122.8
0° and below	33	3.0	2.0	0.1	0.0	0.0	0.0	0.0	0.0	0.0	0.0	0.0	0.8	5.9
AVG. STATION PRESS.(mb)	21	988.8	989.2	987.4	986.5	986.7	987.0	987.9	989.2	989.7	990.0	989.2	989.4	988.4
RELATIVE HUMIDITY (%)														
Hour 01	33	75	76	75	73	77	80	81	83	82	77	75	76	78
Hour 07 (Local Time)	33	78	78	78	76	78	79	81	85	85	80	77	77	79
Hour 13	33	69	68	63	57	57	57	57	60	60	59	65	70	62
Hour 19	33	72	72	69	61	60	61	61	66	70	69	71	74	67
PRECIPITATION (inches):														
Water Equivalent														
-Normal		2.04	2.19	2.91	3.14	3.49	3.70	3.52	3.40	3.44	2.54	3.17	3.09	36.63
-Maximum Monthly	52	7.01	4.70	6.07	6.61	9.14	9.06	9.12	8.96	7.33	9.50	8.80	8.59	9.50
-Year		1950	1990	1954	1961	1989	1972	1992	1975	1990	1954	1985	1990	OCT 1954
-Minimum Monthly	52	0.36	0.48	0.78	1.18	1.00	0.65	1.21	0.53	0.74	0.61	0.80	0.71	0.36
-Year		1961	1978	1958	1946	1963	1988	1982	1969	1964	1952	1976	1958	JAN 1961
-Maximum in 24 hrs	52	2.33	2.33	2.76	2.24	3.73	4.00	2.87	3.07	3.30	3.44	2.73	2.81	4.00
-Year		1959	1959	1948	1961	1955	1972	1969	1947	1990	1954	1985	1992	JUN 1972
Snow,Ice pellets,hail														
-Maximum Monthly	52	42.8	39.1	26.3	14.5	2.1	T	T	0.0	T	8.0	22.3	30.3	42.8
-Year		1978	1993	1954	1943	1974	1993	1993		1993	1962	1950	1962	JAN 1978
-Maximum in 24 hrs	52	10.5	14.8	16.0	11.6	2.1	T	T	0.0	T	6.7	15.0	12.2	16.0
-Year		1978	1993	1987	1982	1974	1993	1993		1993	1962	1950	1974	MAR 1987
WIND:														
Mean Speed (mph)	52	12.2	11.8	12.1	11.5	10.0	9.3	8.6	8.3	9.0	9.9	11.8	12.1	10.5
Prevailing Direction through 1963		SW	S	W	S	S	S	S	S	S	S	S	S	S
Fastest Obs. 1 Min.														
-Direction (!!)	16	22	23	27	23	20	18	23	31	34	23	21	21	22
-Speed (MPH)	16	53	39	41	44	42	37	36	36	31	37	39	43	53
-Year		1978	1988	1986	1982	1983	1992	1983	1988	1986	1983	1988	1982	JAN 1978
Peak Gust														
-Direction (!!)	10	SW	SW	SW	SW	NW	W	NW	NW	W	W	S	SW	SW
-Speed (mph)	10	55	58	63	69	54	56	67	49	45	52	59	63	69
-Date		1985	1988	1986	1984	1988	1988	1992	1988	1986	1988	1988	1987	APR 1984

PRECIPITATION (inches) CLEVELAND, OHIO

YEAR	JAN	FEB	MAR	APR	MAY	JUNE	JULY	AUG	SEP	OCT	NOV	DEC	ANNUAL
1964	1.45	1.49	5.21	4.87	3.02	2.06	3.37	3.82	0.74	1.78	0.92	2.67	31.40
1965	4.45	3.00	1.66	1.83	2.29	3.05	3.01	3.58	2.53	2.55	1.89	2.07	31.91
1966	1.53	2.31	2.26	3.61	2.21	1.83	3.89	3.48	1.66	1.18	5.16	2.84	31.96
1967	0.97	2.35	2.08	3.12	3.82	1.17	1.90	1.85	2.08	2.11	2.88	2.46	26.79
1968	3.27	0.79	2.07	2.25	4.08	2.32	3.58	1.82	3.36	2.90	4.35	3.94	34.73
1969	2.84	0.75	1.82	4.49	5.73	4.61	6.47	0.53	4.92	1.90	2.86	2.46	39.38
1970	1.28	1.35	2.32	2.64	2.95	4.98	4.14	0.92	3.16	3.98	3.69	2.25	33.66
1971	1.35	3.69	2.01	1.24	3.29	3.79	3.72	0.91	4.27	1.61	2.02	3.90	31.80
1972	1.95	2.01	2.97	3.40	3.74	9.06	4.44	6.38	4.91	1.64	4.58	3.26	48.34
1973	1.62	2.40	3.48	3.40	4.79	6.72	2.94	3.11	2.69	3.95	2.62	3.53	41.25
1974	2.56	2.43	3.88	3.64	4.78	3.57	1.90	3.29	3.06	1.19	4.72	4.86	39.88
1975	3.06	3.20	3.47	1.31	3.23	4.10	2.54	8.96	3.35	1.73	2.09	3.77	40.81
1976	3.38	3.97	3.11	2.17	2.94	3.64	3.48	3.50	3.71	2.54	0.80	1.57	34.81
1977	1.29	1.38	4.49	3.56	1.02	4.91	3.94	3.92	2.52	1.93	3.62	3.51	36.09
1978	3.67	0.48	2.17	3.02	3.01	3.30	2.40	3.58	3.68	3.23	1.19	2.96	32.69
1979	2.61	2.74	2.33	3.09	4.77	3.47	3.76	4.46	3.66	1.79	3.16	4.00	39.84
1980	1.18	1.27	3.66	2.65	3.13	2.69	4.77	4.38	3.11	2.38	1.29	2.10	32.61
1981	0.76	2.72	1.61	4.62	2.19	4.68	5.31	2.61	6.75	2.33	1.99	3.44	39.01
1982	4.00	1.41	3.77	1.62	2.65	5.01	1.21	2.66	4.82	0.93	5.17	3.68	36.93
1983	1.08	0.77	3.54	4.48	4.17	3.45	4.16	3.15	2.87	4.14	5.89	2.92	40.62
1984	1.25	3.82	3.80	2.29	5.95	3.40	3.35	5.51	2.43	2.20	3.95	3.38	41.33
1985	1.78	2.60	4.97	1.38	3.45	2.93	3.23	4.01	2.05	3.45	8.80	2.63	41.28
1986	2.23	3.08	2.44	3.90	4.34	2.97	3.10	3.58	6.41	2.83	3.01	2.82	40.71
1987	1.98	0.49	3.84	2.97	2.40	7.94	3.36	5.51	2.07	3.41	1.02	2.96	37.95
1988	1.03	2.84	2.20	3.47	1.33	0.65	3.42	3.35	1.77	2.51	4.63	2.49	29.69
1989	2.07	1.73	3.46	3.73	9.14	5.22	3.02	1.09	4.61	4.50	3.61	1.72	43.90
1990	2.35	4.70	0.86	4.57	6.10	1.72	5.62	4.79	7.33	4.92	2.28	8.59	53.83
1991	2.18	2.31	3.64	4.22	3.24	1.37	1.69	2.79	3.40	2.65	2.92	2.26	32.67
1992	3.32	2.65	3.05	3.77	3.01	2.66	9.12	4.58	3.25	2.27	6.54	4.31	48.53
1993	4.44	2.61	3.85	3.16	1.56	5.18	2.58	1.52	5.94	3.52	4.06	2.21	40.63
Record Mean	2.49	2.34	2.90	2.86	3.22	3.37	3.44	3.06	3.22	2.63	2.78	2.56	34.87

AVERAGE TEMPERATURE (deg. F) CLEVELAND, OHIO

YEAR	JAN	FEB	MAR	APR	MAY	JUNE	JULY	AUG	SEP	OCT	NOV	DEC	ANNUAL
1964	29.8	25.5	37.1	48.8	60.7	67.5	72.2	67.5	63.5	49.2	44.2	32.2	49.8
1965	27.2	28.0	31.7	45.6	63.2	66.9	69.0	68.7	67.4	51.3	42.5	37.4	49.9
1966	21.9	26.7	37.8	46.2	54.2	68.9	72.7	68.8	60.9	50.3	42.8	30.5	48.5
1967	32.4	25.9	37.2	49.7	52.3	71.7	69.7	68.8	61.7	54.0	38.6	34.7	49.7
1968	23.0	22.6	37.6	49.4	54.4	66.5	70.0	71.7	63.9	52.1	42.4	30.0	48.6
1969	25.4	27.9	34.3	49.4	58.6	65.5	72.4	71.8	63.6	51.9	40.2	27.2	49.0
1970	18.9	27.2	33.8	50.2	62.7	69.8	71.9	69.9	66.0	54.4	41.6	32.2	49.9
1971	21.4	27.9	31.6	43.2	56.5	71.0	69.5	68.9	67.7	59.9	41.4	38.1	49.8
1972	27.3	25.7	34.8	46.0	58.6	62.7	71.4	68.9	63.8	49.1	39.6	34.5	48.5
1973	30.4	27.9	46.5	50.2	56.7	70.4	72.6	73.2	66.3	57.7	44.6	34.3	52.6
1974	32.0	27.8	39.6	51.3	56.4	66.2	72.2	70.4	59.9	51.2	42.9	31.7	50.1
1975	31.9	30.4	34.6	41.8	62.3	69.8	71.3	72.3	58.6	53.8	47.0	32.0	50.5
1976	21.6	36.0	45.0	49.1	55.3	69.5	71.6	68.4	61.1	48.1	33.7	23.3	48.6
1977	11.0	25.0	42.7	51.4	61.8	63.3	73.1	69.6	65.6	52.6	45.4	29.2	49.3
1978	20.1	16.8	32.4	47.0	59.4	69.0	72.2	73.0	69.2	53.2	44.2	33.7	49.2
1979	22.0	19.1	42.9	46.6	56.9	66.9	71.1	71.5	65.0	52.4	42.3	33.7	49.2
1980	25.5	21.9	33.6	46.1	58.5	64.0	72.3	73.2	64.7	47.9	39.4	28.5	48.0
1981	20.1	31.5	36.0	50.6	55.7	68.2	71.3	70.0	62.4	50.0	42.6	30.6	49.1
1982	19.8	25.2	37.1	44.6	64.9	64.1	73.6	67.9	62.7	55.3	45.4	40.5	50.1
1983	30.7	33.9	40.8	47.1	55.7	69.0	75.2	73.7	65.1	53.4	43.9	23.2	51.0
1984	20.7	34.5	28.4	46.8	54.0	69.5	68.7	70.6	61.1	56.3	40.9	36.5	49.0
1985	20.8	25.2	40.3	53.6	60.4	62.7	71.1	68.9	64.9	54.0	46.0	24.3	49.4
1986	26.7	28.8	39.5	49.8	60.8	67.2	73.1	69.0	67.0	54.3	40.3	32.6	50.8
1987	27.4	30.5	39.0	49.1	63.0	70.2	75.2	70.8	63.5	47.5	46.1	34.8	51.4
1988	25.6	25.8	37.5	47.9	59.7	68.9	75.9	74.2	64.0	47.1	43.8	31.3	50.1
1989	35.0	26.1	38.1	45.3	57.6	68.3	73.4	71.0	64.0	54.0	41.0	19.2	49.4
1990	35.8	34.1	42.0	49.4	56.3	67.6	71.2	69.8	63.4	53.7	45.3	35.6	52.0
1991	27.3	32.8	40.7	52.6	66.9	71.1	74.7	72.7	64.5	55.7	40.2	34.8	52.8
1992	30.2	32.7	36.6	47.9	57.9	64.1	70.7	67.6	63.3	49.9	42.0	34.1	49.8
1993	32.3	25.6	33.6	47.6	59.2	67.9	75.0	73.2	62.2	51.3	41.8	30.5	50.0
Record Mean	27.1	27.8	36.2	47.3	58.4	67.8	72.3	70.7	64.6	53.6	41.7	31.3	49.9
Max	34.0	35.1	44.0	56.0	67.4	76.7	80.8	79.1	73.2	61.9	48.6	37.5	57.8
Min	20.2	20.5	28.3	38.5	49.3	59.0	63.7	62.2	55.9	45.2	34.8	25.0	41.9

REFERENCE NOTES CLEVELAND, OHIO

GENERAL
T - TRACE AMOUNT.
BLANK ENTRIES DENOTE MISSING/UNREPORTED DATA.
# INDICATES A STATION OR INSTRUMENT RELOCATION.

SPECIFIC

PM - INCLUDES LAST DAY OF PREVIOUS MONTH
ASOS - AUTOMATED SURFACE OBSERVING SYSTEM IN OPERATION DURING THESE MONTHS.

(a) - LENGTH OF RECORD IN YEARS, ALTHOUGH INDIVIDUAL MONTHS MAY BE MISSING.
0.* OR * - THE VALUE IS BETWEEN 0.0 AND 0.05
NORMALS - BASED ON THE 1961-1990 RECORD PERIOD.
EXTREMES - DATES ARE THE MOST RECENT OCCURRENCE.
WIND DIR. - NUMERALS SHOW TENS OF DEGRESS CLOCKWISE FROM TRUE NORTH. "00" INDICATES CALM.
RESULTANT DIRECTIONS ARE GIVEN TO WHOLE DEGREES.
BOLD VALUES INDICATE EXTREME VALUES WHICH OCCURRED AFTER THE ASOS SYSTEM WAS COMMISSIONED.

RECORD = PERIOD OF RECORD
RECORD MEAN PRECIPITATION IS THE MEAN OF ALL DAILY PRECIPITATION AMOUNTS DURING THE PERIOD OF RECORD.
RECORD MAX(MIN) TEMPERATURE IS THE MEAN OF ALL DAILY MAX(MIN) TEMPERATURES DURING THE PERIOD OF RECORD.
RECORD MEAN TEMPERATURE IS THE SUM OF THE RECORD MAX AND RECORD MIN DIVIDED BY 2.
AVERAGE TEMPERATURE IS THE SUM OF THE MEAN DAILY MAX AND MIN TEMPERATURE DIVIDED BY 2.

EXCEPTIONS

RECORD MEANS ARE THROUGH THE CURRENT YEAR, BEGINNING IN 1871 FOR TEMPERATURE
1871 FOR PRECIPITATION
1942 FOR SNOWFALL

HEATING DEGREE DAYS Base 65 deg. F CLEVELAND, OHIO

SEASON	JULY	AUG	SEP	OCT	NOV	DEC	JAN	FEB	MAR	APR	MAY	JUNE	TOTAL
1964-65	3	46	117	483	617	1009	1165	1032	1025	576	130	64	6267
1965-66	24	49	67	418	671	852	1328	1067	837	562	346	53	6274
1966-67	6	15	162	452	655	1063	1000	1087	858	461	393	17	6169
1967-68	21	19	137	351	784	934	1295	1224	845	459	328	59	6456
1968-69	26	34	93	414	672	1080	1220	1032	946	471	234	100	6322
1969-70	1	7	121	406	736	1166	1425	1052	960	462	154	39	6529
1970-71	9	12	86	332	696	1009	1344	1032	1031	650	277	16	6494
1971-72	9	13	63	168	704	828	1160	1133	930	564	196	124	5892
1972-73	32	27	95	485	752	937	1067	1033	569	450	254	3	5704
1973-74	3	9	73	234	605	946	1015	1035	777	419	280	49	5445
1974-75	2	5	176	423	660	1026	1021	962	934	691	154	38	6092
1975-76	5	4	187	345	532	1015	1336	836	614	493	309	25	5701
1976-77	0	25	150	519	932	1286	1672	1113	689	423	166	115	7090
1977-78	4	26	60	378	592	1103	1387	1343	1005	534	218	43	6693
1978-79	7	2	43	362	620	965	1328	1281	680	552	290	60	6190
1979-80	20	11	87	403	670	967	1218	1244	967	561	223	103	6474
1980-81	3	2	97	521	763	1125	1385	935	894	430	298	30	6483
1981-82	11	11	145	458	664	1059	1393	1109	860	608	78	75	6471
1982-83	5	42	136	310	586	760	1056	864	742	533	294	56	5384
1983-84	7	0	116	362	628	1291	1366	878	1126	544	347	19	6684
1984-85	16	17	174	270	716	877	1364	1110	757	370	187	99	5957
1985-86	2	7	118	338	565	1255	1180	1009	785	459	172	52	5942
1986-87	3	40	63	332	736	999	1158	958	795	473	170	23	5750
1987-88	3	22	90	535	562	929	1213	1129	848	506	208	60	6105
1988-89	8	5	83	557	629	1040	922	1084	831	585	272	33	6049
1989-90	0	6	108	350	716	1416	898	858	718	492	270	56	5888
1990-91	7	3	121	350	585	906	1163	897	748	379	111	11	5281
1991-92	0	0	123	310	738	930	1074	929	872	513	243	90	5822
1992-93	8	26	118	462	682	952	1009	1097	967	519	192	56	6088
1993-94	0	3	134	420	691	1063							

COOLING DEGREE DAYS Base 65 deg. F CLEVELAND, OHIO

YEAR	JAN	FEB	MAR	APR	MAY	JUNE	JULY	AUG	SEP	OCT	NOV	DEC	TOTAL
1969	0	0	0	10	41	120	237	223	84	10	0	0	725
1970	0	0	0	22	89	189	230	171	121	10	0	0	832
1971	0	0	0	0	22	198	158	143	152	19	0	0	692
1972	0	0	0	1	5	63	239	157	64	0	0	0	529
1973	0	0	0	13	7	168	244	273	119	17	0	0	841
1974	0	0	0	14	18	91	231	180	30	3	2	0	569
1975	0	0	0	0	75	187	206	241	6	5	0	0	720
1976	0	0	3	23	14	167	214	138	39	2	0	0	600
1977	0	0	4	22	74	73	262	175	84	0	9	0	703
1978	0	0	0	0	53	170	237	256	177	3	0	0	896
1979	0	0	0	6	42	122	213	218	93	21	0	0	715
1980	0	0	0	0	27	83	235	263	97	0	0	0	705
1981	0	0	0	4	16	132	214	175	73	0	0	0	614
1982	0	0	0	3	84	54	278	140	73	17	6	6	661
1983	0	0	0	5	12	185	327	277	127	12	0	0	945
1984	0	0	0	3	13	159	139	197	60	5	0	0	576
1985	0	0	0	38	52	34	201	131	122	4	2	0	584
1986	0	0	1	9	48	128	259	168	131	8	0	0	752
1987	0	0	0	0	114	183	322	209	53	0	3	0	884
1988	0	0	0	0	47	185	348	297	58	9	0	0	944
1989	0	0	4	0	46	138	268	199	83	14	0	0	752
1990	0	0	10	31	8	141	208	158	80	8	1	0	645
1991	0	0	1	14	176	200	307	245	114	26	0	0	1083
1992	0	0	0	8	28	68	191	114	74	0	0	0	483
1993	0	0	0	0	17	147	316	262	62	2	0	0	806

SNOWFALL (inches) CLEVELAND, OHIO

SEASON	JULY	AUG	SEP	OCT	NOV	DEC	JAN	FEB	MAR	APR	MAY	JUNE	TOTAL
1964-65	0.0	0.0	0.0	T	1.0	8.7	13.6	15.6	12.9	0.4	0.0	0.0	52.2
1965-66	0.0	0.0	0.0	T	1.2	1.2	15.3	10.1	7.0	2.5	T	0.0	37.3
1966-67	0.0	0.0	0.0	0.0	8.8	10.9	2.0	18.5	7.3	0.1	0.0	0.0	47.6
1967-68	0.0	0.0	0.0	0.1	9.1	2.8	14.5	8.9	7.7	0.2	T	0.0	43.3
1968-69	0.0	0.0	0.0	T	6.8	8.3	5.8	5.6	9.0	1.5	T	0.0	37.0
1969-70	0.0	0.0	0.0	0.6	6.6	17.4	10.5	6.6	11.5	0.2	T	0.0	53.4
1970-71	0.0	0.0	T	T	5.2	6.0	8.6	14.3	16.6	0.7	0.0	0.0	51.4
1971-72	0.0	0.0	0.0	0.0	5.3	1.9	15.0	14.8	6.3	2.3	0.0	0.0	45.6
1972-73	0.0	0.0	0.0	5.5	7.8	15.2	9.8	20.4	8.3	0.9	0.6	0.0	68.5
1973-74	0.0	0.0	0.0	T	3.3	13.8	8.9	16.9	7.1	6.4	2.1	0.0	58.5
1974-75	0.0	0.0	0.0	1.6	5.3	24.1	9.7	9.9	15.2	1.2	0.0	0.0	67.0
1975-76	0.0	0.0	0.0	0.0	5.6	13.1	21.5	6.8	5.8	1.6	T	0.0	54.4
1976-77	0.0	0.0	T	1.6	8.9	16.3	21.1	9.6	4.2	1.7	0.0	0.0	63.4
1977-78	0.0	0.0	0.0	T	9.7	23.1	42.8	10.8	3.5	0.2	0.0	0.0	90.1
1978-79	0.0	0.0	0.0	0.0	1.9	2.5	15.1	16.0	2.4	0.4	0.0	0.0	38.3
1979-80	0.0	0.0	0.0	0.2	0.5	4.0	11.3	19.2	3.5	T	T	0.0	38.7
1980-81	0.0	0.0	0.0	T	5.4	13.5	15.0	9.7	16.9	T	0.0	0.0	60.5
1981-82	0.0	0.0	0.0	4.0	2.9	27.1	28.1	7.6	17.6	13.2	0.0	0.0	100.5
1982-83	0.0	0.0	0.0	T	2.2	6.3	6.5	8.3	11.3	3.4	0.0	0.0	38.0
1983-84	0.0	0.0	0.0	0.0	7.1	13.0	12.9	27.1	19.3	T	0.0	0.0	79.4
1984-85	0.0	0.0	0.0	0.0	4.0	8.9	25.5	18.2	1.2	5.9	0.0	0.0	63.7
1985-86	0.0	0.0	0.0	0.0	T	23.4	17.2	10.8	6.7	0.2	0.0	0.0	58.3
1986-87	0.0	0.0	0.0	0.0	3.1	1.1	16.4	5.0	26.2	4.0	0.0	0.0	55.8
1987-88	0.0	0.0	0.0	T	1.0	16.4	8.7	22.9	20.4	1.9	0.0	0.0	71.3
1988-89	0.0	0.0	0.0	T	1.7	17.9	6.6	13.8	9.9	4.9	T	0.0	54.8
1989-90	0.0	0.0	0.0	T	9.1	24.0	10.5	9.9	4.4	4.7	0.0	0.0	62.6
1990-91	0.0	0.0	0.0	T	T	7.4	16.6	18.9	4.2	T	0.0	0.0	47.1
1991-92	0.0	0.0	0.0	0.0	3.5	9.4	23.8	6.2	18.4	4.4	0.0	0.0	65.7
1992-93	0.0	0.0	0.0	T	7.1	7.1	8.7	39.1	25.4	1.1	0.0	T	88.5
1993-94	T	0.0	T	0.2	3.0	19.0							
Record Mean	T	0.0	T	0.6	5.0	11.9	12.6	12.3	10.6	2.3	0.1	T	55.4

COLUMBUS, OHIO

Columbus is located in the center of the state and in the drainage area of the Ohio River. The airport is located at the eastern boundary of the city approximately 7 miles from the center of the business district.

Four nearly parallel streams run through or adjacent to the city. The Scioto River is the principal stream and flows from the northwest into the center of the city and then flows straight south toward the Ohio River. The Olentangy River runs almost due south and empties into the Scioto just west of the business district. Two minor streams run through portions of Columbus or skirt the eastern and southern fringes of the area. They are Alum Creek and Big Walnut Creek. Alum Creek empties into the Big Walnut southeast of the city and the Big Walnut empties into the Scioto a few miles downstream. The Scioto and Olentangy are gorge-like in character with very little flood plain and the two creeks have only a little more flood plain or bottomland.

The narrow valleys associated with the streams flowing through the city supply the only variation in the micro-climate of the area. The city proper shows the typical metropolitan effect with shrubs and flowers blossoming earlier than in the immediate surroundings and in retarding light frost on clear quiet nights. Many small areas to the southeast and to the north and northeast show marked effects of air drainage as evidenced by the frequent formation of shallow ground fog at daybreak during the summer and fall months and the higher frequency of frost in the spring and fall.

The average occurrence of the last freezing temperature in the spring within the city proper is mid-April, and the first freeze in the fall is very late October, but in the immediate surroundings there is much variation. For example, at Valley Crossing located at the southeastern outskirts of the city, the average occurrence of the last 32 degree temperature in the spring is very early May, while the first 32 degree temperature in the fall is mid-October.

The records show a high frequency of calm or very low wind speeds during the late evening and early morning hours, from June through September. The rolling landscape is conducive to air drainage and from the Weather Service location at the airport the air drainage is toward the northwest with the wind direction indicated as southeast. Air drainage takes place at speeds generally 4 mph or less and frequently provides the only perceptible breeze during the night.

Columbus is located in the area of changeable weather. Air masses from central and northwest Canada frequently invade this region. Air from the Gulf of Mexico often reachs central Ohio during the summer and to a much lesser extent in the fall and winter. There are also occasional weather changes brought about by cool outbreaks from the Hudson Bay region of Canada, especially during the spring months. At infrequent intervals the general circulation will bring showers or snow to Columbus from the Atlantic. Although Columbus does not have a wet or dry season as such, the month of October usually has the least amount of precipitation.

NORMALS, MEANS, AND EXTREMES

COLUMBUS, OHIO

LATITUDE: 40°00'N LONGITUDE: 82°53' W ELEVATION: FT. GRND 813 BARO 816 TIME ZONE: EASTERN WBAN: 14821

	(a)	JAN	FEB	MAR	APR	MAY	JUNE	JULY	AUG	SEP	OCT	NOV	DEC	YEAR
TEMPERATURE °F:														
Normals														
-Daily Maximum		34.1	38.0	50.5	62.0	72.3	80.4	83.7	82.1	76.2	64.5	51.4	39.2	61.2
-Daily Minimum		18.5	21.2	31.2	40.0	50.1	58.0	62.7	60.8	54.8	42.9	34.3	24.6	41.6
-Monthly		26.4	29.6	40.9	51.0	61.2	69.2	73.2	71.5	65.5	53.7	42.9	31.9	51.4
Extremes														
-Record Highest	54	74	73	85	89	94	102	100	101	100	90	80	76	102
-Year		1950	1957	1945	1948	1941	1944	1988	1983	1951	1951	1987	1982	JUN 1944
-Record Lowest	54	-19	-13	-6	14	25	35	43	39	31	20	5	-17	-19
-Year		1985	1977	1984	1982	1966	1972	1972	1965	1963	1962	1976	1989	JAN 1985
NORMAL DEGREE DAYS:														
Heating (base 65°F)		1197	991	747	420	187	23	0	12	81	361	663	1026	5708
Cooling (base 65°F)		0	0	0	0	69	149	258	214	96	11	0	0	797
% OF POSSIBLE SUNSHINE	42	36	42	43	51	56	60	60	60	60	55	37	31	49
MEAN SKY COVER (tenths)														
Sunrise - Sunset	44	7.7	7.6	7.4	7.0	6.6	6.2	6.1	5.9	5.7	5.8	7.4	7.9	6.8
MEAN NUMBER OF DAYS:														
Sunrise to Sunset														
-Clear	44	4.3	4.0	4.9	5.3	6.1	6.2	6.6	7.1	9.0	10.1	4.8	4.0	72.4
-Partly Cloudy	44	6.4	6.1	6.7	7.8	10.2	11.3	13.2	12.9	9.2	7.4	6.5	5.8	103.4
-Cloudy	44	20.4	18.1	19.4	16.9	14.8	12.6	11.2	11.0	11.8	13.5	18.8	21.2	189.5
Precipitation														
.01 inches or more	54	13.3	11.4	13.9	13.0	12.6	10.9	10.8	9.3	8.5	8.9	11.6	12.8	137.1
Snow, Ice pellets, hail														
1.0 inches or more	45	2.6	2.2	1.4	0.2	0.0	0.0	0.0	0.0	0.0	0.1	0.7	1.8	8.9
Thunderstorms	54	0.4	0.6	2.1	4.0	6.3	7.9	8.2	6.1	3.1	1.3	1.0	0.3	41.3
Heavy Fog Visibility														
1/4 mile or less	44	1.8	1.5	1.1	0.6	0.9	1.0	1.0	1.6	1.8	1.4	1.2	1.6	15.4
Temperature °F														
-Maximum														
90° and above	34	0.0	0.0	0.0	0.0	0.6	3.4	5.9	3.8	1.3	0.0	0.0	0.0	15.1
32° and below	34	12.8	9.2	2.7	0.1	0.0	0.0	0.0	0.0	0.0	0.0	1.1	8.7	34.6
-Minimum														
32° and below	34	26.9	23.5	18.4	6.5	0.6	0.0	0.0	0.0	0.1	4.0	13.6	23.6	117.3
0° and below	34	2.7	1.4	0.1	0.0	0.0	0.0	0.0	0.0	0.0	0.0	0.0	0.9	5.1
AVG. STATION PRESS. (mb)	21	988.9	988.6	986.5	985.7	985.7	986.1	987.2	988.2	988.9	989.4	988.9	989.2	987.8
RELATIVE HUMIDITY (%)														
Hour 01	34	75	74	70	71	77	81	83	84	84	79	77	77	78
Hour 07 (Local Time)	34	77	77	75	75	79	81	84	87	87	82	80	79	80
Hour 13	34	67	64	57	53	55	55	56	58	57	55	63	69	59
Hour 19	34	69	67	60	55	57	58	60	63	65	64	69	73	63
PRECIPITATION (inches):														
Water Equivalent														
-Normal		2.18	2.24	3.27	3.21	3.93	4.04	4.31	3.72	2.96	2.15	3.22	2.86	38.09
-Maximum Monthly	54	8.29	5.15	9.59	6.36	9.11	9.75	12.36	8.63	6.76	5.24	10.67	6.98	12.36
-Year		1950	1990	1964	1964	1968	1958	1992	1979	1979	1954	1985	1990	JUL 1992
-Minimum Monthly	54	0.53	0.29	0.61	0.67	0.95	0.71	0.48	0.58	0.51	0.11	0.60	0.46	0.11
-Year		1944	1978	1941	1971	1977	1984	1940	1951	1963	1963	1976	1955	OCT 1963
-Maximum in 24 hrs	46	4.81	2.15	3.40	2.37	2.72	2.93	5.16	3.79	4.86	2.21	2.47	1.83	5.16
-Year		1959	1975	1964	1957	1968	1958	1992	1972	1979	1986	1985	1990	JUL 1992
Snow, Ice pellets, hail														
-Maximum Monthly	46	34.4	16.4	13.5	12.6	0.8	T	0.0	0.0	T	4.6	15.2	17.3	34.4
-Year		1978	1979	1962	1987	1989	1990			1967	1993	1950	1960	JAN 1978
-Maximum in 24 hrs	46	7.5	8.9	8.6	12.3	0.8	T	0.0	0.0	T	4.6	8.2	8.7	12.3
-Year		1978	1971	1962	1987	1989	1990			1967	1993	1950	1960	APR 1987
WIND:														
Mean Speed (mph)	44	9.9	9.7	10.2	9.7	8.1	7.3	6.6	6.2	6.5	7.4	9.1	9.6	8.4
Prevailing Direction through 1963		SSW	NW	SSW	WNW	S	SSW	SSW	NNW	S	S	S	W	S
Fastest Obs. 1 Min.														
-Direction (!!!)	12	23	18	28	22	25	26	23	27	20	25	27	23	25
-Speed (MPH)	12	37	30	35	40	52	31	35	38	27	35	39	35	52
-Year		1985	1988	1985	1982	1982	1983	1981	1984	1983	1981	1992	1987	MAY 1982
Peak Gust														
-Direction (!!!)	10	W	SW	W	S	W	SW	W	W	S	W	W	SW	W
-Speed (mph)	10	51	51	53	52	52	40	47	56	38	39	53	55	56
-Date		1985	1988	1985	1985	1985	1989	1991	1984	1985	1993	1988	1987	AUG 1984

PRECIPITATION (inches)

COLUMBUS, OHIO

YEAR	JAN	FEB	MAR	APR	MAY	JUNE	JULY	AUG	SEP	OCT	NOV	DEC	ANNUAL
1964	1.82	1.68	9.59	6.36	1.95	5.71	2.97	3.19	1.66	0.38	1.81	4.09	41.21
1965	2.70	3.76	2.90	5.90	4.00	2.42	3.76	4.62	6.18	3.98	1.19	1.24	42.65
1966	2.87	2.59	1.04	4.89	3.13	1.28	5.91	4.90	3.56	0.79	4.05	3.33	38.34
1967	0.78	2.46	4.40	3.29	4.59	2.92	4.22	1.51	2.63	1.39	3.22	2.55	33.96
1968	2.22	0.38	3.01	2.20	9.11	2.96	2.80	3.08	1.77	2.59	4.26	3.40	37.78
1969	3.40	1.17	1.32	3.10	3.04	8.19	7.65	3.25	1.40	1.52	3.87	2.30	40.21
1970	1.60	1.68	3.04	5.52	5.37	5.65	3.73	3.94	3.95	2.07	2.88	2.50	41.93
1971	1.57	3.16	2.70	0.67	3.66	4.16	4.22	2.81	3.08	1.32	1.73	4.61	33.69
1972	1.40	1.74	2.86	3.74	6.56	3.98	2.60	7.96	5.13	1.74	4.40	3.49	45.60
1973	2.46	1.29	3.43	3.72	3.36	8.77	4.07	4.97	2.82	3.29	5.37	2.70	46.25
1974	2.40	2.30	4.38	2.66	3.29	5.04	1.14	4.88	3.32	1.51	3.39	2.68	36.99
1975	3.21	3.47	4.10	2.71	3.17	3.53	2.04	4.51	5.46	2.29	1.54	3.01	39.04
1976	3.15	2.03	2.17	1.44	1.41	4.52	5.12	5.08	2.54	2.86	0.60	0.93	31.85
1977	1.57	1.02	3.88	4.04	0.95	4.02	2.52	4.76	3.48	2.57	3.77	3.54	36.12
1978	5.89	0.29	2.98	3.02	4.15	3.65	1.81	5.23	1.16	2.39	1.56	5.01	37.14
1979	3.32	2.88	1.01	4.01	3.27	4.23	8.06	8.63	6.76	1.26	3.91	1.83	49.17
1980	1.69	1.38	3.77	1.59	4.56	5.17	4.58	6.26	1.86	2.53	2.07	1.96	37.42
1981	0.70	4.60	1.11	5.38	6.50	5.73	4.14	1.41	2.28	1.40	1.65	2.88	37.78
1982	4.77	1.49	3.99	1.90	4.68	3.37	3.90	1.02	4.25	0.92	5.19	3.84	39.32
1983	1.20	0.74	1.69	5.58	5.06	4.59	2.80	2.23	1.91	4.45	5.00	3.16	38.41
1984	1.07	1.97	3.89	3.10	4.93	0.71	3.15	2.96	1.48	2.91	4.41	2.84	33.42
1985	1.31	1.67	3.78	0.73	4.96	1.41	6.88	2.34	1.18	1.93	10.67	1.81	38.67
1986	1.54	2.96	2.61	1.31	2.47	5.53	3.60	1.61	3.44	4.16	3.00	2.81	35.04
1987	1.14	0.59	2.04	2.02	2.85	3.60	3.89	2.96	1.53	1.57	1.63	2.88	26.70
1988	2.14	4.26	2.54	2.24	2.27	1.34	7.80	2.68	3.52	1.70	3.59	2.49	36.57
1989	1.97	3.10	4.16	3.30	4.69	6.36	6.79	4.30	2.16	2.49	2.65	1.79	43.76
1990	2.43	5.15	1.32	2.82	7.01	5.25	8.00	1.86	5.26	5.05	2.03	6.98	53.16
1991	1.97	2.30	3.97	4.15	2.47	2.81	2.14	2.02	4.05	1.76	1.31	3.79	32.74
1992	1.79	0.85	3.40	2.83	3.40	2.33	12.36	3.75	2.14	1.40	4.03	1.32	39.60
1993	4.14	1.82	3.50	4.49	2.47	3.33	5.95	0.74	1.75	3.05	4.45	2.16	37.85
Record Mean	2.84	2.45	3.90	3.14	3.70	3.71	3.91	3.20	2.63	2.25	2.80	2.62	37.16

AVERAGE TEMPERATURE (deg. F)

COLUMBUS, OHIO

YEAR	JAN	FEB	MAR	APR	MAY	JUNE	JULY	AUG	SEP	OCT	NOV	DEC	ANNUAL
1964	30.1	27.7	41.1	53.6	63.4	70.3	74.4	71.2	65.0	51.1	45.1	33.9	52.3
1965	28.3	29.1	34.8	51.4	68.1	69.4	72.0	71.0	68.3	54.4	43.3	36.6	52.2
1966	22.2	29.4	41.8	49.4	56.1	71.4	75.6	71.3	63.2	50.7	42.4	32.6	50.5
1967	34.8	25.8	40.4	52.6	55.4	71.8	71.7	68.3	60.4	51.9	36.7	34.3	50.3
1968	23.6	25.7	43.3	53.1	58.5	71.5	74.1	73.0	65.5	54.1	44.0	31.1	51.5
1969	27.0	31.4	35.3	51.5	61.7	68.5	74.1	71.2	63.8	53.9	39.3	26.8	50.4
1970	20.6	28.5	36.9	53.4	64.9	69.8	73.5	72.4	68.9	55.6	42.4	34.3	51.8
1971	24.3	30.8	36.6	49.0	58.0	73.5	70.5	68.9	67.8	59.8	40.4	38.4	51.5
1972	28.2	27.7	37.0	48.8	60.8	63.6	71.9	70.1	64.6	49.6	40.5	36.1	49.9
1973	31.2	31.4	50.4	51.1	59.5	72.6	74.3	74.2	68.9	58.6	45.1	33.7	54.3
1974	33.2	31.2	44.6	54.3	60.8	67.7	74.4	74.0	62.2	52.8	44.5	34.0	52.8
1975	32.5	33.4	37.3	46.7	66.6	72.4	75.1	77.3	62.7	54.7	47.5	33.5	53.3
1976	24.0	37.4	46.5	50.9	58.1	70.5	72.0	68.3	61.7	47.5	33.9	24.8	49.6
1977	11.4	26.5	45.6	54.8	66.8	67.5	76.2	72.0	68.2	52.0	45.1	29.5	51.3
1978	19.0	16.6	34.5	50.6	59.6	70.4	73.5	73.2	69.8	51.5	44.4	34.4	49.8
1979	21.4	19.3	44.3	50.1	60.5	69.6	71.8	71.9	65.1	53.3	43.6	35.1	50.5
1980	29.3	25.2	37.2	49.5	62.4	67.4	75.9	75.9	68.3	50.8	40.8	32.5	51.3
1981	23.3	34.0	40.2	55.8	59.5	70.9	71.9	70.4	62.3	51.1	40.9	30.6	50.9
1982	21.2	29.2	40.4	46.4	66.8	65.8	74.4	69.2	63.5	56.2	45.4	40.4	51.6
1983	29.9	34.0	43.3	48.4	57.6	69.4	76.7	76.2	67.1	54.5	44.0	24.8	52.2
1984	23.3	37.4	32.3	50.0	57.6	73.1	71.2	72.9	63.1	59.4	40.6	39.5	51.7
1985	21.7	26.0	43.7	56.3	62.6	66.9	72.7	71.2	66.6	57.3	48.2	26.0	51.6
1986	30.1	32.7	42.5	54.5	64.3	70.6	75.7	71.0	69.2	56.3	41.3	33.3	53.5
1987	29.9	34.9	44.3	52.1	66.0	72.7	76.6	74.3	66.9	49.1	47.6	35.7	54.2
1988	26.5	29.3	40.2	50.3	62.6	69.6	77.5	75.3	65.2	47.4	43.9	31.6	51.6
1989	36.6	28.7	42.0	48.2	57.2	68.8	73.9	71.2	65.2	54.2	42.1	19.8	50.7
1990	37.7	37.5	45.3	50.7	59.1	70.3	73.6	72.5	66.4	55.1	46.2	37.2	54.3
1991	29.7	35.7	43.9	56.1	70.9	75.0	77.6	75.0	66.2	55.9	41.0	36.4	55.3
1992	32.2	36.8	40.7	51.8	59.9	67.3	73.5	69.4	64.7	51.9	44.8	34.7	52.3
1993	34.3	27.8	38.6	50.3	62.2	69.8	76.2	75.7	64.9	53.0	43.5	32.9	52.4
Record Mean	29.0	30.8	40.2	51.2	61.9	70.7	74.8	72.8	66.4	54.7	42.4	32.4	52.3
Max	36.5	38.8	49.2	61.4	72.5	81.2	85.2	83.2	77.1	65.0	50.6	39.5	61.7
Min	21.4	22.7	31.1	40.9	51.3	60.2	64.3	62.4	55.8	44.4	34.2	25.3	42.8

REFERENCE NOTES

COLUMBUS, OHIO

GENERAL
T - TRACE AMOUNT.
BLANK ENTRIES DENOTE MISSING/UNREPORTED DATA.
# INDICATES A STATION OR INSTRUMENT RELOCATION.

SPECIFIC

PM - INCLUDES LAST DAY OF PREVIOUS MONTH
ASOS - AUTOMATED SURFACE OBSERVING SYSTEM IN OPERATION DURING THESE MONTHS.

(a) - LENGTH OF RECORD IN YEARS, ALTHOUGH INDIVIDUAL MONTHS MAY BE MISSING.
0.* OR * - THE VALUE IS BETWEEN 0.0 AND 0.05
NORMALS - BASED ON THE 1961-1990 RECORD PERIOD.
EXTREMES - DATES ARE THE MOST RECENT OCCURRENCE
WIND DIR.- NUMERALS SHOW TENS OF DEGRESS CLOCKWISE FROM TRUE NORTH. "00" INDICATES CALM.
RESULTANT DIRECTIONS ARE GIVEN TO WHOLE DEGREES.
BOLD VALUES INDICATE EXTREME VALUES WHICH OCCURRED AFTER THE ASOS SYSTEM WAS COMMISSIONED.

RECORD = PERIOD OF RECORD
RECORD MEAN PRECIPITATION IS THE MEAN OF ALL DAILY PRECIPITATION AMOUNTS DURING THE PERIOD OF RECORD.
RECORD MAX(MIN) TEMPERATURE IS THE MEAN OF ALL DAILY MAX(MIN) TEMPERATURES DURING THE PERIOD OF RECORD.
RECORD MEAN TEMPERATURE IS THE SUM OF THE RECORD MAX AND RECORD MIN DIVIDED BY 2.
AVERAGE TEMPERATURE IS THE SUM OF THE MEAN DAILY MAX AND MIN TEMPERATURE DIVIDED BY 2.

EXCEPTIONS

FASTEST MILE WINDS ARE THROUGH JUNE 1981.

RECORD MEANS ARE THROUGH THE CURRENT YEAR, BEGINNING IN 1879 FOR TEMPERATURE
1879 FOR PRECIPITATION
1948 FOR SNOWFALL

HEATING DEGREE DAYS Base 65 deg. F COLUMBUS, OHIO

SEASON	JULY	AUG	SEP	OCT	NOV	DEC	JAN	FEB	MAR	APR	MAY	JUNE	TOTAL
1964-65	0	28	90	424	589	957	1132	999	932	403	39	26	5619
1965-66	4	28	53	330	645	873	1325	989	710	465	286	29	5737
1966-67	1	5	106	440	671	999	929	1093	756	374	300	9	5683
1967-68	7	23	164	406	843	942	1276	1133	667	351	205	6	6023
1968-69	6	20	57	362	624	1043	1173	933	916	402	143	54	5733
1969-70	0	2	107	359	763	1175	1369	1017	861	365	119	13	6150
1970-71	10	0	58	297	674	944	1256	950	871	475	230	4	5769
1971-72	3	5	52	181	733	815	1133	1077	860	482	146	101	5588
1972-73	22	18	77	473	727	889	1041	934	444	427	184	0	5236
1973-74	0	3	35	219	589	963	977	940	628	332	178	31	4895
1974-75	0	0	130	374	609	954	999	878	850	542	73	18	5427
1975-76	0	0	110	321	520	973	1263	791	570	440	229	4	5221
1976-77	1	25	118	537	925	1241	1659	1071	601	324	91	64	6657
1977-78	1	17	36	394	594	1091	1420	1346	938	424	223	23	6507
1978-79	0	0	38	411	610	943	1346	1270	637	449	185	18	5907
1979-80	11	16	83	376	632	920	1099	1148	855	458	133	53	5784
1980-81	0	0	46	435	717	1000	1286	864	761	287	195	14	5605
1981-82	8	5	141	429	713	1061	1351	997	758	556	45	33	6097
1982-83	3	19	107	304	585	759	1081	863	669	493	239	30	5152
1983-84	6	0	83	325	626	1236	1284	796	1007	447	254	3	6067
1984-85	6	3	143	182	727	782	1339	1086	654	286	134	35	5377
1985-86	0	2	96	249	500	1202	1076	901	694	328	113	19	5180
1986-87	0	26	41	287	702	974	1083	838	637	393	103	9	5093
1987-88	0	4	53	489	521	900	1187	1029	762	433	119	49	5546
1988-89	3	7	57	547	624	1032	873	1009	711	499	274	28	5664
1989-90	0	11	90	345	680	1394	840	766	613	444	190	26	5399
1990-91	0	3	83	310	558	857	1089	817	649	282	42	0	4690
1991-92	0	0	105	296	714	878	1011	814	747	402	190	35	5192
1992-93	0	8	101	403	600	932	942	1034	811	434	130	51	5446
1993-94	0	1	84	366	637	989							

COOLING DEGREE DAYS Base 65 deg. F COLUMBUS, OHIO

YEAR	JAN	FEB	MAR	APR	MAY	JUNE	JULY	AUG	SEP	OCT	NOV	DEC	TOTAL
1969	0	0	0	3	45	165	290	203	76	18	0	0	800
1970	0	0	0	22	125	166	281	237	179	13	0	0	1023
1971	0	0	0	0	21	266	181	135	144	24	0	0	771
1972	0	0	0	1	24	67	245	183	71	0	0	0	591
1973	0	0	3	14	17	236	295	292	160	25	0	0	1042
1974	0	0	4	20	58	117	296	286	52	3	0	0	836
1975	0	0	0	1	130	248	320	389	48	10	1	0	1147
1976	0	0	3	23	23	174	223	135	25	2	0	0	608
1977	0	0	8	24	151	148	354	242	139	0	7	0	1073
1978	0	0	0	0	59	190	270	261	188	0	0	0	968
1979	0	0	0	7	54	163	230	239	93	22	0	0	808
1980	0	0	0	0	61	132	343	344	151	3	0	0	1034
1981	0	0	0	16	32	198	231	181	64	4	0	0	726
1982	0	0	0	4	111	66	301	154	67	39	7	4	753
1983	0	0	1	2	17	167	377	355	152	9	0	0	1080
1984	0	0	0	8	30	253	205	254	94	14	0	0	858
1985	0	0	2	32	64	97	245	201	152	19	2	0	814
1986	0	0	2	19	95	194	339	221	171	25	0	0	1066
1987	0	0	0	11	142	246	366	299	116	0	5	0	1185
1988	0	0	0	0	54	194	396	333	70	5	0	0	1052
1989	0	0	5	2	40	149	282	211	106	12	0	0	807
1990	0	0	11	21	13	191	273	244	133	9	3	0	898
1991	0	0	0	21	232	307	402	317	147	23	0	0	1449
1992	0	0	0	13	37	115	272	152	99	2	0	0	690
1993	0	0	0	0	48	204	352	343	89	2	0	0	1038

SNOWFALL (inches) COLUMBUS, OHIO

SEASON	JULY	AUG	SEP	OCT	NOV	DEC	JAN	FEB	MAR	APR	MAY	JUNE	TOTAL
1964-65	0.0	0.0	0.0	0.0	1.0	3.2	10.2	7.6	8.6	T	0.0	0.0	30.6
1965-66	0.0	0.0	0.0	0.0	0.2	1.1	7.6	6.7	1.2	0.7	T	0.0	17.5
1966-67	0.0	0.0	0.0	T	10.4	6.4	2.8	15.6	11.4	0.0	0.0	0.0	46.6
1967-68	0.0	0.0	T	T	6.5	5.2	11.6	2.8	6.1	T	0.0	0.0	32.2
1968-69	0.0	0.0	0.0	0.0	1.0	6.8	2.5	2.5	1.9	0.5	0.0	0.0	15.2
1969-70	0.0	0.0	0.0	0.0	1.8	9.7	18.4	3.2	10.3	0.9	0.0	0.0	44.3
1970-71	0.0	0.0	0.0	0.0	0.9	1.4	6.5	12.3	12.3	T	0.0	0.0	33.4
1971-72	0.0	0.0	0.0	0.0	5.0	0.6	5.8	6.6	5.0	0.6	0.0	0.0	23.6
1972-73	0.0	0.0	0.0	T	6.3	2.8	4.4	1.8	2.1	7.1	0.0	0.0	24.5
1973-74	0.0	0.0	0.0	0.0	T	6.4	2.3	5.0	4.5	0.3	0.0	0.0	18.5
1974-75	0.0	0.0	0.0	T	0.3	7.4	8.1	3.7	2.6	T	0.0	0.0	22.1
1975-76	0.0	0.0	0.0	0.0	1.1	2.9	12.4	1.8	1.0	T	0.0	0.0	19.2
1976-77	0.0	0.0	0.0	T	3.1	4.6	18.1	6.7	0.3	0.1	0.0	0.0	32.9
1977-78	0.0	0.0	0.0	0.0	2.2	7.5	34.4	4.5	5.5	T	0.0	0.0	54.1
1978-79	0.0	0.0	0.0	T	1.3	1.8	17.3	16.4	0.8	0.3	0.0	0.0	37.9
1979-80	0.0	0.0	0.0	0.0	0.1	0.2	7.0	8.1	1.2	T	0.0	0.0	16.6
1980-81	0.0	0.0	0.0	T	8.0	7.3	7.8	3.7	3.3	0.0	0.0	0.0	30.1
1981-82	0.0	0.0	0.0	0.0	1.9	9.8	11.8	3.7	3.2	4.7	0.0	0.0	35.1
1982-83	0.0	0.0	0.0	0.0	T	1.5	2.6	4.5	2.8	0.1	0.0	0.0	11.5
1983-84	0.0	0.0	0.0	0.0	0.5	5.7	9.0	10.8	9.8	0.3	0.0	0.0	36.1
1984-85	0.0	0.0	0.0	0.0	0.9	7.3	21.9	12.5	T	0.8	0.0	0.0	43.4
1985-86	0.0	0.0	0.0	0.0	0.0	8.6	4.8	9.8	1.8	T	0.0	0.0	25.0
1986-87	0.0	0.0	0.0	0.0	0.4	0.4	2.7	1.2	5.9	12.6	0.0	0.0	23.2
1987-88	0.0	0.0	0.0	T	0.6	4.6	8.4	6.5	3.8	T	0.0	0.0	23.9
1988-89	0.0	0.0	0.0	T	0.8	5.9	0.6	3.9	6.6	0.1	0.8	0.0	18.7
1989-90	0.0	0.0	0.0	0.4	0.3	9.4	3.3	6.0	1.4	0.4	0.0	T	21.2
1990-91	0.0	0.0	0.0	0.0	0.0	3.7	3.4	4.5	4.0	T	0.0	0.0	15.6
1991-92	0.0	0.0	0.0	T	0.6	1.6	12.2	1.8	1.6	1.1	0.0	0.0	18.9
1992-93	0.0	0.0	0.0	T	3.0	2.4	1.5	14.6	8.9	0.2	0.0	0.0	30.6
1993-94	0.0	0.0	0.0	4.6	0.8	4.2							
Record Mean	0.0	0.0	T	0.1	2.3	5.5	8.1	6.2	4.5	0.9	T	T	27.7

DAYTON, OHIO

Dayton is located near the center of the Miami River Valley, which is a nearly flat plain, 50 to 200 feet below the general elevation of the adjacent rolling country. Three Miami River tributaries, the Mad River, the Stillwater River, and Wolf Creek converge, fanwise, from the north to join the master stream within the city limits of Dayton. Heavy rains in March 1913 caused the worst flood disaster in the history of the Miami Valley. During the flood more than 400 people lost their lives and property damage amounted to $100 million. After the 1913 flood, dams were built on the streams north of Dayton, forming retarding basins. No floods have occurred at Dayton since the construction of these dams.

The elevation of the city of Dayton is about 750 feet. Terrain north of the city slopes gradually upward to about 1,100 feet at Indian Lake. Ten miles southeast of Indian Lake, near Bellefontaine, is the highest point in the state, with an elevation of about 1,550 feet. South of the city, the terrain slopes gradually downward to about 450 feet where the Miami River empties into the Ohio River.

Precipitation, which is rather evenly distributed throughout the year, and moderate temperatures help to make the Miami Valley a rich agricultural region. High relative humidities during much of the year cause some discomfort to people with allergies. Temperatures of zero or below will be experienced in about four years out of five, while 100 degrees or higher will be recorded in about one year out of five. Extreme temperatures are usually of short duration. The downward slope of about 700 feet in the 163 miles of the Miami River may have some moderating influence on the winter temperatures in the Miami Valley.

The average last occurrence in the spring of freezing temperatures is mid-April, and the average first occurrence in the autumn is late October.

Cold, polar air, flowing across the Great Lakes, causes much cloudiness during the winter, and is accompanied by frequent snow flurries. These add little to the total snowfall.

NORMALS, MEANS, AND EXTREMES

DAYTON, OHIO

LATITUDE: 39°54'N LONGITUDE: 84°12' W ELEVATION: FT. GRND 995 BARO 1005 TIME ZONE: EASTERN WBAN: 93815

	(a)	JAN	FEB	MAR	APR	MAY	JUNE	JULY	AUG	SEP	OCT	NOV	DEC	YEAR
TEMPERATURE °F:														
Normals														
-Daily Maximum		34.1	38.0	50.0	61.9	72.5	81.6	84.9	83.0	76.5	64.5	51.3	39.1	61.5
-Daily Minimum		17.9	20.8	31.0	40.5	51.0	59.2	63.4	61.3	55.1	43.6	34.4	24.0	41.9
-Monthly		26.0	29.4	40.5	51.2	61.7	70.4	74.2	72.2	65.8	54.1	42.9	31.6	51.7
Extremes														
-Record Highest	50	71	71	82	89	93	102	102	102	101	89	79	72	102
-Year		1950	1976	1986	1962	1962	1988	1988	1988	1954	1951	1975	1982	JUN 1988
-Record Lowest	50	-24	-16	-7	15	27	40	44	40	32	21	-2	-20	-24
-Year		1985	1951	1980	1972	1947	1990	1972	1965	1974	1962	1958	1989	JAN 1985
NORMAL DEGREE DAYS:														
Heating (base 65°F)		1209	997	760	414	185	11	0	6	73	355	663	1035	5708
Cooling (base 65°F)		0	0	0	0	83	173	285	230	97	18	0	0	886
% OF POSSIBLE SUNSHINE	50	41	45	48	53	59	66	67	67	65	59	40	36	54
MEAN SKY COVER (tenths)														
Sunrise - Sunset	50	7.6	7.3	7.4	7.1	6.7	6.3	5.9	5.7	5.7	5.7	7.4	7.7	6.7
MEAN NUMBER OF DAYS:														
Sunrise to Sunset														
-Clear	50	4.8	5.1	4.5	5.4	5.9	6.4	7.5	8.4	9.2	10.1	4.9	4.2	76.4
-Partly Cloudy	50	6.1	5.7	7.5	7.3	9.4	10.3	12.3	12.0	9.0	7.9	6.6	6.2	100.2
-Cloudy	50	20.1	17.4	19.0	17.4	15.7	13.3	11.2	10.6	11.8	13.1	18.5	20.6	188.6
Precipitation														
.01 inches or more	50	12.9	10.9	13.0	12.9	12.1	10.3	10.3	9.4	8.2	8.7	11.5	12.3	132.5
Snow,Ice pellets,hail														
1.0 inches or more	50	2.4	1.9	1.6	0.2	0.0	0.0	0.0	0.0	0.0	0.1	0.5	1.8	8.6
Thunderstorms	50	0.5	0.5	2.4	4.3	6.2	7.2	7.4	6.1	3.2	1.5	0.8	0.3	40.5
Heavy Fog Visibility														
1/4 mile or less	50	3.5	2.6	1.8	0.7	1.2	0.9	1.2	1.7	1.7	1.4	1.8	3.3	21.8
Temperature °F														
-Maximum														
90° and above	30	0.0	0.0	0.0	0.0	0.3	3.6	7.4	4.2	1.2	0.0	0.0	0.0	16.8
32° and below	30	13.0	9.2	2.5	0.1	0.0	0.0	0.0	0.0	0.0	0.0	1.1	7.7	33.6
-Minimum														
32° and below	30	27.3	23.2	18.1	6.0	0.3	0.0	0.0	0.0	0.*	3.8	13.8	23.7	116.2
0° and below	30	2.9	1.7	0.2	0.0	0.0	0.0	0.0	0.0	0.0	0.0	0.0	1.0	5.8
AVG. STATION PRESS.(mb)	21	982.2	982.0	979.9	979.2	979.3	979.8	980.9	982.0	982.5	982.9	982.2	982.5	981.3
RELATIVE HUMIDITY (%)														
Hour 01	30	76	75	74	72	74	77	79	82	81	77	77	78	77
Hour 07 (Local Time)	30	78	78	78	76	77	79	82	86	87	82	81	80	80
Hour 13	30	69	66	61	55	55	54	55	57	57	57	66	71	60
Hour 19	30	72	69	65	59	58	57	59	63	66	65	72	75	65
PRECIPITATION (inches):														
Water Equivalent														
-Normal		2.13	2.17	3.42	3.46	3.88	3.82	3.54	3.20	2.54	2.48	3.07	2.93	36.64
-Maximum Monthly	50	9.86	5.77	7.65	6.78	8.55	10.89	8.55	8.03	5.69	6.25	8.07	10.04	10.89
-Year		1950	1990	1964	1993	1989	1958	1990	1974	1965	1986	1985	1990	JUN 1958
-Minimum Monthly	50	0.30	0.14	1.07	0.56	1.55	0.32	0.47	0.34	0.27	0.10	0.48	0.36	0.10
-Year		1981	1947	1966	1962	1964	1962	1974	1967	1963	1944	1949	1955	OCT 1944
-Maximum in 24 hrs	50	4.30	2.79	2.87	3.10	3.64	3.76	4.54	3.62	2.60	3.75	2.93	2.86	4.54
-Year		1959	1959	1964	1977	1989	1981	1990	1974	1981	1986	1955	1990	JUL 1990
Snow,Ice pellets,hail														
-Maximum Monthly	50	40.2	17.5	13.8	4.9	T	0.0	T	0.0	0.0	5.8	12.7	15.6	40.2
-Year		1978	1979	1984	1974	1989		1992			1989	1950	1960	JAN 1978
-Maximum in 24 hrs	50	12.2	7.7	11.3	4.7	T	0.0	T	0.0	0.0	5.0	10.0	7.6	12.2
-Year		1978	1984	1968	1974	1989		1992			1989	1950	1974	JAN 1978
WIND:														
Mean Speed (mph)	50	11.4	11.3	11.8	11.3	9.5	8.8	7.9	7.3	8.1	8.9	10.9	11.2	9.9
Prevailing Direction through 1963		S	WNW	WNW	SSW	SSW	SSW	SSW	SSW	SSW	SSW	S	SSW	SSW
Fastest Obs 1 Min.														
-Direction (!!)	8	24	21	23	21	18	33	32	36	25	20	30	24	24
-Speed (MPH)	8	33	32	37	35	32	32	44	32	28	29	30	45	45
-Year		1990	1988	1986	1985	1991	1987	1990	1988	1993	1993	1992	1987	DEC 1987
Peak Gust														
-Direction (!!)	10	W	NW	SW	SW	NW	SW	W	N	SW	S	W	SW	SW
-Speed (mph)	10	49	52	67	62	55	60	58	51	48	43	54	62	67
-Date		1990	1987	1986	1985	1993	1984	1992	1992	1986	1991	1988	1987	MAR 1986

PRECIPITATION (inches)

DAYTON, OHIO

YEAR	JAN	FEB	MAR	APR	MAY	JUNE	JULY	AUG	SEP	OCT	NOV	DEC	ANNUAL
1964	1.60	1.52	7.65	5.98	1.55	3.15	2.15	1.12	1.04	0.53	1.95	3.33	31.57
1965	2.32	2.80	2.33	4.54	2.33	1.44	2.24	1.35	5.69	2.76	1.20	1.46	30.46
1966	2.33	2.64	1.07	2.63	2.37	2.63	3.95	3.49	3.63	1.28	3.44	2.81	32.27
1967	0.69	1.38	3.77	4.24	6.72	3.98	2.42	0.34	1.09	2.23	2.99	3.58	33.43
1968	1.67	0.27	2.22	1.44	7.33	3.66	3.61	2.89	2.74	1.38	3.66	3.48	34.35
1969	3.75	0.73	1.41	2.74	4.20	5.90	5.71	3.39	1.13	0.98	3.14	1.83	34.91
1970	1.23	1.16	2.08	5.60	2.68	3.01	2.64	1.04	1.29	2.91	1.63	1.95	27.22
1971	1.64	3.66	1.84	1.00	4.20	2.39	4.09	3.11	3.78	2.17	1.55	3.80	33.23
1972	1.47	0.85	2.46	3.77	4.34	3.04	2.08	3.13	4.64	2.22	5.00	2.81	35.81
1973	1.63	1.28	4.64	3.45	3.10	5.72	3.76	3.93	0.69	3.28	3.86	3.14	38.48
1974	2.67	2.04	3.59	2.97	5.20	4.50	0.47	8.03	3.68	0.98	3.44	2.86	40.43
1975	3.59	3.95	3.34	3.99	2.28	2.23	5.50	4.75	3.30	2.45	2.18	3.53	41.09
1976	3.01	1.73	2.97	1.80	1.90	5.32	0.95	1.88	1.71	2.82	0.87	0.67	25.63
1977	1.64	1.78	3.50	5.13	2.02	2.46	1.73	3.59	2.73	3.85	2.67	4.47	35.57
1978	4.72	0.24	2.45	3.89	2.85	4.66	3.83	6.98	0.43	2.47	2.03	4.45	39.00
1979	3.29	2.85	1.35	3.62	2.90	4.34	4.43	7.95	3.51	2.03	4.91	2.12	43.30
1980	2.16	1.69	4.43	3.55	5.06	9.54	2.98	4.60	1.45	2.25	1.81	1.44	40.96
1981	0.30	3.37	1.18	5.06	4.76	6.32	5.08	3.51	5.06	2.79	2.80	3.46	43.69
1982	6.03	1.82	5.54	1.95	4.80	4.05	1.46	6.42	1.40	1.42	4.10	3.72	42.71
1983	1.39	0.65	2.67	4.73	4.43	5.73	3.57	1.16	0.88	5.54	4.21	2.89	37.85
1984	1.15	2.67	3.63	3.92	4.29	1.87	2.37	2.06	3.30	3.52	3.38	3.83	35.99
1985	1.56	2.26	4.85	1.56	4.43	2.27	2.69	2.50	0.98	2.39	8.07	2.25	35.81
1986	1.68	3.67	4.02	2.68	2.29	6.66	4.62	1.99	3.15	6.25	3.04	2.87	42.92
1987	1.06	1.01	2.22	3.11	2.61	3.16	4.36	0.65	0.28	1.26	1.98	3.00	24.70
1988	1.46	3.81	3.04	2.01	1.62	1.41	3.76	2.86	4.73	3.00	6.22	2.65	36.57
1989	2.72	2.65	5.99	6.52	8.55	4.76	3.56	1.89	5.66	1.56	3.66	1.85	49.37
1990	2.28	5.77	3.70	3.00	8.40	3.21	8.55	3.76	2.60	5.98	2.46	10.04	59.75
1991	2.53	2.55	5.34	4.72	4.25	2.52	2.58	2.13	3.25	1.52	2.12	3.64	37.15
1992	3.06	0.96	2.99	5.44	3.17	3.23	6.83	4.48	2.27	1.51	5.07	1.58	40.59
1993	3.88	2.77	3.85	6.78	3.98	6.25	4.68	3.32	4.69	3.06	6.19	2.78	52.23
Record Mean	2.85	2.21	3.42	3.49	3.79	3.85	3.47	3.07	2.73	2.50	2.91	2.67	36.96

AVERAGE TEMPERATURE (deg. F)

DAYTON, OHIO

YEAR	JAN	FEB	MAR	APR	MAY	JUNE	JULY	AUG	SEP	OCT	NOV	DEC	ANNUAL
1964	31.5	30.0	41.0	54.2	65.4	72.7	75.1	72.8	67.1	52.3	46.1	33.5	53.5
1965	29.2	30.7	34.7	51.3	69.4	71.8	73.8	73.1	68.3	53.1	42.3	36.2	52.8
1966	21.8	30.3	42.2	49.7	57.8	73.8	77.5	72.3	64.3	51.3	42.2	33.0	51.4
1967	33.9	25.1	40.2	52.3	56.5	74.3	73.5	70.9	63.7	54.7	37.6	34.7	51.5
1968	24.7	25.5	41.9	52.9	58.3	71.3	73.9	73.3	65.5	53.9	43.8	30.5	51.3
1969	26.7	31.6	35.5	52.7	62.4	68.1	75.2	73.0	66.4	55.4	39.6	28.9	51.3
1970	20.7	29.3	37.6	54.2	65.1	72.2	75.0	73.6	69.3	56.0	42.0	35.5	52.6
1971	23.9	30.0	37.8	51.4	59.5	74.6	72.7	71.1	68.2	62.9	42.8	38.8	52.8
1972	27.5	28.2	36.7	50.3	62.0	66.8	74.7	72.2	65.8	50.4	39.1	33.6	50.6
1973	31.3	31.2	48.6	50.5	58.7	73.5	75.8	74.5	70.2	58.7	45.0	31.0	54.1
1974	32.4	29.5	42.5	52.7	60.1	67.3	75.5	73.7	60.5	52.5	44.4	33.0	52.0
1975	33.3	33.9	37.9	47.1	65.9	71.6	73.1	74.5	62.0	55.8	48.6	33.8	53.1
1976	24.1	38.6	46.4	52.3	59.1	70.5	73.1	68.8	61.7	47.6	35.2	26.4	50.3
1977	11.6	27.5	45.6	55.7	68.8	69.2	77.5	72.8	68.3	51.8	44.8	29.0	51.9
1978	18.7	16.9	34.8	51.7	60.4	71.6	73.3	71.1	69.4	52.0	44.9	34.4	50.0
1979	20.6	19.1	44.3	50.3	61.5	70.3	73.0	71.6	64.7	52.6	43.2	35.1	50.5
1980	28.0	22.9	35.7	48.3	62.3	68.1	76.8	77.0	68.8	49.3	39.3	31.2	50.7
1981	23.3	32.9	39.3	54.9	57.3	71.9	75.1	72.8	63.4	52.1	43.6	29.7	51.4
1982	20.9	27.6	40.0	46.6	67.9	66.7	75.0	70.7	63.9	55.4	43.9	39.6	51.6
1983	29.1	34.4	42.0	46.9	56.7	69.9	76.4	75.8	66.2	54.1	43.1	21.9	51.4
1984	21.2	36.3	30.9	48.6	57.2	72.4	69.9	71.5	62.5	58.8	39.6	38.1	50.6
1985	19.4	25.3	43.6	56.2	62.8	67.0	73.0	70.8	66.3	57.4	47.6	23.8	51.1
1986	29.1	31.3	42.4	54.0	63.8	71.3	75.6	69.8	68.8	55.3	40.4	32.8	52.9
1987	28.2	34.1	43.3	51.3	66.4	72.5	75.5	73.2	67.4	48.5	46.6	35.2	53.5
1988	26.1	27.1	40.3	50.8	63.5	71.4	78.3	77.2	65.4	47.1	43.6	30.9	51.8
1989	36.3	27.5	42.4	50.2	58.8	70.6	76.1	72.2	64.1	54.1	41.0	19.0	51.0
1990	37.1	37.3	44.9	50.7	58.5	70.2	73.2	71.6	65.1	53.4	47.1	36.3	53.8
1991	28.1	34.9	42.9	55.0	69.3	74.2	76.3	73.7	66.0	56.7	39.2	34.6	54.2
1992	30.0	36.3	41.0	50.9	60.5	68.1	74.2	68.9	64.3	52.2	42.8	33.0	51.9
1993	32.1	26.3	37.7	50.0	62.0	69.7	77.0	75.1	62.7	51.6	42.5	31.1	51.5
Record Mean	28.4	30.9	40.2	51.4	61.8	71.0	74.9	73.1	66.4	55.0	42.5	32.0	52.3
Max	36.0	39.0	49.3	61.5	72.1	81.3	85.2	83.3	76.9	65.2	50.6	39.2	61.6
Min	20.7	22.8	31.1	41.2	51.5	60.7	64.7	62.9	55.9	44.9	34.4	24.8	43.0

REFERENCE NOTES

DAYTON, OHIO

GENERAL
T - TRACE AMOUNT.
BLANK ENTRIES DENOTE MISSING/UNREPORTED DATA.
# INDICATES A STATION OR INSTRUMENT RELOCATION.

SPECIFIC

(a) - LENGTH OF RECORD IN YEARS, ALTHOUGH INDIVIDUAL MONTHS MAY BE MISSING.
0.* OR * - THE VALUE IS BETWEEN 0.0 AND 0.05
NORMALS - BASED ON THE 1961-1990 RECORD PERIOD.
EXTREMES - DATES ARE THE MOST RECENT OCCURRENCE
WIND DIR.- NUMERALS SHOW TENS OF DEGRESS CLOCKWISE FROM TRUE NORTH. "00" INDICATES CALM.
RESULTANT DIRECTIONS ARE GIVEN TO WHOLE DEGREES.
BOLD VALUES INDICATE EXTREME VALUES WHICH OCCURRED AFTER THE ASOS SYSTEM WAS COMMISSIONED.

RECORD = PERIOD OF RECORD
RECORD MEAN PRECIPITATION IS THE MEAN OF ALL DAILY PRECIPITATION AMOUNTS DURING THE PERIOD OF RECORD.
RECORD MAX(MIN) TEMPERATURE IS THE MEAN OF ALL DAILY MAX(MIN) TEMPERATURES DURING THE PERIOD OF RECORD.
RECORD MEAN TEMPERATURE IS THE SUM OF THE RECORD MAX AND RECORD MIN DIVIDED BY 2.
AVERAGE TEMPERATURE IS THE SUM OF THE MEAN DAILY MAX AND MIN TEMPERATURE DIVIDED BY 2.

EXCEPTIONS

FASTEST MILE WINDS ARE THROUGH APRIL 1983.

RECORD MEANS ARE THROUGH THE CURRENT YEAR, BEGINNING IN 1912 FOR TEMPERATURE
1912 FOR PRECIPITATION
1944 FOR SNOWFALL

HEATING DEGREE DAYS Base 65 deg. F DAYTON, OHIO

SEASON	JULY	AUG	SEP	OCT	NOV	DEC	JAN	FEB	MAR	APR	MAY	JUNE	TOTAL
1964-65	0	24	75	386	559	968	1104	957	930	405	31	4	5443
1965-66	0	19	62	371	677	886	1331	965	701	456	240	16	5724
1966-67	0	3	88	420	674	984	959	1112	763	378	277	1	5659
1967-68	3	13	115	336	816	934	1241	1139	708	356	218	10	5889
1968-69	6	14	56	366	629	1062	1181	933	908	371	139	55	5720
1969-70	0	1	71	314	755	1115	1369	995	843	337	115	11	5926
1970-71	5	0	48	278	685	907	1265	975	836	403	193	1	5596
1971-72	0	0	50	97	661	806	1155	1058	870	438	130	67	5332
1972-73	15	6	63	443	771	967	1038	941	502	444	204	0	5394
1973-74	0	2	31	227	593	1045	1004	986	691	379	198	42	5198
1974-75	0	0	164	384	613	988	977	868	835	530	84	22	5465
1975-76	2	0	130	292	492	957	1259	760	570	403	203	2	5070
1976-77	0	25	121	535	886	1191	1651	1042	600	309	63	36	6459
1977-78	0	7	45	403	605	1108	1428	1339	927	395	205	15	6477
1978-79	1	11	47	396	597	942	1372	1281	634	445	166	15	5907
1979-80	7	23	91	397	649	923	1145	1217	901	495	139	49	6036
1980-81	0	0	44	487	764	1043	1289	894	792	305	252	9	5879
1981-82	0	2	119	398	638	1089	1362	1039	766	546	27	22	6008
1982-83	1	9	102	320	630	782	1105	851	706	538	258	28	5330
1983-84	8	0	104	345	649	1331	1351	827	1051	493	263	2	6424
1984-85	7	8	142	191	756	824	1406	1104	660	294	128	34	5554
1985-86	0	1	107	255	516	1271	1109	939	699	343	118	15	5373
1986-87	0	32	42	317	732	991	1134	858	664	411	98	6	5285
1987-88	2	10	44	505	551	916	1200	1091	759	425	106	35	5644
1988-89	2	3	56	549	635	1052	881	1043	700	445	241	14	5621
1989-90	0	5	114	347	713	1419	858	772	628	450	199	29	5534
1990-91	3	1	101	359	534	886	1138	835	678	304	59	1	4899
1991-92	0	0	111	281	765	935	1078	828	735	427	187	26	5373
1992-93	2	10	110	392	658	988	1011	1076	837	444	129	47	5704
1993-94	0	2	121	413	672	1046							

COOLING DEGREE DAYS Base 65 deg. F DAYTON, OHIO

YEAR	JAN	FEB	MAR	APR	MAY	JUNE	JULY	AUG	SEP	OCT	NOV	DEC	TOTAL
1969	0	0	0	6	64	154	327	254	118	22	0	0	945
1970	0	0	0	22	125	234	322	275	185	5	0	0	1168
1971	0	0	0	3	28	298	247	196	153	39	0	0	964
1972	0	0	0	4	42	130	324	237	94	0	0	0	831
1973	0	0	0	15	18	260	343	306	191	39	0	0	1172
1974	0	0	0	15	52	118	331	277	36	3	3	0	835
1975	0	0	0	2	119	228	257	302	46	13	7	0	974
1976	0	0	2	30	27	175	259	147	28	1	0	0	669
1977	0	0	4	36	188	167	393	255	153	0	3	0	1199
1978	0	0	0	1	66	219	265	210	184	0	0	0	945
1979	0	0	1	12	63	179	260	234	87	18	0	0	854
1980	0	0	0	0	61	148	373	378	164	6	0	0	1130
1981	0	0	1	11	17	221	321	251	76	5	0	0	903
1982	0	0	0	0	123	78	316	191	75	29	3	1	816
1983	0	0	0	1	8	180	369	339	148	10	0	0	1055
1984	0	0	0	5	25	233	169	218	72	7	0	0	729
1985	0	0	3	36	63	101	253	190	152	25	1	0	824
1986	0	0	4	23	88	211	335	186	161	21	0	0	1029
1987	0	0	0	5	148	238	331	270	122	0	5	0	1119
1988	0	0	1	4	64	232	423	387	73	3	0	0	1187
1989	0	0	5	7	55	189	350	235	94	17	0	0	952
1990	0	0	12	27	5	192	263	215	111	7	4	0	836
1991	0	0	0	11	199	284	355	277	149	28	0	0	1303
1992	0	0	0	11	51	127	293	138	94	0	0	0	714
1993	0	0	0	0	44	195	376	323	58	4	0	0	1000

SNOWFALL (inches) DAYTON, OHIO

SEASON	JULY	AUG	SEP	OCT	NOV	DEC	JAN	FEB	MAR	APR	MAY	JUNE	TOTAL
1964-65	0.0	0.0	0.0	0.0	2.0	2.1	8.7	5.3	9.8	T	0.0	0.0	27.9
1965-66	0.0	0.0	0.0	0.0	T	2.0	5.2	8.4	2.8	1.7	T	0.0	20.1
1966-67	0.0	0.0	0.0	0.0	9.8	7.5	3.6	8.7	7.8	T	0.0	0.0	37.4
1967-68	0.0	0.0	0.0	T	6.2	4.0	12.1	2.4	13.8	T	0.0	0.0	38.5
1968-69	0.0	0.0	0.0	T	1.4	3.7	2.5	0.9	3.8	0.3	0.0	0.0	12.6
1969-70	0.0	0.0	0.0	0.0	2.3	10.1	12.0	6.4	10.5	T	0.0	0.0	41.3
1970-71	0.0	0.0	0.0	0.0	3.6	0.7	4.6	9.8	11.0	T	0.0	0.0	29.7
1971-72	0.0	0.0	0.0	0.0	3.4	1.0	6.6	9.6	1.2	1.2	0.0	0.0	23.0
1972-73	0.0	0.0	0.0	T	2.5	5.2	2.4	3.6	4.4	3.6	0.0	0.0	21.7
1973-74	0.0	0.0	0.0	0.0	T	11.6	2.0	6.3	3.5	4.9	0.0	0.0	28.3
1974-75	0.0	0.0	0.0	T	6.6	8.8	8.2	6.4	11.1	T	0.0	0.0	41.1
1975-76	0.0	0.0	0.0	0.0	3.0	4.7	11.6	1.3	1.2	T	0.0	0.0	21.8
1976-77	0.0	0.0	0.0	T	2.6	3.9	20.2	10.6	1.2	0.3	0.0	0.0	38.8
1977-78	0.0	0.0	0.0	T	2.4	8.5	40.2	3.6	8.0	0.0	0.0	0.0	62.7
1978-79	0.0	0.0	0.0	0.0	T	1.1	20.5	17.5	0.2	T	0.0	0.0	39.3
1979-80	0.0	0.0	0.0	T	0.4	0.2	7.3	14.6	2.4	T	0.0	0.0	24.9
1980-81	0.0	0.0	0.0	T	5.3	3.3	5.7	2.0	3.3	0.0	0.0	0.0	19.6
1981-82	0.0	0.0	0.0	T	0.8	14.7	13.3	5.1	4.9	4.1	0.0	0.0	42.9
1982-83	0.0	0.0	0.0	0.0	0.1	0.3	1.5	2.5	0.9	0.2	0.0	0.0	5.5
1983-84	0.0	0.0	0.0	0.0	T	5.9	9.2	12.2	13.8	T	0.0	0.0	41.1
1984-85	0.0	0.0	0.0	0.0	3.2	5.3	14.0	12.6	0.3	2.3	0.0	0.0	37.7
1985-86	0.0	0.0	0.0	0.0	0.0	7.5	4.2	10.4	1.6	0.3	0.0	0.0	24.0
1986-87	0.0	0.0	0.0	0.0	1.1	1.7	3.6	3.4	7.9	2.2	0.0	0.0	19.9
1987-88	0.0	0.0	0.0	0.0	0.5	2.0	3.2	7.2	2.1	T	0.0	0.0	15.0
1988-89	0.0	0.0	0.0	T	1.0	4.8	0.1	4.0	5.7	0.1	T	0.0	15.7
1989-90	0.0	0.0	0.0	5.8	0.4	8.6	3.8	3.8	2.2	0.2	0.0	0.0	24.8
1990-91	0.0	0.0	0.0	0.0	0.0	3.5	5.3	1.5	2.4	0.0	0.0	0.0	12.7
1991-92	0.0	0.0	0.0	0.0	1.9	1.0	9.1	1.1	2.9	1.0	0.0	0.0	17.0
1992-93	T	0.0	0.0	0.1	0.8	1.8	2.8	16.4	7.2	T	0.0	0.0	29.1
1993-94	0.0	0.0	0.0	4.1	1.2	6.5							
Record Mean	T	0.0	0.0	0.3	2.1	5.5	7.7	6.2	5.2	0.7	T	0.0	27.6

MANSFIELD, OHIO

Mansfield is in the north central highlands at the geographical and climatological junction of central Ohio, northwest Ohio, and northeast Ohio. The station is on a plateau 3 miles north of the city of Mansfield and surrounded by rolling open farmland. The general elevation ranges from around 1,300 to 1,400 feet above sea level with the 1,000-foot contour east to west some 15 miles to the north. The climate is continental, with the modifying effects of Lake Erie most pronounced in winter. Lake Erie is just 38 miles due north.

The lake influence, plus the elevation, produce cloudy skies and considerable snow shower activity from late November into April with any wind flow from northwest through northeast. Because of this, any windshift with a cold frontal passage in winter does not bring the clearing skies, indeed, more snow is often measured from the flurry activity behind the front than from the pre-frontal conditions. A frozen Lake Erie will allow clearing skies, but an open lake dictates overcast and snow flurries. Usually the lake is open enough to set off the flurries and cloudy conditions. The major snow producer will be an intense storm moving out of the southwest with the Gulf of Mexico moisture available. Snow cover is almost constant from December through March due to almost daily snow flurries, but the depth of cover is rarely more than 8 inches. Daytime winter temperatures are not above the freezing mark too often.

Spring is a short period of rapid transition from hard winter to summer conditions. April usually brings abundant shower activity and the crops and vegetation get a quick start.

Summer is a pleasant season with low humidities and no extremely high temperatures. Rarely does the temperature climb above the 90 degree point. Thunderstorms average about once every three days during the season from June through September. Highest winds are associated with the heavier thunderstorms, and while hail does not occur often, it is of major concern to the applegrowers in the area. Flooding problems are confined to the flash-flood type on the small streams in the area.

The growing season is normally about 153 days. Autumn usually produces many clear warm days and cool invigorating nights. Ground fog is at a maximum incidence during the autumn. Little rainfall occurs to interfere with harvest time and county fair time.

NORMALS, MEANS, AND EXTREMES

MANSFIELD, OHIO

LATITUDE: 40°49'N LONGITUDE: 82°31'W ELEVATION: FT. GRND 1295 BARO 1300 TIME ZONE: EASTERN WBAN: 14891

	(a)	JAN	FEB	MAR	APR	MAY	JUNE	JULY	AUG	SEP	OCT	NOV	DEC	YEAR
TEMPERATURE °F:														
Normals														
-Daily Maximum		32.1	35.0	46.6	58.6	69.3	78.2	82.1	80.1	73.7	62.3	49.2	36.8	58.7
-Daily Minimum		16.8	18.9	28.6	38.2	48.3	57.3	62.0	60.4	54.0	43.1	33.9	22.7	40.4
-Monthly		24.5	27.0	37.6	48.4	58.8	67.8	72.1	70.3	63.9	52.7	41.6	29.8	49.5
Extremes														
-Record H ghest	34	63	67	82	86	92	101	100	97	93	85	78	73	101
-Year		1972	1961	1986	1990	1962	1988	1988	1988	1964	1963	1968	1982	JUN 1988
-Record Lowest	34	-22	-11	-6	8	25	37	43	40	33	20	2	-17	-22
-Year		1985	1982	1980	1982	1966	1990	1988	1965	1991	1988	1976	1989	JAN 1985
NORMAL DEGREE DAYS:														
Heating (base 65°F)		1256	1064	849	498	229	43	0	20	110	396	702	1091	6258
Cooling (base 65°F)		0	0	0	0	37	127	225	185	77	15	0	0	666
% OF POSSIBLE SUNSHINE														
MEAN SKY COVER (tenths)														
Sunrise - Sunset	30	7.7	7.6	7.4	6.9	6.5	6.0	5.7	5.8	5.9	6.1	7.7	8.1	6.8
MEAN NUMBER OF DAYS:														
Sunrise to Sunset														
-Clear	30	4.0	3.8	4.8	5.6	6.9	7.3	7.8	8.4	8.7	8.8	4.0	3.5	73.4
-Partly Cloudy	30	6.4	6.4	7.0	8.0	9.5	11.6	13.3	11.3	8.8	7.9	5.9	4.9	101.1
-Cloudy	30	20.6	18.0	19.2	16.4	14.6	11.2	9.9	11.3	12.5	14.3	20.1	22.6	190.7
Precipitation														
.01 inches or more	34	13.1	11.9	14.6	13.6	12.8	10.7	9.8	10.1	9.1	9.9	12.9	14.1	142.9
Snow,Ice pellets,hail														
1.0 inches or more	34	3.6	3.4	2.3	0.5	0.1	0.0	0.0	0.0	0.0	0.1	1.0	3.0	14.1
Thunderstorms	34	0.2	0.4	2.0	3.6	4.8	6.2	6.6	5.9	3.2	0.9	0.8	0.1	34.8
Heavy Fog Visibility														
1/4 mile or less	34	2.4	3.1	3.4	1.7	2.3	1.3	1.4	2.6	2.1	2.2	2.2	3.9	28.5
Temperature °F														
-Maximum														
90° and above	28	0.0	0.0	0.0	0.0	0.*	1.1	3.3	1.4	0.5	0.0	0.0	0.0	6.3
32° and below	28	15.9	11.9	4.3	0.3	0.0	0.0	0.0	0.0	0.0	0.0	1.8	11.1	45.2
-Minimum														
32° and below	28	28.0	24.3	20.3	8.6	0.7	0.0	0.0	0.0	0.0	4.2	14.7	25.2	126.0
0° and below	28	3.4	2.1	0.3	0.0	0.0	0.0	0.0	0.0	0.0	0.0	0.0	1.1	6.9
AVG. STATION PRESS.(mb)														
RELATIVE HUMIDITY (%)														
Hour 01	10	78	77	76	72	76	81	82	83	82	77	79	79	79
Hour 07 (Local Time)	27	81	81	80	77	79	80	83	87	87	82	81	82	82
Hour 13	27	72	69	62	56	57	58	58	60	60	59	67	74	63
Hour 19	13	75	72	68	61	61	64	64	68	72	69	74	77	69
PRECIPITATION (inches):														
Water Equivalent														
-Normal		1.98	2.02	3.30	3.64	4.35	3.95	4.04	4.08	3.38	2.34	3.51	3.07	39.66
-Maximum Monthly	34	5.54	4.74	7.04	6.58	8.83	10.00	13.23	7.60	7.76	5.66	12.82	11.19	13.23
-Year		1993	1990	1964	1964	1989	1981	1992	1974	1986	1990	1985	1990	JUL 1992
-Minimum Monthly	34	0.41	0.29	1.16	0.75	1.50	0.56	0.94	0.63	0.73	0.43	0.70	0.74	0.29
-Year		1981	1978	1960	1971	1961	1988	1975	1970	1963	1963	1976	1976	FEB 1978
-Maximum in 24 hrs	34	2.10	2.79	2.45	2.34	2.62	3.72	5.06	3.85	4.10	1.84	3.17	3.55	5.06
-Year		1992	1961	1964	1979	1989	1978	1969	1972	1979	1983	1993	1990	JUL 1969
Snow,Ice pellets,hail														
-Maximum Monthly	34	42.1	19.1	17.1	13.4	1.4	0.0	0.0	0.0	T	10.3	7.4	18.0	42.1
-Year		1978	1984	1993	1982	1966				1970	1993	1980	1962	JAN 1978
-Maximum in 24 hrs	34	12.0	10.0	8.0	11.8	1.4	0.0	0.0	0.0	T	8.6	3.2	12.3	12.3
-Year		1968	1985	1987	1987	1966				1970	1993	1966	1974	DEC 1974
WIND:														
Mean Speed (mph)	11	13.4	12.6	12.5	12.3	10.3	9.9	8.4	8.4	9.0	10.6	11.9	12.7	11.0
Prevailing Direction														
Fastest Obs. 1 Min.														
-Direction (!!)	27	24	24	25	33	24	23	25	17	24	27	18	18	33
-Speed (MPH)	27	46	44	37	46	35	40	37	41	30	31	37	46	46
-Year		1971	1967	1982	1981	1990	1973	1983	1988	1986	1988	1988	1971	APR 1981
Peak Gust														
-Direction (!!)	10	SW	S		SW		W	W	N	SW	W		SW	
-Speed (mph)	10	59	58	62	68	55	60	64	68	45	52	69	64	69
-Date		1990	1988	1991	1986	1991	1987	1990	1987	1984	1988	1992	1987	NOV 1992

PRECIPITATION (inches)

MANSFIELD, OHIO

YEAR	JAN	FEB	MAR	APR	MAY	JUNE	JULY	AUG	SEP	OCT	NOV	DEC	ANNUAL
1964	1.45	1.33	7.04	6.58	4.16	3.03	2.16	2.44	1.13	1.09	1.49	2.88	34.78
1965	3.01	2.64	2.28	3.02	3.69	1.84	1.42	3.30	2.11	2.98	2.47	1.88	30.64
1966	1.55	2.17	1.44	2.73	2.84	1.88	3.82	4.66	1.95	1.23	4.33	2.57	31.17
1967	0.74	1.33	3.21	2.61	4.50	1.63	3.29	1.08	4.61	1.75	2.38	2.76	29.89
1968	2.27	0.48	2.40	1.84	6.27	3.94	3.13	1.56	2.26	0.78	4.01	3.17	32.11
1969	2.41	0.74	1.25	5.50	6.15	3.68	8.06	2.36	2.70	1.82	3.25	1.55	39.47
1970	1.02	0.93	1.27	5.27	5.17	1.86	5.67	0.63	4.65	2.69	3.24	2.39	34.79
1971	1.40	2.58	1.78	0.75	4.24	1.29	3.79	2.18	2.41	1.15	1.11	3.42	26.10
1972	0.95	1.15	2.49	5.98	3.07	3.75	3.11	5.74	6.85	1.52	4.30	2.79	41.70
1973	1.83	1.39	4.21	3.35	4.15	7.07	4.35	4.00	1.38	3.59	3.19	2.85	41.36
1974	3.59	1.59	4.14	2.47	5.65	1.90	2.93	7.60	3.41	0.79	5.15	2.64	41.86
1975	3.33	3.74	2.69	1.95	4.01	2.36	0.94	5.63	4.97	2.19	1.59	2.92	36.32
1976	2.69	2.99	3.96	1.37	2.35	4.34	2.55	3.16	3.61	2.43	0.70	0.74	30.89
1977	0.79	1.17	4.26	3.25	1.99	3.69	4.93	3.40	4.56	1.75	3.04	4.15	36.98
1978	3.76	0.29	2.29	4.70	4.20	4.98	1.70	3.20	1.10	4.89	1.96	4.78	37.85
1979	3.29	1.74	1.81	5.63	5.25	4.82	2.55	6.78	5.64	1.39	4.93	2.28	46.11
1980	1.30	1.44	5.72	3.34	4.11	6.31	5.01	4.41	2.45	1.98	2.01	2.15	40.23
1981	0.41	3.41	1.71	5.62	4.49	10.00	2.04	5.08	4.41	2.62	1.53	3.50	44.82
1982	4.53	1.76	5.18	1.93	3.73	6.09	4.08	5.09	2.52	0.99	6.82	4.68	47.40
1983	1.67	1.16	2.43	5.80	5.68	2.92	4.61	4.51	5.16	4.60	6.49	3.37	48.40
1984	1.22	2.91	4.68	5.48	6.23	1.25	3.55	4.43	3.78	3.71	4.15	4.41	45.80
1985	1.49	2.09	6.17	1.42	6.57	3.13	3.00	6.90	1.63	2.46	12.82	3.17	50.85
1986	1.62	4.34	3.92	3.51	3.72	6.67	8.56	4.00	7.76	4.57	3.94	3.71	56.32
1987	1.35	0.82	3.06	4.00	3.76	8.87	6.92	7.44	0.81	2.42	1.60	3.01	44.06
1988	2.04	3.23	2.63	3.89	2.84	0.56	4.90	5.34	2.60	2.29	5.41	2.44	38.17
1989	2.82	1.94	4.29	3.74	8.83	6.40	3.96	0.64	3.10	3.20	4.42	2.06	45.40
1990	2.39	4.74	1.74	4.58	7.79	5.21	9.59	6.95	4.94	5.66	2.44	11.19	67.22
1991	2.65	1.55	3.56	4.70	4.05	1.02	1.97	3.50	3.09	1.86	1.81	2.78	32.54
1992	3.59	2.09	3.21	4.17	3.00	3.05	13.23	2.62	4.07	2.68	7.63	2.35	51.69
1993	5.54	2.02	5.70	6.12	2.52	6.02	1.92	0.94	5.50	3.19	7.23	2.53	49.23
Record Mean	2.17	2.03	3.32	3.71	4.22	3.91	4.16	3.89	3.39	2.34	3.65	2.97	39.75

AVERAGE TEMPERATURE (deg. F)

MANSFIELD, OHIO

YEAR	JAN	FEB	MAR	APR	MAY	JUNE	JULY	AUG	SEP	OCT	NOV	DEC	ANNUAL
1964	28.9	25.3	36.6	49.1	61.9	68.0	72.4	68.3	63.6	49.9	43.9	30.6	49.9
#1965	25.8	25.5	30.6	45.9	62.9	65.2	68.0	67.2	63.9	48.8	40.6	35.2	48.3
1966	20.3	27.1	37.5	45.5	52.6	68.2	72.0	68.1	59.7	49.3	40.8	29.0	47.5
1967	30.7	23.8	37.4	50.2	53.3	70.9	70.2	67.6	60.1	52.6	36.0	33.1	48.8
1968	22.8	23.4	39.7	51.3	55.5	68.9	72.3	73.0	66.4	54.4	42.9	29.7	50.0
1969	26.2	29.8	35.0	51.9	61.4	67.4	74.6	73.0	64.8	54.2	38.7	26.0	50.3
1970	20.1	27.6	34.4	50.6	62.0	69.9	73.7	73.6	69.1	56.4	42.3	32.3	51.0
1971	22.4	28.9	34.1	46.0	56.4	71.3	70.3	68.6	67.8	60.0	40.7	37.6	50.4
1972	27.9	27.3	36.2	47.7	61.4	66.0	73.5	70.5	65.7	49.6	38.6	33.6	49.8
1973	30.0	27.8	47.5	50.0	57.9	71.8	73.5	73.6	67.4	58.2	45.1	32.7	53.0
1974	32.0	28.2	37.4	51.9	58.2	66.3	72.7	71.7	59.9	51.5	41.7	30.2	50.2
1975	29.8	29.7	33.9	42.2	62.4	68.5	72.9	72.6	58.7	54.1	45.8	29.9	50.0
1976	20.8	34.6	42.3	49.1	55.3	67.6	69.2	66.0	59.2	45.3	31.3	20.7	46.8
1977	8.7	23.6	42.0	51.3	63.3	63.5	74.1	70.1	65.8	50.5	43.4	27.4	48.6
1978	18.3	15.4	31.6	47.2	58.1	68.4	71.2	71.8	68.5	50.3	41.9	31.1	47.8
1979	18.6	16.4	40.8	46.2	56.6	66.6	68.6	68.2	62.8	50.8	40.9	31.9	47.4
1980	25.7	21.4	33.9	46.9	58.3	62.8	72.7	73.9	65.2	49.0	39.5	30.4	48.3
1981	22.3	32.4	37.0	51.8	57.1	69.7	72.8	70.5	62.8	51.3	43.4	28.5	50.0
1982	18.6	24.8	36.1	43.2	64.1	62.4	70.9	67.2	60.9	54.6	44.1	39.5	48.9
1983	29.2	33.1	40.6	47.5	56.5	69.7	75.6	74.9	65.5	53.9	44.4	22.3	51.1
1984	20.2	34.5	28.7	48.5	55.9	72.2	70.1	71.1	61.0	58.3	41.0	38.2	50.0
1985	21.0	25.2	42.0	54.4	60.7	63.0	70.7	68.6	64.8	54.4	45.2	23.1	49.4
1986	26.4	28.2	39.7	51.8	61.1	67.4	72.7	67.5	66.2	53.8	38.8	31.2	50.4
1987	26.2	31.1	39.6	49.0	62.8	69.7	73.9	70.1	63.7	46.7	45.1	32.8	50.9
1988	23.9	24.7	36.7	47.0	60.7	68.4	75.6	73.3	62.9	45.1	42.4	29.1	49.2
1989	34.5	24.3	38.8	45.4	55.8	67.5	72.5	69.5	62.7	52.6	38.8	17.0	48.3
1990	34.5	34.1	42.0	48.5	55.8	67.1	70.0	68.9	62.9	52.8	44.0	34.7	51.3
1991	26.2	32.3	41.1	52.8	67.5	71.1	74.1	71.8	63.4	55.2	38.0	32.5	52.2
1992	28.2	32.1	36.2	47.6	57.2	64.2	70.5	66.5	61.5	48.6	41.0	32.3	48.8
1993	30.6	24.2	33.7	47.6	59.4	67.2	73.9	71.3	61.0	49.9	40.7	29.4	49.1
Record Mean	24.8	27.1	37.1	48.5	59.0	67.7	71.9	70.2	63.6	52.5	41.4	29.6	49.5
Max	32.1	35.0	45.9	58.6	69.4	78.0	81.9	80.1	73.4	62.0	49.0	36.5	58.5
Min	17.4	19.2	28.2	38.4	48.5	57.3	61.9	60.3	53.8	42.9	33.8	22.6	40.4

REFERENCE NOTES

MANSFIELD, OHIO

GENERAL
T - TRACE AMOUNT.
BLANK ENTRIES DENOTE MISSING/UNREPORTED DATA.
# INDICATES A STATION OR INSTRUMENT RELOCATION.

SPECIFIC

(a) - LENGTH OF RECORD IN YEARS, ALTHOUGH INDIVIDUAL MONTHS MAY BE MISSING.
0.* OR * - THE VALUE IS BETWEEN 0.0 AND 0.05
NORMALS - BASED ON THE 1961-1990 RECORD PERIOD.
EXTREMES - DATES ARE THE MOST RECENT OCCURRENCE
WIND DIR.- NUMERALS SHOW TENS OF DEGRESS CLOCKWISE FROM TRUE NORTH. "00" INDICATES CALM.
RESULTANT DIRECTIONS ARE GIVEN TO WHOLE DEGREES.
BOLD VALUES INDICATE EXTREME VALUES WHICH OCCURRED AFTER THE ASOS SYSTEM WAS COMMISSIONED.

RECORD = PERIOD OF RECORD
RECORD MEAN PRECIPITATION IS THE MEAN OF ALL DAILY PRECIPITATION AMOUNTS DURING THE PERIOD OF RECORD.
RECORD MAX(MIN) TEMPERATURE IS THE MEAN OF ALL DAILY MAX(MIN) TEMPERATURES DURING THE PERIOD OF RECORD.
RECORD MEAN TEMPERATURE IS THE SUM OF THE RECORD MAX AND RECORD MIN DIVIDED BY 2.
AVERAGE TEMPERATURE IS THE SUM OF THE MEAN DAILY MAX AND MIN TEMPERATURE DIVIDED BY 2.

EXCEPTIONS

MEAN WIND SPEED IS THROUGH 1979.

RECORD MEANS ARE THROUGH THE CURRENT YEAR,
BEGINNING IN 1960 FOR TEMPERATURE
1960 FOR PRECIPITATION
1960 FOR SNOWFALL

HEATING DEGREE DAYS Base 65 deg. F MANSFIELD, OHIO

SEASON	JULY	AUG	SEP	OCT	NOV	DEC	JAN	FEB	MAR	APR	MAY	JUNE	TOTAL
#1964-65	2	37	127	462	627	1061	1208	1098	1059	567	119	76	6443
1965-66	23	59	115	500	728	921	1380	1054	845	579	381	54	6639
1966-67	7	20	189	480	723	1108	1060	1148	852	442	364	13	6406
1967-68	21	28	171	391	865	982	1300	1201	778	406	292	25	6460
1968-69	9	17	48	354	655	1086	1196	982	923	396	168	70	5904
1969-70	0	0	89	347	784	1204	1386	1043	940	444	159	27	6423
1970-71	6	0	53	275	674	1006	1316	1004	955	563	275	16	6143
1971-72	3	12	54	170	722	840	1143	1085	886	515	143	71	5644
1972-73	9	16	70	471	783	966	1078	1037	534	453	219	1	5637
1973-74	1	8	51	223	588	995	1017	1022	847	402	234	56	5444
1974-75	0	0	175	413	695	1073	1085	983	956	676	142	45	6243
1975-76	2	4	188	343	570	1078	1367	870	699	494	307	30	5952
1976-77	8	56	185	606	1006	1369	1743	1155	708	421	128	112	7497
1977-78	9	24	63	442	643	1159	1442	1380	1028	530	259	39	7018
1978-79	12	6	63	449	686	1044	1433	1353	744	557	290	54	6691
1979-80	38	39	118	453	719	1018	1214	1256	956	535	217	115	6678
1980-81	4	3	86	490	757	1067	1318	908	859	389	259	12	6152
1981-82	5	7	128	418	637	1127	1434	1120	887	645	79	106	6593
1982-83	8	41	161	325	624	786	1102	886	751	521	266	39	5510
1983-84	8	0	106	347	613	1319	1383	880	1123	493	298	5	6575
1984-85	9	9	172	214	713	825	1361	1108	706	344	177	88	5726
1985-86	2	11	128	323	587	1293	1191	1025	783	403	174	52	5972
1986-87	4	56	72	354	783	1043	1196	945	782	476	168	29	5908
1987-88	5	24	88	562	593	993	1266	1165	870	534	178	71	6349
1988-89	7	13	97	617	672	1111	940	1135	812	578	312	37	6331
1989-90	3	12	132	394	779	1483	939	859	716	517	282	61	6177
1990-91	13	9	135	375	623	931	1197	909	735	377	92	15	5411
1991-92	0	1	143	321	803	999	1131	948	886	524	259	82	6097
1992-93	14	42	158	501	716	1008	1060	1135	961	516	192	62	6365
1993-94	0	8	158	460	722	1096							

COOLING DEGREE DAYS Base 65 deg. F MANSFIELD, OHIO

YEAR	JAN	FEB	MAR	APR	MAY	JUNE	JULY	AUG	SEP	OCT	NOV	DEC	TOTAL
1969	0	0	0	10	63	148	303	255	91	19	0	0	889
1970	0	0	0	17	71	181	284	271	184	16	0	0	1024
1971	0	0	0	0	16	211	175	131	147	25	0	0	705
1972	0	0	0	0	35	109	279	198	102	0	0	0	723
1973	0	0	0	11	8	210	272	282	129	19	0	0	931
1974	0	0	0	16	30	103	246	215	27	2	1	0	640
1975	0	0	0	0	68	160	254	247	8	9	0	0	746
1976	0	0	0	21	12	117	145	91	16	0	0	0	402
1977	0	0	0	18	84	72	299	190	96	0	1	0	760
1978	0	0	0	0	50	146	209	226	174	0	0	0	805
1979	0	0	0	4	38	109	157	146	60	17	0	0	531
1980	0	0	0	0	17	57	249	285	102	1	0	0	711
1981	0	0	0	2	20	159	255	186	68	3	0	0	693
1982	0	0	0	1	59	33	197	118	47	11	3	3	472
1983	0	0	0	4	12	185	344	313	127	9	0	0	994
1984	0	0	0	6	22	228	176	205	57	14	0	0	708
1985	0	0	0	36	53	36	187	129	129	0	0	0	570
1986	0	0	6	15	60	129	247	139	118	11	0	0	725
1987	0	0	0	3	108	176	287	189	56	0	2	0	821
1988	0	0	0	0	49	182	343	281	40	6	0	0	901
1989	0	0	4	0	33	116	243	160	71	15	0	0	642
1990	0	0	10	28	6	131	175	139	80	5	0	0	574
1991	0	0	0	19	177	205	287	219	104	25	0	0	1036
1992	0	0	0	7	25	66	191	94	60	0	0	0	443
1993	0	0	0	0	24	136	285	211	44	2	0	0	702

SNOWFALL (inches) MANSFIELD, OHIO

SEASON	JULY	AUG	SEP	OCT	NOV	DEC	JAN	FEB	MAR	APR	MAY	JUNE	TOTAL
1964-65	0.0	0.0	0.0	T	1.9	8.2	11.3	10.8	10.2	T	0.0	0.0	42.4
1965-66	0.0	0.0	0.0	T	T	0.8	8.0	5.8	3.6	2.7	1.4	0.0	22.3
1966-67	0.0	0.0	0.0	0.0	6.2	6.3	1.4	12.8	11.2	T	0.0	0.0	37.9
1967-68	0.0	0.0	0.0	T	3.5	8.2	21.3	8.1	8.1	0.3	0.0	0.0	49.5
1968-69	0.0	0.0	0.0	T	1.9	7.5	4.0	7.4	5.5	2.1	T	0.0	28.4
1969-70	0.0	0.0	0.0	T	3.7	11.7	10.7	5.0	4.3	T	0.0	0.0	35.4
1970-71	0.0	0.0	T	0.0	1.7	5.4	7.2	15.9	13.5	T	0.0	0.0	43.7
1971-72	0.0	0.0	0.0	0.0	2.8	1.7	7.1	9.9	5.9	1.4	0.0	0.0	28.8
1972-73	0.0	0.0	0.0	T	4.4	9.1	3.2	9.0	3.3	1.3	T	0.0	30.3
1973-74	0.0	0.0	0.0	0.0	0.2	11.0	8.1	11.0	6.3	2.7	0.0	0.0	39.3
1974-75	0.0	0.0	0.0	T	5.2	16.8	10.1	12.4	8.9	0.7	0.0	0.0	54.1
1975-76	0.0	0.0	0.0	0.0	2.7	6.2	14.0	10.8	2.2	1.7	T	0.0	37.6
1976-77	0.0	0.0	0.0	T	4.7	12.8	24.8	10.0	2.0	1.4	0.0	0.0	55.7
1977-78	0.0	0.0	0.0	T	4.2	16.8	42.1	9.6	4.9	T	0.0	0.0	77.6
1978-79	0.0	0.0	0.0	T	0.3	1.3	20.5	10.9	1.3	0.3	0.0	0.0	34.6
1979-80	0.0	0.0	0.0	T	0.4	5.5	6.8	12.6	1.7	0.6	T	0.0	27.6
1980-81	0.0	0.0	0.0	T	7.4	11.5	9.9	7.1	7.5	0.0	0.0	0.0	43.4
1981-82	0.0	0.0	0.0	T	3.3	16.1	16.5	6.6	11.0	13.4	0.0	0.0	66.9
1982-83	0.0	0.0	0.0	0.0	1.0	3.3	1.6	5.9	3.9	0.9	0.0	0.0	16.6
1983-84	0.0	0.0	0.0	0.0	3.7	9.7	11.6	19.1	13.4	0.2	0.0	0.0	57.7
1984-85	0.0	0.0	0.0	0.0	0.7	7.1	20.3	18.2	0.5	2.2	0.0	0.0	49.0
1985-86	0.0	0.0	0.0	0.0	T	13.3	8.6	7.4	5.7	2.8	0.0	0.0	37.8
1986-87	0.0	0.0	0.0	0.0	5.7	0.2	10.5	2.6	10.6	13.0	0.0	0.0	42.6
1987-88	0.0	0.0	0.0	T	0.9	13.8	7.5	11.1	14.9	0.3	0.0	0.0	48.5
1988-89	0.0	0.0	0.0	0.5	2.1	8.1	3.4	6.3	6.0	0.7	1.1	0.0	28.2
1989-90	0.0	0.0	0.0	3.1	0.9	11.9	7.3	5.8	2.4	1.3	0.0	0.0	32.7
1990-91	0.0	0.0	0.0	0.0	T	6.0	7.0	9.1	2.7	T	0.0	0.0	24.8
1991-92	0.0	0.0	0.0	0.0	3.1	4.2	19.1	2.2	4.6	4.2	0.0	0.0	37.4
1992-93	0.0	0.0	0.0	1.1	2.6	7.3	5.5	15.1	17.1	0.1	0.0	0.0	48.8
1993-94	0.0	0.0	0.0	10.3	2.2	8.0							
Record Mean	0.0	0.0	T	0.5	2.4	8.7	11.0	10.1	7.1	1.9	0.1	0.0	41.7

TOLEDO, OHIO

Toledo is located on the western end of Lake Erie at the mouth of the Maumee River. Except for a bank up from the river about 30 feet, the terrain is generally level with only a slight slope toward the river and Lake Erie. The city has quite a diversified industrial section and excellent harbor facilities, making it a large transportation center for rail, water, and motor freight. Generally rich agricultural land is found in the surrounding area, especially up the Maumee Valley toward the Indiana state line.

Rainfall is usually sufficient for general agriculture. The terrain is level and drainage rather poor, therefore, a little less than the normal precipitation during the growing season is better than excessive amounts. Snowfall is generally light in this area, distributed throughout the winter from November to March with frequent thaws.

The nearness of Lake Erie and the other Great Lakes has a moderating effect on the temperature, and extremes are seldom recorded. On average, only fifteen days a year experience temperatures of 90 degrees or higher, and only eight days when it drops to zero or lower. The growing season averages 160 days, but has ranged from over 220 to less than 125 days.

Humidity is rather high throughout the year in this area, and there is an excessive amount of cloudiness. In the winter months the sun shines during only about 30 percent of the daylight hours. December and January, the cloudiest months, sometimes have as little as 16 percent of the possible hours of sunshine.

Severe windstorms, causing more than minor damage, occur infrequently. There are on the average twenty-three days per year having a sustained wind velocity of 32 mph or more.

Flooding in the Toledo area is produced by several factors. Heavy rains of 1 inch or more will cause a sudden rise in creeks and drainage ditches to the point of overflow. The western shores of Lake Erie are subject to flooding when the lake level is high and prolonged periods of east to northeast winds prevail.

NORMALS, MEANS, AND EXTREMES

TOLEDO, OHIO

LATITUDE: 41°36'N LONGITUDE: 83°48' W ELEVATION: FT. GRND 669 BARO 694 TIME ZONE: EASTERN WBAN: 94830

	(a)	JAN	FEB	MAR	APR	MAY	JUNE	JULY	AUG	SEP	OCT	NOV	DEC	YEAR
TEMPERATURE °F:														
Normals														
-Daily Maximum		30.2	33.4	45.5	58.8	70.5	79.8	83.4	81.3	74.4	62.4	48.5	35.2	58.6
-Daily Minimum		14.9	17.0	26.8	36.4	46.7	56.0	60.6	58.4	51.5	40.0	31.5	20.5	38.4
-Monthly		22.5	25.2	36.2	47.6	58.6	67.9	72.1	69.9	63.0	51.2	40.0	27.9	48.5
Extremes														
-Record Highest	38	62	68	80	88	95	104	103	99	98	91	78	68	104
-Year		1989	1957	1986	1990	1962	1988	1988	1993	1978	1963	1987	1982	JUN 1988
-Record Lowest	38	-20	-14	-6	8	25	32	40	34	26	15	2	-19	-20
-Year		1984	1982	1984	1982	1974	1972	1988	1982	1974	1976	1958	1989	JAN 1984
NORMAL DEGREE DAYS:														
Heating (base 65°F)		1318	1114	893	522	238	33	0	16	109	436	750	1150	6579
Cooling (base 65°F)		0	0	0	0	40	120	225	168	49	8	0	0	610
% OF POSSIBLE SUNSHINE	38	42	47	49	52	60	64	65	63	60	54	37	33	52
MEAN SKY COVER (tenths)														
Sunrise - Sunset	38	7.5	7.3	7.3	6.9	6.4	6.0	5.7	5.7	5.9	6.1	7.7	7.9	6.7
MEAN NUMBER OF DAYS:														
Sunrise to Sunset														
-Clear	38	4.6	4.6	5.1	5.8	6.7	7.0	7.6	8.4	8.1	8.1	3.7	3.3	72.7
-Partly Cloudy	38	6.7	6.7	7.1	7.8	10.4	11.5	13.4	12.2	9.8	8.9	7.0	6.2	107.7
-Cloudy	38	19.7	17.0	18.8	16.4	13.9	11.5	10.1	10.5	12.1	14.0	19.4	21.5	184.8
Precipitation														
.01 inches or more	38	13.1	10.8	13.3	12.8	11.9	10.0	9.6	9.1	10.2	9.3	12.0	14.2	136.3
Snow,Ice pellets,hail														
1.0 inches or more	38	2.8	2.5	2.0	0.5	0.0	0.0	0.0	0.0	0.0	0.1	1.1	2.7	11.6
Thunderstorms	38	0.2	0.5	2.0	3.7	4.9	6.9	6.9	6.2	3.8	1.3	0.7	0.2	37.3
Heavy Fog Visibility														
1/4 mile or less	38	1.4	1.9	1.7	0.9	0.7	0.9	0.8	1.8	1.8	1.9	1.5	2.2	17.6
Temperature °F														
-Maximum														
90° and above	38	0.0	0.0	0.0	0.0	0.8	3.5	5.1	3.5	1.1	0.*	0.0	0.0	14.1
32° and below	38	16.9	12.7	4.4	0.2	0.0	0.0	0.0	0.0	0.0	0.0	2.1	11.6	47.9
-Minimum														
32° and below	38	29.1	25.8	22.8	10.5	1.4	0.*	0.0	0.0	0.4	6.8	17.3	26.4	140.6
0° and below	38	4.2	2.7	0.2	0.0	0.0	0.0	0.0	0.0	0.0	0.0	0.0	1.8	8.9
AVG. STATION PRESS.(mb)	21	992.6	993.2	991.1	990.0	990.1	990.2	991.3	992.6	993.1	993.4	992.6	993.0	991.9
RELATIVE HUMIDITY (%)														
Hour 01	38	76	76	75	76	78	82	84	88	88	82	80	82	81
Hour 07 (Local Time)	38	80	80	81	80	80	82	86	91	91	86	83	83	84
Hour 13	38	70	66	61	55	52	54	55	58	58	57	66	73	60
Hour 19	38	74	71	66	59	56	58	61	67	71	70	74	78	67
PRECIPITATION (inches):														
Water Equivalent														
-Normal		1.75	1.73	2.66	2.96	2.91	3.75	3.27	3.25	2.85	2.10	2.81	2.93	32.97
-Maximum Monthly	38	4.61	5.39	5.70	6.10	5.13	8.48	6.75	8.47	8.10	5.53	6.86	6.81	8.48
-Year		1965	1990	1985	1977	1968	1981	1969	1965	1972	1991	1982	1967	JUN 1981
-Minimum Monthly	38	0.27	0.27	0.58	0.88	0.96	0.27	0.52	0.40	0.58	0.28	0.55	0.54	0.27
-Year		1961	1969	1958	1962	1964	1988	1991	1976	1963	1964	1976	1958	JUN 1988
-Maximum in 24 hrs	38	1.78	2.59	2.60	3.43	2.34	3.21	4.39	2.42	3.97	3.21	3.17	3.53	4.39
-Year		1959	1990	1985	1977	1991	1978	1969	1972	1972	1988	1982	1967	JUL 1969
Snow,Ice pellets,hail														
-Maximum Monthly	38	30.8	14.4	17.7	12.0	1.3	0.0	T	T	T	2.0	17.9	24.2	30.8
-Year		1978	1967	1993	1957	1989		1992	1993	1993	1989	1966	1977	JAN 1978
-Maximum in 24 hrs	38	10.4	7.7	9.7	9.8	1.3	0.0	T	T	T	1.8	8.3	13.9	13.9
-Year		1978	1981	1993	1957	1989		1992	1993	1993	1989	1966	1974	DEC 1974
WIND:														
Mean Speed (mph)	38	10.9	10.6	11.1	10.9	9.5	8.4	7.4	7.1	7.6	8.8	10.2	10.5	9.4
Prevailing Direction through 1963		WSW	WSW	WSW	E	WSW	SW	WSW	SW	SSW	WSW	WSW	SW	WSW
Fastest Mile														
-Direction (!!!)	38	W	SW	W	SW	W	W	NW	W	NW	SW	SW	SW	SW
-Speed (MPH)	38	47	56	56	72	45	50	54	47	47	40	65	45	72
-Year		1972	1967	1957	1956	1957	1969	1970	1965	1969	1956	1957	1971	APR 1956
Peak Gust														
-Direction (!!!)	10	SW	NW	SW	SW	W	SW	SW	SE	NW	SW	NW	SW	SE
-Speed (mph)	10	62	52	64	59	58	55	66	75	54	49	51	56	75
-Date		1990	1990	1991	1991	1989	1992	1992	1988	1986	1990	1989	1987	AUG 1988

PRECIPITATION (inches) — TOLEDO, OHIO

YEAR	JAN	FEB	MAR	APR	MAY	JUNE	JULY	AUG	SEP	OCT	NOV	DEC	ANNUAL
1964	1.87	0.95	4.88	3.49	0.96	1.89	1.58	3.80	1.61	0.28	0.77	2.20	24.28
1965	4.61	1.96	1.77	2.07	3.80	2.57	2.03	8.47	4.93	3.28	1.75	3.61	40.85
1966	0.46	1.46	1.82	2.81	1.88	3.42	3.73	4.60	1.17	0.97	4.63	5.12	32.07
1967	1.29	2.12	1.72	2.77	2.28	1.92	3.95	0.81	2.14	3.07	2.85	6.81	31.73
1968	1.91	1.29	2.26	3.01	5.13	3.40	4.50	1.45	1.52	1.11	3.52	3.97	33.07
1969	3.70	0.27	1.54	3.64	3.74	4.82	6.75	1.15	2.70	1.58	3.81	2.10	35.80
1970	1.09	0.89	2.61	4.26	4.05	4.59	5.99	3.00	5.78	2.00	2.09	1.49	37.84
1971	0.82	2.59	1.34	1.08	2.33	2.64	2.77	1.10	1.84	1.77	1.17	3.73	23.18
1972	1.42	0.77	2.33	3.74	2.63	4.09	2.77	4.47	8.10	1.46	3.55	3.08	38.41
1973	1.63	1.05	4.20	1.79	2.85	6.51	3.17	1.18	1.09	2.76	3.27	3.17	32.67
1974	2.27	2.00	2.93	2.55	4.18	3.31	0.68	1.61	1.41	0.70	3.57	3.41	28.62
1975	2.57	2.57	1.90	2.34	3.83	4.21	4.99	5.52	2.70	2.42	2.17	3.35	38.57
1976	2.80	4.43	3.56	2.79	1.72	3.70	2.08	0.40	3.68	2.14	0.55	0.93	28.78
1977	1.29	1.99	4.43	6.10	1.53	3.48	1.83	5.79	4.27	1.77	2.72	3.56	38.76
1978	3.14	0.54	2.34	3.74	2.48	5.34	1.86	1.67	3.19	1.65	2.48	3.31	31.74
1979	1.24	0.70	2.55	4.03	3.15	4.23	3.96	4.71	2.90	2.02	4.25	2.46	36.20
1980	0.74	0.96	3.65	3.13	2.93	3.26	4.49	5.89	1.63	1.79	0.97	2.48	31.92
1981	0.48	3.27	0.63	3.54	2.38	8.48	3.72	2.28	6.05	3.79	0.84	2.93	38.39
1982	3.61	1.15	3.74	1.53	2.61	2.01	1.97	1.38	2.03	1.14	6.86	3.48	31.51
1983	0.88	0.59	1.86	4.28	3.98	4.06	3.39	2.15	1.42	3.59	5.56	3.91	35.67
1984	0.99	1.18	2.95	5.15	3.48	1.49	2.30	3.87	2.02	1.75	2.74	3.22	31.14
1985	2.02	3.23	5.70	1.40	1.85	2.90	3.86	4.30	2.53	3.05	5.89	1.62	38.35
1986	0.99	2.46	2.16	2.81	2.72	5.32	3.37	5.93	4.75	4.78	1.66	1.87	38.82
1987	1.87	0.53	1.78	1.72	2.32	5.62	1.51	4.45	2.31	2.21	2.59	3.80	30.71
1988	1.17	1.33	1.69	1.45	1.37	0.27	3.76	5.11	1.80	4.37	4.27	1.96	28.55
1989	1.80	0.74	2.03	3.50	4.87	6.74	6.31	3.59	3.30	1.36	1.89	1.29	37.42
1990	2.18	5.39	3.46	2.09	4.63	3.14	1.89	3.32	1.72	2.63	2.27	5.69	38.41
1991	1.41	1.42	1.42	4.29	4.82	1.51	0.52	1.94	0.73	5.53	2.15	1.51	27.25
1992	1.70	1.68	3.05	3.41	3.18	1.28	6.51	2.40	4.01	1.77	4.45	3.60	37.04
1993	3.17	1.71	3.46	3.06	1.13	4.60	1.60	1.15	4.50	1.51	2.73	1.25	29.87
Record Mean	2.12	1.91	2.59	2.85	3.15	3.47	3.00	2.93	2.67	2.32	2.45	2.42	31.89

AVERAGE TEMPERATURE (deg. F) — TOLEDO, OHIO

YEAR	JAN	FEB	MAR	APR	MAY	JUNE	JULY	AUG	SEP	OCT	NOV	DEC	ANNUAL
1964	28.5	25.9	35.0	48.6	62.6	68.8	73.8	67.8	62.5	47.5	42.3	27.8	49.2
1965	24.7	25.8	29.8	46.8	63.5	66.9	68.5	67.2	63.7	49.1	39.5	34.1	48.3
1966	20.2	27.0	38.1	45.6	53.8	70.8	74.1	69.1	61.7	50.4	40.4	27.7	48.2
1967	29.1	23.0	34.8	49.0	52.4	71.1	69.1	65.4	58.4	51.2	34.8	31.6	47.5
1968	22.1	24.5	38.6	49.5	55.9	69.6	71.9	74.0	65.2	52.0	40.6	26.6	49.2
1969	21.6	27.4	33.2	49.3	58.4	64.6	71.8	71.4	62.9	50.5	37.6	25.7	47.9
1970	16.2	24.3	31.8	48.9	61.1	67.2	71.1	69.6	64.4	54.0	39.8	28.9	48.1
1971	20.3	27.9	32.9	46.1	56.4	71.3	69.0	69.0	66.8	59.0	37.6	33.6	49.1
1972	23.4	24.4	34.1	46.1	60.4	63.9	71.4	68.4	62.2	47.2	37.7	30.3	47.5
1973	28.2	25.2	44.1	48.3	55.7	70.1	72.3	71.3	64.4	55.7	41.9	27.5	50.4
1974	26.1	23.2	36.2	48.8	56.0	65.4	72.5	71.5	59.6	49.5	40.4	28.9	48.2
1975	29.2	28.3	33.3	42.7	62.5	69.0	70.8	72.0	57.4	51.9	45.3	28.9	49.3
1976	19.8	32.8	41.6	49.3	56.0	69.3	72.2	68.2	60.5	45.6	32.3	19.9	47.3
1977	9.6	24.3	41.6	53.3	63.6	65.0	74.6	69.3	65.0	49.3	41.0	24.7	48.4
1978	16.7	11.8	28.7	45.8	58.9	67.6	70.9	70.4	68.0	49.8	40.3	30.1	46.6
1979	17.6	15.1	38.7	45.5	57.9	67.7	70.1	68.8	63.0	51.3	40.6	32.1	47.4
1980	24.3	21.4	32.4	46.8	59.5	65.5	73.6	73.3	63.8	46.8	37.4	26.0	47.6
1981	17.6	28.5	36.5	49.9	55.4	68.4	71.7	69.8	61.3	47.7	39.6	27.4	47.8
1982	15.8	20.2	33.4	42.7	64.4	64.3	72.6	67.5	61.9	52.7	41.8	36.6	47.8
1983	27.6	30.5	37.9	44.2	54.8	67.9	74.7	73.8	64.2	51.9	41.3	20.0	49.1
1984	16.6	33.0	27.6	46.8	54.4	71.2	69.8	71.2	60.8	55.2	38.7	34.0	48.3
1985	19.5	22.6	39.3	53.5	61.6	64.8	73.2	69.1	64.0	53.3	43.9	22.3	48.9
1986	25.6	25.0	39.2	50.0	60.3	66.8	73.8	67.0	65.3	53.2	37.2	31.6	49.6
1987	25.8	30.0	39.7	50.3	62.5	70.8	74.9	71.0	63.8	45.4	44.4	33.0	51.0
1988	23.8	23.3	37.5	48.1	61.0	69.3	75.9	73.9	62.5	45.2	41.8	28.0	49.2
1989	33.1	24.5	36.7	45.5	57.2	68.2	73.2	69.8	61.8	52.2	38.5	16.8	48.1
1990	34.3	32.4	41.1	49.4	56.6	69.1	71.8	70.0	63.7	51.8	44.3	33.1	51.5
1991	25.2	31.6	40.3	52.6	67.0	72.6	74.6	73.0	62.9	55.0	37.9	33.0	52.1
1992	28.8	31.9	36.1	47.4	57.9	65.1	70.1	67.8	61.9	49.6	40.8	32.9	49.2
1993	30.2	24.7	34.3	48.3	60.6	68.1	76.1	74.3	61.1	49.8	39.7	29.5	49.7
Record Mean	25.6	26.9	36.0	47.7	59.0	68.7	73.2	71.1	64.3	52.7	40.4	29.5	49.6
Max	32.7	34.4	44.3	57.2	69.1	78.7	83.2	81.0	74.2	62.2	47.9	36.2	58.4
Min	18.4	19.3	27.6	38.1	48.9	58.7	63.1	61.1	54.3	43.2	32.9	22.8	40.7

REFERENCE NOTES — TOLEDO, OHIO

GENERAL
T - TRACE AMOUNT.
BLANK ENTRIES DENOTE MISSING/UNREPORTED DATA.
# INDICATES A STATION OR INSTRUMENT RELOCATION.

SPECIFIC

PM - INCLUDES LAST DAY OF PREVIOUS MONTH
ASOS - AUTOMATED SURFACE OBSERVING SYSTEM IN OPERATION DURING THESE MONTHS.

(a) - LENGTH OF RECORD IN YEARS, ALTHOUGH INDIVIDUAL MONTHS MAY BE MISSING.
0.* OR * - THE VALUE IS BETWEEN 0.0 AND 0.05
NORMALS - BASED ON THE 1961-1990 RECORD PERIOD.
EXTREMES - DATES ARE THE MOST RECENT OCCURRENCE
WIND DIR.- NUMERALS SHOW TENS OF DEGRESS CLOCKWISE FROM TRUE NORTH. "00" INDICATES CALM.
RESULTANT DIRECTIONS ARE GIVEN TO WHOLE DEGREES.
BOLD VALUES INDICATE EXTREME VALUES WHICH OCCURRED AFTER THE ASOS SYSTEM WAS COMMISSIONED.

RECORD = PERIOD OF RECORD
RECORD MEAN PRECIPITATION IS THE MEAN OF ALL DAILY PRECIPITATION AMOUNTS DURING THE PERIOD OF RECORD.
RECORD MAX(MIN) TEMPERATURE IS THE MEAN OF ALL DAILY MAX(MIN) TEMPERATURES DURING THE PERIOD OF RECORD.
RECORD MEAN TEMPERATURE IS THE SUM OF THE RECORD MAX AND RECORD MIN DIVIDED BY 2.
AVERAGE TEMPERATURE IS THE SUM OF THE MEAN DAILY MAX AND MIN TEMPERATURE DIVIDED BY 2.

EXCEPTIONS

RECORD MEANS ARE THROUGH THE CURRENT YEAR, BEGINNING IN 1874 FOR TEMPERATURE
1871 FOR PRECIPITATION
1956 FOR SNOWFALL

HEATING DEGREE DAYS Base 65 deg. F TOLEDO, OHIO

SEASON	JULY	AUG	SEP	OCT	NOV	DEC	JAN	FEB	MAR	APR	MAY	JUNE	TOTAL
1964-65	1	55	150	536	676	1149	1242	1088	1084	538	123	51	6693
1965-66	27	57	129	486	759	953	1381	1058	828	579	348	30	6635
1966-67	1	15	147	451	729	1150	1105	1170	929	479	390	16	6582
1967-68	27	57	206	437	900	1031	1324	1168	814	458	284	27	6733
1968-69	8	17	71	424	726	1184	1340	1047	976	470	239	107	6609
1969-70	3	7	126	446	818	1213	1507	1130	1022	495	176	70	7013
1970-71	14	11	118	345	749	1111	1379	1035	987	561	272	22	6604
1971-72	18	12	78	197	813	966	1283	1169	952	560	158	95	6301
1972-73	28	36	134	543	810	1073	1135	1106	639	499	285	3	6291
1973-74	3	16	114	289	686	1157	1197	1166	885	483	295	71	6362
1974-75	2	0	190	478	730	1108	1104	1021	974	664	148	45	6464
1975-76	7	6	227	406	585	1110	1393	927	717	497	277	16	6168
1976-77	1	33	162	596	976	1393	1708	1135	718	381	135	91	7329
1977-78	3	29	71	481	713	1241	1490	1484	1121	573	243	43	7492
1978-79	11	11	74	466	732	1076	1461	1390	808	577	259	42	6907
1979-80	16	33	121	440	724	1009	1258	1256	1005	542	199	83	6686
1980-81	0	3	113	560	822	1206	1464	1015	879	450	309	24	6845
1981-82	7	15	169	529	754	1160	1522	1250	972	665	81	76	7200
1982-83	3	47	148	386	690	871	1154	958	833	624	311	55	6080
1983-84	8	0	127	407	705	1389	1494	920	1151	545	341	9	7096
1984-85	11	15	173	297	782	951	1404	1182	791	368	158	58	6190
1985-86	0	16	138	356	626	1316	1216	1113	793	449	185	54	6262
1986-87	2	54	87	365	828	1027	1209	972	778	439	173	20	5954
1987-88	5	34	89	601	611	986	1269	1202	845	498	159	53	6352
1988-89	4	5	104	613	691	1141	979	1127	869	578	270	29	6410
1989-90	0	14	159	396	789	1488	947	907	742	492	262	31	6227
1990-91	4	3	125	415	612	981	1228	928	758	377	115	7	5553
1991-92	0	0	167	315	806	986	1116	953	889	525	245	62	6064
1992-93	7	25	146	473	719	987	1072	1123	943	493	156	48	6192
1993-94	0	3	151	465	756	1095							

COOLING DEGREE DAYS Base 65 deg. F TOLEDO, OHIO

YEAR	JAN	FEB	MAR	APR	MAY	JUNE	JULY	AUG	SEP	OCT	NOV	DEC	TOTAL
1969	0	0	0	7	43	101	220	215	69	1	0	0	656
1970	0	0	0	19	62	142	210	159	107	10	0	0	709
1971	0	0	0	1	13	219	148	143	138	18	0	0	680
1972	0	0	0	0	22	67	236	148	55	0	0	0	528
1973	0	0	0	5	3	163	237	222	103	9	0	0	742
1974	0	0	0	4	25	91	243	206	34	5	0	0	608
1975	0	0	0	0	79	172	197	230	7	7	0	0	692
1976	0	0	0	31	10	155	230	137	34	2	0	0	599
1977	0	0	0	37	95	99	309	167	77	0	0	0	784
1978	0	0	0	0	58	128	200	184	170	1	0	0	741
1979	0	0	0	0	46	127	182	158	67	22	0	0	602
1980	0	0	0	3	35	106	275	265	84	4	0	0	772
1981	0	0	1	2	17	132	220	170	64	0	0	0	606
1982	0	0	0	0	68	61	245	132	62	11	0	0	579
1983	0	0	0	4	2	148	311	279	109	11	0	0	864
1984	0	0	0	5	17	203	168	214	51	1	0	0	659
1985	0	0	0	29	60	58	263	147	116	0	0	0	673
1986	0	0	1	4	48	113	282	125	103	4	0	0	680
1987	0	0	0	5	105	202	318	225	59	0	4	0	918
1988	0	0	0	0	43	190	350	286	39	5	0	0	913
1989	0	0	2	0	34	132	259	168	69	5	0	0	669
1990	0	0	7	32	11	164	222	164	91	14	0	0	705
1991	0	0	0	14	185	244	305	256	111	13	0	0	1128
1992	0	0	0	3	32	66	170	120	59	2	0	0	452
1993	0	0	0	0	26	148	351	297	41	1	0	0	864

SNOWFALL (inches) TOLEDO, OHIO

SEASON	JULY	AUG	SEP	OCT	NOV	DEC	JAN	FEB	MAR	APR	MAY	JUNE	TOTAL
1964-65	0.0	0.0	0.0	0.0	3.6	7.4	9.0	12.6	10.2	1.4	0.0	0.0	44.2
1965-66	0.0	0.0	0.0	0.0	0.1	0.9	5.3	4.5	7.9	1.1	T	0.0	19.8
1966-67	0.0	0.0	0.0	0.0	17.9	13.6	4.1	14.4	9.8	0.8	0.0	0.0	60.6
1967-68	0.0	0.0	T	T	1.9	5.1	10.4	5.6	11.2	0.2	0.0	0.0	34.4
1968-69	0.0	0.0	0.0	0.0	1.8	8.2	9.2	2.5	4.9	1.5	0.0	0.0	28.1
1969-70	0.0	0.0	0.0	T	5.7	19.0	14.2	7.7	8.3	4.5	0.0	0.0	59.4
1970-71	0.0	0.0	0.0	0.0	3.6	8.1	8.5	8.0	5.2	T	0.0	0.0	33.4
1971-72	0.0	0.0	0.0	0.0	5.7	1.4	10.1	7.6	3.3	1.8	0.0	0.0	29.9
1972-73	0.0	0.0	0.0	0.2	5.0	7.7	3.0	11.6	4.0	T	0.0	0.0	31.5
1973-74	0.0	0.0	0.0	0.0	0.2	13.8	7.5	11.6	2.9	1.1	T	0.0	37.1
1974-75	0.0	0.0	0.0	T	2.8	23.9	5.4	5.5	5.3	1.8	0.0	0.0	44.7
1975-76	0.0	0.0	0.0	0.0	5.7	12.2	14.5	8.4	4.0	1.3	0.0	0.0	46.1
1976-77	0.0	0.0	0.0	T	1.3	11.1	17.2	8.7	15.0	0.6	0.0	0.0	53.9
1977-78	0.0	0.0	0.0	0.0	6.6	24.2	30.8	9.0	2.5	T	0.0	0.0	73.1
1978-79	0.0	0.0	0.0	0.0	2.8	2.3	7.6	5.1	1.2	4.0	0.0	0.0	23.0
1979-80	0.0	0.0	0.0	T	1.6	1.5	4.1	6.4	3.4	0.5	T	0.0	17.5
1980-81	0.0	0.0	0.0	0.9	3.5	11.6	6.9	11.2	3.6	0.0	0.0	0.0	37.7
1981-82	0.0	0.0	0.0	T	0.8	14.9	18.4	14.3	10.7	9.1	0.0	0.0	68.2
1982-83	0.0	0.0	0.0	T	2.2	1.2	0.7	4.1	3.6	0.7	0.0	0.0	12.5
1983-84	0.0	0.0	0.0	0.0	3.4	13.4	12.2	6.3	9.8	T	T	0.0	45.1
1984-85	0.0	0.0	0.0	0.0	2.4	5.1	14.0	12.4	2.6	2.0	0.0	0.0	38.5
1985-86	0.0	0.0	0.0	0.0	2.5	8.7	6.6	10.2	2.2	0.2	0.0	0.0	30.4
1986-87	0.0	0.0	0.0	T	4.5	1.3	20.5	0.5	10.0	2.4	0.0	0.0	39.2
1987-88	0.0	0.0	0.0	T	0.1	11.1	8.3	14.3	4.2	T	0.0	0.0	38.0
1988-89	0.0	0.0	0.0	T	2.3	6.6	2.4	4.8	2.6	0.7	1.3	0.0	20.7
1989-90	0.0	0.0	0.0	2.0	2.3	6.5	2.5	10.4	3.5	0.3	0.0	0.0	27.5
1990-91	0.0	0.0	0.0	0.0	T	8.2	5.0	10.1	T	T	0.0	0.0	23.3
1991-92	0.0	0.0	0.0	T	1.7	2.5	10.5	3.0	12.5	0.1	0.0	0.0	30.3
1992-93	T	0.0	0.0	1.0	0.2	5.2	6.3	10.2	17.7	0.8	0.0	0.0	41.4
1993-94	0.0	T	T	0.8	1.1	6.9							
Record Mean	T	T	T	0.1	2.9	8.5	9.3	8.2	6.3	1.5	T	0.0	36.9

YOUNGSTOWN, OHIO

The Youngstown Municipal Airport is located in northeastern Ohio approximately 8 miles north of the city of Youngstown in Trumbull County. Airport elevation is 1,178 feet, about 200 feet higher than most communities in the Mahoning and Shenango River Valleys. There are numerous natural and man-made lakes in the region, including Lake Erie, 45 miles to the north. Drainage from the area flows southward through the Mahoning and Shenango Rivers which join to form the Beaver River at New Castle, Pennsylvania. The Beaver empties into the Ohio River at Rochester, Pennsylvania.

This entire area experiences frequent outbreaks of cold Canadian air masses which may be modified by passage over Lake Erie. This effect produces widespread cloudiness especially during the cool months of the year. The winter months are characterized by persistent cloudiness and intermittent snow flurries. The daily temperature range during most winter days is quite small. During most winters, the bulk of the snow falls as flurries of 2 inches or less per occurrence, although several snowstorms per year will produce amounts in the 4- to 10-inch range.

Destructive storms seldom occur, and tornadoes are not common. During recent years flood control projects have all but eliminated the threat of serious river flooding. Flash flooding of small streams and creeks rarely affects residential areas. Certain communities have well known areas of urban flooding during periods of prolonged heavy thunderstorms.

The climate of the Youngstown district has had an important role in the growth and development of this industrial area. Temperatures seldom reach extreme values especially during the summer months. However, high humidity during most days of the year tends to accentuate the temperature. Rainfall, reasonably well distributed throughout the year, provides a more than adequate supply of water for agriculture, industrial, and residential use.

Based on the 1951–1980 period, the average first occurrence of 32 degrees Fahrenheit in the fall is October 14 and the average last occurrence in the spring is May 6.

NORMALS, MEANS, AND EXTREMES

YOUNGSTOWN, OHIO

LATITUDE: 41°16'N LONGITUDE: 80°40' W ELEVATION: FT. GRND 1178 BARO 1199 TIME ZONE: EASTERN WBAN: 14852

	(a)	JAN	FEB	MAR	APR	MAY	JUNE	JULY	AUG	SEP	OCT	NOV	DEC	YEAR
TEMPERATURE °F:														
Normals														
-Daily Maximum		30.7	33.9	45.3	57.7	68.7	77.4	81.3	79.6	72.6	60.9	48.4	36.0	57.7
-Daily Minimum		16.4	17.9	27.3	36.8	46.2	54.9	59.2	57.9	51.6	41.5	33.6	22.9	38.9
-Monthly		23.6	25.9	36.3	47.3	57.5	66.2	70.3	68.8	62.2	51.2	41.0	29.5	48.3
Extremes														
-Record Highest	50	71	67	82	88	92	99	100	97	99	87	80	76	100
-Year		1950	1961	1986	1990	1962	1988	1988	1988	1954	1953	1961	1982	JUL 1988
-Record Lowest	50	-20	-14	-10	11	24	30	42	32	29	20	1	-12	-20
-Year		1985	1979	1980	1950	1970	1972	1990	1982	1991	1988	1976	1989	JAN 1985
NORMAL DEGREE DAYS:														
Heating (base 65°F)		1283	1095	890	531	268	62	8	22	129	435	720	1101	6544
Cooling (base 65°F)		0	0	0	0	35	98	172	140	45	7	0	0	497
% OF POSSIBLE SUNSHINE														
MEAN SKY COVER (tenths)														
Sunrise - Sunset	47	8.3	7.9	7.6	7.2	6.7	6.2	6.1	6.0	6.2	6.4	8.0	8.4	7.1
MEAN NUMBER OF DAYS:														
Sunrise to Sunset														
-Clear	50	2.9	3.1	4.4	4.8	6.0	6.4	6.9	7.3	7.3	8.1	3.2	2.6	63.1
-Partly Cloudy	50	5.2	5.4	6.2	7.5	8.5	11.1	12.8	12.0	10.1	7.6	5.5	5.2	97.2
-Cloudy	50	23.0	19.8	20.3	17.7	16.4	12.5	11.3	11.7	12.6	15.3	21.3	23.3	205.1
Precipitation														
.01 inches or more	50	16.7	14.8	15.5	14.4	12.8	11.5	10.6	9.9	10.4	11.0	14.7	17.3	159.7
Snow,Ice pellets,hail														
1.0 inches or more	50	4.2	3.8	3.2	0.7	0.*	0.0	0.0	0.0	0.0	0.2	1.5	4.4	18.0
Thunderstorms	50	0.3	0.3	1.7	3.3	4.5	6.8	6.7	5.4	3.4	1.2	0.7	0.2	34.6
Heavy Fog Visibility														
1/4 mile or less	50	2.1	2.2	1.9	1.7	2.0	2.2	2.5	3.4	3.2	2.1	1.9	2.8	28.3
Temperature °F														
-Maximum														
90° and above	50	0.0	0.0	0.0	0.0	0.1	1.3	3.1	2.1	0.5	0.0	0.0	0.0	7.2
32° and below	50	15.5	11.8	5.4	0.2	0.0	0.0	0.0	0.0	0.0	0.*	2.3	12.3	47.5
-Minimum														
32° and below	50	28.1	25.1	22.4	10.9	1.5	0.*	0.0	0.*	0.1	4.6	15.8	25.8	134.3
0° and below	50	2.8	2.1	0.2	0.0	0.0	0.0	0.0	0.0	0.0	0.0	0.0	1.0	6.0
AVG. STATION PRESS.(mb)	21	974.2	974.6	973.2	972.5	973.0	973.5	974.6	975.8	976.2	976.2	975.1	974.9	974.5
RELATIVE HUMIDITY (%)														
Hour 01	45	79	78	77	75	79	84	85	87	86	81	78	80	81
Hour 07 (Local Time)	46	81	81	80	77	79	82	85	88	89	85	81	82	83
Hour 13	46	72	68	63	56	54	55	56	57	59	58	67	73	62
Hour 19	46	75	72	67	61	60	63	64	68	74	71	73	77	69
PRECIPITATION (inches):														
Water Equivalent														
-Normal		2.13	2.03	3.11	3.06	3.52	3.94	4.07	3.32	3.48	2.62	3.11	2.93	37.32
-Maximum Monthly	50	7.64	5.26	6.20	6.43	9.87	10.66	9.68	7.86	6.17	8.59	9.11	6.53	10.66
-Year		1950	1950	1964	1957	1946	1986	1992	1956	1945	1954	1985	1990	JUN 1986
-Minimum Monthly	50	0.73	0.55	1.07	1.01	0.78	0.71	1.57	0.51	0.27	0.43	0.94	0.88	0.27
-Year		1985	1987	1990	1982	1977	1988	1957	1969	1960	1953	1976	1958	SEP 1960
-Maximum in 24 hrs	50	2.79	2.76	2.47	1.75	2.85	3.57	3.82	2.86	4.02	4.31	3.00	2.28	4.31
-Year		1959	1959	1954	1957	1946	1986	1967	1980	1979	1954	1985	1979	OCT 1954
Snow,Ice pellets,hail														
-Maximum Monthly	50	36.0	22.7	30.9	12.4	5.4	0.0	T	0.0	T	7.7	30.6	29.5	36.0
-Year		1978	1967	1993	1987	1966		1991		1990	1993	1950	1987	JAN 1978
-Maximum in 24 hrs	50	17.5	13.4	15.3	12.4	5.4	0.0	T	0.0	T	5.2	20.7	14.8	20.7
-Year		1948	1984	1993	1987	1966		1991		1990	1993	1950	1944	NOV 1950
WIND:														
Mean Speed (mph)	44	11.6	11.2	11.4	10.8	9.4	8.4	7.7	7.4	8.1	9.2	10.9	11.3	9.8
Prevailing Direction through 1963		SW	W	W	SW	SW	SW	SW	SW	SSW	SSW	SW	SW	SW
Fastest Obs. 1 Min.														
-Direction (!!!)	44	25	27	25	25	26	23	27	27	36	23	25	28	27
-Speed (MPH)	44	48	58	55	49	40	45	58	44	40	44	52	40	58
-Year		1959	1956	1959	1957	1988	1949	1959	1956	1960	1959	1957	1970	JUL 1959
Peak Gust														
-Direction (!!!)	10	SW	NW	W	W	SW	NW	SW	N	NW	W	W	SW	SW
-Speed (mph)	10	51	54	58	64	56	58	66	44	52	46	53	58	66
-Date		1990	1990	1986	1985	1989	1984	1993	1990	1984	1988	1992	1985	JUL 1993

PRECIPITATION (inches)

YOUNGSTOWN, OHIO

YEAR	JAN	FEB	MAR	APR	MAY	JUNE	JULY	AUG	SEP	OCT	NOV	DEC	ANNUAL
1964	2.03	1.36	6.20	5.97	2.96	3.55	4.18	3.20	0.94	1.83	1.76	4.58	38.56
1965	4.80	3.21	3.31	2.56	2.98	2.95	3.42	4.10	4.02	4.14	2.65	2.33	40.47
1966	2.73	1.93	1.75	4.34	1.81	1.88	2.52	5.05	1.59	1.40	5.52	2.49	33.01
1967	1.26	2.53	3.36	2.52	3.77	2.14	5.51	2.02	3.64	3.26	2.94	1.92	34.87
1968	3.75	0.71	3.24	2.34	5.25	3.16	2.06	5.08	2.88	4.94	4.14	4.00	41.55
1969	2.56	0.83	1.83	3.91	4.07	3.57	5.79	0.51	1.95	2.80	2.67	2.84	33.33
1970	1.39	1.93	2.22	2.85	3.22	3.52	3.48	1.08	1.80	5.20	3.24	2.90	32.83
1971	2.08	3.56	2.36	1.57	2.58	2.46	2.19	2.78	2.74	1.40	3.26	5.52	32.50
1972	1.35	2.28	3.22	3.97	2.41	4.43	3.66	1.11	4.38	0.81	4.29	3.11	35.02
1973	1.80	1.79	3.72	4.39	5.10	3.51	2.97	2.83	2.96	2.89	1.71	3.04	36.71
1974	2.78	1.61	4.48	3.05	4.67	3.95	3.20	7.21	3.36	1.67	3.98	3.03	42.99
1975	3.14	3.12	3.22	1.61	5.94	2.75	2.35	6.06	5.04	2.59	1.91	2.89	40.62
1976	3.20	3.24	4.01	1.64	1.61	4.07	7.16	2.41	5.16	2.43	0.94	1.66	37.53
1977	1.53	1.19	4.46	4.06	0.78	5.75	5.72	4.64	5.05	2.25	4.55	3.75	43.73
1978	4.52	0.60	1.64	3.01	4.49	3.88	3.98	2.72	4.23	4.34	1.36	4.22	38.99
1979	2.95	2.03	1.94	4.02	4.40	2.16	3.60	4.72	5.57	1.63	2.67	3.95	39.64
1980	1.73	1.26	5.17	2.16	2.74	3.77	6.20	7.74	4.37	2.08	1.88	1.52	40.62
1981	0.75	3.85	2.04	4.84	3.99	3.21	3.39	2.01	4.12	2.34	1.70	2.77	35.01
1982	4.27	1.50	3.24	1.01	4.37	5.20	1.73	1.75	2.16	0.59	4.93	2.96	33.71
1983	1.20	1.19	3.57	5.28	3.00	4.42	2.59	2.31	3.98	3.55	3.90	3.70	38.69
1984	1.20	2.40	2.90	2.43	5.84	3.72	4.18	3.84	2.53	2.28	3.55	2.92	37.79
1985	0.73	1.22	5.77	1.76	3.06	2.88	6.01	3.07	2.40	1.34	9.11	2.07	39.42
1986	0.87	2.29	1.38	2.03	3.62	10.66	8.31	1.72	5.83	3.05	2.92	3.42	46.10
1987	1.89	0.55	3.25	2.97	2.54	6.57	3.19	6.77	3.71	2.62	1.55	2.73	38.34
1988	1.05	2.60	1.62	2.43	3.39	0.71	5.31	3.43	3.50	2.62	3.04	1.93	31.63
1989	1.96	1.90	3.64	1.56	6.24	10.09	3.33	1.68	6.11	2.90	2.47	1.45	43.33
1990	2.06	4.56	1.07	3.21	4.60	3.05	6.93	3.04	5.19	4.23	2.24	6.53	46.71
1991	2.13	1.51	2.56	3.02	1.57	1.51	4.04	0.86	3.00	1.11	1.40	2.53	25.24
1992	2.22	1.66	3.60	3.94	2.03	0.99	9.68	3.86	4.60	2.48	4.21	3.50	42.77
1993	3.75	1.94	4.31	2.74	0.90	4.04	3.47	1.79	3.49	2.85	4.93	2.58	36.79
Record Mean	2.63	2.26	3.24	3.37	3.64	3.75	4.12	3.27	3.31	2.62	3.07	2.82	38.10

AVERAGE TEMPERATURE (deg. F)

YOUNGSTOWN, OHIO

YEAR	JAN	FEB	MAR	APR	MAY	JUNE	JULY	AUG	SEP	OCT	NOV	DEC	ANNUAL
1964	27.9	23.4	36.5	48.8	60.9	66.8	71.1	66.3	62.1	49.0	44.6	31.1	49.0
1965	23.9	25.5	30.8	45.0	63.1	65.4	67.8	67.3	65.0	49.4	41.7	35.4	48.3
1966	20.7	27.3	37.6	44.6	52.2	67.7	72.4	68.2	59.1	49.0	40.5	28.3	47.3
1967	29.5	22.0	35.1	48.0	50.2	70.2	68.0	65.7	58.1	50.6	34.6	31.8	47.0
1968	20.3	19.5	36.9	48.9	52.4	65.0	70.1	70.5	64.5	52.4	41.1	27.6	47.5
1969	24.7	26.8	31.5	48.5	56.8	64.3	69.8	69.1	61.4	50.9	38.0	24.6	47.2
1970	17.8	24.5	31.5	48.4	61.2	65.8	70.0	68.9	65.1	53.7	40.9	30.1	48.1
1971	20.7	27.1	30.3	43.4	55.2	69.6	68.1	67.5	66.0	57.9	38.5	35.8	48.3
1972	26.5	23.5	31.6	44.4	58.3	60.8	69.1	67.6	60.9	46.1	36.7	33.3	46.6
1973	27.4	24.9	44.5	47.8	54.4	69.2	71.1	71.0	63.5	54.8	42.8	31.4	50.2
1974	30.0	25.4	36.5	49.8	56.1	65.3	69.7	69.1	58.3	49.3	41.7	30.5	48.5
1975	30.2	29.2	33.4	40.9	62.0	68.6	71.1	71.2	58.7	52.5	46.2	30.5	49.6
1976	19.9	34.6	43.2	50.2	55.4	68.9	68.7	66.1	59.5	45.9	32.4	22.1	47.3
1977	10.3	24.8	42.3	50.7	62.6	63.4	72.3	69.4	65.6	51.0	43.2	28.0	48.6
1978	19.5	15.6	31.5	45.5	57.3	65.4	68.4	70.4	65.9	50.0	42.0	32.4	47.0
1979	20.7	17.2	39.6	46.1	56.0	65.3	68.4	68.1	62.0	50.9	43.0	33.0	47.5
1980	24.4	20.9	32.2	44.9	58.0	61.5	69.4	71.0	63.1	46.1	37.2	25.5	46.2
1981	18.4	29.4	35.9	51.3	58.2	70.2	71.2	68.1	60.3	48.9	40.7	29.2	48.5
1982	19.2	25.6	35.2	43.9	64.2	62.4	71.1	65.9	60.8	55.5	45.5	40.3	49.1
1983	29.9	33.2	40.4	46.9	53.3	66.3	74.1	74.2	63.8	51.8	42.8	24.0	50.1
1984	19.9	35.5	29.4	48.7	54.9	69.2	68.8	71.0	60.4	56.2	39.3	36.2	49.1
1985	19.2	24.1	39.1	53.4	59.6	62.2	68.3	67.9	64.0	52.9	45.2	24.4	48.4
1986	25.9	28.6	39.1	50.5	60.2	65.2	70.8	66.9	64.2	52.6	39.1	31.4	49.5
1987	26.0	29.1	39.0	49.4	61.6	68.9	74.1	68.2	61.9	45.6	44.5	33.5	50.2
1988	24.4	26.0	36.4	47.1	59.0	66.2	73.9	71.7	61.2	45.1	42.4	29.7	48.6
1989	33.0	24.8	37.9	44.2	56.1	66.0	71.8	69.1	62.8	52.6	39.1	18.1	48.0
1990	35.0	34.0	41.5	49.7	55.1	66.0	69.3	68.3	61.9	53.5	44.6	35.1	51.2
1991	26.6	31.8	40.8	52.1	66.2	69.4	72.1	71.7	62.3	54.0	39.2	32.5	51.6
1992	28.8	31.6	35.1	47.8	56.8	62.9	69.8	65.5	60.8	47.9	40.8	32.2	48.3
1993	31.1	24.8	34.2	47.3	57.7	66.1	73.2	71.2	59.8	49.3	40.4	29.3	48.7
Record Mean	25.1	27.0	36.1	47.6	57.8	66.5	70.6	69.2	62.3	51.7	40.6	29.5	48.7
Max	32.5	35.0	45.2	58.2	69.1	77.9	81.8	80.3	73.2	61.8	48.2	36.3	58.3
Min	17.8	18.9	27.0	37.0	46.4	55.1	59.4	58.1	51.5	41.6	33.0	22.7	39.1

REFERENCE NOTES

YOUNGSTOWN, OHIO

GENERAL
T - TRACE AMOUNT.
BLANK ENTRIES DENOTE MISSING/UNREPORTED DATA.
# INDICATES A STATION OR INSTRUMENT RELOCATION.

SPECIFIC

(a) - LENGTH OF RECORD IN YEARS, ALTHOUGH INDIVIDUAL MONTHS MAY BE MISSING.
0.* OR * - THE VALUE IS BETWEEN 0.0 AND 0.05
NORMALS - BASED ON THE 1961-1990 RECORD PERIOD.
EXTREMES - DATES ARE THE MOST RECENT OCCURRENCE.
WIND DIR.- NUMERALS SHOW TENS OF DEGRESS CLOCKWISE FROM TRUE NORTH. "00" INDICATES CALM.
RESULTANT DIRECTIONS ARE GIVEN TO WHOLE DEGREES.
BOLD VALUES INDICATE EXTREME VALUES WHICH OCCURRED AFTER THE ASOS SYSTEM WAS COMMISSIONED.

RECORD = PERIOD OF RECORD
RECORD MEAN PRECIPITATION IS THE MEAN OF ALL DAILY PRECIPITATION AMOUNTS DURING THE PERIOD OF RECORD.
RECORD MAXIMIN) TEMPERATURE IS THE MEAN OF ALL DAILY MAXIMIN) TEMPERATURES DURING THE PERIOD OF RECORD.
RECORD MEAN TEMPERATURE IS THE SUM OF THE RECORD MAX AND RECORD MIN DIVIDED BY 2.
AVERAGE TEMPERATURE IS THE SUM OF THE MEAN DAILY MAX AND MIN TEMPERATURE DIVIDED BY 2.

EXCEPTIONS

RECORD MEANS ARE THROUGH THE CURRENT YEAR, BEGINNING IN 1943 FOR TEMPERATURE
1943 FOR PRECIPITATION
1943 FOR SNOWFALL

HEATING DEGREE DAYS Base 65 deg. F YOUNGSTOWN, OHIO

SEASON	JULY	AUG	SEP	OCT	NOV	DEC	JAN	FEB	MAR	APR	MAY	JUNE	TOTAL
1964-65	8	57	140	490	603	1044	1266	1098	1053	589	123	80	6551
1965-66	24	66	100	481	694	908	1367	1049	844	605	398	65	6601
1966-67	5	19	211	494	728	1130	1093	1200	919	508	453	18	6778
1967-68	34	48	214	445	908	1023	1378	1315	865	476	384	91	7181
1968-69	32	39	73	400	710	1150	1243	1066	1032	491	275	116	6627
1969-70	13	17	167	440	803	1242	1457	1128	1036	509	180	77	7069
1970-71	20	19	101	347	716	1077	1367	1055	1070	642	314	18	6746
1971-72	16	26	76	221	788	898	1186	1198	1028	611	208	157	6413
1972-73	51	39	154	580	842	972	1157	1116	630	512	324	6	6383
1973-74	8	17	120	308	659	1037	1077	1103	878	465	288	60	6020
1974-75	8	10	216	480	692	1064	1069	995	972	717	149	48	6420
1975-76	8	8	189	389	559	1063	1391	873	672	468	313	22	5955
1976-77	18	59	183	584	969	1321	1692	1119	703	449	148	127	7372
1977-78	9	28	65	429	652	1142	1403	1378	1030	582	273	86	7077
1978-79	31	2	82	456	680	1004	1363	1332	781	568	310	74	6683
1979-80	35	37	141	445	652	987	1250	1270	1008	596	233	139	6793
1980-81	11	12	117	576	826	1220	1438	989	894	408	232	16	6739
1981-82	12	21	189	491	721	1101	1414	1098	920	628	81	99	6775
1982-83	11	65	161	302	587	763	1083	885	757	543	359	81	5597
1983-84	13	0	143	406	662	1265	1393	850	1101	486	323	18	6660
1984-85	16	13	182	271	763	885	1417	1137	796	383	200	111	6174
1985-86	10	17	141	365	589	1255	1207	1014	799	437	186	86	6106
1986-87	9	64	105	381	772	1036	1201	999	802	464	185	43	6061
1987-88	6	44	118	594	611	971	1252	1123	878	532	206	91	6426
1988-89	11	29	134	618	675	1086	983	1119	837	618	306	48	6464
1989-90	0	20	130	382	772	1446	925	863	731	488	306	75	6138
1990-91	20	6	141	357	603	918	1183	927	745	399	102	26	5427
1991-92	0	3	165	360	766	999	1115	960	919	515	264	108	6174
1992-93	12	55	168	523	721	1005	1044	1121	950	526	239	70	6434
1993-94	3	8	192	481	732	1099							

COOLING DEGREE DAYS Base 65 deg. F YOUNGSTOWN, OHIO

YEAR	JAN	FEB	MAR	APR	MAY	JUNE	JULY	AUG	SEP	OCT	NOV	DEC	TOTAL
1969	0	0	0	1	26	101	170	151	67	8	0	0	524
1970	0	0	0	16	71	110	182	148	112	4	0	0	643
1971	0	0	0	0	16	164	119	109	114	10	0	0	532
1972	0	0	0	0	8	38	184	125	39	0	0	0	394
1973	0	0	0	4	2	137	204	210	81	0	0	0	638
1974	0	0	0	15	21	74	162	143	21	0	1	0	437
1975	0	0	0	0	65	163	204	207	8	9	0	0	656
1976	0	0	0	30	21	145	137	100	28	0	0	0	461
1977	0	0	4	28	80	86	246	171	89	0	4	0	708
1978	0	0	0	0	37	102	142	176	117	0	0	0	574
1979	0	0	0	7	38	95	147	139	61	14	0	0	501
1980	0	0	0	0	23	41	153	203	66	0	0	0	486
1981	0	0	0	6	30	179	212	124	55	0	0	0	606
1982	0	0	0	2	66	28	209	98	42	16	10	4	475
1983	0	0	0	5	5	127	299	292	117	4	0	0	849
1984	0	0	0	6	18	151	141	207	53	6	0	0	582
1985	0	0	0	41	39	33	119	111	118	0	1	0	462
1986	0	0	2	9	44	97	196	136	90	5	0	0	579
1987	0	0	0	5	84	167	294	153	33	0	1	0	737
1988	0	0	0	0	27	136	295	246	25	7	0	0	736
1989	0	0	4	0	37	84	219	153	71	8	0	0	576
1990	0	0	10	35	9	112	159	117	56	8	0	0	506
1991	0	0	0	18	144	165	227	216	89	23	0	0	882
1992	0	0	0	5	16	52	166	75	49	0	0	0	363
1993	0	0	0	0	18	111	261	209	41	1	0	0	641

SNOWFALL (inches) YOUNGSTOWN, OHIO

SEASON	JULY	AUG	SEP	OCT	NOV	DEC	JAN	FEB	MAR	APR	MAY	JUNE	TOTAL
1964-65	0.0	0.0	0.0	T	3.8	14.2	16.7	12.2	23.2	0.5	0.0	0.0	70.6
1965-66	0.0	0.0	0.0	0.4	3.1	3.9	23.4	6.7	6.1	7.2	5.4	0.0	56.2
1966-67	0.0	0.0	0.0	T	7.4	12.5	8.1	22.7	13.9	2.4	T	0.0	67.0
1967-68	0.0	0.0	0.0	0.6	12.7	8.0	14.0	10.1	4.9	T	0.0	0.0	50.3
1968-69	0.0	0.0	0.0	T	5.8	21.6	8.6	11.8	9.3	T	T	0.0	57.1
1969-70	0.0	0.0	0.0	0.4	6.0	21.2	15.2	13.4	10.0	0.6	0.0	0.0	66.8
1970-71	0.0	0.0	T	T	1.7	15.6	16.0	18.0	22.0	0.4	0.0	0.0	73.7
1971-72	0.0	0.0	0.0	0.0	20.1	7.2	9.0	19.0	9.3	2.1	0.0	0.0	66.7
1972-73	0.0	0.0	0.0	0.4	8.2	8.3	5.3	11.8	8.5	1.8	T	0.0	44.3
1973-74	0.0	0.0	0.0	0.0	3.3	7.0	9.8	9.2	9.2	4.7	T	0.0	43.2
1974-75	0.0	0.0	0.0	0.9	5.3	20.4	6.6	10.2	13.5	1.8	0.0	0.0	58.7
1975-76	0.0	0.0	0.0	0.0	3.7	10.2	19.7	7.9	5.0	0.6	T	0.0	47.1
1976-77	0.0	0.0	0.0	T	8.9	10.9	22.0	10.5	3.3	0.6	0.0	0.0	56.2
1977-78	0.0	0.0	0.0	T	5.6	18.6	36.0	9.6	3.6	T	0.0	0.0	73.4
1978-79	0.0	0.0	0.0	0.0	2.0	5.7	16.2	9.6	2.8	T	0.0	0.0	36.3
1979-80	0.0	0.0	0.0	T	1.6	7.4	6.2	9.2	8.4	T	T	0.0	32.8
1980-81	0.0	0.0	0.0	0.1	5.9	11.0	13.8	8.2	10.1	T	0.0	0.0	49.1
1981-82	0.0	0.0	0.0	0.4	1.5	17.6	13.5	5.4	13.1	10.6	0.0	0.0	62.1
1982-83	0.0	0.0	0.0	T	4.1	10.7	3.4	7.6	11.9	1.7	0.0	0.0	39.4
1983-84	0.0	0.0	T	0.0	2.8	7.9	12.4	22.3	18.0	T	0.0	0.0	63.4
1984-85	0.0	0.0	0.0	0.0	4.6	9.4	21.2	9.1	1.7	5.8	0.0	0.0	51.8
1985-86	0.0	0.0	0.0	0.0	T	16.4	7.0	9.9	7.4	3.2	T	0.0	43.9
1986-87	0.0	0.0	0.0	0.0	6.8	3.2	20.3	3.7	20.6	12.4	0.0	0.0	67.0
1987-88	0.0	0.0	0.0	0.2	4.6	29.5	10.4	17.4	14.4	1.2	0.0	0.0	77.7
1988-89	0.0	0.0	0.0	0.9	2.4	13.0	3.7	9.6	11.1	1.9	0.3	0.0	42.9
1989-90	0.0	0.0	0.0	2.3	3.0	14.6	11.7	8.4	2.3	3.5	T	0.0	45.8
1990-91	0.0	0.0	T	0.0	0.5	6.6	9.3	12.2	1.6	0.4	0.0	0.0	30.6
1991-92	T	0.0	0.0	0.0	2.3	8.3	18.3	3.0	21.1	1.5	0.0	0.0	54.5
1992-93	0.0	0.0	0.0	4.0	6.8	14.0	8.7	17.6	30.9	0.5	0.0	0.0	82.5
1993-94	0.0	0.0	0.0	7.7	1.4	17.4							
Record Mean	T	0.0	T	0.6	5.6	12.6	12.9	10.9	11.0	2.5	0.1	0.0	56.2

1993 OKLAHOMA CITY, OKLAHOMA (OKC)

Oklahoma City is located along the North Canadian River, a frequently nearly-dry stream, at the geographic center of the state. It is not quite 1,000 miles south of the Canadian Border and a little less than 500 miles north of the Gulf of Mexico. The surrounding country is gently rolling with the nearest hills or low mountains, the Arbuckles, 80 miles south. The elevation ranges around 1,250 feet above sea level.

Although some influence is exerted at times by warm, moist air currents from the Gulf of Mexico, the climate of Oklahoma City falls mainly under continental controls characteristic of the Great Plains Region. The continental effect produces pronounced daily and seasonal temperature changes and considerable variation in seasonal and annual precipitation. Summers are long and usually hot. Winters are comparatively mild and short.

During the year, temperatures of 100 degrees or more occur on an average of 10 days, but have occurred on as many as 50 days or more. While summers are usually hot, the discomforting effect of extreme heat is considerably mitigated by low humidity and the prevalence of a moderate southerly breeze. Approximately one winter in three has temperatures of zero or lower.

The length of the growing season varies from 180 to 251 days. Average date of last freeze is early April and average date of first freeze is early November. Freezes have occurred in early October.

During an average year, skies are clear approximately 40 percent of the time, partly cloudy 25 percent, and cloudy 35 percent of the time. The city is almost smoke-free as a result of favorable atmospheric conditions and the almost exclusive use of natural gas for heating. Flying conditions are generally very good with flight by visual flight rules possible about 96 percent of the time.

Summer rainfall comes mainly from showers and thunderstorms. Winter precipitation is generally associated with frontal passages. Measurable precipitation has occurred on as many as 122 days and as few as 55 days during the year. The seasonal distribution of precipitation is normally 12 percent in winter, 34 percent in spring, 30 percent in summer, and 24 percent in fall. The The period with the least number of days with precipitation is November through January, and the month with the most rainy days is May. Thunderstorms occur most often in late spring and early summer. Large hail and/or destructive winds on occasion accompany these thunderstorms.

Snowfall averages less than 10 inches per year and seldom remains on the ground very long. Occasional brief periods of freezing rain and sleet storms occur.

Heavy fogs are infrequent. Prevailing winds are southerly except in January and February when northerly breezes predominate.

NORMALS, MEANS, AND EXTREMES FOR 1993

OKLAHOMA CITY, OK (OKC)

LATITUDE: 35° 23' 35" N LONGTITUDE: 97° 36' 02" W ELEVATION (FT): GRND: 1285 BARO: 1283 TIME ZONE: CENTRAL (UTC+06) WBAN: 13967

	ELEMENT	POR	JAN	FEB	MAR	APR	MAY	JUN	JUL	AUG	SEP	OCT	NOV	DEC	YEAR
TEMPERATURE °F	NORMAL DAILY MAXIMUM	30	46.7	52.1	62.0	71.9	79.1	87.3	93.4	92.5	83.8	73.6	60.4	49.9	71.1
	MEAN DAILY MAXIMUM	55	46.8	52.2	61.3	72.8	80.3	88.9	94.6	94.3	86.0	75.4	60.9	51.2	72.0
	HIGHEST DAILY MAXIMUM	40	80	84	93	100	104	105	109	110	102	96	87	86	110
	YEAR OF OCCURRENCE		1986	1981	1967	1972	1985	1980	1986	1980	1985	1972	1980	1955	AUG 1980
	MEAN OF EXTREME MAXS.	46	69.0	74.5	82.2	87.7	91.4	96.8	101.0	101.6	96.7	87.3	77.7	70.7	86.4
	NORMAL DAILY MINIMUM	30	25.2	29.6	38.5	48.8	57.7	66.1	70.6	69.6	62.2	50.4	38.6	28.6	48.8
	MEAN DAILY MINIMUM	55	26.1	30.5	38.0	49.8	58.9	67.7	72.0	71.1	63.2	51.9	38.6	30.1	49.8
	LOWEST DAILY MINIMUM	40	-4	-3	3	20	37	47	53	51	36	16	11	-8	-8
	YEAR OF OCCURRENCE		1988	1979	1960	1957	1981	1954	1971	1956	1989	1993	1991	1989	DEC 1989
	MEAN OF EXTREME MINS.	46	7.2	12.4	19.8	32.3	44.5	55.7	62.3	59.9	46.5	33.1	21.5	11.8	33.9
	NORMAL DRY BULB	30	35.9	40.9	50.3	60.4	68.4	76.7	82.0	81.1	73.0	62.0	49.6	39.3	60.0
	MEAN DRY BULB	55	36.5	41.4	49.6	61.3	69.6	78.3	83.4	82.7	74.6	63.7	49.8	40.7	61.0
	MEAN WET BULB	47	31.7	35.8	42.7	52.2	60.9	68.3	70.9	70.0	64.0	54.3	42.9	34.7	52.4
	MEAN DEW POINT	47	25.1	28.9	35.0	45.2	56.2	64.3	66.0	64.8	58.9	48.1	36.0	28.0	46.4
	NORMAL NO. DAYS WITH:														
	MAXIMUM ≥ 90°	30	0.0	0.0	0.1	0.5	2.2	12.0	22.4	22.4	8.8	1.2	0.0	0.0	69.6
	MAXIMUM ≤ 32°	30	5.2	2.4	0.2	0.0	0.0	0.0	0.0	0.0	0.0	0.0	0.1	2.9	10.8
	MINIMUM ≤ 32°	30	23.1	17.0	8.4	0.9	0.0	0.0	0.0	0.0	0.0	0.4	7.9	20.0	77.7
	MINIMUM ≤ 0°	30	0.5	0.1	0.0	0.0	0.0	0.0	0.0	0.0	0.0	0.0	0.0	0.2	0.8
H/C	NORMAL HEATING DEG. DAYS	30	902	675	464	176	31	0	0	0	15	137	462	797	3659
	NORMAL COOLING DEG. DAYS	30	0	0	9	38	136	351	527	499	255	44	0	0	1859
RH	NORMAL (PERCENT)	30	67	66	61	61	68	67	61	62	67	64	67	68	65
	HOUR 00 LST	30	72	72	68	69	76	77	70	70	76	72	74	73	72
	HOUR 06 LST	30	77	77	75	76	83	84	80	80	83	79	79	78	79
	HOUR 12 LST	30	58	58	52	51	57	56	49	50	55	52	56	58	54
	HOUR 18 LST	30	58	54	48	48	54	53	46	46	53	54	60	62	53
S	PERCENT POSSIBLE SUNSHINE	39	60	61	65	67	67	75	80	79	73	70	62	59	68
W/O	MEAN NO. DAYS WITH:														
	HEAVY FOG(VISBY ≤ 1/4 MI)	45	3.8	3.2	1.8	1.0	0.7	0.4	0.3	0.4	0.8	1.5	2.1	3.3	19.3
	THUNDERSTORMS	54	0.5	1.4	3.2	5.5	9.1	8.7	6.2	6.4	5.0	3.1	1.3	0.6	51.0
CLOUDINESS	MEAN:														
	SUNRISE-SUNSET (TENTHS)	1		8.0	7.0	6.0	7.0		2.0	4.0	3.0	4.0	4.0	5.0	
	MIDNIGHT-MIDNIGHT (TENTHS)	1		8.0	6.0	6.0	7.0		2.0	3.0	3.0	4.0	5.0	5.0	
	MEAN NO. DAYS WITH:														
	CLEAR	1	3.0	9.0	11.0	9.0	8.0	7.0	15.0	17.0	14.0	14.0	9.0	14.0	130.0
	PARTLY CLOUDY	1	3.0	4.0	4.0	2.0	6.0	6.0	3.0	4.0	6.0	5.0	5.0	7.0	55.0
	CLOUDY	1	13.0	12.0	10.0	8.0	7.0	5.0	2.0	5.0	4.0	5.0	5.0	9.0	85.0
PR	MEAN STATION PRESSURE(IN.)	21	28.74	28.70	28.59	28.58	28.55	28.58	28.63	28.64	28.66	28.69	28.68	28.73	28.65
	MEAN SEA-LEVEL PRES. (IN.)	47	30.14	30.09	29.98	29.94	29.91	29.90	29.96	29.96	29.99	30.04	30.08	30.12	30.01
WINDS	MEAN SPEED (MPH)	30	11.2	12.1	13.5	13.1	11.3	10.6	10.2	9.7	10.1	10.7	11.5	11.4	11.3
	PREVAIL.DIR(TENS OF DEGS.)	30	36	36	17	16	16	17	18	16	17	17	16	36	17
	MAXIMUM 2-MINUTE:														
	SPEED (MPH)	1	39	32	36	40	44	41	39	34	32	38	34	38	44
	DIR. (TENS OF DEGS.)	1	32	31	33	32	21	25	20	24	21	02	36	32	21
	YEAR OF OCCURRENCE		1993	1993	1993	1993	1993	1993	1993	1993	1993	1993	1993	1993	MAY 1993
	MAXIMUM 5-SECOND:														
	SPEED (MPH)	1	43	37	43	47	57	51	47	40	40	44	41	47	57
	DIR. (TENS OF DEGS.)	1	33	19	33	33	22	26	20	25	22	01	36	32	22
	YEAR OF OCCURRENCE		1993	1993	1993	1993	1993	1993	1993	1993	1993	1993	1993	1993	MAY 1993
PRECIPITATION	NORMAL TOTAL (IN.)	30	1.13	1.56	2.71	2.77	5.22	4.31	2.61	2.60	3.84	3.23	1.98	1.40	33.36
	MAXIMUM TOTAL (IN.)	54	5.68	4.63	7.85	10.78	12.07	14.66	8.44	6.77	11.85	13.18	5.46	8.14	14.66
	YEAR OF OCCURRENCE		1949	1990	1988	1947	1982	1989	1959	1966	1991	1983	1964	1984	JUN 1989
	MINIMUM TOTAL (IN.)	54	0.00	T	T	0.17	0.33	0.63	T	0.25	T	T	T	0.03	0.0
	YEAR OF OCCURRENCE		1985	1947	1940	1989	1942	1952	1983	1978	1948	1958	1949	1955	JAN 1985
	MAXIMUM IN 24 HOURS (IN.)	54	3.10	2.21	3.44	3.80	7.56	4.56	5.75	3.56	7.68	8.95	2.21	2.89	8.95
	YEAR OF OCCURRENCE		1982	1978	1944	1970	1993	1989	1981	1989	1970	1983	1986	1991	OCT 1983
	NORMAL NO. DAYS WITH:														
	PRECIPITATION ≥ 0.01	30	5.1	6.1	7.2	7.4	9.6	8.1	6.1	6.7	7.7	6.5	5.7	5.6	81.8
	PRECIPITATION ≥ 1.00	30	0.3	0.3	0.8	0.8	1.5	1.4	0.7	0.9	1.1	0.9	0.6	0.4	9.7
SNOWFALL	NORMAL TOTAL (IN.)	30	2.8	2.7	1.2	0.*	0.0	0.0	0.0	0.0	0.0	0.0	0.6	1.8	9.1
	MAXIMUM TOTAL (IN.)	54	17.3	12.0	13.9	0.7	T	T	0.0	0.0	T	0.1	7.5	8.3	17.3
	YEAR OF OCCURRENCE		1949	1978	1968	1957	1992	1992			1992	1993	1972	1987	JAN 1949
	MAXIMUM IN 24 HOURS (IN.)	54	8.9	6.5	8.4	0.7	T	T	0.0	0.0	T	0.1	5.5	8.3	8.9
	YEAR OF OCCURRENCE		1988	1986	1948	1957	1992	1992	1993	1993	1992	1993	1972	1987	JAN 1988
	MAXIMUM SNOW DEPTH (IN.)	43	12	8	8	T	0	0	0	0	0	T	3	7	12
	YEAR OF OCCURRENCE		1988	1951	1948	1973						1993	1980	1987	JAN 1988
	NORMAL NO. DAYS WITH:														
	SNOWFALL ≥ 1.0	30	0.9	1.1	0.3	0.0	0.0	0.0	0.0	0.0	0.0	0.0	0.2	0.6	3.1

PRECIPITATION (inches) 1993 OKLAHOMA CITY, OKLAHOMA (OKC)

YEAR	JAN	FEB	MAR	APR	MAY	JUN	JUL	AUG	SEP	OCT	NOV	DEC	ANNUAL
1964	0.83	2.17	1.30	2.06	5.21	0.77	2.01	4.91	2.96	0.84	5.46	0.62	29.14
1965	0.98	0.85	0.86	3.24	2.14	3.65	1.57	3.37	3.94	1.00	0.06	2.51	24.17
1966	1.05	2.39	1.30	3.68	0.88	2.63	2.38	6.77	2.82	0.37	0.84	0.45	25.56
1967	0.77	0.20	2.49	5.71	4.25	2.27	1.21	1.40	3.15	2.92	0.40	1.04	25.81
1968	2.19	1.02	2.84	3.03	8.40	2.39	1.41	3.75	2.64	2.40	4.11	1.33	35.51
1969	0.20	1.93	3.01	1.66	3.99	4.92	1.42	2.38	6.51	1.58	0.06	1.44	29.10
1970	0.32	0.29	2.09	5.33	6.53	2.45	1.30	0.80	9.64	3.29	1.03	0.26	33.33
1971	0.75	1.95	0.07	0.62	2.68	5.15	4.13	2.13	4.25	2.62	0.29	2.79	27.43
1972	0.21	0.43	1.13	3.10	4.03	1.36	3.22	1.82	2.04	7.17	2.28	0.84	27.63
1973	3.39	0.31	6.76	2.32	3.61	6.31	3.38	1.36	8.00	3.05	2.81	0.47	41.77
1974	0.10	2.68	3.12	4.66	5.01	3.36	0.48	4.42	6.24	5.57	2.34	1.47	39.45
1975	1.99	1.90	1.72	1.92	8.76	4.82	7.71	0.60	1.92	0.84	1.77	1.30	35.25
1976	T	0.33	3.09	2.94	4.36	0.88	1.38	1.46	1.53	1.78	0.12	0.19	18.06
1977	0.32	1.40	1.30	2.88	7.97	2.00	4.10	3.08	1.20	2.41	1.59	0.34	28.59
1978	1.26	3.23	1.32	1.65	10.12	4.04	3.75	0.25	0.96	1.02	2.88	0.70	31.18
1979	1.55	0.63	2.73	2.78	7.29	9.94	5.62	3.78	0.72	1.58	1.93	2.57	41.12
1980	1.69	1.29	1.38	2.16	9.00	2.52	0.42	0.60	2.21	0.99	0.51	1.58	24.35
1981	0.19	1.15	2.87	2.97	2.73	7.49	6.45	3.61	1.48	7.70	2.11	0.20	38.95
1982	3.68	0.98	1.63	1.92	12.07	4.06	2.11	1.13	2.86	1.03	2.78	1.94	36.19
1983	2.62	1.71	2.51	2.34	6.88	3.18	T	3.18	0.90	13.18	1.90	0.70	39.10
1984	0.35	1.16	4.70	1.79	1.62	3.48	0.30	2.35	1.01	6.64	2.05	8.14	33.59
1985	0.92	3.71	6.60	5.35	1.49	8.34	1.33	2.63	4.59	5.23	3.73	0.26	44.18
1986	0.00	0.68	1.75	4.42	8.21	3.11	0.38	3.29	9.54	8.00	4.63	1.16	45.17
1987	2.45	4.05	2.33	0.41	11.86	6.50	2.99	1.83	4.58	1.82	1.92	3.75	44.49
1988	1.24	0.41	7.85	3.19	1.07	3.59	1.92	1.60	5.19	2.04	2.45	1.39	31.94
1989	1.17	2.20	2.72	0.17	4.33	14.66	1.91	5.55	4.51	3.26	0.09	0.32	40.89
1990	1.85	4.63	4.43	5.11	5.79	1.25	2.65	3.16	7.35	1.27	1.59	1.46	40.54
1991	0.89	0.03	1.59	2.10	6.39	3.85	1.98	3.24	11.85	3.98	1.94	5.90	43.74
1992	1.15	1.28	1.08	3.64	4.88	6.35	4.01	5.82	2.92	1.13	4.51	3.08	39.85
1993	1.90	3.21	2.82	2.50	10.90	2.65	1.24	1.86	7.05	0.47	1.34	1.27	37.21
POR= 55 YRS	1.21	1.49	2.33	3.05	5.49	4.41	2.67	2.56	3.63	2.92	1.72	1.47	32.95

AVERAGE TEMPERATURE (˚F) 1993 OKLAHOMA CITY, OKLAHOMA (OKC)

YEAR	JAN	FEB	MAR	APR	MAY	JUN	JUL	AUG	SEP	OCT	NOV	DEC	ANNUAL
1964	40.1	38.5	47.1	64.1	70.0	77.4	85.3	80.8	72.3	60.3	49.9	37.9	60.3
1965	38.8	39.5	40.7	65.5	71.3	78.0	84.6	80.6	73.9	62.8	56.3	48.8	61.7
1966	33.8	38.6	52.8	57.8	68.7	77.9	86.4	78.8	70.5	60.9	54.3	37.8	59.9
1967	41.8	41.7	56.4	65.4	66.9	77.4	79.7	79.3	70.9	62.6	49.4	39.8	60.9
1968	36.6	36.4	50.7	58.1	64.7	74.8	79.8	80.0	70.5	62.1	46.2	38.1	58.2
1969	38.8	42.3	42.0	60.4	67.7	75.0	83.9	80.2	73.3	57.6	48.7	40.4	59.2
1970	31.8	42.7	45.0	60.2	69.2	76.3	82.1	83.6	75.1	57.9	46.0	43.9	59.5
1971	36.9	39.1	49.1	60.4	67.3	78.6	80.7	77.2	73.3	63.4	49.0	42.2	59.8
1972	34.9	42.1	53.4	63.2	67.6	79.0	79.8	80.4	75.8	61.1	43.4	34.4	59.6
1973	33.3	39.8	52.5	56.0	66.8	75.2	79.8	79.7	70.6	64.3	53.1	39.3	59.2
1974	35.0	44.4	54.8	60.0	71.5	74.1	82.7	78.5	65.5	63.5	49.3	39.6	59.9
1975	40.3	36.5	46.1	58.7	67.4	75.1	78.0	80.1	68.3	63.4	50.7	41.8	58.9
1976	39.0	52.2	52.4	61.6	63.6	74.8	79.8	81.3	72.6	56.5	43.9	38.8	59.7
1977	29.2	45.9	54.1	62.5	70.0	79.6	83.0	80.7	78.0	62.7	50.9	40.0	61.4
1978	26.3	29.4	49.1	64.5	68.1	77.3	87.0	82.6	79.7	64.7	50.4	36.9	59.7
1979	25.4	31.5	51.2	58.1	65.8	75.2	81.0	80.0	73.1	65.7	46.5	43.3	58.1
1980	38.2	38.2	46.3	56.7	69.0	81.4	88.3	88.0	76.3	61.1	50.3	41.9	61.3
1981	37.7	43.9	51.9	65.6	65.7	78.4	84.2	78.8	74.1	60.1	50.3	39.1	60.8
1982	35.3	37.7	52.7	57.5	68.2	72.2	81.0	84.1	74.5	62.7	48.6	43.2	59.8
1983	38.6	42.6	48.8	54.0	64.6	73.4	81.6	84.0	74.9	62.7	50.4	25.8	58.5
1984	34.0	45.4	46.4	56.5	68.4	78.6	81.6	82.6	71.5	61.6	49.7	43.0	59.9
1985	30.6	37.2	53.0	62.7	70.0	76.0	80.9	81.3	73.1	61.2	46.1	35.1	58.9
1986	43.6	44.8	55.5	62.8	69.0	79.0	85.9	80.0	74.8	61.6	44.8	40.8	61.9
1987	35.1	45.9	50.3	61.8	72.6	77.1	80.1	82.2	72.4	60.0	50.5	40.6	60.7
1988	34.2	40.3	49.5	58.9	70.3	78.4	81.6	82.8	73.5	59.3	51.2	43.9	60.3
1989	42.8	33.1	51.1	63.4	69.4	74.3	79.6	78.3	67.8	63.1	52.2	32.7	59.0
1990	45.9	46.0	52.6	59.2	68.6	82.0	80.7	81.6	77.0	60.9	54.9	37.1	62.2
1991	34.9	49.0	54.3	62.5	72.3	78.0	82.2	81.2	70.9	62.6	45.0	44.1	61.4
1992	42.0	49.9	54.1	61.3	66.5	74.1	81.1	74.8	72.5	62.3	45.9	39.8	60.4
1993	36.5	38.8	48.0	56.2	66.0	76.8	83.6	82.3	69.8	57.1	44.2	42.0	58.4
POR= 55 YRS	36.5	41.4	49.7	60.2	68.4	76.9	81.8	81.2	73.2	62.6	48.9	39.9	60.1

REFERENCE NOTES:

H/C INDICATES HEATING AND COOLING DEGREE DAYS.
RH INDICATES RELATIVE HUMIDITY
W/O INDICATES WEATHER AND OBSTRUCTIONS
S INDICATES SUNSHINE.
PR INDICATES PRESSURE.
CLOUDINESS IS THE SUM OF THE CEILOMETER AND SATELLITE DATA NOT TO EXCEED TEN TENTHS.
MAXIMUM SNOW DEPTH IS THE DEPTH MEASURED AT 12 UTC.

GENERAL:
T INDICATES TRACE PRECIPITATION, AN AMOUNT GREATER THAN ZERO BUT LESS THAN THE LOWEST REPORTABLE VALUE.
+ INDICATES THE VALUE ALSO OCCURS ON EARLIER DATES.
BLANK ENTRIES DENOTE MISSING OR UNREPORTED DATA.
NORMALS ARE 30-YEAR AVERAGES (1961 - 1990).
ASOS INDICATES AUTOMATED SURFACE OBSERVING SYSTEM.
PM INDICATES THE LAST DAY OF THE PREVIOUS MONTH.
POR (PERIOD OF RECORD) INDICATES THE NUMBER OF YEARS USED TO COMPUTE THE MEAN. INDIVIDUAL MONTHS WITHIN THE POR MAY BE MISSING.
WHEN THE POR FOR A NORMAL IS LESS THAN 30 YEARS, THE NORMAL IS PROVISIONAL AND IS BASED ON THE NUMBER OF YEARS INDICATED.
0.* OR * INDICATES THE VALUE OR MEAN DAYS WITH IS BETWEEN 0.00 AND 0.05.
CLOUDINESS FOR ASOS STATIONS DIFFERS FROM THE NON-ASOS OBSERVATION TAKEN BY A HUMAN OBSERVER. ASOS STATION CLOUDINESS IS BASED ON TIME-AVERAGED CEILOMETER DATA FOR CLOUDS AT OR BELOW 12,000 FEET AND ON SATELLITE DATA FOR CLOUDS ABOVE 12,000 FEET.
THE NUMBER OF DAYS WITH CLEAR, PARTLY CLOUDY, AND CLOUDY CONDITIONS FOR ASOS STATIONS IS THE SUM OF THE CEILOMETER AND SATELLITE DATA FOR THE SUNRISE TO SUNSET PERIOD.

GENERAL CONTINUED:
CLEAR INDICATES 0 - 3 TENTHS, PARTLY CLOUDY INDICATES 4 - 7 TENTHS, AND CLOUDY INDICATES 8 OR MORE TENTHS. WHEN AT LEAST ONE OF THE ELEMENTS (CEILOMETER OR SATELLITE) IS MISSING, THE DAILY CLOUDINESS IS NOT COMPUTED.
WIND DIRECTION IS RECORDED IN TENS OF DEGREES (2 DIGITS) CLOCKWISE FROM TRUE NORTH. "00" INDICATES CALM. "36" INDICATES TRUE NORTH.
RESULTANT WIND IS THE VECTOR AVERAGE OF THE SPEED AND DIRECTION.
AVERAGE TEMPERATURE IS THE SUM OF THE MEAN DAILY MAXIMUM AND MINIMUM TEMPERATURE DIVIDED BY 2.
SNOWFALL DATA COMPRISE ALL FORMS OF FROZEN PRECIPITATION, INCLUDING HAIL.
A HEATING (COOLING) DEGREE DAY IS THE DIFFERENCE BETWEEN THE AVERAGE DAILY TEMPERATURE AND 65˚ F.
DRY BULB IS THE TEMPERATURE OF THE AMBIENT AIR.
DEW POINT IS THE TEMPERATURE TO WHICH THE AIR MUST BE COOLED TO ACHIEVE 100 PERCENT RELATIVE HUMIDITY.
WET BULB IS THE TEMPERATURE THE AIR WOULD HAVE IF THE MOISTURE CONTENT WAS INCREASED TO 100 PERCENT RELATIVE HUMIDITY.

HEATING DEGREE DAYS (base 65°F) 1993 OKLAHOMA CITY, OKLAHOMA (OKC)

YEAR	JUL	AUG	SEP	OCT	NOV	DEC	JAN	FEB	MAR	APR	MAY	JUN	TOTAL
1964-65	0	0	24	150	454	831	808	706	745	71	2	0	3791
1965-66	0	0	28	129	262	496	961	734	388	223	62	0	3283
1966-67	0	0	6	166	338	837	713	647	307	71	77	0	3162
1967-68	0	0	27	155	464	773	872	826	444	215	71	0	3847
1968-69	0	0	0	152	561	829	808	629	708	158	38	2	3885
1969-70	0	0	0	274	481	752	1022	620	615	187	31	12	3994
1970-71	0	0	18	254	559	651	866	718	492	163	36	0	3757
1971-72	0	0	59	88	475	702	923	660	365	144	46	0	3462
1972-73	0	0	23	225	640	940	975	701	380	283	55	0	4222
1973-74	0	0	37	99	362	787	922	573	330	168	8	0	3286
1974-75	0	0	56	88	463	784	763	792	583	235	29	0	3793
1975-76	0	0	64	126	430	713	801	367	406	128	100	0	3135
1976-77	0	0	19	306	629	805	1103	529	338	107	7	0	3843
1977-78	0	0	0	115	420	766	1192	990	493	90	64	0	4130
1978-79	0	0	2	89	437	866	1221	932	434	217	81	0	4279
1979-80	0	0	2	92	551	669	823	771	572	249	24	0	3753
1980-81	0	0	23	180	444	710	839	587	400	69	69	0	3321
1981-82	0	0	22	189	434	797	913	759	382	248	25	13	3782
1982-83	0	0	14	156	490	671	809	622	496	345	96	9	3708
1983-84	0	0	25	117	439	1207	955	561	572	263	45	0	4184
1984-85	0	0	75	162	462	676	1059	773	377	108	10	0	3702
1985-86	0	0	63	146	562	921	656	562	308	122	17	0	3357
1986-87	0	0	2	137	599	742	918	528	450	177	3	0	3556
1987-88	0	0	1	165	442	748	948	712	473	204	14	0	3707
1988-89	0	0	8	196	408	644	679	887	441	140	38	0	3441
1989-90	0	0	78	135	386	993	583	525	387	202	52	0	3341
1990-91	0	0	9	169	307	860	925	444	339	110	25	0	3188
1991-92	0	0	37	150	594	642	704	430	332	154	59	2	3104
1992-93	0	1	5	115	563	774	878	725	525	265	53	0	3904
1993-	0	1	27	269	619	706							
POR= 50 YRS	0	0	20	146	479	776	879	661	474	185	45	1	3666

COOLING DEGREE DAYS (base 65°F) 1993 OKLAHOMA CITY, OKLAHOMA (OKC)

YEAR	JAN	FEB	MAR	APR	MAY	JUN	JUL	AUG	SEP	OCT	NOV	DEC	ANNUAL
1969	0	0	0	29	128	310	593	477	255	52	0	0	1844
1970	0	0	1	47	169	357	536	582	328	38	0	0	2058
1971	0	0	4	31	117	416	493	388	313	45	3	0	1810
1972	0	2	11	97	133	429	470	483	351	109	0	0	2085
1973	0	0	0	19	119	312	465	462	216	83	11	0	1687
1974	0	0	22	26	217	280	553	426	80	47	0	0	1651
1975	0	0	1	53	108	310	410	476	170	83	4	0	1615
1976	0	1	23	33	62	300	468	512	253	50	0	0	1702
1977	0	1	8	37	170	445	565	491	395	49	2	0	2163
1978	0	0	8	80	165	378	690	553	450	87	7	0	2418
1979	0	0	10	18	112	314	505	471	252	121	2	0	1805
1980	0	0	0	7	155	498	729	721	366	65	11	2	2554
1981	0	4	0	94	98	409	603	435	304	47	0	0	1994
1982	0	0	9	28	130	234	503	598	305	90	3	1	1901
1983	0	0	0	20	91	266	523	599	329	54	8	0	1890
1984	0	0	0	16	159	414	521	551	279	64	5	0	2009
1985	0	0	12	43	172	336	501	512	313	38	0	0	1927
1986	0	2	21	63	147	425	653	473	301	40	0	0	2125
1987	0	0	0	88	242	371	475	543	230	18	12	0	1979
1988	0	0	1	29	186	410	525	558	270	25	1	0	2005
1989	0	0	16	100	179	285	459	419	170	83	8	0	1719
1990	0	0	12	33	169	517	495	522	378	48	13	0	2187
1991	0	0	15	45	257	398	542	507	219	85	1	0	2069
1992	0	0	3	51	114	283	508	312	239	36	0	1	1547
1993	0	0	4	9	89	362	584	545	177	32	0	0	1802
POR= 25 YRS	0	0	6	43	146	362	534	503	276	58	2	0	1930

SNOWFALL (inches) 1993 OKLAHOMA CITY, OKLAHOMA (OKC)

YEAR	JUL	AUG	SEP	OCT	NOV	DEC	JAN	FEB	MAR	APR	MAY	JUN	TOTAL
1964-65	0.0	0.0	0.0	0.0	T	T	2.0	0.9	0.1	0.0	0.0	0.0	3.0
1965-66	0.0	0.0	0.0	0.0	0.0	T	5.8	2.3	0.0	0.0	0.0	0.0	8.1
1966-67	0.0	0.0	0.0	0.0	T	0.6	T	0.1	0.4	0.0	0.0	0.0	1.1
1967-68	0.0	0.0	0.0	T	1.2	1.2	0.4	7.7	13.9	0.0	0.0	0.0	24.4
1968-69	0.0	0.0	0.0	0.0	1.4	1.3	T	1.3	8.2	0.0	0.0	0.0	12.2
1969-70	0.0	0.0	0.0	0.0	0.0	4.2	0.4	T	2.3	T	0.0	0.0	6.9
1970-71	0.0	0.0	0.0	0.0	T	T	T	5.1	0.7	0.0	0.0	0.0	5.8
1971-72	0.0	0.0	0.0	0.0	0.5	5.2	0.8	4.9	0.0	0.0	0.0	0.0	11.4
1972-73	0.0	0.0	0.0	0.0	7.5	1.4	8.3	0.2	T	T	0.0	0.0	17.4
1973-74	0.0	0.0	0.0	0.0	0.0	0.6	0.7	1.0	0.5	0.0	0.0	0.0	2.8
1974-75	0.0	0.0	0.0	0.0	1.0	2.0	0.6	0.9	0.1	0.0	0.0	0.0	4.6
1975-76	0.0	0.0	0.0	0.0	0.7	3.9	T	0.3	0.0	0.0	0.0	0.0	4.9
1976-77	0.0	0.0	0.0	0.0	0.3	T	2.8	0.4	0.0	0.0	0.0	0.0	3.5
1977-78	0.0	0.0	0.0	0.0	T	0.0	8.4	12.0	T	0.0	0.0	0.0	20.4
1978-79	0.0	0.0	0.0	0.0	0.0	3.3	4.0	6.1	0.0	0.0	0.0	0.0	13.4
1979-80	0.0	0.0	0.0	0.0	T	T	T	1.8	T	0.0	0.0	0.0	1.8
1980-81	0.0	0.0	0.0	0.0	4.0	0.0	T	T	0.0	0.0	0.0	0.0	4.0
1981-82	0.0	0.0	0.0	0.0	0.0	T	1.0	3.9	2.5	0.0	0.0	0.0	7.4
1982-83	0.0	0.0	0.0	0.0	T	T	5.1	4.3	T	0.0	0.0	0.0	9.4
1983-84	0.0	0.0	0.0	0.0	T	1.9	5.6	2.0	T	0.0	0.0	0.0	9.5
1984-85	0.0	0.0	0.0	0.0	T	6.1	1.5	2.3	0.0	0.0	0.0	0.0	9.9
1985-86	0.0	0.0	0.0	0.0	T	2.9	0.0	10.9	0.0	0.0	0.0	0.0	13.8
1986-87	0.0	0.0	0.0	0.0	0.0	T	10.0	1.0	T	0.0	0.0	0.0	11.0
1987-88	0.0	0.0	0.0	0.0	2.0	8.3	12.1	0.2	0.9	0.0	0.0	0.0	23.5
1988-89	0.0	0.0	0.0	0.0	0.6	2.0	4.8	T	4.0	0.6	T	0.0	12.0
1989-90	0.0	0.0	0.0	0.0	T	1.7	0.0	1.7	0.1	0.0	0.0	0.0	3.5
1990-91	0.0	0.0	0.0	0.0	0.0	4.2	T	0.0	T	0.0	0.0	0.0	4.2
1991-92	0.0	0.0	0.0	T	2.1	1.0	5.0	0.0	T	0.0	T	T	8.1
1992-93	0.0	0.0	T		T	3.3	0.4	1.8	T	T		0.0	
1993-	0.0	0.0	0.0	0.1	0.0	T							
POR= 52 YRS	0.0	0.0	T	T	0.5	1.8	3.0	2.4	1.2	T	T	T	8.9

1993 TULSA, OKLAHOMA (TUL)

The city of Tulsa lies along the Arkansas River at an elevation of 700 feet above sea level. The surrounding terrain is gently rolling.

At latitude 36 degrees, Tulsa is far enough north to escape the long periods of heat in summer, yet far enough south to miss the extreme cold of winter. The influence of warm moist air from the Gulf of Mexico is often noted, due to the high humidity, but the climate is essentially continental characterized by rapid changes in temperature. Generally the winter months are mild. Temperatures occasionally fall below zero but only last a very short time. Temperatures of 100 degrees or higher are often experienced from late July to early September, but are usually accompanied by low relative humidity and a good southerly breeze. The fall season is long with a great number of pleasant, sunny days and cool, bracing nights.

Rainfall is ample for most agricultural pursuits and is distributed favorably throughout the year. Spring is the wettest season, having an abundance of rain in the form of showers and thunderstorms. The steady rains of fall are a contrast to the spring and summer showers and provide a good supply of moisture and more ideal conditions for the growth of winter grains and pastures. The greatest amounts of snow are received in January and early March. The snow is usually light and only remains on the ground for brief periods.

The average date of the last 32 degree temperature occurrence is late March and the average date of the first 32 degree occurrence is early November. The average growing season is 216 days.

The Tulsa area is occasionally subjected to large hail and violent windstorms which occur mostly during spring and early summer, although occurrences have been noted throughout the year.

Prevailing surface winds are southerly during most of the year. Heavy fogs are infrequent. Sunshine is abundant. The prevalence of good flying weather throughout the year has contributed to the development of Tulsa as an aviation center.

NORMALS, MEANS, AND EXTREMES FOR 1993

TULSA, OK (TUL)

LATITUDE: 36° 11' 54" N   LONGTITUDE: 95° 53' 17" W   ELEVATION (FT): GRND: 650 BARO: 669   TIME ZONE: CENTRAL (UTC+06)   WBAN: 13968

	ELEMENT	POR	JAN	FEB	MAR	APR	MAY	JUN	JUL	AUG	SEP	OCT	NOV	DEC	YEAR
TEMPERATURE °F	NORMAL DAILY MAXIMUM	30	45.4	51.0	62.1	73.0	79.7	87.7	93.7	92.5	83.6	73.8	60.3	48.8	71.0
	MEAN DAILY MAXIMUM	58	46.6	52.0	61.2	73.3	80.6	87.6	93.3	92.7	84.7	74.7	60.1	50.9	71.5
	HIGHEST DAILY MAXIMUM	55	79	86	96	102	96	103	112	110	109	98	87	80	112
	YEAR OF OCCURRENCE		1950	1962	1974	1972	1985	1953	1954	1970	1939	1979	1945	1966	JUL 1954
	MEAN OF EXTREME MAXS.	46	68.7	74.4	83.2	88.3	90.6	96.4	101.9	102.0	96.5	87.6	78.9	71.0	86.6
	NORMAL DAILY MINIMUM	30	24.9	29.5	39.1	49.9	58.8	67.7	72.8	70.6	63.0	50.7	39.5	28.9	49.6
	MEAN DAILY MINIMUM	58	26.3	30.4	38.4	50.6	59.6	67.5	72.1	70.4	62.3	50.7	38.3	30.5	49.8
	LOWEST DAILY MINIMUM	55	-8	-7	-3	22	35	49	51	52	35	18	10	-8	-8
	YEAR OF OCCURRENCE		1947	1979	1948	1957	1961	1954	1971	1988	1984	1993	1976	1989	DEC 1989
	MEAN OF EXTREME MINS.	46	6.4	11.9	20.1	33.4	44.2	55.7	62.3	59.3	46.3	33.3	21.7	11.0	33.8
	NORMAL DRY BULB	30	35.2	40.3	50.6	61.5	69.3	77.7	83.3	81.5	73.3	62.2	49.9	38.9	60.3
	MEAN DRY BULB	58	36.5	41.2	49.8	62.0	70.1	77.6	82.7	81.6	73.5	62.7	49.2	40.7	60.6
	MEAN WET BULB	44	31.6	35.7	43.0	52.7	62.0	69.7	72.6	71.2	65.0	54.6	43.4	35.0	53.0
	MEAN DEW POINT	44	24.8	28.6	35.1	45.6	57.5	65.8	68.2	66.6	60.3	48.6	36.5	28.3	47.2
	NORMAL NO. DAYS WITH:														
	MAXIMUM ≥ 90°	30	0.0	0.0	0.3	0.8	2.1	13.3	24.0	22.0	9.4	1.6	0.0	0.0	73.5
	MAXIMUM ≤ 32°	30	5.8	2.6	0.3	0.0	0.0	0.0	0.0	0.0	0.0	0.0	0.2	3.1	12.0
	MINIMUM ≤ 32°	30	23.9	17.2	8.5	0.5	0.0	0.0	0.0	0.0	0.0	0.3	7.6	20.1	78.1
	MINIMUM ≤ 0°	30	0.6	0.1	0.0	0.0	0.0	0.0	0.0	0.0	0.0	0.0	0.0	0.4	1.1
H/C	NORMAL HEATING DEG. DAYS	30	924	692	457	151	41	0	0	0	20	144	453	809	3691
	NORMAL COOLING DEG. DAYS	30	0	0	11	46	174	381	567	512	269	57	0	0	2017
RH	NORMAL (PERCENT)	30	67	65	62	61	69	69	64	64	70	66	67	68	66
	HOUR 00 LST	30	72	70	68	68	78	78	72	73	79	75	74	73	73
	HOUR 06 LST	30	78	77	76	78	85	86	82	84	86	82	80	79	81
	HOUR 12 LST	30	59	57	53	51	58	58	53	53	58	53	57	60	56
	HOUR 18 LST	30	58	55	49	48	56	56	49	50	57	55	59	61	54
S	PERCENT POSSIBLE SUNSHINE	51	54	55	58	59	59	68	75	74	67	65	56	53	61
W/O	MEAN NO. DAYS WITH:														
	HEAVY FOG(VISBY ≤ 1/4 MI)	55	2.0	1.6	0.9	0.2	0.4	0.3	0.2	0.1	0.6	1.1	1.2	1.6	10.2
	THUNDERSTORMS	55	0.7	1.3	3.3	6.0	8.8	8.2	5.6	6.0	5.2	3.1	1.5	0.8	50.5
CLOUDINESS	MEAN:														
	SUNRISE-SUNSET (TENTHS)	1		8.0	8.0	8.0	8.0		4.0	4.0	5.0	6.0	6.0	5.0	
	MIDNIGHT-MIDNIGHT (TENTHS)	1		8.0	8.0	8.0	8.0		3.0	4.0	5.0	5.0	6.0	6.0	
	MEAN NO. DAYS WITH:														
	CLEAR	1	5.0	9.0	8.0	8.0	6.0	3.0	13.0	16.0	14.0	11.0	9.0	14.0	116.0
	PARTLY CLOUDY	1	1.0	4.0	2.0	1.0	3.0	10.0	5.0	5.0	5.0	3.0	2.0	7.0	48.0
	CLOUDY	1	13.0	11.0	14.0	11.0	10.0	6.0	3.0	7.0	6.0	8.0	9.0	8.0	106.0
PR	MEAN STATION PRESSURE(IN.)	21	29.43	29.38	29.25	29.23	29.20	29.22	29.26	29.28	29.31	29.34	29.34	29.41	29.30
	MEAN SEA-LEVEL PRES. (IN.)	44	30.15	30.10	30.00	29.94	29.92	29.92	29.96	29.97	30.00	30.05	30.08	30.12	30.02
WINDS	MEAN SPEED (MPH)	30	9.6	10.3	11.6	11.5	10.2	9.8	9.4	8.9	8.9	9.2	10.0	9.8	9.9
	PREVAIL.DIR(TENS OF DEGS.)	30	18	18	18	18	18	18	18	18	18	18	18	18	18
	MAXIMUM 2-MINUTE:														
	SPEED (MPH)	1	31	36	32	55	37	48	51	31	34	30	31	31	55
	DIR. (TENS OF DEGS.)	1	35	24	27	34	31	14	19	33	36	01	36	19	34
	YEAR OF OCCURRENCE		1993	1993	1993	1993	1993	1993	1993	1993	1993	1993	1993	1993	APR 1993
	MAXIMUM 5-SECOND:														
	SPEED (MPH)	1	34	46	44	63	44	55	55	36	40	37	39	41	63
	DIR. (TENS OF DEGS.)	1	35	25	27	34	30	25	19	33	35	02	36	32	34
	YEAR OF OCCURRENCE		1993	1993	1993	1993	1993	1993	1993	1993	1993	1993	1993	1993	APR 1993
PRECIPITATION	NORMAL TOTAL (IN.)	30	1.54	1.97	3.46	3.72	5.60	4.44	3.09	3.12	4.70	3.66	3.13	2.16	40.59
	MAXIMUM TOTAL (IN.)	55	6.65	5.73	11.94	9.23	18.00	11.17	10.88	7.47	18.81	16.51	7.57	8.70	18.81
	YEAR OF OCCURRENCE		1949	1985	1973	1947	1943	1948	1961	1942	1971	1941	1946	1984	SEP 1971
	MINIMUM TOTAL (IN.)	55	0.00	0.38	0.08	0.34	1.17	0.53	0.03	0.21	T	T	0.01	0.16	0.0
	YEAR OF OCCURRENCE		1993	1991	1971	1989	1988	1963	1954	1945	1948	1952	1949	1950	JAN 1993
	MAXIMUM IN 24 HOURS (IN.)	55	2.25	4.34	2.67	4.58	9.27	5.01	7.54	5.37	6.39	5.80	5.14	3.27	9.27
	YEAR OF OCCURRENCE		1946	1985	1969	1964	1984	1941	1963	1989	1940	1983	1974	1984	MAY 1984
	NORMAL NO. DAYS WITH:														
	PRECIPITATION ≥ 0.01	30	6.1	6.7	8.4	8.8	10.1	8.5	5.9	6.8	8.0	6.5	6.5	6.7	89.0
	PRECIPITATION ≥ 1.00	30	0.2	0.4	1.1	1.0	1.6	1.3	0.9	1.1	1.5	1.3	1.0	0.6	12.0
SNOWFALL	NORMAL TOTAL (IN.)	30	2.8	2.5	1.5	0.*	0.0	0.0	0.0	0.0	0.0	0.0	0.4	1.9	9.1
	MAXIMUM TOTAL (IN.)	55	12.7	10.1	11.8	1.7	T	0.0	0.0	0.0	T	0.3	5.6	9.9	12.7
	YEAR OF OCCURRENCE		1979	1960	1968	1957	1991				1990	1993	1972	1958	JAN 1979
	MAXIMUM IN 24 HOURS (IN.)	55	9.0	6.5	9.8	1.7	T	0.0	0.0	0.0	T	0.3	4.0	8.8	9.8
	YEAR OF OCCURRENCE		1944	1993	1968	1957	1991	1993	1993	1993	1990	1993	1972	1954	MAR 1968
	MAXIMUM SNOW DEPTH (IN.)	44	11	6	10	T	0	0	0	0	0	0	2	8	11
	YEAR OF OCCURRENCE		1988	1949	1968	1993							1971	1954	JAN 1988
	NORMAL NO. DAYS WITH:														
	SNOWFALL ≥ 1.0	30	1.0	0.9	0.3	0.0	0.0	0.0	0.0	0.0	0.0	0.0	0.2	0.8	3.2

PRECIPITATION (inches) 1993 TULSA, OKLAHOMA (TUL)

YEAR	JAN	FEB	MAR	APR	MAY	JUN	JUL	AUG	SEP	OCT	NOV	DEC	ANNUAL
1964	0.63	2.17	3.96	5.87	4.77	5.79	1.80	6.14	3.33	1.24	6.90	1.67	44.27
1965	1.56	1.45	0.73	3.00	3.91	3.76	3.39	3.72	4.59	0.26	0.03	4.29	30.69
1966	0.69	2.35	0.86	4.84	1.86	2.56	2.00	4.59	2.68	1.39	0.51	2.53	26.86
1967	1.51	0.65	1.42	5.09	5.34	4.60	6.88	0.57	4.89	3.75	1.09	1.12	36.91
1968	3.26	1.08	3.49	4.40	3.56	4.08	1.37	1.90	2.80	2.64	5.19	2.01	35.78
1969	1.63	1.34	3.25	1.56	1.98	6.40	1.08	3.24	1.67	5.86	0.32	1.62	29.95
1970	0.41	0.57	2.05	5.66	4.20	4.60	0.13	1.85	6.73	5.83	0.84	1.15	34.02
1971	1.37	4.18	0.08	1.37	6.59	3.27	3.34	1.86	18.81	7.99	1.21	6.34	56.41
1972	0.17	0.49	0.91	4.45	2.43	2.69	2.68	5.16	2.95	7.58	5.00	1.03	35.54
1973	3.39	0.74	11.94	7.22	5.30	7.69	6.47	4.70	6.56	6.16	6.32	3.39	69.88
1974	0.79	3.17	2.62	3.65	6.94	7.88	0.55	5.30	11.78	6.40	7.30	2.88	59.26
1975	2.61	3.44	5.45	2.20	7.22	6.75	2.14	3.52	3.34	1.47	3.53	3.04	44.71
1976	0.21	0.84	3.95	8.27	6.75	1.87	4.37	1.17	2.60	2.65	0.68	0.55	33.91
1977	1.43	1.57	5.58	2.05	5.72	6.69	2.00	4.86	5.57	2.75	2.31	0.93	41.46
1978	0.81	2.84	2.99	7.14	9.28	6.06	0.36	1.37	0.13	0.95	5.48	0.78	38.19
1979	2.07	0.81	3.97	4.47	6.15	8.90	2.68	4.77	0.28	2.20	5.60	0.45	42.35
1980	2.07	1.32	3.59	3.44	7.23	5.57	0.09	2.34	3.47	2.05	0.79	1.37	33.33
1981	0.69	1.63	1.67	1.90	6.70	3.31	6.22	2.47	3.11	6.73	2.25	0.20	36.88
1982	3.58	0.67	1.04	1.28	9.30	4.13	1.65	1.42	2.95	1.22	4.61	3.39	35.24
1983	2.95	1.98	2.19	3.88	6.85	1.47	0.58	0.65	2.11	9.33	2.14	0.61	34.74
1984	1.00	1.95	6.72	2.44	11.25	1.72	0.48	1.96	2.77	6.98	2.80	8.70	48.77
1985	1.24	5.74	5.39	5.62	4.19	7.63	2.38	1.91	3.29	6.26	6.27	1.39	51.31
1986	0.00	1.22	2.28	5.10	6.97	4.23	1.15	3.96	8.36	5.53	2.99	0.97	42.76
1987	2.21	4.72	2.20	0.70	10.02	2.31	4.20	3.72	3.52	1.27	5.17	5.87	45.91
1988	1.11	1.03	6.52	3.18	1.17	0.58	4.20	2.43	5.37	1.43	4.38	1.82	33.22
1989	2.94	2.26	3.14	0.34	3.95	5.16	4.09	6.69	3.32	2.80	0.15	0.26	35.10
1990	2.93	4.14	6.51	5.31	5.21	1.08	0.24	1.83	4.19	2.15	2.41	2.94	38.94
1991	1.47	0.38	1.02	2.58	5.11	3.64	0.35	1.17	6.15	5.12	1.98	4.57	33.54
1992	0.48	1.32	1.37	4.75	5.65	8.41	2.12	3.09	2.66	3.53	4.83	5.21	43.42
1993	0.00	2.86	2.76	4.59	6.86	3.79	2.42	2.29	6.90	1.13	1.69	1.76	37.05
POR= 57 YRS	1.59	1.85	2.83	3.96	5.53	4.61	3.06	3.04	4.18	3.47	2.55	2.01	38.67

AVERAGE TEMPERATURE (°F) 1993 TULSA, OKLAHOMA (TUL)

YEAR	JAN	FEB	MAR	APR	MAY	JUN	JUL	AUG	SEP	OCT	NOV	DEC	ANNUAL
1964	40.9	39.4	46.9	65.7	71.7	78.2	84.9	80.7	73.0	60.1	52.7	39.1	61.1
1965	39.8	39.4	40.5	66.2	72.6	78.1	83.2	81.7	75.3	63.3	55.7	46.8	61.9
1966	32.8	39.0	52.7	57.9	67.6	77.3	87.3	78.5	70.3	59.5	53.4	37.2	59.5
1967	39.6	38.9	55.1	64.3	65.2	76.5	77.5	76.0	69.2	61.2	48.2	39.7	59.3
1968	36.0	37.0	49.8	59.3	65.5	76.8	80.6	80.9	71.9	62.2	46.7	36.8	58.6
1969	36.9	41.4	42.7	61.6	69.9	75.0	85.9	80.8	74.9	59.5	48.1	39.3	59.7
1970	29.7	41.9	44.6	60.7	70.7	76.9	82.8	84.8	74.5	58.9	45.6	42.5	59.5
1971	36.5	39.0	49.9	60.2	66.7	79.4	80.0	79.0	73.0	65.0	50.1	43.7	60.2
1972	34.8	41.8	53.0	62.8	68.0	79.5	80.4	81.7	75.5	60.9	43.6	33.7	59.6
1973	34.0	39.8	54.3	58.2	67.5	76.7	81.2	79.3	72.3	65.0	53.4	38.1	60.0
1974	34.1	43.7	55.2	61.8	72.1	73.8	85.4	78.3	64.7	63.0	49.1	39.7	60.1
1975	39.9	36.9	45.3	60.4	69.1	76.0	81.2	82.2	69.2	63.2	50.8	40.1	59.5
1976	37.2	51.1	51.9	61.5	63.0	75.0	81.4	79.7	72.8	56.1	43.1	37.1	59.2
1977	26.9	46.6	55.0	64.4	72.6	81.0	84.8	81.7	75.6	62.2	51.1	39.0	61.7
1978	24.9	29.4	47.5	63.5	68.3	77.6	87.8	84.3	80.6	63.5	51.6	38.0	59.7
1979	23.1	30.2	52.4	61.0	68.7	77.8	83.4	81.8	74.7	66.2	47.5	44.4	59.3
1980	38.6	37.1	48.3	61.1	70.6	82.5	91.7	89.7	78.3	61.5	50.5	42.3	62.7
1981	37.6	43.6	53.3	68.0	65.9	80.0	85.9	79.4	73.9	60.9	51.4	38.5	61.5
1982	33.6	38.2	55.3	59.3	72.9	74.7	84.2	85.3	74.6	63.4	50.6	44.4	61.4
1983	39.1	42.9	49.0	55.4	67.0	76.6	84.7	88.1	77.4	64.5	52.9	26.7	60.4
1984	34.4	46.4	48.3	58.0	67.5	80.1	82.0	82.7	71.5	63.8	50.4	44.7	60.8
1985	30.2	35.9	54.7	63.3	70.6	75.8	82.9	81.7	74.6	63.1	47.8	34.5	59.6
1986	42.8	43.2	55.0	62.6	69.4	79.7	86.6	78.2	74.7	61.0	43.6	40.0	61.4
1987	36.0	45.4	51.5	63.2	74.1	78.9	81.9	83.1	72.4	59.3	51.6	41.4	61.6
1988	34.8	39.3	49.3	59.5	71.0	79.9	82.6	83.0	73.2	58.5	51.7	43.4	60.5
1989	43.4	31.9	49.3	63.3	69.2	74.8	80.2	80.4	68.7	64.0	52.7	31.6	59.1
1990	46.1	46.1	53.2	59.6	67.4	82.1	83.2	83.5	78.3	61.2	56.4	38.5	63.0
1991	34.7	48.3	55.1	63.8	73.7	80.0	84.9	82.9	72.7	64.3	45.8	44.6	62.6
1992	42.8	50.1	54.9	61.6	67.6	74.7	81.8	76.6	72.8	60.8	45.9	38.6	60.7
1993	35.7	37.8	46.8	55.8	66.0	76.8	84.4	83.5	68.6	56.2	44.5	42.3	58.2
POR= 58 YRS	36.5	41.2	49.8	60.9	68.9	77.6	82.7	81.5	73.5	62.7	49.2	40.0	60.4

REFERENCE NOTES:

H/C INDICATES HEATING AND COOLING DEGREE DAYS.
RH INDICATES RELATIVE HUMIDITY
W/O INDICATES WEATHER AND OBSTRUCTIONS
S INDICATES SUNSHINE.
PR INDICATES PRESSURE.
CLOUDINESS IS THE SUM OF THE CEILOMETER AND SATELLITE DATA NOT TO EXCEED TEN TENTHS.
MAXIMUM SNOW DEPTH IS THE DEPTH MEASURED AT 12 UTC.

GENERAL:
T INDICATES TRACE PRECIPITATION, AN AMOUNT GREATER THAN ZERO BUT LESS THAN THE LOWEST REPORTABLE VALUE.
+ INDICATES THE VALUE ALSO OCCURS ON EARLIER DATES.
BLANK ENTRIES DENOTE MISSING OR UNREPORTED DATA.
NORMALS ARE 30-YEAR AVERAGES (1961 - 1990).
ASOS INDICATES AUTOMATED SURFACE OBSERVING SYSTEM.
PM INDICATES THE LAST DAY OF THE PREVIOUS MONTH.
POR (PERIOD OF RECORD) INDICATES THE NUMBER OF YEARS USED TO COMPUTE THE MEAN. INDIVIDUAL MONTHS WITHIN THE POR MAY BE MISSING.
WHEN THE POR FOR A NORMAL IS LESS THAN 30 YEARS, THE NORMAL IS PROVISIONAL AND IS BASED ON THE NUMBER OF YEARS INDICATED.
0.* OR * INDICATES THE VALUE OR MEAN DAYS WITH IS BETWEEN 0.00 AND 0.05.
CLOUDINESS FOR ASOS STATIONS DIFFERS FROM THE NON-ASOS OBSERVATION TAKEN BY A HUMAN OBSERVER. ASOS STATION CLOUDINESS IS BASED ON TIME-AVERAGED CEILOMETER DATA FOR CLOUDS AT OR BELOW 12,000 FEET AND ON SATELLITE DATA FOR CLOUDS ABOVE 12,000 FEET.
THE NUMBER OF DAYS WITH CLEAR, PARTLY CLOUDY, AND CLOUDY CONDITIONS FOR ASOS STATIONS IS THE SUM OF THE CEILOMETER AND SATELLITE DATA FOR THE SUNRISE TO SUNSET PERIOD.

GENERAL CONTINUED:
CLEAR INDICATES 0 - 3 TENTHS, PARTLY CLOUDY INDICATES 4 - 7 TENTHS, AND CLOUDY INDICATES 8 OR MORE TENTHS. WHEN AT LEAST ONE OF THE ELEMENTS (CEILOMETER OR SATELLITE) IS MISSING, THE DAILY CLOUDINESS IS NOT COMPUTED.
WIND DIRECTION IS RECORDED IN TENS OF DEGREES (2 DIGITS) CLOCKWISE FROM TRUE NORTH. "00" INDICATES CALM. "36" INDICATES TRUE NORTH.
RESULTANT WIND IS THE VECTOR AVERAGE OF THE SPEED AND DIRECTION.
AVERAGE TEMPERATURE IS THE SUM OF THE MEAN DAILY MAXIMUM AND MINIMUM TEMPERATURE DIVIDED BY 2.
SNOWFALL DATA COMPRISE ALL FORMS OF FROZEN PRECIPITATION, INCLUDING HAIL.
A HEATING (COOLING) DEGREE DAY IS THE DIFFERENCE BETWEEN THE AVERAGE DAILY TEMPERATURE AND 65° F.
DRY BULB IS THE TEMPERATURE OF THE AMBIENT AIR.
DEW POINT IS THE TEMPERATURE TO WHICH THE AIR MUST BE COOLED TO ACHIEVE 100 PERCENT RELATIVE HUMIDITY.
WET BULB IS THE TEMPERATURE THE AIR WOULD HAVE IF THE MOISTURE CONTENT WAS INCREASED TO 100 PERCENT RELATIVE HUMIDITY.

HEATING DEGREE DAYS (base 65°F) 1993 TULSA, OKLAHOMA (TUL)

YEAR	JUL	AUG	SEP	OCT	NOV	DEC	JAN	FEB	MAR	APR	MAY	JUN	TOTAL
1964-65	0	0	18	162	385	797	775	711	752	73	1	0	3674
1965-66	0	0	15	122	283	556	992	722	395	227	67	0	3379
1966-67	0	0	13	202	376	859	776	724	362	100	101	0	3513
1967-68	0	0	37	184	498	781	892	803	479	184	67	0	3925
1968-69	0	0	1	160	543	864	863	652	680	127	24	2	3916
1969-70	0	0	0	241	498	789	1088	642	625	174	26	18	4101
1970-71	0	0	18	217	577	692	878	721	463	176	37	0	3779
1971-72	0	0	53	60	446	653	932	670	373	161	47	0	3395
1972-73	0	0	19	183	634	964	954	700	321	233	42	0	4050
1973-74	0	0	24	95	343	824	951	591	341	137	5	0	3311
1974-75	0	0	74	94	473	777	773	780	610	205	19	0	3805
1975-76	0	0	57	146	429	762	855	402	407	126	109	0	3293
1976-77	0	0	16	317	648	858	1173	511	309	99	1	0	3932
1977-78	0	0	1	118	412	801	1236	989	541	110	67	0	4275
1978-79	0	0	0	121	406	834	1293	972	391	164	47	0	4228
1979-80	0	0	0	90	525	632	812	801	513	154	22	0	3549
1980-81	0	0	13	172	438	703	843	598	360	48	58	0	3233
1981-82	0	0	23	178	402	817	967	747	322	208	11	5	3680
1982-83	0	0	23	146	437	635	794	611	492	321	50	0	3509
1983-84	0	0	19	89	378	1179	941	533	509	229	47	0	3924
1984-85	0	0	73	130	438	628	1073	809	330	103	7	0	3591
1985-86	0	0	46	111	510	936	680	602	322	127	13	0	3347
1986-87	0	0	5	148	632	771	893	544	413	149	0	0	3555
1987-88	0	0	1	189	416	727	928	739	483	187	9	0	3679
1988-89	0	0	8	218	393	662	663	921	487	155	53	0	3560
1989-90	0	0	67	126	375	1029	580	527	376	194	54	0	3328
1990-91	0	0	8	172	271	813	933	459	327	83	17	0	3083
1991-92	0	0	35	121	570	628	682	423	311	156	53	0	2979
1992-93	0	0	9	151	565	812	903	755	556	280	57	0	4088
1993-	0	0	40	294	611	695							
POR= 57 YRS	0	0	19	141	471	761	879	666	473	169	40	0	3619

COOLING DEGREE DAYS (base 65°F) 1993 TULSA, OKLAHOMA (TUL)

YEAR	JAN	FEB	MAR	APR	MAY	JUN	JUL	AUG	SEP	OCT	NOV	DEC	ANNUAL
1969	0	0	0	29	185	312	656	496	307	77	0	0	2062
1970	0	0	0	50	209	382	555	619	307	39	0	0	2161
1971	0	0	3	40	97	442	471	444	298	65	7	0	1867
1972	0	6	11	99	144	446	487	524	339	64	0	0	2120
1973	0	0	0	35	124	357	508	452	249	101	5	0	1831
1974	0	0	47	48	232	270	641	419	71	40	2	0	1770
1975	0	0	9	77	156	335	509	542	192	97	12	0	1929
1976	0	6	7	28	52	307	520	461	256	48	0	0	1685
1977	0	1	6	84	248	486	619	525	327	38	0	0	2334
1978	0	0	7	73	180	388	713	605	476	79	14	0	2535
1979	0	0	9	48	167	388	577	527	298	137	6	0	2157
1980	0	0	0	43	200	533	833	774	419	69	6	4	2881
1981	0	5	4	145	96	456	658	452	296	57	1	0	2170
1982	0	0	28	44	266	300	601	637	319	106	10	5	2316
1983	0	0	3	40	120	353	615	725	396	80	20	0	2352
1984	0	0	0	25	132	464	534	556	272	100	9	2	2094
1985	0	0	19	59	185	333	564	523	340	57	0	0	2080
1986	0	0	20	60	157	448	676	415	303	31	0	0	2110
1987	0	0	2	102	290	421	532	567	230	18	19	0	2181
1988	0	0	2	30	200	454	555	564	262	23	1	0	2091
1989	0	0	6	107	191	300	475	483	183	105	14	0	1864
1990	0	0	17	38	137	521	571	581	416	63	21	0	2365
1991	0	0	29	53	293	458	622	562	274	108	3	0	2402
1992	0	0	5	63	140	298	526	369	251	29	0	0	1681
1993	0	0	1	7	95	360	609	579	153	27	0	0	1831
POR= 25 YRS	0	0	8	56	171	391	585	535	289	65	5	0	2105

SNOWFALL (inches) 1993 TULSA, OKLAHOMA (TUL)

YEAR	JUL	AUG	SEP	OCT	NOV	DEC	JAN	FEB	MAR	APR	MAY	JUN	TOTAL
1964-65	0.0	0.0	0.0	0.0	0.0	T	0.3	1.5	0.6	0.0	0.0	0.0	2.4
1965-66	0.0	0.0	0.0	0.0	0.0	T	4.3	5.1	T	0.0	0.0	0.0	9.4
1966-67	0.0	0.0	0.0	0.0	T	3.1	0.7	0.9	1.1	0.0	0.0	0.0	5.8
1967-68	0.0	0.0	0.0	T	0.9	1.6	0.6	2.1	11.8	0.0	0.0	0.0	17.0
1968-69	0.0	0.0	0.0	0.0	T	1.4	T	5.3	1.3	0.0	0.0	0.0	8.0
1969-70	0.0	0.0	0.0	0.0	0.0	5.8	4.7	T	9.9	T	0.0	0.0	20.4
1970-71	0.0	0.0	0.0	0.0	T	0.0	T	6.5	T	0.0	0.0	0.0	6.5
1971-72	0.0	0.0	0.0	0.0	2.0	1.0	0.8	4.9	T	0.0	0.0	0.0	8.7
1972-73	0.0	0.0	0.0	0.0	5.6	1.7	4.3	2.2	0.0	0.3	0.0	0.0	14.1
1973-74	0.0	0.0	0.0	0.0	T	1.8	T	T	T	T	0.0	0.0	1.8
1974-75	0.0	0.0	0.0	0.0	1.7	T	T	3.0	1.8	T	0.0	0.0	6.5
1975-76	0.0	0.0	0.0	0.0	0.8	1.3	T	T	T	0.0	0.0	0.0	2.1
1976-77	0.0	0.0	0.0	0.0	0.5	T	10.5	0.3	0.0	0.0	0.0	0.0	11.3
1977-78	0.0	0.0	0.0	0.0	T	0.0	5.4	6.3	T	0.0	0.0	0.0	11.7
1978-79	0.0	0.0	0.0	0.0	0.0	2.8	12.7	3.4	0.0	T	0.0	0.0	18.9
1979-80	0.0	0.0	0.0	0.0	T	0.0	0.4	3.8	T	0.0	0.0	0.0	4.2
1980-81	0.0	0.0	0.0	0.0	T	0.0	T	0.9	T	0.0	0.0	0.0	.9
1981-82	0.0	0.0	0.0	0.0	0.0	T	0.3	5.6	T	0.0	0.0	0.0	5.9
1982-83	0.0	0.0	0.0	0.0	T	T	3.8	1.4	T	0.0	0.0	0.0	5.2
1983-84	0.0	0.0	0.0	0.0	T	3.0	4.6	0.2	T	0.0	0.0	0.0	7.8
1984-85	0.0	0.0	0.0	0.0	0.0	6.6	3.3	4.3	0.0	0.0	0.0	0.0	14.2
1985-86	0.0	0.0	0.0	0.0	T	2.5	0.0	4.9	0.0	0.0	0.0	0.0	7.4
1986-87	0.0	0.0	0.0	0.0	0.0	0.0	8.7	4.6	0.0	0.0	0.0	0.0	13.3
1987-88	0.0	0.0	0.0	0.0	T	6.7	11.0	T	0.5	0.0	0.0	0.0	18.2
1988-89	0.0	0.0	0.0	0.0	0.4	2.7	3.4	0.3	9.7	0.0	0.0	0.0	16.5
1989-90	0.0	0.0	0.0	0.0	0.0	2.0	0.0	0.0	0.2	0.0	0.0	0.0	2.2
1990-91	0.0	0.0	0.0	0.0	0.0	4.6	T	1.4	T	0.0	T	0.0	6.0
1991-92	0.0	0.0	0.0	0.0	0.2	0.1	0.8	0.0	T	T	0.0	0.0	1.1
1992-93	0.0	0.0	0.0		3.5	1.1		6.7	T	T	0.0		
1993-	0.0	0.0	0.0	0.3	0.0	0.0							
POR= 58 YRS	0.0	0.0	0.0	T	0.4	1.6	3.1	2.3	1.4	T	T	0.0	8.8

ASTORIA, OREGON

Astoria is ringed by low mountains on the north, east, and south. On the west, the area is open to the Pacific Ocean at the mouth of the Columbia River. North of the station, 8 to 12 miles distant, the Washington hills rise to 1,000 to 1,200 feet. Maximum visibility is 19 miles north–northeastward to the Willapa Hills. East–northeastward 2 to 4 miles, the Astoria hills rise to 600 feet. East–southeastward 4 to 14 miles, consecutively, rise other ridges of the Coast Ranges, and southeastward is the most prominent landmark, Saddle Mountain, 3,283 feet high. Forests cover most of the uplands. From Seaside northward to the south bank of the Columbia are 18 miles of sandy beaches, and a 2 to 3 mile wide stretch of dune lands.

The airport sits by the south bank of the Columbia estuary, west of Youngs Bay, on the flood plain or tidal flats. Low dikes prevent flooding and increase the bog–like characteristics of the area. When air temperature falls below water temperature, fog forms easily, or rolls in from the ocean, river, or bay. This usually begins from late afternoon to early morning, and may persist well into the following day. During the summer months, sea breezes commonly blow up the river by noon and stop the diurnal rise in temperature. In winter, cold air may funnel down the Columbia from the interior.

Weather hazards occasionally occur. For flying, the greatest are fog and gales. Even with moderate surface velocities, wind and turbulence at 800 feet may be severe enough to upset a heavy plane. Even in fair weather, wind and wave may combine to produce a type of breaker known as the widow–maker and swamp a boat. Heavy rains inundate lowlands, and high tides aggravated by gales may push seawater across highways or up beaches. Rains may cause earthslides, mostly in highway cuts. Lightning strikes are rare. Showers of ice pellets may briefly whiten the ground during many of the months. Occasionally in winter there may be rather brief periods of freezing temperatures, with snow or ice.

The climate is generally healthful, except for dampness and a lack of sunshine in winter. Heat waves are uncommon and usually brief. Soil leaching necessitates supplementary mineral diets for both animals and plants.

NORMALS, MEANS, AND EXTREMES

ASTORIA, OREGON

LATITUDE: 46°09'N LONGITUDE: 123°53' W ELEVATION: FT. GRND 8 BARO 13 TIME ZONE: PACIFIC WBAN: 94224

	(a)	JAN	FEB	MAR	APR	MAY	JUNE	JULY	AUG	SEP	OCT	NOV	DEC	YEAR
TEMPERATURE °F:														
Normals														
-Daily Maximum		47.8	51.0	53.2	56.0	60.1	64.1	67.5	68.8	67.8	61.1	53.5	48.2	58.3
-Daily Minimum		35.9	37.3	38.1	40.3	44.8	49.5	52.4	52.6	49.0	44.0	40.2	36.6	43.4
-Monthly		41.9	44.2	45.7	48.2	52.5	56.9	60.0	60.7	58.4	52.6	46.9	42.4	50.9
Extremes														
-Record Highest	40	67	72	73	83	87	93	100	96	95	85	71	64	100
-Year		1986	1968	1979	1987	1985	1955	1961	1981	1972	1987	1970	1980	JUL 1961
-Record Lowest	40	11	9	22	29	30	37	39	39	33	26	15	6	6
-Year		1980	1989	1971	1968	1954	1980	1971	1973	1983	1971	1955	1990	DEC 1990
NORMAL DEGREE DAYS:														
Heating (base 65°F)		716	582	598	504	388	243	157	141	201	384	543	701	5158
Cooling (base 65°F)		0	0	0	0	0	0	0	8	0	0	0	0	8
% OF POSSIBLE SUNSHINE														
MEAN SKY COVER (tenths)														
Sunrise - Sunset	40	8.3	8.1	8.1	8.0	7.7	7.7	6.7	6.7	6.3	7.2	8.1	8.4	7.6
MEAN NUMBER OF DAYS:														
Sunrise to Sunset														
-Clear	40	3.0	3.4	2.8	3.0	3.1	3.3	5.8	6.2	8.1	5.3	2.8	2.7	49.5
-Partly Cloudy	40	3.7	3.3	5.3	5.9	8.2	7.4	9.8	9.6	7.5	6.7	4.9	4.2	76.5
-Cloudy	40	24.2	21.5	22.9	21.0	19.7	19.3	15.3	15.3	14.5	19.0	22.3	24.2	239.2
Precipitation														
.01 inches or more	40	21.6	19.0	20.3	18.3	15.1	12.7	7.7	7.8	9.9	15.6	20.4	22.0	190.3
Snow, Ice pellets, hail														
1.0 inches or more	40	0.6	0.1	0.3	0.*	0.0	0.0	0.0	0.0	0.0	0.0	0.1	0.3	1.4
Thunderstorms	40	0.7	0.3	0.4	0.6	0.3	0.3	0.4	0.3	0.8	0.9	1.2	0.9	7.1
Heavy Fog Visibility 1/4 mile or less	40	3.7	3.0	2.7	2.2	1.5	1.4	2.1	4.2	5.7	6.9	4.1	3.9	41.4
Temperature °F														
-Maximum														
90° and above	40	0.0	0.0	0.0	0.0	0.0	0.1	0.1	0.1	0.1	0.0	0.0	0.0	0.5
32° and below	40	0.6	0.1	0.0	0.0	0.0	0.0	0.0	0.0	0.0	0.0	0.2	0.7	1.5
-Minimum														
32° and below	40	9.6	6.6	6.0	1.9	0.1	0.0	0.0	0.0	0.0	0.5	4.5	8.2	37.4
0° and below	40	0.0	0.0	0.0	0.0	0.0	0.0	0.0	0.0	0.0	0.0	0.0	0.0	0.0
AVG. STATION PRESS. (mb)	21	1017.4	1015.9	1015.4	1017.5	1017.6	1017.9	1018.3	1017.1	1016.6	1017.4	1016.5	1017.7	1017.1
RELATIVE HUMIDITY (%)														
Hour 04	40	86	87	88	89	90	90	90	91	91	90	88	87	89
Hour 10 (Local Time)	40	84	82	78	74	74	75	75	77	75	81	83	85	79
Hour 16	40	78	74	71	69	70	71	69	70	69	73	77	80	73
Hour 22	40	85	86	86	85	85	85	86	88	88	88	87	86	86
PRECIPITATION (inches):														
Water Equivalent														
-Normal		10.00	7.59	7.07	4.60	3.02	2.40	1.15	1.33	2.91	5.73	10.05	10.55	66.40
-Maximum Monthly	40	18.94	21.89	13.47	9.47	6.60	5.48	4.39	5.22	6.93	12.56	16.75	16.57	21.89
-Year		1954	1961	1956	1991	1960	1954	1983	1968	1978	1975	1983	1955	FEB 1961
-Minimum Monthly	40	0.69	1.35	0.93	1.33	0.37	**0.55**	0.01	0.08	0.04	0.52	1.45	2.67	0.01
-Year		1985	1993	1965	1956	1982	**1992**	1960	1970	1975	1987	1976	1985	JUL 1960
-Maximum in 24 hrs	40	5.14	3.39	2.66	3.00	1.82	2.42	1.98	1.65	2.63	3.71	4.19	3.61	5.14
-Year		1990	1990	1956	1991	1979	1968	1974	1968	1953	1982	1986	1974	JAN 1990
Snow, Ice pellets, hail														
-Maximum Monthly	40	26.3	4.0	6.7	1.1	T	T	**0.0**	**0.0**	T	**T**	4.6	19.0	26.3
-Year		1969	1962	1966	1975	1991	1989			1972	**1992**	1985	1964	JAN 1969
-Maximum in 24 hrs	40	10.8	4.0	5.9	1.0	T	T	**0.0**	**0.0**	T	**T**	4.3	7.2	10.8
-Year		1971	1962	1960	1975	1991	1989			1972	**1992**	1985	1964	JAN 1971
WIND:														
Mean Speed (mph)	40	9.2	9.0	8.8	8.7	8.5	8.6	8.7	8.1	7.5	7.5	8.7	9.0	8.5
Prevailing Direction through 1963		E	ESE	SE	WNW	NW	NW	NW	NW	SE	SE	SE	ESE	SE
Fastest Obs. 1 Min.														
-Direction (!!!)	40	17	19	19	20	22	**23**	19	20	20	20	20	25	17
-Speed (MPH)	40	55	47	44	52	37	**29**	29	30	35	44	46	52	55
-Year		1971	1979	1964	1962	1975	**1993**	1988	1991	1959	1962	1962	1961	JAN 1971
Peak Gust														
-Direction (!!!)	10	SW	S	S	S	**21**	**23**	SW	**19**	S	SW	S	**18**	**18**
-Speed (mph)	10	75	70	67	63	**46**	**41**	41	**46**	56	49	66	**83**	**83**
-Date		1986	1987	1989	1988	**1993**	**1993**	1988	**1993**	1988	1990	1991	**1993**	**DEC 1993**

PRECIPITATION (inches)

ASTORIA, OREGON

YEAR	JAN	FEB	MAR	APR	MAY	JUNE	JULY	AUG	SEP	OCT	NOV	DEC	ANNUAL
1964	18.50	4.06	7.41	3.59	2.27	2.70	2.59	2.21	2.73	2.61	11.15	13.67	73.49
1965	16.59	6.77	0.93	5.47	2.74	0.75	0.46	1.95	0.51	3.97	11.82	11.78	63.74
1966	8.61	5.53	8.79	2.90	2.18	2.13	0.54	1.01	2.18	5.83	10.00	14.07	63.77
1967	14.95	6.07	8.38	5.52	1.37	1.14	0.22	0.19	3.07	11.06	5.94	9.04	66.95
1968	9.57	9.57	10.42	4.22	3.91	4.81	1.23	5.22	4.60	8.03	11.96	13.85	87.39
1969	12.02	5.67	3.16	3.84	3.92	3.63	0.56	0.62	6.55	5.28	5.77	11.69	62.71
1970	14.46	5.29	4.28	7.74	1.92	1.19	0.31	0.08	3.65	5.80	9.86	15.93	70.51
1971	16.69	6.67	9.96	4.09	2.30	2.97	1.55	1.14	4.65	6.34	9.08	13.83	79.27
1972	10.62	8.58	10.04	6.82	1.22	0.92	2.01	0.37	4.72	1.96	6.90	13.28	67.44
1973	5.72	2.60	5.71	2.38	3.16	4.26	0.07	0.46	4.19	5.92	14.93	15.75	65.15
1974	12.47	8.38	10.73	4.88	4.37	2.33	4.20	0.29	0.67	1.85	8.95	13.84	72.96
1975	15.21	8.03	5.66	3.90	2.41	1.99	0.22	2.82	0.04	12.56	12.28	15.66	80.78
1976	11.67	7.86	7.17	3.55	2.20	1.27	2.46	2.55	1.58	2.96	1.45	4.20	48.92
1977	3.20	5.22	9.74	1.65	6.00	1.36	0.44	3.85	5.44	4.38	12.37	14.34	67.99
1978	8.66	5.43	4.40	6.35	4.75	3.07	0.90	2.61	6.93	1.01	8.43	4.99	57.53
1979	3.83	11.76	4.52	4.38	4.19	1.82	0.92	0.81	4.35	8.46	7.87	13.18	66.09
1980	7.21	9.60	6.31	4.85	1.45	1.57	0.64	1.24	2.51	2.79	12.02	12.44	62.63
1981	2.63	8.69	5.80	7.30	2.97	5.47	1.06	0.62	2.77	8.67	10.66	11.80	68.44
1982	13.98	10.87	7.19	6.52	0.37	1.22	0.75	0.63	3.72	8.31	9.62	12.14	75.32
1983	13.52	8.66	8.84	4.26	3.59	4.53	4.39	1.14	1.83	1.87	16.75	9.44	78.82
1984	6.60	8.34	5.90	5.02	5.34	3.90	0.05	0.52	3.16	8.10	15.19	6.51	68.63
1985	0.69	4.09	7.00	2.95	1.90	3.09	0.78	1.11	3.23	8.11	5.96	2.67	41.58
1986	11.19	8.93	6.11	3.58	3.30	0.94	1.69	0.14	3.62	5.45	11.42	7.34	63.71
1987	10.38	5.08	8.52	3.02	3.97	0.65	1.10	0.16	0.95	0.52	4.33	8.85	47.53
1988	6.57	3.60	7.86	3.99	4.09	3.50	0.96	0.88	1.23	2.14	13.06	7.32	55.20
1989	8.20	6.61	10.09	2.27	3.01	2.58	1.64	0.84	0.50	5.30	6.73	7.40	55.17
1990	16.09	11.83	5.15	4.44	4.00	3.47	0.54	1.57	0.67	8.44	11.28	5.11	72.59
1991	6.76	8.57	5.65	9.47	2.68	1.86	0.33	2.31	0.07	2.44	10.53	6.60	57.27
1992	9.34	5.69	1.19	7.49	0.52	0.55	0.24	0.77	2.66	4.10	10.11	5.99	48.65
1993	6.27	1.35	6.93	9.01	4.74	3.70	1.81	0.57	0.12	2.23	6.68	9.59	53.00
Record Mean	10.06	7.57	7.04	4.99	2.93	2.57	1.13	1.32	2.73	5.89	10.08	10.33	66.63

AVERAGE TEMPERATURE (deg. F)

ASTORIA, OREGON

YEAR	JAN	FEB	MAR	APR	MAY	JUNE	JULY	AUG	SEP	OCT	NOV	DEC	ANNUAL
1964	43.7	42.3	44.3	45.8	49.5	55.2	59.2	59.3	57.4	53.8	44.8	40.0	49.6
1965	42.3	44.2	47.2	49.3	50.9	56.2	60.2	61.7	57.0	55.3	48.8	40.3	51.1
1966	42.3	42.6	43.7	49.0	50.5	56.2	59.6	60.3	59.2	51.7	47.8	45.5	50.7
1967	43.8	43.2	42.2	44.0	51.7	58.3	61.1	62.8	60.5	52.9	48.1	41.2	50.8
1968	42.0	47.5	46.7	46.1	52.2	56.1	61.0	60.1	57.9	50.6	47.1	39.5	50.6
1969	34.3	40.3	44.3	46.8	53.5	59.6	58.6	58.7	57.2	51.0	46.5	44.1	49.6
1970	41.8	46.3	44.9	45.3	51.4	56.6	59.3	60.2	56.0	51.0	47.1	40.8	50.1
1971	40.7	42.2	42.0	46.5	51.4	54.6	59.4	62.4	57.9	51.3	46.8	41.2	49.7
1972	40.0	43.2	47.8	46.6	54.7	58.3	62.2	63.4	57.6	52.3	47.3	38.6	51.0
1973	40.5	44.3	44.4	48.9	53.4	56.6	59.7	57.6	57.9	51.2	44.3	45.2	50.3
1974	38.9	42.0	44.7	48.0	50.5	56.1	59.3	62.1	60.9	51.9	48.0	45.6	50.7
1975	42.9	43.1	43.8	44.9	52.8	55.6	60.7	59.7	59.2	50.9	46.2	44.7	50.4
1976	44.1	42.6	43.7	48.7	52.8	55.9	61.1	61.7	60.8	53.2	48.0	43.5	51.3
1977	39.9	46.8	44.6	48.9	50.2	56.8	58.7	62.7	57.0	51.7	44.6	43.9	50.5
1978	44.0	46.0	47.0	48.8	52.3	59.4	60.9	61.5	58.1	54.3	41.7	37.6	50.9
1979	35.3	41.8	47.3	49.2	53.8	56.8	62.4	61.8	61.4	55.0	46.9	47.4	51.6
1980	38.7	46.4	44.8	49.5	52.3	55.4	60.7	58.8	58.9	55.0	50.1	47.5	51.5
1981	48.2	46.6	48.1	49.5	53.4	57.7	60.4	62.4	59.1	52.7	49.0	45.3	52.7
1982	41.7	44.0	45.2	46.8	52.2	57.8	59.3	61.3	59.7	55.5	46.6	43.5	51.2
1983	47.3	50.2	50.9	50.2	55.7	59.8	61.2	61.7	56.5	50.4	48.6	37.9	52.5
1984	43.5	45.4	48.4	46.7	50.8	55.6	59.6	59.9	58.4	50.3	46.7	39.6	50.4
1985	39.3	40.3	42.8	48.4	52.5	56.8	60.6	59.4	56.2	50.7	39.2	39.3	48.8
1986	46.9	44.1	49.3	47.5	52.8	58.4	58.7	61.1	55.7	54.9	47.9	44.4	51.8
1987	43.7	46.3	48.0	50.4	55.0	57.4	59.2	60.4	58.4	54.5	49.6	40.9	52.0
1988	41.8	45.8	45.4	48.9	52.5	55.7	59.9	59.2	56.6	55.5	47.4	43.3	51.0
1989	42.1	37.1	45.1	51.5	53.1	58.7	59.8	60.5	60.4	52.8	49.2	44.0	51.2
1990	44.7	43.1	47.6	50.3	52.9	57.6	62.3	62.7	60.8	51.1	47.8	36.9	51.5
1991	41.3	48.0	43.3	46.8	50.8	54.5	59.6	59.7	60.6	51.2	47.3	44.5	50.6
1992	45.2	48.3	49.0	50.9	54.8	58.6	60.7	60.6	56.3	52.6	46.3	40.7	52.0
1993	38.9	41.9	47.4	49.2	56.2	57.5	59.0	60.7	56.6	54.1	42.3	42.7	50.5
Record Mean	41.8	44.2	45.4	48.0	52.5	56.9	60.0	60.6	58.3	52.7	46.7	42.6	50.9
Max	47.7	51.1	53.0	55.8	60.3	64.2	67.6	68.7	67.6	61.2	53.4	48.5	58.3
Min	35.9	37.3	37.7	40.2	44.7	49.6	52.4	52.6	49.0	44.2	39.9	36.7	43.4

REFERENCE NOTES

ASTORIA, OREGON

GENERAL
T - TRACE AMOUNT.
BLANK ENTRIES DENOTE MISSING/UNREPORTED DATA.
# INDICATES A STATION OR INSTRUMENT RELOCATION.

SPECIFIC

(a) - LENGTH OF RECORD IN YEARS, ALTHOUGH INDIVIDUAL MONTHS MAY BE MISSING.
0.* OR * - THE VALUE IS BETWEEN 0.0 AND 0.05
NORMALS - BASED ON THE 1961-1990 RECORD PERIOD.
EXTREMES - DATES ARE THE MOST RECENT OCCURRENCE
WIND DIR. - NUMERALS SHOW TENS OF DEGRESS CLOCKWISE FROM TRUE NORTH. "00" INDICATES CALM.
RESULTANT DIRECTIONS ARE GIVEN TO WHOLE DEGREES.
BOLD VALUES INDICATE EXTREME VALUES WHICH OCCURRED AFTER THE ASOS SYSTEM WAS COMMISSIONED.

RECORD = PERIOD OF RECORD
RECORD MEAN PRECIPITATION IS THE MEAN OF ALL DAILY PRECIPITATION AMOUNTS DURING THE PERIOD OF RECORD.
RECORD MAX(MIN) TEMPERATURE IS THE MEAN OF ALL DAILY MAX(MIN) TEMPERATURES DURING THE PERIOD OF RECORD.
RECORD MEAN TEMPERATURE IS THE SUM OF THE RECORD MAX AND RECORD MIN DIVIDED BY 2.
AVERAGE TEMPERATURE IS THE SUM OF THE MEAN DAILY MAX AND MIN TEMPERATURE DIVIDED BY 2.

EXCEPTIONS

RECORD MEANS ARE THROUGH THE CURRENT YEAR, BEGINNING IN 1954 FOR TEMPERATURE
1954 FOR PRECIPITATION
1954 FOR SNOWFALL

HEATING DEGREE DAYS Base 65 deg. F — ASTORIA, OREGON

SEASON	JULY	AUG	SEP	OCT	NOV	DEC	JAN	FEB	MAR	APR	MAY	JUNE	TOTAL
1964-65	174	179	222	345	598	768	696	576	542	466	432	259	5257
1965-66	144	98	235	295	476	760	697	621	658	472	443	267	5166
1966-67	160	142	169	405	514	595	651	603	697	623	404	194	5157
1967-68	118	68	135	369	500	731	705	500	557	560	387	260	4890
1968-69	124	151	216	438	531	782	944	687	633	540	351	161	5558
1969-70	193	188	226	423	551	637	710	518	617	584	416	250	5313
1970-71	170	142	264	428	531	743	746	632	703	549	415	310	5633
1971-72	165	85	212	420	538	728	768	623	526	545	313	196	5119
1972-73	99	65	230	388	525	810	755	574	635	473	357	246	5157
1973-74	164	225	213	421	614	611	803	639	623	503	445	257	5518
1974-75	174	97	130	398	502	596	678	606	653	598	373	273	5078
1975-76	133	155	183	429	558	624	642	640	651	483	374	269	5141
1976-77	117	105	128	361	503	660	774	504	628	475	453	238	4946
1977-78	190	82	231	403	606	646	643	524	551	482	389	161	4908
1978-79	125	110	202	324	693	843	910	641	543	467	338	243	5439
1979-80	91	92	111	305	537	540	807	529	617	459	388	279	4755
1980-81	140	186	183	309	439	533	513	507	518	458	354	211	4351
1981-82	139	107	173	372	474	604	714	581	610	538	391	212	4915
1982-83	171	112	152	288	546	661	543	407	430	438	282	149	4179
1983-84	109	98	249	445	485	834	658	559	509	541	433	276	5196
1984-85	160	155	199	452	542	780	788	686	683	493	383	252	5573
1985-86	131	172	259	438	769	788	551	577	481	520	373	194	5253
1986-87	188	120	273	308	505	635	654	517	519	431	306	229	4685
1987-88	171	139	194	321	455	741	713	552	601	476	382	273	5018
1988-89	159	173	250	289	520	663	703	773	612	399	363	199	5103
1989-90	157	135	150	371	464	643	621	606	533	435	365	221	4701
1990-91	82	74	127	424	510	863	727	469	667	537	435	309	5224
1991-92	170	155	147	420	523	631	607	479	491	416	309	188	4536
1992-93	129	144	253	379	555	747	803	641	541	469	268	218	5147
1993-94	182	144	247	331	674	688							

COOLING DEGREE DAYS Base 65 deg. F — ASTORIA, OREGON

YEAR	JAN	FEB	MAR	APR	MAY	JUNE	JULY	AUG	SEP	OCT	NOV	DEC	TOTAL
1969	0	0	0	0	1	4	0	0	0	0	0	0	5
1970	0	0	0	0	0	7	0	0	1	0	0	0	8
1971	0	0	0	0	0	2	0	11	4	2	0	0	19
1972	0	0	0	0	0	1	19	23	16	0	0	0	59
1973	0	0	0	0	0	0	6	0	7	0	0	0	13
1974	0	0	0	0	0	0	5	11	13	0	0	0	29
1975	0	0	0	0	1	0	5	0	17	0	0	0	23
1976	0	0	0	0	0	3	4	8	8	1	0	0	24
1977	0	0	0	0	0	0	0	18	0	0	0	0	18
1978	0	0	0	0	0	0	6	8	0	0	0	0	14
1979	0	0	0	0	0	0	15	0	11	1	0	0	27
1980	0	0	0	0	0	0	13	0	7	8	0	0	28
1981	0	0	0	0	0	0	4	36	2	0	0	0	42
1982	0	0	0	0	0	6	0	4	1	1	0	0	12
1983	0	0	0	0	0	0	0	4	1	0	0	0	5
1984	0	0	0	0	0	0	3	1	5	0	0	0	9
1985	0	0	0	0	3	11	1	6	0	0	0	0	21
1986	0	0	0	0	2	0	0	5	0	0	0	0	7
1987	0	0	0	0	2	5	0	5	5	3	0	0	20
1988	0	0	0	0	0	0	10	3	6	0	0	0	19
1989	0	0	0	1	0	15	0	1	17	0	0	0	34
1990	0	0	0	0	0	5	5	12	10	0	0	0	32
1991	0	0	0	0	0	0	7	0	24	1	0	0	32
1992	0	0	0	0	0	6	1	18	1	0	0	0	26
1993	0	0	0	0	2	0	0	16	2	0	0	0	20

SNOWFALL (inches) — ASTORIA, OREGON

SEASON	JULY	AUG	SEP	OCT	NOV	DEC	JAN	FEB	MAR	APR	MAY	JUNE	TOTAL
1964-65	0.0	0.0	0.0	0.0	T	19.0	10.7	0.0	0.0	0.0	0.0	0.0	29.7
1965-66	0.0	0.0	0.0	0.0	0.0	3.4	0.7	T	6.7	T	0.0	0.0	10.8
1966-67	0.0	0.0	0.0	0.0	0.0	0.0	T	T	0.7	T	0.0	0.0	0.7
1967-68	0.0	0.0	0.0	0.0	0.0	0.8	3.7	0.0	0.0	0.0	0.0	0.0	4.5
1968-69	0.0	0.0	0.0	0.0	0.0	4.5	26.3	1.0	0.0	0.0	0.0	0.0	31.8
1969-70	0.0	0.0	0.0	0.0	0.0	0.0	0.0	0.0	0.0	T	T	0.0	T
1970-71	0.0	0.0	0.0	T	T	1.8	16.1	0.2	1.5	0.2	T	0.0	19.8
1971-72	0.0	0.0	0.0	T	0.0	3.4	1.7	T	T	0.2	0.0	0.0	5.3
1972-73	0.0	0.0	T	0.0	0.0	5.4	T	0.0	T	T	T	0.0	5.4
1973-74	0.0	0.0	0.0	0.0	T	T	0.4	T	1.7	0.0	T	0.0	2.1
1974-75	0.0	0.0	0.0	0.0	T	T	0.3	0.2	T	1.1	T	0.0	1.6
1975-76	0.0	0.0	0.0	T	1.1	T	T	0.1	T	T	0.0	0.0	1.2
1976-77	0.0	0.0	0.0	0.0	0.0	T	T	T	T	0.0	T	0.0	T
1977-78	0.0	0.0	0.0	0.0	0.1	T	T	T	0.0	T	0.0	0.0	0.1
1978-79	0.0	0.0	0.0	T	0.1	2.5	0.3	T	T	T	0.0	0.0	2.9
1979-80	0.0	0.0	0.0	T	0.3	0.2	0.5	1.7	0.5	T	0.0	0.0	3.2
1980-81	0.0	0.0	0.0	0.0	T	T	0.0	0.0	0.0	T	0.0	0.0	T
1981-82	0.0	0.0	0.0	0.0	T	T	6.1	T	T	T	0.0	0.0	6.1
1982-83	0.0	0.0	0.0	0.0	T	0.0	T	0.0	T	T	T	0.0	T
1983-84	0.0	0.0	0.0	0.0	T	T	0.0	T	T	T	T	0.0	T
1984-85	0.0	0.0	0.0	T	T	T	0.0	1.1	T	T	T	0.0	1.1
1985-86	0.0	0.0	0.0	0.0	4.6	0.4	T	T	T	T	0.0	0.0	5.0
1986-87	0.0	0.0	0.0	0.0	T	0.0	T	0.0	T	T	0.0	0.0	T
1987-88	0.0	0.0	0.0	0.0	T	T	T	0.0	T	T	0.0	0.0	T
1988-89	0.0	0.0	0.0	0.0	T	T	T	1.5	T	T	T	T	1.5
1989-90	0.0	0.0	0.0	T	T	0.0	T	1.8	T	T	0.0	0.0	1.8
1990-91	0.0	0.0	0.0	0.0	T	1.4	0.0	0.0	T	T	T	0.0	1.4
1991-92	0.0	0.0	0.0	0.0	T	0.0	T	T	0.0	T	0.0	0.0	T
1992-93	0.0	0.0	0.0	T	0.0	T	5.2	3.4	0.0	0.0	0.0	0.0	8.6
1993-94	0.0	0.0	0.0	0.0	T	0.0							
Record Mean	0.0	0.0	T	T	0.2	1.1	2.2	0.4	0.6	T	T	T	4.5

EUGENE, OREGON

Eugene is located at the upper or southern end of the fertile Willamette Valley. Mahlon Sweet Field, location of the National Weather Service Office, is 9 miles northwest of the city center. The Cascade Mountains to the east and the Coast Range to the west bound the valley, and low hills to the south nearly close it, but northward the level valley floor broadens rapidly. Hills of the rolling, wooded Coast Range begin about 5 miles west of the airport and rise to elevations of 1,500 to 2,500 feet midway between Eugene and the Pacific Ocean lying 50 miles to the west. About 10 miles east, the Coburg Hills, rising to an elevation of 2,500 feet, obscure the snow-covered peaks of the Cascade Range, which reach elevations of 10,000 feet about 75 miles away. Abundant moisture and moderate temperatures result in rapid growth of evergreen timber and lumbering is a major industry.

The Willamette River passes about 5 miles east of the airport. The Fern Ridge flood control reservoir, with a normal area of 9,360 acres, begins about 2 miles southwest. These two water areas are the main sources of local fog, but numerous small creeks and low places, which fill with water in the wet season, also produce considerable fog. The Coast Range acts as a barrier to coastal fog, but active storms cross these ridges with little hindrance. The Cascade Range blocks westward passage of all but the strongest continental air masses, but when air does flow into the valley from the east, dry, hot weather develops in summer causing an extreme fire hazard. In winter this situation causes clear, sunny days and cool, frosty nights.

The centers of low barometric pressure, with which rain is associated, generally pass inland north of Eugene and as a result southwest winds with speeds of 10 to 20 mph usually accompany rainfall. Heavier storms bring winds of 30 to 40 mph and occasional southwest winds exceeding 50 mph are experienced. Fair weather in both summer and winter is most often accompanied by calm nights and daytime northerly winds increasing to speeds of 5 to 15 mph in the afternoon.

The first fall rains usually arrive during the second or third week of September, after which rain gradually increases until about the first of January and then slowly decreases to the latter part of June. July and August are normally very dry, occasionally passing without rainfall. When snow occurs, it frequently melts on contact with the ground or within a few hours, but occasionally an accumulation of a few inches will persist as a ground covering for several days. Snowfall for a winter season exceeds 5 inches in about one-third of the years.

Temperatures are so largely controlled by maritime air from the Pacific that long periods of extremely hot or severely cold weather never occur. Temperatures of 95 degrees or higher have occurred only in the months of June, July, August, and September, and average three days a year.

Based on the 1951-1980 period, the average first occurrence of 32 degrees Fahrenheit in the fall is October 25 and the average last occurrence in the spring is April 24.

NORMALS, MEANS, AND EXTREMES

EUGENE, OREGON

LATITUDE: 44°07'N LONGITUDE: 123°13'W ELEVATION: FT. GRND 359 BARO 364 TIME ZONE: PACIFIC WBAN: 24221

	(a)	JAN	FEB	MAR	APR	MAY	JUNE	JULY	AUG	SEP	OCT	NOV	DEC	YEAR
TEMPERATURE °F:														
Normals														
-Daily Maximum		46.4	51.4	55.9	60.5	67.1	74.2	81.7	81.8	76.2	64.6	52.4	46.2	63.2
-Daily Minimum		35.2	37.0	38.9	40.6	44.5	49.7	52.8	53.2	49.3	43.5	39.7	35.9	43.4
-Monthly		40.8	44.2	47.4	50.6	55.8	62.0	67.3	67.6	62.8	54.1	46.1	41.1	53.3
Extremes														
-Record Highest	51	67	71	77	86	93	102	105	108	103	94	76	68	108
-Year		1975	1968	1978	1957	1987	1992	1961	1981	1988	1980	1975	1979	AUG 1981
-Record Lowest	51	-4	-3	20	27	28	32	39	38	32	19	12	-12	-12
-Year		1957	1950	1956	1983	1954	1976	1986	1969	1983	1971	1978	1972	DEC 1972
NORMAL DEGREE DAYS:														
Heating (base 65°F)		750	582	546	432	285	121	38	30	116	338	567	741	4546
Cooling (base 65°F)		0	0	0	0	0	31	109	110	50	0	0	0	300
% OF POSSIBLE SUNSHINE														
MEAN SKY COVER (tenths)														
Sunrise - Sunset	49	8.5	8.3	8.0	7.5	6.8	6.2	3.9	4.5	5.0	7.0	8.5	8.9	6.9
MEAN NUMBER OF DAYS:														
Sunrise to Sunset														
-Clear	51	2.1	2.5	3.0	3.8	5.8	7.6	16.0	14.0	12.0	5.0	1.5	1.5	74.8
-Partly Cloudy	51	4.2	4.6	6.1	6.8	8.6	8.1	7.7	8.6	8.5	8.9	5.6	3.8	81.7
-Cloudy	51	24.7	21.1	21.9	19.4	16.6	14.3	7.3	8.4	9.5	17.1	22.8	25.7	208.7
Precipitation														
.01 inches or more	51	17.6	14.9	16.6	13.0	10.0	7.0	2.6	3.9	5.6	11.0	16.3	18.0	136.6
Snow,Ice pellets,hail														
1.0 inches or more	51	1.0	0.3	0.2	0.0	0.0	0.0	0.0	0.0	0.0	0.0	0.1	0.5	2.2
Thunderstorms	50	0.2	0.1	0.2	0.4	0.9	0.7	0.5	0.7	0.5	0.2	0.1	0.2	4.7
Heavy Fog Visibility														
1/4 mile or less	50	8.9	6.9	3.9	2.2	1.2	0.8	0.4	0.9	4.2	10.9	9.8	9.7	59.7
Temperature °F														
-Maximum														
90° and above	51	0.0	0.0	0.0	0.0	0.1	1.3	6.0	5.1	2.6	0.1	0.0	0.0	15.1
32° and below	51	1.7	0.3	0.0	0.0	0.0	0.0	0.0	0.0	0.0	0.0	0.2	1.0	3.1
-Minimum														
32° and below	51	14.8	9.7	6.8	2.6	0.4	0.*	0.0	0.0	0.1	2.0	7.4	11.5	55.3
0° and below	51	0.1	0.*	0.0	0.0	0.0	0.0	0.0	0.0	0.0	0.0	0.0	0.1	0.2
AVG. STATION PRESS.(mb)	19	1006.3	1004.6	1003.4	1005.0	1004.4	1004.1	1004.2	1003.2	1003.4	1005.0	1005.3	1006.3	1004.6
RELATIVE HUMIDITY (%)														
Hour 04	36	92	92	92	91	91	90	88	88	89	94	93	93	91
Hour 10 (Local Time)	49	88	85	79	71	66	63	57	60	64	80	87	89	74
Hour 16	49	80	72	64	58	54	49	38	39	43	62	79	84	60
Hour 22	38	91	90	87	83	81	79	73	73	77	88	92	92	84
PRECIPITATION (inches):														
Water Equivalent														
-Normal		7.91	5.64	5.52	3.11	2.16	1.43	0.51	1.08	1.67	3.41	8.32	8.61	49.37
-Maximum Monthly	51	15.09	14.22	12.46	7.85	6.92	4.76	3.00	5.79	4.65	12.66	20.48	20.99	20.99
-Year		1990	1986	1974	1993	1993	1952	1987	1968	1986	1950	1973	1964	DEC 1964
-Minimum Monthly	51	0.31	0.86	0.79	0.49	0.14	T	0.00	0.00	T	0.11	1.20	1.24	0.00
-Year		1985	1964	1965	1985	1992	1951	1967	1967	1993	1988	1956	1976	JUL 1967
-Maximum in 24 hrs	51	4.88	4.81	2.44	2.38	2.37	2.36	2.44	1.92	1.68	3.85	4.53	5.15	5.15
-Year		1974	1984	1963	1992	1972	1952	1987	1983	1981	1955	1960	1981	DEC 1981
Snow,Ice pellets,hail														
-Maximum Monthly	51	47.1	8.8	10.8	T	T	T	0.0	0.0	T	T	6.0	10.2	47.1
-Year		1969	1990	1951	1992	1986	1981			1986	1984	1955	1964	JAN 1969
-Maximum in 24 hrs	51	22.9	5.7	4.9	T	T	T	0.0	0.0	T	T	5.0	6.3	22.9
-Year		1969	1993	1951	1992	1986	1981			1986	1984	1955	1972	JAN 1969
WIND:														
Mean Speed (mph)	41	7.9	7.8	8.3	7.8	7.4	7.5	8.0	7.6	7.4	6.7	7.4	7.6	7.6
Prevailing Direction														
Fastest Obs. 1 Min.														
-Direction (!!)	37	20	20	18	18	25	27	32	11	20	18	23	25	18
-Speed (MPH)	37	58	54	48	44	46	29	37	32	32	63	46	40	63
-Year		1961	1961	1963	1972	1961	1988	1986	1979	1959	1962	1957	1961	OCT 1962
Peak Gust														
-Direction (!!)	10	SW	S	W	W	SW	W	NW	N	N	S	S	S	SW
-Speed (mph)	10	58	44	45	41	45	41	51	39	36	36	46	49	58
-Date		1990	1984	1985	1984	1985	1988	1986	1985	1988	1986	1988	1987	JAN 1990

PRECIPITATION (inches)

EUGENE, OREGON

YEAR	JAN	FEB	MAR	APR	MAY	JUNE	JULY	AUG	SEP	OCT	NOV	DEC	ANNUAL
1964	14.83	0.86	4.53	1.28	0.90	1.29	0.54	0.27	0.73	1.03	10.70	20.99	57.95
1965	9.92	1.60	0.79	2.92	0.90	0.44	0.24	0.90	T	2.43	7.43	7.69	35.26
1966	10.97	1.83	5.95	0.52	0.54	0.23	0.45	0.02	1.44	1.93	9.10	8.31	41.29
1967	10.33	2.13	2.90	3.02	1.59	1.70	0.00	0.00	1.83	5.10	3.22	4.99	36.81
1968	7.53	6.32	3.91	1.06	2.86	0.96	0.02	5.79	2.57	5.99	7.02	12.52	56.55
1969	12.67	3.21	2.74	2.92	2.38	3.13	T	T	2.05	4.22	2.74	11.68	47.74
1970	14.38	3.37	2.45	3.04	0.80	0.76	T	T	1.19	4.30	8.23	11.57	50.09
1971	10.83	5.11	7.99	4.49	2.52	3.10	T	1.33	3.04	3.32	9.48	9.46	60.67
1972	12.48	6.52	7.63	5.80	2.94	1.91	0.02	1.70	2.71	1.31	3.76	10.78	57.56
1973	6.20	2.09	5.18	1.67	0.86	1.35	T	0.80	2.48	2.37	20.48	11.82	55.30
1974	12.80	8.42	12.46	2.47	1.12	0.37	1.37	0.42	0.08	1.59	6.42	9.26	56.78
1975	6.93	6.79	7.56	2.90	2.16	0.88	1.18	2.09	T	5.69	8.47	7.12	51.77
1976	9.82	7.66	6.23	1.89	0.94	0.21	0.42	2.04	1.13	1.87	1.33	1.24	34.78
1977	1.11	5.05	4.66	1.47	2.84	0.97	0.11	1.70	2.39	2.87	9.14	14.60	46.91
1978	9.05	3.25	1.68	6.56	2.12	0.74	0.72	2.17	3.45	0.29	6.61	2.86	39.50
1979	2.98	9.52	3.12	4.71	2.61	0.56	0.41	3.46	2.32	8.12	6.09	7.38	51.28
1980	7.45	4.68	5.11	4.20	1.39	2.06	0.39	0.02	0.75	1.90	8.66	14.73	51.34
1981	2.13	4.35	4.16	2.69	3.27	3.51	0.08	T	3.15	5.42	9.51	17.63	55.90
1982	9.31	8.14	4.88	6.89	0.26	1.92	0.54	0.72	2.81	3.95	7.07	13.53	60.02
1983	6.75	12.28	10.58	3.35	1.81	1.78	1.77	3.19	0.54	1.36	13.13	7.47	64.01
1984	2.11	9.58	6.36	5.41	3.91	3.88	0.27	0.03	0.94	6.05	18.67	4.56	61.77
1985	0.31	5.15	5.65	0.49	1.53	2.51	1.37	0.04	2.13	4.83	6.31	3.51	33.83
1986	6.97	14.22	4.41	1.85	3.21	0.33	0.42	0.04	4.65	2.46	11.04	3.30	52.90
1987	9.66	4.47	2.81	2.04	2.00	0.07	3.00	0.26	0.20	0.24	4.65	15.40	44.80
1988	8.72	1.59	4.78	5.65	3.71	2.37	0.11	0.04	1.22	0.11	14.27	5.18	47.75
1989	6.68	3.45	10.93	1.73	3.85	1.03	0.49	1.21	0.64	2.95	5.00	2.70	40.66
1990	15.09	6.32	3.43	2.44	2.99	2.01	0.88	2.38	0.28	7.59	7.59	4.47	55.47
1991	3.73	3.58	9.79	5.22	6.03	0.99	0.64	0.68	0.10	1.93	10.47	5.28	48.44
1992	4.57	5.15	2.10	7.82	0.14	1.68	1.32	0.10	0.99	4.60	6.54	12.59	47.60
1993	6.85	2.64	8.58	7.85	6.92	3.70	1.14	1.80	T	1.45	1.98	10.82	53.73
Record Mean	7.46	5.39	5.29	3.03	2.22	1.43	0.47	0.84	1.40	3.68	7.33	7.86	46.41

AVERAGE TEMPERATURE (deg. F)

EUGENE, OREGON

YEAR	JAN	FEB	MAR	APR	MAY	JUNE	JULY	AUG	SEP	OCT	NOV	DEC	ANNUAL
1964	42.4	41.6	45.1	47.9	53.2	59.8	66.4	65.6	61.0	55.7	43.2	40.5	51.9
1965	40.8	44.3	47.8	51.8	53.6	60.7	68.0	67.4	60.8	55.5	49.4	39.2	53.3
1966	41.3	42.1	46.6	52.5	55.2	62.6	66.7	67.2	64.6	53.5	47.5	45.1	53.7
1967	44.4	44.6	45.0	45.9	55.7	64.8	68.8	72.0	66.0	54.4	46.9	42.4	54.2
1968	41.5	48.9	49.1	48.5	55.5	61.6	68.7	66.4	62.2	52.9	48.2	41.0	53.7
1969	36.9	40.7	47.6	50.2	60.1	65.3	66.5	64.6	61.3	50.6	45.6	43.1	52.7
1970	42.5	44.5	46.8	46.9	56.1	65.4	69.9	67.6	60.2	52.6	45.6	40.0	53.2
1971	40.1	41.1	43.1	48.3	54.5	57.8	66.7	68.2	59.2	49.5	43.5	38.7	50.9
1972	37.9	43.5	48.6	46.6	56.9	60.9	69.0	69.1	59.0	51.1	47.0	34.3	52.0
1973	38.3	44.5	44.9	49.9	55.9	61.7	67.7	64.0	62.7	52.7	44.3	43.4	52.5
1974	39.3	43.2	49.5	52.6	56.1	63.9	67.9	68.9	67.6	55.0	48.7	45.7	54.9
1975	45.9	43.5	44.7	46.4	55.5	61.1	68.5	64.5	66.1	55.0	46.2	43.0	53.4
1976	42.0	41.9	44.3	48.8	54.6	58.1	67.7	66.1	63.8	54.0	47.6	37.7	52.2
1977	38.9	44.5	44.9	50.7	51.3	61.0	64.2	69.7	59.4	52.5	43.5	43.0	52.0
1978	42.5	44.2	48.4	48.1	52.7	62.5	66.4	66.1	59.6	53.0	39.3	35.2	51.5
1979	31.5	42.9	48.8	50.4	55.5	60.0	67.3	65.5	63.6	56.5	42.4	43.0	52.3
1980	35.7	42.9	45.5	50.8	53.2	57.5	66.8	63.5	61.9	54.1	46.4	42.3	51.7
1981	40.4	43.4	45.9	49.9	54.4	59.5	64.8	68.2	62.0	50.2	45.4	41.8	52.2
1982	36.2	41.3	44.1	46.4	53.9	62.2	64.4	66.5	60.9	52.2	40.8	39.3	50.7
1983	41.7	45.1	48.3	49.1	56.1	58.6	63.0	66.1	59.1	52.1	46.6	36.3	51.9
1984	40.5	44.2	48.5	47.9	53.2	58.3	65.9	66.0	60.8	50.8	43.8	37.3	51.4
1985	34.2	40.6	42.8	52.0	54.9	61.4	69.9	65.8	58.3	51.8	37.6	33.0	50.2
1986	43.5	44.7	50.5	48.5	55.3	63.6	63.1	69.9	58.8	54.4	46.5	39.9	53.2
1987	39.6	43.7	47.3	52.4	57.9	63.3	64.4	67.6	62.7	56.5	46.8	39.8	53.5
1988	39.3	44.1	46.8	51.7	54.3	60.7	67.9	66.5	63.0	57.2	47.0	40.6	53.3
1989	42.1	35.8	47.2	55.6	56.5	63.1	65.3	66.7	64.6	53.9	47.4	41.1	53.3
1990	43.1	41.5	48.8	53.4	55.2	62.1	70.4	69.6	64.8	52.1	46.9	34.1	53.5
1991	38.9	48.7	44.4	48.8	52.7	57.5	67.5	67.7	65.1	54.3	47.6	41.3	52.9
1992	44.0	47.5	50.3	54.1	60.1	65.1	68.3	68.9	62.9	54.5	45.9	39.5	55.1
1993	35.3	38.5	49.6	51.2	58.6	60.5	62.5	66.8	62.7	56.1	38.5	39.3	51.6
Record Mean	39.4	43.2	46.1	50.0	55.5	61.1	66.5	66.4	62.0	53.3	45.1	40.5	52.5
Max	45.9	51.2	55.4	60.8	67.5	73.9	81.9	81.6	76.5	64.5	52.5	46.5	63.2
Min	32.9	35.2	36.8	39.3	43.4	48.3	51.1	51.3	47.4	42.0	37.7	34.5	41.7

REFERENCE NOTES

EUGENE, OREGON

GENERAL
T - TRACE AMOUNT.
BLANK ENTRIES DENOTE MISSING/UNREPORTED DATA.
# INDICATES A STATION OR INSTRUMENT RELOCATION.

SPECIFIC

EXCEPTIONS

RECORD MEANS ARE THROUGH THE CURRENT YEAR,
BEGINNING IN 1943 FOR TEMPERATURE
1943 FOR PRECIPITATION
1943 FOR SNOWFALL

(a) - LENGTH OF RECORD IN YEARS, ALTHOUGH INDIVIDUAL MONTHS MAY BE MISSING.
0.* OR * - THE VALUE IS BETWEEN 0.0 AND 0.05
NORMALS - BASED ON THE 1961-1990 RECORD PERIOD.
EXTREMES - DATES ARE THE MOST RECENT OCCURRENCE
WIND DIR. - NUMERALS SHOW TENS OF DEGREES CLOCKWISE FROM TRUE NORTH. "00" INDICATES CALM.
RESULTANT DIRECTIONS ARE GIVEN TO WHOLE DEGREES.
BOLD VALUES INDICATE EXTREME VALUES WHICH OCCURRED AFTER THE ASOS SYSTEM WAS COMMISSIONED.

RECORD = PERIOD OF RECORD
RECORD MEAN PRECIPITATION IS THE MEAN OF ALL DAILY PRECIPITATION AMOUNTS DURING THE PERIOD OF RECORD.
RECORD MAX(MIN) TEMPERATURE IS THE MEAN OF ALL DAILY MAX(MIN) TEMPERATURES DURING THE PERIOD OF RECORD.
RECORD MEAN TEMPERATURE IS THE SUM OF THE RECORD MAX AND RECORD MIN DIVIDED BY 2.
AVERAGE TEMPERATURE IS THE SUM OF THE MEAN DAILY MAX AND MIN TEMPERATURE DIVIDED BY 2.

HEATING DEGREE DAYS Base 65 deg. F EUGENE, OREGON

SEASON	JULY	AUG	SEP	OCT	NOV	DEC	JAN	FEB	MAR	APR	MAY	JUNE	TOTAL
1964-65	40	45	120	282	646	753	744	572	522	389	350	138	4601
1965-66	31	17	132	288	461	791	728	634	565	368	295	108	4418
1966-67	25	31	53	352	517	611	633	566	614	564	285	64	4315
1967-68	5	1	28	322	538	693	722	461	486	487	287	123	4153
1968-69	9	39	100	366	497	738	867	673	536	438	160	44	4467
1969-70	17	44	134	440	576	673	692	568	557	534	272	75	4582
1970-71	11	11	135	374	576	769	767	664	671	496	318	212	5004
1971-72	60	23	169	474	638	811	834	617	503	547	256	123	5055
1972-73	23	27	202	424	536	945	825	567	615	444	276	132	5016
1973-74	29	77	90	374	614	666	790	604	472	364	270	84	4434
1974-75	29	16	25	305	480	589	585	595	623	548	294	138	4227
1975-76	18	56	40	302	557	676	707	664	636	478	316	210	4660
1976-77	10	38	58	340	518	840	801	567	615	422	417	126	4752
1977-78	68	20	178	385	637	675	690	576	508	503	375	102	4717
1978-79	52	48	165	366	763	919	1029	614	496	432	288	159	5331
1979-80	33	19	50	257	673	671	901	634	599	421	361	219	4838
1980-81	29	65	101	344	551	697	752	599	585	446	321	164	4654
1981-82	64	33	121	450	580	710	887	657	642	551	335	124	5154
1982-83	67	20	137	392	719	791	716	553	511	470	287	189	4852
1983-84	78	15	171	393	544	879	753	596	502	504	359	207	5001
1984-85	33	19	145	440	628	851	948	674	678	383	311	128	5238
1985-86	5	43	197	405	817	983	662	562	444	489	321	93	5021
1986-87	74	3	213	319	548	773	784	589	539	370	245	105	4562
1987-88	60	17	94	262	537	774	790	600	561	395	327	136	4553
1988-89	29	31	108	244	533	749	705	811	544	278	271	87	4390
1989-90	31	21	65	336	522	733	674	654	498	340	296	106	4276
1990-91	16	10	39	394	532	952	803	449	632	479	376	220	4902
1991-92	17	25	59	325	515	729	643	502	448	326	165	66	3820
1992-93	7	14	91	319	566	782	912	732	472	406	191	143	4635
1993-94	76	36	108	274	787	792							

COOLING DEGREE DAYS Base 65 deg. F EUGENE, OREGON

YEAR	JAN	FEB	MAR	APR	MAY	JUNE	JULY	AUG	SEP	OCT	NOV	DEC	TOTAL
1969	0	0	0	0	12	60	72	37	33	0	0	0	214
1970	0	0	0	0	4	93	173	99	1	0	0	0	370
1971	0	0	0	0	0	3	119	129	2	0	0	0	253
1972	0	0	0	0	11	6	156	159	26	0	0	0	358
1973	0	0	0	0	2	38	119	50	26	0	0	0	235
1974	0	0	0	0	3	57	127	143	110	0	0	0	440
1975	0	0	0	0	3	25	135	47	82	1	0	0	293
1976	0	0	0	0	3	12	101	82	30	3	0	0	231
1977	0	0	0	0	0	10	52	177	12	0	0	0	251
1978	0	0	0	0	2	34	101	88	10	0	0	0	235
1979	0	0	0	0	0	16	111	40	15	0	0	0	182
1980	0	0	0	0	0	0	91	23	16	12	0	0	142
1981	0	0	0	1	0	4	63	138	38	0	0	0	244
1982	0	0	0	0	0	51	59	73	20	0	0	0	203
1983	0	0	0	0	17	3	23	56	0	0	0	0	99
1984	0	0	0	0	2	14	69	55	23	5	0	0	168
1985	0	0	0	0	4	28	166	73	2	0	0	0	273
1986	0	0	0	0	26	55	20	165	34	0	0	0	300
1987	0	0	0	0	32	59	48	104	30	4	0	0	277
1988	0	0	0	0	2	15	127	85	53	8	0	0	290
1989	0	0	0	1	14	38	47	81	60	0	0	0	241
1990	0	0	0	0	0	26	192	157	42	0	0	0	417
1991	0	0	0	0	0	2	101	117	71	1	0	0	292
1992	0	0	0	2	21	75	115	144	34	0	0	0	391
1993	0	0	0	0	0	12	9	95	46	5	0	0	167

SNOWFALL (inches) EUGENE, OREGON

SEASON	JULY	AUG	SEP	OCT	NOV	DEC	JAN	FEB	MAR	APR	MAY	JUNE	TOTAL
1964-65	0.0	0.0	0.0	0.0	T	10.2	4.9	T	T	0.0	0.0	0.0	15.1
1965-66	0.0	0.0	0.0	0.0	0.0	4.8	0.1	0.3	2.5	0.0	0.0	0.0	7.7
1966-67	0.0	0.0	0.0	0.0	0.0	0.0	T	0.7	T	T	0.0	0.0	0.7
1967-68	0.0	0.0	0.0	0.0	0.0	1.1	3.6	0.0	0.0	0.0	0.0	0.0	4.7
1968-69	0.0	0.0	0.0	0.0	T	2.3	47.1	T	0.0	0.0	0.0	0.0	49.4
1969-70	0.0	0.0	0.0	0.0	0.0	0.0	0.2	T	T	T	0.0	0.0	0.2
1970-71	0.0	0.0	0.0	0.0	T	0.1	18.9	4.5	0.8	T	0.0	0.0	24.3
1971-72	0.0	0.0	T	T	0.0	1.2	1.2	T	0.5	T	0.0	0.0	2.9
1972-73	0.0	0.0	0.0	0.0	0.0	9.9	1.1	0.0	T	T	0.0	0.0	11.0
1973-74	0.0	0.0	0.0	0.0	1.5	T	T	0.4	T	0.0	T	0.0	1.9
1974-75	0.0	0.0	0.0	0.0	0.0	T	4.0	T	T	T	T	0.0	4.0
1975-76	0.0	0.0	0.0	0.0	T	2.8	T	T	1.3	T	0.0	0.0	4.1
1976-77	0.0	0.0	0.0	0.0	0.0	0.0	0.0	T	T	0.0	T	0.0	T
1977-78	0.0	0.0	0.0	0.0	0.2	0.0	T	T	0.0	T	0.0	0.0	0.2
1978-79	0.0	0.0	0.0	0.0	1.0	T	T	T	0.0	T	0.0	0.0	1.0
1979-80	0.0	0.0	0.0	0.0	0.0	T	T	0.0	T	0.0	0.0	0.0	T
1980-81	0.0	0.0	0.0	0.0	0.0	T	0.0	0.0	T	T	T	T	T
1981-82	0.0	0.0	0.0	0.0	T	T	5.0	2.1	T	T	0.0	0.0	7.1
1982-83	0.0	0.0	0.0	T	0.0	T	0.0	0.0	0.0	T	0.0	0.0	T
1983-84	0.0	0.0	0.0	0.0	T	3.7	T	T	T	T	T	0.0	3.7
1984-85	0.0	0.0	0.0	T	0.0	T	T	0.2	T	T	0.0	0.0	0.2
1985-86	0.0	0.0	0.0	0.0	2.3	4.7	0.0	2.8	0.0	T	T	0.0	9.8
1986-87	0.0	0.0	T	0.0	0.0	0.0	0.0	0.0	T	0.0	0.0	0.0	T
1987-88	0.0	0.0	0.0	0.0	0.0	6.0	0.7	1.0	0.0	T	0.0	0.0	7.7
1988-89	0.0	0.0	0.0	0.0	0.0	0.3	T	5.6	T	0.0	0.0	0.0	5.9
1989-90	0.0	0.0	0.0	0.0	T	0.0	T	8.8	0.0	0.0	0.0	0.0	8.8
1990-91	0.0	0.0	0.0	0.0	0.0	3.6	T	0.0	T	T	0.0	0.0	3.6
1991-92	0.0	0.0	0.0	0.0	0.0	0.0	0.0	0.0	0.0	T	0.0	0.0	T
1992-93	0.0	0.0	0.0	0.0	0.0	3.3	6.7	8.2	0.0	0.0	0.0	0.0	18.2
1993-94	0.0	0.0	0.0	0.0	T	0.0							
Record Mean	0.0	0.0	T	T	0.3	1.3	3.7	0.8	0.5	T	T	T	6.5

MEDFORD, OREGON

Medford is located in a mountain valley formed by the famous Rogue River and one of its tributaries, Bear Creek. The major portion of the valley ranges in elevation from 1,300 to 1,400 feet above sea level. Mountains surround the valley on all sides, to the east the Cascades, ranging up to 9,500 feet, to the south the Siskiyous, ranging up to 7,600 feet, and to the west and north, the Coast Range and Umpqua Divide, ranging up to 5,500 feet above sea level. The valley exits to the ocean 80 miles westward through the narrow canyon of the Rogue River.

Medford has a moderate climate of marked seasonal characteristics. Late fall, winter, and early spring months are damp, cloudy, and cool under the influence of marine air. Late spring, summer, and early fall are warm, dry, and sunny, due to the dry continental nature of the prevailing winds aloft that cross this area.

The rain shadow afforded by the Siskiyous and Coast Range results in a relatively light annual rainfall, most of which falls during the winter season. Summertime rainfall is brought by thunderstorm activity. Snowfall is quite heavy in the surrounding mountains during the winter, providing excellent skiing. The mountains provide irrigation water storage which is necessary for production of most commercial crops during the dry summer. Valley snowfall is light. Individual accumulations of snow seldom last more than 24 hours and present little hindrance to transportation on the valley floor.

Few extremes of temperature occur. High temperatures in the summer months average slightly below 90 degrees. High temperatures are always accompanied by low humidity, and hot days give way to cool nights as cool air drains down the mountain slopes into the valley. The length of the growing season is 170 days, from late April to mid-October. The last date of 32 degrees in the spring normally occurs in mid-June and the first date of 32 degrees in the fall occurs in mid-September.

Valley winds are usually very light, prevailing from the north or northwest much of the year. Winds exceeding 10 mph during the winter months nearly always come from the southerly quadrant. Highest velocities are reached when a well developed storm off the northern California coast causes a foehn or chinook wind off the Siskiyou Mountains to the south, speeds to 50 mph are common, and gusts to 70 mph have been recorded occasionally. Summer thunderstorms produce gusty winds to 40 or 50 mph which may come from any direction.

Fog often fills the lower portion of the valley during the winter and early spring months, when rapid clearing of the sky after a storm allows nocturnal cooling of the entrapped, moist air to the saturation point. Duration of the fog is seldom more than three days. Geographical and meteorological conditions contribute to a smoke problem during the fall, winter and early spring months. Smoke, from local sources, occasionally reduces visibility to 1 to 3 miles under stable conditions.

NORMALS, MEANS, AND EXTREMES

MEDFORD, OREGON

LATITUDE: 42°22'N LONGITUDE: 122°52'W ELEVATION: FT. GRND 1298 BARO 1304 TIME ZONE: PACIFIC WBAN: 24225

	(a)	JAN	FEB	MAR	APR	MAY	JUNE	JULY	AUG	SEP	OCT	NOV	DEC	YEAR
TEMPERATURE °F:														
Normals														
-Daily Maximum		45.7	53.3	58.5	64.6	72.9	82.1	90.5	89.9	82.8	69.4	52.6	44.3	67.2
-Daily Minimum		30.4	32.2	35.4	38.0	43.4	50.7	55.2	55.1	48.2	40.4	35.5	31.2	41.3
-Monthly		38.1	42.8	47.0	51.3	58.2	66.5	72.9	72.5	65.5	54.9	44.1	37.8	54.3
Extremes														
-Record Highest	64	71	79	86	93	103	111	115	114	110	99	75	72	115
-Year		1981	1992	1930	1987	1986	1992	1946	1981	1988	1980	1991	1962	JUL 1946
-Record Lowest	64	-3	6	16	21	28	31	38	39	29	18	10	-6	-6
-Year		1930	1950	1956	1936	1968	1952	1976	1962	1950	1971	1978	1972	DEC 1972
NORMAL DEGREE DAYS:														
Heating (base 65°F)		834	622	558	416	219	60	8	11	93	320	627	843	4611
Cooling (base 65°F)		0	0	0	0	8	105	253	244	108	7	0	0	725
% OF POSSIBLE SUNSHINE														
MEAN SKY COVER (tenths)														
Sunrise - Sunset	64	8.2	7.6	7.3	6.7	5.8	4.7	2.1	2.5	3.3	5.5	7.8	8.6	5.8
MEAN NUMBER OF DAYS:														
Sunrise to Sunset														
-Clear	64	2.7	3.6	5.2	5.8	9.3	12.9	22.9	21.5	17.8	10.5	3.4	2.0	117.6
-Partly Cloudy	64	4.9	5.8	6.6	8.3	8.9	8.0	5.4	5.8	6.5	8.2	6.2	4.1	78.7
-Cloudy	64	23.4	18.8	19.2	15.8	12.8	9.1	2.6	3.8	5.7	12.3	20.5	24.9	169.0
Precipitation														
.01 inches or more	64	13.4	11.3	11.7	9.2	8.2	5.2	1.7	2.1	4.0	7.5	12.2	14.0	100.6
Snow,Ice pellets,hail														
1.0 inches or more	64	1.1	0.5	0.3	0.1	0.0	0.0	0.0	0.0	0.0	0.*	0.1	0.5	2.6
Thunderstorms	64	0.*	0.1	0.2	0.8	1.7	1.8	1.5	1.2	0.9	0.3	0.*	0.*	8.6
Heavy Fog Visibility														
1/4 mile or less	64	11.8	5.8	1.6	0.5	0.2	0.2	0.*	0.1	0.5	4.4	10.7	13.8	49.6
Temperature °F														
-Maximum														
90° and above	32	0.0	0.0	0.0	0.1	2.0	7.9	17.3	17.3	9.0	1.2	0.0	0.0	54.7
32° and below	32	0.9	0.*	0.0	0.0	0.0	0.0	0.0	0.0	0.0	0.0	0.*	1.8	2.8
-Minimum														
32° and below	32	19.8	14.9	10.9	6.1	0.8	0.0	0.0	0.0	0.2	3.8	11.1	17.0	84.7
0° and below	32	0.*	0.0	0.0	0.0	0.0	0.0	0.0	0.0	0.0	0.0	0.0	0.2	0.2
AVG. STATION PRESS.(mb)	21	972.0	970.2	968.6	970.0	969.1	968.6	968.5	967.9	968.4	970.5	971.8	972.6	969.9
RELATIVE HUMIDITY (%)														
Hour 04	32	90	88	86	84	83	79	74	75	79	86	92	91	84
Hour 10 (Local Time)	32	88	83	73	63	56	48	45	47	53	70	87	90	67
Hour 16	32	71	57	50	44	39	33	26	27	29	43	67	76	47
Hour 22	32	87	82	75	69	65	59	51	53	60	76	88	89	71
PRECIPITATION (inches):														
Water Equivalent														
-Normal		2.69	1.93	1.82	1.16	1.00	0.58	0.26	0.52	0.86	1.49	3.23	3.32	18.86
-Maximum Monthly	64	6.67	5.67	5.54	3.07	4.58	3.49	1.63	2.83	4.22	9.16	8.62	12.72	12.72
-Year		1936	1983	1957	1965	1945	1931	1966	1976	1977	1950	1942	1964	DEC 1964
-Minimum Monthly	64	0.19	0.20	0.29	0.16	T	0.00	0.00	0.00	0.00	T	0.01	0.36	0.00
-Year		1984	1988	1969	1949	1982	1951	1970	1981	1974	1987	1936	1976	AUG 1981
-Maximum in 24 hrs	64	3.17	2.96	1.61	1.05	1.67	1.96	1.07	1.13	3.09	2.92	2.99	3.75	3.75
-Year		1943	1956	1972	1965	1956	1931	1966	1945	1977	1950	1953	1964	DEC 1964
Snow,Ice pellets,hail														
-Maximum Monthly	64	22.6	11.6	8.1	4.2	0.1	T	T	0.0	0.0	1.3	11.4	12.2	22.6
-Year		1930	1956	1956	1953	1988	1991	1992			1956	1955	1972	JAN 1930
-Maximum in 24 hrs	64	9.3	5.2	7.9	4.2	0.1	T	T	0.0	0.0	1.3	8.5	4.2	9.3
-Year		1971	1956	1956	1953	1988	1991	1992			1956	1977	1964	JAN 1971
WIND:														
Mean Speed (mph)	44	4.1	4.5	5.2	5.7	5.7	5.9	5.8	5.3	4.5	3.7	3.6	3.7	4.8
Prevailing Direction through 1963		SSE	S	NNW	WNW	WNW	WNW	WNW	WNW	WNW	S	N	N	WNW
Fastest Obs. 1 Min.														
-Direction (!!)	44	23	25	16	14	12	17	07	16	13	20	19	14	16
-Speed (MPH)	44	50	46	55	35	38	37	44	48	40	40	40	44	55
-Year		1950	1958	1952	1965	1966	1986	1958	1956	1982	1962	1981	1965	MAR 1952
Peak Gust														
-Direction (!!)	10	SE	S	SE	NW	SW	S	N	SE	W	SE	SE	SE	SE
-Speed (mph)	10	47	45	53	38	39	53	46	36	36	35	52	56	56
-Date		1992	1986	1987	1991	1985	1986	1991	1991	1984	1989	1984	1985	DEC 1985

PRECIPITATION (inches)

MEDFORD, OREGON

YEAR	JAN	FEB	MAR	APR	MAY	JUNE	JULY	AUG	SEP	OCT	NOV	DEC	ANNUAL
1964	5.60	0.21	2.71	0.37	0.82	0.79	0.97	0.10	0.15	0.90	3.75	12.72	29.09
1965	4.30	0.70	0.41	3.07	0.31	1.05	0.03	1.52	T	0.46	2.56	3.71	18.12
1966	4.80	0.37	1.70	0.45	0.20	0.37	1.63	0.19	1.88	0.76	5.89	2.80	21.04
1967	5.44	1.14	2.08	1.72	0.96	0.27	T	T	0.28	2.34	1.04	3.40	18.67
1968	1.86	2.95	0.90	0.38	1.05	0.06	T	1.33	0.32	0.62	3.04	2.78	15.29
1969	6.16	1.46	0.29	0.60	1.62	1.31	0.02	0.00	0.62	2.46	0.49	5.44	20.47
1970	6.19	1.70	1.13	1.44	0.34	0.59	0.00	0.34	0.22	1.39	6.57	3.36	23.27
1971	3.68	1.43	2.72	1.34	1.13	0.97	0.07	0.28	1.24	0.61	3.43	2.45	19.35
1972	3.55	2.49	3.62	0.94	1.61	1.59	T	0.36	0.52	1.21	1.50	3.23	20.62
1973	1.98	0.54	1.58	0.76	0.45	0.06	0.04	0.03	0.64	2.79	7.01	3.02	18.90
1974	4.32	2.78	3.26	1.70	0.22	T	0.10	T	0.00	1.17	1.13	3.91	18.59
1975	2.64	2.64	3.97	1.27	0.24	0.38	0.22	0.54	0.65	2.21	1.85	2.74	19.35
1976	1.62	2.21	1.13	1.67	0.11	0.04	0.84	2.83	0.90	0.18	0.43	0.36	12.32
1977	1.17	0.67	1.12	0.81	2.37	0.53	0.23	0.36	4.22	0.96	4.91	4.81	22.16
1978	1.53	2.45	2.03	1.26	1.59	1.02	0.54	1.46	1.68	0.01	1.50	0.66	15.73
1979	2.81	1.54	0.83	2.24	1.42	0.55	0.02	0.63	0.32	3.98	3.17	2.73	20.24
1980	2.59	1.78	1.27	1.75	0.69	1.22	0.02	0.00	0.18	1.52	2.28	2.59	15.89
1981	0.54	1.72	1.23	0.55	1.17	0.47	0.41	0.00	0.52	1.23	6.05	8.02	21.91
1982	1.43	3.64	2.30	0.87	T	0.85	0.07	0.03	0.97	1.60	2.17	5.31	19.24
1983	0.92	5.67	3.21	1.12	0.81	0.66	0.59	2.21	2.05	1.21	4.97	6.73	30.15
1984	0.19	2.50	2.05	1.11	0.39	0.79	0.16	0.40	0.51	1.93	6.56	1.96	18.55
1985	0.23	1.58	1.22	0.39	1.00	0.37	T	0.02	1.53	1.50	2.02	0.83	10.69
1986	1.99	5.22	1.02	0.23	1.19	0.45	T	T	2.31	1.49	2.45	0.72	17.07
1987	2.89	2.24	1.34	0.45	0.95	0.12	1.34	T	0.00	T	1.68	3.77	14.78
1988	2.53	0.20	0.57	1.07	1.51	1.04	0.00	0.02	0.22	0.12	5.14	1.28	13.70
1989	2.33	0.78	3.94	2.42	1.01	0.16	T	0.41	1.94	0.71	0.71	0.68	15.09
1990	2.94	1.06	1.49	0.82	1.86	0.17	0.11	0.99	0.13	1.29	1.52	1.12	13.50
1991	1.55	1.73	2.42	1.07	1.84	0.68	1.10	0.22	T	0.39	2.42	1.08	14.50
1992	0.84	0.63	0.42	1.10	1.30	2.62	0.58	T	0.06	2.37	1.54	3.52	14.98
1993	2.65	1.37	1.25	1.83	2.63	1.23	0.66	1.21	T	0.66	0.68	2.43	16.60
Record Mean	2.70	2.08	1.65	1.17	1.20	0.82	0.27	0.32	0.71	1.58	2.80	3.10	18.42

AVERAGE TEMPERATURE (deg. F)

MEDFORD, OREGON

YEAR	JAN	FEB	MAR	APR	MAY	JUNE	JULY	AUG	SEP	OCT	NOV	DEC	ANNUAL
1964	35.2	37.9	43.6	47.7	54.9	63.3	70.9	69.6	61.1	55.9	41.1	41.5	51.9
1965	39.0	42.7	48.7	53.4	56.5	64.5	72.9	70.5	62.2	57.4	46.6	35.0	54.1
1966	38.2	39.9	46.8	53.3	59.4	63.8	68.7	71.3	65.2	53.5	45.7	40.6	53.9
1967	39.6	40.5	44.4	45.8	58.9	68.4	75.5	78.0	69.2	53.6	45.5	35.3	54.6
1968	36.6	47.6	48.1	49.3	57.9	68.8	74.1	69.8	68.2	56.7	44.8	38.2	55.0
1969	35.7	40.9	45.5	49.5	62.4	67.9	72.3	69.9	64.9	49.6	42.5	39.7	53.4
1970	42.5	44.0	47.0	46.2	57.9	68.7	75.8	73.7	62.7	53.1	47.4	38.1	54.8
1971	36.1	40.8	44.9	48.4	55.8	61.4	73.1	74.1	62.6	51.1	41.5	36.1	52.2
1972	36.0	43.2	49.8	48.0	57.9	65.4	74.9	73.0	60.8	52.4	44.8	31.5	53.2
1973	37.9	46.2	45.0	52.1	62.4	67.5	75.2	72.0	66.9	53.4	45.8	43.6	55.7
1974	38.7	39.7	44.8	47.9	56.4	67.7	71.3	73.0	69.7	54.0	44.4	38.9	53.9
1975	36.6	41.8	43.3	45.1	57.4	63.9	72.3	68.4	68.9	51.5	41.3	39.3	52.5
1976	38.8	40.2	42.8	48.0	57.3	61.7	71.5	67.3	65.5	54.1	43.6	31.7	51.9
1977	34.0	44.0	43.0	53.0	53.3	69.7	72.0	76.0	63.8	53.1	42.3	42.4	53.9
1978	44.0	44.5	50.5	49.7	56.4	66.4	72.6	70.9	59.1	56.9	39.2	33.4	53.6
1979	35.5	41.9	49.2	51.1	57.6	65.9	72.6	70.3	68.3	59.1	42.4	39.8	54.5
1980	38.7	45.6	44.8	52.2	56.3	61.3	73.1	68.9	65.4	55.8	43.8	38.1	53.7
1981	41.7	42.1	46.9	52.8	57.2	65.5	71.2	74.9	66.6	52.2	45.9	42.8	55.0
1982	34.6	41.2	44.5	48.8	58.0	67.1	71.6	71.9	62.8	54.9	42.1	38.9	53.0
1983	39.7	44.6	48.9	49.1	59.6	64.2	68.6	73.0	63.6	54.3	43.7	40.4	54.2
1984	38.8	43.0	48.9	49.3	58.3	63.3	73.8	71.8	65.2	51.5	44.5	36.3	53.7
1985	37.6	42.1	44.1	56.2	57.9	68.2	76.4	71.0	61.0	53.1	38.1	36.1	53.5
1986	44.0	46.4	51.6	51.4	59.1	70.4	70.5	77.1	60.8	55.6	44.7	37.8	55.8
1987	38.3	44.1	47.9	56.6	62.3	70.4	70.0	74.2	66.6	61.2	46.7	39.4	56.5
1988	39.7	44.8	47.2	53.8	56.6	65.5	75.5	73.6	67.4	61.9	45.2	38.3	55.8
1989	37.8	38.3	47.4	57.3	58.4	67.6	70.2	70.7	65.8	54.6	43.0	35.5	53.9
1990	39.9	41.1	49.0	57.5	57.0	66.4	76.3	73.5	69.5	53.5	42.2	31.7	54.8
1991	37.8	47.7	45.9	49.6	54.8	62.9	74.2	73.6	70.7	58.0	46.8	38.3	55.0
1992	40.9	48.4	51.9	56.5	66.3	69.7	74.2	75.7	67.1	57.3	45.4	37.6	57.6
1993	38.5	41.8	51.4	51.5	60.5	63.7	64.8	70.4	67.3	58.5	39.9	38.9	53.9
Record Mean	37.8	42.6	46.6	51.6	58.1	65.3	72.2	71.5	64.8	54.4	43.8	38.0	53.9
Max	45.4	52.9	58.4	65.0	72.8	81.0	90.0	89.4	82.3	69.0	53.3	44.7	67.0
Min	30.2	32.3	34.8	38.2	43.4	49.6	54.3	53.5	47.2	39.7	34.3	31.3	40.7

REFERENCE NOTES

MEDFORD, OREGON

GENERAL
T - TRACE AMOUNT.
BLANK ENTRIES DENOTE MISSING/UNREPORTED DATA.
# INDICATES A STATION OR INSTRUMENT RELOCATION.

SPECIFIC

(a) - LENGTH OF RECORD IN YEARS, ALTHOUGH INDIVIDUAL MONTHS MAY BE MISSING.
0.* OR * - THE VALUE IS BETWEEN 0.0 AND 0.05
NORMALS - BASED ON THE 1961-1990 RECORD PERIOD.
EXTREMES - DATES ARE THE MOST RECENT OCCURRENCE
WIND DIR. - NUMERALS SHOW TENS OF DEGRESS CLOCKWISE FROM TRUE NORTH. "00" INDICATES CALM.
RESULTANT DIRECTIONS ARE GIVEN TO WHOLE DEGREES.
BOLD VALUES INDICATE EXTREME VALUES WHICH OCCURRED AFTER THE ASOS SYSTEM WAS COMMISSIONED.

RECORD = PERIOD OF RECORD
RECORD MEAN PRECIPITATION IS THE MEAN OF ALL DAILY PRECIPITATION AMOUNTS DURING THE PERIOD OF RECORD.
RECORD MAX(MIN) TEMPERATURE IS THE MEAN OF ALL DAILY MAX(MIN) TEMPERATURES DURING THE PERIOD OF RECORD.
RECORD MEAN TEMPERATURE IS THE SUM OF THE RECORD MAX AND RECORD MIN DIVIDED BY 2.
AVERAGE TEMPERATURE IS THE SUM OF THE MEAN DAILY MAX AND MIN TEMPERATURE DIVIDED BY 2.

EXCEPTIONS

RECORD MEANS ARE THROUGH THE CURRENT YEAR, BEGINNING IN 1912 FOR TEMPERATURE
1912 FOR PRECIPITATION
1930 FOR SNOWFALL

HEATING DEGREE DAYS Base 65 deg. F MEDFORD, OREGON

SEASON	JULY	AUG	SEP	OCT	NOV	DEC	JAN	FEB	MAR	APR	MAY	JUNE	TOTAL
1964-65	8	23	120	284	710	723	802	621	498	346	267	78	4480
1965-66	4	7	106	231	544	923	823	696	557	345	182	93	4511
1966-67	26	11	53	352	571	749	781	678	634	569	206	50	4680
1967-68	0	0	10	344	578	913	873	500	515	466	218	48	4465
1968-69	0	31	56	252	599	826	900	670	599	459	133	37	4562
1969-70	0	9	68	471	670	780	691	583	549	558	227	68	4674
1970-71	0	0	89	363	520	827	890	674	617	489	280	121	4870
1971-72	16	12	108	424	698	889	891	624	465	505	237	54	4923
1972-73	2	11	178	385	599	1031	835	519	616	379	139	48	4742
1973-74	0	12	28	352	569	654	809	703	620	504	267	30	4548
1974-75	20	6	9	333	611	803	873	642	663	590	249	86	4885
1975-76	18	18	14	416	706	789	807	712	679	503	238	127	5027
1976-77	9	31	48	338	636	1023	952	581	675	358	359	17	5027
1977-78	14	4	105	360	673	694	646	571	444	451	262	44	4268
1978-79	4	15	180	244	768	973	905	639	483	410	223	58	4902
1979-80	8	2	5	203	668	776	807	556	620	376	266	121	4408
1980-81	1	9	56	313	630	827	715	635	553	369	244	57	4409
1981-82	20	0	89	392	566	682	935	660	628	479	223	63	4737
1982-83	13	8	114	310	682	803	775	566	491	474	235	68	4539
1983-84	26	3	77	322	632	757	806	633	491	467	219	108	4541
1984-85	0	0	79	416	608	881	842	635	642	257	225	37	4622
1985-86	0	4	135	368	801	890	644	515	407	407	251	29	4451
1986-87	11	0	197	286	602	833	825	579	524	255	144	32	4288
1987-88	21	0	26	136	540	786	780	579	544	330	268	106	4116
1988-89	1	0	62	115	586	822	837	742	538	228	222	24	4177
1989-90	7	7	40	313	651	906	772	663	489	221	251	55	4375
1990-91	3	5	0	351	680	1022	837	477	588	456	313	82	4814
1991-92	1	1	11	246	541	820	738	476	399	253	48	43	3577
1992-93	0	1	35	233	583	844	816	643	411	397	150	90	4203
1993-94	42	15	65	216	747	804							

COOLING DEGREE DAYS Base 65 deg. F MEDFORD, OREGON

YEAR	JAN	FEB	MAR	APR	MAY	JUNE	JULY	AUG	SEP	OCT	NOV	DEC	TOTAL
1969	0	0	0	0	58	131	235	168	70	0	0	0	662
1970	0	0	0	0	15	185	343	278	28	3	0	0	852
1971	0	0	0	0	2	16	275	302	44	2	0	0	641
1972	0	0	0	0	27	70	315	265	58	0	0	0	735
1973	0	0	0	0	64	132	323	233	92	0	0	0	844
1974	0	0	0	0	7	120	224	262	155	0	0	0	768
1975	0	0	0	0	21	60	250	129	140	2	0	0	602
1976	0	0	0	0	6	34	215	110	66	5	0	0	436
1977	0	0	0	2	2	165	239	352	77	0	0	0	837
1978	0	0	0	0	4	93	244	205	11	0	0	0	557
1979	0	0	0	0	2	95	251	172	108	30	0	0	658
1980	0	0	0	0	2	16	257	134	78	35	0	0	522
1981	0	0	0	11	8	78	216	313	144	0	0	0	770
1982	0	0	0	0	14	136	223	228	56	1	0	0	658
1983	0	0	0	0	74	49	144	259	43	0	0	0	569
1984	0	0	0	0	20	64	280	219	90	5	0	0	678
1985	0	0	0	0	10	141	360	196	21	5	0	0	733
1986	0	0	0	5	74	198	189	383	80	0	0	0	929
1987	0	0	0	11	66	203	185	292	81	24	0	0	862
1988	0	0	0	0	12	128	334	276	143	27	0	0	920
1989	0	0	0	4	24	109	176	190	71	0	0	0	574
1990	0	0	0	2	13	99	358	275	142	1	0	0	890
1991	0	0	0	0	0	26	292	271	189	37	0	0	815
1992	0	0	0	5	96	189	292	338	102	2	0	0	1024
1993	0	0	0	0	17	58	42	188	140	20	0	0	465

SNOWFALL (inches) MEDFORD, OREGON

SEASON	JULY	AUG	SEP	OCT	NOV	DEC	JAN	FEB	MAR	APR	MAY	JUNE	TOTAL
1964-65	0.0	0.0	0.0	0.0	0.8	9.3	13.2	T	0.0	0.0	T	0.0	23.3
1965-66	0.0	0.0	0.0	0.0	T	9.0	3.4	T	2.0	0.0	0.0	0.0	14.4
1966-67	0.0	0.0	0.0	0.0	T	T	1.1	3.4	0.4	1.6	T	0.0	6.5
1967-68	0.0	0.0	0.0	0.0	0.0	4.3	6.4	0.0	0.0	T	T	0.0	10.7
1968-69	0.0	0.0	0.0	0.0	T	2.0	13.7	3.4	T	0.0	0.0	0.0	19.1
1969-70	0.0	0.0	0.0	0.0	T	0.4	0.2	1.5	T	0.8	T	0.0	2.9
1970-71	0.0	0.0	0.0	0.0	T	0.1	10.5	2.3	5.0	T	0.0	0.0	17.9
1971-72	0.0	0.0	0.0	T	T	2.8	0.3	0.9	2.6	T	T	0.0	6.6
1972-73	0.0	0.0	0.0	0.0	0.0	12.2	2.7	0.0	T	T	0.0	0.0	14.9
1973-74	0.0	0.0	0.0	0.0	T	T	T	5.2	0.3	0.0	T	0.0	5.5
1974-75	0.0	0.0	0.0	0.0	0.0	5.4	6.6	1.9	T	0.4	T	0.0	14.3
1975-76	0.0	0.0	0.0	0.0	0.4	2.3	2.1	1.0	0.2	T	0.0	0.0	6.0
1976-77	0.0	0.0	0.0	0.0	0.0	0.1	1.0	T	T	T	T	0.0	1.1
1977-78	0.0	0.0	0.0	0.0	8.5	0.2	0.0	0.1	0.0	T	0.0	0.0	8.8
1978-79	0.0	0.0	0.0	0.0	T	T	0.6	T	T	T	T	0.0	0.6
1979-80	0.0	0.0	0.0	0.0	T	0.1	0.3	T	0.1	T	0.0	0.0	0.5
1980-81	0.0	0.0	0.0	0.0	0.0	0.4	T	T	T	0.0	0.0	0.0	0.4
1981-82	0.0	0.0	0.0	0.0	0.0	3.0	4.1	T	1.8	T	0.0	0.0	8.9
1982-83	0.0	0.0	0.0	0.0	T	T	T	0.0	0.0	T	0.0	0.0	T
1983-84	0.0	0.0	0.0	0.0	T	3.4	T	T	0.0	T	T	0.0	3.4
1984-85	0.0	0.0	0.0	T	0.1	0.9	1.8	0.4	2.0	T	0.0	0.0	5.2
1985-86	0.0	0.0	0.0	0.0	0.5	T	0.0	0.1	0.0	T	0.0	0.0	0.6
1986-87	0.0	0.0	0.0	0.0	0.0	T	T	T	T	T	0.0	0.0	T
1987-88	0.0	0.0	0.0	0.0	T	3.8	8.0	T	0.0	T	0.1	0.0	11.9
1988-89	0.0	0.0	0.0	0.0	T	3.4	3.6	1.5	2.2	0.0	0.0	0.0	10.7
1989-90	0.0	0.0	0.0	0.0	T	0.8	0.5	2.2	0.3	0.0	0.0	0.0	3.8
1990-91	0.0	0.0	0.0	0.0	0.0	2.6	T	0.0	0.6	T	T	T	3.2
1991-92	0.0	0.0	0.0	0.0	0.0	T	T	0.0	0.0	0.0	0.0	0.0	T
1992-93	T	0.0	0.0	0.0	0.0	1.0	1.2	1.9	0.0	0.0	0.0	0.0	4.1
1993-94	0.0	0.0	0.0	0.0	0.1	T							
Record Mean	T	0.0	0.0	T	0.5	1.4	3.3	1.2	0.7	0.2	T	T	7.3

PENDLETON, OREGON

Pendleton is located in the southeastern part of the Columbia Basin, that low country of northern Oregon and central and eastern Washington which is almost entirely surrounded by mountains. This Basin is bounded on the south by the high country of central Oregon, on the north by the mountains of western Canada, on the west by the Cascade Range and on the east by the Blue Mountains and the north Idaho plateau. The gorge in the Cascades through which the Columbia River reaches the Pacific is the most important break in the barriers surrounding this basin. These physical features have important influences on the general climate of Pendleton and the surrounding territory.

The Weather Service Office at Pendleton Airport is located in rolling country which slopes generally upward toward the Blue Mountains about 15 miles to the east and southeast. The Columbia River approaches the area from the northwest to its junction with the Walla Walla River at an elevation of 351 feet and some 25 miles north of Pendleton, then turns southwestward to be joined a few miles below by the Umatilla River. Both the Walla Walla and Umatilla Rivers have their sources in the Blue Mountains and flow westward to the Columbia. The observation station is at an elevation of nearly 1,500 feet, about 3 miles northwest of downtown Pendleton. The city of Pendleton lies in the shallow east-west valley of the Umatilla River, approximately 400 feet lower than the airport.

Precipitation in the Pendleton area is definitely seasonal in occurrence with an average of only 10 percent of the annual total occurring in the three-month period, July-September. Most precipitation reaching this area accompanies cyclonic storms moving in from the Pacific Ocean. These storms reach their greatest intensity and frequency from October through April. The Cascade Range west of the Columbia Basin reduces the amount of precipitation received from the Pacific cyclonic storms. This influence is felt, particularly, in the desert area of the central part of the Basin. A gradual rise in elevation from the Columbia River to the foothills of the Blue Mountains again results in increased precipitation. This increase supplies sufficient moisture for productive wheat, pea, and stock raising activity in the area surrounding Pendleton.

The lighter summertime precipitation usually accompanies thunderstorms which often move into the area from the south or southwest. On occasion, these storms are quite intense, causing flash flooding with resultant heavy property damage and even loss of life.

Seasonal temperature extremes are usually quite moderate for the latitude. The last occurrence in spring of temperatures as low as 32 degrees is mid-April, and the average last occurrence in the fall of 32 degrees is late October. At the city station, where cool air settles in the valley on still nights, temperatures of 32 degrees have been recorded later in the spring and earlier in the fall. Under usual atmospheric conditions, air from the Pacific, with moderate temperature characteristics, moves across the Cascades or through the Columbia Gorge resulting in mild temperatures in the Pendleton area. When this flow of air from the west is impeded by slow-moving high pressure systems over the interior of the continent, temperature conditions sometimes become rather severe, hot in summer and cold in winter. During the summer or early fall, if a stagnant high predominates to the north or east of Pendleton, the hot, dry conditions may prove detrimental to crops during late May and June, and cause fire danger in the forest and grassland areas during late summer and early fall. During winter, coldest temperatures occur when air from a cold high pressure system in central Canada moves southwestward across the Rockies and flows down into the Columbia Basin. Under this condition the heavy cold air sometimes remains at low levels in the Basin for several days while warmer air from the Pacific flows above it, causing comparatively mild temperatures at higher elevations. Extreme winter temperatures are not particularly common in the Pendleton area. Below zero readings are recorded in approximately 60 percent of winters. Maximum temperatures usually reach 100 degrees or slightly higher on a few days during the summer.

NORMALS, MEANS, AND EXTREMES

PENDLETON, OREGON

LATITUDE: 45°41'N LONGITUDE: 118°51' W ELEVATION: FT. GRND 1482 BARO 1507 TIME ZONE: PACIFIC WBAN: 24155

	(a)	JAN	FEB	MAR	APR	MAY	JUNE	JULY	AUG	SEP	OCT	NOV	DEC	YEAR
TEMPERATURE °F:														
Normals														
-Daily Maximum		39.7	46.9	54.2	61.3	70.0	79.5	87.8	86.2	76.3	63.7	48.9	40.5	62.9
-Daily Minimum		27.2	31.6	35.4	39.4	45.8	52.9	58.0	57.7	49.9	41.0	34.1	27.9	41.7
-Monthly		33.5	39.2	44.8	50.3	57.9	66.2	72.9	72.0	63.1	52.4	41.5	34.3	52.3
Extremes														
-Record Highest	58	68	72	79	91	100	108	110	113	102	92	77	67	113
-Year		1974	1986	1964	1977	1986	1961	1939	1961	1955	1980	1975	1980	AUG 1961
-Record Lowest	58	-22	-18	1	18	25	35	42	40	30	11	-12	-19	-22
-Year		1957	1950	1993	1936	1954	1991	1971	1980	1970	1935	1985	1983	JAN 1957
NORMAL DEGREE DAYS:														
Heating (base 65°F)		977	722	626	441	226	71	15	23	145	391	705	952	5294
Cooling (base 65°F)		0	0	0	0	6	107	260	240	88	0	0	0	701
% OF POSSIBLE SUNSHINE														
MEAN SKY COVER (tenths)														
Sunrise - Sunset	48	8.4	8.0	7.3	6.8	6.1	5.4	3.1	3.5	4.1	5.7	7.8	8.4	6.2
MEAN NUMBER OF DAYS:														
Sunrise to Sunset														
-Clear	58	2.3	2.7	4.8	5.3	7.5	9.7	19.3	18.0	15.2	10.1	3.7	2.6	101.1
-Partly Cloudy	58	5.2	5.6	7.5	9.5	10.8	10.1	7.7	7.9	7.8	7.9	6.4	4.5	90.9
-Cloudy	58	23.5	19.9	18.7	15.3	12.7	10.2	4.1	5.2	7.0	13.0	19.9	23.8	173.2
Precipitation														
.01 inches or more	58	12.2	10.5	10.8	8.9	7.8	6.5	2.8	3.1	4.2	7.1	11.4	12.4	97.8
Snow,Ice pellets,hail														
1.0 inches or more	58	2.7	1.1	0.4	0.1	0.0	0.0	0.0	0.0	0.0	0.1	0.5	1.4	6.1
Thunderstorms	56	0.0	0.*	0.2	0.9	1.8	1.9	1.9	2.0	1.1	0.3	0.1	0.*	10.2
Heavy Fog Visibility														
1/4 mile or less	56	7.4	4.9	1.9	0.3	0.2	0.1	0.0	0.*	0.2	1.0	6.0	8.6	30.7
Temperature °F														
-Maximum														
90° and above	58	0.0	0.0	0.0	0.*	0.9	4.6	14.0	10.6	2.7	0.*	0.0	0.0	32.8
32° and below	58	9.3	2.9	0.3	0.0	0.0	0.0	0.0	0.0	0.0	0.*	2.1	7.4	22.0
-Minimum														
32° and below	58	21.4	15.7	9.5	2.4	0.1	0.0	0.0	0.0	0.1	2.5	12.5	19.5	83.6
0° and below	58	1.6	0.7	0.0	0.0	0.0	0.0	0.0	0.0	0.0	0.0	0.1	0.6	3.0
AVG. STATION PRESS. (mb)	20	966.2	964.6	961.6	962.5	961.7	961.4	961.8	961.4	963.0	964.8	964.7	966.4	963.4
RELATIVE HUMIDITY (%)														
Hour 04	52	80	79	74	71	69	65	54	54	61	72	79	82	70
Hour 10 (Local Time)	54	77	72	59	51	47	42	34	36	42	55	72	78	55
Hour 16	54	75	65	50	42	37	32	23	26	32	46	69	78	48
Hour 22	51	80	77	69	63	58	52	38	41	51	66	78	81	63
PRECIPITATION (inches):														
Water Equivalent														
-Normal		1.51	1.14	1.16	1.04	0.99	0.64	0.35	0.53	0.59	0.86	1.58	1.63	12.02
-Maximum Monthly	58	3.92	3.03	2.82	2.78	3.18	2.70	1.45	2.58	2.34	2.79	3.76	4.68	4.68
-Year		1970	1940	1983	1978	1991	1947	1993	1977	1941	1947	1973	1973	DEC 1973
-Minimum Monthly	58	0.21	0.07	0.24	0.01	0.03	0.03	T	0.00	T	T	0.04	0.21	0.00
-Year		1949	1964	1941	1956	1964	1986	1967	1969	1993	1987	1939	1989	AUG 1969
-Maximum in 24 hrs	58	1.29	1.09	1.33	1.24	1.52	1.49	1.19	2.19	1.23	1.88	1.35	1.25	2.19
-Year		1956	1959	1983	1990	1972	1947	1948	1993	1981	1982	1971	1978	AUG 1993
Snow,Ice pellets,hail														
-Maximum Monthly	58	41.6	15.8	4.9	2.2	T	0.0	T	0.0	0.0	3.2	14.9	26.6	41.6
-Year		1950	1936	1971	1975	1993		1993			1973	1985	1983	JAN 1950
-Maximum in 24 hrs	58	13.3	9.7	4.0	2.2	T	0.0	T	0.0	0.0	3.2	8.0	9.9	13.3
-Year		1950	1949	1970	1975	1993		1993			1973	1977	1948	JAN 1950
WIND:														
Mean Speed (mph)	40	7.8	8.2	9.2	9.9	9.6	9.6	9.0	8.6	8.4	7.6	7.7	7.8	8.6
Prevailing Direction through 1963		SE	SE	W	W	W	W	WNW	SE	SE	SE	SE	SE	SE
Fastest Obs. 1 Min.														
-Direction (!!)	38	23	25	29	27	27	29	28	27	27	25	27	29	27
-Speed (MPH)	38	49	54	63	77	48	62	46	40	47	49	62	63	77
-Year		1990	1955	1956	1960	1959	1956	1968	1961	1954	1959	1959	1959	APR 1960
Peak Gust														
-Direction (!!)	10	SW	SW	SW	SW	W	W	W	W	W	NW	W	W	SW
-Speed (mph)	10	76	52	74	61	60	59	62	55	56	51	58	62	76
-Date		1990	1988	1991	1987	1988	1992	1990	1990	1984	1991	1989	1990	JAN 1990

PRECIPITATION (inches)

PENDLETON, OREGON

YEAR	JAN	FEB	MAR	APR	MAY	JUNE	JULY	AUG	SEP	OCT	NOV	DEC	ANNUAL
1964	1.07	0.07	0.66	0.34	0.03	1.01	0.64	0.21	0.15	0.80	1.93	3.23	10.14
1965	3.08	0.37	0.29	0.65	0.57	1.10	0.51	1.21	0.23	0.19	1.95	0.27	10.42
1966	2.19	0.83	0.96	0.08	0.07	0.55	0.79	0.17	0.43	0.75	2.09	2.65	11.56
1967	1.59	0.15	0.89	1.05	0.56	0.41	T	T	0.40	0.64	0.63	0.45	6.77
1968	0.59	1.82	0.47	0.17	0.66	0.89	0.17	0.61	0.57	1.03	2.06	2.19	11.23
1969	2.88	0.88	0.57	2.05	1.40	0.86	0.02	0.00	0.42	1.13	0.36	1.88	12.45
1970	3.92	1.48	0.99	0.63	0.32	0.57	0.08	0.03	0.78	0.81	1.78	0.80	12.19
1971	0.84	0.69	1.11	1.15	1.41	1.73	0.32	0.14	1.03	0.70	2.73	2.59	14.44
1972	0.96	1.08	1.47	0.68	1.97	0.80	0.58	0.36	0.16	0.58	0.70	2.31	11.65
1973	0.50	1.09	0.43	0.27	0.67	0.15	0.01	0.08	1.34	1.71	3.76	4.68	14.69
1974	0.79	1.57	0.81	2.13	0.26	0.19	0.90	T	T	0.29	1.00	1.59	9.53
1975	3.53	1.30	0.65	0.97	0.30	0.28	0.73	0.67	0.00	1.80	0.84	1.98	13.05
1976	1.77	1.00	1.65	1.09	0.92	0.33	0.16	1.77	0.18	0.54	0.19	0.44	10.04
1977	0.48	0.64	1.51	0.18	1.87	0.37	0.06	2.58	1.17	0.51	2.00	2.42	13.79
1978	2.82	1.60	1.03	2.78	0.63	0.76	0.77	2.21	0.92	T	2.37	1.86	17.75
1979	1.43	1.72	1.18	1.17	0.39	0.21	0.09	1.40	0.30	1.68	1.83	0.62	12.02
1980	2.48	1.39	1.60	0.59	2.14	1.12	0.77	0.03	0.59	1.22	0.84	1.20	13.97
1981	0.89	1.35	1.43	1.20	1.59	1.53	0.94	0.03	1.31	0.86	1.91	2.31	15.35
1982	1.54	0.77	1.22	0.84	0.31	0.63	0.51	0.24	1.47	2.67	0.34	2.20	12.74
1983	0.86	1.57	2.82	0.70	0.73	1.44	0.52	0.56	0.46	0.84	1.67	3.42	15.59
1984	0.53	1.74	1.83	1.70	1.02	1.13	0.06	0.44	0.39	1.02	2.14	0.92	12.92
1985	0.44	1.33	1.13	0.37	0.44	0.69	0.34	0.26	2.10	0.89	2.11	1.27	11.37
1986	1.66	2.58	1.13	0.43	1.18	0.03	0.48	0.02	1.28	0.80	2.12	0.82	12.53
1987	1.48	0.64	1.39	0.47	0.85	0.38	0.34	0.05	0.03	T	0.76	1.23	7.62
1988	1.86	0.12	0.95	2.47	1.56	0.31	0.01	T	0.31	0.10	2.16	0.37	10.22
1989	1.86	1.36	1.72	1.57	1.47	0.57	0.09	1.25	0.12	0.84	1.27	0.21	12.33
1990	0.77	0.28	1.14	1.54	1.83	0.58	0.18	0.62	T	0.78	0.87	0.84	9.43
1991	0.98	0.57	1.00	0.71	3.18	2.14	0.24	0.42	T	0.92	2.68	0.67	13.51
1992	0.41	1.04	0.26	1.21	0.07	0.94	0.70	0.43	0.42	1.32	1.15	0.73	8.68
1993	1.79	0.80	1.49	1.85	1.51	0.71	1.45	2.19	T	0.22	0.93	0.92	13.86
Record Mean	1.55	1.29	1.22	1.06	1.13	0.94	0.34	0.46	0.69	1.06	1.53	1.56	12.84

AVERAGE TEMPERATURE (deg. F)

PENDLETON, OREGON

YEAR	JAN	FEB	MAR	APR	MAY	JUNE	JULY	AUG	SEP	OCT	NOV	DEC	ANNUAL
1964	40.3	41.1	43.5	48.5	56.8	64.9	72.2	68.4	62.7	51.9	40.1	32.9	51.9
1965	35.4	41.1	40.6	52.9	58.0	66.0	73.8	71.5	60.1	57.5	44.7	36.9	53.2
1966	38.3	39.7	45.4	52.0	59.9	64.6	71.0	72.4	66.6	54.0	45.1	41.1	54.2
1967	42.4	42.5	43.5	45.8	58.5	69.7	76.4	79.7	69.9	54.9	41.1	36.8	55.1
1968	37.3	42.8	48.5	49.6	59.7	67.5	77.2	70.9	65.9	50.8	42.6	33.3	53.8
1969	22.0	35.5	44.6	49.8	61.7	69.8	72.5	70.1	64.7	49.3	43.2	34.9	51.5
1970	32.3	39.5	43.5	45.8	57.8	68.7	74.8	72.5	56.3	47.4	40.4	35.7	51.2
1971	40.0	39.8	40.5	49.4	60.5	63.3	76.1	76.8	59.1	51.4	43.7	36.9	53.1
1972	34.0	37.4	47.8	47.6	60.9	68.3	74.7	76.2	61.1	51.1	42.6	27.1	52.4
1973	31.3	38.4	45.8	50.3	61.3	66.9	75.3	71.7	64.0	52.8	42.6	41.5	53.5
1974	30.4	43.8	46.4	51.7	57.3	71.1	73.3	75.5	67.5	54.8	44.8	40.6	54.8
1975	37.1	39.0	45.2	47.5	59.3	65.8	78.4	70.1	67.0	54.3	42.3	40.5	53.9
1976	39.2	37.9	42.8	50.2	58.8	63.7	73.4	67.7	67.3	53.1	42.7	35.9	52.7
1977	26.3	41.5	44.2	55.3	55.1	69.1	70.3	74.9	58.5	50.0	38.3	34.8	51.5
1978	32.2	39.3	45.7	48.0	54.4	66.3	72.2	69.4	60.5	51.7	33.5	29.5	50.2
1979	15.3	37.7	46.0	50.4	59.5	66.6	72.8	70.6	65.5	54.3	34.7	38.2	51.0
1980	25.6	36.1	41.3	51.9	56.4	60.4	72.1	66.9	63.3	51.3	42.0	39.2	50.6
1981	36.2	38.9	45.7	50.4	56.0	61.6	69.2	74.3	63.8	50.6	44.2	37.2	52.3
1982	35.0	38.1	43.5	47.6	56.8	67.6	71.1	71.5	60.7	50.7	37.3	35.7	51.3
1983	40.8	43.8	47.8	49.0	58.9	62.7	68.4	72.7	58.9	52.5	45.9	23.2	52.1
1984	34.6	39.7	46.8	48.2	54.7	62.1	72.9	72.2	60.4	49.1	41.8	30.4	51.1
1985	26.3	33.5	43.2	53.1	58.5	65.6	77.4	68.1	57.0	50.3	26.5	19.5	48.3
1986	35.9	39.0	48.8	50.0	58.6	70.0	67.6	75.8	58.9	54.0	42.2	31.5	52.7
1987	30.4	39.1	46.4	53.9	59.7	67.2	68.9	70.6	66.2	54.1	42.6	32.7	52.7
1988	32.4	41.1	44.1	51.9	56.8	63.9	72.0	70.0	63.4	58.4	44.3	33.9	52.7
1989	38.3	25.1	42.5	52.9	55.9	65.9	70.3	68.8	63.6	51.8	44.6	33.2	51.1
1990	39.6	37.9	45.7	54.8	56.4	64.7	75.2	72.2	68.2	51.3	45.4	25.8	53.1
1991	31.2	44.7	42.6	49.6	53.9	59.6	71.5	73.3	64.9	51.6	41.2	36.9	51.8
1992	38.7	42.4	47.7	52.8	62.4	70.4	71.9	72.6	61.9	54.2	39.8	32.0	53.9
1993	25.1	28.2	41.9	50.3	61.8	62.9	65.4	68.0	64.3	54.6	34.6	35.6	49.4
Record Mean	32.5	37.9	44.9	51.3	58.4	65.6	72.9	71.4	63.1	52.5	41.3	35.0	52.2
Max	39.4	46.0	55.0	63.4	71.5	79.5	89.2	87.2	77.4	64.7	49.5	41.6	63.7
Min	25.5	29.8	34.8	39.2	45.3	51.6	56.7	55.6	48.7	40.3	33.0	28.5	40.7

REFERENCE NOTES

PENDLETON, OREGON

GENERAL
T - TRACE AMOUNT.
BLANK ENTRIES DENOTE MISSING/UNREPORTED DATA.
# INDICATES A STATION OR INSTRUMENT RELOCATION.

SPECIFIC

PM - INCLUDES LAST DAY OF PREVIOUS MONTH
ASOS - AUTOMATED SURFACE OBSERVING SYSTEM IN OPERATION DURING THESE MONTHS.

(a) - LENGTH OF RECORD IN YEARS, ALTHOUGH INDIVIDUAL MONTHS MAY BE MISSING.
0.* OR * - THE VALUE IS BETWEEN 0.0 AND 0.05
NORMALS - BASED ON THE 1961-1990 RECORD PERIOD.
EXTREMES - DATES ARE THE MOST RECENT OCCURRENCE
WIND DIR.- NUMERALS SHOW TENS OF DEGRESS CLOCKWISE FROM TRUE NORTH. "00" INDICATES CALM.
RESULTANT DIRECTIONS ARE GIVEN TO WHOLE DEGREES.
BOLD VALUES INDICATE EXTREME VALUES WHICH OCCURRED AFTER THE ASOS SYSTEM WAS COMMISSIONED.

RECORD = PERIOD OF RECORD
RECORD MEAN PRECIPITATION IS THE MEAN OF ALL DAILY PRECIPITATION AMOUNTS DURING THE PERIOD OF RECORD.
RECORD MAX(MIN) TEMPERATURE IS THE MEAN OF ALL DAILY MAX(MIN) TEMPERATURES DURING THE PERIOD OF RECORD.
RECORD MEAN TEMPERATURE IS THE SUM OF THE RECORD MAX AND RECORD MIN DIVIDED BY 2.
AVERAGE TEMPERATURE IS THE SUM OF THE MEAN DAILY MAX AND MIN TEMPERATURE DIVIDED BY 2.

EXCEPTIONS

RECORD MEANS ARE THROUGH THE CURRENT YEAR,
BEGINNING IN 1900 FOR TEMPERATURE
1900 FOR PRECIPITATION
1936 FOR SNOWFALL
BEGINNING IN MAY 1989, SNOWFALL DATA COMPRISE ALL FORMS OF SOLID PRECIPITATION INCLUDING HAIL.

HEATING DEGREE DAYS Base 65 deg. F PENDLETON, OREGON

SEASON	JULY	AUG	SEP	OCT	NOV	DEC	JAN	FEB	MAR	APR	MAY	JUNE	TOTAL
1964-65	6	30	92	397	742	990	912	664	750	357	223	44	5207
1965-66	15	19	153	226	602	865	820	702	597	382	194	75	4650
1966-67	19	10	48	333	590	736	691	621	659	571	218	21	4517
1967-68	0	0	24	306	711	866	850	638	505	462	174	46	4582
1968-69	0	15	73	434	664	977	1327	820	623	450	140	32	5555
1969-70	0	11	79	480	646	926	1007	707	660	568	228	83	5395
1970-71	0	1	260	540	731	903	767	698	755	460	169	95	5379
1971-72	11	9	182	428	633	868	955	793	528	515	171	29	5122
1972-73	5	4	165	422	663	1170	1036	738	588	434	169	73	5467
1973-74	1	16	97	372	666	721	1064	589	573	391	241	29	4760
1974-75	8	0	39	313	600	750	857	721	609	517	194	57	4665
1975-76	0	12	43	332	673	751	791	782	679	436	206	89	4794
1976-77	4	42	31	363	660	896	1192	653	639	299	301	26	5106
1977-78	20	35	200	461	792	927	1011	714	593	504	322	46	5625
1978-79	7	41	146	403	936	1094	1533	757	582	432	184	62	6177
1979-80	12	0	43	326	902	823	1210	829	728	388	267	141	5669
1980-81	4	33	88	438	681	794	886	724	593	435	275	126	5077
1981-82	20	1	128	440	617	855	919	747	662	515	256	72	5232
1982-83	22	7	171	435	825	901	741	588	528	470	242	95	5025
1983-84	42	1	180	381	569	1292	935	729	558	496	316	134	5633
1984-85	4	0	182	490	692	1065	1196	876	665	351	224	65	5810
1985-86	4	22	242	452	1149	1402	898	722	497	446	277	25	6136
1986-87	33	0	213	335	675	1031	1065	717	571	332	201	71	5244
1987-88	25	12	65	334	668	995	1004	689	637	387	264	126	5206
1988-89	22	4	120	208	616	957	821	1113	691	354	279	42	5227
1989-90	11	17	76	403	607	978	781	752	591	299	262	89	4866
1990-91	9	13	11	419	583	1211	1039	564	689	454	338	162	5492
1991-92	4	2	52	418	707	865	810	649	527	362	127	36	4559
1992-93	11	28	129	333	752	1015	1231	1025	709	432	153	98	5916
1993-94	27	35	114	318	908	903							

COOLING DEGREE DAYS Base 65 deg. F PENDLETON, OREGON

YEAR	JAN	FEB	MAR	APR	MAY	JUNE	JULY	AUG	SEP	OCT	NOV	DEC	TOTAL
1969	0	0	0	0	45	183	238	177	76	0	0	0	719
1970	0	0	0	0	11	201	313	243	4	0	0	0	772
1971	0	0	0	0	36	50	363	379	12	13	0	0	853
1972	0	0	0	0	50	134	314	358	55	0	0	0	911
1973	0	0	0	0	63	137	327	232	72	0	0	0	831
1974	0	0	0	0	9	219	272	332	122	4	0	0	958
1975	0	0	0	0	27	88	423	179	109	8	0	0	834
1976	0	0	0	0	20	53	270	129	103	3	0	0	578
1977	0	0	0	16	3	152	190	348	16	0	0	0	725
1978	0	0	0	0	1	93	236	182	16	0	0	0	528
1979	0	0	0	0	21	114	261	186	65	3	0	0	650
1980	0	0	0	2	5	13	232	101	44	20	0	0	417
1981	0	0	0	4	2	28	155	297	101	0	0	0	587
1982	0	0	0	0	7	158	219	215	47	0	0	0	646
1983	0	0	0	0	60	32	155	246	6	0	0	0	499
1984	0	0	0	0	7	55	256	231	51	3	0	0	603
1985	0	0	0	0	28	91	394	127	7	0	0	0	647
1986	0	0	0	2	88	184	121	341	35	1	0	0	772
1987	0	0	0	8	41	145	152	194	108	4	0	0	652
1988	0	0	0	0	16	98	246	164	78	9	0	0	611
1989	0	0	0	0	5	76	182	143	41	0	0	0	447
1990	0	0	0	0	4	92	330	245	114	3	0	0	788
1991	0	0	0	0	0	8	214	267	56	9	0	0	554
1992	0	0	0	1	52	204	229	275	45	4	0	0	810
1993	0	0	0	0	59	42	47	136	99	4	0	0	387

SNOWFALL (inches) PENDLETON, OREGON

SEASON	JULY	AUG	SEP	OCT	NOV	DEC	JAN	FEB	MAR	APR	MAY	JUNE	TOTAL
1964-65	0.0	0.0	0.0	0.0	0.2	4.9	7.0	0.5	2.5	0.0	0.0	0.0	15.1
1965-66	0.0	0.0	0.0	0.0	T	0.1	11.5	1.6	0.2	T	0.0	0.0	13.4
1966-67	0.0	0.0	0.0	T	0.4	2.9	2.3	0.4	3.3	T	0.0	0.0	9.3
1967-68	0.0	0.0	0.0	0.0	0.8	2.5	1.3	T	T	T	0.0	0.0	4.6
1968-69	0.0	0.0	0.0	0.0	T	11.9	27.4	2.7	T	0.0	0.0	0.0	42.0
1969-70	0.0	0.0	0.0	0.0	T	3.5	9.9	3.8	1.3	T	T	0.0	18.5
1970-71	0.0	0.0	0.0	0.0	1.6	2.3	4.0	0.6	4.9	T	T	0.0	13.4
1971-72	0.0	0.0	0.0	1.9	T	11.8	3.6	6.2	0.1	1.1	0.0	0.0	24.7
1972-73	0.0	0.0	0.0	T	T	12.6	2.2	5.9	T	0.1	0.0	0.0	20.8
1973-74	0.0	0.0	0.0	3.2	9.1	5.3	2.6	0.5	T	T	0.0	0.0	20.7
1974-75	0.0	0.0	0.0	0.0	T	T	16.6	3.3	T	2.2	T	0.0	22.1
1975-76	0.0	0.0	0.0	0.0	5.2	3.0	0.3	0.3	0.1	0.0	0.0	0.0	8.9
1976-77	0.0	0.0	0.0	0.0	0.0	1.0	3.1	0.5	0.4	0.0	0.0	0.0	5.0
1977-78	0.0	0.0	0.0	0.0	8.5	11.5	6.1	T	3.9	0.0	T	0.0	30.0
1978-79	0.0	0.0	0.0	0.0	9.0	7.4	14.7	2.2	T	0.0	0.0	0.0	33.3
1979-80	0.0	0.0	0.0	0.0	4.3	T	16.6	0.9	3.9	0.0	0.0	0.0	25.7
1980-81	0.0	0.0	0.0	0.0	2.0	2.7	3.6	1.2	0.0	0.0	0.0	0.0	9.5
1981-82	0.0	0.0	0.0	0.0	0.6	5.1	5.7	1.5	1.9	T	0.0	0.0	14.8
1982-83	0.0	0.0	0.0	0.0	T	1.6	0.2	0.9	0.0	0.0	0.0	0.0	2.7
1983-84	0.0	0.0	0.0	0.0	T	26.6	1.0	1.2	T	T	0.0	0.0	28.8
1984-85	0.0	0.0	0.0	0.0	T	6.2	0.8	12.7	0.6	T	0.0	0.0	20.3
1985-86	0.0	0.0	0.0	0.0	14.9	9.1	T	7.6	0.0	0.0	T	0.0	31.6
1986-87	0.0	0.0	0.0	0.0	1.2	6.8	5.8	0.0	T	0.0	0.0	0.0	13.8
1987-88	0.0	0.0	0.0	0.0	0.3	2.3	10.6	0.0	1.5	0.0	0.0	0.0	14.7
1988-89	0.0	0.0	0.0	0.0	T	T	4.3	4.9	4.0	0.0	T	0.0	13.2
1989-90	0.0	0.0	0.0	0.0	0.0	1.0	T	2.0	1.3	0.0	0.0	0.0	4.3
1990-91	0.0	0.0	0.0	0.0	T	6.4	1.6	0.0	0.6	T	0.0	0.0	8.6
1991-92	0.0	0.0	0.0	2.3	1.0	T	0.8	T	0.0	0.0	0.0	0.0	4.1
1992-93	0.0	0.0	0.0	0.0	0.2	7.6	25.1	14.8	1.8	T	T	0.0	49.5
1993-94	T	0.0	0.0	0.0	0.7	0.4							
Record Mean	T	0.0	0.0	0.1	1.7	4.0	7.3	3.5	1.0	0.1	T	0.0	17.7

PORTLAND, OREGON

The Portland Weather Service Office is located 6 miles north-northeast of downtown Portland. Portland is situated about 65 miles inland from the Pacific Coast and midway between the northerly oriented low coast range on the west and the higher Cascade range on the east, each about 30 miles distant. The airport lies on the south bank of the Columbia River. The coast range provides limited shielding from the Pacific Ocean. The Cascade range provides a steep slope for orographic lift of moisture-laden westerly winds and consequent moderate rainfall, and also forms a barrier from continental air masses originating over the interior Columbia Basin. Airflow is usually northwesterly in Portland in spring and summer and southeasterly in fall and winter. The Portland Airport location is drier than most surrounding localities.

Portland has a very definite winter rainfall climate. Approximately 88 percent of the annual total occurs in the months of October through May, 9 percent in June and September, while only 3 percent comes in July and August. Precipitation is mostly rain, as on the average there are only five days each year with measurable snow. Snowfalls are seldom more than a couple of inches, and generally last only a few days.

The winter season is marked by relatively mild temperatures, cloudy skies and rain with southeasterly surface winds predominating. Summer produces pleasantly mild temperatures, northwesterly winds and very little precipitation. Fall and spring are transitional in nature. Fall and early winter are times with most frequent fog.

At all times, incursions of marine air are a frequent moderating influence. Outbreaks of continental high pressure from east of the Cascade Mountains produce strong easterly flow through the Columbia Gorge into the Portland area. In winter this brings the coldest weather with the extremes of low temperature registered in the cold air mass. Freezing rain and ice glaze are sometimes transitional effects. Temperatures below zero are very infrequent. In summer, hot, dry continental air brings the highest temperatures. Temperatures above 100 degrees are infrequent, but 90 degrees or higher are reached every year, but seldom persist for more than two or three days.

Destructive storms are infrequent in the Portland area. Surface winds seldom exceed gale force and rarely in the period of record have winds reached higher than 75 mph. Thunderstorms occur about once a month through the spring and summer months. Heavy downpours are infrequent but gentle rains occur almost daily during winter months.

Most rural areas around Portland are farmed for berries, green beans, and vegetables for fresh market and processing. The long growing season with mild temperatures and ample moisture favors local nursery and seed industries.

Based on the 1951-1980 period, the average first occurrence of 32 degrees Fahrenheit in the fall is November 7 and the average last occurrence in the spring is April 3.

NORMALS, MEANS, AND EXTREMES

PORTLAND, OREGON

LATITUDE: 45°36'N LONGITUDE: 122°36'W ELEVATION: FT. GRND 21 BARO 27 TIME ZONE: PACIFIC WBAN: 24229

	(a)	JAN	FEB	MAR	APR	MAY	JUNE	JULY	AUG	SEP	OCT	NOV	DEC	YEAR
TEMPERATURE °F:														
Normals														
-Daily Maximum		45.4	51.0	56.0	60.6	67.1	74.0	79.9	80.3	74.6	64.0	52.6	45.6	62.6
-Daily Minimum		33.7	36.1	38.6	41.3	47.0	52.9	56.5	56.9	52.0	44.9	39.5	34.8	44.5
-Monthly		39.6	43.6	47.3	51.0	57.1	63.5	68.2	68.6	63.3	54.5	46.1	40.2	53.6
Extremes														
-Record Highest	53	63	71	80	87	100	100	107	107	105	92	73	65	107
-Year		1986	1988	1947	1957	1983	1992	1965	1981	1988	1987	1975	1993	AUG 1981
-Record Lowest	53	-2	-3	19	29	29	39	43	44	34	26	13	6	-3
-Year		1950	1950	1989	1955	1954	1966	1955	1980	1965	1971	1985	1964	FEB 1950
NORMAL DEGREE DAYS:														
Heating (base 65°F)		787	599	549	420	249	91	28	35	102	326	567	769	4522
Cooling (base 65°F)		0	0	0	0	0	46	127	147	51	0	0	0	371
% OF POSSIBLE SUNSHINE	44	28	38	47	52	57	56	69	66	62	44	28	23	48
MEAN SKY COVER (tenths)														
Sunrise - Sunset	45	8.4	8.2	8.1	7.8	7.2	6.7	4.8	5.2	5.5	7.1	8.2	8.7	7.1
MEAN NUMBER OF DAYS:														
Sunrise to Sunset														
-Clear	45	2.9	2.8	3.1	3.5	4.8	6.2	12.6	11.4	10.3	5.5	2.8	2.1	68.0
-Partly Cloudy	45	3.6	3.8	5.0	5.7	7.2	7.6	8.6	9.5	8.1	7.4	4.3	3.3	74.4
-Cloudy	45	24.5	21.6	23.0	20.7	19.0	16.1	9.8	10.1	11.5	18.1	22.9	25.6	222.8
Precipitation														
.01 inches or more	53	18.1	15.7	17.0	14.2	11.9	9.1	3.9	4.8	7.5	12.3	17.8	18.6	150.9
Snow,Ice pellets,hail														
1.0 inches or more	53	1.1	0.3	0.1	0.0	0.*	0.0	0.0	0.0	0.0	0.0	0.1	0.5	2.2
Thunderstorms	53	0.*	0.1	0.5	0.9	1.5	0.9	0.8	1.0	0.7	0.4	0.3	0.1	7.2
Heavy Fog Visibility														
1/4 mile or less	51	4.2	3.7	2.4	1.1	0.2	0.1	0.1	0.2	2.7	7.3	6.1	5.0	33.1
Temperature °F														
-Maximum														
90° and above	53	0.0	0.0	0.0	0.0	0.3	1.2	3.6	3.8	1.8	0.1	0.0	0.0	10.8
32° and below	53	2.2	0.3	0.*	0.0	0.0	0.0	0.0	0.0	0.0	0.0	0.2	0.9	3.6
-Minimum														
32° and below	53	13.3	8.2	4.7	1.0	0.1	0.0	0.0	0.0	0.0	0.6	5.2	9.6	42.6
0° and below	53	0.*	0.*	0.0	0.0	0.0	0.0	0.0	0.0	0.0	0.0	0.0	0.0	*
AVG. STATION PRESS.(mb)	21	1018.4	1016.6	1015.2	1016.8	1016.3	1016.1	1016.2	1015.2	1015.5	1017.1	1017.3	1018.6	1016.6
RELATIVE HUMIDITY (%)														
Hour 04	53	86	86	86	86	85	84	82	84	87	90	88	87	86
Hour 10 (Local Time)	53	82	80	74	69	66	65	62	64	67	78	82	84	73
Hour 16	53	75	67	60	55	53	49	45	46	48	62	74	78	59
Hour 22	53	83	81	78	75	74	71	68	70	75	84	84	85	77
PRECIPITATION (inches):														
Water Equivalent														
-Normal		5.35	3.85	3.56	2.39	2.06	1.48	0.63	1.09	1.75	2.67	5.34	6.13	36.30
-Maximum Monthly	53	12.83	9.46	7.52	5.26	4.57	4.06	2.68	4.53	4.30	8.04	11.57	11.12	12.83
-Year		1953	1949	1957	1993	1945	1984	1983	1968	1986	1947	1942	1968	JAN 1953
-Minimum Monthly	53	0.06	0.72	1.10	0.53	0.10	0.03	0.00	T	T	0.19	0.77	1.38	0.00
-Year		1985	1993	1965	1956	1992	1951	1967	1970	1993	1988	1976	1976	JUL 1967
-Maximum in 24 hrs	53	2.61	2.36	1.83	1.47	1.47	1.82	1.09	1.54	2.38	2.18	2.62	2.59	2.62
-Year		1974	1987	1943	1962	1968	1958	1978	1977	1982	1941	1973	1977	NOV 1973
Snow,Ice pellets,hail														
-Maximum Monthly	53	41.4	13.2	12.9	T	0.6	T	0.0	T	T	0.2	8.2	15.7	41.4
-Year		1950	1949	1951	1993	1953	1981		1989	1949	1950	1955	1968	JAN 1950
-Maximum in 24 hrs	53	10.6	6.4	7.7	T	0.5	T	0.0	T	T	0.2	7.4	8.0	10.6
-Year		1950	1993	1951	1993	1953	1981		1989	1949	1950	1977	1964	JAN 1950
WIND:														
Mean Speed (mph)	45	10.0	9.2	8.3	7.4	7.1	7.2	7.6	7.1	6.5	6.5	8.6	9.5	7.9
Prevailing Direction through 1963		ESE	ESE	ESE	NW	NW	NW	NW	NW	NW	ESE	ESE	ESE	ESE
Fastest Mile														
-Direction (!!!)	41	S	SW	S	S	SW	SW	SW	SW	S	S	SW	S	S
-Speed (MPH)	41	54	61	57	60	42	40	33	29	61	88	56	57	88
-Year		1951	1958	1963	1957	1960	1958	1983	1961	1963	1962	1961	1951	OCT 1962
Peak Gust														
-Direction (!!!)	10	SW	SE	SW	SW	W	W	NW	E	E	SW	SW	SW	SW
-Speed (mph)	10	63	61	52	45	46	32	32	31	39	41	52	60	63
-Date		1990	1989	1991	1992	1993	1989	1992	1993	1985	1990	1984	1992	JAN 1990

PRECIPITATION (inches) PORTLAND, OREGON

YEAR	JAN	FEB	MAR	APR	MAY	JUNE	JULY	AUG	SEP	OCT	NOV	DEC	ANNUAL
1964	9.51	0.78	2.30	1.56	1.04	1.96	0.68	0.90	1.61	0.84	6.78	9.92	37.88
1965	7.44	2.22	1.10	2.20	1.31	0.83	0.44	0.73	0.01	2.03	5.64	7.34	31.29
1966	5.74	1.70	4.71	0.85	0.91	1.02	1.19	0.59	1.70	3.06	5.50	6.89	33.86
1967	6.21	2.02	4.31	2.17	1.02	1.01	0.00	T	0.76	4.72	2.27	4.75	29.24
1968	4.58	6.64	2.68	1.91	3.63	2.20	0.14	4.53	2.20	5.03	6.23	11.12	50.89
1969	7.60	3.14	1.13	2.28	1.61	2.99	0.14	0.04	3.86	3.02	3.18	8.12	37.11
1970	11.81	4.77	2.58	2.94	1.55	0.49	0.05	T	1.10	2.85	5.72	7.49	41.35
1971	7.09	3.36	4.87	2.72	1.00	1.76	0.26	0.95	3.53	2.37	5.76	8.05	41.72
1972	5.71	4.08	5.41	2.98	2.23	0.68	0.56	0.67	3.06	0.87	3.78	8.79	38.82
1973	3.69	1.94	2.45	1.33	1.43	1.45	0.06	1.41	3.29	3.14	11.55	9.93	41.67
1974	8.51	4.61	5.65	1.76	1.74	0.80	2.01	0.07	0.21	2.14	6.73	6.05	40.28
1975	8.43	4.75	3.45	1.88	1.35	1.13	0.43	2.10	T	4.76	4.10	6.68	39.06
1976	5.14	4.92	2.93	2.34	2.29	0.78	0.66	3.29	0.73	1.48	0.77	1.38	26.71
1977	1.07	2.49	3.50	1.04	4.30	0.83	0.39	3.26	3.33	2.28	5.56	8.98	37.03
1978	4.85	3.28	1.49	3.96	3.17	1.69	1.36	2.05	2.07	0.36	3.83	2.51	30.62
1979	2.55	6.53	2.51	2.47	2.41	0.64	0.25	1.18	1.75	4.85	3.38	7.23	35.75
1980	8.51	4.01	3.11	2.58	2.19	2.50	0.19	0.39	1.56	1.18	6.47	9.72	42.41
1981	1.47	3.86	2.33	1.79	2.25	3.23	0.24	0.15	1.86	4.12	4.62	8.37	34.29
1982	6.31	5.98	2.38	3.56	0.46	1.66	0.94	1.66	3.98	4.44	3.51	8.16	43.04
1983	6.23	7.78	6.80	1.87	1.30	1.95	2.68	2.29	0.39	1.95	8.65	5.30	47.19
1984	2.01	3.93	3.19	3.20	3.41	4.06	T	0.09	1.46	3.85	9.74	2.56	37.50
1985	0.06	1.79	3.08	1.07	1.52	2.34	0.55	0.48	2.76	2.75	3.89	2.19	22.48
1986	4.65	5.31	2.60	1.91	2.19	0.23	1.20	0.10	4.30	1.99	6.26	4.30	35.04
1987	6.93	2.45	4.91	1.94	1.63	0.14	1.03	0.35	0.30	0.27	1.96	8.00	29.91
1988	4.95	1.17	3.13	4.57	2.53	2.34	0.69	0.10	1.76	0.19	7.92	2.37	31.72
1989	3.30	2.84	6.73	2.08	2.87	0.78	0.91	1.07	1.48	1.73	3.18	3.08	30.05
1990	7.95	3.43	2.52	2.31	2.37	1.94	0.32	0.95	0.34	4.65	3.68	2.40	32.86
1991	2.56	3.65	4.64	4.05	3.34	2.31	0.07	0.70	0.02	1.51	6.36	4.34	33.55
1992	4.31	4.12	1.87	3.82	0.10	0.60	0.67	0.49	1.12	2.87	4.55	4.98	29.50
1993	3.06	0.72	4.39	5.26	4.36	1.69	2.41	0.37	T	1.59	1.50	5.01	30.36
Record Mean	5.39	3.94	3.64	2.40	2.14	1.57	0.60	0.91	1.64	3.07	5.40	5.88	36.59

AVERAGE TEMPERATURE (deg. F) PORTLAND, OREGON

YEAR	JAN	FEB	MAR	APR	MAY	JUNE	JULY	AUG	SEP	OCT	NOV	DEC	ANNUAL
1964	40.9	40.1	43.8	46.7	52.6	58.7	64.5	63.7	58.4	53.2	41.0	37.0	50.1
1965	40.2	43.4	47.6	51.8	54.3	61.8	69.7	68.6	60.4	57.5	49.7	39.5	53.7
1966	40.3	42.8	47.2	50.6	57.0	62.4	66.4	67.5	64.4	53.7	47.1	44.2	53.7
1967	43.6	43.7	44.0	46.9	57.1	65.9	69.3	72.9	66.4	54.8	46.2	40.4	54.3
1968	39.3	48.2	48.2	48.0	56.3	62.2	68.6	65.8	61.5	52.1	46.5	37.3	52.8
1969	31.9	39.7	46.5	50.1	59.5	66.5	66.6	66.1	63.4	53.3	47.2	42.9	52.8
1970	40.6	46.0	46.9	48.4	57.0	65.9	69.2	68.2	60.7	53.0	47.1	40.0	53.6
1971	40.4	42.8	43.7	49.6	56.8	60.2	69.2	71.6	60.8	52.4	45.5	40.4	52.8
1972	39.2	43.8	49.8	48.0	60.2	64.0	70.9	71.7	61.2	53.0	48.2	37.4	53.9
1973	39.0	44.9	47.9	52.3	59.4	63.9	70.3	65.9	64.4	54.3	44.2	44.7	54.2
1974	38.0	43.0	47.2	51.3	55.7	64.4	67.1	68.9	67.3	55.2	48.1	44.1	54.2
1975	41.5	41.2	45.0	47.3	57.5	61.9	69.0	65.3	65.7	53.5	46.0	42.7	53.1
1976	42.2	42.1	44.4	50.3	56.6	60.4	67.2	65.5	64.2	54.7	47.0	39.5	52.8
1977	35.7	44.6	45.5	52.9	53.8	63.9	66.3	71.7	60.8	53.8	43.3	42.0	52.9
1978	40.1	44.7	49.1	50.5	54.7	65.1	68.4	67.6	60.9	54.7	39.1	35.3	52.5
1979	30.7	42.9	50.8	53.1	60.1	65.1	70.5	68.6	66.3	58.1	45.0	44.4	54.6
1980	35.1	42.5	46.3	53.8	57.3	60.7	68.9	66.4	63.8	56.0	48.5	44.0	53.6
1981	43.9	44.0	48.8	52.5	57.5	61.8	67.5	72.2	64.9	53.3	48.8	42.7	54.8
1982	39.7	43.6	48.5	49.0	57.6	66.0	67.5	68.6	63.2	54.9	44.4	41.7	53.7
1983	44.4	47.3	50.7	52.7	60.4	62.8	66.5	69.1	61.5	54.2	49.3	36.4	54.6
1984	42.2	45.9	51.1	50.4	56.4	62.2	69.1	69.4	63.7	52.9	46.7	38.3	54.0
1985	36.1	41.1	45.8	53.9	58.3	64.4	74.1	69.3	60.8	52.7	37.3	33.0	52.2
1986	42.5	43.7	51.3	50.2	57.6	66.3	65.3	72.3	61.5	57.0	47.7	40.6	54.7
1987	39.6	45.2	48.7	54.2	60.4	66.5	67.2	70.5	65.5	58.2	48.8	39.1	55.3
1988	39.0	44.7	47.2	52.2	56.4	62.4	68.4	68.0	64.0	58.3	47.5	42.0	54.2
1989	42.2	36.0	45.6	56.0	58.0	64.3	65.5	66.1	65.3	54.9	48.6	40.3	53.6
1990	43.4	41.9	49.4	54.5	56.7	63.6	71.2	70.9	67.0	53.9	48.4	34.7	54.6
1991	38.9	48.8	46.2	50.8	54.7	59.9	69.9	70.1	67.4	55.4	47.5	42.7	54.4
1992	44.5	48.1	52.3	55.4	63.1	67.4	70.2	69.8	62.3	55.9	46.3	39.3	56.2
1993	36.5	40.1	47.9	52.6	61.1	62.5	64.3	68.5	65.7	57.4	41.5	41.4	53.3
Record Mean	39.0	43.3	46.7	51.2	57.3	61.4	66.6	66.6	63.0	54.5	45.8	40.6	53.0
Max	44.7	50.5	55.3	60.8	67.4	71.4	78.0	77.9	74.6	63.8	52.4	46.0	61.9
Min	33.2	36.0	38.1	41.6	47.1	51.3	55.2	55.3	51.5	45.1	39.1	35.2	44.1

REFERENCE NOTES PORTLAND, OREGON

GENERAL
T - TRACE AMOUNT.
BLANK ENTRIES DENOTE MISSING/UNREPORTED DATA.
# INDICATES A STATION OR INSTRUMENT RELOCATION.

SPECIFIC

PM - INCLUDES LAST DAY OF PREVIOUS MONTH
ASOS - AUTOMATED SURFACE OBSERVING SYSTEM IN OPERATION DURING THESE MONTHS.

(*) - LENGTH OF RECORD IN YEARS, ALTHOUGH INDIVIDUAL MONTHS MAY BE MISSING.
0.* OR * - THE VALUE IS BETWEEN 0.0 AND 0.05
NORMALS - BASED ON THE 1961-1990 RECORD PERIOD.
EXTREMES - DATES ARE THE MOST RECENT OCCURRENCE.
WIND DIR.- NUMERALS SHOW TENS OF DEGRESS CLOCKWISE FROM TRUE NORTH. "00" INDICATES CALM.
RESULTANT DIRECTIONS ARE GIVEN TO WHOLE DEGREES.
BOLD VALUES INDICATE EXTREME VALUES WHICH OCCURRED AFTER THE ASOS SYSTEM WAS COMMISSIONED.

RECORD = PERIOD OF RECORD
RECORD MEAN PRECIPITATION IS THE MEAN OF ALL DAILY PRECIPITATION AMOUNTS DURING THE PERIOD OF RECORD.
RECORD MAX(MIN) TEMPERATURE IS THE MEAN OF ALL DAILY MAX(MIN) TEMPERATURES DURING THE PERIOD OF RECORD.
RECORD MEAN TEMPERATURE IS THE SUM OF THE RECORD MAX AND RECORD MIN DIVIDED BY 2.
AVERAGE TEMPERATURE IS THE SUM OF THE MEAN DAILY MAX AND MIN TEMPERATURE. DIVIDED BY 2.

EXCEPTIONS

RECORD MEANS ARE THROUGH THE CURRENT YEAR, BEGINNING IN 1941 FOR TEMPERATURE
1941 FOR PRECIPITATION
1941 FOR SNOWFALL

HEATING DEGREE DAYS Base 65 deg. F PORTLAND, OREGON

SEASON	JULY	AUG	SEP	OCT	NOV	DEC	JAN	FEB	MAR	APR	MAY	JUNE	TOTAL
1964-65	67	81	191	358	711	860	761	599	533	388	325	113	4987
1965-66	22	16	139	227	451	786	759	620	545	425	249	99	4338
1966-67	27	16	56	345	531	635	655	591	647	535	246	50	4334
1967-68	3	0	29	306	558	758	789	482	515	500	261	110	4311
1968-69	17	43	123	395	544	852	1022	703	570	442	178	51	4940
1969-70	17	22	85	357	526	678	751	526	553	493	246	71	4325
1970-71	14	14	130	369	530	771	757	615	653	454	253	149	4709
1971-72	33	5	123	388	578	756	793	607	466	501	174	61	4485
1972-73	10	6	153	363	497	848	799	560	525	378	202	89	4430
1973-74	6	47	59	326	618	624	832	610	545	403	282	72	4424
1974-75	32	16	29	301	500	640	722	660	615	523	240	127	4405
1975-76	24	41	48	354	565	686	698	658	632	437	258	155	4556
1976-77	15	41	47	319	536	783	901	564	596	358	340	68	4568
1977-78	40	19	131	339	644	707	764	561	485	430	317	58	4495
1978-79	29	26	134	312	772	915	1058	615	434	351	162	57	4865
1979-80	8	2	19	214	592	631	920	647	575	329	232	125	4294
1980-81	15	25	64	284	485	644	650	583	494	372	229	108	3953
1981-82	23	5	76	355	478	687	780	596	502	472	229	71	4274
1982-83	22	10	99	307	614	715	635	492	435	363	184	81	3957
1983-84	27	2	109	325	463	880	701	546	425	430	269	115	4292
1984-85	9	2	80	377	539	820	893	664	588	327	213	62	4574
1985-86	0	7	124	373	826	982	691	591	417	437	265	43	4756
1986-87	37	0	148	242	510	750	780	550	495	321	173	51	4057
1987-88	22	2	54	214	479	798	801	581	544	380	272	109	4256
1988-89	33	15	91	208	518	705	699	805	594	263	219	77	4227
1989-90	32	27	44	306	486	759	664	641	476	308	251	78	4072
1990-91	10	5	14	336	492	933	802	446	575	420	310	156	4499
1991-92	4	13	25	295	517	685	627	483	387	282	108	35	3461
1992-93	5	9	107	274	556	789	877	692	522	366	135	94	4426
1993-94	37	17	57	231	698	726							

COOLING DEGREE DAYS Base 65 deg. F PORTLAND, OREGON

YEAR	JAN	FEB	MAR	APR	MAY	JUNE	JULY	AUG	SEP	OCT	NOV	DEC	TOTAL
1969	0	0	0	0	13	102	74	65	43	0	0	0	297
1970	0	0	0	0	8	106	150	120	8	2	0	0	394
1971	0	0	0	0	5	14	170	217	7	3	0	0	416
1972	0	0	0	0	27	39	200	221	44	0	0	0	531
1973	0	0	0	0	34	65	178	81	45	0	0	0	403
1974	0	0	0	0	1	60	102	144	102	0	0	0	409
1975	0	0	0	0	12	39	157	57	75	2	0	0	342
1976	0	0	0	0	4	23	89	66	30	4	0	0	216
1977	0	0	0	0	0	42	90	233	10	0	0	0	375
1978	0	0	0	0	3	69	141	112	18	0	0	0	343
1979	0	0	0	0	18	65	183	124	65	7	0	0	462
1980	0	0	0	1	0	2	141	75	35	12	0	0	266
1981	0	0	0	3	4	16	109	232	82	0	0	0	446
1982	0	0	0	0	4	107	103	127	50	0	0	0	391
1983	0	0	0	0	48	23	80	137	12	0	0	0	300
1984	0	0	0	0	10	34	140	144	47	6	0	0	381
1985	0	0	0	0	11	53	291	145	5	0	0	0	505
1986	0	0	0	0	40	87	52	235	50	0	0	0	464
1987	0	0	0	4	37	102	95	177	77	12	0	0	504
1988	0	0	0	0	10	39	147	115	67	8	0	0	386
1989	0	0	0	0	9	62	53	66	60	0	0	0	250
1990	0	0	0	2	3	45	206	193	83	0	0	0	532
1991	0	0	0	0	0	7	164	176	102	2	0	0	451
1992	0	0	0	1	57	114	174	164	31	0	0	0	541
1993	0	0	0	0	21	24	22	132	80	3	0	0	282

SNOWFALL (inches) PORTLAND, OREGON

SEASON	JULY	AUG	SEP	OCT	NOV	DEC	JAN	FEB	MAR	APR	MAY	JUNE	TOTAL
1964-65	0.0	0.0	0.0	0.0	T	11.0	T	0.0	0.3	0.0	T	0.0	11.3
1965-66	0.0	0.0	0.0	0.0	0.0	T	T	T	0.6	0.0	0.0	0.0	0.6
1966-67	0.0	0.0	0.0	0.0	0.0	0.0	T	T	T	T	0.0	0.0	T
1967-68	0.0	0.0	0.0	0.0	0.0	5.7	5.2	0.0	0.0	T	0.0	0.0	10.9
1968-69	0.0	0.0	0.0	0.0	0.0	15.7	18.3	T	0.0	0.0	0.0	0.0	34.0
1969-70	0.0	0.0	0.0	0.0	0.0	0.0	T	T	T	T	0.0	0.0	T
1970-71	0.0	0.0	0.0	0.0	T	1.4	6.9	1.7	T	T	0.0	0.0	10.0
1971-72	0.0	0.0	0.0	T	0.0	4.6	0.4	T	T	T	0.0	0.0	5.0
1972-73	0.0	0.0	0.0	0.0	0.0	6.1	0.4	T	T	T	0.0	0.0	6.5
1973-74	0.0	0.0	0.0	0.0	T	0.0	T	T	T	0.0	T	0.0	T
1974-75	0.0	0.0	0.0	0.0	0.0	T	T	0.1	T	T	T	0.0	0.1
1975-76	0.0	0.0	0.0	0.0	T	T	T	T	T	T	T	0.0	T
1976-77	0.0	0.0	0.0	0.0	0.0	0.0	T	0.0	T	0.0	T	0.0	T
1977-78	0.0	0.0	0.0	0.0	7.6	T	0.0	0.0	0.1	T	T	0.0	7.7
1978-79	0.0	0.0	0.0	0.0	3.0	2.4	1.9	1.1	T	T	0.0	0.0	8.4
1979-80	0.0	0.0	0.0	0.0	0.0	T	12.4	T	T	T	0.0	T	12.4
1980-81	0.0	0.0	0.0	0.0	T	T	0.0	T	T	T	T	T	T
1981-82	0.0	0.0	0.0	0.0	0.0	2.0	2.1	T	T	T	0.0	0.0	4.1
1982-83	0.0	0.0	0.0	T	T	0.0	0.0	0.0	T	T	T	0.0	T
1983-84	0.0	0.0	0.0	0.0	0.0	2.3	0.1	T	T	0.0	T	0.0	2.4
1984-85	0.0	0.0	0.0	T	0.0	2.8	T	4.8	T	T	0.0	0.0	7.6
1985-86	0.0	0.0	0.0	0.0	3.4	1.6	T	5.8	T	T	T	0.0	10.8
1986-87	0.0	0.0	0.0	0.0	T	0.1	0.0	T	T	T	T	0.0	0.1
1987-88	0.0	0.0	0.0	0.0	0.0	2.9	0.6	0.0	0.0	T	T	0.0	3.5
1988-89	0.0	0.0	0.0	0.0	T	T	0.9	0.3	2.0	T	0.0	0.0	3.2
1989-90	0.0	T	0.0	T	0.0	0.0	T	8.3	T	0.0	0.0	0.0	8.3
1990-91	0.0	0.0	0.0	0.0	0.0	1.3	0.6	0.0	T	T	0.0	0.0	1.9
1991-92	0.0	0.0	0.0	0.0	T	0.0	0.0	0.0	0.0	T	0.0	0.0	T
1992-93	0.0	0.0	0.0	0.0	T	4.6	2.9	6.6	0.0	T	T	0.0	14.1
1993-94	0.0	0.0	0.0	0.0	T	T							
Record Mean	0.0	T	T	T	0.4	1.4	3.3	1.0	0.4	T	T	T	6.6

SALEM, OREGON

Salem is located in the middle Willamette Valley some 60 airline miles east of the Pacific Ocean. The valley here is approximately 50 miles wide with the city about equidistant from the valley walls formed by the Coast Range on the west and the Cascade Range on the east.

The usual movement of very moist maritime air masses from the Pacific Ocean inland over the Coast Range produces, near its crest, some of the heaviest yearly rainfall in the United States. Annual totals of nearly 170 inches have been recorded in the mountains. From the ridge crest of the Coast Range, approximately 3,000 feet above sea level, there is a gradual decrease of rainfall downslope to the valley floor where annual totals are between 35 and 45 inches. As these marine conditioned air masses continue to move farther inland they are forced to ascend the west slopes of the Cascades to approximately 5,000 feet above sea level and again rainfall amounts substantially increase with elevation.

Most of this precipitation in both the valley and its bordering mountain ranges occurs during the winter. At Salem, 70 percent of the annual total occurs during the five months of November through March while only 6 percent occurs during the three summer months, with practically all of it falling in the form of rain. In the immediate area, there are only three or four days a year with measurable amounts of snow. Its depth on the ground rarely exceeds 2 or 3 inches, and it usually melts in a day or two. The few thunderstorms that occur each year are not generally severe and seldom do they, or the hail that occasionally accompanies them, cause any serious damage. A tornado in the immediate metropolitan area has never been recorded.

The seasonal difference in temperatures is much less marked than that of precipitation. There is a range of about 28 degrees between the temperature for January, the coldest month, and July, the warmest. Highs of 100 degrees or more seldom occur, and only in a few years since records began in 1892, have 0 degree or lower temperatures been observed. There is an average growing season of six and a half months.

The mild temperatures, long growing season, and plentiful supply of moisture are ideal for a wide variety of crops. In dollar value of agricultural returns, this is the most productive area in Oregon. Large orchards of sweet cherries are grown and processed here for maraschino cherries. Hops, filberts, walnuts, cane, and strawberries each contribute many millions of dollars to the annual farm income. A wide variety of vegetables is raised for both the fresh market and to support a large number of processing plants located in Salem. This climate is also suitable for the production of a number of specialty crops including mint, several seed crops, and nursery stock, particularly roses and ornamental shrubs.

Based on the 1951–1980 period, the average first occurrence of 32 degrees Fahrenheit in the fall is October 22 and the average last occurrence in the spring is May 5.

NORMALS, MEANS, AND EXTREMES

SALEM, OREGON

LATITUDE: 44°55'N LONGITUDE: 123°00' W ELEVATION: FT. GRND 196 BARO 199 TIME ZONE: PACIFIC WBAN: 24232

	(a)	JAN	FEB	MAR	APR	MAY	JUNE	JULY	AUG	SEP	OCT	NOV	DEC	YEAR
TEMPERATURE °F:														
Normals														
-Daily Maximum		46.4	51.5	55.7	60.4	67.0	74.5	81.6	82.0	76.0	64.2	52.4	46.4	63.2
-Daily Minimum		32.7	34.1	35.6	37.7	42.2	48.3	50.9	51.4	47.1	41.1	37.5	33.6	41.0
-Monthly		39.6	42.8	45.7	49.1	54.6	61.4	66.3	66.7	61.6	52.7	45.0	40.0	52.1
Extremes														
-Record Highest	56	65	72	80	88	100	105	108	108	104	93	72	68	108
-Year		1984	1968	1947	1957	1983	1992	1941	1981	1988	1970	1970	1993	AUG 1981
-Record Lowest	56	-10	-4	12	23	25	32	37	36	26	23	9	-12	-12
-Year		1950	1950	1971	1968	1954	1976	1962	1980	1972	1971	1955	1972	DEC 1972
NORMAL DEGREE DAYS:														
Heating (base 65°F)		787	622	598	477	322	141	41	52	131	381	600	775	4927
Cooling (base 65°F)		0	0	0	0	0	33	81	104	29	0	0	0	247
% OF POSSIBLE SUNSHINE														
MEAN SKY COVER (tenths)														
Sunrise - Sunset	49	8.3	8.1	7.9	7.5	6.9	6.4	4.2	4.6	5.1	6.8	8.2	8.7	6.9
MEAN NUMBER OF DAYS:														
Sunrise to Sunset														
-Clear	56	2.8	3.1	3.5	4.0	5.6	7.1	14.7	13.6	11.5	5.9	2.7	2.0	76.4
-Partly Cloudy	56	4.6	4.6	6.2	7.1	7.6	8.1	8.4	8.6	8.3	8.3	5.1	3.7	80.4
-Cloudy	56	23.6	20.6	21.4	18.9	17.8	14.8	7.9	8.8	10.3	16.8	22.3	25.3	208.4
Precipitation														
.01 inches or more	56	18.1	15.9	16.8	13.8	10.9	7.8	3.2	4.0	6.7	12.1	17.7	18.8	145.6
Snow, Ice pellets, hail														
1.0 inches or more	56	1.0	0.3	0.2	0.0	0.0	0.0	0.0	0.0	0.0	0.0	0.1	0.5	2.1
Thunderstorms	56	0.1	0.1	0.2	0.4	0.8	0.6	0.8	0.8	0.8	0.4	0.2	0.1	5.2
Heavy Fog Visibility														
1/4 mile or less	56	6.5	4.2	2.1	0.7	0.5	0.1	0.*	0.3	1.9	6.8	6.6	7.4	37.0
Temperature °F														
-Maximum														
90° and above	31	0.0	0.0	0.0	0.0	0.3	1.9	5.8	6.0	2.5	0.1	0.0	0.0	16.6
32° and below	31	1.1	0.1	0.0	0.0	0.0	0.0	0.0	0.0	0.0	0.0	0.2	1.2	2.7
-Minimum														
32° and below	31	14.4	11.3	9.5	5.7	0.9	0.*	0.0	0.0	0.2	3.0	7.7	12.9	65.6
0° and below	31	0.0	0.*	0.0	0.0	0.0	0.0	0.0	0.0	0.0	0.0	0.0	0.2	0.2
AVG. STATION PRESS. (mb)	21	1012.1	1010.3	1009.2	1010.8	1010.5	1010.2	1010.3	1009.2	1009.4	1011.0	1011.3	1012.5	1010.6
RELATIVE HUMIDITY (%)														
Hour 04	31	87	88	87	87	87	87	86	86	87	90	90	88	88
Hour 10 (Local Time)	31	84	81	76	70	66	62	57	58	63	76	84	86	72
Hour 16	31	75	68	61	57	53	49	41	40	45	59	76	80	59
Hour 22	31	86	84	82	79	76	74	68	70	76	85	88	87	80
PRECIPITATION (inches):														
Water Equivalent														
-Normal		5.92	4.50	4.17	2.42	1.88	1.34	0.56	0.76	1.55	2.98	6.28	6.80	39.16
-Maximum Monthly	56	15.40	12.31	8.56	5.64	4.58	4.19	2.63	4.17	3.98	11.17	15.23	12.40	15.40
-Year		1953	1949	1983	1993	1942	1984	1983	1968	1971	1947	1973	1964	JAN 1953
-Minimum Monthly	56	0.24	0.75	0.87	0.39	0.05	0.01	0.00	0.00	0.00	0.12	0.84	1.26	0.00
-Year		1985	1988	1965	1939	1992	1951	1967	1988	1975	1988	1939	1976	AUG 1988
-Maximum in 24 hrs	56	3.07	3.16	3.03	2.22	2.23	1.81	1.92	1.25	1.86	2.84	2.82	3.12	3.16
-Year		1972	1949	1943	1971	1991	1993	1987	1943	1951	1955	1950	1987	FEB 1949
Snow, Ice pellets, hail														
-Maximum Monthly	56	32.8	14.1	10.9	0.1	T	0.0	0.0	0.0	T	T	6.1	14.6	32.8
-Year		1950	1993	1951	1972	1988				1981	1990	1977	1972	JAN 1950
-Maximum in 24 hrs	56	10.8	12.0	8.5	0.1	T	0.0	0.0	0.0	T	T	6.1	9.4	12.0
-Year		1943	1993	1960	1972	1988				1981	1990	1977	1972	FEB 1993
WIND:														
Mean Speed (mph)	45	8.1	7.6	7.8	7.2	6.7	6.6	6.7	6.3	6.1	6.0	7.4	7.9	7.0
Prevailing Direction through 1963		S	S	S	S	S	S	N	NW	S	S	S	S	S
Fastest Obs. 1 Min.														
-Direction (!!)	44	18	18	19	18	25	18	24	29	18	18	18	23	18
-Speed (MPH)	44	41	46	40	44	28	25	26	24	31	58	44	45	58
-Year		1990	1958	1971	1962	1975	1974	1979	1969	1969	1962	1981	1953	OCT 1962
Peak Gust														
-Direction (!!)	10	S	S	S	S	SW	SW	NW	N	S	S	S	S	S
-Speed (mph)	10	61	47	52	41	45	36	36	30	39	40	58	54	61
-Date		1990	1984	1991	1993	1993	1988	1991	1988	1986	1985	1988	1993	JAN 1990

PRECIPITATION (inches) SALEM, OREGON

YEAR	JAN	FEB	MAR	APR	MAY	JUNE	JULY	AUG	SEP	OCT	NOV	DEC	ANNUAL
1964	11.19	0.78	3.55	1.28	0.59	1.73	0.45	0.41	0.74	0.93	8.44	12.40	42.49
1965	8.15	1.57	0.87	2.41	1.16	1.11	0.19	0.99	0.13	2.20	7.00	7.95	33.73
1966	6.60	2.24	6.08	1.07	0.78	0.58	0.53	0.40	1.66	2.06	5.88	7.32	35.20
1967	7.29	2.06	3.84	2.02	1.87	0.69	0.00	T	0.84	5.08	3.30	5.45	32.44
1968	6.37	7.73	3.32	1.47	3.46	1.29	0.39	4.17	2.48	6.14	6.49	11.05	54.36
1969	8.61	3.24	1.63	2.51	0.89	2.94	0.05	0.05	3.58	4.44	3.21	9.23	40.38
1970	13.47	4.46	1.92	2.63	1.36	0.85	0.01	T	1.81	3.25	7.18	9.74	46.68
1971	6.49	4.34	6.93	4.05	1.89	2.47	0.01	1.49	3.98	3.09	6.27	8.18	49.19
1972	7.98	4.68	4.96	3.79	2.40	0.69	0.12	0.14	2.07	0.70	3.77	8.70	40.00
1973	5.64	1.62	3.50	1.69	1.11	1.48	T	0.80	2.80	2.79	15.23	11.08	47.74
1974	10.89	5.56	7.95	1.48	0.90	0.41	1.80	0.11	0.28	2.15	7.42	6.94	45.89
1975	4.96	4.68	4.22	2.20	1.66	0.81	0.51	1.96	0.00	5.51	6.06	6.07	38.64
1976	5.47	6.92	3.66	2.00	1.33	1.04	0.67	1.89	1.13	1.51	1.13	1.26	28.01
1977	0.88	2.83	3.33	0.62	3.76	0.73	0.26	1.70	2.36	2.37	6.19	8.73	33.76
1978	5.67	3.54	1.23	3.50	2.97	0.48	1.07	2.56	2.64	0.37	4.50	2.64	31.17
1979	2.84	7.19	2.17	2.82	2.20	0.65	0.30	0.70	2.19	6.06	3.83	6.95	37.90
1980	6.58	4.04	3.48	3.58	1.53	1.99	0.22	0.04	1.05	1.45	5.24	10.44	39.64
1981	2.09	3.29	3.45	2.06	2.19	3.33	0.15	T	2.78	4.63	7.54	9.73	41.24
1982	6.11	6.02	2.81	3.80	0.73	1.36	0.33	0.41	1.83	3.04	4.92	9.26	40.62
1983	6.00	10.57	8.56	2.72	2.12	2.48	2.63	2.09	0.32	1.31	10.06	6.49	55.35
1984	2.34	5.50	4.40	3.74	3.32	4.19	0.03	T	1.19	4.31	12.70	3.71	45.43
1985	0.24	3.44	3.80	0.93	0.65	2.42	0.32	0.20	1.25	3.17	4.81	2.51	23.74
1986	6.25	8.26	2.89	1.30	2.22	0.42	1.23	T	3.12	2.71	7.24	3.86	39.50
1987	7.67	3.52	3.98	2.36	1.52	0.26	2.51	0.15	0.14	0.47	3.00	10.92	36.50
1988	6.78	0.75	3.34	3.56	2.39	1.97	0.03	0.00	0.98	0.12	9.50	3.23	32.65
1989	3.57	2.79	6.47	1.06	1.08	1.03	0.77	0.48	1.03	2.50	3.09	3.95	27.82
1990	8.51	5.31	2.39	1.67	1.94	1.09	0.47	0.85	0.42	5.63	4.91	2.82	36.01
1991	2.42	3.72	5.18	3.84	4.51	1.55	0.11	0.67	0.04	1.70	5.94	3.73	33.41
1992	4.76	4.46	0.89	4.35	0.05	1.12	0.50	0.38	0.70	3.52	4.84	7.83	33.40
1993	4.77	1.61	5.12	5.64	3.59	3.17	1.05	0.40	0.01	1.11	1.26	6.84	34.57
Record Mean	6.09	4.87	4.32	2.43	2.01	1.39	0.48	0.63	1.43	3.39	5.92	6.64	39.60

AVERAGE TEMPERATURE (deg. F) SALEM, OREGON

YEAR	JAN	FEB	MAR	APR	MAY	JUNE	JULY	AUG	SEP	OCT	NOV	DEC	ANNUAL
1964	41.7	40.3	43.7	48.1	53.7	60.3	66.4	63.8	58.8	53.2	42.0	39.6	51.0
1965	40.5	43.1	46.6	50.3	52.2	58.9	67.2	66.7	59.6	55.3	47.9	37.8	52.2
1966	39.6	40.6	44.6	50.5	54.9	61.8	65.2	66.9	63.2	52.6	47.1	44.2	52.6
1967	44.1	42.6	43.1	44.6	54.4	64.2	67.6	71.6	64.9	54.3	46.4	42.0	53.3
1968	39.7	48.1	47.3	46.9	54.5	59.6	66.1	65.2	60.5	51.5	45.3	37.0	51.8
1969	33.0	40.3	46.3	49.1	58.9	65.4	64.9	63.5	61.1	49.9	44.1	41.9	51.5
1970	41.6	44.9	45.5	45.8	54.2	64.4	67.2	65.4	58.3	50.7	43.6	39.0	51.7
1971	39.1	39.4	41.9	47.1	53.3	58.2	67.3	69.1	59.1	50.6	44.5	39.5	50.8
1972	38.3	43.2	47.9	46.0	56.5	60.0	67.6	68.2	57.6	50.1	46.4	35.2	51.4
1973	38.8	43.9	44.6	49.1	55.8	61.1	67.5	63.6	62.1	51.9	44.0	43.4	52.2
1974	37.6	41.2	46.1	49.3	52.9	61.8	65.8	67.4	65.8	53.0	46.4	43.0	52.5
1975	42.6	41.9	43.6	45.3	54.3	59.9	67.1	63.1	63.5	51.6	43.9	41.9	51.6
1976	40.7	40.2	42.5	47.2	53.2	56.9	65.3	64.2	62.0	52.1	45.9	38.8	50.8
1977	37.7	44.5	44.5	50.6	51.8	61.8	64.5	70.1	59.0	52.1	44.6	43.2	52.0
1978	41.4	45.3	48.2	48.9	54.8	65.2	68.8	68.3	61.0	54.1	39.4	34.8	52.5
1979	31.2	42.7	49.0	50.5	56.8	61.8	67.9	65.7	62.9	54.7	41.8	44.8	52.5
1980	35.6	43.2	44.6	50.7	53.7	57.5	65.5	62.8	62.5	53.6	47.3	42.9	51.6
1981	40.6	43.1	46.6	50.6	54.5	59.0	65.1	68.7	61.9	50.4	44.6	42.9	52.3
1982	38.6	41.6	44.5	45.9	54.7	65.1	65.7	66.4	61.4	53.4	42.4	41.1	51.7
1983	43.3	46.1	48.9	50.1	57.3	60.3	64.3	67.8	60.3	52.7	49.3	38.4	53.3
1984	43.2	44.2	48.7	47.8	52.9	57.4	65.4	66.8	61.3	51.4	44.4	37.5	51.8
1985	34.3	39.6	42.8	51.4	55.2	61.2	70.2	65.4	58.4	51.2	37.3	33.3	50.0
1986	43.2	44.4	50.2	48.4	55.3	64.2	63.5	69.9	59.0	53.7	46.7	39.6	53.2
1987	39.5	43.8	47.3	51.9	58.1	64.0	64.9	67.8	62.7	56.1	46.9	38.9	53.5
1988	39.4	43.3	45.6	51.1	53.7	60.5	66.8	65.8	62.3	56.6	46.8	40.5	52.7
1989	41.6	34.8	46.4	54.6	56.2	63.3	64.6	64.9	63.1	53.0	47.4	39.6	52.5
1990	42.8	41.3	47.8	52.8	55.2	62.3	69.6	69.1	64.9	51.2	46.9	34.4	53.2
1991	39.5	48.7	44.6	49.2	53.1	57.9	68.0	67.8	65.4	54.3	47.3	41.6	53.1
1992	44.1	47.6	50.5	53.8	60.8	65.8	68.9	68.2	62.0	55.1	46.1	39.2	55.2
1993	36.4	39.4	49.1	52.3	60.5	61.3	63.2	67.2	63.3	56.5	40.5	41.1	52.6
Record Mean	39.2	42.8	45.8	49.9	55.6	61.3	66.6	66.4	62.0	53.3	45.0	40.5	52.4
Max	45.9	51.2	55.6	61.2	68.0	74.2	82.0	81.7	76.5	64.5	52.6	46.8	63.3
Min	32.4	34.4	35.9	38.6	43.2	48.3	51.1	51.2	47.4	42.0	37.3	34.3	41.4

REFERENCE NOTES SALEM, OREGON

GENERAL
T - TRACE AMOUNT.
BLANK ENTRIES DENOTE MISSING/UNREPORTED DATA.
# INDICATES A STATION OR INSTRUMENT RELOCATION.

SPECIFIC

(a) - LENGTH OF RECORD IN YEARS, ALTHOUGH INDIVIDUAL MONTHS MAY BE MISSING.
0.* OR * - THE VALUE IS BETWEEN 0.0 AND 0.05
NORMALS - BASED ON THE 1961-1990 RECORD PERIOD.
EXTREMES - DATES ARE THE MOST RECENT OCCURRENCE
WIND DIR. - NUMERALS SHOW TENS OF DEGRESS CLOCKWISE FROM TRUE NORTH. "00" INDICATES CALM.
RESULTANT DIRECTIONS ARE GIVEN TO WHOLE DEGREES.
BOLD VALUES INDICATE EXTREME VALUES WHICH OCCURRED AFTER THE ASOS SYSTEM WAS COMMISSIONED.

RECORD = PERIOD OF RECORD
RECORD MEAN PRECIPITATION IS THE MEAN OF ALL DAILY PRECIPITATION AMOUNTS DURING THE PERIOD OF RECORD.
RECORD MAX(MIN) TEMPERATURE IS THE MEAN OF ALL DAILY MAX(MIN) TEMPERATURES DURING THE PERIOD OF RECORD.
RECORD MEAN TEMPERATURE IS THE SUM OF THE RECORD MAX AND RECORD MIN DIVIDED BY 2.
AVERAGE TEMPERATURE IS THE SUM OF THE MEAN DAILY MAX AND MIN TEMPERATURE DIVIDED BY 2.

EXCEPTIONS

RECORD MEANS ARE THROUGH THE CURRENT YEAR, BEGINNING IN 1938 FOR TEMPERATURE
1938 FOR PRECIPITATION
1938 FOR SNOWFALL

HEATING DEGREE DAYS Base 65 deg. F — SALEM, OREGON

SEASON	JULY	AUG	SEP	OCT	NOV	DEC	JAN	FEB	MAR	APR	MAY	JUNE	TOTAL
1964-65	41	79	181	357	684	783	751	607	565	433	390	188	5059
1965-66	46	26	166	293	505	836	781	679	629	425	308	114	4808
1966-67	44	24	74	375	530	637	642	620	669	604	323	73	4615
1967-68	11	2	47	323	549	706	778	481	541	536	318	167	4459
1968-69	47	49	142	413	581	860	987	685	575	469	195	49	5052
1969-70	45	70	137	459	621	710	719	555	596	571	329	94	4906
1970-71	33	45	195	438	635	797	796	710	706	528	355	201	5439
1971-72	61	18	169	440	607	784	821	625	523	560	267	148	5023
1972-73	34	42	239	454	552	919	803	586	626	468	288	141	5152
1973-74	30	85	106	398	624	665	845	658	580	463	371	123	4948
1974-75	48	28	47	364	550	674	688	641	656	584	327	164	4771
1975-76	37	84	79	409	624	708	743	709	693	529	356	243	5214
1976-77	43	63	98	395	565	807	842	569	629	424	400	108	4943
1977-78	57	23	181	392	606	670	727	547	513	476	316	51	4559
1978-79	17	20	129	328	763	928	1038	619	486	430	249	121	5128
1979-80	24	21	76	315	691	620	905	627	627	423	344	217	4890
1980-81	51	82	87	357	525	682	746	610	566	426	318	176	4626
1981-82	59	33	124	447	607	676	814	649	629	568	317	85	5008
1982-83	42	27	128	356	671	732	668	524	491	439	253	135	4466
1983-84	59	3	135	376	463	821	667	597	501	511	367	231	4731
1984-85	44	12	124	418	610	846	947	706	684	403	299	132	5225
1985-86	5	51	196	423	824	975	668	574	452	494	324	82	5068
1986-87	64	3	210	345	542	779	781	586	544	389	237	89	4569
1987-88	54	16	92	272	536	803	788	622	595	411	346	145	4680
1988-89	47	41	125	264	538	752	716	841	570	308	267	90	4559
1989-90	44	39	81	367	523	779	682	658	527	363	297	103	4463
1990-91	14	16	43	419	535	939	784	449	624	464	360	210	4857
1991-92	10	32	61	329	524	717	641	498	443	332	152	56	3795
1992-93	12	22	108	303	558	792	881	711	486	373	150	117	4513
1993-94	60	32	93	257	729	734							

COOLING DEGREE DAYS Base 65 deg. F — SALEM, OREGON

YEAR	JAN	FEB	MAR	APR	MAY	JUNE	JULY	AUG	SEP	OCT	NOV	DEC	TOTAL
1969	0	0	0	0	12	66	50	28	27	0	0	0	183
1970	0	0	0	0	2	82	106	66	0	1	0	0	257
1971	0	0	0	0	0	5	140	152	2	0	0	0	299
1972	0	0	0	0	10	6	122	148	21	0	0	0	307
1973	0	0	0	0	6	34	115	47	29	0	0	0	231
1974	0	0	0	0	0	34	78	108	78	0	0	0	298
1975	0	0	0	0	1	16	109	30	41	0	0	0	197
1976	0	0	0	0	0	6	59	48	16	3	0	0	132
1977	0	0	0	0	0	19	50	187	5	0	0	0	261
1978	0	0	0	0	6	63	143	130	16	0	0	0	358
1979	0	0	0	0	0	29	121	48	20	1	0	0	219
1980	0	0	0	0	0	0	73	21	17	9	0	0	120
1981	0	0	0	1	0	2	69	156	38	0	0	0	266
1982	0	0	0	0	5	91	68	82	27	0	0	0	273
1983	0	0	0	0	21	4	46	94	3	0	0	0	168
1984	0	0	0	0	1	10	64	73	19	5	0	0	172
1985	0	0	0	0	1	26	171	71	1	0	0	0	270
1986	0	0	0	0	28	64	25	159	34	0	0	0	310
1987	0	0	0	0	27	64	59	110	31	4	0	0	295
1988	0	0	0	0	3	16	108	75	48	10	0	0	260
1989	0	0	0	0	4	46	39	46	32	0	0	0	167
1990	0	0	0	0	1	31	165	153	46	0	0	0	396
1991	0	0	0	0	0	1	111	127	78	0	0	0	317
1992	0	0	0	1	30	83	138	128	25	0	0	0	405
1993	0	0	0	0	17	11	10	108	51	1	0	0	198

SNOWFALL (inches) — SALEM, OREGON

SEASON	JULY	AUG	SEP	OCT	NOV	DEC	JAN	FEB	MAR	APR	MAY	JUNE	TOTAL
1964-65	0.0	0.0	0.0	0.0	T	5.7	T	0.0	T	0.0	0.0	0.0	5.7
1965-66	0.0	0.0	0.0	0.0	0.0	7.4	1.7	0.2	0.6	0.0	0.0	0.0	9.9
1966-67	0.0	0.0	0.0	0.0	0.0	T	T	T	T	T	0.0	0.0	T
1967-68	0.0	0.0	0.0	0.0	0.0	1.0	9.3	0.0	0.0	T	0.0	0.0	10.3
1968-69	0.0	0.0	0.0	0.0	T	12.6	21.9	0.4	0.0	0.0	0.0	0.0	34.9
1969-70	0.0	0.0	0.0	0.0	0.0	0.0	0.2	0.0	T	T	0.0	0.0	0.2
1970-71	0.0	0.0	0.0	T	T	0.9	11.1	8.1	1.4	T	T	0.0	21.5
1971-72	0.0	0.0	0.0	0.0	0.0	2.2	0.4	T	0.3	0.1	0.0	0.0	3.0
1972-73	0.0	0.0	0.0	0.0	0.0	14.6	0.7	0.0	T	T	0.0	0.0	15.3
1973-74	0.0	0.0	0.0	0.0	0.5	0.0	T	T	T	0.0	T	0.0	0.5
1974-75	0.0	0.0	0.0	T	0.0	0.0	1.6	T	T	T	T	0.0	1.6
1975-76	0.0	0.0	0.0	0.0	T	1.5	0.0	T	0.3	T	0.0	0.0	1.8
1976-77	0.0	0.0	0.0	0.0	0.0	0.0	T	0.0	T	0.0	T	0.0	T
1977-78	0.0	0.0	0.0	0.0	6.1	0.0	0.0	0.0	0.0	T	T	0.0	6.1
1978-79	0.0	0.0	0.0	0.0	1.5	2.9	2.2	0.6	0.0	T	0.0	0.0	7.2
1979-80	0.0	0.0	0.0	0.0	0.0	T	1.8	T	T	T	T	0.0	1.8
1980-81	0.0	0.0	0.0	0.0	0.0	T	0.0	0.0	0.0	0.0	0.0	0.0	T
1981-82	0.0	0.0	T	0.0	T	T	4.4	0.6	T	T	0.0	0.0	5.0
1982-83	0.0	0.0	0.0	0.0	0.0	0.0	0.0	0.0	T	T	T	0.0	T
1983-84	0.0	0.0	0.0	0.0	T	4.3	T	0.0	0.0	T	T	0.0	4.3
1984-85	0.0	0.0	0.0	0.0	0.0	0.3	T	4.6	T	T	0.0	0.0	4.9
1985-86	0.0	0.0	0.0	0.0	2.0	6.5	0.0	3.6	0.0	0.0	0.0	0.0	12.1
1986-87	0.0	0.0	0.0	0.0	0.0	0.0	0.0	0.0	0.0	T	T	0.0	T
1987-88	0.0	0.0	0.0	0.0	0.0	5.8	1.0	0.0	0.0	T	T	0.0	6.8
1988-89	0.0	0.0	0.0	0.0	0.0	T	T	8.9	0.7	T	0.0	0.0	9.6
1989-90	0.0	0.0	0.0	0.0	0.0	0.0	T	9.6	0.0	0.0	0.0	0.0	9.6
1990-91	0.0	0.0	0.0	T	0.0	1.2	T	0.0	0.0	T	0.0	0.0	1.2
1991-92	0.0	0.0	0.0	0.0	0.0	0.0	0.0	0.0	T	T	0.0	0.0	T
1992-93	0.0	0.0	0.0	0.0	0.0	7.9	10.0	14.1	0.0	0.0	0.0	0.0	32.0
1993-94	0.0	0.0	0.0	0.0	0.4	T							
Record Mean	0.0	0.0	T	T	0.3	1.5	3.0	1.3	0.6	T	T	0.0	6.7

CHUUK, EASTERN CAROLINE IS., PACIFIC

Truk Atoll, located in the Western Pacific about 500 miles north of the equator, is comprised of numerous small coral and high volcanic islands scattered over a large lagoon and surrounded by a 125 mile-long barrier reef. Moen Island, roughly triangular in shape and 7.3 square miles in area, is the second largest of the 19 volcanic islands. Its highest point is Mount Teroken, which rises to an elevation of 1,214 feet, and has dense, tropical forests covering its upper reaches and small streams descending its slopes.

The Weather Station is situated on northwestern Moen about 300 feet east of Truk Lagoon and immediately south of the approach end of the airport runway.

A steep-sided 754 foot hill lies a short distance northeast of the Weather Station, and a northeast to southwest trending mountain ridge, of which Mount Teroken is part, to the east. In funneling between these high points, the northeasterly trade winds are deflected to north-northeasterly, and at times generate local eddies with a southerly component near the Weather Station.

From about November to June the climate of Truk is chiefly influenced by the northeasterly trade winds, with average monthly speeds of 8 to 12 miles per hour. By about April, however, the trades begin to weaken, and by July have given way to the lighter and more variable winds of the doldrums. Between July and November the island is frequently under the influence of the Intertropical Convergence Zone which has moved northward into the area. This is also the season when moist southerly winds and tropical disturbances, many of them associated with the ITCZ, are most frequent and when humidities are often oppressively high.

Rainfall at Truk averages about 140 inches a year. A relatively dry period occurs from January to March,when monthly averages run below 8.50 inches. February is the driest month. Rainfall varies widely from year to year and seasonally. Annual totals at Truk have been as low as 120 inches and as high as 180 inches, while even during the ordinarily drier season of January to March, the monthly rainfall has been as much as 24 inches in individual years. However, it has also been less than 1 inch, so that extended dry spells are not uncommon.

The temperature is remarkably uniform throughout the year, with less than one-half degree separating the averages of the warmest and coolest months. With highs in the mid 80s and lows in the mid 70s, the daily range is about 10 degrees. Temperatures below 70 degrees are rare. Every month has had a high temperature of at least 90 degrees. Humidities are high throughout the year.

Although the major typhoon tracks of the Western Pacific lie well to the north and west of Truk, several of the storms have passed close to or over the island within recent years. The storms have caused widespread damage to buildings and to coconut palms and other crops, as well as wave damage to shorelines and coastal structures.

NORMALS, MEANS, AND EXTREMES

CHUUK, EASTERN CAROLINE IS., PACIFIC

LATITUDE: 7°27′N LONGITUDE: 151°50′E ELEVATION: FT. GRND 5 BARO 8 TIME ZONE: 150E MER WBAN: 40505

	(a)	JAN	FEB	MAR	APR	MAY	JUNE	JULY	AUG	SEP	OCT	NOV	DEC	YEAR
TEMPERATURE °F:														
Normals														
-Daily Maximum		85.8	85.9	86.3	86.7	87.2	87.2	87.1	87.3	87.3	87.3	87.2	86.4	86.8
-Daily Minimum		76.9	77.0	77.1	77.0	76.8	76.2	75.5	75.5	75.6	75.7	76.3	77.0	76.4
-Monthly		81.4	81.5	81.7	81.9	82.0	81.8	81.3	81.4	81.5	81.6	81.8	81.7	81.6
Extremes														
-Record Highest	43	91	91	94	92	94	93	92	92	93	92	91	91	94
-Year		1969	1946	1946	1982	1946	1957	1984	1981	1981	1981	1993	1981	MAR 1946
-Record Lowest	43	69	70	71	71	70	70	70	70	68	66	70	70	66
-Year		1990	1986	1968	1967	1980	1965	1974	1968	1973	1980	1990	1980	OCT 1980
NORMAL DEGREE DAYS:														
Heating (base 65°F)		0	0	0	0	0	0	0	0	0	0	0	0	0
Cooling (base 65°F)		508	462	518	507	527	504	505	508	495	515	504	518	6071
% OF POSSIBLE SUNSHINE	33	52	57	57	52	49	46	49	49	46	43	45	45	49
MEAN SKY COVER (tenths)														
Sunrise - Sunset	42	9.1	9.3	9.2	9.2	9.1	9.2	9.1	9.1	9.1	9.1	9.1	9.3	9.2
MEAN NUMBER OF DAYS:														
Sunrise to Sunset														
-Clear	42	0.3	0.2	0.3	0.2	0.4	0.2	0.2	0.1	0.3	0.3	0.1	0.1	2.6
-Partly Cloudy	42	3.6	2.9	3.3	3.7	4.0	3.6	4.1	4.3	4.0	4.1	3.8	3.2	44.8
-Cloudy	42	27.1	25.1	27.4	26.0	26.6	26.2	26.7	26.6	25.7	26.6	26.0	27.7	317.9
Precipitation														
.01 inches or more	42	19.3	16.0	18.7	20.4	24.4	24.0	24.4	24.5	22.3	23.5	23.7	23.0	264.3
Snow,Ice pellets,hail														
1.0 inches or more	42	0.0	0.0	0.0	0.0	0.0	0.0	0.0	0.0	0.0	0.0	0.0	0.0	0.0
Thunderstorms	42	0.9	0.2	1.0	1.2	1.7	1.6	1.7	1.3	1.8	2.0	2.2	1.5	17.2
Heavy Fog Visibility														
1/4 mile or less	42	0.0	0.0	0.0	0.0	0.0	0.0	0.0	0.0	0.0	0.0	0.0	0.0	0.0
Temperature °F														
-Maximum														
90° and above	42	0.1	0.*	0.2	0.6	2.2	2.0	2.6	2.8	3.0	2.8	1.7	0.4	18.3
32° and below	42	0.0	0.0	0.0	0.0	0.0	0.0	0.0	0.0	0.0	0.0	0.0	0.0	0.0
-Minimum														
32° and below	42	0.0	0.0	0.0	0.0	0.0	0.0	0.0	0.0	0.0	0.0	0.0	0.0	0.0
0° and below	42	0.0	0.0	0.0	0.0	0.0	0.0	0.0	0.0	0.0	0.0	0.0	0.0	0.0
AVG. STATION PRESS.(mb)	15	1008.9	1009.7	1009.6	1009.2	1009.2	1009.4	1009.2	1009.4	1009.3	1008.7	1008.2	1008.6	1009.1
RELATIVE HUMIDITY (%)														
Hour 04	23	82	81	82	84	86	87	89	89	88	88	87	84	86
Hour 10 (Local Time)	42	76	76	77	79	80	80	80	80	80	80	80	79	79
Hour 16	42	76	75	75	77	78	78	78	77	77	78	78	78	77
Hour 22	42	81	80	81	83	85	85	87	87	86	86	85	83	84
PRECIPITATION (inches):														
Water Equivalent														
-Normal		8.98	6.42	9.05	11.46	13.94	11.84	14.37	13.77	12.07	14.23	11.10	11.55	138.78
-Maximum Monthly	44	19.19	15.95	24.02	23.38	28.39	21.72	32.99	25.96	21.17	24.71	26.12	34.89	34.89
-Year		1981	1991	1967	1956	1976	1950	1962	1979	1955	1979	1962	1959	DEC 1959
-Minimum Monthly	44	0.96	0.56	1.65	3.28	3.80	6.10	2.65	5.37	5.24	4.17	1.88	3.12	0.56
-Year		1959	1983	1992	1983	1983	1966	1984	1949	1989	1972	1982	1990	FEB 1983
-Maximum in 24 hrs	44	6.78	7.89	8.21	7.16	11.13	7.61	10.07	4.91	6.24	6.55	10.41	14.92	14.92
-Year		1985	1993	1972	1989	1976	1972	1962	1963	1978	1968	1962	1959	DEC 1959
Snow,Ice pellets,hail														
-Maximum Monthly		0.0	0.0	0.0	0.0	0.0	0.0	0.0	0.0	0.0	0.0	0.0	0.0	
-Year														
-Maximum in 24 hrs	44	0.0	0.0	0.0	0.0	0.0	0.0	0.0	0.0	0.0	0.0	0.0	0.0	
-Year														
WIND:														
Mean Speed (mph)	28	10.8	11.3	10.7	9.7	8.6	7.3	7.3	7.3	7.6	7.8	8.0	9.8	8.9
Prevailing Direction through 1963		NNE	NNE	NNE	NNE	NNE	NNE	SE	S	SW	S	NNE	NNE	NNE
Fastest Mile														
-Direction (!!)	27	NW	S	NE	NE	S	SW	NW	W	SW	NW	N	W	S
-Speed (MPH)	27	37	31	34	40	78	40	41	38	50	41	45	39	78
-Year		1985	1962	1978	1971	1971	1972	1962	1979	1972	1979	1962	1979	MAY 1971
Peak Gust														
-Direction (!!)	10	SE	NE	NE	NE	E	SE	SE	SW	SW	E	S	NE	S
-Speed (mph)	10	53	48	54	55	52	49	46	46	43	58	94	47	94
-Date		1987	1990	1991	1990	1985	1986	1989	1992	1986	1986	1987	1993	NOV 1987

PRECIPITATION (inches)

CHUUK, EASTERN CAROLINE IS., PACIFIC

YEAR	JAN	FEB	MAR	APR	MAY	JUNE	JULY	AUG	SEP	OCT	NOV	DEC	ANNUAL
1964	2.00	10.80	2.44	12.29	18.45	9.99	13.55	12.47	16.88	15.80	7.85	17.73	140.25
1965	13.86	6.70	15.30	8.17	10.79	12.17	25.19	9.13	16.97	9.53	5.54	4.26	137.61
1966	4.61	1.70	7.57	7.53	13.62	6.10	20.11	12.39	12.49	8.88	8.44	18.21	121.65
1967	8.04	6.38	24.02	17.21	15.17	14.13	19.90	17.36	7.82	15.80	15.55	14.17	175.55
1968	8.67	9.23	13.91	20.50	10.00	14.20	15.75	7.77	12.37	11.78	6.91	24.51	155.60
1969	1.22	1.44	3.82	11.28	19.26	14.91	16.38	14.29	12.26	10.38	16.00	10.39	131.63
1970	14.80	11.80	2.40	10.43	18.60	13.99	7.49	12.98	10.75	19.04	8.43	13.57	144.28
1971	8.25	8.63	9.85	10.20	15.33	13.92	13.18	10.63	14.54	16.40	5.20	8.04	134.17
1972	9.83	10.65	18.49	15.79	16.68	14.73	16.68	11.58	13.60	4.17	7.49	9.26	148.95
1973	1.36	2.30	4.59	8.83	8.96	10.31	13.78	13.18	11.34	21.16	8.59	17.60	122.00
1974	10.23	13.44	19.75	11.59	13.47	14.83	12.49	10.72	14.33	20.14	14.91	8.84	164.74
1975	3.71	3.86	11.17	4.25	17.91	16.12	7.35	13.72	12.02	12.24	17.44	9.99	129.78
1976	10.57	9.37	5.70	17.80	28.39	12.26	11.55	14.74	15.14	15.22	16.09	6.41	163.24
1977	6.44	1.98	8.31	11.47	11.67	7.07	9.11	14.20	13.94	16.21	12.45	3.24	116.09
1978	5.73	2.29	4.85	8.17	13.25	10.10	8.40	14.37	14.98	21.21	12.99	12.47	128.81
1979	7.69	4.39	7.83	20.32	13.91	19.02	9.02	25.96	10.44	24.71	20.97	7.36	171.62
1980	13.91	5.96	8.14	6.55	18.26	12.88	17.26	12.78	8.50	18.91	2.98	18.44	144.57
1981	19.19	4.41	7.24	12.54	6.04	14.46	8.44	10.76	11.35	14.48	9.62	18.20	136.73
1982	7.04	5.92	11.21	8.67	14.68	11.99	11.55	10.70	9.38	6.76	1.88	4.61	104.39
1983	5.16	0.56	1.95	3.28	3.80	9.28	23.09	12.84	9.75	15.32	12.08	15.17	112.28
1984	12.92	10.10	8.27	7.03	11.06	7.47	2.65	14.88	5.35	18.06	14.58	6.84	119.21
1985	16.99	7.85	3.64	16.39	11.67	10.21	12.24	9.73	15.81	7.47	8.70	12.91	133.61
1986	14.40	14.82	18.35	9.45	10.86	8.81	14.92	8.31	10.74	5.60	12.67	11.45	140.38
1987	9.51	0.58	7.69	19.80	5.89	13.34	14.68	15.70	5.36	7.12	14.14	10.89	124.70
1988	3.46	6.65	2.94	3.49	16.18	8.45	11.41	10.44	11.21	21.17	8.84	11.86	116.10
1989	9.32	3.85	7.13	22.51	13.47	10.50	18.20	14.43	5.24	15.72	8.94	11.27	140.58
1990	12.16	2.02	10.72	12.65	11.54	11.22	10.57	24.39	10.89	9.25	12.59	3.12	131.12
1991	11.42	15.95	15.67	18.92	16.81	11.25	9.91	9.60	18.57	6.14	10.21	4.45	148.90
1992	4.80	2.70	1.65	5.09	7.52	12.60	10.72	18.96	12.07	13.24	7.93	7.72	105.00
1993	14.64	10.36	9.43	16.64	15.49	17.29	10.80	16.36	14.50	12.92	9.04	21.00	168.47
Record Mean	8.87	6.39	8.59	12.29	14.41	12.33	13.76	13.77	12.69	13.65	11.41	12.51	140.67

AVERAGE TEMPERATURE (deg. F)

CHUUK, EASTERN CAROLINE IS., PACIFIC

YEAR	JAN	FEB	MAR	APR	MAY	JUNE	JULY	AUG	SEP	OCT	NOV	DEC	ANNUAL
1964	81.8	80.7	81.7	81.4	81.1	81.2	80.5	80.2	80.2	80.8	81.4	80.2	80.9
1965	80.2	80.3	80.0	80.4	80.8	80.1	79.0	80.7	80.5	80.6	80.9	81.3	80.4
1966	80.1	81.3	81.4	82.0	81.3	81.1	81.1	81.3	81.2	81.4	81.5	80.8	81.2
1967	81.3	81.3	80.7	80.5	81.5	80.8	80.4	80.1	80.8	81.0	81.2	81.6	80.9
1968	81.0	81.0	80.8	80.7	81.4	81.5	80.2	81.1	80.9	81.2	80.8	80.8	80.9
1969	80.7	80.2	81.4	81.0	81.7	81.6	80.7	81.0	81.0	81.4	81.9	81.8	81.2
1970	81.5	81.9	82.6	82.4	81.8	81.8	81.8	81.6	81.7	81.1	82.1	81.9	81.8
1971	81.5	81.3	81.5	81.6	81.2	81.0	80.4	81.5	81.1	81.3	82.1	81.7	81.3
1972	80.9	80.5	81.1	81.0	81.5	81.4	80.7	80.7	81.3	81.6	82.1	81.4	81.2
1973	81.6	81.2	82.2	81.8	82.2	82.4	81.6	81.8	81.7	81.0	82.2	81.8	81.8
1974	81.2	81.2	81.6	81.8	82.0	81.4	81.2	81.3	81.5	81.4	81.8	81.7	81.5
1975	81.6	81.7	81.7	82.2	81.4	81.2	81.0	81.1	81.1	81.0	80.5	81.2	81.3
1976	81.2	80.8	81.3	80.9	80.9	80.9	81.3	80.7	80.5	81.7	81.6	81.7	81.1
1977	81.3	81.6	81.4	82.1	82.0	82.5	81.7	81.9	81.7	82.1	81.2	82.2	81.8
1978	81.3	81.7	82.4	82.1	82.1	82.0	82.4	82.1	81.9	81.8	81.9	82.0	82.0
1979	81.9	82.2	81.8	81.8	81.9	82.3	81.8	81.3	82.4	81.3	81.6	81.5	81.8
1980	81.2	81.5	81.8	82.5	82.0	81.8	80.8	81.1	81.5	80.4	81.3	81.2	81.4
1981	81.0	82.0	82.2	82.6	83.0	82.0	82.7	82.5	82.4	82.3	82.4	82.4	82.3
1982	82.7	82.2	81.8	82.7	82.4	82.3	82.1	81.9	82.2	82.0	82.6	82.1	82.2
1983	80.8	81.5	81.8	82.8	84.0	83.4	81.8	82.5	82.6	82.5	82.7	82.4	82.4
1984	81.4	81.7	82.5	83.3	83.6	81.5	83.8	81.6	82.6	81.9	82.7	83.1	82.5
1985	81.5	82.6	82.9	82.1	82.8	82.7	81.7	81.9	81.6	82.4	82.9	82.7	82.3
1986	81.9	81.0	81.7	82.9	83.0	82.6	82.0	82.6	81.6	82.1	82.0	81.8	82.1
1987	81.8	82.9	81.8	81.8	83.9	82.7	81.2	81.3	82.6	83.0	82.8	82.7	82.4
1988	82.3	81.9	83.9	83.7	82.5	82.8	82.7	82.8	82.5	82.1	83.0	81.9	82.7
1989	82.7	82.9	82.1	81.5	82.0	81.8	81.5	81.5	82.2	82.0	82.1	81.9	82.0
1990	82.0	82.2	81.9	82.3	82.3	82.1	81.4	80.8	81.9	82.0	82.1	82.2	81.9
1991	81.4	80.8	81.6	82.2	82.4	82.3	81.9	82.0	81.0	82.0	81.7	82.2	81.8
1992	81.3	81.1	82.2	82.8	83.2	82.6	82.3	81.4	82.0	82.4	81.8	81.2	82.0
1993	80.9	81.4	81.1	81.7	82.2	81.3	82.3	81.1	81.2	81.5	82.1	81.3	81.5
Record Mean	81.3	81.4	81.7	81.8	81.8	81.7	81.3	81.2	81.4	81.4	81.7	81.6	81.5
Max	85.6	85.8	86.2	86.5	87.0	87.0	87.0	87.0	87.2	87.2	87.0	86.2	86.6
Min	76.9	76.9	77.1	77.1	76.7	76.3	75.6	75.4	75.6	75.7	76.3	77.0	76.4

REFERENCE NOTES

CHUUK, EASTERN CAROLINE IS., PACIFIC

GENERAL
T - TRACE AMOUNT.
BLANK ENTRIES DENOTE MISSING/UNREPORTED DATA.
# INDICATES A STATION OR INSTRUMENT RELOCATION.

SPECIFIC

(a) - LENGTH OF RECORD IN YEARS, ALTHOUGH INDIVIDUAL MONTHS MAY BE MISSING.
0.* OR * - THE VALUE IS BETWEEN 0.0 AND 0.05
NORMALS - BASED ON THE 1961-1990 RECORD PERIOD.
EXTREMES - DATES ARE THE MOST RECENT OCCURRENCE.
WIND DIR.- NUMERALS SHOW TENS OF DEGRESS CLOCKWISE FROM TRUE NORTH. "00" INDICATES CALM.
RESULTANT DIRECTIONS ARE GIVEN TO WHOLE DEGREES.
BOLD VALUES INDICATE EXTREME VALUES WHICH OCCURRED AFTER THE ASOS SYSTEM WAS COMMISSIONED.

RECORD = PERIOD OF RECORD
RECORD MEAN PRECIPITATION IS THE MEAN OF ALL DAILY PRECIPITATION AMOUNTS DURING THE PERIOD OF RECORD.
RECORD MAX(MIN) TEMPERATURE IS THE MEAN OF ALL DAILY MAX(MIN) TEMPERATURES DURING THE PERIOD OF RECORD.
RECORD MEAN TEMPERATURE IS THE SUM OF THE RECORD MAX AND RECORD MIN DIVIDED BY 2.
AVERAGE TEMPERATURE IS THE SUM OF THE MEAN DAILY MAX AND MIN TEMPERATURE DIVIDED BY 2.

EXCEPTIONS

THUNDERSTORMS AND HEAVY FOG ARE FOR APRIL 1973 TO DECEMBER 1978 AND SEPETEMBER 1979 TO DATE. THESE DATA MAY BE INCOMPLETE, DUE TO PART-TIME OPERATIONS.

RECORD MEANS ARE THROUGH THE CURRENT YEAR, BEGINNING IN 1952 FOR TEMPERATURE
1948 FOR PRECIPITATION

HEATING DEGREE DAYS Base 65 deg. F — CHUUK, EASTERN CAROLINE IS., PACIFIC

SEASON	JULY	AUG	SEP	OCT	NOV	DEC	JAN	FEB	MAR	APR	MAY	JUNE	TOTAL
1983-84	0	0	0	0	0	0	0	0	0	0	0	0	0
1984-85	0	0	0	0	0	0	0	0	0	0	0	0	0
1985-86	0	0	0	0	0	0	0	0	0	0	0	0	0
1986-87	0	0	0	0	0	0	0	0	0	0	0	0	0
1987-88	0	0	0	0	0	0	0	0	0	0	0	0	0
1988-89	0	0	0	0	0	0	0	0	0	0	0	0	0
1989-90	0	0	0	0	0	0	0	0	0	0	0	0	0
1990-91	0	0	0	0	0	0	0	0	0	0	0	0	0
1991-92	0	0	0	0	0	0	0	0	0	0	0	0	0
1992-93	0	0	0	0	0	0	0	0	0	0	0	0	0
1993-94	0	0	0	0	0	0							

COOLING DEGREE DAYS Base 65 deg. F — CHUUK, EASTERN CAROLINE IS., PACIFIC

YEAR	JAN	FEB	MAR	APR	MAY	JUNE	JULY	AUG	SEP	OCT	NOV	DEC	TOTAL
1969	493	432	514	486	525	503	492	504	489	518	511	530	5997
1970	516	478	551	528	527	509	531	525	509	506	517	532	6229
1971	517	465	515	507	507	487	482	519	491	509	517	525	6041
1972	500	456	506	487	521	499	492	493	496	521	519	518	6008
1973	520	463	540	509	540	529	523	529	507	503	523	527	6213
1974	508	459	525	511	536	499	507	512	502	513	508	526	6106
1975	522	475	524	522	514	494	501	506	489	501	473	506	6027
1976	507	466	509	483	500	483	512	493	469	524	505	524	5975
1977	515	471	516	517	534	531	529	529	508	538	491	540	6219
1978	514	475	545	520	537	515	546	543	517	528	511	534	6285
1979	533	489	531	511	531	527	529	512	531	513	504	517	6228
1980	507	483	527	531	534	512	496	505	500	483	495	509	6082
1981	501	481	540	534	565	518	557	547	529	543	530	546	6391
1982	557	488	527	538	547	526	540	534	523	533	533	541	6387
1983	497	469	527	540	594	558	528	550	535	550	538	548	6434
1984	513	487	548	556	582	504	590	521	534	530	539	566	6470
1985	517	502	560	517	556	538	526	532	507	550	543	553	6401
1986	530	453	526	542	563	534	538	552	507	537	516	528	6326
1987	530	505	528	512	591	538	509	512	536	565	540	554	6420
1988	542	496	593	567	552	536	556	559	534	536	545	530	6546
1989	555	506	538	503	533	510	520	516	524	536	521	529	6291
1990	533	487	531	527	543	524	516	498	515	531	516	539	6260
1991	516	450	521	524	547	528	529	534	486	532	508	537	6212
1992	512	472	542	540	571	535	542	517	518	546	510	510	6315
1993	498	464	506	507	538	497	545	505	494	519	518	514	6105

SNOWFALL (inches) — CHUUK, EASTERN CAROLINE IS., PACIFIC

SEASON	JULY	AUG	SEP	OCT	NOV	DEC	JAN	FEB	MAR	APR	MAY	JUNE	TOTAL
1971-72	0.0	0.0	0.0	0.0	0.0	0.0	0.0	0.0	0.0	0.0	0.0	0.0	0.0
1972-73	0.0	0.0	0.0	0.0	0.0	0.0	0.0	0.0	0.0	0.0	0.0	0.0	0.0
1973-74	0.0	0.0	0.0	0.0	0.0	0.0	0.0	0.0	0.0	0.0	0.0	0.0	0.0
1974-75	0.0	0.0	0.0	0.0	0.0	0.0	0.0	0.0	0.0	0.0	0.0	0.0	0.0
1975-76	0.0	0.0	0.0	0.0	0.0	0.0	0.0	0.0	0.0	0.0	0.0	0.0	0.0
1976-77	0.0	0.0	0.0	0.0	0.0	0.0	0.0	0.0	0.0	0.0	0.0	0.0	0.0
1977-78	0.0	0.0	0.0	0.0	0.0	0.0	0.0	0.0	0.0	0.0	0.0	0.0	0.0
1978-79	0.0	0.0	0.0	0.0	0.0	0.0	0.0	0.0	0.0	0.0	0.0	0.0	0.0
1979-80	0.0	0.0	0.0	0.0	0.0	0.0	0.0	0.0	0.0	0.0	0.0	0.0	0.0
1980-81	0.0	0.0	0.0	0.0	0.0	0.0	0.0	0.0	0.0	0.0	0.0	0.0	0.0
1981-82	0.0	0.0	0.0	0.0	0.0	0.0	0.0	0.0	0.0	0.0	0.0	0.0	0.0
1982-83	0.0	0.0	0.0	0.0	0.0	0.0	0.0	0.0	0.0	0.0	0.0	0.0	0.0
1983-84	0.0	0.0	0.0	0.0	0.0	0.0	0.0	0.0	0.0	0.0	0.0	0.0	0.0
1984-85	0.0	0.0	0.0	0.0	0.0	0.0	0.0	0.0	0.0	0.0	0.0	0.0	0.0
1985-86	0.0	0.0	0.0	0.0	0.0	0.0	0.0	0.0	0.0	0.0	0.0	0.0	0.0
1986-87	0.0	0.0	0.0	0.0	0.0	0.0	0.0	0.0	0.0	0.0	0.0	0.0	0.0
1987-88	0.0	0.0	0.0	0.0	0.0	0.0	0.0	0.0	0.0	0.0	0.0	0.0	0.0
1988-89	0.0	0.0	0.0	0.0	0.0	0.0	0.0	0.0	0.0	0.0	0.0	0.0	0.0
1989-90	0.0	0.0	0.0	0.0	0.0	0.0	0.0	0.0	0.0	0.0	0.0	0.0	0.0
1990-91	0.0	0.0	0.0	0.0	0.0	0.0	0.0	0.0	0.0	0.0	0.0	0.0	0.0
1991-92	0.0	0.0	0.0	0.0	0.0	0.0	0.0	0.0	0.0	0.0	0.0	0.0	0.0
1992-93	0.0	0.0	0.0	0.0	0.0	0.0	0.0	0.0	0.0	0.0	0.0	0.0	0.0
1993-94	0.0	0.0	0.0	0.0	0.0	0.0							
Record Mean	0.0	0.0	0.0	0.0	0.0	0.0	0.0	0.0	0.0	0.0	0.0	0.0	0.0

GUAM, PACIFIC

Guam is the largest and southernmost of the Mariana Islands. The Philippine Sea lies to the west and the Pacific Ocean to the east. The island is 28 miles long, 4 to 8 miles wide, and is oriented north northeast and south southwest. Located 1,500 miles east of Manila and 3,000 miles west of Honolulu, Guam serves as an important stopping place for aircraft and ships. It also has long been an important American military base. Outside of the activities of the Federal and Territorial governments, the most important single industry is agriculture.

Guam is shaped like a bow tie, and in correspondence with this shape, there are three topographic regions. The northern portion of the island is a limestone plateau that is bounded by steep cliffs that either fall directly to the sea or to narrow beaches. The surface of the plateau is 300 to 600 feet above sea level. The southern portion of the island is mountainous with several peaks that rise above 1,000 feet. The highest of these is Mount Lamlam which reaches 1,334 feet. The third major region, the narrow waist between the northern and southern regions, is quite low, generally less than 200 feet above sea level.

The National Weather Service station on Guam is located on the western side of the northern plateau. The ocean is 1 1/2 miles to the west, 9 miles to the north and east, and 5 miles to the southeast. The weather instruments at the station are well exposed in the center of an open field that is 40 acres in area. The trade winds reach the station after rising sharply up the 500-foot cliffs on the eastern side of the island and flowing 9 miles on an easy downslope grade across the surface of the northern plateau.

The climate of Guam is almost uniformly warm and humid throughout the year. Afternoon temperatures are typically in the middle or high 80s and nighttime temperatures typically fall to the low 70s or high 60s. Relative humidity commonly ranges from around 65 to 75 percent in the afternoon to 85 to 100 percent at night. Though temperature and humidity vary only slightly throughout the year, rainfall and wind conditions vary markedly, and it is these latter variations that really define the seasons.

There are two primary seasons and two secondary seasons on Guam. The primary seasons are the four-month dry season, which extends from January through April, and the four-month rainy season which extends from mid-July to mid-November. The secondary seasons are May to mid-July and mid-November through December. These are transitional seasons that may be either rainy or dry depending upon the nature of the particular year. On the average, about 15 percent of the annual rainfall occurs during the dry season and 55 percent during the rainy season.

At all times of the year the dominant winds on Guam are the trade winds which blow from the east or northeast. The trades are strongest and most constant during the dry season, when wind speeds of 15 to 25 mph are very common. During the rainy season there is often a breakdown of the trades, and on some days the weather may be dominated by westerly-moving storm systems that bring heavy showers or steady, and sometimes torrential, rain. Occasionally there are typhoons, and these bring not only tremendous rains, but also violent winds that may cause a surge of water onto low-lying coastal areas. Typhoons have passed sufficiently close to Guam to produce high winds and heavy rains in every month, but their most frequent occurrence is during the latter half of the year.

NORMALS, MEANS, AND EXTREMES

GUAM, PACIFIC

LATITUDE: 13°33'N LONGITUDE: 144°50'E ELEVATION: FT. GRND 361 BARO 365 TIME ZONE: 150E MER WBAN: 41415

	(a)	JAN	FEB	MAR	APR	MAY	JUNE	JULY	AUG	SEP	OCT	NOV	DEC	YEAR
TEMPERATURE °F:														
Normals														
-Daily Maximum		83.6	83.5	84.5	85.8	86.8	86.9	86.3	85.9	86.1	85.9	85.3	84.2	85.4
-Daily Minimum		71.0	70.9	71.1	72.1	72.6	72.9	72.4	72.2	72.2	72.4	73.1	72.7	72.1
-Monthly		77.3	77.2	77.8	79.0	79.7	79.9	79.4	79.1	79.2	79.1	79.2	78.5	78.8
Extremes														
-Record Highest	36	87	89	90	90	92	93	94	91	95	91	89	89	95
-Year		1992	1989	1989	1993	1993	1993	1983	1991	1957	1957	1991	1966	SEP 1957
-Record Lowest	36	56	59	54	59	62	63	64	63	61	64	62	61	54
-Year		1978	1959	1965	1965	1960	1983	1986	1983	1958	1987	1957	1980	MAR 1965
NORMAL DEGREE DAYS:														
Heating (base 65°F)		0	0	0	0	0	0	0	0	0	0	0	0	0
Cooling (base 65°F)		381	342	397	420	456	447	446	437	426	437	426	419	5034
% OF POSSIBLE SUNSHINE	31	48	51	57	57	55	50	39	34	36	36	39	37	45
MEAN SKY COVER (tenths)														
Sunrise - Sunset	17	6.9	7.4	6.9	7.4	7.4	8.1	8.5	9.0	8.7	8.3	7.2	7.2	7.8
MEAN NUMBER OF DAYS:														
Sunrise to Sunset														
-Clear	17	3.4	2.5	3.1	1.3	0.8	0.6	0.3	0.0	0.2	0.3	2.4	2.9	17.9
-Partly Cloudy	17	12.5	9.3	13.6	13.7	14.5	9.9	8.3	4.3	6.1	9.5	12.2	12.0	125.9
-Cloudy	17	15.1	16.5	14.2	15.0	15.7	19.6	22.5	26.7	23.7	21.1	15.4	16.1	221.6
Precipitation														
.01 inches or more	37	20.5	18.3	18.8	19.8	20.1	23.9	26.0	25.8	25.4	25.5	25.4	23.4	272.9
Snow,Ice pellets,hail														
1.0 inches or more	37	0.0	0.0	0.0	0.0	0.0	0.0	0.0	0.0	0.0	0.0	0.0	0.0	0.0
Thunderstorms	12	0.6	0.0	0.0	0.5	1.1	1.8	4.4	5.8	6.6	3.5	2.2	0.3	26.8
Heavy Fog Visibility														
1/4 mile or less	12	0.0	0.0	0.0	0.0	0.0	0.0	0.0	0.0	0.0	0.0	0.0	0.2	0.2
Temperature °F														
-Maximum														
90° and above	36	0.0	0.0	0.*	0.1	2.4	3.1	1.5	0.4	0.4	0.3	0.0	0.0	8.1
32° and below	36	0.0	0.0	0.0	0.0	0.0	0.0	0.0	0.0	0.0	0.0	0.0	0.0	0.0
-Minimum														
32° and below	36	0.0	0.0	0.0	0.0	0.0	0.0	0.0	0.0	0.0	0.0	0.0	0.0	0.0
0° and below	36	0.0	0.0	0.0	0.0	0.0	0.0	0.0	0.0	0.0	0.0	0.0	0.0	0.0
AVG. STATION PRESS.(mb)														
RELATIVE HUMIDITY (%)														
Hour 04	3	88	88	88	90	92	92	94	96	96	95	91	88	92
Hour 10 (Local Time)	26	77	76	75	74	73	75	79	81	81	80	80	78	77
Hour 16	26	75	74	72	71	72	74	78	80	80	80	80	77	76
Hour 22	25	86	85	85	85	87	88	91	92	92	92	90	88	88
PRECIPITATION (inches):														
Water Equivalent														
-Normal		5.55	5.11	4.45	4.71	7.10	6.49	11.78	14.59	15.02	12.74	9.06	6.44	103.04
-Maximum Monthly	37	20.39	14.79	16.94	19.55	40.13	14.61	20.00	38.13	27.13	26.05	18.14	16.19	40.13
-Year		1976	1980	1971	1963	1976	1985	1972	1992	1982	1979	1957	1963	MAY 1976
-Minimum Monthly	37	1.17	0.67	0.59	0.50	0.64	0.80	4.74	3.87	5.95	6.63	2.08	2.17	0.50
-Year		1993	1960	1965	1965	1987	1983	1957	1965	1992	1976	1973	1992	APR 1965
-Maximum in 24 hrs	37	11.09	9.24	3.55	6.37	27.00	4.55	7.86	15.36	7.48	12.07	7.26	6.09	27.00
-Year		1990	1980	1972	1974	1976	1958	1986	1992	1965	1986	1957	1963	MAY 1976
Snow,Ice pellets,hail														
-Maximum Monthly		0.0	0.0	0.0	0.0	0.0	0.0	0.0	0.0	0.0	0.0	0.0	0.0	
-Year														
-Maximum in 24 hrs	37	0.0	0.0	0.0	0.0	0.0	0.0	0.0	0.0	0.0	0.0	0.0	0.0	
-Year														
WIND:														
Mean Speed (mph)	6	8.2	10.2	9.0	8.9	8.3	6.3	5.1	4.8	4.7	6.2	7.8	9.1	7.4
Prevailing Direction through 1963		E	NE	E	E	E	E	E	E	E	E	E	E	E
Fastest Mile														
-Direction (!!)	32	W	NE	NE	SW	NE	E	E	SE	E	W	NE	NE	NE
-Speed (MPH)	32	64	36	32	64	76	32	34	43	35	44	80	67	80
-Year		1988	1962	1977	1963	1976	1976	1963	1979	1976	1968	1962	1990	NOV 1962
Peak Gust														
-Direction (!!)	10	W	NE	NE	NE	NE	SE	SW	NE	NE	E	E	NE	NE
-Speed (mph)	10	87	37	36	58	39	36	41	117	43	61	79	89	117
-Date		1988	1989	1991	1989	1991	1985	1986	1992	1993	1992	1991	1990	AUG 1992

PRECIPITATION (inches) GUAM, PACIFIC

YEAR	JAN	FEB	MAR	APR	MAY	JUNE	JULY	AUG	SEP	OCT	NOV	DEC	ANNUAL
1964	1.99	4.11	4.44	7.48	20.69	7.49	7.67	8.92	19.30	14.62	6.05	6.92	109.68
1965	10.29	1.02	0.59	0.50	1.69	5.35	15.67	3.87	22.27	6.89	6.97	3.13	78.24
1966	2.13	1.69	1.71	1.31	2.51	5.72	6.74	11.39	22.28	6.96	7.26	4.76	74.46
1967	5.26	4.45	11.28	8.97	3.76	8.37	12.62	23.07	20.28	14.20	11.75	2.51	126.52
1968	5.75	8.48	3.71	1.69	3.16	6.17	12.36	14.61	11.82	13.31	11.92	3.36	96.34
1969	2.79	1.12	0.97	1.36	2.72	1.52	13.83	9.30	6.79	25.32	10.10	10.11	85.93
1970	11.93	3.60	2.68	1.22	2.96	5.60	8.99	10.42	13.70	11.65	9.77	5.71	88.23
1971	5.25	5.78	16.94	5.74	22.68	5.06	16.06	20.92	12.21	8.63	6.75	3.14	129.16
1972	4.86	6.40	11.29	3.35	3.02	7.95	20.00	18.09	16.09	7.09	4.83	6.35	109.32
1973	2.10	2.94	1.57	1.90	3.01	4.85	9.55	6.65	9.05	17.82	2.08	5.35	66.87
1974	4.27	3.13	12.25	14.83	12.00	10.46	9.32	25.66	9.96	10.00	8.03	6.20	126.11
1975	9.68	1.15	2.25	3.67	1.33	2.46	11.13	22.54	8.86	10.68	12.35	6.42	92.52
1976	20.39	13.56	9.04	4.37	40.13	6.40	13.38	24.40	10.79	6.63	7.91	8.91	165.91
1977	3.12	3.01	4.98	2.29	6.88	5.10	7.25	6.09	17.32	16.67	12.38	2.22	87.31
1978	2.44	4.99	0.85	2.65	3.48	7.89	8.91	19.63	10.01	10.49	14.10	4.80	90.24
1979	5.60	1.85	4.20	1.89	1.48	4.28	11.83	12.75	8.34	26.05	6.78	5.67	90.72
1980	2.07	14.79	3.42	2.92	8.60	7.96	10.93	9.42	24.34	12.02	7.76	6.14	110.37
1981	9.05	2.44	3.88	6.61	5.68	6.11	13.18	17.29	9.73	14.92	15.57	7.20	111.66
1982	2.92	8.77	2.46	0.99	6.35	8.77	13.61	6.32	27.13	11.86	9.34	6.33	104.85
1983	1.31	1.21	3.34	1.83	1.10	0.80	6.15	10.88	14.62	10.17	10.52	5.13	67.06
1984	3.19	4.19	4.02	1.56	3.10	6.69	6.58	24.05	17.41	9.40	12.93	5.89	99.01
1985	8.16	3.69	5.53	5.62	11.95	14.61	13.23	15.97	18.06	8.33	4.96	8.40	118.51
1986	2.01	8.71	5.60	8.36	7.77	9.08	17.41	24.87	8.02	19.71	6.05	12.43	130.02
1987	2.63	5.94	2.36	1.35	0.64	1.61	12.30	8.50	14.41	12.18	7.91	6.64	76.47
1988	8.71	1.30	1.50	2.95	2.33	7.93	14.37	9.21	10.67	14.56	5.54	4.33	83.40
1989	3.31	9.95	1.01	10.93	4.30	9.27	11.64	13.27	14.38	12.74	10.52	4.07	105.39
1990	17.01	3.72	2.64	2.52	4.95	6.28	14.49	17.25	22.00	8.46	16.41	7.35	123.08
1991	4.85	4.58	1.79	6.33	4.40	7.28	13.33	18.62	11.27	14.49	13.72	3.29	103.95
1992	9.81	1.75	2.28	2.57	6.23	3.40	10.08	38.13	5.95	15.01	12.91	2.17	110.29
1993	1.17	4.62	1.51	1.13	1.85	3.03	6.97	14.74	12.04	9.68	8.03	5.63	70.40
Record Mean	5.75	4.68	3.99	4.37	6.42	6.30	11.23	15.16	14.51	13.19	9.55	6.06	101.22

AVERAGE TEMPERATURE (deg. F) GUAM, PACIFIC

YEAR	JAN	FEB	MAR	APR	MAY	JUNE	JULY	AUG	SEP	OCT	NOV	DEC	ANNUAL
1964	78.0	77.4	77.8	79.2	79.3	80.0	79.0	78.7	78.7	79.0	79.2	78.0	78.7
1965	76.9	77.4	76.4	78.1	80.5	79.4	78.7	78.8	78.7	78.5	78.7	78.2	78.3
1966	77.0	77.0	77.5	80.0	79.9	80.4	80.6	79.2	78.9	79.1	79.1	78.7	79.0
1967	77.5	76.6	75.9	78.0	79.3	79.5	79.1	78.8	78.9	77.9	79.4	76.8	78.1
1968	77.3	75.6	77.9	77.2	78.6	79.1	79.4	79.5	78.5	78.4	78.0	77.7	78.1
1969	76.9	76.2	77.6	79.9	80.2	80.7	80.1	79.0	79.8	78.7	80.0	79.3	79.1
1970	76.9	77.7	78.7	79.0	79.8	80.2	78.9	79.0	79.0	79.5	79.6	79.0	79.0
1971	78.1	78.1	78.5	79.7	79.6	79.6	78.6	78.1	79.0	78.8	79.1	79.2	78.9
1972	76.7	76.5	77.3	77.7	79.0	79.4	79.3	78.8	79.4	79.2	79.4	78.3	78.4
1973	75.5	76.9	78.3	79.6	80.1	80.2	79.9	79.5	79.4	78.7	79.5	79.0	78.9
1974	77.4	78.7	77.6	79.1	79.6	79.0	78.5	78.9	79.1	79.7	79.5	78.7	78.8
1975	77.9	77.5	77.2	79.2	79.4	80.9	78.4	78.0	78.8	78.8	78.3	78.2	78.5
1976	76.1	76.6	77.4	78.4	79.1	79.5	79.4	79.0	78.6	79.5	78.8	77.6	78.3
1977	76.4	77.0	77.7	78.3	78.9	80.1	80.0	80.0	79.0	78.9	79.2	78.6	78.7
1978	76.7	76.6	78.1	79.8	81.0	79.9	79.7	79.1	79.7	79.7	78.9	78.1	79.0
1979	77.4	77.0	77.7	78.9	80.2	81.1	79.7	78.8	78.9	79.2	78.9	78.0	78.8
1980	76.4	77.4	77.6	79.3	79.5	79.7	79.6	80.2	79.4	79.8	80.4	78.0	78.9
1981	78.3	77.0	78.0		80.1	80.1	79.4	79.0	79.7	79.8	79.3	79.1	
1982	77.4	76.8	77.4	78.7	79.6	79.7	78.8	78.6	78.6	77.5	78.5	77.3	78.2
1983	75.1	75.1	75.6	77.7	79.4	80.7	80.8	78.5	79.9	80.1	79.0	78.0	78.3
1984	76.8	76.2	77.6	79.1	79.9	79.4	78.8	78.1	78.4	79.6	78.9	78.6	78.5
1985	77.7	78.3	78.8	78.9	78.6	78.8	78.9	78.5	78.9	78.9	79.3	78.5	78.7
1986	78.1	76.9	78.5	79.5	79.3	79.9	79.0	79.7	79.4	79.5	80.1	78.2	79.0
1987	76.9	76.9	77.7	79.2	79.8	81.2	79.5	80.0	80.5	79.7	79.5	79.9	79.2
1988	78.8	78.5	79.8	79.9	81.0	79.9	79.4	79.9	79.7	79.4	80.1	79.8	79.7
1989	79.6	78.2	77.7	79.7	80.7	79.6	79.5	79.4	79.1	79.2	79.4	79.5	79.3
1990	78.7	78.2	78.2	79.5	80.6	80.9	80.2	79.6	79.2	79.9	79.5	78.6	79.4
1991	77.8	77.8	78.7	80.3	80.7	81.0	80.6	80.5	80.0	79.6	79.6	78.3	79.6
1992	78.0	77.5	78.7	78.8	79.8	81.4	80.5	79.4	80.6	80.0	79.6	78.8	79.4
1993	78.3	77.5	79.0	81.3	82.0	81.7	81.4	79.5	79.8	79.6	80.4	79.9	80.0
Record Mean	77.3	77.3	77.8	79.2	79.9	80.1	79.5	79.2	79.3	79.2	79.3	78.6	78.9
Max	83.5	83.5	84.5	85.9	86.9	87.1	86.5	86.0	86.2	85.9	85.4	84.3	85.5
Min	71.2	71.0	71.2	72.4	72.8	73.0	72.5	72.3	72.3	72.4	73.2	72.8	72.3

REFERENCE NOTES GUAM, PACIFIC

GENERAL
T - TRACE AMOUNT.
BLANK ENTRIES DENOTE MISSING/UNREPORTED DATA.
# INDICATES A STATION OR INSTRUMENT RELOCATION.

SPECIFIC

(a) - LENGTH OF RECORD IN YEARS, ALTHOUGH INDIVIDUAL MONTHS MAY BE MISSING.
0.* OR * - THE VALUE IS BETWEEN 0.0 AND 0.05
NORMALS - BASED ON THE 1961-1990 RECORD PERIOD.
EXTREMES - DATES ARE THE MOST RECENT OCCURRENCE
WIND DIR.- NUMERALS SHOW TENS OF DEGRESS CLOCKWISE FROM TRUE NORTH. "00" INDICATES CALM.
RESULTANT DIRECTIONS ARE GIVEN TO WHOLE DEGREES.
BOLD VALUES INDICATE EXTREME VALUES WHICH OCCURRED AFTER THE ASOS SYSTEM WAS COMMISSIONED.

RECORD = PERIOD OF RECORD
RECORD MEAN PRECIPITATION IS THE MEAN OF ALL DAILY PRECIPITATION AMOUNTS DURING THE PERIOD OF RECORD.
RECORD MAX(MIN) TEMPERATURE IS THE MEAN OF ALL DAILY MAX(MIN) TEMPERATURES DURING THE PERIOD OF RECORD.
RECORD MEAN TEMPERATURE IS THE SUM OF THE RECORD MAX AND RECORD MIN DIVIDED BY 2.
AVERAGE TEMPERATURE IS THE SUM OF THE MEAN DAILY MAX AND MIN TEMPERATURE DIVIDED BY 2.

EXCEPTIONS

1. THUNDERSTORMS AND HEAVY FOG ARE FOR 1957-1964 AND 1970-1972 AND MAY BE INCOMPLETE, DUE TO PART-TIME OPERATIONS.
2. MEAN WIND SPEED IS THROUGH 1964 AND 1970-1971
3. MEAN SKY COVER, AND DAYS CLEAR-PARTLY CLOUDY-CLOUDY ARE 1968-1970 AND 1976 TO DATE.
4. RELATIVE HUMIDITY HOUR 04 IS 1970-1972.

RECORD MEANS ARE THROUGH THE CURRENT YEAR, BEGINNING IN 1957 FOR TEMPERATURE
1957 FOR PRECIPITATION

HEATING DEGREE DAYS Base 65 deg. F — GUAM, PACIFIC

SEASON	JULY	AUG	SEP	OCT	NOV	DEC	JAN	FEB	MAR	APR	MAY	JUNE	TOTAL
1983-84	0	0	0	0	0	0	0	0	0	0	0	0	0
1984-85	0	0	0	0	0	0	0	0	0	0	0	0	0
1985-86	0	0	0	0	0	0	0	0	0	0	0	0	0
1986-87	0	0	0	0	0	0	0	0	0	0	0	0	0
1987-88	0	0	0	0	0	0	0	0	0	0	0	0	0
1988-89	0	0	0	0	0	0	0	0	0	0	0	0	0
1989-90	0	0	0	0	0	0	0	0	0	0	0	0	0
1990-91	0	0	0	0	0	0	0	0	0	0	0	0	0
1991-92	0	0	0	0	0	0	0	0	0	0	0	0	0
1992-93	0	0	0	0	0	0	0	0	0	0	0	0	0
1993-94	0	0	0	0	0	0							

COOLING DEGREE DAYS Base 65 deg. F — GUAM, PACIFIC

YEAR	JAN	FEB	MAR	APR	MAY	JUNE	JULY	AUG	SEP	OCT	NOV	DEC	TOTAL
1969	375	321	400	454	479	475	476	442	451	431	454	451	5209
1970	373	362	431	425	465	463	437	441	428	460	445	437	5167
1971	411	372	427	448	461	443	429	412	429	437	429	449	5147
1972	370	343	387	390	440	437	449	436	438	448	440	419	4997
1973	331	341	417	444	478	463	470	456	439	434	442	442	5157
1974	393	389	396	430	461	427	427	434	427	465	444	430	5123
1975	409	357	383	434	454	487	422	409	422	435	405	418	5035
1976	350	346	392	411	445	442	454	444	414	456	421	398	4973
1977	363	341	401	405	438	459	474	474	428	442	435	424	5084
1978	368	329	413	451	504	454	466	445	449	463	421	413	5176
1979	391	344	403	423	480	491	461	435	425	447	422	408	5130
1980	358	367	397	434	456	445	457	476	437	465	466	412	5170
1981	416	342	410		475	461	454	440	448	467	437	442	
1982	393	335	390	420	459	446	432	430	414	393	412	389	4913
1983	320	289	337	386	451	480	494	425	455	474	427	411	4949
1984	374	333	396	428	467	437	433	413	408	461	424	429	5003
1985	401	378	436	424	430	423	437	425	427	436	434	427	5078
1986	416	338	425	439	448	454	440	464	437	457	457	415	5190
1987	373	338	399	432	462	496	455	470	474	464	441	468	5272
1988	434	396	467	454	504	455	454	471	449	457	461	464	5466
1989	460	376	401	448	495	446	456	456	428	450	440	456	5312
1990	434	379	417	440	493	485	480	460	431	469	440	431	5359
1991	403	364	431	465	497	489	494	489	456	458	443	422	5411
1992	411	370	430	420	466	497	488	453	475	475	444	437	5366
1993	418	356	442	498	534	507	514	458	454	459	470	469	5579

SNOWFALL (inches) — GUAM, PACIFIC

SEASON	JULY	AUG	SEP	OCT	NOV	DEC	JAN	FEB	MAR	APR	MAY	JUNE	TOTAL
1971-72	0.0	0.0	0.0	0.0	0.0	0.0	0.0	0.0	0.0	0.0	0.0	0.0	0.0
1972-73	0.0	0.0	0.0	0.0	0.0	0.0	0.0	0.0	0.0	0.0	0.0	0.0	0.0
1973-74	0.0	0.0	0.0	0.0	0.0	0.0	0.0	0.0	0.0	0.0	0.0	0.0	0.0
1974-75	0.0	0.0	0.0	0.0	0.0	0.0	0.0	0.0	0.0	0.0	0.0	0.0	0.0
1975-76	0.0	0.0	0.0	0.0	0.0	0.0	0.0	0.0	0.0	0.0	0.0	0.0	0.0
1976-77	0.0	0.0	0.0	0.0	0.0	0.0	0.0	0.0	0.0	0.0	0.0	0.0	0.0
1977-78	0.0	0.0	0.0	0.0	0.0	0.0	0.0	0.0	0.0	0.0	0.0	0.0	0.0
1978-79	0.0	0.0	0.0	0.0	0.0	0.0	0.0	0.0	0.0	0.0	0.0	0.0	0.0
1979-80	0.0	0.0	0.0	0.0	0.0	0.0	0.0	0.0	0.0	0.0	0.0	0.0	0.0
1980-81	0.0	0.0	0.0	0.0	0.0	0.0	0.0	0.0	0.0	0.0	0.0	0.0	0.0
1981-82	0.0	0.0	0.0	0.0	0.0	0.0	0.0	0.0	0.0	0.0	0.0	0.0	0.0
1982-83	0.0	0.0	0.0	0.0	0.0	0.0	0.0	0.0	0.0	0.0	0.0	0.0	0.0
1983-84	0.0	0.0	0.0	0.0	0.0	0.0	0.0	0.0	0.0	0.0	0.0	0.0	0.0
1984-85	0.0	0.0	0.0	0.0	0.0	0.0	0.0	0.0	0.0	0.0	0.0	0.0	0.0
1985-86	0.0	0.0	0.0	0.0	0.0	0.0	0.0	0.0	0.0	0.0	0.0	0.0	0.0
1986-87	0.0	0.0	0.0	0.0	0.0	0.0	0.0	0.0	0.0	0.0	0.0	0.0	0.0
1987-88	0.0	0.0	0.0	0.0	0.0	0.0	0.0	0.0	0.0	0.0	0.0	0.0	0.0
1988-89	0.0	0.0	0.0	0.0	0.0	0.0	0.0	0.0	0.0	0.0	0.0	0.0	0.0
1989-90	0.0	0.0	0.0	0.0	0.0	0.0	0.0	0.0	0.0	0.0	0.0	0.0	0.0
1990-91	0.0	0.0	0.0	0.0	0.0	0.0	0.0	0.0	0.0	0.0	0.0	0.0	0.0
1991-92	0.0	0.0	0.0	0.0	0.0	0.0	0.0	0.0	0.0	0.0	0.0	0.0	0.0
1992-93	0.0	0.0	0.0	0.0	0.0	0.0	0.0	0.0	0.0	0.0	0.0	0.0	0.0
1993-94	0.0	0.0	0.0	0.0	0.0	0.0							
Record Mean	0.0	0.0	0.0	0.0	0.0	0.0	0.0	0.0	0.0	0.0	0.0	0.0	0.0

KOROR ISLAND, PACIFIC

Koror is one of the islands of the Palau Group, which lies in the extreme western Carolines and along the eastern side of the Philippine Sea. It is the administrative center of the Republic of Palau of the U.S. Trust Territory of the Pacific Islands. Like most of the other islands of the group, Koror is hilly and is surrounded by a lagoon whose outer border is a barrier reef. Koror is not an isolated island, but is immediately adjacent to the islands of Malakal and Arakabesan, to which it is joined by causeways. Babelthuap, the largest of the Palau Islands, lies 1 mile north–northeast and is connected to Korror by bridge. The highest hill within 3 miles of the Weather Station on Koror has an elevation of 610 feet.

Precipitation is heavy at Koror. Rainfall of 150 inches or more in a year is not uncommon. Rainfall is variable for each month and from year to year. Normal monthly precipitation exceeds 10 inches and during some years, each month has received at at least 15 inches.

During the period December through March, the northeast trades prevail. Winds are generally light to moderate. Precipitation, heavy during December and January, decreases sharply when the Intertropical Convergence Zone moves well south of the island. February, March, and April are the driest months of the year. During April, the frequency of northeast winds decreases, and there is an increase in the frequency of east winds. In May the winds are predominantly from southeast, to northeast.

Usually during June the Intertropical Convergence Zone moves northward across Koror, bringing with it heavy rainfall and thunderstorms that may yield 1 inch of rain in 15 to 30 minutes. The convergence zone remains in the vicinity of Koror, though most commonly toward the north, from July through January, with heavy rainfall persisting. During November, Koror is usually near the heart of the zone. Calm to light variable winds and continued heavy showers are the rule.

NORMALS, MEANS, AND EXTREMES

KOROR ISLAND, PACIFIC

LATITUDE: 7°20'N LONGITUDE: 134°29' E ELEVATION: FT. GRND 94 BARO 98 TIME ZONE: 135E MER WBAN: 40309

	(a)	JAN	FEB	MAR	APR	MAY	JUNE	JULY	AUG	SEP	OCT	NOV	DEC	YEAR
TEMPERATURE °F:														
Normals														
-Daily Maximum		87.0	87.0	87.7	88.4	88.6	87.8	87.1	87.2	87.7	87.9	88.5	87.9	87.7
-Daily Minimum		75.1	75.0	75.3	75.9	76.1	75.5	75.4	75.7	76.1	75.9	76.0	75.6	75.6
-Monthly		81.1	81.0	81.5	82.2	82.4	81.7	81.3	81.5	81.9	81.9	82.3	81.8	81.7
Extremes														
-Record Highest	44	91	92	91	93	93	95	93	93	92	92	93	93	95
-Year		1990	1954	1993	1988	1972	1976	1968	1947	1988	1990	1993	1987	JUN 1976
-Record Lowest	44	70	71	69	69	72	70	70	70	70	71	70	71	69
-Year		1974	1993	1953	1979	1989	1948	1991	1989	1978	1981	1990	1984	APR 1979
NORMAL DEGREE DAYS:														
Heating (base 65°F)		0	0	0	0	0	0	0	0	0	0	0	0	0
Cooling (base 65°F)		499	448	512	516	539	501	505	512	507	524	519	521	6103
% OF POSSIBLE SUNSHINE	33	54	57	65	63	55	44	47	45	53	48	52	49	53
MEAN SKY COVER (tenths)														
Sunrise - Sunset	41	8.8	8.7	8.4	8.2	8.4	9.1	9.0	9.1	8.8	8.7	8.6	8.8	8.7
MEAN NUMBER OF DAYS:														
Sunrise to Sunset														
-Clear	41	0.4	0.3	0.8	1.2	0.7	0.2	0.2	0.2	0.3	0.6	0.3	0.3	5.4
-Partly Cloudy	41	5.8	5.8	7.6	8.0	7.7	3.9	4.4	4.1	5.4	6.4	6.7	5.8	71.7
-Cloudy	41	24.9	22.1	22.6	20.7	22.6	25.9	26.4	26.7	24.3	24.0	23.0	25.0	288.1
Precipitation														
.01 inches or more	42	22.5	19.1	20.2	19.0	23.6	24.9	24.3	22.5	20.8	22.7	22.1	24.0	265.7
Snow, Ice pellets, hail														
1.0 inches or more	42	0.0	0.0	0.0	0.0	0.0	0.0	0.0	0.0	0.0	0.0	0.0	0.0	0.0
Thunderstorms	42	1.0	0.9	1.0	2.0	3.5	3.6	3.1	3.5	4.0	4.1	3.5	2.5	32.6
Heavy Fog Visibility														
1/4 mile or less	42	0.0	0.0	0.0	0.0	0.0	0.0	0.0	0.0	0.0	0.0	0.0	0.0	0.0
Temperature °F														
-Maximum														
90° and above	42	1.7	1.4	2.6	7.7	12.2	7.1	3.3	2.6	4.8	7.6	10.4	5.4	66.9
32° and below	42	0.0	0.0	0.0	0.0	0.0	0.0	0.0	0.0	0.0	0.0	0.0	0.0	0.0
-Minimum														
32° and below	42	0.0	0.0	0.0	0.0	0.0	0.0	0.0	0.0	0.0	0.0	0.0	0.0	0.0
0° and below	42	0.0	0.0	0.0	0.0	0.0	0.0	0.0	0.0	0.0	0.0	0.0	0.0	0.0
AVG. STATION PRESS. (mb)	15	1005.9	1006.8	1006.8	1006.1	1005.8	1005.8	1005.7	1005.9	1005.9	1005.3	1005.0	1005.6	1005.9
RELATIVE HUMIDITY (%)														
Hour 03	23	90	89	90	90	92	92	91	90	89	90	91	91	90
Hour 09 (Local Time)	42	81	80	79	78	80	81	81	81	79	80	79	80	80
Hour 15	42	76	75	74	74	77	78	78	77	77	77	77	77	76
Hour 21	42	87	86	87	87	89	89	89	87	87	88	89	88	88
PRECIPITATION (inches):														
Water Equivalent														
-Normal		10.70	9.12	8.20	8.67	11.99	17.27	18.04	14.95	11.86	13.87	11.32	11.98	147.97
-Maximum Monthly	44	28.13	22.46	21.98	27.69	27.46	33.83	34.82	33.11	23.16	22.47	22.06	21.10	34.82
-Year		1974	1978	1972	1979	1954	1990	1962	1987	1985	1974	1958	1975	JUL 1962
-Minimum Monthly	44	2.11	0.64	1.71	1.65	5.73	5.91	4.14	6.89	1.04	6.77	4.70	1.49	0.64
-Year		1973	1983	1983	1948	1983	1976	1964	1981	1982	1951	1957	1990	FEB 1983
-Maximum in 24 hrs	44	13.86	8.42	13.78	16.95	9.86	13.83	8.27	8.18	8.47	6.18	9.88	6.47	16.95
-Year		1974	1980	1991	1979	1982	1990	1981	1962	1949	1957	1990	1974	APR 1979
Snow, Ice pellets, hail														
-Maximum Monthly		0.0	0.0	0.0	0.0	0.0	0.0	0.0	0.0	0.0	0.0	0.0	0.0	
-Year														
-Maximum in 24 hrs	44	0.0	0.0	0.0	0.0	0.0	0.0	0.0	0.0	0.0	0.0	0.0	0.0	
-Year														
WIND:														
Mean Speed (mph)	28	8.3	8.5	8.3	7.4	6.3	5.9	6.4	6.8	6.8	7.1	6.3	7.2	7.1
Prevailing Direction through 1963		NE	ENE	NE	ENE	E	E	NW	SW	W	W	NE	ENE	NE
Fastest Mile														
-Direction (!!)	28	NW	SE	S	SW	SW	SW	SW	SW	N	SW	SE	N	S
-Speed (MPH)	28	43	30	73	60	35	40	34	34	34	36	59	50	73
-Year		1975	1968	1967	1976	1976	1967	1986	1978	1969	1968	1964	1972	MAR 1967
Peak Gust														
-Direction (!!)	10	NW	E	NE	E	S	W	SW	NW	W	SW	SW	SE	SW
-Speed (mph)	10	45	39	44	35	46	58	54	52	52	51	83	39	83
-Date		1985	1993	1991	1986	1989	1990	1986	1986	1991	1991	1990	1984	NOV 1990

PRECIPITATION (inches)

KOROR ISLAND, PACIFIC

YEAR	JAN	FEB	MAR	APR	MAY	JUNE	JULY	AUG	SEP	OCT	NOV	DEC	ANNUAL
1964	7.27	16.10	6.98	7.46	18.32	12.12	4.14	15.73	7.06	10.19	13.75	11.45	130.57
1965	6.40	13.03	14.60	7.11	9.78	19.85	30.57	16.76	12.98	9.68	6.16	12.67	159.59
1966	8.23	3.31	10.02	15.27	9.08	10.44	23.68	9.69	7.20	16.66	11.76	14.93	140.27
1967	18.86	3.74	7.20	4.63	9.91	17.19	12.20	17.11	6.75	17.06	11.79	12.23	138.67
1968	8.02	15.54	8.59	7.56	9.51	7.01	16.43	11.00	12.25	11.46	8.15	15.95	131.47
1969	6.14	2.93	5.11	6.83	11.72	16.79	28.21	12.39	14.00	9.81	9.06	6.32	129.31
1970	6.23	5.78	4.83	3.21	8.25	12.38	12.61	15.77	8.51	12.99	9.17	14.82	114.55
1971	13.54	10.68	11.09	8.32	16.31	19.61	14.49	10.40	13.98	19.59	10.26	11.06	159.33
1972	10.78	10.83	21.98	7.08	9.49	20.68	11.15	15.88	14.90	9.85	9.96	7.54	150.12
1973	2.11	1.24	2.95	11.29	10.16	13.78	12.79	11.35	12.18	19.14	16.56	9.87	123.42
1974	28.13	7.98	13.75	10.86	8.10	9.72	21.16	13.75	14.80	22.47	15.73	18.54	184.99
1975	17.29	2.82	6.69	10.00	9.01	16.24	22.86	8.28	17.24	11.52	11.18	21.10	154.23
1976	7.80	7.27	8.05	20.09	8.66	5.91	8.08	16.64	7.72	12.49	6.30	16.54	125.55
1977	5.18	5.30	3.60	4.48	11.36	11.15	20.72	19.20	12.65	10.63	7.38	7.79	119.44
1978	10.34	22.46	6.02	8.98	12.52	16.04	9.13	20.36	10.85	20.06	17.66	10.33	164.75
1979	6.98	6.47	7.96	27.69	11.26	22.84	17.79	11.69	12.29	11.97	11.57	11.57	160.08
1980	8.72	16.01	5.53	18.80	10.02	19.50	12.40	15.26	13.60	17.11	12.17	19.95	169.07
1981	11.32	15.00	4.49	3.00	9.66	29.17	21.14	6.89	16.70	14.30	11.37	9.81	152.85
1982	5.79	6.81	9.90	9.45	19.12	22.41	19.40	10.94	1.04	8.82	9.92	13.71	137.31
1983	3.44	0.64	1.71	3.12	5.73	18.48	21.20	17.96	11.73	14.23	11.40	10.48	120.12
1984	18.57	10.81	13.58	7.23	10.85	16.49	12.82	17.47	10.39	15.94	9.19	9.42	152.76
1985	13.22	13.88	5.38	11.82	10.41	25.25	17.55	14.32	23.16	8.64	13.61	6.28	163.52
1986	11.10	16.60	8.33	3.49	12.24	17.16	26.15	13.09	10.21	15.92	17.97	9.06	161.32
1987	9.12	5.94	6.10	4.74	15.32	19.60	28.23	33.11	4.21	12.75	10.88	8.72	158.72
1988	7.14	7.83	6.22	6.18	14.30	21.88	14.90	14.39	12.34	20.05	15.05	21.04	161.32
1989	12.99	12.66	11.31	7.41	16.39	17.33	21.97	17.81	5.54	13.07	9.54	5.96	151.98
1990	5.09	3.83	11.74	7.00	11.72	33.83	19.84	10.34	15.18	11.96	20.71	1.49	152.73
1991	24.16	5.46	17.72	10.34	9.37	12.79	31.58	13.73	21.16	7.81	7.79	12.25	174.16
1992	8.28	2.74	4.88	2.46	9.76	15.73	14.94	17.46	8.15	15.48	5.34	17.25	122.47
1993	9.25	10.49	9.46	14.67	6.72	18.47	11.85	8.61	11.81	8.68	9.24	9.63	128.88
Record Mean	10.86	8.16	8.43	9.22	12.77	16.50	17.47	15.29	12.69	13.56	11.23	12.27	148.45

AVERAGE TEMPERATURE (deg. F)

KOROR ISLAND, PACIFIC

YEAR	JAN	FEB	MAR	APR	MAY	JUNE	JULY	AUG	SEP	OCT	NOV	DEC	ANNUAL
1964	80.8	79.7	81.3	81.7	82.1	81.5	82.0	80.5	82.0	81.5	81.0	81.0	81.3
1965	80.9	80.0	80.1	81.4	81.4	80.6	78.8	80.9	81.4	81.8	82.3	80.9	80.9
1966	79.6	80.1	80.9	81.8	82.0	81.6	81.1	81.7	82.6	82.1	82.1	81.6	81.4
1967	80.1	81.4	81.2	82.4	82.5	82.0	81.2	80.6	82.5	81.6	82.1	81.0	81.5
1968	81.0	80.0	81.3	82.2	82.7	83.2	82.0	81.9	82.1	82.5	82.5	81.2	81.9
1969	80.7	81.1	81.7	82.4	82.7	82.5	80.8	81.9	81.3	82.2	82.1	81.8	81.8
1970	81.5	81.9	82.0	82.6	82.4	82.4	82.1	82.0	82.3	82.2	82.4	81.6	82.1
1971	81.3	81.3	82.1	82.9	82.3	81.1	80.7	81.5	82.1	80.9	82.0	82.0	81.7
1972	80.8	80.5	80.9	81.8	82.7	81.7	82.2	81.6	81.8	82.5	82.3	81.9	81.7
1973	81.2	81.6	81.9	82.3	82.7	82.7	81.9	82.5	82.2	81.9	82.2	82.3	82.1
1974	81.0	81.7	81.1	81.6	82.7	81.8	81.4	82.1	82.3	82.0	82.1	81.6	81.8
1975	80.8	81.4	81.7	82.0	82.2	81.2	80.8	81.6	81.6	81.8	82.3	81.0	81.5
1976	81.9	80.9	81.4	81.7	82.5	82.0	82.0	80.7	81.8	82.4	82.8	81.6	81.8
1977	81.5	81.2	81.8	82.5	83.0	82.5	81.1	81.2	82.2	82.7	83.0	82.5	82.1
1978	81.3	80.1	82.0	81.9	82.6	81.8	82.3	80.8	80.9	81.4	81.7	81.8	81.6
1979	81.4	81.3	81.3	81.3	82.4	81.0	81.1	81.6	82.1	82.0	82.2	81.5	81.6
1980	80.8	80.2	81.4	81.8	82.5	81.8	81.1	80.9	81.6	82.3	83.0	82.3	81.7
1981	81.1	80.4	81.2	82.0	82.6	80.9	81.2	81.7	81.6	81.2	82.5	82.1	81.5
1982	81.7	81.7	81.1	81.7	81.8	81.5	81.1	81.1	82.4	81.8	82.3	81.7	81.7
1983	80.5	81.2	81.7	82.2	83.0	81.9	81.1	82.0	82.0	82.0	82.4	82.0	81.9
1984	81.1	80.9	81.7	82.3	82.7	81.2	81.5	81.0	81.7	80.9	82.7	82.4	81.7
1985	80.9	81.3	82.1	81.6	82.4	80.5	80.0	80.7	80.5	81.8	82.5	82.4	81.4
1986	81.7	81.0	82.1	82.5	83.3	82.0	81.0	82.5	82.4	82.2	82.3	81.9	82.1
1987	81.3	80.7	81.6	82.4	82.2	82.3	81.6	81.1	83.4	82.7	83.3	83.1	82.1
1988	82.1	81.7	82.6	83.4	82.3	81.9	82.1	82.1	83.0	82.1	82.5	81.6	82.3
1989	82.0	82.4	82.3	82.8	82.4	81.7	81.6	81.9	82.5	82.0	82.6	82.3	82.2
1990	82.1	82.1	82.1	82.9	83.3	82.0	82.1	82.6	81.7	82.4	81.9	82.3	82.3
1991	81.0	80.5	81.2	82.5	82.7	82.4	81.0	82.1	81.2	81.9	82.5	81.1	81.7
1992	80.6	81.2	81.1	82.6	82.5	81.9	81.2	81.2	82.1	81.3	82.1	80.6	81.5
1993	80.6	80.1	81.2	81.5	82.9	81.6	81.8	81.1	81.2	82.2	82.7	82.4	81.6
Record Mean	81.0	80.9	81.4	82.1	82.3	81.7	81.3	81.4	81.8	81.9	82.2	81.6	81.7
Max	86.9	86.9	87.6	88.3	88.6	87.8	87.2	87.1	87.7	87.9	88.5	87.7	87.7
Min	75.0	74.9	75.2	75.8	76.1	75.6	75.3	75.7	75.9	75.9	75.9	75.5	75.6

REFERENCE NOTES

KOROR ISLAND, PACIFIC

GENERAL
T - TRACE AMOUNT.
BLANK ENTRIES DENOTE MISSING/UNREPORTED DATA.
# INDICATES A STATION OR INSTRUMENT RELOCATION.

SPECIFIC

(a) - LENGTH OF RECORD IN YEARS, ALTHOUGH INDIVIDUAL MONTHS MAY BE MISSING.
0.* OR * - THE VALUE IS BETWEEN 0.0 AND 0.05
NORMALS - BASED ON THE 1961-1990 RECORD PERIOD.
EXTREMES - DATES ARE THE MOST RECENT OCCURRENCE
WIND DIR.- NUMERALS SHOW TENS OF DEGRESS CLOCKWISE FROM TRUE NORTH. "00" INDICATES CALM.
RESULTANT DIRECTIONS ARE GIVEN TO WHOLE DEGREES.
BOLD VALUES INDICATE EXTREME VALUES WHICH OCCURRED AFTER THE ASOS SYSTEM WAS COMMISSIONED.

RECORD = PERIOD OF RECORD
RECORD MEAN PRECIPITATION IS THE MEAN OF ALL DAILY PRECIPITATION AMOUNTS DURING THE PERIOD OF RECORD.
RECORD MAX(MIN) TEMPERATURE IS THE MEAN OF ALL DAILY MAX(MIN) TEMPERATURES DURING THE PERIOD OF RECORD.
RECORD MEAN TEMPERATURE IS THE SUM OF THE RECORD MAX AND RECORD MIN DIVIDED BY 2.
AVERAGE TEMPERATURE IS THE SUM OF THE MEAN DAILY MAX AND MIN TEMPERATURE DIVIDED BY 2.

EXCEPTIONS

THUNDERSTORMS AND HEAVY FOG, APRIL 1973 TO DATE, MAY BE INCOMPLETE, DUE TO PART-TIME OPERATIONS.

RECORD MEANS ARE THROUGH THE CURRENT YEAR, BEGINNING IN 1952 FOR TEMPERATURE
1948 FOR PRECIPITATION

HEATING DEGREE DAYS Base 65 deg. F KOROR ISLAND, PACIFIC

SEASON	JULY	AUG	SEP	OCT	NOV	DEC	JAN	FEB	MAR	APR	MAY	JUNE	TOTAL
1983-84	0	0	0	0	0	0	0	0	0	0	0	0	0
1984-85	0	0	0	0	0	0	0	0	0	0	0	0	0
1985-86	0	0	0	0	0	0	0	0	0	0	0	0	0
1986-87	0	0	0	0	0	0	0	0	0	0	0	0	0
1987-88	0	0	0	0	0	0	0	0	0	0	0	0	0
1988-89	0	0	0	0	0	0	0	0	0	0	0	0	0
1989-90	0	0	0	0	0	0	0	0	0	0	0	0	0
1990-91	0	0	0	0	0	0	0	0	0	0	0	0	0
1991-92	0	0	0	0	0	0	0	0	0	0	0	0	0
1992-93	0	0	0	0	0	0	0	0	0	0	0	0	0
1993-94	0	0	0	0	0	0							

COOLING DEGREE DAYS Base 65 deg. F KOROR ISLAND, PACIFIC

YEAR	JAN	FEB	MAR	APR	MAY	JUNE	JULY	AUG	SEP	OCT	NOV	DEC	TOTAL
1969	493	457	522	530	557	532	496	529	495	541	521	527	6200
1970	519	481	536	536	546	526	537	534	527	538	527	523	6330
1971	514	463	537	539	539	488	495	517	521	499	519	536	6167
1972	499	456	502	512	553	509	539	523	510	551	528	530	6212
1973	506	470	530	525	555	536	531	548	523	529	522	547	6322
1974	504	476	504	505	555	513	517	540	526	535	524	519	6218
1975	498	467	524	515	541	493	493	524	504	527	527	503	6116
1976	530	467	515	504	549	517	533	495	510	547	543	523	6233
1977	516	460	527	530	563	532	507	509	524	551	550	551	6320
1978	513	429	535	513	556	510	542	495	484	511	508	529	6125
1979	516	463	511	497	547	488	506	524	520	533	522	520	6147
1980	496	449	514	509	546	509	508	500	503	545	548	543	6170
1981	504	437	507	514	553	485	506	523	504	508	532	536	6109
1982	527	474	505	508	525	501	507	507	530	525	527	526	6162
1983	489	461	524	525	564	511	505	537	516	535	530	535	6232
1984	503	471	523	525	554	493	517	502	510	500	539	547	6184
1985	500	466	536	507	547	471	473	493	473	526	530	545	6067
1986	527	455	540	533	575	513	501	548	528	537	527	530	6314
1987	512	446	525	529	540	527	525	505	556	557	555	568	6345
1988	537	489	554	558	541	515	539	540	544	535	530	520	6402
1989	536	493	543	538	545	504	522	530	533	536	534	545	6359
1990	537	485	537	544	577	520	536	552	508	546	513	541	6396
1991	506	440	506	534	556	529	504	533	492	532	533	507	6172
1992	493	475	506	531	549	515	509	506	519	514	521	492	6130
1993	491	429	508	500	564	506	529	508	493	542	539	546	6155

SNOWFALL (inches) KOROR ISLAND, PACIFIC

SEASON	JULY	AUG	SEP	OCT	NOV	DEC	JAN	FEB	MAR	APR	MAY	JUNE	TOTAL
1971-72	0.0	0.0	0.0	0.0	0.0	0.0	0.0	0.0	0.0	0.0	0.0	0.0	0.0
1972-73	0.0	0.0	0.0	0.0	0.0	0.0	0.0	0.0	0.0	0.0	0.0	0.0	0.0
1973-74	0.0	0.0	0.0	0.0	0.0	0.0	0.0	0.0	0.0	0.0	0.0	0.0	0.0
1974-75	0.0	0.0	0.0	0.0	0.0	0.0	0.0	0.0	0.0	0.0	0.0	0.0	0.0
1975-76	0.0	0.0	0.0	0.0	0.0	0.0	0.0	0.0	0.0	0.0	0.0	0.0	0.0
1976-77	0.0	0.0	0.0	0.0	0.0	0.0	0.0	0.0	0.0	0.0	0.0	0.0	0.0
1977-78	0.0	0.0	0.0	0.0	0.0	0.0	0.0	0.0	0.0	0.0	0.0	0.0	0.0
1978-79	0.0	0.0	0.0	0.0	0.0	0.0	0.0	0.0	0.0	0.0	0.0	0.0	0.0
1979-80	0.0	0.0	0.0	0.0	0.0	0.0	0.0	0.0	0.0	0.0	0.0	0.0	0.0
1980-81	0.0	0.0	0.0	0.0	0.0	0.0	0.0	0.0	0.0	0.0	0.0	0.0	0.0
1981-82	0.0	0.0	0.0	0.0	0.0	0.0	0.0	0.0	0.0	0.0	0.0	0.0	0.0
1982-83	0.0	0.0	0.0	0.0	0.0	0.0	0.0	0.0	0.0	0.0	0.0	0.0	0.0
1983-84	0.0	0.0	0.0	0.0	0.0	0.0	0.0	0.0	0.0	0.0	0.0	0.0	0.0
1984-85	0.0	0.0	0.0	0.0	0.0	0.0	0.0	0.0	0.0	0.0	0.0	0.0	0.0
1985-86	0.0	0.0	0.0	0.0	0.0	0.0	0.0	0.0	0.0	0.0	0.0	0.0	0.0
1986-87	0.0	0.0	0.0	0.0	0.0	0.0	0.0	0.0	0.0	0.0	0.0	0.0	0.0
1987-88	0.0	0.0	0.0	0.0	0.0	0.0	0.0	0.0	0.0	0.0	0.0	0.0	0.0
1988-89	0.0	0.0	0.0	0.0	0.0	0.0	0.0	0.0	0.0	0.0	0.0	0.0	0.0
1989-90	0.0	0.0	0.0	0.0	0.0	0.0	0.0	0.0	0.0	0.0	0.0	0.0	0.0
1990-91	0.0	0.0	0.0	0.0	0.0	0.0	0.0	0.0	0.0	0.0	0.0	0.0	0.0
1991-92	0.0	0.0	0.0	0.0	0.0	0.0	0.0	0.0	0.0	0.0	0.0	0.0	0.0
1992-93	0.0	0.0	0.0	0.0	0.0	0.0	0.0	0.0	0.0	0.0	0.0	0.0	0.0
1993-94	0.0	0.0	0.0	0.0	0.0	0.0							
Record Mean	0.0	0.0	0.0	0.0	0.0	0.0	0.0	0.0	0.0	0.0	0.0	0.0	0.0

KWAJALEIN, MARSHALL ISLANDS, PACIFIC

Kwajalein Island, although only 3 miles in length and 1/2 mile wide, is the largest of the fringing reef islands composing Kwajalein Atoll. Kwajalein Atoll, spanning some 70 miles, is one of the largest coral atolls in the world. The land surface of the island, which has very little effect on the climate of the locality, has an average elevation of less than 10 feet above sea level. The highest points on the island are 12 to 15 feet above sea level.

Kwajalein, located less than 700 miles north of the equator, has a tropical marine climate characterized by relatively high annual rainfall and warm to hot, humid weather throughout the year.

Temperatures are very uniform from day to day and month to month. Because of the low latitude, there are only slight seasonal variations in the length of daylight period and the altitude of the sun at Kwajalein. As a result, the variation of the amount of solar energy received is small. The small variation in solar energy and the marine influence are the principal reasons for the uniform temperatures in the area. The range of normal temperature between the coldest month and the warmest month averages about 2 degrees.

The principal rainfall season extends from May through November. Light, easterly winds, almost constant cloudiness, and frequent moderate to heavy showers prevail during the wet season.

The dry season includes the period December through April, and is characterized not so much by lack of showers as by light showers of short duration. In this season the trade winds are persistent, blowing from the northeast 15 to 20 knots almost continuously. Cloudiness is at a minimum, and the sky is less than one-half covered most of the time, but clear skies are rare.

Severe storms with attendant damaging winds are rare in the vicinity of Kwajalein. During the wet season, however, small, weak depressions may form near the island. Some of these intensify and a few eventually develop into typhoons after moving westward away from the island. These depressions cause heavy rainfall in the Kwajalein Atoll.

The relative humidity is uniformly high throughout the year, and is slightly higher in the wet season than in the dry season. The combination of high humidity and proximity of the salt water ocean presents a corrosion problem.

NORMALS, MEANS, AND EXTREMES

KWAJALEIN, MARSHALL ISLANDS, PACIFIC

LATITUDE: 8°44'N LONGITUDE: 167°44'E ELEVATION: FT. GRND 8 BARO 11 TIME ZONE: 180E MER WBAN: 40604

	(a)	JAN	FEB	MAR	APR	MAY	JUNE	JULY	AUG	SEP	OCT	NOV	DEC	YEAR
TEMPERATURE °F:														
Normals														
-Daily Maximum		85.8	86.5	87.1	86.7	86.7	86.7	86.7	87.1	87.0	87.0	86.7	86.1	86.7
-Daily Minimum		77.0	77.1	77.4	77.4	77.3	77.2	77.1	77.2	77.1	77.3	77.3	77.3	77.2
-Monthly		81.4	81.8	82.3	82.1	82.0	82.0	81.9	82.2	82.1	82.2	82.0	81.7	82.0
Extremes														
-Record Highest	41	90	90	90	90	91	90	91	91	92	92	91	89	92
-Year		1971	1985	1982	1988	1963	1984	1972	1969	1980	1977	1989	1984	SEP 1980
-Record Lowest	41	68	71	70	71	71	71	70	71	68	71	70	69	68
-Year		1979	1963	1985	1964	1968	1976	1954	1971	1984	1967	1975	1963	SEP 1984
NORMAL DEGREE DAYS:														
Heating (base 65°F)		0	0	0	0	0	0	0	0	0	0	0	0	0
Cooling (base 65°F)		508	470	536	513	527	510	524	533	513	533	510	518	6195
% OF POSSIBLE SUNSHINE														
MEAN SKY COVER (tenths)														
Sunrise - Sunset	32	8.0	8.0	8.0	8.5	8.7	8.6	8.7	8.5	8.6	8.8	8.6	8.4	8.5
MEAN NUMBER OF DAYS:														
Sunrise to Sunset														
-Clear	32	2.6	3.0	2.6	1.8	0.8	1.2	1.1	1.5	1.5	1.0	1.0	2.0	20.3
-Partly Cloudy	32	7.7	6.3	7.4	5.3	5.9	5.8	5.4	6.1	5.1	4.8	6.1	5.8	71.6
-Cloudy	32	20.7	18.9	21.0	23.0	24.2	23.0	24.5	23.3	23.4	25.2	22.9	23.2	273.4
Precipitation														
.01 inches or more	41	15.4	13.3	15.0	17.1	20.7	22.6	23.2	23.2	22.5	23.1	23.1	19.2	238.4
Snow, Ice pellets, hail														
1.0 inches or more	41	0.0	0.0	0.0	0.0	0.0	0.0	0.0	0.0	0.0	0.0	0.0	0.0	0.0
Thunderstorms	33	0.2	0.1	0.1	0.4	0.7	0.8	1.3	1.2	2.0	1.4	1.4	0.4	9.9
Heavy Fog Visibility														
1/4 mile or less	33	0.0	0.0	0.0	0.0	0.0	0.0	0.0	0.0	0.0	0.0	0.0	0.0	0.0
Temperature °F														
-Maximum														
90° and above	41	0.1	0.2	0.7	0.6	1.1	0.8	1.3	2.0	2.6	2.1	0.5	0.*	12.0
32° and below	41	0.0	0.0	0.0	0.0	0.0	0.0	0.0	0.0	0.0	0.0	0.0	0.0	0.0
-Minimum														
32° and below	41	0.0	0.0	0.0	0.0	0.0	0.0	0.0	0.0	0.0	0.0	0.0	0.0	0.0
0° and below	41	0.0	0.0	0.0	0.0	0.0	0.0	0.0	0.0	0.0	0.0	0.0	0.0	0.0
AVG. STATION PRESS. (mb)	17	1008.8	1009.2	1009.4	1009.3	1009.5	1009.3	1009.0	1009.0	1009.0	1008.8	1008.1	1008.3	1009.0
RELATIVE HUMIDITY (%)														
Hour 00	33	79	79	80	82	84	84	84	83	83	82	83	81	82
Hour 06 (Local Time)	33	80	79	80	82	84	84	85	84	84	83	83	81	82
Hour 12	33	71	69	70	73	76	77	76	76	76	75	76	74	74
Hour 18	33	75	73	74	77	79	79	78	78	78	79	79	77	77
PRECIPITATION (inches):														
Water Equivalent														
-Normal		4.56	3.23	4.10	7.55	9.98	9.62	10.44	10.11	11.83	11.91	10.66	8.10	102.09
-Maximum Monthly	49	15.66	9.05	24.33	20.29	26.86	19.61	17.33	17.46	21.16	20.05	19.51	30.38	30.38
-Year		1951	1976	1951	1971	1980	1955	1958	1979	1972	1964	1957	1950	DEC 1950
-Minimum Monthly	49	0.48	0.04	0.16	0.20	0.53	3.56	3.53	5.38	5.32	5.04	3.51	1.90	0.04
-Year		1977	1977	1975	1983	1984	1984	1984	1981	1965	1969	1973	1971	FEB 1977
-Maximum in 24 hrs	33	6.46	4.60	3.81	5.41	8.35	4.69	5.68	5.35	4.69	6.53	7.24	17.15	17.15
-Year		1979	1985	1990	1971	1980	1971	1983	1979	1972	1968	1975	1972	DEC 1972
Snow, Ice pellets, hail														
-Maximum Monthly		0.0	0.0	0.0	0.0	0.0	0.0	0.0	0.0	0.0	0.0	0.0	0.0	
-Year														
-Maximum in 24 hrs	49	0.0	0.0	0.0	0.0	0.0	0.0	0.0	0.0	0.0	0.0	0.0	0.0	
-Year														
WIND:														
Mean Speed (mph)	27	16.9	16.4	16.2	14.8	13.8	12.9	11.0	9.6	9.1	9.9	12.2	16.3	13.3
Prevailing Direction through 1963		ENE	ENE	ENE	ENE	ENE	ENE	E	ENE	E	E	ENE	ENE	ENE
Fastest Obs. 1 Min.														
-Direction (!!)	33	22	18	07	08	11	12	09	08	19	20	12	10	12
-Speed (MPH)	33	55	35	36	37	44	41	41	43	35	40	60	40	60
-Year		1969	1963	1981	1964	1967	1986	1962	1970	1967	1968	1991	1990	NOV 1991
Peak Gust														
-Direction (!!)	10	E	E	E	SE	E	SE	SE	SE	E	NE	NE	E	NE
-Speed (mph)	10	66	43	44	45	45	51	46	43	43	53	82	52	82
-Date		1988	1987	1987	1985	1989	1986	1988	1988	1985	1985	1991	1990	NOV 1991

PRECIPITATION (inches)

KWAJALEIN, MARSHALL ISLANDS, PACIFIC

YEAR	JAN	FEB	MAR	APR	MAY	JUNE	JULY	AUG	SEP	OCT	NOV	DEC	ANNUAL
1964	0.90	3.07	3.44	14.13	24.18	9.59	10.67	11.71	19.76	20.05	9.27	12.60	139.37
1965	3.85	1.01	0.18	0.24	3.20	12.55	12.03	8.05	5.32	9.95	9.17	4.49	70.04
1966	2.97	0.40	4.00	11.65	13.72	10.42	7.38	15.03	12.55	7.80	9.19	21.08	116.19
1967	5.14	4.42	17.19	4.61	10.35	14.14	9.64	6.11	10.91	11.18	7.71	7.17	108.57
1968	5.79	3.12	10.28	11.63	11.80	11.56	12.01	12.15	7.25	14.75	10.34	9.57	120.25
1969	6.11	0.43	1.40	6.07	8.13	12.56	12.81	5.74	9.91	5.04	8.46	2.68	79.34
1970	0.82	0.43	0.73	2.99	6.74	8.06	6.84	11.94	13.83	19.12	6.89	5.16	83.55
1971	5.35	3.63	4.28	20.29	18.66	11.54	16.43	11.19	12.52	15.65	4.86	1.90	126.30
1972	10.02	6.20	3.88	8.13	14.26	11.84	12.33	9.44	21.16	11.75	3.69	22.70	135.40
1973	0.52	0.52	1.20	4.53	7.92	8.60	11.29	7.51	8.33	11.45	3.51	5.21	70.59
1974	8.74	7.90	5.82	18.98	11.84	9.05	5.80	17.21	10.28	12.62	13.84	8.50	130.58
1975	1.05	0.24	0.16	6.45	8.17	13.99	11.74	11.22	11.60	15.62	15.61	5.69	101.54
1976	4.18	9.05	6.73	12.31	11.03	11.24	8.79	8.94	13.50	10.43	5.81	2.46	104.47
1977	0.48	0.04	4.71	5.79	12.36	7.93	13.01	7.58	8.17	14.09	8.76	7.16	90.08
1978	13.86	6.31	4.59	5.30	15.81	7.11	8.52	5.48	8.75	11.47	16.16	3.14	106.50
1979	10.73	2.81	1.08	12.15	2.74	6.19	9.29	17.46	8.44	12.52	17.61	8.27	109.29
1980	8.45	2.82	1.20	8.55	26.86	9.29	13.75	13.02	14.42	11.91	9.58	4.27	124.12
1981	1.91	1.29	7.94	5.10	1.74	8.36	11.97	5.38	12.22	10.76	15.74	5.56	87.97
1982	2.16	2.20	1.93	6.50	8.88	13.60	16.14	13.73	13.55	7.10	11.10	4.28	101.17
1983	0.89	0.68	0.36	0.20	1.76	3.83	13.27	10.74	14.44	10.49	18.88	4.24	79.78
1984	3.20	4.07	1.06	2.12	0.53	3.56	3.53	7.54	7.10	8.48	11.84	6.32	59.35
1985	2.28	6.97	5.75	12.46	8.05	7.66	7.61	11.18	7.64	9.99	11.45	5.06	96.10
1986	3.12	5.64	6.87	4.82	9.37	15.08	12.43	7.53	12.43	13.48	16.36	23.28	130.41
1987	3.95	1.86	1.49	3.11	6.43	11.08	10.51	14.42	11.42	9.67	9.67	6.16	89.77
1988	5.08	0.67	1.24	5.67	3.21	5.17	7.54	5.54	16.01	16.68	11.17	8.45	86.43
1989	2.90	4.63	1.67	12.84	7.34	5.77	8.06	8.09	17.32	8.20	6.40	7.13	90.35
1990	9.32	3.36	11.00	4.40	10.75	11.61	9.25	6.36	13.28	9.14	10.15	13.08	111.70
1991	1.88	7.16	10.67	15.83	10.63	10.00	11.40	12.84	19.72	7.09	16.21	3.84	127.27
1992	4.19	0.42	0.20	0.24	3.68	6.86	13.90	7.99	11.44	7.88	14.94	8.87	80.61
1993	2.97	2.72	6.87	2.48	5.94	7.72	7.40	13.07	7.07	10.31	10.26	10.78	87.59
Record Mean	4.17	2.90	5.15	6.78	9.39	9.49	10.13	10.18	11.50	11.59	11.46	8.50	101.24

AVERAGE TEMPERATURE (deg. F)

KWAJALEIN, MARSHALL ISLANDS, PACIFIC

YEAR	JAN	FEB	MAR	APR	MAY	JUNE	JULY	AUG	SEP	OCT	NOV	DEC	ANNUAL
1964	80.8	80.8	81.5	80.7	81.0	81.5	81.1	81.0	80.5	80.6	80.7	80.4	80.9
1965	80.5	81.0	82.6	82.8	82.2	81.1	80.9	81.9	82.2	81.9	80.9	81.3	81.6
1966	80.7	81.8	82.1	81.3	81.6	81.6	82.4	82.4	83.1	82.6	82.0	80.3	81.8
1967	81.0	81.5	80.4	81.9	81.7	81.4	81.8	82.1	82.3	81.6	82.0	81.2	81.6
1968	81.0	81.2	80.4	80.8	80.4	81.3	81.1	82.1	82.9	82.9	82.4	82.3	81.6
1969	80.8	82.2	82.7	82.2	82.0	82.3	81.8	83.0	82.4	83.1	83.0	82.6	82.4
1970	82.4	83.3	83.5	83.5	82.6	82.5	82.0	81.9	81.6	80.8	81.8	82.1	82.3
1971	81.6	81.7	82.1	80.4	80.8	80.9	80.8	81.1	81.0	81.0	81.5	81.6	81.2
1972	80.8	81.3	82.3	81.4	82.4	81.9	82.1	82.5	82.0	81.9	82.4	81.3	81.9
1973	81.1	81.8	83.4	82.7	82.0	82.1	81.8	81.9	81.9	81.5	82.5	81.9	82.0
1974	81.6	81.7	82.1	80.8	81.7	81.6	82.4	81.5	81.7	81.7	81.9	81.7	81.7
1975	81.8	82.4	83.3	82.8	81.3	81.1	80.7	80.8	80.9	80.3	80.8	81.0	81.4
1976	80.4	80.5	81.2	81.3	81.1	81.2	81.6	81.7	81.4	81.5	81.7	81.5	81.3
1977	81.2	82.3	82.1	81.9	81.7	82.1	81.8	82.9	83.3	82.4	82.7	82.4	82.2
1978	81.6	81.5	82.5	82.0	81.2	81.7	81.6	82.3	82.6	82.5	81.6	82.0	81.9
1979	81.1	81.4	82.8	81.5	82.4	82.4	82.5	82.1	83.0	82.9	82.4	82.0	82.2
1980	81.9	82.0	83.3	82.0	81.4	82.4	82.3	82.9	82.3	82.8	82.0	82.5	82.3
1981	82.3	82.3	82.3	82.3	83.3	82.2	81.9	82.6	81.9	83.0	82.2	81.9	82.3
1982	81.6	82.1	82.6	82.7	82.5	82.4	81.8	82.3	81.8	82.3	82.7	80.9	82.2
1983	80.1	81.1	82.0	82.7	83.6	83.5	82.6	82.4	82.2	82.5	81.5	82.1	82.2
1984	82.0	82.2	83.0	83.1	83.6	82.9	82.7	81.8	82.3	82.6	82.0	82.5	82.6
1985	82.2	82.7	82.0	81.4	82.2	82.0	82.2	81.9	82.8	82.9	82.0	82.2	82.2
1986	82.1	82.2	82.1	82.6	82.2	82.3	82.5	83.0	82.1	82.7	82.0	81.5	82.3
1987	81.1	81.1	82.1	82.5	82.9	82.5	82.7	83.1	82.9	83.7	82.9	82.2	82.5
1988	81.7	82.6	83.0	83.1	83.6	82.6	82.3	83.1	82.2	81.9	82.2	81.5	82.5
1989	82.0	82.1	82.8	82.2	82.1	82.2	82.5	82.6	82.4	82.5	82.9	83.0	82.4
1990	82.2	82.2	82.2	83.5	82.3	82.9	82.9	83.3	82.6	83.2	83.3	82.4	82.8
1991	82.6	82.5	81.8	81.6	82.6	82.5	82.3	82.7	82.7	82.8	81.9	81.6	82.3
1992	80.8	81.2	81.8	82.7	82.8	82.5	82.2	83.0	83.5	83.4	81.8	81.7	82.3
1993	81.1	81.5	81.7	82.3	82.5	82.9	82.9	82.5	83.1	83.2	82.4	82.1	82.4
Record Mean	81.4	81.7	82.1	82.1	82.1	82.2	82.2	82.4	82.4	82.4	82.1	81.8	82.1
Max	85.5	86.0	86.5	86.4	86.5	86.6	86.8	87.2	87.3	87.1	86.5	85.9	86.5
Min	77.3	77.4	77.7	77.8	77.7	77.7	77.5	77.6	77.5	77.7	77.6	77.7	77.6

REFERENCE NOTES

KWAJALEIN, MARSHALL ISLANDS, PACIFIC

GENERAL
T - TRACE AMOUNT.
BLANK ENTRIES DENOTE MISSING/UNREPORTED DATA.
# INDICATES A STATION OR INSTRUMENT RELOCATION.

SPECIFIC

(a) - LENGTH OF RECORD IN YEARS, ALTHOUGH INDIVIDUAL MONTHS MAY BE MISSING.
0.* OR * - THE VALUE IS BETWEEN 0.0 AND 0.05
NORMALS - BASED ON THE 1961-1990 RECORD PERIOD.
EXTREMES - DATES ARE THE MOST RECENT OCCURRENCE
WIND DIR.- NUMERALS SHOW TENS OF DEGRESS CLOCKWISE FROM TRUE NORTH. "00" INDICATES CALM.
RESULTANT DIRECTIONS ARE GIVEN TO WHOLE DEGREES.
BOLD VALUES INDICATE EXTREME VALUES WHICH OCCURRED AFTER THE ASOS SYSTEM WAS COMMISSIONED.

RECORD = PERIOD OF RECORD
RECORD MEAN PRECIPITATION IS THE MEAN OF ALL DAILY PRECIPITATION AMOUNTS DURING THE PERIOD OF RECORD.
RECORD MAX(MIN) TEMPERATURE IS THE MEAN OF ALL DAILY MAX(MIN) TEMPERATURES DURING THE PERIOD OF RECORD.
RECORD MEAN TEMPERATURE IS THE SUM OF THE RECORD MAX AND RECORD MIN DIVIDED BY 2.
AVERAGE TEMPERATURE IS THE SUM OF THE MEAN DAILY MAX AND MIN TEMPERATURE DIVIDED BY 2.

EXCEPTIONS

RECORD MEANS ARE THROUGH THE CURRENT YEAR, BEGINNING IN 1948 FOR TEMPERATURE
1945 FOR PRECIPITATION
YEARS 1953 - 1958 ARE EXCLUDED FROM COMPUTATION OF EXTREME RECORD HIGHEST TEMPERATURES

HEATING DEGREE DAYS Base 65 deg. F KWAJALEIN, MARSHALL ISLANDS, PACIFIC

SEASON	JULY	AUG	SEP	OCT	NOV	DEC	JAN	FEB	MAR	APR	MAY	JUNE	TOTAL
1983-84	0	0	0	0	0	0	0	0	0	0	0	0	0
1984-85	0	0	0	0	0	0	0	0	0	0	0	0	0
1985-86	0	0	0	0	0	0	0	0	0	0	0	0	0
1986-87	0	0	0	0	0	0	0	0	0	0	0	0	0
1987-88	0	0	0	0	0	0	0	0	0	0	0	0	0
1988-89	0	0	0	0	0	0	0	0	0	0	0	0	0
1989-90	0	0	0	0	0	0	0	0	0	0	0	0	0
1990-91	0	0	0	0	0	0	0	0	0	0	0	0	0
1991-92	0	0	0	0	0	0	0	0	0	0	0	0	0
1992-93	0	0	0	0	0	0	0	0	0	0	0	0	0
1993-94	0	0	0	0	0	0							

COOLING DEGREE DAYS Base 65 deg. F KWAJALEIN, MARSHALL ISLANDS, PACIFIC

YEAR	JAN	FEB	MAR	APR	MAY	JUNE	JULY	AUG	SEP	OCT	NOV	DEC	TOTAL
1969	498	487	556	524	532	523	528	564	528	566	542	552	6400
1970	547	518	580	562	552	532	534	530	506	497	510	536	6404
1971	519	470	538	469	498	484	494	509	488	505	500	521	5995
1972	497	477	545	496	548	515	536	549	518	530	528	513	6252
1973	509	477	574	538	533	520	526	533	513	520	534	529	6306
1974	521	476	538	480	526	503	545	518	509	523	517	525	6181
1975	526	495	572	539	516	493	496	498	482	480	479	502	6078
1976	483	458	511	496	507	491	522	525	499	521	507	518	6038
1977	507	491	537	514	526	518	526	564	555	547	537	549	6371
1978	525	465	548	519	507	509	521	542	536	552	507	535	6266
1979	502	467	561	504	544	526	546	536	547	559	530	532	6354
1980	528	501	576	515	517	529	542	557	526	560	519	551	6421
1981	545	491	541	527	573	525	532	551	515	565	523	530	6418
1982	521	486	554	535	548	528	530	544	513	541	535	498	6333
1983	478	454	531	538	582	561	552	546	521	548	502	535	6348
1984	535	504	565	551	582	546	557	526	527	554	518	549	6514
1985	541	500	533	501	542	513	542	529	540	564	516	539	6360
1986	539	488	537	536	542	524	550	564	519	558	516	516	6389
1987	507	458	540	530	559	530	557	569	541	585	542	540	6458
1988	526	516	565	551	584	535	545	568	521	529	525	519	6484
1989	532	482	559	520	535	527	549	554	529	550	544	567	6448
1990	538	490	539	561	542	545	561	575	533	572	555	545	6556
1991	553	495	528	505	552	533	546	556	535	556	515	521	6395
1992	494	477	526	535	558	532	541	565	562	577	514	524	6405
1993	507	468	522	524	548	542	562	552	548	571	529	541	6414

SNOWFALL (inches) KWAJALEIN, MARSHALL ISLANDS, PACIFIC

SEASON	JULY	AUG	SEP	OCT	NOV	DEC	JAN	FEB	MAR	APR	MAY	JUNE	TOTAL
1971-72	0.0	0.0	0.0	0.0	0.0	0.0	0.0	0.0	0.0	0.0	0.0	0.0	0.0
1972-73	0.0	0.0	0.0	0.0	0.0	0.0	0.0	0.0	0.0	0.0	0.0	0.0	0.0
1973-74	0.0	0.0	0.0	0.0	0.0	0.0	0.0	0.0	0.0	0.0	0.0	0.0	0.0
1974-75	0.0	0.0	0.0	0.0	0.0	0.0	0.0	0.0	0.0	0.0	0.0	0.0	0.0
1975-76	0.0	0.0	0.0	0.0	0.0	0.0	0.0	0.0	0.0	0.0	0.0	0.0	0.0
1976-77	0.0	0.0	0.0	0.0	0.0	0.0	0.0	0.0	0.0	0.0	0.0	0.0	0.0
1977-78	0.0	0.0	0.0	0.0	0.0	0.0	0.0	0.0	0.0	0.0	0.0	0.0	0.0
1978-79	0.0	0.0	0.0	0.0	0.0	0.0	0.0	0.0	0.0	0.0	0.0	0.0	0.0
1979-80	0.0	0.0	0.0	0.0	0.0	0.0	0.0	0.0	0.0	0.0	0.0	0.0	0.0
1980-81	0.0	0.0	0.0	0.0	0.0	0.0	0.0	0.0	0.0	0.0	0.0	0.0	0.0
1981-82	0.0	0.0	0.0	0.0	0.0	0.0	0.0	0.0	0.0	0.0	0.0	0.0	0.0
1982-83	0.0	0.0	0.0	0.0	0.0	0.0	0.0	0.0	0.0	0.0	0.0	0.0	0.0
1983-84	0.0	0.0	0.0	0.0	0.0	0.0	0.0	0.0	0.0	0.0	0.0	0.0	0.0
1984-85	0.0	0.0	0.0	0.0	0.0	0.0	0.0	0.0	0.0	0.0	0.0	0.0	0.0
1985-86	0.0	0.0	0.0	0.0	0.0	0.0	0.0	0.0	0.0	0.0	0.0	0.0	0.0
1986-87	0.0	0.0	0.0	0.0	0.0	0.0	0.0	0.0	0.0	0.0	0.0	0.0	0.0
1987-88	0.0	0.0	0.0	0.0	0.0	0.0	0.0	0.0	0.0	0.0	0.0	0.0	0.0
1988-89	0.0	0.0	0.0	0.0	0.0	0.0	0.0	0.0	0.0	0.0	0.0	0.0	0.0
1989-90	0.0	0.0	0.0	0.0	0.0	0.0	0.0	0.0	0.0	0.0	0.0	0.0	0.0
1990-91	0.0	0.0	0.0	0.0	0.0	0.0	0.0	0.0	0.0	0.0	0.0	0.0	0.0
1991-92	0.0	0.0	0.0	0.0	0.0	0.0	0.0	0.0	0.0	0.0	0.0	0.0	0.0
1992-93	0.0	0.0	0.0	0.0	0.0	0.0	0.0	0.0	0.0	0.0	0.0	0.0	0.0
1993-94	0.0	0.0	0.0	0.0	0.0	0.0							
Record Mean	0.0	0.0	0.0	0.0	0.0	0.0	0.0	0.0	0.0	0.0	0.0	0.0	0.0

MAJURO, MARSHALL ISLANDS, PACIFIC

The station at Majuro is located on the southeastern end of the Majuro Atoll. This atoll is approximately 160 square miles in area with a lagoon of about 150 square miles. The lagoon is oblong, 22 miles long and about 4 miles wide. Dalap Island, on which the station is located, is oriented roughly east-west.

The climate of Majuro is predominately a trade-wind climate with the trade winds prevailing throughout the year. Tropical storms are very rare.

Minor storms of the easterly wave type are quite common from March to April and October to November. The trades are frequently locally interrupted during the summer months by the movement of the zone of intertropical convergence across the area.

Rainfall is heavy, with the wettest months being October and November. Precipitation is generally of the shower type, however, continuous rain is not uncommon.

One of the outstanding features of the climate is the extremely consistent temperature regime. The range between the coolest and the warmest months averages less than 1 degree. The average daily range is less than 9 degrees. Nighttime minima are generally 2-4 degrees warmer than the average daily minimum because lowest temperatures usually occur during heavy showers in the daytime.

Skies at Majuro are quite cloudy. Cumuliform clouds are predominant but altostratus-altocumulus and cirriform clouds are also present most of the time.

NORMALS, MEANS, AND EXTREMES

MAJURO, MARSHALL ISLANDS, PACIFIC

LATITUDE: 7°05'N LONGITUDE: 171°23'E ELEVATION: FT. GRND 10 BARO 8 TIME ZONE: 180E MER WBAN: 40710

	(a)	JAN	FEB	MAR	APR	MAY	JUNE	JULY	AUG	SEP	OCT	NOV	DEC	YEAR
TEMPERATURE °F:														
Normals														
-Daily Maximum		85.0	85.4	85.6	85.6	85.8	85.8	85.7	86.2	86.3	86.3	86.0	85.3	85.8
-Daily Minimum		76.5	76.7	76.6	76.5	76.6	76.3	76.1	76.3	76.3	76.2	76.3	76.3	76.4
-Monthly		80.7	81.1	81.1	81.1	81.2	81.1	80.9	81.3	81.4	81.3	81.2	80.8	81.1
Extremes														
-Record Highest	38	89	89	89	89	90	89	90	91	90	91	91	90	91
-Year		1979	1991	1990	1992	1992	1992	1980	1969	1992	1958	1990	1990	NOV 1990
-Record Lowest	38	69	70	70	70	70	70	70	71	70	70	68	70	68
-Year		1958	1985	1993	1985	1985	1958	1989	1990	1991	1984	1991	1984	NOV 1991
NORMAL DEGREE DAYS:														
Heating (base 65°F)		0	0	0	0	0	0	0	0	0	0	0	0	0
Cooling (base 65°F)		487	451	499	483	502	483	493	505	492	505	486	490	5876
% OF POSSIBLE SUNSHINE	33	62	65	66	60	59	56	56	61	59	55	53	54	59
MEAN SKY COVER (tenths)														
Sunrise - Sunset	37	8.6	8.3	8.5	8.6	8.6	8.7	8.7	8.4	8.5	8.6	8.7	8.7	8.6
MEAN NUMBER OF DAYS:														
Sunrise to Sunset														
-Clear	37	0.8	0.9	1.2	0.7	0.6	0.4	0.5	0.8	0.9	0.8	0.4	0.6	8.7
-Partly Cloudy	37	6.3	7.5	7.0	6.6	6.6	6.4	6.1	7.9	6.4	6.6	6.3	6.2	79.9
-Cloudy	37	23.9	19.8	22.9	22.6	23.8	23.2	24.4	22.3	22.8	23.5	23.2	24.1	276.6
Precipitation														
.01 inches or more	39	17.4	15.5	18.3	20.7	23.5	24.1	24.5	23.3	22.5	23.3	23.3	21.8	258.3
Snow, Ice pellets, hail														
1.0 inches or more	39	0.0	0.0	0.0	0.0	0.0	0.0	0.0	0.0	0.0	0.0	0.0	0.0	0.0
Thunderstorms	23	0.4	0.5	0.6	0.6	0.9	1.7	1.9	1.5	2.7	2.2	2.1	1.1	16.2
Heavy Fog Visibility														
1/4 mile or less	24	0.0	0.0	0.0	0.0	0.0	0.0	0.0	0.0	0.0	0.0	0.0	0.0	0.0
Temperature °F														
-Maximum														
90° and above	39	0.0	0.0	0.0	0.0	0.1	0.0	0.*	0.4	0.5	0.6	0.3	0.1	1.9
32° and below	39	0.0	0.0	0.0	0.0	0.0	0.0	0.0	0.0	0.0	0.0	0.0	0.0	0.0
-Minimum														
32° and below	39	0.0	0.0	0.0	0.0	0.0	0.0	0.0	0.0	0.0	0.0	0.0	0.0	0.0
0° and below	39	0.0	0.0	0.0	0.0	0.0	0.0	0.0	0.0	0.0	0.0	0.0	0.0	0.0
AVG. STATION PRESS. (mb)	16	1008.8	1009.4	1009.6	1009.4	1009.7	1009.4	1009.3	1009.5	1009.4	1008.9	1008.2	1008.4	1009.2
RELATIVE HUMIDITY (%)														
Hour 00	37	80	80	81	82	84	84	83	82	82	82	82	82	82
Hour 06 (Local Time)	38	81	80	81	84	85	84	84	84	83	83	83	82	83
Hour 12	38	75	74	75	77	78	78	77	76	76	76	77	77	76
Hour 18	37	78	77	78	80	81	80	79	78	78	79	80	80	79
PRECIPITATION (inches):														
Water Equivalent														
-Normal		8.43	6.15	8.28	10.28	11.18	11.59	13.00	11.52	12.42	13.84	12.80	11.85	131.34
-Maximum Monthly	39	21.97	18.34	29.54	31.10	22.23	17.63	21.17	19.98	21.11	24.26	23.56	24.80	31.10
-Year		1961	1957	1991	1971	1956	1975	1987	1986	1964	1955	1978	1968	APR 1971
-Minimum Monthly	39	0.78	0.20	0.15	0.36	1.49	5.40	5.34	5.33	5.90	6.17	4.53	2.28	0.15
-Year		1973	1992	1992	1992	1983	1984	1961	1959	1992	1990	1972	1957	MAR 1992
-Maximum in 24 hrs	39	9.57	6.65	15.26	6.63	6.26	7.39	5.86	5.29	5.76	8.74	10.01	17.88	17.88
-Year		1961	1991	1991	1973	1992	1983	1987	1986	1982	1974	1957	1972	DEC 1972
Snow, Ice pellets, hail														
-Maximum Monthly		0.0	0.0	0.0	0.0	0.0	0.0	0.0	0.0	0.0	0.0	0.0	0.0	
-Year														
-Maximum in 24 hrs	39	0.0	0.0	0.0	0.0	0.0	0.0	0.0	0.0	0.0	0.0	0.0	0.0	
-Year														
WIND:														
Mean Speed (mph)	29	13.0	13.4	13.1	12.1	11.2	10.0	8.5	7.3	7.0	7.4	8.7	12.2	10.3
Prevailing Direction through 1963		ENE	ENE	ENE	ENE	ENE	ENE	ENE	ENE	E	E	E	ENE	ENE
Fastest Mile														
-Direction (!!)	31	W	E	NE	E	E	NE	E	NW	E	E	SW	E	W
-Speed (MPH)	31	47	35	36	35	38	38	34	33	36	38	45	38	47
-Year		1988	1962	1959	1963	1962	1964	1973	1986	1973	1985	1982	1973	JAN 1988
Peak Gust														
-Direction (!!)	10	NE	E	E	E	E	E	E	E	E	E	SW	E	NE
-Speed (mph)	10	53	39	40	39	37	41	40	40	39	47	44	45	53
-Date		1992	1984	1986	1991	1992	1993	1989	1993	1984	1985	1991	1993	JAN 1992

PRECIPITATION (inches)

MAJURO, MARSHALL ISLANDS, PACIFIC

YEAR	JAN	FEB	MAR	APR	MAY	JUNE	JULY	AUG	SEP	OCT	NOV	DEC	ANNUAL
1964	1.40	6.99	7.23	11.46	22.02	11.16	18.69	15.58	21.11	22.79	16.85	7.42	162.70
1965	9.85	5.32	1.98	4.69	7.93	11.45	14.85	6.92	15.46	14.71	12.12	9.55	114.83
1966	3.79	4.42	5.80	16.03	8.64	9.40	14.94	6.52	13.95	13.53	12.24	19.44	128.70
1967	11.88	9.72	12.46	7.64	4.93	10.98	13.87	7.99	13.78	15.16	11.16	6.48	126.05
1968	5.38	3.49	11.12	8.86	9.33	16.07	11.39	11.50	9.77	12.06	11.97	24.80	135.74
1969	8.22	2.35	16.17	17.21	8.78	13.01	16.65	10.24	15.65	7.11	11.68	7.21	134.28
1970	5.62	0.40	1.73	2.87	9.23	10.66	7.73	11.24	11.75	12.64	6.68	8.40	88.95
1971	8.21	5.74	9.80	31.10	19.86	13.42	15.49	14.92	7.93	18.06	9.46	8.40	162.39
1972	9.58	7.11	15.45	9.17	14.96	14.88	14.76	10.84	18.96	14.06	4.53	23.36	157.66
1973	0.78	1.84	11.05	14.59	14.33	12.23	7.29	13.86	12.78	13.79	14.21	7.24	123.99
1974	11.09	8.07	7.18	15.67	12.84	13.66	12.48	13.69	10.44	19.90	9.29	14.49	148.80
1975	5.20	3.21	7.77	12.76	10.58	17.63	14.23	16.35	16.51	18.29	15.28	13.95	151.76
1976	8.57	9.42	15.68	19.41	15.28	9.43	16.78	8.36	17.66	8.95	12.70	2.77	145.01
1977	2.39	0.77	2.60	10.62	17.21	8.37	10.88	11.15	9.72	17.59	11.85	18.88	122.03
1978	3.60	5.25	3.39	12.65	13.90	10.70	16.25	8.86	9.73	20.56	23.56	14.35	142.80
1979	6.78	2.77	7.14	11.75	7.91	13.23	6.67	13.03	6.54	15.04	11.33	7.10	109.29
1980	8.11	9.70	5.05	7.03	11.34	6.73	8.48	13.89	12.85	9.25	5.35	10.56	108.34
1981	0.90	4.34	17.40	10.20	9.04	5.43	16.53	12.24	6.71	7.28	14.61	14.47	119.15
1982	12.63	9.72	13.29	4.68	11.46	16.98	14.66	11.72	18.94	8.17	19.08	3.17	144.50
1983	0.83	0.98	0.66	1.97	1.49	14.45	12.58	6.05	11.25	13.47	9.84	12.74	86.31
1984	16.12	16.83	1.29	3.87	4.18	5.40	9.35	9.20	6.42	14.77	13.31	14.95	115.69
1985	8.70	16.56	4.59	15.38	9.67	14.67	13.18	16.77	8.03	18.06	12.81	11.30	149.72
1986	10.51	3.91	14.75	12.23	14.94	15.89	12.09	19.98	10.52	7.32	9.37	17.10	148.61
1987	6.24	10.38	4.90	2.14	9.22	14.76	21.17	8.36	11.09	11.29	15.45	7.48	122.48
1988	14.65	1.52	6.76	5.92	6.85	9.11	14.33	10.59	13.86	17.87	7.19	13.65	122.30
1989	7.75	8.30	4.76	8.54	11.18	7.20	17.44	10.34	14.55	16.41	19.84	8.52	134.83
1990	7.01	4.21	10.36	9.43	16.56	7.28	9.09	14.39	7.57	6.17	15.87	10.36	118.30
1991	9.87	11.68	29.54	20.46	13.24	16.60	16.41	11.04	19.73	10.44	15.35	3.48	177.84
1992	7.73	0.20	0.15	0.36	14.21	8.41	10.36	12.62	5.90	13.57	10.05	3.52	87.08
1993	5.82	8.51	13.49	14.10	11.31	8.17	13.92	13.32	9.80	20.48	14.41	19.24	152.57
Record Mean	8.11	6.74	9.10	10.74	11.49	11.78	12.99	11.68	12.43	14.33	13.43	11.06	133.87

AVERAGE TEMPERATURE (deg. F)

MAJURO, MARSHALL ISLANDS, PACIFIC

YEAR	JAN	FEB	MAR	APR	MAY	JUNE	JULY	AUG	SEP	OCT	NOV	DEC	ANNUAL
1964	81.8	81.6	81.3	81.4	81.1	80.7	80.8	80.8	80.5	80.7	80.7	80.8	81.0
1965	80.2	80.5	81.5	81.3	81.0	81.2	80.7	82.1	81.3	81.4	81.0	80.8	81.1
1966	81.0	81.3	81.3	80.7	81.7	81.7	81.9	82.5	81.8	81.9	81.2	80.9	81.5
1967	81.0	80.8	80.2	81.2	82.0	81.3	81.4	82.2	82.0	81.4	81.2	81.5	81.4
1968	81.1	81.4	80.3	80.6	80.8	81.1	80.9	81.4	81.8	81.1	81.1	80.6	81.0
1969	80.1	81.1	81.0	80.7	81.5	81.2	80.5	81.6	81.5	82.4	82.0	81.3	81.2
1970	81.2	82.0	82.0	82.1	81.5	80.7	81.1	80.8	81.1	80.6	81.2	80.6	81.3
1971	80.6	80.9	80.9	79.5	80.0	80.2	80.5	80.1	80.8	80.5	81.1	80.5	80.5
1972	80.2	80.9	80.8	80.8	81.2	81.5	80.9	81.2	81.3	80.9	81.7	80.8	81.0
1973	80.9	81.8	81.6	81.3	80.6	80.8	80.9	80.8	80.2	80.5	80.7	81.1	80.9
1974	79.9	80.8	80.8	80.5	80.8	80.6	80.7	81.0	80.9	81.0	80.9	80.3	80.7
1975	80.4	81.0	80.7	80.2	80.5	79.7	79.7	79.9	80.0	78.8	79.4	79.5	80.0
1976	79.4	79.4	79.6	79.5	80.0	80.0	80.0	80.7	80.4	81.4	80.2	80.1	80.1
1977	80.3	81.3	81.5	80.5	80.1	81.2	80.9	81.2	82.4	81.1	81.2	81.1	81.1
1978	81.2	81.1	81.5	80.9	80.5	80.9	80.4	81.5	81.6	81.2	80.6	80.3	81.0
1979	81.2	81.0	81.6	79.7	80.9	81.5	81.5	80.9	82.0	82.1	81.8	81.7	81.3
1980	81.4	81.3	81.3	81.6	81.5	81.9	81.5	81.5	81.7	82.2	81.9	81.0	81.6
1981	81.4	81.5	81.0	80.9	81.4	82.0	80.8	81.4	82.0	82.0	81.0	80.6	81.3
1982	80.5	80.7	80.5	81.8	81.3	81.3	81.1	81.3	81.4	82.0	81.4	80.2	81.1
1983	80.1	80.5	81.4	82.2	83.0	81.4	81.3	82.2	81.8	80.9	81.0	80.3	81.3
1984	80.6	80.6	82.0	81.9	81.5	80.3	80.6	81.1	81.2	80.8	80.6	80.9	81.0
1985	80.8	80.3	80.7	79.8	81.1	80.4	80.7	80.4	81.5	81.4	81.5	80.9	80.8
1986	81.1	81.9	80.3	81.1	81.7	81.1	81.7	81.8	81.8	82.1	81.9	80.5	81.4
1987	80.6	80.7	81.0	81.8	81.5	81.1	80.9	82.1	81.8	82.2	81.8	81.2	81.4
1988	80.9	81.9	81.8	82.1	81.8	81.3	80.2	80.9	80.6	80.3	81.2	80.4	81.1
1989	80.5	80.3	80.6	80.5	80.8	81.0	80.5	80.9	81.1	81.6	81.4	82.0	80.9
1990	81.2	81.6	82.0	81.9	81.1	81.6	81.4	81.7	81.8	81.7	81.8	80.4	81.5
1991	80.1	81.4	79.6	80.3	81.1	80.9	81.3	81.4	81.1	81.8	81.0	81.0	80.9
1992	79.6	81.3	81.6	82.7	81.9	82.0	81.6	82.0	82.6	82.2	81.5	81.6	81.7
1993	80.2	80.4	80.3	80.7	81.1	81.3	81.1	82.0	82.3	81.4	81.5	81.0	81.1
Record Mean	80.7	81.1	81.1	81.0	81.2	81.1	80.9	81.3	81.4	81.4	81.2	80.9	81.1
Max	84.9	85.4	85.6	85.6	85.8	85.8	85.7	86.2	86.4	86.4	86.0	85.3	85.8
Min	76.4	76.7	76.5	76.4	76.6	76.3	76.1	76.4	76.4	76.3	76.4	76.4	76.4

REFERENCE NOTES

MAJURO, MARSHALL ISLANDS, PACIFIC

GENERAL
T - TRACE AMOUNT.
BLANK ENTRIES DENOTE MISSING/UNREPORTED DATA.
# INDICATES A STATION OR INSTRUMENT RELOCATION.

SPECIFIC

(a) - LENGTH OF RECORD IN YEARS, ALTHOUGH INDIVIDUAL MONTHS MAY BE MISSING.
0.* OR * - THE VALUE IS BETWEEN 0.0 AND 0.05
NORMALS - BASED ON THE 1961-1990 RECORD PERIOD.
EXTREMES - DATES ARE THE MOST RECENT OCCURRENCE.
WIND DIR.- NUMERALS SHOW TENS OF DEGRESS CLOCKWISE FROM TRUE NORTH. "00" INDICATES CALM.
RESULTANT DIRECTIONS ARE GIVEN TO WHOLE DEGREES.
BOLD VALUES INDICATE EXTREME VALUES WHICH OCCURRED AFTER THE ASOS SYSTEM WAS COMMISSIONED.

RECORD = PERIOD OF RECORD
RECORD MEAN PRECIPITATION IS THE MEAN OF ALL DAILY PRECIPITATION AMOUNTS DURING THE PERIOD OF RECORD.
RECORD MAX(MIN) TEMPERATURE IS THE MEAN OF ALL DAILY MAX(MIN) TEMPERATURES DURING THE PERIOD OF RECORD.
RECORD MEAN TEMPERATURE IS THE SUM OF THE RECORD MAX AND RECORD MIN DIVIDED BY 2.
AVERAGE TEMPERATURE IS THE SUM OF THE MEAN DAILY MAX AND MIN TEMPERATURE DIVIDED BY 2.

EXCEPTIONS

CHANGED WIND MEASUREMENT TO FASTEST 1-MINUTE WIND AND BEGAN PRESSURE MEASUREMENT ON JULY 1, 1988.

THUNDERSTORMS, AND HEAVY FOG ARE THROUGH 1972

RECORD MEANS ARE THROUGH THE CURRENT YEAR, BEGINNING IN 1955 FOR TEMPERATURE
1955 FOR PRECIPITATION

HEATING DEGREE DAYS Base 65 deg. F MAJURO, MARSHALL ISLANDS, PACIFIC

SEASON	JULY	AUG	SEP	OCT	NOV	DEC	JAN	FEB	MAR	APR	MAY	JUNE	TOTAL
1983-84	0	0	0	0	0	0	0	0	0	0	0	0	0
1984-85	0	0	0	0	0	0	0	0	0	0	0	0	0
1985-86	0	0	0	0	0	0	0	0	0	0	0	0	0
1986-87	0	0	0	0	0	0	0	0	0	0	0	0	0
1987-88	0	0	0	0	0	0	0	0	0	0	0	0	0
1988-89	0	0	0	0	0	0	0	0	0	0	0	0	0
1989-90	0	0	0	0	0	0	0	0	0	0	0	0	0
1990-91	0	0	0	0	0	0	0	0	0	0	0	0	0
1991-92	0	0	0	0	0	0	0	0	0	0	0	0	0
1992-93	0	0	0	0	0	0	0	0	0	0	0	0	0
1993-94	0	0	0	0	0	0							

COOLING DEGREE DAYS Base 65 deg. F MAJURO, MARSHALL ISLANDS, PACIFIC

YEAR	JAN	FEB	MAR	APR	MAY	JUNE	JULY	AUG	SEP	OCT	NOV	DEC	TOTAL
1969	475	458	502	477	519	496	486	520	503	544	515	513	6008
1970	512	483	536	521	518	474	507	497	491	489	491	488	6007
1971	494	454	500	443	473	462	486	476	480	487	492	489	5736
1972	480	466	498	483	509	504	502	508	499	500	506	498	5953
1973	498	476	523	494	490	480	500	493	464	484	479	505	5886
1974	469	449	497	468	498	475	492	503	486	505	487	481	5810
1975	483	456	494	464	490	447	460	468	458	436	439	458	5553
1976	451	424	459	442	470	460	473	491	469	514	465	477	5595
1977	478	464	518	475	476	495	501	510	528	506	492	507	5950
1978	508	456	517	482	486	482	483	518	506	509	475	480	5902
1979	507	454	519	447	501	504	520	500	512	537	510	527	6038
1980	519	478	513	507	519	516	516	518	509	542	515	502	6154
1981	515	470	501	483	516	517	496	512	517	534	489	488	6038
1982	485	447	487	511	511	497	507	511	498	534	498	478	5964
1983	476	439	519	522	566	500	511	537	511	499	487	482	6049
1984	492	459	533	512	518	464	490	507	494	495	475	499	5938
1985	499	434	492	453	507	469	495	486	503	515	500	501	5854
1986	504	480	480	491	525	491	525	530	510	538	513	489	6076
1987	491	447	501	510	520	489	499	537	510	540	510	508	6062
1988	500	498	529	519	528	494	480	499	474	482	493	487	5983
1989	488	436	492	473	499	488	488	499	493	521	500	534	5911
1990	509	473	532	513	502	506	517	526	511	524	510	487	6110
1991	475	467	460	466	508	487	512	518	491	527	487	502	5900
1992	458	480	522	538	529	516	522	535	533	539	502	521	6195
1993	481	437	480	477	507	497	506	531	525	514	503	503	5961

SNOWFALL (inches) MAJURO, MARSHALL ISLANDS, PACIFIC

SEASON	JULY	AUG	SEP	OCT	NOV	DEC	JAN	FEB	MAR	APR	MAY	JUNE	TOTAL
1971-72	0.0	0.0	0.0	0.0	0.0	0.0	0.0	0.0	0.0	0.0	0.0	0.0	0.0
1972-73	0.0	0.0	0.0	0.0	0.0	0.0	0.0	0.0	0.0	0.0	0.0	0.0	0.0
1973-74	0.0	0.0	0.0	0.0	0.0	0.0	0.0	0.0	0.0	0.0	0.0	0.0	0.0
1974-75	0.0	0.0	0.0	0.0	0.0	0.0	0.0	0.0	0.0	0.0	0.0	0.0	0.0
1975-76	0.0	0.0	0.0	0.0	0.0	0.0	0.0	0.0	0.0	0.0	0.0	0.0	0.0
1976-77	0.0	0.0	0.0	0.0	0.0	0.0	0.0	0.0	0.0	0.0	0.0	0.0	0.0
1977-78	0.0	0.0	0.0	0.0	0.0	0.0	0.0	0.0	0.0	0.0	0.0	0.0	0.0
1978-79	0.0	0.0	0.0	0.0	0.0	0.0	0.0	0.0	0.0	0.0	0.0	0.0	0.0
1979-80	0.0	0.0	0.0	0.0	0.0	0.0	0.0	0.0	0.0	0.0	0.0	0.0	0.0
1980-81	0.0	0.0	0.0	0.0	0.0	0.0	0.0	0.0	0.0	0.0	0.0	0.0	0.0
1981-82	0.0	0.0	0.0	0.0	0.0	0.0	0.0	0.0	0.0	0.0	0.0	0.0	0.0
1982-83	0.0	0.0	0.0	0.0	0.0	0.0	0.0	0.0	0.0	0.0	0.0	0.0	0.0
1983-84	0.0	0.0	0.0	0.0	0.0	0.0	0.0	0.0	0.0	0.0	0.0	0.0	0.0
1984-85	0.0	0.0	0.0	0.0	0.0	0.0	0.0	0.0	0.0	0.0	0.0	0.0	0.0
1985-86	0.0	0.0	0.0	0.0	0.0	0.0	0.0	0.0	0.0	0.0	0.0	0.0	0.0
1986-87	0.0	0.0	0.0	0.0	0.0	0.0	0.0	0.0	0.0	0.0	0.0	0.0	0.0
1987-88	0.0	0.0	0.0	0.0	0.0	0.0	0.0	0.0	0.0	0.0	0.0	0.0	0.0
1988-89	0.0	0.0	0.0	0.0	0.0	0.0	0.0	0.0	0.0	0.0	0.0	0.0	0.0
1989-90	0.0	0.0	0.0	0.0	0.0	0.0	0.0	0.0	0.0	0.0	0.0	0.0	0.0
1990-91	0.0	0.0	0.0	0.0	0.0	0.0	0.0	0.0	0.0	0.0	0.0	0.0	0.0
1991-92	0.0	0.0	0.0	0.0	0.0	0.0	0.0	0.0	0.0	0.0	0.0	0.0	0.0
1992-93	0.0	0.0	0.0	0.0	0.0	0.0	0.0	0.0	0.0	0.0	0.0	0.0	0.0
1993-94	0.0	0.0	0.0	0.0	0.0	0.0							
Record Mean	0.0	0.0	0.0	0.0	0.0	0.0	0.0	0.0	0.0	0.0	0.0	0.0	0.0

PAGO PAGO, AMERICAN SAMOA

Pago Pago Airport is located on the southeastern coast of the island of Tutuila in the American Samoa group, approximately 2,600 miles south-southwest of Hawaii, 1,600 miles north-northeast of New Zealand, and 4,500 miles southwest of California. Tutuila is a long, narrow island lying southwest-northeast, with a land area of 76 square miles, a greatest length of just over 20 miles, and a width ranging from 1 to 2 miles in the eastern half and from 2 to 5 miles in the western. It is volcanic in origin, extremely mountainous, and nearly surrounded by a coral reef. The principal ridge extends the length of the island, reaching a maximum height of 2,141 feet, at Matafao peak, near the central portion of the long axis. Vegetation is moderately dense, with many coconut, banana, and other tropical fruit trees, grass, and low-growing brush. The orientation of Tutuila is such that winds from the east-northeast clockwise to south approach Pago Pago Airport directly from the ocean without being deflected by the terrain, while winds from other directions may be considerably disturbed by topography.

Samoa has a maritime climate with abundant rain and warm, humid days and nights. Rainfall, usually falling as showers, is about 125 inches a year at the airport, but varies greatly over small distances because of topography. Thus, Pago Pago, less than 4 miles north of the airport and at the head of a hill-encircled harbor open to the prevailing wind, receives nearly 200 inches a year. The crest of the range receives well above 250 inches. In most years, the airport records about 300 days with a trace or more of rain and about 175 with .1 inch or more.

The driest months are June through September (southern winter) and the wettest, December through March (southern summer). However, the seasonal rainfall may vary widely in individual years, and heavy showers and long rainy periods can occur in any month. Flooding rains are not unknown. Some of these have been associated with hurricanes and tropical storms, but they have occurred at other times as well.

June, July and August are the coolest months and January, February, and March, the warmest. Afternoon temperatures ordinarily reach the upper 80s in summer and the mid 80s in winter, while nighttime temperatures fall to the mid 70s in the summer and low 70s in winter. The highest temperatures recorded at the airport are in the low 90s and lowest near 60.

The prevailing winds throughout the year are the easterly trades. These tend to be more directly from the east in December through March, but predominantly from east-southeast and southeast during the rest of the year. The trade winds are also less prevalent in summer than in winter. As the foregoing suggests, the trades are interrupted more often in summer than in winter. These interruptions are sometimes associated with the proximity of small tropical storms, of bands of converging winds, or of low pressure systems higher in the atmosphere, all of which help make summer the rainy season. At other times, the absence of the trades is marked by periods of light and variable winds and by land and sea breezes. Westerly to northerly winds, in particular, are more frequent then. These are strong at times, but are often quite light, and may then reflect the nighttime drainage of cooled air from the mountains west and north of the airport.

Thunderstorms are less frequent than might be expected, considering the moistness and instability of the tropical air mass which usually overlies Samoa.

NORMALS, MEANS, AND EXTREMES

PAGO PAGO, AMERICAN SAMOA

LATITUDE: 14 °20'S LONGITUDE: 170 °43' W ELEVATION: FT. GRND 12 BARO 15 TIME ZONE: 165W MER WBAN: 61705

	(a)	JAN	FEB	MAR	APR	MAY	JUNE	JULY	AUG	SEP	OCT	NOV	DEC	YEAR
TEMPERATURE °F:														
Normals														
-Daily Maximum		86.4	86.6	86.8	86.3	85.1	84.2	83.1	83.3	84.4	84.8	85.3	86.0	85.2
-Daily Minimum		75.4	75.4	75.6	75.2	75.2	75.3	74.5	74.3	74.7	75.3	75.3	75.4	75.1
-Monthly		80.9	81.1	81.2	80.8	80.2	79.8	78.8	78.9	79.6	80.0	80.3	80.7	80.2
Extremes														
-Record Highest	34	92	92	92	92	91	90	89	90	89	92	92	92	92
-Year		1969	1983	1966	1980	1991	1991	1961	1991	1992	1990	1991	1986	NOV 1991
-Record Lowest	34	67	67	67	68	66	64	62	64	63	67	67	67	62
-Year		1965	1965	1965	1979	1974	1965	1964	1979	1970	1973	1964	1964	JUL 1964
NORMAL DEGREE DAYS:														
Heating (base 65°F)		0	0	0	0	0	0	0	0	0	0	0	0	0
Cooling (base 65°F)		493	451	502	474	471	444	428	431	438	465	459	487	5543
% OF POSSIBLE SUNSHINE	25	42	42	46	37	35	34	40	45	53	41	42	40	41
MEAN SKY COVER (tenths)														
Sunrise - Sunset	27	8.2	8.2	7.8	7.8	7.6	7.1	7.2	6.9	6.9	7.6	7.8	8.0	7.6
MEAN NUMBER OF DAYS:														
Sunrise to Sunset														
-Clear	27	0.6	0.6	0.9	1.0	0.9	1.7	1.7	2.1	2.4	1.1	0.8	1.1	14.8
-Partly Cloudy	27	9.2	8.4	11.4	11.1	12.6	14.0	14.3	15.4	14.5	12.8	11.0	9.4	144.2
-Cloudy	27	21.2	19.3	18.7	17.9	17.6	14.2	14.9	13.5	13.1	17.1	18.3	20.4	206.0
Precipitation														
.01 inches or more	27	24.6	21.6	23.2	22.2	20.8	18.8	18.7	17.6	17.0	21.6	20.7	23.1	250.0
Snow,Ice pellets,hail														
1.0 inches or more	27	0.0	0.0	0.0	0.0	0.0	0.0	0.0	0.0	0.0	0.0	0.0	0.0	0.0
Thunderstorms	27	2.8	2.9	3.0	3.4	2.6	1.1	0.5	0.6	0.8	2.4	3.1	3.2	26.3
Heavy Fog Visibility														
1/4 mile or less	27	0.0	0.0	0.0	0.0	0.0	0.0	0.0	0.0	0.0	0.0	0.0	0.0	0.0
Temperature °F														
-Maximum														
90° and above	34	1.7	2.1	2.4	1.5	0.2	0.1	0.0	0.1	0.0	0.1	0.5	1.7	10.3
32° and below	34	0.0	0.0	0.0	0.0	0.0	0.0	0.0	0.0	0.0	0.0	0.0	0.0	0.0
-Minimum														
32° and below	34	0.0	0.0	0.0	0.0	0.0	0.0	0.0	0.0	0.0	0.0	0.0	0.0	0.0
0° and below	34	0.0	0.0	0.0	0.0	0.0	0.0	0.0	0.0	0.0	0.0	0.0	0.0	0.0
AVG. STATION PRESS.(mb)	21	1007.9	1008.5	1009.4	1010.0	1011.2	1011.9	1012.4	1012.7	1012.6	1011.6	1009.9	1008.2	1010.5
RELATIVE HUMIDITY (%)														
Hour 01	25	88	89	88	89	87	85	83	83	85	86	86	88	86
Hour 07 (Local Time)	25	88	89	89	89	87	85	84	84	84	84	84	86	86
Hour 13	25	76	76	76	77	77	76	75	74	73	76	76	75	76
Hour 19	25	82	82	83	85	84	82	81	81	81	83	82	82	82
PRECIPITATION (inches):														
Water Equivalent														
-Normal		12.59	12.76	11.26	12.04	9.92	7.38	6.28	6.71	6.69	10.79	10.84	14.54	121.80
-Maximum Monthly	34	24.88	32.66	31.84	28.35	20.48	12.82	19.59	16.46	25.29	21.48	25.67	26.95	32.66
-Year		1986	1968	1961	1992	1966	1961	1962	1978	1972	1980	1978	1984	FEB 1968
-Minimum Monthly	34	4.65	5.91	4.01	4.13	1.61	1.97	0.72	0.29	0.63	2.10	1.36	2.84	0.29
-Year		1968	1966	1976	1982	1983	1992	1974	1989	1987	1975	1965	1982	AUG 1989
-Maximum in 24 hrs	34	8.74	9.01	6.48	10.79	6.68	5.94	8.65	6.18	10.28	7.50	6.51	8.97	10.79
-Year		1973	1968	1981	1975	1976	1961	1962	1978	1972	1967	1992	1970	APR 1975
Snow,Ice pellets,hail														
-Maximum Monthly		0.0	0.0	0.0	0.0	0.0	0.0	0.0	0.0	0.0	0.0	0.0	0.0	
-Year														
-Maximum in 24 hrs	34	0.0	0.0	0.0	0.0	0.0	0.0	0.0	0.0	0.0	0.0	0.0	0.0	
-Year														
WIND:														
Mean Speed (mph)	26	9.1	9.0	8.8	8.7	10.6	12.2	13.1	12.8	12.3	12.2	10.7	9.3	10.7
Prevailing Direction														
Fastest Obs. 1 Min.														
-Direction (!!)	14	34	36	02	07	08	07	09	09	06	18	09	21	21
-Speed (MPH)	14	46	63	33	35	35	36	31	33	38	35	33	81	81
-Year		1985	1990	1982	1985	1988	1989	1988	1987	1988	1986	1988	1991	DEC 1991
Peak Gust														
-Direction (!!)	10	N	NE	N	NW	E	E	E	SE	E	S	E	SW	SW
-Speed (mph)	10	71	107	53	51	44	43	49	43	49	43	47	109	109
-Date		1993	1990	1993	1993	1988	1992	1992	1985	1988	1986	1991	1991	DEC 1991

PRECIPITATION (inches)

PAGO PAGO, AMERICAN SAMOA

YEAR	JAN	FEB	MAR	APR	MAY	JUNE	JULY	AUG	SEP	OCT	NOV	DEC	ANNUAL
1964	14.92		10.95	11.72	10.53	9.47	10.34	7.69	14.92	7.54	16.68	16.42	
1965	9.84	12.73	6.01		17.69	12.12	2.06	3.75	1.95	19.67	1.36	12.95	
1966	16.38	5.91	10.83	10.42	20.48	9.17	6.40	7.62	8.46	18.63	5.76	16.21	136.27
1967	6.97	8.60	4.95	24.46	9.97	11.74	9.00	13.02	4.34	15.92	15.30	16.36	140.63
1968	4.65	32.66	11.69	12.60	7.00	2.71	3.56	4.57	8.01	18.88	4.27	6.24	116.84
1969	17.95	11.71	11.93	15.95	4.91	3.96	11.42	7.35	2.26	10.05	9.35	16.68	123.52
1970	9.43	8.87	21.62	13.58	8.78	12.16	4.68	8.10	7.69	5.69	7.98	26.38	134.96
1971	11.48	14.33	7.69	11.70	4.77	12.03	3.16	5.24	2.61	7.25	9.24	6.67	96.17
1972	17.77	11.17	9.45	6.50	6.55	3.67	7.35	5.61	25.29	10.44	6.48	20.30	130.58
1973	22.14	8.44	4.94	9.74	7.93	6.09	8.12	8.58	5.15	16.19	18.93	14.90	131.15
1974	9.26	9.46	9.67	8.80	8.07	5.40	0.72	1.19	0.99	2.44	12.73	9.16	77.89
1975	22.13	9.88	6.57	22.77	17.88	5.21	8.17	2.95	5.57	2.10	7.69	14.15	125.07
1976	5.69	13.36	4.01	10.31	18.98	7.11	8.73	4.09	5.17	3.50	12.19	21.38	114.52
1977	7.29	6.30	11.64	6.98	20.21	5.19	2.92	2.40	3.12	10.06	14.81	6.86	97.78
1978	21.36	9.52	25.64	9.24	8.71	2.62	3.48	16.46	4.71	10.04	25.67	15.62	153.07
1979	12.70	14.43	9.03	9.53	4.23	7.86	8.83	7.19	9.12	11.73	7.89	8.41	110.95
1980	9.73	7.47	14.84	6.52	19.50	11.90	4.14	13.97	15.47	21.48	7.43	13.44	145.89
1981	14.25	14.23	25.37	22.00	9.25	9.47	10.19	10.64	4.04	16.38	10.08	19.58	165.48
1982	9.09	30.25	7.68	4.13	7.90	4.22	6.96	16.20	6.30	4.38	6.35	2.84	106.30
1983	9.45	12.09	6.07	10.97	1.61	2.71	1.12	2.11	7.78	7.17	13.23	13.05	87.36
1984	9.70	8.03	19.34	6.70	4.79	7.44	1.86	4.25	5.23	15.21	8.08	26.95	117.58
1985	16.71	8.27	5.00	18.66	10.41	12.60	3.84	5.76	9.56	10.39	9.24	6.27	116.71
1986	24.88	9.47	6.88	18.83	12.33	5.69	8.79	5.05	17.67	8.87	8.82	22.60	149.88
1987	17.05	15.01	10.09	8.05	5.51	4.02	3.86	6.29	0.63	3.87	3.83	16.42	94.63
1988	8.34	9.32	12.76	10.57	9.76	6.15	9.42	3.98	6.78	9.21	16.10	24.39	126.78
1989	18.37	14.69	8.20	13.60	7.51	5.43	9.56	0.29	1.35	9.78	17.88	14.12	120.78
1990	8.87	22.54	9.90	13.01	3.52	6.31	4.82	2.45	3.68	9.68	11.25	9.30	105.33
1991	13.67	21.01	9.02	9.45	11.25	7.86	9.16	4.36	3.40	9.63	10.20	22.55	131.56
1992	15.40	12.44	13.76	28.35	10.67	1.97	5.22	6.33	6.95	6.59	22.79	8.94	139.41
1993	23.47	7.27	17.26	12.51	8.82	3.89	4.12	8.61	9.28	9.32	8.34	12.62	125.51
Record Mean	13.51	12.80	11.72	12.59	9.93	7.19	6.50	6.79	6.56	10.57	11.14	14.52	123.82

AVERAGE TEMPERATURE (deg. F)

PAGO PAGO, AMERICAN SAMOA

YEAR	JAN	FEB	MAR	APR	MAY	JUNE	JULY	AUG	SEP	OCT	NOV	DEC	ANNUAL
1964	81.9	82.0	82.4	82.0	79.7	79.5	78.0	79.1	78.7	80.1	80.4	79.1	80.3
1965	79.9	79.1	79.7	79.6	79.3	78.0	78.2	78.2	78.7	79.7	80.6	80.8	79.3
1966	81.3	82.1	82.3	81.6	80.1	79.8	79.6	78.4	79.5	80.0	81.2	80.5	80.5
1967	80.4	80.0	81.7	80.0	79.1	77.9	77.4	77.6	79.6	78.8	79.0	80.0	79.3
1968	81.0	80.1	79.7	78.4	79.3	79.8	78.5	77.4	79.1	79.5	80.8	81.7	79.6
1969	81.7	81.1	81.2	80.6	79.8	80.5	78.4	78.3	78.7	80.4	80.1	80.5	80.1
1970	81.0	81.7	80.1	81.0	81.2	80.0	79.6	78.3	78.4	79.8	79.5	78.7	79.9
1971	80.0	79.4	79.5	79.3	79.2	78.5	78.1	78.6	78.4	79.3	80.0	80.1	79.2
1972	79.2	79.7	80.9	81.2	79.9	80.1	78.4	79.5	79.3	80.2	80.8	82.6	80.2
1973	81.6	82.6	82.9	82.2	81.3	80.4	79.9	79.3	80.3	78.8	79.6	80.0	80.7
1974	80.6	78.9	79.9	79.6	78.4	79.2	79.2	78.1	80.5	80.8	80.2	80.0	79.6
1975	80.5	80.6	81.1	79.9	80.0	79.2	79.3	79.3	79.6	81.1	80.3	79.4	80.0
1976	79.6	79.8	80.6	80.5	79.1	78.8	78.5	78.8	78.9	80.8	80.9	80.1	79.7
1977	81.9	82.3	81.4	80.9	79.4	80.5	79.1	79.1	78.5	80.0	80.8	82.1	80.5
1978	80.9	81.6	80.3	80.3	80.5	79.8	78.9	79.5	79.5	80.1	79.3	80.9	80.2
1979	81.0	81.4	82.3	80.1	81.2	81.1	78.4	78.8	80.5	81.0	80.4	80.5	80.5
1980	81.1	81.8	82.2	82.4	79.6	80.6	79.2	79.4	80.2	79.7	81.8	81.6	80.8
1981	81.7	80.8	80.6	80.2	80.6	79.1	79.0	79.2	80.6	79.8	80.8	80.9	80.3
1982	82.0	80.4	82.3	81.8	80.8	80.4	78.7	79.0	79.2	80.8	80.2	81.1	80.5
1983	82.1	83.5	81.9	81.1	81.1	79.7	78.7	78.3	80.2	79.6	80.4	81.4	80.7
1984	80.7	81.5	81.3	81.5	81.8	80.6	79.0	79.8	79.6	79.6	81.0	80.3	80.6
1985	80.3	81.6	82.2	80.8	80.4	80.0	79.5	79.9	79.7	80.9	80.5	81.7	80.6
1986	81.2	81.6	82.1	81.2	80.6	80.6	79.3	78.3	79.9	81.1	81.8	81.7	80.8
1987	81.7	82.5	82.3	82.1	80.4	78.5	78.3	78.2	79.7	80.4	81.8	82.2	80.7
1988	82.6	82.8	82.9	81.4	81.7	81.4	79.6	80.5	80.2	80.1	79.7	79.9	81.1
1989	79.7	80.5	81.3	80.7	80.5	80.1	78.7	79.7	81.7	81.1	80.5	80.4	80.4
1990	81.5	81.4	81.6	80.6	81.9	80.4	80.2	80.1	80.4	81.5	81.6	81.7	81.1
1991	81.5	82.0	82.6	82.6	81.0	80.6	81.1	80.8	80.5	81.1	82.1	81.5	81.5
1992	81.9	82.2	81.5	81.5	82.0	81.3	81.7	80.3	81.5	81.4	80.4	82.0	81.5
1993	81.7	81.9	81.5	81.9	81.2	79.9	78.3	80.0	79.7	80.3	82.5	83.5	81.0
Record Mean	81.1	81.2	81.3	80.9	80.3	79.8	79.0	79.0	79.7	80.2	80.6	80.9	80.4
Max	86.7	86.8	87.0	86.5	85.2	84.3	83.3	83.5	84.5	84.9	85.6	86.2	85.4
Min	75.4	75.5	75.6	75.3	75.4	75.3	74.7	74.5	74.8	75.4	75.6	75.6	75.3

REFERENCE NOTES

PAGO PAGO, AMERICAN SAMOA

GENERAL
T - TRACE AMOUNT.
BLANK ENTRIES DENOTE MISSING/UNREPORTED DATA.
# INDICATES A STATION OR INSTRUMENT RELOCATION.

SPECIFIC

(a) - LENGTH OF RECORD IN YEARS, ALTHOUGH INDIVIDUAL MONTHS MAY BE MISSING.
0.* OR * - THE VALUE IS BETWEEN 0.0 AND 0.05
NORMALS - BASED ON THE 1961-1990 RECORD PERIOD.
EXTREMES - DATES ARE THE MOST RECENT OCCURRENCE
WIND DIR. - NUMERALS SHOW TENS OF DEGRESS CLOCKWISE FROM TRUE NORTH. "00" INDICATES CALM.
RESULTANT DIRECTIONS ARE GIVEN TO WHOLE DEGREES.
BOLD VALUES INDICATE EXTREME VALUES WHICH OCCURRED AFTER THE ASOS SYSTEM WAS COMMISSIONED.

RECORD = PERIOD OF RECORD
RECORD MEAN PRECIPITATION IS THE MEAN OF ALL DAILY PRECIPITATION AMOUNTS DURING THE PERIOD OF RECORD.
RECORD MAX(MIN) TEMPERATURE IS THE MEAN OF ALL DAILY MAX(MIN) TEMPERATURES DURING THE PERIOD OF RECORD.
RECORD MEAN TEMPERATURE IS THE SUM OF THE RECORD MAX AND RECORD MIN DIVIDED BY 2.
AVERAGE TEMPERATURE IS THE SUM OF THE MEAN DAILY MAX AND MIN TEMPERATURE DIVIDED BY 2.

EXCEPTIONS

TEMPERATURES ARE BASED ON DAILY EXTREMES OF HOURLY VALUES APRIL 1964 THROUGH MARCH 1966.

RECORD MEANS ARE THROUGH THE CURRENT YEAR, BEGINNING IN 1960 FOR TEMPERATURE
1960 FOR PRECIPITATION

HEATING DEGREE DAYS Base 65 deg. F — PAGO PAGO, AMERICAN SAMOA

SEASON	JULY	AUG	SEP	OCT	NOV	DEC	JAN	FEB	MAR	APR	MAY	JUNE	TOTAL
1983-84	0	0	0	0	0	0	0	0	0	0	0	0	0
1984-85	0	0	0	0	0	0	0	0	0	0	0	0	0
1985-86	0	0	0	0	0	0	0	0	0	0	0	0	0
1986-87	0	0	0	0	0	0	0	0	0	0	0	0	0
1987-88	0	0	0	0	0	0	0	0	0	0	0	0	0
1988-89	0	0	0	0	0	0	0	0	0	0	0	0	0
1989-90	0	0	0	0	0	0	0	0	0	0	0	0	0
1990-91	0	0	0	0	0	0	0	0	0	0	0	0	0
1991-92	0	0	0	0	0	0	0	0	0	0	0	0	0
1992-93	0	0	0	0	0	0	0	0	0	0	0	0	0
1993-94	0	0	0	0	0	0							

COOLING DEGREE DAYS Base 65 deg. F — PAGO PAGO, AMERICAN SAMOA

YEAR	JAN	FEB	MAR	APR	MAY	JUNE	JULY	AUG	SEP	OCT	NOV	DEC	TOTAL
1969	526	459	511	476	464	472	421	416	419	487	458	487	5596
1970	500	474	474	486	508	456	458	423	408	467	441	430	5525
1971	474	410	454	438	448	410	412	429	410	448	458	475	5266
1972	446	433	503	490	468	461	421	456	436	480	480	551	5625
1973	525	500	561	522	513	464	469	450	465	436	446	473	5824
1974	488	396	469	447	421	435	448	414	472	497	462	471	5420
1975	485	445	506	452	473	431	452	450	444	507	464	454	5563
1976	462	437	492	470	444	422	427	435	426	496	483	477	5471
1977	531	491	516	487	454	474	445	445	411	472	480	537	5743
1978	501	472	483	467	486	448	439	455	445	474	437	500	5607
1979	502	465	540	458	509	487	424	436	469	502	470	487	5749
1980	507	495	543	530	457	475	446	453	463	465	513	523	5870
1981	526	450	493	465	490	432	439	447	473	467	481	502	5665
1982	533	438	543	512	497	469	433	442	434	496	463	506	5766
1983	539	524	529	493	504	449	429	419	463	456	467	516	5788
1984	495	487	514	504	528	472	442	462	445	459	487	482	5777
1985	481	470	543	480	487	458	457	468	449	500	474	526	5793
1986	510	474	538	490	489	476	452	421	454	506	511	527	5848
1987	521	497	541	520	485	412	420	419	449	485	514	541	5804
1988	550	522	563	499	525	496	457	484	459	473	450	467	5945
1989	468	438	513	478	488	461	431	463	506	506	471	484	5707
1990	519	466	522	471	531	470	478	473	467	519	505	524	5945
1991	514	482	553	533	503	473	505	498	470	506	518	520	6075
1992	529	503	519	499	533	497	525	482	504	514	468	535	6108
1993	524	477	521	513	510	454	421	468	448	483	534	584	5937

SNOWFALL (inches) — PAGO PAGO, AMERICAN SAMOA

SEASON	JULY	AUG	SEP	OCT	NOV	DEC	JAN	FEB	MAR	APR	MAY	JUNE	TOTAL
1971-72	0.0	0.0	0.0	0.0	0.0	0.0	0.0	0.0	0.0	0.0	0.0	0.0	0.0
1972-73	0.0	0.0	0.0	0.0	0.0	0.0	0.0	0.0	0.0	0.0	0.0	0.0	0.0
1973-74	0.0	0.0	0.0	0.0	0.0	0.0	0.0	0.0	0.0	0.0	0.0	0.0	0.0
1974-75	0.0	0.0	0.0	0.0	0.0	0.0	0.0	0.0	0.0	0.0	0.0	0.0	0.0
1975-76	0.0	0.0	0.0	0.0	0.0	0.0	0.0	0.0	0.0	0.0	0.0	0.0	0.0
1976-77	0.0	0.0	0.0	0.0	0.0	0.0	0.0	0.0	0.0	0.0	0.0	0.0	0.0
1977-78	0.0	0.0	0.0	0.0	0.0	0.0	0.0	0.0	0.0	0.0	0.0	0.0	0.0
1978-79	0.0	0.0	0.0	0.0	0.0	0.0	0.0	0.0	0.0	0.0	0.0	0.0	0.0
1979-80	0.0	0.0	0.0	0.0	0.0	0.0	0.0	0.0	0.0	0.0	0.0	0.0	0.0
1980-81	0.0	0.0	0.0	0.0	0.0	0.0	0.0	0.0	0.0	0.0	0.0	0.0	0.0
1981-82	0.0	0.0	0.0	0.0	0.0	0.0	0.0	0.0	0.0	0.0	0.0	0.0	0.0
1982-83	0.0	0.0	0.0	0.0	0.0	0.0	0.0	0.0	0.0	0.0	0.0	0.0	0.0
1983-84	0.0	0.0	0.0	0.0	0.0	0.0	0.0	0.0	0.0	0.0	0.0	0.0	0.0
1984-85	0.0	0.0	0.0	0.0	0.0	0.0	0.0	0.0	0.0	0.0	0.0	0.0	0.0
1985-86	0.0	0.0	0.0	0.0	0.0	0.0	0.0	0.0	0.0	0.0	0.0	0.0	0.0
1986-87	0.0	0.0	0.0	0.0	0.0	0.0	0.0	0.0	0.0	0.0	0.0	0.0	0.0
1987-88	0.0	0.0	0.0	0.0	0.0	0.0	0.0	0.0	0.0	0.0	0.0	0.0	0.0
1988-89	0.0	0.0	0.0	0.0	0.0	0.0	0.0	0.0	0.0	0.0	0.0	0.0	0.0
1989-90	0.0	0.0	0.0	0.0	0.0	0.0	0.0	0.0	0.0	0.0	0.0	0.0	0.0
1990-91	0.0	0.0	0.0	0.0	0.0	0.0	0.0	0.0	0.0	0.0	0.0	0.0	0.0
1991-92	0.0	0.0	0.0	0.0	0.0	0.0	0.0	0.0	0.0	0.0	0.0	0.0	0.0
1992-93	0.0	0.0	0.0	0.0	0.0	0.0	0.0	0.0	0.0	0.0	0.0	0.0	0.0
1993-94	0.0	0.0	0.0	0.0	0.0	0.0							
Record Mean	0.0	0.0	0.0	0.0	0.0	0.0	0.0	0.0	0.0	0.0	0.0	0.0	0.0

POHNPEI (PONAPE), EASTERN CAROLINE ISLANDS, PACIFIC

Pohnpei, about 129 square miles in area, is a nearly circular island of volcanic origin, encircled by coral barrier reefs, and covered with lush, tropical vegetation. The island, located less than 500 miles north of the equator, rises from the Pacific Ocean to an elevation of 2,595 feet, the highest point in the Caroline Islands. The topography is a complicated system of ridges and valleys, interlaced with small rivers and intermittent streams, and covered with tall grasses, tropical trees and flowers, and the coconut palms which are the backbone of the island economy. The interior of the island is covered by a rain forest which acts as a watershed area supplying fresh water the year-round.

The Weather Station lies a little north of the center of a bowl-shaped valley, about 3 miles south of the Pacific Ocean. Encircling it on a radius of 1 to 3 miles are volcanic outcroppings which rise rapidly from the ocean to an average elevation of about 2,100 feet in the east, south, and southwest, but to only about 700 feet to the west and northwest. With the exception of the cliff area in the northwest, the vegetation is lush and extremely dense. The valley is relatively level compared to the rest of the island. A small stream, oriented northeast-southwest lies about one-eighth of a mile to the east of the station.

From about November to June the climate of Pohnpei is chiefly influenced by the northeasterly trade winds. Wind speeds in the vicinity of the weather station, Pohnpei, are reduced somewhat due to the surrounding terrain. By about April the trades begin to diminish in strength, and by July have given way to the lighter and more variable winds of the doldrums. Between July and November the island is frequently under the influence of the Intertropical Convergence Zone (ITCZ – also called the Intertropical Front) which has moved northward into the area. This is also the season when moist southerly winds and tropical disturbances, many of them associated with the ITCZ, are most frequent and when humidities are often oppressively high.

Rainfall at Pohnpei is heavy and frequent throughout the year, averaging 192 inches annually. The wettest period is April and May. Measurable rain (.01 inch or more) falls on about 300 days a year.

The temperature is remarkably uniform throughout the year, with only slightly more than 1 degree separating the averages of the warmest and coolest months. High temperatures normally range in the mid to upper 80s and lows in the low to mid 70s. Temperature extremes above 90 degrees and below 70 degrees have occurred in every month of the year. Humidities are usually high throughout the year.

On most days, cumulus clouds predominate and usually cover more than eight-tenths of the sky. Days are normally cloudier than nights. High clouds, such as cirrus or cirrostratus often form and are obscurred by clouds at lower altitudes. Clouds at middle heights, usually altocumulus, sometimes combined with altostratus, occur quite frequently, especially if there are tropical disturbances in the vicinity.

Although Pohnpei is located within the spawning grounds of typhoons, the major typhoon tracks of the Western Pacific lie well to the north and west. Typhoons have caused extensive damage to crops and buildings on the island on several occasions.

The steep slopes surrounding the Weather Station reduce the wind speed. They can cause gentle up and downslope air currents which augment the land and sea breezes. The temperture and humidity are also greatly influenced by the reduced circulation and dense vegetation.

NORMALS, MEANS, AND EXTREMES

POHNPEI (PONAPE), EASTERN CAROLINE ISLANDS, PACIFIC

LATITUDE: 6°58'N LONGITUDE: 158°13'E ELEVATION: FT. GRND 123 BARO 126 TIME ZONE: 165E MER WBAN: 40504

	(a)	JAN	FEB	MAR	APR	MAY	JUNE	JULY	AUG	SEP	OCT	NOV	DEC	YEAR
TEMPERATURE °F:														
Normals														
-Daily Maximum		86.6	86.7	87.2	87.4	87.6	87.6	87.8	88.2	88.1	88.2	88.1	87.2	87.6
-Daily Minimum		75.5	75.6	75.6	75.0	74.8	74.1	73.1	73.0	72.9	72.9	73.4	74.9	74.2
-Monthly		81.0	81.2	81.4	81.2	81.2	80.9	80.4	80.6	80.5	80.6	80.8	81.1	80.9
Extremes														
-Record Highest	43	92	92	95	92	93	94	93	94	96	95	95	93	96
-Year		1980	1986	1950	1983	1993	1991	1991	1991	1950	1991	1989	1991	SEP 1950
-Record Lowest	43	66	67	66	68	69	69	68	68	68	68	65	66	65
-Year		1974	1976	1987	1972	1964	1990	1950	1974	1978	1955	1991	1977	NOV 1991
NORMAL DEGREE DAYS:														
Heating (base 65°F)		0	0	0	0	0	0	0	0	0	0	0	0	0
Cooling (base 65°F)		496	454	508	486	502	477	477	484	465	484	474	499	5806
% OF POSSIBLE SUNSHINE	35	39	41	45	42	41	41	43	44	45	40	40	36	41
MEAN SKY COVER (tenths)														
Sunrise - Sunset	41	9.0	8.9	9.0	9.0	8.8	8.9	8.6	8.6	8.6	8.6	8.7	8.9	8.8
MEAN NUMBER OF DAYS:														
Sunrise to Sunset														
-Clear	41	0.2	0.2	0.2	0.4	0.3	0.3	0.5	0.5	0.8	0.5	0.4	0.2	4.6
-Partly Cloudy	41	4.7	4.7	4.6	4.6	5.1	5.0	6.1	6.5	5.7	6.4	5.8	5.3	64.6
-Cloudy	41	26.1	23.4	26.2	25.0	25.5	24.7	24.4	23.9	23.5	24.1	23.8	25.4	296.0
Precipitation .01 inches or more	42	21.9	19.7	22.6	24.4	27.2	27.1	27.0	26.5	24.3	25.2	24.9	24.7	295.5
Snow,Ice pellets,hail 1.0 inches or more	42	0.0	0.0	0.0	0.0	0.0	0.0	0.0	0.0	0.0	0.0	0.0	0.0	0.0
Thunderstorms	42	0.8	0.8	1.2	1.2	2.0	2.2	2.7	2.3	2.8	3.1	2.9	1.7	23.8
Heavy Fog Visibility 1/4 mile or less	42	0.0	0.0	0.0	0.0	0.0	0.0	0.0	0.0	0.0	0.0	0.0	0.0	0.0
Temperature °F														
-Maximum														
90° and above	42	0.9	1.0	2.7	3.4	5.8	5.0	7.2	9.6	10.0	11.5	8.4	3.4	68.8
32° and below	42	0.0	0.0	0.0	0.0	0.0	0.0	0.0	0.0	0.0	0.0	0.0	0.0	0.0
-Minimum														
32° and below	42	0.0	0.0	0.0	0.0	0.0	0.0	0.0	0.0	0.0	0.0	0.0	0.0	0.0
0° and below	42	0.0	0.0	0.0	0.0	0.0	0.0	0.0	0.0	0.0	0.0	0.0	0.0	0.0
AVG. STATION PRESS.(mb)	5	1003.1	1004.1	1003.6	1003.6	1003.7	1003.9	1003.8	1004.1	1003.9	1003.2	1002.3	1002.8	1003.5
RELATIVE HUMIDITY (%)														
Hour 05	23	84	84	86	88	91	92	94	95	95	95	93	88	90
Hour 11 (Local Time)	42	76	78	78	79	81	81	81	81	80	80	81	79	80
Hour 17	42	78	76	77	79	81	80	79	79	79	79	80	79	79
Hour 23	42	83	82	84	86	89	90	92	93	93	93	91	85	88
PRECIPITATION (inches):														
Water Equivalent														
-Normal		12.07	10.80	13.54	16.44	19.12	17.14	18.39	16.53	16.06	16.71	15.74	15.22	187.76
-Maximum Monthly	43	26.67	19.76	35.30	38.65	38.43	27.97	37.20	32.74	38.81	28.26	31.79	33.35	38.81
-Year		1962	1964	1991	1959	1980	1987	1965	1976	1991	1989	1957	1975	SEP 1991
-Minimum Monthly	43	1.89	1.05	1.52	2.03	2.21	9.60	8.37	10.06	6.57	7.67	4.55	2.40	1.05
-Year		1983	1977	1983	1983	1983	1963	1990	1965	1979	1986	1963	1957	FEB 1977
-Maximum in 24 hrs	43	13.16	8.91	9.15	8.86	7.72	6.26	7.07	13.25	11.11	6.55	22.48	8.04	22.48
-Year		1992	1968	1987	1977	1980	1980	1965	1976	1990	1985	1957	1975	NOV 1957
Snow,Ice pellets,hail														
-Maximum Monthly		0.0	0.0	0.0	0.0	0.0	0.0	0.0	0.0	0.0	0.0	0.0	0.0	
-Year														
-Maximum in 24 hrs	43	0.0	0.0	0.0	0.0	0.0	0.0	0.0	0.0	0.0	0.0	0.0	0.0	
-Year														
WIND:														
Mean Speed (mph)	19	8.4	9.1	8.1	7.2	6.6	5.7	5.0	4.5	4.7	4.8	5.5	7.4	6.4
Prevailing Direction through 1963		NE	NE	NE	NE	NE	NE	E	ESE	S	ESE	NE	NE	NE
Fastest Obs. 1 Min.														
-Direction (!!!)	5	29	08	23	10	04	08	05	21	22	28	24	27	24
-Speed (MPH)	5	28	18	20	16	17	17	14	14	21	21	35	23	35
-Year		1992	1993	1991	1993	1992	1992	1993	1993	1992	1993	1991	1990	NOV 1991
Peak Gust														
-Direction (!!!)	10	NW	NE	SE	E	NW	SE	SE	SE	SW	W	SW	NW	SW
-Speed (mph)	10	55	41	45	48	53	43	35	35	36	46	62	56	62
-Date		1992	1985	1986	1985	1986	1985	1988	1984	1992	1993	1991	1990	NOV 1991

PRECIPITATION (inches)

POHNPEI (PONAPE), EASTERN CAROLINE ISLANDS, PACIFIC

YEAR	JAN	FEB	MAR	APR	MAY	JUNE	JULY	AUG	SEP	OCT	NOV	DEC	ANNUAL
1964	3.59	19.76	14.03	16.02	12.69	13.16	14.33	16.47	15.44	11.02	12.66	18.22	167.39
1965	11.74	11.12	6.36	14.28	14.18	18.73	37.20	10.06	24.67	15.27	15.50	14.20	193.31
1966	15.84	1.71	14.77	6.07	21.27	11.87	24.22	10.18	10.65	16.61	18.74	18.12	170.05
1967	10.21	18.83	21.82	20.90	13.46	11.48	14.02	20.32	19.51	16.37	22.44	12.60	201.96
1968	9.88	13.60	24.47	21.09	16.99	10.54	22.32	17.36	17.58	14.65	9.32	13.33	191.13
1969	6.39	9.47	10.71	22.35	17.75	24.88	28.90	13.28	18.01	12.86	16.20	15.00	195.80
1970	7.98	7.14	7.35	14.58	15.73	16.80	12.40	16.10	15.74	20.40	15.62	19.93	169.77
1971	16.64	13.12	19.98	17.37	24.59	23.62	22.69	16.67	9.89	19.11	12.02	8.67	204.37
1972	10.52	11.79	14.66	20.39	33.46	9.73	36.31	12.38	29.53	7.98	7.30	7.71	201.76
1973	3.31	3.70	7.71	24.38	16.28	23.40	9.40	17.85	18.15	21.32	10.38	18.16	174.04
1974	10.66	16.56	18.81	22.24	16.74	21.27	21.57	18.77	8.84	17.30	20.79	12.32	205.87
1975	6.61	4.26	15.99	16.79	17.50	18.83	15.60	11.26	12.61	22.25	17.22	33.35	192.27
1976	6.02	12.76	25.30	20.18	24.39	20.99	13.04	32.74	24.11	16.94	26.34	13.48	236.29
1977	4.45	1.05	12.65	15.93	26.17	14.70	16.97	18.72	10.88	20.00	16.14	4.95	162.61
1978	16.38	6.18	6.17	18.91	12.82	19.27	10.28	13.62	11.44	16.97	13.19	14.00	159.23
1979	8.16	6.65	11.98	23.38	18.76	23.85	17.07	22.35	6.57	17.57	20.09	19.58	196.01
1980	14.01	18.63	8.11	15.10	38.43	23.21	15.87	15.79	15.32	11.70	7.96	10.07	194.20
1981	14.59	13.75	8.55	10.47	23.00	17.92	16.20	13.41	15.04	15.43	16.46	17.47	182.29
1982	14.06	16.35	12.94	16.59	22.67	17.28	23.00	17.99	25.58	8.42	9.94	16.05	200.87
1983	1.89	1.72	1.52	2.03	2.21	15.91	24.55	14.29	12.36	20.05	20.99	16.10	133.62
1984	23.24	13.24	9.43	5.85	6.73	17.49	10.49	11.49	12.51	16.75	17.79	13.89	158.90
1985	14.83	12.91	5.15	24.22	13.04	13.80	14.62	14.02	12.48	18.81	19.95	20.11	183.94
1986	10.14	9.02	19.80	11.03	28.35	14.16	21.47	13.22	16.88	7.67	15.48	6.33	173.55
1987	17.77	5.86	25.91	20.18	9.22	27.97	25.80	16.52	11.96	16.61	19.46	15.09	212.35
1988	8.62	5.74	8.93	14.85	21.92	14.06	10.54	13.15	13.45	21.58	15.16	17.06	165.06
1989	18.72	10.66	14.37	24.26	19.83	14.59	21.77	23.21	9.24	28.26	8.72	26.21	219.84
1990	11.72	8.24	13.20	17.87	21.56	15.13	8.37	20.37	26.72	15.94	15.12	15.34	189.58
1991	6.18	8.73	35.30	21.43	22.91	14.68	17.30	18.37	38.81	12.51	23.26	9.06	228.54
1992	21.77	2.28	2.60	5.95	9.98	19.16	19.06	15.89	15.36	19.94	14.85	9.45	156.29
1993	15.83	10.07	19.44	18.91	20.41	21.56	19.60	17.07	9.07	15.40	12.03	25.55	204.94
Record Mean	11.99	10.40	14.26	17.82	19.29	17.22	17.70	16.57	16.35	16.36	16.62	15.74	190.33

AVERAGE TEMPERATURE (deg. F)

POHNPEI (PONAPE), EASTERN CAROLINE ISLANDS, PACIFIC

YEAR	JAN	FEB	MAR	APR	MAY	JUNE	JULY	AUG	SEP	OCT	NOV	DEC	ANNUAL
1964	81.8	81.6	80.9	80.8	80.3	79.7	79.6	79.7	79.9	79.6	80.0	79.6	80.3
1965	80.4	80.6	80.9	80.3	80.7	80.4	78.9	80.0	79.9	80.4	80.0	80.6	80.3
1966	80.1	81.4	80.7	81.7	80.6	81.0	80.4	81.1	80.5	80.5	80.3	80.4	80.7
1967	81.5	80.6	80.4	80.0	81.1	80.4	80.1	80.0	79.9	79.8	80.3	81.2	80.5
1968	81.1	80.6	80.2	79.9	80.0	80.1	79.5	80.4	79.9	79.9	80.1	80.6	80.2
1969	80.2	79.1	80.2	80.8	81.1	80.7	79.7	79.6	79.8	80.0	79.5	80.7	80.1
1970	80.8	82.0	81.9	81.6	81.6	80.7	80.1	80.1	79.4	79.3	80.0	81.3	80.7
1971	80.5	79.4	79.5	79.7	79.8	79.5	79.5	79.4	80.3	79.9	80.6	81.2	80.0
1972	80.6	80.8	80.8	80.0	80.0	81.0	79.6	79.9	79.9	80.1	81.1	81.5	80.5
1973	81.0	80.7	82.2	81.0	81.5	81.3	81.5	80.9	80.6	80.4	81.6	81.1	81.2
1974	79.8	81.5	80.9	80.8	81.1	79.6	79.8	80.5	80.2	80.1	80.7	80.8	80.5
1975	81.8	82.3	81.0	81.5	80.8	80.2	79.8	80.2	80.5	79.6	80.2	79.6	80.6
1976	81.6	80.2	81.1	80.1	80.2	80.4	80.3	79.8	79.6	80.9	80.8	80.8	80.5
1977	81.0	82.6	81.9	82.0	81.3	81.5	81.2	81.6	81.5	81.0	81.0	81.7	81.5
1978	81.1	81.3	82.6	81.4	82.0	80.6	80.8	81.3	80.9	80.9	80.8	81.6	81.3
1979	81.7	81.5	81.7	80.6	81.6	81.7	81.4	80.7	81.8	81.6	81.7	81.1	81.4
1980	81.9	81.7	82.8	82.4	80.9	81.7	81.3	81.3	81.0	81.3	81.9	82.3	81.7
1981	81.7	81.7	82.2	82.0	82.3	81.5	81.2	81.6	81.9	81.8	81.8	81.9	81.8
1982	81.4	81.7	81.7	81.8	81.2	81.6	80.3	80.6	80.4	80.7	81.0	82.1	81.2
1983	81.4	81.7	82.4	83.4	84.4	82.8	81.5	81.5	81.7	81.3	81.9	81.5	82.1
1984	80.8	81.5	82.7	83.0	83.3	81.4	81.6	81.1	81.2	81.0	80.8	81.8	81.7
1985	81.3	81.9	82.3	81.0	82.1	81.1	80.5	80.8	81.3	80.9	81.4	81.5	81.3
1986	82.1	81.7	81.3	81.8	80.6	81.6	80.9	81.5	80.9	81.4	81.5	81.4	81.4
1987	81.6	81.3	81.2	81.3	82.0	81.2	80.1	80.9	81.0	81.7	81.2	81.7	81.3
1988	81.7	82.2	82.7	82.1	81.4	81.0	81.0	81.2	80.6	80.2	80.3	80.5	81.2
1989	80.8	81.0	80.8	81.3	80.7	81.0	81.3	80.7	81.0	81.4	81.9	80.7	81.1
1990	81.7	80.7	82.1	81.6	81.3	81.5	81.2	80.9	80.5	81.4	81.1	80.2	81.2
1991	80.5	80.8	79.9	81.0	80.9	81.1	81.5	81.3	80.4	80.2	79.0	79.4	80.5
1992	79.0	80.0	80.0	81.1	81.9	81.6	80.9	81.5	81.3	81.1	80.2	80.5	80.8
1993	79.5	80.2	79.8	80.0	80.7	80.8	80.7	79.9	80.6	80.2	80.3	79.3	80.2
Record Mean	80.9	81.0	81.2	81.1	81.1	80.8	80.4	80.5	80.4	80.6	80.7	80.9	80.8
Max	86.3	86.5	86.9	87.1	87.4	87.5	87.7	88.1	88.2	88.3	88.0	87.0	87.4
Min	75.4	75.5	75.4	75.0	74.8	74.1	73.1	72.8	72.7	72.8	73.3	74.8	74.1

REFERENCE NOTES

POHNPEI (PONAPE), EASTERN CAROLINE ISLANDS, PACIFIC

GENERAL
T - TRACE AMOUNT.
BLANK ENTRIES DENOTE MISSING/UNREPORTED DATA.
# INDICATES A STATION OR INSTRUMENT RELOCATION.

SPECIFIC

(a) - LENGTH OF RECORD IN YEARS, ALTHOUGH INDIVIDUAL MONTHS MAY BE MISSING.
0.* OR * - THE VALUE IS BETWEEN 0.0 AND 0.05
NORMALS - BASED ON THE 1961-1990 RECORD PERIOD.
EXTREMES - DATES ARE THE MOST RECENT OCCURRENCE
WIND DIR.- NUMERALS SHOW TENS OF DEGRESS CLOCKWISE FROM TRUE NORTH. "00" INDICATES CALM.
RESULTANT DIRECTIONS ARE GIVEN TO WHOLE DEGREES.
BOLD VALUES INDICATE EXTREME VALUES WHICH OCCURRED AFTER THE ASOS SYSTEM WAS COMMISSIONED.

RECORD = PERIOD OF RECORD
RECORD MEAN PRECIPITATION IS THE MEAN OF ALL DAILY PRECIPITATION AMOUNTS DURING THE PERIOD OF RECORD.
RECORD MAX(MIN) TEMPERATURE IS THE MEAN OF ALL DAILY MAX(MIN) TEMPERATURES DURING THE PERIOD OF RECORD.
RECORD MEAN TEMPERATURE IS THE SUM OF THE RECORD MAX AND RECORD MIN DIVIDED BY 2.
AVERAGE TEMPERATURE IS THE SUM OF THE MEAN DAILY MAX AND MIN TEMPERATURE DIVIDED BY 2.

EXCEPTIONS

1. MEAN WIND SPEED IS THROUGH 1972.
2. THUNDERSTORMS AND HEAVY FOG ARE APRIL 1973 TO DATE AND MAY BE INCOMPLETE, DUE TO PART-TIME OPERATIONS.

RECORD MEANS ARE THROUGH THE CURRENT YEAR, BEGINNING IN 1952 FOR TEMPERATURE
1950 FOR PRECIPITATION

HEATING DEGREE DAYS Base 65 deg. F POHNPEI (PONAPE), EASTERN CAROLINE ISLANDS, PACIFIC

SEASON	JULY	AUG	SEP	OCT	NOV	DEC	JAN	FEB	MAR	APR	MAY	JUNE	TOTAL
1983-84	0	0	0	0	0	0	0	0	0	0	0	0	0
1984-85	0	0	0	0	0	0	0	0	0	0	0	0	0
1985-86	0	0	0	0	0	0	0	0	0	0	0	0	0
1986-87	0	0	0	0	0	0	0	0	0	0	0	0	0
1987-88	0	0	0	0	0	0	0	0	0	0	0	0	0
1988-89	0	0	0	0	0	0	0	0	0	0	0	0	0
1989-90	0	0	0	0	0	0	0	0	0	0	0	0	0
1990-91	0	0	0	0	0	0	0	0	0	0	0	0	0
1991-92	0	0	0	0	0	0	0	0	0	0	0	0	0
1992-93	0	0	0	0	0	0	0	0	0	0	0	0	0
1993-94	0	0	0	0	0	0							

COOLING DEGREE DAYS Base 65 deg. F POHNPEI (PONAPE), EASTERN CAROLINE ISLANDS, PACIFIC

YEAR	JAN	FEB	MAR	APR	MAY	JUNE	JULY	AUG	SEP	OCT	NOV	DEC	TOTAL
1969	478	401	477	480	506	479	462	460	450	474	444	492	5603
1970	494	480	534	508	521	477	472	476	442	452	455	513	5824
1971	491	408	454	450	465	442	457	452	467	468	476	506	5536
1972	492	465	499	458	472	488	461	469	455	477	493	521	5750
1973	504	447	541	486	520	495	519	500	473	485	505	507	5982
1974	464	467	501	480	506	446	464	489	463	476	476	497	5729
1975	525	491	501	503	498	464	467	481	471	461	459	458	5779
1976	522	445	507	459	477	470	479	464	444	499	481	498	5745
1977	504	498	530	518	511	504	507	522	499	505	490	525	6113
1978	508	459	554	501	536	476	497	513	484	498	479	523	6028
1979	524	465	525	477	520	506	516	494	512	524	508	503	6074
1980	531	488	557	530	502	507	514	512	487	515	511	544	6198
1981	526	474	540	519	546	503	509	521	513	530	511	531	6223
1982	512	472	527	508	509	507	481	491	469	495	486	537	5994
1983	513	473	548	562	609	544	519	519	509	511	512	516	6335
1984	497	488	559	547	576	499	520	504	490	504	479	528	6191
1985	513	481	544	487	535	489	490	496	495	501	500	519	6050
1986	539	474	508	511	493	504	502	516	485	517	500	519	6068
1987	523	463	506	498	532	492	477	502	486	524	493	524	6020
1988	523	507	556	516	517	487	504	507	475	477	468	486	6023
1989	502	454	497	493	494	484	513	495	488	515	514	495	5944
1990	526	448	536	504	513	500	507	500	468	511	489	479	5981
1991	488	449	465	490	498	491	518	514	467	481	428	456	5745
1992	441	440	471	488	534	503	501	516	494	504	465	488	5845
1993	460	430	467	458	493	479	493	469	474	478	470	450	5621

SNOWFALL (inches) POHNPEI (PONAPE), EASTERN CAROLINE ISLANDS, PACIFIC

SEASON	JULY	AUG	SEP	OCT	NOV	DEC	JAN	FEB	MAR	APR	MAY	JUNE	TOTAL
1971-72	0.0	0.0	0.0	0.0	0.0	0.0	0.0	0.0	0.0	0.0	0.0	0.0	0.0
1972-73	0.0	0.0	0.0	0.0	0.0	0.0	0.0	0.0	0.0	0.0	0.0	0.0	0.0
1973-74	0.0	0.0	0.0	0.0	0.0	0.0	0.0	0.0	0.0	0.0	0.0	0.0	0.0
1974-75	0.0	0.0	0.0	0.0	0.0	0.0	0.0	0.0	0.0	0.0	0.0	0.0	0.0
1975-76	0.0	0.0	0.0	0.0	0.0	0.0	0.0	0.0	0.0	0.0	0.0	0.0	0.0
1976-77	0.0	0.0	0.0	0.0	0.0	0.0	0.0	0.0	0.0	0.0	0.0	0.0	0.0
1977-78	0.0	0.0	0.0	0.0	0.0	0.0	0.0	0.0	0.0	0.0	0.0	0.0	0.0
1978-79	0.0	0.0	0.0	0.0	0.0	0.0	0.0	0.0	0.0	0.0	0.0	0.0	0.0
1979-80	0.0	0.0	0.0	0.0	0.0	0.0	0.0	0.0	0.0	0.0	0.0	0.0	0.0
1980-81	0.0	0.0	0.0	0.0	0.0	0.0	0.0	0.0	0.0	0.0	0.0	0.0	0.0
1981-82	0.0	0.0	0.0	0.0	0.0	0.0	0.0	0.0	0.0	0.0	0.0	0.0	0.0
1982-83	0.0	0.0	0.0	0.0	0.0	0.0	0.0	0.0	0.0	0.0	0.0	0.0	0.0
1983-84	0.0	0.0	0.0	0.0	0.0	0.0	0.0	0.0	0.0	0.0	0.0	0.0	0.0
1984-85	0.0	0.0	0.0	0.0	0.0	0.0	0.0	0.0	0.0	0.0	0.0	0.0	0.0
1985-86	0.0	0.0	0.0	0.0	0.0	0.0	0.0	0.0	0.0	0.0	0.0	0.0	0.0
1986-87	0.0	0.0	0.0	0.0	0.0	0.0	0.0	0.0	0.0	0.0	0.0	0.0	0.0
1987-88	0.0	0.0	0.0	0.0	0.0	0.0	0.0	0.0	0.0	0.0	0.0	0.0	0.0
1988-89	0.0	0.0	0.0	0.0	0.0	0.0	0.0	0.0	0.0	0.0	0.0	0.0	0.0
1989-90	0.0	0.0	0.0	0.0	0.0	0.0	0.0	0.0	0.0	0.0	0.0	0.0	0.0
1990-91	0.0	0.0	0.0	0.0	0.0	0.0	0.0	0.0	0.0	0.0	0.0	0.0	0.0
1991-92	0.0	0.0	0.0	0.0	0.0	0.0	0.0	0.0	0.0	0.0	0.0	0.0	0.0
1992-93	0.0	0.0	0.0	0.0	0.0	0.0	0.0	0.0	0.0	0.0	0.0	0.0	0.0
1993-94	0.0	0.0	0.0	0.0	0.0	0.0							
Record Mean	0.0	0.0	0.0	0.0	0.0	0.0	0.0	0.0	0.0	0.0	0.0	0.0	0.0

WAKE ISLAND, PACIFIC

Wake is a coral atoll, a coral and shell accumulation atop the rim of an extinct underwater volcano. The central lagoon is the former crater and the rim is three small islands, Wake, Peale, and Wilkes, totalling less than three square miles in area and approximately nine miles in outside perimeter. The average height above sea level is only about 12 feet. The highest terrain, 15 to 20 feet, consists of coral rock and shell ridges parallel to the shore and inland some 10 to 50 feet from the sand beaches. The terrain slopes gently from the ridges to the lagoon, and includes several tidal flats. There are no other islands within several hundred miles. The terrain does not appreciably affect the climate.

The climate of Wake is maritime and is chiefly controlled by the easterly trade winds which dominate the island throughout the year. Occasionally during late fall, winter, and early spring, polar outbreaks reach the island and are marked by temperature drops of several degrees, increased cloudiness, and light to moderate showers of short duration. The winds during these outbreaks swing to the northerly directions, and may reach gust velocities of 40 mph. The unusual weather lasts only a few days but may produce temperatures in the low 60s.

Frequent tropical disturbances (low pressure systems, aloft or at the surface) approach the island from the southeast quadrant during the late summer and early fall months. These systems bring periods of light wind, high temperatures, high humidities, and moderate to heavy rain showers. When the system is vigorous and close to the island, winds strengthen and showers may be prolonged to several hours duration. Although typhoons in the area are not unusual, only rarely have typhoon-force winds occurred at Wake since observations began in 1935.

Clouds are predominately cumulus types with little difference in amounts from day to night. High cloudiness, rare during winter months, occurs in association with fall and summer tropical disturbances.

Showers account for most of the precipitation and occur most frequently between midnight and sunrise. The sky is seldom completely overcast or completely clear. Thunderstorms are infrequent, occurring mainly with tropical disturbances. The occurrence of hail and fog are virtually unknown to Wake. Although rare, the visibility can be reduced by volcanic haze, carried from the Hawaiian Islands by the Easterly Trades.

NORMALS, MEANS, AND EXTREMES

WAKE ISLAND, PACIFIC

LATITUDE: 19°17'N LONGITUDE: 166°39'E ELEVATION: FT. GRND 11 BARO 16 TIME ZONE: 180E MER WBAN: 41606

	(a)	JAN	FEB	MAR	APR	MAY	JUNE	JULY	AUG	SEP	OCT	NOV	DEC	YEAR
TEMPERATURE °F:														
Normals														
-Daily Maximum		82.3	82.0	83.1	83.9	85.7	87.6	88.2	88.2	88.2	87.4	85.5	83.5	85.5
-Daily Minimum		72.7	72.1	72.9	73.4	75.1	76.7	77.3	77.4	77.7	77.2	76.1	74.2	75.2
-Monthly		77.5	77.1	78.1	78.7	80.4	82.2	82.8	82.8	83.0	82.3	80.8	78.9	80.4
Extremes														
-Record Highest	46	89	88	90	91	92	93	94	95	95	92	90	91	95
-Year		1982	1985	1985	1970	1984	1988	1985	1984	1984	1985	1984	1980	AUG 1984
-Record Lowest	46	65	65	65	65	69	71	69	68	69	68	65	64	64
-Year		1958	1972	1977	1985	1992	1978	1980	1958	1975	1961	1950	1954	DEC 1954
NORMAL DEGREE DAYS:														
Heating (base 65°F)		0	0	0	0	0	0	0	0	0	0	0	0	0
Cooling (base 65°F)		388	339	406	411	477	516	552	552	540	536	474	431	5622
% OF POSSIBLE SUNSHINE	24	68	72	77	75	74	75	71	66	67	68	64	65	70
MEAN SKY COVER (tenths)														
Sunrise - Sunset	42	4.6	4.6	4.6	4.9	5.1	5.2	6.4	6.6	6.4	5.9	4.8	4.3	5.3
MEAN NUMBER OF DAYS:														
Sunrise to Sunset														
-Clear	42	12.5	11.7	12.2	10.9	10.3	9.8	6.9	6.0	6.6	8.5	11.8	14.1	121.3
-Partly Cloudy	42	13.0	11.3	13.6	13.5	14.2	13.2	10.9	11.3	10.9	12.1	12.7	12.6	149.3
-Cloudy	42	5.4	5.1	5.1	5.6	6.4	6.9	13.3	13.7	12.3	10.3	5.5	4.3	93.9
Precipitation .01 inches or more	46	10.8	9.6	12.2	13.9	14.6	15.5	19.1	18.8	19.0	19.1	15.1	12.8	180.5
Snow,Ice pellets,hail 1.0 inches or more	46	0.0	0.0	0.0	0.0	0.0	0.0	0.0	0.0	0.0	0.0	0.0	0.0	0.0
Thunderstorms	45	0.0	0.*	0.1	0.2	0.2	0.5	1.0	1.2	1.1	1.2	0.3	0.2	6.0
Heavy Fog Visibility 1/4 mile or less	46	0.0	0.0	0.0	0.0	0.0	0.0	0.0	0.0	0.0	0.0	0.0	0.0	0.0
Temperature °F														
-Maximum														
90° and above	46	0.0	0.0	0.*	0.2	1.0	2.5	5.8	7.3	6.4	3.3	0.1	0.*	26.7
32° and below	46	0.0	0.0	0.0	0.0	0.0	0.0	0.0	0.0	0.0	0.0	0.0	0.0	0.0
-Minimum														
32° and below	46	0.0	0.0	0.0	0.0	0.0	0.0	0.0	0.0	0.0	0.0	0.0	0.0	0.0
0° and below	46	0.0	0.0	0.0	0.0	0.0	0.0	0.0	0.0	0.0	0.0	0.0	0.0	0.0
AVG. STATION PRESS.(mb)	16	1014.0	1014.4	1014.7	1015.4	1014.7	1014.1	1013.2	1012.4	1012.3	1012.6	1013.3	1014.4	1013.8
RELATIVE HUMIDITY (%)														
Hour 00	45	76	77	79	79	81	81	81	82	82	81	79	77	80
Hour 06 (Local Time)	41	77	78	80	81	82	83	82	83	83	82	80	78	81
Hour 12	46	65	65	67	68	68	68	70	71	71	71	70	67	68
Hour 18	46	70	69	71	72	73	72	73	75	75	76	76	73	73
PRECIPITATION (inches):														
Water Equivalent														
-Normal		1.16	1.60	2.23	2.51	1.74	2.29	4.02	6.16	5.07	4.33	2.79	1.78	35.68
-Maximum Monthly	46	4.25	6.53	6.91	8.36	4.55	5.42	10.74	27.71	17.66	15.45	8.54	6.87	27.71
-Year		1974	1990	1976	1985	1970	1978	1970	1981	1952	1948	1950	1975	AUG 1981
-Minimum Monthly	46	0.06	0.15	0.20	0.31	0.39	0.49	0.41	1.16	0.80	1.04	0.32	0.28	0.06
-Year		1965	1968	1988	1970	1979	1993	1987	1974	1983	1982	1966	1980	JAN 1965
-Maximum in 24 hrs	46	2.31	3.24	3.38	5.67	2.58	3.69	5.89	7.83	15.00	7.21	6.24	2.63	15.00
-Year		1974	1990	1986	1985	1970	1953	1990	1981	1952	1992	1954	1975	SEP 1952
Snow,Ice pellets,hail														
-Maximum Monthly		0.0	0.0	0.0	0.0	0.0	0.0	0.0	0.0	0.0	0.0	0.0	0.0	
-Year														
-Maximum in 24 hrs	46	0.0	0.0	0.0	0.0	0.0	0.0	0.0	0.0	0.0	0.0	0.0	0.0	
-Year														
WIND:														
Mean Speed (mph)	40	13.6	13.5	14.5	15.6	14.3	12.6	12.6	12.1	12.4	14.0	16.0	14.9	13.8
Prevailing Direction through 1963		ENE	ENE	ENE	ENE	ENE	E	E	E	E	ENE	ENE	ENE	ENE
Fastest Mile														
-Direction (!!)	12	N	NE	NE	NE	SE	SE	E	E	N	E	NE	E	N
-Speed (MPH)	12	44	45	36	36	38	34	41	45	104	47	52	43	104
-Year		1978	1976	1972	1976	1968	1968	1972	1968	1967	1978	1979	1972	SEP 1967
Peak Gust														
-Direction (!!)	9	E			NE		E	E	SW	NW	E	E		E
-Speed (mph)	9	52	47	48	45	40	39	44	59	61	104	52	44	104
-Date		1986	1989	1989	1984	1987	1988	1989	1986	1992	1992	1985	1986	OCT 1992

PRECIPITATION (inches)

WAKE ISLAND, PACIFIC

YEAR	JAN	FEB	MAR	APR	MAY	JUNE	JULY	AUG	SEP	OCT	NOV	DEC	ANNUAL
1964	0.92	0.61	1.08	3.68	0.42	5.25	4.46	5.85	6.92	2.80	2.53	1.67	36.19
1965	0.06	1.08	1.38	3.31	2.23	1.41	0.78	4.02	1.44	8.31	1.92	0.51	26.45
1966	0.67	1.34	0.64	1.30	3.20	1.59	1.70	4.85	5.15	2.53	0.32	0.62	23.91
1967	1.17	1.90	6.14	3.80	2.25	4.18	10.56	8.71	16.60	3.12	2.08	1.85	62.36
1968	1.20	0.15	0.74	6.22	3.65	2.26	3.48	6.10	3.69	7.12	4.29	2.03	40.93
1969	1.15	0.74	0.91	1.62	1.38	1.64	1.15	1.63	4.18	2.41	2.92	4.74	24.47
1970	0.95	1.50	1.66	0.31	4.55	0.98	10.74	5.44	2.88	3.26	1.61	1.91	35.79
1971	1.79	3.34	1.62	3.01	2.38	3.52	2.40	6.51	17.08	5.00	5.01	0.98	52.64
1972	1.75	0.75	1.77	1.21	0.67	1.30	3.38	1.44	8.05	3.36	1.77	4.15	29.60
1973	2.17	1.71	1.41	0.89	1.99	1.95	1.23	7.42	3.12	2.73	2.25	0.93	27.80
1974	4.25	2.13	0.59	1.93	0.96	1.74	9.25	1.16	4.94	3.18	1.88	1.73	33.74
1975	0.88	2.12	1.44	0.91	2.19	0.96	3.08	4.48	5.73	3.48	2.82	6.87	34.96
1976	1.19	3.49	6.91	1.53	1.29	1.81	3.18	2.69	2.24	1.40	2.44	0.64	28.81
1977	0.43	0.87	5.55	1.63	1.60	1.83	2.40	12.39	4.37	10.33	2.83	1.36	45.59
1978	1.00	0.33	1.90	4.14	2.35	5.42	2.57	5.09	7.58	3.09	4.21	1.16	38.84
1979	2.28	1.45	1.53	1.48	0.39	1.50	1.77	14.20	4.58	3.95	3.09	1.17	37.39
1980	0.38	1.54	1.11	2.02	1.50	1.00	2.71	3.92	3.85	4.81	3.08	0.28	26.20
1981	0.84	1.39	5.19	1.26	1.19	2.23	7.23	27.71	2.63	3.11	2.67	1.38	56.83
1982	0.14	3.15	3.14	4.04	2.20	2.20	1.28	2.72	4.50	1.04	2.01	0.45	26.87
1983	0.65	0.27	0.50	4.28	0.62	0.99	2.21	1.72	0.80	3.03	2.26	1.11	18.44
1984	2.11	2.51	1.20	3.18	1.61	4.24	5.51	1.62	2.60	8.33	4.00	1.00	37.91
1985	0.82	1.12	1.54	8.36	2.01	1.90	3.57	7.06	2.05	1.09	2.18	0.52	32.22
1986	1.82	1.61	6.68	0.88	0.45	0.68	4.30	12.72	1.88	1.64	3.71	1.64	38.01
1987	0.50	0.67	0.59	1.55	1.99	1.21	0.41	2.09	5.07	6.35	2.66	1.22	24.31
1988	0.66	1.15	0.20	3.20	0.66	1.77	1.55	2.87	3.81	4.25	0.53	2.02	22.67
1989	1.94	2.11	2.63	1.10	2.44	1.87	5.34	6.79	4.60	2.53	5.79	1.73	38.87
1990	1.17	6.53	3.40	3.94	0.73	3.76	9.54	2.70	2.47	2.33	5.06	4.76	46.39
1991	1.57	0.26	3.32	1.03	0.75	1.83	3.37	6.31	2.34	2.42	1.74	1.08	26.02
1992	0.34	0.60	0.68	1.40	1.28	1.44	1.99	2.79	16.07	10.42	2.12	0.81	39.94
1993	0.56	0.89	0.41	0.56	1.19	0.49	1.53	2.46	3.49	3.91	5.08	1.79	22.36
Record Mean	1.14	1.47	1.93	2.14	1.79	2.08	3.99	6.25	5.40	4.77	2.93	1.83	35.72

AVERAGE TEMPERATURE (deg. F)

WAKE ISLAND, PACIFIC

YEAR	JAN	FEB	MAR	APR	MAY	JUNE	JULY	AUG	SEP	OCT	NOV	DEC	ANNUAL
1964	78.1	77.5	78.3	77.8	80.2	82.6	81.9	82.1	82.2	80.4	79.9	76.3	79.8
1965	75.6	76.3	76.8	76.5	77.6	79.4	81.4	81.5	82.3	80.8	78.3	76.6	78.6
1966	75.6	75.8	76.6	78.0	78.8	81.7	82.9	82.1	82.5	81.7	78.9	78.1	79.4
1967	77.4	76.3	76.1	77.4	80.1	82.2	81.4	82.4	81.0	82.3	81.3	79.7	79.8
1968	77.3	75.9	77.2	77.6	80.1	82.5	82.8	83.4	82.9	82.7	81.9	78.0	80.2
1969	77.0	77.0	79.2	80.4	80.8	81.8	83.4	84.7	83.7	83.4	80.6	79.1	80.9
1970	77.7	78.3	80.9	82.6	82.9	84.2	82.2	82.7	83.5	83.0	81.4	79.4	81.6
1971	77.5	76.3	78.6	79.4	81.5	82.5	83.1	82.4	80.7	81.5	80.2	76.6	80.0
1972	74.5	74.5	75.6	75.6	79.7	81.9	82.4	83.3	82.9	82.0	80.6	77.4	79.2
1973	76.5	75.8	78.3	78.4	80.1	81.6	83.2	81.0	82.2	81.1	79.9	78.7	79.7
1974	76.6	76.5	75.9	77.0	78.6	80.3	80.0	81.4	80.4	81.3	79.7	78.3	78.9
1975	76.3	75.5	77.0	77.3	79.7	82.3	81.9	82.2	81.5	81.8	80.3	77.8	79.5
1976	77.4	76.1	77.5	78.3	79.3	82.3	82.4	82.7	82.9	82.5	81.0	79.1	80.1
1977	76.3	76.9	76.6	77.6	80.4	82.3	82.8	82.0	82.8	81.0	80.0	77.7	79.7
1978	76.8	76.8	77.0	79.3	80.1	81.6	82.3	84.1	84.1	83.8	81.2	79.7	80.5
1979	77.4	77.4	79.1	80.0	81.8	83.3	85.2	81.5	83.1	83.4	79.4	78.5	80.9
1980	77.6	76.1	78.3	79.7	80.7	82.8	84.3	83.5	84.6	82.1	81.6	80.2	81.0
1981	79.3	78.6	80.2	81.4	82.1	84.0	83.9	82.7	84.3	82.8	82.8	81.7	82.0
1982	79.4	78.1	79.2	77.3	79.7	81.2	84.0	84.4	84.5	83.8	80.9	78.4	80.9
1983	75.7	76.7	77.3	77.4	79.7	81.3	82.8	83.7	84.3	84.4	82.5	81.2	80.6
1984	79.5	80.1	81.0	82.1	83.7	83.8	84.1	85.8	86.1	83.2	83.1	80.5	82.8
1985	79.2	79.7	79.6	80.0	82.6	83.1	84.2	83.3	84.9	84.1	80.8	79.0	81.7
1986	77.8	77.6	78.2	79.2	79.6	82.2	83.6	82.5	83.5	83.1	80.9	79.3	80.6
1987	78.5	77.4	78.2	78.6	79.8	81.5	83.0	83.9	83.6	82.7	80.9	79.1	80.6
1988	78.0	77.6	80.0	80.5	81.5	83.2	83.8	83.8	83.3	83.1	82.5	80.7	81.5
1989	79.5	77.0	76.6	79.6	81.1	81.9	82.1	82.0	81.9	81.7	80.5	79.4	80.3
1990	78.4	76.3	77.4	78.1	80.5	82.0	82.1	83.5	83.3	82.0	80.9	79.9	80.4
1991	78.2	77.0	78.5	79.6	81.9	82.8	84.6	83.2		82.2	81.5	79.7	
1992	77.9	77.8	77.7	79.4	79.6	82.4	83.5	84.7	82.9	83.3	80.1	78.4	80.6
1993	76.8	76.9	78.6	80.8	81.3	83.7	84.4	84.5	84.7	83.4	81.2	79.8	81.3
Record Mean	77.4	77.0	77.9	78.6	80.2	82.1	82.6	82.8	82.9	82.1	80.7	78.9	80.3
Max	81.9	81.8	82.8	83.6	85.2	87.2	87.8	88.0	88.0	87.0	85.2	83.3	85.1
Min	72.8	72.2	73.0	73.6	75.2	76.9	77.4	77.6	77.8	77.2	76.2	74.4	75.4

REFERENCE NOTES

WAKE ISLAND, PACIFIC

GENERAL
T - TRACE AMOUNT.
BLANK ENTRIES DENOTE MISSING/UNREPORTED DATA.
# INDICATES A STATION OR INSTRUMENT RELOCATION.

SPECIFIC

EXCEPTIONS

RECORD MEANS ARE THROUGH THE CURRENT YEAR,
BEGINNING IN 1948 FOR TEMPERATURE
1936 FOR PRECIPITATION

(a) - LENGTH OF RECORD IN YEARS, ALTHOUGH INDIVIDUAL MONTHS MAY BE MISSING.
0.* OR * - THE VALUE IS BETWEEN 0.0 AND 0.05
NORMALS - BASED ON THE 1961-1990 RECORD PERIOD.
EXTREMES - DATES ARE THE MOST RECENT OCCURRENCE.
WIND DIR. - NUMERALS SHOW TENS OF DEGRESS CLOCKWISE FROM TRUE NORTH. "00" INDICATES CALM.
RESULTANT DIRECTIONS ARE GIVEN TO WHOLE DEGREES.
BOLD VALUES INDICATE EXTREME VALUES WHICH OCCURRED AFTER THE ASOS SYSTEM WAS COMMISSIONED.

RECORD = PERIOD OF RECORD
RECORD MEAN PRECIPITATION IS THE MEAN OF ALL DAILY PRECIPITATION AMOUNTS DURING THE PERIOD OF RECORD.
RECORD MAX(MIN) TEMPERATURE IS THE MEAN OF ALL DAILY MAX(MIN) TEMPERATURES DURING THE PERIOD OF RECORD.
RECORD MEAN TEMPERATURE IS THE SUM OF THE RECORD MAX AND RECORD MIN DIVIDED BY 2.
AVERAGE TEMPERATURE IS THE SUM OF THE MEAN DAILY MAX AND MIN TEMPERATURE DIVIDED BY 2.

HEATING DEGREE DAYS Base 65 deg. F — WAKE ISLAND, PACIFIC

SEASON	JULY	AUG	SEP	OCT	NOV	DEC	JAN	FEB	MAR	APR	MAY	JUNE	TOTAL
1983-84	0	0	0	0	0	0	0	0	0	0	0	0	0
1984-85	0	0	0	0	0	0	0	0	0	0	0	0	0
1985-86	0	0	0	0	0	0	0	0	0	0	0	0	0
1986-87	0	0	0	0	0	0	0	0	0	0	0	0	0
1987-88	0	0	0	0	0	0	0	0	0	0	0	0	0
1988-89	0	0	0	0	0	0	0	0	0	0	0	0	0
1989-90	0	0	0	0	0	0	0	0	0	0	0	0	0
1990-91	0	0	0	0	0	0	0	0	0	0	0	0	0
1991-92	0	0		0	0	0	0	0	0	0	0	0	
1992-93	0	0	0	0	0	0	0	0	0	0	0	0	0
1993-94	0	0	0	0	0	0							

COOLING DEGREE DAYS Base 65 deg. F — WAKE ISLAND, PACIFIC

YEAR	JAN	FEB	MAR	APR	MAY	JUNE	JULY	AUG	SEP	OCT	NOV	DEC	TOTAL
1969	379	338	448	467	495	510	572	616	568	578	472	445	5888
1970	401	378	497	535	562	585	540	554	561	565	500	452	6130
1971	394	324	426	440	517	531	568	547	479	518	463	368	5575
1972	302	279	335	323	463	513	545	571	545	535	477	391	5279
1973	367	309	421	407	475	502	572	504	520	505	455	432	5469
1974	368	329	343	370	431	464	474	516	470	512	445	416	5138
1975	357	300	378	375	460	525	530	539	505	528	463	402	5362
1976	388	331	391	407	450	525	545	554	542	549	486	445	5613
1977	359	341	366	384	484	524	560	534	541	502	457	399	5451
1978	374	338	378	436	476	505	544	599	577	588	495	464	5774
1979	392	354	443	458	527	556	632	518	550	576	436	426	5868
1980	400	326	418	446	493	542	603	580	593	540	506	477	5924
1981	451	386	479	500	535	578	594	557	587	562	540	523	6292
1982	453	378	444	373	465	492	596	606	590	591	484	423	5895
1983	341	334	390	381	462	496	557	586	587	608	534	511	5787
1984	457	444	502	516	586	572	599	655	639	570	552	489	6581
1985	450	415	462	455	554	548	601	574	606	600	482	442	6189
1986	406	358	417	429	456	523	584	549	564	566	484	451	5787
1987	428	352	418	411	464	499	566	592	568	555	482	445	5780
1988	410	372	468	472	515	552	593	590	554	569	534	496	6125
1989	455	339	365	443	506	515	537	536	515	527	472	451	5661
1990	422	323	390	400	486	519	538	583	554	534	484	469	5702
1991	416	347	423	446	532	538	612	570		542	501	462	
1992	407	375	400	440	462	528	579	618	544	575	459	421	5808
1993	374	338	431	480	515	567	608	611	596	574	494	467	6055

SNOWFALL (inches) — WAKE ISLAND, PACIFIC

SEASON	JULY	AUG	SEP	OCT	NOV	DEC	JAN	FEB	MAR	APR	MAY	JUNE	TOTAL
1971-72	0.0	0.0	0.0	0.0	0.0	0.0	0.0	0.0	0.0	0.0	0.0	0.0	0.0
1972-73	0.0	0.0	0.0	0.0	0.0	0.0	0.0	0.0	0.0	0.0	0.0	0.0	0.0
1973-74	0.0	0.0	0.0	0.0	0.0	0.0	0.0	0.0	0.0	0.0	0.0	0.0	0.0
1974-75	0.0	0.0	0.0	0.0	0.0	0.0	0.0	0.0	0.0	0.0	0.0	0.0	0.0
1975-76	0.0	0.0	0.0	0.0	0.0	0.0	0.0	0.0	0.0	0.0	0.0	0.0	0.0
1976-77	0.0	0.0	0.0	0.0	0.0	0.0	0.0	0.0	0.0	0.0	0.0	0.0	0.0
1977-78	0.0	0.0	0.0	0.0	0.0	0.0	0.0	0.0	0.0	0.0	0.0	0.0	0.0
1978-79	0.0	0.0	0.0	0.0	0.0	0.0	0.0	0.0	0.0	0.0	0.0	0.0	0.0
1979-80	0.0	0.0	0.0	0.0	0.0	0.0	0.0	0.0	0.0	0.0	0.0	0.0	0.0
1980-81	0.0	0.0	0.0	0.0	0.0	0.0	0.0	0.0	0.0	0.0	0.0	0.0	0.0
1981-82	0.0	0.0	0.0	0.0	0.0	0.0	0.0	0.0	0.0	0.0	0.0	0.0	0.0
1982-83	0.0	0.0	0.0	0.0	0.0	0.0	0.0	0.0	0.0	0.0	0.0	0.0	0.0
1983-84	0.0	0.0	0.0	0.0	0.0	0.0	0.0	0.0	0.0	0.0	0.0	0.0	0.0
1984-85	0.0	0.0	0.0	0.0	0.0	0.0	0.0	0.0	0.0	0.0	0.0	0.0	0.0
1985-86	0.0	0.0	0.0	0.0	0.0	0.0	0.0	0.0	0.0	0.0	0.0	0.0	0.0
1986-87	0.0	0.0	0.0	0.0	0.0	0.0	0.0	0.0	0.0	0.0	0.0	0.0	0.0
1987-88	0.0	0.0	0.0	0.0	0.0	0.0	0.0	0.0	0.0	0.0	0.0	0.0	0.0
1988-89	0.0	0.0	0.0	0.0	0.0	0.0	0.0	0.0	0.0	0.0	0.0	0.0	0.0
1989-90	0.0	0.0	0.0	0.0	0.0	0.0	0.0	0.0	0.0	0.0	0.0	0.0	0.0
1990-91	0.0	0.0	0.0	0.0	0.0	0.0	0.0	0.0	0.0	0.0	0.0	0.0	0.0
1991-92	0.0	0.0	0.0	0.0	0.0	0.0	0.0	0.0	0.0	0.0	0.0	0.0	0.0
1992-93	0.0	0.0	0.0	0.0	0.0	0.0	0.0	0.0	0.0	0.0	0.0	0.0	0.0
1993-94	0.0	0.0	0.0	0.0	0.0	0.0							
Record Mean	0.0	0.0	0.0	0.0	0.0	0.0	0.0	0.0	0.0	0.0	0.0	0.0	0.0

YAP ISLAND, PACIFIC

The Yap group consists of four large islands and ten small islands surrounded by a coral reef. These islands were formed by land upheaval and are not, therefore, of volcanic or of coral origin. The soil is clay-like and contains considerable rock. The islands are mostly low, rolling grass-covered hills.

The lowlands, which occupy the southwesterly end of Yap, are covered with dense jungle growth and are marsh-like, except during the dry spells that occur from time to time, particularly during the early months of the year. The terrain around the station is level for about 1/2 mile. A ridge about 3 1/4 miles to the northwest rises 224 feet above sea level and slopes northeastward toward the highest hill on the island, elevation 585 feet. Other small hills lie approximately 3 miles northeast and 3 miles southwest of the station. Lush vegetation, interspersed with sparse pandanas and a few coconut trees, is visible in all directions. The ocean itself cannot be seen from the station, although it lies only about 3/4 of a mile away to the east and south and about 1 mile to the west.

During northern summer, the Intertropical Convergence Zone lies near Yap, particularly as it moves northward in July and southward again in October. At such times showers and light variable winds predominate, interspersed with heavier showers or thunderstorms, occasionally accompanied by strong, shifting winds. Thunderstorms are relatively infrequent, averaging two per month from August through December and fifteen for the year as a whole.

Tropical cyclones affect the area much less often than they do the Pacific further to the northwest. June to December are the months of greatest frequency. Fully-developed typhoons are uncommon near Yap. Most of them pass to the north and then move westward to northwestward away from the island.

Yap is under the influence of the northeast trade winds for eight months of the year, November through June. From July through October the prevailing wind is southwesterly, with frequent periods of calm and light variable winds. This is also the wettest season with monthly rainfall exceeding 13 inches. The nearest approach to a dry season is February through April, when the monthly rainfall is less than 7 inches.

Temperature varies much less seasonally than between day and night. Thus, the warmest and coolest months differ by less than 2 degrees in average temperature, as compared with a difference of nearly 12 degrees between the warmest and coolest times of day.

Humidity is higher and clear skies more frequent during the night and early morning than during the day. Cloudless days are rare. A common daily sequence from May through December is to have the late morning fair weather clouds build up in late afternoon into towering cumulus that give rise to evening and early morning showers. Visibility in such showers is seldom less than 5 miles.

Despite their relatively small size and low relief, the islands nevertheless appear to be large and high enough to cause local differences in temperature, wind, humidity, and, perhaps, rainfall.

NORMALS, MEANS, AND EXTREMES

YAP ISLAND, PACIFIC

LATITUDE: 9°29'N LONGITUDE: 138°05'E ELEVATION: FT. GRND 44 BARO 47 TIME ZONE: 135E MER WBAN: 40308

	(a)	JAN	FEB	MAR	APR	MAY	JUNE	JULY	AUG	SEP	OCT	NOV	DEC	YEAR
TEMPERATURE °F:														
Normals														
-Daily Maximum		85.9	86.2	86.9	87.7	87.8	87.3	86.8	86.8	87.1	87.2	87.2	86.5	87.0
-Daily Minimum		74.6	74.8	74.9	75.3	75.5	75.0	74.6	74.5	74.6	74.7	74.9	75.0	74.9
-Monthly		80.3	80.5	80.9	81.5	81.6	81.2	80.7	80.7	80.9	81.0	81.1	80.8	80.9
Extremes														
-Record Highest	45	90	92	90	97	93	94	96	93	93	93	93	94	97
-Year		1963	1950	1988	1951	1984	1958	1950	1957	1956	1957	1956	1955	APR 1951
-Record Lowest	45	67	67	64	69	69	65	70	67	68	69	69	65	64
-Year		1993	1992	1993	1993	1992	1975	1990	1975	1993	1991	1991	1973	MAR 1993
NORMAL DEGREE DAYS:														
Heating (base 65°F)		0	0	0	0	0	0	0	0	0	0	0	0	0
Cooling (base 65°F)		474	434	493	495	515	486	487	487	477	496	483	490	5817
% OF POSSIBLE SUNSHINE	34	58	63	68	68	62	51	48	44	49	45	53	52	55
MEAN SKY COVER (tenths)														
Sunrise - Sunset	45	8.6	8.6	8.4	8.2	8.2	8.9	9.1	9.4	9.1	8.8	8.7	8.6	8.7
MEAN NUMBER OF DAYS:														
Sunrise to Sunset														
-Clear	44	0.8	0.5	0.7	0.9	0.9	0.3	0.5	0.1	0.3	0.5	0.5	0.6	6.6
-Partly Cloudy	44	7.4	7.4	8.2	9.5	10.1	5.4	4.0	3.0	5.1	6.3	7.4	8.2	82.1
-Cloudy	44	22.8	20.3	22.1	19.5	20.1	24.3	26.5	27.9	24.5	24.2	22.1	22.2	276.5
Precipitation														
.01 inches or more	45	20.8	17.6	17.8	17.4	20.8	23.6	24.5	23.7	22.6	23.8	22.8	21.8	257.2
Snow, Ice pellets, hail														
1.0 inches or more	45	0.0	0.0	0.0	0.0	0.0	0.0	0.0	0.0	0.0	0.0	0.0	0.0	0.0
Thunderstorms	45	0.4	0.3	0.3	0.8	1.1	1.6	2.0	2.0	2.7	2.5	2.0	1.5	17.3
Heavy Fog Visibility														
1/4 mile or less	45	0.0	0.0	0.0	0.0	0.0	0.0	0.0	0.0	0.0	0.0	0.0	0.0	0.0
Temperature °F														
-Maximum														
90° and above	45	0.2	0.1	0.3	3.2	5.2	4.1	2.8	3.0	3.3	3.6	2.4	0.6	28.9
32° and below	45	0.0	0.0	0.0	0.0	0.0	0.0	0.0	0.0	0.0	0.0	0.0	0.0	0.0
-Minimum														
32° and below	45	0.0	0.0	0.0	0.0	0.0	0.0	0.0	0.0	0.0	0.0	0.0	0.0	0.0
0° and below	45	0.0	0.0	0.0	0.0	0.0	0.0	0.0	0.0	0.0	0.0	0.0	0.0	0.0
AVG. STATION PRESS.(mb)	14	1008.1	1008.9	1008.7	1008.2	1007.6	1007.5	1007.1	1007.1	1007.2	1006.8	1006.4	1007.4	1007.6
RELATIVE HUMIDITY (%)														
Hour 03	24	87	85	86	87	89	91	91	91	91	92	90	88	89
Hour 09 (Local Time)	45	78	77	75	75	77	79	80	81	80	80	79	79	78
Hour 15	45	76	74	73	73	76	77	77	78	78	78	78	77	76
Hour 21	45	85	84	84	85	87	89	90	89	89	90	89	87	87
PRECIPITATION (inches):														
Water Equivalent														
-Normal		7.33	5.98	5.96	5.76	9.06	12.69	14.54	15.20	13.51	11.97	9.07	8.99	120.06
-Maximum Monthly	45	23.08	13.36	16.46	18.15	18.23	32.01	34.71	29.44	19.57	22.43	20.66	17.05	34.71
-Year		1955	1962	1950	1956	1964	1982	1969	1953	1949	1992	1960	1954	JUL 1969
-Minimum Monthly	45	1.25	0.27	1.38	0.91	1.47	3.40	4.99	5.13	5.32	2.59	1.96	2.22	0.27
-Year		1983	1983	1958	1977	1993	1951	1949	1973	1987	1976	1957	1990	FEB 1983
-Maximum in 24 hrs	45	10.45	5.94	5.09	6.57	10.06	18.75	7.44	7.81	8.35	6.72	8.91	6.66	18.75
-Year		1958	1962	1963	1962	1967	1982	1969	1987	1978	1988	1960	1985	JUN 1982
Snow, Ice pellets, hail														
-Maximum Monthly		0.0	0.0	0.0	0.0	0.0	0.0	0.0	0.0	0.0	0.0	0.0	0.0	
-Year														
-Maximum in 24 hrs	45	0.0	0.0	0.0	0.0	0.0	0.0	0.0	0.0	0.0	0.0	0.0	0.0	
-Year														
WIND:														
Mean Speed (mph)	24	10.0	10.4	9.9	9.0	7.9	6.8	6.4	6.8	6.8	6.5	7.4	8.9	8.1
Prevailing Direction through 1963		NE	NE	NE	NE	NE	NE	SW	SW	SW	W	NE	NE	NE
Fastest Mile														
-Direction (!!!)	30	S	NE	NE	NW	E	08	W	W	SW	SW	SW	SW	SW
-Speed (MPH)	30	35	68	50	34	33	40	45	55	46	56	72	72	72
-Year		1985	1970	1961	1962	1962	1990	1979	1962	1977	1979	1960	1960	NOV 1960
Peak Gust														
-Direction (!!!)	10	NE	E	NE	NE	SE	E	SW	W	SW	SW	E	W	E
-Speed (mph)	10	39	38	45	44	33	63	40	45	39	43	60	54	63
-Date		1989	1986	1991	1989	1989	1990	1986	1993	1986	1985	1990	1993	JUN 1990

PRECIPITATION (inches)

YAP ISLAND, PACIFIC

YEAR	JAN	FEB	MAR	APR	MAY	JUNE	JULY	AUG	SEP	OCT	NOV	DEC	ANNUAL
1964	2.37	6.47	4.01	7.61	18.23	6.74	9.44	16.72	12.55	11.69	6.19	10.98	113.00
1965	3.32	6.00	7.63	4.25	8.12	10.88	26.47	12.39	17.73	8.42	12.02	3.69	120.92
1966	4.98	1.29	2.31	1.86	6.71	12.52	17.98	9.02	9.59	7.11	8.84	9.97	92.18
1967	12.02	6.25	5.37	11.76	16.00	16.71	14.14	16.45	11.72	12.80	10.44	7.48	141.14
#1968	10.77	8.04	3.72	1.82	3.94	5.76	14.24	10.90	10.66	11.21	3.59	8.34	92.99
1969	4.10	1.24	2.08	3.03	7.69	8.78	34.71	11.58	17.03	11.48	9.76	8.32	119.80
1970	4.64	6.17	4.67	3.04	9.76	8.76	8.80	25.45	11.04	12.31	9.56	8.15	112.35
1971	10.42	10.11	13.48	12.25	12.84	13.94	14.12	12.15	13.87	15.15	10.26	9.71	148.30
1972	6.03	10.42	14.21	8.97	5.33	10.18	9.20	11.09	17.60	5.64	9.35	5.14	113.16
1973	2.14	1.00	1.54	5.62	5.98	12.35	10.11	5.13	17.64	14.92	10.57	7.03	94.03
1974	11.84	4.27	9.99	10.07	9.77	14.30	14.40	12.33	9.48	19.11	18.85	13.30	147.71
1975	19.48	1.20	3.12	10.73	9.09	10.67	8.38	11.90	11.25	12.67	6.79	10.93	116.21
1976	7.36	3.19	8.76	6.77	12.52	13.30	11.43	16.29	13.44	2.59	8.88	9.97	114.50
1977	3.94	2.18	2.42	0.91	10.36	7.49	17.21	13.99	18.73	5.76	9.47	11.64	104.10
1978	4.22	5.25	2.04	5.38	4.87	12.89	8.67	18.52	19.17	18.10	11.09	8.98	119.18
1979	3.88	3.16	7.06	3.98	8.82	21.07	14.44	19.57	9.59	12.18	7.34	13.40	124.49
1980	2.32	4.60	6.42	7.72	10.57	13.52	17.84	9.52	12.71	13.41	7.20	14.52	120.35
1981	12.90	8.00	2.89	1.10	5.05	10.77	18.54	13.61	19.03	14.22	10.12	11.01	127.24
1982	7.30	12.58	7.50	2.62	10.49	32.01	13.04	14.26	13.93	9.34	4.95	7.01	135.03
1983	1.25	0.27	2.76	1.36	3.59	6.98	16.14	16.59	12.59	8.37	13.56	5.38	88.84
1984	5.33	9.59	3.90	2.21	1.77	12.38	9.59	15.33	6.41	17.29	12.03	5.44	101.27
1985	14.46	3.27	6.70	8.83	6.81	18.65	11.52	15.49	17.34	10.31	5.79	14.32	133.49
1986	7.53	10.61	10.90	6.94	9.59	13.08	15.36	11.25	12.31	7.62	14.07	5.88	125.14
1987	5.96	4.91	1.96	4.80	3.96	11.04	15.10	27.87	5.32	6.70	6.47	2.74	96.83
1988	3.58	3.63	2.80	3.93	7.25	10.49	13.79	5.35	14.60	22.12	9.02	8.94	105.60
1989	10.09	12.25	7.16	4.66	10.66	13.56	13.55	17.75	11.72	13.45	7.14	8.70	130.69
1990	6.21	2.33	3.18	5.83	12.52	22.96	12.29	22.76	14.56	7.56	9.44	2.22	121.86
1991	8.46	2.09	2.46	3.65	4.92	13.42	17.21	13.49	15.45	17.33	6.78	3.73	108.99
1992	4.20	1.50	2.97	1.31	1.48	7.59	12.98	16.21	10.32	22.43	2.77	6.22	89.98
1993	6.34	6.68	11.53	5.83	1.47	11.53	15.69	14.30	12.08	10.27	9.43	13.64	118.79
Record Mean	7.70	5.54	5.90	5.76	8.68	12.22	13.91	14.70	13.23	12.69	9.52	9.24	119.10

AVERAGE TEMPERATURE (deg. F)

YAP ISLAND, PACIFIC

YEAR	JAN	FEB	MAR	APR	MAY	JUNE	JULY	AUG	SEP	OCT	NOV	DEC	ANNUAL
1964	80.5	80.0	80.7	81.4	81.1	81.0	81.4	80.7	81.7	81.5	81.6	80.7	81.0
1965	81.1	80.5	79.9	81.2	82.0	81.0	79.0	80.7	80.8	81.3	81.0	81.2	80.8
1966	79.6	80.3	80.5	82.1	82.2	81.6	81.1	81.2	81.5	82.2	81.6	80.4	81.2
1967	79.8	80.3	80.1	80.7	81.5	80.9	80.7	80.6	81.7	81.3	80.9	80.5	80.8
#1968	79.9	80.0	81.1	81.0	81.8	81.2	80.6	80.5	80.8	80.4	81.2	80.4	80.7
1969	80.1	79.8	80.5	81.6	81.4	81.9	80.5	80.4	80.1	80.9	80.7	80.8	80.7
1970	80.8	81.0	81.5	82.1	81.7	81.8	81.7	80.6	80.9	81.1	81.4	81.2	81.4
1971	80.3	80.7	81.5	81.1	81.0	80.5	79.9	80.2	80.8	80.0	80.6	80.9	80.6
1972	80.2	80.3	80.4	81.1	81.0	81.4	82.9	81.0	81.0	81.3	81.0	80.8	81.0
1973	79.7	81.1	81.7	82.6	82.2	81.7	80.9	81.0	80.6	80.0	81.2	80.5	81.1
1974	79.3	81.0	80.7	81.2	80.8	80.3	80.7	81.2	80.6	80.5	80.7	80.6	80.6
1975	80.1	80.8	80.8	80.8	80.7	79.9	80.2	79.6	79.7	80.6	80.7	79.9	80.3
1976	79.2	79.9	79.9	79.8	81.0	80.0	79.5	79.5	79.5	81.2	80.6	80.1	80.0
1977	80.0	80.5	81.4	82.3	81.2	81.1	80.1	80.7	80.3	81.3	81.2	80.8	80.9
1978	80.2	79.7	81.2	81.6	82.2	81.3	81.4	80.4	80.2	79.8	81.0	81.1	80.8
1979	80.5	80.2	80.8	81.7	81.5	81.4	80.1	80.2	81.4	81.0	81.3	80.4	80.9
1980	80.7	80.2	81.0	81.7	81.7	80.9	80.6	81.4	80.3	80.9	80.8	80.5	80.9
1981	79.7	80.0	80.6	81.9	82.5	80.6	80.0	80.6	81.5	81.2	81.5	81.7	81.0
1982	80.9	80.6	80.9	82.1	81.3	80.5	80.5	80.3	80.1	80.4	81.1	80.3	80.8
1983	79.3	80.0	80.4	81.0	82.2	81.8	80.8	81.0	81.1	81.8	81.3	81.5	81.0
1984	80.7	80.5	81.3	82.6	83.8	81.8	81.3	80.6	81.4	80.6	81.1	81.8	81.5
1985	80.6	81.4	81.5	81.7	82.1	81.2	80.6	80.5	81.0	81.1	81.7	81.0	81.2
1986	81.3	80.8	80.9	81.2	81.7	81.4	81.3	82.9	81.2	81.7	81.2	81.1	81.4
1987	80.4	80.3	81.0	82.2	82.4	81.9	81.5	80.5	82.4	82.0	81.9	81.8	81.5
1988	81.3	81.0	82.3	82.5	82.4	81.4	81.3	81.2	81.7	81.1	81.1	80.6	81.5
1989	80.9	80.7	80.8	81.6	81.2	80.7	80.6	80.9	80.8	80.5	81.0	80.7	80.9
1990	80.1	80.2	80.3	81.2	81.0	80.2	80.2	80.1	79.9	80.5	79.4	80.1	80.3
1991	79.4	79.5	79.2	81.2	81.4	80.4	79.9	80.1	79.5	79.4	79.9	79.6	80.0
1992	78.7	78.9	80.0	81.2	82.6	82.3	81.0	81.0	80.7	80.4	80.5	79.8	80.6
1993	78.6	78.9	78.9	80.4	82.1	81.2	80.9	79.7	79.3	80.4	80.2	79.8	80.0
Record Mean	80.3	80.5	80.9	81.7	81.8	81.3	80.9	80.8	81.0	81.1	81.2	80.8	81.0
Max	85.8	86.1	86.8	87.7	87.9	87.5	87.2	87.0	87.3	87.4	87.3	86.5	87.0
Min	74.8	74.8	75.0	75.6	75.7	75.2	74.7	74.7	74.7	74.7	75.1	75.1	75.0

REFERENCE NOTES

YAP ISLAND, PACIFIC

GENERAL
T - TRACE AMOUNT.
BLANK ENTRIES DENOTE MISSING/UNREPORTED DATA.
# INDICATES A STATION OR INSTRUMENT RELOCATION.

SPECIFIC

(a) - LENGTH OF RECORD IN YEARS, ALTHOUGH INDIVIDUAL MONTHS MAY BE MISSING.
0.* OR * - THE VALUE IS BETWEEN 0.0 AND 0.05
NORMALS - BASED ON THE 1961-1990 RECORD PERIOD.
EXTREMES - DATES ARE THE MOST RECENT OCCURRENCE
WIND DIR.- NUMERALS SHOW TENS OF DEGRESS CLOCKWISE FROM TRUE NORTH. "00" INDICATES CALM.
RESULTANT DIRECTIONS ARE GIVEN TO WHOLE DEGREES.
BOLD VALUES INDICATE EXTREME VALUES WHICH OCCURRED AFTER THE ASOS SYSTEM WAS COMMISSIONED.

RECORD = PERIOD OF RECORD
RECORD MEAN PRECIPITATION IS THE MEAN OF ALL DAILY PRECIPITATION AMOUNTS DURING THE PERIOD OF RECORD.
RECORD MAX(MIN) TEMPERATURE IS THE MEAN OF ALL DAILY MAX(MIN) TEMPERATURES DURING THE PERIOD OF RECORD.
RECORD MEAN TEMPERATURE IS THE SUM OF THE RECORD MAX AND RECORD MIN DIVIDED BY 2.
AVERAGE TEMPERATURE IS THE SUM OF THE MEAN DAILY MAX AND MIN TEMPERATURE DIVIDED BY 2.

EXCEPTIONS

JUNE 1970 TO DATE, THUNDERSTORMS AND HEAVY FOG MAY BE INCOMPLETE, DUE TO PART-TIME OPERATIONS.

RECORD MEANS ARE THROUGH THE CURRENT YEAR, BEGINNING IN 1949 FOR TEMPERATURE
1949 FOR PRECIPITATION

HEATING DEGREE DAYS Base 65 deg. F — YAP ISLAND, PACIFIC

SEASON	JULY	AUG	SEP	OCT	NOV	DEC	JAN	FEB	MAR	APR	MAY	JUNE	TOTAL
1983-84	0	0	0	0	0	0	0	0	0	0	0	0	0
1984-85	0	0	0	0	0	0	0	0	0	0	0	0	0
1985-86	0	0	0	0	0	0	0	0	0	0	0	0	0
1986-87	0	0	0	0	0	0	0	0	0	0	0	0	0
1987-88	0	0	0	0	0	0	0	0	0	0	0	0	0
1988-89	0	0	0	0	0	0	0	0	0	0	0	0	0
1989-90	0	0	0	0	0	0	0	0	0	0	0	0	0
1990-91	0	0	0	0	0	0	0	0	0	0	0	0	0
1991-92	0	0	0	0	0	0	0	0	0	0	0	0	0
1992-93	0	0	0	0	0	0	0	0	0	0	0	0	0
1993-94	0	0	0	0	0	0							

COOLING DEGREE DAYS Base 65 deg. F — YAP ISLAND, PACIFIC

YEAR	JAN	FEB	MAR	APR	MAY	JUNE	JULY	AUG	SEP	OCT	NOV	DEC	TOTAL
1969	476	422	487	502	517	510	491	480	460	500	480	498	5823
1970	496	454	521	519	526	512	528	492	485	507	499	510	6049
1971	480	443	520	490	504	469	468	478	481	470	474	498	5775
1972	478	449	483	491	503	503	561	505	490	509	488	496	5956
1973	461	457	525	535	538	509	500	502	476	472	495	487	5957
1974	452	455	497	493	501	466	494	511	474	489	477	490	5799
1975	474	451	498	482	491	451	479	459	448	489	477	469	5668
1976	449	437	469	450	502	454	453	455	439	508	474	472	5562
1977	474	440	516	528	508	490	478	493	466	512	493	496	5894
1978	476	417	508	508	538	498	513	486	462	467	488	507	5868
1979	490	434	499	510	518	500	475	477	501	505	493	485	5887
1980	496	449	503	510	522	483	489	513	467	498	480	486	5896
1981	458	428	490	514	548	475	472	491	500	509	501	525	5911
1982	500	442	502	520	511	471	489	479	463	484	490	480	5831
1983	451	426	485	484	540	510	496	506	490	530	495	519	5932
1984	495	454	514	534	591	508	511	490	500	489	491	529	6106
1985	486	465	518	506	536	494	488	485	491	506	510	504	5989
1986	516	448	498	493	525	499	512	563	492	523	492	505	6066
1987	486	437	500	521	549	513	517	491	531	533	513	528	6119
1988	513	470	544	529	546	499	516	508	510	506	488	490	6119
1989	502	447	496	504	509	478	491	499	480	486	485	496	5873
1990	473	431	481	492	502	464	479	474	454	489	438	475	5652
1991	453	413	445	493	515	473	470	476	443	454	453	463	5551
1992	433	410	474	493	552	523	502	505	476	484	470	465	5787
1993	432	397	441	469	537	491	500	461	436	482	463	468	5577

SNOWFALL (inches) — YAP ISLAND, PACIFIC

SEASON	JULY	AUG	SEP	OCT	NOV	DEC	JAN	FEB	MAR	APR	MAY	JUNE	TOTAL
1971-72	0.0	0.0	0.0	0.0	0.0	0.0	0.0	0.0	0.0	0.0	0.0	0.0	0.0
1972-73	0.0	0.0	0.0	0.0	0.0	0.0	0.0	0.0	0.0	0.0	0.0	0.0	0.0
1973-74	0.0	0.0	0.0	0.0	0.0	0.0	0.0	0.0	0.0	0.0	0.0	0.0	0.0
1974-75	0.0	0.0	0.0	0.0	0.0	0.0	0.0	0.0	0.0	0.0	0.0	0.0	0.0
1975-76	0.0	0.0	0.0	0.0	0.0	0.0	0.0	0.0	0.0	0.0	0.0	0.0	0.0
1976-77	0.0	0.0	0.0	0.0	0.0	0.0	0.0	0.0	0.0	0.0	0.0	0.0	0.0
1977-78	0.0	0.0	0.0	0.0	0.0	0.0	0.0	0.0	0.0	0.0	0.0	0.0	0.0
1978-79	0.0	0.0	0.0	0.0	0.0	0.0	0.0	0.0	0.0	0.0	0.0	0.0	0.0
1979-80	0.0	0.0	0.0	0.0	0.0	0.0	0.0	0.0	0.0	0.0	0.0	0.0	0.0
1980-81	0.0	0.0	0.0	0.0	0.0	0.0	0.0	0.0	0.0	0.0	0.0	0.0	0.0
1981-82	0.0	0.0	0.0	0.0	0.0	0.0	0.0	0.0	0.0	0.0	0.0	0.0	0.0
1982-83	0.0	0.0	0.0	0.0	0.0	0.0	0.0	0.0	0.0	0.0	0.0	0.0	0.0
1983-84	0.0	0.0	0.0	0.0	0.0	0.0	0.0	0.0	0.0	0.0	0.0	0.0	0.0
1984-85	0.0	0.0	0.0	0.0	0.0	0.0	0.0	0.0	0.0	0.0	0.0	0.0	0.0
1985-86	0.0	0.0	0.0	0.0	0.0	0.0	0.0	0.0	0.0	0.0	0.0	0.0	0.0
1986-87	0.0	0.0	0.0	0.0	0.0	0.0	0.0	0.0	0.0	0.0	0.0	0.0	0.0
1987-88	0.0	0.0	0.0	0.0	0.0	0.0	0.0	0.0	0.0	0.0	0.0	0.0	0.0
1988-89	0.0	0.0	0.0	0.0	0.0	0.0	0.0	0.0	0.0	0.0	0.0	0.0	0.0
1989-90	0.0	0.0	0.0	0.0	0.0	0.0	0.0	0.0	0.0	0.0	0.0	0.0	0.0
1990-91	0.0	0.0	0.0	0.0	0.0	0.0	0.0	0.0	0.0	0.0	0.0	0.0	0.0
1991-92	0.0	0.0	0.0	0.0	0.0	0.0	0.0	0.0	0.0	0.0	0.0	0.0	0.0
1992-93	0.0	0.0	0.0	0.0	0.0	0.0	0.0	0.0	0.0	0.0	0.0	0.0	0.0
1993-94	0.0	0.0	0.0	0.0	0.0	0.0							
Record Mean	0.0	0.0	0.0	0.0	0.0	0.0	0.0	0.0	0.0	0.0	0.0	0.0	0.0

ALLENTOWN, PENNSYLVANIA

Allentown is located in the east central section of the state and in the Lehigh River valley. Twelve miles to the north is Blue Mountain, a ridge from 1,000 to 1,800 feet in height. The South Mountain, 500 to 1,000 feet high, fringes the southern edge of the city. Otherwise the country is generally rolling with numerous small streams. A modified climate prevails. Temperatures are usually moderate and precipitation generally ample and dependable with the largest amounts occurring during the summer months when precipitation is generally showery. General climatological features of the area are slightly modified by the mountain ranges so that at times during the winter there is a temperature difference of 10 to 15 degrees between Allentown and Philadelphia, only 50 miles to the south.

The growing season averages 177 days, and generally ranges from 170 to 185 days. It begins late in April and ends late in October. The average occurrence of the last temperature of 32 degrees in the spring is late April, and the average first fall minimum of 32 degrees is mid-October.

Maximum temperatures during most years are not excessively high and temperatures above 100 degrees are seldom recorded. However, the average humidity in the valley is quite high, and combined with the normal summer temperatures, causes periods of discomfort.

Winters in the valley are comparatively mild. Minimum temperatures during December, January, and February are usually below freezing, but below zero temperatures are seldom recorded.

Seasonal snowfall is quite variable. Freezing rain is a common problem throughout the Lehigh Valley. Snowstorms producing 10 inches or more occur an average of once in two years. The accumulation of snowfall over the drainage area of the Lehigh River to the north of Allentown, combined with spring rains, frequently presents a flood threat to the city and surrounding area. The valley is also subject to torrential rains that cause quick rises in the river and feeder creeks.

The area is seldom subject to destructive storms of large extent. Heavy thunderstorms and tornadoes occasionally cause damage over limited areas. An exception to the usual weather was the storm that battered the east coast on November 25, 1950, when gusts of 88 mph were observed at the station.

NORMALS, MEANS, AND EXTREMES

ALLENTOWN, PENNSYLVANIA

LATITUDE: 40°39'N LONGITUDE: 75°26' W ELEVATION: FT. GRND 387 BARO 379 TIME ZONE: EASTERN WBAN: 14737

	(a)	JAN	FEB	MAR	APR	MAY	JUNE	JULY	AUG	SEP	OCT	NOV	DEC	YEAR
TEMPERATURE °F:														
Normals														
-Daily Maximum		34.3	37.7	48.8	60.4	71.3	80.0	84.5	82.3	75.1	63.8	51.8	39.2	60.8
-Daily Minimum		18.8	20.9	29.9	38.8	49.3	58.8	63.6	62.0	54.2	42.5	34.2	24.4	41.5
-Monthly		26.6	29.3	39.4	49.7	60.3	69.4	74.1	72.2	64.7	53.2	43.1	31.8	51.2
Extremes														
-Record Highest	50	72	76	86	93	97	100	105	100	99	90	81	72	105
-Year		1950	1985	1945	1976	1962	1966	1966	1955	1980	1951	1950	1984	JUL 1966
-Record Lowest	50	-12	-7	-1	16	28	39	48	41	30	21	11	-8	-12
-Year		1961	1967	1967	1982	1947	1972	1976	1986	1947	1988	1976	1950	JAN 1961
NORMAL DEGREE DAYS:														
Heating (base 65°F)		1190	1000	794	459	183	10	0	8	82	373	657	1029	5785
Cooling (base 65°F)		0	0	0	0	38	142	282	231	73	7	0	0	773
% OF POSSIBLE SUNSHINE	9	47	49	53	48	54	63	55	61	58	57	51	45	53
MEAN SKY COVER (tenths)														
Sunrise - Sunset	49	6.7	6.4	6.4	6.4	6.4	6.0	5.9	5.8	5.8	5.6	6.6	6.7	6.2
MEAN NUMBER OF DAYS:														
Sunrise to Sunset														
-Clear	50	6.9	7.1	7.7	6.9	6.7	7.6	8.3	8.8	9.3	10.4	6.8	6.9	93.2
-Partly Cloudy	50	8.0	7.9	8.5	8.8	10.7	11.1	11.9	11.1	8.9	8.9	8.0	7.4	111.1
-Cloudy	50	16.1	13.3	14.8	14.4	13.7	11.3	10.8	11.1	11.9	11.7	15.3	16.6	161.0
Precipitation														
.01 inches or more	50	10.8	9.7	11.4	11.5	12.1	10.5	10.4	9.9	9.1	8.1	10.2	10.9	124.6
Snow,Ice pellets,hail														
1.0 inches or more	50	2.4	2.2	1.5	0.2	0.0	0.0	0.0	0.0	0.0	0.*	0.4	1.9	8.6
Thunderstorms	50	0.2	0.3	0.8	2.1	4.3	5.9	6.9	5.9	3.0	1.0	0.7	0.1	31.2
Heavy Fog Visibility														
1/4 mile or less	50	2.7	2.5	2.5	1.4	1.8	1.2	1.0	1.7	2.7	3.0	2.6	2.9	26.0
Temperature °F														
-Maximum														
90° and above	50	0.0	0.0	0.0	0.1	0.6	3.5	6.8	3.9	1.1	0.*	0.0	0.0	16.2
32° and below	50	11.0	7.4	1.4	0.*	0.0	0.0	0.0	0.0	0.0	0.0	0.3	7.2	27.4
-Minimum														
32° and below	50	27.9	24.9	20.1	6.2	0.3	0.0	0.0	0.0	0.1	3.8	14.1	25.7	122.9
0° and below	50	1.3	0.5	0.*	0.0	0.0	0.0	0.0	0.0	0.0	0.0	0.0	0.4	2.2
AVG. STATION PRESS.(mb)	21	1003.9	1004.0	1002.6	1001.1	1001.6	1001.6	1002.3	1003.8	1004.9	1005.3	1004.8	1004.8	1003.4
RELATIVE HUMIDITY (%)														
Hour 01	43	74	73	71	72	77	81	82	85	86	83	78	76	78
Hour 07 (Local Time)	43	76	76	75	75	77	79	82	86	88	86	81	79	80
Hour 13	43	61	58	53	50	52	53	53	56	57	56	59	63	56
Hour 19	43	68	64	60	57	59	60	61	66	70	68	69	70	64
PRECIPITATION (inches):														
Water Equivalent														
-Normal		3.16	2.95	3.28	3.52	4.20	3.75	4.14	4.28	3.93	2.94	3.88	3.49	43.52
-Maximum Monthly	50	8.42	5.44	7.21	10.09	10.62	8.58	10.42	12.10	8.87	6.84	9.69	7.89	12.10
-Year		1979	1971	1953	1952	1984	1972	1969	1955	1987	1955	1972	1973	AUG 1955
-Minimum Monthly	50	0.67	1.01	0.97	0.61	0.09	0.34	0.42	0.93	0.94	0.15	0.68	0.39	0.09
-Year		1981	1980	1981	1985	1964	1949	1955	1980	1967	1963	1976	1955	MAY 1964
-Maximum in 24 hrs	50	2.47	2.05	3.08	2.52	3.36	3.55	4.54	5.84	7.85	2.96	3.40	3.04	7.85
-Year		1949	1966	1952	1968	1984	1967	1969	1982	1985	1955	1972	1993	SEP 1985
Snow,Ice pellets,hail														
-Maximum Monthly	50	24.1	29.5	30.5	13.4	T	0.0	0.0	0.0	0.0	1.4	7.8	28.4	30.5
-Year		1966	1983	1958	1982	1977					1972	1967	1966	MAR 1958
-Maximum in 24 hrs	50	16.0	25.2	17.5	11.4	T	0.0	0.0	0.0	0.0	1.4	6.4	13.3	25.2
-Year		1961	1983	1958	1982	1977					1972	1968	1966	FEB 1983
WIND:														
Mean Speed (mph)	44	10.5	10.9	11.4	10.8	9.0	8.2	7.2	6.9	7.2	8.2	9.6	10.0	9.1
Prevailing Direction through 1963		W	WNW	WNW	W	WSW	SW	WSW	SW	SW	WSW	W	W	W
Fastest Obs. 1 Min.														
-Direction (!!)	45	29	25	29	29	30	27	27	23	25	14	30	29	27
-Speed (MPH)	45	55	58	58	60	58	81	55	58	46	49	58	52	81
-Year		1959	1956	1955	1963	1964	1964	1951	1949	1956	1954	1989	1956	JUN 1964
Peak Gust														
-Direction (!!)	10	W	W	NW	NW	SE	NW	W	NW	W	SE	NW	NW	NW
-Speed (mph)	10	54	60	55	61	48	58	66	48	60	46	78	62	78
-Date		1988	1991	1993	1991	1989	1993	1984	1988	1986	1990	1989	1988	NOV 1989

PRECIPITATION (inches)

ALLENTOWN, PENNSYLVANIA

YEAR	JAN	FEB	MAR	APR	MAY	JUNE	JULY	AUG	SEP	OCT	NOV	DEC	ANNUAL
1964	4.67	3.29	1.95	7.01	0.09	3.82	2.29	1.26	2.17	1.37	2.34	4.48	34.74
1965	2.70	3.21	2.69	2.10	1.34	1.81	1.58	5.92	2.29	3.46	1.66	1.79	30.55
1966	4.20	4.84	1.44	3.75	2.53	0.89	1.91	2.68	6.23	2.77	3.05	3.99	38.28
1967	1.76	1.98	6.68	2.61	4.44	4.10	5.85	5.00	0.94	2.30	3.00	5.13	43.79
1968	2.54	1.31	3.15	3.14	6.74	3.59	1.60	2.74	2.64	2.99	4.32	3.67	38.43
1969	1.47	1.61	1.93	3.52	1.90	3.20	10.42	3.33	3.87	1.62	2.59	6.43	41.89
1970	0.69	2.40	3.98	4.65	2.97	2.91	5.40	1.69	2.96	5.06	6.08	2.80	41.59
1971	2.73	5.44	3.52	1.65	5.30	2.66	3.04	9.25	3.71	3.57	6.31	1.32	48.50
1972	2.79	4.05	3.37	2.78	6.20	8.58	5.85	2.26	1.27	3.59	9.69	5.42	55.85
1973	3.58	2.65	2.80	5.94	5.51	5.36	2.16	2.36	6.18	2.07	1.67	7.89	48.17
1974	3.66	1.60	5.38	3.98	3.85	4.31	1.40	9.42	6.67	1.58	1.82	4.52	48.19
1975	5.17	3.57	3.49	3.50	4.25	5.09	7.71	4.23	7.60	4.84	3.37	2.72	55.54
1976	5.03	2.80	2.64	2.16	3.12	3.57	2.11	5.09	4.60	5.70	0.68	2.40	39.90
1977	1.42	2.54	5.90	4.23	1.75	3.66	4.33	5.79	4.10	4.21	6.13	5.54	49.60
1978	7.42	1.60	3.94	1.55	5.52	3.15	3.64	8.95	2.28	1.96	2.02	3.96	45.99
1979	8.42	4.49	2.35	3.74	4.85	1.20	5.57	2.07	7.19	4.36	2.99	2.48	49.71
1980	0.81	1.01	5.84	4.39	3.38	2.90	1.50	0.93	1.99	3.09	3.23	0.75	29.82
1981	0.67	4.79	0.97	3.43	4.82	5.72	2.71	1.60	2.74	2.82	1.63	3.18	35.08
1982	3.48	2.63	1.99	4.73	4.02	7.16	2.86	8.43	1.76	1.56	3.40	1.38	43.40
1983	2.84	3.21	5.35	7.87	4.21	4.68	2.02	2.38	1.41	4.83	6.29	7.61	52.70
1984	1.53	3.64	3.82	5.08	10.62	4.20	9.26	4.17	1.10	4.26	1.66	3.07	52.41
1985	0.84	2.28	1.56	0.61	5.47	5.06	7.61	3.46	8.72	1.97	7.74	1.71	47.03
1986	4.27	4.29	2.07	4.31	1.83	2.51	4.46	3.77	3.07	1.91	6.32	4.39	43.20
1987	3.89	1.11	2.10	4.42	2.28	4.42	4.41	7.09	8.87	2.64	3.98	1.49	46.70
1988	2.75	3.28	2.56	1.36	7.43	1.79	6.17	2.72	3.21	2.18	4.98	0.99	39.42
1989	1.79	2.18	3.82	1.31	9.52	5.81	4.09	3.05	5.93	4.30	1.90	1.31	45.01
1990	4.57	2.38	1.52	2.31	6.66	2.86	3.20	6.47	1.97	3.12	3.53	5.68	44.27
1991	2.77	1.14	3.71	3.11	2.90	1.06	5.28	2.54	3.32	2.57	2.48	3.50	34.38
1992	1.73	1.49	3.32	0.88	4.88	3.36	5.36	4.08	3.42	1.86	5.06	3.27	38.71
1993	1.98	2.81	6.59	6.81	1.43	4.09	2.99	5.39	7.77	5.31	2.83	4.96	52.96
Record Mean	3.13	2.84	3.56	3.74	4.18	3.71	4.35	4.28	4.03	2.87	3.74	3.55	43.98

AVERAGE TEMPERATURE (deg. F)

ALLENTOWN, PENNSYLVANIA

YEAR	JAN	FEB	MAR	APR	MAY	JUNE	JULY	AUG	SEP	OCT	NOV	DEC	ANNUAL
1964	29.2	27.5	39.1	47.4	62.2	68.8	73.9	69.6	65.8	50.8	44.8	32.8	51.0
1965	24.3	30.2	36.2	47.2	64.3	68.3	72.4	71.1	65.8	50.9	41.3	33.9	50.5
1966	26.2	28.4	39.0	45.7	56.5	71.1	77.0	74.0	62.4	49.8	43.5	30.3	50.3
1967	31.3	24.3	35.1	48.8	52.8	70.6	72.7	69.1	61.3	51.1	37.3	32.6	48.9
1968	21.7	25.5	40.6	52.0	56.2	67.2	74.5	73.8	66.5	55.6	42.3	29.6	50.4
1969	28.0	30.2	37.3	52.6	62.4	70.0	72.2	72.0	63.6	51.5	40.9	28.4	50.8
1970	19.2	27.9	34.4	48.4	60.8	67.5	73.8	72.9	67.3	55.5	44.0	30.7	50.2
1971	22.0	30.0	35.7	47.7	56.9	70.2	72.6	70.8	67.7	58.9	40.9	36.5	50.8
1972	30.4	26.8	37.2	47.1	61.4	66.6	74.9	71.1	64.8	48.2	39.9	35.0	50.3
1973	30.7	29.3	44.3	50.0	56.2	71.6	75.2	75.3	65.9	55.6	45.0	35.1	52.8
1974	31.3	30.2	40.3	52.5	59.1	66.6	73.3	72.4	62.5	49.6	43.5	35.2	51.4
1975	31.6	32.0	37.6	45.5	64.1	69.2	73.7	72.5	61.3	56.9	48.8	32.7	52.1
1976	24.1	36.6	41.8	52.3	58.3	72.6	71.6	71.0	63.4	50.2	37.6	27.0	50.5
1977	18.6	30.8	45.4	52.4	63.9	68.7	75.5	73.0	66.7	53.3	44.2	30.0	51.9
1978	24.9	22.3	36.7	50.5	60.8	70.8	72.4	73.3	64.2	52.9	44.3	33.5	50.6
1979	28.0	20.5	44.3	49.5	62.1	68.0	73.2	72.8	65.3	52.5	47.8	36.2	51.7
1980	30.7	27.3	38.6	52.9	63.8	68.2	77.3	78.2	70.3	54.4	41.4	29.6	52.7
1981	23.2	35.3	38.4	51.7	61.2	71.6	74.7	70.6	63.6	49.8	43.3	30.6	51.2
1982	19.9	29.9	37.3	47.2	62.5	65.7	73.5	68.3	64.1	54.0	45.5	38.5	50.5
1983	30.2	32.0	41.9	49.2	58.3	70.0	76.4	75.5	66.5	54.3	43.5	28.6	52.2
1984	23.0	36.7	33.7	49.3	58.7	72.0	73.2	74.1	62.4	59.5	43.4	38.4	52.0
1985	24.9	32.2	42.3	54.2	63.6	67.3	73.3	71.9	66.2	54.2	46.1	29.4	52.1
1986	29.3	28.1	41.1	50.5	64.6	69.1	73.9	69.6	64.1	54.0	39.5	33.8	51.5
1987	27.8	29.2	42.6	50.8	61.6	71.4	75.9	70.5	64.3	49.1	43.4	34.9	51.8
1988	22.7	29.7	40.7	48.5	61.6	68.7	77.9	75.0	62.8	48.2	44.5	31.7	51.0
1989	33.4	30.4	39.0	48.5	59.3	70.2	72.7	71.5	65.5	55.0	42.1	21.2	50.7
1990	37.1	36.3	43.3	51.9	58.8	70.0	74.2	71.8	63.1	57.5	45.6	37.2	53.9
1991	29.5	35.9	42.3	52.6	67.2	71.6	74.7	74.1	64.7	54.7	43.8	35.7	53.9
1992	32.2	33.9	37.4	48.8	59.3	68.5	72.4	69.4	63.7	49.3	43.1	34.0	51.0
1993	33.3	26.9	35.6	50.6	63.0	70.5	76.9	74.4	64.1	51.7	43.1	33.1	51.9
Record Mean	27.7	30.0	39.0	49.9	60.4	69.4	74.1	72.1	64.6	53.6	42.8	31.7	51.3
Max	35.5	38.4	48.5	60.8	71.5	80.3	84.8	82.5	75.1	64.5	51.8	39.4	61.1
Min	19.9	21.5	29.4	39.0	49.2	58.4	63.4	61.6	54.0	42.7	33.8	23.9	41.4

REFERENCE NOTES

ALLENTOWN, PENNSYLVANIA

GENERAL
T - TRACE AMOUNT.
BLANK ENTRIES DENOTE MISSING/UNREPORTED DATA.
# INDICATES A STATION OR INSTRUMENT RELOCATION.

SPECIFIC

(a) - LENGTH OF RECORD IN YEARS, ALTHOUGH INDIVIDUAL MONTHS MAY BE MISSING.
0.* OR * - THE VALUE IS BETWEEN 0.0 AND 0.05
NORMALS - BASED ON THE 1961-1990 RECORD PERIOD.
EXTREMES - DATES ARE THE MOST RECENT OCCURRENCE
WIND DIR.- NUMERALS SHOW TENS OF DEGRESS CLOCKWISE FROM TRUE NORTH. "00" INDICATES CALM.
RESULTANT DIRECTIONS ARE GIVEN TO WHOLE DEGREES.
BOLD VALUES INDICATE EXTREME VALUES WHICH OCCURRED AFTER THE ASOS SYSTEM WAS COMMISSIONED.

RECORD = PERIOD OF RECORD
RECORD MEAN PRECIPITATION IS THE MEAN OF ALL DAILY PRECIPITATION AMOUNTS DURING THE PERIOD OF RECORD.
RECORD MAX(MIN) TEMPERATURE IS THE MEAN OF ALL DAILY MAX(MIN) TEMPERATURES DURING THE PERIOD OF RECORD.
RECORD MEAN TEMPERATURE IS THE SUM OF THE RECORD MAX AND RECORD MIN DIVIDED BY 2.
AVERAGE TEMPERATURE IS THE SUM OF THE MEAN DAILY MAX AND MIN TEMPERATURE DIVIDED BY 2.

EXCEPTIONS

RECORD MEANS ARE THROUGH THE CURRENT YEAR, BEGINNING IN 1944 FOR TEMPERATURE
1944 FOR PRECIPITATION
1944 FOR SNOWFALL

HEATING DEGREE DAYS Base 65 deg. F — ALLENTOWN, PENNSYLVANIA

SEASON	JULY	AUG	SEP	OCT	NOV	DEC	JAN	FEB	MAR	APR	MAY	JUNE	TOTAL
1964-65	1	15	78	436	599	990	1256	969	889	526	92	44	5895
1965-66	0	23	81	432	705	957	1197	1017	801	573	272	29	6087
1966-67	0	0	136	463	640	1070	1038	1134	921	484	375	12	6273
1967-68	0	20	146	431	825	1000	1336	1139	750	386	266	34	6333
1968-69	0	7	33	293	670	1091	1141	967	852	369	133	14	5570
1969-70	3	5	125	420	717	1129	1409	1032	940	490	169	35	6474
1970-71	0	0	64	292	621	1056	1326	969	900	512	250	24	6014
1971-72	0	8	52	191	719	877	1064	1102	853	531	127	54	5578
1972-73	4	11	69	513	745	924	1055	995	633	446	275	6	5676
1973-74	0	2	72	290	593	922	1036	968	755	385	216	32	5271
1974-75	2	0	123	471	638	917	1027	919	841	577	98	19	5632
1975-76	0	7	122	257	482	995	1262	820	713	414	219	28	5319
1976-77	1	9	101	455	817	1174	1428	950	604	384	127	35	6085
1977-78	0	10	56	356	623	1077	1238	1190	871	430	189	15	6055
1978-79	9	0	94	368	614	971	1142	1241	633	460	125	29	5686
1979-80	13	22	76	386	509	884	1058	1087	808	356	87	41	5327
1980-81	0	0	26	326	702	1092	1290	823	818	392	163	6	5638
1981-82	0	3	100	465	643	1058	1389	975	851	528	110	50	6172
1982-83	5	28	82	347	575	815	1071	920	708	477	222	23	5273
1983-84	0	1	96	336	639	1121	1295	813	963	464	211	12	5951
1984-85	0	1	135	180	642	820	1240	914	702	346	106	28	5114
1985-86	0	3	85	324	559	1093	1098	1025	735	427	115	20	5484
1986-87	0	35	92	361	759	960	1143	997	688	423	178	5	5641
1987-88	0	16	73	487	637	926	1306	1016	747	486	145	52	5891
1988-89	5	6	97	515	607	1026	974	963	798	489	204	12	5696
1989-90	4	7	92	309	680	1350	858	800	670	416	188	17	5391
1990-91	7	12	129	269	574	853	1092	806	696	374	92	10	4914
1991-92	0	0	113	326	631	902	1007	895	845	478	197	23	5417
1992-93	4	9	110	482	649	955	974	1060	904	424	103	27	5701
1993-94	0	3	121	407	652	985							

COOLING DEGREE DAYS Base 65 deg. F — ALLENTOWN, PENNSYLVANIA

YEAR	JAN	FEB	MAR	APR	MAY	JUNE	JULY	AUG	SEP	OCT	NOV	DEC	TOTAL
1969	0	0	0	5	58	172	232	230	88	6	0	0	791
1970	0	0	0	2	44	116	278	249	141	6	0	0	836
1971	0	0	0	0	4	184	239	194	140	10	4	0	775
1972	0	0	0	0	18	106	318	208	69	0	0	0	719
1973	0	0	0	3	7	212	322	331	103	4	0	0	982
1974	0	0	0	16	40	86	266	236	52	0	0	0	696
1975	0	0	0	0	79	154	279	247	18	13	1	0	791
1976	0	0	0	37	20	262	214	203	59	2	0	0	797
1977	0	0	4	13	99	155	331	263	115	1	5	0	986
1978	0	0	0	0	65	196	247	264	76	0	0	0	848
1979	0	0	0	1	42	126	274	268	92	7	1	0	811
1980	0	0	0	0	59	147	388	413	194	6	0	0	1207
1981	0	0	0	0	50	210	307	183	61	0	0	0	811
1982	0	0	0	1	37	76	274	137	63	13	0	0	601
1983	0	0	0	11	21	179	361	337	148	13	0	0	1070
1984	0	0	0	0	25	226	260	291	65	18	0	0	885
1985	0	0	2	27	66	100	262	222	125	1	0	0	805
1986	0	0	0	0	111	152	282	187	72	23	0	0	827
1987	0	0	0	3	78	204	344	194	58	0	0	0	881
1988	0	0	0	0	48	172	412	324	36	3	0	0	995
1989	0	0	0	0	37	173	248	218	116	6	1	0	799
1990	0	0	5	31	6	171	299	228	78	42	0	0	860
1991	0	0	0	9	167	215	306	290	110	15	0	0	1112
1992	0	0	0	0	27	135	241	151	79	0	0	0	633
1993	0	0	0	1	49	198	375	302	103	0	0	0	1028

SNOWFALL (inches) — ALLENTOWN, PENNSYLVANIA

SEASON	JULY	AUG	SEP	OCT	NOV	DEC	JAN	FEB	MAR	APR	MAY	JUNE	TOTAL
1964-65	0.0	0.0	0.0	0.0	T	2.2	11.7	1.9	5.1	1.7	0.0	0.0	22.6
1965-66	0.0	0.0	0.0	T	T	T	24.1	15.5	0.1	0.0	T	0.0	39.7
1966-67	0.0	0.0	0.0	0.0	T	28.4	0.7	19.7	17.4	1.0	0.0	0.0	67.2
1967-68	0.0	0.0	0.0	0.0	7.8	8.4	3.0	0.6	0.1	0.0	0.0	0.0	19.9
1968-69	0.0	0.0	0.0	0.0	7.2	3.7	1.4	11.8	8.4	0.0	0.0	0.0	32.5
1969-70	0.0	0.0	0.0	0.0	T	20.0	7.8	4.3	14.8	T	0.0	0.0	46.9
1970-71	0.0	0.0	0.0	T	T	6.8	11.2	4.4	6.8	0.5	0.0	0.0	29.7
1971-72	0.0	0.0	0.0	0.0	6.1	1.4	4.5	14.8	1.4	0.4	0.0	0.0	28.6
1972-73	0.0	0.0	0.0	1.4	2.3	0.6	1.9	1.0	0.2	T	0.0	0.0	7.4
1973-74	0.0	0.0	0.0	0.0	T	7.2	7.2	4.3	0.6	0.1	0.0	0.0	19.4
1974-75	0.0	0.0	0.0	0.0	T	3.2	8.3	9.6	2.1	T	0.0	0.0	23.2
1975-76	0.0	0.0	0.0	0.0	T	1.6	6.3	4.3	6.7	T	0.0	0.0	18.9
1976-77	0.0	0.0	0.0	0.0	0.4	6.0	11.6	8.4	1.5	T	T	0.0	27.9
1977-78	0.0	0.0	0.0	T	2.5	3.9	20.9	19.9	8.1	0.3	0.0	0.0	55.6
1978-79	0.0	0.0	0.0	0.0	2.9	0.9	9.1	21.0	T	T	0.0	0.0	33.9
1979-80	0.0	0.0	0.0	1.0	T	4.5	1.0	5.9	9.5	T	0.0	0.0	21.9
1980-81	0.0	0.0	0.0	0.0	4.2	3.6	9.5	1.3	6.9	0.0	0.0	0.0	25.5
1981-82	0.0	0.0	0.0	0.0	T	8.9	12.4	5.2	4.0	13.4	0.0	0.0	43.9
1982-83	0.0	0.0	0.0	0.0	T	2.3	8.8	29.5	0.6	4.6	0.0	0.0	45.8
1983-84	0.0	0.0	0.0	0.0	T	5.0	12.8	0.6	15.9	T	0.0	0.0	34.3
1984-85	0.0	0.0	0.0	0.0	T	5.7	6.7	11.5	0.3	T	0.0	0.0	24.2
1985-86	0.0	0.0	0.0	0.0	T	2.7	5.7	16.5	T	1.1	0.0	0.0	26.0
1986-87	0.0	0.0	0.0	0.0	0.5	2.4	21.9	9.6	1.1	T	0.0	0.0	35.5
1987-88	0.0	0.0	0.0	0.0	3.7	2.2	21.0	11.6	1.7	T	0.0	0.0	40.2
1988-89	0.0	0.0	0.0	0.0	0.0	0.1	3.9	2.8	5.4	0.0	0.0	0.0	12.2
1989-90	0.0	0.0	0.0	0.0	3.0	7.3	8.8	2.0	0.9	0.1	0.0	0.0	22.1
1990-91	0.0	0.0	0.0	0.0	T	8.8	5.0	0.9	2.1	T	0.0	0.0	16.8
1991-92	0.0	0.0	0.0	0.0	T	0.8	2.8	2.2	9.8	T	0.0	0.0	15.6
1992-93	0.0	0.0	0.0	0.0	T	0.9	1.6	13.4	21.6	0.0	0.0	0.0	37.5
1993-94	0.0	0.0	0.0	0.0	T	1.9							
Record Mean	0.0	0.0	0.0	0.1	1.3	6.1	8.4	8.9	5.8	0.7	T	0.0	31.3

AVOCA, WILKES-BARRE - SCRANTON PENNSYLVANIA

The Wilkes-Barre Scranton National Weather Service Office is located about midway between the two cities, at the southwest end of the crescent-shaped Lackawanna River Valley. The river flows through this valley and empties into the Susquehanna River and the Wyoming Valley a few miles west of the airport. The surrounding mountains protect both cities and the airport from high winds. They influence the temperature and precipitation during both summer and winter, causing wide departures in both within a few miles of the station. Because of the proximity of the mountains, the climate is relatively cool in summer with frequent shower and thunderstorm activity, usually of brief duration. The winter temperatures in the valley are not severe. The occurrence of sub-zero temperatures and severe snowstorms is infrequent. A high percentage of the winter precipitation occurs as rain.

Although severe snowstorms are infrequent, when they do occur they approach blizzard conditions. High winds cause huge drifts and normal routines are disrupted for several days.

While the incidence of tornadoes is very low, Wilkes-Barre has occasionally been hit with these storms which caused loss of life and great property damage.

The area has felt the effects of tropical storms. Considerable wind damage has occasionally occurred, but the most devastating damage has come from flooding caused by the large amounts of precipitation deposited by the storms. The worst natural disaster to hit the region was the result of the flooding caused by a hurricane.

NORMALS, MEANS, AND EXTREMES

AVOCA, WILKES-BARRE - SCRANTON PENNSYLVANIA

LATITUDE: 41°20'N LONGITUDE: 75°44' W ELEVATION: FT. GRND 930 BARO 959 TIME ZONE: EASTERN WBAN: 14777

	(a)	JAN	FEB	MAR	APR	MAY	JUNE	JULY	AUG	SEP	OCT	NOV	DEC	YEAR
TEMPERATURE °F:														
Normals														
-Daily Maximum		31.8	34.5	45.5	57.8	69.3	77.5	81.8	79.7	72.4	61.0	48.8	36.6	58.1
-Daily Minimum		17.5	19.0	28.3	38.1	48.3	56.8	61.6	60.0	52.8	42.1	33.9	23.4	40.2
-Monthly		24.7	26.8	36.9	48.0	58.8	67.2	71.7	69.9	62.6	51.5	41.4	30.0	49.1
Extremes														
-Record Highest	38	67	71	83	92	93	97	101	94	95	84	80	67	101
-Year		1967	1985	1977	1976	1962	1964	1988	1983	1983	1959	1982	1984	JUL 1988
-Record Lowest	38	-14	-16	-4	14	27	34	43	38	30	19	9	-9	-16
-Year		1985	1979	1967	1982	1974	1972	1979	1982	1974	1972	1976	1989	FEB 1979
NORMAL DEGREE DAYS:														
Heating (base 65°F)		1249	1070	871	510	217	38	0	11	109	423	708	1085	6291
Cooling (base 65°F)		0	0	0	0	25	104	210	163	37	0	0	0	539
% OF POSSIBLE SUNSHINE	38	42	47	50	53	56	61	62	60	54	52	36	34	51
MEAN SKY COVER (tenths)														
Sunrise - Sunset	38	7.5	7.2	7.1	6.8	6.7	6.2	6.2	6.1	6.2	6.2	7.6	7.8	6.8
MEAN NUMBER OF DAYS:														
Sunrise to Sunset														
-Clear	38	4.4	4.5	5.4	6.2	5.9	6.9	6.3	7.1	7.1	8.3	3.9	3.8	69.7
-Partly Cloudy	38	7.2	6.9	7.7	7.3	9.6	10.8	12.6	11.7	9.7	8.5	6.8	6.7	105.4
-Cloudy	38	19.4	16.8	17.9	16.5	15.5	12.3	12.1	12.2	13.3	14.2	19.2	20.6	190.1
Precipitation														
.01 inches or more	38	12.1	11.1	12.8	12.4	12.7	11.8	11.3	11.1	10.0	9.8	11.8	12.8	139.6
Snow,Ice pellets,hail														
1.0 inches or more	38	3.1	2.9	2.7	0.7	0.*	0.0	0.0	0.0	0.0	0.1	0.9	2.6	12.9
Thunderstorms	38	0.1	0.2	0.8	2.0	3.5	5.8	6.8	5.1	2.7	1.0	0.4	0.2	28.6
Heavy Fog Visibility														
1/4 mile or less	38	1.8	2.0	1.8	1.3	1.2	1.2	1.6	1.9	2.8	2.1	1.7	2.3	21.8
Temperature °F														
-Maximum														
90° and above	38	0.0	0.0	0.0	0.1	0.3	1.6	3.2	1.8	0.4	0.0	0.0	0.0	7.5
32° and below	38	15.1	11.6	3.9	0.2	0.0	0.0	0.0	0.0	0.0	0.0	1.2	10.6	42.5
-Minimum														
32° and below	38	28.2	24.8	21.6	8.4	0.6	0.0	0.0	0.0	0.2	4.4	13.9	25.4	127.5
0° and below	38	2.1	1.3	0.1	0.0	0.0	0.0	0.0	0.0	0.0	0.0	0.0	0.7	4.1
AVG. STATION PRESS.(mb)	21	982.5	983.0	981.8	980.6	981.4	981.6	982.4	984.0	984.9	984.9	984.0	983.6	982.9
RELATIVE HUMIDITY (%)														
Hour 01	38	72	71	69	67	72	80	81	83	84	79	75	75	76
Hour 07 (Local Time)	38	75	75	74	72	76	82	83	87	88	84	79	77	79
Hour 13	38	65	63	58	52	52	55	56	58	61	59	64	67	59
Hour 19	38	67	64	59	54	56	60	62	66	70	66	68	70	64
PRECIPITATION (inches):														
Water Equivalent														
-Normal		2.10	2.15	2.55	2.97	3.65	3.98	3.79	3.32	3.31	2.79	3.06	2.51	36.18
-Maximum Monthly	38	6.48	8.06	4.83	9.56	8.02	7.22	7.25	5.69	8.15	8.12	7.69	6.58	9.56
-Year		1979	1981	1977	1983	1989	1982	1986	1990	1987	1976	1972	1983	APR 1983
-Minimum Monthly	38	0.39	0.30	0.49	0.97	0.77	0.27	1.04	1.23	0.82	0.03	0.80	0.35	0.03
-Year		1980	1968	1981	1989	1959	1966	1993	1980	1964	1963	1976	1958	OCT 1963
-Maximum in 24 hrs	38	1.89	3.11	3.02	3.80	2.58	3.61	2.45	3.18	6.52	3.27	2.91	2.86	6.52
-Year		1978	1981	1986	1983	1972	1973	1990	1966	1985	1976	1972	1983	SEP 1985
Snow,Ice pellets,hail														
-Maximum Monthly	38	29.6	22.0	32.0	26.7	2.4	0.0	0.0	T	T	4.4	22.5	33.9	33.9
-Year		1987	1964	1993	1983	1977			1993	1993	1962	1971	1969	DEC 1969
-Maximum in 24 hrs	38	20.1	13.3	20.4	12.2	2.4	0.0	0.0	T	T	4.4	20.5	12.4	20.5
-Year		1964	1961	1993	1983	1977			1993	1993	1962	1971	1969	NOV 1971
WIND:														
Mean Speed (mph)	38	8.9	9.1	9.4	9.5	8.5	7.9	7.3	7.0	7.4	7.9	8.6	8.8	8.4
Prevailing Direction through 1963		SW	SW	NW	SW	WSW	SW	WSW	SW	SW	WSW	WSW	SW	SW
Fastest Mile														
-Direction (!!!)	38	SW	W	S	NW	SW	W	NW	NE	SW	E	NW	SW	W
-Speed (MPH)	38	52	60	49	47	46	43	43	50	47	40	49	47	60
-Year		1992	1956	1970	1957	1980	1956	1988	1956	1989	1980	1989	1957	FEB 1956
Peak Gust														
-Direction (!!!)	10	SW	W	E	SW	NW	SW	NW	NW	SW	S	SE	SW	SW
-Speed (mph)	10	74	53	56	64	55	58	58	51	52	52	61	64	74
-Date		1992	1990	1993	1985	1992	1993	1988	1990	1989	1990	1992	1991	JAN 1992

PRECIPITATION (inches)

AVOCA, WILKES-BARRE - SCRANTON PENNSYLVANIA

YEAR	JAN	FEB	MAR	APR	MAY	JUNE	JULY	AUG	SEP	OCT	NOV	DEC	ANNUAL
1964	3.40	2.03	3.54	3.82	0.98	5.00	1.23	2.85	0.82	1.13	1.86	3.67	30.33
1965	2.07	1.90	1.83	2.63	2.51	1.22	1.30	5.23	3.13	1.80	1.43	1.30	26.35
1966	1.66	2.31	1.60	2.91	3.32	0.27	1.89	4.76	2.70	2.04	2.97	2.00	28.43
1967	1.11	0.89	3.91	2.09	4.41	4.48	3.61	5.20	2.13	2.58	2.31	2.45	35.17
1968	2.03	0.30	2.73	2.37	4.64	4.82	1.23	1.43	4.20	1.65	3.40	1.95	30.75
1969	0.64	0.96	1.45	3.04	2.42	4.58	6.81	4.47	1.92	2.20	3.77	3.42	35.68
1970	0.52	2.41	2.33	3.07	2.75	2.56	5.19	2.72	3.03	2.24	2.87	1.85	31.54
1971	1.54	3.92	1.93	1.29	3.38	2.44	5.73	4.88	1.93	3.00	3.55	2.08	35.67
1972	2.05	2.42	4.00	3.31	7.33	7.04	1.23	1.64	1.57	3.30	7.69	3.61	45.19
1973	2.13	1.28	1.79	4.38	3.80	5.99	3.87	2.61	3.62	1.97	1.50	6.07	39.01
1974	2.66	1.48	4.75	2.71	1.89	3.85	2.80	3.50	6.85	1.07	2.26	3.40	37.22
1975	2.78	3.26	2.52	1.17	4.01	5.64	3.85	2.78	6.10	3.29	3.00	1.84	40.24
1976	3.25	2.14	2.18	2.27	3.24	5.43	3.20	2.57	3.81	8.12	0.80	1.50	38.51
1977	0.88	1.82	4.83	3.98	1.72	3.16	3.44	4.23	5.97	5.27	3.98	3.44	42.72
1978	5.33	0.93	2.30	1.67	4.30	2.48	2.16	3.28	3.06	3.35	1.02	3.09	32.97
1979	6.48	2.44	1.52	3.69	5.16	2.54	2.97	2.05	5.84	3.68	3.17	1.70	41.24
1980	0.39	0.69	3.72	2.35	2.37	4.36	3.76	1.23	1.43	2.17	2.83	1.24	26.54
1981	0.63	8.06	0.49	3.54	3.00	3.45	4.27	1.75	2.74	3.50	1.84	2.13	35.40
1982	2.71	2.28	2.55	3.48	3.52	7.22	3.32	3.42	1.10	0.84	3.44	1.52	35.40
1983	1.17	1.46	3.28	9.56	3.28	4.81	2.76	1.77	2.12	2.73	3.71	6.58	43.23
1984	1.11	2.92	2.42	4.09	6.70	4.75	5.12	2.81	1.36	2.30	2.63	2.36	38.57
1985	0.61	1.58	2.24	2.00	6.10	3.00	6.09	2.62	7.83	1.92	4.47	1.96	40.42
1986	2.59	2.58	4.25	2.98	2.24	6.77	7.25	3.94	3.07	2.61	3.94	2.04	44.26
1987	2.60	0.68	1.18	4.38	2.22	4.35	5.80	4.16	8.15	2.77	2.24	0.99	39.52
1988	1.41	2.32	1.97	2.65	4.24	0.82	6.26	5.03	1.89	1.93	3.33	1.08	32.93
1989	1.02	1.73	2.23	0.97	8.02	6.10	2.76	2.92	3.92	4.73	3.57	0.96	38.93
1990	3.81	2.70	1.88	2.48	5.27	4.78	4.36	5.69	3.16	4.33	3.33	4.30	46.09
1991	1.54	1.35	2.91	2.69	2.84	1.72	2.45	3.28	2.55	2.41	3.81	2.73	30.28
1992	1.23	1.41	2.65	1.67	4.21	1.45	3.83	2.13	2.65	1.91	3.47	2.91	29.52
1993	1.97	1.24	4.02	7.47	1.38	1.85	1.04	3.82	6.55	3.99	3.32	3.09	39.74
Record Mean	2.30	2.25	2.82	3.12	3.33	3.75	4.03	3.59	3.23	2.90	2.82	2.61	36.74

AVERAGE TEMPERATURE (deg. F)

AVOCA, WILKES-BARRE - SCRANTON PENNSYLVANIA

YEAR	JAN	FEB	MAR	APR	MAY	JUNE	JULY	AUG	SEP	OCT	NOV	DEC	ANNUAL
1964	27.9	23.5	36.9	46.6	62.0	67.4	74.1	68.2	63.8	49.4	44.1	31.6	49.6
1965	23.0	28.1	34.2	45.3	63.7	68.0	71.0	69.5	64.5	49.6	40.4	33.5	49.2
1966	23.8	27.2	39.0	46.0	55.7	70.1	75.0	72.4	60.2	50.2	43.4	30.5	49.4
1967	32.6	23.5	34.9	49.1	52.7	71.1	71.7	68.9	61.6	51.2	36.6	31.7	48.8
1968	20.0	24.1	40.4	52.9	56.3	67.5	73.7	71.0	64.4	53.4	40.7	27.2	49.3
1969	26.0	26.6	34.4	50.5	59.9	68.2	70.8	69.9	62.7	50.6	39.2	25.7	48.7
1970	17.7	25.7	31.8	48.4	60.4	64.9	70.8	69.6	64.2	52.8	41.5	27.3	47.9
1971	19.5	27.2	33.0	45.0	55.9	68.6	69.9	68.3	65.5	56.4	38.8	34.1	48.5
1972	27.6	24.3	32.8	43.5	60.0	62.9	71.5	69.6	61.9	45.3	36.0	32.9	47.4
1973	28.5	24.5	41.9	47.5	53.4	68.3	71.5	71.6	62.4	53.2	41.5	31.3	49.7
1974	27.8	24.2	34.7	49.1	56.2	64.4	70.8	68.5	61.2	48.5	43.0	34.4	48.6
1975	31.8	32.1	35.8	43.9	64.9	68.8	73.7	71.2	59.9	55.8	47.8	31.1	51.4
1976	22.1	35.0	41.0	51.3	56.7	70.4	69.2	69.1	60.9	48.0	37.0	22.9	48.6
1977	15.0	26.9	40.5	48.9	60.0	63.8	71.1	68.7	62.7	48.5	43.1	29.6	48.2
1978	24.4	19.2	33.1	46.0	58.7	65.2	69.4	71.0	60.5	50.7	40.5	29.1	47.3
1979	24.2	16.0	40.7	46.4	58.6	65.1	70.7	70.7	63.2	51.8	45.9	35.6	49.1
1980	27.8	24.2	35.9	51.0	61.8	65.3	73.2	75.2	66.1	49.8	37.6	25.7	49.5
1981	19.5	34.9	36.2	51.0	59.8	68.5	72.2	70.1	62.1	49.1	41.1	29.1	49.5
1982	18.7	27.8	36.0	46.3	61.2	64.3	71.0	66.2	62.4	52.3	44.3	36.7	48.9
1983	27.3	29.4	38.8	45.9	55.7	67.6	72.4	71.6	64.8	52.5	42.8	27.1	49.7
1984	23.0	35.9	30.9	48.3	57.5	69.3	71.6	72.6	61.7	58.0	40.8	37.3	50.6
1985	21.5	29.8	39.1	51.4	60.6	63.8	70.1	69.0	64.0	52.7	44.5	26.7	49.4
1986	27.2	26.1	39.8	49.3	62.7	66.2	71.0	67.4	61.6	51.2	37.3	32.8	49.4
1987	24.7	25.1	39.7	50.7	60.0	68.7	73.5	68.3	61.8	47.4	41.3	32.7	49.5
1988	22.0	27.4	38.1	46.6	60.1	65.5	75.8	72.9	60.4	46.3	43.3	29.8	49.0
1989	31.3	28.0	37.2	45.6	57.6	67.0	70.4	68.5	62.2	52.8	39.7	18.6	48.2
1990	35.3	33.3	40.6	50.2	56.1	67.6	71.7	69.5	61.6	55.3	44.2	36.0	51.8
1991	27.5	33.4	40.4	51.7	65.7	69.5	73.4	72.2	61.8	53.3	41.4	32.9	51.9
1992	28.6	30.3	34.4	47.4	58.7	66.3	71.1	68.4	62.8	49.0	42.2	32.8	49.3
1993	32.0	23.8	33.3	49.9	61.7	68.8	75.5	72.7	61.2	50.0	41.4	30.0	50.0
Record Mean	26.8	27.4	36.8	48.1	59.2	67.4	72.1	69.9	63.0	52.1	41.2	30.3	49.5
Max	34.0	35.3	45.3	57.9	69.8	78.0	82.5	80.1	73.1	61.9	48.7	37.0	58.6
Min	19.5	19.5	28.3	38.2	48.5	56.8	61.7	59.8	52.8	42.3	33.6	23.6	40.4

REFERENCE NOTES

AVOCA, WILKES-BARRE - SCRANTON PENNSYLVANIA

GENERAL
T - TRACE AMOUNT.
BLANK ENTRIES DENOTE MISSING/UNREPORTED DATA.
# INDICATES A STATION OR INSTRUMENT RELOCATION.

SPECIFIC

PM - INCLUDES LAST DAY OF PREVIOUS MONTH
ASOS - AUTOMATED SURFACE OBSERVING SYSTEM IN OPERATION DURING THESE MONTHS.

(*) - LENGTH OF RECORD IN YEARS, ALTHOUGH INDIVIDUAL MONTHS MAY BE MISSING.
0.* OR * - THE VALUE IS BETWEEN 0.0 AND 0.05
NORMALS - BASED ON THE 1961-1990 RECORD PERIOD.
EXTREMES - DATES ARE THE MOST RECENT OCCURRENCE
WIND DIR.- NUMERALS SHOW TENS OF DEGRESS CLOCKWISE FROM TRUE NORTH. "00" INDICATES CALM.
RESULTANT DIRECTIONS ARE GIVEN TO WHOLE DEGREES.
BOLD VALUES INDICATE EXTREME VALUES WHICH OCCURRED AFTER THE ASOS SYSTEM WAS COMMISSIONED.

RECORD = PERIOD OF RECORD
RECORD MEAN PRECIPITATION IS THE MEAN OF ALL DAILY PRECIPITATION AMOUNTS DURING THE PERIOD OF RECORD.
RECORD MAX(MIN) TEMPERATURE IS THE MEAN OF ALL DAILY MAX(MIN) TEMPERATURES DURING THE PERIOD OF RECORD.
RECORD MEAN TEMPERATURE IS THE SUM OF THE RECORD MAX AND RECORD MIN DIVIDED BY 2.
AVERAGE TEMPERATURE IS THE SUM OF THE MEAN DAILY MAX AND MIN TEMPERATURE DIVIDED BY 2.

EXCEPTIONS

RECORD MEANS ARE THROUGH THE CURRENT YEAR,
BEGINNING IN 1901 FOR TEMPERATURE
1901 FOR PRECIPITATION
1956 FOR SNOWFALL

HEATING DEGREE DAYS Base 65 deg. F

AVOCA, WILKES-BARRE - SCRANTON PENNSYLVANIA

SEASON	JULY	AUG	SEP	OCT	NOV	DEC	JAN	FEB	MAR	APR	MAY	JUNE	TOTAL
1964-65	0	31	110	475	621	1028	1293	1027	951	582	100	55	6273
1965-66	7	35	99	471	731	971	1270	1053	799	565	300	41	6342
1966-67	1	0	170	451	642	1063	998	1154	923	475	379	12	6268
1967-68	5	17	139	427	844	1023	1388	1181	754	357	261	42	6438
1968-69	0	28	55	359	725	1165	1204	1068	943	434	188	38	6207
1969-70	6	15	134	440	765	1212	1459	1094	1022	500	176	76	6899
1970-71	2	5	109	371	695	1162	1404	1051	985	593	292	32	6701
1971-72	5	22	90	263	783	951	1152	1175	992	634	168	109	6344
1972-73	21	20	125	603	860	988	1124	1131	704	521	354	25	6476
1973-74	2	11	140	368	699	1036	1145	1135	934	480	291	65	6306
1974-75	5	2	155	503	655	941	1024	913	902	627	88	25	5840
1975-76	0	8	158	291	509	1043	1322	864	737	451	265	31	5679
1976-77	8	25	155	519	834	1297	1546	1058	756	487	206	90	6981
1977-78	14	37	119	505	653	1090	1252	1279	984	562	240	73	6808
1978-79	38	2	153	436	728	1103	1257	1370	747	552	221	66	6673
1979-80	34	31	120	420	568	900	1144	1175	895	414	137	94	5932
1980-81	1	0	82	466	813	1211	1407	835	886	416	195	19	6331
1981-82	2	5	132	485	706	1105	1426	1034	896	554	147	68	6560
1982-83	17	55	112	390	619	870	1158	992	805	569	292	41	5920
1983-84	7	11	119	392	659	1169	1297	837	1052	493	247	34	6317
1984-85	7	6	148	219	719	852	1342	981	799	421	162	78	5734
1985-86	4	11	127	376	610	1181	1163	1083	777	467	140	61	6000
1986-87	16	50	139	428	823	990	1243	1111	779	425	208	20	6232
1987-88	2	34	119	539	706	995	1326	1082	823	546	176	91	6439
1988-89	13	12	156	574	643	1083	1037	1031	853	575	251	39	6267
1989-90	6	31	133	377	750	1433	915	881	757	465	269	44	6061
1990-91	10	13	152	320	619	894	1153	877	757	402	113	23	5333
1991-92	0	0	160	370	699	989	1121	1001	942	521	209	40	6052
1992-93	4	9	132	488	676	993	1017	1148	975	447	136	39	6064
1993-94	0	4	174	460	703	1077							

COOLING DEGREE DAYS Base 65 deg. F

AVOCA, WILKES-BARRE - SCRANTON PENNSYLVANIA

YEAR	JAN	FEB	MAR	APR	MAY	JUNE	JULY	AUG	SEP	OCT	NOV	DEC	TOTAL
1969	0	0	0	7	34	141	193	178	73	1	0	0	627
1970	0	0	0	10	40	81	189	159	92	2	0	0	573
1971	0	0	0	0	16	145	160	132	113	4	3	0	573
1972	0	0	0	0	20	53	232	167	42	0	0	0	514
1973	0	0	0	4	0	132	212	223	72	7	0	0	650
1974	0	0	0	10	28	52	194	117	46	0	0	0	447
1975	0	0	0	0	91	146	278	207	13	14	0	0	749
1976	0	0	0	46	16	198	145	159	37	0	0	0	601
1977	0	0	2	13	57	62	208	162	59	0	2	0	565
1978	0	0	0	0	52	84	181	194	26	0	0	0	537
1979	0	0	0	2	32	78	218	214	75	15	0	0	634
1980	0	0	0	0	42	107	263	322	122	3	0	0	859
1981	0	0	0	4	42	131	231	172	55	0	0	0	635
1982	0	0	0	1	34	55	208	98	41	3	5	0	445
1983	0	0	0	4	12	125	243	224	118	9	0	0	735
1984	0	0	0	0	20	165	218	248	58	12	0	0	721
1985	0	0	0	20	32	47	169	142	104	0	0	0	514
1986	0	0	2	0	76	104	205	131	44	8	0	0	570
1987	0	0	0	3	62	138	273	141	30	0	0	0	647
1988	0	0	0	0	34	111	356	266	23	3	0	0	793
1989	0	0	0	0	31	109	179	148	57	4	0	0	528
1990	0	0	8	29	2	127	225	160	56	24	0	0	631
1991	0	0	0	12	142	164	264	230	69	14	0	0	895
1992	0	0	0	0	21	85	201	123	71	0	0	0	501
1993	0	0	0	5	41	158	334	249	66	0	0	0	853

SNOWFALL (inches)

AVOCA, WILKES-BARRE - SCRANTON PENNSYLVANIA

SEASON	JULY	AUG	SEP	OCT	NOV	DEC	JAN	FEB	MAR	APR	MAY	JUNE	TOTAL
1964-65	0.0	0.0	0.0	0.2	0.1	4.1	11.5	2.3	9.9	3.8	0.0	0.0	31.9
1965-66	0.0	0.0	0.0	0.6	0.5	1.3	20.7	17.5	3.8	0.7	0.6	0.0	45.7
1966-67	0.0	0.0	0.0	T	T	17.7	8.3	14.4	29.7	4.8	0.4	0.0	75.3
1967-68	0.0	0.0	0.0	T	7.2	13.8	6.0	2.9	2.7	0.0	0.0	0.0	32.6
1968-69	0.0	0.0	0.0	T	10.9	5.8	2.7	14.8	2.5	0.0	0.0	0.0	36.7
1969-70	0.0	0.0	0.0	T	2.7	33.9	9.5	9.5	20.6	0.6	T	0.0	76.8
1970-71	0.0	0.0	0.0	0.3	T	12.3	15.2	12.1	15.6	1.6	0.0	0.0	57.1
1971-72	0.0	0.0	0.0	0.0	22.5	4.1	5.9	19.7	6.6	3.8	0.0	0.0	62.6
1972-73	0.0	0.0	0.0	0.8	7.9	3.9	5.0	3.1	1.9	0.2	0.4	0.0	23.2
1973-74	0.0	0.0	0.0	T	0.4	16.0	12.8	4.5	15.7	2.8	0.0	0.0	52.2
1974-75	0.0	0.0	0.0	0.2	2.2	5.2	13.7	15.2	5.5	1.2	0.0	0.0	43.2
1975-76	0.0	0.0	0.0	0.0	1.3	3.7	13.0	7.5	10.2	0.5	T	0.0	36.2
1976-77	0.0	0.0	0.0	0.0	6.0	6.7	15.7	13.0	11.2	1.4	2.4	0.0	56.4
1977-78	0.0	0.0	0.0	0.6	8.7	9.8	28.8	18.2	6.5	0.9	0.0	0.0	73.5
1978-79	0.0	0.0	0.0	T	4.1	7.9	12.7	14.3	1.1	4.4	0.0	0.0	44.5
1979-80	0.0	0.0	0.0	T	T	5.5	1.4	8.1	10.5	T	0.0	0.0	25.5
1980-81	0.0	0.0	0.0	T	8.6	8.0	11.1	7.0	5.8	T	0.0	0.0	40.5
1981-82	0.0	0.0	0.0	T	1.0	14.2	14.1	13.5	8.7	8.1	0.0	0.0	59.6
1982-83	0.0	0.0	0.0	0.0	0.5	7.4	8.4	12.3	3.8	26.7	0.0	0.0	59.1
1983-84	0.0	0.0	0.0	0.0	3.1	2.7	11.2	4.0	18.4	0.0	0.0	0.0	39.4
1984-85	0.0	0.0	0.0	0.0	3.0	9.2	10.8	9.1	1.4	1.8	0.0	0.0	35.3
1985-86	0.0	0.0	0.0	0.0	1.7	13.4	12.9	11.6	1.1	8.6	0.0	0.0	49.3
1986-87	0.0	0.0	0.0	0.0	8.6	1.4	29.6	6.4	0.9	0.6	0.0	0.0	47.5
1987-88	0.0	0.0	0.0	T	6.4	6.4	13.0	14.9	4.3	0.7	0.0	0.0	45.7
1988-89	0.0	0.0	0.0	T	T	1.1	2.1	3.0	1.1	T	0.0	0.0	7.3
1989-90	0.0	0.0	0.0	T	2.6	8.3	10.8	7.3	6.2	2.1	0.0	0.0	37.3
1990-91	0.0	0.0	T	T	0.4	8.4	6.6	7.6	7.2	1.1	0.0	0.0	31.3
1991-92	0.0	0.0	0.0	0.0	0.1	3.5	5.4	6.3	8.5	0.7	0.0	0.0	24.5
1992-93	0.0	0.0	T	T	0.5	10.0	3.8	12.1	32.0	1.9	0.0	0.0	60.3
1993-94	0.0	T	T	1.9	0.7	8.0							
Record Mean	0.0	T	T	0.2	3.3	8.8	11.2	10.8	9.4	3.1	0.1	0.0	47.0

ERIE, PENNSYLVANIA

Erie is located on the southeast shore of Lake Erie and observations are made at Erie International Airport, which is 6 miles southwest of the center of the city and about 1 mile from the lake shore. The terrain rises gradually in a series of ridges paralleling the shoreline to 500 feet above the lake level 3 to 4 miles inland and to 1,000 feet about 15 miles inland. Snowfall from instability showers moving southward off the lake usually increases due to the upslope terrain. Snowfall is somewhat higher south of the city than along the lake shore.

During the winter months, the many cold air masses moving south from Canada are modified by the relatively warm waters of Lake Erie. However, the temperature difference between air and water produces an excess of cloudiness and frequent snow from November through March.

Spring weather is quite variable in Erie, but generally cloudy and cool. Proximity to the lake frequently prevents killing frosts that occur inland. This has led to the establishment of numerous vineyards and orchards in a narrow belt along the shore. Summer heat waves are tempered by cool lake breezes that may reach several miles inland, and days with temperatures above 90 degrees are infrequent. Summer thunderstorms are usually less destructive in Erie than inland areas because of the stabilizing effects of Lake Erie.

Autumn, with long dry periods and an abundance of sunshine, is usually the most pleasant period of the year in Erie. The growing season is extended by the influence of the warmer waters of the lake. Precipitation is well distributed throughout the year, although the number of days with measurable amounts varies considerably from a low average of about one day in three for the period June through September to about one-half of the days from November through March, when snow flurries and squalls move in from the lake.

NORMALS, MEANS, AND EXTREMES

ERIE, PENNSYLVANIA

LATITUDE: 42°05'N LONGITUDE: 80°11'W ELEVATION: FT. GRND 731 BARO 744 TIME ZONE: EASTERN WBAN: 14860

	(a)	JAN	FEB	MAR	APR	MAY	JUNE	JULY	AUG	SEP	OCT	NOV	DEC	YEAR
TEMPERATURE °F:														
Normals														
-Daily Maximum		32.5	33.7	43.6	54.5	65.9	75.3	79.9	78.5	72.1	61.0	49.4	37.8	57.0
-Daily Minimum		18.2	17.9	28.1	37.7	47.8	57.6	62.6	61.8	55.9	45.6	37.0	25.2	41.3
-Monthly		25.4	25.8	35.9	46.1	56.9	66.5	71.3	70.1	64.0	53.3	43.2	31.5	49.2
Extremes														
-Record Highest	40	65	67	81	89	90	100	99	93	94	88	80	75	100
-Year		1985	1957	1990	1990	1991	1988	1990	1988	1959	1963	1961	1982	JUN 1988
-Record Lowest	40	-16	-17	-9	12	26	32	44	37	33	24	7	-6	-17
-Year		1985	1979	1980	1982	1970	1972	1963	1982	1974	1975	1976	1983	FEB 1979
NORMAL DEGREE DAYS:														
Heating (base 65°F)		1228	1098	902	567	268	52	0	12	89	370	654	1039	6279
Cooling (base 65°F)		0	0	0	0	17	97	199	171	59	7	0	0	550
% OF POSSIBLE SUNSHINE														
MEAN SKY COVER (tenths)														
Sunrise - Sunset	38	8.6	8.0	7.4	6.9	6.4	5.8	5.6	5.8	6.3	6.8	8.4	9.0	7.1
MEAN NUMBER OF DAYS:														
Sunrise to Sunset														
-Clear	38	1.7	2.5	4.8	5.7	6.8	8.1	8.3	8.6	6.7	6.2	2.5	1.3	63.0
-Partly Cloudy	38	4.4	5.9	7.1	7.9	10.3	11.2	12.9	11.4	9.7	8.0	4.5	3.4	96.8
-Cloudy	38	24.9	19.8	19.1	16.4	13.9	10.8	9.8	11.0	13.6	16.9	23.0	26.2	205.4
Precipitation .01 inches or more	40	18.4	15.0	14.9	14.4	11.8	10.3	9.6	10.7	11.0	12.9	16.5	19.0	164.6
Snow, Ice pellets, hail 1.0 inches or more	38	7.4	5.0	3.2	0.9	0.0	0.0	0.0	0.0	0.0	0.2	2.4	6.8	25.9
Thunderstorms	38	0.1	0.4	1.6	2.8	3.9	6.1	6.6	6.7	4.4	2.3	1.5	0.3	36.8
Heavy Fog Visibility 1/4 mile or less	38	1.1	1.7	2.8	1.9	1.9	0.9	0.4	0.4	0.3	0.4	0.9	0.8	13.5
Temperature °F														
-Maximum														
90° and above	28	0.0	0.0	0.0	0.0	0.*	0.5	0.9	0.6	0.1	0.0	0.0	0.0	2.2
32° and below	28	15.8	14.0	6.9	0.4	0.0	0.0	0.0	0.0	0.0	0.0	1.4	9.8	48.4
-Minimum														
32° and below	28	27.8	25.1	23.0	10.9	1.0	0.*	0.0	0.0	0.0	1.7	11.3	24.3	125.1
0° and below	28	1.9	1.7	0.2	0.0	0.0	0.0	0.0	0.0	0.0	0.0	0.0	0.3	4.1
AVG. STATION PRESS. (mb)	21	990.2	991.0	989.4	988.4	988.7	988.8	989.8	991.1	991.6	991.7	990.7	990.8	990.2
RELATIVE HUMIDITY (%)														
Hour 01 (Local Time)	28	75	77	74	73	77	79	80	82	80	74	74	75	77
Hour 07	28	77	78	77	75	76	79	80	82	82	77	75	76	78
Hour 13	28	72	71	66	61	62	63	64	64	65	63	68	72	66
Hour 19	28	74	76	70	64	63	65	65	69	73	72	73	75	70
PRECIPITATION (inches):														
Water Equivalent														
-Normal		2.22	2.28	3.00	3.24	3.44	4.09	3.43	4.06	4.39	3.77	4.02	3.59	41.53
-Maximum Monthly	40	4.59	5.73	6.78	7.11	6.14	7.74	7.70	11.06	10.65	9.87	10.40	6.94	11.06
-Year		1959	1990	1976	1961	1989	1957	1970	1977	1977	1954	1985	1990	AUG 1977
-Minimum Monthly	40	0.87	0.57	0.63	1.63	1.00	0.75	0.65	0.58	1.45	1.13	1.52	1.38	0.57
-Year		1981	1978	1960	1975	1991	1991	1978	1959	1960	1963	1978	1960	FEB 1978
-Maximum in 24 hrs	38	1.51	2.16	2.38	2.53	2.23	2.80	3.22	3.53	6.11	4.35	3.67	2.39	6.11
-Year		1959	1961	1987	1977	1969	1957	1970	1980	1979	1954	1985	1979	SEP 1979
Snow, Ice pellets, hail														
-Maximum Monthly	39	62.4	32.1	27.3	17.2	0.4	T	0.0	T	T	4.0	36.3	66.9	66.9
-Year		1978	1972	1993	1957	1989	1990		1992	1993	1954	1967	1989	DEC 1989
-Maximum in 24 hrs	39	12.9	17.8	13.2	10.0	0.4	T	0.0	T	T	2.3	23.0	19.2	23.0
-Year		1986	1979	1993	1957	1989	1990		1992	1993	1974	1956	1989	NOV 1956
WIND:														
Mean Speed (mph)	39	13.3	12.2	12.1	11.5	10.0	9.6	9.1	9.0	10.0	11.3	13.0	13.5	11.2
Prevailing Direction through 1963		WSW	WSW	NNE	WSW	WSW	S	S	S	S	SSE	SSW	SSW	S
Fastest Obs. 1 Min.														
-Direction (!!!)	36	20	29	14	21	25	36	32	24	17	24	31	26	14
-Speed (MPH)	36	53	52	55	46	37	37	46	35	45	43	41	38	55
-Year		1978	1972	1960	1965	1964	1975	1959	1970	1965	1983	1989	1991	MAR 1960
Peak Gust														
-Direction (!!!)	10	SW	S	W	SW	SW	S	W	SW	NW	NW	W	W	W
-Speed (mph)	10	55	62	64	55	54	52	69	53	61	61	69	60	69
-Date		1990	1988	1991	1985	1990	1990	1991	1984	1986	1986	1992	1985	NOV 1992

PRECIPITATION (inches)

ERIE, PENNSYLVANIA

YEAR	JAN	FEB	MAR	APR	MAY	JUNE	JULY	AUG	SEP	OCT	NOV	DEC	ANNUAL
1964	2.35	1.39	3.81	3.67	3.75	2.26	3.42	4.60	1.97	2.96	3.08	3.21	36.47
1965	3.86	3.00	3.55	1.66	3.40	3.75	3.48	2.68	2.24	3.84	4.27	2.68	38.41
1966	2.01	1.92	3.47	4.26	2.03	3.82	2.08	3.52	3.24	1.73	5.34	4.32	37.74
1967	0.90	1.67	1.67	4.21	3.15	2.49	4.79	4.48	4.25	2.77	5.19	2.21	37.78
1968	2.98	1.13	1.69	2.64	3.06	2.63	2.34	2.76	2.93	3.40	4.71	4.06	34.33
1969	2.95	0.73	1.43	5.27	5.49	4.88	3.65	1.75	1.99	3.02	2.99	2.43	36.58
1970	1.44	2.09	1.61	2.48	3.23	2.65	7.70	2.03	7.08	3.52	4.87	2.80	41.50
1971	1.63	2.05	1.85	1.81	2.20	2.83	2.67	3.46	3.51	3.48	4.51	4.06	34.06
1972	1.94	2.77	3.95	2.64	4.69	7.50	2.91	3.01	5.37	1.72	3.36	3.69	43.55
1973	1.72	2.00	3.18	2.71	4.57	6.28	1.55	3.96	2.22	3.56	2.83	3.46	38.04
1974	2.45	1.93	5.02	4.89	4.44	5.33	1.11	2.97	3.34	1.57	5.20	3.58	41.83
1975	2.99	3.30	3.58	1.63	2.36	4.36	3.22	8.93	2.52	2.42	3.17	4.55	43.03
1976	3.07	5.01	6.78	2.71	4.18	3.98	4.35	1.40	3.87	4.14	2.12	2.24	43.85
1977	1.05	1.44	3.45	5.21	2.53	4.39	6.67	11.06	10.65	3.37	6.25	5.63	61.70
1978	3.61	0.57	1.79	2.45	2.76	2.68	0.65	4.56	4.83	5.81	1.52	3.81	35.04
1979	3.50	2.15	2.50	4.71	3.70	3.58	4.27	5.09	8.44	6.63	5.84	4.90	55.31
1980	1.57	1.29	4.11	3.79	2.23	4.83	5.42	6.76	5.48	6.51	2.56	2.49	47.04
1981	0.87	5.21	1.58	6.09	2.13	4.84	3.04	3.85	4.26	5.04	2.22	2.84	41.97
1982	3.85	1.24	3.50	1.81	3.06	6.02	4.40	2.20	4.07	2.74	5.33	3.34	41.56
1983	1.49	1.07	3.63	2.93	3.91	3.84	5.52	4.74	5.27	3.77	6.11	3.97	46.25
1984	1.65	2.42	1.91	2.63	5.83	4.49	1.94	2.09	5.29	1.82	3.62	4.10	37.79
1985	2.56	2.75	5.08	1.76	2.94	3.50	4.97	1.66	2.22	5.20	10.40	2.83	45.87
1986	2.33	2.72	2.10	2.88	5.24	7.71	2.54	1.83	7.97	4.86	2.99	4.13	47.30
1987	2.15	1.05	4.28	1.87	1.78	5.15	3.91	7.82	5.45	5.76	2.25	3.39	44.86
1988	1.50	2.47	2.44	3.00	3.21	1.26	4.14	3.78	3.21	8.25	2.99	2.62	38.87
1989	1.95	2.41	4.70	2.02	6.14	5.14	1.35	3.96	3.76	3.33	3.87	3.25	41.88
1990	2.30	5.73	1.29	3.52	5.74	2.84	2.53	6.49	7.74	4.15	2.69	6.94	51.96
1991	2.16	1.62	3.38	3.64	1.00	0.75	3.49	3.07	3.25	3.00	3.18	3.17	31.71
1992	2.60	1.91	2.11	4.04	1.78	1.95	6.06	4.11	6.81	4.01	5.37	3.34	44.09
1993	3.36	2.03	3.59	2.34	1.28	3.94	2.80	2.85	4.52	4.41	4.10	2.88	38.10
Record Mean	2.64	2.40	2.81	3.09	3.35	3.48	3.30	3.34	3.76	3.60	3.55	2.94	38.27

AVERAGE TEMPERATURE (deg. F)

ERIE, PENNSYLVANIA

YEAR	JAN	FEB	MAR	APR	MAY	JUNE	JULY	AUG	SEP	OCT	NOV	DEC	ANNUAL
1964	30.0	24.8	36.6	47.2	60.1	66.6	72.5	66.6	62.7	50.0	45.0	33.2	49.6
#1965	25.8	26.6	29.5	41.6	61.0	64.5	67.5	68.2	66.3	50.5	42.4	36.0	48.4
1966	22.8	26.6	37.6	44.5	51.6	67.5	71.8	69.3	61.1	51.3	43.3	32.3	48.3
1967	32.8	24.2	34.6	47.6	49.6	71.1	68.8	66.6	60.0	53.1	36.9	34.0	48.3
1968	23.5	20.7	36.4	48.3	52.6	65.0	70.1	70.6	65.9	55.1	42.4	30.5	48.4
1969	27.0	25.7	31.8	47.2	54.7	62.6	69.0	69.6	61.7	50.0	38.3	25.8	47.0
1970	16.8	23.1	28.4	45.1	58.2	63.9	68.9	68.0	62.8	53.0	40.4	30.8	46.6
1971	22.0	27.3	30.0	40.6	51.9	66.0	66.9	64.7	64.1	57.0	39.3	35.9	47.2
1972	26.9	22.9	29.8	40.0	55.6	59.8	68.4	66.1	61.1	47.2	37.3	33.3	45.7
1973	29.6	23.8	42.2	46.0	52.5	66.2	69.7	69.6	62.6	54.8	42.8	31.7	49.3
1974	30.0	24.6	35.4	47.6	53.0	63.2	67.7	68.1	58.7	48.3	41.0	31.0	47.4
1975	29.9	28.4	32.1	38.8	60.3	67.4	71.8	69.4	58.5	52.7	46.6	30.7	48.9
1976	20.9	32.8	38.9	45.8	52.5	66.4	66.4	67.0	60.5	48.1	34.2	23.9	46.4
1977	12.5	25.0	40.6	47.8	58.4	61.9	70.3	67.1	62.5	49.4	43.7	28.3	47.3
1978	19.8	14.5	27.9	41.8	56.0	65.2	69.5	71.1	62.3	49.3	40.6	31.0	45.7
1979	19.5	14.1	36.5	42.2	53.9	63.5	67.6	69.5	64.3	53.6	43.7	34.5	46.9
1980	26.6	20.9	31.0	43.3	55.7	61.3	69.6	72.2	63.7	48.1	38.9	27.5	46.6
1981	19.6	30.3	33.1	47.4	53.9	66.0	72.1	70.6	61.9	49.7	41.6	31.4	48.1
1982	19.2	23.4	34.4	42.5	60.9	61.3	70.9	66.2	62.5	54.2	44.8	40.3	48.4
1983	30.6	31.6	38.5	45.3	55.0	67.0	73.0	72.4	65.7	54.4	45.2	25.9	50.4
1984	21.6	35.5	28.5	47.0	53.8	68.6	69.6	72.0	61.8	56.4	42.4	37.3	49.5
1985	21.7	25.9	37.4	51.4	59.8	63.4	70.2	70.7	66.9	54.6	46.3	27.2	49.6
1986	27.5	27.0	38.7	48.2	59.6	65.2	72.0	69.2	65.0	53.6	40.0	32.9	49.9
1987	28.0	27.3	37.5	48.3	60.5	69.2	74.6	70.0	64.5	48.2	45.1	35.8	50.8
1988	27.2	25.9	36.3	46.0	58.2	65.6	74.5	72.9	63.0	48.2	44.6	32.4	49.6
1989	33.4	24.8	35.6	43.1	56.1	66.4	71.9	69.4	63.2	54.8	41.2	21.7	48.5
1990	36.0	33.3	40.4	49.4	55.2	67.0	70.9	69.9	63.4	54.8	46.0	36.7	51.9
1991	28.1	32.7	39.0	51.5	64.8	70.3	73.2	72.6	63.8	55.8	41.3	34.6	52.3
1992	30.4	30.4	34.7	46.6	57.5	63.9	69.4	67.8	64.0	51.1	42.4	34.7	49.4
1993	32.5	24.1	31.7	48.1	57.8	68.4	75.4	73.4	61.2	51.1	41.7	30.5	49.7
Record Mean	27.2	26.6	34.5	45.5	56.7	66.4	71.3	69.8	63.8	53.2	42.0	31.7	49.1
Max	33.7	33.8	42.1	53.6	65.1	74.5	79.0	77.5	71.6	60.6	48.2	37.5	56.4
Min	20.6	19.4	27.0	37.4	48.2	58.3	63.5	62.2	56.1	45.7	35.7	25.9	41.7

REFERENCE NOTES

ERIE, PENNSYLVANIA

GENERAL
T - TRACE AMOUNT.
BLANK ENTRIES DENOTE MISSING/UNREPORTED DATA.
# INDICATES A STATION OR INSTRUMENT RELOCATION.

SPECIFIC

(a) - LENGTH OF RECORD IN YEARS, ALTHOUGH INDIVIDUAL MONTHS MAY BE MISSING.
0.* OR * - THE VALUE IS BETWEEN 0.0 AND 0.05
NORMALS - BASED ON THE 1961-1990 RECORD PERIOD.
EXTREMES - DATES ARE THE MOST RECENT OCCURRENCE.
WIND DIR. - NUMERALS SHOW TENS OF DEGRESS CLOCKWISE FROM TRUE NORTH. "00" INDICATES CALM.
RESULTANT DIRECTIONS ARE GIVEN TO WHOLE DEGREES.
BOLD VALUES INDICATE EXTREME VALUES WHICH OCCURRED AFTER THE ASOS SYSTEM WAS COMMISSIONED.

RECORD = PERIOD OF RECORD
RECORD MEAN PRECIPITATION IS THE MEAN OF ALL DAILY PRECIPITATION AMOUNTS DURING THE PERIOD OF RECORD.
RECORD MAX(MIN) TEMPERATURE IS THE MEAN OF ALL DAILY MAX(MIN) TEMPERATURES DURING THE PERIOD OF RECORD.
RECORD MEAN TEMPERATURE IS THE SUM OF THE RECORD MAX AND RECORD MIN DIVIDED BY 2.
AVERAGE TEMPERATURE IS THE SUM OF THE MEAN DAILY MAX AND MIN TEMPERATURE DIVIDED BY 2.

EXCEPTIONS

RECORD MEANS ARE THROUGH THE CURRENT YEAR, BEGINNING IN 1847 FOR TEMPERATURE
1847 FOR PRECIPITATION
1955 FOR SNOWFALL

HEATING DEGREE DAYS Base 65 deg. F ERIE, PENNSYLVANIA

SEASON	JULY	AUG	SEP	OCT	NOV	DEC	JAN	FEB	MAR	APR	MAY	JUNE	TOTAL
1964-65	3	53	140	461	596	978	1209	1072	1093	696	174	105	6580
#1965-66	26	44	76	441	673	892	1302	1068	842	609	417	64	6454
1966-67	5	10	151	420	644	1006	993	1137	941	518	471	18	6314
1967-68	25	32	164	374	835	955	1280	1277	881	494	377	81	6775
1968-69	20	26	55	333	671	1063	1169	1093	1020	528	324	145	6447
1969-70	20	20	169	462	796	1211	1488	1169	1129	605	235	103	7407
1970-71	18	18	125	366	731	1056	1324	1048	1078	723	404	68	6959
1971-72	32	55	97	242	763	895	1172	1217	1082	742	287	184	6768
1972-73	55	47	144	546	820	976	1090	1146	697	568	379	40	6508
1973-74	9	18	141	309	658	1022	1077	1128	915	521	373	101	6272
1974-75	32	9	204	511	713	1046	1081	1016	1013	781	203	51	6660
1975-76	3	14	197	381	546	1057	1361	931	804	580	386	51	6311
1976-77	28	42	163	519	920	1269	1620	1116	751	519	249	126	7322
1977-78	9	51	110	477	636	1131	1395	1408	1143	688	303	84	7435
1978-79	15	2	125	481	723	1046	1406	1424	874	678	369	106	7249
1979-80	36	17	84	378	634	939	1188	1275	1045	644	299	161	6700
1980-81	12	2	108	520	775	1156	1400	967	982	523	345	42	6832
1981-82	3	6	153	467	694	1034	1413	1159	942	672	158	123	6824
1982-83	11	47	126	336	605	762	1062	933	814	585	312	67	5660
1983-84	10	0	89	336	591	1207	1338	847	1123	539	360	25	6465
1984-85	11	5	141	262	673	853	1337	1088	846	423	202	87	5928
1985-86	5	3	75	316	558	1164	1152	1056	811	502	207	74	5923
1986-87	5	32	84	350	742	989	1138	1048	844	495	225	31	5983
1987-88	0	21	67	513	592	902	1166	1128	883	565	236	96	6169
1988-89	5	10	102	528	605	1006	972	1121	905	651	301	45	6251
1989-90	2	16	128	320	706	1335	892	879	773	500	309	58	5918
1990-91	9	0	113	323	564	866	1137	898	800	412	148	15	5285
1991-92	0	0	130	304	704	934	1065	996	934	556	258	94	5975
1992-93	5	23	110	423	673	928	1002	1137	1025	507	227	51	6111
1993-94	0	0	162	430	692	1061							

COOLING DEGREE DAYS Base 65 deg. F ERIE, PENNSYLVANIA

YEAR	JAN	FEB	MAR	APR	MAY	JUNE	JULY	AUG	SEP	OCT	NOV	DEC	TOTAL
1969	0	0	0	4	10	81	150	170	76	5	0	0	496
1970	0	0	0	13	28	76	145	119	62	3	0	0	446
1971	0	0	0	0	7	105	96	55	75	3	0	0	341
1972	0	0	0	0	2	32	165	88	35	0	0	0	322
1973	0	0	0	6	0	84	163	167	76	1	0	0	497
1974	0	0	0	5	9	55	120	114	23	0	0	0	326
1975	0	0	0	0	64	132	221	156	9	6	0	0	588
1976	0	0	0	10	5	98	81	110	33	2	0	0	339
1977	0	0	2	9	52	42	179	121	44	0	1	0	450
1978	0	0	0	0	32	94	162	198	48	0	0	0	534
1979	0	0	0	3	29	69	125	163	71	32	0	0	492
1980	0	0	0	0	20	56	162	234	75	4	0	0	551
1981	0	0	0	2	8	81	230	187	65	0	0	0	573
1982	0	0	0	4	35	19	201	90	59	8	5	4	425
1983	0	0	0	0	7	135	261	235	117	15	0	0	770
1984	0	0	0	7	18	145	161	227	52	3	0	0	613
1985	0	0	0	21	45	45	172	187	140	4	3	0	617
1986	0	0	2	6	46	86	228	169	88	3	0	0	628
1987	0	0	0	0	92	164	304	182	61	0	2	0	805
1988	0	0	0	0	29	119	305	263	52	10	0	0	778
1989	0	0	0	0	35	94	225	159	80	7	0	0	600
1990	0	0	16	40	13	123	198	159	73	13	2	0	637
1991	0	0	0	14	148	181	259	244	102	25	0	0	973
1992	0	0	0	9	30	70	148	117	85	0	0	0	459
1993	0	0	0	4	13	161	331	269	55	7	0	0	840

SNOWFALL (inches) ERIE, PENNSYLVANIA

SEASON	JULY	AUG	SEP	OCT	NOV	DEC	JAN	FEB	MAR	APR	MAY	JUNE	TOTAL
1964-65	0.0	0.0	0.0	T	19.9	13.5	16.8	13.7	26.8	2.2	0.0	0.0	92.9
1965-66	0.0	0.0	0.0	T	18.4	5.3	29.9	8.7	17.7	4.8	T	0.0	84.8
1966-67	0.0	0.0	0.0	0.0	7.2	31.4	6.0	13.5	8.8	1.7	T	0.0	68.6
1967-68	0.0	0.0	0.0	T	36.3	13.3	24.2	24.6	9.1	0.4	0.0	0.0	107.9
1968-69	0.0	0.0	0.0	T	9.8	18.8	26.1	13.6	9.4	2.0	0.0	0.0	79.7
1969-70	0.0	0.0	0.0	1.2	11.4	21.9	19.7	21.9	8.6	0.9	0.0	0.0	85.6
1970-71	0.0	0.0	0.0	T	18.6	23.6	26.8	21.4	26.8	2.8	0.0	0.0	120.0
1971-72	0.0	0.0	0.0	T	14.6	9.3	27.3	32.1	7.9	1.1	0.0	0.0	92.3
1972-73	0.0	0.0	0.0	0.6	6.6	18.5	7.9	12.6	5.7	1.8	T	0.0	53.7
1973-74	0.0	0.0	0.0	T	0.4	10.8	21.2	17.9	13.2	5.1	T	0.0	68.6
1974-75	0.0	0.0	0.0	2.3	2.9	20.1	17.7	12.6	10.0	1.1	0.0	0.0	66.7
1975-76	0.0	0.0	0.0	0.0	2.1	24.7	24.8	6.8	7.5	0.1	T	0.0	66.0
1976-77	0.0	0.0	0.0	3.1	33.9	27.6	29.0	13.0	4.5	0.4	0.0	0.0	111.5
1977-78	0.0	0.0	0.0	0.0	13.1	51.0	62.4	12.6	3.6	0.1	0.0	0.0	142.8
1978-79	0.0	0.0	0.0	0.0	1.8	10.6	33.4	27.8	2.2	0.7	0.0	0.0	76.5
1979-80	0.0	0.0	0.0	0.1	22.2	4.8	9.2	12.2	6.7	T	0.0	0.0	55.2
1980-81	0.0	0.0	0.0	T	4.9	21.0	27.9	22.5	13.1	T	0.0	0.0	89.4
1981-82	0.0	0.0	0.0	T	0.4	17.3	27.1	9.0	7.3	10.2	0.0	0.0	71.3
1982-83	0.0	0.0	0.0	T	9.2	8.9	6.7	9.2	5.6	1.6	0.0	0.0	41.2
1983-84	0.0	0.0	0.0	0.0	1.4	41.1	18.7	27.2	21.6	T	0.0	0.0	110.0
1984-85	0.0	0.0	0.0	0.0	4.7	16.7	57.2	19.0	3.5	5.2	0.0	0.0	106.3
1985-86	0.0	0.0	0.0	0.0	1.9	59.9	30.6	22.9	4.2	5.4	0.0	0.0	124.9
1986-87	0.0	0.0	0.0	0.0	4.6	8.0	31.3	10.1	11.6	2.6	0.0	0.0	68.2
1987-88	0.0	0.0	0.0	T	T	24.3	30.8	31.2	16.8	0.4	0.0	0.0	103.5
1988-89	0.0	0.0	0.0	1.8	0.5	28.1	10.2	21.5	10.5	3.5	0.4	0.0	76.5
1989-90	0.0	0.0	0.0	T	19.6	66.9	13.7	8.3	2.3	4.1	0.0	T	114.9
1990-91	0.0	T	T	T	2.0	15.4	24.3	15.4	2.1	0.4	0.0	0.0	59.6
1991-92	0.0	0.0	0.0	T	13.7	30.0	32.6	8.0	12.8	7.7	0.0	0.0	104.8
1992-93	0.0	T	0.0	T	23.0	15.6	10.0	31.7	27.3	0.9	0.0	0.0	108.5
1993-94	0.0	0.0	T	0.3	3.5	25.7							
Record Mean	0.0	T	T	0.3	10.3	22.9	22.8	16.1	10.4	2.7	T	T	85.5

MIDDLETOWN/HARRISBURG INTL APT
PENNSYLVANIA

No narrative available

NORMALS, MEANS, AND EXTREMES

MIDDLETOWN/HARRISBURG INTL APT PENNSYLVANIA

LATITUDE: 40°12'N LONGITUDE: 76°46' W ELEVATION: FT. GRND 303 BARO 324 TIME ZONE: EASTERN WBAN: 14711

	(a)	JAN	FEB	MAR	APR	MAY	JUNE	JULY	AUG	SEP	OCT	NOV	DEC	YEAR
TEMPERATURE °F:														
Normals														
-Daily Maximum		35.9	39.2	50.3	62.0	72.5	81.2	85.8	83.8	76.3	64.7	52.6	40.6	62.1
-Daily Minimum		21.2	23.3	32.0	41.2	51.1	60.6	65.6	64.3	56.5	44.6	36.1	26.6	43.6
-Monthly		28.6	31.3	41.2	51.6	61.8	70.9	75.7	74.1	66.4	54.7	44.4	33.6	52.9
Extremes														
-Record Highest	55	73	75	86	93	97	100	107	101	102	97	84	75	107
-Year		1950	1985	1945	1985	1942	1966	1966	1944	1953	1941	1950	1984	JUL 1966
-Record Lowest	55	-9	-5	5	19	31	40	49	45	30	23	13	-8	-9
-Year		1985	1979	1984	1982	1966	1980	1945	1976	1963	1969	1955	1960	JAN 1985
NORMAL DEGREE DAYS:														
Heating (base 65°F)		1128	944	738	402	147	9	0	0	59	329	618	973	5347
Cooling (base 65°F)		0	0	0	0	48	186	332	285	101	10	0	0	962
% OF POSSIBLE SUNSHINE	55	49	54	57	58	60	65	68	67	61	57	47	44	57
MEAN SKY COVER (tenths)														
Sunrise - Sunset	43	6.7	6.6	6.6	6.5	6.5	6.1	6.0	5.9	5.8	5.7	6.7	7.0	6.3
MEAN NUMBER OF DAYS:														
Sunrise to Sunset														
-Clear	55	6.7	6.7	7.2	6.4	6.1	6.7	7.2	8.0	8.7	10.0	6.3	6.1	86.0
-Partly Cloudy	55	7.5	7.2	8.2	8.6	10.4	11.5	11.5	11.1	9.5	8.1	8.2	7.5	109.4
-Cloudy	55	16.7	14.4	15.6	15.0	14.5	11.8	12.3	11.9	11.8	12.9	15.6	17.4	169.9
Precipitation														
.01 inches or more	15	10.5	10.1	11.5	12.4	12.8	10.5	10.0	8.9	9.3	8.6	10.1	10.0	124.7
Snow, Ice pellets, hail														
1.0 inches or more	15	2.5	2.3	1.4	0.3	0.0	0.0	0.0	0.0	0.0	0.0	0.4	1.6	8.4
Thunderstorms	50	0.2	0.2	1.1	2.2	5.1	6.2	6.9	5.2	2.9	0.8	0.5	0.2	31.5
Heavy Fog Visibility 1/4 mile or less	50	2.3	2.3	1.7	1.0	0.9	0.6	0.6	0.8	1.5	2.8	1.8	2.4	18.7
Temperature °F														
-Maximum														
90° and above	55	0.0	0.0	0.0	0.3	1.0	4.6	9.0	6.1	1.8	0.1	0.0	0.0	22.9
32° and below	55	9.3	5.9	1.3	0.0	0.0	0.0	0.0	0.0	0.0	0.0	0.2	5.5	22.2
-Minimum														
32° and below	55	26.2	23.1	17.1	3.6	0.1	0.0	0.0	0.0	0.1	2.0	10.9	22.9	105.9
0° and below	55	0.6	0.2	0.0	0.0	0.0	0.0	0.0	0.0	0.0	0.0	0.0	0.1	0.9
AVG. STATION PRESS. (mb)	21	1006.3	1006.2	1004.7	1003.0	1003.3	1003.3	1003.9	1005.4	1006.7	1007.3	1006.8	1007.0	1005.3
RELATIVE HUMIDITY (%)														
Hour 01	51	70	68	67	67	74	79	79	82	81	79	74	71	74
Hour 07 (Local Time)	51	72	71	72	71	74	77	79	83	85	82	77	73	76
Hour 13	50	58	55	52	50	52	53	53	55	56	55	57	58	55
Hour 19	51	63	61	57	54	58	60	60	65	69	67	66	64	62
PRECIPITATION (inches):														
Water Equivalent														
-Normal		2.84	2.93	3.28	3.24	4.26	3.85	3.59	3.31	3.51	2.93	3.52	3.24	40.50
-Maximum Monthly	15	8.01	5.93	6.32	7.96	9.71	8.12	7.20	6.26	8.41	5.59	6.23	7.57	9.71
-Year		1979	1981	1993	1983	1989	1982	1989	1986	1987	1989	1985	1983	MAY 1989
-Minimum Monthly	15	0.43	0.82	1.02	0.45	1.86	1.00	0.97	1.45	0.65	1.34	0.96	0.77	0.43
-Year		1981	1980	1981	1985	1981	1988	1983	1992	1986	1985	1981	1980	JAN 1981
-Maximum in 24 hrs	13	2.09	1.84	1.94	2.06	2.91	2.32	2.58	2.78	3.03	2.19	3.27	2.58	3.27
-Year		1979	1985	1992	1992	1984	1987	1989	1991	1979	1980	1993	1993	NOV 1993
Snow, Ice pellets, hail														
-Maximum Monthly	15	31.5	28.8	22.8	10.2	0.0	T	0.0	T	0.0	T	9.7	12.5	31.5
-Year		1987	1983	1993	1982		1993		1993		1982	1987	1981	JAN 1987
-Maximum in 24 hrs	13	11.2	14.2	20.4	2.6	0.0	T	0.0	T	0.0	T	7.9	9.1	20.4
-Year		1987	1979	1993	1985		1993		1993		1979	1987	1990	MAR 1993
WIND:														
Mean Speed (mph)	51	8.3	9.0	9.5	9.2	7.6	6.9	6.3	5.9	6.1	6.6	7.8	8.1	7.6
Prevailing Direction through 1963		WNW	WNW	WNW	WNW	W	W	W	W	WNW	W	WNW	WNW	WNW
Fastest Obs. 1 Min.														
-Direction (!!)	11	27	31	31	28	29	33	29	35	28	32	31	31	33
-Speed (MPH)	11	44	35	40	35	47	58	35	46	32	37	40	46	58
-Year		1978	1992	1992	1991	1980	1980	1987	1979	1993	1992	1989	1978	JUN 1980
Peak Gust														
-Direction (!!)														
-Speed (mph)														
-Date														

PRECIPITATION (inches)

MIDDLETOWN/HARRISBURG INTL APT PENNSYLVANIA

YEAR	JAN	FEB	MAR	APR	MAY	JUNE	JULY	AUG	SEP	OCT	NOV	DEC	ANNUAL
1964	4.78	3.12	2.94	4.91	0.51	4.20	2.25	3.07	1.77	1.92	1.87	3.11	34.45
1965	2.70	3.29	3.61	1.25	2.38	2.60	3.10	3.99	2.12	3.65	1.63	0.87	31.19
1966	3.57	4.44	1.88	3.44	0.98	0.07	0.81	1.53	6.12	2.12	3.56	3.08	31.60
1967	1.81	1.54	5.26	2.58	4.32	1.90	5.96	5.61	1.80	3.15	2.89	4.27	41.09
1968	1.32	0.53	3.40	2.43	6.55	2.25	1.94	1.77	5.18	2.34	3.38	2.12	33.21
1969	1.06	1.70	2.20	2.13	1.56	2.54	9.72	2.07	2.32	1.63	3.29	6.46	36.68
1970	0.88	3.25	3.64	5.03	2.39	5.80	6.34	2.97	2.12	3.20	4.59	3.50	43.71
1971	2.70	5.62	2.67	1.04	5.30	1.80	2.84	7.77	1.94	2.85	4.96	1.93	41.42
1972	2.65	5.00	2.68	4.10	5.56	18.55	2.26	2.52	1.41	2.03	7.20	5.31	59.27
1973	3.24	2.50	2.00	6.23	6.37	3.34	2.18	2.19	5.73	2.47	1.04	6.52	43.81
1974	3.82	1.36	4.64	3.21	4.38	3.69	2.79	4.13	6.79	1.25	2.30	4.59	42.95
1975	4.12	3.10	3.78	2.80	5.25	6.51	3.13	1.83	14.97	2.62	2.92	3.19	54.22
1976	4.34	1.88	3.43	1.63	5.42	2.42	5.50	3.28	4.79	9.87	0.79	1.96	45.31
1977	1.44	1.75	6.10	4.48	1.00	3.17	3.01	0.93	3.73	3.66	5.61	4.82	39.70
1978	7.44	1.35	3.94	1.97	5.67	5.16	4.35	3.60	1.64	2.51	2.13	3.95	43.71
#1979	8.01	4.74	1.93	3.60	4.66	2.62	3.14	3.24	6.62	3.91	2.67	1.46	46.60
1980	0.90	0.82	5.47	4.27	4.58	2.50	1.59	1.51	1.06	2.94	3.65	0.77	30.06
1981	0.43	5.93	1.02	2.77	1.86	4.66	4.67	4.11	2.20	3.76	0.96	2.41	34.78
1982	3.63	1.92	2.20	4.17	4.89	8.12	2.90	2.47	2.87	1.82	3.37	1.56	39.92
1983	2.26	3.38	4.86	7.96	5.36	2.81	0.97	2.50	1.40	4.21	5.29	7.57	48.57
1984	1.12	4.51	5.36	4.46	6.20	6.36	3.76	2.75	1.49	1.98	3.78	2.28	44.05
1985	1.06	2.91	2.78	0.45	6.29	3.07	2.50	2.14	3.76	1.34	6.23	1.28	33.81
1986	2.24	4.50	3.16	4.10	2.29	1.48	5.17	6.26	0.65	2.59	4.58	4.90	41.92
1987	3.69	1.59	1.43	2.93	3.73	3.46	1.96	2.89	8.41	2.63	4.96	1.84	39.52
1988	2.18	3.28	1.98	2.65	5.79	1.00	4.40	2.67	2.42	1.81	3.67	0.90	32.75
1989	2.29	1.90	3.60	1.10	9.71	6.02	7.20	3.03	2.63	5.59	2.17	1.27	46.51
1990	3.77	2.73	1.76	2.60	7.20	1.10	3.62	6.14	1.65	4.92	2.58	6.05	44.12
#1991	2.61	1.39	3.54	2.00	3.15	1.08	1.99	5.29	1.35	3.15	2.08	3.49	31.12
1992	1.62	1.56	5.13	2.62	3.17	1.90	3.54	1.45	5.65	1.64	4.82	2.42	35.52
1993	2.39	2.32	6.32	6.49	1.96	3.20	3.65	3.45	7.84	2.66	4.17	3.95	48.40
Record Mean	2.79	2.68	3.26	3.07	3.77	3.58	3.61	3.67	3.20	2.91	2.81	2.95	38.29

AVERAGE TEMPERATURE (deg. F)

MIDDLETOWN/HARRISBURG INTL APT PENNSYLVANIA

YEAR	JAN	FEB	MAR	APR	MAY	JUNE	JULY	AUG	SEP	OCT	NOV	DEC	ANNUAL
1964	30.2	28.9	40.5	49.6	64.3	71.4	76.1	71.5	66.2	52.0	45.7	33.9	52.5
1965	26.4	32.5	37.5	50.1	66.5	71.7	75.6	74.6	69.8	53.4	42.2	35.7	53.0
1966	26.3	31.5	43.8	49.0	61.4	74.6	79.9	78.3	66.7	54.7	45.1	31.7	53.6
1967	33.6	26.9	38.8	53.0	55.2	73.9	74.2	71.8	65.3	53.3	39.8	34.8	51.7
1968	24.0	29.2	43.7	53.7	58.0	70.3	77.3	76.7	67.6	56.1	44.9	31.3	52.7
1969	29.9	33.3	38.6	53.4	63.2	72.4	75.3	73.4	65.6	53.8	42.7	31.4	52.8
1970	22.9	30.4	36.7	51.4	65.1	70.7	76.2	75.6	71.6	58.7	46.0	34.8	53.4
1971	26.4	33.2	38.6	50.4	60.0	72.5	75.5	72.2	69.9	60.6	43.5	40.3	53.6
1972	34.5	30.2	40.0	49.6	62.3	67.7	76.2	74.0	68.4	52.1	41.8	36.8	52.8
1973	33.7	31.7	45.2	50.9	57.8	73.4	75.5	76.1	68.6	57.4	46.9	36.2	54.5
1974	34.8	32.3	42.8	56.2	63.0	70.5	77.6	77.1	64.5	51.6	44.9	35.4	54.2
1975	33.3	32.4	38.3	47.4	64.8	70.7	75.3	76.2	63.3	57.7	50.0	34.2	53.7
1976	27.6	39.6	44.2	54.5	59.4	73.4	72.3	72.4	64.8	51.5	39.6	30.0	52.4
1977	20.1	30.4	46.0	54.1	64.5	68.6	75.9	74.3	68.3	52.7	46.1	31.6	52.7
1978	26.2	22.8	38.6	51.0	61.5	69.5	73.1	76.9	67.6	54.4	46.7	36.5	52.1
1979	29.2	22.6	45.1	50.4	62.2	68.5	73.4	72.9	65.5	52.3	46.8	37.6	52.2
1980	30.3	29.1	38.9	52.8	63.3	67.8	76.3	76.1	67.7	51.5	39.4	29.6	51.9
1981	23.7	34.6	38.7	53.7	61.9	71.7	75.7	72.2	63.9	50.7	44.7	31.9	51.9
1982	22.8	30.9	38.6	47.6	62.2	65.1	74.4	70.5	65.3	55.1	47.6	41.4	51.8
1983	33.0	33.4	42.7	49.3	58.4	69.1	75.9	74.9	66.2	53.6	43.9	28.7	52.4
1984	24.8	36.6	33.7	48.0	58.2	72.9	74.3	75.8	64.4	61.5	43.9	41.5	53.0
1985	27.9	34.4	44.5	56.9	65.2	69.4	75.9	74.1	69.2	57.2	47.9	31.0	54.5
1986	31.4	30.0	43.5	53.5	65.6	71.4	76.3	72.0	66.2	56.0	41.2	36.1	53.6
1987	30.0	32.3	44.1	52.3	63.2	72.6	78.2	73.1	65.7	49.4	43.9	36.6	53.5
1988	24.4	31.8	42.4	50.0	62.3	70.8	78.8	76.2	63.5	49.5	43.7	33.3	52.2
1989	34.8	32.4	40.4	50.4	60.0	70.6	73.7	72.6	66.0	55.6	42.7	22.6	51.8
1990	38.2	38.2	44.9	53.1	59.4	71.2	75.2	72.7	65.0	58.2	46.9	38.4	55.1
#1991	31.7	37.4	43.7	53.7	69.1	74.2	78.7	76.2	65.4	56.9	43.7	36.5	55.6
1992	33.0	35.7	40.3	51.8	60.2	69.3	75.0	71.6	65.5	51.3	44.3	35.3	52.8
1993	34.8	28.8	36.6	51.3	64.6	72.1	78.0	75.9	65.6	52.8	43.9	34.0	53.2
Record Mean	30.2	31.3	40.4	50.7	62.1	70.8	75.3	73.4	66.5	55.0	43.6	33.2	52.7
Max	37.1	38.7	48.8	59.6	72.2	80.6	84.9	82.8	75.9	64.3	51.5	40.0	61.4
Min	23.3	23.8	32.0	41.8	52.1	60.9	65.7	63.9	57.0	45.6	35.7	26.4	44.0

REFERENCE NOTES

MIDDLETOWN/HARRISBURG INTL APT PENNSYLVANIA

GENERAL
T - TRACE AMOUNT.
BLANK ENTRIES DENOTE MISSING/UNREPORTED DATA.
# INDICATES A STATION OR INSTRUMENT RELOCATION.

SPECIFIC

PM - INCLUDES LAST DAY OF PREVIOUS MONTH
ASOS - AUTOMATED SURFACE OBSERVING SYSTEM IN OPERATION DURING THESE MONTHS.

(a) - LENGTH OF RECORD IN YEARS, ALTHOUGH INDIVIDUAL MONTHS MAY BE MISSING.
0.* OR * - THE VALUE IS BETWEEN 0.0 AND 0.05
NORMALS - BASED ON THE 1961-1990 RECORD PERIOD.
EXTREMES - DATES ARE THE MOST RECENT OCCURRENCE
WIND DIR.- NUMERALS SHOW TENS OF DEGRESS CLOCKWISE FROM TRUE NORTH. "00" INDICATES CALM.
RESULTANT DIRECTIONS ARE GIVEN TO WHOLE DEGREES.
BOLD VALUES INDICATE EXTREME VALUES WHICH OCCURRED AFTER THE ASOS SYSTEM WAS COMMISSIONED.

RECORD = PERIOD OF RECORD
RECORD MEAN PRECIPITATION IS THE MEAN OF ALL DAILY PRECIPITATION AMOUNTS DURING THE PERIOD OF RECORD.
RECORD MAX(MIN) TEMPERATURE IS THE MEAN OF ALL DAILY MAX(MIN) TEMPERATURES DURING THE PERIOD OF RECORD.
RECORD MEAN TEMPERATURE IS THE SUM OF THE RECORD MAX AND RECORD MIN DIVIDED BY 2.
AVERAGE TEMPERATURE IS THE SUM OF THE MEAN DAILY MAX AND MIN TEMPERATURE DIVIDED BY 2.

HEATING DEGREE DAYS Base 65 deg. F — MIDDLETOWN/HARRISBURG INTL APT PENNSYLVANIA

SEASON	JULY	AUG	SEP	OCT	NOV	DEC	JAN	FEB	MAR	APR	MAY	JUNE	TOTAL
1964-65	0	4	71	399	570	958	1192	902	847	441	66	27	5477
1965-66	0	7	44	357	677	900	1191	932	648	472	166	12	5406
1966-67	0	0	69	312	592	1024	966	1061	806	364	306	3	5503
1967-68	0	5	85	369	750	931	1266	1032	654	331	216	12	5651
1968-69	0	3	18	277	593	1038	1083	882	809	345	123	8	5179
1969-70	0	0	89	352	662	1036	1299	962	872	407	93	13	5785
1970-71	0	0	39	213	564	931	1192	884	812	432	165	12	5244
1971-72	0	3	40	146	643	759	940	1001	766	457	108	44	4907
1972-73	0	1	25	395	686	865	964	926	607	422	227	7	5125
1973-74	0	1	34	238	534	887	931	910	683	289	133	2	4642
1974-75	0	0	94	414	600	911	977	903	818	520	97	13	5347
1975-76	0	0	87	232	445	951	1150	730	639	354	193	23	4804
1976-77	0	5	75	418	756	1075	1387	966	588	340	106	32	5748
1977-78	0	5	35	377	562	1029	1196	1175	810	414	173	17	5793
1978-79	14	0	48	321	544	876	1104	1182	611	435	123	26	5284
1979-80	12	14	71	393	536	844	1070	1033	799	361	103	48	5284
1980-81	0	0	57	411	761	1091	1277	844	809	339	147	6	5742
1981-82	0	1	94	437	599	1021	1304	948	812	518	128	61	5923
1982-83	7	12	67	318	520	725	985	876	686	468	221	25	4910
1983-84	0	2	103	362	628	1117	1238	817	962	502	240	11	5982
1984-85	0	0	105	131	627	724	1143	849	627	292	87	16	4601
1985-86	0	0	41	237	508	1049	1038	974	664	349	89	9	4958
1986-87	2	17	46	300	705	890	1080	907	643	380	142	2	5114
1987-88	0	8	51	477	627	873	1252	961	693	445	131	41	5559
1988-89	4	5	88	475	633	975	931	912	760	433	196	9	5421
1989-90	1	6	81	292	663	1306	824	744	629	385	175	13	5119
1990-91	5	8	96	248	535	816	1026	769	651	345	69	2	4570
#1991-92	0	0	103	279	634	877	986	842	761	392	170	16	5060
#1992-93	0	3	86	420	615	914	931	1005	875	405	67	15	5336
1993-94	0	2	99	371	628	954							

COOLING DEGREE DAYS Base 65 deg. F — MIDDLETOWN/HARRISBURG INTL APT PENNSYLVANIA

YEAR	JAN	FEB	MAR	APR	MAY	JUNE	JULY	AUG	SEP	OCT	NOV	DEC	TOTAL
1969	0	0	0	3	71	235	327	267	113	11	0	0	1027
1970	0	0	0	7	102	191	353	333	246	24	0	0	1256
1971	0	0	0	0	17	241	331	236	195	17	6	0	1043
1972	0	0	0	0	33	133	357	287	137	3	0	0	950
1973	0	0	0	8	13	266	334	352	148	9	0	0	1130
1974	0	0	0	34	79	176	401	381	88	1	3	0	1163
1975	0	0	0	0	97	192	325	354	44	12	2	0	1026
1976	0	0	0	47	22	283	233	240	73	3	0	0	901
1977	0	0	4	19	95	147	347	297	140	1	3	0	1053
1978	0	0	0	0	69	162	273	377	133	1	0	0	1015
1979	0	0	0	5	43	138	279	264	92	7	0	0	828
1980	0	0	0	0	57	138	355	350	145	0	0	0	1045
1981	0	0	0	6	60	213	339	232	66	0	0	0	916
1982	0	0	0	2	48	70	307	191	83	19	4	0	724
1983	0	0	0	6	22	154	343	315	146	13	0	0	999
1984	0	0	0	0	34	256	292	342	95	35	0	1	1055
1985	0	0	0	55	99	157	345	290	173	4	0	0	1123
1986	0	0	5	12	116	205	360	238	88	27	0	0	1051
1987	0	0	0	9	94	237	418	266	76	0	0	0	1100
1988	0	0	0	0	53	219	439	355	52	4	0	0	1122
1989	0	0	5	0	49	182	279	249	114	9	1	0	888
1990	0	0	14	34	12	205	330	254	102	43	0	0	994
#1991	0	0	0	16	202	284	429	355	118	31	0	0	1435
1992	0	0	0	1	29	151	314	214	108	0	0	0	817
1993	0	0	0	0	63	236	410	348	127	0	0	0	1184

SNOWFALL (inches) — MIDDLETOWN/HARRISBURG INTL APT PENNSYLVANIA

SEASON	JULY	AUG	SEP	OCT	NOV	DEC	JAN	FEB	MAR	APR	MAY	JUNE	TOTAL
1964-65	0.0	0.0	0.0	0.0	T	1.4	13.4	1.7	15.1	0.2	0.0	0.0	31.8
1965-66	0.0	0.0	0.0	T	T	T	24.8	16.8	1.0	T	T	0.0	42.6
1966-67	0.0	0.0	0.0	0.0	0.2	19.9	1.6	16.5	10.2	T	0.0	0.0	48.4
1967-68	0.0	0.0	0.0	0.0	9.7	13.0	3.0	1.8	3.5	0.0	0.0	0.0	31.0
1968-69	0.0	0.0	0.0	0.0	3.0	0.2	1.2	16.8	3.8	0.0	0.0	0.0	25.0
1969-70	0.0	0.0	0.0	T	2.1	28.3	9.8	7.5	12.9	T	0.0	0.0	60.6
1970-71	0.0	0.0	0.0	0.0	0.0	10.9	11.6	5.1	5.3	T	0.0	0.0	32.9
1971-72	0.0	0.0	0.0	0.0	8.8	T	2.6	21.6	0.5	1.1	0.0	0.0	34.6
1972-73	0.0	0.0	0.0	1.2	5.9	0.5	T	5.7	T	T	0.0	0.0	13.3
1973-74	0.0	0.0	0.0	0.0	T	15.3	7.0	5.5	T	T	0.0	0.0	27.8
1974-75	0.0	0.0	0.0	0.0	T	0.5	11.3	13.1	6.1	T	0.0	0.0	31.0
1975-76	0.0	0.0	0.0	0.0	T	2.2	3.6	2.5	10.0	0.0	0.0	0.0	18.3
1976-77	0.0	0.0	0.0	0.0	1.4	5.1	12.2	4.5	0.2	T	T	0.0	23.4
1977-78	0.0	0.0	0.0	T	1.5	5.5	33.5	21.1	9.0	0.0	0.0	0.0	70.6
#1978-79	0.0	0.0	0.0	0.0	4.0	0.3	9.2	26.0	T	T	0.0	0.0	39.5
1979-80	0.0	0.0	0.0	T	T	0.2	3.8	2.7	7.9	0.0	0.0	0.0	14.6
1980-81	0.0	0.0	0.0	0.0	4.0	4.3	5.5	4.4	6.7	0.0	0.0	0.0	24.9
1981-82	0.0	0.0	0.0	0.0	0.8	12.5	18.8	8.4	7.8	10.2	0.0	0.0	58.5
1982-83	0.0	0.0	0.0	T	T	1.1	4.4	28.8	0.4	1.3	0.0	0.0	36.0
1983-84	0.0	0.0	0.0	0.0	T	4.6	9.7	2.3	14.9	T	0.0	0.0	31.5
1984-85	0.0	0.0	0.0	0.0	1.9	2.6	10.6	11.4	T	3.6	0.0	0.0	30.1
1985-86	0.0	0.0	0.0	0.0	T	5.6	7.8	23.1	T	T	0.0	0.0	36.5
1986-87	0.0	0.0	0.0	0.0	T	1.9	31.5	10.1	1.4	1.0	0.0	0.0	45.9
1987-88	0.0	0.0	0.0	0.0	9.7	3.6	9.6	2.8	1.0	T	0.0	0.0	26.7
1988-89	0.0	0.0	0.0	0.0	T	T	6.4	2.2	11.3	0.0	0.0	0.0	19.9
1989-90	0.0	0.0	0.0	0.0	1.8	6.7	4.9	1.3	3.5	1.1	0.0	0.0	19.3
1990-91	0.0	0.0	0.0	0.0	0.0	9.3	5.2	0.7	5.9	T	0.0	0.0	21.1
#1991-92	0.0	0.0	0.0	0.0	0.0	T	1.7	4.6	6.6	0.0	0.0	T	12.9
1992-93	0.0	0.0	0.0	0.0	T	3.9	1.4	18.5	22.8	0.6	0.0	T	47.2
1993-94	0.0	T	0.0	0.0	T	0.9							
Record Mean	0.0	T	0.0	T	1.9	6.6	9.4	9.3	6.5	0.5	T	T	34.3

PHILADELPHIA, PENNSYLVANIA

The Appalachian Mountains to the west and the Atlantic Ocean to the east have a moderating effect on climate. Periods of very high or very low temperatures seldom last for more than three or four days. Temperatures below zero or above 100 degrees are a rarity. On occasion, the area becomes engulfed with maritime air during the summer months, and high humidity adds to the discomfort of seasonably warm temperatures.

Precipitation is fairly evenly distributed throughout the year with maximum amounts during the late summer months. Much of the summer rainfall is from local thunderstorms and amounts vary in different areas of the city. This is due, in part, to the higher elevations to the west and north. Snowfall amounts are often considerably larger in the northern suburbs than in the central and southern parts of the city. In many cases, the precipitation will change from snow to rain within the city. Single storms of 10 inches or more occur about every five years.

The prevailing wind direction for the summer months is from the southwest, while northwesterly winds prevail during the winter. The annual prevailing direction is from the west-southwest. Destructive velocities are comparatively rare and occur mostly in gustiness during summer thunderstorms. High winds occurring in the winter months, as a rule, come with the advance of cold air after the passage of a deep low pressure system. Only rarely have hurricanes in the vicinity caused widespread damage, primarily because of flooding.

Flood stages in the Schuylkill River normally occur about twice a year. Flood stages seldom last over 12 hours and usually occur after excessive thunderstorms. Flooding rarely occurs on the Delaware River.

NORMALS, MEANS, AND EXTREMES

PHILADELPHIA, PENNSYLVANIA

LATITUDE: 39°53'N LONGITUDE: 75°15'W ELEVATION: FT. GRND 5 BARO 63 TIME ZONE: EASTERN WBAN: 13739

	(a)	JAN	FEB	MAR	APR	MAY	JUNE	JULY	AUG	SEP	OCT	NOV	DEC	YEAR
TEMPERATURE °F:														
Normals														
-Daily Maximum		37.9	41.0	51.6	62.6	73.1	81.7	86.1	84.6	77.6	66.3	55.1	43.4	63.4
-Daily Minimum		22.8	24.8	33.2	42.1	52.7	61.8	67.2	66.3	58.7	46.4	37.6	28.1	45.1
-Monthly		30.4	33.0	42.4	52.4	62.9	71.8	76.7	75.5	68.2	56.4	46.4	35.8	54.3
Extremes														
-Record Highest	52	74	74	87	94	97	100	104	101	100	96	81	72	104
-Year		1950	1985	1945	1976	1991	1991	1966	1955	1953	1941	1993	1984	JUL 1966
-Record Lowest	52	-7	-4	7	19	28	44	51	44	35	25	15	1	-7
-Year		1984	1961	1984	1982	1966	1984	1966	1986	1963	1969	1976	1983	JAN 1984
NORMAL DEGREE DAYS:														
Heating (base 65°F)		1073	896	701	378	123	5	0	0	32	283	558	905	4954
Cooling (base 65°F)		0	0	0	0	58	209	363	326	128	17	0	0	1101
% OF POSSIBLE SUNSHINE	51	50	53	55	55	56	62	62	61	59	59	52	49	56
MEAN SKY COVER (tenths)														
Sunrise - Sunset	53	6.6	6.4	6.4	6.4	6.5	6.1	6.0	5.8	5.7	5.5	6.3	6.5	6.2
MEAN NUMBER OF DAYS:														
Sunrise to Sunset														
-Clear	53	7.2	7.2	7.7	7.1	6.2	7.1	7.3	8.3	9.6	10.5	7.2	7.4	92.9
-Partly Cloudy	53	7.9	7.4	8.1	8.8	10.6	11.5	11.7	11.3	9.0	8.7	8.9	8.3	112.2
-Cloudy	53	15.9	13.7	15.2	14.1	14.2	11.4	12.1	11.4	11.4	11.8	13.9	15.2	160.2
Precipitation .01 inches or more	53	10.8	9.3	10.9	10.7	11.1	10.0	9.3	9.1	8.2	7.6	9.4	10.1	116.6
Snow,Ice pellets,hail 1.0 inches or more	53	1.9	1.6	1.0	0.1	0.0	0.0	0.0	0.0	0.0	0.*	0.2	0.8	5.7
Thunderstorms	53	0.2	0.3	1.0	2.0	4.2	5.4	5.7	4.9	2.4	0.8	0.6	0.2	27.5
Heavy Fog Visibility 1/4 mile or less	53	2.7	2.2	1.7	1.2	1.3	1.0	0.8	1.0	1.5	3.3	2.4	2.5	21.6
Temperature °F														
-Maximum														
90° and above	34	0.0	0.0	0.0	0.4	1.1	4.2	8.9	6.3	1.9	0.0	0.0	0.0	22.8
32° and below	34	8.5	5.4	0.9	0.0	0.0	0.0	0.0	0.0	0.0	0.0	0.1	4.2	19.1
-Minimum														
32° and below	34	25.8	22.4	14.2	2.4	0.*	0.0	0.0	0.0	0.0	1.4	8.3	20.9	95.4
0° and below	34	0.4	0.1	0.0	0.0	0.0	0.0	0.0	0.0	0.0	0.0	0.0	0.0	0.5
AVG. STATION PRESS.(mb)	21	1017.6	1017.5	1015.9	1014.2	1014.4	1014.3	1014.8	1016.3	1017.6	1018.4	1018.1	1018.3	1016.5
RELATIVE HUMIDITY (%)														
Hour 01	34	71	68	69	70	77	79	81	82	82	80	75	72	76
Hour 07 (Local Time)	34	73	71	72	71	75	76	79	81	83	83	78	74	76
Hour 13	34	59	55	52	49	52	53	54	55	56	54	56	59	55
Hour 19	34	65	60	58	55	59	60	63	65	69	69	66	66	63
PRECIPITATION (inches):														
Water Equivalent														
-Normal		3.21	2.79	3.46	3.62	3.75	3.74	4.28	3.80	3.42	2.62	3.34	3.38	41.41
-Maximum Monthly	51	8.86	6.44	7.01	8.12	7.41	7.88	9.44	9.70	8.78	5.21	9.06	7.37	9.70
-Year		1978	1979	1980	1983	1948	1973	1989	1955	1960	1943	1972	1983	AUG 1955
-Minimum Monthly	51	0.45	0.75	0.68	0.52	0.47	0.11	0.64	0.49	0.44	0.09	0.32	0.25	0.09
-Year		1955	1991	1966	1985	1964	1949	1957	1964	1968	1963	1976	1955	OCT 1963
-Maximum in 24 hrs	47	2.70	1.96	2.39	2.76	3.18	4.62	4.49	5.68	5.45	3.85	3.99	3.03	5.68
-Year		1979	1966	1968	1970	1984	1973	1989	1971	1960	1980	1977	1992	AUG 1971
Snow,Ice pellets,hail														
-Maximum Monthly	51	23.4	27.6	13.4	4.3	T	T	0.0	0.0	0.0	2.1	8.8	18.8	27.6
-Year		1978	1979	1958	1971	1963	1993				1979	1953	1966	FEB 1979
-Maximum in 24 hrs	51	13.2	21.3	12.0	4.3	T	T	0.0	0.0	0.0	2.1	8.7	14.6	21.3
-Year		1961	1983	1993	1971	1963	1993				1979	1953	1960	FEB 1983
WIND:														
Mean Speed (mph)	53	10.3	10.9	11.4	10.9	9.5	8.8	8.2	7.9	8.3	8.9	9.6	10.1	9.6
Prevailing Direction through 1963		WNW	NW	N	SW	WSW	WSW	WSW	SW	SW	WSW	WSW	WNW	WSW
Fastest Mile														
-Direction (!!!)	53	NE	NW	NW	SW	SW	NW	SW	E	NE	SW	SW	NW	NW
-Speed (MPH)	53	61	59	56	59	56	73	49	67	49	66	60	48	73
-Year		1958	1956	1989	1958	1957	1958	1980	1955	1960	1954	1958	1988	JUN 1958
Peak Gust														
-Direction (!!!)	10	NW	NW	NW	NW	NW	NW	W	N	W	SW	NW	W	NW
-Speed (mph)	10	59	44	69	43	67	54	60	47	53	55	61	63	69
-Date		1992	1992	1989	1993	1984	1993	1992	1988	1985	1990	1989	1991	MAR 1989

PRECIPITATION (inches)

PHILADELPHIA, PENNSYLVANIA

YEAR	JAN	FEB	MAR	APR	MAY	JUNE	JULY	AUG	SEP	OCT	NOV	DEC	ANNUAL
1964	3.92	2.83	1.94	5.27	0.47	0.21	3.83	0.49	2.42	1.73	1.64	5.13	29.88
1965	2.35	2.18	3.19	2.33	1.23	2.85	3.22	4.05	3.02	2.02	1.05	1.85	29.34
1966	2.82	4.30	0.68	4.35	2.95	0.41	2.35	1.63	8.70	5.12	2.36	4.33	40.00
1967	1.67	1.82	4.53	2.17	3.49	4.12	7.11	7.08	2.96	2.00	1.99	5.88	44.82
1968	2.90	1.40	4.98	1.57	5.17	5.89	2.00	1.24	0.44	3.15	4.17	2.54	35.45
1969	1.57	1.88	1.92	1.68	3.30	7.31	8.33	2.66	4.38	1.13	1.97	7.23	43.36
1970	0.74	2.08	3.83	6.12	2.57	4.60	2.75	3.99	0.82	3.66	4.71	3.27	39.14
1971	2.13	5.43	2.58	1.84	4.10	1.01	4.84	9.61	5.83	3.84	5.37	1.21	47.79
1972	2.34	5.09	2.69	4.08	4.11	5.79	2.62	3.76	1.12	3.77	9.06	5.20	49.63
1973	3.93	2.96	3.52	6.68	4.14	7.88	2.39	2.03	3.39	2.16	0.64	6.34	46.06
1974	2.95	2.14	4.91	2.77	3.21	4.43	2.08	3.83	4.68	1.93	0.81	4.04	37.78
1975	4.00	2.91	4.68	2.97	4.99	7.57	6.32	2.21	7.21	3.24	3.14	2.89	52.13
1976	4.50	1.66	2.38	2.06	4.35	3.42	4.04	2.17	2.44	4.30	0.32	1.63	33.27
1977	2.61	1.33	4.19	5.59	0.70	5.33	1.47	8.70	3.44	3.11	7.76	5.19	49.42
1978	8.86	1.35	4.31	1.76	6.01	1.75	5.27	6.04	1.59	1.20	2.20	5.61	45.95
1979	8.74	6.44	2.43	4.08	3.98	4.34	3.95	5.95	4.89	3.84	2.48	1.67	52.79
1980	2.27	0.96	7.01	4.79	3.22	1.73	6.58	0.80	2.79	5.03	2.85	0.77	38.80
1981	0.50	2.94	1.61	3.60	4.53	4.40	4.54	5.11	2.83	2.68	0.95	4.14	37.83
1982	4.45	3.16	2.66	6.06	4.47	5.76	1.94	2.20	2.32	1.94	3.67	1.80	40.43
1983	2.81	3.53	6.70	8.12	7.03	2.75	0.68	2.57	3.45	3.69	5.71	7.37	54.41
1984	2.22	2.81	6.14	4.25	6.87	2.85	6.99	3.28	1.96	2.56	1.56	2.17	43.66
1985	1.55	2.44	1.95	0.52	4.99	1.88	4.66	2.82	5.78	1.54	6.09	0.98	35.20
1986	4.13	3.38	1.25	4.46	0.70	1.99	4.10	3.70	2.33	2.22	6.27	5.89	40.42
1987	4.58	1.17	1.16	3.63	3.15	2.01	4.82	3.72	2.78	2.62	2.08	1.68	33.40
1988	2.72	4.11	2.24	2.92	3.67	0.57	8.07	3.16	2.62	2.16	5.17	1.00	38.41
1989	2.41	3.25	4.41	2.27	6.76	4.73	9.44	3.92	5.03	3.44	1.79	1.21	48.66
1990	4.09	1.44	2.59	3.16	6.08	3.39	2.62	4.07	1.71	1.68	1.17	3.79	35.79
1991	4.10	0.75	4.13	2.81	1.82	3.36	4.79	3.86	3.58	1.61	1.55	3.86	36.22
1992	0.88	1.31	3.19	1.26	2.74	1.84	5.05	2.00	3.04	1.23	3.26	4.61	30.41
1993	1.97	3.03	6.61	4.20	2.42	1.52	1.98	5.18	6.66	2.69	2.23	3.69	42.18
Record Mean	3.21	3.02	3.53	3.34	3.45	3.55	4.16	4.41	3.39	2.76	3.11	3.20	41.14

AVERAGE TEMPERATURE (deg. F)

PHILADELPHIA, PENNSYLVANIA

YEAR	JAN	FEB	MAR	APR	MAY	JUNE	JULY	AUG	SEP	OCT	NOV	DEC	ANNUAL
1964	33.0	31.8	42.7	50.8	65.1	72.4	76.6	72.2	67.2	52.6	47.1	37.5	54.1
1965	29.2	33.3	37.6	49.0	65.5	70.0	74.1	73.1	69.2	53.7	44.2	37.0	53.0
1966	29.1	31.5	42.5	47.8	59.5	72.1	77.9	74.8	65.2	53.1	46.8	35.5	53.0
1967	36.0	29.0	38.5	51.7	55.9	72.1	76.6	75.1	67.0	56.8	42.8	38.5	53.3
1968	28.9	30.4	44.4	54.6	59.7	71.2	77.1	77.8	69.4	58.1	45.6	32.3	54.1
1969	29.8	32.0	39.7	55.3	64.6	73.4	75.1	75.2	67.2	55.0	44.4	33.5	53.8
1970	24.5	33.1	38.3	51.5	64.9	71.6	76.9	76.7	72.0	60.1	48.2	35.8	54.5
1971	27.8	36.1	40.7	51.6	60.9	74.3	77.4	75.3	71.6	63.5	46.1	41.6	55.6
1972	35.1	32.4	40.7	49.7	63.6	68.7	77.1	76.0	69.2	52.7	43.6	39.9	54.1
1973	34.4	33.6	47.2	53.4	60.3	74.6	77.9	78.8	70.7	59.2	48.0	38.6	56.4
1974	35.9	31.7	43.3	55.8	62.4	70.3	76.9	76.8	68.1	54.8	48.5	39.4	55.3
1975	37.3	35.8	41.2	48.7	66.6	72.2	76.6	77.1	66.6	61.2	52.7	36.9	56.1
1976	28.7	40.9	46.3	56.6	62.7	75.2	75.3	74.8	67.3	52.5	39.9	30.3	54.2
1977	20.0	33.6	48.8	57.2	65.8	68.6	77.8	76.2	69.9	54.3	46.4	32.6	54.3
1978	28.0	24.7	39.0	50.6	61.4	72.6	75.6	79.2	68.5	55.5	47.9	38.6	53.5
1979	32.5	23.0	47.0	52.3	66.4	69.1	76.2	75.5	68.5	54.9	50.1	38.2	54.5
1980	31.8	29.7	40.2	54.7	65.4	70.6	78.5	80.0	72.2	54.9	43.2	32.5	54.5
1981	25.3	37.9	40.0	54.7	62.6	72.0	76.9	74.9	66.8	53.1	45.6	34.6	53.7
1982	24.7	34.4	41.7	50.2	65.9	68.7	76.9	73.5	67.6	56.9	48.4	41.3	54.2
1983	34.1	34.0	43.7	51.0	62.1	72.0	77.9	77.1	69.0	56.6	46.7	33.2	54.8
1984	26.2	38.7	35.5	50.2	60.2	73.0	73.9	75.2	64.7	61.2	44.4	41.9	53.8
1985	27.3	35.3	44.6	55.5	64.5	68.8	75.4	74.1	69.1	59.3	51.3	33.3	54.9
1986	32.8	32.1	44.5	53.3	66.8	73.8	78.1	74.0	68.3	57.8	44.5	37.9	55.3
1987	31.9	32.5	45.7	53.1	63.9	74.6	79.5	75.4	68.8	52.5	48.0	39.2	55.4
1988	27.3	34.6	44.7	51.3	63.6	72.3	80.7	78.3	66.7	51.8	47.7	35.4	54.5
1989	36.5	34.8	42.3	52.4	62.4	74.7	76.3	75.6	69.7	58.3	44.9	25.5	54.5
1990	40.3	41.2	46.1	53.3	61.3	72.2	78.0	75.8	68.0	61.9	49.7	42.1	57.5
1991	35.2	40.0	46.1	55.5	70.8	75.7	79.0	79.0	69.5	58.9	47.3	39.6	58.1
1992	35.7	37.5	41.6	53.2	62.5	71.3	77.1	73.5	68.4	55.2	48.0	38.8	55.2
1993	38.2	31.9	39.8	54.2	66.4	74.4	81.4	78.9	69.7	58.0	49.0	38.4	56.7
Record Mean	32.8	33.8	41.9	52.4	63.2	71.8	76.7	74.9	68.4	57.3	46.3	36.2	54.6
Max	39.9	41.4	50.3	61.8	72.8	81.2	85.5	83.3	77.0	66.1	54.1	43.1	63.0
Min	25.6	26.2	33.4	42.9	53.5	62.5	67.9	66.4	59.8	48.5	38.5	29.2	46.2

REFERENCE NOTES

PHILADELPHIA, PENNSYLVANIA

GENERAL
T - TRACE AMOUNT.
BLANK ENTRIES DENOTE MISSING/UNREPORTED DATA.
# INDICATES A STATION OR INSTRUMENT RELOCATION.

SPECIFIC

PM - INCLUDES LAST DAY OF PREVIOUS MONTH
ASOS - AUTOMATED SURFACE OBSERVING SYSTEM IN OPERATION DURING THESE MONTHS.

(*) - LENGTH OF RECORD IN YEARS, ALTHOUGH INDIVIDUAL MONTHS MAY BE MISSING.
0.* OR * - THE VALUE IS BETWEEN 0.0 AND 0.05
NORMALS - BASED ON THE 1961-1990 RECORD PERIOD.
EXTREMES - DATES ARE THE MOST RECENT OCCURRENCE
WIND DIR.- NUMERALS SHOW TENS OF DEGRESS CLOCKWISE FROM TRUE NORTH. "00" INDICATES CALM.
RESULTANT DIRECTIONS ARE GIVEN TO WHOLE DEGREES.
BOLD VALUES INDICATE EXTREME VALUES WHICH OCCURRED AFTER THE ASOS SYSTEM WAS COMMISSIONED.

RECORD = PERIOD OF RECORD
RECORD MEAN PRECIPITATION IS THE MEAN OF ALL DAILY PRECIPITATION AMOUNTS DURING THE PERIOD OF RECORD.
RECORD MAX(MIN) TEMPERATURE IS THE MEAN OF ALL DAILY MAX(MIN) TEMPERATURES DURING THE PERIOD OF RECORD.
RECORD MEAN TEMPERATURE IS THE SUM OF THE RECORD MAX AND RECORD MIN DIVIDED BY 2.
AVERAGE TEMPERATURE IS THE SUM OF THE MEAN DAILY MAX AND MIN TEMPERATURE DIVIDED BY 2.

EXCEPTIONS

RECORD MEANS ARE THROUGH THE CURRENT YEAR,
BEGINNING IN 1874 FOR TEMPERATURE
1872 FOR PRECIPITATION
1943 FOR SNOWFALL

HEATING DEGREE DAYS Base 65 deg. F PHILADELPHIA, PENNSYLVANIA

SEASON	JULY	AUG	SEP	OCT	NOV	DEC	JAN	FEB	MAR	APR	MAY	JUNE	TOTAL
1964-65	0	2	51	377	532	847	1107	883	839	475	66	26	5205
1965-66	0	18	41	342	614	862	1110	931	693	509	207	21	5348
1966-67	0	0	83	362	538	908	893	1001	817	396	280	6	5284
1967-68	0	0	55	271	660	814	1112	995	633	305	170	7	5022
1968-69	0	0	14	234	576	1008	1084	918	782	290	84	2	4992
1969-70	0	0	54	316	611	970	1247	890	821	399	92	0	5400
1970-71	0	0	29	191	499	899	1145	802	746	394	140	3	4848
1971-72	0	0	17	79	576	719	920	941	748	450	86	26	4562
1972-73	0	0	22	378	635	775	940	874	547	359	176	1	4707
1973-74	0	0	18	196	507	810	897	926	667	292	128	11	4452
1974-75	0	0	46	313	500	786	852	812	732	483	66	4	4594
1975-76	0	0	45	152	372	866	1120	692	572	307	119	13	4258
1976-77	0	2	42	387	743	1069	1390	873	505	258	73	36	5378
1977-78	0	0	24	328	558	998	1139	1121	797	423	161	10	5559
1978-79	5	0	41	296	507	811	999	1170	556	378	38	17	4818
1979-80	4	7	28	324	439	823	1021	1016	763	301	72	17	4815
1980-81	0	0	22	320	646	999	1222	752	768	309	129	4	5171
1981-82	0	0	58	364	576	936	1243	850	714	440	50	25	5256
1982-83	0	8	31	277	497	730	951	861	653	423	128	2	4561
1983-84	0	0	70	283	540	981	1196	756	911	438	181	13	5369
1984-85	0	0	92	138	613	709	1161	824	627	306	89	9	4568
1985-86	0	0	38	187	407	975	990	914	628	345	77	6	4567
1986-87	0	21	23	255	609	838	1017	904	591	359	129	1	4747
1987-88	0	0	20	379	504	796	1162	876	624	404	105	32	4902
1988-89	0	0	35	408	513	908	876	840	700	371	138	0	4789
1989-90	0	0	43	220	594	1219	757	662	588	375	127	6	4591
1990-91	2	1	55	171	453	701	920	694	576	296	44	0	3913
1991-92	0	0	42	215	527	778	903	789	720	356	124	9	4463
1992-93	0	1	49	310	504	804	824	920	773	314	47	6	4552
1993-94	0	0	34	219	479	820							

COOLING DEGREE DAYS Base 65 deg. F PHILADELPHIA, PENNSYLVANIA

YEAR	JAN	FEB	MAR	APR	MAY	JUNE	JULY	AUG	SEP	OCT	NOV	DEC	TOTAL
1969	0	0	0	9	77	259	319	323	126	15	0	0	1128
1970	0	0	0	3	100	204	376	367	247	46	0	0	1343
1971	0	0	0	0	19	292	394	326	223	37	14	0	1305
1972	0	0	3	0	47	143	381	344	153	3	0	0	1074
1973	0	0	0	16	35	294	404	435	193	23	0	0	1400
1974	0	0	0	24	55	179	373	372	145	5	12	0	1165
1975	0	0	0	0	121	224	366	380	98	42	12	0	1243
1976	0	0	0	64	58	326	326	315	115	7	0	0	1211
1977	0	0	10	32	104	150	402	355	175	3	6	0	1237
1978	0	0	0	0	57	244	338	447	153	8	0	0	1247
1979	0	0	6	5	90	146	357	339	137	16	1	0	1097
1980	0	0	0	0	89	194	428	470	244	10	0	0	1435
1981	0	0	0	9	62	224	373	315	119	1	0	0	1103
1982	0	0	0	3	85	142	376	280	115	31	5	0	1037
1983	0	0	0	11	43	217	409	380	199	27	0	0	1286
1984	0	0	0	0	39	260	283	324	90	30	0	0	1026
1985	0	0	0	27	81	133	330	291	166	19	0	0	1047
1986	0	0	0	0	139	278	413	307	129	40	0	0	1306
1987	0	0	0	7	101	295	456	332	142	0	0	0	1333
1988	0	0	0	1	70	259	495	418	93	7	0	0	1343
1989	0	0	1	1	62	298	357	332	192	18	1	0	1262
1990	0	0	9	29	20	226	413	341	152	83	1	0	1274
1991	0	0	1	21	230	327	443	437	183	30	0	0	1672
1992	0	0	0	10	54	205	380	268	158	13	0	0	1088
1993	0	0	0	0	95	298	517	438	181	11	6	0	1546

SNOWFALL (inches) PHILADELPHIA, PENNSYLVANIA

SEASON	JULY	AUG	SEP	OCT	NOV	DEC	JAN	FEB	MAR	APR	MAY	JUNE	TOTAL
1964-65	0.0	0.0	0.0	0.0	T	2.6	11.9	2.2	6.5	3.0	0.0	0.0	26.2
1965-66	0.0	0.0	0.0	T	0.0	T	16.0	11.4	T	T	0.0	0.0	27.4
1966-67	0.0	0.0	0.0	0.0	T	18.8	0.6	18.4	6.4	0.1	0.0	0.0	44.3
1967-68	0.0	0.0	0.0	0.0	4.9	5.6	1.5	1.7	2.2	0.0	0.0	0.0	15.9
1968-69	0.0	0.0	0.0	0.0	0.4	3.1	1.9	9.5	8.8	0.0	0.0	0.0	23.7
1969-70	0.0	0.0	0.0	0.0	0.2	7.5	7.5	2.7	2.4	T	0.0	0.0	20.3
1970-71	0.0	0.0	0.0	T	0.0	1.1	7.7	0.8	4.4	4.3	0.0	0.0	18.3
1971-72	0.0	0.0	0.0	0.0	T	0.1	3.2	8.2	0.3	0.4	0.0	0.0	12.2
1972-73	0.0	0.0	0.0	T	T	T	T	T	T	T	0.0	0.0	T
1973-74	0.0	0.0	0.0	0.0	T	4.6	4.1	12.1	T	T	0.0	0.0	20.8
1974-75	0.0	0.0	0.0	0.0	T	0.8	3.9	6.6	2.3	T	0.0	0.0	13.6
1975-76	0.0	0.0	0.0	0.0	0.0	1.1	6.4	3.1	6.9	0.0	0.0	0.0	17.5
1976-77	0.0	0.0	0.0	0.0	T	2.8	15.7	0.2	T	0.0	0.0	0.0	18.7
1977-78	0.0	0.0	0.0	0.0	0.2	0.2	23.4	19.0	12.1	T	0.0	0.0	54.9
1978-79	0.0	0.0	0.0	0.0	2.5	T	10.1	27.6	T	T	0.0	0.0	40.2
1979-80	0.0	0.0	0.0	2.1	T	4.9	6.1	0.4	7.4	T	0.0	0.0	20.9
1980-81	0.0	0.0	0.0	0.0	0.2	1.4	5.0	T	8.8	0.0	0.0	0.0	15.4
1981-82	0.0	0.0	0.0	0.0	T	2.8	14.0	3.5	1.1	4.0	0.0	0.0	25.4
1982-83	0.0	0.0	0.0	0.0	0.0	6.8	0.2	26.1	0.9	1.9	0.0	0.0	35.9
1983-84	0.0	0.0	0.0	0.0	0.8	T	10.5	T	10.3	T	0.0	0.0	21.6
1984-85	0.0	0.0	0.0	0.0	T	0.2	11.9	4.4	T	T	0.0	0.0	16.5
1985-86	0.0	0.0	0.0	0.0	0.0	1.5	3.4	11.5	T	T	0.0	0.0	16.4
1986-87	0.0	0.0	0.0	0.0	T	0.4	15.2	10.1	T	T	0.0	0.0	25.7
1987-88	0.0	0.0	0.0	0.0	1.4	1.5	10.6	1.5	T	0.0	0.0	0.0	15.0
1988-89	0.0	0.0	0.0	0.0	0.0	0.4	6.0	2.4	2.4	0.0	0.0	0.0	11.2
1989-90	0.0	0.0	0.0	0.0	4.6	5.3	1.4	0.9	2.4	2.4	0.0	0.0	17.0
1990-91	0.0	0.0	0.0	0.0	0.0	6.4	6.5	1.0	0.7	0.0	0.0	0.0	14.6
1991-92	0.0	0.0	0.0	0.0	T	T	1.2	1.0	2.5	T	0.0	0.0	4.7
1992-93	0.0	0.0	0.0	0.0	T	T	1.0	10.9	12.4	T	0.0	T	24.3
1993-94	0.0	0.0	0.0	0.0	T	0.9							
Record Mean	0.0	0.0	0.0	T	0.6	3.4	6.4	6.4	3.7	0.3	T	T	20.9

PITTSBURGH, GRTR. PITT. AIRPORT
PENNSYLVANIA

Pittsburgh lies at the foothills of the Allegheny Mountains at the confluence of the Allegheny and Monongahela Rivers which form the Ohio. The city is a little over 100 miles southeast of Lake Erie. It has a humid continental type of climate modified only slightly by its nearness to the Atlantic Seaboard and the Great Lakes.

The predominant winter air masses influencing the climate of Pittsburgh have a polar continental source in Canada and move in from the Hudson Bay region or the Canadian Rockies. During the summer, frequent invasions of air from the Gulf of Mexico bring warm humid weather. Occasionally, Gulf air reaches as far north as Pittsburgh during the winter and produces intermittent periods of thawing. The last spring temperature of 32 degrees usually occurs in late April and the first in late October. The average growing season is about 180 days. There is a wide variation in the time of the first and last frosts over a radius of 25 miles from the center of Pittsburgh due to terrain differences.

Precipitation is distributed well throughout the year. During the winter months about a fourth of the precipitation occurs as snow and there is about a 50 percent chance of measurable precipitation on any day. Thunderstorms occur normally during all months, except midwinter, and have a maximum frequency in midsummer. The first appreciable snowfall generally occurs in late November and usually the last occurs early in April. Snow lies on the ground in the suburbs on an average of about 33 days during the year.

Seven months of the year, April through October, have sunshine more than 50 percent of the possible time. During the remaining five months cloudiness is heavier because the track of migratory storms from west to east is closer to the area and because of the frequent periods of cloudy, showery weather associated with northwest winds from across the Great Lakes. Cold air drainage induced by the many hills leads to the frequent formation of early morning fog which may be quite persistent in the river valleys during the colder months.

The Allegheny River flowing south and the Monongahela River flowing north meet to form the Ohio River at Pittsburgh. Heavier rainfall and steeper topography cause the Monongahela River to flood more frequently than the Allegheny River.

Both rivers combine to cause the Ohio River at Pittsburgh to reach the 25 foot flood stage approximately once every four years. The serious flood level of 30 feet is reached much less frequently.

NORMALS, MEANS, AND EXTREMES

PITTSBURGH, GRTR. PITT. AIRPORT PENNSYLVANIA

LATITUDE: 40°30'N LONGITUDE: 80°13'W ELEVATION: FT. GRND 1137 BARO 1213 TIME ZONE: EASTERN WBAN: 94823

	(a)	JAN	FEB	MAR	APR	MAY	JUNE	JULY	AUG	SEP	OCT	NOV	DEC	YEAR
TEMPERATURE °F:														
Normals														
-Daily Maximum		33.7	36.9	49.0	60.3	70.6	78.9	82.6	80.8	74.3	62.5	50.4	38.6	59.9
-Daily Minimum		18.5	20.3	29.8	38.8	48.4	56.9	61.6	60.2	53.5	42.3	34.1	24.4	40.7
-Monthly		26.1	28.7	39.4	49.6	59.5	67.9	72.1	70.5	63.9	52.4	42.3	31.5	50.3
Extremes														
-Record Highest	41	69	69	82	89	91	98	103	100	97	87	82	74	103
-Year		1985	1954	1986	1990	1987	1988	1988	1988	1954	1959	1961	1982	JUL 1988
-Record Lowest	41	-18	-12	-1	14	26	34	42	39	31	16	-1	-12	-18
-Year		1985	1979	1980	1982	1970	1972	1963	1982	1959	1965	1958	1989	JAN 1985
NORMAL DEGREE DAYS:														
Heating (base 65°F)		1206	1016	794	462	214	36	6	14	100	400	681	1039	5968
Cooling (base 65°F)		0	0	0	0	44	123	227	184	67	9	0	0	654
% OF POSSIBLE SUNSHINE	41	32	36	43	46	50	56	57	55	55	51	36	28	45
MEAN SKY COVER (tenths)														
Sunrise - Sunset	41	8.1	7.9	7.5	7.2	6.9	6.5	6.4	6.3	6.2	6.3	7.7	8.2	7.1
MEAN NUMBER OF DAYS:														
Sunrise to Sunset														
-Clear	41	3.0	3.2	4.3	4.5	5.2	5.0	5.3	6.5	7.1	7.7	3.7	2.7	58.3
-Partly Cloudy	41	6.0	6.1	6.9	8.2	9.1	11.7	12.9	11.4	10.2	8.8	6.3	5.6	103.3
-Cloudy	41	22.0	19.0	19.8	17.3	16.6	13.3	12.8	13.0	12.6	14.5	20.0	22.7	203.6
Precipitation														
.01 inches or more	41	16.2	14.0	15.7	13.5	12.6	11.4	10.8	9.6	9.6	10.5	13.1	16.5	153.3
Snow, Ice pellets, hail														
1.0 inches or more	41	3.5	2.8	2.3	0.5	0.*	0.0	0.0	0.0	0.0	0.1	1.0	2.5	12.8
Thunderstorms	41	0.2	0.4	1.7	3.2	5.2	6.7	6.9	5.5	3.1	1.2	0.6	0.3	35.0
Heavy Fog Visibility														
1/4 mile or less	41	1.3	1.2	1.0	0.8	1.1	1.1	1.6	2.2	2.4	1.9	1.4	1.8	17.7
Temperature °F														
-Maximum														
90° and above	34	0.0	0.0	0.0	0.0	0.3	1.6	3.5	1.9	0.6	0.0	0.0	0.0	8.0
32° and below	34	13.7	10.2	3.6	0.1	0.0	0.0	0.0	0.0	0.0	0.0	1.3	9.9	38.8
-Minimum														
32° and below	34	27.0	23.9	19.5	8.2	0.8	0.0	0.0	0.0	0.0	4.2	14.1	24.4	122.0
0° and below	34	2.3	1.4	0.1	0.0	0.0	0.0	0.0	0.0	0.0	0.0	0.*	0.8	4.7
AVG. STATION PRESS. (mb)	21	973.3	973.4	971.9	971.2	971.7	972.3	973.4	974.6	975.1	975.1	974.3	974.0	973.4
RELATIVE HUMIDITY (%)														
Hour 01	33	73	71	70	67	73	78	80	82	82	77	75	75	75
Hour 07 (Local Time)	33	76	75	75	73	76	79	82	86	86	82	79	77	79
Hour 13	33	65	62	57	50	52	52	54	56	57	54	62	67	57
Hour 19	33	67	64	59	53	55	57	60	62	66	62	68	70	62
PRECIPITATION (inches):														
Water Equivalent														
-Normal		2.54	2.39	3.41	3.15	3.59	3.71	3.75	3.21	2.97	2.36	2.85	2.92	36.85
-Maximum Monthly	41	6.25	5.98	6.10	7.61	6.56	10.29	8.71	7.86	6.00	8.20	11.05	8.51	11.05
-Year		1978	1956	1967	1964	1989	1989	1992	1987	1990	1954	1985	1990	NOV 1985
-Minimum Monthly	41	0.77	0.51	1.14	0.48	1.21	0.64	1.62	0.78	0.28	0.16	0.90	0.40	0.16
-Year		1981	1969	1969	1971	1965	1992	1989	1957	1985	1963	1976	1955	OCT 1963
-Maximum in 24 hrs	41	1.69	2.30	2.00	2.15	2.44	2.96	2.97	3.06	2.59	3.56	1.97	2.76	3.56
-Year		1986	1975	1964	1964	1971	1987	1971	1956	1990	1954	1985	1990	OCT 1954
Snow, Ice pellets, hail														
-Maximum Monthly	41	40.2	24.2	34.1	8.1	3.1	T	T	0.0	T	8.5	11.0	21.2	40.2
-Year		1978	1972	1993	1987	1966	1990	1991		1989	1993	1958	1974	JAN 1978
-Maximum in 24 hrs	41	14.0	12.3	23.8	7.7	3.1	T	T	0.0	T	6.6	10.5	12.5	23.8
-Year		1966	1960	1993	1987	1966	1990	1991		1989	1993	1958	1974	MAR 1993
WIND:														
Mean Speed (mph)	41	10.6	10.5	10.7	10.3	8.9	8.0	7.3	6.9	7.4	8.4	9.8	10.4	9.1
Prevailing Direction through 1963		WSW	WSW	WSW	WSW	WSW	WSW	WSW	WSW	WSW	WSW	WSW	WSW	WSW
Fastest Obs. 1 Min.														
-Direction (!!)	41	23	26	25	27	23	27	25	29	26	25	29	25	26
-Speed (MPH)	41	52	58	48	46	44	40	51	46	36	35	45	48	58
-Year		1978	1967	1954	1974	1988	1957	1956	1963	1990	1986	1969	1968	FEB 1967
Peak Gust														
-Direction (!!)	10	SW	W	W	W	SW	W	W	W	W	SW	W	NE	W
-Speed (mph)	10	51	59	60	60	61	52	83	56	48	49	62	55	83
-Date		1992	1990	1985	1991	1988	1990	1992	1986	1990	1990	1992	1990	JUL 1992

PRECIPITATION (inches)

PITTSBURGH, GRTR. PITT. AIRPORT PENNSYLVANIA

YEAR	JAN	FEB	MAR	APR	MAY	JUNE	JULY	AUG	SEP	OCT	NOV	DEC	ANNUAL
1964	2.55	1.73	4.96	7.61	1.77	3.84	4.48	1.79	0.74	1.42	2.74	4.26	37.89
1965	3.84	2.98	3.16	1.79	1.21	2.31	1.82	3.26	4.07	2.82	2.35	0.63	30.24
1966	4.52	3.23	1.88	3.73	2.76	1.72	2.70	5.13	1.92	1.38	3.39	1.70	34.06
1967	1.06	2.54	6.10	4.41	5.21	0.90	4.54	2.67	1.61	2.05	3.07	2.22	36.38
1968	2.83	0.79	4.53	2.33	6.36	2.38	2.36	3.97	3.08	2.13	2.07	3.24	36.07
1969	2.02	0.51	1.14	2.91	1.89	3.74	4.52	2.96	0.91	2.59	2.44	3.95	29.58
1970	1.61	1.92	3.35	3.09	4.36	4.61	3.89	1.55	2.77	4.80	2.64	3.29	37.88
1971	2.29	4.04	3.20	0.48	3.87	1.41	6.82	1.23	3.86	0.84	1.94	3.24	33.22
1972	1.84	3.64	3.68	4.37	1.38	5.08	2.98	1.79	5.42	2.15	4.70	3.04	40.07
1973	2.03	1.80	3.86	4.69	5.87	3.12	2.16	3.40	3.56	4.45	2.65	2.15	39.74
1974	3.47	2.10	3.72	3.26	5.35	5.08	3.30	2.93	4.42	1.12	3.06	4.02	41.83
1975	3.34	4.64	4.62	2.27	1.84	4.58	4.38	7.56	5.06	3.46	1.77	2.90	46.42
1976	3.25	1.74	4.45	1.24	1.99	3.37	4.72	1.25	3.30	3.76	0.90	1.81	31.78
1977	2.06	0.87	4.12	3.26	2.57	2.85	3.38	2.66	3.13	2.44	2.59	3.27	33.20
1978	6.25	0.54	1.65	2.25	4.26	4.11	2.15	3.65	2.64	3.42	1.62	5.24	37.78
1979	4.80	3.12	1.32	3.17	4.49	1.73	4.31	6.84	3.60	2.46	2.43	2.29	40.56
1980	1.56	1.32	5.65	2.94	4.32	4.34	6.76	5.10	1.29	2.42	2.38	1.38	39.46
1981	0.77	4.20	2.12	4.92	2.04	8.20	3.82	0.98	4.13	1.82	1.50	3.00	37.50
1982	4.44	1.93	3.52	1.44	3.98	3.05	2.36	1.97	2.80	0.40	3.33	2.79	32.01
1983	1.19	1.58	3.50	4.33	5.24	4.82	3.32	3.13	2.42	3.67	3.94	4.27	41.41
1984	1.40	2.05	2.32	3.72	5.22	1.98	3.01	5.15	0.84	3.45	3.14	3.04	35.32
1985	1.43	1.45	3.37	1.64	5.80	2.26	4.06	2.64	0.28	2.27	11.05	2.26	38.51
1986	2.49	3.43	1.38	1.94	1.67	5.24	5.66	3.04	2.33	2.83	3.92	3.47	37.40
1987	2.23	0.71	2.65	5.30	2.41	6.30	2.42	7.86	3.97	0.92	2.02	2.41	39.20
1988	1.49	3.46	2.56	1.97	2.78	1.26	2.82	2.04	2.34	1.40	2.80	2.17	27.09
1989	1.99	3.42	5.52	1.43	6.56	10.29	1.62	1.12	4.57	2.04	1.56	2.39	42.51
1990	3.30	3.31	1.47	3.48	6.19	4.24	6.59	3.59	6.00	3.51	2.05	8.51	52.24
1991	2.55	1.88	2.92	2.56	3.29	3.82	3.74	1.63	3.45	0.55	1.97	3.66	32.02
1992	2.13	1.73	3.54	2.30	2.31	0.64	8.71	4.77	2.91	1.47	3.31	2.83	36.65
1993	2.99	2.92	4.14	3.66	2.85	3.35	2.85	2.44	3.87	2.77	4.30	2.12	38.26
Record Mean	2.84	2.46	3.27	3.07	3.39	3.75	3.99	3.21	2.69	2.46	2.47	2.78	36.39

AVERAGE TEMPERATURE (deg. F)

PITTSBURGH, GRTR. PITT. AIRPORT PENNSYLVANIA

YEAR	JAN	FEB	MAR	APR	MAY	JUNE	JULY	AUG	SEP	OCT	NOV	DEC	ANNUAL
1964	31.4	27.0	40.0	51.7	62.7	67.9	72.3	67.1	63.7	50.4	45.5	34.0	51.1
1965	28.2	28.4	35.2	49.0	65.9	66.9	69.9	69.1	64.7	48.1	41.3	37.5	50.4
1966	23.1	30.3	40.9	47.9	56.1	70.4	75.6	71.1	61.3	50.8	42.8	31.4	50.1
1967	32.3	25.6	40.2	52.2	54.3	73.0	71.5	68.8	61.1	52.5	36.8	34.8	50.3
1968	23.4	22.2	40.4	51.2	54.7	66.9	72.4	71.8	64.8	52.2	41.3	27.6	49.1
1969	26.7	29.5	34.3	51.7	60.2	69.3	72.7	69.7	63.0	52.9	39.2	26.7	49.7
1970	20.7	27.7	35.5	52.5	63.9	68.2	71.6	71.6	67.8	54.9	42.2	32.1	50.7
1971	23.7	30.4	34.3	46.0	56.6	71.4	70.2	69.6	68.5	59.5	40.4	38.8	50.8
1972	29.6	26.5	36.4	48.5	61.8	63.8	71.2	70.6	65.3	48.4	39.3	37.2	49.9
1973	29.7	28.8	48.3	49.3	56.4	70.9	73.2	73.2	66.5	56.1	44.1	33.3	52.5
1974	34.0	29.9	41.2	51.8	58.3	65.2	73.1	72.8	62.2	52.4	43.9	32.5	51.4
1975	32.6	32.1	36.3	44.3	63.0	67.8	72.8	73.0	58.8	53.3	46.3	32.9	51.1
1976	23.5	37.2	45.2	50.6	55.6	68.4	67.4	65.3	59.9	45.9	33.1	23.9	48.0
1977	11.4	26.9	43.7	50.8	63.0	63.8	71.8	68.1	64.7	50.5	45.6	31.1	49.3
1978	22.6	20.9	36.9	51.0	60.2	69.4	73.0	71.4	66.2	49.1	43.0	32.7	49.7
1979	21.4	18.0	43.1	49.7	59.1	67.7	70.3	69.6	63.4	50.9	44.7	34.6	49.4
1980	26.9	24.2	35.6	48.1	60.3	66.2	75.0	74.5	67.1	49.5	38.6	28.6	49.5
1981	20.5	31.4	35.6	51.9	58.4	68.8	72.1	69.7	61.9	49.4	40.3	29.4	49.1
1982	20.9	28.4	38.4	45.3	64.7	63.7	72.4	68.2	63.4	54.4	44.7	39.9	50.4
1983	30.0	32.6	40.7	47.1	55.8	67.8	73.0	72.8	64.4	53.0	43.5	25.4	50.5
1984	23.2	36.4	32.2	49.2	55.3	69.7	68.5	70.8	61.4	58.3	40.2	39.3	50.4
1985	22.1	27.7	42.1	55.0	60.6	64.2	70.5	69.6	65.3	55.2	47.1	27.4	50.6
1986	28.3	31.3	41.1	53.1	62.0	68.3	73.3	68.6	66.6	54.2	40.4	33.1	51.7
1987	28.0	32.6	41.9	50.0	63.0	70.9	75.7	71.8	65.1	47.8	46.2	35.1	52.3
1988	26.6	29.0	39.3	49.4	61.4	68.5	76.9	75.1	63.5	46.6	44.2	31.9	51.0
1989	35.5	27.8	41.1	47.0	58.0	69.0	74.1	71.6	64.8	53.3	40.6	19.2	50.2
1990	36.8	36.9	44.0	51.3	57.7	68.3	71.7	70.5	63.7	55.0	45.5	38.0	53.3
1991	29.7	35.4	43.0	54.5	68.7	72.6	75.4	74.4	64.6	55.7	41.6	35.3	54.2
1992	30.5	34.3	38.8	51.3	59.1	66.0	72.4	67.9	63.7	50.1	42.9	33.9	50.9
1993	35.1	27.8	37.6	50.0	61.9	68.9	75.7	75.4	63.3	51.7	43.0	31.7	51.8
Record Mean	30.0	31.1	39.9	51.0	61.7	70.1	74.3	72.5	66.2	54.7	43.1	33.3	52.3
Max	37.6	39.3	49.0	61.1	72.3	80.4	84.3	82.4	76.3	64.4	50.9	40.3	61.5
Min	22.4	22.9	30.8	40.8	51.1	59.8	64.2	62.5	56.1	44.9	35.3	26.3	43.1

REFERENCE NOTES

PITTSBURGH, GRTR. PITT. AIRPORT PENNSYLVANIA

GENERAL
T - TRACE AMOUNT.
BLANK ENTRIES DENOTE MISSING/UNREPORTED DATA.
# INDICATES A STATION OR INSTRUMENT RELOCATION.

(a) - LENGTH OF RECORD IN YEARS, ALTHOUGH INDIVIDUAL MONTHS MAY BE MISSING.
0.* OR * - THE VALUE IS BETWEEN 0.0 AND 0.05
NORMALS - BASED ON THE 1961-1990 RECORD PERIOD.
EXTREMES - DATES ARE THE MOST RECENT OCCURRENCE
WIND DIR.- NUMERALS SHOW TENS OF DEGRESS CLOCKWISE FROM TRUE NORTH. "00" INDICATES CALM.
RESULTANT DIRECTIONS ARE GIVEN TO WHOLE DEGREES.
BOLD VALUES INDICATE EXTREME VALUES WHICH OCCURRED AFTER THE ASOS SYSTEM WAS COMMISSIONED.

RECORD = PERIOD OF RECORD
RECORD MEAN PRECIPITATION IS THE MEAN OF ALL DAILY PRECIPITATION AMOUNTS DURING THE PERIOD OF RECORD.
RECORD MAX(MIN) TEMPERATURE IS THE MEAN OF ALL DAILY MAX(MIN) TEMPERATURES DURING THE PERIOD OF RECORD.
RECORD MEAN TEMPERATURE IS THE SUM OF THE RECORD MAX AND RECORD MIN DIVIDED BY 2.
AVERAGE TEMPERATURE IS THE SUM OF THE MEAN DAILY MAX AND MIN TEMPERATURE DIVIDED BY 2.

EXCEPTIONS

TEMPERATURE DATA MAY BE SUSPECT NOVEMBER 1977 THROUGH JULY 1978 DUE TO INTERMITTENT INSTRUMENT MALFUNCTION.

RECORD MEANS ARE THROUGH THE CURRENT YEAR, BEGINNING IN 1875 FOR TEMPERATURE
1872 FOR PRECIPITATION
1953 FOR SNOWFALL

HEATING DEGREE DAYS Base 65 deg. F PITTSBURGH, GRTR. PITT. AIRPORT PENNSYLVANIA

SEASON	JULY	AUG	SEP	OCT	NOV	DEC	JAN	FEB	MAR	APR	MAY	JUNE	TOTAL
1964-65	1	51	99	447	577	954	1134	1018	920	476	63	53	5793
1965-66	9	40	99	518	702	848	1293	963	741	510	289	34	6046
1966-67	2	6	156	435	659	1035	1007	1097	765	387	332	4	5885
1967-68	10	13	146	391	840	931	1284	1232	758	406	313	60	6384
1968-69	8	31	54	400	703	1152	1181	988	944	394	182	35	6072
1969-70	0	8	127	383	770	1183	1370	1039	908	390	127	31	6336
1970-71	5	1	69	318	678	1013	1277	961	949	562	264	6	6103
1971-72	1	6	41	184	729	807	1093	1112	881	489	128	96	5567
1972-73	20	11	63	508	767	853	1087	1006	508	474	264	2	5563
1973-74	2	8	55	274	621	978	957	978	729	403	223	54	5282
1974-75	0	0	124	384	630	1001	997	916	881	617	116	48	5714
1975-76	0	0	192	362	554	989	1278	801	605	453	301	24	5559
1976-77	15	59	159	587	953	1268	1655	1060	658	436	138	102	7090
1977-78	11	41	78	442	583	1043	1307	1229	860	412	209	38	6253
1978-79	4	3	80	485	656	993	1346	1311	671	458	219	38	6264
1979-80	23	26	111	438	601	935	1175	1177	906	500	172	71	6135
1980-81	0	5	48	476	787	1117	1372	936	904	391	223	18	6277
1981-82	3	10	159	475	736	1098	1361	1017	819	586	82	67	6413
1982-83	9	23	119	336	605	770	1080	904	746	535	280	44	5451
1983-84	10	2	126	365	639	1223	1293	823	1008	471	305	16	6281
1984-85	12	7	165	214	734	790	1322	1038	701	334	163	65	5545
1985-86	3	9	116	300	531	1160	1131	936	737	368	148	37	5476
1986-87	1	40	65	346	733	983	1139	904	710	451	145	22	5539
1987-88	4	20	61	529	560	920	1181	1040	792	461	149	64	5781
1988-89	5	3	83	570	619	1018	905	1033	739	532	260	25	5792
1989-90	1	14	102	364	723	1414	869	781	657	439	229	49	5642
1990-91	4	1	116	314	577	829	1085	820	674	337	63	5	4825
1991-92	0	0	127	308	698	913	1063	880	805	417	210	50	5471
1992-93	1	17	116	457	657	956	920	1037	841	445	135	43	5625
1993-94	0	0	118	407	654	1028							

COOLING DEGREE DAYS Base 65 deg. F PITTSBURGH, GRTR. PITT. AIRPORT PENNSYLVANIA

YEAR	JAN	FEB	MAR	APR	MAY	JUNE	JULY	AUG	SEP	OCT	NOV	DEC	TOTAL
1969	0	0	0	2	42	170	245	162	72	14	0	0	707
1970	0	0	0	21	100	133	215	213	162	11	0	0	855
1971	0	0	0	0	13	204	171	158	153	17	0	0	716
1972	0	0	0	0	34	68	219	192	76	0	0	0	589
1973	0	0	0	10	5	185	264	269	108	5	0	0	846
1974	0	0	0	13	19	66	258	247	45	5	4	0	657
1975	0	0	0	0	60	137	248	257	12	7	0	0	721
1976	0	0	1	25	14	134	99	73	12	0	0	0	358
1977	0	0	3	14	83	75	231	141	72	0	4	0	623
1978	0	0	0	0	69	178	260	207	122	0	0	0	836
1979	0	0	0	9	41	125	193	175	70	7	0	0	620
1980	0	0	0	0	34	115	317	306	118	0	0	0	890
1981	0	0	0	5	25	139	230	160	72	0	0	0	631
1982	0	0	0	0	79	33	246	127	77	15	3	0	580
1983	0	0	0	3	3	135	263	251	115	0	0	0	770
1984	0	0	0	3	12	165	127	194	63	13	0	0	577
1985	0	0	0	41	33	49	181	160	130	4	0	0	598
1986	0	0	3	20	65	144	265	157	121	20	0	0	795
1987	0	0	0	6	93	204	342	240	72	0	1	0	958
1988	0	0	0	0	44	174	381	322	47	7	0	0	975
1989	0	0	5	0	49	154	291	225	100	9	0	0	833
1990	0	0	14	37	9	153	218	179	83	13	0	0	706
1991	0	0	0	27	184	239	333	298	124	27	0	0	1232
1992	0	0	0	9	33	87	233	112	86	3	0	0	563
1993	0	0	0	2	45	166	337	327	73	1	1	0	952

SNOWFALL (inches) PITTSBURGH, GRTR. PITT. AIRPORT PENNSYLVANIA

SEASON	JULY	AUG	SEP	OCT	NOV	DEC	JAN	FEB	MAR	APR	MAY	JUNE	TOTAL
1964-65	0.0	0.0	0.0	T	1.6	6.1	10.6	10.4	13.3	0.2	0.0	0.0	42.2
1965-66	0.0	0.0	0.0	0.2	0.2	1.8	24.6	6.9	8.5	2.7	3.1	0.0	48.0
1966-67	0.0	0.0	0.0	0.0	5.1	7.8	4.5	21.7	20.0	0.5	0.0	0.0	59.6
1967-68	0.0	0.0	0.0	T	10.1	7.9	15.4	6.1	11.0	T	0.0	0.0	50.5
1968-69	0.0	0.0	0.0	T	2.7	13.3	6.5	4.0	3.9	0.0	T	0.0	30.4
1969-70	0.0	0.0	0.0	0.4	7.9	20.6	12.6	13.0	16.1	0.1	0.0	0.0	70.7
1970-71	0.0	0.0	0.0	T	0.1	10.1	12.1	20.6	16.8	0.2	0.0	0.0	59.9
1971-72	0.0	0.0	0.0	0.0	10.5	0.7	4.9	24.2	9.8	1.8	0.0	0.0	51.9
1972-73	0.0	0.0	0.0	1.8	6.1	2.9	3.4	6.1	4.6	1.4	T	0.0	26.3
1973-74	0.0	0.0	0.0	0.0	0.8	4.8	4.9	2.2	2.3	1.6	T	0.0	16.6
1974-75	0.0	0.0	0.0	T	2.6	21.2	10.1	13.9	9.8	1.1	0.0	0.0	58.7
1975-76	0.0	0.0	0.0	0.0	1.9	3.8	21.8	3.3	4.3	0.5	0.0	0.0	35.6
1976-77	0.0	0.0	0.0	T	6.6	7.9	26.5	6.4	0.9	1.3	T	0.0	49.6
1977-78	0.0	0.0	0.0	T	3.3	9.1	40.2	5.4	4.0	0.2	0.0	0.0	62.2
1978-79	0.0	0.0	0.0	0.0	2.3	3.2	18.2	13.7	2.0	1.4	0.0	0.0	40.8
1979-80	0.0	0.0	0.0	T	1.1	1.1	7.8	6.2	7.9	T	0.0	0.0	24.1
1980-81	0.0	0.0	0.0	T	9.7	6.3	12.5	11.9	7.6	T	0.0	0.0	48.0
1981-82	0.0	0.0	0.0	T	0.6	11.5	13.4	3.6	12.2	3.8	0.0	0.0	45.1
1982-83	0.0	0.0	0.0	T	0.1	8.8	3.9	12.0	4.3	1.0	0.0	0.0	30.1
1983-84	0.0	0.0	0.0	0.0	6.1	10.5	10.8	11.4	10.4	T	0.0	0.0	49.2
1984-85	0.0	0.0	0.0	0.0	1.5	4.8	14.6	8.1	0.2	7.2	0.0	0.0	36.4
1985-86	0.0	0.0	0.0	0.0	T	15.3	11.1	12.4	4.8	2.7	0.0	0.0	46.3
1986-87	0.0	0.0	0.0	0.0	1.0	0.9	11.6	1.1	7.3	8.1	0.0	0.0	30.0
1987-88	0.0	0.0	0.0	T	4.1	7.9	5.5	6.9	9.8	0.9	0.0	0.0	35.1
1988-89	0.0	0.0	0.0	0.2	1.1	4.0	4.2	4.1	7.5	0.6	T	0.0	21.7
1989-90	0.0	0.0	T	0.2	1.6	12.5	7.7	2.5	0.6	3.3	0.0	T	28.4
1990-91	0.0	0.0	0.0	0.0	T	4.6	4.8	3.6	4.2	T	0.0	0.0	17.2
1991-92	T	0.0	0.0	0.0	2.1	3.1	12.9	2.4	10.6	2.8	T	0.0	33.9
1992-93	0.0	0.0	0.0	1.3	1.5	14.1	2.1	18.5	34.1	0.5	0.0	0.0	72.1
1993-94	0.0	0.0	0.0	8.5	2.6	10.4							
Record Mean	T	0.0	T	0.4	3.3	8.3	11.3	9.3	8.7	1.7	0.1	T	43.1

WILLIAMSPORT, PENNSYLVANIA

The climate of the Lycoming valley is favorably influenced by the lower elevation of the area compared to the surrounding terrain. Since the prevailing winds reach the area from the southwest to the north, there is a slight moderating effect on winter extremes of cold. Radiation cooling on clear nights is somewhat more frequent than in adjacent areas. Deep valley fogs occasionally persist until nearly midday. Cold air drainage from the surrounding hills is experienced during several nights but the cool temperatures are often modified by the proximity of the river and adjacent damp areas. The winters are milder than those experienced to the west and cold spells are frequently interrupted by incursions of warmer coastal weather. In summer the air frequently becomes trapped in the valley and higher temperatures and humidities result, generally benefitting the local agriculture.

The long irregular range south of the river forms an effective barrier to free air movement. Moderate or strong south to southwest winds are deflected to southeast or south winds in crossing the range and the air becomes quite turbulent with distinct wave effects. Banner clouds with one to three rolls are frequently observed with low overcasts.

An average growing season of 168 days extends from April 29 to October 14. Snowfall in the valley is generally uniform but varies considerably with the rise in terrain, particularly to the north and south. Snow depth on the ridge 2 miles south of the observation point is frequently double the amount at the station.

NORMALS, MEANS, AND EXTREMES

WILLIAMSPORT, PENNSYLVANIA

LATITUDE: 41°15'N LONGITUDE: 76°55' W ELEVATION: FT. GRND 524 BARO 543 TIME ZONE: EASTERN WBAN: 14778

	(a)	JAN	FEB	MAR	APR	MAY	JUNE	JULY	AUG	SEP	OCT	NOV	DEC	YEAR
TEMPERATURE °F:														
Normals														
-Daily Maximum		33.3	36.6	47.7	60.1	71.1	79.0	83.1	81.2	73.8	62.4	49.9	37.9	59.7
-Daily Minimum		17.1	19.2	28.5	38.1	47.9	56.6	61.5	60.5	53.1	41.6	33.6	23.9	40.1
-Monthly		25.2	28.0	38.1	49.1	59.5	67.8	72.3	70.8	63.5	52.0	41.8	30.9	49.9
Extremes														
-Record Highest	49	69	71	87	92	95	102	103	100	102	91	83	67	103
-Year		1967	1985	1986	1990	1969	1952	1988	1955	1953	1951	1950	1984	JUL 1988
-Record Lowest	49	-17	-13	-1	15	28	36	43	38	28	20	8	-15	-17
-Year		1977	1971	1984	1982	1966	1986	1965	1965	1947	1972	1976	1950	JAN 1977
NORMAL DEGREE DAYS:														
Heating (base 65°F)		1234	1036	834	477	200	30	0	9	106	408	696	1057	6087
Cooling (base 65°F)		0	0	0	0	29	114	229	189	61	0	0	0	622
% OF POSSIBLE SUNSHINE														
MEAN SKY COVER (tenths)														
Sunrise - Sunset	49	7.3	7.0	6.9	6.8	6.8	6.3	6.2	6.3	6.6	6.4	7.5	7.6	6.8
MEAN NUMBER OF DAYS:														
Sunrise to Sunset														
-Clear	49	4.9	5.4	6.1	6.1	5.6	6.1	6.2	5.7	5.6	7.1	4.0	4.2	67.1
-Partly Cloudy	49	7.3	6.8	7.9	7.6	10.1	11.6	12.8	13.1	10.6	9.1	6.7	6.8	110.4
-Cloudy	49	18.8	16.1	17.1	16.3	15.3	12.3	11.9	12.2	13.7	14.7	19.2	20.0	187.7
Precipitation .01 inches or more	49	11.9	11.0	12.7	13.1	13.3	11.7	11.5	10.8	10.0	10.0	12.2	12.4	140.6
Snow,Ice pellets,hail 1.0 inches or more	49	3.1	2.9	2.0	0.3	0.0	0.0	0.0	0.0	0.0	0.*	0.9	2.3	11.6
Thunderstorms	24	0.2	0.3	1.0	1.8	4.0	6.9	7.7	6.2	3.2	0.9	0.5	0.3	33.0
Heavy Fog Visibility 1/4 mile or less	24	1.3	1.3	1.3	1.2	2.3	2.9	3.5	4.6	7.9	6.1	3.0	1.9	37.2
Temperature °F														
-Maximum														
90° and above	49	0.0	0.0	0.0	0.2	0.9	2.9	5.9	3.2	0.8	0.*	0.0	0.0	13.9
32° and below	49	12.3	8.0	2.0	0.*	0.0	0.0	0.0	0.0	0.0	0.0	0.6	7.9	30.8
-Minimum														
32° and below	49	28.0	24.8	21.2	7.6	0.7	0.0	0.0	0.0	0.2	4.7	14.9	25.5	127.7
0° and below	49	2.2	1.2	0.1	0.0	0.0	0.0	0.0	0.0	0.0	0.0	0.0	0.8	4.3
AVG. STATION PRESS. (mb)	21	999.0	999.3	997.8	996.3	996.7	996.7	997.5	999.0	1000.2	1000.5	1000.0	999.9	998.6
RELATIVE HUMIDITY (%)														
Hour 01	42	75	74	73	73	81	87	88	90	90	86	80	77	81
Hour 07 (Local Time)	48	76	76	77	75	80	84	87	90	92	88	82	78	82
Hour 13	48	62	58	54	49	51	54	55	58	59	57	61	63	57
Hour 19	47	68	64	60	54	58	61	64	70	75	72	71	71	66
PRECIPITATION (inches):														
Water Equivalent														
-Normal		2.54	2.76	3.19	3.23	3.86	4.32	3.98	3.39	3.39	3.30	3.73	3.03	40.72
-Maximum Monthly	49	8.25	8.42	5.96	7.55	9.45	16.80	9.65	7.67	10.02	9.60	8.09	7.36	16.80
-Year		1978	1981	1980	1993	1946	1972	1992	1988	1975	1990	1972	1973	JUN 1972
-Minimum Monthly	49	0.52	0.57	0.86	0.70	0.80	0.66	0.99	0.96	0.50	0.19	0.83	0.68	0.19
-Year		1985	1968	1981	1989	1964	1966	1955	1951	1964	1963	1976	1989	OCT 1963
-Maximum in 24 hrs	49	2.46	2.72	2.52	2.71	4.15	8.66	2.82	5.53	4.60	4.38	3.46	3.29	8.66
-Year		1978	1971	1964	1977	1946	1972	1990	1988	1975	1955	1956	1983	JUN 1972
Snow,Ice pellets,hail														
-Maximum Monthly	49	40.1	34.3	29.5	13.8	0.2	T	0.0	T	0.0	1.0	13.7	35.5	40.1
-Year		1987	1972	1967	1982	1977	1991		1991		1977	1953	1969	JAN 1987
-Maximum in 24 hrs	49	23.1	20.4	15.0	8.8	0.2	T	0.0	T	0.0	1.0	12.1	16.5	23.1
-Year		1964	1972	1993	1982	1977	1991		1991		1977	1953	1969	JAN 1964
WIND:														
Mean Speed (mph)	32	9.0	9.1	9.3	9.2	7.9	7.0	6.4	6.0	6.2	6.8	8.2	8.6	7.8
Prevailing Direction through 1963		W	WNW	W	W	W	W	W	W	WNW	W	W	W	W
Fastest Obs. 1 Min.														
-Direction (!!)	40	27	14	11	18	18	29	20	29	16	11	09	16	20
-Speed (MPH)	40	66	60	58	62	55	62	78	60	59	75	77	58	78
-Year		1951	1951	1954	1950	1953	1951	1951	1949	1951	1954	1950	1953	JUL 1951
Peak Gust														
-Direction (!!)	9	W	W	W	W	W	NW	W	W	SE	S	NW	W	W
-Speed (mph)	9	54	49	60	60	51	56	58	61	54	44	63	64	64
-Date		1988	1985	1986	1985	1990	1993	1988	1989	1989	1987	1989	1991	DEC 1991

PRECIPITATION (inches) WILLIAMSPORT, PENNSYLVANIA

YEAR	JAN	FEB	MAR	APR	MAY	JUNE	JULY	AUG	SEP	OCT	NOV	DEC	ANNUAL
1964	4.95	3.11	3.87	5.56	0.80	2.60	3.37	4.18	0.50	1.10	2.24	2.66	34.94
1965	2.88	2.27	3.21	3.07	1.88	1.75	1.70	5.03	4.04	3.98	1.69	0.93	32.43
1966	3.32	4.58	1.90	2.83	4.16	0.66	3.90	2.45	5.22	1.43	5.77	1.92	38.14
1967	1.35	1.49	5.36	2.88	5.50	3.24	6.03	4.97	2.20	4.50	3.63	2.71	43.86
1968	2.49	0.57	3.10	1.05	4.93	5.20	1.39	1.47	6.41	3.72	4.50	2.96	37.79
1969	1.24	0.72	1.49	2.78	1.83	4.38	6.46	3.09	1.81	3.04	5.08	5.63	37.55
1970	0.95	3.76	2.95	3.28	3.23	3.32	5.45	3.00	1.91	5.43	5.08	4.12	42.48
1971	2.54	6.50	2.69	1.17	2.88	1.89	5.68	2.48	3.33	3.88	3.18	2.87	39.09
1972	2.49	4.84	3.96	4.22	6.79	16.80	3.79	1.71	1.61	2.18	8.09	4.79	61.27
1973	2.73	2.74	3.47	5.56	5.18	5.86	3.99	1.95	5.50	3.17	2.24	7.36	49.75
1974	3.03	1.90	5.78	2.99	3.04	5.51	2.67	3.05	4.84	0.86	2.57	4.50	40.74
1975	3.53	4.34	3.07	1.53	6.54	5.25	3.66	4.43	10.02	2.98	2.83	3.24	51.42
1976	2.73	1.90	2.75	1.90	3.81	5.31	4.80	3.18	3.21	8.14	0.83	1.74	40.30
1977	1.35	1.74	5.74	6.79	1.02	3.51	3.59	2.29	7.30	4.22	5.45	4.22	47.22
1978	8.25	0.86	2.70	2.01	6.36	4.60	3.99	3.56	2.84	2.76	1.78	3.68	43.39
1979	6.36	3.20	3.18	2.73	4.00	2.23	4.24	5.54	5.83	4.08	4.07	1.90	47.36
1980	0.88	0.77	5.96	6.72	1.71	3.50	1.82	1.82	3.32	3.19	2.68	1.18	33.55
1981	0.68	8.42	0.86	2.93	2.07	5.51	4.09	1.02	2.10	4.03	1.46	1.95	35.12
1982	3.37	1.93	2.13	2.96	3.53	9.23	1.81	1.00	2.59	0.59	3.00	1.35	33.49
1983	1.79	1.87	3.05	7.03	4.51	5.55	3.38	3.68	1.88	3.01	5.89	6.54	48.18
1984	1.09	4.83	3.63	4.58	4.89	5.15	6.42	3.60	0.62	2.78	4.76	2.34	44.69
1985	0.52	1.12	2.81	1.18	3.98	2.49	3.47	4.28	2.58	1.83	5.99	1.42	31.67
1986	2.18	2.49	2.65	4.25	3.93	4.71	5.78	3.31	1.78	2.00	6.13	3.99	43.20
1987	2.68	0.78	1.77	2.63	2.21	2.30	3.69	4.04	4.99	2.70	3.26	1.63	32.68
1988	1.39	3.09	1.09	1.28	4.27	0.99	4.71	7.67	2.62	2.30	3.67	1.14	34.22
1989	1.58	1.75	3.32	0.70	7.29	5.74	3.42	2.94	4.28	4.26	2.06	0.68	38.02
1990	2.37	2.37	1.51	2.57	5.01	2.48	5.07	5.45	1.98	9.60	2.68	5.73	46.82
1991	2.02	1.70	3.84	2.44	1.79	1.36	3.98	3.54	3.35	2.24	3.29	3.20	32.75
1992	2.01	1.40	4.33	2.10	3.47	2.90	9.65	2.47	4.16	1.51	4.74	2.35	41.09
1993	1.81	1.28	4.30	7.55	2.43	4.01	1.84	4.62	4.71	3.08	4.00	2.66	42.29
Record Mean	2.56	2.69	3.39	3.43	3.80	3.84	4.14	3.46	3.39	3.15	3.76	3.06	40.67

AVERAGE TEMPERATURE (deg. F) WILLIAMSPORT, PENNSYLVANIA

YEAR	JAN	FEB	MAR	APR	MAY	JUNE	JULY	AUG	SEP	OCT	NOV	DEC	ANNUAL
1964	27.9	26.8	38.5	47.0	62.5	68.4	74.7	68.8	64.5	51.4	43.5	31.5	50.5
1965	22.3	27.3	33.5	44.5	61.5	65.2	68.9	68.1	63.4	49.1	41.4	34.7	48.3
1966	25.7	29.0	39.5	46.6	55.9	66.3	72.2	70.9	59.4	47.6	41.3	30.1	48.7
1967	31.4	23.5	34.3	50.4	52.8	71.2	71.4	69.4	63.3	51.8	37.9	33.5	49.3
1968	23.0	26.4	41.0	52.0	56.3	66.3	72.4	71.9	64.5	53.9	42.9	30.5	50.1
1969	27.3	28.2	35.5	50.0	60.9	69.5	71.7	71.1	64.2	51.7	41.8	28.5	50.0
1970	18.5	26.5	34.2	49.9	62.6	67.4	73.2	72.4	67.4	56.4	45.1	32.8	50.6
1971	23.0	30.4	36.6	48.4	58.5	70.6	71.8	70.4	68.2	58.8	39.6	35.1	51.0
1972	28.6	24.2	34.9	46.0	61.0	64.0	72.1	70.1	63.1	46.6	37.7	34.3	48.6
1973	28.6	25.9	43.4	48.9	54.9	70.4	71.2	71.1	63.6	53.4	42.4	31.7	50.5
1974	29.7	27.5	38.1	50.9	57.2	65.7	71.8	71.3	61.5	49.0	43.1	34.1	50.0
1975	30.6	31.2	37.2	43.2	61.5	66.9	72.1	70.9	59.5	54.4	47.2	31.8	50.6
1976	22.7	35.4	42.1	52.9	58.3	71.6	70.5	70.2	62.3	49.4	36.3	26.1	49.8
1977	14.9	28.9	44.7	53.2	63.8	67.0	73.8	72.1	67.2	50.4	45.3	28.4	50.8
1978	22.8	20.5	34.5	48.3	60.7	68.4	72.1	74.8	65.0	52.5	43.7	32.5	49.7
1979	26.7	19.7	42.3	48.3	60.9	67.0	72.6	71.0	62.8	50.8	44.8	35.0	50.2
1980	28.0	25.7	36.9	51.3	61.4	65.7	73.8	75.4	66.8	49.6	37.1	25.0	49.7
1981	20.3	33.2	36.5	51.9	62.5	70.8	74.4	72.4	64.5	51.3	42.2	30.2	50.9
1982	19.3	28.4	35.7	46.1	62.1	64.5	71.1	66.6	63.1	52.0	43.8	36.9	49.1
1983	29.3	31.5	41.0	47.9	56.3	66.3	72.2	71.1	62.0	49.9	40.1	26.7	49.5
1984	21.0	34.7	30.4	47.5	55.9	68.7	70.6	72.2	60.6	56.7	40.3	35.7	49.5
1985	23.4	29.9	40.1	53.3	61.0	64.4	71.0	70.1	65.3	52.9	45.1	27.9	50.4
1986	27.3	27.6	40.7	51.1	63.7	67.5	72.6	68.3	62.9	52.3	37.4	33.4	50.4
1987	25.3	27.2	40.5	52.2	61.2	70.6	75.5	69.9	63.0	47.5	41.2	33.7	50.7
1988	22.5	28.5	39.9	48.2	61.1	67.8	77.4	74.4	61.3	47.7	42.5	30.5	50.2
1989	31.3	28.9	37.8	47.7	57.9	69.2	72.9	70.8	63.7	53.6	39.9	19.8	49.5
1990	35.5	34.3	41.3	51.6	57.6	69.1	72.6	71.4	62.4	55.7	44.0	35.3	52.6
1991	28.2	33.8	41.7	52.1	67.1	71.6	75.2	74.0	62.6	54.3	42.0	34.0	53.1
1992	30.6	32.2	37.3	48.7	58.7	66.8	71.4	69.5	63.7	49.3	43.1	33.0	50.4
1993	32.3	25.3	35.2	50.4	62.1	69.2	76.6	73.3	62.7	50.0	40.6	30.6	50.7
Record Mean	26.5	28.6	38.0	49.5	59.7	68.2	72.7	71.0	63.5	52.4	41.5	30.6	50.2
Max	34.4	37.3	47.6	60.5	71.4	79.7	83.9	81.9	74.1	63.2	49.8	38.0	60.2
Min	18.6	20.0	28.4	38.5	48.0	56.6	61.5	60.1	52.9	41.6	33.1	23.3	40.2

REFERENCE NOTES WILLIAMSPORT, PENNSYLVANIA

GENERAL
T - TRACE AMOUNT.
BLANK ENTRIES DENOTE MISSING/UNREPORTED DATA.
# INDICATES A STATION OR INSTRUMENT RELOCATION.

SPECIFIC

EXCEPTIONS

RECORD MEANS ARE THROUGH THE CURRENT YEAR,
BEGINNING IN 1945 FOR TEMPERATURE
1945 FOR PRECIPITATION
1945 FOR SNOWFALL

(a) - LENGTH OF RECORD IN YEARS, ALTHOUGH
INDIVIDUAL MONTHS MAY BE MISSING.
0.* OR * - THE VALUE IS BETWEEN 0.0 AND 0.05
NORMALS - BASED ON THE 1961-1990 RECORD PERIOD.
EXTREMES - DATES ARE THE MOST RECENT OCCURRENCE.
WIND DIR.- NUMERALS SHOW TENS OF DEGRESS CLOCKWISE
FROM TRUE NORTH. "00" INDICATES CALM.
RESULTANT DIRECTIONS ARE GIVEN TO WHOLE DEGREES.
BOLD VALUES INDICATE EXTREME VALUES WHICH OCCURRED
AFTER THE ASOS SYSTEM WAS COMMISSIONED.

RECORD = PERIOD OF RECORD
RECORD MEAN PRECIPITATION IS THE MEAN OF ALL DAILY
PRECIPITATION AMOUNTS DURING THE PERIOD OF RECORD.
RECORD MAX(MIN) TEMPERATURE IS THE MEAN OF ALL DAILY
MAX(MIN) TEMPERATURES DURING THE PERIOD OF RECORD.
RECORD MEAN TEMPERATURE IS THE SUM OF THE RECORD
MAX AND RECORD MIN DIVIDED BY 2.
AVERAGE TEMPERATURE IS THE SUM OF THE MEAN DAILY
MAX AND MIN TEMPERATURE DIVIDED BY 2.

HEATING DEGREE DAYS Base 65 deg. F WILLIAMSPORT, PENNSYLVANIA

SEASON	JULY	AUG	SEP	OCT	NOV	DEC	JAN	FEB	MAR	APR	MAY	JUNE	TOTAL
1964-65	0	20	89	417	639	1032	1317	1047	970	611	137	80	6359
1965-66	17	41	121	484	702	937	1213	1002	783	548	283	60	6191
1966-67	5	4	196	531	703	1075	1035	1157	946	437	371	5	6465
1967-68	5	12	109	406	807	973	1296	1113	739	385	265	50	6160
1968-69	4	22	46	337	656	1063	1162	1022	909	444	171	24	5860
1969-70	2	5	106	412	690	1125	1433	1072	945	453	127	30	6400
1970-71	0	0	56	271	591	990	1297	961	871	495	212	16	5760
1971-72	0	7	58	201	757	921	1122	1176	923	562	141	90	5958
1972-73	12	19	91	565	810	942	1125	1089	664	479	306	16	6118
1973-74	5	16	119	354	672	1027	1087	1047	828	427	258	36	5876
1974-75	1	0	142	488	648	950	1059	940	852	648	155	44	5927
1975-76	1	9	168	324	529	1023	1305	850	704	393	217	19	5542
1976-77	0	12	121	478	855	1201	1547	1008	624	366	119	38	6369
1977-78	4	15	46	444	588	1126	1300	1239	938	495	190	28	6413
1978-79	9	0	76	381	633	998	1179	1264	696	497	167	40	5940
1979-80	15	19	131	438	601	919	1141	1134	863	405	139	77	5882
1980-81	0	0	65	472	830	1232	1378	886	877	391	130	7	6268
1981-82	0	0	97	416	677	1070	1408	1020	903	558	123	57	6329
1982-83	15	45	108	393	632	865	1101	934	736	511	271	50	5661
1983-84	14	8	159	461	741	1182	1360	873	1069	517	288	30	6702
1984-85	6	6	171	257	733	901	1282	978	765	377	151	59	5686
1985-86	4	6	108	369	590	1142	1162	1043	749	412	117	41	5743
1986-87	9	41	108	403	821	969	1225	1051	754	379	179	10	5949
1987-88	2	22	93	534	707	963	1306	1054	774	497	162	60	6174
1988-89	5	6	133	530	667	1063	1038	1005	835	515	244	18	6059
1989-90	3	16	109	344	746	1396	908	853	730	425	226	31	5787
1990-91	6	9	133	305	623	913	1136	870	714	394	104	7	5214
1991-92	0	0	143	342	684	953	1060	945	854	482	201	30	5694
1992-93	2	6	114	478	649	988	1006	1106	915	434	124	33	5855
1993-94	0	4	145	458	726	1059							

COOLING DEGREE DAYS Base 65 deg. F WILLIAMSPORT, PENNSYLVANIA

YEAR	JAN	FEB	MAR	APR	MAY	JUNE	JULY	AUG	SEP	OCT	NOV	DEC	TOTAL
1969	0	0	0	0	50	167	216	201	90	5	0	0	729
1970	0	0	0	8	58	109	262	237	136	12	0	0	822
1971	0	0	0	1	19	189	221	178	160	15	2	0	785
1972	0	0	0	0	25	67	236	183	40	0	0	0	551
1973	0	0	0	4	0	186	205	213	83	1	0	0	692
1974	0	0	0	11	26	63	217	204	43	0	0	0	564
1975	0	0	0	0	54	108	231	197	10	4	0	0	604
1976	0	0	0	35	20	223	173	177	47	0	0	0	675
1977	0	0	0	18	91	106	282	241	119	0	5	0	862
1978	0	0	0	0	63	134	237	310	83	0	0	0	827
1979	0	0	0	3	48	105	262	212	73	4	0	0	707
1980	0	0	0	0	34	104	279	333	128	0	0	0	878
1981	0	0	0	5	60	186	301	236	90	0	0	0	878
1982	0	0	0	0	42	48	213	103	56	2	5	0	469
1983	0	0	0	3	8	99	242	205	76	0	0	0	633
1984	0	0	0	0	9	147	186	236	44	7	0	0	629
1985	0	0	2	30	36	48	194	170	121	1	0	0	602
1986	0	0	1	1	83	123	250	151	53	15	0	0	677
1987	0	0	0	2	67	185	332	181	39	0	0	0	806
1988	0	0	0	0	48	151	398	306	28	2	0	0	933
1989	0	0	0	2	34	150	253	199	77	1	0	0	716
1990	0	0	1	30	4	159	248	216	67	26	0	0	751
1991	0	0	0	12	176	211	322	288	81	18	0	0	1108
1992	0	0	0	1	16	91	207	149	82	0	0	0	546
1993	0	0	0	4	39	165	365	267	83	0	0	0	923

SNOWFALL (inches) WILLIAMSPORT, PENNSYLVANIA

SEASON	JULY	AUG	SEP	OCT	NOV	DEC	JAN	FEB	MAR	APR	MAY	JUNE	TOTAL
1964-65	0.0	0.0	0.0	T	T	1.8	9.4	0.8	9.9	3.0	0.0	0.0	24.9
1965-66	0.0	0.0	0.0	T	T	0.7	25.6	11.7	1.0	0.1	T	0.0	39.1
1966-67	0.0	0.0	0.0	0.0	0.3	11.6	3.0	17.6	29.5	2.7	T	0.0	64.7
1967-68	0.0	0.0	0.0	0.0	9.8	10.4	2.6	3.0	3.2	0.0	0.0	0.0	29.0
1968-69	0.0	0.0	0.0	0.0	7.5	2.1	0.8	6.7	1.2	0.0	0.0	0.0	18.3
1969-70	0.0	0.0	0.0	0.0	1.2	35.5	12.2	13.5	20.0	0.2	T	0.0	82.6
1970-71	0.0	0.0	0.0	T	0.2	14.8	14.2	15.5	16.7	T	0.0	0.0	61.4
1971-72	0.0	0.0	0.0	0.0	9.3	1.6	6.0	34.3	6.7	1.6	0.0	0.0	59.5
1972-73	0.0	0.0	0.0	0.3	12.8	5.1	1.5	8.3	2.6	T	0.0	0.0	30.6
1973-74	0.0	0.0	0.0	0.0	0.4	15.6	7.8	4.8	5.7	6.3	0.0	0.0	40.6
1974-75	0.0	0.0	0.0	T	T	4.6	15.4	14.9	3.8	0.1	0.0	0.0	38.8
1975-76	0.0	0.0	0.0	0.0	0.2	4.6	11.4	5.2	7.5	T	0.0	0.0	28.9
1976-77	0.0	0.0	0.0	T	3.8	6.0	16.0	7.3	8.5	0.3	0.2	0.0	42.1
1977-78	0.0	0.0	0.0	1.0	4.6	17.6	38.1	14.3	6.8	1.2	0.0	0.0	83.6
1978-79	0.0	0.0	0.0	T	3.9	4.8	10.5	17.7	T	0.4	0.0	0.0	37.3
1979-80	0.0	0.0	0.0	T	T	4.8	1.9	5.2	8.6	T	0.0	0.0	20.5
1980-81	0.0	0.0	0.0	T	12.0	6.8	10.8	5.5	6.5	T	0.0	0.0	41.6
1981-82	0.0	0.0	0.0	T	T	8.1	14.3	10.4	7.9	13.8	0.0	0.0	54.5
1982-83	0.0	0.0	0.0	T	T	1.4	6.3	7.6	0.5	1.8	0.0	0.0	17.6
1983-84	0.0	0.0	0.0	0.0	T	5.3	11.7	1.6	22.1	T	0.0	0.0	40.7
1984-85	0.0	0.0	0.0	0.0	T	8.9	8.7	6.4	0.6	0.4	0.0	0.0	25.0
1985-86	0.0	0.0	0.0	0.0	T	6.5	10.2	13.6	0.9	T	0.0	0.0	31.2
1986-87	0.0	0.0	0.0	0.0	7.7	T	40.1	5.3	0.4	T	0.0	0.0	53.5
1987-88	0.0	0.0	0.0	T	9.4	6.0	10.1	9.7	1.6	T	0.0	0.0	36.8
1988-89	0.0	0.0	0.0	T	T	0.7	2.3	1.5	2.5	T	0.0	0.0	7.0
1989-90	0.0	T	0.0	T	2.8	8.3	9.5	3.2	6.8	0.7	0.0	0.0	31.3
1990-91	0.0	0.0	0.0	0.0	T	9.8	5.2	4.1	8.5	0.3	0.0	T	27.9
1991-92	0.0	T	0.0	0.0	T	0.8	4.3	2.8	10.5	1.6	0.0	0.0	20.0
1992-93	0.0	0.0	0.0	T	T	9.2	2.2	21.0	22.6	0.8	0.0	0.0	55.8
1993-94	0.0	0.0	0.0	0.8	0.1	4.1							
Record Mean	0.0	T	0.0	0.1	3.0	8.2	10.4	10.1	8.2	1.2	T	T	41.1

SAN JUAN, PUERTO RICO

San Juan, located on the north coast of the island of Puerto Rico, is surrounded by the waters of the Atlantic Ocean and San Juan Bay. Local custom assigns the name San Juan to the old city which lies right on the coast, but the modern metropolitan area extends inland about 12 miles. These inland sections have a temperature and rainfall regime significantly different from the coastal area. Isla Verde Airport, where weather observations are made, lies on the coast about 7 miles east of old San Juan. The surrounding terrain is level with a gradual upslope inland. Mountain ranges, with peak elevations of 4,000 feet, extend east and west through the central portion of Puerto Rico, and are located 15 to 20 miles east and south of San Juan. These mountain ranges have a decided influence on the rainfall of the San Juan metropolitan area, and on the entire island in general.

The climate is tropical maritime, characteristic of all tropical islands. The predominant easterly trade winds, modified by local effects such as the land and sea breeze and the particular island topography, are a primary feature of the climate of San Juan and have a significant influence on the temperature and rainfall. During daylight hours the wind blows almost constantly off the ocean. Usually, after sunset the wind shifts to the south or southeast, off land. This daily wind variation is a contributing factor to the delightful climate of the city. The annual temperature range is small with about a 5-6 degree difference between the temperatures of the warmest and coldest months. The inland sectors have warmer afternoons and cooler nights. In the interior mountain and valley regions even greater daily and annual ranges of temperature occur. The highest temperatures recorded in Puerto Rico have exceeded 105 degrees and the lowest have been near 40. Sea water temperatures range from 78 degrees in March to about 83 degrees in September.

Although rainfall in San Juan is nearly 60 inches, the geographical distribution of rainfall over the island shows the heaviest rainfall, of about 180 inches per year, in the Luquillo Range, only 23 miles distant from San Juan. The driest area, with annual rainfall of 30 to 35 inches, is located in the southwest corner of the island. Rain showers occur mostly in the afternoon and at night. The nocturnal showers, usually light, are a characteristic feature of the San Juan rainfall pattern. Rainfall is generally of the brief showery type except for the continuous rains occuring with the passage of tropical disturbances, or when the trailing edge of a cold front out of the United States reaches Puerto Rico. This normally occurs from about November to April.

Puerto Rico is in the tropical hurricane region of the eastern Caribbean. The hurricane season begins June 1 and ends November 30. Only a few hurricanes have passed close enough to San Juan to produce hurricane force winds or damage.

Mild temperatures, refreshing sea breezes in the daytime, plenty of sunshine, and adequate rainfall make the climate of San Juan most enjoyable for tourists and residents alike.

NORMALS, MEANS, AND EXTREMES

SAN JUAN, PUERTO RICO

LATITUDE: 18°26'N LONGITUDE: 66°00'W ELEVATION: FT. GRND 13 BARO 69 TIME ZONE: ATLANTIC WBAN: 11641

	(a)	JAN	FEB	MAR	APR	MAY	JUNE	JULY	AUG	SEP	OCT	NOV	DEC	YEAR
TEMPERATURE °F:														
Normals														
-Daily Maximum		83.2	83.6	84.4	85.8	87.2	88.6	88.5	88.7	88.8	88.3	85.9	83.8	86.4
-Daily Minimum		70.8	70.6	71.6	72.9	74.5	76.1	76.8	76.7	76.2	75.5	74.0	72.4	74.0
-Monthly		77.0	77.1	78.0	79.4	80.9	82.3	82.6	82.7	82.5	81.9	80.0	78.1	80.2
Extremes														
-Record Highest	39	92	96	96	97	96	97	95	97	97	98	96	94	98
-Year		1983	1983	1983	1983	1980	1988	1981	1980	1981	1981	1981	1989	OCT 1981
-Record Lowest	39	61	62	60	64	66	69	69	70	69	67	66	63	60
-Year		1962	1968	1957	1968	1962	1957	1959	1956	1960	1959	1969	1964	MAR 1957
NORMAL DEGREE DAYS:														
Heating (base 65°F)		0	0	0	0	0	0	0	0	0	0	0	0	0
Cooling (base 65°F)		372	339	403	432	493	519	546	549	525	524	450	406	5558
% OF POSSIBLE SUNSHINE	38	68	71	76	71	63	64	68	67	62	63	61	60	66
MEAN SKY COVER (tenths)														
Sunrise - Sunset	38	4.9	5.0	5.0	5.4	6.6	6.2	5.9	5.8	6.2	6.1	5.7	5.5	5.7
MEAN NUMBER OF DAYS:														
Sunrise to Sunset														
-Clear	38	8.7	7.2	9.1	7.0	3.5	4.0	4.9	5.3	3.8	4.4	5.3	6.4	69.7
-Partly Cloudy	38	18.1	16.6	17.6	16.7	15.6	15.9	17.4	17.4	16.9	17.0	17.8	18.4	205.4
-Cloudy	38	4.2	4.4	4.4	6.3	11.9	10.1	8.7	8.2	9.3	9.6	6.9	6.1	90.2
Precipitation														
.01 inches or more	38	16.9	13.0	12.4	12.7	16.5	15.4	18.7	18.2	17.3	17.1	18.3	19.1	195.7
Snow, Ice pellets, hail														
1.0 inches or more	38	0.0	0.0	0.0	0.0	0.0	0.0	0.0	0.0	0.0	0.0	0.0	0.0	0.0
Thunderstorms	38	0.2	0.3	0.3	1.1	4.5	4.8	5.2	6.1	8.3	7.4	3.1	0.8	41.9
Heavy Fog Visibility														
1/4 mile or less	38	0.*	0.0	0.0	0.0	0.0	0.0	0.0	0.0	0.0	0.0	0.0	0.0	*
Temperature °F														
-Maximum														
90° and above	38	0.3	0.8	2.0	3.4	5.7	9.4	8.1	10.1	10.7	9.2	1.8	0.5	62.1
32° and below	38	0.0	0.0	0.0	0.0	0.0	0.0	0.0	0.0	0.0	0.0	0.0	0.0	0.0
-Minimum														
32° and below	38	0.0	0.0	0.0	0.0	0.0	0.0	0.0	0.0	0.0	0.0	0.0	0.0	0.0
0° and below	38	0.0	0.0	0.0	0.0	0.0	0.0	0.0	0.0	0.0	0.0	0.0	0.0	0.0
AVG. STATION PRESS. (mb)	21	1014.9	1014.7	1014.2	1013.1	1013.0	1014.4	1015.0	1013.8	1012.3	1011.4	1011.9	1013.9	1013.5
RELATIVE HUMIDITY (%)														
Hour 02	38	82	80	80	80	84	84	84	84	85	85	84	82	83
Hour 18 (Local Time)	38	82	80	77	75	77	78	79	79	79	80	81	81	79
Hour 14	38	64	62	60	62	66	66	67	66	67	66	67	66	65
Hour 20	38	76	74	73	74	77	77	78	78	78	79	78	77	77
PRECIPITATION (inches):														
Water Equivalent														
-Normal		2.81	2.15	2.35	3.76	5.93	4.00	4.37	5.32	5.28	5.71	5.94	4.72	52.34
-Maximum Monthly	39	7.60	6.69	5.41	10.37	14.99	10.96	9.35	11.76	14.83	15.06	15.96	16.81	16.81
-Year		1977	1982	1958	1988	1965	1965	1961	1955	1989	1970	1979	1981	DEC 1981
-Minimum Monthly	39	0.61	0.20	0.72	0.28	0.44	0.29	1.12	1.93	1.73	1.17	1.91	0.68	0.20
-Year		1978	1983	1970	1984	1972	1985	1974	1982	1987	1979	1980	1963	FEB 1983
-Maximum in 24 hrs	39	5.08	2.75	3.91	7.20	4.74	3.55	2.91	5.08	8.84	5.04	7.07	8.40	8.84
-Year		1969	1989	1969	1988	1986	1965	1993	1955	1989	1985	1979	1981	SEP 1989
Snow, Ice pellets, hail														
-Maximum Monthly	39	0.0	0.0	0.0	0.0	0.0	0.0	0.0	0.0	T	0.0	0.0	0.0	T
-Year										1989				SEP 1989
-Maximum in 24 hrs	39	0.0	0.0	0.0	0.0	0.0	0.0	0.0	0.0	T	0.0	0.0	0.0	T
-Year										1989				SEP 1989
WIND:														
Mean Speed (mph)	38	8.5	8.8	9.2	8.9	8.4	8.9	9.6	8.8	7.5	6.8	7.5	8.2	8.4
Prevailing Direction through 1963		ENE	ENE	ENE	ENE	ENE	ENE	ENE	ENE	ENE	ENE	ENE	ENE	ENE
Fastest Obs. 1 Min.														
-Direction (!!!)	8	09	08	07	06	01	12	08	09	34	07	09	11	34
-Speed (MPH)	8	24	23	23	23	25	35	24	23	77	23	25	29	77
-Year		1989	1992	1984	1993	1992	1987	1987	1992	1989	1993	1988	1984	SEP 1989
Peak Gust														
-Direction (!!!)	10	E	ESE	E	SE	N	SW	E	E	NW	SE	E	E	NW
-Speed (mph)	10	36	37	36	38	39	38	36	37	92	35	39	41	92
-Date		1993	1989	1989	1989	1992	1984	1991	1989	1989	1986	1985	1984	SEP 1989

PRECIPITATION (inches)

SAN JUAN, PUERTO RICO

YEAR	JAN	FEB	MAR	APR	MAY	JUNE	JULY	AUG	SEP	OCT	NOV	DEC	ANNUAL
1964	2.02	1.70	1.27	6.38	3.96	4.50	7.03	6.71	5.10	3.13	3.39	2.35	47.54
1965	2.62	0.79	0.86	2.19	14.99	10.96	5.88	8.66	4.80	5.08	3.93	5.05	65.81
1966	1.34	1.64	4.63	5.62	5.69	3.26	4.21	3.41	7.20	8.99	7.99	6.21	60.19
1967	3.07	2.93	1.46	0.85	4.15	3.38	4.79	4.20	5.12	4.47	5.00	3.13	42.55
1968	2.15	1.60	1.79	0.50	6.31	5.98	5.25	7.36	5.26	2.33	11.11	3.56	53.20
1969	7.49	3.97	2.89	2.42	5.79	4.04	7.49	6.89	4.86	6.99	6.70	2.28	61.81
1970	2.94	1.33	0.72	1.15	7.98	9.26	3.58	4.66	5.66	15.06	8.00	5.98	66.32
1971	2.18	3.67	1.78	2.93	3.87	1.24	1.69	5.18	2.19	4.61	2.31	3.93	35.58
1972	2.76	2.00	3.40	2.79	0.44	1.58	2.24	3.06	3.68	5.46	2.78	7.53	37.72
1973	2.27	0.92	4.66	8.48	0.48	4.71	2.44	7.00	3.13	3.29	3.01	4.16	44.55
1974	2.92	0.82	1.92	1.20	2.42	2.34	1.12	6.57	3.67	8.23	6.55	3.92	41.68
1975	2.69	0.71	1.13	1.01	1.04	2.64	3.35	4.08	9.29	6.60	10.90	7.82	51.26
1976	1.50	2.18	2.05	3.94	2.96	2.96	2.48	5.12	11.44	7.69	2.77	2.11	47.20
1977	7.60	1.02	1.73	0.96	4.04	1.49	4.64	4.42	4.71	5.94	12.44	3.82	52.81
1978	0.61	1.56	3.52	8.27	7.14	2.86	3.46	3.21	6.34	4.88	5.40	2.61	49.86
1979	1.29	1.80	2.25	4.28	12.13	5.76	6.61	9.38	10.11	1.17	15.96	3.81	74.55
1980	1.75	1.67	1.47	2.55	5.19	1.31	2.19	3.17	4.85	6.71	1.91	3.18	35.95
1981	2.55	2.72	4.39	2.89	11.02	5.48	7.04	3.32	2.98	9.32	4.94	16.81	73.46
1982	2.53	6.69	0.98	1.01	10.26	5.24	2.33	1.93	2.87	2.06	4.34	4.76	45.00
1983	0.69	0.20	1.47	8.54	3.85	1.91	6.53	5.15	2.75	4.06	3.25	3.50	41.90
1984	1.96	3.13	0.82	0.28	3.75	6.85	2.66	6.04	3.16	5.10	5.65	4.69	44.09
1985	2.80	2.40	1.84	1.02	5.95	0.29	2.85	4.33	5.44	11.10	4.54	2.80	45.36
1986	2.18	1.13	1.61	8.93	12.80	1.52	1.94	5.19	1.98	8.54	5.87	3.59	55.28
1987	2.16	1.20	5.17	8.88	12.17	7.07	3.26	2.48	1.73	2.70	7.49	7.69	62.00
1988	3.83	2.27	1.76	10.37	6.06	1.45	4.02	11.31	5.49	4.12	5.68	4.07	60.43
1989	2.96	6.05	3.39	2.63	4.88	2.97	5.54	7.88	14.83	2.09	4.95	2.50	60.67
1990	4.56	3.02	3.14	1.05	2.44	4.32	5.76	3.42	2.23	8.65	5.33	5.03	48.95
1991	2.57	2.26	1.99	1.76	3.23	2.77	3.30	1.94	5.00	1.84	6.16	2.71	35.53
1992	4.03	1.19	1.47	2.12	8.76	5.55	4.38	4.00	5.35	1.74	11.98	4.72	55.29
1993	2.35	0.51	0.78	6.55	4.48	5.46	7.34	3.01	4.36	2.78	4.34	3.00	44.96
Record Mean	2.96	2.27	2.25	3.79	6.29	4.55	4.77	5.65	5.60	5.55	5.86	4.81	54.35

AVERAGE TEMPERATURE (deg. F)

SAN JUAN, PUERTO RICO

YEAR	JAN	FEB	MAR	APR	MAY	JUNE	JULY	AUG	SEP	OCT	NOV	DEC	ANNUAL
1964	76.6	77.5	78.8	78.5	80.9	81.6	81.8	82.2	82.2	80.6	79.0	75.7	79.6
1965	74.7	75.9	77.7	77.7	78.9	80.3	81.7	81.5	82.2	81.5	79.8	77.4	79.1
1966	77.4	76.7	78.1	78.5	79.5	81.0	82.1	82.6	81.1	80.0	77.5	77.1	79.3
1967	76.7	76.3	75.9	78.0	79.7	81.8	82.1	82.2	81.8	81.4	80.4	77.5	79.5
1968	76.4	76.5	76.2	77.0	79.9	80.4	81.2	82.0	81.8	81.6	79.5	77.5	79.2
1969	75.4	75.0	77.8	79.6	80.7	82.0	80.9	81.1	81.2	80.9	78.7	77.1	79.2
1970	76.5	76.2	77.9	80.0	80.6	81.6	82.4	82.3	82.2	81.9	80.2	78.2	80.0
1971	77.6	77.4	78.9	79.7	81.3	83.0	82.9	83.2	83.9	81.9	80.6	78.8	80.8
1972	77.7	77.8	78.7	80.1	81.7	83.8	83.9	83.4	82.9	82.6	81.1	79.2	81.1
1973	78.9	78.3	78.6	80.0	82.9	83.2	83.4	83.2	82.7	83.5	80.7	77.9	81.1
1974	77.3	78.0	78.5	79.6	81.4	83.4	83.7	83.2	82.9	82.5	80.0	78.3	80.7
1975	76.9	77.2	78.4	79.6	81.0	82.9	83.0	82.7	82.1	81.6	79.8	77.5	80.2
1976	76.3	76.6	76.8	78.5	80.5	81.7	83.1	83.5	83.3	82.5	80.9	78.3	80.2
1977	77.0	77.6	78.3	80.2	80.9	81.9	80.6	81.0	81.1	81.9	80.3	79.3	80.0
1978	78.3	79.2	79.8	80.1	80.8	82.3	83.1	83.2	83.9	82.7	81.8	79.6	81.2
1979	78.5	78.7	77.8	79.4	81.3	83.8	83.5	83.0	81.8	83.0	80.6	78.7	80.8
1980	78.2	78.6	79.2	80.8	83.8	85.1	85.2	85.1	84.5	84.4	82.2	80.7	82.3
1981	79.8	79.4	80.9	80.3	83.4	84.2	85.0	83.8	84.3	83.3	81.6	78.7	82.1
1982	78.5	77.8	78.0	80.4	81.0	82.9	83.3	84.6	84.6	83.8	80.2	78.0	81.1
1983	78.5	79.9	82.2	81.8	83.2	85.4	84.2	84.0	84.3	83.6	81.4	79.9	82.4
1984	78.1	77.8	80.2	81.8	81.1	82.4	82.6	82.6	82.2	81.3	78.7	77.3	80.5
1985	76.1	77.3	76.7	78.3	80.4	83.0	83.5	83.4	81.7	79.6	79.0	76.8	79.7
1986	75.5	76.0	77.4	79.0	79.1	81.6	82.0	82.2	82.5	81.4	79.1	77.7	79.5
1987	76.7	77.4	78.0	81.2	81.6	81.7	82.7	83.7	83.8	83.3	80.6	79.7	80.9
1988	76.9	76.6	77.8	80.4	82.4	84.3	83.6	82.8	82.2	81.6	79.8	76.9	80.4
1989	76.2	76.0	76.1	78.8	80.3	81.2	81.8	82.3	82.3	81.7	80.0	79.2	79.7
1990	77.1	76.4	76.5	79.1	81.8	82.4	82.6	82.9	83.4	82.1	80.7	77.5	80.2
1991	77.0	77.8	78.6	79.2	81.0	83.2	83.1	83.9	83.3	82.4	79.9	77.0	80.5
1992	77.1	78.2	78.9	80.8	81.3	83.4	82.4	82.7	82.3	84.0	80.8	79.3	80.9
1993	77.3	78.5	79.3	80.3	82.1	83.3	82.8	84.2	83.0	82.7	80.9	78.9	81.1
Record Mean	76.6	76.7	77.6	79.0	80.6	82.0	82.2	82.5	82.2	81.6	79.7	77.8	79.8
Max	82.9	83.2	84.2	85.5	86.9	88.3	88.2	88.6	88.6	88.2	85.8	83.6	86.2
Min	70.3	70.2	71.0	72.5	74.2	75.6	76.2	76.3	75.7	75.0	73.6	71.9	73.5

REFERENCE NOTES

SAN JUAN, PUERTO RICO

GENERAL
T - TRACE AMOUNT.
BLANK ENTRIES DENOTE MISSING/UNREPORTED DATA.
# INDICATES A STATION OR INSTRUMENT RELOCATION.

SPECIFIC

PM - INCLUDES LAST DAY OF PREVIOUS MONTH
ASOS - AUTOMATED SURFACE OBSERVING SYSTEM IN OPERATION DURING THESE MONTHS.

(a) - LENGTH OF RECORD IN YEARS, ALTHOUGH INDIVIDUAL MONTHS MAY BE MISSING.
0.* OR * - THE VALUE IS BETWEEN 0.0 AND 0.05
NORMALS - BASED ON THE 1961-1990 RECORD PERIOD.
EXTREMES - DATES ARE THE MOST RECENT OCCURRENCE
WIND DIR.- NUMERALS SHOW TENS OF DEGRESS CLOCKWISE FROM TRUE NORTH. "00" INDICATES CALM.
RESULTANT DIRECTIONS ARE GIVEN TO WHOLE DEGREES.
BOLD VALUES INDICATE EXTREME VALUES WHICH OCCURRED AFTER THE ASOS SYSTEM WAS COMMISSIONED.

RECORD = PERIOD OF RECORD
RECORD MEAN PRECIPITATION IS THE MEAN OF ALL DAILY PRECIPITATION AMOUNTS DURING THE PERIOD OF RECORD.
RECORD MAX(MIN) TEMPERATURE IS THE MEAN OF ALL DAILY MAX(MIN) TEMPERATURES DURING THE PERIOD OF RECORD.
RECORD MEAN TEMPERATURE IS THE SUM OF THE RECORD MAX AND RECORD MIN DIVIDED BY 2.
AVERAGE TEMPERATURE IS THE SUM OF THE MEAN DAILY MAX AND MIN TEMPERATURE DIVIDED BY 2.

EXCEPTIONS

FASTEST MILE WINDS ARE THROUGH MAY 1983.

RECORD MEANS ARE THROUGH THE CURRENT YEAR, BEGINNING IN 1951 FOR TEMPERATURE
1951 FOR PRECIPITATION

HEATING DEGREE DAYS Base 65 deg. F — SAN JUAN, PUERTO RICO

SEASON	JULY	AUG	SEP	OCT	NOV	DEC	JAN	FEB	MAR	APR	MAY	JUNE	TOTAL
1983-84	0	0	0	0	0	0	0	0	0	0	0	0	0
1984-85	0	0	0	0	0	0	0	0	0	0	0	0	0
1985-86	0	0	0	0	0	0	0	0	0	0	0	0	0
1986-87	0	0	0	0	0	0	0	0	0	0	0	0	0
1987-88	0	0	0	0	0	0	0	0	0	0	0	0	0
1988-89	0	0	0	0	0	0	0	0	0	0	0	0	0
1989-90	0	0	0	0	0	0	0	0	0	0	0	0	0
1990-91	0	0	0	0	0	0	0	0	0	0	0	0	0
1991-92	0	0	0	0	0	0	0	0	0	0	0	0	0
1992-93	0	0	0	0	0	0	0	0	0	0	0	0	0
1993-94	0	0	0	0	0	0							

COOLING DEGREE DAYS Base 65 deg. F — SAN JUAN, PUERTO RICO

YEAR	JAN	FEB	MAR	APR	MAY	JUNE	JULY	AUG	SEP	OCT	NOV	DEC	TOTAL
1969	328	287	401	447	496	517	501	506	493	500	417	382	5275
1970	362	321	409	460	490	506	548	540	523	528	461	416	5564
1971	398	353	438	445	511	545	562	573	577	531	473	438	5844
1972	400	376	431	458	525	569	592	578	543	554	490	446	5962
1973	438	379	428	458	562	553	578	570	539	579	475	406	5965
1974	390	368	424	442	511	558	586	575	544	549	455	418	5820
1975	375	346	420	446	501	543	565	556	521	519	450	395	5637
1976	357	342	375	413	485	506	566	579	557	548	485	419	5632
1977	382	360	419	463	499	514	491	502	487	532	469	452	5570
1978	420	404	464	462	499	526	566	570	575	555	513	459	6013
1979	426	393	404	436	512	571	582	564	511	569	477	432	5877
1980	414	402	446	479	590	610	633	628	591	608	522	494	6417
1981	467	407	499	468	578	584	626	587	588	574	505	429	6312
1982	423	364	412	470	504	543	573	615	593	590	466	405	5958
1983	426	424	541	509	573	621	604	593	583	582	502	468	6426
1984	415	377	479	508	503	528	553	553	522	517	417	389	5761
1985	352	349	368	403	483	547	579	577	508	459	429	371	5425
1986	329	315	390	428	444	501	535	538	534	517	429	401	5361
1987	366	354	407	490	521	508	556	588	571	575	477	464	5877
1988	377	341	403	470	547	587	581	558	523	522	453	375	5737
1989	354	313	353	421	482	495	529	543	525	526	456	443	5440
1990	383	325	362	430	526	526	551	561	559	539	478	392	5632
1991	379	363	428	430	503	551	571	593	558	547	455	380	5758
1992	383	393	435	479	512	556	547	555	524	596	482	454	5916
1993	391	383	450	469	535	557	561	603	550	557	483	440	5979

SNOWFALL (inches) — SAN JUAN, PUERTO RICO

SEASON	JULY	AUG	SEP	OCT	NOV	DEC	JAN	FEB	MAR	APR	MAY	JUNE	TOTAL
1971-72	0.0	0.0	0.0	0.0	0.0	0.0	0.0	0.0	0.0	0.0	0.0	0.0	0.0
1972-73	0.0	0.0	0.0	0.0	0.0	0.0	0.0	0.0	0.0	0.0	0.0	0.0	0.0
1973-74	0.0	0.0	0.0	0.0	0.0	0.0	0.0	0.0	0.0	0.0	0.0	0.0	0.0
1974-75	0.0	0.0	0.0	0.0	0.0	0.0	0.0	0.0	0.0	0.0	0.0	0.0	0.0
1975-76	0.0	0.0	0.0	0.0	0.0	0.0	0.0	0.0	0.0	0.0	0.0	0.0	0.0
1976-77	0.0	0.0	0.0	0.0	0.0	0.0	0.0	0.0	0.0	0.0	0.0	0.0	0.0
1977-78	0.0	0.0	0.0	0.0	0.0	0.0	0.0	0.0	0.0	0.0	0.0	0.0	0.0
1978-79	0.0	0.0	0.0	0.0	0.0	0.0	0.0	0.0	0.0	0.0	0.0	0.0	0.0
1979-80	0.0	0.0	0.0	0.0	0.0	0.0	0.0	0.0	0.0	0.0	0.0	0.0	0.0
1980-81	0.0	0.0	0.0	0.0	0.0	0.0	0.0	0.0	0.0	0.0	0.0	0.0	0.0
1981-82	0.0	0.0	0.0	0.0	0.0	0.0	0.0	0.0	0.0	0.0	0.0	0.0	0.0
1982-83	0.0	0.0	0.0	0.0	0.0	0.0	0.0	0.0	0.0	0.0	0.0	0.0	0.0
1983-84	0.0	0.0	0.0	0.0	0.0	0.0	0.0	0.0	0.0	0.0	0.0	0.0	0.0
1984-85	0.0	0.0	0.0	0.0	0.0	0.0	0.0	0.0	0.0	0.0	0.0	0.0	0.0
1985-86	0.0	0.0	0.0	0.0	0.0	0.0	0.0	0.0	0.0	0.0	0.0	0.0	0.0
1986-87	0.0	0.0	0.0	0.0	0.0	0.0	0.0	0.0	0.0	0.0	0.0	0.0	0.0
1987-88	0.0	0.0	0.0	0.0	0.0	0.0	0.0	0.0	0.0	0.0	0.0	0.0	0.0
1988-89	0.0	0.0	0.0	0.0	0.0	0.0	0.0	0.0	0.0	0.0	0.0	0.0	0.0
1989-90	0.0	0.0	T	0.0	0.0	0.0	0.0	0.0	0.0	0.0	0.0	0.0	T
1990-91	0.0	0.0	0.0	0.0	0.0	0.0	0.0	0.0	0.0	0.0	0.0	0.0	0.0
1991-92	0.0	0.0	0.0	0.0	0.0	0.0	0.0	0.0	0.0	0.0	0.0	0.0	0.0
1992-93	0.0	0.0	0.0	0.0	0.0	0.0	0.0	0.0	0.0	0.0	0.0	0.0	0.0
1993-94	0.0	0.0	0.0	0.0	0.0	0.0							
Record Mean	0.0	0.0	T	0.0	0.0	0.0	0.0	0.0	0.0	0.0	0.0	0.0	T

PROVIDENCE, RHODE ISLAND

The proximity to Narragansett Bay and the Atlantic Ocean plays an important part in determining the climate for Providence and vicinity. In winter, the temperatures are modified considerably, and many major snowstorms change to rain before reaching the area. In summer, many days that could be uncomfortably warm are cooled by refreshing sea breezes. At other times of the year, sea fog may be advected in over land by onshore winds. In fact, most cases of dense fog are produced this way, but the number of such days is few, averaging two or three days per month. In early fall, severe coastal storms of tropical origin sometimes bring destructive winds to this area. Even at other times of the year, it is usually coastal storms which produce the severest weather.

The temperature for the entire year averages around 50 degrees with 70 degree temperatures common from near the end of May to the latter part of September. During this period, there may be several days reaching 90 degrees or more. Temperatures of 100 degrees and more are rare.

Freezing temperatures occur on the average about 125 days per year. They become a common daily occurrence in the latter part of November, and become less frequent near the end of March. The average date for the last freeze in spring is mid-April, while the average date for the first freeze in fall is late October, making the growing season about 195 days in length. Sub-zero weather in winter seldom occurs, averaging less than one day for December and one or two days each for January and February.

Measurable precipitation occurs on about one day out of every three, and is fairly evenly distributed throughout the year. There is usually no definite dry season, but occasionally droughts do occur.

Thunderstorms are responsible for much of the rainfall from May through August. They usually produce heavy, and sometimes even excessive amounts of rainfall. However, since their duration is relatively short, damage is ordinarily light. The thunderstorms of summer are frequently accompanied by extremely gusty winds, which may result in some damage to property.

The first measurable snowfall of winter usually comes toward the end of November, and the last in spring is about the middle of March. Winters with over 50 inches of snow are not common. The area normally receives less than 25 inches. The month of greatest snowfall is usually February, but January and March are close seconds. It is unusual for the ground to remain well covered with snow for any long period of time.

NORMALS, MEANS, AND EXTREMES

PROVIDENCE, RHODE ISLAND

LATITUDE: 41°44'N LONGITUDE: 71°26' W ELEVATION: FT. GRND 51 BARO 58 TIME ZONE: EASTERN WBAN: 14765

	(a)	JAN	FEB	MAR	APR	MAY	JUNE	JULY	AUG	SEP	OCT	NOV	DEC	YEAR
TEMPERATURE °F:														
Normals														
-Daily Maximum		36.6	38.3	46.1	57.0	67.3	76.9	82.1	80.7	74.3	64.1	53.0	41.2	59.8
-Daily Minimum		19.1	20.9	28.8	37.7	47.3	56.8	63.2	61.9	53.8	43.0	34.9	24.4	41.0
-Monthly		27.9	29.7	37.4	47.4	57.3	66.9	72.7	71.3	64.1	53.6	44.0	32.8	50.4
Extremes														
-Record Highest	40	66	72	80	98	94	97	102	104	100	86	78	70	104
-Year		1974	1985	1989	1976	1992	1988	1991	1975	1983	1979	1993	1984	AUG 1975
-Record Lowest	40	-13	-7	1	14	29	41	48	40	33	20	6	-10	-13
-Year		1976	1979	1967	1954	1956	1980	1988	1965	1980	1976	1989	1980	JAN 1976
NORMAL DEGREE DAYS:														
Heating (base 65°F)		1150	988	856	528	246	31	0	8	90	359	630	998	5884
Cooling (base 65°F)		0	0	0	0	7	88	239	203	63	6	0	0	606
% OF POSSIBLE SUNSHINE	40	57	57	58	56	58	61	63	61	62	60	50	52	58
MEAN SKY COVER (tenths)														
Sunrise - Sunset	40	6.2	6.3	6.5	6.6	6.7	6.4	6.4	6.2	5.9	5.5	6.4	6.2	6.3
MEAN NUMBER OF DAYS:														
Sunrise to Sunset														
-Clear	40	9.6	7.9	8.6	7.3	6.4	6.7	6.8	8.3	9.5	10.9	8.2	8.4	98.5
-Partly Cloudy	40	6.8	7.1	7.7	8.1	9.9	10.3	11.9	10.4	8.1	7.9	6.8	7.8	102.9
-Cloudy	40	14.6	13.3	14.8	14.5	14.6	13.1	12.3	12.3	12.4	12.2	14.9	14.8	163.7
Precipitation														
.01 inches or more	40	10.9	10.0	11.9	11.1	11.4	10.7	9.1	9.4	8.4	8.7	10.8	12.0	124.3
Snow,Ice pellets,hail														
1.0 inches or more	40	2.8	2.5	2.0	0.3	0.*	0.0	0.0	0.0	0.0	0.1	0.3	2.1	10.0
Thunderstorms	40	0.3	0.2	0.6	1.2	2.6	3.7	4.4	3.6	1.6	1.0	0.8	0.2	20.2
Heavy Fog Visibility 1/4 mile or less	40	2.0	2.0	2.1	2.0	2.1	2.4	1.9	1.5	1.7	3.0	2.0	2.0	24.8
Temperature °F														
-Maximum														
90° and above	30	0.0	0.0	0.0	0.1	0.7	1.9	3.8	2.5	0.8	0.0	0.0	0.0	9.8
32° and below	30	11.0	7.6	1.4	0.*	0.0	0.0	0.0	0.0	0.0	0.0	0.2	6.0	26.2
-Minimum														
32° and below	30	27.9	24.4	19.5	5.5	0.2	0.0	0.0	0.0	0.0	3.6	12.7	24.4	118.2
0° and below	30	1.3	0.6	0.0	0.0	0.0	0.0	0.0	0.0	0.0	0.0	0.0	0.3	2.2
AVG. STATION PRESS.(mb)	21	1014.3	1014.6	1013.6	1012.2	1013.1	1012.4	1013.0	1014.8	1016.2	1016.4	1015.6	1015.4	1014.3
RELATIVE HUMIDITY (%)														
Hour 01	30	69	68	70	71	78	82	82	84	84	80	75	72	76
Hour 07 (Local Time)	30	71	70	71	70	73	75	77	80	82	80	77	74	75
Hour 13	30	56	54	53	49	52	56	56	56	56	54	57	58	55
Hour 19	30	63	61	61	60	64	67	68	71	74	71	69	66	66
PRECIPITATION (inches):														
Water Equivalent														
-Normal		3.88	3.61	4.05	4.11	3.76	3.33	3.18	3.63	3.48	3.69	4.43	4.38	45.53
-Maximum Monthly	40	11.66	7.20	8.84	12.74	8.38	11.08	8.08	11.12	7.92	11.89	11.01	10.75	12.74
-Year		1979	1984	1983	1983	1984	1982	1976	1955	1961	1962	1983	1969	APR 1983
-Minimum Monthly	40	0.50	0.39	0.56	1.48	0.71	0.39	1.00	0.71	0.77	1.53	0.81	0.58	0.39
-Year		1970	1987	1981	1966	1964	1957	1970	1984	1959	1992	1976	1955	FEB 1987
-Maximum in 24 hrs	40	3.34	3.14	4.53	4.45	5.17	5.03	4.83	6.71	4.89	6.63	4.18	3.85	6.71
-Year		1962	1978	1968	1983	1984	1984	1976	1979	1961	1962	1983	1969	AUG 1979
Snow,Ice pellets,hail														
-Maximum Monthly	40	28.7	30.9	31.6	7.6	7.0	0.0	0.0	0.0	0.0	2.5	8.0	19.8	31.6
-Year		1965	1962	1956	1982	1977					1979	1989	1963	MAR 1956
-Maximum in 24 hrs	40	10.8	27.6	16.9	7.6	7.0	0.0	0.0	0.0	0.0	2.5	8.0	11.9	27.6
-Year		1978	1978	1960	1982	1977					1979	1989	1981	FEB 1978
WIND:														
Mean Speed (mph)	40	11.1	11.5	12.0	12.0	10.7	9.9	9.4	9.3	9.4	9.6	10.4	10.8	10.5
Prevailing Direction through 1963		NW	NNW	WNW	SW	S	SW	SW	SSW	SW	NW	SW	WNW	SW
Fastest Obs. 1 Min.														
-Direction (!!)	40	20	16	18	20	20	20	34	11	18	14	18	14	11
-Speed (MPH)	40	46	46	60	51	42	40	35	90	58	41	52	48	90
-Year		1978	1972	1959	1956	1956	1957	1964	1954	1960	1954	1957	1957	AUG 1954
Peak Gust														
-Direction (!!)	10	S	SW	SW	SE	S	SE	SW	E	S	NE	SW	NW	S
-Speed (mph)	10	53	55	60	54	49	54	43	63	81	58	54	51	81
-Date		1992	1989	1986	1984	1990	1989	1984	1991	1985	1991	1992	1988	SEP 1985

PRECIPITATION (inches)

PROVIDENCE, RHODE ISLAND

YEAR	JAN	FEB	MAR	APR	MAY	JUNE	JULY	AUG	SEP	OCT	NOV	DEC	ANNUAL
1964	5.65	3.15	2.26	5.34	0.71	2.34	2.63	2.38	3.95	2.11	2.43	5.46	38.41
1965	3.46	3.77	1.72	2.43	1.08	1.91	1.28	1.90	1.64	2.75	2.08	1.42	25.44
1966	3.40	4.30	2.40	1.48	3.85	2.31	2.77	3.37	5.23	2.60	3.93	3.04	38.68
1967	1.60	2.51	5.49	4.19	7.27	2.72	3.95	3.24	3.17	2.25	2.75	7.36	46.50
1968	3.50	1.31	7.83	1.49	3.54	4.74	1.49	1.61	1.14	1.79	6.22	6.70	41.36
1969	2.23	4.30	3.10	3.95	2.41	1.23	2.98	2.58	3.09	1.62	6.35	10.75	44.59
1970	0.50	5.34	4.75	3.91	3.03	4.25	1.00	6.59	1.79	4.41	5.31	4.54	45.42
1971	2.01	5.36	3.81	2.31	3.83	1.64	3.48	3.03	2.54	2.88	5.16	2.37	38.42
1972	1.85	5.19	6.70	3.71	5.73	6.83	4.25	2.98	7.31	4.36	8.45	7.70	65.06
1973	3.06	3.55	2.78	7.16	3.99	3.48	2.92	5.17	3.04	3.17	2.29	7.63	48.24
1974	4.45	3.04	4.51	2.86	2.74	3.28	1.64	3.10	6.15	2.79	1.56	4.54	40.66
1975	6.78	3.29	3.07	2.99	2.06	4.73	3.51	2.19	6.15	4.66	6.29	5.11	50.83
1976	6.38	2.91	3.44	2.00	2.53	1.60	8.08	7.01	1.57	6.52	0.81	3.47	46.32
1977	3.90	2.87	5.62	3.35	3.43	3.92	2.04	2.12	5.60	6.90	3.24	5.85	48.84
1978	9.01	3.20	3.10	2.53	5.27	1.97	2.63	6.46	1.82	3.22	2.61	5.19	47.01
1979	11.66	4.08	2.21	5.12	7.62	1.44	1.65	10.09	4.08	3.94	4.49	1.81	58.19
1980	1.40	1.16	8.11	6.18	1.78	3.85	2.03	1.99	0.90	3.41	3.73	1.57	36.11
1981	0.77	4.79	0.56	4.10	1.92	2.31	3.75	2.65	2.58	3.38	3.20	6.36	36.37
1982	6.09	3.08	3.76	3.64	1.61	11.08	3.51	3.67	3.61	3.08	4.32	1.81	49.26
1983	4.32	4.81	8.84	12.74	4.67	1.91	2.14	2.71	2.16	4.50	11.01	7.71	67.52
1984	2.00	7.20	5.77	4.30	8.38	4.09	5.16	0.71	1.77	4.25	1.95	3.16	48.74
1985	1.18	1.57	3.08	1.65	4.76	4.70	2.88	8.57	1.69	1.78	7.14	1.42	40.42
1986	5.88	3.18	2.86	2.10	2.29	3.27	5.95	3.29	0.97	2.48	5.77	8.09	46.13
1987	4.73	0.39	5.62	6.91	1.80	2.00	1.20	2.58	7.47	2.28	3.40	2.29	40.67
1988	2.69	5.29	4.09	3.11	2.83	0.91	5.73	0.94	2.38	1.77	7.60	1.03	38.37
1989	1.17	2.69	4.13	5.30	6.07	5.84	5.59	6.14	4.75	8.37	4.35	1.66	56.06
1990	5.01	2.93	2.01	5.57	5.70	1.13	3.52	3.74	2.28	4.96	2.45	5.48	44.78
1991	3.44	2.31	6.61	4.80	3.30	0.93	2.76	5.98	5.09	2.65	4.65	3.17	45.69
1992	4.82	2.10	4.04	2.34	1.42	4.61	3.59	6.06	5.09	1.53	5.05	6.83	47.48
1993	2.42	5.06	6.99	5.02	1.12	1.40	2.18	1.23	4.08	3.55	3.35	5.76	42.16
Record Mean	3.73	3.29	3.86	3.75	3.25	3.02	3.17	3.70	3.34	3.20	3.87	3.89	42.08

AVERAGE TEMPERATURE (deg. F)

PROVIDENCE, RHODE ISLAND

YEAR	JAN	FEB	MAR	APR	MAY	JUNE	JULY	AUG	SEP	OCT	NOV	DEC	ANNUAL
1964	30.6	27.9	37.8	46.1	60.3	66.3	71.7	66.6	62.5	52.6	44.6	33.2	50.0
1965	25.0	28.4	36.1	45.4	60.0	67.0	71.6	71.6	64.1	52.1	41.1	35.4	49.8
1966	28.8	29.9	38.8	43.8	54.4	67.4	72.7	70.7	61.3	51.4	45.5	32.8	49.8
1967	33.7	25.7	33.3	44.8	51.2	66.8	72.8	70.6	62.7	53.7	39.5	34.5	49.1
1968	24.5	24.6	38.1	49.9	55.7	65.0	73.1	70.6	64.8	55.9	42.4	30.2	49.6
1969	28.8	28.7	35.0	49.7	57.7	68.2	71.5	74.3	64.0	53.3	42.4	30.5	50.4
1970	19.6	29.3	35.0	47.8	58.2	65.6	74.1	72.3	64.4	54.2	44.7	28.5	49.5
1971	22.9	30.9	36.7	45.9	58.1	69.0	74.3	73.0	68.7	59.2	40.4	35.0	51.2
1972	30.8	28.0	36.4	44.3	57.7	64.9	72.6	70.6	65.1	49.6	40.9	34.3	49.6
1973	31.1	29.6	43.7	50.0	56.7	70.3	73.6	75.0	63.4	54.2	43.8	38.3	52.5
1974	31.6	29.0	38.7	50.6	55.6	65.3	72.6	72.6	63.2	48.2	43.7	35.7	50.5
1975	34.1	30.4	35.5	44.5	61.4	66.0	74.3	71.4	61.0	55.3	48.0	32.1	51.2
1976	23.5	35.5	39.0	52.6	58.0	70.0	70.0	70.0	61.6	48.7	37.9	25.4	49.4
1977	20.9	29.8	43.6	50.6	61.2	66.7	74.3	73.0	64.1	52.9	45.9	31.6	51.2
1978	25.1	22.1	33.8	46.7	57.8	68.0	71.7	71.3	59.5	51.5	42.3	33.4	48.6
1979	30.0	19.7	40.4	46.8	60.3	65.1	73.5	70.2	64.0	53.0	48.3	37.3	50.7
1980	29.7	26.8	37.1	49.5	60.3	64.5	74.8	73.4	64.9	49.8	41.0	28.5	50.0
1981	20.3	37.4	38.7	51.4	58.5	69.4	75.6	70.0	62.4	49.1	43.0	31.1	50.6
1982	21.5	31.5	38.8	47.8	58.9	63.9	73.6	69.2	64.1	53.2	47.5	38.6	50.7
1983	31.4	32.9	40.4	49.9	56.9	70.2	76.6	74.3	69.6	55.3	46.0	32.5	53.0
1984	26.4	37.1	33.8	47.6	57.4	69.1	71.5	73.5	62.1	56.3	43.6	37.9	51.4
1985	22.5	32.1	40.8	51.0	60.2	64.8	73.0	71.1	65.2	54.6	45.9	30.4	51.0
1986	31.1	29.0	39.9	49.4	59.4	66.4	71.0	69.3	62.3	53.0	41.6	35.4	50.7
1987	29.0	28.6	39.8	48.4	59.3	68.4	72.2	69.6	64.3	51.4	43.0	35.1	50.8
1988	26.8	31.8	39.4	47.0	58.0	66.9	74.3	75.3	63.0	48.9	45.2	32.4	50.8
1989	33.8	29.9	37.5	46.2	59.3	68.7	72.3	72.1	65.3	54.1	42.5	21.8	50.3
1990	36.3	34.3	40.1	48.1	56.0	67.7	73.0	73.5	63.7	58.6	46.5	39.5	53.1
1991	29.6	35.1	41.3	51.8	63.9	69.3	74.2	73.6	63.1	56.1	45.4	36.3	53.3
1992	31.4	33.0	36.6	46.5	57.6	67.3	70.3	70.1	64.0	51.7	43.0	34.2	50.5
1993	31.4	26.3	35.8	49.6	61.8	69.3	74.5	73.8	65.1	51.5	44.0	33.7	51.4
Record Mean	29.1	29.6	37.8	47.8	58.1	67.0	72.8	71.1	63.8	53.7	43.5	32.7	50.6
Max	37.0	37.7	46.1	57.0	67.8	76.7	82.0	80.2	73.2	63.2	51.6	40.4	59.4
Min	21.1	21.4	29.5	38.6	48.3	57.3	63.6	62.0	54.5	44.2	35.3	25.0	41.7

REFERENCE NOTES

PROVIDENCE, RHODE ISLAND

GENERAL
T - TRACE AMOUNT.
BLANK ENTRIES DENOTE MISSING/UNREPORTED DATA.
# INDICATES A STATION OR INSTRUMENT RELOCATION.

SPECIFIC

PM - INCLUDES LAST DAY OF PREVIOUS MONTH
ASOS - AUTOMATED SURFACE OBSERVING SYSTEM IN OPERATION DURING THESE MONTHS.

(a) - LENGTH OF RECORD IN YEARS, ALTHOUGH INDIVIDUAL MONTHS MAY BE MISSING.
0.* OR * - THE VALUE IS BETWEEN 0.0 AND 0.05
NORMALS - BASED ON THE 1961-1990 RECORD PERIOD.
EXTREMES - DATES ARE THE MOST RECENT OCCURRENCE
WIND DIR.- NUMERALS SHOW TENS OF DEGRESS CLOCKWISE FROM TRUE NORTH. "00" INDICATES CALM.
RESULTANT DIRECTIONS ARE GIVEN TO WHOLE DEGREES.
BOLD VALUES INDICATE EXTREME VALUES WHICH OCCURRED AFTER THE ASOS SYSTEM WAS COMMISSIONED.

RECORD = PERIOD OF RECORD
RECORD MEAN PRECIPITATION IS THE MEAN OF ALL DAILY PRECIPITATION AMOUNTS DURING THE PERIOD OF RECORD.
RECORD MAX(MIN) TEMPERATURE IS THE MEAN OF ALL DAILY MAX(MIN) TEMPERATURES DURING THE PERIOD OF RECORD.
RECORD MEAN TEMPERATURE IS THE SUM OF THE RECORD MAX AND RECORD MIN DIVIDED BY 2.
AVERAGE TEMPERATURE IS THE SUM OF THE MEAN DAILY MAX AND MIN TEMPERATURE DIVIDED BY 2.

EXCEPTIONS

RECORD MEANS ARE THROUGH THE CURRENT YEAR,
BEGINNING IN 1905 FOR TEMPERATURE
1905 FOR PRECIPITATION
1954 FOR SNOWFALL

HEATING DEGREE DAYS Base 65 deg. F PROVIDENCE, RHODE ISLAND

SEASON	JULY	AUG	SEP	OCT	NOV	DEC	JAN	FEB	MAR	APR	MAY	JUNE	TOTAL
1964-65	9	24	125	377	605	981	1231	1018	891	581	182	67	6091
1965-66	3	29	99	395	711	907	1115	975	806	630	326	57	6053
1966-67	1	1	135	417	577	994	963	1093	976	598	424	48	6227
1967-68	0	7	103	356	761	937	1246	1166	827	447	281	74	6205
1968-69	2	16	59	295	672	1072	1117	1010	923	452	241	22	5881
1969-70	2	4	119	365	673	1065	1399	996	924	509	214	60	6330
1970-71	0	0	102	342	602	1124	1298	949	868	566	212	34	6097
1971-72	0	7	42	181	736	922	1054	1064	879	616	226	55	5782
1972-73	8	10	64	473	717	945	1044	984	653	451	268	16	5633
1973-74	2	3	125	331	632	819	1028	1003	808	433	313	62	5559
1974-75	0	0	114	512	634	899	951	962	907	606	160	64	5809
1975-76	0	13	132	298	506	1013	1283	850	798	403	223	39	5558
1976-77	2	23	124	501	806	1219	1361	983	653	434	176	51	6333
1977-78	0	6	103	368	568	1030	1231	1192	964	542	238	26	6268
1978-79	8	8	180	412	673	970	1075	1261	755	540	162	52	6096
1979-80	11	25	94	380	496	849	1088	1104	857	459	158	93	5614
1980-81	0	1	120	465	715	1125	1379	769	808	405	228	13	6028
1981-82	0	20	119	486	651	1044	1343	932	802	510	190	91	6188
1982-83	1	26	78	363	518	809	1038	892	755	449	254	13	5196
1983-84	0	4	62	323	563	1001	1190	802	961	513	236	36	5691
1984-85	1	0	125	270	637	832	1309	914	743	417	177	63	5488
1985-86	0	6	78	321	567	1065	1045	999	772	460	216	57	5586
1986-87	14	25	113	380	697	911	1111	1014	772	494	228	23	5782
1987-88	2	25	70	414	653	921	1177	954	787	532	238	67	5840
1988-89	8	10	89	491	587	1003	960	975	847	557	181	22	5730
1989-90	2	9	89	332	668	1329	882	854	761	511	275	24	5736
1990-91	6	0	107	242	549	781	1090	829	726	400	121	29	4880
1991-92	1	0	125	275	581	884	1034	919	876	549	246	27	5517
1992-93	11	4	100	404	652	951	1036	1077	901	455	118	32	5741
1993-94	1	0	102	413	623	966							

COOLING DEGREE DAYS Base 65 deg. F PROVIDENCE, RHODE ISLAND

YEAR	JAN	FEB	MAR	APR	MAY	JUNE	JULY	AUG	SEP	OCT	NOV	DEC	TOTAL
1969	0	0	0	0	23	125	211	299	94	9	0	0	761
1970	0	0	0	0	13	86	289	237	91	14	0	0	730
1971	0	0	0	0	3	157	296	263	158	8	5	0	890
1972	0	0	0	1	7	60	248	190	76	1	0	0	583
1973	0	0	0	8	17	181	272	318	84	3	0	0	883
1974	0	0	0	7	27	79	242	244	66	0	1	0	666
1975	0	0	0	0	55	100	300	218	16	4	1	0	694
1976	0	0	0	40	13	196	163	183	33	3	0	0	631
1977	0	0	0	5	68	108	295	260	85	0	0	0	821
1978	0	0	0	0	25	126	224	211	24	0	0	0	610
1979	0	0	0	0	26	59	279	190	74	12	0	0	640
1980	0	0	0	0	21	84	312	272	122	0	0	0	811
1981	0	0	0	2	33	152	335	183	47	0	0	0	752
1982	0	0	0	0	11	64	276	165	59	3	2	0	580
1983	0	0	0	1	8	177	367	298	206	30	0	0	1087
1984	0	0	0	0	6	164	206	272	47	7	0	0	702
1985	0	0	0	5	34	65	256	203	90	5	0	0	658
1986	0	0	0	0	51	105	207	164	38	14	0	0	579
1987	0	0	0	0	57	130	231	177	53	0	0	0	648
1988	0	0	0	0	26	131	302	336	37	2	0	0	834
1989	0	0	0	0	10	141	237	237	103	0	0	0	728
1990	0	0	0	8	1	114	262	272	74	49	2	0	782
1991	0	0	0	12	96	166	295	276	73	8	0	0	926
1992	0	0	0	0	27	103	183	169	75	1	0	0	558
1993	0	0	0	0	26	167	303	281	110	1	1	0	889

SNOWFALL (inches) PROVIDENCE, RHODE ISLAND

SEASON	JULY	AUG	SEP	OCT	NOV	DEC	JAN	FEB	MAR	APR	MAY	JUNE	TOTAL
1964-65	0.0	0.0	0.0	0.0	T	6.9	28.7	2.7	6.6	T	0.0	0.0	44.9
1965-66	0.0	0.0	0.0	0.0	T	1.9	16.1	9.1	8.1	T	0.0	0.0	35.2
1966-67	0.0	0.0	0.0	0.0	T	7.2	1.3	23.1	24.9	1.6	0.0	0.0	58.1
1967-68	0.0	0.0	0.0	0.0	0.8	19.0	13.5	4.6	5.1	0.0	0.0	0.0	43.0
1968-69	0.0	0.0	0.0	0.0	0.1	2.7	0.5	26.7	6.0	0.0	0.0	0.0	36.0
1969-70	0.0	0.0	0.0	0.0	T	15.4	6.5	6.8	15.7	1.1	0.0	0.0	45.5
1970-71	0.0	0.0	0.0	T	0.0	17.8	11.0	5.0	6.1	1.9	0.0	0.0	41.8
1971-72	0.0	0.0	0.0	0.0	T	4.7	2.2	13.7	8.7	0.7	0.0	0.0	30.0
1972-73	0.0	0.0	0.0	0.7	0.3	4.1	2.0	3.2	0.6	0.4	0.0	0.0	11.3
1973-74	0.0	0.0	0.0	0.0	T	T	15.1	11.1	0.4	1.3	0.0	0.0	27.9
1974-75	0.0	0.0	0.0	0.0	0.3	2.1	2.0	18.2	2.2	0.2	0.0	0.0	25.0
1975-76	0.0	0.0	0.0	T	1.2	7.5	15.6	3.7	9.5	0.0	0.0	0.0	37.5
1976-77	0.0	0.0	0.0	0.0	T	9.8	14.0	11.4	4.4	T	7.0	0.0	46.6
1977-78	0.0	0.0	0.0	0.0	1.3	3.9	20.5	28.6	15.9	T	0.0	0.0	70.2
1978-79	0.0	0.0	0.0	0.0	2.3	2.4	6.0	5.5	T	1.1	0.0	0.0	17.3
1979-80	0.0	0.0	0.0	2.5	0.0	T	0.6	3.8	5.3	0.0	0.0	0.0	12.2
1980-81	0.0	0.0	0.0	0.0	4.1	3.6	12.9	0.6	0.3	0.0	0.0	0.0	21.5
1981-82	0.0	0.0	0.0	T	T	16.4	13.4	4.3	5.7	7.6	0.0	0.0	47.4
1982-83	0.0	0.0	0.0	0.0	0.0	7.3	3.8	21.3	T	T	0.0	0.0	32.4
1983-84	0.0	0.0	0.0	0.0	T	4.5	17.9	T	13.7	T	0.0	0.0	36.1
1984-85	0.0	0.0	0.0	0.0	T	2.0	9.8	10.0	0.6	T	0.0	0.0	22.4
1985-86	0.0	0.0	0.0	0.0	1.8	2.6	0.7	13.0	0.5	T	T	0.0	18.6
1986-87	0.0	0.0	0.0	0.0	4.4	8.0	21.5	4.7	1.6	1.1	0.0	0.0	41.3
1987-88	0.0	0.0	0.0	0.0	8.0	7.8	13.5	6.7	2.7	T	0.0	0.0	38.7
1988-89	0.0	0.0	0.0	0.0	T	1.2	0.2	7.3	1.9	0.3	0.0	0.0	10.9
1989-90	0.0	0.0	0.0	0.0	8.0	15.8	10.8	10.5	9.3	1.8	0.0	0.0	56.2
1990-91	0.0	0.0	0.0	0.0	T	6.9	6.4	6.0	5.3	0.0	0.0	0.0	24.6
1991-92	0.0	0.0	0.0	0.0	T	4.8	2.4	4.9	8.2	2.0	0.0	0.0	22.3
1992-93	0.0	0.0	0.0	T	T	3.6	5.4	12.7	17.8	0.2	0.0	0.0	39.7
1993-94	0.0	0.0	0.0	0.0	T	10.1							
Record Mean	0.0	0.0	0.0	0.1	1.0	7.0	9.4	9.6	7.6	0.7	0.2	0.0	35.7

CHARLESTON, SOUTH CAROLINA

Charleston is a peninsula city bounded on the west and south by the Ashley River, on the east by the Cooper River, and on the southeast by a spacious harbor. Weather records for the airport are from a site some 10 miles inland. The terrain is generally level, ranging in elevation from sea level to 20 feet on the peninsula, with gradual increases in elevation toward inland areas. The soil is sandy to sandy loam with lesser amounts of loam. The drainage varies from good to poor. Because of the very low elevation, a considerable portion of this community and the nearby coastal islands are vulnerable to tidal flooding.

The climate is temperate, modified considerably by the nearness to the ocean. The marine influence is noticeable during winter when the low temperatures are sometimes 10–15 degrees higher on the peninsula than at the airport. By the same token, high temperatures are generally a few degrees lower on the peninsula. The prevailing winds are northerly in the fall and winter, southerly in the spring and summer.

Summer is warm and humid. Temperatures of 100 degrees or more are infrequent. High temperatures are generally several degrees lower along the coast than inland due to the cooling effect of the sea breeze. Summer is the rainiest season with 41 percent of the annual total. The rain, except during occasional tropical storms, generally occurs as showers or thunderstorms.

The fall season passes through the warm Indian Summer period to the pre–winter cold spells which begin late in November. From late September to early November the weather is mostly sunny and temperature extremes are rare. Late summer and early fall is the period of maximum threat to the South Carolina coast from hurricanes.

The winter months, December through February, are mild with periods of rain. However, the winter rainfall is generally of a more uniform type. There is some chance of a snow flurry, with the best probability of its occurrence in January, but a significant amount is rarely measured. An average winter would experience less than one cold wave and severe freeze. Temperatures of 20 degrees or less on the peninsula and along the coast are very unusual.

The most spectacular time of the year, weatherwise, is spring with its rapid changes from windy and cold in March to warm and pleasant in May. Severe local storms are more likely to occur in spring than in summer.

The average occurrence of the first freeze in the fall is early December, and the average last freeze is late February, giving an average growing season of about 294 days.

NORMALS, MEANS, AND EXTREMES

CHARLESTON, SOUTH CAROLINA

LATITUDE: 32°54'N LONGITUDE: 80°02' W ELEVATION: FT. GRND 40 BARO 47 TIME ZONE: EASTERN WBAN: 13880

	(a)	JAN	FEB	MAR	APR	MAY	JUNE	JULY	AUG	SEP	OCT	NOV	DEC	YEAR
TEMPERATURE °F:														
Normals														
-Daily Maximum		57.8	61.0	68.6	75.8	82.7	87.6	90.2	89.0	84.9	77.2	69.5	61.6	75.5
-Daily Minimum		37.7	40.0	47.5	53.9	62.9	69.1	72.7	72.2	67.9	56.3	47.2	40.7	55.7
-Monthly		47.8	50.5	58.1	64.9	72.8	78.3	81.5	80.6	76.4	66.8	58.4	51.2	65.6
Extremes														
-Record Highest	51	83	87	90	94	98	103	104	102	99	94	88	83	104
-Year		1950	1989	1974	1989	1989	1944	1986	1954	1944	1986	1961	1972	JUL 1986
-Record Lowest	51	6	12	15	29	36	50	58	56	42	27	15	8	6
-Year		1985	1973	1980	1944	1963	1972	1952	1979	1967	1976	1950	1962	JAN 1985
NORMAL DEGREE DAYS:														
Heating (base 65°F)		548	414	239	66	0	0	0	0	0	74	233	439	2013
Cooling (base 65°F)		15	8	25	63	242	399	512	484	342	130	35	11	2266
% OF POSSIBLE SUNSHINE	34	56	60	65	70	69	66	67	63	61	63	59	56	63
MEAN SKY COVER (tenths)														
Sunrise - Sunset	44	6.3	6.1	6.0	5.3	6.0	6.3	6.5	6.4	6.2	5.2	5.3	5.9	5.9
MEAN NUMBER OF DAYS:														
Sunrise to Sunset														
-Clear	45	8.9	8.7	9.2	11.2	8.0	6.0	5.1	5.5	6.8	11.6	11.8	9.6	102.3
-Partly Cloudy	45	6.3	6.4	8.0	7.9	11.0	11.4	12.3	12.9	10.5	8.3	6.6	7.0	108.6
-Cloudy	45	15.8	13.2	13.8	11.0	12.1	12.6	13.6	12.6	12.7	11.2	11.6	14.4	154.3
Precipitation														
.01 inches or more	51	9.9	8.8	10.0	7.6	9.0	10.8	13.3	12.8	9.5	6.0	6.9	8.3	112.9
Snow,Ice pellets,hail														
1.0 inches or more	51	0.0	0.1	0.*	0.0	0.0	0.0	0.0	0.0	0.0	0.0	0.0	0.1	0.2
Thunderstorms	51	0.8	1.1	2.3	2.9	6.7	9.9	13.3	11.7	5.3	1.4	0.7	0.5	56.5
Heavy Fog Visibility														
1/4 mile or less	44	4.2	2.0	2.4	2.1	2.1	1.5	0.7	1.2	1.8	2.5	3.6	3.7	27.8
Temperature °F														
-Maximum														
90° and above	51	0.0	0.0	0.*	0.8	3.9	11.3	16.9	14.7	5.4	0.4	0.0	0.0	53.4
32° and below	51	0.2	0.1	0.*	0.0	0.0	0.0	0.0	0.0	0.0	0.0	0.0	0.1	0.3
-Minimum														
32° and below	51	10.6	7.5	2.7	0.2	0.0	0.0	0.0	0.0	0.0	0.1	3.0	8.9	33.1
0° and below	51	0.0	0.0	0.0	0.0	0.0	0.0	0.0	0.0	0.0	0.0	0.0	0.0	0.0
AVG. STATION PRESS. (mb)	21	1018.8	1017.9	1016.2	1015.3	1014.7	1014.8	1015.8	1016.1	1016.1	1017.5	1018.6	1019.5	1016.8
RELATIVE HUMIDITY (%)														
Hour 01	51	80	78	81	83	88	89	90	91	90	87	85	81	85
Hour 07 (Local Time)	51	83	81	83	84	85	86	88	90	90	88	86	83	86
Hour 13	51	56	52	51	49	53	58	62	63	62	55	53	55	56
Hour 19	51	71	67	66	65	70	74	77	79	80	78	76	73	73
PRECIPITATION (inches):														
Water Equivalent														
-Normal		3.45	3.30	4.34	2.67	4.01	6.43	6.84	7.22	4.73	2.90	2.49	3.15	51.53
-Maximum Monthly	51	8.92	6.35	11.11	9.50	9.28	27.24	18.46	16.99	17.31	9.12	7.35	7.09	27.24
-Year		1993	1983	1983	1958	1957	1973	1964	1974	1945	1959	1972	1953	JUN 1973
-Minimum Monthly	51	0.63	0.33	0.99	0.01	0.68	0.96	1.76	0.73	0.18	0.08	0.48	0.66	0.01
-Year		1950	1947	1963	1972	1944	1970	1972	1980	1990	1943	1966	1984	APR 1972
-Maximum in 24 hrs	51	3.90	3.28	6.63	4.10	6.23	10.10	5.81	5.77	8.84	5.77	5.24	3.40	10.10
-Year		1993	1944	1959	1958	1967	1973	1960	1964	1945	1944	1969	1978	JUN 1973
Snow,Ice pellets,hail														
-Maximum Monthly	51	1.0	7.1	2.0	T	0.0	T	T	0.0	0.0	0.0	T	8.0	8.0
-Year		1977	1973	1969	1985		1989	1993				1950	1989	DEC 1989
-Maximum in 24 hrs	51	0.8	5.9	2.0	T	0.0	T	T	0.0	0.0	0.0	T	6.6	6.6
-Year		1966	1973	1969	1985		1989	1993				1950	1989	DEC 1989
WIND:														
Mean Speed (mph)	44	9.1	9.8	10.0	9.7	8.6	8.3	7.9	7.4	7.8	8.0	8.1	8.5	8.6
Prevailing Direction through 1963		SW	NNE	SSW	SSW	S	S	SW	SW	NNE	NNE	N	NNE	NNE
Fastest Obs. 1 Min.														
-Direction (!!)	18	20	30	21	19	25	32	33	08	21	18	15	24	21
-Speed (MPH)	18	40	38	46	36	32	40	37	35	52	30	37	39	52
-Year		1978	1991	1993	1983	1978	1991	1986	1988	1989	1976	1985	1975	SEP 1989
Peak Gust														
-Direction (!!)	10	SW	NW	SW	S	W	NW	W	N	SW	NW	SE	SW	SW
-Speed (mph)	10	51	48	69	48	49	55	59	64	98	41	48	44	98
-Date		1992	1991	1993	1988	1984	1990	1992	1986	1989	1986	1985	1993	SEP 1989

PRECIPITATION (inches)

CHARLESTON, SOUTH CAROLINA

YEAR	JAN	FEB	MAR	APR	MAY	JUNE	JULY	AUG	SEP	OCT	NOV	DEC	ANNUAL
1964	6.53	6.32	4.40	2.72	4.77	6.95	18.46	7.67	4.11	7.53	0.52	3.01	72.99
1965	1.69	5.49	7.99	3.10	1.32	6.68	8.90	5.90	4.66	6.23	0.87	1.20	54.03
1966	6.68	4.61	2.65	2.83	7.71	6.03	11.48	3.45	2.20	1.70	0.48	3.76	53.58
1967	4.93	3.12	2.81	0.84	8.91	4.06	9.19	3.74	1.39	0.52	1.35	2.79	43.65
1968	2.25	1.27	1.35	1.99	2.54	9.41	7.77	4.42	1.66	6.86	2.67	3.65	45.84
1969	1.19	2.05	5.14	3.46	2.34	4.88	5.34	11.84	5.37	1.50	5.49	3.52	52.12
1970	2.51	2.86	7.72	1.34	3.78	0.96	5.93	10.64	2.53	4.08	0.67	2.90	45.92
1971	5.45	4.71	4.05	4.11	4.15	4.07	6.04	16.32	0.53	7.22	1.61	2.28	60.54
1972	4.13	5.18	2.52	0.01	5.67	5.29	1.76	4.52	1.82	0.25	7.35	4.36	42.86
1973	4.59	5.57	6.15	2.55	1.83	27.24	3.60	6.66	7.93	0.63	0.84	4.58	72.17
1974	1.42	2.96	3.04	0.86	4.82	9.45	3.09	16.99	4.80	0.40	3.78	3.00	54.61
1975	4.92	3.54	4.54	3.74	5.06	5.96	9.34	7.18	5.16	1.97	1.43	3.35	56.19
1976	1.62	0.95	2.33	0.62	8.87	5.59	4.48	5.22	6.03	4.10	3.57	5.12	48.50
1977	2.72	1.38	5.31	0.45	4.66	2.12	3.86	8.13	2.48	2.49	1.76	5.88	41.24
1978	4.31	1.82	3.25	1.97	4.68	3.42	6.19	4.01	5.06	0.18	1.87	4.13	40.89
1979	3.43	3.04	3.01	3.81	8.09	2.23	8.35	0.88	15.36	3.87	3.29	2.62	57.98
1980	3.99	1.25	7.99	3.43	5.85	3.15	6.97	0.73	2.60	1.52	2.19	1.25	40.92
1981	0.93	2.23	2.38	1.87	4.02	6.04	12.66	9.30	1.27	1.95	1.06	5.73	49.44
1982	2.18	3.64	1.26	6.51	3.04	9.16	5.40	4.10	3.92	2.42	1.19	4.20	47.02
1983	4.86	6.35	11.11	3.57	0.75	2.37	8.89	2.90	3.50	2.36	3.08	4.35	54.09
1984	5.12	3.51	5.63	6.30	6.89	2.96	4.87	1.96	5.27	1.67	1.39	0.66	46.23
1985	0.87	2.70	1.50	1.12	2.79	7.02	12.06	8.48	2.53	4.58	5.49	1.21	50.35
1986	2.05	4.17	2.67	0.83	0.93	2.51	5.07	13.41	4.60	2.95	4.03	5.21	48.43
1987	7.17	4.58	5.55	1.31	2.29	5.64	2.92	6.97	14.49	0.56	3.65	1.57	56.70
1988	2.76	2.38	1.78	3.21	1.86	2.32	4.13	11.88	9.72	0.73	1.08	0.72	42.57
1989	2.31	1.17	2.87	4.84	2.14	7.26	1.93	9.18	13.35	4.08	1.85	4.74	55.72
1990	3.96	1.68	6.63	1.65	1.91	3.12	5.95	6.32	0.18	7.29	3.75	2.69	45.13
1991	7.78	0.94	4.66	4.59	5.37	4.54	7.38	8.09	2.29	0.77	1.64	1.62	49.67
1992	4.93	2.23	3.59	2.75	5.07	6.22	4.36	9.55	3.04	4.87	5.76	1.50	53.87
1993	8.92	3.08	5.80	2.72	2.67	3.70	4.21	7.69	5.01	3.00	3.59	2.30	52.69
Record Mean	3.14	3.23	3.75	2.77	3.54	5.17	7.07	6.77	5.09	3.10	2.33	2.87	48.84

AVERAGE TEMPERATURE (deg. F)

CHARLESTON, SOUTH CAROLINA

YEAR	JAN	FEB	MAR	APR	MAY	JUNE	JULY	AUG	SEP	OCT	NOV	DEC	ANNUAL
1964	48.5	47.5	57.6	66.0	73.1	80.0	78.0	79.6	75.4	62.5	60.5	52.9	65.1
1965	48.4	51.2	54.5	64.5	75.3	76.1	79.0	79.6	75.9	65.2	57.6	49.3	64.7
1966	43.9	49.6	53.9	63.2	70.6	73.7	79.6	80.0	75.8	67.4	56.3	48.9	63.6
1967	50.6	48.5	60.0	67.1	71.3	75.6	79.9	79.5	69.9	62.6	53.8	53.4	64.4
1968	43.5	41.8	55.7	65.7	71.7	77.8	81.5	81.5	75.1	67.1	53.7	46.5	63.5
1969	44.6	45.2	50.6	63.3	69.9	80.3	82.7	77.5	73.7	68.2	53.4	46.5	63.0
1970	41.0	48.5	57.1	66.3	72.7	78.4	82.5	80.6	77.3	67.5	54.1	52.1	64.8
1971	49.9	49.8	51.9	62.7	71.1	80.4	80.3	79.6	77.1	70.8	57.1	58.8	65.8
1972	54.8	48.9	56.9	64.0	69.8	74.0	80.1	80.4	76.4	67.3	56.8	55.7	65.5
1973	48.0	46.1	61.0	61.5	71.7	78.4	82.6	81.0	79.6	69.6	61.4	51.3	66.0
1974	61.8	51.5	62.0	63.6	74.0	75.4	78.2	79.3	75.0	61.8	55.5	51.0	65.8
1975	53.8	54.7	56.9	62.3	75.1	78.5	79.2	81.5	76.8	69.0	59.3	49.7	66.4
1976	44.8	55.9	62.3	64.0	70.3	75.8	81.0	77.2	73.9	61.4	50.9	48.8	63.9
1977	38.7	46.3	60.6	66.4	72.8	81.2	83.8	81.4	78.7	63.5	61.3	5.0	65.4
1978	43.5	42.7	55.2	66.5	72.0	78.6	81.1	81.3	77.4	65.7	63.3	52.7	65.0
1979	45.4	46.8	57.4	64.9	72.4	75.9	82.0	81.4	76.5	66.0	59.4	48.7	64.7
1980	48.7	45.9	54.6	64.3	71.4	78.4	82.4	82.1	79.8	65.0	55.4	47.5	64.6
1981	41.6	50.8	54.3	67.5	70.8	82.7	83.5	80.3	74.8	64.1	55.4	46.2	64.3
1982	45.1	51.5	59.2	61.8	72.2	78.8	81.2	80.0	74.5	65.1	60.9	57.0	65.6
1983	45.6	49.0	56.4	61.0	71.7	76.9	82.8	82.9	75.5	68.9	57.4	48.8	64.8
1984	46.1	52.8	57.6	64.0	71.7	78.8	79.9	81.1	73.0	71.3	54.2	57.2	65.6
1985	42.6	50.5	60.7	67.8	73.6	79.6	80.9	79.9	75.8	72.2	67.3	47.9	66.6
1986	45.8	55.5	58.0	66.1	74.3	81.4	86.1	79.9	78.6	68.8	63.1	52.8	67.5
1987	47.2	48.8	56.8	62.6	73.3	80.1	83.0	83.5	77.8	61.0	60.1	53.5	65.6
1988	43.2	49.2	57.4	64.4	71.9	76.7	81.8	82.0	76.3	62.7	61.0	50.6	64.8
1989	55.6	55.0	59.7	65.3	72.3	80.4	82.8	80.7	76.6	68.7	60.6	43.2	66.7
1990	55.4	59.2	62.5	66.0	74.4	81.0	83.6	82.5	79.2	70.5	60.4	56.4	69.3
1991	50.8	54.9	60.5	67.6	76.3	78.9	83.6	82.0	77.4	67.6	56.0	54.3	67.5
1992	49.5	55.0	57.9	63.6	70.3	77.5	83.9	80.6	76.2	65.1	60.4	51.0	65.9
1993	53.7	49.7	55.9	61.6	72.5	79.5	85.5	82.0	78.3	67.3	59.3	48.5	66.2
Record Mean	49.7	51.5	57.5	64.6	72.6	78.7	81.3	80.6	76.5	67.2	58.0	51.0	65.8
Max	58.6	60.6	66.7	73.8	81.1	86.5	88.8	87.9	83.8	75.7	67.3	60.1	74.2
Min	40.8	42.4	48.4	55.3	64.0	70.8	73.8	73.2	69.2	58.6	48.6	42.0	57.3

REFERENCE NOTES

CHARLESTON, SOUTH CAROLINA

GENERAL
T - TRACE AMOUNT.
BLANK ENTRIES DENOTE MISSING/UNREPORTED DATA.
# INDICATES A STATION OR INSTRUMENT RELOCATION.

SPFCIFIC

PM - INCLUDES LAST DAY OF PREVIOUS MONTH
ASOS - AUTOMATED SURFACE OBSERVING SYSTEM IN OPERA'ION DURING THESE MONTHS.

(a) - LENGTH OF RECORD IN YEARS, ALTHOUGH INDIVIDUAL MONTHS MAY BE MISSING.
0.* OR * - THE VALUE IS BETWEEN 0.0 AND 0.05
NORMALS - BASED ON THE 1961-1990 RECORD PERIOD.
EXTREMES - DATES ARE THE MOST RECENT OCCURRENCE
WIND DIR. - NUMERALS SHOW TENS OF DEGREES CLOCKWISE FROM TRUE NORTH. "00" INDICATES CALM.
RESULTANT DIRECTIONS ARE GIVEN TO WHOLE DEGREES.
BOLD VALUES INDICATE EXTREME VALUES WHICH OCCURRED AFTER THE ASOS SYSTEM WAS COMMISSIONED.

RECORD = PERIOD OF RECORD
RECORD MEAN PRECIPITATION IS THE MEAN OF ALL DAILY PRECIPITATION AMOUNTS DURING THE PERIOD OF RECORD.
RECORD MAX(MIN) TEMPERATURE IS THE MEAN OF ALL DAILY MAX(MIN) TEMPERATURES DURING THE PERIOD OF RECORD.
RECORD MEAN TEMPERATURE IS THE SUM OF THE RECORD MAX AND RECORD MIN DIVIDED BY 2.
AVERAGE TEMPERATURE IS THE SUM OF THE MEAN DAILY MAX AND MIN TEMPERATURE DIVIDED BY 2.

EXCEPTIONS

RECORD MEANS ARE THROUGH THE CURRENT YEAR,
BEGINNING IN 1874 FOR TEMPERATURE
1871 FOR PRECIPITATION
1943 FOR SNOWFALL

HEATING DEGREE DAYS Base 65 deg. F CHARLESTON, SOUTH CAROLINA

SEASON	JULY	AUG	SEP	OCT	NOV	DEC	JAN	FEB	MAR	APR	MAY	JUNE	TOTAL
1964-65	0	0	0	135	153	385	507	392	340	95	0	0	2007
1965-66	0	0	0	95	225	481	644	427	338	119	15	2	2346
1966-67	0	0	0	51	280	498	442	460	194	50	29	4	2008
1967-68	0	0	15	112	334	367	657	667	307	67	4	0	2530
1968-69	0	0	0	80	334	564	624	551	444	90	6	0	2693
1969-70	0	0	1	50	349	567	735	454	249	82	9	0	2496
1970-71	0	0	11	42	324	392	465	424	404	127	16	0	2205
1971-72	0	0	0	13	261	220	317	463	249	113	8	0	1644
1972-73	0	0	0	33	268	302	520	524	167	141	18	0	1973
1973-74	0	0	0	34	158	428	131	378	150	114	2	0	1395
1974-75	0	0	5	136	299	432	350	294	273	152	0	0	1941
1975-76	0	0	0	40	221	466	624	265	146	94	15	3	1874
1976-77	0	0	0	159	418	501	808	516	186	58	17	0	2663
1977-78	0	0	0	112	175	459	663	616	309	52	18	0	2404
1978-79	0	0	0	57	83	399	602	505	241	70	2	0	1959
1979-80	0	0	0	68	203	500	495	555	321	82	17	0	2241
1980-81	0	0	0	80	287	537	719	393	333	55	16	0	2420
1981-82	0	0	3	88	291	577	611	372	214	132	3	0	2291
1982-83	0	0	0	102	154	276	596	440	264	146	2	0	1980
1983-84	0	0	4	24	230	500	578	347	240	92	16	0	2031
1984-85	0	0	9	13	337	249	692	418	183	47	4	0	1952
1985-86	0	0	2	16	54	526	586	261	244	74	4	0	1767
1986-87	0	6	0	56	128	376	545	446	272	131	7	0	1967
1987-88	0	0	0	135	188	358	669	458	239	85	7	2	2141
1988-89	0	0	0	107	145	442	286	312	220	121	14	0	1647
1989-90	0	0	1	50	169	669	294	189	137	67	0	0	1576
1990-91	0	0	0	65	152	280	432	293	185	34	0	0	1441
1991-92	0	0	0	41	281	349	472	293	234	119	28	0	1817
1992-93	0	0	3	70	197	430	353	421	279	132	1	0	1886
1993-94	0	0	3	49	206	504							

COOLING DEGREE DAYS Base 65 deg. F CHARLESTON, SOUTH CAROLINA

YEAR	JAN	FEB	MAR	APR	MAY	JUNE	JULY	AUG	SEP	OCT	NOV	DEC	TOTAL
1969	0	0	1	46	165	465	555	394	268	158	6	0	2058
1970	0	0	9	126	253	410	552	487	384	130	3	1	2355
1971	2	5	5	65	215	469	480	457	369	199	31	32	2329
1972	7	3	5	93	163	275	475	488	351	110	30	22	2022
1973	1	0	50	42	233	412	554	501	445	184	56	10	2488
1974	41	7	63	80	288	319	417	450	312	46	18	3	2044
1975	8	13	26	74	318	414	449	516	361	171	58	0	2408
1976	2	9	73	70	187	329	502	384	274	52	2	1	1885
1977	0	1	54	107	263	493	588	518	417	71	71	1	2584
1978	0	0	13	106	242	414	505	514	378	86	40	21	2319
1979	0	2	9	71	241	335	533	514	354	105	40	0	2204
1980	0	9	7	69	221	407	549	539	451	87	5	1	2345
1981	0	0	9	138	199	539	582	481	307	66	9	0	2330
1982	0	2	42	42	232	420	510	475	293	111	36	34	2197
1983	0	0	6	32	217	362	559	567	322	149	10	4	2228
1984	0	1	19	67	228	420	471	509	254	212	21	10	2212
1985	7	17	57	136	276	445	501	470	332	245	129	5	2620
1986	0	2	36	114	300	499	662	474	414	182	78	5	2766
1987	0	0	26	62	269	459	567	580	389	18	48	9	2427
1988	0	2	12	74	229	359	529	534	349	43	30	4	2165
1989	5	37	64	136	246	470	561	493	358	173	44	0	2587
1990	4	34	65	105	302	487	583	548	430	238	24	21	2841
1991	0	20	52	118	356	426	584	531	379	128	18	25	2637
1992	0	7	20	83	199	381	592	495	345	79	68	5	2274
1993	7	0	4	37	241	440	644	536	410	127	42	0	2488

SNOWFALL (inches) CHARLESTON, SOUTH CAROLINA

SEASON	JULY	AUG	SEP	OCT	NOV	DEC	JAN	FEB	MAR	APR	MAY	JUNE	TOTAL
1970-71	0.0	0.0	0.0	0.0	0.0	0.0	T	T	T	0.0	0.0	0.0	T
1971-72	0.0	0.0	0.0	0.0	0.0	0.0	0.0	0.0	0.0	0.0	0.0	0.0	0.0
1972-73	0.0	0.0	0.0	0.0	0.0	0.0	T	7.1	0.0	0.0	0.0	0.0	7.1
1973-74	0.0	0.0	0.0	0.0	0.0	T	0.0	0.0	0.0	0.0	0.0	0.0	T
1974-75	0.0	0.0	0.0	0.0	0.0	0.0	0.0	0.0	0.0	0.0	0.0	0.0	0.0
1975-76	0.0	0.0	0.0	0.0	0.0	0.0	0.4	0.0	0.0	0.0	0.0	0.0	0.4
1976-77	0.0	0.0	0.0	0.0	0.0	0.0	1.0	0.3	0.0	0.0	0.0	0.0	1.3
1977-78	0.0	0.0	0.0	0.0	0.0	0.0	T	0.4	T	0.0	0.0	0.0	0.4
1978-79	0.0	0.0	0.0	0.0	0.0	0.0	0.0	1.8	0.0	0.0	0.0	0.0	1.8
1979-80	0.0	0.0	0.0	0.0	0.0	0.0	0.0	T	1.3	0.0	0.0	0.0	1.3
1980-81	0.0	0.0	0.0	0.0	0.0	3.8	0.0	0.0	0.0	0.0	0.0	0.0	3.8
1981-82	0.0	0.0	0.0	0.0	0.0	0.0	T	0.0	0.0	0.0	0.0	0.0	T
1982-83	0.0	0.0	0.0	0.0	0.0	0.0	0.0	0.0	T	0.0	0.0	0.0	T
1983-84	0.0	0.0	0.0	0.0	0.0	0.0	T	0.0	0.0	0.0	0.0	0.0	T
1984-85	0.0	0.0	0.0	0.0	0.0	0.0	T	0.0	0.0	T	0.0	0.0	T
1985-86	0.0	0.0	0.0	0.0	0.0	0.0	0.5	0.0	0.0	0.0	0.0	0.0	0.5
1986-87	0.0	0.0	0.0	0.0	0.0	0.0	T	0.0	T	0.0	0.0	0.0	T
1987-88	0.0	0.0	0.0	0.0	0.0	0.0	0.4	0.0	0.0	0.0	0.0	0.0	0.4
1988-89	0.0	0.0	0.0	0.0	0.0	T	0.0	0.9	T	0.0	0.0	T	0.9
1989-90	0.0	0.0	0.0	0.0	0.0	8.0	0.0	0.0	0.0	0.0	0.0	0.0	8.0
1990-91	0.0	0.0	0.0	0.0	0.0	0.0	0.0	T	0.0	0.0	0.0	0.0	T
1991-92	0.0	0.0	0.0	0.0	0.0	0.0	0.0	0.0	0.0	0.0	0.0	0.0	0.0
1992-93	0.0	0.0	0.0	0.0	0.0	0.0	0.0	0.0	T	0.0	0.0	0.0	T
1993-94	T	0.0	0.0	0.0	0.0	T							
Record Mean	T	0.0	0.0	0.0	T	0.3	0.1	0.2	0.1	T	0.0	T	0.7

COLUMBIA,
SOUTH CAROLINA

Columbia is centrally located within the state of South Carolina and lies on the Congaree River near the confluence of the Broad and Saluda Rivers. The surrounding terrain is rolling, sloping from about 350 feet above sea level in northern Columbia to about 200 feet in the southeastern part of the city.

The climate in the Columbia area is relatively temperate. The Appalachian Mountain chain, some 150 miles to the northwest, frequently retards the approach of unseasonable cold weather in the winter. The terrain offers little moderating effect on the summer heat.

Long summers are prevalent with warm weather usually lasting from sometime in May into September. In summer the Bermuda high is the greatest single weather factor influencing the area. This permanent high more or less blocks the entry of cold fronts so that many stall before reaching central South Carolina. Also, the southwestern flow around the offshore Bermuda high pressure supplies moisture for the many summer thunderstorms. There are relatively few breaks in the heat during midsummer. The typical summer has about six days with 100 degrees or more. Thunderstorm activity usually shows a decided increase during June, decreasing about the first of September. About once or twice a year, passing tropical storms produce strong winds and heavy rains. The incidence of these storms is greatest in September, although they represent a possible threat from midsummer to late fall. Damage from tropical storms is usually minor in the Columbia area.

Fall is the most pleasant time of the year. Rainfall during the late fall is at an annual minimum, while the sunshine is at a relative maximum. Winters are mild with the cold weather usually lasting from late November to mid–March. The winter weather at Columbia is largely made up of polar air outbreaks that reach this area in a much modified form. On rare occasions in winter, Arctic air masses push southward as far as central South Carolina and cause some of the coldest temperatures. Disruption of activities from snowfall is unusual, in fact, more than three days of sustained snow cover is rare.

Spring is the most changeable season of the year. The temperature varies from an occasional cold snap in March to generally warm and pleasant in May. While tornadoes are infrequent, they occur most often in the spring. Hailstorms are not frequent, with the annual incidence at a maximum in spring and early summer. The average occurrence of the last spring freeze is very late March, and the first fall freeze is early November, for a growing period of about 218 days.

NORMALS, MEANS, AND EXTREMES

COLUMBIA, SOUTH CAROLINA

LATITUDE: 33°57'N LONGITUDE: 81°07'W ELEVATION: FT. GRND 213 BARO 245 TIME ZONE: EASTERN WBAN: 13883

	(a)	JAN	FEB	MAR	APR	MAY	JUNE	JULY	AUG	SEP	OCT	NOV	DEC	YEAR
TEMPERATURE °F:														
Normals														
-Daily Maximum		55.3	59.3	68.2	76.5	83.5	88.8	91.6	90.1	85.1	76.3	67.8	58.8	75.1
-Daily Minimum		32.1	34.2	42.2	49.4	58.2	66.0	70.0	69.2	63.2	50.1	41.5	34.9	50.9
-Monthly		43.8	46.8	55.2	63.0	70.9	77.4	80.8	79.7	74.2	63.3	54.6	46.9	63.1
Extremes														
-Record Highest	46	84	84	91	94	101	107	107	107	101	101	90	83	107
-Year		1975	1990	1974	1986	1953	1954	1952	1983	1954	1954	1961	1978	AUG 1983
-Record Lowest	46	-1	5	4	26	34	44	54	53	40	23	12	4	-1
-Year		1985	1973	1980	1983	1963	1984	1951	1969	1967	1952	1970	1958	JAN 1985
NORMAL DEGREE DAYS:														
Heating (base 65°F)		665	514	317	108	18	0	0	0	5	138	323	561	2649
Cooling (base 65°F)		8	0	13	48	201	372	490	456	281	86	11	0	1966
% OF POSSIBLE SUNSHINE	40	55	59	64	69	68	67	66	66	64	66	63	58	64
MEAN SKY COVER (tenths)														
Sunrise - Sunset	45	6.2	6.0	5.9	5.3	5.7	5.8	6.1	5.8	5.6	4.8	5.1	5.9	5.7
MEAN NUMBER OF DAYS:														
Sunrise to Sunset														
-Clear	46	9.2	8.8	9.2	11.3	9.3	7.9	6.8	8.0	9.6	13.4	12.3	10.2	115.9
-Partly Cloudy	46	5.8	6.3	7.8	7.7	10.3	11.4	12.7	12.7	8.9	7.1	6.1	6.1	102.8
-Cloudy	46	16.0	13.1	14.0	11.0	11.5	10.7	11.6	10.4	11.5	10.5	11.6	14.7	146.6
Precipitation														
.01 inches or more	46	10.3	9.6	10.5	8.1	8.8	9.4	11.5	10.8	7.6	6.3	7.0	9.2	109.2
Snow,Ice pellets,hail														
1.0 inches or more	46	0.2	0.2	0.1	0.0	0.0	0.0	0.0	0.0	0.0	0.0	0.0	0.*	0.5
Thunderstorms	46	0.7	1.5	2.7	3.8	6.3	9.1	12.8	9.8	3.9	1.4	0.8	0.3	53.2
Heavy Fog Visibility														
1/4 mile or less	45	2.8	2.3	1.7	1.3	1.5	1.4	1.6	2.3	2.8	2.6	3.0	3.2	26.6
Temperature °F														
-Maximum														
90° and above	27	0.0	0.0	0.1	1.9	5.4	15.5	22.3	17.9	9.1	0.6	0.0	0.0	72.9
32° and below	27	0.6	0.1	0.*	0.0	0.0	0.0	0.0	0.0	0.0	0.0	0.0	0.1	0.8
-Minimum														
32° and below	27	16.3	13.4	6.2	1.0	0.0	0.0	0.0	0.0	0.0	1.0	7.7	14.1	59.7
0° and below	27	0.*	0.0	0.0	0.0	0.0	0.0	0.0	0.0	0.0	0.0	0.0	0.0	*
AVG. STATION PRESS. (mb)	21	1012.1	1011.1	1009.2	1008.3	1007.9	1008.1	1009.0	1009.5	1009.9	1011.3	1012.0	1012.8	1010.1
RELATIVE HUMIDITY (%)														
Hour 01	27	78	76	77	77	85	86	87	90	90	88	85	80	83
Hour 07 (Local Time)	27	83	81	84	84	87	86	89	92	92	91	89	84	87
Hour 13	27	54	49	48	43	49	51	54	57	56	51	51	53	51
Hour 19	27	65	57	54	50	58	61	66	71	73	74	72	69	64
PRECIPITATION (inches):														
Water Equivalent														
-Normal		4.42	4.12	4.82	3.28	3.68	4.80	5.50	6.09	3.67	3.04	2.90	3.59	49.91
-Maximum Monthly	46	9.26	8.68	10.89	6.85	8.85	14.81	17.46	16.72	8.78	12.09	7.20	8.54	17.46
-Year		1978	1961	1973	1979	1967	1973	1991	1949	1953	1959	1957	1981	JUL 1991
-Minimum Monthly	46	0.84	0.87	0.56	0.35	0.29	0.74	0.57	1.02	0.07	T	0.41	0.32	T
-Year		1981	1976	1985	1986	1951	1993	1977	1976	1985	1963	1973	1955	OCT 1963
-Maximum in 24 hrs	46	3.15	3.69	3.59	3.66	5.57	5.44	5.81	7.66	6.23	5.46	2.60	3.18	7.66
-Year		1993	1962	1960	1956	1967	1973	1959	1949	1953	1964	1986	1970	AUG 1949
Snow,Ice pellets,hail														
-Maximum Monthly	46	4.3	16.0	4.1	T	0.0	0.0	0.0	T	0.0	0.0	T	9.1	16.0
-Year		1988	1973	1980	1992				1993			1976	1958	FEB 1973
-Maximum in 24 hrs	46	4.3	15.7	4.1	T	0.0	0.0	0.0	T	0.0	0.0	T	8.8	15.7
-Year		1988	1973	1980	1992				1993			1976	1958	FEB 1973
WIND:														
Mean Speed (mph)	45	7.2	7.7	8.3	8.3	7.0	6.6	6.3	5.8	6.1	6.1	6.4	6.7	6.9
Prevailing Direction through 1963		SW	SW	SW	SW	SW	SW	SW	SW	NE	NE	SW	WSW	SW
Fastest Obs. 1 Min.														
-Direction (!!)	40	28	20	27	33	23	22	32	16	30	21	35	28	27
-Speed (MPH)	40	46	40	60	44	46	46	40	44	48	27	35	35	60
-Year		1964	1966	1954	1988	1958	1989	1990	1961	1989	1968	1967	1975	MAR 1954
Peak Gust														
-Direction (!!)	10	W	W	W	NW	W	SW	W	NW	NW	N	SW	W	SW
-Speed (mph)	10	54	69	69	61	59	78	64	54	70	54	51	49	78
-Date		1992	1989	1992	1988	1989	1989	1990	1993	1989	1985	1989	1984	JUN 1989

PRECIPITATION (inches)

COLUMBIA, SOUTH CAROLINA

YEAR	JAN	FEB	MAR	APR	MAY	JUNE	JULY	AUG	SEP	OCT	NOV	DEC	ANNUAL
1964	6.34	5.33	6.16	3.60	2.63	2.97	10.32	9.97	6.93	10.34	1.36	4.58	70.53
1965	1.43	5.33	7.68	3.99	1.46	8.20	4.33	9.39	5.99	2.34	1.77	0.64	52.55
1966	7.22	4.54	2.23	3.58	6.14	3.66	2.87	3.22	2.02	2.47	1.05	3.31	42.31
1967	2.79	4.36	3.08	3.72	8.85	4.18	7.27	11.16	2.38	0.62	3.71	2.59	54.71
1968	5.94	1.14	1.92	4.52	4.17	5.41	9.28	1.11	2.40	4.31	5.21	3.26	48.67
1969	2.64	3.03	5.16	4.57	3.28	4.70	4.31	2.93	3.17	1.17	1.20	4.51	40.67
1970	3.28	2.58	8.42	0.91	4.50	2.05	4.74	7.13	3.72	8.18	1.43	4.55	51.49
1971	4.55	5.23	9.53	4.31	2.71	7.46	11.13	10.68	5.03	3.44	2.35	2.90	69.32
1972	7.62	3.58	3.79	1.16	6.41	6.10	9.31	2.87	2.51	1.15	5.62	5.39	55.51
1973	5.25	5.75	10.89	4.47	4.04	14.81	3.19	6.92	4.47	0.71	0.41	6.66	67.57
1974	6.16	4.49	2.36	2.97	3.40	4.50	4.40	6.20	4.44	0.02	4.47	4.61	48.02
1975	4.26	6.43	5.41	4.59	7.88	2.85	9.91	3.16	3.32	0.88	2.23	5.03	55.95
1976	3.58	0.87	5.24	0.81	4.63	11.67	6.55	1.02	5.74	5.21	5.13	7.54	57.99
1977	4.20	1.22	6.34	0.91	0.89	2.20	0.57	10.73	1.51	4.81	2.10	3.69	39.17
1978	9.26	1.28	3.49	4.28	3.09	4.73	2.10	4.45	4.09	0.79	2.98	1.82	42.36
1979	5.19	8.10	3.53	6.85	6.47	5.48	7.28	4.05	7.86	1.76	3.89	1.51	61.97
1980	4.72	1.88	10.72	2.02	4.51	2.27	1.24	3.29	7.25	1.58	1.72	1.33	42.53
1981	0.84	4.08	2.25	1.87	3.38	5.28	5.42	4.65	0.39	1.90	1.47	8.54	40.07
1982	3.74	4.39	1.65	6.44	2.92	4.23	9.98	5.88	3.32	1.47	2.62	3.72	50.36
1983	3.66	5.38	7.35	5.68	0.70	2.85	0.73	3.36	3.25	2.22	3.63	6.58	45.39
1984	3.99	4.88	5.54	3.75	4.29	6.47	8.69	3.23	0.67	1.03	0.78	1.75	45.07
1985	3.27	7.15	0.56	1.29	3.13	3.96	7.47	5.65	0.07	8.44	5.98	0.88	47.85
1986	1.05	1.46	3.21	0.35	1.13	0.88	1.25	9.55	0.56	6.04	6.26	2.52	34.26
1987	8.36	5.39	5.38	0.40	1.12	6.49	3.95	10.81	5.27	0.99	4.55	1.50	54.21
1988	4.10	2.02	1.98	3.01	2.08	1.66	3.24	11.75	7.53	3.68	1.59	0.75	43.39
1989	1.90	3.30	4.89	4.27	4.44	5.99	9.41	3.19	5.16	2.25	1.85	5.28	51.93
1990	2.44	2.56	2.28	1.26	4.03	1.27	5.14	6.51	2.64	11.66	2.04	1.64	43.47
1991	5.48	1.87	7.57	4.69	6.96	3.56	17.46	7.77	2.45	0.54	1.46	2.62	62.43
1992	3.14	4.16	3.38	3.16	1.93	6.37	2.15	9.61	4.60	4.22	4.02	3.26	50.00
1993	7.49	3.29	6.01	1.63	2.98	0.74	2.02	2.34	3.90	4.29	1.94	2.39	39.02
Record Mean	3.47	3.83	3.98	3.18	3.28	4.14	5.59	5.67	3.78	2.60	2.86	3.19	45.58

AVERAGE TEMPERATURE (deg. F)

COLUMBIA, SOUTH CAROLINA

YEAR	JAN	FEB	MAR	APR	MAY	JUNE	JULY	AUG	SEP	OCT	NOV	DEC	ANNUAL
1964	44.1	43.7	54.8	64.2	72.4	80.5	78.8	78.4	73.8	59.1	58.2	49.8	63.2
1965	45.7	48.3	52.2	65.1	76.1	75.5	79.9	80.5	75.4	62.8	54.7	47.3	63.6
#1966	40.3	46.9	51.8	63.1	70.2	75.0	81.6	79.2	73.8	62.8	53.4	45.7	62.0
1967	46.9	43.0	57.1	65.6	67.0	73.5	77.6	75.7	67.3	60.2	49.1	49.6	61.0
1968	41.1	40.1	54.8	63.8	69.6	77.9	80.5	82.9	72.4	64.0	53.6	42.7	62.0
1969	42.7	43.5	51.1	65.2	70.6	78.3	83.1	77.5	72.8	65.3	50.8	44.1	62.1
1970	38.3	47.5	57.1	67.6	73.0	78.8	83.4	81.3	78.6	64.7	50.7	50.1	64.3
1971	45.4	46.3	50.0	61.3	69.8	80.3	80.0	80.1	77.1	69.8	53.8	56.4	64.2
1972	50.9	45.7	53.9	63.1	68.4	73.9	79.2	79.1	75.2	62.7	53.5	51.3	63.1
1973	44.9	44.0	58.6	60.0	67.8	77.2	80.7	80.6	78.6	66.7	59.0	49.5	64.0
1974	59.2	50.9	61.7	65.3	74.7	77.7	81.3	78.7	72.5	61.1	54.3	48.0	65.5
1975	51.6	52.0	54.0	63.3	74.8	77.8	79.0	80.5	75.3	67.2	56.1	46.3	64.8
1976	43.5	55.5	61.0	64.4	69.1	74.6	79.6	79.0	72.4	58.6	47.7	44.0	62.5
1977	35.8	46.1	59.6	66.2	72.5	78.3	82.5	78.6	75.4	59.5	56.4	44.4	62.9
1978	37.3	38.4	51.4	63.0	69.2	77.0	79.9	80.3	77.3	63.9	59.9	49.5	62.3
1979	43.4	42.9	55.8	62.4	69.0	73.0	78.5	79.7	73.7	62.0	55.0	44.9	61.7
1980	44.2	41.4	49.8	61.5	69.0	75.7	81.9	80.5	76.8	60.2	51.7	44.7	61.5
1981	38.7	46.7	51.2	66.9	68.0	81.4	80.9	76.1	71.4	61.8	53.2	43.3	61.7
1982	40.7	49.8	56.5	60.2	72.0	77.4	80.3	77.9	71.8	61.9	55.0	51.3	62.9
1983	40.9	45.3	53.6	57.8	69.9	75.3	83.3	82.7	73.2	63.7	52.3	43.3	61.8
1984	41.7	48.0	53.6	59.6	70.1	79.1	78.7	79.2	71.7	70.4	50.6	53.4	63.0
1985	39.5	47.0	57.5	65.3	71.5	78.7	79.9	77.7	72.4	68.4	62.5	42.9	63.6
1986	41.2	51.4	55.0	63.8	72.2	82.0	86.1	79.8	76.4	65.0	59.3	46.7	64.9
1987	43.5	45.7	53.4	60.4	74.3	79.4	83.1	83.2	75.2	56.5	54.9	49.0	63.2
1988	39.6	44.8	54.5	62.5	69.4	77.0	82.0	82.6	75.5	58.7	56.0	46.0	62.4
1989	49.6	49.8	56.5	62.3	69.0	79.8	80.9	80.4	74.8	64.7	55.8	41.0	63.7
1990	52.1	56.2	59.4	63.4	71.9	79.9	83.6	81.7	76.0	66.4	55.7	51.7	66.5
1991	46.3	51.7	58.3	67.1	75.4	79.0	83.0	80.2	75.0	64.7	52.2	50.0	65.2
1992	46.3	51.0	54.4	62.3	67.6	75.9	83.8	78.8	74.9	61.6	55.1	45.8	63.1
1993	48.5	45.4	51.2	58.9	71.5	79.9	86.2	81.9	76.6	63.6	54.5	45.0	63.6
Record Mean	45.9	48.0	55.2	63.3	71.7	78.4	80.9	79.8	75.0	64.2	54.4	47.1	63.7
Max	56.1	58.8	66.5	75.2	83.2	89.2	91.1	89.7	85.0	75.7	65.9	57.5	74.5
Min	35.7	37.1	43.8	51.4	60.1	67.6	70.7	70.0	64.9	52.6	42.9	36.7	52.8

REFERENCE NOTES

COLUMBIA, SOUTH CAROLINA

GENERAL
T - TRACE AMOUNT.
BLANK ENTRIES DENOTE MISSING/UNREPORTED DATA.
# INDICATES A STATION OR INSTRUMENT RELOCATION.

SPECIFIC

(a) - LENGTH OF RECORD IN YEARS, ALTHOUGH INDIVIDUAL MONTHS MAY BE MISSING.
0.* OR * - THE VALUE IS BETWEEN 0.0 AND 0.05
NORMALS - BASED ON THE 1961-1990 RECORD PERIOD.
EXTREMES - DATES ARE THE MOST RECENT OCCURRENCE.
WIND DIR.- NUMERALS SHOW TENS OF DEGRESS CLOCKWISE FROM TRUE NORTH. "00" INDICATES CALM.
RESULTANT DIRECTIONS ARE GIVEN TO WHOLE DEGREES.
BOLD VALUES INDICATE EXTREME VALUES WHICH OCCURRED AFTER THE ASOS SYSTEM WAS COMMISSIONED.

RECORD = PERIOD OF RECORD
RECORD MEAN PRECIPITATION IS THE MEAN OF ALL DAILY PRECIPITATION AMOUNTS DURING THE PERIOD OF RECORD.
RECORD MAX(MIN) TEMPERATURE IS THE MEAN OF ALL DAILY MAX(MIN) TEMPERATURES DURING THE PERIOD OF RECORD.
RECORD MEAN TEMPERATURE IS THE SUM OF THE RECORD MAX AND RECORD MIN DIVIDED BY 2.
AVERAGE TEMPERATURE IS THE SUM OF THE MEAN DAILY MAX AND MIN TEMPERATURE DIVIDED BY 2.

EXCEPTIONS

RECORD MEANS ARE THROUGH THE CURRENT YEAR, BEGINNING IN 1880 FOR TEMPERATURE
1880 FOR PRECIPITATION
1948 FOR SNOWFALL

HEATING DEGREE DAYS Base 65 deg. F COLUMBIA, SOUTH CAROLINA

SEASON	JULY	AUG	SEP	OCT	NOV	DEC	JAN	FEB	MAR	APR	MAY	JUNE	TOTAL
1964-65	0	0	3	204	216	473	592	466	398	93	0	0	2445
1965-66	0	0	2	138	303	543	759	502	408	128	19	1	2803
#1966-67	0	0	0	127	351	596	554	605	270	67	71	10	2651
1967-68	0	0	32	164	469	480	732	717	327	106	20	0	3047
1968-69	0	0	0	121	339	684	683	594	430	71	13	0	2935
1969-70	0	0	2	86	419	640	823	484	252	73	12	0	2791
1970-71	0	0	6	85	420	462	602	519	462	155	27	0	2738
1971-72	0	0	0	13	354	292	429	555	341	136	12	1	2133
1972-73	0	0	0	98	350	423	618	583	219	179	47	0	2517
1973-74	0	0	0	57	205	477	199	394	161	99	7	0	1599
1974-75	0	0	18	163	342	521	417	365	344	136	0	0	2306
1975-76	0	0	0	56	296	577	662	280	176	97	26	4	2174
1976-77	0	0	0	219	512	645	901	525	211	69	20	0	3102
1977-78	0	0	1	192	277	629	850	741	418	109	37	0	3254
1978-79	0	0	0	89	162	492	664	613	294	107	27	2	2450
1979-80	0	0	6	138	314	616	641	680	466	147	27	0	3035
1980-81	0	0	14	178	394	622	809	506	424	59	43	0	3049
1981-82	0	0	13	131	349	665	748	425	283	179	5	0	2798
1982-83	0	0	18	162	306	440	739	546	352	230	24	1	2818
1983-84	0	0	21	89	373	666	717	485	355	190	31	2	2929
1984-85	0	0	12	22	444	355	792	503	262	79	9	0	2478
1985-86	0	0	16	45	110	676	731	374	319	115	21	0	2407
1986-87	0	2	0	110	207	565	657	532	362	187	5	0	2627
1987-88	0	0	0	259	298	495	780	579	322	114	20	1	2868
1988-89	0	0	2	218	274	583	469	435	289	165	50	0	2485
1989-90	0	0	9	110	283	738	393	265	220	115	6	0	2139
1990-91	0	0	8	95	277	414	571	376	233	51	4	0	2029
1991-92	0	0	6	79	392	477	574	402	323	150	47	2	2452
1992-93	0	0	17	133	305	587	509	543	422	200	9	0	2725
1993-94	0	0	14	110	334	614							

COOLING DEGREE DAYS Base 65 deg. F COLUMBIA, SOUTH CAROLINA

YEAR	JAN	FEB	MAR	APR	MAY	JUNE	JULY	AUG	SEP	OCT	NOV	DEC	TOTAL
1969	0	0	5	81	191	406	570	392	242	104	1	0	1992
1970	2	0	12	158	266	420	576	513	421	82	0	6	2456
1971	0	2	6	51	185	465	470	473	369	168	25	32	2246
1972	1	0	5	86	126	274	450	445	313	35	14	4	1753
1973	0	0	28	33	143	375	494	490	413	116	32	6	2130
1974	29	8	65	116	316	388	511	432	249	51	29	1	2195
1975	8	6	11	94	308	390	442	488	317	129	37	0	2230
1976	2	10	58	87	160	295	460	442	231	30	0	0	1775
1977	0	3	53	110	257	407	549	429	318	29	25	0	2180
1978	0	0	1	56	172	369	466	478	376	62	15	17	2012
1979	0	0	13	36	157	253	425	461	276	50	22	0	1693
1980	0	0	0	47	159	328	530	487	376	35	2	0	1964
1981	0	0	6	119	143	497	499	349	213	41	0	2	1869
1982	0	4	26	39	229	377	484	406	226	74	11	20	1896
1983	0	0	6	21	184	319	573	558	270	58	0	0	1989
1984	0	0	10	33	194	430	436	448	219	199	18	3	1990
1985	9	2	38	94	215	417	468	402	244	155	45	0	2089
1986	0	0	17	85	249	516	660	471	347	118	44	5	2512
1987	0	0	10	58	300	440	571	571	315	2	4	6	2277
1988	0	0	5	47	162	367	536	555	323	29	11	0	2035
1989	0	19	37	91	180	451	500	482	309	107	14	0	2190
1990	0	25	54	74	228	454	583	524	343	146	6	9	2446
1991	0	12	32	119	331	427	565	476	313	77	15	18	2385
1992	0	3	3	75	134	337	591	435	320	33	14	0	1945
1993	2	0	0	22	221	452	663	532	369	72	27	0	2360

SNOWFALL (inches) COLUMBIA, SOUTH CAROLINA

SEASON	JULY	AUG	SEP	OCT	NOV	DEC	JAN	FEB	MAR	APR	MAY	JUNE	TOTAL
1964-65	0.0	0.0	0.0	0.0	0.0	0.0	1.1	0.0	T	0.0	0.0	0.0	1.1
1965-66	0.0	0.0	0.0	0.0	0.0	0.0	1.0	T	0.0	0.0	0.0	0.0	1.0
1966-67	0.0	0.0	0.0	0.0	0.0	T	0.0	3.4	0.0	0.0	0.0	0.0	3.4
1967-68	0.0	0.0	0.0	0.0	0.0	0.0	1.7	1.0	0.0	0.0	0.0	0.0	2.7
1968-69	0.0	0.0	0.0	0.0	T	0.0	T	0.8	0.0	0.0	0.0	0.0	0.8
1969-70	0.0	0.0	0.0	0.0	0.0	T	1.8	T	0.0	0.0	0.0	0.0	1.8
1970-71	0.0	0.0	0.0	0.0	0.0	T	T	T	1.7	0.0	0.0	0.0	1.7
1971-72	0.0	0.0	0.0	0.0	0.0	0.8	0.0	0.0	0.0	0.0	0.0	0.0	0.8
1972-73	0.0	0.0	0.0	0.0	0.0	0.0	2.2	16.0	0.0	0.0	0.0	0.0	18.2
1973-74	0.0	0.0	0.0	0.0	0.0	T	0.0	0.0	0.0	0.0	0.0	0.0	T
1974-75	0.0	0.0	0.0	0.0	0.0	0.0	0.0	T	0.0	0.0	0.0	0.0	T
1975-76	0.0	0.0	0.0	0.0	0.0	0.0	T	0.0	0.0	0.0	0.0	0.0	T
1976-77	0.0	0.0	0.0	0.0	T	0.0	T	0.0	0.0	0.0	0.0	0.0	T
1977-78	0.0	0.0	0.0	0.0	0.0	0.0	T	0.5	0.0	0.0	0.0	0.0	0.5
1978-79	0.0	0.0	0.0	0.0	0.0	T	0.0	5.5	0.0	0.0	0.0	0.0	5.5
1979-80	0.0	0.0	0.0	0.0	0.0	0.0	0.0	3.0	4.1	0.0	0.0	0.0	7.1
1980-81	0.0	0.0	0.0	0.0	0.0	0.3	T	T	0.0	0.0	0.0	0.0	0.3
1981-82	0.0	0.0	0.0	0.0	0.0	T	3.5	0.8	0.0	0.0	0.0	0.0	4.3
1982-83	0.0	0.0	0.0	0.0	0.0	0.0	0.1	T	0.4	0.0	0.0	0.0	0.5
1983-84	0.0	0.0	0.0	0.0	0.0	0.0	T	T	0.0	0.0	0.0	0.0	T
1984-85	0.0	0.0	0.0	0.0	0.0	0.0	T	0.0	0.0	0.0	0.0	0.0	T
1985-86	0.0	0.0	0.0	0.0	0.0	0.0	0.0	0.0	0.0	0.0	0.0	0.0	0.0
1986-87	0.0	0.0	0.0	0.0	0.0	0.0	0.9	0.0	0.1	0.0	0.0	0.0	1.0
1987-88	0.0	0.0	0.0	0.0	0.0	0.0	4.3	0.0	0.0	0.0	0.0	0.0	4.3
1988-89	0.0	0.0	0.0	0.0	0.0	0.0	0.0	4.3	0.0	0.0	0.0	0.0	4.3
1989-90	0.0	0.0	0.0	0.0	0.0	T	0.0	0.0	0.0	T	0.0	0.0	T
1990-91	0.0	0.0	0.0	0.0	0.0	0.0	0.4	0.0	0.0	0.0	0.0	0.0	0.4
1991-92	0.0	0.0	0.0	0.0	0.0	0.0	0.8	0.0	T	T	0.0	0.0	0.8
1992-93	0.0	0.0	0.0	0.0	0.0	0.0	0.0	T	1.2	0.0	0.0	0.0	1.2
1993-94	0.0	T	0.0	0.0	0.0	2.3							
Record Mean	0.0	T	0.0	0.0	T	0.3	0.5	0.8	0.3	T	0.0	0.0	1.8

GREENVILLE–SPARTANBURG (GREER), SOUTH CAROLINA

This station, three miles south of Greer, South Carolina, is located in the Piedmont section, on the eastern slope of the Southern Appalachian Mountains. It is rolling country with the first ridge of the mountains about 20 miles to the northwest and the main ridge about 55 miles to the northwest. These mountains usually protect this area from the full force of the cold air masses which move southeastward from central Canada during the winter months.

At present, the National Weather Service Office is located at the Greenville–Spartanburg Jet Age Airport, on a level with, or slightly higher than, most of the surrounding countryside. No bodies of water are nearby. Temperatures are quite consistent with those in Greer, Greenville, and Spartanburg.

The elevation of the area, ranging from 800 to 1,100 feet is conducive to cool nights, especially during the summer months. Winters are quite pleasant, with the temperature remaining below freezing throughout the daylight hours only a few times during a normal year. There are usually two freezing rainstorms each winter and two or three small snowstorms.

Rainfall in this section is usually abundant and spread quite evenly through the months. Droughts have been experienced, but are usually of short duration.

The mountain ridges, which lie in a northeast–southwest direction, appear to have a definite overall influence on the direction of the wind. The prevailing directions are northeast and southwest, divided almost evenly, with fall and winter favoring northeast and spring and summer favoring southwest. Destructive winds occur occasionally, while tornadoes are infrequent in this vicinity.

In the southern two–thirds of Greenville and Spartanburg Counties, including the cities of the same names, the average occurrence of the last temperature of 32 degrees in spring is late March and the average occurrence of the first in fall is early November, giving an average growing season of 225 days. In a normal year some flowering shrubs bloom through the winter. In the higher elevations in the northern thirds of these counties, the growing season begins about one month later and ends about one month earlier.

NORMALS, MEANS, AND EXTREMES

GREENVILLE-SPARTANBURG (GREER), SOUTH CAROLINA

LATITUDE: 34°54'N LONGITUDE: 82°13'W ELEVATION: FT. GRND 957 BARO 951 TIME ZONE: EASTERN WBAN: 03870

	(a)	JAN	FEB	MAR	APR	MAY	JUNE	JULY	AUG	SEP	OCT	NOV	DEC	YEAR
TEMPERATURE °F:														
Normals														
-Daily Maximum		50.1	54.3	63.5	72.0	79.4	85.5	88.2	86.8	81.1	71.9	62.5	53.2	70.7
-Daily Minimum		30.0	32.4	39.9	47.6	56.4	64.4	68.3	67.4	61.2	49.0	40.5	33.4	49.2
-Monthly		40.1	43.4	51.7	59.8	67.9	75.0	78.2	77.1	71.2	60.5	51.6	43.3	60.0
Extremes														
-Record Highest	31	79	79	88	93	97	100	103	103	96	92	85	76	103
-Year		1975	1982	1967	1986	1967	1985	1986	1983	1975	1986	1974	1984	JUL 1986
-Record Lowest	31	-6	8	11	25	31	40	54	52	36	25	12	5	-6
-Year		1966	1967	1993	1983	1989	1972	1979	1968	1967	1976	1970	1985	JAN 1966
NORMAL DEGREE DAYS:														
Heating (base 65°F)		772	605	412	177	49	0	0	0	7	175	402	673	3272
Cooling (base 65°F)		0	0	0	21	139	300	409	375	193	36	0	0	1473
% OF POSSIBLE SUNSHINE	31	55	59	63	67	62	63	62	61	62	67	59	55	61
MEAN SKY COVER (tenths) Sunrise - Sunset	31	5.9	5.8	5.7	5.3	5.9	5.8	6.0	5.8	5.5	4.6	5.2	5.8	5.6
MEAN NUMBER OF DAYS:														
Sunrise to Sunset														
-Clear	31	10.5	9.9	10.2	11.0	8.5	8.5	7.2	7.9	10.0	14.2	12.4	11.0	121.3
-Partly Cloudy	31	5.8	5.4	7.6	8.4	10.0	10.8	12.5	12.5	8.6	6.7	5.8	5.8	99.9
-Cloudy	31	14.7	12.9	13.1	10.6	12.6	10.8	11.3	10.6	11.4	10.1	11.7	14.2	144.1
Precipitation .01 inches or more	31	10.9	9.2	11.0	8.8	10.9	9.7	11.7	10.4	8.5	7.2	8.7	10.1	117.1
Snow,Ice pellets,hail 1.0 inches or more	31	0.7	0.6	0.3	0.0	0.0	0.0	0.0	0.0	0.0	0.0	0.*	0.1	1.7
Thunderstorms	31	0.6	0.9	2.5	3.1	6.2	6.6	10.0	7.0	3:2	0.8	0.8	0.5	42.3
Heavy Fog Visibility 1/4 mile or less	31	4.4	3.7	3.1	2.0	1.6	1.3	2.0	2.5	2.3	1.8	3.8	4.9	33.4
Temperature °F														
-Maximum														
90° and above	31	0.0	0.0	0.0	0.2	1.5	7.9	13.9	9.9	2.6	0.1	0.0	0.0	36.2
32° and below	31	1.2	0.2	0.*	0.0	0.0	0.0	0.0	0.0	0.0	0.0	0.0	0.4	1.8
-Minimum														
32° and below	31	18.5	15.1	7.3	1.1	0.*	0.0	0.0	0.0	0.0	0.8	7.2	15.6	65.7
0° and below	31	0.1	0.0	0.0	0.0	0.0	0.0	0.0	0.0	0.0	0.0	0.0	0.0	0.1
AVG. STATION PRESS.(mb)	21	984.6	983.9	982.2	981.5	981.5	982.1	983.1	983.7	984.2	985.1	985.2	985.5	983.5
RELATIVE HUMIDITY (%)														
Hour 01	31	73	70	70	71	81	83	85	87	87	82	77	75	78
Hour 07 (Local Time)	31	77	75	77	78	84	85	87	90	89	85	82	78	82
Hour 13	31	55	51	50	48	53	54	57	59	59	52	53	55	54
Hour 19	31	62	58	56	52	62	63	67	70	74	70	67	66	64
PRECIPITATION (inches):														
Water Equivalent														
-Normal		4.10	4.41	5.39	3.86	4.42	4.77	4.63	3.95	3.96	3.99	3.65	4.14	51.27
-Maximum Monthly	31	7.19	7.43	11.37	11.30	8.89	9.59	13.57	9.02	11.65	10.24	7.85	8.45	13.57
-Year		1993	1971	1980	1964	1972	1969	1984	1991	1975	1964	1992	1983	JUL 1984
-Minimum Monthly	31	0.29	0.53	1.13	0.69	1.09	0.17	0.75	0.87	0.27	0.24	1.34	0.37	0.17
-Year		1981	1978	1985	1976	1965	1993	1993	1993	1978	1991	1973	1965	JUN 1993
-Maximum in 24 hrs	31	3.30	3.57	4.45	3.76	3.58	4.80	3.89	4.49	6.21	4.93	2.83	3.00	6.21
-Year		1982	1984	1963	1963	1972	1980	1964	1967	1973	1990	1964	1972	SEP 1973
Snow,Ice pellets,hail														
-Maximum Monthly	31	12.0	12.3	9.8	0.3	T	0.0	T	0.0	0.0	0.0	1.9	11.4	12.3
-Year		1988	1979	1993	1987	1992		1989				1968	1971	FEB 1979
-Maximum in 24 hrs	31	12.0	8.2	9.8	0.3	T	0.0	T	0.0	0.0	0.0	1.9	11.4	12.0
-Year		1988	1979	1993	1987	1992		1989				1968	1971	JAN 1988
WIND:														
Mean Speed (mph)	31	7.5	8.0	8.2	8.0	6.9	6.5	6.0	5.8	6.1	6.6	6.8	7.4	7.0
Prevailing Direction through 1963		NE	NE	SW	SW	NE	NE	WSW	NE	NE	NE	SW	NE	NE
Fastest Mile														
-Direction (!!)	28	SW	SW	W	SW	SW	NW	NE	23	NE	06	NE	NE	NE
-Speed (MPH)	28	44	44	38	44	36	35	52	40	31	33	34	47	52
-Year		1967	1966	1963	1970	1967	1969	1966	1990	1964	1990	1974	1963	JUL 1966
Peak Gust														
-Direction (!!)	10	SW	N	SW	SW	W	NW	SW	SW	N	NE	SW	N	SW
-Speed (mph)	10	47	47	51	71	53	60	66	58	45	48	47	41	71
-Date		1987	1987	1984	1989	1989	1987	1992	1990	1989	1990	1989	1987	APR 1989

PRECIPITATION (inches)

GREENVILLE-SPARTANBURG (GREER), SOUTH CAROLINA

YEAR	JAN	FEB	MAR	APR	MAY	JUNE	JULY	AUG	SEP	OCT	NOV	DEC	ANNUAL
1964	5.44	4.67	7.11	11.30	1.59	8.07	7.44	6.64	0.93	10.24	3.36	3.62	70.41
1965	2.39	5.22	7.60	4.93	1.09	8.62	3.13	3.57	2.32	3.60	2.82	0.37	45.66
1966	4.64	6.78	3.26	2.53	3.06	3.84	2.98	5.01	7.98	3.78	1.93	3.15	48.94
1967	3.97	3.32	1.98	2.36	4.97	4.87	3.86	7.51	2.05	2.35	3.50	7.40	48.14
1968	4.12	1.00	3.68	2.40	3.93	5.71	6.92	1.31	3.04	2.82	5.07	3.18	43.18
1969	3.94	5.24	4.56	7.18	1.93	9.59	3.17	6.53	3.68	2.38	2.24	4.60	55.04
1970	1.74	3.74	3.45	2.94	3.13	3.60	2.31	3.59	1.34	7.02	1.77	2.88	37.51
1971	3.33	7.43	5.52	3.09	5.72	2.19	5.64	2.44	3.28	9.51	4.22	3.79	56.16
1972	6.14	3.04	4.59	2.28	8.89	8.16	4.18	3.21	2.20	3.44	5.31	6.68	58.12
1973	4.33	4.88	8.73	4.04	5.59	3.87	3.70	2.03	7.56	0.98	1.34	7.55	54.60
1974	4.24	4.90	3.26	4.06	5.45	3.78	3.23	4.03	3.76	0.24	4.81	2.50	44.26
1975	5.42	5.78	8.64	1.14	7.81	5.39	4.79	3.21	11.65	7.45	3.98	3.07	68.33
1976	4.49	2.15	7.30	0.69	8.10	2.81	5.75	2.09	8.28	8.49	2.75	6.21	59.11
1977	3.53	2.00	8.47	3.23	2.71	2.88	0.80	4.99	9.44	6.39	4.43	3.55	52.42
1978	6.93	0.53	6.09	2.97	4.84	3.51	6.77	2.98	0.27	0.81	1.93	3.39	41.02
1979	7.19	6.11	4.19	10.15	5.69	3.74	8.66	4.34	7.50	3.33	3.91	1.25	66.06
1980	4.28	1.19	11.37	3.47	5.92	6.72	1.05	3.33	5.82	2.83	4.11	0.64	50.73
1981	0.29	3.86	3.22	0.88	4.15	1.29	5.30	1.17	2.08	4.40	1.66	7.19	35.49
1982	6.27	5.21	2.77	4.57	6.18	3.32	12.52	1.66	1.44	3.07	4.17	5.02	56.20
1983	2.70	5.26	6.26	4.66	5.80	4.67	1.13	3.27	3.59	3.05	5.29	8.45	54.13
1984	3.04	7.04	5.67	4.76	8.30	3.07	13.57	4.00	1.34	2.28	2.60	2.22	57.89
1985	4.94	4.29	1.13	1.31	2.42	2.85	6.96	5.93	1.62	4.55	7.52	1.44	44.96
1986	1.10	1.46	2.64	1.10	6.34	0.93	1.63	5.93	2.56	6.11	5.37	4.17	39.34
1987	4.65	7.33	5.01	2.30	1.31	6.68	3.58	2.79	3.33	0.37	2.81	4.62	44.78
1988	3.91	1.79	3.67	3.41	1.96	3.25	2.18	3.93	4.57	3.38	4.26	1.90	38.21
1989	1.51	4.93	4.48	3.15	3.64	6.00	5.11	4.71	5.42	3.10	3.74	4.76	50.55
1990	4.37	5.97	6.67	2.22	2.70	0.90	3.61	6.21	2.12	9.45	1.93	3.26	49.41
1991	4.72	2.24	5.82	5.65	6.37	1.72	5.74	9.02	1.44	0.24	1.39	2.90	47.25
1992	2.50	6.12	5.45	4.81	5.03	4.97	2.66	5.54	4.30	6.27	7.85	5.08	60.58
1993	7.19	3.56	10.27	2.91	3.08	0.17	0.75	0.87	1.71	2.07	3.73	2.94	39.25
Record Mean	4.10	4.20	5.57	3.76	4.54	4.25	4.57	3.97	3.91	4.01	3.68	3.92	50.48

AVERAGE TEMPERATURE (deg. F)

GREENVILLE-SPARTANBURG (GREER), SOUTH CAROLINA

YEAR	JAN	FEB	MAR	APR	MAY	JUNE	JULY	AUG	SEP	OCT	NOV	DEC	ANNUAL
1964	41.7	40.7	51.9	61.2	71.3	77.6	77.6	76.7	72.4	57.8	55.9	45.3	60.8
1965	42.8	44.2	47.8	61.7	73.1	72.7	77.5	78.1	72.4	59.6	52.3	45.0	60.6
1966	36.5	42.8	50.1	59.1	66.9	74.2	77.9	75.7	69.9	58.7	51.6	43.0	58.8
1967	44.0	41.0	56.5	64.6	66.1	72.6	75.2	74.2	66.8	59.7	48.2	47.2	59.7
1968	39.0	38.3	53.3	60.2	65.4	74.2	77.0	78.7	69.8	60.9	49.5	39.0	58.8
1969	39.5	40.9	45.9	60.7	67.6	76.3	80.6	74.6	69.8	60.9	48.8	41.4	58.9
1970	35.4	43.3	51.7	63.3	68.8	75.2	79.3	76.9	74.8	63.0	48.7	44.7	60.5
1971	39.5	41.9	46.3	59.2	65.3	76.2	76.7	75.8	72.9	64.9	48.9	50.4	59.8
1972	44.6	40.9	50.7	59.5	64.9	70.9	77.0	76.3	73.1	59.2	49.3	48.1	59.5
1973	41.1	42.0	54.5	57.8	65.8	74.6	78.2	76.9	74.5	61.8	51.9	41.8	60.1
1974	51.4	43.9	55.9	58.8	68.0	72.2	78.0	77.2	69.9	58.8	49.6	43.4	60.6
1975	45.8	46.8	48.5	58.7	70.5	74.4	76.5	78.6	70.0	62.6	53.0	42.7	60.7
1976	38.0	50.6	54.0	59.2	63.3	71.4	77.2	75.3	68.6	55.6	44.6	39.9	58.1
1977	30.7	42.4	54.5	62.4	71.5	77.5	81.9	78.7	74.4	59.9	54.8	42.8	61.0
1978	35.3	36.7	49.2	60.8	65.5	75.4	77.8	77.9	73.9	60.4	55.6	43.5	59.3
1979	37.3	39.2	54.2	61.4	69.6	72.0	74.7	76.6	70.5	59.2	52.7	45.4	59.4
1980	44.0	39.3	47.6	59.1	68.2	74.5	81.5	79.4	73.6	57.4	49.3	43.4	59.8
1981	38.1	43.3	49.1	64.2	66.3	80.3	80.1	75.4	69.3	58.0	50.8	39.3	59.5
1982	36.1	45.8	53.3	56.7	70.6	74.2	77.6	75.2	69.8	61.0	52.7	48.3	60.1
1983	39.2	42.1	51.4	54.7	66.3	73.1	79.7	79.5	69.0	59.4	49.2	39.2	58.6
1984	38.8	45.6	50.3	56.2	66.9	77.7	75.6	76.3	69.0	67.5	47.8	50.1	60.2
1985	36.2	44.6	54.2	61.8	67.9	76.7	76.4	75.1	69.6	64.3	58.6	38.9	60.4
1986	39.6	47.8	52.4	61.6	68.9	78.6	83.1	76.0	72.7	61.9	52.9	42.8	61.5
1987	39.5	43.0	50.5	58.3	71.1	76.6	80.4	79.8	71.4	55.2	52.9	45.5	60.4
1988	36.0	42.0	51.5	59.3	67.2	75.4	78.5	80.4	71.0	55.3	52.3	43.0	59.3
1989	45.8	45.4	52.9	58.6	64.4	75.5	78.3	76.8	70.3	61.6	50.8	36.9	59.8
1990	46.6	50.6	54.6	59.0	67.1	76.3	78.7	78.5	72.1	62.1	54.0	48.0	62.3
1991	41.8	46.4	53.5	62.1	71.5	75.2	80.5	76.7	71.8	61.6	49.3	47.4	61.5
1992	44.0	48.0	50.9	59.2	64.0	71.9	79.5	74.2	70.7	57.5	49.7	41.7	59.3
1993	44.2	41.9	46.9	55.6	67.3	76.3	83.2	78.8	71.9	60.3	50.4	41.4	59.9
Record Mean	40.4	43.3	51.5	59.9	67.8	75.0	78.5	77.0	71.2	60.3	51.2	43.5	60.0
Max	50.4	54.2	63.3	72.2	79.1	85.6	88.6	86.7	81.2	71.8	62.3	53.4	70.7
Min	30.3	32.3	39.8	47.6	56.4	64.4	68.4	67.3	61.1	48.8	40.1	33.5	49.2

REFERENCE NOTES

GREENVILLE-SPARTANBURG (GREER), SOUTH CAROLINA

GENERAL
T - TRACE AMOUNT.
BLANK ENTRIES DENOTE MISSING/UNREPORTED DATA.
# INDICATES A STATION OR INSTRUMENT RELOCATION.

SPECIFIC

(a) - LENGTH OF RECORD IN YEARS, ALTHOUGH
INDIVIDUAL MONTHS MAY BE MISSING.
0.* OR * - THE VALUE IS BETWEEN 0.0 AND 0.05
NORMALS - BASED ON THE 1961-1990 RECORD PERIOD.
EXTREMES - DATES ARE THE MOST RECENT OCCURRENCE
WIND DIR.- NUMERALS SHOW TENS OF DEGRESS CLOCKWISE
FROM TRUE NORTH. "00" INDICATES CALM.
RESULTANT DIRECTIONS ARE GIVEN TO WHOLE DEGREES.
BOLD VALUES INDICATE EXTREME VALUES WHICH OCCURRED
AFTER THE ASOS SYSTEM WAS COMMISSIONED.

RECORD = PERIOD OF RECORD
RECORD MEAN PRECIPITATION IS THE MEAN OF ALL DAILY
PRECIPITATION AMOUNTS DURING THE PERIOD OF RECORD.
RECORD MAX(MIN) TEMPERATURE IS THE MEAN OF ALL DAILY
MAX(MIN) TEMPERATURES DURING THE PERIOD OF RECORD.
RECORD MEAN TEMPERATURE IS THE SUM OF THE RECORD
MAX AND RECORD MIN DIVIDED BY 2.
AVERAGE TEMPERATURE IS THE SUM OF THE MEAN DAILY
MAX AND MIN TEMPERATURE DIVIDED BY 2.

EXCEPTIONS

RECORD MEANS ARE THROUGH THE CURRENT YEAR,
BEGINNING IN 1963 FOR TEMPERATURE
1963 FOR PRECIPITATION
1963 FOR SNOWFALL

HEATING DEGREE DAYS Base 65 deg. F GREENVILLE-SPARTANBURG (GREER), SOUTH CAROLINA

SEASON	JULY	AUG	SEP	OCT	NOV	DEC	JAN	FEB	MAR	APR	MAY	JUNE	TOTAL
1964-65	0	2	0	237	281	604	681	572	527	147	0	6	3057
1965-66	0	0	6	186	375	615	877	616	454	194	42	5	3370
1966-67	0	0	6	202	394	675	643	663	276	88	85	14	3046
1967-68	0	0	41	181	498	547	799	769	368	169	54	0	3426
1968-69	0	1	1	168	461	801	785	668	585	142	44	0	3656
1969-70	0	0	19	165	481	725	910	600	405	104	28	0	3437
1970-71	0	0	10	120	480	620	782	643	572	188	73	0	3488
1971-72	0	0	2	74	492	449	625	694	434	205	48	14	3037
1972-73	0	0	4	189	476	519	734	641	323	229	61	0	3176
1973-74	0	0	1	132	387	712	419	584	290	205	54	0	2784
1974-75	0	0	39	205	464	663	590	505	507	227	9	0	3209
1975-76	0	0	24	118	361	683	828	415	345	190	85	15	3064
1976-77	0	0	11	297	605	773	1057	626	319	117	19	0	3824
1977-78	0	0	2	184	322	681	916	787	486	143	88	0	3609
1978-79	0	0	6	152	276	658	851	715	336	129	15	3	3141
1979-80	1	0	9	193	362	600	647	737	532	199	27	0	3307
1980-81	0	0	27	246	462	663	829	602	491	91	50	0	3461
1981-82	0	0	26	229	422	791	887	533	365	249	12	0	3514
1982-83	0	0	24	177	370	515	796	631	414	307	43	1	3278
1983-84	0	0	59	179	467	792	807	553	452	271	67	1	3648
1984-85	0	0	37	32	508	454	884	565	341	141	32	0	2994
1985-86	0	0	38	101	194	800	783	475	392	154	26	0	2963
1986-87	0	7	3	160	361	682	786	610	441	225	15	0	3290
1987-88	0	0	4	296	356	596	892	660	417	180	34	3	3438
1988-89	0	0	2	308	373	674	589	543	386	229	104	0	3208
1989-90	0	0	35	152	420	863	564	395	333	199	41	0	3002
1990-91	0	0	23	131	328	518	713	515	358	125	20	0	2731
1991-92	0	0	22	138	471	547	646	487	433	205	94	5	3048
1992-93	0	0	19	226	455	713	639	640	555	276	34	0	3557
1993-94	0	0	27	179	440	721							

COOLING DEGREE DAYS Base 65 deg. F GREENVILLE-SPARTANBURG (GREER), SOUTH CAROLINA

YEAR	JAN	FEB	MAR	APR	MAY	JUNE	JULY	AUG	SEP	OCT	NOV	DEC	TOTAL
1969	0	0	0	19	133	344	493	305	169	46	0	0	1509
1970	0	0	0	59	154	313	454	378	310	67	0	0	1735
1971	0	0	0	20	88	342	369	345	246	77	15	3	1505
1972	0	0	0	48	49	197	380	356	253	15	9	1	1308
1973	0	0	6	21	92	297	414	375	291	36	0	0	1532
1974	1	0	17	26	153	222	408	386	193	19	10	0	1435
1975	0	0	0	45	185	290	363	427	180	50	7	0	1547
1976	0	2	11	22	40	213	387	322	126	12	0	0	1135
1977	0	0	1	46	226	383	528	431	289	34	20	0	1958
1978	0	0	0	23	109	319	403	404	282	19	0	0	1559
1979	0	0	8	28	164	215	312	368	180	20	1	0	1296
1980	0	0	0	27	135	292	519	454	294	16	0	0	1737
1981	0	0	4	75	96	467	475	327	158	19	0	0	1621
1982	0	1	8	6	193	282	396	324	174	58	5	3	1450
1983	0	0	1	3	91	252	465	457	185	15	0	0	1469
1984	0	0	0	13	132	387	335	356	164	116	2	0	1505
1985	0	0	13	50	132	360	359	320	183	88	9	0	1514
1986	0	0	8	62	154	416	567	355	237	74	8	0	1881
1987	0	0	0	33	213	355	485	465	200	0	0	1	1752
1988	0	0	2	13	109	320	423	481	187	14	0	0	1549
1989	0	1	16	44	94	321	419	371	200	57	1	0	1524
1990	0	0	18	25	108	346	431	429	242	46	4	0	1649
1991	0	0	8	45	228	315	489	369	235	42	6	9	1746
1992	0	0	1	37	72	218	458	292	196	0	0	0	1274
1993	0	0	0	0	113	344	572	434	242	38	10	0	1753

SNOWFALL (inches) GREENVILLE-SPARTANBURG (GREER), SOUTH CAROLINA

SEASON	JULY	AUG	SEP	OCT	NOV	DEC	JAN	FEB	MAR	APR	MAY	JUNE	TOTAL
1964-65	0.0	0.0	0.0	0.0	0.0	0.0	7.7	1.0	1.3	0.0	0.0	0.0	10.0
1965-66	0.0	0.0	0.0	0.0	0.0	0.0	9.1	3.4	0.0	0.0	0.0	0.0	12.5
1966-67	0.0	0.0	0.0	0.0	0.0	T	0.5	1.5	0.0	0.0	0.0	0.0	2.0
1967-68	0.0	0.0	0.0	0.0	0.0	T	0.9	1.8	0.0	0.0	0.0	0.0	2.7
1968-69	0.0	0.0	0.0	0.0	1.9	T	T	6.9	1.9	0.0	0.0	0.0	10.7
1969-70	0.0	0.0	0.0	0.0	0.0	T	2.0	0.0	0.0	0.0	0.0	0.0	2.0
1970-71	0.0	0.0	0.0	0.0	0.0	2.9	T	1.0	6.6	0.0	0.0	0.0	10.5
1971-72	0.0	0.0	0.0	0.0	T	11.4	0.0	3.9	1.8	0.0	0.0	0.0	17.1
1972-73	0.0	0.0	0.0	0.0	T	0.0	4.6	0.0	0.0	0.0	0.0	0.0	4.6
1973-74	0.0	0.0	0.0	0.0	0.0	T	0.0	0.0	0.4	0.0	0.0	0.0	0.4
1974-75	0.0	0.0	0.0	0.0	0.0	T	T	1.3	T	0.0	0.0	0.0	1.3
1975-76	0.0	0.0	0.0	0.0	T	T	T	0.0	0.0	0.0	0.0	0.0	T
1976-77	0.0	0.0	0.0	0.0	0.1	0.0	6.9	0.0	0.0	0.0	0.0	0.0	7.0
1977-78	0.0	0.0	0.0	0.0	0.0	T	T	1.1	2.8	0.0	0.0	0.0	3.9
1978-79	0.0	0.0	0.0	0.0	0.0	0.0	0.9	12.3	0.0	0.0	0.0	0.0	13.2
1979-80	0.0	0.0	0.0	0.0	0.0	0.0	3.2	7.6	3.6	0.0	0.0	0.0	14.4
1980-81	0.0	0.0	0.0	0.0	0.0	T	1.0	0.0	T	0.0	0.0	0.0	1.0
1981-82	0.0	0.0	0.0	0.0	0.0	T	7.0	5.1	0.0	0.0	0.0	0.0	12.1
1982-83	0.0	0.0	0.0	0.0	0.0	0.0	4.4	3.0	9.3	0.1	0.0	0.0	16.8
1983-84	0.0	0.0	0.0	0.0	0.0	T	T	0.6	0.0	0.0	0.0	0.0	0.6
1984-85	0.0	0.0	0.0	0.0	0.0	0.0	1.2	0.3	0.0	0.0	0.0	0.0	1.5
1985-86	0.0	0.0	0.0	0.0	0.0	0.0	T	1.4	T	0.0	0.0	0.0	1.4
1986-87	0.0	0.0	0.0	0.0	T	0.0	10.8	0.9	T	0.3	0.0	0.0	12.0
1987-88	0.0	0.0	0.0	0.0	0.5	T	12.0	0.0	0.0	0.0	0.0	0.0	12.5
1988-89	0.0	0.0	0.0	0.0	0.0	0.0	T	2.5	0.0	0.0	0.0	0.0	2.5
1989-90	T	0.0	0.0	0.0	T	0.9	0.0	0.0	0.0	T	0.0	0.0	0.9
1990-91	0.0	0.0	0.0	0.0	0.0	T	3.8	0.0	0.0	0.0	0.0	0.0	3.8
1991-92	0.0	0.0	0.0	0.0	T	T	T	T	0.0	0.0	T	0.0	T
1992-93	0.0	0.0	0.0	0.0	0.0	T	0.0	0.4	9.8	0.0	0.0	0.0	10.2
1993-94	0.0	0.0	0.0	0.0	0.0	0.7							
Record Mean	T	0.0	0.0	0.0	0.1	0.6	2.5	1.9	1.2	T	T	0.0	6.3

ABERDEEN, SOUTH DAKOTA

Aberdeen is located in the northeast quarter of South Dakota, approximately 200 miles south of the geographical center of the North American continent. The surrounding area, extensively cultivated, is the bed of glacial Lake Dakota, which is by far the largest flat area in South Dakota. The lake bed slopes gently to the south. The elevation of Aberdeen at the northern end of the lake bed is 1,300 feet. The elevation at the southern end, some 30 miles distant is 1,280 feet. Low hills rim the area on the east and west. These hills effect ceilings, visibility, and precipitation, which are a hazard to private aircraft operating in the area during periods of marginal weather. Principal drainage for the area is through the southward flowing, meandering James River with its associated meandering rivers and creeks.

Located near the center of the North American land mass, the climate is continental with distinct seasons. Frequent and rapid weather changes occur during all seasons of the year as migratory storms sweep through the area. The winters are cold and dry. Sub-zero minimum temperatures may set in as early as late November, although temperatures of zero and below are generally not recorded until mid-December. Lowest temperatures of the winter generally occur in the period from mid-January to mid-February. During the coldest periods the days are generally sunny with light winds, and these conditions partially moderate the discomfort experienced at such low temperatures. Some days of the winter will be extremely unpleasant with temperatures near or below zero and brisk winds. Heavy snowfalls rarely occur during the first two-thirds of the winter season, with heaviest snowfalls developing during late February and early March as temperatures moderate. Blizzards are infrequent, many winters will pass without a single occurrence of this type of weather phenomenon. However, difficult driving conditions occur several times during most winters during periods of weather termed ground blizzards by residents of the state. These ground blizzards may develop at wind speeds as low as 12 mph, depending on the condition of the snow pack, wind direction with respect to the highway, etc. Depending on the wind speed, the ground blizzard may restrict visibility only a few inches to as much as 5 feet above the highway surface, completely obscuring the highway surface, but with excellent visibility prevailing above the low blanket of wind-driven snow.

Spring is a very short and transitional period, the shortest season of the four distinct seasons, and one marked by very rapid weather changes. Cool to quite cold nights prevail into mid-May, although afternoon temperatures may be quite warm, as high as the mid-80s. By mid-May temperatures below the freezing point rarely occur and frost is rarely experienced after the end of May. Precipitation increases markedly during the spring, 42 percent of the total annual precipitation normally being recorded in the three month period from April through June.

Summers are pleasant with a maximum of sunshine, warm days, and generally cool and comfortable nights. Temperatures of 100 degrees or above may occur several times during the summer season, but low humidities, brisk winds during the heat of the day, and rapid cooling during the evening hours, which generally occur during the periods of elevated temperatures, markedly moderate the physical discomfort normally experienced at these high temperatures. Thunderstorms occur frequently. In June and August, thunderstorms are more likely to occur during the early evening and nighttime hours. During July, thunderstorms are approximately equally distributed throughout the 24 hours of the day. Hail, occurring in connection with thunderstorm activity, is most likely in late May and early June.

Autumn is most pleasant with mild days, cool nights, ample sunshine, and declining occurrences and amounts of precipitation. The first frost may be expected by late September, although it may occur as early as late August. By mid-October, the temperatures during the night will be near or below freezing, and late October temperatures are normally in the 20 degree range. The growing season is about 132 days.

NORMALS, MEANS, AND EXTREMES

ABERDEEN, SOUTH DAKOTA

LATITUDE: 45 °27'N LONGITUDE: 98 °26' W ELEVATION: FT. GRND 1296 BARO 1315 TIME ZONE: CENTRAL WBAN: 14929

	(a)	JAN	FEB	MAR	APR	MAY	JUNE	JULY	AUG	SEP	OCT	NOV	DEC	YEAR
TEMPERATURE °F:														
Normals														
-Daily Maximum		20.9	26.9	39.8	57.3	69.7	78.8	85.9	84.4	72.9	60.4	40.5	25.4	55.2
-Daily Minimum		-0.6	6.5	19.8	33.0	44.5	54.3	59.6	56.8	46.1	34.2	19.9	5.3	31.6
-Monthly		10.1	16.7	29.8	45.2	57.1	66.6	72.8	70.6	59.6	47.3	30.3	15.3	43.5
Extremes														
-Record Highest	32	60	62	82	98	96	108	110	112	103	96	78	62	112
-Year		1981	1987	1963	1992	1992	1988	1966	1965	1970	1963	1975	1969	AUG 1965
-Record Lowest	32	-35	-37	-29	-2	19	33	39	32	20	8	-27	-39	-39
-Year		1972	1971	1962	1975	1961	1964	1971	1987	1965	1991	1964	1967	DEC 1967
NORMAL DEGREE DAYS:														
Heating (base 65°F)		1702	1352	1091	594	269	68	15	29	195	549	1041	1541	8446
Cooling (base 65°F)		0	0	0	0	24	116	257	203	33	0	0	0	633
% OF POSSIBLE SUNSHINE														
MEAN SKY COVER (tenths)														
Sunrise - Sunset	27	6.5	6.7	6.9	6.6	6.3	5.8	4.9	4.9	5.3	5.9	7.0	6.7	6.1
MEAN NUMBER OF DAYS:														
Sunrise to Sunset														
-Clear	27	7.7	6.6	6.3	7.0	7.4	8.7	11.1	11.7	10.7	9.9	6.0	7.4	100.7
-Partly Cloudy	27	7.6	6.4	7.4	7.3	9.6	9.9	12.7	11.0	8.4	8.0	6.3	6.7	101.2
-Cloudy	27	15.7	15.2	17.4	15.7	14.0	11.4	7.2	8.3	10.9	13.1	17.7	16.9	163.4
Precipitation														
.01 inches or more	62	6.4	6.0	7.1	8.2	9.4	10.3	8.6	8.2	6.3	4.9	5.8	6.2	87.4
Snow,Ice pellets,hail														
1.0 inches or more	62	2.4	2.1	2.0	0.9	0.*	0.0	0.0	0.0	0.0	0.3	1.7	1.9	11.4
Thunderstorms	14	0.0	0.0	0.4	1.9	4.9	9.6	9.0	6.1	2.9	1.4	0.0	0.0	36.1
Heavy Fog Visibility														
1/4 mile or less	14	1.6	2.1	2.4	1.4	0.5	0.9	0.9	0.9	1.1	1.1	2.5	2.9	18.3
Temperature °F														
-Maximum														
90° and above	32	0.0	0.0	0.0	0.2	0.5	3.1	9.1	8.3	2.1	0.3	0.0	0.0	23.5
32° and below	32	22.9	17.0	9.3	0.3	0.0	0.0	0.0	0.0	0.0	0.2	8.2	20.6	78.5
-Minimum														
32° and below	32	30.9	27.8	26.8	15.1	3.0	0.0	0.0	0.*	2.2	14.0	27.5	30.8	178.2
0° and below	32	15.5	9.6	2.9	0.*	0.0	0.0	0.0	0.0	0.0	0.0	2.0	11.0	40.9
AVG. STATION PRESS.(mb)	7	971.0	970.8	966.7	967.4	965.7	965.4	967.2	966.8	968.5	968.8	969.6	969.8	968.1
RELATIVE HUMIDITY (%)														
Hour 00	13	73	76	79	75	72	77	75	72	74	74	80	80	76
Hour 06 (Local Time)	25	77	79	83	81	81	85	85	86	85	81	83	81	82
Hour 12	27	70	71	67	54	51	57	53	51	53	54	68	72	60
Hour 18	27	73	72	65	50	47	53	48	46	49	54	71	76	59
PRECIPITATION (inches):														
Water Equivalent														
-Normal		0.37	0.47	1.34	1.95	2.41	3.15	2.75	2.13	1.86	1.12	0.59	0.41	18.55
-Maximum Monthly	62	2.23	2.06	3.45	7.88	7.36	8.88	7.71	6.62	5.31	5.14	2.36	1.86	8.88
-Year		1937	1952	1977	1986	1991	1939	1972	1942	1988	1982	1977	1935	JUN 1939
-Minimum Monthly	62	0.01	0.00	0.04	0.13	0.28	0.37	0.30	0.06	0.05	0.00	T	T	0.00
-Year		1961	1932	1971	1988	1948	1974	1975	1947	1979	1952	1980	1986	OCT 1952
-Maximum in 24 hrs	62	1.12	1.02	3.00	2.28	3.81	5.20	3.46	3.50	3.49	2.38	1.30	0.91	5.20
-Year		1939	1958	1937	1938	1949	1978	1983	1990	1988	1982	1977	1988	JUN 1978
Snow,Ice pellets,hail														
-Maximum Monthly	62	26.2	25.1	27.9	24.4	2.0	T	T	T	0.1	5.5	30.1	18.5	30.1
-Year		1937	1969	1975	1970	1943	1993	1992	1993	1965	1970	1993	1935	NOV 1993
-Maximum in 24 hrs	62	10.0	14.3	13.0	15.0	2.0	T	T	T	0.1	5.0	12.6	9.1	15.0
-Year		1937	1951	1937	1970	1943	1993	1992	1993	1965	1932	1993	1988	APR 1970
WIND:														
Mean Speed (mph)	18	11.3	11.3	12.5	13.0	12.4	10.5	9.5	10.2	10.6	11.1	10.9	10.7	11.2
Prevailing Direction														
Fastest Obs. 1 Min.														
-Direction (!!)	24	34	06	35	36	16	35	32	33	32	35	34	36	32
-Speed (MPH)	24	58	52	52	49	46	45	58	46	41	38	41	40	58
-Year		1967	1967	1966	1968	1965	1990	1973	1987	1977	1987	1982	1972	JUL 1973
Peak Gust														
-Direction (!!)	10	NW	N	NW	S	SE	N	SE	NW	NW	N	N	NW	SE
-Speed (mph)	10	55	56	56	60	62	61	64	63	53	52	54	59	64
-Date		1990	1984	1988	1985	1988	1990	1988	1987	1992	1987	1989	1985	JUL 1988

PRECIPITATION (inches)

ABERDEEN, SOUTH DAKOTA

YEAR	JAN	FEB	MAR	APR	MAY	JUNE	JULY	AUG	SEP	OCT	NOV	DEC	ANNUAL
1964	0.36	0.31	0.89	3.19	2.68	2.69	3.47	4.04	0.56	T	0.24	0.39	18.82
1965	0.15	0.09	0.71	2.93	3.01	2.11	2.22	1.26	3.25	0.37	0.65	0.38	17.13
1966	0.16	0.52	2.34	0.92	0.91	4.53	4.66	2.77	1.25	1.86	0.59	0.03	20.54
1967	0.50	0.42	0.09	3.38	0.58	5.13	1.06	1.40	3.41	0.65	0.15	0.47	17.24
1968	0.15	0.10	2.74	4.81	1.66	3.24	2.20	1.07	2.82	0.60	0.83	0.94	21.16
1969	1.11	1.54	0.36	0.92	3.38	3.37	4.07	0.74	0.38	0.57	0.14	0.71	17.29
1970	0.21	0.17	1.33	3.42	1.49	2.11	2.04	0.33	1.67	1.24	1.39	0.19	15.59
1971	0.35	0.41	0.04	1.86	1.70	4.29	3.75	1.39	2.30	2.80	1.39	0.47	20.75
1972	0.39	0.39	1.02	1.05	4.86	1.80	7.71	0.52	0.25	1.26	0.26	0.88	20.39
1973	0.21	0.15	1.29	1.02	1.00	1.37	1.77	1.17	3.51	1.28	0.66	0.44	13.87
1974	0.05	0.26	0.74	2.10	3.77	0.37	1.72	0.94	0.15	0.54	0.12	0.05	10.81
1975	0.82	0.29	2.71	2.60	2.26	5.28	0.30	1.50	1.73	1.23	0.29	0.30	19.31
1976	0.81	0.52	0.70	1.27	0.52	1.41	0.50	0.66	0.86	0.32	0.01	0.30	7.88
1977	0.34	0.91	3.45	0.90	2.82	1.99	2.16	2.48	4.36	1.24	2.36	0.83	23.84
1978	0.15	0.23	0.46	2.25	4.23	7.30	2.17	4.04	0.60	0.08	0.77	0.11	22.39
1979	1.01	0.73	1.60	2.93	1.93	4.99	2.56	1.24	0.05	1.18	0.02	0.14	18.38
1980	0.51	0.44	0.88	1.15	1.64	2.53	0.80	5.93	0.92	1.44	T	0.14	16.38
1981	0.12	0.20	2.00	0.12	1.60	2.10	3.97	2.91					13.02
1982										5.14	0.59	0.09	
1983	0.16	0.26	2.65	0.69	1.66	3.47	6.46	2.21	1.55	0.81	0.60	0.55	21.07
1984	0.47	0.70	1.94	2.39	1.13	5.65	2.64	2.23	0.84	2.93	0.06	0.61	21.59
1985	0.23	0.08	1.82	0.63	3.41	1.76	2.38	2.71	2.71	0.87	1.60	0.57	18.77
1986	0.43	0.71	0.58	7.88	3.32	2.48	3.78	2.85	2.82	0.19	0.77	T	25.81
1987	0.09	1.12	1.91	0.41	2.01	0.77	2.13	1.87	1.33	0.20	0.79	0.09	12.72
1988	0.35	0.31	0.37	0.13	3.43	0.93	3.14	2.80	5.31	0.11	0.73	1.37	18.98
1989	0.52	0.37	1.46	3.42	1.20	2.05	2.00	3.83	2.23	0.58	0.74	0.15	18.55
1990	0.13	0.39	0.81	1.87	1.41	7.72	1.98	4.85	3.01	0.44	0.11	0.37	23.09
1991	0.11	0.70	0.77	3.70	7.36	4.76	1.32	2.28	0.57	1.06	0.32	0.13	23.08
1992	0.66	0.47	0.54	0.40	0.78	5.61	2.97	1.55	1.63	0.83	1.19	0.20	16.83
1993	0.61	0.49	0.42	1.51	3.11	6.20	7.37	4.42	1.21	0.35	1.88	0.56	28.13
Record Mean	0.52	0.58	1.12	2.04	2.48	3.56	2.65	2.23	1.62	1.08	0.69	0.50	19.07

AVERAGE TEMPERATURE (deg. F)

ABERDEEN, SOUTH DAKOTA

YEAR	JAN	FEB	MAR	APR	MAY	JUNE	JULY	AUG	SEP	OCT	NOV	DEC	ANNUAL
1964	16.3	24.3	23.0	46.4	58.5	66.7	75.1	66.4	56.8	46.7	26.9	7.2	42.9
1965	2.1	11.2	19.3	43.7	56.2	65.1	71.7	69.5	49.7	50.2	29.2	25.4	41.1
1966	-0.1	9.2	34.0	38.9	53.8	67.1	76.4	66.6	59.0	46.7	25.5	19.3	41.4
1967	15.1	11.7	33.1	43.2	52.7	64.2	69.2	68.0	61.7	46.3	31.2	16.5	42.7
1968	10.4	15.8	36.3	44.0	51.3	65.0	69.9	68.2	58.9	46.7	32.4	13.8	42.7
1969	4.4	17.9	16.2	46.4	57.5	59.9	70.0	73.1	63.2	42.6	33.6	18.5	42.0
1970	2.0	15.6	23.1	41.2	56.7	68.2	72.5	72.1	62.8	48.3	30.5	13.8	42.3
1971	5.4	15.2	31.4	47.9	55.6	70.3	68.2	73.0	59.7	48.5	32.2	12.6	43.4
1972	6.3	9.6	29.2	44.8	60.7	67.2	69.5	71.6	60.3	44.2	30.3	10.5	42.0
1973	17.0	24.3	39.5	45.6	57.5	68.2	72.9	75.6	57.9	50.6	27.0	12.2	45.7
1974	6.3	18.2	30.6	46.0	53.9	65.7	77.4	68.1	57.2	50.2	32.5	22.9	44.1
1975	17.1	14.7	24.9	40.6	59.8	68.1	78.3	72.3	58.9	51.9	34.0	17.6	44.9
1976	12.0	26.8	30.5	49.7	57.0	69.9	75.0	75.9	62.1	43.0	26.5	14.6	45.2
1977	1.4	24.2	33.6	51.9	64.9	67.7	74.1	65.7	59.6	46.5	27.2	11.1	44.0
1978	-1.7	6.0	25.3	42.9	58.2	65.6	70.1	70.7	65.1	47.4	24.2	10.8	40.4
1979	-1.2	2.5	24.6	40.7	52.6	65.3	70.5	68.3	63.2	45.9	29.2	25.6	40.6
1980	13.0	14.7	25.6	49.9	59.5	67.0	72.9	68.9	59.5	45.3	36.2	20.0	44.4
1981	19.4	23.9	35.6	49.9	56.3	65.3	74.4	72.1					33.1
1982										46.6	27.7	25.0	
1983	23.1	28.0	33.3	41.5	53.6	65.2	75.2	76.5	60.8	47.3	32.2	-0.6	44.7
1984	15.7	30.3	27.9	47.0	54.4	65.6	71.3	72.7	55.4	48.6	32.5	13.0	44.5
1985	8.5	16.3	36.0	50.3	61.2	61.3	72.6	67.3	56.5	45.3	16.4	7.8	41.6
1986	17.3	12.1	35.8	44.1	57.6	68.0	72.5	66.2	56.7	47.3	26.3	24.2	44.0
1987	22.7	30.5	33.3	51.5	62.3	69.3	75.7	67.8	61.1	43.3	35.1	24.5	48.1
1988	9.9	14.6	33.0	45.4	63.5	75.0	76.4	72.9	59.2	44.4	29.1	18.8	45.2
1989	15.6	5.0	22.9	45.1	57.4	64.7	75.3	72.0	59.8	46.8	28.6	9.3	41.9
1990	25.3	21.3	35.8	44.0	54.9	66.2	69.7	70.9	63.6	45.8	34.7	12.2	45.4
1991	11.3	27.8	33.8	48.3	60.8	70.8	72.2	72.7	60.3	44.8	24.9	24.1	46.0
1992	21.6	28.8	36.2	44.1	59.5	63.7	64.3	64.3	58.5	46.2	29.6	14.7	44.3
1993	9.7	12.0	31.1	44.7	57.6	63.0	68.6	70.2	56.1	45.6	27.9	18.3	42.1
Record Mean	10.6	16.3	29.2	45.0	57.3	66.7	73.0	71.0	60.0	47.6	30.1	16.8	43.6
Max	21.3	26.7	39.2	57.3	70.2	79.1	86.5	84.9	73.8	61.0	40.4	26.7	55.6
Min	0.0	5.9	19.1	32.7	44.3	54.2	59.4	57.0	46.1	34.2	19.7	6.9	31.6

REFERENCE NOTES

ABERDEEN, SOUTH DAKOTA

GENERAL
T - TRACE AMOUNT.
BLANK ENTRIES DENOTE MISSING/UNREPORTED DATA.
# INDICATES A STATION OR INSTRUMENT RELOCATION.

SPECIFIC

(#) - LENGTH OF RECORD IN YEARS, ALTHOUGH INDIVIDUAL MONTHS MAY BE MISSING.
0.* OR * - THE VALUE IS BETWEEN 0.0 AND 0.05
NORMALS - BASED ON THE 1961-1990 RECORD PERIOD.
EXTREMES - DATES ARE THE MOST RECENT OCCURRENCE
WIND DIR.- NUMERALS SHOW TENS OF DEGRESS CLOCKWISE FROM TRUE NORTH. "00" INDICATES CALM.
RESULTANT DIRECTIONS ARE GIVEN TO WHOLE DEGREES.
BOLD VALUES INDICATE EXTREME VALUES WHICH OCCURRED AFTER THE ASOS SYSTEM WAS COMMISSIONED.

RECORD = PERIOD OF RECORD
RECORD MEAN PRECIPITATION IS THE MEAN OF ALL DAILY PRECIPITATION AMOUNTS DURING THE PERIOD OF RECORD.
RECORD MAX(MIN) TEMPERATURE IS THE MEAN OF ALL DAILY MAX(MIN) TEMPERATURES DURING THE PERIOD OF RECORD.
RECORD MEAN TEMPERATURE IS THE SUM OF THE RECORD MAX AND RECORD MIN DIVIDED BY 2.
AVERAGE TEMPERATURE IS THE SUM OF THE MEAN DAILY MAX AND MIN TEMPERATURE DIVIDED BY 2.

EXCEPTIONS

1. PRIOR TO 1948, MAXIMUM 24-HOUR PRECIPITATION IS FOR OBSERVATIONAL DAY (A 24-HOUR PERIOD OTHER THAN MIDNIGHT TO MIDNIGHT) INSTEAD OF ANY CONSECUTIVE 24 HOURS.
2. MEAN WIND SPEED, THUNDERSTORMS, AND HEAVY FOG ARE THROUGH 1977.
3. TOTAL BREAK IN RECORD SEPTEMBER 1981 THROUGH SEPTEMBER 1982.
4. FASTEST OBSERVED WINDS ARE THROUGH JUNE 1978 AND OCTOBER 1982 TO DATE.

RECORD MEANS ARE THROUGH THE CURRENT YEAR, BEGINNING IN 1930 FOR TEMPERATURE
1930 FOR PRECIPITATION
1930 FOR SNOWFALL

HEATING DEGREE DAYS Base 65 deg. F — ABERDEEN, SOUTH DAKOTA

SEASON	JULY	AUG	SEP	OCT	NOV	DEC	JAN	FEB	MAR	APR	MAY	JUNE	TOTAL
1964-65	2	77	274	560	1139	1789	1951	1505	1413	635	279	52	9676
1965-66	5	38	456	447	1069	1217	2017	1560	950	774	355	67	8955
1966-67	0	60	212	560	1181	1409	1544	1491	982	647	395	56	8537
1967-68	35	39	134	575	1008	1500	1690	1421	883	623	423	77	8408
1968-69	21	60	207	562	970	1580	1882	1312	1505	550	273	169	9091
1969-70	7	4	126	688	935	1438	1954	1379	1294	706	266	43	8840
1970-71	11	9	187	513	1029	1583	1846	1388	1035	515	287	20	8423
1971-72	32	15	220	504	977	1621	1821	1603	1100	597	201	39	8730
1972-73	19	35	176	638	1033	1688	1484	1135	783	580	238	27	7836
1973-74	2	0	239	442	1132	1630	1819	1305	1058	563	350	73	8613
1974-75	0	43	250	455	967	1302	1477	1401	1234	725	184	44	8082
1975-76	2	6	208	411	921	1465	1639	1101	1063	454	245	39	7554
1976-77	3	2	165	676	1149	1557	1973	1135	968	395	83	23	8129
1977-78	2	39	172	569	1128	1669	2070	1650	1228	659	226	74	9486
1978-79	13	23	142	540	1221	1678	2051	1750	1245	727	389	68	9847
1979-80	6	41	124	585	1068	1215	1608	1452	1213	463	221	31	8027
1980-81	1	24	192	601	860	1388	1408	1146	907	447	273	43	7290
1981-82	13	1	0	0	0	0							
1982-83				562	1111	1232	1291	1032	976	697	351	89	
1983-84	4	0	198	540	975	2030	1522	1001	1147	533	332	44	8326
1984-85	12	15	312	514	967	1608	1747	1361	891	444	158	145	8174
1985-86	2	42	292	608	1453	1770	1473	1477	897	622	245	32	8913
1986-87	0	59	246	540	1152	1258	1303	961	978	409	134	31	7071
1987-88	4	60	140	666	892	1247	1707	1461	983	584	120	0	7864
1988-89	3	30	193	631	1069	1426	1527	1682	1298	597	244	81	8781
1989-90	0	5	203	561	1085	1724	1222	1221	900	639	308	68	7936
1990-91	7	12	155	586	901	1633	1661	1035	961	491	205	6	7653
1991-92	5	2	206	619	1195	1264	1338	1041	884	631	224	97	7506
1992-93	61	97	206	582	1054	1556	1713	1481	1043	602	240	110	8745
1993-94	17	18	277	599	1109	1441							

COOLING DEGREE DAYS Base 65 deg. F — ABERDEEN, SOUTH DAKOTA

YEAR	JAN	FEB	MAR	APR	MAY	JUNE	JULY	AUG	SEP	OCT	NOV	DEC	TOTAL
1969	0	0	0	0	48	17	170	265	75	0	0	0	575
1970	0	0	0	0	14	141	253	235	127	3	0	0	773
1971	0	0	0	7	1	186	138	269	67	0	0	0	668
1972	0	0	0	0	73	112	166	245	43	0	0	0	639
1973	0	0	0	2	10	129	252	335	33	0	0	0	761
1974	0	0	0	0	13	99	393	145	20	3	0	0	673
1975	0	0	0	0	27	142	420	241	34	13	0	0	877
1976	0	0	0	0	5	193	322	345	85	1	0	0	951
1977	0	0	0	9	85	109	291	67	19	0	0	0	580
1978	0	0	0	0	21	100	179	206	152	0	0	0	658
1979	0	0	0	0	11	83	183	150	75	0	0	0	502
1980	0	0	0	16	55	98	251	150	33	0	0	0	603
1981	0	0	0	2	11	57	314	230			0	0	614
1982										0	0	0	
1983	0	0	0	0	3	100	329	363	79	0	0	0	874
1984	0	0	0	0	12	70	210	259	31	15	0	0	597
1985	0	0	0	9	48	41	244	121	46	0	0	0	509
1986	0	0	0	1	22	126	240	105	6	0	0	0	500
1987	0	0	0	11	57	166	341	153	30	0	0	0	758
1988	0	0	0	3	81	308	362	279	28	0	0	0	1061
1989	0	0	0	5	14	77	326	228	54	2	0	0	706
1990	0	0	0	16	3	108	159	204	120	0	0	0	610
1991	0	0	0	0	82	187	238	244	72	3	0	0	826
1992	0	0	0	11	59	65	47	85	17	5	0	0	289
1993	0	0	0	0	19	54	137	187	16	2	0	0	415

SNOWFALL (inches) — ABERDEEN, SOUTH DAKOTA

SEASON	JULY	AUG	SEP	OCT	NOV	DEC	JAN	FEB	MAR	APR	MAY	JUNE	TOTAL
1964-65	0.0	0.0	0.0	T	4.5	8.6	4.1	1.9	10.4	10.9	T	0.0	40.4
1965-66	0.0	0.0	0.1	0.0	5.8	0.7	3.3	5.7	18.9	7.2	0.3	0.0	42.0
1966-67	0.0	0.0	0.0	T	4.5	0.7	7.9	8.8	1.2	1.3	0.5	0.0	24.9
1967-68	0.0	0.0	0.0	1.4	1.3	9.1	3.1	1.4	3.2	15.5	T	0.0	35.0
1968-69	0.0	0.0	0.0	0.3	3.7	12.7	13.7	25.1	2.8	T	0.0	0.0	58.3
1969-70	0.0	0.0	0.0	0.3	1.1	12.0	4.8	3.7	6.8	24.4	0.1	0.0	53.2
1970-71	0.0	0.0	0.0	5.5	8.9	3.9	10.2	6.2	1.3	3.5	0.0	0.0	39.5
1971-72	0.0	0.0	T	2.9	9.1	9.9	8.2	7.0	6.0	0.4	0.0	0.0	43.5
1972-73	0.0	0.0	T	1.3	4.2	15.0	4.3	2.0	T	2.7	0.0	0.0	29.5
1973-74	0.0	0.0	0.0	T	4.2	10.7	1.6	6.6	4.2	T	0.0	0.0	27.3
1974-75	0.0	0.0	0.0	0.0	1.1	1.1	14.6	4.7	27.9	0.5	0.0	0.0	49.9
1975-76	0.0	0.0	0.0	1.9	1.5	5.1	13.6	8.5	6.8	0.4	0.3	0.0	38.1
1976-77	0.0	0.0	0.0	2.4	0.2	6.3	5.3	11.0	9.0	1.0	0.0	0.0	35.2
1977-78	0.0	0.0	0.0	0.2	20.2	12.5	2.9	5.0	5.8	1.5	0.0	0.0	48.1
1978-79	0.0	0.0	0.0	0.0	8.8	2.1	15.1	10.5	7.2	7.0	0.9	0.0	51.6
1979-80	0.0	0.0	0.0	3.2	0.7	1.0	5.3	6.4	10.7	1.5	0.0	0.0	28.8
1980-81	0.0	0.0	0.0	0.4	0.0	2.3	1.8	2.2	1.6	T	0.0	0.0	8.3
1981-82	0.0	0.0	0.0	0.0	0.0	0.0							
1982-83				0.0	0.0	0.0	0.9	2.2	7.9	2.3	0.0	0.0	
1983-84	0.0	0.0	0.0	T	5.5	5.2	2.0	3.8	17.3	0.1	0.0	0.0	33.9
1984-85	0.0	0.0	T	1.0	0.2	5.6	3.7	1.0	18.4	0.1	0.0	0.0	30.0
1985-86	0.0	0.0	0.0	T	22.4	9.6	4.0	10.5	1.9	9.2	0.0	0.0	57.6
1986-87	0.0	0.0	0.0	0.0	3.4	0.1	1.1	10.5	5.7	T	0.0	0.0	20.8
1987-88	0.0	0.0	0.0	T	5.6	0.9	4.4	4.2	2.5	1.0	0.0	0.0	18.6
1988-89					5.0	16.4	5.1	5.8	13.1	1.6	T	0.0	
1989-90	T	T	0.0	0.1	6.4	1.5	0.8	9.2	4.8	4.7	T	T	27.5
1990-91	0.0	0.0	0.0	1.0	T	7.0	2.2	13.0	5.1	5.5	0.0	0.0	33.8
1991-92	0.0	0.0	0.0	0.4	4.0	0.3	7.9	2.4	0.8	1.1	0.0	0.0	16.9
1992-93	T	0.0	T	3.6	9.3	2.7	10.5	7.1	2.5	3.2	0.0	T	38.9
1993-94	0.0	T	0.0	T	30.1	9.7							
Record Mean	T	T	T	0.7	5.2	6.1	6.6	6.8	6.7	3.5	0.2	T	35.8

HURON, SOUTH DAKOTA

Located on the west bank of the James River at about the middle of the river valley, Huron has a climate that can be classified as continental with frequent daily temperature fluctuations and distinct seasons. The seasons have varied quite markedly from year to year. Agriculture is the main industry in the area.

Winter is characteristically cold and dry with storms of short duration. Normal temperatures for the season are in the middle teens and precipitation is mainly in the form of snow. Seasonal snowfall has varied from under 9 inches to over 75 inches. Wintertime storms of blizzard proportions are infrequent, but they do occur. These blizzards are characterized by strong winds, low temperatures, snow, and very poor visibility. Many mild days can be expected during the winter since about 39 percent of the daily maximum temperatures are above the freezing mark.

Spring is characterized by marked upward trends in both precipitation and temperature with the moisture amounts increasing some three to four fold over winter. Nearly one–third, over 5 inches, of the annual total precipitation usually falls during the spring months. In early spring some of the precipitation falls as snow. As a consequence, the month of March has a slightly higher snowfall average than any of the winter months.

Based on the 1951–1980 period, the average first occurrence of 32 degrees Fahrenheit in the fall is September 29 and the average last occurrence in the spring is May 11.

Summers are hot but not extreme. Temperatures of 100 degrees or higher usually occur three or four times a year, but nights are normally cool and comfortable. Summertime precipitation is mainly in the form of showers and thunderstorms. Hail occurs about three times a year in the summertime thunderstorms. Sunshine also reaches its maximum during the summer months, when the sun shines during more than 70 percent of the daylight hours.

Autumn, with a relatively slow drop in temperature and steadily lessening amounts of rainfall, is a delightful season with mildly warm days, cool nights, and plentiful sunshine.

The terrain around Huron is exceptionally level and flat. Even the James River has a slope of only 4 to 6 inches per mile. However, floods in the Huron area have generally been of minor importance because the flood plain is agricultural land, and is not developed for homes or businesses. Heavy rains from summertime thunderstorms may, at times, cause some local flooding problems in the city. Moderate to fresh winds occur quite frequently during the daytime in all seasons of the year. Winds are normally from the north in the winter and from the south in the summer. The unusually level terrain helps to accentuate this tendency toward windy days.

NORMALS, MEANS, AND EXTREMES

HURON, SOUTH DAKOTA

LATITUDE: 44°23'N LONGITUDE: 98°13'W ELEVATION: FT. GRND 1281 BARO 1285 TIME ZONE: CENTRAL WBAN: 14936

	(a)	JAN	FEB	MAR	APR	MAY	JUNE	JULY	AUG	SEP	OCT	NOV	DEC	YEAR
TEMPERATURE °F:														
Normals														
-Daily Maximum		24.1	29.7	42.1	58.6	70.4	80.3	87.1	84.8	74.2	61.5	43.0	28.3	57.0
-Daily Minimum		2.3	9.1	21.7	34.0	44.8	55.5	61.7	58.8	47.3	35.4	21.8	7.8	33.4
-Monthly		13.2	19.4	31.9	46.3	57.6	67.9	74.4	71.8	60.8	48.5	32.4	18.1	45.2
Extremes														
-Record Highest	52	63	71	89	97	99	109	112	110	106	102	78	62	112
-Year		1944	1958	1943	1980	1959	1988	1966	1965	1970	1963	1990	1969	JUL 1966
-Record Lowest	52	-37	-39	-24	-2	17	32	37	36	19	9	-21	-30	-39
-Year		1988	1962	1960	1975	1976	1964	1971	1964	1974	1976	1964	1990	FEB 1962
NORMAL DEGREE DAYS:														
Heating (base 65°F)		1606	1277	1026	561	256	48	10	22	173	512	978	1454	7923
Cooling (base 65°F)		0	0	0	0	27	135	302	232	47	0	0	0	743
% OF POSSIBLE SUNSHINE	52	57	60	59	62	66	70	77	75	69	64	51	50	63
MEAN SKY COVER (tenths)														
Sunrise - Sunset	50	6.5	6.6	7.0	6.6	6.2	5.6	4.7	4.7	5.0	5.5	6.7	6.7	6.0
MEAN NUMBER OF DAYS:														
Sunrise to Sunset														
-Clear	54	7.9	6.6	5.9	6.6	7.5	8.8	12.0	12.4	11.8	10.7	6.8	7.5	104.4
-Partly Cloudy	54	7.7	7.5	7.6	8.6	9.7	11.0	12.2	11.2	8.3	8.5	7.4	7.2	107.0
-Cloudy	54	15.4	14.2	17.5	14.8	13.8	10.2	6.8	7.4	9.9	11.8	15.8	16.3	153.8
Precipitation														
.01 inches or more	54	5.9	6.7	8.3	9.0	10.1	10.7	8.8	8.3	7.0	5.5	5.9	5.7	91.9
Snow, Ice pellets, hail														
1.0 inches or more	50	2.1	2.7	2.6	0.8	0.1	0.0	0.0	0.0	0.0	0.2	1.6	1.9	12.0
Thunderstorms	54	0.*	0.*	0.4	2.4	5.5	8.7	9.3	7.9	4.0	1.4	0.1	0.*	39.7
Heavy Fog Visibility														
1/4 mile or less	54	1.7	1.9	2.1	0.7	0.7	0.4	0.8	1.1	1.1	1.1	1.4	2.1	15.2
Temperature °F														
-Maximum														
90° and above	34	0.0	0.0	0.0	0.3	0.6	4.3	11.5	9.4	2.9	0.2	0.0	0.0	29.1
32° and below	34	20.2	15.1	7.0	0.3	0.0	0.0	0.0	0.0	0.0	0.2	6.6	18.0	67.4
-Minimum														
32° and below	34	30.8	27.5	25.9	13.5	2.6	0.*	0.0	0.0	1.5	12.1	26.2	30.8	170.8
0° and below	34	13.7	8.5	2.3	0.*	0.0	0.0	0.0	0.0	0.0	0.0	1.2	8.7	34.4
AVG. STATION PRESS. (mb)	21	971.2	971.6	968.0	967.1	966.3	965.9	967.4	968.0	968.7	969.6	969.5	970.7	968.7
RELATIVE HUMIDITY (%)														
Hour 00	34	75	79	80	76	76	80	78	79	79	75	79	78	78
Hour 06 (Local Time)	34	75	79	83	83	84	86	87	89	87	81	82	79	83
Hour 12	34	67	68	65	54	54	56	53	53	54	53	64	69	59
Hour 18	34	71	71	64	51	51	53	48	48	52	56	69	74	59
PRECIPITATION (inches):														
Water Equivalent														
-Normal		0.41	0.68	1.66	2.09	2.87	3.35	2.67	1.97	1.72	1.47	0.72	0.47	20.08
-Maximum Monthly	54	1.93	3.87	5.89	5.59	7.69	11.49	6.69	5.47	4.03	6.44	3.01	1.53	11.49
-Year		1975	1962	1977	1991	1962	1984	1993	1956	1986	1946	1947	1968	JUN 1984
-Minimum Monthly	54	0.02	0.03	0.12	0.24	0.33	0.67	0.42	0.14	0.10	T	T	T	T
-Year		1990	1968	1939	1981	1992	1950	1941	1976	1952	1952	1990	1986	NOV 1990
-Maximum in 24 hrs	54	1.58	2.16	2.82	2.40	3.74	5.48	4.98	4.15	2.60	4.20	2.00	1.24	5.48
-Year		1944	1962	1985	1970	1990	1967	1986	1956	1950	1961	1972	1949	JUN 1967
Snow, Ice pellets, hail														
-Maximum Monthly	54	27.7	39.9	33.9	12.8	1.6	T	T	T	T	7.3	32.7	26.0	39.9
-Year		1975	1962	1975	1970	1954	1993	1992	1993	1965	1976	1985	1968	FEB 1962
-Maximum in 24 hrs	54	12.3	17.5	18.3	8.9	1.6	T	T	T	T	5.0	10.0	10.7	18.3
-Year		1982	1962	1985	1957	1954	1993	1992	1993	1965	1970	1953	1955	MAR 1985
WIND:														
Mean Speed (mph)	54	11.4	11.3	12.4	13.4	12.3	11.2	10.3	10.5	11.2	11.2	11.6	11.1	11.5
Prevailing Direction through 1963		SSE	NW	SSE	SSE	SSE	SSE	SSE	SSE	SSE	SSE	SSE	SSE	SSE
Fastest Mile														
-Direction (!!)	53	NW	NW	NW	SE	NW	SE	NW	NW	NW	W	NW	NW	NW
-Speed (MPH)	53	57	56	68	73	70	65	77	72	64	72	73	59	77
-Year		1967	1955	1953	1955	1942	1960	1957	1968	1951	1949	1954	1963	JUL 1957
Peak Gust														
-Direction (!!)	10	W	NW	NW	S	SE	NW	NW	NW	NE	NW	W	NW	NW
-Speed (mph)	10	53	52	55	64	60	63	82	76	54	52	58	60	82
-Date		1986	1987	1993	1985	1988	1993	1992	1987	1984	1990	1986	1992	JUL 1992

PRECIPITATION (inches)

HURON, SOUTH DAKOTA

YEAR	JAN	FEB	MAR	APR	MAY	JUNE	JULY	AUG	SEP	OCT	NOV	DEC	ANNUAL
1964	0.14	0.20	1.06	2.78	1.29	1.61	0.69	1.28	0.86	0.06	0.19	0.69	10.85
1965	0.22	0.51	1.21	2.42	4.33	3.49	0.68	1.72	3.57	0.35	0.45	0.57	19.52
1966	0.32	0.82	1.35	1.95	1.13	2.42	4.24	4.05	2.69	3.12	0.40	0.06	22.55
1967	0.39	0.77	0.19	0.87	0.56	8.30	0.69	1.03	1.55	0.29	0.06	0.47	15.17
1968	0.29	0.03	0.16	5.39	1.02	7.23	2.05	3.45	2.66	1.93	0.43	1.53	26.17
1969	0.65	1.94	0.28	1.45	3.84	2.38	4.52	1.57	0.67	2.11	0.09	0.59	20.09
1970	0.29	0.08	2.10	4.69	2.62	3.33	3.45	0.61	1.49	1.44	1.66	0.44	22.20
1971	0.18	0.99	0.15	2.22	1.99	2.87	0.91	1.85	1.78	3.16	2.04	0.41	18.55
1972	0.43	0.65	1.21	3.80	6.86	1.25	4.93	0.69	0.30	2.66	2.44	1.24	26.46
1973	0.74	0.42	2.16	1.17	2.04	0.82	2.11	1.61	2.69	2.12	0.94	0.56	17.38
1974	0.10	0.68	1.62	1.07	4.26	2.00	1.29	1.26	0.15	0.49	0.08	0.03	13.03
1975	1.93	0.21	2.71	3.31	1.98	3.44	0.46	2.50	0.82	0.81	0.15	0.06	18.38
1976	0.64	0.63	0.73	1.10	0.62	1.97	2.05	0.14	1.43	1.26	0.01	0.39	10.97
1977	0.18	1.73	5.89	2.13	1.83	1.31	1.50	2.57	2.29	1.59	1.60	0.52	23.14
1978	0.13	0.49	0.35	3.11	3.66	2.20	2.49	1.81	0.99	0.13	0.27	0.26	15.89
1979	0.89	0.29	1.96	2.23	1.64	2.37	2.61	1.50	0.23	1.97	0.22	0.04	15.95
1980	0.25	0.49	0.81	0.82	1.68	5.31	3.21	3.94	0.53	0.91	0.08	0.12	18.15
1981	0.12	0.12	1.90	0.24	0.68	1.48	2.24	4.63	1.39	1.28	1.20	0.51	15.79
1982	1.36	0.17	1.60	1.29	4.58	1.25	5.10	2.85	2.32	3.49	0.60	0.43	25.04
1983	0.08	0.21	2.34	1.14	1.37	4.62	2.22	0.45	2.55	1.27	1.81	0.48	18.54
1984	0.36	0.69	1.50	2.83	2.52	11.49	1.81	3.55	0.94	3.36	0.13	0.55	29.73
1985	0.40	0.10	3.86	1.57	2.03	2.36	4.67	2.57	3.28	1.06	2.46	0.59	24.95
1986	0.41	0.43	1.56	5.08	3.15	2.50	6.44	2.84	4.03	0.42	0.52	T	27.38
1987	0.19	1.16	4.78	0.42	2.34	1.54	2.19	2.98	1.36	0.42	0.74	1.28	19.40
1988	0.39	0.53	0.88	2.06	5.58	0.77	1.44	2.13	3.69	0.09	0.99	0.65	19.20
1989	0.21	0.81	2.17	1.59	1.00	3.19	1.14	1.25	1.62	0.26	1.05	0.21	14.50
1990	0.02	0.61	1.31	1.92	6.28	6.34	3.51	0.72	0.57	1.27	T	0.43	22.98
1991	0.08	0.92	0.62	5.59	3.72	5.72	1.76	1.18	1.94	0.98	0.81	0.08	23.40
1992	0.71	1.18	0.86	2.10	0.33	5.43	4.86	2.78	2.57	0.39	0.97	0.31	22.49
1993	0.32	0.55	1.57	3.18	3.32	7.12	6.69	2.39	1.74	0.63	1.78	0.72	30.01
Record Mean	0.50	0.60	1.16	2.13	2.72	3.53	2.62	2.17	1.62	1.30	0.68	0.52	19.56

AVERAGE TEMPERATURE (deg. F)

HURON, SOUTH DAKOTA

YEAR	JAN	FEB	MAR	APR	MAY	JUNE	JULY	AUG	SEP	OCT	NOV	DEC	ANNUAL
1964	20.9	26.2	25.9	49.5	60.4	69.2	78.6	69.4	61.2	49.9	29.3	10.1	45.9
1965	5.4	14.3	21.1	45.6	57.9	65.5	73.0	71.0	50.5	50.7	32.4	28.4	43.0
1966	4.5	15.1	35.9	41.3	55.6	68.7	78.3	66.9	58.9	47.5	28.6	20.8	43.5
1967	19.5	14.8	36.3	46.3	53.0	66.2	70.6	70.3	62.7	48.2	33.8	20.9	45.2
1968	15.4	19.3	38.6	45.5	52.2	67.7	71.4	70.2	59.7	48.4	33.9	16.0	44.8
1969	7.8	20.6	20.7	49.0	57.0	60.4	71.7	71.7	62.7	42.7	34.8	21.7	43.4
1970	6.5	22.5	27.4	43.6	58.1	68.2	74.8	73.7	63.2	48.1	33.0	15.1	44.5
1971	7.6	15.7	28.5	45.4	54.3	71.4	69.4	74.0	60.0	49.4	33.3	16.8	43.8
1972	9.6	12.9	31.5	44.8	60.0	66.1	69.7	70.6	60.5	44.8	31.5	15.4	43.1
1973	18.9	23.9	39.8	45.5	56.2	67.0	72.1	75.5	58.0	52.6	31.6	16.5	46.4
1974	13.1	24.4	34.2	47.9	54.2	66.0	79.2	69.0	58.4	51.5	33.0	23.1	46.2
1975	16.7	15.7	24.9	39.8	59.0	67.0	78.6	73.6	57.9	50.8	32.6	21.4	44.8
1976	14.5	28.3	31.1	48.1	55.0	69.2	74.7	75.2	61.5	42.1	25.4	16.1	45.1
1977	5.3	25.4	33.9	52.8	65.4	70.2	76.7	67.5	61.9	48.4	31.3	15.5	46.2
1978	0.3	7.1	29.3	44.0	57.5	65.9	70.7	71.2	66.1	47.4	29.1	14.5	42.0
1979	0.3	6.1	28.3	42.4	53.1	68.9	75.2	69.8	63.9	46.2	31.6	28.8	42.9
1980	16.9	18.5	29.7	49.1	60.3	68.5	73.4	70.6	61.7	46.5	37.0	22.0	46.2
1981	22.1	25.6	36.3	52.6	57.8	69.4	77.5	72.2	63.6	49.3	39.7	17.9	48.7
1982	3.5	18.9	33.2	44.5	59.6	63.6	75.4	72.7	61.2	49.0	31.4	26.3	44.9
1983	25.6	30.1	35.0	42.4	54.3	66.3	76.5	78.9	63.1	49.5	34.0	2.8	46.5
1984	18.8	31.1	29.3	47.1	55.8	67.2	74.0	74.5	56.9	49.2	35.4	18.1	46.5
1985	13.2	20.2	37.2	51.6	62.0	63.2	72.5	66.9	58.3	47.0	18.7	8.6	43.3
1986	19.3	14.6	37.3	45.3	57.2	69.1	73.6	67.0	58.9	48.8	28.4	26.9	45.5
1987	24.9	33.3	36.1	53.6	64.6	71.1	75.7	68.7	62.2	44.3	37.1	24.5	49.7
1988	8.7	14.9	35.0	46.2	63.5	75.3	76.7	73.4	61.2	45.3	33.3	22.0	46.3
1989	23.3	7.7	26.6	47.4	56.5	65.7	76.2	73.2	61.6	49.0	30.7	11.0	44.1
1990	28.3	24.2	37.4	45.4	56.8	68.6	71.4	72.7	66.5	47.4	36.3	14.4	47.5
1991	12.8	29.7	35.8	48.9	61.0	73.4	73.5	74.1	61.5	46.4	27.0	26.3	47.5
1992	26.4	29.8	37.2	44.6	59.3	65.3	65.1	65.6	59.7	47.3	30.9	17.6	45.7
1993	14.5	14.2	31.3	44.7	57.1	64.1	69.9	70.3	56.2	45.7	29.6	20.5	43.2
Record Mean	12.9	17.3	30.5	46.0	57.3	67.3	73.3	71.3	61.3	48.6	31.9	19.1	44.7
Max	23.5	27.7	40.7	58.1	69.6	79.3	86.1	84.2	74.5	61.5	42.6	29.1	56.4
Min	2.3	6.8	20.2	34.0	44.9	55.3	60.6	58.3	48.0	35.7	21.2	9.1	33.0

REFERENCE NOTES

HURON, SOUTH DAKOTA

GENERAL
T - TRACE AMOUNT.
BLANK ENTRIES DENOTE MISSING/UNREPORTED DATA.
# INDICATES A STATION OR INSTRUMENT RELOCATION.

SPECIFIC

(a) - LENGTH OF RECORD IN YEARS, ALTHOUGH INDIVIDUAL MONTHS MAY BE MISSING.
0.* OR * - THE VALUE IS BETWEEN 0.0 AND 0.05
NORMALS - BASED ON THE 1961-1990 RECORD PERIOD.
EXTREMES - DATES ARE THE MOST RECENT OCCURRENCE
WIND DIR.- NUMERALS SHOW TENS OF DEGRESS CLOCKWISE FROM TRUE NORTH. "00" INDICATES CALM.
RESULTANT DIRECTIONS ARE GIVEN TO WHOLE DEGREES.
BOLD VALUES INDICATE EXTREME VALUES WHICH OCCURRED AFTER THE ASOS SYSTEM WAS COMMISSIONED.

RECORD = PERIOD OF RECORD
RECORD MEAN PRECIPITATION IS THE MEAN OF ALL DAILY PRECIPITATION AMOUNTS DURING THE PERIOD OF RECORD.
RECORD MAX(MIN) TEMPERATURE IS THE MEAN OF ALL DAILY MAX(MIN) TEMPERATURES DURING THE PERIOD OF RECORD.
RECORD MEAN TEMPERATURE IS THE SUM OF THE RECORD MAX AND RECORD MIN DIVIDED BY 2.
AVERAGE TEMPERATURE IS THE SUM OF THE MEAN DAILY MAX AND MIN TEMPERATURE DIVIDED BY 2.

EXCEPTIONS

RECORD MEANS ARE THROUGH THE CURRENT YEAR, BEGINNING IN 1882 FOR TEMPERATURE
1882 FOR PRECIPITATION
1940 FOR SNOWFALL

HEATING DEGREE DAYS Base 65 deg. F HURON, SOUTH DAKOTA

SEASON	JULY	AUG	SEP	OCT	NOV	DEC	JAN	FEB	MAR	APR	MAY	JUNE	TOTAL
1964-65	1	55	179	463	1068	1699	1846	1415	1352	579	229	40	8926
1965-66	0	29	431	436	971	1127	1875	1394	894	707	307	51	8222
1966-67	0	62	208	535	1083	1362	1403	1404	884	556	393	38	7928
1967-68	38	24	118	521	933	1361	1533	1318	812	577	392	56	7683
1968-69	13	35	190	511	926	1515	1769	1239	1366	473	280	162	8479
1969-70	0	6	118	684	899	1340	1814	1186	1158	635	225	40	8105
1970-71	7	2	181	522	952	1541	1777	1375	1127	579	325	16	8404
1971-72	32	8	211	478	946	1486	1714	1510	1032	600	210	57	8284
1972-73	26	38	171	621	997	1530	1428	1144	776	578	271	31	7611
1973-74	5	0	227	380	995	1499	1607	1131	948	506	346	73	7717
1974-75	1	36	218	414	954	1293	1489	1373	1239	756	195	67	8035
1975-76	4	3	236	447	966	1346	1561	1057	1043	502	306	39	7510
1976-77	5	4	173	706	1181	1514	1849	1101	958	374	61	12	7938
1977-78	0	27	129	506	1002	1530	2007	1620	1099	624	247	89	8880
1978-79	12	27	132	539	1070	1560	2006	1647	1130	668	374	35	9200
1979-80	0	32	122	574	997	1117	1485	1343	1088	489	193	29	7469
1980-81	1	19	160	578	835	1326	1322	1098	882	376	243	12	6852
1981-82	5	1	105	480	752	1456	1908	1287	976	609	173	85	7837
1982-83	0	19	168	489	1002	1193	1215	970	923	673	329	83	7064
1983-84	2	0	166	473	924	1926	1430	978	1101	528	294	36	7858
1984-85	4	7	277	493	883	1448	1598	1250	856	403	141	123	7483
1985-86	7	46	267	553	1383	1746	1412	1405	855	584	244	20	8522
1986-87	0	55	191	496	1091	1174	1235	881	889	354	96	20	6482
1987-88	7	53	118	635	828	1248	1744	1449	924	556	117	1	7680
1988-89	3	26	151	603	946	1326	1284	1601	1182	529	262	83	7996
1989-90	0	11	182	491	1024	1674	1131	1138	848	611	258	37	7405
1990-91	7	9	124	543	855	1564	1614	981	898	477	203	C	7275
1991-92	4	0	192	580	1131	1192	1191	1015	855	617	217	62	7056
1992-93	43	79	177	544	1021	1463	1561	1421	1037	602	243	86	8277
1993-94	3	25	274	593	1054	1371							

COOLING DEGREE DAYS Base 65 deg. F HURON, SOUTH DAKOTA

YEAR	JAN	FEB	MAR	APR	MAY	JUNE	JULY	AUG	SEP	OCT	NOV	DEC	TOTAL
1969	0	0	0	0	36	31	214	219	56	0	0	0	556
1970	0	0	0	0	18	143	320	279	136	4	0	0	900
1971	0	0	0	0	1	217	175	293	69	0	0	0	755
1972	0	0	0	0	62	94	181	222	46	0	0	0	605
1973	0	0	0	0	4	98	233	331	22	3	0	0	691
1974	0	0	0	0	18	109	448	166	29	1	0	0	771
1975	0	0	0	4	17	136	432	276	33	12	0	0	910
1976	0	0	0	0	2	170	313	328	75	3	0	0	891
1977	0	0	0	14	79	176	369	113	44	0	0	0	795
1978	0	0	0	0	21	119	196	227	173	0	0	0	736
1979	0	0	0	0	13	162	324	188	96	0	0	0	783
1980	0	0	0	18	56	139	269	199	66	10	0	0	757
1981	0	0	0	9	25	153	399	232	70	0	0	0	888
1982	0	0	0	0	14	50	329	263	57	0	0	0	713
1983	0	0	0	0	4	130	366	437	113	0	0	0	1050
1984	0	0	0	0	13	107	291	306	41	9	0	0	767
1985	0	0	0	7	55	73	246	112	71	0	0	0	564
1986	0	0	0	0	8	152	272	122	16	0	0	0	570
1987	0	0	0	20	92	209	342	173	41	0	0	0	877
1988	0	0	0	3	78	316	374	292	45	0	0	0	1108
1989	0	0	0	9	4	110	354	274	87	4	0	0	842
1990	0	0	0	28	9	154	211	257	174	6	0	0	839
1991	0	0	0	1	85	260	275	289	93	11	0	0	1014
1992	0	0	0	10	48	77	52	105	25	3	0	0	320
1993	0	0	0	0	7	66	163	197	14	1	0	0	448

SNOWFALL (inches) HURON, SOUTH DAKOTA

SEASON	JULY	AUG	SEP	OCT	NOV	DEC	JAN	FEB	MAR	APR	MAY	JUNE	TOTAL
1964-65	0.0	0.0	0.0	0.0	3.6	10.1	3.6	10.7	17.2	1.2	T	0.0	46.4
1965-66	0.0	0.0	T	0.0	2.3	0.2	6.7	8.4	14.4	6.3	0.5	0.0	38.8
1966-67	0.0	0.0	0.0	T	4.3	1.7	6.7	17.5	2.1	0.6	0.7	0.0	33.6
1967-68	0.0	0.0	0.0	T	0.8	4.3	4.5	0.7	2.5	5.3	0.3	0.0	18.4
1968-69	0.0	0.0	0.0	T	3.4	26.0	13.2	22.3	2.5	T	0.0	0.0	67.4
1969-70	0.0	0.0	0.0	1.4	1.4	12.2	7.3	1.6	15.0	12.8	T	0.0	51.7
1970-71	0.0	0.0	0.0	5.3	1.9	7.8	4.6	16.0	1.9	T	0.0	0.0	37.5
1971-72	0.0	0.0	0.0	3.8	11.8	10.4	5.3	13.5	0.9	1.0	0.0	0.0	46.7
1972-73	0.0	0.0	0.0	1.3	10.9	13.8	10.7	6.2	0.1	T	T	0.0	43.0
1973-74	0.0	0.0	0.0	T	6.1	10.6	1.4	8.0	6.0	T	0.0	0.0	32.1
1974-75	0.0	0.0	0.0	0.0	2.0	1.3	27.7	5.9	33.9	0.7	0.0	0.0	71.5
1975-76	0.0	0.0	0.0	T	2.9	1.6	10.9	6.6	13.0	T	T	0.0	35.0
1976-77	0.0	0.0	0.0	7.3	0.4	5.8	3.5	9.8	9.1	0.3	0.0	0.0	36.2
1977-78	0.0	0.0	0.0	T	7.8	6.5	4.2	9.0	4.8	0.8	0.0	0.0	33.1
1978-79	0.0	0.0	0.0	0.0	3.5	5.9	23.8	5.9	2.6	4.0	T	0.0	45.7
1979-80	0.0	0.0	0.0	2.5	3.2	0.4	1.5	6.0	8.6	T	0.0	0.0	22.2
1980-81	0.0	0.0	0.0	1.5	0.8	3.0	1.6	2.3	1.2	T	0.0	0.0	10.4
1981-82	0.0	0.0	0.0	0.1	8.0	7.4	23.8	1.8	9.5	9.1	0.0	0.0	59.7
1982-83	0.0	0.0	0.0	2.7	2.1	4.4	0.7	0.5	11.8	5.1	0.0	0.0	27.3
1983-84	0.0	0.0	0.0	0.0	16.9	6.9	4.3	6.8	19.1	T	0.0	0.0	54.0
1984-85	0.0	0.0	0.0	T	0.2	8.1	5.5	1.8	26.7	2.0	0.0	0.0	44.3
1985-86	0.0	0.0	0.0	T	32.7	11.8	6.3	4.8	7.0	4.0	0.0	0.0	66.6
1986-87	0.0	0.0	0.0	0.0	5.6	T	2.4	6.7	7.0	T	0.0	0.0	21.7
1987-88	0.0	0.0	0.0	T	1.0	20.0	7.3	7.3	11.1	3.2	0.0	0.0	49.9
1988-89	0.0	0.0	0.0	0.0	5.0	11.2	3.0	14.9	26.6	0.4	0.2	0.0	61.3
1989-90	0.0	0.0	0.0	0.0	3.7	3.6	0.2	14.1	1.1	2.3	0.0	T	25.0
1990-91	0.0	0.0	0.0	1.8	T	7.4	1.9	13.0	6.3	1.2	T	0.0	31.6
1991-92	0.0	0.0	0.0	T	4.9	0.1	4.0	12.1	0.7	1.2	0.0	T	23.0
1992-93	T	0.0	0.0	0.4	7.9	4.7	5.8	11.7	8.1	2.7	0.0	T	41.3
1993-94	0.0	T	0.0	T	13.4	12.1							
Record Mean	T	T	T	0.7	5.2	6.4	6.6	8.8	9.3	2.2	0.1	T	39.3

RAPID CITY, SOUTH DAKOTA

Rapid City, which is not far from the geographical center of North America, experiences the large temperature ranges, both daily and seasonal, that are typical of semi-arid continental climates.

The city is surrounded by contrasting landforms, with the forested Black Hills rising immediately west of the city, and rolling prairie extending out in the other directions. From 40 to 70 miles southeast lie the eroded Badlands. The Black Hills, many of which are more than 5,000 feet above sea level, with a number of peaks above 7,000 feet, exert a pronounced influence on the climate of this area. The rolling land to the east of the city is cut by the valleys of the Box Elder and Rapid Creeks, which flow generally east-southeastward. The station is located on the north slope of the irrigated Rapid Valley. An east-west ridge 200 to 300 feet higher than the airport separates the station from the Box Elder Creek Valley.

The principal agricultural products in the area are cattle and wheat, and ranchers and farmers are dependent on the current weather forecasts, which are at times of vital interest in the protection of livestock.

Although the annual precipitation is light at lower elevations, the distribution is beneficial to agriculture with the greatest amounts occurring during the growing season. The heaviest snows are expected in the spring, which helps to furnish moisture for the early maturing crops such as wheat, while heavy winter snows at the higher elevations provide irrigation water for the fertile valleys.

Summer days are normally warm with cool, comfortable nights. Nearly all of the summer precipitation occurs as thunderstorms. Hail is often associated with the more severe thunderstorms, with resultant damage to vegetation as well as other fragile material in the path of the storms. Autumn, which begins soon after the first of September, is characterized by mild, balmy days, and cool, invigorating mornings and evenings. Autumn weather usually extends into November and often into December.

Temperatures for the winter months of December, January, and February are among the warmest in South Dakota due to the protection of the Black Hills, the frequent occurrence of Chinook winds, and the fact that the winter tracks of arctic air masses usually pass east of Rapid City. Rapid City has become the retirement home for many farmers and ranchers from the western half of the state because of the cool summer nights and the relatively mild winters.

Snowfall is normally light with the greatest monthly average of about 8 inches occurring in March. Cold waves can be expected occasionally, and one or more blizzards may occur each winter.

Spring is characterized by unsettled conditions. Wide variations usually occur in temperatures, and snows may fall as late as May.

Based on the 1951-1980 period, the average first occurrence of 32 degrees Fahrenheit in the fall is September 29 and the average last occurrence in the spring is May 7.

NORMALS, MEANS, AND EXTREMES

RAPID CITY, SOUTH DAKOTA

LATITUDE: 44 °03'N LONGITUDE: 103 °04' W ELEVATION: FT. GRND 3162 BARO 3169 TIME ZONE: MOUNTAIN WBAN: 24090

	(a)	JAN	FEB	MAR	APR	MAY	JUNE	JULY	AUG	SEP	OCT	NOV	DEC	YEAR
TEMPERATURE °F:														
Normals														
-Daily Maximum		33.8	38.2	45.9	57.9	68.1	77.8	86.2	85.1	74.4	62.5	46.7	35.6	59.4
-Daily Minimum		10.7	15.2	22.2	32.2	42.3	51.7	58.2	56.1	45.5	34.9	22.8	12.7	33.7
-Monthly		22.3	26.7	34.1	45.1	55.2	64.8	72.2	70.6	60.0	48.7	34.8	24.2	46.6
Extremes														
-Record Highest	51	76	75	82	93	98	106	110	106	104	94	78	75	110
-Year		1987	1988	1993	1989	1969	1988	1989	1988	1978	1993	1990	1965	JUL 1989
-Record Lowest	51	-27	-24	-17	1	18	31	39	38	18	-2	-19	-30	-30
-Year		1950	1989	1962	1975	1950	1951	1987	1992	1985	1991	1959	1990	DEC 1990
NORMAL DEGREE DAYS:														
Heating (base 65°F)		1324	1072	958	597	311	103	18	31	211	505	906	1265	7301
Cooling (base 65°F)		0	0	0	0	7	97	241	205	61	0	0	0	611
% OF POSSIBLE SUNSHINE	51	56	60	63	62	60	65	72	74	69	66	55	55	63
MEAN SKY COVER (tenths)														
Sunrise - Sunset	51	6.3	6.4	6.6	6.6	6.3	5.5	4.4	4.3	4.5	5.1	6.2	6.1	5.7
MEAN NUMBER OF DAYS:														
Sunrise to Sunset														
-Clear	51	8.0	6.5	6.4	5.9	6.8	9.1	12.8	13.7	13.3	12.1	7.7	8.4	110.6
-Partly Cloudy	51	7.8	8.2	9.3	9.5	10.8	11.4	12.8	11.7	8.7	8.2	8.5	8.3	115.2
-Cloudy	51	15.2	13.6	15.4	14.6	13.3	9.5	5.5	5.6	8.0	10.7	13.8	14.3	139.5
Precipitation														
.01 inches or more	51	6.6	7.2	8.7	9.3	11.8	12.4	9.4	7.8	6.4	5.1	5.9	6.0	96.5
Snow,Ice pellets,hail														
1.0 inches or more	43	1.3	2.5	2.6	2.0	0.2	0.*	0.0	0.0	0.1	0.4	1.7	1.8	12.6
Thunderstorms	51	0.0	0.0	0.1	1.2	5.9	10.8	11.7	8.7	3.4	0.4	0.*	0.*	42.3
Heavy Fog Visibility														
1/4 mile or less	51	1.5	2.1	2.5	1.7	0.9	1.2	0.5	0.6	0.6	0.6	2.0	1.9	16.0
Temperature °F														
-Maximum														
90° and above	51	0.0	0.0	0.0	0.1	0.5	3.3	11.4	11.5	3.6	0.3	0.0	0.0	30.6
32° and below	51	12.9	9.9	6.6	0.7	0.*	0.0	0.0	0.0	0.0	0.4	5.1	10.8	46.4
-Minimum														
32° and below	51	30.2	26.9	27.0	14.9	2.7	0.1	0.0	0.0	1.6	10.4	24.7	29.9	168.5
0° and below	51	8.0	4.4	2.0	0.0	0.0	0.0	0.0	0.0	0.0	0.*	1.1	4.4	19.8
AVG. STATION PRESS.(mb)	21	904.5	904.7	902.1	902.8	902.7	903.4	905.2	905.4	905.8	905.8	904.4	904.3	904.3
RELATIVE HUMIDITY (%)														
Hour 05	43	68	72	74	73	75	77	73	71	68	66	69	69	71
Hour 11 (Local Time)	43	59	59	55	48	49	51	45	42	41	42	52	59	50
Hour 17	43	63	61	54	46	47	48	41	37	38	45	59	64	50
Hour 23	43	68	71	72	68	71	73	65	62	60	62	68	68	67
PRECIPITATION (inches):														
Water Equivalent														
-Normal		0.39	0.52	1.03	1.89	2.68	3.06	2.04	1.67	1.23	1.10	0.56	0.47	16.64
-Maximum Monthly	51	1.77	2.46	3.02	5.16	7.35	7.00	6.13	4.83	3.94	3.82	2.22	1.65	7.35
-Year		1944	1953	1945	1967	1946	1968	1969	1982	1946	1982	1985	1975	MAY 1946
-Minimum Monthly	51	0.01	0.06	0.12	0.02	0.33	0.64	0.38	0.10	0.03	T	0.03	0.01	T
-Year		1952	1985	1981	1987	1966	1973	1988	1943	1975	1960	1945	1986	OCT 1960
-Maximum in 24 hrs	51	1.26	1.00	2.19	3.01	3.40	4.01	2.51	2.60	2.13	2.49	1.09	1.04	4.01
-Year		1944	1953	1945	1946	1965	1963	1944	1982	1966	1982	1944	1975	JUN 1963
Snow,Ice pellets,hail														
-Maximum Monthly	51	24.0	23.7	30.7	30.6	11.6	3.6	T	T	2.0	10.2	33.6	17.9	33.6
-Year		1949	1953	1950	1970	1950	1951	1993	1990	1970	1971	1985	1975	NOV 1985
-Maximum in 24 hrs	51	16.3	10.0	14.9	16.0	13.4	3.6	T	T	2.0	7.6	9.4	9.8	16.3
-Year		1944	1953	1973	1970	1967	1951	1993	1990	1970	1971	1977	1975	JAN 1944
WIND:														
Mean Speed (mph)	43	10.8	11.1	12.7	13.3	12.4	10.9	10.2	10.3	11.0	11.2	10.9	10.6	11.3
Prevailing Direction through 1963		NNW	NNW	NNW	NNW	NNW	NNW	NNW	NNW	NNW	NNW	NNW	NNW	NNW
Fastest Obs. 1 Min.														
-Direction (!!)	10	33	32	34	32	34	25	34	23	32	33	34	33	25
-Speed (MPH)	10	52	45	53	51	46	54	51	41	52	44	45	46	54
-Year		1987	1988	1988	1992	1988	1990	1991	1987	1991	1988	1993	1993	JUN 1990
Peak Gust														
-Direction (!!)	10	N	NW	N	NW	NW	W	NW	NW	S	NW	NW	NW	NW
-Speed (mph)	10	68	61	66	67	61	66	72	69	70	67	63	66	72
-Date		1988	1984	1988	1992	1987	1990	1984	1985	1985	1988	1993	1993	JUL 1984

PRECIPITATION (inches)

RAPID CITY, SOUTH DAKOTA

YEAR	JAN	FEB	MAR	APR	MAY	JUNE	JULY	AUG	SEP	OCT	NOV	DEC	ANNUAL
1964	0.35	0.83	0.63	1.24	2.52	4.69	0.77	1.87	0.69	0.50	0.30	0.78	15.17
1965	0.61	0.22	0.46	1.50	6.97	3.56	0.60	1.46	1.46	0.57	0.15	0.12	17.68
1966	0.24	1.00	1.78	2.50	0.33	1.31	3.93	3.24	2.84	1.50	0.95	0.79	20.41
1967	0.47	0.59	0.82	5.16	3.20	6.78	1.07	0.95	2.10	0.28	0.28	0.89	22.59
1968	0.43	0.35	0.21	1.82	1.68	7.00	2.44	2.46	0.89	0.14	0.46	0.58	18.46
1969	0.11	0.73	0.66	1.60	2.20	2.04	6.13	0.31	0.35	0.85	0.35	0.57	15.90
1970	0.73	0.54	1.48	4.63	2.41	2.16	1.04	0.67	1.57	1.23	0.80	0.61	17.87
1971	1.18	1.00	1.25	2.86	3.70	1.92	1.46	0.52	2.32	2.02	0.79	0.15	19.17
1972	0.22	0.44	0.47	2.78	3.28	4.11	1.67	2.49	0.24	0.77	0.38	0.34	17.19
1973	0.11	0.31	2.71	2.69	2.37	0.64	1.46	0.74	1.44	1.38	0.73	0.54	15.12
1974	0.16	0.30	0.34	1.55	1.32	1.10	0.68	1.37	0.88	1.18	0.12	0.12	9.12
1975	1.05	0.35	2.45	1.37	1.23	5.63	1.57	0.87	0.03	0.69	0.57	1.65	17.46
1976	0.28	0.47	0.33	2.70	2.74	4.81	1.05	1.31	0.28	0.21	0.61	0.41	15.20
1977	0.83	0.25	2.63	1.57	2.49	1.76	2.98	1.79	2.86	1.06	0.82	0.36	19.40
1978	0.19	0.84	0.40	2.19	3.12	2.01	4.08	1.42	0.18	0.26	0.63	0.25	15.57
1979	0.49	0.33	0.47	0.31	1.17	3.60	4.11	2.32	0.07	0.90	0.15	0.07	13.99
1980	0.20	0.51	0.86	1.13	1.58	4.75	1.78	2.38	0.48	2.28	0.57	0.66	17.18
1981	0.14	0.09	0.12	0.32	2.81	1.89	4.47	1.74	0.16	1.81	0.23	0.35	14.13
1982	0.39	0.37	1.35	0.69	6.50	2.89	1.81	4.83	2.69	3.82	0.27	0.36	25.97
1983	0.34	0.18	0.84	1.00	2.18	3.01	1.94	2.39	0.33	1.74	1.07	0.47	15.49
1984	0.10	0.18	0.69	3.10	1.57	4.72	1.57	1.00	0.74	0.67	0.51	0.38	15.23
1985	0.46	0.06	1.55	0.32	1.24	1.58	1.03	1.86	1.57	0.98	2.22	0.77	13.64
1986	0.49	0.92	0.88	4.74	1.43	4.56	0.91	1.32	3.14	1.64	1.40	0.01	21.44
1987	0.04	1.71	1.14	0.02	3.39	1.37	0.83	2.37	0.68	0.26	0.30	0.31	12.42
1988	0.17	0.34	0.52	0.60	3.25	1.09	0.38	1.98	0.56	0.76	0.81	0.46	10.92
1989	0.02	0.34	0.96	1.46	1.40	1.04	0.82	1.70	3.09	1.49	0.43	0.82	13.57
1990	0.22	0.37	1.17	0.77	4.87	1.42	1.94	1.87	2.44	0.61	0.44	0.33	16.45
1991	0.32	0.77	0.63	2.99	4.40	3.27	1.97	0.58	0.59	1.00	0.73	0.04	17.29
1992	0.29	0.16	1.92	0.71	2.47	2.17	3.25	0.47	0.42	0.68	0.39	0.57	13.50
1993	0.68	0.61	0.82	3.05	2.16	3.39	4.31	1.18	1.46	0.90	0.70	0.53	19.79
Record Mean	0.41	0.47	1.00	1.91	3.06	3.21	2.27	1.61	1.25	1.02	0.61	0.51	17.33

AVERAGE TEMPERATURE (deg. F)

RAPID CITY, SOUTH DAKOTA

YEAR	JAN	FEB	MAR	APR	MAY	JUNE	JULY	AUG	SEP	OCT	NOV	DEC	ANNUAL
1964	28.4	28.1	29.9	46.3	58.5	64.9	76.7	69.5	59.0	51.7	33.7	18.9	47.1
#1965	25.4	25.6	21.2	47.0	54.7	64.5	72.1	70.7	49.7	54.8	38.2	33.3	46.4
1966	14.2	19.3	35.0	38.7	56.0	65.3	76.6	65.9	61.3	48.6	34.6	26.7	45.2
1967	27.9	27.5	35.2	44.1	49.8	60.0	70.6	70.6	63.4	50.1	34.6	22.5	46.3
1968	23.4	27.4	40.4	42.6	50.8	63.0	69.2	67.0	59.8	51.0	35.5	18.5	45.7
1969	16.0	25.4	27.8	49.6	57.5	59.8	71.0	73.8	64.8	40.6	38.7	28.2	46.1
1970	17.8	30.8	27.5	39.2	55.6	65.6	72.7	73.8	59.1	44.1	34.2	21.9	45.2
1971	17.5	22.6	32.6	46.0	53.6	66.4	68.0	73.9	57.4	46.1	34.5	23.3	45.2
1972	17.3	24.0	36.5	43.6	55.2	64.7	65.6	69.1	59.4	43.8	31.0	18.6	44.1
1973	26.8	28.9	37.5	42.9	53.7	64.7	70.7	74.3	57.1	51.3	32.3	25.4	47.1
1974	21.9	32.5	37.6	47.2	53.8	66.6	77.3	67.7	57.6	51.7	36.4	28.9	48.3
1975	23.7	17.6	27.6	40.5	54.0	62.2	74.7	70.3	58.5	49.2	33.8	29.4	45.1
1976	23.9	34.6	34.4	46.9	55.4	64.3	72.8	72.6	63.3	45.4	31.7	26.8	47.7
1977	12.7	34.6	36.4	49.5	60.3	69.0	73.4	66.0	61.0	48.8	33.3	21.9	47.3
1978	11.0	15.3	35.4	45.2	55.1	65.1	71.1	69.7	66.2	50.5	29.2	17.1	44.2
1979	7.4	16.8	35.4	44.4	53.6	65.4	70.4	68.5	66.3	50.9	33.0	33.2	45.4
1980	21.0	27.1	31.5	48.9	57.4	67.1	74.9	68.6	61.6	48.9	39.3	30.3	48.0
1981	32.6	29.4	40.0	51.5	54.9	64.9	72.0	70.0	63.5	47.7	40.4	25.8	49.4
1982	11.9	23.9	33.0	42.1	53.3	59.7	70.7	70.2	58.7	47.2	32.7	28.6	44.3
1983	32.1	37.3	36.4	40.7	52.0	63.1	73.6	78.0	60.8	49.4	34.9	8.1	47.2
1984	28.0	36.1	34.3	43.8	53.6	62.8	72.2	74.8	57.2	47.2	37.5	21.4	47.4
1985	21.6	23.8	35.9	52.0	61.8	62.1	74.6	69.0	55.6	47.3	16.0	21.0	45.1
1986	29.8	21.5	43.0	44.2	54.9	67.6	70.9	69.6	55.0	48.7	30.6	30.5	47.2
1987	31.1	32.4	32.6	51.6	59.5	67.1	75.4	68.1	61.4	47.1	40.3	28.9	49.6
1988	21.7	26.9	35.6	47.1	60.0	75.6	76.1	72.5	60.4	49.7	36.4	28.4	49.2
1989	28.7	14.4	31.1	45.8	55.4	64.0	77.0	73.1	61.0	48.8	36.4	19.4	46.3
1990	32.4	28.9	36.4	45.0	53.2	66.6	71.7	73.8	65.9	48.2	40.5	17.8	48.4
1991	18.4	36.0	38.5	46.3	56.6	67.2	72.9	74.0	61.1	46.4	30.0	32.7	48.3
1992	33.6	36.3	40.9	47.1	57.4	62.6	64.3	65.9	62.6	49.3	32.4	19.0	47.6
1993	15.0	15.2	37.6	43.5	56.1	60.3	65.1	68.6	56.6	48.2	32.4	31.5	44.3
Record Mean	22.9	25.5	33.4	45.0	55.1	64.6	72.2	70.8	60.6	49.2	35.7	26.3	46.8
Max	34.5	36.9	44.9	57.0	66.9	76.6	85.4	84.4	74.2	62.1	47.3	37.5	59.0
Min	11.3	14.1	21.9	33.1	43.3	52.6	59.0	57.3	47.1	36.3	24.1	15.1	34.6

REFERENCE NOTES

RAPID CITY, SOUTH DAKOTA

GENERAL
T - TRACE AMOUNT.
BLANK ENTRIES DENOTE MISSING/UNREPORTED DATA.
# INDICATES A STATION OR INSTRUMENT RELOCATION.

SPECIFIC

PM - INCLUDES LAST DAY OF PREVIOUS MONTH
ASOS - AUTOMATED SURFACE OBSERVING SYSTEM IN OPERATION DURING THESE MONTHS.

(a) - LENGTH OF RECORD IN YEARS, ALTHOUGH INDIVIDUAL MONTHS MAY BE MISSING.
0.* OR * - THE VALUE IS BETWEEN 0.0 AND 0.05
NORMALS - BASED ON THE 1961-1990 RECORD PERIOD.
EXTREMES - DATES ARE THE MOST RECENT OCCURRENCE
WIND DIR.- NUMERALS SHOW TENS OF DEGRESS CLOCKWISE FROM TRUE NORTH. "00" INDICATES CALM.
RESULTANT DIRECTIONS ARE GIVEN TO WHOLE DEGREES.
BOLD VALUES INDICATE EXTREME VALUES WHICH OCCURRED AFTER THE ASOS SYSTEM WAS COMMISSIONED.

RECORD = PERIOD OF RECORD
RECORD MEAN PRECIPITATION IS THE MEAN OF ALL DAILY PRECIPITATION AMOUNTS DURING THE PERIOD OF RECORD.
RECORD MAX(MIN) TEMPERATURE IS THE MEAN OF ALL DAILY MAX(MIN) TEMPERATURES DURING THE PERIOD OF RECORD.
RECORD MEAN TEMPERATURE IS THE SUM OF THE RECORD MAX AND RECORD MIN DIVIDED BY 2.
AVERAGE TEMPERATURE IS THE SUM OF THE MEAN DAILY MAX AND MIN TEMPERATURE DIVIDED BY 2.

EXCEPTIONS

RECORD MEANS ARE THROUGH THE CURRENT YEAR, BEGINNING IN 1900 FOR TEMPERATURE
1900 FOR PRECIPITATION
1943 FOR SNOWFALL

HEATING DEGREE DAYS Base 65 deg. F RAPID CITY, SOUTH DAKOTA

SEASON	JULY	AUG	SEP	OCT	NOV	DEC	JAN	FEB	MAR	APR	MAY	JUNE	TOTAL
1964-65	0	64	198	410	931	1419	1218	1098	1351	534	327	55	7605
#1965-66	3	27	461	312	797	978	1573	1273	923	784	285	103	7519
1966-67	0	83	158	504	905	1180	1143	1042	917	622	473	162	7189
1967-68	31	19	129	464	902	1311	1285	1083	755	669	432	107	7187
1968-69	26	46	176	428	881	1438	1516	1102	1145	458	264	176	7656
1969-70	0	0	62	750	781	1135	1461	950	1157	767	293	74	7430
1970-71	2	0	245	644	918	1330	1463	1182	996	563	344	35	7722
1971-72	43	2	267	578	909	1284	1473	1182	877	634	322	74	7645
1972-73	74	43	193	649	1010	1436	1181	1008	847	659	350	83	7533
1973-74	16	0	246	416	972	1223	1329	905	840	530	343	87	6907
1974-75	1	42	242	407	849	1112	1274	1318	1151	728	343	119	7586
1975-76	3	16	206	493	929	1096	1269	878	940	535	295	98	6758
1976-77	3	6	132	606	991	1177	1616	846	877	459	165	17	6895
1977-78	1	48	163	494	944	1330	1669	1383	912	588	312	91	7935
1978-79	17	40	111	443	1068	1480	1781	1348	910	614	362	82	8256
1979-80	3	25	64	433	952	982	1359	1094	1032	483	251	54	6732
1980-81	1	18	144	510	763	1070	998	993	765	402	311	65	6040
1981-82	21	7	108	531	730	1209	1646	1146	985	682	358	170	7593
1982-83	7	21	226	545	962	1119	1012	772	880	723	407	113	6787
1983-84	8	0	208	474	896	1762	1139	832	948	626	366	101	7360
1984-85	0	0	268	546	820	1344	1341	1148	895	393	146	144	7045
1985-86	8	27	327	544	1466	1358	1083	1211	672	617	317	35	7665
1986-87	5	12	296	497	1025	1059	1045	907	997	408	199	46	6496
1987-88	10	49	147	545	736	1111	1340	1103	905	533	195	17	6691
1988-89	3	18	163	470	850	1127	1120	1414	1047	586	303	116	7217
1989-90	3	6	182	495	847	1410	1004	1004	880	597	363	68	6859
1990-91	10	5	112	514	730	1462	1440	807	815	556	269	16	6736
1991-92	2	6	183	581	1045	997	964	828	742	539	262	107	6256
1992-93	67	83	138	493	973	1418	1515	1390	841	636	269	158	7981
1993-94	56	26	256	522	972	1035							

COOLING DEGREE DAYS Base 65 deg. F RAPID CITY, SOUTH DAKOTA

YEAR	JAN	FEB	MAR	APR	MAY	JUNE	JULY	AUG	SEP	OCT	NOV	DEC	TOTAL
1969	0	0	0	0	35	29	193	280	63	0	0	0	600
1970	0	0	0	0	7	96	248	279	78	5	0	0	713
1971	0	0	0	0	0	81	142	284	43	0	0	0	550
1972	0	0	0	0	21	70	98	178	34	0	0	0	401
1973	0	0	0	0	7	80	202	295	15	0	0	0	599
1974	0	0	0	1	3	143	390	132	28	0	0	0	697
1975	0	0	0	0	9	41	314	188	19	12	0	0	583
1976	0	0	0	0	4	83	251	248	87	3	0	0	676
1977	0	0	0	0	25	143	270	85	51	0	0	0	574
1978	0	0	0	0	9	99	214	193	1.54	0	0	0	669
1979	0	0	0	4	15	99	179	141	110	2	0	0	550
1980	0	0	0	6	25	123	315	136	48	14	0	0	667
1981	0	0	0	3	5	67	243	170	74	0	0	0	562
1982	0	0	0	0	3	18	189	190	41	0	0	0	441
1983	0	0	0	0	9	62	282	407	88	0	0	0	848
1984	0	0	0	0	18	41	234	309	42	2	0	0	646
1985	0	0	0	10	53	64	312	158	51	0	0	0	648
1986	0	0	0	0	11	124	192	164	0	0	0	0	491
1987	0	0	0	13	33	118	341	152	45	0	0	0	702
1988	0	0	0	2	46	341	355	255	33	2	0	0	1034
1989	0	0	0	15	9	95	380	265	70	0	0	0	834
1990	0	0	0	2	4	120	226	282	147	3	0	0	784
1991	0	0	0	3	16	89	256	294	76	15	0	0	749
1992	0	0	0	9	36	41	51	118	72	11	0	0	338
1993	0	0	0	0	0	25	66	146	9	7	0	0	253

SNOWFALL (inches) RAPID CITY, SOUTH DAKOTA

SEASON	JULY	AUG	SEP	OCT	NOV	DEC	JAN	FEB	MAR	APR	MAY	JUNE	TOTAL
1964-65	0.0	0.0	0.0	0.0	2.6	7.8	4.3	1.9	4.6	5.5	8.8	0.0	35.5
1965-66	0.0	0.0	1.5	0.0	0.8	0.7	2.4	10.0	16.5	10.0	0.4	0.0	42.3
1966-67	0.0	0.0	T	2.1	6.4	8.5	4.0	7.4	7.6	14.0	3.1	0.0	53.1
1967-68	0.0	0.0	0.0	0.0	2.2	9.1	4.0	3.4	1.3	13.7	T	0.0	33.7
1968-69	0.0	0.0	0.0	0.0	2.7	7.5	1.2	7.9	6.3	5.2	T	0.5	31.3
1969-70	0.0	0.0	0.0	4.1	2.9	5.7	5.5	6.0	14.8	30.6	T	0.0	69.6
1970-71	0.0	0.0	2.0	5.7	4.5	7.8	13.2	15.7	13.8	5.3	0.0	0.0	68.0
1971-72	0.0	0.0	0.0	10.2	6.9	2.0	2.6	8.0	1.2	1.0	T	0.0	31.9
1972-73	0.0	0.0	0.0	0.7	3.6	6.0	2.0	3.1	16.9	1.9	0.0	0.0	34.2
1973-74	0.0	0.0	0.0	1.4	9.7	5.6	2.0	4.0	1.7	4.7	0.0	0.0	29.1
1974-75	0.0	0.0	0.0	T	1.5	1.4	13.7	6.1	27.4	1.3	0.0	0.0	51.4
1975-76	0.0	0.0	0.0	4.6	8.4	17.9	2.0	6.5	4.3	3.5	0.0	0.0	47.2
1976-77	0.0	0.0	0.0	0.5	7.0	6.1	11.1	2.6	26.0	3.4	0.0	0.0	56.7
1977-78	0.0	0.0	0.0	0.8	10.6	7.4	3.1	15.0	3.0	2.6	T	0.0	42.5
1978-79	0.0	0.0	0.0	T	9.8	4.0	6.1	4.4	2.6	1.8	2.8	0.0	31.5
1979-80	0.0	0.0	0.0	T	1.8	0.3	3.0	10.1	8.6	5.4	0.0	0.0	29.2
1980-81	0.0	0.0	0.0	1.4	6.9	6.1	1.2	1.3	T	T	0.0	0.0	16.9
1981-82	0.0	0.0	0.0	1.6	1.2	3.8	6.2	5.0	11.5	5.5	0.0	0.0	34.8
1982-83	0.0	0.0	0.0	1.4	1.2	4.0	2.9	0.3	6.5	4.3	4.3	0.0	24.9
1983-84	0.0	0.0	0.3	0.9	6.9	7.1	1.9	2.5	6.1	22.1	0.2	0.0	48.0
1984-85	0.0	0.0	1.3	0.7	2.0	4.9	3.8	0.7	16.2	0.4	0.0	T	30.0
1985-86	0.0	0.0	1.4	0.6	33.6	10.2	5.7	10.7	6.0	12.7	0.0	0.0	80.9
1986-87	0.0	0.0	0.0	T	12.6	0.1	0.5	21.5	10.9	0.3	0.0	0.0	45.9
1987-88	0.0	0.0	0.0	1.7	T	4.7	2.7	3.6	10.6	6.1	0.0	0.0	29.4
1988-89	0.0	0.0	0.0	0.0	2.2	9.0	0.4	7.3	10.7	6.4	0.0	0.0	36.0
1989-90	0.0	T	0.0	3.9	4.6	10.9	3.1	5.0	9.2	3.0	0.5	T	40.2
1990-91	0.0	T	0.0	1.5	1.0	6.2	6.3	8.2	3.6	11.1	4.9	T	42.8
1991-92	T	0.0	T	4.0	6.7	0.8	5.2	1.2	8.2	3.5	0.0	0.0	29.6
1992-93	T	0.0	0.0	0.9	5.0	12.4	13.6	9.0	1.0	12.2	0.0	0.0	54.1
1993-94	T	0.0	T	2.4	11.4	5.5							
Record Mean	T	T	0.1	1.6	5.1	5.1	5.0	6.4	8.8	6.1	0.8	0.1	39.1

SIOUX FALLS, SOUTH DAKOTA

Sioux Falls is located in the Big Sioux River Valley in southeast South Dakota. The surrounding terrain is gently rolling. The land slopes upward for about 100 miles north and northwest to an elevation about 400 feet higher than the city. To the southeast, the land slopes downward 200 to 300 feet over the same distance. Little change in elevation occurs in the remaining directions.

The climate is of the continental type. There are frequent weather changes from day to day or week to week as the locality is visited by differing air masses. Cold air masses arrive from the interior of Canada, cool, dry air from the northern Pacific, warm, moist air from the Gulf of Mexico, or hot, dry air from the southwest.

Temperatures fluctuate frequently as cold air masses move in very rapidly. During the late fall and winter, cold fronts accompanied by strong, gusty winds drop temperatures by 20 to 30 degrees in a 24-hour period. Severe cold spells usually last only a few days. The winter months of December through February have experienced cold spells with average temperatures under 8 degrees and more than 60 consecutive days below 32 degrees.

Temperatures of 100 degrees and above occur about one in every three years, and will most likely happen in July. Summer nights are usually comfortable with temperatures below 70 degrees.

Rainfall is heavier during the spring and summer with lighter amounts in winter. Nearly 64 percent of the normal yearly precipitation falls during the growing season of April through August.

One or two very heavy snows usually fall each winter. Eight to 12 inches of snow may fall in 24 hours. There have been a few snows in excess of 15 inches and almost 30 inches have fallen during a severe winter storm. Strong winds often cause drifting snow, and blizzard conditions may block highways for a day or so.

Southerly winds prevail from late spring to early fall with northwest winds the remainder of the year. Strong winds of 70 mph with gusts to 90 mph have occurred.

Thunderstorms are frequent during the late spring and summer with June and July the most active months. The thunderstorms usually occur during the late afternoon and evening with a secondary peak of activity between 2 and 5 in the morning. Some of the most severe thunderstorms with damaging winds, hail and an occasional tornado, occur most frequently June.

There is occasional flooding in the lower areas of Sioux Falls along the Big Sioux River and Skunk Creek. Runoff from the melting snow in the spring often causes substantial rises in the rivers. A diversion canal around Sioux Falls has reduced the threat of damaging floods.

Based on the 1951-1980 period, the average first occurrence of 32 degrees Fahrenheit in the fall is October 1 and the average last occurrence in the spring is May 10.

NORMALS, MEANS, AND EXTREMES

SIOUX FALLS, SOUTH DAKOTA

LATITUDE: 43°35'N LONGITUDE: 96°44'W ELEVATION: FT. GRND 1418 BARO 1432 TIME ZONE: CENTRAL WBAN: 14944

	(a)	JAN	FEB	MAR	APR	MAY	JUNE	JULY	AUG	SEP	OCT	NOV	DEC	YEAR
TEMPERATURE °F:														
Normals														
-Daily Maximum		24.3	29.6	42.3	59.0	70.7	80.5	86.3	83.3	73.1	61.2	43.4	28.0	56.8
-Daily Minimum		3.3	9.7	22.6	34.8	45.9	56.1	62.3	59.4	48.7	36.0	22.6	8.6	34.2
-Monthly		13.8	19.7	32.5	46.9	58.4	68.3	74.3	71.4	60.9	48.6	33.0	18.3	45.5
Extremes														
-Record Highest	48	66	70	87	94	100	110	108	108	104	94	76	61	110
-Year		1981	1982	1968	1962	1967	1988	1989	1973	1976	1963	1978	1984	JUN 1988
-Record Lowest	48	-36	-31	-23	5	17	33	38	34	22	9	-17	-28	-36
-Year		1970	1962	1948	1982	1967	1969	1971	1950	1974	1972	1964	1990	JAN 1970
NORMAL DEGREE DAYS:														
Heating (base 65°F)		1587	1268	1008	543	240	50	10	22	165	508	960	1448	7809
Cooling (base 65°F)		0	0	0	0	35	149	298	220	42	0	0	0	744
% OF POSSIBLE SUNSHINE														
MEAN SKY COVER (tenths)														
Sunrise - Sunset	48	6.4	6.5	6.9	6.5	6.3	5.6	4.9	5.0	5.1	5.5	6.7	6.6	6.0
MEAN NUMBER OF DAYS:														
Sunrise to Sunset														
-Clear	48	8.1	7.0	6.3	7.2	7.3	8.7	11.7	11.8	11.8	11.1	6.9	7.6	105.6
-Partly Cloudy	48	7.7	6.6	7.6	8.2	9.9	10.9	11.7	10.6	8.0	7.9	7.3	7.0	103.4
-Cloudy	48	15.1	14.6	17.0	14.7	13.8	10.4	7.6	8.6	10.3	12.0	15.8	16.4	156.3
Precipitation														
.01 inches or more	48	6.1	6.5	8.7	9.4	10.5	10.9	9.7	9.0	8.1	6.1	6.4	6.2	97.6
Snow, Ice pellets, hail														
1.0 inches or more	48	2.0	2.1	2.5	0.6	0.0	0.0	0.0	0.0	0.0	0.1	1.6	2.0	11.0
Thunderstorms	48	0.*	0.1	0.9	3.0	6.2	9.0	9.5	8.0	5.4	1.8	0.4	0.1	44.3
Heavy Fog Visibility 1/4 mile or less	48	2.5	3.0	2.8	0.9	0.8	0.5	0.7	1.2	1.4	1.5	2.7	3.6	21.4
Temperature °F														
-Maximum														
90° and above	30	0.0	0.0	0.0	0.2	0.7	4.0	10.2	7.3	1.7	0.0	0.0	0.0	24.1
32° and below	30	20.7	15.3	7.0	0.4	0.0	0.0	0.0	0.0	0.0	0.2	6.8	18.6	68.9
-Minimum														
32° and below	30	30.9	27.6	25.6	12.2	2.0	0.0	0.0	0.0	1.3	11.9	25.7	30.7	167.8
0° and below	30	12.7	7.8	1.5	0.0	0.0	0.0	0.0	0.0	0.0	0.0	1.4	7.8	31.2
AVG. STATION PRESS. (mb)	21	966.4	966.8	963.3	962.7	962.3	962.3	963.9	964.7	965.0	965.4	964.9	966.0	964.5
RELATIVE HUMIDITY (%)														
Hour 00	30	76	78	79	73	72	75	76	79	78	75	80	80	77
Hour 06 (Local Time)	30	77	80	82	81	81	82	83	86	86	81	83	81	82
Hour 12	30	68	68	64	54	53	55	54	55	57	55	66	71	60
Hour 18	30	72	70	63	51	49	51	51	54	56	57	71	76	60
PRECIPITATION (inches):														
Water Equivalent														
-Normal		0.51	0.64	1.64	2.52	3.03	3.40	2.68	2.85	3.02	1.78	1.09	0.70	23.86
-Maximum Monthly	48	1.71	4.05	3.60	5.79	8.26	8.43	8.41	9.09	9.26	5.73	2.95	2.62	9.26
-Year		1969	1962	1977	1984	1993	1984	1992	1975	1986	1973	1983	1968	SEP 1986
-Minimum Monthly	48	0.05	0.05	0.14	0.17	0.61	0.91	0.25	0.53	0.29	T	0.02	T	T
-Year		1958	1986	1967	1969	1981	1988	1947	1970	1956	1952	1980	1986	DEC 1986
-Maximum in 24 hrs	48	1.61	2.00	1.96	2.64	3.92	4.32	3.39	4.59	4.02	4.54	1.62	1.44	4.59
-Year		1960	1962	1956	1953	1972	1957	1992	1975	1966	1973	1972	1955	AUG 1975
Snow, Ice pellets, hail														
-Maximum Monthly	48	19.6	48.4	31.5	18.4	0.2	T	T	0.0	0.9	10.0	21.9	41.1	48.4
-Year		1969	1962	1951	1983	1954	1993	1993		1985	1991	1985	1968	FEB 1962
-Maximum in 24 hrs	48	11.8	26.0	18.9	9.0	0.2	T	T	0.0	0.9	8.8	11.8	16.6	26.0
-Year		1960	1962	1956	1957	1954	1993	1993		1985	1991	1979	1968	FEB 1962
WIND:														
Mean Speed (mph)	45	11.0	11.0	12.5	13.2	11.9	10.7	9.8	9.8	10.3	10.7	11.5	10.8	11.1
Prevailing Direction through 1963		NW	NW	NW	NW	S	S	S	S	S	S	NW	NW	S
Fastest Obs. 1 Min.														
-Direction (!!)	45	32	30	02	25	11	23	36	29	16	27	36	36	23
-Speed (MPH)	45	47	44	60	48	46	70	69	52	50	60	52	46	70
-Year		1990	1972	1950	1955	1950	1952	1982	1956	1953	1949	1975	1982	JUN 1952
Peak Gust														
-Direction (!!)	10	NW	NW	NW	SW	SE	E	NW	E	N	NW	W	NW	E
-Speed (mph)	10	67	55	56	64	53	71	66	47	53	53	58	53	71
-Date		1990	1984	1991	1985	1992	1990	1991	1986	1990	1984	1986	1992	JUN 1990

PRECIPITATION (inches)

SIOUX FALLS, SOUTH DAKOTA

YEAR	JAN	FEB	MAR	APR	MAY	JUNE	JULY	AUG	SEP	OCT	NOV	DEC	ANNUAL
1964	0.34	0.08	2.12	4.03	1.29	1.68	4.03	3.87	4.06	0.09	0.38	0.59	22.56
1965	0.24	1.46	1.09	3.35	7.29	4.91	1.49	1.29	4.91	1.05	0.26	0.60	27.94
1966	0.54	0.99	0.70	1.71	1.94	2.68	1.54	2.12	6.34	1.43	0.20	0.65	20.84
1967	0.75	0.44	0.14	3.90	0.72	4.26	0.53	3.46	0.87	0.39	0.03	0.91	16.40
1968	0.33	0.09	0.61	4.34	2.69	4.10	2.37	1.70	4.01	4.57	0.39	2.62	27.82
1969	1.71	2.55	1.09	0.17	2.43	4.85	2.73	5.07	2.41	2.05	0.34	1.18	26.58
1970	0.37	0.10	2.03	3.75	4.83	3.81	2.98	0.53	3.14	3.13	2.17	0.54	27.38
1971	0.13	0.90	0.85	1.59	1.06	6.10	2.92	0.71	3.23	3.06	2.45	0.64	23.64
1972	0.18	0.40	0.97	2.73	7.25	2.09	3.49	2.65	1.75	1.78	1.89	1.25	26.43
1973	0.43	0.43	3.52	2.12	1.93	2.38	3.50	1.05	5.61	5.73	1.01	0.48	28.19
1974	0.13	0.30	1.65	1.33	3.11	2.79	1.27	5.16	0.58	0.34	0.27	0.10	17.03
1975	1.35	0.22	1.95	2.45	1.66	4.48	0.62	9.09	1.35	0.49	2.25	0.19	26.10
1976	0.41	0.48	1.60	2.15	1.02	1.02	1.53	1.31	0.76	0.71	0.07	0.36	11.42
1977	0.19	0.83	3.60	2.17	3.17	1.73	3.64	5.63	5.63	2.36	1.80	0.87	31.62
1978	0.47	0.33	0.56	3.98	3.47	2.91	4.79	3.08	2.45	0.14	0.48	0.51	23.17
1979	1.14	0.41	3.47	2.75	4.90	3.01	3.13	4.35	4.03	3.30	1.72	0.04	32.25
1980	0.18	0.47	0.70	0.77	2.52	2.17	1.63	2.92	0.79	1.36	0.02	0.29	13.82
1981	0.12	0.33	1.86	0.58	0.61	3.90	3.89	2.28	0.50	2.45	1.21	0.38	18.11
1982	0.76	0.13	1.17	1.87	4.72	1.18	4.60	5.23	3.49	5.18	2.94	1.99	33.26
1983	0.52	0.22	3.35	2.88	2.92	6.75	1.82	2.00	1.92	0.71	2.95	0.73	26.77
1984	0.37	1.10	1.83	5.79	2.95	8.43	1.63	0.76	1.62	4.11	0.03	1.02	29.64
1985	0.45	0.05	2.37	5.18	3.29	2.52	2.70	4.07	3.34	0.75	1.97	0.47	27.16
1986	0.72	0.05	1.50	5.15	2.42	3.93	2.59	2.77	9.26	1.22	0.89	T	30.50
1987	0.19	0.26	3.27	0.28	2.94	1.78	3.16	1.36	2.05	0.31	1.66	1.40	18.66
1988	1.54	0.25	0.63	3.00	1.54	0.91	0.49	4.02	4.39	0.02	1.98	0.37	19.14
1989	0.23	0.51	1.07	1.59	1.42	2.50	1.37	2.46	3.38	0.10	0.91	0.25	15.79
1990	0.08	0.31	1.57	1.86	4.07	4.86	1.77	1.17	0.47	1.82	0.61	0.61	19.20
1991	0.22	0.34	0.86	2.21	6.20	6.36	2.26	1.41	3.95	1.65	1.78	0.20	27.44
1992	0.75	1.76	2.36	2.01	1.80	2.44	8.41	5.29	3.06	2.72	1.04	0.83	32.47
1993	0.70	0.81	2.04	2.61	8.26	6.43	7.86	3.10	1.88	0.62	1.50	0.30	36.11
Record Mean	0.60	0.76	1.45	2.49	3.53	4.13	3.04	3.08	2.74	1.58	1.07	0.71	25.17

AVERAGE TEMPERATURE (deg. F)

SIOUX FALLS, SOUTH DAKOTA

YEAR	JAN	FEB	MAR	APR	MAY	JUNE	JULY	AUG	SEP	OCT	NOV	DEC	ANNUAL
1964	23.0	26.9	26.6	48.1	61.6	70.1	77.3	68.3	59.3	48.7	32.7	15.5	46.5
1965	12.0	13.0	21.7	45.9	59.0	65.5	70.8	69.4	51.5	51.7	32.8	28.4	43.5
1966	5.1	17.3	37.7	41.1	54.3	68.5	78.7	68.8	59.9	48.3	31.0	19.4	44.2
1967	17.6	15.3	37.1	46.0	53.1	66.3	71.7	69.9	61.3	47.2	32.7	21.9	45.0
1968	16.8	18.9	39.8	47.8	52.6	69.6	71.8	72.7	61.0	49.4	33.5	16.2	45.9
1969	8.8	19.0	20.5	48.4	59.6	61.4	73.5	73.3	62.3	43.4	33.4	19.5	43.6
1970	4.7	19.4	26.4	44.8	59.9	69.0	73.3	72.5	61.9	46.2	32.0	15.6	43.8
1971	8.3	18.8	31.2	47.8	55.3	71.3	69.3	73.0	61.5	51.7	33.3	17.4	44.9
1972	8.9	11.5	31.5	43.8	59.1	66.7	70.2	70.5	59.5	44.4	31.9	14.8	42.7
1973	18.5	23.1	39.7	46.4	56.9	68.9	74.4	76.7	60.1	52.9	34.7	17.1	47.5
1974	12.8	24.3	34.5	48.9	55.7	65.9	79.2	67.6	57.7	51.4	33.9	23.6	46.3
1975	17.1	16.5	25.5	41.4	61.1	67.6	78.5	72.7	57.0	52.1	33.5	20.4	45.3
1976	15.6	29.5	34.8	51.0	57.1	71.5	76.7	75.2	62.9	44.3	27.6	16.6	46.9
1977	4.7	25.6	36.0	53.8	66.8	71.1	77.0	68.1	61.9	47.7	31.3	16.3	46.7
1978	1.9	8.6	30.5	44.3	57.7	66.6	70.9	71.1	66.7	47.8	31.3	14.6	42.7
1979	1.8	7.5	28.2	44.1	55.2	67.8	74.0	69.9	64.3	48.2	30.2	27.1	43.2
1980	18.1	18.1	31.7	50.1	58.4	68.8	74.2	71.1	62.0	45.5	36.7	21.2	46.3
1981	22.3	26.0	38.4	53.5	58.3	70.4	75.5	71.1	63.0	48.7	39.5	18.0	48.8
1982	3.8	20.2	32.8	43.9	59.6	62.6	74.4	71.3	60.0	48.8	30.3	24.5	44.3
1983	20.1	26.2	33.1	41.5	55.1	66.7	77.1	78.3	63.5	49.6	34.9	2.1	45.7
1984	17.4	27.8	24.6	45.8	55.9	68.1	73.6	74.2	57.2	50.4	36.2	20.4	46.0
1985	13.4	19.9	37.8	52.8	62.7	64.2	71.5	66.3	58.2	46.6	20.7	9.5	43.6
1986	20.7	16.9	36.9	48.2	58.8	69.6	75.0	66.5	59.6	48.6	28.7	25.2	46.2
1987	24.0	32.8	37.8	52.6	64.9	71.4	77.0	68.8	62.8	43.9	38.1	24.4	49.9
1988	9.6	15.0	36.4	46.4	65.1	76.3	77.4	74.8	62.6	44.6	34.2	22.1	47.0
1989	25.5	9.2	29.8	47.8	58.0	66.9	77.3	72.0	60.2	49.5	29.5	11.5	44.8
1990	28.2	24.9	36.7	46.4	56.4	70.1	71.2	72.9	66.5	48.0	35.6	15.3	47.7
1991	13.7	29.8	37.0	50.0	62.5	73.5	73.1	72.8	62.0	46.1	25.7	26.0	47.7
1992	26.5	30.0	36.7	43.7	60.5	67.0	65.5	66.1	60.3	48.7	30.4	19.2	46.2
1993	14.0	15.0	29.0	44.1	57.1	65.1	71.4	71.4	56.8	47.1	29.9	21.9	43.6
Record Mean	15.3	20.6	32.3	47.1	58.9	68.5	74.2	71.7	61.7	49.8	33.1	20.3	46.1
Max	25.3	30.6	42.2	58.9	71.1	80.3	86.1	83.5	73.6	61.7	42.9	29.6	57.1
Min	5.2	10.6	22.4	35.3	46.7	56.6	62.2	59.9	49.8	37.8	23.3	11.0	35.1

REFERENCE NOTES

SIOUX FALLS, SOUTH DAKOTA

GENERAL
T - TRACE AMOUNT.
BLANK ENTRIES DENOTE MISSING/UNREPORTED DATA.
# INDICATES A STATION OR INSTRUMENT RELOCATION.

SPECIFIC

PM - INCLUDES LAST DAY OF PREVIOUS MONTH
ASOS - AUTOMATED SURFACE OBSERVING SYSTEM IN OPERATION DURING THESE MONTHS.

(a) - LENGTH OF RECORD IN YEARS, ALTHOUGH INDIVIDUAL MONTHS MAY BE MISSING.
0.* OR * - THE VALUE IS BETWEEN 0.0 AND 0.05
NORMALS - BASED ON THE 1961-1990 RECORD PERIOD.
EXTREMES - DATES ARE THE MOST RECENT OCCURRENCE
WIND DIR.- NUMERALS SHOW TENS OF DEGRESS CLOCKWISE FROM TRUE NORTH. "00" INDICATES CALM.
RESULTANT DIRECTIONS ARE GIVEN TO WHOLE DEGREES.
BOLD VALUES INDICATE EXTREME VALUES WHICH OCCURRED AFTER THE ASOS SYSTEM WAS COMMISSIONED.

RECORD = PERIOD OF RECORD
RECORD MEAN PRECIPITATION IS THE MEAN OF ALL DAILY PRECIPITATION AMOUNTS DURING THE PERIOD OF RECORD.
RECORD MAX(MIN) TEMPERATURE IS THE MEAN OF ALL DAILY MAX(MIN) TEMPERATURES DURING THE PERIOD OF RECORD.
RECORD MEAN TEMPERATURE IS THE SUM OF THE RECORD MAX AND RECORD MIN DIVIDED BY 2.
AVERAGE TEMPERATURE IS THE SUM OF THE MEAN DAILY MAX AND MIN TEMPERATURE DIVIDED BY 2.

EXCEPTIONS

RECORD MEANS ARE THROUGH THE CURRENT YEAR, BEGINNING IN 1921 FOR TEMPERATURE
1891 FOR PRECIPITATION
1946 FOR SNOWFALL

HEATING DEGREE DAYS Base 65 deg. F SIOUX FALLS, SOUTH DAKOTA

SEASON	JULY	AUG	SEP	OCT	NOV	DEC	JAN	FEB	MAR	APR	MAY	JUNE	TOTAL
1964-65	0	55	210	499	965	1528	1639	1452	1340	568	195	41	8492
1965-66	2	37	405	408	964	1129	1856	1331	838	709	343	49	8071
1966-67	0	41	188	508	1016	1406	1462	1385	858	563	399	37	7863
1967-68	36	32	141	549	963	1328	1490	1331	776	511	380	53	7590
1968-69	14	14	152	489	938	1511	1742	1283	1373	493	217	143	8369
1969-70	2	2	117	668	941	1404	1867	1274	1192	603	206	46	8322
1970-71	16	3	195	577	986	1526	1754	1288	1041	511	297	17	8211
1971-72	29	12	177	409	945	1470	1737	1551	1031	628	236	50	8275
1972-73	29	43	194	631	984	1555	1440	1166	775	549	256	19	7641
1973-74	1	0	178	373	902	1481	1616	1133	939	477	303	85	7488
1974-75	2	39	240	419	927	1279	1476	1351	1217	701	157	51	7859
1975-76	3	5	271	407	937	1377	1528	1023	929	418	261	23	7182
1976-77	1	0	142	643	1114	1495	1867	1097	891	341	38	4	7633
1977-78	0	22	125	530	1000	1503	1957	1576	1063	613	252	78	8719
1978-79	11	18	104	525	1004	1554	1960	1607	1134	622	321	40	8900
1979-80	0	28	103	512	1038	1168	1448	1349	1027	464	243	32	7412
1980-81	1	14	157	602	841	1354	1318	1087	816	353	230	5	6778
1981-82	5	5	114	497	758	1452	1899	1247	991	632	174	105	7879
1982-83	0	28	188	494	1035	1249	1385	1082	982	699	317	67	7526
1983-84	1	0	160	476	894	1947	1468	1072	1247	569	285	23	8142
1984-85	2	12	265	451	857	1376	1594	1260	836	378	124	101	7256
1985-86	5	44	269	562	1327	1721	1363	1341	865	498	204	17	8216
1986-87	0	54	180	504	1082	1227	1265	896	835	387	96	23	6549
1987-88	5	54	110	649	801	1252	1715	1448	878	554	83	1	7550
1988-89	0	22	126	628	916	1321	1219	1559	1083	514	235	61	7684
1989-90	0	5	192	481	1056	1655	1131	1117	871	586	268	39	7401
1990-91	11	7	109	527	875	1538	1587	977	859	455	192	7	7144
1991-92	6	3	180	583	1171	1200	1186	1009	871	636	187	35	7067
1992-93	35	63	159	500	1030	1414	1576	1397	1109	621	244	74	8222
1993-94	2	19	257	557	1046	1331							

COOLING DEGREE DAYS Base 65 deg. F SIOUX FALLS, SOUTH DAKOTA

YEAR	JAN	FEB	MAR	APR	MAY	JUNE	JULY	AUG	SEP	OCT	NOV	DEC	TOTAL
1969	0	0	0	0	58	43	270	265	41	4	0	0	681
1970	0	0	0	6	54	172	280	241	108	2	0	0	863
1971	0	0	0	4	1	211	167	268	79	2	0	0	732
1972	0	0	0	0	60	108	197	219	35	0	0	0	619
1973	0	0	0	0	13	143	300	370	38	8	0	0	872
1974	0	0	0	2	23	118	450	126	26	6	0	0	751
1975	0	0	0	0	44	134	428	252	36	16	0	0	910
1976	0	0	0	2	24	226	370	324	85	9	0	0	1040
1977	0	0	0	11	101	197	382	123	40	0	0	0	854
1978	0	0	0	0	32	133	202	213	163	0	0	0	743
1979	0	0	0	3	25	131	289	189	87	0	0	0	724
1980	0	0	0	23	46	153	290	207	74	6	0	0	799
1981	0	0	0	14	27	175	342	203	61	0	0	0	822
1982	0	0	0	4	12	42	298	230	45	0	0	0	631
1983	0	0	0	0	15	122	381	420	120	2	0	0	1060
1984	0	0	0	0	10	120	276	305	36	4	0	0	751
1985	0	0	0	19	59	85	212	93	72	0	0	0	540
1986	0	0	0	0	20	163	318	108	25	0	0	0	634
1987	0	0	0	20	100	219	381	178	53	0	0	0	951
1988	0	0	0	1	94	349	393	330	61	0	0	0	1228
1989	0	0	0	6	25	125	387	228	56	7	0	0	834
1990	0	0	0	31	11	198	209	258	160	7	0	0	874
1991	0	0	0	11	119	265	264	249	100	5	0	0	1013
1992	0	0	0	4	53	103	58	103	25	5	0	0	351
1993	0	0	0	0	5	85	208	223	16	9	0	0	546

SNOWFALL (inches) SIOUX FALLS, SOUTH DAKOTA

SEASON	JULY	AUG	SEP	OCT	NOV	DEC	JAN	FEB	MAR	APR	MAY	JUNE	TOTAL
1964-65	0.0	0.0	0.0	0.0	3.9	6.5	3.3	21.1	13.3	0.4	0.0	0.0	48.5
1965-66	0.0	0.0	0.0	0.0	1.6	0.1	6.2	1.8	4.6	0.9	T	0.0	15.2
1966-67	0.0	0.0	0.0	0.4	2.3	9.3	6.0	11.8	1.0	T	0.1	0.0	30.9
1967-68	0.0	0.0	0.0	0.5	0.4	2.5	4.6	0.9	T	0.1	0.1	0.0	9.1
1968-69	0.0	0.0	0.0	0.7	1.9	41.1	19.6	28.5	2.9	0.0	0.0	0.0	94.7
1969-70	0.0	0.0	0.0	T	2.9	15.2	5.9	1.5	19.3	3.0	T	0.0	47.8
1970-71	0.0	0.0	0.0	5.1	0.5	8.5	1.9	1.7	4.7	0.3	0.0	0.0	22.7
1971-72	0.0	0.0	0.0	T	4.4	10.6	1.8	10.6	5.0	0.8	0.0	0.0	33.2
1972-73	0.0	0.0	0.0	0.2	2.5	6.2	5.5	7.4	0.3	T	0.0	0.0	22.1
1973-74	0.0	0.0	0.0	T	0.1	11.6	1.9	2.9	5.1	3.3	0.0	0.0	24.9
1974-75	0.0	0.0	0.0	0.0	2.5	1.1	18.3	3.1	17.9	T	0.0	0.0	42.9
1975-76	0.0	0.0	0.0	T	13.2	2.5	6.5	6.6	7.9	T	0.1	0.0	36.8
1976-77	0.0	0.0	0.0	3.4	0.7	6.7	2.7	1.9	13.5	T	0.0	0.0	28.9
1977-78	0.0	0.0	0.0	0.9	8.6	8.0	7.6	6.5	5.5	0.6	0.0	0.0	37.7
1978-79	0.0	0.0	0.0	0.0	2.9	9.5	19.0	7.3	13.2	1.5	0.0	0.0	53.4
1979-80	0.0	0.0	0.0	T	15.2	0.3	2.9	5.5	5.8	T	0.0	0.0	29.7
1980-81	0.0	0.0	0.0	0.8	0.4	4.6	1.4	3.1	T	0.5	0.0	0.0	10.8
1981-82	0.0	0.0	0.0	0.8	8.9	5.5	16.9	1.0	1.8	7.5	0.0	0.0	42.4
1982-83	0.0	0.0	0.0	3.3	4.1	17.6	4.7	3.6	18.8	18.4	0.0	0.0	70.5
1983-84	0.0	0.0	0.0	T	19.0	13.7	5.0	11.9	19.4	6.0	0.0	0.0	75.0
1984-85	0.0	0.0	T	T	T	4.7	7.4	0.7	16.1	2.5	0.0	0.0	31.4
1985-86	0.0	0.0	0.9	0.2	21.9	9.1	9.1	0.9	8.2	0.3	0.0	0.0	50.6
1986-87	0.0	0.0	0.0	T	5.4	T	2.4	0.3	T	T	0.0	0.0	8.1
1987-88	0.0	0.0	0.0	0.1	7.5	13.3	17.9	7.3	2.5	11.3	0.0	0.0	59.9
1988-89	0.0	0.0	0.0	T	10.9	2.2	2.0	10.0	16.0	0.7	T	0.0	41.8
1989-90	0.0	0.0	0.0	0.0	2.4	4.0	0.2	5.8	0.7	T	0.0	T	13.1
1990-91	0.0	0.0	0.0	1.2	8.4	8.8	5.4	5.4	3.2	T	T	T	32.4
1991-92	0.0	0.0	0.0	10.0	10.8	0.2	3.8	11.2	8.4	3.5	T	T	47.9
1992-93	0.0	0.0	0.0	T	4.7	8.4	8.6	13.1	14.9	2.2	T	T	51.9
1993-94	T	0.0	T	T	8.8	4.6							
Record Mean	T	0.0	T	0.7	5.4	7.2	6.4	8.1	9.6	2.2	T	T	39.7

BRISTOL, JOHNSON CITY, KINGSPORT, TENNESSEE

The Weather Service Office is located an almost equal distance of 15 miles in the middle of a geographical triangle between the cities of Bristol, Tennessee–Virginia, Kingsport and Johnson City, Tennessee, and is more commonly known as the Tri–City Area. This location is situated in the extreme upper East Tennessee Valley. The terrain immediately surrounding the station ranges from gently rolling on the east and south to very hilly on the west and north. Mountain ranges begin about 10 miles to the southeast and about 15 miles to the west and north, with many peaks and ridges rising to 4,000 feet, and some to 6,000 feet toward the southeast.

This section does not lie directly within any of the principal storm tracks that cross the country, but comes under the influence of storm centers that pass along the Gulf Coast and then up the Atlantic Coast toward the northeast. Being quite varied, the topography has considerable influence on the weather. Moist air from the east is forced up the slopes of the mountains causing much of the moisture to be precipitated before the air mass reaches the Bristol area. The same process occurs to a lesser extent when air masses move over the smaller mountain ranges to the west and north. The maximum monthly precipitation occurs in July, usually from afternoon and early evening thunderstorms. A second maximum of precipitation occurs in the late winter months, due mainly to moist air associated with stórm centers to the south or northeast. Annual precipitation amounts recorded in mountainous sections to the east and southeast are almost double what they are in the immediate vicinity.

Lowest temperatures normally occur during the early morning hours, but rise rapidly during the morning hours. Periods of cold weather are generally associated with air flow from winter storm centers near the northeast coast. Periods of unusually high temperatures occur most frequently when Gulf air associated with the Bermuda high pressure system dominates the area.

Snowfall seldom occurs before November and rarely remains on the ground for more than a few days. However, mountains to the east and south of the station are frequently well blanketed with snow for much longer periods of time.

Agricultural activies within this area include such staple crops as tobacco, beans, and hay which are raised in such amounts as to be important commercially. The last freezing temperature in spring normally occurs in late April, and the first in autumn around mid–October. The growing season of 180 days, usually coupled with ample sunshine and rainfall, permits a second planting and harvesting of some staple crops.

NORMALS, MEANS, AND EXTREMES

BRISTOL, JOHNSON CITY, KINGSPORT, TENNESSEE

LATITUDE: 36°29'N LONGITUDE: 82°24' W ELEVATION: FT. GRND 1507 BARO 1558 TIME ZONE: EASTERN WBAN: 13877

	(a)	JAN	FEB	MAR	APR	MAY	JUNE	JULY	AUG	SEP	OCT	NOV	DEC	YEAR
TEMPERATURE °F:														
Normals														
-Daily Maximum		43.7	48.0	58.9	67.4	75.2	82.2	84.6	84.1	79.1	69.1	58.2	48.1	66.6
-Daily Minimum		24.3	26.8	35.4	43.0	51.6	59.9	64.1	63.1	56.6	44.2	35.9	28.2	44.4
-Monthly		34.0	37.4	47.2	55.2	63.4	71.1	74.4	73.6	67.9	56.7	47.0	38.2	55.5
Extremes														
-Record Highest	48	79	80	85	89	92	97	102	101	100	90	81	78	102
-Year		1950	1977	1954	1986	1969	1952	1952	1988	1954	1954	1993	1951	JUL 1952
-Record Lowest	48	-21	-5	-2	21	30	38	45	43	34	20	5	-9	-21
-Year		1985	1958	1980	1982	1989	1966	1947	1986	1983	1962	1950	1962	JAN 1985
NORMAL DEGREE DAYS:														
Heating (base 65°F)		961	773	552	298	121	6	0	0	44	280	540	831	4406
Cooling (base 65°F)		0	0	0	0	71	189	291	267	131	23	0	0	972
% OF POSSIBLE SUNSHINE														
MEAN SKY COVER (tenths)														
Sunrise - Sunset	45	7.2	6.9	6.8	6.3	6.3	6.0	6.2	6.0	5.6	5.1	6.3	6.9	6.3
MEAN NUMBER OF DAYS:														
Sunrise to Sunset														
-Clear	56	5.9	5.9	6.6	7.6	6.8	6.2	6.0	6.8	9.6	12.2	8.4	6.7	88.6
-Partly Cloudy	56	6.8	6.8	7.7	8.7	10.6	12.4	12.5	13.5	9.8	8.3	7.3	7.1	111.5
-Cloudy	56	18.3	15.6	16.7	13.7	13.6	11.4	12.5	10.8	10.6	10.5	14.3	17.2	165.2
Precipitation														
.01 inches or more	48	13.6	11.8	13.0	11.3	11.7	11.1	12.0	10.5	8.2	8.1	10.4	11.5	133.4
Snow,Ice pellets,hail														
1.0 inches or more	50	1.7	1.2	0.6	0.1	0.0	0.0	0.0	0.0	0.0	0.*	0.2	0.9	4.7
Thunderstorms	50	0.3	0.9	2.0	3.4	6.4	8.1	9.3	7.4	3.5	0.8	0.4	0.2	42.8
Heavy Fog Visibility														
1/4 mile or less	50	3.2	2.4	1.4	1.6	3.5	3.6	4.9	7.1	5.6	5.0	2.8	3.1	44.3
Temperature °F														
-Maximum														
90° and above	32	0.0	0.0	0.0	0.0	0.3	2.5	5.8	3.5	1.7	0.0	0.0	0.0	13.7
32° and below	32	5.1	2.8	0.4	0.*	0.0	0.0	0.0	0.0	0.0	0.0	0.2	2.6	11.2
-Minimum														
32° and below	32	23.1	19.7	13.3	3.6	0.2	0.0	0.0	0.0	0.0	2.9	12.5	20.4	95.8
0° and below	32	0.9	0.3	0.*	0.0	0.0	0.0	0.0	0.0	0.0	0.0	0.0	0.2	1.4
AVG. STATION PRESS.(mb)	21	964.5	963.8	962.2	961.9	962.3	963.3	964.4	965.0	965.3	965.7	965.3	965.2	964.1
RELATIVE HUMIDITY (%)														
Hour 01	32	77	74	72	72	83	87	88	89	88	84	79	78	81
Hour 07 (Local Time)	32	80	79	79	81	88	90	92	93	93	89	84	81	86
Hour 13	32	62	58	52	49	55	57	60	60	58	53	56	62	57
Hour 19	32	65	60	55	51	61	63	67	69	70	64	65	68	63
PRECIPITATION (inches):														
Water Equivalent														
-Normal		3.23	3.44	3.70	3.30	3.84	3.54	4.32	3.17	3.26	2.59	2.94	3.39	40.72
-Maximum Monthly	48	9.18	7.29	9.56	5.85	9.71	6.97	9.73	7.07	7.09	5.65	5.90	6.75	9.73
-Year		1957	1956	1955	1970	1950	1989	1949	1966	1972	1959	1948	1961	JUL 1949
-Minimum Monthly	48	1.37	0.75	1.31	0.21	1.31	0.75	0.79	0.55	0.50	0.07	1.07	0.21	0.07
-Year		1981	1968	1985	1976	1966	1986	1957	1987	1985	1963	1953	1965	OCT 1963
-Maximum in 24 hrs	48	2.34	2.02	3.35	2.66	3.26	3.10	2.90	3.07	3.61	3.65	2.55	2.95	3.65
-Year		1950	1991	1973	1977	1984	1954	1946	1982	1972	1964	1957	1969	OCT 1964
Snow,Ice pellets,hail														
-Maximum Monthly	56	22.1	20.4	27.9	14.8	T	T	0.0	0.0	0.0	1.3	18.1	12.9	27.9
-Year		1966	1979	1960	1987	1992	1993				1993	1952	1963	MAR 1960
-Maximum in 24 hrs	50	9.7	10.7	14.2	10.0	T	T	0.0	0.0	0.0	1.3	16.2	9.6	16.2
-Year		1955	1969	1993	1987	1992	1993				1993	1952	1969	NOV 1952
WIND:														
Mean Speed (mph)	39	6.5	6.7	7.2	6.9	5.2	4.8	4.2	3.9	4.3	4.7	5.6	5.9	5.5
Prevailing Direction through 1963		WSW	NE	WNW	WSW	WSW	NE	WSW	NE	NE	NE	W	WSW	NE
Fastest Obs. 1 Min.														
-Direction (!!)	38	25	25	25	25	32	27	23	34	31	28	26	24	32
-Speed (MPH)	38	40	46	40	41	50	39	40	46	29	35	37	40	50
-Year		1965	1961	1952	1977	1951	1989	1961	1962	1967	1965	1988	1968	MAY 1951
Peak Gust														
-Direction (!!)	10	W	W	W	W	W	SW	SW	S	SW	W	W	W	W
-Speed (mph)	10	55	49	52	63	46	59	47	48	54	40	53	48	63
-Date		1987	1990	1988	1991	1990	1990	1990	1993	1984	1990	1988	1984	APR 1991

PRECIPITATION (inches)

BRISTOL, JOHNSON CITY, KINGSPORT, TENNESSEE

YEAR	JAN	FEB	MAR	APR	MAY	JUNE	JULY	AUG	SEP	OCT	NOV	DEC	ANNUAL
1964	3.25	4.11	4.66	4.65	2.27	1.14	3.73	3.79	3.14	4.95	1.88	2.48	40.05
1965	3.23	2.77	5.82	3.42	2.75	3.45	7.17	2.65	1.78	2.16	1.85	0.21	37.26
1966	3.30	3.50	2.05	5.04	1.31	1.77	6.21	7.07	4.73	3.50	3.71	3.07	45.26
1967	2.00	4.14	3.02	2.80	6.29	2.14	4.87	3.68	2.59	1.99	3.73	5.62	42.87
1968	3.58	0.75	4.95	4.12	3.83	2.89	2.89	2.32	0.93	2.51	1.89	2.13	32.79
1969	2.72	5.72	2.14	2.11	2.06	5.24	8.18	2.11	4.31	1.54	1.86	5.78	43.77
1970	2.89	2.75	1.60	5.85	1.60	2.48	2.95	4.28	1.67	3.29	1.33	3.40	34.09
1971	3.83	3.59	3.42	3.38	4.38	5.41	5.94	2.32	2.91	3.18	2.45	2.43	43.24
1972	4.87	4.81	2.86	2.92	5.92	2.97	3.18	2.70	7.09	4.57	2.85	6.43	51.17
1973	1.80	2.04	7.18	3.58	6.46	2.99	5.66	2.60	1.68	2.29	4.52	5.07	45.87
1974	5.33	4.07	4.34	3.96	8.66	4.01	1.78	1.95	3.12	2.34	3.29	3.53	46.38
1975	4.19	3.13	9.22	3.67	4.29	3.19	2.46	3.87	5.02	1.54	3.14	3.33	47.05
1976	2.60	3.07	3.24	0.21	3.31	5.62	2.14	2.27	4.44	5.30	1.41	3.80	37.41
1977	1.90	1.01	4.86	5.43	1.91	3.77	2.13	3.09	2.36	4.89	4.91	2.72	38.98
1978	4.22	0.89	3.18	2.44	4.50	5.67	4.78	3.71	3.04	0.58	2.64	4.89	40.54
1979	5.29	3.58	3.16	3.68	3.35	3.55	6.12	2.80	3.89	2.19	4.44	1.66	43.71
1980	3.91	1.39	5.68	3.56	2.69	1.10	3.82	2.54	3.01	2.09	2.10	1.38	33.27
1981	1.37	2.59	1.94	5.10	4.51	4.28	6.24	3.05	4.17	2.51	1.95	3.20	40.91
1982	4.07	5.07	3.35	2.30	2.55	5.52	9.14	4.70	5.53	2.54	4.12	2.89	51.78
1983	1.67	2.14	1.73	4.44	4.83	4.60	3.29	5.05	1.88	2.18	2.74	4.15	38.70
1984	1.79	4.50	2.73	2.85	7.42	3.86	4.63	1.23	1.43	1.14	2.61	1.76	35.95
1985	3.21	3.40	1.31	2.08	2.85	4.35	4.38	3.09	0.50	3.02	5.87	1.17	35.23
1986	1.55	4.11	1.56	0.51	4.16	0.75	5.50	3.40	3.93	1.69	2.67	3.66	33.49
1987	4.11	4.13	2.80	5.23	1.62	2.64	1.91	0.55	4.57	0.62	2.10	3.00	33.28
1988	2.74	3.20	1.54	2.69	2.48	0.89	3.20	2.78	3.20	1.79	3.44	2.73	30.68
1989	3.69	4.07	3.76	2.97	4.10	6.97	3.81	3.41	6.95	1.77	3.18	3.16	47.84
1990	3.23	5.06	4.00	2.44	6.57	2.90	3.78	3.51	1.47	5.23	1.32	4.85	44.36
1991	2.01	5.43	6.30	3.39	2.10	4.51	3.81	3.98	2.44	0.31	3.42	6.73	44.43
1992	2.42	3.52	2.62	1.53	5.46	3.44	5.11	2.41	1.49	3.42	2.25	5.05	38.72
1993	3.21	2.12	5.51	2.43	3.42	0.98	4.59	5.01	2.50	2.19	3.58	5.85	41.39
Record Mean	3.38	3.52	3.82	3.13	3.65	3.50	4.76	3.47	2.99	2.28	2.87	3.47	40.87

AVERAGE TEMPERATURE (deg. F)

BRISTOL, JOHNSON CITY, KINGSPORT, TENNESSEE

YEAR	JAN	FEB	MAR	APR	MAY	JUNE	JULY	AUG	SEP	OCT	NOV	DEC	ANNUAL
1964	34.1	34.6	44.7	56.4	65.9	74.2	75.2	73.2	68.6	54.0	50.2	41.6	56.1
1965	36.5	37.3	43.6	57.9	68.1	71.2	74.5	73.9	71.3	55.7	47.6	38.9	56.4
1966	29.5	37.9	47.1	54.9	62.9	71.9	75.4	73.3	66.4	55.0	47.0	37.7	54.9
1967	38.2	33.7	50.2	57.7	58.7	68.9	69.8	70.7	62.6	56.1	42.7	40.8	54.2
1968	33.2	30.1	46.9	55.8	61.3	70.8	75.8	75.5	67.1	57.6	45.2	35.2	54.6
1969	34.3	35.8	38.6	54.7	63.0	73.1	76.4	73.0	66.2	55.7	41.4	34.5	53.9
1970	27.6	36.8	47.0	59.0	66.1	72.1	76.8	75.0	72.9	61.5	46.8	41.2	56.9
1971	35.6	36.9	43.0	55.1	61.5	73.9	74.0	73.7	72.3	63.4	46.5	47.5	57.0
1972	40.8	35.7	45.1	54.9	62.7	67.4	73.6	73.7	69.8	55.9	46.1	43.7	55.8
1973	35.8	36.9	53.9	53.9	60.7	73.1	74.7	73.4	71.4	60.3	48.8	39.3	56.8
1974	47.0	39.8	51.2	56.4	64.3	67.7	74.0	73.0	64.8	54.5	47.2	39.5	56.6
1975	40.4	42.8	44.7	54.2	67.1	71.1	74.8	76.1	67.8	59.0	49.8	40.0	57.3
1976	33.5	47.3	51.9	55.3	61.2	70.7	73.2	71.6	65.1	51.6	38.8	33.3	54.5
1977	22.1	34.5	50.0	59.4	66.8	71.2	76.8	74.8	69.6	53.3	49.9	35.3	55.3
1978	25.8	28.8	44.8	55.5	62.3	71.5	74.4	74.5	72.3	56.3	51.0	39.5	54.7
1979	30.4	31.9	48.3	55.4	63.8	68.2	70.9	73.1	66.8	54.6	48.7	38.8	54.2
1980	37.9	30.6	43.2	55.4	64.2	70.6	76.4	76.9	70.9	54.1	45.3	36.3	55.1
1981	29.3	37.9	42.6	60.4	61.3	75.1	75.3	72.1	65.0	53.7	45.8	34.5	54.4
1982	31.2	40.2	49.8	53.1	68.2	71.4	75.1	72.3	65.7	57.6	48.3	43.3	56.3
1983	35.1	38.2	46.4	50.5	60.9	69.5	74.2	74.7	65.9	57.1	45.3	34.8	54.4
1984	32.7	40.3	44.5	53.0	59.4	72.4	71.3	72.9	64.5	63.6	43.0	45.5	55.3
1985	27.6	35.5	48.2	57.0	63.8	69.5	73.2	71.2	66.1	61.4	55.3	33.9	55.2
1986	32.0	40.4	46.2	57.2	64.8	74.1	77.2	73.0	69.7	58.7	51.4	38.8	57.0
1987	34.7	39.9	47.8	52.0	69.3	73.5	77.0	77.5	68.2	50.8	48.5	41.6	56.7
1988	31.5	37.3	47.8	54.1	62.3	70.5	76.7	77.1	68.1	50.2	47.6	37.6	55.1
1989	42.3	39.5	51.2	54.7	59.8	71.9	75.1	73.3	67.8	56.6	45.6	28.1	55.5
1990	41.6	45.4	51.2	55.3	63.7	72.0	75.2	74.4	68.7	58.0	49.1	44.3	58.2
1991	38.8	41.2	49.2	59.7	69.6	72.0	76.6	73.2	67.7	58.1	46.1	40.6	57.7
1992	38.6	42.7	44.8	55.5	61.2	69.2	74.9	71.1	68.4	54.0	47.0	38.3	55.5
1993	41.7	37.4	43.6	53.2	64.7	72.7	79.6	73.8	67.1	55.3	46.8	38.3	56.2
Record Mean	35.8	38.9	46.6	55.5	64.4	71.9	75.0	73.9	68.2	57.3	46.5	38.3	56.0
Max	45.3	49.5	58.1	68.2	76.3	83.3	85.6	84.8	79.8	69.8	57.7	48.1	67.2
Min	26.2	28.2	35.1	42.9	52.4	60.4	64.4	63.0	56.5	44.7	35.2	28.5	44.8

REFERENCE NOTES

BRISTOL, JOHNSON CITY, KINGSPORT, TENNESSEE

GENERAL
T - TRACE AMOUNT.
BLANK ENTRIES DENOTE MISSING/UNREPORTED DATA.
# INDICATES A STATION OR INSTRUMENT RELOCATION.

SPECIFIC

EXCEPTIONS

RECORD MEANS ARE THROUGH THE CURRENT YEAR,
BEGINNING IN 1938 FOR TEMPERATURE
1938 FOR PRECIPITATION
1938 FOR SNOWFALL

(a) - LENGTH OF RECORD IN YEARS, ALTHOUGH INDIVIDUAL MONTHS MAY BE MISSING.
0.* OR * - THE VALUE IS BETWEEN 0.0 AND 0.05
NORMALS - BASED ON THE 1961-1990 RECORD PERIOD.
EXTREMES - DATES ARE THE MOST RECENT OCCURRENCE
WIND DIR.- NUMERALS SHOW TENS OF DEGRESS CLOCKWISE FROM TRUE NORTH. "00" INDICATES CALM.
RESULTANT DIRECTIONS ARE GIVEN TO WHOLE DEGREES.
BOLD VALUES INDICATE EXTREME VALUES WHICH OCCURRED AFTER THE ASOS SYSTEM WAS COMMISSIONED.

RECORD = PERIOD OF RECORD
RECORD MEAN PRECIPITATION IS THE MEAN OF ALL DAILY PRECIPITATION AMOUNTS DURING THE PERIOD OF RECORD.
RECORD MAX(MIN) TEMPERATURE IS THE MEAN OF ALL DAILY MAX(MIN) TEMPERATURES DURING THE PERIOD OF RECORD.
RECORD MEAN TEMPERATURE IS THE SUM OF THE RECORD MAX AND RECORD MIN DIVIDED BY 2.
AVERAGE TEMPERATURE IS THE SUM OF THE MEAN DAILY MAX AND MIN TEMPERATURE DIVIDED BY 2.

HEATING DEGREE DAYS Base 65 deg. F BRISTOL, JOHNSON CITY, KINGSPORT, TENNESSEE

SEASON	JULY	AUG	SEP	OCT	NOV	DEC	JAN	FEB	MAR	APR	MAY	JUNE	TOTAL
1964-65	0	12	26	349	436	719	874	772	656	223	17	5	4089
1965-66	0	3	21	292	518	803	1096	755	547	307	109	20	4471
1966-67	0	0	23	303	533	843	821	868	455	218	216	29	4309
1967-68	12	2	107	271	662	745	979	1004	553	276	137	1	4749
1968-69	0	4	19	245	590	920	947	813	814	300	113	9	4774
1969-70	0	0	49	293	702	940	1155	784	553	201	66	1	4744
1970-71	0	0	26	147	538	737	903	780	676	296	135	0	4238
1971-72	0	0	0	87	560	537	743	844	610	307	86	36	3810
1972-73	8	0	10	281	559	652	898	779	334	333	152	0	4006
1973-74	0	0	12	169	483	792	552	700	424	257	86	14	3489
1974-75	0	0	72	326	527	785	755	616	621	328	29	1	4060
1975-76	0	0	46	188	451	768	970	509	402	297	134	9	3774
1976-77	0	0	46	409	776	973	1321	845	459	184	61	21	5095
1977-78	0	0	19	359	455	913	1207	1005	620	286	130	4	4998
1978-79	0	0	0	268	411	786	1065	920	511	283	88	11	4343
1979-80	3	3	23	323	483	804	832	990	670	285	76	4	4496
1980-81	0	0	31	337	586	882	1099	752	686	161	137	0	4671
1981-82	0	0	83	346	570	942	1042	687	465	351	32	0	4518
1982-83	0	0	66	256	494	667	922	745	570	436	140	19	4315
1983-84	0	0	98	242	583	928	995	709	627	357	199	8	4746
1984-85	0	0	91	73	652	599	1154	819	518	243	88	19	4256
1985-86	0	2	76	124	283	960	1015	681	577	240	80	0	4038
1986-87	0	19	2	227	402	804	932	698	527	393	30	0	4034
1987-88	0	0	24	436	489	718	1031	798	527	322	113	45	4503
1988-89	0	0	22	455	512	844	698	709	428	325	198	1	4192
1989-90	0	6	55	262	575	1139	718	540	423	303	93	0	4114
1990-91	0	0	48	225	474	636	806	659	483	175	32	2	3540
1991-92	0	0	68	224	562	750	813	638	617	293	148	22	4135
1992-93	0	3	32	337	532	820	714	767	657	346	65	10	4283
1993-94	0	0	59	298	537	823							

COOLING DEGREE DAYS Base 65 deg. F BRISTOL, JOHNSON CITY, KINGSPORT, TENNESSEE

YEAR	JAN	FEB	MAR	APR	MAY	JUNE	JULY	AUG	SEP	OCT	NOV	DEC	TOTAL
1969	0	0	0	0	60	261	363	255	93	14	0	0	1046
1970	0	0	0	28	106	220	373	314	270	45	0	0	1356
1971	0	0	0	5	32	273	285	276	228	44	10	2	1155
1972	0	0	0	11	22	113	279	274	160	3	0	0	862
1973	0	0	0	5	26	249	309	267	212	32	1	0	1101
1974	0	0	0	9	73	102	286	254	73	5	1	0	803
1975	0	0	0	12	101	190	310	353	136	7	0	0	1109
1976	0	0	2	14	24	185	262	212	54	0	0	0	753
1977	0	0	1	23	124	213	374	309	163	2	8	0	1217
1978	0	0	0	6	53	204	299	302	224	4	0	0	1092
1979	0	0	0	1	59	113	194	259	85	9	0	0	720
1980	0	0	0	6	61	179	359	377	215	5	0	0	1202
1981	0	0	0	27	30	309	325	226	87	1	0	0	1005
1982	0	0	0	0	136	199	322	232	93	36	0	0	1018
1983	0	0	2	5	21	163	294	307	131	3	0	0	926
1984	0	0	0	2	35	236	202	253	83	38	0	0	849
1985	0	0	5	11	53	159	262	201	114	22	0	0	827
1986	0	0	0	14	78	278	385	275	149	38	3	0	1220
1987	0	0	0	11	171	259	378	395	126	0	0	0	1340
1988	0	0	0	4	32	214	369	383	120	2	0	0	1124
1989	0	0	5	24	42	218	322	268	144	11	0	0	1034
1990	0	0	3	20	62	217	325	299	161	17	0	0	1104
1991	0	0	0	23	180	218	367	258	157	20	2	0	1225
1992	0	0	0	14	36	155	312	199	143	0	0	0	859
1993	0	0	0	1	62	250	460	281	131	6	0	0	1191

SNOWFALL (inches) BRISTOL, JOHNSON CITY, KINGSPORT, TENNESSEE

SEASON	JULY	AUG	SEP	OCT	NOV	DEC	JAN	FEB	MAR	APR	MAY	JUNE	TOTAL
1964-65	0.0	0.0	0.0	0.0	0.7	T	7.5	9.0	5.5	0.0	0.0	0.0	22.7
1965-66	0.0	0.0	0.0	T	T	0.1	22.1	0.3	T	T	0.0	0.0	22.5
1966-67	0.0	0.0	0.0	0.0	1.9	5.8	4.6	4.8	T	0.0	0.0	0.0	17.1
1967-68	0.0	0.0	0.0	0.0	T	4.1	12.1	6.3	1.0	0.0	0.0	0.0	23.5
1968-69	0.0	0.0	0.0	T	2.9	1.0	5.9	15.1	4.4	0.0	0.0	0.0	29.3
1969-70	0.0	0.0	0.0	0.0	1.0	9.7	10.1	7.8	0.6	0.0	0.0	0.0	29.2
1970-71	0.0	0.0	0.0	0.0	0.4	4.2	4.5	9.9	6.3	T	0.0	0.0	25.3
1971-72	0.0	0.0	0.0	0.0	0.7	4.3	T	5.3	2.4	T	0.0	0.0	12.7
1972-73	0.0	0.0	0.0	0.0	2.1	0.1	4.2	1.2	T	T	0.0	0.0	7.6
1973-74	0.0	0.0	0.0	0.0	0.0	3.3	0.0	0.9	T	T	0.0	0.0	4.2
1974-75	0.0	0.0	0.0	0.0	0.5	3.1	1.8	0.3	3.4	T	0.0	0.0	9.1
1975-76	0.0	0.0	0.0	0.0	T	T	0.5	3.2	T	0.0	0.0	0.0	3.7
1976-77	0.0	0.0	0.0	0.0	1.2	6.2	12.7	2.4	0.4	T	0.0	0.0	22.9
1977-78	0.0	0.0	0.0	T	1.6	0.4	13.6	6.7	3.9	T	0.0	0.0	26.2
1978-79	0.0	0.0	0.0	0.0	0.0	T	9.2	20.4	0.8	0.0	0.0	0.0	30.4
1979-80	0.0	0.0	0.0	0.0	T	0.4	7.8	5.4	3.6	0.0	0.0	0.0	17.2
1980-81	0.0	0.0	0.0	0.0	T	T	7.3	1.1	1.1	0.0	0.0	0.0	9.5
1981-82	0.0	0.0	0.0	0.0	T	7.3	3.6	2.3	2.6	0.6	0.0	0.0	16.4
1982-83	0.0	0.0	0.0	0.0	0.0	6.3	3.4	5.5	1.1	5.6	0.0	0.0	21.9
1983-84	0.0	0.0	0.0	0.0	T	0.3	4.6	3.8	1.2	T	0.0	0.0	9.9
1984-85	0.0	0.0	0.0	0.0	0.0	0.3	9.7	6.0	T	T	0.0	0.0	16.0
1985-86	0.0	0.0	0.0	0.0	0.0	3.2	10.5	7.7	T	T	0.0	0.0	21.4
1986-87	0.0	0.0	0.0	0.0	T	0.2	12.2	4.0	0.3	14.8	0.0	0.0	31.5
1987-88	0.0	0.0	0.0	0.0	0.8	0.2	7.6	T	0.4	0.0	0.0	0.0	9.0
1988-89	0.0	0.0	0.0	T	T	2.3	3.7	11.0	T	T	T	0.0	17.0
1989-90	0.0	0.0	0.0	0.1	0.9	6.9	0.7	0.4	0.5	T	0.0	0.0	9.5
1990-91	0.0	0.0	0.0	0.0	0.0	T	T	1.4	T	0.0	0.0	0.0	1.4
1991-92	0.0	0.0	0.0	0.0	T	T	0.2	T	0.1	0.4	T	0.0	0.7
1992-93	0.0	0.0	0.0	0.0	T	3.5	T	2.3	14.2	T	0.0	T	20.0
1993-94	0.0	0.0	0.0	1.3	T	1.9							
Record Mean	0.0	0.0	0.0	T	1.0	2.6	5.1	4.3	2.3	0.4	T	T	15.8

CHATTANOOGA, TENNESSEE

Chattanooga is located in the southern portion of the Great Valley of Tennessee, an area of the Tennessee River between the Cumberland Mountains to the west and the Appalachian Mountains to the east. Local topography is complex with a number of minor valleys and ridges giving a local relief of as much as 500 feet. The Tennessee River approaches Chattanooga from the northeast and forms a loop southwest to west to northwest of the city at an elevation of about 630 feet above mean sea level. Most of the city lies on the south side of the river. On the north and southwest sides, the terrain rises abruptly to about 1,200 feet above the river. This complex topography results in marked variations in air drainage, wind, and minimum temperatures within short distances. In winter the Cumberland Mountains have a moderating influence on the local climate by retarding the flow of cold air from the north and west.

Chattanooga enjoys a moderate climate, characterized by cool winters and quite warm summers. Because of the sheltering effect of the mountains, winter temperatures average about 3.0 degrees warmer than at stations on the southern Cumberland Plateau section of the state. Winter weather is changeable and alternates between cool spells with an occasional cold period, but extreme cold is rare. Temperatures fall as low as the freezing point on a little over one-half of the winter days, but temperatures below zero rarely occur. Snowfall from year to year is highly variable with some winters having little or none. Heavy snowfalls have occurred, but any accumulation of snow seldom remains on the ground more than a few days. Ice storms of freezing rain or glaze are not uncommon, occasionally midwinter icing becomes severe enough to do some damage in the area.

Summer temperatures are either in the high 80s or low 90s and temperatures over 100 degrees are unusual. Most afternoon temperatures are modified by thunderstorms. Temperatures frequently plunge 10 to 15 degrees in a matter of minutes during one of these showers.

Precipitation in the Chattanooga area is well distributed throughout the year with the greater amounts in wintertime when cyclonic storms from the Gulf of Mexico reach the area with greater intensity and frequency. A second peak rainfall period generally occurs in July, principally from thunderstorms that move into the area from the south and southwest. During any year there are usually a few of these storms that can be classified as severe, with hail and damaging winds.

The growing season averages 228 days. The average occurrence of last freezing temperature in spring is early April and the average first freezing temperature in the fall is early November.

Spring and autumn are very enjoyable seasons in Chattanooga, with many days being nearly ideal in temperature. To many, the fall months of September, October, and November are the most pleasant. Rainfall is at a minimum, sunshine at a relative maximum and the temperature range is comfortable.

NORMALS, MEANS, AND EXTREMES

CHATTANOOGA, TENNESSEE

LATITUDE: 35°02'N LONGITUDE: 85°12' W ELEVATION: FT. GRND 665 BARO 681 TIME ZONE: EASTERN WBAN: 13882

	(a)	JAN	FEB	MAR	APR	MAY	JUNE	JULY	AUG	SEP	OCT	NOV	DEC	YEAR
TEMPERATURE °F:														
Normals														
-Daily Maximum		46.8	52.4	61.8	71.4	78.5	86.4	89.0	88.2	82.1	72.0	61.1	50.8	70.0
-Daily Minimum		28.0	31.1	39.0	46.8	55.4	63.9	68.3	67.7	62.6	48.9	39.9	31.2	48.6
-Monthly		37.4	41.8	50.4	59.1	66.9	75.2	78.7	77.9	72.4	60.4	50.5	41.1	59.3
Extremes														
-Record Highest	54	78	79	87	93	99	104	106	105	102	94	84	78	106
-Year		1949	1989	1963	1942	1941	1952	1952	1947	1954	1954	1961	1951	JUL 1952
-Record Lowest	54	-10	1	8	25	34	41	51	50	36	22	4	-2	-10
-Year		1985	1958	1960	1992	1971	1972	1972	1946	1967	1952	1950	1983	JAN 1985
NORMAL DEGREE DAYS:														
Heating (base 65°F)		856	650	453	196	57	0	0	0	9	190	435	741	3587
Cooling (base 65°F)		0	0	0	19	116	306	425	400	231	47	0	0	1544
% OF POSSIBLE SUNSHINE	63	43	49	52	61	65	65	62	63	64	63	53	44	57
MEAN SKY COVER (tenths)														
Sunrise - Sunset	63	6.8	6.5	6.4	5.9	5.9	5.7	6.0	5.7	5.5	4.9	5.8	6.6	6.0
MEAN NUMBER OF DAYS:														
Sunrise to Sunset														
-Clear	63	7.2	7.5	7.9	9.1	8.6	8.0	7.0	8.3	9.8	13.2	10.2	7.8	104.6
-Partly Cloudy	63	6.6	5.9	7.5	8.0	10.5	11.6	12.7	12.8	9.6	7.6	6.7	6.2	105.7
-Cloudy	63	17.2	14.8	15.6	13.0	12.0	10.3	11.3	9.9	10.6	10.2	13.1	17.0	155.0
Precipitation														
.01 inches or more	63	11.8	10.4	12.0	9.6	10.3	10.3	11.7	9.8	7.9	6.9	8.9	10.9	120.4
Snow,Ice pellets,hail														
1.0 inches or more	63	0.7	0.5	0.2	0.*	0.0	0.0	0.0	0.0	0.0	0.0	0.*	0.3	1.8
Thunderstorms	63	1.2	1.9	3.7	4.7	7.3	9.3	11.3	9.2	4.1	1.3	1.3	0.6	55.8
Heavy Fog Visibility														
1/4 mile or less	63	3.0	1.6	1.7	1.6	2.3	2.0	2.1	2.6	3.7	5.9	3.9	3.2	33.4
Temperature °F														
-Maximum														
90° and above	53	0.0	0.0	0.0	0.2	2.5	10.5	16.5	14.5	4.9	0.3	0.0	0.0	49.4
32° and below	53	2.0	0.9	0.1	0.0	0.0	0.0	0.0	0.0	0.0	0.0	0.*	0.8	3.8
-Minimum														
32° and below	53	19.1	15.1	8.7	1.5	0.0	0.0	0.0	0.0	0.0	1.1	10.0	17.6	73.1
0° and below	53	0.2	0.0	0.0	0.0	0.0	0.0	0.0	0.0	0.0	0.0	0.0	0.*	0.2
AVG. STATION PRESS.(mb)	21	995.7	994.6	992.3	991.7	991.2	991.7	992.7	993.1	993.6	994.9	995.3	996.1	993.6
RELATIVE HUMIDITY (%)														
Hour 01	53	79	77	76	77	86	88	88	89	89	88	83	81	83
Hour 07 (Local Time)	63	81	81	81	82	86	87	89	91	91	89	85	83	86
Hour 13	63	62	57	53	49	53	55	57	57	56	53	56	62	56
Hour 19	63	66	59	55	51	57	60	63	65	66	67	66	69	62
PRECIPITATION (inches):														
Water Equivalent														
-Normal		4.89	4.81	6.03	4.31	4.37	3.52	4.85	3.53	4.15	3.22	4.61	5.17	53.46
-Maximum Monthly	54	12.28	11.03	16.32	11.92	9.21	9.40	11.78	7.54	14.18	9.91	13.59	13.68	16.32
-Year		1947	1944	1980	1964	1979	1949	1979	1975	1977	1949	1948	1961	MAR 1980
-Minimum Monthly	54	0.90	0.62	1.17	0.44	0.54	0.63	0.20	0.56	0.34	0.22	0.93	0.86	0.20
-Year		1986	1941	1967	1942	1941	1988	1957	1963	1941	1991	1953	1965	JUL 1957
-Maximum in 24 hrs	54	4.44	4.27	6.53	3.36	3.47	4.85	5.77	3.70	6.62	4.04	4.56	5.25	6.62
-Year		1949	1990	1973	1983	1979	1949	1979	1941	1977	1986	1948	1942	SEP 1977
Snow,Ice pellets,hail														
-Maximum Monthly	63	10.2	10.4	20.0	2.8	T	0.0	0.0	0.0	0.0	T	2.8	9.1	20.0
-Year		1988	1960	1993	1987	1944					1993	1950	1963	MAR 1993
-Maximum in 24 hrs	63	10.2	8.7	20.0	2.8	T	0.0	0.0	0.0	0.0	T	2.8	8.9	20.0
-Year		1988	1960	1993	1987	1944					1993	1950	1963	MAR 1993
WIND:														
Mean Speed (mph)	53	6.9	7.3	7.8	7.4	5.9	5.3	5.0	4.6	4.9	4.9	6.0	6.4	6.0
Prevailing Direction through 1963		S	S	S	S	S	S	S	S	S	S	S	S	S
Fastest Obs. 1 Min.														
-Direction (!!)	18	31	25	31	18	18	34	23	05	32	31	31	28	05
-Speed (MPH)	18	30	37	31	32	35	30	29	37	28	24	32	28	37
-Year		1978	1976	1993	1977	1990	1981	1992	1979	1987	1981	1988	1990	AUG 1979
Peak Gust														
-Direction (!!)	10	SW	NW	N	NW	S	NW	W	NW	S	SE	W	S	NW
-Speed (mph)	10	41	48	44	45	54	56	44	40	40	39	45	41	56
-Date		1992	1991	1993	1991	1990	1987	1992	1989	1991	1984	1988	1990	JUN 1987

PRECIPITATION (inches)

CHATTANOOGA, TENNESSEE

YEAR	JAN	FEB	MAR	APR	MAY	JUNE	JULY	AUG	SEP	OCT	NOV	DEC	ANNUAL
1964	5.98	4.51	10.44	11.92	4.28	4.06	4.99	4.87	2.06	3.53	4.03	3.59	64.26
1965	3.51	4.46	9.43	2.87	1.65	4.92	3.13	3.55	4.47	1.03	4.48	0.86	44.36
1966	3.67	6.50	2.99	2.93	4.00	1.03	4.51	4.16	6.40	4.77	2.65	3.58	47.19
1967	3.02	4.10	1.17	3.39	6.02	1.63	10.56	2.67	2.08	4.91	4.49	7.89	51.93
1968	4.98	0.94	3.70	3.72	3.93	0.87	1.92	2.90	2.82	2.52	2.81	3.98	35.09
1969	6.84	5.82	2.96	4.20	2.36	3.61	2.32	6.20	4.70	1.72	2.11	7.98	50.82
1970	2.45	1.80	4.08	7.36	2.76	4.69	3.88	2.92	2.92	5.56	2.03	4.30	44.75
1971	5.19	8.50	6.71	2.26	2.17	2.71	7.28	1.91	3.68	2.11	1.93	6.56	51.01
1972	8.64	4.53	5.72	3.46	6.47	3.95	7.53	5.31	3.19	4.80	3.81	7.09	64.50
1973	5.09	3.76	13.80	5.17	6.37	4.58	6.93	2.58	5.38	3.69	5.87	8.34	71.56
1974	8.00	5.68	3.44	3.09	6.53	1.19	4.74	6.10	1.30	1.71	5.04	4.51	51.33
1975	5.63	6.34	10.44	1.03	4.45	3.92	2.18	7.54	10.31	6.28	5.00	4.98	68.10
1976	4.32	2.10	5.41	1.02	5.86	6.76	5.57	1.38	3.35	3.55	2.92	5.54	47.78
1977	3.14	2.98	8.39	9.10	2.55	2.44	1.01	3.53	14.18	6.21	8.66	2.37	64.56
1978	5.79	0.72	4.71	4.78	4.17	2.59	0.80	4.17	1.62	0.32	2.84	7.57	40.08
1979	5.44	3.85	4.77	7.49	9.21	2.32	11.78	4.50	7.75	2.45	7.35	1.66	68.57
1980	4.69	2.16	16.32	4.78	4.25	2.67	1.28	0.96	4.62	1.56	4.68	0.90	48.87
1981	1.91	5.07	4.44	3.81	3.38	3.70	3.99	4.21	2.74	2.80	4.85	5.02	45.92
1982	8.29	5.95	4.80	4.48	1.44	2.76	4.79	6.42	1.34	1.98	7.65	8.05	57.95
1983	2.65	4.52	3.66	6.40	6.65	3.29	2.52	1.34	1.15	2.64	9.57	8.34	52.73
1984	3.25	4.40	3.75	6.28	6.57	2.52	6.32	2.48	0.51	6.38	3.44	1.84	47.74
1985	3.45	4.95	2.27	2.81	4.36	2.22	3.10	4.93	0.93	4.83	3.34	2.36	39.55
1986	0.90	3.78	1.42	1.28	3.14	1.64	2.61	2.78	5.86	6.66	7.60	4.82	42.49
1987	6.29	7.16	5.13	1.66	4.97	5.12	1.84	3.53	4.07	1.13	2.40	3.25	46.55
1988	7.10	1.87	1.83	3.30	2.20	0.63	5.85	4.07	6.76	2.10	4.64	3.54	43.89
1989	5.30	7.16	5.86	3.99	4.56	9.19	9.93	3.46	11.22	1.77	5.45	3.71	71.60
1990	7.46	9.74	9.47	2.86	6.49	3.35	4.69	2.29	3.69	4.45	3.73	10.34	68.56
1991	2.71	8.12	6.58	6.78	4.58	4.11	3.59	4.00	2.37	0.22	4.88	8.02	55.96
1992	4.33	4.63	4.89	2.08	1.99	8.87	5.33	3.07	4.71	4.48	5.88	5.60	55.86
1993	5.16	3.69	7.40	3.34	3.37	1.64	1.30	2.12	1.68	1.80	3.33	5.26	40.09
Record Mean	5.16	4.98	5.89	4.55	3.98	3.91	4.62	3.72	3.42	3.03	3.88	5.11	52.24

AVERAGE TEMPERATURE (deg. F)

CHATTANOOGA, TENNESSEE

YEAR	JAN	FEB	MAR	APR	MAY	JUNE	JULY	AUG	SEP	OCT	NOV	DEC	ANNUAL
1964	39.9	39.9	50.2	62.9	69.9	77.0	76.8	76.7	72.6	57.2	52.4	43.9	60.0
1965	40.9	40.7	45.9	62.5	70.4	73.3	76.8	76.9	71.9	58.3	51.4	41.5	59.2
1966	33.1	41.3	49.0	58.3	64.8	73.8	78.9	75.5	69.5	56.8	49.5	40.4	57.6
1967	41.0	38.2	53.9	63.5	64.9	73.0	72.8	73.6	67.5	59.4	46.9	45.8	58.4
1968	37.1	34.4	49.8	57.9	65.8	75.3	79.3	80.9	72.5	61.9	49.7	38.9	58.6
1969	38.3	42.2	44.6	61.2	67.1	75.1	78.6	74.3	68.3	56.5	43.6	35.8	57.1
1970	32.0	38.9	46.8	60.2	67.6	75.1	79.7	78.1	75.6	62.4	47.0	44.2	59.0
1971	38.8	39.9	45.2	58.9	65.1	78.3	77.1	78.0	75.5	66.4	48.7	49.4	60.1
1972	43.4	40.0	48.8	59.2	63.9	70.2	74.5	75.0	71.9	56.8	46.0	43.1	57.7
1973	36.0	37.5	55.4	57.3	64.8	76.4	79.0	77.6	75.8	62.7	52.3	41.5	59.7
1974	49.3	43.3	54.5	57.1	65.8	68.4	75.7	75.8	68.3	57.0	47.6	40.7	58.6
1975	42.7	44.0	45.6	56.1	68.0	72.8	78.0	77.1	68.4	60.7	49.0	39.4	58.5
1976	34.6	48.6	53.8	59.4	63.5	72.8	75.8	76.6	69.1	56.3	42.7	37.4	57.6
1977	28.5	40.5	53.6	63.1	70.5	77.6	83.4	81.8	75.3	57.8	53.0	40.6	60.5
1978	30.3	35.5	49.7	62.6	68.1	76.5	81.2	78.7	75.6	59.0	54.9	44.3	59.7
1979	33.7	38.8	52.1	60.1	67.2	74.1	76.5	78.0	72.4	60.2	50.2	41.3	58.7
1980	40.7	36.8	47.8	57.8	67.2	73.8	82.8	82.2	74.0	57.2	49.8	41.2	59.3
1981	35.5	43.4	48.1	64.7	64.6	78.0	81.4	77.3	69.7	58.6	50.8	38.2	59.2
1982	36.0	43.4	53.4	55.9	70.2	74.2	79.6	76.7	70.0	60.5	50.9	48.0	59.9
1983	38.8	42.8	50.5	54.3	65.9	73.5	80.4	81.9	71.2	60.8	48.5	37.0	58.8
1984	35.5	43.2	48.2	56.8	64.4	77.4	76.2	76.6	68.0	66.7	46.5	49.0	59.0
1985	32.8	40.2	52.9	60.8	67.6	74.9	78.4	76.6	69.4	64.7	58.8	36.7	59.5
1986	38.4	47.1	51.8	61.0	70.2	78.4	82.8	78.0	74.0	62.0	54.8	41.5	61.7
1987	39.3	44.7	51.9	58.0	73.1	77.1	81.2	80.5	71.4	54.9	52.0	45.5	60.8
1988	35.1	41.0	52.0	60.3	67.0	76.9	79.0	80.5	72.4	55.3	51.7	42.3	59.5
1989	45.3	43.2	54.5	59.6	65.4	74.8	78.9	78.4	71.7	60.9	50.7	35.0	59.9
1990	46.8	51.5	54.6	60.5	67.5	76.9	79.4	80.2	74.6	61.3	53.4	46.4	62.8
1991	42.6	45.8	54.1	63.7	72.9	77.1	81.5	79.6	74.0	61.4	47.7	45.2	62.1
1992	42.1	46.9	49.9	59.2	64.3	71.4	78.2	73.5	71.3	58.3	49.3	42.5	58.9
1993	45.8	43.0	48.8	58.5	68.9	78.2	85.2	82.5	74.4	61.4	50.5	43.0	61.7
Record Mean	40.9	43.6	51.2	60.2	68.3	75.9	78.9	78.0	72.6	61.2	50.2	42.5	60.3
Max	49.7	53.2	61.5	71.4	79.5	86.4	88.8	87.9	82.9	72.5	60.5	51.4	70.5
Min	32.1	33.9	40.8	49.0	57.1	65.3	68.9	68.0	62.2	49.9	39.8	33.6	50.0

REFERENCE NOTES

CHATTANOOGA, TENNESSEE

GENERAL
T - TRACE AMOUNT.
BLANK ENTRIES DENOTE MISSING/UNREPORTED DATA.
# INDICATES A STATION OR INSTRUMENT RELOCATION.

SPECIFIC

(a) - LENGTH OF RECORD IN YEARS, ALTHOUGH INDIVIDUAL MONTHS MAY BE MISSING.
0.* OR * - THE VALUE IS BETWEEN 0.0 AND 0.05
NORMALS - BASED ON THE 1961-1990 RECORD PERIOD.
EXTREMES - DATES ARE THE MOST RECENT OCCURRENCE.
WIND DIR.- NUMERALS SHOW TENS OF DEGRESS CLOCKWISE FROM TRUE NORTH. "00" INDICATES CALM.
RESULTANT DIRECTIONS ARE GIVEN TO WHOLE DEGREES.
BOLD VALUES INDICATE EXTREME VALUES WHICH OCCURRED AFTER THE ASOS SYSTEM WAS COMMISSIONED.

RECORD = PERIOD OF RECORD
RECORD MEAN PRECIPITATION IS THE MEAN OF ALL DAILY PRECIPITATION AMOUNTS DURING THE PERIOD OF RECORD.
RECORD MAX(MIN) TEMPERATURE IS THE MEAN OF ALL DAILY MAX(MIN) TEMPERATURES DURING THE PERIOD OF RECORD.
RECORD MEAN TEMPERATURE IS THE SUM OF THE RECORD MAX AND RECORD MIN DIVIDED BY 2.
AVERAGE TEMPERATURE IS THE SUM OF THE MEAN DAILY MAX AND MIN TEMPERATURE DIVIDED BY 2.

EXCEPTIONS

RECORD MEANS ARE THROUGH THE CURRENT YEAR, BEGINNING IN 1930 FOR TEMPERATURE
1930 FOR PRECIPITATION
1930 FOR SNOWFALL

HEATING DEGREE DAYS Base 65 deg. F CHATTANOOGA, TENNESSEE

SEASON	JULY	AUG	SEP	OCT	NOV	DEC	JAN	FEB	MAR	APR	MAY	JUNE	TOTAL
1964-65	0	1	5	255	378	646	741	674	587	130	2	0	3419
1965-66	0	0	15	215	399	721	982	656	486	231	63	10	3778
1966-67	0	0	7	252	459	756	737	743	351	94	93	8	3500
1967-68	1	0	45	200	538	587	857	881	465	227	58	0	3859
1968-69	0	0	1	169	452	805	821	634	628	141	48	1	3700
1969-70	0	0	20	263	636	897	1017	729	556	194	47	1	4360
1970-71	0	0	13	119	535	636	807	698	606	206	73	0	3693
1971-72	0	0	0	45	496	478	661	721	496	215	62	10	3184
1972-73	0	0	10	253	567	674	890	767	294	246	80	0	3781
1973-74	0	0	3	143	374	722	480	604	322	242	65	16	2971
1974-75	0	0	28	250	517	746	686	581	593	290	14	0	3705
1975-76	0	0	53	152	474	785	934	470	340	180	87	2	3477
1976-77	0	0	11	269	661	851	1123	679	353	111	17	0	4075
1977-78	0	0	5	233	359	753	1072	821	467	114	60	0	3884
1978-79	0	0	0	192	299	637	962	725	390	152	42	0	3399
1979-80	0	0	0	167	438	728	745	813	526	215	38	0	3670
1980-81	0	0	24	246	450	736	907	600	515	70	81	0	3629
1981-82	0	0	31	211	422	823	890	597	371	275	13	0	3633
1982-83	0	0	29	202	418	528	807	616	447	321	42	0	3410
1983-84	0	0	38	146	487	863	910	626	515	246	100	2	3933
1984-85	0	0	27	40	550	488	990	687	381	158	28	4	3353
1985-86	0	0	32	67	198	869	818	497	402	148	23	0	3054
1986-87	0	1	0	148	306	722	789	562	399	238	0	0	3165
1987-88	0	0	4	309	384	597	919	688	396	164	26	1	3488
1988-89	0	0	1	312	391	697	602	608	333	220	92	0	3256
1989-90	0	0	29	164	423	924	561	371	325	171	48	0	3016
1990-91	0	0	19	154	344	571	689	534	349	75	7	0	2742
1991-92	0	0	17	142	513	610	701	520	457	207	98	3	3268
1992-93	0	0	8	202	465	689	586	612	496	197	21	0	3276
1993-94	0	0	15	155	439	675							

COOLING DEGREE DAYS Base 65 deg. F CHATTANOOGA, TENNESSEE

YEAR	JAN	FEB	MAR	APR	MAY	JUNE	JULY	AUG	SEP	OCT	NOV	DEC	TOTAL
1969	0	0	0	33	119	311	429	298	125	7	0	0	1322
1970	0	0	0	57	133	313	462	412	337	44	0	0	1758
1971	0	0	0	27	81	406	384	411	324	96	16	2	1747
1972	0	0	0	51	35	177	303	316	224	8	4	0	1118
1973	0	0	1	22	79	350	443	400	334	79	1	0	1709
1974	0	0	3	11	97	124	338	343	132	8	2	0	1058
1975	0	0	0	29	115	238	411	382	162	27	1	0	1365
1976	0	0	1	19	49	241	342	363	142	8	0	0	1165
1977	0	0	9	60	194	384	578	532	320	15	3	0	2095
1978	0	0	0	48	164	350	507	432	325	14	2	5	1847
1979	0	0	0	12	119	279	361	410	227	24	0	0	1432
1980	0	0	0	8	111	272	562	540	300	10	1	0	1804
1981	0	0	0	68	77	400	513	385	178	21	2	0	1644
1982	0	0	19	9	184	283	460	368	187	68	1	7	1586
1983	0	0	2	6	78	264	485	529	230	22	0	0	1616
1984	0	0	0	8	86	379	356	367	124	102	0	0	1422
1985	0	0	12	38	118	310	422	366	171	64	16	0	1517
1986	0	0	0	33	192	409	558	414	276	63	7	0	1952
1987	0	0	0	34	261	372	508	488	202	0	0	0	1865
1988	0	0	2	27	95	364	442	486	230	17	0	0	1663
1989	0	2	16	68	109	300	437	424	236	43	0	0	1635
1990	0	0	13	42	133	363	455	480	312	45	3	0	1846
1991	0	0	20	43	258	370	518	460	295	37	0	2	2003
1992	0	0	0	40	81	202	414	273	202	0	0	0	1212
1993	0	0	4	12	152	401	632	547	305	49	10	0	2112

SNOWFALL (inches) CHATTANOOGA, TENNESSEE

SEASON	JULY	AUG	SEP	OCT	NOV	DEC	JAN	FEB	MAR	APR	MAY	JUNE	TOTAL
1964-65	0.0	0.0	0.0	0.0	T	0.0	1.8	0.9	0.7	0.0	0.0	0.0	3.4
1965-66	0.0	0.0	0.0	0.0	0.0	0.0	8.4	T	T	0.0	0.0	0.0	8.4
1966-67	0.0	0.0	0.0	0.0	T	T	T	1.8	T	0.0	0.0	0.0	1.8
1967-68	0.0	0.0	0.0	0.0	0.0	T	5.6	4.9	T	0.0	0.0	0.0	10.5
1968-69	0.0	0.0	0.0	0.0	T	T	7.3	0.8	0.2	0.0	0.0	0.0	8.3
1969-70	0.0	0.0	0.0	0.0	T	5.1	7.9	T	T	0.0	0.0	0.0	13.0
1970-71	0.0	0.0	0.0	0.0	0.0	3.4	0.6	0.6	4.5	1.0	0.0	0.0	10.1
1971-72	0.0	0.0	0.0	0.0	T	T	T	T	1.0	0.0	0.0	0.0	1.0
1972-73	0.0	0.0	0.0	0.0	T	0.0	2.3	T	T	T	0.0	0.0	2.3
1973-74	0.0	0.0	0.0	0.0	0.0	0.2	0.0	T	T	0.0	0.0	0.0	0.2
1974-75	0.0	0.0	0.0	0.0	0.0	T	T	T	T	0.0	0.0	0.0	T
1975-76	0.0	0.0	0.0	0.0	T	T	0.1	T	0.0	0.0	0.0	0.0	0.1
1976-77	0.0	0.0	0.0	0.0	T	T	3.7	T	0.0	0.0	0.0	0.0	3.7
1977-78	0.0	0.0	0.0	0.0	T	T	2.5	0.7	0.4	0.0	0.0	0.0	3.6
1978-79	0.0	0.0	0.0	0.0	0.0	0.0	0.5	8.7	T	0.0	0.0	0.0	9.2
1979-80	0.0	0.0	0.0	0.0	T	0.0	T	7.4	0.4	0.0	0.0	0.0	7.8
1980-81	0.0	0.0	0.0	0.0	T	0.0	T	T	0.0	0.0	0.0	0.0	T
1981-82	0.0	0.0	0.0	0.0	0.0	0.1	2.9	2.3	T	0.0	0.0	0.0	5.3
1982-83	0.0	0.0	0.0	0.0	0.0	T	1.0	0.9	0.6	0.0	0.0	0.0	2.5
1983-84	0.0	0.0	0.0	0.0	0.0	T	0.6	1.8	T	0.0	0.0	0.0	2.4
1984-85	0.0	0.0	0.0	0.0	0.0	T	3.8	1.1	0.0	0.0	0.0	0.0	4.9
1985-86	0.0	0.0	0.0	0.0	0.0	0.3	2.0	T	0.0	0.0	0.0	0.0	2.3
1986-87	0.0	0.0	0.0	0.0	T	T	6.5	0.0	0.9	2.8	0.0	0.0	10.2
1987-88	0.0	0.0	0.0	0.0	0.0	T	10.2	T	T	0.0	0.0	0.0	10.2
1988-89	0.0	0.0	0.0	0.0	0.0	0.8	1.5	0.1	0.0	0.0	0.0	0.0	2.4
1989-90	0.0	0.0	0.0	0.0	0.3	0.3	T	0.0	0.0	0.0	0.0	0.0	0.6
1990-91	0.0	0.0	0.0	0.0	0.0	T	T	0.2	2.0	0.0	0.0	0.0	2.2
1991-92	0.0	0.0	0.0	0.0	T	0.0	T	0.0	T	0.0	0.0	0.0	T
1992-93	0.0	0.0	0.0	0.0	0.0	T	T	2.7	20.0	T	0.0	0.0	22.7
1993-94	0.0	0.0	0.0	T	0.0	T							
Record Mean	0.0	0.0	0.0	T	T	0.6	1.8	1.2	0.7	0.1	T	0.0	4.4

KNOXVILLE, TENNESSEE

Knoxville is located in a broad valley between the Cumberland Mountains, which lie northwest of the city, and the Great Smoky Mountains, which lie southeast of the city. These two mountain ranges exercise a marked influence upon the climate of the valley. The Cumberland Mountains, to the northwest, serve to retard and weaken the force of the cold winter air which frequently penetrates far south of the latitude of Knoxville over the plains areas to the west of the mountains.

The mountains also serve to modify the hot summer winds which are common to the plains to the west. In addition, they serve as a fixed incline plane which lifts the warm, moist air flowing northward from the Gulf of Mexico and thereby increases the frequency of afternoon thunderstorms. Relief from extremely high temperatures which such thunderstorms produce serves to reduce the number of extremely warm days in the valley.

July is usually the warmest month of the year. The coldest weather usually occurs during the month of January. Sudden great temperature changes occur infrequently. This again is due mainly to the retarding effect of the mountains. Summer nights are nearly always comfortable.

Rainfall is ample for agricultural purposes and is favorably distributed during the year for most crops. Precipitation is greatest in the wintertime. Another peak period occurs during the late spring and summer months. The period of lowest rainfall occurs during the fall. A cumulative total of approximately 12 inches of snow falls annually. However, this usually comes in amounts of less than 4 inches at one time. It is unusual for snow to remain on the ground in measurable amounts longer than one week.

The topography also has a pronounced effect upon the prevailing wind direction. Daytime winds usually have a southwesterly component, while nighttime winds usually move from the northeast. The winds are relatively light and tornadoes are extremely rare.

NORMALS, MEANS, AND EXTREMES

KNOXVILLE, TENNESSEE

LATITUDE: 35°49'N LONGITUDE: 83°59' W ELEVATION: FT. GRND 979 BARO 994 TIME ZONE: EASTERN WBAN: 13891

	(a)	JAN	FEB	MAR	APR	MAY	JUNE	JULY	AUG	SEP	OCT	NOV	DEC	YEAR
TEMPERATURE °F:														
Normals														
-Daily Maximum		45.9	50.9	61.3	70.4	77.6	84.5	87.1	86.7	81.2	70.6	59.9	50.1	68.9
-Daily Minimum		26.0	29.1	36.6	44.6	53.1	61.8	66.0	65.3	59.0	46.0	37.5	30.0	46.3
-Monthly		36.0	40.1	49.0	57.5	65.4	73.2	76.6	76.0	70.1	58.4	48.7	40.1	57.6
Extremes														
-Record Highest	52	77	83	86	92	94	102	103	102	103	91	84	80	103
-Year		1950	1977	1963	1942	1962	1988	1952	1944	1954	1953	1948	1982	SEP 1954
-Record Lowest	52	-24	-2	1	22	32	43	49	49	36	25	5	-6	-24
-Year		1985	1958	1980	1987	1986	1956	1988	1946	1967	1987	1950	1983	JAN 1985
NORMAL DEGREE DAYS:														
Heating (base 65°F)		899	697	496	234	94	0	0	0	18	238	489	772	3937
Cooling (base 65°F)		0	0	0	9	106	246	360	341	171	33	0	0	1266
% OF POSSIBLE SUNSHINE	51	40	47	53	62	64	65	63	63	61	61	49	40	56
MEAN SKY COVER (tenths)														
Sunrise - Sunset	51	7.1	6.7	6.7	6.1	6.1	5.8	5.9	5.6	5.6	5.0	6.2	6.8	6.1
MEAN NUMBER OF DAYS:														
Sunrise to Sunset														
-Clear	51	6.2	6.7	7.2	8.2	7.8	7.5	7.2	8.4	9.5	12.3	8.7	7.0	96.6
-Partly Cloudy	51	6.6	6.0	7.3	8.6	10.1	12.3	12.8	12.3	9.5	7.9	6.9	6.6	106.9
-Cloudy	51	18.2	15.6	16.5	13.2	13.1	10.1	11.0	10.3	11.0	10.8	14.4	17.4	161.7
Precipitation														
.01 inches or more	51	12.3	11.3	12.7	10.9	11.0	10.2	11.3	9.5	8.4	7.8	9.9	11.0	126.3
Snow,Ice pellets,hail														
1.0 inches or more	51	1.3	1.0	0.5	0.1	0.0	0.0	0.0	0.0	0.0	0.0	0.1	0.5	3.5
Thunderstorms	51	0.7	1.4	3.3	4.3	6.7	8.0	9.6	7.0	3.2	1.3	1.0	0.5	47.0
Heavy Fog Visibility														
1/4 mile or less	51	2.8	1.8	1.6	1.1	2.4	2.1	2.1	3.5	3.9	4.7	3.2	2.6	31.7
Temperature °F														
-Maximum														
90° and above	33	0.0	0.0	0.0	0.*	0.8	5.1	11.1	8.5	2.8	0.0	0.0	0.0	28.3
32° and below	33	3.3	1.3	0.2	0.0	0.0	0.0	0.0	0.0	0.0	0.0	0.1	1.2	6.0
-Minimum														
32° and below	33	20.1	16.5	8.6	1.8	0.*	0.0	0.0	0.0	0.0	1.0	8.0	17.5	73.4
0° and below	33	0.4	0.*	0.0	0.0	0.0	0.0	0.0	0.0	0.0	0.0	0.0	0.1	0.6
AVG. STATION PRESS.(mb)	21	984.7	983.7	981.6	981.1	980.9	981.5	982.6	983.0	983.5	984.5	984.6	985.2	983.1
RELATIVE HUMIDITY (%)														
Hour 01	33	77	74	72	72	83	86	87	88	88	85	80	78	81
Hour 07 (Local Time)	33	81	80	80	81	87	88	90	92	92	90	84	83	86
Hour 13	33	63	59	55	51	57	59	61	61	60	55	59	64	59
Hour 19	33	66	60	56	53	61	63	65	67	68	65	66	69	63
PRECIPITATION (inches):														
Water Equivalent														
-Normal		4.17	4.06	5.09	3.72	4.13	3.97	4.67	3.13	3.07	2.84	3.75	4.54	47.14
-Maximum Monthly	52	11.74	9.38	10.42	7.20	10.98	8.21	10.09	8.88	9.19	6.67	10.36	11.63	11.74
-Year		1954	1944	1975	1970	1974	1989	1967	1942	1989	1949	1948	1961	JAN 1954
-Minimum Monthly	52	0.95	0.74	1.69	0.39	0.74	0.20	0.70	0.77	0.42	T	0.97	0.45	T
-Year		1986	1968	1986	1976	1970	1944	1957	1954	1985	1963	1942	1965	OCT 1963
-Maximum in 24 hrs	52	3.89	3.42	4.85	3.65	3.40	3.57	4.69	3.25	5.08	2.44	4.06	4.89	5.08
-Year		1946	1991	1973	1977	1984	1972	1942	1959	1944	1961	1948	1969	SEP 1944
Snow,Ice pellets,hail														
-Maximum Monthly	52	15.1	23.3	20.2	10.7	T	0.0	0.0	0.0	0.0	T	18.2	12.2	23.3
-Year		1962	1960	1960	1987	1945					1993	1952	1963	FEB 1960
-Maximum in 24 hrs	52	12.0	17.5	14.1	10.7	T	0.0	0.0	0.0	0.0	T	18.2	8.9	18.2
-Year		1962	1960	1993	1987	1945					1993	1952	1969	NOV 1952
WIND:														
Mean Speed (mph)	51	7.8	8.1	8.6	8.5	6.8	6.4	6.0	5.5	5.7	5.7	6.8	7.2	6.9
Prevailing Direction through 1963		NE	NE	NE	WSW	SW	SW	WSW	NE	NE	NE	NE	NE	NE
Fastest Obs. 1 Min.														
-Direction (!!!)	19	27	23	22	29	30	07	36	32	36	27	23	20	23
-Speed (MPH)	19	35	37	36	39	31	35	35	29	26	28	40	39	40
-Year		1978	1990	1977	1988	1989	1975	1985	1983	1989	1988	1988	1988	NOV 1988
Peak Gust														
-Direction (!!!)	9	SW	SW	W	W	W	NW	NE	N	W	SW	W	SW	W
-Speed (mph)	9	46	53	53	59	48	58	46	44	39	39	64	58	64
-Date		1989	1990	1989	1985	1989	1989	1985	1989	1989	1988	1988	1988	NOV 1988

PRECIPITATION (inches)

KNOXVILLE, TENNESSEE

YEAR	JAN	FEB	MAR	APR	MAY	JUNE	JULY	AUG	SEP	OCT	NOV	DEC	ANNUAL
1964	4.71	4.09	5.75	6.98	3.96	0.97	3.70	5.75	1.10	2.10	2.80	4.02	45.93
1965	3.94	3.25	9.31	4.16	3.13	4.84	3.49	2.18	2.48	1.08	2.50	0.45	40.81
1966	3.88	4.68	2.72	2.88	2.92	2.27	5.44	4.09	4.41	4.80	5.12	2.66	45.87
1967	2.67	4.58	4.08	2.00	4.10	6.53	10.09	4.06	2.70	2.33	5.57	6.95	55.66
1968	4.13	0.74	4.78	4.12	3.01	3.97	2.57	1.29	2.53	3.40	2.00	3.22	35.76
1969	4.11	5.54	2.89	2.41	1.33	7.58	3.51	6.72	3.03	1.56	2.56	7.74	48.98
1970	3.04	2.86	3.18	7.20	0.74	4.26	3.11	4.89	2.75	5.33	1.40	4.67	43.43
1971	5.03	4.93	4.21	3.87	3.78	3.73	8.76	3.05	3.41	1.98	2.21	5.48	50.44
1972	7.35	4.19	4.98	2.54	4.49	5.02	6.76	1.61	4.70	5.99	3.36	7.02	58.01
1973	3.24	2.59	10.24	5.15	5.71	5.26	4.38	2.31	3.28	3.48	5.01	7.38	58.03
1974	7.05	5.24	6.15	5.77	10.98	2.70	2.92	3.14	3.33	2.35	5.18	4.52	59.33
1975	4.66	4.68	10.42	2.43	2.98	2.43	2.25	1.61	3.28	4.02	2.92	3.59	45.27
1976	3.86	2.18	5.22	0.39	5.53	3.46	3.75	1.98	2.87	5.33	3.45	4.42	42.44
1977	2.55	1.52	6.08	6.96	1.16	6.49	1.08	5.78	6.91	4.04	5.06	3.30	50.93
1978	5.22	1.01	4.42	4.10	3.44	5.27	5.06	2.44	1.26	0.82	3.62	5.91	42.57
1979	6.18	4.17	4.21	4.30	7.21	3.80	9.47	2.29	2.64	1.97	5.73	1.92	53.89
1980	5.54	1.78	8.72	3.30	3.80	1.94	3.57	2.34	2.38	1.53	3.78	1.78	40.46
1981	1.05	3.62	2.83	4.84	3.02	5.53	2.03	3.48	6.09	4.15	3.01	4.14	43.79
1982	6.03	4.88	6.36	3.26	5.52	3.93	6.60	2.68	2.68	2.66	5.21	4.89	54.70
1983	1.58	2.90	1.99	5.88	5.42	3.26	3.18	3.89	0.95	3.34	4.40	5.69	42.48
1984	2.26	4.42	3.79	3.37	10.14	4.34	9.03	1.72	0.85	3.26	2.87	2.49	48.54
1985	3.17	4.11	1.98	2.86	1.60	4.77	2.63	4.07	0.42	3.04	5.39	2.36	36.40
1986	0.95	3.90	1.69	2.25	2.40	0.69	1.89	3.37	3.59	3.84	3.83	4.08	32.48
1987	4.68	4.63	2.91	2.18	4.62	2.66	4.67	1.08	1.93	0.60	1.21	3.49	34.66
1988	4.29	2.94	2.42	2.34	2.35	0.51	3.60	3.20	2.68	1.52	4.82	3.99	34.66
1989	4.96	6.26	3.82	3.50	5.31	8.21	2.68	3.16	9.19	1.47	4.92	2.74	56.22
1990	5.88	6.90	5.73	2.56	4.71	1.72	7.56	3.02	2.68	3.64	1.77	8.99	55.16
1991	2.53	6.99	6.36	3.97	3.10	6.02	3.45	6.13	3.14	1.10	5.24	10.23	58.26
1992	3.87	3.36	3.99	2.11	3.20	3.07	4.62	4.22	1.77	3.05	4.29	6.68	44.23
1993	4.09	2.20	6.16	2.50	3.78	0.90	2.03	6.04	4.77	2.25	3.33	7.04	45.09
Record Mean	4.54	4.53	5.08	4.04	3.82	4.00	4.48	3.66	2.87	2.64	3.43	4.47	47.57

AVERAGE TEMPERATURE (deg. F)

KNOXVILLE, TENNESSEE

YEAR	JAN	FEB	MAR	APR	MAY	JUNE	JULY	AUG	SEP	OCT	NOV	DEC	ANNUAL
1964	36.3	36.1	47.7	60.5	66.9	75.1	75.7	73.7	70.0	55.5	52.3	43.5	57.8
1965	40.1	39.7	46.8	62.4	72.3	73.9	78.0	77.5	73.5	59.0	51.7	43.0	59.9
1966	34.0	41.6	50.4	58.9	66.1	74.3	79.3	75.0	68.9	57.5	49.1	40.0	57.9
1967	41.9	38.1	54.7	62.7	63.5	73.2	71.8	73.2	65.6	58.2	44.2	44.6	57.6
1968	37.1	33.3	49.9	59.0	65.3	73.6	77.2	78.4	69.1	58.8	48.1	37.0	57.3
1969	36.5	39.4	42.1	59.5	67.3	74.5	78.9	74.5	69.3	59.1	45.5	37.7	57.0
1970	30.6	39.3	48.7	61.2	68.6	73.4	78.5	77.1	75.3	62.8	48.1	44.1	59.0
1971	38.5	39.4	45.6	58.5	63.9	75.9	75.4	76.0	73.3	65.3	48.7	50.4	59.2
1972	42.5	39.4	48.2	59.5	64.9	70.8	75.5	75.9	71.9	56.8	47.5	44.8	58.1
1973	38.0	39.9	55.8	56.4	63.5	74.8	76.7	76.2	73.8	62.4	52.4	41.1	59.2
1974	49.3	43.1	55.2	59.8	68.1	70.5	77.7	76.5	69.8	58.6	50.8	41.9	60.1
1975	43.3	45.5	46.6	57.5	70.1	74.6	77.6	78.8	68.9	60.1	50.2	41.1	59.5
1976	34.3	48.2	52.6	58.9	63.4	73.9	76.2	74.9	67.7	54.5	43.2	37.5	57.1
1977	27.2	41.4	55.0	62.9	69.8	75.3	80.3	78.9	74.0	57.4	52.6	40.2	59.6
1978	29.4	34.6	49.0	61.1	66.6	75.4	78.7	77.1	74.8	58.2	54.4	42.8	58.5
1979	33.0	36.7	53.1	59.9	66.9	72.8	75.0	77.6	72.2	58.3	51.4	42.2	58.3
1980	41.6	36.6	47.6	59.1	68.1	75.2	82.1	81.7	73.7	56.0	47.3	40.1	59.1
1981	33.4	41.9	46.8	63.5	63.9	77.1	81.4	78.3	70.2	58.9	49.4	38.7	58.6
1982	33.8	43.1	53.0	55.2	69.3	72.6	77.6	76.7	70.0	59.5	50.9	46.4	59.0
1983	38.0	41.1	49.7	53.2	65.4	72.5	78.5	79.6	70.3	60.2	47.4	36.4	57.7
1984	35.3	43.4	47.7	57.3	62.8	74.8	73.3	74.1	67.1	67.6	46.3	47.5	58.1
1985	29.4	37.1	50.2	58.6	65.7	72.5	76.5	75.1	68.5	63.7	56.8	34.8	57.4
1986	35.0	43.9	48.7	58.6	68.1	77.2	81.4	76.7	72.8	60.2	52.4	39.1	59.5
1987	36.8	42.2	49.2	55.0	71.3	75.5	78.5	79.6	70.1	52.8	50.2	43.5	58.7
1988	33.9	39.4	50.1	57.2	65.0	74.4	78.7	79.6	71.3	52.7	49.7	40.4	57.7
1989	42.8	41.3	53.5	57.8	62.4	73.3	77.5	76.8	70.9	59.4	48.1	32.2	58.0
1990	44.7	48.9	53.6	57.9	65.9	75.4	78.0	77.9	72.1	59.9	52.1	45.1	61.0
1991	41.2	43.8	52.0	62.8	72.0	74.9	79.3	76.5	71.4	61.0	47.1	43.5	60.5
1992	40.5	45.8	48.3	59.0	64.0	72.2	77.9	73.7	71.3	57.4	48.3	40.0	58.2
1993	43.5	39.5	45.7	56.2	67.0	76.0	83.0	78.7	70.6	57.9	48.4	40.2	58.9
Record Mean	39.0	41.8	49.5	58.7	67.2	74.8	77.8	76.8	71.4	59.7	48.4	40.7	58.8
Max	47.6	51.1	59.8	69.7	78.1	85.1	87.7	86.7	81.8	70.9	58.5	49.2	68.8
Min	30.4	32.4	39.1	47.7	56.3	64.4	68.0	66.9	60.9	48.5	38.3	32.1	48.8

REFERENCE NOTES

KNOXVILLE, TENNESSEE

GENERAL
T - TRACE AMOUNT.
BLANK ENTRIES DEMOTE MISSING/UNREPORTED DATA.
# INDICATES A STATION OR INSTRUMENT RELOCATION.

SPECIFIC

PM - INCLUDES LAST DAY OF PREVIOUS MONTH
ASOS - AUTOMATED SURFACE OBSERVING SYSTEM IN OPERATION DURING THESE MONTHS.

(a) - LENGTH OF RECORD IN YEARS, ALTHOUGH INDIVIDUAL MONTHS MAY BE MISSING.
0.* OR * - THE VALUE IS BETWEEN 0.0 AND 0.05
NORMALS - BASED ON THE 1961-1990 RECORD PERIOD.
EXTREMES - DATES ARE THE MOST RECENT OCCURRENCE
WIND DIR.- NUMERALS SHOW TENS OF DEGRESS CLOCKWISE FROM TRUE NORTH. "00" INDICATES CALM.
RESULTANT DIRECTIONS ARE GIVEN TO WHOLE DEGREES.
BOLD VALUES INDICATE EXTREME VALUES WHICH OCCURRED AFTER THE ASOS SYSTEM WAS COMMISSIONED.

RECORD = PERIOD OF RECORD
RECORD MEAN PRECIPITATION IS THE MEAN OF ALL DAILY PRECIPITATION AMOUNTS DURING THE PERIOD OF RECORD.
RECORD MAX(MIN) TEMPERATURE IS THE MEAN OF ALL DAILY MAX(MIN) TEMPERATURES DURING THE PERIOD OF RECORD.
RECORD MEAN TEMPERATURE IS THE SUM OF THE RECORD MAX AND RECORD MIN DIVIDED BY 2.
AVERAGE TEMPERATURE IS THE SUM OF THE MEAN DAILY MAX AND MIN TEMPERATURE DIVIDED BY 2.

EXCEPTIONS

RECORD MEANS ARE THROUGH THE CURRENT YEAR, BEGINNING IN 1871 FOR TEMPERATURE
1871 FOR PRECIPITATION
1943 FOR SNOWFALL

HEATING DEGREE DAYS Base 65 deg. F — KNOXVILLE, TENNESSEE

SEASON	JULY	AUG	SEP	OCT	NOV	DEC	JAN	FEB	MAR	APR	MAY	JUNE	TOTAL
1964-65	0	8	9	308	374	657	765	701	559	125	0	0	3506
1965-66	0	0	13	200	393	674	956	650	452	217	59	9	3623
1966-67	0	0	9	232	472	767	709	748	330	112	121	8	3508
1967-68	1	0	63	229	617	626	860	912	467	191	66	0	4032
1968-69	0	0	5	215	501	859	879	710	704	175	50	4	4102
1969-70	0	0	11	200	579	840	1063	712	496	139	45	0	4085
1970-71	0	0	13	109	497	640	813	710	596	209	76	0	3663
1971-72	0	0	0	53	496	450	692	735	516	211	47	11	3211
1972-73	0	0	9	251	522	622	828	697	283	272	97	0	3581
1973-74	0	0	10	135	373	734	481	606	304	190	41	10	2884
1974-75	0	0	23	215	439	713	666	542	563	257	9	0	3427
1975-76	0	0	42	172	437	733	944	485	380	203	91	0	3487
1976-77	0	0	19	320	651	845	1166	658	319	110	33	2	4123
1977-78	0	0	3	242	374	764	1097	846	487	148	74	0	4035
1978-79	0	0	0	210	310	681	985	786	376	167	46	0	3561
1979-80	0	0	0	220	407	700	715	815	529	185	32	1	3604
1980-81	0	0	23	284	523	761	974	641	556	104	94	0	3960
1981-82	0	0	32	196	461	809	959	606	384	295	20	0	3762
1982-83	0	0	30	228	416	577	829	662	472	356	50	3	3623
1983-84	0	0	51	163	523	878	912	619	530	240	139	5	4060
1984-85	0	0	44	30	556	536	1095	776	459	208	51	13	3768
1985-86	0	0	44	86	250	927	922	582	499	206	51	0	3567
1986-87	0	2	0	197	377	797	863	631	483	313	13	0	3676
1987-88	0	0	15	370	437	660	956	734	458	247	66	9	3952
1988-89	0	0	3	384	454	755	681	660	360	258	148	1	3704
1989-90	0	1	36	204	499	1011	622	443	358	239	68	0	3481
1990-91	0	0	26	182	382	612	730	589	405	93	16	0	3035
1991-92	0	0	46	159	538	660	753	549	512	225	113	6	3561
1992-93	0	0	13	229	494	769	659	708	592	270	38	2	3774
1993-94	0	0	37	232	495	761							

COOLING DEGREE DAYS Base 65 deg. F — KNOXVILLE, TENNESSEE

YEAR	JAN	FEB	MAR	APR	MAY	JUNE	JULY	AUG	SEP	OCT	NOV	DEC	TOTAL
1969	0	0	0	15	125	294	438	304	147	23	0	0	1346
1970	0	0	0	33	162	258	427	384	326	46	0	0	1636
1971	0	0	0	20	50	333	328	350	252	69	13	3	1418
1972	0	0	0	53	51	193	332	347	223	4	2	0	1205
1973	0	0	4	19	56	298	369	355	282	60	2	0	1445
1974	0	0	9	39	142	179	399	365	171	19	17	0	1340
1975	0	0	0	39	173	293	397	434	168	26	0	0	1530
1976	0	3	1	26	48	273	356	314	107	5	0	0	1133
1977	0	5	14	57	188	317	483	440	283	14	10	0	1811
1978	0	0	0	37	130	319	432	384	302	8	0	0	1612
1979	0	0	15	20	111	242	317	399	224	23	4	0	1355
1980	0	0	0	16	136	315	538	525	290	12	0	0	1832
1981	0	0	0	62	65	373	512	421	193	14	1	0	1641
1982	0	0	17	6	157	233	396	372	187	65	2	8	1443
1983	0	0	4	6	71	235	425	462	217	23	0	0	1443
1984	0	0	0	12	77	306	263	290	116	120	2	0	1186
1985	0	0	11	20	81	247	363	322	155	53	11	0	1263
1986	0	0	0	21	156	374	517	373	241	55	6	0	1743
1987	0	0	0	18	215	321	427	460	172	0	0	0	1613
1988	0	0	3	19	74	297	431	458	200	9	0	0	1491
1989	0	2	10	47	76	257	395	374	219	35	0	0	1415
1990	0	1	13	32	101	316	410	406	245	32	1	0	1557
1991	0	0	9	32	239	306	453	364	244	42	6	2	1697
1992	0	0	0	49	91	230	410	279	206	0	0	0	1265
1993	0	0	0	10	106	339	566	432	212	21	3	0	1689

SNOWFALL (inches) — KNOXVILLE, TENNESSEE

SEASON	JULY	AUG	SEP	OCT	NOV	DEC	JAN	FEB	MAR	APR	MAY	JUNE	TOTAL
1964-65	0.0	0.0	0.0	0.0	T	T	6.3	6.9	4.4	0.0	0.0	0.0	17.6
1965-66	0.0	0.0	0.0	0.0	T	T	14.2	T	0.5	0.0	0.0	0.0	14.7
1966-67	0.0	0.0	0.0	0.0	T	3.1	2.2	5.7	0.0	0.0	0.0	0.0	11.0
1967-68	0.0	0.0	0.0	0.0	0.0	T	5.9	4.7	1.9	0.0	0.0	0.0	12.5
1968-69	0.0	0.0	0.0	0.0	0.3	T	0.5	7.7	5.2	0.0	0.0	0.0	13.7
1969-70	0.0	0.0	0.0	0.0	T	8.9	12.9	4.6	T	0.0	0.0	0.0	26.4
1970-71	0.0	0.0	0.0	0.0	T	6.3	1.3	6.4	3.6	7.0	0.0	0.0	24.6
1971-72	0.0	0.0	0.0	0.0	T	T	0.4	4.2	6.7	0.0	0.0	0.0	11.3
1972-73	0.0	0.0	0.0	0.0	T	T	9.0	T	1.9	T	0.0	0.0	10.9
1973-74	0.0	0.0	0.0	0.0	0.0	2.0	T	T	1.8	0.0	0.0	0.0	3.8
1974-75	0.0	0.0	0.0	0.0	T	1.6	1.3	T	2.5	0.0	0.0	0.0	5.4
1975-76	0.0	0.0	0.0	0.0	T	T	0.3	2.8	0.0	0.0	0.0	0.0	3.1
1976-77	0.0	0.0	0.0	0.0	1.0	2.1	7.9	0.2	0.0	0.0	0.0	0.0	11.2
1977-78	0.0	0.0	0.0	0.0	0.1	0.6	11.3	5.5	1.8	0.0	0.0	0.0	19.3
1978-79	0.0	0.0	0.0	0.0	0.0	T	4.7	18.4	T	0.0	0.0	0.0	23.1
1979-80	0.0	0.0	0.0	0.0	T	T	0.5	11.0	3.5	T	0.0	0.0	15.0
1980-81	0.0	0.0	0.0	0.0	T	T	5.0	2.5	T	0.0	0.0	0.0	7.5
1981-82	0.0	0.0	0.0	0.0	T	0.1	4.4	0.1	1.1	T	0.0	0.0	5.7
1982-83	0.0	0.0	0.0	0.0	0.0	3.0	1.1	3.4	0.7	2.0	0.0	0.0	10.2
1983-84	0.0	0.0	0.0	0.0	0.0	0.7	2.8	3.5	T	0.0	0.0	0.0	7.0
1984-85	0.0	0.0	0.0	0.0	0.0	T	14.2	8.3	0.0	0.0	0.0	0.0	22.5
1985-86	0.0	0.0	0.0	0.0	0.0	0.4	3.6	5.0	T	0.0	0.0	0.0	9.0
1986-87	0.0	0.0	0.0	0.0	T	T	7.5	T	1.6	10.7	0.0	0.0	19.8
1987-88	0.0	0.0	0.0	0.0	T	T	9.2	0.9	T	0.0	0.0	0.0	10.1
1988-89	0.0	0.0	0.0	0.0	T	2.6	1.8	2.8	0.0	0.0	0.0	0.0	7.2
1989-90	0.0	0.0	0.0	T	T	1.1	T	T	T	0.0	0.0	0.0	1.1
1990-91	0.0	0.0	0.0	0.0	0.0	T	T	1.2	0.7	T	0.0	0.0	1.9
1991-92	0.0	0.0	0.0	0.0	0.2	T	T	0.0	T	T	0.0	0.0	0.2
1992-93	0.0	0.0	0.0	0.0	0.0	0.3	T	1.0	15.1	T	0.0	0.0	16.4
1993-94	0.0	0.0	0.0	T	0.0	0.3							
Record Mean	0.0	0.0	0.0	T	0.6	1.6	3.9	3.6	1.8	0.4	T	0.0	11.8

MEMPHIS, TENNESSEE

Topography varies from the level alluvial area in east–central Arkansas to the slightly rolling area in northwestern Mississippi and southwestern Tennessee.

Agricultural interests are varied, with major crops being cotton, corn, hay, soybeans, peaches, apples, and a considerable number of vegetables. The climate is quite favorable for dairy interests, and for the raising of cattle and hogs.

The growing season is about 230 days in length. The average date for the last occurrence of temperatures as low as 32 degrees is late March. The average date of the first temperature of 32 degrees or below is early November.

Precipitation of nearly 50 inches per year is fairly well distributed. Crops and pastures receive, on the average, an adequate supply of moisture during the growing season, with lesser amounts during the fall harvesting period.

Sunshine averages slightly over 70 percent of the possible amount during the growing season. Relative humidity averages about 70 percent for the year.

Memphis, although not in the normal paths of storms coming from the Gulf or from western Canada, is affected by both, and thereby has comparatively frequent changes in weather. Extremely high or low temperatures, however, are relatively rare.

NORMALS, MEANS, AND EXTREMES

MEMPHIS, TENNESSEE

LATITUDE: 35°03'N LONGITUDE: 90°00'W ELEVATION: FT. GRND 258 BARO 271 TIME ZONE: CENTRAL WBAN: 13893

	(a)	JAN	FEB	MAR	APR	MAY	JUNE	JULY	AUG	SEP	OCT	NOV	DEC	YEAR
TEMPERATURE °F:														
Normals														
-Daily Maximum		48.5	53.5	63.2	73.3	81.0	89.3	92.3	90.8	83.9	74.3	62.3	52.5	72.1
-Daily Minimum		30.9	34.8	43.0	52.4	61.2	68.9	72.9	71.1	64.5	51.9	42.7	34.8	52.4
-Monthly		39.7	44.2	53.1	62.9	71.2	79.1	82.6	81.0	74.2	63.1	52.5	43.7	62.3
Extremes														
-Record Highest	52	78	81	85	94	99	104	108	105	103	95	85	81	108
-Year		1972	1962	1986	1987	1977	1954	1980	1943	1954	1954	1955	1982	JUL 1980
-Record Lowest	52	-4	-11	12	29	38	48	52	48	36	25	9	-13	-13
-Year		1985	1951	1943	1987	1944	1966	1947	1946	1949	1952	1950	1963	DEC 1963
NORMAL DEGREE DAYS:														
Heating (base 65°F)		784	582	383	127	25	0	0	0	10	131	380	660	3082
Cooling (base 65°F)		0	0	14	64	217	423	546	496	286	72	0	0	2118
% OF POSSIBLE SUNSHINE	35	50	54	56	64	69	74	74	75	69	70	58	50	64
MEAN SKY COVER (tenths)														
Sunrise - Sunset	43	6.8	6.4	6.5	6.0	5.9	5.4	5.3	5.0	5.1	4.6	5.6	6.4	5.8
MEAN NUMBER OF DAYS:														
Sunrise to Sunset														
-Clear	41	7.9	7.8	7.8	8.5	8.5	9.7	10.3	11.6	12.3	14.5	10.2	9.0	118.0
-Partly Cloudy	41	5.8	5.8	6.6	7.3	9.6	11.1	12.0	11.8	7.5	7.0	6.2	5.6	96.3
-Cloudy	41	17.3	14.7	16.7	14.2	12.9	9.2	8.7	7.6	10.1	9.5	13.6	16.4	151.0
Precipitation														
.01 inches or more	43	10.0	9.5	10.9	10.2	9.4	8.5	8.5	7.6	7.4	6.4	8.7	9.8	106.7
Snow,Ice pellets,hail														
1.0 inches or more	43	0.8	0.5	0.3	0.0	0.0	0.0	0.0	0.0	0.0	0.0	0.*	0.1	1.7
Thunderstorms	43	1.9	2.4	4.4	6.4	6.8	7.4	8.3	6.2	3.5	2.1	2.3	1.6	53.5
Heavy Fog Visibility														
1/4 mile or less	43	2.0	1.3	0.7	0.3	0.2	0.2	0.3	0.3	0.6	1.0	1.3	1.7	10.0
Temperature °F														
-Maximum														
90° and above	52	0.0	0.0	0.0	0.1	3.1	14.5	21.7	18.5	8.0	0.6	0.0	0.0	66.5
32° and below	52	3.1	1.3	0.2	0.0	0.0	0.0	0.0	0.0	0.0	0.0	0.1	1.4	6.0
-Minimum														
32° and below	52	17.7	12.3	5.3	0.3	0.0	0.0	0.0	0.0	0.0	0.3	5.8	14.4	56.2
0° and below	52	0.1	0.*	0.0	0.0	0.0	0.0	0.0	0.0	0.0	0.0	0.0	0.1	0.2
AVG. STATION PRESS.(mb)	21	1011.4	1009.9	1006.5	1005.6	1004.7	1005.2	1006.3	1006.7	1007.4	1009.1	1009.5	1011.0	1007.8
RELATIVE HUMIDITY (%)														
Hour 00	54	75	73	70	71	76	78	79	80	80	77	74	74	76
Hour 06 (Local Time)	54	79	78	76	78	82	82	84	86	86	83	80	78	81
Hour 12	54	63	60	56	53	55	56	57	57	56	51	56	61	57
Hour 18	54	67	62	57	54	57	57	58	60	62	60	63	67	60
PRECIPITATION (inches):														
Water Equivalent														
-Normal		3.73	4.35	5.41	5.46	4.98	3.57	3.79	3.43	3.53	3.01	5.10	5.74	52.10
-Maximum Monthly	43	12.21	10.51	12.08	17.13	11.58	7.50	8.84	9.65	7.61	7.75	10.52	13.81	17.13
-Year		1951	1989	1975	1991	1953	1992	1959	1978	1958	1984	1988	1982	APR 1991
-Minimum Monthly	43	0.57	1.12	1.50	1.39	0.83	0.04	0.43	0.43	0.19	T	0.75	1.05	T
-Year		1986	1980	1966	1992	1977	1953	1954	1953	1953	1963	1965	1955	OCT 1963
-Maximum in 24 hrs	43	3.89	4.24	5.95	4.35	4.94	4.76	4.71	4.04	4.63	3.40	5.65	5.42	5.95
-Year		1974	1989	1975	1985	1958	1980	1980	1978	1957	1981	1988	1978	MAR 1975
Snow,Ice pellets,hail														
-Maximum Monthly	43	12.4	8.3	17.3	T	T	0.0	0.0	0.0	0.0	T	1.5	14.3	17.3
-Year		1985	1985	1968	1993	1993					1993	1976	1963	MAR 1968
-Maximum in 24 hrs	43	8.1	5.8	16.1	T	T	0.0	0.0	0.0	0.0	T	1.2	14.3	16.1
-Year		1985	1960	1968	1993	1993					1993	1976	1963	MAR 1968
WIND:														
Mean Speed (mph)	45	10.0	10.1	10.8	10.3	8.8	7.9	7.5	6.9	7.4	7.7	9.1	9.7	8.9
Prevailing Direction through 1963		S	S	S	S	S	S	S	S	E	S	S	S	S
Fastest Obs. 1 Min.														
-Direction (!!)	17	34	24	16	24	34	02	34	20	36	28	23	30	24
-Speed (MPH)	17	35	38	40	46	40	40	37	35	39	40	40	36	46
-Year		1976	1993	1973	1979	1979	1977	1990	1980	1972	1984	1979	1972	APR 1979
Peak Gust														
-Direction (!!)	5	N	NW	NW	SW	SE	SE	W	SE	N	W	NW	NW	SE
-Speed (mph)	5	40	49	49	69	48	71	45	39	43	59	43	49	71
-Date		1984	1984	1984	1991	1990	1993	1992	1993	1993	1984	1992	1992	JUN 1993

PRECIPITATION (inches)

MEMPHIS, TENNESSEE

YEAR	JAN	FEB	MAR	APR	MAY	JUNE	JULY	AUG	SEP	OCT	NOV	DEC	ANNUAL
1964	3.73	3.50	7.34	11.03	3.28	1.39	6.14	5.76	2.74	2.21	2.59	7.97	57.68
1965	4.79	6.78	5.35	2.05	7.42	0.98	1.60	3.98	7.38	0.54	0.75	1.17	42.79
1966	2.84	6.88	1.50	5.42	5.69	0.52	2.18	4.28	3.23	1.92	1.57	5.21	41.24
1967	2.23	2.33	4.65	4.46	6.38	1.70	6.01	5.17	1.86	2.38	1.90	7.37	46.44
1968	5.57	1.98	6.52	5.15	5.21	3.76	2.69	1.61	5.58	2.87	4.89	6.04	51.87
1969	3.14	3.20	2.63	8.29	1.34	1.60	1.92	6.62	0.90	1.24	4.19	7.05	42.12
1970	1.16	3.87	5.32	7.08	3.70	5.76	4.99	1.78	3.80	6.20	2.62	3.71	49.99
1971	2.15	7.21	3.64	2.89	3.90	3.82	2.90	6.00	3.42	0.06	1.49	6.71	44.19
1972	4.73	2.23	4.80	3.51	4.55	5.50	4.89	1.94	5.46	3.92	8.05	9.37	58.95
1973	4.62	3.62	7.63	9.44	6.23	1.00	4.49	4.88	5.06	3.37	8.49	5.35	64.18
1974	8.90	4.65	3.40	6.34	7.76	6.30	6.33	4.78	3.45	2.67	4.96	5.03	64.57
1975	4.65	5.53	12.08	4.98	8.72	2.42	2.26	2.03	2.62	2.69	7.77	2.93	58.68
1976	2.85	4.41	7.68	2.41	4.73	4.06	3.82	0.86	5.40	5.66	1.83	1.79	45.50
1977	2.57	1.99	4.13	5.42	0.83	3.38	3.41	1.62	6.43	2.02	6.01	3.39	41.20
1978	8.13	1.31	4.05	2.14	8.14	4.45	3.89	9.65	1.52	1.82	5.56	13.12	63.78
1979	5.98	5.66	6.60	11.47	7.78	4.93	3.12	5.92	4.49	2.60	7.42	4.92	70.89
1980	3.23	1.12	10.86	7.53	4.43	5.75	4.73	1.23	5.32	3.14	5.23	1.86	54.43
1981	1.38	3.66	4.98	3.67	7.06	2.93	1.71	4.21	0.61	5.83	2.12	1.84	40.00
1982	6.61	4.16	4.47	6.76	5.50	6.68	4.13	3.11	1.92	5.23	6.43	13.81	68.81
1983	2.32	2.61	3.66	8.84	9.58	3.50	3.83	0.61	1.52	2.94	9.56	8.68	57.65
1984	1.88	4.37	6.07	5.24	9.06	1.12	4.59	5.00	1.96	7.75	5.85	4.35	57.24
1985	3.78	4.10	4.96	6.51	2.23	4.55	3.50	3.50	4.03	3.36	3.87	3.27	47.66
1986	0.57	2.50	1.90	3.72	4.63	3.80	1.21	2.74	1.21	3.75	8.67	3.92	38.62
1987	1.76	5.81	3.38	3.78	2.96	3.66	2.06	4.12	2.01	1.96	10.45	11.39	53.34
1988	4.25	3.49	4.20	2.85	2.38	2.15	5.21	0.85	4.73	3.62	10.52	5.99	50.24
1989	7.91	10.51	5.50	2.13	2.36	7.20	7.55	1.43	6.08	2.37	3.65	2.20	58.89
1990	3.97	8.99	5.65	6.93	4.55	2.68	2.21	1.18	5.21	4.37	3.44	10.61	59.79
1991	2.90	6.46	3.68	17.13	5.10	1.42	1.92	2.06	1.47	4.39	5.54	7.04	59.11
1992	1.78	2.18	7.07	1.39	3.68	7.50	5.38	2.44	3.62	4.01	4.82	3.28	47.15
1993	3.59	2.46	3.14	6.20	4.56	4.20	0.86	3.69	3.73	1.91	4.07	5.59	44.00
Record Mean	4.82	4.37	5.20	5.16	4.37	3.69	3.41	3.21	3.00	2.89	4.45	4.81	49.40

AVERAGE TEMPERATURE (deg. F)

MEMPHIS, TENNESSEE

YEAR	JAN	FEB	MAR	APR	MAY	JUNE	JULY	AUG	SEP	OCT	NOV	DEC	ANNUAL
1964	41.1	40.2	51.7	64.0	71.9	79.3	80.6	78.6	72.9	58.9	54.4	44.3	61.5
1965	43.4	42.9	44.0	66.4	74.7	78.1	81.8	80.2	73.7	60.8	55.8	45.8	62.3
1966	34.2	42.2	52.8	60.6	67.9	76.6	84.7	76.9	70.8	58.1	53.7	41.4	60.0
1967	42.2	39.2	56.7	66.5	68.8	78.3	77.7	76.0	70.3	62.4	49.1	45.0	61.0
1968	38.9	37.3	50.9	62.7	69.7	79.5	80.8	82.0	71.3	62.2	50.9	41.7	60.7
1969	41.2	43.3	45.0	62.9	72.1	78.8	84.9	79.3	72.9	62.9	49.0	40.1	61.1
1970	35.3	41.7	48.7	65.0	72.3	77.3	79.7	81.2	77.9	61.7	49.8	46.4	61.4
1971	39.6	43.5	48.4	60.5	66.6	80.6	80.6	78.6	76.1	69.3	50.9	50.7	62.1
1972	42.3	44.7	52.2	63.1	69.7	77.5	79.4	79.7	75.9	61.4	45.5	40.0	60.9
1973	38.6	40.5	57.3	59.8	68.3	81.0	83.2	79.6	76.1	67.6	57.3	43.3	62.7
1974	45.7	45.6	58.7	61.8	72.1	74.7	82.5	79.2	68.5	62.4	53.3	45.2	62.5
1975	45.9	46.2	49.9	61.9	73.5	78.8	81.1	81.2	70.9	65.8	53.8	44.1	62.8
1976	39.5	53.8	58.5	63.6	65.6	76.4	81.5	78.9	73.0	58.9	45.5	41.9	61.5
1977	30.7	45.1	58.6	66.9	76.4	81.9	84.7	82.6	79.0	62.2	55.1	44.1	64.0
1978	32.7	35.0	50.3	66.3	70.9	79.8	83.8	80.9	77.7	62.5	57.7	44.0	61.8
1979	30.9	38.5	54.3	63.0	70.0	77.9	82.6	80.9	73.4	65.8	50.7	45.4	61.1
1980	43.2	39.5	49.4	60.9	72.5	80.9	88.8	87.2	80.5	62.7	53.3	45.9	63.8
1981	40.9	47.3	54.3	70.2	70.0	82.5	84.6	81.8	74.0	62.5	53.8	40.9	63.6
1982	36.6	40.5	55.5	58.5	74.5	78.0	85.0	82.9	73.3	63.7	53.4	49.5	62.6
1983	40.4	45.3	51.9	56.8	68.2	77.2	83.6	84.9	75.2	66.2	53.1	34.7	61.5
1984	35.9	47.6	51.1	61.0	69.5	80.9	80.9	79.8	71.9	68.4	50.9	53.8	62.6
1985	32.4	40.6	57.7	65.0	71.8	78.9	82.2	80.2	73.6	67.2	57.6	36.9	62.0
1986	41.9	48.0	55.2	64.4	72.5	81.0	86.5	79.2	79.1	64.2	51.1	42.6	63.8
1987	39.6	47.1	54.7	62.2	76.5	80.0	82.5	83.6	75.3	59.2	54.4	46.7	63.5
1988	36.8	42.2	52.3	62.8	71.8	80.3	81.6	83.7	76.3	59.1	54.6	44.9	62.2
1989	47.3	40.0	53.8	62.6	69.5	77.6	80.7	81.2	72.4	64.2	54.2	33.6	61.4
1990	48.5	52.0	55.3	61.4	68.3	80.7	82.5	82.0	77.9	61.3	57.1	45.8	64.4
1991	41.2	48.0	55.4	65.0	74.6	80.3	83.3	80.9	75.3	65.5	49.2	46.8	63.8
1992	42.2	49.9	53.7	63.0	70.4	76.9	82.5	76.8	73.2	63.8	50.7	43.8	62.2
1993	42.7	42.6	51.1	59.3	69.9	79.2	86.4	83.6	73.5	62.2	50.2	44.8	62.1
Record Mean	40.9	44.1	52.5	62.3	70.6	78.4	81.5	80.2	74.3	63.5	51.8	43.5	62.0
Max	48.9	52.6	61.5	71.6	79.8	87.4	90.3	89.1	83.6	73.6	60.9	51.5	70.9
Min	32.9	35.6	43.5	52.9	61.3	69.3	72.7	71.3	64.9	53.3	42.7	35.6	53.0

REFERENCE NOTES

MEMPHIS, TENNESSEE

GENERAL
T - TRACE AMOUNT.
BLANK ENTRIES DENOTE MISSING/UNREPORTED DATA.
# INDICATES A STATION OR INSTRUMENT RELOCATION.

SPECIFIC

PM - INCLUDES LAST DAY OF PREVIOUS MONTH
ASOS - AUTOMATED SURFACE OBSERVING SYSTEM IN OPERATION DURING THESE MONTHS.

(*) - LENGTH OF RECORD IN YEARS, ALTHOUGH INDIVIDUAL MONTHS MAY BE MISSING.
0.* OR * - THE VALUE IS BETWEEN 0.0 AND 0.05
NORMALS - BASED ON THE 1961-1990 RECORD PERIOD.
EXTREMES - DATES ARE THE MOST RECENT OCCURRENCE
WIND DIR. - NUMERALS SHOW TENS OF DEGRESS CLOCKWISE FROM TRUE NORTH. "00" INDICATES CALM.
RESULTANT DIRECTIONS ARE GIVEN TO WHOLE DEGREES.
BOLD VALUES INDICATE EXTREME VALUES WHICH OCCURRED AFTER THE ASOS SYSTEM WAS COMMISSIONED.

RECORD = PERIOD OF RECORD
RECORD MEAN PRECIPITATION IS THE MEAN OF ALL DAILY PRECIPITATION AMOUNTS DURING THE PERIOD OF RECORD.
RECORD MAX(MIN) TEMPERATURE IS THE MEAN OF ALL DAILY MAX(MIN) TEMPERATURES DURING THE PERIOD OF RECORD.
RECORD MEAN TEMPERATURE IS THE SUM OF THE RECORD MAX AND RECORD MIN DIVIDED BY 2.
AVERAGE TEMPERATURE IS THE SUM OF THE MEAN DAILY MAX AND MIN TEMPERATURE DIVIDED BY 2.

EXCEPTIONS

RECORD MEANS ARE THROUGH THE CURRENT YEAR, BEGINNING IN 1875 FOR TEMPERATURE
1872 FOR PRECIPITATION
1951 FOR SNOWFALL

HEATING DEGREE DAYS Base 65 deg. F — MEMPHIS, TENNESSEE

SEASON	JULY	AUG	SEP	OCT	NOV	DEC	JAN	FEB	MAR	APR	MAY	JUNE	TOTAL
1964-65	0	0	13	202	335	634	666	614	645	72	0	0	3181
1965-66	0	0	22	162	275	590	947	634	382	175	47	5	3239
1966-67	0	0	9	227	344	732	707	715	301	60	35	0	3130
1967-68	0	0	34	144	469	613	803	795	444	114	21	0	3437
1968-69	0	0	1	149	423	716	733	601	608	103	12	0	3346
1969-70	0	0	0	151	473	768	917	648	500	97	20	0	3574
1970-71	0	0	7	150	455	571	781	593	509	171	50	0	3287
1971-72	0	0	9	13	432	435	698	582	391	146	27	0	2733
1972-73	0	0	12	172	583	766	809	679	237	200	32	0	3490
1973-74	0	0	8	67	244	665	599	535	235	150	1	0	2504
1974-75	0	0	28	121	367	607	591	521	463	180	2	0	2880
1975-76	0	0	40	90	352	643	783	326	238	100	58	0	2630
1976-77	0	0	0	231	581	708	1056	547	212	61	4	0	3400
1977-78	0	0	0	123	313	640	995	835	454	74	47	0	3481
1978-79	0	0	0	116	230	643	1049	734	345	121	23	0	3261
1979-80	0	0	0	76	426	598	669	733	478	156	7	0	3143
1980-81	0	0	5	146	362	586	739	492	342	18	23	0	2713
1981-82	0	0	9	153	331	739	873	680	324	215	2	0	3326
1982-83	0	0	20	134	352	500	759	543	406	273	25	0	3012
1983-84	0	0	27	73	368	935	894	499	426	162	24	0	3408
1984-85	0	0	37	48	423	367	1004	683	254	100	6	0	2922
1985-86	0	0	17	54	257	864	708	475	307	102	8	0	2792
1986-87	0	0	0	102	413	687	782	492	322	154	0	0	2952
1987-88	0	0	0	186	324	559	867	657	393	108	0	0	3094
1988-89	0	0	1	202	314	619	544	694	369	174	41	0	2958
1989-90	0	0	24	102	337	966	503	363	321	181	38	0	2835
1990-91	0	0	11	182	249	593	728	471	329	57	7	0	2627
1991-92	0	0	23	90	481	556	699	433	347	144	30	0	2803
1992-93	0	0	15	87	428	650	685	617	433	200	15	2	3132
1993-94	0	0	14	164	447	618							

COOLING DEGREE DAYS Base 65 deg. F — MEMPHIS, TENNESSEE

YEAR	JAN	FEB	MAR	APR	MAY	JUNE	JULY	AUG	SEP	OCT	NOV	DEC	TOTAL
1969	0	0	0	48	240	422	627	449	243	91	0	0	2120
1970	3	0	0	104	251	375	463	509	400	51	6	2	2164
1971	0	0	3	46	107	474	489	426	349	154	14	0	2062
1972	0	0	3	95	179	383	449	464	346	66	6	0	1991
1973	0	0	4	48	143	486	571	458	350	156	19	0	2235
1974	6	0	46	59	228	299	550	445	138	46	23	0	1840
1975	8	0	3	93	272	421	507	510	224	121	23	2	2184
1976	0	7	44	64	84	349	519	438	247	48	0	0	1800
1977	0	0	23	123	362	516	619	551	426	41	20	0	2681
1978	0	0	6	122	235	452	590	501	387	46	18	0	2357
1979	0	0	19	68	184	394	553	499	259	108	4	0	2088
1980	0	0	0	40	249	480	744	695	476	80	18	2	2784
1981	0	5	20	181	184	532	614	527	285	80	2	0	2430
1982	0	1	36	26	305	399	623	563	275	100	14	25	2367
1983	0	0	7	32	131	373	584	622	338	116	17	0	2220
1984	0	1	4	51	169	482	502	462	249	162	5	23	2110
1985	0	6	30	107	224	425	540	478	285	129	42	0	2266
1986	0	3	12	91	247	487	673	448	427	81	4	0	2473
1987	0	0	9	78	366	458	549	584	315	12	13	0	2384
1988	0	0	7	52	221	469	518	586	347	24	8	0	2232
1989	0	2	27	109	186	386	496	510	254	84	20	0	2074
1990	0	7	27	83	146	477	550	534	404	71	18	2	2319
1991	0	0	34	62	309	464	576	500	339	113	13	0	2410
1992	0	0	5	92	204	363	551	375	268	55	4	0	1917
1993	0	0	8	38	173	436	669	584	275	84	13	0	2280

SNOWFALL (inches) — MEMPHIS, TENNESSEE

SEASON	JULY	AUG	SEP	OCT	NOV	DEC	JAN	FEB	MAR	APR	MAY	JUNE	TOTAL
1964-65	0.0	0.0	0.0	0.0	T	T	T	3.3	4.6	0.0	0.0	0.0	7.9
1965-66	0.0	0.0	0.0	0.0	0.0	0.0	12.2	T	T	0.0	0.0	0.0	12.2
1966-67	0.0	0.0	0.0	0.0	T	0.7	0.6	T	0.3	0.0	0.0	0.0	1.6
1967-68	0.0	0.0	0.0	0.0	0.0	T	2.0	4.5	17.3	0.0	0.0	0.0	23.8
1968-69	0.0	0.0	0.0	0.0	0.0	0.0	T	T	T	0.0	0.0	0.0	T
1969-70	0.0	0.0	0.0	0.0	T	0.1	3.3	0.2	T	0.0	0.0	0.0	3.6
1970-71	0.0	0.0	0.0	0.0	T	1.0	T	6.7	1.6	T	0.0	0.0	9.3
1971-72	0.0	0.0	0.0	0.0	0.8	0.0	0.3	0.1	T	0.0	0.0	0.0	1.2
1972-73	0.0	0.0	0.0	0.0	T	T	1.4	T	0.0	0.0	0.0	0.0	1.4
1973-74	0.0	0.0	0.0	0.0	0.0	0.2	0.9	0.5	T	0.0	0.0	0.0	1.6
1974-75	0.0	0.0	0.0	0.0	T	0.2	3.9	0.5	1.4	0.0	0.0	0.0	6.0
1975-76	0.0	0.0	0.0	0.0	0.0	0.1	0.3	T	0.0	0.0	0.0	0.0	0.4
1976-77	0.0	0.0	0.0	0.0	1.5	0.3	3.5	T	0.0	0.0	0.0	0.0	5.3
1977-78	0.0	0.0	0.0	0.0	0.0	T	4.3	3.2	T	0.0	0.0	0.0	7.5
1978-79	0.0	0.0	0.0	0.0	0.0	T	3.0	7.4	0.0	0.0	0.0	0.0	10.4
1979-80	0.0	0.0	0.0	0.0	T	0.0	1.3	1.5	0.8	0.0	0.0	0.0	3.6
1980-81	0.0	0.0	0.0	0.0	T	T	T	T	0.0	0.0	0.0	0.0	T
1981-82	0.0	0.0	0.0	0.0	0.0	T	4.5	0.7	1.2	0.0	0.0	0.0	6.4
1982-83	0.0	0.0	0.0	0.0	0.0	T	7.3	T	0.2	0.0	0.0	0.0	7.5
1983-84	0.0	0.0	0.0	0.0	0.0	0.8	2.0	T	0.5	0.0	0.0	0.0	3.3
1984-85	0.0	0.0	0.0	0.0	0.0	T	12.4	8.3	0.0	0.0	0.0	0.0	20.7
1985-86	0.0	0.0	0.0	0.0	0.0	T	T	2.0	0.0	0.0	0.0	0.0	2.0
1986-87	0.0	0.0	0.0	0.0	0.0		T	T	0.4	0.0	0.0	0.0	
1987-88	0.0	0.0	0.0	0.0	0.0	T	8.2	3.0	T	0.0	0.0	0.0	11.2
1988-89	0.0	0.0	0.0	0.0	0.0	T	T	0.3	T	0.0	T	0.0	0.3
1989-90	0.0	0.0	0.0	T	T	0.4	0.0	0.0	0.0	T	0.0	0.0	0.4
1990-91	0.0	0.0	0.0	0.0	0.0	0.4	T	T	1.0	0.0	0.0	0.0	1.4
1991-92	0.0	0.0	0.0	0.0	0.6	0.0	T	0.0	T	0.0	0.0	0.0	0.6
1992-93	0.0	0.0	0.0	0.0	T	T	T	T	T	T	T	0.0	T
1993-94	0.0	0.0	0.0	T	0.0	0.0							
Record Mean	0.0	0.0	0.0	T	0.1	0.6	2.3	1.3	0.8	T	T	0.0	5.2

NASHVILLE, TENNESSEE

The city of Nashville is located on the Cumberland River, in the northwestern corner of the Central Basin of middle Tennessee near the escarpment of the Highland Rim. The Rim, as it is called, rises to the height of 300 to 400 feet above the mean elevation of the basin, forming an amphitheater about the city from the southwest to the southeast, with the south being more or less open but undulating.

Temperatures are moderate, with great extremes of either heat or cold rarely occurring, yet there are changes of sufficient amplitude and frequency to give variety.

Based on the 1951–1980 period, the average first occurrence of 32 degrees Fahrenheit in the fall is October 29 and the average last occurrence in the spring is April 5.

Humidity is an important phase of climate in relation to bodily health and comfort. The Nashville records show that the average relative humidity is moderate as compared with the general conditions east of the Mississippi River and south of the Ohio.

Nashville is not in the most frequented path of general storms that cross the country, however, it is in the zone of moderate frequency of thunderstorms. The thunderstorm season usually begins in the latter part of March and continues through September.

NORMALS, MEANS, AND EXTREMES

NASHVILLE, TENNESSEE

LATITUDE: 36 °07'N LONGITUDE: 86 °41' W ELEVATION: FT. GRND 590 BARO 630 TIME ZONE: CENTRAL WBAN: 13897

	(a)	JAN	FEB	MAR	APR	MAY	JUNE	JULY	AUG	SEP	OCT	NOV	DEC	YEAR
TEMPERATURE °F:														
Normals														
-Daily Maximum		45.9	50.8	61.2	70.8	78.8	86.5	89.5	88.4	82.5	72.5	60.4	50.2	69.8
-Daily Minimum		26.5	29.9	39.1	47.5	56.6	64.7	68.9	67.7	61.1	48.3	39.6	30.9	48.4
-Monthly		36.2	40.4	50.2	59.2	67.7	75.6	79.3	78.1	71.8	60.4	50.0	40.5	59.1
Extremes														
-Record Highest	54	78	84	86	91	97	106	107	104	105	94	84	79	107
-Year		1972	1962	1982	1989	1941	1952	1952	1954	1954	1953	1971	1982	JUL 1952
-Record Lowest	54	-17	-13	2	23	34	42	51	47	36	26	-1	-10	-17
-Year		1985	1951	1980	1982	1976	1966	1947	1946	1983	1987	1950	1989	JAN 1985
NORMAL DEGREE DAYS:														
Heating (base 65°F)		893	689	469	193	59	0	0	0	21	195	450	760	3729
Cooling (base 65°F)		0	0	10	19	143	318	443	406	225	52	0	0	1616
% OF POSSIBLE SUNSHINE	51	41	47	52	59	60	65	63	63	63	62	50	42	56
MEAN SKY COVER (tenths)														
Sunrise - Sunset	53	7.0	6.7	6.6	6.1	6.0	5.6	5.6	5.3	5.3	4.9	6.1	6.8	6.0
MEAN NUMBER OF DAYS:														
Sunrise to Sunset														
-Clear	52	6.4	7.0	7.4	8.3	8.1	8.2	8.2	10.0	10.6	12.8	8.8	7.1	102.9
-Partly Cloudy	52	6.3	5.8	7.1	8.5	9.9	12.5	13.1	12.0	9.0	8.0	6.8	7.0	105.9
-Cloudy	52	18.4	15.4	16.4	13.3	13.0	9.4	9.7	9.0	10.4	10.2	14.4	16.9	156.4
Precipitation														
.01 inches or more	52	10.9	10.6	11.8	10.7	10.9	9.3	10.2	8.7	8.1	7.0	9.5	11.0	118.6
Snow,Ice pellets,hail														
1.0 inches or more	52	1.2	1.2	0.5	0.*	0.0	0.0	0.0	0.0	0.0	0.0	0.1	0.5	3.5
Thunderstorms	52	1.2	1.7	4.1	5.1	7.3	8.1	9.5	7.7	3.7	1.6	1.7	1.1	52.8
Heavy Fog Visibility 1/4 mile or less	52	2.4	1.3	1.0	0.5	1.0	0.9	1.1	1.6	1.8	2.1	1.7	1.8	17.1
Temperature °F														
-Maximum														
90° and above	28	0.0	0.0	0.0	0.1	1.0	9.5	17.4	12.4	5.4	0.1	0.0	0.0	45.9
32° and below	28	4.8	2.2	0.1	0.0	0.0	0.0	0.0	0.0	0.0	0.0	0.1	1.8	9.1
-Minimum														
32° and below	28	21.6	16.8	9.3	1.7	0.0	0.0	0.0	0.0	0.0	1.1	8.2	17.1	75.8
0° and below	28	0.7	0.2	0.0	0.0	0.0	0.0	0.0	0.0	0.0	0.0	0.0	0.2	1.1
AVG. STATION PRESS. (mb)	21	999.1	997.9	995.2	994.5	994.0	994.4	995.4	995.9	996.6	998.0	998.1	999.1	996.5
RELATIVE HUMIDITY (%)														
Hour 00	28	75	73	71	72	82	83	85	85	85	81	77	76	79
Hour 06 (Local Time)	28	80	80	78	81	86	87	89	90	90	86	82	80	84
Hour 12	28	63	59	53	51	56	55	57	57	58	53	59	63	57
Hour 18	28	65	60	54	52	59	59	62	62	64	60	64	67	61
PRECIPITATION (inches):														
Water Equivalent														
-Normal		3.58	3.81	4.85	4.37	4.88	3.57	3.97	3.46	3.46	2.62	4.12	4.61	47.30
-Maximum Monthly	54	13.92	10.31	12.35	8.41	11.04	9.37	7.75	8.31	11.44	6.13	9.04	13.63	13.92
-Year		1950	1956	1975	1984	1983	1960	1950	1942	1979	1959	1945	1978	JAN 1950
-Minimum Monthly	54	0.19	0.64	1.18	0.52	0.69	0.45	0.71	0.69	0.28	T	0.54	0.98	T
-Year		1986	1968	1987	1986	1941	1988	1954	1968	1956	1963	1949	1985	OCT 1963
-Maximum in 24 hrs	54	4.40	4.73	4.66	3.29	4.27	4.91	4.32	5.34	6.68	3.75	3.74	5.12	6.68
-Year		1946	1989	1975	1979	1984	1960	1992	1963	1979	1975	1973	1978	SEP 1979
Snow,Ice pellets,hail														
-Maximum Monthly	54	18.8	18.9	16.1	1.1	0.0	T	0.0	T	0.0	0.4	9.2	13.2	18.9
-Year		1948	1979	1960	1971		1993		1989		1993	1950	1963	FEB 1979
-Maximum in 24 hrs	54	8.1	8.3	8.8	1.1	0.0	T	0.0	i	0.0	0.4	9.2	10.2	10.2
-Year		1988	1979	1951	1971		1993		1989		1993	1950	1963	DEC 1963
WIND:														
Mean Speed (mph)	52	9.1	9.3	9.9	9.3	7.7	7.1	6.6	6.2	6.5	6.8	8.4	8.9	8.0
Prevailing Direction through 1963		S	S	S	S	S	S	S	S	S	S	S	S	S
Fastest Obs. 1 Min.														
-Direction (!!)	18	34	32	13	18	36	36	34	02	34	36	15	23	13
-Speed (MPH)	18	32	35	41	35	41	35	33	40	33	32	39	41	41
-Year		1985	1980	1987	1993	1984	1990	1978	1983	1977	1986	1984	1987	MAR 1987
Peak Gust														
-Direction (!!)	10	W	SW	SE	W	NW	SW	N	NW	N	N	W	SW	NW
-Speed (mph)	10	46	47	56	67	55	49	51	70	47	48	60	54	70
-Date		1984	1988	1987	1991	1984	1988	1988	1990	1989	1986	1985	1987	AUG 1990

PRECIPITATION (inches)

NASHVILLE, TENNESSEE

YEAR	JAN	FEB	MAR	APR	MAY	JUNE	JULY	AUG	SEP	OCT	NOV	DEC	ANNUAL
1964	3.70	3.26	5.92	5.86	5.04	1.21	2.16	4.56	2.65	1.83	3.67	5.15	45.01
1965	2.98	4.71	6.13	5.72	3.12	2.74	3.32	2.53	5.02	0.57	1.82	1.01	39.67
1966	3.93	3.63	1.39	5.08	3.99	1.09	2.70	5.29	3.87	2.50	2.76	5.69	41.92
1967	1.62	1.78	4.44	3.40	6.98	4.23	7.46	2.06	1.93	1.57	3.87	5.88	45.22
1968	3.50	0.64	4.47	3.57	6.28	2.26	6.87	0.69	2.76	3.92	5.39	3.58	43.93
1969	4.96	4.48	2.12	6.03	4.81	3.34	5.33	2.27	2.06	2.01	1.83	8.03	47.27
1970	1.16	4.36	3.87	6.81	5.90	6.73	3.61	2.99	2.76	2.94	2.20	3.60	46.93
1971	2.66	4.70	2.95	3.34	2.93	3.47	5.00	5.87	2.11	1.27	1.18	5.17	40.65
1972	5.15	3.45	4.34	3.58	3.52	2.54	6.40	4.30	3.71	4.06	5.22	8.14	54.41
1973	3.40	3.63	9.88	7.00	5.72	4.80	7.67	1.79	1.56	3.32	7.78	3.23	59.78
1974	9.45	3.01	5.25	3.97	5.04	6.80	2.10	4.13	10.44	1.47	6.23	2.81	60.70
1975	4.67	5.22	12.35	3.55	6.52	2.22	2.96	4.69	5.42	5.86	3.00	4.12	60.58
1976	4.11	2.28	5.32	1.53	6.19	4.72	4.01	8.05	5.08	5.17	1.30	1.81	49.57
1977	2.53	3.27	5.83	7.87	1.65	4.29	1.15	4.65	5.04	4.22	5.96	4.25	50.71
1978	5.95	1.57	4.88	2.42	8.03	1.46	4.03	3.81	1.37	2.28	4.01	13.63	53.44
1979	7.13	4.01	4.92	7.80	8.18	2.79	4.27	4.59	11.44	3.97	5.98	5.04	70.12
1980	2.59	1.38	7.27	3.67	6.14	2.89	3.53	1.24	1.09	1.17	2.55	1.40	34.92
1981	1.60	3.83	3.38	4.78	3.05	8.05	3.49	3.10	1.37	2.82	3.83	2.38	41.68
1982	6.50	4.80	3.00	4.36	4.19	2.28	5.47	3.46	3.23	1.91	3.87	6.36	49.43
1983	2.56	2.93	3.44	6.80	11.04	3.93	1.71	1.36	0.45	2.77	6.98	7.75	51.72
1984	1.79	2.38	5.14	8.41	9.68	4.49	6.63	2.42	0.97	6.00	6.20	2.38	56.49
1985	3.02	3.30	2.70	2.91	2.65	1.53	2.00	3.91	2.52	1.59	3.81	0.98	30.92
1986	0.19	3.59	2.29	0.52	3.36	2.38	0.77	3.38	2.19	2.19	7.43	3.31	31.60
1987	1.61	4.87	1.18	1.03	4.41	2.82	2.56	0.73	1.95	0.21	3.40	5.46	30.23
1988	3.73	2.02	2.18	2.09	1.86	0.45	3.26	2.39	2.45	1.54	5.49	3.95	31.41
1989	4.52	9.36	5.31	2.68	4.61	7.87	3.18	3.67	6.30	3.62	3.94	1.97	57.03
1990	2.76	4.73	3.26	1.60	2.80	2.37	4.86	3.12	2.13	4.41	4.29	10.76	47.09
1991	2.92	5.44	4.25	3.35	5.63	1.25	2.82	1.79	5.47	3.88	2.87	7.27	46.94
1992	2.97	2.60	4.50	0.77	3.12	4.31	5.89	3.25	3.45	1.62	4.48	2.88	39.84
1993	2.76	3.33	5.50	3.33	4.50	5.31	3.64	1.76	2.90	2.20	2.53	6.62	44.38
Record Mean	4.55	4.12	5.06	4.13	4.07	3.77	3.93	3.35	3.28	2.51	3.61	4.06	46.44

AVERAGE TEMPERATURE (deg. F)

NASHVILLE, TENNESSEE

YEAR	JAN	FEB	MAR	APR	MAY	JUNE	JULY	AUG	SEP	OCT	NOV	DEC	ANNUAL
1964	38.9	37.2	49.0	62.6	69.7	77.7	78.7	77.2	71.3	56.7	51.7	42.7	59.4
#1965	40.0	39.8	41.9	60.9	71.7	75.1	78.6	79.0	74.1	59.3	51.9	44.8	59.7
1966	32.3	41.3	50.5	58.8	65.3	75.1	82.3	76.5	69.5	57.0	50.7	40.2	58.3
1967	42.3	37.5	57.0	63.9	66.3	76.3	75.7	72.6	66.8	59.3	44.3	42.6	58.7
1968	34.0	32.4	47.8	58.7	66.3	74.9	77.8	79.5	69.9	59.5	48.7	38.1	57.3
1969	37.2	39.8	42.6	60.9	68.9	77.2	82.7	78.1	70.8	60.6	46.5	37.3	58.6
1970	32.4	38.4	46.8	61.3	68.4	73.6	77.2	79.2	76.9	61.0	47.5	43.2	58.8
1971	35.6	38.7	44.6	57.9	63.3	77.5	76.8	76.4	74.4	66.7	49.9	49.2	59.3
1972	41.8	41.8	50.2	60.5	67.4	73.0	77.3	77.2	75.7	60.2	47.3	42.8	59.6
1973	38.0	39.7	56.8	56.4	64.2	76.1	78.7	78.1	76.2	66.2	54.7	40.5	60.5
1974	45.4	41.8	54.9	58.6	70.0	71.4	78.0	77.6	67.5	59.4	50.0	42.6	59.8
1975	43.4	44.6	47.3	58.5	70.4	76.0	78.5	79.1	67.9	62.4	52.1	42.8	60.2
1976	36.7	50.5	55.9	59.9	64.0	73.3	76.4	74.4	66.8	53.9	40.9	36.6	57.5
1977	24.5	40.6	53.9	63.0	71.9	77.2	82.2	79.5	74.0	57.0	50.8	38.6	59.4
1978	27.6	29.2	46.9	61.0	66.7	76.2	80.5	78.7	75.5	57.4	53.6	42.4	58.0
1979	29.7	33.4	50.7	57.8	66.3	73.7	77.6	77.0	70.5	60.3	48.6	41.5	57.3
1980	39.7	35.7	46.3	57.3	67.8	75.5	82.8	81.7	76.0	57.8	48.5	41.0	59.2
1981	35.5	42.6	47.5	64.0	64.2	77.5	79.8	76.6	69.1	60.4	49.9	38.3	58.7
1982	34.0	39.5	52.5	54.6	71.0	73.3	79.8	76.1	69.6	61.1	51.4	48.2	59.3
1983	38.8	42.7	50.3	54.5	64.8	75.5	80.5	83.2	73.7	62.4	49.9	34.0	59.2
1984	32.2	43.4	46.1	58.2	64.2	77.4	76.1	76.5	68.6	66.7	46.0	49.6	58.8
1985	27.8	36.5	53.2	61.9	68.4	75.7	80.2	77.2	70.8	64.4	56.9	34.2	58.9
1986	37.2	44.7	50.8	60.8	68.6	76.5	82.4	76.7	74.9	61.0	49.9	39.9	60.3
1987	36.1	43.1	51.8	57.7	73.4	77.5	80.2	81.1	72.2	54.6	52.4	44.1	60.4
1988	34.4	38.7	49.3	57.1	67.3	77.3	81.4	81.9	72.8	54.2	51.1	42.4	59.0
1989	44.9	39.0	52.6	59.3	65.7	74.7	79.1	78.0	70.5	61.0	51.4	29.5	58.8
1990	45.8	49.9	53.6	58.4	66.4	78.2	80.4	79.6	74.7	60.1	54.3	43.7	62.1
1991	39.2	43.9	52.5	63.8	74.2	78.2	81.1	78.3	72.3	61.2	47.2	44.5	61.4
1992	40.0	45.9	50.1	59.6	65.8	72.4	79.9	74.9	70.9	59.4	49.5	41.2	59.1
1993	41.6	39.3	47.1	56.7	67.6	75.9	83.3	81.0	71.0	58.6	47.4	40.3	59.2
Record Mean	38.5	41.0	49.7	59.4	68.1	76.2	79.5	78.3	72.2	60.8	49.1	41.0	59.5
Max	47.2	50.2	59.6	69.9	78.5	86.3	89.3	88.2	82.6	72.0	58.7	49.6	69.3
Min	29.9	31.8	39.7	48.9	57.7	66.1	69.6	68.3	61.7	49.6	39.4	32.4	49.6

REFERENCE NOTES

NASHVILLE, TENNESSEE

GENERAL
T - TRACE AMOUNT.
BLANK ENTRIES DENOTE MISSING/UNREPORTED DATA.
# INDICATES A STATION OR INSTRUMENT RELOCATION.

SPECIFIC

PM - INCLUDES LAST DAY OF PREVIOUS MONTH
ASOS - AUTOMATED SURFACE OBSERVING SYSTEM IN OPERATION DURING THESE MONTHS.

(a) - LENGTH OF RECORD IN YEARS, ALTHOUGH INDIVIDUAL MONTHS MAY BE MISSING.
0.* OR * - THE VALUE IS BETWEEN 0.0 AND 0.05
NORMALS - BASED ON THE 1961-1990 RECORD PERIOD.
EXTREMES - DATES ARE THE MOST RECENT OCCURRENCE
WIND DIR.- NUMERALS SHOW TENS OF DEGRESS CLOCKWISE FROM TRUE NORTH. "00" INDICATES CALM.
RESULTANT DIRECTIONS ARE GIVEN TO WHOLE DEGREES.
BOLD VALUES INDICATE EXTREME VALUES WHICH OCCURRED AFTER THE ASOS SYSTEM WAS COMMISSIONED.

RECORD = PERIOD OF RECORD
RECORD MEAN PRECIPITATION IS THE MEAN OF ALL DAILY PRECIPITATION AMOUNTS DURING THE PERIOD OF RECORD.
RECORD MAX(MIN) TEMPERATURE IS THE MEAN OF ALL DAILY MAX(MIN) TEMPERATURES DURING THE PERIOD OF RECORD.
RECORD MEAN TEMPERATURE IS THE SUM OF THE RECORD MAX AND RECORD MIN DIVIDED BY 2.
AVERAGE TEMPERATURE IS THE SUM OF THE MEAN DAILY MAX AND MIN TEMPERATURE DIVIDED BY 2.

EXCEPTIONS

RECORD MEANS ARE THROUGH THE CURRENT YEAR,
BEGINNING IN 1871 FOR TEMPERATURE
1871 FOR PRECIPITATION
1942 FOR SNOWFALL

HEATING DEGREE DAYS Base 65 deg. F NASHVILLE, TENNESSEE

SEASON	JULY	AUG	SEP	OCT	NOV	DEC	JAN	FEB	MAR	APR	MAY	JUNE	TOTAL
#1964-65	0	3	18	265	398	685	768	698	711	168	3	0	3717
1965-66	0	0	26	198	386	618	1007	657	452	234	78	7	3663
1966-67	0	0	13	255	423	763	697	763	300	106	69	1	3390
1967-68	0	3	58	216	615	688	952	941	531	205	64	1	4274
1968-69	0	0	4	220	484	825	855	700	692	149	38	3	3970
1969-70	0	0	10	201	551	854	1005	737	556	156	51	0	4121
1970-71	0	0	13	159	522	671	902	733	624	227	101	0	3952
1971-72	0	0	0	39	462	483	713	667	454	193	36	6	3053
1972-73	0	0	10	168	533	682	830	702	261	275	83	0	3544
1973-74	0	0	8	84	316	753	601	641	320	227	28	3	2981
1974-75	0	0	48	196	464	685	665	567	547	241	6	0	3419
1975-76	0	0	68	138	398	683	870	417	303	183	94	0	3154
1976-77	0	0	31	349	718	872	1250	679	350	129	28	1	4407
1977-78	0	0	3	255	425	813	1152	996	556	164	92	0	4456
1978-79	0	0	1	240	338	695	1088	877	449	213	57	0	3958
1979-80	0	0	5	180	487	723	777	848	571	240	38	0	3869
1980-81	0	0	9	259	487	739	909	621	537	97	96	0	3754
1981-82	0	0	42	175	445	820	956	707	416	309	8	0	3878
1982-83	0	0	30	194	413	537	806	620	458	322	71	0	3451
1983-84	0	0	45	121	447	956	1009	621	578	220	106	0	4103
1984-85	0	0	59	63	564	473	1146	794	383	145	25	6	3658
1985-86	0	0	30	91	264	948	854	561	432	171	55	0	3406
1986-87	0	3	0	175	447	773	889	608	401	242	6	0	3544
1987-88	0	0	7	317	376	640	941	756	485	242	43	2	3809
1988-89	0	0	5	343	408	693	618	721	397	258	90	0	3533
1989-90	0	0	36	158	408	1095	590	422	373	245	65	1	3393
1990-91	0	0	21	195	323	654	791	586	402	80	9	0	3061
1991-92	0	0	42	166	535	628	768	544	456	217	85	2	3443
1992-93	0	0	26	181	461	731	717	713	552	252	32	4	3669
1993-94	0	0	27	227	528	759							

COOLING DEGREE DAYS Base 65 deg. F NASHVILLE, TENNESSEE

YEAR	JAN	FEB	MAR	APR	MAY	JUNE	JULY	AUG	SEP	OCT	NOV	DEC	TOTAL
1969	0	0	0	33	165	378	554	416	191	74	0	0	1811
1970	0	0	0	56	163	265	386	446	374	40	1	1	1732
1971	0	0	0	23	55	380	374	360	293	101	16	1	1603
1972	0	0	1	62	117	250	385	387	341	24	6	0	1573
1973	0	0	14	25	61	339	432	412	351	128	8	0	1770
1974	0	0	16	39	191	203	410	399	130	30	22	0	1440
1975	3	0	3	55	183	341	424	444	164	62	19	0	1698
1976	0	1	28	36	68	257	363	299	92	10	0	0	1154
1977	0	0	11	74	253	371	543	458	281	13	4	0	2008
1978	0	0	1	50	152	344	489	432	324	13	2	0	1807
1979	0	0	11	5	103	264	393	381	175	44	0	0	1376
1980	0	0	0	17	131	322	562	527	344	44	1	0	1948
1981	0	0	1	71	81	383	464	366	145	42	0	0	1553
1982	0	0	37	4	199	256	470	352	177	84	12	21	1612
1983	0	0	9	12	69	320	488	568	315	49	2	0	1832
1984	0	0	0	21	87	382	352	364	173	121	0	1	1501
1985	0	2	24	59	137	335	479	386	206	79	29	0	1736
1986	0	0	1	52	174	352	551	371	304	59	0	0	1864
1987	0	0	0	31	272	381	479	507	227	3	7	0	1907
1988	0	0	5	17	120	380	515	531	246	17	0	0	1831
1989	0	0	21	93	120	298	446	408	208	39	8	0	1641
1990	0	4	26	51	115	401	485	458	315	52	10	0	1917
1991	0	0	22	50	300	403	507	419	268	57	4	0	2030
1992	0	0	0	60	115	233	471	311	208	15	2	0	1415
1993	0	0	1	9	121	336	573	506	215	33	7	0	1801

SNOWFALL (inches) NASHVILLE, TENNESSEE

SEASON	JULY	AUG	SEP	OCT	NOV	DEC	JAN	FEB	MAR	APR	MAY	JUNE	TOTAL
1964-65	0.0	0.0	0.0	0.0	T	0.0	1.2	2.9	3.4	0.0	0.0	0.0	7.5
1965-66	0.0	0.0	0.0	0.0	T	0.0	11.4	T	T	0.0	0.0	0.0	11.4
1966-67	0.0	0.0	0.0	0.0	7.2	4.3	1.2	T	T	0.0	0.0	0.0	12.7
1967-68	0.0	0.0	0.0	0.0	0.0	8.4	7.2	2.9	8.5	0.0	0.0	0.0	27.0
1968-69	0.0	0.0	0.0	0.0	T	T	5.2	6.9	4.8	0.0	0.0	0.0	16.9
1969-70	0.0	0.0	0.0	0.0	0.3	3.8	5.6	3.0	T	0.0	0.0	0.0	12.7
1970-71	0.0	0.0	0.0	0.0	T	1.2	1.2	6.5	3.0	1.1	0.0	0.0	13.0
1971-72	0.0	0.0	0.0	0.0	0.1	T	0.4	0.5	0.9	T	0.0	0.0	1.9
1972-73	0.0	0.0	0.0	0.0	0.1	0.6	4.8	T	0.2	0.1	0.0	0.0	5.8
1973-74	0.0	0.0	0.0	0.0	0.0	2.4	T	0.3	T	0.0	0.0	0.0	2.7
1974-75	0.0	0.0	0.0	0.0	T	2.1	4.2	T	T	T	0.0	0.0	6.3
1975-76	0.0	0.0	0.0	0.0	T	T	1.1	2.3	T	0.0	0.0	0.0	3.4
1976-77	0.0	0.0	0.0	0.0	1.2	1.8	18.5	T	0.0	T	0.0	0.0	21.5
1977-78	0.0	0.0	0.0	0.0	T	0.1	12.9	9.8	2.4	0.0	0.0	0.0	25.2
1978-79	0.0	0.0	0.0	0.0	0.0	T	8.0	18.9	0.6	0.0	0.0	0.0	27.5
1979-80	0.0	0.0	0.0	0.0	T	T	0.3	6.6	3.1	0.0	0.0	0.0	10.0
1980-81	0.0	0.0	0.0	0.0	T	T	1.2	1.7	T	0.0	0.0	0.0	2.9
1981-82	0.0	0.0	0.0	0.0	0.0	0.2	4.8	3.7	1.0	0.0	0.0	0.0	9.7
1982-83	0.0	0.0	0.0	0.0	0.0	0.4	0.3	0.8	T	0.0	0.0	0.0	1.5
1983-84	0.0	0.0	0.0	0.0	0.0	0.7	5.3	3.7	T	0.0	0.0	0.0	9.7
1984-85	0.0	0.0	0.0	0.0	0.0	0.8	9.8	8.0	0.0	0.0	0.0	0.0	18.6
1985-86	0.0	0.0	0.0	0.0	0.0	0.5	0.4	2.1	T	0.0	0.0	0.0	3.0
1986-87	0.0	0.0	0.0	0.0	0.0	T	1.4	1.3	1.6	T	0.0	0.0	4.3
1987-88	0.0	0.0	0.0	0.0	T	T	8.6	1.4	T	0.0	0.0	0.0	10.0
1988-89	0.0	0.0	0.0	0.0	0.0	1.6	T	5.2	0.0	0.0	0.0	0.0	6.8
1989-90	0.0	T	0.0	T	T	0.4	T	T	0.4	0.0	0.0	0.0	0.8
1990-91	0.0	0.0	0.0	0.0	0.0	0.3	T	0.6	1.1	0.0	0.0	0.0	2.0
1991-92	0.0	0.0	0.0	0.0	T	0.0	T	0.0	1.0	0.0	0.0	0.0	1.0
1992-93	0.0	0.0	0.0	0.0	T	0.3	T	5.9	2.8	T	0.0	T	9.0
1993-94	0.0	0.0	0.0	0.4	T	0.3							
Record Mean	0.0	T	0.0	T	0.4	1.5	3.8	3.1	1.4	T	0.0	T	10.3

OAK RIDGE, TENNESSEE

Oak Ridge is located in a broad valley between the Cumberland Mountains, which lie to the northwest of the area, and the Great Smoky Mountains, to the southeast. These mountain ranges are oriented northeast–southwest and the valley between is corrugated by broken ridges 300 to 500 feet high and oriented parallel to the main valley. During periods of light winds, daytime winds are usually southwesterly, nighttime winds northeasterly. Wind velocities are somewhat decreased by the ridges. Tornadoes rarely occur in the valley between the Cumberlands and the Great Smokies. In winter the Cumberland Mountains have a moderating influence on the local climate by retarding the flow of cold air from the north and west.

Temperatures of 100 degrees or more have occurred during less than one–half of the years of the period of record, and temperatures of zero or below are rare. Summer nights are seldom oppressively hot and humid.

Precipitation is more than adequate for agriculture and is normally well distributed through the year for agricultural purposes. Occasionally there is sufficient dry weather in late summer or early fall to cause small damage to crops and pastures and to create conditions favorable for destructive forest fires. Winter and early spring are the seasons of heaviest precipitation. A few of the larger monthly precipitation amounts recorded have occurred in the normally drier fall months.

Light snow usually occurs in all of the months from November through March, but the total monthly snowfall is often only a trace. Snowfalls sufficiently heavy to interfere with traffic and outdoor activities occur infrequently.

Based on the 1951–1980 period, the average first occurrence of 32 degrees Fahrenheit in the fall is October 27 and the average last occurrence in the spring is April 11.

NORMALS, MEANS, AND EXTREMES

OAK RIDGE, TENNESSEE

LATITUDE: 36 °00'N LONGITUDE: 84 °15' W ELEVATION: FT. GRND 880 BARO TIME ZONE: EASTERN WBAN: 03841

	(a)	JAN	FEB	MAR	APR	MAY	JUNE	JULY	AUG	SEP	OCT	NOV	DEC	YEAR
TEMPERATURE °F:														
Normals														
-Daily Maximum		44.8	49.8	60.1	69.9	77.4	84.2	86.7	86.0	80.1	69.9	58.4	48.3	68.0
-Daily Minimum		25.1	27.7	35.7	43.5	52.2	60.5	64.9	64.3	58.0	44.8	36.2	28.7	45.1
-Monthly		35.0	38.8	47.9	56.8	64.9	72.4	75.8	75.2	69.1	57.4	47.3	38.6	56.6
Extremes														
-Record Highest	46	75	79	85	92	93	101	105	103	102	90	83	78	105
-Year		1952	1977	1982	1986	1962	1988	1952	1983	1954	1954	1961	1982	JUL 1952
-Record Lowest	46	-17	1	1	20	30	39	49	50	33	21	0	-7	-17
-Year		1985	1965	1980	1987	1976	1977	1988	1989	1967	1952	1950	1983	JAN 1985
NORMAL DEGREE DAYS:														
Heating (base 65°F)		930	734	530	252	96	0	0	0	32	260	531	818	4183
Cooling (base 65°F)		0	0	0	6	93	226	335	316	155	25	0	0	1156
% OF POSSIBLE SUNSHINE														
MEAN SKY COVER (tenths)														
Sunrise - Sunset	39	6.7	6.5	6.5	5.9	5.9	5.5	5.8	5.5	5.6	4.9	6.0	6.5	5.9
MEAN NUMBER OF DAYS:														
Sunrise to Sunset														
-Clear	41	7.2	7.6	7.9	9.7	9.2	9.4	8.2	9.3	9.9	13.7	9.2	8.1	109.4
-Partly Cloudy	41	6.6	5.6	7.1	6.9	9.0	10.4	11.5	11.4	8.9	7.0	7.2	6.3	97.9
-Cloudy	41	17.2	15.1	16.0	13.4	12.8	10.2	11.3	10.2	11.2	10.2	13.6	16.6	157.9
Precipitation														
.01 inches or more	45	12.4	11.2	12.5	10.6	10.9	10.5	11.8	10.3	8.4	8.1	9.8	11.3	127.9
Snow, Ice pellets, hail														
1.0 inches or more	41	1.2	1.1	0.3	0.*	0.0	0.0	0.0	0.0	0.0	0.0	0.0	0.6	3.3
Thunderstorms	17	0.7	1.7	2.6	4.2	7.4	8.0	10.8	9.1	3.4	1.4	1.2	0.8	51.3
Heavy Fog Visibility														
1/4 mile or less	15	1.3	1.1	0.8	0.9	1.5	1.4	3.1	4.2	3.8	7.6	5.7	2.2	33.7
Temperature °F														
-Maximum														
90° and above	45	0.0	0.0	0.0	0.1	1.2	6.3	12.0	9.2	3.2	0.1	0.0	0.0	32.0
32° and below	45	2.9	1.3	0.2	0.0	0.0	0.0	0.0	0.0	0.0	0.0	0.1	1.6	6.0
-Minimum														
32° and below	45	21.0	17.4	12.1	2.8	0.1	0.0	0.0	0.0	0.0	2.1	12.2	19.6	87.3
0° and below	45	0.4	0.0	0.0	0.0	0.0	0.0	0.0	0.0	0.0	0.0	0.*	0.1	0.5
AVG. STATION PRESS. (mb)														
RELATIVE HUMIDITY (%)														
Hour 01														
Hour 07 (Local Time)														
Hour 13														
Hour 19														
PRECIPITATION (inches):														
Water Equivalent														
-Normal		4.57	4.34	5.68	4.08	4.68	4.34	5.45	3.70	3.86	3.18	4.59	5.30	53.77
-Maximum Monthly	46	13.27	10.47	12.24	9.71	10.70	11.14	19.27	10.46	9.10	6.95	12.22	12.64	19.27
-Year		1954	1956	1975	1956	1984	1989	1967	1960	1957	1972	1948	1990	JUL 1967
-Minimum Monthly	46	0.93	0.84	2.13	0.88	0.80	0.53	1.55	0.54	0.41	T	1.37	0.67	T
-Year		1981	1968	1957	1976	1970	1988	1970	1953	1961	1963	1949	1965	OCT 1963
-Maximum in 24 hrs	46	4.25	3.11	4.74	6.24	4.41	3.70	4.91	7.48	3.76	2.66	5.29	5.12	7.48
-Year		1954	1991	1973	1977	1973	1969	1967	1960	1982	1976	1973	1969	AUG 1960
Snow, Ice pellets, hail														
-Maximum Monthly	46	9.6	17.2	21.0	5.9	T	0.0	T	T	0.0	T	6.5	14.8	21.0
-Year		1966	1979	1960	1987	1993		1993	1990		1993	1950	1963	MAR 1960
-Maximum in 24 hrs	46	8.3	9.1	12.0	5.4	T	0.0	T	T	0.0	T	6.5	10.8	12.0
-Year		1962	1960	1960	1987	1993		1993	1990		1993	1950	1963	MAR 1960
WIND:														
Mean Speed (mph)	16	4.8	5.0	5.3	5.7	4.5	4.2	3.9	3.7	3.8	3.6	4.1	4.5	4.4
Prevailing Direction through 1961		SW	ENE	SW	SW	SW	SW	SW	E	E	E	E	SW	SW
Fastest Mile														
-Direction (!!)														
-Speed (MPH)	22	59	47	48	50	46	50	50	53	38	39	45	50	59
-Year		1959	1967	1962	1959	1973	1975	1961	1964	1959	1967	1968	1964	JAN 1959
Peak Gust														
-Direction (!!)														
-Speed (mph)														
-Date														

PRECIPITATION (inches)

OAK RIDGE, TENNESSEE

YEAR	JAN	FEB	MAR	APR	MAY	JUNE	JULY	AUG	SEP	OCT	NOV	DEC	ANNUAL
1964	4.81	4.33	6.32	7.13	2.84	0.86	3.41	5.05	3.14	2.61	3.82	5.62	49.94
1965	4.17	2.89	9.92	3.86	3.27	5.75	7.54	2.76	3.01	1.08	2.92	0.67	47.84
1966	3.72	5.73	2.29	5.50	3.52	2.65	2.57	4.73	4.20	3.85	4.74	3.72	47.22
1967	3.78	3.77	6.11	2.62	4.77	6.40	19.27	2.22	3.27	3.61	5.01	7.94	68.77
1968	4.47	0.84	5.26	5.21	4.01	3.56	2.80	1.25	3.47	2.23	2.24	4.28	39.62
1969	4.30	4.84	2.24	2.86	2.76	6.07	5.27	4.17	4.08	1.83	2.62	8.54	49.58
1970	2.96	3.74	4.06	9.24	0.80	5.90	1.55	9.10	2.54	6.07	1.78	4.47	52.21
1971	4.88	4.68	4.61	4.73	7.17	2.25	8.22	2.57	3.32	1.75	2.25	6.90	53.33
1972	7.32	4.83	5.99	2.94	5.81	5.56	5.22	1.77	4.28	6.95	5.02	9.20	64.89
1973	4.21	3.42	11.43	5.66	10.43	6.94	5.54	2.25	3.41	3.36	10.78	8.90	76.33
1974	9.62	4.72	7.22	3.78	7.98	2.00	1.99	4.31	4.18	1.78	5.04	4.96	57.58
1975	5.88	5.91	12.24	2.36	4.02	5.60	2.88	2.67	5.79	5.55	3.42	4.36	60.68
1976	5.47	2.60	6.00	0.88	7.24	4.80	4.03	4.89	4.56	5.91	2.27	4.68	53.33
1977	2.52	1.89	5.42	8.50	1.67	6.68	4.11	4.40	8.67	4.97	9.12	4.82	62.77
1978	5.99	1.13	4.69	3.21	4.50	3.97	4.89	5.75	1.70	0.39	5.54	6.65	48.41
1979	7.60	4.30	5.01	5.25	9.32	3.73	12.92	5.49	3.74	1.93	5.77	2.24	67.30
1980	6.24	1.51	9.20	3.89	3.01	0.85	2.39	2.07	3.31	1.08	4.55	2.02	40.12
1981	0.93	4.69	3.59	4.58	2.65	4.50	2.42	3.11	3.90	4.89	3.20	4.12	42.58
1982	6.70	5.42	6.20	2.79	2.88	2.15	7.01	4.61	5.31	1.95	7.81	7.22	60.05
1983	1.75	4.38	2.57	6.40	6.90	2.53	2.41	1.28	2.07	4.58	5.85	6.95	47.67
1984	2.62	3.92	4.81	4.16	10.70	4.70	8.72	1.87	1.87	6.19	4.37	2.59	56.52
1985	2.88	3.73	2.60	2.23	3.83	5.08	6.43	8.50	1.56	3.08	4.49	2.11	46.52
1986	1.16	5.15	2.70	1.73	2.74	1.45	2.84	2.84	4.70	4.51	3.67	5.34	38.83
1987	4.37	5.64	2.82	2.97	2.02	4.26	3.94	1.92	5.64	0.69	2.11	3.43	40.31
1988	5.44	3.43	3.80	3.42	2.65	0.53	7.60	2.39	5.63	1.97	6.56	5.53	48.95
1989	6.94	5.07	6.03	2.76	6.14	11.14	3.62	3.90	8.86	2.46	6.06	3.03	66.01
1990	5.29	8.01	5.09	2.57	6.59	1.53	5.06	5.09	1.44	4.07	2.40	12.64	59.78
1991	2.97	7.94	7.44	3.84	2.55	7.05	4.06	4.62	2.86	1.76	4.15	10.92	60.16
1992	3.37	3.25	3.79	1.64	2.27	4.00	5.10	3.38	3.11	2.62	5.67	5.79	43.99
1993	3.98	4.13	6.42	3.70	4.29	1.60	1.99	3.25	5.12	2.15	4.09	8.29	49.01
Record Mean	5.04	4.82	5.66	3.99	4.26	4.06	5.21	3.75	3.73	2.97	4.62	5.71	53.82

AVERAGE TEMPERATURE (deg. F)

OAK RIDGE, TENNESSEE

YEAR	JAN	FEB	MAR	APR	MAY	JUNE	JULY	AUG	SEP	OCT	NOV	DEC	ANNUAL
1964	36.9	35.4	47.5	60.7	67.6	75.4	74.9	74.3	68.8	55.1	51.3	41.7	57.5
1965	38.5	38.9	45.1	60.7	68.6	71.4	75.5	75.8	71.1	56.1	49.5	41.4	57.7
1966	33.1	39.7	48.8	57.0	64.6	72.5	78.4	75.0	68.5	55.5	47.8	38.3	56.6
1967	40.5	36.3	52.9	61.3	62.7	72.0	72.1	72.8	64.9	57.8	44.3	44.2	56.8
1968	36.8	32.7	48.9	58.2	65.2	73.9	77.3	78.3	68.1	58.7	47.3	36.4	56.8
1969	35.7	38.7	41.2	59.4	67.3	74.6	79.1	74.8	68.5	58.1	44.8	36.5	56.6
1970	29.8	38.0	47.0	60.3	67.7	72.1	76.7	76.5	73.9	61.2	46.9	42.5	57.7
1971	36.8	39.1	44.4	57.6	63.5	75.6	75.0	74.8	72.6	64.0	46.7	47.8	58.2
1972	40.6	37.4	46.5	58.0	63.9	69.8	75.2	75.5	72.1	57.3	47.0	44.2	57.3
1973	37.4	38.2	55.0	55.4	62.0	74.0	76.8	75.3	72.8	62.1	50.9	39.3	58.2
1974	47.1	41.6	53.6	57.7	65.9	67.7	75.4	74.2	65.6	54.1	45.8	39.6	57.4
1975	41.1	43.0	44.6	54.9	68.2	72.0	74.7	76.2	66.0	58.5	48.4	39.4	57.3
1976	32.6	45.7	49.9	56.2	60.5	70.9	73.5	72.3	65.0	52.3	39.7	34.6	54.4
1977	24.8	36.7	50.8	59.2	66.4	72.0	77.3	75.3	70.3	53.6	49.9	36.1	56.0
1978	27.3	31.1	46.1	57.2	64.1	72.5	75.7	75.6	72.8	55.7	51.7	39.7	55.8
1979	31.1	35.3	50.0	57.2	64.4	71.5	73.9	74.9	69.7	56.5	47.9	39.5	56.0
1980	38.8	34.3	44.5	55.4	65.7	72.8	80.6	78.9	73.7	55.4	46.6	38.3	57.1
1981	32.3	39.9	44.1	61.2	62.3	75.4	77.8	74.1	66.8	55.0	46.0	34.3	55.8
1982	31.6	41.2	50.7	52.6	69.3	71.6	77.5	73.8	66.8	57.6	47.4	43.0	57.0
1983	35.4	38.6	47.7	51.2	63.8	71.5	77.9	79.2	69.4	59.4	45.7	34.4	56.2
1984	33.2	41.3	44.9	54.7	60.7	74.0	73.5	75.1	66.5	65.5	43.1	45.8	56.5
1985	28.0	35.3	47.8	57.3	64.2	71.0	74.8	73.1	67.0	61.9	54.5	30.9	55.5
1986	35.2	43.2	48.4	58.3	66.7	75.7	80.3	75.6	72.1	59.9	51.1	37.7	58.7
1987	36.0	40.8	48.3	56.7	71.6	75.0	78.1	78.1	69.7	53.4	50.0	42.8	58.4
1988	32.9	38.5	49.5	58.3	66.5	74.1	77.6	78.7	70.9	51.8	48.2	38.9	57.2
1989	42.3	40.0	52.8	57.9	62.2	73.2	77.1	75.8	70.8	58.8	46.8	30.5	57.4
1990	42.8	47.7	52.8	56.2	64.6	75.0	77.6	77.6	71.8	59.1	50.4	43.1	59.9
1991	40.0	42.6	51.3	62.8	72.2	74.9	79.1	76.8	71.4	60.7	45.5	42.1	60.0
1992	39.3	44.8	47.5	58.5	63.7	71.5	77.4	73.3	70.9	57.4	47.0	38.9	57.5
1993	42.2	38.7	45.6	55.2	67.4	74.7	82.1	78.7	71.7	60.1	46.2	39.6	58.5
Record Mean	37.0	40.4	48.1	58.0	66.1	73.4	76.9	75.8	69.9	58.4	47.3	39.2	57.5
Max	46.2	51.1	59.8	70.7	78.3	84.9	87.6	86.6	81.1	70.7	58.4	48.6	68.7
Min	27.7	29.7	36.4	45.3	53.8	61.9	66.2	65.1	58.8	46.1	36.1	29.8	46.4

REFERENCE NOTES

OAK RIDGE, TENNESSEE

GENERAL
T - TRACE AMOUNT.
BLANK ENTRIES DENOTE MISSING/UNREPORTED DATA.
# INDICATES A STATION OR INSTRUMENT RELOCATION.

SPECIFIC

(a) - LENGTH OF RECORD IN YEARS, ALTHOUGH INDIVIDUAL MONTHS MAY BE MISSING.
0.* OR * - THE VALUE IS BETWEEN 0.0 AND 0.05
NORMALS - BASED ON THE 1961-1990 RECORD PERIOD.
EXTREMES - DATES ARE THE MOST RECENT OCCURRENCE
WIND DIR. - NUMERALS SHOW TENS OF DEGRESS CLOCKWISE FROM TRUE NORTH. "00" INDICATES CALM.
RESULTANT DIRECTIONS ARE GIVEN TO WHOLE DEGREES.
BOLD VALUES INDICATE EXTREME VALUES WHICH OCCURRED AFTER THE ASOS SYSTEM WAS COMMISSIONED.

RECORD = PERIOD OF RECORD
RECORD MEAN PRECIPITATION IS THE MEAN OF ALL DAILY PRECIPITATION AMOUNTS DURING THE PERIOD OF RECORD.
RECORD MAX(MIN) TEMPERATURE IS THE MEAN OF ALL DAILY MAX(MIN) TEMPERATURES DURING THE PERIOD OF RECORD.
RECORD MEAN TEMPERATURE IS THE SUM OF THE RECORD MAX AND RECORD MIN DIVIDED BY 2.
AVERAGE TEMPERATURE IS THE SUM OF THE MEAN DAILY MAX AND MIN TEMPERATURE DIVIDED BY 2.

EXCEPTIONS

1. PEAK GUST WINDS ARE THROUGH 1979.
2. THUNDERSTORMS AND HEAVY FOG ARE THROUGH 1984 AND MAY BE INCOMPLETE, DUE TO PART-TIME OPERATIONS.
3. MEAN WIND SPEED IS THROUGH 1984.

RECORD MEANS ARE THROUGH THE CURRENT YEAR, BEGINNING IN 1948 FOR TEMPERATURE
1948 FOR PRECIPITATION
1949 FOR SNOWFALL

HEATING DEGREE DAYS Base 65 deg. F OAK RIDGE, TENNESSEE

SEASON	JULY	AUG	SEP	OCT	NOV	DEC	JAN	FEB	MAR	APR	MAY	JUNE	TOTAL
1964-65	0	6	29	310	405	715	815	724	610	166	9	2	3791
1965-66	0	0	22	279	460	726	984	702	496	267	79	12	4027
1966-67	0	0	15	288	511	820	754	797	378	138	135	10	3846
1967-68	2	0	70	243	616	638	870	930	494	216	70	0	4149
1968-69	0	0	11	219	523	878	902	731	728	181	48	7	4228
1969-70	0	0	22	227	602	876	1088	751	550	172	52	2	4342
1970-71	0	0	19	134	536	689	870	717	633	233	87	0	3918
1971-72	0	0	0	73	551	526	748	792	568	244	67	17	3586
1972-73	0	0	10	241	536	642	848	745	304	301	122	0	3749
1973-74	0	0	11	139	415	789	548	651	347	228	65	26	3219
1974-75	0	0	61	330	568	781	736	610	629	318	24	0	4057
1975-76	0	0	79	203	490	786	1000	556	459	262	148	5	3988
1976-77	1	0	45	389	752	936	1242	788	437	169	60	23	4842
1977-78	0	0	10	352	448	890	1161	946	577	232	110	2	4728
1978-79	0	0	0	281	393	780	1043	826	462	230	86	2	4103
1979-80	0	0	7	267	507	783	804	883	629	288	63	3	4234
1980-81	0	0	23	304	545	821	1009	695	643	145	120	0	4305
1981-82	0	0	68	307	562	944	1030	662	450	367	29	0	4419
1982-83	0	0	62	263	522	678	910	734	529	413	78	11	4200
1983-84	0	0	57	181	572	942	977	682	616	304	173	5	4509
1984-85	0	0	58	47	648	588	1140	824	525	239	84	20	4173
1985-86	0	0	61	123	311	1049	917	603	508	214	62	0	3848
1986-87	0	3	0	199	417	841	892	673	515	282	11	0	3833
1987-88	0	0	9	353	445	682	988	764	477	217	45	11	3991
1988-89	0	0	6	413	498	805	697	695	379	260	154	0	3907
1989-90	0	1	40	220	539	1066	682	481	384	282	83	0	3778
1990-91	0	0	26	200	436	675	770	624	426	96	14	0	3267
1991-92	0	0	48	167	580	705	794	583	538	236	121	7	3779
1992-93	0	0	14	230	536	803	701	729	595	292	37	3	3940
1993-94	0	0	26	182	559	781							

COOLING DEGREE DAYS Base 65 deg. F OAK RIDGE, TENNESSEE

YEAR	JAN	FEB	MAR	APR	MAY	JUNE	JULY	AUG	SEP	OCT	NOV	DEC	TOTAL
1969	0	0	0	23	129	300	443	311	136	22	0	0	1364
1970	0	0	0	37	140	219	371	361	291	25	0	0	1444
1971	0	0	0	16	46	323	316	312	232	47	9	0	1301
1972	0	0	0	41	40	168	322	330	229	5	2	0	1137
1973	0	0	2	16	34	277	374	325	254	56	0	0	1338
1974	0	0	1	20	103	116	330	295	86	2	0	0	953
1975	0	0	0	23	133	216	306	353	114	11	0	0	1156
1976	0	1	0	6	17	187	273	234	53	0	0	0	771
1977	0	0	3	4	109	241	387	324	177	7	4	0	1256
1978	0	0	0	3	89	234	342	335	240	1	0	0	1244
1979	0	0	2	5	74	202	285	314	151	11	0	0	1044
1980	0	0	0	6	92	244	491	435	291	13	0	0	1572
1981	0	0	0	39	40	319	404	286	126	3	0	0	1217
1982	0	0	15	2	167	207	392	279	120	44	0	3	1229
1983	0	0	0	5	47	211	405	445	196	12	0	0	1321
1984	0	0	0	3	47	283	275	318	106	69	0	0	1101
1985	0	0	2	16	63	207	313	258	125	31	3	0	1018
1986	0	0	0	22	121	329	477	337	219	47	5	0	1557
1987	0	0	0	36	222	306	412	409	152	0	0	0	1537
1988	0	0	1	23	94	291	400	429	187	9	0	0	1434
1989	0	1	3	51	72	249	379	341	219	31	0	0	1346
1990	0	0	11	25	78	305	393	397	235	26	3	0	1473
1991	0	0	5	36	242	300	441	370	245	39	0	0	1678
1992	0	0	0	43	86	204	387	261	198	1	0	0	1180
1993	0	0	0	3	115	298	532	429	234	36	0	0	1647

SNOWFALL (inches) OAK RIDGE, TENNESSEE

SEASON	JULY	AUG	SEP	OCT	NOV	DEC	JAN	FEB	MAR	APR	MAY	JUNE	TOTAL
1964-65	0.0	0.0	0.0	0.0	T	T	4.3	5.8	1.2	0.0	0.0	0.0	11.3
1965-66	0.0	0.0	0.0	0.0	0.0	T	9.6	T	0.3	0.0	0.0	0.0	9.9
1966-67	0.0	0.0	0.0	0.0	1.0	4.4	4.2	8.2	T	0.0	0.0	0.0	17.8
1967-68	0.0	0.0	0.0	0.0	0.0	0.2	3.9	2.7	0.5	0.0	0.0	0.0	7.3
1968-69	0.0	0.0	0.0	0.0	0.4	0.5	2.8	10.2	0.9	0.0	0.0	0.0	14.8
1969-70	0.0	0.0	0.0	0.0	T	8.1	8.6	4.6	0.3	0.0	0.0	0.0	21.6
1970-71	0.0	0.0	0.0	0.0	T	5.4	1.7	1.9	2.6	0.3	0.0	0.0	11.9
1971-72	0.0	0.0	0.0	0.0	0.2	0.7	0.4	5.0	4.0	0.0	0.0	0.0	10.3
1972-73	0.0	0.0	0.0	0.0	0.4	T	8.0	0.2	0.6	0.2	0.0	0.0	9.4
1973-74	0.0	0.0	0.0	0.0	0.0	0.6	0.0	0.3	0.4	0.0	0.0	0.0	1.3
1974-75	0.0	0.0	0.0	0.0	0.4	1.5	1.1	T	3.1	T	0.0	0.0	6.1
1975-76	0.0	0.0	0.0	0.0	0.1	T	2.2	3.9	0.0	0.0	0.0	0.0	6.2
1976-77	0.0	0.0	0.0	0.0	T	1.3	8.7	T	T	0.0	0.0	0.0	10.0
1977-78	0.0	0.0	0.0	0.0	T	1.0	8.6	6.2	3.4	T	0.0	0.0	19.2
1978-79	0.0	0.0	0.0	0.0	0.0	T	4.2	17.2	0.5	0.0	0.0	0.0	21.9
1979-80	0.0	0.0	0.0	0.0	T	T	1.0	7.9	3.2	T	0.0	0.0	12.1
1980-81	0.0	0.0	0.0	0.0	T	T	5.3	0.2	0.1	0.0	0.0	0.0	5.6
1981-82	0.0	0.0	0.0	0.0	T	0.9	4.4	0.5	1.8	T	0.0	0.0	7.6
1982-83	0.0	0.0	0.0	0.0	0.0	2.8	1.4	4.8	T	0.2	0.0	0.0	9.2
1983-84	0.0	0.0	0.0	0.0	T	0.3	4.6	5.1	0.2	0.0	0.0	0.0	10.2
1984-85	0.0	0.0	0.0	0.0	0.0	0.1	9.4	3.6	0.0	0.0	0.0	0.0	13.1
1985-86	0.0	0.0	0.0	0.0	0.0	T	2.5	5.0	T	0.0	0.0	0.0	7.5
1986-87	0.0	0.0	0.0	0.0	0.0	T	4.7	T	T	5.9	0.0	0.0	10.6
1987-88	0.0	0.0	0.0	0.0	T	T	6.9	0.1	T	0.0	0.0	0.0	7.0
1988-89	0.0	0.0	0.0	0.0	T	4.2	T	4.1	0.0	T	0.0	0.0	8.3
1989-90	0.0	0.0	0.0	T	T	9.5	T	T	T	0.0	T	0.0	9.5
1990-91	0.0	T	0.0	0.0	0.0	T	T	T	T	0.0	0.0	0.0	T
1991-92	T	0.0	0.0	0.0	T	T	T	0.0	1.0	T	0.0	0.0	1.0
1992-93	0.0	0.0	0.0	0.0	0.0	0.6	0.0	2.8	12.0	T	T	0.0	15.4
1993-94	T	0.0	0.0	T	0.0	0.5							
Record Mean	T	T	0.0	T	0.3	1.7	3.2	3.0	1.4	0.1	T	0.0	9.8

ABILENE, TEXAS

Abilene is located in north central Texas. The station elevation is 1,750 feet above sea level. Topography of the area includes rolling plains, treeless except for mesquite, broken by low hills to the south and west. The land rises gently to the east and southeast. Regional agricultural products are mainly cattle, dry-land cotton, and feed crops.

Abilene is on the boundary between the humid east Texas climate and the semi-arid west and north Texas climate. The rainfall pattern is typical of the Great Plains. Most precipitation occurs from April to October and is usually associated with thunderstorms. Severe storms are infrequent, occurring mostly in the spring.

The large range of high and low temperatures, characteristic of the Great Plains, extends south to the Abilene area. High daytime temperatures prevail in the summer, but are normally broken by thunderstorms about five times a month. Rapid cooling after sunset results in pleasant nights with low summertime temperatures in the upper 60s and low 70s. High summer temperatures are usually associated with fair skies, southwesterly winds, and low humidities.

Rapid wintertime temperature changes occur when cold, dry, arctic air replaces warm moist tropical air. Drops in temperature of 20 to 30 degrees in one hour are not unusual. However, cold weather periods are short lived. Fair, mild weather is typical.

South is the prevailing wind direction, and southerly winds are frequently high and persist for several days. Strong northerly winds often occur during the passage of cold fronts. Dusty conditions are infrequent, occurring mostly with westerly winds. Dust storm frequency and intensity depend on soil conditions in eastern New Mexico, west Texas, and the Texas Panhandle.

Based on the 1951-1980 period, the average first occurrence of 32 degrees Fahrenheit in the fall is November 13 and the average last occurrence in the spring is March 25.

NORMALS, MEANS, AND EXTREMES

ABILENE, TEXAS

LATITUDE: 32°25'N LONGITUDE: 99°41' W ELEVATION: FT. GRND 1784 BARO 1789 TIME ZONE: CENTRAL WBAN: 13962

	(a)	JAN	FEB	MAR	APR	MAY	JUNE	JULY	AUG	SEP	OCT	NOV	DEC	YEAR
TEMPERATURE °F:														
Normals														
-Daily Maximum		54.8	59.7	68.9	77.8	84.4	91.4	95.2	94.5	86.7	77.9	66.3	57.0	76.2
-Daily Minimum		30.8	35.1	43.3	52.9	61.1	68.9	72.7	71.7	65.3	54.8	43.4	33.9	52.8
-Monthly		42.8	47.4	56.1	65.4	72.7	80.2	84.0	83.2	76.0	66.4	54.9	45.5	64.6
Extremes														
-Record Highest	54	89	90	97	99	107	109	110	109	106	103	92	89	110
-Year		1943	1940	1974	1948	1967	1980	1978	1943	1952	1979	1980	1955	JUL 1978
-Record Lowest	54	-9	-7	7	25	36	47	55	50	35	23	14	-7	-9
-Year		1947	1985	1943	1973	1979	1964	1940	1992	1942	1993	1976	1989	JAN 1947
NORMAL DEGREE DAYS:														
Heating (base 65°F)		688	493	299	94	7	0	0	0	6	75	317	605	2584
Cooling (base 65°F)		0	0	23	106	245	456	589	564	336	118	14	0	2451
% OF POSSIBLE SUNSHINE	45	61	64	70	71	70	78	79	77	70	72	67	62	70
MEAN SKY COVER (tenths)														
Sunrise - Sunset	54	5.7	5.6	5.3	5.2	5.4	4.5	4.4	4.4	4.6	4.3	4.8	5.3	5.0
MEAN NUMBER OF DAYS:														
Sunrise to Sunset														
-Clear	54	11.0	10.0	11.5	11.6	10.1	12.7	14.4	14.4	13.7	15.1	13.4	11.9	149.8
-Partly Cloudy	54	6.2	6.1	7.4	7.7	10.1	10.3	9.8	9.6	8.0	6.9	6.0	6.3	94.4
-Cloudy	54	13.8	12.2	12.1	10.7	10.9	7.0	6.8	7.0	8.3	9.0	10.6	12.8	121.1
Precipitation														
.01 inches or more	54	5.2	5.4	4.8	6.1	7.9	6.3	4.8	5.6	6.1	5.5	4.4	4.7	66.8
Snow,Ice pellets,hail														
1.0 inches or more	54	0.7	0.5	0.2	0.0	0.0	0.0	0.0	0.0	0.0	0.0	0.1	0.3	1.8
Thunderstorms	54	0.6	1.4	2.9	5.1	8.0	6.2	4.7	5.3	3.5	2.9	1.4	0.7	42.7
Heavy Fog Visibility														
1/4 mile or less	54	1.3	1.4	0.6	0.5	0.4	0.1	0.*	0.1	0.4	0.7	1.0	1.0	7.5
Temperature °F														
-Maximum														
90° and above	30	0.0	0.0	0.5	2.6	7.3	19.2	26.2	24.4	11.7	2.1	0.1	0.0	94.1
32° and below	30	1.9	1.0	0.*	0.0	0.0	0.0	0.0	0.0	0.0	0.0	0.2	1.1	4.2
-Minimum														
32° and below	30	17.0	11.1	4.2	0.4	0.0	0.0	0.0	0.0	0.0	0.3	5.4	13.7	52.1
0° and below	30	0.*	0.*	0.0	0.0	0.0	0.0	0.0	0.0	0.0	0.0	0.0	0.1	0.1
AVG. STATION PRESS.(mb)	21	956.1	954.6	951.1	950.5	949.6	950.8	952.8	952.9	953.3	954.3	954.4	955.7	953.0
RELATIVE HUMIDITY (%)														
Hour 00	30	67	67	61	63	69	66	58	61	68	67	68	67	65
Hour 06 (Local Time)	30	73	73	70	72	79	78	72	74	78	76	74	73	74
Hour 12	30	54	54	48	46	51	50	45	47	53	51	52	53	50
Hour 18	30	50	47	40	40	44	43	38	40	48	48	51	52	45
PRECIPITATION (inches):														
Water Equivalent														
-Normal		1.03	1.16	1.36	1.90	2.97	2.86	2.09	2.80	3.21	2.51	1.48	1.03	24.40
-Maximum Monthly	54	4.35	3.60	5.16	6.80	13.19	9.60	7.15	8.18	11.03	10.68	4.60	6.28	13.19
-Year		1968	1992	1979	1966	1957	1961	1968	1969	1974	1981	1968	1991	MAY 1957
-Minimum Monthly	54	T	0.04	0.03	T	0.15	0.03	T	T	T	0.00	0.00	T	0.00
-Year		1967	1962	1963	1961	1956	1954	1970	1943	1956	1952	1949	1972	OCT 1952
-Maximum in 24 hrs	54	2.18	1.94	2.23	3.75	4.76	3.66	3.74	6.30	6.70	6.08	2.43	2.62	6.70
-Year		1961	1992	1977	1957	1990	1959	1960	1978	1961	1981	1975	1991	SEP 1961
Snow,Ice pellets,hail														
-Maximum Monthly	54	13.5	8.4	7.3	T	T	T	T	0.0	0.0	0.3	8.1	7.8	13.5
-Year		1973	1956	1970	1991	1993	1993	1992			1993	1968	1983	JAN 1973
-Maximum in 24 hrs	54	7.5	4.5	6.1	T	T	T	T	0.0	0.0	0.3	5.3	4.2	7.5
-Year		1973	1979	1970	1991	1993	1993	1992			1993	1976	1946	JAN 1973
WIND:														
Mean Speed (mph)	49	11.7	12.5	13.9	13.8	12.9	12.6	10.8	10.2	10.3	11.0	11.6	11.8	11.9
Prevailing Direction through 1963		S	S	S	SSE	SSE	SSE	SSE	SSE	SSE	SSE	S	SSW	SSE
Fastest Obs. 1 Min.														
-Direction (!!)	13	36	35	27	34	20	34	36	26	29	31	19	32	20
-Speed (MPH)	13	35	36	41	41	46	40	39	35	35	46	35	40	46
-Year		1993	1984	1984	1990	1990	1989	1981	1987	1993	1984	1982	1993	MAY 1990
Peak Gust														
-Direction (!!)	10	W	N	W	N	S	N	N	W	N	NW	S	NW	S
-Speed (mph)	10	49	52	61	56	71	63	49	52	59	58	47	59	71
-Date		1988	1990	1984	1990	1993	1989	1992	1987	1987	1984	1991	1993	MAY 1993

PRECIPITATION (inches)

ABILENE, TEXAS

YEAR	JAN	FEB	MAR	APR	MAY	JUNE	JULY	AUG	SEP	OCT	NOV	DEC	ANNUAL
1964	2.36	2.47	1.26	2.76	1.53	1.55	2.58	5.87	2.11	0.50	2.19	0.17	25.35
1965	1.13	1.48	0.59	2.78	5.04	3.70	0.69	1.59	1.86	2.67	1.85	0.97	24.35
1966	0.85	0.40	0.40	6.80	0.46	1.16	0.12	5.08	4.08	2.09	0.28	0.05	21.77
1967	T	0.24	1.01	0.49	4.39	2.85	1.32	0.54	5.44	1.74	2.89	2.43	23.34
1968	4.35	1.93	2.12	3.32	2.40	1.09	7.15	0.89	0.53	0.14	4.60	0.12	28.64
1969	0.81	1.28	2.08	2.49	7.76	2.42	2.15	8.18	4.10	1.98	1.11	2.48	36.84
1970	0.04	1.84	2.41	2.56	3.68	1.94	T	1.54	2.41	1.18	0.08	0.24	17.92
1971	0.01	0.57	0.04	2.44	2.17	1.78	1.85	6.92	5.33	2.43	0.76	1.81	26.11
1972	0.29	0.11	0.23	1.04	2.47	3.67	1.22	4.92	3.15	4.49	0.37	T	21.96
1973	3.52	1.61	3.28	1.10	1.38	2.21	2.68	0.30	4.65	3.15	0.22	0.01	24.11
1974	0.18	0.57	0.72	3.44	0.72	1.03	2.20	5.61	11.03	5.06	1.75	1.14	33.45
1975	0.83	1.86	1.49	0.52	4.43	1.68	1.93	2.47	2.27	0.84	2.67	1.22	22.21
1976	0.02	0.08	0.25	3.70	1.04	0.68	4.27	1.47	3.97	6.24	0.63	0.18	22.53
1977	1.22	0.08	2.35	3.11	0.44	1.71	1.82	2.54	0.09	2.15	0.57	0.19	16.27
1978	0.62	1.26	0.17	1.00	1.50	1.24	0.72	6.70	2.36	1.49	0.94	0.28	18.28
1979	0.72	1.10	5.16	1.72	1.86	2.89	1.55	1.55	0.01	0.53	0.65	2.62	20.36
1980	0.77	0.72	0.69	0.17	5.00	1.14	0.24	1.62	6.30	0.71	1.62	1.69	20.67
1981	1.20	1.06	2.36	3.52	1.48	2.73	1.69	0.63	1.74	10.68	0.43	0.25	27.77
1982	1.00	1.20	0.43	0.38	6.87	3.98	1.50	1.12	1.09	0.74	1.72	1.28	21.31
1983	2.28	0.12	1.95	0.54	1.42	3.86	2.57	0.10	0.87	3.25	1.77	0.77	19.50
1984	0.98	0.46	0.47	0.20	0.42	1.70	0.97	3.24	3.55	5.15	2.11	3.08	22.33
1985	0.53	1.50	2.86	0.90	4.03	1.78	1.71	3.66	1.28	2.43	1.56	0.03	22.27
1986	0.01	1.51	0.50	0.99	4.03	4.49	3.54	2.45	1.71	7.46	3.24	2.05	31.98
1987	0.54	3.55	1.55	0.87	4.58	1.86	1.91	4.17	2.12	0.21	1.16	4.55	27.07
1988	0.05	0.37	0.54	2.09	2.93	3.75	1.76	1.18	1.74	1.01	T	0.80	16.22
1989	0.54	3.08	1.72	0.71	3.07	9.53	1.01	2.53	4.09	0.97	0.20	0.21	27.66
1990	1.94	2.36	2.43	4.57	5.06	0.36	4.27	0.77	4.79	1.73	2.26	0.81	31.35
1991	2.71	0.14	0.51	1.65	1.07	8.05	2.24	1.85	5.99	2.47	0.32	6.28	33.28
1992	1.87	3.60	0.83	2.02	3.51	4.81	2.70	1.99	1.79	1.19	2.57	1.00	27.88
1993	0.91	1.23	1.03	0.84	2.01	3.80	0.25	1.82	2.42	3.19	0.40	0.90	18.80
Record Mean	0.95	1.06	1.21	2.33	3.75	2.79	2.04	2.19	2.67	2.63	1.29	1.25	24.15

AVERAGE TEMPERATURE (deg. F)

ABILENE, TEXAS

YEAR	JAN	FEB	MAR	APR	MAY	JUNE	JULY	AUG	SEP	OCT	NOV	DEC	ANNUAL
1964	46.2	44.9	57.2	68.8	75.3	81.3	87.0	85.3	76.1	66.0	56.7	48.6	66.1
1965	48.3	45.8	48.1	68.1	72.3	79.3	85.7	84.7	79.2	66.5	62.3	51.4	66.0
1966	39.3	44.4	56.9	63.2	71.0	79.9	86.9	80.4	75.2	64.1	59.5	44.6	63.8
1967	48.0	46.6	63.5	71.9	72.5	81.9	82.8	80.2	71.1	63.7	54.2	43.7	65.0
1968	43.9	43.6	53.1	61.7	71.4	79.3	80.7	82.5	74.4	69.5	52.6	45.5	63.2
1969	48.0	48.0	47.5	65.7	70.4	78.4	87.4	85.6	75.9	61.9	53.4	47.7	64.2
1970	40.8	48.8	51.4	65.1	70.3	79.2	84.4	83.8	78.1	63.4	51.2	51.5	64.0
1971	46.4	48.7	55.1	64.5	72.4	80.4	84.4	76.9	73.3	66.0	55.0	48.9	64.3
1972	43.7	49.2	59.5	69.4	70.6	80.6	80.8	78.9	75.3	63.7	46.4	42.1	63.3
1973	37.5	44.4	57.2	58.6	70.6	77.4	81.2	82.5	73.8	66.5	58.2	46.8	62.9
1974	42.8	51.2	62.6	65.6	76.0	80.8	84.4	78.7	66.3	64.6	51.7	44.4	64.1
1975	44.8	45.1	54.1	62.9	70.2	78.5	79.7	81.3	70.8	65.6	54.3	46.1	62.8
1976	42.4	56.0	56.5	64.8	68.2	78.9	77.7	81.6	73.5	56.9	45.6	42.5	62.1
1977	35.7	50.9	56.8	62.4	73.4	81.3	83.8	83.4	82.6	66.9	54.6	48.7	65.0
1978	34.5	38.7	55.0	70.3	76.3	83.3	89.0	82.3	77.3	66.9	55.3	43.3	64.3
1979	35.6	44.9	56.7	64.2	69.8	79.0	83.7	82.2	77.7	71.6	50.9	47.7	63.7
1980	45.2	47.9	54.8	63.9	72.1	84.4	89.4	86.1	77.7	64.9	52.6	49.3	65.7
1981	44.8	49.4	54.9	68.3	71.6	80.3	85.8	83.3	78.0	66.2	57.4	48.8	65.8
1982	44.8	44.7	58.3	62.9	72.1	78.8	84.1	85.6	77.6	65.9	54.4	45.8	64.6
1983	42.6	47.0	54.9	60.6	70.8	76.7	82.8	85.2	77.9	69.1	57.0	34.2	63.2
1984	39.5	49.9	55.3	63.6	75.3	83.1	83.7	83.1	73.1	63.8	52.7	48.8	64.3
1985	37.2	44.4	57.5	66.7	73.8	77.9	81.6	84.9	75.7	65.7	54.6	41.4	63.5
1986	47.9	50.8	60.6	68.6	73.9	79.3	84.3	81.8	77.4	63.5	51.2	43.9	65.3
1987	43.4	49.4	52.1	62.7	72.1	77.5	81.6	83.5	74.0	67.0	54.0	44.8	63.5
1988	41.7	45.2	55.4	64.0	71.1	78.9	81.7	83.9	76.3	66.4	57.9	47.4	64.2
1989	49.4	40.2	56.8	66.5	75.9	75.9	82.9	81.9	72.6	67.7	56.4	38.4	63.7
1990	50.0	53.0	56.4	62.8	72.1	84.9	79.7	82.1	76.8	64.5	57.1	42.9	65.2
1991	41.9	52.1	58.9	68.0	75.4	78.2	82.3	79.2	71.3	66.1	49.9	46.7	64.2
1992	43.1	50.6	58.8	64.9	69.4	78.2	81.1	78.0	76.0	68.7	51.5	47.8	64.0
1993	42.9	46.6	54.8	63.6	70.4	79.7	85.7	83.9	74.1	62.8	49.6	47.8	63.5
Record Mean	44.0	47.9	56.0	65.0	72.3	80.0	83.3	82.8	76.0	65.9	53.9	46.0	64.5
Max	55.5	59.8	68.5	77.2	83.7	91.1	94.5	94.0	86.9	77.4	65.2	57.1	75.9
Min	32.6	36.0	43.6	52.8	61.0	68.8	72.2	71.7	65.0	54.4	42.6	34.9	53.0

REFERENCE NOTES

ABILENE, TEXAS

GENERAL
T - TRACE AMOUNT.
BLANK ENTRIES DENOTE MISSING/UNREPORTED DATA.
# INDICATES A STATION OR INSTRUMENT RELOCATION.

SPECIFIC

(a) - LENGTH OF RECORD IN YEARS, ALTHOUGH INDIVIDUAL MONTHS MAY BE MISSING.
0.* OR * - THE VALUE IS BETWEEN 0.0 AND 0.05
NORMALS - BASED ON THE 1961-1990 RECORD PERIOD.
EXTREMES - DATES ARE THE MOST RECENT OCCURRENCE
WIND DIR.- NUMERALS SHOW TENS OF DEGRESS CLOCKWISE FROM TRUE NORTH. "00" INDICATES CALM.
RESULTANT DIRECTIONS ARE GIVEN TO WHOLE DEGREES.
BOLD VALUES INDICATE EXTREME VALUES WHICH OCCURRED AFTER THE ASOS SYSTEM WAS COMMISSIONED.

RECORD = PERIOD OF RECORD
RECORD MEAN PRECIPITATION IS THE MEAN OF ALL DAILY PRECIPITATION AMOUNTS DURING THE PERIOD OF RECORD.
RECORD MAX(MIN) TEMPERATURE IS THE MEAN OF ALL DAILY MAX(MIN) TEMPERATURES DURING THE PERIOD OF RECORD.
RECORD MEAN TEMPERATURE IS THE SUM OF THE RECORD MAX AND RECORD MIN DIVIDED BY 2.
AVERAGE TEMPERATURE IS THE SUM OF THE MEAN DAILY MAX AND MIN TEMPERATURE DIVIDED BY 2.

EXCEPTIONS

FASTEST MILE WINDS ARE THROUGH JANUARY 1980.

RECORD MEANS ARE THROUGH THE CURRENT YEAR, BEGINNING IN 1886 FOR TEMPERATURE
1886 FOR PRECIPITATION
1940 FOR SNOWFALL

HEATING DEGREE DAYS Base 65 deg. F ABILENE, TEXAS

SEASON	JULY	AUG	SEP	OCT	NOV	DEC	JAN	FEB	MAR	APR	MAY	JUNE	TOTAL
1964-65	0	0	2	44	293	505	519	538	522	59	0	0	2482
1965-66	0	0	4	67	117	424	791	570	281	122	42	0	2418
1966-67	0	0	0	119	198	633	532	508	154	14	23	0	2181
1967-68	0	0	17	118	321	654	648	611	366	134	15	0	2884
1968-69	0	0	0	42	387	599	520	468	541	53	15	1	2626
1969-70	0	0	0	192	350	528	748	449	422	92	31	1	2813
1970-71	0	0	15	147	414	411	569	455	326	96	22	0	2455
1971-72	0	0	44	44	312	492	651	464	207	49	12	0	2275
1972-73	0	0	13	158	554	702	844	574	235	228	43	0	3351
1973-74	0	0	6	60	227	557	680	389	181	78	7	0	2185
1974-75	0	0	60	67	400	631	620	553	351	144	7	0	2833
1975-76	0	0	55	97	327	578	692	276	296	68	39	0	2428
1976-77	0	0	11	276	575	690	901	393	262	89	0	0	3197
1977-78	0	0	0	56	314	500	938	730	337	37	34	0	2946
1978-79	0	0	5	56	308	667	906	567	272	84	55	0	2920
1979-80	0	0	0	43	429	530	605	496	316	118	10	0	2547
1980-81	0	0	16	116	389	485	619	439	312	51	20	0	2447
1981-82	0	0	7	116	237	496	617	570	253	149	22	0	2467
1982-83	0	0	0	111	340	589	686	497	325	209	23	4	2784
1983-84	0	0	9	37	261	947	781	431	310	107	12	0	2895
1984-85	0	0	65	108	375	501	853	572	259	52	3	0	2788
1985-86	0	0	28	70	328	725	522	409	162	38	9	0	2291
1986-87	0	0	1	102	413	647	663	429	396	159	1	0	2811
1987-88	0	0	0	45	347	618	715	567	324	103	10	0	2729
1988-89	0	0	3	33	248	539	477	690	310	116	4	0	2420
1989-90	0	0	40	92	286	820	460	334	293	123	58	0	2506
1990-91	0	0	3	104	258	680	711	358	224	22	12	0	2372
1991-92	0	0	33	110	448	562	672	408	200	88	30	0	2551
1992-93	0	0	0	20	413	530	677	509	328	125	24	0	2626
1993-94	0	0	10	174	459	524							

COOLING DEGREE DAYS Base 65 deg. F ABILENE, TEXAS

YEAR	JAN	FEB	MAR	APR	MAY	JUNE	JULY	AUG	SEP	OCT	NOV	DEC	TOTAL
1969	0	0	5	83	190	412	699	644	332	105	12	0	2482
1970	2	0	5	104	202	433	607	590	411	106	6	1	2467
1971	0	4	26	88	256	469	607	376	297	83	20	0	2226
1972	0	12	41	188	189	474	497	436	330	128	1	0	2296
1973	0	0	0	44	224	379	508	550	277	111	32	0	2125
1974	0	7	113	102	354	480	607	430	104	62	8	0	2267
1975	0	0	20	89	175	411	462	513	236	120	13	0	2039
1976	0	21	40	67	145	424	401	521	275	31	0	0	1925
1977	0	4	13	17	267	496	590	579	534	122	7	1	2630
1978	0	0	35	201	392	555	751	542	379	121	23	1	3000
1979	0	9	20	67	209	428	586	539	390	253	11	0	2512
1980	0	7	6	91	237	588	762	660	405	119	21	5	2901
1981	0	6	8	154	230	464	653	575	403	162	15	1	2671
1982	1	8	52	92	249	422	600	645	386	147	28	1	2631
1983	0	0	21	84	208	365	558	634	403	171	30	0	2474
1984	0	0	20	71	339	550	589	570	315	80	13	4	2551
1985	0	0	32	107	281	393	519	621	356	97	25	0	2431
1986	0	18	34	155	294	435	610	524	382	64	3	0	2519
1987	0	0	3	97	228	379	525	579	277	116	25	0	2229
1988	0	0	34	80	206	422	524	592	350	85	37	0	2330
1989	0	0	64	169	347	334	563	530	275	183	33	0	2498
1990	2	6	33	63	285	603	461	537	363	94	30	0	2477
1991	0	0	40	118	340	404	544	449	228	151	2	1	2277
1992	0	0	13	92	173	402	501	410	336	142	13	6	2088
1993	0	0	19	88	198	448	648	593	289	116	4	0	2403

SNOWFALL (inches) ABILENE, TEXAS

SEASON	JULY	AUG	SEP	OCT	NOV	DEC	JAN	FEB	MAR	APR	MAY	JUNE	TOTAL
1964-65	0.0	0.0	0.0	0.0	0.0	T	0.0	T	T	0.0	0.0	0.0	T
1965-66	0.0	0.0	0.0	0.0	0.0	0.0	6.6	1.5	0.0	0.0	0.0	0.0	8.1
1966-67	0.0	0.0	0.0	0.0	0.0	T	0.0	T	T	0.0	0.0	0.0	T
1967-68	0.0	0.0	0.0	T	0.0	4.0	T	1.6	2.5	0.0	0.0	0.0	8.1
1968-69	0.0	0.0	0.0	0.0	8.1	0.0	0.0	T	0.5	0.0	0.0	0.0	8.6
1969-70	0.0	0.0	0.0	0.0	T	3.1	0.2	0.0	7.3	0.0	0.0	0.0	10.6
1970-71	0.0	0.0	0.0	0.0	0.0	0.0	0.0	1.2	0.4	0.0	0.0	0.0	1.6
1971-72	0.0	0.0	0.0	0.0	0.0	T	1.0	0.0	T	0.0	0.0	0.0	1.0
1972-73	0.0	0.0	0.0	0.0	0.1	0.0	13.5	3.0	0.0	0.0	0.0	0.0	16.6
1973-74	0.0	0.0	0.0	0.0	0.0	T	T	0.2	T	0.0	0.0	0.0	0.2
1974-75	0.0	0.0	0.0	0.0	0.0	T	6.0	1.6	T	0.0	0.0	0.0	7.6
1975-76	0.0	0.0	0.0	0.0	0.0	2.6	0.0	0.0	T	0.0	0.0	0.0	2.6
1976-77	0.0	0.0	0.0	0.0	5.6	0.0	3.9	0.0	0.0	0.0	0.0	0.0	9.5
1977-78	0.0	0.0	0.0	0.0	0.0	0.0	1.7	1.4	1.0	0.0	0.0	0.0	4.1
1978-79	0.0	0.0	0.0	0.0	0.0	0.9	0.5	5.9	0.0	0.0	0.0	0.0	7.3
1979-80	0.0	0.0	0.0	0.0	0.0	0.3	T	2.2	T	T	0.0	0.0	2.5
1980-81	0.0	0.0	0.0	T	3.7	0.0	1.9	T	0.0	0.0	0.0	0.0	5.6
1981-82	0.0	0.0	0.0	0.0	0.0	0.0	5.0	1.2	0.0	0.0	0.0	0.0	6.2
1982-83	0.0	0.0	0.0	0.0	T	1.4	10.8	T	T	0.0	0.0	0.0	12.2
1983-84	0.0	0.0	0.0	0.0	0.0	7.8	T	0.4	0.8	0.0	0.0	0.0	9.0
1984-85	0.0	0.0	0.0	0.0	0.0	0.0	6.1	1.9	0.0	0.0	0.0	0.0	8.0
1985-86	0.0	0.0	0.0	0.0	0.0	0.2	0.0	T	0.0	0.0	0.0	0.0	0.2
1986-87	0.0	0.0	0.0	0.0	0.0	0.6	2.4	3.9	T	0.0	0.0	0.0	6.9
1987-88	0.0	0.0	0.0	0.0	0.0	3.3	T	2.0	0.0	0.0	0.0	0.0	5.3
1988-89	0.0	0.0	0.0	0.0	0.0	T	0.2	T	5.2	0.0	T	T	5.4
1989-90	0.0	0.0	0.0	0.0	0.0	T	T	T	T	T	0.0	0.0	T
1990-91	0.0	0.0	0.0	0.0	0.0	2.6	1.6	0.0	T	T	0.0	T	4.2
1991-92	0.0	0.0	0.0	T	0.0	T	0.5	0.5	T	0.0	0.0	0.0	1.0
1992-93	T	0.0	0.0	0.0	2.8	T	T	T	0.3	0.0	T	T	3.1
1993-94	0.0	0.0	0.0	0.3	T	1.7							
Record Mean	T	0.0	0.0	T	0.4	0.7	1.8	1.1	0.7	T	T	T	4.7

1993 AMARILLO, TEXAS (AMA)

The station is located 7 statute miles east northeast of the downtown post office in a region of rather flat topography. The Canadian River flows eastward 18 miles north of the station, with its bed about 800 feet below the plains. The Prairie Dog Town Fork of the Red River flows southeastward about 15 miles south of the station where it enters the Palo Duro Canyon, which is about 1,000 feet deep. There are numerous shallow Playa lakes, often dry, over the area, and the nearly treeless grasslands slope downward to the east. The terrain gradually rises to the west and northwest.

Three-fourths of the total annual precipitation falls from April through September, occurring from thunderstorm activity. Snow usually melts within a few days after it falls. Heavier snowfalls of 10 inches or more, usually with near blizzard conditions, average once every 5 years and last 2 to 3 days.

The Amarillo area is subject to rapid and large temperature changes, especially during the winter months when cold fronts from the northern Rocky Mountain and Plains states sweep across the area. Temperature drops of 50 to 60 degrees within a 12-hour period are not uncommon. Temperature drops of 40 degrees have occurred within a few minutes.

Humidity averages are low, occasionally dropping below 20 percent in the spring. Low humidity moderates the effect of high summer afternoon temperatures, permits evaporative cooling systems to be very effective, and provides many pleasant evenings and nights.

Severe local storms are infrequent, although a few thunderstorms with damaging hail, lightning, and wind in a very localized area occur most years, usually in spring and summer. These storms are often accompanied by very heavy rain, which produces local flooding, particularly of roads and streets. Tornadoes are rare.

Based on the 1951-1980 period, the average first occurrence of 32 degrees Fahrenheit in the fall is October 29 and the average last occurrence in the spring is April 14.

NORMALS, MEANS, AND EXTREMES FOR 1993

AMARILLO, TX (AMA)

LATITUDE: 35° 13' 10" N LONGTITUDE: 101° 42' 20" W ELEVATION (FT): GRND: 3604 BARO: 3591 TIME ZONE: CENTRAL (UTC+06) WBAN: 23047

	ELEMENT	POR	JAN	FEB	MAR	APR	MAY	JUN	JUL	AUG	SEP	OCT	NOV	DEC	YEAR
TEMPERATURE °F	NORMAL DAILY MAXIMUM	30	49.0	52.8	61.6	71.5	79.1	87.6	91.7	89.1	81.8	72.5	59.7	50.1	70.5
	MEAN DAILY MAXIMUM	54	48.6	53.1	61.1	70.9	78.7	87.7	91.1	91.1	83.9	73.8	60.1	51.8	71.0
	HIGHEST DAILY MAXIMUM	53	81	88	94	98	102	108	105	106	102	95	87	81	108
	YEAR OF OCCURRENCE		1950	1963	1971	1989	1953	1990	1981	1944	1983	1954	1980	1955	JUN 1990
	MEAN OF EXTREME MAXS.	46	71.3	75.3	81.9	88.9	93.8	99.8	99.6	98.4	94.5	89.1	78.3	72.0	86.9
	NORMAL DAILY MINIMUM	30	21.2	25.5	32.7	42.1	51.6	60.7	65.5	63.8	56.4	44.5	32.3	23.7	43.3
	MEAN DAILY MINIMUM	54	21.8	26.1	32.0	42.0	51.5	61.1	65.6	65.3	57.6	46.1	32.5	25.0	43.9
	LOWEST DAILY MINIMUM	53	-11	-14	-3	14	28	42	51	49	30	12	0	-8	-14
	YEAR OF OCCURRENCE		1984	1951	1948	1945	1954	1955	1990	1956	1984	1993	1976	1989	FEB 1951
	MEAN OF EXTREME MINS.	46	3.0	8.2	14.4	27.1	37.8	50.0	58.1	56.4	42.8	30.1	15.6	7.6	29.2
	NORMAL DRY BULB	30	35.1	39.2	47.1	56.8	65.4	74.1	78.6	76.5	69.1	58.5	46.0	36.9	56.9
	MEAN DRY BULB	54	35.2	39.6	46.6	56.5	65.1	74.4	78.4	78.2	70.8	60.0	46.3	38.4	57.5
	MEAN WET BULB	44	28.7	32.3	37.3	45.2	53.9	62.1	65.1	64.4	58.5	48.5	37.4	30.5	47.0
	MEAN DEW POINT	44	18.7	22.4	25.2	33.0	44.7	54.6	58.1	58.0	51.5	39.5	27.8	20.8	37.9
	NORMAL NO. DAYS WITH:														
	MAXIMUM ≥ 90°	30	0.0	0.0	0.1	1.0	4.7	12.8	21.1	16.5	6.4	1.0	0.0	0.0	63.6
	MAXIMUM ≤ 32°	30	4.5	2.8	0.9	*	0.0	0.0	0.0	0.0	0.0	*	0.8	3.6	12.6
	MINIMUM ≤ 32°	30	27.5	21.8	14.7	3.5	0.1	0.0	0.0	0.0	0.2	1.8	14.5	26.7	110.8
	MINIMUM ≤ 0°	30	1.1	0.4	0.0	0.0	0.0	0.0	0.0	0.0	0.0	0.0	*	0.7	2.2
H/C	NORMAL HEATING DEG. DAYS	30	927	722	555	266	89	6	0	0	26	226	570	871	4258
	NORMAL COOLING DEG. DAYS	30	0	0	0	20	102	279	422	357	149	25	0	0	1354
RH	NORMAL (PERCENT)	30	58	59	52	48	54	56	53	58	61	56	59	59	56
	HOUR 00 LST	30	64	65	58	56	63	65	61	67	70	64	66	66	64
	HOUR 06 LST	30	70	72	68	68	74	77	74	78	80	73	72	70	73
	HOUR 12 LST	30	49	50	42	38	43	45	42	46	49	43	47	49	45
	HOUR 18 LST	30	46	43	35	31	37	39	38	43	45	42	48	50	41
S	PERCENT POSSIBLE SUNSHINE	52	69	68	73	75	73	78	79	78	74	76	73	68	73
W/O	MEAN NO. DAYS WITH:														
	HEAVY FOG(VISBY ≤ 1/4 MI)	52	3.3	4.2	3.3	1.9	2.1	0.7	0.5	0.7	2.0	2.5	2.7	2.8	26.7
	THUNDERSTORMS	52	0.2	0.5	1.6	3.5	8.4	9.6	9.2	9.0	4.2	2.4	0.6	0.2	49.4
CLOUDINESS	MEAN:														
	SUNRISE-SUNSET (TENTHS)	1		7.0	5.0	5.0	4.0		3.0	3.0	2.0	4.0	4.0	3.0	
	MIDNIGHT-MIDNIGHT (TENTHS)	1		7.0	5.0	5.0	4.0		3.0	4.0	2.0	4.0	4.0	3.0	
	MEAN NO. DAYS WITH:														
	CLEAR	1	5.0	10.0	15.0	8.0	12.0	12.0	14.0	16.0	19.0	13.0	10.0	19.0	153.0
	PARTLY CLOUDY	1	6.0	3.0	3.0	4.0	8.0	6.0	2.0	3.0	2.0	2.0	6.0	7.0	52.0
	CLOUDY	1	8.0	12.0	7.0	5.0	3.0		4.0	3.0	3.0	6.0	5.0	4.0	
PR	MEAN STATION PRESSURE(IN.)	21	26.35	26.32	26.23	26.25	26.25	26.30	26.37	26.38	26.37	26.37	26.33	26.35	26.32
	MEAN SEA-LEVEL PRES. (IN.)	44	30.07	30.02	29.91	29.87	29.84	29.84	29.90	29.92	29.95	29.99	30.03	30.06	29.95
WINDS	MEAN SPEED (MPH)	30	12.7	13.7	15.4	15.4	14.4	14.1	12.9	12.2	12.7	12.8	12.8	12.8	13.5
	PREVAIL.DIR(TENS OF DEGS.)	30	23	36	21	18	18	19	18	18	18	20	20	23	18
	MAXIMUM 2-MINUTE:														
	SPEED (MPH)	1	31	48	44	43	43	46	45	43	40	38	39	51	51
	DIR. (TENS OF DEGS.)	1	03	25	06	26	14	20	20	05	32	02	24	32	32
	YEAR OF OCCURRENCE		1993	1993	1993	1993	1993	1993	1993	1993	1993	1993	1993	1993	DEC 1993
	MAXIMUM 5-SECOND:														
	SPEED (MPH)	1	36	55	51	52	48	52	52	53	51	43	46	57	57
	DIR. (TENS OF DEGS.)	1	02	24	06	25	14	20	19	31	31	02	25	31	31
	YEAR OF OCCURRENCE		1993	1993	1993	1993	1993	1993	1993	1993	1993	1993	1993	1993	DEC 1993
PRECIPITATION	NORMAL TOTAL (IN.)	30	0.50	0.61	0.96	0.99	2.48	3.70	2.62	3.22	1.99	1.37	0.69	0.43	19.56
	MAXIMUM TOTAL (IN.)	53	2.33	1.83	3.99	3.74	9.81	10.73	7.59	7.55	5.02	7.64	2.26	4.52	10.73
	YEAR OF OCCURRENCE		1968	1948	1973	1942	1951	1965	1960	1974	1950	1941	1961	1959	JUN 1965
	MINIMUM TOTAL (IN.)	53	0.00	T	T	T	0.04	0.01	0.12	0.28	0.03	0.00	0.00	T	0.0
	YEAR OF OCCURRENCE		1986	1991	1950	1964	1984	1953	1946	1983	1977	1952	1989	1976	NOV 1989
	MAXIMUM IN 24 HOURS (IN.)	53	1.74	1.28	2.27	1.99	6.75	6.15	4.74	4.26	3.42	3.45	1.53	3.11	6.75
	YEAR OF OCCURRENCE		1968	1971	1973	1985	1951	1960	1982	1945	1941	1948	1980	1943	MAY 1951
	NORMAL NO. DAYS WITH:														
	PRECIPITATION ≥ 0.01	30	3.9	4.4	5.2	4.9	7.9	8.9	7.7	8.7	6.5	4.7	3.6	4.1	70.5
	PRECIPITATION ≥ 1.00	30	*	0.1	0.1	0.1	0.5	0.9	0.7	0.9	0.5	0.2	*	0.0	4.0
SNOWFALL	NORMAL TOTAL (IN.)	30	3.8	4.6	2.9	0.6	0.*	0.0	0.0	0.0	0.*	0.3	1.9	2.8	16.9
	MAXIMUM TOTAL (IN.)	53	14.5	17.3	14.7	6.4	0.5	T	0.0	0.0	0.3	3.9	13.6	15.3	17.3
	YEAR OF OCCURRENCE		1983	1971	1961	1947	1978	1992			1984	1976	1952	1987	FEB 1971
	MAXIMUM IN 24 HOURS (IN.)	53	9.8	13.5	9.8	5.1	0.5	T	0.0	0.0	0.3	3.2	12.2	11.3	13.5
	YEAR OF OCCURRENCE		1983	1971	1957	1947	1978	1992	1993	1993	1984	1976	1952	1987	FEB 1971
	MAXIMUM SNOW DEPTH (IN.)	44	10	14	10	6	T	T	0	0	T	1	9	13	14
	YEAR OF OCCURRENCE		1987	1983	1957	1973	1988	1949			1984	1970	1952	1987	FEB 1983
	NORMAL NO. DAYS WITH:														
	SNOWFALL ≥ 1.0	30	1.4	1.3	1.0	0.2	0.0	0.0	0.0	0.0	0.0	0.1	0.6	0.8	5.4

PRECIPITATION (inches) 1993 AMARILLO, TEXAS (AMA)

YEAR	JAN	FEB	MAR	APR	MAY	JUN	JUL	AUG	SEP	OCT	NOV	DEC	ANNUAL
1964	T	1.37	0.03	T	1.69	1.90	0.94	5.69	3.95	0.08	1.53	0.79	17.97
1965	0.55	0.47	0.72	0.23	1.88	10.73	1.54	1.71	0.79	1.02	0.07	0.38	20.09
1966	0.43	0.69	0.01	0.87	0.19	4.62	1.37	3.77	2.40	0.29	0.08	0.19	14.91
1967	T	0.15	0.42	1.95	1.40	2.55	3.70	1.81	2.47	1.61	0.28	0.51	16.85
1968	2.33	0.73	0.45	0.93	2.84	1.68	2.96	3.35	0.62	0.90	0.92	0.26	17.97
1969	0.02	0.50	1.15	0.30	2.93	4.09	2.55	4.51	2.77	2.56	0.34	0.83	22.55
1970	0.02	0.02	2.10	1.33	0.23	1.54	1.39	1.27	0.34	1.06	0.26	T	9.56
1971	0.10	1.65	0.10	0.77	0.91	4.17	1.75	3.33	4.70	2.59	2.08	0.89	23.04
1972	0.21	0.11	0.11	0.03	2.81	3.87	2.59	1.73	0.71	1.66	1.19	0.32	15.34
1973	0.56	0.42	3.99	1.88	1.43	0.84	4.08	2.31	1.22	1.05	0.10	0.17	18.05
1974	0.33	0.24	0.60	0.04	4.06	3.33	1.31	7.55	1.65	3.44	0.12	0.42	23.09
1975	0.28	1.33	0.51	1.02	2.47	4.15	5.19	3.97	0.76	0.33	0.92	0.15	21.08
1976	T	0.10	0.79	1.65	1.36	2.94	1.77	1.78	4.28	1.14	0.43	T	16.24
1977	0.64	0.53	0.24	2.74	4.01	2.06	3.14	4.94	0.03	0.26	0.32	0.27	19.18
1978	0.63	0.80	0.21	0.55	5.76	6.50	1.82	1.61	2.42	0.97	0.47	0.27	22.01
1979	0.92	0.28	1.46	1.29	3.94	3.19	2.03	5.08	0.52	1.28	0.40	0.07	20.46
1980	0.85	0.55	1.38	0.82	2.88	1.30	0.65	1.80	1.55	0.42	0.84	0.35	13.39
1981	0.11	0.23	1.87	0.90	2.11	1.04	2.73	5.22	3.47	1.79	1.50	0.03	21.00
1982	0.15	0.39	0.52	0.43	1.96	4.75	6.23	0.55	1.37	0.71	0.75	0.79	18.60
1983	1.78	1.19	0.98	0.83	2.85	1.76	0.74	0.28	0.37	3.23	0.33	0.64	14.98
1984	0.56	0.37	0.98	1.18	0.04	6.76	0.83	2.28	0.95	3.19	1.09	1.00	19.23
1985	0.99	0.77	1.49	2.79	0.86	3.08	2.07	1.67	4.96	3.07	0.39	0.26	22.40
1986	0.00	1.02	0.60	0.30	3.28	3.70	3.52	7.04	1.45	1.94	1.82	0.66	25.33
1987	1.26	0.84	0.92	0.57	4.28	3.29	0.83	3.28	3.40	1.17	0.43	1.75	22.02
1988	0.33	0.04	1.19	2.22	6.02	3.68	3.30	3.59	3.15	0.71	0.29	0.17	24.69
1989	0.16	0.55	0.52	0.75	2.51	6.07	2.74	3.22	1.80	0.74	0.00	0.49	19.55
1990	1.22	1.61	2.56	1.10	0.90	0.14	3.28	2.79	2.72	0.46	0.50	0.23	17.51
1991	0.86	T	0.41	0.04	3.08	2.47	2.20	1.28	2.04	0.64	0.66	2.24	15.92
1992	0.50	0.30	1.11	1.60	3.10	7.57	2.36	2.27	0.16	0.31	0.80	0.55	20.63
1993	0.76	0.36	1.29	0.35	1.92	2.76	3.36	4.64	1.00	0.53	0.51	0.95	18.43
POR= 54 YRS	0.55	0.57	0.87	1.16	2.80	3.42	2.73	3.03	1.89	1.59	0.64	0.64	19.90

AVERAGE TEMPERATURE (°F) 1993 AMARILLO, TEXAS (AMA)

YEAR	JAN	FEB	MAR	APR	MAY	JUN	JUL	AUG	SEP	OCT	NOV	DEC	ANNUAL
1964	37.2	32.1	44.8	57.9	68.9	76.5	81.9	78.1	69.6	60.6	47.5	39.1	57.8
1965	41.0	37.6	38.4	59.4	66.9	72.1	78.6	76.9	68.5	59.5	52.5	42.9	57.9
1966	27.8	34.1	50.0	54.5	66.0	74.2	82.9	73.5	68.5	57.7	51.6	35.5	56.4
1967	40.7	40.6	53.4	60.6	63.4	73.7	77.2	75.0	68.3	60.9	46.6	35.4	58.0
1968	38.3	38.0	48.7	55.3	63.4	74.9	77.0	77.0	69.1	62.1	45.4	36.9	57.2
1969	41.4	40.4	38.8	59.9	67.2	72.8	82.6	80.2	71.1	54.5	46.7	38.4	57.8
1970	33.2	43.2	41.7	56.0	68.8	74.5	80.7	79.1	70.9	54.0	46.6	43.4	57.7
1971	37.8	38.8	48.3	55.7	64.9	75.9	76.9	72.2	67.1	57.2	45.6	37.6	56.5
1972	35.7	40.9	51.7	59.5	63.2	73.6	74.5	73.9	69.2	57.9	36.9	32.9	55.8
1973	33.4	39.5	47.1	50.3	62.9	75.1	78.1	78.8	68.7	62.0	50.8	38.6	57.1
1974	35.0	42.2	53.1	60.2	71.5	75.0	79.3	73.6	62.9	59.0	45.7	36.1	57.8
1975	37.1	35.3	45.0	54.9	64.5	73.3	75.5	76.7	65.8	61.1	45.8	40.4	56.3
1976	36.9	47.8	46.5	56.8	60.2	72.2	74.8	75.1	66.5	49.9	38.5	37.5	55.2
1977	30.1	43.8	48.7	57.8	67.3	78.1	80.7	77.5	74.3	60.7	47.4	39.9	58.9
1978	29.0	30.1	47.2	61.4	63.7	75.4	80.7	76.1	70.7	59.4	45.6	32.7	56.0
1979	24.9	40.0	46.6	54.5	62.2	70.9	77.1	73.6	69.8	60.1	40.5	38.8	54.9
1980	34.9	37.7	43.9	52.4	61.9	78.3	82.9	78.5	70.7	56.6	42.7	41.4	56.8
1981	37.9	42.2	48.8	63.1	65.6	78.5	81.3	74.4	69.0	56.6	49.1	40.1	58.9
1982	37.1	35.9	47.3	53.8	63.3	72.2	78.8	78.7	71.1	57.3	45.6	36.1	56.4
1983	33.4	36.2	45.3	50.9	60.3	70.4	80.0	81.0	73.6	60.5	47.9	24.7	55.4
1984	31.6	40.3	44.0	51.8	66.9	74.6	75.6	75.3	65.7	56.9	47.0	40.5	55.9
1985	31.5	37.2	49.5	60.0	68.1	75.1	80.1	79.6	68.9	56.9	43.5	34.0	57.0
1986	42.5	40.4	52.7	59.4	65.1	73.3	80.6	74.4	68.8	55.7	42.7	37.1	57.7
1987	34.0	42.0	44.6	54.8	64.5	72.1	77.2	75.3	67.6	57.9	45.4	34.6	55.8
1988	32.2	38.3	44.3	54.1	63.6	73.3	75.8	76.4	67.7	59.0	47.6	38.5	55.9
1989	40.7	32.3	51.2	59.9	67.3	69.4	76.2	76.0	66.3	60.4	47.9	31.4	56.6
1990	39.2	40.8	47.0	55.9	63.6	81.3	76.4	76.3	72.0	57.7	49.7	33.2	57.8
1991	32.6	45.9	49.7	58.2	68.5	74.7	76.5	76.2	67.1	58.5	41.2	39.0	57.3
1992	38.0	45.0	50.6	58.7	63.3	70.8	76.6	73.5	70.2	60.1	39.3	34.2	56.7
1993	33.1	36.0	46.1	54.4	64.5	74.0	79.1	74.9	67.9	54.6	40.9	38.7	55.4
POR= 54 YRS	35.3	39.6	46.6	56.5	65.2	74.4	78.4	76.9	69.5	58.9	45.5	37.7	57.0

REFERENCE NOTES:

THE TEMPERATURE GRAPH SHOWS NORMAL MAXIMUM AND NORMAL MINIMUM DAILY TEMPERATURES (SOLID CURVES) AND THE ACTUAL DAILY HIGH AND LOW TEMPERATURES (VERTICAL BARS).

H/C INDICATES HEATING AND COOLING DEGREE DAYS.
RH INDICATES RELATIVE HUMIDITY
W/O INDICATES WEATHER AND OBSTRUCTIONS
S INDICATES SUNSHINE.
PR INDICATES PRESSURE.
CLOUDINESS IS THE SUM OF THE CEILOMETER AND SATELLITE DATA NOT TO EXCEED TEN TENTHS.
MAXIMUM SNOW DEPTH IS THE DEPTH MEASURED AT 12 UTC.

GENERAL:
T INDICATES TRACE PRECIPITATION, AN AMOUNT GREATER THAN ZERO BUT LESS THAN THE LOWEST REPORTABLE VALUE.
+ INDICATES THE VALUE ALSO OCCURS ON EARLIER DATES.
BLANK ENTRIES DENOTE MISSING OR UNREPORTED DATA.
NORMALS ARE 30-YEAR AVERAGES (1961 - 1990).
ASOS INDICATES AUTOMATED SURFACE OBSERVING SYSTEM.
PM INDICATES THE LAST DAY OF THE PREVIOUS MONTH.
POR (PERIOD OF RECORD) INDICATES THE NUMBER OF YEARS USED TO COMPUTE THE MEAN. INDIVIDUAL MONTHS WITHIN THE POR MAY BE MISSING.
WHEN THE POR FOR A NORMAL IS LESS THAN 30 YEARS, THE NORMAL IS PROVISIONAL AND IS BASED ON THE NUMBER OF YEARS INDICATED.
0.* OR * INDICATES THE VALUE OR MEAN DAYS WITH IS BETWEEN 0.00 AND 0.05.
CLOUDINESS FOR ASOS STATIONS DIFFERS FROM THE NON-ASOS OBSERVATION TAKEN BY A HUMAN OBSERVER. ASOS STATION CLOUDINESS IS BASED ON TIME-AVERAGED CEILOMETER DATA FOR CLOUDS AT OR BELOW 12,000 FEET AND ON SATELLITE DATA FOR CLOUDS ABOVE 12,000 FEET.
THE NUMBER OF DAYS WITH CLEAR, PARTLY CLOUDY, AND CLOUDY CONDITIONS FOR ASOS STATIONS IS THE SUM OF THE CEILOMETER AND SATELLITE DATA FOR THE SUNRISE TO SUNSET PERIOD.

GENERAL CONTINUED:
CLEAR INDICATES 0 - 3 TENTHS, PARTLY CLOUDY INDICATES 4 - 7 TENTHS, AND CLOUDY INDICATES 8 OR MORE TENTHS. WHEN AT LEAST ONE OF THE ELEMENTS (CEILOMETER OR SATELLITE) IS MISSING, THE DAILY CLOUDINESS IS NOT COMPUTED.
WIND DIRECTION IS RECORDED IN TENS OF DEGREES (2 DIGITS) CLOCKWISE FROM TRUE NORTH. "00" INDICATES CALM. "36" INDICATES TRUE NORTH.
RESULTANT WIND IS THE VECTOR AVERAGE OF THE SPEED AND DIRECTION.
AVERAGE TEMPERATURE IS THE SUM OF THE MEAN DAILY MAXIMUM AND MINIMUM TEMPERATURE DIVIDED BY 2.
SNOWFALL DATA COMPRISE ALL FORMS OF FROZEN PRECIPITATION, INCLUDING HAIL.
A HEATING (COOLING) DEGREE DAY IS THE DIFFERENCE BETWEEN THE AVERAGE DAILY TEMPERATURE AND 65° F.
DRY BULB IS THE TEMPERATURE OF THE AMBIENT AIR.
DEW POINT IS THE TEMPERATURE TO WHICH THE AIR MUST BE COOLED TO ACHIEVE 100 PERCENT RELATIVE HUMIDITY.
WET BULB IS THE TEMPERATURE THE AIR WOULD HAVE IF THE MOISTURE CONTENT WAS INCREASED TO 100 PERCENT RELATIVE HUMIDITY.

HEATING DEGREE DAYS (base 65°F) 1993 AMARILLO, TEXAS (AMA)

YEAR	JUL	AUG	SEP	OCT	NOV	DEC	JAN	FEB	MAR	APR	MAY	JUN	TOTAL
1964-65	0	0	48	149	517	795	740	762	820	194	56	7	4088
1965-66	0	0	73	191	376	676	1145	857	456	312	99	1	4186
1966-67	0	23	22	250	395	908	743	675	356	159	135	9	3675
1967-68	0	1	23	185	544	911	820	777	504	290	109	0	4164
1968-69	0	0	6	159	583	865	726	681	808	176	69	24	4097
1969-70	0	0	1	374	543	817	981	602	713	280	50	33	4394
1970-71	0	0	47	359	544	661	837	726	524	290	89	0	4077
1971-72	1	0	120	245	575	843	900	692	409	203	106	0	4094
1972-73	13	2	48	274	833	987	972	706	548	438	113	0	4934
1973-74	0	0	56	154	420	813	922	633	368	190	17	0	3573
1974-75	0	0	119	199	571	890	853	826	612	323	67	16	4476
1975-76	0	0	104	158	569	757	861	492	567	253	171	0	3932
1976-77	0	0	59	464	790	846	1075	592	499	215	22	0	4562
1977-78	0	0	1	150	522	768	1107	972	551	136	139	6	4352
1978-79	0	0	33	197	574	992	1236	697	561	319	144	30	4783
1979-80	0	2	28	186	727	806	926	788	649	373	146	0	4631
1980-81	0	0	35	280	662	723	832	630	496	111	65	0	3834
1981-82	0	0	26	271	469	765	849	803	534	340	96	2	4155
1982-83	0	0	23	252	575	888	972	800	603	421	171	32	4737
1983-84	0	0	40	175	506	1241	1028	709	642	390	56	0	4787
1984-85	0	0	125	262	531	752	1034	769	474	169	37	5	4158
1985-86	0	0	111	249	640	957	691	681	379	203	72	0	3983
1986-87	0	2	26	290	665	858	954	634	624	315	70	9	4447
1987-88	0	8	18	226	584	936	1010	765	633	323	102	4	4609
1988-89	0	7	32	197	517	815	747	909	429	219	59	27	3958
1989-90	0	0	91	185	507	1037	795	672	551	276	140	0	4254
1990-91	0	0	11	234	454	981	994	529	467	208	53	1	3932
1991-92	0	0	71	258	709	798	833	576	437	203	98	10	3993
1992-93	0	0	21	171	765	949	984	808	577	317	82	5	4679
1993-	0	2	39	332	717	810							
POR= 53 YRS	0	1	41	217	576	843	911	711	564	269	92	10	4235

COOLING DEGREE DAYS (base 65°F) 1993 AMARILLO, TEXAS (AMA)

YEAR	JAN	FEB	MAR	APR	MAY	JUN	JUL	AUG	SEP	OCT	NOV	DEC	ANNUAL
1969	0	0	0	30	147	264	550	478	190	54	0	0	1713
1970	0	0	0	14	176	327	495	446	231	24	0	0	1713
1971	0	0	13	18	94	334	381	232	189	10	0	0	1271
1972	0	0	4	44	59	265	313	285	182	60	0	0	1212
1973	0	0	0	3	53	310	414	435	173	67	1	0	1456
1974	0	0	6	52	228	306	449	274	61	20	0	0	1396
1975	0	0	0	24	61	273	330	367	134	46	0	0	1235
1976	0	0	0	13	32	223	312	318	110	5	0	0	1013
1977	0	0	0	10	100	399	493	389	287	22	0	0	1700
1978	0	0	6	33	108	324	497	351	209	28	0	0	1556
1979	0	0	0	12	64	218	380	276	177	41	0	0	1168
1980	0	0	0	4	58	408	562	429	211	26	0	0	1698
1981	0	0	1	59	87	410	512	299	154	16	0	0	1538
1982	0	0	0	9	52	225	437	432	213	22	0	0	1390
1983	0	0	0	2	31	201	473	502	306	41	3	0	1559
1984	0	0	0	1	121	298	331	325	151	17	0	0	1244
1985	0	0	2	27	143	315	473	458	235	6	0	0	1659
1986	0	0	3	45	87	256	489	299	144	10	0	0	1333
1987	0	0	0	19	64	227	386	334	105	11	0	0	1146
1988	0	0	0	3	67	263	340	366	120	19	0	0	1178
1989	0	0	10	75	136	164	354	347	138	51	0	0	1275
1990	0	0	0	9	101	494	359	358	229	17	1	0	1568
1991	0	0	0	12	166	298	364	358	144	66	0	0	1408
1992	0	0	0	19	52	188	366	271	186	25	0	0	1107
1993	0	0	0	6	77	283	443	315	134	15	0	0	1273
POR= 25 YRS	0	0	0	21	94	290	419	357	176	28	0	0	1385

SNOWFALL (inches) 1993 AMARILLO, TEXAS (AMA)

YEAR	JUL	AUG	SEP	OCT	NOV	DEC	JAN	FEB	MAR	APR	MAY	JUN	TOTAL
1964-65	0.0	0.0	0.0	0.0	T	0.5	5.6	3.8	2.3	0.0	0.0	0.0	12.2
1965-66	0.0	0.0	0.0	0.0	0.0	1.4	7.2	0.4	0.1	0.0	0.0	0.0	9.1
1966-67	0.0	0.0	0.0	T	T	1.8	T	1.1	0.4	0.0	0.0	0.0	3.3
1967-68	0.0	0.0	0.0	T	2.1	6.1	0.6	8.0	2.9	0.0	0.0	0.0	19.7
1968-69	0.0	0.0	0.0	0.0	0.3	1.8	0.2	2.9	10.3	0.0	0.0	0.0	15.5
1969-70	0.0	0.0	0.0	0.0	T	7.1	0.1	T	14.1	1.8	0.0	0.0	23.1
1970-71	0.0	0.0	0.0	3.9	2.6	T	0.8	17.3	1.0	T	0.0	0.0	25.6
1971-72	0.0	0.0	0.0	0.0	T	7.9	1.6	2.0	0.6	0.0	0.0	0.0	12.1
1972-73	0.0	0.0	0.0	0.4	9.9	2.8	7.0	4.2	0.5	5.7	0.0	0.0	30.5
1973-74	0.0	0.0	0.0	0.0	0.4	1.5	3.6	1.2	T	0.0	0.0	0.0	6.7
1974-75	0.0	0.0	0.0	0.0	T	1.5	3.8	11.7	1.2	T	0.0	0.0	18.2
1975-76	0.0	0.0	0.0	0.0	0.4	1.8	T	T	4.2	0.0	0.0	0.0	6.4
1976-77	0.0	0.0	0.0	3.9	4.3	T	6.4	3.7	T	T	0.0	0.0	18.3
1977-78	0.0	0.0	0.0	0.0	1.3	0.2	8.2	9.4	1.5	T	0.5	0.0	21.1
1978-79	0.0	0.0	0.0	0.0	2.0	3.3	5.9	1.9	T	T	0.0	0.0	13.1
1979-80	0.0	0.0	0.0	1.6	2.2	0.3	1.4	5.2	1.0	T	0.0	0.0	11.7
1980-81	0.0	0.0	0.0	0.0	8.6	T	1.1	0.1	0.1	0.0	0.0	0.0	9.9
1981-82	0.0	0.0	0.0	0.0	T	0.3	1.1	4.7	1.7	T	0.0	0.0	7.8
1982-83	0.0	0.0	0.0	0.0	4.5	1.9	14.5	13.0	8.1	5.9	0.0	0.0	47.9
1983-84	0.0	0.0	0.0	0.0	T	5.2	7.0	3.5	2.5	0.0	0.0	0.0	18.2
1984-85	0.0	0.0	0.3	0.0	0.1	0.3	7.4	0.3	0.2	0.0	0.0	0.0	8.6
1985-86	0.0	0.0	T	0.0	T	2.8	0.0	10.9	1.0	0.0	0.0	0.0	14.7
1986-87	0.0	0.0	0.0	T	0.2	3.3	12.1	3.1	6.4	0.1	0.0	0.0	25.2
1987-88	0.0	0.0	0.0	0.0	0.7	15.3	4.3	0.5	8.5	4.2	0.0	0.0	33.5
1988-89	0.0	0.0	0.0	0.0	2.4	2.2	T	0.1	4.2	0.1	T	T	9.0
1989-90	0.0	0.0	T	0.0	0.0	5.4	8.5	3.0	0.0	0.0	0.0	0.0	16.9
1990-91	0.0	0.0	0.0	0.0	2.2	3.8	5.0	T	1.0	0.0	T	T	12.0
1991-92	0.0	0.0	T	3.2	T	2.0	1.0	0.2	T	0.3	T	T	6.7
1992-93	0.0	0.0	0.0	T	9.0	10.0	2.5	2.2	2.1	T	T	0.0	25.8
1993-	0.0	0.0	0.0	T	T	0.5							
POR= 54 YRS	0.0	0.0	T	0.2	1.8	2.8	3.9	3.6	2.5	0.6	T	T	15.4

AUSTIN, TEXAS

Austin, capital of Texas, is located on the Colorado River where the stream crosses the Balcones escarpment separating the Texas Hill Country from the Blackland Prairies to the east. Elevations within the city vary from 400 feet to nearly 1,000 feet above sea level. Native trees include cedar, oak, walnut, mesquite, and pecan.

The climate of Austin is humid subtropical with hot summers. Winters are mild, with below freezing temperatures occurring on an average of about 25 days each year. Rather strong northerly winds, accompanied by sharp drops in temperature, frequently occur during the winter months in connection with cold fronts, but cold spells are usually of short duration, seldom lasting more than two days. Daytime temperatures in summer are hot, but summer nights are usually pleasant.

Precipitation is fairly evenly distributed throughout the year, with heaviest amounts occurring in late spring. A secondary rainfall peak occurs in September, primarily because of tropical cyclones that migrate out of the Gulf of Mexico. Precipitation from April through September usually results from thunderstorms, with fairly large amounts of rain falling within short periods of time. While thunderstorms and heavy rains may occur in all months of the year, most of the winter precipitation consists of light rain. Snow is insignificant as a source of moisture, and usually melts as rapidly as it falls. The city may experience several seasons in succession with no measurable snowfall.

Prevailing winds are southerly, however in winter, northerly winds are about as frequent as those from the south. Destructive winds and damaging hailstorms are infrequent. On rare occasions dissipating tropical storms produce strong winds and heavy rains in the area. Blowing dust occurs occasionally in spring, but visibility rarely drops substantially, and then only for a few hours.

The average length of the warm season (freeze-free period) is 273 days. The average occurrence of the last temperature of 32 degrees in spring is early March and the average occurrence of the first temperature of 32 degrees is late November.

NORMALS, MEANS, AND EXTREMES

AUSTIN, TEXAS

LATITUDE: 30°18'N LONGITUDE: 97°42' W ELEVATION: FT. GRND 587 BARO 590 TIME ZONE: CENTRAL WBAN: 13958

	(a)	JAN	FEB	MAR	APR	MAY	JUNE	JULY	AUG	SEP	OCT	NOV	DEC	YEAR
TEMPERATURE °F:														
Normals														
-Daily Maximum		58.9	63.4	71.9	79.4	84.7	91.1	95.0	95.5	90.5	82.1	71.8	62.0	78.9
-Daily Minimum		38.6	42.1	51.1	59.8	66.5	71.5	73.9	73.9	69.8	60.0	49.9	41.2	58.2
-Monthly		48.8	52.8	61.5	69.6	75.6	81.3	84.5	84.8	80.2	71.1	60.9	51.6	68.6
Extremes														
-Record Highest	52	90	97	98	98	100	105	109	106	104	98	91	90	109
-Year		1971	1986	1971	1982	1984	1980	1954	1990	1985	1991	1951	1955	JUL 1954
-Record Lowest	52	-2	7	18	35	43	53	64	61	41	30	20	4	-2
-Year		1949	1951	1948	1973	1954	1970	1970	1967	1942	1993	1976	1989	JAN 1949
NORMAL DEGREE DAYS:														
Heating (base 65°F)		514	353	175	28	0	0	0	0	0	17	175	426	1688
Cooling (base 65°F)		12	11	66	166	329	489	605	614	456	206	52	10	3016
% OF POSSIBLE SUNSHINE	52	48	51	55	54	56	69	75	75	66	64	55	49	60
MEAN SKY COVER (tenths)														
Sunrise - Sunset	52	6.3	6.1	6.0	6.2	6.2	5.2	4.7	4.6	5.0	4.7	5.4	6.0	5.5
MEAN NUMBER OF DAYS:														
Sunrise to Sunset														
-Clear	52	8.9	8.3	8.8	7.8	6.5	8.2	11.7	11.7	10.6	12.6	10.9	9.9	116.0
-Partly Cloudy	52	5.7	6.4	7.8	7.7	11.4	14.9	13.4	13.8	10.8	9.3	6.9	5.8	114.0
-Cloudy	52	16.3	13.6	14.4	14.6	13.1	6.9	5.9	5.5	8.5	9.1	12.2	15.3	135.3
Precipitation														
.01 inches or more	52	8.1	7.8	7.3	7.4	9.1	6.6	4.8	5.1	7.1	6.4	7.1	7.6	84.3
Snow, Ice pellets, hail														
1.0 inches or more	52	0.2	0.1	0.*	0.0	0.0	0.0	0.0	0.0	0.0	0.0	0.1	0.0	0.4
Thunderstorms	52	1.0	2.2	3.3	4.7	7.1	4.9	3.9	4.8	4.0	2.8	1.7	1.1	41.4
Heavy Fog Visibility														
1/4 mile or less	52	4.4	2.9	2.5	1.3	1.0	0.4	0.3	0.3	0.8	2.0	2.7	4.0	22.6
Temperature °F														
-Maximum														
90° and above	32	0.*	0.1	0.6	1.5	6.3	20.9	28.1	28.3	17.3	4.2	0.0	0.0	107.3
32° and below	32	0.6	0.2	0.0	0.0	0.0	0.0	0.0	0.0	0.0	0.0	0.0	0.3	1.1
-Minimum														
32° and below	32	8.0	4.3	1.0	0.0	0.0	0.0	0.0	0.0	0.0	0.*	1.1	5.3	19.7
0° and below	32	0.0	0.0	0.0	0.0	0.0	0.0	0.0	0.0	0.0	0.0	0.0	0.0	0.0
AVG. STATION PRESS. (mb)	21	998.3	996.7	992.9	992.0	990.7	991.6	993.5	993.2	993.4	995.4	996.2	997.8	994.3
RELATIVE HUMIDITY (%)														
Hour 00	32	73	72	71	75	81	80	75	73	77	75	76	73	75
Hour 06 (Local Time)	32	79	79	79	82	88	89	88	87	86	84	82	80	84
Hour 12	32	61	59	56	57	61	57	51	50	55	55	57	60	57
Hour 18	32	58	52	49	53	58	54	47	46	54	54	58	59	54
PRECIPITATION (inches):														
Water Equivalent														
-Normal		1.71	2.17	1.87	2.56	4.78	3.72	2.04	2.05	3.30	3.43	2.37	1.88	31.88
-Maximum Monthly	52	9.21	6.56	6.03	9.93	9.98	14.96	10.54	8.90	8.11	12.31	7.91	14.16	14.96
-Year		1991	1992	1983	1957	1965	1981	1979	1974	1942	1960	1946	1991	JUN 1981
-Minimum Monthly	52	0.04	0.28	T	0.06	0.81	T	0.00	0.00	0.07	T	T	T	0.00
-Year		1971	1954	1972	1984	1960	1967	1962	1952	1947	1952	1970	1950	JUL 1962
-Maximum in 24 hrs	52	4.41	3.73	2.69	3.86	5.66	6.50	5.46	4.68	6.74	7.22	5.09	6.19	7.22
-Year		1991	1958	1980	1942	1979	1964	1961	1945	1973	1960	1974	1991	OCT 1960
Snow, Ice pellets, hail														
-Maximum Monthly	52	7.5	6.0	2.0	0.0	T	0.0	0.0	0.0	0.0	0.0	2.0	T	7.5
-Year		1985	1966	1965		1993						1980	1993	JAN 1985
-Maximum in 24 hrs	52	7.0	6.0	2.0	T	T	0.0	0.0	0.0	0.0	0.0	2.0	T	7.0
-Year		1944	1966	1965	1989	1993						1980	1993	JAN 1944
WIND:														
Mean Speed (mph)	52	9.6	10.1	10.8	10.4	9.5	9.1	8.3	7.9	7.9	8.1	9.0	9.1	9.1
Prevailing Direction through 1963		S	S	S	SSE	SSE	S	S	S	S	S	S	S	S
Fastest Obs. 1 Min.														
-Direction (!!!)	14	35	34	33	33	32	34	28	03	02	35	30	31	02
-Speed (MPH)	14	37	39	36	40	41	41	40	35	52	33	36	44	52
-Year		1993	1984	1984	1983	1992	1988	1980	1988	1987	1985	1983	1987	SEP 1987
Peak Gust														
-Direction (!!!)	10	W	NW	NW	W	NW	N	SW	NE	N	NW	NW	NW	N
-Speed (mph)	10	52	55	56	51	63	54	44	47	81	46	49	63	81
-Date		1992	1984	1984	1989	1992	1990	1988	1988	1987	1988	1988	1987	SEP 1987

PRECIPITATION (inches)

AUSTIN, TEXAS

YEAR	JAN	FEB	MAR	APR	MAY	JUNE	JULY	AUG	SEP	OCT	NOV	DEC	ANNUAL
1964	2.57	1.47	1.95	1.47	1.87	7.54	0.65	2.09	6.29	3.74	2.45	0.88	32.97
1965	4.09	5.06	1.30	1.91	9.98	0.89	0.37	1.32	5.46	3.26	2.65	4.28	40.57
1966	1.58	3.23	0.50	3.74	3.13	1.53	0.47	6.21	3.22	0.60	0.11	0.87	25.19
1967	0.25	1.52	1.09	4.44	3.35	T	1.15	3.71	5.71	4.55	4.36	3.41	33.54
1968	7.94	1.64	2.09	1.87	8.75	3.10	3.11	0.74	3.42	0.60	4.91	0.55	38.72
1969	0.40	4.18	3.26	5.04	3.25	2.66	0.12	5.78	1.17	2.65	0.79	4.29	33.59
1970	1.83	5.70	2.47	1.36	8.18	0.29	0.66	1.00	3.82	5.22	T	0.11	30.64
1971	0.04	0.69	0.79	1.07	1.37	1.68	1.23	5.69	2.13	3.02	3.02	4.22	24.95
1972	1.48	0.31	T	1.46	7.88	2.20	2.55	2.53	1.55	2.96	2.62	0.53	26.07
1973	3.42	2.05	2.92	3.09	1.38	4.70	2.95	0.06	7.44	11.11	0.58	0.76	40.46
1974	2.74	0.36	1.34	1.79	5.88	0.21	0.61	8.90	1.58	3.45	7.35	2.00	36.21
1975	1.11	2.30	0.80	3.86	8.16	7.07	2.25	2.54	3.62	2.54	0.52	2.04	36.81
1976	1.16	1.11	2.11	8.13	6.05	3.19	4.71	0.80	3.80	5.93	1.78	2.48	41.25
1977	2.25	2.58	2.18	6.08	1.24	1.22	0.21	0.06	3.10	1.19	1.69	0.34	22.14
1978	0.88	1.95	0.84	1.72	5.78	2.98	1.19	1.49	4.44	1.38	5.48	2.84	30.97
1979	2.11	3.54	3.76	2.98	7.29	0.83	10.54	0.61	1.40	0.45	0.59	3.40	37.50
1980	0.85	2.33	3.20	2.20	5.43	0.31	0.28	1.18	5.66	1.29	3.41	1.24	27.38
1981	1.61	1.18	3.05	0.81	9.02	14.96	3.39	0.91	2.65	7.04	0.72	0.39	45.73
1982	0.85	0.80	1.39	4.17	5.68	2.99	0.13	0.77	1.88	2.66	3.19	2.12	26.63
1983	1.88	2.84	6.03	0.16	5.33	3.84	2.85	2.21	2.83	2.82	2.66	0.53	33.98
1984	1.66	1.00	2.49	0.06	1.27	1.69	1.44	0.45	0.79	10.34	1.88	3.23	26.30
1985	1.34	2.10	1.84	2.39	1.65	5.64	1.53	0.37	3.98	5.84	4.75	1.06	32.49
1986	0.45	1.14	0.41	1.46	7.36	2.20	0.45	1.21	4.77	7.98	1.81	5.77	35.01
1987	0.92	2.87	1.36	0.45	6.75	10.85	3.46	0.27	5.03	0.31	3.08	1.31	36.66
1988	0.27	0.32	2.66	2.02	3.33	2.60	2.77	1.67	1.43	0.66	0.34	1.14	19.21
1989	3.79	0.85	2.12	2.43	6.90	3.10	0.09	2.72	0.27	2.20	1.26	0.14	25.87
1990	1.28	3.55	2.08	3.12	3.65	1.55	3.14	0.33	1.76	3.39	3.87	0.72	28.44
1991	9.21	2.99	0.90	4.91	3.98	4.40	1.16	4.28	2.25	3.06	0.91	14.16	52.21
1992	4.83	6.56	5.43	1.90	9.05	4.96	0.96	1.95	1.98	1.38	3.76	3.29	46.05
1993	3.39	3.14	2.09	2.94	5.30	3.99	T	0.75	0.34	2.42	1.00	1.14	26.50
Record Mean	2.08	2.41	2.24	3.39	4.38	2.98	2.18	2.11	3.51	3.15	2.38	2.50	33.30

AVERAGE TEMPERATURE (deg. F)

AUSTIN, TEXAS

YEAR	JAN	FEB	MAR	APR	MAY	JUNE	JULY	AUG	SEP	OCT	NOV	DEC	ANNUAL
1964	50.3	48.4	60.7	70.8	76.6	81.3	85.2	86.0	79.9	67.6	62.5	52.2	68.5
1965	53.4	49.4	53.4	71.4	74.7	80.7	84.2	84.3	80.5	66.6	64.1	56.0	68.2
1966	45.3	48.9	60.1	69.1	73.9	79.9	84.7	82.2	78.2	68.6	64.2	51.0	67.2
1967	50.9	51.4	67.3	75.9	75.3	83.7	85.2	82.6	75.2	67.7	58.9	49.4	68.6
1968	47.6	47.2	57.3	68.1	74.2	79.5	82.2	84.1	76.0	72.0	55.8	50.4	66.2
1969	53.6	52.6	53.9	69.0	73.7	80.7	86.3	84.9	79.7	68.6	57.9	53.8	67.9
1970	44.0	53.1	55.2	68.7	71.9	79.9	83.8	85.8	79.2	66.7	57.4	58.7	67.1
1971	54.6	54.9	62.1	68.2	76.2	83.7	85.9	81.2	79.3	72.2	60.5	56.4	69.6
1972	51.0	55.7	66.1	73.5	73.5	81.7	82.3	83.1	82.4	70.9	52.8	48.3	68.5
1973	45.9	50.3	63.7	63.1	73.7	78.4	82.8	82.5	79.7	71.1	65.1	52.4	67.4
1974	49.0	56.3	66.5	68.9	77.7	80.1	84.4	81.3	72.3	69.5	56.9	50.7	67.8
1975	53.4	52.2	59.7	67.8	73.7	79.9	82.0	82.6	75.8	70.0	60.4	52.4	67.5
1976	49.5	61.3	62.0	68.0	70.7	79.8	80.2	83.4	78.1	61.3	51.6	48.5	66.2
1977	41.6	54.1	61.3	67.0	75.3	82.2	85.1	86.8	83.8	71.6	61.7	53.6	68.7
1978	40.7	45.0	58.5	69.8	77.3	82.3	86.4	84.8	79.0	69.8	60.6	49.7	67.0
1979	40.4	48.5	61.1	67.0	71.5	79.8	83.1	82.5	77.6	73.5	56.5	53.0	66.2
1980	51.2	52.8	61.4	66.7	75.1	84.6	87.9	86.0	81.7	68.4	57.5	53.7	68.9
1981	50.6	53.4	59.3	72.4	74.6	81.4	84.9	85.5	79.7	72.2	63.7	53.7	69.3
1982	51.9	51.0	64.7	67.3	75.6	81.6	86.5	87.4	91.8	70.7	58.5	52.3	69.1
1983	47.4	51.3	57.8	64.6	72.7	78.4	82.7	84.4	78.3	72.0	62.4	41.9	66.2
1984	46.3	55.5	63.4	71.1	78.0	82.6	85.1	86.0	76.3	70.5	58.7	58.4	69.3
1985	44.1	49.7	64.3	70.0	76.3	80.7	82.9	87.0	79.3	71.7	63.0	49.1	68.2
1986	54.0	57.6	63.7	71.8	73.8	81.4	85.9	84.7	82.0	68.8	59.0	50.8	69.5
1987	50.9	55.5	57.9	67.4	76.1	80.2	83.4	85.4	78.8	69.9	60.0	53.3	68.2
1988	47.3	53.3	60.4	68.5	74.7	80.2	84.6	86.7	81.6	72.5	64.9	55.5	69.2
1989	55.3	48.6	60.8	69.2	78.3	81.3	85.7	84.7	79.3	72.0	62.7	44.3	68.5
1990	57.9	59.3	61.7	69.8	78.5	85.9	83.4	86.1	81.2	70.0	63.6	51.3	70.7
1991	47.3	56.6	63.5	71.4	76.7	81.5	83.9	85.1	77.2	72.7	56.8	54.5	68.9
1992	50.6	58.9	62.7	68.0	73.2	81.6	83.9	82.0	80.8	72.7	56.9	54.8	68.8
1993	49.9	54.5	60.9	66.3	73.7	81.4	85.3	87.4	81.8	70.5	55.7	55.0	68.5
Record Mean	50.0	53.5	60.7	68.1	74.8	81.6	84.2	84.3	79.1	70.0	59.3	51.7	68.1
Max	60.2	64.1	71.6	78.6	84.7	91.6	94.7	95.2	89.5	81.0	69.9	61.9	78.6
Min	39.8	42.9	49.7	57.5	64.9	71.5	73.7	73.5	68.7	58.9	48.7	41.5	57.6

REFERENCE NOTES

AUSTIN, TEXAS

GENERAL
T - TRACE AMOUNT.
BLANK ENTRIES DENOTE MISSING/UNREPORTED DATA.
# INDICATES A STATION OR INSTRUMENT RELOCATION.

SPECIFIC

(a) - LENGTH OF RECORD IN YEARS, ALTHOUGH INDIVIDUAL MONTHS MAY BE MISSING.
0.* OR * - THE VALUE IS BETWEEN 0.0 AND 0.05
NORMALS - BASED ON THE 1961-1990 RECORD PERIOD.
EXTREMES - DATES ARE THE MOST RECENT OCCURRENCE
WIND DIR.- NUMERALS SHOW TENS OF DEGRESS CLOCKWISE FROM TRUE NORTH. "00" INDICATES CALM.
RESULTANT DIRECTIONS ARE GIVEN TO WHOLE DEGREES.
BOLD VALUES INDICATE EXTREME VALUES WHICH OCCURRED AFTER THE ASOS SYSTEM WAS COMMISSIONED.

RECORD = PERIOD OF RECORD
RECORD MEAN PRECIPITATION IS THE MEAN OF ALL DAILY PRECIPITATION AMOUNTS DURING THE PERIOD OF RECORD.
RECORD MAX(MIN) TEMPERATURE IS THE MEAN OF ALL DAILY MAX(MIN) TEMPERATURES DURING THE PERIOD OF RECORD.
RECORD MEAN TEMPERATURE IS THE SUM OF THE RECORD MAX AND RECORD MIN DIVIDED BY 2.
AVERAGE TEMPERATURE IS THE SUM OF THE MEAN DAILY MAX AND MIN TEMPERATURE DIVIDED BY 2.

EXCEPTIONS

RECORD MEANS ARE THROUGH THE CURRENT YEAR, BEGINNING IN 1898 FOR TEMPERATURE
1856 FOR PRECIPITATION
1942 FOR SNOWFALL

HEATING DEGREE DAYS Base 65 deg. F AUSTIN, TEXAS

SEASON	JULY	AUG	SEP	OCT	NOV	DEC	JAN	FEB	MAR	APR	MAY	JUNE	TOTAL
1964-65	0	0	0	17	158	415	370	431	367	12	0	0	1770
1965-66	0	0	2	58	76	288	611	449	168	24	11	0	1687
1966-67	0	0	0	31	109	453	446	380	77	0	2	0	1498
1967-68	0	0	7	41	192	477	538	510	265	37	1	0	2068
1968-69	0	0	0	7	294	446	368	345	346	6	1	0	1813
1969-70	0	0	0	69	253	339	653	330	309	47	10	0	2010
1970-71	0	0	3	75	240	231	328	300	171	52	0	0	1400
1971-72	0	0	3	0	184	271	434	298	66	6	0	0	1262
1972-73	0	0	0	46	375	513	586	405	72	128	3	0	2128
1973-74	0	0	0	8	83	387	496	258	108	36	0	0	1376
1974-75	0	0	5	11	271	441	377	356	204	53	0	0	1718
1975-76	0	0	1	27	204	404	476	158	173	19	3	0	1465
1976-77	0	0	0	170	402	506	719	308	148	20	0	0	2273
1977-78	0	0	0	16	136	354	750	561	225	15	7	0	2064
1978-79	0	0	0	5	186	475	754	460	144	37	10	0	2071
1979-80	0	0	0	16	278	372	425	357	168	47	0	0	1663
1980-81	0	0	0	69	262	361	443	340	196	6	0	0	1677
1981-82	0	0	0	44	86	349	424	400	147	81	2	0	1533
1982-83	0	0	0	37	245	408	538	376	229	93	4	0	1930
1983-84	0	0	4	7	154	721	573	280	123	16	0	0	1878
1984-85	0	0	18	34	221	228	643	428	99	19	0	0	1690
1985-86	0	0	7	12	134	490	338	267	87	14	2	0	1351
1986-87	0	0	0	18	214	431	434	266	228	83	0	0	1674
1987-88	0	0	0	11	203	369	548	357	193	34	0	0	1715
1988-89	0	0	0	3	116	307	307	464	198	64	0	0	1459
1989-90	0	0	0	39	157	638	239	171	154	26	0	0	1424
1990-91	0	0	0	44	121	430	540	242	110	8	1	0	1496
1991-92	0	0	8	34	280	332	440	188	105	39	4	0	1430
1992-93	0	0	0	0	267	327	462	296	163	54	0	0	1569
1993-94	0	0	0	70	298	324							

COOLING DEGREE DAYS Base 65 deg. F AUSTIN, TEXAS

YEAR	JAN	FEB	MAR	APR	MAY	JUNE	JULY	AUG	SEP	OCT	NOV	DEC	TOTAL
1969	25	5	8	132	276	476	668	620	450	186	46	1	2893
1970	8	1	10	165	235	451	589	651	435	134	22	43	2744
1971	14	20	86	157	355	570	656	511	436	229	53	13	3100
1972	6	34	111	269	272	510	547	571	529	239	16	3	3107
1973	1	0	34	78	277	407	557	550	447	205	93	4	2653
1974	6	21	160	158	398	463	606	513	229	157	34	6	2751
1975	21	0	47	141	277	454	536	552	329	187	72	22	2638
1976	0	57	85	115	186	451	479	577	401	61	6	0	2418
1977	0	7	38	90	324	526	629	681	572	227	40	7	3141
1978	5	6	31	167	396	526	670	621	427	161	60	7	3077
1979	2	7	29	102	216	449	570	549	385	285	29	8	2631
1980	5	11	62	106	322	598	718	659	510	182	45	19	3237
1981	2	23	27	233	304	501	627	641	446	272	54	5	3135
1982	23	14	144	157	337	506	672	701	509	219	61	18	3361
1983	0	0	13	90	250	406	556	608	409	227	80	11	2650
1984	0	13	80	204	413	535	629	659	364	206	40	30	3173
1985	0	9	85	174	358	480	562	687	443	228	82	4	3112
1986	2	64	53	227	278	499	654	618	519	141	39	0	3092
1987	5	4	17	158	352	463	576	638	421	170	58	14	2876
1988	7	23	57	146	305	463	614	679	503	243	123	19	3182
1989	16	10	76	197	417	498	650	620	435	263	93	2	3277
1990	28	16	57	178	425	634	579	662	495	204	86	11	3375
1991	0	11	72	207	372	501	594	633	382	280	39	15	3106
1992	0	15	41	137	264	505	595	538	479	246	28	17	2865
1993	0	7	44	99	276	498	641	700	509	248	27	19	3068

SNOWFALL (inches) AUSTIN, TEXAS

SEASON	JULY	AUG	SEP	OCT	NOV	DEC	JAN	FEB	MAR	APR	MAY	JUNE	TOTAL
1970-71	0.0	0.0	0.0	0.0	0.0	0.0	0.0	T	0.0	0.0	0.0	0.0	T
1971-72	0.0	0.0	0.0	0.0	0.0	T	T	0.0	0.0	0.0	0.0	0.0	T
1972-73	0.0	0.0	0.0	0.0	0.0	0.0	0.8	0.9	0.0	0.0	0.0	0.0	1.7
1973-74	0.0	0.0	0.0	0.0	0.0	0.0	0.0	0.0	0.0	0.0	0.0	0.0	0.0
1974-75	0.0	0.0	0.0	0.0	0.0	0.0	T	T	0.0	0.0	0.0	0.0	T
1975-76	0.0	0.0	0.0	0.0	0.0	0.0	0.0	0.0	0.0	0.0	0.0	0.0	0.0
1976-77	0.0	0.0	0.0	0.0	T	0.0	0.0	T	0.0	0.0	0.0	0.0	T
1977-78	0.0	0.0	0.0	0.0	0.0	0.0	T	T	T	0.0	0.0	0.0	T
1978-79	0.0	0.0	0.0	0.0	0.0	T	T	T	0.0	0.0	0.0	0.0	T
1979-80	0.0	0.0	0.0	0.0	T	0.0	T	T	0.0	0.0	0.0	0.0	T
1980-81	0.0	0.0	0.0	0.0	2.0	0.0	T	T	0.0	0.0	0.0	0.0	2.0
1981-82	0.0	0.0	0.0	0.0	0.0	0.0	2.0	T	T	0.0	0.0	0.0	2.0
1982-83	0.0	0.0	0.0	0.0	0.0	0.0	0.0	0.0	0.0	0.0	0.0	0.0	0.0
1983-84	0.0	0.0	0.0	0.0	0.0	T	0.0	0.0	0.0	0.0	0.0	0.0	T
1984-85	0.0	0.0	0.0	0.0	0.0	0.0	7.5	1.2	0.0	0.0	0.0	0.0	8.7
1985-86	0.0	0.0	0.0	0.0	0.0	0.0	T	0.0	0.0	0.0	0.0	0.0	T
1986-87	0.0	0.0	0.0	0.0		0.0	0.0	0.0	T	0.0	0.0	0.0	
1987-88	0.0	0.0	0.0	0.0	0.0	0.0	0.0	T	0.0	0.0	0.0	0.0	T
1988-89	0.0	0.0	0.0	0.0	0.0	0.0	0.0	T	T	0.0	T	0.0	T
1989-90	0.0	0.0	0.0	0.0	0.0	T	0.0	0.0	0.0	0.0	0.0	0.0	T
1990-91	0.0	0.0	0.0	0.0	0.0	T	0.0	0.0	0.0	0.0	0.0	0.0	T
1991-92	0.0	0.0	0.0	0.0	0.0	0.0	0.0	0.0	0.0	0.0	0.0	0.0	0.0
1992-93	0.0	0.0	0.0	0.0	T	0.0	T	0.0	T	0.0	T	0.0	T
1993-94	0.0	0.0	0.0	0.0	0.0	T							
Record Mean	0.0	0.0	0.0	0.0	0.1	T	0.5	0.3	T	0.0	T	0.0	0.9

BROWNSVILLE, TEXAS

Brownsville is located at the southern tip of Texas. It is the largest city in the four county area referred to as the Lower Rio Grande Valley or just the Valley.

The Gulf of Mexico, located about 18 miles east, is the dominant influence on local weather. Prevailing southeast breezes off the Gulf provide a humid but generally mild climate. Winds are frequently strong and gusty in the spring.

Brownsville weather is generally favorable for outdoor activities and the Valley is a popular tourist area, especially for Winter Texans who come to enjoy the mild winters. High temperatures range mostly in the 70s and 80s from October through April, with lows in the 50s and 60s during the same period. For the remainder of the year highs are frequently in the 90s with lows in the 70s.

Temperature extremes are rare but do occur. Temperatures in the 90s have occurred in every month of the year, with 100 degree readings noted as early as March and as late as September. Temperatures of 100 degrees or more are associated with west winds bringing hot dry air out of Mexico. Very hot temperatures are often moderated by a cooling sea breeze from the Gulf during the afternoon hours.

Located about 150 miles north of the tropics, cold weather in Brownsville is infrequent and of short duration. Some winters pass without a single day with freezing temperatures. This climate permits year around gardening and cultivation of citrus and other cold sensitive tropical and sub-tropical plants. Damaging cold comes from frigid air masses, called northers or arctic outbreaks, plunging south from Canada or the Arctic. The worst of these can drop temperatures well below freezing for several hours, and a few have produced readings in the teens. Fortunately such events are very rare since they are disasterous to the local economy.

Rainfall is not well distributed. Heaviest rains occur in May through June and mid August through mid October. Extended periods of cool rainy weather, called overrunning, can occur in winter. Torrential rains of 10 to 20 inches or more may accompany tropical storms or hurricanes that occasionally move over the area in summer or fall. Rainy spells may be followed by long dry periods. Irrigation is required to ensure production of corps such as cotton, grains, and vegetables. Snow and freezing rain or drizzle are so rare that years may pass between occurrences.

Brownsville is blessed by having little severe weather. Damaging hail or winds from heavy thunderstorms are generally limited to the Spring season and many years may elapse between occurrences. Tornadoes are even more rare. Tropical storms and hurricanes from the Gulf are a threat each summer and fall, but again, damaging storms are quite rare.

NORMALS, MEANS, AND EXTREMES

BROWNSVILLE, TEXAS

LATITUDE: 25°54'N LONGITUDE: 97°26' W ELEVATION: FT. GRND 19 BARO 22 TIME ZONE: CENTRAL WBAN: 12919

	(a)	JAN	FEB	MAR	APR	MAY	JUNE	JULY	AUG	SEP	OCT	NOV	DEC	YEAR
TEMPERATURE °F:														
Normals														
-Daily Maximum		68.9	72.2	78.4	84.0	87.8	91.0	93.3	93.6	90.4	85.3	78.3	71.7	82.9
-Daily Minimum		49.9	52.5	59.1	66.5	72.0	74.9	75.7	75.4	73.2	66.1	59.0	52.4	64.7
-Monthly		59.4	62.4	68.8	75.3	79.9	83.0	84.5	84.5	81.8	75.7	68.7	62.1	73.8
Extremes														
-Record Highest	55	93	94	106	102	102	102	102	102	104	96	97	94	106
-Year		1971	1986	1984	1984	1974	1989	1939	1962	1947	1986	1988	1977	MAR 1984
-Record Lowest	55	19	22	32	38	52	60	68	63	55	35	31	16	16
-Year		1962	1951	1989	1980	1970	1975	1989	1967	1942	1993	1993	1989	DEC 1989
NORMAL DEGREE DAYS:														
Heating (base 65°F)		232	135	41	0	0	0	0	0	0	0	50	177	635
Cooling (base 65°F)		59	62	159	312	462	540	605	605	504	332	161	87	3888
% OF POSSIBLE SUNSHINE	51	41	48	53	58	64	73	80	76	68	65	51	42	60
MEAN SKY COVER (tenths)														
Sunrise - Sunset	51	7.0	6.6	6.7	6.6	6.1	5.2	4.8	5.0	5.4	4.9	5.8	6.8	5.9
MEAN NUMBER OF DAYS:														
Sunrise to Sunset														
-Clear	51	6.2	6.9	6.7	5.1	5.9	8.4	11.0	10.5	9.0	11.2	8.8	6.6	96.4
-Partly Cloudy	51	6.7	6.1	7.9	10.4	14.1	15.4	14.3	13.8	12.9	12.3	9.5	7.5	130.8
-Cloudy	51	18.1	15.2	16.4	14.6	11.0	6.2	5.7	6.7	8.1	7.5	11.7	16.9	138.0
Precipitation														
.01 inches or more	51	7.8	6.1	4.1	3.8	5.0	5.9	4.8	6.6	9.9	6.4	6.0	6.6	73.0
Snow, Ice pellets, hail														
1.0 inches or more	51	0.0	0.0	0.0	0.0	0.0	0.0	0.0	0.0	0.0	0.0	0.0	0.0	0.0
Thunderstorms	51	0.6	0.8	0.6	2.4	3.6	3.0	2.8	4.5	4.9	2.0	1.0	0.5	26.7
Heavy Fog Visibility														
1/4 mile or less	51	6.0	4.6	3.5	2.3	0.9	0.2	0.1	0.2	0.3	0.8	2.9	5.3	27.1
Temperature °F														
-Maximum														
90° and above	27	0.1	0.2	1.6	3.3	10.0	23.6	27.2	28.0	18.7	6.2	0.5	0.*	119.4
32° and below	27	0.0	0.0	0.0	0.0	0.0	0.0	0.0	0.0	0.0	0.0	0.0	0.*	*
-Minimum														
32° and below	27	1.0	0.4	0.1	0.0	0.0	0.0	0.0	0.0	0.0	0.0	0.*	0.8	2.4
0° and below	27	0.0	0.0	0.0	0.0	0.0	0.0	0.0	0.0	0.0	0.0	0.0	0.0	0.0
AVG. STATION PRESS. (mb)	21	1018.8	1017.5	1013.7	1012.5	1011.0	1012.2	1014.3	1013.8	1013.3	1015.6	1016.7	1018.5	1014.8
RELATIVE HUMIDITY (%)														
Hour 00	27	87	86	85	86	87	87	86	86	87	86	85	86	86
Hour 06 (Local Time)	27	88	89	88	88	90	91	91	91	90	89	87	87	89
Hour 12	27	68	63	59	59	61	59	55	55	60	59	61	65	60
Hour 18	27	75	69	66	68	69	66	63	63	68	69	72	75	69
PRECIPITATION (inches):														
Water Equivalent														
-Normal		1.56	1.06	0.53	1.56	2.94	2.73	1.90	2.77	6.00	2.80	1.51	1.25	26.61
-Maximum Monthly	54	5.11	10.25	4.27	10.35	9.12	13.06	9.43	9.56	20.18	17.12	7.69	9.45	20.18
-Year		1945	1958	1941	1991	1982	1942	1976	1975	1984	1958	1986	1940	SEP 1984
-Minimum Monthly	54	T	T	T	T	T	0.01	T	0.02	0.07	0.34	0.01	T	T
-Year		1956	1954	1986	1988	1978	1955	1993	1974	1959	1961	1949	1969	JUL 1993
-Maximum in 24 hrs	54	3.00	4.98	2.59	9.37	4.56	8.18	4.25	5.48	12.19	6.67	4.08	5.69	12.19
-Year		1988	1958	1981	1991	1969	1942	1976	1980	1967	1954	1986	1940	SEP 1967
Snow, Ice pellets, hail														
-Maximum Monthly	54	T	T	T	0.0	0.0	0.0	0.0	T	0.0	0.0	T	T	T
-Year		1993	1973	1993					1992			1991	1966	JAN 1993
-Maximum in 24 hrs	54	T	T	T	0.0	0.0	0.0	0.0	T	0.0	0.0	T	T	T
-Year		1993	1973	1993					1992			1991	1966	JAN 1993
WIND:														
Mean Speed (mph)	51	11.2	12.0	13.3	13.7	13.0	11.9	11.3	10.3	9.4	9.5	10.6	10.7	11.4
Prevailing Direction through 1963		SSE	SSE	SE	SE	SE	SE	SE	SE	SE	SE	SSE	NNW	SE
Fastest Obs. 1 Min.														
-Direction (!!)	14	19	23	28	15	18	30	16	19	11	36	17	17	19
-Speed (MPH)	14	33	36	44	35	33	41	32	48	41	35	39	35	48
-Year		1990	1983	1993	1993	1989	1988	1986	1980	1988	1988	1983	1981	AUG 1980
Peak Gust														
-Direction (!!)	10	N	S	W	SE	NW	NW	S	E	NE	N	S	S	NE
-Speed (mph)	10	46	49	52	54	60	62	51	54	62	49	44	44	62
-Date		1985	1990	1993	1985	1985	1988	1986	1992	1988	1988	1990	1988	SEP 1988

PRECIPITATION (inches)

BROWNSVILLE, TEXAS

YEAR	JAN	FEB	MAR	APR	MAY	JUNE	JULY	AUG	SEP	OCT	NOV	DEC	ANNUAL
1964	0.35	1.32	0.26	1.05	3.12	0.59	0.22	0.11	4.94	0.40	0.60	2.76	15.72
1965	0.51	1.57	0.13	0.10	1.16	0.07	1.83	3.12	5.40	0.74	2.60	3.98	21.21
1966	3.45	0.83	0.53	0.80	6.05	2.18	1.58	0.58	0.52	7.45	0.02	0.68	24.67
1967	1.69	1.05	0.44	T	0.64	1.51	0.58	5.95	19.26	2.82	0.63	1.10	35.67
1968	3.08	0.80	1.08	1.27	5.90	4.71	1.20	2.78	4.48	3.39	0.85	0.10	29.64
1969	0.51	1.81	0.82	0.24	6.69	0.08	0.23	7.74	2.80	3.25	3.18	T	27.35
1970	4.12	0.22	0.34	2.22	3.28	1.64	2.35	1.79	6.95	3.28	0.14	0.12	26.45
1971	0.22	0.30	T	1.82	2.47	3.44	1.08	2.64	10.78	3.11	0.74	1.08	27.68
1972	1.30	2.72	1.14	1.54	2.02	8.52	5.16	0.90	4.22	3.33	1.25	0.74	32.84
1973	2.07	4.74	0.13	0.69	1.24	7.57	0.59	2.80	4.61	3.77	0.88	0.53	29.62
1974	0.65	0.01	0.20	0.52	1.82	4.59	1.02	0.02	4.93	2.60	0.65	0.78	17.79
1975	0.60	0.09	0.01	0.01	2.22	2.19	4.78	9.56	4.77	0.51	1.66	2.17	28.57
1976	0.48	0.03	1.28	5.71	4.95	0.80	9.43	3.35	2.85	8.45	2.49	1.32	41.14
1977	1.24	1.37	0.12	6.62	0.76	4.73	0.27	1.27	2.84	2.87	4.07	0.14	26.30
1978	1.94	1.29	0.01	2.39	T	2.25	0.39	3.20	8.28	4.45	0.82	1.86	26.88
1979	1.43	1.10	0.14	3.91	0.59	1.52	2.10	5.25	8.84	1.18	0.12	2.04	28.22
1980	1.05	1.74	0.28	0.01	1.78	0.02	1.46	7.29	1.48	2.26	2.50	1.90	21.77
1981	1.79	0.76	3.47	0.34	5.88	2.29	2.65	4.47	5.05	2.47	0.33	0.75	30.25
1982	0.04	0.75	0.19	4.08	9.12	0.18	T	1.04	2.42	1.63	3.11	2.70	25.26
1983	1.10	2.62	0.61	T	1.41	1.78	6.11	2.34	8.61	2.53	0.52	0.48	28.11
1984	4.79	0.42	0.13	T	6.18	2.44	1.59	1.80	20.18	0.93	0.02	1.85	40.33
1985	1.49	0.54	0.40	1.91	4.21	6.47	4.18	2.10	6.04	4.04	1.02	0.42	32.82
1986	1.07	0.21	T	0.87	2.89	3.72	0.35	2.14	1.71	4.61	7.69	2.42	27.68
1987	2.46	2.26	0.58	1.39	1.52	4.78	1.64	0.73	4.70	4.44	3.83	0.42	28.75
1988	3.97	1.53	1.42	T	0.25	2.86	1.00	2.56	7.48	1.80	0.14	0.07	23.08
1989	1.94	0.08	0.17	3.83	1.23	2.35	2.13	1.25	2.46	3.06	0.93	1.73	21.16
1990	0.58	0.56	0.81	1.55	2.72	1.08	1.53	2.87	3.90	2.29	0.91	0.05	18.85
1991	0.47	2.50	0.02	10.35	2.97	1.93	2.36	0.89	5.57	3.33	0.15	1.18	31.72
1992	3.50	1.99	0.12	4.15	5.55	1.50	0.40	3.71	3.62	0.85	5.61	0.85	31.85
1993	1.79	2.86	1.68	0.34	3.64	6.72	T	0.04	1.93	4.69	1.25	2.29	27.23
Record Mean	1.40	1.31	0.97	1.54	2.61	2.74	1.81	2.57	5.52	3.18	1.78	1.51	26.94

AVERAGE TEMPERATURE (deg. F)

BROWNSVILLE, TEXAS

YEAR	JAN	FEB	MAR	APR	MAY	JUNE	JULY	AUG	SEP	OCT	NOV	DEC	ANNUAL
1964	59.6	58.5	67.5	77.3	80.3	81.6	84.9	86.0	83.2	72.2	70.7	60.3	73.5
1965	63.0	61.3	64.2	76.3	79.0	82.7	83.5	81.8	81.8	71.5	72.1	64.9	73.5
#1966	54.3	56.7	65.5	74.1	76.2	80.0	83.3	84.4	81.2	72.9	68.9	61.1	71.6
1967	58.5	60.9	70.1	80.1	80.9	83.2	85.3	82.0	77.7	72.8	69.9	61.3	73.6
1968	57.8	57.5	64.0	74.2	80.2	83.3	83.6	84.3	81.6	77.7	68.0	64.7	73.1
1969	64.2	65.7	63.4	75.9	78.8	83.8	87.2	84.9	81.8	77.3	66.9	65.3	74.6
1970	57.6	64.0	66.0	76.6	76.5	82.2	84.0	85.7	82.7	75.2	66.2	69.8	73.9
1971	66.5	67.2	72.1	74.3	79.5	82.8	83.4	83.0	81.4	76.9	69.3	67.3	75.3
1972	66.0	64.7	72.4	77.4	77.6	80.6	81.3	82.2	82.0	77.5	63.1	59.5	73.7
1973	54.5	58.7	70.4	71.9	77.8	81.1	83.6	81.5	81.6	76.4	74.6	63.0	72.9
1974	61.4	63.9	73.1	75.6	81.6	80.9	81.9	84.9	80.0	73.6	66.5	61.2	73.7
1975	61.7	64.0	71.0	76.7	81.8	82.6	82.4	82.4	78.3	74.8	68.2	61.3	73.7
1976	59.1	65.6	71.0	74.2	75.8	81.6	81.3	82.3	81.7	70.2	60.8	57.4	71.8
1977	55.0	61.9	69.0	73.7	80.2	82.6	84.6	85.9	84.3	77.1	69.8	64.4	74.0
1978	54.6	55.9	66.8	74.8	83.3	85.0	86.8	85.4	82.3	75.9	71.2	62.9	73.8
1979	56.3	59.5	69.0	76.1	77.0	82.1	85.5	84.5	77.7	76.6	65.3	59.7	72.5
1980	63.9	60.6	69.6	72.0	82.0	86.9	87.5	84.9	84.9	73.6	63.1	61.6	74.2
1981	59.7	62.9	68.1	77.0	80.3	84.3	85.3	85.6	81.7	77.6	70.9	64.8	74.9
1982	62.7	61.1	71.7	76.0	80.1	85.7	86.9	86.3	83.4	77.0	68.4	62.4	75.1
1983	59.8	62.8	68.2	73.0	79.8	83.4	84.5	85.6	81.7	76.2	71.5	55.4	73.5
1984	55.9	62.4	69.7	76.3	79.3	82.8	84.2	84.5	79.4	79.3	68.8	70.3	74.4
1985	54.4	59.2	71.4	75.9	80.6	82.5	82.9	84.9	82.1	76.7	73.0	60.4	73.7
1986	61.0	65.9	69.0	77.0	79.4	83.1	84.8	84.8	84.4	76.3	66.7	60.2	74.4
1987	59.4	64.0	64.3	69.6	79.6	82.6	84.3	86.2	83.0	74.1	67.8	64.3	73.3
1988	55.6	61.4	66.9	73.4	78.3	81.8	85.7	84.5	80.8	76.0	72.3	64.0	73.4
1989	66.2	61.2	67.1	74.3	82.0	84.3	84.0	84.4	81.0	74.5	70.6	51.8	73.5
1990	65.1	67.0	70.2	75.5	81.0	85.4	84.5	85.4	81.3	74.9	70.8	61.8	75.2
1991	58.5	64.5	73.4	78.0	81.6	84.7	84.0	86.0	79.6	77.1	64.6	64.3	74.7
1992	58.0	65.5	70.3	72.8	76.8	84.1	85.4	84.3	82.0	76.4	66.4	64.7	73.9
1993	61.2	65.7	68.6	73.4	77.7	82.3	84.7	85.6	82.9	75.0	65.8	63.8	73.9
Record Mean	60.2	63.1	68.5	74.4	79.1	82.7	83.9	84.2	81.3	75.5	67.9	62.1	73.6
Max	69.3	72.5	77.4	82.8	87.0	90.7	92.4	93.0	89.8	84.7	77.1	71.3	82.3
Min	51.0	53.7	59.5	66.0	71.1	74.6	75.3	75.4	72.8	66.3	58.8	52.8	64.8

REFERENCE NOTES

BROWNSVILLE, TEXAS

GENERAL
T - TRACE AMOUNT.
BLANK ENTRIES DENOTE MISSING/UNREPORTED DATA.
# INDICATES A STATION OR INSTRUMENT RELOCATION.

SPECIFIC

(a) - LENGTH OF RECORD IN YEARS, ALTHOUGH INDIVIDUAL MONTHS MAY BE MISSING.
0.* OR * - THE VALUE IS BETWEEN 0.0 AND 0.05
NORMALS - BASED ON THE 1961-1990 RECORD PERIOD.
EXTREMES - DATES ARE THE MOST RECENT OCCURRENCE.
WIND DIR. - NUMERALS SHOW TENS OF DEGRESS CLOCKWISE FROM TRUE NORTH. "00" INDICATES CALM.
RESULTANT DIRECTIONS ARE GIVEN TO WHOLE DEGREES.
BOLD VALUES INDICATE EXTREME VALUES WHICH OCCURRED AFTER THE ASOS SYSTEM WAS COMMISSIONED.

RECORD = PERIOD OF RECORD
RECORD MEAN PRECIPITATION IS THE MEAN OF ALL DAILY PRECIPITATION AMOUNTS DURING THE PERIOD OF RECORD.
RECORD MAX(MIN) TEMPERATURE IS THE MEAN OF ALL DAILY MAX(MIN) TEMPERATURES DURING THE PERIOD OF RECORD.
RECORD MEAN TEMPERATURE IS THE SUM OF THE RECORD MAX AND RECORD MIN DIVIDED BY 2.
AVERAGE TEMPERATURE IS THE SUM OF THE MEAN DAILY MAX AND MIN TEMPERATURE DIVIDED BY 2.

EXCEPTIONS

RECORD MEANS ARE THROUGH THE CURRENT YEAR, BEGINNING IN 1882 FOR TEMPERATURE
1870 FOR PRECIPITATION
1940 FOR SNOWFALL

HEATING DEGREE DAYS Base 65 deg. F BROWNSVILLE, TEXAS

SEASON	JULY	AUG	SEP	OCT	NOV	DEC	JAN	FEB	MAR	APR	MAY	JUNE	TOTAL
1964-65	0	0	0	5	52	215	127	136	138	0	0	0	673
1965-66	0	0	0	8	8	76	333	239	65	1	0	0	730
#1966-67	0	0	0	4	48	185	238	151	32	0	0	0	658
1967-68	0	0	1	4	8	174	262	218	126	4	0	0	797
1968-69	0	0	0	0	73	96	141	45	115	0	0	0	470
1969-70	0	0	0	0	92	69	256	72	76	4	1	0	570
1970-71	0	0	1	5	86	46	130	76	38	6	0	0	388
1971-72	0	0	0	0	16	67	112	93	11	2	0	0	301
1972-73	0	0	0	0	144	217	339	190	4	27	0	0	921
1973-74	0	0	0	0	16	130	190	102	40	5	0	0	483
1974-75	0	0	0	3	77	169	177	90	31	5	0	0	552
1975-76	0	0	0	5	74	180	215	74	22	0	0	0	570
1976-77	0	0	0	21	177	256	319	123	43	3	0	0	942
1977-78	0	0	0	0	25	101	342	286	66	6	0	0	826
1978-79	0	0	0	0	35	152	305	203	41	0	0	0	736
1979-80	0	0	0	0	99	204	108	180	51	13	0	0	655
1980-81	0	0	0	34	141	166	181	117	38	0	0	0	677
1981-82	0	0	0	14	23	101	183	187	54	17	0	0	579
1982-83	0	0	0	0	82	145	187	85	18	11	0	0	528
1983-84	0	0	0	0	22	365	299	138	39	2	0	0	865
1984-85	0	0	0	2	51	55	349	213	21	0	0	0	691
1985-86	0	0	0	1	21	184	153	97	25	0	0	0	481
1986-87	0	0	0	12	117	177	192	79	80	36	0	0	693
1987-88	0	0	0	0	53	119	298	170	65	4	0	0	709
1988-89	0	0	0	0	21	123	100	206	104	9	0	0	563
1989-90	0	0	0	16	43	415	95	45	25	2	2	0	643
1990-91	0	0	0	3	26	194	217	69	5	0	0	0	514
1991-92	0	0	1	0	121	134	223	63	26	10	0	0	578
1992-93	0	0	0	0	82	93	152	58	48	1	0	0	434
1993-94	0	0	0	36	110	134							

COOLING DEGREE DAYS Base 65 deg. F BROWNSVILLE, TEXAS

YEAR	JAN	FEB	MAR	APR	MAY	JUNE	JULY	AUG	SEP	OCT	NOV	DEC	TOTAL
1969	123	71	72	335	437	570	695	624	511	390	156	87	4071
1970	35	48	116	356	364	522	597	648	540	330	128	202	3886
1971	185	142	266	294	459	541	578	566	498	375	153	143	4200
1972	148	91	246	382	398	475	512	542	517	395	95	53	3854
1973	19	21	178	242	404	490	585	519	504	365	309	74	3710
1974	83	75	299	329	520	485	534	624	458	275	130	59	3871
1975	79	68	223	363	526	536	548	546	408	315	176	69	3857
1976	41	97	216	283	343	507	511	547	509	190	59	24	3327
1977	15	44	174	272	476	535	615	658	585	384	177	88	4023
1978	32	38	129	308	572	605	680	638	524	342	228	92	4188
1979	45	56	173	339	380	517	645	614	388	368	118	46	3689
1980	83	62	198	230	536	661	707	623	602	309	91	67	4169
1981	22	64	139	365	482	586	635	646	507	414	204	102	4166
1982	119	85	267	357	474	626	684	667	555	381	190	73	4478
1983	29	29	124	254	466	560	607	645	506	354	222	73	3869
1984	24.	70	190	349	451	541	604	611	441	451	175	224	4131
1985	28	57	226	334	492	533	560	623	518	370	269	47	4057
1986	36	126	155	367	453	550	620	620	584	367	174	39	4055
1987	26	58	65	181	457	533	605	663	544	289	144	105	3670
1988	15	73	131	263	416	512	649	612	481	349	245	99	3845
1989	145	108	176	296	534	586	599	607	490	318	218	12	4089
1990	105	107	194	324	503	619	612	641	494	318	209	100	4226
1991	23	62	273	395	522	599	599	657	446	383	117	120	4196
1992	13	81	198	249	375	579	640	603	518	361	129	89	3835
1993	44	82	166	260	401	522	620	647	543	349	142	103	3879

SNOWFALL (inches) BROWNSVILLE, TEXAS

SEASON	JULY	AUG	SEP	OCT	NOV	DEC	JAN	FEB	MAR	APR	MAY	JUNE	TOTAL
1970-71	0.0	0.0	0.0	0.0	0.0	0.0	0.0	0.0	0.0	0.0	0.0	0.0	0.0
1971-72	0.0	0.0	0.0	0.0	0.0	0.0	0.0	0.0	0.0	0.0	0.0	0.0	0.0
1972-73	0.0	0.0	0.0	0.0	0.0	0.0	0.0	T	0.0	0.0	0.0	0.0	T
1973-74	0.0	0.0	0.0	0.0	0.0	0.0	0.0	0.0	0.0	0.0	0.0	0.0	0.0
1974-75	0.0	0.0	0.0	0.0	0.0	0.0	0.0	0.0	0.0	0.0	0.0	0.0	0.0
1975-76	0.0	0.0	0.0	0.0	0.0	0.0	0.0	0.0	0.0	0.0	0.0	0.0	0.0
1976-77	0.0	0.0	0.0	0.0	T	0.0	0.0	0.0	0.0	0.0	0.0	0.0	T
1977-78	0.0	0.0	0.0	0.0	0.0	0.0	0.0	0.0	0.0	0.0	0.0	0.0	0.0
1978-79	0.0	0.0	0.0	0.0	0.0	0.0	0.0	0.0	0.0	0.0	0.0	0.0	0.0
1979-80	0.0	0.0	0.0	0.0	0.0	0.0	0.0	0.0	0.0	0.0	0.0	0.0	0.0
1980-81	0.0	0.0	0.0	0.0	0.0	0.0	0.0	0.0	0.0	0.0	0.0	0.0	0.0
1981-82	0.0	0.0	0.0	0.0	0.0	0.0	0.0	0.0	0.0	0.0	0.0	0.0	0.0
1982-83	0.0	0.0	0.0	0.0	0.0	0.0	0.0	0.0	0.0	0.0	0.0	0.0	0.0
1983-84	0.0	0.0	0.0	0.0	0.0	0.0	0.0	0.0	0.0	0.0	0.0	0.0	0.0
1984-85	0.0	0.0	0.0	0.0	0.0	0.0	T	0.0	0.0	0.0	0.0	0.0	T
1985-86	0.0	0.0	0.0	0.0	0.0	0.0	0.0	0.0	0.0	0.0	0.0	0.0	0.0
1986-87	0.0	0.0	0.0	0.0	0.0	0.0	0.0	0.0	0.0	0.0	0.0	0.0	0.0
1987-88	0.0	0.0	0.0	0.0	0.0	0.0	0.0	0.0	0.0	0.0	0.0	0.0	0.0
1988-89	0.0	0.0	0.0	0.0	0.0	0.0	0.0	0.0	0.0	0.0	0.0	0.0	0.0
1989-90	0.0	0.0	0.0	0.0	0.0	0.0	T	0.0	0.0	0.0	0.0	0.0	T
1990-91	0.0	0.0	0.0	0.0	0.0	0.0	0.0	0.0	0.0	0.0	0.0	0.0	0.0
1991-92	0.0	0.0	0.0	0.0	T	0.0	0.0	0.0	0.0	0.0	0.0	0.0	T
1992-93	0.0	T	0.0	0.0	0.0	0.0	T	0.0	T	0.0	0.0	0.0	T
1993-94	0.0	0.0	0.0	0.0	0.0	0.0							
Record Mean	0.0	T	0.0	0.0	T	T	T	T	T	0.0	0.0	0.0	T

CORPUS CHRISTI, TEXAS

Corpus Christi is located on Corpus Christi Bay, an inlet of the Gulf of Mexico, in south Texas. The climatic conditions vary between the humid subtropical region to the northeast along the Texas coast and the semi-arid region to the west and southwest. Temperatures at the International Airport, which is about 7 miles west of downtown Corpus Christi, may be substantially different than those in the city during calm winter mornings and during summer afternoon sea breezes.

Peak rainfall months are May and September. Winter months have the least amounts of rainfall. The hurricane season from June to November can greatly effect the rainfall totals. Dry periods frequently occur. Several months during the years of record have had no rainfall, or only a trace. Snow falls on an average of about one day every two years.

There is little change in the day-to-day weather of the summer months, except for an occasional rainshower or a tropical storm in the area. High temperatures range in the high 80s to mid 90s, except for brief periods in the high 90s. The sea breeze during the afternoon and evening hours moderates the summer heat. Low temperatures are usually in the mid 70s. Mornings are generally warm. Summertime temperatures rarely reach 100 degrees near the bay, but occasionally do in most other parts of the city. Temperatures above 100 degrees are frequent about 30 to 60 miles to the west and southwest. Summertime afternoons are more pleasant than mornings because they are usually clear and windy. In the summer season the region receives nearly 80 percent of the possible sunshine.

The fall months of September and October are essentially an extension of the summer months. November is a transition to the conditions of the coming winter months, with greater temperature extremes, stronger winds, and the first occurrences of northers. The winter months are relatively mild, but with temperatures sufficiently low to be stimulating. Temperatures below 32 degrees seldom occur near the bay, but are more frequent inland. January is the coldest month with a prevailing northerly wind. The most extreme cold weather, in which daytime highs do not exceed 32 degrees, does not occur more than once every three or four years. The earliest occurrence of a temperature below 32 degrees is in early November and the latest occurrence in the spring is mid to late March.

Relative humidity, because of the nearness of the Gulf of Mexico, is high throughout the year. However, during the afternoons the humidity usually drops to between 50 and 60 percent.

Severe tropical storms average about one every ten years. Lesser strength storms average about one every five years. The city of Corpus Christi has a feature not found in most other coastal cities. A bluff rises 30 to 40 feet above the level of the lowlands area near the bay. This serves as a natural protection from high water. Protection for the main city is now furnished by sea walls.

Chief hurricane months are August and September, although tropical storms have occurred as early as June and as late as October. The majority of the storms pass either to the south or east of the city. Tornadoes are an infrequent occurrence in the area, and hail occurs only about once a year.

NORMALS, MEANS, AND EXTREMES

CORPUS CHRISTI, TEXAS

LATITUDE: 27°46'N LONGITUDE: 97°30' W ELEVATION: FT. GRND 41 BARO 56 TIME ZONE: CENTRAL WBAN: 12924

	(a)	JAN	FEB	MAR	APR	MAY	JUNE	JULY	AUG	SEP	OCT	NOV	DEC	YEAR
TEMPERATURE °F:														
Normals														
-Daily Maximum		65.0	69.0	75.7	81.7	86.2	90.4	93.3	93.4	89.7	83.9	75.8	68.3	81.0
-Daily Minimum		45.3	48.0	55.3	63.2	69.5	73.4	74.8	75.0	72.3	63.9	55.6	48.4	62.1
-Monthly		55.1	58.5	65.6	72.5	77.9	81.9	84.1	84.2	81.0	73.9	65.7	58.3	71.6
Extremes														
-Record Highest	55	91	98	102	102	103	101	104	103	103	98	98	91	104
-Year		1971	1940	1989	1984	1984	1980	1939	1962	1977	1950	1988	1977	JUL 1939
-Record Lowest	55	14	18	24	33	47	58	64	64	50	28	28	13	13
-Year		1962	1951	1980	1987	1970	1975	1967	1967	1942	1993	1993	1989	DEC 1989
NORMAL DEGREE DAYS:														
Heating (base 65°F)		338	213	100	9	0	0	0	0	0	0	95	261	1016
Cooling (base 65°F)		31	31	119	234	400	507	592	595	480	281	116	53	3439
% OF POSSIBLE SUNSHINE	51	44	49	55	56	59	72	80	77	68	68	54	44	61
MEAN SKY COVER (tenths)														
Sunrise - Sunset	51	6.9	6.4	6.6	6.7	6.4	5.2	4.8	4.8	5.2	4.7	5.8	6.6	5.9
MEAN NUMBER OF DAYS:														
Sunrise to Sunset														
-Clear	51	6.8	7.3	7.3	5.6	5.5	8.7	11.1	11.3	9.8	12.5	8.9	7.6	102.6
-Partly Cloudy	51	6.7	6.0	7.5	9.2	12.2	14.6	14.2	13.1	12.0	10.4	8.6	6.4	120.9
-Cloudy	51	17.5	14.9	16.3	15.2	13.3	6.6	5.7	6.6	8.2	8.1	12.4	17.0	141.8
Precipitation														
.01 inches or more	54	8.3	6.9	5.4	5.0	6.7	6.0	4.8	5.8	9.1	6.5	5.9	6.8	77.2
Snow,Ice pellets,hail														
1.0 inches or more	54	0.*	0.*	0.0	0.0	0.0	0.0	0.0	0.0	0.0	0.0	0.0	0.0	0.1
Thunderstorms	54	0.8	1.2	1.5	2.4	4.7	3.3	2.8	3.8	4.7	2.3	1.1	0.7	29.1
Heavy Fog Visibility														
1/4 mile or less	51	6.0	4.9	4.0	2.5	1.1	0.2	0.2	0.2	0.9	1.2	3.4	5.3	29.2
Temperature °F														
-Maximum														
90° and above	29	0.1	0.2	0.9	1.8	4.7	18.9	27.4	27.1	17.4	5.2	0.3	0.1	104.3
32° and below	29	0.0	0.1	0.0	0.0	0.0	0.0	0.0	0.0	0.0	0.0	0.0	0.1	0.2
-Minimum														
32° and below	29	2.9	1.3	0.3	0.0	0.0	0.0	0.0	0.0	0.0	0.*	0.3	1.6	6.4
0° and below	29	0.0	0.0	0.0	0.0	0.0	0.0	0.0	0.0	0.0	0.0	0.0	0.0	0.0
AVG. STATION PRESS.(mb)	21	1018.6	1017.1	1013.1	1012.0	1010.4	1011.5	1013.5	1013.1	1012.8	1015.1	1016.3	1018.1	1014.3
RELATIVE HUMIDITY (%)														
Hour 00	29	85	84	84	87	90	89	88	87	86	85	84	83	86
Hour 06 (Local Time)	29	87	88	87	90	92	93	93	92	90	89	87	86	90
Hour 12	29	69	65	61	63	66	63	57	58	61	59	61	64	62
Hour 18	29	71	67	65	69	71	68	63	64	68	68	71	71	68
PRECIPITATION (inches):														
Water Equivalent														
-Normal		1.71	1.96	0.94	1.72	3.33	3.38	2.39	3.31	5.52	3.02	1.59	1.26	30.13
-Maximum Monthly	55	10.78	8.11	4.80	8.04	9.38	13.35	11.92	14.79	20.33	11.02	8.53	9.80	20.33
-Year		1958	1982	1974	1956	1968	1973	1976	1980	1967	1981	1947	1991	SEP 1967
-Minimum Monthly	55	0.03	T	T	T	T	0.03	0.00	0.10	0.49	0.00	T	0.01	0.00
-Year		1971	1976	1971	1984	1961	1980	1957	1952	1981	1952	1949	1950	JUL 1957
-Maximum in 24 hrs	51	6.38	4.85	2.99	7.19	4.65	5.65	4.61	8.92	8.76	7.25	3.44	6.81	8.92
-Year		1958	1982	1990	1956	1968	1978	1981	1980	1967	1960	1947	1991	AUG 1980
Snow,Ice pellets,hail														
-Maximum Monthly	55	1.2	1.1	T	T	T	0.0	0.0	0.0	0.0	T	T	T	1.2
-Year		1940	1973	1990	1993	1993					1993	1979	1989	JAN 1940
-Maximum in 24 hrs	55	1.1	1.1	T	T	T	0.0	0.0	0.0	0.0	T	T	T	1.1
-Year		1940	1973	1990	1993	1993					1993	1979	1989	FEB 1973
WIND:														
Mean Speed (mph)	51	12.0	12.9	14.0	14.2	12.8	11.7	11.6	11.0	10.4	10.4	11.7	11.6	12.0
Prevailing Direction through 1963		SSE	SSE	SSE	SE	SE	SE	SSE	SSE	SE	SE	SSE	SSE	SSE
Fastest Obs. 1 Min.														
-Direction (!!)	17	30	15	14	30	32	36	02	11	13	36	17	18	11
-Speed (MPH)	17	38	41	38	45	44	37	46	55	40	35	39	38	55
-Year		1992	1981	1983	1988	1985	1988	1980	1980	1993	1988	1983	1982	AUG 1980
Peak Gust														
-Direction (!!)	10	NW	NW	NW	NW	NW	N	S	E	E	E	NW	SE	E
-Speed (mph)	10	52	60	54	56	60	61	49	48	61	53	60	47	61
-Date		1992	1987	1991	1988	1985	1988	1986	1986	1988	1984	1987	1987	SEP 1988

PRECIPITATION (inches)

CORPUS CHRISTI, TEXAS

YEAR	JAN	FEB	MAR	APR	MAY	JUNE	JULY	AUG	SEP	OCT	NOV	DEC	ANNUAL
1964	1.61	1.53	1.14	0.08	4.39	0.38	2.25	0.50	6.98	0.19	0.21	2.45	21.71
1965	0.86	4.41	0.78	0.80	4.01	1.99	1.25	2.64	2.09	1.36	1.96	3.14	25.29
1966	2.12	1.15	0.69	5.03	7.23	4.35	1.23	4.15	2.84	0.85	0.07	0.18	29.89
1967	2.63	2.38	0.08	0.23	1.83	0.35	1.05	5.36	20.33	2.86	0.28	0.84	38.22
1968	2.11	2.42	0.90	0.82	9.38	8.36	5.43	0.62	6.34	3.68	1.34	0.13	41.53
1969	0.35	2.92	0.49	2.89	2.07	0.13	0.03	2.83	2.05	2.85	5.09	1.87	23.57
1970	1.79	1.01	1.55	0.15	3.92	9.16	1.72	7.32	8.51	3.13	0.81	0.40	39.47
1971	0.03	0.22	T	2.29	4.55	1.24	0.31	8.32	12.17	3.96	0.44	3.42	36.95
1972	1.23	3.41	1.44	1.53	5.99	3.65	2.82	3.74	9.49	0.46	2.48	0.17	36.41
1973	2.18	1.42	0.16	1.73	0.58	13.35	0.52	5.63	7.58	9.95	0.31	0.12	43.53
1974	1.98	T	4.80	0.08	4.24	2.43	0.19	1.05	3.88	3.57	1.76	0.83	24.81
1975	1.94	0.42	0.05	0.08	1.67	1.31	4.05	4.84	6.70	2.02	0.90	1.21	25.19
1976	0.15	T	0.15	3.68	5.95	0.76	11.92	0.86	2.54	6.81	4.27	2.30	39.39
1977	3.11	1.72	0.96	6.00	1.96	3.56	1.15	0.39	0.87	4.73	1.74	0.06	26.25
1978	2.01	0.84	0.03	2.20	1.68	12.04	3.92	0.81	10.83	2.46	0.50	1.82	39.14
1979	3.93	0.83	1.55	3.69	4.28	3.23	3.52	2.53	13.77	0.41	0.28	1.02	39.04
1980	1.24	1.01	0.31	0.34	2.82	0.03	1.47	14.79	6.01	1.18	3.16	0.33	32.69
1981	2.55	1.91	2.37	0.98	8.64	3.02	5.98	5.79	0.49	11.02	0.12	1.15	44.02
1982	0.07	8.11	0.46	1.01	4.17	0.72	0.01	0.64	0.55	1.70	4.33	0.70	22.47
1983	0.75	3.27	3.03	T	2.77	2.50	8.78	2.67	7.04	3.99	1.53	0.58	36.91
1984	5.91	0.39	0.19	T	2.22	0.23	0.25	0.90	3.03	6.49	1.71	0.92	22.24
1985	2.68	2.86	1.82	3.54	2.87	3.99	1.04	2.88	8.39	3.40	1.62	1.61	36.70
1986	1.70	1.07	0.14	0.66	5.13	3.10	0.25	4.94	1.86	5.02	3.74	4.54	32.15
1987	2.22	6.01	0.42	1.13	4.15	4.92	3.17	3.49	0.99	1.44	1.79	0.93	30.66
1988	0.85	1.13	0.91	0.52	0.94	1.64	1.79	1.52	6.27	2.60	0.13	0.98	19.28
1989	1.96	0.95	0.21	3.59	0.10	3.17	1.02	2.36	2.05	0.11	1.83	1.50	18.85
1990	0.41	3.96	2.97	3.40	1.26	0.89	1.74	0.69	2.66	1.35	1.34	0.43	21.10
1991	1.69	2.06	1.59	4.00	6.25	6.97	0.55	4.63	8.60	1.57	0.36	9.80	48.07
1992	5.11	4.48	4.09	3.73	7.41	1.53	0.92	3.44	4.26	0.70	4.77	0.98	41.42
1993	0.44	1.69	2.83	3.30	8.07	12.02	T	0.68	1.12	1.90	0.81	4.82	37.68
Record Mean	1.57	1.68	1.38	1.91	3.25	2.92	1.93	2.50	4.69	2.65	1.82	1.73	28.03

AVERAGE TEMPERATURE (deg. F)

CORPUS CHRISTI, TEXAS

YEAR	JAN	FEB	MAR	APR	MAY	JUNE	JULY	AUG	SEP	OCT	NOV	DEC	ANNUAL
#1964	56.9	53.7	64.1	74.6	78.8	81.3	84.2	85.6	80.9	70.3	67.8	57.6	71.3
1965	59.1	56.4	59.8	73.6	78.4	82.2	84.3	83.1	82.6	70.1	71.2	62.6	72.0
1966	50.5	54.4	64.5	71.9	75.6	79.5	83.6	84.7	80.7	72.6	67.0	58.2	70.3
1967	56.1	57.4	69.2	77.3	78.5	82.8	84.6	81.4	77.2	71.0	65.7	56.1	71.5
1968	53.2	52.4	60.1	71.7	76.9	79.9	82.0	83.6	79.1	74.5	62.6	58.9	69.6
1969	59.1	60.2	57.8	70.8	75.5	82.1	86.6	85.3	81.3	74.2	62.9	61.2	71.4
1970	51.2	59.0	60.9	73.3	75.4	80.4	82.6	84.2	80.9	72.6	63.5	66.9	70.9
1971	62.3	62.0	68.7	70.9	78.4	82.9	84.9	82.7	81.1	76.6	67.1	64.9	73.5
1972	61.6	61.1	70.2	76.7	77.0	82.0	83.9	83.8	83.6	76.4	60.7	56.2	72.8
1973	51.7	56.0	68.8	69.3	77.1	81.1	84.9	82.8	81.8	76.2	72.9	59.7	71.9
1974	57.9	61.5	70.7	73.6	80.9	81.0	83.8	85.3	78.5	74.0	64.2	57.5	72.4
1975	59.7	60.1	67.2	75.0	80.5	83.4	84.5	83.3	79.0	74.4	66.5	58.5	72.7
1976	56.7	64.4	68.8	73.5	74.4	81.6	81.2	83.0	91.4	67.3	57.3	54.8	70.4
1977	50.3	58.6	66.0	71.5	78.8	82.9	84.6	86.7	85.5	75.7	67.5	61.9	72.5
1978	49.2	51.7	63.9	72.3	81.0	83.5	85.6	84.9	81.9	73.2	68.8	59.1	71.3
1979	51.3	57.0	67.9	74.3	76.6	81.7	85.5	84.9	78.8	75.4	63.3	58.8	71.3
1980	60.0	58.1	67.6	69.4	77.8	83.8	85.8	83.2	82.4	71.3	59.9	57.7	71.4
1981	55.3	58.4	63.7	74.4	77.0	82.5	83.4	83.7	60.7	75.2	68.3	59.4	71.9
1982	57.4	55.9	66.8	71.1	76.7	82.7	84.6	84.9	82.3	73.9	64.5	58.4	71.6
1983	55.1	57.9	63.7	69.0	76.0	81.2	83.1	84.2	79.4	74.1	67.7	49.3	70.1
1984	51.4	59.2	66.4	73.2	77.1	81.6	83.3	83.7	78.3	76.5	64.9	66.3	71.8
1985	49.2	54.1	67.6	72.3	77.6	80.2	82.0	84.1	80.7	75.2	69.0	55.5	70.6
1986	57.4	62.2	66.3	74.7	77.1	81.8	84.1	83.4	83.2	73.8	64.8	55.8	72.1
1987	55.2	60.0	62.3	68.2	78.2	81.1	83.2	84.2	81.4	73.2	63.9	60.3	70.9
1988	53.1	58.3	63.5	71.3	76.6	81.2	85.2	85.5	81.0	74.7	69.5	61.0	71.7
1989	61.7	57.4	64.8	71.7	81.9	83.4	84.1	84.5	80.2	74.4	68.0	48.0	71.7
1990	61.8	64.1	66.8	72.8	79.3	84.4	83.5	85.6	81.8	73.1	68.5	58.1	73.3
1991	54.6	61.4	69.6	75.8	79.1	83.2	84.0	85.1	79.4	75.5	61.6	61.9	72.6
1992	55.4	62.6	67.6	71.1	75.0	82.0	83.8	82.9	81.7	75.2	63.3	61.9	71.9
1993	56.8	61.6	65.3	69.8	75.1	81.6	83.5	84.6	82.6	73.3	62.0	60.5	71.4
Record Mean	56.3	59.1	65.1	71.5	76.9	81.2	83.3	83.5	80.6	73.9	65.0	58.4	71.3
Max	64.5	67.5	73.1	78.6	83.4	88.2	90.6	90.9	87.8	82.0	73.3	66.8	78.9
Min	48.0	50.7	57.1	64.4	70.4	74.2	76.0	76.1	73.4	65.8	56.6	50.0	63.6

REFERENCE NOTES

CORPUS CHRISTI, TEXAS

GENERAL
T - TRACE AMOUNT.
BLANK ENTRIES DENOTE MISSING/UNREPORTED DATA.
# INDICATES A STATION OR INSTRUMENT RELOCATION.

SPECIFIC

(a) - LENGTH OF RECORD IN YEARS, ALTHOUGH INDIVIDUAL MONTHS MAY BE MISSING.
0.* OR * - THE VALUE IS BETWEEN 0.0 AND 0.05
NORMALS - BASED ON THE 1961-1990 RECORD PERIOD.
EXTREMES - DATES ARE THE MOST RECENT OCCURRENCE
WIND DIR.- NUMERALS SHOW TENS OF DEGRESS CLOCKWISE FROM TRUE NORTH. "00" INDICATES CALM.
RESULTANT DIRECTIONS ARE GIVEN TO WHOLE DEGREES.
BOLD VALUES INDICATE EXTREME VALUES WHICH OCCURRED AFTER THE ASOS SYSTEM WAS COMMISSIONED.

RECORD = PERIOD OF RECORD
RECORD MEAN PRECIPITATION IS THE MEAN OF ALL DAILY PRECIPITATION AMOUNTS DURING THE PERIOD OF RECORD.
RECORD MAX(MIN) TEMPERATURE IS THE MEAN OF ALL DAILY MAX(MIN) TEMPERATURES DURING THE PERIOD OF RECORD.
RECORD MEAN TEMPERATURE IS THE SUM OF THE RECORD MAX AND RECORD MIN DIVIDED BY 2.
AVERAGE TEMPERATURE IS THE SUM OF THE MEAN DAILY MAX AND MIN TEMPERATURE DIVIDED BY 2.

EXCEPTIONS

RECORD MEANS ARE THROUGH THE CURRENT YEAR, BEGINNING IN
1887 FOR TEMPERATURE
1887 FOR PRECIPITATION
1940 FOR SNOWFALL

HEATING DEGREE DAYS Base 65 deg. F CORPUS CHRISTI, TEXAS

SEASON	JULY	AUG	SEP	OCT	NOV	DEC	JAN	FEB	MAR	APR	MAY	JUNE	TOTAL
1964-65	0	0	0	18	70	289	219	252	222	0	0	0	1070
1965-66	0	0	0	22	14	125	450	300	79	12	1	0	1003
1966-67	0	0	0	5	68	267	304	222	42	0	0	0	908
1967-68	0	0	0	17	63	299	373	362	202	9	0	0	1325
1968-69	0	0	0	3	144	219	237	146	226	0	0	0	975
1969-70	0	0	0	10	160	152	442	168	166	29	2	0	1129
1970-71	0	0	2	12	125	93	174	151	64	26	0	0	647
1971-72	0	0	0	0	51	112	188	166	22	2	0	0	541
1972-73	0	0	0	8	188	295	415	262	23	44	0	0	1235
1973-74	0	0	0	0	20	204	264	151	56	13	0	0	708
1974-75	0	0	0	4	117	260	227	167	77	10	0	0	862
1975-76	0	0	0	4	107	247	275	94	45	1	0	0	773
1976-77	0	0	0	59	253	311	455	192	65	6	0	0	1341
1977-78	0	0	0	4	56	160	502	382	101	17	0	0	1222
1978-79	0	0	0	1	57	236	445	256	38	1	0	0	1034
1979-80	0	0	0	6	131	233	195	247	74	30	0	0	916
1980-81	0	0	0	43	204	254	294	221	90	0	0	0	1106
1981-82	0	0	0	24	34	207	292	275	92	40	0	0	964
1982-83	0	0	0	10	134	245	308	195	88	40	0	0	1020
1983-84	0	0	0	6	66	511	415	199	69	7	0	0	1273
1984-85	0	0	3	5	106	93	485	320	48	6	0	0	1066
1985-86	0	0	2	3	50	304	235	144	44	0	0	0	782
1986-87	0	0	0	11	139	294	298	156	122	61	0	0	1081
1987-88	0	0	0	0	126	190	374	225	126	13	0	0	1054
1988-89	0	0	0	0	59	181	167	280	141	25	0	0	853
1989-90	0	0	0	17	74	521	160	84	59	8	3	0	926
1990-91	0	0	0	15	42	274	321	123	19	4	0	0	798
1991-92	0	0	0	9	170	172	293	102	43	24	0	0	813
1992-93	0	0	0	0	139	146	264	132	83	11	0	0	775
1993-94	0	0	0	41	157	205							

COOLING DEGREE DAYS Base 65 deg. F CORPUS CHRISTI, TEXAS

YEAR	JAN	FEB	MAR	APR	MAY	JUNE	JULY	AUG	SEP	OCT	NOV	DEC	TOTAL
1969	62	16	11	180	332	521	677	636	496	300	103	40	3374
1970	20	8	47	287	329	466	552	599	485	255	89	160	3297
1971	98	74	187	207	423	542	625	556	489	367	121	116	3805
1972	93	61	187	360	380	520	594	588	561	368	68	28	3808
1973	13	14	149	179	381	489	629	560	511	353	265	48	3591
1974	52	60	243	276	498	486	587	637	410	289	97	32	3667
1975	70	35	154	319	486	557	608	573	422	300	161	52	3737
1976	25	84	170	265	299	505	510	563	498	136	31	3	3089
1977	4	18	104	206	435	545	616	677	621	342	136	71	3775
1978	18	16	74	243	502	561	647	622	511	264	177	56	3691
1979	26	39	140	289	366	507	639	624	421	333	89	47	3520
1980	45	52	159	166	405	573	649	570	530	247	56	34	3486
1981	4	41	57	288	378	535	580	589	476	349	141	41	3479
1982	63	24	158	229	370	540	613	627	525	289	128	48	3614
1983	10	6	55	166	349	494	568	605	438	295	152	31	3169
1984	0	39	119	257	382	502	573	588	411	367	108	141	3487
1985	5	19	136	230	399	461	532	595	480	327	179	19	3382
1986	6	72	89	298	382	510	596	575	552	290	140	16	3520
1987	4	21	47	165	414	488	572	603	499	260	103	54	3230
1988	11	37	86	207	366	491	631	643	487	309	203	64	3535
1989	77	71	143	235	533	558	597	611	462	313	172	2	3774
1990	67	68	123	249	452	589	580	650	512	273	154	69	3786
1991	6	28	168	335	445	551	597	629	438	341	77	85	3700
1992	1	42	128	214	316	517	590	564	506	323	95	58	3354
1993	17	43	98	163	318	500	578	611	535	308	72	73	3316

SNOWFALL (inches) CORPUS CHRISTI, TEXAS

SEASON	JULY	AUG	SEP	OCT	NOV	DEC	JAN	FEB	MAR	APR	MAY	JUNE	TOTAL
1970-71	0.0	0.0	0.0	0.0	0.0	0.0	0.0	0.0	0.0	0.0	0.0	0.0	0.0
1971-72	0.0	0.0	0.0	0.0	0.0	0.0	0.0	0.0	0.0	0.0	0.0	0.0	0.0
1972-73	0.0	0.0	0.0	0.0	0.0	0.0	0.2	1.1	0.0	0.0	0.0	0.0	1.3
1973-74	0.0	0.0	0.0	0.0	0.0	0.0	0.0	0.0	0.0	0.0	0.0	0.0	0.0
1974-75	0.0	0.0	0.0	0.0	0.0	0.0	0.0	0.0	0.0	0.0	0.0	0.0	0.0
1975-76	0.0	0.0	0.0	0.0	0.0	0.0	0.0	0.0	0.0	0.0	0.0	0.0	0.0
1976-77	0.0	0.0	0.0	0.0	T	0.0	0.0	0.0	0.0	0.0	0.0	0.0	T
1977-78	0.0	0.0	0.0	0.0	0.0	0.0	T	0.0	0.0	0.0	0.0	0.0	T
1978-79	0.0	0.0	0.0	0.0	0.0	0.0	0.0	0.0	0.0	0.0	0.0	0.0	0.0
1979-80	0.0	0.0	0.0	0.0	T	0.0	0.0	0.0	0.0	0.0	0.0	0.0	T
1980-81	0.0	0.0	0.0	0.0	0.0	0.0	T	0.0	0.0	0.0	0.0	0.0	T
1981-82	0.0	0.0	0.0	0.0	0.0	0.0	0.0	0.0	0.0	0.0	0.0	0.0	0.0
1982-83	0.0	0.0	0.0	0.0	0.0	0.0	0.0	0.0	0.0	0.0	0.0	0.0	0.0
1983-84	0.0	0.0	0.0	0.0	0.0	T	0.0	0.0	0.0	0.0	0.0	0.0	T
1984-85	0.0	0.0	0.0	0.0	0.0	0.0	T	T	0.0	0.0	0.0	0.0	T
1985-86	0.0	0.0	0.0	0.0	0.0	0.0	0.0	0.0	0.0	0.0	0.0	0.0	0.0
1986-87	0.0	0.0	0.0	0.0	0.0	0.0	0.0	0.0	0.0	0.0	0.0	0.0	0.0
1987-88	0.0	0.0	0.0	0.0	0.0	0.0	0.0	T	0.0	0.0	0.0	0.0	T
1988-89	0.0	0.0	0.0	0.0	0.0	0.0	0.0	0.0	T	0.0	0.0	0.0	T
1989-90	0.0	0.0	0.0	0.0	0.0	T	0.0	0.0	T	0.0	0.0	0.0	T
1990-91	0.0	0.0	0.0	0.0	0.0	0.0	0.0	0.0	0.0	0.0	0.0	0.0	0.0
1991-92	0.0	0.0	0.0	0.0	0.0	0.0	T	0.0	0.0	0.0	0.0	0.0	T
1992-93	0.0	0.0	0.0	0.0	0.0	0.0	0.0	0.0	0.0	T	T	0.0	T
1993-94	0.0	0.0	0.0	T	0.0	0.0							
Record Mean	0.0	0.0	0.0	T	T	T	0.1	T	T	T	T	0.0	0.1

DALLAS – FORT WORTH, TEXAS

The Dallas–Fort Worth Metroplex is located in North Central Texas, approximately 250 miles north of the Gulf of Mexico. It is near the headwaters of the Trinity River, which lie in the upper margins of the Coastal Plain. The rolling hills in the area range from 500 to 800 feet in elevation.

The Dallas–Fort Worth climate is humid subtropical with hot summers. It is also continental, characterized by a wide annual temperature range. Precipitation also varies considerably, ranging from less than 20 to more than 50 inches.

Winters are mild, but northers occur about three times each month, and often are accompanied by sudden drops in temperature. Periods of extreme cold that occasionally occur are short–lived, so that even in January mild weather occurs frequently.

The highest temperatures of summer are associated with fair skies, westerly winds and low humidities. Characteristically, hot spells in summer are broken into three–to–five day periods by thunderstorm activity. There are only a few nights each summer when the low temperature exceeds 80 degrees. Summer daytime temperatures frequently exceed 100 degrees. Air conditioners are recommended for maximum comfort indoors and while traveling via automobile.

Throughout the year, rainfall occurs more frequently during the night. Usually, periods of rainy weather last for only a day or two, and are followed by several days with fair skies. A large part of the annual precipitation results from thunderstorm activity, with occasional heavy rainfall over brief periods of time. Thunderstorms occur throughout the year, but are most frequent in the spring. Hail falls on about two or three days a year, ordinarily with only slight and scattered damage. Windstorms occurring during thunderstorm activity are sometimes destructive. Snowfall is rare.

The average length of the warm season (freeze–free period) in the Dallas–Fort Worth Metroplex is about 249 days. The average last occurrence of 32 degrees or below is mid March and the average first occurrence of 32 degrees or below is in late November.

NORMALS, MEANS, AND EXTREMES

DALLAS - FORT WORTH, TEXAS

LATITUDE: 32°54'N LONGITUDE: 97°02' W ELEVATION: FT. GRND 551 BARO 575 TIME ZONE: CENTRAL WBAN: 03927

	(a)	JAN	FEB	MAR	APR	MAY	JUNE	JULY	AUG	SEP	OCT	NOV	DEC	YEAR
TEMPERATURE °F:														
Normals														
-Daily Maximum		54.1	58.9	67.8	76.3	82.9	91.9	96.5	96.2	87.8	78.5	66.8	57.5	76.3
-Daily Minimum		32.7	36.9	45.6	54.7	62.6	70.0	74.1	73.6	66.9	55.8	45.4	36.3	54.6
-Monthly		43.4	47.9	56.7	65.5	72.8	81.0	85.3	84.9	77.4	67.2	56.2	46.9	65.4
Extremes														
-Record Highest	40	88	88	96	95	103	113	110	108	106	102	89	88	113
-Year		1969	1986	1991	1990	1985	1980	1980	1964	1985	1979	1989	1955	JUN 1980
-Record Lowest	40	4	7	15	29	41	51	59	56	43	29	20	-1	-1
-Year		1964	1985	1980	1989	1978	1964	1972	1967	1984	1993	1959	1989	DEC 1989
NORMAL DEGREE DAYS:														
Heating (base 65°F)		670	484	286	75	0	0	0	0	0	51	275	566	2407
Cooling (base 65°F)		0	5	29	90	246	480	629	617	372	119	11	5	2603
% OF POSSIBLE SUNSHINE	15	53	54	60	62	59	67	76	73	69	62	57	52	62
MEAN SKY COVER (tenths)														
Sunrise - Sunset	40	6.2	5.8	5.8	5.9	5.9	5.0	4.3	4.3	4.7	4.7	5.3	5.7	5.3
MEAN NUMBER OF DAYS:														
Sunrise to Sunset														
-Clear	40	9.7	9.6	9.9	8.8	8.1	10.7	14.9	14.7	12.6	13.8	11.9	11.0	135.5
-Partly Cloudy	40	5.8	5.6	7.6	8.2	10.4	11.5	9.6	10.1	8.9	7.3	5.9	6.1	96.9
-Cloudy	40	15.5	13.1	13.5	13.1	12.6	7.8	6.5	6.2	8.5	9.9	12.3	13.9	132.8
Precipitation														
.01 inches or more	40	7.3	6.5	7.2	7.9	8.8	6.5	4.7	4.8	6.8	6.1	5.8	6.5	78.9
Snow, Ice pellets, hail														
1.0 inches or more	40	0.5	0.3	0.1	0.0	0.0	0.0	0.0	0.0	0.0	0.0	0.*	0.1	1.1
Thunderstorms	40	1.1	1.8	4.3	5.9	7.5	6.3	4.8	4.5	3.6	3.0	1.9	1.1	45.9
Heavy Fog Visibility 1/4 mile or less	40	2.5	1.5	1.0	0.6	0.3	0.1	0.0	0.*	0.1	0.9	1.5	2.4	10.9
Temperature °F														
-Maximum														
90° and above	30	0.0	0.0	0.2	0.8	4.0	19.7	27.7	26.3	14.5	2.7	0.0	0.0	95.9
32° and below	30	1.5	0.7	0.1	0.0	0.0	0.0	0.0	0.0	0.0	0.0	0.0	0.8	3.1
-Minimum														
32° and below	30	14.5	8.8	2.6	0.2	0.0	0.0	0.0	0.0	0.0	0.1	2.9	9.7	38.8
0° and below	30	0.0	0.0	0.0	0.0	0.0	0.0	0.0	0.0	0.0	0.0	0.0	0.*	*
AVG. STATION PRESS. (mb)	21	999.6	997.8	993.9	993.1	991.8	992.6	994.2	994.1	994.8	996.5	997.2	998.9	995.4
RELATIVE HUMIDITY (%)														
Hour 00	30	73	72	70	73	79	74	67	67	74	73	74	73	72
Hour 06 (Local Time)	30	80	79	80	82	87	86	81	80	85	83	81	79	82
Hour 12	30	60	58	56	56	60	55	49	49	55	54	57	59	56
Hour 18	30	58	54	51	52	57	51	44	45	53	55	58	59	53
PRECIPITATION (inches):														
Water Equivalent														
-Normal		1.83	2.18	2.77	3.50	4.88	2.98	2.31	2.21	3.39	3.52	2.29	1.84	33.70
-Maximum Monthly	40	4.54	6.20	6.39	12.19	13.66	8.75	11.13	6.85	9.52	14.18	6.23	8.75	14.18
-Year		1990	1965	1968	1957	1982	1989	1973	1970	1964	1981	1964	1991	OCT 1981
-Minimum Monthly	40	T	0.15	0.10	0.11	0.99	0.40	0.00	T	0.09	T	0.20	0.17	0.00
-Year		1986	1963	1972	1987	1977	1964	1993	1980	1984	1975	1970	1981	JUL 1993
-Maximum in 24 hrs	40	3.11	4.06	4.39	4.55	5.34	3.15	3.76	4.05	4.76	5.91	2.83	4.22	5.91
-Year		1990	1965	1977	1957	1989	1989	1975	1976	1965	1959	1964	1991	OCT 1959
Snow, Ice pellets, hail														
-Maximum Monthly	40	12.1	13.5	2.5	T	T	0.0	0.0	0.0	0.0	T	5.0	2.6	13.5
-Year		1964	1978	1962	1990	1990					1993	1976	1963	FEB 1978
-Maximum in 24 hrs	40	12.1	7.5	2.5	T	T	0.0	0.0	0.0	0.0	T	4.8	2.5	12.1
-Year		1964	1978	1962	1990	1990					1993	1976	1963	JAN 1964
WIND:														
Mean Speed (mph)	40	10.9	11.7	12.7	12.4	11.0	10.5	9.7	8.9	9.2	9.7	10.7	10.8	10.7
Prevailing Direction through 1963		S	S	S	S	S	S	S	S	S	S	S	S	S
Fastest Obs. 1 Min.														
-Direction (!!)	40	36	36	29	32	14	32	36	36	11	30	34	32	36
-Speed (MPH)	40	55	51	55	55	55	52	65	73	53	44	50	53	73
-Year		1985	1962	1954	1970	1955	1955	1961	1959	1961	1992	1957	1968	AUG 1959
Peak Gust														
-Direction (!!)	10	N	S	W	SE	N	NW	NE	NW	S	W	W	S	NW
-Speed (mph)	10	66	54	53	58	54	58	54	81	51	63	68	55	81
-Date		1985	1985	1992	1992	1993	1989	1989	1985	1986	1992	1992	1988	AUG 1985

PRECIPITATION (inches)

DALLAS – FORT WORTH, TEXAS

YEAR	JAN	FEB	MAR	APR	MAY	JUNE	JULY	AUG	SEP	OCT	NOV	DEC	ANNUAL
1964	3.53	1.17	3.35	2.71	2.85	0.40	0.25	2.43	9.52	0.62	6.23	1.25	34.31
1965	2.77	6.20	1.45	2.15	8.97	1.50	0.09	2.26	5.04	1.97	2.43	1.73	36.56
1966	1.68	2.84	1.38	10.74	3.13	5.47	3.26	3.38	4.23	1.48	0.53	1.17	39.29
1967	0.28	0.32	2.09	3.84	4.02	0.72	2.20	0.48	5.94	4.19	0.92	2.30	27.30
1968	3.60	1.48	6.39	2.41	6.02	3.50	1.88	2.71	2.53	2.18	4.58	1.20	38.48
1969	1.26	1.99	3.62	3.40	7.12	0.63	0.77	2.56	4.55	5.82	1.22	2.75	35.69
1970	0.72	4.78	3.49	4.68	3.62	0.61	0.94	6.85	6.25	2.95	0.20	1.01	36.10
1971	0.19	1.32	0.34	2.76	1.88	0.83	3.60	5.70	3.24	7.64	1.77	6.99	36.26
1972	1.09	0.26	0.10	3.25	2.35	1.50	0.59	0.81	2.42	6.89	2.36	0.61	22.23
1973	3.26	1.92	2.28	6.06	3.18	5.88	11.13	0.01	7.16	6.85	2.06	0.83	50.62
1974	1.79	1.01	0.80	2.51	6.00	5.44	0.67	4.19	6.04	5.93	3.32	1.93	39.63
1975	3.34	3.72	1.67	3.40	6.88	1.95	5.06	0.30	0.87	T	0.42	1.49	29.10
1976	0.13	0.52	2.29	5.71	6.03	1.40	3.83	4.75	5.02	3.46	0.50	1.99	35.63
1977	2.39	1.68	5.88	4.31	0.99	0.69	2.20	2.33	1.72	2.96	1.79	0.25	27.19
1978	1.41	3.33	2.66	1.34	8.01	0.77	0.33	1.53	0.93	0.55	2.73	0.78	24.37
1979	3.35	1.52	6.33	2.03	5.90	1.36	1.94	2.47	0.99	3.38	0.43	2.72	32.42
1980	2.52	0.84	1.24	2.23	3.01	1.25	0.71	T	6.54	1.08	1.23	1.43	22.08
1981	0.58	1.44	3.39	2.69	6.24	7.85	1.81	2.32	2.40	14.18	1.53	0.17	44.60
1982	2.33	1.89	1.71	2.71	13.66	4.28	2.73	0.52	0.58	3.36	4.22	2.76	40.75
1983	2.55	1.25	4.36	0.59	5.83	2.07	1.56	5.55	0.22	4.04	2.22	0.83	31.07
1984	1.07	3.11	4.92	1.41	3.04	2.79	0.43	1.47	0.09	6.50	2.97	6.09	33.89
1985	0.81	2.62	3.70	3.75	2.13	3.78	2.40	0.53	3.35	3.91	3.11	0.61	30.70
1986	T	2.49	1.08	5.30	5.52	3.92	0.41	1.63	4.60	1.81	3.25	2.44	32.45
1987	1.22	3.67	1.70	0.11	5.95	3.45	1.77	0.81	1.38	0.12	4.17	2.90	27.25
1988	0.88	1.23	2.03	2.21	2.11	3.23	2.47	0.44	4.04	1.64	2.28	2.48	25.04
1989	2.56	3.70	3.72	1.86	9.62	8.75	2.61	1.89	2.40	2.02	0.49	0.33	39.95
1990	4.54	4.72	5.89	6.90	7.16	1.89	2.60	2.37	1.12	2.81	3.81	1.46	45.27
1991	2.72	2.60	1.35	3.63	6.97	4.26	3.99	4.30	4.61	9.32	1.04	8.75	53.54
1992	3.25	2.40	3.24	2.46	6.93	5.23	2.48	2.08	3.25	3.05	3.56	4.26	42.19
1993	1.74	5.78	3.03	3.49	1.75	3.75	0.00	0.75	3.28	5.10	1.62	2.54	32.83
Record Mean	1.83	2.10	2.43	3.75	4.79	3.04	2.19	2.18	2.87	3.08	2.27	2.05	32.57

AVERAGE TEMPERATURE (deg. F)

DALLAS – FORT WORTH, TEXAS

YEAR	JAN	FEB	MAR	APR	MAY	JUNE	JULY	AUG	SEP	OCT	NOV	DEC	ANNUAL
1964	43.8	43.8	55.6	66.8	73.2	81.0	87.1	85.3	76.9	63.6	57.6	47.1	65.2
1965	47.1	45.8	47.0	68.4	72.9	79.9	86.3	84.1	79.4	66.6	62.9	52.8	66.1
1966	40.3	45.4	56.3	63.8	70.8	79.6	86.3	82.7	75.8	65.0	60.7	45.3	64.4
1967	48.3	46.8	63.3	71.1	71.4	81.4	82.9	83.1	74.1	65.4	55.6	47.0	65.9
1968	44.4	44.2	54.6	63.4	72.4	79.5	81.0	83.5	74.6	67.8	53.9	47.4	63.9
1969	49.0	50.0	49.8	65.4	71.9	79.8	87.9	84.2	77.1	65.3	55.1	49.9	65.5
1970	40.6	48.6	52.1	66.2	71.7	79.1	84.0	85.8	78.2	65.1	54.7	53.6	65.0
1971	46.7	49.2	55.6	64.0	70.5	82.9	84.4	79.5	77.1	70.1	57.0	52.2	65.8
1972	45.0	51.5	62.1	70.1	72.7	81.4	83.1	84.7	80.8	67.5	50.1	44.0	66.1
1973	42.5	47.9	60.0	60.7	71.7	79.3	83.9	82.9	76.1	68.3	59.8	48.4	65.1
1974	43.6	52.3	62.9	65.8	75.7	78.7	86.1	82.9	70.9	69.2	55.5	47.2	65.9
1975	49.0	46.6	53.8	64.7	72.4	80.9	83.6	84.8	75.6	69.8	57.3	49.0	65.6
1976	45.0	58.4	59.4	64.9	68.6	78.8	82.1	84.2	76.1	60.2	49.5	45.0	64.3
1977	34.7	49.4	57.2	66.8	77.4	84.1	87.1	84.9	81.6	66.7	56.4	47.6	66.2
1978	33.8	36.7	54.1	67.1	73.1	82.3	88.4	84.6	80.2	68.9	57.7	46.1	64.4
1979	35.4	42.2	56.7	64.4	69.7	81.0	84.5	82.5	77.0	70.8	52.9	49.4	63.9
1980	45.5	46.6	54.2	63.1	75.0	87.0	92.0	88.5	80.3	65.4	54.9	49.4	66.8
1981	44.6	48.9	55.7	69.2	70.5	80.3	85.9	83.4	76.2	66.1	57.5	47.3	65.4
1982	44.6	44.5	59.8	62.5	72.5	79.2	84.6	86.7	78.1	67.0	55.6	49.2	65.4
1983	43.4	48.5	54.5	60.6	69.5	77.3	83.6	84.9	77.1	67.8	57.3	34.8	63.3
1984	39.3	50.9	56.3	63.7	73.7	82.5	85.5	85.8	76.1	67.0	54.6	52.6	65.7
1985	37.8	45.0	60.8	67.2	74.0	80.2	84.4	87.6	77.7	67.6	56.3	42.3	65.1
1986	48.8	51.2	60.2	67.2	71.5	80.8	86.4	83.4	80.2	65.7	52.4	46.1	66.2
1987	44.5	50.8	53.9	65.0	75.1	79.6	83.4	86.5	77.1	66.5	55.7	47.3	65.5
1988	42.2	47.1	56.0	64.5	72.8	80.4	85.3	87.9	79.2	65.7	58.1	49.1	65.7
1989	50.0	42.2	56.7	66.4	74.3	77.9	82.8	82.3	74.7	69.0	58.2	39.0	64.5
1990	51.8	53.9	57.7	64.0	73.4	84.0	82.5	84.6	80.0	66.4	59.8	44.0	66.8
1991	42.8	53.7	59.8	67.4	75.4	81.0	85.0	82.5	75.2	68.1	51.7	50.3	66.1
1992	46.9	54.4	59.1	66.0	71.1	79.4	84.3	80.2	77.7	69.5	52.7	49.9	65.9
1993	45.1	49.0	56.1	63.3	71.9	81.7	87.3	87.5	78.2	63.8	51.6	49.5	65.4
Record Mean	45.1	48.8	56.9	65.2	72.7	80.9	84.6	84.7	77.8	67.6	56.0	47.5	65.6
Max	55.4	59.4	68.0	75.9	82.7	91.1	95.0	95.3	88.2	78.5	66.5	57.6	76.1
Min	34.8	38.1	45.8	54.5	62.7	70.7	74.2	74.0	67.3	56.6	45.4	37.3	55.1

REFERENCE NOTES

DALLAS – FORT WORTH, TEXAS

GENERAL
T - TRACE AMOUNT.
BLANK ENTRIES DENOTE MISSING/UNREPORTED DATA.
# INDICATES A STATION OR INSTRUMENT RELOCATION.

SPECIFIC

PM - INCLUDES LAST DAY OF PREVIOUS MONTH
ASOS - AUTOMATED SURFACE OBSERVING SYSTEM IN OPERATION DURING THESE MONTHS.

(•) - LENGTH OF RECORD IN YEARS, ALTHOUGH INDIVIDUAL MONTHS MAY BE MISSING.
0.* OR * - THE VALUE IS BETWEEN 0.0 AND 0.05
NORMALS - BASED ON THE 1961-1990 RECORD PERIOD.
EXTREMES - DATES ARE THE MOST RECENT OCCURRENCE
WIND DIR.- NUMERALS SHOW TENS OF DEGRESS CLOCKWISE FROM TRUE NORTH. "00" INDICATES CALM.
RESULTANT DIRECTIONS ARE GIVEN TO WHOLE DEGREES.
BOLD VALUES INDICATE EXTREME VALUES WHICH OCCURRED AFTER THE ASOS SYSTEM WAS COMMISSIONED.

RECORD = PERIOD OF RECORD
RECORD MEAN PRECIPITATION IS THE MEAN OF ALL DAILY PRECIPITATION AMOUNTS DURING THE PERIOD OF RECORD.
RECORD MAX(MIN) TEMPERATURE IS THE MEAN OF ALL DAILY MAX(MIN) TEMPERATURES DURING THE PERIOD OF RECORD.
RECORD MEAN TEMPERATURE IS THE SUM OF THE RECORD MAX AND RECORD MIN DIVIDED BY 2.
AVERAGE TEMPERATURE IS THE SUM OF THE MEAN DAILY MAX AND MIN TEMPERATURE DIVIDED BY 2.

EXCEPTIONS

RECORD MEANS ARE THROUGH THE CURRENT YEAR, BEGINNING IN 1899 FOR TEMPERATURE
1899 FOR PRECIPITATION
1954 FOR SNOWFALL

HEATING DEGREE DAYS Base 65 deg. F DALLAS - FORT WORTH, TEXAS

SEASON	JULY	AUG	SEP	OCT	NOV	DEC	JAN	FEB	MAR	APR	MAY	JUNE	TOTAL
1964-65	0	0	6	81	260	550	550	530	551	36	0	0	2564
1965-66	0	0	2	60	103	376	760	542	274	84	26	0	2227
1966-67	0	0	0	79	182	627	514	503	146	15	21	0	2087
1967-68	0	0	13	80	282	548	631	598	330	100	2	0	2584
1968-69	0	0	0	47	348	540	492	416	468	49	6	4	2370
1969-70	0	0	0	116	306	463	756	455	404	63	21	1	2585
1970-71	0	0	7	105	316	369	564	440	307	97	19	0	2224
1971-72	0	0	12	7	270	389	615	398	143	26	1	0	1861
1972-73	0	0	3	96	446	644	690	475	155	182	12	0	2703
1973-74	0	0	1	36	182	509	656	352	173	70	1	0	1980
1974-75	0	0	20	16	296	546	489	508	355	112	0	0	2342
1975-76	0	0	4	33	266	500	616	217	222	48	20	0	1926
1976-77	0	0	0	214	459	614	931	431	241	37	0	0	2927
1977-78	0	0	0	55	257	536	962	786	346	54	41	0	3037
1978-79	0	0	0	27	247	578	911	635	261	78	29	0	2766
1979-80	0	0	0	34	370	478	597	530	339	102	6	0	2456
1980-81	0	0	18	99	330	486	625	448	284	26	23	0	2339
1981-82	0	0	10	116	228	541	625	569	232	140	9	0	2470
1982-83	0	0	1	94	316	495	663	454	324	186	21	2	2556
1983-84	0	0	12	52	269	933	789	401	281	89	11	0	2837
1984-85	0	0	38	66	322	389	837	558	171	37	0	0	2418
1985-86	0	0	19	53	285	696	495	400	164	41	5	0	2158
1986-87	0	0	0	61	376	580	632	387	342	109	0	0	2487
1987-88	0	0	0	55	297	540	703	512	301	70	0	0	2478
1988-89	0	0	0	51	240	487	460	630	294	102	4	0	2268
1989-90	0	0	14	80	251	799	401	306	251	102	19	0	2223
1990-91	0	0	0	100	190	646	681	314	198	37	7	0	2173
1991-92	0	0	12	69	405	456	555	302	182	73	14	0	2068
1992-93	0	0	0	14	366	474	612	445	290	120	4	0	2325
1993-94	0	0	2	145	414	476							

COOLING DEGREE DAYS Base 65 deg. F DALLAS - FORT WORTH, TEXAS

YEAR	JAN	FEB	MAR	APR	MAY	JUNE	JULY	AUG	SEP	OCT	NOV	DEC	TOTAL
1969	3	3	3	67	228	453	715	602	372	133	17	0	2596
1970	5	0	7	108	236	433	595	653	409	115	12	22	2595
1971	0	6	21	71	195	546	606	456	382	171	36	1	2491
1972	2	14	57	185	249	498	569	618	480	183	4	0	2859
1973	1	0	6	60	230	435	593	559	342	146	33	0	2405
1974	0	2	115	101	341	419	660	563	202	153	20	2	2578
1975	0	0	15	107	236	483	580	620	331	189	39	9	2609
1976	0	32	59	52	138	421	537	602	338	72	0	0	2251
1977	0	0	7	94	391	581	693	626	505	112	6	2	3017
1978	0	0	16	125	301	524	733	614	462	153	37	0	2965
1979	0	1	9	67	179	489	613	551	366	220	14	0	2509
1980	0	0	11	52	320	668	844	737	485	117	35	10	3279
1981	0	5	5	158	200	467	654	577	352	155	8	0	2581
1982	1	2	77	71	252	433	614	679	403	160	40	10	2742
1983	0	0	7	61	171	382	582	626	381	145	46	0	2401
1984	0	0	20	60	288	531	644	652	376	135	16	12	2734
1985	0	5	51	108	287	460	608	706	408	139	29	0	2801
1986	0	19	24	112	212	480	673	578	464	91	3	0	2656
1987	0	0	6	114	318	442	576	674	370	111	23	0	2634
1988	4	0	28	61	247	467	639	714	433	78	39	1	2711
1989	1	0	45	154	297	393	561	542	314	208	52	0	2567
1990	1	2	30	79	286	575	551	617	457	152	41	2	2793
1991	0	3	42	115	335	484	624	550	324	174	14	5	2670
1992	0	2	4	109	213	437	606	480	386	161	4	13	2415
1993	0	0	19	75	223	507	697	701	406	115	20	4	2767

SNOWFALL (inches) DALLAS - FORT WORTH, TEXAS

SEASON	JULY	AUG	SEP	OCT	NOV	DEC	JAN	FEB	MAR	APR	MAY	JUNE	TOTAL
1964-65	0.0	0.0	0.0	0.0	0.0	T	0.0	T	T	0.0	0.0	0.0	T
1965-66	0.0	0.0	0.0	0.0	0.0	0.0	4.4	2.9	0.0	0.0	0.0	0.0	7.3
1966-67	0.0	0.0	0.0	0.0	0.0	0.0	0.0	T	T	0.0	0.0	0.0	T
1967-68	0.0	0.0	0.0	0.0	0.0	0.0	0.4	2.6	T	0.0	0.0	0.0	3.0
1968-69	0.0	0.0	0.0	0.0	T	0.0	0.0	T	0.0	0.0	0.0	0.0	T
1969-70	0.0	0.0	0.0	0.0	0.0	T	T	0.0	0.8	0.0	0.0	0.0	0.8
1970-71	0.0	0.0	0.0	0.0	0.0	0.0	0.0	T	1.6	0.0	0.0	0.0	1.6
1971-72	0.0	0.0	0.0	0.0	T	0.0	T	T	0.0	0.0	0.0	0.0	T
1972-73	0.0	0.0	0.0	0.0	T	1.4	2.3	T	0.0	0.0	0.0	0.0	3.7
1973-74	0.0	0.0	0.0	0.0	0.0	0.0	T	0.0	0.0	0.0	0.0	0.0	T
1974-75	0.0	0.0	0.0	0.0	T	T	T	3.7	T	0.0	0.0	0.0	3.7
1975-76	0.0	0.0	0.0	0.0	0.0	0.4	0.0	0.0	T	0.0	0.0	0.0	0.4
1976-77	0.0	0.0	0.0	0.0	5.0	0.0	5.4	0.0	0.0	0.0	0.0	0.0	10.4
1977-78	0.0	0.0	0.0	0.0	0.0	0.0	4.1	13.5	T	0.0	0.0	0.0	17.6
1978-79	0.0	0.0	0.0	0.0	0.0	0.8	1.8	0.7	0.0	0.0	0.0	0.0	3.3
1979-80	0.0	0.0	0.0	0.0	0.0	T	0.0	1.6	0.0	0.0	0.0	0.0	1.6
1980-81	0.0	0.0	0.0	0.0	0.0	0.0	T	T	0.0	0.0	0.0	0.0	T
1981-82	0.0	0.0	0.0	0.0	0.0	0.0	0.8	T	0.0	0.0	0.0	0.0	0.8
1982-83	0.0	0.0	0.0	0.0	0.0	T	T	T	0.0	0.0	0.0	0.0	T
1983-84	0.0	0.0	0.0	0.0	0.0	2.0	0.0	0.0	0.0	0.0	0.0	0.0	2.0
1984-85	0.0	0.0	0.0	0.0	0.0	0.0	3.4	1.7	0.0	0.0	0.0	0.0	5.1
1985-86	0.0	0.0	0.0	0.0	0.0	T	0.0	0.8	0.0	0.0	0.0	0.0	0.8
1986-87	0.0	0.0	0.0	0.0	0.0	1.7	T	T	0.5	0.0	0.0	0.0	2.2
1987-88	0.0	0.0	0.0	0.0	0.0	T	0.8	2.7	0.0	0.0	0.0	0.0	3.5
1988-89	0.0	0.0	0.0	0.0	0.0	T	T	0.7	1.1	0.0	T	0.0	1.8
1989-90	0.0	0.0	0.0	0.0	0.0	T	0.0	0.0	0.0	T	T	0.0	T
1990-91	0.0	0.0	0.0	0.0	0.0	0.3	0.3	0.0	0.0	0.0	0.0	0.0	0.6
1991-92	0.0	0.0	0.0	0.0	0.0	0.0	T	0.0	0.0	0.0	0.0	0.0	T
1992-93	0.0	0.0	0.0	0.0	0.0	T	0.0	0.0	0.0	0.0	0.0	0.0	T
1993-94	0.0	0.0	0.0	T	0.3	T							
Record Mean	0.0	0.0	0.0	T	0.1	0.2	1.2	1.0	0.2	T	T	0.0	2.8

DEL RIO, TEXAS

Del Rio is located on the Rio Grande River, on the western tip of the Balcones escarpment, in southwest Texas. Elevation is near 1,000 feet and varies little within the city but rises to 2,300 feet in the northern part of the county. Regional agriculture is chiefly wool and mohair production to the north and west of Del Rio and garden crops to the southeast. Lake Amistad, a reservoir of 65,000 surface acres, lies 10 miles west of Del Rio.

The climate of Del Rio is semi-arid continental. Annual precipitation is insufficient for dry farming. However, San Felipe Springs and the Rio Grande provide adequate water for irrigation farming. Over 80 percent of the average annual precipitation occurs from April through October. During this period, rainfall is chiefly in the form of showers and thunderstorms, often as heavy downpours, resulting in flash flooding. The small amount of precipitation for November through March usually falls as steady light rain.

Hail occurs in the vicinity of Del Rio about once per year and reaches severe proportions about once every five years. Sleet or snow falls on an average of once a year, but frequently melts as it falls. A snowfall heavy enough to blanket the ground only occurs about once every four or five years, and seldom remains more than 24 hours.

Temperature averages indicate mild winters and quite warm summers. Cold periods in winter are ushered in by strong, dry, dusty north, and northwest winds known as northers, and temperature drops of as much as 25 degrees in a few hours are not uncommon. Cold weather periods usually do not last more than two or three days. Temperatures as low as 32 degrees have occurred as early as October and as late as March. Normal occurrences of the earliest freezing temperature in autumn and the latest in spring are early December and mid February, which results in an average growing season of 300 days. Hot weather is rather persistent from late May to mid September and temperatures above 100 degrees have been recorded as early as March and as late as October. Low humidity and fresh breezes tend to alleviate uncomfortable conditions usually associated with high temperatures. The mean early morning humidity is about 79 percent, and the mean afternoon humidity is near 44 percent.

Clear to partly cloudy skies predominate, and even in the more cloudy winter months, the mean number of cloudy days are less than the number of clear days.

NORMALS, MEANS, AND EXTREMES

DEL RIO, TEXAS

LATITUDE: 29°22'N LONGITUDE: 100°50' W ELEVATION: FT. GRND 1026 BARO 1030 TIME ZONE: CENTRAL WBAN: 22010

	(a)	JAN	FEB	MAR	APR	MAY	JUNE	JULY	AUG	SEP	OCT	NOV	DEC	YEAR
TEMPERATURE °F:														
Normals														
-Daily Maximum		61.9	67.1	75.7	83.4	88.2	93.6	96.2	96.0	90.4	81.7	71.7	63.5	80.8
-Daily Minimum		38.5	42.9	50.9	59.2	66.3	71.7	74.1	73.6	69.1	59.7	49.0	40.6	58.0
-Monthly		50.2	55.1	63.3	71.3	77.3	82.7	85.2	84.8	79.8	70.7	60.4	52.1	69.4
Extremes														
-Record Highest	31	89	97	101	106	106	112	108	109	105	106	96	90	112
-Year		1972	1986	1971	1984	1979	1988	1980	1969	1983	1979	1988	1977	JUN 1988
-Record Lowest	31	15	14	21	33	45	55	64	64	48	28	22	10	10
-Year		1982	1985	1980	1987	1970	1970	1976	1966	1970	1993	1976	1989	DEC 1989
NORMAL DEGREE DAYS:														
Heating (base 65°F)		465	291	139	14	0	0	0	0	0	19	171	407	1506
Cooling (base 65°F)		6	14	87	203	381	531	626	614	444	196	33	7	3142
% OF POSSIBLE SUNSHINE														
MEAN SKY COVER (tenths)														
Sunrise - Sunset	16	5.8	5.5	5.5	6.1	6.4	5.5	4.9	5.0	5.4	4.9	5.2	5.6	5.5
MEAN NUMBER OF DAYS:														
Sunrise to Sunset														
-Clear	16	10.1	10.3	10.6	8.4	6.5	7.9	11.8	11.0	9.2	12.4	11.8	11.3	121.2
-Partly Cloudy	16	6.6	6.7	7.6	7.4	8.9	13.3	11.3	12.1	11.0	8.4	6.9	5.6	105.8
-Cloudy	16	14.3	11.3	12.8	14.2	15.6	8.8	7.9	7.9	9.8	10.1	11.4	14.2	138.3
Precipitation														
.01 inches or more	30	5.2	5.1	4.9	5.5	7.4	5.4	4.2	4.1	6.5	4.9	4.5	5.4	63.2
Snow,Ice pellets,hail														
1.0 inches or more	30	0.2	0.1	0.*	0.0	0.0	0.0	0.0	0.0	0.0	0.*	0.0	0.0	0.3
Thunderstorms	18	0.3	0.5	2.0	4.5	7.9	4.5	3.8	4.0	3.9	2.0	0.9	0.3	34.6
Heavy Fog Visibility														
1/4 mile or less	18	3.1	1.5	0.9	0.3	0.2	0.0	0.0	0.0	0.2	0.9	3.0	4.1	14.1
Temperature °F														
-Maximum														
90° and above	30	0.0	0.3	1.8	6.2	13.2	24.6	28.0	28.3	18.8	4.2	0.2	0.*	125.7
32° and below	30	0.1	0.*	0.0	0.0	0.0	0.0	0.0	0.0	0.0	0.0	0.0	0.1	0.3
-Minimum														
32° and below	30	6.5	3.2	0.8	0.0	0.0	0.0	0.0	0.0	0.0	0.1	1.3	4.7	16.5
0° and below	30	0.0	0.0	0.0	0.0	0.0	0.0	0.0	0.0	0.0	0.0	0.0	0.0	0.0
AVG. STATION PRESS.(mb)	7	982.9	981.7	976.7	976.3	974.6	976.5	977.7	977.9	977.8	980.1	981.5	982.2	978.8
RELATIVE HUMIDITY (%)														
Hour 00	16	66	62	57	62	69	65	59	62	70	71	71	68	65
Hour 06 (Local Time)	16	76	74	72	77	83	82	79	80	84	82	80	76	79
Hour 12	16	55	52	48	52	57	55	52	54	57	57	57	54	54
Hour 18	16	46	40	36	40	46	42	39	41	49	50	52	49	44
PRECIPITATION (inches):														
Water Equivalent														
-Normal		0.56	0.95	0.69	1.98	2.03	2.11	1.85	1.47	2.83	2.24	0.92	0.61	18.24
-Maximum Monthly	31	1.92	3.80	3.20	7.51	5.15	5.74	13.18	6.10	15.79	11.33	3.36	3.06	15.79
-Year		1992	1992	1990	1981	1980	1987	1976	1971	1964	1969	1969	1991	SEP 1964
-Minimum Monthly	31	T	0.00	T	0.12	0.20	T	0.04	T	0.36	0.00	T	T	0.00
-Year		1971	1974	1971	1988	1973	1990	1970	1985	1981	1979	1988	1973	OCT 1979
-Maximum in 24 hrs	27	1.33	2.57	1.73	4.57	1.86	2.67	6.34	3.81	5.52	7.60	2.74	1.37	7.60
-Year		1986	1987	1979	1969	1964	1993	1975	1969	1970	1969	1978	1974	OCT 1969
Snow,Ice pellets,hail														
-Maximum Monthly	30	9.8	2.7	3.0	T	T	0.0	0.0	0.0	0.0	1.2	T	T	9.8
-Year		1985	1973	1965	1990	1991					1993	1976	1989	JAN 1985
-Maximum in 24 hrs	27	8.6	2.7	3.0	T	T	0.0	0.0	0.0	0.0	1.2	T	T	8.6
-Year		1985	1973	1965	1990	1991					1993	1976	1989	JAN 1985
WIND:														
Mean Speed (mph)	16	8.8	9.5	10.9	11.0	10.7	11.4	10.9	10.2	9.2	9.1	8.5	8.4	9.9
Prevailing Direction														
Fastest Obs. 1 Min.														
-Direction (!!!)	16	30	30	33	31	32	36	32	14	31	33	35	29	14
-Speed (MPH)	16	37	37	52	37	48	38	38	60	35	46	35	43	60
-Year		1973	1974	1964	1969	1970	1968	1969	1970	1976	1967	1972	1969	AUG 1970
Peak Gust														
-Direction (!!!)														
-Speed (mph)														
-Date														

PRECIPITATION (inches)

DEL RIO, TEXAS

YEAR	JAN	FEB	MAR	APR	MAY	JUNE	JULY	AUG	SEP	OCT	NOV	DEC	ANNUAL
1964	0.42	0.76	0.45	1.20	2.28	0.03	0.34	1.16	15.79	2.47	T	0.25	25.15
1965	0.16	1.72	0.33	3.13	2.34	2.03	0.38	1.64	2.61	0.43	1.26	1.09	17.12
1966	0.94	1.11	0.48	4.36	2.23	3.37	0.05	2.10	1.45	1.26	T	0.04	17.39
1967	0.02	0.29	0.58	1.11	0.65	0.36	0.12	1.00	7.02	1.39	0.66	0.76	13.96
1968	0.66	1.53	1.15	2.53	1.03	1.72	3.07	0.25	1.20	1.84	2.19	T	17.17
1969	1.04	1.17	0.60	5.46	2.36	0.20	1.47	4.27	0.78	11.33	3.36	1.18	33.22
1970	0.42	1.61	0.57	0.14	2.85	1.38	0.04	1.43	9.87	0.06	T	0.12	18.49
1971	T	0.29	T	2.16	0.59	4.87	0.45	6.10	0.50	2.36	0.83	0.25	18.40
1972	0.63	0.08	0.52	1.58	2.61	1.82	2.60	5.74	0.92	0.38	0.71	0.01	17.60
1973	0.76	1.36	1.19	2.08	0.20	3.22	2.43	1.08	2.86	4.78	T	T	19.96
1974	0.05	0.00	1.39	1.39	1.42	T	0.09	3.37	4.98	2.81	0.62	1.47	17.59
1975	0.51	1.92	0.06	1.06	2.17	1.12	8.26	0.27	1.24	1.36	0.10	0.53	18.60
1976	0.15	0.01	0.14	2.45	2.82	0.46	13.18	0.43	1.10	3.20	0.89	1.51	26.34
1977	0.89	1.23	0.87	2.66	2.27	1.23	0.06	0.10	1.41	4.96	0.67	0.01	16.36
1978	0.07	0.36	0.15	2.09	3.46	2.02	2.15	1.14	2.76	0.95	3.35	0.76	19.26
1979	0.29	1.57	2.66	1.53	0.62	4.33	0.25	0.72	0.68	0.00	0.60	0.78	14.03
1980	0.22	0.44	0.30	0.44	5.15	0.32	0.28	2.06	1.86	0.06	2.00	1.02	14.15
1981	0.67	0.21	1.64	7.51	1.94	5.54	0.56	2.18	0.36	6.64	0.10	0.06	27.41
1982	0.12	2.45	0.13	0.47	1.59	2.37	0.86	0.12	1.10	0.17	1.77	0.79	11.94
1983	0.70	1.01	1.07	0.13	1.19	1.07	0.32	0.79	1.35	5.09	1.42	0.06	14.20
1984	1.44	0.24	0.06	0.55	2.09	0.45	0.90	0.51	2.78	3.02	1.03	2.44	15.51
1985	1.25	0.51	0.81	2.51	2.20	3.91	1.71	T	2.81	0.63	1.05	0.03	17.42
1986	1.69	0.17	0.02	0.78	3.56	4.47	1.20	0.81	3.70	5.44	0.42	2.36	24.62
1987	0.16	3.47	0.77	3.21	3.89	5.74	1.97	2.05	0.58	0.24	0.45	0.59	23.12
1988	0.05	0.18	0.15	0.12	1.45	1.78	4.38	2.17	4.48	0.27	T	0.38	15.41
1989	1.06	1.04	0.92	0.18	1.94	0.26	0.26	0.23	1.24	0.49	1.32	0.30	9.24
1990	0.59	1.25	3.20	4.82	2.13	T	4.13	0.43	6.19	1.51	1.16	0.22	25.63
1991	0.57	0.75	0.04	1.88	2.37	2.73	1.23	0.75	7.07	0.17	1.42	3.06	22.04
1992	1.92	3.80	2.57	1.27	4.13	1.82	4.28	1.15	0.41	0.80	0.69	0.42	23.26
1993	0.60	0.77	0.47	0.72	1.29	5.28	0.97	2.54	0.77	0.68	0.13	0.18	14.40
Record Mean	0.64	0.91	0.80	1.66	2.36	2.20	1.82	1.47	2.61	2.22	0.93	0.72	18.34

AVERAGE TEMPERATURE (deg. F)

DEL RIO, TEXAS

YEAR	JAN	FEB	MAR	APR	MAY	JUNE	JULY	AUG	SEP	OCT	NOV	DEC	ANNUAL
1964	49.2	50.7	63.8	74.4	79.2	84.9	88.2	87.0	79.8	67.5	62.4	51.3	69.9
1965	54.5	50.7	57.4	73.6	75.9	81.9	85.8	84.5	81.3	69.1	65.4	56.4	69.7
1966	46.8	51.0	62.3	71.0	75.3	80.2	86.9	83.2	79.7	68.5	63.7	50.9	68.3
1967	51.5	54.2	68.8	76.1	79.1	85.2	86.8	83.3	76.2	68.2	60.8	49.5	70.0
1968	49.8	50.4	58.0	69.4	77.4	82.3	83.5	85.5	77.4	73.1	58.5	51.3	68.0
1969	54.5	56.6	57.3	71.1	75.5	84.4	88.4	86.0	79.9	68.5	57.2	54.0	69.5
1970	47.1	54.9	57.5	70.1	73.8	81.2	85.4	85.3	78.7	66.9	57.3	57.7	68.0
1971	56.1	56.8	64.9	70.2	78.3	80.1	82.0	78.6	78.5	71.1	61.7	56.5	69.6
1972	53.6	58.1	68.8	75.3	74.9	82.2	83.0	80.8	80.5	71.9	55.9	50.9	69.7
1973	47.7	51.2	65.9	68.1	77.9	80.1	82.5	83.7	80.2	71.4	64.2	53.0	68.8
1974	53.1	56.9	68.8	72.2	80.6	83.7	85.9	83.6	73.9	70.1	58.0	50.9	69.8
1975	53.1	55.2	63.8	70.9	76.4	82.6	81.2	83.4	76.3	70.8	61.4	53.5	69.1
1976	50.1	63.0	66.5	70.3	72.9	82.5	78.0	81.5	78.8	63.0	53.2	49.2	67.4
1977	46.4	55.4	62.8	68.4	75.6	82.9	85.8	88.3	85.8	72.4	61.7	55.5	70.1
1978	46.5	49.9	63.4	73.9	80.0	83.8	87.7	84.6	78.5	69.5	61.5	51.1	69.2
1979	44.2	52.2	62.9	70.2	75.8	79.6	87.4	85.0	80.8	75.4	57.8	53.9	68.8
1980	54.1	56.1	63.6	70.8	78.2	87.2	90.1	85.0	83.6	70.7	57.6	53.5	70.9
1981	52.2	55.3	61.6	71.4	76.3	80.5	84.3	84.9	80.4	72.1	63.0	54.5	69.7
1982	50.9	52.2	64.9	71.0	76.0	84.3	86.2	87.3	81.2	72.2	59.6	51.8	69.8
1983	51.1	55.1	62.5	69.0	78.0	81.9	86.3	86.2	81.1	72.9	62.4	44.6	69.3
1984	47.4	57.0	65.7	72.8	79.7	84.3	85.8	86.7	78.8	71.8	59.0	57.9	70.6
1985	45.4	51.8	64.2	71.1	79.1	81.1	83.6	87.3	80.7	71.6	64.6	49.7	69.2
1986	52.8	60.5	66.8	75.6	77.4	81.1	85.7	85.3	81.1	69.1	58.9	51.2	70.5
1987	51.4	55.8	57.5	65.4	74.9	79.1	82.7	85.2	79.1	71.6	59.3	53.0	67.9
1988	47.7	54.8	63.5	71.5	76.8	82.9	84.5	84.4	79.8	71.6	63.2	53.6	69.5
1989	55.3	52.4	62.7	71.5	82.0	84.8	87.3	85.1	78.6	70.8	60.4	44.3	69.6
1990	55.4	59.5	62.4	69.7	77.5	87.8	82.7	83.5	78.9	69.5	62.6	52.4	70.2
1991	51.5	58.3	66.7	74.1	80.7	83.5	84.6	86.7	77.2	72.8	57.4	55.5	70.8
1992	51.7	58.2	64.4	69.3	74.1	82.9	84.5	83.2	81.8	74.0	58.0	55.5	69.8
1993	52.2	57.4	65.0	71.7	77.4	82.7	86.1	87.0	79.4	71.3	57.8	57.0	70.4
Record Mean	51.4	56.0	63.4	70.9	77.3	83.0	85.1	85.1	79.9	71.0	59.7	52.4	69.6
Max	62.4	67.5	75.4	82.5	88.0	93.4	95.6	95.8	90.2	81.5	70.5	63.3	80.5
Min	40.3	44.4	51.3	59.3	66.5	72.5	74.6	74.4	69.7	60.4	48.9	41.5	58.7

REFERENCE NOTES

DEL RIO, TEXAS

GENERAL
T - TRACE AMOUNT.
BLANK ENTRIES DENOTE MISSING/UNREPORTED DATA.
# INDICATES A STATION OR INSTRUMENT RELOCATION.

SPECIFIC

(a) - LENGTH OF RECORD IN YEARS, ALTHOUGH INDIVIDUAL MONTHS MAY BE MISSING.
0.* OR * - THE VALUE IS BETWEEN 0.0 AND 0.05
NORMALS - BASED ON THE 1961-1990 RECORD PERIOD.
EXTREMES - DATES ARE THE MOST RECENT OCCURRENCE.
WIND DIR. - NUMERALS SHOW TENS OF DEGRESS CLOCKWISE FROM TRUE NORTH. "00" INDICATES CALM.
RESULTANT DIRECTIONS ARE GIVEN TO WHOLE DEGREES.
BOLD VALUES INDICATE EXTREME VALUES WHICH OCCURRED AFTER THE ASOS SYSTEM WAS COMMISSIONED.

RECORD = PERIOD OF RECORD
RECORD MEAN PRECIPITATION IS THE MEAN OF ALL DAILY PRECIPITATION AMOUNTS DURING THE PERIOD OF RECORD.
RECORD MAX(MIN) TEMPERATURE IS THE MEAN OF ALL DAILY MAX(MIN) TEMPERATURES DURING THE PERIOD OF RECORD.
RECORD MEAN TEMPERATURE IS THE SUM OF THE RECORD MAX AND RECORD MIN DIVIDED BY 2.
AVERAGE TEMPERATURE IS THE SUM OF THE MEAN DAILY MAX AND MIN TEMPERATURE DIVIDED BY 2.

EXCEPTIONS

MAXIMUM 24-HOUR PRECIPITATION AND SNOW, WINDS, MEAN SKY COVER, DAYS CLEAR-PARTLY CLOUDY-CLOUDY, THUNDERSTORMS, AND HEAVY FOG ARE THROUGH 1979.

RECORD MEANS ARE THROUGH THE CURRENT YEAR, BEGINNING IN 1906 FOR TEMPERATURE
1906 FOR PRECIPITATION
1964 FOR SNOWFALL

HEATING DEGREE DAYS Base 65 deg. F　　DEL RIO, TEXAS

SEASON	JULY	AUG	SEP	OCT	NOV	DEC	JAN	FEB	MAR	APR	MAY	JUNE	TOTAL
1964-65	0	0	0	21	152	417	326	397	268	4	0	0	1585
1965-66	0	0	3	16	41	263	554	388	128	8	9	0	1410
1966-67	0	0	0	26	93	434	417	299	54	0	0	0	1323
1967-68	0	0	1	36	159	475	466	419	246	28	0	0	1830
1968-69	0	0	0	0	225	419	324	240	266	1	0	0	1475
1969-70	0	0	0	62	273	330	551	277	242	40	7	0	1782
1970-71	0	0	24	69	233	224	278	239	107	35	0	0	1209
1971-72	0	0	9	6	135	263	350	228	34	2	0	0	1027
1972-73	0	0	0	21	283	429	531	381	31	72	0	0	1748
1973-74	0	0	0	1	83	362	363	231	53	11	0	0	1104
1974-75	0	0	6	9	223	431	375	274	114	22	0	0	1454
1975-76	0	0	0	16	180	361	456	125	76	3	5	0	1222
1976-77	0	0	0	127	348	481	568	269	122	13	0	0	1928
1977-78	0	0	0	6	126	294	564	422	111	3	0	0	1526
1978-79	0	0	1	8	147	426	639	363	125	8	0	0	1717
1979-80	0	0	0	10	233	346	336	279	119	26	0	0	1349
1980-81	0	0	0	51	268	353	388	280	119	9	0	0	1468
1981-82	0	0	0	33	92	319	436	359	125	42	0	0	1406
1982-83	0	0	0	20	221	409	423	269	103	66	0	0	1511
1983-84	0	0	0	2	141	625	539	240	82	9	0	0	1638
1984-85	0	0	10	22	214	222	603	367	96	28	0	0	1562
1985-86	0	0	8	9	85	466	370	173	41	4	0	0	1156
1986-87	0	0	0	19	205	419	414	252	235	98	0	0	1642
1987-88	0	0	0	2	220	362	533	307	142	12	0	0	1578
1988-89	0	0	0	0	131	346	299	362	162	39	0	0	1339
1989-90	0	0	2	40	171	634	300	169	133	30	8	0	1487
1990-91	0	0	0	34	128	388	412	195	58	2	0	0	1217
1991-92	0	0	13	24	247	294	405	193	70	36	0	0	1282
1992-93	0	0	0	0	230	291	394	218	102	5	0	0	1240
1993-94	0	0	0	69	221	258							

COOLING DEGREE DAYS Base 65 deg. F　　DEL RIO, TEXAS

YEAR	JAN	FEB	MAR	APR	MAY	JUNE	JULY	AUG	SEP	OCT	NOV	DEC	TOTAL
1969	6	13	32	193	330	590	735	660	455	175	46	0	3235
1970	3	0	15	200	286	496	642	637	443	135	11	0	2868
1971	10	19	114	196	421	461	534	429	423	201	43	8	2859
1972	3	37	161	317	315	521	566	496	473	241	18	1	3149
1973	0	0	68	170	410	460	550	586	463	207	69	0	2983
1974	2	10	178	237	489	565	654	584	278	177	23	2	3199
1975	13	1	83	205	363	534	508	582	345	202	78	13	2927
1976	0	74	130	168	255	534	410	521	419	73	3	0	2587
1977	0	7	63	121	334	543	649	733	630	239	34	6	3359
1978	0	7	71	274	473	571	710	614	415	159	49	1	3344
1979	0	9	66	172	340	444	701	625	483	337	24	9	3210
1980	6	25	81	206	417	671	785	629	567	233	53	5	3678
1981	0	15	24	210	356	472	603	624	470	258	40	0	3072
1982	7	9	129	228	348	587	662	698	493	253	68	5	3487
1983	0	0	33	193	409	511	669	664	489	253	71	0	3292
1984	0	12	111	249	463	585	651	682	431	241	40	12	3477
1985	0	6	78	218	443	491	580	698	481	222	81	0	3298
1986	0	51	103	329	389	493	650	637	491	153	30	0	3326
1987	0	0	8	120	314	431	556	634	428	216	55	0	2762
1988	0	19	104	217	374	543	610	609	449	210	84	2	3221
1989	7	15	98	243	534	599	698	631	418	228	40	0	3511
1990	13	20	58	177	401	694	557	581	423	177	61	4	3166
1991	0	11	115	283	494	560	614	677	386	273	27	7	3447
1992	0	1	58	173	286	545	609	574	510	287	26	3	3072
1993	0	12	110	212	391	536	660	690	442	271	13	14	3351

SNOWFALL (inches)　　DEL RIO, TEXAS

SEASON	JULY	AUG	SEP	OCT	NOV	DEC	JAN	FEB	MAR	APR	MAY	JUNE	TOTAL
1964-65	0.0	0.0	0.0	0.0	0.0	T	0.0	T	3.0	0.0	0.0	0.0	3.0
1965-66	0.0	0.0	0.0	0.0	0.0	0.0	T	2.2	0.0	0.0	0.0	0.0	2.2
1966-67	0.0	0.0	0.0	0.0	0.0	0.0	0.0	0.6	0.0	0.0	0.0	0.0	0.6
1967-68	0.0	0.0	0.0	T	0.0	T	0.0	0.0	T	0.0	0.0	0.0	T
1968-69	0.0	0.0	0.0	0.0	0.0	0.0	0.0	0.0	0.0	0.0	0.0	0.0	0.0
1969-70	0.0	0.0	0.0	0.0	0.0	T	0.0	T	0.0	0.0	0.0	0.0	T
1970-71	0.0	0.0	0.0	0.0	0.0	0.0	0.0	0.0	0.0	0.0	0.0	0.0	0.0
1971-72	0.0	0.0	0.0	0.0	0.0	0.0	T	0.0	0.0	0.0	0.0	0.0	T
1972-73	0.0	0.0	0.0	0.0	0.0	0.0	T	2.7	0.0	0.0	0.0	0.0	2.7
1973-74	0.0	0.0	0.0	0.0	0.0	0.0	0.0	0.0	0.0	0.0	0.0	0.0	0.0
1974-75	0.0	0.0	0.0	0.0	0.0	0.0	0.0	0.0	0.0	0.0	0.0	0.0	0.0
1975-76	0.0	0.0	0.0	0.0	0.0	0.0	T	0.0	0.0	0.0	0.0	0.0	T
1976-77	0.0	0.0	0.0	0.0	T	0.0	0.0	0.0	0.0	0.0	0.0	0.0	T
1977-78	0.0	0.0	0.0	0.0	0.0	0.0	0.0	0.0	0.0	0.0	0.0	0.0	0.0
1978-79	0.0	0.0	0.0	0.0	0.0	0.0	0.0	0.0	0.0	0.0	0.0	0.0	0.0
1979-80	0.0	0.0	0.0	0.0	0.0	0.0	0.0	T	0.0	0.0	0.0	0.0	T
1980-81	0.0	0.0	0.0	0.0	0.0	0.0	0.2	0.0	0.0	0.0	0.0	0.0	0.2
1981-82	0.0	0.0	0.0	0.0	0.0	0.0	1.0	T	0.0	0.0	0.0	0.0	1.0
1982-83	0.0	0.0	0.0	0.0	0.0	0.0	0.0	T		0.0	0.0	0.0	
1983-84	0.0	0.0	0.0	0.0	0.0	0.0	T	0.0	0.0	0.0	0.0	0.0	T
1984-85	0.0	0.0	0.0	0.0	0.0	0.0	9.8	T	0.0	0.0	0.0	0.0	9.8
1985-86	0.0	0.0	0.0	0.0	0.0	0.0	8.2	0.0	0.0	0.0	0.0	0.0	8.2
1986-87	0.0	0.0	0.0	0.0	0.0	0.0	0.0	0.0	0.0	0.0	0.0	0.0	0.0
1987-88	0.0	0.0	0.0	0.0	0.0	0.0	0.0	0.1	0.0	0.0	0.0	0.0	0.1
1988-89	0.0	0.0	0.0	0.0	0.0	0.0	0.0	T	0.0	0.0	0.0	0.0	T
1989-90	0.0	0.0	0.0	0.0	0.0	T	0.0	0.0	0.0	T	0.0	0.0	T
1990-91	0.0	0.0	0.0	0.0	0.0	0.0	0.0	0.0	0.0	0.0	T	0.0	T
1991-92	0.0	0.0	0.0	0.0	0.0	0.0	0.0	0.0	T	0.0	0.0	0.0	T
1992-93	0.0	0.0	0.0	0.0	0.0	0.0	0.0	0.0	0.0	0.0	0.0	0.0	0.0
1993-94	0.0	0.0	0.0	1.2	0.0	0.0							
Record Mean	0.0	0.0	0.0	T	T	T	0.7	0.2	0.1	T	T	0.0	1.0

EL PASO, TEXAS

The city of El Paso is located in the extreme west point of Texas at an elevation of about 3,700 feet . The National Weather Service station is located on a mesa about 200 feet higher than the city. The climate of the region is characterized by an abundance of sunshine throughout the year, high daytime summer temperatures, very low humidity, scanty rainfall, and a relatively mild winter season. The Franklin Mountains begin within the city limits and extend northward for about 16 miles. Peaks of these mountains range from 4,687 to 7,152 feet above sea level.

Rainfall throughout the year is light, insufficient for any growth except desert vegetation. Irrigation is necessary for crops, gardens, and lawns. Dry periods lasting several months are not unusual. Almost half of the precipitation occurs in the three-month period, July through September, from brief but often heavy thunderstorms. Small amounts of snow fall nearly every winter, but snow cover rarely amounts to more than an inch and seldom remains on the ground for more than a few hours.

Daytime summer temperatures are high, frequently above 90 degrees and occasionally above 100 degrees. Summer nights are usually comfortable, with temperatures in the 60s. It should be noted that when temperatures are high the relative humidity is generally quite low. A 20-year tabulation of observations with temperatures above 90 degrees shows that in April, May, and June the humidity averaged from 10 to 14 percent, while in July, August, and September it averaged 22 to 24 percent. This low humidity aids the efficiency of evaporative air coolers, which are widely used in homes and public buildings and are quite effective in cooling the air to comfortable temperatures.

Winter daytime temperatures are mild. At night they drop below freezing about half the time in December and January. The flat, irrigated land of the Rio Grande Valley in the vicinity of El Paso is noticeably cooler, particularly at night, than the airport or the city proper, both in summer and winter. This results in more comfortable temperatures in summer but increases the severity of freezes in winter. The cooler air in the Valley also causes marked short-period fluctuations of temperature and dewpoint at the airport with changes in wind direction, especially during the early morning hours.

Dust and sandstorms are the most unpleasant features of the weather in El Paso. While wind velocities are not excessively high, the soil surface is dry and loose and natural vegetation is sparse, so moderately strong winds raise considerable dust and sand. A tabulation of duststorms for a period of 20 years shows that they are most frequent in March and April, and comparatively rare in the period July through December. prevailing winds are from the north in winter and the south in summer.

NORMALS, MEANS, AND EXTREMES

EL PASO, TEXAS

LATITUDE: 31°48'N LONGITUDE: 106°24' W ELEVATION: FT. GRND 3918 BARO 3932 TIME ZONE: MOUNTAIN WBAN: 23044

	(a)	JAN	FEB	MAR	APR	MAY	JUNE	JULY	AUG	SEP	OCT	NOV	DEC	YEAR
TEMPERATURE °F:														
Normals														
-Daily Maximum		56.1	62.2	69.9	78.7	87.1	96.5	96.1	93.5	87.1	78.4	66.4	57.5	77.5
-Daily Minimum		29.4	33.9	40.2	48.0	56.5	64.3	68.4	66.6	61.6	49.6	38.4	30.7	49.0
-Monthly		42.8	48.1	55.1	63.4	71.8	80.4	82.3	80.1	74.4	64.0	52.4	44.1	63.2
Extremes														
-Record Highest	54	80	83	89	98	104	111	112	108	104	96	87	80	112
-Year		1970	1986	1989	1989	1951	1978	1979	1980	1982	1979	1983	1973	JUL 1979
-Record Lowest	54	-8	8	14	23	31	46	57	56	41	25	1	5	-8
-Year		1962	1985	1971	1983	1967	1988	1988	1973	1945	1970	1976	1953	JAN 1962
NORMAL DEGREE DAYS:														
Heating (base 65°F)		688	473	316	110	7	0	0	0	0	88	378	648	2708
Cooling (base 65°F)		0	0	10	62	217	462	536	468	282	57	0	0	2094
% OF POSSIBLE SUNSHINE	51	77	82	86	88	89	90	81	81	82	84	83	77	83
MEAN SKY COVER (tenths)														
Sunrise - Sunset	51	4.6	4.3	4.2	3.6	3.2	2.8	4.5	4.3	3.4	3.2	3.5	4.3	3.8
MEAN NUMBER OF DAYS:														
Sunrise to Sunset														
-Clear	51	13.9	13.6	15.3	16.6	18.7	19.8	12.2	13.8	17.7	19.0	17.5	15.1	193.2
-Partly Cloudy	51	7.4	7.4	8.2	8.1	8.1	7.5	13.4	12.1	7.2	6.8	6.2	7.3	99.5
-Cloudy	51	9.7	7.3	7.5	5.3	4.3	2.7	5.5	5.1	5.1	5.3	6.2	8.7	72.5
Precipitation .01 inches or more	54	4.2	3.0	2.4	1.7	2.5	3.3	8.0	8.1	5.5	4.1	2.8	3.9	49.4
Snow,Ice pellets,hail 1.0 inches or more	54	0.5	0.3	0.*	0.1	0.0	0.0	0.0	0.0	0.0	0.*	0.3	0.5	1.8
Thunderstorms	54	0.2	0.4	0.5	1.0	2.9	4.5	10.4	10.2	4.1	1.8	0.3	0.2	36.5
Heavy Fog Visibility 1/4 mile or less	54	0.7	0.2	0.1	0.*	0.*	0.0	0.0	0.0	0.1	0.1	0.3	0.6	2.1
Temperature °F														
-Maximum														
90° and above	33	0.0	0.0	0.0	2.0	12.7	25.9	26.8	23.6	11.9	1.8	0.0	0.0	104.9
32° and below	33	0.3	0.1	0.0	0.0	0.0	0.0	0.0	0.0	0.0	0.0	0.1	0.2	0.6
-Minimum														
32° and below	33	18.9	11.8	4.9	0.8	0.*	0.0	0.0	0.0	0.0	0.4	8.1	18.2	63.2
0° and below	33	0.1	0.0	0.0	0.0	0.0	0.0	0.0	0.0	0.0	0.0	0.0	0.0	0.1
AVG. STATION PRESS.(mb)	21	883.8	882.7	880.1	879.9	879.4	880.2	882.3	882.9	882.8	883.4	883.3	884.0	882.1
RELATIVE HUMIDITY (%)														
Hour 05	33	66	57	47	40	42	46	63	67	68	64	62	66	57
Hour 11 (Local Time)	33	45	37	29	23	23	25	38	42	43	38	38	45	36
Hour 17	33	35	27	21	16	17	18	30	33	34	30	32	38	28
Hour 23	33	56	45	35	28	29	32	47	53	55	52	52	57	45
PRECIPITATION (inches):														
Water Equivalent														
-Normal		0.40	0.41	0.29	0.20	0.25	0.67	1.54	1.58	1.70	0.76	0.44	0.57	8.81
-Maximum Monthly	54	1.84	1.69	2.26	1.42	4.22	3.18	5.53	5.57	6.68	4.31	1.63	3.29	6.68
-Year		1949	1973	1958	1983	1992	1984	1968	1984	1974	1945	1961	1991	SEP 1974
-Minimum Monthly	54	0.00	0.00	T	0.00	0.00	T	0.04	T	T	0.00	0.00	0.00	0.00
-Year		1967	1943	1982	1978	1962	1990	1978	1962	1959	1952	1964	1955	APR 1978
-Maximum in 24 hrs	54	0.61	0.87	1.72	1.08	2.40	1.56	2.63	2.30	2.52	1.77	1.19	1.76	2.63
-Year		1960	1956	1941	1966	1992	1986	1968	1984	1958	1945	1943	1987	JUL 1968
Snow,Ice pellets,hail														
-Maximum Monthly	54	8.3	8.9	7.3	16.5	T	T	T	0.0	T	1.0	12.7	25.9	25.9
-Year		1949	1956	1958	1983	1992	1992	1990		1993	1980	1976	1987	DEC 1987
-Maximum in 24 hrs	54	5.2	7.2	7.3	8.8	T	T	T	0.0	T	1.0	7.8	16.8	16.8
-Year		1992	1956	1958	1983	1992	1992	1990		1993	1980	1961	1987	DEC 1987
WIND:														
Mean Speed (mph)	51	8.3	9.1	11.0	11.0	10.3	9.3	8.3	7.7	7.6	7.5	8.0	7.9	8.8
Prevailing Direction through 1963		N	N	WSW	WSW	WSW	S	SSE	S	S	N	N	N	N
Fastest Obs. 1 Min.														
-Direction (!!!)	18	28	26	30	30	13	22	13	13	11	25	24	24	26
-Speed (MPH)	18	40	48	48	42	35	40	37	35	29	35	40	42	48
-Year		1976	1977	1977	1975	1987	1976	1988	1979	1993	1991	1975	1975	FEB 1977
Peak Gust														
-Direction (!!!)	10	W	W	SW	W	SW	N	S	SE	SE	W	W	W	W
-Speed (mph)	10	47	60	59	55	48	48	55	48	48	52	49	54	60
-Date		1988	1987	1991	1984	1985	1985	1989	1991	1993	1993	1988	1993	FEB 1987

PRECIPITATION (inches)

EL PASO, TEXAS

YEAR	JAN	FEB	MAR	APR	MAY	JUNE	JULY	AUG	SEP	OCT	NOV	DEC	ANNUAL
1964	T	T	0.99	0.08	0.02	T	0.18	0.76	2.40	0.40	0.00	0.52	5.35
1965	0.19	0.59	0.03	0.01	0.11	0.66	0.17	0.49	2.12	0.18	0.12	0.74	5.41
1966	0.38	0.20	T	1.08	0.04	2.67	1.17	1.85	1.79	0.01	0.01	0.04	9.24
1967	0.00	0.04	0.17	0.03	0.05	1.41	0.84	0.54	1.54	0.09	0.23	0.78	5.72
1968	0.47	1.11	0.85	0.10	T	0.03	5.53	1.71	0.53	0.11	1.35	0.23	12.02
1969	0.05	0.08	0.17	T	0.28	T	1.14	0.28	0.43	0.59	0.63	0.69	4.34
1970	0.03	0.55	0.47	T	0.71	0.73	1.41	0.41	1.01	0.68	T	0.06	6.06
1971	0.17	0.04	0.00	0.42	T	0.01	2.34	1.59	0.96	1.07	0.14	0.50	7.24
1972	0.44	T	T	0.00	0.04	1.62	0.71	2.59	1.60	1.25	0.33	0.42	9.00
1973	1.23	1.69	0.60	0.00	0.29	0.71	2.12	0.73	0.01	0.07	0.08	T	7.53
1974	0.27	T	0.36	0.12	0.05	0.36	2.21	0.63	6.68	1.90	0.50	0.87	13.95
1975	0.70	0.59	0.19	T	0.03	T	1.11	0.45	2.18	0.25	T	0.71	6.21
1976	0.26	0.52	T	0.30	0.74	0.50	3.17	0.23	1.70	1.20	1.20	0.32	10.14
1977	0.57	T	0.17	0.09	0.06	0.04	1.09	1.36	0.16	1.65	0.05	0.26	5.50
1978	0.44	0.47	0.07	0.00	0.57	1.46	0.04	2.18	4.14	2.28	0.45	0.47	12.57
1979	0.77	0.68	T	0.28	0.24	0.03	0.98	2.16	0.41	T	0.04	0.25	5.84
1980	0.54	0.73	0.25	0.31	0.08	T	0.21	1.76	1.90	0.95	0.54	0.04	7.31
1981	1.10	0.36	0.39	0.65	0.72	0.64	2.08	5.26	0.52	0.53	0.30	0.08	12.63
1982	0.34	0.55	T	0.05	0.19	0.18	1.00	0.48	5.28	T	0.29	2.61	10.97
1983	0.35	0.60	0.45	1.42	0.05	0.23	0.43	0.97	1.51	1.48	0.34	0.16	7.99
1984	0.31	0.00	0.44	0.01	0.59	3.18	0.69	5.57	0.58	3.12	0.51	1.17	16.17
1985	0.95	0.19	0.59	0.07	0.01	0.10	1.32	1.46	1.47	1.82	0.13	0.05	8.16
1986	0.01	0.39	0.39	T	0.83	3.05	2.66	0.70	0.85	0.45	1.42	1.42	12.17
1987	0.29	0.30	0.49	0.32	0.24	2.24	0.64	2.22	0.89	0.15	0.29	2.87	10.94
1988	0.25	0.70	0.10	0.23	0.15	0.03	3.35	3.46	1.52	0.59	0.24	0.44	11.06
1989	0.11	0.72	0.62	T	0.65	T	1.23	3.06	0.48	0.23	T	0.16	7.26
1990	0.29	0.14	0.41	0.25	0.10	T	3.96	1.98	3.46	0.58	1.34	0.34	12.85
1991	0.82	0.66	0.10	T	0.23	0.01	2.69	2.06	1.82	0.20	0.50	3.29	12.38
1992	1.14	0.16	0.50	0.30	4.22	0.27	0.65	2.11	0.15	0.27	0.28	1.35	11.40
1993	1.34	0.32	0.01	0.12	T	1.47	0.95	2.73	1.32	0.17	0.49	0.71	9.63
Record Mean	0.45	0.41	0.33	0.25	0.37	0.63	1.67	1.54	1.31	0.80	0.44	0.56	8.75

AVERAGE TEMPERATURE (deg. F)

EL PASO, TEXAS

YEAR	JAN	FEB	MAR	APR	MAY	JUNE	JULY	AUG	SEP	OCT	NOV	DEC	ANNUAL
1964	39.3	40.8	53.6	63.2	73.9	81.5	84.5	82.6	75.2	63.4	51.8	44.0	62.8
1965	48.0	46.7	52.1	65.3	71.8	78.2	84.2	81.1	74.0	63.0	56.8	45.7	63.9
1966	40.1	42.9	56.4	65.1	74.5	79.5	83.6	78.7	73.4	62.1	54.5	42.4	62.8
1967	41.6	48.0	59.6	65.1	70.9	79.1	83.0	78.6	73.3	63.7	53.1	41.5	63.1
1968	42.4	50.4	53.0	61.4	73.0	81.0	79.1	76.5	72.4	65.0	51.0	41.3	62.2
1969	48.6	48.4	49.4	65.8	72.1	81.5	84.9	85.7	77.1	67.7	52.5	48.6	65.2
1970	46.9	52.3	55.6	63.5	72.2	79.7	82.8	81.4	74.2	59.5	51.9	48.0	64.0
1971	44.6	48.4	58.1	62.6	72.1	81.1	82.3	77.0	73.5	62.5	52.1	44.7	63.2
1972	45.2	52.3	61.2	65.2	69.8	78.3	82.2	77.7	72.9	65.7	48.8	46.6	63.8
1973	42.9	47.0	52.4	57.7	68.7	76.6	79.7	79.2	75.9	63.3	53.8	45.7	61.9
1974	44.3	44.8	59.6	65.0	75.6	82.8	79.6	77.0	69.4	63.0	49.9	41.3	62.7
1975	43.1	48.9	55.1	59.7	69.8	80.9	79.8	80.9	72.2	64.0	51.5	44.2	62.5
1976	42.3	52.6	55.9	64.1	69.5	79.4	78.2	78.7	70.5	58.5	44.8	41.9	61.4
1977	44.7	47.3	49.7	61.5	70.7	81.5	82.2	82.5	77.4	64.3	53.7	49.6	63.8
#1978	45.4	48.8	58.6	66.1	73.6	83.4	84.5	80.0	72.3	63.6	55.8	44.6	64.7
1979	41.0	47.1	53.1	63.6	70.1	78.4	85.1	78.8	74.2	66.2	47.9	43.1	62.4
1980	46.8	50.6	54.1	60.6	70.5	86.3	87.2	82.4	75.6	60.3	49.2	48.5	64.3
1981	45.2	50.3	57.2	64.8	73.6	82.6	83.6	79.5	75.9	64.6	54.3	48.7	65.1
1982	42.3	48.7	57.7	64.4	69.6	80.9	84.2	83.5	77.1	64.6	53.2	43.3	64.1
1983	41.6	49.5	54.6	56.3	68.9	77.3	82.9	81.8	78.7	66.6	54.3	45.5	63.2
1984	44.4	47.0	55.7	62.0	75.0	79.5	81.1	80.4	72.9	61.4	51.6	45.7	63.1
1985	40.0	45.6	55.2	64.2	72.1	79.0	79.4	80.6	72.8	61.4	52.9	43.1	62.2
1986	44.7	52.1	55.7	67.3	71.5	77.7	80.0	80.4	74.1	62.4	49.4	42.6	63.2
1987	41.3	46.2	51.2	59.7	68.6	78.1	81.6	79.1	72.1	67.0	51.4	40.5	61.4
1988	42.6	48.4	53.4	61.1	70.3	79.0	80.3	77.6	72.4	66.7	54.0	42.6	62.4
1989	43.6	51.4	58.7	67.4	74.2	81.3	81.8	79.1	73.4	63.2	53.3	41.9	64.1
1990	44.1	49.1	56.1	66.4	73.1	87.1	80.2	76.8	73.9	63.4	52.9	44.8	64.0
1991	44.3	51.1	53.9	64.0	72.6	79.0	78.6	79.5	70.7	65.4	50.3	46.3	63.0
1992	44.0	50.2	58.1	67.9	70.6	81.5	85.1	81.3	77.8	66.5	48.0	44.7	64.6
1993	48.5	51.4	57.9	66.7	74.6	82.7	85.2	82.5	75.8	64.4	52.3	46.9	65.7
Record Mean	44.5	49.3	55.6	63.8	72.1	80.8	82.0	80.3	74.8	64.5	52.3	45.0	63.7
Max	57.2	62.6	69.3	77.8	86.1	94.7	94.1	92.1	87.0	77.9	65.7	57.5	76.8
Min	31.9	35.9	41.8	49.7	58.1	66.8	69.8	68.4	62.5	51.1	38.9	32.6	50.6

REFERENCE NOTES

EL PASO, TEXAS

GENERAL
T - TRACE AMOUNT.
BLANK ENTRIES DENOTE MISSING/UNREPORTED DATA.
# INDICATES A STATION OR INSTRUMENT RELOCATION.

SPECIFIC

PM - INCLUDES LAST DAY OF PREVIOUS MONTH
ASOS - AUTOMATED SURFACE OBSERVING SYSTEM IN OPERATION DURING THESE MONTHS.

(*) - LENGTH OF RECORD IN YEARS, ALTHOUGH INDIVIDUAL MONTHS MAY BE MISSING.
0.* OR * - THE VALUE IS BETWEEN 0.0 AND 0.05
NORMALS - BASED ON THE 1961-1990 RECORD PERIOD.
EXTREMES - DATES ARE THE MOST RECENT OCCURRENCE
WIND DIR.- NUMERALS SHOW TENS OF DEGRESS CLOCKWISE FROM TRUE NORTH. "00" INDICATES CALM.
RESULTANT DIRECTIONS ARE GIVEN TO WHOLE DEGREES.
BOLD VALUES INDICATE EXTREME VALUES WHICH OCCURRED AFTER THE ASOS SYSTEM WAS COMMISSIONED.

RECORD = PERIOD OF RECORD
RECORD MEAN PRECIPITATION IS THE MEAN OF ALL DAILY PRECIPITATION AMOUNTS DURING THE PERIOD OF RECORD.
RECORD MAX(MIN) TEMPERATURE IS THE MEAN OF ALL DAILY MAX(MIN) TEMPERATURES DURING THE PERIOD OF RECORD.
RECORD MEAN TEMPERATURE IS THE SUM OF THE RECORD MAX AND RECORD MIN DIVIDED BY 2.
AVERAGE TEMPERATURE IS THE SUM OF THE MEAN DAILY MAX AND MIN TEMPERATURE DIVIDED BY 2.

EXCEPTIONS

RECORD MEANS ARE THROUGH THE CURRENT YEAR,
BEGINNING IN 1887 FOR TEMPERATURE
1879 FOR PRECIPITATION
1940 FOR SNOWFALL

HEATING DEGREE DAYS Base 65 deg. F EL PASO, TEXAS

SEASON	JULY	AUG	SEP	OCT	NOV	DEC	JAN	FEB	MAR	APR	MAY	JUNE	TOTAL
1964-65	0	0	0	76	391	643	521	504	397	54	7	2	2595
1965-66	0	0	4	107	240	592	769	612	264	59	4	0	2651
1966-67	0	0	2	126	307	695	718	469	173	56	25	0	2571
1967-68	0	0	2	106	352	720	691	415	377	128	0	0	2791
1968-69	0	0	0	61	414	728	503	464	477	43	24	0	2714
1969-70	0	0	0	62	371	504	556	348	286	94	33	0	2254
1970-71	0	0	39	180	388	519	625	457	254	110	6	0	2578
1971-72	0	0	31	112	381	624	607	364	126	56	3	0	2304
1972-73	0	0	3	87	480	563	679	499	384	218	31	0	2944
1973-74	0	0	7	90	336	592	636	558	178	79	5	0	2481
1974-75	0	0	41	107	445	728	672	445	309	188	13	0	2948
1975-76	0	0	20	66	399	640	696	351	278	91	26	0	2567
1976-77	0	0	7	214	601	709	623	492	469	138	3	0	3256
1977-78	0	0	0	56	328	472	603	449	200	57	22	0	2187
#1978-79	0	0	16	106	272	625	735	494	362	118	26	1	2755
1979-80	0	0	24	56	505	670	555	410	331	157	19	0	2727
1980-81	0	0	2	203	467	503	607	405	233	82	2	0	2504
1981-82	0	0	0	93	313	499	697	449	237	82	17	0	2387
1982-83	0	0	0	88	344	668	720	430	316	284	23	0	2873
1983-84	0	0	0	52	317	599	633	514	285	126	8	0	2534
1984-85	0	0	18	144	404	592	768	537	302	71	5	0	2841
1985-86	0	0	10	125	358	670	621	356	283	47	22	0	2492
1986-87	0	0	1	116	460	687	725	521	420	173	5	0	3108
1987-88	0	0	0	13	405	750	686	474	360	121	16	0	2825
1988-89	0	2	2	17	337	684	661	377	208	52	9	0	2349
1989-90	0	0	2	108	344	708	640	439	271	45	20	0	2577
1990-91	0	2	8	88	354	617	636	383	336	78	3	0	2505
1991-92	0	0	26	66	434	576	644	423	209	49	3	0	2430
1992-93	0	0	0	21	502	624	504	373	222	53	4	0	2303
1993-94	0	0	0	145	375	551							

COOLING DEGREE DAYS Base 65 deg. F EL PASO, TEXAS

YEAR	JAN	FEB	MAR	APR	MAY	JUNE	JULY	AUG	SEP	OCT	NOV	DEC	TOTAL
1969	0	0	1	71	250	501	627	647	372	155	2	0	2626
1970	0	0	2	57	263	448	559	514	321	19	0	0	2183
1971	0	0	45	45	235	492	543	375	293	43	0	0	2071
1972	0	4	15	70	159	404	543	401	247	120	0	0	1963
1973	0	0	0	7	152	355	459	448	340	44	7	0	1812
1974	0	0	19	84	338	540	459	378	181	54	0	0	2053
1975	0	0	9	35	170	482	469	502	241	41	1	0	1950
1976	0	0	4	71	170	441	418	434	179	18	0	0	1735
1977	0	0	0	39	186	500	540	552	380	43	0	0	2240
#1978	0	0	8	98	295	559	612	474	238	70	2	0	2356
1979	0	0	0	84	190	414	630	432	308	99	0	0	2157
1980	0	0	0	34	198	646	693	546	329	63	0	0	2509
1981	0	2	2	84	275	534	586	455	333	89	0	2	2362
1982	0	0	14	70	167	484	602	583	371	82	0	0	2373
1983	0	0	0	32	151	374	564	527	417	108	6	0	2179
1984	0	0	2	44	324	441	507	482	260	38	8	0	2106
1985	0	0	9	55	233	428	457	488	252	18	0	0	1940
1986	0	3	0	122	228	391	474	485	281	41	0	0	2025
1987	0	0	0	22	121	399	521	446	222	83	2	0	1816
1988	0	0	9	15	186	429	478	399	226	76	16	0	1834
1989	0	2	20	130	300	494	530	442	261	60	0	0	2239
1990	0	1	0	96	278	667	480	374	282	43	0	0	2221
1991	0	0	0	55	245	424	428	455	204	88	0	0	1899
1992	0	0	3	144	182	500	628	511	389	75	0	0	2432
1993	0	0	11	110	306	538	632	550	331	132	2	0	2612

SNOWFALL (inches) EL PASO, TEXAS

SEASON	JULY	AUG	SEP	OCT	NOV	DEC	JAN	FEB	MAR	APR	MAY	JUNE	TOTAL
1964-65	0.0	0.0	0.0	0.0	0.0	T	0.0	0.3	T	0.0	0.0	0.0	0.3
1965-66	0.0	0.0	0.0	0.0	0.0	T	0.4	T	T	0.0	0.0	0.0	0.4
1966-67	0.0	0.0	0.0	0.0	0.0	T	0.0	T	0.0	T	0.0	0.0	T
1967-68	0.0	0.0	0.0	0.0	0.0	5.6	2.1	2.3	0.8	0.0	0.0	0.0	10.8
1968-69	0.0	0.0	0.0	0.0	7.0	T	0.0	0.0	T	0.0	0.0	0.0	7.0
1969-70	0.0	0.0	0.0	0.0	6.0	1.9	0.3	1.7	T	0.0	0.0	0.0	9.9
1970-71	0.0	0.0	0.0	0.0	0.0	0.0	3.8	T	0.0	T	0.0	0.0	3.8
1971-72	0.0	0.0	0.0	0.0	0.0	1.0	2.5	0.0	0.0	0.0	0.0	0.0	3.5
1972-73	0.0	0.0	0.0	0.0	0.0	T	5.3	4.6	T	0.0	0.0	0.0	9.9
1973-74	0.0	0.0	0.0	0.0	T	0.0	T	T	0.0	0.0	0.0	0.0	T
1974-75	0.0	0.0	0.0	0.0	0.0	5.3	2.2	0.2	2.0	0.0	0.0	0.0	9.7
1975-76	0.0	0.0	0.0	0.0	0.0	T	1.0	0.0	T	T	0.0	0.0	1.0
1976-77	0.0	0.0	0.0	T	12.7	2.0	T	T	0.0	T	0.0	0.0	14.7
1977-78	0.0	0.0	0.0	0.0	0.0	0.0	T	0.0	0.0	0.0	0.0	0.0	T
1978-79	0.0	0.0	0.0	0.0	T	T	T	1.2	0.0	0.0	0.0	0.0	1.2
1979-80	0.0	0.0	0.0	0.0	0.0	0.0	T	3.6	0.0	2.0	0.0	0.0	5.6
1980-81	0.0	0.0	0.0	1.0	4.0	0.0	4.8	0.0	0.0	0.0	0.0	0.0	9.8
1981-82	0.0	0.0	0.0	0.0	0.0	0.0	T	T	0.0	0.0	0.0	0.0	T
1982-83	0.0	0.0	0.0	0.0	0.3	18.2	T	0.0	T	16.5	0.0	0.0	35.0
1983-84	0.0	0.0	0.0	0.0	T	T	T	0.0	6.1	0.0	0.0	0.0	6.1
1984-85	0.0	0.0	0.0	0.0	T	2.9	5.4	1.1	0.0	0.0	0.0	0.0	9.4
1985-86	0.0	0.0	0.0	0.0	0.0	0.9	T	0.9	T	0.0	0.0	0.0	1.8
1986-87	0.0	0.0	0.0	0.0	0.0	0.6	3.4	2.8	0.6	0.0	0.0	0.0	7.4
1987-88	0.0	0.0	0.0	0.0	0.0	25.9	T	6.6	0.0	0.0	0.0	0.0	32.5
1988-89	0.0	0.0	0.0	0.0	T	0.3	0.0	0.0	T	0.0	T	0.0	0.3
1989-90	0.0	0.0	0.0	0.0	0.0	T	T	T	0.0	T	T	0.0	T
1990-91	T	0.0	0.0	0.0	4.3	T	1.5	T	T	0.0	T	0.0	5.8
1991-92	0.0	0.0	0.0	0.0	0.0	0.5	6.9	0.0	T	0.0	T	T	7.4
1992-93	0.0	0.0	0.0	0.0	0.0	1.6	0.0	0.0	0.0	0.0	0.0	0.0	1.6
1993-94	0.0	0.0	T	0.8	0.0	1.6							
Record Mean	T	0.0	T	T	0.9	1.7	1.4	0.8	0.4	0.3	T	T	5.6

GALVESTON, TEXAS

The city of Galveston is located on Galveston Island off the southeast coast of Texas. The island is about 2 3/4 miles across at the widest point and 29 miles long. It is bounded on the southeast by the Gulf of Mexico and on the northwest by Galveston Bay, which is about 3 miles wide at this point. The climate of the Galveston area is predominantly marine, with periods of modified continental influence during the colder months, when cold fronts from the northwest sometimes reach the coast.

Because of its coastal location and relatively low latitude, cold fronts which do reach the area are very seldom severe and temperatures below 32 degrees are recorded on an average only four times a year. Normal monthly high temperatures range from about 60 degrees in January to nearly 88 degrees in August. Lows range from about 48 degrees in January to the upper 70s during the summer season.

High humidities prevail throughout the year. Annual precipitation averages about 42 inches. Rainfall during the summer months may vary greatly on different parts of the island, as most of the rain in this season is from local thunderstorm activity. Hail is rare because the necessary strong vertical lifting is usually absent. There have been several instances when a monthly rainfall total amounted to only a trace, but these have been offset in the means by many monthly totals in excess of 15 inches. Winter precipitation comes mainly from frontal activity and from low stratus clouds, which produce slow, steady rains.

The island has been subject at infrequent intervals to major tropical storms of hurricane force.

NORMALS, MEANS, AND EXTREMES

GALVESTON, TEXAS

LATITUDE: 29°18'N LONGITUDE: 94°48' W ELEVATION: FT. GRND 7 BARO 69 TIME ZONE: CENTRAL WBAN: 12944

	(a)	JAN	FEB	MAR	APR	MAY	JUNE	JULY	AUG	SEP	OCT	NOV	DEC	YEAR
TEMPERATURE °F:														
Normals														
-Daily Maximum		58.3	60.5	66.7	73.5	79.8	85.0	87.3	87.7	84.5	77.5	69.3	61.9	74.3
-Daily Minimum		47.1	49.9	56.7	65.0	71.8	77.2	79.2	79.2	75.4	68.0	59.0	50.9	65.0
-Monthly		52.7	55.2	61.7	69.3	75.8	81.1	83.3	83.5	80.0	72.8	64.2	56.4	69.7
Extremes														
-Record Highest	123	78	83	85	92	94	99	101	100	96	94	85	80	101
-Year		1989	1932	1879	1953	1984	1918	1932	1924	1989	1952	1988	1918	JUL 1932
-Record Lowest	123	11	8	26	38	52	57	66	67	52	39	26	14	8
-Year		1886	1899	1980	1938	1954	1903	1910	1966	1942	1993	1911	1989	FEB 1899
NORMAL DEGREE DAYS:														
Heating (base 65°F)		396	287	145	16	0	0	0	0	0	7	123	289	1263
Cooling (base 65°F)		15	13	43	145	335	483	567	574	450	248	99	22	2994
% OF POSSIBLE SUNSHINE	102	47	51	56	61	67	75	72	71	68	71	59	48	62
MEAN SKY COVER (tenths)														
Sunrise - Sunset														
MEAN NUMBER OF DAYS:														
Sunrise to Sunset														
-Clear														
-Partly Cloudy														
-Cloudy														
Precipitation .01 inches or more	122	10.0	8.5	7.7	6.2	6.3	6.7	8.5	8.9	9.2	6.4	7.7	9.7	95.8
Snow,Ice pellets,hail 1.0 inches or more	122	0.*	0.*	0.0	0.0	0.0	0.0	0.0	0.0	0.0	0.0	0.0	0.*	*
Thunderstorms	1	3.0	2.0	2.5	3.0	5.5	3.0	2.5	2.5	5.0	3.0	3.0	1.0	36.0
Heavy Fog Visibility 1/4 mile or less	1	4.0	1.0	3.5	1.5	0.0	0.0	0.0	0.0	0.0	0.0	1.5	6.5	18.0
Temperature °F														
-Maximum														
90° and above	108	0.0	0.0	0.0	0.*	0.1	1.3	3.8	5.5	1.7	0.1	0.0	0.0	12.5
32° and below	123	0.1	0.1	0.0	0.0	0.0	0.0	0.0	0.0	0.0	0.0	0.0	0.*	0.2
-Minimum														
32° and below	123	1.9	0.8	0.1	0.0	0.0	0.0	0.0	0.0	0.0	0.0	0.*	0.8	3.6
0° and below	123	0.0	0.0	0.0	0.0	0.0	0.0	0.0	0.0	0.0	0.0	0.0	0.0	0.0
AVG. STATION PRESS.(mb)														
RELATIVE HUMIDITY (%)														
Hour 00	44	83	82	84	85	83	80	80	78	78	75	81	82	81
Hour 06 (Local Time)	96	85	84	85	86	84	81	81	81	81	80	83	85	83
Hour 12	66	77	74	74	75	73	70	70	69	68	65	72	76	72
Hour 18	96	80	77	79	80	77	73	73	73	74	71	77	79	76
PRECIPITATION (inches):														
Water Equivalent														
-Normal		3.26	2.26	2.23	2.43	3.59	4.44	3.96	4.47	5.93	2.84	3.37	3.50	42.28
-Maximum Monthly	123	10.75	8.34	9.49	11.04	11.03	15.49	18.74	19.08	26.01	17.78	16.18	10.28	26.01
-Year		1991	1992	1973	1904	1991	1919	1900	1915	1885	1871	1940	1887	SEP 1885
-Minimum Monthly	123	0.02	0.09	0.06	T	T	T	T	0.00	0,04	T	0.03	0.23	0.00
-Year		1909	1954	1953	1984	1978	1907	1962	1902	1924	1952	1903	1889	AUG 1902
-Maximum in 24 hrs	123	5.38	6.55	8.10	9.23	7.71	12.56	14.35	10.86	11.65	14.10	9.01	5.43	14.35
-Year		1923	1952	1973	1904	1975	1961	1900	1981	1961	1901	1940	1964	JUL 1900
Snow,Ice pellets,hail														
-Maximum Monthly	123	2.5	15.4	T	T	T	0.0	0.0	0.0	0.0	0.0	0.0	1.0	15.4
-Year		1973	1895	1989	1992	1991							1989	FEB 1895
-Maximum in 24 hrs	123	2.5	15.4	T	T	T	0.0	0.0	0.0	0.0	0.0	0.0	1.0	15.4
-Year		1973	1895	1989	1992	1991							1989	FEB 1895
WIND:														
Mean Speed (mph)	93	11.6	11.8	11.9	12.1	11.5	10.7	9.8	9.4	10.1	10.3	11.2	11.3	11.0
Prevailing Direction														
Fastest Mile														
-Direction (!!)	122	S	N	SE	NW	W	SE	NW	E	NE	SE	NW	NW	NE
-Speed (MPH)	122	53	60	50	68	66	62	68	91	100	66	72	50	100
-Year		1915	1927	1952	1983	1959	1921	1943	1915	1900	1949	1987	1954	SEP 1900
Peak Gust														
-Direction (!!)														
-Speed (mph)														
-Date														

PRECIPITATION (inches) GALVESTON, TEXAS

YEAR	JAN	FEB	MAR	APR	MAY	JUNE	JULY	AUG	SEP	OCT	NOV	DEC	ANNUAL
1964	4.13	3.39	2.53	0.37	1.72	1.29	5.79	5.67	4.38	1.00	1.49	7.12	38.88
1965	1.65	1.77	0.36	0.55	2.57	2.45	1.24	4.88	4.02	1.91	4.26	9.00	34.66
1966	5.52	4.50	0.70	5.41	10.34	4.86	1.25	5.56	7.99	4.03	0.93	1.99	53.08
1967	1.60	2.26	1.55	1.69	3.33	0.81	8.73	5.74	2.21	3.18	1.64	2.50	35.24
1968	6.14	1.70	1.93	3.93	4.82	13.03	2.26	3.17	3.95	2.28	4.73	3.22	51.16
1969	1.59	3.54	2.30	6.03	6.97	3.45	4.76	5.75	0.28	1.90	0.91	4.31	41.79
1970	1.47	1.33	6.40	3.47	4.33	5.14	3.07	1.37	14.31	5.14	1.48	0.96	48.47
1971	0.18	4.12	0.64	1.77	1.31	0.82	1.98	3.61	10.21	2.37	2.29	6.67	35.97
1972	3.68	1.87	1.11	0.76	6.37	0.50	3.44	1.40	7.85	3.03	6.30	3.64	39.95
1973	3.19	2.62	9.49	10.41	1.61	4.01	2.44	10.25	6.80	6.44	0.57	2.64	60.47
1974	3.28	0.83	2.79	0.91	7.84	2.03	2.18	8.08	3.31	3.57	4.10	4.34	43.26
1975	3.36	1.79	0.96	2.19	10.79	3.63	1.28	6.58	4.61	4.83	4.56	3.96	48.54
1976	1.00	0.31	0.85	2.53	2.16	4.28	5.63	1.43	8.27	3.85	4.02	7.73	42.06
1977	3.41	1.07	1.39	6.74	0.67	2.83	0.68	6.45	4.91	2.95	9.51	1.46	42.07
1978	8.88	1.81	0.38	1.62	T	3.55	4.23	0.63	3.55	0.30	2.99	1.34	29.28
1979	3.48	2.80	3.34	4.91	3.45	0.79	17.48	4.47	10.86	2.75	2.05	2.97	59.35
1980	6.35	0.83	3.93	0.57	7.44	0.44	3.85	2.35	5.05	0.28	1.25	1.74	34.58
1981	1.38	1.30	0.52	0.33	4.31	10.66	4.87	13.38	2.49	3.03	1.76	2.25	46.78
1982	1.80	4.00	1.24	1.81	3.38	4.51	0.24	1.80	1.29	1.20	6.61	6.38	34.26
1983	4.24	3.37	2.91	0.24	1.52	4.94	7.28	11.18	11.58	1.24	2.76	2.64	53.90
1984	3.19	1.19	1.36	T	5.07	0.86	1.82	3.08	4.97	8.97	2.41	2.72	35.64
1985	2.88	5.42	5.63	1.20	2.01	3.41	7.08	2.46	2.93	3.41	2.24	2.57	41.24
1986	1.35	1.05	1.30	1.03	3.10	4.45	0.65	3.39	2.98	9.04	3.96	4.04	36.34
1987	3.94	4.78	0.43	0.32	3.77	6.86	4.77	2.76	4.47	0.05	2.89	1.80	36.84
1988	2.88	0.96	4.41	5.22	0.81	2.42	5.82	3.24	9.82	0.48	0.93	2.89	39.88
1989	6.25	0.17	2.37	0.69	2.70	12.35	2.60	7.01	1.32	1.87	2.76	0.50	40.59
1990	3.37	3.79	4.07	3.19	4.26	1.17	4.76	0.18	7.16	2.49	1.94	1.81	38.19
1991	10.75	6.30	1.21	2.31	11.03	4.38	4.64	4.35	5.50	2.25	4.01	2.35	59.08
1992	8.46	8.34	2.17	5.59	4.42	2.57	3.43	1.37	4.16	1.15	7.17	3.80	52.63
1993	9.26	3.75	4.92	5.42	3.29	3.32	0.06	0.54	4.88	7.07	2.18	2.43	47.12
Record Mean	3.55	2.81	2.62	2.94	3.48	3.88	4.02	4.38	5.62	3.74	3.58	3.79	44.40

AVERAGE TEMPERATURE (deg. F) GALVESTON, TEXAS

YEAR	JAN	FEB	MAR	APR	MAY	JUNE	JULY	AUG	SEP	OCT	NOV	DEC	ANNUAL
1964	52.1	51.5	59.9	69.4	76.7	80.6	82.3	83.8	80.5	69.8	65.8	55.5	69.0
1965	56.9	55.0	57.0	71.1	76.3	81.7	83.7	82.8	80.7	70.8	70.3	60.7	70.6
1966	50.5	53.3	60.6	68.8	75.2	79.6	83.4	81.6	79.2	70.8	66.3	56.0	68.7
1967	53.3	54.6	65.4	75.0	75.6	81.7	81.5	80.7	77.8	72.3	64.5	57.5	70.0
1968	52.6	50.8	57.4	69.7	75.7	80.3	82.3	84.4	79.0	74.2	62.0	55.8	68.7
1969	55.4	56.5	55.8	69.6	75.5	81.4	84.3	84.4	81.2	73.9	63.2	59.1	70.0
1970	48.9	56.1	58.9	69.7	74.1	81.0	82.7	83.6	80.0	69.7	60.6	61.8	69.0
1971	57.0	57.1	60.6	68.1	75.0	81.5	83.4	82.5	79.8	75.7	65.2	62.2	70.7
1972	59.0	58.1	66.0	72.5	76.3	81.8	82.2	84.1	82.8	73.4	58.8	55.7	70.9
1973	51.1	53.0	64.8	65.5	74.3	79.9	83.4	82.0	80.6	76.2	70.4	57.3	69.9
1974	57.0	58.8	65.8	70.4	77.2	80.7	83.0	82.6	76.0	72.8	63.0	56.9	70.3
1975	58.7	58.1	62.6	68.2	77.0	80.9	83.8	84.1	77.4	72.9	64.1	55.2	70.2
1976	53.3	60.5	63.5	70.0	73.1	80.0	81.4	82.5	79.7	65.3	54.6	52.8	68.0
1977	46.6	55.1	62.2	70.0	76.1	81.0	82.7	82.6	81.4	72.8	65.1	57.8	69.4
1978	45.3	47.0	58.3	68.4	76.7	82.6	83.9	84.5	81.4	73.4	66.7	57.4	68.8
1979	47.0	51.9	61.6	68.6	73.5	81.0	82.3	82.4	76.0	73.2	58.9	55.0	67.7
1980	56.3	52.9	60.0	67.1	75.7	81.8	84.3	84.2	82.8	70.8	59.0	55.0	69.2
1981	52.5	53.7	61.0	72.1	74.8	82.0	84.7	84.8	80.4	74.3	66.0	57.2	70.3
1982	53.2	53.1	63.3	67.4	76.3	82.4	84.4	84.5	81.0	71.5	63.4	59.1	69.9
1983	53.2	55.9	61.3	65.9	75.1	81.3	84.7	84.4	79.2	74.8	67.4	49.8	69.4
1984	49.3	57.2	61.5	68.8	75.6	81.1	84.5	82.3	76.8	75.1	63.2	62.4	69.8
1985	48.1	50.0	63.2	70.3	76.3	79.9	81.7	83.7	79.9	73.0	68.1	54.3	69.0
1986	56.2	58.7	64.9	70.8	75.7	81.8	84.3	83.1	83.1	72.5	64.8	54.2	70.8
1987	53.7	57.7	61.3	68.0	77.1	81.2	83.9	85.1	80.8	72.3	63.9	58.6	70.3
1988	50.6	55.0	61.5	69.0	75.4	80.3	83.3	85.4	81.0	74.0	68.5	58.5	70.2
1989	58.8	53.9	59.7	69.2	78.0	80.6	83.1	83.6	79.0	72.9	65.8	47.9	69.4
1990	57.6	60.9	63.9	69.1	77.2	84.2	83.6	85.5	82.0	72.0	66.4	57.4	71.7
1991	53.0	59.2	64.8	72.4	77.4	82.2	84.3	84.1	79.3	74.8	59.3	59.0	70.8
1992	54.0	60.0	63.8	68.5	74.1	81.2	84.1	82.2	80.6	74.3	60.6	58.7	70.2
1993	56.4	58.5	62.2	66.7	74.5	81.5	85.1	85.7	81.6	71.8	61.1	59.2	70.4
Record Mean	53.8	56.3	62.0	69.0	75.7	81.3	83.3	83.4	80.1	73.0	63.3	56.8	69.8
Max	59.3	61.6	67.1	73.5	80.0	85.6	87.7	87.9	84.7	77.9	68.7	62.1	74.7
Min	48.3	50.9	57.0	64.5	71.3	77.1	78.9	78.8	75.4	68.0	58.0	51.4	65.0

REFERENCE NOTES GALVESTON, TEXAS

GENERAL
T - TRACE AMOUNT.
BLANK ENTRIES DENOTE MISSING/UNREPORTED DATA.
# INDICATES A STATION OR INSTRUMENT RELOCATION.

SPECIFIC

(a) - LENGTH OF RECORD IN YEARS, ALTHOUGH INDIVIDUAL MONTHS MAY BE MISSING.
0.* OR * - THE VALUE IS BETWEEN 0.0 AND 0.05
NORMALS - BASED ON THE 1961-1990 RECORD PERIOD.
EXTREMES - DATES ARE THE MOST RECENT OCCURRENCE
WIND DIR. - NUMERALS SHOW TENS OF DEGRESS CLOCKWISE FROM TRUE NORTH. "00" INDICATES CALM.
RESULTANT DIRECTIONS ARE GIVEN TO WHOLE DEGREES.
BOLD VALUES INDICATE EXTREME VALUES WHICH OCCURRED AFTER THE ASOS SYSTEM WAS COMMISSIONED.

RECORD = PERIOD OF RECORD
RECORD MEAN PRECIPITATION IS THE MEAN OF ALL DAILY PRECIPITATION AMOUNTS DURING THE PERIOD OF RECORD.
RECORD MAX(MIN) TEMPERATURE IS THE MEAN OF ALL DAILY MAX(MIN) TEMPERATURES DURING THE PERIOD OF RECORD.
RECORD MEAN TEMPERATURE IS THE SUM OF THE RECORD MAX AND RECORD MIN DIVIDED BY 2.
AVERAGE TEMPERATURE IS THE SUM OF THE MEAN DAILY MAX AND MIN TEMPERATURE DIVIDED BY 2.

EXCEPTIONS

1. FASTEST MILE WIND OF 100 M.P.H. WAS RECORDED AT 1815H SEPTEMBER 8, 1900 JUST BEFORE ANEMOMETER BLEW AWAY, MAXIMUM VELOCITY ESTIMATED 120 M.P.H. NE BETWEEN 1930H-2030H.
2. MEAN WIND SPEED IS THROUGH 1964.
3. RELATIVE HUMIDITY IS THROUGH 1983.

RECORD MEANS ARE THROUGH THE CURRENT YEAR, BEGINNING IN 1874 FOR TEMPERATURE
1871 FOR PRECIPITATION
1871 FOR SNOWFALL

HEATING DEGREE DAYS Base 65 deg. F — GALVESTON, TEXAS

SEASON	JULY	AUG	SEP	OCT	NOV	DEC	JAN	FEB	MAR	APR	MAY	JUNE	TOTAL
1964-65	0	0	0	5	102	293	254	276	249	2	0	0	1181
1965-66	0	0	0	5	12	152	447	320	138	9	0	0	1083
1966-67	0	0	0	16	63	298	356	287	61	0	0	0	1081
1967-68	0	0	3	12	78	243	381	404	233	8	0	0	1362
1968-69	0	0	0	4	142	283	296	236	281	0	0	0	1242
1969-70	0	0	0	9	127	189	490	244	185	25	3	0	1272
1970-71	0	0	0	31	161	130	250	223	143	30	0	0	968
1971-72	0	0	0	0	91	131	210	211	53	2	0	0	698
1972-73	0	0	0	23	225	289	424	329	39	73	0	0	1402
1973-74	0	0	0	1	23	249	251	181	61	10	0	0	776
1974-75	0	0	0	2	121	254	208	201	120	28	0	0	934
1975-76	0	0	0	4	126	309	355	128	82	0	0	0	1004
1976-77	0	0	0	96	313	373	563	274	102	3	0	0	1724
1977-78	0	0	0	6	63	238	606	501	204	16	2	0	1636
1978-79	0	0	0	0	72	258	550	361	120	12	2	0	1375
1979-80	0	0	0	7	200	302	268	348	158	28	0	0	1311
1980-81	0	0	0	43	222	306	382	314	127	0	0	0	1394
1981-82	0	0	0	30	45	242	358	333	109	51	0	0	1168
1982-83	0	0	0	23	109	201	359	249	129	49	0	0	1119
1983-84	0	0	2	2	56	491	478	228	108	17	0	0	1382
1984-85	0	0	2	1	120	109	515	412	81	2	0	0	1242
1985-86	0	0	1	11	41	333	266	178	53	5	0	0	888
1986-87	0	0	0	11	115	333	345	199	132	61	0	0	1196
1987-88	0	0	0	0	110	207	441	286	135	17	0	0	1196
1988-89	0	0	0	0	61	209	204	322	200	38	0	0	1034
1989-90	0	0	0	30	88	528	224	116	71	16	0	0	1073
1990-91	0	0	0	24	49	268	364	165	65	2	0	0	937
1991-92	0	0	0	2	193	199	334	150	74	25	0	0	977
1992-93	0	0	0	0	171	193	266	186	105	30	0	0	951
1993-94	0	0	0	47	149	203							

COOLING DEGREE DAYS Base 65 deg. F — GALVESTON, TEXAS

YEAR	JAN	FEB	MAR	APR	MAY	JUNE	JULY	AUG	SEP	OCT	NOV	DEC	TOTAL
1969	3	4	4	145	333	500	602	609	493	289	81	11	3074
1970	0	0	3	174	292	488	557	583	456	181	36	39	2809
1971	8	5	15	130	316	500	574	549	448	340	103	49	3037
1972	30	20	91	234	360	510	540	597	540	294	48	8	3272
1973	1	0	41	93	296	454	576	533	476	354	191	15	3030
1974	10	14	94	178	385	481	561	553	335	253	68	7	2939
1975	19	11	53	132	379	483	588	599	378	254	106	11	3013
1976	0	6	40	156	259	457	516	550	445	113	7	0	2549
1977	0	4	21	160	350	488	553	555	499	254	75	21	2980
1978	1	2	5	124	373	535	595	613	499	273	128	25	3173
1979	0	0	22	128	271	488	544	547	339	268	22	0	2629
1980	2	1	12	99	339	512	604	602	542	228	52	0	2993
1981	0	0	12	218	311	520	617	620	465	327	80	7	3177
1982	1	6	63	132	357	524	608	611	486	230	66	29	3113
1983	3	0	19	83	320	496	617	609	435	314	135	25	3056
1984	0	7	7	138	338	489	610	545	364	324	71	38	2931
1985	0	0	31	168	358	456	523	587	457	269	137	9	2995
1986	0	7	56	186	340	513	605	568	551	250	116	6	3198
1987	1	1	22	157	384	493	595	631	481	233	84	15	3097
1988	2	2	31	140	330	466	573	638	486	289	173	11	3141
1989	18	16	44	168	410	478	569	584	426	283	118	3	3117
1990	0	7	46	148	384	585	580	643	516	248	100	38	3295
1991	0	9	63	233	390	521	604	600	435	311	28	18	3212
1992	0	10	43	134	289	494	598	539	474	298	46	6	2931
1993	7	10	27	89	301	503	633	648	503	262	39	30	3052

SNOWFALL (inches) — GALVESTON, TEXAS

SEASON	JULY	AUG	SEP	OCT	NOV	DEC	JAN	FEB	MAR	APR	MAY	JUNE	TOTAL
1970-71	0.0	0.0	0.0	0.0	0.0	0.0	T	0.0	0.0	0.0	0.0	0.0	T
1971-72	0.0	0.0	0.0	0.0	0.0	0.0	0.0	0.0	0.0	0.0	0.0	0.0	0.0
1972-73	0.0	0.0	0.0	0.0	0.0	0.0	2.5	1.6	0.0	0.0	0.0	0.0	4.1
1973-74	0.0	0.0	0.0	0.0	0.0	0.0	0.0	0.0	0.0	0.0	0.0	0.0	0.0
1974-75	0.0	0.0	0.0	0.0	0.0	0.0	0.0	0.0	0.0	0.0	0.0	0.0	0.0
1975-76	0.0	0.0	0.0	0.0	0.0	0.0	0.0	0.0	0.0	0.0	0.0	0.0	0.0
1976-77	0.0	0.0	0.0	0.0	0.0	0.0	0.0	0.0	0.0	0.0	0.0	0.0	0.0
1977-78	0.0	0.0	0.0	0.0	0.0	0.0	T	0.0	0.0	0.0	0.0	0.0	T
1978-79	0.0	0.0	0.0	0.0	0.0	0.0	0.0	0.0	0.0	0.0	0.0	0.0	0.0
1979-80	0.0	0.0	0.0	0.0	0.0	0.0	0.0	T	0.0	0.0	0.0	0.0	T
1980-81	0.0	0.0	0.0	0.0	0.0	0.0	0.0	0.0	0.0	0.0	0.0	0.0	0.0
1981-82	0.0	0.0	0.0	0.0	0.0	0.0	T	0.0	0.0	0.0	0.0	0.0	T
1982-83	0.0	0.0	0.0	0.0	0.0	0.0	0.0	0.0	0.0	0.0	0.0	0.0	0.0
1983-84	0.0	0.0	0.0	0.0	0.0	0.0	0.0	0.0	0.0	0.0	0.0	0.0	0.0
1984-85	0.0	0.0	0.0	0.0	0.0	0.0	T	0.0	0.0	0.0	0.0	0.0	T
1985-86	0.0	0.0	0.0	0.0	0.0	0.0	0.0	0.0	0.0	0.0	0.0	0.0	0.0
1986-87	0.0	0.0	0.0	0.0	0.0	0.0	0.0	0.0	0.0	0.0	0.0	0.0	0.0
1987-88	0.0	0.0	0.0	0.0	0.0	0.0	0.0	T	0.0	0.0	0.0	0.0	T
1988-89	0.0	0.0	0.0	0.0	0.0	0.0	0.0	0.0	T	0.0	0.0	0.0	T
1989-90	0.0	0.0	0.0	0.0	0.0	1.0	0.0	0.0	0.0	0.0	0.0	0.0	1.0
1990-91	0.0	0.0	0.0	0.0	0.0	T	0.0	0.0	0.0	0.0	T	0.0	T
1991-92	0.0	0.0	0.0	0.0	0.0	0.0	0.0	0.0	0.0	T	0.0	0.0	T
1992-93	0.0	0.0	0.0	0.0	0.0	0.0	T	0.0	0.0	0.0	0.0	0.0	T
1993-94	0.0	0.0	0.0	0.0	0.0	0.0							
Record Mean	0.0	0.0	0.0	0.0	0.0	T	T	0.2	T	T	T	0.0	0.2

HOUSTON, TEXAS

Houston, the largest city in Texas, is located in the flat Coastal Plains, about 50 miles from the Gulf of Mexico and about 25 miles from Galveston Bay. The climate is predominantly marine. The terrain includes numerous small streams and bayous which, together with the nearness to Galveston Bay, favor the development of both ground and advective fogs. Prevailing winds are from the southeast and south, except in January, when frequent passages of high pressure areas bring invasions of polar air and prevailing northerly winds.

Temperatures are moderated by the influence of winds from the Gulf, which result in mild winters. Another effect of the nearness of the Gulf is abundant rainfall, except for rare extended dry periods. Polar air penetrates the area frequently enough to provide variability in the weather.

Records of sky cover for daylight hours indicate about one-fourth of the days per year as clear, with a high number of clear days in October and November. Cloudy days are relatively frequent from December to May and partly cloudy days are the more frequent for June through September. Sunshine averages nearly 60 percent of the possible amount for the year ranging from 42 percent in January to 67 percent in June.

Heavy fog occurs on an average of 16 days a year and light fog occurs about 62 days a year in the city. The frequency of heavy fog is considerably higher at William P. Hobby Airport and at Intercontinental Airport.

Destructive windstorms are fairly infrequent, but both thundersqualls and tropical storms occasionally pass through the area.

NORMALS, MEANS, AND EXTREMES

HOUSTON, TEXAS

LATITUDE: 29°58'N LONGITUDE: 95°21'W ELEVATION: FT. GRND 96 BARO 122 TIME ZONE: CENTRAL WBAN: 12960

	(a)	JAN	FEB	MAR	APR	MAY	JUNE	JULY	AUG	SEP	OCT	NOV	DEC	YEAR
TEMPERATURE °F:														
Normals														
-Daily Maximum		61.0	65.3	71.1	78.4	84.6	90.1	92.7	92.5	88.4	81.6	72.4	64.7	78.6
-Daily Minimum		39.7	42.6	50.0	58.1	64.4	70.6	72.4	72.0	67.9	57.6	49.6	42.2	57.3
-Monthly		50.4	53.9	60.6	68.3	74.5	80.4	82.6	82.3	78.2	69.6	61.0	53.5	67.9
Extremes														
-Record Highest	24	84	91	91	95	97	103	104	107	102	96	89	83	107
-Year		1975	1986	1989	1987	1990	1980	1980	1980	1985	1991	1989	1978	AUG 1980
-Record Lowest	24	12	20	22	31	44	52	62	60	48	29	19	7	7
-Year		1982	1985	1980	1987	1978	1970	1990	1992	1975	1993	1976	1989	DEC 1989
NORMAL DEGREE DAYS:														
Heating (base 65°F)		468	322	187	36	0	0	0	0	0	31	181	374	1599
Cooling (base 65°F)		16	11	50	135	295	462	546	536	396	174	61	18	2700
% OF POSSIBLE SUNSHINE	24	43	49	52	56	60	66	68	67	65	63	51	50	58
MEAN SKY COVER (tenths)														
Sunrise - Sunset	24	7.0	6.5	6.6	6.6	6.4	5.8	5.7	5.6	5.6	5.2	5.9	6.8	6.1
MEAN NUMBER OF DAYS:														
Sunrise to Sunset														
-Clear	24	7.2	7.3	7.1	7.2	6.1	7.3	7.0	6.1	8.3	11.0	9.1	7.5	91.0
-Partly Cloudy	24	5.3	5.6	6.5	7.1	10.7	13.3	15.7	17.0	11.6	9.1	7.1	5.3	114.5
-Cloudy	24	18.5	15.4	17.4	15.7	14.2	9.4	8.3	7.8	10.1	10.9	13.8	18.2	159.7
Precipitation														
.01 inches or more	24	10.7	8.5	9.2	7.3	8.5	9.3	9.4	8.8	9.3	7.4	8.5	9.2	105.8
Snow,Ice pellets,hail														
1.0 inches or more	24	0.1	0.1	0.0	0.0	0.0	0.0	0.0	0.0	0.0	0.0	0.0	0.*	0.3
Thunderstorms	24	2.2	2.0	3.8	3.9	7.2	8.1	10.6	10.2	7.5	3.9	2.8	1.8	63.9
Heavy Fog Visibility														
1/4 mile or less	24	5.0	3.6	3.0	2.8	1.7	0.6	0.2	0.4	1.3	2.9	3.6	4.3	29.5
Temperature °F														
-Maximum														
90° and above	24	0.0	0.*	0.1	0.8	5.1	19.8	26.6	25.6	15.0	2.9	0.0	0.0	96.0
32° and below	24	0.2	0.2	0.0	0.0	0.0	0.0	0.0	0.0	0.0	0.0	0.0	0.3	0.7
-Minimum														
32° and below	24	7.2	4.3	1.3	0.1	0.0	0.0	0.0	0.0	0.0	0.1	1.7	5.1	19.7
0° and below	24	0.0	0.0	0.0	0.0	0.0	0.0	0.0	0.0	0.0	0.0	0.0	0.0	0.0
AVG. STATION PRESS.(mb)	21	1017.1	1015.7	1012.1	1011.1	1009.7	1010.4	1012.3	1011.9	1011.7	1014.0	1015.0	1016.6	1013.1
RELATIVE HUMIDITY (%)														
Hour 00 (Local Time)	24	82	83	83	85	87	87	86	87	89	88	85	84	86
Hour 06 (Local Time)	24	86	86	87	89	92	92	93	93	93	91	89	87	90
Hour 12 (Local Time)	24	64	61	59	58	60	60	57	57	60	56	60	62	60
Hour 18 (Local Time)	24	67	61	60	61	63	63	62	62	66	68	71	71	65
PRECIPITATION (inches):														
Water Equivalent														
-Normal		3.29	2.96	2.92	3.21	5.24	4.96	3.60	3.49	4.89	4.27	3.79	3.45	46.07
-Maximum Monthly	24	9.78	5.99	8.52	10.92	14.39	16.28	8.10	9.42	11.35	16.05	8.91	9.34	16.28
-Year		1991	1992	1972	1976	1970	1989	1979	1983	1976	1984	1982	1991	JUN 1989
-Minimum Monthly	24	0.36	0.38	0.88	0.43	0.79	0.26	0.47	0.31	0.80	0.05	0.41	0.64	0.05
-Year		1971	1976	1987	1983	1977	1970	1993	1990	1975	1978	1988	1973	OCT 1978
-Maximum in 24 hrs	24	2.66	2.22	7.47	8.16	10.36	10.35	3.99	6.83	7.98	9.31	4.19	4.04	10.36
-Year		1992	1985	1972	1976	1989	1989	1973	1981	1976	1984	1986	1991	MAY 1989
Snow,Ice pellets,hail														
-Maximum Monthly	24	2.0	2.8	T	T	T	T	0.0	0.0	0.0	0.0	T	1.7	2.8
-Year		1973	1973	1992	1993	1993	1990					1979	1989	FEB 1973
-Maximum in 24 hrs	24	2.0	1.4	T	T	T	T	0.0	0.0	0.0	0.0	T	1.7	2.0
-Year		1973	1980	1992	1993	1993	1990					1979	1989	JAN 1973
WIND:														
Mean Speed (mph)	24	8.2	8.8	9.3	9.2	8.2	7.8	7.0	6.3	6.9	7.0	8.0	8.0	7.9
Prevailing Direction through 1963		NNW	SSE	SSE	SSE	SSE	SSE	S	SSE	SSE	ESE	SSE	SSE	SSE
Fastest Obs. 1 Min.														
-Direction (!!)	24	33	26	22	14	23	30	10	08	05	27	33	31	08
-Speed (MPH)	24	32	46	35	45	46	45	46	51	37	41	37	35	51
-Year		1993	1984	1989	1978	1983	1973	1969	1983	1982	1988	1972	1973	AUG 1983
Peak Gust														
-Direction (!!)	10	W	W	W	W	W	NE	SE	E	E	W	NW	NW	E
-Speed (mph)	10	44	61	51	56	52	68	52	78	44	58	45	56	78
-Date		1992	1984	1990	1990	1993	1990	1990	1983	1990	1988	1988	1990	AUG 1983

PRECIPITATION (inches)

HOUSTON, TEXAS

YEAR	JAN	FEB	MAR	APR	MAY	JUNE	JULY	AUG	SEP	OCT	NOV	DEC	ANNUAL
1964	2.89	4.97	2.24	1.63	2.25	1.89	1.68	2.61	6.76	2.35	4.28	5.57	39.12
1965	1.87	3.27	0.81	0.95	6.53	3.06	1.57	2.29	3.56	3.09	4.82	6.15	37.97
1966	4.46	7.75	2.20	7.98	11.21	4.42	1.45	7.11	4.01	5.45	1.56	1.53	59.13
1967	2.41	2.17	1.83	4.42	2.54	0.17	7.77	1.60	4.84	3.18	0.50	5.02	36.45
1968	8.02	1.99	2.92	3.02	13.24	11.18	6.49	2.90	3.87	3.91	2.71	1.19	61.44
#1969	2.74	5.31	3.18	3.34	4.73	1.51	3.89	2.67	6.08	3.30	2.13	4.38	43.26
1970	1.93	2.52	5.08	2.21	14.39	0.26	2.28	2.03	6.22	9.09	1.54	0.64	48.19
1971	0.36	2.11	1.21	2.14	3.41	2.42	1.42	6.95	5.17	3.49	1.82	7.33	37.83
1972	3.30	1.20	8.52	2.85	6.99	3.02	2.76	3.90	6.23	3.34	6.49	2.20	50.80
1973	5.00	3.40	3.68	7.15	4.22	13.46	6.77	3.73	9.38	9.31	1.59	2.47	70.16
1974	7.68	0.55	4.20	1.68	5.61	0.59	1.75	6.94	4.51	4.53	7.90	3.35	49.29
1975	1.97	2.63	3.19	4.80	7.57	7.50	5.48	5.72	0.80	5.62	2.08	3.61	50.97
1976	1.39	0.38	1.53	10.92	5.80	2.63	3.93	1.59	11.35	5.83	3.05	6.22	54.62
1977	2.67	1.70	1.95	4.34	0.79	3.55	2.69	4.45	3.92	0.82	5.17	2.89	34.94
1978	7.15	3.07	1.70	0.57	4.15	9.37	2.35	3.66	4.27	0.05	5.99	2.60	44.93
1979	6.30	5.23	2.88	7.79	3.78	1.88	8.10	4.57	9.83	2.80	1.78	4.03	58.97
1980	6.09	2.54	5.39	2.05	5.63	0.92	1.57	1.40	6.00	4.03	2.12	1.25	38.99
1981	2.32	2.21	1.74	2.69	8.75	9.65	4.43	7.01	2.91	6.96	5.26	2.05	55.98
1982	1.82	1.59	1.55	2.28	6.87	1.10	4.32	1.90	0.98	6.64	8.91	4.91	42.87
1983	2.00	3.97	3.85	0.43	7.29	5.37	5.23	9.42	7.23	1.56	3.17	3.69	53.21
1984	3.99	4.37	2.41	0.56	3.13	1.99	3.43	3.52	3.87	16.05	2.28	2.59	48.19
1985	2.10	5.38	4.52	4.31	1.57	5.29	4.93	1.14	4.67	6.54	4.84	3.85	49.14
1986	0.71	2.74	1.44	2.63	4.29	6.34	0.61	3.27	3.70	6.83	6.66	5.71	44.93
1987	2.42	4.26	0.88	0.47	5.39	9.31	4.79	1.48	3.46	0.17	3.41	4.56	40.60
1988	1.27	1.29	4.88	1.26	1.32	2.00	3.23	3.52	1.20	1.29	0.41	1.26	22.93
1989	4.80	0.90	3.96	1.48	13.56	16.28	1.92	2.74	2.69	1.76	1.84	0.80	52.73
1990	3.96	4.54	5.11	6.21	2.23	2.98	4.85	0.31	1.57	3.79	3.01	1.81	40.37
1991	9.78	5.79	1.77	8.06	4.02	7.69	1.31	2.97	2.76	2.57	5.03	9.34	61.09
1992	7.70	5.99	6.28	3.74	7.05	3.38	3.85	2.78	1.08	1.03	5.99	3.46	52.33
1993	5.79	2.67	6.41	7.88	8.50	12.08	0.47	1.82	1.10	5.32	3.27	2.68	57.99
Record Mean	3.79	3.16	2.86	3.48	4.92	4.81	3.96	3.91	4.40	3.91	4.01	4.00	47.20

AVERAGE TEMPERATURE (deg. F)

HOUSTON, TEXAS

YEAR	JAN	FEB	MAR	APR	MAY	JUNE	JULY	AUG	SEP	OCT	NOV	DEC	ANNUAL
1964	51.6	49.4	60.4	70.2	75.8	80.4	83.7	84.4	79.3	67.8	65.0	55.9	68.7
1965	56.0	55.1	58.7	73.5	77.1	83.0	85.1	83.5	81.2	69.4	69.1	58.9	70.9
1966	48.4	52.8	61.5	70.6	75.9	80.0	84.3	82.4	79.6	69.7	64.7	54.2	68.7
1967	54.9	54.5	67.6	75.7	75.9	82.6	82.1	81.1	77.6	70.5	64.1	55.4	70.2
1968	52.7	50.4	59.0	71.1	76.3	80.5	82.5	84.0	78.6	73.5	59.7	55.7	68.7
#1969	56.7	56.9	56.1	70.3	75.4	80.0	84.4	83.2	78.2	71.0	58.5	55.2	68.8
1970	46.7	53.9	56.9	69.3	72.1	78.4	81.4	83.1	78.9	66.7	56.9	60.5	67.1
1971	56.7	55.7	59.4	66.6	74.1	80.3	83.9	80.4	78.3	72.0	60.0	59.9	68.9
1972	56.5	55.2	64.3	71.2	73.7	80.8	80.3	80.3	79.6	69.8	54.7	52.0	68.2
1973	47.4	51.4	63.7	64.6	72.8	79.2	83.0	79.5	78.1	71.8	67.3	53.5	67.7
1974	55.0	56.2	66.8	67.2	76.9	79.9	82.8	81.6	74.6	70.6	60.2	54.6	68.9
1975	56.9	55.4	61.1	68.3	75.9	80.0	81.5	81.1	74.9	69.5	59.7	52.7	68.1
1976	50.6	60.1	62.2	67.7	70.5	78.4	80.5	81.4	76.2	60.6	51.8	49.2	65.8
1977	42.7	53.8	60.9	66.9	74.5	81.0	82.4	83.1	80.0	69.2	61.8	53.7	67.5
1978	40.8	45.1	57.3	67.6	76.0	80.4	83.7	83.1	79.3	68.9	64.7	52.9	66.7
1979	44.1	51.7	62.4	68.7	73.1	79.8	82.6	81.5	75.6	70.7	55.7	52.4	66.5
1980	55.0	53.7	60.9	66.2	77.3	85.1	87.5	86.6	83.2	67.8	58.0	55.2	69.7
1981	51.4	55.4	60.9	74.3	75.3	82.7	84.4	84.4	78.6	72.3	64.4	54.5	69.9
1982	52.9	52.1	64.9	67.8	75.3	83.0	85.4	84.1	79.3	69.5	60.9	55.4	69.2
1983	50.1	52.5	58.3	64.0	73.4	79.0	82.2	82.6	76.6	70.1	63.1	45.7	66.5
1984	47.0	54.0	61.9	67.8	74.9	78.6	81.8	82.9	77.4	74.2	60.0	63.4	68.7
1985	45.7	49.6	64.7	70.0	75.6	81.0	81.6	84.2	79.8	72.5	67.0	51.0	68.6
1986	54.4	59.9	63.3	71.7	75.8	82.0	85.9	82.6	81.8	68.9	62.0	51.7	70.0
1987	51.4	56.1	58.9	67.2	77.1	81.3	83.5	86.2	78.9	68.7	60.5	55.6	68.8
1988	48.1	54.1	61.3	67.6	73.6	80.5	84.4	85.3	80.8	72.0	65.7	55.4	69.1
1989	57.5	52.7	61.3	69.4	77.8	79.9	82.4	81.7	77.0	70.2	62.9	44.4	68.1
1990	57.0	59.1	62.9	69.4	78.1	84.8	82.1	85.1	80.1	68.7	63.4	53.6	70.4
1991	50.4	57.4	63.5	72.2	78.0	82.0	84.0	83.0	77.4	72.3	56.7	56.2	69.4
1992	51.0	58.5	64.0	68.7	73.7	81.7	83.6	80.1	79.3	71.4	56.8	56.7	68.8
1993	53.6	56.7	61.1	65.9	73.4	81.6	85.8	86.5	80.2	69.5	56.9	54.6	68.8
Record Mean	50.9	54.6	61.8	68.3	74.9	80.8	83.2	82.9	78.5	70.0	60.4	54.0	68.3
Max	61.0	65.5	72.7	78.9	84.8	90.7	93.4	93.3	88.6	81.5	71.5	64.7	78.9
Min	40.8	43.7	50.8	57.8	65.0	71.0	73.0	72.5	68.4	58.4	49.3	43.3	57.8

REFERENCE NOTES

HOUSTON, TEXAS

GENERAL
T - TRACE AMOUNT.
BLANK ENTRIES DENOTE MISSING/UNREPORTED DATA.
# INDICATES A STATION OR INSTRUMENT RELOCATION.

SPECIFIC

PM - INCLUDES LAST DAY OF PREVIOUS MONTH
ASOS - AUTOMATED SURFACE OBSERVING SYSTEM IN OPERATION DURING THESE MONTHS.

(a) - LENGTH OF RECORD IN YEARS, ALTHOUGH INDIVIDUAL MONTHS MAY BE MISSING.
0.* OR * - THE VALUE IS BETWEEN 0.0 AND 0.05
NORMALS - BASED ON THE 1961-1990 RECORD PERIOD.
EXTREMES - DATES ARE THE MOST RECENT OCCURRENCE
WIND DIR.- NUMERALS SHOW TENS OF DEGRESS CLOCKWISE FROM TRUE NORTH. "00" INDICATES CALM.
RESULTANT DIRECTIONS ARE GIVEN TO WHOLE DEGREES.
BOLD VALUES INDICATE EXTREME VALUES WHICH OCCURRED AFTER THE ASOS SYSTEM WAS COMMISSIONED.

RECORD = PERIOD OF RECORD
RECORD MEAN PRECIPITATION IS THE MEAN OF ALL DAILY PRECIPITATION AMOUNTS DURING THE PERIOD OF RECORD.
RECORD MAX(MIN) TEMPERATURE IS THE MEAN OF ALL DAILY MAX(MIN) TEMPERATURES DURING THE PERIOD OF RECORD.
RECORD MEAN TEMPERATURE IS THE SUM OF THE RECORD MAX AND RECORD MIN DIVIDED BY 2.
AVERAGE TEMPERATURE IS THE SUM OF THE MEAN DAILY MAX AND MIN TEMPERATURE DIVIDED BY 2.

EXCEPTIONS

RECORD MEANS ARE THROUGH THE CURRENT YEAR, BEGINNING IN 1969 FOR TEMPERATURE
1933 FOR PRECIPITATION
1935 FOR SNOWFALL

HEATING DEGREE DAYS Base 65 deg. F HOUSTON, TEXAS

SEASON	JULY	AUG	SEP	OCT	NOV	DEC	JAN	FEB	MAR	APR	MAY	JUNE	TOTAL
1964-65	0	0	0	31	114	315	300	284	250	0	0	0	1294
1965-66	0	0	2	20	23	212	516	334	144	12	0	0	1263
1966-67	0	0	0	27	99	359	342	298	68	0	0	0	1193
1967-68	0	0	3	18	99	312	390	415	229	17	0	0	1483
#1968-69	0	0	0	5	199	297	284	234	281	1	0	2	1303
1969-70	0	0	0	29	238	304	579	309	252	51	12	0	1774
1970-71	0	0	0	72	274	209	298	273	219	72	3	0	1420
1971-72	0	0	2	6	195	194	315	295	85	17	0	0	1109
1972-73	0	0	2	50	320	410	540	379	75	117	5	0	1898
1973-74	0	0	0	8	74	364	330	273	95	60	0	0	1204
1974-75	0	0	0	15	196	336	290	270	179	48	0	0	1334
1975-76	0	0	0	26	217	399	441	178	155	26	7	0	1449
1976-77	0	0	0	173	398	484	687	312	166	25	0	0	2245
1977-78	0	0	0	34	150	365	752	553	250	33	17	0	2154
1978-79	0	0	0	22	111	393	646	376	135	23	2	0	1708
1979-80	0	0	0	27	297	389	308	350	169	45	0	0	1585
1980-81	0	0	0	67	255	323	416	291	144	6	1	0	1503
1981-82	0	0	0	50	82	326	409	363	143	79	1	0	1453
1982-83	0	0	0	53	175	328	457	346	219	96	0	0	1674
1983-84	0	0	6	27	138	606	549	325	150	45	2	0	1848
1984-85	0	0	6	12	204	144	591	432	91	22	0	0	1502
1985-86	0	0	5	17	76	434	326	209	99	11	0	0	1177
1986-87	0	0	0	28	175	411	421	245	196	82	0	0	1558
1987-88	0	0	0	16	185	301	525	331	171	35	0	0	1564
1988-89	0	0	0	5	120	309	260	379	210	56	0	0	1339
1989-90	0	0	0	47	160	637	264	177	122	34	0	0	1441
1990-91	0	0	0	61	129	395	448	222	115	8	0	0	1378
1991-92	0	0	0	15	289	303	428	197	95	37	4	0	1368
1992-93	0	0	0	1	270	268	351	235	157	62	0	0	1344
1993-94	0	0	0	76	269	343							

COOLING DEGREE DAYS Base 65 deg. F HOUSTON, TEXAS

YEAR	JAN	FEB	MAR	APR	MAY	JUNE	JULY	AUG	SEP	OCT	NOV	DEC	TOTAL
#1969	35	13	11	167	328	456	608	569	402	222	48	9	2868
1970	18	3	7	188	238	409	513	567	423	131	38	76	2611
1971	48	18	55	126	292	466	594	485	409	229	52	44	2818
1972	58	20	71	208	275	480	480	482	447	206	17	12	2756
1973	1	4	41	111	253	434	564	458	401	225	151	12	2655
1974	24	33	158	132	374	454	558	519	295	196	60	18	2821
1975	47	8	61	155	342	455	514	505	303	174	68	24	2656
1976	5	43	75	110	182	408	490	520	341	42	9	0	2225
1977	0	5	44	91	302	487	547	565	456	173	58	23	2751
1978	10	5	19	120	369	471	584	568	437	150	108	25	2866
1979	7	13	62	142	261	454	552	519	324	211	26	6	2577
1980	4	31	49	86	388	610	705	677	553	162	52	26	3343
1981	1	28	23	295	330	538	606	609	413	285	71	7	3206
1982	39	11	147	170	329	547	641	599	437	199	60	40	3219
1983	0	0	18	76	268	427	541	554	362	196	87	18	2547
1984	0	13	64	135	315	415	527	562	384	302	62	100	2879
1985	0	6	87	180	335	487	521	602	456	257	143	9	3083
1986	4	71	52	220	341	518	654	553	510	157	92	4	3172
1987	4	4	14	154	383	497	580	661	423	137	54	15	2926
1988	7	20	65	121	274	472	609	637	478	229	144	20	3076
1989	33	44	105	194	405	454	547	526	363	218	105	5	2999
1990	20	19	65	174	413	603	536	630	456	181	87	47	3231
1991	0	14	76	231	408	514	593	565	376	248	47	37	3109
1992	0	17	74	155	281	508	584	476	437	210	32	20	2794
1993	4	10	42	92	267	506	652	674	463	221	31	27	2989

SNOWFALL (inches) HOUSTON, TEXAS

SEASON	JULY	AUG	SEP	OCT	NOV	DEC	JAN	FEB	MAR	APR	MAY	JUNE	TOTAL
1970-71	0.0	0.0	0.0	0.0	0.0	0.0	T	0.0	0.0	0.0	0.0	0.0	T
1971-72	0.0	0.0	0.0	0.0	0.0	0.0	T	0.0	0.0	0.0	0.0	0.0	T
1972-73	0.0	0.0	0.0	0.0	0.0	0.0	2.0	2.8	0.0	0.0	0.0	0.0	4.8
1973-74	0.0	0.0	0.0	0.0	0.0	0.0	0.0	0.0	0.0	0.0	0.0	0.0	0.0
1974-75	0.0	0.0	0.0	0.0	0.0	0.0	T	0.0	0.0	0.0	0.0	0.0	T
1975-76	0.0	0.0	0.0	0.0	0.0	0.0	0.0	0.0	0.0	0.0	0.0	0.0	0.0
1976-77	0.0	0.0	0.0	0.0	T	0.0	0.0	0.0	0.0	0.0	0.0	0.0	T
1977-78	0.0	0.0	0.0	0.0	0.0	0.0	0.4	0.0	0.0	0.0	0.0	0.0	0.4
1978-79	0.0	0.0	0.0	0.0	0.0	0.0	T	0.0	0.0	0.0	0.0	0.0	T
1979-80	0.0	0.0	0.0	0.0	T	0.0	0.0	1.4	0.0	0.0	0.0	0.0	1.4
1980-81	0.0	0.0	0.0	0.0	0.0	0.0	T	T	0.0	0.0	0.0	0.0	T
1981-82	0.0	0.0	0.0	0.0	0.0	0.0	T	0.0	0.0	0.0	0.0	0.0	T
1982-83	0.0	0.0	0.0	0.0	0.0	0.0	0.0	0.0	0.0	0.0	0.0	0.0	0.0
1983-84	0.0	0.0	0.0	0.0	0.0	0.0	0.0	0.0	0.0	0.0	0.0	0.0	0.0
1984-85	0.0	0.0	0.0	0.0	0.0	0.0	1.4	0.3	0.0	0.0	0.0	0.0	1.7
1985-86	0.0	0.0	0.0	0.0	0.0	0.0	0.0	0.0	0.0	0.0	0.0	0.0	0.0
1986-87	0.0	0.0	0.0	0.0	0.0	0.0	0.0	0.0	0.0	0.0	0.0	0.0	0.0
1987-88	0.0	0.0	0.0	0.0	0.0	0.0	0.0	T	0.0	0.0	0.0	0.0	T
1988-89	0.0	0.0	0.0	0.0	0.0	0.0	0.0	T	0.0	0.0	0.0	0.0	T
1989-90	0.0	0.0	0.0	0.0	0.0	1.7	0.0	0.0	0.0	0.0	0.0	T	1.7
1990-91	0.0	0.0	0.0	0.0	0.0	0.0	0.0	0.0	0.0	0.0	0.0	0.0	0.0
1991-92	0.0	0.0	0.0	0.0	0.0	0.0	0.0	0.0	T	0.0	T	0.0	T
1992-93	0.0	0.0	0.0	0.0	0.0	0.0	0.0	0.0	0.0	T	T	0.0	T
1993-94	0.0	0.0	0.0	0.0	0.0	0.0							
Record Mean	0.0	0.0	0.0	0.0	T	T	0.2	0.2	T	T	T	T	0.4

LUBBOCK, TEXAS

Lubbock is located on a plateau area of Northwestern Texas that is referred to locally as the South Plains Region. The general elevation of the area is about 3,250 feet. The Region is a major part of the Llano Estacado (staked plains). The latter, which includes a large portion of Northwest Texas, is bounded on the east and southeast by an erosional escarpment that is usually referred to as the Cap Rock. The Llano Estacado extends southwestward into the upper Pecos Valley and westward into eastern New Mexico.

The South Plains are predominately flat, but contain numerous small playas (or clay lined depressions) and small stream valleys. During the rainy months the playas collect run-off water and form small lakes or ponds. The stream valleys drain into the major rivers of West Texas, but throughout most of the year these streams carry only very light flows.

The escarpment, or Cap Rock, is the primary terrain feature that causes a noticeable distortion of the smooth wind flow patterns across the South Plains. The most noticeable influence is on southeasterly winds as they are deflected upward along the face of the escarpment.

The Lubbock area is the heart of the largest cotton-producing section of Texas. Grain sorghum production and cattle feeding make significant contributions to the agroeconomy of the area. Irrigation from underground sources is often used as a supplement to natural rainfall to improve crop yields. The soils of the region are sandy clay loams which consist of limy clays, silts, and sands of a reddish hue.

The area is semi-arid, transitional between the desert conditions on the west and the humid climates to the east and southeast. The greatest monthly rainfall totals occur from May through September when warm moist tropical air may be carried into the area from the Gulf of Mexico. This air mass often brings moderate to heavy afternoon and evening thunderstorms, accompanied by hail. Precipitation across the area is characterized by its variability. The monthly precipitation extremes range from trace amounts in several isolated months to 14 inches.

Snow may occur from late October until April. Each snowfall is generally light and seldom remains on the ground for more than two or three days at any one period.

High winds are associated primarily with intense thunderstorms and at times may cause significant damage to structures. Winds in excess of 25 mph occasionally occur for periods of 12 hours or longer. These prolonged winds are generally associated with late winter and springtime low-pressure centers. Spring winds often bring widespread dust causing discomfort to residents for periods of several hours.

Overall, the climate of the region is rated as pleasant. Most periods of disagreeable weather are of short duration. They generally occur from the winter months into the early summer months.

The summer heat is not considered oppressive. One moderating factor is a variable, but usually gentle, wind. Intrusions of dry air from the west often reduce any discomfort from the summer heat and lower temperatures into the 60s.

The average first occurrence of temperatures below 32 degrees Fahrenheit in the fall is the first of November and the average last occurrence in the spring is in mid April.

NORMALS, MEANS, AND EXTREMES

LUBBOCK, TEXAS

LATITUDE: 33°39'N LONGITUDE: 101°49' W ELEVATION: FT. GRND 3254 BARO 3258 TIME ZONE: CENTRAL WBAN: 23042

	(a)	JAN	FEB	MAR	APR	MAY	JUNE	JULY	AUG	SEP	OCT	NOV	DEC	YEAR
TEMPERATURE °F:														
Normals														
-Daily Maximum		52.9	57.6	66.0	75.4	83.1	90.0	91.9	89.6	82.9	74.7	63.2	54.1	73.5
-Daily Minimum		24.6	28.6	36.4	46.7	55.8	64.3	68.0	66.2	59.4	48.1	36.5	27.2	46.8
-Monthly		38.8	43.1	51.2	61.1	69.4	77.2	80.0	77.9	71.1	61.4	49.8	40.6	60.1
Extremes														
-Record Highest	47	83	87	95	100	104	110	108	106	103	98	86	81	110
-Year		1972	1979	1989	1989	1989	1990	1983	1966	1948	1979	1980	1958	JUN 1990
-Record Lowest	47	-16	-8	2	22	30	44	51	52	33	18	-1	-2	-16
-Year		1963	1960	1948	1948	1967	1947	1952	1956	1983	1993	1957	1989	JAN 1963
NORMAL DEGREE DAYS:														
Heating (base 65°F)		812	613	437	161	25	0	0	0	17	154	456	756	3431
Cooling (base 65°F)		0	0	10	44	162	366	465	400	200	42	0	0	1689
% OF POSSIBLE SUNSHINE	21	65	66	73	74	71	75	77	76	71	75	68	65	71
MEAN SKY COVER (tenths)														
Sunrise - Sunset	45	5.3	5.2	5.0	4.8	5.0	4.2	4.4	4.3	4.4	3.8	4.4	4.9	4.6
MEAN NUMBER OF DAYS:														
Sunrise to Sunset														
-Clear	47	12.5	10.7	11.9	12.5	11.2	13.6	14.0	14.9	14.4	16.8	14.7	13.3	160.3
-Partly Cloudy	47	6.4	7.2	8.7	8.7	11.1	10.6	11.0	10.0	7.7	6.6	6.7	7.1	101.7
-Cloudy	47	12.1	10.3	10.4	8.9	8.7	5.8	6.1	6.1	7.9	7.7	8.6	10.7	103.3
Precipitation														
.01 inches or more	47	3.9	4.2	4.1	4.4	7.5	7.1	6.9	6.8	6.0	4.7	3.4	4.1	63.2
Snow,Ice pellets,hail														
1.0 inches or more	47	0.8	0.9	0.5	0.*	0.0	0.0	0.0	0.0	0.0	0.*	0.3	0.6	3.3
Thunderstorms	47	0.2	0.5	1.9	3.6	8.6	9.4	7.6	7.4	4.4	2.6	0.8	0.3	47.3
Heavy Fog Visibility														
1/4 mile or less	47	3.1	2.9	1.6	1.0	1.1	0.4	0.1	0.4	1.2	1.7	2.3	1.9	18.0
Temperature °F														
-Maximum														
90° and above	46	0.0	0.0	0.1	1.8	8.2	18.2	22.2	19.5	8.5	1.0	0.0	0.0	79.5
32° and below	46	3.1	1.3	0.4	0.0	0.0	0.0	0.0	0.0	0.0	0.*	0.2	1.6	6.7
-Minimum														
32° and below	46	25.5	18.8	10.6	1.8	0.1	0.0	0.0	0.0	0.0	1.0	12.1	23.5	93.3
0° and below	46	0.3	0.1	0.0	0.0	0.0	0.0	0.0	0.0	0.0	0.0	0.*	0.1	0.5
AVG. STATION PRESS.(mb)	21	905.5	904.5	901.4	901.6	901.2	902.6	904.8	905.2	905.2	905.5	904.8	905.4	904.0
RELATIVE HUMIDITY (%)														
Hour 00	46	66	64	56	55	62	63	62	65	69	67	65	65	63
Hour 06 (Local Time)	46	73	73	68	69	76	78	75	78	81	78	74	72	75
Hour 12	46	50	50	41	39	43	45	47	49	52	48	46	48	47
Hour 18	46	47	42	33	31	36	37	39	42	45	44	46	48	41
PRECIPITATION (inches):														
Water Equivalent														
-Normal		0.39	0.68	0.89	0.97	2.35	2.75	2.37	2.51	2.60	1.86	0.75	0.53	18.65
-Maximum Monthly	47	4.05	2.51	3.23	3.48	7.80	7.95	7.20	8.85	6.90	10.80	2.67	2.24	10.80
-Year		1949	1961	1958	1957	1949	1967	1976	1966	1986	1983	1968	1991	OCT 1983
-Minimum Monthly	47	0.00	T	T	0.04	0.10	T	T	0.05	T	0.00	0.00	T	0.00
-Year		1967	1955	1972	1989	1962	1990	1970	1960	1954	1952	1960	1973	JAN 1967
-Maximum in 24 hrs	47	1.56	2.15	1.80	2.18	5.14	5.70	3.25	3.78	3.61	5.82	1.57	1.12	5.82
-Year		1983	1961	1973	1982	1949	1967	1985	1966	1991	1983	1968	1959	OCT 1983
Snow,Ice pellets,hail														
-Maximum Monthly	45	25.3	16.8	14.3	5.3	0.7	T	T	0.0	0.0	7.5	21.4	9.9	25.3
-Year		1983	1956	1958	1983	1991	1992	1990			1976	1980	1960	JAN 1983
-Maximum in 24 hrs	45	16.3	12.1	10.0	4.5	0.7	T	T	0.0	0.0	4.7	10.8	6.3	16.3
-Year		1983	1961	1969	1983	1991	1992	1990			1976	1980	1960	JAN 1983
WIND:														
Mean Speed (mph)	44	12.0	13.2	14.7	14.8	14.2	13.6	11.3	10.0	10.6	11.2	11.7	11.9	12.4
Prevailing Direction through 1963		SW	SW	SW	SW	S	S	S	S	S	S	WSW	WSW	S
Fastest Obs. 1 Min.														
-Direction (!!)	45	28	25	34	25	36	05	25	30	36	25	25	25	36
-Speed (MPH)	45	59	58	69	58	70	63	64	46	45	65	59	58	70
-Year		1965	1960	1957	1956	1952	1955	1950	1989	1953	1957	1955	1957	MAY 1952
Peak Gust														
-Direction (!!)	10	W	N	W	S	NW	NE	N	NW	NW	NW	W	NW	NE
-Speed (mph)	10	54	64	77	71	69	85	72	59	58	52	60	64	85
-Date		1989	1984	1993	1985	1992	1987	1990	1989	1988	1992	1988	1993	JUN 1987

PRECIPITATION (inches)

LUBBOCK, TEXAS

YEAR	JAN	FEB	MAR	APR	MAY	JUNE	JULY	AUG	SEP	OCT	NOV	DEC	ANNUAL
1964	0.45	0.16	0.64	0.11	1.67	5.00	0.82	1.14	2.46	0.30	0.57	0.90	14.22
1965	0.08	0.35	0.22	0.41	1.63	1.44	2.14	0.62	5.68	1.06	0.02	0.50	14.15
1966	0.52	0.06	0.13	3.03	0.67	2.27	0.57	8.85	2.18	T	0.11	0.03	18.42
1967	0.00	0.14	2.09	0.95	3.45	7.95	3.29	0.71	0.98	0.45	0.11	0.52	20.64
1968	0.94	0.82	2.77	0.58	2.01	1.81	3.14	2.72	0.67	0.81	2.67	0.48	19.42
1969	T	1.13	1.77	1.14	3.88	1.41	2.99	2.59	4.93	7.76	0.77	0.82	29.19
1970	T	0.11	2.15	0.26	4.30	1.36	T	1.18	1.80	1.34	0.05	0.08	12.63
1971	T	0.81	0.21	1.36	2.44	2.25	0.76	4.15	5.22	1.79	0.43	0.81	20.23
1972	0.16	0.13	T	0.35	3.20	5.37	4.47	5.40	2.95	1.75	0.97	0.32	25.07
1973	1.44	1.26	1.90	1.40	0.43	0.32	4.16	0.36	0.73	0.89	T	T	12.89
1974	0.08	0.01	1.56	0.82	1.23	1.11	2.22	5.14	6.62	3.89	0.89	0.44	24.01
1975	0.41	1.53	0.04	0.45	2.74	1.80	4.32	2.21	2.61	0.06	1.18	0.34	17.69
1976	T	0.03	0.24	1.76	1.19	2.46	7.20	1.99	3.28	1.39	0.56	0.01	20.11
1977	0.24	0.38	0.82	2.90	2.46	2.28	1.13	4.31	0.49	1.11	0.02	0.01	16.15
1978	0.59	1.39	0.23	0.21	3.20	1.93	0.15	0.34	3.29	1.06	1.11	0.17	13.67
1979	0.33	0.85	2.95	1.17	4.00	3.69	1.84	3.81	0.21	0.59	0.09	1.29	20.82
1980	0.54	0.38	0.19	1.13	3.46	1.78	0.20	1.64	3.55	0.19	2.29	0.51	15.86
1981	0.32	0.67	1.19	2.05	1.25	0.79	3.35	5.41	1.78	5.34	0.64	0.20	22.99
1982	0.05	0.39	0.44	2.53	4.54	4.99	2.08	1.08	1.29	0.48	1.18	1.95	21.00
1983	2.75	0.32	0.55	0.77	1.23	1.79	0.41	0.32	0.39	10.80	0.54	0.36	20.23
1984	0.03	0.17	0.23	0.23	0.45	4.32	0.53	3.72	0.15	1.74	1.87	1.18	14.62
1985	0.38	0.27	1.19	0.48	2.97	4.51	3.94	0.63	4.73	3.60	0.27	0.18	23.15
1986	0.00	0.94	0.39	0.72	1.82	4.92	1.41	3.60	6.90	2.89	1.73	1.29	26.61
1987	0.54	1.47	0.41	0.09	3.30	2.40	4.29	1.68	2.67	0.77	0.11	1.09	18.82
1988	0.22	0.45	0.79	1.08	2.64	1.03	2.93	0.92	2.29	0.02	0.19	0.56	13.12
1989	0.50	1.04	0.70	0.04	0.39	4.98	0.26	3.05	3.74	T	T	0.31	15.01
1990	0.37	2.14	0.87	1.44	1.15	T	3.13	1.87	1.24	1.91	1.29	0.42	15.83
1991	1.15	0.54	0.08	0.07	1.87	5.15	2.14	2.39	6.77	0.57	1.07	2.24	24.04
1992	1.32	2.01	1.36	1.26	5.25	4.40	1.71	1.56	0.69	T	1.47	1.36	22.39
1993	1.03	0.39	0.37	1.16	2.06	3.78	0.82	1.78	0.24	0.49	0.30	0.33	12.75
Record Mean	0.54	0.63	0.79	1.03	2.71	2.90	2.22	2.09	2.23	1.83	0.63	0.54	18.15

AVERAGE TEMPERATURE (deg. F)

LUBBOCK, TEXAS

YEAR	JAN	FEB	MAR	APR	MAY	JUNE	JULY	AUG	SEP	OCT	NOV	DEC	ANNUAL
1964	38.8	36.5	49.2	61.5	71.3	76.7	82.0	79.9	70.9	62.1	50.5	42.8	60.2
#1965	44.3	40.4	42.8	63.6	71.1	77.5	81.7	78.0	72.9	63.5	57.3	47.0	61.7
1966	34.2	40.5	55.7	60.4	69.8	78.5	85.4	76.8	71.6	60.9	55.6	38.0	60.6
1967	40.7	41.5	55.8	64.1	66.0	75.2	77.0	74.9	68.6	60.5	49.3	36.8	59.2
1968	40.7	38.5	48.5	55.3	65.5	74.5	75.3	74.2	66.7	61.9	46.6	39.9	57.3
1969	44.9	43.3	41.5	61.8	68.4	76.9	83.3	79.6	70.2	56.4	48.6	43.1	59.9
1970	37.5	45.5	46.2	58.5	68.7	75.7	80.8	79.0	71.5	56.7	48.7	45.1	59.5
1971	41.4	42.7	51.0	59.7	68.8	78.4	80.0	73.4	70.4	61.9	50.8	43.7	60.2
1972	41.3	45.9	56.3	65.7	67.4	76.9	76.5	74.4	71.0	59.6	42.4	38.0	59.6
1973	34.9	40.2	51.4	55.1	67.8	76.9	77.3	77.7	70.3	63.9	53.5	42.2	59.3
1974	41.3	45.6	58.5	63.1	75.1	78.6	80.6	74.4	64.0	59.9	48.1	40.6	60.8
1975	40.8	41.0	49.5	58.7	67.6	77.4	75.4	77.5	67.5	62.5	50.3	42.8	59.2
1976	39.6	51.2	52.1	62.4	66.0	77.1	75.0	77.9	70.0	54.1	42.6	40.3	59.0
1977	34.5	46.2	51.9	60.4	71.8	79.3	80.3	79.5	77.4	63.3	52.3	45.0	61.8
1978	32.1	33.9	51.7	65.1	70.1	79.0	82.9	78.2	72.0	62.1	49.9	38.1	59.6
1979	31.7	41.9	52.9	61.8	69.2	76.7	81.8	77.4	73.2	64.6	45.7	42.5	59.9
1980	40.3	44.3	50.8	59.0	68.7	83.1	84.3	80.6	73.0	60.2	46.0	45.7	61.3
1981	41.7	44.7	51.3	64.0	68.1	79.6	81.7	76.2	70.5	59.7	53.4	44.0	61.2
1982	39.7	42.4	53.7	59.3	67.8	73.5	80.1	81.0	74.4	60.6	48.6	38.3	60.0
1983	32.5	42.8	51.0	54.8	65.9	74.0	80.6	80.4	74.7	63.6	52.4	31.7	58.7
1984	37.9	45.3	49.4	58.0	71.2	76.9	78.3	78.2	69.3	59.5	49.5	43.9	59.8
1985	35.6	41.9	52.3	63.0	70.0	75.5	79.7	81.5	71.4	61.2	49.8	38.0	60.0
1986	44.4	45.6	56.5	64.2	69.9	75.9	81.9	77.6	71.5	59.7	47.0	40.5	61.2
1987	38.2	44.9	48.1	58.9	68.8	76.1	79.5	79.4	70.9	63.1	49.8	39.6	59.8
1988	36.3	42.8	50.1	59.9	68.8	78.0	79.3	79.7	71.4	62.9	51.9	41.9	60.3
1989	44.7	38.4	54.5	64.8	73.0	74.5	81.4	79.0	69.1	64.1	51.1	35.4	60.8
1990	43.2	46.9	51.7	61.6	69.5	84.4	77.7	78.4	73.7	61.6	53.2	38.9	61.7
1991	37.8	49.0	54.0	63.8	73.7	77.6	79.6	77.9	68.0	62.2	45.2	43.1	61.0
1992	39.6	48.6	55.0	62.4	67.2	75.6	80.8	76.5	73.3	64.3	46.4	41.7	61.0
1993	39.3	42.2	51.1	60.3	69.3	78.4	82.2	79.9	72.0	59.4	45.5	43.2	60.2
Record Mean	38.7	43.1	50.5	60.5	69.0	77.4	79.8	78.2	71.3	61.3	48.7	41.1	60.0
Max	52.6	57.4	65.4	75.1	82.8	90.6	92.1	90.4	83.7	75.0	62.4	54.8	73.5
Min	24.8	28.7	35.6	45.9	55.2	64.2	67.6	65.9	58.8	47.6	34.9	27.4	46.4

REFERENCE NOTES

LUBBOCK, TEXAS

GENERAL
T - TRACE AMOUNT.
BLANK ENTRIES DENOTE MISSING/UNREPORTED DATA.
# INDICATES A STATION OR INSTRUMENT RELOCATION.

SPECIFIC

(a) - LENGTH OF RECORD IN YEARS, ALTHOUGH INDIVIDUAL MONTHS MAY BE MISSING.
0.* OR * - THE VALUE IS BETWEEN 0.0 AND 0.05
NORMALS - BASED ON THE 1961-1990 RECORD PERIOD.
EXTREMES - DATES ARE THE MOST RECENT OCCURRENCE
WIND DIR. - NUMERALS SHOW TENS OF DEGRESS CLOCKWISE FROM TRUE NORTH. "00" INDICATES CALM.
RESULTANT DIRECTIONS ARE GIVEN TO WHOLE DEGREES.
BOLD VALUES INDICATE EXTREME VALUES WHICH OCCURRED AFTER THE ASOS SYSTEM WAS COMMISSIONED.

RECORD = PERIOD OF RECORD
RECORD MEAN PRECIPITATION IS THE MEAN OF ALL DAILY PRECIPITATION AMOUNTS DURING THE PERIOD OF RECORD.
RECORD MAX(MIN) TEMPERATURE IS THE MEAN OF ALL DAILY MAX(MIN) TEMPERATURES DURING THE PERIOD OF RECORD.
RECORD MEAN TEMPERATURE IS THE SUM OF THE RECORD MAX AND RECORD MIN DIVIDED BY 2.
AVERAGE TEMPERATURE IS THE SUM OF THE MEAN DAILY MAX AND MIN TEMPERATURE DIVIDED BY 2.

EXCEPTIONS

RECORD MEANS ARE THROUGH THE CURRENT YEAR, BEGINNING IN 1947 FOR TEMPERATURE
1947 FOR PRECIPITATION
1949 FOR SNOWFALL

HEATING DEGREE DAYS Base 65 deg. F LUBBOCK, TEXAS

SEASON	JULY	AUG	SEP	OCT	NOV	DEC	JAN	FEB	MAR	APR	MAY	JUNE	TOTAL
1964-65	0	0	35	111	433	677	635	683	682	107	12	0	3375
#1965-66	0	0	22	94	235	551	946	677	307	166	48	0	3046
1966-67	0	8	3	170	283	827	748	652	289	85	84	3	3152
1967-68	0	4	17	190	463	866	748	762	507	293	73	4	3927
1968-69	0	1	31	151	543	770	614	601	723	126	45	7	3612
1969-70	0	0	3	312	486	673	846	539	577	198	47	21	3702
1970-71	0	0	50	286	484	608	723	618	438	186	30	0	3423
1971-72	0	0	83	124	421	656	728	549	275	94	40	0	2970
1972-73	2	0	23	220	672	831	928	688	416	312	58	0	4150
1973-74	0	0	31	99	340	703	726	536	223	127	11	0	2796
1974-75	0	0	105	166	500	753	744	664	474	233	18	4	3661
1975-76	0	0	76	122	436	680	780	397	392	116	57	0	3056
1976-77	0	0	30	333	665	760	939	520	397	153	1	0	3798
1977-78	0	0	0	80	373	611	1014	864	419	75	64	0	3500
1978-79	0	0	31	129	447	827	1023	640	369	134	45	9	3654
1979-80	0	0	9	104	570	690	756	592	436	205	48	0	3410
1980-81	0	0	21	190	565	590	712	563	420	108	33	0	3202
1981-82	0	0	16	202	341	643	777	635	351	198	45	2	3210
1982-83	0	0	0	189	485	821	1001	613	430	326	68	13	3946
1983-84	0	0	18	107	371	1025	836	566	477	216	20	0	3636
1984-85	0	0	80	194	460	646	908	639	392	104	26	3	3452
1985-86	0	0	65	133	448	832	631	540	272	100	32	0	3053
1986-87	0	0	7	185	535	752	822	555	519	224	25	0	3624
1987-88	0	0	2	96	456	784	881	638	459	168	32	0	3516
1988-89	0	6	11	95	388	707	620	739	341	132	17	1	3057
1989-90	0	0	55	115	409	912	672	500	411	147	86	0	3307
1990-91	0	0	5	138	357	801	836	444	340	82	18	0	3021
1991-92	0	0	69	163	587	671	782	465	303	134	47	3	3224
1992-93	0	0	0	90	554	716	788	632	421	176	40	0	3417
1993-94	0	3	13	222	580	669							

COOLING DEGREE DAYS Base 65 deg. F LUBBOCK, TEXAS

YEAR	JAN	FEB	MAR	APR	MAY	JUNE	JULY	AUG	SEP	OCT	NOV	DEC	TOTAL
1969	0	0	0	36	160	370	575	461	165	56	0	0	1823
1970	0	0	0	11	167	348	498	440	249	32	2	0	1747
1971	0	0	14	35	157	408	476	269	252	34	4	0	1649
1972	0	0	14	120	119	362	366	297	211	63	0	0	1552
1973	0	0	0	19	153	364	387	401	197	71	2	0	1594
1974	0	0	30	78	330	413	491	300	80	18	0	0	1740
1975	0	0	0	49	107	383	332	392	155	49	0	0	1467
1976	0	1	0	45	93	369	317	406	187	4	0	0	1422
1977	0	0	0	19	221	435	482	455	379	36	0	0	2027
1978	0	0	12	85	230	426	562	417	248	46	0	0	2026
1979	0	0	0	44	183	371	524	394	264	98	0	0	1878
1980	0	0	0	32	172	549	605	488	272	48	3	0	2169
1981	0	0	3	84	137	443	524	356	187	42	0	0	1776
1982	0	5	11	34	141	265	476	504	289	58	0	0	1783
1983	0	0	2	27	101	291	490	486	313	70	2	0	1782
1984	0	0	0	13	219	363	420	416	217	30	1	0	1679
1985	0	0	6	52	187	324	462	518	264	22	0	0	1835
1986	0	3	15	79	190	334	528	400	209	26	0	0	1784
1987	0	0	0	48	149	338	456	452	187	44	7	0	1681
1988	0	0	6	24	158	394	450	467	210	37	3	0	1749
1989	0	0	26	134	274	292	517	442	182	93	0	0	1960
1990	0	0	5	50	231	590	401	423	270	41	9	0	2020
1991	0	0	6	52	295	386	461	404	167	81	0	0	1852
1992	0	0	1	61	121	328	495	364	258	72	0	0	1700
1993	0	0	0	42	181	408	541	470	229	51	0	0	1922

SNOWFALL (inches) LUBBOCK, TEXAS

SEASON	JULY	AUG	SEP	OCT	NOV	DEC	JAN	FEB	MAR	APR	MAY	JUNE	TOTAL
1964-65	0.0	0.0	0.0	0.0	0.0	0.1	0.2	2.0	1.2	0.0	0.0	0.0	3.5
1965-66	0.0	0.0	0.0	0.0	0.0	0.7	4.9	0.4	0.0	0.0	0.0	0.0	6.0
1966-67	0.0	0.0	0.0	0.0	T	T	0.0	0.7	T	0.0	0.0	0.0	0.7
1967-68	0.0	0.0	0.0	0.2	T	3.0	0.2	3.8	9.7	0.0	0.0	0.0	16.9
1968-69	0.0	0.0	0.0	0.0	1.2	0.6	0.0	0.1	11.7	0.0	0.0	0.0	13.6
1969-70	0.0	0.0	0.0	0.0	3.0	4.8	T	0.4	6.2	0.0	0.0	0.0	14.4
1970-71	0.0	0.0	0.0	0.0	0.1	0.0	T	5.2	2.1	0.0	0.0	0.0	7.4
1971-72	0.0	0.0	0.0	0.0	T	6.5	2.2	1.0	T	0.0	0.0	0.0	9.7
1972-73	0.0	0.0	0.0	T	5.9	0.4	9.4	9.6	T	0.3	0.0	0.0	25.6
1973-74	0.0	0.0	0.0	0.0	T	T	T	0.2	T	0.0	0.0	0.0	0.2
1974-75	0.0	0.0	0.0	0.0	0.0	1.1	1.7	3.8	0.7	T	0.0	0.0	7.3
1975-76	0.0	0.0	0.0	0.0	0.0	3.4	T	T	T	0.0	0.0	0.0	3.4
1976-77	0.0	0.0	0.0	7.5	9.1	0.1	2.0	1.7	T	0.0	0.0	0.0	20.4
1977-78	0.0	0.0	0.0	0.0	T	T	5.7	10.2	0.7	0.0	0.0	0.0	16.6
1978-79	0.0	0.0	0.0	0.0	0.5	1.7	0.8	8.6	0.6	0.0	0.0	0.0	12.2
1979-80	0.0	0.0	0.0	0.0	0.1	5.0	3.6	2.9	T	1.4	0.0	0.0	13.0
1980-81	0.0	0.0	0.0	0.0	21.4	T	3.5	0.1	T	0.0	0.0	0.0	25.0
1981-82	0.0	0.0	0.0	0.0	0.0	0.0	0.5	2.7	T	0.0	0.0	0.0	3.2
1982-83	0.0	0.0	0.0	0.0	0.4	6.8	25.3	3.4	T	5.3	0.0	0.0	41.2
1983-84	0.0	0.0	0.0	0.0	0.0	2.3	0.5	1.7	T	0.0	0.0	0.0	4.5
1984-85	0.0	0.0	0.0	0.0	T	0.1	2.0	T	T	0.0	0.0	0.0	2.1
1985-86	0.0	0.0	0.0	0.0	0.0	1.7	0.0	6.7	0.0	0.0	0.0	0.0	8.4
1986-87	0.0	0.0	0.0	T	T	5.8	4.9	3.3	1.3	0.1	0.0	0.0	15.4
1987-88	0.0	0.0	0.0	0.0	T	4.3	0.9	4.7	0.4	T	0.0	0.0	10.3
1988-89	0.0	0.0	0.0	0.0	1.4	3.6	1.4	T	3.9	0.0	T	T	10.3
1989-90	0.0	0.0	0.0	0.0	0.0	0.9	0.5	T	0.2	0.0	T	0.0	1.6
1990-91	T	0.0	0.0	0.0	0.3	0.5	0.6	T	0.4	0.0	0.7	T	2.5
1991-92	0.0	0.0	0.0	0.4	0.4	2.5	7.4	T	0.0	T	T	T	10.7
1992-93	0.0	0.0	0.0	0.0	2.9	5.3	0.6	0.2	0.8	T	T	0.0	9.8
1993-94	0.0	0.0	0.0	0.1	T	T							
Record Mean	T	0.0	0.0	0.2	1.2	1.9	2.4	2.9	1.5	0.2	T	T	10.2

MIDLAND – ODESSA, TEXAS

The Midland–Odessa region is on the southern extension of the South Plains of Texas. The terrain is level with only slight occasional undulations.

The climate is typical of a semi–arid region. The vegetation of the area consists mostly of native grasses and a few trees, mostly of the mesquite variety.

Most of the annual precipitation in the area comes as a result of very violent spring and early summer thunderstorms. These are usually accompanied by excessive rainfall, over limited areas, and sometimes hail. Due to the flat nature of the countryside, local flooding occurs, but is of short duration. Tornadoes are occasionally sighted.

During the late winter and early spring months, blowing dust occurs frequently. The flat plains of the area with only grass as vegetation offer little resistance to the strong winds. The sky is occasionally obscured by dust but in most storms visibilities range from 1 to 3 miles.

Daytime temperatures are quite hot in the summer, but there is a large diurnal range of temperature and most nights are comfortable. The temperature drops below 32 degrees in the fall about mid–November and the last temperature below 32 degrees in spring comes early in April.

Winters are characterized by frequent cold periods followed by rapid warming. Cold frontal passages are followed by chilly weather for two or three days. Cloudiness is at a minimum. Summers are hot and dry with numerous small convective showers.

The prevailing wind direction in this area is from the southeast. This, together with the upslope of the terrain from the same direction, causes occasional low cloudiness and drizzle during winter and spring months. Snow is infrequent. Maximum temperatures during the summer month frequently are from 2 to 6 degrees cooler than those at places 100 miles southeast, due to the cooling effect of the upslope winds.

Very low humidities are conducive to personal comfort, because even though summer afternoon temperatures are frequently above 90 degrees, the low humidity with resultant rapid evaporation, has a cooling effect. The climate of the area is generally quite pleasant with the most disagreeable weather concentrated in the late winter and spring months.

NORMALS, MEANS, AND EXTREMES

MIDLAND - ODESSA, TEXAS

LATITUDE: 31°57'N LONGITUDE: 102°11' W ELEVATION: FT. GRND 2862 BARO 2865 TIME ZONE: CENTRAL WBAN: 23023

	(a)	JAN	FEB	MAR	APR	MAY	JUNE	JULY	AUG	SEP	OCT	NOV	DEC	YEAR
TEMPERATURE °F:														
Normals														
-Daily Maximum		56.5	61.6	71.2	79.8	87.4	93.4	95.4	94.1	85.4	77.3	66.2	58.4	77.2
-Daily Minimum		28.5	32.6	40.2	49.4	58.1	65.7	68.5	67.5	61.1	50.6	38.9	30.8	49.3
-Monthly		42.5	47.1	55.7	64.6	72.8	79.6	82.0	80.8	73.3	64.0	52.6	44.6	63.3
Extremes														
-Record Highest	46	84	90	95	101	108	109	112	107	107	100	89	85	112
-Year		1974	1986	1989	1989	1989	1990	1989	1964	1953	1979	1988	1954	JUL 1989
-Record Lowest	46	-8	-11	9	20	34	47	53	54	36	24	13	-1	-11
-Year		1962	1985	1980	1973	1970	1983	1978	1989	1989	1993	1993	1989	FEB 1985
NORMAL DEGREE DAYS:														
Heating (base 65°F)		698	501	309	108	10	0	0	0	10	106	377	632	2751
Cooling (base 65°F)		0	0	21	96	252	438	527	490	259	75	5	0	2163
% OF POSSIBLE SUNSHINE	13	66	68	76	77	79	79	78	74	76	72	72	64	73
MEAN SKY COVER (tenths)														
Sunrise - Sunset	45	5.2	5.1	4.8	4.7	4.7	4.0	4.5	4.4	4.4	3.9	4.3	4.8	4.5
MEAN NUMBER OF DAYS:														
Sunrise to Sunset														
-Clear	45	12.4	11.2	13.3	13.2	13.2	15.0	13.1	14.1	14.3	16.8	14.7	13.9	165.3
-Partly Cloudy	45	6.3	7.0	7.6	7.9	9.5	9.4	11.0	10.5	7.8	6.3	6.4	6.0	95.7
-Cloudy	45	12.3	10.0	10.1	8.9	8.3	5.6	6.9	6.4	7.9	7.9	8.9	11.1	104.3
Precipitation														
.01 inches or more	46	3.8	3.9	2.6	3.4	5.9	4.9	5.2	5.7	5.8	4.5	3.0	3.4	52.2
Snow,Ice pellets,hail														
1.0 inches or more	46	0.7	0.5	0.1	0.0	0.0	0.0	0.0	0.0	0.0	0.0	0.2	0.4	1.8
Thunderstorms	45	0.3	0.6	1.4	3.2	6.8	5.8	5.9	6.2	3.8	2.5	0.6	0.4	37.3
Heavy Fog Visibility														
1/4 mile or less	45	3.5	2.8	0.8	0.6	0.4	0.1	0.*	0.1	0.6	1.3	2.5	3.1	15.8
Temperature °F														
-Maximum														
90° and above	30	0.0	0.*	0.4	3.5	11.3	20.7	25.7	22.3	10.4	2.1	0.0	0.0	96.4
32° and below	30	1.4	0.4	0.1	0.0	0.0	0.0	0.0	0.0	0.0	0.0	0.1	0.9	2.9
-Minimum														
32° and below	30	19.9	13.4	5.6	0.7	0.0	0.0	0.0	0.0	0.0	0.3	7.1	17.5	64.4
0° and below	30	0.*	0.1	0.0	0.0	0.0	0.0	0.0	0.0	0.0	0.0	0.0	0.*	0.1
AVG. STATION PRESS.(mb)	21	918.6	917.4	914.2	914.0	913.3	914.6	916.7	917.0	917.2	917.9	917.6	918.4	916.4
RELATIVE HUMIDITY (%)														
Hour 00	30	65	64	54	53	61	62	58	62	70	70	68	65	63
Hour 06 (Local Time)	30	73	73	66	67	76	78	73	76	81	80	76	72	74
Hour 12	30	48	45	36	34	38	42	42	44	51	46	45	45	43
Hour 18	30	43	36	28	27	31	34	35	37	44	43	44	44	37
PRECIPITATION (inches):														
Water Equivalent														
-Normal		0.40	0.62	0.58	0.83	1.98	1.55	1.70	1.69	2.62	1.74	0.69	0.56	14.96
-Maximum Monthly	46	3.66	2.55	2.86	2.85	7.63	3.93	8.50	4.43	9.70	7.45	2.32	3.30	9.70
-Year		1949	1992	1970	1949	1992	1949	1991	1974	1980	1986	1968	1986	SEP 1980
-Minimum Monthly	46	0.00	0.01	T	0.00	0.08	0.01	T	0.17	0.08	0.00	0.00	T	0.00
-Year		1967	1971	1991	1964	1953	1990	1983	1967	1979	1952	1950	1958	JAN 1967
-Maximum in 24 hrs	40	1.15	1.22	2.20	1.62	4.75	3.07	5.99	2.41	4.37	3.59	2.16	1.58	5.99
-Year		1958	1965	1970	1979	1968	1993	1961	1965	1986	1985	1975	1979	JUL 1961
Snow,Ice pellets,hail														
-Maximum Monthly	46	9.0	3.9	5.9	0.5	T	T	0.0	T	0.0	0.6	7.2	8.5	9.0
-Year		1985	1973	1970	1983	1992	1993		1992		1993	1980	1986	JAN 1985
-Maximum in 24 hrs	40	6.8	3.9	5.0	0.5	T	T	0.0	T	0.0	0.6	5.7	4.3	6.8
-Year		1974	1985	1970	1983	1992	1993		1992		1993	1980	1986	JAN 1974
WIND:														
Mean Speed (mph)	40	10.2	11.2	12.7	12.8	12.4	12.2	10.8	10.0	10.1	10.1	10.4	10.1	11.1
Prevailing Direction through 1963		S	SW	S	SSE	SSE	SSE	SSE	SE	SSE	S	S	SW	SSE
Fastest Obs. 1 Min.														
-Direction (!!)	40	27	25	27	30	20	24	05	24	23	14	25	26	25
-Speed (MPH)	40	41	67	53	53	52	58	58	54	53	46	48	47	67
-Year		1965	1960	1984	1983	1958	1966	1979	1987	1985	1983	1993	1977	FEB 1960
Peak Gust														
-Direction (!!)	10	W	SW	W	W	NW	S	SW	SW	SW	N	W	N	SW
-Speed (mph)	10	58	55	74	67	69	71	82	63	82	69	59	47	82
-Date		1988	1993	1984	1984	1993	1991	1987	1987	1985	1986	1988	1990	JUL 1987

PRECIPITATION (inches)

MIDLAND - ODESSA, TEXAS

YEAR	JAN	FEB	MAR	APR	MAY	JUNE	JULY	AUG	SEP	OCT	NOV	DEC	ANNUAL
1964	0.40	0.27	0.64	0.00	2.14	0.40	0.24	0.58	2.19	0.19	0.20	0.40	7.65
1965	0.06	1.71	0.06	0.33	3.04	0.90	0.06	3.23	0.78	1.41	0.29	0.29	12.16
1966	0.48	0.20	0.41	2.44	2.86	2.91	0.17	3.37	3.29	0.80	0.04	0.02	16.99
1967	0.00	0.03	1.45	0.44	0.56	1.17	1.64	0.17	1.59	0.20	0.44	0.78	8.47
1968	1.07	1.06	1.62	1.59	4.96	1.02	1.01	2.10	1.39	0.19	2.32	0.18	18.51
1969	0.02	1.79	0.51	2.10	3.28	2.68	0.74	0.32	1.53	2.52	0.85	0.60	16.94
1970	0.01	1.40	2.86	0.29	0.18	0.99	0.23	0.87	2.00	0.78	T	0.11	9.72
1971	T	0.01	T	0.81	2.77	1.07	0.79	3.27	2.55	0.67	0.42	0.24	12.60
1972	0.37	0.04	T	0.29	1.20	1.70	0.58	3.80	1.05	2.26	0.22	0.15	11.66
1973	0.93	1.63	1.73	1.38	0.87	0.34	2.47	0.25	1.09	0.59	T	0.00	11.28
1974	0.54	0.28	0.18	1.17	0.34	0.75	0.26	4.43	6.16	5.42	0.72	0.25	20.50
1975	0.81	0.69	0.08	0.09	3.44	2.05	7.73	1.68	3.44	0.70	2.16	0.37	23.24
1976	0.03	0.18	0.30	2.02	1.08	2.14	3.56	0.84	1.25	1.84	0.26	0.08	13.58
1977	0.63	0.55	0.61	1.22	0.60	0.81	0.40	0.37	0.09	1.27	0.02	0.27	6.84
1978	0.22	0.34	T	0.09	1.97	1.15	2.51	1.01	5.02	2.51	2.27	0.20	17.29
1979	0.16	0.26	0.81	1.63	1.14	2.98	3.05	2.17	0.08	0.97	T	2.80	16.05
1980	0.49	0.29	T	0.86	1.85	1.59	T	0.93	9.70	0.12	0.78	1.15	17.76
1981	0.56	0.67	0.56	2.15	2.21	0.46	0.40	3.33	1.68	5.53	T	0.05	17.60
1982	0.42	0.16	0.01	1.08	3.16	1.45	3.24	0.67	1.43	1.01	0.62	1.40	14.65
1983	1.14	0.26	0.21	0.06	0.50	0.19	T	0.43	0.62	4.64	1.62	0.33	10.00
1984	0.41	0.17	T	0.01	3.40	2.15	0.84	2.45	1.89	1.99	2.31	0.83	16.45
1985	0.84	0.59	0.63	0.21	0.99	2.23	0.90	0.81	3.15	5.93	0.08	0.05	16.41
1986	0.23	0.17	0.03	0.01	3.89	3.42	1.26	3.94	6.98	7.45	1.45	3.30	32.13
1987	0.10	1.93	1.23	0.63	4.13	2.98	0.33	0.51	1.37	0.44	0.02	0.38	14.05
1988	0.01	0.44	0.74	0.08	2.26	0.84	6.66	1.66	5.07	T	T	0.53	18.29
1989	0.25	1.28	0.34	0.05	0.95	0.61	0.34	1.19	2.78	0.17	0.04	0.14	8.14
1990	0.21	1.09	0.94	1.97	0.35	0.01	1.65	2.15	2.89	1.46	0.75	0.69	14.16
1991	1.80	0.25	T	0.01	0.54	3.50	8.50	1.58	3.76	0.58	0.71	2.23	23.46
1992	1.20	2.55	0.97	1.03	7.63	2.12	1.63	2.75	0.16	0.03	0.70	0.61	21.38
1993	1.28	0.82	0.26	0.57	1.45	3.11	0.89	4.37	2.11	1.11	0.06	0.34	16.37
Record Mean	0.64	0.64	0.47	0.81	2.09	1.67	1.85	1.66	2.16	1.65	0.59	0.63	14.87

AVERAGE TEMPERATURE (deg. F)

MIDLAND - ODESSA, TEXAS

YEAR	JAN	FEB	MAR	APR	MAY	JUNE	JULY	AUG	SEP	OCT	NOV	DEC	ANNUAL
1964	43.6	42.7	55.6	67.1	75.7	83.5	86.9	87.2	76.5	66.4	56.6	49.0	65.9
1965	47.6	43.2	47.7	67.4	71.7	79.0	82.8	79.6	74.9	63.9	57.4	49.7	63.7
1966	38.0	42.7	57.2	61.9	70.4	78.0	84.1	78.9	72.8	61.2	56.8	43.4	62.1
1967	43.6	46.4	60.5	68.7	72.3	79.9	80.7	77.7	71.1	64.0	54.0	41.9	63.4
1968	44.3	44.7	53.5	60.8	70.8	77.5	78.5	79.2	72.6	67.0	49.6	43.3	61.8
1969	47.6	48.1	47.5	65.2	69.3	78.2	84.1	82.7	74.2	61.0	51.4	47.6	63.1
1970	42.0	47.6	49.8	61.5	69.9	77.5	82.2	79.9	73.4	61.9	53.1	50.7	62.4
1971	47.4	47.2	55.8	62.8	72.2	78.4	80.7	74.4	71.6	64.0	53.3	47.8	63.0
1972	44.2	49.3	60.0	69.7	70.4	78.9	79.2	75.7	72.5	62.5	46.4	43.5	62.7
1973	38.9	43.8	54.4	57.1	68.8	77.0	78.1	78.9	72.0	65.0	56.6	46.7	61.4
1974	45.2	49.5	62.7	66.1	77.5	81.1	83.4	79.5	67.8	64.3	53.6	45.6	64.7
1975	45.9	48.0	54.9	63.0	71.8	80.3	77.8	79.0	70.6	64.9	53.5	46.7	63.1
1976	43.9	56.2	57.0	64.6	68.5	78.1	75.8	78.8	71.7	58.4	46.6	44.3	62.0
1977	40.5	50.2	54.8	61.6	76.4	82.1	83.6	85.4	82.1	67.0	55.0	49.4	65.7
1978	38.2	43.8	58.1	67.8	72.1	78.1	81.9	78.9	71.4	61.3	52.3	42.6	62.2
1979	35.9	45.3	55.4	62.6	70.7	77.4	82.1	79.0	75.1	68.1	49.5	46.4	62.3
1980	44.9	46.3	52.4	59.1	69.1	81.9	84.1	80.8	72.7	60.2	47.2	46.2	62.1
1981	43.2	48.6	52.4	63.9	70.2	79.1	83.6	80.2	75.7	65.6	57.0	47.8	64.0
1982	43.6	45.9	57.5	63.8	70.7	79.1	82.3	82.8	77.4	64.2	52.2	43.4	63.6
1983	41.8	48.0	55.9	59.5	71.5	79.3	83.7	84.1	77.8	67.8	55.0	37.5	63.5
1984	40.1	48.2	55.0	63.6	74.1	79.8	80.9	80.2	71.5	61.7	50.7	48.0	62.8
1985	37.5	44.3	56.7	66.9	74.8	77.3	81.2	84.1	74.2	63.7	54.6	41.1	63.0
1986	46.3	51.3	58.9	68.4	72.8	77.2	80.8	79.6	74.2	61.7	50.7	43.7	63.8
1987	42.6	47.5	49.5	59.2	69.2	75.7	80.5	81.3	73.1	66.3	50.8	44.2	61.7
1988	39.3	46.2	54.3	63.1	70.8	78.5	77.7	79.3	72.5	64.7	54.9	43.7	62.1
1989	45.9	43.3	56.9	66.8	76.9	78.7	83.1	80.5	71.7	65.2	53.6	38.7	63.4
1990	46.4	50.5	54.9	64.7	73.0	86.5	79.0	78.6	75.0	63.8	55.6	43.4	64.3
1991	41.7	50.6	56.9	65.9	76.6	79.5	79.2	77.6	69.9	64.6	47.7	45.8	63.0
1992	41.3	50.9	57.3	63.2	68.0	77.0	79.9	76.9	74.8	66.4	49.2	46.5	62.6
1993	43.4	47.1	54.6	63.3	71.2	78.9	83.5	81.1	71.9	61.9	49.5	46.9	62.8
Record Mean	43.3	47.9	55.0	64.0	72.1	79.6	81.4	80.7	74.2	64.8	52.4	45.3	63.4
Max	56.7	61.9	69.8	78.7	85.9	92.7	94.0	93.2	86.4	77.9	65.8	58.7	76.8
Min	29.9	33.8	40.1	49.3	58.3	66.4	68.8	68.1	61.9	51.6	38.9	31.9	49.9

REFERENCE NOTES

MIDLAND - ODESSA, TEXAS

GENERAL
T - TRACE AMOUNT.
BLANK ENTRIES DENOTE MISSING/UNREPORTED DATA.
# INDICATES A STATION OR INSTRUMENT RELOCATION.

SPECIFIC

EXCEPTIONS

RECORD MEANS ARE THROUGH THE CURRENT YEAR, BEGINNING IN 1930 FOR TEMPERATURE
1930 FOR PRECIPITATION
1949 FOR SNOWFALL

(a) - LENGTH OF RECORD IN YEARS, ALTHOUGH INDIVIDUAL MONTHS MAY BE MISSING.
0.* OR * - THE VALUE IS BETWEEN 0.0 AND 0.05
NORMALS - BASED ON THE 1961-1990 RECORD PERIOD.
EXTREMES - DATES ARE THE MOST RECENT OCCURRENCE.
WIND DIR. - NUMERALS SHOW TENS OF DEGRESS CLOCKWISE FROM TRUE NORTH. "00" INDICATES CALM.
RESULTANT DIRECTIONS ARE GIVEN TO WHOLE DEGREES.
BOLD VALUES INDICATE EXTREME VALUES WHICH OCCURRED AFTER THE ASOS SYSTEM WAS COMMISSIONED.

RECORD = PERIOD OF RECORD
RECORD MEAN PRECIPITATION IS THE MEAN OF ALL DAILY PRECIPITATION AMOUNTS DURING THE PERIOD OF RECORD.
RECORD MAX(MIN) TEMPERATURE IS THE MEAN OF ALL DAILY MAX(MIN) TEMPERATURES DURING THE PERIOD OF RECORD.
RECORD MEAN TEMPERATURE IS THE SUM OF THE RECORD MAX AND RECORD MIN DIVIDED BY 2.
AVERAGE TEMPERATURE IS THE SUM OF THE MEAN DAILY MAX AND MIN TEMPERATURE DIVIDED BY 2.

HEATING DEGREE DAYS Base 65 deg. F MIDLAND - ODESSA, TEXAS

SEASON	JULY	AUG	SEP	OCT	NOV	DEC	JAN	FEB	MAR	APR	MAY	JUNE	TOTAL
1964-65	0	0	6	34	283	493	535	604	529	52	5	0	2541
1965-66	0	0	11	96	232	468	831	621	258	124	46	0	2687
1966-67	0	3	4	159	242	663	654	516	176	23	22	0	2462
1967-68	0	0	9	95	324	708	634	582	360	143	14	0	2869
1968-69	0	0	0	60	456	665	535	467	534	52	30	0	2799
1969-70	0	0	0	216	405	532	706	482	463	137	48	5	2994
1970-71	0	0	41	150	348	441	540	493	297	119	15	0	2444
1971-72	0	0	51	71	350	528	639	446	176	37	12	0	2310
1972-73	0	0	19	158	551	660	804	587	320	249	55	0	3403
1973-74	0	0	15	81	252	561	607	429	121	68	1	0	2135
1974-75	0	0	44	65	344	594	582	468	314	135	2	3	2551
1975-76	0	0	40	82	340	560	652	258	261	73	38	0	2304
1976-77	0	0	10	220	545	635	755	409	320	117	0	0	3011
1977-78	0	0	0	47	295	478	823	586	245	49	30	0	2553
1978-79	0	0	36	138	374	689	893	546	297	106	34	3	3116
1979-80	0	0	3	63	459	566	619	535	384	184	34	0	2847
1980-81	0	0	20	173	531	575	669	452	383	99	20	0	2922
1981-82	0	0	0	109	234	527	655	535	245	112	21	0	2438
1982-83	0	0	0	128	378	661	711	469	281	224	19	3	2874
1983-84	0	0	9	56	304	844	768	481	313	100	7	0	2882
1984-85	0	0	66	135	426	521	843	573	277	53	2	1	2897
1985-86	0	0	30	87	308	734	575	390	192	39	11	0	2366
1986-87	0	0	2	137	423	653	689	484	473	205	14	0	3080
1987-88	0	0	2	38	427	636	788	540	338	100	11	0	2880
1988-89	0	0	10	54	321	654	586	601	279	102	2	0	2609
1989-90	0	0	36	98	341	806	569	401	324	105	58	0	2738
1990-91	0	0	2	101	295	662	715	398	260	47	8	0	2488
1991-92	0	0	47	107	513	587	728	404	235	113	29	0	2763
1992-93	0	0	1	32	475	564	665	494	322	118	16	0	2687
1993-94	0	3	10	185	457	556							

COOLING DEGREE DAYS Base 65 deg. F MIDLAND - ODESSA, TEXAS

YEAR	JAN	FEB	MAR	APR	MAY	JUNE	JULY	AUG	SEP	OCT	NOV	DEC	TOTAL
1969	0	0	0	66	170	402	597	559	286	97	5	0	2182
1970	0	0	0	37	205	389	542	467	301	59	0	1	2001
1971	0	0	19	59	245	410	491	301	256	47	3	0	1831
1972	0	0	31	187	183	423	446	338	249	88	0	0	1945
1973	0	0	0	18	178	368	414	437	229	88	5	0	1737
1974	0	0	58	109	394	488	575	455	136	50	5	0	2270
1975	0	0	10	80	220	466	404	439	217	86	5	0	1927
1976	0	10	21	68	155	398	344	434	219	22	0	0	1671
1977	0	0	9	21	359	522	584	638	518	112	1	0	2764
1978	0	0	38	139	260	401	531	438	233	30	0	0	2070
1979	0	0	5	43	216	383	539	442	314	167	0	0	2109
1980	0	0	0	17	166	514	600	498	260	33	2	0	2090
1981	0	0	0	71	188	428	583	480	327	135	0	0	2212
1982	0	3	20	84	204	429	543	556	374	111	1	0	2325
1983	0	0	5	64	227	439	588	599	401	151	11	0	2485
1984	0	0	10	69	296	447	500	482	267	39	2	0	2112
1985	0	1	26	115	311	376	505	599	310	56	3	0	2302
1986	0	13	13	147	259	371	496	460	286	38	0	0	2083
1987	0	0	0	39	151	328	487	511	251	86	7	0	1860
1988	0	0	13	51	198	414	397	452	242	52	22	0	1841
1989	0	0	36	164	376	416	569	487	244	111	4	0	2407
1990	0	0	15	103	312	651	441	428	310	68	19	0	2347
1991	0	1	17	80	374	440	447	399	200	101	0	0	2059
1992	0	0	1	68	131	367	468	377	303	83	4	0	1802
1993	0	0	6	75	213	422	579	510	224	95	0	0	2124

SNOWFALL (inches) MIDLAND - ODESSA, TEXAS

SEASON	JULY	AUG	SEP	OCT	NOV	DEC	JAN	FEB	MAR	APR	MAY	JUNE	TOTAL
1964-65	0.0	0.0	0.0	0.0	0.0	T	0.0	3.5	1.0	0.0	0.0	0.0	4.5
1965-66	0.0	0.0	0.0	0.0	0.0	T	3.6	T	0.0	0.0	0.0	0.0	3.6
1966-67	0.0	0.0	0.0	0.0	0.0	0.0	0.0	T	0.0	0.0	0.0	0.0	T
1967-68	0.0	0.0	0.0	0.0	0.0	1.2	0.8	1.7	0.2	0.0	0.0	0.0	3.9
1968-69	0.0	0.0	0.0	0.0	4.5	T	0.0	0.0	1.5	0.0	0.0	0.0	6.0
1969-70	0.0	0.0	0.0	0.0	1.3	0.5	0.1	1.7	5.9	0.0	0.0	0.0	9.5
1970-71	0.0	0.0	0.0	0.0	0.0	0.0	T	0.1	T	0.0	0.0	0.0	0.1
1971-72	0.0	0.0	0.0	0.0	0.0	T	2.4	T	T	0.0	0.0	0.0	2.4
1972-73	0.0	0.0	0.0	0.0	0.6	T	2.9	3.9	0.0	T	0.0	0.0	7.4
1973-74	0.0	0.0	0.0	0.0	0.0	0.0	7.0	0.7	0.0	0.0	0.0	0.0	7.7
1974-75	0.0	0.0	0.0	0.0	0.0	0.1	0.4	1.6	T	0.0	0.0	0.0	2.1
1975-76	0.0	0.0	0.0	0.0	0.0	1.3	0.3	T	0.0	0.0	0.0	0.0	1.6
1976-77	0.0	0.0	0.0	0.0	5.1	T	0.7	0.0	0.0	0.0	0.0	0.0	5.8
1977-78	0.0	0.0	0.0	0.0	0.0	T	1.1	T	T	0.0	0.0	0.0	1.1
1978-79	0.0	0.0	0.0	0.0	0.0	1.1	T	2.0	0.0	0.0	0.0	0.0	3.1
1979-80	0.0	0.0	0.0	0.0	0.0	2.1	0.1	3.2	0.0	T	0.0	0.0	5.4
1980-81	0.0	0.0	0.0	T	7.2	T	4.4	0.0	0.0	0.0	0.0	0.0	11.6
1981-82	0.0	0.0	0.0	0.0	0.0	0.0	3.8	T	0.0	0.0	0.0	0.0	3.8
1982-83	0.0	0.0	0.0	0.0	0.1	6.4	5.3	T	T	0.5	0.0	0.0	12.3
1983-84	0.0	0.0	0.0	0.0	0.0	2.1	T	T	T	0.0	0.0	0.0	2.1
1984-85	0.0	0.0	0.0	0.0	0.0	T	9.0	0.7	0.0	0.0	0.0	0.0	9.7
1985-86	0.0	0.0	0.0	0.0	0.0	0.5	2.0	0.2	0.0	0.0	0.0	0.0	2.7
1986-87	0.0	0.0	0.0	0.0	T	8.5	1.1	2.0	0.1	0.0	0.0	0.0	11.7
1987-88	0.0	0.0	0.0	0.0	0.0	1.2	T	3.2	0.1	0.0	0.0	0.0	4.5
1988-89	0.0	0.0	0.0	0.0	0.0	1.5	0.6	T	5.5	0.0	0.0	T	7.6
1989-90	0.0	0.0	0.0	0.0	0.4	0.3	0.0	0.0	0.0	0.3	0.0	0.0	1.0
1990-91	0.0	0.0	0.0	0.0	T	0.2	5.8	T	0.0	0.0	0.0	T	6.0
1991-92	0.0	0.0	0.0	0.1	0.0	T	5.8	2.1	T	0.0	T	T	8.0
1992-93	0.0	T	0.0	0.0	T	4.2	T	T	T	0.0	0.0	T	4.2
1993-94	0.0	0.0	0.0	0.6	0.0	0.9							
Record Mean	0.0	T	0.0	T	0.5	0.8	1.8	0.9	0.4	T	T	T	4.4

PORT ARTHUR, TEXAS

Port Arthur is located on the flat Coastal Plain in the extreme southeast corner of Texas. The climate is a mixture of tropical and temperate zone conditions.

Sea breezes prevent extremely high temperatures in the summer, except on rare occasions. The area lies far enough south so that cold air masses modify in severity but still provide freezing temperatures up to six times a year.

High humidity is the result of fairly evenly distributed high normal rainfall and prevailing southerly winds from the Gulf of Mexico.

Cloudy, rainy weather is most common in the winter. Only slightly more than half the winters record even a trace of sleet or snow. Heavy rainfall in summer occurs in short duration thunderstorms and in infrequent tropical storms.

Slow moving systems in the spring and fall often result in three to five days of stormy weather and heavy rain. The lightest precipitation usually occurs in March and October. Funnel clouds and waterspouts are common near the coast. The area enjoys approximately 60 percent of possible sunshine.

Fog, most frequent in midwinter and early spring, is rare in summer. It usually dissipates before noon, but occasionally under stagnant conditions lasts a day or two. Along the immediate coast, fog usually does not form until daybreak, but inland it may form before midnight.

The average wind movement is near 11 mph. Except for severe storms and tropical disturbances, wind seldom exceeds 45 mph. It exceeds 30 mph on only about 40 days in any one year.

The climate is favorable for outdoor activities throughout the year. The abundant rainfall, moderate temperatures, and the short period of temperatures below freezing are particularly favorable for farming and livestock production. Heaviest rain usually falls in the summer when needed for rice. The comparatively dry harvest season simplifies the gathering of rice and feed crops. Cattle on the open range of the coastal marshes need little supplemental feeding or protection. Improved pastures are easily provided because of the moderate temperatures and abundant rainfall.

NORMALS, MEANS, AND EXTREMES

PORT ARTHUR, TEXAS

LATITUDE: 29°57'N LONGITUDE: 94°01'W ELEVATION: FT. GRND 16 BARO 21 TIME ZONE: CENTRAL WBAN: 12917

	(a)	JAN	FEB	MAR	APR	MAY	JUNE	JULY	AUG	SEP	OCT	NOV	DEC	YEAR
TEMPERATURE °F:														
Normals														
-Daily Maximum		60.3	64.3	71.5	78.3	84.1	89.4	91.9	91.7	87.3	80.2	71.3	64.2	77.9
-Daily Minimum		41.5	44.4	51.3	59.5	66.3	72.0	73.7	73.3	69.7	59.2	51.2	44.3	58.9
-Monthly		50.9	54.4	61.4	68.9	75.2	80.7	82.8	82.6	78.6	69.7	61.2	54.3	68.4
Extremes														
-Record Highest	40	81	85	87	94	97	100	103	107	100	95	88	84	107
-Year		1989	1986	1974	1987	1977	1954	1980	1962	1980	1977	1989	1978	AUG 1962
-Record Lowest	40	14	20	23	32	46	56	61	60	45	30	22	12	12
-Year		1962	1981	1989	1987	1954	1984	1990	1989	1967	1993	1976	1989	DEC 1989
NORMAL DEGREE DAYS:														
Heating (base 65°F)		450	307	155	32	0	0	0	0	0	26	177	352	1499
Cooling (base 65°F)		13	10	43	149	316	471	552	546	408	172	63	21	2764
% OF POSSIBLE SUNSHINE	26	42	52	52	52	64	69	65	63	62	67	57	47	58
MEAN SKY COVER (tenths)														
Sunrise - Sunset	40	7.0	6.4	6.6	6.6	6.1	5.4	5.8	5.7	5.5	4.8	5.8	6.6	6.0
MEAN NUMBER OF DAYS:														
Sunrise to Sunset														
-Clear	40	6.6	7.6	7.3	6.6	6.7	8.4	6.7	7.1	8.8	12.4	9.6	8.1	95.8
-Partly Cloudy	40	6.1	5.8	7.2	7.7	11.9	13.7	15.0	15.0	11.6	9.9	7.3	5.9	117.1
-Cloudy	40	18.4	14.9	16.5	15.7	12.4	7.9	9.3	8.9	9.6	8.8	13.1	16.9	152.4
Precipitation														
.01 inches or more	40	9.9	8.7	7.9	6.7	7.5	8.4	11.3	11.9	9.8	6.3	7.8	9.3	105.4
Snow,Ice pellets,hail														
1.0 inches or more	40	0.*	0.1	0.*	0.0	0.0	0.0	0.0	0.0	0.0	0.0	0.0	0.0	0.1
Thunderstorms	40	2.7	2.5	3.4	3.9	6.2	8.2	13.8	12.5	7.4	3.2	2.8	2.2	68.8
Heavy Fog Visibility														
1/4 mile or less	40	7.1	5.4	5.8	3.0	1.4	0.2	0.2	0.2	0.7	3.0	4.5	6.1	37.5
Temperature °F														
-Maximum														
90° and above	33	0.0	0.0	0.0	0.2	2.6	17.0	24.5	23.7	13.0	1.8	0.0	0.0	82.7
32° and below	33	0.2	0.*	0.0	0.0	0.0	0.0	0.0	0.0	0.0	0.0	0.0	0.2	0.3
-Minimum														
32° and below	33	6.0	3.4	1.0	0.*	0.0	0.0	0.0	0.0	0.0	0.1	1.0	4.1	15.5
0° and below	33	0.0	0.0	0.0	0.0	0.0	0.0	0.0	0.0	0.0	0.0	0.0	0.0	0.0
AVG. STATION PRESS.(mb)	20	1020.4	1018.9	1015.7	1014.8	1013.3	1014.2	1015.9	1015.3	1015.1	1017.4	1018.5	1020.1	1016.6
RELATIVE HUMIDITY (%)														
Hour 00	31	86	85	86	88	90	91	92	92	90	88	87	86	88
Hour 06 (Local Time)	33	88	87	88	90	92	93	94	94	92	91	89	89	91
Hour 12	33	68	63	62	62	64	64	65	64	64	58	62	67	64
Hour 18	33	75	70	68	70	70	70	71	72	73	73	75	77	72
PRECIPITATION (inches):														
Water Equivalent														
-Normal		4.77	3.38	3.24	3.51	5.71	5.59	5.38	5.34	6.31	4.29	4.85	4.81	57.18
-Maximum Monthly	40	14.87	11.76	9.35	15.30	13.18	18.90	18.71	17.26	21.96	15.09	10.84	17.98	21.96
-Year		1991	1959	1979	1973	1991	1989	1959	1966	1980	1970	1977	1982	SEP 1980
-Minimum Monthly	40	0.60	0.21	0.06	0.26	0.10	0.76	0.63	0.98	0.50	0.00	0.15	1.32	0.00
-Year		1971	1989	1955	1987	1978	1980	1956	1968	1953	1963	1967	1954	OCT 1963
-Maximum in 24 hrs	40	4.92	5.05	6.04	10.09	9.89	10.20	10.56	8.45	17.16	8.06	7.26	9.98	17.16
-Year		1961	1965	1979	1973	1989	1961	1979	1966	1980	1970	1961	1982	SEP 1980
Snow,Ice pellets,hail														
-Maximum Monthly	40	3.0	4.4	1.4	0.0	T	0.0	T	0.0	0.0	0.0	T	0.7	4.4
-Year		1973	1960	1968		1991		1990				1976	1989	FEB 1960
-Maximum in 24 hrs	40	3.0	4.4	1.4	0.0	T	0.0	T	0.0	0.0	0.0	T	0.7	4.4
-Year		1973	1960	1968		1991		1990				1976	1989	FEB 1960
WIND:														
Mean Speed (mph)	40	10.7	11.3	11.6	11.7	10.2	8.8	7.6	7.2	8.4	8.8	10.1	10.4	9.7
Prevailing Direction through 1963		N	S	S	S	S	S	S	S	NE	N	N	N	S
Fastest Obs. 1 Min.														
-Direction (!!)	14	06	24	18	30	29	14	21	01	08	31	27	32	14
-Speed (MPH)	14	39	35	38	46	52	55	40	44	29	32	38	32	55
-Year		1991	1984	1992	1981	1983	1986	1991	1988	1980	1985	1987	1988	JUN 1986
Peak Gust														
-Direction (!!)	10	NE	SW	NW	W	S	SE	SW	N	SE	NW	NW	S	SE
-Speed (mph)	10	52	54	53	59	62	76	69	55	39	47	54	47	76
-Date		1991	1984	1991	1990	1988	1986	1991	1988	1985	1985	1991	1984	JUN 1986

PRECIPITATION (inches) PORT ARTHUR, TEXAS

YEAR	JAN	FEB	MAR	APR	MAY	JUNE	JULY	AUG	SEP	OCT	NOV	DEC	ANNUAL
1964	4.28	3.10	3.64	2.11	6.02	1.62	6.16	4.96	5.57	0.07	4.19	5.06	46.78
1965	1.59	5.41	2.96	0.35	3.15	2.41	3.21	1.91	4.62	0.78	2.29	5.78	34.46
1966	6.22	5.68	0.65	5.86	9.63	3.50	3.97	17.26	3.17	6.49	4.85	3.39	70.67
1967	1.92	2.06	0.91	5.86	5.85	1.06	4.20	8.63	3.35	2.13	0.15	4.55	40.67
1968	5.48	2.56	2.51	7.38	3.20	12.17	6.15	0.98	6.01	3.11	5.69	2.74	57.98
1969	1.19	4.14	3.67	5.38	5.69	2.08	10.14	3.06	1.50	1.74	1.52	8.33	48.44
1970	1.78	2.88	4.32	2.47	8.54	0.96	1.55	3.07	7.37	15.09	1.57	2.03	51.63
1971	0.60	4.31	1.15	1.76	3.47	1.86	4.31	10.56	5.07	2.16	2.64	7.53	45.42
1972	8.33	1.83	3.93	5.42	5.92	3.43	4.86	5.80	6.50	2.88	6.11	3.95	58.96
1973	4.76	3.00	7.24	15.30	5.47	5.80	8.72	7.16	11.44	5.11	2.01	2.90	78.91
1974	8.81	2.37	4.88	5.33	7.61	1.43	3.75	4.39	2.83	3.69	5.84	4.53	55.46
1975	5.46	1.60	1.75	3.77	5.30	4.97	7.56	6.86	2.80	4.68	3.26	3.60	51.61
1976	3.64	0.44	2.69	1.89	6.39	6.16	3.99	2.57	4.47	5.16	5.54	6.11	49.05
1977	4.93	1.06	2.23	4.79	3.41	3.37	2.11	7.20	2.87	4.06	10.84	2.01	48.88
1978	6.66	2.32	0.44	0.36	0.10	7.47	3.41	5.11	3.29	T	4.80	3.72	37.68
1979	5.41	4.21	9.35	7.35	4.39	3.96	15.68	4.07	9.45	6.62	3.20	2.51	76.20
1980	4.66	1.92	6.83	0.93	7.76	0.76	2.24	1.18	21.96	7.27	4.82	2.22	62.55
1981	3.33	3.26	2.09	3.16	2.78	9.14	10.75	1.63	2.72	5.68	2.16	3.85	50.55
1982	2.04	3.38	3.12	5.63	9.36	9.30	5.88	2.47	1.00	4.15	7.46	17.98	71.77
1983	5.78	3.97	4.57	0.51	11.39	8.10	6.53	14.35	11.89	0.33	4.24	6.49	78.15
1984	6.69	4.27	1.28	1.03	10.43	1.05	3.83	3.91	4.14	14.94	6.44	2.03	60.04
1985	3.44	6.85	4.74	1.13	5.15	3.17	6.91	11.25	7.39	11.60	3.85	2.74	68.22
1986	1.81	1.11	2.52	1.44	10.56	11.23	2.25	3.97	3.41	7.37	9.08	9.59	64.34
1987	8.05	7.68	0.94	0.26	4.44	12.31	3.39	3.86	11.49	0.27	6.74	6.14	65.57
1988	3.23	5.05	5.54	2.64	0.61	4.63	3.84	5.54	10.32	1.70	1.12	5.80	50.02
1989	6.76	0.21	3.40	2.75	12.69	18.90	7.74	5.03	1.20	1.17	3.14	3.21	66.20
1990	8.50	7.69	6.53	4.22	8.62	4.21	6.18	1.67	5.30	4.37	5.54	3.82	66.65
1991	14.87	4.56	2.44	6.78	13.18	7.27	3.11	3.04	7.14	3.61	3.96	11.59	81.55
1992	11.04	6.51	3.80	5.78	1.39	10.66	5.37	2.26	2.07	2.96	6.89	6.45	65.18
1993	9.92	4.09	5.56	6.38	4.62	9.32	5.41	6.19	2.39	5.99	5.31	1.94	67.12
Record Mean	4.79	3.79	3.27	3.71	5.01	5.05	6.04	5.27	5.29	3.64	4.14	5.06	55.08

AVERAGE TEMPERATURE (deg. F) PORT ARTHUR, TEXAS

YEAR	JAN	FEB	MAR	APR	MAY	JUNE	JULY	AUG	SEP	OCT	NOV	DEC	ANNUAL
1964	50.4	49.2	59.9	69.2	76.6	81.1	82.1	83.4	79.1	66.6	64.1	54.7	68.0
1965	55.0	54.4	57.1	72.7	77.3	81.2	83.8	82.4	80.7	68.4	68.1	57.4	69.9
1966	48.0	52.9	59.8	69.9	75.8	80.2	84.3	81.7	78.9	68.3	64.2	53.5	68.1
1967	51.8	53.4	63.9	73.2	74.7	81.9	81.7	80.9	76.4	69.0	62.7	55.5	68.8
1968	51.0	48.0	58.2	69.9	74.6	79.5	81.3	83.2	76.6	70.4	58.9	53.8	67.1
1969	54.6	54.6	54.7	69.2	74.9	81.0	84.9	83.1	78.4	72.0	59.7	55.9	68.6
1970	47.5	54.4	59.7	71.5	74.7	80.9	84.7	85.5	82.0	70.1	59.3	62.1	69.4
1971	57.5	56.6	60.7	69.0	75.2	81.9	81.7	79.3	77.8	71.8	59.3	59.8	69.2
1972	55.3	55.3	63.0	69.6	73.3	80.2	79.3	80.2	79.4	69.6	54.0	52.5	67.6
1973	47.5	51.0	63.5	63.7	72.6	79.4	81.7	79.6	78.9	72.1	66.5	52.4	67.4
1974	56.4	55.1	66.2	68.6	76.5	78.9	81.7	81.1	74.9	68.7	59.0	53.0	68.3
1975	56.2	55.2	60.9	66.9	75.9	79.9	81.4	81.2	75.9	70.3	60.2	52.0	68.0
1976	49.8	59.9	62.8	68.7	72.1	78.7	81.2	81.2	78.3	61.9	52.5	50.6	66.5
1977	44.5	55.3	63.2	68.4	77.4	83.6	85.0	84.9	83.1	71.8	63.8	55.3	69.7
1978	43.8	47.2	59.8	70.2	79.0	83.3	85.1	85.3	81.8	72.0	67.4	56.8	69.3
1979	45.8	52.5	62.5	69.5	72.9	80.3	81.8	81.7	76.5	70.6	56.4	52.8	66.9
1980	55.2	52.7	61.4	65.6	76.4	82.9	85.1	84.2	82.6	67.5	57.8	54.4	68.8
1981	50.3	53.8	61.2	72.9	73.7	82.0	83.2	82.7	76.9	71.0	63.4	55.1	68.8
1982	52.8	52.4	65.0	68.0	75.6	81.6	82.7	82.7	78.6	70.0	61.8	57.6	69.0
1983	50.4	54.0	58.9	63.8	73.3	79.1	82.4	82.4	76.7	70.3	63.7	48.7	67.0
1984	48.7	56.0	62.2	68.6	74.7	79.7	81.2	81.7	77.0	74.8	60.8	63.7	69.1
1985	48.1	53.1	67.0	72.1	76.2	81.0	81.7	83.3	78.7	72.4	66.6	50.8	69.3
1986	53.5	59.3	62.2	70.7	75.4	81.3	84.5	82.4	82.2	69.4	63.6	51.5	69.7
1987	50.3	56.2	59.5	66.7	76.6	80.6	83.3	84.2	78.1	67.2	60.3	56.9	68.3
1988	48.4	53.8	60.9	67.9	73.8	79.9	83.5	84.4	79.2	69.1	63.9	54.4	68.3
1989	57.4	52.6	60.0	67.3	76.7	79.6	82.5	81.8	77.7	70.6	63.1	45.8	67.9
1990	57.5	60.4	63.4	68.9	77.2	83.5	82.8	83.8	80.6	68.3	63.2	55.3	70.4
1991	51.1	57.8	63.7	72.2	77.8	81.6	83.7	82.5	78.1	73.0	56.9	57.4	69.7
1992	51.6	58.3	63.3	67.9	73.7	81.0	82.8	80.7	79.4	70.9	56.4	57.3	68.6
1993	54.3	56.3	60.9	66.0	74.1	81.7	84.3	85.0	80.8	70.0	58.1	55.5	68.9
Record Mean	52.3	55.7	61.4	68.7	75.3	81.1	82.8	82.8	79.0	70.7	60.7	54.7	68.7
Max	60.9	64.5	70.3	77.1	83.4	89.0	91.0	91.2	87.5	80.5	70.1	63.5	77.4
Min	43.7	46.8	52.5	60.2	67.1	73.1	74.5	74.3	70.4	60.8	51.2	45.8	60.0

REFERENCE NOTES PORT ARTHUR, TEXAS

GENERAL
T - TRACE AMOUNT.
BLANK ENTRIES DENOTE MISSING/UNREPORTED DATA.
# INDICATES A STATION OR INSTRUMENT RELOCATION.

SPECIFIC

(a) - LENGTH OF RECORD IN YEARS, ALTHOUGH INDIVIDUAL MONTHS MAY BE MISSING.
0.* OR * - THE VALUE IS BETWEEN 0.0 AND 0.05
NORMALS - BASED ON THE 1961-1990 RECORD PERIOD.
EXTREMES - DATES ARE THE MOST RECENT OCCURRENCE
WIND DIR.- NUMERALS SHOW TENS OF DEGRESS CLOCKWISE FROM TRUE NORTH. "00" INDICATES CALM.
RESULTANT DIRECTIONS ARE GIVEN TO WHOLE DEGREES.
BOLD VALUES INDICATE EXTREME VALUES WHICH OCCURRED AFTER THE ASOS SYSTEM WAS COMMISSIONED.

RECORD = PERIOD OF RECORD
RECORD MEAN PRECIPITATION IS THE MEAN OF ALL DAILY PRECIPITATION AMOUNTS DURING THE PERIOD OF RECORD.
RECORD MAX(MIN) TEMPERATURE IS THE MEAN OF ALL DAILY MAX(MIN) TEMPERATURES DURING THE PERIOD OF RECORD.
RECORD MEAN TEMPERATURE IS THE SUM OF THE RECORD MAX AND RECORD MIN DIVIDED BY 2.
AVERAGE TEMPERATURE IS THE SUM OF THE MEAN DAILY MAX AND MIN TEMPERATURE DIVIDED BY 2.

EXCEPTIONS

PERCENT OF POSSIBLE SUNSHINE IS THROUGH 1979.

RECORD MEANS ARE THROUGH THE CURRENT YEAR, BEGINNING IN 1911 FOR TEMPERATURE
1911 FOR PRECIPITATION
1954 FOR SNOWFALL

HEATING DEGREE DAYS Base 65 deg. F — PORT ARTHUR, TEXAS

SEASON	JULY	AUG	SEP	OCT	NOV	DEC	JAN	FEB	MAR	APR	MAY	JUNE	TOTAL
1964-65	0	0	0	52	135	337	322	298	270	3	0	0	1417
1965-66	0	0	0	43	34	250	529	336	180	19	0	0	1391
1966-67	0	0	0	46	115	383	417	325	111	0	2	0	1399
1967-68	0	0	10	33	133	313	435	485	244	18	0	0	1671
1968-69	0	0	0	20	207	343	331	289	317	4	0	0	1511
1969-70	0	0	0	22	212	281	545	293	180	22	3	0	1558
1970-71	0	0	0	28	213	168	269	254	167	43	0	0	1142
1971-72	0	0	0	6	208	199	317	291	109	20	0	0	1150
1972-73	0	0	0	57	340	390	532	389	77	118	2	0	1905
1973-74	0	0	0	13	83	384	288	288	85	36	0	0	1177
1974-75	0	0	0	20	230	384	297	276	180	67	0	0	1454
1975-76	0	0	0	16	209	413	464	170	128	16	0	0	1416
1976-77	0	0	0	149	375	439	626	268	110	16	0	0	1983
1977-78	0	0	0	16	112	321	662	497	184	14	0	0	1806
1978-79	0	0	0	4	75	296	591	360	125	12	0	0	1463
1979-80	0	0	0	23	278	382	306	377	151	44	0	0	1561
1980-81	0	0	0	70	248	339	448	315	140	3	3	0	1566
1981-82	0	0	3	59	94	315	402	357	124	59	0	0	1413
1982-83	0	0	0	43	154	266	446	301	203	91	1	0	1505
1983-84	0	0	1	27	122	524	500	265	129	31	1	0	1600
1984-85	0	0	5	8	176	125	514	342	34	6	0	0	1210
1985-86	0	0	3	14	78	438	354	196	120	10	0	0	1213
1986-87	0	0	0	26	140	415	451	241	182	77	0	0	1532
1987-88	0	0	0	29	182	267	510	329	171	29	0	0	1517
1988-89	0	0	0	10	135	328	255	371	216	60	0	0	1375
1989-90	0	0	0	45	149	592	245	151	108	26	0	0	1316
1990-91	0	0	0	67	121	338	422	208	113	9	0	0	1278
1991-92	0	0	0	6	287	265	409	198	93	34	7	0	1299
1992-93	0	0	0	2	273	251	328	238	152	55	0	0	1299
1993-94	0	0	0	58	240	312							

COOLING DEGREE DAYS Base 65 deg. F — PORT ARTHUR, TEXAS

YEAR	JAN	FEB	MAR	APR	MAY	JUNE	JULY	AUG	SEP	OCT	NOV	DEC	TOTAL
1969	17	4	5	139	316	484	621	569	408	247	60	6	2876
1970	13	2	23	225	309	484	617	644	518	194	49	85	3163
1971	42	24	42	167	325	517	527	451	388	223	42	43	2791
1972	21	18	52	164	262	461	449	479	440	209	17	9	2581
1973	0	2	38	84	244	440	524	459	426	242	137	6	2602
1974	27	15	130	153	362	423	524	505	305	143	58	20	2665
1975	28	9	61	130	344	454	516	508	335	186	71	18	2660
1976	2	29	66	133	226	418	512	512	404	61	8	0	2371
1977	0	4	62	125	391	562	627	621	551	235	82	28	3288
1978	11	8	29	177	439	555	627	635	510	227	154	49	3421
1979	4	17	54	156	250	464	529	523	352	203	27	11	2590
1980	8	27	47	69	359	546	633	603	534	152	37	15	3030
1981	0	8	28	245	278	515	570	557	367	250	53	14	2885
1982	27	9	129	155	335	504	555	554	416	205	67	46	3002
1983	0	0	20	64	264	429	547	546	361	200	88	21	2540
1984	0	9	50	147	309	449	508	524	369	317	56	94	2832
1985	0	13	104	228	354	489	527	573	422	253	134	5	3102
1986	1	44	39	185	331	495	611	546	522	169	104	4	3050
1987	2	2	16	136	367	476	574	604	402	105	48	22	2754
1988	4	11	49	121	283	452	581	606	432	148	108	6	2801
1989	28	32	69	137	370	444	548	531	387	226	99	5	2876
1990	19	29	68	151	384	563	559	591	471	173	74	42	3124
1991	0	9	77	233	406	504	584	548	401	260	49	38	3109
1992	0	9	49	126	284	486	558	495	439	188	23	17	2674
1993	3	2	32	90	290	506	605	627	482	221	38	26	2922

SNOWFALL (inches) — PORT ARTHUR, TEXAS

SEASON	JULY	AUG	SEP	OCT	NOV	DEC	JAN	FEB	MAR	APR	MAY	JUNE	TOTAL
1970-71	0.0	0.0	0.0	0.0	0.0	0.0	0.0	0.0	0.0	0.0	0.0	0.0	0.0
1971-72	0.0	0.0	0.0	0.0	0.0	0.0	0.0	0.0	0.0	0.0	0.0	0.0	0.0
1972-73	0.0	0.0	0.0	0.0	0.0	0.0	3.0	0.4	0.0	0.0	0.0	0.0	3.4
1973-74	0.0	0.0	0.0	0.0	0.0	0.0	0.0	0.0	0.0	0.0	0.0	0.0	0.0
1974-75	0.0	0.0	0.0	0.0	0.0	0.0	0.0	0.0	0.0	0.0	0.0	0.0	0.0
1975-76	0.0	0.0	0.0	0.0	0.0	0.0	T	0.0	0.0	0.0	0.0	0.0	T
1976-77	0.0	0.0	0.0	0.0	T	0.0	0.0	0.0	0.0	0.0	0.0	0.0	T
1977-78	0.0	0.0	0.0	0.0	0.0	0.0	T	T	0.0	0.0	0.0	0.0	T
1978-79	0.0	0.0	0.0	0.0	0.0	0.0	T	0.0	0.0	0.0	0.0	0.0	T
1979-80	0.0	0.0	0.0	0.0	0.0	0.0	0.0	0.7	0.0	0.0	0.0	0.0	0.7
1980-81	0.0	0.0	0.0	0.0	0.0	0.0	0.0	T	0.0	0.0	0.0	0.0	T
1981-82	0.0	0.0	0.0	0.0	0.0	0.0	0.2	0.0	0.0	0.0	0.0	0.0	0.2
1982-83	0.0	0.0	0.0	0.0	0.0	0.0	0.0	0.0	0.0	0.0	0.0	0.0	0.0
1983-84	0.0	0.0	0.0	0.0	0.0	T	0.0	0.0	0.0	0.0	0.0	0.0	T
1984-85	0.0	0.0	0.0	0.0	0.0	0.0	0.4	T	0.0	0.0	0.0	0.0	0.4
1985-86	0.0	0.0	0.0	0.0	0.0	0.0	0.0	0.0	0.0	0.0	0.0	0.0	0.0
1986-87	0.0	0.0	0.0	0.0	0.0	0.0	T	0.0	T	0.0	0.0	0.0	T
1987-88	0.0	0.0	0.0	0.0	0.0	0.0	0.0	0.2	0.0	0.0	0.0	0.0	0.2
1988-89	0.0	0.0	0.0	0.0	0.0	0.0	0.0	0.0	0.0	0.0	0.0	0.0	0.0
1989-90	0.0	0.0	0.0	0.0	0.0	0.7	0.0	0.0	0.0	0.0	0.0	0.0	0.7
1990-91	T	0.0	0.0	0.0	0.0	0.0	0.0	0.0	0.0	0.0	T	0.0	T
1991-92	0.0	0.0	0.0	0.0	0.0	0.0	0.0	0.0	0.0	0.0	0.0	0.0	0.0
1992-93	0.0	0.0	0.0	0.0	0.0	0.0	0.0	0.0	0.0	0.0	0.0	0.0	0.0
1993-94	0.0	0.0	0.0	0.0	0.0	0.0							
Record Mean	T	0.0	0.0	0.0	T	T	0.1	0.2	T	0.0	T	0.0	0.4

SAN ANGELO, TEXAS

San Angelo is located near the center of Texas at the northern edge of the Edwards Plateau. Ground elevation ranges from about 1,700 to 2,700 feet above sea level. Topography varies from level and slightly rolling to broken. The climate is generally classified as semi-arid or steppe, but has some humid temperate characteristics. Warm, dry weather predominates, although changes may be rapid and frequent with the passage of cold fronts or northers.

High temperatures of summer are associated with fair skies, south to southwest winds and dry air. Low humidities, however, are conducive to personal comfort because of rapid evaporation. Rapid temperature drops occur after sunset, and most nights are pleasant with lows in the upper 60s and lower 70s. Rapid temperature drops occur in the winter as cold polar air invades the region. Temperature drops of 20 to 30 degrees in a short time are not uncommon. Cold polar outbreaks have produced record low temperatures of zero or below throughout the area.

The rainfall is typical of the Great Plains. Much of the rainfall occurs from thunderstorm activity, and wide variations in annual precipitation occur from year to year. Heavy rainfall occurs in April, May, June, September and October. Also, in the late summer months, heavy precipitation may occur when tropical disturbances move inland over south Texas and pass near the San Angelo area.

The prevailing wind direction is from the south, and winds are frequently high and persistent for several days. Dusty conditions are infrequent and occur in early spring when west or northwest winds predominate. The frequency and intensity of the dust storms are dependent on soil conditions in the Texas Panhandle and in New Mexico.

Agriculture in the region consists of cattle, sheep, and goat raising. Cotton, from dry-land and irrigated fields, maize, corn, melons, truck farming, and pecan production are also important crops.

NORMALS, MEANS, AND EXTREMES

SAN ANGELO, TEXAS

LATITUDE: 31°22'N LONGITUDE: 100°30' W ELEVATION: FT. GRND 1903 BARO 1909 TIME ZONE: CENTRAL WBAN: 23034

	(a)	JAN	FEB	MAR	APR	MAY	JUNE	JULY	AUG	SEP	OCT	NOV	DEC	YEAR
TEMPERATURE °F:														
Normals														
-Daily Maximum		56.8	62.0	72.6	81.2	87.4	92.7	96.2	95.3	86.8	78.8	68.2	59.0	78.1
-Daily Minimum		30.6	34.7	43.5	52.7	61.1	66.4	69.1	68.4	64.0	53.6	42.6	33.0	51.6
-Monthly		43.7	48.4	58.1	67.0	74.2	79.5	82.7	81.9	75.4	66.2	55.4	46.0	64.9
Extremes														
-Record Highest	46	90	91	97	103	107	110	111	109	107	100	93	91	111
-Year		1969	1986	1974	1972	1989	1969	1960	1986	1952	1951	1980	1954	JUL 1960
-Record Lowest	46	5	-1	8	25	35	48	56	54	37	26	13	-4	-4
-Year		1982	1985	1980	1973	1967	1964	1990	1992	1989	1993	1979	1989	DEC 1989
NORMAL DEGREE DAYS:														
Heating (base 65°F)		660	465	247	70	7	0	0	0	6	71	299	589	2414
Cooling (base 65°F)		0	0	33	130	292	435	549	524	318	108	11	0	2400
% OF POSSIBLE SUNSHINE														
MEAN SKY COVER (tenths)														
Sunrise - Sunset	45	5.5	5.3	5.1	5.1	5.2	4.4	4.3	4.2	4.6	4.2	4.6	5.2	4.8
MEAN NUMBER OF DAYS:														
Sunrise to Sunset														
-Clear	45	11.7	10.8	12.2	11.2	10.9	13.6	14.5	15.1	13.3	15.3	13.5	12.3	154.5
-Partly Cloudy	45	6.2	6.3	7.5	8.0	10.0	10.3	9.9	10.0	8.5	7.2	6.4	6.6	96.9
-Cloudy	45	13.1	11.1	11.4	10.8	10.1	6.1	6.6	5.9	8.2	8.4	10.1	12.1	113.9
Precipitation .01 inches or more	46	4.9	4.5	4.1	4.9	7.1	5.3	4.5	5.0	6.1	4.8	3.8	4.1	59.1
Snow,Ice pellets,hail 1.0 inches or more	46	0.5	0.3	0.1	0.0	0.0	0.0	0.0	0.0	0 0	0.0	0.2	0.1	1.2
Thunderstorms	46	0.6	1.1	2.4	4.3	7.2	5.3	4.1	4.8	3.8	2.6	1.0	0.5	37.8
Heavy Fog Visibility 1/4 mile or less	46	1.3	1.2	0.5	0.3	0.2	0.1	0.*	0.*	0.2	0.7	1.3	1.3	7.2
Temperature °F														
-Maximum														
90° and above	33	0.*	0.1	0.9	5.0	11.5	20.9	27.0	26.0	12.4	2.8	0.2	0.0	106.8
32° and below	33	1.4	0.5	0.0	0.0	0.0	0.0	0.0	0.0	0.0	0.0	0.1	0.6	2.5
-Minimum														
32° and below	33	17.4	10.5	4.3	0.6	0.0	0.0	0.0	0.0	0.0	0.2	5.5	14.2	52.7
0° and below	33	0.0	0.*	0.0	0.0	0.0	0.0	0.0	0.0	0.0	0.0	0.0	0.1	0.1
AVG. STATION PRESS. (mb)	21	952.1	950.6	947.2	946.5	945.5	946.7	948.6	948.8	949.2	950.3	950.6	951.8	949.0
RELATIVE HUMIDITY (%)														
Hour 00	33	70	68	62	63	69	67	61	64	73	74	73	72	68
Hour 06 (Local Time)	33	77	76	72	74	81	82	78	79	84	83	80	78	79
Hour 12	33	53	50	44	43	49	50	44	45	54	52	51	52	49
Hour 18	33	48	43	36	36	41	42	37	40	49	50	51	51	44
PRECIPITATION (inches):														
Water Equivalent														
-Normal		0.80	1.07	0.91	1.67	3.00	2.33	1.06	1.93	3.41	2.40	1.08	0.79	20.45
-Maximum Monthly	46	3.65	4.45	5.00	5.10	11.24	6.01	7.21	8.13	11.00	8.68	3.55	3.98	11.24
-Year		1961	1987	1953	1977	1987	1982	1959	1971	1980	1981	1968	1991	MAY 1987
-Minimum Monthly	46	0.00	0.01	T	0.07	0.26	0.05	T	T	T	0.00	0.00	T	0.00
-Year		1967	1974	1972	1986	1962	1990	1970	1959	1983	1952	1950	1973	JAN 1967
-Maximum in 24 hrs	46	2.49	3.16	4.65	3.32	3.12	2.86	2.95	3.00	6.25	5.11	2.16	2.71	6.25
-Year		1961	1987	1953	1971	1987	1961	1959	1971	1980	1959	1975	1984	SEP 1980
Snow,Ice pellets,hail														
-Maximum Monthly	46	9.0	5.8	3.1	T	T	T	T	0.0	T	T	8.8	3.7	9.0
-Year		1978	1973	1962	1990	1993	1992	1992		1992	1993	1968	1986	JAN 1978
-Maximum in 24 hrs	46	7.4	4.1	3.1	T	T	T	T	0.0	T	T	5.8	3.3	7.4
-Year		1978	1966	1962	1990	1993	1992	1992		1992	1993	1968	1986	JAN 1978
WIND:														
Mean Speed (mph)	44	10.2	10.8	12.3	12.0	11.2	11.0	9.8	9.1	9.0	9.3	10.0	10.0	10.4
Prevailing Direction through 1963		SW	SSW	SSW	S	S	S	S	S	S	S	SW	W	S
Fastest Obs. 1 Min.														
-Direction (!!)	45	27	29	27	28	02	02	17	02	34	29	29	30	28
-Speed (MPH)	45	44	48	58	75	60	57	46	44	52	60	66	43	75
-Year		1960	1960	1961	1969	1963	1955	1993	1986	1981	1960	1958	1965	APR 1969
Peak Gust														
-Direction (!!)	10	N	W	W	SW	NE	NW	S	SW	NE	W	W	NW	NE
-Speed (mph)	10	55	59	61	67	76	71	56	67	52	68	47	53	76
-Date		1993	1989	1987	1993	1988	1991	1993	1992	1992	1992	1992	1993	MAY 1988

PRECIPITATION (inches)

SAN ANGELO, TEXAS

YEAR	JAN	FEB	MAR	APR	MAY	JUNE	JULY	AUG	SEP	OCT	NOV	DEC	ANNUAL
1964	0.95	1.27	0.83	1.90	0.41	0.16	0.47	1.35	2.80	0.63	1.14	0.27	12.18
1965	0.53	2.54	0.12	0.12	3.65	2.01	0.26	2.71	1.18	2.11	0.39	0.63	16.25
1966	0.49	0.58	0.27	2.16	2.10	0.77	0.09	5.05	2.41	1.90	T	T	15.82
1967	0.00	0.17	0.78	0.10	1.91	2.42	0.59	1.43	7.21	1.82	2.41	1.14	19.98
1968	2.32	1.58	1.88	2.26	1.98	4.71	1.23	1.87	1.81	0.11	3.55	T	23.30
1969	0.02	1.07	1.76	4.38	3.20	0.26	0.62	2.96	5.74	6.59	1.20	2.24	30.04
1970	0.14	1.31	2.36	1.10	1.83	0.39	T	0.47	4.28	0.94	T	0.06	12.88
1971	T	0.42	T	3.89	1.42	1.72	1.84	8.13	3.59	2.09	0.27	0.88	24.25
1972	0.48	0.09	T	0.57	4.89	4.21	0.17	3.66	5.30	3.22	0.31	0.03	22.93
1973	1.86	1.82	1.18	1.67	1.43	2.75	1.06	0.36	4.22	2.03	0.03	T	18.41
1974	0.12	0.01	0.48	1.86	2.43	0.16	1.54	4.77	6.11	4.82	1.40	1.40	25.10
1975	0.37	1.28	0.09	0.86	5.26	2.32	2.66	0.38	2.25	3.37	2.19	0.55	21.58
1976	0.04	0.42	0.30	3.41	1.77	0.41	3.51	1.41	4.66	5.11	0.58	0.18	21.80
1977	0.61	0.26	0.99	5.10	2.14	0.47	0.52	0.38	0.27	1.38	0.68	0.15	12.95
1978	0.61	1.17	0.41	0.73	1.83	2.81	0.41	2.93	1.35	0.42	1.75	0.25	14.67
1979	0.19	1.83	2.25	1.90	0.67	2.15	1.30	2.18	0.06	0.93	T	2.70	16.16
1980	0.84	0.79	0.71	0.52	4.72	3.14	0.33	3.30	11.00	0.01	2.53	2.20	30.09
1981	1.17	0.68	2.81	3.51	4.70	1.97	2.68	1.23	2.72	8.68	T	0.02	30.17
1982	1.06	1.53	0.40	0.83	4.17	6.01	0.35	0.46	0.09	1.26	1.11	0.91	18.18
1983	2.06	0.42	1.20	0.81	0.52	3.72	1.18	0.03	T	3.47	1.78	0.07	15.26
1984	2.38	0.54	0.49	0.23	0.54	2.82	0.60	0.26	2.99	3.74	1.09	3.48	19.16
1985	0.67	0.38	1.69	0.42	4.78	3.55	1.13	0.24	2.84	5.64	0.48	0.01	21.83
1986	0.30	0.65	0.52	0.07	7.28	3.30	0.74	2.50	7.53	5.72	1.83	2.48	32.92
1987	0.65	4.45	1.77	1.33	11.24	3.28	0.22	1.91	3.86	0.34	0.80	2.05	31.90
1988	0.01	0.42	0.69	1.36	3.31	2.14	1.22	0.92	3.20	T	0.00	0.79	14.06
1989	0.68	3.01	1.95	1.04	1.06	2.83	0.35	2.78	2.82	0.41	0.48	0.23	17.64
1990	1.60	1.65	0.85	4.14	4.02	0.05	4.09	1.05	6.23	2.40	2.51	0.21	28.80
1991	2.08	0.26	0.61	0.48	1.15	4.22	1.80	1.36	4.77	3.34	0.24	3.98	24.29
1992	1.90	3.79	1.05	1.00	1.49	4.74	1.98	1.78	0.27	1.33	0.95	0.75	21.03
1993	0.94	0.81	0.28	1.46	3.87	0.94	0.82	2.64	1.99	1.07	0.05	0.76	15.63
Record Mean	0.89	0.96	0.92	1.84	3.02	2.06	1.66	1.98	2.93	2.25	1.12	1.04	20.66

AVERAGE TEMPERATURE (deg. F)

SAN ANGELO, TEXAS

YEAR	JAN	FEB	MAR	APR	MAY	JUNE	JULY	AUG	SEP	OCT	NOV	DEC	ANNUAL
1964	45.1	45.4	60.6	70.3	77.3	83.1	86.9	86.6	77.1	64.7	56.9	48.9	66.9
1965	49.7	46.6	49.7	70.5	74.0	81.4	85.7	82.7	78.5	65.8	61.2	51.5	66.5
1966	40.7	45.5	59.0	66.3	73.6	81.0	87.0	81.0	74.1	62.8	59.3	45.2	64.6
1967	46.5	48.5	64.8	72.5	74.4	83.3	84.5	83.0	72.6	66.1	57.0	45.2	66.5
1968	46.3	45.8	55.0	62.2	74.5	79.3	82.5	84.1	74.4	69.4	53.3	47.2	64.5
1969	52.1	51.6	51.6	67.0	71.4	82.0	87.5	84.9	75.2	64.5	54.4	50.5	66.1
1970	43.9	50.8	52.9	66.2	71.3	78.9	83.3	85.4	78.3	64.0	54.4	54.6	65.4
1971	50.2	52.0	59.6	66.7	75.2	80.9	82.8	75.4	74.3	66.8	56.0	50.2	65.8
1972	47.4	52.4	62.4	71.7	70.8	81.1	83.3	79.9	78.2	66.2	50.4	46.2	65.8
1973	41.0	46.4	59.0	60.6	73.0	78.9	83.3	82.6	75.1	67.0	60.7	49.4	64.7
1974	47.0	52.4	64.9	67.9	76.0	80.9	84.0	79.6	68.9	66.4	55.0	46.8	65.8
1975	46.5	48.0	56.1	64.8	71.8	79.7	78.6	80.6	71.6	65.3	54.7	48.0	63.8
1976	43.9	56.7	58.1	65.5	68.8	79.7	76.2	80.7	73.8	58.6	47.5	45.6	62.9
1977	40.4	51.5	57.9	65.0	75.5	82.3	84.8	86.2	83.5	69.2	57.1	52.3	67.2
1978	38.5	43.8	56.3	70.6	75.7	80.3	85.1	79.5	74.6	64.7	55.5	44.2	64.1
1979	36.2	46.7	57.2	65.4	72.1	78.0	83.4	80.4	75.2	69.9	50.8	47.4	63.5
1980	46.1	49.2	55.4	63.7	71.8	82.6	86.9	84.4	76.9	64.5	53.0	49.9	65.4
1981	45.8	50.2	54.2	66.4	70.7	77.8	82.1	80.0	74.3	64.6	55.6	48.7	64.2
1982	46.0	45.9	59.1	63.5	69.6	77.9	83.2	84.6	77.4	66.0	54.3	45.8	64.5
1983	45.1	48.8	56.6	62.2	72.5	76.6	82.2	82.1	77.4	70.2	58.4	36.3	64.0
1984	40.4	50.3	57.4	65.1	76.0	82.3	81.5	82.8	72.5	64.5	53.0	50.2	64.7
1985	38.8	46.2	59.3	68.1	75.1	77.5	80.5	84.5	75.7	66.9	57.9	43.6	64.5
1986	48.4	52.9	60.9	70.8	73.7	78.0	83.3	81.4	77.9	63.8	53.0	45.3	65.8
1987	44.9	49.9	51.3	61.5	71.1	76.0	80.2	82.4	72.9	66.5	53.5	46.2	63.0
1988	42.6	47.9	56.8	64.7	70.7	79.2	80.7	81.9	75.4	66.1	57.9	48.0	64.3
1989	50.2	44.7	58.4	66.6	78.2	77.9	83.2	81.9	72.2	67.3	56.5	39.5	64.7
1990	50.6	53.4	57.7	64.7	73.2	85.1	79.1	80.0	75.0	63.6	57.1	45.1	65.4
1991	43.7	51.9	58.9	66.7	76.6	78.5	81.5	80.3	72.0	65.9	50.0	49.3	64.6
1992	45.7	52.9	58.4	64.8	70.5	79.2	83.1	79.3	77.6	68.3	53.0	49.6	65.2
1993	44.9	49.2	56.9	66.3	72.4	81.2	85.8	84.4	74.1	64.1	51.9	50.1	65.1
Record Mean	45.5	49.6	57.2	65.8	73.3	80.7	83.5	83.0	75.9	66.2	54.3	47.3	65.2
Max	58.6	63.2	71.6	79.9	86.3	93.0	96.0	95.5	88.1	79.1	67.4	60.3	78.2
Min	32.3	35.9	42.7	51.7	60.3	68.3	70.9	70.4	63.7	53.3	41.1	34.3	52.1

REFERENCE NOTES

SAN ANGELO, TEXAS

GENERAL
T - TRACE AMOUNT.
BLANK ENTRIES DENOTE MISSING/UNREPORTED DATA.
# INDICATES A STATION OR INSTRUMENT RELOCATION.

SPECIFIC

(a) - LENGTH OF RECORD IN YEARS, ALTHOUGH INDIVIDUAL MONTHS MAY BE MISSING.
0.* OR * - THE VALUE IS BETWEEN 0.0 AND 0.05
NORMALS - BASED ON THE 1961-1990 RECORD PERIOD.
EXTREMES - DATES ARE THE MOST RECENT OCCURRENCE
WIND DIR.- NUMERALS SHOW TENS OF DEGRESS CLOCKWISE FROM TRUE NORTH. "00" INDICATES CALM.
RESULTANT DIRECTIONS ARE GIVEN TO WHOLE DEGREES.
BOLD VALUES INDICATE EXTREME VALUES WHICH OCCURRED AFTER THE ASOS SYSTEM WAS COMMISSIONED.

RECORD = PERIOD OF RECORD
RECORD MEAN PRECIPITATION IS THE MEAN OF ALL DAILY PRECIPITATION AMOUNTS DURING THE PERIOD OF RECORD.
RECORD MAX(MIN) TEMPERATURE IS THE MEAN OF ALL DAILY MAX(MIN) TEMPERATURES DURING THE PERIOD OF RECORD.
RECORD MEAN TEMPERATURE IS THE SUM OF THE RECORD MAX AND RECORD MIN DIVIDED BY 2.
AVERAGE TEMPERATURE IS THE SUM OF THE MEAN DAILY MAX AND MIN TEMPERATURE DIVIDED BY 2.

EXCEPTIONS

RECORD MEANS ARE THROUGH THE CURRENT YEAR, BEGINNING IN 1907 FOR TEMPERATURE
1867 FOR PRECIPITATION
1948 FOR SNOWFALL

HEATING DEGREE DAYS Base 65 deg. F SAN ANGELO, TEXAS

SEASON	JULY	AUG	SEP	OCT	NOV	DEC	JAN	FEB	MAR	APR	MAY	JUNE	TOTAL
1964-65	0	0	2	65	277	492	469	514	471	31	0	0	2321
1965-66	0	0	0	76	144	416	747	538	221	55	30	0	2227
1966-67	0	0	0	127	192	614	568	459	113	2	19	0	2094
1967-68	0	0	5	76	249	608	577	550	304	127	1	0	2497
1968-69	0	0	0	38	362	542	399	372	416	33	9	0	2171
1969-70	0	0	0	122	335	441	654	391	376	92	44	0	2455
1970-71	0	0	26	126	320	326	453	359	211	81	9	0	1911
1971-72	0	0	23	33	289	450	539	382	138	26	3	0	1883
1972-73	0	0	3	97	438	577	738	514	183	191	20	0	2761
1973-74	0	0	1	33	170	476	554	352	115	53	1	0	1755
1974-75	0	0	33	37	308	560	568	470	288	101	1	0	2366
1975-76	0	0	32	75	320	519	645	251	237	48	30	0	2157
1976-77	0	0	3	220	518	595	757	377	227	45	0	0	2742
1977-78	0	0	0	44	249	390	818	588	286	31	21	0	2427
1978-79	0	0	3	74	294	642	886	510	251	60	34	0	2754
1979-80	0	0	0	57	430	542	578	453	300	105	11	0	2476
1980-81	0	0	5	101	374	464	588	411	330	62	17	0	2352
1981-82	0	0	8	133	281	496	582	532	220	124	20	0	2396
1982-83	0	0	0	105	338	586	611	448	268	163	13	3	2535
1983-84	0	0	7	20	245	883	754	422	258	74	5	0	2668
1984-85	0	0	59	90	365	455	804	522	223	52	1	0	2571
1985-86	0	0	22	50	232	655	506	354	158	15	7	0	1999
1986-87	0	0	0	96	354	604	615	418	417	167	0	0	2671
1987-88	0	0	0	49	354	577	689	491	280	96	9	0	2545
1988-89	0	0	2	35	245	520	449	567	264	101	1	0	2184
1989-90	0	0	37	84	284	784	444	324	259	95	52	0	2363
1990-91	0	0	3	124	257	610	651	364	214	38	8	0	2269
1991-92	0	0	33	90	443	482	592	344	199	88	20	0	2291
1992-93	0	0	0	27	370	473	619	436	273	89	4	0	2291
1993-94	0	0	7	159	401	457							

COOLING DEGREE DAYS Base 65 deg. F SAN ANGELO, TEXAS

YEAR	JAN	FEB	MAR	APR	MAY	JUNE	JULY	AUG	SEP	OCT	NOV	DEC	TOTAL
1969	6	1	6	99	213	513	707	620	311	113	23	0	2612
1970	6	1	7	135	248	422	575	639	432	101	10	9	2585
1971	0	4	49	139	332	482	557	328	308	96	23	0	2318
1972	0	20	65	232	191	488	574	468	406	140	7	2	2593
1973	0	0	6	67	274	425	573	554	314	104	49	0	2366
1974	3	4	120	147	349	481	597	460	159	87	15	0	2422
1975	3	0	18	103	220	448	427	493	238	90	18	0	2058
1976	0	12	29	66	155	449	354	491	272	27	0	0	1855
1977	0	6	14	53	333	523	621	662	562	183	17	1	2975
1978	0	0	22	204	360	467	634	454	302	72	13	0	2528
1979	0	2	17	77	263	394	579	487	309	217	12	0	2357
1980	0	0	9	73	229	535	685	610	368	90	19	2	2620
1981	0	1	2	112	202	390	536	474	290	129	6	0	2142
1982	0	4	43	88	170	392	572	616	378	142	24	0	2429
1983	0	0	13	88	253	359	539	535	390	191	55	0	2423
1984	0	2	30	86	352	528	517	558	291	82	10	4	2460
1985	0	1	53	153	318	381	486	610	352	113	26	0	2493
1986	0	19	38	196	284	397	572	515	396	69	4	0	2490
1987	0	0	1	65	195	338	475	548	246	100	17	0	1985
1988	0	0	33	91	193	430	492	528	322	78	40	1	2208
1989	0	3	65	158	417	396	571	531	261	165	33	0	2600
1990	4	5	40	94	312	611	443	469	311	88	26	0	2403
1991	0	0	30	98	371	413	517	481	248	123	0	2	2283
1992	0	0	3	86	195	432	567	450	386	134	16	3	2272
1993	0	0	28	131	239	494	653	608	283	138	14	0	2588

SNOWFALL (inches) SAN ANGELO, TEXAS

SEASON	JULY	AUG	SEP	OCT	NOV	DEC	JAN	FEB	MAR	APR	MAY	JUNE	TOTAL
1970-71	0.0	0.0	0.0	0.0	0.0	0.0	0.0	T	T	0.0	0.0	0.0	T
1971-72	0.0	0.0	0.0	0.0	0.0	0.0	3.3	0.0	T	0.0	0.0	0.0	3.3
1972-73	0.0	0.0	0.0	0.0	T	T	7.7	5.8	0.0	0.0	0.0	0.0	13.5
1973-74	0.0	0.0	0.0	0.0	0.0	0.0	T	T	0.2	0.0	0.0	0.0	0.2
1974-75	0.0	0.0	0.0	0.0	0.0	0.1	1.5	T	0.0	0.0	0.0	0.0	1.6
1975-76	0.0	0.0	0.0	0.0	0.0	0.0	T	0.0	0.0	0.0	0.0	0.0	T
1976-77	0.0	0.0	0.0	0.0	3.0	0.0	0.1	0.0	0.0	0.0	0.0	0.0	3.1
1977-78	0.0	0.0	0.0	0.0	0.0	0.0	9.0	T	0.6	0.0	0.0	0.0	9.6
1978-79	0.0	0.0	0.0	0.0	0.0	1.0	T	T	0.0	0.0	0.0	0.0	1.0
1979-80	0.0	0.0	0.0	0.0	0.0	0.0	T	1.0	0.0	T	0.0	0.0	1.0
1980-81	0.0	0.0	0.0	0.0	2.3	0.0	1.8	T	0.0	0.0	0.0	0.0	4.1
1981-82	0.0	0.0	0.0	0.0	0.0	0.0	6.8	T	0.0	0.0	0.0	0.0	6.8
1982-83	0.0	0.0	0.0	0.0	0.0	T	2.9	0.0	0.0	0.0	0.0	0.0	2.9
1983-84	0.0	0.0	0.0	0.0	0.0	0.3	T	0.9	T	0.0	0.0	0.0	1.2
1984-85	0.0	0.0	0.0	0.0	T	T	8.4	0.4	0.0	0.0	0.0	0.0	8.8
1985-86	0.0	0.0	0.0	0.0	0.0	T	3.3	T	0.0	0.0	0.0	0.0	3.3
1986-87	0.0	0.0	0.0	0.0	0.0	3.7	0.7	T	T	0.0	0.0	0.0	4.4
1987-88	0.0	0.0	0.0	0.0	0.0	T	T	2.7	0.0	0.0	0.0	0.0	2.7
1988-89	0.0	0.0	0.0	0.0	0.0	T	1.2	T	2.6	0.0	T	T	3.8
1989-90	0.0	0.0	0.0	0.0	T	T	T	0.0	T	T	T	0.0	T
1990-91	0.0	0.0	0.0	0.0	0.0	1.1	3.3	0.0	T	0.0	T	0.0	4.4
1991-92	0.0	0.0	0.0	T	0.0	0.0	T	T	T	0.0	0.0	T	T
1992-93	T	0.0	T	T	T	T	0.1	0.0	T	0.0	T	0.0	0.1
1993-94	0.0	0.0	0.0	T	0.0	T							
Record Mean	T	0.0	T	T	0.5	0.2	1.5	0.7	0.2	T	T	T	3.1

SAN ANTONIO, TEXAS

The city of San Antonio is located in the south-central portion of Texas on the Balcones escarpment. Northwest of the city, the terrain slopes upward to the Edwards Plateau, and to the southeast it slopes downward to the Gulf Coastal Plains. Soils are blackland clay and silty loam on the Plains and thin limestone soils on the Edwards Plateau. With its location on the NW edge of the Gulf Coastal Plain, San Antonio experiences a modified subtropical climate. Average monthly temperatures range from the 50s in the winter to the 80s in the summer.

During the winter months the area is alternately influenced by a continental climate, when winds blow from the north and west; and by a modified maritime climate, when south and southeast winds blow from the Gulf of Mexico. Mild weather prevails during most of the winter. Below freezing temperatures occur on average about 20 days each year. When cold arctic outbreaks do occur, they cause relatively cold temperatures for this part of the U.S., as the cold north winds block any moderating affects from the Gulf. Some winters may see a mostly cloudy weather pattern that modify daily temperatures, that is cooler days and warmer nights. During sunny winters, the South Texas sun warms daytime temperatures to pleasant levels. On some occasions dry west winds in a mild airmass allow winter temperatures to climb to spring or summerlike levels.

During the summer the climate becomes even more tropical with prevailing south and southeast winds. The moderating effects of the Gulf of Mexico prevent extremely high temperatures; however, summers are usually long and hot with daily maximum temperatures above 90 degrees over 80 percent of the time. In many years summer conditions continue into September and sometimes even October. Occasionally, cool fronts may move through the area, dropping overnight lows into the 50s; however, the cooling unfortunately is brief and in a day or two the minimum temperatures are back up to the 70s. The drier air that is brought in by these cool fronts also allows daily maximum temperatures to rise above normal.

San Antonio is situated, climatically, between a semi-arid area to the west and the tropical area to the east. Such a location allows for large variations in monthly and annual precipitation amounts. The normal annual precipitation for San Antonio is about 30 inches although it may range from 10 to 50 inches from one year to another. Heavy rain may occur with or without thunder in any season. During some of these events, rain has exceeded 5 inches in several hours and caused flash flooding. A year of normal precipitation is sufficient for the production of most crops, although during the drier years irrigation is essential. On average the heaviest rains fall in May and September, while the driest months are December through March, and July. Since rainfall is sporadic, the wettest and driest month in any one year may occur in any season and vary widely from year to year. Precipitation from April through September usually occurs as a result of thunderstorms; however, thunderstorms may occur in any month.

Hail of damaging intensity seldom occurs; however, small hail is frequent with the springtime thunderstorms and has been observed during other seasons. Measurable snow usually occurs only once in three or four years. Snowfall of 2 to 4 inches occurs about every ten years.

Since San Antonio is located only 140 miles from the Gulf of Mexico, tropical storms occasionally affect the city with strong winds and heavy rains. One of the fastest winds recorded, 74 mph, occurred as a tropical storm moved inland east of the city in August 1942.

Relative humidity is above 80 percent during the early morning hours most of the year, dropping to near 50 percent in late afternoon.

San Antonio receives about 50 percent of possible sunshine during the winter months and more than 70 percent during the summer. Stratus clouds frequently develop at night during the summer as Gulf moisture is lifted over the higher terrain. These clouds usually dissipate before noon and lead to partly cloudy to clear skies during the afternoon. During the winter these stratus clouds may be accompanied by fog and drizzle.

Normally the first freezing temperatures occur around December the 1st while the average last freeze occurs in late February. Extreme freeze dates are late October and early April.

NORMALS, MEANS, AND EXTREMES

SAN ANTONIO, TEXAS

LATITUDE: 29°32'N LONGITUDE: 98°28'W ELEVATION: FT. GRND 788 BARO 796 TIME ZONE: CENTRAL WBAN: 12921

	(a)	JAN	FEB	MAR	APR	MAY	JUNE	JULY	AUG	SEP	OCT	NOV	DEC	YEAR
TEMPERATURE °F:														
Normals														
-Daily Maximum		60.8	65.7	73.5	80.3	85.3	91.8	95.0	95.3	89.3	81.7	71.9	63.5	79.5
-Daily Minimum		37.9	41.3	49.7	58.4	65.7	72.6	75.0	74.5	69.2	58.8	48.8	40.8	57.7
-Monthly		49.3	53.5	61.7	69.3	75.5	82.2	85.0	84.9	79.3	70.2	60.4	52.2	68.6
Extremes														
-Record Highest	52	89	97	100	100	103	105	106	108	103	99	94	90	108
-Year		1971	1986	1991	1984	1989	1980	1989	1986	1985	1991	1988	1955	AUG 1986
-Record Lowest	52	0	6	19	31	43	53	62	61	41	27	21	6	0
-Year		1949	1951	1980	1987	1984	1964	1967	1992	1942	1993	1976	1989	JAN 1949
NORMAL DEGREE DAYS:														
Heating (base 65°F)		494	332	167	32	0	0	0	0	0	30	180	409	1644
Cooling (base 65°F)		8	10	64	161	326	516	620	617	429	191	42	12	2996
% OF POSSIBLE SUNSHINE	51	47	51	57	56	56	67	75	74	67	64	54	48	60
MEAN SKY COVER (tenths)														
Sunrise - Sunset	51	6.3	6.2	6.1	6.4	6.4	5.5	5.1	4.9	5.2	4.9	5.5	6.1	5.7
MEAN NUMBER OF DAYS:														
Sunrise to Sunset														
-Clear	51	8.9	8.2	8.7	7.4	6.0	6.9	9.0	10.2	9.3	11.6	10.3	9.6	106.1
-Partly Cloudy	51	5.9	5.9	7.2	7.6	11.3	15.3	15.4	14.8	12.3	10.0	7.0	5.9	118.5
-Cloudy	51	16.2	14.1	15.1	15.0	13.8	7.8	6.6	6.0	8.5	9.4	12.7	15.5	140.7
Precipitation														
.01 inches or more	51	8.1	7.7	7.3	7.3	8.6	6.4	4.4	5.1	7.0	6.2	6.5	7.5	82.1
Snow,Ice pellets,hail														
1.0 inches or more	51	0.1	0.1	0.0	0.0	0.0	0.0	0.0	0.0	0.0	0.0	0.0	0.0	0.2
Thunderstorms	51	1.0	1.6	2.6	4.0	6.6	4.6	3.5	4.0	4.1	2.7	1.7	0.9	37.3
Heavy Fog Visibility														
1/4 mile or less	51	5.1	2.9	2.4	1.4	0.7	0.1	0.1	0.*	0.2	1.4	2.8	4.4	21.5
Temperature °F														
-Maximum														
90° and above	51	0.0	0.1	0.8	2.3	8.5	21.8	28.3	28.2	17.5	4.5	0.1	0.*	112.1
32° and below	51	0.3	0.1	0.0	0.0	0.0	0.0	0.0	0.0	0.0	0.0	0.0	0.1	0.5
-Minimum														
32° and below	51	8.1	4.5	1.5	0.*	0.0	0.0	0.0	0.0	0.0	0.*	2.0	5.8	22.1
0° and below	51	0.*	0.0	0.0	0.0	0.0	0.0	0.0	0.0	0.0	0.0	0.0	0.0	*
AVG. STATION PRESS. (mb)	21	991.8	990.3	986.7	985.8	984.5	985.6	987.5	987.3	987.3	989.2	989.9	991.4	988.1
RELATIVE HUMIDITY (%)														
Hour 00	51	76	75	72	76	81	80	75	74	77	77	77	76	76
Hour 06 (Local Time)	51	80	80	79	83	88	88	87	86	86	84	81	80	84
Hour 12	51	59	57	53	56	59	56	52	51	54	53	55	57	55
Hour 18	51	57	52	47	51	55	52	45	45	51	52	56	57	52
PRECIPITATION (inches):														
Water Equivalent														
-Normal		1.71	1.81	1.52	2.50	4.22	3.81	2.16	2.54	3.41	3.17	2.62	1.51	30.98
-Maximum Monthly	51	8.52	6.43	6.12	9.32	12.85	11.95	8.29	11.14	15.78	9.56	6.01	13.96	15.78
-Year		1968	1965	1992	1957	1987	1986	1990	1974	1946	1942	1977	1991	SEP 1946
-Minimum Monthly	51	0.04	0.03	0.03	0.11	0.17	0.01	T	0.00	0.06	T	T	0.03	0.00
-Year		1971	1954	1961	1984	1961	1967	1993	1952	1947	1952	1966	1950	AUG 1952
-Maximum in 24 hrs	51	3.18	2.44	3.59	4.88	6.53	6.30	6.97	5.57	7.28	5.29	4.87	6.90	7.28
-Year		1968	1986	1992	1977	1972	1986	1958	1950	1973	1942	1977	1991	SEP 1973
Snow,Ice pellets,hail														
-Maximum Monthly	51	15.9	3.5	T	T	T	T	0.0	0.0	0.0	T	0.3	0.2	15.9
-Year		1985	1966	1992	1993	1993	1989				1993	1957	1964	JAN 1985
-Maximum in 24 hrs	51	13.2	3.5	T	T	T	T	0.0	0.0	0.0	T	0.3	0.2	13.2
-Year		1985	1966	1992	1993	1993	1989				1993	1957	1964	JAN 1985
WIND:														
Mean Speed (mph)	51	8.9	9.4	10.2	10.1	9.8	9.8	9.1	8.4	8.4	8.3	8.7	8.4	9.1
Prevailing Direction through 1963		N	NE	SE	SE	SE	SE	SSE	SE	SE	N	N	N	SE
Fastest Obs. 1 Min.														
-Direction (!!)	17	31	31	36	35	02	34	09	20	18	02	33	32	09
-Speed (MPH)	17	35	42	35	39	43	35	48	37	42	31	37	30	48
-Year		1979	1984	1980	1979	1983	1984	1979	1984	1977	1977	1983	1987	JUL 1979
Peak Gust														
-Direction (!!)	10	N	NW	NW	NW	NW	NW	SE	SW	N	N	NW	W	NW
-Speed (mph)	10	51	56	46	47	55	51	41	49	51	43	41	48	56
-Date		1985	1984	1984	1988	1987	1984	1989	1984	1987	1985	1987	1987	FEB 1984

PRECIPITATION (inches)

SAN ANTONIO, TEXAS

YEAR	JAN	FEB	MAR	APR	MAY	JUNE	JULY	AUG	SEP	OCT	NOV	DEC	ANNUAL
1964	3.40	1.89	1.73	1.16	1.79	4.88	0.02	5.19	4.15	1.64	4.81	1.22	31.88
1965	2.40	6.43	2.30	1.97	8.18	2.42	0.08	1.65	3.13	2.69	0.89	4.51	36.65
1966	1.47	2.30	1.14	3.20	3.53	1.78	0.06	4.28	2.13	1.11	T	0.44	21.44
1967	0.18	0.48	2.18	0.94	2.22	0.01	2.12	3.17	11.16	2.00	3.42	1.38	29.26
1968	8.52	1.85	1.27	1.92	2.82	2.63	1.53	0.94	2.99	0.69	4.58	0.66	30.40
1969	1.76	2.90	2.35	2.46	4.61	2.32	0.36	4.19	1.32	5.85	1.02	2.28	31.42
1970	1.10	2.66	1.98	1.13	7.30	0.89	0.91	0.95	4.35	1.31	0.01	0.15	22.74
1971	0.04	0.81	0.04	1.39	1.52	2.74	1.05	9.42	4.57	4.62	2.74	2.86	31.80
1972	1.35	0.40	0.13	1.94	11.24	2.86	3.13	4.24	1.40	1.99	2.37	0.44	31.49
1973	2.77	2.76	1.58	5.41	2.73	10.44	6.91	1.29	13.09	4.85	0.29	0.16	52.28
1974	1.36	0.04	0.94	2.18	4.28	1.02	1.28	11.14	3.85	4.09	5.39	1.43	37.00
1975	1.04	3.30	0.52	2.69	6.91	4.60	1.06	1.28	0.51	2.25	0.03	1.48	25.67
1976	0.56	0.13	1.20	5.67	5.80	1.61	5.39	2.09	3.79	8.48	2.46	1.95	39.13
1977	3.10	0.91	0.88	8.80	1.62	2.26	0.10	0.06	2.11	3.47	6.01	0.32	29.64
1978	0.68	1.76	1.71	3.62	2.45	3.96	1.43	4.97	8.86	0.55	4.91	1.09	35.99
1979	4.07	1.38	3.55	5.34	1.98	5.59	7.38	2.09	0.86	0.11	1.43	2.86	36.64
1980	0.72	0.74	0.98	1.67	6.42	0.52	0.26	2.64	5.05	1.09	3.53	0.61	24.23
1981	2.06	0.96	1.96	2.21	6.43	8.71	0.25	2.41	1.36	8.61	0.72	0.69	36.37
1982	0.72	1.28	0.69	1.23	6.42	1.37	0.14	0.55	0.87	2.84	4.54	2.31	22.96
1983	1.48	1.54	3.89	0.18	4.37	1.27	2.43	2.00	3.86	1.64	3.06	0.39	26.11
1984	1.87	0.54	1.91	0.11	3.76	1.40	T	3.04	1.06	5.94	2.91	3.41	25.95
1985	2.68	1.91	2.85	3.27	2.47	8.20	5.80	0.45	4.80	3.91	5.00	0.09	41.43
1986	0.76	2.52	0.35	0.60	6.29	11.95	0.05	1.86	2.83	6.58	1.83	7.11	42.73
1987	1.13	4.78	1.10	1.48	12.85	7.69	1.21	0.33	2.24	0.44	2.53	2.18	37.96
1988	0.39	0.92	0.86	1.23	0.41	5.50	5.58	1.98	0.83	0.62	0.02	0.67	19.01
1989	2.96	0.29	1.24	2.55	0.33	3.96	0.69	0.48	1.54	5.81	1.93	0.36	22.14
1990	1.17	2.68	5.17	4.52	3.28	1.18	8.29	1.30	3.70	3.71	3.11	0.20	38.31
1991	5.08	2.34	1.06	4.91	5.30	2.28	2.23	2.84	1.42	0.87	0.47	13.96	42.76
1992	5.64	6.37	6.12	3.03	8.15	5.67	1.28	2.56	1.12	0.92	3.47	2.16	46.49
1993	1.31	3.72	1.56	1.81	12.47	6.43	T	0.01	0.52	3.07	0.66	0.44	32.00
Record Mean	1.66	1.74	1.70	2.89	3.66	3.10	2.06	2.32	3.11	2.55	1.96	1.76	28.52

AVERAGE TEMPERATURE (deg. F)

SAN ANTONIO, TEXAS

YEAR	JAN	FEB	MAR	APR	MAY	JUNE	JULY	AUG	SEP	OCT	NOV	DEC	ANNUAL
1964	51.0	49.8	61.5	70.5	77.6	82.4	86.3	86.2	80.0	66.4	62.6	52.3	68.9
1965	54.4	49.8	54.9	71.6	75.0	81.6	84.9	84.0	80.7	66.8	64.5	55.5	68.6
1966	45.4	49.8	60.0	68.6	73.5	78.8	84.2	81.9	77.5	67.0	63.0	50.7	66.7
1967	50.2	51.8	66.9	76.6	76.6	84.5	85.3	82.7	75.5	66.9	60.5	51.0	69.0
1968	49.8	48.3	58.0	68.1	75.3	80.5	82.7	84.2	76.0	72.2	56.4	50.7	66.8
1969	52.5	53.6	54.9	69.0	73.5	81.2	86.8	85.7	79.6	69.8	58.1	55.1	68.3
1970	45.6	54.8	56.8	70.2	72.9	80.7	84.0	85.7	81.1	67.7	58.0	60.1	68.1
1971	56.0	57.4	64.6	69.4	78.1	83.6	85.9	81.2	80.1	73.9	63.2	57.2	70.9
1972	52.8	56.7	66.3	73.7	72.8	80.3	82.2	82.1	82.0	71.9	54.0	50.3	68.8
1973	47.2	51.9	66.1	66.0	74.7	79.2	83.2	82.1	79.3	72.5	65.8	52.2	68.4
1974	51.0	56.5	67.9	69.7	77.3	79.4	83.0	81.2	72.3	68.2	57.3	50.9	67.9
1975	53.2	53.5	61.4	68.4	73.5	80.0	80.9	81.7	76.0	71.1	60.3	53.1	67.8
1976	49.6	61.2	63.8	68.9	71.3	79.8	79.8	81.6	77.5	61.1	52.1	49.9	66.4
1977	44.1	52.8	61.8	66.9	74.8	81.5	84.9	84.7	82.3	71.2	61.4	53.4	68.3
1978	43.4	46.4	59.6	68.9	77.1	82.7	86.1	83.1	78.5	69.3	62.4	51.7	67.5
1979	43.7	52.4	63.3	69.7	73.9	80.9	84.7	83.1	78.7	74.7	58.2	55.4	68.2
1980	52.6	53.7	61.5	67.6	76.1	85.1	88.1	85.3	83.7	70.7	58.3	55.0	69.8
1981	50.8	53.7	60.7	72.9	75.3	81.5	84.2	84.7	79.9	71.9	62.4	53.0	69.2
1982	50.8	49.7	63.1	66.9	74.5	81.6	85.5	86.0	80.1	69.3	59.4	52.4	68.3
1983	48.9	52.1	58.7	65.2	73.6	79.2	82.9	84.5	78.5	70.9	62.5	43.0	66.7
1984	46.7	54.1	64.2	69.7	77.1	82.8	85.0	84.7	77.6	71.2	58.8	59.6	69.3
1985	44.2	50.5	64.1	69.4	76.7	80.2	82.2	85.5	79.4	71.7	64.4	49.9	68.2
1986	53.4	58.0	62.9	72.6	74.6	81.5	85.8	85.7	83.7	69.7	59.4	51.6	69.9
1987	50.7	55.9	57.8	66.1	75:8	80.5	83.8	86.0	79.2	71.2	60.6	54.2	68.5
1988	47.6	54.3	61.3	69.1	76.1	81.2	84.6	86.4	80.7	73.2	65.1	56.0	69.6
1989	56.2	51.6	61.9	70.4	81.7	83.3	86.6	86.0	79.1	71.3	61.8	43.4	69.4
1990	56.4	58.9	61.5	69.7	79.3	87.5	83.4	85.3	80.0	69.3	63.0	51.9	70.5
1991	48.9	56.6	64.0	72.4	77.7	82.8	84.5	85.8	77.8	73.3	57.4	55.5	69.7
1992	50.8	59.1	63.3	69.0	73.7	82.5	84.7	82.2	81.7	73.4	57.3	56.3	69.5
1993	51.2	55.5	61.5	67.3	73.9	81.6	86.1	87.3	81.5	70.7	56.3	55.1	69.0
Record Mean	51.6	55.2	62.2	69.3	75.5	81.6	84.0	84.2	79.4	70.9	60.5	53.5	69.0
Max	62.1	66.1	73.4	80.0	85.5	91.6	94.4	94.9	89.6	81.8	71.1	64.0	79.5
Min	41.1	44.2	50.9	58.6	65.5	71.6	73.6	73.5	69.2	59.9	49.8	43.1	58.4

REFERENCE NOTES

SAN ANTONIO, TEXAS

GENERAL
T - TRACE AMOUNT.
BLANK ENTRIES DENOTE MISSING/UNREPORTED DATA.
# INDICATES A STATION OR INSTRUMENT RELOCATION.

SPECIFIC

PM - INCLUDES LAST DAY OF PREVIOUS MONTH
ASOS - AUTOMATED SURFACE OBSERVING SYSTEM IN OPERATION DURING THESE MONTHS.

(a) - LENGTH OF RECORD IN YEARS, ALTHOUGH INDIVIDUAL MONTHS MAY BE MISSING.
0.* OR * - THE VALUE IS BETWEEN 0.0 AND 0.05
NORMALS - BASED ON THE 1961-1990 RECORD PERIOD.
EXTREMES - DATES ARE THE MOST RECENT OCCURRENCE
WIND DIR.- NUMERALS SHOW TENS OF DEGRESS CLOCKWISE FROM TRUE NORTH. "00" INDICATES CALM.
RESULTANT DIRECTIONS ARE GIVEN TO WHOLE DEGREES.
BOLD VALUES INDICATE EXTREME VALUES WHICH OCCURRED AFTER THE ASOS SYSTEM WAS COMMISSIONED.

RECORD = PERIOD OF RECORD
RECORD MEAN PRECIPITATION IS THE MEAN OF ALL DAILY PRECIPITATION AMOUNTS DURING THE PERIOD OF RECORD.
RECORD MAX(MIN) TEMPERATURE IS THE MEAN OF ALL DAILY MAX(MIN) TEMPERATURES DURING THE PERIOD OF RECORD.
RECORD MEAN TEMPERATURE IS THE SUM OF THE RECORD MAX AND RECORD MIN DIVIDED BY 2.
AVERAGE TEMPERATURE IS THE SUM OF THE MEAN DAILY MAX AND MIN TEMPERATURE DIVIDED BY 2.

EXCEPTIONS

RECORD MEANS ARE THROUGH THE CURRENT YEAR, BEGINNING IN 1885 FOR TEMPERATURE
1885 FOR PRECIPITATION
1943 FOR SNOWFALL

HEATING DEGREE DAYS Base 65 deg. F SAN ANTONIO, TEXAS

SEASON	JULY	AUG	SEP	OCT	NOV	DEC	JAN	FEB	MAR	APR	MAY	JUNE	TOTAL
1964-65	0	0	0	41	155	414	346	419	327	13	0	0	1715
1965-66	0	0	2	62	64	301	607	426	182	39	5	0	1688
1966-67	0	0	0	57	131	456	470	366	80	0	0	0	1560
1967-68	0	0	8	48	164	429	477	478	254	39	0	0	1897
1968-69	0	0	0	9	278	437	394	319	315	5	3	0	1760
1969-70	0	0	0	52	253	299	599	282	266	45	7	0	1803
1970-71	0	0	1	72	247	201	282	239	134	52	1	0	1229
1971-72	0	0	1	0	129	266	382	263	61	7	0	0	1109
1972-73	0	0	0	29	334	457	551	362	29	94	1	0	1857
1973-74	0	0	0	4	85	391	437	257	74	39	0	0	1287
1974-75	0	0	2	19	260	433	389	316	152	41	0	0	1612
1975-76	0	0	1	21	214	394	472	166	143	11	2	0	1424
1976-77	0	0	0	160	382	461	643	336	144	32	0	0	2158
1977-78	0	0	0	19	138	360	667	521	192	27	4	0	1928
1978-79	0	0	0	12	152	413	657	356	109	20	4	0	1723
1979-80	0	0	0	15	243	306	386	333	163	42	0	0	1488
1980-81	0	0	0	62	245	331	437	332	157	10	0	0	1574
1981-82	0	0	2	52	112	368	445	430	171	77	2	0	1659
1982-83	0	0	0	49	237	404	490	356	208	99	1	0	1844
1983-84	0	0	5	20	154	681	563	315	120	21	2	0	1881
1984-85	0	0	9	28	228	203	635	406	109	26	0	0	1644
1985-86	0	0	10	9	112	467	354	232	106	8	1	0	1299
1986-87	0	0	0	14	204	413	443	254	233	98	0	0	1659
1987-88	0	0	0	1	194	339	538	323	179	38	0	0	1612
1988-89	0	0	0	0	122	291	292	392	187	55	0	0	1339
1989-90	0	0	0	42	165	663	283	190	154	32	0	0	1529
1990-91	0	0	0	50	142	422	494	240	96	7	0	0	1451
1991-92	0	0	5	30	271	306	435	188	91	31	5	0	1362
1992-93	0	0	0	0	260	287	421	269	147	42	0	0	1426
1993-94	0	0	0	85	287	323							

COOLING DEGREE DAYS Base 65 deg. F SAN ANTONIO, TEXAS

YEAR	JAN	FEB	MAR	APR	MAY	JUNE	JULY	AUG	SEP	OCT	NOV	DEC	TOTAL
1969	11	5	8	133	273	494	683	652	448	207	53	2	2969
1970	3	3	20	208	259	478	592	651	493	163	40	57	2967
1971	14	36	130	189	414	564	658	509	459	281	81	31	3366
1972	11	31	105	276	252	465	542	539	515	249	12	6	3003
1973	8	0	69	129	310	431	570	536	437	242	114	0	2846
1974	11	22	171	188	387	439	568	506	229	124	34	5	2684
1975	29	1	51	151	273	457	502	524	337	217	80	30	2652
1976	3	62	113	136	202	451	467	521	383	45	0	0	2383
1977	0	3	52	98	311	502	620	618	525	218	38	5	2990
1978	3	7	30	152	384	537	660	567	410	154	79	11	2994
1979	3	13	65	166	285	482	619	570	418	322	42	13	2998
1980	11	14	61	127	355	614	725	635	567	245	51	26	3431
1981	3	24	30	255	324	502	603	619	424	273	41	3	3101
1982	11	5	117	142	304	504	645	659	459	191	72	19	3128
1983	0	0	21	111	276	435	560	611	417	207	84	8	2730
1984	0	8	101	169	383	541	625	618	394	230	46	44	3159
1985	0	8	85	165	368	462	539	641	450	223	101	5	3047
1986	2	45	49	244	304	500	652	646	568	166	40	4	3218
1987	4	5	17	135	340	471	589	658	434	199	67	12	2931
1988	6	19	71	166	352	492	617	671	480	264	131	20	3289
1989	24	23	99	222	524	557	678	656	429	244	75	0	3531
1990	22	26	53	177	450	681	578	635	459	192	91	23	3387
1991	0	10	70	234	402	541	612	654	396	295	49	20	3283
1992	0	23	47	158	281	531	618	542	508	267	34	24	3033
1993	1	7	45	117	283	503	660	698	503	267	31	22	3137

SNOWFALL (inches) SAN ANTONIO, TEXAS

SEASON	JULY	AUG	SEP	OCT	NOV	DEC	JAN	FEB	MAR	APR	MAY	JUNE	TOTAL
1964-65	0.0	0.0	0.0	0.0	0.0	0.2	0.0	T	T	0.0	0.0	0.0	0.2
1965-66	0.0	0.0	0.0	0.0	0.0	0.0	T	3.5	0.0	0.0	0.0	0.0	3.5
1966-67	0.0	0.0	0.0	0.0	0.0	0.0	T	T	0.0	0.0	0.0	0.0	T
1967-68	0.0	0.0	0.0	0.0	0.0	0.0	0.0	T	0.0	0.0	0.0	0.0	T
1968-69	0.0	0.0	0.0	0.0	0.0	0.0	0.0	0.0	0.0	0.0	0.0	0.0	0.0
1969-70	0.0	0.0	0.0	0.0	0.0	T	T	T	0.0	0.0	0.0	0.0	T
1970-71	0.0	0.0	0.0	0.0	0.0	0.0	0.0	0.0	0.0	0.0	0.0	0.0	0.0
1971-72	0.0	0.0	0.0	0.0	0.0	0.0	T	0.0	0.0	0.0	0.0	0.0	T
1972-73	0.0	0.0	0.0	0.0	0.0	T	0.8	2.1	0.0	0.0	0.0	0.0	2.9
1973-74	0.0	0.0	0.0	0.0	0.0	0.0	0.0	0.0	0.0	0.0	0.0	0.0	0.0
1974-75	0.0	0.0	0.0	0.0	0.0	T	T	0.0	0.0	0.0	0.0	0.0	T
1975-76	0.0	0.0	0.0	0.0	0.0	0.0	T	0.0	0.0	0.0	0.0	0.0	T
1976-77	0.0	0.0	0.0	0.0	T	T	0.0	0.0	0.0	0.0	0.0	0.0	T
1977-78	0.0	0.0	0.0	0.0	0.0	0.0	0.0	0.0	T	0.0	0.0	0.0	T
1978-79	0.0	0.0	0.0	0.0	0.0	0.0	T	0.0	0.0	0.0	0.0	0.0	T
1979-80	0.0	0.0	0.0	0.0	T	0.0	0.0	T	0.0	0.0	0.0	0.0	T
1980-81	0.0	0.0	0.0	0.0	T	0.0	T	0.0	0.0	0.0	0.0	0.0	T
1981-82	0.0	0.0	0.0	0.0	0.0	0.0	0.5	0.0	0.0	0.0	0.0	0.0	0.5
1982-83	0.0	0.0	0.0	0.0	0.0	0.0	0.0	0.0	0.0	0.0	0.0	0.0	0.0
1983-84	0.0	0.0	0.0	0.0	0.0	0.0	0.0	0.0	0.0	0.0	0.0	0.0	0.0
1984-85	0.0	0.0	0.0	0.0	0.0	0.0	15.9	T	0.0	0.0	0.0	0.0	15.9
1985-86	0.0	0.0	0.0	0.0	0.0	0.0	T	0.0	0.0	0.0	0.0	0.0	T
1986-87	0.0	0.0	0.0	0.0	0.0	0.0	1.3	0.0	0.0	0.0	0.0	0.0	1.3
1987-88	0.0	0.0	0.0	0.0	0.0	0.0	0.0	0.1	0.0	0.0	0.0	0.0	0.1
1988-89	0.0	0.0	0.0	0.0	0.0	0.0	0.0	0.0	T	0.0	0.0	T	T
1989-90	0.0	0.0	0.0	0.0	0.0	T	0.0	T	T	T	0.0	0.0	T
1990-91	0.0	0.0	0.0	0.0	0.0	T	0.0	T	0.0	T	T	0.0	T
1991-92	0.0	0.0	0.0	0.0	0.0	0.0	0.0	T	T	0.0	0.0	0.0	T
1992-93	0.0	0.0	0.0	0.0	T	0.0	T	0.0	0.0	T	T	0.0	T
1993-94	0.0	0.0	0.0	T	0.0	0.0							
Record Mean	0.0	0.0	0.0	T	T	T	0.5	0.2	T	T	T	T	0.7

VICTORIA, TEXAS

The city of Victoria is located in the south-central Texas Coastal Plain. The climate is classified as humid subtropical. Summers are hot with about 100 days with temperatures of 90 degrees or above. However, pleasant sea breezes from the nearby Gulf of Mexico make the high temperatures bearable.

Spring is characterized by mild days, brisk winds, and occasional showers and thunderstorms. Strong southeast winds begin in March, diminish in April and May, and become pleasant sea breezes in the first half of June. Thunderstorm activity increases through March and April, reaching a peak in May. Considerable cloudiness is the rule, with almost 50 percent of the days in the spring having overcast or nearly overcast skies.

The sea breeze diminishes during the summer, and at times fails altogether, and some hot nights are experienced in late June, July, and early August. High summer humidity gives way to clear, drier air in late August. Nighttime temperatures drop to pleasant levels. Thunderstorms continue, and lawns and fields remain green.

The first norther usually arrives near the beginning of fall, in late September. October and November are ideal fall months with long periods of clear days with mild temperatures and cool nights. The amount of rainfall decreases.

The winter season weather conditions alternate between clear, cold, dry periods and cloudy, mild, drizzly days as fronts move down from the north. The temperature drops below 32 degrees on an average of about a dozen mornings per year.

The normal rainfall of about 36 inches is well distributed throughout the year, with the heaviest falls coming during the growing season. Some of the smaller streams dry up in the late summer, and during occasional periods of general drought some of the larger streams may reach pool stage.

The area is subject to occasional tropical disturbances during summer and fall. Destructive winds and torrential rains may occur in these storms. Approximately 50 days per year have thunderstorms, but hail is infrequent. Destructive storms with tornados are rare.

NORMALS, MEANS, AND EXTREMES

VICTORIA, TEXAS

LATITUDE: 28°51'N LONGITUDE: 96°55'W ELEVATION: FT. GRND 104 BARO 107 TIME ZONE: CENTRAL WBAN: 12912

	(a)	JAN	FEB	MAR	APR	MAY	JUNE	JULY	AUG	SEP	OCT	NOV	DEC	YEAR
TEMPERATURE °F:														
Normals														
-Daily Maximum		62.8	66.8	73.7	80.2	85.3	90.7	93.5	93.9	88.9	82.4	73.4	66.0	79.8
-Daily Minimum		42.5	45.4	52.8	61.0	67.7	72.7	74.6	74.2	70.3	60.9	52.4	45.2	60.0
-Monthly		52.7	56.1	63.3	70.6	76.6	81.7	84.1	84.1	79.6	71.7	62.9	55.6	69.9
Extremes														
-Record Highest	33	88	95	97	98	101	101	104	107	102	99	93	88	107
-Year		1971	1986	1989	1963	1964	1990	1964	1962	1989	1991	1988	1964	AUG 1962
-Record Lowest	33	14	19	21	33	49	59	62	62	49	31	24	9	9
-Year		1982	1985	1980	1987	1978	1984	1967	1992	1984	1993	1976	1989	DEC 1989
NORMAL DEGREE DAYS:														
Heating (base 65°F)		403	266	128	15	0	0	0	0	0	15	147	322	1296
Cooling (base 65°F)		21	17	76	183	360	501	592	592	438	223	84	31	3118
% OF POSSIBLE SUNSHINE														
MEAN SKY COVER (tenths)														
Sunrise - Sunset	32	7.0	6.6	6.8	7.1	6.7	5.8	5.6	5.7	5.7	5.1	5.9	7.0	6.2
MEAN NUMBER OF DAYS:														
Sunrise to Sunset														
-Clear	32	6.5	7.3	6.9	5.3	4.8	6.2	7.4	7.0	7.6	11.3	9.1	6.7	86.1
-Partly Cloudy	32	6.2	5.6	6.7	7.3	10.8	14.7	15.2	15.4	12.6	10.1	7.3	5.7	117.5
-Cloudy	32	18.3	15.3	17.4	17.3	15.4	9.1	8.5	8.6	9.8	9.6	13.7	18.7	161.6
Precipitation .01 inches or more	32	8.4	7.1	6.7	6.1	7.6	7.7	7.7	8.3	10.0	6.2	6.7	8.1	90.4
Snow,Ice pellets,hail 1.0 inches or more	32	0.1	0.*	0.0	0.0	0.0	0.0	0.0	0.0	0.0	0.0	0.0	0.0	0.1
Thunderstorms	32	1.2	1.8	2.6	3.6	6.5	6.1	6.9	8.8	7.9	3.7	1.8	1.0	52.0
Heavy Fog Visibility 1/4 mile or less	32	7.0	5.0	5.2	4.0	2.1	0.6	0.2	0.2	1.1	3.0	5.6	6.1	40.1
Temperature °F														
-Maximum														
90° and above	32	0.0	0.1	0.3	0.9	5.7	20.1	27.6	28.1	17.5	4.7	0.2	0.0	105.0
32° and below	32	0.1	0.1	0.0	0.0	0.0	0.0	0.0	0.0	0.0	0.0	0.0	0.2	0.5
-Minimum														
32° and below	32	4.9	2.3	0.5	0.0	0.0	0.0	0.0	0.0	0.0	0.*	0.7	3.1	11.5
0° and below	32	0.0	0.0	0.0	0.0	0.0	0.0	0.0	0.0	0.0	0.0	0.0	0.0	0.0
AVG. STATION PRESS.(mb)	21	1016.7	1015.2	1011.5	1010.4	1008.9	1009.9	1011.9	1011.4	1011.2	1013.5	1014.6	1016.3	1012.6
RELATIVE HUMIDITY (%)														
Hour 00	29	84	83	83	85	87	88	87	87	87	86	86	84	86
Hour 06 (Local Time)	32	87	87	86	88	91	92	92	92	92	90	88	87	89
Hour 12	32	65	61	58	59	61	60	56	56	60	56	59	63	60
Hour 18	32	68	63	60	63	66	65	60	61	65	65	68	70	65
PRECIPITATION (inches):														
Water Equivalent														
-Normal		2.16	2.00	1.55	2.41	4.50	4.89	3.34	3.01	5.60	3.46	2.45	2.04	37.41
-Maximum Monthly	33	7.76	9.08	5.51	11.09	14.66	12.68	13.59	7.30	19.05	10.16	8.68	6.97	19.05
-Year		1991	1992	1985	1991	1993	1973	1990	1974	1978	1981	1982	1975	SEP 1978
-Minimum Monthly	33	0.02	0.23	0.18	T	0.69	T	0.07	0.34	1.11	0.34	0.02	0.36	T
-Year		1971	1988	1971	1987	1989	1980	1982	1965	1982	1987	1981	1972	APR 1987
-Maximum in 24 hrs	33	4.70	3.21	2.78	9.87	8.45	9.30	8.41	6.14	8.51	5.05	6.63	6.12	9.87
-Year		1991	1992	1991	1991	1972	1977	1990	1964	1967	1981	1982	1975	APR 1991
Snow,Ice pellets,hail														
-Maximum Monthly	33	2.1	1.0	T	0.0	T	0.0	0.0	0.0	0.0	0.0	0.2	T	2.1
-Year		1985	1973	1990		1993						1976	1990	JAN 1985
-Maximum in 24 hrs	33	2.1	1.0	T	0.0	T	0.0	0.0	0.0	0.0	0.0	0.2	T	2.1
-Year		1985	1973	1990		1993						1976	1990	JAN 1985
WIND:														
Mean Speed (mph)	32	10.4	11.0	11.6	11.7	10.7	9.6	8.9	8.4	8.7	8.9	9.8	10.2	10.0
Prevailing Direction														
Fastest Obs. 1 Min.														
-Direction (!!)	8	36	10	15	16	30	35	21	05	17	34	34	02	21
-Speed (MPH)	8	35	36	39	37	40	35	44	25	32	35	44	35	44
-Year		1991	1993	1990	1993	1993	1989	1989	1988	1993	1993	1987	1991	JUL 1989
Peak Gust														
-Direction (!!)	10	N	N	N	W	N	NE	E	E	S	N	N	N	N
-Speed (mph)	10	49	54	54	54	68	56	54	48	44	54	55	45	68
-Date		1993	1984	1993	1989	1986	1985	1993	1986	1993	1985	1987	1991	MAY 1986

PRECIPITATION (inches)

VICTORIA, TEXAS

YEAR	JAN	FEB	MAR	APR	MAY	JUNE	JULY	AUG	SEP	OCT	NOV	DEC	ANNUAL
1964	2.03	2.49	2.10	0.50	3.09	3.77	1.81	6.84	7.64	0.34	0.90	1.81	33.32
1965	2.07	2.86	0.80	0.91	3.34	4.88	0.29	0.34	4.34	4.04	2.07	4.91	30.85
1966	3.71	2.30	0.39	4.02	4.77	3.60	5.14	6.05	2.39	0.99	0.87	1.24	35.47
1967	2.31	1.41	0.49	1.72	2.38	T	1.26	3.03	14.52	4.92	1.31	0.55	33.90
1968	3.27	2.03	1.41	0.81	14.08	10.67	3.75	1.92	4.28	1.91	4.24	0.95	49.32
1969	0.43	5.42	2.90	9.43	2.09	2.33	1.74	3.49	6.71	3.57	3.05	3.48	44.64
1970	2.74	1.68	4.44	2.38	9.93	2.87	3.51	1.44	7.41	2.63	0.10	0.65	39.78
1971	0.02	1.10	0.18	1.75	1.00	5.23	0.27	4.01	12.03	6.87	1.37	2.23	36.06
1972	1.74	0.72	1.55	0.35	11.24	3.17	7.30	4.38	5.97	3.44	2.19	0.36	42.41
1973	2.40	2.75	1.04	4.73	1.22	12.68	2.89	2.55	7.20	6.26	0.80	1.13	45.65
1974	2.89	0.27	1.75	0.90	11.16	3.33	0.99	7.30	5.84	2.88	3.43	2.60	43.34
1975	0.96	0.46	0.36	0.89	6.73	7.68	3.71	2.38	1.93	3.88	1.01	6.97	36.96
1976	0.77	0.39	1.45	5.90	2.61	1.32	5.75	2.76	7.61	6.18	3.05	5.46	43.25
1977	2.39	2.56	1.10	3.90	2.26	12.21	0.76	2.53	3.20	4.21	3.64	0.45	39.21
1978	3.43	2.87	0.54	1.95	1.05	4.88	2.51	2.08	19.05	0.57	2.88	1.27	43.08
1979	5.21	2.32	1.69	5.16	6.66	4.03	6.94	2.07	10.50	1.65	0.89	2.18	49.30
1980	4.52	1.78	1.97	0.48	8.16	T	0.41	5.65	6.07	0.90	1.81	0.79	32.54
1981	2.22	1.01	1.40	1.42	8.39	9.29	4.37	4.23	1.22	10.16	0.02	1.37	45.10
1982	0.39	5.38	0.23	1.40	8.61	0.06	0.07	1.78	1.11	4.07	8.68	0.75	32.53
1983	1.64	3.79	4.21	0.24	1.76	2.96	10.47	1.88	4.80	7.00	3.14	0.52	42.41
1984	3.02	1.34	1.74	0.09	4.02	2.05	1.02	4.16	1.87	8.52	2.16	3.93	33.92
1985	3.37	1.97	5.51	8.56	1.03	6.97	1.26	1.88	3.29	2.03	1.74	2.38	39.99
1986	1.12	0.50	1.03	0.50	6.77	7.45	0.81	3.62	3.56	6.79	2.79	4.25	39.19
1987	2.42	4.24	0.43	T	4.96	11.70	4.98	3.07	3.20	0.34	5.89	1.86	43.09
1988	0.30	0.23	1.68	1.10	1.03	1.73	2.79	1.12	2.77	0.77	0.15	2.24	15.91
1989	3.91	0.47	1.72	1.10	0.69	4.35	2.47	1.73	2.43	3.89	1.90	1.13	25.79
1990	1.73	2.04	3.00	3.63	1.19	0.82	13.59	1.47	3.59	1.56	2.34	0.81	35.77
1991	7.76	3.05	3.23	11.09	2.16	5.77	7.90	2.42	2.94	3.27	0.87	6.26	56.72
1992	5.00	9.08	1.18	7.81	12.26	1.33	2.30	1.65	3.63	1.17	3.88	2.09	51.38
1993	3.21	3.61	2.80	4.43	14.66	11.36	1.16	0.72	1.78	2.49	1.17	4.01	51.40
Record Mean	2.30	2.21	2.02	2.76	4.28	3.70	3.37	2.85	4.28	3.44	2.46	2.47	36.15

AVERAGE TEMPERATURE (deg. F)

VICTORIA, TEXAS

YEAR	JAN	FEB	MAR	APR	MAY	JUNE	JULY	AUG	SEP	OCT	NOV	DEC	ANNUAL
1964	54.3	51.8	62.3	72.5	79.8	83.1	85.5	84.5	80.0	68.6	66.0	56.1	70.4
1965	56.9	54.5	58.3	73.5	77.3	82.0	85.0	84.7	81.8	68.8	68.7	59.3	70.9
1966	48.9	52.6	62.8	71.5	75.3	80.0	83.8	83.6	80.0	70.7	66.1	55.3	69.2
1967	54.5	55.5	67.8	76.1	76.8	83.5	84.6	82.7	77.4	70.8	65.4	54.8	70.8
1968	52.1	50.7	60.0	71.2	76.8	80.8	82.7	84.1	78.4	74.1	60.2	56.0	68.9
1969	57.0	57.7	56.9	70.9	74.9	81.9	87.0	86.0	80.5	72.7	62.0	58.7	70.5
1970	49.3	56.7	59.0	71.4	73.3	80.0	82.8	84.7	79.9	69.3	60.3	62.7	69.1
1971	59.7	58.0	64.2	69.1	77.0	82.8	85.2	82.5	79.7	74.5	64.2	61.8	71.6
1972	57.9	57.9	67.5	74.7	76.1	82.1	82.9	83.5	82.9	73.5	57.3	53.0	70.8
1973	49.6	54.4	67.7	67.2	76.3	80.7	85.3	82.8	80.7	74.9	70.6	57.1	70.6
1974	55.5	59.6	69.3	71.7	79.5	81.2	84.6	83.9	76.3	73.3	62.2	56.1	71.1
1975	58.8	58.2	65.6	71.7	78.4	82.5	84.1	83.9	78.1	72.3	63.2	55.5	71.0
1976	53.2	62.5	65.2	70.9	73.3	80.7	81.0	82.5	78.5	64.0	54.4	51.9	68.2
1977	46.0	56.2	64.4	69.3	76.9	82.0	84.1	85.0	83.2	72.8	64.9	57.5	70.2
1978	45.2	48.4	60.7	70.0	78.6	82.4	84.8	84.1	80.6	71.4	64.6	54.3	68.8
1979	45.9	52.7	64.7	70.7	73.5	80.6	83.0	82.8	76.9	72.7	57.6	54.3	68.0
1980	55.7	54.9	62.8	67.2	77.1	84.1	86.9	84.1	82.4	69.4	58.5	56.0	69.9
1981	53.1	56.1	62.0	73.7	75.6	82.2	84.0	83.3	78.8	72.6	65.4	56.2	70.2
1982	53.8	53.2	65.2	69.5	75.9	83.2	85.9	86.1	81.8	71.9	62.6	57.4	70.5
1983	52.4	55.1	61.1	67.0	75.2	80.7	82.8	84.0	77.7	72.1	65.3	46.6	68.4
1984	50.3	57.4	64.9	71.6	76.5	82.2	85.2	84.2	78.3	75.3	62.7	65.0	71.1
1985	47.3	52.5	66.7	71.4	78.5	82.0	83.5	86.2	81.1	73.8	67.4	52.7	70.3
1986	56.0	60.6	64.8	73.6	76.0	82.0	85.0	84.5	82.5	71.3	62.6	53.6	71.0
1987	53.3	57.5	61.1	68.3	77.5	81.0	83.9	85.6	80.6	71.2	62.4	58.0	70.0
1988	50.0	55.8	62.6	69.3	75.3	81.2	85.2	86.7	81.6	74.4	67.7	57.7	70.6
1989	59.5	54.4	63.1	70.8	81.1	82.3	84.2	84.0	78.7	72.6	65.0	46.0	70.1
1990	59.1	61.1	63.7	70.6	78.7	85.3	83.3	85.3	81.4	70.7	66.2	55.1	71.7
1991	51.6	58.8	66.0	74.1	78.0	82.2	84.0	84.5	78.6	74.2	59.5	58.7	70.9
1992	53.4	61.4	66.0	70.6	75.0	83.8	85.3	83.0	81.6	74.1	59.7	59.3	71.1
1993	54.3	57.9	64.0	67.9	74.2	81.2	84.7	85.9	81.7	71.7	58.5	57.2	69.9
Record Mean	54.5	57.5	63.9	71.8	76.6	82.2	84.4	84.7	80.3	72.6	63.0	56.2	70.7
Max	64.9	68.4	74.7	80.7	86.2	91.7	94.3	95.1	90.6	84.1	74.1	66.8	81.0
Min	44.0	46.6	53.1	62.8	67.0	72.6	74.4	74.3	70.1	61.0	51.8	45.6	60.3

REFERENCE NOTES

VICTORIA, TEXAS

GENERAL
T - TRACE AMOUNT.
BLANK ENTRIES DENOTE MISSING/UNREPORTED DATA.
# INDICATES A STATION OR INSTRUMENT RELOCATION.

SPECIFIC

(a) - LENGTH OF RECORD IN YEARS, ALTHOUGH INDIVIDUAL MONTHS MAY BE MISSING.
0.* OR * - THE VALUE IS BETWEEN 0.0 AND 0.05
NORMALS - BASED ON THE 1961-1990 RECORD PERIOD.
EXTREMES - DATES ARE THE MOST RECENT OCCURRENCE
WIND DIR. - NUMERALS SHOW TENS OF DEGRESS CLOCKWISE FROM TRUE NORTH. "00" INDICATES CALM.
RESULTANT DIRECTIONS ARE GIVEN TO WHOLE DEGREES.
BOLD VALUES INDICATE EXTREME VALUES WHICH OCCURRED AFTER THE ASOS SYSTEM WAS COMMISSIONED.

RECORD = PERIOD OF RECORD
RECORD MEAN PRECIPITATION IS THE MEAN OF ALL DAILY PRECIPITATION AMOUNTS DURING THE PERIOD OF RECORD.
RECORD MAX(MIN) TEMPERATURE IS THE MEAN OF ALL DAILY MAX(MIN) TEMPERATURES DURING THE PERIOD OF RECORD.
RECORD MEAN TEMPERATURE IS THE SUM OF THE RECORD MAX AND RECORD MIN DIVIDED BY 2.
AVERAGE TEMPERATURE IS THE SUM OF THE MEAN DAILY MAX AND MIN TEMPERATURE DIVIDED BY 2.

EXCEPTIONS

THUNDERSTORMS AND HEAVY FOG MAY BE INCOMPLETE, DUE TO PART-TIME OPERATIONS.

RECORD MEANS ARE THROUGH THE CURRENT YEAR, BEGINNING IN 1904 FOR TEMPERATURE
1893 FOR PRECIPITATION
1962 FOR SNOWFALL

HEATING DEGREE DAYS Base 65 deg. F VICTORIA, TEXAS

SEASON	JULY	AUG	SEP	OCT	NOV	DEC	JAN	FEB	MAR	APR	MAY	JUNE	TOTAL
1964-65	0	0	0	22	102	321	275	294	255	0	0	0	1269
1965-66	0	0	2	28	25	202	499	341	122	16	3	0	1238
1966-67	0	0	0	17	83	327	350	267	50	0	0	0	1094
1967-68	0	0	4	15	83	332	414	409	208	16	0	0	1481
1968-69	0	0	0	6	188	287	281	208	261	1	0	0	1232
1969-70	0	0	0	14	174	202	498	230	202	34	5	0	1359
1970-71	0	0	0	44	189	149	228	222	120	45	0	0	997
1971-72	0	0	2	0	96	152	270	229	50	2	0	0	801
1972-73	0	0	0	14	264	376	476	301	20	64	0	0	1515
1973-74	0	0	0	0	26	254	310	183	70	17	0	0	860
1974-75	0	0	0	7	155	286	244	206	92	14	0	0	1004
1975-76	0	0	0	6	165	316	364	120	95	2	0	0	1068
1976-77	0	0	0	111	325	398	585	246	82	8	0	0	1755
1977-78	0	0	0	8	86	261	620	467	166	23	1	0	1632
1978-79	0	0	0	3	121	349	591	352	79	13	6	0	1514
1979-80	0	0	0	20	243	329	299	309	135	39	0	0	1374
1980-81	0	0	0	59	231	303	362	272	120	2	0	0	1349
1981-82	0	0	0	46	64	284	375	337	126	47	0	0	1279
1982-83	0	0	0	21	155	270	390	274	151	60	0	0	1321
1983-84	0	0	3	12	99	580	451	230	88	14	0	0	1477
1984-85	0	0	8	6	139	123	544	358	60	5	0	0	1243
1985-86	0	0	4	6	69	385	275	183	67	3	0	0	992
1986-87	0	0	0	12	162	353	360	207	148	64	0	0	1306
1987-88	0	0	0	2	148	243	462	288	144	31	0	0	1318
1988-89	0	0	0	0	91	252	211	335	172	37	0	0	1098
1989-90	0	0	0	28	109	584	206	133	100	23	1	0	1184
1990-91	0	0	0	32	73	349	408	186	61	4	0	0	1113
1991-92	0	0	0	18	204	239	353	133	67	24	0	0	1038
1992-93	0	0	0	0	202	206	330	206	101	26	0	0	1071
1993-94	0	0	0	61	227	267							

COOLING DEGREE DAYS Base 65 deg. F VICTORIA, TEXAS

YEAR	JAN	FEB	MAR	APR	MAY	JUNE	JULY	AUG	SEP	OCT	NOV	DEC	TOTAL
1969	38	10	15	184	313	512	687	659	472	263	89	14	3256
1970	19	6	23	234	269	458	561	617	452	184	55	84	2962
1971	70	35	104	172	380	540	630	548	452	303	78	58	3370
1972	58	28	134	299	350	519	562	582	541	285	37	13	3408
1973	3	10	109	138	358	477	638	556	476	313	201	15	3294
1974	20	38	207	225	458	494	613	594	346	271	75	17	3358
1975	61	19	118	223	421	535	597	591	401	240	116	29	3351
1976	7	54	107	186	263	478	505	550	414	88	16	1	2669
1977	4	8	74	144	376	517	599	625	551	259	89	37	3283
1978	10	8	39	177	428	528	622	598	476	205	118	23	3232
1979	7	18	78	190	277	476	563	559	362	266	28	5	2829
1980	15	24	73	114	383	582	684	599	526	202	41	29	3272
1981	0	32	35	268	333	523	595	576	421	285	83	18	3169
1982	40	13	140	191	344	552	655	661	513	244	93	41	3487
1983	7	0	36	126	323	479	557	598	393	238	112	20	2889
1984	1	21	95	219	363	524	632	602	413	332	74	128	3404
1985	4	15	117	204	422	517	581	667	493	287	146	12	3465
1986	6	66	67	269	350	516	626	609	530	214	96	8	3351
1987	4	4	29	170	395	485	589	649	475	204	79	34	3117
1988	3	32	76	166	326	493	634	679	506	298	179	32	3424
1989	45	47	121	220	504	523	603	599	418	270	116	1	3467
1990	32	28	69	195	432	616	575	638	499	216	117	47	3464
1991	0	18	97	280	409	523	595	608	415	309	46	49	3349
1992	0	37	106	202	315	568	636	567	506	291	51	35	3314
1993	8	13	80	121	291	494	618	655	508	279	39	31	3137

SNOWFALL (inches) VICTORIA, TEXAS

SEASON	JULY	AUG	SEP	OCT	NOV	DEC	JAN	FEB	MAR	APR	MAY	JUNE	TOTAL
1970-71	0.0	0.0	0.0	0.0	0.0	0.0	0.0	0.0	0.0	0.0	0.0	0.0	0.0
1971-72	0.0	0.0	0.0	0.0	0.0	0.0	0.0	0.0	0.0	0.0	0.0	0.0	0.0
1972-73	0.0	0.0	0.0	0.0	0.0	0.0	1.2	1.0	0.0	0.0	0.0	0.0	2.2
1973-74	0.0	0.0	0.0	0.0	0.0	0.0	0.0	0.0	0.0	0.0	0.0	0.0	0.0
1974-75	0.0	0.0	0.0	0.0	0.0	0.0	T	0.0	0.0	0.0	0.0	0.0	T
1975-76	0.0	0.0	0.0	0.0	0.0	0.0	0.0	0.0	0.0	0.0	0.0	0.0	0.0
1976-77	0.0	0.0	0.0	0.0	0.2	0.0	0.0	0.0	0.0	0.0	0.0	0.0	0.2
1977-78	0.0	0.0	0.0	0.0	0.0	0.0	0.0	0.0	0.0	0.0	0.0	0.0	0.0
1978-79	0.0	0.0	0.0	0.0	0.0	0.0	0.0	0.0	0.0	0.0	0.0	0.0	0.0
1979-80	0.0	0.0	0.0	0.0	T	0.0	0.0	0.0	0.0	0.0	0.0	0.0	T
1980-81	0.0	0.0	0.0	0.0	0.0	0.0	T	0.0	0.0	0.0	0.0	0.0	T
1981-82	0.0	0.0	0.0	0.0	0.0	0.0	T	0.0	0.0	0.0	0.0	0.0	T
1982-83	0.0	0.0	0.0	0.0	0.0	0.0	0.0	0.0	0.0	0.0	0.0	0.0	0.0
1983-84	0.0	0.0	0.0	0.0	0.0	0.0	T	0.0	0.0	0.0	0.0	0.0	T
1984-85	0.0	0.0	0.0	0.0	0.0	0.0	2.1	T	0.0	0.0	0.0	0.0	2.1
1985-86	0.0	0.0	0.0	0.0	0.0	0.0	0.0	0.0	0.0	0.0	0.0	0.0	0.0
1986-87	0.0	0.0	0.0	0.0	0.0	0.0	T	0.0	0.0	0.0	0.0	0.0	T
1987-88	0.0	0.0	0.0	0.0	0.0	0.0	0.0	T	0.0	0.0	0.0	0.0	T
1988-89	0.0	0.0	0.0	0.0	0.0	0.0	0.0	0.0	T	0.0	0.0	0.0	T
1989-90	0.0	0.0	0.0	0.0	0.0	T	0.0	0.0	T	0.0	0.0	0.0	T
1990-91	0.0	0.0	0.0	0.0	0.0	T	0.0	0.0	0.0	0.0	0.0	0.0	T
1991-92	0.0	0.0	0.0	0.0	0.0	0.0	0.0	0.0	0.0	0.0	0.0	0.0	0.0
1992-93	0.0	0.0	0.0	0.0	0.0	0.0	0.0	0.0	0.0	0.0	T	0.0	T
1993-94	0.0	0.0	0.0	0.0	0.0	0.0							
Record Mean	0.0	0.0	0.0	0.0	T	T	0.1	T	T	0.0	T	0.0	0.1

WACO, TEXAS

One of the major cities of Texas, Waco is located in the rich agricultural region of the Brazos River Valley in North Central Texas. The city lies on the edge of the gently rolling Blackland Prairies. To the west lies the rolling to hilly Grand Prairie. Waco is a commercial hub with an economy based on industry, education and agriculture. Baylor University, founded in 1845, is located here. Regional agriculture includes chiefly cattle, poultry, sorghum, cotton and corn. Soils are black waxy, loam and sandy types. Lake Waco, a reservoir of 7,260 surface acres, lies within the Waco city limits, with the north shoreline approximately 0.8 mile south of the Municipal Airport.

The climate of Waco is humid subtropical with hot summers. It is a continental type climate characterized by extreme variations in temperature. Tropical maritime air masses predominate throughout the late spring, summer and early fall months, while Polar air masses frequent the area in winter. In an average year, April and May are the wettest months, while the July–August period is the driest. Most warm season rainfall occurs from thunderstorm activity. Consequently, considerable spatial variation in amounts occur.

Winters are mild. Cold fronts moving down from the High Plains often are accompanied by strong, gusty, northerly winds and sharp drops in temperature. Cold spells are of short duration, rarely lasting longer than 2 or 3 days before a rapid warming occurs. Winter precipitation is closely associated with frontal activity, and may fall as rain, freezing rain, sleet or snow. During most years, snowfall is of little or no consequence.

Daytime temperatures are hot in summer, particularly in July and August. The highest temperatures are associated with fair skies, light winds, and comparatively low humidities. There is little variety in the day–to–day weather during July and August. Air conditioning is recommended for maximum comfort indoors or while traveling.

The spring and fall seasons are very pleasant at Waco. Temperatures are comfortable. Cloudiness and showers are more frequent in the spring than in the fall. The average first occurrence of 32 degrees Fahrenheit is late November and the average last occurrence is in mid March.

NORMALS, MEANS, AND EXTREMES

WACO, TEXAS

LATITUDE: 31°37'N LONGITUDE: 97°13' W ELEVATION: FT. GRND 501 BARO 499 TIME ZONE: CENTRAL WBAN: 13959

	(a)	JAN	FEB	MAR	APR	MAY	JUNE	JULY	AUG	SEP	OCT	NOV	DEC	YEAR
TEMPERATURE °F:														
Normals														
-Daily Maximum		56.1	60.8	69.6	78.0	84.4	92.0	96.8	97.2	89.6	80.3	68.8	59.3	77.7
-Daily Minimum		34.2	38.0	46.8	56.2	64.2	70.9	74.4	74.0	67.6	56.8	46.6	37.3	55.6
-Monthly		45.2	49.4	58.2	67.1	74.3	81.5	85.6	85.6	78.6	68.5	57.7	48.3	66.7
Extremes														
-Record Highest	51	88	92	100	101	102	109	108	112	106	101	92	91	112
-Year		1971	1986	1971	1963	1985	1980	1986	1969	1985	1989	1988	1955	AUG 1969
-Record Lowest	51	-5	4	15	27	37	52	60	**53**	40	**25**	17	-4	-5
-Year		1949	1985	1948	1975	1981	1964	1990	**1992**	1983	**1993**	1976	1989	JAN 1949
NORMAL DEGREE DAYS:														
Heating (base 65°F)		620	441	247	66	0	0	0	0	0	38	243	524	2179
Cooling (base 65°F)		6	0	36	129	288	495	639	639	408	146	24	6	2816
% OF POSSIBLE SUNSHINE														
MEAN SKY COVER (tenths)														
Sunrise - Sunset	48	6.3	6.1	5.9	6.1	6.1	5.0	4.4	4.3	4.8	4.6	5.3	5.8	5.4
MEAN NUMBER OF DAYS:														
Sunrise to Sunset														
-Clear	50	8.8	8.7	9.9	8.7	7.9	10.4	14.1	14.0	12.4	13.5	11.6	10.6	130.4
-Partly Cloudy	50	6.2	5.8	6.7	7.2	10.1	12.0	10.4	11.1	9.1	8.1	5.9	5.8	98.5
-Cloudy	50	16.0	13.8	14.5	14.1	12.9	7.6	6.5	5.9	8.5	9.5	12.5	14.7	136.4
Precipitation														
.01 inches or more	50	7.4	7.4	7.3	7.2	8.7	6.6	4.4	4.9	6.4	5.7	6.4	6.3	78.7
Snow, Ice pellets, hail														
1.0 inches or more	50	0.3	0.2	0.1	0.0	0.0	0.0	0.0	0.0	0.0	0.0	0.*	0.*	0.6
Thunderstorms	50	1.3	2.5	3.8	5.5	7.9	5.9	4.1	4.3	4.0	3.1	2.1	1.4	45.9
Heavy Fog Visibility														
1/4 mile or less	50	3.0	2.1	1.2	0.8	0.6	0.2	0.*	0.*	0.2	0.8	1.8	2.5	13.2
Temperature °F														
-Maximum														
90° and above	30	0.0	0.*	0.4	1.3	6.9	22.3	28.6	28.3	17.5	4.0	0.1	0.0	109.5
32° and below	30	0.9	0.3	0.*	0.0	0.0	0.0	0.0	0.0	0.0	0.0	0.0	0.5	1.7
-Minimum														
32° and below	30	12.7	7.2	2.1	0.2	0.0	0.0	0.0	0.0	0.0	0.1	2.6	8.9	33.8
0° and below	30	0.0	0.0	0.0	0.0	0.0	0.0	0.0	0.0	0.0	0.0	0.0	0.*	*
AVG. STATION PRESS. (mb)	21	1002.6	1000.9	996.9	996.1	994.7	995.6	997.3	997.1	997.6	999.5	1000.3	1002.0	998.4
RELATIVE HUMIDITY (%)														
Hour 00	30	78	76	74	76	80	76	69	68	75	76	78	78	75
Hour 06 (Local Time)	30	83	83	82	84	88	86	82	82	86	85	84	83	84
Hour 12	30	64	61	58	59	61	55	48	47	54	55	59	62	57
Hour 18	30	62	57	52	54	58	52	44	44	53	57	62	64	55
PRECIPITATION (inches):														
Water Equivalent														
-Normal		1.65	2.09	2.33	3.19	4.58	3.28	1.99	1.68	3.52	3.36	2.43	1.86	31.96
-Maximum Monthly	51	5.83	6.29	6.84	13.37	15.00	12.06	8.58	8.91	7.29	10.51	6.24	8.44	15.00
-Year		1961	1992	1945	1957	1965	1961	1971	1974	1970	1984	1952	1991	MAY 1965
-Minimum Monthly	51	0.03	0.17	0.04	0.12	0.65	0.27	**0.00**	T	0.00	0.00	0.13	0.04	**0.00**
-Year		1971	1972	1956	1983	1988	1953	**1993**	1952	1956	1952	1970	1950	JUL 1993
-Maximum in 24 hrs	51	2.24	3.96	3.17	5.09	7.18	4.21	4.49	4.80	4.57	5.72	4.26	4.56	7.18
-Year		1961	1986	1990	1957	1953	1947	1973	1958	1957	1974	1952	1991	MAY 1953
Snow, Ice pellets, hail														
-Maximum Monthly	51	7.0	4.8	1.0	T	T	0.0	**0.0**	T	**0.0**	**T**	**2.4**	2.0	7.0
-Year		1949	1966	1987	1993	1989			1991		**1993**	**1993**	1946	JAN 1949
-Maximum in 24 hrs	51	7.0	4.8	1.0	T	T	0.0	**0.0**	T	**0.0**	**T**	**2.4**	2.0	7.0
-Year		1949	1966	1987	1993	1989			1991		**1993**	**1993**	1946	JAN 1949
WIND:														
Mean Speed (mph)	44	11.4	12.0	13.0	12.8	11.7	11.3	10.8	10.0	9.7	10.1	10.9	11.1	11.2
Prevailing Direction through 1963		S	S	S	S	S	S	S	S	S	S	S	S	S
Fastest Obs. 1 Min.														
-Direction (!!)	45	32	36	27	36	36	09	36	05	32	34	29	32	09
-Speed (MPH)	45	49	58	65	62	60	69	60	60	60	52	62	52	69
-Year		1952	1954	1952	1953	1952	1961	1953	1951	1952	1960	1953	1954	JUN 1961
Peak Gust														
-Direction (!!)	9	N	W	S	SW	W	N	SE	NW	**13**	**NW**	**25**	**29**	S
-Speed (mph)	9	54	60	75	55	61	51	49	54	**51**	**60**	**54**	**53**	75
-Date		1985	1990	1990	1990	1987	1992	1985	1985	**1993**	**1992**	**1993**	**1993**	MAR 1990

PRECIPITATION (inches) WACO, TEXAS

YEAR	JAN	FEB	MAR	APR	MAY	JUNE	JULY	AUG	SEP	OCT	NOV	DEC	ANNUAL
1964	3.15	1.29	2.85	2.85	1.48	4.32	0.23	2.30	5.49	0.79	3.67	0.90	29.32
1965	2.97	4.41	3.13	1.35	15.00	0.82	0.12	0.64	3.80	2.91	4.64	2.31	42.10
1966	1.15	2.00	0.77	9.37	2.55	2.34	0.46	4.09	4.73	0.01	0.13	0.45	28.05
1967	0.37	0.77	1.93	8.80	4.17	0.36	1.59	0.43	4.31	3.80	3.53	2.91	32.97
1968	3.47	1.78	2.15	2.94	7.12	5.11	0.98	0.87	5.61	1.46	4.77	0.45	36.71
1969	0.45	2.10	3.62	3.98	3.98	0.44	0.05	2.23	4.70	5.53	1.93	2.49	31.50
1970	0.76	3.82	4.33	1.80	2.24	0.92	0.10	2.47	7.29	3.37	0.13	0.83	28.06
1971	0.03	1.33	0.64	2.79	0.75	0.99	8.58	2.16	0.74	3.26	1.64	4.30	27.21
1972	2.54	0.17	0.09	2.04	1.95	3.05	4.14	1.79	2.57	6.62	2.87	0.78	28.61
1973	3.48	1.52	3.04	5.41	3.16	7.10	4.89	0.36	5.24	9.36	0.66	0.76	44.98
1974	1.33	1.21	0.61	3.30	2.21	1.07	0.73	8.91	6.80	6.76	2.54	1.92	37.39
1975	1.40	2.94	1.09	4.58	13.21	2.84	2.77	0.72	2.27	2.40	0.40	1.83	36.45
1976	1.74	0.33	1.60	6.54	4.99	3.23	7.40	0.24	5.66	5.19	0.66	2.50	40.08
1977	1.83	3.67	2.90	7.38	1.71	0.78	0.21	1.96	0.26	1.50	2.43	0.15	24.78
1978	1.03	2.83	2.46	2.37	2.88	1.54	0.26	1.42	0.53	1.87	4.57	2.01	23.77
1979	1.99	1.97	4.31	1.38	9.68	4.70	5.01	4.58	2.51	1.94	0.33	3.97	42.37
1980	2.26	1.83	2.13	4.03	4.47	0.34	0.01	0.16	2.03	0.76	2.29	2.70	23.01
1981	0.85	1.80	2.97	1.21	3.69	7.23	0.16	2.14	3.61	8.41	0.99	0.50	33.56
1982	1.86	1.41	3.79	2.46	5.54	3.93	2.61	0.18	0.15	1.43	3.57	2.11	29.04
1983	1.28	2.97	3.46	0.12	3.83	0.95	3.41	2.54	0.71	1.55	2.90	0.46	24.18
1984	0.82	1.00	3.45	0.54	3.71	1.75	1.82	1.33	1.32	10.51	3.05	5.07	34.37
1985	0.76	1.91	2.57	3.37	1.59	5.62	0.68	0.68	6.38	3.84	3.53	2.02	32.95
1986	0.07	4.51	0.43	2.74	6.89	3.95	0.37	0.68	3.94	5.09	3.56	3.48	35.71
1987	1.29	2.60	1.67	1.63	5.16	5.84	1.42	0.76	5.69	1.19	4.44	3.31	35.00
1988	0.44	1.68	3.09	1.28	0.65	5.57	1.15	0.56	5.52	0.98	1.76	2.01	24.69
1989	2.45	2.27	2.42	1.60	9.34	4.16	1.99	2.21	0.27	0.91	0.32	0.54	28.48
1990	2.72	2.80	5.07	3.11	6.12	0.97	0.69	1.05	6.57	4.48	3.50	1.35	38.43
1991	3.13	1.43	0.93	2.68	5.89	3.16	3.65	7.21	3.81	3.58	1.68	8.44	45.59
1992	4.49	6.29	3.57	1.27	6.36	2.66	2.39	0.49	2.22	1.51	4.05	3.31	38.61
1993	2.45	3.84	4.31	3.63	2.06	2.15	0.00	0.81	4.44	5.50	2.11	1.48	32.78
Record Mean	2.02	2.31	2.68	3.84	4.59	3.08	1.98	2.09	3.03	2.82	2.51	2.53	33.48

AVERAGE TEMPERATURE (deg. F) WACO, TEXAS

YEAR	JAN	FEB	MAR	APR	MAY	JUNE	JULY	AUG	SEP	OCT	NOV	DEC	ANNUAL
1964	46.7	45.5	57.4	68.7	75.5	82.6	87.9	86.8	78.7	65.8	60.9	50.6	67.3
1965	51.4	48.2	51.2	71.9	75.6	82.6	87.2	86.0	79.9	66.0	63.2	52.4	68.0
1966	42.1	46.2	55.6	65.6	73.6	81.4	88.6	84.2	78.4	68.1	63.4	47.6	66.2
1967	48.5	48.7	65.3	74.0	73.2	84.9	86.9	85.2	76.8	68.4	58.8	49.4	68.4
1968	46.7	47.0	56.1	64.8	73.1	79.9	82.9	85.6	76.3	70.7	55.9	49.1	65.7
1969	51.1	51.3	51.8	68.0	75.2	82.7	90.5	88.1	80.5	68.8	56.8	50.5	67.9
1970	41.7	50.0	52.7	68.0	72.8	80.8	85.1	88.1	79.6	66.3	55.9	56.6	66.5
1971	51.7	52.9	60.8	67.8	75.1	84.0	85.9	80.5	79.1	71.2	58.5	53.4	68.4
1972	47.3	52.9	63.6	71.0	73.3	82.7	82.2	84.4	81.3	67.7	50.8	44.1	66.8
1973	42.3	47.2	60.4	61.1	71.6	77.3	83.8	84.4	79.1	69.9	63.2	50.0	65.8
1974	45.3	53.4	64.6	67.4	77.2	80.1	86.8	84.4	71.8	69.8	56.7	46.2	67.0
1975	48.5	47.0	54.5	64.1	72.0	79.7	82.0	84.6	77.2	70.5	59.8	50.5	65.9
1976	46.8	59.0	60.5	66.2	70.7	80.4	82.2	85.6	77.9	60.5	50.1	45.5	65.4
1977	38.0	52.7	59.8	67.2	76.5	84.6	89.1	87.9	85.7	71.4	60.2	51.7	68.7
1978	37.8	42.0	56.4	70.2	76.5	83.7	89.2	86.3	81.4	69.5	57.3	46.5	66.4
1979	36.4	44.5	58.4	64.9	69.5	78.9	82.3	81.8	75.5	70.4	52.4	49.3	63.7
1980	46.7	48.7	56.0	62.5	73.5	85.2	89.2	87.5	81.2	65.7	55.9	51.4	67.0
1981	46.6	50.1	55.7	69.9	72.6	81.7	85.8	83.9	77.5	68.9	58.8	48.7	66.7
1982	46.8	47.5	61.8	64.4	75.0	80.5	86.3	88.2	80.4	68.8	57.9	50.4	67.3
1983	45.5	48.9	55.4	61.8	71.5	79.2	84.3	85.6	77.4	69.8	59.3	38.3	64.7
1984	41.8	51.5	58.8	65.5	76.2	82.4	86.0	86.6	76.6	69.1	56.0	54.9	67.1
1985	40.4	46.2	61.5	68.4	76.8	82.1	85.3	87.4	77.5	69.4	59.7	44.9	66.6
1986	50.5	53.0	60.8	69.0	73.5	81.5	87.5	84.6	83.1	67.2	55.1	47.3	67.8
1987	45.9	51.9	55.3	64.8	75.3	80.4	84.4	87.3	77.3	67.7	57.0	49.4	66.4
1988	44.2	49.4	57.9	66.6	73.6	79.8	86.0	88.2	80.2	68.5	60.8	50.8	67.2
1989	51.7	44.8	58.3	68.2	76.5	78.9	84.2	83.9	75.6	70.4	59.6	40.5	66.1
1990	54.0	54.9	60.0	66.8	75.7	86.3	84.8	87.0	80.6	67.6	61.5	47.0	68.9
1991	44.3	53.8	60.7	69.3	75.8	81.4	85.3	83.5	76.7	69.9	53.9	52.2	67.2
1992	48.6	55.0	61.2	67.3	72.4	81.4	86.3	81.5	80.2	70.9	54.2	51.2	67.5
1993	46.4	50.4	57.0	63.8	72.1	82.1	85.7	86.8	78.0	64.6	51.5	49.7	65.7
Record Mean	47.3	50.2	58.7	66.9	74.7	82.3	85.8	85.7	79.2	68.9	57.5	49.3	67.2
Max	57.7	60.5	70.1	77.8	84.9	92.6	96.4	96.6	89.9	80.3	68.3	59.7	77.9
Min	36.8	39.8	47.3	56.0	64.4	71.9	75.1	74.8	68.4	57.5	46.6	38.8	56.5

REFERENCE NOTES WACO, TEXAS

GENERAL
T - TRACE AMOUNT.
BLANK ENTRIES DENOTE MISSING/UNREPORTED DATA.
# INDICATES A STATION OR INSTRUMENT RELOCATION.

SPECIFIC

(a) - LENGTH OF RECORD IN YEARS, ALTHOUGH INDIVIDUAL MONTHS MAY BE MISSING.
0.* OR * - THE VALUE IS BETWEEN 0.0 AND 0.05
NORMALS - BASED ON THE 1961-1990 RECORD PERIOD.
EXTREMES - DATES ARE THE MOST RECENT OCCURRENCE
WIND DIR.- NUMERALS SHOW TENS OF DEGRESS CLOCKWISE FROM TRUE NORTH. "00" INDICATES CALM.
RESULTANT DIRECTIONS ARE GIVEN TO WHOLE DEGREES.
BOLD VALUES INDICATE EXTREME VALUES WHICH OCCURRED AFTER THE ASOS SYSTEM WAS COMMISSIONED.

RECORD = PERIOD OF RECORD
RECORD MEAN PRECIPITATION IS THE MEAN OF ALL DAILY PRECIPITATION AMOUNTS DURING THE PERIOD OF RECORD.
RECORD MAX(MIN) TEMPERATURE IS THE MEAN OF ALL DAILY MAX(MIN) TEMPERATURES DURING THE PERIOD OF RECORD.
RECORD MEAN TEMPERATURE IS THE SUM OF THE RECORD MAX AND RECORD MIN DIVIDED BY 2.
AVERAGE TEMPERATURE IS THE SUM OF THE MEAN DAILY MAX AND MIN TEMPERATURE DIVIDED BY 2.

EXCEPTIONS

RECORD MEANS ARE THROUGH THE CURRENT YEAR, BEGINNING IN 1897 FOR TEMPERATURE
1897 FOR PRECIPITATION
1944 FOR SNOWFALL

HEATING DEGREE DAYS Base 65 deg. F — WACO, TEXAS

SEASON	JULY	AUG	SEP	OCT	NOV	DEC	JAN	FEB	MAR	APR	MAY	JUNE	TOTAL
1964-65	0	0	0	45	196	458	427	465	424	15	0	0	2030
1965-66	0	0	2	72	100	393	706	522	291	75	13	0	2174
1966-67	0	0	0	52	126	557	511	450	114	4	14	0	1828
1967-68	0	0	6	46	196	480	560	519	288	79	0	0	2174
1968-69	0	0	0	22	294	484	438	382	405	15	1	0	2041
1969-70	0	0	0	75	278	444	719	414	372	60	11	0	2373
1970-71	0	0	4	87	286	288	411	349	188	50	5	0	1668
1971-72	0	0	6	5	230	360	543	368	109	19	2	0	1642
1972-73	0	0	4	89	426	638	696	492	149	170	13	0	2677
1973-74	0	0	0	22	118	460	605	322	145	52	0	0	1724
1974-75	0	0	13	14	274	578	507	495	343	120	0	0	2344
1975-76	0	0	2	27	230	462	558	200	204	38	6	0	1727
1976-77	0	0	0	205	446	596	827	345	175	21	2	0	2617
1977-78	0	0	0	22	173	417	839	639	284	19	19	0	2412
1978-79	0	0	0	23	256	573	880	570	217	59	30	0	2608
1979-80	0	0	0	43	384	483	562	475	297	119	6	0	2369
1980-81	0	0	2	98	305	427	561	417	285	27	13	0	2135
1981-82	0	0	6	75	194	498	566	486	202	103	3	0	2133
1982-83	0	0	1	73	265	453	599	443	298	159	9	0	2300
1983-84	0	0	11	38	220	823	710	390	221	73	6	0	2492
1984-85	0	0	23	41	293	319	752	527	157	41	0	0	2153
1985-86	0	0	14	38	207	619	443	353	151	39	3	0	1867
1986-87	0	0	0	50	305	540	583	360	308	111	0	0	2257
1987-88	0	0	0	44	270	480	644	451	256	49	0	0	2194
1988-89	0	0	3	20	188	434	403	562	255	82	0	0	1947
1989-90	0	0	12	67	227	753	341	287	195	64	7	0	1953
1990-91	0	0	0	90	157	555	633	312	177	15	7	0	1946
1991-92	0	0	10	57	350	401	502	283	144	59	10	0	1816
1992-93	0	0	1	7	331	435	570	405	262	111	6	0	2128
1993-94	0	0	0	155	419	477							

COOLING DEGREE DAYS Base 65 deg. F — WACO, TEXAS

YEAR	JAN	FEB	MAR	APR	MAY	JUNE	JULY	AUG	SEP	OCT	NOV	DEC	TOTAL
1969	12	2	1	111	325	537	796	723	470	200	40	0	3217
1970	4	0	0	155	260	483	629	723	447	136	20	35	2892
1971	6	14	66	142	325	576	655	489	436	206	40	7	2962
1972	3	24	72	203	266	539	541	608	499	178	6	0	2939
1973	0	0	12	60	225	375	587	611	429	182	72	0	2553
1974	0	5	138	128	382	461	683	610	225	169	31	2	2834
1975	2	0	26	100	227	448	533	618	376	206	83	19	2638
1976	0	33	72	81	193	469	542	644	395	73	5	0	2507
1977	0	7	20	94	366	593	754	719	628	224	34	13	3452
1978	2	0	21	181	383	568	757	666	496	169	30	6	3279
1979	0	1	21	61	177	426	544	529	320	220	17	1	2317
1980	0	6	25	49	278	611	758	705	498	126	39	12	3107
1981	0	5	4	184	256	509	648	594	384	207	15	0	2806
1982	8	0	112	93	321	470	666	725	471	197	62	9	3134
1983	0	0	7	68	217	431	604	648	390	193	56	3	2617
1984	0	4	39	95	360	529	658	677	379	174	28	14	2957
1985	0	7	56	150	370	518	637	699	397	177	55	3	3069
1986	1	24	27	165	273	502	703	616	551	124	16	0	3001
1987	0	0	14	112	327	469	609	697	374	134	35	3	2774
1988	5	5	43	105	273	451	655	726	465	135	68	2	2933
1989	0	2	56	186	367	427	601	594	335	244	72	0	2884
1990	6	9	49	125	344	649	619	687	474	177	59	5	3203
1991	0	4	49	150	351	500	638	581	366	219	25	11	2894
1992	0	1	33	137	247	497	666	518	465	194	14	13	2785
1993	0	0	24	82	234	519	649	681	398	149	21	5	2762

SNOWFALL (inches) — WACO, TEXAS

SEASON	JULY	AUG	SEP	OCT	NOV	DEC	JAN	FEB	MAR	APR	MAY	JUNE	TOTAL
1964-65	0.0	0.0	0.0	0.0	0.0	T	0.0	2.0	0.2	0.0	0.0	0.0	2.2
1965-66	0.0	0.0	0.0	0.0	0.0	0.0	2.6	4.8	0.0	0.0	0.0	0.0	7.4
1966-67	0.0	0.0	0.0	0.0	0.0	0.0	0.0	T	0.0	0.0	0.0	0.0	T
1967-68	0.0	0.0	0.0	0.0	0.0	T	T	1.0	T	0.0	0.0	0.0	1.0
1968-69	0.0	0.0	0.0	0.0	T	0.0	T	T	0.0	0.0	0.0	0.0	T
1969-70	0.0	0.0	0.0	0.0	T	T	T	0.0	0.0	0.0	0.0	0.0	T
1970-71	0.0	0.0	0.0	0.0	0.0	0.0	T	T	T	0.0	0.0	0.0	T
1971-72	0.0	0.0	0.0	0.0	0.0	0.0	T	0.0	0.0	0.0	0.0	0.0	T
1972-73	0.0	0.0	0.0	0.0	T	T	2.6	T	0.0	0.0	0.0	0.0	2.6
1973-74	0.0	0.0	0.0	0.0	0.0	T	T	0.0	0.0	0.0	0.0	0.0	T
1974-75	0.0	0.0	0.0	0.0	0.0	T	1.0	0.8	0.0	0.0	0.0	0.0	1.8
1975-76	0.0	0.0	0.0	0.0	0.0	T	0.0	0.0	0.0	0.0	0.0	0.0	T
1976-77	0.0	0.0	0.0	0.0	0.8	0.0	1.8	0.0	0.0	0.0	0.0	0.0	2.6
1977-78	0.0	0.0	0.0	0.0	T	0.0	2.2	0.2	T	0.0	0.0	0.0	2.4
1978-79	0.0	0.0	0.0	0.0	0.0	0.2	T	T	0.0	0.0	0.0	0.0	0.2
1979-80	0.0	0.0	0.0	0.0	T	0.0	0.0	T	0.0	0.0	0.0	0.0	T
1980-81	0.0	0.0	0.0	0.0	0.8	0.0	T	1.0	0.0	0.0	0.0	0.0	1.8
1981-82	0.0	0.0	0.0	0.0	0.0	0.0	6.0	T	T	0.0	0.0	0.0	6.0
1982-83	0.0	0.0	0.0	0.0	0.0	0.0	T	T	0.0	0.0	0.0	0.0	T
1983-84	0.0	0.0	0.0	0.0	0.0	T	T	0.0	T	0.0	0.0	0.0	T
1984-85	0.0	0.0	0.0	0.0	0.0	0.0	1.0	1.9	0.0	0.0	0.0	0.0	2.9
1985-86	0.0	0.0	0.0	0.0	0.0	0.0	T	T	0.0	0.0	0.0	0.0	T
1986-87	0.0	0.0	0.0	0.0	T	T	T	0.0	1.0	0.0	0.0	0.0	1.0
1987-88	0.0	0.0	0.0	0.0	0.0	0.0	0.1	1.4	0.0	0.0	0.0	0.0	1.5
1988-89	0.0	0.0	0.0	0.0	0.0	0.0	0.0	0.2	0.8	0.0	T	0.0	1.0
1989-90	0.0	0.0	0.0	0.0	0.0	T	0.0	T	0.0	0.0	0.0	0.0	T
1990-91	0.0	0.0	0.0	0.0	0.0	T	T	0.0	0.0	0.0	0.0	0.0	T
1991-92	0.0	T	0.0	T	0.0	0.0	T	0.0	T	T	0.0	0.0	T
1992-93	0.0	0.0	0.0	0.0	T	0.0	T	0.0	T	T	0.0	0.0	T
1993-94	0.0	0.0	0.0	T	2.4	T							
Record Mean	0.0	T	0.0	T	0.1	0.1	0.8	0.3	0.1	T	T	0.0	1.4

WICHITA FALLS, TEXAS

Wichita Falls is located in the West Cross Timbers subdivision of the North Central Plains of Texas, about 10 miles south of the Red River and 400 miles northwest of the nearest portion of the Gulf of Mexico. The topography is gently rolling mesquite plain, and the elevation of the area is about 1,000 feet.

This region lies between the humid subtropical climate of east Texas and a continental climate to the north and west. The climate of Wichita Falls is classified as continental. It is characterized by rapid changes in temperature, large daily and annual temperature extremes, and by rather erratic rainfall.

The area lies in the path of polar air masses which move down from the north during the winter season. With the passage of cold fronts or northers in the fall and winter, abrupt drops in temperature of as much as 20 to 30 degrees within an hour sometimes occur. While the area is subject to a wide range of temperature, winters are on the whole relatively mild. January, the coldest month, has an average temperature around 40 degrees. Sub–zero temperatures occur about once every five years.

The summers in Wichita Falls are generally of the continental climate type, characterized by low humidity and windy conditions. Temperatures over 100 degrees are frequent during the common periods of hot weather. July and August, the hottest months, have average temperatures in the middle 80s.

The normal rainfall is nearly 27 inches per year, but the distribution is erratic to such an extent that prolonged dry periods are common. Several lakes in the area provide water for domestic, industrial, and irrigation purposes. The greater part of the rainfall comes in the form of showers rather than general rains. Over 75 percent of the annual moisture occurs during the period from late March to mid November, but dry periods of three to four weeks are to be expected during this time almost every year. While the dry conditions materially affect agriculture in this region, complete crop failure seldom results. Moderate flooding along Holliday Creek and the Wichita River, which run through the city, occur about once in each ten–year period. Snowfall, measuring an inch or more, occurs on average only two days a year.

Wind speeds average over 11 mph, and southerly winds prevail. Rather strong winds are observed in all months. Even though strong, gusty winds occur frequently, severe duststorms are rare. Most severe dust observed in the area is blown in from the north and west.

The area around Wichita Falls enjoys excellent aviation weather. Flying activities are possible on all but a very few days of the year. Approximately 95 percent of the time the ceiling is 1,000 feet or more with visibility of 3 miles or more.

NORMALS, MEANS, AND EXTREMES

WICHITA FALLS, TEXAS

LATITUDE: 33°58'N LONGITUDE: 98°29'W ELEVATION: FT. GRND 994 BARO 1007 TIME ZONE: CENTRAL WBAN: 13966

	(a)	JAN	FEB	MAR	APR	MAY	JUNE	JULY	AUG	SEP	OCT	NOV	DEC	YEAR
TEMPERATURE °F:														
Normals														
-Daily Maximum		52.0	57.2	66.4	75.8	83.3	91.5	97.2	95.9	86.8	76.9	64.2	54.9	75.2
-Daily Minimum		27.6	32.1	40.6	50.3	59.1	68.0	72.7	71.4	63.9	52.2	40.6	30.8	50.8
-Monthly		39.8	44.7	53.5	63.1	71.2	79.8	85.0	83.6	75.4	64.6	52.4	42.8	63.0
Extremes														
-Record Highest	47	87	92	100	102	105	117	114	113	108	102	89	88	117
-Year		1969	1979	1971	1972	1985	1980	1980	1964	1977	1979	1988	1954	JUN 1980
-Record Lowest	47	-5	-8	8	24	36	51	54	**53**	38	25	14	-7	-8
-Year		1966	1985	1989	1975	1979	1983	1970	**1992**	1989	1957	1950	1989	FEB 1985
NORMAL DEGREE DAYS:														
Heating (base 65°F)		781	568	377	125	10	0	0	0	6	104	383	688	3042
Cooling (base 65°F)		0	0	20	68	202	444	620	577	318	91	0	0	2340
% OF POSSIBLE SUNSHINE														
MEAN SKY COVER (tenths)														
Sunrise - Sunset	48	5.6	5.6	5.5	5.3	5.4	4.5	4.2	4.1	4.3	4.2	4.9	5.3	4.9
MEAN NUMBER OF DAYS:														
Sunrise to Sunset														
-Clear	50	11.0	9.7	10.9	11.0	10.6	12.8	14.9	15.4	14.8	15.2	12.9	12.2	151.4
-Partly Cloudy	50	6.4	6.4	7.7	7.7	9.0	10.5	9.4	9.3	7.3	7.2	6.4	6.0	93.4
-Cloudy	50	13.6	12.2	12.4	11.4	11.4	6.7	6.7	6.2	7.9	8.6	10.7	12.8	120.5
Precipitation														
.01 inches or more	50	5.0	5.6	6.0	6.6	8.9	6.6	5.2	5.6	6.3	5.8	4.7	5.0	71.3
Snow, Ice pellets, hail														
1.0 inches or more	50	0.8	0.6	0.3	0.0	0.0	0.0	0.0	0.0	0.0	0.*	0.1	0.3	2.1
Thunderstorms	50	0.9	1.4	3.4	5.3	9.2	7.1	5.3	5.4	4.1	3.2	1.6	0.9	47.9
Heavy Fog Visibility														
1/4 mile or less	50	2.1	2.1	1.0	0.7	0.6	0.2	0.1	0.1	0.5	1.0	1.4	2.0	11.8
Temperature °F														
-Maximum														
90° and above	33	0.0	0.*	0.8	2.3	8.0	20.3	28.1	26.4	14.2	3.7	0.0	0.0	104.0
32° and below	33	3.4	1.6	0.1	0.0	0.0	0.0	0.0	0.0	0.0	0.0	0.1	1.6	6.8
-Minimum														
32° and below	33	21.3	14.0	6.1	0.6	0.0	0.0	0.0	0.0	0.0	0.3	6.8	17.4	66.6
0° and below	33	0.1	0.*	0.0	0.0	0.0	0.0	0.0	0.0	0.0	0.0	0.0	0.1	0.2
AVG. STATION PRESS. (mb)	21	983.5	981.8	977.6	976.9	975.7	976.6	978.3	978.5	979.4	980.7	981.1	982.8	979.4
RELATIVE HUMIDITY (%)														
Hour 00	33	73	73	70	72	77	75	66	68	76	74	75	75	73
Hour 06 (Local Time)	33	80	80	79	80	86	85	78	80	86	84	83	81	82
Hour 12	33	57	55	50	49	53	51	44	45	53	51	54	56	52
Hour 18	33	56	52	46	47	51	47	39	42	51	54	58	61	50
PRECIPITATION (inches):														
Water Equivalent														
-Normal		1.04	1.46	2.21	3.01	4.07	3.52	1.72	2.48	3.82	2.74	1.54	1.29	28.90
-Maximum Monthly	50	4.48	4.55	5.38	8.50	13.22	8.60	11.86	7.61	10.23	7.86	5.69	6.93	13.22
-Year		1968	1990	1990	1957	1982	1989	1950	1971	1980	1972	1957	1991	MAY 1982
-Minimum Monthly	50	0.00	T	T	0.32	0.01	0.26	T	T	T	T	0.00	0.02	0.00
-Year		1986	1991	1956	1989	1966	1980	1943	1943	1983	1952	1949	1950	JAN 1986
-Maximum in 24 hrs	50	2.02	3.00	4.32	4.09	5.70	5.36	3.93	4.62	6.22	5.61	2.58	2.98	6.22
-Year		1968	1981	1988	1967	1975	1985	1950	1971	1980	1959	1968	1991	SEP 1980
Snow, Ice pellets, hail														
-Maximum Monthly	50	11.9	11.8	10.9	0.8	T	**T**	**0.0**	**0.0**	**0.0**	**1.0**	3.9	7.1	11.9
-Year		1966	1978	1989	1973	1990	**1992**				**1993**	1957	1983	JAN 1966
-Maximum in 24 hrs	50	8.1	4.5	9.7	0.8	T	**T**	**0.0**	**0.0**	**0.0**	**1.0**	3.9	5.6	9.7
-Year		1985	1958	1989	1973	1990	**1992**				**1993**	1957	1983	MAR 1989
WIND:														
Mean Speed (mph)	45	11.2	11.9	13.2	13.1	12.2	12.0	11.1	10.3	10.4	10.7	11.4	11.2	11.5
Prevailing Direction through 1963		N	N	S	S	SSE	S	S	S	SE	S	S	S	S
Fastest Obs. 1 Min.														
-Direction (!!)	45	32	29	27	14	34	27	34	34	01	29	32	29	27
-Speed (MPH)	45	49	57	59	52	58	60	46	55	53	60	56	55	60
-Year		1949	1952	1950	1950	1949	1954	1958	1949	1973	1949	1949	1949	JUN 1954
Peak Gust														
-Direction (!!)	10	N	N	NW	W	**15**	NW	**18**	**10**	**W**	**01**	W	**33**	**15**
-Speed (mph)	10	59	61	61	66	**90**	76	**54**	**61**	**69**	**60**	59	**56**	**90**
-Date		1985	1984	1984	1984	**1993**	1989	**1993**	**1993**	**1992**	**1993**	1988	**1993**	**MAY 1993**

PRECIPITATION (inches)

WICHITA FALLS, TEXAS

YEAR	JAN	FEB	MAR	APR	MAY	JUNE	JULY	AUG	SEP	OCT	NOV	DEC	ANNUAL
1964	1.35	1.30	1.32	3.22	8.43	2.16	0.38	3.74	6.26	0.35	2.74	0.60	31.85
1965	1.73	0.83	0.66	2.45	5.92	3.88	0.35	2.29	2.58	3.11	0.01	0.62	24.43
1966	1.10	1.41	1.76	7.97	0.01	0.28	3.13	5.05	7.45	0.59	0.60	0.27	29.62
1967	0.03	0.20	1.30	5.84	6.50	1.84	3.18	0.22	3.93	1.98	0.36	1.44	26.82
1968	4.48	1.08	2.58	1.80	4.41	1.19	3.29	2.35	2.21	2.40	4.05	0.72	30.56
1969	0.57	2.82	3.18	1.87	5.04	3.11	0.34	3.95	4.61	3.16	0.54	2.42	31.61
1970	0.02	1.10	2.95	3.00	0.94	1.48	0.10	0.36	4.10	1.39	0.22	0.41	16.07
1971	0.54	1.37	0.12	0.74	1.06	0.63	1.83	7.61	5.13	4.21	0.91	3.37	27.52
1972	0.17	0.44	0.40	2.39	2.90	1.75	1.09	1.78	2.20	7.86	1.86	0.52	23.36
1973	2.74	1.08	3.89	3.51	0.74	5.17	4.51	0.05	5.31	1.93	1.68	0.08	30.69
1974	0.29	1.50	1.71	6.15	1.81	1.40	0.86	3.28	5.92	3.47	0.53	0.88	27.80
1975	0.91	1.71	1.57	2.94	9.66	1.69	3.41	3.46	3.65	0.82	1.31	1.86	32.99
1976	T	0.10	1.90	5.39	2.28	2.18	3.98	2.42	8.06	6.02	0.30	1.00	33.63
1977	1.09	1.28	2.08	2.51	3.78	2.53	1.49	4.10	0.39	0.69	0.93	0.06	20.93
1978	0.41	1.96	3.53	0.61	3.51	3.10	0.27	4.16	2.16	1.07	2.14	0.65	23.57
1979	2.04	0.71	3.43	2.71	4.69	6.07	1.74	3.57	0.06	1.16	2.20	2.00	30.38
1980	1.57	0.63	0.76	0.35	6.67	0.26	0.03	0.26	10.23	1.65	1.57	1.94	25.92
1981	0.14	3.30	1.86	3.95	3.54	4.59	1.33	2.29	1.49	7.83	0.86	0.30	31.48
1982	1.66	1.03	2.04	3.38	13.22	7.41	0.92	0.71	2.06	1.99	2.73	2.22	39.37
1983	1.87	0.90	2.06	2.19	3.21	5.05	0.19	0.19	T	7.79	0.91	0.89	25.25
1984	0.17	0.79	1.48	0.62	1.44	1.78	0.92	3.07	0.80	6.24	3.32	5.03	25.66
1985	1.08	2.61	3.77	6.15	1.69	7.07	0.21	1.75	1.46	3.69	1.11	0.11	30.70
1986	0.00	1.10	0.84	3.24	3.87	7.61	0.77	2.10	6.77	4.37	3.03	0.91	34.61
1987	1.78	4.16	1.89	0.32	10.17	3.74	1.47	2.43	2.20	0.11	1.66	4.84	34.77
1988	1.17	0.60	5.24	2.16	0.93	2.45	0.95	0.58	7.04	0.76	0.51	1.11	23.50
1989	1.04	3.50	1.54	0.32	4.54	8.60	3.19	6.17	5.01	2.25	0.03	0.28	36.47
1990	2.30	4.55	5.38	6.95	5.01	2.73	2.21	2.08	1.78	1.33	2.49	0.97	37.78
1991	2.66	T	0.68	0.99	4.24	2.76	1.91	2.75	8.28	3.92	0.81	6.93	35.93
1992	1.78	1.15	1.27	1.08	3.56	7.94	1.78	1.86	1.73	0.17	4.57	2.21	29.10
1993	1.19	3.37	2.80	3.21	5.25	3.21	0.72	1.21	2.83	2.09	0.58	2.64	29.10
Record Mean	1.14	1.39	1.84	2.68	4.22	3.46	1.94	2.10	3.10	2.80	1.47	1.49	27.64

AVERAGE TEMPERATURE (deg. F)

WICHITA FALLS, TEXAS

YEAR	JAN	FEB	MAR	APR	MAY	JUNE	JULY	AUG	SEP	OCT	NOV	DEC	ANNUAL
1964	43.6	42.9	53.2	66.8	73.1	80.4	88.7	86.0	75.1	63.0	54.3	42.1	64.1
1965	42.8	44.7	44.0	67.2	72.6	77.8	86.2	83.4	77.2	63.4	57.0	48.9	63.8
1966	34.8	42.1	57.1	61.7	71.8	81.3	88.2	81.3	72.8	63.2	58.0	41.4	62.8
1967	45.6	45.5	62.9	69.8	69.6	80.8	83.0	82.7	73.1	65.1	52.6	43.0	64.5
1968	42.2	41.0	53.7	61.3	69.7	79.6	82.7	83.5	73.7	66.4	50.7	42.9	62.3
1969	43.9	45.4	45.0	63.6	71.0	79.2	88.8	83.8	75.4	59.8	51.2	44.3	62.6
1970	36.1	46.0	47.6	63.2	71.6	79.7	84.8	85.0	76.3	60.6	49.0	47.7	62.3
1971	40.6	44.8	52.5	63.0	71.6	83.8	86.4	79.9	75.5	67.9	53.0	46.1	63.7
1972	39.4	45.2	56.8	67.0	70.0	82.0	83.3	82.3	77.1	63.4	45.8	39.9	62.7
1973	37.7	44.1	58.0	57.7	68.3	76.5	82.8	83.2	74.4	67.7	57.7	44.8	62.7
1974	40.5	49.9	60.6	64.5	76.2	79.9	86.6	81.3	68.0	65.4	51.7	43.1	64.0
1975	43.9	42.4	51.2	62.3	71.6	79.3	81.8	82.7	71.0	65.7	54.0	45.3	62.6
1976	42.5	55.3	55.9	64.2	67.5	78.0	81.0	84.2	73.9	57.6	46.4	41.1	62.3
1977	31.7	48.4	55.8	63.3	72.9	82.7	85.8	83.2	81.5	65.5	53.3	44.0	64.0
1978	30.1	33.3	52.8	66.5	72.1	80.8	89.9	82.5	79.8	65.5	53.3	40.3	62.3
1979	30.7	38.1	54.4	62.1	70.0	79.5	85.1	82.4	75.6	68.6	49.8	46.7	61.9
1980	42.8	42.4	49.9	61.1	70.9	84.8	91.9	88.8	77.9	62.8	51.2	46.5	64.3
1981	44.0	48.6	54.0	68.3	69.6	80.8	86.9	82.7	77.4	63.1	53.2	44.5	64.4
1982	40.8	40.3	54.9	59.3	71.5	76.9	82.3	85.7	76.5	65.2	52.4	45.1	62.6
1983	41.2	44.7	52.3	57.8	67.9	75.7	84.6	86.4	78.1	66.3	54.1	30.5	61.6
1984	37.2	47.1	51.4	60.1	72.7	82.9	84.8	84.4	73.9	63.3	51.9	46.2	63.0
1985	34.8	40.8	55.9	64.2	72.2	78.1	83.0	85.4	75.9	64.7	50.7	38.6	62.0
1986	45.7	47.6	58.4	65.6	71.1	79.6	86.9	82.5	78.1	63.7	49.0	43.3	64.3
1987	39.2	48.7	51.2	62.1	73.1	78.3	82.2	85.1	74.4	63.8	53.2	42.8	62.8
1988	38.2	43.1	53.5	61.4	72.1	79.5	84.7	86.2	75.5	63.5	54.7	45.8	63.2
1989	46.4	37.2	53.8	65.4	72.5	76.1	82.9	81.0	72.0	66.2	54.6	35.0	61.9
1990	47.9	49.4	55.4	61.6	71.7	84.5	83.0	84.1	79.1	64.1	56.5	39.2	64.7
1991	39.3	51.5	57.0	65.4	73.9	79.8	85.1	81.1	72.0	65.2	48.3	46.8	63.8
1992	43.2	51.3	56.6	63.4	69.4	78.3	83.0	78.9	75.4	66.8	49.9	45.2	63.5
1993	40.9	43.4	52.5	60.5	68.1	78.7	86.1	84.0	73.7		47.2	45.2	
Record Mean	41.2	46.1	54.2	64.0	71.8	80.7	85.2	84.7	76.6	66.0	52.6	44.0	63.9
Max	52.8	58.2	67.1	76.8	83.8	92.7	97.6	97.2	88.7	78.6	64.7	55.4	76.1
Min	29.5	34.0	41.2	51.1	59.8	68.7	72.8	72.1	64.5	53.4	40.5	32.5	51.7

REFERENCE NOTES

WICHITA FALLS, TEXAS

GENERAL
T - TRACE AMOUNT.
BLANK ENTRIES DENOTE MISSING/UNREPORTED DATA.
# INDICATES A STATION OR INSTRUMENT RELOCATION.

SPECIFIC

EXCEPTIONS

RECORD MEANS ARE THROUGH THE CURRENT YEAR,
BEGINNING IN 1924 FOR TEMPERATURE
1924 FOR PRECIPITATION
1944 FOR SNOWFALL

(a) - LENGTH OF RECORD IN YEARS, ALTHOUGH INDIVIDUAL MONTHS MAY BE MISSING.
0.* OR * - THE VALUE IS BETWEEN 0.0 AND 0.05
NORMALS - BASED ON THE 1961-1990 RECORD PERIOD.
EXTREMES - DATES ARE THE MOST RECENT OCCURRENCE
WIND DIR. - NUMERALS SHOW TENS OF DEGRESS CLOCKWISE FROM TRUE NORTH. "00" INDICATES CALM.
RESULTANT DIRECTIONS ARE GIVEN TO WHOLE DEGREES.
BOLD VALUES INDICATE EXTREME VALUES WHICH OCCURRED AFTER THE ASOS SYSTEM WAS COMMISSIONED.

RECORD = PERIOD OF RECORD
RECORD MEAN PRECIPITATION IS THE MEAN OF ALL DAILY PRECIPITATION AMOUNTS DURING THE PERIOD OF RECORD.
RECORD MAX(MIN) TEMPERATURE IS THE MEAN OF ALL DAILY MAX(MIN) TEMPERATURES DURING THE PERIOD OF RECORD.
RECORD MEAN TEMPERATURE IS THE SUM OF THE RECORD MAX AND RECORD MIN DIVIDED BY 2.
AVERAGE TEMPERATURE IS THE SUM OF THE MEAN DAILY MAX AND MIN TEMPERATURE DIVIDED BY 2.

HEATING DEGREE DAYS Base 65 deg. F WICHITA FALLS, TEXAS

SEASON	JULY	AUG	SEP	OCT	NOV	DEC	JAN	FEB	MAR	APR	MAY	JUNE	TOTAL
1964-65	0	0	12	86	352	706	679	563	647	47	2	0	3094
1965-66	0	0	13	120	241	494	929	633	275	139	33	0	2877
1966-67	0	2	0	119	247	723	593	538	167	30	41	0	2460
1967-68	0	0	14	93	366	674	699	687	357	133	23	0	3046
1968-69	0	0	0	68	431	676	646	543	609	91	14	4	3082
1969-70	0	0	1	235	411	634	891	526	534	131	19	4	3386
1970-71	0	0	20	205	474	529	749	560	395	131	15	0	3078
1971-72	0	0	37	25	364	578	787	574	273	77	12	0	2727
1972-73	0	0	8	193	571	771	839	577	213	234	47	0	3453
1973-74	0	0	10	42	248	621	752	418	212	88	1	0	2392
1974-75	0	0	36	54	398	673	644	628	430	157	0	0	3020
1975-76	0	0	52	87	338	606	690	295	318	73	40	0	2499
1976-77	0	0	7	265	552	731	1023	459	281	80	1	0	3399
1977-78	0	0	0	64	345	645	1078	881	394	73	50	0	3530
1978-79	0	0	1	76	364	758	1056	750	343	130	47	0	3525
1979-80	0	0	0	57	464	563	682	649	460	162	18	0	3055
1980-81	0	0	32	139	416	567	644	455	339	50	33	0	2675
1981-82	0	0	11	144	355	629	745	689	331	220	25	0	3149
1982-83	0	0	4	116	387	612	732	564	387	259	43	4	3108
1983-84	0	0	13	60	342	1059	855	509	420	181	17	0	3456
1984-85	0	0	65	129	400	582	930	673	301	67	1	0	3148
1985-86	0	0	31	83	431	812	592	482	226	76	12	0	2745
1986-87	0	0	0	98	473	665	791	452	419	162	1	0	3061
1987-88	0	0	0	92	370	682	824	630	376	138	7	0	3119
1988-89	0	0	3	89	318	587	568	772	391	123	16	0	2867
1989-90	0	0	41	100	318	923	525	428	316	154	44	0	2849
1990-91	0	0	3	120	267	793	791	376	265	63	15	0	2693
1991-92	0	0	25	116	496	558	671	391	262	107	35	0	2661
1992-93	0	0	1	47	457	614	737	597	394	169	30	0	3046
1993-94	0	0	15		530	607							

COOLING DEGREE DAYS Base 65 deg. F WICHITA FALLS, TEXAS

YEAR	JAN	FEB	MAR	APR	MAY	JUNE	JULY	AUG	SEP	OCT	NOV	DEC	TOTAL
1969	0	0	0	57	206	438	742	591	320	79	3	0	2436
1970	0	0	4	85	233	452	622	627	365	76	3	0	2467
1971	0	0	14	78	226	573	672	470	357	124	11	0	2525
1972	0	4	26	145	174	517	572	543	377	147	2	0	2507
1973	0	0	3	21	155	351	559	570	298	132	35	0	2124
1974	0	0	85	80	356	454	676	512	135	73	6	0	2377
1975	0	0	10	81	210	436	529	558	237	116	13	0	2190
1976	0	21	39	54	124	395	500	601	282	44	0	0	2060
1977	0	2	2	34	254	537	652	570	502	88	3	0	2644
1978	0	0	27	128	281	481	777	550	449	98	17	0	2808
1979	0	2	18	49	208	444	628	551	324	176	14	1	2415
1980	0	0	0	52	204	603	843	745	427	80	7	2	2963
1981	0	5	6	159	179	480	687	558	389	90	5	0	2558
1982	0	3	24	59	234	364	542	648	354	130	18	3	2379
1983	0	0	1	49	140	331	614	672	411	107	18	0	2343
1984	0	0	2	42	262	545	620	609	338	87	13	3	2521
1985	0	0	27	53	230	401	562	638	365	81	11	0	2368
1986	0	2	30	102	204	448	686	552	399	66	0	0	2489
1987	0	0	0	80	258	404	542	629	288	63	21	0	2285
1988	0	0	26	37	234	443	620	664	322	51	15	0	2412
1989	0	0	50	138	253	336	561	506	258	143	12	0	2257
1990	0	0	27	59	259	593	567	596	435	98	17	0	2651
1991	0	1	24	78	296	454	627	506	242	128	3	1	2360
1992	0	0	7	68	178	405	565	436	321	112	11	6	2109
1993	0	0	13	42	137	415	663	595	285		3	0	

SNOWFALL (inches) WICHITA FALLS, TEXAS

SEASON	JULY	AUG	SEP	OCT	NOV	DEC	JAN	FEB	MAR	APR	MAY	JUNE	TOTAL
1964-65	0.0	0.0	0.0	0.0	0.0	T	T	T	1.4	0.0	0.0	0.0	1.4
1965-66	0.0	0.0	0.0	0.0	0.0	T	11.9	0.8	0.0	0.0	0.0	0.0	12.7
1966-67	0.0	0.0	0.0	0.0	0.0	T	0.0	T	0.6	0.0	0.0	0.0	0.6
1967-68	0.0	0.0	0.0	T	T	T	T	8.6	3.8	0.0	0.0	0.0	12.4
1968-69	0.0	0.0	0.0	0.0	T	T	T	1.7	3.0	0.0	0.0	0.0	4.7
1969-70	0.0	0.0	0.0	0.0	T	5.4	T	T	2.3	T	0.0	0.0	7.7
1970-71	0.0	0.0	0.0	0.0	0.0	0.0	0.0	T	3.0	0.0	0.0	0.0	3.0
1971-72	0.0	0.0	0.0	0.0	T	4.2	0.4	1.9	0.0	0.0	0.0	0.0	6.5
1972-73	0.0	0.0	0.0	0.0	2.1	T	3.1	4.6	0.0	0.8	0.0	0.0	10.6
1973-74	0.0	0.0	0.0	0.0	0.0	0.5	T	T	0.0	0.0	0.0	0.0	0.5
1974-75	0.0	0.0	0.0	0.0	0.6	0.0	T	1.9	0.2	0.0	0.0	0.0	2.7
1975-76	0.0	0.0	0.0	0.0	0.5	2.3	T	T	0.0	0.0	0.0	0.0	2.8
1976-77	0.0	0.0	0.0	0.0	3.7	T	5.9	0.0	0.0	0.0	0.0	0.0	9.6
1977-78	0.0	0.0	0.0	0.0	T	0.0	1.9	11.8	T	0.0	0.0	0.0	13.7
1978-79	0.0	0.0	0.0	0.0	0.0	3.0	6.0	1.6	0.0	0.0	0.0	0.0	10.6
1979-80	0.0	0.0	0.0	0.0	0.0	T	T	3.6	T	0.0	0.0	0.0	3.6
1980-81	0.0	0.0	0.0	0.0	2.7	0.0	T	T	0.0	0.0	0.0	0.0	2.7
1981-82	0.0	0.0	0.0	0.0	0.0	0.0	1.3	3.2	T	0.0	0.0	0.0	4.5
1982-83	0.0	0.0	0.0	0.0	T	1.1	7.3	T	0.0	0.0	0.0	0.0	8.4
1983-84	0.0	0.0	0.0	0.0	0.0	7.1	0.3	0.0	0.0	0.0	0.0	0.0	7.4
1984-85	0.0	0.0	0.0	0.0	0.0	0.4	8.7	0.2	0.0	0.0	0.0	0.0	9.3
1985-86	0.0	0.0	0.0	0.0	0.0	0.9	0.0	2.3	0.0	0.0	0.0	0.0	3.2
1986-87	0.0	0.0	0.0	0.0	0.0	T	2.4	T	0.0	0.0	0.0	0.0	2.4
1987-88	0.0	0.0	0.0	0.0	0.0	4.1	4.9	3.0	T	0.0	0.0	0.0	12.0
1988-89	0.0	0.0	0.0	0.0	T	T	2.3	T	10.9	0.0	0.0	0.0	13.2
1989-90	0.0	0.0	0.0	T	0.0	T	T	0.0	T	T	T	0.0	T
1990-91	0.0	0.0	0.0	0.0	0.0	1.1	T	0.0	T	0.0	0.0	0.0	1.1
1991-92	0.0	0.0	0.0	T	T	T	8.4	T	0.0	0.0	0.0	T	8.4
1992-93	0.0	0.0	0.0	T	T	T	T	T	T	T			
1993-94	0.0	0.0	0.0	1.0	0.0								
Record Mean	0.0	0.0	0.0	T	0.3	1.0	2.1	1.6	0.8	T	T	T	5.8

SALT LAKE CITY, UTAH

Salt Lake City is located in a northern Utah valley surrounded by mountains on three sides and the Great Salt Lake to the northwest. The city varies in altitude from near 4,200 to 5,000 feet above sea level.

The Wasatch Mountains to the east have peaks to nearly 12,000 feet above sea level. Their orographic effects cause more precipitation in the eastern part of the city than over the western part.

The Oquirrh Mountains to the southwest of the city have several peaks to above 10,000 feet above sea level. The Traverse Mountain Range at the south end of the Salt Lake Valley rises to above 6,000 feet above sea level. These mountain ranges help to shelter the valleys from storms from the southwest in the winter, but are instrumental in developing thunderstorms which can drift over the valley in the summer.

Besides the mountain ranges, the most influential natural condition affecting the climate of Salt Lake City is the Great Salt Lake. This large inland body of water, which never freezes over due to its high salt content, can moderate the temperatures of cold winter winds blowing from the northwest and helps drive a lake/valley wind system. The warmer lake water during the winter and spring also contributes to increased precipitation in the valley downwind from the lake. The combination of the Great Salt Lake and the Wasatch Mountains often enhances storm precipitation in the valley.

Salt Lake City normally has a semi-arid continental climate with four well-defined seasons. Summers are characterized by hot, dry weather, but the high temperatures are usually not oppressive since the relative humidity is generally low and the nights usually cool. July is the hottest month with temperature readings in the 90s.

The mean diurnal temperature range is about 30 degrees in the summer and 18 degrees during the winter. Temperatures above 102 degrees in the summer or colder than -10 degrees in the winter are likely to occur one season out of four.

Winters are cold, but usually not severe. Mountains to the north and east act as a barrier to frequent invasions of cold continental air. The average annual snowfall is under 60 inches at the airport but much higher amounts fall in higher bench locations. Heavy fog can develop under temperature inversions in the winter and persist for several days.

Precipitation, generally light during the summer and early fall, is heavy in the spring when storms from the Pacific Ocean are moving through the area more frequently than at any other season of the year.

Winds are usually light, although occasional high winds have occurred in every month of the year, particularly in March.

The growing season is over five months in length. Yard and garden foilage generally are making good growth by mid-April. The last freezing temperature in the spring averages late April and the first freeze of the fall is mid-October.

NORMALS, MEANS, AND EXTREMES

SALT LAKE CITY, UTAH

LATITUDE: 40°47'N LONGITUDE: 111°57'W ELEVATION: FT. GRND 4221 BARO 4224 TIME ZONE: MOUNTAIN WBAN: 24127

	(a)	JAN	FEB	MAR	APR	MAY	JUNE	JULY	AUG	SEP	OCT	NOV	DEC	YEAR
TEMPERATURE °F:														
Normals														
-Daily Maximum		36.4	43.6	52.2	61.3	71.9	82.8	92.2	89.4	79.2	66.1	50.8	37.8	63.6
-Daily Minimum		19.3	24.6	31.4	37.9	45.6	55.4	63.7	61.8	51.0	40.2	30.9	21.6	40.3
-Monthly		27.9	34.1	41.8	49.7	58.8	69.1	77.9	75.6	65.2	53.2	40.8	29.7	52.0
Extremes														
-Record Highest	65	62	69	78	86	93	104	107	104	100	89	75	67	107
-Year		1982	1972	1960	1992	1984	1979	1960	1979	1979	1963	1967	1969	JUL 1960
-Record Lowest	65	-22	-30	2	14	25	35	40	37	27	16	-14	-21	-30
-Year		1949	1933	1966	1936	1965	1962	1968	1965	1965	1971	1955	1932	FEB 1933
NORMAL DEGREE DAYS:														
Heating (base 65°F)		1150	865	719	464	215	51	0	0	108	373	726	1094	5765
Cooling (base 65°F)		0	0	0	0	23	174	400	329	114	7	0	0	1047
% OF POSSIBLE SUNSHINE	55	45	54	63	68	72	80	83	82	82	72	53	42	66
MEAN SKY COVER (tenths)														
Sunrise - Sunset	58	7.3	7.1	6.7	6.4	5.7	4.3	3.6	3.7	3.6	4.6	6.3	7.2	5.6
MEAN NUMBER OF DAYS:														
Sunrise to Sunset														
-Clear	65	5.6	5.2	6.9	6.8	9.1	13.7	16.8	15.8	16.5	13.9	8.5	6.3	125.0
-Partly Cloudy	65	6.6	6.9	8.2	9.4	10.2	9.9	9.7	10.7	8.3	7.8	7.2	6.5	101.2
-Cloudy	65	18.9	16.1	15.9	13.8	11.7	6.4	4.5	4.6	5.2	9.3	14.4	18.2	139.0
Precipitation														
.01 inches or more	65	10.0	8.8	9.8	9.4	8.2	5.4	4.5	5.7	5.3	6.3	7.8	9.1	90.6
Snow, Ice pellets, hail														
1.0 inches or more	65	4.1	3.2	2.8	1.4	0.2	0.0	0.0	0.0	0.*	0.4	2.0	3.8	17.8
Thunderstorms	65	0.3	0.7	1.4	2.2	5.2	5.4	6.8	7.8	4.2	1.9	0.5	0.3	36.6
Heavy Fog Visibility														
1/4 mile or less	65	4.5	2.3	0.3	0.1	0.*	0.0	0.0	0.0	0.0	0.*	0.9	3.6	11.9
Temperature °F														
-Maximum														
90° and above	34	0.0	0.0	0.0	0.0	0.6	8.9	23.1	19.1	3.7	0.0	0.0	0.0	55.5
32° and below	34	10.9	3.7	0.6	0.0	0.0	0.0	0.0	0.0	0.0	0.*	0.7	8.7	24.7
-Minimum														
32° and below	34	27.7	22.7	15.6	6.3	0.7	0.0	0.0	0.0	0.3	4.7	18.2	27.6	123.9
0° and below	34	1.7	0.4	0.0	0.0	0.0	0.0	0.0	0.0	0.0	0.0	0.0	0.8	2.9
AVG. STATION PRESS. (mb)	21	874.8	873.4	869.6	869.7	869.0	869.9	871.2	871.6	872.2	873.6	873.6	874.9	872.0
RELATIVE HUMIDITY (%)														
Hour 05	34	79	78	71	67	65	60	52	55	61	68	75	79	68
Hour 11 (Local Time)	34	71	64	52	44	39	31	27	29	35	43	58	70	47
Hour 17	34	70	60	47	39	33	26	22	23	29	41	59	71	43
Hour 23	34	79	77	68	61	57	50	43	45	54	66	74	78	63
PRECIPITATION (inches):														
Water Equivalent														
-Normal		1.11	1.23	1.91	2.12	1.80	0.93	0.81	0.86	1.28	1.44	1.29	1.40	16.18
-Maximum Monthly	65	3.23	3.22	3.97	4.90	4.76	2.93	2.57	3.66	7.04	3.91	2.63	4.37	7.04
-Year		1993	1936	1983	1944	1977	1947	1982	1968	1982	1981	1985	1983	SEP 1982
-Minimum Monthly	65	0.09	0.12	0.10	0.45	T	0.01	T	T	T	0.00	0.01	0.08	0.00
-Year		1961	1946	1956	1981	1934	1946	1963	1944	1951	1952	1939	1976	OCT 1952
-Maximum in 24 hrs	65	1.36	1.05	1.83	2.41	2.03	1.88	2.35	1.96	2.30	1.76	1.13	1.82	2.41
-Year		1953	1958	1944	1957	1942	1948	1962	1932	1982	1984	1954	1972	APR 1957
Snow, Ice pellets, hail														
-Maximum Monthly	65	50.3	27.9	41.9	26.4	7.5	T	T	T	4.0	20.4	27.2	35.2	50.3
-Year		1993	1969	1977	1974	1975	1993	1991	1993	1971	1984	1985	1972	JAN 1993
-Maximum in 24 hrs	65	10.7	11.9	15.4	16.2	6.4	T	T	T	4.0	18.4	11.0	18.1	18.4
-Year		1980	1989	1944	1974	1975	1993	1991	1993	1971	1984	1930	1972	OCT 1984
WIND:														
Mean Speed (mph)	64	7.5	8.1	9.3	9.6	9.5	9.4	9.5	9.7	9.1	8.5	7.9	7.5	8.8
Prevailing Direction through 1963		SSE	SE	SSE	SE	SE	SSE	SSE	SSE	SE	SE	SSE	SSE	SSE
Fastest Mile														
-Direction (!!!)	58	NW	SE	NW	NW	NW	W	NW	SW	W	NW	NW	S	NW
-Speed (MPH)	58	59	56	71	57	57	63	51	58	61	67	63	54	71
-Year		1980	1954	1954	1964	1953	1963	1986	1946	1952	1950	1937	1955	MAR 1954
Peak Gust														
-Direction (!!!)	10	N	S	NW	NW	SW	SW	W	SW	NW	NW	SW	S	SW
-Speed (mph)	10	59	54	59	54	69	58	59	67	61	63	54	49	69
-Date		1988	1989	1989	1984	1989	1993	1991	1989	1992	1985	1992	1992	MAY 1989

PRECIPITATION (inches)

SALT LAKE CITY, UTAH

YEAR	JAN	FEB	MAR	APR	MAY	JUNE	JULY	AUG	SEP	OCT	NOV	DEC	ANNUAL
1964	0.94	0.35	2.26	2.69	2.77	2.61	0.26	0.17	0.13	0.45	1.42	3.82	17.87
1965	2.13	1.13	0.14	2.30	2.02	1.87	1.50	2.08	1.93	0.39	1.13	1.81	18.43
1966	0.41	1.19	1.21	1.43	0.51	0.07	0.33	0.22	0.83	1.18	0.75	0.86	8.99
1967	2.05	0.67	1.94	2.08	2.15	2.73	1.14	0.07	0.73	0.66	0.66	1.64	16.52
1968	0.46	2.32	2.21	2.82	2.18	1.58	0.09	3.66	0.56	1.64	1.32	2.27	21.11
1969	1.69	2.84	0.57	1.38	0.18	2.83	1.51	0.34	0.18	1.96	0.92	1.69	16.09
1970	1.24	0.94	1.01	3.25	0.89	1.63	0.86	0.57	2.80	1.61	2.27	2.80	19.87
1971	1.06	2.13	1.01	2.16	1.34	0.64	0.94	2.15	1.75	3.23	1.03	1.35	18.79
1972	1.22	0.48	1.18	3.62	0.14	0.15	0.06	0.21	1.36	2.74	1.36	3.22	15.74
1973	1.49	0.91	2.67	1.64	1.74	0.19	1.07	1.16	4.07	0.67	2.52	2.26	20.39
1974	1.80	1.65	0.97	4.57	0.39	0.28	0.18	0.32	0.03	2.03	0.90	1.34	14.46
1975	1.28	1.24	3.44	2.46	2.58	1.81	0.28	0.10	0.08	1.91	1.71	1.03	17.92
1976	0.63	1.90	1.90	2.47	0.99	1.24	1.55	0.82	0.16	0.57	0.03	0.08	12.34
1977	0.76	0.64	3.10	0.59	4.76	0.06	0.61	1.85	1.85	0.83	1.20	1.42	17.67
1978	2.33	1.96	3.47	2.90	1.57	0.06	0.06	0.92	2.51	T	1.73	0.58	18.09
1979	0.72	1.05	0.80	1.04	0.84	0.35	0.40	0.63	0.05	1.29	0.98	0.55	8.70
1980	2.87	2.25	2.46	0.89	2.70	0.42	1.34	0.26	0.72	1.74	1.17	0.37	17.19
1981	0.64	0.81	2.11	0.45	3.68	1.03	0.33	0.23	0.48	3.91	1.03	1.89	16.59
1982	1.08	0.53	2.39	1.63	1.86	0.66	2.57	0.56	7.04	1.87	0.75	1.92	22.86
1983	1.19	1.36	3.97	1.63	2.58	0.62	1.02	2.64	1.03	1.62	2.23	4.37	24.26
1984	0.50	0.95	1.76	4.43	1.17	1.86	1.72	1.49	1.72	3.70	1.45	0.80	21.55
1985	0.91	0.85	1.80	0.64	2.95	1.30	0.85	0.03	1.98	1.61	2.63	1.42	16.97
1986	0.86	1.28	2.32	4.55	3.39	0.42	0.85	1.32	2.75	0.39	1.17	0.10	19.40
1987	1.53	1.41	1.52	0.79	2.41	0.19	0.79	0.36	0.05	1.18	1.17	1.10	12.50
1988	1.06	0.13	0.94	1.84	2.16	0.03	0.04	0.22	0.07	0.01	2.17	0.62	9.29
1989	0.56	1.57	1.77	0.46	1.83	0.22	0.39	0.90	0.49	1.82	0.73	0.13	10.87
1990	0.57	0.35	2.17	1.14	1.65	0.66	0.64	0.46	0.56	0.69	1.24	0.56	10.69
1991	1.11	0.61	1.11	2.71	2.76	1.09	0.32	0.86	2.55	2.10	2.17	0.40	17.79
1992	0.78	1.24	1.11	0.96	1.86	0.45	0.29	0.35	0.47	1.03	2.46	1.07	12.07
1993	3.23	1.35	1.37		3.99	1.14	1.38	0.46	0.22	2.77	0.54	0.88	
Record Mean	1.28	1.32	1.84	1.98	1.82	0.87	0.63	0.86	0.95	1.43	1.37	1.34	15.69

AVERAGE TEMPERATURE (deg. F)

SALT LAKE CITY, UTAH

YEAR	JAN	FEB	MAR	APR	MAY	JUNE	JULY	AUG	SEP	OCT	NOV	DEC	ANNUAL
1964	21.9	25.8	32.0	45.6	55.8	63.2	77.5	71.9	61.5	53.0	37.8	33.3	48.3
1965	31.0	33.0	36.8	51.0	54.7	64.8	75.0	70.9	57.5	54.5	46.1	30.2	50.5
1966	30.6	29.4	41.6	49.6	62.7	69.2	80.1	74.1	67.5	49.9	43.1	29.2	52.3
1967	29.4	37.5	44.2	46.1	56.3	64.6	78.4	78.6	66.7	52.4	43.0	25.1	51.8
1968	24.5	38.2	44.7	45.4	56.4	67.5	78.3	69.4	61.4	51.7	38.5	26.8	50.2
1969	32.2	28.7	38.4	50.4	64.0	64.8	76.6	77.6	69.7	47.7	39.5	32.4	51.9
1970	34.6	40.4	40.6	44.2	58.8	67.6	76.6	77.7	59.0	47.1	42.6	29.2	51.5
1971	32.4	34.9	40.4	48.2	56.6	67.5	76.4	76.9	59.8	47.5	37.6	26.9	50.4
1972	29.8	37.8	46.9	48.1	60.5	71.9	77.2	75.8	63.9	53.6	39.4	22.7	52.3
1973	19.6	32.3	41.8	47.6	61.6	70.2	76.6	76.6	61.5	54.1	40.5	33.4	51.3
1974	26.7	31.4	45.2	48.1	58.8	73.4	79.2	74.2	66.5	54.7	43.4	31.7	52.8
1975	27.4	35.5	41.1	44.3	54.3	64.8	78.8	73.4	65.4	53.4	37.3	32.9	50.7
1976	27.9	34.1	38.1	49.3	62.2	67.6	78.7	72.3	66.4	51.0	41.8	29.4	51.6
1977	26.8	35.8	37.7	54.1	55.0	73.2	77.3	75.0	66.4	55.6	42.5	37.9	53.2
1978	36.3	39.8	48.0	50.2	56.0	69.2	78.0	74.0	64.0	55.5	41.0	26.8	53.3
1979	22.1	32.5	43.2	51.1	60.2	70.1	78.9	75.6	71.4	56.7	36.5	32.9	52.6
1980	33.7	36.0	41.5	52.7	57.0	67.5	77.6	74.1	66.3	52.6	41.3	33.6	52.8
1981	32.1	38.3	44.1	53.4	57.6	69.6	78.2	78.0	68.5	50.5	44.3	36.4	54.3
1982	29.8	32.3	43.3	46.5	56.7	68.0	75.5	78.4	64.0	48.8	38.1	29.9	51.0
1983	35.2	39.4	44.6	45.9	55.8	67.7	76.6	77.8	67.8	56.0	43.0	31.9	53.5
1984	23.8	25.8	40.1	48.5	61.6	67.3	78.5	77.2	66.5	49.5	42.7	29.9	51.0
1985	24.2	25.6	40.8	55.7	63.9	72.5	80.7	76.5	62.7	53.1	37.4	27.7	51.7
1986	29.0	41.4	47.7	48.8	57.2	73.5	74.2	77.9	60.2	51.3	40.9	29.8	52.7
1987	26.5	36.1	42.8	55.9	62.7	71.6	75.7	74.7	66.5	56.4	40.8	30.5	53.4
1988	25.0	34.8	41.4	52.0	59.6	75.7	80.9	76.5	63.8	60.0	41.1	28.1	53.2
1989	22.3	25.3	45.8	54.8	59.9	69.2	81.1	75.1	66.4	53.4	40.5	31.4	52.1
1990	33.4	32.8	45.0	54.9	57.8	72.0	78.9	76.2	72.0	54.0	41.4	21.0	53.3
1991	24.4	36.5	42.8	47.9	55.6	68.3	79.1	78.4	64.4	53.6	40.0	30.0	51.8
1992	25.6	39.3	49.3	57.1	65.6	70.4	75.4	77.3	66.5	56.0	34.1	27.1	53.6
1993	24.9	29.5	45.5	48.5	63.4	63.7	69.9	72.5	65.5	52.6	34.8	31.5	50.2
Record Mean	28.0	33.3	41.1	49.4	58.4	68.4	77.3	75.5	65.3	53.2	40.4	31.2	51.8
Max	36.1	41.7	50.7	60.3	70.5	81.8	91.2	89.1	78.6	65.1	50.0	39.1	62.8
Min	19.9	24.9	31.5	38.5	46.3	54.9	63.3	61.9	51.9	41.3	30.8	23.3	40.7

REFERENCE NOTES

SALT LAKE CITY, UTAH

GENERAL
T - TRACE AMOUNT.
BLANK ENTRIES DENOTE MISSING/UNREPORTED DATA.
# INDICATES A STATION OR INSTRUMENT RELOCATION.

SPECIFIC

PM - INCLUDES LAST DAY OF PREVIOUS MONTH
ASOS - AUTOMATED SURFACE OBSERVING SYSTEM IN OPERATION DURING THESE MONTHS.

(a) - LENGTH OF RECORD IN YEARS, ALTHOUGH INDIVIDUAL MONTHS MAY BE MISSING.
0.* OR * - THE VALUE IS BETWEEN 0.0 AND 0.05
NORMALS - BASED ON THE 1961-1990 RECORD PERIOD.
EXTREMES - DATES ARE THE MOST RECENT OCCURRENCE
WIND DIR. - NUMERALS SHOW TENS OF DEGRESS CLOCKWISE FROM TRUE NORTH. "00" INDICATES CALM.
RESULTANT DIRECTIONS ARE GIVEN TO WHOLE DEGREES.
BOLD VALUES INDICATE EXTREME VALUES WHICH OCCURRED AFTER THE ASOS SYSTEM WAS COMMISSIONED.

RECORD = PERIOD OF RECORD
RECORD MEAN PRECIPITATION IS THE MEAN OF ALL DAILY PRECIPITATION AMOUNTS DURING THE PERIOD OF RECORD.
RECORD MAX(MIN) TEMPERATURE IS THE MEAN OF ALL DAILY MAX(MIN) TEMPERATURES DURING THE PERIOD OF RECORD.
RECORD MEAN TEMPERATURE IS THE SUM OF THE RECORD MAX AND RECORD MIN DIVIDED BY 2.
AVERAGE TEMPERATURE IS THE SUM OF THE MEAN DAILY MAX AND MIN TEMPERATURE DIVIDED BY 2.

EXCEPTIONS

RECORD MEANS ARE THROUGH THE CURRENT YEAR, BEGINNING IN 1874 FOR TEMPERATURE
1874 FOR PRECIPITATION
1929 FOR SNOWFALL

HEATING DEGREE DAYS Base 65 deg. F — SALT LAKE CITY, UTAH

SEASON	JULY	AUG	SEP	OCT	NOV	DEC	JAN	FEB	MAR	APR	MAY	JUNE	TOTAL
1964-65	0	44	134	365	808	975	1046	889	869	414	316	61	5921
1965-66	0	20	239	317	564	1069	1058	989	717	456	140	40	5609
1966-67	0	4	57	460	649	1101	1097	763	638	564	287	76	5696
1967-68	0	0	57	387	653	1228	1246	772	622	583	276	57	5881
1968-69	3	49	166	407	786	1174	1009	1010	818	433	75	67	5997
1969-70	1	0	17	530	759	1003	935	681	754	619	218	69	5586
1970-71	0	0	218	550	667	1103	1002	836	754	499	258	55	5942
1971-72	0	0	201	535	817	1176	1085	783	556	499	168	2	5822
1972-73	0	0	110	347	761	1307	1400	909	711	515	135	67	6262
1973-74	1	0	140	333	732	975	1181	935	603	502	214	41	5657
1974-75	0	5	54	316	638	1025	1157	819	734	613	334	92	5787
1975-76	0	1	62	365	825	989	1144	890	826	464	112	67	5745
1976-77	0	7	37	432	689	1096	1175	813	838	333	304	0	5724
1977-78	0	11	73	282	670	835	880	697	522	433	293	36	4732
1978-79	0	12	144	284	714	1178	1327	902	666	414	196	57	5894
1979-80	0	0	7	270	846	987	964	835	723	371	250	77	5330
1980-81	0	10	57	379	704	965	1013	742	641	346	233	46	5136
1981-82	0	0	34	444	614	879	1087	909	668	548	259	62	5504
1982-83	7	0	134	495	800	1080	916	710	624	569	314	36	5685
1983-84	6	0	49	276	650	1018	1269	1130	763	493	157	76	5887
1984-85	0	0	98	480	662	1084	1260	1097	740	285	109	17	5832
1985-86	0	0	140	360	821	1151	1110	655	527	477	283	14	5538
1986-87	6	0	203	416	720	1085	1186	803	679	291	123	17	5529
1987-88	0	0	51	260	719	1060	1235	870	723	381	222	3	5524
1988-89	0	0	142	158	711	1138	1318	1105	587	313	193	35	5700
1989-90	0	15	44	355	729	1036	971	895	612	297	232	30	5216
1990-91	0	0	17	347	704	1359	1249	792	682	508	289	32	5979
1991-92	0	0	78	347	742	1078	1220	741	481	247	57	33	5024
1992-93	11	14	42	285	921	1169	1237	986	596	489	151	108	6009
1993-94	19	3	82	391	902	1031							

COOLING DEGREE DAYS Base 65 deg. F — SALT LAKE CITY, UTAH

YEAR	JAN	FEB	MAR	APR	MAY	JUNE	JULY	AUG	SEP	OCT	NOV	DEC	TOTAL
1969	0	0	0	1	53	68	366	398	164	0	0	0	1050
1970	0	0	0	0	32	152	365	398	46	0	0	0	993
1971	0	0	0	0	5	136	361	374	50	0	0	0	926
1972	0	0	0	0	34	213	386	340	85	0	0	0	1058
1973	0	0	0	0	38	226	370	367	44	3	0	0	1048
1974	0	0	0	2	31	303	446	298	108	3	0	0	1191
1975	0	0	0	0	9	89	439	269	80	14	0	0	900
1976	0	0	0	0	34	151	431	237	87	3	0	0	943
1977	0	0	0	12	2	254	389	328	123	0	0	0	1108
1978	0	0	0	0	21	167	411	299	120	0	0	0	1018
1979	0	0	0	2	54	214	439	336	208	21	0	0	1274
1980	0	0	0	9	10	159	399	301	99	1	0	0	978
1981	0	0	0	3	12	190	412	409	145	2	0	0	1173
1982	0	0	0	0	11	158	338	423	109	0	0	0	1039
1983	0	0	0	0	37	123	370	405	138	4	0	0	1077
1984	0	0	0	3	58	153	426	383	147	4	0	0	1174
1985	0	0	0	11	78	249	493	364	79	0	0	0	1274
1986	0	0	0	0	47	277	296	407	66	0	0	0	1093
1987	0	0	0	25	60	222	338	309	103	0	0	0	1057
1988	0	0	0	0	61	334	501	363	112	9	0	0	1380
1989	0	0	0	13	43	171	506	337	92	0	0	0	1162
1990	0	0	0	2	19	247	438	351	235	11	0	0	1303
1991	0	0	0	1	3	138	442	425	63	4	0	0	1076
1992	0	0	0	15	81	204	340	401	93	14	0	0	1148
1993	0	0	0	0	108	75	178	246	102	13	0	0	722

SNOWFALL (inches) — SALT LAKE CITY, UTAH

SEASON	JULY	AUG	SEP	OCT	NOV	DEC	JAN	FEB	MAR	APR	MAY	JUNE	TOTAL
1964-65	0.0	0.0	0.0	0.0	6.7	6.2	15.7	9.5	1.1	2.4	5.3	0.0	46.9
1965-66	0.0	0.0	2.2	0.0	2.6	12.8	5.9	18.1	17.4	2.8	0.0	0.0	61.8
1966-67	0.0	0.0	0.0	3.6	3.2	8.7	30.4	4.5	11.8	11.4	1.0	0.0	74.6
1967-68	0.0	0.0	0.0	0.0	4.2	27.1	6.8	13.6	8.4	14.2	T	T	74.3
1968-69	0.0	0.0	0.0	T	8.7	33.3	13.7	27.9	5.4	0.2	0.0	0.0	89.2
1969-70	0.0	0.0	0.0	0.1	5.6	16.0	2.8	3.9	5.2	23.6	0.0	0.0	57.2
1970-71	0.0	0.0	0.0	0.3	0.7	25.8	13.6	8.7	8.9	1.7	1.4	0.0	61.1
1971-72	0.0	0.0	4.0	16.6	5.4	17.7	10.5	7.6	1.4	15.0	0.0	0.0	78.2
1972-73	0.0	0.0	0.0	6.0	1.1	35.2	20.9	3.6	17.8	2.6	0.0	0.0	87.2
1973-74	0.0	0.0	0.0	1.3	19.5	19.6	20.1	17.2	6.7	26.4	T	T	110.8
1974-75	0.0	0.0	0.0	T	T	8.8	12.5	7.9	22.8	13.1	7.5	0.0	72.6
1975-76	0.0	0.0	0.0	0.1	18.0	11.8	8.6	15.8	18.7	3.5	0.0	T	76.5
1976-77	0.0	0.0	0.0	0.0	T	1.2	8.6	3.2	41.9	4.8	0.6	0.0	60.3
1977-78	0.0	0.0	0.0	0.2	8.5	8.2	15.6	15.5	6.2	2.5	4.6	0.0	61.3
1978-79	0.0	0.0	1.0	0.0	17.4	8.7	13.8	12.4	3.6	7.7	T	0.0	64.6
1979-80	0.0	0.0	0.0	0.0	4.6	8.5	24.5	2.9	19.9	1.2	T	0.0	61.6
1980-81	0.0	0.0	0.0	T	3.9	3.3	8.9	2.7	11.1	0.3	T	T	30.2
1981-82	0.0	0.0	0.0	4.4	2.4	11.5	15.3	4.5	10.2	9.5	T	T	57.8
1982-83	0.0	0.0	0.0	0.2	1.0	20.1	6.2	1.0	13.3	9.0	5.0	0.0	55.8
1983-84	0.0	0.0	0.0	0.0	5.9	34.2	7.6	18.5	6.7	25.1	T	T	98.0
1984-85	0.0	0.0	T	20.4	6.6	12.9	12.7	11.4	8.0	0.7	0.0	0.0	72.7
1985-86	0.0	0.0	0.0	T	27.2	14.7	3.9	1.7	1.0	5.5	T	0.0	54.0
1986-87	0.0	0.0	T	0.0	4.4	1.7	16.4	9.9	3.0	2.1	0.0	0.0	37.5
1987-88	0.0	0.0	0.0	0.0	0.6	11.0	16.3	0.4	6.1	T	0.9	0.0	35.3
1988-89	0.0	0.0	T	0.0	8.5	12.5	9.4	27.5	2.1	T	0.0	0.0	60.0
1989-90	0.0	0.0	0.0	2.7	2.4	1.7	8.2	8.5	11.8	0.7	T	T	36.0
1990-91	0.0	0.0	T	0.0	4.8	14.3	10.9	1.2	1.9	13.7	T	0.0	46.8
1991-92	T	T	T	2.8	11.4	6.8	12.2	4.7	0.2	0.4	0.0	0.0	38.5
1992-93	0.0	T	0.0	0.0	14.2	16.9	50.3	13.2	T	2.4	1.7	T	98.7
1993-94	0.0	T	0.0	0.0	6.1	8.3							
Record Mean	T	T	0.1	1.3	6.6	12.0	13.5	9.4	9.4	5.0	0.6	T	57.9

BURLINGTON, VERMONT

Burlington is located on the eastern shore of Lake Champlain at the widest part of the lake. About 35 miles to the west lie the highest peaks of the Adirondacks, while the foothills of the Green Mountains begin 10 miles to the east and southeast.

Its northerly latitude assures the variety and vigor of a true New England climate, while thanks to the modifying influence of the lake, the many rapid and marked weather changes are tempered in severity. Due to its location in the path of the St. Lawrence Valley storm track and the lake effects, the city is one of the cloudiest in the United States.

Lake Champlain exercises a tempering influence on the local temperature. During the winter months and prior to the lake freezing, temperatures along the lake shore are often 5-10 degrees warmer than at the airport 3 1/2 miles inland. At the airport the average occurrence of the last freeze in spring is around May 10th and that of the first in fall is early October, giving a growing season of 145 days. This location is justly proud of its delightful summer weather. On average, there are few days a year with maxima of 90 degrees or higher. This moderate summer heat gives way to a cooler, but none the less pleasant fall period, usually extending well into October. High pressure systems moving down rapidly from central Canada or Hudson Bay produce the coldest temperatures during the winter months, but extended periods of very cold weather are rare.

Precipitation, although generally plentiful and well distributed throughout the year, is less in the Champlain Valley than in other areas of Vermont due to the shielding effect of the mountain barriers to the east and west. The heaviest rainfall usually occurs during summer thunderstorms, but excessively heavy rainfall is quite uncommon. Droughts are infrequent.

Because of the trend of the Champlain Valley between the Adirondack and Green Mountain ranges, most winds have a northerly or southerly component. The prevailing direction most of the year is from the south. Winds of damaging force are very uncommon.

Smoke pollution is nearly non-existent since there is no concentration of heavy industry here, however, haze has been on the increase over the years due to the large increase in industry to the north and south. During the spring and fall months, fog occasionally forms along the Winooski River to the north and east and may drift over the airport with favorable winds. In spite of the high percentage of cloudiness, periods of low aircraft ceilings and visibilities are usually of short duration, allowing this area to have one of the highest percentages of flying weather in New England.

NORMALS, MEANS, AND EXTREMES

BURLINGTON, VERMONT

LATITUDE: 44 °28'N LONGITUDE: 73 °09' W ELEVATION: FT. GRND 332 BARO 350 TIME ZONE: EASTERN WBAN: 14742

	(a)	JAN	FEB	MAR	APR	MAY	JUNE	JULY	AUG	SEP	OCT	NOV	DEC	YEAR
TEMPERATURE °F:														
Normals														
-Daily Maximum		25.1	27.5	39.3	53.6	67.2	75.8	81.2	77.9	69.0	57.0	44.0	30.4	54.0
-Daily Minimum		7.5	8.9	22.0	34.2	45.4	54.6	59.7	57.9	48.8	38.6	29.6	15.5	35.2
-Monthly		16.3	18.2	30.7	43.9	56.3	65.2	70.5	67.9	58.9	47.8	36.8	23.0	44.6
Extremes														
-Record Highest	50	63	62	84	91	93	97	99	101	94	85	75	65	101
-Year		1950	1981	1946	1976	1977	1988	1977	1944	1945	1949	1948	1982	AUG 1944
-Record Lowest	50	-30	-30	-20	2	24	33	39	35	25	15	-2	-26	-30
-Year		1957	1979	1948	1972	1966	1986	1962	1976	1963	1972	1958	1980	FEB 1979
NORMAL DEGREE DAYS:														
Heating (base 65°F)		1510	1310	1063	633	282	58	6	29	199	533	846	1302	7771
Cooling (base 65°F)		0	0	0	0	13	64	176	119	16	0	0	0	388
% OF POSSIBLE SUNSHINE	50	41	48	51	49	54	58	64	60	54	47	31	33	49
MEAN SKY COVER (tenths)														
Sunrise - Sunset	50	7.5	7.3	7.1	7.1	7.0	6.8	6.4	6.4	6.5	7.0	8.2	8.0	7.1
MEAN NUMBER OF DAYS:														
Sunrise to Sunset														
-Clear	50	4.5	4.4	5.8	5.1	4.9	4.7	5.2	5.9	5.9	5.9	2.5	3.0	57.9
-Partly Cloudy	50	6.4	6.8	7.0	7.4	9.2	10.8	12.9	11.8	10.1	7.8	5.4	5.8	101.4
-Cloudy	50	20.1	17.0	18.2	17.4	16.8	14.5	12.8	13.3	14.0	17.4	22.1	22.1	205.9
Precipitation														
.01 inches or more	50	14.4	11.6	13.1	12.4	13.6	12.4	11.8	12.5	11.6	11.9	14.1	14.7	154.0
Snow,Ice pellets,hail														
1.0 inches or more	50	5.2	4.8	3.7	1.2	0.1	0.0	0.0	0.0	0.0	0.1	1.9	5.1	22.0
Thunderstorms	50	0.*	0.0	0.4	0.9	2.5	5.0	6.0	5.2	2.0	0.6	0.3	0.*	23.0
Heavy Fog Visibility														
1/4 mile or less	50	0.9	1.1	1.3	1.2	0.9	1.1	0.8	1.4	2.4	2.0	1.1	1.1	15.4
Temperature °F														
-Maximum														
90° and above	29	0.0	0.0	0.0	0.1	0.6	1.1	2.9	1.2	0.1	0.0	0.0	0.0	6.0
32° and below	29	21.1	17.9	8.7	0.5	0.0	0.0	0.0	0.0	0.0	0.*	4.3	16.2	68.7
-Minimum														
32° and below	29	29.8	26.5	25.7	14.8	2.6	0.0	0.0	0.0	0.6	8.8	18.7	27.6	155.1
0° and below	29	9.8	8.3	2.0	0.0	0.0	0.0	0.0	0.0	0.0	0.0	0.0	4.7	24.8
AVG. STATION PRESS.(mb)	21	1003.8	1004.8	1003.6	1001.9	1002.3	1001.6	1002.4	1004.3	1005.5	1005.8	1004.7	1004.9	1003.8
RELATIVE HUMIDITY (%)														
Hour 01	28	71	72	73	74	77	81	83	85	85	79	76	75	78
Hour 07 (Local Time)	28	72	74	75	74	74	77	79	83	86	81	78	76	77
Hour 13	28	64	62	58	53	51	55	53	57	61	61	66	68	59
Hour 19	28	67	65	62	58	58	61	61	67	73	71	72	72	66
PRECIPITATION (inches):														
Water Equivalent														
-Normal		1.82	1.63	2.23	2.76	3.12	3.47	3.65	4.06	3.30	2.88	3.13	2.42	34.47
-Maximum Monthly	50	4.69	5.38	3.58	6.55	6.31	7.69	6.12	11.54	8.18	6.22	6.85	5.95	11.54
-Year		1978	1981	1972	1983	1983	1973	1972	1955	1945	1959	1983	1973	AUG 1955
-Minimum Monthly	50	0.42	0.21	0.38	0.93	0.29	1.09	1.23	0.72	0.87	0.50	0.63	0.62	0.21
-Year		1989	1978	1965	1966	1977	1949	1979	1957	1948	1963	1952	1960	FEB 1978
-Maximum in 24 hrs	50	1.53	1.93	1.62	2.16	2.26	2.83	2.69	3.59	3.26	2.17	2.48	2.60	3.59
-Year		1978	1981	1971	1968	1955	1972	1985	1955	1983	1983	1990	1950	AUG 1955
Snow,Ice pellets,hail														
-Maximum Monthly	50	42.4	34.3	39.9	21.3	3.9	0.0	T	0.0	0.1	5.1	19.2	56.7	56.7
-Year		1978	1958	1993	1983	1966		1989		1992	1969	1971	1970	DEC 1970
-Maximum in 24 hrs	50	14.5	16.5	22.4	15.6	3.5	0.0	T	0.0	0.1	5.1	10.1	17.0	22.4
-Year		1961	1966	1993	1983	1966		1989		1992	1969	1958	1978	MAR 1993
WIND:														
Mean Speed (mph)	50	9.7	9.4	9.5	9.5	9.0	8.5	8.0	7.6	8.3	8.8	9.7	9.9	9.0
Prevailing Direction through 1963		S	S	N	S	S	S	S	S	S	S	S	S	S
Fastest Obs. 1 Min.														
-Direction (!!)	10	16	17	16	17	29	16	18	17	17	17	16	16	17
-Speed (MPH)	10	38	39	33	30	35	32	35	26	32	31	35	33	39
-Year		1989	1990	1988	1992	1993	1989	1989	1987	1989	1991	1989	1985	FEB 1990
Peak Gust														
-Direction (!!)	10	SE	S	SW	S	S	S	S	W	S	S	SE	W	SE
-Speed (mph)	10	49	54	54	40	54	44	60	41	52	47	62	51	62
-Date		1989	1990	1991	1993	1990	1989	1991	1993	1989	1990	1989	1993	NOV 1989

PRECIPITATION (inches) — BURLINGTON, VERMONT

YEAR	JAN	FEB	MAR	APR	MAY	JUNE	JULY	AUG	SEP	OCT	NOV	DEC	ANNUAL
1964	2.27	0.63	2.64	2.11	4.67	3.00	2.87	4.10	1.49	2.20	2.10	1.63	29.71
1965	0.60	0.93	0.38	2.16	1.05	4.08	2.91	6.27	3.19	3.32	2.65	1.47	29.01
1966	2.02	2.49	2.63	0.93	2.49	2.63	1.92	4.46	3.33	1.41	1.41	2.82	28.54
1967	1.65	0.77	0.51	3.77	3.19	3.12	4.60	3.79	3.06	3.03	2.12	2.61	32.22
1968	1.26	1.28	3.23	3.54	2.43	3.66	2.70	2.36	2.06	2.73	4.37	3.12	32.74
1969	2.43	0.94	1.93	2.93	3.10	4.01	2.40	3.71	1.88	1.62	4.98	4.59	34.52
1970	0.65	1.95	2.01	2.78	3.14	4.38	1.92	3.44	3.93	2.66	2.35	3.77	32.98
1971	1.24	2.98	2.71	2.65	2.97	2.29	4.29	4.85	1.63	2.16	2.29	1.93	31.99
1972	0.93	1.69	3.58	2.26	2.83	6.52	6.12	2.35	1.69	2.60	4.10	3.43	38.10
1973	2.13	1.55	2.09	3.80	5.38	7.69	3.02	5.41	5.02	1.93	2.31	5.95	46.28
1974	1.90	1.54	2.73	3.47	4.61	4.45	3.70	2.60	3.23	0.78	3.60	2.08	34.69
1975	2.20	2.01	2.86	1.71	1.17	2.47	3.77	2.85	4.12	3.85	3.14	2.36	32.51
1976	2.99	2.85	2.35	2.54	5.86	4.04	3.05	4.69	3.77	4.34	1.63	1.97	40.08
1977	1.61	1.78	2.97	3.13	0.29	2.06	3.34	6.27	6.33	5.02	4.22	3.42	40.44
1978	4.69	0.21	2.98	2.51	2.16	4.36	3.50	1.82	2.07	3.72	0.95	2.11	31.08
1979	4.50	0.60	2.15	3.61	3.12	1.39	1.23	3.42	3.84	2.31	3.89	1.50	31.56
1980	0.61	0.67	2.44	2.39	1.61	1.92	6.11	3.83	4.41	2.48	2.92	1.50	30.89
1981	0.49	5.38	1.32	3.05	3.76	3.07	3.22	5.58	6.24	5.26	2.73	2.03	42.13
1982	2.74	1.43	2.31	2.63	1.95	4.95	3.07	3.55	2.12	2.31	3.59	1.69	32.34
1983	3.09	1.66	2.60	6.55	6.31	1.49	3.92	4.31	3.77	4.38	6.85	5.23	50.16
1984	0.81	2.73	1.72	4.25	5.27	1.70	5.11	3.30	2.81	1.89	3.08	3.14	35.81
1985	1.46	1.26	2.46	1.90	3.53	3.76	4.42	2.67	3.30	3.31	3.68	1.59	33.34
1986	3.69	1.68	3.17	0.95	4.11	4.40	4.53	5.82	4.86	2.50	2.99	1.32	40.02
1987	1.91	0.49	1.33	1.42	2.69	4.42	2.79	2.09	3.58	3.28	2.24	1.17	27.41
1988	0.69	1.69	1.55	1.91	1.80	3.26	2.55	4.27	1.50	2.05	4.51	0.90	26.68
1989	0.42	0.67	2.60	1.89	3.19	3.68	3.65	7.30	5 98	2.98	2.41	1.26	36.03
1990	2.36	2.82	1.81	2.97	3.66	3.08	5.12	4.85	2.03	5.99	3.91	3.58	42.18
1991	1.65	0.51	2.55	3.41	3.15	1.28	2.83	4.00	5.14	5.07	1.58	1.35	32.52
1992	1.65	1.56	2.13	2.58	2.38	1.72	4.58	1.89	4.73	3.00	3.67	0.96	30.85
1993	2.17	1.90	1.54	3.76	2.19	3.35	3.34	4.46	3.38	2.93	2.27	1.57	32.86
Record Mean	1.82	1.66	2.16	2.53	3.02	3.45	3.62	3.54	3.39	2.96	2.79	2.07	33.01

AVERAGE TEMPERATURE (deg. F) — BURLINGTON, VERMONT

YEAR	JAN	FEB	MAR	APR	MAY	JUNE	JULY	AUG	SEP	OCT	NOV	DEC	ANNUAL
#1964	21.9	17.2	31.1	43.3	58.6	63.0	69.6	62.9	55.8	45.3	35.7	24.8	44.1
1965	13.7	20.0	27.4	39.3	57.2	62.9	65.1	66.0	58.7	46.8	32.9	28.6	43.2
1966	15.5	17.9	30.3	41.3	51.2	65.3	69.6	67.4	55.7	47.0	40.6	23.4	43.7
1967	23.9	11.7	25.3	40.7	47.6	67.6	70.1	67.0	58.1	48.8	32.9	25.5	43.3
1968	8.4	11.0	29.6	46.2	51.7	60.8	68.7	63.7	61.0	49.8	32.2	17.9	41.8
1969	16.5	18.5	24.3	41.4	51.3	64.1	67.8	68.6	57.9	45.8	36.3	18.6	42.6
1970	3.6	16.9	25.8	42.6	54.1	63.7	70.6	68.4	60.1	50.5	39.0	14.3	42.5
1971	9.7	20.0	24.1	37.3	54.5	64.9	68.9	67.1	63.5	53.8	33.5	24.3	43.5
1972	21.1	17.0	24.8	35.6	56.2	63.1	69.5	65.3	58.7	42.4	32.1	22.5	42.4
1973	21.5	14.6	37.1	44.6	53.6	66.9	70.6	72.1	58.5	49.3	37.2	27.4	46.1
1974	18.7	15.6	29.2	44.4	51.3	66.5	70.2	69.1	58.2	43.4	36.2	28.5	44.3
1975	23.6	20.7	28.0	37.1	62.3	66.4	74.6	69.1	58.0	50.4	42.1	20.1	46.1
1976	11.1	24.6	33.4	47.4	54.7	69.2	68.5	65.7	57.0	43.7	33.0	16.3	43.7
1977	11.1	20.5	37.6	45.3	60.0	64.7	69.6	67.5	58.7	46.6	40.0	22.4	45.3
1978	15.1	9.5	26.0	38.7	60.1	63.9	69.4	68.3	55.2	46.4	34.8	25.2	42.7
1979	18.0	7.5	36.9	43.5	58.5	65.3	72.2	65.9	58.6	48.1	41.3	29.0	45.4
1980	21.2	17.6	31.1	46.5	58.9	64.4	70.6	70.7	57.9	45.0	32.3	15.0	44.2
1981	8.9	32.9	33.5	46.7	58.2	66.1	71.1	67.1	59.3	44.9	36.9	25.3	45.9
1982	9.6	19.1	30.3	43.4	57.3	60.7	69.5	65.9	62.3	50.1	42.3	31.9	45.2
1983	21.0	22.3	33.0	42.3	52.9	66.3	71.3	68.6	62.9	48.2	38.1	22.4	45.8
1984	16.5	28.7	21.9	44.7	52.3	66.0	70.3	71.1	57.2	50.0	38.4	30.3	45.6
1985	13.4	22.5	31.6	44.3	55.8	61.7	69.6	67.5	60.3	49.1	36.9	21.3	44.5
1986	18.5	16.2	33.7	48.5	58.3	62.3	68.5	66.1	58.1	46.9	34.5	27.8	45.0
1987	18.1	15.0	33.3	48.6	55.5	66.3	71.5	66.7	59.5	45.9	37.0	28.5	45.5
1988	19.9	21.4	29.7	44.3	57.9	63.4	73.2	70.7	58.2	44.4	39.6	22.9	45.5
1989	23.7	19.7	28.4	41.6	59.6	67.2	71.7	67.7	61.4	50.3	36.4	7.6	44.6
1990	29.8	23.5	33.8	46.2	52.9	65.9	70.2	69.8	59.4	49.4	39.5	30.1	47.5
1991	18.9	26.5	34.0	49.1	59.3	65.9	70.4	70.5	57.8	50.4	37.6	24.0	47.0
1992	18.6	19.1	26.5	42.3	56.5	64.4	66.1	67.6	60.2	45.5	36.7	28.1	44.3
1993	21.7	10.6	27.4	45.3	56.6	64.7	72.2	70.8	59.1	46.1	36.8	24.7	44.7
Record Mean	17.9	18.6	29.6	43.0	55.5	64.8	69.8	67.4	59.5	48.5	36.7	23.4	44.5
Max	26.5	27.5	38.0	52.2	65.8	74.9	79.9	77.2	68.9	57.2	43.5	30.7	53.5
Min	9.3	9.7	21.2	33.7	45.2	54.7	59.7	57.7	50 1	39.8	29.9	16.0	35.6

REFERENCE NOTES — BURLINGTON, VERMONT

GENERAL
T - TRACE AMOUNT.
BLANK ENTRIES DENOTE MISSING/UNREPORTED DATA.
# INDICATES A STATION OR INSTRUMENT RELOCATION.

SPECIFIC

PM - INCLUDES LAST DAY OF PREVIOUS MONTH
ASOS - AUTOMATED SURFACE OBSERVING SYSTEM IN OPERATION DURING THESE MONTHS.

(a) - LENGTH OF RECORD IN YEARS, ALTHOUGH INDIVIDUAL MONTHS MAY BE MISSING.
0.* OR * - THE VALUE IS BETWEEN 0.0 AND 0.05
NORMALS - BASED ON THE 1961-1990 RECORD PERIOD.
EXTREMES - DATES ARE THE MOST RECENT OCCURRENCE
WIND DIR.- NUMERALS SHOW TENS OF DEGRESS CLOCKWISE FROM TRUE NORTH. "00" INDICATES CALM.
RESULTANT DIRECTIONS ARE GIVEN TO WHOLE DEGREES.
BOLD VALUES INDICATE EXTREME VALUES WHICH OCCURRED AFTER THE ASOS SYSTEM WAS COMMISSIONED.

RECORD = PERIOD OF RECORD
RECORD MEAN PRECIPITATION IS THE MEAN OF ALL DAILY PRECIPITATION AMOUNTS DURING THE PERIOD OF RECORD.
RECORD MAX(MIN) TEMPERATURE IS THE MEAN OF ALL DAILY MAX(MIN) TEMPERATURES DURING THE PERIOD OF RECORD.
RECORD MEAN TEMPERATURE IS THE SUM OF THE RECORD MAX AND RECORD MIN DIVIDED BY 2.
AVERAGE TEMPERATURE IS THE SUM OF THE MEAN DAILY MAX AND MIN TEMPERATURE DIVIDED BY 2.

EXCEPTIONS

FASTEST MILE WIND IS THROUGH NOVEMBER 1983.

RECORD MEANS ARE THROUGH THE CURRENT YEAR, BEGINNING IN 1893 FOR TEMPERATURE
1884 FOR PRECIPITATION
1944 FOR SNOWFALL

HEATING DEGREE DAYS Base 65 deg. F BURLINGTON, VERMONT

SEASON	JULY	AUG	SEP	OCT	NOV	DEC	JAN	FEB	MAR	APR	MAY	JUNE	TOTAL
#1964-65	12	102	280	601	872	1238	1585	1256	1159	766	257	136	8264
1965-66	43	80	236	558	956	1125	1531	1313	1069	707	432	91	8141
1966-67	17	26	280	551	725	1285	1269	1490	1225	722	533	29	8152
1967-68	11	35	223	495	957	1216	1751	1561	1089	562	407	140	8447
1968-69	32	104	127	472	979	1451	1496	1298	1256	700	422	107	8444
1969-70	41	41	244	589	856	1434	1906	1342	1208	663	341	105	8770
1970-71	10	36	174	444	773	1567	1710	1257	1263	821	336	83	8474
1971-72	12	49	131	344	938	1254	1357	1387	1239	872	281	113	7977
1972-73	26	69	212	694	982	1310	1344	1410	855	608	345	86	7941
1973-74	10	17	256	480	825	1160	1431	1378	1101	618	430	37	7743
1974-75	2	6	224	665	858	1128	1276	1236	1141	831	152	82	7601
1975-76	0	45	208	448	681	1385	1669	1168	973	545	331	50	7503
1976-77	20	68	254	654	954	1505	1667	1240	842	590	223	89	8106
1977-78	24	53	207	564	740	1314	1539	1547	1202	781	225	90	8286
1978-79	49	38	295	571	897	1227	1452	1610	866	641	224	90	7960
1979-80	23	65	213	528	703	1107	1350	1371	1043	550	204	91	7248
1980-81	10	3	240	611	976	1545	1738	894	969	544	239	43	7812
1981-82	13	36	204	617	837	1224	1716	1277	1069	643	255	133	8024
1982-83	30	54	124	455	676	1021	1356	1188	983	675	367	77	7006
1983-84	19	36	148	518	803	1317	1500	1044	1331	602	395	68	7781
1984-85	6	24	241	460	792	1068	1592	1185	1029	615	296	118	7426
1985-86	11	42	169	489	835	1344	1436	1361	966	492	219	113	7477
1986-87	40	60	215	553	906	1144	1446	1397	975	488	328	48	7600
1987-88	19	66	185	584	833	1125	1389	1260	1088	614	236	136	7535
1988-89	15	52	212	635	755	1298	1273	1265	1128	691	188	45	7557
1989-90	2	43	164	451	849	1776	1084	1156	961	577	370	63	7496
1990-91	19	10	180	480	758	1074	1424	1072	954	475	206	59	6711
1991-92	7	11	240	451	813	1266	1434	1327	1187	674	277	83	7770
1992-93	49	33	197	597	843	1137	1335	1517	1159	584	256	80	7787
1993-94	3	12	211	579	839	1243							

COOLING DEGREE DAYS Base 65 deg. F BURLINGTON, VERMONT

YEAR	JAN	FEB	MAR	APR	MAY	JUNE	JULY	AUG	SEP	OCT	NOV	DEC	TOTAL
1969	0	0	0	0	2	86	134	160	38	0	0	0	420
1970	0	0	0	0	11	75	189	150	36	1	0	0	462
1971	0	0	0	0	17	87	138	118	90	4	0	0	454
1972	0	0	0	0	14	64	169	81	30	0	0	0	358
1973	0	0	0	3	0	149	187	243	68	0	0	0	650
1974	0	0	0	5	9	89	171	140	27	1	0	0	442
1975	0	0	0	0	75	131	306	181	5	1	0	0	699
1976	0	0	0	24	19	185	135	97	23	0	0	0	483
1977	0	0	0	7	75	86	174	138	27	0	0	0	507
1978	0	0	0	0	79	64	194	146	6	0	0	0	489
1979	0	0	0	2	29	106	253	101	27	13	0	0	531
1980	0	0	0	0	24	78	189	184	34	0	0	0	509
1981	0	0	0	2	35	85	211	110	39	0	0	0	482
1982	0	0	0	1	24	11	179	90	51	0	0	0	356
1983	0	0	0	0	0	121	223	155	92	6	0	0	597
1984	0	0	0	0	7	106	175	217	15	3	0	0	523
1985	0	0	0	0	15	25	160	123	34	0	0	0	357
1986	0	0	0	4	19	38	156	104	14	0	0	0	335
1987	0	0	0	3	42	92	228	126	30	0	0	0	521
1988	0	0	0	0	19	96	274	238	15	3	0	0	645
1989	0	0	0	0	28	117	216	134	63	0	0	0	558
1990	0	0	0	16	1	95	189	165	18	6	0	0	490
1991	0	0	0	5	35	92	182	186	32	6	0	0	538
1992	0	0	0	3	21	71	91	121	61	0	0	0	368
1993	0	0	0	0	2	79	235	198	39	0	0	0	553

SNOWFALL (inches) BURLINGTON, VERMONT

SEASON	JULY	AUG	SEP	OCT	NOV	DEC	JAN	FEB	MAR	APR	MAY	JUNE	TOTAL
1964-65	0.0	0.0	0.0	0.1	1.2	23.0	11.8	4.3	7.9	1.1	0.0	0.0	49.4
1965-66	0.0	0.0	0.0	0.4	12.4	11.9	41.3	28.5	8.3	4.9	3.9	0.0	111.6
1966-67	0.0	0.0	0.0	T	2.4	36.2	20 5	12.6	6.1	4.7	2.6	0.0	85.1
1967-68	0.0	0.0	0.0	T	10.3	17.1	18.4	24.8	14.5	T	0.0	0.0	85.1
1968-69	0.0	0.0	0.0	T	18.8	28.6	15.8	17.0	12.4	3.7	0.0	0.0	96.3
1969-70	0.0	0.0	0.0	5.1	10.5	50.8	11.1	13.8	10.5	2.4	0.4	0.0	104.6
1970-71	0.0	0.0	0.0	0.1	2.7	56.7	17.1	23.1	33.1	12.6	0.0	0.0	145.4
1971-72	0.0	0.0	0.0	0.0	19.2	19.3	14.3	25.1	21.8	9.2	0.0	0.0	108.9
1972-73	0.0	0.0	0.0	T	12.2	39.0	11.4	18.5	2.3	6.3	0.0	0.0	89.7
1973-74	0.0	0.0	0.0	0.1	2.6	24.1	21.5	9.9	20.9	16.8	0.0	0.0	95.9
1974-75	0.0	0.0	0.0	0.1	11.5	16.8	14.8	22.0	12.4	13.3	0.0	0.0	90.9
1975-76	0.0	0.0	0.0	T	5.3	16.0	28.3	20.4	18.8	0.9	T	0.0	89.7
1976-77	0.0	0.0	0.0	0.9	13.3	11.5	24.2	16.4	9.6	1.8	T	0.0	77.7
1977-78	0.0	0.0	0.0	0.0	16.0	22.6	42.4	4.0	12.5	1.9	T	0.0	99.4
1978-79	0.0	0.0	0.0	T	5.7	24.1	37.9	6.6	1.6	8.4	0.0	0.0	84.3
1979-80	0.0	0.0	0.0	1.5	0.4	6.0	3.0	11.6	16.8	0.3	0.0	0.0	39.6
1980-81	0.0	0.0	0.0	T	12.2	17.5	8.7	11.9	13.3	1.1	0.0	0.0	64.7
1981-82	0.0	0.0	0.0	T	3.9	32.8	19.4	8.3	13.0	4.1	0.0	0.0	81.5
1982-83	0.0	0.0	0.0	T	0.8	5.0	22.5	18.3	11.9	21.3	0.7	0.0	80.5
1983-84	0.0	0.0	0.0	T	4.7	14.4	15.2	13.7	16.1	0.4	T	0.0	64.5
1984-85	0.0	0.0	0.0	0.0	6.0	29.3	25.9	10.9	16.6	2.7	0.0	0.0	91.4
1985-86	0.0	0.0	0.0	T	4.6	21.3	33.6	18.3	8.4	T	T	0.0	86.2
1986-87	0.0	0.0	0.0	T	10.5	7.7	34.4	7.0	6.0	2.1	0.0	0.0	67.7
1987-88	0.0	0.0	0.0	0.6	6.5	12.4	9.2	26.9	6.4	2.4	0.0	0.0	64.4
1988-89	0.0	0.0	0.0	0.3	0.6	12.4	6.6	8.5	9.7	2.3	0.0	0.0	40.4
1989-90	T	0.0	0.0	0.0	5.6	20.7	17.6	20.5	10.2	2.1	0.0	0.0	76.7
1990-91	0.0	0.0	T	T	7.3	10.3	17.8	3.9	3.2	T	0.0	0.0	42.5
1991-92	0.0	0.0	T	T	2.3	14.9	12.2	27.2	14.0	8.6	0.0	0.0	79.2
1992-93	0.0	0.0	0.1	T	2.9	2.6	24.8	33.8	39.9	12.8	0.0	0.0	116.9
1993-94	0.0	0.0	0.0	1.3	7.9	9.1							
Record Mean	T	0.0	T	0.2	6.6	18.1	18.8	16.8	12.4	3.8	0.2	0.0	77.1

LYNCHBURG, VIRGINIA

Lynchburg is situated in the valley of the James River, and on the eastern edge of the Blue Ridge Mountains. The terrain is definitely hilly, with sheltered valleys which are visited by early autumn and late spring frosts. The climate is usually a pleasant one, being neither too hot in the summer, nor too cold in the winter. Rainfall is fairly evenly distributed throughout the year, but there is a distinct summertime rainfall, occasioned by afternoon thunderstorms.

Spring makes itself felt in March, when the mean monthly temperature increases about 7 degrees over the February temperature. Autumn rapidly comes in October, which shows about a 10 degree drop below the September mean. The approaching autumn season brings periods of two to three days of cloudy, cool weather, with high humidity and light rain or drizzle. In midwinter, however, after the passage of a cold front, dry invigorating air, with clear skies, is the rule in Lynchburg. There are occasional snow showers, but the mountains to the immediate west act as a barrier and shelter the area from many storms and high winds.

The mountains also act as a barrier to extremely cold weather. Temperatures have fallen below zero only on a few days, and 100 degree heat is almost as rare, although this mark has been exceeded in the months of May through September.

Great variation in temperature is quite frequently noted during clear, still nights in the winter months. On some such nights, differences of as much as 10–15 degrees occur between the low valleys and the higher terrain.

Based on the 1951–1980 period, the average first occurrence of 32 degrees Fahrenheit in the fall is October 23 and the average last occurrence in the spring is April 13.

NORMALS, MEANS, AND EXTREMES

LYNCHBURG, VIRGINIA

LATITUDE: 37°20'N LONGITUDE: 79°12' W ELEVATION: FT. GRND 921 BARO 971 TIME ZONE: EASTERN WBAN: 13733

	(a)	JAN	FEB	MAR	APR	MAY	JUNE	JULY	AUG	SEP	OCT	NOV	DEC	YEAR
TEMPERATURE °F:														
Normals														
-Daily Maximum		43.6	47.1	57.3	67.6	75.6	82.6	86.0	84.9	78.5	68.3	58.1	47.6	66.4
-Daily Minimum		24.7	27.4	35.4	43.8	52.6	60.8	65.1	64.4	57.7	45.7	37.3	29.1	45.3
-Monthly		34.2	37.2	46.4	55.7	64.1	71.7	75.6	74.6	68.1	57.0	47.7	38.4	55.9
Extremes														
-Record Highest	49	76	79	87	92	93	100	103	102	101	93	83	78	103
-Year		1952	1985	1986	1985	1969	1945	1954	1988	1954	1951	1974	1984	JUL 1954
-Record Lowest	49	-10	0	7	20	31	40	49	45	36	21	8	-4	-10
-Year		1985	1965	1993	1985	1966	1977	1988	1965	1983	1969	1970	1983	JAN 1985
NORMAL DEGREE DAYS:														
Heating (base 65°F)		955	778	577	283	95	6	0	0	32	270	519	825	4340
Cooling (base 65°F)		0	0	0	0	67	207	329	298	125	22	0	0	1048
% OF POSSIBLE SUNSHINE	49	52	56	58	61	62	66	62	61	60	62	56	53	59
MEAN SKY COVER (tenths)														
Sunrise - Sunset	47	6.2	6.1	6.0	5.9	6.0	5.7	5.8	5.7	5.5	4.9	5.6	5.9	5.8
MEAN NUMBER OF DAYS:														
Sunrise to Sunset														
-Clear	47	8.9	8.5	8.8	8.6	8.1	8.1	8.0	8.8	10.2	13.3	10.3	10.1	111.7
-Partly Cloudy	47	7.3	6.9	8.8	9.2	10.4	11.7	11.5	11.2	8.8	7.1	7.3	6.7	106.9
-Cloudy	47	14.8	12.9	13.4	12.3	12.5	10.2	11.4	11.0	11.0	10.6	12.4	14.2	146.6
Precipitation														
.01 inches or more	49	10.8	9.5	11.1	9.9	11.6	9.7	11.3	10.0	8.3	7.7	9.0	9.5	118.5
Snow,Ice pellets,hail														
1.0 inches or more	49	1.5	1.5	0.8	0.1	0.0	0.0	0.0	0.0	0.0	0.*	0.2	1.0	5.2
Thunderstorms	23	0.3	0.4	1.2	2.9	6.4	7.1	9.3	7.8	3.5	0.9	0.5	0.1	40.5
Heavy Fog Visibility														
1/4 mile or less	23	3.9	3.9	2.7	2.6	3.5	2.3	2.6	3.4	3.7	2.6	3.8	3.9	39.0
Temperature °F														
-Maximum														
90° and above	30	0.0	0.0	0.0	0.4	0.5	4.4	9.4	7.1	2.4	0.*	0.0	0.0	24.2
32° and below	30	4.7	2.3	0.2	0.0	0.0	0.0	0.0	0.0	0.0	0.0	0.1	2.0	9.3
-Minimum														
32° and below	30	22.9	20.5	12.0	2.9	0.1	0.0	0.0	0.0	0.0	2.4	10.0	19.3	90.2
0° and below	30	0.4	0.*	0.0	0.0	0.0	0.0	0.0	0.0	0.0	0.0	0.0	0.*	0.5
AVG. STATION PRESS.(mb)	10	984.0	984.3	982.8	982.8	981.8	983.2	983.9	985.2	984.9	985.6	985.7	984.9	984.1
RELATIVE HUMIDITY (%)														
Hour 01														
Hour 07 (Local Time)	30	72	72	73	73	81	83	86	89	88	85	79	75	80
Hour 13	30	53	50	49	45	53	55	57	58	57	52	52	54	53
Hour 19	15	60	55	55	51	62	67	70	73	73	71	61	63	63
PRECIPITATION (inches):														
Water Equivalent														
-Normal		2.86	3.04	3.47	3.09	3.91	3.45	4.16	3.59	3.24	3.70	3.14	3.23	40.88
-Maximum Monthly	49	7.97	5.70	9.24	7.95	9.07	9.97	10.30	11.36	9.90	11.40	8.77	7.15	11.40
-Year		1978	1972	1975	1987	1971	1989	1984	1952	1989	1976	1985	1973	OCT 1976
-Minimum Monthly	49	0.49	0.54	0.74	0.28	1.36	0.47	1.15	0.93	0.02	0.38	0.90	0.33	0.02
-Year		1981	1978	1966	1976	1957	1986	1977	1963	1978	1974	1960	1965	SEP 1978
-Maximum in 24 hrs	49	2.28	2.87	2.47	3.67	3.47	6.27	4.82	4.47	4.72	4.98	2.65	3.03	6.27
-Year		1978	1984	1975	1978	1960	1972	1984	1967	1989	1954	1951	1948	JUN 1972
Snow,Ice pellets,hail														
-Maximum Monthly	49	31.8	19.2	24.9	4.8	0.0	0.0	0.0	0.0	0.0	2.4	11.6	17.9	31.8
-Year		1966	1979	1960	1971						1979	1968	1966	JAN 1966
-Maximum in 24 hrs	49	12.3	14.6	13.4	4.8	0.0	0.0	0.0	0.0	0.0	2.4	6.7	12.7	14.6
-Year		1987	1983	1969	1971						1979	1968	1969	FEB 1983
WIND:														
Mean Speed (mph)	28	8.6	8.6	9.1	9.0	7.5	6.9	6.5	6.3	6.9	7.3	7.9	7.9	7.7
Prevailing Direction through 1963		SW	SW	SW	SW	SW	SW	SW	N	N	N	SW	SW	SW
Fastest Mile														
-Direction (!!)	49	W	S	S	NE	N	SW	NW	W	NE	N	NW	SE	N
-Speed (MPH)	49	45	50	43	43	56	56	43	48	40	41	43	45	56
-Year		1971	1961	1950	1978	1958	1951	1954	1977	1956	1954	1949	1950	MAY 1958
Peak Gust														
-Direction (!!)	10	NW	W	NE	SW	W	W	W	SE	SE	W	S	NW	W
-Speed (mph)	10	48	46	56	53	59	74	64	46	41	48	64	47	74
-Date		1987	1989	1993	1993	1989	1993	1989	1992	1989	1990	1989	1985	JUN 1993

PRECIPITATION (inches)

LYNCHBURG, VIRGINIA

YEAR	JAN	FEB	MAR	APR	MAY	JUNE	JULY	AUG	SEP	OCT	NOV	DEC	ANNUAL
1964	4.82	4.46	1.99	2.14	1.64	0.67	6.07	2.02	1.26	2.06	3.08	3.52	33.73
1965	2.18	3.35	4.35	1.42	2.21	2.04	1.83	2.87	2.31	4.39	1.00	0.33	28.28
1966	3.62	4.51	0.74	1.70	3.16	1.36	2.97	2.58	6.22	4.69	2.65	3.18	37.38
1967	2.18	2.13	2.61	1.45	3.69	2.54	5.41	11.27	1.17	1.83	1.28	5.95	41.51
1968	2.35	0.64	3.28	2.48	3.58	1.49	2.58	1.53	1.74	3.65	3.32	1.78	28.42
1969	2.30	2.56	4.32	2.21	2.37	3.27	4.06	3.03	2.29	0.87	1.24	5.37	33.89
1970	1.16	2.39	2.36	2.82	2.12	0.74	4.72	3.62	0.78	5.25	4.01	2.90	32.87
1971	1.73	4.45	2.64	2.29	9.07	2.85	4.48	3.02	1.85	7.40	3.11	1.38	44.27
1972	3.25	5.70	1.76	2.81	8.04	8.58	8.45	1.21	3.10	6.71	5.78	4.32	59.71
1973	3.05	2.94	6.44	4.57	3.81	4.17	6.64	4.44	2.53	3.95	1.04	7.15	50.73
1974	3.94	2.05	2.96	3.11	4.86	2.93	2.84	6.00	5.55	0.38	1.74	3.68	40.04
1975	3.82	2.66	9.24	2.31	6.54	4.78	4.42	4.68	7.42	2.43	2.96	4.34	55.60
1976	2.70	1.98	4.07	0.28	5.32	7.34	1.29	1.07	5.06	11.40	1.38	2.83	44.72
1977	1.76	0.61	2.71	4.84	1.50	3.39	1.15	5.73	2.25	5.77	5.70	3.29	38.70
1978	7.97	0.54	3.84	5.33	6.69	3.36	4.83	3.89	0.02	0.84	3.22	4.06	44.59
1979	6.65	4.99	3.78	3.26	3.10	5.28	4.50	3.42	9.22	3.77	3.18	1.13	52.28
1980	4.63	1.07	5.03	3.99	3.03	0.65	3.61	1.34	1.79	2.35	2.85	0.56	30.90
1981	0.49	3.81	1.81	2.44	1.66	5.24	5.02	4.21	3.61	2.80	0.93	3.90	35.92
1982	3.59	4.41	2.41	3.28	3.67	6.45	5.07	2.60	2.25	2.75	2.85	2.26	41.59
1983	1.14	3.70	4.13	6.67	2.83	1.64	1.46	2.62	1.66	5.99	5.14	5.48	42.46
1984	1.54	5.39	6.45	4.27	4.21	2.74	10.30	5.98	2.15	2.18	1.99	2.01	49.21
1985	2.77	3.41	0.96	1.83	5.97	1.45	3.61	7.10	0.06	3.30	8.77	1.11	40.34
1986	0.83	2.85	0.97	1.21	3.10	0.47	3.20	3.57	1.63	2.38	3.27	4.86	28.34
1987	4.35	3.33	2.35	7.95	3.15	0.86	2.98	0.99	7.36	0.67	3.92	3.38	41.29
1988	2.09	1.26	1.55	3.72	4.20	2.04	3.04	3.68	1.73	1.84	4.25	1.20	30.60
1989	1.37	2.91	3.55	5.03	5.57	9.97	8.30	4.39	9.90	3.73	1.38	3.01	59.11
1990	3.19	3.20	4.02	3.10	5.70	4.79	2.78	3.27	2.03	10.81	1.68	5.14	49.71
1991	3.00	2.20	6.66	2.80	2.66	1.14	4.51	1.37	0.74	0.62	3.25	5.52	34.47
1992	2.31	2.94	3.20	4.26	5.99	1.88	3.27	1.50	4.13	1.76	5.60	3.45	40.29
1993	4.14	3.59	7.01	3.59	2.23	2.33	5.06	1.61	1.67	1.77	3.16	4.89	41.05
Record Mean	3.24	3.00	3.61	3.14	3.56	3.76	4.11	3.97	3.29	3.14	2.73	3.20	40.75

AVERAGE TEMPERATURE (deg. F)

LYNCHBURG, VIRGINIA

YEAR	JAN	FEB	MAR	APR	MAY	JUNE	JULY	AUG	SEP	OCT	NOV	DEC	ANNUAL
1964	36.9	35.9	45.9	56.2	66.0	73.6	74.8	72.4	66.1	52.7	50.2	40.8	55.9
1965	34.9	36.0	41.5	54.3	68.8	69.4	73.9	73.9	69.4	54.7	47.3	41.6	55.5
1966	31.3	36.4	47.0	52.2	64.0	71.8	77.9	73.6	66.8	55.4	48.2	37.5	55.2
1967	40.5	34.3	46.0	57.7	58.7	71.0	72.8	72.1	63.1	54.3	41.9	40.7	54.4
1968	32.0	32.4	49.9	56.0	62.1	72.1	76.0	77.0	68.0	57.8	47.4	34.0	55.4
1969	32.2	36.4	40.7	57.1	65.4	72.3	76.2	72.8	65.9	56.0	44.4	34.0	54.5
1970	29.4	36.3	41.1	56.0	65.9	73.0	75.6	73.6	73.1	60.0	47.2	39.3	55.9
1971	33.5	38.7	43.1	54.8	62.1	73.4	74.6	72.6	70.0	62.6	44.9	47.3	56.5
1972	39.1	36.3	46.4	56.2	63.1	67.7	75.2	74.1	69.4	54.8	45.1	43.1	55.9
1973	36.0	36.0	50.7	54.0	60.9	73.4	74.7	75.4	70.8	59.4	49.8	37.3	56.6
1974	44.2	38.6	48.9	56.9	63.8	68.0	74.4	74.3	65.7	55.1	47.2	39.0	56.3
1975	38.7	39.7	42.3	52.9	66.9	71.4	73.9	76.5	66.2	59.2	50.5	37.3	56.3
1976	32.7	45.8	50.5	57.6	61.8	70.4	74.1	73.2	66.4	53.2	40.8	34.3	55.1
1977	23.0	36.8	51.5	58.8	67.1	70.1	78.4	76.0	70.0	54.6	49.6	36.2	56.0
1978	29.2	30.1	44.1	57.2	63.6	72.4	74.5	77.3	71.0	55.6	50.1	39.7	55.4
1979	31.9	27.9	49.3	56.3	63.9	69.2	74.3	75.0	67.9	55.2	50.1	40.7	55.2
1980	36.0	33.7	43.8	56.9	65.8	69.4	77.8	78.6	72.6	55.2	45.2	38.7	56.1
1981	31.3	40.8	43.3	59.4	62.5	75.7	76.4	73.3	66.3	53.8	47.0	34.0	55.3
1982	27.8	38.6	45.5	53.8	67.5	70.8	75.7	73.5	68.8	59.0	50.0	44.4	56.3
1983	35.9	37.7	46.9	52.4	62.1	72.1	77.4	78.4	68.7	57.5	48.0	34.9	56.0
1984	32.8	43.3	43.6	53.7	63.3	74.5	73.3	74.9	65.6	65.5	45.4	47.4	56.9
1985	32.3	39.2	49.6	61.2	65.9	71.5	74.9	73.4	68.5	60.3	54.6	35.6	57.3
1986	35.6	38.9	47.8	58.5	64.4	74.6	78.8	72.4	68.9	59.3	47.4	38.8	57.1
1987	33.9	36.9	46.7	53.8	67.0	74.7	79.4	77.5	69.9	52.2	49.7	41.3	56.9
1988	31.5	37.8	47.6	55.6	64.1	71.4	78.2	78.4	66.7	51.4	47.5	38.1	55.7
1989	41.2	38.4	47.3	54.9	62.0	73.9	75.7	73.7	67.8	58.5	47.1	29.4	55.8
1990	43.9	45.6	50.9	55.9	64.0	71.0	75.5	73.3	67.3	59.2	51.1	43.5	58.4
1991	38.9	42.2	49.6	57.8	69.4	73.5	78.5	75.4	69.8	59.6	46.6	42.4	58.6
1992	39.7	41.4	46.7	55.7	61.7	70.6	78.1	72.8	68.9	56.4	48.0	39.9	56.7
1993	40.1	36.4	43.8	56.1	67.6	73.9	80.7	77.8	70.8	56.8	48.1	37.7	57.5
Record Mean	37.0	38.9	46.8	56.2	65.5	73.0	76.8	75.2	69.1	58.0	47.6	39.0	56.9
Max	46.0	48.4	57.1	67.5	76.7	83.6	87.0	85.1	79.3	69.1	57.7	48.0	67.1
Min	28.0	29.4	36.4	44.8	54.3	62.4	66.6	65.3	58.8	46.9	37.4	30.1	46.7

REFERENCE NOTES

LYNCHBURG, VIRGINIA

GENERAL
T - TRACE AMOUNT.
BLANK ENTRIES DENOTE MISSING/UNREPORTED DATA.
# INDICATES A STATION OR INSTRUMENT RELOCATION.

SPECIFIC

(a) - LENGTH OF RECORD IN YEARS, ALTHOUGH INDIVIDUAL MONTHS MAY BE MISSING.
0.* OR * - THE VALUE IS BETWEEN 0.0 AND 0.05
NORMALS - BASED ON THE 1961-1990 RECORD PERIOD.
EXTREMES - DATES ARE THE MOST RECENT OCCURRENCE
WIND DIR.- NUMERALS SHOW TENS OF DEGRESS CLOCKWISE FROM TRUE NORTH. "00" INDICATES CALM.
RESULTANT DIRECTIONS ARE GIVEN TO WHOLE DEGREES.
BOLD VALUES INDICATE EXTREME VALUES WHICH OCCURRED AFTER THE ASOS SYSTEM WAS COMMISSIONED.

RECORD = PERIOD OF RECORD
RECORD MEAN PRECIPITATION IS THE MEAN OF ALL DAILY PRECIPITATION AMOUNTS DURING THE PERIOD OF RECORD.
RECORD MAX(MIN) TEMPERATURE IS THE MEAN OF ALL DAILY MAX(MIN) TEMPERATURES DURING THE PERIOD OF RECORD.
RECORD MEAN TEMPERATURE IS THE SUM OF THE RECORD MAX AND RECORD MIN DIVIDED BY 2.
AVERAGE TEMPERATURE IS THE SUM OF THE MEAN DAILY MAX AND MIN TEMPERATURE DIVIDED BY 2.

EXCEPTIONS

THUNDERSTORMS AND HEAVY FOG ARE THROUGH 1966 AND MAY BE INCOMPLETE, DUE TO PART-TIME OPERATIONS PRIOR TO AUGUST 1962.

RECORD MEANS ARE THROUGH THE CURRENT YEAR, BEGINNING IN 1875 FOR TEMPERATURE
1872 FOR PRECIPITATION
1945 FOR SNOWFALL

HEATING DEGREE DAYS Base 65 deg. F — LYNCHBURG, VIRGINIA

SEASON	JULY	AUG	SEP	OCT	NOV	DEC	JAN	FEB	MAR	APR	MAY	JUNE	TOTAL
1964-65	0	5	48	374	438	743	926	803	724	321	15	27	4424
1965-66	0	9	33	317	522	718	1038	795	554	392	102	22	4502
1966-67	0	0	46	294	496	844	752	855	583	237	212	17	4336
1967-68	0	0	97	330	688	744	1018	941	468	268	117	1	4672
1968-69	0	2	7	233	523	953	1009	793	745	239	74	15	4593
1969-70	0	0	58	294	612	954	1095	798	732	278	79	0	4900
1970-71	0	0	21	183	528	789	968	729	673	303	119	3	4316
1971-72	0	0	13	110	613	546	796	826	571	282	73	36	3866
1972-73	2	0	20	310	593	671	893	809	435	339	158	0	4230
1973-74	0	0	9	192	448	851	641	736	495	265	105	16	3758
1974-75	0	0	74	304	537	799	808	701	696	362	45	11	4337
1975-76	0	0	60	185	432	853	994	550	443	273	132	23	3945
1976-77	0	2	34	370	719	947	1294	783	417	217	58	26	4867
1977-78	0	0	12	325	464	887	1103	972	637	237	125	6	4768
1978-79	0	0	25	289	439	777	1018	1033	489	265	83	11	4429
1979-80	0	6	27	320	444	746	892	901	652	255	65	19	4327
1980-81	0	0	28	312	589	807	1039	670	666	197	131	2	4441
1981-82	0	0	55	353	535	954	1144	731	597	331	29	1	4730
1982-83	0	1	33	229	446	637	893	758	554	380	127	9	4067
1983-84	0	0	73	243	502	927	991	622	656	348	111	5	4478
1984-85	0	0	91	59	580	539	1006	716	482	170	49	11	3703
1985-86	1	0	53	167	307	901	904	723	529	208	95	2	3890
1986-87	0	18	26	225	520	805	957	782	564	341	55	0	4293
1987-88	0	0	11	389	450	729	1030	782	538	282	83	37	4331
1988-89	0	0	39	417	517	828	732	737	556	317	155	0	4298
1989-90	0	2	58	221	525	1100	645	536	455	301	87	8	3938
1990-91	0	0	54	206	408	659	804	631	469	231	37	4	3503
1991-92	0	0	41	195	546	697	780	678	562	294	144	5	3942
1992-93	0	0	48	266	504	773	766	796	653	270	29	3	4108
1993-94	0	0	33	263	510	840							

COOLING DEGREE DAYS Base 65 deg. F — LYNCHBURG, VIRGINIA

YEAR	JAN	FEB	MAR	APR	MAY	JUNE	JULY	AUG	SEP	OCT	NOV	DEC	TOTAL
1969	0	0	0	9	94	239	355	247	91	18	0	0	1053
1970	0	0	0	17	116	249	338	272	271	37	0	0	1300
1971	0	0	0	3	36	263	304	242	170	44	16	5	1083
1972	0	0	0	25	20	126	324	290	159	2	3	0	949
1973	0	0	3	16	34	260	308	330	188	24	0	0	1163
1974	0	0	3	30	75	113	298	296	103	7	10	0	935
1975	0	0	0	6	108	209	282	365	103	14	4	0	1091
1976	0	0	3	57	37	190	289	259	80	13	0	0	928
1977	0	0	5	38	129	184	424	349	167	9	8	0	1313
1978	0	0	0	9	88	237	305	389	212	5	0	0	1245
1979	0	0	11	12	54	143	298	322	121	21	0	0	982
1980	0	0	0	15	99	159	403	432	262	18	0	0	1388
1981	0	0	0	35	59	331	361	265	101	11	0	0	1163
1982	0	0	0	3	113	183	339	272	153	49	3	4	1119
1983	0	0	0	10	45	227	394	422	193	17	0	0	1308
1984	0	0	0	17	65	296	265	314	115	81	0	1	1154
1985	0	1	13	65	83	215	316	268	163	29	2	0	1155
1986	0	0	3	23	82	295	434	254	149	52	0	0	1292
1987	0	0	0	12	122	297	454	395	166	0	0	0	1446
1988	0	0	1	7	65	234	420	421	94	3	0	0	1245
1989	0	0	13	22	70	276	339	281	150	26	0	0	1177
1990	0	0	25	35	62	193	335	266	127	31	0	0	1074
1991	0	0	2	22	182	267	424	330	194	30	1	0	1452
1992	0	0	0	18	50	181	415	252	171	8	0	0	1095
1993	0	0	0	9	118	277	489	403	213	17	9	0	1535

SNOWFALL (inches) — LYNCHBURG, VIRGINIA

SEASON	JULY	AUG	SEP	OCT	NOV	DEC	JAN	FEB	MAR	APR	MAY	JUNE	TOTAL
1964-65	0.0	0.0	0.0	0.0	0.4	T	8.3	3.7	2.8	0.0	0.0	0.0	15.2
1965-66	0.0	0.0	0.0	0.0	0.0	0.1	31.8	6.3	T	0.0	0.0	0.0	38.2
1966-67	0.0	0.0	0.0	0.0	T	17.9	4.6	14.7	T	0.0	0.0	0.0	37.2
1967-68	0.0	0.0	0.0	0.0	T	4.0	3.9	3.6	T	0.0	0.0	0.0	11.5
1968-69	0.0	0.0	0.0	0.0	11.6	T	1.0	4.1	18.2	0.0	0.0	0.0	34.9
1969-70	0.0	0.0	0.0	0.0	T	12.7	5.8	1.6	T	T	0.0	0.0	20.1
1970-71	0.0	0.0	0.0	0.0	1.8	9.3	1.7	2.3	7.8	4.8	0.0	0.0	27.7
1971-72	0.0	0.0	0.0	0.0	5.9	T	T	15.2	T	T	0.0	0.0	21.1
1972-73	0.0	0.0	0.0	0.0	0.5	0.0	2.3	1.7	5.3	0.2	0.0	0.0	10.0
1973-74	0.0	0.0	0.0	0.0	0.0	6.6	0.0	5.9	T	0.0	0.0	0.0	12.5
1974-75	0.0	0.0	0.0	0.0	0.6	0.6	6.5	4.4	3.8	0.0	0.0	0.0	15.9
1975-76	0.0	0.0	0.0	0.0	0.0	0.1	T	1.2	0.6	0.0	0.0	0.0	1.9
1976-77	0.0	0.0	0.0	0.0	T	2.8	10.0	0.3	0.0	0.0	0.0	0.0	13.1
1977-78	0.0	0.0	0.0	0.0	0.7	0.3	10.8	5.9	8.1	0.0	0.0	0.0	25.8
1978-79	0.0	0.0	0.0	0.0	T	T	4.0	19.2	0.0	0.0	0.0	0.0	23.2
1979-80	0.0	0.0	0.0	2.4	0.0	T	14.2	7.5	8.9	0.0	0.0	0.0	33.0
1980-81	0.0	0.0	0.0	0.0	0.0	T	3.8	0.0	3.6	0.0	0.0	0.0	7.4
1981-82	0.0	0.0	0.0	0.0	3.7	5.9	7.6	15.1	0.2	2.5	0.0	0.0	35.0
1982-83	0.0	0.0	0.0	0.0	0.0	6.8	0.5	18.6	T	T	0.0	0.0	25.9
1983-84	0.0	0.0	0.0	0.0	0.0	T	3.8	0.1	0.4	T	0.0	0.0	4.3
1984-85	0.0	0.0	0.0	0.0	T	0.0	6.4	T	0.5	0.0	0.0	0.0	6.9
1985-86	0.0	0.0	0.0	0.0	0.0	1.2	2.5	8.2	T	0.0	0.0	0.0	11.9
1986-87	0.0	0.0	0.0	0.0	T	T	28.3	13.5	3.1	0.2	0.0	0.0	45.1
1987-88	0.0	0.0	0.0	0.0	1.2	0.0	8.3	0.0	0.0	T	0.0	0.0	9.5
1988-89	0.0	0.0	0.0	0.0	0.0	3.5	0.3	10.1	1.0	T	0.0	0.0	14.9
1989-90	0.0	0.0	0.0	0.0	1.0	11.2	T	0.2	5.5	0.0	0.0	0.0	17.9
1990-91	0.0	0.0	0.0	0.0	0.0	1.0	0.0	1.4	1.7	0.0	0.0	0.0	4.1
1991-92	0.0	0.0	0.0	0.0	T	T	T	0.8	0.9	1.4	0.0	0.0	3.1
1992-93	0.0	0.0	0.0	0.0	T	T	T	5.4	13.0	T	0.0	0.0	18.4
1993-94	0.0	0.0	0.0	T	0.0	5.4							
Record Mean	0.0	0.0	0.0	T	0.7	2.9	5.3	5.3	3.5	0.3	0.0	0.0	18.1

NORFOLK, VIRGINIA

The city of Norfolk, Virginia, is located near the coast and the southern border of the state. It is almost surrounded by water, with the Chesapeake Bay immediately to the north, Hampton Roads to the west, and the Atlantic Ocean only 18 miles to the east. It is traversed by numerous rivers and waterways and its average elevation above sea level is 13 feet. There are no nearby hilly areas and the land is low and level throughout the city. The climate is generally marine. The geographic location of the city with respect to the principal storm tracks, is especially favorable, being south of the average path of storms originating in the higher latitudes and north of the usual tracks of hurricanes and other tropical storms.

The winters are usually mild, while the autumn and spring seasons usually are delightful. Summers, though warm and long, frequently are tempered by cool periods, often associated with northeasterly winds off the Atlantic. Temperatures of 100 degrees or higher occur infrequently. Extreme cold waves seldom penetrate the area and temperatures of zero or below are almost nonexistent. Winters pass, on occasion, without a measurable amount of snowfall. Most of the snowfall in Norfolk is light and generally melts within 24 hours.

Based on the 1951–1980 period, the average first occurrence of 32 degrees Fahrenheit in the fall is November 17 and the average last occurrence in the spring is March 23.

NORMALS, MEANS, AND EXTREMES

NORFOLK, VIRGINIA

LATITUDE: 36°54'N LONGITUDE: 76°12'W ELEVATION: FT. GRND 24 BARO 44 TIME ZONE: EASTERN WBAN: 13737

	(a)	JAN	FEB	MAR	APR	MAY	JUNE	JULY	AUG	SEP	OCT	NOV	DEC	YEAR
TEMPERATURE °F:														
Normals														
-Daily Maximum		47.3	49.7	57.9	66.9	75.3	82.9	86.4	85.1	79.6	69.5	61.2	52.2	67.8
-Daily Minimum		30.9	32.3	39.3	47.1	56.8	65.2	70.0	69.4	64.2	52.9	43.8	35.4	50.6
-Monthly		39.1	41.0	48.6	57.0	66.1	74.1	78.2	77.2	71.9	61.2	52.5	43.8	59.2
Extremes														
-Record Highest	45	78	81	88	97	100	101	103	104	99	95	86	80	104
-Year		1970	1989	1990	1960	1991	1964	1993	1980	1983	1954	1974	1991	AUG 1980
-Record Lowest	45	-3	8	18	28	36	45	54	49	45	27	20	7	-3
-Year		1985	1965	1980	1982	1966	1967	1979	1982	1967	1976	1950	1983	JAN 1985
NORMAL DEGREE DAYS:														
Heating (base 65°F)		803	672	508	249	51	0	0	0	11	164	380	657	3495
Cooling (base 65°F)		0	0	0	9	85	277	409	378	218	46	0	0	1422
% OF POSSIBLE SUNSHINE	29	54	57	60	63	63	67	63	62	62	59	57	56	60
MEAN SKY COVER (tenths)														
Sunrise - Sunset	45	6.3	6.2	6.1	5.8	6.1	5.8	6.0	5.9	5.7	5.4	5.5	6.1	5.9
MEAN NUMBER OF DAYS:														
Sunrise to Sunset														
-Clear	45	8.9	8.1	8.8	8.7	7.8	7.5	7.4	7.7	9.0	11.6	10.4	9.2	105.2
-Partly Cloudy	45	6.5	6.3	7.6	9.3	10.0	11.9	11.9	12.0	9.6	7.1	8.0	7.0	107.1
-Cloudy	45	15.6	13.8	14.7	12.0	13.2	10.6	11.7	11.2	11.4	12.3	11.7	14.8	153.0
Precipitation														
.01 inches or more	45	10.6	10.2	11.0	10.0	9.9	9.1	11.1	10.4	7.9	7.6	8.0	9.2	115.1
Snow, Ice pellets, hail														
1.0 inches or more	45	0.8	0.7	0.2	0.*	0.0	0.0	0.0	0.0	0.0	0.0	0.0	0.3	2.0
Thunderstorms	45	0.4	0.7	1.9	2.7	4.8	5.8	8.3	6.9	2.7	1.3	0.5	0.4	36.3
Heavy Fog Visibility														
1/4 mile or less	45	2.1	2.6	1.9	1.4	1.9	1.0	0.5	1.0	1.2	2.2	1.8	2.2	19.9
Temperature °F														
-Maximum														
90° and above	45	0.0	0.0	0.0	0.4	1.6	6.7	11.9	8.9	2.9	0.1	0.0	0.0	32.5
32° and below	45	2.5	1.2	0.1	0.0	0.0	0.0	0.0	0.0	0.0	0.0	0.0	1.0	4.8
-Minimum														
32° and below	45	16.2	14.2	6.2	0.3	0.0	0.0	0.0	0.0	0.0	0.1	3.0	13.3	53.5
0° and below	45	0.*	0.0	0.0	0.0	0.0	0.0	0.0	0.0	0.0	0.0	0.0	0.0	*
AVG. STATION PRESS. (mb)	21	1018.4	1018.0	1016.4	1015.0	1015.0	1015.0	1015.6	1016.7	1017.6	1018.6	1018.9	1019.2	1017.0
RELATIVE HUMIDITY (%)														
Hour 01	45	73	72	72	73	81	83	84	86	84	82	76	73	78
Hour 07 (Local Time)	45	74	74	74	73	77	79	81	84	83	83	79	75	78
Hour 13	45	59	56	54	50	56	56	59	61	61	59	56	58	57
Hour 19	45	67	65	62	61	66	67	70	74	75	74	70	68	68
PRECIPITATION (inches):														
Water Equivalent														
-Normal		3.78	3.47	3.70	3.06	3.81	3.82	5.06	4.81	3.90	3.15	2.85	3.23	44.64
-Maximum Monthly	45	9.93	6.23	8.50	7.25	10.12	9.72	13.73	14.32	13.80	10.12	7.01	6.10	14.32
-Year		1987	1983	1989	1984	1979	1963	1975	1992	1979	1971	1951	1983	AUG 1992
-Minimum Monthly	45	1.05	0.84	0.75	0.43	0.64	0.37	0.36	0.74	0.26	0.57	0.49	0.67	0.26
-Year		1981	1991	1986	1985	1991	1954	1993	1975	1986	1984	1965	1988	SEP 1986
-Maximum in 24 hrs	45	3.80	2.71	3.18	5.90	3.41	6.85	5.64	11.40	6.79	4.38	3.35	2.76	11.40
-Year		1967	1983	1958	1991	1980	1963	1969	1964	1959	1971	1952	1983	AUG 1964
Snow, Ice pellets, hail														
-Maximum Monthly	45	14.2	24.4	13.7	1.2	0.0	T	0.0	T	0.0	0.0	0.6	14.7	24.4
-Year		1966	1989	1980	1964		1990		1991			1950	1958	FEB 1989
-Maximum in 24 hrs	45	9.1	14.2	9.9	1.2	0.0	T	0.0	T	0.0	0.0	0.6	11.4	14.2
-Year		1973	1989	1980	1964		1990		1991			1950	1958	FEB 1989
WIND:														
Mean Speed (mph)	45	11.5	11.9	12.4	11.8	10.5	9.8	9.0	8.8	9.6	10.3	10.6	11.1	10.6
Prevailing Direction through 1963		SW	NNE	SW	SW	SW	SW	SW	SW	NE	NE	SW	SW	SW
Fastest Obs. 1 Min.														
-Direction (!!)	21	23	36	22	02	28	30	34	35	30	04	21	01	04
-Speed (MPH)	21	39	44	46	41	38	46	46	46	46	48	40	39	48
-Year		1978	1973	1973	1990	1989	1977	1973	1979	1985	1982	1989	1989	OCT 1982
Peak Gust														
-Direction (!!)	10	N	E	SW	N	NW	SW	N	E	NW	N	E	N	N
-Speed (mph)	10	58	56	66	56	66	69	63	63	67	69	55	53	69
-Date		1987	1984	1993	1990	1984	1987	1986	1986	1985	1990	1985	1989	OCT 1990

PRECIPITATION (inches)

NORFOLK, VIRGINIA

YEAR	JAN	FEB	MAR	APR	MAY	JUNE	JULY	AUG	SEP	OCT	NOV	DEC	ANNUAL
1964	4.56	4.56	2.26	2.38	1.56	2.58	7.33	10.58	12.26	5.55	1.14	2.95	57.71
1965	2.73	2.53	2.83	2.24	1.48	4.69	3.46	3.08	0.77	1.29	0.49	1.08	26.67
1966	4.86	3.83	1.50	1.68	5.95	1.82	4.26	5.24	3.39	1.25	1.05	3.13	37.96
1967	5.44	3.56	1.34	1.31	3.25	1.37	7.21	11.19	3.02	0.93	1.75	4.84	45.21
1968	3.62	2.01	4.76	3.17	2.16	3.07	4.23	2.04	1.51	4.44	3.56	3.14	37.71
1969	2.26	2.16	4.88	2.07	2.05	4.13	12.70	5.28	2.72	3.18	2.97	3.93	48.33
1970	2.27	3.97	3.37	3.19	2.58	4.10	5.33	2.04	1.72	1.30	2.34	3.01	35.22
1971	4.03	3.59	3.88	2.18	4.46	2.16	4.81	4.63	5.46	10.12	0.97	1.44	47.73
1972	2.94	3.50	2.55	2.15	3.35	4.93	4.65	1.60	6.91	4.09	5.44	4.12	46.23
1973	2.54	3.21	4.69	3.44	3.62	5.93	4.19	7.92	0.86	1.37	1.90	5.83	45.50
1974	3.52	2.98	5.16	3.34	3.74	4.76	5.47	8.33	4.40	1.23	1.22	3.81	47.96
1975	4.18	4.18	5.72	4.19	3.37	1.16	13.73	0.74	4.82	3.19	1.63	3.62	50.53
1976	2.51	1.50	2.21	0.99	3.74	1.59	5.19	2.62	3.51	2.90	2.38	3.22	32.36
1977	3.33	2.23	4.05	2.20	3.86	2.41	2.70	4.57	3.00	6.09	5.41	3.92	43.77
1978	6.32	1.91	7.80	2.90	5.64	7.84	4.19	1.66	1.17	1.50	4.40	2.31	47.64
1979	6.47	5.01	5.13	7.00	10.12	2.97	4.69	1.79	13.80	1.74	5.26	0.98	64.96
1980	4.54	2.91	4.40	3.25	5.17	1.39	1.85	4.54	1.47	4.21	2.01	2.64	38.38
1981	1.05	2.26	1.88	2.26	2.75	5.00	5.10	6.87	3.18	3.28	1.78	5.77	41.18
1982	3.35	5.81	3.04	1.71	3.07	4.22	5.83	6.51	3.63	4.25	3.43	4.30	49.15
1983	2.21	6.23	4.55	6.13	3.52	3.84	0.77	3.07	4.52	5.29	3.24	6.10	49.47
1984	2.77	4.66	5.09	7.25	6.23	1.50	7.66	2.25	1.94	0.57	2.68	2.22	44.82
1985	3.98	3.53	2.02	0.43	3.23	6.81	6.14	1.89	6.36	3.92	5.71	0.79	44.81
1986	2.52	2.71	0.75	3.31	1.41	1.51	2.59	4.80	0.26	1.67	1.21	3.74	26.48
1987	9.93	3.11	2.30	3.83	2.65	2.98	3.20	2.04	7.00	1.81	3.51	2.33	44.69
1988	3.12	2.70	2.11	3.53	5.49	3.83	2.93	5.69	1.74	2.85	4.02	0.67	38.68
1989	2.70	5.80	8.50	3.62	2.97	5.10	4.86	7.49	5.10	2.94	3.69	3.86	56.63
1990	3.26	2.93	3.49	3.55	3.79	3.51	4.06	11.85	1.00	3.73	1.68	2.67	45.52
1991	4.74	0.84	4.70	6.39	0.64	4.54	6.46	3.77	2.04	4.65	1.72	2.43	42.92
1992	4.48	2.07	2.63	1.26	3.46	2.22	4.52	14.32	2.06	2.85	4.26	3.15	47.28
1993	4.89	2.36	5.91	3.59	2.88	2.79	0.36	1.45	4.14	3.40	0.97	3.29	36.03
Record Mean	3.37	3.34	3.76	3.25	3.68	3.98	5.54	5.35	3.83	3.13	2.66	3.19	45.06

AVERAGE TEMPERATURE (deg. F)

NORFOLK, VIRGINIA

YEAR	JAN	FEB	MAR	APR	MAY	JUNE	JULY	AUG	SEP	OCT	NOV	DEC	ANNUAL
1964	42.3	39.9	49.5	55.2	66.4	74.5	77.3	74.4	70.8	57.7	54.6	46.5	59.1
1965	39.7	41.0	44.4	54.5	69.7	72.8	76.7	77.1	73.6	59.0	51.6	43.5	58.6
1966	35.8	38.6	47.9	54.5	63.6	71.6	78.0	74.9	69.9	59.3	50.5	41.4	57.1
1967	45.7	39.9	47.7	58.0	61.2	71.2	76.3	75.2	66.5	58.7	45.9	44.0	57.5
1968	34.8	34.0	50.0	55.1	64.6	74.9	78.0	80.5	71.5	63.1	52.9	41.4	58.4
1969	38.5	39.8	44.7	59.8	67.4	77.1	79.2	76.2	71.1	62.2	49.1	40.9	58.9
1970	33.9	39.3	44.7	56.7	67.5	74.9	76.9	78.0	74.7	63.7	51.7	47.0	59.1
1971	38.6	44.7	46.9	55.9	65.0	76.0	77.2	75.7	73.2	66.7	52.4	52.3	60.4
1972	46.4	43.2	49.0	56.4	63.5	70.5	77.6	75.8	71.9	59.2	51.6	49.2	59.5
1973	40.5	39.7	52.3	58.5	66.9	76.9	78.3	78.5	75.0	64.2	53.5	46.2	60.9
1974	48.6	43.4	53.1	60.8	66.8	72.8	78.3	77.4	71.4	58.7	53.5	46.0	60.9
1975	46.0	45.4	47.4	52.7	68.3	77.0	78.6	79.6	72.3	63.4	55.7	43.2	60.8
1976	38.9	49.9	53.4	61.9	66.3	75.9	78.2	75.9	71.1	57.7	45.9	41.4	59.7
1977	29.2	41.5	54.7	61.9	68.2	74.3	81.4	81.0	76.3	60.5	54.8	43.5	60.6
1978	37.0	32.6	46.1	57.2	65.6	74.1	76.1	80.5	73.2	60.8	56.0	45.3	58.7
1979	39.4	33.3	49.1	58.1	66.7	70.4	77.1	78.5	72.8	60.4	56.4	44.9	58.9
1980	40.3	34.7	46.5	58.6	67.8	73.9	80.9	80.9	76.1	60.4	49.9	42.3	59.4
1981	32.7	43.1	45.4	61.2	65.1	78.3	79.8	75.1	70.7	59.6	50.7	41.0	58.6
1982	35.4	42.0	48.8	55.0	69.4	73.4	78.6	75.3	70.0	60.2	54.4	48.8	59.3
1983	40.2	40.8	51.0	55.7	65.8	73.0	80.3	79.0	72.8	62.7	52.6	41.6	59.6
1984	35.5	46.7	45.4	55.6	67.8	76.2	76.7	78.4	70.5	66.9	49.9	50.9	60.0
1985	34.9	40.4	51.8	62.0	68.8	74.2	78.2	77.2	73.4	65.9	60.3	41.2	60.7
1986	39.3	42.1	49.9	57.3	67.6	76.1	82.1	76.6	72.4	65.4	54.9	44.8	60.7
1987	39.6	38.7	47.5	54.6	68.3	77.0	82.4	79.6	74.3	56.6	54.3	46.0	59.9
1988	37.3	42.5	49.5	56.5	65.8	73.6	80.1	80.8	70.5	56.9	54.2	42.4	59.2
1989	45.3	43.6	50.1	56.5	65.6	78.5	79.2	77.7	73.9	62.7	53.3	34.8	60.1
1990	47.3	50.2	53.2	58.7	66.6	75.5	80.6	78.0	71.6	65.9	55.0	50.5	62.8
1991	43.5	46.0	52.7	61.6	72.8	76.2	82.0	79.5	72.3	61.9	52.7	47.6	62.4
1992	42.7	45.0	48.8	58.2	62.4	72.0	81.8	75.6	73.5	59.7	53.8	44.6	59.8
1993	44.2	39.1	46.8	57.3	68.5	76.1	83.4	78.9	74.6	61.3	53.5	41.5	60.4
Record Mean	41.2	42.2	48.9	57.5	66.8	74.7	78.8	77.6	72.5	62.1	52.3	43.7	59.8
Max	49.0	50.5	57.8	66.9	75.8	83.3	87.0	85.1	79.8	69.9	60.3	51.4	68.1
Min	33.3	33.9	40.0	48.1	57.7	66.1	70.6	70.0	65.1	54.3	44.2	35.9	51.6

REFERENCE NOTES

NORFOLK, VIRGINIA

GENERAL
T - TRACE AMOUNT.
BLANK ENTRIES DENOTE MISSING/UNREPORTED DATA.
# INDICATES A STATION OR INSTRUMENT RELOCATION.

SPECIFIC

PM - INCLUDES LAST DAY OF PREVIOUS MONTH
ASOS - AUTOMATED SURFACE OBSERVING SYSTEM IN OPERATION DURING THESE MONTHS.

(a) - LENGTH OF RECORD IN YEARS, ALTHOUGH INDIVIDUAL MONTHS MAY BE MISSING.
0.* OR * - THE VALUE IS BETWEEN 0.0 AND 0.05
NORMALS - BASED ON THE 1961-1990 RECORD PERIOD.
EXTREMES - DATES ARE THE MOST RECENT OCCURRENCE
WIND DIR.- NUMERALS SHOW TENS OF DEGRESS CLOCKWISE FROM TRUE NORTH. "00" INDICATES CALM.
RESULTANT DIRECTIONS ARE GIVEN TO WHOLE DEGREES.
BOLD VALUES INDICATE EXTREME VALUES WHICH OCCURRED AFTER THE ASOS SYSTEM WAS COMMISSIONED.

RECORD = PERIOD OF RECORD
RECORD MEAN PRECIPITATION IS THE MEAN OF ALL DAILY PRECIPITATION AMOUNTS DURING THE PERIOD OF RECORD.
RECORD MAX(MIN) TEMPERATURE IS THE MEAN OF ALL DAILY MAX(MIN) TEMPERATURES DURING THE PERIOD OF RECORD.
RECORD MEAN TEMPERATURE IS THE SUM OF THE RECORD MAX AND RECORD MIN DIVIDED BY 2.
AVERAGE TEMPERATURE IS THE SUM OF THE MEAN DAILY MAX AND MIN TEMPERATURE DIVIDED BY 2.

EXCEPTIONS

PERCENT OF POSSIBLE SUNSHINE IS THROUGH 1980.

RECORD MEANS ARE THROUGH THE CURRENT YEAR, BEGINNING IN 1875 FOR TEMPERATURE
1871 FOR PRECIPITATION
1949 FOR SNOWFALL

HEATING DEGREE DAYS Base 65 deg. F — NORFOLK, VIRGINIA

SEASON	JULY	AUG	SEP	OCT	NOV	DEC	JAN	FEB	MAR	APR	MAY	JUNE	TOTAL
1964-65	0	0	4	232	312	575	780	667	635	320	29	15	3569
1965-66	0	6	1	195	398	657	897	734	527	330	121	21	3887
1966-67	0	0	22	191	437	725	588	699	533	244	157	21	3617
1967-68	0	0	36	211	566	644	928	895	471	294	88	0	4133
1968-69	0	0	0	124	361	726	814	697	624	192	44	0	3582
1969-70	0	0	8	131	469	741	960	714	622	263	57	0	3965
1970-71	0	0	16	93	393	552	812	567	555	269	69	0	3326
1971-72	0	0	3	27	391	390	572	628	494	272	81	11	2869
1972-73	0	0	4	197	406	486	752	703	403	217	47	0	3215
1973-74	0	0	0	83	353	575	504	599	377	183	63	0	2737
1974-75	0	0	16	213	371	584	584	547	541	382	47	0	3285
1975-76	0	0	6	98	290	671	804	443	362	186	62	6	2928
1976-77	0	0	0	245	566	726	1104	657	330	150	40	1	3819
1977-78	0	0	0	158	321	661	860	902	580	235	72	3	3792
1978-79	0	0	3	162	268	614	787	879	499	213	52	5	3482
1979-80	0	0	0	190	272	616	759	872	564	196	58	2	3529
1980-81	0	0	11	181	449	699	994	610	605	159	96	0	3804
1981-82	0	0	12	189	423	739	907	636	495	303	21	0	3725
1982-83	0	4	6	177	334	498	762	674	426	295	85	3	3264
1983-84	0	0	27	126	370	718	908	522	601	281	54	3	3610
1984-85	0	0	16	37	450	432	928	686	421	172	21	0	3163
1985-86	0	0	6	61	162	731	790	637	465	228	69	1	3150
1986-87	0	1	8	88	311	620	779	730	538	306	58	0	3439
1987-88	0	0	0	252	320	582	851	646	474	266	86	15	3492
1988-89	0	0	2	265	324	692	602	601	486	282	80	0	3334
1989-90	0	0	12	134	356	928	541	417	410	234	39	3	3074
1990-91	0	0	13	102	301	444	657	527	386	166	19	2	2617
1991-92	0	0	22	132	377	542	686	575	496	262	125	1	3218
1992-93	0	0	9	179	337	623	638	722	555	246	22	5	3336
1993-94	0	0	10	140	353	726							

COOLING DEGREE DAYS Base 65 deg. F — NORFOLK, VIRGINIA

YEAR	JAN	FEB	MAR	APR	MAY	JUNE	JULY	AUG	SEP	OCT	NOV	DEC	TOTAL
1969	0	0	0	42	125	369	446	357	199	49	0	0	1587
1970	1	0	0	19	140	303	374	412	311	60	0	0	1620
1971	0	3	0	3	76	336	383	343	259	87	23	5	1518
1972	0	0	8	20	40	183	398	343	217	22	10	2	1243
1973	0	0	16	27	112	363	420	424	307	64	17	1	1751
1974	3	0	16	64	124	244	419	390	213	26	32	0	1531
1975	2	3	0	22	157	366	429	460	233	55	17	0	1744
1976	1	13	11	102	110	337	417	347	193	27	0	0	1558
1977	0	4	16	66	145	289	515	502	347	24	22	0	1930
1978	0	0	0	9	96	286	352	487	257	36	3	9	1535
1979	0	0	11	13	112	171	385	426	239	54	22	0	1433
1980	0	0	0	11	153	274	499	497	351	45	1	1	1832
1981	0	0	0	51	103	407	468	320	189	29	0	0	1567
1982	0	0	1	8	166	257	428	331	164	39	21	4	1419
1983	0	0	0	21	115	250	481	440	265	62	4	0	1638
1984	0	0	0	5	146	345	368	426	188	102	5	2	1587
1985	0	5	20	91	146	284	419	382	267	97	28	0	1739
1986	0	0	2	2	153	343	537	367	237	109	15	0	1765
1987	0	0	0	2	168	364	544	461	285	0	7	0	1831
1988	0	0	1	18	118	280	477	498	173	17	10	0	1592
1989	0	9	30	31	106	412	447	399	286	69	10	0	1799
1990	0	8	52	51	98	324	489	407	218	137	8	5	1797
1991	0	0	13	71	269	347	534	456	248	43	18	11	2010
1992	0	0	0	63	51	220	527	336	270	23	9	0	1499
1993	0	0	0	23	136	345	579	438	305	31	18	0	1875

SNOWFALL (inches) — NORFOLK, VIRGINIA

SEASON	JULY	AUG	SEP	OCT	NOV	DEC	JAN	FEB	MAR	APR	MAY	JUNE	TOTAL
1964-65	0.0	0.0	0.0	0.0	T	0.0	10.6	3.9	T	0.0	0.0	0.0	14.5
1965-66	0.0	0.0	0.0	0.0	0.0	T	14.2	0.5	0.0	0.0	0.0	0.0	14.7
1966-67	0.0	0.0	0.0	0.0	T	1.0	4.2	5.1	T	0.0	0.0	0.0	10.3
1967-68	0.0	0.0	0.0	0.0	T	2.0	1.5	2.9	0.9	0.0	0.0	0.0	7.3
1968-69	0.0	0.0	0.0	0.0	T	3.8	T	0.8	1.9	0.0	0.0	0.0	6.5
1969-70	0.0	0.0	0.0	0.0	0.0	T	3.0	2.8	T	0.0	0.0	0.0	5.8
1970-71	0.0	0.0	0.0	0.0	T	T	T	2.4	4.2	T	0.0	0.0	6.6
1971-72	0.0	0.0	0.0	0.0	T	T	0.0	1.8	T	T	0.0	0.0	1.8
1972-73	0.0	0.0	0.0	0.0	T	0.0	9.1	4.7	T	0.0	0.0	0.0	13.8
1973-74	0.0	0.0	0.0	0.0	0.0	1.4	T	0.9	7.5	0.0	0.0	0.0	9.8
1974-75	0.0	0.0	0.0	0.0	0.0	T	0.3	T	0.8	T	0.0	0.0	1.1
1975-76	0.0	0.0	0.0	0.0	0.0	T	T	T	0.0	0.0	0.0	0.0	T
1976-77	0.0	0.0	0.0	0.0	T	1.0	4.7	1.4	0.0	0.0	0.0	0.0	7.1
1977-78	0.0	0.0	0.0	0.0	0.0	T	1.3	9.2	2.3	0.0	0.0	0.0	12.8
1978-79	0.0	0.0	0.0	0.0	0.0	0.0	1.0	12.7	T	0.0	0.0	0.0	13.7
1979-80	0.0	0.0	0.0	0.0	0.0	0.0	9.3	18.9	13.7	0.0	0.0	0.0	41.9
1980-81	0.0	0.0	0.0	0.0	0.0	T	T	0.0	0.3	0.0	0.0	0.0	0.3
1981-82	0.0	0.0	0.0	0.0	0.0	1.8	4.2	0.1	T	T	0.0	0.0	6.1
1982-83	0.0	0.0	0.0	0.0	0.0	0.4	T	3.0	T	T	0.0	0.0	3.4
1983-84	0.0	0.0	0.0	0.0	0.0	T	T	5.2	T	0.0	0.0	0.0	5.2
1984-85	0.0	0.0	0.0	0.0	0.0	T	4.3	0.0	T	0.0	0.0	0.0	4.3
1985-86	0.0	0.0	0.0	0.0	0.0	T	3.6	1.1	T	T	0.0	0.0	4.7
1986-87	0.0	0.0	0.0	0.0	0.0	T	1.6	1.0	1.2	T	0.0	0.0	3.8
1987-88	0.0	0.0	0.0	0.0	0.3	T	4.4	T	T	0.0	0.0	0.0	4.7
1988-89	0.0	0.0	0.0	0.0	0.0	T	T	24.4	T	0.5	0.0	0.0	24.9
1989-90	0.0	0.0	0.0	0.0	0.0	0.5	0.0	0.0	T	0.0	0.0	T	0.5
1990-91	0.0	0.0	0.0	0.0	0.0	T	0.0	T	T	0.0	0.0	0.0	T
1991-92	0.0	T	0.0	0.0	0.0	T	T	0.0	T	T	0.0	0.0	T
1992-93	0.0	0.0	0.0	0.0	0.0	0.0	0.3	1.9	1.7	0.0	0.0	0.0	3.9
1993-94	0.0	0.0	0.0	0.0	0.0	3.9							
Record Mean	0.0	T	0.0	0.0	T	0.9	2.6	2.9	1.0	T	0.0	T	7.5

RICHMOND, VIRGINIA

Richmond is located in east-central Virginia at the head of navigation on the James River and along a line separating the Coastal Plains (Tidewater Virginia) from the Piedmont. The Blue Ridge Mountains lie about 90 miles to the west and the Chesapeake Bay 60 miles to the east. Elevations range from a few feet above sea level along the river to a little over 300 feet in parts of the western section of the city.

The climate might be classified as modified continental. Summers are warm and humid and winters generally mild. The mountains to the west act as a partial barrier to outbreaks of cold, continental air in winter. The cold winter air is delayed long enough to be modified, then further warmed as it subsides in its approach to Richmond. The open waters of the Chesapeake Bay and Atlantic Ocean contribute to the humid summers and mild winters. The coldest weather normally occurs in late December and January, when low temperatures usually average in the upper 20s, and the high temperatures in the upper 40s. Temperatures seldom lower to zero, but there have been several occurrences of below zero temperatures. Summertime high temperatures above 100 degrees are not uncommon, but do not occur every year.

Precipitation is rather uniformly distributed throughout the year. However, dry periods lasting several weeks do occur, especially in autumn when long periods of pleasant, mild weather are most common. There is considerable variability in total monthly amounts from year to year. Snow usually remains on the ground only one or two days at a time. Ice storms (freezing rain or glaze) are not uncommon, but they are seldom severe enough to do any considerable damage. A notable exception was the spectacular glaze storm of January 27–28, 1943, when nearly 1 inch of ice accumulation caused heavy damage to trees and overhead transmission lines.

The James River reaches tidewater at Richmond where flooding may occur in every month of the year, most frequently in March and least in July. Hurricanes and tropical storms have been responsible for most of the flooding during the summer and early fall months. Hurricanes passing near Richmond have produced record rainfalls. In 1955, three hurricanes brought record rainfall to Richmond within a six-week period. The most noteworthy of these were Hurricanes Connie and Diane that brought heavy rains five days apart.

Damaging storms occur mainly from snow and freezing rain in winter and from hurricanes, tornadoes, and severe thunderstorms in other seasons. Damage may be from wind, flooding, or rain, or from any combination of these. Tornadoes are infrequent but some notable occurrences have been observed within the Richmond area.

Based on the 1951–1980 period, the average first occurrence of 32 degrees Fahrenheit in the fall is October 26 and the average last occurrence in the spring is April 10.

NORMALS, MEANS, AND EXTREMES

RICHMOND, VIRGINIA

LATITUDE: 37°30'N LONGITUDE: 77°20'W ELEVATION: FT. GRND 164 BARO 178 TIME ZONE: EASTERN WBAN: 13740

	(a)	JAN	FEB	MAR	APR	MAY	JUNE	JULY	AUG	SEP	OCT	NOV	DEC	YEAR
TEMPERATURE °F:														
Normals														
-Daily Maximum		45.7	49.2	59.5	70.0	77.8	85.1	88.4	87.1	80.9	70.7	61.3	50.2	68.8
-Daily Minimum		25.7	28.1	36.3	44.6	54.2	62.7	67.5	66.4	59.0	46.5	37.9	29.9	46.6
-Monthly		35.7	38.7	48.0	57.3	66.0	73.9	78.0	76.8	70.0	58.6	49.6	40.1	57.7
Extremes														
-Record Highest	64	80	83	93	96	100	104	105	102	103	99	86	80	105
-Year		1950	1932	1938	1990	1941	1952	1977	1983	1954	1941	1993	1971	JUL 1977
-Record Lowest	64	-12	-10	11	23	31	40	51	46	35	21	10	-1	-12
-Year		1940	1936	1960	1985	1956	1967	1965	1934	1974	1962	1933	1942	JAN 1940
NORMAL DEGREE DAYS:														
Heating (base 65°F)		908	736	527	241	61	0	0	0	23	233	462	772	3963
Cooling (base 65°F)		0	0	0	10	92	270	403	366	173	34	0	0	1348
% OF POSSIBLE SUNSHINE	43	54	58	62	66	66	70	68	66	64	63	59	54	63
MEAN SKY COVER (tenths)														
Sunrise - Sunset	48	6.4	6.3	6.2	6.1	6.3	6.0	6.1	6.1	5.8	5.4	5.8	6.2	6.1
MEAN NUMBER OF DAYS:														
Sunrise to Sunset														
-Clear	48	8.3	8.2	8.2	7.8	7.0	6.8	6.9	7.2	9.2	11.5	9.3	9.4	99.8
-Partly Cloudy	48	6.7	6.4	8.2	9.2	10.1	11.9	11.6	11.6	8.6	7.3	7.7	6.4	105.8
-Cloudy	48	16.1	13.6	14.6	13.0	13.9	11.3	12.5	12.2	12.2	12.1	13.0	15.2	159.6
Precipitation														
.01 inches or more	56	10.3	9.2	10.8	9.4	10.7	9.4	11.1	9.7	8.1	7.3	8.3	9.0	113.3
Snow,Ice pellets,hail														
1.0 inches or more	56	1.3	1.1	0.6	0.1	0.0	0.0	0.0	0.0	0.0	0.0	0.1	0.6	3.8
Thunderstorms	56	0.2	0.4	1.6	2.4	5.4	6.7	8.7	6.6	2.9	1.0	0.6	0.3	36.6
Heavy Fog Visibility														
1/4 mile or less	64	2.7	2.0	1.7	1.6	1.9	1.5	2.1	2.5	3.0	3.3	2.2	2.8	27.2
Temperature °F														
-Maximum														
90° and above	64	0.0	0.0	0.1	0.8	2.7	9.4	14.1	11.0	4.5	0.4	0.0	0.0	42.9
32° and below	64	3.2	1.4	0.2	0.0	0.0	0.0	0.0	0.0	0.0	0.0	0.*	1.7	6.5
-Minimum														
32° and below	64	21.3	18.8	10.4	2.1	0.1	0.0	0.0	0.0	0.0	1.8	10.0	20.1	84.5
0° and below	64	0.4	0.1	0.0	0.0	0.0	0.0	0.0	0.0	0.0	0.0	0.0	0.*	0.5
AVG. STATION PRESS.(mb)	21	1012.9	1012.5	1010.8	1009.3	1009.4	1009.5	1010.1	1011.3	1012.4	1013.3	1013.5	1013.7	1011.6
RELATIVE HUMIDITY (%)														
Hour 01	59	77	74	73	74	83	86	88	90	90	87	80	77	82
Hour 07 (Local Time)	59	80	79	78	76	80	82	85	89	90	89	84	81	83
Hour 13	59	57	53	49	46	51	53	56	57	56	53	51	55	53
Hour 19	59	68	63	59	55	65	67	71	75	78	76	69	70	68
PRECIPITATION (inches):														
Water Equivalent														
-Normal		3.24	3.16	3.61	2.96	3.84	3.62	5.03	4.40	3.34	3.53	3.17	3.26	43.16
-Maximum Monthly	56	7.97	5.97	8.65	7.31	8.87	9.24	18.87	14.10	10.98	9.39	7.64	7.07	18.87
-Year		1978	1979	1984	1987	1972	1938	1945	1955	1975	1971	1959	1973	JUL 1945
-Minimum Monthly	56	0.64	0.48	0.94	0.64	0.87	0.38	0.51	0.52	0.26	0.30	0.36	0.40	0.26
-Year		1981	1978	1966	1963	1965	1980	1983	1943	1978	1963	1965	1980	SEP 1978
-Maximum in 24 hrs	56	3.31	2.67	3.43	2.97	3.08	4.61	5.73	8.79	4.02	6.50	4.07	3.16	8.79
-Year		1962	1979	1992	1987	1981	1963	1969	1955	1985	1961	1956	1958	AUG 1955
Snow,Ice pellets,hail														
-Maximum Monthly	56	28.5	21.4	19.7	2.0	T	0.0	0.0	0.0	0.0	T	7.3	12.5	28.5
-Year		1940	1983	1960	1940	1989					1979	1953	1958	JAN 1940
-Maximum in 24 hrs	56	21.6	16.8	12.1	2.0	T	0.0	0.0	0.0	0.0	T	7.3	7.5	21.6
-Year		1940	1983	1962	1940	1989					1979	1953	1966	JAN 1940
WIND:														
Mean Speed (mph)	45	8.1	8.7	9.1	9.0	7.8	7.4	6.9	6.4	6.7	7.0	7.6	7.8	7.7
Prevailing Direction through 1963		S	NNE	W	S	SSW	S	SSW	S	S	NNE	S	SW	S
Fastest Obs. 1 Min.														
-Direction (!!)	8	23	32	27	21	23	30	23	18	36	36	27	21	23
-Speed (MPH)	8	30	32	41	29	46	32	40	30	37	29	32	32	46
-Year		1992	1987	1989	1990	1989	1987	1992	1991	1989	1990	1989	1988	MAY 1989
Peak Gust														
-Direction (!!)	9	NW	NW	W	N	SW	NW	N	SW	N	N	SW	SW	SW
-Speed (mph)	9	48	48	67	49	79	53	61	48	49	43	54	49	79
-Date		1985	1987	1989	1992	1989	1987	1986	1992	1989	1990	1988	1988	MAY 1989

PRECIPITATION (inches) **RICHMOND, VIRGINIA**

YEAR	JAN	FEB	MAR	APR	MAY	JUNE	JULY	AUG	SEP	OCT	NOV	DEC	ANNUAL
1964	4.16	4.46	2.61	2.71	1.14	2.40	6.46	9.88	2.56	3.62	1.98	3.05	45.03
1965	2.51	2.77	3.68	2.13	0.87	3.39	6.33	0.81	4.81	1.38	0.36	0.72	29.76
1966	4.58	3.80	0.94	2.18	2.58	2.54	4.07	1.31	5.06	4.81	1.31	3.07	36.25
1967	1.50	3.35	2.34	1.32	3.71	3.58	5.00	6.65	0.95	1.00	1.76	6.48	37.64
1968	2.53	0.98	4.00	2.93	3.13	2.89	3.41	3.71	1.78	1.59	3.87	2.28	33.10
1969	2.04	3.95	3.95	2.60	3.32	4.36	13.90	9.31	3.89	1.88	1.87	5.26	56.33
1970	1.32	2.37	3.70	2.84	1.84	1.12	4.74	1.69	1.02	1.55	3.10	3.00	28.29
1971	1.84	4.37	2.68	1.76	6.82	4.10	4.40	3.73	2.35	9.39	2.76	0.75	44.95
1972	1.43	5.15	2.11	3.35	8.87	8.82	5.80	3.84	3.35	7.89	5.82	2.91	59.34
1973	2.66	3.11	3.44	4.58	3.56	2.45	3.64	4.34	1.82	2.56	1.27	7.07	40.50
1974	3.21	2.54	3.79	1.58	3.02	1.80	2.25	6.84	4.83	0.39	1.23	4.22	35.70
1975	5.71	2.96	8.04	2.78	2.59	4.00	12.29	2.31	10.98	3.10	2.04	4.51	61.31
1976	3.39	1.35	2.14	1.08	3.76	2.85	2.63	1.35	4.78	6.99	1.88	2.56	34.76
1977	2.22	1.34	2.67	2.33	3.99	1.25	4.20	6.15	2.16	7.88	4.32	5.57	44.08
1978	7.97	0.48	5.67	4.31	3.92	5.26	4.24	5.93	0.26	1.21	4.57	3.80	47.62
1979	6.16	5.97	2.59	3.97	3.80	2.42	4.36	7.08	9.76	3.87	5.50	1.64	57.12
1980	6.05	1.01	5.49	4.28	4.68	0.38	5.18	2.15	2.37	6.96	2.18	0.40	41.13
1981	0.64	2.76	1.52	2.96	6.62	3.69	4.01	2.89	2.70	2.36	0.68	5.04	35.87
1982	2.76	4.44	3.74	2.97	3.48	3.97	9.21	4.39	2.55	2.90	2.70	3.37	46.48
1983	1.59	3.95	6.04	5.21	2.50	5.46	0.51	0.97	3.05	4.02	5.63	4.50	43.43
1984	3.98	3.97	8.65	5.92	4.52	2.01	3.55	4.58	1.86	2.14	3.34	1.52	46.04
1985	3.54	3.20	1.80	0.65	2.36	4.01	5.31	10.58	4.97	5.09	6.99	0.58	49.08
1986	2.69	2.67	1.16	1.16	3.15	1.30	7.01	6.75	0.63	2.43	2.46	5.15	36.56
1987	5.53	2.57	1.65	7.31	2.94	6.29	1.20	1.11	4.43	1.25	3.13	2.86	40.27
1988	2.53	3.08	1.98	2.55	4.81	2.25	7.50	2.95	1.74	2.74	4.34	0.79	37.26
1989	1.88	4.34	5.00	4.27	5.02	5.85	4.00	4.89	5.33	3.54	3.00	2.62	49.74
1990	2.84	2.38	2.54	2.81	6.85	0.97	6.74	5.76	1.92	3.90	1.70	3.52	41.93
1991	3.62	1.09	5.87	0.87	0.91	6.24	3.47	3.32	2.69	2.50	0.67	4.53	35.78
1992	1.57	2.89	5.87	2.21	4.95	2.28	5.68	6.40	2.35	1.94	2.62	2.79	41.55
1993	4.48	2.88	7.24	3.23	4.66	1.75	1.91	3.89	2.97	2.23	3.24	3.77	42.25
Record Mean	3.13	2.99	3.65	2.96	3.77	3.64	5.35	4.87	3.48	3.29	3.17	3.16	43.47

AVERAGE TEMPERATURE (deg. F) **RICHMOND, VIRGINIA**

YEAR	JAN	FEB	MAR	APR	MAY	JUNE	JULY	AUG	SEP	OCT	NOV	DEC	ANNUAL
1964	38.1	37.2	47.6	55.4	66.4	73.1	75.8	73.1	67.1	53.4	51.5	42.9	56.8
1965	35.6	38.8	43.0	53.9	69.6	70.7	74.9	75.9	70.7	56.1	48.2	41.3	56.6
1966	31.1	37.7	47.5	52.8	63.1	71.4	76.4	74.6	67.2	55.5	49.5	38.0	55.4
1967	40.9	34.8	46.6	58.8	60.7	72.1	76.6	75.5	65.7	57.2	44.0	41.9	56.2
1968	33.9	34.2	52.0	58.8	64.7	74.7	78.9	78.9	70.9	61.9	51.3	37.0	58.1
1969	33.9	36.8	42.3	57.6	65.5	75.7	78.3	75.1	68.1	58.5	46.8	35.5	56.2
1970	30.1	37.1	42.9	58.2	69.1	75.7	78.3	78.0	74.8	62.9	49.9	40.4	58.1
1971	33.8	39.5	44.5	55.0	63.3	74.7	76.6	75.3	71.4	64.6	48.5	48.0	57.9
1972	40.7	37.6	47.2	56.2	64.6	70.1	77.1	75.2	70.1	55.8	47.9	45.9	57.4
1973	37.6	38.5	52.6	57.9	65.1	76.0	77.4	77.5	72.3	60.6	51.3	40.8	59.0
1974	45.8	40.1	50.4	59.9	65.8	70.6	76.9	75.7	67.4	55.4	48.5	41.7	58.2
1975	40.7	41.4	45.3	52.9	67.7	73.6	76.0	78.8	69.3	62.5	53.6	40.0	58.5
1976	35.1	48.5	52.6	60.5	65.2	74.6	77.3	75.7	68.7	54.4	42.7	36.7	57.7
1977	25.3	40.5	53.7	61.1	68.2	73.0	81.4	79.8	74.2	57.3	52.3	39.5	58.9
1978	33.4	30.3	44.5	57.3	65.5	74.7	77.5	80.1	72.9	58.3	52.5	42.5	57.5
1979	36.4	28.6	51.1	58.4	67.1	70.8	76.9	77.8	71.0	58.3	53.3	42.3	57.7
1980	38.8	36.0	47.4	61.1	68.3	72.8	80.0	80.7	74.7	56.9	46.2	38.6	58.5
1981	31.2	42.2	44.6	60.6	64.1	77.9	79.6	75.1	69.4	56.4	49.1	38.0	57.3
1982	31.6	41.7	49.1	55.9	70.4	73.4	78.6	75.0	69.8	59.2	51.9	46.1	58.6
1983	37.8	39.1	50.4	56.1	66.1	75.6	79.4	77.7	68.8	58.1	49.0	36.2	57.9
1984	32.6	44.5	43.6	55.8	65.4	77.7	76.0	77.0	67.5	66.1	46.6	47.7	58.4
1985	32.6	40.2	49.7	62.0	68.0	74.3	79.0	77.5	70.8	62.6	56.6	37.8	59.3
1986	36.2	39.3	50.0	59.2	66.9	76.1	80.9	74.2	70.8	61.8	49.1	40.9	58.8
1987	34.7	37.0	47.1	54.3	67.3	75.8	81.3	78.5	72.3	52.9	51.3	43.0	58.0
1988	32.3	39.1	47.9	56.0	65.8	72.8	79.9	79.8	68.4	53.6	50.6	39,2	57.1
1989	42.3	39.6	47.9	55.8	64.1	76.1	77.7	75.3	70.8	60.2	49.3	31.3	57.5
1990	46.3	48.0	52.1	57.9	65.8	75.0	79.9	76.3	69.5	63.1	52.6	46.3	61.1
1991	40.2	44.0	50.8	60.2	72.1	75.3	80.7	78.6	71.6	59.9	49.6	44.3	60.6
1992	40.7	42.7	47.2	57.8	61.7	70.8	79.5	73.7	70.3	56.3	50.6	40.9	57.7
1993	41.4	37.8	45.9	55.9	67.7	75.0	82.5	78.5	72.5	58.6	50.9	39.0	58.8
Record Mean	37.5	39.5	47.3	57.2	66.3	74.1	78.1	76.5	70.1	58.8	49.1	39.9	57.9
Max	47.3	50.1	58.7	69.6	78.1	85.3	88.4	86.6	80.9	70.6	60.6	50.0	68.9
Min	27.6	28.9	35.9	44.7	54.4	62.9	67.7	66.3	59.3	46.9	37.6	29.8	46.8

REFERENCE NOTES **RICHMOND, VIRGINIA**

GENERAL
T - TRACE AMOUNT.
BLANK ENTRIES DENOTE MISSING/UNREPORTED DATA.
# INDICATES A STATION OR INSTRUMENT RELOCATION.

SPECIFIC

PM - INCLUDES LAST DAY OF PREVIOUS MONTH
ASOS - AUTOMATED SURFACE OBSERVING SYSTEM IN OPERATION DURING THESE MONTHS.

(*) - LENGTH OF RECORD IN YEARS, ALTHOUGH INDIVIDUAL MONTHS MAY BE MISSING.
0.* OR * - THE VALUE IS BETWEEN 0.0 AND 0.05
NORMALS - BASED ON THE 1961-1990 RECORD PERIOD.
EXTREMES - DATES ARE THE MOST RECENT OCCURRENCE
WIND DIR.- NUMERALS SHOW TENS OF DEGRESS CLOCKWISE FROM TRUE NORTH. "00" INDICATES CALM.
RESULTANT DIRECTIONS ARE GIVEN TO WHOLE DEGREES.
BOLD VALUES INDICATE EXTREME VALUES WHICH OCCURRED AFTER THE ASOS SYSTEM WAS COMMISSIONED.

RECORD = PERIOD OF RECORD
RECORD MEAN PRECIPITATION IS THE MEAN OF ALL DAILY PRECIPITATION AMOUNTS DURING THE PERIOD OF RECORD.
RECORD MAX(MIN) TEMPERATURE IS THE MEAN OF ALL DAILY MAX(MIN) TEMPERATURES DURING THE PERIOD OF RECORD.
RECORD MEAN TEMPERATURE IS THE SUM OF THE RECORD MAX AND RECORD MIN DIVIDED BY 2.
AVERAGE TEMPERATURE IS THE SUM OF THE MEAN DAILY MAX AND MIN TEMPERATURE DIVIDED BY 2.

EXCEPTIONS

RECORD MEANS ARE THROUGH THE CURRENT YEAR,
BEGINNING IN 1930 FOR TEMPERATURE
1930 FOR PRECIPITATION
1938 FOR SNOWFALL

HEATING DEGREE DAYS Base 65 deg. F RICHMOND, VIRGINIA

SEASON	JULY	AUG	SEP	OCT	NOV	DEC	JAN	FEB	MAR	APR	MAY	JUNE	TOTAL
1964-65	0	0	32	352	402	676	909	726	674	339	17	34	4161
1965-66	0	6	25	275	498	726	1043	759	538	371	133	27	4401
1966-67	0	0	47	293	466	833	738	841	560	230	171	17	4196
1967-68	0	0	64	256	623	708	956	887	416	191	86	0	4187
1968-69	0	0	0	161	403	864	957	783	695	237	66	0	4166
1969-70	0	0	45	221	541	907	1076	778	677	231	51	0	4527
1970-71	0	0	12	124	445	756	960	709	627	295	104	3	4035
1971-72	0	0	11	69	512	526	748	788	554	286	58	21	3573
1972-73	0	0	17	285	513	588	843	735	394	247	79	0	3701
1973-74	0	0	5	163	414	744	589	691	455	204	75	5	3345
1974-75	0	0	62	310	513	715	746	654	604	368	44	1	4017
1975-76	0	0	27	121	356	770	917	480	386	227	78	11	3373
1976-77	0	1	15	332	660	869	1227	680	366	176	42	7	4375
1977-78	0	0	4	259	401	784	974	964	627	235	88	5	4341
1978-79	0	0	16	214	366	694	876	1011	439	218	44	4	3882
1979-80	0	0	8	242	353	698	806	835	541	135	47	2	3667
1980-81	0	0	14	267	557	813	1042	633	626	171	107	0	4230
1981-82	0	1	29	273	473	834	1029	645	486	280	6	1	4057
1982-83	0	6	10	213	399	585	836	718	445	282	69	2	3565
1983-84	0	1	86	236	475	887	994	589	657	282	93	3	4303
1984-85	0	0	73	57	546	531	997	692	484	177	35	5	3597
1985-86	0	0	31	114	257	838	886	713	465	187	78	3	3572
1986-87	0	16	24	172	476	741	931	777	550	317	57	0	4061
1987-88	0	0	5	370	409	677	1008	746	527	279	79	32	4132
1988-89	0	0	27	361	425	794	696	709	546	293	108	0	3959
1989-90	0	3	38	181	468	1036	574	472	436	258	50	3	3519
1990-91	0	0	33	146	365	574	762	582	443	193	27	1	3126
1991-92	0	0	25	190	458	637	749	640	545	261	142	10	3657
1992-93	0	0	37	275	427	739	724	757	585	280	24	0	3848
1993-94	0	0	26	213	436	799							

COOLING DEGREE DAYS Base 65 deg. F RICHMOND, VIRGINIA

YEAR	JAN	FEB	MAR	APR	MAY	JUNE	JULY	AUG	SEP	OCT	NOV	DEC	TOTAL
1969	0	0	0	21	90	328	417	321	147	26	0	0	1350
1970	0	0	0	35	185	328	418	410	313	67	0	0	1756
1971	0	0	0	0	56	297	367	327	209	62	22	5	1345
1972	0	0	7	30	52	180	381	326	178	9	8	0	1171
1973	0	0	13	42	91	338	391	395	231	32	9	2	1544
1974	0	0	10	58	106	180	377	340	141	21	26	0	1259
1975	0	0	0	16	135	267	348	433	165	51	18	0	1433
1976	0	8	9	99	91	307	389	337	133	12	0	0	1385
1977	0	0	22	66	148	258	513	467	289	24	27	0	1814
1978	0	0	0	12	112	302	393	475	263	15	0	1	1573
1979	0	0	16	30	117	188	374	404	195	42	9	0	1375
1980	0	0	1	25	157	243	472	494	313	23	1	0	1729
1981	0	0	1	45	89	395	458	319	169	16	0	0	1492
1982	0	0	0	13	181	259	428	323	157	43	13	7	1424
1983	0	0	0	23	108	325	452	405	207	27	0	0	1547
1984	0	0	0	10	114	392	346	381	154	100	0	2	1499
1985	0	4	20	94	139	290	441	392	213	51	10	0	1654
1986	0	0	8	19	142	344	498	308	205	79	6	0	1609
1987	0	0	0	2	136	329	513	427	227	0	3	0	1637
1988	0	0	3	16	108	269	466	465	137	12	0	0	1476
1989	0	3	24	27	88	341	403	331	218	42	4	0	1481
1990	0	1	43	51	81	312	470	356	177	96	0	2	1589
1991	0	0	10	53	254	316	494	428	227	40	3	0	1825
1992	0	0	0	51	48	192	457	275	202	12	3	0	1240
1993	0	0	0	14	112	307	547	426	261	20	18	0	1705

SNOWFALL (inches) RICHMOND, VIRGINIA

SEASON	JULY	AUG	SEP	OCT	NOV	DEC	JAN	FEB	MAR	APR	MAY	JUNE	TOTAL
1964-65	0.0	0.0	0.0	0.0	0.4	0.0	12.4	6.6	1.0	0.0	0.0	0.0	20.4
1965-66	0.0	0.0	0.0	0.0	0.0	T	26.2	3.0	0.0	0.0	0.0	0.0	29.2
1966-67	0.0	0.0	0.0	0.0	0.2	12.2	6.3	17.1	T	0.0	0.0	0.0	35.8
1967-68	0.0	0.0	0.0	0.0	T	5.6	2.3	2.4	2.8	0.0	0.0	0.0	13.1
1968-69	0.0	0.0	0.0	0.0	1.2	2.8	T	T	11.9	0.0	0.0	0.0	15.9
1969-70	0.0	0.0	0.0	0.0	0.0	1.8	5.4	0.4	0.0	0.0	0.0	0.0	7.6
1970-71	0.0	0.0	0.0	0.0	0.0	0.9	3.3	2.0	8.4	0.6	0.0	0.0	15.2
1971-72	0.0	0.0	0.0	0.0	T	0.0	T	13.7	0.0	T	0.0	0.0	13.7
1972-73	0.0	0.0	0.0	T	0.6	0.0	4.3	0.4	1.4	0.0	0.0	0.0	6.7
1973-74	0.0	0.0	0.0	0.0	0.0	9.9	T	5.0	T	0.0	0.0	0.0	14.9
1974-75	0.0	0.0	0.0	0.0	0.0	T	2.7	2.9	0.4	0.0	0.0	0.0	6.0
1975-76	0.0	0.0	0.0	0.0	0.0	T	0.2	T	1.0	0.0	0.0	0.0	1.2
1976-77	0.0	0.0	0.0	0.0	1.0	1.7	11.1	T	0.0	0.0	0.0	0.0	13.8
1977-78	0.0	0.0	0.0	0.0	T	T	1.3	5.1	5.0	0.0	0.0	0.0	11.4
1978-79	0.0	0.0	0.0	0.0	0.0	0.0	0.7	19.5	T	0.0	0.0	0.0	20.2
1979-80	0.0	0.0	0.0	T	0.0	T	16.6	7.0	15.0	0.0	0.0	0.0	38.6
1980-81	0.0	0.0	0.0	0.0	0.0	0.2	0.6	0.0	0.2	0.0	0.0	0.0	1.0
1981-82	0.0	0.0	0.0	0.0	T	1.9	8.3	10.8	T	0.2	0.0	0.0	21.2
1982-83	0.0	0.0	0.0	0.0	0.0	7.9	0.1	21.4	0.0	T	0.0	0.0	29.4
1983-84	0.0	0.0	0.0	0.0	0.1	T	1.1	2.8	0.3	0.0	0.0	0.0	4.3
1984-85	0.0	0.0	0.0	0.0	0.0	T	8.3	T	T	0.0	0.0	0.0	8.3
1985-86	0.0	0.0	0.0	0.0	0.0	1.3	3.3	4.4	T	T	0.0	0.0	9.0
1986-87	0.0	0.0	0.0	0.0	0.0	0.0	15.8	5.3	0.7	T	0.0	0.0	21.8
1987-88	0.0	0.0	0.0	0.0	4.5	T	8.1	T	T	T	0.0	0.0	12.6
1988-89	0.0	0.0	0.0	0.0	0.0	1.8	T	13.6	T	T	T	0.0	15.4
1989-90	0.0	0.0	0.0	0.0	1.1	9.9	0.0	T	T	0.2	0.0	0.0	11.2
1990-91	0.0	0.0	0.0	0.0	0.0	T	T	1.9	T	0.0	0.0	0.0	1.9
1991-92	0.0	0.0	0.0	0.0	0.1	T	0.0	0.8	T	T	0.0	0.0	0.9
1992-93	0.0	0.0	0.0	0.0	0.0	0.0	0.5	5.3	3.5	T	0.0	0.0	9.3
1993-94	0.0	0.0	0.0	0.0	0.0	5.1							
Record Mean	0.0	0.0	0.0	T	0.4	2.0	4.9	4.1	2.4	0.1	T	0.0	14.0

ROANOKE, VIRGINIA

The climate of Roanoke is relatively mild. Roanoke is nestled among mountains which interrupt the Great Valley, extending from northernmost Virginia southwestward into east Tennessee. This location, at a point where the valley is pinched between the Blue Ridges and the Alleghenies, offers a natural barrier to the winter cold as it moves southward. It is also far enough inland that hurricanes lose much of their destructive force before reaching Roanoke. Finally, the rough terrain is an inhospitable breeding ground for tornadic activity. The elevation in the vicinity usually produces cool summer nights that make a light cover comfortable for sleeping. Although past records show extremes over 100 degrees and below zero, many years pass without either extreme being threatened.

Roanoke is located near the headwaters of the Roanoke River, which flows in a general southeasterly direction. Numerous creeks and small streams from nearby mountainous areas empty into the Roanoke River. The usual low water stage is 1 to 1.5 feet, and flood stage is 10 feet. Some low-lying streets in Roanoke and nearby Salem have to be blocked off during 7 to 8 foot stages, but damage is minor until the river overflows its banks. The highest stage on record exceeds 19 feet. Damage has been widespread on occasion and has amounted to several million dollars in the city of Roanoke alone.

The growing season averages 190 days. The average date of the last freezing temperature in spring is mid-April and the average date of the first freezing date in the fall is late October.

Rainfall is well apportioned throughout the year. Droughts are so infrequent that quoting actual records would be difficult. Snow usually falls each winter, ranging from only a trace to more than 60 inches.

NORMALS, MEANS, AND EXTREMES

ROANOKE, VIRGINIA

LATITUDE: 37°19'N LONGITUDE: 79°58'W ELEVATION: FT. GRND 1149 BARO 1193 TIME ZONE: EASTERN WBAN: 13741

	(a)	JAN	FEB	MAR	APR	MAY	JUNE	JULY	AUG	SEP	OCT	NOV	DEC	YEAR
TEMPERATURE °F:														
Normals														
-Daily Maximum		43.8	47.3	57.8	67.3	75.7	82.9	86.4	85.3	78.5	68.1	58.0	47.6	66.6
-Daily Minimum		25.0	27.2	35.7	43.8	52.5	60.2	64.8	63.8	56.8	44.8	37.0	28.9	45.0
-Monthly		34.5	37.3	46.8	55.6	64.1	71.5	75.6	74.6	67.7	56.5	47.5	38.3	55.8
Extremes														
-Record Highest	46	78	80	87	95	96	100	104	105	101	93	83	76	105
-Year		1952	1985	1986	1957	1962	1959	1954	1983	1954	1951	1950	1984	AUG 1983
-Record Lowest	46	-11	1	9	20	31	39	47	42	34	22	9	-4	-11
-Year		1985	1970	1993	1985	1966	1977	1988	1986	1983	1976	1950	1983	JAN 1985
NORMAL DEGREE DAYS:														
Heating (base 65°F)		946	776	564	289	105	5	0	0	39	283	525	828	4360
Cooling (base 65°F)		0	0	0	7	78	200	329	298	120	20	0	0	1052
% OF POSSIBLE SUNSHINE														
MEAN SKY COVER (tenths)														
Sunrise - Sunset	45	6.3	6.3	6.3	6.0	6.2	6.0	6.0	5.9	5.7	5.0	5.9	6.2	6.0
MEAN NUMBER OF DAYS:														
Sunrise to Sunset														
-Clear	46	8.2	7.8	8.0	8.6	7.4	7.2	6.9	7.7	9.6	12.8	8.8	8.6	101.6
-Partly Cloudy	46	7.8	7.0	8.6	8.7	10.5	11.8	13.0	12.5	8.9	7.4	8.4	8.0	112.6
-Cloudy	46	15.0	13.5	14.4	12.8	13.1	11.0	11.1	10.8	11.5	10.8	12.8	14.4	151.0
Precipitation .01 inches or more	46	10.0	9.7	11.2	10.1	12.0	9.9	11.6	10.5	8.6	7.6	8.7	8.9	118.8
Snow,Ice pellets,hail 1.0 inches or more	46	1.6	1.9	1.0	0.1	0.0	0.0	0.0	0.0	0.0	0.*	0.5	1.1	6.2
Thunderstorms	46	0.1	0.3	1.1	3.1	6.1	6.3	8.5	6.6	2.6	0.9	0.4	0.1	36.0
Heavy Fog Visibility 1/4 mile or less	46	2.5	2.8	2.0	1.0	1.7	1.0	1.3	1.5	2.5	1.9	2.2	2.4	22.8
Temperature °F														
-Maximum														
90° and above	29	0.0	0.0	0.0	0.4	0.7	5.2	10.6	7.9	2.2	0.0	0.0	0.0	27.1
32° and below	29	4.7	2.7	0.2	0.0	0.0	0.0	0.0	0.0	0.0	0.0	0.1	2.3	10.0
-Minimum														
32° and below	29	22.9	20.1	12.4	2.6	0.1	0.0	0.0	0.0	0.0	2.7	10.1	19.5	90.4
0° and below	29	0.3	0.0	0.0	0.0	0.0	0.0	0.0	0.0	0.0	0.0	0.0	0.2	0.5
AVG. STATION PRESS.(mb)	21	976.1	975.8	974.4	973.5	974.0	974.8	975.7	976.7	977.3	977.7	977.3	977.0	975.9
RELATIVE HUMIDITY (%)														
Hour 01	29	67	65	65	65	78	83	84	86	87	80	72	69	75
Hour 07 (Local Time)	29	70	69	71	71	80	81	83	87	89	84	76	72	78
Hour 13	29	52	50	48	47	53	54	54	56	57	52	52	54	52
Hour 19	29	58	55	51	49	59	62	64	67	69	64	60	61	60
PRECIPITATION (inches):														
Water Equivalent														
-Normal		2.62	3.04	3.48	3.25	3.98	3.19	3.91	4.15	3.50	3.85	3.19	2.97	41.13
-Maximum Monthly	46	6.12	7.17	7.91	11.35	8.42	7.76	10.09	9.54	11.09	9.89	12.36	7.10	12.36
-Year		1978	1960	1993	1987	1950	1989	1989	1984	1987	1990	1985	1948	NOV 1985
-Minimum Monthly	46	0.29	0.56	0.43	0.48	1.27	0.62	0.45	1.04	0.15	0.04	0.44	0.18	0.04
-Year		1981	1968	1966	1976	1951	1986	1977	1993	1991	1991	1960	1965	OCT 1991
-Maximum in 24 hrs	46	2.71	2.62	3.02	5.57	3.99	3.98	2.74	5.22	6.60	6.41	6.63	3.40	6.63
-Year		1968	1984	1983	1978	1973	1972	1989	1985	1987	1968	1985	1948	NOV 1985
Snow,Ice pellets,hail														
-Maximum Monthly	46	41.2	27.6	30.3	7.3	T	T	0.0	0.0	T	1.0	13.8	22.6	41.2
-Year		1966	1960	1960	1971	1990	1989			1953	1957	1968	1966	JAN 1966
-Maximum in 24 hrs	46	13.7	18.4	17.4	7.3	T	T	0.0	0.0	T	1.0	10.0	16.4	18.4
-Year		1966	1983	1960	1971	1990	1989			1953	1957	1968	1969	FEB 1983
WIND:														
Mean Speed (mph)	45	9.4	9.7	10.0	9.6	7.8	6.9	6.5	6.1	6.1	6.9	8.2	8.7	8.0
Prevailing Direction through 1963		WNW	SE	WNW	SE	SE	SE	W	SE	SE	SE	NW	NW	SE
Fastest Obs. 1 Min.														
-Direction (!!)	32	30	31	32	32	36	28	34	30	15	34	34	30	32
-Speed (MPH)	32	53	40	52	58	46	46	46	44	35	35	52	40	58
-Year		1964	1972	1967	1963	1962	1966	1980	1975	1989	1963	1963	1970	APR 1963
Peak Gust														
-Direction (!!)	9	31	NW	NW	NW	NW	SW	NW	NW	SE	W	W	NW	NW
-Speed (mph)	9	56	59	52	77	52	72	38	39	54	39	46	53	77
-Date		1989	1987	1985	1989	1990	1993	1991	1993	1989	1989	1986	1991	APR 1989

PRECIPITATION (inches)

ROANOKE, VIRGINIA

YEAR	JAN	FEB	MAR	APR	MAY	JUNE	JULY	AUG	SEP	OCT	NOV	DEC	ANNUAL
1964	5.20	5.33	1.81	3.54	1.79	1.97	3.37	3.54	2.70	2.80	3.09	2.73	37.87
1965	3.72	3.53	3.93	1.70	3.83	1.83	5.50	1.16	1.97	3.42	0.95	0.18	31.72
1966	4.26	4.78	0.43	2.59	3.49	1.54	3.00	4.73	7.25	4.04	1.73	3.25	41.09
1967	1.25	2.51	4.51	1.67	3.95	3.19	4.05	6.36	1.78	2.42	1.30	4.84	37.83
1968	3.33	0.56	3.03	2.73	1.89	2.11	2.95	5.36	0.44	8.06	2.82	1.80	35.08
1969	1.86	2.74	3.35	1.40	1.58	4.95	5.35	5.41	2.79	2.22	1.40	5.54	38.59
1970	1.31	2.36	1.95	2.80	1.51	4.87	3.28	6.40	1.99	7.51	3.67	2.81	40.46
1971	1.21	5.13	2.28	2.55	7.50	4.84	5.23	4.46	3.87	6.75	2.18	0.83	46.83
1972	2.49	4.80	1.76	3.31	6.00	7.55	4.89	2.62	4.79	3.18	5.63	4.62	51.64
1973	2.60	2.95	5.92	5.39	5.58	3.65	5.10	3.34	1.84	4.28	1.79	5.60	48.04
1974	3.33	2.13	3.12	1.86	3.76	2.93	3.71	4.93	3.04	0.77	1.28	3.16	34.02
1975	3.59	3.05	7.80	2.04	6.65	1.54	5.15	5.68	6.46	3.01	1.77	3.67	50.41
1976	2.16	1.27	4.54	0.48	6.13	5.20	1.24	2.20	3.17	9.72	1.31	2.59	40.01
1977	1.46	0.73	2.61	3.50	1.52	2.41	0.45	2.29	2.71	4.70	6.46	2.49	31.33
1978	6.12	0.65	5.92	7.54	4.85	2.05	4.83	6.33	0.52	0.78	2.55	3.15	45.29
1979	5.27	5.37	3.38	3.99	2.65	5.78	3.97	3.37	9.18	3.56	3.77	1.13	51.42
1980	4.10	0.67	5.41	5.51	2.66	1.81	5.18	2.87	1.66	2.30	1.78	0.60	34.55
1981	0.29	2.43	2.30	1.75	4.56	2.49	2.86	1.32	4.52	3.90	0.68	3.79	30.89
1982	3.76	4.75	2.33	2.01	4.83	4.99	3.98	5.20	2.67	4.13	3.65	2.53	44.83
1983	1.28	4.12	6.41	7.95	3.17	2.38	1.67	2.23	1.52	7.73	4.26	5.61	48.33
1984	1.35	4.85	4.30	3.97	4.49	2.34	4.17	9.54	2.69	1.42	2.67	1.84	43.63
1985	2.45	3.64	1.80	1.75	6.89	2.08	4.18	8.67	1.26	3.77	12.36	0.85	49.70
1986	0.93	2.87	1.36	1.67	4.15	0.62	2.83	4.31	3.04	2.76	3.73	5.48	33.75
1987	4.53	4.55	4.11	11.35	2.68	0.71	3.21	1.08	11.09	1.10	5.00	2.16	51.57
1988	1.87	1.07	0.88	3.40	2.76	3.66	3.75	4.30	3.01	1.26	2.42	1.28	29.66
1989	1.31	2.04	2.96	2.54	6.46	7.76	10.09	1.65	8.94	4.13	3.86	2.60	54.34
1990	2.33	2.76	3.42	2.07	7.45	0.83	3.80	4.42	1.86	9.89	1.08	3.79	43.70
1991	3.55	2.10	7.58	2.49	2.88	2.42	7.22	2.31	0.15	0.04	2.51	3.81	37.06
1992	2.51	3.75	2.54	4.89	6.06	6.87	1.73	2.38	2.07	1.90	5.60	2.62	42.92
1993	3.77	3.44	7.91	2.77	2.37	2.49	1.24	1.04	4.58	2.15	2.60	5.27	39.63
Record Mean	2.72	3.15	3.66	3.30	3.95	3.37	3.72	4.04	3.34	3.41	2.99	3.09	40.75

AVERAGE TEMPERATURE (deg. F)

ROANOKE, VIRGINIA

YEAR	JAN	FEB	MAR	APR	MAY	JUNE	JULY	AUG	SEP	OCT	NOV	DEC	ANNUAL
#1964	37.3	35.6	46.6	56.9	67.4	74.6	75.9	74.0	66.8	54.0	51.1	40.0	56.7
1965	36.0	36.4	41.9	54.9	69.2	69.7	73.9	74.2	69.7	56.0	47.5	41.5	55.9
1966	30.6	35.1	46.7	51.5	63.3	71.1	77.9	73.0	66.1	54.6	47.2	37.8	54.6
1967	40.8	34.8	48.5	59.0	59.0	70.3	73.4	72.6	63.2	54.7	43.4	41.3	55.1
1968	33.0	33.8	51.3	56.5	62.4	71.3	75.6	75.2	66.7	57.4	48.1	33.7	55.4
1969	32.3	37.2	40.7	57.3	65.0	73.1	76.0	73.5	66.7	56.6	44.5	33.6	54.7
1970	29.8	36.9	42.5	56.5	66.8	72.9	76.1	74.1	72.6	59.7	46.3	40.0	56.2
1971	33.9	39.1	43.6	55.5	61.5	72.6	73.5	72.6	69.8	62.1	45.4	45.3	56.3
1972	39.6	36.0	45.6	55.5	62.0	67.7	74.3	73.2	67.6	52.2	45.4	44.5	55.3
1973	37.3	35.8	51.1	53.8	61.1	73.8	75.5	76.0	70.8	59.2	49.3	38.2	56.8
1974	45.3	39.0	50.8	57.2	64.3	68.1	74.5	73.5	64.8	55.0	47.3	38.9	56.6
1975	39.6	40.4	42.2	52.9	66.5	71.5	74.5	77.0	66.6	60.0	51.1	38.8	56.8
1976	33.4	46.7	50.8	57.4	62.3	70.0	74.0	72.1	65.2	50.3	40.2	34.3	54.7
1977	23.6	36.6	52.5	58.9	68.1	71.1	79.7	77.8	70.4	53.6	48.5	35.9	56.4
1978	27.8	29.5	44.2	56.3	63.8	72.7	76.0	77.5	71.4	54.0	49.5	39.5	55.2
1979	31.4	29.4	48.6	55.7	63.8	69.3	73.5	74.0	66.3	54.6	49.6	40.9	54.8
1980	37.5	34.0	43.5	56.7	64.8	70.0	78.1	76.8	71.5	55.5	45.4	38.7	56.1
1981	32.0	38.2	43.1	58.6	60.9	74.0	76.1	73.4	66.1	53.7	45.2	32.8	54.5
1982	28.7	38.1	44.8	50.8	67.5	70.1	75.5	72.3	65.8	57.4	47.8	42.4	55.1
1983	35.6	36.5	47.0	52.4	61.0	70.1	77.0	77.9	68.1	57.1	47.1	34.9	55.4
1984	33.6	43.6	43.3	54.2	63.1	74.2	73.2	74.4	64.1	64.3	46.0	47.1	56.8
1985	31.3	38.6	50.3	61.9	67.1	72.7	76.5	73.8	68.4	60.3	55.4	35.1	57.6
1986	35.2	38.3	47.2	58.8	63.9	74.2	78.8	72.4	68.9	59.0	47.4	38.5	56.9
1987	34.2	36.9	46.6	53.2	67.1	75.0	79.3	77.8	68.9	51.5	49.6	41.4	56.8
1988	30.8	37.9	47.5	55.4	63.4	70.9	77.1	77.5	66.0	51.2	46.7	38.9	55.3
1989	41.3	38.2	47.4	54.3	61.2	73.7	75.9	74.1	68.1	58.5	46.0	29.6	55.7
1990	43.7	45.9	51.5	56.0	64.3	72.6	76.8	74.6	68.4	59.2	52.2	43.8	59.1
1991	39.3	42.7	49.2	57.5	69.6	73.5	77.9	74.8	69.3	59.4	46.8	42.4	58.5
1992	39.6	41.5	46.5	55.5	60.2	68.4	77.0	72.1	68.0	55.5	46.3	38.5	55.8
1993	39.7	36.0	42.1	54.5	65.6	71.9	80.3	76.8	68.6	55.3	47.3	36.2	56.2
Record Mean	35.9	38.3	46.1	56.6	64.7	72.1	76.3	74.8	67.8	57.1	47.1	38.4	56.3
Max	45.1	48.3	56.8	67.7	76.2	83.5	87.2	85.6	78.7	68.6	57.4	47.5	66.9
Min	26.7	28.4	35.4	45.5	53.1	60.6	65.3	64.0	56.9	45.6	36.7	29.2	45.6

REFERENCE NOTES

ROANOKE, VIRGINIA

GENERAL
T - TRACE AMOUNT.
BLANK ENTRIES DENOTE MISSING/UNREPORTED DATA.
# INDICATES A STATION OR INSTRUMENT RELOCATION.

SPECIFIC

PM - INCLUDES LAST DAY OF PREVIOUS MONTH
ASOS - AUTOMATED SURFACE OBSERVING SYSTEM IN OPERATION DURING THESE MONTHS.

(a) - LENGTH OF RECORD IN YEARS, ALTHOUGH INDIVIDUAL MONTHS MAY BE MISSING.
0.* OR * - THE VALUE IS BETWEEN 0.0 AND 0.05
NORMALS - BASED ON THE 1961-1990 RECORD PERIOD.
EXTREMES - DATES ARE THE MOST RECENT OCCURRENCE
WIND DIR.- NUMERALS SHOW TENS OF DEGRESS CLOCKWISE FROM TRUE NORTH. "00" INDICATES CALM.
RESULTANT DIRECTIONS ARE GIVEN TO WHOLE DEGREES.
BOLD VALUES INDICATE EXTREME VALUES WHICH OCCURRED AFTER THE ASOS SYSTEM WAS COMMISSIONED.

RECORD = PERIOD OF RECORD
RECORD MEAN PRECIPITATION IS THE MEAN OF ALL DAILY PRECIPITATION AMOUNTS DURING THE PERIOD OF RECORD.
RECORD MAX(MIN) TEMPERATURE IS THE MEAN OF ALL DAILY MAX(MIN) TEMPERATURES DURING THE PERIOD OF RECORD.
RECORD MEAN TEMPERATURE IS THE SUM OF THE RECORD MAX AND RECORD MIN DIVIDED BY 2.
AVERAGE TEMPERATURE IS THE SUM OF THE MEAN DAILY MAX AND MIN TEMPERATURE DIVIDED BY 2.

EXCEPTIONS

RECORD MEANS ARE THROUGH THE CURRENT YEAR, BEGINNING IN 1948 FOR TEMPERATURE
1948 FOR PRECIPITATION
1948 FOR SNOWFALL

HEATING DEGREE DAYS Base 65 deg. F — ROANOKE, VIRGINIA

SEASON	JULY	AUG	SEP	OCT	NOV	DEC	JAN	FEB	MAR	APR	MAY	JUNE	TOTAL
#1964-65	0	9	41	338	409	770	892	794	710	305	21	19	4308
1965-66	0	9	38	287	520	722	1058	831	565	409	118	25	4582
1966-67	0	0	51	317	527	837	740	838	505	202	206	28	4251
1967-68	0	0	105	321	643	728	984	902	423	255	111	3	4475
1968-69	0	12	11	244	502	963	1005	772	746	227	77	9	4568
1969-70	0	0	53	274	607	967	1087	783	691	269	66	2	4799
1970-71	1	1	24	194	553	769	959	719	658	285	143	3	4309
1971-72	0	0	12	117	595	603	780	837	594	305	99	31	3973
1972-73	10	1	33	391	582	628	852	812	428	344	144	1	4226
1973-74	0	0	12	196	461	826	607	722	440	255	96	14	3629
1974-75	0	0	84	308	539	801	783	683	699	361	60	12	4330
1975-76	0	0	59	173	415	809	973	523	438	271	126	22	3809
1976-77	0	3	47	452	735	945	1275	786	385	203	56	23	4910
1977-78	0	0	18	350	496	896	1147	989	637	261	112	4	4910
1978-79	0	0	29	335	461	784	1037	992	512	279	88	15	4532
1979-80	3	9	49	329	458	738	848	893	656	266	78	14	4341
1980-81	0	0	30	301	582	807	1016	744	672	212	158	3	4525
1981-82	0	0	58	357	589	991	1121	746	618	421	51	3	4955
1982-83	0	6	61	264	509	695	904	792	551	381	157	14	4334
1983-84	1	0	87	246	531	924	966	614	664	336	123	6	4498
1984-85	0	0	120	74	565	549	1041	734	471	168	39	11	3772
1985-86	0	0	55	171	282	918	917	739	551	206	115	1	3955
1986-87	0	22	22	231	523	813	950	782	562	352	58	0	4315
1987-88	0	0	18	412	455	723	1055	778	535	289	101	46	4412
1988-89	1	0	53	423	543	802	726	743	556	340	172	0	4359
1989-90	0	3	63	234	560	1091	654	528	441	297	77	3	3951
1990-91	0	0	38	200	381	652	787	618	486	245	36	4	3447
1991-92	0	0	49	194	539	695	777	677	567	290	172	18	3978
1992-93	0	2	55	290	552	813	777	807	698	310	52	12	4368
1993-94	0	0	52	300	529	886							

COOLING DEGREE DAYS Base 65 deg. F — ROANOKE, VIRGINIA

YEAR	JAN	FEB	MAR	APR	MAY	JUNE	JULY	AUG	SEP	OCT	NOV	DEC	TOTAL
1969	0	0	0	2	86	256	347	272	107	21	0	0	1091
1970	0	0	0	22	128	246	351	291	261	35	0	0	1334
1971	0	0	0	4	40	239	270	245	163	35	13	1	1010
1972	0	0	0	26	13	118	305	262	118	0	2	0	844
1973	0	0	3	15	29	269	331	346	192	23	0	0	1208
1974	0	0	7	27	82	117	303	267	83	3	12	0	901
1975	0	0	0	5	108	213	301	379	114	27	5	0	1152
1976	0	0	8	53	49	179	285	230	58	1	0	0	863
1977	0	0	7	28	158	214	461	400	185	4	7	0	1464
1978	0	0	0	7	79	243	347	391	228	2	0	0	1297
1979	0	0	10	7	56	150	273	296	97	12	1	0	902
1980	0	0	0	21	78	171	412	374	231	13	0	0	1300
1981	0	0	2	26	39	278	350	267	97	13	0	0	1072
1982	0	0	0	0	137	165	332	242	90	35	0	0	1001
1983	0	0	0	11	41	172	382	407	188	10	0	0	1211
1984	0	0	0	20	71	290	260	301	101	56	0	0	1099
1985	0	1	18	78	112	247	365	280	163	34	3	0	1301
1986	0	0	4	24	84	282	438	257	145	48	0	0	1282
1987	0	0	0	4	130	306	450	403	140	0	0	0	1433
1988	0	0	0	7	58	232	386	395	90	1	0	0	1169
1989	0	0	17	27	61	265	344	293	163	38	0	0	1208
1990	0	0	29	34	61	238	377	305	145	27	3	0	1219
1991	0	0	2	29	188	267	407	311	185	28	1	0	1418
1992	0	0	2	12	29	125	377	230	152	4	0	0	931
1993	0	0	0	2	78	226	481	374	166	8	6	0	1341

SNOWFALL (inches) — ROANOKE, VIRGINIA

SEASON	JULY	AUG	SEP	OCT	NOV	DEC	JAN	FEB	MAR	APR	MAY	JUNE	TOTAL
1964-65	0.0	0.0	0.0	0.0	0.5	T	12.1	4.0	4.1	0.0	0.0	0.0	20.7
1965-66	0.0	0.0	0.0	0.0	T	0.3	41.2	8.4	T	0.0	0.0	0.0	49.9
1966-67	0.0	0.0	0.0	0.0	T	22.6	2.7	15.6	0.8	0.0	0.0	0.0	41.7
1967-68	0.0	0.0	0.0	0.0	T	14.8	15.4	3.8	T	0.0	0.0	0.0	34.0
1968-69	0.0	0.0	0.0	T	13.8	T	0.1	11.4	13.8	0.0	0.0	0.0	39.1
1969-70	0.0	0.0	0.0	0.0	T	16.8	4.4	5.4	T	0.2	0.0	0.0	26.8
1970-71	0.0	0.0	0.0	0.0	2.0	10.8	1.1	3.6	7.3	7.3	0.0	0.0	32.1
1971-72	0.0	0.0	0.0	0.0	10.2	0.2	T	12.8	T	T	0.0	0.0	23.2
1972-73	0.0	0.0	0.0	0.0	T	T	2.4	1.4	5.5	0.2	0.0	0.0	9.5
1973-74	0.0	0.0	0.0	T	T	7.9	T	9.7	T	0.0	0.0	0.0	17.6
1974-75	0.0	0.0	0.0	T	3.3	6.6	4.6	6.3	6.2	T	0.0	0.0	27.0
1975-76	0.0	0.0	0.0	0.0	T	0.1	T	T	2.2	0.0	0.0	0.0	2.3
1976-77	0.0	0.0	0.0	0.0	2.4	6.4	8.9	1.5	T	T	0.0	0.0	19.2
1977-78	0.0	0.0	0.0	T	2.4	0.4	14.5	9.6	10.4	0.0	0.0	0.0	37.3
1978-79	0.0	0.0	0.0	0.0	1.6	T	3.0	19.3	T	0.0	0.0	0.0	23.9
1979-80	0.0	0.0	0.0	0.3	T	0.2	15.1	4.2	12.0	T	0.0	0.0	31.8
1980-81	0.0	0.0	0.0	T	T	T	1.4	T	10.4	0.0	0.0	0.0	11.8
1981-82	0.0	0.0	0.0	0.0	4.0	3.9	8.4	12.2	0.5	1.9	0.0	0.0	30.9
1982-83	0.0	0.0	0.0	T	0.0	6.2	3.8	24.3	0.3	0.4	0.0	0.0	35.0
1983-84	0.0	0.0	0.0	0.0	T	0.6	7.2	1.5	0.3	0.2	0.0	0.0	9.8
1984-85	0.0	0.0	0.0	0.0	T	0.8	2.7	1.2	1.3	T	0.0	0.0	6.0
1985-86	0.0	0.0	0.0	0.0	0.0	1.2	1.7	7.1	T	T	0.0	0.0	10.0
1986-87	0.0	0.0	0.0	0.0	T	T	27.9	19.1	2.7	6.3	0.0	0.0	56.0
1987-88	0.0	0.0	0.0	0.0	1.7	T	6.7	T	T	T	0.0	0.0	8.4
1988-89	0.0	0.0	0.0	T	0.0	3.8	0.7	9.1	0.2	0.2	T	T	14.0
1989-90	0.0	0.0	0.0	0.0	3.3	11.1	T	0.2	1.5	0.0	T	0.0	16.1
1990-91	0.0	0.0	0.0	0.0	0.0	0.8	T	T	0.4	0.0	0.0	0.0	1.2
1991-92	0.0	0.0	0.0	0.0	T	T	0.4	2.2	0.5	1.9	0.0	0.0	5.0
1992-93	0.0	0.0	0.0	0.0	T	0.9	0.1	10.7	16.0	0.3	0.0	0.0	28.0
1993-94	0.0	0.0	0.0	T	T	6.2							
Record Mean	0.0	0.0	T	T	1.5	3.8	6.2	7.0	3.8	0.5	T	T	22.9

OLYMPIA
WASHINGTON

The climate of Olympia and vicinity is characterized by warm, generally dry summers and wet, mild winters.

Fall rains usually begin about mid–October, borne inland to the Cascade Mountains by frequent maritime disturbances originating in the Pacific Ocean. These rains continue with few interruptions through spring. Daytime temperatures will be in the 40s and low 50s, with nighttime temperatures in the 30s. The progression of the wet and mild Pacific disturbances is usually broken once or twice each winter by the formation of large anticyclones, which originate in Alaska and northwestern Canada, and move southward over the state of Washington. These southerly migrations of polar–continental air from the interior will normally lower daytime temperatures to about freezing and nighttime temperatures to 10 – 20 degrees. In exceptional cases, temperatures slightly below zero have been recorded in this vicinity. Often the onset of the cold weather is accompanied by a little snow, but during these cold snaps the air is dry and skies are clear. Snow in an amount sufficient to seriously hinder highway travel occurs only once every two or three years.

During the spring months, the track of the Pacific storms moves gradually farther north, and the semi–permanent Pacific anticyclone also moves northward. The effects of the maritime disturbances lessen, and the periods of improving weather between storms lengthen. During the spring and early fall, clearing skies at night will usually be followed by fog or low stratus clouds in the early morning, which normally dissipate by noon.

Daily high temperatures average 70 to 80 degrees during July, August, and September. The temperature will equal or exceed 90 degrees about 6 days each summer, but as the warm weather is usually accompanied by lowering humidity, it is seldom uncomfortably hot. Rainfall is near 1 inch per month during July and August and about 2 inches per month during the transitional period of May, June, and September. About two–thirds of the days are sunny in July, August, and September, and about half are sunny during May and June.

Olympia and vicinity are quite well protected by the Coast Range from the strong south and southwest winds accompanying many of the Pacific storms during the fall and winter. Winds which reach hurricane force along the coast, only 45 miles away, will reach only 50 or 55 mph in gusts in this vicinity. Some damage to utility lines occurs every fall and winter from trees and limbs broken or felled by the wind, but damage rarely occurs to dwellings or buildings. The prevailing wind in Olympia is southerly during most of the year, but during the fair weather in the summers the wind is gentle and from the north to east.

The length of the growing season in the vicinity of Olympia varies with distance from the waterfront and elevation above sea level. The average length of the growing season is 166 days.

NORMALS, MEANS, AND EXTREMES

OLYMPIA WASHINGTON

LATITUDE: 46°58'N LONGITUDE: 122°54' W ELEVATION: FT. GRND 195 BARO 196 TIME ZONE: PACIFIC WBAN: 24227

	(a)	JAN	FEB	MAR	APR	MAY	JUNE	JULY	AUG	SEP	OCT	NOV	DEC	YEAR
TEMPERATURE °F:														
Normals														
-Daily Maximum		44.4	49.5	53.9	58.8	65.3	70.9	76.5	77.1	71.0	60.5	50.0	44.2	60.2
-Daily Minimum		31.6	32.7	33.6	36.2	41.0	46.4	49.2	49.5	44.7	38.8	35.0	32.1	39.2
-Monthly		38.0	41.1	43.8	47.5	53.2	58.7	62.9	63.3	57.9	49.7	42.5	38.2	49.7
Extremes														
-Record Highest	52	63	73	76	87	96	101	103	104	98	90	74	64	104
-Year		1942	1986	1969	1987	1983	1942	1941	1981	1988	1987	1949	1958	AUG 1981
-Record Lowest	52	-8	-1	9	23	25	30	35	33	25	20	-1	-7	-8
-Year		1979	1972	1989	1975	1954	1976	1962	1973	1972	1971	1955	1983	JAN 1979
NORMAL DEGREE DAYS:														
Heating (base 65°F)		837	669	657	525	366	201	96	102	222	474	675	831	5655
Cooling (base 65°F)		0	0	0	0	0	12	31	49	9	0	0	0	101
% OF POSSIBLE SUNSHINE														
MEAN SKY COVER (tenths)														
Sunrise - Sunset	49	8.7	8.4	8.1	7.8	7.2	6.9	5.4	5.8	6.0	7.7	8.6	8.8	7.5
MEAN NUMBER OF DAYS:														
Sunrise to Sunset														
-Clear	52	2.2	2.2	2.5	2.8	4.2	5.3	10.0	8.9	7.9	3.2	1.5	1.7	52.2
-Partly Cloudy	52	3.5	4.3	6.5	7.3	8.6	8.3	9.8	9.9	9.5	7.8	5.1	3.5	84.3
-Cloudy	52	25.3	21.8	22.0	19.9	18.3	16.4	11.2	12.2	12.6	20.0	23.4	25.8	228.8
Precipitation														
.01 inches or more	52	19.5	17.2	17.8	14.9	11.1	9.1	5.0	5.6	8.2	13.7	19.2	20.7	161.9
Snow,Ice pellets,hail														
1.0 inches or more	52	2.3	1.0	0.6	0.*	0.0	0.0	0.0	0.0	0.0	0.0	0.3	1.1	5.3
Thunderstorms	50	0.1	0.2	0.3	0.6	0.8	0.6	0.5	0.8	0.7	0.4	0.3	0.1	5.3
Heavy Fog Visibility														
1/4 mile or less	50	9.4	8.4	8.1	5.2	3.2	2.2	3.2	5.5	9.3	14.0	11.1	10.2	89.7
Temperature °F														
-Maximum														
90° and above	34	0.0	0.0	0.0	0.0	0.1	0.7	2.3	2.6	0.6	0.*	0.0	0.0	6.3
32° and below	34	1.1	0.1	0.0	0.0	0.0	0.0	0.0	0.0	0.0	0.0	0.3	1.3	2.8
-Minimum														
32° and below	34	16.0	13.7	13.7	8.4	1.9	0.1	0.0	0.0	0.7	5.3	10.6	15.4	85.9
0° and below	34	0.1	0.*	0.0	0.0	0.0	0.0	0.0	0.0	0.0	0.0	0.1	0.2	0.4
AVG. STATION PRESS.(mb)	21	1010.9	1009.4	1008.4	1010.1	1009.9	1010.1	1010.6	1009.5	1009.8	1010.8	1009.9	1011.1	1010.0
RELATIVE HUMIDITY (%)														
Hour 04	34	92	91	91	91	90	91	91	91	93	94	93	92	92
Hour 10 (Local Time)	34	89	87	80	71	67	67	65	67	73	85	89	90	78
Hour 16	34	80	71	62	57	55	54	50	50	54	68	80	84	64
Hour 22	30	90	89	87	83	80	79	78	80	86	92	92	91	86
PRECIPITATION (inches):														
Water Equivalent														
-Normal		8.01	5.77	4.95	3.29	2.09	1.63	0.82	1.29	2.26	4.31	8.05	8.12	50.59
-Maximum Monthly	52	19.84	13.18	10.13	7.80	5.83	6.48	3.00	5.45	7.59	10.08	15.51	14.32	19.84
-Year		1953	1961	1950	1991	1948	1946	1983	1968	1978	1967	1962	1970	JAN 1953
-Minimum Monthly	52	0.29	0.22	0.48	0.37	0.15	0.04	T	0.00	T	0.39	1.37	2.28	0.00
-Year		1985	1993	1965	1956	1947	1945	1984	1946	1993	1987	1976	1944	AUG 1946
-Maximum in 24 hrs	52	4.52	4.93	3.92	3.46	1.54	1.91	1.50	1.39	2.48	3.60	5.90	3.83	5.90
-Year		1990	1951	1972	1991	1948	1985	1979	1977	1978	1981	1990	1956	NOV 1990
Snow,Ice pellets,hail														
-Maximum Monthly	52	58.7	27.4	20.6	2.2	T	T	0.0	0.0	T	T	14.8	21.4	58.7
-Year		1969	1990	1951	1972	1991	1989			1972	1975	1978	1968	JAN 1969
-Maximum in 24 hrs	52	20.5	11.8	9.1	1.8	T	T	0.0	0.0	T	T	14.5	11.9	20.5
-Year		1972	1989	1966	1972	1991	1989			1972	1975	1978	1974	JAN 1972
WIND:														
Mean Speed (mph)	41	7.1	7.0	7.3	7.4	6.9	6.7	6.2	6.0	5.7	5.8	6.9	7.2	6.7
Prevailing Direction through 1963		SSW	SSW	SSW	SW	SW	SSW	SW	SW	SW	SSW	SW	SSW	SSW
Fastest Obs. 1 Min.														
-Direction (!!)	45	18	18	23	23	29	25	18	27	18	23	18	18	18
-Speed (MPH)	45	55	45	40	46	39	32	29	26	35	58	60	41	60
-Year		1956	1958	1956	1962	1960	1949	1957	1964	1967	1962	1958	1969	NOV 1958
Peak Gust														
-Direction (!!)	10	SW	S	SW	S	W	S	W	S	SW	SW	SW	S	SW
-Speed (mph)	10	58	43	45	43	38	38	32	32	38	45	47	49	58
-Date		1986	1984	1986	1991	1993	1986	1987	1991	1990	1990	1991	1993	JAN 1986

PRECIPITATION (inches)

OLYMPIA WASHINGTON

YEAR	JAN	FEB	MAR	APR	MAY	JUNE	JULY	AUG	SEP	OCT	NOV	DEC	ANNUAL
1964	15.13	2.54	4.47	1.58	0.98	2.35	1.07	1.47	2.26	1.79	9.18	9.11	51.93
1965	9.37	4.93	0.48	3.61	1.89	0.33	0.48	2.05	0.60	3.30	5.84	7.81	40.69
1966	7.89	3.38	7.28	1.71	1.30	1.28	1.34	0.68	1.95	4.83	8.16	11.53	51.33
1967	12.21	3.58	4.31	2.88	0.25	1.49	0.02	T	1.36	10.08	3.90	5.94	46.02
1968	9.04	7.83	6.53	3.02	2.57	2.43	0.89	5.45	2.51	6.07	7.96	9.95	64.25
1969	9.45	3.41	2.90	3.44	2.07	1.68	0.50	0.18	5.23	2.69	3.60	7.24	42.39
1970	12.48	4.30	3.07	4.76	1.21	0.14	0.16	0.15	3.20	2.71	7.40	14.32	53.90
1971	11.15	4.41	9.11	2.78	1.50	3.00	0.78	0.71	3.06	4.43	7.59	9.18	57.70
1972	12.43	11.06	10.01	5.87	0.83	1.07	1.72	0.70	5.04	0.85	4.17	10.66	64.41
1973	5.66	1.71	3.02	2.23	2.66	2.60	0.05	0.59	2.18	4.60	12.95	11.61	49.86
1974	10.57	5.68	6.65	4.77	2.65	1.57	2.29	0.07	0.50	1.38	7.44	8.86	52.43
1975	9.70	5.61	4.32	1.88	1.51	0.72	0.25	3.97	T	8.38	9.54	11.42	57.30
1976	9.40	7.25	4.24	2.79	2.66	1.07	1.26	2.71	1.22	2.64	1.37	3.00	39.61
1977	1.55	3.70	4.44	1.27	5.21	0.64	0.32	4.17	4.58	3.40	9.30	12.36	50.94
1978	6.61	4.39	3.45	3.97	2.90	1.54	1.49	1.47	7.59	0.78	6.81	2.95	43.95
1979	2.67	9.25	2.73	2.21	1.40	1.11	1.58	1.56	2.72	6.20	2.38	13.01	46.82
1980	6.27	6.18	3.98	4.14	0.93	2.46	0.41	0.36	2.21	1.84	9.77	10.92	49.47
1981	2.54	9.19	3.86	4.77	1.81	2.96	0.36	0.85	2.44	8.17	7.22	8.60	52.77
1982	7.79	9.91	4.67	3.89	0.51	1.15	0.54	0.55	1.89	6.12	5.74	10.64	53.40
1983	9.99	7.09	6.63	2.26	1.51	2.80	3.00	2.18	1.89	1.66	12.84	7.28	59.13
1984	6.97	5.49	6.42	3.67	5.48	3.74	T	0.21	1.76	5.13	12.19	4.97	56.03
1985	0.29	3.54	4.10	2.65	0.94	2.48	0.37	0.70	2.65	9.86	4.99	2.50	35.07
1986	12.14	6.84	2.29	2.87	3.20	0.92	1.11	0.01	3.38	4.12	11.09	5.20	53.17
1987	8.38	3.55	7.14	3.11	2.71	0.32	0.84	0.24	0.29	0.39	3.65	9.14	39.76
1988	5.18	2.35	5.66	4.94	3.36	2.06	0.42	0.43	1.93	2.26	10.14	5.16	43.89
1989	5.41	4.19	7.88	2.49	1.99	1.47	0.70	0.55	0.49	2.42	8.50	5.66	41.75
1990	14.53	8.52	3.54	3.29	2.06	2.86	0.32	1.79	0.03	6.34	15.06	5.05	63.39
1991	5.35	5.85	4.32	7.80	1.69	1.30	0.33	2.31	T	2.20	7.00	4.28	42.43
1992	9.45	4.19	1.50	5.15	0.19	1.15	0.36	0.82	2.44	2.80	6.82	5.28	40.15
1993	5.36	0.22	4.95	6.67	4.57	1.77	1.35	0.15	T	1.63	2.86	6.02	35.55
Record Mean	7.73	5.90	4.94	3.41	1.99	1.66	0.78	1.14	2.06	4.66	7.94	7.86	50.08

AVERAGE TEMPERATURE (deg. F)

OLYMPIA WASHINGTON

YEAR	JAN	FEB	MAR	APR	MAY	JUNE	JULY	AUG	SEP	OCT	NOV	DEC	ANNUAL
1964	39.8	39.0	42.1	44.6	49.9	56.1	61.1	60.1	55.9	50.3	41.0	34.4	47.8
1965	38.9	41.5	43.8	49.0	51.6	59.1	65.9	65.2	54.8	53.3	46.6	38.3	50.7
1966	39.4	40.6	43.8	49.1	52.6	56.3	60.9	61.8	59.4	49.1	44.0	42.1	49.9
1967	40.7	41.3	41.0	44.8	53.6	62.2	63.9	68.8	62.3	52.5	43.8	39.4	51.2
1968	38.3	43.9	46.2	47.0	54.0	58.3	64.8	61.7	57.7	48.6	44.5	35.6	50.1
1969	30.9	38.2	44.4	48.1	56.2	64.0	62.1	60.6	58.9	48.9	44.3	40.3	49.7
1970	38.1	42.7	43.4	45.1	51.8	60.8	64.1	62.1	55.1	47.6	42.4	36.8	49.2
1971	38.6	39.2	40.0	46.0	53.7	54.8	63.5	65.4	55.3	48.6	43.6	36.0	48.7
1972	34.0	38.8	45.3	43.8	54.9	58.0	62.8	64.2	55.1	46.5	43.0	35.5	48.5
1973	36.5	40.4	43.5	47.2	53.9	57.9	62.7	59.5	58.8	49.8	39.8	40.5	49.2
1974	37.4	41.9	44.7	46.6	51.8	59.6	62.4	64.7	63.2	49.5	43.6	40.1	50.5
1975	38.4	39.4	40.5	44.0	52.5	57.5	63.9	60.7	57.8	49.1	41.3	38.6	48.6
1976	40.1	39.3	40.3	46.0	51.5	54.9	61.7	60.7	58.9	49.6	43.2	39.1	48.8
1977	35.7	45.2	43.4	50.2	51.4	59.8	59.7	64.6	56.7	48.5	40.4	38.0	49.5
1978	41.8	43.6	45.9	48.1	52.9	63.1	65.2	63.5	56.8	50.6	37.9	34.1	50.3
1979	30.9	40.3	47.5	49.6	54.7	58.7	65.1	62.8	59.3	51.7	40.2	41.3	50.2
1980	31.6	39.8	42.0	48.9	52.8	56.3	62.9	62.3	59.6	51.9	46.5	43.7	49.9
1981	42.1	43.5	47.1	49.0	53.9	57.8	62.7	67.4	60.0	49.7	45.3	39.7	51.5
1982	38.0	40.6	42.6	45.8	54.1	62.9	63.1	64.4	59.5	50.8	40.2	38.6	50.1
1983	42.6	45.3	47.5	48.9	56.5	58.7	62.0	64.1	56.3	49.2	46.0	33.9	50.9
1984	41.7	43.0	46.7	46.7	51.6	57.8	63.3	63.8	58.2	48.4	43.3	35.4	50.0
1985	35.7	37.4	42.1	48.9	54.5	59.0	67.5	63.5	56.6	49.2	33.8	34.0	48.5
1986	42.0	40.2	47.2	46.3	54.4	60.8	59.7	66.5	57.2	51.4	43.4	38.7	50.7
1987	38.7	43.2	46.0	50.0	55.5	60.2	62.4	63.6	60.1	52.4	46.1	37.6	51.3
1988	37.9	42.5	43.8	48.4	53.3	56.7	62.6	62.1	57.3	53.8	43.0	39.3	50.1
1989	39.4	33.2	41.1	50.9	53.9	60.3	61.1	62.4	59.4	49.7	44.7	41.1	49.8
1990	40.3	37.4	44.2	49.6	52.5	57.7	65.8	66.0	61.1	48.5	44.0	32.6	50.0
1991	37.1	45.1	41.4	46.4	51.9	55.7	63.8	65.1	59.7	49.3	44.6	41.0	50.1
1992	41.2	45.2	47.9	50.4	57.5	63.0	64.7	64.8	58.2	51.4	43.7	37.0	52.1
1993	34.9	38.5	45.1	49.1	58.6	59.3	60.3	64.1	59.7	53.1	39.1	39.9	50.1
Record													
Mean	37.5	40.8	43.3	47.8	53.8	58.7	63.2	63.2	58.4	50.3	42.9	38.7	49.9
Max	44.0	49.1	53.2	59.0	66.0	70.9	77.0	77.0	71.5	60.8	50.4	44.8	60.3
Min	30.9	32.6	33.4	36.5	41.5	46.5	49.3	49.4	45.2	39.8	35.4	32.6	39.4

REFERENCE NOTES

OLYMPIA WASHINGTON

GENERAL
T - TRACE AMOUNT.
BLANK ENTRIES DENOTE MISSING/UNREPORTED DATA.
\# INDICATES A STATION OR INSTRUMENT RELOCATION.

SPECIFIC

(a) - LENGTH OF RECORD IN YEARS, ALTHOUGH INDIVIDUAL MONTHS MAY BE MISSING.
0.* OR * - THE VALUE IS BETWEEN 0.0 AND 0.05
NORMALS - BASED ON THE 1961-1990 RECORD PERIOD.
EXTREMES - DATES ARE THE MOST RECENT OCCURRENCE
WIND DIR. - NUMERALS SHOW TENS OF DEGREES CLOCKWISE FROM TRUE NORTH. "00" INDICATES CALM.
RESULTANT DIRECTIONS ARE GIVEN TO WHOLE DEGREES.
BOLD VALUES INDICATE EXTREME VALUES WHICH OCCURRED AFTER THE ASOS SYSTEM WAS COMMISSIONED.

RECORD = PERIOD OF RECORD
RECORD MEAN PRECIPITATION IS THE MEAN OF ALL DAILY PRECIPITATION AMOUNTS DURING THE PERIOD OF RECORD.
RECORD MAX(MIN) TEMPERATURE IS THE MEAN OF ALL DAILY MAX(MIN) TEMPERATURES DURING THE PERIOD OF RECORD.
RECORD MEAN TEMPERATURE IS THE SUM OF THE RECORD MAX AND RECORD MIN DIVIDED BY 2.
AVERAGE TEMPERATURE IS THE SUM OF THE MEAN DAILY MAX AND MIN TEMPERATURE DIVIDED BY 2.

EXCEPTIONS

MEAN WIND SPEED, THUNDERSTORMS, AND HEAVY FOG ARE THROUGH 1969 AND 1972 TO DATE.

RECORD MEANS ARE THROUGH THE CURRENT YEAR, BEGINNING IN 1942 FOR TEMPERATURE
1942 FOR PRECIPITATION
1942 FOR SNOWFALL

HEATING DEGREE DAYS Base 65 deg. F OLYMPIA WASHINGTON

SEASON	JULY	AUG	SEP	OCT	NOV	DEC	JAN	FEB	MAR	APR	MAY	JUNE	TOTAL
1964-65	130	150	266	450	715	945	801	652	650	470	408	177	5814
1965-66	52	52	298	356	546	820	788	676	649	471	377	258	5343
1966-67	129	99	161	485	624	700	745	653	739	599	350	96	5380
1967-68	51	13	94	379	631	786	820	604	575	534	334	201	5022
1968-69	64	113	214	503	610	903	1048	746	633	498	274	76	5682
1969-70	95	137	183	494	616	759	826	615	665	591	403	149	5533
1970-71	65	102	293	535	672	868	810	716	771	560	344	304	6040
1971-72	113	49	286	500	639	891	954	752	602	632	315	204	5937
1972-73	85	69	295	570	651	909	874	681	659	528	345	223	5889
1973-74	97	170	184	465	748	749	852	644	625	544	403	161	5642
1974-75	103	51	78	472	634	763	817	711	751	624	377	225	5606
1975-76	73	138	210	486	703	811	766	739	760	563	410	296	5955
1976-77	104	127	177	470	649	798	902	547	660	438	415	158	5445
1977-78	166	90	242	502	734	829	711	596	587	502	367	91	5417
1978-79	55	78	240	442	803	952	1048	683	536	454	314	188	5793
1979-80	59	71	165	404	737	726	1027	724	702	476	371	254	5716
1980-81	81	96	158	399	549	653	701	597	551	474	336	212	4807
1981-82	100	50	161	465	585	777	831	678	686	570	330	103	5336
1982-83	78	63	170	435	738	813	687	545	535	476	275	182	4997
1983-84	98	38	255	482	566	958	712	631	560	545	408	211	5464
1984-85	80	64	205	507	644	911	903	767	704	476	320	181	5762
1985-86	15	68	245	484	932	954	704	689	542	555	332	128	5648
1986-87	158	25	244	414	642	803	812	604	583	442	293	162	5182
1987-88	100	75	153	382	562	841	834	645	650	492	360	250	5344
1988-89	108	106	242	340	656	790	789	882	735	420	336	153	5557
1989-90	128	88	164	467	604	735	760	768	640	454	380	220	5408
1990-91	39	31	116	505	622	1000	854	551	725	549	398	273	5663
1991-92	73	59	159	478	604	736	729	567	525	434	235	107	4706
1992-93	49	57	205	417	630	862	927	737	609	468	197	171	5329
1993-94	139	66	175	361	772	770							

COOLING DEGREE DAYS Base 65 deg. F OLYMPIA WASHINGTON

YEAR	JAN	FEB	MAR	APR	MAY	JUNE	JULY	AUG	SEP	OCT	NOV	DEC	TOTAL
1969	0	0	0	0	7	51	13	6	8	0	0	0	85
1970	0	0	0	0	0	29	44	16	0	0	0	0	89
1971	0	0	0	0	0	3	71	67	0	0	0	0	141
1972	0	0	0	0	8	0	25	50	5	0	0	0	88
1973	0	0	0	0	6	14	35	7	5	0	0	0	67
1974	0	0	0	0	0	6	30	50	32	0	0	0	118
1975	0	0	0	0	0	4	45	9	1	0	0	0	59
1976	0	0	0	0	0	2	9	3	0	0	0	0	14
1977	0	0	0	0	0	14	7	87	1	0	0	0	109
1978	0	0	0	0	1	43	64	38	0	0	0	0	146
1979	0	0	0	0	0	6	73	12	2	0	0	0	93
1980	0	0	0	0	0	0	21	19	3	2	0	0	45
1981	0	0	0	0	0	2	35	131	18	0	0	0	186
1982	0	0	0	0	0	46	24	52	10	0	0	0	132
1983	0	0	0	0	18	1	10	18	0	0	0	0	47
1984	0	0	0	0	0	1	34	37	6	1	0	0	79
1985	0	0	0	0	3	7	97	30	0	0	0	0	137
1986	0	0	0	0	12	9	0	78	17	0	0	0	116
1987	0	0	0	0	8	25	26	40	12	1	0	0	112
1988	0	0	0	0	4	7	39	23	17	0	0	0	90
1989	0	0	0	0	0	20	13	14	3	0	0	0	50
1990	0	0	0	0	0	5	71	70	6	0	0	0	152
1991	0	0	0	0	0	0	42	69	7	0	0	0	118
1992	0	0	0	0	9	50	48	57	8	0	0	0	172
1993	0	0	0	0	5	6	0	49	22	0	0	0	82

SNOWFALL (inches) OLYMPIA WASHINGTON

SEASON	JULY	AUG	SEP	OCT	NOV	DEC	JAN	FEB	MAR	APR	MAY	JUNE	TOTAL
1964-65	0.0	0.0	0.0	0.0	1.3	16.5	14.8	T	0.2	0.0	0.0	0.0	32.8
1965-66	0.0	0.0	0.0	0.0	0.0	14.0	10.7	2.0	15.5	0.1	0.0	0.0	42.3
1966-67	0.0	0.0	0.0	0.0	0.0	1.7	3.3	T	1.1	T	0.0	0.0	6.1
1967-68	0.0	0.0	0.0	0.0	0.0	7.9	16.9	0.0	0.0	T	0.0	0.0	24.8
1968-69	0.0	0.0	0.0	0.0	0.0	21.4	58.7	1.4	0.0	0.0	0.0	0.0	81.5
1969-70	0.0	0.0	0.0	0.0	0.0	T	3.0	0.0	T	0.1	T	0.0	3.1
1970-71	0.0	0.0	0.0	T	0.4	9.0	15.4	5.1	6.1	0.1	T	T	36.1
1971-72	0.0	0.0	0.0	T	T	17.2	29.2	2.4	3.6	2.2	0.0	0.0	54.6
1972-73	0.0	0.0	T	0.0	0.0	5.8	2.9	0.0	T	0.0	0.0	0.0	8.7
1973-74	0.0	0.0	0.0	0.0	1.9	2.6	4.1	1.1	T	T	T	0.0	9.7
1974-75	0.0	0.0	0.0	0.0	T	13.5	7.2	1.5	0.3	T	T	0.0	22.5
1975-76	0.0	0.0	0.0	T	4.5	3.7	0.6	7.2	T	T	T	T	16.0
1976-77	0.0	0.0	0.0	0.0	0.0	0.0	1.8	T	0.8	T	T	0.0	2.6
1977-78	0.0	0.0	0.0	0.0	3.8	1.6	0.0	0.0	T	0.0	T	0.0	5.4
1978-79	0.0	0.0	0.0	0.0	14.8	2.5	0.1	1.5	0.0	0.0	0.0	0.0	18.9
1979-80	0.0	0.0	0.0	0.0	T	T	19.7	11.9	4.1	T	0.0	0.0	35.7
1980-81	0.0	0.0	0.0	0.0	T	3.0	0.0	0.9	0.0	0.4	0.0	0.0	4.3
1981-82	0.0	0.0	0.0	0.0	0.0	2.8	11.8	5.1	1.9	T	0.0	0.0	21.6
1982-83	0.0	0.0	0.0	0.0	T	T	0.0	0.0	0.0	T	0.0	0.0	T
1983-84	0.0	0.0	0.0	0.0	0.0	4.9	T	0.0	T	0.0	T	0.0	4.9
1984-85	0.0	0.0	0.0	0.0	T	4.7	T	13.8	T	T	T	0.0	18.5
1985-86	0.0	0.0	0.0	0.0	12.3	7.3	0.0	5.3	0.0	T	T	0.0	24.9
1986-87	0.0	0.0	0.0	0.0	0.0	0.0	0.0	0.0	0.0	T	0.0	0.0	T
1987-88	0.0	0.0	0.0	0.0	T	1.0	T	0.1	0.0	0.2	T	0.0	1.3
1988-89	0.0	0.0	0.0	0.0	T	T	6.8	13.8	9.0	T	0.0	T	29.6
1989-90	0.0	0.0	0.0	0.0	0.0	0.0	1.7	27.4	0.2	0.0	0.0	0.0	29.3
1990-91	0.0	0.0	0.0	0.0	T	2.2	3.0	0.0	T	T	T	0.0	5.2
1991-92	0.0	0.0	0.0	0.0	0.0	0.0	T	0.0	0.0	T	0.0	0.0	T
1992-93	0.0	0.0	0.0	0.0	0.0	9.4	3.8	0.2	0.0	0.0	0.0	0.0	13.4
1993-94	0.0	0.0	0.0	0.0	0.5	0.0							
Record Mean	0.0	0.0	T	T	1.4	3.5	7.2	3.3	1.8	0.1	T	T	17.2

QUILLAYUTE AIRPORT
WASHINGTON

Quillayute Airport, located on the coastal plain between the Pacific Ocean and the Olympic Mountains, is 3 miles inland from the coast, and 10 miles west of the city of Forks. The terrain is slightly rolling with a gradual increase in elevation from sea level to 180 feet at the station, to 350 feet in the vicinity of Forks. Foothills of the Olympic Mountains begin near the eastern edge of Forks, and within 10 to 15 miles, the higher ridges reach elevations of 3,000 to 6,000 feet.

Timber is the primary economic product of this northwestern section of the Olympic Peninsula. Only small areas of the Quillayute plains and a few other localities are devoted to cattle raising and agriculture. Logging operations continue throughout the year in the lower elevations with little delay due to normal rain or snow. In the foothills and mountains, heavy snowfall, excessive precipitation, and high winds during the winter season result in shutdowns, but seldom for more than one or two days. Forests are closed to logging and recreation for short periods of time almost every summer when the relative humidity is low and the fire danger is high.

Maritime air from over the Pacific has an influence on the climate throughout the year. In the late fall and winter, the low pressure center in the Gulf of Alaska intensifies and is of major importance in controlling weather systems entering the Pacific Northwest. At this season of the year, storm systems crossing the Pacific follow a more southerly path striking the coast at frequent intervals. The prevailing flow of air is from the southwest and west. Air reaching this area is moist and near the temperature of the ocean water along the coast which ranges from 45 degrees in February to 57 degrees in August. The wet season begins in September or October. From October through January, rain may be expected on about 26 days per month, from February through March, on 20 days, from April to June, on 15 days, and from July to September, on 10 days. As the weather systems move inland, rainfall is usually of moderate intensity and continuous, rather than heavy downpours for brief periods. Gale force winds are not unusual. Most of the winter precipitation over the coastal plains falls as rain, however, snow can be expected each year. Snow seldom reaches depths in excess of 10 inches or remains on the ground longer than two weeks.

Annual precipitation increases from approximately 90 inches near the coast, to amounts in excess of 120 inches over the coastal plains, to 200 inches or more on the wettest slopes of the Olympic Mountains.

During the rainy season, temperatures show little diurnal or day-to-day change. Maximums are in the 40s and minimums in the mid-30s. A few brief outbreaks of cold air from the interior of Canada can be expected each winter. Clear, dry, cold weather generally prevails during periods of easterly winds.

In the late spring and summer, a clockwise circulation of air around the large high pressure center over the north Pacific brings a prevailing northwesterly and westerly flow of cool, comparatively dry, stable air into the northwest Olympic Peninsula. The dry season begins in May with the driest period between mid-July and mid-August. The total rainfall for July is less than .5 inch in one summer out of ten. It also exceeds 5 inches in one summer out of ten. During the warmest months, afternoon temperatures are in the upper 60s and lower 70s, reaching the upper 70s and the lower 80s on a few days. Occasionally, hot, dry air from the east of the Cascade Mountains reaches this area and temperatures are in the mid- or upper-90s for one to three days.

In summer and early fall, fog or low clouds form over the ocean and frequently move inland at night, but generally disappear by midday. In winter, under the influence of a surface high pressure system, centered off the coast, fog, low clouds, and drizzle are a daily occurrence as long as this type of pressure continues.

NORMALS, MEANS, AND EXTREMES

QUILLAYUTE AIRPORT WASHINGTON

LATITUDE: 47°57'N LONGITUDE: 124°33' W ELEVATION: FT. GRND 179 BARO 1816 TIME ZONE: PACIFIC WBAN: 94240

	(a)	JAN	FEB	MAR	APR	MAY	JUNE	JULY	AUG	SEP	OCT	NOV	DEC	YEAR
TEMPERATURE °F:														
Normals														
-Daily Maximum		46.0	49.1	51.4	55.0	59.9	64.2	68.2	69.1	67.0	59.4	50.9	46.2	57.2
-Daily Minimum		33.7	34.9	35.2	37.3	41.8	46.5	49.4	49.7	46.2	41.4	37.2	34.4	40.6
-Monthly		39.9	42.0	43.3	46.2	50.9	55.4	58.8	59.4	56.6	50.4	44.1	40.3	48.9
Extremes														
-Record Highest	27	65	73	71	83	92	96	97	99	97	83	69	64	99
-Year		1986	1992	1987	1987	1987	1982	1988	1981	1988	1988	1976	1969	AUG 1981
-Record Lowest	27	7	11	19	24	29	33	38	36	28	24	5	7	5
-Year		1969	1989	1989	1975	1977	1976	1977	1985	1972	1984	1985	1972	NOV 1985
NORMAL DEGREE DAYS:														
Heating (base 65°F)		778	644	673	564	437	288	192	180	256	453	627	766	5858
Cooling (base 65°F)		0	0	0	0	0	0	0	6	0	0	0	0	6
% OF POSSIBLE SUNSHINE	27	21	29	33	36	36	35	43	44	47	34	21	18	33
MEAN SKY COVER (tenths)														
Sunrise - Sunset	27	8.4	8.0	7.9	8.0	7.8	7.7	6.7	6.7	6.2	7.3	8.2	8.2	7.6
MEAN NUMBER OF DAYS:														
Sunrise to Sunset														
-Clear	27	3.3	3.5	3.3	2.9	2.6	3.4	6.4	6.3	8.0	5.1	2.8	3.6	51.2
-Partly Cloudy	27	3.1	3.7	5.7	5.8	7.8	7.0	8.1	9.0	7.7	7.2	5.1	3.9	74.3
-Cloudy	27	24.6	21.1	21.9	21.3	20.6	19.5	16.4	15.7	14.3	18.7	22.1	23.6	239.8
Precipitation														
.01 inches or more	27	22.2	19.1	20.9	19.6	16.8	13.9	11.7	10.4	11.8	17.7	22.0	22.7	208.6
Snow,Ice pellets,hail														
1.0 inches or more	27	1.6	1.0	0.5	0.1	0.0	0.0	0.0	0.0	0.0	0.0	0.4	0.9	4.4
Thunderstorms	26	0.7	0.3	0.4	0.4	0.2	0.2	0.6	0.3	0.5	1.0	1.7	0.9	7.0
Heavy Fog Visibility														
1/4 mile or less	26	4.1	3.3	3.6	2.3	2.9	2.8	4.7	8.4	7.0	7.0	4.0	3.9	54.0
Temperature °F														
-Maximum														
90° and above	27	0.0	0.0	0.0	0.0	0.1	0.2	0.4	0.4	0.2	0.0	0.0	0.0	1.3
32° and below	27	0.6	0.1	0.0	0.0	0.0	0.0	0.0	0.0	0.0	0.0	0.3	0.9	1.9
-Minimum														
32° and below	27	13.3	10.4	9.6	5.2	0.5	0.0	0.0	0.0	0.2	3.3	7.5	13.0	62.9
0° and below	27	0.0	0.0	0.0	0.0	0.0	0.0	0.0	0.0	0.0	0.0	0.0	0.0	0.0
AVG. STATION PRESS.(mb)	21	1009.4	1008.0	1007.6	1009.8	1010.1	1010.6	1011.4	1010.2	1009.9	1010.0	1008.4	1009.6	1009.6
RELATIVE HUMIDITY (%)														
Hour 04	27	91	91	92	93	93	94	94	95	94	94	93	92	93
Hour 10 (Local Time)	27	90	86	82	75	73	74	74	76	76	85	89	91	81
Hour 16	27	83	76	72	68	66	66	65	67	67	75	83	86	73
Hour 22	27	90	90	90	89	88	88	88	90	90	92	92	91	90
PRECIPITATION (inches):														
Water Equivalent														
-Normal		14.37	12.59	11.48	7.51	5.40	3.12	2.57	2.54	4.88	10.52	14.74	15.46	105.18
-Maximum Monthly	27	23.99	20.60	21.86	13.90	12.45	8.50	11.02	15.07	10.93	27.17	29.14	27.82	29.14
-Year		1992	1982	1974	1992	1974	1981	1983	1991	1969	1975	1983	1979	NOV 1983
-Minimum Monthly	27	1.22	0.87	1.84	2.94	1.05	0.40	0.36	0.30	0.12	1.37	4.41	3.63	0.12
-Year		1985	1993	1992	1973	1972	1967	1985	1986	1991	1987	1976	1985	SEP 1991
-Maximum in 24 hrs	27	8.32	5.07	4.23	2.77	3.54	2.34	6.45	4.70	4.13	5.54	5.70	6.76	8.32
-Year		1968	1982	1974	1968	1973	1990	1972	1991	1968	1975	1991	1972	JAN 1968
Snow,Ice pellets,hail														
-Maximum Monthly	27	40.1	16.1	10.2	2.8	T	T	0.0	0.0	T	T	15.6	11.6	40.1
-Year		1969	1990	1971	1975	1992	1988			1972	1991	1985	1972	JAN 1969
-Maximum in 24 hrs	27	8.2	8.4	7.5	2.4	T	T	0.0	0.0	T	T	7.7	7.3	8.4
-Year		1969	1971	1989	1975	1992	1988			1972	1991	1985	1981	FEB 1971
WIND:														
Mean Speed (mph)	27	6.7	6.8	6.7	6.4	6.1	5.9	5.5	5.1	5.0	5.4	6.4	6.6	6.0
Prevailing Direction														
Fastest Mile														
-Direction (!!!)	27	W	SE	SE	SW	W	SE	NE	SE	SE	SE	SE	SW	SE
-Speed (MPH)	27	34	46	33	32	28	23	23	27	33	42	37	39	46
-Year		1986	1977	1974	1970	1989	1988	1972	1969	1972	1982	1973	1969	FEB 1977
Peak Gust														
-Direction (!!!)	10	S	S	SE	S	S	SE	SE	SE	SE	S	SE	S	S
-Speed (mph)	10	55	56	46	40	41	44	26	48	36	44	54	58	58
-Date		1992	1991	1990	1993	1985	1988	1989	1991	1988	1990	1986	1990	DEC 1990

PRECIPITATION (inches)

QUILLAYUTE AIRPORT WASHINGTON

YEAR	JAN	FEB	MAR	APR	MAY	JUNE	JULY	AUG	SEP	OCT	NOV	DEC	ANNUAL
1966								1.79	3.50	10.52	16.07	23.47	
1967	20.62	16.80	11.73	5.90	2.50	0.40	1.10	0.62	7.75	24.86	8.47	19.27	120.02
1968	22.59	12.89	15.17	8.15	4.33	4.58	1.74	4.61	8.46	17.46	13.54	15.90	129.42
1969	13.82	9.11	10.58	11.17	5.49	2.04	1.59	3.43	10.93	6.85	8.40	12.87	96.28
1970	13.67	5.09	7.43	13.89	2.82	1.20	2.45	0.55	7.21	9.09	11.32	16.55	91.27
1971	23.34	12.81	14.83	6.21	3.90	5.04	1.04	3.52	5.93	13.21	14.33	15.42	119.58
1972	12.92	15.51	15.38	11.46	1.05	1.49	9.33	0.45	6.91	2.30	10.88	22.20	109.88
1973	17.29	5.90	9.72	2.94	7.43	5.47	1.14	1.03	3.63	10.16	17.17	19.89	101.77
1974	19.91	17.45	21.86	9.26	12.45	3.14	5.42	0.51	2.71	3.13	12.90	19.02	127.76
1975	14.25	11.89	9.61	5.14	5.68	3.02	0.43	10.12	0.36	27.17	24.28	19.69	131.64
1976	15.70	14.91	14.23	4.32	5.35	2.75	3.43	4.12	2.39	6.07	4.41	9.09	86.77
1977	8.26	11.22	10.82	3.88	8.87	2.10	1.92	3.15	6.28	9.21	18.10	15.03	98.84
1978	8.20	6.52	7.37	5.54	5.66	3.47	0.67	6.89	9.79	3.08	6.67	7.68	71.54
1979	3.64	19.13	8.39	5.41	4.00	2.58	2.18	0.84	8.64	8.86	7.53	27.82	99.02
1980	6.30	13.10	10.52	8.11	3.15	1.85	4.69	2.00	7.74	4.41	22.17	18.31	102.35
1981	4.49	11.03	11.66	12.65	4.80	8.50	1.43	1.24	5.18	15.32	14.04	14.72	105.06
1982	19.34	20.60	9.10	9.72	1.55	2.39	1.53	2.54	3.81	13.61	12.71	12.25	109.15
1983	13.77	20.11	12.97	3.91	3.81	4.08	11.02	2.09	5.73	5.44	29.14	8.93	121.00
1984	18.27	13.77	9.79	8.89	10.56	2.83	0.55	1.61	3.69	13.91	14.65	13.89	112.41
1985	1.22	6.90	9.33	6.19	1.86	2.43	0.36	1.09	4.75	14.69	7.79	3.63	60.24
1986	16.11	12.74	13.09	6.79	11.02	1.48	3.45	0.30	3.68	5.36	13.29	12.42	99.73
1987	14.05	8.20	13.37	6.95	8.52	1.74	1.39	0.55	1.75	1.37	11.34	11.71	80.94
1988	9.39	8.45	12.83	10.87	10.68	2.28	1.80	1.30	5.73	8.07	15.51	12.12	99.03
1989	15.65	6.22	10.03	5.57	3.31	4.52	2.36	1.99	0.13	9.80	16.57	8.82	84.97
1990	17.09	15.85	8.68	5.51	4.59	6.47	0.95	2.30	0.38	15.85	22.59	16.79	117.05
1991	12.88	16.45	6.89	7.88	4.31	2.27	1.28	15.07	0.12	2.66	18.49	10.71	99.01
1992	23.99	7.19	1.84	13.90	1.27	1.72	0.43	1.85	5.34	11.19	11.86	9.79	90.37
1993	7.88	0.87	10.96	12.25	6.83	3.67	0.98	1.25	0.23	5.46	6.62	12.85	69.85
Record Mean	13.88	11.88	11.04	7.87	5.40	3.09	2.39	2.74	4.74	9.97	13.96	14.67	101.64

AVERAGE TEMPERATURE (deg. F)

QUILLAYUTE AIRPORT WASHINGTON

YEAR	JAN	FEB	MAR	APR	MAY	JUNE	JULY	AUG	SEP	OCT	NOV	DEC	ANNUAL
1966								58.7	57.8	49.4	45.2	43.6	
1967	41.9	42.3	40.4	42.9	50.9	57.6	60.1	62.6	59.7	52.0	45.8	40.3	49.7
1968	40.2	44.7	45.6	45.1	51.5	54.8	60.6	58.6	56.2	49.0	45.6	36.7	49.0
1969	31.4	38.7	43.4	45.5	53.2	59.2	58.0	56.8	56.1	49.8	45.6	43.6	48.4
1970	40.3	45.6	43.7	43.8	49.7	57.0	58.3	58.4	54.8	49.4	43.8	39.0	48.7
1971	39.5	40.7	39.3	45.5	50.5	52.5	58.4	61.3	54.5	48.8	44.3	36.4	47.7
1972	35.7	40.1	43.5	42.7	53.7	55.8	60.5	61.3	52.5	48.4	44.5	37.5	48.0
1973	39.0	42.0	43.0	46.7	51.7	54.9	58.0	55.4	56.4	48.7	40.5	42.2	48.2
1974	38.1	40.4	42.0	46.5	48.2	55.0	57.9	59.4	60.4	50.6	45.0	43.6	48.9
1975	39.0	40.3	41.0	43.3	51.4	53.5	59.2	57.4	57.7	49.1	43.8	40.6	48.0
1976	41.9	40.0	39.9	46.7	49.9	52.9	58.4	58.1	57.9	51.0	45.8	43.8	48.8
1977	39.4	44.9	42.2	47.0	48.7	55.4	57.1	60.7	55.1	49.3	42.1	40.1	48.5
1978	42.9	43.9	45.4	47.5	50.7	58.7	59.9	60.4	55.9	51.7	40.4	35.9	49.4
1979	35.3	39.7	45.5	47.2	52.4	54.9	60.0	60.4	60.0	52.5	43.9	44.5	49.7
1980	35.7	44.4	43.4	48.1	50.5	54.2	59.7	56.7	56.6	52.3	46.3	43.9	49.3
1981	45.3	43.7	46.6	46.8	51.1	54.2	58.1	61.3	56.8	49.3	45.8	41.0	50.0
1982	37.8	40.6	41.5	43.1	49.8	56.6	57.1	59.5	57.7	50.3	41.2	39.4	47.9
1983	44.4	44.5	46.2	47.7	53.5	55.8	59.6	60.2	54.4	47.8	45.8	35.2	49.6
1984	42.0	44.0	46.4	45.7	49.4	53.2	58.6	58.9	55.1	47.9	43.5	36.7	48.5
1985	38.9	39.4	40.4	45.9	51.7	54.9	60.3	58.6	54.9	48.7	35.0	38.3	47.3
1986	44.6	40.3	47.3	44.8	51.3	57.4	56.8	60.7	55.1	52.4	44.6	42.8	49.8
1987	41.8	44.4	46.0	49.1	52.7	55.3	58.4	59.5	57.7	52.6	47.8	39.2	50.4
1988	39.7	45.0	44.3	47.7	51.4	54.4	59.5	59.9	55.4	53.0	44.8	41.9	49.8
1989	39.6	35.0	41.9	50.4	51.6	57.4	58.6	59.4	59.3	51.1	45.9	43.6	49.5
1990	42.5	39.8	45.1	48.9	51.4	55.9	61.4	61.5	59.9	48.3	45.3	36.3	49.7
1991	39.5	46.4	41.8	46.0	50.3	53.9	59.2	58.6	60.1	49.3	45.6	43.6	49.5
1992	44.1	46.5	48.5	49.0	54.3	58.4	60.7	60.7	54.8	51.1	44.8	38.1	50.9
1993	36.8	40.9	45.4	48.1	55.5	56.8	57.8	60.9	55.6	52.7	41.6	41.6	49.5
Record Mean	39.9	42.2	43.7	46.3	51.4	55.5	59.0	59.5	56.7	50.2	44.0	40.3	49.0
Max	46.0	49.4	51.7	55.1	60.4	64.2	68.2	69.1	67.0	59.2	50.8	46.2	57.3
Min	33.8	34.9	35.6	37.6	42.3	46.9	49.7	49.9	46.4	41.2	37.3	34.4	40.8

REFERENCE NOTES

QUILLAYUTE AIRPORT WASHINGTON

GENERAL
T - TRACE AMOUNT.
BLANK ENTRIES DENOTE MISSING/UNREPORTED DATA.
# INDICATES A STATION OR INSTRUMENT RELOCATION.

SPECIFIC

(a) - LENGTH OF RECORD IN YEARS, ALTHOUGH INDIVIDUAL MONTHS MAY BE MISSING.
0.* OR * - THE VALUE IS BETWEEN 0.0 AND 0.05
NORMALS - BASED ON THE 1961-1990 RECORD PERIOD.
EXTREMES - DATES ARE THE MOST RECENT OCCURRENCE
WIND DIR.- NUMERALS SHOW TENS OF DEGRESS CLOCKWISE FROM TRUE NORTH. "00" INDICATES CALM.
RESULTANT DIRECTIONS ARE GIVEN TO WHOLE DEGREES.
BOLD VALUES INDICATE EXTREME VALUES WHICH OCCURRED AFTER THE ASOS SYSTEM WAS COMMISSIONED.

RECORD = PERIOD OF RECORD
RECORD MEAN PRECIPITATION IS THE MEAN OF ALL DAILY PRECIPITATION AMOUNTS DURING THE PERIOD OF RECORD.
RECORD MAX(MIN) TEMPERATURE IS THE MEAN OF ALL DAILY MAX(MIN) TEMPERATURES DURING THE PERIOD OF RECORD.
RECORD MEAN TEMPERATURE IS THE SUM OF THE RECORD MAX AND RECORD MIN DIVIDED BY 2.
AVERAGE TEMPERATURE IS THE SUM OF THE MEAN DAILY MAX AND MIN TEMPERATURE DIVIDED BY 2.

EXCEPTIONS

RECORD MEANS ARE THROUGH THE CURRENT YEAR,
BEGINNING IN 1966 FOR TEMPERATURE
1966 FOR PRECIPITATION
1966 FOR SNOWFALL

HEATING DEGREE DAYS Base 65 deg. F QUILLAYUTE AIRPORT WASHINGTON

SEASON	JULY	AUG	SEP	OCT	NOV	DEC	JAN	FEB	MAR	APR	MAY	JUNE	TOTAL
1966-67		191	208	479	586	656	709	628	755	655	431	214	
1967-68	153	83	163	396	570	760	761	580	596	589	412	301	5364
1968-69	147	195	261	487	576	871	1036	729	661	578	358	188	6087
1969-70	208	248	259	462	574	656	758	536	655	628	467	239	5690
1970-71	208	197	298	475	628	799	783	673	787	578	445	372	6243
1971-72	201	126	305	493	614	879	903	716	660	663	354	269	6183
1972-73	160	131	379	508	608	845	800	638	675	543	409	299	5995
1973-74	217	291	250	499	728	701	828	681	705	548	513	295	6256
1974-75	223	168	159	438	595	656	797	684	738	643	417	338	5856
1975-76	178	230	214	486	628	750	711	718	770	542	460	357	6044
1976-77	196	207	207	427	569	647	784	554	703	530	499	283	5606
1977-78	240	142	290	479	679	767	680	585	601	516	434	187	5600
1978-79	170	154	266	404	732	892	913	703	600	526	383	294	6037
1979-80	157	137	147	382	625	628	905	591	661	500	443	319	5495
1980-81	167	250	244	386	554	647	603	591	565	542	422	314	5285
1981-82	210	148	248	480	570	739	838	679	722	648	465	257	6004
1982-83	237	170	215	450	706	786	635	565	577	514	353	266	5474
1983-84	163	146	313	526	570	915	707	601	569	571	478	347	5906
1984-85	199	184	291	525	637	871	802	712	753	564	408	301	6247
1985-86	145	190	296	498	894	820	622	685	542	599	422	223	5936
1986-87	248	134	290	384	608	682	712	572	582	471	380	285	5348
1987-88	197	177	217	381	510	791	777	574	632	512	417	311	5496
1988-89	183	155	293	366	599	710	783	830	709	434	407	245	5714
1989-90	194	168	173	426	566	659	689	698	610	477	415	273	5348
1990-91	118	114	156	510	584	883	782	515	710	563	447	326	5708
1991-92	185	189	159	481	575	657	644	531	505	476	325	199	4926
1992-93	132	152	299	424	602	825	868	667	598	500	288	239	5594
1993-94	215	145	275	379	695	719							

COOLING DEGREE DAYS Base 65 deg. F QUILLAYUTE AIRPORT WASHINGTON

YEAR	JAN	FEB	MAR	APR	MAY	JUNE	JULY	AUG	SEP	OCT	NOV	DEC	TOTAL
1969	0	0	0	0	2	18	0	1	0	0	0	0	21
1970	0	0	0	0	0	7	8	1	0	0	0	0	16
1971	0	0	0	0	0	3	5	18	0	0	0	0	26
1972	0	0	0	0	11	0	31	21	11	0	0	0	74
1973	0	0	0	0	2	2	7	0	1	0	0	0	12
1974	0	0	0	0	0	0	10	3	27	0	0	0	40
1975	0	0	0	0	0	0	5	0	1	0	0	0	6
1976	0	0	0	0	0	0	0	0	0	0	0	0	0
1977	0	0	0	0	0	1	1	15	0	0	0	0	17
1978	0	0	0	0	0	7	18	17	0	0	0	0	42
1979	0	0	0	0	0	0	11	1	3	0	0	0	15
1980	0	0	0	0	0	0	11	0	0	0	0	0	11
1981	0	0	0	0	0	0	1	38	7	0	0	0	46
1982	0	0	0	0	0	10	0	4	2	0	0	0	16
1983	0	0	0	0	4	0	0	6	0	0	0	0	10
1984	0	0	0	0	0	0	5	1	0	0	0	0	6
1985	0	0	0	0	1	6	6	1	0	0	0	0	14
1986	0	0	0	0	4	1	0	7	0	0	0	0	12
1987	0	0	0	0	2	2	0	12	2	0	0	0	18
1988	0	0	0	0	0	0	19	4	13	0	0	0	36
1989	0	0	0	0	0	24	0	0	9	0	0	0	33
1990	0	0	0	0	0	7	16	10	11	0	0	0	44
1991	0	0	0	0	0	0	11	0	20	0	0	0	31
1992	0	0	0	0	2	9	6	25	0	0	0	0	42
1993	0	0	0	0	2	0	1	25	0	0	0	0	28

SNOWFALL (inches) QUILLAYUTE AIRPORT WASHINGTON

SEASON	JULY	AUG	SEP	OCT	NOV	DEC	JAN	FEB	MAR	APR	MAY	JUNE	TOTAL
1966-67		0.0	0.0	T	0.0	0.1	T	T	2.7	T	0.0	0.0	
1967-68	0.0	0.0	0.0	0.0	T	1.8	6.8	0.0	T	0.6	0.0	0.0	9.2
1968-69	0.0	0.0	0.0	0.0	T	4.0	40.1	6.2	T	0.4	0.0	0.0	50.7
1969-70	0.0	0.0	0.0	0.0	0.1	0.0	T	0.0	0.4	1.1	0.0	0.0	1.6
1970-71	0.0	0.0	0.0	0.0	1.7	7.0	26.4	12.8	10.2	0.6	T	0.0	58.7
1971-72	0.0	0.0	0.0	T	0.0	8.3	13.7	1.0	0.2	2.5	0.0	0.0	25.7
1972-73	0.0	0.0	T	T	0.0	11.6	1.2	0.0	0.4	T	T	T	13.2
1973-74	0.0	0.0	0.0	0.0	4.4	T	2.4	6.7	8.6	0.0	T	0.0	22.1
1974-75	0.0	0.0	0.0	0.0	T	0.8	7.7	3.0	T	2.8	0.0	0.0	14.3
1975-76	0.0	0.0	0.0	0.0	3.8	2.3	0.3	13.3	8.7	0.1	0.0	0.0	28.5
1976-77	0.0	0.0	0.0	0.0	0.0	0.0	0.0	T	T	T	T	0.0	T
1977-78	0.0	0.0	0.0	0.0	T	4.5	1.4	T	T	T	0.0	0.0	5.9
1978-79	0.0	0.0	0.0	0.0	T	1.6	T	0.5	T	0.0	0.0	0.0	2.1
1979-80	0.0	0.0	0.0	0.0	0.0	0.0	11.0	T	2.1	T	T	0.0	13.1
1980-81	0.0	0.0	0.0	0.0	0.0	7.5	0.0	T	T	0.6	0.0	0.0	8.1
1981-82	0.0	0.0	0.0	0.0	T	8.3	13.1	7.2	0.6	0.8	T	0.0	30.0
1982-83	0.0	0.0	0.0	T	T	T	T	T	T	T	T	0.0	T
1983-84	0.0	0.0	0.0	0.0	T	0.9	0.0	T	T	T	T	0.0	0.9
1984-85	0.0	0.0	0.0	T	T	6.7	0.0	4.7	0.4	T	T	0.0	11.8
1985-86	0.0	0.0	0.0	T	15.6	T	T	3.4	T	T	T	0.0	19.0
1986-87	0.0	0.0	0.0	0.0	T	T	0.1	0.9	0.3	0.1	0.0	0.0	1.4
1987-88	0.0	0.0	0.0	0.0	T	0.9	T	T	T	T	T	T	0.9
1988-89	0.0	0.0	0.0	0.0	1.1	T	3.2	0.3	7.5	T	0.0	0.0	12.1
1989-90	0.0	0.0	0.0	T	T	T	5.2	16.1	0.1	T	T	0.0	21.4
1990-91	0.0	0.0	0.0	T	1.5	5.1	T	0.0	0.7	T	T	0.0	7.3
1991-92	0.0	0.0	0.0	T	T	T	0.0	T	0.0	0.2	T	0.0	0.2
1992-93	0.0	0.0	0.0	0.0	0.0	6.8	0.3	2.1	T	T	0.0	0.0	9.2
1993-94	0.0	0.0	0.0	0.0	T	T							
Record Mean	0.0	0.0	T	T	1.0	2.8	4.9	2.9	1.6	0.4	T	T	13.6

SEATTLE, WASHINGTON
SEATTLE – TACOMA AIRPORT

The Seattle–Tacoma International Airport is located 6 miles south of the Seattle city limits and 14 miles north of Tacoma. It is situated on a low ridge lying between Puget Sound on the west and the Green River valley on the east with terrain sloping moderately to the shores of Puget Sound some 2 miles to the west. The Olympic Mountains, rising sharply from Puget Sound, are about 50 miles to the northwest. Rather steep bluffs border the Green River Valley about 2.5 miles to the east and the foothills of the Cascade Range begin 10 to 15 miles to the east of the airport.

The mild climate of the Pacific Coast is modified by the Cascade Mountains and, to a lesser extent, by the Olympic Mountains. The climate is characterized by mild temperatures, a pronounced though not sharply defined rainy season, and considerable cloudiness, particularly during the winter months. The Cascades are very effective in shielding the Seattle–Tacoma area from the cold, dry continental air during the winter and the hot, dry continental air during the summer months. The extremes of temperature that occur in western Washington are the result of the occasional pressure distributions that force the continental air into the Puget Sound area. But the prevailing southwesterly circulation keeps the average winter daytime temperatures in the 40s and the nighttime readings in the 30s. During the summer, daytime temperatures are usually in the 70s with nighttime lows in the 50s. Extremes of temperatures, both in the winter and summer, are usually of short duration. The dry season is centered around July and early August with July being the driest month of the year. The rainy season extends from October to March with December normally the wettest month, however, precipitation is rather evenly distributed through the winter and early spring months with more than 75 percent of the yearly precipitation falling during the winter wet season. Most of the rainfall in the Seattle area comes from storms common to the middle latitudes. These disturbances are most vigorous during the winter as they move through western Washington. The storm track shifts to the north during the summer and those that reach the State are not the wind and rain producers of the winter months. Local summer afternoon showers and a few thunderstorms occur in the Seattle–Tacoma area but they do not contribute materially to the precipitation.

The occurrence of snow in the Seattle–Tacoma area is extremely variable and usually melts before accumulating measurable depths. There are winters on record with only a trace of snow, but at the other extreme, over 21 inches has fallen in a 24–hour period. Usually, winter storms do not produce snow unless the storm moves in such a way to bring cold air out of Canada directly or with only a short over water trajectory.

The highest winds recorded in the Seattle–Tacoma area were associated with strong storms crossing the state from the southwest. Prevailing winds are from the southwest but occasional severe winter storms will produce strong northerly winds. Winds during the summer months are relatively light with occasional land–sea breeze effects creating afternoon northerly winds of 8 to 15 miles an hour. Fog or low clouds that form over the southern Puget Sound area in the late summer, fall, and early winter months, often dominate the weather conditions during the late night and early morning hours with visibilities occasionally lower for a few hours near sunrise. Most of the summer clouds form along the coast and move into the Seattle area from the southwest.

Based on the 1951–1980 period, the average first occurrence of 32 degrees Fahrenheit in the fall is November 11 and the average last occurrence in the spring is March 24.

NORMALS, MEANS, AND EXTREMES

SEATTLE, WASHINGTON SEATTLE – TACOMA AIRPORT

LATITUDE: 47°27'N LONGITUDE: 122°18' W ELEVATION: FT. GRND 400 BARO 451 TIME ZONE: PACIFIC WBAN: 24233

	(a)	JAN	FEB	MAR	APR	MAY	JUNE	JULY	AUG	SEP	OCT	NOV	DEC	YEAR
TEMPERATURE °F:														
Normals														
-Daily Maximum		45.0	49.5	52.7	57.2	63.9	69.9	75.2	75.2	69.3	59.7	50.5	45.1	59.4
-Daily Minimum		35.2	37.4	38.5	41.2	46.3	51.9	55.2	55.7	51.9	45.8	40.1	35.8	44.6
-Monthly		40.1	43.5	45.6	49.2	55.1	60.9	65.2	65.5	60.6	52.8	45.3	40.5	52.0
Extremes														
-Record Highest	49	64	70	75	85	93	96	99	99	98	89	74	64	99
-Year		1981	1968	1987	1976	1963	1955	1991	1981	1988	1987	1949	1993	JUL 1991
-Record Lowest	49	0	1	11	29	28	38	43	44	35	28	6	6	0
-Year		1950	1950	1955	1975	1954	1952	1954	1955	1972	1949	1955	1968	JAN 1950
NORMAL DEGREE DAYS:														
Heating (base 65°F)		772	602	601	474	307	144	58	65	156	378	591	760	4908
Cooling (base 65°F)		0	0	0	0	0	21	64	81	24	0	0	0	190
% OF POSSIBLE SUNSHINE	27	26	39	49	52	55	56	64	65	61	43	28	22	47
MEAN SKY COVER (tenths)														
Sunrise - Sunset	49	8.4	8.1	7.9	7.8	7.2	7.0	5.4	5.7	6.0	7.5	8.4	8.6	7.3
MEAN NUMBER OF DAYS:														
Sunrise to Sunset														
-Clear	49	2.8	2.9	3.2	2.8	4.4	5.1	10.3	9.1	8.2	4.1	2.5	2.2	57.6
-Partly Cloudy	49	3.9	4.2	5.8	7.2	8.8	7.8	9.6	9.9	8.6	7.4	4.3	3.8	81.2
-Cloudy	49	24.3	21.2	22.0	20.0	17.8	17.1	10.8	12.1	13.2	19.5	23.2	25.1	226.1
Precipitation														
.01 inches or more	49	18.4	15.6	16.9	14.0	10.4	9.0	5.1	6.2	8.8	13.0	17.9	19.1	154.5
Snow, Ice pellets, hail														
1.0 inches or more	49	1.5	0.6	0.5	0.*	0.0	0.0	0.0	0.0	0.0	0.*	0.3	0.9	3.8
Thunderstorms	49	0.2	0.3	0.7	0.9	0.9	0.7	0.7	0.7	0.7	0.4	0.7	0.3	7.2
Heavy Fog Visibility														
1/4 mile or less	49	5.3	3.5	2.4	1.2	0.7	0.7	1.5	2.6	5.1	7.4	5.6	6.2	42.2
Temperature °F														
-Maximum														
90° and above	34	0.0	0.0	0.0	0.0	0.2	0.4	1.1	1.2	0.3	0.0	0.0	0.0	3.2
32° and below	34	1.0	0.2	0.0	0.0	0.0	0.0	0.0	0.0	0.0	0.0	0.2	1.1	2.6
-Minimum														
32° and below	34	10.0	5.4	2.9	0.2	0.0	0.0	0.0	0.0	0.0	0.2	3.6	8.9	31.2
0° and below	34	0.0	0.0	0.0	0.0	0.0	0.0	0.0	0.0	0.0	0.0	0.0	0.0	0.0
AVG. STATION PRESS. (mb)	21	1001.3	1000.0	999.0	1000.8	1000.8	1000.9	1001.5	1000.5	1000.7	1001.6	1000.4	1001.6	1000.8
RELATIVE HUMIDITY (%)														
Hour 04	34	81	81	83	84	83	82	82	84	86	87	84	83	83
Hour 10 (Local Time)	34	79	77	75	71	68	66	65	68	73	79	80	81	74
Hour 16	34	74	67	62	58	54	53	49	50	56	67	75	78	62
Hour 22	34	78	76	76	74	72	70	68	71	75	81	81	81	75
PRECIPITATION (inches):														
Water Equivalent														
-Normal		5.38	3.99	3.54	2.33	1.70	1.50	0.76	1.14	1.88	3.23	5.83	5.91	37.19
-Maximum Monthly	49	12.92	9.11	8.40	6.53	4.76	3.90	2.39	4.59	5.95	8.95	10.71	11.85	12.92
-Year		1953	1961	1950	1991	1948	1946	1983	1975	1978	1947	1990	1979	JAN 1953
-Minimum Monthly	49	0.58	0.35	0.57	0.33	0.12	0.13	T	0.01	T	0.31	0.74	1.37	T
-Year		1985	1993	1965	1956	1992	1951	1960	1974	1991	1987	1976	1978	SEP 1991
-Maximum in 24 hrs	49	3.22	3.41	2.86	3.32	1.83	2.08	0.85	1.75	2.23	3.74	3.58	2.61	3.74
-Year		1986	1951	1972	1991	1969	1985	1981	1968	1978	1981	1990	1979	OCT 1981
Snow, Ice pellets, hail														
-Maximum Monthly	49	57.2	13.1	18.2	2.3	T	0.0	T	0.0	T	2.0	17.5	22.1	57.2
-Year		1950	1949	1951	1972	1993		1980		1972	1971	1985	1968	JAN 1950
-Maximum in 24 hrs	49	21.4	9.8	7.4	2.3	T	0.0	T	0.0	T	2.0	9.4	13.0	21.4
-Year		1950	1990	1989	1972	1993		1980		1972	1971	1946	1968	JAN 1950
WIND:														
Mean Speed (mph)	45	9.7	9.5	9.7	9.5	8.9	8.7	8.3	7.9	8.1	8.5	9.3	9.5	9.0
Prevailing Direction through 1963		SSW	SW	SSW	SW	SW	SW	SW	SW	N	S	S	SSW	SW
Fastest Mile														
-Direction (!!!)	25	S	S	SW	SW	SW	SW	SW	SW	S	SW	S	S	S
-Speed (MPH)	25	47	51	44	38	32	29	26	29	35	38	66	49	66
-Year		1993	1981	1984	1972	1968	1974	1981	1977	1990	1982	1981	1982	NOV 1981
Peak Gust														
-Direction (!!!)	10	S	S	S	SW	W	S	S	S	S	S	S	S	S
-Speed (mph)	10	64	46	43	44	39	32	29	38	39	45	52	52	64
-Date		1993	1988	1991	1990	1993	1990	1988	1991	1990	1990	1991	1993	JAN 1993

PRECIPITATION (inches)

SEATTLE, WASHINGTON SEATTLE – TACOMA AIRPORT

YEAR	JAN	FEB	MAR	APR	MAY	JUNE	JULY	AUG	SEP	OCT	NOV	DEC	ANNUAL
1964	9.76	1.66	2.96	1.56	0.91	3.82	0.99	1.23	2.27	1.00	9.65	5.53	41.34
1965	5.27	3.88	0.57	3.73	1.63	0.59	0.38	2.18	0.49	2.76	4.98	7.10	33.56
1966	5.43	2.31	4.38	1.99	1.35	1.15	1.35	0.42	1.77	2.92	6.85	8.31	38.23
1967	9.32	2.72	3.71	2.50	0.38	2.04	0.01	0.02	0.94	6.66	2.56	4.72	35.58
1968	6.90	6.08	5.08	1.33	1.67	3.02	0.83	4.58	1.93	4.32	5.86	8.55	50.15
1969	5.71	3.16	2.20	3.45	2.93	0.91	0.27	0.45	5.57	1.19	2.21	5.68	33.73
1970	8.22	2.26	3.16	3.31	1.17	0.43	0.48	0.32	2.23	2.52	5.03	8.28	37.41
1971	5.32	4.36	7.12	2.39	1.43	2.28	0.68	0.57	3.51	3.57	5.31	6.67	43.21
1972	7.24	8.11	6.74	4.12	0.69	1.81	1.34	1.13	4.10	0.72	3.38	8.98	48.36
1973	4.29	1.89	1.62	1.35	1.60	2.50	0.08	0.27	1.81	3.31	7.99	8.33	35.04
1974	7.78	4.01	5.84	2.39	1.37	1.25	1.51	0.01	0.21	1.99	5.06	6.45	37.87
1975	6.01	5.80	2.87	2.49	1.13	0.84	0.27	4.59	T	7.75	5.07	7.66	44.48
1976	5.55	4.74	2.71	1.67	1.61	0.63	1.17	2.71	1.25	2.06	0.74	1.86	26.70
1977	1.77	1.58	3.80	0.55	3.70	0.54	0.42	3.59	2.55	2.60	5.27	6.47	32.84
1978	4.30	3.59	2.43	4.19	1.79	0.75	1.40	1.19	5.95	0.98	6.05	1.37	33.99
1979	2.25	5.32	1.55	0.81	0.88	0.46	0.73	1.02	2.07	3.38	1.94	11.85	32.26
1980	4.09	5.04	2.10	3.23	0.97	1.77	0.46	0.64	1.43	1.32	7.16	7.39	35.60
1981	2.42	4.45	2.23	1.58	1.33	2.31	1.38	0.25	3.42	6.40	4.07	5.56	35.40
1982	5.35	7.57	3.73	2.07	0.63	1.03	0.59	0.62	1.49	4.07	5.31	6.86	39.32
1983	7.07	4.57	3.81	1.06	2.10	1.85	2.39	1.90	1.85	1.34	7.97	5.02	40.93
1984	3.62	3.91	3.91	2.87	3.38	2.81	0.17	0.13	1.01	2.14	8.09	4.95	36.99
1985	0.58	2.63	2.56	1.30	0.85	2.80	0.10	0.55	1.98	5.74	4.26	1.78	25.13
1986	8.54	4.41	2.67	1.38	1.71	0.68	1.10	0.10	1.89	4.21	7.98	3.67	38.34
1987	5.98	2.05	5.53	2.61	2.38	0.16	0.39	0.29	0.91	0.31	3.21	6.11	29.93
1988	4.07	0.71	3.75	3.20	3.01	1.56	0.50	0.28	1.75	2.24	8.43	3.48	32.98
1989	2.78	3.43	5.79	2.80	2.78	1.14	0.64	0.89	0.54	2.98	6.13	4.79	34.69
1990	9.41	3.72	2.58	2.54	1.98	3.05	0.58	0.71	0.05	5.79	10.71	3.63	44.75
1991	4.46	4.69	4.66	6.53	1.39	1.29	0.28	2.17	T	1.31	5.33	3.31	35.42
1992	7.82	3.09	1.68	4.12	0.12	1.14	0.89	0.66	1.15	2.45	5.57	4.09	32.78
1993	4.09	0.35	4.80	4.54	2.86	2.48	1.27	0.16	0.03	1.54	2.20	4.48	28.80
Record Mean	5.63	4.19	3.70	2.54	1.68	1.51	0.77	1.07	1.88	3.51	5.79	5.82	38.09

AVERAGE TEMPERATURE (deg. F)

SEATTLE, WASHINGTON SEATTLE – TACOMA AIRPORT

YEAR	JAN	FEB	MAR	APR	MAY	JUNE	JULY	AUG	SEP	OCT	NOV	DEC	ANNUAL
1964	40.0	41.3	44.1	46.8	53.2	57.9	63.5	62.6	58.5	53.5	42.1	36.4	50.0
1965	40.2	43.0	47.0	49.5	51.9	60.8	67.8	65.7	58.4	56.4	49.4	40.4	52.6
1966	41.1	43.9	45.1	50.0	54.5	58.7	62.1	64.5	61.5	51.4	45.4	43.5	51.8
1967	42.4	42.8	42.2	46.6	55.4	62.7	66.5	71.1	65.7	54.8	47.3	41.6	53.2
1968	40.9	48.5	48.6	48.7	57.3	60.7	67.0	63.7	59.1	51.5	46.8	36.6	52.5
1969	33.1	42.3	46.9	48.9	58.0	64.3	64.7	64.0	61.0	52.4	46.6	45.2	52.3
1970	41.2	47.0	46.0	46.1	54.7	62.7	64.9	64.5	58.6	50.8	46.5	39.0	51.8
1971	39.7	42.3	41.3	48.9	54.5	55.9	65.5	67.7	57.6	51.0	45.7	37.5	50.6
1972	37.0	41.4	46.9	47.0	58.3	60.1	66.0	66.7	55.4	50.1	46.7	38.1	51.1
1973	38.7	43.9	44.1	48.6	56.5	59.3	64.7	61.6	61.9	52.2	43.7	44.4	51.6
1974	38.7	43.2	46.3	50.3	54.9	62.6	64.0	64.6	64.4	52.5	45.1	42.4	52.4
1975	38.8	40.8	42.9	45.8	54.6	60.7	67.5	63.2	63.0	51.4	44.9	41.5	51.3
1976	41.8	40.9	41.3	49.5	56.4	60.0	65.9	64.1	62.6	54.9	47.8	44.7	52.5
1977	39.4	48.7	45.7	53.6	54.5	63.0	65.1	68.5	58.9	52.2	43.9	42.2	53.0
1978	44.4	46.0	48.6	49.9	54.5	64.3	65.8	65.5	58.8	54.3	41.2	37.5	52.6
1979	37.8	42.3	49.3	50.8	57.2	62.5	67.4	64.0	62.6	54.2	43.9	44.1	53.0
1980	34.8	43.8	44.3	51.6	54.2	57.5	63.8	61.9	59.6	53.9	46.7	44.1	51.4
1981	44.4	44.2	48.8	49.6	54.7	57.5	63.3	68.1	61.1	50.9	47.2	41.7	52.7
1982	39.3	42.1	44.1	47.4	54.7	63.1	62.8	65.1	60.6	52.7	43.2	40.8	51.3
1983	45.0	46.9	49.4	50.7	57.7	59.9	63.3	65.6	58.3	51.7	47.8	36.1	52.7
1984	43.2	44.8	48.5	48.7	52.9	58.8	65.0	64.9	59.9	49.7	44.6	36.8	51.5
1985	37.1	39.0	43.3	49.2	54.8	60.0	68.6	65.2	58.1	51.4	35.8	36.2	49.9
1986	44.9	42.8	49.2	48.1	55.7	62.7	61.7	68.4	59.1	54.3	45.3	42.0	52.9
1987	40.5	46.3	48.9	52.0	56.9	62.6	64.2	66.1	62.6	55.8	48.5	39.2	53.6
1988	40.1	44.4	45.6	50.3	54.9	59.6	65.3	65.4	60.5	55.4	45.4	41.9	52.4
1989	40.5	35.9	43.7	53.4	56.0	63.2	64.5	65.3	64.1	53.1	47.0	42.9	52.5
1990	42.5	40.0	47.1	52.1	54.7	59.8	68.0	67.3	63.4	51.2	46.6	35.3	52.3
1991	40.0	47.7	44.1	49.1	54.3	58.9	66.8	66.6	62.9	52.9	47.3	43.7	52.9
1992	43.9	47.3	50.3	53.1	59.8	65.0	66.7	66.8	60.0	54.4	45.5	38.8	54.3
1993	37.9	42.3	48.1	50.6	59.6	60.6	61.2	65.5	61.9	55.4	42.0	41.4	52.2
Record Mean	39.3	42.7	44.9	49.0	55.3	60.3	64.6	64.6	60.0	52.3	44.7	40.5	51.5
Max	44.3	48.8	52.0	57.2	64.4	69.5	75.0	74.6	69.1	59.4	50.2	45.3	59.1
Min	34.2	36.5	37.7	40.8	46.2	51.1	54.2	54.5	50.9	45.1	39.2	35.7	43.8

REFERENCE NOTES

SEATTLE, WASHINGTON SEATTLE – TACOMA AIRPORT

GENERAL
T - TRACE AMOUNT.
BLANK ENTRIES DENOTE MISSING/UNREPORTED DATA.
# INDICATES A STATION OR INSTRUMENT RELOCATION.

SPECIFIC

PM - INCLUDES LAST DAY OF PREVIOUS MONTH
ASOS - AUTOMATED SURFACE OBSERVING SYSTEM IN OPERATION DURING THESE MONTHS.

(a) - LENGTH OF RECORD IN YEARS, ALTHOUGH INDIVIDUAL MONTHS MAY BE MISSING.
0.* OR * - THE VALUE IS BETWEEN 0.0 AND 0.05
NORMALS - BASED ON THE 1961-1990 RECORD PERIOD.
EXTREMES - DATES ARE THE MOST RECENT OCCURRENCE.
WIND DIR. - NUMERALS SHOW TENS OF DEGRESS CLOCKWISE FROM TRUE NORTH. "00" INDICATES CALM.
RESULTANT DIRECTIONS ARE GIVEN TO WHOLE DEGREES.
BOLD VALUES INDICATE EXTREME VALUES WHICH OCCURRED AFTER THE ASOS SYSTEM WAS COMMISSIONED.

RECORD = PERIOD OF RECORD
RECORD MEAN PRECIPITATION IS THE MEAN OF ALL DAILY PRECIPITATION AMOUNTS DURING THE PERIOD OF RECORD.
RECORD MAX(MIN) TEMPERATURE IS THE MEAN OF ALL DAILY MAX(MIN) TEMPERATURES DURING THE PERIOD OF RECORD.
RECORD MEAN TEMPERATURE IS THE SUM OF THE RECORD MAX AND RECORD MIN DIVIDED BY 2.
AVERAGE TEMPERATURE IS THE SUM OF THE MEAN DAILY MAX AND MIN TEMPERATURE DIVIDED BY 2.

EXCEPTIONS

RECORD MEANS ARE THROUGH THE CURRENT YEAR, BEGINNING IN 1945 FOR TEMPERATURE
1945 FOR PRECIPITATION
1945 FOR SNOWFALL

HEATING DEGREE DAYS Base 65 deg. F — SEATTLE, WASHINGTON SEATTLE - TACOMA AIRPORT

SEASON	JULY	AUG	SEP	OCT	NOV	DEC	JAN	FEB	MAR	APR	MAY	JUNE	TOTAL
1964-65	76	91	189	349	679	882	761	611	553	459	400	136	5186
1965-66	24	44	194	261	462	754	732	584	610	442	321	190	4618
1966-67	95	54	106	414	585	658	695	614	700	548	292	92	4853
1967-68	16	0	44	310	524	718	737	472	503	485	232	139	4180
1968-69	33	70	179	415	538	871	983	627	554	478	230	71	5049
1969-70	49	49	144	381	547	607	731	499	586	563	314	122	4592
1970-71	53	44	190	435	548	801	778	628	728	472	321	267	5265
1971-72	82	17	214	429	570	843	863	678	557	531	222	144	5150
1972-73	48	32	295	455	544	825	807	586	639	484	272	183	5170
1973-74	70	114	111	388	633	632	809	606	573	433	306	99	4774
1974-75	60	66	74	380	591	690	804	671	678	570	317	144	5045
1975-76	23	73	93	413	594	723	712	693	731	465	265	157	4942
1976-77	24	52	81	307	510	625	786	451	591	335	320	79	4161
1977-78	34	43	178	390	625	701	631	525	498	447	323	78	4473
1978-79	44	42	180	324	706	846	837	630	479	420	235	96	4839
1979-80	27	40	86	327	628	642	929	610	634	395	329	218	4865
1980-81	66	104	158	343	543	639	633	577	494	455	316	220	4548
1981-82	80	28	138	430	530	715	790	636	640	521	312	103	4923
1982-83	93	42	141	373	647	745	613	502	479	422	244	149	4450
1983-84	72	19	196	406	511	890	672	577	507	482	372	183	4887
1984-85	54	42	159	467	603	867	857	719	666	469	310	160	5373
1985-86	8	48	199	413	870	888	618	616	479	502	305	90	5036
1986-87	105	12	196	323	586	707	754	522	491	384	253	105	4438
1987-88	58	37	102	284	485	792	767	590	593	435	316	165	4624
1988-89	60	38	162	291	583	708	749	807	654	340	273	93	4758
1989-90	41	29	68	362	534	677	689	696	547	379	312	158	4492
1990-91	29	23	61	420	546	913	767	475	639	472	322	179	4846
1991-92	24	40	90	368	526	654	645	507	447	351	171	63	3886
1992-93	26	22	151	322	580	805	834	629	517	427	170	135	4618
1993-94	113	48	120	290	685	723							

COOLING DEGREE DAYS Base 65 deg. F — SEATTLE, WASHINGTON SEATTLE - TACOMA AIRPORT

YEAR	JAN	FEB	MAR	APR	MAY	JUNE	JULY	AUG	SEP	OCT	NOV	DEC	TOTAL
1969	0	0	0	0	19	55	44	25	28	0	0	0	171
1970	0	0	0	0	1	60	58	36	6	0	0	0	161
1971	0	0	0	0	4	2	106	107	0	0	0	0	219
1972	0	0	0	0	22	3	85	91	11	0	0	0	212
1973	0	0	0	0	16	19	67	17	21	0	0	0	140
1974	0	0	0	0	0	36	38	62	60	0	0	0	196
1975	0	0	0	0	0	21	108	29	39	0	0	0	197
1976	0	0	0	8	4	14	59	29	15	0	0	0	129
1977	0	0	0	0	0	26	44	158	4	0	0	0	232
1978	0	0	0	0	4	66	76	64	0	0	0	0	210
1979	0	0	0	0	2	27	106	15	21	0	0	0	171
1980	0	0	0	0	0	0	34	15	3	2	0	0	54
1981	0	0	0	0	1	3	35	131	24	0	0	0	194
1982	0	0	0	0	0	53	31	55	15	0	0	0	154
1983	0	0	0	0	24	2	24	44	0	0	0	0	94
1984	0	0	0	0	1	5	62	45	11	0	0	0	124
1985	0	0	0	0	3	17	125	59	0	0	0	0	204
1986	0	0	0	0	22	27	10	124	26	0	0	0	209
1987	0	0	0	0	11	42	39	80	35	5	0	0	212
1988	0	0	0	0	7	10	79	56	36	1	0	0	189
1989	0	0	0	0	2	47	32	45	46	0	0	0	172
1990	0	0	0	0	0	10	129	100	21	0	0	0	260
1991	0	0	0	0	0	0	85	96	34	0	0	0	215
1992	0	0	0	0	18	68	83	85	8	0	0	0	262
1993	0	0	0	0	10	8	1	72	32	0	0	0	123

SNOWFALL (inches) — SEATTLE, WASHINGTON SEATTLE - TACOMA AIRPORT

SEASON	JULY	AUG	SEP	OCT	NOV	DEC	JAN	FEB	MAR	APR	MAY	JUNE	TOTAL
1964-65	0.0	0.0	0.0	0.0	3.3	7.6	7.3	T	T	0.0	T	0.0	18.2
1965-66	0.0	0.0	0.0	0.0	0.0	15.3	2.1	T	5.5	T	0.0	0.0	22.9
1966-67	0.0	0.0	0.0	0.0	0.0	2.0	5.9	T	T	T	0.0	0.0	7.9
1967-68	0.0	0.0	0.0	0.0	0.0	3.6	7.5	0.0	0.0	0.5	0.0	0.0	11.6
1968-69	0.0	0.0	0.0	0.0	0.0	22.1	45.4	T	0.0	0.0	0.0	0.0	67.5
1969-70	0.0	0.0	0.0	0.0	T	0.0	T	0.0	T	T	0.0	0.0	T
1970-71	0.0	0.0	0.0	0.0	T	2.9	9.1	2.2	1.9	T	0.0	0.0	16.1
1971-72	0.0	0.0	0.0	2.0	T	10.6	14.0	0.3	T	2.3	0.0	0.0	29.2
1972-73	0.0	0.0	T	0.0	T	5.6	2.7	T	0.8	T	0.0	0.0	9.1
1973-74	0.0	0.0	0.0	0.0	0.2	0.3	3.7	T	T	0.0	T	0.0	4.2
1974-75	0.0	0.0	0.0	0.0	0.0	9.8	1.3	T	T	0.2	0.0	0.0	11.3
1975-76	0.0	0.0	0.0	0.0	1.6	2.6	T	0.5	0.2	T	0.0	0.0	4.9
1976-77	0.0	0.0	0.0	0.0	0.0	T	1.0	T	0.9	0.0	0.0	0.0	1.9
1977-78	0.0	0.0	0.0	0.0	3.5	T	T	0.0	T	T	0.0	0.0	3.5
1978-79	0.0	0.0	0.0	0.0	4.9	0.2	0.5	0.4	0.0	0.0	0.0	0.0	6.0
1979-80	0.0	0.0	0.0	0.0	0.0	1.2	8.8	2.5	0.1	T	0.0	0.0	12.6
1980-81	T	0.0	0.0	0.0	T	0.3	0.0	1.1	0.0	0.0	0.0	0.0	1.4
1981-82	0.0	0.0	0.0	0.0	0.0	T	7.0	T	2.0	T	0.0	0.0	9.0
1982-83	0.0	0.0	0.0	T	0.0	T	0.0	0.0	0.0	0.0	0.0	0.0	T
1983-84	0.0	0.0	0.0	0.0	T	0.3	T	0.0	0.0	T	0.0	0.0	0.3
1984-85	0.0	0.0	0.0	T	T	2.4	T	5.7	T	T	0.0	0.0	8.1
1985-86	0.0	0.0	0.0	T	17.5	1.7	0.0	1.1	0.0	T	0.0	0.0	20.3
1986-87	0.0	0.0	0.0	0.0	T	0.0	1.4	0.0	0.0	0.0	0.0	0.0	1.4
1987-88	0.0	0.0	0.0	0.0	0.0	T	T	0.0	T	0.0	0.0	0.0	T
1988-89	0.0	0.0	0.0	0.0	T	T	1.0	5.8	7.4	T	T	0.0	14.2
1989-90	0.0	0.0	0.0	0.0	0.0	0.0	T	9.8	T	0.0	T	0.0	9.8
1990-91	0.0	0.0	0.0	0.0	0.0	3.8	0.4	0.0	2.5	T	0.0	0.0	6.7
1991-92	0.0	0.0	0.0	0.0	0.0	0.0	0.0	0.0	0.0	0.0	0.0	0.0	0.0
1992-93	0.0	0.0	0.0	0.0	T	6.7	1.3	1.4	0.0	0.0	T	0.0	9.4
1993-94	0.0	0.0	0.0	0.0	T	0.0							
Record Mean	T	0.0	T	T	1.2	2.5	5.0	1.6	1.4	0.1	T	0.0	11.8

SEATTLE, WASHINGTON
URBAN CLIMATOLOGY STATION

Seattle is located on a hill between the salt waters of Puget Sound to the west and the fresh waters of Lake Washington to the east. The lake shore roughly parallels the shore of Puget Sound at distances varying from about 2-1/2 to 6 miles. Hills rise abruptly from both shorelines and reach elevations of more than 300 feet in the central area to more than 500 feet in the northern and the southwestern sections. The north-south orientation of the city is matched on the east by the Cascade Mountains and the Olympic Mountains to the west and northwest. The Seattle Urban Climatological Station was relocated to the Portage Bay site in November 1972. Portage Bay is an arm of Lake Washington near the University of Washington. The observational site is on a grassy plot about ten yards from the edge of the Bay.

The climate of Seattle is mild and moist, the result of the prevailing westerly winds off the Pacific Ocean approximately 90 miles to the west, and the Cascade Mountains which tend to shield the city from the cold continental air from the east. Winters are comparatively warm and summers cool because of the steady influx of marine air. The daily temperature range is small and extremes of temperature, both hot and cold, are moderate and usually of short duration.

The warmest summer and the coldest winter days come with north winds from British Columbia or east winds from eastern Washington. An average year will have less than three days during the summer with a high temperature of 90 degrees or more with the maximum temperature rarely reaching 100 degrees. Nighttime temperatures during the warmest months seldom remain above 65 degrees. Daily highs during the winter fail to rise above 32 degrees on an average of about two days per year, while the number of days with minimum temperatures of 32 degrees or less averages only 15 days per year. Low temperatures may vary by several degrees throughout the city and depend upon the wind direction, distance from water, and site elevation.

The growing season often lasts from about the mid-March to late November and grass is usually green throughout the winter.

The city lies within the lee side dry area caused by the Olympic Mountains. As a result, the normal precipitation of less than 36 inches is relatively light when compared to the 50 inches or more that falls on the nearby Cascade foothills. The western slopes of these hills and mountains lift the moist marine air, causing very heavy precipitation on the seaward slopes and significantly less at the summits. The winter wet season, usually from October to March, is the result of the air flowing around the Aleutian low pressure system, but, in the summer the Eastern Pacific high pressure system moves north and forces the moist marine air to the north of Washington and brings relatively dry and cool air to the state. The warmest temperatures of the summer usually occur when the Pacific high extends into southwest Canada creating a hot and dry flow of continental air across the Cascade Mountains into the Puget Sound area. Less than 20 percent of the annual rainfall occurs during the summer dry season, April through September.

The average winter snowfall is about 9 inches but the snow seldom remains on the ground for more than two days at a time. Annual totals range from as little as a trace in several instances to over 36 inches in one season. Fog is a frequent occurrence during the late fall and winter months. Severe weather is relatively infrequent over Seattle with an average of just six thunderstorms per year and no tornado ever reported within the city.

NORMALS, MEANS, AND EXTREMES

SEATTLE, WASHINGTON URBAN CLIMATOLOGY STATION

LATITUDE: 47°39'N LONGITUDE: 122°18' W ELEVATION: FT. GRND 22 BARO 25 TIME ZONE: PACIFIC WBAN: 24281

	(a)	JAN	FEB	MAR	APR	MAY	JUNE	JULY	AUG	SEP	OCT	NOV	DEC	YEAR
TEMPERATURE °F:														
Normals														
-Daily Maximum		46.1	50.6	53.7	58.1	64.2	69.6	74.1	74.1	68.8	60.0	51.5	46.2	59.8
-Daily Minimum		36.4	38.0	39.5	42.7	47.9	53.2	56.4	57.2	52.9	46.9	41.0	37.0	45.8
-Monthly		41.3	44.3	46.6	50.4	56.1	61.4	65.3	65.7	60.8	53.5	46.3	41.6	52.8
Extremes														
-Record Highest	60	66	74	75	87	92	100	100	97	92	82	73	65	100
-Year		1935	1968	1941	1947	1963	1955	1941	1960	1967	1991	1970	1980	JUN 1955
-Record Lowest	60	11	11	22	31	35	42	47	48	40	30	13	9	9
-Year		1950	1989	1955	1936	1954	1991	1979	1980	1972	1935	1985	1990	DEC 1990
NORMAL DEGREE DAYS:														
Heating (base 65°F)		735	580	570	438	276	123	44	58	144	357	561	725	4611
Cooling (base 65°F)		0	0	0	0	0	15	54	80	18	0	0	0	167
% OF POSSIBLE SUNSHINE	31	28	34	42	47	52	49	63	56	53	37	28	23	43
MEAN SKY COVER (tenths)														
Sunrise - Sunset	24	8.0	7.7	7.4	6.9	6.4	6.4	4.9	5.3	5.6	7.2	8.0	8.1	6.8
MEAN NUMBER OF DAYS:														
Sunrise to Sunset														
-Clear	24	3.0	3.0	4.0	5.0	7.0	7.0	12.0	10.0	9.0	5.0	3.0	3.0	71.0
-Partly Cloudy	24	5.0	6.0	8.0	9.0	10.0	8.0	10.0	10.0	8.0	8.0	6.0	5.0	93.0
-Cloudy	24	23.0	19.0	19.0	16.0	14.0	15.0	9.0	11.0	13.0	18.0	21.0	23.0	201.0
Precipitation														
.01 inches or more	43	18.3	15.6	16.7	13.9	10.7	8.9	5.4	5.9	8.3	10.5	17.8	18.4	150.4
Snow, Ice pellets, hail														
1.0 inches or more	43	1.2	0.3	0.1	0.*	0.0	0.0	0.0	0.0	0.0	0.0	0.2	0.6	2.4
Thunderstorms	25	0.3	0.3	0.3	0.4	0.6	0.6	0.7	0.8	0.7	0.6	0.4	0.3	6.0
Heavy Fog Visibility 1/4 mile or less														
Temperature °F														
-Maximum														
90° and above	43	0.0	0.0	0.0	0.0	0.1	0.3	0.9	0.5	0.1	0.0	0.0	0.0	1.9
32° and below	43	0.8	0.1	0.0	0.0	0.0	0.0	0.0	0.0	0.0	0.0	0.3	0.7	2.0
-Minimum														
32° and below	43	6.6	3.4	1.3	0.0	0.0	0.0	0.0	0.0	0.0	0.*	2.3	5.0	18.6
0° and below	43	0.0	0.0	0.0	0.0	0.0	0.0	0.0	0.0	0.0	0.0	0.0	0.0	0.0
AVG. STATION PRESS. (mb)														
RELATIVE HUMIDITY (%)														
Hour 04														
Hour 10 (Local Time)														
Hour 16														
Hour 22														
PRECIPITATION (inches):														
Water Equivalent														
-Normal		5.35	4.03	3.77	2.51	1.84	1.59	0.85	1.22	1.94	3.25	5.65	6.00	38.00
-Maximum Monthly	60	10.93	7.75	7.23	5.25	4.67	3.68	2.16	5.49	5.62	8.04	11.20	10.41	11.20
-Year		1953	1961	1950	1991	1948	1964	1978	1977	1978	1975	1983	1939	NOV 1983
-Minimum Monthly	60	0.60	0.54	0.44	0.16	0.34	0.12	T	T	T	0.29	0.50	1.00	T
-Year		1985	1993	1965	1939	1972	1987	1960	1967	1993	1987	1976	1944	SEP 1993
-Maximum in 24 hrs	60	4.48	2.69	2.32	2.34	1.35	2.07	1.22	1.93	1.91	3.48	3.20	3.31	4.48
-Year		1986	1945	1950	1991	1948	1985	1954	1977	1953	1981	1937	1937	JAN 1986
Snow, Ice pellets, hail														
-Maximum Monthly	60	31.0	10.4	7.5	1.0	T	0.0	0.0	0.0	0.0	T	9.6	13.5	31.0
-Year		1950	1949	1951	1972	1993					1984	1985	1968	JAN 1950
-Maximum in 24 hrs	60	11.5	7.6	7.1	1.0	T	0.0	0.0	0.0	0.0	T	6.0	10.0	11.5
-Year		1943	1937	1989	1972	1993					1984	1985	1968	JAN 1943
WIND:														
Mean Speed (mph)														
Prevailing Direction														
Fastest Mile														
-Direction (!!)	11	S	SSW	WSW	WSW	WNW	SW	S	WNW	W	S	S	SSE	SSW
-Speed (MPH)	11	44	48	44	39	37	35	36	37	39	37	45	46	48
-Year		1978	1979	1980	1980	1983	1980	1976	1975	1981	1977	1981	1973	FEB 1979
Peak Gust														
-Direction (!!)	10	SSE	SSW	WSW	SW	WSW	WSW	S	S	S	SSW	SSW	SW	SSW
-Speed (mph)	10	51	40	54	44	46	33	30	33	33	41	63	46	63
-Date		1993	1990	1985	1992	1993	1991	1988	1991	1992	1990	1991	1990	NOV 1991

PRECIPITATION (inches)

SEATTLE, WASHINGTON URBAN CLIMATOLOGY STATION

YEAR	JAN	FEB	MAR	APR	MAY	JUNE	JULY	AUG	SEP	OCT	NOV	DEC	ANNUAL
1964	8.16	1.55	3.20	1.29	1.07	3.68	0.84	1.46	2.01	0.83	8.11	4.86	37.06
1965	5.83	4.28	0.44	3.79	1.25	0.47	0.48	1.61	0.75	2.03	5.01	7.06	33.00
1966	6.40	2.36	4.76	2.02	1.34	0.75	1.39	0.17	1.57	2.21	7.18	7.73	37.88
1967	9.02	2.18	3.61	2.76	0.52	1.35	0.05	T	1.21	5.42	2.01	5.33	33.46
1968	7.39	4.87	5.11	2.06	1.34	2.52	0.48	4.28	1.63	3.52	4.87	10.07	48.14
1969	5.83	3.58	1.97	3.39	2.27	1.22	0.23	0.15	5.41	1.72	2.73	6.89	35.39
1970	8.11	2.01	3.07	2.73	1.00	0.78	0.53	0.74	2.01	3.42	4.30	8.34	37.04
1971	4.25	3.97	7.16	2.33	1.51	1.85	0.58	0.46	3.08	3.06	4.25	5.29	37.79
#1972	5.15	5.61	6.01	4.20	0.34	1.98	0.67	1.06	3.27	0.48	3.19	8.36	40.32
1973	4.61	2.00	1.55	0.93	1.11	1.73	0.25	0.57	1.56	2.74	8.40	9.56	35.01
1974	8.53	4.19	5.69	1.83	1.48	1.07	1.95	0.11	0.23	1.78	5.56	6.52	38.94
1975	5.03	4.91	4.82	2.51	1.51	1.26	0.11	2.81	0.03	8.04	6.40	7.57	45.00
1976	4.33	4.83	3.64	2.10	1.87	1.07	1.02	3.05	1.43	1.94	0.50	2.00	27.78
1977	2.13	2.11	5.24	1.35	4.24	0.59	0.80	5.49	2.82	2.72	5.03	5.81	38.33
1978	5.57	3.72	2.61	3.53	2.02	0.57	2.16	1.65	5.62	0.87	6.52	1.42	36.26
1979	2.20	6.32	1.32	2.69	1.07	0.45	0.97	0.61	1.86	4.81	2.28	10.29	34.87
1980	4.96	5.64	3.27	3.65	1.51	2.88	0.49	1.74	1.43	1.42	6.12	7.19	40.30
1981	2.58	4.07	2.54	2.09	2.25	2.07	1.52	0.34	3.25	6.45	5.21	7.40	39.77
1982	4.68	7.50	3.86	2.48	0.57	1.07	0.80	0.63	1.68	4.20	5.30	8.99	41.76
1983	6.84	5.29	4.37	0.91	2.18	1.99	2.07	2.08	2.55	1.11	11.20	5.52	46.11
1984	2.68	4.18	3.72	2.81	2.99	3.17	0.46	0.22	1.51	2.34	9.26	5.27	38.61
1985	0.60	2.88	2.92	2.59	1.99	3.05	0.22	0.72	1.77	6.81	3.77	1.22	28.54
1986	10.38	3.44	2.55	2.37	2.09	0.94	1.36	0.13	2.03	4.30	6.71	4.58	40.88
1987	6.51	2.25	5.81	3.20	2.05	0.12	0.45	0.45	0.52	0.29	2.84	7.10	31.59
1988	4.33	0.78	3.94	3.03	2.94	1.93	0.23	0.35	1.87	1.73	7.56	3.93	32.62
1989	3.31	3.45	5.71	1.69	2.98	1.10	0.62	0.88	0.15	4.17	4.12	4.52	32.70
1990	8.72	4.19	3.01	2.20	1.50	2.63	0.98	0.60	0.03	5.00	6.40	4.51	39.77
1991	3.97	4.69	5.21	5.25	1.35	1.12	0.47	2.52	0.07	1.82	5.36	2.64	34.47
1992	7.57	3.69	1.74	3.44	0.58	0.97	1.35	0.66	2.43	2.06	6.11	4.05	34.65
1993	2.61	0.54	3.99	4.89	2.72	2.45	1.63	0.58	T	1.62	2.42	6.09	29.54
Record Mean	5.54	3.79	3.62	2.45	1.65	1.48	0.77	1.06	1.69	3.09	5.32	5.62	36.08

AVERAGE TEMPERATURE (deg. F)

SEATTLE, WASHINGTON URBAN CLIMATOLOGY STATION

YEAR	JAN	FEB	MAR	APR	MAY	JUNE	JULY	AUG	SEP	OCT	NOV	DEC	ANNUAL
1964	43.4	44.2	46.3	49.6	55.7	60.3	65.3	64.4	60.6	55.9	45.2	39.4	52.5
1965	42.8	44.6	48.9	52.5	54.6	62.5	67.1	66.0	59.7	57.3	51.5	43.4	54.2
1966	43.8	46.1	48.4	52.6	57.1	61.2	64.5	66.4	63.7	54.6	49.1	47.2	54.5
1967	45.0	45.9	45.1	49.0	57.7	65.3	67.3	70.6	66.4	55.9	50.0	43.7	54.7
1968	43.0	50.7	50.2	50.6	58.6	61.9	68.5	64.9	61.3	54.3	49.6	39.5	54.4
1969	34.8	43.2	48.4	50.7	59.9	65.0	64.9	63.5	61.1	54.0	47.7	46.0	53.3
1970	42.9	48.4	48.2	49.7	57.3	64.4	66.2	65.5	60.0	52.4	47.2	40.9	53.6
1971	41.6	42.6	43.5	50.9	56.1	58.1	66.1	66.9	60.6	51.9	47.0	39.1	52.1
#1972	38.2	43.2	48.3	48.1	58.9	60.1	65.9	66.3	57.6	51.6	46.8	39.1	52.1
1973	39.5	44.8	46.8	50.5	57.4	60.6	65.7	62.3	61.4	52.8	43.4	44.2	52.4
1974	39.3	43.5	45.4	49.9	53.8	61.1	64.3	66.4	64.8	54.1	47.9	44.4	52.9
1975	41.7	42.1	44.3	47.3	55.8	59.6	66.1	62.8	62.3	52.7	45.9	43.3	52.0
1976	44.0	42.9	44.0	51.2	56.0	59.1	65.0	63.1	61.7	54.0	47.9	44.4	52.8
1977	39.7	47.5	44.8	52.3	53.7	61.4	63.0	68.1	58.6	52.2	44.4	42.0	52.3
1978	43.6	46.2	49.1	50.6	54.9	63.8	65.4	65.0	59.0	54.8	42.0	38.3	52.7
1979	36.2	42.2	49.3	51.1	57.8	61.3	66.3	65.3	63.2	55.0	45.1	44.8	53.2
1980	36.6	45.2	46.9	53.4	56.1	59.2	64.5	63.2	61.5	55.0	48.4	45.7	53.0
1981	45.5	45.5	50.4	52.2	56.7	60.1	65.1	69.4	62.2	52.5	48.5	42.5	54.2
1982	40.9	43.3	45.6	48.7	56.1	64.2	64.0	65.1	61.3	53.6	43.7	42.1	52.4
1983	46.0	47.9	50.2	52.6	59.1	61.6	64.6	66.6	59.2	53.2	48.9	37.1	53.9
1984	44.6	46.3	49.9	50.6	55.1	60.3	65.7	65.5	60.3	51.5	46.1	38.4	52.9
1985	39.1	40.5	44.8	50.6	56.8	61.5	69.5	65.6	58.8	52.6	37.4	38.1	51.3
1986	45.3	43.5	50.4	49.7	57.4	63.7	63.1	69.3	60.8	55.8	47.4	43.5	54.2
1987	42.8	47.7	50.1	54.6	59.3	63.5	64.5	65.7	62.1	55.5	49.1	40.3	54.6
1988	41.3	45.6	46.6	51.4	56.5	60.3	65.4	64.8	60.3	55.6	46.4	42.4	53.1
1989	41.8	36.6	44.4	53.7	56.1	63.0	63.8	64.6	62.6	53.4	47.8	44.0	52.7
1990	43.5	41.3	47.9	52.9	56.6	61.2	68.0	67.5	63.0	52.0	48.1	36.4	53.2
1991	40.7	48.3	44.6	49.6	55.5	58.6	66.3	66.8	62.0	52.3	47.4	42.5	52.9
1992	43.2	46.9	50.1	53.8	59.5	65.0	66.4	67.0	58.9	53.6	45.0	39.5	54.1
1993	39.2	42.3	48.9	51.7	61.2	62.1	62.9	66.2	62.7	56.1	43.1	41.9	53.2
Record Mean	41.7	44.6	46.8	51.1	57.0	61.5	65.6	65.6	61.4	54.1	46.7	42.4	53.2
Max	46.2	50.3	53.4	58.6	65.0	69.5	74.6	74.2	69.2	60.3	51.5	46.7	60.0
Min	37.1	38.9	40.2	43.6	49.0	53.4	56.6	57.0	53.6	47.9	41.8	38.1	46.4

REFERENCE NOTES

SEATTLE, WASHINGTON URBAN CLIMATOLOGY STATION

GENERAL
T - TRACE AMOUNT.
BLANK ENTRIES DENOTE MISSING/UNREPORTED DATA.
# INDICATES A STATION OR INSTRUMENT RELOCATION.

SPECIFIC

(a) - LENGTH OF RECORD IN YEARS, ALTHOUGH INDIVIDUAL MONTHS MAY BE MISSING.
0.* OR * - THE VALUE IS BETWEEN 0.0 AND 0.05
NORMALS - BASED ON THE 1961-1990 RECORD PERIOD.
EXTREMES - DATES ARE THE MOST RECENT OCCURRENCE
WIND DIR.- NUMERALS SHOW TENS OF DEGRESS CLOCKWISE FROM TRUE NORTH. "00" INDICATES CALM.
RESULTANT DIRECTIONS ARE GIVEN TO WHOLE DEGREES.
BOLD VALUES INDICATE EXTREME VALUES WHICH OCCURRED AFTER THE ASOS SYSTEM WAS COMMISSIONED.

RECORD = PERIOD OF RECORD
RECORD MEAN PRECIPITATION IS THE MEAN OF ALL DAILY PRECIPITATION AMOUNTS DURING THE PERIOD OF RECORD.
RECORD MAX(MIN) TEMPERATURE IS THE MEAN OF ALL DAILY MAX(MIN) TEMPERATURES DURING THE PERIOD OF RECORD.
RECORD MEAN TEMPERATURE IS THE SUM OF THE RECORD MAX AND RECORD MIN DIVIDED BY 2.
AVERAGE TEMPERATURE IS THE SUM OF THE MEAN DAILY MAX AND MIN TEMPERATURE DIVIDED BY 2.

EXCEPTIONS

1. MEAN SKY COVER, DAYS CLEAR-PARTLY CLOUDY-CLOUDY, AND THUNDERSTORMS ARE 1934-1957.
2. PERCENT OF POSSIBLE SUNSHINE IS 1934-1964.

RECORD MEANS ARE THROUGH THE CURRENT YEAR, BEGINNING IN 1951 FOR TEMPERATURE
1951 FOR PRECIPITATION
1951 FOR SNOWFALL

HEATING DEGREE DAYS Base 65 deg. F — SEATTLE, WASHINGTON URBAN CLIMATOLOGY STATION

SEASON	JULY	AUG	SEP	OCT	NOV	DEC	JAN	FEB	MAR	APR	MAY	JUNE	TOTAL
1964-65	39	53	127	276	586	788	683	564	492	372	315	90	4385
1965-66	33	35	154	233	402	663	649	523	510	365	242	123	3932
1966-67	51	20	54	316	473	545	612	528	612	471	226	47	3955
1967-68	12	3	26	276	446	652	673	410	452	426	191	102	3669
1968-69	16	50	120	324	458	782	929	604	507	422	173	60	4445
1969-70	39	54	133	338	511	584	679	460	513	455	235	82	4083
1970-71	27	30	151	381	527	741	720	624	662	418	275	203	4759
1971-72	65	7	127	401	531	795	822	626	509	499	206	141	4729
#1972-73	49	30	225	407	538	797	783	558	560	429	235	148	4759
1973-74	42	100	116	371	642	637	790	595	600	448	338	124	4803
1974-75	58	26	55	331	504	634	714	635	638	522	279	165	4561
1975-76	28	81	93	374	566	666	644	633	645	411	271	176	4588
1976-77	32	66	96	336	507	635	780	483	617	374	346	105	4377
1977-78	78	41	185	387	609	704	657	522	487	423	307	70	4470
1978-79	45	45	170	310	685	822	886	630	479	411	215	118	4816
1979-80	33	12	68	304	591	620	872	569	554	341	270	169	4403
1980-81	47	69	107	307	491	588	597	540	443	376	255	148	3968
1981-82	39	7	103	383	489	691	743	604	593	483	271	77	4483
1982-83	55	31	119	347	633	699	583	474	452	367	194	95	4049
1983-84	38	6	167	360	477	859	628	537	459	427	297	145	4400
1984-85	32	29	145	413	558	816	797	678	618	427	253	114	4880
1985-86	3	27	180	378	822	827	603	594	445	454	252	60	4645
1986-87	60	1	146	280	520	660	681	480	454	303	183	81	3849
1987-88	44	33	97	291	471	760	728	556	563	398	263	147	4351
1988-89	43	37	161	283	550	693	711	788	633	336	267	90	4592
1989-90	52	26	81	354	510	645	660	656	523	358	254	121	4240
1990-91	18	7	66	397	501	881	747	462	627	454	289	185	4634
1991-92	19	34	105	387	521	689	669	519	457	328	170	57	3955
1992-93	25	19	181	346	591	783	793	629	494	391	130	94	4476
1993-94	64	35	102	269	651	709							

COOLING DEGREE DAYS Base 65 deg. F — SEATTLE, WASHINGTON URBAN CLIMATOLOGY STATION

YEAR	JAN	FEB	MAR	APR	MAY	JUNE	JULY	AUG	SEP	OCT	NOV	DEC	TOTAL
1973	0	0	0	0	7	20	69	22	13	0	0	0	131
1974	0	0	0	0	0	12	42	76	54	0	0	0	184
1975	0	0	0	0	2	10	71	20	17	0	0	0	120
1976	0	0	0	4	1	6	41	14	7	0	0	0	73
1977	0	0	0	0	0	6	24	144	1	0	0	0	175
1978	0	0	0	0	2	44	66	49	0	0	0	0	161
1979	0	0	0	0	1	12	79	27	20	0	0	0	139
1980	0	0	0	0	0	0	38	22	8	5	0	0	73
1981	0	0	0	0	3	5	47	151	24	0	0	0	230
1982	0	0	0	0	1	61	33	43	16	0	0	0	154
1983	0	0	0	0	20	1	33	64	1	0	0	0	119
1984	0	0	0	0	2	13	61	50	10	2	0	0	138
1985	0	0	0	0	6	20	147	53	0	0	0	0	226
1986	0	0	0	0	22	27	10	140	28	0	0	0	227
1987	0	0	0	0	15	44	36	60	21	1	0	0	177
1988	0	0	0	0	3	14	61	37	26	0	0	0	141
1989	0	0	0	0	1	39	20	23	14	0	0	0	97
1990	0	0	0	0	0	16	116	89	13	0	0	0	234
1991	0	0	0	0	0	1	66	98	21	1	0	0	187
1992	0	0	0	0	8	65	79	87	2	0	0	0	241
1993	0	0	0	0	22	10	5	79	41	0	0	0	157

SNOWFALL (inches) — SEATTLE, WASHINGTON URBAN CLIMATOLOGY STATION

SEASON	JULY	AUG	SEP	OCT	NOV	DEC	JAN	FEB	MAR	APR	MAY	JUNE	TOTAL
1964-65	0.0	0.0	0.0	0.0	1.5	6.4	4.3	0.0	0.0	0.0	0.0	0.0	12.2
1965-66	0.0	0.0	0.0	0.0	0.0	7.6	T	T	0.8	0.0	0.0	0.0	8.4
1966-67	0.0	0.0	0.0	0.0	0.0	1.0	5.0	0.0	0.0	0.0	0.0	0.0	6.0
1967-68	0.0	0.0	0.0	0.0	0.0	2.5	6.0	0.0	0.0	T	0.0	0.0	8.5
1968-69	0.0	0.0	0.0	0.0	0.0	13.5	22.7	0.0	0.0	0.0	0.0	0.0	36.2
1969-70	0.0	0.0	0.0	0.0	0.0	0.0	T	0.0	0.0	T	0.0	0.0	T
1970-71	0.0	0.0	0.0	0.0	0.3	T	9.1	5.2	1.0	0.0	0.0	0.0	15.6
1971-72	0.0	0.0	0.0	T	0.0	6.3	8.0	0.0	0.0	1.0	0.0	0.0	15.3
#1972-73	0.0	0.0	0.0	0.0	0.0	5.0	3.7	0.0	0.0	0.0	0.0	0.0	8.7
1973-74	0.0	0.0	0.0	0.0	T	0.0	2.0	0.0	1.0	0.0	0.0	0.0	3.0
1974-75	0.0	0.0	0.0	0.0	0.0	6.0	1.0	0.5	T	T	0.0	0.0	7.5
1975-76	0.0	0.0	0.0	0.0	0.5	0.5	0.0	0.0	T	0.0	0.0	0.0	1.0
1976-77	0.0	0.0	0.0	0.0	0.0	0.0	1.0	0.0	T	0.0	0.0	0.0	1.0
1977-78	0.0	0.0	0.0	0.0	0.3	T	T	0.0	T	0.0	0.0	0.0	0.3
1978-79	0.0	0.0	0.0	0.0	6.0	T	T	T	0.0	0.0	0.0	0.0	6.0
1979-80	0.0	0.0	0.0	0.0	0.0	T	11.9	2.5	0.0	0.0	0.0	0.0	14.4
1980-81	0.0	0.0	0.0	0.0	0.0	1.0	0.0	0.5	0.0	T	0.0	0.0	1.5
1981-82	0.0	0.0	0.0	0.0	0.0	T	5.2	T	T	0.0	0.0	0.0	5.2
1982-83	0.0	0.0	0.0	0.0	T	0.0	0.0	0.0	0.0	0.0	0.0	0.0	T
1983-84	0.0	0.0	0.0	0.0	0.0	1.4	T	0.0	0.0	0.0	0.0	0.0	1.4
1984-85	0.0	0.0	0.0	T	0.0	3.4	T	3.5	T	0.0	0.0	0.0	6.9
1985-86	0.0	0.0	0.0	0.0	9.6	1.0	0.0	0.5	0.0	0.0	0.0	0.0	11.1
1986-87	0.0	0.0	0.0	0.0	0.0	0.0	0.3	0.0	0.0	0.0	0.0	0.0	0.3
1987-88	0.0	0.0	0.0	0.0	0.0	0.0	0.0	0.0	0.0	0.0	0.0	0.0	0.0
1988-89	0.0	0.0	0.0	0.0	0.0	0.0	6.2	2.9	7.1	0.0	0.0	0.0	16.2
1989-90	0.0	0.0	0.0	0.0	0.0	0.0	T	6.0	T	0.0	0.0	0.0	6.0
1990-91	0.0	0.0	0.0	0.0	0.0	9.6	T	0.0	0.2	0.0	0.0	0.0	9.8
1991-92	0.0	0.0	0.0	0.0	0.0	0.0	0.0	0.0	0.0	0.0	0.0	0.0	0.0
1992-93	0.0	0.0	0.0	0.0	0.0	3.0	T	T	0.0	0.0	T	0.0	3.0
1993-94	0.0	0.0	0.0	0.0	1.0	0.0							
Record Mean	0.0	0.0	0.0	T	0.7	1.9	3.0	0.9	0.6	T	T	0.0	7.1

SPOKANE
WASHINGTON

Spokane lies on the eastern edge of the broad Columbia Basin area of Washington which is bounded by the Cascade Range on the west and the Rocky Mountains on the east. The elevations in eastern Washington vary from less than 400 feet above sea level near Pasco where the Columbia River flows out of Washington to over 5,000 feet in the mountain areas of the extreme eastern edge of the State. Spokane is located on the upper plateau area where the long gradual slope from the Columbia River meets the sharp rise of the Rocky Mountain Ranges.

Much of the urban area of Spokane lies along both sides of the Spokane River at an elevation of approximately 2,000 feet, but the residential areas have spread to the crests of the plateaus on either side of the river with elevations up to 2,500 feet above sea level. Spokane International Airport is situated on the plateau area 6 miles west-southwest and some 400 feet higher than the downtown business district.

The climate of Spokane combines some of the characteristics of damp coastal type weather and arid interior conditions. Most of the air masses which reach Spokane are brought in by the prevailing westerly and southwesterly circulations. Frequently, much of the moisture in the storms that move eastward and southeastward from the Gulf of Alaska and the eastern Pacific Ocean is precipitated out as the storms are lifted across the Coast and Cascade Ranges. Annual precipitation totals in the Spokane area are generally less than 20 inches and less than 50 percent of the amounts received west of the Cascades. However, the precipitation and total cloudiness in the Spokane vicinity is greater than that of the desert areas of south-central Washington. The lifting action of the air masses as they move up the east slope of the Columbia Basin frequently produces the cooling and condensation necessary for formation of clouds and precipitation.

Infrequently, the Spokane area comes under the influence of dry continental air masses from the north or east. On occasions when these air masses penetrate into eastern Washington the result is high temperatures and very low humidity in the summer and sub-zero temperatures in the winter. In the winter most of the severe arctic outbursts of cold air move southward on the east side of the Continental Divide and do not affect Spokane.

In general, Spokane weather has the characteristics of a mild, arid climate during the summer months and a cold, coastal type in the winter. Approximately 70 percent of the total annual precipitation falls between the first of October and the end of March and about half of that falls as snow. The growing season usually extends over nearly six months from mid-April to mid-October. Irrigation is required for all crops except dry-land type grains. The summer weather is ideal for full enjoyment of the many mountain and lake recreational areas in the immediate vicinity. Winter weather includes many cloudy or foggy days and below freezing temperatures with occasional snowfall of several inches in depth. Sub-zero temperatures and traffic-stopping snowfalls are infrequent.

Based on the 1951-1980 period, the average first occurrence of 32 degrees Fahrenheit in the fall is October 6 and the average last occurrence in the spring is May 4.

NORMALS, MEANS, AND EXTREMES

SPOKANE WASHINGTON

LATITUDE: 47°38'N LONGITUDE: 117°32' W ELEVATION: FT. GRND 2357 BARO 2360 TIME ZONE: PACIFIC WBAN: 24157

	(a)	JAN	FEB	MAR	APR	MAY	JUNE	JULY	AUG	SEP	OCT	NOV	DEC	YEAR
TEMPERATURE °F:														
Normals														
-Daily Maximum		33.2	40.6	47.7	57.0	65.8	74.7	83.1	82.5	72.0	58.6	41.4	33.8	57.5
-Daily Minimum		20.8	25.9	29.6	34.7	41.9	49.2	54.4	54.3	45.8	36.0	28.8	21.7	36.9
-Monthly		27.1	33.3	38.7	45.9	53.9	62.0	68.8	68.4	58.9	47.3	35.1	27.8	47.3
Extremes														
-Record Highest	46	59	61	71	90	96	101	103	108	98	86	67	56	108
-Year		1971	1958	1960	1977	1986	1992	1967	1961	1988	1980	1975	1980	AUG 1961
-Record Lowest	46	-22	-17	-7	17	24	33	37	35	24	10	-21	-25	-25
-Year		1979	1979	1989	1966	1954	1984	1981	1965	1985	1991	1985	1968	DEC 1968
NORMAL DEGREE DAYS:														
Heating (base 65°F)		1175	888	815	573	344	139	30	56	223	549	897	1153	6842
Cooling (base 65°F)		0	0	0	0	0	49	148	161	40	0	0	0	398
% OF POSSIBLE SUNSHINE	45	28	41	54	61	64	66	80	78	72	55	29	23	54
MEAN SKY COVER (tenths)														
Sunrise - Sunset	46	8.3	7.9	7.4	7.1	6.6	6.1	3.9	4.2	4.8	6.3	8.1	8.4	6.6
MEAN NUMBER OF DAYS:														
Sunrise to Sunset														
-Clear	46	3.0	3.4	4.3	4.5	5.7	7.2	16.1	15.1	12.3	8.0	3.3	2.8	86.0
-Partly Cloudy	46	4.2	4.9	7.7	8.2	10.2	10.3	8.4	8.5	8.3	7.7	5.0	3.9	87.3
-Cloudy	46	23.8	20.0	19.0	17.2	15.1	12.5	6.5	7.3	9.3	15.3	21.7	24.3	191.9
Precipitation														
.01 inches or more	46	14.0	11.3	11.3	8.9	9.3	7.8	4.7	4.8	5.6	7.5	12.7	14.7	112.5
Snow, Ice pellets, hail														
1.0 inches or more	46	5.2	2.8	1.6	0.2	0.*	0.0	0.0	0.0	0.0	0.1	2.1	5.0	17.0
Thunderstorms	46	0.*	0.*	0.3	0.7	1.6	2.9	2.5	2.2	0.7	0.3	0.1	0.0	11.1
Heavy Fog Visibility														
1/4 mile or less	46	9.4	7.3	3.1	1.2	0.8	0.5	0.2	0.3	0.8	4.0	8.5	11.9	48.0
Temperature °F														
-Maximum														
90° and above	34	0.0	0.0	0.0	0.*	0.3	2.1	8.3	7.2	0.9	0.0	0.0	0.0	18.8
32° and below	34	14.6	4.6	0.9	0.0	0.0	0.0	0.0	0.0	0.0	0.1	4.1	13.9	38.2
-Minimum														
32° and below	34	26.5	22.4	20.6	10.4	1.7	0.0	0.0	0.0	0.8	9.3	20.1	26.7	138.6
0° and below	34	2.4	0.6	0.*	0.0	0.0	0.0	0.0	0.0	0.0	0.0	0.3	1.9	5.1
AVG. STATION PRESS.(mb)	20	934.0	933.0	930.0	931.1	930.6	930.8	931.8	931.5	932.8	933.9	932.9	934.2	932.2
RELATIVE HUMIDITY (%)														
Hour 04	34	85	85	81	77	77	74	65	63	71	79	87	88	78
Hour 10 (Local Time)	34	84	80	69	57	53	49	41	43	51	65	83	86	63
Hour 16	34	79	69	55	44	41	36	28	28	34	48	76	83	52
Hour 22	34	84	81	74	66	63	58	46	46	56	70	85	87	68
PRECIPITATION (inches):														
Water Equivalent														
-Normal		1.98	1.49	1.49	1.18	1.41	1.26	0.67	0.72	0.73	0.99	2.15	2.42	16.49
-Maximum Monthly	46	4.96	3.94	3.75	3.08	5.71	3.06	2.33	1.83	2.05	4.05	5.10	5.13	5.71
-Year		1959	1961	1950	1948	1948	1964	1990	1976	1959	1950	1973	1964	MAY 1948
-Minimum Monthly	46	0.38	0.35	0.31	0.08	0.20	0.16	T	T	T	0.03	0.22	0.60	T
-Year		1985	1988	1965	1956	1982	1960	1973	1988	1990	1987	1976	1976	SEP 1990
-Maximum in 24 hrs	46	1.48	1.11	0.96	1.01	1.67	2.07	1.80	1.09	1.12	0.98	1.41	1.60	2.07
-Year		1954	1963	1989	1982	1948	1964	1990	1959	1973	1955	1960	1951	JUN 1964
Snow, Ice pellets, hail														
-Maximum Monthly	46	56.9	28.5	15.3	6.6	3.5	T	0.0	0.0	T	6.1	24.7	42.0	56.9
-Year		1950	1975	1962	1964	1967	1993			1991	1957	1955	1964	JAN 1950
-Maximum in 24 hrs	46	13.0	11.0	6.1	4.9	3.5	T	0.0	0.0	T	6.1	9.0	12.1	13.0
-Year		1950	1993	1989	1964	1967	1993			1991	1957	1973	1951	JAN 1950
WIND:														
Mean Speed (mph)	46	8.8	9.2	9.6	10.0	9.3	9.2	8.6	8.2	8.3	8.1	8.6	8.6	8.9
Prevailing Direction through 1963		NE	SSW	SSW	SW	SSW	SSW	SW	SW	NE	SSW	NE	NE	SSW
Fastest Mile														
-Direction (!!!)	46	SW	SW	SW	SW	W	SW	SW	SW	SW	SW	SW	SW	SW
-Speed (MPH)	46	59	54	54	52	49	44	43	50	38	56	54	51	59
-Year		1972	1949	1971	1987	1957	1986	1970	1982	1961	1950	1949	1956	JAN 1972
Peak Gust														
-Direction (!!!)	10	SW	S	W	SW	W	SW	SW	W	SW	SW	SW	SW	SW
-Speed (mph)	10	56	51	52	62	53	49	51	47	47	62	56	51	62
-Date		1986	1987	1988	1987	1986	1989	1989	1993	1987	1991	1990	1991	OCT 1991

PRECIPITATION (inches)

SPOKANE WASHINGTON

YEAR	JAN	FEB	MAR	APR	MAY	JUNE	JULY	AUG	SEP	OCT	NOV	DEC	ANNUAL
1964	3.15	0.98	1.53	0.98	0.45	3.06	0.39	1.46	1.03	0.46	2.89	5.13	21.51
1965	2.82	1.13	0.31	2.35	1.02	0.74	0.69	1.73	0.28	0.05	1.71	1.63	14.46
1966	1.94	0.50	2.43	0.13	0.49	0.70	0.95	0.15	0.51	0.36	3.01	2.96	14.13
1967	2.44	0.40	1.72	1.71	1.31	1.99	0.06	T	0.24	1.18	0.82	2.02	13.89
1968	1.57	2.12	0.71	0.10	1.16	0.87	0.23	1.35	0.63	2.24	2.35	2.93	16.26
1969	4.08	1.21	0.53	2.16	0.54	1.17	0.03	T	0.71	0.45	0.37	2.45	13.70
1970	4.15	1.83	1.30	0.93	0.94	1.60	0.59	0.10	0.48	2.13	2.04	1.43	17.52
1971	2.11	0.88	2.11	1.85	1.39	2.46	0.50	0.59	1.37	0.82	1.51	2.89	18.48
1972	1.74	1.13	1.05	1.09	1.99	1.56	0.25	0.87	0.86	0.19	0.88	1.92	13.53
1973	2.05	0.48	0.77	0.42	1.34	0.57	T	0.19	1.44	0.97	5.10	3.78	17.11
1974	3.79	1.79	2.22	0.80	1.03	0.23	0.71	0.04	0.18	0.12	2.59	2.54	16.04
1975	2.53	3.12	1.83	1.78	1.41	1.45	1.60	0.93	0.03	2.23	1.94	2.42	21.27
1976	1.28	2.04	0.83	0.97	1.24	0.78	0.79	1.83	0.05	0.59	0.22	0.60	11.22
1977	0.75	0.52	1.15	0.13	1.71	1.45	0.11	1.25	1.42	0.44	2.12	4.52	15.57
1978	2.53	1.64	0.77	2.62	2.81	1.22	1.76	1.71	0.93	0.13	2.02	1.05	19.19
1979	1.11	2.19	1.03	0.69	1.60	0.78	0.85	1.01	0.78	1.22	1.15	1.94	14.35
1980	1.96	1.90	0.91	1.06	2.34	0.99	0.21	0.79	0.84	0.64	1.67	3.72	17.03
1981	1.00	1.41	1.57	0.85	2.02	1.92	0.51	0.04	0.59	1.53	0.96	2.51	14.91
1982	1.61	1.67	1.49	2.23	0.20	0.85	1.05	0.25	1.77	1.48	1.86	2.79	17.25
1983	1.89	2.07	2.20	0.61	0.92	2.84	1.85	0.96	0.79	1.33	4.80	2.38	22.64
1984	0.99	1.37	1.80	1.75	2.01	1.89	0.07	0.27	0.56	0.76	4.26	2.28	18.01
1985	0.38	0.93	1.39	0.28	1.13	0.67	0.26	0.19	1.64	1.40	2.23	0.71	11.21
1986	3.08	2.02	1.58	1.33	1.08	0.48	0.44	0.15	1.65	0.46	2.25	1.03	15.55
1987	1.59	0.88	2.18	1.12	0.90	0.59	2.27	1.81	0.01	0.03	1.37	4.93	17.68
1988	1.76	0.35	1.57	2.15	1.50	1.12	0.23	T	1.63	0.11	4.35	1.75	16.52
1989	0.82	1.34	2.87	0.72	2.17	0.41	0.40	1.61	0.18	1.58	1.66	0.95	14.71
1990	2.45	1.01	0.85	1.34	3.11	1.91	2.33	1.03	T	3.05	0.84	1.69	19.61
1991	1.72	0.81	2.31	1.35	1.72	1.13	0.58	0.17	0.01	0.34	3.08	1.23	14.45
1992	2.12	1.76	0.43	0.65	0.28	1.51	1.09	0.33	0.36	0.81	3.02	2.16	14.52
1993	1.40	0.86	1.13	1.90	1.36	0.48	2.08	1.24	0.28	0.42	0.68	1.80	13.63
Record Mean	2.05	1.57	1.36	1.10	1.36	1.27	0.57	0.62	0.81	1.20	2.06	2.21	16.17

AVERAGE TEMPERATURE (deg. F)

SPOKANE WASHINGTON

YEAR	JAN	FEB	MAR	APR	MAY	JUNE	JULY	AUG	SEP	OCT	NOV	DEC	ANNUAL
1964	29.3	29.2	35.7	44.5	52.9	60.4	68.3	62.8	55.2	47.9	32.2	24.0	45.2
1965	28.6	32.1	34.4	47.2	52.4	61.3	70.1	67.9	53.8	52.7	38.0	30.0	47.4
1966	29.7	32.8	38.7	46.0	55.9	58.8	68.2	68.2	64.6	47.3	36.9	33.4	48.4
1967	33.9	36.2	37.1	42.3	52.8	63.5	70.6	74.5	65.3	48.4	35.4	27.8	49.0
1968	27.8	37.8	42.1	43.0	53.8	61.2	71.1	65.1	58.9	45.2	34.9	24.6	47.1
1969	16.3	26.1	35.6	46.2	57.4	65.2	67.4	67.1	59.8	43.7	36.3	29.4	45.9
1970	25.9	36.3	37.0	41.6	54.9	66.2	72.5	70.2	54.2	44.9	36.0	27.7	47.3
1971	31.8	33.6	35.2	45.3	56.3	58.2	69.7	74.1	55.2	44.2	35.4	25.8	47.0
1972	22.6	30.7	41.4	42.0	56.9	62.0	68.1	71.1	55.4	47.2	38.3	25.4	46.8
1973	27.0	34.9	41.1	46.2	56.5	62.0	71.2	69.1	59.7	47.2	33.7	33.3	48.5
1974	24.1	35.4	38.5	46.4	50.2	66.0	67.8	68.2	60.5	48.0	36.4	30.5	47.7
1975	23.6	24.7	34.0	41.7	52.7	59.2	72.4	64.1	61.0	46.9	33.8	30.9	45.4
1976	29.6	32.1	35.1	45.2	54.5	58.5	68.8	65.4	63.4	46.8	35.8	29.6	47.1
1977	22.0	35.1	38.2	50.9	51.6	65.0	67.0	71.2	55.1	46.5	34.0	26.2	46.9
1978	27.6	34.0	42.2	45.7	51.4	62.7	68.3	65.9	56.6	46.6	28.7	19.0	45.7
1979	10.5	28.8	40.4	45.5	54.7	62.7	70.4	70.0	63.1	51.1	30.5	35.2	46.9
1980	20.7	34.5	38.6	51.7	55.8	57.8	69.2	64.1	58.4	47.4	36.3	33.2	47.3
1981	32.8	33.9	40.9	45.7	52.0	57.0	65.1	71.5	59.7	45.9	39.9	29.7	47.8
1982	26.0	32.1	40.3	43.5	54.2	66.5	67.6	69.8	59.5	46.1	31.7	27.3	47.0
1983	35.8	38.1	43.0	46.3	57.1	61.9	65.5	72.3	57.1	49.7	39.3	16.2	48.5
1984	30.5	34.5	41.7	44.0	50.1	59.2	69.1	70.1	56.7	43.4	35.8	20.4	46.3
1985	21.4	24.9	35.9	48.0	56.2	61.8	75.0	64.9	53.3	44.7	19.5	19.3	43.7
1986	30.1	31.6	42.8	44.9	55.3	66.2	64.0	72.6	54.8	49.0	34.8	26.3	47.7
1987	26.5	35.1	41.8	51.1	57.2	65.1	66.6	66.2	62.8	49.5	38.1	25.9	48.8
1988	24.7	35.4	39.7	48.9	54.6	61.1	68.7	68.4	58.9	53.3	36.3	27.0	48.1
1989	28.8	21.8	36.6	48.9	53.1	64.3	68.7	64.8	60.1	47.0	38.0	31.0	46.9
1990	33.4	30.2	40.9	49.7	52.8	60.7	70.4	68.5	65.3	45.1	39.0	21.1	48.1
1991	25.7	39.2	36.8	45.8	51.6	56.6	68.7	70.2	61.8	46.2	34.2	32.8	47.5
1992	31.8	38.9	45.5	48.8	58.9	68.0	67.8	69.6	57.4	49.5	34.3	22.9	49.5
1993	21.9	25.4	37.8	45.5	59.8	60.2	60.2	64.2	58.7	50.0	29.4	30.9	45.3
Record Mean	26.8	31.8	39.5	47.6	55.6	62.4	69.8	68.6	59.4	48.5	36.5	29.7	48.0
Max	32.6	38.9	48.4	58.5	67.3	74.5	83.9	82.7	72.3	59.3	43.0	35.0	58.0
Min	20.9	24.7	30.5	36.7	43.8	50.3	55.7	54.5	46.4	37.7	30.0	24.4	38.0

REFERENCE NOTES

SPOKANE WASHINGTON

GENERAL
T - TRACE AMOUNT.
BLANK ENTRIES DENOTE MISSING/UNREPORTED DATA.
# INDICATES A STATION OR INSTRUMENT RELOCATION.

SPECIFIC

PM - INCLUDES LAST DAY OF PREVIOUS MONTH
ASOS - AUTOMATED SURFACE OBSERVING SYSTEM IN OPERATION DURING THESE MONTHS.

(a) - LENGTH OF RECORD IN YEARS, ALTHOUGH INDIVIDUAL MONTHS MAY BE MISSING.
0.* OR * - THE VALUE IS BETWEEN 0.0 AND 0.05
NORMALS - BASED ON THE 1961-1990 RECORD PERIOD.
EXTREMES - DATES ARE THE MOST RECENT OCCURRENCE
WIND DIR. - NUMERALS SHOW TENS OF DEGRESS CLOCKWISE FROM TRUE NORTH. "00" INDICATES CALM.
RESULTANT DIRECTIONS ARE GIVEN TO WHOLE DEGREES.
BOLD VALUES INDICATE EXTREME VALUES WHICH OCCURRED AFTER THE ASOS SYSTEM WAS COMMISSIONED.

RECORD = PERIOD OF RECORD
RECORD MEAN PRECIPITATION IS THE MEAN OF ALL DAILY PRECIPITATION AMOUNTS DURING THE PERIOD OF RECORD.
RECORD MAX(MIN) TEMPERATURE IS THE MEAN OF ALL DAILY MAX(MIN) TEMPERATURES DURING THE PERIOD OF RECORD.
RECORD MEAN TEMPERATURE IS THE SUM OF THE RECORD MAX AND RECORD MIN DIVIDED BY 2.
AVERAGE TEMPERATURE IS THE SUM OF THE MEAN DAILY MAX AND MIN TEMPERATURE DIVIDED BY 2.

EXCEPTIONS

RECORD MEANS ARE THROUGH THE CURRENT YEAR, BEGINNING IN 1882 FOR TEMPERATURE
1882 FOR PRECIPITATION
1948 FOR SNOWFALL

HEATING DEGREE DAYS Base 65 deg. F SPOKANE WASHINGTON

SEASON	JULY	AUG	SEP	OCT	NOV	DEC	JAN	FEB	MAR	APR	MAY	JUNE	TOTAL
1964-65	31	118	290	524	976	1268	1121	915	942	528	387	129	7229
1965-66	31	62	330	377	804	1078	1088	896	808	561	291	190	6516
1966-67	30	42	67	544	838	975	956	799	859	677	370	96	6253
1967-68	8	2	71	508	882	1146	1149	783	702	654	343	138	6386
1968-69	19	89	199	607	897	1245	1504	1080	905	559	236	88	7428
1969-70	40	44	192	655	855	1097	1208	797	859	696	305	101	6849
1970-71	13	5	321	614	864	1146	1022	873	918	584	270	215	6845
1971-72	64	19	297	641	882	1208	1308	991	726	684	274	127	7221
1972-73	36	18	292	545	795	1219	1171	838	734	558	286	152	6644
1973-74	17	47	193	546	933	978	1265	824	814	554	455	97	6723
1974-75	41	22	134	519	852	1062	1276	1122	953	694	375	173	7223
1975-76	22	75	136	554	933	1048	1091	946	922	588	317	213	6845
1976-77	20	71	74	556	871	1089	1324	832	824	436	409	66	6572
1977-78	57	56	289	563	921	1197	1154	862	701	576	412	101	6889
1978-79	37	97	252	562	1083	1424	1684	1011	756	577	313	134	7930
1979-80	41	4	91	423	1029	918	1365	880	809	392	283	211	6446
1980-81	19	77	195	543	854	977	992	867	741	570	395	243	6473
1981-82	73	7	209	584	747	1088	1202	912	761	639	328	76	6626
1982-83	62	17	193	582	996	1163	897	747	672	558	285	113	6285
1983-84	55	2	230	468	765	1508	1065	880	715	621	460	194	6963
1984-85	21	18	264	662	870	1381	1345	1117	895	501	280	128	7482
1985-86	0	64	343	622	1363	1409	1076	927	680	595	357	67	7503
1986-87	81	4	311	488	902	1193	1186	831	710	417	253	86	6462
1987-88	51	50	116	474	799	1206	1240	850	775	477	330	173	6541
1988-89	47	16	240	361	856	1171	1113	1205	873	473	364	65	6784
1989-90	22	76	149	554	805	1048	976	968	739	454	373	166	6330
1990-91	37	42	54	610	774	1356	1212	716	866	568	406	248	6889
1991-92	15	16	108	574	918	992	1024	750	598	477	206	61	5739
1992-93	32	60	232	481	916	1297	1331	1102	834	578	192	165	7220
1993-94	151	83	217	457	1063	1051							

COOLING DEGREE DAYS Base 65 deg. F SPOKANE WASHINGTON

YEAR	JAN	FEB	MAR	APR	MAY	JUNE	JULY	AUG	SEP	OCT	NOV	DEC	TOTAL
1969	0	0	0	0	7	99	121	112	40	0	0	0	379
1970	0	0	0	0	3	143	253	175	2	0	0	0	576
1971	0	0	0	0	10	17	216	306	9	0	0	0	558
1972	0	0	0	0	28	41	138	213	10	0	0	0	430
1973	0	0	0	0	31	67	216	177	39	0	0	0	530
1974	0	0	0	0	0	137	134	127	7	0	0	0	405
1975	0	0	0	0	0	7	256	57	20	0	0	0	340
1976	0	0	0	0	0	24	143	93	33	0	0	0	293
1977	0	0	0	18	2	72	126	254	0	0	0	0	472
1978	0	0	0	0	0	42	144	131	9	0	0	0	326
1979	0	0	0	0	1	73	217	166	39	0	0	0	496
1980	0	0	0	1	3	2	156	56	6	3	0	0	227
1981	0	0	0	0	0	9	82	213	60	0	0	0	364
1982	0	0	0	0	2	128	148	171	32	0	0	0	481
1983	0	0	0	0	46	26	77	235	1	0	0	0	385
1984	0	0	0	0	3	28	155	181	23	1	0	0	391
1985	0	0	0	0	15	36	317	68	0	0	0	0	436
1986	0	0	0	0	65	109	57	247	8	0	0	0	486
1987	0	0	0	8	20	94	110	97	53	1	0	0	383
1988	0	0	0	0	12	63	169	128	67	0	0	0	439
1989	0	0	0	0	0	49	145	78	9	0	0	0	281
1990	0	0	0	0	0	42	213	157	68	0	0	0	480
1991	0	0	0	0	0	0	139	187	20	0	0	0	346
1992	0	0	0	0	25	159	124	209	11	8	0	0	536
1993	0	0	0	0	36	27	11	64	34	0	0	0	172

SNOWFALL (inches) SPOKANE WASHINGTON

SEASON	JULY	AUG	SEP	OCT	NOV	DEC	JAN	FEB	MAR	APR	MAY	JUNE	TOTAL
1964-65	0.0	0.0	0.0	T	15.2	42.0	20.1	2.3	2.0	0.1	0.0	0.0	81.7
1965-66	0.0	0.0	0.0	T	6.3	15.4	13.9	1.6	7.2	T	T	0.0	44.4
1966-67	0.0	0.0	0.0	T	0.9	9.0	6.5	3.1	6.4	0.8	3.5	0.0	30.2
1967-68	0.0	0.0	0.0	T	4.8	12.7	11.8	0.4	T	T	T	0.0	29.7
1968-69	0.0	0.0	0.0	0.0	1.2	19.8	48.7	5.4	2.0	0.4	0.0	0.0	77.5
1969-70	0.0	0.0	0.0	0.0	T	10.4	19.4	2.8	6.9	0.3	0.1	0.0	39.9
1970-71	0.0	0.0	0.0	T	6.8	12.0	6.1	5.5	1.5	T	0.0	0.0	31.9
1971-72	0.0	0.0	0.0	3.1	4.0	34.2	17.2	5.9	2.5	0.2	0.0	0.0	67.1
1972-73	0.0	0.0	0.0	0.8	T	4.7	6.5	3.5	0.5	T	0.0	0.0	16.0
1973-74	0.0	0.0	0.0	0.8	23.6	9.1	15.0	4.4	2.5	0.4	0.4	0.0	56.2
1974-75	0.0	0.0	0.0	0.0	0.3	16.6	30.9	28.5	7.6	5.1	T	0.0	89.0
1975-76	0.0	0.0	0.0	3.9	11.4	6.9	15.3	6.3	4.6	0.4	0.0	0.0	48.8
1976-77	0.0	0.0	0.0	0.0	0.1	4.2	6.8	2.5	2.7	T	T	0.0	16.3
1977-78	0.0	0.0	0.0	0.0	11.2	30.3	19.1	6.6	2.2	T	T	0.0	69.4
1978-79	0.0	0.0	0.0	0.0	15.4	14.8	16.5	10.6	3.4	T	T	0.0	60.7
1979-80	0.0	0.0	0.0	0.0	3.9	10.4	16.6	5.9	1.1	0.4	0.0	0.0	38.3
1980-81	0.0	0.0	0.0	0.0	1.2	6.8	2.6	3.3	T	T	0.3	0.0	14.2
1981-82	0.0	0.0	0.0	T	0.8	13.0	23.3	2.2	2.1	6.0	T	0.0	47.4
1982-83	0.0	0.0	0.0	T	5.4	17.4	8.1	5.5	T	0.2	T	0.0	36.6
1983-84	0.0	0.0	0.0	0.0	5.7	24.8	5.3	8.0	1.9	1.3	0.8	0.0	47.8
1984-85	0.0	0.0	0.0	1.1	12.0	24.7	4.6	14.8	9.6	T	T	0.0	66.8
1985-86	0.0	0.0	0.0	0.4	23.7	8.3	14.7	13.8	T	0.2	T	0.0	61.1
1986-87	0.0	0.0	0.0	0.0	5.0	7.9	11.7	1.1	T	T	T	0.0	25.7
1987-88	0.0	0.0	0.0	0.0	1.5	20.3	9.1	1.2	1.6	T	T	0.0	33.7
1988-89	0.0	0.0	0.0	0.0	10.9	16.3	10.5	19.0	9.4	T	T	0.0	66.1
1989-90	0.0	0.0	0.0	T	5.2	1.1	10.3	18.0	2.6	3.5	T	0.0	40.7
1990-91	0.0	0.0	0.0	0.0	1.2	14.3	15.9	1.1	9.5	0.2	0.0	0.0	42.2
1991-92	0.0	0.0	T	0.8	4.9	2.4	9.0	1.4	0.0	T	0.0	0.0	18.5
1992-93	0.0	0.0	0.0	0.0	11.1	40.2	18.8	15.1	2.1	T	T	T	87.3
1993-94	0.0	0.0	0.0	T	3.7	6.4							
Record Mean	0.0	0.0	T	0.4	6.2	15.1	16.2	7.7	4.1	0.6	0.1	T	50.5

YAKIMA
WASHINGTON

Yakima is located in a small east–west valley in the upper (northwestern) part of the Yakima Valley. Local topography is complex with a number of minor valleys and ridges giving a local relief of as much as 1,000 feet. This complex topography results in marked variations in air drainage, winds, and low temperatures within short distances.

The climate of the Yakima Valley is relatively mild and dry. It has characteristics of both maritime and continental climates, modified by the Cascade and the Rocky Mountains, respectively. Summers are dry and rather hot, and winters cool with only light snowfall. The maritime influence is strongest in winter when the prevailing westerlies are the strongest and most steady. The Selkirk and Rocky Mountains in British Columbia and Idaho shield the area from most of the very cold air masses that sweep down from Canada into the Great Plains and eastern United States. Sometimes a strong polar high pressure area over western Canada will occur at the same time that a low pressure area covers the southwestern United States. On these occasions, the cold arctic air will pour through the passes and down the river valleys of British Columbia, bringing very cold temperatures to Yakima. However, over one–half of the winters remain above zero.

The modifying influence of the Pacific Ocean is much less in summer. Afternoons are hot, but the dry air results in a rapid temperature fall after sunset, and nights are pleasantly cool with summertime low temperatures, usually in the 50s. Spells of 4 to 11 days of 100 degrees or more have occurred.

The length of the growing season varies depending on the immediate topography and the crop grown. Temperatures below 32 degrees are infrequent during the period from mid–May through September. Temperatures below 40 degrees during July and August have occurred in about half of the years.

Precipitation follows the pattern of a West Coast marine climate with the typical late fall and early winter high. However, since Yakima lies in the rain shadow of the Cascades, total amounts are small. The three months, November to January, total nearly half of the annual fall. Late June, July, and August are very dry.

Irrigation is necessary for nearly all crops. Ample water supplies are available from the snowmelt in the Cascade Mountains which is collected in storage reservoirs for summer use.

Snowfall in the Yakima area is light, averaging 20 to 25 inches.

Summers are sunny, with about 85 percent of the possible sunshine. Winters are generally cloudy, with only a third of the possible sunshine.

Winds are mostly light, averaging about 7 mph for the year, being somewhat stronger in late spring and weaker in winter. Speeds of 30 to 35 mph are reached at least once in about half the months and speeds over 40 mph occur in about 1 out of 5 months. The most common wind direction in downtown Yakima is northwest, while at the airport the wind is from the west in winter and the west–northwest in summer.

NORMALS, MEANS, AND EXTREMES

YAKIMA WASHINGTON

LATITUDE: 46°34'N LONGITUDE: 120°32' W ELEVATION: FT. GRND 1052 BARO 1068 TIME ZONE: PACIFIC WBAN: 24243

	(a)	JAN	FEB	MAR	APR	MAY	JUNE	JULY	AUG	SEP	OCT	NOV	DEC	YEAR
TEMPERATURE °F:														
Normals														
-Daily Maximum		37.5	46.4	55.2	63.2	71.6	79.9	86.7	85.7	76.8	64.4	48.3	37.5	62.8
-Daily Minimum		21.8	26.4	30.8	35.5	42.3	49.2	53.1	52.3	44.6	35.3	29.0	22.1	36.9
-Monthly		29.7	36.4	43.0	49.4	57.0	64.6	69.9	69.0	60.7	49.9	38.6	29.8	49.8
Extremes														
-Record Highest	47	68	69	80	92	102	105	108	110	100	88	73	67	110
-Year		1977	1947	1960	1977	1986	1992	1971	1971	1988	1992	1989	1980	AUG 1971
-Record Lowest	47	-21	-25	-1	20	25	30	34	35	24	11	-13	-17	-25
-Year		1950	1950	1960	1985	1954	1984	1971	1960	1985	1971	1985	1964	FEB 1950
NORMAL DEGREE DAYS:														
Heating (base 65°F)		1094	801	682	468	255	90	19	38	169	468	792	1091	5967
Cooling (base 65°F)		0	0	0	0	7	78	171	162	40	0	0	0	458
% OF POSSIBLE SUNSHINE														
MEAN SKY COVER (tenths)														
Sunrise - Sunset	47	7.9	7.4	6.8	6.5	5.9	5.3	3.2	3.5	4.0	5.8	7.4	7.9	6.0
MEAN NUMBER OF DAYS:														
Sunrise to Sunset														
-Clear	47	4.0	4.5	6.1	6.3	8.4	10.3	18.6	17.4	15.0	9.5	5.1	4.0	109.0
-Partly Cloudy	47	5.4	5.9	8.1	9.3	10.6	9.7	8.0	7.9	7.9	8.2	6.1	5.3	92.4
-Cloudy	47	21.6	17.9	16.8	14.4	12.0	9.9	4.4	5.7	7.1	13.3	18.8	21.8	163.8
Precipitation .01 inches or more	47	9.4	7.1	6.6	4.7	5.0	4.9	2.2	2.8	3.1	5.0	8.4	9.6	68.8
Snow,Ice pellets,hail 1.0 inches or more	45	2.7	1.2	0.6	0.0	0.0	0.0	0.0	0.0	0.0	0.1	0.7	2.7	7.9
Thunderstorms	47	0.0	0.*	0.1	0.4	1.1	1.7	1.4	1.3	0.6	0.1	0.0	0.0	6.8
Heavy Fog Visibility 1/4 mile or less	47	4.5	2.4	0.6	0.*	0.1	0.0	0.*	0.0	0.1	0.7	3.4	6.7	18.6
Temperature °F														
-Maximum														
90° and above	47	0.0	0.0	0.0	0.*	1.4	4.6	13.2	10.7	2.3	0.0	0.0	0.0	32.3
32° and below	47	10.2	2.5	0.2	0.0	0.0	0.0	0.0	0.0	0.0	0.0	1.7	8.5	23.1
-Minimum														
32° and below	47	28.0	23.7	20.6	12.0	2.8	0.1	0.0	0.0	1.1	10.9	21.3	27.9	148.4
0° and below	47	2.3	0.5	0.*	0.0	0.0	0.0	0.0	0.0	0.0	0.0	0.2	0.9	4.0
AVG. STATION PRESS.(mb)	19	982.1	979.5	976.8	977.4	976.3	976.1	976.3	975.8	977.8	979.8	980.1	982.1	978.3
RELATIVE HUMIDITY (%)														
Hour 04	45	83	83	77	72	70	70	68	71	77	81	84	85	77
Hour 10 (Local Time)	47	78	71	54	42	39	39	36	39	44	55	73	81	54
Hour 16	47	71	58	41	33	31	31	26	28	32	43	63	75	44
Hour 22	44	81	79	69	59	56	55	51	55	65	74	81	84	67
PRECIPITATION (inches):														
Water Equivalent														
-Normal		1.21	0.74	0.67	0.50	0.45	0.53	0.16	0.40	0.40	0.47	1.03	1.41	7.97
-Maximum Monthly	47	3.66	2.46	2.63	1.62	2.76	2.53	0.71	2.10	2.07	2.22	2.83	4.19	4.19
-Year		1970	1961	1957	1963	1948	1991	1966	1975	1986	1950	1973	1964	DEC 1964
-Minimum Monthly	47	0.09	T	0.01	T	0.03	0.01	T	0.00	0.00	0.00	T	0.07	0.00
-Year		1985	1988	1973	1985	1964	1970	1988	1955	1986	1978	1990	1976	SEP 1986
-Maximum in 24 hrs	47	1.37	0.87	0.74	1.25	0.90	1.56	0.66	1.74	1.49	1.05	1.08	1.58	1.74
-Year		1963	1961	1987	1974	1986	1982	1963	1990	1986	1982	1955	1977	AUG 1990
Snow,Ice pellets,hail														
-Maximum Monthly	47	26.6	16.5	10.8	0.2	T	0.0	0.0	0.0	0.0	2.9	21.2	37.5	37.5
-Year		1950	1949	1971	1993	1986					1991	1955	1964	DEC 1964
-Maximum in 24 hrs	47	13.6	5.8	7.4	0.2	T	0.0	0.0	0.0	0.0	2.4	11.2	14.0	14.0
-Year		1963	1956	1951	1993	1986					1991	1984	1964	DEC 1964
WIND:														
Mean Speed (mph)	40	5.6	6.4	7.9	8.6	8.5	8.2	7.8	7.4	7.3	6.6	5.8	5.2	7.1
Prevailing Direction through 1963		W	W	W	WNW	WNW	NW	WNW	WNW	WNW	WNW	W	W	WNW
Fastest Obs. 1 Min.														
-Direction (!!)	39	25	28	23	29	18	20	24	29	30	31	29	23	28
-Speed (MPH)	39	44	48	48	46	46	47	43	35	38	41	45	48	48
-Year		1962	1967	1956	1961	1961	1955	1968	1988	1992	1988	1955	1955	FEB 1967
Peak Gust														
-Direction (!!)	10	W	W	W	S	NE	SE	SW	S	NW	SW	NW	W	NE
-Speed (mph)	10	55	56	51	52	69	51	54	44	55	54	58	61	69
-Date		1988	1985	1988	1989	1985	1987	1990	1993	1992	1990	1989	1991	MAY 1985

PRECIPITATION (inches)

YAKIMA WASHINGTON

YEAR	JAN	FEB	MAR	APR	MAY	JUNE	JULY	AUG	SEP	OCT	NOV	DEC	ANNUAL
1964	0.60	T	0.14	0.25	0.03	1.18	0.08	0.20	0.03	0.15	0.70	4.19	7.55
1965	1.33	0.08	0.10	0.48	0.05	0.51	0.27	0.21	0.04	0.06	1.43	1.39	5.95
1966	1.73	0.11	0.81	T	0.10	0.17	0.71	T	0.87	0.41	2.14	0.95	8.00
1967	0.60	T	0.45	1.03	0.16	1.12	T	0.01	0.09	0.21	0.30	0.55	4.52
1968	1.76	0.88	0.11	T	0.47	0.02	0.02	1.71	0.32	0.94	1.32	1.91	9.46
1969	1.52	0.91	0.16	0.27	0.54	0.61	T	0.01	0.32	0.24	0.08	2.28	6.94
1970	3.66	0.49	0.22	0.16	0.06	0.01	0.13	T	0.07	0.54	1.25	1.41	8.00
1971	1.48	T	1.56	0.47	0.54	0.20	0.04	0.14	0.73	0.27	0.97	1.45	7.85
1972	0.88	0.31	1.05	0.09	0.60	1.50	0.04	0.65	0.06	0.12	0.72	1.31	7.33
1973	1.19	0.24	0.01	0.04	0.08	0.02	T	0.01	0.81	1.52	2.83	2.22	8.97
1974	1.67	0.85	1.21	1.46	0.80	0.12	0.18	T	0.02	0.45	0.30	1.14	8.20
1975	2.28	1.16	0.49	0.40	0.23	0.22	0.18	2.10	T	0.79	0.43	0.55	8.83
1976	0.56	0.78	0.70	0.33	0.09	0.69	0.26	0.50	0.13	0.07	T	0.07	4.18
1977	0.13	0.69	0.23	0.01	0.68	0.46	T	1.16	0.89	0.17	0.70	2.80	7.92
1978	2.30	1.30	0.52	0.91	0.28	0.32	0.29	0.38	0.64	0.00	0.94	0.14	8.02
1979	0.91	0.54	0.23	0.14	0.04	0.57	0.04	0.42	0.36	0.74	1.53	1.33	6.85
1980	2.23	1.30	0.29	0.80	0.84	1.12	T	0.29	0.48	0.23	1.00	2.69	11.27
1981	0.95	0.65	0.10	0.01	0.68	0.39	0.29	0.09	0.59	1.16	1.36	2.38	8.65
1982	0.58	1.48	0.34	0.30	0.37	1.70	0.12	0.39	1.08	1.46	0.90	2.15	10.87
1983	1.97	1.59	1.95	0.66	0.30	0.77	0.29	0.44	0.33	0.23	2.77	1.92	13.22
1984	0.13	0.92	1.04	1.05	0.51	1.45	0.13	0.04	0.46	0.16	2.62	0.51	9.02
1985	0.09	0.68	0.62	T	0.46	0.37	0.12	0.03	0.84	0.75	0.92	1.02	5.90
1986	1.82	1.26	0.54	0.05	0.94	0.08	0.25	0.11	2.07	0.38	0.64	0.89	9.03
1987	1.46	0.25	1.44	0.57	0.10	0.05	0.40	T	0.00	0.02	0.68	3.30	8.27
1988	0.68	T	0.21	1.41	0.18	1.00	T	T	0.13	0.05	1.12	0.67	5.45
1989	0.19	1.29	1.71	0.85	0.63	0.05	0.07	0.41	0.09	0.67	0.72	0.21	6.89
1990	1.47	0.11	0.21	0.18	1.13	0.31	0.02	2.00	0.04	0.45	T	0.24	6.16
1991	0.34	0.23	1.16	0.61	0.17	2.53	0.18	0.06	0.08	0.73	1.24	0.28	7.61
1992	0.27	0.62	0.41	0.96	0.04	1.25	0.44	0.25	0.28	0.68	0.98	2.33	8.51
1993	0.91	0.66	0.62	0.50	0.43	0.72	0.58	0.17	0.03	0.07	0.21	1.02	5.92
Record Mean	1.11	0.76	0.56	0.46	0.49	0.62	0.18	0.26	0.40	0.53	0.99	1.19	7.57

AVERAGE TEMPERATURE (deg. F)

YAKIMA WASHINGTON

YEAR	JAN	FEB	MAR	APR	MAY	JUNE	JULY	AUG	SEP	OCT	NOV	DEC	ANNUAL
1964	33.2	36.1	40.4	45.5	53.6	62.0	68.0	62.8	57.1	50.1	38.2	27.2	47.9
1965	28.9	37.5	41.2	50.5	56.4	64.5	70.0	68.4	57.9	52.5	42.4	31.0	50.1
1966	29.2	37.4	43.4	52.0	58.3	61.4	67.6	71.2	66.5	52.1	43.8	36.4	51.6
1967	38.6	41.3	41.7	44.9	57.6	67.8	71.4	73.3	65.7	51.8	40.3	31.5	52.2
1968	30.9	40.0	46.0	47.9	56.8	63.9	70.9	65.3	61.0	46.3	39.7	28.4	49.8
1969	18.2	28.5	41.8	48.3	59.9	68.9	68.4	66.2	61.2	47.3	39.8	33.5	48.5
1970	28.4	40.0	43.7	46.2	58.6	68.8	72.5	70.1	57.5	48.6	38.6	28.0	50.1
1971	34.8	37.8	38.7	47.8	60.7	60.7	71.3	72.3	55.5	47.1	38.5	27.8	49.4
1972	29.4	32.2	44.2	45.5	59.9	65.6	69.9	71.6	56.8	49.1	40.9	26.8	49.4
1973	27.6	37.7	45.0	51.7	58.5	64.4	72.2	69.3	61.9	49.2	37.6	35.9	50.9
1974	27.4	40.1	43.7	50.4	54.3	68.3	68.3	70.7	63.0	50.2	41.3	35.2	51.1
1975	30.1	33.0	40.9	46.7	57.7	63.8	75.0	66.3	62.1	49.4	37.8	34.7	49.8
1976	32.5	35.1	39.1	48.1	56.1	59.9	69.0	66.3	64.4	49.6	39.1	29.9	49.1
1977	24.5	38.7	42.8	54.1	52.7	68.8	68.8	74.3	57.7	48.9	35.9	33.5	50.1
1978	32.8	37.9	45.8	49.4	54.8	64.8	71.0	67.5	58.8	49.5	32.3	27.5	49.3
1979	14.8	32.6	44.0	49.6	59.5	65.4	72.1	70.2	64.0	52.9	34.1	34.8	49.5
1980	20.8	33.7	43.7	53.7	58.9	62.1	70.4	66.4	62.6	51.5	40.4	34.8	49.9
1981	39.9	39.3	47.7	49.7	56.8	61.8	69.0	73.1	61.6	48.8	41.2	30.2	51.6
1982	26.2	36.8	43.0	45.8	56.0	66.8	68.7	68.1	60.2	49.0	35.8	29.0	48.8
1983	37.6	40.2	47.8	49.8	60.9	62.9	67.6	70.9	58.1	50.6	42.4	22.2	50.9
1984	31.2	37.8	45.3	47.1	52.4	61.1	70.5	68.8	59.0	47.1	36.1	20.1	48.0
1985	26.4	28.9	41.9	50.8	58.8	64.6	74.9	66.0	55.3	46.7	23.0	19.1	46.4
1986	30.1	35.2	45.6	47.3	57.6	67.4	65.7	72.6	56.9	51.1	38.5	28.5	49.7
1987	28.1	37.6	44.0	52.8	59.6	65.9	67.9	68.5	62.9	50.3	40.1	29.0	50.6
1988	28.8	38.2	42.2	51.1	55.1	62.1	69.2	67.7	59.5	55.1	39.5	28.1	49.7
1989	33.7	26.4	40.0	52.0	56.2	65.8	68.6	66.9	61.0	48.8	41.7	31.4	49.4
1990	36.5	35.2	44.3	53.4	55.0	63.7	73.2	70.1	65.4	46.8	41.6	22.8	50.7
1991	27.7	41.5	40.6	49.2	55.0	60.1	70.6	71.3	63.8	48.6	38.2	34.9	50.1
1992	34.9	40.3	48.1	50.7	60.7	70.0	69.5	70.1	59.1	51.2	38.2	24.5	51.4
1993	23.6	28.8	40.8	47.8	61.5	62.0	63.0	66.6	60.9	51.7	31.4	33.0	47.6
Record Mean	28.6	35.5	43.6	50.9	58.6	65.5	71.7	70.2	61.6	50.9	38.5	30.8	50.5
Max	36.6	45.2	56.0	64.9	73.1	80.2	88.0	86.4	77.4	65.0	48.3	38.2	63.3
Min	20.6	25.9	31.2	36.9	44.1	50.8	55.4	53.9	45.8	36.7	28.7	23.4	37.8

REFERENCE NOTES

YAKIMA WASHINGTON

GENERAL
T - TRACE AMOUNT.
BLANK ENTRIES DENOTE MISSING/UNREPORTED DATA.
# INDICATES A STATION OR INSTRUMENT RELOCATION.

SPECIFIC

(a) - LENGTH OF RECORD IN YEARS, ALTHOUGH INDIVIDUAL MONTHS MAY BE MISSING.
0.* OR * - THE VALUE IS BETWEEN 0.0 AND 0.05
NORMALS - BASED ON THE 1961-1990 RECORD PERIOD.
EXTREMES - DATES ARE THE MOST RECENT OCCURRENCE.
WIND DIR.- NUMERALS SHOW TENS OF DEGRESS CLOCKWISE FROM TRUE NORTH. "00" INDICATES CALM.
RESULTANT DIRECTIONS ARE GIVEN TO WHOLE DEGREES.
BOLD VALUES INDICATE EXTREME VALUES WHICH OCCURRED AFTER THE ASOS SYSTEM WAS COMMISSIONED.

RECORD = PERIOD OF RECORD
RECORD MEAN PRECIPITATION IS THE MEAN OF ALL DAILY PRECIPITATION AMOUNTS DURING THE PERIOD OF RECORD.
RECORD MAX(MIN) TEMPERATURE IS THE MEAN OF ALL DAILY MAX(MIN) TEMPERATURES DURING THE PERIOD OF RECORD.
RECORD MEAN TEMPERATURE IS THE SUM OF THE RECORD MAX AND RECORD MIN DIVIDED BY 2.
AVERAGE TEMPERATURE IS THE SUM OF THE MEAN DAILY MAX AND MIN TEMPERATURE DIVIDED BY 2.

EXCEPTIONS

RECORD MEANS ARE THROUGH THE CURRENT YEAR, BEGINNING IN 1910 FOR TEMPERATURE
1910 FOR PRECIPITATION
1947 FOR SNOWFALL

HEATING DEGREE DAYS Base 65 deg. F YAKIMA WASHINGTON

SEASON	JULY	AUG	SEP	OCT	NOV	DEC	JAN	FEB	MAR	APR	MAY	JUNE	TOTAL
1964-65	32	94	229	456	798	1167	1114	763	731	430	264	49	6127
1965-66	29	38	207	378	671	1048	1104	767	662	385	218	129	5636
1966-67	30	11	28	394	629	880	814	659	716	593	228	30	5012
1967-68	4	0	49	401	732	1031	1049	721	583	508	253	79	5410
1968-69	20	64	132	572	751	1130	1447	1016	711	495	161	43	6542
1969-70	20	39	147	542	753	966	1127	693	655	556	200	54	5752
1970-71	13	5	221	504	785	1141	932	754	813	508	158	140	5974
1971-72	52	21	282	549	790	1146	1096	941	633	578	192	55	6335
1972-73	26	11	258	486	717	1179	1153	760	610	390	230	95	5915
1973-74	13	42	128	485	816	895	1157	692	652	433	329	72	5714
1974-75	44	8	79	452	706	917	1072	889	741	540	232	80	5760
1975-76	0	37	100	477	809	931	1004	859	795	500	267	173	5952
1976-77	26	53	61	472	769	1081	1248	730	684	323	371	34	5852
1977-78	32	24	217	491	862	968	992	753	588	464	311	73	5775
1978-79	6	53	184	470	975	1155	1549	901	641	456	178	72	6640
1979-80	29	4	59	371	919	929	1364	902	654	336	190	109	5866
1980-81	24	41	98	412	731	929	773	712	530	459	255	117	5081
1981-82	40	7	163	497	707	1072	1195	783	675	569	276	72	6056
1982-83	43	30	164	486	868	1111	840	690	527	452	191	90	5492
1983-84	32	2	206	440	672	1320	1043	781	602	531	384	142	6155
1984-85	13	23	199	546	860	1389	1189	1002	709	421	219	61	6631
1985-86	0	32	286	561	1255	1416	1079	828	595	525	291	45	6913
1986-87	58	2	255	424	791	1125	1133	761	642	367	189	72	5819
1987-88	31	19	112	447	739	1108	1116	770	699	413	311	139	5904
1988-89	33	14	197	303	762	1136	963	1073	767	383	269	52	5952
1989-90	26	28	116	497	694	1035	876	827	634	342	307	111	5493
1990-91	15	21	36	558	697	1302	1149	652	749	466	306	153	6104
1991-92	5	12	63	499	795	926	926	710	515	419	159	33	5062
1992-93	9	34	184	427	799	1249	1274	1005	742	505	153	123	6504
1993-94	80	53	172	406	1002	983							

COOLING DEGREE DAYS Base 65 deg. F YAKIMA WASHINGTON

YEAR	JAN	FEB	MAR	APR	MAY	JUNE	JULY	AUG	SEP	OCT	NOV	DEC	TOTAL
1969	0	0	0	0	8	166	132	81	37	0	0	0	424
1970	0	0	0	0	6	174	252	168	3	0	0	0	603
1971	0	0	0	0	34	16	257	256	1	0	0	0	564
1972	0	0	0	0	42	80	187	221	19	0	0	0	549
1973	0	0	0	0	34	84	245	181	41	0	0	0	585
1974	0	0	0	0	3	174	153	192	24	0	0	0	546
1975	0	0	0	0	10	52	314	86	22	0	0	0	484
1976	0	0	0	0	0	27	156	103	50	0	0	0	336
1977	0	0	0	5	0	153	158	319	4	0	0	0	639
1978	0	0	0	0	0	73	198	136	5	0	0	0	412
1979	0	0	0	0	12	91	255	174	37	2	0	0	571
1980	0	0	0	2	8	29	198	92	34	2	0	0	365
1981	0	0	0	6	9	28	169	263	66	0	0	0	541
1982	0	0	0	0	6	132	165	131	28	0	0	0	462
1983	0	0	0	0	72	35	119	191	6	0	0	0	423
1984	0	0	0	0	3	33	192	150	25	0	0	0	403
1985	0	0	0	0	32	57	312	71	2	0	0	0	474
1986	0	0	0	0	69	122	87	245	20	0	0	0	543
1987	0	0	0	14	29	110	128	133	56	0	0	0	470
1988	0	0	0	2	11	59	168	105	39	1	0	0	385
1989	0	0	0	0	4	82	146	95	4	0	0	0	331
1990	0	0	0	1	1	79	277	186	53	0	0	0	597
1991	0	0	0	0	1	13	184	215	32	0	0	0	445
1992	0	0	0	0	31	189	156	199	14	5	0	0	594
1993	0	0	0	0	49	42	26	111	54	0	0	0	282

SNOWFALL (inches) YAKIMA WASHINGTON

SEASON	JULY	AUG	SEP	OCT	NOV	DEC	JAN	FEB	MAR	APR	MAY	JUNE	TOTAL
1964-65	0.0	0.0	0.0	0.0	1.6	37.5	12.1	0.1	1.0	0.0	T	0.0	52.3
1965-66	0.0	0.0	0.0	0.0	1.4	14.0	14.6	0.8	1.1	0.0	0.0	0.0	31.9
1966-67	0.0	0.0	0.0	0.0	T	5.4	1.0	0.0	T	T	0.0	0.0	6.4
1967-68	0.0	0.0	0.0	0.0	2.0	4.7	11.1	0.4	0.0	T	0.0	0.0	18.2
1968-69	0.0	0.0	0.0	0.0	T	15.3	20.2	5.5	T	0.0	0.0	0.0	41.0
1969-70	0.0	0.0	0.0	0.0	T	9.0	15.0	0.5	T	0.0	0.0	0.0	24.5
1970-71	0.0	0.0	0.0	0.0	2.9	14.3	10.8	T	10.8	0.0	0.0	0.0	38.8
1971-72	0.0	0.0	0.0	1.1	0.0	14.7	10.1	3.6	5.5	T	0.0	0.0	35.0
1972-73	0.0	0.0	0.0	0.0	T	4.8	8.7	2.7	T	0.0	0.0	0.0	16.2
1973-74	0.0	0.0	0.0	2.4	7.5	18.6	6.1	0.0	0.6	0.0	T	0.0	35.2
1974-75	0.0	0.0	0.0	0.0	0.0	1.6	16.4	10.8	2.3	T	0.0	0.0	31.1
1975-76	0.0	0.0	0.0	0.0	2.4	4.7	5.4	5.2	T	0.0	0.0	0.0	17.7
1976-77	0.0	0.0	0.0	0.0	T	0.7	3.2	T	T	0.0	0.0	0.0	3.9
1977-78	0.0	0.0	0.0	0.0	3.6	6.6	3.0	2.8	0.9	T	T	0.0	16.9
1978-79	0.0	0.0	0.0	0.0	5.8	0.8	13.1	1.9	T	0.0	0.0	0.0	21.6
1979-80	0.0	0.0	0.0	0.0	9.3	11.4	17.6	7.6	1.7	T	0.0	0.0	47.6
1980-81	0.0	0.0	0.0	0.0	0.4	7.5	1.0	3.1	0.0	0.0	0.0	0.0	12.0
1981-82	0.0	0.0	0.0	0.0	0.0	20.6	5.6	0.1	1.9	T	0.0	0.0	28.2
1982-83	0.0	0.0	0.0	0.0	0.6	13.2	5.0	3.1	0.0	T	0.0	0.0	21.9
1983-84	0.0	0.0	0.0	0.0	3.7	16.3	1.2	T	0.0	0.0	T	0.0	21.2
1984-85	0.0	0.0	0.0	T	11.5	7.2	0.2	8.6	0.8	0.0	0.0	0.0	28.3
1985-86	0.0	0.0	0.0	T	10.0	11.5	9.8	7.3	0.0	0.0	T	0.0	38.6
1986-87	0.0	0.0	0.0	0.0	0.3	9.6	7.8	T	0.0	0.0	0.0	0.0	17.7
1987-88	0.0	0.0	0.0	0.0	2.0	10.6	9.2	T	0.0	0.0	0.0	0.0	21.8
1988-89	0.0	0.0	0.0	0.0	T	7.2	0.4	11.4	5.0	0.0	0.0	0.0	24.0
1989-90	0.0	0.0	0.0	0.0	T	0.2	0.4	1.9	T	0.0	0.0	0.0	2.5
1990-91	0.0	0.0	0.0	0.0	0.0	1.0	1.4	T	0.6	T	0.0	0.0	3.0
1991-92	0.0	0.0	0.0	2.9	T	0.6	0.3	0.9	0.0	T	0.0	0.0	4.7
1992-93	0.0	0.0	0.0	0.0	4.2	30.4	11.6	8.3	5.1	0.2	0.0	0.0	59.8
1993-94	0.0	0.0	0.0	T	1.7	3.0							
Record Mean	0.0	0.0	0.0	0.1	2.2	8.3	8.4	3.3	1.5	T	T	0.0	24.0

BECKLEY, WEST VIRGINIA

The city of Beckley is located in the Appalachian Mountains about 30 miles northwest of the high ridges through eastern West Virginia. The Raleigh County Memorial Airport is on a plateau about 2.5 miles east of the city. Beckley is almost surrounded by distant peaks. The entire area is on a broad plateau composed of rough, hilly ground and lush valleys. The generalized 2,000-foot contour line is about 25 miles to the northwest and runs from southwest to northeast. The generalized 3,000-foot contour is about 10 miles to the southeast and follows the general configuration of the Appalachians.

One of the important results of this location is that only the air reaching Beckley from the northwest has had a trajectory which is primarily upslope. When a northwesterly circulation persists following the passage of a cold front, clearing often is delayed because of the upslope conditions causing clouds. The downslope effect is noticeable with an easterly circulation. When this occurs, the weather often remains clear over Beckley even though the low stratus to the east may obscure the ridges and be solid to the east of these ridges. To a lesser degree, the upslope effect from the northwest is largely responsible for the formation of heavy fogs at the Beckley Airport. Likewise, downslope motion resists formation of fog when the flow is from an easterly direction. The exception to this is a light southeasterly flow which frequently produces fog that is advected over the airport after forming over the large lake in New River Valley. Smoke is no problem due to the lack of any source region.

Beckley has a climate characterized by sharp temperature contrasts, both seasonal and day to day. The months of May through September are generally warm, those of November through March moderately cold, with April and October months of fairly rapid transition. Cold waves occur on an average of two or three times during the winter, but severe cold spells are seldom of more than two or three days duration. Below-zero temperatures, as well as temperatures in the 70s have been recorded during winter months. A low of -13 degrees may be expected once every 10 years and -16 degrees once every 25 years. Summer highs near 90 degrees have occurred, contrasting with lows in the 30s during the same months. Highs seldom reach above the mid-80s. Cool nights are common throughout the summer, with lowest temperatures usually ranging from the 50s to the low 60s.

Ample precipitation is well distributed throughout the year. July has the highest monthly average while October has the lowest average. Summer rainfall occurs mostly during thunderstorms or showery precipitation, while the heaviest winter precipitation usually is associated with storms originating to the southwest and moving northeastward over the Ohio Valley. The formation of storms in the eastern Gulf of Mexico which move up the east coast will sometimes bring heavy snow to the Beckley area. Snowfall occurs chiefly from November through March and occasionally in October and April. The average seasonal snowfall is greater than snowfall at stations to the west at lower elevations, and considerably less than the totals for stations at higher elevations to the east and northeast.

NORMALS, MEANS, AND EXTREMES

BECKLEY, WEST VIRGINIA

LATITUDE: 37°47'N LONGITUDE: 81°07' W ELEVATION: FT. GRND 2504 BARO 2512 TIME ZONE: EASTERN WBAN: 03872

	(a)	JAN	FEB	MAR	APR	MAY	JUNE	JULY	AUG	SEP	OCT	NOV	DEC	YEAR
TEMPERATURE °F:														
Normals														
-Daily Maximum		37.9	41.2	51.8	61.8	70.1	76.4	79.2	78.4	72.4	62.6	52.3	42.4	60.5
-Daily Minimum		20.4	22.9	32.1	40.4	48.9	56.1	60.1	59.1	53.0	42.1	34.4	25.6	41.3
-Monthly		29.2	32.1	42.0	51.2	59.5	66.2	69.7	68.8	62.8	52.4	43.4	34.0	50.9
Extremes														
-Record Highest	30	69	74	81	86	87	90	94	96	89	81	78	73	96
-Year		1985	1977	1989	1976	1991	1964	1988	1988	1973	1989	1993	1971	AUG 1988
-Record Lowest	30	-22	-10	-5	11	23	32	41	36	30	18	4	-18	-22
-Year		1985	1970	1980	1985	1966	1972	1988	1986	1983	1976	1970	1989	JAN 1985
NORMAL DEGREE DAYS:														
Heating (base 65°F)		1110	921	713	414	198	47	10	19	117	400	648	961	5558
Cooling (base 65°F)		0	0	0	0	28	83	156	136	51	9	0	0	463
% OF POSSIBLE SUNSHINE														
MEAN SKY COVER (tenths)														
Sunrise - Sunset	30	7.7	7.6	7.5	7.1	7.1	7.1	7.2	7.0	6.7	6.1	7.2	7.7	7.2
MEAN NUMBER OF DAYS:														
Sunrise to Sunset														
-Clear	30	4.8	4.5	5.0	5.1	4.9	3.5	3.3	3.9	6.2	8.5	5.3	4.5	59.6
-Partly Cloudy	30	5.3	5.5	5.7	7.3	8.7	10.5	11.8	11.3	8.7	8.0	6.9	5.8	95.4
-Cloudy	30	21.0	18.3	20.3	17.7	17.4	16.0	15.9	15.7	15.1	14.5	17.8	20.7	210.3
Precipitation														
.01 inches or more	30	16.1	14.7	15.3	14.6	13.8	12.1	13.2	11.3	11.1	10.2	12.6	15.0	160.1
Snow, Ice pellets, hail														
1.0 inches or more	30	5.7	5.1	2.9	0.6	0.*	0.0	0.0	0.0	0.0	0.2	1.1	3.6	19.2
Thunderstorms	30	0.3	0.7	2.0	3.7	6.2	8.3	10.0	7.5	3.5	0.8	0.5	0.2	43.8
Heavy Fog Visibility 1/4 mile or less	30	4.2	3.6	3.7	2.5	3.3	4.0	4.9	6.0	5.9	3.8	3.2	3.4	48.6
Temperature °F														
-Maximum														
90° and above	30	0.0	0.0	0.0	0.0	0.0	0.*	0.5	0.5	0.0	0.0	0.0	0.0	1.1
32° and below	30	10.0	7.0	2.3	0.2	0.0	0.0	0.0	0.0	0.0	0.*	1.8	6.0	27.4
-Minimum														
32° and below	30	25.3	21.8	16.5	6.8	0.9	0.*	0.0	0.0	0.1	5.4	13.8	22.0	112.7
0° and below	30	2.2	1.1	0.1	0.0	0.0	0.0	0.0	0.0	0.0	0.0	0.0	0.6	4.0
AVG. STATION PRESS. (mb)	21	928.4	928.2	927.4	927.5	928.7	930.1	931.4	932.2	932.1	931.6	930.4	929.5	929.8
RELATIVE HUMIDITY (%)														
Hour 01	30	77	75	73	70	79	87	89	90	90	82	76	77	80
Hour 07 (Local Time)	30	80	79	78	75	81	86	89	92	92	86	80	79	83
Hour 13	30	67	64	58	52	55	59	62	62	63	57	60	66	60
Hour 19	30	71	68	62	55	61	68	72	75	79	70	68	73	69
PRECIPITATION (inches):														
Water Equivalent														
-Normal		2.92	2.94	3.40	3.43	3.98	3.84	4.70	3.38	3.33	2.89	2.99	3.23	41.03
-Maximum Monthly	30	6.36	6.01	9.18	7.63	7.13	7.05	9.61	5.93	8.27	5.91	6.33	6.43	9.61
-Year		1974	1972	1975	1987	1985	1969	1982	1977	1964	1990	1985	1991	JUL 1982
-Minimum Monthly	30	0.53	0.94	1.74	0.28	1.03	1.48	1.65	1.74	0.54	0.14	0.96	0.57	0.14
-Year		1983	1968	1966	1976	1977	1978	1993	1976	1985	1963	1965	1965	OCT 1963
-Maximum in 24 hrs	30	2.27	2.04	2.22	3.77	2.50	2.47	2.69	2.55	5.34	3.31	2.54	2.22	5.34
-Year		1974	1984	1967	1977	1985	1983	1984	1980	1964	1989	1985	1991	SEP 1964
Snow, Ice pellets, hail														
-Maximum Monthly	30	35.5	30.8	35.5	18.6	1.3	T	0.0	0.0	T	6.5	13.1	26.5	35.5
-Year		1977	1964	1993	1987	1989	1993			1967	1993	1974	1993	MAR 1993
-Maximum in 24 hrs	30	9.5	14.4	28.2	13.1	1.3	T	0.0	0.0	T	6.5	7.0	13.8	28.2
-Year		1971	1983	1993	1987	1989	1993			1967	1993	1968	1967	MAR 1993
WIND:														
Mean Speed (mph)	30	10.3	10.3	10.8	10.3	8.7	7.5	6.8	6.5	7.2	8.3	9.6	10.2	8.9
Prevailing Direction														
Fastest Obs. 1 Min.														
-Direction (!!)	30	24	26	27	27	24	27	36	32	30	30	16	28	27
-Speed (MPH)	30	46	40	58	44	41	40	46	40	46	30	44	41	58
-Year		1965	1965	1977	1970	1978	1973	1980	1965	1967	1981	1967	1964	MAR 1977
Peak Gust														
-Direction (!!)	10	SW	SE	W	W	NW	W	SW	W	SE	SE	SW	SE	SE
-Speed (mph)	10	51	62	49	49	56	53	48	46	48	40	60	53	62
-Date		1992	1984	1988	1989	1993	1985	1988	1991	1989	1986	1988	1986	FEB 1984

PRECIPITATION (inches)

BECKLEY, WEST VIRGINIA

YEAR	JAN	FEB	MAR	APR	MAY	JUNE	JULY	AUG	SEP	OCT	NOV	DEC	ANNUAL
1964	3.44	3.83	4.75	3.46	1.31	1.98	2.98	2.38	8.27	3.39	2.35	3.41	41.55
1965	5.31	2.11	4.70	3.98	1.71	3.13	3.03	2.26	2.19	2.67	0.96	0.57	32.62
1966	3.54	3.69	1.74	4.60	1.45	1.58	3.66	3.99	8.08	2.45	3.39	4.08	42.25
1967	1.40	2.85	6.50	2.84	6.25	2.25	7.62	3.81	2.83	2.42	3.91	4.96	47.64
1968	2.64	0.94	2.93	3.36	4.62	3.08	3.17	3.50	2.20	2.63	3.17	2.24	34.48
1969	1.91	1.61	2.48	2.55	2.81	7.05	5.14	4.66	2.19	1.18	2.45	6.14	40.17
1970	2.11	2.99	2.55	3.62	1.36	3.58	3.34	5.06	3.03	2.50	1.93	3.27	35.34
1971	3.74	3.67	2.87	2.97	6.24	4.49	4.23	1.74	4.51	3.01	2.03	1.91	41.41
1972	5.45	6.01	1.86	4.62	4.73	5.46	5.78	3.56	3.65	2.84	3.94	4.29	52.19
1973	1.75	2.66	4.49	5.12	4.63	2.11	5.21	2.79	2.33	3.22	4.52	4.62	43.45
1974	6.36	2.03	4.91	3.36	4.67	4.42	3.93	2.76	2.89	2.90	4.07	3.45	45.75
1975	4.54	4.34	9.18	3.92	5.65	4.61	7.26	4.21	4.62	2.75	2.51	3.51	57.10
1976	3.08	2.96	3.32	0.28	3.79	2.98	3.56	1.74	5.51	5.88	1.30	2.77	37.17
1977	2.02	1.58	2.48	6.27	1.03	5.65	3.28	5.93	4.06	4.09	2.94	2.11	41.44
1978	4.40	0.98	3.14	3.82	4.16	1.48	3.10	4.66	1.49	0.97	2.42	5.88	36.50
1979	5.31	2.54	2.60	3.11	4.99	6.40	4.89	3.40	4.59	2.59	3.68	1.48	45.58
1980	3.35	1.78	3.71	4.85	2.38	2.69	6.80	5.54	1.60	1.81	2.15	1.06	37.72
1981	0.57	2.11	1.79	3.40	5.16	5.65	3.29	2.50	5.00	3.15	1.25	3.42	37.29
1982	2.07	3.68	3.49	1.44	6.22	3.23	9.61	2.87	1.74	2.64	3.46	1.69	42.14
1983	0.53	1.43	1.83	3.43	3.98	7.00	3.72	2.52	1.60	4.55	1.84	2.27	34.70
1984	1.50	4.07	1.85	3.21	5.27	2.87	8.53	2.65	1.76	2.18	4.13	3.41	41.43
1985	2.25	2.02	2.14	0.87	7.13	4.11	3.14	4.32	0.54	1.57	6.33	1.71	36.13
1986	1.44	2.67	2.07	2.23	4.59	2.16	6.13	2.90	3.75	3.11	4.50	2.69	38.24
1987	3.24	3.59	1.93	7.63	2.47	4.16	1.66	3.77	2.80	1.29	1.69	4.27	38.50
1988	2.11	1.79	2.50	2.70	2.98	2.31	4.56	1.90	5.89	2.10	2.55	2.22	33.61
1989	2.97	3.27	3.39	4.29	4.55	3.56	4.27	4.46	3.36	5.25	2.05	2.39	43.81
1990	2.94	4.71	2.26	3.31	3.53	3.09	3.61	3.50	1.94	5.91	1.66	5.26	41.72
1991	4.20	2.46	6.57	2.54	2.16	1.64	3.16	3.63	2.55	0.82	4.00	6.43	40.16
1992	1.94	2.35	4.39	4.06	3.35	3.70	5.28	3.97	1.60	1.47	1.92	3.94	37.97
1993	1.56	2.84	6.37	2.54	4.10	3.15	1.65	3.50	4.06	3.29	2.58	4.49	40.13
Record Mean	3.03	3.06	3.78	3.45	3.83	3.88	4.59	3.67	3.36	2.67	2.90	3.26	41.47

AVERAGE TEMPERATURE (deg. F)

BECKLEY, WEST VIRGINIA

YEAR	JAN	FEB	MAR	APR	MAY	JUNE	JULY	AUG	SEP	OCT	NOV	DEC	ANNUAL
1964	33.1	28.1	41.4	53.7	61.1	67.6	68.9	67.3	63.4	48.4	46.4	36.8	51.4
1965	31.2	32.8	36.0	52.2	64.6	65.0	68.7	68.0	65.9	49.7	44.0	37.1	51.3
1966	24.6	30.9	42.8	48.6	59.1	66.5	71.3	67.3	59.7	49.7	44.0	32.7	49.8
1967	35.5	28.6	44.8	53.8	55.2	66.6	66.5	65.4	57.3	51.9	38.1	36.9	50.0
1968	28.3	22.9	44.3	52.1	56.9	66.0	69.6	69.6	61.2	52.1	42.8	29.9	49.7
1969	28.6	30.7	33.8	53.2	61.3	67.7	71.5	67.8	60.8	52.2	38.0	28.1	49.5
1970	24.0	31.3	39.1	53.7	61.6	66.4	69.5	69.0	67.3	55.4	43.2	36.1	51.4
1971	27.7	33.3	36.0	49.4	56.2	69.1	68.3	67.5	66.0	58.3	41.2	42.5	51.3
1972	35.3	30.7	39.7	49.9	58.3	61.3	69.0	68.7	63.8	49.4	40.3	39.8	50.5
1973	31.3	31.4	48.5	49.1	56.8	68.8	70.2	69.9	65.6	55.3	44.3	34.1	52.1
1974	41.4	34.1	45.7	53.1	59.5	62.6	68.2	68.2	59.3	50.4	42.8	34.4	51.6
1975	34.3	36.5	37.3	47.7	62.6	66.6	69.5	71.9	61.1	55.5	47.1	35.3	52.1
1976	27.8	42.2	47.7	52.0	57.2	66.6	68.1	66.0	59.9	46.4	34.3	29.5	49.8
1977	16.5	29.9	46.7	54.8	63.0	63.5	71.9	69.8	65.1	49.3	46.1	32.6	50.7
1978	22.2	21.6	38.9	52.8	58.7	66.1	69.1	70.3	66.0	51.5	47.4	36.9	50.1
1979	25.6	26.6	45.5	51.8	60.1	64.0	67.7	68.6	62.2	50.5	44.4	36.2	50.3
1980	30.8	25.8	38.6	50.5	59.5	64.2	72.2	71.4	65.6	49.9	39.9	33.5	50.1
1981	24.2	34.4	36.2	55.6	56.6	68.6	70.6	67.3	60.7	50.1	42.4	29.9	49.7
1982	25.8	34.3	44.2	48.5	64.4	64.6	70.2	66.6	61.2	53.4	45.3	41.0	51.6
1983	30.2	34.2	42.8	48.6	57.8	66.0	70.3	70.9	63.0	51.8	41.5	28.5	50.5
1984	28.1	37.9	38.1	49.0	57.0	69.7	67.5	68.7	60.1	61.1	41.5	45.1	52.0
1985	24.2	31.6	43.2	54.8	60.5	65.1	69.5	67.5	63.0	57.2	52.2	28.8	51.5
1986	29.3	35.8	42.3	54.4	60.0	68.1	72.3	67.7	64.7	54.2	44.8	34.5	52.3
1987	29.6	33.4	42.6	47.6	64.3	68.8	72.7	72.8	64.0	48.4	46.4	37.6	52.4
1988	27.8	32.3	44.0	50.7	59.6	66.5	72.7	73.4	63.0	46.1	44.5	34.0	51.2
1989	38.6	32.9	45.8	50.3	56.1	68.3	71.5	69.1	63.2	54.1	43.1	22.9	51.3
1990	38.6	41.7	46.3	51.1	59.5	67.2	70.6	69.9	63.6	54.7	47.6	41.0	54.3
1991	33.9	37.5	44.3	55.6	66.6	69.0	73.4	70.4	63.9	55.7	42.9	38.6	54.3
1992	34.7	39.5	41.9	51.8	58.2	66.1	72.2	67.8	64.4	52.3	43.7	34.6	52.3
1993	36.8	31.5	37.9	51.1	60.4	67.2	74.5	71.5	64.1	52.6	43.9	32.8	52.0
Record Mean	30.4	33.3	41.3	51.5	59.8	66.2	70.0	68.8	62.7	52.2	42.8	34.0	51.0
Max	39.6	43.1	51.6	63.1	71.3	77.1	80.3	79.3	73.5	63.4	52.6	43.0	61.5
Min	21.3	23.6	31.0	39.9	48.3	55.2	59.6	58.3	51.9	40.9	32.9	25.0	40.6

REFERENCE NOTES

BECKLEY, WEST VIRGINIA

GENERAL
T - TRACE AMOUNT.
BLANK ENTRIES DENOTE MISSING/UNREPORTED DATA.
# INDICATES A STATION OR INSTRUMENT RELOCATION.

SPECIFIC

(a) - LENGTH OF RECORD IN YEARS, ALTHOUGH INDIVIDUAL MONTHS MAY BE MISSING.
0.* OR * - THE VALUE IS BETWEEN 0.0 AND 0.05
NORMALS - BASED ON THE 1961-1990 RECORD PERIOD.
EXTREMES - DATES ARE THE MOST RECENT OCCURRENCE
WIND DIR.- NUMERALS SHOW TENS OF DEGRESS CLOCKWISE FROM TRUE NORTH. "00" INDICATES CALM.
RESULTANT DIRECTIONS ARE GIVEN TO WHOLE DEGREES.
BOLD VALUES INDICATE EXTREME VALUES WHICH OCCURRED AFTER THE ASOS SYSTEM WAS COMMISSIONED.

RECORD = PERIOD OF RECORD
RECORD MEAN PRECIPITATION IS THE MEAN OF ALL DAILY PRECIPITATION AMOUNTS DURING THE PERIOD OF RECORD.
RECORD MAX(MIN) TEMPERATURE IS THE MEAN OF ALL DAILY MAX(MIN) TEMPERATURES DURING THE PERIOD OF RECORD.
RECORD MEAN TEMPERATURE IS THE SUM OF THE RECORD MAX AND RECORD MIN DIVIDED BY 2.
AVERAGE TEMPERATURE IS THE SUM OF THE MEAN DAILY MAX AND MIN TEMPERATURE DIVIDED BY 2.

EXCEPTIONS

RECORD MEANS ARE THROUGH THE CURRENT YEAR, BEGINNING IN 1951 FOR TEMPERATURE
1951 FOR PRECIPITATION
1964 FOR SNOWFALL

HEATING DEGREE DAYS Base 65 deg. F BECKLEY, WEST VIRGINIA

SEASON	JULY	AUG	SEP	OCT	NOV	DEC	JAN	FEB	MAR	APR	MAY	JUNE	TOTAL
1964-65	8	48	93	509	550	868	1037	895	891	379	60	70	5408
1965-66	7	40	77	468	621	855	1243	946	680	490	199	53	5679
1966-67	5	27	172	469	625	993	905	1012	620	335	306	43	5512
1967-68	37	49	228	400	802	865	1133	1215	634	381	249	58	6051
1968-69	10	53	124	392	661	1078	1124	955	961	351	140	50	5899
1969-70	0	4	166	400	804	1139	1269	938	798	351	150	38	6057
1970-71	17	12	58	299	645	890	1149	883	890	461	280	4	5588
1971-72	6	10	46	212	714	692	916	987	775	446	204	139	5147
1972-73	35	10	74	475	737	776	1036	936	506	477	258	11	5331
1973-74	5	13	55	295	613	951	724	859	594	362	198	113	4782
1974-75	5	9	188	447	660	943	946	794	849	514	116	40	5511
1975-76	0	2	146	289	530	914	1148	655	533	421	241	37	4916
1976-77	24	49	151	566	914	1094	1496	978	565	310	113	97	6357
1977-78	8	11	67	479	564	997	1319	1211	801	366	216	56	6095
1978-79	14	3	58	411	521	864	1213	1070	597	389	175	69	5384
1979-80	28	37	116	448	611	890	1053	1130	809	435	186	81	5824
1980-81	0	1	69	463	746	970	1257	849	887	287	264	15	5808
1981-82	6	18	164	453	670	1084	1211	856	636	487	71	43	5699
1982-83	9	23	144	363	585	737	1071	856	680	490	230	46	5234
1983-84	14	12	138	400	699	1126	1138	781	827	483	266	18	5902
1984-85	24	14	189	137	698	608	1259	929	672	324	151	56	5061
1985-86	4	19	128	247	378	1116	1099	812	696	332	173	26	5030
1986-87	2	46	75	347	598	939	1089	878	686	518	92	19	5289
1987-88	5	8	82	507	552	842	1148	943	644	422	182	84	5419
1988-89	14	5	85	580	606	953	812	894	597	441	298	16	5301
1989-90	1	22	118	334	652	1299	813	648	574	427	190	29	5107
1990-91	2	9	122	320	515	736	957	765	636	294	70	23	4449
1991-92	0	4	117	293	655	812	936	732	710	395	229	49	4932
1992-93	1	17	98	390	630	937	865	932	833	414	147	41	5305
1993-94	0	4	97	378	631	993							

COOLING DEGREE DAYS Base 65 deg. F BECKLEY, WEST VIRGINIA

YEAR	JAN	FEB	MAR	APR	MAY	JUNE	JULY	AUG	SEP	OCT	NOV	DEC	TOTAL
1969	0	0	0	4	32	139	208	100	45	11	0	0	539
1970	0	0	0	16	53	88	165	142	131	11	0	0	606
1971	0	0	0	0	14	132	119	91	86	11	5	0	458
1972	0	0	0	2	4	34	165	132	42	0	0	0	379
1973	0	0	1	6	9	130	172	173	81	2	0	0	574
1974	0	0	1	12	35	51	112	112	22	0	0	0	345
1975	0	0	0	1	51	92	145	221	37	1	0	0	548
1976	0	0	2	38	7	90	127	86	6	0	0	0	356
1977	0	0	1	11	58	60	229	169	77	0	2	0	607
1978	0	0	0	5	33	96	147	175	96	0	0	0	552
1979	0	0	0	1	31	49	121	157	39	6	0	0	404
1980	0	0	0	4	24	66	235	207	94	2	0	0	632
1981	0	0	0	11	8	130	187	99	40	0	0	0	475
1982	0	0	0	0	61	37	176	83	37	12	0	0	406
1983	0	0	0	4	16	84	189	201	86	0	0	0	580
1984	0	0	0	8	26	166	107	134	52	24	0	0	517
1985	0	0	5	25	20	68	150	100	75	9	0	0	452
1986	0	0	0	21	27	124	235	136	74	20	0	0	637
1987	0	0	0	2	78	143	250	257	59	0	0	0	789
1988	0	0	0	1	22	135	258	268	37	1	0	0	722
1989	0	0	10	5	29	121	208	155	71	3	0	0	602
1990	0	0	1	16	25	100	183	169	83	7	0	0	584
1991	0	0	0	17	126	150	266	178	94	13	1	0	845
1992	0	0	0	7	23	85	230	111	86	2	0	0	544
1993	0	0	0	1	13	113	304	213	79	4	3	0	730

SNOWFALL (inches) BECKLEY, WEST VIRGINIA

SEASON	JULY	AUG	SEP	OCT	NOV	DEC	JAN	FEB	MAR	APR	MAY	JUNE	TOTAL
1964-65	0.0	0.0	0.0	0.0	3.3	1.4	21.3	11.0	10.5	T	0.0	0.0	47.5
1965-66	0.0	0.0	0.0	0.2	0.6	6.9	25.2	15.9	5.9	6.6	T	0.0	61.3
1966-67	0.0	0.0	0.0	T	4.0	18.1	9.2	26.6	6.3	T	T	0.0	64.2
1967-68	0.0	0.0	T	T	0.9	19.3	21.2	11.8	7.2	0.0	0.0	0.0	60.4
1968-69	0.0	0.0	0.0	T	11.0	12.0	4.0	10.9	15.4	T	T	0.0	53.3
1969-70	0.0	0.0	0.0	0.0	6.5	20.7	18.9	12.6	4.0	0.2	0.0	0.0	62.9
1970-71	0.0	0.0	0.0	0.0	4.5	10.3	14.0	13.4	19.8	8.2	T	0.0	70.2
1971-72	0.0	0.0	0.0	0.0	7.6	0.9	2.4	22.1	3.4	1.9	0.0	0.0	38.3
1972-73	0.0	0.0	0.0	0.2	3.6	2.3	5.5	10.9	5.0	1.4	0.0	0.0	28.9
1973-74	0.0	0.0	0.0	3.2	0.7	19.2	1.4	10.4	2.5	1.2	0.0	0.0	38.6
1974-75	0.0	0.0	0.0	1.4	13.1	16.5	11.7	7.8	17.3	1.0	0.0	0.0	68.8
1975-76	0.0	0.0	0.0	0.0	3.1	5.0	16.4	7.3	3.8	T	T	0.0	35.6
1976-77	0.0	0.0	0.0	T	12.7	19.3	35.5	15.4	3.5	2.2	0.0	0.0	88.6
1977-78	0.0	0.0	0.0	T	5.4	7.9	31.6	20.3	15.1	0.2	0.0	0.0	80.5
1978-79	0.0	0.0	0.0	0.0	2.4	3.8	27.7	29.5	5.4	T	0.0	0.0	68.8
1979-80	0.0	0.0	0.0	2.7	1.4	4.7	21.1	26.1	12.8	1.0	0.0	0.0	69.8
1980-81	0.0	0.0	0.0	T	1.4	4.4	21.4	8.4	20.4	T	0.0	0.0	56.0
1981-82	0.0	0.0	0.0	0.5	3.6	22.0	17.9	18.2	12.8	4.0	0.0	0.0	79.0
1982-83	0.0	0.0	0.0	0.0	0.3	14.8	10.9	25.6	6.8	0.7	0.0	0.0	59.1
1983-84	0.0	0.0	0.0	0.0	3.8	8.5	22.3	22.9	9.8	0.3	0.0	0.0	67.6
1984-85	0.0	0.0	0.0	0.0	0.5	4.8	34.1	18.5	2.2	6.9	0.0	0.0	67.0
1985-86	0.0	0.0	0.0	0.0	T	14.4	24.4	18.7	9.4	1.7	0.0	0.0	68.6
1986-87	0.0	0.0	0.0	0.0	1.1	0.3	29.4	21.9	1.3	18.6	0.0	0.0	72.6
1987-88	0.0	0.0	0.0	T	7.3	15.9	16.8	9.2	7.5	0.4	0.0	0.0	57.1
1988-89	0.0	0.0	0.0	T	0.2	12.8	9.5	10.9	0.4	5.7	1.3	0.0	40.8
1989-90	0.0	0.0	0.0	1.2	7.4	24.6	15.0	15.3	4.4	6.1	T	0.0	74.0
1990-91	0.0	0.0	0.0	T	T	4.6	4.7	9.9	3.7	T	0.0	0.0	22.9
1991-92	0.0	0.0	0.0	0.0	1.4	4.5	9.5	5.2	10.7	11.3	T	0.0	42.6
1992-93	0.0	0.0	0.0	0.2	2.5	14.8	0.4	20.1	35.5	1.0	T	T	74.5
1993-94	0.0	0.0	0.0	6.5	0.7	26.5							
Record Mean	0.0	0.0	T	0.5	3.7	11.4	16.5	16.3	9.0	2.7	T	T	60.1

CHARLESTON, WEST VIRGINIA

Charleston lies at the junction of the Kanawha and Elk Rivers in the western foothills of the Appalachian Mountains. The main urban and business areas have developed along the two river valleys, while some residential areas are in nearby valleys and on the surrounding hills. The hilltops are around 1,100 feet above sea level, about 500 feet higher than the valleys. The Kanawha Airport is just over 2 miles northeast of the center-city area, on an artificial plateau constructed from several hilltops.

Weather records are maintained at the Kanawha Airport by National Weather Service personnel. This site tends to be slightly cooler than the river valleys during the afternoons. Conversely, the valleys can become cooler than the hilltops during clear, calm nights. The weather at Charleston is highly changeable, especially from mid-autumn through the spring.

Winters can vary greatly from one season to the next. Snow does not favor any given winter month, heavy snowstorms are infrequent, and most snowfalls are in the 4-inch or less category. Snow and ice usually do not persist on valley roads, but can linger longer on nearby hills and outlying rural roads.

Afternoon temperatures in the 40s and morning readings in the 20s are common during the winter. Yet, every winter typically has two or three extended cold spells when temperatures stay below freezing for a few consecutive days. Northwesterly winds are associated with the cold weather. Air reaching Charleston from the northwest can cause cloudiness and flurries, even when there is no nearby organized storm system. Winter conditions are much more severe over the higher mountains less than 50 miles to the northeast through the southeast. Temperatures warm rapidly in the spring and are accompanied by low daytime humidities.

Summer and early autumn have more day-to-day consistency in the weather. Sunshine is more abundant than in winter. Summer precipitation falls mostly in brief, but sometimes heavy, showers. Flash flooding can occur along small streams, but flooding is rare on the dam-controlled Kanawha and Elk Rivers.

Afternoon summer temperatures are mostly in the 80s. Readings above 95 degrees are rare. However, during a hot spell, haze and humidity can add to the unpleasantness and indoor air conditioning is recommended. Cooler and less humid air often penetrates the area from the north to end a hot spell.

Early morning fog is common from late June into October. Industrial and vehicular pollutants can contribute to limited visibility any time of the year, especially when cooler air becomes trapped in the valleys. Autumn foliage is generally at its peak during the second and third weeks of October. By the end of October, the first 32 degree temperature has usually arrived.

Ample precipitation is well distributed throughout the year. July is quite often the wettest month of the year, while October averages the least rain. Droughts severe enough to limit water use are scarce. Any dry spells during the spring or autumn can cause conditions favorable for brush fires in outlying areas.

NORMALS, MEANS, AND EXTREMES

CHARLESTON, WEST VIRGINIA

LATITUDE: 38°22'N LONGITUDE: 81°36' W ELEVATION: FT. GRND 1016 BARO 1019 TIME ZONE: EASTERN WBAN: 13866

	(a)	JAN	FEB	MAR	APR	MAY	JUNE	JULY	AUG	SEP	OCT	NOV	DEC	YEAR
TEMPERATURE °F:														
Normals														
-Daily Maximum		41.2	45.3	56.7	66.8	75.5	83.1	85.7	84.4	78.8	68.2	57.3	46.0	65.8
-Daily Minimum		23.0	25.7	35.0	42.8	51.5	59.8	64.4	63.4	56.5	44.2	36.3	28.0	44.2
-Monthly		32.1	35.5	45.9	54.8	63.5	71.4	75.1	73.9	67.7	56.2	46.8	37.0	55.0
Extremes														
-Record Highest	46	79	78	89	94	93	98	104	101	102	92	85	80	104
-Year		1950	1977	1990	1990	1991	1988	1988	1988	1953	1951	1993	1982	JUL 1988
-Record Lowest	46	-15	-6	0	19	26	33	46	41	34	17	6	-12	-15
-Year		1985	1968	1980	1982	1966	1972	1963	1965	1983	1962	1950	1989	JAN 1985
NORMAL DEGREE DAYS:														
Heating (base 65°F)		1020	826	592	312	129	10	0	0	44	299	546	868	4646
Cooling (base 65°F)		0	0	0	6	83	202	313	276	125	26	0	0	1031
% OF POSSIBLE SUNSHINE														
MEAN SKY COVER (tenths)														
Sunrise - Sunset	46	7.7	7.5	7.3	6.9	6.6	6.4	6.5	6.3	6.2	6.0	7.2	7.6	6.9
MEAN NUMBER OF DAYS:														
Sunrise to Sunset														
-Clear	46	3.9	4.3	4.8	5.8	6.1	4.9	4.6	5.2	6.7	8.8	5.2	4.7	65.0
-Partly Cloudy	46	6.3	5.9	7.5	7.8	9.9	12.8	13.5	14.2	11.1	9.3	7.0	6.3	111.5
-Cloudy	46	20.8	18.0	18.7	16.4	15.0	12.2	12.9	11.6	12.2	12.9	17.8	20.1	188.8
Precipitation														
.01 inches or more	46	15.4	13.7	15.0	13.9	13.2	11.3	12.8	10.8	9.6	9.5	12.0	14.0	151.2
Snow,Ice pellets,hail														
1.0 inches or more	46	3.3	2.6	1.5	0.2	0.0	0.0	0.0	0.0	0.0	0.*	0.8	1.8	10.2
Thunderstorms	46	0.5	0.8	2.3	4.1	6.7	7.7	9.7	7.1	3.2	1.1	0.7	0.3	44.2
Heavy Fog Visibility														
1/4 mile or less	46	3.9	3.0	2.8	2.9	7.7	11.8	14.7	18.3	16.1	10.6	4.5	3.7	100.1
Temperature °F														
-Maximum														
90° and above	46	0.0	0.0	0.0	0.3	1.1	4.8	8.3	5.6	2.2	0.*	0.0	0.0	22.3
32° and below	46	7.1	4.5	0.9	0.*	0.0	0.0	0.0	0.0	0.0	0.0	0.5	4.3	17.3
-Minimum														
32° and below	46	22.9	19.9	14.6	4.7	0.4	0.0	0.0	0.0	0.0	3.4	12.7	20.8	99.3
0° and below	46	0.9	0.3	0.*	0.0	0.0	0.0	0.0	0.0	0.0	0.0	0.0	0.3	1.5
AVG. STATION PRESS.(mb)	21	984.2	983.7	981.7	980.9	981.0	981.7	982.7	983.7	984.3	984.9	984.5	984.6	983.2
RELATIVE HUMIDITY (%)														
Hour 01	46	74	72	68	67	80	87	90	91	89	84	75	75	79
Hour 07 (Local Time)	46	77	77	75	75	83	87	90	92	91	88	80	78	83
Hour 13	46	63	59	53	47	51	54	59	58	56	54	56	62	56
Hour 19	46	65	61	54	49	56	61	66	69	71	66	62	66	62
PRECIPITATION (inches):														
Water Equivalent														
-Normal		2.91	3.04	3.63	3.31	3.94	3.59	4.99	4.01	3.24	2.89	3.59	3.39	42.53
-Maximum Monthly	46	9.11	6.89	6.80	6.46	6.79	7.54	13.54	10.45	7.61	6.49	8.45	8.02	13.54
-Year		1950	1956	1967	1965	1989	1989	1961	1958	1971	1983	1985	1978	JUL 1961
-Minimum Monthly	46	1.09	0.64	1.30	0.50	0.84	0.70	1.98	0.66	0.65	0.09	0.64	0.45	0.09
-Year		1981	1968	1987	1976	1977	1966	1993	1957	1959	1963	1965	1965	OCT 1963
-Maximum in 24 hrs	46	1.91	2.45	2.86	2.72	3.31	2.24	5.60	4.17	2.40	2.48	2.88	2.47	5.60
-Year		1961	1951	1967	1948	1982	1962	1961	1958	1956	1961	1991	1978	JUL 1961
Snow,Ice pellets,hail														
-Maximum Monthly	46	39.5	21.8	20.4	20.7	0.6	T	T	T	T	2.8	25.8	18.6	39.5
-Year		1978	1964	1993	1987	1989	1993	1990	1989	1989	1961	1950	1962	JAN 1978
-Maximum in 24 hrs	46	15.8	11.2	17.1	11.3	0.6	T	T	T	T	2.8	15.1	11.2	17.1
-Year		1978	1983	1993	1987	1989	1993	1990	1989	1989	1961	1950	1967	MAR 1993
WIND:														
Mean Speed (mph)	46	7.4	7.5	8.1	7.5	6.0	5.4	5.0	4.4	4.7	5.1	6.6	7.0	6.2
Prevailing Direction through 1963		WSW	WSW	WSW	SW	SW	SW	S	S	S	S	SW	SW	SW
Fastest Obs. 1 Min.														
-Direction (!!!)	44	25	19	32	27	25	32	29	29	20	25	29	25	25
-Speed (MPH)	44	45	40	46	45	55	50	46	50	35	45	40	55	55
-Year		1951	1981	1955	1953	1953	1951	1957	1952	1956	1950	1954	1953	MAY 1953
Peak Gust														
-Direction (!!!)	10	SW	S	NW	SW	W	NW	NW	SW	W	SW	W	W	SW
-Speed (mph)	10	66	56	53	66	49	49	56	47	44	38	49	62	66
-Date		1992	1986	1985	1991	1988	1993	1986	1991	1990	1990	1992	1987	JAN 1992

PRECIPITATION (inches)

CHARLESTON, WEST VIRGINIA

YEAR	JAN	FEB	MAR	APR	MAY	JUNE	JULY	AUG	SEP	OCT	NOV	DEC	ANNUAL
1964	2.58	3.58	3.65	3.20	0.95	3.82	2.76	3.23	2.53	0.59	2.95	3.14	32.98
1965	4.65	2.02	4.52	6.46	1.90	2.18	2.46	4.38	3.21	2.09	0.64	0.45	34.96
1966	3.57	2.78	1.51	5.06	1.52	0.70	2.94	3.31	3.74	1.72	3.05	2.53	32.43
1967	1.21	2.95	6.80	3.21	6.45	1.83	4.59	1.85	1.68	2.09	4.30	4.72	41.68
1968	2.01	0.64	4.79	2.58	6.59	2.83	4.02	5.42	3.32	3.16	2.47	2.38	40.21
1969	1.50	1.27	1.43	2.35	1.95	2.43	6.13	8.20	3.27	1.52	2.22	4.85	37.12
1970	1.15	3.51	4.23	3.19	1.13	2.35	3.53	4.84	3.55	5.19	2.34	3.81	38.82
1971	2.35	3.40	1.97	1.19	5.17	2.58	6.59	2.12	7.61	1.30	2.83	1.71	38.82
1972	5.47	5.51	2.17	5.16	2.55	4.33	4.13	4.13	3.61	2.48	5.26	6.35	51.15
1973	1.52	2.41	3.40	5.44	5.36	4.48	6.88	2.07	3.91	4.75	5.42	3.68	49.32
1974	4.67	2.50	4.54	3.05	6.06	5.07	2.16	4.22	2.64	1.64	3.72	3.19	43.46
1975	4.84	3.10	6.01	4.03	6.44	4.25	2.71	5.14	4.99	3.08	2.66	3.74	50.99
1976	2.89	2.11	4.21	0.50	3.66	4.24	6.93	2.23	5.37	5.44	1.02	2.18	40.78
1977	1.90	1.08	3.16	4.06	0.84	5.93	4.92	6.58	1.14	4.16	3.78	2.07	39.62
1978	5.59	1.31	2.67	3.31	3.99	2.96	9.83	8.21	1.45	2.68	2.26	8.02	52.28
#1979	6.48	3.76	3.00	3.82	3.87	3.54	5.17	4.78	3.95	3.67	4.02	2.81	48.87
1980	2.85	2.25	5.32	4.49	2.67	2.17	8.47	10.32	2.37	2.03	3.02	1.85	47.81
1981	1.09	4.59	1.80	4.04	3.78	6.46	3.02	2.24	2.36	2.43	1.29	2.71	35.81
1982	3.74	3.23	4.96	1.14	6.19	7.00	2.68	2.65	2.58	1.65	4.65	2.71	43.18
1983	1.24	2.72	3.15	3.96	5.98	2.77	4.19	2.54	1.33	6.49	4.80	3.19	42.36
1984	1.67	2.56	2.72	4.00	3.71	2.56	4.37	4.57	2.95	3.28	4.73	3.78	40.90
1985	3.07	2.32	4.23	1.84	5.88	3.07	3.22	2.02	0.71	3.65	8.45	2.71	41.17
1986	2.12	4.35	1.87	1.39	4.86	2.36	7.61	4.71	3.51	2.20	6.88	3.89	45.75
1987	3.23	3.34	1.30	4.05	2.49	3.38	4.23	3.56	3.89	1.10	2.71	4.13	37.41
1988	1.62	2.50	2.71	2.17	2.59	0.94	3.00	2.86	3.46	1.87	5.02	2.66	31.40
1989	2.92	6.05	5.81	4.13	6.79	7.54	3.04	5.62	7.28	4.09	2.87	1.83	57.97
1990	2.86	3.74	1.94	2.89	4.87	3.01	5.35	2.54	4.26	3.51	2.07	7.01	44.05
1991	2.68	2.98	6.07	3.49	1.47	2.49	2.84	2.95	5.51	1.10	5.00	5.89	42.47
1992	1.94	2.72	4.79	2.93	4.66	3.21	6.41	4.41	1.38	0.94	3.15	3.50	40.04
1993	1.87	2.98	6.68	1.78	1.98	5.01	1.98	2.71	5.99	3.50	3.95	3.23	41.66
Record Mean	3.57	3.27	4.04	3.57	3.78	3.87	4.90	4.11	3.07	2.76	3.22	3.36	43.51

AVERAGE TEMPERATURE (deg. F)

CHARLESTON, WEST VIRGINIA

YEAR	JAN	FEB	MAR	APR	MAY	JUNE	JULY	AUG	SEP	OCT	NOV	DEC	ANNUAL
1964	35.3	33.2	46.7	58.5	67.0	72.9	75.0	73.1	67.5	53.2	49.0	39.2	55.9
1965	33.3	35.6	39.6	55.1	69.2	70.0	73.2	72.5	69.3	52.6	46.1	39.8	54.7
1966	27.1	36.6	47.7	54.1	62.1	72.0	77.3	73.5	66.5	55.0	48.2	35.5	54.6
1967	38.5	31.6	48.5	56.3	58.3	71.6	71.1	70.2	62.1	55.9	41.3	38.9	53.7
1968	29.1	27.1	47.5	56.4	61.0	70.6	74.2	74.1	66.3	56.0	46.8	34.2	53.6
1969	32.7	35.2	38.7	56.2	64.1	73.7	76.6	72.1	64.1	53.9	41.3	32.3	53.4
1970	28.5	36.2	43.3	58.2	66.9	73.1	74.6	73.6	70.0	57.4	46.0	38.4	55.5
1971	29.4	35.2	39.7	51.9	59.9	73.5	72.4	71.9	70.7	63.8	45.9	47.3	55.2
1972	38.7	35.2	43.8	54.5	63.7	64.9	73.4	72.7	68.7	52.5	43.4	42.1	54.5
1973	34.3	35.1	52.2	53.8	60.8	73.3	74.5	74.6	70.1	58.9	46.8	38.0	56.0
1974	43.5	37.1	49.1	56.9	64.2	67.8	74.2	73.6	63.2	52.4	45.2	36.8	55.3
1975	35.7	38.0	39.8	50.2	66.1	71.7	74.2	76.7	63.9	57.6	50.2	38.6	55.2
1976	31.7	45.8	51.6	55.2	61.9	71.9	72.4	70.2	63.3	49.5	37.6	31.0	53.5
1977	18.6	33.2	49.8	58.4	67.0	68.6	76.9	73.6	70.2	53.4	49.1	35.1	54.5
1978	24.4	24.2	42.6	56.7	63.1	70.9	73.8	74.9	71.0	53.8	49.4	39.0	53.6
1979	28.1	27.9	50.5	55.1	63.0	68.8	72.9	73.4	67.0	54.6	47.5	38.3	54.0
1980	34.1	29.7	42.0	53.3	63.6	68.8	76.6	76.3	69.8	53.6	43.2	36.3	54.0
1981	28.0	37.2	41.4	59.1	60.4	73.2	75.7	72.8	66.5	54.2	45.3	34.6	54.0
1982	29.8	36.1	47.2	51.5	68.6	68.8	76.2	71.1	65.9	57.7	49.0	44.8	55.6
1983	34.0	37.7	47.0	52.1	61.1	71.6	77.0	78.0	68.4	58.1	47.4	32.0	55.4
1984	30.6	41.5	41.1	54.2	61.4	75.3	73.2	74.9	65.4	64.4	44.6	46.9	56.1
1985	27.2	34.0	49.4	60.8	66.3	71.0	75.8	74.0	69.6	62.3	55.5	33.8	56.6
1986	34.1	40.5	47.1	57.9	65.3	72.2	77.2	71.9	69.5	57.9	46.4	36.4	56.4
1987	33.0	37.2	47.0	52.7	68.4	73.5	77.1	77.0	67.1	50.3	49.0	39.8	56.0
1988	31.1	35.2	46.1	54.1	63.2	71.0	78.6	77.4	66.6	49.3	47.0	37.4	54.8
1989	41.1	34.9	47.8	52.8	59.4	71.7	75.7	73.2	67.0	56.7	46.4	26.0	54.4
1990	42.3	45.2	51.7	55.1	62.9	72.3	75.8	74.1	68.7	58.2	50.7	43.6	58.4
1991	36.5	40.3	47.4	60.3	71.7	73.9	77.9	75.2	68.4	59.1	45.8	41.3	58.2
1992	35.6	41.8	45.4	55.7	61.2	68.8	76.0	70.9	67.4	54.1	47.4	37.9	55.2
1993	40.2	34.0	41.8	55.0	64.9	71.4	79.0	76.4	66.6	55.1	47.1	36.6	55.7
Record Mean	35.7	37.7	46.0	55.9	64.7	72.1	76.0	74.7	69.0	57.5	46.7	38.0	56.1
Max	45.6	48.4	57.8	68.7	77.4	84.0	87.3	86.0	81.0	70.0	57.9	47.8	67.6
Min	25.7	26.9	34.3	43.1	52.0	60.1	64.6	63.4	56.9	44.9	35.5	28.2	44.6

REFERENCE NOTES

CHARLESTON, WEST VIRGINIA

GENERAL
T - TRACE AMOUNT.
BLANK ENTRIES DENOTE MISSING/UNREPORTED DATA.
# INDICATES A STATION OR INSTRUMENT RELOCATION.

SPECIFIC

PM - INCLUDES LAST DAY OF PREVIOUS MONTH
ASOS - AUTOMATED SURFACE OBSERVING SYSTEM IN OPERATION DURING THESE MONTHS.

(*) - LENGTH OF RECORD IN YEARS, ALTHOUGH INDIVIDUAL MONTHS MAY BE MISSING.
0.* OR * - THE VALUE IS BETWEEN 0.0 AND 0.05
NORMALS - BASED ON THE 1961-1990 RECORD PERIOD.
EXTREMES - DATES ARE THE MOST RECENT OCCURRENCE.
WIND DIR.- NUMERALS SHOW TENS OF DEGRESS CLOCKWISE FROM TRUE NORTH. "00" INDICATES CALM.
RESULTANT DIRECTIONS ARE GIVEN TO WHOLE DEGREES.
BOLD VALUES INDICATE EXTREME VALUES WHICH OCCURRED AFTER THE ASOS SYSTEM WAS COMMISSIONED.

RECORD = PERIOD OF RECORD
RECORD MEAN PRECIPITATION IS THE MEAN OF ALL DAILY PRECIPITATION AMOUNTS DURING THE PERIOD OF RECORD.
RECORD MAX(MIN) TEMPERATURE IS THE MEAN OF ALL DAILY MAX(MIN) TEMPERATURES DURING THE PERIOD OF RECORD.
RECORD MEAN TEMPERATURE IS THE SUM OF THE RECORD MAX AND RECORD MIN DIVIDED BY 2.
AVERAGE TEMPERATURE IS THE SUM OF THE MEAN DAILY MAX AND MIN TEMPERATURE DIVIDED BY 2.

EXCEPTIONS

RECORD MEANS ARE THROUGH THE CURRENT YEAR, BEGINNING IN 1902 FOR TEMPERATURE
1901 FOR PRECIPITATION
1948 FOR SNOWFALL

HEATING DEGREE DAYS Base 65 deg. F CHARLESTON, WEST VIRGINIA

SEASON	JULY	AUG	SEP	OCT	NOV	DEC	JAN	FEB	MAR	APR	MAY	JUNE	TOTAL
1964-65	0	13	40	364	474	791	974	820	780	299	24	15	4594
1965-66	0	15	58	382	561	776	1166	793	538	354	147	25	4815
1966-67	0	1	44	315	507	910	815	932	517	275	217	17	4550
1967-68	4	6	124	297	704	802	1104	1095	539	263	141	15	5094
1968-69	2	11	35	298	541	946	994	828	807	264	98	9	4833
1969-70	0	0	95	352	703	1007	1125	801	666	239	82	1	5071
1970-71	3	0	40	246	563	817	1097	828	778	387	184	2	4945
1971-72	0	1	13	78	578	543	809	856	649	333	90	81	4031
1972-73	16	2	14	378	642	701	945	832	394	351	157	4	4436
1973-74	0	1	19	202	541	833	659	775	500	277	115	25	3947
1974-75	0	0	110	388	590	869	899	749	772	445	59	4	4885
1975-76	0	0	106	227	441	813	1025	549	427	342	142	4	4076
1976-77	0	9	84	475	814	1047	1432	888	482	242	81	52	5606
1977-78	0	2	19	357	482	919	1249	1138	691	258	137	23	5275
1978-79	0	0	18	344	462	797	1137	1031	456	308	125	19	4697
1979-80	5	10	39	331	519	820	951	1017	707	349	106	27	4881
1980-81	0	0	33	356	650	882	1138	774	727	207	175	2	4944
1981-82	0	1	76	335	585	936	1086	801	545	405	36	2	4808
1982-83	1	2	69	268	480	626	955	757	554	388	153	16	4269
1983-84	4	0	66	227	521	1019	1059	674	734	346	171	5	4826
1984-85	1	0	98	74	613	563	1164	860	488	192	54	18	4125
1985-86	0	0	51	127	294	960	954	679	554	249	83	7	3958
1986-87	0	23	23	255	550	880	989	770	549	374	63	4	4480
1987-88	0	0	37	447	473	774	1043	859	577	326	112	38	4686
1988-89	2	0	37	484	534	849	735	837	536	367	221	2	4604
1989-90	0	7	72	270	553	1203	697	549	446	323	111	8	4239
1990-91	0	0	59	230	428	655	876	685	558	192	21	1	3705
1991-92	0	0	80	229	576	729	904	670	602	319	170	24	4303
1992-93	0	1	67	335	522	834	760	862	713	306	73	24	4497
1993-94	0	0	64	307	540	873							

COOLING DEGREE DAYS Base 65 deg. F CHARLESTON, WEST VIRGINIA

YEAR	JAN	FEB	MAR	APR	MAY	JUNE	JULY	AUG	SEP	OCT	NOV	DEC	TOTAL
1969	0	0	0	7	79	277	368	227	75	15	0	0	1048
1970	0	0	0	40	147	251	310	273	197	18	0	0	1236
1971	0	0	0	0	32	265	237	219	190	49	11	5	1008
1972	0	0	0	24	56	85	283	247	132	0	0	0	827
1973	0	0	4	22	34	256	304	305	181	22	2	0	1130
1974	0	0	14	43	99	118	292	275	62	6	1	0	910
1975	0	0	0	7	99	212	291	372	82	7	4	0	1074
1976	0	1	17	58	54	218	238	175	39	1	0	0	801
1977	0	0	18	50	148	165	373	277	180	4	12	0	1227
1978	0	0	0	16	88	207	279	314	205	4	1	0	1114
1979	0	0	13	18	69	138	257	277	105	17	0	0	894
1980	0	0	0	6	71	147	370	358	182	9	0	0	1143
1981	0	0	2	38	41	256	340	251	126	5	0	0	1059
1982	0	0	0	6	154	122	355	196	101	47	5	6	992
1983	0	0	2	6	39	222	385	407	177	18	0	0	1256
1984	0	0	0	27	64	318	261	312	116	65	7	8	1178
1985	0	0	9	72	105	204	339	285	194	52	14	0	1274
1986	0	0	4	41	100	227	384	244	167	43	0	0	1210
1987	0	0	0	13	177	268	381	379	108	0	2	0	1328
1988	0	0	3	9	64	225	430	392	91	4	3	0	1221
1989	0	0	11	6	55	211	339	273	140	23	2	0	1060
1990	0	0	41	33	54	232	342	286	174	28	7	0	1197
1991	0	0	17	60	236	273	408	324	190	53	5	0	1566
1992	0	0	2	47	57	143	347	192	145	4	0	0	937
1993	0	0	0	14	76	222	444	364	119	7	8	0	1254

SNOWFALL (inches) CHARLESTON, WEST VIRGINIA

SEASON	JULY	AUG	SEP	OCT	NOV	DEC	JAN	FEB	MAR	APR	MAY	JUNE	TOTAL
1964-65	0.0	0.0	0.0	0.0	1.3	0.9	13.7	6.7	8.5	T	0.0	0.0	31.1
1965-66	0.0	0.0	0.0	T	T	2.7	19.8	8.0	1.7	T	0.0	0.0	32.2
1966-67	0.0	0.0	0.0	T	3.8	6.0	3.6	20.6	2.2	0.0	0.0	0.0	36.2
1967-68	0.0	0.0	0.0	0.0	0.2	16.6	13.3	9.2	2.0	0.0	0.0	0.0	41.3
1968-69	0.0	0.0	0.0	T	4.3	4.7	2.0	1.3	4.8	0.0	0.0	0.0	17.1
1969-70	0.0	0.0	0.0	0.0	4.3	9.5	12.6	14.2	2.9	T	0.0	0.0	43.5
1970-71	0.0	0.0	0.0	0.0	0.2	10.9	11.6	10.5	11.7	T	0.0	0.0	44.9
1971-72	0.0	0.0	0.0	0.0	4.2	0.8	4.0	9.8	3.5	T	0.0	0.0	22.3
1972-73	0.0	0.0	0.0	0.9	6.9	1.4	3.4	2.9	4.8	0.9	0.0	0.0	21.2
1973-74	0.0	0.0	0.0	0.0	T	6.5	0.3	13.0	0.9	1.2	0.0	0.0	21.9
1974-75	0.0	0.0	0.0	0.6	2.7	8.3	18.1	2.4	8.1	0.1	0.0	0.0	40.3
1975-76	0.0	0.0	0.0	0.0	T	4.0	14.3	2.6	5.4	0.0	0.0	0.0	26.3
1976-77	0.0	0.0	0.0	T	4.7	7.4	22.2	11.1	1.0	1.5	0.0	0.0	47.9
1977-78	0.0	0.0	0.0	0.0	4.4	5.4	39.5	15.6	11.7	0.0	0.0	0.0	76.6
#1978-79	0.0	0.0	0.0	0.0	T	1.5	27.5	20.1	5.5	T	0.0	0.0	54.6
1979-80	0.0	0.0	0.0	T	0.9	0.4	11.7	12.7	10.5	T	0.0	0.0	36.2
1980-81	0.0	0.0	0.0	T	0.5	2.6	9.1	6.8	7.5	T	0.0	0.0	26.5
1981-82	0.0	0.0	0.0	T	0.5	8.2	12.1	6.4	7.4	1.0	0.0	0.0	35.6
1982-83	0.0	0.0	0.0	0.0	T	2.9	5.8	15.0	5.2	0.1	0.0	0.0	29.0
1983-84	0.0	0.0	0.0	0.0	0.3	3.8	12.8	9.7	2.4	0.0	0.0	0.0	29.0
1984-85	0.0	0.0	0.0	0.0	T	3.7	17.6	20.1	0.9	1.7	0.0	0.0	44.0
1985-86	0.0	0.0	0.0	0.0	0.0	8.9	13.1	17.7	3.7	T	0.0	0.0	43.4
1986-87	0.0	0.0	0.0	0.0	0.2	0.1	16.3	9.7	3.9	20.7	0.0	0.0	50.9
1987-88	0.0	0.0	0.0	T	2.4	5.7	8.3	7.8	4.6	T	0.0	0.0	28.8
1988-89	0.0	0.0	0.0	T	T	6.9	1.7	4.6	T	1.4	0.6	0.0	15.2
1989-90	0.0	T	T	T	2.0	14.1	11.0	3.8	6.6	1.1	0.0	0.0	38.6
1990-91	T	0.0	0.0	0.0	0.0	1.2	3.5	6.5	5.3	T	0.0	0.0	16.5
1991-92	0.0	0.0	0.0	0.0	4.1	0.7	5.6	1.1	8.6	3.6	T	T	23.7
1992-93	0.0	0.0	0.0	T	2.5	3.7	0.4	12.0	20.4	T	0.0	T	39.0
1993-94	0.0	0.0	0.0	1.5	0.4	12.4							
Record Mean	T	T	T	0.2	2.2	5.1	10.1	8.7	5.1	0.9	T	T	32.2

ELKINS, WEST VIRGINIA

Elkins, West Virginia, is located near the principal storm tracks and is therefore subjected to frequent weather changes throughout the year. While changes may be rigorous, they bring relief from summer heat waves and winter cold waves. The airport and city are located near the middle of a valley with a narrow floor and ridges at or near 3,000 feet. The ridges are oriented north northeast to south southwest, 3 to 4 miles to the east and west. The valley is located on the general northwest slope of the Appalachian Mountains which crest about 20 miles to the southeast at about 4,500 feet, with some higher peaks.

The seasonal climates vary greatly from year to year. When the Atlantic High extends westward, warm weather with high humidities occur in both summer and winter. Conversely, if the Atlantic High is displaced eastward and the circulation is principally from the northwest, weather is colder than normal. Hottest weather is from warm westerlies which have been over land for a long time and have a path over the south central or southwestern areas of the United States.

Summers are characterized by warm, humid, showery weather, but the heat is moderated by elevation and orographically induced cloudiness. A daily high temperature of 90 or above may occasionally be expected during the summer months. Winters are moderately severe with rapid changes. Snowfall may be frequent, and at times, heavy. However, it seldom remains on the ground for extended periods. Snows often fall upon warm ground thereby causing preliminary melting, then freezing, resulting in slippery road conditions. Glaze formation upon the ground or upon wires and trees is rare. Cold spells alternate frequently with thaws, and snow is subject to frequent complete melting during the winter. Severe cold spells occur occasionally but they seldom last more than two or three days. A daily low of zero degrees or below can be expected several times annually.

Significant climatic characteristics are associated with air currents rising and descending over the mountains. Orographic lifting of air delays post-frontal clearing after the passage of a cold front, especially during the winter when low clouds and snow flurries sometimes persist for 24 hours or more after the front has passed. While this upslope effect prevails with winds from the north-west quadrant, a foehn effect prevails with easterly and southerly winds, tending to diminish existing low cloud layers and to keep ceilings higher than otherwise anticipated. Nocturnal radiation fog is common during the summer and the autumn but it usually dissipates rapidly after sunrise.

Tornadoes are rare in this area, and severe thunderstorms are very infrequent. However, occasionally intense local rainfall from warm-season thunderstorms causes flash flooding in the narrow valleys of the area. Due to the remote location of the city with respect to concentrated industry, the air is usually relatively unpolluted. There are no important smoke sources in the locality, and smoke or haze seldom reduces the visibility below four miles.

The average last occurrence in the spring of temperatures as low as 32 is early to mid-May, and the first occurrence in the autumn is early October. The length of the growing season averages about 148 days.

NORMALS, MEANS, AND EXTREMES

ELKINS, WEST VIRGINIA

LATITUDE: 38°53'N LONGITUDE: 79°51' W ELEVATION: FT. GRND 1948 BARO 1985 TIME ZONE: EASTERN WBAN: 13729

	(a)	JAN	FEB	MAR	APR	MAY	JUNE	JULY	AUG	SEP	OCT	NOV	DEC	YEAR
TEMPERATURE °F:														
Normals														
-Daily Maximum		37.9	41.3	52.1	61.6	70.9	77.6	80.4	79.3	73.6	63.6	52.9	42.6	61.2
-Daily Minimum		16.2	18.3	27.1	34.8	44.0	52.0	56.9	56.2	49.9	37.2	29.7	21.3	37.0
-Monthly		27.1	29.8	39.6	48.2	57.4	64.8	68.7	67.8	61.8	50.4	41.3	32.0	49.1
Extremes														
-Record Highest	49	76	72	84	89	90	93	99	95	97	86	80	76	99
-Year		1950	1985	1954	1986	1991	1952	1988	1948	1953	1951	1958	1951	JUL 1988
-Record Lowest	49	-24	-22	-15	3	20	25	32	34	27	11	0	-24	-24
-Year		1984	1977	1978	1985	1978	1977	1988	1965	1963	1952	1958	1989	DEC 1989
NORMAL DEGREE DAYS:														
Heating (base 65°F)		1175	986	787	504	249	62	11	23	131	458	711	1023	6120
Cooling (base 65°F)		0	0	0	0	13	56	126	110	35	6	0	0	346
% OF POSSIBLE SUNSHINE	10	30	31	39	44	44	48	44	44	45	46	37	28	40
MEAN SKY COVER (tenths)														
Sunrise - Sunset	48	8.0	7.8	7.6	7.3	7.1	7.1	7.2	7.0	6.8	6.5	7.4	7.9	7.3
MEAN NUMBER OF DAYS:														
Sunrise to Sunset														
-Clear	48	3.5	3.5	4.2	4.1	4.4	3.1	2.5	2.6	4.3	7.3	4.6	4.2	48.2
-Partly Cloudy	48	5.6	5.5	6.4	7.8	9.3	11.0	12.6	13.7	11.3	8.3	6.4	5.4	103.3
-Cloudy	48	21.9	19.3	20.4	18.1	17.3	15.9	15.9	14.8	14.6	13.3	19.0	21.5	211.8
Precipitation														
.01 inches or more	49	18.0	15.7	16.9	15.2	14.3	13.4	13.7	11.9	11.0	10.7	13.5	16.7	171.2
Snow, Ice pellets, hail														
1.0 inches or more	31	6.2	6.0	3.7	1.4	0.0	0.0	0.0	0.0	0.0	0.2	2.1	4.9	24.3
Thunderstorms	25	0.2	0.5	2.0	4.1	6.2	7.8	10.0	7.7	3.3	1.3	0.7	0.4	44.3
Heavy Fog Visibility														
1/4 mile or less	25	2.0	1.8	1.7	1.9	5.2	10.8	13.1	16.9	14.7	9.6	3.1	2.0	82.8
Temperature °F														
-Maximum														
90° and above	49	0.0	0.0	0.0	0.0	0.*	0.3	1.1	0.9	0.2	0.0	0.0	0.0	2.6
32° and below	49	9.7	6.9	2.7	0.2	0.0	0.0	0.0	0.0	0.0	0.*	1.9	6.7	28.1
-Minimum														
32° and below	49	26.6	23.8	21.9	11.9	2.7	0.1	0.*	0.0	0.8	10.5	19.7	25.7	143.8
0° and below	49	3.4	2.2	0.3	0.0	0.0	0.0	0.0	0.0	0.0	0.0	0.*	1.4	7.4
AVG. STATION PRESS. (mb)	10	945.7	945.9	944.7	945.1	945.4	947.3	948.3	949.6	949.2	948.9	947.9	947.0	947.1
RELATIVE HUMIDITY (%)														
Hour 01	6	81	78	82	81	85	95	97	97	96	91	84	85	88
Hour 07 (Local Time)	31	80	80	82	83	87	91	94	96	95	90	84	82	87
Hour 13	31	65	61	56	52	54	58	61	62	62	54	59	66	59
Hour 19	29	73	69	62	56	61	69	73	79	84	75	74	76	71
PRECIPITATION (inches):														
Water Equivalent														
-Normal		3.08	3.00	3.83	3.82	4.12	4.46	4.53	4.35	3.76	3.08	3.33	3.48	44.84
-Maximum Monthly	49	6.09	5.62	8.85	6.95	7.67	8.33	9.30	10.40	7.52	8.43	11.08	6.73	11.08
-Year		1949	1972	1963	1972	1967	1981	1958	1980	1988	1954	1985	1978	NOV 1985
-Minimum Monthly	49	1.05	0.79	1.39	1.02	1.45	1.66	1.31	1.09	0.32	0.31	1.19	0.90	0.31
-Year		1967	1978	1957	1971	1970	1988	1987	1976	1985	1963	1976	1965	OCT 1963
-Maximum in 24 hrs	31	1.85	1.73	2.94	2.02	2.63	2.63	2.70	3.21	2.88	3.62	5.10	2.22	5.10
-Year		1971	1984	1963	1966	1990	1986	1985	1969	1988	1985	1985	1970	NOV 1985
Snow, Ice pellets, hail														
-Maximum Monthly	31	54.1	32.0	33.5	24.8	0.7	T	0.0	0.0	0.0	3.9	14.8	36.7	54.1
-Year		1985	1986	1993	1987	1963	1993				1979	1976	1993	JAN 1985
-Maximum in 24 hrs	31	18.7	12.8	16.9	11.6	0.7	T	0.0	0.0	0.0	3.6	12.4	17.8	18.7
-Year		1971	1983	1993	1987	1963	1993				1979	1970	1967	JAN 1971
WIND:														
Mean Speed (mph)	24	7.3	8.0	8.2	7.9	6.7	4.9	4.3	4.1	4.4	5.0	6.9	6.9	6.2
Prevailing Direction through 1963		W	NW	WNW	NW	SSE	NW	NW	NW	NW	NNW	WNW	W	NW
Fastest Mile														
-Direction (!!)	10	W	W	N	W	NW	W	NW	NW	NW	NW	NW	NW	W
-Speed (MPH)	10	49	32	43	52	52	27	31	40	26	40	34	47	52
-Year		1987	1990	1993	1991	1991	1986	1984	1988	1984	1991	1993	1988	APR 1991
Peak Gust														
-Direction (!!)	10	W	W	W	W	NW	W	NW	NW	SW	W	NW	NW	W
-Speed (mph)	10	56	49	48	69	60	69	47	46	44	41	59	52	69
-Date		1993	1992	1992	1991	1991	1993	1984	1988	1989	1991	1992	1988	JUN 1993

PRECIPITATION (inches)

ELKINS, WEST VIRGINIA

YEAR	JAN	FEB	MAR	APR	MAY	JUNE	JULY	AUG	SEP	OCT	NOV	DEC	ANNUAL
1964	3.20	3.07	3.95	5.61	1.47	5.01	5.19	3.55	4.20	0.94	2.00	3.39	41.58
1965	5.31	1.94	4.51	6.90	1.45	2.21	2.27	1.74	2.89	2.51	1.43	0.90	34.06
1966	3.49	3.10	1.44	6.63	1.68	1.98	3.24	4.84	5.14	2.46	2.63	2.45	39.08
1967	1.05	3.03	7.69	3.24	7.67	2.83	4.55	5.10	3.33	3.44	3.55	4.47	49.95
1968	1.88	1.80	4.28	1.54	7.30	2.05	3.46	4.07	3.01	3.21	3.20	3.04	38.84
1969	3.01	2.13	2.26	3.34	2.76	3.83	6.82	5.60	5.94	1.60	1.81	5.84	44.94
1970	1.53	1.92	2.80	4.46	1.45	4.65	5.88	4.83	3.28	1.95	2.43	5.51	40.69
1971	4.33	3.16	2.90	1.02	4.23	2.34	4.10	3.87	6.29	1.85	2.76	1.65	38.50
1972	5.46	5.62	3.16	6.95	4.72	7.03	4.56	2.09	2.67	4.64	5.82	6.02	58.74
1973	2.05	2.52	2.13	5.87	3.75	3.53	3.82	3.47	3.66	4.22	3.28	4.53	42.83
1974	4.04	3.41	3.48	2.93	5.23	7.38	1.93	6.84	3.38	2.03	2.18	3.20	46.03
1975	4.17	2.84	4.64	4.38	7.28	4.87	2.56	3.55	3.71	2.49	2.04	3.76	46.29
1976	3.10	3.05	3.17	1.13	1.94	3.23	6.17	1.09	4.71	6.28	1.19	2.66	37.72
1977	1.73	1.13	3.40	2.83	1.96	5.08	4.49	5.60	2.14	3.74	3.30	2.59	37.99
1978	4.26	0.79	2.25	1.99	4.24	6.84	7.28	4.85	1.47	1.88	2.90	6.73	45.48
1979	5.88	3.63	2.03	3.40	5.71	4.94	6.22	5.06	4.61	5.16	3.54	2.33	52.51
1980	2.75	1.90	4.80	5.46	3.66	5.04	7.80	10.40	3.94	1.99	3.76	2.08	53.58
1981	1.15	4.22	2.76	3.72	5.79	8.33	3.43	1.72	6.26	3.48	1.54	3.17	45.57
1982	3.53	4.14	6.34	2.15	2.26	6.46	6.01	4.61	4.35	1.42	3.92	2.55	47.74
1983	1.59	1.50	3.69	5.26	5.49	4.05	2.03	3.59	1.98	4.13	3.65	3.57	40.53
1984	1.69	4.42	4.89	3.96	2.64	2.52	5.66	7.21	2.23	4.08	3.94	3.87	47.11
1985	3.26	1.99	4.68	2.44	6.95	3.56	5.58	2.78	0.32	6.00	11.08	2.59	51.23
1986	2.09	5.27	3.44	3.38	3.04	7.84	6.42	5.03	4.29	2.91	5.68	2.61	52.00
1987	4.00	2.95	1.50	3.94	3.88	2.97	1.31	4.29	5.69	1.51	3.18	3.54	38.76
1988	2.81	2.50	2.92	3.14	4.84	1.66	3.60	4.05	7.52	2.03	3.60	2.35	41.02
1989	3.78	3.82	5.52	4.10	6.09	5.94	5.31	7.98	2.17	3.75	3.65	3.03	55.14
1990	3.62	2.87	2.25	3.72	7.03	4.01	4.01	3.25	4.11	3.31	1.66	6.12	45.96
1991	4.10	2.50	5.94	3.12	2.32	2.59	8.34	3.52	1.92	2.37	2.66	6.09	45.47
1992	2.66	3.72	4.33	2.58	3.20	1.95	8.26	2.49	2.63	0.80	2.66	4.47	39.75
1993	1.71	2.87	6.30	3.90	2.69	2.12	2.71	2.18	6.24	2.49	4.54	4.97	42.72
Record Mean	3.60	3.10	4.05	3.73	4.23	4.91	5.17	4.12	3.44	2.98	2.96	3.48	45.76

AVERAGE TEMPERATURE (deg. F)

ELKINS, WEST VIRGINIA

YEAR	JAN	FEB	MAR	APR	MAY	JUNE	JULY	AUG	SEP	OCT	NOV	DEC	ANNUAL
1964	31.6	26.4	39.7	51.2	59.4	65.7	68.6	65.8	61.3	47.0	44.0	35.2	49.7
1965	28.1	31.6	35.3	49.2	62.3	63.8	67.9	67.3	64.9	48.7	41.9	36.0	49.8
1966	24.2	30.8	40.9	47.4	56.8	64.3	69.9	66.6	58.4	47.7	42.1	31.7	48.4
1967	33.7	27.6	42.5	50.5	52.7	66.5	66.7	65.4	57.5	50.4	37.2	35.8	48.9
1968	25.0	22.0	41.8	49.9	55.8	65.3	69.2	69.9	61.3	51.3	42.2	28.6	48.5
1969	27.8	30.4	32.6	50.4	59.0	66.9	70.7	67.2	61.0	50.1	37.1	25.7	48.2
1970	20.0	29.5	35.9	49.6	59.8	64.4	67.9	67.2	64.1	53.1	41.4	33.0	48.8
1971	24.2	31.5	34.5	45.4	55.4	68.8	68.1	67.0	66.2	57.7	39.7	41.1	50.0
1972	33.8	28.8	38.6	48.3	57.8	60.8	68.0	68.0	63.5	48.5	39.9	39.2	49.6
1973	30.3	30.2	48.0	47.9	55.4	67.2	69.1	68.5	62.5	52.8	41.5	32.6	50.5
1974	39.0	30.6	42.5	49.1	57.1	62.0	66.5	67.1	59.4	46.5	38.6	30.2	49.0
1975	31.3	33.7	35.7	43.7	60.1	65.6	68.4	69.8	60.3	53.8	44.9	34.5	50.2
1976	26.8	39.0	44.7	48.3	55.8	66.7	67.7	65.7	59.9	46.7	32.5	27.5	48.4
1977	15.0	28.3	45.5	52.5	60.8	62.0	69.2	68.8	65.9	48.6	44.6	28.9	49.2
1978	19.6	16.6	36.5	49.3	56.6	65.0	69.1	70.4	65.0	47.5	44.2	33.8	47.8
1979	25.7	22.4	42.9	49.2	58.0	63.5	68.0	68.1	61.6	49.0	43.0	34.0	48.8
1980	30.0	23.2	36.7	48.1	58.5	62.9	70.1	71.1	64.6	47.5	37.3	30.9	48.4
1981	20.4	31.7	35.0	52.2	53.7	66.0	68.1	65.1	59.6	47.0	37.4	27.6	47.0
1982	24.1	32.5	41.3	44.8	60.8	63.3	69.8	64.9	60.7	51.8	44.5	39.0	49.8
1983	29.3	30.8	38.3	46.0	53.9	63.5	67.8	69.0	59.8	51.9	40.4	27.5	48.2
1984	24.2	35.6	34.5	47.8	54.8	65.9	65.9	67.9	58.6	58.8	37.7	41.3	49.4
1985	21.6	28.4	41.0	50.2	59.1	63.2	68.9	66.7	61.6	54.8	50.6	27.6	49.5
1986	26.5	33.1	38.0	50.1	58.4	65.1	71.8	66.6	63.1	52.6	42.9	33.0	50.1
1987	27.7	31.3	40.4	47.0	61.3	67.0	71.6	69.5	62.1	45.2	43.2	34.4	50.1
1988	23.9	29.5	38.7	48.0	57.2	62.7	70.7	71.6	61.4	43.7	41.8	31.5	48.4
1989	35.5	29.5	42.5	45.6	54.8	67.0	71.2	68.0	63.1	51.4	39.6	18.4	48.9
1990	35.7	37.7	44.3	48.3	56.8	65.7	69.9	70.0	63.0	52.2	42.9	38.5	52.1
1991	31.3	34.4	41.2	52.4	65.3	67.2	71.8	69.2	62.0	51.6	39.5	36.0	51.8
1992	30.7	35.2	39.4	48.3	56.3	63.7	70.4	66.3	62.8	48.3	43.0	32.6	49.8
1993	36.3	28.0	35.2	47.9	58.1	66.3	73.3	72.0	63.4	49.6	41.4	30.7	50.2
Record Mean	30.3	31.2	39.7	48.9	58.2	65.8	69.7	68.5	62.6	51.7	41.0	32.5	50.0
Max	40.5	41.8	51.1	61.4	70.8	78.0	81.2	79.9	74.9	64.6	52.1	42.4	61.6
Min	20.1	20.6	28.3	36.3	45.6	53.7	58.1	57.0	50.3	38.7	29.8	22.5	38.4

REFERENCE NOTES

ELKINS, WEST VIRGINIA

GENERAL
T - TRACE AMOUNT.
BLANK ENTRIES DENOTE MISSING/UNREPORTED DATA.
# INDICATES A STATION OR INSTRUMENT RELOCATION.

SPECIFIC

(a) - LENGTH OF RECORD IN YEARS, ALTHOUGH INDIVIDUAL MONTHS MAY BE MISSING.
0.* OR * - THE VALUE IS BETWEEN 0.0 AND 0.05
NORMALS - BASED ON THE 1961-1990 RECORD PERIOD.
EXTREMES - DATES ARE THE MOST RECENT OCCURRENCE
WIND DIR.- NUMERALS SHOW TENS OF DEGRESS CLOCKWISE FROM TRUE NORTH. "00" INDICATES CALM.
RESULTANT DIRECTIONS ARE GIVEN TO WHOLE DEGREES.
BOLD VALUES INDICATE EXTREME VALUES WHICH OCCURRED AFTER THE ASOS SYSTEM WAS COMMISSIONED.

RECORD = PERIOD OF RECORD
RECORD MEAN PRECIPITATION IS THE MEAN OF ALL DAILY PRECIPITATION AMOUNTS DURING THE PERIOD OF RECORD.
RECORD MAX(MIN) TEMPERATURE IS THE MEAN OF ALL DAILY MAX(MIN) TEMPERATURES DURING THE PERIOD OF RECORD.
RECORD MEAN TEMPERATURE IS THE SUM OF THE RECORD MAX AND RECORD MIN DIVIDED BY 2.
AVERAGE TEMPERATURE IS THE SUM OF THE MEAN DAILY MAX AND MIN TEMPERATURE DIVIDED BY 2.

EXCEPTIONS

1. FASTEST OBSERVED WINDS ARE 1955-1958 AND 1963-1968.
2. MEAN WIND SPEED, THUNDERSTORMS, AND HEAVY FOG ARE THROUGH 1968.

RECORD MEANS ARE THROUGH THE CURRENT YEAR, BEGINNING IN 1899 FOR TEMPERATURE
1899 FOR PRECIPITATION
1963 FOR SNOWFALL

HEATING DEGREE DAYS Base 65 deg. F — ELKINS, WEST VIRGINIA

SEASON	JULY	AUG	SEP	OCT	NOV	DEC	JAN	FEB	MAR	APR	MAY	JUNE	TOTAL
1964-65	23	55	124	553	624	914	1134	930	912	466	106	92	5933
1965-66	12	50	81	497	688	893	1258	951	741	525	254	81	6031
1966-67	15	29	207	531	679	1027	963	1039	692	430	378	47	6037
1967-68	33	37	226	443	826	900	1233	1239	711	444	279	67	6438
1968-69	16	46	122	425	673	1121	1147	964	996	430	194	57	6191
1969-70	0	8	160	462	828	1212	1390	986	892	456	182	65	6641
1970-71	26	14	97	362	702	987	1258	930	938	580	296	5	6195
1971-72	0	18	54	233	756	731	961	1047	815	494	217	156	5482
1972-73	30	13	76	504	745	793	1068	966	522	509	304	28	5558
1973-74	12	12	110	373	698	996	799	959	692	472	255	119	5497
1974-75	29	11	188	564	784	1074	1037	871	901	632	174	46	6311
1975-76	2	7	160	341	596	940	1175	746	625	492	292	44	5420
1976-77	24	44	152	564	968	1159	1548	1021	600	377	151	136	6744
1977-78	22	17	59	503	608	1111	1401	1349	877	465	265	70	6747
1978-79	13	6	75	534	617	961	1213	1187	680	467	231	83	6067
1979-80	39	46	130	491	652	954	1080	1203	872	500	210	113	6290
1980-81	4	5	72	538	823	1050	1375	928	924	384	339	42	6484
1981-82	21	54	187	553	822	1150	1265	904	728	600	166	76	6526
1982-83	13	57	149	403	609	798	1099	951	817	563	336	85	5880
1983-84	30	21	184	404	729	1153	1258	847	937	511	318	50	6442
1984-85	39	22	199	190	811	727	1338	1020	738	437	194	101	5816
1985-86	1	27	160	312	427	1153	1185	888	830	443	218	68	5712
1986-87	9	50	98	397	657	987	1149	938	753	533	163	43	5777
1987-88	12	26	122	605	645	939	1264	1024	808	506	243	131	6325
1988-89	27	6	119	653	686	1031	908	987	689	573	321	20	6020
1989-90	0	29	120	416	755	1438	903	758	637	496	249	49	5850
1990-91	1	4	131	394	654	814	1038	852	730	383	80	43	5124
1991-92	0	6	144	406	760	893	1056	856	785	495	267	82	5750
1992-93	4	32	122	510	654	996	884	1028	918	508	210	49	5915
1993-94	0	2	115	472	705	1056							

COOLING DEGREE DAYS Base 65 deg. F — ELKINS, WEST VIRGINIA

YEAR	JAN	FEB	MAR	APR	MAY	JUNE	JULY	AUG	SEP	OCT	NOV	DEC	TOTAL
1969	0	0	0	0	18	122	184	83	45	7	0	0	459
1970	0	0	0	2	28	52	123	86	75	1	0	0	367
1971	0	0	0	0	6	128	105	88	99	11	3	0	440
1972	0	0	0	0	3	36	131	111	36	0	0	0	317
1973	0	0	0	3	11	98	147	125	41	3	0	0	428
1974	0	0	0	1	15	35	81	85	27	0	0	0	244
1975	0	0	0	0	25	70	114	161	27	0	0	0	397
1976	0	0	0	1	16	105	117	73	2	0	0	0	314
1977	0	0	1	6	26	55	160	142	91	0	4	0	485
1978	0	0	0	1	13	76	147	179	82	0	0	0	498
1979	0	0	0	2	21	45	142	149	33	0	0	0	392
1980	0	0	0	0	16	57	169	202	69	0	0	0	513
1981	0	0	0	6	0	76	126	65	32	0	0	0	305
1982	0	0	0	0	43	33	170	60	30	1	0	0	337
1983	0	0	0	0	0	46	124	153	33	4	0	0	360
1984	0	0	0	2	9	82	72	117	16	6	0	0	304
1985	0	0	0	0	17	54	129	87	63	5	0	0	355
1986	0	0	0	1	20	78	224	104	46	18	0	0	491
1987	0	0	0	0	57	109	224	171	42	0	0	0	603
1988	0	0	0	0	6	67	211	220	18	1	0	0	523
1989	0	0	0	0	10	90	198	130	70	1	0	0	499
1990	0	0	3	1	3	78	158	167	74	4	0	0	488
1991	0	0	0	13	97	116	217	144	61	0	0	0	648
1992	0	0	0	0	3	49	178	79	62	0	0	0	371
1993	0	0	0	2	5	97	263	225	69	2	2	0	665

SNOWFALL (inches) — ELKINS, WEST VIRGINIA

SEASON	JULY	AUG	SEP	OCT	NOV	DEC	JAN	FEB	MAR	APR	MAY	JUNE	TOTAL
1964-65	0.0	0.0	0.0	0.0	4.6	5.7	21.6	11.2	14.0	0.2	0.0	0.0	57.3
1965-66	0.0	0.0	0.0	0.4	1.6	4.6	21.1	21.0	8.1	5.0	T	0.0	61.8
1966-67	0.0	0.0	0.0	0.0	7.8	14.2	7.2	28.1	7.0	0.5	T	0.0	64.8
1967-68	0.0	0.0	0.0	T	9.0	23.2	14.1	24.1	10.3	0.1	0.0	0.0	80.8
1968-69	0.0	0.0	0.0	0.5	11.9	19.7	3.1	13.6	14.6	0.7	T	0.0	64.1
1969-70	0.0	0.0	0.0	T	8.9	34.9	20.0	13.8	8.3	0.7	0.0	0.0	86.6
1970-71	0.0	0.0	0.0	0.0	13.2	16.9	25.2	14.1	33.5	4.8	T	0.0	107.7
1971-72	0.0	0.0	0.0	0.0	11.0	1.6	4.1	26.9	7.2	4.6	0.0	0.0	55.4
1972-73	0.0	0.0	0.0	2.2	7.4	6.7	4.6	9.9	7.6	7.4	T	0.0	45.8
1973-74	0.0	0.0	0.0	2.5	3.2	25.0	0.8	15.1	1.9	2.0	0.0	0.0	50.5
1974-75	0.0	0.0	0.0	T	6.8	17.7	16.9	10.0	11.1	3.6	0.0	0.0	66.1
1975-76	0.0	0.0	0.0	0.0	3.9	6.5	21.0	8.1	8.6	1.8	T	0.0	49.9
1976-77	0.0	0.0	0.0	T	14.8	21.1	37.3	14.3	1.3	2.6	0.0	0.0	91.4
1977-78	0.0	0.0	0.0	T	5.3	12.6	32.8	13.2	13.8	T	0.0	0.0	77.7
1978-79	0.0	0.0	0.0	T	2.0	5.1	30.8	18.7	4.2	3.0	0.0	0.0	63.8
1979-80	0.0	0.0	0.0	3.9	3.8	10.2	23.0	19.7	14.1	3.6	0.0	0.0	78.3
1980-81	0.0	0.0	0.0	0.4	4.0	8.0	23.6	12.1	14.9	0.6	0.0	0.0	63.6
1981-82	0.0	0.0	0.0	0.4	8.4	20.6	18.2	10.1	13.4	7.1	0.0	0.0	78.2
1982-83	0.0	0.0	0.0	T	1.3	10.5	17.0	27.0	10.5	1.8	0.0	0.0	68.1
1983-84	0.0	0.0	0.0	T	11.1	7.0	26.4	23.7	21.6	2.9	0.0	0.0	92.7
1984-85	0.0	0.0	0.0	0.0	4.0	4.0	54.1	21.3	0.5	8.0	0.0	0.0	91.9
1985-86	0.0	0.0	0.0	0.0	0.0	22.6	33.5	32.0	16.8	18.8	0.0	0.0	123.7
1986-87	0.0	0.0	0.0	0.0	1.0	4.0	42.6	19.5	4.9	24.8	0.0	0.0	96.8
1987-88	0.0	0.0	0.0	T	12.4	20.3	21.5	16.1	15.9	1.6	0.0	0.0	87.8
1988-89	0.0	0.0	0.0	T	2.1	14.3	11.5	14.1	1.8	7.9	0.5	0.0	52.2
1989-90	0.0	0.0	0.0	0.9	9.0	32.2	21.7	9.0	12.3	13.8	0.0	0.0	98.9
1990-91	0.0	0.0	0.0	T	T	8.9	19.9	20.0	18.9	0.3	0.0	0.0	68.0
1991-92	0.0	0.0	0.0	0.0	5.8	6.9	29.5	11.8	19.3	5.8	0.0	0.0	79.1
1992-93	0.0	0.0	0.0	2.4	3.4	27.5	4.4	27.4	33.5	2.5	0.0	T	101.1
1993-94	0.0	0.0	0.0	3.1	3.5	36.7							
Record Mean	0.0	0.0	0.0	0.5	6.0	15.0	20.4	18.0	11.7	4.4	T	T	76.1

HUNTINGTON, WEST VIRGINIA

The Tri-State Airport is near the confluence of the Ohio and Big Sandy Rivers, located on a man-made plateau constructed by cutting the tops off several hills and filling intervening valleys. The elevation of the ground at the National Weather Service Office is 260 feet higher than at the Federal Building in downtown Huntington.

The temperature record for the valley locations is not compatible with that for the airport, which is generally cooler throughout the year. The summer season is moderately warm and humid, with the valley locations considerably warmer and more humid than the Tri-State Airport site. The winter months are moderately cold, with an occasional severe cold wave lasting a few days. The four seasons are nearly equal in length and autumn is the most pleasant, with warm days and cool nights.

The heaviest rainfall occurs in July and August, mostly in thunderstorms, and flash floods are common in the area. The winter rainfall occurs mostly prior to and with a frontal passage and frequently lasts from two to four days, causing frequent general flooding on all streams.

Snow seldom remains on the ground more than two days in the valleys. However, at higher elevations surrounding the airport, roads are frequently blocked for several days during the winter months.

NORMALS, MEANS, AND EXTREMES

HUNTINGTON, WEST VIRGINIA

LATITUDE: 38°22'N LONGITUDE: 82°33' W ELEVATION: FT. GRND 827 BARO 837 TIME ZONE: EASTERN WBAN: 03860

	(a)	JAN	FEB	MAR	APR	MAY	JUNE	JULY	AUG	SEP	OCT	NOV	DEC	YEAR
TEMPERATURE °F:														
Normals														
-Daily Maximum		40.7	44.2	56.5	66.8	75.2	81.3	84.3	83.1	78.0	67.2	55.9	45.3	64.9
-Daily Minimum		23.2	26.1	35.5	43.8	52.4	60.4	65.0	63.9	57.1	45.0	37.0	28.4	44.8
-Monthly		32.0	35.1	46.0	55.3	63.8	70.9	74.7	73.5	67.6	56.1	46.5	36.9	54.9
Extremes														
-Record Highest	33	74	79	86	92	93	100	102	100	97	86	82	80	102
-Year		1967	1977	1989	1985	1991	1988	1988	1988	1983	1991	1993	1982	JUL 1988
-Record Lowest	33	-16	-6	-2	20	27	40	46	43	31	16	8	-13	-16
-Year		1985	1970	1980	1985	1966	1977	1968	1986	1983	1962	1964	1989	JAN 1985
NORMAL DEGREE DAYS:														
Heating (base 65°F)		1023	837	589	298	121	11	0	7	53	300	555	871	4665
Cooling (base 65°F)		0	0	0	7	83	188	301	271	131	24	0	0	1005
% OF POSSIBLE SUNSHINE														
MEAN SKY COVER (tenths)														
Sunrise - Sunset	32	7.7	7.6	7.4	7.1	6.9	6.8	6.7	6.7	6.5	6.1	7.4	7.8	7.1
MEAN NUMBER OF DAYS:														
Sunrise to Sunset														
-Clear	32	4.5	4.4	4.6	5.2	5.8	4.6	4.5	4.6	6.4	8.7	4.9	4.4	62.6
-Partly Cloudy	32	6.2	5.3	6.8	7.8	8.7	11.1	12.0	12.4	9.6	7.9	6.1	5.3	99.3
-Cloudy	32	20.4	18.5	19.5	17.0	16.6	14.3	14.5	14.0	13.9	14.4	19.0	21.3	203.3
Precipitation														
.01 inches or more	32	13.6	12.8	14.1	12.9	12.5	10.8	11.7	9.4	8.8	9.3	11.3	12.9	140.3
Snow,Ice pellets,hail														
1.0 inches or more	32	2.7	2.3	1.2	0.1	0.0	0.0	0.0	0.0	0.0	0.*	0.3	1.2	7.8
Thunderstorms	31	0.4	0.6	2.8	4.2	6.2	6.8	9.1	6.7	2.5	1.1	0.8	0.3	41.5
Heavy Fog Visibility														
1/4 mile or less	31	2.6	2.7	2.1	1.5	4.4	6.5	9.4	10.4	9.7	5.9	3.3	2.8	61.3
Temperature °F														
-Maximum														
90° and above	32	0.0	0.0	0.0	0.2	0.7	4.1	7.8	6.2	1.6	0.0	0.0	0.0	20.4
32° and below	32	8.5	5.4	0.9	0.0	0.0	0.0	0.0	0.0	0.0	0.0	0.3	4.8	20.0
-Minimum														
32° and below	32	23.7	20.5	13.5	3.6	0.3	0.0	0.0	0.0	0.*	3.4	11.1	20.0	96.2
0° and below	32	1.4	0.3	0.*	0.0	0.0	0.0	0.0	0.0	0.0	0.0	0.0	0.4	2.1
AVG. STATION PRESS.(mb)	21	989.0	988.4	986.3	985.4	985.4	986.0	987.0	987.9	988.5	989.2	989.0	989.4	987.6
RELATIVE HUMIDITY (%)														
Hour 01	31	73	72	68	66	79	86	88	89	89	81	74	74	78
Hour 07 (Local Time)	32	77	77	76	76	85	88	90	92	92	87	80	78	83
Hour 13	32	65	61	55	49	54	57	60	59	60	55	59	65	58
Hour 19	32	64	59	53	48	57	62	66	67	69	61	62	66	61
PRECIPITATION (inches):														
Water Equivalent														
-Normal		2.83	2.90	3.68	3.43	4.24	3.51	4.65	3.83	2.93	2.83	3.30	3.36	41.49
-Maximum Monthly	33	6.37	8.67	7.71	6.56	9.26	7.63	8.57	6.93	6.32	5.71	7.40	8.69	9.26
-Year		1978	1989	1991	1966	1974	1979	1962	1989	1989	1983	1985	1978	MAY 1974
-Minimum Monthly	33	0.64	0.53	1.12	0.74	0.93	0.41	1.37	0.68	0.35	T	0.73	0.31	T
-Year		1981	1968	1966	1976	1965	1966	1974	1962	1985	1963	1976	1965	OCT 1963
-Maximum in 24 hrs	33	2.63	2.69	3.43	2.26	3.73	3.42	4.27	2.90	2.74	2.95	2.75	3.36	4.27
-Year		1974	1989	1967	1978	1990	1979	1962	1964	1964	1985	1988	1978	JUL 1962
Snow,Ice pellets,hail														
-Maximum Monthly	32	30.3	23.9	23.6	14.4	0.2	T	0.0	T	0.0	1.5	4.6	13.2	30.3
-Year		1978	1986	1993	1987	1989	1990		1991		1993	1987	1967	JAN 1978
-Maximum in 24 hrs	32	11.6	11.2	21.1	8.1	0.2	T	0.0	T	0.0	1.5	4.4	6.7	21.1
-Year		1978	1985	1993	1987	1989	1990		1991		1993	1969	1967	MAR 1993
WIND:														
Mean Speed (mph)	31	7.6	7.6	8.1	7.8	6.2	5.7	5.3	5.1	5.3	5.9	7.1	7.5	6.6
Prevailing Direction														
Fastest Obs. 1 Min.														
-Direction (!!)	31	27	26	25	18	29	24	31	24	34	27	23	30	29
-Speed (MPH)	31	38	41	37	44	47	35	32	35	28	29	35	35	47
-Year		1971	1967	1971	1968	1967	1973	1976	1965	1963	1969	1966	1983	MAY 1967
Peak Gust														
-Direction (!!)	10	W	NW	NW	W	W	E	N	W	W	W	W	SW	E
-Speed (mph)	10	51	53	52	52	46	56	53	49	41	35	55	51	56
-Date		1987	1993	1993	1993	1990	1993	1993	1991	1993	1990	1988	1987	JUN 1993

PRECIPITATION (inches)

HUNTINGTON, WEST VIRGINIA

YEAR	JAN	FEB	MAR	APR	MAY	JUNE	JULY	AUG	SEP	OCT	NOV	DEC	ANNUAL
1964	2.11	3.07	4.54	3.09	1.25	1.97	2.18	6.21	4.53	0.80	3.37	3.90	37.02
1965	3.36	2.17	4.81	5.68	0.93	4.35	4.72	4.24	3.49	2.25	0.96	0.31	37.27
1966	3.26	3.97	1.12	6.56	1.58	0.41	6.37	4.15	5.64	2.28	3.06	2.81	41.21
1967	1.43	2.53	7.52	3.48	6.81	1.11	5.82	1.48	1.21	2.27	4.71	3.55	41.92
1968	1.72	0.53	6.05	2.81	6.10	1.67	2.59	6.03	1.14	2.89	2.41	2.20	36.14
1969	2.34	0.94	1.24	2.85	3.30	2.45	3.34	3.93	3.91	1.85	2.14	4.80	33.09
1970	1.12	3.25	3.33	3.68	3.03	3.28	3.75	4.28	2.80	5.23	2.11	4.11	39.97
1971	2.57	2.71	1.96	1.20	6.04	4.80	7.57	1.05	2.61	1.48	1.96	1.50	35.45
1972	4.79	4.90	2.76	6.07	3.82	4.39	2.69	2.06	3.32	2.02	4.38	5.52	46.72
1973	1.48	2.05	3.27	4.91	4.89	3.14	3.52	1.69	2.55	3.39	5.17	2.23	38.29
1974	5.57	1.85	4.45	2.19	9.26	4.86	1.37	4.94	3.63	1.64	4.01	2.87	46.64
1975	4.14	3.11	6.32	5.55	3.20	4.06	4.55	3.96	4.84	3.47	2.64	3.64	49.48
1976	2.84	2.38	3.90	0.74	2.41	4.25	5.48	5.72	4.27	5.54	0.73	2.17	40.43
1977	2.42	0.76	3.10	3.58	2.32	4.70	3.22	5.93	3.26	4.36	3.57	2.33	39.55
1978	6.37	1.06	3.11	4.06	4.30	1.66	4.49	4.22	1.37	3.11	2.74	8.69	45.18
1979	5.28	4.28	2.24	2.83	4.65	7.63	6.18	6.86	5.29	2.67	3.19	2.77	53.87
1980	2.46	1.71	5.04	3.11	2.64	1.95	7.94	5.91	1.89	1.57	2.96	1.82	39.00
1981	0.64	4.23	1.65	5.35	5.19	5.91	3.64	1.03	1.60	2.53	1.23	2.61	35.61
1982	4.49	2.36	4.40	1.49	6.28	4.53	3.76	4.45	2.00	1.80	4.33	3.34	43.23
1983	1.40	1.92	1.89	3.99	6.82	3.02	2.03	3.56	0.63	5.71	2.87	3.10	36.94
1984	1.82	2.09	2.45	4.46	4.45	1.75	3.37	4.66	2.67	4.87	3.85	4.36	40.80
1985	3.13	2.92	3.41	0.83	5.40	3.10	6.13	4.68	0.35	5.17	7.40	2.10	44.62
1986	1.31	3.60	1.55	0.95	3.57	3.52	5.44	3.53	3.73	1.88	6.67	3.51	39.26
1987	2.52	3.32	2.44	4.78	3.59	2.56	3.12	1.19	2.08	0.55	2.51	4.82	33.48
1988	1.96	2.42	3.01	3.10	2.02	0.77	6.88	2.69	2.65	1.91	5.52	3.17	36.10
1989	3.60	8.67	5.83	3.73	3.96	5.49	6.48	6.93	6.32	4.30	2.39	2.28	59.98
1990	3.15	3.35	2.44	3.21	8.29	4.18	4.19	3.71	5.35	3.47	2.51	8.07	51.92
1991	2.22	3.47	7.71	3.91	1.20	1.63	2.79	4.05	2.27	1.01	3.08	5.36	38.70
1992	1.67	2.32	3.78	2.85	4.97	3.15	7.93	2.54	1.19	1.08	3.18	2.83	37.49
1993	2.23	3.68	5.53	2.18	2.52	5.65	2.80	2.74	4.35	3.48	3.70	3.28	42.14
Record Mean	3.13	2.96	3.86	3.37	4.04	3.72	4.68	3.56	3.00	2.47	3.12	3.32	41.21

AVERAGE TEMPERATURE (deg. F)

HUNTINGTON, WEST VIRGINIA

YEAR	JAN	FEB	MAR	APR	MAY	JUNE	JULY	AUG	SEP	OCT	NOV	DEC	ANNUAL
1964	34.2	32.1	46.4	58.0	66.0	71.7	76.0	73.0	65.7	52.2	48.0	39.4	55.2
1965	33.9	35.9	40.9	56.2	69.3	71.9	74.5	73.2	69.2	54.2	47.8	41.6	55.7
1966	26.6	34.3	46.4	52.5	61.2	72.4	76.4	72.0	65.1	53.0	47.0	36.2	53.6
1967	39.3	32.0	48.1	56.3	59.2	73.1	71.4	70.4	62.9	55.2	40.6	38.9	54.0
1968	29.4	27.4	45.8	55.7	60.3	70.3	75.2	73.6	67.0	56.7	47.1	33.8	53.5
1969	31.7	35.6	38.6	56.1	64.1	72.6	77.9	75.0	66.3	55.3	42.1	30.7	53.9
1970	27.3	33.9	40.9	56.7	65.7	71.4	73.5	73.4	72.7	58.6	47.2	39.0	55.0
1971	29.6	37.3	41.9	54.0	60.6	73.3	71.9	71.8	71.0	63.9	44.8	45.6	55.5
1972	35.8	34.6	43.5	54.1	62.2	66.3	74.6	72.3	69.2	51.0	42.9	40.4	53.9
1973	34.2	33.8	53.2	52.5	61.1	74.3	76.4	75.9	71.2	61.0	48.1	37.6	56.6
1974	41.9	37.9	49.4	58.1	64.5	68.4	75.4	73.8	63.4	54.1	46.4	38.6	56.0
1975	37.3	39.6	42.4	52.7	68.1	73.2	75.0	78.4	65.1	58.0	50.1	38.6	56.6
1976	31.6	46.4	51.9	56.0	62.6	72.2	72.7	72.1	63.2	50.1	39.2	32.3	54.2
1977	19.7	35.1	50.4	60.3	68.2	69.1	78.6	75.4	68.8	54.2	50.2	35.1	55.4
1978	25.1	24.7	42.3	56.9	62.2	73.1	76.4	75.9	72.1	54.1	48.0	38.5	54.1
1979	26.3	28.6	49.5	55.6	63.7	70.1	74.0	74.4	67.4	55.4	48.4	39.1	54.4
1980	34.6	29.2	41.9	53.4	64.7	70.5	77.7	77.8	70.7	53.9	43.7	36.5	54.6
1981	28.2	37.1	42.7	60.5	60.8	73.6	75.6	74.1	67.2	55.2	45.4	34.2	54.6
1982	27.5	34.5	45.9	50.6	68.0	67.5	74.4	69.8	63.9	57.6	49.7	45.5	54.6
1983	34.7	37.9	48.0	52.3	60.8	71.7	78.6	79.8	69.1	57.3	46.9	30.3	55.6
1984	28.4	41.7	41.8	54.8	61.6	75.2	73.4	75.6	65.8	62.9	42.9	45.7	55.8
1985	25.9	32.1	48.5	59.6	64.9	70.2	74.3	72.1	67.9	61.0	53.3	32.6	55.2
1986	33.7	39.4	47.2	58.7	66.6	73.2	77.9	72.9	70.8	58.3	45.8	36.2	56.7
1987	33.2	37.5	47.1	53.4	68.7	73.5	76.6	77.4	68.9	51.0	49.7	39.2	56.4
1988	31.2	35.1	46.1	54.3	63.6	71.5	78.0	77.0	66.9	49.4	46.9	37.4	54.8
1989	40.7	34.6	47.6	54.1	59.9	71.4	75.6	73.8	67.3	56.4	46.3	25.6	54.4
1990	42.1	43.9	51.3	54.9	62.8	72.2	75.7	73.7	68.2	57.1	50.4	43.1	58.0
1991	35.9	40.4	48.0	60.0	71.8	75.0	78.9	76.2	69.1	59.3	44.9	40.4	58.3
1992	35.6	41.4	45.6	55.8	61.7	69.1	75.6	71.5	67.3	54.9	47.3	37.9	55.3
1993	39.3	33.7	41.7	54.3	65.2	72.2	79.4	76.7	66.8	55.2	46.6	36.3	55.6
Record Mean	33.8	36.5	45.4	55.9	64.5	72.3	75.9	74.7	68.3	57.1	46.3	37.0	55.6
Max	42.6	46.2	56.2	67.9	76.5	83.7	86.7	85.6	79.6	68.9	56.3	45.8	66.3
Min	25.0	26.8	34.6	43.9	52.5	60.8	65.2	63.7	56.9	45.3	36.2	28.3	44.9

REFERENCE NOTES

HUNTINGTON, WEST VIRGINIA

GENERAL
T - TRACE AMOUNT.
BLANK ENTRIES DENOTE MISSING/UNREPORTED DATA.
# INDICATES A STATION OR INSTRUMENT RELOCATION.

SPECIFIC

(a) - LENGTH OF RECORD IN YEARS, ALTHOUGH INDIVIDUAL MONTHS MAY BE MISSING.
0.* OR * - THE VALUE IS BETWEEN 0.0 AND 0.05
NORMALS - BASED ON THE 1961-1990 RECORD PERIOD.
EXTREMES - DATES ARE THE MOST RECENT OCCURRENCE.
WIND DIR. - NUMERALS SHOW TENS OF DEGRESS CLOCKWISE FROM TRUE NORTH. "00" INDICATES CALM.
RESULTANT DIRECTIONS ARE GIVEN TO WHOLE DEGREES.
BOLD VALUES INDICATE EXTREME VALUES WHICH OCCURRED AFTER THE ASOS SYSTEM WAS COMMISSIONED.

RECORD = PERIOD OF RECORD
RECORD MEAN PRECIPITATION IS THE MEAN OF ALL DAILY PRECIPITATION AMOUNTS DURING THE PERIOD OF RECORD.
RECORD MAX(MIN) TEMPERATURE IS THE MEAN OF ALL DAILY MAX(MIN) TEMPERATURES DURING THE PERIOD OF RECORD.
RECORD MEAN TEMPERATURE IS THE SUM OF THE RECORD MAX AND RECORD MIN DIVIDED BY 2.
AVERAGE TEMPERATURE IS THE SUM OF THE MEAN DAILY MAX AND MIN TEMPERATURE DIVIDED BY 2.

EXCEPTIONS

RECORD MEANS ARE THROUGH THE CURRENT YEAR, BEGINNING IN 1941 FOR TEMPERATURE
1941 FOR PRECIPITATION
1962 FOR SNOWFALL

HEATING DEGREE DAYS Base 65 deg. F — HUNTINGTON, WEST VIRGINIA

SEASON	JULY	AUG	SEP	OCT	NOV	DEC	JAN	FEB	MAR	APR	MAY	JUNE	TOTAL
1964-65	0	17	69	396	504	786	956	806	741	264	17	7	4563
1965-66	0	10	59	337	510	719	1182	853	583	384	167	20	4824
1966-67	0	3	60	369	535	889	790	919	528	269	194	13	4569
1967-68	7	4	111	326	724	805	1095	1085	590	279	160	18	5204
1968-69	0	11	26	290	532	960	1028	819	810	268	95	18	4857
1969-70	0	0	72	320	681	1057	1163	866	737	264	103	3	5266
1970-71	4	0	28	215	529	800	1089	769	711	322	165	2	4634
1971-72	2	0	11	84	606	597	899	877	661	335	110	57	4239
1972-73	8	2	14	429	664	757	950	866	363	385	158	1	4597
1973-74	0	0	18	159	503	845	710	754	501	248	109	23	3870
1974-75	0	0	113	337	560	812	851	704	696	376	43	4	4496
1975-76	0	0	91	228	445	811	1032	535	422	314	125	3	4006
1976-77	0	1	76	460	765	1003	1397	829	475	200	67	45	5318
1977-78	0	0	35	332	455	919	1232	1122	698	254	164	12	5223
1978-79	0	0	15	334	504	815	1194	1014	484	292	110	12	4774
1979-80	2	5	36	312	494	798	936	1033	708	351	93	16	4784
1980-81	0	0	26	348	632	878	1132	777	690	183	167	0	4833
1981-82	0	0	67	307	577	947	1157	845	588	429	29	8	4954
1982-83	1	8	99	263	464	611	933	750	529	377	156	16	4207
1983-84	0	0	67	248	537	1070	1127	667	713	324	171	3	4927
1984-85	1	0	98	91	660	597	1205	918	507	206	72	18	4373
1985-86	0	0	63	164	348	1001	964	710	550	233	65	3	4101
1986-87	0	19	15	246	567	886	980	765	554	352	53	3	4440
1987-88	0	0	25	426	458	793	1043	858	584	325	104	28	4644
1988-89	0	0	38	485	538	849	744	845	548	340	213	4	4604
1989-90	0	7	70	283	557	1217	702	585	457	331	107	7	4323
1990-91	0	2	65	261	440	670	894	682	533	188	25	0	3760
1991-92	0	0	75	225	597	756	904	678	595	311	160	17	4318
1992-93	0	1	59	312	527	837	789	871	714	323	71	24	4528
1993-94	0	0	55	307	548	885							

COOLING DEGREE DAYS Base 65 deg. F — HUNTINGTON, WEST VIRGINIA

YEAR	JAN	FEB	MAR	APR	MAY	JUNE	JULY	AUG	SEP	OCT	NOV	DEC	TOTAL
1969	0	0	0	8	73	255	405	316	115	26	0	0	1198
1970	0	0	0	21	130	201	275	268	265	25	0	0	1185
1971	0	0	0	0	35	258	225	215	197	56	7	2	995
1972	0	0	1	16	30	102	308	234	147	0	3	0	841
1973	0	0	4	14	40	284	357	344	212	41	0	0	1296
1974	0	0	21	45	100	130	332	279	69	8	7	0	991
1975	0	0	0	15	143	255	319	423	100	17	2	0	1274
1976	0	1	24	51	59	224	244	228	32	4	0	0	867
1977	0	0	25	66	173	173	427	330	156	4	19	0	1373
1978	0	0	0	18	85	261	361	344	234	5	0	0	1308
1979	0	0	13	17	76	173	286	301	113	22	3	0	1004
1980	0	0	0	10	89	190	403	401	206	11	0	0	1310
1981	0	0	3	55	45	262	334	289	140	10	0	0	1138
1982	0	0	0	7	131	88	299	162	72	42	10	11	822
1983	0	0	7	3	35	222	426	466	198	17	0	0	1374
1984	0	0	0	26	71	314	269	336	132	34	5	6	1193
1985	0	0	6	53	75	182	297	227	155	45	7	0	1047
1986	0	0	8	51	125	253	406	273	199	45	0	0	1360
1987	0	0	0	13	176	267	367	391	147	0	5	0	1366
1988	0	0	7	10	68	230	409	379	102	10	2	0	1217
1989	0	0	12	19	61	201	331	285	147	24	0	0	1080
1990	0	0	35	33	47	231	340	281	165	25	8	0	1165
1991	0	0	15	45	245	308	438	355	206	54	4	0	1670
1992	0	0	1	41	66	149	334	210	136	5	0	0	942
1993	0	0	0	6	82	249	453	371	116	10	7	0	1294

SNOWFALL (inches) — HUNTINGTON, WEST VIRGINIA

SEASON	JULY	AUG	SEP	OCT	NOV	DEC	JAN	FEB	MAR	APR	MAY	JUNE	TOTAL
1964-65	0.0	0.0	0.0	0.0	1.0	1.1	10.0	2.7	7.3	0.0	0.0	0.0	22.1
1965-66	0.0	0.0	0.0	0.0	T	1.0	12.9	2.0	0.5	T	0.0	0.0	16.4
1966-67	0.0	0.0	0.0	T	4.1	5.0	3.4	11.7	3.0	0.0	0.0	0.0	27.2
1967-68	0.0	0.0	0.0	0.0	0.5	13.2	9.0	4.1	2.2	0.0	0.0	0.0	29.0
1968-69	0.0	0.0	0.0	0.0	1.5	2.5	2.1	2.8	4.1	0.0	0.0	0.0	13.0
1969-70	0.0	0.0	0.0	0.0	4.6	9.9	9.2	7.9	7.2	T	0.0	0.0	38.8
1970-71	0.0	0.0	0.0	0.0	0.1	4.7	7.1	9.8	11.0	0.0	0.0	0.0	32.7
1971-72	0.0	0.0	0.0	0.0	4.0	0.2	6.0	6.6	4.5	T	0.0	0.0	21.3
1972-73	0.0	0.0	0.0	0.2	3.6	0.8	2.6	1.5	3.4	0.7	0.0	0.0	12.8
1973-74	0.0	0.0	0.0	0.0	T	2.0	0.3	7.5	1.0	0.8	0.0	0.0	11.6
1974-75	0.0	0.0	0.0	0.4	T	5.9	14.7	1.9	6.3	T	0.0	0.0	29.2
1975-76	0.0	0.0	0.0	0.0	0.3	1.6	10.2	1.2	6.2	0.0	0.0	0.0	19.5
1976-77	0.0	0.0	0.0	T	2.2	4.8	23.7	5.3	0.8	T	0.0	0.0	36.8
1977-78	0.0	0.0	0.0	T	3.5	2.0	30.3	9.7	10.2	0.0	0.0	0.0	55.7
1978-79	0.0	0.0	0.0	0.0	T	0.6	14.7	18.1	0.3	T	0.0	0.0	33.7
1979-80	0.0	0.0	0.0	0.0	0.4	0.3	9.2	9.3	5.5	0.2	0.0	0.0	24.9
1980-81	0.0	0.0	0.0	T	0.7	1.1	4.6	2.4	3.3	0.0	0.0	0.0	12.1
1981-82	0.0	0.0	0.0	0.0	T	6.4	7.6	5.1	4.2	T	0.0	0.0	23.3
1982-83	0.0	0.0	0.0	0.0	T	0.7	3.3	11.1	3.9	T	0.0	0.0	19.0
1983-84	0.0	0.0	0.0	0.0	T	2.0	11.7	8.7	2.2	0.0	0.0	0.0	24.6
1984-85	0.0	0.0	0.0	0.0	0.1	2.5	19.5	21.2	0.2	0.4	0.0	0.0	43.9
1985-86	0.0	0.0	0.0	0.0	0.0	5.8	10.4	23.9	1.5	T	0.0	0.0	41.6
1986-87	0.0	0.0	0.0	0.0	0.7	T	9.1	6.1	2.8	14.4	0.0	0.0	33.1
1987-88	0.0	0.0	0.0	T	4.6	3.7	7.6	3.4	1.9	T	0.0	0.0	21.2
1988-89	0.0	0.0	0.0	0.0	T	2.5	0.3	2.0	T	T	0.2	0.0	5.0
1989-90	0.0	0.0	0.0	T	2.7	9.9	0.8	2.9	5.3	0.7	0.0	T	22.3
1990-91	0.0	0.0	0.0	0.0	0.0	2.6	2.8	5.1	4.8	0.0	0.0	0.0	15.3
1991-92	0.0	T	0.0	0.0	0.4	0.5	2.3	0.4	4.6	3.0	T	0.0	11.2
1992-93	0.0	0.0	0.0	T	1.3	5.9	0.5	13.6	23.6	0.1	T	0.0	45.0
1993-94	0.0	0.0	0.0	1.5	T	10.7							
Record Mean	0.0	T	0.0	0.1	1.2	3.9	8.3	7.3	4.3	0.6	T	T	25.7

GREEN BAY, WISCONSIN

The Green Bay climate is modified by surrounding topography. The modification is caused by the Bay of Green Bay, Lakes Michigan, and Superior, and to a lesser extent, the slightly higher surrounding terrain terminating in the Fox River Valley. The city of Green Bay is located at the mouth of the Fox River, one of the largest rivers flowing northward in the United States. It empties into the south end of the Bay.

The modified continental climate of Green Bay is shown by the few occurrences of 90 degree temperatures in the summer season and the few occurrences of sub-zero temperatures in the winter season. The narrow temperature range stems from the lake effects and the limited hours of sunshine caused by cloudiness.

Precipitation normally falls in the five-month period May through September. Three-fifths of the annual total is in the growing season, most often falling during thunderstorms. During the winter months, snowfall is less than in nearby communities where the ground is slightly higher.

The comparatively low range in temperature along with the greater portion of the precipitation falling during the growing season is conducive to the development of the dairy industry. Cherry and apple orchards are important crops in nearby lake communities. The growing of potatoes and canning vegetables are predominant inland. Paper products are the major manufacturing industry.

High winds, excessive precipitation, and electrical storms cause occasional damage. Snowstorms are the principal winter hazard. While the winters are long in Green Bay, the extremes are never as severe as the northern latitude location would indicate.

Based on the 1951-1980 period, the average first occurrence of 32 degrees Fahrenheit in the fall is October 2 and the average last occurrence in the spring is May 12.

NORMALS, MEANS, AND EXTREMES

GREEN BAY, WISCONSIN

LATITUDE: 44 °29′N LONGITUDE: 88 °08′ W ELEVATION: FT. GRND 682 BARO 694 TIME ZONE: CENTRAL WBAN: 14898

	(a)	JAN	FEB	MAR	APR	MAY	JUNE	JULY	AUG	SEP	OCT	NOV	DEC	YEAR
TEMPERATURE °F:														
Normals														
-Daily Maximum		22.8	27.1	38.5	54.0	67.2	75.5	80.5	77.5	69.1	57.4	42.0	27.7	53.3
-Daily Minimum		5.8	9.5	21.4	33.9	43.7	53.5	58.9	56.8	48.8	38.5	26.8	12.5	34.2
-Monthly		14.3	18.3	30.0	44.0	55.5	64.5	69.7	67.1	59.0	48.0	34.4	20.2	43.8
Extremes														
-Record Highest	44	50	55	77	89	91	98	99	99	95	88	72	62	99
-Year		1961	1981	1986	1980	1959	1988	1977	1988	1955	1963	1990	1970	AUG 1988
-Record Lowest	44	-31	-26	-29	7	21	32	40	38	24	15	-9	-27	-31
-Year		1951	1971	1962	1954	1966	1958	1965	1967	1949	1966	1976	1983	JAN 1951
NORMAL DEGREE DAYS:														
Heating (base 65°F)		1572	1308	1085	630	318	93	17	43	189	527	918	1389	8089
Cooling (base 65°F)		0	0	0	0	23	78	163	108	9	0	0	0	381
% OF POSSIBLE SUNSHINE	44	49	53	54	54	61	64	65	63	55	47	38	40	54
MEAN SKY COVER (tenths)														
Sunrise - Sunset	44	6.6	6.5	6.7	6.8	6.4	6.0	5.7	5.8	6.1	6.5	7.4	7.1	6.5
MEAN NUMBER OF DAYS:														
Sunrise to Sunset														
-Clear	44	7.8	7.1	7.1	6.3	7.0	7.5	8.2	8.5	8.1	7.2	4.8	6.3	85.9
-Partly Cloudy	44	6.5	6.6	7.6	7.7	9.5	11.0	12.0	10.9	9.3	8.5	6.5	6.0	102.1
-Cloudy	44	16.8	14.5	16.3	16.0	14.4	11.5	10.8	11.6	12.5	15.3	18.8	18.6	177.2
Precipitation														
.01 inches or more	44	10.0	8.0	10.8	11.0	10.8	10.4	9.8	10.4	10.2	9.1	9.8	10.6	121.0
Snow,Ice pellets,hail														
1.0 inches or more	44	3.6	2.8	2.8	0.9	0.*	0.0	0.0	0.0	0.0	0.1	1.4	3.2	14.8
Thunderstorms	44	0.1	0.1	1.2	2.3	3.9	6.5	6.5	5.9	4.0	1.9	0.5	0.2	33.1
Heavy Fog Visibility														
1/4 mile or less	44	1.7	2.4	2.7	2.2	1.4	1.3	1.1	2.3	1.9	2.4	2.3	2.3	24.1
Temperature °F														
-Maximum														
90° and above	32	0.0	0.0	0.0	0.0	0.1	1.9	3.1	1.7	0.2	0.0	0.0	0.0	6.8
32° and below	32	22.9	18.5	8.1	0.4	0.0	0.0	0.0	0.0	0.0	0.0	4.8	19.2	74.0
-Minimum														
32° and below	32	30.6	27.6	26.5	13.8	2.6	0.0	0.0	0.0	0.8	8.5	22.3	29.4	162.0
0° and below	32	11.1	7.2	1.2	0.0	0.0	0.0	0.0	0.0	0.0	0.0	0.2	6.0	25.7
AVG. STATION PRESS.(mb)	21	991.1	992.4	990.1	989.3	989.1	988.6	989.9	991.1	991.3	991.3	990.4	991.3	990.5
RELATIVE HUMIDITY (%)														
Hour 00	32	77	77	78	75	76	79	83	86	86	81	81	80	80
Hour 06 (Local Time)	32	78	79	81	79	79	81	86	90	89	85	83	81	83
Hour 12	32	70	68	65	58	54	57	58	61	62	62	69	73	63
Hour 18	32	73	70	68	60	57	58	61	65	70	71	76	77	67
PRECIPITATION (inches):														
Water Equivalent														
-Normal		1.15	1.03	2.05	2.40	2.82	3.39	3.10	3.50	3.47	2.23	2.16	1.53	28.83
-Maximum Monthly	44	2.64	3.56	4.68	5.52	8.21	10.29	6.83	9.04	7.80	5.00	5.32	3.15	10.29
-Year		1950	1953	1977	1953	1973	1990	1993	1975	1965	1954	1992	1971	JUN 1990
-Minimum Monthly	44	0.12	0.04	0.31	0.49	0.06	0.31	0.83	0.90	0.28	T	0.16	0.10	T
-Year		1981	1969	1978	1989	1988	1976	1981	1955	1976	1952	1976	1960	OCT 1952
-Maximum in 24 hrs	44	1.14	1.78	1.64	2.00	3.28	4.90	2.95	4.60	2.99	3.68	2.30	1.55	4.90
-Year		1980	1966	1990	1981	1973	1990	1959	1975	1964	1954	1985	1959	JUN 1990
Snow,Ice pellets,hail														
-Maximum Monthly	44	28.0	20.6	24.2	11.8	4.3	T	0.0	T	T	1.7	16.5	27.0	28.0
-Year		1982	1962	1989	1977	1990	1992		1993	1965	1959	1985	1977	JAN 1982
-Maximum in 24 hrs	44	8.8	9.2	12.7	10.2	4.3	T	0.0	T	T	1.6	8.2	14.4	14.4
-Year		1988	1959	1989	1977	1990	1992		1993	1965	1989	1977	1990	DEC 1990
WIND:														
Mean Speed (mph)	44	10.9	10.5	10.8	11.2	10.1	9.2	8.2	7.9	8.9	9.9	10.9	10.5	9.9
Prevailing Direction through 1963		SW	SW	NE	NE	NE	SW	SW	SW	SW	SW	SW	SW	SW
Fastest Obs. 1 Min.														
-Direction (!!)	8	22	35	21	22	29	25	30	28	22	22	33	35	29
-Speed (MPH)	8	30	31	35	32	46	29	35	35	30	30	35	32	46
-Year		1993	1987	1990	1993	1989	1985	1988	1985	1993	1991	1989	1989	MAY 1989
Peak Gust														
-Direction (!!)	10	W	NW	S	NE	W	NW	NW	W	SW	SW	NW	NE	W
-Speed (mph)	10	46	46	55	49	81	49	56	53	47	44	49	53	81
-Date		1990	1987	1990	1993	1989	1984	1988	1985	1985	1984	1989	1990	MAY 1989

PRECIPITATION (inches)

GREEN BAY, WISCONSIN

YEAR	JAN	FEB	MAR	APR	MAY	JUNE	JULY	AUG	SEP	OCT	NOV	DEC	ANNUAL
1964	1.14	0.26	1.76	2.55	4.14	1.05	4.55	2.72	6.74	0.44	2.07	0.70	28.12
1965	0.93	0.85	2.38	3.62	3.95	1.89	1.96	3.38	7.80	1.32	2.19	2.31	32.58
1966	1.18	2.25	2.46	1.38	1.28	1.09	4.19	2.65	1.21	0.72	1.58	1.65	21.64
1967	2.52	0.84	1.13	2.77	2.45	8.47	1.96	2.43	0.46	4.71	1.66	1.17	30.57
1968	0.94	0.45	0.97	4.84	3.10	6.97	2.00	2.66	3.31	1.01	1.01	2.69	29.95
1969	2.60	0.04	1.04	2.86	2.66	7.62	2.51	1.19	2.03	3.46	0.43	1.43	27.87
1970	0.73	0.23	1.07	1.61	5.76	1.11	4.02	1.25	6.11	2.98	2.68	1.24	28.79
1971	1.60	2.03	2.04	1.05	1.67	1.87	3.44	2.99	3.36	2.01	3.21	3.15	28.42
1972	0.65	0.96	2.19	1.45	0.82	2.25	1.85	5.86	5.76	1.84	1.15	2.49	27.27
1973	1.86	0.72	2.43	3.23	8.21	3.20	1.93	2.57	2.91	3.96	1.45	2.41	34.88
1974	1.71	1.17	1.07	2.62	4.46	4.91	4.25	1.61	1.05	1.72	2.09	1.67	28.33
1975	1.52	1.48	3.44	2.35	2.79	5.27	1.78	9.04	3.18	0.36	3.42	0.84	35.47
1976	1.72	1.33	3.65	2.44	2.42	0.31	2.96	1.15	0.28	0.82	0.16	0.61	17.85
1977	0.67	1.38	4.68	3.33	2.47	2.27	2.13	2.37	2.44	1.36	2.70	2.31	28.11
1978	1.33	0.35	0.31	3.44	3.38	2.72	6.03	4.36	4.82	2.33	2.93	1.30	33.30
1979	1.78	1.17	4.49	1.93	3.01	2.21	3.55	5.97	0.76	2.72	2.49	1.28	31.36
1980	1.92	0.35	1.00	2.73	1.77	3.82	1.87	7.31	3.42	1.79	1.25	1.35	28.58
1981	0.12	2.76	0.42	4.22	0.56	2.63	0.83	3.37	3.25	3.44	1.08	1.10	23.78
1982	1.34	0.14	1.95	2.66	2.74	2.67	5.10	2.91	1.43	1.20	4.51	2.50	29.15
1983	0.72	1.46	1.52	1.39	4.80	1.82	3.76	5.27	3.59	2.24	2.63	1.18	30.38
1984	0.59	1.59	1.64	3.33	1.65	5.60	3.17	3.78	5.66	4.92	2.55	1.72	36.20
1985	0.86	2.55	2.70	2.24	2.58	2.21	4.03	8.03	3.65	2.72	4.96	1.83	38.36
1986	0.60	0.83	2.48	2.26	1.15	4.06	4.95	3.85	7.51	1.89	1.27	0.48	31.33
1987	0.47	0.39	1.53	2.33	2.58	1.83	2.18	3.41	1.57	1.76	3.07	2.04	23.16
1988	1.79	0.73	1.10	2.53	0.06	0.67	2.34	3.47	4.11	1.96	4.43	0.84	24.03
1989	0.41	0.38	2.88	0.49	4.22	1.56	2.27	1.05	0.58	4.76	1.25	0.55	20.40
1990	0.64	0.58	3.25	1.28	3.99	10.29	2.93	2.51	5.13	2.34	1.61	2.10	36.65
1991	0.57	0.37	2.87	2.77	2.42	1.08	4.16	2.11	2.55	3.50	2.72	1.42	26.54
1992	0.72	0.55	2.48	3.01	1.54	1.61	4.18	2.10	5.61	0.92	5.32	2.27	30.31
1993	1.42	0.34	0.76	3.99	4.28	6.82	6.83	2.30	2.78	2.29	1.56	0.44	33.81
Record Mean	1.29	1.25	1.94	2.57	3.09	3.40	3.14	3.05	3.20	2.15	2.03	1.43	28.54

AVERAGE TEMPERATURE (deg. F)

GREEN BAY, WISCONSIN

YEAR	JAN	FEB	MAR	APR	MAY	JUNE	JULY	AUG	SEP	OCT	NOV	DEC	ANNUAL
1964	20.9	21.4	27.4	44.1	59.5	65.4	71.6	65.4	58.5	47.6	39.1	19.9	45.1
1965	14.4	15.3	24.4	40.4	60.0	63.6	68.8	67.0	57.5	49.3	35.3	28.7	43.7
1966	7.9	21.8	33.6	41.6	50.9	67.0	70.9	64.7	56.9	47.7	34.2	20.3	43.2
1967	20.1	13.2	29.5	44.2	49.8	66.2	68.3	63.9	58.0	46.7	30.9	22.5	42.8
1968	16.4	16.7	38.0	46.5	52.6	64.7	68.3	67.5	61.6	50.5	34.8	21.9	45.0
1969	14.0	20.7	24.4	43.5	54.3	57.2	68.0	70.4	58.7	45.0	31.8	20.9	42.4
1970	7.2	15.5	28.6	46.8	54.9	66.2	71.7	69.1	60.3	49.9	34.8	19.8	43.7
1971	6.9	15.2	25.9	43.7	53.0	69.8	67.8	65.3	62.3	54.6	34.7	23.5	43.6
1972	10.2	14.4	24.4	39.4	59.6	64.2	69.8	68.1	59.5	45.3	35.0	17.3	42.3
1973	21.2	22.9	39.5	43.8	53.0	68.2	71.7	71.8	59.5	54.5	34.9	20.9	46.8
1974	16.9	17.1	29.7	45.5	51.6	61.5	70.2	66.3	54.2	45.8	34.4	24.5	43.2
1975	20.3	18.4	24.1	38.9	60.0	65.3	71.0	68.0	54.6	49.9	39.1	21.2	44.2
1976	12.8	25.8	30.8	45.8	52.2	68.1	71.8	66.9	57.7	43.3	26.4	9.1	42.6
1977	3.1	20.0	37.0	48.2	63.4	63.8	73.3	65.5	59.5	47.3	33.3	18.0	44.4
1978	11.4	11.2	25.2	40.3	57.1	63.6	67.3	68.5	62.4	47.1	33.0	18.3	42.1
1979	5.9	9.0	28.8	41.7	52.4	64.6	70.4	66.5	60.0	45.8	33.8	27.7	42.2
1980	17.7	17.2	27.2	45.1	58.3	62.6	70.6	69.1	59.1	43.4	35.0	19.7	43.8
1981	15.3	22.9	34.7	45.7	53.6	65.6	69.3	68.2	56.8	45.0	36.9	22.9	44.7
1982	6.7	15.8	28.4	40.3	60.8	59.3	70.8	65.1	57.6	49.3	33.3	28.0	43.0
1983	21.4	26.3	31.1	40.5	48.8	64.5	72.9	70.8	59.8	48.5	36.5	10.6	44.3
1984	12.7	28.3	25.3	44.5	51.7	67.4	68.6	69.5	57.5	51.0	34.8	24.0	44.6
1985	12.2	17.6	34.2	47.6	58.5	62.3	69.1	66.4	60.8	48.3	30.9	9.4	43.1
1986	16.7	18.1	32.5	48.1	57.7	63.9	71.5	64.3	59.3	48.0	29.7	24.8	44.6
1987	21.6	27.9	35.1	49.1	58.8	69.0	73.0	67.6	61.1	43.2	37.6	27.1	47.6
1988	12.5	15.0	32.1	44.1	59.8	68.3	73.4	72.3	61.0	42.4	37.1	20.7	44.9
1989	25.5	13.0	25.3	41.9	54.9	63.4	70.9	68.7	59.0	49.6	31.4	11.2	42.9
1990	26.5	22.3	34.0	47.6	52.5	66.0	68.6	67.6	61.8	47.2	40.0	21.1	46.3
1991	13.8	23.5	33.6	47.9	61.4	69.2	69.8	69.9	58.5	48.2	30.2	24.5	45.9
1992	22.9	27.1	30.4	42.0	56.6	62.3	64.9	64.3	58.5	46.8	32.8	23.7	44.4
1993	19.6	19.0	30.2	40.4	56.2	62.8	70.3	70.3	55.4	45.9	33.7	25.3	44.1
Record Mean	16.0	18.3	29.4	43.5	55.2	65.2	70.3	67.9	60.0	48.6	34.5	21.8	44.3
Max	24.0	26.6	37.4	52.7	66.0	75.6	80.8	78.2	69.8	57.6	41.5	28.8	53.3
Min	8.0	9.9	21.3	34.2	44.3	54.7	59.7	57.7	50.2	39.6	27.4	14.8	35.2

REFERENCE NOTES

GREEN BAY, WISCONSIN

GENERAL
T - TRACE AMOUNT.
BLANK ENTRIES DENOTE MISSING/UNREPORTED DATA.
# INDICATES A STATION OR INSTRUMENT RELOCATION.

SPECIFIC

(a) - LENGTH OF RECORD IN YEARS, ALTHOUGH INDIVIDUAL MONTHS MAY BE MISSING.
0.* OR * - THE VALUE IS BETWEEN 0.0 AND 0.05
NORMALS - BASED ON THE 1961-1990 RECORD PERIOD.
EXTREMES - DATES ARE THE MOST RECENT OCCURRENCE
WIND DIR.- NUMERALS SHOW TENS OF DEGRESS CLOCKWISE FROM TRUE NORTH. "00" INDICATES CALM.
RESULTANT DIRECTIONS ARE GIVEN TO WHOLE DEGREES.
BOLD VALUES INDICATE EXTREME VALUES WHICH OCCURRED AFTER THE ASOS SYSTEM WAS COMMISSIONED.

RECORD = PERIOD OF RECORD
RECORD MEAN PRECIPITATION IS THE MEAN OF ALL DAILY PRECIPITATION AMOUNTS DURING THE PERIOD OF RECORD.
RECORD MAX(MIN) TEMPERATURE IS THE MEAN OF ALL DAILY MAX(MIN) TEMPERATURES DURING THE PERIOD OF RECORD.
RECORD MEAN TEMPERATURE IS THE SUM OF THE RECORD MAX AND RECORD MIN DIVIDED BY 2.
AVERAGE TEMPERATURE IS THE SUM OF THE MEAN DAILY MAX AND MIN TEMPERATURE DIVIDED BY 2.

EXCEPTIONS

RECORD MEANS ARE THROUGH THE CURRENT YEAR,
BEGINNING IN 1886 FOR TEMPERATURE
1886 FOR PRECIPITATION
1950 FOR SNOWFALL

HEATING DEGREE DAYS Base 65 deg. F — GREEN BAY, WISCONSIN

SEASON	JULY	AUG	SEP	OCT	NOV	DEC	JAN	FEB	MAR	APR	MAY	JUNE	TOTAL
1964-65	15	74	229	531	771	1392	1564	1388	1249	733	197	86	8229
1965-66	26	51	245	485	885	1118	1763	1207	967	695	437	79	7958
1966-67	10	74	253	531	917	1377	1386	1444	1093	620	467	44	8216
1967-68	40	100	235	563	1017	1313	1506	1395	832	551	378	97	8027
1968-69	39	82	130	467	899	1327	1580	1234	1252	641	346	247	8244
1969-70	38	11	223	619	991	1362	1788	1381	1121	550	333	73	8490
1970-71	17	17	185	466	898	1398	1797	1391	1206	636	365	43	8419
1971-72	32	57	155	328	902	1282	1694	1463	1253	762	215	92	8235
1972-73	37	54	180	606	895	1469	1350	1172	785	632	367	9	7556
1973-74	4	11	221	330	897	1363	1487	1334	1087	590	413	132	7869
1974-75	6	39	331	588	912	1248	1381	1297	1260	779	205	90	8136
1975-76	29	27	307	465	769	1351	1614	1130	1057	579	394	29	7751
1976-77	2	56	255	667	1152	1730	1918	1254	861	500	131	113	8639
1977-78	3	64	162	544	942	1454	1658	1500	1227	733	279	107	8673
1978-79	35	18	152	549	953	1442	1830	1564	1117	691	389	70	8810
1979-80	14	36	166	590	928	1151	1461	1380	1165	597	227	130	7845
1980-81	11	11	189	661	893	1398	1538	1175	932	573	350	45	7776
1981-82	26	21	248	614	839	1300	1805	1373	1127	733	152	178	8416
1982-83	3	75	250	483	946	1140	1344	1077	1046	727	495	100	7686
1983-84	17	2	210	507	847	1682	1617	1055	1223	611	406	13	8190
1984-85	18	20	237	430	899	1262	1632	1324	949	533	204	114	7622
1985-86	9	34	196	508	1016	1719	1491	1307	1002	508	244	90	8124
1986-87	12	65	191	519	1052	1240	1341	1033	918	478	240	36	7125
1987-88	18	44	132	673	815	1167	1623	1447	1012	624	201	74	7830
1988-89	4	23	146	694	830	1365	1216	1451	1224	687	315	98	8053
1989-90	7	19	200	475	1000	1666	1189	1191	952	547	380	55	7681
1990-91	24	28	157	547	744	1357	1579	1154	967	516	220	23	7316
1991-92	17	17	250	515	1038	1248	1296	1093	1067	683	275	122	7621
1992-93	44	87	209	560	960	1274	1398	1281	1071	731	277	108	8000
1993-94	1	16	292	585	930	1226							

COOLING DEGREE DAYS Base 65 deg. F — GREEN BAY, WISCONSIN

YEAR	JAN	FEB	MAR	APR	MAY	JUNE	JULY	AUG	SEP	OCT	NOV	DEC	TOTAL
1969	0	0	0	0	21	17	138	184	40	3	0	0	403
1970	0	0	0	10	26	114	231	150	50	5	0	0	586
1971	0	0	0	0	1	195	125	71	77	15	0	0	484
1972	0	0	0	0	52	75	190	156	22	0	0	0	495
1973	0	0	0	2	0	112	220	228	63	13	0	0	638
1974	0	0	0	7	6	35	175	87	13	0	0	0	323
1975	0	0	0	0	56	106	221	126	3	2	0	0	514
1976	0	0	0	8	1	129	219	122	41	0	0	0	520
1977	0	0	0	5	87	81	269	87	5	0	0	0	534
1978	0	0	0	0	43	71	115	131	80	0	0	0	440
1979	0	0	0	0	6	68	191	91	23	1	0	0	380
1980	0	0	0	5	27	64	192	146	18	0	0	0	452
1981	0	0	0	0	5	71	168	127	9	0	0	0	380
1982	0	0	0	0	30	16	187	85	35	4	0	0	357
1983	0	0	0	0	0	95	270	188	60	4	0	0	617
1984	0	0	0	0	3	94	136	165	17	0	0	0	415
1985	0	0	0	16	11	41	141	85	79	0	0	0	373
1986	0	0	0	8	25	65	220	48	27	0	0	0	393
1987	0	0	0	7	56	161	274	133	23	0	0	0	654
1988	0	0	0	0	46	182	270	255	33	0	0	0	786
1989	0	0	0	0	7	55	199	141	27	0	0	0	429
1990	0	0	0	34	0	92	140	116	70	0	0	0	452
1991	0	0	0	8	115	155	171	177	60	0	0	0	686
1992	0	0	0	0	22	49	49	69	23	1	0	0	213
1993	0	0	0	0	13	48	173	187	11	0	0	0	432

SNOWFALL (inches) — GREEN BAY, WISCONSIN

SEASON	JULY	AUG	SEP	OCT	NOV	DEC	JAN	FEB	MAR	APR	MAY	JUNE	TOTAL
1964-65	0.0	0.0	0.0	T	0.1	6.3	12.5	6.3	17.0	5.5	0.0	0.0	47.7
1965-66	0.0	0.0	T	0.0	0.3	2.8	14.6	1.1	4.5	0.9	T	0.0	24.2
1966-67	0.0	0.0	0.0	T	2.9	11.9	16.4	9.9	4.0	0.2	T	0.0	45.3
1967-68	0.0	0.0	0.0	0.5	2.9	3.7	9.2	3.0	0.6	0.7	0.0	0.0	20.6
1968-69	0.0	0.0	0.0	0.0	2.4	26.7	11.8	0.5	2.2	T	0.0	0.0	43.6
1969-70	0.0	0.0	0.0	0.1	1.7	16.5	9.8	2.9	8.8	T	0.0	0.0	39.8
1970-71	0.0	0.0	0.0	0.0	5.9	10.7	20.8	11.0	14.8	0.2	T	0.0	63.4
1971-72	0.0	0.0	0.0	0.0	12.6	13.2	8.2	12.1	22.2	2.3	0.0	0.0	70.6
1972-73	0.0	0.0	0.0	T	0.7	17.3	6.8	5.7	T	5.6	T	0.0	36.1
1973-74	0.0	0.0	0.0	0.0	2.4	15.5	4.4	18.5	11.2	2.0	T	0.0	54.0
1974-75	0.0	0.0	0.0	T	1.5	10.4	8.6	19.5	13.4	T	0.0	0.0	53.4
1975-76	0.0	0.0	0.0	0.0	4.4	9.3	24.9	12.4	11.5	0.3	0.1	0.0	62.9
1976-77	0.0	0.0	0.0	1.4	1.7	12.1	12.1	2.4	13.7	11.8	0.0	0.0	55.2
1977-78	0.0	0.0	0.0	0.0	14.1	27.0	18.0	7.1	1.8	T	0.0	0.0	68.0
1978-79	0.0	0.0	0.0	0.0	9.6	15.0	24.0	11.8	6.2	5.0	0.0	0.0	71.6
1979-80	0.0	0.0	0.0	T	6.2	1.4	7.3	4.4	11.4	7.4	0.0	0.0	38.1
1980-81	0.0	0.0	0.0	T	3.4	14.5	2.3	7.5	2.5	T	0.0	0.0	30.2
1981-82	0.0	0.0	0.0	1.0	2.7	11.4	28.0	2.0	6.0	2.9	0.0	0.0	54.0
1982-83	0.0	0.0	0.0	T	2.6	1.3	8.5	15.3	11.4	0.6	0.0	0.0	39.7
1983-84	0.0	0.0	0.0	0.0	7.2	12.7	10.5	1.8	6.5	T	0.0	0.0	38.7
1984-85	0.0	0.0	0.0	0.0	2.0	15.9	13.8	15.1	17.7	6.2	0.0	0.0	70.7
1985-86	0.0	0.0	0.0	0.0	16.5	22.7	6.8	9.9	6.8	0.5	0.0	0.0	63.2
1986-87	0.0	0.0	0.0	T	9.1	5.2	7.8	1.6	10.8	1.6	0.0	0.0	36.1
1987-88	0.0	0.0	0.0	T	1.8	15.8	20.6	8.9	1.8	0.4	0.0	0.0	49.3
1988-89	0.0	0.0	0.0	T	11.5	5.8	1.5	8.6	24.2	0.3	T	0.0	51.9
1989-90	0.0	0.0	0.0	1.6	1.5	15.3	10.7	6.9	2.7	2.7	4.3	0.0	45.7
1990-91	0.0	0.0	0.0	T	1.2	26.9	8.8	6.9	5.3	7.5	0.0	0.0	56.6
1991-92	0.0	0.0	0.0	T	8.7	10.7	4.3	7.4	9.0	2.8	0.0	T	42.9
1992-93	0.0	T	0.0	1.2	8.5	15.5	11.8	5.6	7.2	7.2	T	0.0	57.0
1993-94	0.0	T	0.0	T	1.6	1.9							
Record Mean	0.0	T	T	0.2	4.7	11.0	10.5	8.2	8.9	2.4	0.2	T	46.0

LA CROSSE, WISCONSIN

The city of La Crosse is situated on the east bank of the Mississippi River at the confluence of the Mississippi, Black, and La Crosse Rivers. The official records are taken at the La Crosse Municipal Airport which is 6 1/2 miles north of the main Post Office, on the north end of French Island. This island is about 6 miles long from north to south and 2 to 4 miles wide with the Mississippi River to the west and the old channel of the Black River to the east. A rather level sandy plain exists on each side of the river extending between the Wisconsin and Minnesota bluffs which rise 450 to 500 feet above the valley floor. The distance from bluff to bluff averages about 5 miles. The Mississippi River bends to the northwest and continues directly southward from the city.

The prevailing winds in the area are from the northwest from January through April and southerly during the remainder of the year. The situation of the city and airport in a natural bowl between the hills results in somewhat colder temperatures at night due to the settling of cooler air. Valley fogs often persist to mid-forenoon. Steepsided hills with narrow valleys are characteristic of most of the surrounding area.

The flow of the Mississippi River is regulated by dams built for the purpose of navigation, but the reservoirs have limited storage capacity. La Crosse is in the area of Pool No. 8 with a mean sea level elevation of 631 feet. When the river reaches an elevation of 639 feet, with open gate operation, there is considerable flooding of land near the river and some industrial sections of the city.

The invigorating continental-type climate results in wide and frequent variations in temperature. General storms moving eastward or northeastward into our area bring warmer weather and supply most of our moisture. These are usually followed by cooler air from Canada. The winters are cold and humid. The summers are warm with moderate humidities, while periods of hot and humid weather occur occasionally, usually lasting from a few days to a week at a time.

Sixty percent of the precipitation falls during the main growing season, extending from May through September. Most of the summer rainfall occurs during scattered thunderstorms. Some damage from heavy rains, high winds, and hail occurs each year, but tornadoes are infrequent and cover very small areas. Snow is frequent and is the predominant form of precipitation in winter. Heavy snow sometimes falls with larger amounts over the ridges. Glaze storms are not numerous since La Crosse is north of the main path of freezing rain.

Farming is diversified with dairying the leading activity. The more important field crops are corn, oats, and hay. Some of the more specialized crops are soybeans, tobacco, small fruits, and cranberries. Commercial apple orchards are numerous across the Mississippi River in Minnesota.

Based on the 1951-1980 period, the average first occurrence of 32 degrees Fahrenheit in the fall is October 13 and the average last occurrence in the spring is April 29.

NORMALS, MEANS, AND EXTREMES

LA CROSSE, WISCONSIN

LATITUDE: 43°49'N LONGITUDE: 091°15' W ELEVATION: FT. GRND 650 BARO TIME ZONE: CENTRAL WBAN: 14920

	(a)	JAN	FEB	MAR	APR	MAY	JUNE	JULY	AUG	SEP	OCT	NOV	DEC	YEAR
TEMPERATURE °F:														
Normals														
-Daily Maximum		23.5	29.6	42.1	58.5	70.9	79.8	84.5	81.6	72.4	60.2	43.4	28.1	56.2
-Daily Minimum		5.3	10.1	23.6	37.0	48.1	57.1	62.3	59.9	51.2	40.2	27.7	12.5	36.3
-Monthly		14.4	19.9	32.8	47.7	59.5	68.5	73.5	70.8	61.8	50.2	35.6	20.3	46.3
Extremes														
-Record Highest	41	57	64	84	93	94	102	104	105	100	93	75	64	105
-Year		1981	1981	1986	1980	1992	1988	1980	1988	1978	1963	1978	1982	AUG 1988
-Record Lowest	42	-37	-36	-28	7	26	37	33	40	28	14	-9	-30	-37
-Year		1951	1971	1962	1982	1989	1978	1982	1965	1967	1988	1977	1983	JAN 1951
NORMAL DEGREE DAYS:														
Heating (base 65°F)		1569	1263	998	519	220	36	7	16	131	464	882	1386	7491
Cooling (base 65°F)		0	0	0	0	50	141	270	196	35	0	0	0	692
% OF POSSIBLE SUNSHINE														
MEAN SKY COVER (tenths)														
Sunrise - Sunset	17	6.7	6.2	6.7	6.7	6.5	6.2	5.6	5.7	5.8	5.7	7.1	7.1	6.3
MEAN NUMBER OF DAYS:														
Sunrise to Sunset														
-Clear	17	7.4	7.9	7.4	6.8	6.9	7.3	9.8	9.5	9.1	10.8	5.8	6.8	95.4
-Partly Cloudy	17	6.9	7.1	7.1	7.1	9.1	10.4	10.9	10.6	8.9	6.9	6.1	5.8	96.8
-Cloudy	17	16.7	13.2	16.5	16.1	15.0	12.3	10.4	10.9	11.9	13.4	18.1	18.5	173.0
Precipitation														
.01 inches or more	41	8.1	6.9	9.8	10.2	11.2	10.8	10.0	9.8	9.7	7.9	8.3	8.9	111.5
Snow,Ice pellets,hail														
1.0 inches or more	41	3.0	2.4	2.6	0.5	0.0	0.0	0.0	0.0	0.0	0.1	1.3	3.0	12.9
Thunderstorms	40	0.1	0.2	1.2	3.0	5.5	7.7	7.5	7.0	4.7	2.0	0.6	0.2	39.9
Heavy Fog Visibility														
1/4 mile or less	40	1.2	1.3	1.7	0.7	0.7	1.1	1.6	3.6	3.5	2.0	1.3	1.1	19.8
Temperature °F														
-Maximum														
90° and above	40	0.0	0.0	0.0	0.1	0.8	3.5	6.7	4.3	1.2	0.*	0.0	0.0	16.7
32° and below	40	22.0	15.5	6.6	0.2	0.0	0.0	0.0	0.0	0.0	0.0	5.2	18.3	67.8
-Minimum														
32° and below	41	30.6	27.2	25.6	9.6	0.7	0.0	0.0	0.0	0.4	6.4	21.3	29.4	151.4
0° and below	41	11.4	7.2	1.2	0.0	0.0	0.0	0.0	0.0	0.0	0.0	0.4	5.8	26.0
AVG. STATION PRESS.(mb)	16	994.1	994.7	991.5	990.5	989.9	989.8	991.4	992.0	992.8	993.4	992.4	994.1	992.2
RELATIVE HUMIDITY (%)														
Hour 00	33	77	77	77	73	75	81	82	88	87	78	80	79	80
Hour 06 (Local Time)	41	77	79	81	79	80	84	87	91	90	84	82	81	83
Hour 12	41	67	65	61	53	52	55	57	60	61	58	67	71	61
Hour 18	41	71	68	63	53	52	56	57	61	65	64	72	75	63
PRECIPITATION (inches):														
Water Equivalent														
-Normal		0.93	0.90	1.98	2.88	3.26	3.90	3.79	3.92	3.79	2.20	1.73	1.27	30.55
-Maximum Monthly	41	2.86	2.58	3.82	7.31	8.83	10.79	9.35	9.84	10.52	5.09	6.23	2.91	10.79
-Year		1967	1959	1951	1973	1960	1993	1987	1980	1965	1984	1991	1990	JUN 1993
-Minimum Monthly	41	0.14	0.05	0.30	0.60	0.94	1.33	0.16	0.54	0.42	0.02	T	0.30	T
-Year		1981	1969	1978	1966	1988	1989	1967	1976	1952	1952	1976	1962	NOV 1976
-Maximum in 24 hrs	27	1.31	1.06	1.64	3.84	2.72	3.94	5.24	3.92	2.85	2.17	2.80	1.42	5.24
-Year		1967	1966	1966	1954	1960	1967	1987	1962	1992	1966	1991	1990	JUL 1987
Snow,Ice pellets,hail														
-Maximum Monthly	41	29.7	31.0	33.5	17.0	0.8	0.0	0.0	0.0	T	1.8	30.3	30.4	33.5
-Year		1979	1959	1959	1973	1960				1985	1992	1991	1990	MAR 1959
-Maximum in 24 hrs	27	8.7	10.9	15.7	7.3	0.8	0.0	0.0	0.0	T	1.8	13.0	14.4	15.7
-Year		1963	1959	1959	1952	1960				1985	1992	1991	1990	MAR 1959
WIND:														
Mean Speed (mph)	41	8.7	8.6	9.3	10.3	9.4	8.4	7.6	7.4	8.2	9.2	9.7	8.8	8.8
Prevailing Direction through 1963		S	NW	NW	NW	S	S	S	S	S	S	S	S	S
Fastest Obs. 1 Min.														
-Direction (!!)	18	32	34	34	25	09	34	27	32	27	34	18	34	34
-Speed (MPH)	18	45	37	40	53	58	63	52	63	40	39	46	43	63
-Year		1962	1963	1960	1964	1952	1963	1961	1963	1961	1964	1958	1957	JUN 1963
Peak Gust														
-Direction (!!)														
-Speed (mph)														
-Date														

PRECIPITATION (inches)

LA CROSSE, WISCONSIN

YEAR	JAN	FEB	MAR	APR	MAY	JUNE	JULY	AUG	SEP	OCT	NOV	DEC	ANNUAL
1964	0.34	0.09	1.30	2.37	3.72	3.13	3.39	3.34	5.48	0.25	1.09	0.86	25.36
1965	0.81	0.72	2.05	4.87	5.11	3.51	4.48	2.85	10.52	1.66	2.98	2.02	41.58
1966	0.77	1.51	3.31	0.60	2.28	3.47	4.60	1.87	0.96	3.32	0.30	1.20	24.19
1967	2.86	1.11	2.14	2.43	1.22	9.33	0.16	2.44	1.38	2.52	0.18	0.51	26.28
1968	0.87	0.06	0.98	4.46	3.20	9.53	4.06	1.96	5.10	2.59	0.61	2.24	35.66
1969	1.95	0.05	0.78	1.83	2.55	6.39	3.00	2.16	2.32	3.73	0.58	1.69	27.03
1970	0.56	0.33	2.17	2.58	5.82	2.46	4.48	1.02	5.30	3.43	2.60	0.97	31.72
1971	1.52	2.06	1.11	1.74	4.89	4.33	4.06	2.22	2.43	1.09	2.20	2.55	30.20
1972	0.62	0.50	1.68	2.38	1.80	2.81	6.03	5.97	6.72	4.01	1.45	2.46	36.43
1973	0.85	1.38	3.69	7.31	6.21	4.47	2.53	8.63	3.68	2.06	1.93	1.37	44.11
1974	0.40	1.65	2.13	1.83	3.50	5.30	1.76	4.20	1.40	1.91	1.11	1.39	26.58
1975	1.50	1.71	2.14	6.07	2.52	2.88	1.42	2.79	1.28	0.36	3.07	0.70	26.44
1976	0.69	0.64	2.82	4.37	2.95	1.82	2.15	0.54	0.54	0.28	T	0.67	17.47
1977	0.88	0.83	2.64	3.62	3.04	3.51	4.32	2.92	4.12	1.87	1.42	1.40	30.57
1978	0.76	0.68	0.30	4.01	3.97	7.02	9.17	1.20	5.12	0.80	2.35	0.93	36.31
1979	2.41	0.65	2.02	1.79	5.68	1.58	3.52	7.55	0.53	4.41	2.44	0.67	33.25
1980	1.61	0.35	0.65	1.74	2.66	3.57	2.26	9.84	8.51	2.36	0.14	0.61	34.30
1981	0.14	2.10	0.60	4.37	1.88	2.60	7.66	9.56	1.77	1.89	1.08	0.86	34.51
1982	1.34	0.17		1.65	4.71	1.35	1.89	2.91	2.67	2.83	3.64	2.03	
1983	0.89	2.27	1.60	2.37	4.50	1.67	3.16	3.06	4.94	3.35	3.72	0.68	32.21
1984	0.28	0.92	1.94	3.65	2.18	7.43	3.00	2.11	2.87	5.09	1.36	2.42	33.25
1985	0.88	1.27	2.66	2.85	1.08	2.82	2.39	3.21	5.63				
1986		0.77	1.91	3.42	1.49	4.04	4.79	2.38	8.10	3.68	0.95	0.38	
1987	1.17	0.30	2.23	2.35	4.58	2.55	9.35	3.71	2.10	0.56	2.67	1.82	33.39
1988	1.09	0.19	1.89	2.01	0.94	3.25	2.39	4.60	5.21	0.64	3.48	0.78	26.47
1989	0.41	0.40	2.36	1.78	3.02	1.33	2.59	4.60	1.95	2.71	1.47	0.49	23.11
1990	0.79	0.68	3.23	2.61	3.74	8.07	4.03	8.02	1.82	1.43	0.74	2.91	38.07
1991	0.93	0.46	2.23	5.96	6.85	1.72	7.67	2.91	3.90	1.76	6.23	1.67	42.29
1992	0.87	0.81	3.64	4.38	1.49	2.32	4.02	2.67	7.01	0.62	3.64	1.58	33.05
1993	1.18	1.10	2.51	5.78	5.41	10.79	3.78	5.31	1.80	0.94	1.38	0.75	40.73
Record Mean	1.10	1.04	1.87	2.69	3.61	4.24	3.84	3.66	3.83	2.23	1.75	1.25	31.12

AVERAGE TEMPERATURE (deg. F)

LA CROSSE, WISCONSIN

YEAR	JAN	FEB	MAR	APR	MAY	JUNE	JULY	AUG	SEP	OCT	NOV	DEC	ANNUAL
#1964	24.2	26.8	30.0	48.5	63.8	70.0	76.3	69.7	61.0	49.6	37.9	17.5	47.9
1965	11.0	14.0	23.0	43.5	60.7	66.3	70.4	67.5	56.6	50.3	35.3	30.5	44.1
1966	7.7	19.5	37.3	44.7	55.2	69.8	75.3	69.8	61.4	50.5	35.6	22.0	45.7
1967	19.4	12.3	33.3	48.4	54.2	68.5	71.3	67.9	61.1	49.3	32.9	25.3	45.3
1968	19.3	19.5	40.5	51.3	56.3	68.0	71.9	72.0	62.1	52.0	35.7	19.7	47.4
1969	12.3	20.7	26.0	49.9	60.6	61.4	72.5	73.9	62.8	47.8	35.2	22.3	45.4
1970	8.7	20.7	30.8	49.1	60.6	70.6	73.4	70.4	61.4	51.8	35.4	19.4	46.0
1971	7.4	17.0	28.1	47.1	55.3	72.1	68.5	68.1	64.3	55.7	36.4	24.5	45.4
1972	11.1	16.1	28.0	43.0	63.4	66.5	70.3	71.1	61.3	45.9	34.4	15.7	43.9
1973	20.1	24.8	41.3	44.5	55.2	70.3	73.9	72.2	62.6	57.1	36.7	18.9	48.1
1974	18.0	19.1	32.1	48.7	55.0	65.0	75.9	68.4	56.9	50.4	37.4	26.6	46.1
1975	18.1	18.7	24.5	41.4	62.3	68.9	74.0	71.7	57.4	53.0	40.0	23.7	46.2
1976	13.1	29.0	33.7	49.7	56.8	70.0	74.4	71.6	61.0	45.0	27.9	12.8	45.4
1977	2.9	22.0	39.9	54.6	67.4	68.0	75.2	66.8	61.7	49.0	33.6	17.0	46.5
1978	6.7	10.8	29.1	45.9	60.0	67.4	70.6	73.5	68.3	51.0	35.8	19.9	44.9
1979	6.0	12.6	32.8	45.7	58.0	67.3	72.1	68.8	62.8	47.7	34.9	28.3	44.7
1980	18.5	17.9	29.7	49.4	62.8	70.3	78.3	74.1	63.7	45.2	36.5	20.8	47.3
1981	19.5	25.3	37.7	51.2	58.4	69.1	73.6	71.4	60.8	47.6	38.0	19.6	47.7
1982	4.3	17.2		44.2	62.8	63.5	74.2	70.3	60.9	50.8	34.1	27.2	
1983	19.9	25.2	33.8	42.6	54.3	69.0	76.7	75.3	62.9	49.6	35.9	6.4	46.0
1984	14.1	30.3	27.3	48.2	55.9	69.6	71.1	72.9	59.0	52.4	35.0	21.5	46.4
1985	10.9	16.5	37.2	52.5	62.0	64.5	72.1	68.4	62.2				
1986		18.4	36.4	52.1	61.9	69.1	75.0	67.5	62.5	50.9	30.2	25.1	
1987	21.7	31.2	39.0	52.9	63.7	72.8	77.1	70.2	63.0	45.3	40.3	27.6	50.4
1988	13.4	15.8	35.0	48.0	65.2	73.4	76.6	76.0	64.6	44.6	36.6	21.8	47.6
1989	24.2	12.8	29.2	46.0	58.9	67.7	75.4	71.4	61.1	51.7	31.5	12.1	45.2
1990	28.1	25.4	38.4	49.3	56.6	69.5	73.2	72.1	65.5	49.6	41.4	19.0	49.0
1991				51.0	63.9	74.3	73.2	72.7	60.5	49.0	27.3	23.8	
1992	23.9	30.3	33.9	45.6	61.2	66.9	68.0	66.6		48.9	33.1	23.6	
1993	16.9	18.6	30.7	44.1	59.0	66.3	71.5	71.9	56.5	48.7	33.8	25.2	45.3
Record Mean	15.7	20.0	32.0	47.5	59.3	68.4	73.1	70.7	62.1	50.6	35.1	22.1	46.4
Max	24.6	29.1	40.8	57.3	69.6	78.6	83.4	80.9	72.2	60.2	42.8	29.5	55.8
Min	6.7	10.8	23.2	37.7	48.9	58.3	62.8	60.5	52.0	40.9	27.4	14.6	37.0

REFERENCE NOTES

LA CROSSE, WISCONSIN

GENERAL
T - TRACE AMOUNT.
BLANK ENTRIES DENOTE MISSING/UNREPORTED DATA.
# INDICATES A STATION OR INSTRUMENT RELOCATION.

SPECIFIC

(a) - LENGTH OF RECORD IN YEARS, ALTHOUGH INDIVIDUAL MONTHS MAY BE MISSING.
0.* OR * - THE VALUE IS BETWEEN 0.0 AND 0.05
NORMALS - BASED ON THE 1961-1990 RECORD PERIOD.
EXTREMES - DATES ARE THE MOST RECENT OCCURRENCE
WIND DIR. - NUMERALS SHOW TENS OF DEGRESS CLOCKWISE FROM TRUE NORTH. "00" INDICATES CALM.
RESULTANT DIRECTIONS ARE GIVEN TO WHOLE DEGREES.
BOLD VALUES INDICATE EXTREME VALUES WHICH OCCURRED AFTER THE ASOS SYSTEM WAS COMMISSIONED.

RECORD = PERIOD OF RECORD
RECORD MEAN PRECIPITATION IS THE MEAN OF ALL DAILY PRECIPITATION AMOUNTS DURING THE PERIOD OF RECORD.
RECORD MAX(MIN) TEMPERATURE IS THE MEAN OF ALL DAILY MAX(MIN) TEMPERATURES DURING THE PERIOD OF RECORD.
RECORD MEAN TEMPERATURE IS THE SUM OF THE RECORD MAX AND RECORD MIN DIVIDED BY 2.
AVERAGE TEMPERATURE IS THE SUM OF THE MEAN DAILY MAX AND MIN TEMPERATURE DIVIDED BY 2.

EXCEPTIONS

1. MEAN SKY COVER, AND DAYS CLEAR-PARTLY CLOUDY-CLOUDY ARE THROUGH 1967.
2. MAXIMUM 24-HOUR PRECIPITATION AND SNOW, AND FASTEST OBSERVED WINDS ARE THROUGH SEPTEMBER 1968.

RECORD MEANS ARE THROUGH THE CURRENT YEAR, BEGINNING IN 1873 FOR TEMPERATURE
1873 FOR PRECIPITATION
1951 FOR SNOWFALL

HEATING DEGREE DAYS Base 65 deg. F — LA CROSSE, WISCONSIN

SEASON	JULY	AUG	SEP	OCT	NOV	DEC	JAN	FEB	MAR	APR	MAY	JUNE	TOTAL
#1964-65	0	37	179	469	806	1470	1672	1427	1299	640	182	25	8206
1965-66	8	61	256	456	884	1061	1774	1267	852	599	312	35	7565
1966-67	1	18	158	450	873	1325	1406	1472	978	494	355	17	7547
1967-68	21	42	153	491	957	1223	1411	1312	754	411	275	54	7104
1968-69	7	15	120	423	869	1398	1626	1236	1200	447	196	135	7672
1969-70	5	1	135	540	886	1315	1746	1236	1052	499	207	30	7652
1970-71	10	5	173	406	883	1406	1781	1340	1137	533	297	18	7989
1971-72	15	22	140	301	850	1244	1668	1414	1140	655	154	62	7665
1972-73	21	32	147	586	913	1523	1389	1119	727	608	299	8	7372
1973-74	1	7	123	267	844	1424	1453	1279	1011	490	316	69	7284
1974-75	0	24	261	445	822	1182	1446	1292	1248	702	162	56	7640
1975-76	20	2	236	384	741	1274	1600	1039	964	460	254	11	6985
1976-77	3	14	176	624	1104	1613	1923	1200	770	344	66	36	7873
1977-78	1	40	122	488	936	1483	1804	1515	1106	565	209	36	8305
1978-79	6	0	78	427	871	1388	1828	1466	992	569	239	29	7893
1979-80	1	37	116	535	894	1133	1438	1358	1085	480	155	15	7247
1980-81	0	3	112	607	849	1362	1402	1107	840	408	214	11	6915
1981-82	5	7	157	535	802	1402	1878	1333		619	112	77	
1982-83	0	41	177	433	919	1164	1391	1109	960	666	325	46	7231
1983-84	6	0	160	480	867	1814	1572	1002	1159	501	285	5	7851
1984-85	9	6	219	384	893	1344	1673	1355	854	409	129	92	7367
1985-86	0	24	203					1298	880	394	141	25	
1986-87	1	35	120	431	1039	1230	1336	942	798	372	137	11	6452
1987-88	0	28	102	602	733	1151	1595	1421	924	503	85	10	7154
1988-89	0	10	78	627	844	1332	1257	1458	1100	563	218	52	7539
1989-90	1	4	159	417	1000	1638	1138	1102	818	504	267	38	7086
1990-91	3	2	112	474	700	1421				438	176	2	
1991-92	5	3	217	490	1123	1275	1268	1001	958	576	173	59	7148
1992-93	22	46		498	952	1274	1486	1295	1057	620	213	55	
1993-94	1	10	265	507	929	1227							

COOLING DEGREE DAYS Base 65 deg. F — LA CROSSE, WISCONSIN

YEAR	JAN	FEB	MAR	APR	MAY	JUNE	JULY	AUG	SEP	OCT	NOV	DEC	TOTAL
1969	0	0	0	0	66	35	245	284	74	13	0	0	717
1970	0	0	0	29	75	208	279	178	71	5	0	0	845
1971	0	0	0	1	3	237	129	128	126	19	0	0	643
1972	0	0	0	0	113	112	191	230	39	0	0	0	685
1973	0	0	0	0	4	174	283	239	59	28	0	0	787
1974	0	0	0	10	15	74	343	137	23	0	0	0	602
1975	0	0	0	0	85	177	306	215	16	15	0	0	814
1976	0	0	0	5	7	167	304	228	60	10	0	0	781
1977	0	0	0	41	145	132	325	100	33	0	0	0	776
1978	0	0	0	0	61	116	187	272	183	1	0	0	820
1979	0	0	0	0	30	104	226	161	60	3	0	0	584
1980	0	0	0	20	96	180	416	293	79	0	0	0	1084
1981	0	0	0	0	19	139	277	211	37	0	0	0	683
1982	0	0		0	54	38	292	212	60	0	0	0	
1983	0	0	0	1	0	172	375	324	104	12	0	0	988
1984	0	0	0	2	8	149	206	257	48	0	0	0	670
1985	0	0	0	41	42	84	229	136	126				
1986		0	1	13	54	156	319	118	53	0	0	0	
1987	0	0	0	16	103	249	383	196	48	0	0	0	995
1988	0	0	0	0	101	269	365	360	73	0	0	0	1168
1989	0	0	1	0	38	138	331	212	48	11	0	0	779
1990	0	0	0	39	13	181	264	230	134	2	0	0	863
1991				24	148	286	267	251	84	0	0	0	
1992	0	0	0	0	59	124	121	102		5	0	0	
1993	0	0	0	0	35	100	209	232	14	10	0	0	600

SNOWFALL (inches) — LA CROSSE, WISCONSIN

SEASON	JULY	AUG	SEP	OCT	NOV	DEC	JAN	FEB	MAR	APR	MAY	JUNE	TOTAL
1964-65	0.0	0.0	0.0	T	4.7	6.7	12.7	6.6	15.0	1.6	0.0	0.0	47.3
1965-66	0.0	0.0	T	0.0	0.2	1.3	10.9	5.1	2.5	T	T	0.0	20.0
1966-67	0.0	0.0	0.0	T	0.2	11.7	18.3	15.2	6.1	T	0.0	0.0	51.5
1967-68	0.0	0.0	0.0	T	0.8	1.5	4.5	0.8	0.1	T	0.0	0.0	7.7
1968-69	0.0	0.0	0.0	0.0	1.0	26.6	17.0	0.8	3.7	0.0	0.0	0.0	49.1
1969-70	0.0	0.0	0.0	0.0	T	19.8	7.7	3.2	6.1	T	T	0.0	36.8
1970-71	0.0	0.0	0.0	T	1.1	13.5	26.5	20.4	5.6	0.5	0.0	0.0	67.6
1971-72	0.0	0.0	0.0	0.0	3.8	12.0	10.2	8.0	11.9	3.0	0.0	0.0	48.9
1972-73	0.0	0.0	0.0	T	0.8	18.6	7.3	9.0	T	17.0	0.0	0.0	52.7
1973-74	0.0	0.0	0.0	0.0	T	9.8	2.0	14.9	7.6	0.7	0.0	0.0	35.0
1974-75	0.0	0.0	T	0.0	2.2	10.1	18.8	20.1	18.1	3.9	0.0	0.0	73.2
1975-76	0.0	0.0	0.0	0.0	1.7	1.0	8.0	2.7	6.0	0.0	T	0.0	19.4
1976-77	0.0	0.0	0.0	0.2	T	6.8	10.1	2.9	6.6	0.5	0.0	0.0	27.1
1977-78	0.0	0.0	0.0	0.0	12.0	18.8	10.4	8.0	3.4	T	0.0	0.0	52.6
1978-79	0.0	0.0	0.0	0.0	6.6	12.0	29.7	7.3	3.0	1.0	0.0	0.0	59.6
1979-80	0.0	0.0	0.0	T	3.2	T	10.7	4.1	10.1	3.9	0.0	0.0	32.0
1980-81	0.0	0.0	0.0	T	0.4	4.6	1.9	13.9	1.0	T	0.0	0.0	21.8
1981-82	0.0	0.0	0.0	0.5	3.6	11.6	14.3	2.2		0.0	0.0	0.0	
1982-83	0.0	0.0	0.0	0.0	T	0.0	9.8	18.5	4.2	2.1	0.0	0.0	34.6
1983-84	0.0	0.0	0.0	0.0	4.0	9.2	3.5	1.7	11.0	T	0.0	0.0	29.4
1984-85	0.0	0.0	0.0	T	2.9	13.3	9.4	5.3	14.5	T	0.0	0.0	45.4
1985-86	0.0	0.0	T					8.2	T	T	0.0	0.0	
1986-87	0.0	0.0	0.0	T	8.7	2.6	12.6	2.3	12.5	T	0.0	0.0	38.7
1987-88	0.0	0.0	0.0	T	T	13.3	19.1	4.2	1.2	T	0.0	0.0	37.8
1988-89	0.0	0.0	0.0	T	7.6	5.2	3.8	7.9	19.3	T	T	0.0	43.8
1989-90	0.0	0.0	0.0	T	8.7	5.9	10.0	7.0	T	T	0.0	0.0	31.6
1990-91	0.0	0.0	0.0	T	0.8	30.4	12.6	2.6	0.3	2.5	0.0	0.0	49.2
1991-92	0.0	0.0	0.0	1.2	30.3	8.7	5.3	7.5	11.1	T	0.0	0.0	64.1
1992-93	0.0	0.0	0.0	1.8	3.8	10.6	10.4	12.6	10.0	8.9	0.0	0.0	58.1
1993-94	0.0	0.0	0.0	0.2	4.5	2.3							
Record Mean	0.0	0.0	T	0.1	4.4	9.1	9.9	7.9	9.2	1.7	T	0.0	42.4

MADISON, WISCONSIN

Madison is set on a narrow isthmus of land between Lakes Mendota and Monona. Lake Mendota (15 square miles) lies northwest of Lake Monona (5 square miles) and the lakes are only two-thirds of a mile apart at one point. Drainage at Madison is southeast through two other lakes into the Rock River, which flows south into Illinois, and then west to the Mississippi. The westward flowing Wisconsin River is only 20 miles northwest of Madison. Madison lakes are normally frozen from mid-December to early April.

Madison has the typical continental climate of interior North America with a large annual temperature range and with frequent short period temperature changes. The range of extreme temperatures is from about 110 to -40 degrees. Winter temperatures (December-February) average near 20 degrees and the summer average (June-August) is in the upper 60s. Daily temperatures average below 32 degrees about 120 days and above 40 degrees for about 210 days of the year.

Madison lies in the path of the frequent cyclones and anticyclones which move eastward over this area during fall, winter and spring. In summer, the cyclones have diminished intensity and tend to pass farther north. The most frequent air masses are of polar origin. Occasional outbreaks of arctic air affect this area during the winter months. Although northward moving tropical air masses contribute considerable cloudiness and precipitation, the true Gulf air mass does not reach this area in winter, and only occasionally at other seasons. Summers are pleasant, with only occasional periods of extreme heat or high humidity.

There are no dry and wet seasons, but about 60 percent of the annual precipitation falls in the five months of May through September. Cold season precipitation is lighter, but lasts longer. Soil moisture is usually adequate in the first part of the growing season. During July, August, and September, the crops depend on current rainfall, which is mostly from thunderstorms and tends to be erratic and variable. Average occurrence of thunderstorms is just under 7 days per month during this period.

March and November are the windiest months. Tornadoes are infrequent. Dane County has about one tornado in every three to five years.

The ground is covered with 1 inch or more of snow about 60 percent of the time from about December 10 to near February 25 in an average winter. The soil is usually frozen from the first of December through most of March with an average frost penetration of 25 to 30 inches. The growing season averages 175 days.

Farming is diversified with the main emphasis on dairying. Field crops are mainly corn, oats, clover, and alfalfa, but barley, wheat, rye, and tobacco are also raised. Canning factories pack peas, sweet corn, and lima beans. Fruits are mainly apples, strawberries, and raspberries.

NORMALS, MEANS, AND EXTREMES

MADISON, WISCONSIN

LATITUDE: 43°08'N LONGITUDE: 89°20'W ELEVATION: FT. GRND 858 BARO 860 TIME ZONE: CENTRAL WBAN: 14837

	(a)	JAN	FEB	MAR	APR	MAY	JUNE	JULY	AUG	SEP	OCT	NOV	DEC	YEAR
TEMPERATURE °F:														
Normals														
-Daily Maximum		24.8	30.1	41.5	56.7	68.9	78.2	82.4	79.6	71.5	59.9	44.0	29.8	55.6
-Daily Minimum		7.2	11.1	23.0	34.1	44.2	54.2	59.5	56.9	48.2	37.7	26.7	13.5	34.7
-Monthly		16.0	20.6	32.3	45.4	56.5	66.2	71.0	68.3	59.8	48.9	35.4	21.7	45.2
Extremes														
-Record Highest	54	56	61	82	94	93	101	104	102	99	90	76	62	104
-Year		1989	1981	1986	1980	1975	1988	1976	1988	1953	1976	1964	1984	JUL 1976
-Record Lowest	54	-37	-28	-29	0	19	31	36	35	25	13	-11	-25	-37
-Year		1951	1985	1962	1982	1978	1972	1965	1968	1974	1988	1947	1983	JAN 1951
NORMAL DEGREE DAYS:														
Heating (base 65°F)		1519	1243	1014	588	294	68	12	38	168	499	888	1342	7673
Cooling (base 65°F)		0	0	0	0	30	104	198	141	12	0	0	0	485
% OF POSSIBLE SUNSHINE	47	48	51	52	52	58	64	67	65	60	54	39	40	54
MEAN SKY COVER (tenths)														
Sunrise - Sunset	45	6.7	6.6	6.9	6.8	6.5	6.1	5.8	5.8	5.8	6.0	7.3	7.1	6.5
MEAN NUMBER OF DAYS:														
Sunrise to Sunset														
-Clear	47	7.4	7.1	6.3	6.3	6.9	7.2	8.7	9.3	9.4	9.4	5.5	6.3	89.9
-Partly Cloudy	47	6.5	6.0	7.4	7.7	9.2	10.0	11.1	10.6	8.1	7.4	6.0	6.2	96.2
-Cloudy	47	17.0	15.2	17.3	16.0	14.8	12.8	11.2	11.2	12.4	14.2	18.6	18.5	179.2
Precipitation														
.01 inches or more	45	10.1	8.0	11.0	11.6	11.4	10.4	9.7	9.4	9.4	9.0	9.8	10.0	119.7
Snow,Ice pellets,hail														
1.0 inches or more	45	2.8	2.4	2.5	0.6	0.*	0.0	0.0	0.0	0.0	0.*	1.3	3.3	13.0
Thunderstorms	45	0.2	0.2	1.9	3.5	5.2	7.0	7.3	6.5	4.8	2.0	0.8	0.4	39.8
Heavy Fog Visibility														
1/4 mile or less	47	2.3	2.0	2.7	1.3	1.3	0.9	1.3	2.0	1.8	1.7	1.9	2.9	22.1
Temperature °F														
-Maximum														
90° and above	34	0.0	0.0	0.0	0.*	0.3	3.0	5.0	2.8	0.6	0.1	0.0	0.0	11.8
32° and below	34	21.0	15.3	6.0	0.4	0.0	0.0	0.0	0.0	0.0	0.*	3.9	16.8	63.4
-Minimum														
32° and below	34	30.1	27.2	25.7	13.7	3.2	0.*	0.0	0.0	1.3	10.0	21.6	28.9	161.7
0° and below	34	10.7	6.5	1.2	0.*	0.0	0.0	0.0	0.0	0.0	0.0	0.1	5.4	23.9
AVG. STATION PRESS.(mb)	21	985.7	986.4	983.8	982.9	982.9	982.9	984.3	985.5	985.8	985.7	984.7	985.6	984.7
RELATIVE HUMIDITY (%)														
Hour 00	34	78	78	78	76	77	80	84	87	87	81	82	81	81
Hour 06 (Local Time)	34	79	80	82	81	80	82	86	91	91	86	85	83	84
Hour 12	34	69	66	62	55	54	56	57	59	61	59	67	72	61
Hour 18	34	73	70	65	56	54	56	59	62	69	68	75	77	65
PRECIPITATION (inches):														
Water Equivalent														
-Normal		1.07	1.08	2.17	2.86	3.14	3.66	3.39	4.04	3.37	2.17	2.09	1.84	30.88
-Maximum Monthly	54	2.45	2.77	5.04	7.11	6.26	9.95	10.93	9.49	9.51	5.63	5.13	4.09	10.93
-Year		1974	1953	1973	1973	1960	1978	1950	1980	1941	1984	1985	1987	JUL 1950
-Minimum Monthly	54	0.14	0.08	0.28	0.96	0.64	0.81	1.38	0.70	0.11	0.06	0.11	0.25	0.06
-Year		1981	1958	1978	1946	1981	1973	1946	1948	1979	1952	1976	1960	OCT 1952
-Maximum in 24 hrs	45	1.27	1.58	2.52	2.83	3.64	3.67	5.25	2.90	3.57	2.78	2.36	2.19	5.25
-Year		1960	1981	1973	1975	1966	1963	1950	1965	1961	1984	1985	1990	JUL 1950
Snow,Ice pellets,hail														
-Maximum Monthly	45	26.9	20.9	25.4	17.4	3.0	T	T	T	T	3.1	18.3	32.8	32.8
-Year		1979	1975	1959	1973	1990	1992	1991	1992	1993	1990	1985	1987	DEC 1987
-Maximum in 24 hrs	45	11.6	10.3	13.6	12.9	3.0	T	T	T	T	3.0	9.0	17.3	17.3
-Year		1971	1950	1971	1973	1990	1992	1991	1992	1993	1990	1985	1990	DEC 1990
WIND:														
Mean Speed (mph)	47	10.5	10.4	11.2	11.4	10.0	9.2	8.1	8.0	8.7	9.6	10.8	10.3	9.8
Prevailing Direction through 1963		WNW	WNW	NW	NW	S	S	S	S	S	S	S	W	S
Fastest Mile														
-Direction (!!)	46	E	W	SW	SW	SW	W	NW	W	W	SW	SE	SW	SW
-Speed (MPH)	46	68	57	70	73	77	59	72	47	52	73	56	65	77
-Year		1947	1948	1954	1947	1950	1947	1951	1955	1948	1951	1947	1949	MAY 1950
Peak Gust														
-Direction (!!)	10	SW	NW	S	W	SW	W	N	N	NW	NW	S	NE	N
-Speed (mph)	10	46	62	67	53	63	70	83	64	64	62	52	58	83
-Date		1988	1987	1990	1984	1988	1984	1991	1989	1985	1990	1988	1987	JUL 1991

PRECIPITATION (inches)

MADISON, WISCONSIN

YEAR	JAN	FEB	MAR	APR	MAY	JUNE	JULY	AUG	SEP	OCT	NOV	DEC	ANNUAL
1964	0.93	0.26	2.12	3.15	3.87	2.28	4.28	2.52	1.85	0.08	1.94	0.34	23.62
1965	1.80	0.74	2.51	2.94	1.86	2.31	3.30	6.77	9.22	1.69	1.96	2.50	37.60
1966	1.07	1.36	2.11	1.54	4.31	2.91	3.24	3.83	0.51	1.65	1.28	2.62	26.43
1967	1.63	1.17	1.49	2.57	3.53	6.46	2.51	2.71	2.68	5.52	1.83	1.89	33.99
1968	0.56	0.49	0.59	4.18	2.02	7.82	2.54	2.58	4.45	0.85	1.74	2.89	30.71
1969	2.26	0.18	1.47	2.72	3.45	7.96	4.28	0.96	1.35	2.65	0.70	1.66	29.64
1970	0.44	0.16	1.17	2.53	6.09	2.26	2.42	0.97	8.82	2.65	1.06	2.12	30.69
1971	1.48	2.59	1.52	2.42	0.98	2.27	1.65	3.96	1.87	1.30	3.48	3.64	27.16
1972	0.40	0.42	2.23	2.02	2.83	1.65	3.49	7.47	5.26	2.42	0.86	1.91	30.96
1973	1.54	1.20	5.04	7.11	5.27	0.81	2.68	2.53	3.59	2.30	1.48	1.98	35.53
1974	2.45	1.17	3.43	4.24	5.77	3.86	2.69	4.60	1.08	3.18	1.79	1.80	36.06
1975	0.98	1.54	3.09	4.19	4.57	4.30	6.05	5.25	0.84	0.64	2.79	0.29	34.53
1976	0.56	1.72	4.75	4.80	1.95	1.38	1.46	1.99	0.50	1.49	0.11	0.37	21.08
1977	0.53	1.44	3.03	2.59	2.52	2.63	6.63	5.19	2.84	1.41	2.12	1.60	32.53
1978	1.03	0.24	0.28	3.50	3.96	9.95	4.54	1.63	5.44	1.11	3.05	1.71	36.44
1979	1.69	0.90	2.67	2.46	2.70	2.53	2.80	4.96	0.11	3.10	2.27	1.93	28.12
1980	1.11	0.64	0.68	2.36	2.08	3.43	2.67	9.49	7.84	1.13	1.33	1.62	34.38
1981	0.14	2.47	0.33	3.42	0.64	4.99	4.81	7.06	3.10	2.68	1.71	0.75	32.10
1982	1.42	0.17	2.11	3.26	4.34	3.40	3.47	2.67	1.42	1.46	4.21	3.65	31.58
1983	0.53	2.26	2.70	2.23	4.21	1.85	1.92	5.05	2.85	2.59	3.18	2.30	31.67
1984	0.36	1.26	1.15	3.86	3.32	7.01	1.96	1.89	2.79	5.63	1.83	2.66	33.72
1985	1.43	1.89	3.13	1.52	3.35	3.06	4.48	2.98	5.00	4.58	5.13	2.39	38.94
1986	1.02	2.72	1.55	2.27	1.97	3.24	4.31	4.38	6.82	1.85	1.03	0.69	31.85
1987	0.68	0.62	1.99	2.46	3.90	1.17	3.26	7.16	3.61	1.24	3.24	4.09	33.42
1988	1.82	0.46	1.20	2.65	0.92	2.06	2.44	2.95	3.33	1.60	3.58	1.56	24.57
1989	0.61	0.57	1.69	1.69	1.72	1.67	4.97	6.46	0.89	1.88	0.98	0.26	23.39
1990	1.60	0.99	4.18	1.90	5.35	4.88	2.61	6.03	1.64	2.25	1.65	3.46	36.54
1991	1.17	0.44	4.24	4.89	2.20	3.75	5.18	2.34	3.96	5.35	3.86	1.71	39.09
1992	0.78	1.34	1.90	3.17	1.12	1.53	5.54	2.48	5.99	1.06	4.83	2.39	32.13
1993	1.60	1.18	3.29	5.33	3.81	6.67	9.34	5.57	3.74	0.91	1.55	0.35	43.34
Record Mean	1.17	1.06	2.13	2.90	3.23	3.95	3.82	3.74	3.26	2.19	2.12	1.64	31.22

AVERAGE TEMPERATURE (deg. F)

MADISON, WISCONSIN

YEAR	JAN	FEB	MAR	APR	MAY	JUNE	JULY	AUG	SEP	OCT	NOV	DEC	ANNUAL
1964	24.9	24.9	30.8	47.5	62.6	68.5	73.1	66.9	59.6	46.9	38.6	18.8	46.9
1965	15.9	18.7	25.0	43.7	59.6	64.2	68.3	66.7	58.3	49.2	35.6	29.8	44.6
1966	10.0	21.3	35.4	42.0	50.3	66.6	71.1	66.2	57.1	47.7	36.0	23.2	43.9
1967	22.2	14.7	33.3	45.6	50.2	66.2	67.1	62.0	57.6	48.0	33.0	25.1	43.7
1968	19.6	19.0	39.2	47.8	54.3	66.2	68.9	68.8	60.1	50.7	35.6	22.6	46.1
1969	14.8	23.6	27.9	47.0	56.4	59.5	69.6	70.3	59.5	46.4	33.1	21.3	44.1
1970	9.9	20.1	31.1	47.9	58.5	66.5	70.6	68.5	60.0	51.0	36.3	22.4	45.2
1971	9.6	19.9	28.6	45.4	55.1	71.7	68.5	68.3	65.1	55.9	35.2	26.8	45.9
1972	12.7	16.5	28.7	41.3	59.2	62.8	68.3	69.2	59.4	45.8	34.6	17.3	43.0
1973	23.4	24.0	41.6	44.9	54.4	67.9	71.6	70.6	60.7	54.1	36.6	21.5	47.6
1974	19.2	18.4	33.1	48.7	54.1	64.0	72.1	66.8	57.4	50.5	37.1	26.8	45.7
1975	21.9	21.3	26.1	41.0	62.5	69.2	72.4	70.6	57.5	52.2	41.9	25.5	46.8
1976	15.7	28.4	36.5	49.3	54.3	68.3	73.4	68.8	58.0	43.7	28.1	13.2	44.8
1977	3.7	22.4	39.8	51.9	65.2	64.9	73.6	64.6	59.7	47.6	34.0	19.6	45.6
1978	10.5	12.4	29.4	44.5	57.9	65.9	69.7	69.5	63.8	47.3	33.5	21.3	43.8
1979	6.9	11.7	32.1	42.4	56.7	66.0	69.8	66.6	61.1	47.5	35.1	28.8	43.7
1980	17.3	15.7	28.0	45.5	57.8	65.3	73.4	70.3	59.9	43.7	35.4	22.6	44.6
1981	20.5	25.3	36.9	48.7	55.3	67.4	70.6	68.7	59.1	46.6	36.7	22.0	46.5
1982	8.0	19.1	30.6	41.7	60.8	59.6	70.9	66.3	59.0	50.6	34.2	28.8	44.1
1983	21.4	26.3	33.1	41.6	51.9	67.5	75.0	72.2	60.1	48.2	37.3	10.8	45.5
1984	14.8	30.2	26.7	45.6	53.4	67.5	70.2	71.3	59.3	52.0	33.9	26.4	45.9
1985	12.2	19.0	37.7	52.2	60.7	63.8	70.0	66.4	61.6	49.4	31.0	11.3	44.6
1986	18.2	19.4	36.2	49.8	58.4	65.9	73.2	64.8	61.6	49.7	31.2	25.5	46.2
1987	22.6	30.5	37.2	49.9	60.8	70.4	74.5	68.7	60.6	43.4	40.0	28.4	48.9
1988	13.8	17.4	34.6	46.0	60.5	69.5	74.1	74.5	63.0	43.5	38.8	24.6	46.7
1989	27.6	14.6	30.1	44.7	56.1	65.7	72.3	68.6	58.7	50.8	33.1	14.2	44.7
1990	28.6	25.8	37.7	48.5	53.6	67.6	70.6	69.9	63.7	48.3	41.0	21.4	48.1
1991	15.1	26.5	36.8	49.4	63.6	71.1	72.3	70.2	60.0	49.4	31.4	26.3	47.7
1992	25.5	29.3	34.8	43.6	58.0	64.9	67.2	65.3	60.4	48.4	34.7	24.9	46.4
1993	21.8	21.2	31.5	43.7	59.6	65.9	72.0	72.1	56.7	47.9	35.2	26.6	46.2
Record Mean	17.3	21.4	32.3	46.3	57.3	66.8	71.5	69.5	60.6	49.9	35.4	22.6	45.9
Max	26.1	30.6	41.7	57.7	69.6	79.0	83.6	81.4	72.3	61.2	43.9	30.6	56.5
Min	8.5	12.2	22.8	34.8	44.9	54.6	59.4	57.5	48.8	38.5	26.8	14.5	35.3

REFERENCE NOTES

MADISON, WISCONSIN

GENERAL
T - TRACE AMOUNT
BLANK ENTRIES DENOTE MISSING/UNREPORTED DATA.
# INDICATES A STATION OR INSTRUMENT RELOCATION.

SPECIFIC

PM - INCLUDES LAST DAY OF PREVIOUS MONTH
ASOS - AUTOMATED SURFACE OBSERVING SYSTEM IN OPERATION DURING THESE MONTHS.

(a) - LENGTH OF RECORD IN YEARS, ALTHOUGH INDIVIDUAL MONTHS MAY BE MISSING.
0.* OR * - THE VALUE IS BETWEEN 0.0 AND 0.05
NORMALS - BASED ON THE 1961-1990 RECORD PERIOD.
EXTREMES - DATES ARE THE MOST RECENT OCCURRENCE
WIND DIR. - NUMERALS SHOW TENS OF DEGRESS CLOCKWISE FROM TRUE NORTH. "00" INDICATES CALM.
RESULTANT DIRECTIONS ARE GIVEN TO WHOLE DEGREES.
BOLD VALUES INDICATE EXTREME VALUES WHICH OCCURRED AFTER THE ASOS SYSTEM WAS COMMISSIONED.

RECORD = PERIOD OF RECORD
RECORD MEAN PRECIPITATION IS THE MEAN OF ALL DAILY PRECIPITATION AMOUNTS DURING THE PERIOD OF RECORD.
RECORD MAX(MIN) TEMPERATURE IS THE MEAN OF ALL DAILY MAX(MIN) TEMPERATURES DURING THE PERIOD OF RECORD.
RECORD MEAN TEMPERATURE IS THE SUM OF THE RECORD MAX AND RECORD MIN DIVIDED BY 2.
AVERAGE TEMPERATURE IS THE SUM OF THE MEAN DAILY MAX AND MIN TEMPERATURE DIVIDED BY 2.

EXCEPTIONS

RECORD MEANS ARE THROUGH THE CURRENT YEAR, BEGINNING IN 1940 FOR TEMPERATURE
1940 FOR PRECIPITATION
1949 FOR SNOWFALL

HEATING DEGREE DAYS Base 65 deg. F — MADISON, WISCONSIN

SEASON	JULY	AUG	SEP	OCT	NOV	DEC	JAN	FEB	MAR	APR	MAY	JUNE	TOTAL
1964-65	13	68	216	556	789	1427	1517	1294	1234	631	208	79	8032
1965-66	27	69	231	486	878	1086	1702	1220	911	684	454	83	7831
1966-67	11	40	252	533	862	1290	1321	1407	978	574	462	46	7776
1967-68	61	120	239	535	955	1229	1401	1327	792	510	330	76	7575
1968-69	34	66	159	460	873	1305	1548	1152	1143	535	282	197	7754
1969-70	13	9	202	579	951	1346	1705	1252	1044	521	244	73	7939
1970-71	28	18	196	431	853	1310	1718	1258	1124	582	312	22	7852
1971-72	28	21	131	293	885	1179	1616	1401	1119	705	212	117	7707
1972-73	44	42	188	587	905	1475	1279	1143	720	596	325	15	7319
1973-74	4	25	180	349	847	1342	1416	1298	979	494	347	90	7371
1974-75	1	37	253	443	829	1179	1329	1220	1198	714	150	43	7396
1975-76	18	11	236	412	687	1217	1520	1056	877	477	333	32	6876
1976-77	4	40	236	656	1102	1602	1898	1188	772	409	110	95	8112
1977-78	6	95	161	533	925	1404	1688	1466	1096	608	269	59	8310
1978-79	19	22	130	543	940	1348	1800	1489	1013	671	283	52	8310
1979-80	14	62	144	546	890	1112	1471	1424	1138	586	255	84	7726
1980-81	2	11	178	651	881	1303	1373	1107	864	482	307	30	7189
1981-82	16	27	193	566	842	1327	1765	1281	1059	688	155	172	8091
1982-83	5	66	230	444	918	1117	1346	1078	978	693	400	57	7332
1983-84	11	6	193	519	823	1678	1550	1006	1181	575	358	20	7920
1984-85	9	21	215	397	927	1191	1632	1287	839	418	155	96	7187
1985-86	12	36	198	475	1012	1661	1444	1272	888	462	220	73	7753
1986-87	7	59	145	471	1007	1218	1309	963	857	452	192	27	6707
1987-88	3	45	150	661	743	1127	1586	1377	938	565	176	53	7424
1988-89	4	18	107	661	777	1242	1153	1404	1076	602	290	68	7402
1989-90	5	22	207	437	952	1568	1122	1092	835	519	349	46	7154
1990-91	7	12	133	511	713	1349	1539	1072	868	467	173	22	6866
1991-92	8	11	222	476	1002	1195	1216	1031	929	634	244	73	7041
1992-93	26	68	176	514	903	1236	1333	1220	1032	633	196	74	7411
1993-94	0	9	260	525	887	1182							

COOLING DEGREE DAYS Base 65 deg. F — MADISON, WISCONSIN

YEAR	JAN	FEB	MAR	APR	MAY	JUNE	JULY	AUG	SEP	OCT	NOV	DEC	TOTAL
1969	0	0	0	0	25	39	161	179	45	11	0	0	460
1970	0	0	0	12	47	125	210	133	53	4	0	0	584
1971	0	0	0	0	13	229	144	131	140	20	0	0	677
1972	0	0	0	0	41	61	156	180	27	0	0	0	465
1973	0	0	0	0	2	112	215	207	58	19	0	0	613
1974	0	0	0	9	17	68	228	102	31	2	0	0	457
1975	0	0	0	0	81	176	256	190	18	21	0	0	742
1976	0	0	0	14	6	136	270	165	34	2	0	0	627
1977	0	0	0	24	123	99	278	88	10	0	0	0	622
1978	0	0	0	0	56	92	171	168	102	0	0	0	589
1979	0	0	0	0	33	88	168	115	33	13	0	0	450
1980	0	0	0	8	39	100	268	183	31	0	0	0	629
1981	0	0	0	0	13	107	198	148	19	0	0	0	485
1982	0	0	0	0	29	16	194	114	53	3	0	0	409
1983	0	0	0	0	0	138	327	237	52	6	0	0	760
1984	0	0	0	1	5	102	177	224	50	0	0	0	559
1985	0	0	0	40	29	66	175	84	102	0	0	0	496
1986	0	0	0	13	24	105	269	59	49	0	0	0	519
1987	0	0	0	8	69	194	304	165	26	0	0	0	766
1988	0	0	0	0	43	194	296	315	54	0	0	0	902
1989	0	0	0	0	21	97	237	141	25	3	0	0	524
1990	0	0	0	32	2	132	191	171	100	1	0	0	629
1991	0	0	0	8	136	210	241	180	80	3	0	0	858
1992	0	0	0	0	33	76	100	86	42	5	0	0	342
1993	0	0	0	0	33	111	223	240	18	5	0	0	630

SNOWFALL (inches) — MADISON, WISCONSIN

SEASON	JULY	AUG	SEP	OCT	NOV	DEC	JAN	FEB	MAR	APR	MAY	JUNE	TOTAL
1964-65	0.0	0.0	0.0	0.5	3.7	3.5	18.3	4.7	19.4	0.8	0.0	0.0	50.9
1965-66	0.0	0.0	T	0.0	T	5.7	9.6	3.8	5.5	0.1	0.7	0.0	25.4
1966-67	0.0	0.0	0.0	T	0.1	10.1	9.4	13.9	4.5	T	T	0.0	38.0
1967-68	0.0	0.0	0.0	0.9	0.7	2.4	3.9	3.9	0.5	0.4	0.0	0.0	12.7
1968-69	0.0	0.0	0.0	0.0	0.5	11.4	9.7	1.7	10.1	T	T	0.0	33.4
1969-70	0.0	0.0	0.0	0.0	1.0	19.5	6.4	1.7	7.0	1.9	0.0	0.0	37.5
1970-71	0.0	0.0	0.0	0.0	0.2	20.8	21.9	3.7	20.1	0.7	0.0	0.0	67.4
1971-72	0.0	0.0	0.0	0.0	8.8	8.9	3.6	6.3	18.8	3.9	0.0	0.0	50.3
1972-73	0.0	0.0	0.0	T	1.3	16.3	1.9	6.0	1.1	17.4	T	0.0	44.0
1973-74	0.0	0.0	0.0	0.0	0.4	10.9	10.5	14.1	6.6	0.4	T	0.0	42.9
1974-75	0.0	0.0	0.0	0.0	3.0	15.4	5.2	20.9	10.0	5.9	0.0	0.0	60.4
1975-76	0.0	0.0	0.0	0.0	5.5	2.8	10.1	10.4	2.0	T	0.0	0.0	30.8
1976-77	0.0	0.0	0.0	T	1.1	5.8	8.7	2.6	5.8	2.3	0.0	0.0	26.3
1977-78	0.0	0.0	0.0	0.0	10.4	24.6	13.5	4.7	3.0	0.5	0.0	0.0	56.7
1978-79	0.0	0.0	0.0	0.0	6.2	23.0	26.9	8.7	4.0	7.3	0.0	0.0	76.1
1979-80	0.0	0.0	0.0	0.2	4.4	1.3	4.9	7.5	5.6	7.1	0.0	0.0	31.0
1980-81	0.0	0.0	0.0	T	3.5	9.2	2.9	9.2	1.7	0.0	0.0	0.0	26.5
1981-82	0.0	0.0	0.0	0.1	2.0	7.2	19.4	2.4	8.6	10.3	0.0	0.0	50.0
1982-83	0.0	0.0	0.0	0.0	0.3	3.3	6.5	13.0	14.1	4.2	0.0	0.0	41.4
1983-84	0.0	0.0	0.0	0.0	2.1	22.6	6.0	0.8	6.8	3.9	0.0	0.0	42.2
1984-85	0.0	0.0	0.0	0.0	0.5	15.8	19.9	7.4	8.2	1.9	0.0	0.0	53.7
1985-86	0.0	0.0	0.0	0.0	18.3	24.0	13.9	13.3	2.7	0.2	0.0	0.0	72.4
1986-87	0.0	0.0	0.0	T	8.6	8.0	8.7	0.3	8.9	T	0.0	0.0	34.5
1987-88	0.0	0.0	0.0	0.4	3.9	32.8	16.3	6.4	1.1	1.3	0.0	0.0	62.2
1988-89	0.0	0.0	0.0	0.2	5.5	8.2	2.6	9.7	9.3	0.2	0.5	T	36.2
1989-90	0.0	0.0	0.0	0.7	4.4	4.3	10.1	11.7	0.1	0.5	3.0	T	34.8
1990-91	0.0	0.0	0.0	3.1	4.5	23.0	14.5	5.0	3.6	1.3	0.0	0.0	55.0
1991-92	T	0.0	0.0	0.5	8.0	10.2	4.5	12.3	6.9	0.1	0.0	T	42.5
1992-93	0.0	T	0.0	2.1	3.9	10.5	12.5	12.1	21.6	8.5	0.0	0.0	71.2
1993-94	0.0	0.0	T	0.2	1.2	2.5							
Record Mean	T	T	T	0.2	3.7	11.0	9.7	7.3	8.4	2.2	0.1	T	42.7

MILWAUKEE, WISCONSIN

Milwaukee possesses a continental climate characterized by a wide range of temperatures between summer and winter. Precipitation is moderate and occurs mostly in the spring, less in the autumn, and very little in the wintertime. Rainfall is well distributed for agricultural purposes, although spring planting is sometimes delayed by wet ground and cold weather.

Milwaukee is in a region of frequently changeable weather and its climate is influenced by general easterly-moving storms which traverse the nations midsection. The most severe winter storms, which produce in excess of 10 inches of snow, develop in the southern Great Plains and move northeast across Illinois and Indiana.

Occasionally during the cold season, frigid air masses from Canada push southeast across the Great Lakes region. These arctic air masses account for the coldest winter temperatures. Very low temperatures, zero degrees or lower, most often occur in air that flows southward to the west of Lake Superior before reaching the Milwaukee area. If northwesterly wind circulation persists, repeated incursions of arctic air will result in a period of bitterly cold weather lasting several days.

Summer temperatures, which reach into the 90s but rarely exceed 100 degrees, occur with brisk southwest winds that carry hot air from the plains and lower Mississippi River Valley across the city. A combination of high temperatures and humidity occasionally develops, usually building up over a period of several days when persistent southerly winds transport moisture from the Gulf of Mexico into the area.

The Gulf is a major source of moisture for Milwaukee in all seasons, but the type of precipitation which results is dependent upon the time of year. Cold-season precipitation (rain, snow, or a mixture) is usually of relatively long duration and low intensity, and occasionally persists for two days or more, whereas in the warm season, relatively short-duration and high-intensity showery rainfall, usually lasting a few hours or less, predominates.

The Great Lakes significantly influence the local climate. Temperature extremes are modified by Lake Michigan and, to a lesser extent, the other Great Lakes. In late autumn and winter, air masses that are initially very cold often reach the city only after being tempered by passage over one or more of the lakes. Similarly, air masses that approach from the northeast in the spring and summer are cooler because of movement over the Great Lakes.

The influence of Lake Michigan is variable and occasionally dramatic, especially when the temperature of the lake water differs strongly from the air temperature. During the spring and early summer, a wind shift from a westerly to an easterly direction frequently causes a sudden 10 to 20 degree temperature drop. When the breeze off the lake is light, this effect reaches inland only a mile or two. With stronger on-shore winds, the entire city is cooled. In the winter the relatively warm water of the lake moderates the temperature during easterly wind situations. Lake-induced snows usually occur a few times each winter, but snow accumulation is rarely heavy.

Topography does not significantly affect air flow, except that lesser frictional drag over Lake Michigan causes winds to be frequently stronger along the lake shore, and often permits air masses approaching from the north to reach shore areas one hour or more before affecting inland portions of the city.

NORMALS, MEANS, AND EXTREMES

MILWAUKEE, WISCONSIN

LATITUDE: 42°57'N LONGITUDE: 87°54' W ELEVATION: FT. GRND 672 BARO 691 TIME ZONE: CENTRAL WBAN: 14839

	(a)	JAN	FEB	MAR	APR	MAY	JUNE	JULY	AUG	SEP	OCT	NOV	DEC	YEAR
TEMPERATURE °F:														
Normals														
-Daily Maximum		26.1	30.1	40.4	52.9	64.3	74.9	79.9	77.8	70.6	58.7	44.7	31.2	54.3
-Daily Minimum		11.6	15.9	26.2	35.8	44.8	55.0	62.0	60.8	52.8	41.8	30.7	17.5	37.9
-Monthly		18.9	23.0	33.3	44.4	54.6	65.0	70.9	69.3	61.7	50.3	37.7	24.4	46.1
Extremes														
-Record Highest	53	62	65	82	91	93	101	101	103	98	89	77	63	103
-Year		1944	1976	1986	1980	1991	1988	1955	1988	1953	1963	1950	1982	AUG 1988
-Record Lowest	53	-26	-19	-10	12	21	33	40	44	28	18	-5	-20	-26
-Year		1982	1951	1962	1982	1966	1945	1965	1982	1974	1981	1950	1983	JAN 1982
NORMAL DEGREE DAYS:														
Heating (base 65°F)		1429	1176	983	618	338	82	14	27	123	456	819	1259	7324
Cooling (base 65°F)		0	0	0	0	16	82	197	160	24	0	0	0	479
% OF POSSIBLE SUNSHINE	53	45	47	49	53	59	65	69	66	59	54	39	38	54
MEAN SKY COVER (tenths)														
Sunrise - Sunset	53	6.8	6.8	7.0	6.7	6.3	6.0	5.4	5.4	5.7	5.9	7.2	7.2	6.4
MEAN NUMBER OF DAYS:														
Sunrise to Sunset														
-Clear	53	7.1	6.4	6.0	6.2	7.2	7.5	9.6	10.0	9.3	9.2	5.4	6.2	90.1
-Partly Cloudy	53	6.4	6.3	7.7	7.8	9.8	10.4	11.3	10.8	9.0	8.5	6.1	6.0	100.2
-Cloudy	53	17.5	15.5	17.3	16.0	14.1	12.0	10.1	10.2	11.6	13.3	18.5	18.8	175.0
Precipitation														
.01 inches or more	53	11.2	9.5	11.8	12.1	11.6	10.7	9.7	9.1	9.2	9.1	10.5	11.0	125.4
Snow,Ice pellets,hail														
1.0 inches or more	53	3.6	2.7	2.5	0.4	0.*	0.0	0.0	0.0	0.0	0.1	0.9	3.1	13.3
Thunderstorms	53	0.3	0.3	1.5	3.4	4.5	6.3	6.5	5.6	4.0	1.6	1.1	0.3	35.3
Heavy Fog Visibility														
1/4 mile or less	53	2.1	2.0	3.2	3.0	3.2	2.5	1.2	1.6	1.3	2.0	2.2	2.1	26.5
Temperature °F														
-Maximum														
90° and above	33	0.0	0.0	0.0	0.*	0.1	2.1	4.0	2.5	0.6	0.0	0.0	0.0	9.3
32° and below	33	19.8	15.1	6.2	0.5	0.0	0.0	0.0	0.0	0.0	0.0	2.6	14.9	59.2
-Minimum														
32° and below	33	29.5	26.2	23.3	9.5	1.0	0.0	0.0	0.0	0.1	4.1	17.7	27.7	139.2
0° and below	33	7.1	3.3	0.2	0.0	0.0	0.0	0.0	0.0	0.0	0.0	0.1	3.0	13.6
AVG. STATION PRESS.(mb)	21	991.9	992.8	990.3	989.5	989.5	989.3	990.6	991.7	992.1	992.1	991.1	992.0	991.1
RELATIVE HUMIDITY (%)														
Hour 00	33	75	74	76	75	75	77	80	83	82	78	78	78	78
Hour 06 (Local Time)	33	76	77	79	78	77	79	82	86	86	82	81	80	80
Hour 12	33	68	67	65	61	60	60	61	62	63	62	67	71	64
Hour 18	33	71	70	68	64	61	62	63	67	71	70	74	75	68
PRECIPITATION (inches):														
Water Equivalent														
-Normal		1.60	1.45	2.67	3.50	2.84	3.24	3.47	3.53	3.38	2.41	2.51	2.33	32.93
-Maximum Monthly	53	4.04	3.94	6.93	7.31	7.56	8.28	7.66	9.05	9.87	7.03	7.11	5.42	9.87
-Year		1960	1986	1976	1973	1990	1954	1964	1987	1941	1991	1985	1987	SEP 1941
-Minimum Monthly	53	0.31	0.05	0.31	0.81	0.50	0.70	0.95	0.46	0.02	0.15	0.62	0.29	0.02
-Year		1981	1969	1968	1942	1988	1988	1946	1948	1979	1956	1949	1976	SEP 1979
-Maximum in 24 hrs	53	1.73	1.67	2.57	3.11	3.11	3.13	4.35	6.84	5.28	2.60	2.18	2.24	6.84
-Year		1985	1960	1960	1976	1978	1950	1959	1986	1941	1959	1943	1982	AUG 1986
Snow,Ice pellets,hail														
-Maximum Monthly	53	33.6	42.0	26.7	15.8	3.2	T	T	T	T	6.3	16.1	27.9	42.0
-Year		1979	1974	1965	1973	1990	1992	1990	1989	1993	1989	1977	1978	FEB 1974
-Maximum in 24 hrs	53	13.8	16.7	11.2	11.6	3.2	T	T	T	T	6.3	10.6	13.1	16.7
-Year		1990	1960	1961	1973	1990	1992	1990	1989	1993	1989	1977	1987	FEB 1960
WIND:														
Mean Speed (mph)	53	12.6	12.4	12.9	12.8	11.6	10.4	9.7	9.5	10.4	11.4	12.5	12.3	11.5
Prevailing Direction through 1963		WNW	WNW	WNW	NNE	NNE	NNE	SW	SW	SSW	SSW	WNW	WNW	WNW
Fastest Obs. 1 Min.														
-Direction (!!)	11	02	02	19	23	36	25	30	30	26	24	24	04	30
-Speed (MPH)	11	35	36	39	45	44	38	54	44	35	35	38	40	54
-Year		1985	1984	1990	1983	1983	1990	1984	1989	1987	1984	1988	1987	JUL 1984
Peak Gust														
-Direction (!!)	10	SW	NE	SW	SW	SW	W	NW	NW	SW	NW	NW	NE	NW
-Speed (mph)	10	54	46	77	64	54	56	81	64	58	53	56	59	81
-Date		1989	1990	1991	1984	1985	1990	1984	1989	1986	1990	1989	1987	JUL 1984

PRECIPITATION (inches)

MILWAUKEE, WISCONSIN

YEAR	JAN	FEB	MAR	APR	MAY	JUNE	JULY	AUG	SEP	OCT	NOV	DEC	ANNUAL
1964	1.18	0.41	3.05	3.81	2.57	1.70	7.66	2.62	1.74	0.17	2.29	0.98	28.18
1965	3.33	1.04	3.61	3.47	2.12	0.85	2.64	6.15	6.85	2.68	2.02	3.73	38.49
1966	2.06	1.27	3.61	2.67	2.00	1.68	3.32	3.27	0.48	1.76	2.70	2.31	27.13
1967	1.49	1.31	1.35	2.70	1.80	7.38	1.35	1.23	1.69	2.70	1.52	1.33	25.85
1968	0.98	0.56	0.31	2.90	3.28	7.79	3.59	2.59	3.36	0.94	2.56	2.65	31.51
1969	1.83	0.05	1.05	3.42	3.05	7.53	6.61	0.53	2.18	4.48	1.14	1.18	33.05
1970	0.41	0.13	1.62	2.71	3.41	3.92	1.93	0.64	6.94	2.09	2.03	3.02	28.85
1971	1.37	2.50	2.83	1.31	0.90	2.67	2.60	2.28	1.30	1.90	2.45	4.34	26.45
1972	0.75	0.86	2.57	2.76	2.33	3.33	4.60	4.82	7.57	3.28	1.34	2.47	36.68
1973	1.12	1.51	2.86	7.31	3.39	1.96	1.55	0.95	4.50	2.97	1.83	3.80	33.75
1974	3.61	3.10	4.29	3.83	4.10	3.48	3.51	2.54	0.50	1.96	1.86	2.10	34.88
1975	2.25	2.53	3.01	4.08	2.01	3.99	1.14	3.89	1.00	0.72	2.83	1.70	29.15
1976	1.16	2.65	6.93	5.01	3.77	2.27	2.12	2.05	1.70	2.82	0.65	0.29	31.42
1977	0.90	0.59	4.56	2.09	0.90	5.78	5.99	3.82	4.11	2.02	2.56	3.27	36.59
1978	2.03	0.55	1.08	4.41	4.66	4.52	5.98	3.43	6.81	2.22	2.13	2.92	40.74
1979	3.00	0.97	4.17	5.43	1.82	2.84	1.06	4.85	0.02	1.77	2.67	2.27	30.87
1980	1.65	1.75	0.77	4.02	1.81	4.67	3.39	5.06	3.57	1.63	1.57	3.52	33.41
1981	0.31	2.88	0.51	4.87	3.05	2.39	4.35	4.26	5.47	2.71	2.05	1.03	33.88
1982	2.92	0.29	3.20	4.47	2.76	3.06	3.88	3.33	0.64	3.17	4.74	4.10	36.56
1983	0.75	2.23	4.12	4.66	5.83	1.41	1.34	4.70	2.79	2.65	4.10	2.89	37.47
1984	0.79	1.20	2.17	5.04	4.21	4.07	3.39	2.93	2.51	5.30	3.74	4.25	39.60
1985	1.94	2.34	4.11	1.93	2.73	1.27	2.18	2.23	3.44	5.39	7.11	2.62	37.29
1986	0.91	3.94	1.85	1.83	2.74	4.51	6.15	8.82	7.26	2.24	0.89	1.03	42.17
1987	1.22	1.22	1.74	4.26	3.76	2.23	4.20	9.05	2.22	1.09	2.73	5.42	39.14
1988	3.25	1.29	1.30	4.12	0.50	0.70	1.53	3.25	4.94	2.97	5.15	1.43	30.43
1989	0.86	0.69	3.03	1.33	2.86	1.89	6.16	5.19	3.25	2.67	1.90	0.47	30.30
1990	2.57	1.90	2.75	2.67	7.56	4.97	3.02	4.68	1.89	2.65	3.54	2.66	40.86
1991	1.55	0.38	4.06	3.70	4.25	2.13	4.34	2.27	4.34	7.03	3.36	1.94	39.35
1992	1.09	1.54	2.61	2.41	0.60	3.13	5.64	3.50	4.13	1.45	5.40	2.45	33.95
1993	2.63	0.98	3.19	6.64	1.56	6.39	4.22	4.20	3.91	0.44	1.98	0.70	36.84
Record Mean	1.80	1.56	2.51	2.93	3.17	3.48	3.03	3.02	3.18	2.35	2.16	1.86	31.06

AVERAGE TEMPERATURE (deg. F)

MILWAUKEE, WISCONSIN

YEAR	JAN	FEB	MAR	APR	MAY	JUNE	JULY	AUG	SEP	OCT	NOV	DEC	ANNUAL
1964	26.1	25.3	31.7	44.8	60.3	66.5	72.3	67.7	61.2	48.2	40.4	23.2	47.3
1965	19.5	21.2	25.3	42.7	58.6	63.1	68.9	66.8	61.7	50.8	38.5	32.9	45.8
1966	13.9	22.3	35.4	42.3	49.6	67.0	73.5	66.8	59.1	48.9	38.0	24.5	45.1
1967	24.5	17.5	33.2	44.9	50.2	66.5	68.6	66.0	60.8	50.4	35.2	28.9	45.6
1968	21.8	20.7	40.3	47.6	53.2	66.3	69.5	71.4	63.5	52.7	38.1	25.6	47.5
1969	18.6	27.3	29.8	44.9	55.3	58.9	67.8	70.9	62.0	47.5	34.3	25.1	45.2
1970	13.3	22.6	30.5	46.2	56.0	65.3	73.3	71.8	62.0	52.5	37.3	24.3	46.2
1971	12.7	21.6	28.6	40.9	51.1	67.5	68.0	67.0	64.9	55.5	37.3	29.5	45.4
1972	15.9	19.8	27.7	39.4	55.0	61.6	69.3	69.1	61.3	47.5	36.2	20.2	43.6
1973	24.7	25.4	39.9	42.7	51.0	69.2	71.4	72.5	63.9	54.6	38.5	25.5	48.3
1974	21.6	22.9	33.8	46.1	50.6	62.4	71.5	67.3	58.0	49.9	38.6	29.1	46.0
1975	24.0	23.5	28.6	37.7	57.1	65.7	71.7	71.2	58.5	54.1	44.5	27.9	47.0
1976	18.5	31.1	38.6	48.7	52.6	68.2	72.6	69.9	62.6	45.9	29.5	16.3	46.2
1977	8.3	23.7	39.2	48.9	61.3	62.5	73.1	67.2	61.9	49.0	37.1	22.8	46.2
1978	15.4	16.4	29.8	42.5	55.2	65.2	68.6	69.9	65.8	49.5	38.2	23.8	45.0
1979	11.6	15.1	33.2	42.1	54.9	64.7	70.9	68.8	64.7	51.1	38.2	31.4	45.6
1980	20.7	20.3	30.5	45.3	57.2	61.3	71.2	69.7	61.1	45.7	37.7	24.4	45.4
1981	18.9	25.3	35.6	46.5	51.5	65.2	67.3	67.8	59.1	45.8	37.4	24.2	45.4
1982	9.7	19.4	31.5	41.2	58.5	59.8	71.1	67.3	60.8	52.7	38.0	33.2	45.3
1983	26.4	29.4	35.0	41.7	50.2	66.3	76.2	74.4	62.9	52.0	39.9	14.4	47.4
1984	18.9	33.4	29.2	45.5	54.9	68.7	71.7	73.3	60.9	52.9	37.7	29.1	48.0
1985	15.2	21.3	37.9	50.8	58.8	63.8	72.4	68.4	64.3	50.7	36.7	15.7	46.3
1986	21.9	23.3	38.1	48.5	56.1	63.3	72.5	67.1	63.8	51.6	35.1	29.2	47.5
1987	24.8	32.0	37.8	48.0	60.1	72.2	74.8	69.9	63.3	46.2	41.9	31.4	50.2
1988	18.3	20.1	34.8	45.7	58.7	70.2	75.4	75.7	63.5	45.8	40.7	26.6	48.0
1989	30.4	18.0	32.4	43.2	54.9	64.4	71.6	68.8	60.2	52.7	35.1	16.7	45.7
1990	31.1	28.9	38.8	49.3	52.8	67.6	70.5	71.2	66.0	51.5	44.5	26.9	49.9
1991	20.1	29.9	38.2	49.7	63.2	70.7	74.4	73.5	63.2	52.3	35.1	29.6	50.0
1992	27.9	31.9	35.4	42.9	56.7	63.0	67.7	67.4	61.6	49.8	37.3	28.2	47.5
1993	25.7	24.4	32.5	43.0	56.9	63.8	73.0	73.6	60.6	51.0	39.5	29.9	47.8
Record Mean	20.8	23.4	33.0	44.5	54.6	64.7	70.9	69.8	62.5	51.2	37.7	25.9	46.6
Max	28.0	30.5	40.0	52.5	63.7	73.8	79.5	77.9	70.7	59.1	44.5	32.5	54.4
Min	13.6	16.3	25.9	36.4	45.5	55.5	62.3	61.6	54.4	43.3	30.8	19.3	38.7

REFERENCE NOTES

MILWAUKEE, WISCONSIN

GENERAL
T - TRACE AMOUNT.
BLANK ENTRIES DENOTE MISSING/UNREPORTED DATA.
# INDICATES A STATION OR INSTRUMENT RELOCATION.

SPECIFIC

PM - INCLUDES LAST DAY OF PREVIOUS MONTH
ASOS - AUTOMATED SURFACE OBSERVING SYSTEM IN OPERATION DURING THESE MONTHS.

(a) - LENGTH OF RECORD IN YEARS, ALTHOUGH INDIVIDUAL MONTHS MAY BE MISSING.
0.* OR * - THE VALUE IS BETWEEN 0.0 AND 0.05
NORMALS - BASED ON THE 1961-1990 RECORD PERIOD.
EXTREMES - DATES ARE THE MOST RECENT OCCURRENCE
WIND DIR.- NUMERALS SHOW TENS OF DEGRESS CLOCKWISE FROM TRUE NORTH. "00" INDICATES CALM.
RESULTANT DIRECTIONS ARE GIVEN TO WHOLE DEGREES.
BOLD VALUES INDICATE EXTREME VALUES WHICH OCCURRED AFTER THE ASOS SYSTEM WAS COMMISSIONED.

RECORD = PERIOD OF RECORD
RECORD MEAN PRECIPITATION IS THE MEAN OF ALL DAILY PRECIPITATION AMOUNTS DURING THE PERIOD OF RECORD.
RECORD MAX(MIN) TEMPERATURE IS THE MEAN OF ALL DAILY MAX(MIN) TEMPERATURES DURING THE PERIOD OF RECORD.
RECORD MEAN TEMPERATURE IS THE SUM OF THE RECORD MAX AND RECORD MIN DIVIDED BY 2.
AVERAGE TEMPERATURE IS THE SUM OF THE MEAN DAILY MAX AND MIN TEMPERATURE DIVIDED BY 2.

EXCEPTIONS

FASTEST MILE WINDS ARE THROUGH JUNE 1982.

RECORD MEANS ARE THROUGH THE CURRENT YEAR, BEGINNING IN 1875 FOR TEMPERATURE
1871 FOR PRECIPITATION
1941 FOR SNOWFALL

HEATING DEGREE DAYS Base 65 deg. F MILWAUKEE, WISCONSIN

SEASON	JULY	AUG	SEP	OCT	NOV	DEC	JAN	FEB	MAR	APR	MAY	JUNE	TOTAL
1964-65	7	47	177	515	730	1290	1404	1222	1226	664	232	123	7637
1965-66	25	51	149	438	793	987	1579	1189	915	674	473	88	7361
1966-67	4	41	198	496	804	1249	1249	1325	978	596	458	47	7445
1967-68	46	53	164	460	888	1112	1333	1277	758	521	363	74	7049
1968-69	31	23	82	403	799	1214	1434	1053	1087	596	317	215	7254
1969-70	34	5	143	539	913	1228	1600	1180	1062	561	301	104	7670
1970-71	7	7	145	383	823	1259	1615	1211	1122	716	421	65	7774
1971-72	20	37	119	308	824	1097	1518	1305	1149	758	305	139	7579
1972-73	40	32	133	534	859	1381	1242	1101	769	659	426	6	7182
1973-74	10	5	111	324	788	1218	1340	1173	959	560	448	106	7042
1974-75	0	20	237	461	786	1103	1260	1157	1122	814	267	69	7296
1975-76	17	4	203	353	610	1144	1438	978	813	507	382	43	6492
1976-77	2	21	124	589	1056	1504	1754	1152	790	490	173	151	7806
1977-78	8	47	106	485	827	1302	1531	1356	1086	667	335	83	7833
1978-79	21	5	76	473	796	1273	1654	1391	980	681	322	91	7763
1979-80	20	25	70	436	797	1036	1368	1290	1063	594	259	154	7112
1980-81	8	9	140	590	812	1250	1423	1106	905	548	417	69	7277
1981-82	44	21	187	590	820	1257	1712	1272	1032	707	215	172	8029
1982-83	3	44	170	381	802	983	1186	990	925	692	453	81	6710
1983-84	10	0	148	405	748	1565	1424	910	1103	579	318	35	7245
1984-85	3	7	179	373	812	1103	1542	1215	831	461	222	96	6844
1985-86	2	13	139	436	843	1523	1328	1161	827	494	302	128	7196
1986-87	13	34	98	407	891	1106	1242	917	839	502	236	19	6304
1987-88	12	28	91	576	686	1037	1442	1294	930	571	245	55	6967
1988-89	3	7	87	587	720	1183	1065	1307	1006	649	324	85	7023
1989-90	0	16	166	381	890	1493	1040	1004	805	502	375	51	6723
1990-91	21	9	93	418	612	1173	1385	980	822	467	201	26	6207
1991-92	0	1	160	394	889	1091	1141	955	914	655	278	104	6582
1992-93	27	35	145	472	825	1132	1211	1132	1004	654	245	104	6986
1993-94	1	2	158	428	757	1080							

COOLING DEGREE DAYS Base 65 deg. F MILWAUKEE, WISCONSIN

YEAR	JAN	FEB	MAR	APR	MAY	JUNE	JULY	AUG	SEP	OCT	NOV	DEC	TOTAL
1969	0	0	0	0	24	38	126	197	59	5	0	0	449
1970	0	0	0	4	29	119	270	227	60	4	0	0	713
1971	0	0	0	0	0	148	120	105	123	20	0	0	516
1972	0	0	0	0	3	42	180	166	26	0	0	0	417
1973	0	0	0	0	0	140	216	247	84	6	0	0	693
1974	0	0	0	3	6	36	210	98	32	1	0	0	386
1975	0	0	0	0	30	98	230	203	16	21	0	0	598
1976	0	0	0	24	5	144	247	181	62	4	0	0	667
1977	0	0	0	12	65	81	264	122	20	0	0	0	564
1978	0	0	0	0	40	97	138	164	109	0	0	0	548
1979	0	0	0	0	16	87	209	147	68	11	0	0	538
1980	0	0	0	9	25	50	207	164	29	0	0	0	484
1981	0	0	0	2	3	84	121	112	16	0	0	0	338
1982	0	0	0	0	21	24	199	121	51	5	0	0	421
1983	0	0	0	0	0	127	364	299	92	9	0	0	891
1984	0	0	0	1	11	152	216	270	63	3	0	0	716
1985	0	0	0	42	35	68	240	127	127	0	0	0	639
1986	0	0	3	7	31	84	251	105	70	0	0	0	551
1987	0	0	0	2	87	244	323	189	47	0	1	0	893
1988	0	0	0	0	57	215	333	344	48	0	0	0	997
1989	0	0	2	0	16	76	214	144	29	4	0	0	485
1990	0	0	2	38	5	135	198	210	132	7	1	0	728
1991	0	0	0	13	150	204	300	268	114	7	0	0	1056
1992	0	0	0	0	25	49	117	119	50	4	0	0	364
1993	0	0	0	0	5	73	258	277	35	3	0	0	651

SNOWFALL (inches) MILWAUKEE, WISCONSIN

SEASON	JULY	AUG	SEP	OCT	NOV	DEC	JAN	FEB	MAR	APR	MAY	JUNE	TOTAL
1964-65	0.0	0.0	0.0	T	1.4	8.1	23.6	10.1	26.7	4.1	0.0	0.0	74.0
1965-66	0.0	0.0	0.0	T	T	14.5	24.6	7.7	2.8	1.3	0.1	0.0	51.0
1966-67	0.0	0.0	0.0	0.0	2.0	9.9	13.1	27.1	7.4	T	T	0.0	59.5
1967-68	0.0	0.0	0.0	0.8	0.4	1.2	4.6	3.5	1.2	0.4	0.0	0.0	12.1
1968-69	0.0	0.0	0.0	0.0	0.3	11.6	11.1	0.7	6.2	0.0	T	0.0	29.9
1969-70	0.0	0.0	0.0	T	0.7	14.9	6.0	2.0	10.7	5.2	0.0	0.0	39.5
1970-71	0.0	0.0	0.0	0.0	0.6	19.6	15.8	2.5	18.1	0.7	T	0.0	57.3
1971-72	0.0	0.0	0.0	0.0	6.1	2.7	6.8	10.2	14.9	1.2	0.0	0.0	41.9
1972-73	0.0	0.0	0.0	T	3.4	13.7	0.2	9.9	1.8	15.8	T	0.0	44.8
1973-74	0.0	0.0	0.0	0.0	T	19.6	14.2	42.0	7.4	T	0.0	0.0	83.2
1974-75	0.0	0.0	0.0	T	2.0	9.1	3.5	12.2	15.1	10.4	0.0	0.0	52.3
1975-76	0.0	0.0	0.0	0.0	8.4	12.2	14.8	7.6	2.1	0.1	T	0.0	45.2
1976-77	0.0	0.0	0.0	4.0	3.6	5.3	15.6	5.6	12.4	2.1	0.0	0.0	48.6
1977-78	0.0	0.0	0.0	T	16.1	20.8	25.7	13.3	4.8	T	0.0	0.0	80.7
1978-79	0.0	0.0	0.0	0.0	5.3	27.9	33.6	9.1	6.2	0.8	0.0	0.0	82.9
1979-80	0.0	0.0	0.0	T	2.1	0.6	11.6	22.8	6.3	3.6	T	0.0	47.0
1980-81	0.0	0.0	0.0	T	2.3	17.5	4.9	15.7	1.5	T	0.0	0.0	41.9
1981-82	0.0	0.0	0.0	T	2.0	8.3	29.2	3.0	13.0	11.7	0.0	0.0	67.2
1982-83	0.0	0.0	0.0	T	0.4	3.1	6.3	13.5	13.8	1.0	0.0	0.0	38.1
1983-84	0.0	0.0	0.0	0.0	0.3	13.3	9.6	1.2	8.2	0.5	T	0.0	33.1
1984-85	0.0	0.0	0.0	0.0	T	19.0	20.8	15.3	9.0	2.5	0.0	0.0	66.6
1985-86	0.0	0.0	0.0	0.0	3.5	13.5	10.4	14.0	0.7	0.3	0.0	0.0	42.4
1986-87	0.0	0.0	0.0	T	2.4	2.5	11.4	T	5.2	0.4	0.0	0.0	21.9
1987-88	0.0	0.0	0.0	0.6	0.4	19.9	10.2	20.7	2.9	T	0.0	0.0	54.7
1988-89	0.0	0.0	0.0	T	2.7	7.1	2.7	13.1	13.3	0.4	0.6	0.0	39.9
1989-90	T	T	0.0	6.3	11.6	7.4	19.9	17.9	0.2	1.2	3.2	0.0	67.7
1990-91	T	0.0	0.0	0.0	0.4	10.5	15.2	2.4	1.1	0.4	0.0	0.0	30.0
1991-92	0.0	0.0	T	T	4.8	14.7	4.3	5.3	11.1	0.8	0.0	T	41.0
1992-93	0.0	0.0	0.0	1.2	0.4	8.4	12.2	11.6	12.7	3.4	T	0.0	49.9
1993-94	0.0	0.0	T	T	8.0	1.2							
Record Mean	T	T	T	0.2	3.0	10.4	12.8	9.6	8.7	1.7	0.1	T	46.6

CASPER, WYOMING

Casper is located in the central portion of Wyoming in the North Platte River Valley at an elevation of about 5,300 feet. The country immediately surrounding Casper is mostly rolling and hilly with considerable flat prairie land in each direction except toward the south where Casper Mountain rises some 3,500 feet above the valley floor. The prairie land is used mainly for grazing.

The National Weather Service Office is located at Natrona County International Airport, some 8 miles west-northwest of the Casper Post Office and about 200 feet higher in elevation.

The climate of Casper is semi-arid. Most of the air masses reaching this area move in from the Pacific and the mountains to the west are effective moisture barriers. About 70 percent of the annual precipitation occurs during the growing season of late spring and summer, mostly in thunderstorms. Monthly snowfall amounts are unusually uniform from November through February, and a bit heavier in March and April. Snow has occurred as early in the season as September and as late as early June.

Casper experiences large diurnal and annual temperature ranges. This is due to the advent of both warm and cold air masses and the relatively high elevation which permits rapid incoming and outgoing radiation. The mean daily temperature averages about 71 degrees in summer and 22 degrees in winter. Temperatures during winter months average a few degrees higher and summer temperatures average several degrees cooler than locations in the Missouri Valley to the east.

Windy days are quite frequent during winter and spring months. Usually the stronger winds are from the southwest and this tends to raise the temperature because the air is moving downslope.

Based on the 1951-1980 period, the average first occurrence of 32 degrees Fahrenheit in the fall is September 22 and the average last occurrence in the spring is May 22.

NORMALS, MEANS, AND EXTREMES

CASPER, WYOMING

LATITUDE: 42°55'N LONGITUDE: 106°28' W ELEVATION: FT. GRND 5338 BARO 5323 TIME ZONE: MOUNTAIN WBAN: 24089

	(a)	JAN	FEB	MAR	APR	MAY	JUNE	JULY	AUG	SEP	OCT	NOV	DEC	YEAR
TEMPERATURE °F:														
Normals														
-Daily Maximum		32.8	37.0	45.2	56.1	66.6	78.6	87.6	85.7	73.8	60.5	44.3	33.9	58.5
-Daily Minimum		12.0	16.0	21.8	29.5	37.9	46.9	54.0	51.8	41.6	32.2	21.8	13.7	31.6
-Monthly		22.4	26.5	33.5	42.8	52.3	62.7	70.8	68.7	57.8	46.4	33.1	23.8	45.1
Extremes														
-Record Highest	43	60	68	74	84	92	102	104	102	96	87	71	61	104
-Year		1971	1982	1986	1992	1984	1990	1954	1979	1983	1993	1983	1980	JUL 1954
-Record Lowest	43	-40	-29	-21	-4	16	28	30	33	16	-3	-21	-41	-41
-Year		1972	1989	1965	1966	1953	1969	1972	1977	1983	1971	1985	1990	DEC 1990
NORMAL DEGREE DAYS:														
Heating (base 65°F)		1321	1078	977	666	394	139	10	36	250	577	957	1277	7682
Cooling (base 65°F)		0	0	0	0	0	70	190	151	34	0	0	0	445
% OF POSSIBLE SUNSHINE														
MEAN SKY COVER (tenths)														
Sunrise - Sunset	43	6.6	6.5	6.8	6.8	6.6	5.3	4.3	4.5	4.6	5.3	6.3	6.2	5.8
MEAN NUMBER OF DAYS:														
Sunrise to Sunset														
-Clear	43	6.9	6.3	5.7	5.5	5.7	9.9	13.8	13.2	13.2	11.3	7.3	8.2	107.0
-Partly Cloudy	43	8.0	8.4	8.8	9.0	10.7	11.2	10.9	10.8	8.7	8.0	8.4	8.1	111.0
-Cloudy	43	16.0	13.6	16.5	15.5	14.7	8.9	6.3	7.0	8.1	11.6	14.3	14.7	147.2
Precipitation														
.01 inches or more	43	7.1	7.9	9.5	10.2	10.9	8.6	8.0	5.7	6.3	6.5	7.1	7.5	95.3
Snow,Ice pellets,hail														
1.0 inches or more	43	3.5	3.6	4.1	3.3	1.0	0.1	0.0	0.0	0.3	1.8	3.3	3.0	24.2
Thunderstorms	43	0.0	0.*	0.3	1.4	6.4	8.5	9.2	6.9	3.0	0.4	0.1	0.0	36.2
Heavy Fog Visibility														
1/4 mile or less	43	0.8	0.7	0.9	1.3	0.9	0.4	0.3	0.3	0.7	0.9	0.9	0.6	8.7
Temperature °F														
-Maximum														
90° and above	29	0.0	0.0	0.0	0.0	0.1	4.2	12.5	10.6	1.3	0.0	0.0	0.0	28.8
32° and below	29	13.2	8.3	4.2	1.1	0.1	0.0	0.0	0.0	0.1	0.8	6.1	12.7	46.7
-Minimum														
32° and below	29	29.4	26.3	26.7	18.8	7.0	0.3	0.*	0.0	3.9	15.5	25.0	28.8	181.6
0° and below	29	6.8	3.5	1.4	0.1	0.0	0.0	0.0	0.0	0.0	0.1	1.9	5.3	19.1
AVG. STATION PRESS.(mb)	21	835.9	835.9	833.5	834.8	835.5	837.3	839.6	839.8	839.4	838.7	836.4	836.0	836.9
RELATIVE HUMIDITY (%)														
Hour 05	29	68	70	73	75	78	76	70	67	67	68	69	68	71
Hour 11 (Local Time)	29	59	58	53	47	44	38	32	31	36	44	54	58	46
Hour 17	29	61	56	47	42	40	33	27	25	30	40	57	62	43
Hour 23	29	69	69	70	68	68	63	55	53	58	63	67	68	64
PRECIPITATION (inches):														
Water Equivalent														
-Normal		0.55	0.60	0.95	1.56	2.13	1.46	1.26	0.67	0.94	0.97	0.77	0.66	12.52
-Maximum Monthly	43	1.42	1.42	2.43	3.92	6.46	4.15	3.05	2.66	3.40	2.63	2.72	3.71	6.46
-Year		1987	1987	1954	1974	1978	1982	1951	1979	1982	1986	1983	1982	MAY 1978
-Minimum Monthly	43	T	0.15	0.25	0.08	0.30	0.03	0.09	0.02	0.07	T	0.07	0.03	T
-Year		1952	1957	1953	1992	1966	1956	1991	1950	1956	1965	1965	1952	OCT 1965
-Maximum in 24 hrs	43	0.82	0.56	1.00	3.00	2.61	2.34	2.07	1.74	2.04	2.49	1.21	1.64	3.00
-Year		1987	1989	1958	1974	1978	1982	1983	1979	1989	1962	1983	1982	APR 1974
Snow,Ice pellets,hail														
-Maximum Monthly	43	24.0	23.8	36.2	56.3	24.6	3.0	T	T	11.5	16.1	37.1	62.8	62.8
-Year		1987	1952	1975	1973	1978	1969	1990	1993	1982	1986	1983	1982	DEC 1982
-Maximum in 24 hrs	43	14.0	11.0	14.6	16.5	14.1	3.0	T	T	6.8	13.3	14.3	31.1	31.1
-Year		1987	1987	1954	1973	1950	1969	1990	1993	1982	1986	1983	1982	DEC 1982
WIND:														
Mean Speed (mph)	43	16.5	15.1	13.8	12.6	11.6	11.0	10.2	10.4	11.0	12.1	14.4	16.0	12.9
Prevailing Direction through 1963		SW	SW	SW	WSW	WSW	WSW	WSW	SW	WSW	SW	SW	SW	SW
Fastest Obs. 1 Min.														
-Direction (!!!)	40	20	23	25	25	32	36	25	25	32	25	25	20	25
-Speed (MPH)	40	58	58	81	54	58	52	52	50	53	55	49	63	81
-Year		1954	1957	1956	1967	1959	1959	1974	1954	1965	1954	1970	1955	MAR 1956
Peak Gust														
-Direction (!!!)	10	SW	SW	NW	NW	SW	W	NW	SW	SW	SW	SW	SW	SW
-Speed (mph)	10	67	64	63	60	64	64	60	62	63	62	60	66	67
-Date		1990	1986	1988	1985	1985	1991	1990	1988	1986	1985	1984	1984	JAN 1990

PRECIPITATION (inches)

CASPER, WYOMING

YEAR	JAN	FEB	MAR	APR	MAY	JUNE	JULY	AUG	SEP	OCT	NOV	DEC	ANNUAL
1964	0.75	0.72	0.66	3.09	2.17	0.59	0.73	0.47	0.34	0.23	0.66	0.47	10.88
1965	0.78	0.37	0.52	0.96	2.64	2.47	1.16	0.21	2.07	T	0.07	0.87	12.12
1966	0.30	0.43	0.49	1.29	0.30	1.12	0.46	1.03	0.39	1.33	0.43	0.57	8.14
1967	0.58	0.65	0.66	1.24	1.62	3.75	1.46	0.41	1.67	1.37	1.09	0.82	15.32
1968	0.33	0.46	0.92	1.18	3.35	1.86	1.02	0.61	0.30	0.58	0.56	0.71	11.88
1969	0.08	0.55	1.10	1.52	0.85	2.39	0.70	0.78	0.16	1.43	0.96	0.16	10.68
1970	0.59	0.27	1.53	1.57	2.03	2.03	0.73	0.16	0.21	1.01	1.01	0.89	12.03
1971	0.19	0.80	1.17	3.40	5.59	0.17	0.11	0.11	0.74	1.73	0.61	0.34	14.96
1972	0.99	0.41	1.41	1.44	0.76	0.83	1.04	1.52	0.37	2.04	0.66	0.41	11.88
1973	0.63	0.57	0.52	3.86	1.02	0.66	2.26	0.34	3.28	0.60	0.80	0.55	15.09
1974	0.69	0.42	0.37	3.92	0.62	0.60	0.75	1.00	0.83	1.15	0.46	0.22	11.03
1975	0.51	0.51	2.01	0.97	3.88	0.94	0.63	0.12	0.11	1.01	0.51	0.72	11.92
1976	0.44	0.89	0.50	1.52	2.69	1.66	1.07	0.31	1.15	1.21	0.09	0.38	11.91
1977	0.35	0.69	1.05	1.07	1.93	0.19	1.88	0.34	0.30	0.71	0.79	0.71	10.01
1978	0.70	0.75	0.93	1.05	6.46	1.40	2.62	0.93	0.26	0.40	0.94	1.20	17.64
1979	0.81	0.39	1.31	0.83	2.25	1.31	1.22	2.66	0.18	0.50	0.75	0.36	12.57
1980	0.81	0.63	1.19	0.35	2.82	0.10	0.85	0.65	0.10	0.64	0.74	0.37	9.25
1981	0.46	0.23	0.77	1.56	3.51	0.37	1.27	0.50	0.23	0.76	0.75	0.43	10.84
1982	0.41	0.33	0.60	1.25	2.10	4.15	1.92	0.88	3.40	1.18	0.55	3.71	20.48
1983	0.42	0.35	2.29	2.28	1.40	3.76	2.61	0.75	0.20	0.88	2.72	0.75	18.41
1984	1.19	0.48	1.59	2.23	1.33	1.34	2.26	0.25	0.50	0.71	0.70	0.78	13.36
1985	0.79	0.61	0.52	1.25	1.37	1.32	1.57	0.09	1.09	0.46	1.56	1.05	11.68
1986	0.36	0.89	0.55	1.89	1.33	4.06	0.88	0.27	1.31	2.63	1.48	0.27	15.92
1987	1.42	1.42	1.43	0.35	1.40	0.70	1.93	1.80	0.59	0.55	0.77	0.63	12.99
1988	0.28	0.79	0.71	0.71	1.24	0.28	0.55	0.16	0.74	0.12	0.48	0.50	6.56
1989	0.16	1.37	0.49	0.72	2.77	1.95	0.28	1.00	3.22	1.17	0.34	0.31	13.78
1990	0.27	0.70	1.13	1.35	1.09	0.66	2.15	1.89	0.52	0.90	1.27	0.61	12.54
1991	0.54	0.38	0.45	1.42	4.75	1.70	0.09	0.61	0.40	0.55	1.18	0.34	12.41
1992	0.17	0.35	1.27	0.08	2.05	0.79	1.51	0.86	0.29	0.22	0.90	0.88	9.37
1993	0.72	1.00	0.52	2.37	2.85	2.79	0.86	1.14	0.97	1.14	0.87	0.68	15.91
Record Mean	0.53	0.57	0.94	1.50	2.12	1.39	1.10	0.67	0.89	0.91	0.76	0.58	11.96

AVERAGE TEMPERATURE (deg. F)

CASPER, WYOMING

YEAR	JAN	FEB	MAR	APR	MAY	JUNE	JULY	AUG	SEP	OCT	NOV	DEC	ANNUAL
#1964	23.0	21.2	25.1	39.7	54.1	61.1	74.1	66.1	56.1	47.1	31.3	23.4	43.5
1965	29.7	22.6	20.5	45.8	49.9	60.1	69.3	66.3	47.7	51.8	39.8	29.1	44.4
1966	21.8	24.3	36.2	38.9	55.8	61.8	74.9	66.6	62.2	45.2	36.1	25.3	45.8
1967	26.7	27.7	36.4	42.7	49.1	58.5	69.8	69.0	61.0	48.2	30.8	17.4	44.8
1968	23.2	30.3	36.7	38.4	48.1	60.5	69.1	64.9	56.5	47.9	32.0	18.8	43.8
1969	27.1	29.0	28.4	46.7	55.4	57.0	71.6	71.8	63.3	37.9	34.4	28.5	45.9
1970	23.3	32.7	27.6	38.2	53.3	62.6	71.2	72.3	54.7	41.2	34.9	25.2	44.8
1971	25.4	24.7	31.7	40.9	50.1	64.3	67.2	71.3	53.9	42.8	30.5	23.9	43.9
1972	17.7	28.5	38.2	43.0	52.1	65.0	65.9	68.0	56.7	44.1	30.6	17.7	44.0
1973	19.2	24.5	33.0	36.1	51.1	63.0	68.4	70.7	55.0	48.7	33.1	26.7	44.1
1974	18.7	27.6	36.2	43.1	51.4	64.9	72.6	64.4	54.4	48.5	34.3	24.3	45.0
1975	21.1	20.9	30.2	37.7	48.9	58.8	72.9	68.0	56.7	47.6	32.5	28.4	43.7
1976	23.0	30.2	32.0	44.0	53.6	60.7	72.8	68.4	59.2	43.8	34.3	28.3	45.9
1977	20.2	30.9	31.3	46.5	55.1	69.4	71.7	66.1	60.3	48.8	33.3	24.9	46.5
1978	17.1	22.3	37.2	44.6	49.6	63.6	69.6	66.8	60.2	47.9	26.9	13.2	43.3
1979	8.8	23.6	35.4	43.7	49.3	61.8	70.8	67.7	62.1	47.8	26.5	28.7	43.8
1980	16.4	27.5	31.0	44.2	51.6	64.4	72.2	66.4	59.7	46.5	33.8	34.3	45.7
1981	31.4	29.9	38.0	48.3	51.4	64.3	71.8	69.2	62.0	44.4	39.0	27.3	48.1
1982	20.1	25.7	36.6	41.1	51.1	60.3	70.7	73.7	56.9	45.2	32.2	23.3	44.7
1983	31.2	31.5	36.3	38.0	48.4	60.4	71.2	74.8	60.3	49.7	31.2	10.9	45.3
1984	21.9	27.5	34.4	39.1	54.5	61.5	70.4	71.2	52.5	40.1	34.8	22.9	44.2
1985	16.1	19.7	33.3	45.5	55.3	61.3	71.6	66.7	53.4	45.3	20.3	21.3	42.5
1986	31.6	28.5	42.8	44.4	50.9	67.4	69.0	69.3	55.5	45.9	32.0	26.2	47.0
1987	23.1	28.4	32.4	49.7	57.5	64.8	70.7	66.2	58.7	46.7	35.5	23.8	46.5
1988	20.3	27.4	32.0	46.0	54.2	72.6	73.5	69.8	58.0	49.8	34.0	26.3	47.0
1989	26.1	11.9	35.6	44.9	54.0	61.2	73.4	69.0	58.2	46.3	37.3	24.1	45.2
1990	29.1	27.9	35.4	44.3	50.5	64.6	69.8	68.5	63.2	46.3	36.9	15.8	46.0
1991	21.3	35.5	36.5	42.0	51.8	65.0	71.5	72.0	59.5	46.0	30.4	27.9	46.6
1992	28.1	36.3	40.7	49.0	56.2	63.7	67.3	68.4	61.1	50.7	32.3	21.9	48.0
1993	21.5	19.6	38.3	43.9	56.2	60.3	66.7	68.0	58.2	47.4	29.4	28.9	44.9
Record Mean	22.7	26.7	33.0	43.0	52.5	62.6	70.8	69.1	58.4	47.1	33.1	25.4	45.4
Max	33.0	37.3	44.5	55.9	66.2	77.7	87.0	85.3	74.0	60.9	43.9	35.3	58.4
Min	12.4	16.1	21.4	30.0	38.8	47.5	54.5	52.9	42.7	33.3	22.2	15.5	32.3

REFERENCE NOTES

CASPER, WYOMING

GENERAL
T - TRACE AMOUNT.
BLANK ENTRIES DENOTE MISSING/UNREPORTED DATA.
# INDICATES A STATION OR INSTRUMENT RELOCATION.

SPECIFIC

(a) - LENGTH OF RECORD IN YEARS, ALTHOUGH INDIVIDUAL MONTHS MAY BE MISSING.
0.* OR * - THE VALUE IS BETWEEN 0.0 AND 0.05
NORMALS - BASED ON THE 1961-1990 RECORD PERIOD.
EXTREMES - DATES ARE THE MOST RECENT OCCURRENCE
WIND DIR. - NUMERALS SHOW TENS OF DEGRESS CLOCKWISE FROM TRUE NORTH. "00" INDICATES CALM.
RESULTANT DIRECTIONS ARE GIVEN TO WHOLE DEGREES.
BOLD VALUES INDICATE EXTREME VALUES WHICH OCCURRED AFTER THE ASOS SYSTEM WAS COMMISSIONED.

RECORD = PERIOD OF RECORD
RECORD MEAN PRECIPITATION IS THE MEAN OF ALL DAILY PRECIPITATION AMOUNTS DURING THE PERIOD OF RECORD.
RECORD MAX(MIN) TEMPERATURE IS THE MEAN OF ALL DAILY MAX(MIN) TEMPERATURES DURING THE PERIOD OF RECORD.
RECORD MEAN TEMPERATURE IS THE SUM OF THE RECORD MAX AND RECORD MIN DIVIDED BY 2.
AVERAGE TEMPERATURE IS THE SUM OF THE MEAN DAILY MAX AND MIN TEMPERATURE DIVIDED BY 2.

EXCEPTIONS

RECORD MEANS ARE THROUGH THE CURRENT YEAR,
BEGINNING IN 1940 FOR TEMPERATURE
1940 FOR PRECIPITATION
1951 FOR SNOWFALL

HEATING DEGREE DAYS Base 65 deg. F — CASPER, WYOMING

SEASON	JULY	AUG	SEP	OCT	NOV	DEC	JAN	FEB	MAR	APR	MAY	JUNE	TOTAL
#1964-65	0	101	272	545	1004	1283	1084	1181	1377	569	463	145	8024
1965-66	4	49	512	401	749	1108	1334	1132	887	775	289	141	7381
1966-67	0	47	122	609	862	1225	1180	1037	878	664	488	193	7305
1967-68	6	21	149	516	1017	1470	1289	999	868	794	518	166	7813
1968-69	30	81	251	520	984	1426	1164	998	1127	541	304	252	7678
1969-70	3	3	71	834	914	1125	1289	897	1149	797	355	143	7580
1970-71	1	0	322	733	894	1225	1222	1119	1025	714	456	71	7782
1971-72	36	0	352	681	1026	1268	1461	1052	824	653	393	48	7794
1972-73	67	42	251	641	1025	1461	1415	1124	984	861	424	117	8412
1973-74	36	0	296	495	951	1181	1432	1044	885	650	416	105	7491
1974-75	3	61	319	504	912	1253	1352	1230	1066	813	493	193	8199
1975-76	1	18	247	532	969	1126	1295	1004	1017	625	346	149	7329
1976-77	3	14	199	654	913	1133	1385	949	1040	549	298	12	7149
1977-78	3	51	164	492	945	1240	1481	1191	854	605	471	107	7604
1978-79	17	60	212	522	1137	1602	1738	1153	910	631	479	151	8612
1979-80	0	41	118	523	1149	1116	1505	1081	1048	617	408	75	7681
1980-81	0	44	176	566	931	943	1036	979	829	494	418	86	6502
1981-82	15	17	129	633	775	1161	1385	1097	873	709	423	173	7390
1982-83	4	0	276	604	977	1287	1040	930	881	802	506	156	7463
1983-84	22	0	190	468	1009	1673	1331	1082	941	769	332	146	7963
1984-85	0	0	386	767	899	1298	1512	1264	973	578	297	151	8125
1985-86	0	51	355	606	1337	1348	1031	1017	678	613	434	37	7507
1986-87	4	7	282	586	986	1196	1293	1018	1001	453	235	67	7128
1987-88	26	59	204	561	880	1271	1377	1084	1016	564	339	21	7402
1988-89	2	15	223	465	924	1191	1201	1484	905	597	336	162	7505
1989-90	0	14	214	574	825	1261	1107	1032	912	614	442	116	7111
1990-91	22	3	111	570	835	1523	1349	819	875	681	403	49	7240
1991-92	5	0	192	587	1031	1140	1140	825	747	478	283	81	6509
1992-93	33	71	136	443	972	1328	1341	1266	821	625	266	168	7470
1993-94	40	25	208	543	1063	1111							

COOLING DEGREE DAYS Base 65 deg. F — CASPER, WYOMING

YEAR	JAN	FEB	MAR	APR	MAY	JUNE	JULY	AUG	SEP	OCT	NOV	DEC	TOTAL
1969	0	0	0	0	12	18	213	222	26	0	0	0	491
1970	0	0	0	0	0	77	200	234	24	0	0	0	535
1971	0	0	0	0	0	58	109	202	26	0	0	0	395
1972	0	0	0	0	0	54	101	142	7	0	0	0	304
1973	0	0	0	0	1	65	147	182	5	0	0	0	400
1974	0	0	0	0	0	109	243	47	7	0	0	0	406
1975	0	0	0	0	0	16	254	119	5	1	0	0	395
1976	0	0	0	0	0	29	250	127	33	0	0	0	439
1977	0	0	0	0	0	150	218	92	31	0	0	0	491
1978	0	0	0	0	2	72	170	124	74	0	0	0	442
1979	0	0	0	0	0	62	187	129	37	0	0	0	415
1980	0	0	0	0	1	64	232	93	20	0	0	0	410
1981	0	0	0	0	0	73	235	155	45	0	0	0	508
1982	0	0	0	0	0	41	188	275	37	0	0	0	541
1983	0	0	0	0	0	24	220	309	56	0	0	0	609
1984	0	0	0	0	14	48	175	197	15	0	0	0	449
1985	0	0	0	0	4	48	210	109	13	0	0	0	384
1986	0	0	0	0	0	117	135	150	0	0	0	0	402
1987	0	0	0	0	8	68	211	100	20	0	0	0	407
1988	0	0	0	0	12	252	273	170	20	0	0	0	727
1989	0	0	0	0	0	56	268	148	18	0	0	0	490
1990	0	0	0	0	0	113	176	120	65	0	0	0	474
1991	0	0	0	0	1	54	214	225	31	4	0	0	529
1992	0	0	0	4	18	50	108	186	27	2	0	0	395
1993	0	0	0	0	1	33	98	123	10	4	0	0	269

SNOWFALL (inches) — CASPER, WYOMING

SEASON	JULY	AUG	SEP	OCT	NOV	DEC	JAN	FEB	MAR	APR	MAY	JUNE	TOTAL
1964-65	0.0	T	0.0	T	12.9	6.7	16.5	8.6	11.4	7.3	7.1	0.0	70.5
1965-66	0.0	0.0	8.8	0.0	0.5	10.0	5.5	6.2	9.2	17.0	2.1	3.0	62.3
1966-67	0.0	0.0	T	11.9	8.0	11.8	9.2	10.6	8.8	7.5	10.2	0.0	78.0
1967-68	0.0	0.0	0.0	4.3	13.1	17.8	5.2	5.0	13.3	14.5	3.4	T	76.6
1968-69	0.0	0.0	T	0.7	4.9	16.4	2.0	6.9	13.1	12.1	T	3.0	59.1
1969-70	0.0	0.0	0.0	12.5	12.0	2.7	9.0	4.8	22.3	10.9	1.7	T	75.9
1970-71	0.0	0.0	0.2	11.2	7.4	11.9	3.7	11.2	14.8	24.2	2.2	0.0	86.8
1971-72	0.0	0.0	2.8	13.1	9.7	4.7	19.2	7.2	14.5	13.9	0.0	0.0	85.1
1972-73	0.0	0.0	0.2	9.8	8.1	9.1	9.6	11.3	10.7	56.3	1.7	0.0	116.8
1973-74	0.0	0.0	T	5.9	13.2	9.6	14.0	8.1	7.5	19.4	2.0	0.0	79.7
1974-75	0.0	0.0	5.1	1.2	8.9	5.1	10.3	8.1	36.2	13.4	21.3	0.0	109.6
1975-76	0.0	0.0	T	11.8	6.8	16.1	6.9	16.1	7.3	4.1	1.4	T	70.5
1976-77	0.0	0.0	0.0	3.8	1.8	6.2	8.5	11.8	26.6	4.8	T	0.0	63.5
1977-78	0.0	0.0	0.0	4.3	18.2	16.8	14.2	14.6	8.0	5.4	24.6	0.0	106.1
1978-79	0.0	0.0	T	3.1	17.1	25.7	16.3	7.3	18.6	9.2	22.7	T	120.0
1979-80	0.0	0.0	0.0	3.6	16.5	8.8	22.1	13.5	17.3	3.7	15.7	0.0	101.2
1980-81	0.0	0.0	0.0	6.1	10.0	7.9	9.9	7.3	5.0	10.3	0.4	0.0	56.9
1981-82	0.0	0.0	0.0	5.8	7.3	6.1	10.0	6.1	9.3	17.8	6.3	T	68.7
1982-83	0.0	0.0	11.5	11.3	2.5	62.8	4.4	6.2	27.7	23.3	1.9	0.0	151.6
1983-84	0.0	0.0	0.3	1.3	37.1	12.0	19.4	7.6	26.1	33.6	5.7	0.0	143.1
1984-85	0.0	0.0	3.1	9.4	6.1	10.2	14.7	10.5	9.7	11.6	0.0	0.0	75.3
1985-86	0.0	0.0	7.0	4.8	28.3	14.4	6.2	13.4	6.1	16.3	9.4	0.0	105.9
1986-87	0.0	0.0	0.0	16.1	17.2	5.5	24.0	22.7	16.4	3.3	T	0.0	105.2
1987-88	0.0	0.0	0.0	5.0	3.4	11.8	5.9	13.7	13.6	8.9	4.9	0.0	67.2
1988-89	0.0	0.0	T	T	7.5	10.1	4.9	21.5	6.3	6.0	T	T	56.3
1989-90	0.0	T	0.0	6.4	3.2	5.6	4.4	12.8	12.4	4.6	2.5	T	51.9
1990-91	T	T	0.0	7.1	13.8	9.5	10.2	5.9	5.9	15.1	0.4	T	67.9
1991-92	0.0	0.0	T	7.6	16.8	4.7	3.6	3.7	10.3	T	T	T	46.7
1992-93	0.0	T	0.0	1.2	16.5	13.2	12.6	12.5	5.4	19.5	0.4	0.0	81.3
1993-94	0.0	T	0.4	5.0	11.2	12.2							
Record Mean	T	T	1.1	5.2	10.7	10.8	10.0	10.4	14.0	12.7	3.7	0.2	78.8

CHEYENNE, WYOMING

The city of Cheyenne is located on a broad plateau between the North and South Platte Rivers in the extreme southeastern corner of Wyoming at an elevation of approximately 6,100 feet. The surrounding country is mostly rolling prairie which is used primarily for grazing. The ground level rises rapidly to a ridge approximately 9,000 feet in elevation about 30 miles west of the city. This ridge is known as the Laramie Mountains, one of the ranges of the Rockies, and extends in a north-south direction. Because of this ridge, winds from the northwest through west to southwest are downslope and produce a marked chinook effect in Cheyenne which is especially noticeable during the winter months. Also, winds from the north through east to south are upslope and may cause fog or low stratus clouds in the Cheyenne area throughout the year. Because of this terrain variation, the wind direction plays an important role in controlling the local temperature and weather.

Cheyenne experiences large diurnal and annual temperature ranges. This is due to the advent of both warm and cold air masses and the relatively high elevation of the city which permits rapid incoming and outgoing radiation. The daily temperature range averages about 30 degrees in the summer and 23 degrees in the winter. Many cold air masses from the north during the winter months miss Cheyenne. Because of the downslope of land to the east and the prevailing westerlies, some of the cold air masses do move over the city, but only about 13 percent of the days in an average January, the coldest month of the year, show temperatures dropping to zero or below. Temperatures during the winter months average a few degrees higher than over the Mississippi and Missouri Valleys at the same latitude.

Windy days are quire frequent during the winter and spring months. Since the wind is usually strongest during the daytime it is a very noticeable weather element. Usually the strong winds are from a westerly direction and this tends to raise the temperature because the air is moving downslope.

Most of the air masses reaching this area move in from the Pacific and since the mountains to the west are quite effective moisture barriers the climate is semi-arid. Fortunately, about 70 percent of normal annual precipitation occurs during the growing season. In the summer months, precipitation is mostly of the shower type and occurs mainly with thunderstorms. Hail is frequent and occasionally destructive in some thunderstorms. Most of the snow falls during the late winter and early spring months. It is not uncommon to have heavy snow in May.

The growing season in Cheyenne averages about 132 days a year and extends from around May 18th to September 27th. Freezing temperatures have occurred as late in the spring as mid-June, and as early in the fall as late August.

Relative humidity averages near 50 percent on an annual basis with large daily variations. Very seldom is the relative humidity above 30 percent when the temperature is above 80 degrees.

NORMALS, MEANS, AND EXTREMES

CHEYENNE, WYOMING

LATITUDE: 41°09'N LONGITUDE: 104°48' W ELEVATION: FT. GRND 6130 BARO 6129 TIME ZONE: MOUNTAIN WBAN: 24018

	(a)	JAN	FEB	MAR	APR	MAY	JUNE	JULY	AUG	SEP	OCT	NOV	DEC	YEAR
TEMPERATURE °F:														
Normals														
-Daily Maximum		37.7	40.5	44.9	54.7	64.6	74.4	82.2	80.0	71.1	60.0	46.8	38.8	58.0
-Daily Minimum		15.2	18.1	22.1	30.1	39.4	48.3	54.6	52.8	43.7	33.9	23.7	16.7	33.2
-Monthly		26.5	29.3	33.6	42.5	52.0	61.3	68.4	66.4	57.4	47.0	35.2	27.8	45.6
Extremes														
-Record Highest	58	66	71	74	83	90	100	100	96	93	83	73	69	100
-Year		1982	1962	1986	1992	1969	1954	1939	1979	1960	1992	1954	1939	JUN 1954
-Record Lowest	58	-29	-34	-21	-8	16	25	38	36	8	-1	-16	-28	-34
-Year		1984	1936	1943	1975	1947	1951	1952	1975	1985	1991	1993	1990	FEB 1936
NORMAL DEGREE DAYS:														
Heating (base 65°F)		1194	1000	973	675	403	150	31	44	251	558	894	1153	7326
Cooling (base 65°F)		0	0	0	0	0	39	136	87	23	0	0	0	285
% OF POSSIBLE SUNSHINE	54	64	66	67	63	61	66	69	68	70	69	61	60	65
MEAN SKY COVER (tenths)														
Sunrise - Sunset	58	5.8	6.2	6.3	6.5	6.6	5.5	5.0	5.1	4.6	4.8	5.6	5.7	5.6
MEAN NUMBER OF DAYS:														
Sunrise to Sunset														
-Clear	58	8.9	7.1	7.0	5.7	4.8	8.5	9.5	9.7	13.1	12.8	9.5	9.5	106.2
-Partly Cloudy	58	9.5	9.0	9.8	10.3	12.0	12.3	15.0	13.2	8.5	8.9	9.2	9.1	126.8
-Cloudy	58	12.9	12.1	14.1	13.9	14.3	9.2	6.5	8.0	8.4	9.3	11.3	12.6	132.6
Precipitation														
.01 inches or more	58	5.7	6.3	9.4	9.6	12.0	10.9	11.0	10.1	7.4	5.8	6.3	5.7	100.2
Snow, Ice pellets, hail														
1.0 inches or more	58	2.1	2.0	3.6	2.4	0.8	0.1	0.0	0.0	0.3	1.3	2.5	2.0	16.9
Thunderstorms	58	0.0	0.1	0.2	2.2	7.8	11.4	13.3	10.7	4.5	0.9	0.*	0.0	51.0
Heavy Fog Visibility														
1/4 mile or less	58	1.0	1.8	3.1	3.0	2.9	1.9	1.2	1.4	1.9	2.0	1.9	1.4	23.4
Temperature °F														
-Maximum														
90° and above	34	0.0	0.0	0.0	0.0	0.*	1.2	5.2	2.2	0.2	0.0	0.0	0.0	8.9
32° and below	34	9.4	7.1	5.2	1.3	0.*	0.0	0.0	0.0	0.1	0.7	4.4	8.8	37.0
-Minimum														
32° and below	34	28.9	26.4	27.4	17.7	3.6	0.0	0.0	0.0	2.0	12.2	24.7	28.6	171.6
0° and below	34	4.3	2.5	1.0	0.1	0.0	0.0	0.0	0.0	0.0	0.1	0.6	3.1	11.7
AVG. STATION PRESS. (mb)	21	809.0	809.2	807.3	809.0	810.3	812.8	815.3	815.5	814.5	813.3	810.1	809.3	811.3
RELATIVE HUMIDITY (%)														
Hour 05	34	57	62	66	68	72	72	70	69	67	61	61	58	65
Hour 11 (Local Time)	34	46	46	47	42	42	40	36	36	37	38	44	46	42
Hour 17	34	50	48	47	42	44	41	38	38	38	41	51	53	44
Hour 23	34	58	61	64	65	67	64	63	63	62	59	60	59	62
PRECIPITATION (inches):														
Water Equivalent														
-Normal		0.40	0.39	1.03	1.37	2.39	2.08	2.09	1.69	1.27	0.74	0.53	0.42	14.40
-Maximum Monthly	58	2.78	2.16	3.65	5.04	5.67	5.32	5.01	6.64	4.52	3.57	2.48	1.68	6.64
-Year		1949	1953	1990	1942	1981	1955	1973	1985	1973	1942	1979	1937	AUG 1985
-Minimum Monthly	58	T	T	0.12	0.34	0.11	0.07	0.58	0.03	0.01	0.03	T	0.03	T
-Year		1952	1983	1966	1992	1974	1980	1969	1944	1992	1964	1965	1959	FEB 1983
-Maximum in 24 hrs	58	1.41	1.60	1.88	1.94	2.07	2.68	3.42	6.06	2.75	1.70	1.66	1.19	6.06
-Year		1949	1953	1946	1984	1991	1955	1973	1985	1973	1947	1979	1979	AUG 1985
Snow, Ice pellets, hail														
-Maximum Monthly	58	35.5	19.9	39.2	31.8	30.4	8.7	T	0.5	7.4	21.3	31.1	21.3	39.2
-Year		1980	1953	1990	1984	1943	1947	1992	1993	1985	1969	1979	1958	MAR 1990
-Maximum in 24 hrs	58	12.7	14.0	15.6	17.4	15.0	8.7	T	0.5	5.8	8.6	19.8	11.7	19.8
-Year		1992	1953	1973	1984	1942	1947	1992	1993	1985	1990	1979	1979	NOV 1979
WIND:														
Mean Speed (mph)	36	15.3	14.6	14.5	14.2	12.6	11.4	10.4	10.4	11.2	12.3	13.5	14.7	12.9
Prevailing Direction through 1963		WNW	W	WNW	WNW	WNW	WNW	WNW	W	W	W	WNW	WNW	WNW
Fastest Obs. 1 Min.														
-Direction (!!!)	12	28	27	29	30	26	29	28	32	30	29	31	27	30
-Speed (MPH)	12	49	48	51	58	48	41	46	42	46	46	46	49	58
-Year		1989	1988	1982	1982	1988	1990	1993	1983	1992	1990	1986	1992	APR 1982
Peak Gust														
-Direction (!!!)	10	W	27	W	NW	W	W	W	NW	NW	SW	W	W	W
-Speed (mph)	10	77	70	69	62	71	58	79	59	62	71	69	71	79
-Date		1990	1986	1985	1992	1986	1993	1993	1989	1992	1985	1986	1992	JUL 1993

PRECIPITATION (inches)

CHEYENNE, WYOMING

YEAR	JAN	FEB	MAR	APR	MAY	JUNE	JULY	AUG	SEP	OCT	NOV	DEC	ANNUAL
1964	0.03	0.24	0.58	1.30	0.84	1.01	1.00	0.28	0.33	0.03	0.14	0.16	5.94
1965	0.57	0.26	0.76	0.57	3.11	4.03	0.89	1.54	1.32	0.75	T	0.22	14.02
1966	0.11	0.22	0.12	0.48	0.21	1.95	3.38	2.78	2.12	0.68	0.29	0.08	12.42
1967	0.45	0.54	0.76	2.15	4.04	2.63	1.71	0.99	0.88	0.46	0.32	0.46	15.39
1968	0.04	0.30	0.31	2.34	3.54	0.87	1.25	1.21	0.20	0.86	0.71	0.28	11.91
1969	0.23	0.25	0.27	0.82	1.77	2.70	0.58	0.99	0.84	2.04	0.23	0.21	10.93
1970	0.10	0.04	1.32	0.85	3.13	2.42	0.82	0.14	1.10	1.30	0.30	0.31	11.83
1971	0.51	0.62	1.08	2.81	2.38	0.97	1.08	1.41	1.76	1.16	0.05	0.07	13.90
1972	0.36	0.02	0.79	0.80	2.76	1.71	1.35	1.83	1.01	0.42	0.59	0.40	12.04
1973	0.23	0.07	1.85	1.75	0.31	1.20	5.01	0.27	4.52	0.06	1.25	1.06	17.58
1974	0.48	0.03	1.24	0.50	0.11	2.81	1.41	1.29	0.50	0.91	0.49	0.10	9.87
1975	0.40	0.17	1.17	0.47	2.27	1.49	2.62	0.39	0.52	0.49	0.20	0.52	10.71
1976	0.32	0.71	0.32	1.79	2.07	0.68	2.39	1.40	0.77	0.15	0.28	0.10	10.98
1977	0.14	0.08	1.21	1.86	2.50	2.44	3.49	1.07	0.19	0.08	0.35	0.24	13.65
1978	0.58	0.78	0.35	0.52	3.98	0.63	0.98	1.38	0.12	0.50	0.45	0.54	10.81
1979	0.27	0.14	1.34	0.77	2.90	3.32	1.83	1.86	0.32	0.46	2.48	1.50	17.19
1980	2.71	0.73	1.36	0.93	2.39	0.07	2.00	1.55	0.97	0.51	0.46	0.08	13.76
1981	0.30	0.20	0.70	0.73	5.67	1.66	2.85	2.90	0.31	0.85	0.09	0.45	16.71
1982	0.41	0.19	0.17	0.53	3.56	4.52	2.71	1.81	2.87	1.20	0.43	0.83	19.23
1983	0.02	T	2.96	4.45	2.31	2.81	2.12	1.95	0.78	0.49	2.34	0.46	20.69
1984	0.54	0.84	1.28	3.71	0.78	2.43	2.57	2.84	0.65	1.55	0.11	0.34	17.64
1985	0.66	0.19	0.36	1.10	1.05	1.59	3.99	6.64	1.78	0.94	0.84	0.80	19.94
1986	0.13	0.50	0.54	2.26	1.03	2.42	1.04	1.55	2.47	1.78	0.66	0.18	14.56
1987	0.09	0.90	1.25	0.68	4.43	1.80	2.04	1.23	0.93	0.33	0.76	0.85	15.29
1988	0.52	0.65	1.34	1.84	3.09	2.03	1.79	1.79	1.66	0.09	0.42	0.53	15.75
1989	0.27	1.26	0.49	0.48	1.37	2.51	1.70	1.79	1.62	0.41	0.14	0.69	12.73
1990	0.35	0.69	3.65	1.66	3.37	1.03	3.64	1.98	0.80	1.35	0.72	0.39	19.63
1991	0.36	0.12	0.43	1.15	3.84	4.56	3.39	1.49	1.87	0.44	0.87	0.13	18.65
1992	1.18	0.11	1.75	0.34	1.86	2.11	1.87	1.90	0.01	0.46	1.69	0.49	13.77
1993	0.35	0.81	0.63	2.49	1.92	3.32	0.64	2.21	3.16	1.85	1.23	0.30	18.91
Record Mean	0.43	0.54	1.05	1.74	2.44	1.86	2.01	1.58	1.19	0.88	0.57	0.45	14.72

AVERAGE TEMPERATURE (deg. F)

CHEYENNE, WYOMING

YEAR	JAN	FEB	MAR	APR	MAY	JUNE	JULY	AUG	SEP	OCT	NOV	DEC	ANNUAL
1964	26.1	22.3	27.4	40.8	54.6	60.5	70.9	64.3	56.9	48.7	35.4	29.5	44.8
1965	32.1	26.0	22.2	45.7	50.8	59.3	68.2	64.7	49.6	52.0	41.4	32.1	45.4
1966	25.8	26.8	37.7	39.6	55.6	61.5	74.5	66.2	59.9	47.8	38.8	31.0	47.1
1967	31.4	31.1	39.0	44.1	49.1	57.7	67.3	66.0	58.4	49.3	35.1	22.5	45.9
1968	28.7	32.6	38.0	38.9	49.2	62.6	68.2	64.5	57.6	49.2	34.0	28.0	46.0
1969	31.4	32.0	28.7	48.1	55.7	57.0	71.5	70.7	61.2	37.1	36.4	31.3	46.8
1970	28.7	34.6	28.9	37.8	54.3	60.9	69.5	70.7	55.0	41.2	35.7	29.1	45.5
1971	29.7	25.8	32.4	41.0	49.2	63.4	65.5	68.7	53.1	44.3	36.0	29.1	44.8
1972	25.3	33.2	39.3	42.5	51.2	63.2	64.3	64.6	56.6	45.5	29.5	21.0	44.7
1973	23.7	29.7	31.9	37.3	51.4	62.9	65.8	68.4	54.1	49.3	35.3	29.4	44.9
1974	24.1	31.9	37.1	43.3	55.0	64.1	70.1	65.2	55.3	48.3	35.6	26.6	46.4
1975	25.5	25.0	30.2	37.5	48.4	57.9	67.3	66.3	56.0	48.1	34.2	32.6	44.1
1976	26.9	32.9	30.9	42.7	51.1	59.7	69.2	65.0	57.7	43.4	33.5	30.6	45.3
1977	22.4	32.9	32.1	44.7	53.4	65.2	68.2	63.5	60.7	48.3	34.4	29.0	46.3
1978	22.1	25.0	37.7	43.9	49.0	61.0	68.7	64.3	59.6	47.9	33.5	21.1	44.5
1979	17.3	31.6	36.0	44.9	49.9	61.5	69.9	65.9	62.9	49.8	29.5	32.9	46.0
1980	22.2	29.4	33.0	42.3	51.2	65.5	71.4	66.5	60.4	46.7	36.8	38.5	47.0
1981	33.4	32.2	36.8	50.1	50.5	63.7	69.0	65.2	61.2	46.1	40.7	30.7	48.3
1982	25.5	28.2	36.1	41.8	50.5	57.5	67.7	69.1	56.7	45.1	31.4	28.3	44.8
1983	32.8	33.5	32.0	34.8	47.4	57.3	67.8	70.2	59.0	48.6	32.3	15.0	44.2
1984	24.3	28.4	32.4	35.7	53.6	59.4	68.4	66.6	53.0	40.0	35.0	27.0	43.7
1985	19.5	23.3	35.2	45.6	54.3	61.1	69.3	66.8	52.8	45.1	26.2	26.1	43.8
1986	37.0	30.1	42.2	43.9	50.6	64.5	68.5	67.0	55.3	44.6	34.3	28.7	47.2
1987	29.1	31.6	32.4	46.7	54.5	63.3	69.2	64.9	58.0	46.9	36.7	25.7	46.6
1988	21.5	28.4	32.3	44.4	53.3	67.4	69.7	68.7	57.5	50.3	35.3	28.7	46.5
1989	30.2	17.3	37.4	44.3	54.3	60.3	70.6	67.0	57.5	46.5	38.7	24.3	45.7
1990	31.5	28.7	31.9	42.8	49.3	64.1	65.0	66.4	62.2	46.7	38.7	20.8	45.7
1991	25.0	36.2	36.4	41.1	52.4	62.5	67.7	67.7	58.0	46.1	31.7	31.6	46.4
1992	30.5	36.7	38.6	48.6	55.0	60.8	64.6	63.8	60.0	49.2	30.9	25.5	47.0
1993	26.1	24.9	37.2	42.1	53.4	58.7	66.1	64.8	54.3	44.1	29.2	30.7	44.3
Record Mean	26.1	28.0	32.8	41.6	51.2	61.1	67.8	66.3	57.4	46.4	35.0	28.4	45.2
Max	37.3	39.2	44.1	53.6	63.5	74.6	81.8	80.2	71.4	59.4	46.6	39.3	57.6
Min	15.0	16.7	21.4	29.6	38.8	47.6	53.8	52.5	43.4	33.4	23.4	17.5	32.8

REFERENCE NOTES

CHEYENNE, WYOMING

GENERAL
T - TRACE AMOUNT.
BLANK ENTRIES DENOTE MISSING/UNREPORTED DATA.
# INDICATES A STATION OR INSTRUMENT RELOCATION.

SPECIFIC

PM - INCLUDES LAST DAY OF PREVIOUS MONTH
ASOS - AUTOMATED SURFACE OBSERVING SYSTEM IN OPERATION DURING THESE MONTHS.

(a) - LENGTH OF RECORD IN YEARS, ALTHOUGH INDIVIDUAL MONTHS MAY BE MISSING.
0.* OR * - THE VALUE IS BETWEEN 0.0 AND 0.05
NORMALS - BASED ON THE 1961-1990 RECORD PERIOD.
EXTREMES - DATES ARE THE MOST RECENT OCCURRENCE
WIND DIR.- NUMERALS SHOW TENS OF DEGRESS CLOCKWISE FROM TRUE NORTH. "00" INDICATES CALM.
RESULTANT DIRECTIONS ARE GIVEN TO WHOLE DEGREES.
BOLD VALUES INDICATE EXTREME VALUES WHICH OCCURRED AFTER THE ASOS SYSTEM WAS COMMISSIONED.

RECORD = PERIOD OF RECORD
RECORD MEAN PRECIPITATION IS THE MEAN OF ALL DAILY PRECIPITATION AMOUNTS DURING THE PERIOD OF RECORD.
RECORD MAX(MIN) TEMPERATURE IS THE MEAN OF ALL DAILY MAX(MIN) TEMPERATURES DURING THE PERIOD OF RECORD.
RECORD MEAN TEMPERATURE IS THE SUM OF THE RECORD MAX AND RECORD MIN DIVIDED BY 2.
AVERAGE TEMPERATURE IS THE SUM OF THE MEAN DAILY MAX AND MIN TEMPERATURE DIVIDED BY 2.

EXCEPTIONS

FASTEST MILE WINDS ARE THROUGH MARCH 1981.

RECORD MEANS ARE THROUGH THE CURRENT YEAR, BEGINNING IN 1871 FOR TEMPERATURE
1871 FOR PRECIPITATION
1936 FOR SNOWFALL

HEATING DEGREE DAYS Base 65 deg. F CHEYENNE, WYOMING

SEASON	JULY	AUG	SEP	OCT	NOV	DEC	JAN	FEB	MAR	APR	MAY	JUNE	TOTAL
1964-65	2	95	239	500	883	1095	1012	1088	1320	569	435	169	7407
1965-66	11	47	454	397	702	1016	1208	1061	841	758	291	143	6929
1966-67	0	51	166	526	781	1049	1036	945	800	622	490	221	6687
1967-68	12	45	202	481	891	1308	1117	932	833	777	487	107	7192
1968-69	40	82	221	481	924	1137	1034	919	1118	498	297	248	6999
1969-70	3	3	113	859	854	1038	1115	846	1113	811	326	168	7249
1970-71	3	5	302	732	874	1108	1087	1093	1007	713	481	99	7504
1971-72	56	6	364	633	863	1108	1226	916	791	668	419	61	7111
1972-73	85	75	248	599	1056	1358	1275	980	1017	824	414	122	8053
1973-74	80	4	323	482	883	1098	1264	922	862	643	304	110	6975
1974-75	4	55	302	509	873	1180	1215	1115	1070	819	506	212	7860
1975-76	11	39	274	515	920	998	1175	924	1048	661	425	158	7148
1976-77	11	44	224	664	937	1059	1314	894	1014	602	352	44	7159
1977-78	21	74	150	511	910	1108	1324	1115	840	627	491	157	7328
1978-79	28	73	200	523	937	1358	1471	933	893	597	459	139	7611
1979-80	2	62	105	468	1058	990	1321	1027	984	673	424	65	7179
1980-81	0	41	151	558	840	812	974	913	868	440	445	95	6137
1981-82	21	50	120	580	722	1058	1216	1025	892	687	446	227	7044
1982-83	29	7	264	608	1002	1131	992	875	1016	898	538	233	7593
1983-84	23	0	202	502	974	1547	1259	1058	1002	870	348	177	7962
1984-85	6	15	365	769	892	1171	1403	1163	917	576	323	162	7762
1985-86	11	37	364	612	1158	1199	862	970	698	629	440	71	7051
1986-87	4	20	286	630	914	1121	1102	928	1002	542	321	78	6948
1987-88	27	79	214	554	841	1213	1343	1056	1008	611	361	42	7349
1988-89	12	18	236	449	884	1116	1075	1332	851	615	334	180	7102
1989-90	4	18	236	566	779	1257	1037	1012	1021	662	480	106	7178
1990-91	75	28	127	558	784	1364	1231	799	879	712	385	95	7037
1991-92	26	17	217	586	993	1031	1062	813	811	488	307	131	6482
1992-93	63	92	161	484	1014	1216	1198	1118	856	680	351	202	7435
1993-94	45	58	322	642	1068	1059							

COOLING DEGREE DAYS Base 65 deg. F CHEYENNE, WYOMING

YEAR	JAN	FEB	MAR	APR	MAY	JUNE	JULY	AUG	SEP	OCT	NOV	DEC	TOTAL
1969	0	0	0	0	14	12	211	185	9	0	0	0	431
1970	0	0	0	0	2	52	149	189	11	0	0	0	403
1971	0	0	0	0	0	58	77	128	13	0	0	0	276
1972	0	0	0	0	0	11	69	69	2	0	0	0	151
1973	0	0	0	0	0	65	112	116	0	0	0	0	293
1974	0	0	0	0	4	88	173	67	17	0	0	0	349
1975	0	0	0	0	0	7	90	86	10	0	0	0	193
1976	0	0	0	0	0	7	145	50	15	0	0	0	217
1977	0	0	0	0	0	59	126	38	29	0	0	0	252
1978	0	0	0	0	1	43	150	59	44	0	0	0	297
1979	0	0	0	0	1	42	160	100	49	0	0	0	352
1980	0	0	0	0	0	88	205	94	21	0	0	0	408
1981	0	0	0	0	0	59	156	64	15	0	0	0	294
1982	0	0	0	0	0	8	120	140	21	0	0	0	289
1983	0	0	0	0	0	10	115	169	28	0	0	0	322
1984	0	0	0	0	1	14	118	72	8	0	0	0	213
1985	0	0	0	0	0	52	150	98	8	0	0	0	308
1986	0	0	0	0	0	62	118	89	0	0	0	0	269
1987	0	0	0	0	0	35	164	83	9	0	0	0	291
1988	0	0	0	0	4	122	166	140	18	0	0	0	450
1989	0	0	0	0	7	46	188	86	19	0	0	0	346
1990	0	0	0	0	0	86	84	79	49	0	0	0	298
1991	0	0	0	0	2	28	119	106	13	4	0	0	272
1992	0	0	0	1	2	15	58	63	16	0	0	0	155
1993	0	0	0	0	0	19	89	60	6	0	0	0	174

SNOWFALL (inches) CHEYENNE, WYOMING

SEASON	JULY	AUG	SEP	OCT	NOV	DEC	JAN	FEB	MAR	APR	MAY	JUNE	TOTAL
1964-65	0.0	0.0	0.0	0.0	2.0	1.6	8.5	5.5	10.0	2.5	1.4	0.0	31.5
1965-66	0.0	0.0	2.0	T	T	2.4	1.3	2.7	1.1	3.6	T	0.0	13.1
1966-67	0.0	0.0	T	4.5	2.3	0.8	5.4	8.0	3.4	15.9	11.0	0.0	51.3
1967-68	0.0	0.0	0.0	0.4	3.6	4.8	0.2	2.5	2.8	15.4	2.2	0.0	31.9
1968-69	0.0	0.0	0.0	0.5	6.0	5.5	3.0	4.7	2.9	3.0	0.0	T	25.6
1969-70	0.0	0.0	0.0	21.3	1.8	2.4	1.2	1.5	19.4	8.3	T	T	55.9
1970-71	0.0	0.0	T	11.0	4.7	4.9	8.6	9.3	15.2	13.0	1.2	0.0	67.9
1971-72	0.0	0.0	7.4	8.1	0.9	1.4	7.4	0.4	10.6	8.5	0.3	0.0	45.0
1972-73	0.0	0.0	T	5.0	8.8	7.6	4.6	0.6	27.0	12.8	1.2	0.0	67.6
1973-74	0.0	0.0	T	0.5	17.4	13.9	5.0	0.8	16.3	8.0	0.0	0.0	61.9
1974-75	0.0	0.0	1.9	1.4	2.4	1.5	5.8	4.3	14.0	5.4	0.8	0.0	37.5
1975-76	0.0	0.0	0.6	5.5	4.7	8.4	6.0	8.5	9.0	6.7	T	T	49.4
1976-77	0.0	0.0	0.0	1.5	3.5	1.2	2.5	0.8	12.9	7.0	T	0.0	29.4
1977-78	0.0	0.0	0.0	0.4	4.3	5.6	7.0	12.0	2.5	1.0	18.3	0.0	51.1
1978-79	0.0	0.0	0.1	2.0	9.3	17.6	6.7	2.2	21.1	4.0	14.1	T	77.1
1979-80	0.0	0.0	0.0	3.6	31.1	15.6	35.5	10.7	17.8	3.4	3.8	0.0	121.5
1980-81	0.0	0.0	0.0	1.1	6.3	2.1	3.4	2.9	9.0	2.0	0.8	0.0	27.6
1981-82	0.0	0.0	0.0	4.6	0.8	5.8	5.6	2.4	1.7	2.0	4.0	0.0	26.9
1982-83	0.0	0.0	0.0	12.2	7.9	13.1	0.1	T	31.9	25.7	10.1	0.0	101.0
1983-84	0.0	0.0	T	0.2	27.2	7.0	7.9	12.1	13.0	31.8	T	0.0	99.2
1984-85	0.0	0.0	1.6	3.8	1.5	5.6	9.8	1.9	3.6	2.6	0.4	0.0	30.8
1985-86	0.0	0.0	7.4	6.0	12.4	13.0	1.3	4.9	5.6	11.3	3.8	0.0	65.7
1986-87	0.0	0.0	0.0	9.2	6.9	2.2	1.1	9.9	13.7	4.4	T	0.0	47.4
1987-88	0.0	0.0	0.0	1.3	4.5	16.1	9.0	7.5	16.0	7.5	2.2	0.0	64.1
1988-89	0.0	0.0	0.2	0.0	4.5	7.2	4.4	17.6	5.0	4.3	T	T	43.2
1989-90	T	T	2.6	3.4	1.9	9.7	5.6	11.6	39.2	5.6	1.3	T	80.9
1990-91	T	T	T	12.3	8.8	6.6	7.7	1.6	3.7	10.5	T	T	51.2
1991-92	T	T	T	7.5	9.1	2.0	18.2	0.8	11.3	0.7	4.4	0.5	54.5
1992-93	T	T	0.0	2.1	25.5	10.9	7.1	14.6	6.6	11.3	T	0.5	78.6
1993-94	0.0	0.5	5.5	3.1	18.0	5.7							
Record Mean	T	T	0.8	3.7	7.4	6.2	6.4	5.9	12.1	9.0	3.4	0.3	55.1

LANDER, WYOMING

Lander, located in the central Wyoming valley of the Popo Agie River, lies at the foot and east of the Wind River Range. Situated on a flat-topped mesa, the airport station is 1 1/2 miles south-southeast and approximately 200 feet above the town.

The terrain to the north, east and south varies from rolling to broken with some grass covered hills 2 to 5 miles distant, rising approximately 400 feet above the station elevation. To the west and southwest the foothills of the Wind River Range begin about 3 miles from the station, sloping upward to over 12,000 feet above sea level along the Continental Divide, 20 miles distant.

Because Lander is in a pocket, winds from all directions except northeast are downslope and produce a Chinook effect, most noticable in winds from westerly quadrants. The airport, on its mesa, receives more wind than the town of Lander, the wind speed averaging 4.7 mph for the 56 years of record kept in the town. Because of light winds, steep temperature inversions are the rule during winter nights and early mornings. Temperatures in the valley will be as much as 15 degrees lower than at the airport on calm, clear nights when there is a snow cover. However, when the wind is calm and the humidities low, the chilling effect is much less than is usual in extreme cold. Winds are often so light that little or no mixing occurs between the cold surface air and the warmer layer 2,000 to 3,000 feet above the valley. For several days each winter, temperatures are 20 to 30 degrees lower than in the surrounding areas where higher wind speeds occur. The sheltered location, however, offers protection from most severe storms that sweep down from Canada.

Lander does not have a true spring season, and snow has been recorded in June.

Usually on 15 to 20 days a year the temperature reaches or exceeds 90 degrees. Even the warmest days are not oppressive, the humidity being low, and the nights being cool. The normal daily range of summer temperature is near 30 degrees.

Mountains block moisture from the Pacific, creating a semi-arid climate. The heaviest and most persistent precipitation comes when the wind in the lower levels is from easterly quadrants, through a combination of low pressure to the south, usually over Colorado, and high pressure to the north over Montana or the western Dakotas. Lander receives 45 percent more precipitation than the area 24 miles to the northeast and 83 percent more than areas 50 miles northeast. More than a third of the annual precipitation occurs in April and May, with another but lesser peak in September and October. Summer moisture comes from occasional showers but is very erratic and spotty. Since about one-third of the annual snowfall comes in March and April, when the temperature is comparatively high, the snow soon melts.

Hardier plants and vegetables do well in this area. Based on the 1951-1980 period, the average first occurrence of 32 degrees Fahrenheit in the fall is September 24 and the average last occurrence in the spring is May 22.

NORMALS, MEANS, AND EXTREMES

LANDER, WYOMING

LATITUDE: 42°49'N LONGITUDE: 108°44' W ELEVATION: FT. GRND 5557 BARO 5562 TIME ZONE: MOUNTAIN WBAN: 24021

	(a)	JAN	FEB	MAR	APR	MAY	JUNE	JULY	AUG	SEP	OCT	NOV	DEC	YEAR
TEMPERATURE °F:														
Normals														
-Daily Maximum		31.2	37.1	45.7	55.8	66.0	77.4	86.1	83.9	72.3	59.8	42.5	32.1	57.5
-Daily Minimum		7.8	13.6	22.1	30.7	39.9	48.8	55.9	53.8	44.2	33.7	19.9	9.4	31.7
-Monthly		19.5	25.4	33.9	43.3	53.0	63.1	71.0	68.9	58.3	46.8	31.2	20.8	44.6
Extremes														
-Record Highest	47	63	68	76	82	91	100	101	101	94	85	70	64	101
-Year		1971	1951	1966	1992	1954	1954	1954	1979	1990	1963	1988	1980	AUG 1979
-Record Lowest	47	-37	-28	-16	-2	18	25	39	35	10	-3	-18	-37	-37
-Year		1963	1949	1960	1973	1954	1951	1993	1962	1965	1991	1985	1983	DEC 1983
NORMAL DEGREE DAYS:														
Heating (base 65°F)		1411	1109	964	651	376	137	9	38	246	564	1014	1370	7889
Cooling (base 65°F)		0	0	0	0	0	80	195	159	45	0	0	0	479
% OF POSSIBLE SUNSHINE	47	65	68	70	67	64	72	75	75	72	68	58	62	68
MEAN SKY COVER (tenths)														
Sunrise - Sunset	47	6.1	6.1	6.3	6.3	6.4	5.1	4.3	4.4	4.4	5.0	5.9	5.6	5.5
MEAN NUMBER OF DAYS:														
Sunrise to Sunset														
-Clear	47	7.9	7.0	6.9	6.0	6.3	10.3	13.5	13.0	13.6	11.6	7.8	9.5	113.4
-Partly Cloudy	47	9.9	9.6	10.0	10.1	10.8	10.8	11.6	11.9	8.7	9.3	9.8	9.7	122.2
-Cloudy	47	13.2	11.7	14.0	13.8	13.7	8.9	5.9	6.1	7.6	10.1	12.5	11.8	129.5
Precipitation														
.01 inches or more	47	4.4	5.3	7.5	8.1	9.1	6.4	6.1	4.7	5.3	5.1	5.3	4.7	72.1
Snow,Ice pellets,hail														
1.0 inches or more	47	2.5	2.8	4.2	4.1	1.4	0.2	0.0	0.0	0.7	2.1	3.2	2.7	23.9
Thunderstorms	47	0.0	0.0	0.2	1.0	4.7	7.4	9.6	7.4	2.9	0.4	0.*	0.0	33.5
Heavy Fog Visibility														
1/4 mile or less	47	1.0	0.8	0.2	0.3	0.1	0.1	0.0	0.*	0.2	0.1	0.7	0.9	4.4
Temperature °F														
-Maximum														
90° and above	47	0.0	0.0	0.0	0.0	0.1	2.5	10.1	6.6	0.8	0.0	0.0	0.0	20.0
32° and below	47	15.4	8.9	4.1	0.6	0.1	0.0	0.0	0.0	0.1	0.7	6.7	14.0	50.6
-Minimum														
32° and below	47	30.6	27.7	28.2	17.9	4.8	0.3	0.0	0.0	2.7	13.6	27.7	30.7	184.0
0° and below	47	8.1	4.2	1.2	0.*	0.0	0.0	0.0	0.0	0.0	0.1	1.9	5.3	20.9
AVG. STATION PRESS.(mb)	21	827.7	827.5	825.2	826.4	827.2	829.1	831.4	831.6	831.1	830.5	828.2	827.9	828.7
RELATIVE HUMIDITY (%)														
Hour 05	47	68	68	67	66	66	62	56	54	59	63	69	68	64
Hour 11 (Local Time)	47	60	57	50	45	44	40	34	33	39	45	57	61	47
Hour 17	47	60	54	46	40	38	32	27	27	32	41	57	62	43
Hour 23	47	67	65	61	57	56	50	44	43	49	56	66	68	57
PRECIPITATION (inches):														
Water Equivalent														
-Normal		0.48	0.57	1.15	2.08	2.32	1.46	0.81	0.53	1.10	1.13	0.80	0.58	13.01
-Maximum Monthly	47	1.65	2.18	3.30	5.46	6.03	6.88	2.50	2.30	4.68	3.58	3.37	1.62	6.88
-Year		1949	1955	1977	1957	1957	1947	1977	1979	1973	1971	1983	1985	JUN 1947
-Minimum Monthly	47	T	T	0.35	0.21	0.27	T	0.05	T	0.01	T	0.01	0.03	T
-Year		1952	1970	1960	1992	1984	1971	1963	1970	1992	1988	1949	1954	OCT 1988
-Maximum in 24 hrs	47	0.81	0.89	1.28	2.16	2.75	3.56	2.13	1.08	2.21	1.71	1.38	1.25	3.56
-Year		1963	1987	1977	1971	1964	1947	1977	1979	1973	1966	1983	1985	JUN 1947
Snow,Ice pellets,hail														
-Maximum Monthly	47	26.5	43.8	52.0	66.0	33.9	18.4	T	T	32.9	39.9	48.7	28.0	66.0
-Year		1962	1955	1977	1973	1975	1947	1993	1992	1982	1971	1983	1985	APR 1973
-Maximum in 24 hrs	47	13.8	21.0	20.3	21.9	20.8	18.4	T	T	16.9	19.4	23.1	20.5	23.1
-Year		1980	1987	1973	1967	1975	1947	1993	1992	1982	1966	1958	1985	NOV 1958
WIND:														
Mean Speed (mph)	47	5.9	5.9	7.0	7.8	7.8	7.7	7.6	7.4	6.9	6.0	5.6	5.6	6.8
Prevailing Direction through 1963		SW	SW	SW	SW	SW	SW	SW	SW	SW	SW	SW	SW	SW
Fastest Mile														
-Direction (!!)	43	SW	SW	SW	SW	SW	SW	W	W	W	SW	N	SW	SW
-Speed (MPH)	43	73	77	80	72	66	61	57	56	56	70	75	73	80
-Year		1967	1957	1972	1955	1980	1960	1959	1962	1966	1950	1958	1964	MAR 1972
Peak Gust														
-Direction (!!)	10	SW	SW	SW	SW	S	SW	W	SW	SW	SW	W	NW	SW
-Speed (mph)	10	86	55	63	63	67	63	66	67	64	53	59	66	86
-Date		1989	1986	1991	1987	1986	1986	1985	1988	1988	1990	1989	1984	JAN 1989

PRECIPITATION (inches)

LANDER, WYOMING

YEAR	JAN	FEB	MAR	APR	MAY	JUNE	JULY	AUG	SEP	OCT	NOV	DEC	ANNUAL
1964	0.43	1.20	0.80	2.97	3.42	2.10	0.36	0.17	0.55	0.34	0.40	0.07	12.81
1965	0.19	0.83	0.43	1.60	3.15	2.07	1.67	0.18	2.62	0.34	0.44	0.44	13.96
1966	0.27	0.33	0.87	2.19	0.80	1.31	0.28	1.77	0.71	2.17	0.51	0.41	11.62
1967	0.25	1.18	0.73	2.42	3.97	5.06	0.91	0.11	1.36	0.21	1.66	1.12	18.98
1968	0.17	0.61	1.65	2.43	2.16	2.14	0.68	0.93	0.48	0.14	0.46	0.70	12.55
1969	0.12	0.13	0.92	2.32	1.02	5.29	0.07	T	0.01	1.89	0.21	0.38	12.36
1970	0.06	T	2.20	2.56	0.49	2.28	0.40	T	0.66	1.49	0.75	0.38	11.27
1971	0.36	0.45	0.85	5.22	4.95	T	0.27	0.27	2.27	3.58	0.41	0.36	18.99
1972	1.08	0.55	0.38	2.16	2.06	0.70	0.44	1.01	0.09	2.36	0.89	1.46	13.18
1973	0.89	0.28	3.02	4.02	0.91	0.22	2.10	0.33	4.68	1.47	0.74	0.64	19.30
1974	0.43	0.99	0.56	2.35	0.31	0.37	0.47	0.66	1.07	2.02	0.19	0.66	10.08
1975	0.74	0.43	1.18	1.68	4.44	1.88	0.29	0.10	0.38	1.37	0.73	0.79	14.01
1976	0.27	0.82	0.43	1.17	2.09	1.97	1.05	0.40	0.54	1.67	0.40	0.20	11.01
1977	0.53	0.11	3.30	1.47	1.11	0.66	2.50	0.47	0.18	0.91	0.65	0.33	12.22
1978	0.59	0.50	0.42	1.28	5.16	0.16	0.82	0.25	1.17	0.65	2.15	1.22	14.37
1979	0.75	0.05	0.56	1.69	3.00	0.83	0.18	2.30	0.01	0.48	0.57	1.09	11.51
1980	0.95	0.54	1.50	1.68	3.32	0.05	0.31	0.27	0.07	1.41	0.70	0.22	11.02
1981	0.67	0.27	1.98	1.11	3.20	0.04	0.88	0.92	0.61	0.61	0.05	0.05	10.39
1982	0.44	0.16	0.44	0.76	1.65	1.35	1.46	0.32	3.83	0.82	0.59	1.44	13.26
1983	0.08	0.49	2.11	3.34	3.02	1.21	0.20	0.32	0.50	0.57	3.37	0.60	15.81
1984	0.95	1.16	0.80	3.61	0.27	1.48	1.54	1.21	1.04	0.58	0.48	0.07	13.19
1985	0.45	0.26	0.48	0.83	0.83	1.70	1.00	0.07	1.74	0.16	1.55	1.62	10.69
1986	0.30	1.04	0.49	2.15	1.79	0.27	0.89	0.64	0.44	2.83	1.10	0.22	12.16
1987	0.79	1.73	2.24	1.36	2.37	1.39	0.49	1.50	0.30	0.94	1.20	0.72	15.03
1988	0.05	0.37	1.38	0.93	1.95	0.20	0.33	0.66	0.91	T	0.13	0.72	7.63
1989	0.04	0.76	0.37	1.13	5.30	2.72	0.27	0.80	2.19	1.21	0.32	0.39	15.50
1990	0.02	0.12	1.55	1.90	0.42	0.36	1.88	0.39	1.50	0.75	2.27	0.41	11.57
1991	0.41	0.75	0.39	2.83	4.50	0.73	0.91	0.14	0.43	1.35	1.98	0.28	14.70
1992	0.51	0.17	1.95	0.21	3.27	1.39	2.22	0.57	0.01	0.49	1.61	0.82	13.22
1993	1.00	0.34	1.47	1.90	1.79	4.96	0.83	0.46	0.42	1.34	1.62	0.41	16.54
Record Mean	0.49	0.67	1.21	2.21	2.48	1.32	0.80	0.55	1.06	1.33	0.83	0.61	13.55

AVERAGE TEMPERATURE (deg. F)

LANDER, WYOMING

YEAR	JAN	FEB	MAR	APR	MAY	JUNE	JULY	AUG	SEP	OCT	NOV	DEC	ANNUAL
1964	20.4	21.7	26.8	41.1	53.5	59.1	74.2	66.7	57.6	48.5	32.6	24.9	44.0
1965	29.7	24.0	23.7	45.4	49.6	60.4	70.0	67.1	48.0	51.9	38.6	25.8	44.5
1966	22.7	24.2	35.0	40.0	56.7	62.2	74.7	67.5	61.4	43.2	34.2	24.2	45.5
1967	26.6	27.8	36.7	41.4	50.0	57.5	69.0	69.2	60.6	48.1	28.1	15.1	44.2
1968	15.5	24.7	36.3	38.7	49.5	61.1	69.7	64.5	56.9	47.9	30.4	17.5	42.7
1969	25.3	28.2	29.5	47.0	57.0	56.8	71.4	73.1	63.8	37.8	33.9	25.6	45.8
1970	24.4	33.6	29.7	36.8	54.3	62.9	71.1	72.8	53.4	41.2	32.4	21.4	44.5
1971	26.5	26.1	32.6	41.0	50.9	63.9	68.3	72.6	53.4	41.0	27.0	21.5	43.7
1972	19.6	27.5	40.9	43.3	51.9	64.1	67.0	67.3	56.8	43.9	27.9	12.5	43.6
1973	9.4	13.6	26.4	34.1	52.5	63.7	68.5	69.9	53.8	48.0	31.9	25.6	41.5
1974	18.2	27.1	36.0	43.7	51.4	66.3	72.5	65.2	56.1	48.1	33.9	19.5	44.8
1975	20.9	22.9	31.9	36.9	48.1	58.1	71.5	66.8	58.0	46.7	28.8	23.6	42.8
1976	19.2	27.1	32.4	44.0	54.6	60.4	72.5	67.0	59.7	43.6	33.7	25.4	45.0
1977	15.9	31.3	30.6	46.7	53.2	68.3	70.7	65.3	59.7	48.0	32.0	26.3	45.7
1978	17.5	22.0	38.3	45.3	49.0	62.5	69.8	66.5	59.3	48.1	20.7	9.1	42.3
1979	1.3	20.3	33.8	44.1	51.4	62.9	70.7	66.5	63.8	49.8	26.0	24.9	43.0
1980	14.0	25.3	32.5	45.0	52.2	64.1	72.2	66.5	59.9	47.1	31.2	34.3	45.3
1981	28.5	27.2	39.3	48.4	52.4	64.8	71.2	69.7	62.3	43.7	37.7	27.1	47.7
1982	22.1	25.4	35.8	41.3	51.9	61.0	70.6	73.3	55.2	44.3	29.5	18.3	44.1
1983	26.3	28.1	36.2	37.9	48.6	61.5	70.0	74.0	60.9	49.0	28.2	6.1	43.9
1984	16.1	20.4	31.1	40.3	55.9	63.2	73.3	72.4	55.1	41.5	32.2	20.5	43.5
1985	16.2	24.5	36.0	48.3	59.1	65.4	73.7	67.1	53.2	45.4	17.8	16.1	43.6
1986	22.0	24.6	44.1	44.9	52.5	68.2	68.7	69.8	55.2	45.9	31.2	18.8	45.5
1987	16.9	28.1	32.2	50.7	57.5	65.9	70.3	66.7	60.7	48.5	30.2	20.6	45.7
1988	19.1	29.7	33.2	48.8	55.6	72.5	75.3	70.9	58.7	53.6	34.7	20.6	47.7
1989	24.8	10.5	38.5	47.5	53.8	61.4	74.0	68.0	58.5	46.0	35.3	22.0	45.0
1990	29.1	28.7	37.0	46.4	52.5	65.1	70.7	70.9	64.8	47.2	33.4	13.3	46.6
1991	18.9	33.9	37.6	41.5	52.2	64.3	72.2	71.5	60.0	46.4	26.7	20.8	45.5
1992	20.6	33.1	41.6	50.7	57.3	63.3	65.7	68.1	60.2	50.1	28.9	15.9	46.3
1993	11.4	11.0	33.3	43.6	55.8	58.7	64.7	66.1	57.2	45.7	24.6	24.3	41.4
Record Mean	19.1	23.8	32.7	42.9	52.3	61.7	69.4	67.6	57.4	45.7	30.8	21.0	43.7
Max	31.6	36.4	45.0	55.5	65.4	76.3	85.1	83.3	72.4	59.6	43.3	33.0	57.2
Min	6.7	11.2	20.3	30.2	39.2	47.1	53.7	51.8	42.4	31.8	18.3	8.9	30.2

REFERENCE NOTES

LANDER, WYOMING

GENERAL
T - TRACE AMOUNT.
BLANK ENTRIES DENOTE MISSING/UNREPORTED DATA.
# INDICATES A STATION OR INSTRUMENT RELOCATION.

SPECIFIC

(a) - LENGTH OF RECORD IN YEARS, ALTHOUGH INDIVIDUAL MONTHS MAY BE MISSING.
0.* OR * - THE VALUE IS BETWEEN 0.0 AND 0.05
NORMALS - BASED ON THE 1961-1990 RECORD PERIOD.
EXTREMES - DATES ARE THE MOST RECENT OCCURRENCE
WIND DIR.- NUMERALS SHOW TENS OF DEGRESS CLOCKWISE FROM TRUE NORTH. "00" INDICATES CALM.
RESULTANT DIRECTIONS ARE GIVEN TO WHOLE DEGREES.
BOLD VALUES INDICATE EXTREME VALUES WHICH OCCURRED AFTER THE ASOS SYSTEM WAS COMMISSIONED.

RECORD = PERIOD OF RECORD
RECORD MEAN PRECIPITATION IS THE MEAN OF ALL DAILY PRECIPITATION AMOUNTS DURING THE PERIOD OF RECORD.
RECORD MAX(MIN) TEMPERATURE IS THE MEAN OF ALL DAILY MAX(MIN) TEMPERATURES DURING THE PERIOD OF RECORD.
RECORD MEAN TEMPERATURE IS THE SUM OF THE RECORD MAX AND RECORD MIN DIVIDED BY 2.
AVERAGE TEMPERATURE IS THE SUM OF THE MEAN DAILY MAX AND MIN TEMPERATURE DIVIDED BY 2.

EXCEPTIONS

RECORD MEANS ARE THROUGH THE CURRENT YEAR,
BEGINNING IN 1892 FOR TEMPERATURE
1892 FOR PRECIPITATION
1947 FOR SNOWFALL

HEATING DEGREE DAYS Base 65 deg. F LANDER, WYOMING

SEASON	JULY	AUG	SEP	OCT	NOV	DEC	JAN	FEB	MAR	APR	MAY	JUNE	TOTAL
1964-65	0	84	230	505	965	1237	1088	1142	1275	582	469	143	7720
1965-66	1	39	507	395	786	1208	1304	1137	924	745	265	138	7449
1966-67	2	47	142	667	919	1261	1188	1036	869	703	466	225	7525
1967-68	0	9	157	515	1097	1541	1531	1165	881	780	474	153	8303
1968-69	18	89	251	522	1031	1467	1223	1025	1096	534	252	258	7766
1969-70	8	0	64	833	928	1212	1253	869	1087	842	323	136	7555
1970-71	3	0	357	733	970	1344	1187	1083	997	716	428	91	7909
1971-72	12	0	366	738	1133	1343	1404	1083	739	643	402	63	7926
1972-73	56	46	245	645	1105	1625	1720	1432	1186	918	386	122	9486
1973-74	35	3	332	521	988	1211	1451	1055	893	633	415	91	7628
1974-75	4	61	274	513	930	1401	1360	1172	1019	836	518	210	8298
1975-76	0	38	219	561	1079	1277	1411	1095	1006	624	316	170	7796
1976-77	1	16	182	657	933	1220	1514	940	1063	543	361	22	7452
1977-78	2	50	183	520	986	1193	1466	1192	823	586	490	127	7618
1978-79	26	54	226	519	1321	1730	1974	1248	962	620	415	137	9232
1979-80	0	61	84	463	1164	1234	1578	1145	1000	595	390	94	7808
1980-81	0	43	166	550	1005	943	1124	1051	789	487	383	83	6624
1981-82	18	15	104	655	812	1170	1324	1100	898	702	400	164	7362
1982-83	12	0	329	632	1058	1440	1194	1027	884	804	508	125	8013
1983-84	26	0	159	490	1096	1826	1510	1287	1046	736	290	135	8601
1984-85	0	0	316	721	977	1372	1507	1132	890	493	181	84	7673
1985-86	4	41	363	602	1413	1507	1329	1126	641	596	385	26	8033
1986-87	10	5	288	584	1006	1424	1486	1026	1008	425	239	46	7547
1987-88	14	50	146	501	1038	1372	1417	1018	981	481	296	15	7329
1988-89	1	7	217	347	905	1371	1239	1525	816	522	342	154	7446
1989-90	0	20	204	581	886	1329	1105	1011	860	552	382	123	7053
1990-91	22	0	95	545	939	1599	1422	865	844	696	391	62	7480
1991-92	7	1	179	572	1143	1361	1369	919	716	425	240	93	7025
1992-93	50	53	150	457	1075	1514	1658	1509	976	635	278	206	8561
1993-94	76	33	233	590	1205	1256							

COOLING DEGREE DAYS Base 65 deg. F LANDER, WYOMING

YEAR	JAN	FEB	MAR	APR	MAY	JUNE	JULY	AUG	SEP	OCT	NOV	DEC	TOTAL
1969	0	0	0	0	13	19	212	258	36	0	0	0	538
1970	0	0	0	0	1	81	196	247	14	0	0	0	539
1971	0	0	0	0	0	66	123	246	25	0	0	0	460
1972	0	0	0	0	1	43	123	127	3	0	0	0	297
1973	0	0	0	0	4	92	152	163	1	0	0	0	412
1974	0	0	0	0	1	138	243	73	12	0	0	0	467
1975	0	0	0	0	0	8	210	100	14	1	0	0	333
1976	0	0	0	0	1	41	238	84	30	0	0	0	394
1977	0	0	0	0	0	124	185	67	34	0	0	0	410
1978	0	0	0	0	0	56	179	109	64	0	0	0	408
1979	0	0	0	0	2	81	185	113	55	0	0	0	436
1980	0	0	0	0	3	74	231	97	19	0	0	0	424
1981	0	0	0	0	0	83	219	166	29	0	0	0	497
1982	0	0	0	0	1	51	188	266	41	0	0	0	547
1983	0	0	0	0	5	27	188	286	42	0	0	0	548
1984	0	0	0	0	13	87	264	237	26	0	0	0	627
1985	0	0	0	0	8	102	284	113	16	0	0	0	523
1986	0	0	0	0	6	131	131	159	1	0	0	0	428
1987	0	0	0	4	15	81	188	105	22	0	0	0	415
1988	0	0	0	0	9	248	329	196	33	0	0	0	815
1989	0	0	0	0	3	51	284	120	16	0	0	0	474
1990	0	0	0	0	0	133	204	189	97	0	0	0	623
1991	0	0	0	0	0	50	238	209	34	0	0	0	531
1992	0	0	0	2	10	49	79	154	13	0	0	0	307
1993	0	0	0	0	1	24	74	74	7	0	0	0	180

SNOWFALL (inches) LANDER, WYOMING

SEASON	JULY	AUG	SEP	OCT	NOV	DEC	JAN	FEB	MAR	APR	MAY	JUNE	TOTAL
1964-65	0.0	0.0	0.0	3.0	8.4	0.4	3.7	15.7	10.2	14.2	16.0	0.0	71.6
1965-66	0.0	0.0	23.6	0.0	5.2	12.0	9.3	7.0	18.7	34.3	1.4	0.0	111.5
1966-67	0.0	0.0	T	26.5	6.6	7.3	5.2	21.4	17.8	37.0	24.5	0.0	146.3
1967-68	0.0	0.0	0.0	1.4	21.0	26.3	2.4	9.9	26.5	39.5	5.2	0.0	132.2
1968-69	0.0	0.0	0.0	0.5	8.2	12.3	2.0	4.7	21.3	19.1	0.0	1.2	69.3
1969-70	0.0	0.0	0.0	32.4	4.2	9.3	1.9	T	40.3	37.2	T	0.0	125.3
1970-71	0.0	0.0	6.7	20.6	7.5	9.0	6.9	12.6	16.2	45.2	1.5	0.0	126.2
1971-72	0.0	0.0	10.1	39.9	9.3	6.6	18.7	10.7	4.6	28.5	0.0	0.0	128.4
1972-73	0.0	0.0	0.0	19.1	15.1	27.2	25.7	9.1	49.5	66.0	7.6	0.0	219.3
1973-74	0.0	0.0	T	16.1	13.5	12.4	7.2	18.3	12.8	21.8	0.2	0.0	102.3
1974-75	0.0	0.0	6.6	3.4	4.0	9.6	10.6	8.1	22.9	34.7	33.9	0.0	133.8
1975-76	0.0	0.0	0.0	15.9	14.9	19.7	4.9	16.7	7.6	1.3	2.4	2.9	86.3
1976-77	0.0	0.0	0.0	13.3	9.8	6.1	9.1	2.6	52.0	16.6	1.1	0.0	110.6
1977-78	0.0	0.0	0.0	3.7	12.8	7.7	9.3	9.1	4.1	T	33.7	0.0	80.4
1978-79	0.0	0.0	3.0	4.7	38.8	25.4	15.9	1.1	9.3	28.6	30.3	2.6	159.7
1979-80	0.0	0.0	0.0	5.3	10.0	20.4	26.2	10.6	25.4	12.6	13.9	0.0	124.4
1980-81	0.0	0.0	0.0	8.9	11.1	4.1	15.0	6.1	15.6	6.8	T	0.0	67.6
1981-82	0.0	0.0	0.0	0.2	1.8	1.7	8.1	2.8	7.4	7.7	12.1	0.0	41.8
1982-83	0.0	0.0	32.9	3.1	11.2	22.6	1.3	4.8	24.1	43.3	22.4	0.0	165.7
1983-84	0.0	0.0	1.9	1.2	48.7	11.9	18.0	17.4	16.3	45.4	1.1	0.0	161.9
1984-85	0.0	0.0	3.5	10.4	8.7	2.3	9.1	4.7	8.6	8.4	0.0	0.0	55.7
1985-86	0.0	0.0	7.6	2.2	32.3	28.0	4.9	13.8	5.7	14.4	11.3	0.0	120.2
1986-87	0.0	0.0	0.0	11.9	17.5	5.3	14.1	32.7	26.7	17.5	T	0.0	125.7
1987-88	0.0	0.0	0.0	1.0	13.7	14.4	1.1	6.2	30.7	5.0	3.9	0.0	76.0
1988-89	0.0	0.0	0.4	0.0	3.2	13.3	0.6	14.3	6.7	11.1	T	0.0	49.6
1989-90	0.0	0.0	0.0	16.9	4.3	8.6	0.6	2.9	20.9	2.8	2.5	0.0	59.5
1990-91	T	T	0.0	6.9	28.5	6.5	6.6	10.3	2.2	36.9	4.8	T	102.7
1991-92	0.0	0.0	T	16.4	26.4	3.0	10.7	2.7	16.0	0.0	T	T	75.2
1992-93	0.0	T	0.0	3.4	27.5	16.1	17.1	5.5	13.1	16.8	0.0	0.5	100.0
1993-94	T	0.0	0.3	6.9	23.1	7.6							
Record Mean	T	T	2.4	8.9	14.3	10.5	9.1	11.0	17.6	20.3	7.0	1.1	102.3

SHERIDAN, WYOMING

Sheridan is located east of the Rocky Mountains at an elevation of a little less than 4,000 feet. To the northwest, east, and southeast are rolling hills, but to the southwest and west the Bighorn Mountains rise abruptly, oriented generally northwest–southeast. The foothills are only about 15 miles from Sheridan, and within 30 miles to the southwest the average elevation is near 10,000 feet, with Cloud Peak rising to 13,175 feet. This mountain range has a marked effect on the climate at Sheridan.

During the winter months, a few days after the outbreak of cold arctic air from Canada, the winds generally shift to the west or southwest and increase in velocity. These downslope winds produce a pronounced warming or Chinook. At other times, a gentle downslope flow will persist for several days and result in a prolonged period of mild weather. The Chinook is very effective in moderating the weather of the winter season, which otherwise would be more severe. On the other hand, winds from the east or northeast blowing toward the mountains are upslope and usually cause cooling, persistent low clouds, and often heavy precipitation. The upslope precipitation occurs at times all through the year, but most frequently during the winter and spring. Sheridan will often receive much heavier snow or rain with an easterly wind condition than the surrounding country farther away from the mountains. In the summer, the mountains act as a breeding ground for thunderstorms that frequently move away from the mountains toward the northeast and give afternoon or evening showers to Sheridan. Because of the close proximity to the Bighorn Mountains, the annual precipitation at Sheridan is greater, on the average, than in the neighboring area to the east and north.

Based on the 1951–1980 period, the average first occurrence of 32 degrees Fahrenheit in the fall is September 20 and the average last occurrence in the spring is May 20. Because of the short growing season and cold periods during winter, only the most hardy fruits can be grown successfully, but most varieties of vegetables will reach maturity.

The climate of Sheridan can be described generally as semi–arid with long cold winters and short hot summers. However, during all of the winter months, more than 50 percent of the possible sunshine is received, while the hot days in the summer are marked by very low humidity and nights are cool. There are few summer nights when the temperature remains above 60 degrees. During July, the warmest month, even though temperatures of 90 degrees or above occur frequently, the nights are cool. January is usually the coldest month. The cold weather comes from outbreaks of Canadian air moving southeastward down the east side of the Rockies, and the initial onslaught of arctic air is usually accompanied by strong northerly winds with drifting snow. The coldest nights, however, come after the skies have cleared and the wind becomes very light.

The yearly precipitation pattern for Sheridan is heavy in the spring and early summer. The three winter months constitute the period with the least moisture. Amounts of snowfall are quite generous during the winter, but the water content of the snow is usually low. This dry snow is ordinarily not injurious to livestock and does not result in serious inconvenience or discomfort to the public. During the spring months of March and April, however, precipitation often begins as rain, gradually turning to rain and snow mixed or to heavy wet snow. These snowstorms are frequently accompanied by strong winds and drifting. As a result, these two months are considered to have the most disagreeable weather of the year and are most likely to cause livestock loss. March has more snow than any other month.

NORMALS, MEANS, AND EXTREMES

SHERIDAN, WYOMING

LATITUDE: 44°46'N LONGITUDE: 106°58' W ELEVATION: FT. GRND 3964 BARO 3946 TIME ZONE: MOUNTAIN WBAN: 24029

	(a)	JAN	FEB	MAR	APR	MAY	JUNE	JULY	AUG	SEP	OCT	NOV	DEC	YEAR
TEMPERATURE °F:														
Normals														
-Daily Maximum		33.0	38.3	46.2	57.1	66.3	76.9	86.1	85.2	72.8	61.7	45.3	35.0	58.7
-Daily Minimum		8.5	14.5	21.5	30.4	39.0	47.3	53.0	51.7	41.4	31.7	19.7	10.2	30.7
-Monthly		20.8	26.4	33.9	43.8	52.7	62.1	69.6	68.4	57.1	46.7	32.5	22.6	44.7
Extremes														
-Record Highest	53	70	76	77	87	95	105	106	106	103	92	78	72	106
-Year		1974	1982	1993	1946	1960	1988	1989	1983	1983	1992	1975	1981	JUL 1989
-Record Lowest	53	-35	-32	-23	-2	13	27	35	33	6	-9	-25	-37	-37
-Year		1963	1989	1965	1975	1954	1951	1971	1993	1984	1991	1959	1983	DEC 1983
NORMAL DEGREE DAYS:														
Heating (base 65°F)		1370	1081	964	636	381	156	28	63	269	567	975	1314	7804
Cooling (base 65°F)		0	0	0	0	0	69	170	168	32	0	0	0	439
% OF POSSIBLE SUNSHINE	53	56	60	62	60	60	65	75	75	68	63	52	54	63
MEAN SKY COVER (tenths)														
Sunrise - Sunset	50	6.9	6.9	7.0	6.8	6.6	5.7	4.3	4.4	4.9	5.7	6.7	6.7	6.0
MEAN NUMBER OF DAYS:														
Sunrise to Sunset														
-Clear	53	5.8	5.3	4.7	5.4	5.6	7.8	13.4	13.6	11.7	10.0	6.0	6.5	95.8
-Partly Cloudy	53	8.0	7.7	9.4	9.0	10.5	11.8	12.2	10.9	9.1	8.7	8.1	7.8	113.1
-Cloudy	53	17.2	15.2	16.9	15.6	14.8	10.5	5.5	6.5	9.2	12.3	15.9	16.8	156.4
Precipitation														
.01 inches or more	53	8.7	8.5	10.7	10.9	11.7	11.0	7.5	6.5	7.3	7.3	8.0	8.8	106.8
Snow,Ice pellets,hail														
1.0 inches or more	50	3.5	3.6	4.4	2.9	0.5	0.*	0.0	0.0	0.4	1.3	3.0	3.9	23.6
Thunderstorms	53	0.0	0.*	0.*	0.8	4.7	9.2	9.5	7.2	2.5	0.3	0.*	0.0	34.4
Heavy Fog Visibility														
1/4 mile or less	53	0.8	0.8	0.8	0.4	0.3	0.3	0.1	0.1	0.2	0.5	0.7	0.6	5.6
Temperature °F														
-Maximum														
90° and above	29	0.0	0.0	0.0	0.0	0.3	2.9	10.1	10.8	2.3	0.*	0.0	0.0	26.4
32° and below	29	13.3	8.1	4.3	0.7	0.0	0.0	0.0	0.0	0.*	0.5	5.3	12.4	44.6
-Minimum														
32° and below	29	30.2	27.1	27.2	17.9	5.1	0.2	0.0	0.0	3.4	16.2	27.6	30.2	185.2
0° and below	29	9.4	4.4	1.4	0.*	0.0	0.0	0.0	0.0	0.0	0.1	2.0	7.0	24.4
AVG. STATION PRESS.(mb)	7	877.8	877.5	874.8	876.2	876.7	877.9	880.1	879.6	880.5	880.0	879.0	877.2	878.1
RELATIVE HUMIDITY (%)														
Hour 05	29	69	71	74	73	76	78	73	69	70	70	73	70	72
Hour 11 (Local Time)	29	62	60	53	48	48	46	37	35	41	47	57	61	50
Hour 17	29	64	59	50	44	46	45	34	31	37	46	61	65	49
Hour 23	26	70	71	70	67	71	71	62	57	63	67	72	71	68
PRECIPITATION (inches):														
Water Equivalent														
-Normal		0.73	0.64	0.97	1.72	2.39	2.25	0.88	0.82	1.37	1.18	0.83	0.70	14.48
-Maximum Monthly	53	1.79	2.68	3.26	4.80	6.80	9.54	3.78	3.02	3.08	3.16	2.41	2.41	9.54
-Year		1972	1955	1946	1963	1978	1944	1958	1968	1951	1971	1991	1989	JUN 1944
-Minimum Monthly	53	0.07	0.05	0.14	0.18	0.30	0.28	0.07	T	0.06	0.02	0.10	0.10	T
-Year		1983	1992	1978	1980	1958	1971	1988	1970	1964	1965	1981	1986	AUG 1970
-Maximum in 24 hrs	53	1.01	1.10	2.25	3.84	2.04	3.44	2.28	1.71	1.57	1.91	1.10	0.86	3.84
-Year		1972	1955	1946	1948	1956	1944	1948	1943	1982	1974	1991	1980	APR 1948
Snow,Ice pellets,hail														
-Maximum Monthly	53	26.3	35.0	36.8	39.6	12.5	4.0	T	0.0	21.0	17.5	25.8	43.5	43.5
-Year		1977	1955	1954	1955	1979	1969	1993		1984	1989	1964	1989	DEC 1989
-Maximum in 24 hrs	53	13.5	11.2	13.3	26.7	10.9	4.0	T	0.0	12.9	14.0	12.0	13.3	26.7
-Year		1972	1990	1946	1955	1979	1969	1993		1984	1989	1942	1989	APR 1955
WIND:														
Mean Speed (mph)	51	7.7	7.8	9.0	9.8	9.0	8.1	7.3	7.4	7.5	7.5	7.7	7.6	8.0
Prevailing Direction through 1963		NW	NW	NW	NW	NW	NW	NW	NW	NW	NW	NW	NW	NW
Fastest Obs. 1 Min.														
-Direction (!!)	8	30	24	29	29	31	29		31	31	29	30	30	30
-Speed (MPH)	8	53	45	46	43	46	40	37	35	46	46	44	46	53
-Year		1990	1988	1990	1992	1990	1992	1992	1985	1992	1985	1985	1988	JAN 1990
Peak Gust														
-Direction (!!)	10	NW	SW	SW	NW	NW	N	NW	NW	NW	W	SW	SW	NW
-Speed (mph)	10	71	69	68	69	63	54	59	56	67	64	64	69	71
-Date		1988	1988	1988	1987	1989	1992	1985	1991	1992	1985	1984	1987	JAN 1988

PRECIPITATION (inches)

SHERIDAN, WYOMING

YEAR	JAN	FEB	MAR	APR	MAY	JUNE	JULY	AUG	SEP	OCT	NOV	DEC	ANNUAL
1964	0.35	0.81	0.51	2.74	1.76	5.11	0.11	1.18	0.06	0.34	1.99	0.66	15.62
1965	1.18	0.72	0.70	0.65	2.12	2.15	0.93	0.68	1.75	0.02	0.22	0.24	11.36
1966	0.36	0.40	1.04	2.25	1.08	1.29	0.29	0.86	0.98	1.02	0.62	0.82	11.01
1967	0.61	1.28	1.18	2.51	1.33	6.11	0.22	0.70	1.56	0.88	0.84	0.98	18.20
1968	1.05	0.75	0.92	0.71	2.29	3.89	0.33	3.02	2.12	0.63	0.91	0.97	17.59
1969	1.11	0.18	0.40	1.84	1.79	2.44	0.47	0.24	0.09	1.47	0.99	0.51	11.53
1970	0.72	0.71	2.31	1.67	5.20	2.10	1.06	T	1.87	0.94	1.43	0.62	18.63
1971	1.43	1.17	0.62	3.83	2.45	0.28	0.21	0.52	0.90	3.16	0.71	0.52	15.80
1972	1.79	0.66	0.97	1.02	1.35	1.96	1.62	0.96	0.94	1.01	0.32	0.90	13.50
1973	0.35	0.39	1.41	4.05	0.51	2.21	0.64	0.47	2.79	0.81	0.70	0.80	15.13
1974	0.67	0.46	0.96	1.58	1.44	0.48	0.65	0.70	1.48	2.96	0.64	0.24	12.26
1975	1.11	0.47	1.18	1.67	3.51	4.56	0.88	0.19	0.19	1.50	0.88	1.15	17.29
1976	0.51	0.66	0.44	2.24	1.11	2.32	0.73	0.80	1.62	1.05	0.93	0.26	12.67
1977	1.45	0.08	1.84	1.16	3.03	2.09	1.66	1.95	0.93	1.27	0.66	1.36	17.48
1978	1.29	0.84	0.14	2.26	6.80	0.46	1.73	1.50	1.67	0.27	1.45	0.95	19.36
1979	0.48	0.38	0.60	1.31	2.95	1.13	0.84	1.10	0.39	2.28	0.56	0.23	12.25
1980	0.60	0.86	1.25	0.18	3.65	1.11	0.33	1.12	1.31	0.88	1.27	0.72	13.28
1981	0.30	0.36	0.57	0.24	5.71	1.94	3.00	0.67	0.63	1.00	0.10	0.51	15.03
1982	0.70	0.25	1.33	0.93	1.46	3.87	0.51	0.58	2.90	0.65	0.26	1.12	14.56
1983	0.07	0.33	0.48	0.88	1.84	1.01	0.12	1.23	0.94	1.25	1.06	0.76	9.97
1984	0.69	0.46	2.11	2.22	1.55	2.34	0.40	0.96	2.28	0.28	0.57	0.35	14.21
1985	0.81	0.28	0.54	0.98	1.55	1.73	0.93	0.44	2.36	0.57	0.70	0.40	11.29
1986	0.26	1.52	1.08	0.88	1.78	3.21	0.63	1.02	2.43	1.21	1.94	0.10	16.06
1987	0.45	0.81	1.45	0.22	2.76	1.23	3.01	1.15	1.49	0.54	1.17	0.29	14.57
1988	0.67	0.78	0.64	2.09	2.43	0.55	0.07	0.26	0.97	1.27	0.60	0.63	10.96
1989	0.36	0.48	1.61	2.26	1.69	1.62	0.54	0.27	0.84	2.70	0.61	2.41	15.39
1990	0.46	0.89	1.01	2.27	1.92	2.00	1.11	0.46	0.24	1.80	0.87	0.41	13.44
1991	0.85	0.61	1.09	3.69	2.67	1.02	0.29	0.45	1.50	0.83	2.41	0.57	15.98
1992	0.21	0.05	0.96	1.43	2.02	3.56	2.42	0.63	0.45	0.90	1.35	1.26	15.24
1993	0.55	0.32	0.78	2.52	2.37	3.61	3.31	0.43	0.58	1.93	0.45	0.49	17.34
Record Mean	0.73	0.66	1.15	2.00	2.55	2.32	1.18	0.82	1.36	1.24	0.84	0.67	15.51

AVERAGE TEMPERATURE (deg. F)

SHERIDAN, WYOMING

YEAR	JAN	FEB	MAR	APR	MAY	JUNE	JULY	AUG	SEP	OCT	NOV	DEC	ANNUAL
#1964	26.4	26.7	28.8	43.4	54.3	60.1	73.4	65.9	55.2	48.0	28.0	20.0	44.2
1965	29.9	23.4	19.3	45.0	49.8	59.9	69.8	67.1	47.4	52.9	38.1	28.8	44.3
1966	18.4	23.4	36.8	38.5	53.8	60.2	72.5	65.3	61.6	45.9	30.6	24.6	44.3
1967	25.4	29.5	32.8	41.1	49.2	56.8	66.6	67.8	59.2	47.2	29.9	19.4	43.8
1968	16.9	29.7	38.5	40.2	49.4	58.8	67.6	64.5	55.9	46.8	33.1	15.8	43.1
1969	10.3	22.6	28.7	47.5	54.0	57.4	67.9	70.7	60.5	38.6	34.2	28.6	43.4
1970	19.2	31.1	28.1	38.4	53.6	62.1	68.8	70.4	53.9	42.8	33.2	21.4	43.6
1971	19.5	24.8	33.5	43.7	53.2	62.8	66.0	73.6	54.5	42.8	32.8	16.3	43.6
1972	14.0	27.4	40.2	44.5	54.1	64.8	64.1	68.5	55.1	44.4	33.4	16.5	43.9
1973	20.3	28.7	35.3	39.3	52.3	62.9	68.2	70.1	55.0	47.8	29.2	27.1	44.7
1974	19.7	33.1	35.7	45.8	49.9	65.5	72.8	63.0	54.6	48.2	34.3	26.7	45.8
1975	22.2	16.0	28.7	36.5	50.3	58.5	71.0	66.0	56.8	46.3	31.3	28.7	42.7
1976	23.5	32.0	31.3	45.2	54.3	60.3	71.5	67.9	59.6	42.3	33.2	27.6	45.7
1977	15.7	31.7	32.8	47.2	54.9	65.4	70.0	64.0	57.5	47.7	31.5	18.1	44.7
1978	11.4	17.5	34.2	46.7	49.9	60.6	66.7	65.4	59.2	47.8	23.6	12.5	41.3
1979	3.4	15.8	33.3	40.5	50.4	60.9	68.1	68.0	62.4	48.8	29.7	31.1	42.7
1980	15.8	24.5	30.5	47.9	55.0	62.1	71.3	65.6	58.8	46.8	36.9	28.0	45.3
1981	32.4	28.5	39.4	47.8	52.0	61.3	70.9	70.1	62.7	45.7	41.3	26.1	48.2
1982	16.4	26.0	34.8	42.0	50.7	59.6	69.8	73.5	58.1	46.9	32.1	22.8	44.4
1983	33.7	36.4	39.4	42.2	51.1	63.4	72.8	77.8	59.6	51.3	34.4	6.8	47.4
1984	25.0	32.5	37.1	43.2	52.8	61.3	71.0	72.8	53.5	41.2	33.1	14.1	44.8
1985	15.3	16.7	32.1	47.7	58.2	62.5	72.2	66.2	52.6	45.2	15.5	25.6	42.5
1986	32.7	22.9	43.8	44.8	52.5	66.9	68.1	69.9	53.5	48.0	29.2	29.0	46.8
1987	26.8	33.7	35.9	51.2	58.3	64.3	68.2	64.8	58.9	47.1	37.4	26.6	47.8
1988	21.6	26.4	36.7	46.6	56.5	74.2	74.5	70.3	58.0	51.0	34.0	28.6	48.2
1989	25.9	14.3	29.5	44.3	53.3	61.5	73.7	68.3	58.9	45.5	36.7	21.0	44.4
1990	29.5	26.6	36.2	43.7	51.1	60.7	69.2	69.9	64.3	45.9	37.9	15.8	45.9
1991	19.3	35.8	37.2	42.0	53.0	62.9	71.4	73.1	59.7	44.5	29.6	28.5	46.4
1992	30.8	37.7	42.8	48.3	56.0	63.5	63.9	65.2	58.8	48.8	32.7	15.2	47.0
1993	15.4	17.7	37.6	44.8	56.7	58.5	61.7	64.4	55.1	45.4	29.0	30.4	43.1
Record Mean	20.5	24.9	32.9	43.8	53.1	62.0	69.8	68.2	57.4	46.4	32.8	23.7	44.6
Max	33.0	37.2	45.2	56.7	66.4	76.1	85.8	84.6	73.0	61.1	45.6	35.9	58.4
Min	8.0	12.5	20.5	30.9	39.8	47.9	53.8	51.7	41.8	31.7	20.1	11.4	30.8

REFERENCE NOTES

SHERIDAN, WYOMING

GENERAL
T - TRACE AMOUNT.
BLANK ENTRIES DENOTE MISSING/UNREPORTED DATA.
# INDICATES A STATION OR INSTRUMENT RELOCATION.

SPECIFIC

(a) - LENGTH OF RECORD IN YEARS, ALTHOUGH INDIVIDUAL MONTHS MAY BE MISSING.
0.* OR * - THE VALUE IS BETWEEN 0.0 AND 0.05
NORMALS - BASED ON THE 1961-1990 RECORD PERIOD.
EXTREMES - DATES ARE THE MOST RECENT OCCURRENCE
WIND DIR.- NUMERALS SHOW TENS OF DEGRESS CLOCKWISE FROM TRUE NORTH. "00" INDICATES CALM.
RESULTANT DIRECTIONS ARE GIVEN TO WHOLE DEGREES.
BOLD VALUES INDICATE EXTREME VALUES WHICH OCCURRED AFTER THE ASOS SYSTEM HAS COMMISSIONED.

RECORD = PERIOD OF RECORD
RECORD MEAN PRECIPITATION IS THE MEAN OF ALL DAILY PRECIPITATION AMOUNTS DURING THE PERIOD OF RECORD.
RECORD MAX(MIN) TEMPERATURE IS THE MEAN OF ALL DAILY MAX(MIN) TEMPERATURES DURING THE PERIOD OF RECORD.
RECORD MEAN TEMPERATURE IS THE SUM OF THE RECORD MAX AND RECORD MIN DIVIDED BY 2.
AVERAGE TEMPERATURE IS THE SUM OF THE MEAN DAILY MAX AND MIN TEMPERATURE DIVIDED BY 2.

EXCEPTIONS

RECORD MEANS ARE THROUGH THE CURRENT YEAR, BEGINNING IN 1908 FOR TEMPERATURE
1908 FOR PRECIPITATION
1941 FOR SNOWFALL

HEATING DEGREE DAYS Base 65 deg. F　　SHERIDAN, WYOMING

SEASON	JULY	AUG	SEP	OCT	NOV	DEC	JAN	FEB	MAR	APR	MAY	JUNE	TOTAL
#1964-65	0	101	292	520	1103	1389	1078	1158	1412	594	464	153	8264
1965-66	7	44	523	372	800	1116	1438	1158	869	787	345	171	7630
1966-67	0	86	138	584	1023	1244	1221	988	991	707	482	238	7702
1967-68	17	21	183	546	1047	1404	1487	1016	813	740	477	199	7950
1968-69	35	75	267	557	947	1519	1692	1181	1121	522	345	236	8497
1969-70	20	6	151	811	918	1121	1414	944	1138	794	348	132	7797
1970-71	10	4	342	686	948	1347	1407	1122	968	631	358	107	7930
1971-72	59	5	328	683	963	1501	1577	1082	761	607	339	56	7961
1972-73	95	28	303	633	942	1502	1380	1010	915	766	387	130	8091
1973-74	23	4	299	523	1070	1169	1400	887	902	572	462	95	7406
1974-75	3	93	309	513	918	1178	1317	1370	1117	852	450	194	8314
1975-76	1	43	246	575	1003	1122	1282	949	1040	587	328	161	7337
1976-77	1	20	186	696	949	1154	1520	926	989	527	307	53	7328
1977-78	12	73	233	528	1001	1452	1654	1323	951	543	463	153	8386
1978-79	45	75	232	525	1235	1623	1911	1376	979	727	447	154	9329
1979-80	6	20	120	497	1051	1044	1523	1171	1061	507	312	111	7423
1980-81	0	50	195	557	836	1140	1004	1018	788	510	396	136	6630
1981-82	15	6	115	593	703	1200	1503	1087	932	685	436	186	7461
1982-83	26	2	257	556	983	1304	962	794	788	677	431	89	6869
1983-84	22	0	215	419	908	1802	1236	935	856	647	380	147	7567
1984-85	7	3	374	731	953	1574	1535	1351	1016	512	215	116	8387
1985-86	5	62	373	605	1483	1215	996	1175	650	598	390	41	7593
1986-87	15	2	338	521	1069	1109	1177	871	896	413	218	76	6705
1987-88	46	76	197	549	824	1183	1341	1114	870	545	271	23	7039
1988-89	3	20	221	424	925	1122	1204	1415	1094	618	359	142	7547
1989-90	0	25	205	598	841	1358	1096	1070	885	632	425	183	7318
1990-91	17	1	109	585	804	1526	1407	811	858	682	367	83	7250
1991-92	4	0	201	638	1056	1120	1052	785	682	496	290	86	6410
1992-93	78	103	199	510	963	1538	1533	1317	844	599	253	206	8143
1993-94	126	72	298	603	1073	1068							

COOLING DEGREE DAYS Base 65 deg. F　　SHERIDAN, WYOMING

YEAR	JAN	FEB	MAR	APR	MAY	JUNE	JULY	AUG	SEP	OCT	NOV	DEC	TOTAL
1969	0	0	0	0	9	16	114	191	25	0	0	0	355
1970	0	0	0	0	1	55	134	176	16	2	0	0	384
1971	0	0	0	0	0	49	96	276	21	0	0	0	442
1972	0	0	0	0	9	55	74	145	11	0	0	0	294
1973	0	0	0	0	1	73	129	168	5	0	0	0	376
1974	0	0	0	0	0	119	248	39	4	0	0	0	410
1975	0	0	0	0	0	6	194	83	10	0	0	0	293
1976	0	0	0	0	0	28	211	115	28	0	0	0	382
1977	0	0	0	0	3	73	176	48	16	0	0	0	316
1978	0	0	0	0	1	30	103	94	62	0	0	0	290
1979	0	0	0	0	1	42	108	121	46	0	0	0	318
1980	0	0	0	2	9	33	202	75	14	3	0	0	338
1981	0	0	0	0	0	31	207	173	56	0	0	0	467
1982	0	0	0	0	0	31	185	273	53	0	0	0	542
1983	0	0	0	0	6	47	272	402	57	0	0	0	784
1984	0	0	0	0	7	45	199	253	38	0	0	0	542
1985	0	0	0	0	13	47	238	104	10	0	0	0	412
1986	0	0	0	0	8	105	116	156	0	0	0	0	385
1987	0	0	0	6	19	62	150	76	24	2	0	0	339
1988	0	0	0	0	17	304	305	192	19	0	0	0	837
1989	0	0	0	6	0	44	276	132	27	0	0	0	485
1990	0	0	0	0	0	64	154	159	93	0	0	0	470
1991	0	0	0	0	1	28	208	257	49	9	0	0	552
1992	0	0	0	1	17	49	51	113	21	13	0	0	265
1993	0	0	0	0	1	16	32	64	8	0	0	0	121

SNOWFALL (inches)　　SHERIDAN, WYOMING

SEASON	JULY	AUG	SEP	OCT	NOV	DEC	JAN	FEB	MAR	APR	MAY	JUNE	TOTAL
1964-65	0.0	0.0	T	0.0	25.8	9.4	11.8	12.5	12.8	3.7	4.7	0.0	80.7
1965-66	0.0	0.0	6.0	0.0	4.5	5.3	8.6	5.3	12.9	12.9	3.4	0.0	58.9
1966-67	0.0	0.0	0.0	4.4	11.1	14.6	12.1	20.4	16.5	23.0	7.5	0.0	109.6
1967-68	0.0	0.0	0.0	0.2	11.5	20.6	13.5	7.3	14.0	7.1	0.6	0.0	74.8
1968-69	0.0	0.0	0.0	T	6.9	15.3	15.0	3.0	4.9	6.7	1.3	4.0	57.1
1969-70	0.0	0.0	0.0	4.9	7.8	5.7	11.5	10.1	26.1	9.4	2.0	0.0	77.5
1970-71	0.0	0.0	9.5	0.9	11.3	10.0	15.6	15.7	7.5	5.2	T	0.0	75.7
1971-72	0.0	0.0	T	14.8	9.3	9.2	25.1	9.4	9.1	7.8	T	0.0	84.7
1972-73	0.0	0.0	3.6	2.6	1.4	15.6	7.8	8.2	16.1	37.5	1.4	0.0	94.2
1973-74	0.0	0.0	3.2	4.9	11.7	10.7	10.0	6.8	9.8	7.7	0.4	0.0	65.2
1974-75	0.0	0.0	0.3	2.0	4.4	4.7	13.8	10.1	17.6	14.5	4.8	0.0	72.2
1975-76	0.0	0.0	0.0	8.5	14.1	18.0	8.5	12.9	8.1	8.7	0.0	0.0	78.8
1976-77	0.0	0.0	0.0	3.5	11.9	5.4	26.3	2.2	25.0	10.2	0.0	0.0	84.5
1977-78	0.0	0.0	0.0	8.4	9.1	24.0	23.2	15.5	1.5	0.7	5.2	0.0	87.6
1978-79	0.0	0.0	0.0	T	18.5	18.3	10.7	6.9	7.6	11.0	12.5	0.0	85.5
1979-80	0.0	0.0	0.0	10.1	8.9	4.2	12.1	18.2	19.3	3.0	0.0	0.0	75.8
1980-81	0.0	0.0	0.4	4.5	14.0	11.1	6.2	5.5	4.8	0.2	T	0.0	46.7
1981-82	0.0	0.0	0.0	5.6	1.0	8.2	12.8	5.2	18.8	7.3	T	0.0	58.9
1982-83	0.0	0.0	6.9	3.4	5.5	19.6	1.5	4.1	5.7	10.7	8.8	0.0	66.2
1983-84	0.0	0.0	3.3	0.0	12.3	17.6	12.4	7.7	20.5	25.6	5.4	0.0	104.8
1984-85	0.0	0.0	21.0	2.5	8.3	7.7	16.3	5.6	7.8	8.2	0.0	0.0	77.4
1985-86	0.0	0.0	5.2	1.8	13.7	6.3	2.9	23.5	11.3	3.5	5.5	0.0	73.7
1986-87	0.0	0.0	0.0	2.0	20.4	2.1	9.2	12.7	19.4	1.4	4.1	0.0	71.3
1987-88	0.0	0.0	0.0	4.2	3.4	6.0	11.6	12.1	10.4	16.1	0.8	0.0	64.6
1988-89	0.0	0.0	0.1	1.0	7.3	12.1	6.7	11.5	23.5	15.4	0.0	T	77.6
1989-90	0.0	0.0	T	17.5	7.7	43.5	6.0	16.4	12.8	10.7	2.2	T	116.8
1990-91	0.0	0.0	0.0	9.5	9.1	9.3	16.0	5.3	12.7	31.2	1.8	0.0	94.9
1991-92	0.0	0.0	0.0	10.0	22.1	7.2	3.1	0.4	5.8	1.3	0.3	T	50.2
1992-93	T	0.0	0.0	8.2	12.9	21.2	6.3	5.2	9.2	9.0	T	T	72.0
1993-94	T	0.0	T	14.7	7.5	6.3							
Record Mean	T	0.0	1.3	4.5	9.3	10.9	10.5	10.5	12.8	9.9	2.0	0.1	71.8